Woldman's
ENGINEERING ALLOYS

7th Edition

Edited by:

John P. Frick, Ph.D.

Library of Congress Catalog Card Number: 90-84697
ISBN: 0-87170-408-0
SAN: 204-7586

ASM International
Materials Park OH 44073
Telephone: (216)338-5151

PRINTED IN THE UNITED STATES OF AMERICA

SEVENTH EDITION

DEDICATED

TO

DR. NORMAN E. WOLDMAN, Ph.D.
Editor of *Engineering Alloys*

(1899 - 1969)

Preface

The first edition of Engineering Alloys by Norman Woldman and Albert Dornblatt was published in 1936. Over half a century and 30,000 alloys later the book is entering its seventh edition. Since the last edition was published in 1979, an additional 5,000 alloys have been added and 12,000 existing entries have been revised. Included is material from 1,500 manufacturers in twenty-three countries. Additionally, obsolete alloys which were listed only by name in the sixth edition have been reintroduced along with engineering data. The alloys are sorted by ASCII code. This means that APEX 470 precedes APEX 50 in the listing.

All of the early editions contained cautionary statements regarding the use of the book. To quote the second edition, "A pertinent admonition to anyone seeking information on alloys is:

> 'A little learning is a dang'rous thing;
> Drink deep, or taste not the Pierian spring.'"

This is still sound advice. Any alloy can vary significantly in properties, depending on variation in the chemical composition, the mechanical and thermal processing sequences, and the size of the part. The data contained in this volume should only be regarded as typical for the alloy. Before selection of an alloy more detailed information should be acquired.

The editor would like to thank Ms. Fran Cverna and ASM International for their assistance in the preparation of this edition. The editor would also like to express appreciation to Aluminium-Verlag, Dusseldorf, The Aluminum Association, American Iron and Steel Institute, American Society for Testing and Materials, Copper Development Association, and Society of Automotive Engineers, Inc. for permission to include some of their materials. The contributions of all individuals who reviewed and revised existing entries and who provided information on new alloys are gratefully acknowledged.

John P. Frick
November 1990

Contents

ABBREVIATIONS

AA
Aluminum Association

ACI
Alloy Casting Institute

AISI
American Iron and Steel Institute

Ann
Anneal or annealed

AOD
Argon-oxygen-decarburization

ASTM
American Society for Testing and Materials

B
Flux density

Bal
Balance or remainder

BOF
Basic oxygen furnace

Br
Remanence flux in gausses

Brin
Brinell hardness number

BS
British Standard

CC
Combined carbon

CD
Cold drawn

CDA
Copper Development Association

CEL
Elastic limit in compression in pounds per square inch

CF
Cold finished

cmf
Circular mil feet

Coef Exp
Coefficient of expansion per $^{\circ}$C

Coef Res
Temperature coefficient of resistance per $^{\circ}$C

CS
Crushing strength in pounds per square inch

CUS
Ultimate strength in compression in pounds per square inch

CYP
Yield point in compression in pounds per square inch

DCEN
Direct current, electrode negative

DCEP
Direct current, electrode positive

DPH
Diamond pyramid hardness (Vickers hardness)

DVM
Double vacuum melted

EFM
Electroflux melting

EI
Elongation in percent

El Res
Electrical resistivity in microhms per centimeter cube (20 $^{\circ}$C)

ESR
Electroslag melting

h
hours

H
Magnetizing field strength in oersteds

Hc
Coercive force in gilberts per centimeter or oersteds

HSLA
High strength low alloy (steels)

Ht
Heat treated

Ht Tr
Heat treated

in.
inches

IS(Charpy)
Charpy Impact Strength

IS(Izod)
Izod Impact Strength

kg-fmm^2
Kilograms force per square millimeter

kg/mm^2
Kilograms per square millimeter

ksi
Thousand pounds per square inch

lb
pound

ME
Modulus of elasticity in pounds per square inch

min
minutes or minimum

MPa
Megapascals

N/mm^2
Newtons per square millimeter

N & T
Normalized and tempered

PM
Permanent mold

ppm
Parts per million

psi
Pounds per square inch

Q & T
Quenched and tempered

RA
Reduction of area in percent

Rock A
Rockwell "A" Hardness

Rock B
Rockwell "B" Hardness

Rock C
Rockwell "C" Hardness

RT
Room temperature

s
seconds

SAE
Society of Automotive Engineers

SC
Sand cast

SS
Shear strength in pounds per square inch

STA
Solution treated and aged

TC
Total carbon content

Tr D
Transverse deflection in inches

Tr S
Transverse strength in pounds

TS
Tensile strength

tsi
Tons per square inch

UTS
Ultimate tensile strength

VAM
Vacuum arc melting

VDH
Vickers diamond hardness

VIM
Vacuum induction melting

W.Nr.
Werkstoff number (German standard)

YP
Yield point in tension

YS
Yield strength

ALLOY DATA

000 EXTRA
Thyssen Edelstahlwerke AG
C 0.9, W 19, Cr 4, V 2.2, bal Fe.
For high speed cutters; high speed steel. *Obsolete*

000 SPEZIAL 31
Thyssen Edelstahlwerke AG
C 0.8, Cr 4, Mo 0.3, V 2.6, W 10, bal Fe.
For turning, shaping and slotting tools; high speed steel.
Obsolete

01420
Russian manufacture
Mg 4-7, Li 1.5-2.6, bal Al.
Wrought: 65,000 psi TS; 43,000 psi YS.

03KH16N15M3
Russian manufacture
C 0-0.03, Mn 0.8, Si 0.6, Cr 15-17, Ni 14-16, Mo 2, bal Fe.
Austenitic stainless steel.

03KH16N15M3B
Russian manufacture
C 0-0.03, Mn 0.8, Si 0.6, Cr 15-17, Ni 14-16, Mo 2, Cb
0.25-0.5, bal Fe.
Austenitic stainless steel.

03KH17N14M2
Russian manufacture
C 0.03, Mn 1-2, Si 0.8, Cr 16-18, Ni 13-15, Mo 2, bal Fe.
Austenitic stainless steel.

03KH18N11
Russian manufacture
C 0-0.03, Mn 2, Si 0.8, Cr 17-19, Ni 10.5-12.5, bal Fe.
Austenitic stainless steel. Similar to AISI 304 L.

03KH18N12
Russian manufacture
C 0-0.03, Mn 0.4, Si 0.4, Cr 17-19, Ni 11.5-13, Ti 0, bal Fe.
Austenitic stainless steel.

03KH18N12T
Russian manufacture
C 0-0.03, Mn 2, Si 0.8, Cr 17-19, Ni 11-13, bal Fe.
Austenitic stainless steel.

03KH21N21M4GB
Russian manufacture
C 0-0.03, Mn 1.8-2.5, Si 0.6, Cr 20-22, Ni 20-22, Mo 3.4-3.7,
Cb, bal Fe.
Austenitic stainless steel.

04KH18N10
Russian manufacture
C 0-0.04, Mn 2, Si 0.8, Cr 17-19, Ni 9-11, ba.
Austenitic stainless steel; similar to AISI 304L.

06KH18N11
Russian manufacture
C 0-0.06, Mn 2, Si 0.8, Cr 17-19, Ni 10-12, bal Fe.
Austenitic stainless steel. Similar to AISI 304.

07KH16N6
Russian manufacture
C 0.05-0.09, Mn 0.8, Si 0.8, Cr 15.5-17.5, Ni 5-8, bal Fe.
Austenitic-martensitic type stainless steel.

07KH21G7AN5
Russian manufacture
C 0-0.07, Mn 6-7.5, Si 0.7, Cr 19.5-21, Ni 5-6, N 0.15-0.25, bal
Fe.
Austenitic stainless steel.

08KH10N20T2
Russian manufacture
C 0-0.08, Mn 2, Si 0.8, Cr 10-12, Ni 18-20, Ti 1, Al 0-1, bal Fe.
Austenitic stainless steel.

08KH13
Russian manufacture
C 0-0.08, Mn 0.8, Si 0.8, Cr 12-14, Al 1-1.8, bal.
Ferritic type stainless steel.

08KH15N24V4TR
Russian manufacture
C 0-0.08, Mn 0.5-1, Si 0.6, Cr 14-16, W 4-5, B 0-0.005, Ce
0-0.025, Ni 22-25, Ti 1.4-1.8, bal Fe.
Austenitic stainless steel.

08KH16N13M2B
Russian manufacture
C 0.06-0.12, Mn 1, Si 0.8, Cr 15-17, Ni 12.5-14.5, Cb 0.9-1.3,
Mo, bal Fe.
Austenitic stainless steel.

08KH17N13M2T
Russian manufacture
C 0-0.08, Mn 2, Si 0.8, Cr 16-18, Ni 12-14, Mo 2, Ti, bal Fe.
Austenitic stainless steel.

08KH17N15M3T
Russian manufacture
C 0-0.08, Mn 2, Si 0.8, Cr 16-18, Ni 14-16, Mo 3, Ti 0.3-0.6,
bal Fe.
Austenitic stainless steel.

08KH17N5M3
Russian manufacture
C 0.06-0.1, Mn 0.8, Si 0.8, Cr 16-17.5, Ni 4.5-5.5, bal Fe.
Austenitic-martensitic type stainless steel.

08KH17T
Russian manufacture
C 0.08, Mn 0.8, Si 0.8, Cr 16-18, Ti, bal Fe.
Ferritic type stainless steel.

08KH18G8N2T
Russian manufacture
C 0-0.08, Mn 7-9, Si 0.8, Cr 17-19, Ni 1.8-2.8, Ti 0.2-0.5, bal
Fe.
Austenitic-ferritic type stainless steel.

08KH18N10
Russian manufacture
C 0-0.08, Mn 2, Si 0.8, Cr 17-19, Ni 9-11, bal Fe.
Austenitic stainless steel; similar to AISI 304.

08KH18N10T
Russian manufacture
C 0-0.08, Mn 2, Si 0.8, Cr 17-19, Ni 9-11, bal Fe.
Austenitic stainless steel; similar to AISI 321.

08KH18N12B
Russian manufacture
C 0-0.08, Mn 2, Si 0.8, Cr 17-19, Ni 11-13, Mo 1, Cb, bal Fe.
Austenitic stainless steel; similar to AISI 347.

08KH20N14S2
Russian manufacture
C 0-0.08, Mn 1.5, Si 2-3, Cr 19-22, Ni 12-15, bal Fe.
Austenitic-ferritic type stainless steel.

08KH21N6M2T
Russian manufacture
C 0-0.08, Mn 0.8, Si 0.8, Cr 20-22, Ni 5.5-6.5, Ti 0.2-0.4, bal
Fe.
Austenitic-ferritic type stainless steel.

08KH22N6T
Russian manufacture
C 0-0.08, Mn 0.8, Si 0.8, Cr 21-23, Ni 5.3-6.3, bal Fe.
Austenitic-ferritic type stainless steel.

09KH14N16B
Russian manufacture
C 0.07-0.12, Mn 1-2, Si 0.6, Cr 13-15, Ni 14-17, Cb 0.9-1.3,
bal Fe.
Austenitic stainless steel.

09KH14N19V2BR
Russian manufacture
C 0.07-0.12, Mn 2, Si 0.6, Cr 13-15, Ni 18-20, Cb 0.9-1.3, B
0-0.005, Ce 0-0.02, W, bal Fe.
Austenitic stainless steel.

09KH14N19VBR1
Russian manufacture
C 0.07-0.12, Mn 2, Si 0.6, Cr 13-15, Ni 18-20, Cb 0.9-1.3, B
0-0.025, Ce 0-0.02, W, bal Fe.
Austenitic stainless steel.

09KH15N8JU
Russian manufacture
C 0-0.09, Mn 0.8, Si 0.8, Cr 14-16, Ni 7-9.4, bal Fe.
Austenitic-martensitic type stainless steel.

09KH16N15M3B
Russian manufacture
C 0-0.09, Mn 0.8, Si 0.8, Cr 15-17, Ni 14-16, Mo 2, Cb 0.6 0.0,
bal Fe.
Austenitic stainless steel.

09KH16N4B
Russian manufacture
C 0.05-0.13, Mn 0.5, Si 0.6, Cr 15-17, Ni 3.5-4.5, Cb 0, bal Fe.
Stainless steel.

09KH17N7JU
Russian manufacture
C 0-0.09, Mn 0.8, Si 0.8, Cr 16-17.5, Ni 7-8, bal Fe.
Austenitic-martensitic type stainless steel.

09KH17N7JU1
Russian manufacture
C 0-0.09, Mn 0.8, Si 0.8, Cr 16.5-18, Ni 6.5-7.5, bal Fe.
Austenitic-martensitic type stainless steel.

1% NEB-BRONZE
New England Brass Co.
Cu 89-93, Sn 0.7-1.3, Pb 0-0.1, Fe 0-0.05, bal Zn.
Good hot and cold working properties; good for welding,
brazing, soldering. For jewelry products, flat springs for
electrical switchgear. Copper alloy No. 413.

10% NICKEL SILVER (C74000)
Criterion Metals Inc.
Copper. Cu 71.25, Mn 0-0.5, Fe 0-0.25, Ni 10, Pb 0-0.1, rem
Zn.
Thin gauge sheet, various tempers. 50-92 ksi TS min; 15-89
ksi YS min. C74000

10% NICKEL SILVER (C74500)
Criterion Metals Inc.
Copper. Cu 65.25, Mn 0-0.5, Fe 0-0.25, Ni 10, Pb 0-0.1, rem
Zn.
Thin gauge sheet, various tempers. 56-95 ksi TS min. C74500

100 CR 6
Thyssen Edelstahlwerke AG
C 1, Cr 1.5, bal Fe.
For ball bearings and races, header dies; water hardened,
wear resistant. *Obsolete*

100 CRMN 6
Thyssen Edelstahlwerke AG
C 1, Mn 1.1, Cr 1.5, bal Fe.
For roller bearings; oil or water hardened. *Obsolete*

101 VAN PUNCH
Flockton, Tompkin & Co., Ltd.
C 0.5, W 5, bal Fe.
For punches, piercing dies; head, impact and abrasion
resistant.

10KH11N20T3R
Russian manufacture
C 0-0.1, Mn 1, Si 1, Cr 10.5-12.5, Ni 18-21, Ti 2, B 0.008-0.02,
Al 0-0.8, bal Fe.
Austenitic stainless steel.

10KH11N23T3MR
Russian manufacture
C 0-0.1, Mn 0.6, Si 0.6, Cr 10-12.5, Ni 21-25, Mo 1, Ti 2.6-3.2, B 0-0.02, Al 0-0.8, bal Fe.
Austenitic stainless steel.

10KH13SJU
Russian manufacture
C 0.7-0.12, Mn 0.8, Si 1.2-2, Cr 12-14, Al 1-1.8, bal Fe.
Ferritic type stainless steel.

10KH14AG15
Russian manufacture
C 0-0.1, Mn 14.5-16.5, Si 0.8, Cr 13-15, bal Fe.
Austenitic stainless steel.

10KH14G14N3
Russian manufacture
C 0.09-0.14, Mn 13-15, Si 0.7, Cr 12.5-14, bal Fe.
Austenitic stainless steel.

10KH14G14N4T
Russian manufacture
C 0-0.1, Mn 13-15, Si 0.8, Cr 13-15, Ni, bal Fe.
Austenitic stainless steel.

10KH17N13M2T
Russian manufacture
C 0-0.1, Mn 2, Si 0.8, Cr 16-18, Ni 12-14, Mo 2, Ti, bal Fe.
Austenitic stainless steel.

10KH17N13M3T
Russian manufacture
C 0-0.1, Mn 2, Si 0.8, Cr 16-18, Ni 12-14, Mo 3, Ti, bal Fe.
Austenitic stainless steel.

10KH23N18
Russian manufacture
C 0-0.1, Mn 2, Si 1, Cr 22-25, Ni 17-20, bal Fe.
Austenitic stainless steel.

10W-TA
Now TA-10W.

115CRV3
Westig (U.K.) Ltd.
Alloy steel. C 1.17, Si 0.22, Mn 0.3, P 0-0.03, S 0-0.03, Cr 0.65, V 0.1, bal Fe.
Alloyed tool steel equivalent to AISI L2.

11A45
Heppenstall Co.
C 0.4, bal Fe.
For gears, shafts. *Obsolete*

11KH11N2V2MF
Russian manufacture
C 0.09-0.13, Mn 0.6, Si 0.6, Cr 10.5-12, Ni 1.5-1.8, W 1.6-2, V 0.18-0.3, Mo, bal Fe.
Martensitic stainless steel; air hardening.

12 MOV
American manufacture
C 0.25, Mn 0.5, Si 0.5, Cr 12, Ni 0.5, Mo 1, V 0.3.
Martensitic stainless. Similar to AISI 420.

12% NICKEL SILVER (C75700)
Criterion Metals Inc.
Copper. Cu 65, Mn 0-0.5, Fe 0-0.25, Ni 12, Pb 0-0.05, rem Zn.
Thin gauge sheet, various tempers. 56-96 ksi TS min. C75700

12% NICKEL SILVER (C76200)
Criterion Metals Inc.
Copper. Cu 59, Mn 0-0.5, Fe 0-0.25, Ni 12, Pb 0-0.1, rem Zn.
Thin gauge sheet, various tempers. 57-114 ksi TS min; 21-102 ksi YS min. C76200

12-2 W, 12-3 W
Now GREEK ASCOLOY.

120WV4
Westig (U.K.) Ltd.
Alloy steel. C 1.2, Si 0.25, Mn 0.3, P 0-0.03, S 0-0.03, Cr 0.2, V 0.1, W 1, bal Fe.
Alloyed tool steel.

12KH13
Russian manufacture
C 0.09-0.15, Mn 0.8, Si 0.8, Cr 12-14, bal Fe.
Stainless steel; martensitic-ferritic types; similar to AISI 410.

12KH17
Russian manufacture
C 0-0.12, Mn 0.8, Si 0.8, Cr 16-18, bal Fe.
Ferritic type stainless steel; similar to AISI 430.

12KH17G9AN4
Russian manufacture
C 0-0.12, Mn 8-10.5, Si 0.8, Cr 16-18, Ni 3.5-4.5, N 0.15-0.25, bal Fe.
Austenitic stainless steel.

12KH18N10E
Russian manufacture
C 0-0.12, Mn 2, Si 0.8, Cr 17-19, Ni 9-11, Se, bal Fe.
Austenitic stainless steel; similar to AISI 303 Se.

12KH18N10T
Russian manufacture
C 0.12, Mn 2, Si 0.8, Cr 17-19, Ni 9-11, bal Fe.
Austenitic stainless steel.

12KH18N12T
Russian manufacture
C 0-0.12, Mn 2, Si 0.8, Cr 17-19, Ni 11-13, bal Fe.
Austenitic stainless steel.

12KH18N9
Russian manufacture
C 0-0.12, Mn 2, Si 0.8, Cr 17-19, Ni 8-10, bal Fe.
Austenitic stainless steel; similar to AISI 302.

12KH18N9T
Russian manufacture
C 0-0.12, Mn 2, Si 0.8, Cr 17-19, Ni 8-9.5, bal Fe.
Austenitic stainless steel.

12KH21N5T
Russian manufacture
C 0.09-0.14, Mn 0.8, Si 0.8, Cr 20-22, Ni 4.8-5.8, Al 0-0.08, Ti, bal Fe.
Austenitic-ferritic type stainless steel.

12KH25N16G7AR
Russian manufacture
C 0-0.12, Mn 5-7, Si 1, Cr 23-26, N 15-18, B 0-0.1, bal Fe.
Austenitic stainless steel.

12KH8VF
Russian manufacture
C 0.08-0.15, Mn 0.5, Si 0.6, Cr 7-8.5, W 0.6-1, bal Fe.
Martensitic stainless steel; air hardening.

13 CRMO 44
Thyssen Edelstahlwerke AG
C 0.13, Cr 0.85, Mo 0.45, Mn 0.55, bal Fe.
For cams, camshafts, bolts, shafts; case hardening, tough. *Obsolete*

13 MN
Kawasaki Steel Corp.
C 0.96, Mn 13, P 0.016, S 0.005, Si 0.53, bal Fe.
Austenitic 13% Mn for high abrasion resistance.

13A
Texas Instruments Inc./Materials Control
Ag 73.7-76, Cu 23-26, Ni 3-7.
Electrical contact material providing increased hardness over fine silver.

13KH14N3V2FR
Russian manufacture
C 0.1-0.16, Mn 0.6, Si 0.6, Cr 13-15, Ni 2.8-3.4, V 0.18-0.28, W 1.6-2.2, B 0-0.004, Ti, bal Fe.
Martensitic stainless steel.

14 NICR 74
Thyssen Edelstahlwerke AG
C 0.13, Cr 0.7, Ni 3.5, Mn 0.5, bal Fe.
Heat treated: 157,000-185,000 TS; 143,000-171,000 YS; 10-13 El. For gears, cams, camshafts, shafts; case hardening steel, tough. *Obsolete*

14-4 PH
American manufacture
C 0-0.07, Cr 13-15, Ni 3-5, Mo 2-3, Cu 2-5.
Precipitation hardening steel (Experimental).

14D
Texas Instruments Inc./Materials Control
Au 68.5-69.5, Ag 24.7-25.3, Pt 5.8-6.2.
Electrical contact material for low noise, stable contact applications.

14KH17N2
Russian manufacture
C 0.11-0.17, Mn 0.8, Si 0.8, Cr 16-18, Ni 1.5-2.5, bal Fe.
Stainless steel; martensitic-ferritic type.

15 CRNI 6
Thyssen Edelstahlwerke AG
C 0.15, Cr 1.5, Ni 1.5, bal Fe.
For gears, shafts, cams, camshafts; case hardening, tough.
Obsolete

15% NICKEL SILVER
Criterion Metals Inc.
Copper. Cu 65, Mn 0-0.5, Fe 0-0.25, Ni 15, Pb 0-0.1, rem Zn.
Thin gauge sheet, various tempers. 57-92 ksi TS min. C75400

15-5 PH
Now ARMCO 15-5 PH, REPUBLIC 15-5 PH, and JOSLYN ST.

15KH11MF
Russian manufacture
C 0.12-0.19, Mn 0.7, Si 0.5, Cr 10-11.5, Mo 0.6-0.8, V, bal Fe.
Martensitic stainless steel; air hardening.

15KH12VNMF
Russian manufacture
C 0.12-0.18, Mn 0.5-0.9, Si 0.4, Cr 11-13, W 0.7-1.1, V 0.15-0.3, Ni 0.4-0.8, Mo, bal Fe.
Stainless steel; martensitic-ferritic type.

15KH17AG14
Russian manufacture
C 0-0.15, Mn 13.5-15.5, Si 0.8, Cr 16-18, Ni, N, bal Fe.
Austenitic stainless steel.

15KH18N12S4TJU
Russian manufacture
C 0.12-0.17, Mn 0.5-1, Si 3.8-4.5, Cr 17-19, Al 0.13-0.35, Ni 11-13, Ti 0.4-0.7, bal Fe.
Austenitic-ferritic type stainless steel.

15KH18SJU
Russian manufacture
C 0-0.15, Mn 0.8, Si 1-1.5, Cr 17-20, Al 0.7-1.2, bal Fe.
Ferritic type stainless steel.

15KH25T
Russian manufacture
C 0-0.15, Mn 0.8, Si 1, Cr 24-27, Ti, bal Fe.
Ferritic type stainless steel.

15KH28
Russian manufacture
C 0-0.15, Mn 0.8, Si 1, Cr 27-30, bal Fe.
Ferritic type stainless steel.

15KH5
Russian manufacture
C 0-0.15, Mn 0.5, Si 0.5, Cr 4.5-6, bal Fe.
Martensitic stainless steel; air hardening.

15KH5M
Russian manufacture
C 0-0.15, Mn 0.5, Si 0.5, Cr 4.5-6, Mo 0.45-0.6, bal Fe.
Martensitic stainless steel; air hardening; similar to AISI 502.

15KH5VF
Russian manufacture
C 0-0.15, Mn 0.5, Si 0.3-0.6, Cr 4.5-6, W 0.4-0.7, bal Fe.
Martensitic stainless steel; air hardening.

15KH6SJU
Russian manufacture
C 0-0.15, Mn 0.5, Si 1.2-1.8, Cr 5.5-7, Al 0.7-1.1, bal Fe.
Stainless steel; martensitic-ferritic type.

16 MNCR 5
Thyssen Edelstahlwerke AG
C 0.16, Mn 1.2, Cr 1, bal Fe.
For gears, shafts, cams, camshafts; case hardened, tough.
Obsolete

16-15-6
Now TIMKEN 16-15-6.

16-25-6
Now TIMKEN 16-25-6.

16-6 PH
Now B & W CROLOY 16-6 PH.

16KH11N2V2MF
Russian manufacture
C 0.14-0.18, Mn 0.6, Si 0.6, Cr 10.5-12, Ni 1.4-1.8, W 1.6-2, V 0.18-0.3, Mo, bal Fe.
Martensitic stainless steel; air hardening.

17 PS TUNGSTEN
Manufacturer not listed.
2 ThO_2, bal W.
Filament wire.

17 W
American manufacture
C 0.5, Ni 19, Fe 63, Cr 13, Mo 1, W 2.
High temperature alloy.

17-10 P
Now ARMCO 17-10.

17-14 CU MO
Armco
C 0.12, Mn 0.75, Si 0.5, Cr 15.9, Ni 14.1, Mo 2.5, Cb 0.45, Cu 3, Ti, bal Fe.
Experimental. *Obsolete*

17-22 A
Now TIMKEN 17-22.

17-4 PH
Now ARMCO 17-4 PH AND JOSLYN STAINLESS TYPE 17-4.

17-4 PH, 17-7 PH
Now ARMCO, CRUCIBLE REPUBLIC, JOSLYN, etc., 17-4.

17-4-6
Now WSS 17-4-6.

17-7 PH WIRE
National Standard Co.. C 0-0.09, Mn 0-1, Cr 16-18, Si 0-1, N 6.5-7.75, Al 0.75-1.5, bal Fe.
As cold drawn (0.062 inch); 242,000 psi TS. Cond. "CH" (1 hr 900°) (0.062 inch); 287,000-327,000 psi TS. For stainless springs.

1741 CVM
Teledyne Vasco
C 1.3, Cr 17.5, Mo 4, V 1, Mn 0-0.3, bal Fe.
For high temperature bearings, pistons, rings and seals, cryogenic applications. Good wear and heat resistance in high temperature service. *Obsolete*

17KH18N9
Russian manufacture
C 0.13-0.21, Mn 2, Si 0.8, Cr 17-19, Ni 8-10, bal Fe.
Austenitic stainless steel.

18 CR-15 MN
Now USS TENELON.

18 CRNI 8
Thyssen Edelstahlwerke AG
C 0.18, Cr 2, Ni 2, bal Fe.
For gears, shafts, cams, camshafts; case hardened, tough.
Obsolete

18 SR
Now ARMCO 18 SR.

18% NICKEL SILVER (C73500)
Criterion Metals Inc.
Copper. Cu 72, Mn 0-0.5, Fe 0-0.25, Ni 18, Pb 0-0.1, rem Zn.
Thin gauge sheet, various tempers. 50-88 ksi TS min; 15-78 ksi YS min. C73500

18% NICKEL SILVER (C75200)
Criterion Metals Inc.
Copper. Cu 64.5-64.55, Mn 0-0.5, Fe 0-0.25, Ni 18, Pb 0-0.1, rem Zn.
Thin gauge sheet, various tempers. 53-96 ksi TS min; 18-95 ksi YS min. C75200

18% NICKEL SILVER (C77000)
Criterion Metals Inc.
Copper. Cu 55, Mn 0-0.5, Fe 0-0.25, Ni 18, Pb 0-0.1, rem Zn.
Thin gauge sheet, various tempers. 61-116 ksi TS min; 23-115 ksi YS min. C77000

18-18-2
Now USS 18-18-2.

18-2 MN
Now ARMCO 18-2 MN and ARMCO NITRONIC 32.

18-5-8
Now USS 18-5-8.

18-9 LW
Now ARMCO 19-9 LW.

18-9 LW 302 HQ
Techalloy Co. Inc.
C 0.1, Mn 2, Si 1, Cr 17-19, Ni 8-10, Mo 3-4, bal Fe.
Low work hardening austenitic stainless steel. For cold headed parts.

1845/4
Johnson Matthey plc
Cu 28, Zn 10, bal Ag.
Silver braze filler metal. 690-735°C MP.

1845/5
Johnson Matthey plc
Cu 37, Zn 19, bal Ag.
Silver braze filler metal. 700-775°C MP. *Obsolete*

18KH11MNFB
Russian manufacture
C 0.15-0.21, Mn 0.6-1, Si 0.6, Cr 10-11.5, Cb 0.2-0.45, Ni 0.5-1, Mo 0.8-1.1, bal Fe.
Martensitic stainless steel; air hardening.

18KH12VMBFR
Russian manufacture
C 0.15-0.22, Mn 0.5, Si 0.5, Cr 11-13, Mo 0.4-0.6, W 0.4-0.7, V 0.15-0.3, Cb 0.2-0.4, B 0-0.003, bal Fe.
Stainless steel; Martensitic-ferritic type.

19-9 DL, 19-9 DX
Now ALTEMP, CARPENTER, UNITEMP 19-9 DL, 19-9 DX.

19D
Texas Instruments Inc./Materials Control
Au 74.5-75.5, Ag 24.5-25.5.
Electrical contact material for low noise, stable contact applications.

2 AS
Acieries du Forez
High speed steel; similar to AISI M2.

20 CB-3
Now CARPENTER STAINLESS NO. 20CB-3.

20 MNCR 5
Thyssen Edelstahlwerke AG
C 0.2, Mn 1.3, Cr 1.2, bal Fe.
For gears, shafts, cams, camshafts; case hardened, tough.
Obsolete

20 PLUS
George Cook & Co., Ltd.
C 0.35, Mn 0.8, Si 0.5, Cr 1.7, Mo 0.4, bal Fe.
Supplied prehardened at 300 Brin. Machinable; for plastic molds, zinc diecasting dies, bolsters. May be nitrided for increased wear resistance.

2021
American manufacture
Cu 6.3, Mn 0.0, Ti 0.06, V 0.1, Zr 0.18, Cd 0.15, Sn 0.05, Al.
Precipitation hardenable wrought aluminum alloy. T 81
Cond: 73,000 TS; 63,000 YS; 9 El.

203 EZ
See AL TECH TYPE 203EZ.

20KH12VNMF
Russian manufacture
C 0.17-0.23, Mn 0.5-0.9, Si 0.6, Cr 10.5-12.5, W 0.7-1.1, V 0.15-0.3, Ni 0.5-0.9, Mo 0.5-0.7, bal Fe.
Martensitic stainless steel; air hardening.

20KH13
Russian manufacture
C 0.16-0.26, Mn 0.0, Si 0.8, Cr 12-14, bal Fe.
Martensitic stainless steel; air hardening; similar to AISI 420.

20KH13N4G9
Russian manufacture
C 0.15-0.3, Mn 8-10, Si 0.8, Cr 12-14, Ni 3.7-4.7, bal Fe.
Austenitic-martensitic type stainless steel.

20KH17N2
Russian manufacture
C 0.17-0.25, Mn 0.8, Sr 0.8, Cr 16-18, Ni 1.5-2.5, bal Fe.
Martensitic stainless steel; air hardening.

20KH20N14S2
Russian manufacture
C 0-0.2, Mn 1.5, Si 2-3, Cr 19-22, Ni 12-15, bal Fe.
Austenitic-ferritic type stainless steel.

20KH23N13
Russian manufacture
C 0-0.2, Mn 2, Si 1, Cr 22-25, Ni 12-15, bal Fe.
Stainless steel.

20KH23N18
Russian manufacture
C 0-0.2, Mn 2, Si 1, Cr 22-25, Ni 17-20, bal Fe.
Austenitic stainless steel.

20KH25N2052

Russian manufacture
C 0-0.2, Mn 1.5, Si 2-3, Cr 24-27, Ni 18-21, bal Fe.
Austenitic stainless steel.

21-2N

American manufacture
C 0.55, Mn 8.25, Si 0-0.25, Cr 20.35, Ni 2.1, N 0.3, S 0, bal Fe.
For diesel engine exhaust valves. SAE EV12.

21-4N

Manufacturer not listed.
C 0.53, Mn 9, Si 0.15, Cr 21, Ni 3.75, S 0.07, N 0.42, bal Fe.
1400°F: 62,000 psi TS; 37,000 psi YS; 18 El. Exhaust valve steel. SAE No. EV8.

21-6-9

Allegheny Ludlum Steel
C 0.04, Mn 9, Cr 20.5, Ni 6.5, N 0.3, bal Fe.
112,000 TS; 68,000 YS; 44 El. Austenitic, corrosion resistant steel. See also Armco 21-6-9 and Carpenter Stainless 21-6-9.

21-6-9

Bishop Tube Co.
Cr 21, Ni 6, Mn 9, bal Fe.
Annealed: 115,000 psi TS; 70,000 YS; 44 El; 94 Rock B. 30% cold worked: 175,000 psi TS; 150,000 psi YS; 12 El; 32 Rock C. Welded tubing.

21/4/N

Now FIRTH-BROWN 21/4/N and FIRTH-VICKERS 21/4.

212MN

Acciaierie Valbruna s.p.a.
Stainless Steel. C 0.55, Si 0.25, Mn 8.25, P 0-0.04, S 0-0.04, Cr 21, Ni 2.4, N 0.35, bal Fe.
Valve steel. W. Nr. 1.4875.

214MN

Acciaierie Valbruna s.p.a.
Stainless Steel. C 0.53, Si 0-0.25, Mn 9, P 0-0.04, S 0-0.035, Cr 21, Ni 3.85, N 0.45, bal Fe.
Valve steel. W. Nr. 1.4871.

214MNC

Acciaierie Valbruna s.p.a.
Stainless Steel. C 0.53, Si 0-0.25, Mn 9, P 0-0.04, S 0-0.035, Cr 21, Ni 3.85, N 0.45, Nb + Ta = 2.50, bal Fe.
Valve Steel.

2155N

Carpenter Technology Corp.
C 0.2, Mn 5, Si 0.5, Cr 21, Ni 4.5, N 0.3, bal Fe.
For exhaust valves. SAE EV-7. See also Carpenter 21-55N. *Obsolete*

22-13-5

Now ARMCO NITRONIC 50.

22-4-9

Now ARMCO 22-4-9.

24 CRMO 5

Thyssen Edelstahlwerke AG
C 0.24, Cr 1.15, Mo 0.25, Mn 0.55, bal Fe.
For bolts, hardware, shafts, rams; water hardened. *Obsolete*

245 ALLOY

Driver Harris Co.
Nickel. Ni 75, Cr 20, Si 1, Al 5.
For heating elements from 2000-2300°F. *Obsolete*

25-20-SI

Now USS 25-20-SI.

25/20 HIGH CARBON STAINLESS WELD FILLER

Telcon Metals Ltd.
C 0.4, Cr 25, Ni 20, bal Fe.
For TIG and MIG welding.

25KH13N2

Russian manufacture
C 0.2-0.3, Mn 0.8-1.2, P 0.08-0.15, S 0.15-0.25, Si 0.5, Cr 12-14, Ni 1.5-2, bal Fe.
Free-machining, martensitic stainless steel; air hardening. Similar to AISI 420F.

29-4

American manufacture
C 0.01, N 0.01, Cr 29, Mo 4, bal Fe.
For high temperature equipment.

2V PERMENDUR

Teledyne Vasco
Fe 49, Co 49, V 2.
Magnetically soft Fe-Co alloy, either air melted or vacuum arc melted. Cores for power transformers, pulse transformers, magnetic amplifiers, etc. Maximum permeability: air melt, 3000-4500; vacuum melt; 11,000.

30 CRNIMO 8

Thyssen Edelstahlwerke AG
C 0.3, Cr 0.2, Mo 0.2, Ni 2, bal Fe.
For gears, shafts, machinery parts; shock resistant, oil hardened. *Obsolete*

300 M

Now REPUBLIC 300 M.

304 N

Now JOSLYN STAINLESS 304.

30KH13

Russian manufacture
C 0.26-0.35, Mn 0.8, Si 0.8, Cr 12-14, bal Fe.
Martensitic stainless steel; air hardening. Similar to AISI 420.

30KH13N7S2

Russian manufacture
C 0.25-0.34, Mn 0.8, Si 2-3, Cr 12-14, Ni 6-7.5, bal Fe.
Stainless steel.

310 STAINLESS ELECTRODE

J.W. Harris Co., Inc.
All purpose stainless steel electrode, AC-DC, reverse polarity. AWS AS.4 E310. For maintenance repairs on steels, stainless, and clad steels.

312 SUPER BLUE

J.W. Harris Co., Inc.
Nickel chrome high strength electrode, AC-DC. AWS AS.4 E312. For repairs of practically all steels.

31A

Texas Instruments Inc./Materials Control
Ag 99.5-99.9, Ni 0.1-0.5.
Electrical contact material providing increased hardness over fine silver.

31KH19N9MVBT

Russian manufacture
C 0.28-0.35, Mn 0.8-1.5, Si 0.8, Cr 18-20, Ni 8-10, Ti 0.2-0.5, Cb 0.2-0.5, V 1-1.5, bal Fe.
Austenitic stainless steel.

34 CR 4

Thyssen Edelstahlwerke AG
C 0.33, Mn 0.65, Cr 1, Si 0.25, bal Fe.
For fasteners, gears, shafts; water or oil hardened. *Obsolete*

34 CRNIMO 6

Thyssen Edelstahlwerke AG
C 0.34, Cr 1.5, Mo 0.15, Ni 1.5, bal Fe.
For gears, shafts, machinery parts; shock resistant, oil hardened. *Obsolete*

34 SP

Creusot-Loire
C 0.32, Mn 0.35, Si 0.25, Ni 22, Cr 2.2, Mo 0.7, bal Fe.
Oil hardenable to 1220 N/mm² psi TS min. For shafts, bolts, levers. AFNOR 32 CND8.

355

Fansteel/Wellman Dynamics
Aluminum. Cu 1-1.5, Si 4.5-5.5, Fe 0.6, Mn 0.5, Zn 0.35, Mg 0.4-0.6, Ti 0.25, bal Al.
Aluminum casting alloy. Cast: 25,000-32,000 psi TS; 18,000-22,000 psi YS; 2 El.

356

Fansteel/Wellman Dynamics
Aluminum. Cu 0.25, Si 6.5-7.5, Fe 0.6, Mn 0.35, Zn 0.35, Mg 0.2-0.4, Ti 0.25, bal Al.
Aluminum casting alloy. Cast: 23,000-30,000 psi TS; 16,000-20,000 psi YS; 3 El.

36 CRNIMO 4

Thyssen Edelstahlwerke AG
C 0.36, Cr 1, Mo 0.15, Ni 1, bal Fe.
For gears, shafts, machinery parts; shock resistant, oil hardened. *Obsolete*

368 SPECIAL

Ackerlind Steel Co., Inc.
C 0.4, Cr 3.3, Mo 1.3, W 2.5, V 1.3, bal Fe.
Hot work tool steel for forming cooper, brass and steel.

3693 A

Creusot-Loire
C 0.28, Cr 1.35, Mo 0.75, V 0.3, B, bal Fe.
Quenched and tempered: 830-930 N/mm² psi TS; 735 N/mm² YS min; El. For high temperature bolts. AFNOR 28 CDV 8-08; ASTM A 193 Gr. B 16.

36KH18N25S2

Russian manufacture
C 0.32-0.4, Mn 1.5, Si 2-3, Cr 17-19, Ni 23-26, bal Fe.
Austenitic stainless steel.

37 MNSI 5

Thyssen Edelstahlwerke AG
C 0.37, Si 1.25, Mn 1.25, bal Fe.
For punches, upsetters, crimpers; oil hardened, tough. *Obsolete*

37KH12N8G8MFB

Russian manufacture
C 0.34-0.4, Mn 7.5-9.5, Si 0.3-0.8, Cr 11.5-13.5, Ni 7-9, Mo 1.1-1.4, Cb 0.25-0.4, V 1.25-1.55, bal Fe.
Austenitic stainless steel.

4 BEST (FRASSE)

Peter A. Frasse & Co.
C, bal Fe.
For tools. *Obsolete*

4-6 CHROME

National Supply Co.
Cr 4-6, C 0.1, Ti = 4 to 6 x C, bal Fe.
Annealed: 66,000 TS; 39,000 YS; 45 El; 143 Brin. For oil stills and refinery tubes; corrosion resistant.

401 COPPER-NICKEL ALLOY

Now MONEL ALLOY 401.

40HK15N7G7F2MS

Russian manufacture
C 0.38-0.47, Mn 6-8, Si 0.9-1.4, Cr 14-16, Ni 6-8, Mo 0.65-0.95, V 1.5-1.9, bal Fe.
Austenitic stainless steel.

40KH10S2M
Russian manufacture
C 0.35-0.45, Mn 0.8, Si 1.9-2.6, Cr 9-10.5, Mo 0.7-0.9, bal Fe.
Martensitic stainless steel; air hardening.

40KH13
Russian manufacture
C 0.36-0.45, Mn 0.8, Si 0.8, Cr 12-14, bal Fe.
Martensitic stainless steel; air hardening.

40KH9S2
Russian manufacture
C 0.35-0.45, Mn 0.8, Si 2-3, Cr 8-10, bal Fe.
Martensitic stainless steel; air hardening.

4140 FLAME DIE
U.S. Metalsource
C 0.36-0.44, Mn 0.7-1, Si 0.15-0.4, Cr 0.8-1.15, Mo 0.15-0.25,
bal Fe.
Annealed: 90-100 ksi TS; 55-70 ksi YP; 17-25 El; 45-55 RA;
179-205 Brin. For tooling and die applications.

42
Now BISHOP 42.

42 CRMO 4
Thyssen Edelstahlwerke AG
C 0.42, Cr 1.1, Mo 0.2, bal Fe.
For gears, bolts, shafts, axles; oil or water hardened.
Obsolete

420 MFQ STAINLESS PLASTIC MOULD STEEL
Eagle & Globe Steel Ltd.
Stainless steel. C 0.35, Mn 0-1, Si 0-1, Cr 13, bal Fe.
Annealed: 200-220 Brin. For injection, compression and
transfer molding and for the production of glass molds.
AS1444 420; AISI 420; BS970 Part 4 420S45.

45 ALUMINUM-55 MOLYBDENUM-TITANIUM
Reading Alloys, Inc.
Reactive/refractive. Al 40-45, C 0.1, Mo 50-55, Ni 0.004, Si
0.3, Ti 2-5.
Master alloy.

45KH14N14V2M
Russian manufacture
C 0.4-0.5, Mn 0.7, Si 0.8, Cr 13-15, Ni 13-15, Mo 0.25-0.4, W
2-2.8, bal Fe.
Austenitic stainless steel.

45KH22N4M3
Russian manufacture
C 0.4-0.5, Mn 0.85-1.25, Si 0.7-1, Cr 21-23, Ni 4-5, Mo 2.5-3,
bal Fe.
Austenitic stainless steel.

47 CR 4
Thyssen Edelstahlwerke AG
C 0.4, Cr 1.1, bal Fe.
For gears, cams, shafts; oil hardened. *Obsolete*

476
Sanderson Kayser Ltd.
C 1.55, Cr 12, Mo 0.85, V 0.28, bal Fe.
Cold work tool steel. B.S. 4659 Type BD2; AISI D2.

495
Now AMPCOLOY 495.

5 CR MO V
Now HALCOMB 218, AISI H11.

5-317
Now CARPENTER NO. 5-317.

50 CRMO 4
Thyssen Edelstahlwerke AG
C 0.5, Cr 1, Mo 0.2, Mn 0.65, bal Fe.
For gears, bolts, shafts, axles; water hardened. *Obsolete*

50 CRV 4
Thyssen Edelstahlwerke AG
C 0.5, Mn 0.85, Cr 1, V 0.1, bal Fe.
For bolts, springs, shafts, gears; oil hardened, shock
resistant. *Obsolete*

52
Now BISHOP 52.

52 CB BEARING STEEL
Now CRUCIBLE 52 CB BEARING STEEL.

52NI-FE
Now WESTINGHOUSE 52NI-FE.

55KH20G9AN4
Russian manufacture
C 0.5-0.6, Mn 8-10, Si 0.45, Cr 20-22, Ni 3.5-4.5, N 0.3-0.6,
bal Fe.

57 HW
Bethlehem Steel Corp.
C 0.35, Cr 3.25, V 0.3, W 9, bal Fe.
For dies, trimmers, pliers, punches, shear blades, forming
dies; hot-work tool steel. *Obsolete*

57 SPECIAL
Now BETHLEHEM 57 SPECIAL.

58 CRV 4
Thyssen Edelstahlwerke AG
C 0.58, Mn 0.95, Cr 1, V 0.09, bal Fe.
For springs, punches, forging dies; oil hardened, tough.

6 X
American manufacture
C 0.025, Cr 20.25, Ni 24.25, Mo 6.25.
High temperature alloy.

60 CASE
Thomson Industries Inc.
C 0.6, bal Fe.
Heat treated: 600 Brin. For shafts, rolls, piston rods, axles.

62 BE-38AL
Now LOCKALLOY.

6542
George Cook & Co., Ltd.
C 0.85, Cr 4, Mo 5, W 6.25, V 2, bal Fe.
Mo-W high speed steel. For drills, lathe tolls, form tools. BS
4659 BM 2; AISI M2.

67 CHISEL
Bethlehem Steel Corp.
C 0.5, W 2.5, Cr 1.15, Si 0.75, bal Fe.
Heat treated: 185,000-337,000 TS; 180,000-255,000 YS; 9-14
El; 420-670 Brin. For chisels, rivet sets, pneumatic tools,
upsetters; Type S1; shock resistant. *Obsolete*

70 LA-2
Chemetron Corp.
Weld metal: 0.07 C, 0.90 Mn, 0.50 Si, bal Fe. As welded:
79,000 psi TS; 69,500 psi YS; 28 El. Lime coated, AC-DC,
reverse polarity electrode for general purpose welding and
repair. AWS Class E7016.

70-30 ALLOY
Harrison Alloys Inc.
Nickel.
Now HAI-NICR 70.

70/30 COPPER
Telcon Metals Ltd.
For MIG and TIG welding.

70/30 CUPRO NICKEL
Criterion Metals Inc.
Copper. Cu 68.9, Zn 0-1, Mn 0.6, Fe 0.5, Ni 30, Pb 0-0.05.
Thin gauge sheet, various tempers: 55-84 ksi TS min. C71500

70/30 HIGH BRASS
Criterion Metals Inc.
Copper. Cu 70, Zn 30, Fe 0-0.05, Pb 0-0.07.
Thin gauge sheet, various tempers: 40-95 Rock B; 45-95 ksi
TS min; 10-86 ksi YS min (0.2% offset). C26000

7031 M
Creusot-Loire
C 0-0.6, Mn 0-0.6, Si 0-0.5, Cu 0-0.6, Fe 0-0.6, bal Ni.
Cast, annealed: 350 N/mm^2 psi TS min. Resists potash and
soda in all concentrations. ACI CZ 100.

71 ALLOY
Bethlehem Steel Corp.
C 0.6, Ni 0.25, Mn 0.9, Cr 0-0.25, Si 2, bal Fe.
For tools, hand and pneumatic chisels, shear blades,
punches; resists shock. *Obsolete*

713 C
Now HAYNES ALLOY NO. 713 C.

7152
Teledyne Vasco
C 1.6-1.8, Cr 17-17.5, bal Fe.
For pump parts, bearings, pivots; corrosion and wear
resistant.

718
Now INCONEL ALLOY 718 and ALLVAC 718. ser Na

718 ALUM
J.W. Harris Co., Inc.
Aluminum brazing rod. For use on light gage aluminum
sheet, tubing, furniture, trucks, trailers. AWS AS.10 ER 4043.

8-N-2
Teledyne Vasco
C 0.72-0.81, W 1.3-1.8, Cr 3.5-4, Mo 8-9.2, V 0.9-1.3.
For cutting tools, reamers, taps, broaches, drills; high speed
steel. Type M1.

8-N-2 COBALT
Teledyne Vasco
C 0.7, Mo 8, V 0.5, Cr 4, W 1, Co 5, bal Fe.
For lathe and planer tools, reamers, drills, taps; high speed
steel. Type M30.

80-20 ALLOY
Now CHROMELA NICHROME V.

80/20 CUPRO NICKEL
Criterion Metals Inc.
Copper. Cu 80, Zn 0-1, Mn 0.4, Fe 0.8, Ni 20, Pb 0-0.05.
Thin gauge sheet, various tempers: 47-76 ksi TS min. C71000

80/20 LOW BRASS
Criterion Metals Inc.
Copper. Cu 80, Zn 20, Fe 0-0.05, Pb 0-0.05.
Thin gauge sheet, various tempers: 38-93 Rock B; 44-89 ksi
TS min; 12-73 ksi YS min (0.2% offset). C24000

85/15 RICH LOW BRASS
Criterion Metals Inc.
Copper. Cu 85, Zn 15, Fe 0-0.05, Pb 0-0.05.
Thin gauge sheet, various tempers: 33-89 Rock B; 39-82 ksi
TS min; 8-73 ksi YS min (0.2% offset). C23000

9% CR MO
Walsingham Steel Co., Ltd.
C 0-0.2, Si 0-1, Mn 0.3-0.7, Ni 0-0.4, Cu 0-0.4, Cr 8-10, Mo
0.9-1.2, bal Fe.
For use at temperatures above 400°C. Meets BS 1463; ASTM
A217-65 Gr.C12.

90 CASTING ALLOY

Johnson Matthey plc
Ag 90.
High silver casting; splint silver alloy.

90/10 COMMERCIAL BRONZE

Criterion Metals Inc.
Copper. Cu 90, Zn 10, Fe 0-0.05, Pb 0-0.05.
Thin gauge sheet, various tempers: 27-83 Rock B; 36-72 ksi
TS min; 8-67 ksi YS min (0.2% offset). C22000

90/10 CUPRO NICKEL

Criterion Metals Inc.
Copper. Cu 88.35, Zn 0-1, Mn 0.4, Fe 1.25, Ni 10, Pb 0-0.05.
Thin gauge sheet, various tempers: 43-78 ksi TS min; 13-78
ksi YS min. C70600

9086 M

Creusot-Loire
C 0-0.12, Cr 15.5-17.5, Mo 16-18, V 0.2-0.4, Fe 4-7, W
3.75-5.25, bal Ni.
Cast, annealed: 500 N/mm^2 psi TS min. Good resistance to
acids and hyperchlorites. ACI CW 12 M (Hastelloy C).

9097 M

Creusot-Loire
C 0-0.12, Cr 0-1, Mo 26-30, V 0.2-0.6, Fe 4-6, bal Ni.
Cast, annealed: 500 N/mm^2 psi TS min. Good resistance to
acids. ACI N-12M. (Hastelloy B).

9098 M

Creusot-Loire
C 0-0.1, Si 8-10, Cu 2-4, Fe 0-2, bal Ni.
Cast stainless alloy. Resists sulfuric acid in all concentrations;
also organic acids. (Hastelloy D).

9128 M

Creusot-Loire
C 0-0.3, Mn 0-1.5, Si 0-1.5, Cu 26-33, bal Ni.
Cast, annealed: 450 N/mm^2 psi TS min. For valves, pumps
for marine applications. ACI M 35 (Monel 400).

92/8

N.C. Ashton Ltd.
Al 8, Fe 0.25, bal Cu.
Weld metal: 25 tsi TS; 50 El. Welding wire for automatic
processes.

930 ALLOY

N.C. Ashton Ltd.
Al 6.25, Fe 0.75, Si 2.25, bal Cu.
Cast or wrought aluminum bronze. DGS 8453.

95/5 GILDING METAL

Criterion Metals Inc.
Copper. Cu 95, Zn 5, Fe 0-0.05, Pb 0-0.03.
Thin gauge sheet, various tempers: 20-76 Rock B; 34-61 ksi
TS min; 5-56 ksi YS min (0.2% offset). C21000

95KH18

Russian manufacture
C 0.9-1, Mn 0.8, Si 0.8, Cr 17-19, bal Fe.
Martensitic stainless steel; air hardening. Similar to AISI 440C.

98M2

Now HAYNES STELLITE 98M2 ALLOY.

9W18CR4V

China Metallurgical Import&Export Corp.
Tool steel. C 0.9-1, W 17.5-19, Mo 0-0.3, Cr 3.8-4.4, V 1-1.4,
Si 0.2-0.4, Mn 0.2-0.4, S 0-0.03, P 0-0.03, bal Fe.
Quenched and tempered: 66-68 Rock C. Suitable for easy-
to-grind, high hardness cutting tools that need medium
toughness for hard-to-cut materials.

9W6MO5CR4V2

China Metallurgical Import&Export Corp.
Tool steel. C 0.95-1.05, W 5.5-6.75, Mo 4.5-5.5, Cr 3.8-4.4, V
1.75-2.2, Si 0.2-0.4, Mn 0.2-0.4, S 0-0.03, P 0-0.03, bal Fe.
Quenched and tempered: 66-68 Rock C. Used for high
hardness cutting tools which need definite toughness for
hard-to-cut materials. AISI M2, high C.

A 1050
Comalco Ltd.
99.50 Al min.
H12-temper: 97 MPa TS; 83 MPa YS; 15 El. H18-temper: 145 MPa TS; 138 MPa YS; 6 El. Chemical and process plant and equipment.

A 1100
Comalco Ltd.
Cu 0.12, 99.0 Al min.
O-temper: 90 MPa TS; 34 MPa YS; 35 El. H18-temper: 165 MPa TS; 152 MPa YS; 5 El; 44 Brin. Spinnings, holloware and general sheet metal work.

A 1145
Comalco Ltd.
99.45 Al min.
General foil uses.

A 1199
Comalco Ltd.
99.99 Al min.
Electrical and electronic foil uses.

A 2011
Comalco Ltd.
Cu 5.5, Pb 0.5, Bi 0.5, bal Al.
T3-temper: 379 MPa TS; 296 MPa YS; 15 El; 110 Vickers. T4-temper: 310 MPa TS; 145 MPa YS; 20 El. Screw machine products not requiring decorative anodizing.

A 2024
Comalco Ltd.
Cu 4.5, Mn 0.6, Mg 1.5, bal Al.
O-temper: 166 MPa TS; 76 MPa YS; 20 El. T42 temper: 469 MPa TS; 310 MPa YS; 20 El; 120 Brin. Aircraft sheeting.

A 25
Now MATTHEY A 25.

A 300
Bergische Stahl Industrie
C 0.4-0.5, Si 0-1, Mn 0-1, Cr 12-14, bal Fe.
Martensitic stainless steel. W.-Nr. 1.4034.

A 3003
Comalco Ltd.
Mn 1.2, Cu 0.12, bal Al.
O-temper: 110 MPa TS; 41 MPa YS; 30 El. H18-temper: 200 MPa TS; 186 MPa YS; 4 El; 55 Brin. Chemical equipment, sheet metal work, rigid foil containers.

A 3004
Comalco Ltd.
Mn 1.2, Mg 1, bal Al.
O-temper: 179 MPa TS; 69 MPa YS; 20 El. H38-temper: 283 MPa TS; 248 MPa YS; 5 El; 70 Brin. Storage tanks, car bodies, seam welded tubing.

A 3203
Comalco Ltd.
Mn 1.2, 0-1.36911e+038, CO 8.27544e+008-0.000489523, H1 131-4.54451e+030, MP 0, fo 0, CO 8.90405e+011-3.37774e+009, ba 0-1.06483e+037.

A 350
Bergische Stahl Industrie
C 0.17-0.22, Si 0-1, Mn 0-1, Cr 12-14, bal Fe.
Martensitic stainless steel. W.-Nr. 1.4021.

A 400
Bergische Stahl Industrie
C 0-0.08, Si 0-1, Mn 0-1, Cr 13-15.
Ferritic type stainless steel. W.-Nr. 1.4001.

A 4543
Comalco Ltd.
Si 6, Mg 0.3, bal Al.
T1-temper: 145 MPa TS; 83 MPa YS; 20 El; 40 Vickers. T8-temper: 207 MPa TS; 179 MPa YS; 12 El; 75 Vickers. Architectural extrusions.

A 5
Now MATTHEY A 5.

A 500 H
Bergische Stahl Industrie
C 0.75-0.85, Si 0.3-0.6, Mn 0.2-0.5, Co 0.9-1.3, Cr 12.5-14.5, bal Fe.
Cold work tool steel. W.-Nr. 1.2883.

A 5005
Comalco Ltd.
Mg 0.8, bal Al.
O-temper: 124 MPa TS; 41 MPa YS; 25 El. H18-temper: 200 MPa TS; 193 MPa YS; 4 El; 56 Vickers. H38-temper: 200 MPa TS; 186 MPa YS; 5 El; 51 Brin. Appliances and utensils, high strength foil.

A 5052
Comalco Ltd.
Mg 2.5, Cr 0.25, bal Al.
O-temper: 193 MPa TS; 90 MPa YS; 25 El; 47 Brin. H38-temper: 290 MPa TS; 255 MPa YS; 7 El; 77 Brin. Sheet metal work, marine applications.

A 5056
Comalco Ltd.
Mg 5.2, Mn 0.1, Cr 0.1, bal Al.
O-temper: 290 MPa TS; 152 MPa YS; 35 El; 70 Vickers. H38-temper: 414 MPa TS; 345 MPa YS; 15 El; 100 Brin. Cable sheathing, rivets, zippers, screen wire.

A 5086
Comalco Ltd.
Mg 4, Mn 0.5, Cr 0.15, bal Al.
O-temper: 262 MPa TS; 117 MPa YS; 22 El; 60 Brin. H38-temper: 359 MPa TS; 303 MPa YS; 7 El. Unfired pressure vessels, TV towers.

A 5252
Comalco Ltd.
Mg 2.5, bal Al.
H24 temper: 221 MPa TS; 159 MPa YS; 12 El. H28 temper: 283 MPa TS; 241 MPa YS; 5 El; 75 Brin. High strength automobile trim.

A 5454
Comalco Ltd.
Mg 2.7, Mn 0.8, Cr 0.1, bal Al.
O-temper: 248 MPa TS; 117 MPa YS; 22 El; 62 Brin. H34-temper: 303 MPa TS; 241 MPa YS; 10 El; 81 Brin. Welded structures, pressure vessels.

A 5457
Comalco Ltd.
Mg 1, Mn 0.2, Cu 0.1, bal Al.
O-temper: 131 MPa TS; 48 MPa YS; 22 El; 32 Brin. H25-temper: 179 MPa TS; 159 MPa YS; 12 El; 48 Brin. Automobile trim.

A 5557
Comalco Ltd.
Mg 0.6, Mn 0.2, Cu 0.1, bal Al.
O-temper: 110 MPa TS; 41 MPa YS; 25 El. H25-temper: 159 MPa TS; 138 MPa YS; 12 El. Automobile trim.

A 60 PB
Westig (U.K.) Ltd.
C 0.7, Si 0.2, Mn 0.55, P 0-0.035, S 0.2, Pb 0.2, bal Fe.
Free cutting steel.

A 600
Bergische Stahl Industrie
C 1-1.1, Si 0.15-0.3, Mn 0.8-1.1, Cr 0.9-1.1, W 1-1.3, bal Fe.
Cold work tool steel. W.-Nr. 1.2419.

A 6061
Comalco Ltd.
Mg 1, Si 0.6, Cu 0.25, Cr 0.2, bal Al.
O-temper: 124 MPa TS; 55 MPa YS; 25 El; 30 Brin. T6 temper: 310 MPa TS; 276 MPa YS; 12 El; 95 Brin. Structural applications, transport.

A 6262
Comalco Ltd.
Mg 1, Si 0.6, Cu 0.25, Cr 0.1, Bi 0.6, Pb 0.6, bal Al.
T6-temper: 310 MPa TS; 276 MPa YS; 12 El; 100 Vickers. T9-temper: 400 MPa TS; 379 MPa YS; 10 El; 125 Vickers. Screw machine products suitable for decorative anodizing.

A 70
Now CRUCIBLE A 70.

A 8
English manufacture
Al 7.5-9, Zn 0.3-1, Mn 0.15-0.4, bal Mg.
Sand cast: 140 MPa TS; 85 MPa YS; 2 El; 55 Brin. General purpose sand or chill casting. BS 2970 MAG 1.

A 8 H.P
English manufacture
Al 7.5-9, Zn 0.3-1, Mn 0.15-0.7, bal Mg.
Sand cast: 140 MPa TS; 85 MPa YS; 2 El; 55 Brin. Higher purity sand or chill casting. BS 2970 MAG 2.

A ALLOY-1
National Physical Laboratory
Aluminum. Al 77, Zn 20, Cu 3.
Hot rolled: 60,000 TS; 41,000 YS; 21 El; 36 RA. For light alloy wrought parts; non-hardenable.

A ALLOY-2
National Physical Laboratory
Aluminum. Cu 4.7, Mg 1.34, Ni 1.85, Pb 1.5, Fe 0.5, bal Al.
For light alloy wrought parts; age-hardenable.

A L 601
Allegheny Ludlum Steel
C 0-0.15, Cr 14, Ni 72, Cb.
Solid solution with excellent high temperature properties. UNS 6601.

A L 625
Allegheny Ludlum Steel
Now ALTEMP 625.

A L 718
Allegheny Ludlum Steel
Now ALTEMP 718.

A L 75
Sterling International Technology Ltd.
Cu 2.8-3.8, Cr 4-6, Mn 0.3-0.6, bal Al.
Chill cast: 36,000-40,000 TS; 14,500-16,800 YS; 8-10 El; 70-80 Brin; 3.5 Izod. Gravity die aluminum castings requiring strength and ductility. BS 1490 LM 22-W.

A L 800
Allegheny Ludlum Steel
C 0.1, Cr 19, Ni 30, Al 0-0.6, Ti 0-0.6, bal Fe.
Resistant to oxidation and carburization at elevated temperatures. N08800

A L 800
AL Tech Specialty Steel Corp.
C 0.1, Cr 19, Ni 30, Al 0-0.6, Ti 0-0.6, bal Fe.
Resistant to oxidation and carburization at elevated temperatures. N08800

A L 825
Allegheny Ludlum Steel
C 0-0.05, Cr 19.5, Ni 38, Mo 0-3.5, bal Fe.
Excellent corrosion resistance. N08825

A L 825
AL Tech Specialty Steel Corp.
C 0-0.05, Cr 19.5, Ni 38, Mo 0-3.5, bal Fe.
Excellent corrosion resistance. N08825

A L-602
Allegheny Ludlum Steel
C 0.5, Mn 0.7, Si 1.7, V 0.12, Mo 0.4, bal Fe.
AISI Type S5 shock resisting tool steel. *Obsolete*

A M 3 DIE STEEL
Edgar Allen Balfour Ltd.
C 0.4, Si 1.05, Cr 5, Mo 1.35, V 1.1, bal Fe.
For die casting dies, extrusion dies, upsetters; air hardened, resists heat checking.

A METAL
Manufacturer not listed.
Cu 5-7, Ni 44, bal Fe.
For sound transmitting devices.

A NICKEL ELECTRON GRADE
Now NICKEL 205.

A PHOSPHOR BRONZE
Criterion Metals Inc.
Copper. Cu 94.82, Zn 0-0.3, Sn 5, P 0.05, Fe 0-0.1, Pb 0-0.05.
Thin gauge sheet, various tempers: 46-100 ksi TS min; 19-98 ksi YS min. ASTM B-103. C51000

A QUALITY
Eagle & Globe Steel Ltd.
Tool material. C 0.73, Mn 0.2, V 0.2, Si 0.2, bal Fe.
Carbon, water hardened for punches, scale pivots. AS1239 W2A-7; AISI W2.

A-10 SPB
AFORA
Alloy steel. C 0.05-0.15, Mn 1-1.5, Si 0-0.06, P 0-0.07, S 0.3-0.4, Pb 0.15-0.3, bal Fe.
Normalized at 900°C (16 mm): 39-52 kgf/mm^2 TS; 22 kgf/mm^2 min YS; 20 min El. Cold drawn (<16 mm): 52-82 kgf/mm^2 TS; 42 kgf/mm^2 min YS; 7 min El. Cold drawn (<40 mm): 47-77 kgf/mm^2 TS; 38 kgf/mm^2 min YS; 8 min El. Cold drawn (<63 mm): 42-72 kgf/mm^2 TS; 31 kgf/mm^2 min YS; 9 min El.

A-10S
AFORA
Alloy steel. C 0.05-0.15, Mn 1-1.5, Si 0-0.06, P 0-0.07, S 0.3-0.4, bal Fe.
Normalized at 900°C (16 mm): 39-52 kgf/mm^2 TS; 22 kgf/mm^2 min YS; 20 min El. Cold drawn (<16 mm): 52-82 kgf/mm^2 TS; 42 kgf/mm^2 min YS; 7 min El. Cold drawn (<40 mm): 47-77 kgf/mm^2 TS; 38 kgf/mm^2 min YS; 8 min El. Cold drawn (<63 mm): 42-72 kgf/mm^2 TS; 31 kgf/mm^2 min YS; 9 min El.

A-110
Heppenstall Co.
Tool material. C 1.2, Mn 0.3, Si 0.3, bal Fe.
Water hardened tool steel; AISI W1.

A-1110
AFORA
Carbon steel. C 0.1-0.2, Mn 0.4-0.7, Si 0.15-0.4, P 0-0.35, S 0-0.35, bal Fe.
Water quenched and tempered: 45-65 kgf/mm^2 TS; 30 kgf/mm^2 min YS; 20 min El.

A-1120
AFORA
Carbon steel. C 0.2-0.3, Mn 0.5-0.8, Si 0.15-0.4, P 0-0.035, S 0-0.035, bal Fe.
Water quenched and tempered: 54-74 kgf/mm^2 TS; 36 kgf/mm^2 min YS; 19 min El.

A-1130
AFORA
Carbon steel. C 0.3-0.4, Mn 0.5-0.8, Si 0.15-0.4, P 0-0.035, S 0-0.035, bal Fe.
Water quenched and tempered: 63-83 kgf/mm^2 TS; 43 kgf/mm^2 min YS; 16 min El.

A-1131
AFORA
Carbon steel. C 0.3-0.35, Mn 0.5-0.8, Si 0.15-0.4, P 0-0.035, S 0-0.035, bal Fe.
Water quenched and tempered: 63-78 kgf/mm^2 TS; 43 kgf/mm^2 min YS; 17 min El.

A-1132
AFORA
Carbon steel. C 0.35-0.4, Mn 0.5-0.8, Si 0.15-0.4, P 0-0.035, S 0-0.035, bal Fe.
Water quenched and tempered: 68-83 kgf/mm^2 TS; 45 kgf/mm^2 min YS; 16 min El.

A-1140
AFORA
Carbon steel. C 0.4-0.5, Mn 0.5-0.8, Si 0.15-0.4, P 0-0.035, S 0-0.035, bal Fe.
Water quenched and tempered: 71-91 kgf/mm^2 TS; 48 kgf/mm^2 min YS; 13 min El.

A-1141
AFORA
Carbon steel. C 0.4-0.45, Mn 0.5-0.8, Si 0.15-0.4, P 0-0.035, S 0-0.035, bal Fe.
Water quenched and tempered: 71-86 kgf/mm^2 TS; 48 kgf/mm^2 min YS; 14 min El.

A-1142
AFORA
Carbon steel. C 0.45-0.5, Mn 0.5-0.8, Si 0.15-0.4, P 0-0.035, S 0-0.035, bal Fe.
Water quenched and tempered: 76-91 kgf/mm^2 TS; 52 kgf/mm^2 min YS; 13 min El.

A-1150
AFORA
Carbon steel. C 0.5-0.6, Mn 0.6-0.9, Si 0.15-0.4, P 0-0.035, S 0-0.035, bal Fe.
Water quenched and tempered: 80-100 kgf/mm^2 TS; 54 kgf/mm^2 min YS; 12 min El.

A-1200
AFORA
Alloy steel. C 0.34-0.41, Mn 0.6-0.9, Si 0.15-0.4, P 0-0.035, S 0-0.035, Cr 0.5-0.8, bal Fe.
85-105 kgf/mm^2 TS; 60 kgf/mm^2 min YS; 12 min El.

A-1201
AFORA
Alloy steel. C 0.34-0.41, Mn 0.6-0.9, Si 0.15-0.4, P 0-0.035, S 0-0.035, Cr 0.9-1.2, bal Fe.
95-115 kgf/mm^2 TS; 75 kgf/mm^2 min YS; 11 min El.

A-1202
AFORA
Alloy steel. C 0.38-0.45, Mn 0.6-0.9, Si 0.15-0.4, P 0-0.035, S 0-0.035, Cr 0.9-1.2, bal Fe.
100-120 kgf/mm^2 TS; 80 kgf/mm^2 min YS; 11 min El.

A-1203
AFORA
Alloy steel. C 0.33-0.4, Mn 1.3-1.65, Si 0.15-0.4, P 0-0.035, S 0-0.035, bal Fe.
85-105 kgf/mm^2 TS; 65 kgf/mm^2 min YS; 12 min El.

A-1204
AFORA
Alloy steel. C 0.37-0.44, Mn 0.7-1, Si 0.15-0.4, P 0-0.035, S 0-0.035, Cr 0.4-0.7, Ni 0.4-0.7, Mo 0.15-0.3, bal Fe.
105-125 kgf/mm^2 TS; 85 kgf/mm^2 min YS; 10 min El.

A-1250
AFORA
Alloy steel. C 0.32-0.38, Mn 0.6-0.9, Si 0.15-0.4, P 0-0.035, S 0-0.035, Cr 0.15-0.85, Mo 0.15-0.25, bal Fe.
100-120 kgf/mm^2 TS; 80 kgf/mm^2 min YS; 11 min El.

A-1251
AFORA
Alloy steel. C 0.27-0.33, Mn 0.6-0.9, Si 0.15-0.4, P 0-0.035, S 0-0.035, Cr 0.85-1.15, Mo 0.15-0.25, bal Fe.
95-115 kgf/mm^2 TS; 75 kgf/mm^2 min YS; 12 min El.

A-1252
AFORA
Alloy steel. C 0.37-0.43, Mn 0.6-0.9, Si 0.15-0.4, P 0-0.035, S 0-0.035, Cr 0.85-1.15, Mo 0.15-0.25, bal Fe.
110-130 kgf/mm^2 TS; 90 kgf/mm^2 min YS; 10 min El.

A-1260
AFORA
Alloy steel. C 0.3-0.37, Mn 0.3-0.6, Si 0.15-0.4, P 0-0.035, S 0-0.035, Cr 1.1-1.4, Ni 3.7-4.2, Mo 0.25-0.4, bal Fe.
115-135 kgf/mm^2 TS; 100 kgf/mm^2 min YS; 9 min El.

A-1262
AFORA
Alloy steel. C 0.3-0.36, Mn 0.6-0.8, Si 0.15-0.4, P 0-0.035, S 0-0.035, Cr 0.7-0.9, Ni 2.75-3.25, Mo 0.3-0.4, bal Fe.
110-130 kgf/mm^2 TS; 95 kgf/mm^2 min YS; 9 min El.

A-1270
AFORA
Alloy steel. C 0.32-0.38, Mn 0.55-0.85, Si 0.15-0.4, P 0-0.035, S 0-0.035, Cr 0.65-0.95, Ni 1.6-2, Mo 0.15-0.8, bal Fe.
115-135 kgf/mm^2 TS; 95 kgf/mm^2 min YS; 11 min El.

A-1272
AFORA
Alloy steel. C 0.37-0.43, Mn 0.55-0.85, Si 0.15-0.4, P 0-0.035, S 0-0.035, Cr 0.65-0.95, Ni 1.6-2, Mo 0.15-0.3, bal Fe.
120-140 kgf/mm^2 TS; 100 kgf/mm^2 min YS; 9 min El.

A-1280
AFORA
Alloy steel. C 0.32-0.38, Mn 0.5-0.8, Si 0.15-0.4, P 0-0.035, S 0-0.035, Cr 0.6-0.9, Ni 0.7-1, Mo 0.15-0.3, bal Fe.
100-120 kgf/mm^2 TS; 80 kgf/mm^2 min YS; 11 min El.

A-1282
AFORA
Alloy steel. C 0.37-0.42, Mn 0.5-0.8, Si 0.15-0.4, P 0-0.035, S 0-0.035, Cr 0.6-0.9, Ni 0.7-1, Mo 0.15-0.3, bal Fe.
105-125 kgf/mm^2 TS; 85 kgf/mm^2 min YS; 10 min El.

A-1310
AFORA
Alloy steel. C 0.95-1.1, Mn 0.2-0.4, Si 0.15-0.35, P 0-0.03, S 0-0.025, Cr 1.35-1.6, bal Fe.

A-1311
AFORA
Alloy steel. C 0.95-1.1, Mn 0.2-0.4, Si 0.15-0.35, P 0-0.03, S 0-0.025, Cr 0.4-0.6, bal Fe.

A-1312
AFORA
Alloy steel. C 0.9-1.05, Mn 0.2-0.4, Si 0.2-0.45, P 0-0.03, S 0-0.025, Cr 1.65-1.95, Mo 0.15-0.3, bal Fe.

A-140C
AFORA
Alloy steel. C 0.55-0.65, Mn 0.65-0.95, Si 1.5-2, P 0-0.03, S 0-0.03, Cr 0.2-0.5, bal Fe.
115-135 kgf/mm^2 TS; 95 kgf/mm^2 min YS; 8 min El.

A-143
AFORA
Alloy steel. C 0.45-0.55, Mn 0.5-0.7, Si 0.15-0.4, P 0-0.03, S 0-0.03, Cr 0.8-1.1, V 0.15-0.25, bal Fe.
115-135 kgf/mm^2 TS; 95 kgf/mm^2 min YS; 8 min El.

A-144
AFORA
Alloy steel. C 0.5-0.6, Mn 0.7-1, Si 1.5-2, P 0-0.03, S 0-0.03, bal Fe.
110-130 kgf/mm^2 TS; 90 kgf/mm^2 min YS; 8 min El.

A-145
AFORA
Alloy steel. C 0.45-0.55, Mn 0.6-0.9, Si 1.5-2, P 0-0.03, S 0-0.03, bal Fe.
95-115 kgf/mm^2 TS; 75 kgf/mm^2 min YS; 11 min El.

A-1510
AFORA
Carbon steel. C 0.07-0.13, Mn 0.3-0.6, Si 0.15-0.4, P 0-0.035, S 0-0.035, bal Fe.
45-75 kgf/mm^2 TS; 30 kgf/mm^2 min YS; 15 min El.

A-1511
AFORA
Carbon steel. C 0.12-0.18, Mn 0.3-0.6, Si 0.15-0.4, P 0-0.035, S 0-0.035, bal Fe.
50-80 kgf/mm^2 TS; 85 kgf/mm^2 min YS; 14 min El.

A-1515
AFORA
Alloy steel. C 0.18-0.23, Mn 1.3-1.6, Si 0.15-0.4, P 0-0.035, S 0-0.035, bal Fe.
80-110 kgf/mm^2 TS; 55 kgf/mm^2 min YS; 9 min El.

A-1516
AFORA
Alloy steel. C 0.13-0.19, Mn 1-1.3, Si 0.15-0.4, P 0-0.035, S 0-0.035, Cr 0.8-1.1, bal Fe.
85-115 kgf/mm^2 TS; 60 kgf/mm^2 min YS; 10 min El.

A-1522
AFORA
Alloy steel. C 0.17-0.22, Mn 0.6-0.9, Si 0.15-0.4, P 0-0.035, S 0-0.035, Cr 0.35-0.65, Ni 0.4-0.7, Mo 0.15-0.25, bal Fe.
85-115 kgf/mm^2 TS; 60 kgf/mm^2 min YS; 9 min El.

A-1523
AFORA
Alloy steel. C 0.18-0.23, Mn 0.6-0.9, Si 0.15-0.4, P 0-0.035, S 0-0.035, Cr 0.3-0.5, Mo 0.4-0.5, bal Fe.
80-115 kgf/mm^2 TS; 60 kgf/mm^2 min YS; 11 min El.

A-1524
AFORA
Alloy steel. C 0.18-0.23, Mn 0.6-0.9, Si 0.15-0.4, P 0-0.035, S 0-0.035, Cr 0.4-0.6, Ni 0.7-0.9, Mo 0.3-0.4, bal Fe.
95-130 kgf/mm^2 TS; 70 kgf/mm^2 min YS; 9 min El.

A-1525
AFORA
Alloy steel. C 0.18-0.23, Mn 0.6-0.9, Si 0.15-0.4, P 0-0.035, S 0-0.035, Cr 0.4-0.8, Ni 1.4-1.7, Mo 0.3-0.4, bal Fe.
105-135 kgf/mm^2 TS; 80 kgf/mm^2 min YS; 8 min El.

A-1540
AFORA
Alloy steel. C 0.1-0.16, Mn 0.35-0.65, Si 0.15-0.4, P 0-0.035, S 0-0.035, Cr 0.6-0.9, Ni 2.75-3.25, bal Fe.
95-125 kgf/mm^2 TS; 70 kgf/mm^2 min YS; 9 min El.

A-1550
AFORA
Alloy steel. C 0.15-0.21, Mn 0.6-0.9, Si 0.15-0.4, P 0-0.035, S 0-0.035, Cr 0.85-1.15, Mo 0.15-0.25, bal Fe.
90-120 kgf/mm^2 TS; 65 kgf/mm^2 min YS; 10 min El.

A-1551
AFORA
Alloy steel. C 0.1-0.15, Mn 0.6-0.9, Si 0.15-0.4, P 0-0.035, S 0-0.035, Cr 0.85-1.15, Mo 0.15-0.25, bal Fe.
80-115 kgf/mm^2 TS; 55 kgf/mm^2 min YS; 10 min El.

A-1560
AFORA
Alloy steel. C 0.11-0.17, Mn 0.3-0.6, Si 0.15-0.4, P 0-0.035, S 0-0.035, Cr 0.8-1.1, Ni 3-3.5, Mo 0.2-0.3, bal Fe.
115-145 kgf/mm^2 TS; 90 kgf/mm^2 min YS; 10 min El.

A-1580
AFORA
Alloy steel. C 0.17-0.22, Mn 0.8-1, Si 0.15-0.4, P 0-0.035, S 0-0.035, Cr 0.8-1.2, Ni 0.8-1.2, bal Fe.
95-125 kgf/mm^2 TS; 70 kgf/mm^2 min YS; 9 min El.

A-1581
AFORA
Alloy steel. C 0.13-0.18, Mn 0.8-1, Si 0.15-0.4, P 0-0.035, S 0-0.035, Cr 0.8-1.2, Ni 0.8-1.2, bal Fe.
85-115 kgf/mm^2 TS; 60 kgf/mm^2 min YS; 9 min El.

A-210.G, 35MNS6
AFORA
Alloy steel. C 0.32-0.39, Mn 1.35-1.65, P 0-0.035, S 0.08-0.13, <0.40 Si, bal Fe.
Quenched and tempered (<16 mm): 76-95 kgf/mm^2 TS; 52 kgf/mm^2 min YS; 12 min El. Quenched and tempered (<40 mm): 68-88 kgf/mm^2 TS; 48 kgf/mm^2 min YS; 14 min El. Quenched and tempered (<63 mm): 63-83 kgf/mm^2 TS; 45 kgf/mm^2 min YS; 15 min El. Quenched and tempered (<100 mm): 63-83 kgf/mm^2 TS; 45 kgf/mm^2 min YS; 15 min El. Quenched, tempered and cold drawn (<16 mm): 80-100 kgf/mm^2 TS; 66 kgf/mm^2 min YS; 8 min El. Quenched, tempered and cold drawn (<40 mm): 73-93 kgf/mm^2 TS; 63 kgf/mm^2 min YS; 10 min El. Quenched, tempered and cold drawn (<63 mm): 68-88 kgf/mm^2 TS; 58 kgf/mm^2 min YS; 12 min El. Quenched, tempered and cold drawn (<100 mm): 68-88 kgf/mm^2 TS; 58 kgf/mm^2 min YS; 12 min El.

A-286
Driver Harris Co.
Stainless steel. Ni 26, Cr 15, Cb 1, Ti 2, bal Fe.
Wire and rod for welding wire and fastener stock. *Obsolete*

A-286
Bishop Tube Co.
Cr 15, Ni 24, Ti 2, Mo 1.5, V 0.3, bal Fe.
As hardened: 135,000-160,000 psi TS; 85,000-115,000 psi YS; 20-30 El; 24-34 Rock C. High strength; corrosion resistant. Precipitation hardened alloy. Welded or seamless tubing.

A-3401
AFORA
Stainless steel. C 0.09-0.15, Mn 0-1, Si 0-1, P 0-0.04, S 0-0.03, Cr 11.5-14, bal Fe.

A-3507
AFORA
Stainless steel. C 0-0.12, Mn 0-2, Si 0-1, P 0-0.045, S 0-0.03, Cr 17.9-19, Ni 8-10, bal Fe.

A-6
Atrax Cemented Carbide
Sintered carbide. 240,000 transverse strength; 14.8-15.0 g/cm^3 density; 91.5-92.5 RA. Industry code: C-2; ISO K10, 15, 20, M20.

A 60
GTE Valenite Corp.
Ceramic coating TiC and Al_2O_3. For light rough and finishing applications on steel and cast iron. High performance coated grade for high speed machining. Code C-2/C-3/C-4/C-6/C-7/C-8.

A-7
Atrax Cemented Carbide
Sintered carbide. 200,000 transverse strength; 14.95-15.2 g/cm^3 density; 91.7-92.7 RA. Industry code: C-3-4, ISO K01, 05.

A-7W
Wallace Murray Corp.
C 2.25, Cr 5.25, V 4.75, W 1, Mo 1, bal Fe.
Air hardening tool steel; AISI A7.

A-ALLOY GR. A
Alloy Engineering & Casting Co.
C 0.4-0.8, Ni 66-68, Cr 19-21, Mn 0.5-1, bal Fe.
Cast: 60,000-80,000 TS; 40,000-50,000 YS; 40 El; 30 RA; 217 Brin. For heat and corrosion resistant parts. Heat and corrosion resistant. *Obsolete*

A-COPPER
Isabellenhuette
Copper. Cu 99.5, Ni 0.5.
220 N/mm^2 TS. For electrical equipment and instruments. Resistance alloy. Maximum working temperature to 200°C.

A-D-70
John A. Crowley Inc.
C 0.7, Mn 0.35, W 0.98, Cr 0.3, V 0.12, bal Fe.
For dies, crimpers, punches; oil hardened.

A-G5
French manufacture
Mg 5, Mn 0.5, bal Al.
For automobile chassis; light weight construction.

A-HT
Bethlehem Steel Corp.
C 1, Cr 3, Mo 1.1, V 0.25, W 1.05, Ti 1, bal Fe.
Air-hardened tool steel. For dies, punches, shear blades, blanking tools. *Obsolete*

A-L 18 NICOMO(250)
Allegheny Ludlum Steel
C 0.02, Mn 0.08, Si 0.08, P 0.003, S 0.007, Ni 18.5, Co 7.5, Mo 4.8, Ti 0.4, Zr 0.01, B 0.003, bal Fe.
Hot rolled: 165,100 TS; 132,600 YS; 10% El; 53.4% RA. Heat treated: 252,000 TS; 245,000 YS; 9.5% El; 46.5% RA. Uses: Rocket cases, aircraft landing gears, crimping dies. Maraging steel. Strengthened by precipitation hardening. *Obsolete*

A-L 18 NICOMO(300)
Allegheny Ludlum Steel
C 0.02, Mn 0.08, Si 0.08, P 0.005, S 0.005, Ni 18.5, Co 9, Mo 5, Ti 0.6, Zr 0.01, B 0.003, bal Fe.
Annealed: 150,000 TS; 120,000 YS; 18% El; .75% RA; Rock C 28. Heat treated: 306,000 TS; 303,000 YS; 12% El; 60 RA. Uses: Rocket cases, aircraft landing gears. Maraging steel. Strengthened by precipitation hardening. *Obsolete*

A-L 20 NI(250)
Allegheny Ludlum Steel
C 0.02, Mn 0.08, Si 0.08, P 0.007, S 0.002, Ni 20, Ti 1.4, Al 0.2, Cb 0.5, bal Fe.
Annealed: 148,500 TS; 110,900 YS; 26.4 % El; 68.9% RA; Rock C 29. Hardened: 269,000 TS; 263,000 YS; 12% El; 59% RA; Rock C 53. Uses: Rocket cases, aircraft landing gears. Maraging steel. Strengthened by precipitation hardening. *Obsolete*

A-L 25 NI(250)
Allegheny Ludlum Steel
C 0.02, Mn 0.08, Si 0.08, P 0.008, S 0.002, Ni 25.3, Ti 1.4, Al 0.2, Cb 0.5, bal Fe.
Annealed: 105,000 TS; 38,000 YS; 34% El; 77.3% RA; Rock B 89. Heat treated: 286,000 TS; 276,000 YS; 12% El; 57% RA; Rock C 54. Uses: Rocket cases, aircraft landing gear. Maraging steel. Strengthened by precipitation hardening. *Obsolete*

A-M52
AFORA
Alloy steel. C 0.12-0.2, Mn 0.9-1.4, Si 0.15-0.4, P 0-0.035, S 0-0.035, bal Fe.
65-85 kgf/mm^2 TS; 55 kgf/mm^2 min YS; 15 min El.

A-R-COL
British Steel plc
C 0-0.17, Si 0-0.5, Mn 0-1.6, P 0-0.03, S 0-0.015, Cr 0.65-0.95, Mo 0.3-0.5, bal Fe.
Roller quenched: 320 Brin min. Abrasion resistant plate. For use as hoppers, chutes, and mining equipment.

A-R-COL 360
British Steel plc
C 0-0.13, Mn 1.35, Cr 0.8, Mo 0.4, Si 0.05, bal Fe.
Abrasion resisting steel for applications such as liner plates, chutes, buckets, storage bins, and mold boards.

A-TU.9
AFORA
Alloy steel. Mn 0.3-0.6, Si 0.5-1, P 0-0.03, S 0-0.03, Cr 8-10, Mo 0.9-1.1, <0.15 C, bal Fe.
65-75 kgf/mm^2 TS; 45 kgf/mm^2 min YS; 20 min El.

A-TU.D
AFORA
Alloy steel. C 0.12-0.2, Mn 0.5-0.8, Si 0.15-0.4, P 0-0.03, S 0-0.03, Mo 0.45-0.65, bal Fe.
50-60 kgf/mm^2 TS; 31 kgf/mm^2 min YS; 25 min El.

A-TU.E
AFORA
Alloy steel. C 0.1-0.2, Mn 0.3-0.6, Si 0.15-0.4, P 0-0.03, S 0-0.03, Cr 0.5-0.8, Mo 0.45-0.65, bal Fe.
50-60 kgf/mm^2 TS; 35 kgf/mm^2 min YS; 20 min El.

A-TU.G
AFORA
Alloy steel. Mn 0.3-0.6, Si 0.5-1, P 0-0.03, S 0-0.03, Cr 1-1.5, Mo 0.45-0.65, <0.15 C, bal Fe.
52-62 kgf/mm^2 TS; 35 kgf/mm^2 min YS; 20 min El.

A-TU.H
AFORA
Alloy steel. Mn 0.3-0.6, Si 0.15-0.5, P 0-0.03, S 0-0.03, Cr 2-2.5, Mo 0.9-1.1, <0.15 C, bal Fe.
52-75 kgf/mm^2 TS; 32 kgf/mm^2 min YS; 20 min El.

A-TU.I
AFORA
Alloy steel. Mn 0.3-0.6, Si 0.15-0.5, P 0-0.03, S 0-0.03, Cr 2.7-3.3, Mo 0.8-1.05, <0.15 C, bal Fe.
55-75 kgf/mm^2 TS; 32 kgf/mm^2 min YS; 20 min El.

A-TU.J
AFORA
Alloy steel. Mn 0.3-0.6, Si 0.15-0.5, P 0-0.03, S 0-0.03, Cr 4-6, Mo 0.45-0.65, <0.15 C, bal Fe.
65-75 kgf/mm^2 TS; 45 kgf/mm^2 min YS; 20 min El.

A-U4G1
Russian manufacture
Cu 4.36, Mg 1.5, Mn 0.6, Si 0.5, bal Al.
Heat treated: 75,000 TS; 58,000 YS; 135 Brin. For aircraft structures, fittings, hardware, rivets, fastenings; similar to Aluminum 2024, age-hardenable, high strength.

A-U4SG
Russian manufacture
Cu 4.35, Si 0.85, Mn 0.8, Mg 0.5, bal Al.
Heat treated: 70,000 TS; 60,000 YS; 12 El; 135 Brin. For hydraulic fittings, structures, hardware, engine components; similar to Aluminum 2014, age-hardenable, high strength.

A. D. S.
T. Inman & Co. Ltd.
C 0.35, Mn 0.4, Si 1, Cr 5, Mo 1.5, V 0.54, bal Fe.
Aluminum die-casting steel; air or oil hardenable.

A. M. C.
A. Milne & Co.
C 0.7, W 18, Cr 4, V 1, bal Fe.
For tools, cutters, broaches; high speed steel.

A.A.A.A
Seitzinger's Inc.
Pb, Sb, bal Sn.
For bearings. Babbitt, anti-friction metal.

A.B.C
Darwin & Milner Inc.
C 0.4, W 0.5, Cr 0.25, Si 0.3, Mn 0.5, bal Fe.
For tools, chisels, hammers, drills, gears, shafts; tough and hard.

A.B.C
Balfour Darwins Ltd.
C 0.4, W 0.5, Cr 0.25, Si 0.3, Mn 0.5, bal Fe.
For tools, chisels, hammers, drills, gears, shafts; tough and hard.

A.B.C.
P.F. McDonald & Co.
High C, alloy, bal Fe.
For chisels; shock resistant.

A.C.T.
Atlantic Steel Corp.
C 0.65, Mn 0.15, W 18, Cr 4.5, V 2, Mo 0.75, Co 10, bal Fe.
For tools, drills, cutters, reamers; high speed steel.

A.E. SUPER NICKEL
IMI Knoch Ltd.
Ni 30, Fe 1, Mn 1, bal Cu.
Annealed: 65,000-90,000 TS; 25,000-53,000 YS; 44-6 El; 36-22 RA; 77-145 Brin. For condensers, coolers, feed heaters; corrosion resistant. *Obsolete*

A.E.B. STAINLESS
Uddeholm Corp.
C 0.95, Mn 1, Cr 13.5, bal Fe.
For razor blades, surgical instruments; hardenable, corrosion resistant.

A.H. CHROME DIE
Joseph Beardshaw & Son Ltd.
C 1, Cr 5, Mo 1, V 0.3, bal Fe.
For dies, gages, mandrels; cold work steel, air hardened, non-deforming.

A.M. 1 SPECIAL
Edgar Allen Balfour Ltd.
C 0.35, Si 1, W 1.35, Cr 5, Mo 1.5, V 0.45, bal Fe.
For extrusion dies, heading dies, mandrels; hot-work steel, resists heat checking.

A.M. 230
Aluminum Company of America
Al 9-11, Zn 0-0.3, Si 0.5-1, 0.1% min Mn, bal Mg.
Die cast: 32,000 TS; 22,000 YS; 1.5 El; 63 Brin. For aircraft motor parts, switch housings, vacuum cleaner parts; die casting alloy. *Obsolete*

A.M. 403
Aluminum Company of America
Mn 1.2-2, Si 0-0.3, bal Mg.
Sand cast: 14,000 TS; 45,000 YS; 5.0 El; 33 Brin. For gasoline tanks, fittings; light casting alloy. *Obsolete*

A.M. 57 S
Aluminum Company of America
Al 6.5, Mn 0.4, Zn 1, bal Mg.
Extruded: 38,000-42,000 TS; 22,000-28,000 YS; 8-14 El; 15-24 RA; 55 Brin. For extrusion and hot pressings; good corrosion resistance. *Obsolete*

A.M. 59S
Aluminum Company of America
Al 9.5-10.5, Zn 0-0.3, Si 0-0.5, 0.10% min Mn, bal Mg.
Heat treated: 56,000 TS; 40,000 YS; 3.5 El; 85 Brin. For light alloy parts. *Obsolete*

A.M.C.O.H. SPECIAL
A. Milne & Co.
C 0.9, Mn 1.2, Cr 0.5, W 0.5, bal Fe.
For die, punches; nondeforming.

A.M.D.
Diehl Steel Co.
C 0.34, Mn 0.4, Cr 4.75, W 1.1, Mo 1.45, bal Fe.
For hot work dies; hot work steel.

A.M.I.
Edgar Allen Balfour Ltd.
C, alloy, bal Fe.
For dies; tough and abrasion resistant.

A.P. 33
Alais Forges et Camargue
Cu 4.5, Mg 0.5, Ti 0.4, bal Al.
36,000-43,000 TS; 29,000-36,000 YS; 1-2 El. For light alloy parts; age-hardenable.

A.R. ALLOY
International Development Corp.
Sn 10, Pb 2, bal Cu.
10-5 El; 50-60 Brin. For bearings for cold rolling mills, landing gears for airplanes; tough, strong.

A.R. STAINLESS
Crucible Steel Castings Co.
C 0.2, Cr 18, Ni 8, bal Fe.
For stainless parts; stainless. *Obsolete*

A.R.C. 2702T
Creusot-Loire
C 0-0.08, Cr 18-20, Ni 8-11, Mn 0-2, bal Fe.
Annealed: 90,000 TS; 45,000 YS; 60 El; 135 Brin. Cold drawn: 180,000 TS; 150,000 YS; 10 El; 330 Brin. For chemical plant equipment, welded structures; Type 304; stainless, austenitic. *Obsolete*

A.S. ACK-LOW
Ackerlind Steel Co., Inc.
C 0.7, Mn 2.1, Cr 1, Mo 1.3, bal Fe.
Air hardening cold work tool steel; AISI A6. *Obsolete*

A.S. BEARCAT
Ackerlind Steel Co., Inc.
C 0.5, Mn 0.7, Cr 3.25, Mo 1.4, bal Fe.
Air or oil hardening tool steel; good resistance to impact; AISI S7. *Obsolete*

A.S. BLUE LABEL
Ackerlind Steel Co., Inc.
C 1-1.1, Mn 0.25, Si 0.25, bal Fe.
For taps, drills, reamers, hobs, broaches; Type W1; water hardened. *Obsolete*

A.S. BRAKE & DIE STEEL
Ackerlind Steel Co., Inc.
C 0.5, Cr 1, Mn 0.9, Mo 0.2, bal Fe.
For press brakes for forming and bending; resists shock and wear. *Obsolete*

A.S. CROMAT V
Ackerlind Steel Co., Inc.
C 0.35, Si 1.05, Cr 5.15, V 0.3, W 1.25, Mo 1.55, bal Fe.
Hot work tool steel for forging and hot forming dies, chromium type; AISI H12. *Obsolete*

A.S. DURAMOLD B
Ackerlind Steel Co., Inc.
C 0.07, Mn 0.3, Si 0.15, Cr 1, Mo 0.25, B, bal Fe.
For cold hubbing and case hardening for plastic molds; AISI P2. *Obsolete*

A.S. HOBBING IRON
Ackerlind Steel Co., Inc.
C 0.05, Mn 0.1, bal Fe.
For tools, hobbing molds; case-hardened. *Obsolete*

A.S. HOLLOW DIE
Ackerlind Steel Co., Inc.
C 1, Mn 0.4, Cr 1.5, bal Fe.
For dies; tough. *Obsolete*

A.S. HOT DIE
Ackerlind Steel Co., Inc.
C 0.45, Si 0.9, Cr 1.15, V 0.25, W 2.5, bal Fe.
For hot header dies, forming tools, piercers; hot work steel, oil hardened. *Obsolete*

A.S. LUSTRE-DIE
Ackerlind Steel Co., Inc.
C 0.5, Mn 1, Si 0.3, Cr 1.1, Mo 0.25, bal Fe.
Oil hardening steel for arbors, shafts, lathe centers. *Obsolete*

A.S. NICKEL-CHROME HOBBING STEEL

Ackerlind Steel Co., Inc.
C 0.1, Mn 0.5, Cr 0.6, Ni 1.25, bal Fe.
For tools, hobbing molds; case-hardened. *Obsolete*

A.S. NO. 121

Ackerlind Steel Co., Inc.
C 0.95, Mn 2, Cr 2.2, Mo 1.1, bal Fe.
Air hardening, medium alloy tool steel; for cold working tools; AISI A4. *Obsolete*

A.S. NO. 15

Ackerlind Steel Co., Inc.
C 0.5, Mn 0.7, Cr 0.8-1.1, V 0.15, bal Fe.
For dies; pressure resisting steel. *Obsolete*

A.S. NO. 27

Ackerlind Steel Co., Inc.
C 0.95-1.1, Mn 0.3, Cr 1.5, V 0.2, bal Fe.
For tools, dies; pressure resistant steel. *Obsolete*

A.S. NO. 42

Ackerlind Steel Co., Inc.
C 0.3, Mn 0.2, Cr 3.5, W 9.75, V 0.5, bal Fe.
For hot dies; hot work steel. *Obsolete*

A.S. NO. 5

Ackerlind Steel Co., Inc.
C 1, Mn 0.7, Cr 5, V 0.25, Mo 1.1, bal Fe.
For tools and dies. Air hardening; AISI A2.

A.S. NO. 66

Ackerlind Steel Co., Inc.
C 0.8, Cr 4, V 2, Mo 4-5, W 6, bal Fe.
For lathe and planer tools, drills, form cutters, hot punches; high speed steel. *Obsolete*

A.S. NO. 670

Ackerlind Steel Co., Inc.
C 0.38, V 1, Cr 5.25, V 1.05, Mo 1.25, bal Fe.
Air hardening hot work tool and die steel, chromium type; AISI H13. *Obsolete*

A.S. NO. 7

Ackerlind Steel Co., Inc.
C 0.45, Si 0.9, Cr 1.15, V 0.2, W 2.5, bal Fe.
For master hobs, shear blades, and punches. Type S1; shock resisting steel.

A.S. NO. 85

Ackerlind Steel Co., Inc.
C 0.55, Mn 0.4, Si 0.3, Cr 1, Ni 3, Mo 0.3, bal Fe.
Oil hardening steel for shafts, lathe centers, tool holders. *Obsolete*

A.S. SPECIAL HOBBING IRON

Ackerlind Steel Co., Inc.
C 0.05, Mn 0.1-0, Si 0.1, bal Fe.
For cold hubbing; can be case hardened for molds; AISI P1. *Obsolete*

A.S. TRI-ACK

Ackerlind Steel Co., Inc.
C 1.5, Mn 0.3, Cr 12, V 0.2, Mo 0.8, bal Fe.
For dies; air hardening, nondeforming. *Obsolete*

A.S. TRI-MO

Ackerlind Steel Co., Inc.
C 1.5, Cr 12, V 0.2, Mo 0.8, bal Fe.
For threading dies, cutting tools, punches, dies; nondeforming. *Obsolete*

A.S. WHITE LABEL

Ackerlind Steel Co., Inc.
C 1.1, V 0.2, bal Fe.
For blanking and forming dies, plug gauges; water hardened. *Obsolete*

A.S.V.

Teledyne Firth Sterling
Steel. C 0.7, bal Fe.
For tools, cutters; water hardened. *Obsolete*

A.T. 2

CCS Braeburn Alloy Steel
C 2.2, Cr 12, V 0.25, Mo 0.9, bal Fe.
Hardened: 63-65 Rock C. For blanking and forming dies, gauges, trimmer and drawing dies, lathe centers, brick mold liners. Cold work Type D4, oil or air hardening. High abrasion resistance. *Obsolete*

A.T. 2 DIE

CCS Braeburn Alloy Steel
C 2.2, Cr 12, V 0.25, Mo 0.9, bal Fe.
For blanking and forming dies, gauges. Oil hardened, non-deforming. *Obsolete*

A.T. STEEL

Carpenter Technology Corp.
C 0.35, Mn 0.5, Si 0.3, Cr 1.5, Ni 3.5, bal Fe.
Air hardening tool steel. *Obsolete*

A.W. 70-90 TYPE A

Alan Wood Steel Co.
C 0-0.25, Mn 0.7, P 0.1, Cu 0.5, bal Fe.
90,000 TS; 70,000 YS; 20 El. For sheets, plates; easy to weld. *Obsolete*

A.W. 70-90 TYPE B

Alan Wood Steel Co.
C 0-0.25, Mn 0.7, P 0.1, Cu 0.5, bal Fe.
Rolled: 65,000 TS; 50,000 YS; 20 El. For structural work, transportation industry, truck and tank truck bodies; weldable, atmospheric corrosion resistant. *Obsolete*

A.W. DYNALLOY

Alan Wood Steel Co.
C 0.12, Mn 0.75, Ni 0.7, Mo 0.07, bal Fe.
65,000-80,000 TS; 50,000 YS; 25 El. For railroad equipment, trucks, buses, smoke chambers, stacks; good weldability. *Obsolete*

A.Z. ALLOY

Atlantic Zinc Works
Zn 33, Pb 1, bal Cu.
For plates for photoengraving. Free-cutting.

A201

Fansteel/Wellman Dynamics
Aluminum. Cu 4-5, Si 0.05, Fe 0.1, Mn 0.2-0.4, Zn 0.03, Mg 0.15-0.35, Ti 0.15-0.35, Be 0.03, Ag 0.4-1, bal Al.
Aluminum casting alloy. Cast: 60,000 psi TS; 50,000 psi YS; 3-5 El.

A206

Fansteel/Wellman Dynamics
Aluminum. Cu 4.2-5, Si 0.05, Fe 0.1, Mn 0.2-0.5, Zn 0.1, Mg 0.15-0.35, Ti 0.15-0.3, Be 0.05, bal Al.
Aluminum casting alloy. Cast: 50,000 psi TS; 30,000-40,000 psi YS; 3-10 El.

A28F

Acieries et Forges d'Anor
C 0.25-0.45, Cr 28-30, Ni 3-5, bal Fe.
Heat resisting steel. AFNOR Z35 CN 28.04.

A319

Bunting Bearings Corp.
Aluminum. 3.5 Cu min, 6.3 Si min, bal Al.
Permanent mold process.

A356

Fansteel/Wellman Dynamics
Aluminum. Cu 0.2, Si 6.5-7.5, Fe 0.2, Mn 0.1, Zn 0.1, Mg 0.25-0.45, Ti 0.2, bal Al.
Aluminum casting alloy. Cast: 33,000-45,000 psi TS; 28,000-34,000 psi YS; 3-5 El.

A356.0

Bunting Bearings Corp.
Aluminum. Cu 0-0.2, Si 6.5-7.5, Fe 0-0.2, Mn 0-0.1, Zn 0-0.1, Mg 0.25-0.45, Ti 0-0.2, bal Al.
MIL-A-21180D specification alloy A356.0. Permanent mold process.

A357

Fansteel/Wellman Dynamics
Aluminum. Cu 0.2, Si 6.5-7.5, Fe 0.2, Mn 0.1, Zn 0.1, Mg 0.4-0.7, Be 0.04-0.07, 0.04 Ti min, bal Al.
Aluminum casting alloy. Cast: 40,000 psi TS; 30,000 psi YS; 2 El.

A36CC

LTV Steel
C 0.16, Mn 0.5, Si 0.04, bal Fe.
36,000 psi YS min. Good weldability and impact (notch) toughness. For automotive frames and cross members. ASTM A-36.

A441 STEEL

Gulf States Steel, Inc.
HSLA steel. C 0.22, Mn 0.85-1.25, P 0.04, S 0.05, Si 0.3, 0.20 Cu min, 0.02 V min, bal Fe.
High strength low alloy steel with strength, toughness and resistance to atmospheric corrosion. 50,000 YS min; 70,000 TS min; 22 El.

A45YO/A45YK/A45YF

Acme Steel Co.
C 0-0.22, Mn 0-1.35, P 0-0.04, S 0-0.05, Si 0-0.3, 0.005 Cb min, 0.01 V min, Cu optional, bal Fe.
Plate: 60,000 psi TS; 45,000 psi YS; 25 El. For applications requiring high strength with good formability and weldability. Meets ASTM A607, A572; SAE J410c.

A50XF

Acme Steel Co.
C 0-0.15, Mn 0-1.65, P 0-0.025, S 0-0.035, Si 0-0.6, 0.005 Cb min, 0.01 V min, bal Fe.
Plate: 60,000 psi TS; 50,000 psi YS; 24 El. For applications requiring high strength with good formability and weldability. Meets ASTM A715.

A50YO/A50YK

Acme Steel Co.
C 0-0.23, Mn 0-1.35, P 0-0.04, S 0-0.05, Si 0-0.3, 0.005 Cb min, 0.01 V min, Cu optional, bal Fe.
Plate: 65,000 psi TS; 50,000 psi YS; 22 El. For applications requiring high strength with good formability and weldability. Meets ASTM A607, A572; SAE J410c.

A55YO/A55YK

Acme Steel Co.
C 0-0.25, Mn 0-1.35, P 0-0.04, S 0-0.05, Si 0-0.3, 0.005 Cb, 0.01 V min, Cu optional, bal Fe.
Plate: 70,000 psi TS; 55,000 psi YS; 20 El. For applications requiring high strength with good formability and weldability. Meets ASTM A607, A572; SAE J410c.

A60

Acieries et Forges d'Anor
C 0.1-0.15, Cr 24-26, Ni 19-22.
Refractory stainless steel. AFNOR Z12 CN 25.20.

A60XF

Acme Steel Co.
C 0-0.15, Mn 0-1.65, P 0-0.025, S 0-0.035, Si 0-0.6, 0.005 Cb min, 0.01 V min, bal Fe.
Plate: 70,000 psi TS; 60,000 psi YS; 22 El. For applications requiring high strength with good formability and weldability. Meets ASTM A715.

A60YQ/A60YK

Acme Steel Co.
C 0-0.26, Mn 0-1.35, P 0-0.04, S 0-0.05, Si 0-0.3, 0.005 Cb min, 0.01 V min, Cu optional, bal Fe.
Plate: 75,000 psi TS; 60,000 psi YS; 18 El. For applications requiring high strength with good formability and weldability. Meets ASTM A607, A572; SAE J410c.

A70XF

Acme Steel Co.
C 0-0.15, Mn 0-1.65, P 0-0.025, S 0-0.035, Si 0-0.6, 0.005 Cb min, 0.01 V min, bal Fe.
Plate: 80,000 psi TS; 70,000 psi YS; 20 El. For applications requiring high strength with good formability and weldability.

A71NB QUALITY

Delta Metal (BW) Ltd.
Cu 61, Pb 0-0.5, 1.0% Sn min, bal Zn.
Extruded: 60,000 TS; 25 El; 25 RA; 120 Brin. Drawn: 66,000 TS; 22 El; 20 RA; 130 Brin. For condenser tubes; naval brass. *Obsolete*

A80L

Acieries et Forges d'Anor
C 0-0.03, Si 0-1.5, Mn 0-1.5, Cr 17-19, Ni 10-12, Mo 2-3, bal Fe.
Austenitic 18/8 plus Mo type stainless steel. AFNOR Z 08 CND 18-08.

A80XF

Acme Steel Co.
C 0-0.15, Mn 0-1.65, P 0-0.025, S 0-0.035, Si 0-0.6, 0.005 Cb min, 0.01 V min, bal Fe.
Plate: 90,000 psi TS; 80,000 psi YS; 18 El. For applications requiring high strength with good formability and weldability. Meets ASTM A715.

A9V

Farbenindustrie Atkiengesellschaft,I.G.
Magnesium. Al 8.5, Zn 0.5, Mn 0.3, bal Mg.
For castings, light metal parts; heat treatable.

AA 7049

Manufacturer not listed.
For production of high quality weapon parts.

AA NICKEL-COBALT

Hanson-Van Winkle-Munning Co.
Ni-Co.
For anodes; electroplating.

AA-1

Glacier Metal Co.
Sn 15, bal Cu.
For bearings. *Obsolete*

AB 2

Acciaierie Valbruna s.p.a.
Tool Material. C 0.7, Mn 0-0.3, Si 0-0.3, bal Fe.
Cold work tool steel. AISI W1-7; W. Nr. 1.1520.

AB 3

Acciaierie Valbruna s.p.a.
Tool Material. C 0.85, Mn 0-0.25, Si 0-0.25, bal Fe.
Cold work tool steel. AISI W1-8.

AB 4

Acciaierie Valbruna s.p.a.
C 0.98, Mn 0-0.4, Si 0-0.35, bal Fe.
Cold work tool steel. AISI W1-9.5; W. Nr. 1.1545.

AB164 QUALITY

Delta Metal (BW) Ltd.
Al, Fe, Ni, bal Cu.
Extruded: 82,000 TS; 42,000 YS; 20 El; 12 RA; 170 Brin.
Drawn: 88,000 TS; 46,000 YS; 18 El; 10 RA; 180 Brin. For engine valve guides; Al bronze, heat resistant. *Obsolete*

AB197 QUALITY

Delta Metal (BW) Ltd.
Al, Fe, Ni, bal Cu.
Extruded: 110,000 TS; 62,000 YS; 18 El; 10 RA; 190 Brin.
Drawn: 115,000 TS; 68,000 YS; 15 El; 10 RA; 210 Brin. For hardware; Al bronze, corrosion resistant. *Obsolete*

AB2

English manufacture
Al 8.5-10.5, Fe 3.5-5.5, Ni 4.6-6.5, Mn 0-1.5, Zn 0-0.5, bal Cu.
For impellers and propellers, gears, pumps, shafts, bushings. B.S. 1400 equivalent; heat treatable, corrosion resistant.

AB49 TRI-FOIL

Johnson Matthey plc
Ag 49, Cu 27.5, Zn 20.5, Mn 2.5, Ni 0.5.
Silver brazing alloys for tungsten carbide; 670-710°C MP; DIN 8513.

AB65

Eagle & Globe Steel Ltd.
Tool material. C 0.65, Mn 0.75, Cr 0.75, Si 1.6, bal Fe.
Cold work tool and die steel. General purpose for shear blades, punches, dies and vise jaws. AS1239 S101A.

ABC-III

German manufacture
C 0.7, Mo 2.5, V 2.5, W 3, bal Fe.
For cutting tools; oil hardened.

ABEX NO. 10B

Abex Corp.
Cu 64, Al 4, Fe 3, Zn 26, Mn 3.
Cast: 90,000 psi TS; 45,000 psi YS; 18 El; 180 Brin. High tensile-90 Manganese bronze; CDA 862.

ABEX NO. 10C

Abex Corp.
Cu 58, Al 4, Fe 1, Zn 39, Mn 1.
Cast: 65,000 psi TS; 25,000 psi YS; 20 El; 130 Brin. Manganese bronze; CDA 865.

ABEX NO. 10D

Abex Corp.
Cu 63, Al 6, Fe 3, Zn 25, Mn 3.
Cast: 110,000 psi TS; 60,000 psi YS; 12 El; 200 Brin. High tensile-110 Manganese bronze; CDA 863.

ABEX NO. 15A

Abex Corp.
Cu 73, Sn 5, Pb 20, Zn 3.
Sand cast: 22,000 psi TS; 14,000 psi YS; 7 El; 48 Brin. Journal bearing bronze; CDA 941.

ABEX NO. 20-A

Abex Corp.
Cu 97, Sn 2, Zn 1.
Sand cast: 25,000 psi TS; 8,000 psi YS; 35 El; 35 Brin; 28% IACS conductivity. Electrode copper.

ABEX NO. 20-B

Abex Corp.
Cu 99.7.
Sand cast: 25,000 psi TS; 7,000 psi YS; 35 El; 35 Brin; 70% IACS min conductivity. Blast furnace copper.

ABEX NO. 20-C

Abex Corp.
Cu 99.9.
Sand cast: 25,000 psi TS; 7,000 psi YS; 35 El; 35 Brin; 80% IACS min conductivity.

ABEX NO. 20-H

Abex Corp.
Cu 99.2, 0.8 Cr (Chrome Cu).
Cast, heat treated: 45,000 psi TS; 35,000 psi YS; 12 El; 95 Brin; 80% IACS min conductivity. CDA 815.

ABEX NO. 20-L

Abex Corp.
99.9 Cu.
Sand cast: 25,000 psi TS; 7,000 psi YS; 35 El; 35 Brin; 90% IACS min conductivity.

ABEX NO. 20-Y

Abex Corp.
Cu 99, 1 Sn (blast furnace Cu).
Sand cast: 25,000 psi TS; 7,000 psi YS; 35 El; 35 Brin; 30% IACS min conductivity.

ABEX NO. 41

Abex Corp.
Cu 85, Sn 5, Pb 5, Zn 5.
Sand cast: 30,000 psi TS; 14,000 psi YS; 20 El; 60 Brin. Red brass; CDA 836.

ABEX NO. 4K

Abex Corp.
Cu 88, Sn 10, Zn 2.
Sand cast: 40,000 psi TS; 18,000 psi YS; 20 El; 75 Brin. Gun bronze; CDA 905.

ABEX NO. 4L

Abex Corp.
Cu 88, Sn 8, Zn 4.
Sand cast: 40,000 psi TS; 18,000 psi YS; 20 El; 70 Brin. Navy "G" bronze; CDA 903.

ABEX NO. 61

Abex Corp.
Cu 75, Sn 6, Pb 16, Zn 3.
Sand cast: 20,000 psi TS; 14,000 psi YS; 7 El; 50 Brin. "TIGER" bronze. Similar to CDA 939.

ABEX NO. 622

Abex Corp.
Cu 88, Sn 6, Pb 1.5, Zn 4.5.
Sand cast: 34,000 psi TS; 16,000 psi YS; 22 El; 60 Brin. Navy "M" bronze; CDA 922.

ABEX NO. 63

Abex Corp.
Cu 88, Sn 10, Pb 2.
Sand cast: 35,000 psi TS; 16,000 psi YS; 10 El; 67 Brin. Leaded tin bronze; CDA 927.

ABEX NO. 64

Abex Corp.
Cu 80, Sn 10, Pb 10, P.
Sand cast: 25,000 psi TS; 12,000 psi YS; 8 El; 60 Brin. Phosphor bronze; CDA 937.

ABEX NO. 640

Abex Corp.
Cu 87, Sn 11, Pb 1, Ni 1.
Sand cast: 35,000 psi TS; 10 El; 67 Brin. Nickel phosphor bronze; CDA 925.

ABEX NO. 65

Abex Corp.
Cu 89, Sn 11, P 0-0.3.
Sand cast: 35,000 psi TS; 18,000 psi YS; 10 El; 70 Brin. Gear bronze; CDA 907.

ABEX NO. 65-N

Abex Corp.
Cu 87, Sn 11.5, Ni 1.5.
Sand cast: 35,000 psi TS; 17,000 psi YS; 10 El; 75 Brin. Centrifuge cast: 50,000 psi TS; 25,000 psi YS; 10 El; 100 Brin. Nickel gear bronze; CDA 917.

ABEX NO. 660

Abex Corp.
Cu 83, Sn 7, Pb 7, Zn 3.
Sand cast: 30,000 psi TS; 14,000 psi YS; 12 El; 60 Brin. Modified red brass; CDA 932.

ABEX NO. 6H

Abex Corp.
Cu 74, Sn 7, Pb 20.
Sand cast: 22,000 psi TS; 14,000 psi YS; 7 El; 50 Brin. High leaded bronze; CDA 945.

ABEX NO. 6R

Abex Corp.
Cu 70, Sn 10, Pb 20.
Sand cast: 25,000 psi TS; 14,000 psi YS; 7 El; 50 Brin. High leaded tin bronze.

ABEX NO. 6X

Abex Corp.
Cu 70, Sn 5, Pb 25.
Centrifuge cast: 25,000 psi TS; 17,000 psi YS; 8 El. Soft bronze; CDA 943.

ABEX NO. 7A

Abex Corp.
Cu 84, Sn 10, Pb 2.5, Ni 3.5.
Sand cast: 40,000 psi TS; 20,000 psi YS; 15 El; 75 Brin. Nickel bronze; CDA 915.

ABEX NO. 964

Abex Corp.
Cu 69.1, Fe 0.9, Ni 30, Mn 0-1.5.
Cast: 60,000 psi TS; 32,000 psi YS; 15 El; 140 Brin. Copper-nickel bronze; similar to CDA 964.

ABEX NO. 9A

Abex Corp.
Cu 89, Al 10, Fe 1.
As cast: 65,000 psi TS; 25,000 psi YS; 20 El; 120 Brin. Heat treated: 80,000 psi TS; 40,000 psi YS; 12 El; 160 Brin. Aluminum bronze; CDA 953.

ABEX NO. 9AF

Abex Corp.
Cu 85, Al 11, Fe 4.
As cast: 75,000 psi TS; 30,000 psi YS; 12 El; 150 Brin. Heat treated: 90,000 psi TS; 45,000 psi YS; 6 El; 180 Brin. Aluminum bronze; CDA 954.

ABEX NO. 9L

Abex Corp.
Cu 81, Al 11, Fe 4, Ni 4.
As cast: 90,000 psi TS; 40,000 psi YS; 6 El; 190 Brin. Heat treated: 110,000 psi TS; 60,000 psi YS; 5 El; 200 Brin. Nickel aluminum bronze; CDA 955.

ABK METAL

Abex Corp.
C 3-3.75, Mn 0.3-0.9, Si 0.3-1.2, Ni 3.5-6.5, Cr 1.5-4, bal Fe.
Cast: 35,000-80,000 TS; 0.10-0.30 El; 500-650 Brin. For castings, crushing rolls, liners; wear resistant, cast iron.

ABK NI-HARD

Abex Corp.
Ni 4.5, C 2.7-3.6, Cr 1.5, Si 0.5-1.5, Mn 0.3-0.7, bal Fe.
Cast: 30,000-40,000 TS; 575-750 Brin. For pump plungers, roller bearing races, chilled rolls; corrosion resistant, tough.

ABRACORR 25

Societe Nouvelle des Acieries de Pompey
C 0.1, Mn 0.5, Ni 0.75, Cr 4, Al 0.65, Mo 0.2, bal Fe.
Weldable and easily formable. For abrasion and oxidation resisting parts.

ABRACORR 30

Societe Nouvelle des Acieries de Pompey
C 0.1, Mn 0.8, Cr 6, Al 0.8, Ni 1.5, Mo + V, bal Fe.
Weldable and easily formable. For abrasion and oxidation resisting parts.

ABRADUR 500

Societe Nouvelle des Acieries de Pompey
C 0.4, Ni 3.5, Cr 1.4, Mo 0.4, bal Fe.
For gears, shafts, crankshafts; oil hardened, shock resistant. Obsolete

ABRADUR C

Societe Nouvelle des Acieries de Pompey
C 0.9, Mn 0.75, Si 0.25, bal Fe.
Hardenable to 60 Rock C on surface. For wear and abrasion resisting parts.

ABRADUR C 65 K

Societe Nouvelle des Acieries de Pompey
C 0.65, Mn 0.8, Si 0.25, Cr 0.3, bal Fe.
Hardenable to 55 Rock C on surface. For wear and abrasion resisting parts.

ABRADUR C 75 K

Societe Nouvelle des Acieries de Pompey
C 0.75, Mn 0.8, Si 0.25, Cr 0.3, bal Fe.
Hardenable to 58 Rock C on surface. For wear and abrasion resisting parts.

ABRADUR C 90 K

Societe Nouvelle des Acieries de Pompey
C 0.9, Mn 0.8, Si 0.25, Cr 0.35, bal Fe.
Hardenable to 61 Rock C on surface. For wear and abrasion resisting parts.

ABRADUR PM 25

Societe Nouvelle des Acieries de Pompey
C 0.2, Mn 1.3, Cr 0.75, bal Fe.
Hardenable to 40 Rock C (approximate); weldable. For wear and abrasion resisting parts.

ABRADUR PM 35

Societe Nouvelle des Acieries de Pompey
C 0.16, Mn 1.6, Cr 1.5, Mn 0.2, V 0.1, bal Fe.
Hardenable to 35 Rock C (approximate); weldable. For wear and abrasion resisting parts.

ABRADUR PM 40

Societe Nouvelle des Acieries de Pompey
C 0.28, Mn 1.6, Ni 0.8, Cr 1.7, Mo 0.45, V 0.1, bal Fe.
Hardenable to 45-50 Rock C (approximate); weldable. For wear and abrasion resisting parts.

ABRADUR S

Societe Nouvelle des Acieries de Pompey
C 0.45, Mn 0.7, Si 1.8, bal Fe.
Heat treatable to 40-55 Rock C. For wear and abrasion resisting parts.

ABRADUR SK

Societe Nouvelle des Acieries de Pompey
C 0.65, Mn 0.85, Si 1.65, Cr 0.85, bal Fe.
Hardenable to 46-60 Rock C. For wear and abrasion resisting parts.

ABRASALLOY

Atlantic Steel Corp.
C 0.35, Cr 1, Mo 0.25, Mn 0.75, Si 0.35, bal Fe.
For crusher hammers, coal screens, dredge pumps; tough, wear resistant.

ABRASION RESISTING

Inland Steel Co.
Alloy steel. C 0.35-0.5, Mn 1.2-1.65, Si 0.1-0.3, bal Fe.
Z10 Brin min. For wear resistant sheet or plate.

ABRASIST

Thomas Foundries, Inc.
C, alloy, bal Fe.
For machinery castings; abrasion resistant. Obsolete

ABRASOCRAFT

E.M.F. Electric Co. Ltd.
C, alloy, bal Fe.
Welded: 350-500 Brin. For hard surfacing electrode; abrasion resistant.

ABRASOHARD

Universal Power Corp.
C alloy, bal Fe.
For welding rods; abrasion resistant.

ABRASOWELD

Lincoln Electric Co.
C 2.1, Mn 1.1, Si 0.75, Mo 0.4, Cr 6.5, bal Fe.
Hard surfacing, arc welding electrodes; resistance to abrasion and impact.

ABRATEC N700

Eutectic Corp.
Electrode for AC-DC metallic arc abrasion resistant coating of steel; 68-72 Rock C (double pass).

ABRAZO 60

British Steel plc
C 0-0.6, Si 0.25, Mn 0.6, bal Fe.
Abrasion resistant steel for applications such as liner plates, chutes, buckets, storage bins, and mold boards. Obsolete

ABROS

English manufacture
Cr 10, Ni 88, Mn 2.
For electrical resistance, heating elements; stainless and corrosion resistant.

ABSCO METAL 25H

Abex Corp.
high C, Si, Mn, bal Fe.
Cast: 25,000 TS; 150-210 Brin. For ingot molds, pig molds; cast iron, Class 25.

ABSCO METAL 30

Abex Corp.
high C, Si, Mn, bal Fe.
Cast: 30,000 TS; 20,000 YS; 0.6 El; 170-240 Brin. For gears, pulleys, general castings; cast iron, Class 30.

ABSCO METAL 30H

Abex Corp.
high C, Si, Mn, alloy, bal Fe.
Cast: 30,000 TS; 220-280 Brin. For heat and wear resistant castings; cast iron.

ABSCO METAL 35

Abex Corp.
high C, S, Mn, bal Fe.
Cast: 35,000 TS; 200-260 Brin. For gears, housings, general castings; cast iron, Class 35.

ABSCO METAL 40

Abex Corp.
high C, Si, Mn, bal Fe.
Cast: 40,000 TS; 210-260 Brin. For meter bodies, water valves, compressor cylinders; cast iron, Class 40.

ABSCO METAL 400

Abex Corp.
high C, Si, Mn, alloy, bal Fe.
Cast: 400 Brin. For ash-sluice pipe chutes, burner nozzles; heat and wear resistant.

ABSCO METAL 40H

Abex Corp.
high C, Si, Mn, alloy, bal Fe.
Cast: 40,000 TS; 230-290 Brin. For heat and wear resistant castings; cast iron.

ABSCO METAL 50

Abex Corp.
high C, Si, Mn, bal Fe.
Cast: 50,000 TS; 220-280 Brin. For press parts, cylinder heads, forming dies; cast iron, Class 50.

ABSCO METAL 55

Abex Corp.
high C, Si, Mn, bal Fe.
Cast: 55,000 TS; 45,000 YS; 0.7 El; 230-310 Brin. For crankshafts, tool shanks, cutter bodies; cast iron, Class 55.

ABYSSINIAN GOLD

English manufacture
Cu 90-92, Zn 8-10, 1 plated Au.
For ornaments, jewelry; sheets plated with Au.

ABYSSINIAN GOLD

English manufacture
Cu 88, Zn 11.5, Au 0.5.
For ornaments, jewelry.

AC-254
Allis-Chalmers Mfg. Co.
C 0.24, Mn 1, Si 0.4, Cr 12, Ni 1, Mo 2.5, W 1, V 0.25, B 0.22, bal Fe.
For steam turbine blading, jet engine components. Stainless, martensitic, creep resistant.

AC-AMSCO NO. 217
Edgcomb Metals Co.
Cr 3, Cr 8, W 15, bal Fe.
600-700 Brin. For hard facing welding rod; maintains hardness up to 1000 F. *Obsolete*

AC-AMSCO NO. 459
Edgcomb Metals Co.
C 3, Cr 4, Mo 4, bal Fe.
500-600 Brin. For hard surfacing welding rod; for severe wear and abrasion. *Obsolete*

AC-COATED 18-8
Edgcomb Metals Co.
Cr 18, Ni 8, C, bal Fe.
Coated. For stainless steel welding rod; coated rods. *Obsolete*

AC-WELD
Edgcomb Metals Co.
C 0.2, bal Fe.
Weld: 65,000-70,000 TS; 28-30 El. For welding rod; A.C. welding. *Obsolete*

ACADAL
Alumetal (Montecatini Edison)
Aluminum.
Aluminum alloy for color anodizing.

ACCOLLOY
American Chain & Cable
Ni 3.5, C 0.3, bal Fe.
For chains. *Obsolete*

ACCOLOY CN-1
Alloy Engineering & Casting Co.
C 0-0.5, Cr 28, Ni 3, bal Fe.
For chemical, mining, and paper industries; bushings, pump casings, valve bodies, and corrosive resistors; ferritic.

ACCOLOY CN-10
Alloy Engineering & Casting Co.
Ni 9, Mo 2-3, C 0-0.08, Cr 18, bal Fe.
Cast. For corrosion and heat resistance. Austenitic; ACI Type CF-8M.

ACCOLOY CN-2
Alloy Engineering & Casting Co.
C 0.4, Cr 28, Ni 10, bal Fe.
Cast: 95,000 TS; 45,000 YS; 20 El; 200 Brin. For furnace parts, heat treating equipment, and pots. Type HE; austenitic; heat resistant.

ACCOLOY CN-3
Alloy Engineering & Casting Co.
C 0.4, Cr 25, Ni 20, bal Fe.
Cast: 75,000 TS; 50,000 YS; 17 El; 170 Brin. For furnace parts, heat treating equipment, and pots. Type HK; austenitic; heat resistant.

ACCOLOY CN-4
Alloy Engineering & Casting Co.
C 0.3, Cr 25, Ni 12, bal Fe.
Cast: 80,000 TS; 50,000 YS; 25 El; 185 Brin. For furnace parts, heat treating equipment, and pots. Type HH2; austenitic; heat resistant.

ACCOLOY CN-5
Alloy Engineering & Casting Co.
C 0.25, Cr 18, Ni 8, bal Fe.
Cast: 75,000 TS; 45,000 YS; 30 El; 175 Brin. For carburizing boxes; corrosion and heat resistant, austenitic. *Obsolete*

ACCOLOY CN-6
Sandvik Hard Materials Ltd.
C 0.2-0.6, Cr 30, Ni 20, bal Fe.
Cast: 82,000 psi TS; 52,000 psi YS; 19 El; 192 Brin. For furnace fixtures, furnace parts, heat treating equipment, radiant tubes. Type HL; heat and corrosion resistant. *Obsolete*

ACCOLOY CN-7
Alloy Engineering & Casting Co.
C 0-0.2, Cr 18, Ni 8, bal Fe.
Cast. For chemical, petrochemical, and food industries; pumps, return bends, valve bodies, and cylinder liners. Heat and corrosion resistant. ACI Type CF-20.

ACCOLOY CN-8
Alloy Engineering & Casting Co.
C 0-0.08, Cr 18, Ni 8, bal Fe.
Cast. For blast furnace, guide rollers, headers, pumps, and corrosion resistant operations. ACI Type CF-8.

ACCOLOY CNC-4B
Alloy Engineering & Casting Co.
C 0-0.1, Cr 24, Ni 12, bal Fe.
For furnace parts, heat treating equipment, pots; corrosion and heat resistant, austenitic. *Obsolete*

ACCOLOY CNC-4D
Alloy Engineering & Casting Co.
C 0-0.2, Cr 24, Ni 12, bal Fe.
For furnace parts, heat treating equipment, pots; heat and corrosion resistant, austenitic. *Obsolete*

ACCOLOY CNC-5A
Alloy Engineering & Casting Co.
C 0.07, Cr 18, Ni 8, bal Fe.
Cast: 85,000 TS; 150 Brin. For castings, baskets, carburizing boxes; Type 304; stainless, austenitic. *Obsolete*

ACCOLOY CNC-5B
Alloy Engineering & Casting Co.
C 0.1, Cr 18, Ni 8, bal Fe.
Cast: 85,000 TS; 150 Brin. For heat treating boxes, pots, baskets; Type 302; stainless, austenitic. *Obsolete*

ACCOLOY CNC-5C
Alloy Engineering & Casting Co.
C 0-0.16, Cr 18, Ni 8, bal Fe.
Cast: 90,000 TS; 160 Brin. For heat treating boxes, chemical plant equipment; Type 302; stainless, austenitic. *Obsolete*

ACCOLOY NC-1
Alloy Engineering & Casting Co.
C 0.35-0.75, Ni 68, Cr 18, bal Fe.
Cast: 65,000 TS; 36,000 YS; 9 El; 176 Brin. For furnace parts, radiant tubes, heating elements, and heat treating equipment. ACI Type HX; austenitic.

ACCOLOY NC-10
Alloy Engineering & Casting Co.
C 0.35-0.75, C 25, Ni 35, bal Fe.
Cast: 71,000 TS; 40,000 YS; 11.5 El. For heat treating, petrochemical and petroleum industries. For ethylene heaters, heat treat fixtures, and refinery tubes. Type HP; austenitic; heat resistant.

ACCOLOY NC-2
Alloy Engineering & Casting Co.
C 0.35-0.75, Ni 60, Cr 12, bal Fe.
Cast: 68,000 TS; 36,000 YS; 4 El; 185 Brin. For pots, electric heat elements, retorts, hearth plates, and muffles. ACI Type HW.

ACCOLOY NC-3
Alloy Engineering & Casting Co.
C 0.35-0.75, Ni 38, Cr 18, bal Fe.
Cast: 70,000 TS, 40,000 YS; 9 El; 170 Brin. For heat treating trays, burner tubes, carburizing retorts, conveyer screws, chains, furnace rolls, and radiant tubes. ACI Type HU; austenitic; heat resistant.

ACCOLOY NC-4
Alloy Engineering & Casting Co.
C 0.35-0.75, Ni 35, Cr 15, bal Fe.
Cast: 70,000 TS; 40,000 YS; 10 El; 180 Brin. For carburizing containers, furnace parts, heat treating equipment and fixtures, radiant tubes, roller rails, and feed screws. Type HT; austenitic; heat resistant.

ACCOLOY NC-5
Alloy Engineering & Casting Co.
Ni 30, Cr 10, C, bal Fe.
For castings; heat resistant *Obsolete*

ACCOLOY NC-6
Alloy Engineering & Casting Co.
C 0.2-0.5, Ni 25, Cr 20, bal Fe.
Cast: 68,000 TS; 38,000 YS; 13 El; 160 Brin. For brazing fixtures, chains, furnace beams and parts, pier caps, radiant tubes, nozzles, and trays. ACI Type HN; austenitic; heat resistant.

ACCOLOY NC-7
Alloy Engineering & Casting Co.
C 0.2-0.4, Cr 21, Ni 9, bal Fe.
Cast: 92,000 TS; 45,000 YS; 38 El; 165 Brin. For annealing boxes and trays, baskets, burner tips, conveyer belts, furnace rails, wear plates, and hearth plates. ACI Type HF; austenitic; heat resistant.

ACCOLOY NC-9MO
Alloy Engineering & Casting Co.
C 0.35-0.75, Cr 22, Ni 46, bal Fe.
Cast to resist stress-corrosion cracking in petrochemical applications and furnace applications over 2000°F.

ACCULOY 280 C
Fugi Iron & Steel Co., Ltd.
Au, bal Sn.
Uses: Brazing alloy for electronic parts. High corrosion and etch resistance. High electrical conductivity. *Obsolete*

ACCUMULATOR METAL
American manufacture
Pb 90, Sn 9.2, Sb 0.8.
Bearings, battery plates, antrifriction.

ACCUNAMEL
Bethlehem Steel Corp.
C 0.03, Mn 0.15, B 0.004, bal Fe.
Low carbon, continuous cast, continuous annealed, cold rolled sheet steel. For porcelain enameling applications; washing machines, dishwashers, gas and electric ranges, sinks, and bathtubs.

ACCURATE METAL
Connecticut Metals Corp.
Ni 29-32, Fe 0.4-0.7, 68.0 Cu min.
Annealed: 57,000 TS; 21,000 YS; 43 El; 44 Rock B. Hard: 82,000 TS; 75,000 YS; 5 El; 86 Rock B. Soft: 54,000 TS; 16,000 YS; 35 El; 40 Rock B. For electronic components, marine hardware, ferrules, pump valves. Corrosion resistant, non-magnetic.

ACCURLOY
Baldwin Steel Co.
C 0.51, Cr 1.05, Mo 0.25, Ni 0.53, V 0.21, Mn 0.97, bal Fe.
Heat treated: 165,000 TS; 150,000 YS; 20 El; 59 RA; 280-300 Brin. For shafts, pins, boring bars, piston rods; heat treated and stress-relieved bars, shock resistant.

ACE OIL HARDENING
Horace T. Potts Co.
C, alloy, bal Fe.
For tools; non-deforming. *Obsolete*

ACENOR 111
ACENOR, S.A.
C 0.13-0.18, Mn 0.7-1.1, Si 0-0.35, Cr 0.8-1.2, Ni 0.8-1.2, bal Fe.
15NiCr4. Quenched and treated: 85-115 kg/mm² TS. Case hardening steel for gears, bolts, etc.

ACENOR 15CR3
ACENOR, S.A.
C 0.12-0.18, Mn 0.4-0.6, Si 0.15-0.4, Cr 0.6-0.8, bal Fe.
Quenched and treated: 70-100 kg/mm^2 TS. Case hardening, carburizing steel for bolts. G51170

ACENOR 18NICRMO5
ACENOR, S.A.
C 0.15-0.21, Mn 0.5-0.9, Si 0.15-0.4, Cr 0.7-1, Ni 1.2-1.5, Mo 0.15-0.25, bal Fe.
Quenched and treated: 110-140 kg/mm^2 TS. Case hardening, carburizing steel. G43200

ACENOR 27CM12
ACENOR, S.A.
C 0.24-0.3, Mn 0.5-0.7, Si 0-0.35, Cr 2.75-3.25, Mo 0.3-0.5, bal Fe.
Quenched and treated: 90-110 kg/mm^2 TS. Nitriding steel for gears, bolts, etc.

ACENOR 27MNCR5
ACENOR, S.A.
C 0.23-0.31, Mn 1.1-1.4, Si 0.1-0.4, Cr 1-1.3, bal Fe.
Quenched and treated: 85-105 kg/mm^2 TS. Steel for superficial hardening. G51300

ACENOR 45SCD6
ACENOR, S.A.
C 0.42-0.5, Mn 0.5-0.8, Si 1.3-1.7, Cr 0.5-0.75, Mo 0.15-0.3.
Quenched and treated: 100-120 kg/mm^2 TS. Spring steel. G92540

ACENOR 55CR3
ACENOR, S.A.
C 0.52-0.57, Mn 0.7-0.9, Si 0.2-0.4, Cr 0.6-0.8, Ni 0-0.2, Mo 0-0.06, bal Fe.
Quenched and treated: 130-150 kg/mm^2 TS. Spring steel. G51550

ACENOR ALS
ACENOR, S.A.
C 0-0.11, Mn 1-1.4, Si 0-0.05, P 0.03-0.1, S 0.28-0.38, bal Fe.
Normalized: 38 kg/mm^2 TS min. Free cutting steel for bolts, screws, etc. G12130

ACENOR ALSP
ACENOR, S.A.
C 0-0.11, Mn 1-1.4, Si 0-0.5, P 0.03-0.1, S 0.28-0.38, Pb 0.2-0.35, bal Fe.
Free cutting leaded steel for screw machine products. G12144

ACENOR B-137
ACENOR, S.A.
C 0.3-0.38, Mn 0.4-0.7, Si 0.15-0.4, Cr 1.45-1.65, Ni 1.4-1.7, Mo 0.15-0.3, bal Fe.
Quenched and treated: 100-130 kg/mm^2 TS; for gears, bolts, etc.

ACENOR CAN
ACENOR, S.A.
C 0.35-0.41, Mn 0.4-0.6, Si 0-0.4, Cr 1.3-1.7, Mo 0.15-0.25, Al 0.75-1.25, bal Fe.
Quenched and treated: 90-110 kg/mm^2 TS. Nitriding steel.

ACENOR CB
ACENOR, S.A.
C 0.95-1.2, Mn 0.2-0.4, Si 0.15-0.35, Cr 1.4-1.8, bal Fe.
Quenched and treated: 60-65 Rock C. Ball and roller bearing steel. G52986

ACENOR CNE
ACENOR, S.A.
C 0.11-0.16, Mn 0.35-0.65, Si 0.15-0.4, Cr 0.6-0.9, Ni 2.5-3, bal Fe.
Quenched and treated: 95-125 kg/mm^2 TS. Case hardening steel for gear and link components.

ACENOR DTA
ACENOR, S.A.
C 0.29-0.35, Mn 0.45-0.75, Si 0.15-0.4, Cr 1.1-1.4, Ni 4-4.5, bal Fe.
Quenched and treated: 110-130 kg/mm^2 TS. Highly stressed crankshafts, seal axle arbors.

ACENOR ELASTIC
ACENOR, S.A.
C 0.29-0.35, Mn 0.45-0.75, Si 0.15-0.4, Cr 0.5-0.8, Ni 2.25-2.75, Mo 0.45-0.55, bal Fe.
Quenched and treated: 100-125 kg/mm^2 TS. Oil hardening; high tensile strength for gears, shafts.

ACENOR F-1130
ACENOR, S.A.
C 0.3-0.4, Mn 0.5-0.8, Si 0.15-0.4, bal Fe.
Quenched and treated: 58-78 kg/mm^2 TS. For structural components subjected to high stress. G10260

ACENOR F-1140
ACENOR, S.A.
C 0.4-0.5, Mn 0.5-0.8, Si 0.15-0.4, bal Fe.
Quenched and treated: 67-87 kg/mm^2 TS. Automobile and motor construction. Steel for superficial hardening. G10430

ACENOR F-1150
ACENOR, S.A.
C 0.5-0.6, Mn 0.6-0.9, Si 0.15-0.4, bal Fe.
Quenched and treated: 79-93 kg/mm^2 TS. Automobile and motor construction. Steel for superficial hardening. G10550

ACENOR F-1202
ACENOR, S.A.
C 0.38-0.45, Mn 0.5-0.9, Si 0.15-0.4, Cr 0.9-1.2, bal Fe.
Quenched and treated: 90-110 kg/mm^2 TS. Axle journals, control components. G41400

ACENOR F-1282
ACENOR, S.A.
C 0.37-0.43, Mn 0.5-0.8, Si 0.15-0.4, Cr 0.6-0.9, Ni 0.7-1, Mo 0.15-0.3, bal Fe.
Quenched and treated: 100-120 kg/mm^2 TS. Crankshafts, eccentric shafts gear, etc. G86400

ACENOR F-1515
ACENOR, S.A.
C 0.18-0.23, Mn 1.3-1.6, Si 0.15-0.4, bal Fe.
Quenched and treated: 70-100 kg/mm^2 TS. Case hardening steel for structural components. G15220

ACENOR F-1516
ACENOR, S.A.
C 0.13-0.19, Mn 1-1.3, Si 0.16-0.4, Cr 0.0-1.1, bal Fe.
Quenched and treated: 80-110 kg/mm^2 TS. Small cog wheels, arbors, cardan joints, control parts. G15200

ACENOR F-1522
ACENOR, S.A.
C 0.17-0.23, Mn 0.6-0.9, Si 0.15-0.4, Cr 0.4-0.6, Ni 0.4-0.6, Mo 0.15-0.25, bal Fe.
Quenched and treated: 85-115 kg/mm^2 TS. Case hardening steel for gear components, bolts, arbors, bushes, coupling boxes. G86200

ACENOR F-1523
ACENOR, S.A.
C 0.18-0.23, Mn 0.6-0.9, Si 0.15-0.4, Cr 0.3-0.5, Mo 0.4-0.5, bal Fe.
Quenched and treated: 75-105 kg/mm^2 TS. Case hardening steel for gear components, gear boxes, driving pinons, plate, wheels. G43200

ACENOR F-1525
ACENOR, S.A.
C 0.18-0.23, Mn 0.6-0.8, Si 0.15-0.4, Cr 0.4-0.6, Ni 1.4-1.7, Mo 0.3-0.4, bal Fe.
Quenched and treated: 105-135 kg/mm^2 TS. Case hardening steel for plate wheels, driving pinions, and highly stressed cog wheels. G43200

ACENOR F-1580
ACENOR, S.A.
C 0.17-0.22, Mn 0.8-1, Si 0.15-0.4, Cr 0.8-1.2, Ni 0.8-1.2, bal Fe.
Quenched and treated: 95-125 kg/mm^2 TS. Case hardening steel for plate wheels, driving pinions, and highly stressed wheels.

ACENOR HSL
ACENOR, S.A.
C 0.11-0.17, Mn 0.25-0.55, Si 0.15-0.4, Cr 0.8-1.2, Ni 3.75-4.25, bal Fe.
Quenched and treated: 115-140 kg/mm^2 TS. Case hardening steel for crankshafts, gear steels, cardan joints for standard stress. G33106

ACENOR K 1110
ACENOR, S.A.
C 0.1-0.2, Mn 0.4-0.7, Si 0.15-0.4, bal Fe.
Quenched and treated: 40-60 kg/mm^2 TS. Small machine components, levers, links, bushes, bolts, pins. G10120

ACENOR MCV
ACENOR, S.A.
C 0.45-0.55, Mn 0.6-0.9, Si 0.1-0.35, Cr 0.85-1.15, V 0.11-0.21, bal Fe.
Quenched and treated: 130-165 kg/mm^2 TS. Spring steel for laminated springs, helical and plate springs, and torsion bar springs (40 mm min). G61500

ACENOR MP 2311
ACENOR, S.A.
Tool steel. C 0.31-0.35, Mn 1-1.25, Si 0.18-0.42, Cr 1.55-1.85, Mo 0.18-0.22, V 0.08-0.12, bal Fe.
Pressure casting molds for light metals.

ACENOR SPT
ACENOR, S.A.
C 0-0.11, Mn 1-1.4, Si 0-0.05, P 0.03-0.1, S 0.28-0.38, Pb 0.2-0.35, Te 0-0.035, bal Fe.
Normalized: 38 kg/mm^2 TS min. Free cutting leaded steel; for screw machine products. G12144

ACENOR TROKER
ACENOR, S.A.
Tool steel. C 0.5-0.6, Mn 0.4-0.6, Si 0.15-0.35, Cr 0.9-1.2, Ni 2.8-3.2, Mo 0.25-0.45, bal Fe.
Cold heading dies of all kinds, hobbing dies, shear blades.

ACEROID
Central Foundry Co.
Zn, bal Cu.
For castings.

ACEROLD
Central Foundry Co.
C, alloy, bal Fe.
For machinery castings.

ACHORN "350" FINISHING
Achorn Steel Co.
C 1.3, W 3.5, bal Fe.
For tools, dies; fast finishing steel.

ACHORN 100 CHROMIUM
Achorn Steel Co.
C 1, Cr 1.5, bal Fe.
Low alloy tool steel, oil hardening. AISI L1.

ACHORN 11W HOT WORK
Achorn Steel Co.
C 0.4, Cr 3, V 0.45, W 12, bal Fe.
Oil hardening hot work tool steel, tungsten type; for forging and hot working dies. AISI H22.

ACHORN 15 W HOT WORK
Achorn Steel Co.
C 0.45, Cr 3.5, V 0.7, W 14, bal Fe.
Air or oil hardening hot working tool and die steel; AISI H24.

ACHORN 18-4-1
Achorn Steel Co.
C 0.65, Cr 3.75, W 18.5, V 1.2, bal Fe.
For tools, cutters; high speed steel.

ACHORN 225 C HIGH PRODUCTION DIE
Achorn Steel Co.
C 2.1, Cr 12, W 0.75, bal Fe.
Oil or air hardening cold work tool steel; chromium type; for punch and trim dies, thread rolling tools, gages. AISI D3.

ACHORN 33 HOT WORK
Achorn Steel Co.
C 0.33, Si 0.85, Cr 5, V 0.2, W 1.25, Mo 1.45, bal Fe.
Hot work tool steel, chromium type; AISI H12.

ACHORN 33A HOT WORK
Achorn Steel Co.
C 0.4, Mn 0.3, Si 0.9, Cr 5, V 0.5, Mo 1.3, bal Fe.
Oil or air hardening tool and die steel; for forging dies and hot forming dies. AISI H11.

ACHORN 33M HOT WORK
Achorn Steel Co.
C 0.4, Si 1, Cr 5, V 1, Mo 1, bal Fe.
Air or oil hardening hot work tool steel; AISI H13.

ACHORN 512
Achorn Steel Co.
C 0.1, Mn 0.48, Cr 1.5, Ni 3.5, bal Fe.
Low carbon steel for cold hubbing and then case hardening for molds.

ACHORN 6-6-4-2
Achorn Steel Co.
C 0.85, Cr 4, V 2, W 6.5, Mo 5, bal Fe.
For tools, cutters; high speed steel.

ACHORN 9W HOT WORK
Achorn Steel Co.
C 0.3, Cr 3, V 0.5, W 9.5, bal Fe.
Hot work tool steel; oil hardening; for forging dies, hot forming dies; AISI H21.

ACHORN AF-33 HOT WORK
Achorn Steel Co.
C 0.3-0.35, Si 0.8-1, Cr 5, W 1.1, Mo 1.5, bal Fe.
For hot work dies; hot work steel.

ACHORN ALLOY PIVOT
Achorn Steel Co.
C 1.1, Mn 0.4, Si 0.25, Cr 1.35, Mo 0.4, bal Fe.
Oil hardening steel for shafts, pivots, lathe centers. AISI L7.

ACHORN BEST CARBON
Achorn Steel Co.
C 0.8-1.2, Si 0.25, Mn 0.25, bal Fe.
Water hardened: 166,000-216,000 TS; 110,000-150,000 YS; 11-15 El; 32-37 RA; 330-600 Brin. For taps, reamers, drills, punches, stamps, knurls, mandrels, cutters. Type W1 water hardening.

ACHORN CARBON DRILL ROD
Achorn Steel Co.
C 1, Mn 0.25, Si 0.28, bal Fe.
Water hardening tool steel; AISI W1.

ACHORN CM AIR HARDENING
Achorn Steel Co.
C 0.7, Mn 2, Si 0.3, Cr 1, Mo 1.35, bal Fe.
Medium alloy, air hardening cold work tool steel; AISI A6.

ACHORN COLD DRAWN TOOL STEEL
Achorn Steel Co.
C 1.05, Mn 0.25, Si 0.28, bal Fe.
Water hardening tool steel, for drills, arbors, lathe centers, bushings. AISI W1.

ACHORN COLD HEADING
Achorn Steel Co.
C 0.8-1.2, Si 0.25, Mn 0.25, bal Fe.
Water hardened: 166,000-216,000 TS; 110,000-150,000 YS; 11-15 El; 32-37 RA; 330-600 Brin. For taps, drills, reamers, punches, stamps, knurls, mandrels, cold heading tools. Type W1 water hardening.

ACHORN COMPOSITE STEEL
Achorn Steel Co.
C 1, Cr 0.4, V 0.4, W 1.25, bal Fe.
Water or oil hardening steel for shafts, arbors, lathe centers, drill bushings. AISI F1.

ACHORN CRM-50
Achorn Steel Co.
C 0.5-0.55, Mn 0.6, Cr 1, Mo 0.2, bal Fe.
For hot work dies; hot work steel.

ACHORN CVM
Forges et Acieries de Voelkingen
C 0.95, Mn 0.7, Cr 5, V 0.25, Mo 1.2, bal Fe.
For punches, blanking and forming dies; air hardened, nondeforming.

ACHORN EXTRA BLADE
Achorn Steel Co.
C 1.1, Mn 0.25, bal Fe.
For tools, cutters; oil hardened.

ACHORN EXTRA CARBON
Achorn Steel Co.
C 1.05, Mn 0.25, Si 0.28, bal Fe.
Water hardening tool steel; AISI W1.

ACHORN EXTRA CHISEL
Achorn Steel Co.
C 0.8-1.1, Si 0.25, Mn 0.25, bal Fe.
Water hardened: 166,000-216,000 TS; 110,000-150,000 YS; 11-15 El; 32-37 RA; 330-600 Brin. For taps, drills, reamers, punches, stamps, knurls, mandrels. Type W1 water hardening.

ACHORN EXTRA SOLID DRILL
Achorn Steel Co.
C 1.15, Mn 0.3, bal Fe.
For tools, drills; water hardened.

ACHORN FAGERSTA BEST
Achorn Steel Co.
C 1.05-1.15, Mn 0.2-0.25, Si 0.25, bal Fe.
For drills, cutters, reamers, taps, broaches; Type W1; water hardened.

ACHORN FAGERSTA CHISEL
Achorn Steel Co.
C 0.8-0.9, Mn 0.3, Si 0.25, bal Fe.
For chisels, screw drivers; water hardened; Type W1.

ACHORN FAGERSTA COLD HEADING
Achorn Steel Co.
C 0.95, Mn 0.3, Si 0.25, bal Fe.
For cold heading dies, form tools; Type W1; water hardened.

ACHORN FAGERSTA ENVELOPE DIE
Achorn Steel Co.
C 1.05-1.15, Mn 0.3, Si 0.25, bal Fe.
For envelope dies, cutters, drills, form tools; Type W1; water hardened.

ACHORN FAGERSTA EXTRA
Achorn Steel Co.
C 1-1.1, Mn 0.3, Si 0.25, bal Fe.
For tools, cutters, drills, taps, reamers; Type W1; water hardened.

ACHORN FAGERSTA EXTRA CUTLERY
Achorn Steel Co.
C 1.07-1.12, Mn 0.25, Si 0.3, bal Fe.
For cutlery tools, drills, taps; Type W1; water hardened.

ACHORN FAGERSTA FINISHING
Achorn Steel Co.
C 1.15, Mn 0.4, Cr 0.4, W 2.5, bal Fe.
For finishing cutters; water hardened, keen cutting edge.

ACHORN FAGERSTA HIGH PRODUCTION
Bruksconcernen AB
C 1.6, Mn 0.8, Cr 12, V 0.2, bal Fe.
For blanking and forming dies; air hardened, non-deforming.

ACHORN FAGERSTA HIGH PRODUCTION
Achorn Steel Co.
C 1.6, Mn 0.8, Cr 12, V 0.2, bal Fe.
For blanking and forming dies; air hardened, non-deforming.

ACHORN FAGERSTA HOT DIE
Achorn Steel Co.
C 0.55, Mn 0.45, W 5, Ni 3, bal Fe.
For upsetting and forging dies, extrusion rams and liners; hot work steel, oil hardened.

ACHORN FAGERSTA SHOE DIE
Bruksconcernen AB
C 0.55, Cr 0.65, Mo 0.35, bal Fe.
For shoe dies, cutting dies for leather and rubber; oil hardening.

ACHORN FAGERSTA SHOE DIE
Achorn Steel Co.
C 0.55, Cr 0.65, Mo 0.35, bal Fe.
For shoe dies, cutting dies for leather and rubber; oil hardening.

ACHORN FAGERSTA SILVER DIE
Achorn Steel Co.
C 1, Mn 0.25, Si 0.25, bal Fe.
For silver dies; Type W1; water hardened.

ACHORN FAGERSTA SMOOTH BOR HOLLOW
Achorn Steel Co.
C 0.8, Mn 0.25, Si 0.25, bal Fe.
For rock drills; Type W1; water hardened.

ACHORN FAGERSTA SOLID DRILL
Achorn Steel Co.
C 0.8-0.9, Mn 0.35, Si 0.25, bal Fe.
For solid drills; Type W1; water hardened.

ACHORN FAGERSTA SPECIAL ALLOY DIE
Achorn Steel Co.
C 0.45-0.55, Cr 0.9, W 1.2, Mo 0.2, bal Fe.
For dies, tools; oil hardened.

ACHORN FAGERSTA STANDARD
Achorn Steel Co.
C 0.9-1, Mn 0.3, Si 0.25, bal Fe.
For tools, drills, taps, hobs, reamers; Type W1; water hardened.

ACHORN FAGERSTA STANDARD CUTLERY
Achorn Steel Co.
C 0.85-0.95, Mn 0.35, Si 0.25, bal Fe.
For cutlery tools, drills; reamers; Type W1; water hardened.

ACHORN FAGERSTA SUPERIOR OIL
Achorn Steel Co.
C 1, Mn 1, Cr 0.4, W 0.4, bal Fe.
For tools, dies, cold headers; water or oil hardened.

ACHORN FAGERSTA UNBREAKABLE CHISEL
Achorn Steel Co.
C 0.45, Cr 0.9, W 1.2, Mo 0.2, bal Fe.
For chisels, upsetters; oil hardened, tough.

ACHORN FAGERSTA WHITE GOLD
Achorn Steel Co.
C 1.15, Mn 0.4, Cr 0.4, W 1.2, bal Fe.
For tools, cutters; oil or water hardened.

ACHORN GRAPHITIC OIL
Achorn Steel Co.
C 1.45, Mn 0.8, Si 1.15, Cr 0.2, Mo 0.25, bal Fe.
Oil hardening cold work tool steel; AISI O6.

ACHORN HEAT TREATED MOLD STEEL
Achorn Steel Co.
C 0.3, Mn 0.8, Si 0.5, Cr 1.7, Mo 0.4, bal Fe.
Hardened, machinable, for molds; AISI P20.

ACHORN HIGH PRODUCTION
Achorn Steel Co.
C 1.6, Cr 12, Mo 0.8, V 0.8, bal Fe.
For punches, drawing dies, form tools; air hardened, non-deforming.

ACHORN HOLLOW DRILL
Achorn Steel Co.
C 0.75, Mn 0.3, bal Fe.
Heat treated: 180,000 TS; 130,000 YS; 12 El; 36 RA; 360 Brin.
For hollow drills, die blocks; water hardened; Type W1.

ACHORN KLOSTER BRILLIANT AX
Achorn Steel Co.
C 0.7, Cr 4, W 18, V 1.75, bal Fe.
For tools, cutters, reamers, hobs, lathe and planer tools; high speed steel; Type T1.

ACHORN KLOSTER BRILLIANT WKE
Achorn Steel Co.
C 0.7, Cr 4.5, W 18, V 1.5, Co 5, bal Fe.
For lathe and planer tools, hobs, reamers, taps, drills; high speed steel; Type T4.

ACHORN KLOSTER PRIOR EXTRA
Achorn Steel Co.
C 0.6-0.7, Cr 4-5, W 18-19, V 1.2, Co 1, bal Fe.
For lathe and planer tools, hobs, reamers, taps, drills; high speed steel, oil hardened.

ACHORN KLOSTER REMA
Achorn Steel Co.
C 0.02-0.05, Mn 0.02-0.07, bal Fe.
For molding dies for plastics; hobbing steel.

ACHORN M-2
Achorn Steel Co.
C 0.8, Mo 5, W 6.5, V 2, Cr 4, bal Fe.
For lathe and planer tools, drills, reamers; high speed steel.

ACHORN M1 HIGH SPEED
Achorn Steel Co.
C 0.8, Cr 4, V 1, W 1.5, Mo 8, bal Fe.
High speed steel, molybdenum type; for drills, lathe tools, milling cutters. AISI M1.

ACHORN M10 HIGH SPEED
Achorn Steel Co.
C 0.85, Cr 4, V 2, Mo 8, bal Fe.
Molybdenum type high speed steel; AISI M10.

ACHORN M2 1/2 (CLASS 1) HIGH SPEED
Achorn Steel Co.
C 1, Cr 4, V 2.5, W 6.25, Mo 6.25, bal Fe.
High speed tool steel, molybdenum-tungsten type; AISI M3 Class 1.

ACHORN M3 (CLASS 2) HIGH SPEED
Achorn Steel Co.
C 1.15, Cr 4, V 3, W 6, Mo 5.5, bal Fe.
High speed steel, tungsten-molybdenum-vanadium-chromium; AISI M3 Class 2.

ACHORN M34 HIGH SPEED
Achorn Steel Co.
C 0.9, Cr 4, V 2, W 1.75, Mo 8.5, Co 8, bal Fe.
High speed steel, molybdenum-cobalt type, AISI M34.

ACHORN M4 HIGH SPEED
Achorn Steel Co.
C 1.28, Cr 4.5, V 4, W 5.5, Mo 4.5, bal Fe.
High speed steel, tungsten-molybdenum-vanadium-chromium; AISI M4.

ACHORN MANGANESE OIL HARDENING
Achorn Steel Co.
C 0.9, Mn 1.5, Si 0.25, Mo 0.3, bal Fe.
Oil hardening in small sections; AISI O2.

ACHORN MOLDALOY
Achorn Steel Co.
C 0.1, Mn 0.5, Cr 0.5, Ni 1.2, bal Fe.
For dies, molds; carburizing grade.

ACHORN NI-CRO-MO
Achorn Steel Co.
C 0.68, Mn 0.6, Si 0.25, Cr 0.65, Ni 1.4, Mo 0.2, bal Fe.
Oil hardening, low alloy tool steel for shafts, arbors, lathe centers, tool holders. AISI L6.

ACHORN OIL HARDENING DRILL ROD
Achorn Steel Co.
C 0.9, Mn 1.1, W 0.5, Cr 0.5, V 0.15, bal Fe.
Oil hardenable in small sections; AISI O1.

ACHORN OILWEAR
Achorn Steel Co.
C 1.25, Mn 0.3, Si 0.35, Cr 0.4, V 0.2, W 1.4, bal Fe.
Oil hardening cold work tool steel, wear resistant type; AISI O7.

ACHORN REMA IRON
Achorn Steel Co.
C 0.06, Mn 0.21, bal Fe.
For plastic mold dies; water hardened.

ACHORN SOLID DRILL
Achorn Steel Co.
C 0.85, Mn 0.35, bal Fe.
Heat treated: 190,000 TS; 145,000 YS; 10 El; 30 RA; 400 Brin.
For drills, punches, taps, reamers, cutters; water hardened; Type W1.

ACHORN SPRING STEEL SHEET
Achorn Steel Co.
C 1, Mn 0.25, Si 0.28, bal Fe.
Water hardened sheet steel, spring temper; AISI W1.

ACHORN STANDARD CARBON
Achorn Steel Co.
C 1-1.1, Mn 0.3, bal Fe.
For cutters, drills, reamers, punches; water hardened; Type W1.

ACHORN SUPERIOR OIL HARDENING
Achorn Steel Co.
C, alloy, bal Fe.
For tools, dies, punches; oil hardened.

ACHORN T15 HIGH SPEED
Achorn Steel Co.
C 1.5, Cr 4.75, V 5, W 12.5, Co 5, bal Fe.
High speed steel, tungsten, cobalt, vanadium type; AISI T15.

ACHORN T2 HIGH SPEED
Achorn Steel Co.
C 0.8, Cr 4, V 2, W 18.5, Mo 0.75, bal Fe.
Tungsten type high speed steel; AISI T2.

ACHORN T4 HIGH SPEED
Achorn Steel Co.
C 0.75, Cr 4, V 1.05, W 18.5, Co 5, bal Fe.
Tungsten-cobalt type high speed steel; AISI T4.

ACHORN T5 HIGH SPEED
Achorn Steel Co.
C 0.8, Cr 4.25, V 2, W 19, Mo 1, Co 8, bal Fe.
Tungsten-cobalt type high speed steel; AISI T5.

ACHORN TOOL STEEL SHEET
Achorn Steel Co.
C 1.05, Mn 0.25, Si 0.28, bal Fe.
Water hardenable sheet steel; AISI W1.

ACHORN UBC
Achorn Steel Co.
C 0.48, Cr 0.9, W 1.1, Mo 0.2, bal Fe.
For hot work dies; hot work steel.

ACHORN USI STEEL
Achorn Steel Co.
Mn 0.85, Si 2, Mo 0.25, Cr 0.25-0.6, V 0.2, bal Fe.
Water or oil hardening tool steel, shock resisting type; AISI S5.

ACHORN V85 STEEL
Achorn Steel Co.
C 0.9, Mn 0.3, Si 0.25, V 0.25, bal Fe.
Water hardening tool steel; AISI W2.

ACHORN VAMO
Achorn Steel Co.
C 0.5, Mn 0.45, Si 1, Cr 5.2, Ni 1.5, V 1, Mo 1.4, bal Fe.
Medium alloy, air hardening cold work tool steel; AISI A9.

ACHORN VBC
Achorn Steel Co.
C 0.5, Cr 0.9, W 1.25, Mo 2, Mn 0.25, bal Fe.
For punches, rivet sets, upsetters; oil hardened, hot work steel.

ACIBAL
Manufacturer not listed.
Al alloy. For light alloy parts.

ACIBEL
British Steel Corp.
C 0.07-0.13, Pb 0.15-0.25, Mn 0.85-1.15, S 0.23-0.33, bal Fe.
Rolled: 70,000 TS. For shafts, gears, machine tool parts; free-cutting, case hardened. *Obsolete*

ACIBRADE
British Steel plc
C 0.6-0.75, Mn 1.25-1.55, Si 0.05, bal Fe.
Resistant to abrasion and tough in the as-rolled condition. For rod mills, coke crushers, mineral dressing, chutes, conveyers, and buckets. *Obsolete*

ACICULAR
Sheepbridge Engineering Ltd.
C 2.9, Si 2, Ni 2.5, Mo 0.9, bal Fe.
Cast: 280-350 Brin. For castings, gears, housings; alloy cast iron.

ACICULAR
Sheepbridge Alloy Castings Ltd.
C 2.9, Si 2, Ni 2.5, Mo 0.9, bal Fe.
Cast: 280-350 Brin. For castings, gears, housings; alloy cast iron.

ACID BRONZE-1
English manufacture
Cu 88, Sn 10, Pb 2.
For bearings, chemical equipment; corrosion resistant.

ACID BRONZE-2
English manufacture
Cu 82, Sn 8, Pb 8, Zn 2.
For chemical equipment; corrosion resistant.

ACID BRONZE-3
English manufacture
Cu 84, Sn 9.5, Pb 6.3.
For chemical equipment; corrosion resistant.

ACID BRONZE-4
English manufacture
Cu 74, Sn 8, Pb 17, Zn 1.5.
For chemical equipment; corrosion resistant.

ACID METAL
American manufacture
Cu 88, Sn 10, Pb 2.
For chemical equipment; corrosion resisting.

ACID RESISTING-1
English manufacture
Cr 10, Ni 30, Fe 40.8, W 6.2, Mn 1.55, Si 0.45.
For chemical apparatus, acid resisting vessels and tanks;
stainless and corrosion resistant.

ACID RESISTING-2
English manufacture
Cr 14.5, Ni 20, Fe 56, Co 5, Cu 4.5.
For chemical apparatus; resists attack of HNO_3.

ACID RESISTING-3
English manufacture
Cr 15, Ni 53, Fe 23, W 4, Mn 1.25, Si 3.75.
For chemical apparatus; heat and corrosion resistant.

ACIDUR
Machinenbau A.G.
Si 16-17, bal Fe.
Cast: for chemical apparatus; corrosion resistant. *Obsolete*

ACIER A3
Creusot-Loire
C 0.5, bal Fe.
Annealed: 96,000 TS; 52,000 YS; 16 El; 23 RA; 170 Brin. For
axles, gears, bolts, bushings, crankshafts; water hardened.
Obsolete

ACIER ACM
Creusot-Loire
C 0.55, Ni 1.4, Cr 0.6, Mo 0.35, bal Fe.
For gears, bolts, forging dies; oil hardened, tough. *Obsolete*

ACIER ACMA
Creusot-Loire
C 0.55, Ni 1.4, Cr 0.6, Mo 0.35, bal Fe.
For gears, forging dies; oil hardened, tough. *Obsolete*

ACIER AF NO. 1
Creusot-Loire
C 0.85, bal Fe.
Heat treated: 200,000 TS; 150,000 YS; 10 El; 30 RA; 400 Brin.
For drills, taps, reamers, hobs, springs; Type W1; water
hardened. *Obsolete*

ACIER AFS
Creusot-Loire
C 0.8, bal Fe.
Heat treated: 188,000 TS; 143,000 YS; 12 El; 35 RA; 388 Brin.
For drills, punches, reamers taps, springs, hobs; Type W1;
water hardened. *Obsolete*

ACIER AM3
Creusot-Loire
C 0.35, Mn 1.1, bal Fe.
For punches, axles, gears, shafts; water hardened. *Obsolete*

ACIER AS
Creusot-Loire
C 0.5, bal Fe.
Annealed: 96,000 TS; 52,000 YS; 16 El; 23 RA; 170 Brin. For
axles, gears, bolts, tie rods; water hardened. *Obsolete*

ACIER BLN
Creusot-Loire
C 0.28, W 10, Cr 3, Mo 0.3, V 0.3, bal Fe.
For extrusion rams and dies, punches; hot work steel, oil
hardened. *Obsolete*

ACIER BTR
Creusot-Loire
C 0.35, Ni 4.6, Cr 0.3, Mo 1.1, bal Fe.
For gears, dies, crankshafts; oil hardened, tough. *Obsolete*

ACIER CMYO
Creusot-Loire
C 1, Cr 1, Mo 0.2, bal Fe.
For bearings, liners, sleeves; water or oil hardened. *Obsolete*

ACIER CNW
Creusot-Loire
C 0.38, W 3, Cr 2.5, Mo 0.7, V 0.3, bal Fe.
For hot work dies and tools; hot work steel. *Obsolete*

ACIER CTN2
Creusot-Loire
C 0.1, Ni 2, bal Fe.
For spindles, gears, cams, camshafts; case hardening steel,
tough. *Obsolete*

ACIER CTN6
Creusot-Loire
C 0.1, Ni 6, bal Fe.
For spindles, gears, cams, shafts; case hardening steel,
tough. *Obsolete*

ACIER DIABOLIQUE SATAN NO. 2
Creusot-Loire
C 1.15, W 1.9, Cr 0.5, bal Fe.
For header dies, fast finishing cutters; water hardened, wear
resistant. *Obsolete*

ACIER FAM
Creusot-Loire
C 0.42, Si 2, Mn 0.6, bal Fe.
For heat resistant parts; heat resistant. *Obsolete*

ACIER FFV NO. 3
Creusot-Loire
C 0.85, Mn 0.3, bal Fe.
Heat treated: 200,000 TS; 150,000 YS; 10 El; 30 RA; 400 Brin.
For drills, taps, reamers, springs, hobs; Type W1; water
hardened. *Obsolete*

ACIER M13AFY
Creusot-Loire
C 1.1, Mn 13, bal Fe.
For wear resistant parts; wear and abrasion resistant.
Obsolete

ACIER NC2
Creusot-Loire
C 0.3, Ni 2.8, Cr 0.5, bal Fe.
For gears, bolts, machine tool parts; oil hardened, tough.
Obsolete

ACIER TRIPLE SATAN
Creusot-Loire
C 0.32, W 9, Cr 3.2, bal Fe.
For extrusion rams and dies, punches; hot work steel, oil
hardened. *Obsolete*

ACIER VDLD
Creusot-Loire
C 0.35, Ni 3.8, Cr 1.5, bal Fe.
For gears, bolts, dies, crankshafts; oil hardened, tough.
Obsolete

ACIER VDLDM
Creusot-Loire
C 0.35, Ni 4, Cr 1.5, Mo 0.5, bal Fe.
For gears, shafts, crankshafts; oil hardened, tough. *Obsolete*

ACIERAL-1
Acieral Co. of America
Aluminum. Cu 6, Zn 0.4, Ni 0.9, Fe 0.1, Si 0.4, bal Al.
Cast: 20,000 TS. For automotive engine parts; castings.

ACIERAL-2
Acieral Co. of America
Aluminum. Cu 2.3-3.8, Fe 0.7-1.4, Mn 1-1.5, bal Al.
Rolled: 22,000 TS; 2 El. For automotive engine parts; high
strength.

ACIMET C-1
Cleveland Brass Corp.
Cu 90, Al 10, C 0-0.06.
Heat treated: 70,000 TS; 28,000 YS; 20 El; 20 RA; 100 Brin.
For acetic acid equipment castings, pumps, impeller shafts
for rubber pumps; corrosion resistant. *Obsolete*

ACIMET C-2
Cleveland Brass Corp.
Cu 85, Sn 5, Pb 5, Zn 5.
30,000-33,000 TS; 18,000-20,000 YS; 20 El; 20 RA; 50-60
Brin. For fittings and pumps for brewery, distilleries, alcohol
plants, etc.; pressure tight. *Obsolete*

ACIMET C-3
Cleveland Brass Corp.
Cu 60, Pb 2, Zn 38.
50,000-75,000 TS; 18,000 YS; 40 El. For close grained forged
oil fittings; free-cutting. *Obsolete*

ACIMET C-4
Cleveland Brass Corp.
Cu 58, Al 0.5, Sn 0.5, Zn 39, Fe 0-0.75, Mn 0-0.75.
70,000-75,000 TS; 35,000 YS; 20 El; 20 RA; 100-110 Brin. For
impeller shaft and gland fittings for rubber and plastic pumps
for handling HCl; corrosion resistant. *Obsolete*

ACIMET HARD LEAD ALLOY
Cleveland Brass Corp.
Pb 94.5-95, Sb 4.5-5.5.
2,800 TS; 1,600 YS; 75 El; 90 RA. For valves, pumps, fittings,
castings, handling corrosive chemicals; MP 495 F. *Obsolete*

ACIPCO 1005
ACIPCO Steel Products Division
C 0-0.08, Mn 0.3-0.6, bal Fe.
Cast, normalized and tempered: 48,000 TS; 25,000 YS; 35 El;
110 Brin. For high magnetic permeability parts. AISI 1005.

ACIPCO 1015
ACIPCO Steel Products Division
C 0.1-0.2, Mn 0.3-0.7, bal Fe.
Cast, normalized and tempered: 55,000 TS; 30,000 YS; 30 El;
130 Brin. Standard mild steel, high ductility. AISI 1015.

ACIPCO 1025
ACIPCO Steel Products Division
C 0.2-0.3, Mn 0.3-0.7, bal Fe.
Cast, normalized and tempered: 65,000 TS; 35,000 YS; 25 El;
150 Brin. Standard structural steel, good weldability. AISI
1025.

ACIPCO 1045
ACIPCO Steel Products Division
C 0.4-0.5, Mn 0.5-0.9, bal Fe.
Cast, normalized and tempered: 85,000 TS; 45,000 YS; 15 El;
180 Brin. For machine parts and for flame hardening. AISI
1045.

ACIPCO 1070
ACIPCO Steel Products Division
C 0.65-0.75, Mn 0.5-9, bal Fe.
Cast, normalized and tempered: 105,000 TS; 50,000 YS; 5 El;
220 Brin. Particularly for parts requiring flame or induction
hardened areas. AISI 1070.

ACIPCO 4130
ACIPCO Steel Products Division
C 0.25-0.35, Mn 0.4-0.7, Cr 0.8-1.15, Mo 0.15-0.25, bal Fe.
Cast, normalized and tempered: 80,000 TS; 45,000 YS; 17 El;
185 Brin. Cast, water quenched and tempered:
100,000-160,000 TS; 65,000-145,000 YS; 5-16 El; 250-450
Brin. Preferred water quenching grade. AISI 4130.

ACIPCO 4140
ACIPCO Steel Products Division
C 0.35-0.45, Mn 0.6-1, Cr 0.8-1.15, Mo 0.15-0.25, bal Fe.
Cast, normalized and tempered: 100,000 TS; 50,000 YS; 16
El; 200 Brin. Cast, oil quenched and tempered:
120,000-180,000 TS; 85,000-165,000 YS; 16-5 El; 250-450
Brin. Oil hardening alloy steel. AISI 4140.

ACIPCO 4330
ACIPCO Steel Products Division
C 0.25-0.35, Mn 0.4-0.7, Ni 1.65-2, Cr 0.6-0.9, Mo 0.2-0.3, bal Fe.
Cast, normalized and tempered: 100,000 TS; 60,000 YS; 16 El; 200 Brin. Cast, quenched and tempered: 120,000-180,000 TS; 85,000-165,000 YS; 5-16 El; 250-450 Brin. Hardenable in heavy sections.

ACIPCO 8620
ACIPCO Steel Products Division
C 0.15-0.25, Mn 0.6-1, Ni 0.4-0.7, Cr 0.4-0.7, Mo 0.15-0.25, bal Fe.
Cast, normalized and tempered: 70,000 TS; 40,000 YS; 20 El; 150 Brin. Low alloy carburizing grade. AISI 8620.

ACIPCO ACICULAR IRONS
ACIPCO Steel Products Division
C 2.5-3, Mn 0.5-1.5, P 0-0.1, S 0-0.12, Ni 1.7-2.3, Mo 0.8-1.2, bal Fe.
Cast: 60,000 TS; 300 Brin. For highly stressed rolls and cylinders; heat treatable.

ACIPCO CA-15
ACIPCO Steel Products Division
C 0-0.15, Mn 0-1, Si 0-1.5, Cr 11.5-14, Ni 0-1, bal Fe.
Cast, normalized and tempered: 100,000 TS; 65,000 YS; 18 El; 228 Brin. Heat treatable chromium stainless. ACI CA 15; AISI 410.

ACIPCO CF-8
ACIPCO Steel Products Division
C 0-0.08, Mn 0-1.5, Si 0-2, Cr 18-21, Ni 8-11, bal Fe.
Cast: 70,000 TS; 30,000 YS; 30 El; 150 Brin. Austenitic stainless castings. ACI CF-8; AISI 304.

ACIPCO CF-8M
ACIPCO Steel Products Division
C 0-0.08, Mn 0-1.5, Si 0-1.5, Cr 18-21, Ni 9-12, Mo 2-3, bal Fe.
Cast: 75,000 TS; 40,000 YS; 35 El; 160 Brin. Austenitic stainless; for special corrosive conditions. ACI CF-8M; AISI 316.

ACIPCO CLASS 30 CAST IRON
ACIPCO Steel Products Division
C 3.1-3.5, Mn 0.4-1, P 0-0.1, S 0-0.12, Si 1.2-2.5, bal Fe.
Cast: 30,000 TS; 200 Brin. Castings for engineering parts.

ACIPCO CLASS 40 CAST IRON
ACIPCO Steel Products Division
C 2.8-3.3, Mn 0.4-1, P 0-0.1, S 0-0.12, Si 1.6-2.2, Si inoculated, bal Fe.
Cast: 40,000 TS; 225 Brin. Standard high strength gray iron.

ACIPCO CLASS 50 CAST IRON
ACIPCO Steel Products Division
C 2.5-3.1, Mn 0.4-1.7, P 0-0.1, S 0-0.12, Si 1.6-2.2, Si inoculated, bal Fe.
Cast: 50,000 TS; 269 Brin. For highly stressed engineering parts.

ACIPCO CR-CU CAST IRON
ACIPCO Steel Products Division
Mn 0.4-1, P 0-0.1, S 0-0.12, Cr 0.3-0.7, Cu 0.8-1.2, (C + Si adjusted), bal Fe.
Cast: 40,000 TS; 235 Brin. Moderate hardness gray iron for rolls, etc.

ACIPCO CR. ALLOY CAST IRON
ACIPCO Steel Products Division
C 3-3.6, Mn 0.4-1, P 0-0.1, S 0-0.12, Cr 0.5-1, bal Fe.
Cast: 240-300 Brin. For parts requiring increased hardness and wear resistance.

ACIPCO HEAT RESISTING IRON
ACIPCO Steel Products Division
C 3.2, bal Fe.
Cast. For retorts for magnesium manufacture; heat and pressure resistant alloy. *Obsolete*

ACIPCO HF
ACIPCO Steel Products Division
C 0.2-0.4, Mn 0-2, Si 0-2, Cr 19-23, Ni 9-12, bal Fe.
Cast: 75,000 TS; 38,000 YS; 28 El; 165 Brin. Austenitic stainless steel. ACI HF; AISI 302B.

ACIPCO HH
ACIPCO Steel Products Division
C 0.2-0.5, Mn 0-2, Si 0-2, Cr 24-28, Ni 11-14, bal Fe.
Cast: 75,000 TS; 38,000 YS; 15 El; 170 Brin. Austenitic stainless; for furnace parts; service to 2000°F. ACI HH; AISI 309.

ACIPCO HK
ACIPCO Steel Products Division
C 0.2-0.6, Mn 0-2, Si 0-2, Cr 24-28, Ni 18-22, bal Fe.
Cast: 70,000 TS; 38,000 YS; 15 El; 187 Brin. Completely austenitic stainless; service to 2000°F. ACI HK; AISI 310.

ACIPCO HT
ACIPCO Steel Products Division
C 0.35-0.75, Mn 0-2, Si 0-2.5, Cr 13-17, Ni 33-37, bal Fe.
Cast: 68,000 TS; 35,000 YS; 5 El; 170 Brin. Good thermal shock resistance. ACI HT; AISI 330.

ACIPCO NI-CR-MO CAST IRON
ACIPCO Steel Products Division
Mn 0.4-1, P 0-0.1, S 0-0.12, Ni 1.3-1.8, Cr 0.2-0.4, Mo 0.4-0.6, (C + Si adjusted), bal Fe.
Cast: 270 Brin. Wear and heat resistant alloy for die parts; heat treatable.

ACIPCO REGULAR CAST IRON
ACIPCO Steel Products Division
C 3-3.6, Mn 0.4-1, P 0-0.1, S 0-0.12, Si 1.2-2.5, bal Fe.
Cast: 25,000 TS. General gray iron castings.

ACK DIE STEEL
Ackerlind Steel Co., Inc.
C 1, Cr 0.4, V 0.2, W 1.6, bal Fe.
For tools, dies; water hardening. *Obsolete*

ACME
H.K. Porter Co., Inc.
Cr 35, Ni 18, bal Fe.
Annealed: 75,000 TS; 40 El; 60 RA. For resistances; now Alray D. *Obsolete*

ACME 18-8 STAINLESS 302
Acme Steel Co.
Stainless steel. C 0.08-0.2, Cr 18, Ni 8, bal Fe.
Annealed: 80,000 TS; 35,000 YS; 45 El; 65 RA; 150 Brin. For chemical plant equipment; Type 302 stainless steel, austenitic. *Obsolete*

ACME 18-8 STAINLESS 304
Acme Steel Co.
Stainless steel. C 0-0.11, Cr 17-19, Ni 7-9, bal Fe.
Annealed: 75,000 TS; 32,000 YS; 45 El; 60 RA; 150 Brin. For chemical plant equipment; Type 304 stainless steel, austenitic. *Obsolete*

ACME COLORSTRIP
Acme Steel Co.
C 0.2, bal Fe.
For ornamental architecture; cold rolled strip steel in colors. *Obsolete*

ACME M
Acme Foundry Co.
C 0.2, Cr 18, Ni 8, bal Fe.
For heat resistant castings. Corrosion and heat resistant.

ACME NICKEL STEEL
English manufacture
Ni 30.5, Cr 14.2, C 0.3, bal Fe.
For tanks and vessels to resist corrosion; stainless, corrosion resistant.

ACME SATIN STRIP
Acme Steel Co.
Stainless steel. C 0.8, Cr 18, Ni 8, bal Fe.
For ornamental architecture; cold rolled stainless strip steel. *Obsolete*

ACME SILCROME 12
Interlake Inc.
C 0-0.12, Cr 12-15, bal Fe.
For corrosion resistant parts. *Obsolete*

ACME SILCROME CC
Interlake Inc.
C 0-0.12, Cr 12-15, Si, Cu, bal Fe.
For corrosion resistant parts. *Obsolete*

ACME SILCROME KA2
Interlake Inc.
C 0-0.2, Cr 16-23, Ni 7-11, bal Fe.
For stainless and corrosion resistant parts. *Obsolete*

ACME SILCROME KA2S
Interlake Inc.
C 0-0.08, Cr 16-23, Ni 7-11, bal Fe.
For stainless and corrosion resistant parts. *Obsolete*

ACME SILCROME L-12
Interlake Inc.
C 0.13-0.2, Cr 12-15, bal Fe.
For corrosion resistant parts. *Obsolete*

ACME SILCROME M-17
Interlake Inc.
C 0.13-0.2, Cr 16-23, bal Fe.
For corrosion resistant parts. *Obsolete*

ACME STAINLESS 410
Acme Steel Co.
Stainless steel. C 0-0.12, Cr 12-14, bal Fe.
Annealed: 75,000 TS; 40,000 YS; 35 El; 70 RA; 155 Brin. Cold drawn: 100,000 TS; 85,000 YS; 17 El; 60 RA; 205 Brin. For springs, hardware, tableware, turbine blades, pistons; Type 410 stainless steel, corrosion resistant. *Obsolete*

ACME STAINLESS 425
Acme Steel Co.
Stainless steel. C 0-0.12, Cr 14-16, bal Fe.
For chemical plant equipment; Type 425 stainless steel, corrosion resistant. *Obsolete*

ACME STAINLESS 430
Acme Steel Co.
Stainless steel. C 0-0.12, Cr 16-18, bal Fe.
Annealed: 70,000 TS; 40,000 YS; 30 El; 55 RA; 140 Brin. Cold drawn: 130,000 TS; 125,000 YS; 2 El; 185 Brin. For automotive trim, kitchen sinks, fasteners, bolts; Type 430 stainless steel, corrosion resistant. *Obsolete*

ACME STAINLESS STEEL TYPE 301X
Acme Steel Co.
Stainless steel. C 0.1, Cr 18, Ni 8, bal Fe.
For stainless parts; corrosion resistant. *Obsolete*

ACMEITE
Madison Foundry Co.
C, alloy, bal Fe.
For die blocks; oil hardened.

ACMELOY
Acme Foundry & Machine Co.
C 2.75-3.25, Si 1-1.75, Mn 0.65-1.75, bal Fe.
Cast: 30,000-60,000 psi TS; 175-275 Brin. Heat treated: 75,000-85,000 psi TS; 400-600 Brin. For cast iron castings, gears, housings, and shafts. High strength; corrosion, heat and wear resistant.

ACMELOY
Acme Electric Welder Co.
Copper, Cu alloy.
For resistance and spot welding electrodes.

ACMITE
Columbia Tool Steel Co.
C 0.7, W 18, Cr 4, V 1, Co 5, bal Fe.
For cutters, hobs, reamers, drills; high speed steel; Type T4.
Obsolete

ACMITE-L
Columbia Tool Steel Co.
C 0.5, Mn 0.25, Cr 4, W 18, V 1, bal Fe.
For hot work dies and tools; hot work steel. *Obsolete*

ACMONITAL
Italian manufacture
C, Cr, Ni, bal Fe.
For coins; stainless.

ACN10
Soc. Alluminio Veneto per Azioni
Aluminum. Cu 9.5-10.5, Si 0.8-1.2, Mg 0.3, Ti 0.2, Ni 1.3-1.7, bal Al.
Heat treated: 45,000-54,000 TS; 37,000-48,000 YS; 0.5-1.0 El; 115-140 Brin. For engine cylinder heads, pistons; heat resistant.

ACN3
Soc. Alluminio Veneto per Azioni
Aluminum. Cu 2.9-3.2, Fe 1.5, Si 0.7, Mg 0.6, Ti 0.2, Ni 0.6, bal Al.
Heat treated: 47,000-60,000 TS; 43,000-51,000 YS; 1-5 El; 115-150 Brin. For light alloy parts; age hardened.

ACORN
Swedish Iron & Steel Corp.
C 1.2, bal Fe.
For tools, drills; oil or water hardened. *Obsolete*

ACORN BRAND
A.W. Cadman Mfg. Co.
Cu 3.7, Sb 7.4, special hardener, bal Sn.
Cast: 12,500 TS; 22.5 El; 24 Brin. For gas and diesel engine bearings; Babbitt, low coefficient of friction.

ACROM 20
ACENOR, S.A.
C 0.15-0.21, Mn 0.6-0.9, Si 0.15-0.4, Cr 0.85-1.15, Mo 0.15-0.25, bal Fe.
Quenched and treated: 85-115 kg/mm^2 TS. Case hardening steel for bolts.

ACRON
Aluminium Industrie Aktiengesellschaft
Si 1, Cu 4, bal Al.
Annealed: 31,000-35,000 TS; 16,000-19,000 YS; 50-60 Brin.
Heat treated: 42,000-60,000 TS; 28,000-40,000 YS; 90-100 Brin. For automotive engine parts; age hardenable, high strength.

ACT
Acciaierie Valbruna s.p.a.
Stainless Steel.
Now AU177.

ACT 2
Adamas Carbide Corp.
TiC coated sintered carbide. Excellent abrasion, heat, crater and deformation resistance; medium toughness. For machining cast iron, stainless steel and non-ferrous alloys. C-2 general purpose and C-3 finishing cuts.

ACT 5
Adamas Carbide Corp.
TiC coated sintered carbide. Good abrasion, heat and crater resistance. Tough. For rough machining carbon, leaded, alloy and tool steels, and 400 series stainless. C-5 roughing and some C-6 general purpose.

ACT 7
Adamas Carbide Corp.
TiC coated sintered carbide. Excellent abrasion, heat and crater resistance. Good toughness and shock resistance. For general purpose and finishing carbon, leaded, and alloy steels.

ACT CARBIDE
Atlantic Steel Corp.
WC + Co.
For tools, cutters; sintered carbide.

ADABRAZE
Adamas Carbide Corp.
Sintered carbide coated with thin layer of pure cobalt. To improve brazing of carbide to base metal shanks or supports.

ADALLOY
Adams Hardfacing Co.
40.0 Cr + Mn + Si + C, bal Fe.
Cast alloy hardfacing rod; coated for AC-DC; bare for oxy-acetylene; 58-60 Rock C. Wear resistant; non-machinable; for plows, drag chains, cultivators.

ADALLOY "A"
Adams Hardfacing Co.
9.0 Cr + Ni + Mn + C + Si, bal Fe.
Hardfacing electrode; coated for AC-DC; bare for oxy-acetylene; 29-33 Rock C. Shock and impact resistant; machinable after slow cool; for gears, sprockets, latch pins.

ADALLOY "B"
Adams Hardfacing Co.
13.0 Cr + Mn + Si + C, bal Fe.
Hardfacing electrode; coated for AC-DC; bare for oxy-acetylene; 45-50 Rock C. Forgeable at red heat; machinable. For general preventive maintence on shovel buckets and lips, crawler running gear.

ADALLOY "C"
Adams Hardfacing Co.
10.0 Cr + Mn + Si + B + C, bal Fe.
Hardfacing electrode; coated for AC-DC; bare for oxy-acetylene; 55-60 Rock C. Forgeable, not readily machinable. For hardfacing heavy equipment subject to impact and abrasion, as pulverizers.

ADALLOY NO. 3
Adirondack Steel Casting Co.
C 0.3, Ni 1, bal Fe.
For steel castings. *Obsolete*

ADAMANT
Richard W. Carr & Co., Ltd.
C 1.45, Cr 0.5, V 0.25, W 4.75, bal Fe.
For rifling tools, burnishing tools; water hardened, keen cutting edge. *Obsolete*

ADAMANT SUPER-GENUINE BABBITT
Magnolia Metal Corp.
Cu, Sb, bal Sn.
12,850 TS; 8,600 YS; 22.8 Brin. For marine, airplane, diesel and other internal combustion engine bearings; Sp. Gr. 7.34; tough.

ADAMANT XL
Firth Brown Ltd.
C 1.25, Ni 1.75, Cr 3, Mo 0.4, bal Fe.
For cold work tools and dies; oil hardened. *Obsolete*

ADAMANTINE
Babcock & Wilcox
C 0.7, Cr 0.7, Mn 0.7, bal Fe.
Heat treated: 145,000 TS; 77,000 YS; 6.5 El; 16.6 RA; 260-300 Brin. For heat treated steel balls for grinding mills; wear resistant. *Obsolete*

ADAMAS 387
Adamas Carbide Corp.
WC 89, Co 11.
Sintered: 365,000 TrS; A 90.2 Rock *Obsolete*

ADAMAS 434
Adamas Carbide Corp.
WC, TaC, TiC, Co.
Sintered: 270,000 transverse strength; 91 Rock A. For rough machining steel. BHMA 5-7-5; ISO P40.

ADAMAS 474
Adamas Carbide Corp.
WC, Co, TaC.
Sintered: 300,000 transverse strength; 90 Rock A. For rough machining steel. BHMA 3-8-2; ISO P40.

ADAMAS 484
Adamas Carbide Corp.
Sintered: 230,000 TrS; A 91.1 Rock. For rough machining steel. BHMA 5-5-4; ISO P40. *Obsolete*

ADAMAS 490
Adamas Carbide Corp.
Co 4.5, WC, TaC, TiC.
Sintered: 10.7 density; 92.9 Rock A; transverse rupture strength, 200 psi. High hardness, good wear resistance. For finishing operations. ISO P10, M10.

ADAMAS 495
Adamas Carbide Corp.
Co 5, WC, TaC, TiC.
Sintered: 12 density; 92.4 Rock A; transverse rupture strength, 250 psi. Carbide grade with extra wear resistance for difficult machining operations. ISO P15.

ADAMAS 499
Adamas Carbide Corp.
Co 9, WC, TaC, TiC.
Sintered: 12.15 density; 91.3 Rock A; transverse rupture strength, 325,000 psi. Tough carbide grade with good shock resistance, wear and heat resistance; for medium to heavy cuts. ISO P25, P30.

ADAMAS 502
Adamas Carbide Corp.
WC 88, Co 12.
Sintered: 390,000 transverse strength; 88 Rock A.

ADAMAS 548
Adamas Carbide Corp.
WC, TaC, TiC, Co.
Sintered: 230,000 transverse strength; 92 Rock A. For finish machining steel. BHMA 8-5-4; ISO P10.

ADAMAS 569
Adamas Carbide Corp.
WC 90, Co 10.
Sintered: 370,000 transverse strength; 88.5 Rock A.

ADAMAS 5X
Adamas Carbide Corp.
WC, TaC, TiC, Co.
Sintered: 230,000 TrS; A 91.0 Rock. For rough machining steel. BHMA 5-5-5; ISO P30. *Obsolete*

ADAMAS 619
Adamas Carbide Corp.
High in TaC.
Sintered: 230,000 TrS; A 91.1 Rock. For general machining at higher heat. *Obsolete*

ADAMAS 6X
Adamas Carbide Corp.
WC, TaC, TiC, Co.
Sintered: 230,000 transverse strength; 91.5 Rock A. For general purpose machining steel. BHMA 6-5-5; ISO P20.

ADAMAS 783
Adamas Carbide Corp.
WC 89, Co 11.
Sintered: 385,000 transverse strength; 88.4 Rock A.

ADAMAS 7X
Adamas Carbide Corp.
WC, TaC, TiC, Co.
Sintered: 200,000 transverse strength; 92.4 Rock A. For finish machining steel. BHMA 9-3-5; ISO P10.

ADAMAS 815
Adamas Carbide Corp.
WC 91, Co 9.
Sintered: 340,000 transverse strength; 89.2 Rock A.

ADAMAS A
Adamas Carbide Corp.
WC 94, Co 6.
Sintered: 300,000 transverse strength; 91.7 Rock A. For general purpose machining ferrous and non-ferrous metals. BHMA. 7-6-0; ISO K20.

ADAMAS AA
Adamas Carbide Corp.
WC 96, Co 4.
Sintered: 240,000 transverse strength; 92 Rock A. For finish machining ferrous and non-ferrous metals. BHMA 8-5-8; ISO K10, M10.

ADAMAS AAA
Adamas Carbide Corp.
WC 97, Co 3.
Sintered: 200,000 transverse strength; 92.6 Rock A. For finish boring ferrous and non-ferrous alloys. BHMA 9-3-0; ISO KO1.

ADAMAS ACT 2, ETC.
Now ACT 2 etc.

ADAMAS AM
Adamas Carbide Corp.
WC, Co, TaC.
Sintered: 240,000 transverse strength; 91.9 Rock A. For general purpose machining steel and cast iron. BHMA 7-5-1; ISO M20.

ADAMAS B
Adamas Carbide Corp.
WC 91, Co 9.
Sintered: 300,000 transverse strength; 90.5 Rock A. For rough machining ferrous and non-ferrous alloys. BHMA 4-8-0; ISO K30.

ADAMAS BB
Adamas Carbide Corp.
WC 87, Co 13.
Sintered: 360,000 transverse strength; 89.4 Rock A. For rough machining ferrous and non-ferrous metals. BHMA 2-9-0; ISO K40.

ADAMAS C
Adamas Carbide Corp.
WC 76, Co 9, 15.0 TiC.
Sintered: 210,000 transverse strength; 92 Rock A. For finish machining steel. BHMA 8-4-6; ISO P10.

ADAMAS CC
Adamas Carbide Corp.
WC 80, Co 4, 16% TiC.
Sintered: 175,000 TrS; A 93.2 Rock.For precision boring steel. BHMA 9-2-6; ISO PO1. *Obsolete*

ADAMAS D
Adamas Carbide Corp.
WC 82, Co 10, 8% TiC.
Sintered: 240,000 TrS; A 91.0 Rock. For general purpose machining steel. BHMA 5-5-4; ISO P20. *Obsolete*

ADAMAS DD
Adamas Carbide Corp.
WC + WTiC + Co.
770 Brin. For cutting tools; sintered carbides. *Obsolete*

ADAMAS GG
Adamas Carbide Corp.
WC, Co, TaC.
Sintered: 290,000 transverse strength; 89.2 Rock A. For rough machining steel. BHMA 2-8-5; ISO P50.

ADAMAS GU-1
Adamas Carbide Corp.
WC base sintered carbide. Density: 13.84; 88.5 Rock A; transverse rupture strength 500,000 psi. Very high strength carbide; for stamping, cold heading, and saw tips.

ADAMAS HD 20
Adamas Carbide Corp.
WC 80, Co 20.
Sintered: 385,000 transverse strength; 85.5 Rock A. For heading dies, swaging dies, nail gripper dies.

ADAMAS HD-15
Adamas Carbide Corp.
WC 85, Co 15.
Sintered: 400,000 transverse strength; 87.4 Rock A. For lamination punches and dies and wear parts.

ADAMAS HD-20T
Adamas Carbide Corp.
WC 75, Co 20, 5.0 TaC.
Sintered: 380,000 transverse strength; 85.3 Rock A. For small swaging dies and gripper dies where anti-galling properties are important.

ADAMAS HD-25
Adamas Carbide Corp.
WC 75, Co 25.
Sintered: 360,000 transverse strength; 83.5 Rock A. For dies where high impact strength coupled with high abrasion resistance are necessary.

ADAMAS HD-25T
Adamas Carbide Corp.
WC 70, Co 25, 5.0 TaC.
Sintered: 380,000 transverse strength; 83.5 Rock A. For dies requiring maximum impact strength and anti-galling properties.

ADAMAS PWX
Adamas Carbide Corp.
WC, Co, TaC.
Sintered: 230,000 transverse strength; 92.3 Rock A. For finish machining steel. BHMA 9-5-1; ISO M10, KO5.

ADAMAS TITAN
Now TITAN.

ADAMAS TITAN 60
Adamas Carbide Corp.
TiC, Mo$_2$C, Ni.
Sintered: 235,000 transverse strength; 91.9 Rock A. For rough machining ferrous and non-ferrous alloys, including superalloys.

ADAMAS TITAN 80
Adamas Carbide Corp.
TiC, Mo$_2$C, Ni.
Sintered: 200,000 transverse strength; 93.0 Rock A. For finish machining and boring ferrous and non-ferrous alloys, including superalloys. BHMA 9-2-9; ISO PO1, P10.

ADAMITE
Gulf & Western Mfg. Co.
Cr 1, Ni 0.75, high C, bal Fe.
For press dies, rolls, mill guides, furnace parts, castings; pearlitic cast iron, wear resistant.

ADAMS HFA
J.D. Adams & Co.
C, alloy, bal Fe.
Cast: 550 Brin. Heat treated: 730 Brin. For hard facing electrode; wear and shock resistant. *Obsolete*

ADAMS HFB
J.D. Adams & Co.
C, alloy, bal Fe.
Cast: 653 Brin. Heat treated: 800 Brin. For hard facing electrode; wear and shock resistant. *Obsolete*

ADAPTALOY
American Smelting & Refining Co.
Cu 1.8, Si 1, Mg 1, bal Zn.
Cast: 22,000-25,000 psi TS; 8000-10,000 psi YS; 11-7 El; 48 Brin. For ornamental grills, valve handles, structural castings; high elongation and impact strength.

ADIC
Darwins Alloy Castings
C 0.37, Si 1.1, Cr 5, V 1, Mo 1.3, bal Fe.
Annealed: 230 Brin. For die casting dies, hot forging dies; hot-work steel; air hardened. *Obsolete*

ADIC
Eagle & Globe Steel Ltd.
Tool material. C 0.4, Si 1, Cr 5, V 1, Mo 1.3, bal Fe.
Annealed: 235 Brin max. Air hardening die steel, may be water cooled. Aluminum zinc and magnesium die casting tools; extrusion dies and mandrels, hot piercing tools, hot blanking and forging dies, shear blades, hot punches and die inserts. AS1239 H13A; AISI H13; BS4659 BH13; Werkstoff 1.2344.

ADIC C.T.U.
Darwins Alloy Castings
C 0.37, Si 1.1, Cr 5, W 1.75, V 1, Mo 1.3, bal Fe.
Annealed: 230 Brin. For die casting dies, shear blades, slitters; hot-work steel, air hardened. *Obsolete*

ADLERSTAH AZ1
Zweigbetrieb der Carp & Hones
C 1.05, Cr 1, W 1.15, Mn 0.9, bal Fe.
For bearings, liners, cutters, forming dies; water hardened.

ADLERSTAHL 03
Zweigbetrieb der Carp & Hones
C 1, Cr 1.1, Si 0.25, Mn 0.07, bal Fe.
For bearings, races, liners, sleeves; water hardened, wear resistant.

ADLERSTAHL 1W
Zweigbetrieb der Carp & Hones
C 1, Si 0.25, Mn 0.07, Cr 1.1, bal Fe.
For bearings, races, liners, sleeves: water hardened, wear resistant.

ADLERSTAHL 3 MH
Zweigbetrieb der Carp & Hones
Mn 0.25, C 1.1, Si 0.25, bal Fe.
Annealed: 105,000 TS; 55,000 YS; 20 El; 40 RA; 205 Brin. For drills, hobs, reamers, dies, cutters; Type W1; water hardened.

ADLERSTAHL 4 ZH
Zweigbetrieb der Carp & Hones
C 1, Si 0-0.25, Mn 0-0.25, bal Fe.
Annealed: 100,000 TS; 53,000 YS; 21 El; 42 RA; 200 Brin. For drills, taps, lathe, and planer tools, cutters; Type W1; water hardened.

ADLERSTAHL 5 ZAH
Zweigbetrieb der Carp & Hones
C 0.85, Si 0.25, Mn 0.25, bal Fe.
Heat treated: 190,000 TS; 145,000 YS; 12 El; 35 RA; 400 Brin. For drills, taps, tools, springs, cutters; Type W1; water hardened.

ADLERSTAHL B11
Zweigbetrieb der Carp & Hones
C 1, Si 0.2, Mn 0.25, V 0.1, bal Fe.
For drills, taps, punches, reamers, hobs; Type W2; water hardened.

ADLERSTAHL C1
Zweigbetrieb der Carp & Hones
C 1.45, Cr 1.4, Mn 0.6, bal Fe.
For bearings, sleeves, liners; water hardened, wear resistant.

ADLERSTAHL C13
Zweigbetrieb der Carp & Hones
C 2.1, Cr 11.5, Mn 0.3, bal Fe.
For blanking and forming dies, punches; oil hardened, nondeforming.

ADLERSTAHL C13TU
Zweigbetrieb der Carp & Hones
C 2.1, Cr 11.5, Mn 0.3, bal Fe.
For blanking and forming dies, punches; oil hardened, nondeforming.

ADLERSTAHL C13W
Zweigbetrieb der Carp & Hones
C 1.65, Cr 11.5, V 0.1, bal Fe.
For blanking and forming dies, punches; oil hardened, nondeforming.

ADLERSTAHL EC2
Zweigbetrieb der Carp & Hones
C 1.3, Si 0-0.25, Mn 0-0.25, bal Fe.
For reamers, taps, form cutters, drills, hobs; Type W1; water hardened.

ADLERSTAHL EC3
Zweigbetrieb der Carp & Hones
C 1.15, Si 0-0.25, Mn 0-0.25, bal Fe.
Annealed: 110,000 TS; 55,000 YS; 18 El; 38 RA; 210 Brin. For reamers, drills, taps, cutters, broaches; Type W1; water hardened.

ADLERSTAHL EC4
Zweigbetrieb der Carp & Hones
C 1, Si 0-0.25, Mn 0-0.25, bal Fe.
Annealed: 100,000 TS; 53,000 YS; 21 El; 42 RA; 200 Brin. For springs, tools, cutters, drills, reamers; Type W1; water hardened.

ADLERSTAHL EC5
Zweigbetrieb der Carp & Hones
C 0.85, Si 0.25, Mn 0-0.25, bal Fe.
Heat treated: 190,000 TS; 145,000 YS; 10 El; 30 RA; 400 Brin. For springs, tools, drills, cutters, taps, hammers; Type W1; water hardened.

ADLERSTAHL EC6
Zweigbetrieb der Carp & Hones
C 0.7, Si 0-0.25, Mn 0-0.25, bal Fe.
Heat treated: 175,000 TS; 128,000 YS; 12 El; 37 RA; 355 Brin. For springs, tools, rails, axes; Type W1; water hardened.

ADLERSTAHL EGH1
Zweigbetrieb der Carp & Hones
C 0.35, Si 0.9, Mn 0.3, Cr 1.05, V 0.18, W 1.85, bal Fe.
For header dies, upsetters, crimpers, punches, oil hardened, tough.

ADLERSTAHL EGH3
Zweigbetrieb der Carp & Hones
C 0.55, Si 0.9, Cr 1.05, V 0.18, W 1.85, bal Fe.
For header dies, upsetters, shears, crimpers; oil hardened, tough.

ADLERSTAHL EXTRA
Zweigbetrieb der Carp & Hones
C 0.86, Cr 4, V 2.5, Mo 0.85, W 12, bal Fe.
For reamers, taps, drills, hobs, broaches; high-speed steel.

ADLERSTAHL EXTRA P12
Zweigbetrieb der Carp & Hones
C 0.76, Co 10, Cr 4.2, Mo 0.8, V 1.8, W 18, bal Fe.
For lathe and planer tools, reamers, hobs, taps; high-speed steel.

ADLERSTAHL GCN1
Zweigbetrieb der Carp & Hones
C 0.55, Cr 0.7, Mo 0.18, Ni 1.65, V 0.1, bal Fe.
For gears, shafts, bolts, studs, crankshafts; oil hardening, shock resistant.

ADLERSTAHL GCN2
Zweigbetrieb der Carp & Hones
C 0.56, Ni, Cr, Mo, V, bal Fe.
For gears, fasteners, bolts, studs; oil hardened, shock resistant.

ADLERSTAHL GDM
Zweigbetrieb der Carp & Hones
C 0.3, Si 1, Cr 1.1, V 0.18, W 3.75, bal Fe.
For header dies, upsetters, crimpers; oil hardened, tough.

ADLERSTAHL KJ
Zweigbetrieb der Carp & Hones
C 0.74, Cr 4.1, V 1.1, W 18.5, bal Fe.
For lathe and planer tools, reamers, broaches; high-speed steel.

ADLERSTAHL KT
Zweigbetrieb der Carp & Hones
C 0.86, Co 2.8, Cr 4.3, Mo 0.85, V 2, W 12, bal Fe.
For lathe and planer tools, drills, taps, reamers; high speed steel.

ADLERSTAHL NL5H
Zweigbetrieb der Carp & Hones
C 0.5, Cr 1.05, Ni 3.25, Mn 0.5, bal Fe.
For gears, bolts, crankshafts, dies, punches; oil hardened, shock resistant.

ADLERSTAHL NWO
Zweigbetrieb der Carp & Hones
C 0.45, Ni, Cr, bal Fe.
For gears, bolts, crankshafts; oil hardened, shock resistant.

ADLERSTAHL P15
Zweigbetrieb der Carp & Hones
C 1.35, Cr, V, W, Co, bal Fe.
For reamers, broaches, cutters, taps; high-speed steel.

ADLERSTAHL P45
Zweigbetrieb der Carp & Hones
C 1.3, Cr 4.3, Mo 0.85, V 3.8, W 12, bal Fe.
For reamers, broaches, cutters, taps; high-speed steel.

ADLERSTAHL PM35
Zweigbetrieb der Carp & Hones
C 0.38, Si, Cr, V, bal Fe.
For gears, springs, punches, bolts, shafts; oil hardened, tough.

ADLERSTAHL PM45
Zweigbetrieb der Carp & Hones
C 0.45, Si, Cr, V, bal Fe.
For springs, bolts, crankshafts, gears; oil hardened, shock resistant.

ADLERSTAHL PM50
Zweigbetrieb der Carp & Hones
C 0.61, Cr 1.18, V 0.1, Mn 0.75, Si 0.85, bal Fe.
For upsetters, heading and forging dies; oil hardened, tough.

ADLERSTAHL PROPHET EXTRA
Zweigbetrieb der Carp & Hones
C 0.86, Co 2.8, Cr 4.3, Mo 0.85, V 2, W 12, bal Fe.
For lathe and planer tools, cutters, reamers; high-speed steel.

ADLERSTAHL PZM
Zweigbetrieb der Carp & Hones
C 1.4, V 0.1, Cr 0.3, Mn 0.3, bal Fe.
For bearings, forming and blanking dies; water or oil hardened, wear resistant.

ADLERSTAHL RO
Zweigbetrieb der Carp & Hones
C 1.42, W, V, bal Fe.
For blanking and forming dies, bearings; water or oil hardened, wear resistant.

ADLERSTAHL SICR
Zweigbetrieb der Carp & Hones
C 1.1, Cr 0.4, Mn 0.3, bal Fe.
Heat treated: 190,000 TS; 120,000 YS; 10 El; 30 RA; 375 Brin.
For springs, taps, cutters, drills, reamers; Type W1; water hardened.

ADLERSTAHL SPEZIAL M25
Zweigbetrieb der Carp & Hones
C 0.95, W, Mo, bal Fe.
For cutters, punches, dies; oil hardened.

ADLERSTAHL V
Zweigbetrieb der Carp & Hones
C 0.8, W 18, Co 4.7, Cr 4.3, Mo 0.75, V 1.5, bal Fe.
For lathe and planer tools, taps, reamers; high-speed steel.

ADLERSTAHL WKS
Zweigbetrieb der Carp & Hones
C 0.55, Cr 1.05, V 0.18, W 1.85, Si 0.9, bal Fe.
For upsetting and heading dies, punches; oil hardened, tough.

ADLERSTAHL WPS
Zweigbetrieb der Carp & Hones
C 0.3, Cr 2.65, V 0.35, W 8.5, bal Fe.
For extrusion rams and dies, hot punches; oil hardened, hot work steel.

ADLERSTAHL WPS-EXTRA
Zweigbetrieb der Carp & Hones
C 0.3, Co 2, Cr 2.4, V 0.25, W 8.5, bal Fe.
For extrusion rams and dies, hot punches; oil hardened, hot work steel.

ADMIRALTY ALLOY-442
Anaconda Co.
Cu 70, Zn 29, Sn 1.
Soft: 48,000 TS; 18,000 YS; 65 El. Hard: 85,000 TS; 10 El. For condenser and heat exchanger tubes, ferrules. *Obsolete*

ADMIRALTY BRASS
Anaconda Co.
Copper. Cu 62, Zn 37, Sn 1.
Cast: 56,000 TS; 27,000 YS; 15 El; 25 RA. Rolled hard: 71,680 TS; 55,500 YS; 20 El; 30 RA. For pump rods, sheet.

ADMIRALTY BRASS
Yorkshire Imperial Metals Ltd.
Cu 70, Sn 1, As 0.04, bal Zn.
Annealed: 40,300-49,300 TS; 18,000-22,400 YS; 60-75 El; 54-73 Brin. Hard: 71,700-87,400 TS; 60,000-76,000 YS; 8-15 El; 145-175 Brin. Good resistance to sea air; for fixtures on marine equipment.

ADMIRALTY BRASS, INHIBITED WITH ANTIMONY
Chase Brass & Copper Co., Inc.
Copper. Cu 71, Zn 28, Sn 1.
Annealed: 53,000 psi TS; 22,000 psi YS; 65 El. Drawn (35%): 85,000 psi TS; 65,000 psi YS; 15 El. Corrosion resistant; for condenser and heat exchanger tubes, ferrules.

ADMIRALTY BRONZE
Anaconda Co.
Cu 88, Sn 10, Zn 2.
Sand cast: 38,500 TS; 23,300 YS; 14 El; 77 Brin. For gears, worm wheels, trolley wheels. *Obsolete*

ADMIRALTY BRONZE MODIFIED
Belmont Metals Inc.
Cu 86, Sn 8, Pb 2, Zn 4.
For valves, fittings; pressure tight.

ADMIRALTY CAG

A. Cohn Ltd.
Cu 0-88, Sn 0-10, Zn 2.
Cast: 36,000 TS; 18,000 YS; 20 El. For marine castings; resists marine corrosion.

ADMIRALTY GUN METAL

British Insulated Callender's Cables
Cu 87.7, Sn 10.3, Zn 1.4, As 0.4, Pb 0.08, Fe 0.05.
Cast: 37,300 TS; 21,300 YS; 9 El; 100 Brin. For valves and pump parts. Corrosion resistant. *Obsolete*

ADMIRALTY METAL

Chase Brass & Copper Co.
Cu 70-73, Sn 0.9-1.2, Pb 0-0.07, Fe 0-0.06, bal Zn.
Annealed: 45,000 TS; 18,000 YS; 60 El; 70 Brin. Hard: 100,000 TS; 98,000 YS; 4 El; 210 Brin. For condenser tubes, ferrules, distiller tubes; reduced dezincification, corrosion resistant. *Obsolete*

ADMIRALTY NICKEL

Belmont Metals Inc.
Cu 70, Zn 29, Ni 1.
For condenser tubes; corrosion resistant.

ADMIRALTY WHITE METAL

English manufacture
Sb 8-9, Cu 2-7, bal Sn.
For bearings.

ADMIRO

Allgemeines Deutsches Metallwerk GmbH
Cu 48, Ni 10, Zn 35, Al 2, Mn 3, Fe 2.
Cast: 57,000-70,000 TS; 20-40 El; 110-130 Brin. Hot pressed: 63,000-76,000 TS; 25-40 El; 150-170 Brin. For turbine bearings, tubes. Corrosion resistant.

ADMIRO V

Allgemeines Deutsches Metallwerk GmbH
Cu 43, Ni 15, Zn 35, Al 2, Mn 3, Fe 2.
Cast: 5-18 El; 160 Brin. Pressed: 8-20 El; 185 Brin. For turbine bearings. Corrosion resistant.

ADMOS

Allgemeines Deutsches Metallwerk GmbH
Cu 40-55, Ni 3-15, Mn 1-3, Fe 1-2, Al 0.5-3, bal Zn.
Cast: 56,000-71,680 TS; 20-35 El; 80-90 Brin. Hot pressed: 69,440-85,120 TS; 20-35 El; 90-120 Brin. For bearing bushes, worm wheels, condenser tubes, turbine blading, gear wheels, valves. Resists corrosion and erosion of superheated steam.

ADNIC

Century Brass Products Inc.
Cu 70, Ni 29, Sn 1.
Hard: 113,000 TS; 107,000 YS; 10 El; 56 RA. Hot rolled: 64,500 TS; 35,600 YS; 46 El; 72 RA. For condenser tubes, auto head lights, cafeteria ware; corrosion resistant. *Obsolete*

ADNIC 280

Aubert & Duval
Nickel. C 0.25, Si 0-0.5, Mn 1.2, Cr 20, Fe 2, bal Ni.
Wrought. Equivalent to Brightray.

ADNIC 482 D

Aubert & Duval
Nickel. C 0.8, Si 0-1, Mn 0-1, Cr 0-1, Mo 34, W 12, Co 0-0.15, Cu 0-0.4, Fe 0-6, bal Ni.
Cast.

ADNIC 56 D

Aubert & Duval
Nickel. C 0.02-0.08, Si 0.5-1, Mn 0-1, Cr 15.5, Mo 16, W 4, Co 2.5, Fe 0-4, V 0-0.35, bal Ni.
Wrought and cast. Equivalent to Hastelloy C and C276. UNS Nr. N10002 and N10276.

ADNIC 562 D.Z

Aubert & Duval
Nickel. C 0.8, Si 0-1, Mn 0-1, Cr 17, Mo 17, W 5, Co 0-0.15, Cu 0-0.4, Fe 0-6, bal Ni.
Cast.

ADRIATICAL

Alfa Romeo
Cu 4-5, bal Al.
For light alloy parts; similar to Duralumin.

ADSPEC 1076 D

N.C. Ashton Ltd.
Al 8.5, Fe 2.5, bal Cu.
34 tsi TS; 30 El. Wrought aluminum bronze.

ADVALOY

Advance Aluminum Casting Corp.
Al alloy.
For castings; permanent mold.

ADVANCE

U.S. Reduction Co.
Cu, Sb, bal Sn.
For bearings; Babbitt metal. *Obsolete*

ADVANCED DUTY OILITE BRONZE

Chrysler Corp.
Cu 87.5-90.5, Fe 0-1, C 0-1.75, Sn 9.5-10.5.
Sintered: 16,000-18,000 TS; 15,000-20,000 YS; 6.8-7.2 density. Oilite bearing material. ASTM B255-61 Type 11; SAE 842; PMPMA BT-0010-S.

AECMA FE PL 52 S

Commentryenne
Now VASCOJET 90.

AEGIS

Republic Steel Corp.
C 0.15-0.45, Mn 0.5-0.8, Si 0.1-0.2, Ni 3.25-3.75, bal Fe.
For machinery parts, gears, shafts, axles. *Obsolete*

AELENER ZINC

Manufacturer not listed.
Sb 3, Cu 3.5, Pb 21.6, Sn 20.5, Zn 51.4.
For solder, bearings.

AEM 120

Plansee Metallwerk Gesellschaft
For permanent magnets. Sintered.

AEM 90

Plansee Metallwerk Gesellschaft
For permanent magnets. Sintered.

AEONITE

Shanks & Co. Ltd.
Ni, bal Cu.
For sanitary appliances. White metal.

AERAL

William Mills & Co. Ltd.
Cu 2-4.5, Mg 0.2-1.5, Si 0.7, Mn 0.3, Cd 0.5-2.5, bal Al.
Cast: 31,500 TS; 25,000 YS; 3 El; 90 Brin. For casings, housings, oil pans. Age-hardenable, good machinability. *Obsolete*

AERAL

Compagnie des Alliages
Cu 3.5-4.5, Mn 0.4-1, Mg 0.3-0.75, bal Al.
56,000-75,000 TS; 30,000-55,000 YS; 10-20 El. For light alloys, airplane and dirigible parts; Duralumin type alloy; age hardening.

AERAL A

Compagnie des Alliages
Cu 2-4.5, Mg 0.2-1.5, Si 0-2.5, Fe 0-0.8, Ti 0-0.4, Cd 0.5-2.5, Mn 0-0.5, bal Al.
Solution treated: 28,000-34,000 TS; 22,000-24,000 YS; 3-5 El. For sand and chill castings; heat treatable.

AERALLOY

Engelhard Corp.
Pt alloy.
For electrical contacts; heat resistant. *Obsolete*

AERALLOY NO. 1

Engelhard Corp.
Ir, bal Pt.
For electrical contacts, vibrators, temperature gages; resists arc erosion *Obsolete*

AERALLOY NO. 2

Engelhard Corp.
Ir, bal Pt.
For electrical contacts, vibrators, thermostats, resists arc erosion. *Obsolete*

AERISWELD

Lincoln Electric Co.
P 0.05, Sn 7, Al 0.1, Pb 0.02, bal Cu.
Bronze arc welding electrode. AWS Class E CuSn-C.

AERMET ALLOY 100

Carpenter Technology Corp.
C 0.23, Cr 3.1, Ni 11.1, Mo 1.2, Co 13.4, bal Fe.
Heat treated and aged, longitudinal: 1689-1758 MPa YS; 1930-2000 MPa TS; 14-16 El; 63-67 RA. High temperature alloy for use in structural components requiring high strength, high fracture toughness and stress corrosion and cracking resistance.

AERO METAL (CAST)

Garford Engineering Co.
Aluminum. Cu 4.2, Zn 27.8, Fe 0.5, Si 0.5, Al 67.
Cast: 25,000 TS. For automotive engine parts; cast.

AERO METAL (SHEET)

Garford Engineering Co.
Aluminum. Cu 0.2-0.6, Mg 2.1-2.9, Al 96, Fe 0.3-1.3.
Rolled: 20,000 TS. For automotive engine parts; sheet.

AERO-12-3

Colt Industries
C 1.45, Cr 12.75, Mo 0.65, bal Fe.
For dies, cutters; non-deforming. *Obsolete*

AEROLITE

British Aluminium Co., Ltd.
Cu 12, Mn 2, bal Al.
Cast: 22,000 TS; 14,000 YS; 60 Brin. For pistons, automobile parts; good machinability. *Obsolete*

AEROLITE

English manufacture
Al 97, Cu 1.2, Fe 1, Si 0.5, Mg 0.4.
For pistons, auto parts.

AEROMIN

German manufacture
Al 92.7, Mn 6.2, Fe 0.8, Si 0.3.
For light alloy parts.

AERON

Degefors Iron & Steel Works
Cu 4.5, Si 0.75, Mn 0.75, bal Al.
Annealed: 23,000-35,000 TS; 7,000-12,000 YS; 12-20 El; 45-55 Brin. Quenched: 45,000-53,000 TS; 15,000-30,000 YS; 15-22 El; 68-85 Brin. For light anti-corrosive alloy parts. Light anti-corrosion alloy; similar to Alcoa No. 25S.

AERON

Metallbank AG
Cu 1.8, Mn 0.8, Si 1, bal Al.
Heat treated: 54,000-60,000 TS; 26,000-34,000 YS; 18-25 El; 90-120 Brin. For light alloy parts; heat treatable.

AERON

Metallgesellschaft A.G.
Cu 1.8, Mn 0.8, Si 1, bal Al.
Heat treated: 54,000-60,000 TS; 26,000-34,000 YS; 18-25 El; 90-120 Brin. For light alloy parts; heat treatable.

AEROSIL 10A
Georgsmarienwerke Selesiastahl GmbH
C 0.15, Cr 19.5, Ni 9.5, bal Fe.
Annealed: 80,000 TS; 35,000 YS; 55 El; 75 RA; 150 Brin. For chemical plant equipment; Type 302; stainless, austenitic.

AEROSIL 10F
Georgsmarienwerke Selesiastahl GmbH
C 0-0.12, Al 1, Cr 18, bal Fe.
Annealed: 80,000 TS; 50,000 YS; 25 El; 50 RA; 160 Brin. For oil refinery equipment; corrosion and heat resistant.

AEROSIL 11FO
Georgsmarienwerke Selesiastahl GmbH
C 0.2, Cr 25, Ni 4, bal Fe.
For furnace parts, heat treating boxes; heat resistant.

AEROSIL 12A
Georgsmarienwerke Selesiastahl GmbH
C 0.15, Cr 24, Ni 19, bal Fe.
For furnace equipment heat treating boxes, pots; Type 310; stainless, austenitic.

AEROSIL 12F
Georgsmarienwerke Selesiastahl GmbH
C 0-0.12, Al 1.5, Cr 24, bal Fe.
For furnace parts, conveyors, boxes, fixtures; heat and creep resistant.

AEROSIL 8F
Georgsmarienwerke Selesiastahl GmbH
C 0-0.12, Al 0.8, Cr 6.5, bal Fe.
For oil refining equipment; creep and heat resistant.

AEROSIL 9F
Georgsmarienwerke Selesiastahl GmbH
C 0-0.12, Al 1, Cr 13, bal Fe.
For oil refinery and chemical plant equipment; corrosion and heat resistant.

AETERNA 600 METAL
Allied Process Brass Corp.
Cu 60, Zn 40, Mn, Si.
85,000-90,000 TS. For gears, levers, cams, cranks; high strength. *Obsolete*

AF 1753
American manufacture
C 0.24, Cr 16.25, Co 7.2, Mo 1.6, W 8.4, Ti 3.1, Al 1.9, Zr 0.06, B 0.008, Fe 9.5, bal Ni.
At 70°F: 187,000 TS; 121,200 SS. At -320°F: 217,100 TS. At -423°F: 219,400 TS, 149,200 SS. For cryogenic fasteners. Good tensile-impact properties.

AF 2-1DA
Universal Cyclops
C 0.35, Cr 12, W 6, Mo 3, Co 10, Ti 3, Al 4.6, B 0.015, Zr 0.1, Ta 1.5, bal Ni.
Nickel base superalloy.

AF 2-1DA
Homogeneous Metals Inc.
C 0.35, Cr 12, W 6, Mo 3, Co 10, Ti 3, Al 4.6, B 0.015, Zr 0.1, Ta 1.5, bal Ni.
Nickel base superalloy.

AF-1410
American manufacture
C 0.16, Ni 10, Cr 2, Co 14, Mo 1, bal Fe.
Heat treatable to 230,000 psi YS. Weldable and corrosion resistant.

AFC 77
Now CRUCIBLE AFC 77.

AFCOLOY 1025
Atlas Foundry Co.
C 0.2-0.3, Mn 0.6-0.9, Si 0-0.6, bal Fe.
Annealed: 60,000 TS; 35,000 YS; 30 El; 50 RA; 170 Brin. For machinery castings, gears, good machinability. *Obsolete*

AFCOLOY 1040
Atlas Foundry Co.
C 0.35-0.45, Mn 0.6-0.8, Si 0-0.6, bal Fe.
Annealed: 75,000 TS; 40,000 YS; 25 El; 45 RA; 190 Brin. For machinery castings, gears; good machinability. *Obsolete*

AFCOLOY 13
Atlas Foundry Co.
C 0-0.15, Mn 0-1, Si 0-1.5, Cr 11.5-14, Ni 0-1, bal Fe.
Heat treated: 95,000-120,000 TS; 78,000-110,000 YS; 18-22 El; 187-241 Brin. For corrosion resistant castings; ACI-CA15; corrosion resistant. *Obsolete*

AFCOLOY 17H
Atlas Foundry Co.
C 0.6-1.2, Mn 0-1.5, Si 0-1.5, Cr 16-18, bal Fe.
Heat treated: 150,000-270,000 TS; 110,000-245,000 YS; 2-10 El; 3-25 RA; 300-545 Brin. For corrosion and wear resistant castings corrosion resistant. *Obsolete*

AFCOLOY 18-8
Atlas Foundry Co.
C 0-0.2, Mn 0-1.5, Si 0-2, Cr 18-21, Ni 8-11, bal Fe.
Annealed: 80,000 TS; 40,000 YS; 45 El; 50 RA; 150 Brin. For stainless castings, chemical plant equipment; ACI-CF 20, stainless, austenitic. *Obsolete*

AFCOLOY 18-8 CB
Atlas Foundry Co.
C 0-0.08, Mn 0-1.5, Si 0-2, Cr 18-21, Ni 9-12, Cb = 8 x C, bal Fe.
Annealed: 80,000 TS; 40,000 YS; 45 El; 50 RA; 150 Brin. For welded castings; ACI-CF 8 C, stainless, austenitic. *Obsolete*

AFCOLOY 18-8 L
Atlas Foundry Co.
C 0-0.08, Mn 0-1.5, Si 0-2, Cr 18-21, Ni 8-11, bal Fe.
Annealed: 80,000 TS; 40,000 YS; 45 El; 50 RA; 150 Brin. For stainless castings, chemical plant equipment; ACI-CF 8, stainless, austenitic. *Obsolete*

AFCOLOY 18-8 MO
Atlas Foundry Co.
C 0-0.08, Mn 0-1.5, Si 0-2, Cr 19-21, Ni 9-12, Mo 2-3, bal Fe.
Annealed: 80,000 TS; 40,000 YS; 45 El; 50 RA; 150 Brin. For acid resistant castings, chemical plant equipment; ACI-CF 8 M, stainless, austenitic. *Obsolete*

AFCOLOY 18-8 SE
Atlas Foundry Co.
C 0-0.16, Mn 0-1.5, Si 0-2, Cr 18-21, Ni 9-12, Se 0.2-0.35, bal Fe.
Annealed: 80,000 TS; 40,000 YS; 45 El; 50 RA; 150 Brin. For stainless castings, chemical plant equipment; ACI-CF 16 F, stainless, austenitic, free-cutting. *Obsolete*

AFCOLOY 18-8 TI
Atlas Foundry Co.
C 0-0.08, Mn 0-1.5, Si 0-2, Cr 18-21, Ni 9-12, Ti = 5 x C, bal Fe.
Annealed: 80,000 TS; 40,000 YS; 45 El; 50 RA; 150 Brin. For welded castings, chemical plant equipment; stainless, austenitic. *Obsolete*

AFCOLOY 20
Atlas Foundry Co.
C 0-0.3, Mn 0-1, Si 0-1, Cr 18-22, Ni 0-2, bal Fe.
Annealed: 100,000 TS; 65,000 YS; 7 El; 8 RA; 200 Brin. Tempered: 75,000 TS; 50,000 YS; 10 El; 170 Brin. For heat and corrosion resistant castings; ACI-CB30; corrosion resistant. *Obsolete*

AFCOLOY 2140
Atlas Foundry Co.
C 0.35-0.45, Mn 0.6-0.8, Ni 1.5, bal Fe.
Annealed: 90,000 TS; 60,000 YS; 25 El; 45 RA; 210 Brin. For machinery parts, gears; shock resistant. *Obsolete*

AFCOLOY 25-12
Atlas Foundry Co.
C 0.2-0.5, Mn 0-2, Si 0-2, Cr 24-28, Ni 11-14, Mo 0-0.5, bal Fe.
Cast and aged: 85,000-95,000 TS; 50,000-55,000 YS; 11-25 El; 15-30 RA; 180-200 Brin. For corrosion and heat resistant castings; ACI-HH; heat and corrosion resistant. *Obsolete*

AFCOLOY 25-20
Atlas Foundry Co.
C 0.2-0.6, Mn 0-2, Si 0-3, Cr 24-28, Ni 18-22, bal Fe.
Cast and aged: 75,000-78,000 TS; 47,000-50,000 YS; 10-17 El; 187 Brin. For corrosion and heat resistant castings; heat and corrosion resistant; ACI-HK. *Obsolete*

AFCOLOY 28
Atlas Foundry Co.
C 0-0.5, Mn 0-1, Si 0-1, Cr 26-30, bal Fe.
Annealed: 105,000 TS; 70,000 YS; 12 El; 12 RA; 200-220 Brin. For heat resistant castings; ACI-CC50; corrosion resistant. *Obsolete*

AFCOLOY 3140
Atlas Foundry Co.
C 0.35-0.45, Mn 0.6-0.8, Cr 0.45-0.75, Ni 1-1.5, bal Fe.
Annealed: 100,000 TS; 65,000 YS; 25 El; 45 RA; 220 Brin. For gears, shafts; shock resistant. *Obsolete*

AFCOLOY 35
Atlas Foundry Co.
C 0.35-0.75, Mn 0-2, Si 0-2.5, Cr 13-17, Ni 33-37, bal Fe.
Cast: 70,000 TS; 40,000 YS; 10 El; 15 RA; 168 Brin. Aged: 75,000 TS; 45,000 YS; 5 El; 8 RA; 187 Brin. For furnace parts; ACI-HT; heat resistant. *Obsolete*

AFCOLOY 35L
Atlas Foundry Co.
C 0-0.1, Mn 0-2, Si 0-2.5, Cr 13-17, Ni 33-37, bal Fe.
Cast: 60,000 TS; 35,000 YS; 150 Brin. For salt pots and fixtures; heat resistant. *Obsolete*

AFCOLOY 4025
Atlas Foundry Co.
C 0.2-0.3, Mn 0.7-0.9, Mo 0.2-0.3, bal Fe.
Annealed: 80,000 TS; 65,000 YS; 25 El; 45 RA; 220 Brin. For strong castings, gears, shafts, housings; impact resistant. *Obsolete*

AFCOLOY 4140
Atlas Foundry Co.
C 0.35-0.45, Mn 0.75-1, Cr 0.8-1.1, Mo 0.15-0.25, bal Fe.
Annealed: 100,000 TS; 65,000 YS; 25 El; 45 RA; 220 Brin. For high strength castings; oil hardening. *Obsolete*

AFCOLOY 4340
Atlas Foundry Co.
C 0.35-0.45, Mn 0.6-0.8, Cr 0.7-0.9, Ni 1.6-2, Mo 0.2-0.3, bal Fe.
Annealed: 120,000 TS; 80,000 YS; 18 El; 35 RA; 250 Brin. For shock resistant castings, gears, crankshafts; tough. *Obsolete*

AFCOLOY 60
Atlas Foundry Co.
C 0.35-0.37, Mn 0-2, Si 2.5, Cr 10-14, Ni 58-62, bal Fe.
Cast: 68,000 TS; 50,000 YS; 4 El; 179 Brin. Aged: 75,000 TS; 54,000 YS; 4 El; 179 Brin. For furnace parts; ACI-HW; heat and corrosion resistant. *Obsolete*

AFCOLOY 60L
Atlas Foundry Co.
C 0-0.1, Mn 0-2, Si 0-2.5, Cr 10-14, Ni 58-62, bal Fe.
Cast: 55,000 TS; 40,000 YS; 145 Brin. For salt pots and fixtures; heat and corrosion resistant. *Obsolete*

AFCOMET 150-12
Atlas Foundry Co.
C 1.4-1.8, Mn 0.5-1.25, Si 0.25-1, Cr 12-14, bal Fe.
Cast: 50,000 TS; 500 Brin. For forging dies, gauges, tools; Tr.S. 4000; Tr.D. 0.10; wear resistant. *Obsolete*

AFCOMET 20
Atlas Foundry Co.
C 3.3-3.5, Mn 0.6-0.8, Si 2.2-2.4, P 0.2-0.3, S 0-0.15, bal Fe.
Annealed: 20,000 TS; 170 Brin. For castings, gears, housings, shafts; Tr.S. 1800; Tr.D. 0.20. *Obsolete*

AFCOMET 20 CR
Atlas Foundry Co.
C 3.3-3.5, Mn 0.6-0.8, Si 2.2-2.4, P 0.2-0.3, Cr 1-1.5, bal Fe.
Annealed: 25,000 TS; 200 Brin. For light machinery castings, gears, shafts; Tr.S. 2000; Tr.D. 0.20. *Obsolete*

AFCOMET 250-12
Atlas Foundry Co.
C 2.25-2.85, Mn 0.5-1.25, Si 0.25-1, Cr 12-14, bal Fe.
Cast: 55,000 TS; 500 Brin. For wear and corrosion resistant castings; Tr.S. 4500; Tr.D. 0.10. *Obsolete*

AFCOMET 250-25
Atlas Foundry Co.
C 2.25-2.85, Mn 0.5-1.25, Si 0.25-1, Cr 24-30, bal Fe.
Cast: 60,000 TS; 500 Brin. For abrasion and corrosion resistant castings; Tr.S. 6000; Tr.D. 0.10. *Obsolete*

AFCOMET 30
Atlas Foundry Co.
C 3.1-3.3, Mn 0.6-0.8, Si 2-2.2, P 0.2-0.3, bal Fe.
Annealed: 30,000 TS; 200 Brin. For light machinery castings; Tr.S. 2200; Tr.D. 0.20. *Obsolete*

AFCOMET 30 CR
Atlas Foundry Co.
C 3.1-3.3, Mn 0.6-0.8, Si 2-2.2, P 0.2-0.3, Cr 1-1.5, bal Fe.
Annealed: 35,000 TS; 220 Brin. For light machinery castings; Tr.S. 2400; Tr.D. 0.20. *Obsolete*

AFCOMET 40
Atlas Foundry Co.
C 2.9-3.1, Mn 0.6-0.8, Si 1.8-2, P 0.2-0.3, Mo 0.3-0.5, bal Fe.
Annealed: 40,000 TS; 230 Brin. For machinery and pressure castings; Tr.S. 2600; Tr.D. 0.25. *Obsolete*

AFCOMET 40 CR
Atlas Foundry Co.
C 2.9-3.1, Mn 0.6-0.8, Si 1.8-2, P 0.2-0.3, Cr 1-1.5, Mo 0.3-0.5, bal Fe.
Annealed: 45,000 TS; 240 Brin. For machinery and pressure castings; Tr.S. 2900; Tr.D. 0.25. *Obsolete*

AFCOMET 50
Atlas Foundry Co.
C 2.8-3, Mn 0.6-0.8, Si 1.5-1.8, P 0.2-0.3, Mo 1-1.25, bal Fe.
Annealed: 50,000 TS; 240 Brin. For heavy castings, gears, housings, shafts; Tr.S. 3400; Tr.D. 0.30. *Obsolete*

AFCOMET 50 CR
Atlas Foundry Co.
C 2.8-3, Mn 0.6-0.8, Si 1.5-1.8, P 0.2-0.3, Cr 1-1.5, Mo 1-1.25, bal Fe.
Annealed: 55,000 TS; 250 Brin. For heavy castings, gears, housings, shafts; Tr.S. 3500; Tr.D. 0.30. *Obsolete*

AFFIMET AL-CU 10 MG
Aluminium Francais
Cu 9-11, Si 0-2.5, Zn 0-0.8, Fe 0-1, Mn 0-0.6, Ni 0-0.5, Mg 0.2-0.4, bal Al.
Cast aluminium alloy. AFNOR A-U 10 G; BS 1490 LM12.

AFFIMET AL-CU 4 NI
Aluminium Francais
Cu 3.5-4.5, Si 0-0.6, Fe 0-0.6, Mn 0-0.6, Ni 1.8-2.3, Mg 1.2-1.7, bal Al.
Cast aluminium alloy. AFNOR A-U 4NT; BS 1490 LM14.

AFFIMET AL-CU 5 TI
Aluminium Francais
Cu 4-5, Si 0-0.25, Fe 0-0.25, Ti 0.05-0.3, bal Al.
Cast aluminium alloy. BS 1490 LM11.

AFFIMET AL-CU 7 SI 3 ZN 3
Aluminium Francais
Cu 6-8, Si 2-4, Zn 2-4, Fe 0-1, Mn 0-0.6, 0.3 others each max, bal Al.
Cast aluminium alloy. AFNOR A-U 8 SZ; BS 1490 LM1.

AFFIMET AL-MG 10
Aluminium Francais
Si 0-0.25, Fe 0-0.35, Mg 9.5-11, bal Al.
Cast aluminium alloy. AFNOR A-G 10; BS 1490 LM10.

AFFIMET AL-MG 5
Aluminium Francais
Si 0-0.3, Fe 0-0.6, Mg 3-6, Mn 0.3-0.7, bal Al.
Cast aluminium alloy. AFNOR A-G3T; BS 1490 LM5.

AFFIMET AL-SI 10 CU
Aluminium Francais
Cu 0.7-2.5, Si 9-11.5, Zn 0-2, Fe 0-1, Mn 0-0.5, Ni 0-0.5, 0.3 others each max, bal Al.
Cast aluminium alloy. BS 1490 LM2.

AFFIMET AL-SI 10 CU 3 MG
Teledyne Allvac
Aluminum. Cu 2-4, Si 8.5-10.5, Zn 0-1, Fe 0-1.2, Mn 0-0.5, Ni 0-1, Mg 0.5-1.5, bal Al.
Cast aluminium alloy. BS 1490 LM26.

AFFIMET AL-SI 10 MG
Aluminium Francais
Si 10-13, Fe 0-0.6, Mn 0.3-0.7, Mg 0.2-0.6, bal Al.
Cast aluminium alloy. AFNOR A-S 9 G; BS 1490 LM9.

AFFIMET AL-SI 12 CU NI
Aluminium Francais
Cu 0.7-1.5, Si 10-12, Zn 0-0.5, Fe 0-1, Mn 0-0.5, Ni 0-1.5, Mg 0.8-1.5, bal Al.
Cast aluminium alloy. AFNOR A-S 1i UN; BS 1490 LM13.

AFFIMET AL-SI 13
Aluminium Francais
Si 10-13, Fe 0-0.6, Mn 0-0.5, bal Al.
Cast aluminium alloy. AFNOR A-S 13; BS 1490 LM6.

AFFIMET AL-SI 17 CU 4 MG
Aluminium Francais
Cu 4-5, Si 16-18, Fe 0-1.1, Mn 0-0.3, Mg 0.4-0.7, bal Al.
Cast aluminium alloy. BS 1490 LM30.

AFFIMET AL-SI 19 CU NI CO
Aluminium Francais
Cu 1.3-1.8, Si 17-20, Fe 0-0.7, Mn 0-0.6, Ni 0.8-1.5, Mg 0.8-1.5, Cr 0-0.6, Co 0-0.5, bal Al.
Cast aluminium alloy. AFNOR A-S 20 U; BS 1490 LM28.

AFFIMET AL-SI 2
Aluminium Francais
Cu 0-0.4, Si 10-13, Zn 0-0.2, Fe 0-1, Mn 0-0.5, bal Al.
Cast aluminium alloy. AFNOR A-S 12; BS 1490 LM20.

AFFIMET AL-SI 2 CU
Aluminium Francais
Cu 1-2.5, Si 1.5-3.5, Fe 0.3-1.4, Mg 0.05-2, Ni 0.5-1.7, Ti 0.05-0.3, bal Al.
Cast aluminium alloy. BS 1490 LM7.

AFFIMET AL-SI 2 CU 2 MG NI
Aluminium Francais
Cu 1.3-3, Si 0.6-2, Fe 0.8-1.4, Ni 0.5-2, Mg 0.5-1.7, Ti 0.05-0.3, bal Al.
Cast aluminium alloy. BS 1490 LM15.

AFFIMET AL-SI 24 CU NI CO
Aluminium Francais
Cu 0.8-1.3, Si 22-25, Fe 0-0.7, Mn 0-0.6, Ni 0.8-1.3, Mg 0.8-1.3, Cr 0.6, Co 0.5, bal Al.
Cast aluminium alloy. AFNOR A-S 22 UNK; BS 1490 LM29.

AFFIMET AL-SI 4 CU 4 ZN
Aluminium Francais
Cu 3-5, Si 5-7, Zn 0-2, Fe 0-1, Mn 0.2-0.6, Mg 0.1-0.3, bal Al.
Cast aluminium alloy. AFNOR A-S 5 UZ; BS 1490 LM21.

AFFIMET AL-SI 4 MG
Aluminium Francais
Si 3.5-6, Fe 0-0.6, Mn 0-0.5, Mg 0.3-0.8, bal Al.
Cast aluminium alloy. AFNOR A-S 4 G; BS 1490 LM8.

AFFIMET AL-SI 5
Aluminium Francais
Si 4.5-6, Fe 0-0.6, Mn 0-0.5, bal Al.
Cast aluminium alloy. BS 1490 LM18.

AFFIMET AL-SI 5 CU
Aluminium Francais
Cu 2.8-3.8, Si 4-6, Zn 0-0.15, Fe 0-0.6, Mn 0.2-0.6, Ni 0-0.15, bal Al.
Cast aluminium alloy. AFNOR A-S 5 U; BS 1490 LM22.

AFFIMET AL-SI 5 CU 3
Aluminium Francais
Cu 2-4, Si 4-6, Zn 0-0.5, Fe 0-0.8, Mn 0.2-0.6, bal Al.
Cast aluminium alloy. AFNOR A-S 5 U; BS 1490 LM4.

AFFIMET AL-SI 5 CU MG
Aluminium Francais
Cu 1-1.5, Si 4.5-5.5, Mg 0.4-0.6, Ti 0.12-0.2, bal Al.
Cast aluminium alloy. AFNOR A-S 5 U G.

AFFIMET AL-SI 5 MG CU
Aluminium Francais
Cu 1-1.5, Si 4.5-5.5, Fe 0-0.6, Mn 0-0.5, Ni 0-0.25, Mg 0.4-0.6, bal Al.
Cast aluminium alloy. AFNOR A-S 5 U1; BS 1490 LM16.

AFFIMET AL-SI 7 MG
Aluminium Francais
Si 6.5-7.5, Fe 0-0.5, MN 0-0.3, Mg 0.2-0.45, bal Al.
Cast aluminium alloy. AFNOR A-S 7 G; BS 1490 LM25.

AFFIMET AL-SI 8 CU 3 FE/A
Aluminium Francais
Cu 3-4, Si 7.5-9.5, Zn 0-1.5, Fe 0-1.3, Mn 0-0.5, Mg 0-0.3, Pb 0-0.3, Sn 0-0.2, Ti 0-0.2, bal Al.
Cast aluminium alloy. AFNOR A-S 9U 3/A; BS 1490 LM24; Type A.

AFFIMET AL-SI 8 CU 3 FE/B
Aluminium Francais
Cu 3-4, Si 7.5-9.5, Zn 0-3, Fe 0-1.3, Mn 0-0.5, Mg 0-0.3, Pb 0-0.3, Sn 0-0.2, Ti 0-0.2, bal Al.
Cast aluminium alloy. AFNOR A-S 9U 3/B; BS 1490 LM24; Type B.

AFFIMET AL-Z 10 CU 3
Aluminium Francais
Cu 2.5-4.5, Si 0-1.3, Zn 9-13, Fe 0-1, Mn 0-0.5, Ni 0-0.5, Mg 0-0.1, 0.3 others each max, bal Al.
Cast aluminium alloy. BS 1490 LM3.

AFFMET AL-SI 7 CU 2
Aluminium Francais
Cu 1.5-2.5, Si 6-8, Zn 0-1, Fe 0-0.8, Mn 0.2-0.6, Mg 0-0.3, bal Al.
Cast aluminium alloy. AFNOR A-S 7 U 3; BS 1490 LM27.

AFFREEZE-III
British Steel Corp.
C 0-0.15, Mn 1.3, Mo 0.25, V 0.07, bal Fe.
For nuclear parts, pressure vessels. Good creep strength, Maximum service life of 790 F. *Obsolete*

AFORA 19 NCB6
AFORA
Alloy steel. C 0.15-0.21, Mn 0.6-0.9, Si 0.15-0.4, P 0-0.035, S 0-0.035, Cr 0.85-1.15, Ni 1-1.2, bal Fe.
125-155 kgf/mm^2 TS; 100 kgf/mm^2 min YS; 8 min El.

AFORA 19 NCDB2
AFORA
Alloy steel. C 0.17-0.23, Mn 0.65-0.95, Si 0.15-0.4, P 0-0.035, S 0-0.035, Cr 0.4-0.65, Ni 0.4-0.7, Mo 0.15-0.25, bal Fe.
125-155 kgf/mm^2 TS; 100 kgf/mm^2 min YS; 8 min El.

AFORA 20 MB5
AFORA
Alloy steel. C 0.16-0.22, Mn 1.1-1.4, Si 0.15-0.4, P 0-0.035, S 0-0.035, bal Fe.
85-105 kgf/mm^2 TS; 70 kgf/mm^2 min YS; 11 min El.

AFORA 21 B3
AFORA
Alloy steel. C 0.18-0.24, Mn 0.6-0.9, Si 0.15-0.4, P 0-0.035, S 0-0.035, bal Fe.
75-95 kgf/mm^2 TS; 60 kgf/mm^2 min YS; 13 min El.

AFORA 38 B3
AFORA
Alloy steel. C 0.34-0.4, Mn 0.6-0.9, Si 0.15-0.4, P 0-0.035, S 0-0.035, bal Fe.
82-97 kgf/mm^2 TS; 67 kgf/mm^2 min YS; 12 min El.

AFORA 38 CB1
AFORA
Alloy steel. C 0.34-0.4, Mn 0.6-0.9, Si 0.15-0.4, P 0-0.035, S 0-0.035, Cr 0.2-0.4, bal Fe.
90-110 kgf/mm^2 TS; 75 kgf/mm^2 min YS; 10 min El.

AFORA 38 MB5
AFORA
Alloy steel. C 0.34-0.4, Mn 1.1-1.4, Si 0.15-0.4, P 0-0.035, S 0-0.035, bal Fe.
95-115 kgf/mm^2 TS; 80 kgf/mm^2 min YS; 10 min El.

AG 15
Otto Fuchs Metallwerke
Aluminum. Mn 1, bal Al.
295 N/mm^2 TS; 100 N/mm^2 YS; 17 El. Corrosion resistant, weldable with good formability. Type AlMn1.

AG/CU EUTETIC
Johnson Matthey plc
Cu 28, bal Ag.
778°C MP. Silver braze alloy.

AGATHON ARMATURE
Republic Steel Corp.
Si 0.5, bal Fe.
For electric motors; high permeability. *Obsolete*

AGATHON BIT STEEL
Republic Steel Corp.
C 0.5-0.7, Mn 0.5-0.8, bal Fe.
For bits, tools. *Obsolete*

AGATHON DYNAMO
Republic Steel Corp.
Si 2.5, bal Fe.
For dynamos; high permeability. *Obsolete*

AGATHON ELECTRIC
Republic Steel Corp.
Si 1, bal Fe.
For electric motors; high permeability. *Obsolete*

AGATHON NICKEL ALLOY
Republic Steel Corp.
Ni 2, C 0.15, bal Fe.
For staybolts and carburizing purposes; carburizing steel. *Obsolete*

AGATHON SPECIAL ELECTRIC
Republic Steel Corp.
Si 1.2, bal Fe.
For electric motors; high permeability. *Obsolete*

AGGERLIT A
Dorrenberg Edelstahl GmbH
C 0.3, Si 2.2, Cr 6, bal Fe.
For oil refinery equipment; heat resistant.

AGGERLIT B
Dorrenberg Edelstahl GmbH
C 0.5, Si 1.5, Cr 17, bal Fe.
For heat treating boxes, furnace parts, fixtures; corrosion and heat resistant.

AGGERLIT C
Dorrenberg Edelstahl GmbH
C 0.6, Si 1.5, Cr 22, bal Fe.
For furnace parts and equipment, fixtures; corrosion and heat resistant.

AGGERLIT D
Dorrenberg Edelstahl GmbH
C 0.6, Si 1.5, Cr 22, bal Fe.
For furnace parts and equipment; fixtures; corrosion and heat resistant.

AGGERLIT E
Dorrenberg Edelstahl GmbH
C 0.6, Si 1.5, Cr 29, bal Fe.
For furnace parts and equipment, fixtures; heat resistant.

AGGERSTAHL
Dorrenberg Edelstahl GmbH
C 1.45, Cr 1.4, Si 0.25, Mn 0.6, bal Fe.
For bearings, sleeves, liners; water hardened, wear resistant.

AGILE ACTARC
Nagle/Sybron Corp.
Pure Fe or low alloy steel.
For welding electrodes; for sheet metal.

AGILE NO. 200
Nagle/Sybron Corp.
Copper. Sn 15, bal Cu.
For welding and brazing rods; coated, MP 1596°F.

AGILE SILVER A-C ELECTRODE
Nagle/Sybron Corp.
C 0.9, Mn 0.7, Cr 0.7, W 1.5, bal Fe.
400-650 Brin. For welding electrodes; deposit should be heat treated.

AGILE SILVER BLACK
Nagle/Sybron Corp.
C 0.6, Cr 12, Mo 8, W 8, B 1.5, bal Fe.
For welding high speed steel cutters; high speed steel.

AGILE SILVER CHISEL
Nagle/Sybron Corp.
C 0.8-1, Si 1, W 1.5, B 0.3-0.5, bal Fe.
400-620 Brin. For welding electrodes for chisels and shock resisting tools; tough, shock resisting.

AGILE SILVER D-C-ELECTRODE
Nagle/Sybron Corp.
C 1.8-2.2, Cr 12-14, V 0.5, Mo 0.8, bal Fe.
400-640 Brin. For welding electrodes for punches and cold working dies; must be heat treated.

AGILE SILVER ROD
Nagle/Sybron Corp.
C 0.75, Mo 4, W 6, Cr 4, B 0.8, bal Fe.
For welding high speed steel tools; high speed steel.

AGILE YELLOW ROD
Nagle/Sybron Corp.
Nickel. Ni 67, Cu 31, B 0.2.
190 Brin. For welding, electrode for cast iron; machinable.

AGMOLITE
Engelhard Corp.
Ag, Mo.
For electrical contracts for circuit breakers; sintered. *Obsolete*

AGNILITE
Engelhard Corp.
Ag, Ni.
For electric contacts for circuit breakers; sintered. *Obsolete*

AGPU 11
Societe Nouvelle des Acieries de Pompey
C, S, bal Fe.
For screw machine products; free-cutting.

AGRAPHITE
Engelhard Corp.
Ag, graphite.
For electrical contacts for circuit breakers; sintered. *Obsolete*

AGRICOLA
Manco Products, Inc.
Copper. Cu 70, Pb 30.
For diesel engine bearings, seals; heavy duty.

AGRICOLA BRONZE
British Metal Corp., Ltd.
Cu 70, Ag 1, Pb 29.
For bearings, bushings.

AGRILITE BEARING METAL
American Injector Co.
Copper. Cu 70, Sn 5, Pb 25.
25,000 TS; 16,350 YS; 50 Brin. For bearings for rolling mill, thrust bearings on steam boats, bearings for washing and ironing machines; self lubricating bearing bronze.

AGRILITE NO. 5
American Injector Co.
Copper. Cu 70.49, Sn 5.39, P 0.005, Pb 24, Ni 0.09.
For connecting rod bearings for aero engines, rolling mill bearings; self lubricating bearing bronze.

AH NO. 2
Texas Instruments Inc./Materials Control
Ag 30, Cu 52.45, Zn 9.75, Ni 7.8.
For brazing; MP 1400-1735°F. *Obsolete*

AH-5
Now BETHLEHEM AH-5.

AICH METAL
English manufacture
Cu 59.37, Zn 39.68, Fe 0.95.
Cast: 60,000 TS; 21,000 YS; 44 El; 45 RA; 90 Brin. Forged: 64,500 TS; 31,000 YS; 44 El; 64 RA; 107 Brin. For hydraulic cylinders and forgings; resists oxidation; similar to "Sterro."

AIDION
John Finn Metal Works
Sb, Pb, bal Sn.
For bearings. Anti-friction metal.

AIKEN METAL
American manufacture
Cu 50, Pb 50.
For plastic metal parts.

AIP
Acciaierie Valbruna s.p.a.
Stainless Steel. C 0-0.08, Si 0-1.5, Mn 0-1, Cr 18, N 12, P 0.04, S 0.03, bal Fe.
Austenitic. W. Nr. 1.4303. AISI 305.

AIR HARDENING
Republic Steel Corp.
C, Cr, W, V, bal Fe.
Water hardened: 120,000-214,000 TS; 102,000-192,000 YS; 12-20 El; 38-57 RA; 255-415 Brin. For tools, dies; air hardening. *Obsolete*

AIR HARDENING
DoAll Co.
C 1, Mn 0.7, Si 0.3, Cr 5.25, V 0.3, Mo 1.15, bal Fe.
AISI Type A2 air hardening tool steel.

AIR SERVICE ALUMINUM ALLOY NO. 1
English manufacture
Al 92, Cu 8.
For light alloy parts, airplane construction; non-hardenable.

AIR SERVICE ALUMINUM ALLOY NO. 2
English manufacture
Al 88.5, Cu 10, Mg 0.25, Fe 1.25.
For aircraft parts, light alloy parts; non-hardenable.

AIR SERVICE ALUMINUM ALLOY NO. 3
English manufacture
Al 95.75, Cu 2.5, Mg 0.5, Fe 1.25.
For aircraft parts; heat treatable.

AIR SERVICE ALUMINUM ALLOY NO. 4
English manufacture
Al 94, Cu 5, Si 1.
For aircraft parts; age-hardenable.

AIR SERVICE ALUMINUM ALLOY NO. 5
English manufacture
Al 93, Cu 4, Si 3.
For aircraft parts; age-hardenable.

AIR SERVICE ALUMINUM ALLOY NO. 6
English manufacture
Cu 4, Mg 1-5, Ni 2, bal Al.
For aircraft parts; age-hardenable.

AIR SHOCK
Peninsular Steel Co.
C 0.5, Mn 0.7, Cr 3.25, Mo 1.4, bal Fe.
Air or oil hardening. For rivet sets, hot gripper dies, punches, chisels, plastic mold, die casting dies. Type 57 shock resisting tool steel.

AIR-4
Bethlehem Steel Corp.
C 0.95, Mn 2, Cr 2.2, Mo 1.1, Pb 0.15-0.35, bal Fe.
General purpose tools, cold forming dies, blanking dies, forming rolls, knurling tools. Free machining type. Type A4; air-hardened tool steel. *Obsolete*

AIR-CHROM
Manufacturer not listed.
C 1, Mn 0.6, Cr 5.25, V 0.25, Mo 1.1, bal Fe.
For tools, dies. AISI A2 air hardening tool steel.

AIR-EEZ
Manufacturer not listed.
C 0.95, Mn 2, Si 0.35, Cr 2.2, Mo 1.1, Pb 0.3, bal Fe.
Free machining grade. AISI A4 low temperature tool steel.

AIR-HARD
Carpenter Technology Corp.
C 1, Cr 5, Mo 1.5, bal Fe.
For tools, dies; wear resistant, air hardening. *Obsolete*

AIR-HARD
Teledyne Vasco
C 1, Si 0.3, Mn 0.7, Cr 5.25, Mo 1.1, V 0.25, bal Fe.
Annealed: 55,000 YS; 18 El; 38 RA; 223 Brin. Hardened: 184,000 YS; 5 El; 14 RA; 46-54 Rock C. For trimming and blanking dies, cold shears, gauges, punches. Air hardening, non-deforming, wear resistant. Type A2.

AIR-TOUGH
Carpenter Technology Corp.
C 0.7, Mn 2, Cr 1, Mo 1.25, bal Fe.
For tools, dies; air hardening, tough. *Obsolete*

AIR-TRUE
Precision-Kidd Steel Co.
C 0.95-1.05, Mn 0.65-0.75, Si 0.25-0.35, Cr 5.25, V 0.2-0.25, M 1-1.2, bal Fe.
Air hardening, non-distorting steel; particularly for drill rod. AISI A2.

AIR-TRUE
Simonds Worden White Co.
C 1, Mn 3, Cr 1, Mo 1, bal Fe.
For dies, templates, gauges, punch and die facings; air hardening, non-deforming.

AIR-WEAR
Carpenter Technology Corp.
C 1.5, Cr 12, V 1, bal Fe.
For tools, dies; wear resistant, air hardening. *Obsolete*

AIRALOY
Now REPUBLIC A-4.

AIRCO
Wilson Welder & Metals Co.
C 0.2, bal Fe.
For welding rod. *Obsolete*

AIRCO 100
Airco Vacuum Metals
Al 9-11, Fe 0-1.5, bal Cu.
Welded: 77,000 psi TS; 35,000 psi YS; 27 El; 120 Brin. Aluminum bronze brazing rod; coated; DC reverse polarity. For welding and build-up on brass, manganese bronze, aluminum bronze and dissimilar metals. AWS-ASTM E Cu A1-A2.

AIRCO 10013-G
Airco Vacuum Metals
AC-DC straight polarity. For welding low alloy high tensile steels. AWS-ASTM E10013-G. Low carbon; low alloy coated electrode.

AIRCO 10016-D2
Airco Vacuum Metals
C 0.09, Mn 1.8, Si 0.58, Mo 0.4, bal Fe.
Welded, stress relieved: 106,000 psi TS; 92,000 psi YS; 25 El. AC-DC reverse polarity coated electrode. For welding higher strength steels. AWS-ASTM E10016-D2.

AIRCO 10016-G
Airco Vacuum Metals
C 0.06, Mn 0.8, Si 0.34, Mo 0.3, Ni 1.63, V 0.13, bal Fe.
Welded, stress relieved: 109,500 psi TS; 102,000 psi YS; 25 El. AC-DC reverse polarity coated electrode. For high strength assemblies. AWS-ASTM E10016-G.

AIRCO 116
Airco Vacuum Metals
Al 11-12, Fe 3-4.25, bal Cu.
Welded: 89,000 psi TS; 47,000 psi YS; 15 El; 160 Brin. Aluminum bronze brazing rod; coated; DC reverse polarity. For overlays on bearing surfaces that are subjected to shock or impact loading, and for joining many dissimilar metals. AWS-ASTM E Cu Al-B.

AIRCO 120
Airco Vacuum Metals
Al 12-13, Fe 3-5, bal Cu.
Welded: 90,000 psi TS; 46,000 psi YS; 4 El; 200 Brin. Aluminum bronze brazing rod; coated; DC reverse polarity. For surfacing and overlays for good wear and strength for severe service. AWS-ASTM E Cu Al-C.

AIRCO 12015-G
Airco Vacuum Metals
Low hydrogen electrode; reverse polarity only. For welding high strength and armor steels. AWS-ASTM E12015-G.

AIRCO 125
Airco Vacuum Metals
Al 13-14, Fe 3-5, bal Cu.
Welded: 77,000 psi TS; 54,000 psi YS; 1.0 El; 260 Brin. Aluminum bronze brazing rod; coated; DC reverse polarity. For surfacing and overlays for rebuilding aluminum bronze dies or finish layer on ferrous dies to reduce galling. AWS-ASTM E Cu Al-D.

AIRCO 130
Airco Vacuum Metals
Al 14-15, Fe 3-5, bal Cu.
Welded: 80,000 psi TS; 68,000 psi YS; 0 El; 300 Brin. Aluminum bronze brazing rod; coated; DC reverse polarity. For overlaying dies, shafts, bearings to minimize scratching and galling. AWS-ASTM E Cu Al-E.

AIRCO 308
Airco Welding Products
Cr 19.
9 Ni stainless welding rod; bare wire, cut lengths. AWS ER 308; AISI 308 *Obsolete*

AIRCO 308 DC
Airco Vacuum Metals
C 0.05, Cr 19, Ni 10, Mn 1.8, Si 0.3, bal Fe.
DC reverse polarity lime coated electrode. For welding austenitic 18-8 type stainless steels. AWS-ASTM E 308-15; AISI 308.

AIRCO 308 ELC
Airco Vacuum Metals
Low carbon grade of 308.
For welding AISI 304L, 321 and 347 stainless steels. AWS ER 308 L; AISI 308 L.

AIRCO 309
Airco Vacuum Metals
Cr 25.
For welding AISI 309 Stainless steel. AWS ER 309; AISI 309. 12.0 Ni stainless welding rod; bare wire, cut lengths.

AIRCO 309 CB AC-DC
Airco Vacuum Metals
C 0.09, Cr 24, Ni 13, Mn 2.1, Si 0.55, Cb 0.85, bal Fe.
Welded: 90,000 psi TS; 70,000 psi YS; 35 El. AC-DC reverse polarity titania coated electrode. AWS-ASTM E 309 Cb-16, AISI 309 Cb.

AIRCO 309 DC
Airco Vacuum Metals
C 0.07, Cr 23, Ni 12.75, Mn 1.9, Si 0.7, bal Fe.
Welded: 89,000 psi TS; 63,500 psi YS; 42 El. DC reverse polarity lime coated electrode. For welding austenitic type 309 stainless. AWS-ASTM E 309-15; AISI 309.

AIRCO 310
Airco Vacuum Metals
Cr 25, Ni 20.
For welding AISI 310 stainless by TIG or oxyacetylene (with flux) methods; for elevated temperature operations. AWS ER 310; AISI 310. Stainless welding rod; bare wire; cut lengths.

AIRCO 310 CB AC-DC
Airco Vacuum Metals
C 0.11, Cr 26, Ni 21, Mn 1.9, Si 0.6, Cb 0.85, bal Fe.
Welded: 93,500 psi TS; 69,500 psi YS; 42 El. AC-DC reverse polarity titania coated electrode. For welding austenitic stainless steels and some dissimilar metals. AWS-ASTM E 310 Cb-16; AISI 310 Cb.

AIRCO 310 DC
Airco Vacuum Metals
C 0.1, Cr 26, Ni 21, Mn 1.8, Si 0.6, bal Fe.
Welded: 91,000 psi TS; 61,000 psi YS; 36 El. DC reverse polarity lime coated electrode. For welding austenitic types 310 and 314 stainless steels. AWS-ASTM E 310-15; AISI 310.

AIRCO 310 MO AC-DC
Airco Vacuum Metals
C 0.1, Cr 26, Ni 21, Mn 1.9, Si 0.55, Mo 2.5, bal Fe.
Welded: 89,000 psi TS; 62,500 psi YS; 37 El. AC-DC reverse polarity titania coated electrode. For welding type 310 and similar steels and for overlay mild steel for paper pulp digestors and similar applications. AWS-ASTM E 310 Mo-16; AISI 310 Mo.

AIRCO 312 AC-DC

Airco Vacuum Metals

C 0.12, Cr 28, Ni 9.25, Mn 2, Si 0.7, bal Fe.
Welded: 120,000 psi TS; 80,000 psi YS; 36 El. AC-DC reverse polarity titania coated electrode. For welding high strength type 312 base materials particularly for jet aircraft industry, and for dissimilar metals. AWS-ASTM E312-16; AISI 312.

AIRCO 316

Airco Vacuum Metals

Cr 19, Ni 12.
For welding AISI 316 stainless steel. AWS ER 316; AISI 316.
2.5 Mo stainless welding rod; bare wire; cut lengths.

AIRCO 316 CB AC-DC

Airco Vacuum Metals

C 0.07, Cr 18.5, Ni 11.5, Mn 1.9, Si 0.55, Mo 2.25, Cb 0.5, bal Fe.
Welded: 90,000 psi TS; 63,000 psi YS; 36 El. AC-DC reverse polarity titania coated electrode. For welding type 316 and other austenitic stainless steels for corrosion-resistant joints.

AIRCO 316 ELC

Airco Vacuum Metals

For welding AISI 316L stainless steel. AWS ER 316L; AISI 316L. Low carbon grade of 316.

AIRCO 317 AC-DC

Airco Vacuum Metals

C 0.07, Cr 19, Ni 13, Mn 2, Si 0.55, Mo 3.5, bal Fe.
Welded: 102,000 psi TS; 87,000 psi YS; 31 El. AC-DC reverse polarity titania coated electrode. For welding type 317 stainless for severe corrosion resistance. AWS-ASTM E317-16; AISI 317.

AIRCO 330 AC-DC

Airco Vacuum Metals

C 0.2, Cr 15, Ni 34.5, Mn 1.75, Si 0.6, bal Fe.
Welded: 89,500 psi TS; 70,000 psi YS; 35.5 El. AC-DC reverse polarity titania coated electrode. For welding high nickel alloy castings for high temperature operation. AWS-ASTM E330-16; AISI 330.

AIRCO 347

Airco Vacuum Metals

Cr 19, Ni 9.
For welding AISI 304L, 321 and 347 stainless steels. AWS ER 347; AISI 347. Cb-stabilized stainless steel welding rod; bare wire; cut lengths.

AIRCO 35-NI 15-CR TITANIA

Airco Vacuum Metals

C 0.07, Mn 1.8, Cr 16, Ni 35, Mo 0.9, bal Fe.
For welding rod. Heat and corrosion resistant.

AIRCO 361

Airco Vacuum Metals

C 0.5, Mn 4.75, Cr 19, Ni 9.5, Si 0.75, bal Fe.
For joining and hardfacing of railroad frogs and switches and similar applications. Impact resistant. Stainless core rod; composite coating.

AIRCO 363

Airco Vacuum Metals

C 0.2, Mn 1, Si 0.5-0.65, Cr 0.5, bal Fe.
20-28 Rock C. AC or DC, either polarity. For surfacing or intermediate layer with some abrasion and impact resistance; for brake drums, gear teeth, sprockets and bushings. Coated electrode.

AIRCO 375

Airco Vacuum Metals

C 1.2, Mn 1, Si 0.45, Fe 2.75, bal Ni.
AC-DC reverse polarity. For welding cast iron. AWS-ASTM E NiCl.

AIRCO 376

Airco Vacuum Metals

Mn 0.38, P 0.007, S 0.004, Si 0.04, C 1.1, bal Fe.
To produce machinable welds in cast iron using either AC or DC reverse polarity. AWS Class E NiFe-Cl.

AIRCO 4043

Airco Vacuum Metals

Si 4.5-6, Cu 0-0.3, Fe 0-0.8, Zn 0-0.1, Ti 0-0.2, bal Al.
Bare aluminum alloy wire for gas or TIG welding of 3003, 3004, 5052, 6061, 6063, and casting alloys 43, 355, 356 and 214. Flows at 1155°F. AWS ER 4043.

AIRCO 410

Airco Welding Products

Cr 12.
Stainless welding rod; bare wire, cut lengths. For welding AISI 403 and 410 stainless steels. AWS ER 410; AISI 410. *Obsolete*

AIRCO 410 DC

Airco Vacuum Metals

C 0.09, Cr 12.5, Ni 0.6, Mn 0.8, Si 0.6, bal Fe.
Welded: 150,000 psi TS; 5 El. Heat treated: 71,500 psi TS; 53,000 psi YS; 20 El. DC reverse polarity lime type coated electrode. For welding types 403 and 410 chrome corrosion resistant steels and for stainless overlay on carbon steels. AWS-ASTM E410-15; AISI 410.

AIRCO 4130

Airco Welding Products

C 0.3, Mn 0.5, Si 0.3, Mo 0.2, Cr 0.95, bal Fe.
Alloy wire for welding AISI 4130 steel by automatic submerged arc. *Obsolete*

AIRCO 43

Airco Vacuum Metals

For arc welding aluminum. AWS-ASTM Al-43. 5% Si-Aluminum-base electrode; coated; DC reverse polarity.

AIRCO 46

Airco Vacuum Metals

C 0.6, Mn 0.2, Si 0.1, bal Fe.
20-25 Rock C. For build-up to resist abrasion and impact. For excavator bucket fingers, locomotive tire flanges, rails and switches. Coated electrode; DC reverse polarity.

AIRCO 502 DC

Airco Vacuum Metals

C 0.08, Cr 5, Ni 0.4, Mn 0.6, Si 0.65, Mo 0.55, bal Fe.
Welded: 145,000 psi TS; 2.8 El. Heat treated: 80,000 psi TS; 67,500 psi YS; 38 El. DC reverse polarity lime type coated electrode. For welding types 501 and 502 stainless and some alloy steels as 4130. AWS-ASTM E502-15; AISI 502.

AIRCO 5183

Airco Vacuum Metals

Mg 4.3-5.2, Mn 0.5-1, Si 0-0.4, Fe 0-0.4, Zn 0-0.25, Ti 0-0.15, Cr 0.05-0.25, Cu 0-0.1, bal Al.
Bare aluminum alloy welding rod for gas or TIG welding of 5052, 5083, 5356, 5454 and 5456 alloys. Flows at 1180°F. AWS ER 5183.

AIRCO 5356

Airco Vacuum Metals

Mg 4.5-5.5, 0.5 max Si + Fe, bal Al.
Bare aluminum alloy; general purpose. For gas or TIG welding of 5050, 5052, 5083, 5086, 5356, 5454, and 5456. Flows at 1180°F. AWS ER 5356.

AIRCO 5554

Airco Welding Products

Mg 2.4-3, Mn 0.5-1, Zn 0-0.25, Ti 0.05-0.2, bal Al.
Bare aluminum alloy rod for gas or TIG welding of 5454 and 5052 alloys. Flows freely at 1095 F. AWS ER 5554 *Obsolete*

AIRCO 5556

Airco Vacuum Metals

Mg 4.7-5.5, Mn 0.5-1, Zn 0-0.25, Cr 0.05-0.2, bal Al.
Bare aluminum wire for gas or TIG welding 5052, 5083, 5086, 5356, 5454, and 5456 for maximum weld strength. Flows at 1180°F. AWS ER 5556.

AIRCO 57

Airco Vacuum Metals

Si 4.5-6, Fe 0.8, Cu 0.3, Zn 0.1, Ti 0.2, bal Al.
Welded: 14,000 psi TS min. General purpose aluminum electrode; coated; DC reverse polarity. For welding aluminum alloys 6061, 6063, 5052, 5154, 5454, 3003, and 2024; AWS-ASTM Al-43.

AIRCO 6010

Airco Vacuum Metals

C 0.07-0.12, Mn 0.2-0.4, Si 0.1-0.3, bal Fe.
Welded: 68,500 psi TS; 58,250 psi YS; 28.5 El. DC reverse polarity coated electrode. AWS-ASTM Class E6010.

AIRCO 6010 ARCRODS (001)

Airco Vacuum Metals

AWS-ASTM Class E6010. Mild steel coated electrode; DC reverse polarity.

AIRCO 6011

Airco Vacuum Metals

C 0.08, Mn 0.35, Si 0.11, bal Fe.
Welded: 70,000 psi TS; 62,000 psi YS; 30 El. AC-DC reverse polarity coated electrode. AWS-ASTM Class E6011.

AIRCO 6011C

Airco Vacuum Metals

C 0.1, Mn 0.58, Si 0.19, bal Fe.
Welded: 72,000 psi TS; 59,000 psi YS; 29.5 El. AC-DC reverse polarity coated electrode. AWS-ASTM Class E6011.

AIRCO 6011LOC

Airco Vacuum Metals

C 0.11, Mn 0.6, Si 0.25, bal Fe.
Welded: 71,000 psi TS; 59,000 psi YS; 22 El. AC-DC reverse polarity coated electrode. AWS-ASTM Class E6011.

AIRCO 6012

Airco Vacuum Metals

C 0.08, Mn 0.45, Si 0.3, bal Fe.
Welded: 83,000 psi TS; 74,000 psi YS; 20 El. AC-DC straight polarity coated electrode. AWS-ASTM E-6012.

AIRCO 6012 C

Airco Vacuum Metals

C 0.08, Mn 0.38, Si 0.25, bal Fe.
Welded: 70,000 psi TS; 60,000 psi YS; 28 El. AC-DC straight polarity coated electrode. AWS-ASTM E6012.

AIRCO 6013

Airco Vacuum Metals

C 0.1, Mn 0.35, S 0-0.035, P 0-0.035, Si 0.4, bal Fe.
Welded: 70,000 psi TS; 62,000 psi YS; 20 El. AC-DC all position, either polarity. AWS Class E6013.

AIRCO 6013 C

Airco Vacuum Metals

C 0.09, Mn 0.52, Si 0.39, bal Fe.
Welded: 74,500 psi TS; 63,500 psi YS; 27 El. AWS-ASTM E6013.

AIRCO 6020
Airco Vacuum Metals
C 0.12, Mn 0.35, Si 0.1, bal Fe.
Welded: 72,000 psi TS; 62,000 psi YS; 28 El. AC-DC either polarity coated electrode. AWS-ASTM E6020.

AIRCO 6020 D
Airco Vacuum Metals
AC or DC; either polarity. AWS-ASTM E6020. Mild steel coated electrode.

AIRCO 6020 F
Airco Vacuum Metals
AC or DC; either polarity. AWS-ASTM E6020. Mild steel coated electrode.

AIRCO 6150
Airco Welding Products
C 0.5, Mn 0.8, Si 0.28, Cr 1, bal Fe.
Alloy wire for welding AISI 6150 steel by automatic submerged arc. *Obsolete*

AIRCO 630
Airco Vacuum Metals
C 4.5, Cr 30, Si 2, Mn 1, bal Fe.
56-61 Rock C. For build-ups having resistance to severe abrasion with moderate impact. For dredging parts, dipper teeth, and sand muller parts. Steel core; composite type coating; AC-DC straight polarity.

AIRCO 660
Airco Vacuum Metals
C 1.8, Cr 4, Mn 0.5, Si 0.5, bal Fe.
45-55 Rock C. AC-DC polarity. For surfacing or build-up for resistance to abrasion and impact; for crusher rolls, hammermill hammers, and impact breaker bars. Low alloy steel rod; composite coating.

AIRCO 70
Airco Vacuum Metals
Sn 4.25, Fe 0.5, Zn 0.15, P 0.1, Si 0.1, bal Cu.
Welded: 40,000 psi TS; 20,000 psi YS; 15 El; 75 Brin.
Phosphor bronze brazing rod; coated; DC reverse polarity. For joining brass, bronze, copper, steel, cast iron, and malleable iron. AWS-ASTM E Cu Sn-A.

AIRCO 7010-A1
Airco Vacuum Metals
C 0.09, Mn 0.43, Si 0.14, Mo 0.48, bal Fe.
Welded, stress relieved: 80,000 psi TS; 67,000 psi YS; 29 El. DC reverse polarity coated electrode. For welding long distance pipe lines. AWS-ASTM E7010-A1.

AIRCO 7016
Airco Vacuum Metals
C 0.09, Mn 0.55, Si 0.5, bal Fe.
Welded, 74,000 psi TS; 62,000 psi YS; 30 El. AC-DC reverse polarity coated electrode; low hydrogen. For assemblies to be enameled. AWS-ASTM E7016.

AIRCO 7016 M
Airco Vacuum Metals
C 0.07, Mn 0.9, Si 0.5, Mo 0.35, Ni 0.25, Cr 0.15, bal Fe.
Welded: 82,000 psi TS; 72,000 psi YS; 29 El. AC-DC reverse polarity coated electrode. Good impact strength after stress anneal. AWS-ASTM E7016.

AIRCO 7020-A1
Airco Vacuum Metals
C 0.09, Mn 0.29, Si 0.12, Mo 0.52, bal Fe.
Welded, stress relieved: 73,000 psi TS; 63,500 psi YS; 29 El. AC-DC either polarity coated electrode. AWS-ASTM E7020-A1.

AIRCO 718
Airco Vacuum Metals
Si 12, bal Al.
Bare aluminum brazing rod. For braze-welding 1060, 1100, 3003, 5005, 6061 6063, and cast alloys A612 and C612. AWS B AlSi-4.

AIRCO 77
Airco Vacuum Metals
C 0.1, Mn 0.22, Si 0.17, bal Fe.
AC-DC reverse polarity coated electrode. For arc welding cast iron cylinder blocks and heads, bearing blocks, machine parts, and large frames. AWS-ASTM E St.

AIRCO 8016-B2
Airco Vacuum Metals
C 0.07, Mn 0.6, Si 0.42, Mo 0.45, Cr 1.3, bal Fe.
Welded, stress relieved: 84,000 psi TS; 68,000 psi YS; 28 El. AC-DC reverse polarity coated electrode. AWS-ASTM 8016-B2.

AIRCO 8016-C1
Airco Vacuum Metals
C 0.05, Mn 0.82, Si 0.4, Ni 2.4, bal Fe.
Welded, stress relieved: 82,000 psi TS; 69,000 psi YS; 31 El; 108 ft·lb IS (at -75°F, Charpy V-notch). AC-DC reverse polarity coated electrode. AWS-ASTM E8016-C1.

AIRCO 8016-C3
Airco Vacuum Metals
AC-DC reverse polarity. For welding low allow high tensile steel. AWS-ASTM E8016-C3. Low carbon; low alloy coated electrode.

AIRCO 9016-B3
Airco Vacuum Metals
C 0.09, Mn 0.57, Si 0.42, Mo 0.98, Cr 2.26, bal Fe.
Welded, stress relieved: 99,000 psi TS; 87,000 psi YS; 22 El. AC-DC reverse polarity coated electrode. For welding equipment for steam power and other assemblies. For use at elevated temperatures. AWS-ASTM E9016-B3.

AIRCO A715
Airco Welding Products
C 0.11, Mn 0.44, Si 0.5, Cr 1.24, Mo 0.5, bal Fe.
Low alloy bare wire for machine welding of low alloy steel; copper coated. *Obsolete*

AIRCO A725
Airco Welding Products
C 0.12, Mn 0.46, Si 0.48, Cr 2.35, Mo 1, bal Fe.
Low alloy bare wire for machine welding of low alloy steel; copper coated. *Obsolete*

AIRCO AIR-MANG "HC"
Airco Vacuum Metals
C 0.7, Mn 14, Cr 4, Ni 3.5, bal Fe.
For joining or build-up of dipper teeth, coal crusher segments, rail frogs and switches. Impact resistant. Steel core rod; composite coating.

AIRCO AIRCOLOY NO. 1
Airco Vacuum Metals
C 2.5, Co 50, Cr 31, W 13, Si 1.5.
50-58 Rock C. For wear resistant and corrosion resistant build-up and surfacing. AWS ASTM E Co Cr-C and R Co Cr-C. Bare or coated cobalt base electrode.

AIRCO AIRCOLOY NO. 6
Airco Vacuum Metals
C 1, Co 60, Cr 28, W 5, Si 1.5.
38-45 Rock C. For build-ups resistant to corrosion and abrasion at high temperatures. AWS-ASTM E Co Cr-A and R Co Cr-A. Bare or coated cobalt base electrode.

AIRCO AIROD
Airco Vacuum Metals
Steel tube containing tungsten carbide particles. For bare filler rod for gas welding, for hardfacing plow shares, cane knives, and conveyor screws. Abrasion resistant.

AIRCO AX-140
Airco Vacuum Metals
C 0.07-0.11, Mn 1.7-2, S 0.25-0.45, Ni 2-2.5, Cr 0.85-1.2, Mo 0.5-0.6, Ti, bal Fe.
Welded: 140,000 psi YS; 157,000 psi TS; 16 El; 57 RA. For welding high strength steel plates. Tough; strong and crack free welding wire.

AIRCO BULK TUNGSTEN CARBIDE
Airco Vacuum Metals
Hard carbide to be applied with a binder and fused into the surface for a hard abrasion resistant skin. Abrasion resistant.

AIRCO CODE-ARC 11018 MR
Airco Vacuum Metals
C 0.04, Mn 1.6, Si 0.4, Mo 0.42, Cr 0.25, Ni 1.8, bal Fe.
Welded, stress relieved: 113,000 psi TS; 105,000 psi YS; 23 El. AC-DC reverse polarity coated electrode. For welding low alloy high strength steels. AWS-ASTM E11018-M.

AIRCO CODE-ARC 12018
Airco Vacuum Metals
C 0.04, Mn 1.8, Si 0.4, Mo 0.48, Cr 0.45, Ni 2, bal Fe.
Welded, stress relieved: 122,000 psi TS; 113,000 psi YS; 22 El. AC-DC reverse polarity coated electrode. For welding low alloy high strength steels. AWS-ASTM E12018-M.

AIRCO CODE-ARC 308 ELC-MR
Airco Vacuum Metals
C 0.03, Cr 20, Ni 10, Mn 1.9, Si 0.4, bal Fe.
AC-DC reverse polarity titania coated electrode. For welding 304 L, 321, 347 and other 18-8 stainless steels. AWS-ASTM E 308 L-16; AISI 308 ELC.

AIRCO CODE-ARC 308-MR
Airco Vacuum Metals
C 0.07, Mn 1.7, Si 0.4, Cr 20, Ni 10, bal Fe.
AC-DC reverse polarity titania coated electrode. For welding austenitic 18-8 type stainless steels. ASW-ASTM E 308-16; AISI 308.

AIRCO CODE-ARC 309 MR
Airco Vacuum Metals
C 0.1, Cr 24, Ni 13, Mn 1.8, Si 0.55, bal Fe.
Welded: 89,000 psi TS; 65,000 psi YS; 42 El. AC-DC reverse polarity titania coated electrode. For welding austenitic type 309 stainless. AWS-ASTM E 309-16; AISI 309.

AIRCO CODE-ARC 310 MR
Airco Vacuum Metals
C 0.11, Cr 26.5, Ni 21, Mn 1.8, Si 0.5, bal Fe.
Welded: 91,000 psi TS; 61,000 psi YS; 36.5 El. AC-DC reverse polarity titania coated electrode. For welding austenitic types 310 and 314 stainless steels. AWS-ASTM E 310-16; AISI 310.

AIRCO CODE-ARC 316 ELC-MR
Airco Vacuum Metals
C 0.03, Cr 18.5, Ni 13, Mn 1.9, Si 0.6, Mo 2.25, bal Fe.
Welded: 88,700 psi TS; 74,350 psi YS, 33.5 El. AC-DC reverse polarity titania coated electrode. For welding extra low carbon molybdenum bearing austenitic stainless. AWS-ASTM E 316L-16; AISI 316 ELC.

AIRCO CODE-ARC 316 MR
Airco Vacuum Metals
C 0.06, Cr 18.5, Ni 13, Mn 1.9, Si 0.6, Mo 2.25, bal Fe.
Welded: 93,000 psi TS; 75,000 psi YS; 35 El. AC-DC reverse polarity titania coated electrode. For welding type 316 for corrosion resistant or elevated temperature applications. AWS-ASTM E316-16; AISI 316.

AIRCO CODE-ARC 347 MR
Airco Vacuum Metals
C 0.06, Cr 19.5, Ni 10, Mn 1.5, Si 0.6, Cb 0.7, bal Fe.
Welded: 109,000 psi TS; 79,500 psi YS; 34.5 El. AC-DC reverse polarity titania coated electrode. For welding types 321 and 347 stainless. AWS-ASTM E347-16; AISI 347.

AIRCO CODE-ARC 7018 MR
Airco Vacuum Metals
C 0.04, Mn 0.85, Si 0.55, bal Fe.
Welded: 78,000 psi TS; 67,000 psi YS; 30 El. AC-DC reverse polarity coated electrode. AWS-ASTM E7018.

AIRCO CODE-ARC 7018-A1 MR
Airco Vacuum Metals
C 0.04, Mn 0.6, Si 0.6, Mo 0.48, bal Fe.
Welded, stress relieved: 81,000 psi TS; 71,000 psi YS; 30 El.
AC-DC reverse polarity coated electrode. AWS-ASTM E
7018-A1.

AIRCO CODE-ARC 8018-C3 MR
Airco Vacuum Metals
C 0.04, Mn 0.9, Si 0.4, Ni 1, Mo 0.2, bal Fe.
Welded: 83,000 psi TS; 72,000 psi YS; 31 El. AC-DC reverse
polarity coated electrode. Good impact properties. AWS-
ASTM E8018-C3.

AIRCO COMMERCIALLY PURE ALUMINUM A1100
Airco Vacuum Metals
Aluminum.
For machine welding of 1100 and 3003 aluminum, window
frames, heat exchangers, food containers, electrical bus bars,
and racks. AWS ER 1100.

AIRCO EASYARC 308 AC-DC
Airco Vacuum Metals
C 0.07, Cr 19.6, Ni 9.5, Mo 1.65, Si 0.68.
Welded: 87,900 psi TS; 57,400 psi YS; 42 El. AC-DC reverse
polarity; titania type coating with metal powders. For welding
18-8 type steel. AWS-ASTM E308-16; AISI 308.

AIRCO EASYARC 316 AC-DC
Airco Vacuum Metals
C 0.06, Cr 18.5, Ni 12.3, Mn 1.8, Si 0.31, Mo 2.28, bal Fe.
Welded: 80,400 psi TS; 56,400 psi YS; 44 El. AC-DC reverse
polarity; titania type coating with metal powders. For welding
type 316 stainless steel. AWS-ASTM E316-16; AISI 316.

AIRCO EASYARC 6027
Airco Vacuum Metals
C 0.07, Mn 0.9, Si 0.14, bal Fe.
Welded: 68,000 psi TS; 57,000 psi YS; 30 El. AC-DC either
polarity coated electrode. AWS-ASTM E6027.

AIRCO EASYARC 620
Airco Vacuum Metals
C 6, Cr 22, Mo 7, W 5, Si 1.5, Mn 0.5, bal Fe.
60-65 Rock C. For build-ups resistant to severe abrasion,
moderate impact and hot wear up to 1100°F. For crusher
parts, coke pusher shoes, and hot slag dipper teeth. Alloy
steel core rod; chrome iron composite coating; AC DC either
polarity.

AIRCO EASYARC 7014
Airco Vacuum Metals
C 0.08, Mn 0.45, Si 0.3, bal Fe.
Welded: 81,000 psi TS; 73,000 psi YS; 26 El. AC-DC either
polarity coated electrode. AWS-ASTM E7014.

AIRCO EASYARC 7018 C
Airco Vacuum Metals
C 0.07, Mn 0.9, Si 0.5, bal Fe.
Welded: 84,000 psi TS; 73,000 psi YS; 30 El. AC-DC reverse
polarity coated electrode. AWS-ASTM E7018.

AIRCO EASYARC 7018 MR
Airco Vacuum Metals
C 0.06, Mn 0.65, Si 0.5, bal Fe.
Welded: 78,000 psi TS; 66,000 psi YS; 28 El. AC-DC reverse
polarity coated electrode. AWS-ASTM E7018.

AIRCO EASYARC 7024
Airco Vacuum Metals
C 0.07, Mn 0.75, Si 0.5, bal Fe.
Welded: 76,000 psi TS; 68,000 psi YS; 20 El. DC reverse
polarity coated electrode. AWS-ASTM E7024.

AIRCO EASYARC 7028
Airco Vacuum Metals
C 0.06, Mn 0.7, Si 0.35, bal Fe.
Welded: 80,000 psi TS; 72,000 psi YS; 30 El. AC-DC reverse
polarity coated electrode. AWS-ASTM E7028.

AIRCO ELECTRODE
Airco Welding Products
C 0.13-0.18, bal Fe.
Weld: 55,000 TS. For welding electrode for low carbon and
mild steels. *Obsolete*

AIRCO KA2S
Airco Welding Products
C 0-0.07, Mn 0.5, P 0.03, S 0.03, Si 0.75, Cr 17-20, Ni 8-12,
bal Fe.
Weld: 87,000 TS; 36 El. For stainless steel welding rod;
shielded arc. *Obsolete*

AIRCO KA2SMO
Airco Welding Products
C 0-0.07, Mn 0.2-0.5, Si 0-0.75, Cr 17-19, Ni 7-11, Mo 2-40,
bal Fe.
Weld: 87,000 TS; 36 El. For stainless steel welding rod;
shielded arc. *Obsolete*

AIRCO NICKEL MANGANESE
Airco Welding Products
C 0.8, Ni 4, Mn 13-16, bal Fe.
Welded: 200 Brin. For hard facing electrodes; austenitic,
impact and abrasion resistant. *Obsolete*

AIRCO NICKEL MANGANESE C
Airco Vacuum Metals
C 0.6, Mn 13, Ni 3, Cr 0.4, Si 0.75.
For joining and build-up of Hadfield's manganese steel
dipper teeth, shovel tracks, and crusher pads. Impact
resistant. Steel core rod; composite coating. AWS-ASTM E Fe
Mn-A.

AIRCO NO. 1
Airco Vacuum Metals
C 0-0.15, Mn 0.45, Si 0.2, Ni 1-1.5, Cr 0-0.3, bal Fe.
Welded: 62,000 psi TS; 22 El. Low alloy steel welding rod. For
gas or tungsten inert gas welding of carbon and low alloy
steels. AWS RG 60.

AIRCO NO. 10
Airco Vacuum Metals
C 3.3, Mn 0.6, Si 2.25, Ni 1.4, Mo 0.35, bal Fe.
Square low alloy cast iron. For welding low alloy gray irons.
AWS-ASTM RC1-A.

AIRCO NO. 1010 SILICON BRONZE ROD
Airco Vacuum Metals
Cu 95.8, Si 3.1, Mn 1.1, Zn 1.
50,000 psi TS min. Bare rod for gas welding of copper,
copper-silicon and copper-zinc base alloys to themselves
and to plain and galvanized steel. Everdue grade. AWS R Cu
Si-A.

AIRCO NO. 11
Airco Welding Products
C 0-0.15, Cr 0-0.2, Mn 0.25-0.65, Ni 0.35-0.75, Mo 0.45-0.7,
bal Fe.
For welding rod; for high pressure and high steam piping.
Obsolete

AIRCO NO. 20 BRONZE ROD
Airco Vacuum Metals
Sn 0.5-1, Cu 60, Fe 0-0.06, bal Zn.
40,000 psi TS min. Fuses at 1625°F. Bare or coated rod for
braze-welding copper, bronze and nickel alloys. AWS RB
CuZn-A.

AIRCO NO. 21 NICKEL SILVER ROD
Airco Vacuum Metals
Cu 46-50, Ni 9-11, bal Zn.
70,000 psi TS min. Bare or coated rod for joining dissimilar
metals; mild steel, tool steel, cast iron, stainless steel, monel,
inconel, bronze, copper, and malleable iron. AWS RB Cu
Zn-D.

AIRCO NO. 22 BRONZE ROD
Airco Vacuum Metals
Cu 59, Fe 0.4, Mn 0.23, Sn 1, Ni 0.6, bal Zn.
45,000 psi TS min. Bare or coated rod for braze-welding
steel, cast iron, malleable iron, copper base alloys and to
build up wearing surfaces and bearings. AWS R Cu Zn-B.

AIRCO NO. 23 A SILICON COPPER ROD
Airco Welding Products
Sn 0.65-0.9, Si 0.1-0.3, Mn 0.1-0.25, P 0-0.15, bal Cu
(deoxidized).
Bare rod for TIG welding; 25,000 TS. AWS R Cu. *Obsolete*

AIRCO NO. 27 LOW FUMING BRONZE ROD
Airco Vacuum Metals
Cu 56-59, Fe 0.4-0.8, Si 0.1, Mn 0.3, Sn 1, bal Zn.
50,000 psi TS min. Bare or coated rod, low fuming. For
general purpose braze-welding of steel cast iron, brass and
bronze. AWS R Cu Zn-C.

AIRCO NO. 28
Airco Welding Products
Cu 56.5-58.5, Si 0.1-0.3, Sn 2.1-2.7, bal Zn.
For bronze welding rod; for wear resistant surfaces. *Obsolete*

AIRCO NO. 383
Airco Welding Products
C 0.11, Si 0.19, Mn 0.34, Mo 0.52, P 0.012, Cr 0.54, bal Fe.
Welded: 89,000 TS; 75,000 YS; 20 El; 55 RA. For steel
welding rod; ASTM E-8011. *Obsolete*

AIRCO NO. 393
Airco Welding Products
C 0.12, Mn 0.46, Si 0.21, Mo 0.5, Cr 0.98, bal Fe.
Welded: 95,000 TS; 86,000 YS; 16 El; 55 RA. For steel
welding electrodes; ASTM E- 9011. *Obsolete*

AIRCO NO. 397
Airco Welding Products
C 0.11, Si 0.51, Mn 0.5, V 0.1, Mo 1.25, bal Fe.
Welded: 125,000 TS; 16 El; 45 RA. For welding high
temperature and high pressure equipment; AS E-10016.
Obsolete

AIRCO NO. 398
Airco Welding Products
C 0.06, Mn 1.48, Si 0.62, V 0.1, bal Fe.
Welded: 86,000 TS; 72,000 YS; 31 El; 65 RA. For repairing
railroad castings, welding electrode; ASTM B-8016. *Obsolete*

AIRCO NO. 4
Airco Vacuum Metals
C 0.08-0.2, Mn 0.9-1.2, Si 0.15-0.3, bal Fe.
Welded: 62,000-67,000 psi TS; 20-25 El. Low alloy steel
welding rod. For gas or tungsten inert gas welding of carbon
and low alloy steels.

AIRCO NO. 48
Airco Welding Products
C 0-0.06, bal Fe.
Weld: 45,000 TS. For welding electrode; shock and fatigue
resistant. *Obsolete*

AIRCO NO. 5
Airco Welding Products
C 0.1-0.3, Ni 3-4, Si 0-0.3, Mn 0.2-0.7, bal Fe.
Welded: 60,000 TS; 18 El. For welding rod; welds to be heat
treated. *Obsolete*

AIRCO NO. 53
Airco Welding Products
C 0.27, Mn 0.36, Si 0.07, bal Fe.
Welded: 250 Brin. For hard facing electrodes; abrasion
resistant. *Obsolete*

AIRCO NO. 55
Airco Welding Products
Ni 67, Cu 28, Fe 2, Mn 1-2.
For welding electrode for cast iron; light coated. *Obsolete*

AIRCO NO. 58
Airco Welding Products
Ni 67, Cu 28, Fe 2, Mn 2.
For Monel welding rod; flux coated. *Obsolete*

AIRCO NO. 6
Airco Welding Products
C 0.2-0.4, Mn 0.8-1.15, V 0.1-0.3, Cr 0.9-1.25, bal Fe.
Welded: 207 Brin. For welding rod; gas welding. *Obsolete*

AIRCO NO. 62
Airco Welding Products
C 0-0.06, bal Fe.
For welding electrode; shielded arc. *Obsolete*

AIRCO NO. 63
Airco Welding Products
C 0.03, Mn 0.34, P 0.01, S 0.015, bal Fe.
Welded: 45,000 TS; 5 El. For steel welding electrodes; ASTM
E-4510. *Obsolete*

AIRCO NO. 7
Airco Vacuum Metals
C 0-0.06, Mn 0-0.25, bal Fe.
Welded: 52,000 psi TS; 23 El. AWS RG 45. Low carbon steel
welding rod, copper coated, for gas welding of low carbon
steel sheet, plate and pipe, and auto body repair.

AIRCO NO. 72
Airco Welding Products
Si 3.5, bal Cu.
Weld: 46,000 TS; 25,000 YS; 22 El; 24 RA. For welding
electrode; ASTM-E Cu Si. *Obsolete*

AIRCO NO. 76
Airco Welding Products
C 0.13-0.18, bal Fe.
Weld: 55,000 TS. For welding electrode; shielded arc.
Obsolete

AIRCO NO. 84
Airco Welding Products
C 0-0.17, Ni 4.5-5.25, bal Fe.
Weld: 85,000 TS. For welding electrode; not machinable,
shielded arc. *Obsolete*

AIRCO NO. 85
Airco Welding Products
C 0.85-1.1, Mn 0.3-0.6, bal Fe.
Weld: 290-425 Brin. For hard surfacing electrode; shielded
arc, not machinable. *Obsolete*

AIRCO NO. 86
Airco Welding Products
C 0.25-0.35, Mn 0.55-0.75, Cu 0.2-0.3, bal Fe.
Weld: 65,000-80,000 TS; 20-30 El. For welding electrode;
shielded arc. *Obsolete*

AIRCO NO. 87
Airco Welding Products
C 0.07, Mn 0.28, bal Fe.
Weld: 60,000-75,000 TS; 67,000 YS; 20-25 El; 38 RA. For
welding electrode; shielded arc; ASTM E 6012. *Obsolete*

AIRCO NO. 9
Airco Vacuum Metals
C 3.25, Mn 0.6, Si 3, bal Fe.
Square cast iron gas welding rod. For general gas welding of
gray cast iron. AWS-ASTM RCI.

AIRCO NO. 92 SILICON BRONZE ROD
Airco Vacuum Metals
Cu 95.8, Si 3, Mn 1, Zn 1.
Tin coated rod for brazing galvanized iron by carbon arc
process or TIG AWS R Cu Si-A.

AIRCO PHOS-COPPER ROD
Airco Vacuum Metals
P 7.25-7.35, bal Cu.
Bare rod for torch or furnace brazing of transformers,
radiators, motors, water coolers, plumbing installations, and
heat exchangers. AWS B Cu P-2.

AIRCO PHOS-SILVER 2 ROD
Airco Vacuum Metals
P 6.75-7.25, Ag 1.75-2.25, bal Cu.
Bare rod for torch or furnace brazing components of copper,
brass and bronze. 1435-1455°F MP.

AIRCO RAILROAD ROD
National Intergroup Inc.
C 0.2-0.4, Mn 0.85-1.15, V 0.1-0.3, Cr 0.9-1.25, bal Fe.
Welded: 100,000 TS. For welding rod; for rail ends.

AIRCO S-10
Airco Welding Products
C 0.1, Mn 0.55, bal Fe.
Mild steel automatic submerged arc welding wire. AWS-EL
12. *Obsolete*

AIRCO S-15
Airco Welding Products
C 0.14, Mn 0.9, bal Fe.
Mild steel automatic submerged arc welding wire. AWS EH
14. *Obsolete*

AIRCO S-16
Airco Welding Products
C 0.14, Mn 1.9, Mo 0.5, bal Fe.
Low alloy steel automatic submerged arc welding wire.
Obsolete

AIRCO S-20
Airco Welding Products
C 0.16, Mn 1.1, Si 0.2, bal Fe.
Mild steel automatic submerged arc welding wire. AWS EM
15 K. *Obsolete*

AIRCO S-308
Airco Welding Products
C 0.06, Mn 1.8, Si 0.5, Cr 20.5, Ni 9.8, bal Fe.
Stainless steel wire for automatic submerged arc welding; for
18-8 stainless. AWS ER 308. *Obsolete*

AIRCO S-309
Airco Welding Products
C 0.07, Mn 1.75, Si 0.5, Cr 23, Ni 13, bal Fe.
Stainless steel wire for automatic submerged arc welding;
designed for AISI 309. AWS ER 309. *Obsolete*

AIRCO S-310
Airco Welding Products
C 0.12, Mn 1.5, Si 0.5, Cr 26.8, Ni 21, bal Fe.
Stainless steel wire for automatic submerged arc welding;
designed for AISI 310. AWS ER 310. *Obsolete*

AIRCO S-316
Airco Welding Products
C 0.06, Mn 1.65, Si 0.45, Cr 19, Ni 12.9, Mo 2.2, bal Fe.
Stainless steel wire for automatic submerged arc welding;
designed for AISI 316. AWS ER 316. *Obsolete*

AIRCO S-347
Airco Welding Products
C 0.06, Mn 1.65, Si 0.5, Cr 19.5, Ni 9.6, Cb 0.85, bal Fe.
Stainless steel wire for automatic submerged arc welding of
stabilized 18-8 stainless steel. AWS ER 347. *Obsolete*

AIRCO S-65
Airco Welding Products
C 0.6, Mn 1.05, Si 0.15, bal Fe.
Higher carbon steel automatic submerged arc welding wire.
Obsolete

AIRCO SELF HARDENING
Airco Vacuum Metals
C 0.85, Cr 5, Mn 0.9, Si 0.55, V 0.35.
50-55 Rock C. For hard surfacing and build-up for resistance
to abrasion and moderate to heavy impact. For dump truck
bodies, brake shoe hangers, and bulldozer trunnions. Steel
core wire; composite coating.

AIRCO SELF-HARDENING GAS
Airco Welding Products
C 1, Mo 1, Cr 2, Mn 1.5, bal Fe.
Welded: 450 Brin. For hard facing electrodes; abrasion and
impact resistant to 400 F. *Obsolete*

AIRCO SEMI-AUTOMATIC AIR-MANG
Airco Welding Products
For welding railroad track segments, dredge parts, clamshell
lips. Impact resistant. Continuous wire version of Airco Air-
Mang. *Obsolete*

AIRCO SILICON BRONZE
Airco Welding Products
Si 3, bal Cu.
For electrodes for welding silicon bronze, copper; corrosion
resistant. *Obsolete*

AIRCO SPECIAL "B" TUNGTUBE
Airco Vacuum Metals
Steel tube containing tungsten carbide particles. Filler rod for
build up on work core and rock drill bits and post hole digger
blades. Abrasion and impact resistant.

AIRCO STAINLESS PLY
Airco Welding Products
C 0.08, Cr 18, Ni 8, bal Fe.
For stainless steel welding, electrode for stainless clad steel;
shielded arc. *Obsolete*

AIRCO TIMANG
Airco Welding Products
Deposit composition: 0.75% C, 1.40% Mn, 3.0% Ni, bal Fe.
For joining or build-up of manganese steel mill hammers,
bucket teeth, crusher jaws. Impact resistant. Bare or coated
electrode. *Obsolete*

AIRCO TRAC-ROD 27
Airco Welding Products
C 0.08, Mn 0.31, Si 0.11, bal Fe.
AC-DC either polarity coated electrode. Weld metal: 66,200
TS; 56,300 YS; 27 El. Welds over rust and scale. AWS-ASTM
E 6027. *Obsolete*

AIRCO TRACKWEAR
Airco Vacuum Metals
C 0.85, Mn 17, V 0.45, bal Fe.
For joining and build-up of railroad frogs, switches and
cross-overs. Impact resistant; steel core rod; composite
coating.

AIRCO TUBE AIRCOLITE
Airco Vacuum Metals
Cr 30, Si 2, Mn 1.5, C 4, bal Fe.
60 Rock C. Bare tubular wire for gas welding and build-up for
resistance to severe abrasion with moderate impact. For plow
shares, cultivator spades, and corn picker runners.

AIRCO TUNGSERTS
Airco Welding Products
C 4, Co 10-15, bal W.
Cast: 800 Brin. For cutting tips, wear resistant inserts, earth
drilling equipment; sintered carbides, abrasion resistant.
Obsolete

AIRCO TUNGTUBE
Airco Vacuum Metals
Steel tube containing tungsten carbide. For filler rod for
oxyacetylene flame for hard facing. Abrasion resistant.

AIRCOMATIC A100 ALUMINUM BRONZE
Airco Vacuum Metals
Al, Fe, bal Cu.
Bare rod for inert gas welding of aluminum bronze or bronze surfacing of steel. AWS E Cu Al-A2.

AIRCOMATIC A102 PHOS BRONZE
Airco Vacuum Metals
Sn, P, bal Cu.
Bare rod for inert gas welding of brass, bronze or for overlays on ferrous metals. AWS E Cu Sn-C.

AIRCOMATIC A104 NI-AL-MN BRONZE
Airco Vacuum Metals
Al, Ni, Mn, bal Cu.
Bare rod for inert gas welding of ship propellers and turbine runners.

AIRCOMATIC A110 SILICON BRONZE
Airco Vacuum Metals
Si, bal Cu.
Bare rod for inert gas welding of silicon bronze or of many dissimilar metals. AWS E Cu Si.

AIRCOMATIC A145 COPPER
Airco Vacuum Metals
Cu.
Bare rod or wire for inert gas welding of copper and copper base alloys and for overlay on steels. AWS E Cu.

AIRCOMATIC A158 ALUMINUM BRONZE
Airco Vacuum Metals
Al, bal Cu.
Iron free bare wire for welding aluminum bronzes, manganese bronze, copper, cast iron, mild steel and alloy steel, and for surfacing. For inert gas processes. AWS E Cu Al-Al.

AIRCOMATIC A160 ALUMINUM BRONZE
Airco Vacuum Metals
Al, bal Cu.
Aluminum bronze bare wire largely for overlay work to produce corrosion resistant and wear resistant surface. 160-210 Brin. AWS E Cu Al-B.

AIRCOMATIC A308
Airco Vacuum Metals
Wire for welding AISI types 301, 305, and 308 stainless steels, and for depositing stainless overlays on mild steel. AWS ER 308.

AIRCOMATIC A308 ELC
Airco Vacuum Metals
Wire for welding AISI 304 ELC and 308 ELC, also AISI 321 and 347. AWS-ER 308 L; low carbon grade of A308.

AIRCOMATIC A308 L SI
Airco Vacuum Metals
Higher silicon version of A308 for improved arc stability.

AIRCOMATIC A309
Airco Vacuum Metals
Wire for welding AISI 309 stainless for high temperature operation and for welding some straight chromium stainless grades. AWS-ER 309.

AIRCOMATIC A309 ELC
Airco Welding Products
Low carbon grade of A309.
For same purpose as A309 where added insurance against carbide precipitation is required. *Obsolete*

AIRCOMATIC A310
Airco Vacuum Metals
Wire for welding AISI 310 stainless steel for high temperatures or corrosion resistant applications. AWS ER 310.

AIRCOMATIC A312
Airco Vacuum Metals
Wire designed for welding stainless steels to carbon steels. AWS ER312.

AIRCOMATIC A316
Airco Vacuum Metals
Wire for welding AISI 316 stainless. AWS ER 316.

AIRCOMATIC A316 ELC
Airco Vacuum Metals
For welding AISI 316 L stainless steel. AWS ER 316 L; low carbon grade of A316.

AIRCOMATIC A316 L SI
Airco Vacuum Metals
Higher silicon version of A316 for improved arc stability.

AIRCOMATIC A347
Airco Vacuum Metals
Wire for welding AISI types 321, 347 and 304 L. For elevated temperature operation or corrosive conditions. AWS ER 347.

AIRCOMATIC A4043 ALUMINUM ALLOY
Airco Vacuum Metals
AA4043 composition (5.2 Si).
For welding 2014, 4043, 6061, 6062, 6063 alloys for truck bodies, pressure vessels, and structural members. AWS ER 4043.

AIRCOMATIC A5183 ALUMINUM ALLOY
Airco Vacuum Metals
Mg, bal Al.
Wire for welding liquid oxygen and liquid nitrogen containers. AWS ER 5183.

AIRCOMATIC A5356 ALUMINUM ALLOY
Airco Vacuum Metals
Mn 0.12, Mg 5, Cr 0.12, bal Al.
Wire for welding 5056, 5083, 5086, 5154, 5356 aluminum alloys for truck frames, diesel engine bases, cargo tanks, gun mount bases. AWS ER 5356.

AIRCOMATIC A5554 ALUMINUM ALLOY
Airco Vacuum Metals
Wire for welding 5052 and 5454 aluminum alloys for truck tankers, dump bodies, and railroad tank cars. AWS-ER 5554.

AIRCOMATIC A5556 ALUMINUM ALLOY
Airco Vacuum Metals
bal Fe.
Wire for welding 5083, 5086, 5456 high tensile aluminum alloys. For truck frames, diesel engine bases, and storage tanks. AWS ER 5556.

AIRCOMATIC A666
Airco Welding Products
C 0.13, Mn 1.3, Si 0.5, Al 0.6, bal Fe.
Bare mild steel welding wire for carbon dioxide and inert gas welding, particularly of rusted and scaled steel. AWS E 70 S-5. *Obsolete*

AIRCOSIL 105
Airco Welding Products
Ag 45, Cu 30, Zn 12, Mn 13.
For brazing alloy; MP 1298 F; eutectic alloy. *Obsolete*

AIRCOSIL 15
Airco Vacuum Metals
Ag 15, Cu 80, P 5.
For silver brazing of Cu alloys; 1185-1500°F MP; self-fluxing. AWS, ASTM BCuP-5.

AIRCOSIL 3
Airco Vacuum Metals
Ag 50, Ni 3, Cd 16, Cu 15.5, bal Zn.
For silver brazing of carbide tools; 1170-1270°F MP. AWS A5.8-62 and ASTM B260-62; BAg-3.

AIRCOSIL 35
Airco Vacuum Metals
Ag 35, Cd 18, Cu 26, Zn 21.
For silver brazing, general purpose; 1125-1295°F MP; high plastic range. AWS, ASTM BAg-2.

AIRCOSIL 45
Airco Vacuum Metals
Ag 45, Cd 24, Cu 15, Zn 16.
For silver brazing; 1125-1145°F MP; free-flowing. AWS, ASTM BAg-1.

AIRCOSIL 5
Airco Vacuum Metals
Ag 5, Cu 88.75, P 6.25.
For brazing alloy; 1190-1480°F MP. AWS, ASTM BCuP-3.

AIRCOSIL 50
Airco Vacuum Metals
Ag 50, Cd 18, Cu 15.5, Zn 16.5.
For silver brazing for all metals; 1160-1175°F MP; high strength. AWS, ASTM BAg-1a.

AIRCOSIL 60
Airco Vacuum Metals
Ag 60, Cu 30, Sn 10.
For brazing alloy; 1115-1325°F MP. AWS, ASTM BAg-18.

AIRCOSIL A
Airco Vacuum Metals
Ag 9, Cu 53, Zn 38.
For brazing alloy; 1410-1565°F MP.

AIRCOSIL AE-100
Airco Vacuum Metals
Ag 92.5, Cu 7.3.
For brazing alloy; 1435-1635°F MP. AWS-ASTM BAg-19.

AIRCOSIL B
Airco Vacuum Metals
Ag 20, Cu 45, Zn 35.
For brazing alloy; 1315-1500°F MP.

AIRCOSIL C
Airco Vacuum Metals
Ag 20, Cu 45, Zn 30, Cd 5.
For brazing alloy; 1140-1500°F MP.

AIRCOSIL D
Airco Vacuum Metals
Ag 30, Cu 38, Zn 32.
For brazing alloy; 1370-1410°F MP.

AIRCOSIL E
Airco Vacuum Metals
Ag 40, Cu 30, Zn 28, Ni 2.
For brazing alloy for carbides; 1240-1435°F MP. AWS, ASTM BAg-4.

AIRCOSIL EASY
Airco Vacuum Metals
Ag 65, Cu 20, Zn 15.
For brazing alloy for sterling silver; 1240-1325°F MP. AWS, ASTM BAg-9.

AIRCOSIL F
Airco Vacuum Metals
Ag 40, Cu 36, Zn 24.
For brazing alloy; 1325-1415°F MP.

AIRCOSIL G
Airco Vacuum Metals
Ag 45, Cu 30, Zn 25.
For brazing alloy in food industry; 1250-1370°F MP. AWS, ASTM BAg-5.

AIRCOSIL H
Airco Vacuum Metals
Ag 50, Cu 34, Zn 16.
For brazing alloy in food industry; 1270-1425°F MP. AWS, ASTM BAg-6.

AIRCOSIL HARD
Airco Vacuum Metals
Ag 75, Cu 22, Zn 3.
For brazing alloy for sterling silver; 1365-1450°F MP. AWS, ASTM BAg-11.

AIRCOSIL J
Airco Vacuum Metals
Ag 56, Cu 22, Zn 17, Sn 5.
For brazing alloy on stainless steels; 1145-1205°F MP. AWS, ASTM BAg-7.

AIRCOSIL K
Airco Vacuum Metals
Ag 60, Cu 25, Zn 15.
For brazing alloy; 1245-1325°F MP.

AIRCOSIL L
Airco Vacuum Metals
Ag 54, Cu 40, Zn 5, Ni 1.
For brazing alloy; 1325-1575°F MP.

AIRCOSIL M
Airco Vacuum Metals
Ag 72, Cu 28.
For brazing alloy in electronic industry; 1435°F MP; eutectic alloy. AWS, ASTM BAg-8.

AIRCOSIL MEDIUM
Airco Vacuum Metals
Ag 70, Cu 20, Zn 10.
For brazing alloy for sterling silver; 1275-1360°F MP. AWS, AWTM BAg-10.

AIRCOSIL N
Airco Vacuum Metals
Ag 80, Cu 16, Zn 4.
For brazing alloy; 1345-1490°F MP.

AIRCOSIL P
Airco Vacuum Metals
Ag 85, Mn 15.
For brazing alloy for high temperature applications; 1760-1778°F MP. AWS, ASTM BAg-Mn.

AIRCOSIL Q
Airco Welding Products
Ag 50, Cu 28, Zn 22.
For brazing alloy; MP 1250-1340 F. *Obsolete*

AIRCOSIL R
Airco Vacuum Metals
Ag 40, Cu 30, Zn 25, Ni 5.
For brazing alloy; 1240-1560°F MP.

AIRCOSIL S
Airco Vacuum Metals
Ag 25, Cu 52.5, Zn 22.5.
For brazing alloy; 1250-1575°F MP.

AIRCRAFT
English manufacture
Al 10-11.5, Fe 4-6, Ni 4-6, bal Cu.
For corrosion resisting parts, aircraft; corrosion resistant.

AIRCRAT
Marshall Steel Co.
C 1, Mn 0.5, Mo 1.1, V 0.2, Cr 5, bal Fe.
For tools, dies, punches, crimpers; air hardened, ground flat stock.

AIRDI 110
Colt Industries
C 1.1, Cr 11.5, V 0.2, Mo 0.8, bal Fe.
For dies, gauges, rolls; nondeforming. *Obsolete*

AIRDI 150
Crucible Materials Corp.
Tool material. C 1.55, Cr 11.5, V 0.9, Mo 0.8, bal Fe.
Hardened: 58-63 Rock C. Air hardening, somewhat corrosion resistant. For blanking and coining dies, gauges, punches, thread rolling dies, shafts. AISI Type D2. Cold work tool steel.

AIRDI 150-S
Crucible Materials Corp.
Tool material. C 1.5, Cr 12, V 0.9, Mo 0.8, S 0.15, bal Fe.
For blanking and forming dies, punches; air hardened, non-deforming.

AIRDI 225
Colt Industries
C 2.25, Cr 12, Mo 1, V 0.25, bal Fe.
For tools, forming and crimping rolls, perforating dies, plug gages. *Obsolete*

AIRESIST 13
Cannon-Muskegon Corp.
Cr 21, W 11, C 0.45, Cb 2, Al 3.5, Y 0.5, bal Co.
For jet engine and gas turbine components. Corrosion resistant. Good high temperature strength. Resists sulfidation and oxidation.

AIRESIST 213
Cannon-Muskegon Corp.
Al 3.5, Y 0.1, C 0.2, Cr 20, Ni 0.5, W 4.5, Ta 6.5, Al 3.5, Fe 0.5, bal Co.
Sheet: 160,000 TS; 100,000 YS; 10 El. At 1800°F: 35,000 TS; 22,000 YS; 70 El. For jet engine and gas turbine components. Resists sulfur attack up to 2200°F. Corrosion and oxidation resistant.

AIRESIST 215
Cannon-Muskegon Corp.
Cr 19, W 4.5, Al 4.3, C 0.35, Ni 0.5, Y 0.1, Ta 7.5, Fe 0.5, Zr 0.1, bal Co.
Resists oxidation and sulfidation.

AIREX
Columbia Tool Steel Co.
C 0.55, Si 2, Mn 0.75, bal Fe.
For chisels, tools; tough, shock resistant. *Obsolete*

AIRITE NO. 21
English manufacture
Ni alloy.
For compressor blades, aircraft rotor discs; corrosion and heat resistant.

AIRKOOL
Crucible Specialty Steel
C 1, Cr 5, V 0.4, Mo 1.1, bal Fe.
For dies, gauges, punches, shears, broaches; air hardened, non-deforming.

AIRKOOL
Colt Industries
C 1, Cr 5, V 0.4, Mo 1.1, bal Fe.
For dies, gauges, punches, shears, broaches; air hardened, non-deforming.

AIRKOOL-S
Crucible Materials Corp.
Tool material. C 1, Cr 5.2, V 0.3, Mo 1.1, S 0.15, bal Fe.
For blanking and forming dies, punches; air hardened, non-deforming. AISI A2.

AIRKOOL-V
Crucible Materials Corp.
Tool material. C 2.25, Cr 5.25, V 4.75, Mo 1, W 1, bal Fe.
Heat treated: 650 Brin. For blanking dies, brick mold liners; oil hardened, non-deforming.

AIRLOY
Allegheny Ludlum Steel
C 1, Mn 3, Cr 1, Mo 1, bal Fe.
For dies, gauges, thread roller dies, blanking and forming punches; air hardening. *Obsolete*

AIRMO
Teledyne Firth Sterling
Steel. C 1, Mn 2, Cr 1, Mo 1, bal Fe.
For cold work dies, crimpers; cold work steel, air hardened. *Obsolete*

AIRMOLD
Crucible Materials Corp.
Tool material. C 0.1, Cr 5, Mo 0.5, bal Fe.
Hardened: 157,000 TS; 110,000 YS; 17 El; 60 RA; 321 Brin.
For cold hubbed plastic molds; deep hardening.

AIROLITE
National Airoil Burner Co.
C, alloy, bal Fe.
For high temperature applications; heat resistant.

AIROMAT 33 SELF-HARDENING
Atlantic Steel Corp.
C 0.33, Cr 2.5, Mn 0.75, Mo 1, Si 0.75, Cu 0.5, bal Fe.
For shock tools, dies, punches, shears; air hardening. *Obsolete*

AIRPLANE BABBITT
Southern Metals Co.
Sn 16, Sb 10, Pb 72, Cu 2, Bi.
For bearings; melting point 600-700°F.

AIRPRO
Bethlehem Steel Corp.
C 1, Mn 6, Cr 5.25, Mo 1.1, V 0.25, bal Fe.
For thread rolling dies, broaches, trimming and blanking dies; cold work steel. *Obsolete*

AIRQUE
CCS Braeburn Alloy Steel
C 1, Mn 0.7, Cr 5.25, V 0.25, Mo 1.15, bal Fe.
For gauges, punches, shear blades, dies. Air hardened, non-deforming. *Obsolete*

AIRQUE SPECIAL
CCS Braeburn Alloy Steel
C 0.8, Mn 0.7, Cr 5.2, V 0.25, Mo 1.15, bal Fe.
For blanking and trimming dies, punches, shears; air hardened, nondeforming, tough. *Obsolete*

AIRQUE-4
CCS Braeburn Alloy Steel
C 1, Mn 2.5, Cr 1.1, Mo 1.1, bal Fe.
Air hardening tool steel; AISI A4. *Obsolete*

AIRQUE-V
CCS Braeburn Alloy Steel
C 1.25, Cr 5.25, Mo 1.15, V 1, bal Fe.
For thread rolling dies, blanking and forming dies. Air hardened, non-deforming. *Obsolete*

AIRTEM
Lehigh Steel Corp.
C 1.3, Mn 0.5, Cr 5, V 0.3, Mo 1.25, Si 1, bal Fe.
For dies, punches, knives; air hardened.

AIRTREAT A.H.
Jamison Steel Corp.
C, alloy, bal Fe.
For dies; air hardened.

AIRTRUE
Now SIMONDS AIRTRUE.

AIRTRUE LC
Wallace Murray Corp.
C 0.55, Cr 5, Mo 1.5, W 1.25, Si 1, bal Fe.
Cold work tool steel; shock resistant; AISI A8.

AIRVAN
Teledyne Firth Sterling
Steel. C 1, Cr 5.25, Mo 1.15, V 0.25, bal Fe.
Air hardened: 615 Brin. For tools, cutters, cold work dies, shear blades, gauges, punches; air hardening, non-shrinking. *Obsolete*

AIS
Acciaierie Valbruna s.p.a.
Stainless Steel. C 0-0.08, Si 0-2, Mn 0-0.75, Cr 19, Ni 10, P 0.045, S 0.03, bal Fe.
Austenitic. W. Nr. 1.4301; AISI 304.

AISC
Acciaierie Valbruna s.p.a.
Stainless Steel. C 0-0.07, Si 0-2, Mn 0-2, Cr 18, Ni 10, P 0.04, S 0.03, Nb + Ta = 10 x C, bal Fe.
Austenitic. W. Nr. 1.4543; AISI 347.

AISL
Acciaierie Valbruna s.p.a.
Stainless Steel. C 0-0.03, Si 0-2, Mn 0-1, Cr 19, Ni 10, P 0.045, S 0.03, bal Fe.
Low carbon austenitic. W. Nr. 1.4306; AISI 304L.

AISRU
Acciaierie Valbruna s.p.a.
Stainless Steel. C 0-0.03, Si 0-0.7, Mn 0-1.2, P 0-0.04, S 0-0.015, Cr 17.5, Ni 10, Cu 3.5, bal Fe.
Austenitic. W. Nr. 1.4567.

AIST
Acciaierie Valbruna s.p.a.
Stainless Steel. C 0-0.08, Si 0-1, Mn 0-2, Cr 18, Ni 10, P 0.04, S 0.03, bal Fe.
Austenitic. W. Nr. 1.4541; AISI 321.

AISU
Acciaierie Valbruna s.p.a.
Stainless Steel. C 0-0.08, Si 0-1, Mn 0-2, P 0-0.04, S 0-0.03, Cr 17.5, Ni 9, Cu 3.5, bal Fe.
Austenitic. W. Nr. 1.4567.

AITCH METAL
Blackwell's Metallurgical Works
Cu 59, Zn 40, Fe 1.
For hydraulic cylinders and forgings; high strength. *Obsolete*

AJAX
H. Kramer & Co. (Ajax Metal Div.)
Cu 78.3, Sn 11, Pb 10, P 0.7.
For bearings. *Obsolete*

AJAX
English manufacture
Fe 30-70, Ni 25-50, Cu 5-20.
For bearings.

AJAX 1
Cytemp Specialty Steel Div.
C 1, Cr 4, V 0.5, Mo 0.4, bal Fe.
For tools, dies, air hardened. *Obsolete*

AJAX 2
Cytemp Specialty Steel Div.
C 1, Cr 4, V 0.5, Mo 0.4, bal Fe.
For tools, forging mandrels; hot work steel. *Obsolete*

AJAX ANTI-ACID METAL
H. Kramer & Co. (Ajax Metal Div.)
Cu alloy.
For chemical equipment; corrosion resistant. *Obsolete*

AJAX BULL
H. Kramer & Co. (Ajax Metal Div.)
Sb 17, Sn 7, bal Pb.
Cast: 30,000 TS; 20,000 YS. For bearings, bearing liners; anti-friction. *Obsolete*

AJAX DRILL ROD
Precision-Kidd Steel Co.
C 0.95-1.05, Mn 0.25-0.35, Si 0.15-0.3, bal Fe.
Water hardening tool steel; AISI W1.

AJAX DRILL ROD
Kidd Drawn Steel Co.
C 1.1, bal Fe.
For drills, taps; water hardening.

AJAX GOLDEN GLOW BRASS
H. Kramer & Co. (Ajax Metal Div.)
Cu 64.5, Pb 2.5, Zn 33.
Chill cast: 40,000 TS; 18,000 YS; 32 El; 39 Brin. For hardware, ornamental fixtures; free-cutting. *Obsolete*

AJAX MANGANESE BRONZE XX
H. Kramer & Co. (Ajax Metal Div.)
Cu 60-68, Fe 2-4, Al 3-7.5, Mn 3.5-5, bal Zn.
Cast: 110,000 TS; 60,000 YS; 12 El; 210 Brin. For propellers, gears, cams, gibs; corrosion resistant, heavy duty bronze. *Obsolete*

AJAX NICKALLOY BRONZE, GRADE 23
H. Kramer & Co. (Ajax Metal Div.)
Cu 61, Zn 13, Ni 23, Al 3.
Chill cast: 80,000 TS; 60,000 YS; 5 El; 130 Brin. For automobile trimming, typewriter parts, milking machine parts; corrosion resistant. *Obsolete*

AJAX PLASTIC BRONZE GR. A
H. Kramer & Co. (Ajax Metal Div.)
Pb 25, Sn 5, Ni 1, bal Cu.
Cast: 22,000 TS; 12,000 YS; 15 El; 50 Brin. For car and locomotive bearings, bushings; heavy duty. *Obsolete*

AJAX PLASTIC BRONZE GR. B
H. Kramer & Co. (Ajax Metal Div.)
Pb 20, Sn 7, Zn 1, bal Cu.
Cast: 30,000 TS; 16,000 YS; 22 El; 54 Brin. For bearings, bushings; resists severe loads and shock. *Obsolete*

AJAX STRUCTURAL BRONZE, GRADE "X"
H. Kramer & Co. (Ajax Metal Div.)
Cu 57, Zn 40.
Three special hardeners. Chill cast: 75,000 TS; 33,000 YS; 25 El; 100 Brin. For structural work; corrosion resistant. *Obsolete*

AJAX T-M
Ajax Steel & Forge Co.
C 0.4, Mn 0.5, Si 1, Cr 3.3, Mo 2.5, V 0.35, bal Fe.
For extrusion and forging dies, hot punches and shears. Hot work die steel, tough, shock resistant.

AJAX TOOL STEEL
American Saw & Mfg. Co.
C 0.9, Cr 3.7, V 0.3, bal Fe.
For tools, compression tools, dies, bolt and rivet headers, shear blades; hot die steel. *Obsolete*

AK 13
Otto Fuchs Metallwerke
Aluminum. Cu 4, Mg 0.8, Mn 0.9, bal Al.
375-410 N/mm^2 TS; 215-275 N/mm^2 YS; 8-14 El. High strength, heat treatable alloy. Type AlCuMg 1. Similar to AA 2017.

AK 15
Otto Fuchs Metallwerke
Aluminum. Cu 3.8, Mg 0.7, Mn 0.5, bal Al.
375-410 N/mm^2 TS; 215-275 N/mm^2 YS; 8-14 El. Special quality for hard drawn tube, rods, bar, and wire. Type AlCuMg1. Similar to AA 2017.

AK 24
Otto Fuchs Metallwerke
Aluminum. Cu 4.2, Mg 1.4, Mn 0.8, bal Al.
410-470 N/mm^2 TS; 245-335 N/mm^2 YS; 6-12 El. High strength heat treatable alloy. Type AlCuMg2. Similar to AA 2024.

AK 34
Otto Fuchs Metallwerke
Aluminum. Cu 4.3, Mg 0.6, Si 0.8, Mn 0.9, bal Al.
400-460 N/mm^2 TS; 335-385 N/mm^2 YS; 6-8 El. High strength heat treatable alloy. Type AlCuSiMn. Similar to AA 2014.

AK 39
Otto Fuchs Metallwerke
Aluminum. Cu 4.3, Mg 0.6, Si 0.8, Mn 0.6, bal Al.
400-430 N/mm^2 TS; 335-385 N/mm^2 YS; 6-8 El. High strength heat treatable alloy, particularly suitable for hand forging. Type AlCuSiMn. Similar to AA 2014.

AKRIT
American manufacture
Co 38, Cr 30, W 16, Ni 10, Mo 4, C 25.
For tools, high speed cutting tips on lathe tools; corrosion and heat resistant.

AKRIT CO 40
Thyssen Edelstahlwerke AG
C 1.1, Si 1.4, Cr 27, W 4.5, Fe 0-2, bal Co.
Welding consumables for welding valve stem ends. Meets DIN 8555.

AKRIT CO 40
Thyssen Edelstahlwerke AG
C 1-1.1, Cr 27-28, W 4-4.5, Fe 0-5, bal Co.
Coated electrodes for overlaying on outlet valves. Meets DIN 8555.

AKRIT CO 42
Thyssen Edelstahlwerke AG
C 1.6, Si 1.3, Cr 26, Co 38, Ni 23, W 12, Fe 0-1.35.
Welding consumables for overlaying valve stem ends. Meets DIN 8555.

AKRIT CO 50
Thyssen Edelstahlwerke AG
C 1.5-2, Cr 27-29, W 8-9.5, Fe 0-5, bal Co.
Coated electrodes for overlaying on outlet valves. Meets DIN 8555.

AKRIT CO 58
Thyssen Edelstahlwerke AG
C 2.7, Cr 31, W 14, Fe 0-2, bal Co.
Welding consumables for gas welding on valve stem ends. Meets DIN 8555.

AKRIT CO 58
Thyssen Edelstahlwerke AG
C 2.5-2.7, Cr 30-31, W 12-14, Fe 0-5, bal Co.
Coated electrodes for overlaying on wear rings. Meets DIN 8555.

AKRIT COMO 35
Thyssen Edelstahlwerke AG
C 0.25-0.3, Cr 27-28, Mo 6, Ni 3, Fe 0-5, bal Co.
Coated electrodes for overlaying on outlet valves. Meets DIN 8555.

AKRO 6
Vereinigte Edelstahlwerke
C 0-0.12, Al 0.8, Cr 6.5, bal Fe.
For oil refinery equipment; creep and heat resistant. *Obsolete*

AKRON
English manufacture
Cu 63, Sn 1, Zn 36.
For fittings, hardware; corrosion resistant.

AL 15-7
Allegheny Ludlum Corp./Special Materials
Stainless steel. C 0.07, Mn 0.5, P 0.015, S 0.01, Si 0.3, Cr 15, Ni 7, Mo 2, Al 1.25, bal Fe.
Precipitation hardening alloy. For applications requiring high strength and a moderate level of corrosion resistance, aerospace and spring type applications. S15700

AL 17-7
Allegheny Ludlum Corp./Special Materials
Stainless steel. C 0.07, Mn 0.5, P 0.015, S 0.01, Si 0.25, Cr 17, Ni 7, Al 1.25, bal Fe.
Precipitation hardening alloy. For applications requiring high strength and a moderate level of corrosion resistance. S17700

AL 200 ALLOY
Allegheny Ludlum Corp./Special Materials
Nickel. Cu 0.02, Fe 0.05, Mn 0.02, C 0-0.15, Si 0.05, S 0.002, bal Ni + Co.
67,000 psi TS; 21,500 psi YS; 47 El. Provides corrosion resistance in neutral to moderately reducing environments. N02200

AL 201 ALLOY
Allegheny Ludlum Corp./Special Materials
Nickel. Cu 0.02, Fe 0.05, Mn 0.02, C 0-0.02, Si 0.05, S 0.002, bal Ni + Co.
58,500 psi TS; 15,000 psi YS; 50 El. Provides corrosion resistance to neutral to moderately reducing environments. N02201

AL 2205
Allegheny Ludlum Corp./Special Materials
Stainless steel. C 0.02, Mn 0.7, P 0.025, S 0.003, Si 0.4, Cr 22, Ni 5.5, Mo 3, N 0.15, bal Fe.
RT: 90,000 psi TS min; 65,000 psi YS min; 25 El min; 32 Rock C min. For welded pipe or tubular components. S31803

AL 29-4-2
Allegheny Ludlum Steel
Stainless steel. Cr 29, Mo 4, Ni 2.1, Cu 0.06, Mn 0.05, P 0.02, S 0.01, Si 0.1, C 0.003, N 0.015, 0.018 C + N, bal Fe.
Sheet, plate, annealed: 95 ksi TS; 75 ksi YS; 25 El; 92 Rock B. For chemical and food processing, power and oil refining.

AL 29-4C
Allegheny Ludlum Steel
Stainless steel. Cr 29, Mo 4, Ni 0.3, Mn 0.5, P 0.03, S 0.01, Si 0.35, Co 0.03, C 0.02, N 0.02, Ti 0.5, bal Fe.
90 ksi TS; 75 ksi YS; 25 El. For power plant surface condenser tubing.

AL 400
Allegheny Ludlum Corp./Special Materials
Nickel. C 0.1, Mn 0.5, P 0.005, S 0.005, Si 0.25, Al 0.02, Cu 32, Fe 1, bal Ni + Co.
Plate: 80,000 psi TS; 45,000 psi YS; 30 El. Applied from mildly oxidizing trough neutral and to moderately reducing conditions, marine environments. N04400

AL 439
Allegheny Ludlum Steel
Stainless steel. C 0-0.07, Cr 17-19, Mn 0-1, Si 0-1, Ni 0-0.5, Ti 0.4-1.1, bal Fe.
Ferritic; corrosion resistant. ASTM XM-8. S43035

AL 439 HP
Allegheny Ludlum Steel
Stainless steel. C 0.01, Mn 0.34, P 0.025, S 0.003, Si 0.35, Cr 18, Ni 0.29, Ti 0.4, N 0.014, bal Fe.
Annealed: 68,000 psi TS; 43,000 psi YS; 34 El; 78 Rock B. For severe tube bending requirements in automotive exhaust manifolds, automotive exhaust system components.

AL 439 MSR
Allegheny Ludlum Steel
Stainless steel. C 0.018, Mn 0.4, P 0.025, S 0.003, Si 0.39, Cr 18, Ni 0.32, Ti 0.5, N 0.014, bal Fe.
Annealed: 71,000 psi TS; 45,000 psi YS; 32 El; 80 Rock B. For finned tubes, automotive exhaust system components.

AL 441 HP
Allegheny Ludlum Steel
Stainless steel. C 0.009, Mn 0.35, P 0.023, S 0.002, Si 0.34, Cr 18, Ni 0.3, Ti 0.29, Cb 0.71, N 0.014, Al 0.05, bal Fe.
Annealed: 72,000 psi TS; 45,000 psi YS; 31 El; 82 Rock B. For the severe tube bending and forming operations typical in automotive exhaust manifold fabrication.

AL 600
Allegheny Ludlum Steel
Nickel. C 0-0.08, Mn 0-1, Si 0-0.5, Cr 14-17, Fe 6-10, Cu 0-0.5, Co 0-0.1, 72.0 Ni min.
93,000 psi TS; 38,000 psi YS; 45 El; 65 RA. Corrosion resistant, non-magnetic; will withstand temperatures to 2200°F. Used in nuclear power, chemical, and aerospace industries.

AL 6X
Allegheny Ludlum Steel
Cr 20, Mn 1.5, Mo 6, Ni 24, bal Fe.
Alloy for service in chloride and other pitting environments.

AL 800
Allegheny Ludlum Corp./Special Materials
Nickel. C 0.02, Mn 1, P 0.02, S 0.01, Si 0.35, Cr 21, Ni 32, Ti 0.4, Al 0.4, Cu 0.3, bal Ni.
Annealed at 1800°F: 87,000 psi TS; 43,000 psi YS; 44 El. Designed to resist oxidation and carburization at elevated temperatures. N08800

AL 800H
Allegheny Ludlum Corp./Special Materials
Stainless steel. C 0.08, Mn 1, P 0.02, S 0.01, Si 0.35, Cr 21, Ni 32, Ti 0.4, Al 0.4, Cu 0.3, bal Ni.
Annealed at 2100°F: 77,000 psi TS; 29,000 psi YS; 52 El. Designed to resist oxidation and carburization at elevated temperatures. N08810

AL ALLOY
Allegheny Ludlum Corp./Special Materials
Stainless steel. C 0-0.03, Si 0-1, Mn 0-2, P 0-0.03, S 0-0.02, Cr 21-23, Ni 4.5-6.5, Mo 2.5-3.5, N 0.08-0.2, bal Fe.
RT: 90,000 psi TS; 65,000 psi YS; 25 El. For tubing and pipe, plate, sheet, and strip and forgings product forms used in ASME B and PVC, Section VIII Division 1 Construction. 331803

AL HX
Allegheny Ludlum Steel
Now ALTEMP HX.

AL TECH 15CR-5NI ALLOY
AL Tech Specialty Steel Corp.
Stainless steel. C 0-0.07, Mn 0-1, Cr 14-15, Ni 3.5-5.5, Cu 2.5-4.5, 0.15-0.45 Cb + Ta, bal Fe.
Condition A: 150 ksi TS; 110 ksi YS; 38 max Rock C. Precipitation hardening. For high-strength fittings, bolts, shafting, instruments, aerospace hardware. S15500

AL TECH 17CR-4NI EZ
AL Tech Specialty Steel Corp.
Stainless steel. C 0-0.07, Mn 0-1, S 0-0.03, Cr 15-17.5, Ni 3-5, Cu 3-5, 0.15-0.45 Cb+Ta, bal Fe.
Condition A: 150 ksi TS; 110 ksi YS; 10 El; 38 max Rock C. Precipitation hardening. For high-strength fittings, valves, bolts, shafting, pump parts, instruments. S17400

AL TECH 21-12N
AL Tech Specialty Steel Corp.
C 0.2, Cr 21, Ni 12, N 0.21, bal Fe.
For valves, valve seats for high temperature applications. SAE EV4.

AL TECH 21-2
AL Tech Specialty Steel Corp.
C 0.55, Mn 8.3, Cr 21, Ni 2.2, N 0.3, bal Fe.
For valves, valve seats. SAE EV12.

AL TECH 21-4
AL Tech Specialty Steel Corp.
C 0.50, Mn 9, Cr 21, Ni 3.75, N 0.42, bal Fe.
For valves and valve seats. SAE EV8.

AL TECH 21-6-9
Now AL TECH XM-11

AL TECH 26-1S
AL Tech Specialty Steel Corp.
Cr 26, Mo 1, C 0-0.03, Ti or Cb stabilized, bal Fe.
Superior corrosion resistant ferritic stainless. For fasteners, screens.

AL TECH 415
AL Tech Specialty Steel Corp.
Cr 12, Ni 2.5, Mo 1.5, C 0.12, N 0.04, bal Fe.
For turbine parts; martensitic stainless steel; heat and corrosion resistant.

AL TECH 6X
AL Tech Specialty Steel Corp.
Cr 21, Ni 25, Mo 6.5, C 0-0.03, bal Fe.
2150°F water quench anneal. Superior corrosion resistance, especially crevice.

AL TECH 88X
AL Tech Specialty Steel Corp.
C 0.2-0.3, Mn 10-12.5, Ni 7-8.5, bal Fe.
For large electrical transformers, switchboard parts; nonmagnetic, high electric resistivity.

AL TECH A-8
AL Tech Specialty Steel Corp.
C 0.55, Si 0.9, Cr 5, W 1.25, Mo 1.3, bal Fe.
Usual hardness 57-60 Rock C. Punch and die steel for hot and cold work applications. For shear blades, chipper knives, forming dies, forging dies, back up rolls, and plastic molds. AISI A-8.

AL TECH AL 173
AL Tech Specialty Steel Corp.
C 0.4, Mn 0.55, Si 1, Cr 3.3, V 0.5, Mn 2.5, bal Fe.
A hot work die steel capable of exceptionally good resistance to softening at elevated temperatures. Air or oil hardening. AISI H10. *Obsolete*

AL TECH AL 46
AL Tech Specialty Steel Corp.
C 1.24, Cr 4, V 3.15, W 2, Mo 8.25, Co 8.25, bal Fe.
High speed steel, molybdenum-cobalt type; AISI M-46. *Obsolete*

AL TECH AL-609
AL Tech Specialty Steel Corp.
C 0.6, Mn 0.8, Si 2, Cr 0.25, V 0.2, Mo 0.25, bal Fe.
AISI Type S5 shock resisting tool steel.

AL TECH AL-7
AL Tech Specialty Steel Corp.
C 0.5, Mn 0.8, Si 0.25, Cr 3.25, Mo 1.45, bal Fe.
Usual hardness 54-57 Rock C. General purpose air hardening die steel with toughness and resistance to softening by heat. For blanking and forming dies, plastic mold dies, shear blades and punches. AISI S-7.

AL TECH ALTEMP 19-19DX
AL Tech Specialty Steel Corp.
C 0.28-0.35, Cr 18-20, Ni 8-11, Mo 1.2-2, W 1-1.7, Ti 0.6, Cu 0-0.5, bal Fe.
Hot rolled: 118,500 TS; 69,000 YS; 58 El; 55 RA; 216 Brin. Heat treated: 103,000 TS; 43,000 YS; 54 El; 57 RA; 189 Brin. For jet engine and gas turbine components; austenitic, heat resistant. *Obsolete*

AL TECH AM
AL Tech Specialty Steel Corp.
C 0-0.1, Mo 2.5-3, Cr 16.5-17.5, Ni 4-4.5, bal Fe.
Heat treated: 161,000-200,000 TS; 45,000-142,000 YS; 22-8 El; 200-390 Brin. For valves, springs, knife blades; for applications up to 1000°F. *Obsolete*

AL TECH AM 355
AL Tech Specialty Steel Corp.
C 0.13, Cr 15.5, Ni 4.35, Mo 2.75, N 1, bal Fe.
SCT-heat treatment: 190,000 TS; 165,000 YS; 10 El; 20 RA. For jet engine and missile components; high heat resistance. *Obsolete*

AL TECH B-47 HOT WORK
AL Tech Specialty Steel Corp.
C 0.4, Mn 0.35, Cr 4.2, W 4.2, Co 4.2, V 2.2, Mo 0.4, bal Fe.
At 800°F: 178,000 TS; 9 El; 36 RA. At 1200°F: 110,00 TS;
16.5 El; 37 RA. W-Cr-Co hotwork tool steel for heavy duty
applications; forging die inserts, hotwork punches, brass
extrusion tooling. AISI H-19.

AL TECH D-5
AL Tech Specialty Steel Corp.
C 1.4, Mn 0.3, Si 0.5, Cr 12.5, Mo 0.8, Co 3.3, Ni 0.5, bal Fe.
Usual hardness 57-62 Rock C. Air hardenable high carbon-
high chromium with Co die steel with improved abrasion
resistance and red hardness. For long run blanking and
forming dies, hot punches and shear blades. AISI D-5.

AL TECH DBL 2
AL Tech Specialty Steel Corp.
C 0.8, Mn 0.3, Si 0.3, Cr 4, V 2, W 6, Mo 5, bal Fe.
62-64 Rock C. General purpose high speed steel for cutting
tools and ultra high strength non-cutting tool applications as
punches and die inserts. AISI M-2.

AL TECH DBL 2 1/2
AL Tech Specialty Steel Corp.
C 1, Cr 4, W 6.25, V 2.5, Mo 6.25, bal Fe.
High speed tool steel, W-Mo type for lathe and planer tools,
form tools, broaches, drills, reamers, hobs, taps, AISI M3.
Obsolete

AL TECH DBL 3
AL Tech Specialty Steel Corp.
C 1.15, Cr 4, W 6, V 3, Mo 5.5, bal Fe.
High speed tool steel, W-Mo-V type for lathe and planer tools,
broaches form tools, hobs, taps, AISI M3.

AL TECH DBL2-EZ
AL Tech Specialty Steel Corp.
C 0.8, Cr 4, V 2, W 6, Mo 5, bal Fe.
For lathe and planer tools, reamers, drills, taps; high speed
steel, oil hardened.

AL TECH HI-CRO 20
AL Tech Specialty Steel Corp.
C 0-0.6, Mn 0-2, Si 0-1, Cr 19-21, Ni 32.5-35, Mo 2-3, Cu 3-4,
Cb 0-1.
Excellent corrosion resistance especially for sulfuric acid.
Used for valves, pumps, process equipment, seamless tube
and pipe.

AL TECH HTB-1
AL Tech Specialty Steel Corp.
C 1.02, Mn 0.35, Si 0.4, Cr 1.45, Al 1, bal Fe.
For jet engine bearings; for high temperature applications.

AL TECH HTB-2
AL Tech Specialty Steel Corp.
C 0.58, Mn 0.3, Si 1.15, Cr 4.75, Mo 5.25, V 0.55, bal Fe.
For aircraft bearings and other high strength components
operating up to 900°F. AISI M-50.

AL TECH HTB-5
AL Tech Specialty Steel Corp.
C 0.12, Cr 4, Mo 4.25, Ni 3.5, V 1.2, bal Fe.
Surface hardness of 58-62 Rock C obtained by carburizing.
Modified M-50 high temperature bearing steel developed by
G.E. Aircraft Engine Div. as M50 NiL. Capable of high fracture
toughness and rolling contact fatigue life.

AL TECH LMW
AL Tech Specialty Steel Corp.
C 0.83, Cr 4, W 1.5, Mo 8.5, V 1.1, bal Fe.
Mo-type high speed steel for taps, drills, reamers, planer
tools, milling cutters, and wood working tools. AISI M-1.

AL TECH LMW-V
AL Tech Specialty Steel Corp.
C 1, Cr 3.75, V 2, W 1.75, Mo 8.75, bal Fe.
Mo-type high speed steel for drills, end mills, reamers, form
tools, milling cutters, planer and boring tools. AISI M-7.

AL TECH LXX
AL Tech Specialty Steel Corp.
C 0.7, W 18, Cr 4, V 1, bal Fe.
Original 18-4-1 general purpose high speed steel for wide
variety of cutting requirements. AISI T-1.

AL TECH M50NIL
AL Tech Specialty Steel Corp.
C 0.13, Mn 0.25, Si 0.2, Cr 4.1, Ni 3.4, Mo 4.25, V 1.2, bal Fe.
Race hardness of 58-62 Rock C. Bearing steel for higher jet
engine speed requirements.

AL TECH ML
AL Tech Specialty Steel Corp.
C 0.8, W 18, Cr 4, V 2, Mo 0.75, bal Fe.
Hardened: 64-66 Rock C. 18-4-2 high speed steel for cutting
applications where added abrasion resistance is needed.

AL TECH TYPE 203EZ
AL Tech Specialty Steel Corp.
Stainless steel. C 0-0.8, Mn 5-6.5, P 0-0.04, Si 0-1, Cr 16-18,
Ni 5-6.5, Cu 1.75-2.25, 0.15 S min, bal Fe.
Annealed: At 70°F: 80 ksi TS; 56 El; 65 RA. Austenitic. For
use in high-speed, automatic machine operations. S20300

AL TECH TYPE 303EZ
AL Tech Specialty Steel Corp.
Stainless steel. C 0-0.15, Mn 0-2, P 0-0.04, Si 0-1, Cr 17-19,
Ni 8-10, 0.15 S min, bal Fe.
Annealed: At 70°F: 85 ksi TS; 52 El; 66 RA. For high speed
automatic machining. S30300

AL TECH TYPE 304EZ
AL Tech Specialty Steel Corp.
Stainless steel. C 0-0.08, Mn 0-2, P 0-0.045, S 0-0.03, Si 0-1,
Cr 18-20, Ni 8-10.5, bal Fe.
Annealed: At 70°F: 82 ksi TS; 62 El; 72 RA. Drawing and
forming operations. S30400

AL TECH TYPE 316EZ
AL Tech Specialty Steel Corp.
Stainless steel. C 0-0.08, Mn 0-2, P 0-0.045, S 0-0.03, Si 0-1,
Cr 16-18, Ni 12-13, Mo 2-3, bal Fe.
Annealed: At 70°F: 82 ksi TS; 65 El; 75 RA. For screws,
fittings, shafts, pump parts, and valves. S31600

AL TECH TYPE 416 EZ
AL Tech Specialty Steel Corp.
Stainless steel. C 0-0.15, Mn 0-1.25, P 0-0.06, S 0.15-0.35, Si
0-1, Cr 12-14, bal Fe.
For high speed automatic machining; used for screws,
fittings, shafts, gears, nuts and bolts, pump parts and
shafting. S41600

AL TECH TYPE 416 EZR
AL Tech Specialty Steel Corp.
Stainless Steel. C 0-0.15, Mn 0-1.25, P 0-0.06, S 0.15-0.35, Si
0-1, Cr 12-14, bal Fe.
For high speed automatic machining; used for screws,
fittings, shafts, gears, nuts and bolts, pump parts and
shafting. S41600

AL TECH TYPE 416-F
AL Tech Specialty Steel Corp.
Stainless steel. C 0-0.15, Mn 0-1.25, P 0-0.06, Si 0-1, Cr
12-14, Mo 0.4-0.6, 0.30 S min, bal Fe.
For screws, fittings, shafts, nuts and bolts, pump parts and
shafting. S41600

AL TECH TYPE A-286
AL Tech Specialty Steel Corp.
Stainless steel. C 0-0.8, Mn 0-2, S 0-0.025, Si 0-1, Cr 13.5-16,
Ni 24-27, Mo 1-1.5, Ti 1.9-2.35, bal Fe.
At 68°F: 146 ksi TS; 102 ksi YS; 25 El. Austenitic precipitation
hardening. Vacuum arc remelted. Designed for service
applications at temperatures up to 1300°F. K66286

AL TECH VLM
AL Tech Specialty Steel Corp.
C 0.85, Mo 8, Cr 4, V 2, bal Fe.
Hardened: 63-66 Rock C. For tools, cutters, reamers, twist
drills, milling cutters. Type M-10 high-speed steel.

AL TECH XM-11
AL Tech Specialty Steel Corp.
Cr 21, Ni 6, Mn 9, C 0-0.04, N 0.35, bal Fe.
Solution strengthened austenitic alloy. High temperature,
high strength, corrosion resistant. Used where non-magnetic,
high strength needed; aircraft hydraulic tubing.

AL TECH XM-19
AL Tech Specialty Steel Corp.
Stainless Steel. C 0-0.06, Mn 4-6, Si 0-1, Cr 20.5-23.5, Ni
11.5-13.5, Mo 1.5-3, N 0.2-0.4, Cb 0.1-0.3, V 0.1-0.3, bal Fe.
High strength austenitic stainless steel. Used for paper and
pulp equipment, petrochemical, and salt water shafting.

AL-0001
Miller Thermal, Inc.
Reactive/refractory. Co 47.6, Cr 30, W 12.5, Fe 3, C 2.4, Si 1,
0.5 Mn (nominal), 3.0 Ni (nominal).
Hard facing alloy; 47-49 Rock C.

AL-0006
Miller Thermal, Inc.
Reactive/refractory. Co 59.4, Cr 29, W 3.8, Ni 2.3, Fe 2.7, C
1.3, Si 1, Mn 0.38, 0.12 Mn (nominal).
Hard facing alloy; 37-42 Rock C.

AL-0012
Miller Thermal, Inc.
Reactive/refractory. Co 51.9, Cr 30, W 8.2, C 1.4, Si 1.5, Mn
1, 3.0 Ni (nominal), 3.0 Fe (nominal).
Hard facing alloy; 40-42 Rock C.

AL-0021
Miller Thermal, Inc.
Reactive/refractory. Co 59.3, Cr 27, C 0.2, Mo 5.5, 3.0 Ni
(nominal), 3.0 Fe (nominal), 1.0 Si (nominal),, 1.0 Mn
(nominal).
Hard facing alloy; 27-29 Rock C.

AL-1001
Miller Thermal, Inc.
Reactive/refractory. C 3.9-4.3, Co 10-12, Fe 0-2, W 81.7-86.1.
Powder for plasma spray application; abrasion resistant.
Particle size: -325 mesh + 5 microns and -45 mesh + 5
microns. Meets AMS-7879.

AL-1002
Miller Thermal, Inc.
Reactive/refractory. C 3.6-4.2, Co 11-13, Fe 0-2, W 81.7-86.1.
Powder for plasma spray application; abrasion resistant.
Particle size: -200 + 325 mesh and -75 + 45 microns.

AL-1003
Miller Thermal, Inc.
Reactive/refractory. Ni 80, Cr 20.
Powder for plasma spray application; abrasion resistant.
Particle size: -325 mesh + 5 microns and -45 + 5 microns.

AL-1005
Miller Thermal, Inc.
Reactive/refractory. Ni 20, Cr 5, 75.0 Cr_3C_2.
Powder for plasma spray application; abrasion resistant.
Particle size: -325 mesh + 5 microns and -45 + 5 microns.

AL-1007
Miller Thermal, Inc.
Reactive/refractory. Ni 20, Cr 5, 75.0 Cr_3C_2.
Powder for plasma spray application; abrasion resistant.
Particle size: -140 mesh + 10 microns and -105 + 10
microns.

AL-1008
Miller Thermal, Inc.
Reactive/refractory. Ni 12, Cr 3, 85.0 Cr_3C_2.
Powder for plasma spray application; abrasion resistant.
Particle size: -140 mesh + 10 microns and -105 + 10
microns.

AL-1009
Miller Thermal, Inc.
Reactive/refractory. C 5.45-5.65, Co 8-10, W 81.7-86.1.
Powder for plasma spray application; abrasion resistant.
Particle size: -325 mesh + 5 microns and -45 + 5 microns.

AL-1015
Miller Thermal, Inc.
Reactive/refractory. Ni 80, Cr 20.
Powder for plasma spray application; abrasion resistant.
Particle size: -120 + 325 mesh and -125 + 45 microns.

AL-1016
Miller Thermal, Inc.
Reactive/refractory. Co 57.5, Cr 25, Ni 10, W 7.5.
Powder for plasma spray application; abrasion resistant.
Particle size: -325 mesh + 5 microns and -45 + 5 microns.

AL-1017
Miller Thermal, Inc.
Reactive/refractory. Ni 80, Cr 20.
Powder for plasma spray application; abrasion resistant.
Particle size: -170 mesh + 10 microns and -90 + 10 microns.

AL-1018
Miller Thermal, Inc.
Reactive/refractory. Co 57.5, Cr 25, Ni 10, W 7.5.
Powder for plasma spray application; abrasion resistant.
Particle size: -170 + 325 mesh and -90 + 45 microns.

AL-1019
Miller Thermal, Inc.
Reactive/refractory. Ni 80, Cr 20.
Powder for plasma spray application; abrasion resistant.
Particle size: -325 mesh + 5 microns and -45 + 5 microns.

AL-1031-C
Miller Thermal, Inc.
Reactive/refractory. Ni 46, WC 35, Cr 11, Fe 2.5, Si 2.5, B 2.5,
C 0.5.
Powder for plasma spray application; abrasion resistant.
Particle size: -120 + 325 mesh and -125 + 45 microns.

AL-1031-F
Miller Thermal, Inc.
Reactive/refractory. Ni 46, WC 35, Cr 11, Fe 2.5, Si 2.5, B 2.5,
C 0.5.
Powder for plasma spray application; abrasion resistant.
Particle size: -120 mesh + 15 microns and -125 + 15
microns.

AL-1032-C
Miller Thermal, Inc.
Reactive/refractory. WC 80, Ni 14, Cr 3.5, Fe 0.8, Si 0.8, B
0.8, C 0.1.
Powder for plasma spray application; abrasion resistant.
Particle size: -120 + 325 mesh and -125 + 45 microns.

AL-1032-F
Miller Thermal, Inc.
Reactive/refractory. WC 80, Ni 14, Cr 3.5, Fe 0.8, Si 0.8, B
0.8, C 0.1.
Powder for plasma spray application; abrasion resistant.
Particle size: -120 mesh + 15 microns and -125 + 15
microns.

AL-1034-F
Miller Thermal, Inc.
Reactive/refractory. WC 50, Ni 33, Cr 9, Fe 3.5, Si 2, B 2, C
0.5.
Powder for plasma spray application; abrasion resistant.
Particle size: -270 mesh + 15 microns and -53 + 15 microns.

AL-1035
Miller Thermal, Inc.
Reactive/refractory. Al 88, Si 12.
Powder for plasma spray application; abrasion resistant.
Particle size: -140 + 325 mesh and -105 + 45 microns.

AL-1037
Miller Thermal, Inc.
Reactive/refractory. Ni 95, Al 5.
Powder for plasma spray application; abrasion resistant.
Particle size: -140 + 325 mesh and -105 + 45 microns.

AL-1039
Miller Thermal, Inc.
Reactive/refractory. WC 50, Cr 6, Fe 1.5, Si 1.5, B 1, C 0.5,
Ni-Al 14.0.
Powder for plasma spray application; abrasion resistant.
Particle size: -170 mesh + 15 microns and -90 + 15 microns.

AL-1044
Miller Thermal, Inc.
Reactive/refractory. Ni 75, Cr 8, Al 7, Fe 5, Mo 5.
Powder for plasma spray application; abrasion resistant.
Particle size: -140 + 325 mesh and -105 + 45 microns.

AL-1047
Miller Thermal, Inc.
Reactive/refractory. Ni 75, Cr 19, Al 6.
Powder for plasma spray application; abrasion resistant.
Particle size: -100 + 325 mesh and -150 + 45 microns.

AL-1052-1
Miller Thermal, Inc.
Reactive/refractory. Ni 75, C 25.
Powder for plasma spray application; abrasion resistant.
Particle size: -170 + 325 mesh and -90 + 45 microns.

AL-1052-2
Miller Thermal, Inc.
Reactive/refractory. Ni 85, C 15.
Powder for plasma spray application; abrasion resistant.
Particle size: -170 + 325 mesh and -90 + 45 microns.

AL-1052-3
Miller Thermal, Inc.
Reactive/refractory. Ni 80, C 20.
Powder for plasma spray application; abrasion resistant.
Particle size: -170 + 325 mesh and -90 + 45 microns.

AL-1054
Miller Thermal, Inc.
Reactive/refractory. Ni 20.9, B 0.4, Fe 0.7, Si 0.7, Cr 2.3, Mo
75.
Powder for plasma spray application; abrasion resistant.
Particle size: -325 mesh + 5 microns and -45 + 5 microns.

AL-1055
Miller Thermal, Inc.
Reactive/refractory. Al 88, Si 12.
Powder for plasma spray application; abrasion resistant.
Particle size: -250 mesh + 5 microns and -58 + 5 microns.

AL-1057
Miller Thermal, Inc.
Reactive/refractory. Ni 41.7, B 0.8, Fe 1.5, Si 1.5, Cr 4.5, Mo
30, W 20.
Powder for plasma spray application; abrasion resistant.
Particle size: -325 mesh + 10 microns and -45 + 10 microns.

AL-1058-A
Miller Thermal, Inc.
Reactive/refractory. Cu 59, Ni 36, In 5.
Powder for plasma spray application; abrasion resistant.
Particle size: -200 + 325 mesh and -74 + 45 microns.

AL-1058-B
Miller Thermal, Inc.
Reactive/refractory. Cu 59, Ni 36, In 5.
Powder for plasma spray application; abrasion resistant.
Particle size: -325 mesh + 5 microns and -45 + 5 microns.

AL-1064
Miller Thermal, Inc.
Reactive/refractory. Ni 5.6, Cr 1.4, 93.0 Cr_3C_2.
Powder for plasma spray application; abrasion resistant.
Particle size: -325 mesh + 5 microns and -45 + 5 microns.

AL-1067
Miller Thermal, Inc.
Reactive/refractory. Ni 8, Cr 2, 90.0 Cr_3C_2.
Powder for plasma spray application; abrasion resistant.
Particle size: -325 mesh + 5 microns and -45 + 5 microns.

AL-1071
Miller Thermal, Inc.
Reactive/refractory. Co 6-10, WC 90-94.
Powder for plasma spray application; abrasion resistant.
Particle size: -200 + 325 mesh and -74 + 45 microns.

AL-1072
Miller Thermal, Inc.
Reactive/refractory. Co 11-13, Fe 0-1.5, 5.15 C min, bal W.
Powder for plasma spray application; abrasion resistant.
Particle size: -325 mesh + 5 microns and -45 + 5 microns.

AL-1073
Miller Thermal, Inc.
Reactive/refractory. C 4.8-5.6, Co 15-18, Fe 0-2, W 76.
Powder for plasma spray application; abrasion resistant.
Particle size: -325 mesh + 5 microns and -45 + 5 microns.

AL-1081
Miller Thermal, Inc.
Reactive/refractory. Fe 71, Cr 19, Ni 10.
Powder for plasma spray application; abrasion resistant.
Particle size: -150 + 325 mesh + 5 microns and -97 + 45
microns.

AL-1082
Miller Thermal, Inc.
Reactive/refractory. Fe 64.6, Cr 17.1, Ni 13, Mo 2.2, 2.0 Mn
(nominal), 1.0 Si (nominal), 0.10 C (nominal).
Powder for plasma spray application; abrasion resistant.
Particle size: -200 + 325 mesh and -74 + 45 microns.

AL-1082-F
Miller Thermal, Inc.
Reactive/refractory. Fe 64.6, Cr 17.1, Ni 13, Mo 2.2, 2.0 Mn
(nominal), 1.0 Si (nominal), 0.10 C (nominal).
Powder for plasma spray application; abrasion resistant.
Particle size: -325 mesh + 10 microns and -45 + 10 microns.

AL-1083
Miller Thermal, Inc.
Reactive/refractory. Fe 85, Cr 13.3, Si 0.85, Ni 0.15, Mn 0.7.
Powder for plasma spray application; abrasion resistant.
Particle size: -100 mesh + 10 microns and -150 + 10
microns.

AL-1083-F
Miller Thermal, Inc.
Reactive/refractory. Fe 85, Cr 13.3, Si 0.85, Ni 0.15, Mn 0.7.
Powder for plasma spray application; abrasion resistant.
Particle size: -325 mesh + 5 microns and -45 + 5 microns.

AL-1085
Miller Thermal, Inc.
Reactive/refractory. C 5.55-6.05, Cr 20.5-21.5, Ni 5.5-6.5, Fe
0-2, bal W.
Powder for plasma spray application; abrasion resistant.
Particle size: -325 mesh + 5 microns and -45 + 5 microns.

AL-1086
Miller Thermal, Inc.
Reactive/refractory. C 3.25-3.75, Co 9.5-11, Cr 3.5-4.5, Fe
0-1.25, bal W.
Powder for plasma spray application; abrasion resistant.
Particle size: -325 mesh + 5 microns and -45 + 5 microns.

AL-1087
Miller Thermal, Inc.
Reactive/refractory. Ni 89.5, Al 5.5, Mo 5.
Powder for plasma spray application; abrasion resistant.
Particle size: -170 + 325 mesh and -90 + 45 microns.

AL-1090
Miller Thermal, Inc.
Reactive/refractory. Cu 90, Al 10.
Powder for plasma spray application; abrasion resistant.
Particle size: -100 + 325 mesh and -150 + 45 microns.

AL-1090-F
Miller Thermal, Inc.
Reactive/refractory. Cu 90, Al 10.
Powder for plasma spray application; abrasion resistant.
Particle size: -200 mesh + 5 microns and -74 + 5 microns.

AL-1094
Miller Thermal, Inc.
Reactive/refractory. Ni 75, Cr 16, Fe 8, Si 0.7.
Powder for plasma spray application; abrasion resistant.
Particle size: -140 + 325 mesh and -105 + 45 microns.

AL-1094-F
Miller Thermal, Inc.
Reactive/refractory. Ni 75, Cr 16, Fe 8, Si 0.7.
Powder for plasma spray application; abrasion resistant.
Particle size: -325 mesh + 5 microns and -45 + 5 microns.

AL-1095
Miller Thermal, Inc.
Reactive/refractory. WC 30, Co 28, Ni 19, Cr 12.5, Mo 4, Si 2.5, B 2, Fe 1.9, C 0.1.
Powder for plasma spray application; abrasion resistant.
Particle size: -120 + 325 mesh and -125 + 45 microns.

AL-1098
Miller Thermal, Inc.
Reactive/refractory. C 3.3, Ni 19.3, Co 8.6, Al 0.4, Cr 3, B 0.6, Si 0.8, W 63.3.
Powder for plasma spray application; abrasion resistant.
Particle size: -120 mesh + 15 microns and -125 + 15 microns.

AL-1101
Miller Thermal, Inc.
Reactive/refractory. C 3.9-4.3, Co 10-12, Fe 0-2, W 81.7-86.1.
Powder for plasma spray application; abrasion resistant.
Particle size: -325 mesh + 15 microns and -45 mesh + 15 microns. Meets AMS-7879.

AL-1102
Miller Thermal, Inc.
Reactive/refractory. C 3.6-4.2, Co 10-13, Fe 0-2, W 81.7-86.1.
Powder for plasma spray application; abrasion resistant.
Particle size: -100 + 200 mesh and -150 + 75 microns.

AL-1109
Miller Thermal, Inc.
Reactive/refractory. C 5.45-5.65, Co 8-10, W 81.7-86.1.
Powder for plasma spray application; abrasion resistant.
Particle size: -325 mesh + 15 microns and -45 + 15 microns.

AL-1171
Miller Thermal, Inc.
Reactive/refractory. Co 6-10, WC 90-94.
Powder for plasma spray application; abrasion resistant.
Particle size: -325 mesh + 5 microns and -45 + 5 microns.

AL-1172
Miller Thermal, Inc.
Reactive/refractory. Co 11-13, Fe 0-1.5, 5.15 C min, bal W.
Powder for plasma spray application; abrasion resistant.
Particle size: -325 mesh + 15 microns and -45 + 15 microns. Meets AMS-7880.

AL-1173
Miller Thermal, Inc.
Reactive/refractory. C 4.8-5.6, Co 15-18, Fe 0-2, W 76.
Powder for plasma spray application; abrasion resistant.
Particle size: -325 mesh + 15 microns and -45 + 15 microns.

AL-1185
Miller Thermal, Inc.
Reactive/refractory. C 5.55-6.05, Cr 20.5-21.5, Ni 5.5-6.5, Fe 0-2, bal W.
Powder for plasma spray application; abrasion resistant.
Particle size: -325 mesh + 15 microns and -45 + 5 microns.

AL-1186
Miller Thermal, Inc.
Reactive/refractory. C 3.25-3.75, Co 9.5-11, Cr 3.5-4.5, Fe 0-1.25, bal W.
Powder for plasma spray application; abrasion resistant.
Particle size: -325 mesh + 15 microns and -45 + 15 microns.

AL-1236
Miller Thermal, Inc.
Reactive/refractory. WC 35, Cr 11, B 2.5, Fe 2.5, Si 2.5, C 0.5, bal Ni.
Powder for plasma spray application; abrasion resistant.
Particle size: -120 + 325 mesh and -125 + 45 microns.

AL-1255
Miller Thermal, Inc.
Reactive/refractory. Co 42.9, Cr 19, W 5, Ni 19, Fe 3.5, C 0.65, Si 3.45, Mo 2.25, B 3, Cu 1.25.
Hard facing alloy; 55-57 Rock C.

AL-1275
Miller Thermal, Inc.
Reactive/refractory. C 3.9, Ni 17.7, Co 12.8, Cr 4.1, B 0.8, Si 1.1, Fe 2.2, W 57.6.
Powder for plasma spray application; abrasion resistant.
Particle size: -120 mesh + 15 microns and -125 + 15 microns.

AL-840
Now ALLEGHENY 334.

AL-L-24
British Aluminium Co., Ltd.
Cu 4, Ni 2, Mn 1.5, Si 0.75, bal Al.
Cast and treated: 36,000 TS; 1.5 El; 90 Brin. Chill cast and treated: 40,000-46,000 TS; 4-7 El. For pistons, valves, cylinder heads, internal combustion engines; resists high temperature. *Obsolete*

AL99 (901) 1100
J. & A. Erbsloh Aluminium
Aluminum. Cu 0.05, Mn 0.05, Zn 0.1, Ti 0.05, 1.00 others, bal Al.
Pure aluminum. Pressed: 75 N/mm^2 TS; 30 N/mm^2 YS; 22 Brin. Werkstoff Nr. 3.0205.08.

AL99.5 (955) 1050
J. & A. Erbsloh Aluminium
Aluminum. Si 0.25, Fe 0.4, Cu 0.05, Mn 0.05, Mg 0.05, Zn 0.07, Ti 0.05, 0.50 others, bal Al.
Pure aluminum. Pressed: 65 N/mm^2 TS; 20 N/mm^2 YS; 20 Brin. Werkstoff Nr. 3.0255.08.

AL99.7 (971) 1030
J. & A. Erbsloh Aluminium
Aluminum. Si 0.2, Fe 0.25, Cu 0.03, Mn 0.03, Mg 0.03, Zn 0.07, Ti 0.03, 0.30 others, bal Al.
Pure aluminum. Pressed: 60 N/mm^2 TS; 20 N/mm^2 YS; 20 Brin. Werkstoff Nr. 3.0275.08.

AL99.8 (981) 1001
J. & A. Erbsloh Aluminium
Aluminum. Si 0.15, Fe 0.15, Cu 0.03, Mn 0.02, Mg 0.02, Zn 0.06, Ti 0.02, 0.20 others, bal Al.
Pure aluminum. Pressed: 55 N/mm^2 TS; 20 N/mm^2 YS; 18 Brin. Werkstoff Nr. 3.0285.08.

AL99.85 (985) 1002
J. & A. Erbsloh Aluminium
Aluminum. Si 0.08, Fe 0.08, Mn 0.01, Zn 0.05, Ti 0.01, 0.15 others, bal Al.
Nonhardenable alloy.

AL99.85MG 1 (512) 5102
J. & A. Erbsloh Aluminium
Aluminum. Si 0.08, Fe 0.08, Mn 0.03, Mg 0.4-0.6, Zn 0.05, Ti 0.02, 0.15 others, bal Al.
Nonhardenable alloy. Pressed: 100 N/mm^2 TS; 30 Brin.
Werkstoff Nr. 3.3317.08.

AL99.85MG0.5 (915) 5052
J. & A. Erbsloh Aluminium
Aluminum. Si 0.08, Fe 0.08, Mn 0.03, Mg 0.4-0.6, Zn 0.05, Ti 0.02, 0.15 others, bal Al.
Nonhardenable alloy. Pressed: 70 N/mm^2 TS; 23 Brin.
Werkstoff Nr. 3.3307.08.

AL99.85MGSI (486) 6042
J. & A. Erbsloh Aluminium
Aluminum. Si 0.4-0.7, Fe 0.08, Cu 0.05-0.15, Mn 0.03, Mg 0.4-0.6, Zn 0.05, Ti 0.02, 0.15 others, bal Al.
Hardenable alloy. Cold hardened: 140 N/mm^2 TS; 70 N/mm^2 YS; 45 Brin. Warm hardened: 240 N/mm^2 TS; 195 N/mm^2 YS; 75 Brin. Werkstoff Nr. 3.2307.51, 3.2307.72.

AL99.85MGSI (487) 6052
J. & A. Erbsloh Aluminium
Aluminum. Si 0.4-0.7, Fe 0.08, Cu 0.05-0.15, Mn 0.03, Mg 0.4-0.6, Zn 0.05, Ti 0.02, 0.15 others, bal Al.
Hardenable alloy. Cold hardened: 155 N/mm^2 TS; 80 N/mm^2 YS; 50 Brin. Warm hardened: 280 N/mm^2 TS; 230 N/mm^2 YS; 85 Brin. Werkstoff Nr. 3.2307.

AL99.85MGSI0.4 (481) 6032
J. & A. Erbsloh Aluminium
Aluminum. Si 0.35-0.6, Fe 0.08, Cu 0.02, Mn 0.03, Mg 0.35-0.6, Zn 0.05, Ti 0.02, 0.15 others, bal Al.
Hardenable alloy. Cold hardened: 120 N/mm^2 TS; 55 N/mm^2 YS; 35 Brin. Warm hardened: 190 N/mm^2 TS; 120 N/mm^2 YS; 60 Brin.

AL99.9 (933) 1003
J. & A. Erbsloh Aluminium
Aluminum. Si 0.06, Fe 0.04, Zn 0.04, Ti 0.006, 0.10 others, bal Al.
Nonhardenable alloy.

AL99.9MG 1 (513) 5103
J. & A. Erbsloh Aluminium
Aluminum. Si 0.06, Fe 0.04, Mn 0.03, Mg 0.9-1.1, Zn 0.04, Ti 0.01, 0.10 others, bal Al.
Nonhardenable alloy. Pressed: 100 N/mm^2 TS; 30 Brin.
Werkstoff Nr. 3.3318.08.

AL99.9MG0.5 (913) 5053
J. & A. Erbsloh Aluminium
Aluminum. Si 0.06, Fe 0.04, Mn 0.03, Mg 0.4-0.6, Zn 0.04, Ti 0.01, 0.10 others, bal Al.
Nonhardenable alloy. Pressed: 70 N/mm^2 TS; 23 Brin.
Werkstoff Nr. 3.3308.08.

AL99.9MGSI (473) 6043
J. & A. Erbsloh Aluminium
Aluminum. Si 0.4-0.7, Fe 0.04, Cu 0.1-0.15, Mn 0.03, Mg 0.4-0.6, Zn 0.05, Ti 0.01, 0.10 others, bal Al.
Hardenable alloy. Cold hardened: 130 N/mm^2 TS; 65 N/mm^2 YS; 45 Brin. Warm hardened: 240 N/mm^2 TS; 195 N/mm^2 YS; 75 Brin. Werkstoff Nr. 3.3208.51, 3.3208.72.

AL99.9MGSI (473) 6053
J. & A. Erbsloh Aluminium
Aluminum. Si 0.4-0.7, Fe 0.04, Cu 0.1-0.15, Mn 0.03, Mg 0.4-0.6, Zn 0.05, Ti 0.01, 0.10 others, bal Al.
Hardenable alloy. Cold hardened: 155 N/mm^2 TS; 80 N/mm^2 YS; 50 Brin. Warm hardened: 280 N/mm^2 TS; 230 N/mm^2 YS; 85 Brin. Werkstoff Nr. 3.3208.

ALA-ZE 63

American Light Alloys Co.
Zn 5-6, Zr 0.4-1, 2-3 rare earth metals, bal Mg.
Sand cast: 45,000 TS; 25,000 YS; 8 El. For general purpose castings, oil pans, cases, housings. Hydrogen heat treated, high strength and good ductility, pressure tight.

ALACRITE 502

Aubert & Duval
Cobalt. Cr 31, W 14, C 1, Fe 0-2, Ni 0-2.5, Si 0.7, bal Co.
For hard facing electrode; corrosion and wear resistant. Equivalent to Stellite 4.

ALACRITE 52 N.C

Aubert & Duval
Cobalt. C 0.25, Si 0.9, Mn 0.5, Ni 10.5, Cr 25, W 7.5, Fe 0-2, bal Co.
Cast. Equivalent to X 45 G.E.

ALACRITE 631 T

Aubert & Duval
Cobalt. C 0.45, Si 0.3, Mn 0.25, Ni 0-1, Cr 21, W 11, Nb 2, Fe 0-2, bal Co.
Cast: 750 MPa TS. Equivalent to Haynes alloy 152.

ALADAR

English manufacture
Cu 0.13, Fe 0.4, Si 12.5, Mn 0.09, bal Al.
For light alloys, castings; same as "Alpax," "Alcoa No. 47," and "Silumen."

ALADDIN

Aladdin Welding Products Inc.
Al, Cu, bal Zn.
For welding rod; for white metal die castings.

ALAIS 4N2

Compagnie des Mines
C 0.07-0.13, Ni 1.3-2.3, Mn 0.5, bal Fe.
Annealed: 64,000 TS; 47,000 YS; 25 El. Hardened: 121,000 TS; 99,500 YS; 12 El. For exhaust valves, gears, shafts; case hardening steel.

ALAIS 5 1/2C1

Compagnie des Mines
C 0.25-0.3, Cr 0.8-1.2, Mn 0.6-0.9, bal Fe.
Annealed: 78,000 TS; 48,400 YS; 16 El. Hardened: 128,000 TS; 85,300 YS; 10 El. For tools, shafts, gears; case hardening steel.

ALAIS 5 1/2NCD1

Compagnie des Mines
C 0.07-0.13, Cr 0.4-0.7, Ni 1-1.3, Mo 0.1-0.2, Mn 0.5-0.9, bal Fe.
Annealed: 78,000 TS. Hardened: 140,000 TS; 100,000 YS; 10 El. For shafts, gears; tough, case hardening.

ALAIS 5C1

Compagnie des Mines
C 0.09-0.15, Cr 0.6-1, Mn 0.6-0.9, bal Fe.
Annealed: 71,100 TS; 42,600 YS; 20 El. Hardened: 114,000 TS; 78,000 YS; 12 El. For shafts, axles; case hardening steel.

ALAIS 5N2

Compagnie des Mines
C 0.18-0.25, Ni 1.3-2.3, Mn 0.5, bal Fe.
Annealed: 78,000 TS; 60,000 YS; 20 El. Hardened: 160,000 TS; 128,000 YS; 8 El. For gears, shafts; case hardening steel.

ALAIS 5NC3

Compagnie des Mines
C 0.08-0.13, Cr 0.6-0.9, Ni 2.5-3, Mn 0.35-0.65, bal Fe.
Annealed: 71,000 TS; 57,000 YS; 25 El. Hardened: 140,000 TS; 102,000 YS; 12 El. For shafts, gears; tough, case hardening.

ALAIS 6 1/2

Compagnie des Mines
C 0.38, bal Fe.
Rolled: 85,000-121,000 TS; 57,000-83,000 YS; 12-16 El. For shafts, tools; water hardening.

ALAIS 6CD1

Compagnie des Mines
C 0.08-0.16, Cr 0.8-1.2, Mo 0.15-0.3, Mn 0.6-0.9, bal Fe.
Annealed: 78,000 TS; 50,000 YS; 25 El. Hardened: 170,000 TS; 114,000 YS; 8 El. For shafts, gears; case hardening steel.

ALAIS AE4

Compagnie des Mines
C 0.1, bal Fe.
Rolled: 50,000-60,000 TS; 35,000 YS; 30 El. For forgings, shafts; weldable.

ALAIS AE4 1/2

Compagnie des Mines
C 0.12, bal Fe.
Rolled: 57,000-72,000 TS; 40,000 YS; 22 El. For forgings, shafts; weldable.

ALAIS AE5

Compagnie des Mines
C 0.18, bal Fe.
Rolled: 64,000-100,000 TS; 37,000-60,000 YS; 15-23 El. For forgings, axles, shafts; case hardening.

ALAIS AE6

Compagnie des Mines
C 0.32, bal Fe.
Rolled: 78,000-114,000 TS; 50,000-79,000 YS; 13-20 El. For shafts, axles; water hardening.

ALAIS AES7

Compagnie des Mines
C 0.4-0.5, Si 1.6-2.1, Mn 0.4-0.8, bal Fe.
Annealed: 99,500 TS; 60,000 YS; 20 El. Hardened: 210,000 TS; 160,000 YS; 5 El. For tools, dies; shock resistant.

ALAIS AES8

Compagnie des Mines
C 0.52-0.58, Si 1.5-1.9, Mn 0.5-0.9, bal Fe.
Annealed: 120,000 TS; 64,000 YS; 15 El. Hardened: 200,000 TS; 160,000 YS; 6 El. For tools, springs, dies; shock resistant.

ALAIS HC3

Compagnie des Mines
C 0.1, bal Fe.
Rolled: 47,000-72,000 TS; 34,000-43,000 YS; 30-34 El. For case hardened parts; carburized.

ALAIS HC4

Compagnie des Mines
C 0.12, bal Fe.
Rolled: 57,000-85,000 TS; 28,000-47,000 YS; 22-30 El. For case hardened parts; carburized.

ALAIS HCC1

Compagnie des Mines
C 0.09-0.15, Cr 0.6-1, Mn 0.6-0.9, bal Fe.
Annealed: 74,000 TS; 50,000 YS; 22 El. Hardened: 93,000 TS; 78,000 YS; 20 El. For gears, shafts; for carburized parts.

ALAIS HCCD1

Compagnie des Mines
C 0.08-0.16, Cr 0.8-1.2, Mo 0.15-0.3, Mn 0.6-0.9, bal Fe.
Annealed: 78,000 TS; 50,000 YS; 25 El. Hardened: 170,000 TS; 114,000 YS; 8 El. For gears, shafts; case hardening steel.

ALAIS HCN2

Compagnie des Mines
C 0.07-0.13, Ni 1.3-2.3, bal Fe.
Annealed: 60,000 TS; 45,500 YS; 30 El. Hardened: 99,500 TS; 71,200 YS; 26 El. For gears, shafts; case hardening steel.

ALAIS HCN5

Compagnie des Mines
C 0.07-0.13, Ni, bal Fe.
Annealed: 72,000 TS; 57,000 YS; 23 El. Hardened: 160,000 TS; 128,000 YS; 10 El. For gears, shafts, case hardening steel.

ALAIS HCNC1 1/2

Compagnie des Mines
C 0.11-0.18, Cr 0.8-1.2, Ni 1.2-1.6, bal Fe.
Hardened: 170,000-200,000 TS; 140,000 YS; 6 El. For gears, shafts; case hardening steel.

ALAIS HCNCD1

Compagnie des Mines
C 0.07-0.13, Cr 0.2, Ni 0.8-1.2, Mo 0.1-0.2, Mn 0.4, bal Fe.
Annealed: 78,000 TS. Hardened: 155,000 TS; 99,500 YS; 10 El. For gears, shafts; case hardening steel.

ALAKRON

William Mills & Co. Ltd.
Si 5, bal Al.
For decorative light alloy castings; corrosion resistant.
Obsolete

ALAKRON

William Mills & Co. Ltd.
Aluminum.
Aluminum alloy AlSi5.

ALAMO

Birmetals Ltd.
C 0.3, Mn 0.6, Al 1-1.5, Mo 0.6-1, bal Fe.
Heat treated: 160,000 TS; 135,000 YS; 18 El; 60 RA; 340 Brin. For gears, pinions, shafts, cams; nitriding steel. *Obsolete*

ALAMO

Alamo Iron Works
C 0.3, Mn 0.6, Al 1-1.5, Mo 0.6-1, bal Fe.
Heat treated: 160,000 TS; 135,000 YS; 18 El; 60 RA; 340 Brin. For gears, pinions, shafts, cams; nitriding steel. *Obsolete*

ALAR 00.12

Birmingham Aluminium Casting Co.
C 0.4, Mg 0.15, Si 10-13, Fe 0.7, Mn 0.5, Ni 0.1, Zn 0.2, Ti 0.2, bal Al.
Sand cast: 23,400 TS; 8000 YS; 3.5 El; 55 Brin. Permanent mold: 27,000 TS; 9000 YS; 5 El; 60 Brin. For general purpose light castings; corrosion resistant, good castability.

ALAR 00.12

Alar, Ltd.
C 0.4, Mg 0.15, Si 10-13, Fe 0.7, Mn 0.5, Ni 0.1, Zn 0.2, Ti 0.2, bal Al.
Sand cast: 23,400 TS; 8000 YS; 3.5 El; 55 Brin. Permanent mold: 27,000 TS; 9000 YS; 5 El; 60 Brin. For general purpose light castings; corrosion resistant, good castability.

ALAR 00.5

Birmingham Aluminium Casting Co.
Cu 0.1, Mg 0.1, Si 4.5-6, Fe 0.6, Mn 0.5, Ni 0.1, bal Al.
Sand cast: 17,000 TS; 8,000 YS; 3 El; 40 Brin. Permanent mold: 20,200 TS; 9000 YS; 4 El; 50 Brin. For general purpose light castings; good castability, corrosion resistant.

ALAR 00.5

Alar, Ltd.
Cu 0.1, Mg 0.1, Si 4.5-6, Fe 0.6, Mn 0.5, Ni 0.1, bal Al.
Sand cast: 17,000 TS; 8,000 YS; 3 El; 40 Brin. Permanent mold: 20,200 TS; 9000 YS; 4 El; 50 Brin. For general purpose light castings; good castability, corrosion resistant.

ALAR 21

Alar, Ltd.
Cu 2-4, Si 4-6, Fe 0.8, Mn 0.4, Zn 0.3, Ti 0.2, bal Al.
Sand cast: 20,200 TS; 89,000 YS; 2 El; 65 Brin. Permanent mold: 22,400 TS; 11,200 YS; 2 El; 70 Brin. For crankcases, gear cases, covers, and housings. Age-hardenable; good castability.

ALARGAN

English manufacture
Ag-Al.

ALASKA WHITE BRASS

American manufacture
Zn, Ni, bal Cu.
For bearings, bushings; heavy duty.

ALBA

Vacuumschmelze GmbH
Pd 30, Ag 60, Au 5.
For fountain pen points, dental industry; corrosion resistant, heat treatable. *Obsolete*

ALBACAST

J.F. Jelenko & Co.
Precious metal. Pd 25, Ag 70.
Quenched: 63,000 psi TS; 38,000 psi YS; 10 El; 130 Brin. Hardened: 68,000 psi TS; 47,000 psi YS; 8 El; 140 Brin. Type III hard dental alloy. For inlays, crowns, and fixed bridgework.

ALBALOY

Hanson-Van Winkle-Munning Co.
Cu 55, Sn 30, Zn 15.
For bright alloy plating; corrosion resistant.

ALBANOID

Henry Wiggin & Co. Ltd.
Ni 20, Cu 62, Zn 18.
For architectural and sanitary cast fittings. *Obsolete*

ALBANY

Allegheny Ludlum Steel
C 0.75, Cr 0.9, V 0.2, Mn 0.6, bal Fe.
For shear blades, dies, punches, knives, collets, cutlery dies, machine tool parts; great toughness and resistance to impact. *Obsolete*

ALBATRA METAL

American manufacture
Cu 57.5, Zn 22.5, Ni 18.75, Pb 1.25.
For automobile trimmings, hardware, fittings, (Albata).

ALBERTSON SPECIAL

Sioux Tools Inc.
C 3.2-3.4, Mo 0.35-0.4, Cr 0.3-0.4, Si 1.75-2, Mn 0.6-0.7, Ni 1.25-1.5, bal Fe.
For valve seat inserts; alloy cast iron. *Obsolete*

ALBIDUR-ALUMINUM

English manufacture
Al alloy.
For light alloy parts; non-hardenable.

ALBIN

English manufacture
Zn, Ni, bal Cu.
For fixtures, hardware, fittings; white brass.

ALBION HIGH SPEED

Hobson, Houghton & Co.
C 0.4, Mn 0.05, Cr 2.3, W 15.54, bal Fe.
For tools, high speed cutters; high speed steel.

ALBION SPECIAL

Hobson, Houghton & Co.
C 0.7, W 18, Cr 4, V 1, bal Fe.
For high speed cutting tools; high speed steel.

ALBION TOOL

Hobson, Houghton & Co.
C 1.27, Mn 0.33, bal Fe.
For tools, drills, reamers; water hardening.

ALBONDUR

Vereinigte Leichtmetallwerke G.m.b.H.
Aluminum. Mn 0-1.5, Mg 0-1, Si 0-1.2, bal Al.
For tanks, furniture, marine parts; good formability and weldability.

ALBONDUR

German manufacture
Cu 3.5-5.5, Si 0.3-0.5, Mn 0.3-1, Mg 0.2-0.7, bal Al.
Heat treated: 74,000-80,000 TS; 57,000-63,000 YS; 8-12 El; 130-150 Brin. For general purpose applications; age hardenable, high strength.

ALBOR

Jessop-Saville Ltd.
C 0.9, Cr 1.2, bal Fe.
For dies for cold stamping white gold and all hard metals; oil hardening.

ALBOR DIE

Ackerlind Steel Co., Inc.
C 0.85, Mn 0.3, Cr 0.7, Mo 0.2, bal Fe.
For cutting tools, dies; oil hardening. *Obsolete*

ALBOR DIE

Jessop-Saville Ltd.
C 0.85, Cr 0.7, Mo 0.2, bal Fe.
For coining and medal dies, cutlery dies; deep hardening.

ALBORIUM

J.F. Jelenko & Co.
Precious metal. Au 15, Pd 25, Ag 45.
Quenched: 83,500 psi TS; 63,000 psi YS; 10 El; 165 Brin. Hardened: 101,000 psi TS; 85,000 psi YS; 6 El; 245 Brin. Type IV extra-hard dental alloy. For hard inlays, thin crowns, fixed bridgework, and partial dentures.

ALBORO H.F.

J.F. Jelenko & Co.
Dental solder; approximate range 1565-1580°F.

ALBORO L.F.

J.F. Jelenko & Co.
Dental solder; approximate range 1310-1370°F.

ALBRO

Bronze Die Casting Co.
Al 10, Fe 1, bal Cu.
For acid resisting castings. Corrosion resistant.

ALCALOY

Alcaloy Inc.
Copper. Cu 87-89, Al 9-11, Fe 1-2, Be 0.1-0.5.
Cast: 85,000 TS; 41,000 YS; 20 El; 83-150 Brin. For safety tools, chains, chain fittings, and hand tools. Non-sparking; corrosion resistant.

ALCALOY

English manufacture
Al 9-11, bal Cu.
Cast: 85,000 TS; 41,000 YS; 20 El; 50-80 Rock B. For safety tools, chains, fittings, hand tools; corrosion resistant, Al-bronze.

ALCAN 100

Aluminum Co. of Canada, Ltd.
Al 99.5, 0.50 others max.
Casting: 8500 TS; 24 Brin. For unstressed castings in food and chemical industries and electrical equipment.

ALCAN 1050A

Alcan-Booth Industries, Ltd.
99.5 Al min.
Same as ALCAN GB-1S. Similar to AA 1050.

ALCAN 1080A

Alcan-Booth Industries, Ltd.
99.8 Al min.
Similar to AA 1080.

ALCAN 117

Aluminum Co. of Canada, Ltd.
Cu 3, Mg 0.1, Mn 0.5, Si 4.5, bal Al.
Cast: 25,000 TS; 16,000 YS; 3-5 El; 50 Brin. Cast: 29,000 TS; 17,000 YS; 4 El; 65 Brin. For sand and permanent mold castings; for medium stresses.

ALCAN 123

Aluminum Co. of Canada, Ltd.
Si 5, Ti 0-0.2, bal Al.
Cast: 19,000 TS; 9000 YS; 6 El; 40 Brin. Cast: 24,000 TS; 9000 YS; 10 El; 40 Brin. For sand and permanent mold castings; for thin sections.

ALCAN 125

Aluminum Co. of Canada, Ltd.
Cu 2-4, Si 4-6, Fe 0-0.8, Mn 0.3-0.7, bal Al. Cast: 23,000 TS; 15,000 YS; 2 El; 85 Brin. Cast and aged 40,000 TS; 115 Brin. General purpose castings for light weight and low cost; gear boxes, clutch housings, valve bodies, radiators; non-solution heat treatable.

ALCAN 160

Aluminum Co. of Canada, Ltd.
Cu 0.1, Mg 0-0.1, Si 10-13, Fe 0-0.6, Mn 0-0.5, Ni 0.1, Zn 0.1, Ti 0.2, bal Al.
Cast: 26,000 TS; 10,500 YS; 6 El; 60 Brin. Not heat treatable; for large castings, general purpose, marine and electrical, valve bodies, radiators, grids and grills.

ALCAN 160N

Aluminum Co. of Canada, Ltd.
Si 12, bal Al.
Sand cast: 25,000 TS; 12,000 YS; 8 El; 50 Brin. For castings; high fluidity.

ALCAN 160X

Aluminum Co. of Canada, Ltd.
Si 12, bal Al.
Die cast: 37,000 TS; 18,000 YS; 2 El. For die castings, instrument cases; for thin wall sections.

ALCAN 162

Aluminum Co. of Canada, Ltd.
Cu 1, Mg 1, Si 12, bal Al.
Permanent mold: 36,000 TS; 28,000 YS; 0.5 El; 105 Brin. For permanent mold castings, pistons; low coefficient expansion.

ALCAN 16S

Aluminum Co. of Canada, Ltd.
Cu 2.5, Fe 0-0.7, Mg 0.3, Si 0-0.7, bal Al.
S-T temper: 43,000 TS. For aircraft and structural rivets; heat treatable.

ALCAN 17S

Aluminum Co. of Canada, Ltd.
Cu 3.8-4.5, Mg 0.5-0.8, Mn 0.5-0.7, Fe 0-0.5, bal Al.
Forgings, TB condition: 67,000 TS; 31,000 YS; 14 El; 31,000 shearing strength; 120 Brin. For stressed parts in aircraft and other structures; heat treatable.

ALCAN 18S

Aluminum Co. of Canada, Ltd.
Cu 4, Mg 0.6, Si 0-0.9, bal Al.
S-T temper: 55,000 TS; 40,000 YS; 10 El; 100 Brin. For forgings for light temperature service; heat treatable.

ALCAN 1S

Aluminum Co. of Canada, Ltd.
Al 99.5, Cu 0-0.02, Si 0-0.15, Fe 0-0.15, Mn 0-0.03, Zn 0-0.06, 0.2 others max.
Wrought, O temper: 7500-13,500 TS; 29-35 El; 18 Brin. H4 temper: 13,500-17,000 TS; 5-8 El; 33 Brin. H8 temper: 18,000 TS; 3-5 El; 44 Brin. For food and chemical plant equipment, food containers, atomic energy, compression accessories for electrical conductors. Available as sheet and forgings. Not heat treatable.

ALCAN 2017A

Alcan-Booth Industries, Ltd.
Si 0.5, Fe 0.7, Cu 4, Mn 0.7, Mg 0.6, bal Al.
Wrought, heat treatable, high strength. Similar to AA 2017.

ALCAN 225

Aluminum Co. of Canada, Ltd.
Cu 4.5, Si 0-1.5, Ti 0-0.2, bal Al.
T-22 temper: 39,000 TS; 28,000 YS; 5 El; 80 Brin. For sand castings; stressed parts, heat treatable.

ALCAN 236

Aluminum Co. of Canada, Ltd.
Cu 0.65-0.8, Fe 0.95-1.4, Si 1-3, Ti 0-0.2, bal Al.
Cast: 22,000 TS; 14,000 YS; 2 El; 65 Brin. For sand castings; good machinability.

ALCAN 24S
Aluminium Co. of Canada, Ltd.
Cu 3.8-4.9, Mg 1.2-1.8, Si 0-0.5, Fe 0-0.5, Mn 0.3-0.9, Zn 0-0.2, bal Al.
Sheet and plate, heat treatable. Sheet: O: 31,000 max TS. TB: 61,000 TS; 40,000 YS; 120 Brin. For all types of stressed components in aircraft and general engineering.

ALCAN 24S
Alcan-Booth Industries, Ltd.
Cu 3.8-4.9, Mg 1.2-1.8, Si 0-0.5, Fe 0-0.5, Mn 0.3-0.9, Zn 0-0.2, bal Al.
Sheet and plate, heat treatable. Sheet: O: 31,000 max TS. TB: 61,000 TS; 40,000 YS; 120 Brin. For all types of stressed components in aircraft and general engineering.

ALCAN 24S ALCLAD
Aluminium Co. of Canada, Ltd.
Cu 3.8-4.9, Mg 1.2-1.8, Si 0-0.5, Fe 0-0.5, Mn 0.3-0.9, Zn 0-0.2, bal Al.
Sheet and plate, heat treatable. Sheet: O: 31,000 max TS. TB: 61,000 TS; 40,000 YS; 15 El. For all types of stressed components in aircraft and general engineering.

ALCAN 24S ALCLAD
Alcan-Booth Industries, Ltd.
Cu 3.8-4.9, Mg 1.2-1.8, Si 0-0.5, Fe 0-0.5, Mn 0.3-0.9, Zn 0-0.2, bal Al.
Sheet and plate, heat treatable. Sheet: O: 31,000 max TS. TB: 61,000 TS; 40,000 YS; 15 El. For all types of stressed components in aircraft and general engineering.

ALCAN 250
Aluminum Co. of Canada, Ltd.
Cu 10, Fe 0-1.5, Mg 0.2, Si 0-0.6, Ti 0-0.2, bal Al.
A62 temper: 25,000 TS; 20,000 YS; 1 El; 75 Brin. T25 temper: 36,000 TS; 30,000 YS; 1 El; 100 Brin. For engine parts; heat treatable.

ALCAN 2618A
Alcan-Booth Industries, Ltd.
Cu 2.5, Mg 1.5, Ni 1.2, Si 0.2, Fe 1, bal Al.
Same as ALCAN GB-D19S. Similar to AA 2618.

ALCAN 26S
Aluminum Co. of Canada, Ltd.
Cu 3.9-5, Mg 0.2-0.8, Si 0.5-1, Fe 0-0.7, Mn 0.4-1.2, Cr 0-0.1, bal Al.
Plate, extrusions and forgings, heat treatable. Plate, TB: 54,700 TS; 34,800 YS; 10-14 El. Heat treated to TF: 62,500 TS; 54,000 YS; 6-9 El. For stressed components of all types in aircraft and general engineering.

ALCAN 28S
Aluminum Co. of Canada, Ltd.
Cu 5.6, Pb 0.2-0.6, Bi 0.2-0.6, Fe 0-0.7, Si 0-0.4, Zn 0-0.3, bal Al.
Extrusions: 41,000-44,000 TS; 27,000-35,000 YS; 6 El; 25,000-26,000 shearing strength (as heat treated). For repetition machined parts; free machining screw machined parts; heat treatable.

ALCAN 2S
Aluminum Co. of Canada, Ltd.
Al 99, Cu 0-0.1, Si 0-0.5, Fe 0-0.7, Mn 0-0.1, Zn 0-0.1.
Sheet, plate, extrusions, tube; not heat treatable. O temper: 10,000-15,000 TS; 20-30 El; 21 Brin. H4 temper: 15,000-20,000 TS; 3-5 El; 36 Brin. H6 temper: 17,800-21,300 TS; 2-4 El; 42 Brin. H8 temper: 20,000 TS; 2-4 El; 46 Brin. For paneling and moldings, lightly stressed and decorative assemblies, food and brewing equipment.

ALCAN 3103
Alcan-Booth Industries, Ltd.
Mn 1.2, Fe 0.5, bal Al.
Same as ALCAN GB-3S.

ALCAN 340
Aluminum Co. of Canada, Ltd.
Mg 8, bal Al.
Die cast: 42,000 TS; 23,000 YS; 7 El. For die castings; corrosion resistant.

ALCAN 350
Aluminum Co. of Canada, Ltd.
Mg 10, bal Al.
Cast, T4 temper: 47,000 TS; 26,000 YS; 17 El; 75 Brin. For sand castings; heat treatable, corrosion resistant.

ALCAN 3S
Aluminum Co. of Canada, Ltd.
Mn 0.8-1.5, Cu 0-0.1, Mg 0-0.1, Fe 0-0.7, Zn 0-0.2, bal Al.
Sheet, O temper: 12,800-18,500 TS; 20-24 El; 27 Brin. H4 temper: 20,000-25,000 TS; 3-6 El; 46 Brin. H6 temper: 22,800-27,700 TS; 2-4 El; 50 Brin. H8 temper: 25,000 TS; 2-4 El; 57 Brin. For building sheet, vehicle paneling and sheet metal work, packaging, hollowware, not heat treatable.

ALCAN 5083
Alcan-Booth Industries, Ltd.
Mg 4.3, Mn 0.6, Cr 0.13, bal Al.
Same as ALCAN BG-D54S.

ALCAN 50S
Aluminum Co. of Canada, Ltd.
Cu 0-0.1, Mg 0.4-0.9, Si 0.3-0.7, Fe 0-0.4, Zn 0-0.2, bal Al.
Extrusions, tube, forgings, heat treatable. Extrusions, heat treated to TF condition: 26,000 TS; 22,700 YS; 7-8 El; 75 Brin. Architectural members as window frames, window-screen sections, road transport.

ALCAN 5251
Alcan-Booth Industries, Ltd.
Mg 2.1, Mn 0.3, bal Al.
Same as ALCAN GB-M57S.

ALCAN 55S
Aluminum Co. of Canada, Ltd.
Mg 1.2, Cr 0.3, bal Al.
For rivets for aircraft; 200,000 shearing strength.

ALCAN 56S
Aluminum Co. of Canada, Ltd.
Mg 5, Mn 0.1, Cr 0.1, bal Al.
S-O temper: 42,000 TS; 20,000 YS; 35 El. S-H temper: 58,000 TS; 48,000 YS; 15 El. For rivets, wire products; resists sea water corrosion.

ALCAN 6082
Alcan-Booth Industries, Ltd.
Mg 0.8, Si 1, Mg 0.7, bal Al.
Same as ALCAN GB-B51S.

ALCAN 61S
Aluminum Co. of Canada, Ltd.
Cu 0-0.3, Mg 0.6, Si 1, Cr 0.3, bal Al.
S-T temper: 44,000 TS; 34,000 YS; 14 El; 90 Brin. For automotive and aircraft forgings; heat treatable.

ALCAN 65S
Aluminum Co. of Canada, Ltd.
Cu 0.3, Mg 1, Si 0.6, Cr 0-0.3, bal Al.
S-O temper: 18,000 TS; 8000 YS; 22 El; 30 Brin. S-W temper: 35,000 TS; 21,000 YS; 22 El; 65 Brin. S-T temper: 45,000 TS; 40,000 YS; 12 El; 95 Brin. For structures; heat treatable, corrosion resistant.

ALCAN 75S
Aluminum Co. of Canada, Ltd.
Cu 1.5, Mg 2.5, Zn 5.6, Cr 0.3, Ti 0-0.2, bal Al.
S-O temper: 33,000 TS; 15,000 YS; 17 El. S-T temper: 82,000 TS; 72,000 YS; 11 El; 150 Brin. For aircraft, structures; heat treatable.

ALCAN 75S ALCLAD
Aluminum Co. of Canada, Ltd.
Cu 1.5, Mg 2.5, Zn 5.6, Cr 0.3, Ti 0-0.2, bal Al.
S-O temper: 32,000 TS; 14,000 YS; 17 El. S-T temper: 76,000 TS; 67,000 YS; 11 El. For aircraft, structures; heat treatable, corrosion resistant.

ALCAN 99.8
Aluminum Co. of Canada, Ltd.
Al 99.8, 0.2 others max.
Sheet, O temper: 13,000 TS max; 16 Brin. H4 temper: 13,500-17,000 TS; 28 Brin. H8 temper: 18,000 TS; 40 Brin. For fully supported roofing and flashing, collapsible tubes, atomic energy equipment; annealed material is very ductile, malleable and durable. Not heat treatable.

ALCAN A 320
Aluminum Co. of Canada, Ltd.
Mg 4, Si 0.5, Ti 0-0.2, bal Al.
Cast: 25,000 TS; 12,000 YS; 9 El; 50 Brin. For architectural usage, ornamental and marine parts, sand casting; corrosion resistant.

ALCAN A-143
Aluminum Co. of Canada, Ltd.
Si 8.5-10.5, Cu 2-4, Mg 0.5-1.5, Ni 0.5-1.5, Fe 0-1, Mn 0-0.5, Zn 0-0.5, Ti 0-0.2, bal Al.
Cast, heat treated: 30,000 TS; 23,000 YS; 105 Brin. Low-thermal expansion, good strength and frictional properties; mainly for automotive pistons.

ALCAN A111
Aluminum Co. of Canada, Ltd.
Cu 1.2, Fe 1, Ni 1, Si 2.5, bal Al.

ALCAN A56S
Aluminum Co. of Canada, Ltd.
Mg 5, Mn 0.3, bal Al.
F-temper: 38,500 TS; 23,000 YS; 28 El; 70 Brin. For shipbuilding, structural members, extrusions; good formability.

ALCAN B26S
Aluminum Co. of Canada, Ltd.
Cu 3.9-5, Mg 0.2-0.8, Si 0.5-1, Fe 0-0.7, Mn 0.4-1.2, Zn 0-0.2, bal Al.
Sheet, O temper: 25,500 TS; 18-20 El. Heat treated to TF condition: 61,000 TS; 53,000 YS; 6-7 El; 145 Brin. For stressed components of all types in aircraft and general engineering. Heat treatable.

ALCAN B26S ALCLAD
Aluminium Co. of Canada, Ltd.
Cu 3.9-5, Mg 0.2-0.8, Si 0.5-1, Fe 0-0.7, Mn 0.4-1.2, Cr 0-0.1, Zn 0-0.2, bal Al.
Sheet, 0: 24,000 TS; 18-20 El (est). HtTr to TF Cond: 57,000 TS; 46,000 YS; 7-8 El. For parts requiring good strength plus good corrosion resistance; heat treatable.

ALCAN B26S ALCLAD
Alcan-Booth Industries, Ltd.
Cu 3.9-5, Mg 0.2-0.8, Si 0.5-1, Fe 0-0.7, Mn 0.4-1.2, Cr 0-0.1, Zn 0-0.2, bal Al.
Sheet, 0: 24,000 TS; 18-20 El (est). HtTr to TF Cond: 57,000 TS; 46,000 YS; 7-8 El. For parts requiring good strength plus good corrosion resistance; heat treatable.

ALCAN B51S
Aluminum Co. of Canada, Ltd.
Cu 0.1, Mg 0.5-1.2, Si 0.7-1.3, Fe 0-0.5, Mn 0.4-1, Zn 0-0.2, Cr 0-0.25, bal Al.
Sheet, plate, extrusions, tube, forgings. Plate, heat treated to TF condition: 42,000 TS; 34,000 YS; 8 El; 85 Brin. Medium strength, good properties, heat treatable. For structural applications as bridges, cranes, transport equipment, roof trusses.

ALCAN B53S
Aluminum Co. of Canada, Ltd.
Cu 0.1, Mg 3.1-3.9, Si 0-0.5, Fe 0-0.5, Mn 0-0.5, Zn 0-0.2, Cr 0-0.25, bal Al.
Sheet, plate, extrusions, tube, forgings. Plate, rolled: 30,000 TS approx; 75 Brin. Not heat treatable, used in as-rolled or as-forged for road and rail transport, pressure vessels; good weldability and corrosion resistance.

ALCAN B54S
Aluminum Co. of Canada, Ltd.
Mg 4.4, Mn 0.3, bal Al.
F-temper: 43,000 TS; 31,000 YS; 16-20 El; 70 Brin. For shipbuilding, structural members; good formability.

ALCAN C1S
Aluminium Co. of Canada, Ltd.
Al 99.5, Cu 0-0.04, Si 0-0.1, Fe 0-0.4.
Extrusions: Not heat treatable. M Temper: 8500 TS; 25 El. H2 Temper: 12,000 TS; 15 El. For electrical conductors; bus bar, overhead lines, windings, insulated cable.

ALCAN C1S
Alcan-Booth Industries, Ltd.
Al 99.5, Cu 0-0.04, Si 0-0.1, Fe 0-0.4.
Extrusions: Not heat treatable. M Temper: 8500 TS; 25 El. H2 Temper: 12,000 TS; 15 El. For electrical conductors; bus bar, overhead lines, windings, insulated cable.

ALCAN C50S
Aluminum Co. of Canada, Ltd.
Mg 0.45-0.85, Si 0.2-0.6, Cu 0.05-0.2, Fe 0-0.15, bal Al.
Extrusions, solution treated, TB condition: 17,800 TS; 16 El. Heat treated to TF Cond: 26,000 TS; 22,700 YS; 10 El. Heat treatable. Good formability, takes good finish. For motor car trim and other applications requiring bright finish even with anodizing.

ALCAN C75S
Aluminum Co. of Canada, Ltd.
Zn 5.2-6.2, Mg 2.2-3.2, Cu 0.3-0.7, Mn 0.3-0.7, Si 0-0.5, Fe 0-0.5, Ti 0-0.3, Ni 0-0.1, bal Al.
Forging, TF: 70,000 TS; 64,000 YS; 6 El; 160 Brin. Heat treatable; for aircraft and other structures where strength/weight ratio is of prime importance.

ALCAN D135
Aluminum Co. of Canada, Ltd.
Cu 0-0.1, Mg 0.2-0.45, Si 6.5-7.5, Fe 0-0.5, Ni 0-0.1, Mn 0-0.3, bal Al.
Cast: 22,700 TS; 12,000 YS; 3 El; 60 Brin. Heat treated to TF: 40,000 TS, 35,000 YS; 2 El; 95 Brin. Castings for engine components, cylinder blocks, chemical and food equipment; heat treatable.

ALCAN D19S
Aluminum Co. of Canada, Ltd.
Cu 2.5, Mg 1.5, Ni 1.3, Fe 1, bal Al.
Plate and forgings, plate, heat treated to TF condition: 61,000 TS; 54,000 YS; 5 El. Heat treatable. Aircraft structural parts operating up to 200°C; good strength at 200°C.

ALCAN D50S
Aluminum Co. of Canada, Ltd.
Cu 0.04, Mg 0.4-0.9, Si 0.3-0.7, Fe 0.5, bal Al.
Extrusions, heat treated to TF condition: 26,000 TS; 22,700 YS; 7-8 El. Best combination of electrical conductor and mechanical properties; 55% conductivity. Heat treatable. For bus bars, electrical conductors and fittings.

ALCAN D54S
Aluminum Co. of Canada, Ltd.
Mg 4-4.9, Si 0-0.4, Fe 0-0.4, Mn 0.5-1, Zn 0-0.2, Cr 0-0.25, bal Al.
Sheet, plate, extrusions, forgings. Rolled plate: 40,000 TS approx; 10-12 El; 65-70 Brin. Not heat treatable, used in road transport equipment and in shipbuilding; weldable, good resistance to marine atmospheres.

ALCAN GB 162 (OR A162)
Alcan-Booth Industries, Ltd.
Al-Si-11.5Ni-2Cu-Mg.
Aluminum alloy casting. Cast, HT to TF: 42,000 TS; 38,000 YS; 130 Brin. For pistons, engine components; low thermal expansion and good high temperature strength. *Obsolete*

ALCAN GB-1S
Alcan-Booth Industries, Ltd.
Al 99.5.
Higher purity aluminum. For deep drawn containers or spun components.

ALCAN GB-24S
Alcan-Booth Industries, Ltd.
Cu 4.2, Mg 1.5, Mn 0.6, bal Al.
TB, solution treated: 440 MN/m^2 TS; 305 MN/m^2 YS; 12 El. High strength alloy with good elongation, used mainly for aircraft structures.

ALCAN GB-2S
Alcan-Booth Industries, Ltd.
Al 99.
Commercially pure aluminum. For paneling, cooking utensils.

ALCAN GB-3S
Alcan-Booth Industries, Ltd.
Mn 1.2, Fe 0.5, bal Al.
O temper: 90 MN/m^2 TS; 20-25 El. For cooking utensils, containers, roofing sheet.

ALCAN GB-50S
Alcan-Booth Industries, Ltd.
Mg 0.6, Si 0.4, bal Al.
Extruded, heat treated: 150-185 MN/m^2 TS; 130-160 MN/m^2 YS. For window frames, architectural sections.

ALCAN GB-51S
Alcan-Booth Industries, Ltd.
Mg 0.8, Si 1, bal Al.
Extruded, TF, solution treated and aged: 300 MN/m^2 TS; 240 MN/m^2 YS; 7-8 El. Medium strength, good corrosion resistance, good surface finish.

ALCAN GB-54S
Alcan-Booth Industries, Ltd.
Mg 3.5, Mn 0.5, bal Al.
H2: 245 MN/m^2 TS; 165 MN/m^2 YS; 5-8 El. Suitable for marine and low temperature applications.

ALCAN GB-65S
Alcan-Booth Industries, Ltd.
Mg 1, Si 0.6, Cu 0.25, Cr 0.25, bal Al.
H4: 160 MN/m^2 TS; 185 MN/m^2 YS; 5 El. HF (solution treated and aged): 295 MN/m^2 TS; 225 MN/m^2 YS; 7-9 El. Values listed are for drawn tube; use is mainly for tubular furniture.

ALCAN GB-B110
Alcan-Booth Industries, Ltd.
Cast aluminum alloy.
Heat treat (TF): 47,000 TS; 36,000 YS; 110 Brin. For decorative castings and those requiring both strength and surface suitable for polishing and anodizing. *Obsolete*

ALCAN GB-B26S
Alcan-Booth Industries, Ltd.
Cu 4.6, Mg 0.7, Si 0.7, Mn 0.7, bal Al.
TF, solution treated and aged: 280-310 MN/m^2 TS; 240-270 MN/m^2 YS; 5-8 El. Strongest general utility alloy; for aircraft and structural applications.

ALCAN GB-B51S
Alcan-Booth Industries, Ltd.
Mg 0.8, Si 1, Mg 0.7, bal Al.
TF, solution treated and aged: 295 MN/m^2 TS; 240 MN/m^2 YS; 8 El. Good strength for structural purposes, as bridges, vehicle structures, cranes.

ALCAN GB-B53S
Alcan-Booth Industries, Ltd.
Mg 2.7, Mn 0.75, Cr 0.12, bal Al.
H4: 280 MN/m^2 TS; 205 MN/m^2 YS; 6 El. Good corrosion resistance; for chemical and transport industries.

ALCAN GB-B57S
Alcan-Booth Industries, Ltd.
Mg 0.8, Fe 0.6, bal Al.
H4: 160 MN/m^2 TS; 140 MN/m^2 YS; 8 El. Medium strength for general metalworking.

ALCAN GB-B79S
Alcan-Booth Industries, Ltd.
Mg 3.2, Cu 0.6, Zn 4.2, bal Al.
Very high strength alloy. Similar to AA 7079.

ALCAN GB-C125
Alcan-Booth Industries, Ltd.
Al-Si-5Cu-1.5Mg.
Aluminum alloy casting. Heat treated, TF: 47,000 TS; 37,000 YS; 115 Brin. For cylinder heads, valve bodies, water-jackets, hose couplings; can be cast into intricate shapes; high strength at moderate cost. *Obsolete*

ALCAN GB-D135
Alcan-Booth Industries, Ltd.
Al-Si7 Mg.
Aluminum alloy casting. As cast (M): 23,000 TS; 12,000 YS; 60 Brin. Heat treated, TF: 40,000 TS; 36,000 YS; 115 Brin. For engine cylinder blocks, chemical and food equipment, many uses.

ALCAN GB-D19S
Alcan-Booth Industries, Ltd.
Cu 2.5, Mg 1.5, Ni 1.2, Si 0.2, Fe 1, bal Al.
TF, solution treated and aged: 402-417 MN/m^2 TS; 309-324 MN/m^2 YS; 6-9 El. High strength alloy for use up to 300°C.

ALCAN GB-D50S
Alcan-Booth Industries, Ltd.
Mg 0.5, Si 0.5, bal Al.
For bus bars, electrical conductors. Similar to AA 6101.

ALCAN GB-D54S
Alcan-Booth Industries, Ltd.
Mg 4.3, Mn 0.6, Cr 0.13, bal Al.
H4: 345 MN/m^2 TS; 270 MN/m^2 YS; 4-8 El. Strongest of common non-heat treatable alloys with good corrosion resistance. For ship building, low temperatures. Weldable.

ALCAN GB-D75S
Alcan-Booth Industries, Ltd.
Cu 0.6, Mg 2.9, Zn 5.8, Mn 0.4, bal Al.
TF, solution treated and aged: 494-540 MN/m^2 TS; 417-463 MN/m^2 YS; 5-7 El. Very high strength alloy; for aircraft and structural applications.

ALCAN GB-E74S
Alcan-Booth Industries, Ltd.
Mg 2.6, Zn 3.9, bal Al.
TF, solution treated and aged: 390 MN/m^2 TS; 320 MN/m^2 YS; 10 El. Weldable, medium strength alloy.

ALCAN GB-L57S
Alcan-Booth Industries, Ltd.
Mg 0.8, bal Al.
For motor car trim and hollow ware. Similar to AA 5657.

ALCAN GB-M57S
Alcan-Booth Industries, Ltd.
Mg 2.1, Mn 0.3, bal Al.
H6: 225 MN/m^2 TS; 175 MN/m^2 YS; 3-5 El. Medium strength alloy; weldable; for paneling on marine and road vehicles, and containers.

ALCAN L57S

Aluminum Co. of Canada, Ltd.
Al 99.8, Mg 0.8.
Sheet, O temper: 22,000 TS max. H4 temper: 22,000-29,000 TS. H6: temper: 26,000-33,000 TS. H8 temper: 30,000 TS. Good formability; takes good finish. For motor car trim and other applications requiring a bright finish; hollowware; not heat treatable.

ALCAN M57S

Aluminum Co. of Canada, Ltd.
Cu 0.1, Mg 1.7-2.4, Si 0-0.5, Fe 0-0.5, Mn 0-0.5, Zn 0-0.2, Cr 0-0.25, bal Al.
Sheet and plate. Plate, O temper: 22,700-28,500 TS; 8500 YS; 18-20 El. H4 temper: 32,000-39,000 TS; 25,000 YS; 3-5 El. Not heat treatable, good weldability and corrosion resistance, used in panelling and structures for boats, road transport, and aircraft.

ALCAN M75S

Aluminum Co. of Canada, Ltd.
Zn 5.3-6.5, Mg 2.2-3.2, Cu 0.3-1.5, Mn 0.3, Ni 0.1, Si 0.5, Fe 0-0.5, Ti 0-0.3, bal Al.
Sheet and plate, heat treatable. Sheet, O temper: 27,000 TS; 18-20 El; 15,600 shear strength. Heat treated to TF: 70,000 TS; 61,000 YS; 8 El; 40,000 shear strength. High strength; for aircraft and other structures where strength/weight ratio is of prime importance.

ALCAN M75S ALCLAD

Aluminum Co. of Canada, Ltd.
Zn 5.3-6.5, Mg 2.2-3.2, Cu 0.3-1.5, Mn 0.3, Ni 0.1, Si 0.5, Fe 0-0.5, Ti 0-0.3, bal Al.
Sheet and plate, heat treatable, coated with corrosion resistant material.

ALCET

Remington Arms Co. Inc.
High temperature refractory material, sintered. For aluminum vapor deposition crucibles and other refractory applications.

ALCHROME 3

Wilbur B. Driver Co.
Stainless steel. Cr 14, Al 3, bal Fe.
For resistors, heating elements; operating below 1500°F.

ALCHROME 6

Wilbur B. Driver Co.
Cr 19-21, Al 5.25-6.5, bal Fe.
Annealed: 100,000 TS; 18 El. For electrical resistors and heating elements; resists sulfur atmospheres. *Obsolete*

ALCHROME 750

Wilbur B. Driver Co.
Stainless steel. Cr 15, Al 4.5, bal Fe.
For resistors, low cost heating elements. Maximum operating temperature 1900°F.

ALCHROME ALLOY 750

Carpenter Technology Corp.
Cr 15, Al 4, bal Fe.
621 MPa TS; 483 MPa YS; 20 El. For controls and heating appliances operating at temperatures up to 1038°C.

ALCHROME ALLOY D

Carpenter Technology Corp.
Cr 15.25, Al 5.25, bal Fe.
621 MPa TS; 483 MPa YS; 35 El. For open elements in high temperature heating applications up to 1260°C including toasters, hair dryers and other small appliances. ASTM B-603 Class IIA.

ALCHROME ALLOY DK

Carpenter Technology Corp.
Cr 21, Al 4.5, bal Fe.
634 MPa TS; 441 MPa YS; 15 El. For open elements in high temperature heating applications up to 1260°C including toasters, hair dryers and other small appliances. ASTM B-603 Class IIB, B-287.

ALCHROME DK

Wilbur B. Driver Co.
Stainless steel. Cr 20, Al 4.5, Co 0.5, bal Fe.
For resistors, heating elements. Maximum operating temperature 2350°F.

ALCHROME-D

Wilbur B. Driver Co.
Stainless steel. Fe 79.5, Cr 15, Al 5.5.
Wire, specific resistivity: 137 micro-ohm/cm; temperature coefficient: +/-20 x 10^{-6} ohm/ohm/·C. For resistance elements; high specific electrical resistance.

ALCLAD 5056

Alcoa/Massena Operations
Aluminum.
Non-heat-treatable wire. High resistance to corrosion; slightly less strength than bare 5056. For insect screen cloth; braided cable armor wire, and chain link fence.

ALCO

Dirigold Corp.
Zn, bal Cu.
For metalware.

ALCO

Cytemp Specialty Steel Div.
C 0.5, Si 1, Cr 1.7, W 2.2, V 0.2, Mo 0.5, bal Fe.
For tools, chisels, punches, pneumatic drills; for hot and cold work. *Obsolete*

ALCO BRONZE

Skandinaviska Armaturfabriken AG
Sn 10, Ni 5, bal Cu.
For ships forgings, propellers, skates, surgical instruments. Tough, hard, corrosion resistant.

ALCO M AND ALCO S

Now CYCLOPS SL.

ALCOA 050

Alcoa of Great Britain
99.5 Al min.
Annealed (O): 13,700 TS max; 22-32 El. Maximum cold work (H8): 19,500 TS min; 3-4 El. Good corrosion resistance and ductile. For formed parts for chemical plant, dairy, brewery and bakery equipment. BS 1470, 1471, 1472, 1474.

ALCOA 051

Alcoa of Great Britain
99.5 Al min, 0.05 Cu + Si + Fe max.
Extruded: 9500 TS; 4500 YS; 25 El. H2 condition: 12,200 TS min; 15 El. Electrical conductivity: 60% IACS min. For electric purposes, bus bars. BS 2898: E1E.

ALCOA 101

Aluminum Company of America
Al 99.45.
Sand cast-F temper: 14,000 TS; 3,000 YS; 20 El; 25 Brin. Permanent mold-F temper: 17,000 TS; 4,000 YS; 22 El; 35 Brin. For electrical conductors, fittings; high electrical conductivity, corrosion resistant. *Obsolete*

ALCOA 102

Alcoa of Great Britain
99.0 Al min.
Annealed: 13,000 TS; 5,000 YS; 45 El. Good corrosion resistance. For equipment for chemical, petroleum, pharmaceutical, brewery and food industry; paneling and fully supported roof cladding. BS 1470, 1471, 1474.

ALCOA 105

Aluminum Company of America
Cu 3-4.5, Fe 0-1, Si 0-2.5, Mn 0-0.5, Zn 0-0.5, Ni 0-0.3, Ti 0-0.2, bal Al.
T 4-temper: 28,000 TS; 18,000 YS; 4.5 El; 60 Brin. T 6-temper: 35,000 TS; 25,000 YS; 3.5 El; 75 Brin. T 75-temper: 28,000 TS; 16,000 YS; 3.0 El; 60 Brin. For castings (sand); heat treatable. *Obsolete*

ALCOA 108

Now AA 208.0.

ALCOA 109

Aluminum Company of America
Cu 12, Al 88.
Sand cast: 20,000-28,000 TS; 18,000 YS; 0.5-1.5 El; 1.0 RA; 75 Brin. For pump parts, automobile engine manifolds; castings (sand). *Obsolete*

ALCOA 1100

Alcoa/Massena Operations
Aluminum.
12,000-15,000 psi TS. Non-heat-treatable wire. For trim, rivets, cold threaded parts, welding rod, and light forgings.

ALCOA 113

Aluminum Company of America
Si 7, Cu 2, Fe 0-1.2, Ti 0-0.2, Zn 0-0.2, bal Al.
Cast-Permanent mold: 28,000 TS; 19,000 YS; 20 El; 70 Brin. Cast-Sand cast: 24,000 TS; 15,000 YS; 1.5 El; 70 Brin. For light weight castings; general purposes. *Obsolete*

ALCOA 12

Aluminum Company of America
Cu 7-8, Fe 0-0.4, Si 0-0.5, bal Al.
Sand cast: 24,000 TS; 15,000 YS; 1.5 El; 70 Brin. For household articles, aircraft and automotive industry, ship building, ornamental work; casting alloy. *Obsolete*

ALCOA 122

Now AA 222.

ALCOA 1260

Aluminum Company of America
99.60% Al min.
O-temper: 10,000 TS; 4,000 YS; 43 El; 19 Brin. H18-temper: 10,000 TS; 12,000 YS; 6 El; 35 Brin. For chemical equipment; corrosion resistant. AA 1260. *Obsolete*

ALCOA 13

Now AA 413.0.

ALCOA 132

Aluminum Company of America
Si 14, Mg 1, Ni 2.5, bal Al.
T-4-temper: 33,000 TS; 1 El; 105 Brin. For aircraft engine pistons; casting alloy. *Obsolete*

ALCOA 1350

Alcoa/Massena Operations
Aluminum.
Non heat-treatable wire. Excellent forming qualities; high resistance to corrosion; anodizes well. For refrigerator trays, Christmas tree wire, and packaging closure wire.

ALCOA 138

Now AA 238.0.

ALCOA 142

Now AA 242.0.

ALCOA 144

Aluminum Company of America
Cu 10, Si 4, Mg 0.2, bal Al.
Cast: 32,000 TS; 24,000 YS; 0.5 El; 85-110 Brin. For permanent mold castings. *Obsolete*

ALCOA 145

Aluminum Company of America
Zn 10, Cu 2.5, Fe 1.2, bal Al.
Sand cast: 25,000-37,000 TS; 12,000 YS; 6.3 El; 65 Brin. For aircraft and auto engine parts; shock resistant. *Obsolete*

ALCOA 15 S

Aluminum Company of America
Zn 12.5, Cu 1.5, bal Al.
For screw machine products; wrought alloy. *Obsolete*

ALCOA 152
Aluminum Company of America
Cu 7, Si 5.5, Mn 0.3, Fe 0-1.4, bal Al.
T-74 Temper: 35,000 TS; 26,000 YS; 0.5 El; 100 Brin. T-524
Temper: 29,000 TS; 16,000 YS; 1.0 El; 95 Brin. For
permanent mold castings, pistons; heat treatable. *Obsolete*

ALCOA 170
Aluminum Company of America
Cu 4-6, Fe 0-1, Si 0.3-0.8, Sn 4-6, bal Al.
Sand cast-F temper: 23,000 TS; 12,000 YS; 6 El; 55 Brin. For
castings requiring bearing qualities; heat treatable. *Obsolete*

ALCOA 172
Aluminum Company of America
Cu 8, Si 2.5, bal Al.
Sand cast: 23,000 TS; 15,000 YS; 1 El; 65 Brin. For match
plates and patterns. *Obsolete*

ALCOA 190
Alcoa of Great Britain
Si 0-0.6, Fe 0-0.7, Mn 0.8-1.5, Cu 0-0.1, Mg 0-0.1, Zn 0-0.2,
bal Al.
Condition O: 13,000-19,000 TS; 20-25 El. Condition H8:
25,400 TS min; 2-4 El. For general purpose non-heat-
treatable parts and assemblies. Good corrosion resistance.
BS 1470: NS3.

ALCOA 195
Now AA 295.0.

ALCOA 196
Aluminum Company of America
Cu 4, Mg 0.2, bal Al.
Sand cast: 33,000 TS; 27,000 YS; 0.2 El; 110 Brin. T-Temp:
45,000 TS; 40,000 YS; 2.5 El; 115 Brin. For special fittings for
high strength; casting alloy; low shock resistance. *Obsolete*

ALCOA 2011
Alcoa/Massena Operations
Aluminum.
Heat-treatable wire. For screw machine products, machine
parts, pipe stems, brake pistons, ball-point pens, radio and
television components, and carburetor parts.

ALCOA 2014
Alcoa/Massena Operations
Aluminum.
Heat-treatable wire. For heavy duty forgings, airplane fittings,
and structural members.

ALCOA 2017
Alcoa/Massena Operations
Aluminum.
Heat-treatable wire. For screw machine products, fittings,
wire, and brake pistons.

ALCOA 2024
Alcoa/Massena Operations
Aluminum.
Heat-treatable wire. For screw machine products, aircraft
parts, ski poles, fastening devices, hose couplings, fittings
and bolts.

ALCOA 2117
Alcoa/Massena Operations
Aluminum.
Heat-treatable wire. Highest strength rivet alloy. For rivets.

ALCOA 213
Aluminum Company of America
Cu 0.2-0.6, Si 0-0.6, Mn 0-0.3, Mg 3.5-4.5, Ni 0-0.2, bal Al.
Sand cast-F temper: 25,000 TS; 12,000 YS; 9 El; 50 Brin. For
cooking utensils, sand castings; corrosion resistant.
Obsolete

ALCOA 214
Now AA 514.0.

ALCOA 216
Aluminum Company of America
Mg 6, bal Al.
Sand cast: 27,000 TS; 16,000 YS; 6 El; 60 Brin. For corrosion
resistant light alloy castings. *Obsolete*

ALCOA 218
Now AA 518.0.

ALCOA 220
Now AA 520.0.

ALCOA 231
Aluminum Company of America
Si 0.8, Mg 1.5, Cr 0.25, bal Al.
For light alloy parts. *Obsolete*

ALCOA 27 S
Aluminum Company of America
Cu 4.5, Si 0.8, Mn 0.8, Sn 0.05, bal Al.
T6-temper: 60,000 TS; 50,000 YS; 11 El; 115 Brin. For
wrought light alloy parts, bridges, trucks; S.S. 37,000; End.
Lim. 13,000. *Obsolete*

ALCOA 2EC
Now AA 6101.

ALCOA 3003
Alcoa/Massena Operations
Aluminum.
Non-heat-treatable wire. Similar to Alcoa 1100 with the
following exceptions: higher strength and lower electrical and
thermal conductivity. Same as Alcoa 1100 where higher
properties are required.

ALCOA 3004
Aluminum Company of America
Mn 1.25, Mg 1, bal Al.
O-Temper: 26,000 TS; 10,000 YS; 20 El; 45 Brin. H34-Temper:
35,000 TS; 29,000 YS; 9 El; 63 Brin. H38-Temper: 41,000 TS;
36,000 YS; 5 El; 77 Brin. For fuel lines, fan blades, sheeting
for roofs, light sockets, underground cable sheeting; high
resistance to weather corrosion. *Obsolete*

ALCOA 304
Aluminum Company of America
Cu 2.5-4.5, Fe 0-1, Si 2.5-4.5, Mn 0-0.5, Ti 0-0.2, bal Al.
Sand cast-F temper: 23,000 TS; 15,000 YS; 2.5 El; 60 Brin. T
533-temper: 26,000 TS; 21,000 YS; 2.0 El; 65 Brin. For sand
castings; age-hardenable. *Obsolete*

ALCOA 314
Aluminum Company of America
Si 13.5, Cu 4, bal Al.
Cast: 31,000 TS; 1 El; 85 Brin. For permanent mold castings.
Obsolete

ALCOA 315
Aluminum Company of America
Cu 0-0.2, Fe 0-0.6, Si 1-2, bal Al.
For pressure castings; pressure-tight. *Obsolete*

ALCOA 319
Now AA 319.0.

ALCOA 333
Now AA 333.0.

ALCOA 348
Aluminum Company of America
Cu 1-2.5, Fe 0-1, Si 4.5-7, Ti 0-0.2, Mn 0-0.6, bal Al.
Sand cast-T 6-temper: 35,000 TS; 25,000 YS; 2.5 El; 80 Brin.
Permanent mold-T 6-temper: 43,000 TS; 27,000 YS; 3.0 El; 90
Brin. For castings, engine castings. *Obsolete*

ALCOA 355
Now AA 355.0.

ALCOA 356
Now AA 356.0.

ALCOA 364
Now AA 364.0.

ALCOA 380
Now AA 380.0.

ALCOA 384
Now AA 384.0.

ALCOA 4043
Alcoa/Massena Operations
Aluminum.
12,000-30,000 psi TS. Non-heat-treatable wire. For welding
rod.

ALCOA 406
Aluminum Company of America
Mn 2, bal Al.
Sand cast: 20,000 TS; 9,000 YS; 12 El; 35 Brin. For cast
fittings; resists corrosion of S gases. *Obsolete*

ALCOA 43
Now AA 443.0.

ALCOA 45
Aluminum Company of America
Si 10, Mn 0-0.5, Cu 0-0.6, Fe 0-0.6, bal Al.
Sand cast: 22,000 TS; 10,000 YS; 4 El; 50 Brin. For light alloy
parts and castings requiring very fluid alloy; sand and
permanent mold casting. *Obsolete*

ALCOA 47
Aluminum Company of America
Cu 0.13, Si 12.5, Mn 0.09, Na 0.05, Fe 0.4, bal Al.
Sand cast: 24,000-31,000 TS; 9,000 YS; 5-15 El; 55 Brin. For
light alloy parts and leak-proof castings; castings. *Obsolete*

ALCOA 5 S
Aluminum Company of America
Al 95, Cu 2, Zn 2, Mn 1.5.
For light alloy parts; wrought alloy. *Obsolete*

ALCOA 505
Aluminum Company of America
Cu 0.3-0.7, Fe 0-0.8, Si 0.3-0.7, Ni 4-5, Ti 0-0.2, bal Al.
Die cast: 23,000 TS; 14,000 YS; 6 El. For stove burners, die
castings. *Obsolete*

ALCOA 505
Alcoa of Great Britain
Cu 0-0.2, Mg 0.5-1.2, Si 0-0.4, Fe 0-0.7, Mn 0-0.5, Zn 0-0.2,
bal Al.
Condition O: 14,000-22,000 TS; 18-22 El. Condition H8:
27,000 TS min; 24,000 min YS; 1-3 El. Medium strength alloy,
particularly suitable for anodizing. BS 4 300/7: NS41 (Al-
Mg-1).

ALCOA 5052
Alcoa/Massena Operations
Aluminum.
Non-heat-treatable wire. Excellent resistance to corrosion;
workable. For transportation and marine applications; home
appliances.

ALCOA 5056
Alcoa/Massena Operations
Aluminum.
Non-heat-treatable wire. For rivets, nails, braided cable armor,
fasteners, wire products, and zippers.

ALCOA 51
Aluminum Company of America
Cu 1.9, Mg 0.6, Si 1, Al 96.5.
O-Temp: 14,000-18,000 TS; 4,000-6,000 YS; 15-30 El; 25-32
Brin. W-Temp: 30,000-40,000 TS; 15,000-20,000 YS; 20-30 El;
55-70 Brin. T-Temp: 45,000-50,000 TS; 30,000-40,000 YS;
10-18 El; 85-100 Brin. For airplane propellers, pistons, crank
cases, connecting rods; strain hardened when cold worked.
Obsolete

ALCOA 510
Alcoa of Great Britain
Cu 0-0.1, Mg 1.7-2.4, Si 0-0.5, Fe 0-0.5, Zn 0-0.2, 0.5 Mn + Cr max, bal Al.
Condition O: 31,000-40,000 TS; 19,000 YS; 12-18 El.
Condition H6: 33,000-40,000 TS; 25,000 min YS; 3-5 El.
Medium strength alloy for architectural and marine applications; paneling containers, domestic appliances, welded assemblies. BS 1470: NS4; BS 1472: NF4, etc.

ALCOA 520
Alcoa of Great Britain
Cu 0-0.1, Mg 3.1-3.9, Si 0-0.5, Fe 0-0.5, Zn 0-0.2, 0.5 Mn + Cr max, bal Al.
Condition O: 31,000-40,000 TS; 12,000 YS min; 12-18 El.
Condition H4: 40,000-48,000 TS; 33,000 YS min; 4-6 El.
Strong non-heat treatable alloy for general fabrication and for welded structures. BS 1470: NS5; BS 1472: NF5, etc.

ALCOA 5356
Alcoa/Massena Operations
Aluminum.
42,000 psi TS. Non-heat-treatable wire. For architectural and structural welding applications.

ALCOA 540
Alcoa of Great Britain
Cu 0-0.15, Mg 0.7-1.2, Si 0-0.1, Fe 0-0.1, Mn 0-0.3, bal Al.
Condition O: 22,500 TS max. Condition H8: 31,000 TS min.
Alloy developed specifically for cars and other bright trim.

ALCOA 550
Alcoa of Great Britain
Cu 0-0.15, Mg 0.3-0.8, Si 0-0.1, Fe 0-0.1, Mn 0-0.2, bal Fe.
Condition O: 18,000 TS max. Condition H4: 18,000-27,000 TS. Condition H8: 27,000 TS min. Aluminum alloy more ductile than Alcoa 540; for cars and other bright trim.

ALCOA 5556
Alcoa/Massena Operations
Aluminum.
46,000 psi TS. Non-heat-treatable wire. For structural and marine applications.

ALCOA 6
Aluminum Company of America
Zn 12.5-14.5, Cu 2.5-3, Fe 0-0.8, Si 0-0.7, bal Al.
Sand cast: 24,600 TS; 14,000 YS; 3 El. For light alloy parts. *Obsolete*

ALCOA 6053
Alcoa/Massena Operations
Aluminum.
Heat-treatable wire. Moderate shear strength; good corrosion resistance. For rivets.

ALCOA 6061
Alcoa/Massena Operations
Aluminum.
Heat-treatable wire. For truck bodies, furniture, nails, structures, valves, automotive transmission pistons, nuts, chain link fence, and armor rods.

ALCOA 6262
Alcoa/Massena Operations
Aluminum.
Heat-treatable wire. For screw machine products, fittings, couplings, camera parts, and nuts.

ALCOA 645
Aluminum Company of America
Cu 2.5, Fe 1.5, Zn 11, bal Al.
Sand cast: 29,000 TS; 22,000 YS; 4 El; 70 Brin. For light alloy parts, sand castings. *Obsolete*

ALCOA 70 S
Aluminum Company of America
Cu 1, Mn 0.7, Mg 0.4, Zn 10, bal Al.
T-6 Temper: 58,000 TS; 45,000 YS; 18 El; 100 Brin. For light alloy parts; non-hardenable; forgings. *Obsolete*

ALCOA 7075
Alcoa/Massena Operations
Aluminum.
Heat-treatable wire. Very high strength and hardness. For aircraft.

ALCOA 750
Now AA 850.0.

ALCOA 79
Aluminum Company of America
Cu 4, Si 7, bal Al.
Die cast: 36,000 TS; 20,000 YS; 2.5 El. For die castings. *Obsolete*

ALCOA 8
Aluminum Company of America
Al 95, Cu 4, Si 0.5, Fe 0.5.
For light alloy parts, heat treated castings. *Obsolete*

ALCOA 8 S
Aluminum Company of America
Al 67, Zn 33.
For strong, tough light alloy; wrought alloy. *Obsolete*

ALCOA 800
Alcoa of Great Britain
Cu 5-6, Si 0-0.4, Fe 0-0.7, Pb 0.2-0.7, Bi 0.2-0.7, bal Al.
Condition TD: 53,000 TS; 45,000 YS; 11 El. Condition TF: 56,000 TS; 43,000 YS; 13 El. Free machining, heat-treatable, aluminum alloy for high speed automatic work. Nearest US Spec AA 2011.

ALCOA 81
Aluminum Company of America
S 3, Cu 7, bal Al.
Cast: 33,000 TS; 25,000 YS; 1 El; 80 Brin. For castings requiring high fluidity; die casting alloy. *Obsolete*

ALCOA 82
Aluminum Company of America
Cu 14, Si 5, bal Al.
Die cast: 44,000 TS; 0.2 El; 100 Brin. For die castings. *Obsolete*

ALCOA 83
Aluminum Company of America
Si 3, Cu 2, bal Al.
Die cast: 25,000-30,000 TS; 14,000 YS; 3-6 El; 55-65 Brin. For pressure die castings for small machine and ornamental parts; die castings. *Obsolete*

ALCOA 85
Aluminum Company of America
Si 5.5, Cu 4, bal Al.
Die cast: 40,000 TS; 24,000 YS; 5 El; 60 Brin. For die castings; high strength. *Obsolete*

ALCOA 85
Chicago Steel Foundry
C 0.5, Ni 1.5, Cr 0.8, Mo 0.2, bal Fe.
140,000 TS; 120,000 YS; 14 El; 10 RA; 450 Brin. For tractors. *Obsolete*

ALCOA 910
Alcoa of Great Britain
Cu 0-0.1, Mg 0.4-0.9, Si 0.3-0.7, Fe 0-0.4, bal Al.
Condition O: 14,500 TS; 30-32 El. Condition TF: 37,000 TS; 32,000 YS; 13-14 El. General purpose, weldable, heat-treatable alloy for glazing bars, window frames. BS 1471: HT9, etc.; nearest US Spec AA 6063.

ALCOA 920
Alcoa of Great Britain
Cu 0-0.1, Mg 0.5-1.2, Si 0.7-1.3, Fe 0-0.5, Mn 0.4-1, Cr 0-0.25, bal Al.
Condition O: 18,000 TS; 9400 YS; 26-28 El. Condition TF: 49,000 TS; 43,000 YS; 10-12 El. Heat treatable, forgeable, weldable alloy for structural sections subject to stresses and shock loading such as lorry body components. BS 1470: HS30, etc.

ALCOA 93
Aluminum Company of America
Si 2, Ni 4, Cu 4, bal Al.
Die cast: 33,000 TS; 28,000 YS; 1-0.5 El; 80 Brin. For alloy parts and castings for small machines; die castings. *Obsolete*

ALCOA 945
Alcoa of Great Britain
Cu 0.15-0.4, Mg 0.8-1.2, Si 0.4-0.8, Fe 0-0.5, Mn 0-0.1, Cr 0-0.1, bal Al.
Condition TB: 31,000 TS; 19,000 YS; 19 El. Condition TF: 45,000 TS; 40,000 YS; 12 El. For structural road transport sections; forgeable, weldable, heat-treatable.

ALCOA 946
Alcoa of Great Britain
Cu 0.15-0.4, Mg 0.8-1.2, Si 0.4-0.8, Fe 0-0.5, Pb 0.4-0.7, Bi 0.4-0.7, bal Al.
Free machining version of Alcoa 945.

ALCOA A 142
Now AA242.0.

ALCOA A 213
Aluminum Company of America
Cu 0.2-0.6, Si 0-0.6, Mn 0-0.3, Mg 3.5-4.5, Zn 1.4-2.2, Ni 0-0.2, bal Al.
Permanent mold-F temper: 27,000 TS; 17,000 YS; 7 El; 60 Brin. For cooking utensils, permanent mold castings; corrosion resistant. *Obsolete*

ALCOA A 254
Aluminum Company of America
Cu 0-0.6, Fe 0-0.8, Si 0-0.7, Mn 1-1.8, Mg 5.5-6.5, Ni 1.2-1.8, bal Al.
For induction motor rotor castings; 18% conductivity. *Obsolete*

ALCOA A 360
Now AA A360.0.

ALCOA A 612
Now AA A712.0.

ALCOA A-108
Now AA A208.0.

ALCOA A-132
Now AA A332.0.

ALCOA A13
Now AA A413.0.

ALCOA A334
Aluminum Company of America
Cu 3, Si 4, Mg 0.3, bal Al.
Sand cast: 26,000 TS; 20,000 YS; 1.8 El; 70 Brin. For sand cast light alloy castings. *Obsolete*

ALCOA A356
Now AA A356.0.

ALCOA A380
Now AA A380.0.

ALCOA A750
Now AA A850.0.

ALCOA AM-C88S
Aluminum Company of America
Al 0.5-10.5, Zn 0.5-2, 0.1% Mn min, bal Mg.
For welding rod; extruded. *Obsolete*

ALCOA AM262
Aluminum Company of America
Al 8.3-9.7, Zn 1.7-2.3, 0.1% Mn min, bal Mg.
Cast: 23,000 TS; 14,000 YS; 1.5 El; 65 Brin. T4-Temper:
39,000 TS; 14,000 YS; 10 El; 63 Brin. T6-Temper: 39,000 TS;
22,000 YS; 2 El; 81 Brin. For castings; heat treatable.
Obsolete

ALCOA AM266
Aluminum Company of America
Al 5.3-6.7, Zn 2.5-3.5, 0.15% Mn min, bal Mg.
Cast: 26,000 TS; 11,000 YS; 3 El; 51 Brin. T4-Temper: 38,000
TS; 14,000 YS; 10 El; 53 Brin. T6-Temper: 38,000 TS; 19,000
YS; 4 El; 71 Brin. For castings; heat treatable. *Obsolete*

ALCOA AMX269
Aluminum Company of America
Al 7-10, Zn 0-2.75, 0.1% Mn min, bal Mg.
Cast: 23,000 TS; 14,000 YS; 1 El; 65 Brin. T4-Temper: 38,000
TS; 14,000 YS; 8 El; 61 Brin. T6-Temper: 38,000 TS; 20,000
YS; 1 El; 85 Brin. For castings; heat treatable. *Obsolete*

ALCOA B 105
Aluminum Company of America
Cu 4-5, Fe 0-1, Si 2-3, Mn 0-0.4, Ti 0-0.2, Zn 0-1, Ni 0-0.2, bal
Al.
T 4-temper: 37,000 TS; 15,000 YS; 7.0 El; 75 Brin. T 6-temper:
39,000 TS; 27,000 YS; 3.5 El; 90 Brin. T 74-temper: 37,000
TS; 19,000 YS; 4.5 El; 80 Brin. For permanent mold castings.
Obsolete

ALCOA B 213
Aluminum Company of America
Cu 0.2-0.6, Si 1.4-2.2, Mn 0.3-0.4, Mg 3.5-4.5, bal Al.
Sand cast-F temper: 20,000 TS; 15,000 YS; 2 El; 50 Brin. For
sand castings; corrosion resistant. *Obsolete*

ALCOA B-113
Aluminum Company of America
Cu 7, Fe 1.2, Si 1.7, bal Al.
Cast permanent mold: 28,000 TS; 19,000 YS; 2 El; 70-90 Brin.
For permanent mold castings; general purpose. *Obsolete*

ALCOA B-17S
Aluminum Company of America
Cu 3.5, Fe 0.5, Si 0.6, Mn 0.56, Mg 0.55, bal Al.
O-Temp: 20,000-25,000 TS; 20-28 El; 30-40 Brin. T-Temp:
42,000-50,000 TS; 20,000-25,000 YS; 20-28 El; 65-85 Brin.
For airplane structures, floats, dirigibles; age-hardening.
Obsolete

ALCOA B-195
Now AA B295.0.

ALCOA B214
Now AA B214.0.

ALCOA B50S
Aluminum Company of America
Cu 0-0.25, Fe 0-0.8, Si 0-0.5, Mg 1-1.8, Cr 0-0.1, bal Al.
O-temper: 21,000 TS; 8,000 YS; 24 El; 35 Brin. H 32-temper:
25,000 TS; 22,000 YS; 9 El; 42 Brin. H 38-temper: 31,000 TS;
28,000 YS; 6 El; 52 Brin. For sheet and plate, light alloy parts;
good corrosion resistance and formability. *Obsolete*

ALCOA B750
Now AA B850.0.

ALCOA C 214
Aluminum Company of America
Cu 0-0.1, Fe 0-0.3, Mn 0.4-0.5, Mg 3.2-4, Zn 0-0.1, Si 0-0.3,
bal Al.
For architectural requirements; corrosion resistant. *Obsolete*

ALCOA C 612
Now AA C712.0.

ALCOA C-113
Now AA 213.0.

ALCOA C-17S
Aluminum Company of America
Cu 4.07, Si 0.6, Mn 0.56, Mg 0.55, Fe 0.5, bal Al.
O-Temp: 25,000-35,000 TS; 7,000-10,000 YS; 12-20 El; 42-55
Brin. T-Temp: 63,000-90,000 TS; 50,000-55,000 YS; 8-14 El;
95-125 Brin. For airplane structures, floats, dirigibles; age-
hardening. *Obsolete*

ALCOA C355
Now AA C355.0.

ALCOA D 132
Aluminum Company of America
Cu 2-4, Fe 0-1, Si 8.5-10.5, Mg 0.5-1.5, Mn 0-0.5, Ni 0.5-1.2,
Ti 0-0.2, bal Al.
Permanent mold-T 5-temper: 36,000 TS; 28,000 YS; 1 El; 100
Brin. For pistons, cylinder heads; low coefficient of thermal
expansion; replaced by F 132. *Obsolete*

ALCOA D-195
Aluminum Company of America
Cu 5.5, Si 0.7, bal Al.
T4 temper: 42,000 TS; 28,000 YS; 6 El; 60-80 Brin. For
permanent mold castings. *Obsolete*

ALCOA EC
Now AA 1350.

ALCOA F 214
Now AA F214.0.

ALCOA F132
Now AA F332.0.

ALCOA NO. 360
Now AA 360.0.

ALCOA NO. 716
Aluminum Company of America
Si 10, Cu 4, bal Al.
For brazing alloy; wire.

ALCOA XA75S
Aluminum Company of America
Cu, Si, bal Al.
For aircraft parts; age-hardenable. *Obsolete*

ALCOA XA80S
Aluminum Company of America
Sn 6, bal Al.
For bearings; wrought. *Obsolete*

ALCODIE
Columbia Tool Steel Co.
Tool material. C 0.35, Cr 5, V 0.4, W 1.5, Mo 1.5, bal Fe.
Oil or air hardening tool steel for die casting dies, hot
blanking and forging dies, shear blades; AISI H12.

ALCOLOY
Olin Corp.
Zn 22.7, Al 3.4, Co 0.4, bal Cu.
Annealed: 78,000-82,000 TS; 47,000-55,000 YS; 36-39 El.
Spring temper: 120,000 TS; 108,000 YS; 1.5 El. For springs,
diaphragms, connectors. Readily deep drawn, good
corrosion and stress-corrosion resistance. *Obsolete*

ALCOMAX I
Permanent Magnet Association
Al, Ni, Co, bal Fe.
For magnets; permanent. *Obsolete*

ALCOMAX III
English manufacture
Ni 13.5, Co 25.5, Co 3, Al 7.5, Cb 0.9, Si 0.17, bal Fe.
For magnets for motors, loud speakers; permanent magnet.

ALCOMAX IV
English manufacture
Al 7.4, Ni 13.3, Co 24.5, Cu 3, Cb 2.5, bal Fe.
For magnets for cycle dynamos; permanent magnet.

ALCOMAX-II
English manufacture
Al 8.1, Ni 11.9, Co 21, Cu 4, bal Fe.
For magnetic and electrical equipment. Permanent magnet,
high coercive force.

ALCON
Wieland Werke AG
Aluminum.
Aluminum alloy AlMgSiO5.

ALCONIT 16
German manufacture
Ni, Al, bal Fe.
For permanent magnets.

ALCONIT-0
German manufacture
Ni, Al, bal Fe.
For permanent magnets.

ALCONIT-10
German manufacture
Ni, Al, bal Fe.
For permanent magnets.

ALCONITE
Fairmount Foundry Inc.
C, alloy, bal Fe.
For castings.

ALCOP BRONZE
Janney Cylinder Co.
Cu 62, Al 2.5, Zn 35.5, Pb 0-15.
Cast: 73,000 TS; 37,000 YS; 35 El; 150 Brin. For centrifugal
castings, bushings, liners, bearing cages; heavy duty.

ALCOP NO. 3
Pressco Casting & Mfg. Corp.
Cu 61.5-63.5, Zn 35-38, Pb 0.75-1.25, Al 0.9-1.3.
Cast: 63,000 TS; 35,000 YS; 30 El; 102 Brin. For hardware,
die castings; leaded bronze, free-cutting.

ALCRES
American manufacture
Cr 12, Al 5, Fe 83.
For electrical resistances; heat resisting.

ALCRESS
English manufacture
Fe 70, Cr 20, Al 5.
For electrical resistors; heat resistant.

ALCROSIL 5
Timken Co.
C 0-0.15, P 0-0.5, Si 0.5-1, Cr 4-6, Mo 0.45-0.65, Al 0.3-0.7,
bal Fe.
Annealed: 60,000 TS; 25,000 YS; 30 El; 163 Brin. For high
temperature applications; corrosion, heat resistant. *Obsolete*

ALCROSIL NO. 3
Timken Co.
C 0.12, Si 1, Al 0.5, Cr 2.8, Ni 0.2, Mo 0.5, bal Fe.
For oil refinery and chemical plant equipment; resists hot
petroleum products. *Obsolete*

ALCUFONT 10
Soc. Alluminio Veneto per Azioni
Aluminum. Cu 9.3-10.7, Fe 0.5-1.3, Mg 0.2-0.4, bal Al.
Heat treated: 40,000-51,200 TS; 34,000-37,000 YS; 0.2-0.5 El;
125-150 Brin. For pistons; high strength.

ALCUFONT 12
United States Bronze Powders Inc.
Cu 11-12.5, Ti 0.2, Fe 1, bal Al.
Heat treated: 38,400-42,700 TS; 22,800-25,600 YS; 1.0-1.5 El;
110-125 Brin. For pistons; high strength.

ALCUFONT 4.5

Soc. Alluminio Veneto per Azioni
Aluminum. Cu 4.5, Fe 0-0.8, Si 0-1, Ti 0-0.2, bal Al.
Heat treated: 36,000-43,000 TS; 28,000-33,000 YS; 2-4 El;
75-95 Brin. For light alloy parts; age hardenable.

ALCUFONT 8

Soc. Alluminio Veneto per Azioni
Aluminum. Cu 7-8.5, Ti 0-0.2, Si 0-0.5, Fe 0-0.6, bal Al.
Heat treated: 20,000-26,000 TS; 11,000-13,500 YS; 2-5 El;
55-70 Brin. For pistons, light alloy parts; age hardened.

ALCULOY

NL Industries
Cu 7, Si 3, Fe 0-2.25, Zn 0-1.5, bal Al.
Die cast: 29,000-32,000 TS; 12,000-17,000 YS; 1-2 El; 75-80
Brin. For housings, instrument cases, engineering parts;
good castability. *Obsolete*

ALCULOY 12

NL Industries
Cu 7, Si 3, bal Al.
Die cast: 30,000 TS; 15,000 YS; 1 El; 80 Brin. For general die
castings; SAE 312. *Obsolete*

ALCULOY 75

NL Industries
Cu 7.5, Si 5.5, bal Al.
Die cast: 35,000 TS; 20,000 YS; 8 El; 85 Brin. For general die
castings. *Obsolete*

ALCUMAN

German manufacture
Cu 4.8-6, Si 0-0.2, Mn 0.5, Mg 1.5, bal Al.
Heat treated: 60,000-71,000 TS; 50,000-57,000 YS; 5-10 El;
120-140 Brin. For screw machine products, construction and
transportation equipment; age-hardenable, high strength.

ALCUMITE

Duriron Co. Inc.
Cu 88.92, Al 7.65, Fe 2.02, Ni 1.41.
Sand cast: 75,000 TS; 25,000 YS; 30 El; 35 RA; 116 Brin. Hot
rolled: 100,000 TS; 52,000 YS; 10 El; 18 RA; 187 Brin.
Annealed: 90,000 TS; 47,000 YS; 25 El; 38 RA; 132 Brin. For
blowers, pumps, ventilator tanks to resist weak acids and
other corrosives. *Obsolete*

ALCUNIC

Century Brass Products Inc.
Cu 80, Ni 1, Al 2, Zn 17.
Hard: 103,000-112,000 TS; 94,000-106,000 YS; 3.0-3.5 El;
200-220 Brin. For condenser tubing, belt buckles; corrosion
resistant. *Obsolete*

ALCUNIC "G"

Century Brass Products Inc.
Cu 70, Ni 1, Al 2, bal Zn.
Soft: 59,000 TS; 20,000 YS; 56 El. Hard: 100,000 TS; 95,000
YS; 3 El. For condenser tubes, belt buckles; corrosion
resistant. *Obsolete*

ALCUPLATE

Texas Instruments Inc./Materials Control
Cu clad to 1100 aluminum alloy. Soft: 17,000 TS; 7800 YS;
33 El. Hard: 30,000 TS; 25,000 YS; 5 El. For electrical
contacts; Al alloys plated by lamination with copper,
80.0Al-20.0Cu.

ALDA

British Oxygen Co., Ltd.
C 0.1-0.2, bal Fe.
For welding rods; copper coated.

ALDAL

Societe du Duralumin
Cu 3.5-4.5, Mn 0.4-1, Mg 0.3-0.75, bal Fe.
60,000 TS; 22 El; 120 Brin. For light alloy parts, aircraft
construction; Duralumin type.

ALDECOR

Republic Steel Corp.
C 0-0.12, Cu 0.35-0.6, Mo 0.16-0.28, Cr 0.25-1.5, Ni 0.25-1.5,
bal Fe.
Hot rolled: 70,000 TS; 50,000 YS; 22 El. For transportation
equipment, bus and truck bodies; good forming and welding
characteristics. *Obsolete*

ALDIVAN

Latrobe Steel Co.
C 0.4, Cr 2.5, V 0.3, bal Fe.
For tool steel for aluminum die castings; hot work steel.
Obsolete

ALDREI

Manufacturer not listed.
Si, Mg, bal Al.
42,500-51,000 TS; 38,000-44,000 YS; 5-9 El. For light alloy
parts.

ALDREY

Aluminium Walzwerke Singen GmbH
Aluminum. Si 0.5-0.6, Mg 0.4-0.5, bal Al.
Heat treated: 43,000-50,000 TS; 38,400-44,000 YS; 5-9 El;
70-80 Brin. For general structural members, aircraft and
aerospace components. Age hardenable, corrosion resistant.

ALDREY

Aluminium Industrie Aktiengesellschaft
Si 0.2-1, Mg 0.4, Fe 0.3, bal Al.
Heat treated: 44,000-49,000 TS; 38,000-43,000 YS; 5-9 El;
90-100 Brin. For automobile engine parts, electric
transmission wires and cables; age hardenable.

ALDREY 14

Aluminium Industries, AG
Si 0.4-0.6, Mg 0.4-0.6, Fe 0-0.3, Ti 0-0.3, bal Al.
Annealed: 11,400-17,000 TS; 4300-7100 YS; 20-3 El. THA16
Temper: 42,700-50,000 TS; 38,400-44,100 YS; 4-9 El. For
marine structures and hardware, architectural applications.
Heat treatable.

ALDREY 14

Alluminio SA
Si 0.4-0.6, Mg 0.4-0.6, Fe 0-0.3, Ti 0-0.3, bal Al.
Annealed: 11,400-17,000 TS; 4300-7100 YS; 20-3 El. THA16
Temper: 42,700-50,000 TS; 38,400-44,100 YS; 4-9 El. For
marine structures and hardware, architectural applications.
Heat treatable.

ALDREY 14

Lavorazione Leghe Leggere SpA
Si 0.4-0.6, Mg 0.4-0.6, Fe 0-0.3, Ti 0-0.3, bal Al.
Annealed: 11,400-17,000 TS; 4300-7100 YS; 20-3 El. THA16
Temper: 42,700-50,000 TS; 38,400-44,100 YS; 4-9 El. For
marine structures and hardware, architectural applications.
Heat treatable.

ALDUR 35

Voest International, Inc.
C 0.09, Si 0.25, Mn 0.6, bal Fe.
For steel plate.

ALDUR 41

Voest International, Inc.
C 0.15, Si 0.25, Mn 0.7, bal Fe.
For steel plate.

ALDUR 44

Voest International, Inc.
C 0.16, Si 0.3, Mn 1, bal Fe.
For steel plate.

ALDUR 45/60

Voest International, Inc.
C 0.18, Si 0.45, Mn 1.4, bal Fe.
For steel plate.

ALDUR 47

Voest International, Inc.
C 0.17, Si 0.35, Mn 1.2, bal Fe.
For steel plate.

ALDUR 50

Voest International, Inc.
C 0.18, Si 0.4, Mn 1.3, bal Fe.
For steel plate.

ALDUR 50/65

Voest International, Inc.
C 0.2, Si 0.45, Mn 1.5, bal Fe.
For steel plate.

ALDUR 55

Voest International, Inc.
C 0.19, Si 0.45, Mn 1.5, bal Fe.
For steel plate.

ALDUR 55/68

Voest International, Inc.
C 0.2, Si 0.5, Mn 1.6, bal Fe.
For steel plate.

ALDUR 58

Voest International, Inc.
C 0.2, Si 0.5, Mn 1.6, bal Fe.
For steel plate.

ALDUR 58/72

Voest International, Inc.
C 0.2, Si 0.5, Mn 1.7, bal Fe.
For steel plate.

ALDUR, ALDUR G, ALDUR D, ALDUR DG

Manufacturer not listed.
Varying quality and use grades of numbered ALDUR steels.

ALDURAL

Alcan-Booth Industries, Ltd.
Al clad Duralumin.
For aircraft structures; clad with 99.7% Al. *Obsolete*

ALDURAL E

Alcan-Booth Industries, Ltd.
Cu 3-4.5, Si 0-1, Mg 0-1, Mn 0-1.2, Fe 0-1, Al clad Al alloy
core, bal Al.
50,000 TS; 38,000 YS; 8 El. For light metal parts; age
hardenable. *Obsolete*

ALDURAL K

Alcan-Booth Industries, Ltd.
C 0.4, Mg 2.7, Mn 0.5, Zn 5.3, bal Al.
Heat treated: 67,000-72,000 TS; 58,000-60,000 YS; 8 El. For
aircraft structures; age hardenable, high strength. *Obsolete*

ALDURAL Q

Alcan-Booth Industries, Ltd.
Cu 3.5-5, Mg 0.4-1.2, Si 0.7, Fe 0.7, Mn 0.4-1.2, Ti 0.2, bal Al.
Heat treated: 53,800 TS; 31,400 YS: 10-12 El. For structural
members; age-hardened, high strength. *Obsolete*

ALDURAL S

Alcan-Booth Industries, Ltd.
Cu 3.5-4.8, Mg 0.8, Si 0.9, Fe 1, Mn 1.2, Ti 0.3, bal Al.
W-temper: 54,800 TS; 31,400 YS: 15 El. WP-temper: 58,300
TS; 44,800 YS: 8 El. For structural members; age-hardened,
high strength. *Obsolete*

ALDURAL Z

Alcan-Booth Industries, Ltd.
Cu 3.5-4.8, Mg 0.8, Si 1.5, Fe 1, Mn 1.2, Ti 0.3, bal Al.
WP-temper: 60,500 TS; 47,000 YS: 8 El. For structural
members; age-hardened, high strength. *Obsolete*

ALDURAL-G

Alcan-Booth Industries, Ltd.
Cu 4.5, Mn 0.6, Mg 1.5, bal Al.
Annealed: 27,000 TS; 11,000 YS; 20 El; 47 Brin. Heat treated: 72,000 TS; 57,000 YS; 13 El; 130 Brin. For aircraft components, fittings, rivets, fasteners. Age hardenable. High fatigue and tensile strength. *Obsolete*

ALDURAL-L

Alcan-Booth Industries, Ltd.
Duralumin-L clad with Al-1% Zn.
Heat treated: 64,000 TS; 58,300 YS; 6 El. For aircraft skins; improved corrosion resistance. *Obsolete*

ALDURAL-LC

Alcan-Booth Industries, Ltd.
Cu 0.4, Mg 2.7, Mn 0.5, Zn 5.3, bal Al.
Age hardened: 70,000 TS; 60,000 YS; 10 El. For high strength application, manifolds, cylinders. DTD 687A spec. Age-hardenable. *Obsolete*

ALDURBRA

IMI Knoch Ltd.
Cu 76, Zn 22, Al 2.
80,000 TS; 44,000 YS; 16 El; 130 Brin. For condenser tubes; corrosion resistant.

ALDURBRA

Charles Clifford Industries Ltd.
Cu 76, Zn 22, Al 2.
80,000 TS; 44,000 YS; 16 El; 130 Brin. For condenser tubes; corrosion resistant.

ALEMITE

Stewart Warner Corp.
Si 10, bal Al.
For light alloy parts, die casting.

ALFA-I

Allegheny Ludlum Steel
Stainless steel. C 0.025, Mn 0.035, Si 0.3, Cr 13, Al 3, Ti 0.4, bal Fe.
RT: 76,000 psi TS; 55,000 psi YS; 28 El; 88 Rock B. For automotive exhaust system, furnace parts, heat exchangers, recuperator parts, kiln linings and refractory reinforcement fibers.

ALFA-II

Allegheny Ludlum Steel
Stainless steel. C 0.025, Mn 0.035, Si 0.3, Cr 13, Al 4, Ti 0.4, bal Fe.
RT: 80,000 psi TS; 60,000 psi YS; 28 El; 89 Rock B. For automotive exhaust system, furnace parts, heat exchangers, recuperator parts, kiln linings and refractory reinforcement fibers.

ALFENIDE

Manufacturer not listed.
Cu 59.1, Zn 30.12, Ni 9.7, Fe 1.
For ornaments, hardware, fittings; nickel silver.

ALFENOL

Colt Industries
Al 16, bal Fe.
For electrical equipment; high permeability. *Obsolete*

ALFENOL 12

Hamilton Technology Inc.
Al 12, bal Fe.
Residual inductance: 4,000; maximum permeability: 20,000; coercive force (oersted): 0.100. Wear resistant, high permeability, magnetically soft alloy.

ALFENOL 16

Hamilton Technology Inc.
Al 16, bal Fe.
Residual inductance: 3,000; maximum permeability: 110,000; coercive force (oersted): 0.015. Wear resistant, high permeability, magnetically soft alloy.

ALFENOL-16

Carpenter Technology Corp.
Al 16, Fe 84-84.2.
For relays, gyros, signal generators, servo mechanisms; soft magnetic material, high permeability. *Obsolete*

ALFER

Manufacturer not listed.
Al 12-14, bal Fe.
For permanent magnets; high permeability.

ALFER

Texas Instruments Inc./Materials Control
Now ALIRON.

ALFERIUM

Creusot-Loire
Cu 3.5-4.5, Mn 0.4-1, Mg 0.3-0.75, bal Al.
Heat treated: 55,000 TS; 310,000 YS; 16 El. For airplane and dirigible parts; similar to "Duralumin." *Obsolete*

ALFERIUM

Schneider & Cie
Aluminum.
Aluminum alloy AlCu4MnMg.

ALFERON

Harrison Alloys Inc.
Stainless steel.
Now HAI-FECRAL-25.

ALGIERS (ALGERS) METAL

Manufacturer not listed.
Sn 94.5, Sb 0.5, Cu 5.
For bearings; anti-friction.

ALGIERS METAL-1

Manufacturer not listed.
Sn 90, Sb 10.
For bearings; corrosion resistant.

ALGIERS METAL-2

Manufacturer not listed.
Sn 75, Sb 25.
For bearings; corrosion resistant.

ALGO-LOY 1315

Algoma Steel Corp. Ltd.
C 0.1-0.2, Mn 1.1-1.65, Cu 0-0.4, bal Fe.
Rolled: 70,000 TS; 50,000 YS; 22 El. For gears, cams, camshafts, bolts; case hardened, tough. *Obsolete*

ALGO-LOY 1317

Algoma Steel Corp. Ltd.
C 0-0.25, Mn 1.65, Mo 0-0.05, Cr 0-0.3, Ni 0-0.3, 0.20% Cu min, bal Fe.
Rolled: 70,000 TS; 50,000 YS; 22 El. For gears, cams, camshafts, crankshafts; case hardened, tough. *Obsolete*

ALGO-TUF 50

Algoma Steel Corp. Ltd.
HSLA steel. C 0.1-0.18, Mn 0.9-1.25, Si 0.15-0.3, Cr 0.4-0.6, Cu 0.2-0.4, V 0.01-0.1, bal Fe.
50,000 psi YS min. For welded, bolted or riveted construction. *Obsolete*

ALGO-TUF 70

Algoma Steel Corp. Ltd.
HSLA steel. C 0-0.2, Mn 0-1.6, Si 0.15-0.35, Cu 0.2-0.5, Cb 0-0.06, V 0.03-0.1, bal Fe.
70,000 psi YS min. For welded, bolted or riveted construction. *Obsolete*

ALGOFORM 35B

Algoma Steel Corp. Ltd.
HSLA steel. C 0.06, Mn 0.35, Cb 0.05, Al 0.04, bal Fe.
0.10 in. sheet: 45,000 psi (310 MPa) TS min; 35,000 (241 MPa) YS min; 28 El min. High strength low alloy sheet.

ALGOFORM 45

Algoma Steel Corp. Ltd.
HSLA steel. C 0-0.06, Mn 0.35, Al 0.04, Cb 0.01, bal Fe.
0.100 in. sheet: 55,000 psi (379 MPa) TS min; 45,000 psi (310 MPa) YS min; 40 El min. High strength low alloy steel plate.

ALGOFORM 45B

Algoma Steel Corp. Ltd.
HSLA steel. C 0.06, Mn 0.35, Cb 0.018, Al 0.04, bal Fe.
0.10 in. sheet: 55,000 psi (379 MPa) TS min; 45,000 psi (310 MPa) YS min; 30 El min. High strength low alloy sheet.

ALGOFORM 50

Algoma Steel Corp. Ltd.
HSLA steel. C 0-0.06, Mn 0.35, Al 0.04, Cb 0.015, bal Fe.
0.100 in. sheet: 60,000 psi (414 MPa) TS min; 50,000 psi (345 MPa) YS min; 36 El min. High strength low alloy steel plate.

ALGOFORM 50B

Algoma Steel Corp. Ltd.
HSLA steel. C 0.06, Mn 0.35, Al 0.04, Cb 0.025, bal Fe.
0.10 in. sheet: 60,000 psi (414 MPa) TS min; 50,000 psi (345 MPa) YS min; 25 El min. High strength low alloy sheet.

ALGOFORM 55B

Algoma Steel Corp. Ltd.
HSLA steel. C 0.06, Mn 0.35, Al 0.04, Cb 0.035, bal Fe.
0.10 in. sheet: 65,000 psi (448 MPa) TS min; 55,000 psi (379 MPa) YS min; 24 El min. High strength low alloy sheet.

ALGOFORM 60

Algoma Steel Corp. Ltd.
HSLA steel. C 0-0.06, Mn 0.35, Al 0.04, Cb 0.035, bal Fe.
0.100 in. sheet: 70,000 psi (483 MPa) TS min; 60,000 psi (414 MPa) YS min; 32 El min. High strength low alloy steel plate.

ALGOFORM 60B

Algoma Steel Corp. Ltd.
HSLA steel. C 0.06, Mn 0.35, Al 0.04, Cb 0.05, bal Fe.
0.10 in. sheet: 70,000 psi (483 MPa) TS min; 60,000 psi (414 MPa) YS min; 24 El min. High strength low alloy sheet.

ALGOFORM 65B

Algoma Steel Corp. Ltd.
HSLA steel. C 0.06, Mn 0.45, Al 0.04, Cb 0.065, bal Fe.
0.10 sheet: 75,000 psi (517 MPa) TS min; 65,000 psi (448 MPa) YS min; 21 El min. High strength low alloy sheet.

ALGOFORM 70

Algoma Steel Corp. Ltd.
HSLA steel. C 0-0.06, Mn 0.55, Al 0.04, Cb 0.065, bal Fe.
0.100 in. sheet: 80,000 psi (552 MPa) TS min; 70,000 (483 MPa) YS min; 28 El min. High strength low alloy steel plate.

ALGOFORM 80

Algoma Steel Corp. Ltd.
HSLA steel. C 0-0.06, Mn 0.75, Al 0.04, Cb 0.1, bal Fe.
0.100 in. sheet: 90,000 psi (620 MPa) TS min; 80,000 psi (552 MPa) YS min; 24 El min. High strength low alloy steel plate.

ALGOFORM 80B

Algoma Steel Corp. Ltd.
HSLA steel. C 0.1, Mn 1.45, Al 0.04, Cb 0.105, bal Fe.
0.13 in. sheet: 90,000 psi (621 MPa) TS min; 80,000 (553 MPa) YS min; 16 El min. High strength low alloy sheet.

ALGOMA 100

Algoma Steel Corp. Ltd.
HSLA steel. C 0-0.2, Mn 0-1.5, Si 0-0.5, Mo 0-0.5, Cb 0-0.06, B 0-0.003, bal Fe.
Quenched and tempered, 2.5 in. (65 mm) plate: 110,000 psi (760 MPa) TS; 100,000 psi (700 MPa) YS; 18 El min; 240-300 Brin. High strength; structural quality; improved notch toughness, weldability, and formability.

ALGOMA 1315

Algoma Steel Corp. Ltd.
C 0.1-0.2, Mn 1.1-1.65, Si 0.15-0.35, Cu 0.4, bal Fe.
Rolled: 70,000 TS; 50,000 YS; 22 El. For compressed gas cylinders, dished heads; good formability and drawability. *Obsolete*

ALGOMA 1317
Algoma Steel Corp. Ltd.
C 0-0.25, Mn 0-1.65, Ni 0-0.3, Cr 0-0.3, Mo 0-0.05, 0.20% min Cu, bal Fe.
Rolled: 67,000 TS; 46,000 YS; 20 El. For highway guard rails, transportation equipment, cars; structural steel. *Obsolete*

ALGOMA 360
Algoma Steel Corp. Ltd.
HSLA steel. C 0-0.25, Mn 0-1.5, Si 0-0.5, Mo 0-0.55, B 0-0.003, bal Fe.
Quenched and tempered, 2.0 in. (50.8 mm) plate: 190,000 psi (1310 MPa) TS; 180,000 psi (1240 MPa) YS; 10-14 El min; 360-415 Brin. Abrasion resistant.

ALGOMA 425
Algoma Steel Corp. Ltd.
HSLA steel. C 0-0.25, Mn 0-1.5, Si 0-0.5, Mo 0-0.55, B 0-0.003, bal Fe.
Quenched and tempered, 2.0 in. (50.8 mm) plate: 200,000 psi (1380 MPa) TS; 190,000 psi (1310 MPa) YS; 9-14 El min; 400-477 Brin. Abrasion resistant.

ALGOMA 500
Algoma Steel Corp. Ltd.
HSLA steel. C 0-0.33, Mn 0-1.5, Si 0-0.7, Cr 0.7, Ni 0.7, Mo 0-0.35, B 0-0.003, bal Fe.
Quenched and tempered, 1.0 in. (25.4 mm) plate: 225,000 psi (1550 MPa) TS; 200,000 psi (1380 MPa) YS; 9-12 El min; 477-545 Brin. Abrasion resistant.

ALGOMA AR
Algoma Steel Corp. Ltd.
HSLA steel. C 0.35-0.42, Mn 1-1.15, Si 0.3-0.6, bal Fe.
90,000 psi TS. For applications where abrasive resistance is required.

ALGONQUIN
Atlas Specialty Steels
C 1, Mn 0.25, Si 0.2, bal Fe.
For dies; water hardened. *Obsolete*

ALGRITAL
Aluminium Press-ung Walzwerk
Aluminum.
Aluminum alloy AlSi5.

ALHEAD
Allegheny Ludlum Steel
C 1, W 1.5, Co 1.5, Mo 0-0.15, bal Fe.
Annealed: 217 Brin. Heat treated: 600 Brin. For cold heading dies; wear and abrasion resistant. *Obsolete*

ALIRON
Texas Instruments Inc./Materials Control
Low C steel clad with Al alloy. For electron tubes.

ALJEP
Manufacturer not listed.
Cu 3.8, Mg 1.3, Ni 1.5, V 0.5, bal Al.

ALKADUR 300
Kreidler Werke G.m.b.H.
Aluminum. Cu 3.5-4.5, Si 0.2-1, Mn 0.3-1.2, Mg 0.4-1.4, bal Al.
Heat treated: 62,700-71,100 TS; 48,000-57,000 YS; 5-15 El; 110-150 Brin. For aircraft components, fittings, fasteners, rivets. Heat treatable. High strength and fatigue.

ALKALI-RESISTING METAL
Manufacturer not listed.
Fe 95, Ni 5.
For corrosion resisting parts, chemical apparatus; corrosion resistant.

ALKROHMIT 140
Vereinigte Edelstahlwerke
C 0.8, Cr 20, Al 5, bal Fe.
For electric resistances; heat and oxidation resistant. *Obsolete*

ALKRONID
SWB Stahlformguss Gesellschaft mbH
C 0.34, Al 1.1, Cr 1.4, bal Fe.
For oil refinery equipment; creep resistant. *Obsolete*

ALKROTHAL
Kanthal Corp.
Cr 15, Al 4.5, bal Fe.
For electrical resistances, for service up to 1050°C; wire, ribbon, strip, foil.

ALKROTHAL
Kanthal A.B.
Cr 15, Al 4.5, bal Fe.
For electrical resistances, for service up to 1050°C; wire, ribbon, strip, foil.

ALKUMAG 302
Kreidler Werke G.m.b.H.
Cu 2.5-5, Mg 0.2-1.8, Mn 0.3-1.5, bal Al.
For aircraft structures and parts; age hardenable. *Obsolete*

ALKUMAG 303
Kreidler Werke G.m.b.H.
Cu 2.5-5, Mg 0.2-1.8, Mn 0.3-1.5, bal Al.
For aircraft structural parts; age hardenable. *Obsolete*

ALKUMAG 311
Kreidler Werke G.m.b.H.
Cu 2.5-5, Mg 0.2-1.8, Mn 0.3-1.5, bal Al.
For aircraft structures and parts; age hardenable. *Obsolete*

ALL-PO
Grand Northern Products Ltd.
C 0.9, Si 0.4, Mn 0.4, Cr 9.5, Mo 0.6, bal Fe.
Coated, hard facing electrode. As arc welded: 185,000 psi TS; 52-55 Rock C. For overlay or buildup on scrapers, loaders, dozer bits, blades, shovels, buckets. Good abrasion resistance.

ALL-STATE AH, HSS, HW, OH AND WH
Now TOOL-ARC AIR HARDENING, HIGH SPEED, HOT WORK, OIL HARDENING AND WATER HARDENING.

ALL-STATE CHAMFER ROD
Chemetron Corp.
Electrode for AC-DC straight polarity for chamfering and gouging.

ALL-STATE CUTTING ELECTRODE
Chemetron Corp.
Electode for AC-DC straight polarity cutting without the use of oxygen or compressed air; for nickel, copper, brass, bronze, aluminum, stainless and alloy steels.

ALL-STATE GALVOVER
Chemetron Corp.
Electrode to restore galvanized surface on welded or damaged galvanized parts. Working temperature: 600°F.

ALL-STATE HS-1
Chemetron Corp.
Cr, Mn, Ni.
Welded: 125,000 TS; 19 El; 260-320 Brin. For hard facing electrode; abrasion and wear resistant, work hardening. *Obsolete*

ALL-STATE LOW FUMING BRONZE
Chemetron Corp.
Bronze rod for torch brazing steel, cast iron, galvanized malleable and copper base alloys. Working temperature: 1600°F; strengths up to 60,000 psi.

ALL-STATE MANGANESE BRONZE
Chemetron Corp.
Alloy for torch brazing various metals; resistant to salt water.

ALL-STATE MONOWELD
Chemetron Corp.
Electrode for AC-DC, reverse polarity, all-position welding of mild and low alloy steels. As welded: up to 62,000 psi YS; 20-22 El.

ALL-STATE NICKEL BRONZE
Chemetron Corp.
Ni-bronze alloy for torch brazing various alloys. Working range: 1200-1750°F; strengths up to 85,000 psi.

ALL-STATE NO 101
Chemetron Corp.
Ag 45, Cu Zn, Cd.
Silver solder; melt range: 1125-1145°F. For joining wide range of metals; strength up to 50,000 psi.

ALL-STATE NO 101FC
Chemetron Corp.
Flux coated grade of ALL-STATE NO 101. (Also known as TRUCOTE NO 101FC.)

ALL-STATE NO 107
Chemetron Corp.
Silver bearing soft solder for torch, furnace or induction soldering of ferrous and non-ferrous alloys. Melt range: 480-600°F; strength up to 20,000 psi.

ALL-STATE NO 11
Chemetron Corp.
Ni 10, 47 Cu 43 Zn.
For brazing; M.P. 1665-1715°F.

ALL-STATE NO 111
Chemetron Corp.
41% Ag, brazing alloy.
Melt range: 1100-1150°F (lowest melting); TS to 50,000 psi.

ALL-STATE NO 13
Chemetron Corp.
Cu 48, Ni 9.6, Si 0.29, Fe 0.18, Mn 0.16, bal Zn.
For brazing alloy; nickel silver.

ALL-STATE NO 13FC
Chemetron Corp.
Flux coated grade of ALL-STATE NO 13.

ALL-STATE NO 155
Chemetron Corp.
Silver solder (Cadmium free). For brazing stainless steel food handling equipment to hot water systems. Melt range: 1150-1200°F; TS to 50,000 psi.

ALL-STATE NO 155FC
Chemetron Corp.
Flux coated grade of ALL-STATE NO 155. (Also known as TRUCOTE NO 155FC.)

ALL-STATE NO 16
Chemetron Corp.
125-150 Brin. Electrode for DC reverse polarity, all position welding of Ni-Cu alloys.

ALL-STATE NO 164
Chemetron Corp.
Silver solder for torch or furnace brazing of "hard-to-solder" applications, particularly carbide tips to steel tools. Working temperature 1500°F.

ALL-STATE NO 18
Chemetron Corp.
Electrode for DC reverse polarity welding of Ni-Cr-Fe alloys includin 9% Ni steels for cryogenic purposes.

ALL-STATE NO 20
Chemetron Corp.
120-150 Brin; strengths up to 91,000 psi. Aluminum bronze electrode for DC reverse polarity welding and build-up o ferrous and non-ferrous metals.

ALL-STATE NO 21
Chemetron Corp.
Working temperature: 1445-1460°F. Phosphor-copper brazing rod for torch brazing copper tubing and pipe.

ALL-STATE NO 23
Chemetron Corp.
Ag, P, bal Cu.
Working temperature: 1435-1450°F; strength up to 45,000 psi. Brazing alloy for torch brazing copper, brass and bronze.

ALL-STATE NO 252
Chemetron Corp.
Electrode for AC-DC, reverse polarity, all-position welding of 310, 31 stainless and other high temperature steels. 85 Rb; YS up to 65,000 psi (room temp.).

ALL-STATE NO 3
Chemetron Corp.
Copper coated cast iron alloy for torch brazing manifolds, cylinder heads and other cast iron. Machinable. 175-225 Brin; tensile up to 45,000 psi; working temperature; 1500°F.

ALL-STATE NO 31
Chemetron Corp.
Working temperature 1075°F; strengths up to 30,000 psi. Alloy for brazing thin sheet aluminum ; (not recommended for heat-treatabl alloys).

ALL-STATE NO 33
Chemetron Corp.
Working temperature: 1050°F. Alloy for torch brazing aluminum castings.

ALL-STATE NO 34
Chemetron Corp.
Aluminum base electrode for DC reverse polarity, all-position welding an repair of aluminum housings. (Also called SMOOTHCOTE NO 34.)

ALL-STATE NO 4
Chemetron Corp.
175-225 Brin. Electrode for AC-DC, reverse polarity welding thin sections of cast iron.

ALL-STATE NO 4 IMP
Chemetron Corp.
Very machinable; strength to 50,000 psi. Electrode fo AC-DC, reverse polarity welding of cast iron.

ALL-STATE NO 4-60
Chemetron Corp.
200-300 Brin. Electrode for AC-DC reverse polarity, all position welding heavy sections of cast iron.

ALL-STATE NO 4-60
Chemetron Corp.
200-300 Brin. Electrode for AC-DC reverse polarity, all position welding heavy sections of cast iron.

ALL-STATE NO 41FC
Chemetron Corp.
Working temperature: 1400-1650°F; strengths up to 60,000 psi. Flux coated bronze for torch brazing; general maintenance and repair steel cast iron, brass bronze.

ALL-STATE NO 55 RUBBON
Chemetron Corp.
Working temperature; 705-720°F; strength up to 24,000 psi. Lead free, self-fluxing aluminum solder.

ALL-STATE NO 6 IMP
Chemetron Corp.
Electrode for AC-DC, reverse polarity welding of cast iron. Strength to 60,000 psi; not readily machinable.

ALL-STATE NO 7
Chemetron Corp.
Solder for sealing cracks in cast iron, steel, copper, brass and bronze. Working temperature: 450-600°F.

ALL-STATE NO 8
Chemetron Corp.
175-225 Brin. Electrode for AC-DC, reverse polarity, all position welding thin sections of cast iron.

ALL-STATE NO 8-60
Chemetron Corp.
200-300 Brin; strength up to 80,000 psi. Electrode for AC-DC, reverse polarity, all position welding of heavy sections of cast iron and ductile iron.

ALL-STATE NO. 100
Chemetron Corp.
Ag 40, Cu 30, Zn 28, Ni 2.
For brazing, silver solder; M.P. 1220-1435°F.

ALL-STATE NO. 2
Chemetron Corp.
C 3, Si 3.2, Mn 0.5, P 0.7, bal Fe.
For cast iron welding rod; machinable, for arc welding.
Obsolete

ALL-STATE NO. 22
Chemetron Corp.
P, Sn, bal Cu.
Welded: 40,000 TS; 80 Brin. For welding electrodes; arc welding, corrosion resistant. *Obsolete*

ALL-STATE NO. 24
Chemetron Corp.
Sn 7-9, P 0.03-0.35, bal Cu.
Welded: 45,000 TS; 80-160 Brin. For P-bronze welding electrodes; for welding steel, cast iron and copper alloys.

ALL-STATE NO. 321
Chemetron Corp.
Sn 30, Zn 30, Pb 40, plus flux.
powder for galvanizing. Working temperature: 450°F.

ALL-STATE NO. 35
Chemetron Corp.
Si 5, bal Fe.
For Al welding rod. *Obsolete*

ALL-STATE NO. 39
Chemetron Corp.
Sn 59, Zn 42-41.

ALL-STATE NO. 41
Chemetron Corp.
Cu 58, Sn 1, Fe 1, bal Zn.
For brazing rod; for gas welding.

ALL-STATE NO. 430
Chemetron Corp.
Low temperature silver bearing solder; cadmium free. Melt temp: 430°F; strength up to 15,000 psi. For food handling equipment.

ALL-STATE NO. 509 STRONGSET
Chemetron Corp.
Cd Zn.
Solder for joining aluminum and other metals (except magnesium). Melt temp: 509°F: strength up to 29,000 psi.

ALL-STATE NO. 53
Chemetron Corp.
Pb 4, Al 3, Cu 2, bal Zn.
For Zn base welding rod for Zn alloys; for gas welding.

ALL-STATE NO. 61
Chemetron Corp.
Al 10, Zn 1, Mg 89.
For magnesium alloy solder. *Obsolete*

ALL-STATE NO. 616
Chemetron Corp.
Electrode for AC-DC, reverse polarity, all-position welding, particula high tensile and sulphur bearing steels. As welded: up to 80,000 psi TS.

ALL-STATE NO.275
Chemetron Corp.
Electrode for AC-DC reverse polarity, all-position welding of high strength steels. As welded: up to 120,000 psi TS; work-hardenable to 180,000 psi.

ALL-STATE SEALCOR
Chemetron Corp.
Flux cored rod for torch brazing of weldable grades of aluminum. Working temperature; 1100°F; strengths up to 30,000 psi.

ALL-STATE SILFLO "0"
Chemetron Corp.
P 7, bal Cu.
Self-fluxing brazing alloy; very fluid. Working temperature: 1350-1550°F.

ALL-STATE SILFLO "15"
Chemetron Corp.
Ag 15, P 5, bal Cu.
Self-fluxing brazing alloy; for ductile joints. Working range: 1300-1500°F.

ALL-STATE SILFLO "5"
Chemetron Corp.
Ag 5, P 6, bal Cu.
Self-fluxing brazing alloy; for wider gaps. Working temperature: 1300-1500°F.

ALL-STATE SILICON BRONZE
Chemetron Corp.
Si, bal Cu.
Electrode for tungsten inert gas or carbon are brazing or welding silicone bronze, cooper alloys and some iron base metals. Strengths up to 58,000 Psi.

ALL-STATE STEELARC
Chemetron Corp.
Electrode for AC-DC, reverse polarity, all-position welding mild steel particularly dirty, rusty sheet metal. As welded; up to 70,000 psi Ts.

ALL-STATE STEELARC PLUS
Chemetron Corp.
Improved grade of ALL-STATE STEELARC.

ALLADIN
Manufacturer not listed.
Zn alloy. For welding rod; for white metal die castings.

ALLAN NO.2 BRONZE
Manufacturer not listed.
Cu 66, Sn 9, Pb 25.
For bearings, hardware; tough.

ALLAN RED BRONZE
Manufacturer not listed.
Cu 62.5, Pb 30, Sn 7.5.
For bearings, hardware; tough.

ALLAN RED METAL
A. Allan & Son
Copper. Pb 50, Cu 50.
13.5 Brin. For bearings, crank-pins for facing pistons; not a true alloy but a mechanical mixture.

ALLAN RED METAL NO. 2
A. Allan & Son
Copper. Cu 60, Pb 40.
For high speed bearings, piston wearing rings, operating conditions not exceeding 600°F.

ALLAUTAL
VAW Vereinigte Aluminium-Werke AG
Aluminum. Cu 4.5-5.5, Si 0.2-0.5, bal Al.
Heat treated: 60,000-71,000 TS; 42,000-57,000 YS; 2-10 El; 120-140 Brin. For aircraft components, fasteners, fittings, rivets. Age hardenable, high tensile and fatigue strength.

ALLAUTAL

German manufacture
surface of pure Al on base of "Laurel.".
Annealed: 20,000 TS; 25 El. Normalized: 60,000 TS; 18 El.
For light alloy parts; similar to "Alclad."

ALLCAST

Sheepbridge Equipment Ltd.
Aluminum. Cu 2.5-3.5, Si 4.5-6, Mg 0-0.15, Fe 0-1, Ni 0-0.2, Mn 0-0.5, bal Al.
Cast: 24,000 TS; 12,000 YS; 2 El; 60 Brin. Heat treated: 30,000 TS; 18,000 YS; 2.5 El; 75 Brin. For sand and permanent mold castings; age hardenable.

ALLCAST 125

Apex International Alloys, Inc.
Mg 9.7-10.5, Ti 0-0.25, bal Al.
T4 Temper: 57,000 TS; 29,000 YS; 30 El; 80 Brin. For aircraft fittings, car frames, marine parts, lever brackets. Age hardenable, corrosion resistant, shock resistant, tough. *Obsolete*

ALLCAST 400

Apex International Alloys, Inc.
Mg 3.6-4.2, Si 0-0.5, Zn 0-0.5, Mn 0-0.6, Fe 0-0.5, Ti 0-0.25, bal Al.
Cast (Sand): 25,000 TS; 12,000 YS; 9 El; 50 Brin. At 300°F: 22,000 TS; 12,000 YS; 7 El. For food handling equipment, cooking utensils, chemical and textile plant equipment, marine hardware. Similar to Aluminum 214, good corrosion resistance. Not heat treatable. *Obsolete*

ALLCAST 405

Apex International Alloys, Inc.
Zn 4-5, Si 0-0.5, Fe 0-1, Ni 0-0.2, Mn 0-0.15, Sb 0.9-1.4, bal Al.
Die cast: 22,000 TS; 11,000 YS; 10 El. For oil pans, instrument cases, housings, ornaments. Good corrosion resistance. *Obsolete*

ALLCAST 417

Apex International Alloys, Inc.
Mg 6.5-7.5, Si 0-0.12, Mn 0.1-0.25, Ti 0.1-0.25, bal Al.
Sand cast: 36,000 TS; 19,000 YS; 10 El; 70 Brin. For automotive and marine castings, cooking utensils, gasoline fuel gages, textile machinery. Similar to Aluminum 2185P and Almag 35. High strength and shock resistant. *Obsolete*

ALLCAST 418

Apex International Alloys, Inc.
Mg 7.5-8.5, Si 0-0.3, Fe 0.9-1.3, Mn 0-0.3, Ni 0-0.1, bal Al.
Die cast: 43,000 TS; 25,000 YS; 5 El; 80 Brin. For cylinder blocks, marine fittings, hardware, brake shoes, wheel flanges. Similar to Aluminum 218. Good strength and corrosion resistance. *Obsolete*

ALLCAST 420

Apex International Alloys, Inc.
Mg 9.7-10.5, Si 0-0.15, Fe 0-0.2, Ti 0-0.25, bal Al.
T4 Temper: 46,000 TS; 25,000 YS; 16 El; 75 Brin. T5 Temper: 35,000 TS; 110 Brin. For aircraft fittings, marine parts, railroad car frames. Sand castings. Similar to Aluminum 220. High strength and corrosion resistance. *Obsolete*

ALLCAST 440

Apex International Alloys, Inc.
Mg 0.5-0.65, Zn 5.2-6, Cr 0.4-0.6, Ti 0.15-0.25, Fe 0-0.4, bal Al.
Cast: 28,000 TS; 16,000 YS; 12 El; 55 Brin. Aged: 36,000 TS; 25,000 YS; 6 El; 75 Brin. For manifolds, aircraft components, transmission housings, tight castings. Self-aging sand casting. Natural aging. Similar to Aluminum 40 E. *Obsolete*

ALLCAST 46

Apex International Alloys, Inc.
Cu 0-0.6, Si 6.5-7.5, Mg 0.2-0.4, Zn 0-0.5, Fe 0-0.75, Ti 0.1-0.2, bal Al.
Cast: 23,000 TS; 12,000 YS; 3.5 El; 55 Brin. T6 Temper: 33,000 TS; 24,000 YS; 4 El; 70 Brin. For instrument cases, gear cases, fittings, crankcases, oil pans. Heat treatable. Similar to Aluminum 356. Corrosion resistant. *Obsolete*

ALLCAST 46-10

Apex International Alloys, Inc.
Cu 0-0.1, Si 6.5-7.5, Mg 0.28-0.4, Zn 0-0.1, Fe 0-0.15, Ti 0-0.2, bal Al.
Permanent Mold Cast: 27,000 TS; 14,000 YS; 8 El, 55 Brin. T61 Temper: 43,000 TS; 28,000 YS; 12 El; 90 Brin. For gear housings, fittings, oil pans, crankcases, instrument housings. Heat treatable, permanent mold. Similar to Aluminum A356. *Obsolete*

ALLCAST 46A

Apex International Alloys, Inc.
Cu 0.2, Si 6.5-7.5, Mg 0.25-0.4, Zn 0-0.3, Fe 0-0.5, Ti 0-0.2, bal Al.
Permanent Mold Cast: 25,000 TS; 13,000 YS; 7 El; 65 Brin. T6 Temper: 40,000 TS; 27,000 YS; 5 El; 90 Brin. For permanent mold castings, oil pans, instrument cases, crankcases. Heat treatable. Similar to Aluminum 356. *Obsolete*

ALLCAST 470

Apex International Alloys, Inc.
Cu 0.4-1, Mg 0.25-0.5, Zn 7-8, Fe 0-0.8, Ti 0-0.25, bal Al.
Sand Cast: 28,000 TS; 15,000 YS; 8 El; 55 Brin. Aged: 34,000 TS; 23,000 YS; 5 El; 75 Brin. For housings, machinery parts, fittings, hardware, automobile parts. Similar to Tenzalloy. Ages at room temperature. *Obsolete*

ALLCAST 50

Apex International Alloys, Inc.
Cu 3.5-4.5, Si 2.5-3.5, Mg 0-0.1, Zn 0-1, Fe 0-1, Ti 0-0.2, bal Al.
Permanent Mold Cast: 27,000 TS; 16,000 YS; 3 El; 65 Brin. T6 Temper: 40,000 TS; 26,000 YS; 4 El; 85 Brin. For castings, housings, oil pans, manifold and valve bodies, machine frames, piano plates and frames. Heat treatable, permanent mold. Similar to Aluminum 108. *Obsolete*

ALLCAST 51

Apex International Alloys, Inc.
Cu 3.5-4.5, Si 5.5-6.5, Mg 0-0.1, Zn 0-1, Fe 0-1, Ti 0-0.2, bal Al.
Cast: 28,000 TS; 16,000 YS; 2 El; 70 Brin. T6: 45,000 TS; 28,000 YS; 4 El; 105 Brin. For permanent mold castings, housings, oil pans, casings, fittings, Heat treatable. Similar to Aluminum A108. *Obsolete*

ALLCAST 52

Apex International Alloys, Inc.
Cu 4-5, Si 0-1.5, Zn 0-0.5, bal Al.
T6 Temper: 37,000 TS; 24,000 YS; 5 El; 75 Brin. At 600°F: 4000 TS; 3000 YS; 75 El. For flywheel housings, aircraft wheels, fittings, crankcases, instrument cases. Heat treatable. Similar to Aluminum 195. Strong and rugged. *Obsolete*

ALLCAST 52A

Apex International Alloys, Inc.
Cu 4-5, Si 0-1.5, Zn 0-0.35, Fe 0-0.8, Ti 0-0.25, bal Al.
T4 Temper: 32,000 TS; 16,000 YS; 8 El; 60 Brin. T62 Temper: 41,000 TS; 32,000 YS; 2 El; 95 Brin. For flywheel housings, crankcases, aircraft and bus wheels. Heat treatable. Strong and rugged. Similar to Aluminum 195. *Obsolete*

ALLCAST 53

Apex International Alloys, Inc.
Cu 4-5, Si 2-3, Zn 0-0.5, Ni 0-0.35, Ti 0-0.2, Fe 0-1, bal Al.
Cast: 30,000 TS; 14,000 YS; 3.5 El; 70 Brin. T6: 40,000 TS; 26,000 YS; 5 El; 90 Brin. For aircraft fittings and wheels, fuel pump bodies, seat frames. Heat treatable, permanent mold. Similar to Aluminum B195. *Obsolete*

ALLCAST 59A

Apex International Alloys, Inc.
Cu 3.5-4.5, Si 0-0.6, Mg 1.3-1.8, Cr 0-0.25, Ni 1.7-2.0, Ti 0-0.25, bal Al.
Cast: 34,000 TS; 24,000 YS; 1 El; 105 Brin. T6: 54,000 TS; 51,000 YS; 0.8 El; 120 Brin. For air cooled cylinder heads, diesel engine pistons, generator housings. Heat treatable, permanent mold. Similar to Aluminum 142. *Obsolete*

ALLCAST 60

Apex International Alloys, Inc.
Si 6, Cu 3.5, bal Al.
F-temper: 27,000 TS; 16,000 YS; 2.5 El; 65 Brin. T6-temper: 35,000 TS; 24,000 YS; 3.5 El; 85 Brin. For pressure tight castings; heat treatable. *Obsolete*

ALLCAST 70

Apex International Alloys, Inc.
Si 6, Cu 3.5, bal Al.
F-temper: 26,000 TS; 16,000 YS; 2.0 El; 70 Brin. T6-temper: 36,000 TS; 24,000 YS; 2.0 El; 80 Brin. For pressure tight castings; heat treatable. *Obsolete*

ALLCAST 700

Apex International Alloys, Inc.
Cu 2-3, Si 0-0.7, Ni 0.9-1.5, Ti 0-0.2, Sn 6-7.5, bal Al.
T5 or T533 Temper: 20,000 TS; 16,000 YS; 7 El; 55 Brin. For bearings, bushings, heat treatable. *Obsolete*

ALLCAST 700A

Apex International Alloys, Inc.
Cu 0.7-1.3, Sn 5.5-7, Si 0-0.7, Ti 0-0.2, Ni 0.7-1.3, bal Al.
T5 or T533 Temper: 20,000 TS; 8500 YS; 10 El; 55 Brin. For bearings. Similar to Aluminum 750. Good imbeddability. *Obsolete*

ALLCAST 78

Apex International Alloys, Inc.
Cu 3-4, Si 7.5-8.5, Zn 0-1, Fe 0-0.6, Ti 0-0.25, bal Al.
Die cast: 43,000 TS; 26,000 YS; 2 El. At 300°F: 37,000 TS; 24,000 YS; 4 El. For motor frames, hand truck wheels, oil pans, ornamental parts. Good high temperature strength. *Obsolete*

ALLCAST D400

Apex International Alloys, Inc.
Si 1.6-2, Mg 3.6-4.5, Fe 0-0.5, Mn 0-0.6, Ti 0-0.25, bal Al.
Sand Cast: 20,000 TS; 13,000 YS; 2 El; 50 Brin. Permanent Mold: 22,000 TS; 13,000 YS; 2 El; 50 Brin. For cooking utensils, marine fittings. Similar to Aluminum B214. Good corrosion resistance. Not heat treatable. *Obsolete*

ALLCAST P400

Apex International Alloys, Inc.
Mg 3.6-4.5, Zn 1.5-2, Si 0-0.5, Fe 0-0.5, Mn 0-0.6, Ti 0-0.25, bal Al.
Permanent Mold: 27,000 TS; 16,000 YS; 7 El; 60 Brin. Die Cast: 40,000 TS; 22,000 YS; 10 El; 65 Brin. For food and dairy handling equipment, chemical and textile plant equipment, marine hardware. Similar to Aluminum A214. Not heat treatable, corrosion resistant. *Obsolete*

ALLCAST SC8

Apex International Alloys, Inc.
Cu 2.5-4, Si 5-6.5, Mg 0-0.15, bal Al.
Permanent mold cast: 28,000-36,000 TS; 15,000-20,000 YS; 2-3 El; 70-8 Brin. For permanent mold castings; light alloy parts. *Obsolete*

ALLCAST SC8

Sheepbridge Equipment Ltd.
Aluminum. Cu 2.5-3.5, Si 4.5-6, bal Al.
Cast: 24,000 TS; 12,000 YS; 2 El; 60 Brin. T6 temper: 30,000 TS; 18,000 YS; 2.5 El; 75 Brin. For light alloy castings; sand and permanent mold, hardenable.

ALLCAST Z50

Apex International Alloys, Inc.
Cu 3-4, Si 2.5-3.5, Mg 0-0.1, Zn 2-3, Fe 0-1, Ti 0-0.2, bal Al.
Permanent Mold Cast: 27,000 TS; 16,000 YS; 3 El; 65 Brin. T6 Temper: 40,000 TS; 26,000 YS; 4 El; 85 Brin. For casings, housings, fittings, crankcases, axle housings. Heat treatable, permanent mold, corrosion resistant. *Obsolete*

ALLCAST Z51

Apex International Alloys, Inc.
Cu 3-4, Si 5.5-6.5, Mg 0-0.1, Zn 2-3, Fe 0-1, Ti 0-0.2, bal Al.
Cast: 39,000 TS; 22,000 YS; 5.5 El; 90 Brin. T6: 45,000 TS; 28,000 YS; 4 El; 105 Brin. For permanent mold castings, casings, housings, fittings, oil pans. Heat treatable. *Obsolete*

ALLEGHENY 12
Allegheny Ludlum Steel
C 0-0.15, Mn 0-1, Cr 11.5-13.5, bal Fe.
Annealed: 60,000 TS; 30,000 YS; 20 El; 50 RA; 200 Brin.
Hardened: 200,000 TS; 181,000 YS; 2 El; 51 RA; 400 Brin. For
valves, pump shafts, screens; corrosion resistant; Type No.
410. *Obsolete*

ALLEGHENY 16-16-1
Allegheny Ludlum Steel
Cr 16, Mn 16, Ni 1, C, bal Fe.
For chemical plant equipment; stainless. *Obsolete*

ALLEGHENY 18-8C
Allegheny Ludlum Steel
C 0.08, Cr 17-19, Ni 9-12, Cb = 12 x C, bal Fe.
Annealed: 80,000 TS; 30,000 YS; 40 El; 50 RA; 180 Brin. For
exhaust manifolds, fire walls; stabilized stainless; Type 307.
Obsolete

ALLEGHENY 18-8SM
Allegheny Ludlum Steel
C 0-0.12, Cr 16-18, Ni 0-14, Mo 2-3, bal Fe.
Annealed: 73,000-85,000 TS; 35,000-45,000 YS; 45-60 El;
45-60 RA. For stainless parts; austenitic. *Obsolete*

ALLEGHENY 19-10M
Allegheny Ludlum Steel
C 0-0.16, Cr 18-20, Ni 0-14, Mo 3-4, bal Fe.
Cast: 70,000-80,000 TS; 35,000-45,000 YS; 45-60 El; 45-60
RA. For stainless steel castings; resists sulfuric, nitric and
acetic acids. *Obsolete*

ALLEGHENY 19-10SM
Allegheny Ludlum Steel
C 0-0.12, Cr 18-20, Ni 0-14, Mo 3-4, bal Fe.
Cast: 70,000-80,000 TS; 40,000-60,000 YS; 45-60 El; 45-65
RA. For stainless parts; stainless steel castings. *Obsolete*

ALLEGHENY 19-9 DL
Allegheny Ludlum Steel
C 0.3, Mn 1, Si 0.5, Cr 19, Ni 9, Mo 1.4, W 1.3, Ti 0.3, 0.4 C +
Ta, bal Fe.
Heat treated: 148,200 TS; 118,000 YS; 19 El. Stress relieved:
118,500 TS; 69,000 YS; 58 El. At 1500°F: 32,500 TS; 29,500
YS; 53 El. Uses: gas turbine and turbo supercharger
component parts as rotors, buckets, fasteners, tail cones.
Stainless. High temperature properties. Heat resistant to
1200°F. Nonhardenable.

ALLEGHENY 19-9DX
Allegheny Ludlum Steel
C 0.3, Mn 1, Si 0.5, Cr 19, Ni 9, Mo 1.6, W 1.3, Ti 0.6, bal Fe.
Heat treated: 148,200 TS; 118,000 YS; 19 El. Stress-relieved:
118,500 TS; 69,000 YS; 58% El. At 1500 F: 32,500 TS; 29,500
YS; 53% El. Uses: gas turbine and turbo supercharger parts
as rotors, buckets, fasteners, tail cones. Stainless. High
temperature properties. Heat resistant to 1200 F. Non-
hardenable. *Obsolete*

ALLEGHENY 20-10
Allegheny Ludlum Steel
Cr 20, Ni 10, C, bal Fe.
For applications requiring maximum resistance to corrosion
coupled with faculty of fabrication; corrosion resistant,
stainless. *Obsolete*

ALLEGHENY 20-10S
Allegheny Ludlum Steel
C 0-0.08, Cr 19-21, Ni 10-12, bal Fe.
Annealed: 80,000 TS; 30,000 YS; 40 El; 50 RA; 200 Brin. For
chemical equipment; Stainless Type 308. *Obsolete*

ALLEGHENY 20-25S
Allegheny Ludlum Steel
C 0-0.11, Cr 19-21, Ni 24-26, bal Fe.
Cast: 60,000-75,000 TS; 40,000-55,000 YS; 40-60 El; 40-60
RA. For castings; stainless steel castings. *Obsolete*

ALLEGHENY 20-25SM
Allegheny Ludlum Steel
C 0-0.11, Cr 19-21, Ni 24-26, Mo 2-3, bal Fe.
Cast: 60,000-75,000 TS; 50,000-70,000 YS; 40-60 El; 40-60
RA. For castings: stainless steel castings. *Obsolete*

ALLEGHENY 21
Allegheny Ludlum Steel
Cr 18-23, C, bal Fe.
For high temperature service. *Obsolete*

ALLEGHENY 216
Allegheny Ludlum Steel
C 0-0.08, Cr 19.75, Mn 8.25, Ni 6, Mo 2.5, N 0.37, bal Fe.
Annealed: 100,000 TS; 55,000 YS; 45 El; 92 Rock B. At
1000°F: 75,400 TS; 36,300 YS; 43.3 El; 67.8 RA. For marine
engineering components, space hardware, woven wire cloth,
bushings, bolts, shafting. Springs, valve seats. Stainless,
improved high temperature strength and endurance limit.
Good weldability.

ALLEGHENY 216L
Allegheny Ludlum Steel
C 0-0.03, Cr 19.75, Mn 8.25, Ni 6, Mo 2.5, N 0.37, bal Fe.
Annealed: 100,000 TS; 55,000 YS; 45 El; B 92 Rock. For
marine engineering components, space hardware, woven
wire cloth, bushings, bolts, shafting, springs. Good
weldability, high fatigue strength. Stainless and corrosion
resistant. *Obsolete*

ALLEGHENY 22
Allegheny Ludlum Steel
C 0.08-0.2, Cr 19-22, Ni 9-12, bal Fe.
For digesters, tanks, agitators, pipes, autoclaves, strainers;
Type 302; stainless, austenitic. *Obsolete*

ALLEGHENY 22 SP
Allegheny Ludlum Steel
Cr 19-22, Ni 9-12, C 0-0.11, bal Fe.
Annealed: 75,000 TS; 50 El; 60 RA; 135 Brin. For chemical
engineering equipment; stainless and corrosion resistant.
Obsolete

ALLEGHENY 25-12
Allegheny Ludlum Steel
C 0-0.2, Mn 0-1.25, Cr 22-24, Ni 12-15, bal Fe.
Annealed: 75,000 TS; 30,000 YS; 40 El; 50 RA; 200 Brin. For
furnace parts, boiler baffles, pump parts; austenitic, resists
heat to 2000 F. *Obsolete*

ALLEGHENY 25-12S
Allegheny Ludlum Steel
C 0-0.11, Cr 22-26, Ni 12-14, bal Fe.
Cast: 75,000-90,000 TS; 40,000-55,000 YS; 30-50 El; 30-50
RA. For stainless steel castings; resists sulfuric, nitric and
acetic acids. *Obsolete*

ALLEGHENY 25-20
Allegheny Ludlum Steel
Cr 24-26, Ni 19-21, C 0-0.25, bal Fe.
Annealed: 75,000-85,000 TS; 30,000 YS; 40-50 El; 50-60 RA;
150-180 Brin. For chemical engineering equipment, furnace
parts; stainless and corrosion resistant. *Obsolete*

ALLEGHENY 25-20S
Allegheny Ludlum Steel
C 0-0.18, Cr 24-26, Ni 19-21, bal Fe.
Cast: 60,000-75,000 TS; 40,000-55,000 YS; 40-60 El; 40-60
RA. For castings; stainless steel castings. *Obsolete*

ALLEGHENY 28
Allegheny Ludlum Steel
C 0-0.35, Cr 23-27, N 0-0.25, bal Fe.
Annealed: 75,000 TS; 40,000 YS; 20 El; 40 RA; 200 Brin. For
furnace parts, boiler baffles; resists oxidation up to 2150 F.
Obsolete

ALLEGHENY 28-4
Allegheny Ludlum Steel
C 0-0.25, N 0.15, Cr 25-30, Ni 3-5, bal Fe.
Cast: 70,000-90,000 TS; 35,000-50,000 YS; 10-25 El; 10-25
RA. For stainless steel castings; resists nitric and acetic acids.
Obsolete

ALLEGHENY 29-9
Allegheny Ludlum Steel
C 0-0.2, Cr 27-31, Ni 8-10, bal Fe.
Cast: 80,000-95,000 TS; 45,000-70,000 YS; 20-30 El; 20-35
RA. For stainless steel castings; resists nitric and acetic acids.
Obsolete

ALLEGHENY 29-9S
Allegheny Ludlum Steel
C 0-0.12, Cr 27-31, Ni 8-10, bal Fe.
Cast: 75,000-95,000 TS; 36,000-46,000 YS; 20-40 El; 20-45
RA. For castings; stainless steel castings. *Obsolete*

ALLEGHENY 303 EZ
Allegheny Ludlum Steel
C 0.08, Cr 18, Ni 9, S 0.3, bal Fe.
Annealed: 80,000 TS; 35,000 YS; 45 El; 175 Brin. Cold drawn:
135,000-208,000 TS; 100,000-185,000 YS. For screw machine
products, fasteners, shafts. Free machining, stainless,
austenitic. Machinability rating of 70%. *Obsolete*

ALLEGHENY 309 S CB
Allegheny Ludlum Steel
C 0-0.08, Cr 23, Ni 13.5, Cb, bal Fe.
Austenitic stainless steel, modified AISI, with low carbon, and
Cb added to improve welding characteristics.

ALLEGHENY 33 FM
Allegheny Ludlum Steel
Cr 12-15, C 0-0.12, S 0-0.4, bal Fe.
Annealed: 73,000 TS; 35 El; 75 RA; 150 Brin. For chemical
engineering equipment; corrosion resistant, free machining.
Obsolete

ALLEGHENY 33 NH NON-HARDENING
Allegheny Ludlum Steel
Cr 11.5-13.5, Ni 0-0.5, C 0-0.08, Al 0.1-0.25, bal Fe.
Annealed: 70,000 TS; 39,000 YS; 30 El; 68 RA; 143 Brin. For
chemical engineering equipment; corrosion resistant.
Obsolete

ALLEGHENY 33 TURBINE
Allegheny Ludlum Steel
Cr 11.5-13, C 0-0.12, bal Fe.
Heat treated: 100,000 TS; 25 El; 70 RA; 200 Brin. For turbine
blades, chemical engineering equipment; corrosion resistant;
IZ 60 min. *Obsolete*

ALLEGHENY 33 W
Allegheny Ludlum Steel
Cr 12-15, C 0-0.12, W 2.5-3.5, bal Fe.
Annealed: 87,000 TS; 64,000 YS; 29 El; 64 RA; 170 Brin. For
chemical engineering equipment; high temperature service;
corrosion resistant. *Obsolete*

ALLEGHENY 34
Allegheny Ludlum Steel
C 0-0.12, Cr 14-16, bal Fe.
Annealed: 78,000 TS; 42,000 YS; 30 El; 65 RA; 156 Brin. For
corrosion resistant parts; corrosion resistant. *Obsolete*

ALLEGHENY 404
Allegheny Ludlum Steel
Cr 13.5, C, bal Fe.
Low residual stainless steel.

ALLEGHENY 410 HC
Allegheny Ludlum Steel
C 0.21, Cr 12.5, bal Fe.
Modified AISI 410 stainless with higher carbon for higher
mechanical properties. Similar to AISI 420.

ALLEGHENY 410 KB
Allegheny Ludlum Steel
C 0.3, Cr 12.5, bal Fe.
Higher carbon than AISI 410 to produce higher mechanical properties of heat treating. AISI 420.

ALLEGHENY 410 S
Allegheny Ludlum Steel
C 0.05, Cr 12.5, bal Fe.
Low carbon grade of AISI 410 for welding purposes. S41008

ALLEGHENY 416 EZ
Allegheny Ludlum Steel
C 0-0.15, Mn 0-1.25, P 0-0.06, Si 0-1, Cr 12-14, 0.15% S min, bal Fe.
Annealed: 73,000 TS; 42,000 YS; 35 El; 170 Brin. Hardened: 135,000 TS; 108,000 YS; 20 El; 285 Brin. For screw machine products, hardware, fasteners, shafts. Free machining, hardenable, corrosion resistant. *Obsolete*

ALLEGHENY 433
Allegheny Ludlum Steel
C 0.05, Cr 16.7, Mo 1, Cu 0.95, Mn 0.5, Si 0.3, bal Fe.
Annealed: 80,000 TS; 55,000 YS; 28 El. For kitchen utensils, automobile trim, hub caps, cowls; good formability and weldability, corrosion resistant. *Obsolete*

ALLEGHENY 436
Allegheny Ludlum Steel
C 0-0.12, Cr 17, Mo 1, Cb, bal Fe.
Modified Type 434 with Cb added. S43600

ALLEGHENY 439
Allegheny Ludlum Steel
Now AL 439.

ALLEGHENY 46 TUNGSTEN
Allegheny Ludlum Steel
C 0-0.35, Cr 5-7, W, bal Fe.
For corrosion resistant parts; corrosion resistant. *Obsolete*

ALLEGHENY 46M
Allegheny Ludlum Steel
C 0-0.25, Cr 4-6, Mo 0.4-0.6, bal Fe.
Annealed: 62,000 TS; 25,000 YS; 25 El; 60 RA; 180 Brin. For corrosion resistant parts, oil cracking stills; corrosion resistant. *Obsolete*

ALLEGHENY 4750
Allegheny Ludlum Steel
Ni 47, bal Fe.
For transformers; high permeability. *Obsolete*

ALLEGHENY 66W
Allegheny Ludlum Steel
Cr 15-18, C 0-0.12, W 2.5-3.5, bal Fe.
Annealed: 80,000 TS; 50,000 YS; 26 El; 50 RA; 170 Brin. For chemical engineering equipment; high temperature service; corrosion resistant. *Obsolete*

ALLEGHENY 67
Allegheny Ludlum Steel
Cr 18-23, C 0-0.35, bal Fe.
Annealed: 82,000 TS; 52,000 YS; 25 El; 55 RA; 163 Brin. For chemical engineering equipment; corrosion resistant. *Obsolete*

ALLEGHENY A-286
Crucible Specialty Steel
C 0-0.08, Mn 1-2, Cr 13.5-16, Ni 24-28, Mo 1.3, Ti 2, V 0.3, Al 0.2, bal Fe.
Rolled: 135,000-160,000 TS; 85,000-115,000 YS; 30-20 El; 55-30 RA. For jet engine and supercharger parts, afterburners; age-hardenable, austenitic, heat resistant.

ALLEGHENY A-286
Allegheny Ludlum Steel
C 0-0.08, Mn 1-2, Cr 13.5-16, Ni 24-28, Mo 1.3, Ti 2, V 0.3, Al 0.2, bal Fe.
Rolled: 135,000-160,000 TS; 85,000-115,000 YS; 30-20 El; 55-30 RA. For jet engine and supercharger parts, afterburners; age-hardenable, austenitic, heat resistant.

ALLEGHENY AF-183
Allegheny Ludlum Steel
C 0.3, Mn 18, Cr 12, Mo 3, V 0.75, N 0.2, bal Fe.
Heat treated: 136,800 TS; 81,600 YS; 33.5 El; 23.4 RA; 258 Brin. For gas turbine parts, afterburners, nozzles; high heat resistance, heat treatable. *Obsolete*

ALLEGHENY AF-71
Allegheny Ludlum Steel
C 0.25, Mn 18, Cr 12.5, Mo 3, B 0.18, N 0.2, V 0.8, bal Fe.
Rolled: 159,000 TS; 103,600 YS; 24 El; 37 RA. For jet and gas turbine components. *Obsolete*

ALLEGHENY AF-71
Allegheny Ludlum Steel
C 0.25, Mn 18, Cr 12.5, Mo 3, B 0.2, N 0.2, bal Fe.
Heat treated: 159,000 TS; 90,000 YS; 24 El; 37 RA. For aircraft gas turbines, missiles, airframes; austenitic, age-hardenable. *Obsolete*

ALLEGHENY ALX
Allegheny Ludlum Co.
Cr, bal Fe.
For cutting tools; cast to shape. *Obsolete*

ALLEGHENY AM 357
Allegheny Ludlum Steel
C 0.21-0.26, Mn 0.5-1.25, Cr 13.5, Ni 4-5, Mo 2.5-3.25, N 0.07-0.13, bal Fe.
Sheet: 309,000 TS; 303,000 YS; 4 El. At 1000 F: 158,000 TS; 141,000 YS; 5.5 El. For valves, pressure tanks, aircraft and jet engine components, missiles. Semi-austenitic, precipitation hardenable stainless steel. *Obsolete*

ALLEGHENY AM-367
Allegheny Ludlum Steel
C 0.02, Cr 14.25, Ni 3.4, Co 15.4, Mo 2, Ti 0.35, Al 0.03, bal Fe.
Sheet, cold rolled 50%, annealed and maraged. At 80 F: 245,7000 TS; 242,300 YS; 3 El. At 100 F: 261,900 TS; 261,000 YS; 3 El. For aircraft and missile industries, missile and gas turbine components. Maraging steel, stainless, martensitic. *Obsolete*

ALLEGHENY ARMATURE
Allegheny Ludlum Steel
Si 0.5, bal Fe.
For motors; good permeability. *Obsolete*

ALLEGHENY AUDIO TRANSFORMER "A"
Allegheny Ludlum Steel
Si 3.5, bal Fe.
For audio transformers; high permeability. *Obsolete*

ALLEGHENY AUDIO TRANSFORMER "B"
Allegheny Ludlum Steel
Si 3.5, bal Fe.
For audio transformers; high permeability. *Obsolete*

ALLEGHENY D979
Allegheny Ludlum Steel
C 0.05, Cr 15, Ni 15, Mo 4, W 4, Al 1, Ti 3, Fe 30, bal Ni.
Rolled: 200,000 TS; 142,000 YS; 17 El; 34 RA. For aircraft and auto engine components; heat resistant to 1600 F. *Obsolete*

ALLEGHENY DYNAMO
Allegheny Ludlum Steel
Si 2.5, bal Fe.
For electric machinery; high permeability. *Obsolete*

ALLEGHENY DYNAMO SPECIAL
Allegheny Ludlum Steel
Si 3.25, bal Fe.
For transformers. *Obsolete*

ALLEGHENY ELECTRIC METAL
Allegheny Ludlum Steel
Ni 47, bal Fe.
For transformers; high permeability. *Obsolete*

ALLEGHENY ELECTRICAL SHEET STEEL
Allegheny Ludlum Steel
Si 1, bal Fe.
For motors. *Obsolete*

ALLEGHENY FIELD
Allegheny Ludlum Steel
Si 0.25, bal Fe.
For electrical equipment. *Obsolete*

ALLEGHENY G-192
Allegheny Ludlum Steel
C 0.6, Cr 22, Mn 8.5, N 0.35, Si 0.5, bal Fe.
At RT: 148,000 TS; 85,000 YS; 55 El. At 1200 F: 92,000 TS; 37,000 YS; 15 El. For high temperature applications, heat treating equipment. Heat and corrosion resistant. *Obsolete*

ALLEGHENY LUDLUM -LMW
Allegheny Ludlum Steel
C 0.8, Cr 3.25-4.25, W 1.25-2, V 0.75-1.5, Mo 7.5-9.5, bal Fe.
For reamers, drills, lathe tools, broaches, hobs, form tools. Type M1 high speed steel. Good red-hardness. *Obsolete*

ALLEGHENY LUDLUM 1251
Allegheny Ludlum Steel
C 0.05, Mn 1.5, Cr 13, Mo 6, Ti 2.5, Al 0.2, B 0.01, Fe 35, Ni 43.
At 70 F: 165,000 TS; 110,000 YS; 14 El. At 1200 F: 135,000 TS; 100,000 YS; 20 El. For high temperature applications; heat and oxidation resistant. *Obsolete*

ALLEGHENY LUDLUM 129
Allegheny Ludlum Steel
C 0.7, Mn 0.3, Si 1, Cr 12, Mo 5.25, bal Fe.
Heat treated: 322,000 TS; 265,000 YS; 4 El; C 58 Rock. For tools and bearings for high temperature applications. Stainless, wear resistant, hardenable. High oxidation resistant. *Obsolete*

ALLEGHENY LUDLUM 21-12N
Allegheny Ludlum Steel
C 0.2, Cr 21, Ni 12, N 0.2, bal Fe.
Water quenched: 226 Brin.; at 1400 F: 145 Brin. For valves and valve seats, valve heads for high temperature applications. Heat and corrosion resistant. *Obsolete*

ALLEGHENY LUDLUM 418
Allegheny Ludlum Steel
C 0.2, Mn 1, Si 0.5, Cr 12.5, W 3, bal Fe.
For high temperature applications; heat resistant. *Obsolete*

ALLEGHENY LUDLUM 419
Allegheny Ludlum Steel
C 0.25, Mn 1, Si 0.3, Cr 11.5, Ni 0.5, Mo 0.5, W 2.5, V 0.4, bal Fe.
Heat treated: 204,000-272,000 TS; 134,000-206,000 YS; 5-33.5 El; 16-40 RA; 480-510 Brin. For high temperature applications; heat resistant, heat treatable. *Obsolete*

ALLEGHENY LUDLUM MF-1
Allegheny Ludlum Steel
Stainless steel. C 0.045, Cr 11, Ti 0.5, bal Fe.
Annealed, at 200°F: 58,900 psi TS; 30,000 psi YS; 29.5 El.
For automotive exhaust system applications.

ALLEGHENY LUDLUM TYPE 316
Allegheny Ludlum Steel
Stainless steel. C 0-0.08, Mn 0-2, Si 0-1, Cr 16-18, Ni 10-14, Mo 2-3, P 0-0.04, S 0-0.03, bal Fe.
85,000 psi TS; 40,000 psi YS; 50 El. Chemically resistant.

ALLEGHENY LUDLUM TYPE 316L
Allegheny Ludlum Steel
Stainless steel. C 0-0.03, Mn 0-2, Si 0-1, Cr 16-18, Ni 10-14, Mo 2-3, P 0-0.04, S 0-0.03, bal Fe.
85,000 psi TS; 40,000 psi YS; 50 El. Chemically resistant.

ALLEGHENY LUDLUM TYPE 317
Allegheny Ludlum Steel
Stainless steel. C 0-0.08, Mn 0-2, Si 0-1, Cr 18-20, Ni 11-15, Mo 3-4, P 0-0.04, S 0-0.03, bal Fe.
85,000 psi TS; 40,000 psi YS; 50 El. Chemically resistant.

ALLEGHENY LUDLUM TYPE 317L
Allegheny Ludlum Steel
Stainless steel. C 0-0.03, Mn 0-2, Si 0-1, Cr 18-20, Ni 11-15, Mo 3-4, P 0-0.04, S 0-0.03, bal Fe.
85,000 psi TS; 40,000 psi YS; 50 El. Chemically resistant.

ALLEGHENY LUDLUM TYPE 321
Allegheny Ludlum Steel
Stainless steel. C 0-0.08, Mn 0-2, Si 0-1, Cr 17-19, Ni 9-12, Mo 2-3, P 0-0.045, S 0-0.03, Ti = 5 x C, bal Fe.
RT: 86,650 psi TS; 33.250 psi YS; 51 El. For exhaust manifolds, cabin heaters, flash boilers, pressure vessels, fire walls and stack liners, boiler casings, and afterburners.

ALLEGHENY LUDLUM TYPE 405
Allegheny Ludlum Steel
Stainless steel. C 0-0.08, Mn 0-1, Si 0-1, Cr 11.5-14.5, Al 0.1-0.3, P 0-0.04, S 0-0.03, Ni 0-0.5, bal Fe.
Hot rolled: 89,000-100,000 psi TS; 78,000-88,000 psi YS; 12-13 El; 170-190 Brin. For annealing boxes, quenching racks, oxidation resistant partitions and baffles.

ALLEGHENY LUDLUM TYPE 408 CB
Allegheny Ludlum Steel
Stainless steel. C 0.02, Mn 0.32, P 0.024, S 0.002, Si 0.37, Cr 12, Ni 0.29, Ti 0.28, Cb 0.7, N 0.02, Al 1, bal Fe.
Annealed: 72,000 psi TS; 47,000 psi YS; 32 El; 81 Rock B. For automotive exhaust system components.

ALLEGHENY LUDLUM TYPE 410
Allegheny Ludlum Steel
Stainless steel. C 0-0.15, Mn 0-1, Si 0-1, Cr 11.5-13.5, P 0-0.04, S 0-0.03, bal Fe.
Annealed, RT: 60,000-75,000 psi TS; 32,000-42,000 psi YS; 20-40 El; 50-75 RA. For valve parts, screens, shafting or screw machine parts, in the chemical, petroleum and other processing industries.

ALLEGHENY LUDLUM TYPE 425 MODIFIED
Allegheny Ludlum Steel
Stainless steel. C 0.54, Mn 0.35, Si 0.35, Cr 13.5, Mo 1, bal Fe.
Annealed: 94,000 psi TS; 55,000 psi YS; 24 El; 93 Rock B. For cutlery and other cutting applications.

ALLEGHENY LUDLUM TYPE 430
Allegheny Ludlum Steel
Stainless steel. C 0-0.12, Mn 0-1, P 0-0.04, S 0-0.03, Si 0-1, Cr 16-18, bal Fe.
Sheet, strip, RT: 65-80 ksi TS; 45-60 ksi YS; 20-32 El. For interior and exterior trim applications.

ALLEGHENY LUDLUM TYPE 440A
Allegheny Ludlum Steel
Stainless steel. C 0.6-0.75, Mn 0-1, Si 0-1, Cr 16-18, Mo 0-0.75, Ni 0-0.5, S 0-0.03, P 0-0.04, bal Fe.
Annealed: 60-62 ksi YS; 96 Rock B. For nozzles, valve parts, hardened steel balls and seats for oil well pumps, cutlery, separating screens and strainers, shears and surgical equipment.

ALLEGHENY LUDLUM TYPE 441
Allegheny Ludlum Steel
Stainless steel. C 0.02, Mn 0.33, P 0.025, S 0.002, Si 0.39, Cr 18, Ni 0.32, Ti 0.27, Cb 0.69, N 0.021, Al 0.05, bal Fe.
Annealed: 74,000 psi TS; 48,000 psi YS; 29 El; 84 Rock B. For automotive exhaust system components.

ALLEGHENY LUDLUM TYPE 446
Allegheny Ludlum Steel
Stainless steel. C 0-0.35, Mn 0-1.5, Si 0-1, Cr 23-27, Ni 0-0.5, P 0-0.04, S 0-0.03, bal Fe.
Annealed: 75,000 psi TS min; 45,000 psi YS min; 20 El min; 40 RA min. For furnace parts, boiler baffles, recuperator parts, kiln linings, glass molds, furnace and stack dampers, pyrometer protection tubes.

ALLEGHENY M-17
Allegheny Ludlum Steel
C 0.65, Mn 0.25, Si 0.4, Cr 17, bal Fe.
Annealed: 95,000 TS; 55,000 YS; 20 El; 40 RA; 240 Brin.
Hardened: 275,000 TS; 240,000 YS; 2 El; 550 Brin. For cutlery; stainless. *Obsolete*

ALLEGHENY M-252
Allegheny Ludlum Steel
C 0.15, Cr 19, Ni 55, Co 10, Mo 10, Ti 2.5, Al 1, B 0.005, Fe 2.
Heat treated: 178,000 TS; 128,000 YS; 21 El. At 1500 F: 92,000 TS; 83,000 YS; 19 El. For high temperature service. For jet engine turbine blades, fasteners, missile and space vehicle components. Precipitation hardening. Oxidation resistant. *Obsolete*

ALLEGHENY METAL "A"
Allegheny Ludlum Steel
Cr 17-19, Ni 7-9, C 0-0.11, bal Fe.
Annealed: 75,000 TS; 50 El; 60 RA; 130 Brin. For chemical engineering equipment; stainless, austenitic. *Obsolete*

ALLEGHENY METAL "A-MO"
Allegheny Ludlum Steel
Cr 17-19, Ni 7-9, Mo 2-4, C 0-0.11, bal Fe.
Annealed: 85,000 TS; 50 El; 60 RA; 145 Brin. For chemical engineering equipment; stainless. *Obsolete*

ALLEGHENY METAL "A-TI"
Allegheny Ludlum Steel
Cr 17-19, Ni 7-9, C 0-0.2, Ti 1, bal Fe.
Annealed: 75,000 TS; 50 El; 60 RA; 130 Brin. For chemical engineering equipment; high temperature service; stainless, austenitic; range 800-1500 F. *Obsolete*

ALLEGHENY METAL "B"
Allegheny Ludlum Steel
Cr 18-20, Ni 8-10, C 0.08-0.2, bal Fe.
Annealed: 80,000 TS; 50 El; 60 RA; 135 Brin. For chemical engineering equipment; stainless, austenitic. *Obsolete*

ALLEGHENY METAL "B-TI"
Allegheny Ludlum Steel
Cr 18-20, Ni 8-10, C 0-0.2, Ti 1, bal Fe.
Annealed: 75,000 TS; 50 El; 60 RA; 130 Brin. For chemical engineering equipment; high temperature service; stainless, austenitic. *Obsolete*

ALLEGHENY METAL "C"
Allegheny Ludlum Steel
Cr 17-19, Ni 7-9, C 0.08-0.2, bal Fe.
Annealed: 80,000 TS; 50 El; 60 RA; 135 Brin. For chemical engineering equipment; stainless. *Obsolete*

ALLEGHENY METAL (TYPE 345)
Allegheny Ludlum Steel
C 0-0.15, Cr 17-19, Ni 8-12, Cb = 6 to 10 x C, bal Fe.
Annealed: 75,000 TS; 50 El; 60 RA; 130 Brin. For corrosion resistant parts; immune to intergranular corrosion. *Obsolete*

ALLEGHENY METAL (TYPE 346)
Allegheny Ludlum Steel
C 0-0.15, Cr 19-20, Ni 8-12, Cb = 6 to 10 x C.
Annealed: 75,000 TS; 50 El; 60 RA; 130 Brin. For corrosion resistant parts; immune to intergranular corrosion. *Obsolete*

ALLEGHENY METAL 12 CAST
Allegheny Ludlum Steel
C 0-0.17, Mn 0-1.5, Cr 11.5-13.5, Ni 0.75-1.25, bal Fe.
For glass molds, paper and pulp industry equipment; corrosion resistant. *Obsolete*

ALLEGHENY METAL 12 EZ
Allegheny Ludlum Steel
C 0-0.15, Cr 12-14, Zr 0-0.6, 0.07% S min, P, Se, Mo, bal Fe.
Annealed: 85,000 TS; 50,000 YS; 15 El; 40 RA; 180 Brin.
Hardened: 145,000 TS; 120,000 YS; 15 El; 65 RA; 280 Brin.
For screw machine products; corrosion resistant; Type 416. *Obsolete*

ALLEGHENY METAL 12 PLURAMELT
Allegheny Ludlum Steel
C 0-0.15, Mn 0-1, Cr 11.5-13.5, bal Fe.
For petroleum refining equipment; corrosion resistant. *Obsolete*

ALLEGHENY METAL 12-2
Allegheny Ludlum Steel
C 0-0.15, Cr 11.5-13.5, Ni 1.2-2.5, Mn 0-1, Si 0-1, bal Fe.
Annealed: 100,000 TS; 65,000 YS; 25 El; 50 RA; 250 Brin.
Heat treated: 220,000 TS; 175,000 YS; 2 El; 440 Brin. For valve parts, tie rods, pumps; AISI Type 414; corrosion resistant. *Obsolete*

ALLEGHENY METAL 12-NH
Allegheny Ludlum Steel
C 0-0.8, Mn 0-1, Cr 11.5-13.5, Al 0.1-0.3, bal Fe.
Annealed: 60,000 TS; 32,000 YS; 20 El; 50 RA; 180 Brin. For annealing boxes, quenching racks; corrosion resistant; Type 405. *Obsolete*

ALLEGHENY METAL 12-TB
Allegheny Ludlum Steel
C 0-0.15, Mn 1, Cr 11.5-13, bal Fe.
Annealed: 60,000 TS; 200 Brin. For turbine blades, buckets; corrosion resistant; Type 403. *Obsolete*

ALLEGHENY METAL 12-W
Allegheny Ludlum Steel
C 0-0.15, W 2.5-3.5, Cr 12-14, bal Fe.
Annealed: 75,000 TS; 40,000 YS; 20 El; 50 RA; 212 Brin.
Hardened: 200,000 TS; 180,000 YS; 10 El; 400 Brin. For oil refinery and chemical equipment; corrosion resistant; Type 418. *Obsolete*

ALLEGHENY METAL 15-35
Allegheny Ludlum Steel
Cr 14-16, Ni 33-36, C 0-0.25, bal Fe.
Casting: 50,000-75,000 TS; 40,000-56,000 YS; 40-60 El; 40-60 RA. For castings; stainless. *Obsolete*

ALLEGHENY METAL 16-1
Allegheny Ludlum Steel
C 0-0.2, Cr 15-17, Ni 1.2-2.5, Mn 0-1, Si 0-1, bal Fe.
Annealed: 105,000 TS; 90,000 YS; 20 El; 60 RA; 270 Brin.
Heat treated: 220,000 TS; 185,000 YS; 10 El; 240 Brin. For tie rods, pumps, valve parts; corrosion resistant. *Obsolete*

ALLEGHENY METAL 17
Allegheny Ludlum Steel
C 0-0.12, Mn 0-1, Cr 1-18, bal Fe.
Annealed: 60,000 TS; 35,000 YS; 20 El; 40 RA; 200 Brin.
Hardened: 150,000 TS; 110,000 YS; 3 El; 300 Brin. For automotive trim, cold headed parts; corrosion and heat resistant; Type 340. *Obsolete*

ALLEGHENY METAL 17 PLURAMELT
Allegheny Ludlum Steel
C 0-0.12, Mn 0-1, Cr 14-18, bal Fe.
For tanks and parts in nitric acid industries; corrosion resistant. *Obsolete*

ALLEGHENY METAL 17EZ
Allegheny Ludlum Steel
C 0-0.12, Cr 14-18, Ni 0-0.5, Mn 0-1.25, P, S or Se, 0.7% min or 0.6% max Mo, bal Fe.
Annealed: 60,000 TS; 35,000 YS; 15 El; 40 RA. Heat treated: 150,000 TS; 110,000 YS; 3 El; 300 Brin. For screw machine products, cold headed parts; AISI Type 430F; free cutting. *Obsolete*

ALLEGHENY METAL 18-11C
Allegheny Ludlum Steel
C 0-0.08, Cr 17-18.5, 10% Ni min, Cb = 8 x 10 C, bal Fe.
Cast: 75,000-90,000 TS; 35-50 El; 40-50 RA. For castings,
stainless parts; stainless steel castings. *Obsolete*

ALLEGHENY METAL 18-8 C PLURAMELT
Allegheny Ludlum Steel
C 0-0.08, Cr 17-19, Ni 9-12, Mn 0-2, Cb = 10 x C, bal Fe.
For chemical industry equipment; austenitic, stainless.
Obsolete

ALLEGHENY METAL 18-8 CAST
Allegheny Ludlum Steel
C 0.09-0.16, Cr 18-20, Ni 8-10, Mn 0-1.5, bal Fe.
Annealed: 73,500 TS; 154 Brin. For pump and valve parts,
chemical plant equipment; stainless, austenitic; Type 302.
Obsolete

ALLEGHENY METAL 18-8 PLURAMELT
Allegheny Ludlum Steel
C 0.08-0.2, Cr 17-19, Ni 8-10, Mn 0-2, bal Fe.
For cooking utensils; austenitic, stainless. *Obsolete*

ALLEGHENY METAL 18-8 S PLURAMELT
Allegheny Ludlum Steel
C 0-0.08, Cr 18-20, Ni 8-11, Mn 0-2, bal Fe.
For pans, trays, tanks, for chemical industry; austenitic,
stainless. *Obsolete*

ALLEGHENY METAL 18-8C CAST
Allegheny Ludlum Steel
C 0-0.1, Mn 0-1.5, Cr 17.5-20, Ni 9-12, Cb = 8 x C, bal Fe.
Annealed: 73,500 TS; 145 Brin. For pump and valve liners,
chemical plant equipment; stainless welded structures; Type
347. *Obsolete*

ALLEGHENY METAL 18-8FS
Allegheny Ludlum Steel
C 0-0.12, Cr 17-19, Ni 10-13, Mn 0-2, Si 0-1, bal Fe.
Annealed: 75,000 TS; 25,000 YS; 50 El; 60 RA; 180 Brin. For
stainless structures; stainless. *Obsolete*

ALLEGHENY METAL 18-8M
Allegheny Ludlum Steel
C 0-0.1, Cr 16-18, Ni 10-14, Mo 2-3, bal Fe.
Annealed: 75,000 TS; 30,000 YS; 40 El; 50 RA; 200 Brin. For
chemical equipment; Stainless Type 316. *Obsolete*

ALLEGHENY METAL 18-8M CAST
Allegheny Ludlum Steel
C 0-0.12, Cr 17.5-19.5, Ni 8-12, Mo 2-3, bal Fe.
Annealed: 83,000 TS; 161 Brin. For chemical plant
equipment, mixers, agitators; acid resistant; Type 316;
austenitic. *Obsolete*

ALLEGHENY METAL 18-8M SPECIAL
Allegheny Ludlum Steel
C 0-0.1, Cr 18-20, Ni 11-14, Mo 3-4, Mn 0-2, Si 0-1, bal Fe.
Annealed: 75,000 TS; 30,000 YS; 40 El; 50 RA; 200 Brin. For
pharmaceuticals production; stainless. *Obsolete*

ALLEGHENY METAL 18-8MC
Allegheny Ludlum Steel
C 0-0.08, Cr 17-19, Ni 10-14, Mo 1.75-2.5, Cb 0.5-0.7, bal Fe.
Annealed: 75,000 TS; 30,000 YS; 40 El; 50 RA; 200 Brin. For
chemical equipment; resists pitting and intergranular attack.
Obsolete

ALLEGHENY METAL 18-8S
Allegheny Ludlum Steel
C 0-0.08, Cr 18-20, Ni 8-10, bal Fe.
Casting: 70,000-85,000 TS; 35,000-45,000 YS; 50-75 El; 45-65
RA; 180 Brin. For castings, stainless parts; stainless, resists
acids. *Obsolete*

ALLEGHENY METAL 18-8S CAST
Allegheny Ludlum Steel
C 0-0.08, Cr 18-21, Ni 8-10, Mn 0-1.5, bal Fe.
Annealed: 79,000 TS; 130 Brin. For pump and valve parts,
chemical plant equipment; stainless, austenitic; Type 304.
Obsolete

ALLEGHENY METAL 25-12 CAST
Allegheny Ludlum Steel
C 0.2-0.3, Mn 0-2, Cr 23-25, Ni 12-14, bal Fe.
Annealed: 80,000 TS; 163 Brin. For furnace parts, chemical
plant equipment; corrosion and heat resistant, austenitic.
Obsolete

ALLEGHENY METAL 25-12C
Allegheny Ludlum Steel
C 0-0.3, Cr 22-24, Ni 12-15, Mn 0-2, Cb = 10 x C, bal Fe.
Annealed: 75,000 TS; 30,000 YS; 40 El; 50 RA; 200 Brin. For
stainless parts; resists intergranular corrosion. *Obsolete*

ALLEGHENY METAL 25-2
Allegheny Ludlum Steel
C 0-0.3, Cr 25-30, Ni 2-5, Mn 0-2, 0.05-0.15% N 2 , bal Fe.
For furnace parts, chemical industry; corrosion resistant.
Obsolete

ALLEGHENY METAL 25-2 CAST
Allegheny Ludlum Steel
C 0-0.3, Mn 0-2, Cr 25-30, Ni 2-5, 0.05-0.15% N 2 , bal Fe.
For furnace parts, chemical plant and pulp equipment;
corrosion resistant. *Obsolete*

ALLEGHENY METAL 25-20 + SI
Allegheny Ludlum Steel
C 0-0.25, Cr 23-26, Ni 19-22, Si 1.5-3, Mn 0-2, bal Fe.
Annealed: 75,000 TS; 30,000 YS; 40 El; 50 RA; 180 Brin. For
stainless and heat resistant parts; stainless, heat resistant.
Obsolete

ALLEGHENY METAL 25-20 H
Allegheny Ludlum Steel
C 0-0.25, Cr 23-26, Ni 17-21, Mn 0-1.5, bal Fe.
Cast: 80,000 TS; 163 Brin. For castings, chemical industry,
furnace parts; corrosion resistant. *Obsolete*

ALLEGHENY METAL 25-20C CAST
Allegheny Ludlum Steel
C 0-0.15, Mn 0-1.5, Cr 23-26, Ni 17-21, bal Fe.
Annealed: 80,000 TS; 163 Brin. For furnace parts, chemical
plant equipment; corrosion and heat resistant, austenitic.
Obsolete

ALLEGHENY METAL 25-20H CAST
Allegheny Ludlum Steel
C 0-0.25, Mn 0-1.5, Cr 23-26, Ni 17-21, bal Fe.
Annealed: 80,000 TS; 163 Brin. For furnace parts, chemical
plant equipment; corrosion and heat resistant, austenitic.
Obsolete

ALLEGHENY METAL 28 PLURAMELT
Allegheny Ludlum Steel
C 0-0.35, Mn 0-1.5, Cr 23-27, 0.25% N 2 max, bal Fe.
For furnace linings and parts; corrosion and heat resistant.
Obsolete

ALLEGHENY METAL 42
Allegheny Ludlum Steel
Ni 42, bal Fe.
Cold drawn: 70,000 TS; 50,000 YS; 28 El; 116 Brin. For
glass-to-metal seals; controlled expansion. *Obsolete*

ALLEGHENY METAL 46SM
Allegheny Ludlum Steel
C 0-0.1, Cr 4-6, Ni 0-0.5, Mn 0-1, Mo 0.4-0.6, bal Fe.
Annealed: 60,000 TS; 25,000 YS; 25 El; 60 RA; 180 Brin. Heat
treated: 150,000 TS; 110,000 YS; 15 El. For oil refinery and
chemical equipment; AISI Type 502 + Mo; corrosion
resistant. *Obsolete*

ALLEGHENY METAL AM-350
Now AM-350.

ALLEGHENY METAL AM-355
Now AM-355.

ALLEGHENY METAL B (SP)
Allegheny Ludlum Steel
C 0-0.11, Cr 18-20, Ni 8-10, bal Fe.
Annealed: 75,000 TS; 50 El; 60 RA; 130 Brin. For general
engineering purposes; welding grade. *Obsolete*

ALLEGHENY METAL FM FREE MACHINING
Allegheny Ludlum Steel
Cr 17-19, Ni 7-9, C 0.08-0.2, S 0.3, bal Fe.
Annealed: 75,000 TS; 25,000 YS; 50 El; 60 RA; 130 Brin. For
chemical engineering equipment; corrosion resistant; free
machining. *Obsolete*

ALLEGHENY METAL H-17
Now AISI 440C.

ALLEGHENY METAL L-12
Allegheny Ludlum Steel
Cr 12-14, Ni 0-0.5, 0.15% C min, bal Fe.
Annealed: 90,000 TS; 50,000 YS; 15 El; 40 RA; 240 Brin. For
cutlery; Stainless Type 420; hardenable. *Obsolete*

ALLEGHENY METAL MF-1
Allegheny Ludlum Steel
C 0-0.045, Cr 11, Ti 0.5, bal Fe.
Sheet: 65,000 TS; 35,000 YS; 30 El; B 80 Rock. For
automobile mufflers, cargo containers, drainage culverts,
transformer cases. Corrosion resistant. Good weldability.
Obsolete

ALLEGHENY METAL S-590
Allegheny Ludlum Steel
C 0.38-0.47, Mn 1-2, Cr 19-21, Ni 19-21, Co 19-21, Mo
3.5-4.5, W 3.5-4.5, Cb 3.5-4.5, bal Fe.
Heat resistant to 1100-1400°F. Corrosion resistant.

ALLEGHENY METAL S-816
Allegheny Ludlum Steel
C 0.4, Mn 1.2, Cr 20, Ni 20, Co 43, W 4, Fe 3, Cb 4.
For turbine blades, rotors, high temperature hardware, jet
propulsion engine parts. High heat and corrosion resistant.
For high temperature service to 1520°F.

ALLEGHENY METALCLAD STEEL
Allegheny Ludlum Steel
20% of 18-8 stainless steel rolled out 80% of plain C, steel.
For sheet or plate to resist corrosion; stainless. *Obsolete*

ALLEGHENY MOLY IRON
Allegheny Ludlum Steel
C 0.02, Mo 3, bal Fe.
Annealed: 50,000 TS; 25,000 YS; 38 El; 62 RA; 107 Brin.
Rolled: 60,000 TS; 45,000 YS; 29 El; 55 RA; 185 Brin. For
D.C. relays, core material; low hysteresis loss. *Obsolete*

ALLEGHENY NO. 33
Allegheny Ludlum Steel
C 0-0.12, Mn 0-0.5, Si 0-0.5, Cr 12-15, bal Fe.
Annealed: 77,300 TS; 40,600 YS; 37 El; 78.7 RA; 153 Brin.
Heat treated: 109,000-178,750 TS; 91,000-161,500 YS; 15-23
El; 51-70 RA; 228-364 Brin. At 1000 C: 70,000 TS; 56.5 El;
92.5 RA. For automotive parts, interval engine and steam
engine parts, tanks, fans, blowers; corrosion resistant.
Obsolete

ALLEGHENY NO. 44
Allegheny Ludlum Steel
C 0-0.2, Mn 0-1.25, Si 0-0.5, Cr 22-26, Ni 11-13, bal Fe.
Hot rolled: 130,500 TS; 107,300 YS; 31 El; 50.8 RA; 202 Brin.
Annealed: 96,000 TS; 45,000 YS; 62 El; 69 RA; 143 Brin. For
furnace and pump parts, boiler baffles, carburizing boxes,
reuperators; high heat resistance; easy to fabricate.
Obsolete

ALLEGHENY NO. 55

Allegheny Ludlum Steel
C 0-0.35, Mn 0-1, Si 0-0.6, Cr 23-30, Ni 0-0.6, bal Fe.
Annealed forgings: 75,000-100,000 TS; 45,000-60,000 YS;
25-35 El; 45-65 RA; 160-200 Brin. Annealed castings:
40,000-80,000 TS; 30,000-70,000 YS; 2-5 El; 2-6 RA; 170-190
Brin. For furnace parts, boiler baffles, furnace dampers,
recuperators; heat resistant; corrosion resistant to acid mine
waters. *Obsolete*

ALLEGHENY NO. 66

Allegheny Ludlum Steel
C 0-0.12, Cr 16-18, bal Fe.
Annealed: 70,000-85,000 TS; 40,000-55,000 YS; 20-35 El;
75-85 RA; 156 Brin. For hardware, steam engine parts, fans,
blowers, condensers, evaporators; corrosion resistant to nitric
acid. *Obsolete*

ALLEGHENY R41

Allegheny Ludlum Steel
C 0.1, Cr 19, Ni 53, Co 11, Mo 10, Ti 3.2, Al 1.6, B 0.005, Fe
2.
Heat treated: 206,000 TS; 154,000 YS; 14 El. At 1600 F:
90,000 TS; 80,000 YS; 19 El. For afterburners, turbine
casings, combustion liners, high speed airframe
components. Precipitation hardening. Heat resistant.
Obsolete

ALLEGHENY RADIO TRANSFORMER

Allegheny Ludlum Steel
Si 3.8, bal Fe.
For electrical equipment, chokes, transformers; low induction.
Obsolete

ALLEGHENY S-816 + B

Allegheny Ludlum Steel
C 0.32-0.42, Cr 19-21, Ni 19-21, W 3.5-5, Mo 3.5-4.5, Mn
0.6-1.8, Fe 0-5, B, 3.0-4.5% Cb + Ta, bal Co.
Heat treated: 140,000 TS; 67,000 YS; 31 El; 31 RA; 269-352
Brin. For high temperature applications; age-hardenable,
heat and corrosion resistant. *Obsolete*

ALLEGHENY SUPER DYNAMO

Allegheny Ludlum Steel
Si 2.5, bal Fe.
For electrical equipment, transformers; low energy losses.
Obsolete

ALLEGHENY TRANSFORMER A

Allegheny Ludlum Steel
Si 4-5, bal Fe.
For electrical equipment, transformers; low energy losses.
Obsolete

ALLEGHENY TRANSFORMER AA

Allegheny Ludlum Steel
Si 4-5, bal Fe.
For power transformers; lowest energy losses. *Obsolete*

ALLEGHENY TRANSFORMER B

Allegheny Ludlum Steel
Si 4-5, bal Fe.
For electrical equipment, transformers; low energy losses.
Obsolete

ALLEGHENY TRANSFORMER C

Allegheny Ludlum Steel
Si 4-5, bal Fe.
For electrical equipment; low energy losses. *Obsolete*

ALLEGHENY TRC

Allegheny Ludlum Steel
C 0-0.1, Cr 15, Mn 16, Ni 0-1, 0.15% N 2 max, bal Fe.
Annealed: 80,000-120,000 TS; 40,000 min YS; 50 min El;
Rockwell B95. For railroad cars and trailer truck bodies;
stainless, austenitic, good formability. *Obsolete*

ALLEGHENY TYPE 201

Now AISI 201.

ALLEGHENY TYPE 203 EZ

Allegheny Ludlum Steel
C 0.06, Cr 17, Ni 6, Mn 6, Cu 2, S 0.3, bal Fe.
Bar annealed: 84,000 TS; 40,000 YS; 58 El, 183 Brin. Bar,
cold drawn: 135,000-208,000 TS. For screw machine
products, fasteners, hardware, shafts. Austenitic, stainless,
free-cutting. Machinability rating 85%. *Obsolete*

ALLEGHENY TYPE 332

Allegheny Ludlum Steel
Stainless steel. C 0-0.08, Cr 21, Ni 32, Ti 0.4, Al 0.3, Mn 1, Si
0.35, Cu 0.3, bal Fe.
Resists oxidation and carburization at elevated temperatures
resistant to chloride stress corrosion cracking.

ALLEGHENY TYPE 334

Allegheny Ludlum Steel
C 0-0.05, Cr 20, Ni 20, Ti 0.5, Al 0.5, bal Fe.
Austenitic stainless steel with good oxidation resistance at
elevated temperatures.

ALLEGHENY TYPE 347

Allegheny Ludlum Steel
Stainless steel. C 0-0.08, Cr 17-19, Ni 9-13, Mn 0-2, Si 0-1, Cb
+ Ta = 10 x C, bal Fe.
Annealed: 75,000 TS; 50 El; 60 RA; 130 Brin. For corrosion
resistant parts; immune to intergranular corrosion.

ALLEGHENY TYPE 348

Allegheny Ludlum Steel
C 0-0.08, Cr 17-19, Ni 9-13, Mn 0-2, Si 0-1, Cb + Ta = 10 x C,
bal Fe.
Annealed: 75,000 TS; 50 El; 60 RA; 130 Brin. For corrosion
resistant parts; immune to intergranular corrosion.

ALLEGHENY TYPE 434

Allegheny Ludlum Steel
Stainless steel. C 0-0.12, Cr 16-18, Mo 0.75-1.25, Mn 0-1, Si
0-1, bal Fe.
74-84 ksi TS; 50-65 ksi YS; 22-31 El. For interior and exterior
trim applications. Corrosion and oxidation resistant. S43400

ALLEGHENY V-36

Allegheny Ludlum Steel
C 0.25, Cr 25, Ni 20, W 2, Mo 4, Co 44, Cb 2, bal Fe.
For high temperature applications; oxidation and corrosion
resistant. *Obsolete*

ALLEGHENY-LUDLUM 88

Allegheny Ludlum Steel
Ni 6.5-14, Mn 5-12, bal Fe.
For electrical apparatus; nonmagnetic. *Obsolete*

ALLEN "B" METAL

Edgar Allen Balfour Ltd.
C 0.22-0.27, Si 0.2-0.35, Mn 0.6-0.8, bal Fe.
For machinery parts, gears; castings.

ALLEN "C" STEEL

Edgar Allen Balfour Ltd.
C 0.3-0.35, Si 0.2-0.35, Mn 0.6-0.8, bal Fe.
For gears, pinions, machinery parts; castings.

ALLEN "D" STEEL

Edgar Allen Balfour Ltd.
C 0.4-0.5, Si 0.2-0.35, Mn 0.7-1, bal Fe.
For gears, machinery parts, housings; castings.

ALLEN "T" STEEL

Edgar Allen Balfour Ltd.
C 0.14-0.16, Si 0.2-0.35, Mn 0.4-0.5, bal Fe.
For gears, shafts, housings; castings.

ALLEN C. CU

Edgar Allen Balfour Ltd.
C 0.3-0.35, Si 0.2-0.35, Mn 0.6-0.8, Cu 0.8-1.1, bal Fe.
For machinery parts; castings.

ALLEN CAST IRON SOLDER

L.B. Allen Co.
For soldering cast iron.

ALLEN CR. MO. "P"

Edgar Allen Balfour Ltd.
C 0.43-0.48, Si 0.2-0.35, Mn 0.5-0.8, Cr 0.7-0.9, Mo 0.25-0.35,
bal Fe.
For gears, shafts, crankshafts, bolts, housings; castings.

ALLEN CR. MO. 329

Edgar Allen Balfour Ltd.
C 0.2-0.35, Si 0.2-0.35, Mn 0.4-0.6, Cr 3-3.25, Mo 0.45-0.55,
bal Fe.
For oil refinery equipment; castings.

ALLEN CR. MO./F. MO

Edgar Allen Balfour Ltd.
C 0.5-0.55, Si 0.2-0.35, Mn 0.5-0.8, Cr 0.7-0.9, Mo 0.25-0.35,
bal Fe.
For gears, shafts, axles, bolts, crankshafts; castings.

ALLEN HI LUSTER

L.B. Allen Co.
For solder for stainless steel; bright finish.

ALLEN MAGNESIUM SOLDER

L.B. Allen Co.
For solder for magnesium.

ALLEN MT. "P"

Edgar Allen Balfour Ltd.
C 0.22-0.27, Si 0.2-0.35, Mn 1.2-1.6, bal Fe.
For machinery parts, gears, camshafts; castings.

ALLEN MT. "R"

Edgar Allen Balfour Ltd.
C 0.25-0.33, Si 0.2-0.35, Mn 1.2-1.6, bal Fe.
For machinery parts, camshafts, gears; castings.

ALLEN MT. MO. "P"

Edgar Allen Balfour Ltd.
C 0.18-0.25, Si 0.2-0.35, Mn 1.2-1.6, Mo 0.2-0.3, bal Fe.
For machinery parts, gears, camshafts; castings.

ALLEN MT. MO. "R"

Edgar Allen Balfour Ltd.
C 0.25-0.33, Si 0.2-0.35, Mn 1.2-1.6, Mo 0.2-0.3, bal Fe.
For gears, shafts, crankshafts, castings.

ALLEN PRESTO SOLDER

L.B. Allen Co.
Sn, Pb, Ag.
For solder.

ALLEN RED BRONZE

Manufacturer not listed.
Cu 55-70, Pb 20-40, Sn 5-10, S 1.
For bearings, hardware; tough.

ALLENITE A.S.

Edgar Allen Balfour Ltd.
WC.
For cutting tools; sintered carbide.

ALLENITE C.H.

Edgar Allen Balfour Ltd.
WC.
For cutting tools for chilled iron rolls; sintered carbide.

ALLENITE D.N.

Edgar Allen Balfour Ltd.
WC.
For drawing dies, wood working tools; sintered carbide.

ALLENITE E.S.

Edgar Allen Balfour Ltd.
WC.
For cutting tools for finish machining; sintered carbide.

ALLENITE L. C. G.

Edgar Allen Balfour Ltd.
WC.
For percussion drilling tools; sintered carbide.

ALLENITE N
Edgar Allen Balfour Ltd.
WC.
For cutting tools for cast iron; sintered carbide.

ALLENITE T.N.
Edgar Allen Balfour Ltd.
WC.
For cutting tools for intermittent cuts; sintered carbide.

ALLENITE T.S.
Edgar Allen Balfour Ltd.
WC.
For cutting tools for rough machining; sintered carbide.

ALLENOY
Allen Manufacturing Co.
C 0.35-0.4, Mn 0.9, Mo 0.2-0.3, bal Fe.
Hardened: 170,000 TS; 150,000 YS; 12 El; 35 RA; 400 Brin.
For set and cap screws; water hardening.

ALLENOY 1038
Allen Manufacturing Co.
C 0.35-0.42, Mn 0.6-0.9, bal Fe.
Heat treated: 145,000-165,000 TS; 1158000-135,000 YS; 11
El; 35 RA; 350-40 Brin. For pipe plugs; similar to AISI 1038.

ALLENOY 4137
Allen Manufacturing Co.
C 0.35-0.4, Mn 0.7-0.9, Cr 0.8-1.1, Mo 0.15-0.25, bal Fe.
Heat treated: 170,000-225,000 TS; 150,000-180,000 YS; 12-8
El; 35-25 RA; 400-500 Brin. For nuts, cap screws, set screws;
similar to AISI 4137.

ALLENOY 8650
Allen Manufacturing Co.
C 0.48-0.53, Mn 0.75-1, Ni 0.4-0.7, Cr 0.4-0.6, Mo 0.15-0.25,
bal Fe.
For wrenches; similar to AISI 8650.

ALLENS IMPERIAL SPECIAL
American manufacture
C 0.68, Mn 0.29, Cr 1.64, W 19.1, V 1.37, bal Fe.
For tools, cutters, dies, punches; high speed steel.

ALLENS OIL HARD
American manufacture
C 0.8, Mn 1.64, bal Fe.
For tools, dies; non-deforming.

ALLIAGE-180
Gilby-Fodor S.A.
Ni 21-23, bal Cu.
For heating elements. Useful operating temperature up to
950°F.

ALLIAGE-30
Gilby-Fodor S.A.
Ni 2, bal Cu.
For heating elements. Useful operating temperature up to
300°C.

ALLIAGE-60
Gilby-Fodor S.A.
Ni 5-7, bal Cu.
For heating elements. Useful operating temperature up to
300°C.

ALLIAGE-90
Gilby-Fodor S.A.
Ni 11-13, bal Cu.
For heating elements. Useful operating temperature up to
400°C.

ALLIS-CHALMERS X-1
Allis-Chalmers Mfg. Co.
C, alloy, bal Fe.
600 Brin. For hard facing electrodes; for abrasion and mild
impact.

ALLIS-CHALMERS X-2
Allis-Chalmers Mfg. Co.
C, Cr, Mo, bal Fe.
For hard surfacing, arc welding electrodes; abrasion resistant;
erosion and mild impact.

ALLIS-CHALMERS X-3
Allis-Chalmers Mfg. Co.
C, Cr, Mo, bal Fe.
For hard surfacing, arc welding electrodes; resists severe
abrasion; erosion and impact.

ALLISITE
Allis-Chalmers Mfg. Co.
C 3, bal Fe.
For machinery castings; high strength cast iron.

ALLITE
Allied Products Corp.
Zn, alloy.
For forming and stamping dies.

ALLOY 10
CCS Braeburn Alloy Steel
C 0.8, bal Fe.
For springs, punches, drills, taps. Type S5. *Obsolete*

ALLOY 100NT-2
English manufacture
C 1, Mn 1.5, Cr 20, Ni 30, Mo 3, W 2.2, Ta 2, Co 20, bal Fe.
Heat treated: 100,000 TS; 1.5 El; 1.0 RA. For high
temperature abrasion resistant parts; corrosion resistant.

ALLOY 110VT-2
English manufacture
C 1.1, Ni 20, Cr 23, Mo 6, Ta 2, bal Co.
For high temperature applications; heat and corrosion
resistant.

ALLOY 11A45
Heppenstall Co.
C 0.4, bal Fe.
For gears, shafts, axles, bolts; water hardened. *Obsolete*

ALLOY 12-2
Climax Performance Materials Corp.
C 1.1-1.4, Mn 12-14, Mo 1.9-2.1, Si 0.4-1, bal Fe.
Heat treated: 130,000 TS; 70,000 YS; 30 El; 30 RA; 215 Brin.
For wear and abrasion resistant parts; wear and abrasion
resistant.

ALLOY 15-3
Climax Performance Materials Corp.
C 3-4, Si 0.1, Mn 0.5-0.9, Cr 12-18, Mo 2.4, bal Fe.
Cast: 550 Brin. Heat treated: 620 Brin. For sand pumps,
chute liners, brick mold liners; tough and wear resistant, cast
iron.

ALLOY 150S
Kaiser Aluminum & Chemical Corp.
Mg 1-1.8, Fe 0-0.8, bal Al.
Annealed: 21,000 TS; 8,500 YS; 25 El. Rolled: 32,000 TS;
29,000 YS; 5 El. For general purpose, trim molding; sheets.
Obsolete

ALLOY 17-W
Cytemp Specialty Steel Div.
C 0.5, Cr 12, Ni 19, W 2.2, Mo 1, bal Fe.
For supercharger wheels for jet engines; heat resistant.
Obsolete

ALLOY 1751
Engelhard Corp.
Now ENGELHARD ALLOY NO. 19182.

ALLOY 19-9 DL
Cytemp Specialty Steel Div.
Now UNITEMP 19-9 DL.

ALLOY 19-9 WMO
Cytemp Specialty Steel Div.
C 0.08-0.12, Ni 8-10, Cr 18-22, Mo 0.2-0.5, W 1-1.5, Cb
0.2-0.6, Ti 0.2-0.6, bal Fe.
For jet engines, gas turbines, buckets, wheels, welding rod;
heat resistant cast alloy. *Obsolete*

ALLOY 2112
American manufacture
Cr 21, Ni 12, Si 0.8, Mn 1.4, C 2, bal Fe.
For exhaust valves; heat resistant, austenitic.

ALLOY 2129
Telcon Metals Ltd.
Ni 45-50, bal Fe.
For magnetic applications; soft. *Obsolete*

ALLOY 21VFS
Light Metal Works
Cu 2.5, Ni 1-2.5, Mg 0.5-2, Si, Fe, V, bal Al.
For pistons, cylinder heads; age-hardened.

ALLOY 22-3
Spang Specialty Metals
Ni 22, Cr 3.1, C 0.1, Mn 0.5, Si 0.25, bal Fe.
Annealed: 49 kg/mm² TS; 28 kg/mm² YS; 35 El; 74 Rock B.
Mean coefficient of thermal expansion at 30-450°C: 19.9 x
10^{-6}. cm/cm·°C. x 10^{-6}. *Obsolete*

ALLOY 223
American Electro Metals Corp.
Cr 81, Mo 16, Si 3.
For high temperature applications; oxidation and creep
resistant.

ALLOY 234-A-5
Manufacturer not listed.
C 0.38, Mn 4.2, Cr 18.5, Ni 4.5, Mo 1.3, W 1.3, Cb 0.57, bal
Fe,
For high temperature applications; heat resistant.

ALLOY 234A5
Manufacturer not listed.
C 0.1, Cr 19, Ni 5, Mn 4, Mo, W, Cb, Ti, bal Fe.
For jet and turbine parts; high heat resistance.

ALLOY 29-17
Spang Specialty Metals
Ni 28.8-29.2, C 0-0.04, Mn 0-0.5, Si 0-0.2, Co 16.7-17.3, bal
Fe.
Controlled expansion alloy in sheet and strip form. Conforms
to ASTM F-15 and MIL I 23011C, Class I with regards to
composition, temper and expansion properties.

ALLOY 3-2-1
Climax Performance Materials Corp.
C 3.3-3.6, Ni 2.75-3.25, Cr 1.5-2, Mo 0.7-1.1, Mn 0.5-0.8, Si
0.3-0.6, bal Fe.
55-62 Rock C. For ball mills, grinding balls, wear shoes,
liners, chutes; abrasion resistant, martensitic, white iron.

ALLOY 3-C DIE
Edgar T. Ward's Sons Co.
C 0.9, Mn 1.2, bal Fe.
55-62 Rock C. For ball mills, grinding balls, wear shoes,
liners, chutes; abrasion resistant, martensitic white iron.

ALLOY 30
Isabellenhuette
Copper. Cu 98, Ni 2.
220 N/mm² TS. For electrical equipment and instruments.
Resistance alloy. Maximum working temperature to 250°C.

ALLOY 300M
Universal Cyclops
C 0.4-0.47, Si 1.4-1.8, Ni 1.6-2, Cr 0.7-0.95, Mo 0.3-0.45, V
0.05-0, bal Fe.
Heat treated: 290,000 TS; 250,000 YS; 20 El; 10 RA; 540 Brin.
For aircraft landing gears, bolts, fittings; fatigue and creep
resistant.

ALLOY 300M

Bethlehem Steel Corp.
C 0.4-0.47, Si 1.4-1.8, Ni 1.6-2, Cr 0.7-0.95, Mo 0.3-0.45, V 0.05-0, bal Fe.
Heat treated: 290,000 TS; 250,000 YS; 20 El; 10 RA; 540 Brin.
For aircraft landing gears, bolts, fittings; fatigue and creep resistant.

ALLOY 325M

Cytemp Specialty Steel Div.
C, alloy, bal Fe.
For aircraft landing gears, bolts; fatigue and creep resistant.
Obsolete

ALLOY 348

German manufacture
Fe 46, Ni 38, Al 16.
For permanent magnet.

ALLOY 348

English manufacture
Fe 46, Ni 38, Al 16.
For permanent magnets; high permeability.

ALLOY 349

English manufacture
Fe 78, Ni 14, Al 8.
For permanent magnets; high permeability.

ALLOY 35

Westinghouse Electric Corp.
Ag 2.5, Cu 0.25, bal Pb.
For solder; MP 304°C. *Obsolete*

ALLOY 36

Spang Specialty Metals
Ni 36, C 0.12, Mn 0.35, Si 0.3, bal Fe.
Annealed: 46 kg/mm^2 TS; 28 kg/mm^2 YS; 35 El; 70 Rock B.
Mean coefficient of thermal expansion at -18 to +175°C: 1.63 x 10^{-6}cm/cm·°. *Obsolete*

ALLOY 39

Spang Specialty Metals
Ni 39, C 0.08, Mn 0.4, Si 0.25, bal Fe.
Annealed: 53 kg/mm^2 TS; 27 kg/mm^2 YS; 30 El; 76 Rock B.
Mean coefficient of thermal expansion at 25-200°C: 2.9 x 10^{-6} cm/cm·°C. *Obsolete*

ALLOY 4-22-19

American manufacture
C 0.35-0.5, Cr 23-29, Ni 13-17, Mo 5.7, Fe 0-2, bal Co.
For turbo supercharger and jet engine parts; corrosion and heat resistant.

ALLOY 4002

Alcan Canada Products Ltd.
Si 3.5-4.5, Fe 0.35, Cu 0.05-0.15, Mn 0.03, Mg 0.05-0.15, Ti 0.02, Cd 0.8-1.4, bal Al.
Coiled sheet, slab or plate for connecting rod bearings, main bearings, flange bearings, crank and camshaft bearings, for automotive and diesel engines. Conforms to AA 4002.

ALLOY 40N

American manufacture
Ni 40, 6 Cr$_3$C$_2$. 54.0 TiC.
For cutting tools, wear parts; sintered carbides.

ALLOY 41SM

French manufacture
Al alloy.
For ornaments; corrosion resistant.

ALLOY 42

Spang Specialty Metals
Ni 40.5-41.5, C 0-0.05, Mn 0-0.8, Si 0-0.3, bal Fe.
Controlled expansion alloy in sheet and strip form. Conforms to ASTM F-30 and MIL I 23011C, Class 5 with regards to composition, temper and expansion properties.

ALLOY 42-6

Spang Specialty Metals
Ni 41.5-42.5, Cr 5.2-6, C 0-0.07, Mn 0-0.25, Si 0-0.3, bal Fe.
Controlled expansion alloy in sheet and strip form. Conforms to ASTM F-31 and MIL I 23011C, Class 6 with regards to composition, temper and expansion properties.

ALLOY 42-C

Heinkel-Herth Co.
C 0.5, Mn 0.6, Si 0.4, Ni 35, Cr 15, Mo 5, W 5, Co 25, bal Fe.
For jet engine and turbine blades; high heat resistance.

ALLOY 422-19

Cabot Corporation
C 0.4, Mn 0.3, Si 0.5, Cr 25, Ni 16, Mo 6, Fe 1, bal Co.
At 70 F: 98,100 TS; 55,100 YS; 5 El; 12 RA. At 1500 F: 64,000 TS; 47,600 YS; 3 El; 3.5 RA. At 1800 F: 37,800 TS; 22 El; 39 RA. For jet engine components; oxidation and heat resistant.
Obsolete

ALLOY 42B

Manufacturer not listed.
Mg, Si, Be, bal Al.
Sand cast: 46,000 TS; 36,000 YS; 5 El. Permanent mold: 50,000 TS; 40,000 YS; 6 El. For gear cases and housings, fittings; heat treatable, high strength.

ALLOY 45-6

Spang Specialty Metals
Ni 45, Cr 6, C 0.02, Mn 0.4, Si 0.15, bal Fe.
Annealed: 56 kg/mm^2 TS; 28 kg/mm^2 YS; 30 El; 80 Rock B.
Mean coefficient of thermal expansion at 30-425°C: 10.5 x 10^{-6} cm/cm·°C. For sealing 0120 glass to metal. *Obsolete*

ALLOY 46

Spang Specialty Metals
Ni 46, C 0.02, Mn 0.4, Si 0.15, bal Fe.
Annealed: 58 kg/mm^2 TS; 24 kg/mm^2 YS; 27 El; 76 Rock B.
Mean coefficient of thermal expansion at 30-500°C: 8.5 x 10^{-6} cm/cm·°C. For sealing soft glasses to metal. *Obsolete*

ALLOY 479

Sigmund Cohn Corp.
Pt 91-93, W 7-8.
Resistance and high tensile strength elements.

ALLOY 48

Spang Specialty Metals
Ni 47-49, Mn 0.8, Si 0.5, C 0.035, bal Fe.
Soft, magnetic and controlled expansion alloy in sheet, strip, bar and rod form. Conforms to ASTM A-753 and MILN 14411C with regards to composition, temper and magnetic properties. Conforms to ASTM F-30 and MIL I 23011C, Class 3 with regards to composition, temper and expansion properties.

ALLOY 48S5

American manufacture
C 0.15, Mn 1.09, V 0.09, Ti 0.009, bal Fe.
For pressure vessels; impact resistant.

ALLOY 4D

British Aluminium Co., Ltd.
Si 11-14, bal Al.
For die castings, housings, cases; corrosion resistant.
Obsolete

ALLOY 52

Spang Specialty Metals
Ni 50-52, C 0-0.05, Mn 0-0.6, Si 0-0.3, bal Fe.
Controlled expansion alloy in sheet and strip form. Conforms to ASTM F-30 and MIL I 23011C, Class 2 with regards to composition, temper and expansion properties.

ALLOY 60

Isabellenhuette
Copper. Cu 94, Ni 6.
Annealed: 250 N/mm^2 TS. For electrical equipment and instruments. Resistance alloy. Maximum working temperature to 300°C.

ALLOY 600

Now INCONEL ALLOY 600.

ALLOY 6059

Cabot Corporation
C 0.4, Ni 33, Fe 1, Cr 25, Mo 6, Mn 0.3, Si 0.65, bal Co.
Cast: 82,500 TS; 46,900 YS; 7 El; 10 RA. At 1500 F: 51,200 TS; 38,200 YS; 10 El; 14 RA. At 1800 F: 33,700 TS; 26 El; 42 RA. For jet engine components; high strength, heat resistant.
Obsolete

ALLOY 6059

General Electric Co.
C 0.4, Ni 33, Fe 1, Cr 25, Mo 6, Mn 0.3, Si 0.65, bal Co.
Cast: 82,500 TS; 46,900 YS; 7 El; 10 RA. At 1500 F: 51,200 TS; 38,200 YS; 10 El; 14 RA. At 1800 F: 33,700 TS; 26 El; 42 RA. For jet engine components; high strength, heat resistant.
Obsolete

ALLOY 61

Cabot Corporation
C 0.42, Cr 24, Ni 1, W 5.5, Fe 1, Mn 0.3, Si 0.6, bal Co.
At 70 F: 105,000 TS; 58,400 YS; 7 El; 11 RA. At 1000 F: 77,000 TS; 41,400 YS; 15 El; 25 RA. At 1800 F: 33,000 TS. For jet engine components; heat and oxidation resistant.
Obsolete

ALLOY 644

Anaconda Co.
Copper. Ni 4.6, Si 1.1, Al 4, bal Cu.
Precipitation hardenable alloy; can be hardened and cold worked to about 150,000 psi YS. Electrical conductivity: 10-13% IACS. For springs, switches, electrical connections.

ALLOY 694C

Manufacturer not listed.
C 0.4, Ni 24, Cr 15, bal Fe.
For furnace equipment, heat treat boxes; heat and corrosion resistant.

ALLOY 713 C

Michigan Steel Casting Co.
C 0.08-0.2, Cr 12-14, Mo 3.8-5.2, Ti 0.5-1, Al 5.5-6.5, Cb 1-3, bal Ni.
Cast: 123,000 TS; 107,000 YS; 8 El; 30-42 Rock C. At 1500°F: 120,000 TS; 95,000 YS; 6 El. For gas turbine blades in jet engines, high temperature bolting. Cast alloy with high rupture strength to 1700°F. High thermal fatigue resistance.

ALLOY 713 C

Cannon-Muskegon Corp.
C 0.08-0.2, Cr 12-14, Mo 3.8-5.2, Ti 0.5-1, Al 5.5-6.5, Cb 1-3, bal Ni.
Cast: 123,000 TS; 107,000 YS; 8 El; 30-42 Rock C. At 1500°F: 120,000 TS; 95,000 YS; 6 El. For gas turbine blades in jet engines, high temperature bolting. Cast alloy with high rupture strength to 1700°F. High thermal fatigue resistance.

ALLOY 713 C

Howmet Corp.
C 0.08-0.2, Cr 12-14, Mo 3.8-5.2, Ti 0.5-1, Al 5.5-6.5, Cb 1-3, bal Ni.
Cast: 123,000 TS; 107,000 YS; 8 El; 30-42 Rock C. At 1500°F: 120,000 TS; 95,000 YS; 6 El. For gas turbine blades in jet engines, high temperature bolting. Cast alloy with high rupture strength to 1700°F. High thermal fatigue resistance.

ALLOY 713LC

Howmet Corp.
C 0.05, B 0.01, Zr 0.1, Cr 12, Mo 4.5, Ti 0.7, Al 6, Cb 2, bal Ni.
Cast: 130,000 TS; 109,000 YS; 15 El; 21 RA. At 1500°F: 123,000 TS; 99,000 YS; 15 El; 17 RA. For turbine wheels high temperature parts, turbine blading. casting alloy. Good thermal fatigue. Combines good low temperature ductility and high temperature strength.

ALLOY 713LC
Cannon-Muskegon Corp.
C 0.05, B 0.01, Zr 0.1, Cr 12, Mo 4.5, Ti 0.7, Al 6, Cb 2, bal Ni.
Cast: 130,000 TS; 109,000 YS; 15 El; 21 RA. At 1500°F:
123,000 TS; 99,000 YS; 15 El; 17 RA. For turbine wheels high
temperature parts, turbine blading. casting alloy. Good
thermal fatigue. Combines good low temperature ductility
and high temperature strength.

ALLOY 713LC
Michigan Steel Casting Co.
C 0.05, B 0.01, Zr 0.1, Cr 12, Mo 4.5, Ti 0.7, Al 6, Cb 2, bal Ni.
Cast: 130,000 TS; 109,000 YS; 15 El; 21 RA. At 1500°F:
123,000 TS; 99,000 YS; 15 El; 17 RA. For turbine wheels high
temperature parts, turbine blading. casting alloy. Good
thermal fatigue. Combines good low temperature ductility
and high temperature strength.

ALLOY 718
Now INCONEL ALLOY 718.

ALLOY 73J
English manufacture
C 0.73, Mn 1, Cr 23, Ni 6, Mo 6, Ta 2, bal Co.
For valves, pump parts; age-hardenable, corrosion resistant.

ALLOY 8081
Alcan Canada Products Ltd.
Si 0.7, Fe 0.7, Cu 0.7-1.3, Mn 0.1, Zn 0.05, Sn 18-22, Ti, bal Al.
Coiled sheet, slab or plate for connecting rod bearings, main
bearings, flange bearings, crank and camshaft bearings, for
automotive and diesel engines. Conforms to AA 8081.

ALLOY 8280
Alcan Canada Products Ltd.
Si 1-2, Fe 0.7, Cu 0.7-1.3, Mn 0.1, Ni 0.2-0.7, Ti 0.1, Zn 0.05,
Sn 5.5-7, bal Al.
Coiled sheet, slab or plate for connecting rod bearings, main
bearings, flange bearings, crank and camshaft bearings, for
automotive and diesel engines. Conforms to AA 8280.

ALLOY 89MC
Jessop Steel Co.
C, alloy, bal Fe.
For machinery parts; oil hardened, tough. *Obsolete*

ALLOY 8J
Wilbur B. Driver Co.
Ni-Fe.
For instruments, electrical equipment; constant modulus.
Obsolete

ALLOY 90
Isabellenhuette
Copper. Cu 90, Ni 10.
290 N/mm² TS. For electrical equipment and instruments.
Resistance alloy. Maximum working temperature to 250°C.

ALLOY 9257
George Cook & Co., Ltd.
C 0.93, Mn 0.23, Cr 0.13, Si 0.2, V 0.2, bal Fe.
For cutters, punches, dies; oil or water hardened.

ALLOY 936
Sigmund Cohn Corp.
W 8, Pt 92.
For high temperature applications; oxidation resistant.
Obsolete

ALLOY 95-M-255
American manufacture
Ni 67, Fe 1, Mo 25, Al 6.
Cast. For high temperature applications; high strength, heat
resistant.

ALLOY 9AL-30MN
Ford Motor Co.
Al 9, Mn 30, Si 1, C 1, bal Fe.
At room temperature: 120,000 TS; 70,000 YS; 70 El. At 1200
F: 80,000 TS; 68,000 YS; 18 El. For high temperature
applications; austenitic, high temperature strength. *Obsolete*

ALLOY "B" FINISHING
Jessop-Saville Ltd.
C 1.4, W 4.5, Cr 0.8, bal Fe.
For finishing tools, draw dies; wear resistant.

ALLOY "C"
Jessop-Saville Ltd.
C 2.2, Cr 13, bal Fe.
For tools, dies, gauges; non-deforming.

ALLOY A226
English manufacture
C, alloy, bal Fe.
Heat treated: 19,000 TS. For jet engine components; age
hardenable, high heat resistance.

ALLOY AP
English manufacture
C 3.45, Si 1.6, Mn 1, P 0.15, Ni 1.2, Cr 0.9, bal Fe.
Cast. For gears, shafts, machine tool housings; cast iron.

ALLOY B
Cytemp Specialty Steel Div.
C 1.05, Mn 0.35, Cr 1.35, bal Fe.
For forming rolls, deep drawing dies; oil hardening. *Obsolete*

ALLOY B-1900
Universal Cyclops
C 0.1, Cr 8, Co 10, Mo 6, Al 6, Ta 4.3, Ti 1, B 0.015, Zr 0.07,
bal Ni.
Hardened: 125,000 TS; 110,000 YS; 5 min El. At 1600°F:
97,000 TS; 83,000 YS; 1.5 min El. For turbine blades and
other high temperature applications. Precipitation hardened.
Vacuum-melted and vacuum-cast.

ALLOY B-1900
Teledyne Vasco
C 0.1, Cr 8, Co 10, Mo 6, Al 6, Ta 4.3, Ti 1, B 0.015, Zr 0.07,
bal Ni.
Hardened: 125,000 TS; 110,000 YS; 5 min El. At 1600°F:
97,000 TS; 83,000 YS; 1.5 min El. For turbine blades and
other high temperature applications. Precipitation hardened.
Vacuum-melted and vacuum-cast.

ALLOY B5
Haywood Tyler of Canada, Ltd.
C 0-0.07, Ni 29, Cr 20, Mo 2-3, Cu 4, bal Fe.
Cast: 65,000-75,000 TS; 28,000-38,000 YS; 35-50 El; 40-50
RA; 120-150 Brin. For chemical plant equipment, tanks;
resists mixed acids, austenitic.

ALLOY BROACHRITE
LaSalle Steel Co.
C 0.4, Mn, bal Fe.
Drawn: 100,000 TS; 90,000 YS; 13 El; 45 RA; 217 Brin. For
heat treated parts; modified S.A.E. X1340. *Obsolete*

ALLOY BSZ
Italian manufacture
For solder; soft.

ALLOY C-422
American manufacture
C 0.2, Cr 13, Mo 1, W 0.8, V 0.25, bal Fe.
For turbine wheels; corrosion and heat resistant.

ALLOY CB-16
Union Carbide Corp.
Cb 67, Ti 10, W 20, V. 3.
For high temperature applications; oxidation resistant to
2000°F.

ALLOY CB-22
Union Carbide Corp.
Cb 94, Al 3, V 3.
For high temperature applications; oxidation resistant to
2000°F.

ALLOY CB-7
Union Carbide Corp.
Zr 0.75, Ti 7.5, bal Cb.
Rolled: 60,800 YS; 32 El. For high temperature applications;
high temperature strength to 2000°F

ALLOY CB-74
Union Carbide Corp.
Cb 85, W 10, Zr 5.
At 2200°F: 44,800 TS; 42,400 YS. For high temperature
applications; oxidation resistant to 2000°F.

ALLOY CH42
Ackerlind Steel Co., Inc.
C 1.65, Mn 0.45, Cr 13, V 0.3, Mo 0.7, bal Fe.
For dies; air hardening. *Obsolete*

ALLOY CM1
American manufacture
Fe 11, Cr 5.7, Ni 5.6, 65.0 TiC, 11.0 CbC, 0.7 TaC.
For cutting tools, wear parts, sintered carbides.

ALLOY CROSS CUTS
Jessop Steel Co.
C 0.9, Mn 0.35, Cr 0.5, bal Fe.
For saws; water hardened. *Obsolete*

ALLOY CSA-39
English manufacture
C 0.1, Ni 27, Fe 40, Cr 19, Mo 9, W 3.
For jet engine components; high strength, heat resistant.

ALLOY CW
Manufacturer not listed.
C 0.1, Mn 1.5, Si 0.5, Cr 21.5, Ni 20.5, Co 20, Mo 3, W 2, Cb
1, N 0.14, bal Fe.
At 70°F: 118,000 TS; 58,000 YS; 49 El. At 1200°F: 73,500 TS;
37,600 YS; 28 El. For jet engine components; oxidation and
heat resistant.

ALLOY D-31
American manufacture
Ti 10, Mo 10, Cb 80.
At 70°F: 100,000 TS. At 2000°F: 34,500 TS. For high
temperature applications; high oxidation resistance.

ALLOY EL 437B
Russian manufacture
C 0.05, Mn 0.23, Cr 20.5, bal Ni.
For high temperature applications; high heat and oxidation
resistance.

ALLOY FDP
English manufacture
C, Cr, Ni, bal Fe.
For aircraft structures; austenitic, stainless.

ALLOY FINISHING
CCS Braeburn Alloy Steel
C 1.4, W 3, bal Fe.
For cutting and finishing tools. Non-deforming. *Obsolete*

ALLOY GAS WELDING 2320
U.S. Steel Corp.
C 0.17-0.22, Mn 0.3-0.6, Si 0.3, Ni 3.25-3.75, bal Fe.
Deposited metal: 70,000-80,000 TS; 15-20 El. For welding
wire; acetylene welding. *Obsolete*

ALLOY GAS WELDING A-2317
U.S. Steel Corp.
C 0.1-0.2, Mn 0.3-0.6, P 0-0.04, S 0-0.05, Si 0.15-0.3, Ni
3.25-3.75, bal Fe.
Deposited metal : 70,000-80,000 TS; 15-20 El. For welding
wire; acetylene welding. *Obsolete*

ALLOY GAS WELDING A-2330
U.S. Steel Corp.
C 0.28-0.33, Mn 0.5-0.8, P 0-0.04, S 0-0.05, Si 0.15-0.3, Ni 3.25-3.75, bal Fe.
Deposited metal: 70,000-80,000 TS; 15-20 El. For welding wire; acetylene welding. *Obsolete*

ALLOY H40
English manufacture
C 0.25, Cr 3, V 0.8, W 0.5, Mo 0.5, Ni 5, bal Fe.
For high temperature bolts, turbine wheels; ferritic, heat resistant.

ALLOY HOLLOW DRILL
Bethlehem Steel Corp.
C 0.28, Mn 0.95, Ni 2.25, Cr 0.65, Mo 0.3, bal Fe.
For mining and quarrying tools, drills; shock resistant, tough. *Obsolete*

ALLOY HTB-1
Allegheny Ludlum Steel
C 1, Cr 1.45, Al 1, bal Fe.
For high temperature bearings; heat and creep resistant. *Obsolete*

ALLOY HTB-1
Teledyne Vasco
C 1, Cr 1.45, Al 1, bal Fe.
For high temperature bearings; heat and creep resistant. *Obsolete*

ALLOY HTB-1
Timken Co.
C 1, Cr 1.45, Al 1, bal Fe.
For high temperature bearings; heat and creep resistant. *Obsolete*

ALLOY HTB-1
Colt Industries
C 1, Cr 1.45, Al 1, bal Fe.
For high temperature bearings; heat and creep resistant. *Obsolete*

ALLOY HUBBING DIE
Bethlehem Steel Corp.
C 0.09, Mn 0.48, Cr 0.58, Ni 1.25, bal Fe.
For hubbing dies; case hardened. *Obsolete*

ALLOY I-1360
American manufacture
C 0.1, Ni 70, Fe 6, Cr 10, Mo 5, Cb 2, Al 6, Be 0.5.
Cast. For high temperature applications; high strength, heat resistant.

ALLOY I-336
American manufacture
C 0.2, Co 50, Ni 15, Fe 1, Cr 20, W 12, Cb 1.
For high temperature applications; high strength, heat resistant.

ALLOY IIIVT2-2
English manufacture
C 1.11, Cr 23, Mo 6, Ta 2, bal Co.
Cast: 120,000 TS; 4 El; 1 RA. For valves, pump parts; corrosion, abrasion and heat resistant.

ALLOY L-251
American manufacture
C 0.4, Co 54, Ni 10, Fe 1, Cr 19, W 14.
Cast. For high temperature applications; high strength, heat resistant.

ALLOY L605
Cytemp Specialty Steel Div.
Refractory.
Now UNITEMP L 605.

ALLOY LM-5
Manufacturer not listed.
Mo 40, Si 40, Cr 8, B 2, Al 10.
Uses: Coating for columbium alloys for high temperature applications. High oxidation resistance.

ALLOY M-252
Universal Cyclops
C 0.2, Co 9-11, Fe 3, Cr 18-20, Mo 9-11, Ti 2.5, Al 1, bal Ni.
Annealed: 175,000 TS; 20 El. At 1000°F: 165,000 TS; 93,000 YS. At 1600°F: 70,000 TS; 60,000 YS; 35 El. For jet engine buckets; high heat and corrosion resistant.

ALLOY M-252
Crucible Specialty Steel
C 0.2, Co 9-11, Fe 3, Cr 18-20, Mo 9-11, Ti 2.5, Al 1, bal Ni.
Annealed: 175,000 TS; 20 El. At 1000°F: 165,000 TS; 93,000 YS. At 1600°F: 70,000 TS; 60,000 YS; 35 El. For jet engine buckets; high heat and corrosion resistant.

ALLOY M-255
Evans Steel Co.
Aluminum. 0.5 Al_2O_3, bal Al.
At 70°F: 22,600 TS; 17,600 YS; 22 El. At 300°F: 15,400 TS; 13,600 YS; 22 El. For high temperature applications; powder metallurgy.

ALLOY M-257
Aluminium Industrie Aktiengesellschaft
7.8 Al_2O_3, bal Al.
Sintered: 40,000 TS; 26,000 YS; 14 El. For high temperature applications; powder metallurgy.

ALLOY M-276
Aluminium Industrie Aktiengesellschaft
16.5 Al_2O_3, bal Al.
Sintered: 54,000 TS; 35,000 YS; 3 El. For high temperature applications; powder metallurgy.

ALLOY M-308
Cytemp Specialty Steel Div.
C 0.08, Cr 14, Ni 33, Mo 4, Al 0.25, Zr 0.25, Ti 2, W 6.5, bal Fe.
For jet engine parts; heat resistant. *Obsolete*

ALLOY M22 VC
Mond Nickel Co. Ltd.
C 0.08-0.16, Cr 5-6.5, Al 5.9-6.6, Mo 1.5-2.5, W 10.5-11.5, Ta 2.6-3.4, Zr 0.4-0.8, Fe 0-1, bal Ni.
High temperature applications. Vacuum cast.

ALLOY MC-102
Mond Nickel Co. Ltd.
C 0.02-0.06, Si 0.1-0.4, Mn 0.1-0.5, Fe 0-4, Cr 19-20.5, Co 0-5, Mo 5.5-6.5, W 2-3, 6.2-6.7 Cb + Ta, bal Ni.
For gas turbine stator blades, turbine rotors, pre-combustion chambers of diesel engines. Good oxidation resistance and strength to 900°C. Age-hardenable.

ALLOY MCS
Cytemp Specialty Steel Div.
C 0.7, Cr 3.5, Mo 5.5, Si 1, bal Fe.
For high temperature bearings; heat resistant. *Obsolete*

ALLOY MN5
Russian manufacture
Ni, bal Cu.
For high temperature applications; heat resistant.

ALLOY N 153
Cytemp Specialty Steel Div.
C 0.1, Mn 1.5, Si 0.5, Cr 17, Ni 15, Co 13, Mo 3, W 2, Cb 1, N 0.14, bal Fe.
For jet engine components; oxidation and heat resistant. *Obsolete*

ALLOY N 155
Cytemp Specialty Steel Div.
Refractory.
Now UNITEMP N-155.

ALLOY NO 4
English manufacture
Ni 42-47, Cr 4-6, bal Fe.
For glass to metal seals; controlled expansion.

ALLOY NO. 1270
Edgcomb Metals Co.
Ag 2, bal Cu.
For silver brazing; M.P. 1220-1270 F. *Obsolete*

ALLOY NO. 188
Now HAYNES ALLOY NO. 188.

ALLOY NO. 20
Janney Cylinder Co.
Ni 2, Sn 5, Pb 15, bal Cu.
For bearings, pump liners, bushings. *Obsolete*

ALLOY NO. 471
Hoskins Mfg. Co.
100 Ni (practically).
60,000 TS; 40-50 El. *Obsolete*

ALLOY NO. 484
Hoskins Mfg. Co.
Mn 2, bal Ni.
65,000 TS; 30-40 El. For spark plug electrodes, lead wire, radio parts; resistance 80 ohms per C.M.F. *Obsolete*

ALLOY NO. 6
Midvale-Heppenstall Co.
C, Cr, bal Fe.
For tools; water hardened. *Obsolete*

ALLOY S 844
American manufacture
C 0.3, Co 44, Ni 20, Fe 2, Cr 25, Mo 3, W 2, Cb 2.
For high temperature applications; high strength, heat resistant.

ALLOY SHOE DIE
Colt Industries
C 0.5, Cr 0.7, Mo 0.15, bal Fe.
For shoe dies; tough. *Obsolete*

ALLOY SSMP
English manufacture
C 2.98, Si 1.14, Mn 1.1, P 0.18, bal Fe.
Cast. For gears, shafts, machine tool housings; cast iron.

ALLOY V814
German manufacture
CC 45, Ni 7, Co 3, 45.0 TiC.
For cutting tools bits, dies; sintered carbides.

ALLOY VCM
English manufacture
C 0.28, Mn 0.7, Cr 1, Mo 1, Ni 0.7, bal Fe.
For gears, shafts, cams; nitriding steel.

ALLOY W
German manufacture
Cu 0.5-0.9, Ni 0.28-0.36, Co 0.25, Mg 0.16, V, W, Cr, Be, bal Al.
For anodized die castings, housings.

ALLOY W.5
Henry Wiggin & Co. Ltd.
Si 4, Mn 0.5, bal Ni.
For spark plug electrodes. Resists high temperature corrosion. *Obsolete*

ALLOY W.6
Henry Wiggin & Co. Ltd.
Si 2, Mn 0.5, bal Ni.
For spark plug electrodes. Resists high temperature corrosion. *Obsolete*

ALLOY W.7
Henry Wiggin & Co. Ltd.
Si 1, Mn 2.5, bal Ni.
For spark plug electrodes. Resists high temperature
corrosion. *Obsolete*

ALLOY W.9
Henry Wiggin & Co. Ltd.
Si 1, Mn 4.5, bal Ni.
For spark plug electrodes. Resists high temperature
corrosion. *Obsolete*

ALLOY WOOD CHISEL
Colt Industries
C 0.75, Cr 0.3, bal Fe.
For wood chisel; water or oil hardening. *Obsolete*

ALLOY Z 132
Now AA 332.

ALLOY Z-FM
Atlantic Steel Corp.
C 0.4, Cr 1, Ni 0.75, Mo 0.35, Pb 0.35, bal Fe.
Heat treated: 143,000 TS; 122,500 YS; 16 El; 52 RA; 300 Brin.
For machinery parts, gears, pins, shafting; oil hardened,
shock resistant.

ALLOY-10
Metals & Controls, Inc.
Ni 36, Fe 64.
Annealed: 70,000 TS; 24,000 YP; 36 El; 143 Brin. For
instruments, geodetic equipment, textile machinery parts.
Controlled expansion, corrosion resistant.

ALLOY-M
Metals & Controls, Inc.
Cr 18, Ni 8, C 0.08, bal Fe.
Annealed: 80,000 TS; 35,000 YS; 80 Rock B. For chemical
plant equipment, mixers, digesters, tanks, agitators.
Stainless, austenitic.

ALLOYED GENUINE WROUGHT IRON
A.M. Byers Co.
Mn 0.05, P 0.1-0.15, Ni 0-5, C 0.05, Cu 0-0.75, bal Fe.
50,000-60,000 TS; 30,000-45,000 YS. For pipes, hull plates;
tough. *Obsolete*

ALLOYMET 1005
Alloy Metal Products Inc.
Cr 4.5-5.5, bal Cu.
For Cr additions to Cu alloys; master alloy.

ALLOYMET 1010
Alloy Metal Products Inc.
Cr 9.5-10.5, bal Cu.
For Cr additions to Cu alloys; master alloy.

ALLOYMET 1110
Alloy Metal Products Inc.
Fe 9-11, bal Cu.
For Cu additions to steel and cast iron; master alloy.

ALLOYMET 1115
Alloy Metal Products Inc.
Fe 4.5-5.5, bal Cu.
For Cu additions to steel and cast iron; master alloy.

ALLOYMET 1130
Alloy Metal Products Inc.
Fe 28-32, bal Cu.
For Cu additions to steel and cast iron; master alloy.

ALLOYMET 1230A
Alloy Metal Products Inc.
Mn 28-32, bal Cu.
For Mn additions for Ni alloys. Hardener; deoxidizer.

ALLOYMET 1230B
Alloy Metal Products Inc.
Mn 23.5-26.5, Fe 4-6, bal Cu.
For Mn additions to bronzes. Hardener; deoxidizer.

ALLOYMET 1315
Alloy Metal Products Inc.
Ni 15, bal Cu.
For alloying additions for bronzes.

ALLOYMET 1330
Alloy Metal Products Inc.
Ni 30, bal Cu.
For alloying additions.

ALLOYMET 1350
Alloy Metal Products Inc.
Ni 50, bal Cu.
For alloying additions.

ALLOYMET 1410
Alloy Metal Products Inc.
Si 9-11, bal Cu.
For high conductivity copper; master alloy.

ALLOYMET 1420
Alloy Metal Products Inc.
Si 18-22, bal Cu.
For high conductivity copper; master alloy.

ALLOYMET 1430
Alloy Metal Products Inc.
Si 28-32, bal Cu.
For high conductivity copper. Hardener; master alloy.

ALLOYMET 2000
Alloy Metal Products Inc.
Ni 99.8.
For anodes for plating and additions to iron and steel.

ALLOYMET 2023
Alloy Metal Products Inc.
Ni 60, Cu 25, Si, Fe and C, bal Mn.
For Ni and Cu additions to steel and cast iron; alloying
master alloy.

ALLOYMET 2030
Alloy Metal Products Inc.
Cu 30, Ni 70.
For Ni and Cu additions to steel and cast iron; alloying
master alloy.

ALLOYMET 2030M
Alloy Metal Products Inc.
Ni 65, Cu 30, Fe 0-2, Mn 0-1, C 0-0.15.
For Ni and Cu additions to steel and cast iron; alloying
master alloys.

ALLOYMET 2041
Alloy Metal Products Inc.
Ni 59, Si 29, bal Fe.
For alloying additions of nickel to ferrous and some non-
ferrous alloys.

ALLOYMET 2115
Alloy Metal Products Inc.
Cr 14-17, Fe 6-10, Mn 1, Cu 0-0.5, 77.0 Ni min.
For alloying additions to stainless steels. For high
temperature castings.

ALLOYMET 2115 WH
Alloy Metal Products Inc.
Ni 33-37, Cr 14-17, Mn 0.5-2, Si 0.8-2.5, bal Fe.
For alloying additions to stainless steels. For high
temperature castings.

ALLOYMET 2120
Alloy Metal Products Inc.
Ni 77-79, Cr 19-20, Fe 1, C, Si, bal Mn.
For alloying additions to stainless steels; alloying master
alloy.

ALLOYMET 2140
Alloy Metal Products Inc.
Cr 14-18, Mn 0-3, Si 0-1.5, 57.0 Ni min, bal Fe.
For alloying additions to stainless steels; alloying master
alloy.

ALLOYMET 2165
Alloy Metal Products Inc.
Ni 35, Cr 10, bal Fe.
For alloying additions to Ni hard iron and to stainless steel.

ALLOYMET 2200
Alloy Metal Products Inc.
Ni 56, Cu 23, Cr 7.5, bal Fe.
For alloying additions for Ni-resist iron.

ALLOYMET 2350
Alloy Metal Products Inc.
Ni 47-50, bal Fe.
For alloying for high permeability; soft magnets.

ALLOYMET 2364
Alloy Metal Products Inc.
Ni 36, bal Fe.
For alloying for low expansion applications; glass seals.

ALLOYMET 2470
Alloy Metal Products Inc.
Ni 30, Cu 15, Mn 0.2, bal Fe.
For glass to metal seals; low expansion.

ALLOYMET AM1
Alloy Metal Products Inc.
30 Ni ferro-nickel.
For alloying additions to alloy iron.

ALLOYMET AM2
Alloy Metal Products Inc.
40 Ni ferro-nickel.
For alloying additions to stainless steel.

ALLOYMET MG
Alloy Metal Products Inc.
Ni 76, Cu 14, bal Fe.
For high permeability alloys; low yield strength.

ALLOYMET NI CR MO 1
Alloy Metal Products Inc.
Ni 35, Cr 10, Mo 5, bal Fe.
For alloying additions to alloy steel.

ALLOYMET NI CU 1
Alloy Metal Products Inc.
Ni 50, Cu 25, bal Fe.
For alloying additions to low alloy steels.

ALLOYMET NI CU 2
Alloy Metal Products Inc.
Ni 45, Cu 15, bal Fe.
For alloying additions to ferrous and some non-ferrous
alloys.

ALLOYMET NI MO 1
Alloy Metal Products Inc.
Ni 35, Mo 5, bal Fe.
For alloy additions to ferrous alloys.

ALLOYMET PERMAG
Alloy Metal Products Inc.
Ni 20, Cu 60, bal Fe.
For permanent magnets.

ALLVAC 18-15B
Teledyne Allvac
C 0.01, Cr 18, Ni 15, B 0.8, bal Fe.
For neutron control rods, atomic energy uses; stainless, austenitic. *Obsolete*

ALLVAC 3-2.5
Teledyne Allvac
Titanium. Al 3, V 2.5, bal Ti.
Titanium alloy. Density 0.162 lb/in.^{3}.

ALLVAC 30
Teledyne Allvac
Titanium. Fe 0.1, 0.10 O_2, bal Ti.
Titanium alloy; CP Grade I. Density 0.163 lb/cu in.

ALLVAC 316 SS
Teledyne Allvac
C 0.05, Cr 17.4, Mo 2.7, Ni 13.8, Mn 1.75, Si 0.6, bal Fe.
Corrosion resistant alloy; nuclear quality only.

ALLVAC 40
Teledyne Allvac
Titanium. Fe 0.1, Pd 0.15, 0.15 O_2, bal Ti.
Titanium alloy; CP Grade II. Density 0.163 lb/in.^{3}.

ALLVAC 5-2.5
Teledyne Allvac
Titanium. Al 5, Sn 2.5, bal Ti.
Titanium alloy. Density 0.161 lb/in.^{3}.

ALLVAC 50
Teledyne Allvac
Titanium. Fe 0.2, 0.20 O_2, bal Ti.
Titanium alloy; CP Grade III.

ALLVAC 500 ZB
Teledyne Allvac
C 0.08, Cr 19, Co 18.8, Mo 4, Ni 54, Ti 3, Al 3, B 0.006, Zr 0.06.
At room temperature: 176,000 TS; 110,000 YS; 16 El. At 1400°F: 151,000 TS; 106,000 YS; 21 El. For missiles, space equipment, jet engine and gas turbine parts. Corrosion and oxidation resistant. Good creep resistance and notch toughness.

ALLVAC 520
Teledyne Allvac
C 0.04, Cr 19, Co 13, Mo 6.3, Ti 3, Al 2, B 0.005, W 1, bal Ni.
Corrosion resistant high temperature alloy.

ALLVAC 6-2-4-2
Teledyne Allvac
Titanium. Al 6, V 2, Zr 4, Mo 2, bal Ti.
Titanium alloy. Density 0.164 lb/in.^{3}.

ALLVAC 6-2-4-6
Teledyne Allvac
Titanium. Al 6, V 2, Zr 4, Mo 6, bal Ti.
Titanium alloy. Density 0.168 lb/in.^{3}.

ALLVAC 6-4
Teledyne Allvac
Titanium. Al 6, V 4, bal Ti.
Titanium alloy. Density 0.160 lb/in.^{3}.

ALLVAC 6-6-2
Teledyne Allvac
Titanium. Al 6, V 6, Sn 2, bal Ti.
Titanium alloy. Density 0.164 lb/in.^{3}.

ALLVAC 70
Teledyne Allvac
Titanium. Fe 0.35, 0.35 O_2, bal Ti.
Titanium alloy; CP Grade IV.

ALLVAC 700
Teledyne Allvac
C 0.13, Cr 15.3, Co 18.5, Mo 5, Ti 3.5, Al 4.4, B 0.014, bal Ni.
For jet engine components and space vehicles.

ALLVAC 718
Teledyne Allvac
C 0.04, Cr 17-21, Mo 2.8-3.3, Ni 50-55, Cb 5-5.5, Ti 0.65-1.15, Al 0.4-0.8, bal Fe.
Heat treated: 190-000-215,000 TS; 160,000-185,00 YS; 15-25 El; 20-30 RA; 40-44 Rock C. For heat resistant parts to 1300°F. Age hardening, high strength. Corrosion and oxidation resistant.

ALLVAC 8-1-1
Teledyne Allvac
Titanium. Al 8, V 1, Mo 1, bal Ti.
Titanium alloy. Density 0.156 lb/in.^{3}.

ALLVAC ASTROLOY
Teledyne Allvac
C 0.06, Cr 15, Co 17, Mo 5, Cb 3.5, Ti 3.5, Al 4, B 0.03, Zr 0.06, bal Ni.
Heat treated: 205,000 TS; 152,000 YS; 12 El; 14 RA. At 1400°F: 160,000 TS; 132,000 YS; 15 El; 17 RA. For turbine wheels and blades, jet engine components, nozzles, compressor discs. Heat and corrosion resistant. High creep and stress-rupture strength.

ALLVAC I-700
Teledyne Allvac
C 0.13, Cr 15, Co 29, Mo 3.75, Ni 46, Ti 2.55, Al 3.25.
At room temperature: 170,000 TS; 103,000 YS; 25 El. At 1400°F: 127,000 TS; 8900 YS; 11 El. For jet engine and gas turbine components, space vehicles and missiles. Oxidation and corrosion resistant. Good creep resistance and notch toughness.

ALLVAC M-252
Teledyne Allvac
C 0.1-0.15, Mn 0-0.5, Cr 18-20, Co 9-11, Mo 9-11, Ti 2.25-2.75, Al 0.5-1.25, B 0.006, Fe 0-5, bal Ni.
Bar: 180,000 TS; 20 El; 40 Rock C; 122,000 TS. At 1200°F: 168,000 TS; 108,000 YS; 11 El; 33 Rock C. For heavy duty jet engine and gas turbine components. Corrosion and heat resistant, age hardenable. Good creep strength.

ALLVAC N-105
Teledyne Allvac
Co 20, Mo 5, Ti 1.3, Al 4.5, B 0.007, C 0.15, Cr 15, bal Ni.
Corrosion resisting and high temperature alloy.

ALLVAC N-115
Teledyne Allvac
C 0.16, Cr 15, Co 15, Mo 4, Ti 4, Al 5, B 0.017, bal Ni.
Corrosion resistant and high temperature alloy.

ALLVAC R-235
Teledyne Allvac
C 0.12, Cr 15.5, Mo 5.5, Ni 65, Ti 2.5, Al 2, Fe 10.
At room temperature: 160,000 TS; 100,000 YS; 30 El. At 1500°F: 105,000 TS; 85,000 YS; 18 El. For high temperature components. Precipitation hardening. Oxidation and corrosion resistant.

ALLVAC RENE 41
Teledyne Allvac
C 0.09, Cr 19, Co 11, Mo 10, Ti 3, Al 1.5, B 0.006, bal Ni.
At 1400°F: 126,000 TS; 88,000 YS; 11 El. At room temperature: 17,000 TS; 103,000 YS; 25 El. For jet engine and gas turbine components, afterburners, high temperature bolts. High temperature, high strength properties. Corrosion and oxidation resistant.

ALLVAC WASPALOY
Teledyne Allvac
C 0-0.1, Co 12-15, Mo 3.5-5, Ti 2.6-3.2, Al 1-1.6, Zr 0, B 0.008, Cr 18-21, bal Ni.
Annealed: 121,000 TS; 70,000 YS; 54 El; B 90 Rock. At 1400°F: 120,000 TS; 99,000 YS; 28 El; 42 RA. For turbine engines, airframe assemblies, missile components, fasteners. High stress-rupture strength and oxidation resistant.

ALLVAC-N
Teledyne Allvac
C 0.05, Cr 7, Mo 16.5, Ni 76.5, 10 ppm B.
At RT: 116,500 TS; 43,000 YS; 62.6 El. At 1200 F: 96,300 TS; 35,500 YS; 54 El. For high temperature applications and for equipment handling hot fluoride salts. High heat, corrosion and oxidation resistant. *Obsolete*

ALMADUR MP6
German manufacture
Al alloy.
For bearings.

ALMADUR MZ3
German manufacture
Al alloy.
For bearings.

ALMAG
Aluminium Francais
Cu 3.5-4.5, Mn 0.4-1, 0.3-0.75 Mag, bal Al.
Rolled: 54,000 TS; 22 El; 100 Brin. For light alloy parts, aircraft construction; Duralumin type.

ALMAG 35
Acme Aluminum Alloys, Inc.
Ti 0.1, Mn 0.1, Be 0.05, Mg 6.5-7.5, bal Al.
Cast: 38,000-42,000 TS; 18,000-21,000 YS; 10-15 El; 70 Brin. For castings, ornaments, and machine tool parts. Not hardenable; corrosion resistant.

ALMANITE S
Meehanite Metal Corp.
C, Al, Mn, bal Fe.
Cast: 90,000 TS; 12 El; 200 Brin. For mill liners, grinding balls, crushing rolls; austenitic, wear and abrasion resistant. *Obsolete*

ALMANITE-C
Meehanite Metal Corp.
C, Al, Mn, bal Fe.
Cast: 90,000 TS; 15 El; 200-400 Brin. For mill liners, grinding balls, mixer blades; work hardens, abrasion and wear resistant. *Obsolete*

ALMAR 15 (280 HT)
Allegheny Ludlum Steel
C 0-0.03, Mn 0-0.1, S 0-0.1, P 0-0.01, S 0-0.01, Ni 15, Mo 4.8, Ti 0.7, Co 9, Zr 0.02, B 0.003, bal Fe.
Heat treated: 299,700 TS; 293,700 YS; 10 El; 41.5 RA, Rock C 54-55. For die casting inserts, pistons, cylinders, hot extrusion dies. High yield strength and ductility. *Obsolete*

ALMAR 18
AL Tech Specialty Steel Corp.
C 0.03, Ni 17-19, Co 7-9.5, Mo 3-5.2, Ti 0.15-0.8, Al 0.05-0.15, Zr 0.02, B 0.003, bal Fe.
Annealed: 148,000 TS; 105,000 YS; 16 El; 68 RA; 30 Rock C.
Aged: 286,000 TS; 280,000 YS; 11 El; 55 RA; 54 Rock C. For rocket motor cases, airframes, airborne equipment, large machinery parts, bolts, cryogenic equipment, high temperature equipment. Maraging steel, shock resistant.

ALMAR 18 (200)
AL Tech Specialty Steel Corp.
C 0-0.03, Ni 17-19, Co 8-9, Mo 3-3.5, Ti 0.15-0.25, Zr 0.02, B 0.003, Al 0.05-0.15, bal Fe.
Maraged: 210,000 TS; 200,000 YS; 15 El; 67 RA. For fuel rocket cases, aircraft landing gears, missiles and aircraft parts Maraging steel, high strength and toughness and ductility. *Obsolete*

ALMAR 18 (250)
AL Tech Specialty Steel Corp.
C 0-0.03, Ni 17-19, Co 7-8.5, Mo 4.6-5.2, Ti 0.3-0.5, B 0.003, Al 0.05-15, bal Fe.
Maraged: 243,000 TS; 232,000 YS; 12 El; 57 RA; 48 Rock C. For fuel rocket cases, aircraft landing gears, missile and aircraft components. Maraging steel, high strength, ductility and toughness.

ALMAR 18 (300)

AL Tech Specialty Steel Corp.
C 0-0.03, Ni 18-19, Co 8.5-9.5, Mo 4.6-5.2, Ti 0.5-8, Zr 0.02, B 0.003, Al 0.05-0.15, bal Fe.
Maraged: 295,000 TS; 292,000 YS; 11.5 El; 59 RA, 48-50 Rock C. For rocket cases, missile landing gears, missile and aircraft components. Maraging steel, high strength, ductility and toughness.

ALMAR 20

AL Tech Specialty Steel Corp.
C 0-0.03, Ni 19-20, Ti 1.3-1.6, Al 0.15-0.3, Cb 0.5, Zr 0.02, B 0.003, bal Fe.
Heat treated: 253,000-265,500 TS; 244,000-260,000 YS; 10-13 El; 52-59 RA; 49-53 Rock C. For rocket motor cases, airframes, airborne equipment, large machinery parts, bolts, cryogenic and high temperature equipment. Maraging steel, shock resistant. *Obsolete*

ALMAR 25

AL Tech Specialty Steel Corp.
C 0.03, Ni 25-26, Ti 1.3-1.6, Al 0.15-0.3, Zr 0.02, Cb 0-0.5, bal Fe.
Heat treated: 240,000-265,000 TS; 212,000-246,000 YS; 3-5 El. For rocket motor cases, airframes, airborne equipment, large machinery parts, bolts, cryogenic equipment, high temperature equipment. Maraging steel, shock resistant. *Obsolete*

ALMAR 362

Allegheny Ludlum Corp./Special Materials
Stainless steel. C 0.03, Mn 0.3, P 0.015, S 0.015, Si 0.2, Cr 14.5, Ni 6.5, Ti 0.55-0.9, bal Fe.
Annealed: 120,000-140,000 psi TS; 105,000-115,000 psi YS; 10-20 El; 22-27 Rock C; 50-70 RA. For chemical processing and industrial equipment industries.

ALMAR 362

AL Tech Specialty Steel Corp.
C 0-0.05, Cr 14-14.5, Ni 6.25-7, Ti 0.6-9, bal Fe.
Annealed: 125,000 TS; 108,000 YS; 16 El; 25 Rock C. Age Hardened: 188,000 TS; 182,000 YS; 13 El, 41 Rock C. For jet engine, aircraft and missile components, hydraulic and pneumatic equipment, valve and pump parts. Age-hardenable, martensitic, corrosion resistant. S36200

ALMAR 363

AL Tech Specialty Steel Corp.
C 0.04, Cr 11.5, Ni 4, Ti 0.4, bal Fe.
Strip Annealed: 123,000 TS; 106,000 YS; 12 El; 28 Rock C. At 600°F: 97,700 YS; 89,800 YS; 5 El. For jet engine, missile and space components. Maraging steel, good fracture toughness. Corrosion resistant. Good fabricability.

ALMASILIUM

Societe du Duralumin
Si 2, Mg 1, bal Al.
50,000 TS; 6 El; 90 Brin. For light alloy parts; corrosion resistant.

ALMECO

Pechiney/Soc. Vente de l'Aluminium
Plated aluminum alloy wire. For insulated building wire.

ALMELEC

Pechiney/Soc. Vente de l'Aluminium
Mg 0.5-0.6, Si 0.5-0.6, bal Al.
Heat treated to 34.5 kg/mm² TS; 4 El. For overhead line conductors.

ALMET 1-8

H.K. Porter Co., Inc.
C 0-0.2, Cr 18, Ni 8, bal Fe.
For stainless wire parts; Type 302 stainless. *Obsolete*

ALMET 12

H.K. Porter Co., Inc.
C 0.12, Cr 12-14, bal Fe.
For corrosion resistant wire parts; Type 410 stainless. *Obsolete*

ALMET 12 M

H.K. Porter Co., Inc.
Cr 12-14, C, bal Fe.
For corrosion resistant wire parts; Type 416 stainless, free machining. *Obsolete*

ALMET 1235

H.K. Porter Co., Inc.
C 0.35, Cr 12-14, bal Fe.
For corrosion resistant wire parts; Type 420 stainless. *Obsolete*

ALMET 17

H.K. Porter Co., Inc.
Cr 14-18, C, bal Fe.
For corrosion resistant wire parts; Type 430 stainless. *Obsolete*

ALMET 17 CU

H.K. Porter Co., Inc.
Cr 14-18, C, Cu, bal Fe.
For corrosion resistant wire parts; Type 434 stainless. *Obsolete*

ALMET 17-7 PH

H.K. Porter Co., Inc.
C 0-0.09, Cr 17, Ni 7, Al 0.75-1.25, bal Fe.
Rolled: 200,000-285,000 TS; 185,000-265,000 YS; 400 Brin. Heat treated: 244,000-345,000 TS; 234,000-340,000 YS; 2-7 El; 480 Brin. For springs, clips, aircraft parts; age-hardened, stainless. *Obsolete*

ALMET 18-12 SMO

H.K. Porter Co., Inc.
Cr 18, Ni 12, Mo 3, C, bal Fe.
For stainless wire parts; Type 316 stainless. *Obsolete*

ALMET 18-8 CB

H.K. Porter Co., Inc.
Cr 18, Ni 10, C, Cb, bal Fe.
For stainless wire parts; Type 347 stainless. *Obsolete*

ALMET 18-8 TI

H.K. Porter Co., Inc.
Cr 18, Ni 10, C, Ti, bal Fe.
For stainless wire parts; Type 321 stainless. *Obsolete*

ALMET 18-8M

H.K. Porter Co., Inc.
Cr 18, Ni 8, C, bal Fe.
For stainless wire parts; Type 303 stainless, free machining *Obsolete*

ALMET 18-8S

H.K. Porter Co., Inc.
C 0-0.08, Cr 18, Ni 8, bal Fe.
For stainless wire parts; Type 304 stainless. *Obsolete*

ALMET 20-10

H.K. Porter Co., Inc.
Cr 20, Ni 10, C, bal Fe.
For stainless wire parts; Type 308 stainless. *Obsolete*

ALMET 21

H.K. Porter Co., Inc.
Cr 21, C, bal Fe.
For corrosion resistant wire parts; Type 442 stainless. *Obsolete*

ALMET 25-12

H.K. Porter Co., Inc.
Cr 25, Ni 12, C, bal Fe.
For stainless and heat resistant wire parts; Type 309 stainless. *Obsolete*

ALMET 25-20

H.K. Porter Co., Inc.
Cr 25, Ni 20, C, bal Fe.
For heat resistant wire parts; Type 310 stainless. *Obsolete*

ALMET 28

H.K. Porter Co., Inc.
Cr 25, C, bal Fe.
For heat resistant wire parts; Type 446 stainless. *Obsolete*

ALMET 302

H.K. Porter Co., Inc.
Stainless steel. C 0-0.2, Cr 17-19, Ni 8-10, bal Fe.
Soft: 105,000 TS; 70,000 YS; 35 El. Hard: 330,000 TS; 310,000 YS; 2 El. For springs, trim, wire parts; stainless, austenitic; Type 302.

ALMET 303

H.K. Porter Co., Inc.
Stainless steel. C 0.15, Cr 17-19, Ni 8-10, 0.07 S or Se min, bal Fe.
Soft: 100,000 TS; 45,000 YS; 25 El; 159 Brin. Hard: 150,000 TS; 130,000 YS; 15 El; 240 Brin. For screw machine products; stainless, austenitic, free-cutting.

ALMET 304

H.K. Porter Co., Inc.
Stainless steel. C 0-0.08, Cr 18-20, Ni 8-10, bal Fe.
Soft: 90,000 TS; 35,000 YS; 55 El; 144 Brin. Hard: 230,000 TS; 220,000 YS; 2 El; 430 Brin. For woven wire, cold headed parts; stainless, austenitic; Type 304.

ALMET 305

H.K. Porter Co., Inc.
Stainless steel. C 0-0.12, Cr 18, Ni 12, bal Fe.
Annealed: 85,000 TS; 47,000 YS; 60 El; 77 RA; 144 Brin. Soft: 100,000 TS; 54,000 YS; 58 El; 74 RA; 156 Brin. For spun, cold drawn and cold headed parts; austenitic, stainless.

ALMET 308

Aluminium Industrie Aktiengesellschaft
C 0-0.08, Cr 19-21, Ni 10-12, bal Fe.
Soft: 85,000-105,000 TS; 35,000-40,000 YS; 25-55 El; 150 Brin. Hard: 160,000-190,000 TS; 150,000-185,000 YS; 4-8 El; 380-430 Brin. For electrical equipment; austenitic, stainless.

ALMET 309

H.K. Porter Co., Inc.
C 0-0.2, Cr 22-24, Ni 12-15, bal Fe.
Soft: 85,000 TS; 35,000 YS; 55 El; 159 Brin. Hard 190,000 TS; 185,000 YS; 4 El; 430 Brin. For heat resistant parts; corrosion and heat resistant.

ALMET 310

H.K. Porter Co., Inc.
C 0-0.25, Cr 24-26, Ni 19-22, bal Fe.
Soft: 85,000 TS; 35,000 YS; 25 El; 139 Brin. Hard: 190,000 TS; 185,000 YS; 4 El; 430 Brin. For heat resistant parts; corrosion and heat resistant.

ALMET 314

H.K. Porter Co., Inc.
C 0-0.25, Cr 23-26, Ni 19-22, Si 2-3, bal Fe.
Soft: 85,000 TS; 35,000 YS; 25 El; 139 Brin. Hard: 190,000 TS; 185,000 YS; 4 El; 430 Brin. For heat resistant parts; corrosion and heat resistant.

ALMET 316

H.K. Porter Co., inc.
Stainless steel. C 0-0.1, Cr 18, Ni 10-14, Mo 2-3, bal Fe.
Soft: 85,000 TS; 35,000 YS; 25 El; 139 Brin. Hard: 190,000 TS; 185,000 YS; 4 El; 430 Brin. For acid and chemical equipment; stainless, austenitic.

ALMET 317

H.K. Porter Co., Inc.
Stainless steel. C 0.1, Cr 18, Ni 12, Mo 3-4, bal Fe.
Soft: 85,000 TS; 35,000 YS; 25 El; 139 Brin. Hard: 190,000 TS; 185,000 YS; 4 El; 430 Brin. For welded stainless structures; stainless, austenitic.

ALMET 321
H.K. Porter Co., Inc.
Stainless steel. C 0-0.08, Cr 17-19, Ni 8-11, Ti = 5 x C, bal Fe.
Soft: 90,000 TS; 35,000 YS; 25 El; 137 Brin. Hard: 230,000 TS; 220,000 YS; 2 El; 430 Brin. For welded stainless parts; austenitic, stainless, stabilized.

ALMET 330
H.K. Porter Co., Inc.
C 0-0.2, Cr 15-17, Ni 34-36, bal Fe.
Soft: 85,000 TS; 35,000 YS; 25 El; 139 Brin. Hard: 190,000 TS; 185,000 YS; 4 El; 430 Brin. For carburizing and heat treating equipment; heat resistant, austenitic.

ALMET 347
H.K. Porter Co., Inc.
Stainless steel. C 0-0.08, Cr 17-19, Ni 9-12, Cb = 10 x C, bal Fe.
Soft: 85,000 TS; 35,000 YS; 25 El; 139 Brin. Hard: 190,000 TS; 185,000 YS; 4 El; 430 Brin. For welded stainless parts; stainless, austenitic.

ALMET 35-15
H.K. Porter Co., Inc.
Cr 15, Ni 35, C, bal Fe.
For heat resistant wire parts; Type 330 stainless. *Obsolete*

ALMET 405
H.K. Porter Co., Inc.
C 0-0.05, Cr 11.5-13.5, Al 0.1-0.3, bal Fe.
Soft: 75,000 TS; 35,000 YS; 20 El; 150 Brin. Hard: 190,000 TS; 145,000 YS; 8 El; 300 Brin. For chemical plant equipment; corrosion resistant. *Obsolete*

ALMET 410
H.K. Porter Co., Inc.
C 0-0.15, Cr 11.5-13.5, bal Fe.
Soft: 75,000 TS; 35,000 YS; 15 El; 150 Brin. Hard: 190,000 TS; 145,000 YS; 8 El; 300 Brin. For valve trim, cold headed parts; corrosion resistant.

ALMET 416
H.K. Porter Co., Inc.
C 0-0.15, Cr 12-14, 0.07 S or Se min, bal Fe.
Soft: 75,000 TS; 35,000 YS; 14 El; 150 Brin. 1/2 H-temper: 115,000 TS; 80,000 YS; 16 El; 185 Brin. For valve trim, cold headed parts; corrosion resistant, free-cutting.

ALMET 420
H.K. Porter Co., Inc.
Cr 12-14, 0.15% C min, bal Fe.
Soft: 95,000 TS; 45,000 YS; 14 El; 176 Brin. Hard: 250,000 TS; 230,000 YS; 5 El; 520 Brin. For springs, dental instruments; hardenable, corrosion resistant. Spring: 75,000 TS; 156 Brin. For costume jewelry, ornamental parts; brass. *Obsolete*

ALMET 430
H.K. Porter Co., Inc.
C 0-0.12, Cr 14-18, bal Fe.
Soft: 75,000 TS; 40,000 YS; 15 El; 156 Brin. Hard: 140,000 TS; 130,000 YS; 4 El; 300 Brin. For screws, rivets, formed parts; corrosion resistant, good formability.

ALMET 434
H.K. Porter Co., Inc.
C 0-0.1, Cr 15-17, Cu 0.75-1.1, bal Fe.
Soft: 75,000 TS; 40,000 YS; 15 El; 156 Brin. Hard: 140,000 TS; 130,000 YS; 4 El; 300 Brin. For sponge wire; corrosion resistant. *Obsolete*

ALMET 442
H.K. Porter Co., Inc.
C 0-0.35, Cr 18-23, bal Fe.
Soft: 100,000 TS; 50,000 YS; 11 El; 156 Brin. Hard: 165,000 TS; 155,000 YS; 1 El; 260 Brin. For heat resistant parts; corrosion and heat resistant. *Obsolete*

ALMET 446
H.K. Porter Co., Inc.
C 0-0.35, Cr 23-27, Ni 0-0.25, bal Fe.
Soft: 110,000 TS; 50,000 YS; 10 El; 156 Brin. Hard: 175,000 TS; 160,000 YS; 1 El; 260 Brin. For heat resistant parts; heat and oxidation resistant. *Obsolete*

ALMET C-20
H.K. Porter Co., Inc.
C 0-0.07, Ni 29, Cr 20, 3.0 Cu min, bal Fe.
Rolled: 85,000 TS; 35,000 YS; 35-50 El; 50-70 RA; 150-180 Brin. For pump and valve parts; corrosion and heat resistant.

ALMG0.5 (910) 5050
J. & A. Erbsloh Aluminium
Aluminum. Si 0.3, Fe 0.4, Cu 0.05, Mn 0.05, Mg 0.3-0.6, Cr 0.03, Zn 0.1, Ti 0.03, 0.15 others, bal Al.
Nonhardenable alloy. Pressed: 85 N/mm^2 TS; 35 N/mm^2 YS; 25 Brin.

ALMG1 (511) 5100
J. & A. Erbsloh Aluminium
Aluminum. Si 0.3, Fe 0.4, Cu 0.05, Mn 0.05, Mg 0.9-1.1, Cr 0.03, Zn 0.1, Ti 0.03, 0.15 others, bal Al.
Nonhardenable alloy. Pressed: 100 N/mm^2 TS; 40 N/mm^2 YS; 30 Brin. Werkstoff Nr. 3.3315.08.

ALMG1.8 (521) 5200
J. & A. Erbsloh Aluminium
Aluminum. Si 0.3, Fe 0.4, Cu 0.05, Mn 0.05, Mg 1.7-2.2, Cr 0.03, Zn 0.1, Ti 0.03, 0.15 others, bal Al.
Nonhardenable alloy. Pressed: 145 N/mm^2 TS; 50 N/mm^2 YS; 40 Brin. Werkstoff Nr. 3.3326.08.

ALMG3 EQ (531) 5300
J. & A. Erbsloh Aluminium
Aluminum. Si 0.3, Fe 0.4, Cu 0.05, Mn 0.05, Mg 2.6-2.9, Cr 0.03, Zn 0.1, Ti 0.03, 0.15 others, bal Al.
Nonhardenable alloy. Pressed: 180 N/mm^2 TS; 80 N/mm^2 YS; 45 Brin. Werkstoff Nr. 3.3535.08.

ALMGSI 0.7 (426) 6070
J. & A. Erbsloh Aluminium
Aluminum. Si 0.5-0.7, Fe 0.25, Cu 0.15-0.2, Mn 0.04-0.08, Mg 0.5-0.6, Cr 0.08-0.12, Zn 0.1, Ti 0.03, 0.15 others, bal Al.
Hardenable alloy. Warm hardened: 270 N/mm^2 TS; 225 N/mm^2 YS; 85 Brin. Werkstoff Nr. 3.3210.71.

ALMGSI 0.8 (420) 6080
J. & A. Erbsloh Aluminium
Aluminum. Si 0.8-1, Fe 0.25, Cu 0.05, Mn 0.15, Mg 0.6-0.7, Cr 0.03, Zn 0.1, Ti 0.05, 0.15 others, bal Al.
Hardenable alloy. Warm hardened: 275 N/mm^2 TS; 200 N/mm^2 YS; 80 Brin.

ALMGSI 1 (410) 6100
J. & A. Erbsloh Aluminium
Aluminum. Si 0.7-1.05, Fe 0.4, Cu 0.05, Mn 0.7-0.8, Mg 0.8-1, Cr 0.1, Zn 0.1, Ti 0.03, 0.15 others, bal Al.
Hardenable alloy. Cold hardened: 205 N/mm^2 TS; 110 N/mm^2 YS; 65 Brin. Warm hardened: 275-310 N/mm^2 TS; 200-260 N/mm^2 YS; 80-95 Brin. Werkstoff Nr. 3.2315.51, 3.2315.71, 3.2315.72.

ALMGSI0.3 (466) 6030
J. & A. Erbsloh Aluminium
Aluminum. Si 0.3, Fe 0.2, Cu 0.03, Mn 0.03, Mg 0.4, Cr 0.03, Zn 0.05, Ti 0.02, 0.15 others, bal Al.
Hardenable alloy. Cold hardened: 120 N/mm^2 TS; 55 N/mm^2 YS; 35 Brin. Warm hardened: 190 N/mm^2 TS; 120 N/mm^2 YS; 60 Brin. Werkstoff Nr. 3.3206.

ALMGSI0.5 (461) 6040
J. & A. Erbsloh Aluminium
Aluminum. Si 0.35-0.5, Fe 0.25, Cu 0.05, Mn 0.05, Mg 0.4-0.5, Cr 0.03, Zn 0.1, Ti 0.03, 0.15 others, bal Al.
Hardenable alloy. Cold hardened: 130 N/mm^2 TS; 65 N/mm^2 YS; 45 Brin. Warm hardened: 215 N/mm^2 TS; 160 N/mm^2 YS; 70 Brin. Werkstoff Nr. 3.3206.51, 3.3206.7

ALMGSI0.5 (471) 6060
J. & A. Erbsloh Aluminium
Aluminum. Si 0.45-0.6, Fe 0.25, Cu 0.05, Mn 0.15, Mg 0.6-0.7, Cr 0.03, Zn 0.1, Ti 0.05, 0.15 others, bal Al.
Hardenable alloy. Cold hardened: 140 N/mm^2 TS; 70 N/mm^2 YS; 45 Brin. Warm hardened: 245 N/mm^2 TS; 195 N/mm^2 YS; 75 Brin. Werkstoff Nr. 3.3206, 3.3206.72.

ALMINAL A2
Imperial Aluminium Co., Ltd.
Aluminum. Cu 0.02, Si 0.15, Fe 0.15, 99.8 Al min.
M-temper: 8000 TS; 30 El. For structural members; corrosion resistant.

ALMINAL A2
Renfrew Foundries Ltd.
Cu 0.02, Si 0.15, Fe 0.15, 99.8 Al min.
M-temper: 8000 TS; 30 El. For structural members; corrosion resistant.

ALMINAL A3
Imperial Aluminium Co., Ltd.
Aluminum. Cu 0.05, Si 0.3, Fe 0.4, Mn 0.05, bal Al.
M-temper: 9000 TS; 25 El. For structural members; corrosion resistant.

ALMINAL A3
Renfrew Foundries Ltd.
Cu 0.05, Si 0.3, Fe 0.4, Mn 0.05, bal Al.
M-temper: 9000 TS; 25 El. For structural members; corrosion resistant.

ALMINAL A4
Imperial Aluminium Co., Ltd.
Aluminum. Cu 0.1, Si 0.5, Fe 0.7, Mn 0.1, bal Al.
M-temper: 9000 TS; 20 El. For structural members; corrosion resistant.

ALMINAL A4
Renfrew Foundries Ltd.
Cu 0.1, Si 0.5, Fe 0.7, Mn 0.1, bal Al.
M-temper: 9000 TS; 20 El. For structural members; corrosion resistant.

ALMINAL C10
Imperial Aluminium Co., Ltd.
Aluminum. Cu 0.1, Mg 9.5-11, Si 0.35, Fe 0.35, Ti 0.2, bal Al.
Sand cast: 40,000 TS; 8 El. Permanent mold: 44,800 TS; 12 El. For structural castings; corrosion resistant.

ALMINAL C10
Renfrew Foundries Ltd.
Cu 0.1, Mg 9.5-11, Si 0.35, Fe 0.35, Ti 0.2, bal Al.
Sand cast: 40,000 TS; 8 El. Permanent mold: 44,800 TS; 12 El. For structural castings; corrosion resistant.

ALMINAL C11
Imperial Aluminium Co., Ltd.
Aluminum. Cu 4-5, Si 0.25, Fe 0.25, Ti 0.2, bal Al.
Sand cast: 40,000 TS; 7 El. Permanent mold: 44,800 TS; 19 El. For aircraft engine castings; age-hardened.

ALMINAL C11
Renfrew Foundries Ltd.
Cu 4-5, Si 0.25, Fe 0.25, Ti 0.2, bal Al.
Sand cast: 40,000 TS; 7 El. Permanent mold: 44,800 TS; 19 El. For aircraft engine castings; age-hardened.

ALMINAL C12
Imperial Aluminium Co., Ltd.
Aluminum. Cu 9-10.5, Mg 0.4, Si 2, Fe 0.5-1.5, Mn 0.6, Ni 0.5, Zn 0.1, bal Al.
Sand cast: 100 Brin. Permanent mold: 150 Brin. For structural members and castings; age-hardened.

ALMINAL C12
Renfrew Foundries Ltd.
Cu 9-10.5, Mg 0.4, Si 2, Fe 0.5-1.5, Mn 0.6, Ni 0.5, Zn 0.1, bal Al.
Sand cast: 100 Brin. Permanent mold: 150 Brin. For structural members and castings; age-hardened.

ALMINAL C13

Imperial Aluminium Co., Ltd.
Aluminum. Cu 0.5-1.3, Mg 0.8-1.5, Si 11-13, Ni 2-3, Fe 0.8, Mn 0.5, bal Al.
Sand cast: 24,500 TS; 100 Brin. Permanent mold: 36,000 TS; 150 Brin. For pressure tight castings; age-hardened, corrosion resistant.

ALMINAL C13

Renfrew Foundries Ltd.
Cu 0.5-1.3, Mg 0.8-1.5, Si 11-13, Ni 2-3, Fe 0.8, Mn 0.5, bal Al.
Sand cast: 24,500 TS; 100 Brin. Permanent mold: 36,000 TS; 150 Brin. For pressure tight castings; age-hardened, corrosion resistant.

ALMINAL C15

Imperial Aluminium Co., Ltd.
Aluminum. Cu 1.3-3, Mg 0.5-1.7, Si 0.6-2, Fe 0.8-1.4, Ni 0.5-2, Ti 0.2, bal Al.
Sand cast: 40,400 TS; 100 Brin. Permanent mold: 47,100 TS, 150 Brin. For structural members; age-hardened, high temperature uses.

ALMINAL C15

Renfrew Foundries Ltd.
Cu 1.3-3, Mg 0.5-1.7, Si 0.6-2, Fe 0.8-1.4, Ni 0.5-2, Ti 0.2, bal Al.
Sand cast: 40,400 TS; 100 Brin. Permanent mold: 47,100 TS; 150 Brin. For structural members; age-hardened, high temperature uses.

ALMINAL C2

Imperial Aluminium Co., Ltd.
Aluminum. Cu 0.7-2.5, Si 9-11, Mg 0.3, Fe 1, Mn 0.5, Ni 1, Zn 1, Ti 0.2, bal Al.
Sand cast: 18,000 TS. Permanent mold: 20,000 TS. For pressure tight castings; corrosion resistant.

ALMINAL C2

Renfrew Foundries Ltd.
Cu 0.7-2.5, Si 9-11, Mg 0.3, Fe 1, Mn 0.5, Ni 1, Zn 1, Ti 0.2, bal Al.
Sand cast: 18,000 TS. Permanent mold: 20,000 TS. For pressure tight castings; corrosion resistant.

ALMINAL C4

Imperial Aluminium Co., Ltd.
Aluminum. Cu 2-4, Si 4-6, Mg 0.15, Fe 0.8, Mn 0.6, Ni 0.3, Zn 0.5, Ti 0.2, bal Al.
Sand cast: 20,000 TS; 2 El. Permanent mold: 22,400 TS; 2 El. For general castings; corrosion resistant.

ALMINAL C4

Renfrew Foundries Ltd.
Cu 2-4, Si 4-6, Mg 0.15, Fe 0.8, Mn 0.6, Ni 0.3, Zn 0.5, Ti 0.2, bal Al.
Sand cast: 20,000 TS; 2 El. Permanent mold: 22,400 TS; 2 El. For general castings; corrosion resistant.

ALMINAL C5

Imperial Aluminium Co., Ltd.
Aluminum. Cu 0.1, Mg 3-6, Si 0.3, Fe 0.6, Mn 0.5, Ni 0.1, Zn 0.1, bal Al.
Sand cast: 20,000 TS; 3 El. Permanent mold: 24,600 TS; 5 El. For structural members and castings; corrosion resistant.

ALMINAL C5

Renfrew Foundries Ltd.
Cu 0.1, Mg 3-6, Si 0.3, Fe 0.6, Mn 0.5, Ni 0.1, Zn 0.1, bal Al.
Sand cast: 20,000 TS; 3 El. Permanent mold: 24,600 TS; 5 El. For structural members and castings; corrosion resistant.

ALMINAL C6

Imperial Aluminium Co., Ltd.
Aluminum. Cu 0.1, Mg 0.1, Si 10-13, Mn 0.5, Ni 0.1, Zn 0.1, bal Al.
Sand cast: 23,500 TS; 5 El. Permanent mold: 27,000 TS; 7 El. For pressure tight castings; corrosion resistant.

ALMINAL C6

Renfrew Foundries Ltd.
Cu 0.1, Mg 0.1, Si 10-13, Mn 0.5, Ni 0.1, Zn 0.1, bal Al.
Sand cast: 23,500 TS; 5 El. Permanent mold: 27,000 TS; 7 El. For pressure tight casting; corrosion resistant.

ALMINAL C7

Imperial Aluminium Co., Ltd.
Aluminum. Cu 1-2.5, Mg 0.2, Si 1.5-3.5, Fe 0.3-1.4, Ni 0.5-1.7, Mn 0.1, bal Al.
Sand cast: 20,000 TS; 2 El. Permanent mold: 22,400 TS; 2 El. For structural members and castings; pressure tight castings.

ALMINAL C7

Renfrew Foundries Ltd.
Cu 1-2.5, Mg 0.2, Si 1.5-3.5, Fe 0.3-1.4, Ni 0.5-1.7, Mn 0.1, bal Al.
Sand cast: 20,000 TS; 2 El. Permanent mold: 22,400 TS; 2 El. For structural members and castings; pressure tight castings.

ALMINAL C8

Imperial Aluminium Co., Ltd.
Aluminum. Cu 0.1, Mg 0.4, Fe 0.6, Si 4.5-6, Mn 0.4, Ni 0.1, Zn 0.1, bal Al.
Sand cast: 24,600 TS; 2.5 El. WP-temper: 38,000 TS; 2.5 El. For pressure tight castings; corrosion resistant, good castability.

ALMINAL C8

Renfrew Foundries Ltd.
Cu 0.1, Mg 0.4, Fe 0.6, Si 4.5-6, Mn 0.4, Ni 0.1, Zn 0.1, bal Al.
Sand cast: 24,600 TS; 2.5 El. WP-temper: 38,000 TS; 2.5 El. For pressure tight castings; corrosion resistant, good castability.

ALMINAL C9

Imperial Aluminium Co., Ltd.
Aluminum. Cu 0.1, Mg 0.4, Fe 0.6, Si 10-13, Mn 0.4, Ni 0.1, Zn 0.1, bal Al.
Sand cast: 24,600 TS; 1.5 El. WP-temper: 42,500 TS. For pressure tight castings; corrosion resistant, good castability.

ALMINAL C9

Renfrew Foundries Ltd.
Cu 0.1, Mg 0.4, Fe 0.6, Si 10-13, Mn 0.4, Ni 0.1, Zn 0.1, bal Al.
Sand cast: 24,600 TS; 1.5 El. WP-temper: 42,500 TS. For pressure tight castings; corrosion resistant, good castability.

ALMINAL W-17

Imperial Aluminium Co., Ltd.
Aluminum. Cu 3.5-4.5, Mg 1.2, Fe 0.6, Si 0.6, Ti 0.3, Ni 1.8-2.3, bal Al.
WP-temper: 49,300 TS; 8 El. For aircraft engine components; age-hardened, high temperature uses.

ALMINAL W-17

Renfrew Foundries Ltd.
Cu 3.5-4.5, Mg 1.2, Fe 0.6, Si 0.6, Ti 0.3, Ni 1.8-2.3, bal Al.
WP-temper: 49,300 TS; 8 El. For aircraft engine components; age-hardened, high temperature uses.

ALMINAL W10

Imperial Aluminium Co., Ltd.
Aluminum. Cu 0.15, Mg 0.4-1.5, Si 0.75-1.3, Fe 0.6, Mn 1, Ti 0.2, bal Al.
W-temper: 27,000 TS; 15,700 YS; 18 El. WP-temper: 40,400 TS; 33,600 YS; 10 El. For structural members; age hardened.

ALMINAL W10

Renfrew Foundries Ltd.
Cu 0.15, Mg 0.4-1.5, Si 0.75-1.3, Fe 0.6, Mn 1, Ti 0.2, bal Al.
W-temper: 27,000 TS; 15,700 YS; 18 El. WP-temper: 40,400 TS; 33,600 YS; 10 El. For structural members; age-hardened.

ALMINAL W11

Imperial Aluminium Co., Ltd.
Aluminum. Cu 1-2, Mg 0.5-1.25, Si 0.75-1.25, Fe 0.75, Mn 1, Ti 0.3, bal Al.
W-temper: 38,100 TS; 22,400 YS; 15 El. WP-temper: 56,000 TS; 42,600 YS; 6 El. For structural members; age-hardened.

ALMINAL W11

Imperial Aluminium Co., Ltd.
Cu 1-2, Mg 0.5-1.25, Si 0.75-1.25, Fe 0.75, Mn 1, Ti 0.3, bal Al.
W-temper: 38,100 TS; 22,400 YS; 15 El. WP-temper: 56,000 TS; 42,600 YS; 6 El. For structural members; age-hardened.

ALMINAL W12

Imperial Aluminium Co., Ltd.
Aluminum. Cu 1.8-2.5, Mg 0.65-1.2, Si 0.55-1.2, Fe 0.6-1.2, Ni 0.6-1.4, Ti 0.2, bal Al.
WP-temper: 60,500 TS; 47,100 YS; 10 El. For aircraft engine components; age-hardened, high strength.

ALMINAL W12

Renfrew Foundries Ltd.
Cu 1.8-2.5, Mg 0.65-1.2, Si 0.55-1.2, Fe 0.6-1.2, Ni 0.6-1.4, Ti 0.2, bal Al.
WP-temper: 60,500 TS; 47,100 YS; 10 El. For aircraft engine components; age-hardened, high strength.

ALMINAL W14

Imperial Aluminium Co., Ltd.
Aluminum. Cu 3.5-5, Mg 0.4-1.2, Si 0.7, Fe 0.7, Mn 0.4-1.2, Ti 0.3, bal Al.
Heat treated: 53,500-56,000 TS; 31,000-33,600 YS; 10-15 El. For structural members; age-hardened.

ALMINAL W14

Renfrew Foundries Ltd.
Cu 3.5-5, Mg 0.4-1.2, Si 0.7, Fe 0.7, Mn 0.4-1.2, Ti 0.3, bal Al.
Heat treated: 53,500-56,000 TS; 31,000-33,600 YS; 10-15 El. For structural members; age-hardened.

ALMINAL W15

Imperial Aluminium Co., Ltd.
Aluminum. Cu 3.5-4.8, Mn 0.6, Si 1.5, Fe 1, Mn 1.2, Ti 0.3, bal Al.
W-temper: 53,500-56,000 TS; 31,400-33,600 YS; 10-15 El. WP-temper: 62,700-67,200 TS; 53,800-58,000 YS; 6-8 El. For structural members; age-hardened.

ALMINAL W15

Renfrew Foundries Ltd.
Cu 3.5-4.8, Mn 0.6, Si 1.5, Fe 1, Mn 1.2, Ti 0.3, bal Al.
W-temper: 53,500-56,000 TS; 31,400-33,600 YS; 10-15 El. WP-temper: 62,700-67,200 TS; 53,800-58,000 YS; 6-8 El. For structural members; age-hardened.

ALMINAL W150

Imperial Aluminium Co., Ltd.
Aluminum. Cu 4.8, Mg 0.85, Si 0.9, Ti 0.3, Mn 1.2, bal Al.
W-temper: 49,000-56,000 TS; 27,000-33,600 YS; 12-15 El. For structural members; age-hardened.

ALMINAL W150

Renfrew Foundries Ltd.
Cu 4.8, Mg 0.85, Si 0.9, Ti 0.3, Mn 1.2, bal Al.
W-temper: 49,000-56,000 TS; 27,000-33,600 YS; 12-15 El. For structural members; age-hardened.

ALMINAL W151

Imperial Aluminium Co., Ltd.
Si 0.4, Fe 0.7, Cu 5-6, Zn 0.3, Bi 0.4, Pb 0.4, bal Al.
Heat treated: 57,000 TS; 47,000 YS; 17 El; 97 Brin. For screw machine products, fasteners, screws, clips. Free-machining. Heat treatable.

ALMINAL W16

Imperial Aluminium Co., Ltd.
Aluminum. Cu 3, Mg 4, Si 0.6, Fe 0.6, Mn 1, Ti 0.2, Zn 4-8.5, bal Al.
WP-temper: 78,000-85,000 TS; 67,000-74,000 YS; 4-5 El. For structural members; age-hardened, high strength.

ALMINAL W16

Renfrew Foundries Ltd.
Cu 3, Mg 4, Si 0.6, Fe 0.6, Mn 1, Ti 0.2, Zn 4-8.5, bal Al.
WP-temper: 78,000-85,000 TS; 67,000-74,000 YS; 4-5 El. For structural members; age-hardened, high strength.

ALMINAL W160
Imperial Aluminium Co., Ltd.
Aluminum. Cu 1.5, Mg 2-3.5, Si 0.5, Fe 0.5, Zn 4.5-6.5, Ti 0.3, Mn 0.8, bal Al.
WP-temper: 72,000-78,500 TS; 60,000-67,000 YS; 7-5 El. For structural members; age-hardened.

ALMINAL W160
Renfrew Foundries Ltd.
Cu 1.5, Mg 2-3.5, Si 0.5, Fe 0.5, Zn 4.5-6.5, Ti 0.3, Mn 0.8, bal Al.
WP-temper: 72,000-78,500 TS; 60,000-67,000 YS; 5-7 El. For structural members; age-hardened.

ALMINAL W18
Imperial Aluminium Co., Ltd.
Aluminum. Cu 1.8-2.5, Mg 1.2-1.8, Si 0.5-1.3, Fe 0.6-1.2, Ni 0.6-1.4, bal Al.
W-temper: 53,800 TS; 6 El. For aircraft engine components; age-hardened, high temperature uses.

ALMINAL W18
Renfrew Foundries Ltd.
Cu 1.8-2.5, Mg 1.2-1.8, Si 0.5-1.3, Fe 0.6-1.2, Ni 0.6-1.4, bal Al.
W-temper: 53,800 TS; 6 El. For aircraft engine components; age-hardened, high temperature uses.

ALMINAL W4
Imperial Aluminium Co., Ltd.
Aluminum. Cu 0.15, Mg 1.75-2.25, Si 0.6, Fe 0.7, Mn 0.5, Ti 0.2, bal Al.
M-temper: 24,600 TS; 18 El. For structural members; corrosion resistant.

ALMINAL W4
Renfrew Foundries Ltd.
Cu 0.15, Mg 1.75-2.25, Si 0.6, Fe 0.7, Mn 0.5, Ti 0.2, bal Al.
M-temper: 24,600 TS; 18 El. For structural members; corrosion resistant.

ALMINAL W5
Imperial Aluminium Co., Ltd.
Aluminum. Mg 3-4, Cu 0.15, Si 0.6, Fe 0.7, Mn 1, Ti 0.2, bal Al.
M-temper: 31,400 TS; 12,000 YS; 18 El. For structural members; corrosion resistant.

ALMINAL W5
Renfrew Foundries Ltd.
Mg 3-4, Cu 0.15, Si 0.6, Fe 0.7, Mn 1, Ti 0.2, bal Al.
M-temper: 31,400 TS; 12,000 YS; 18 El. For structural members; corrosion resistant.

ALMINAL W6
Imperial Aluminium Co., Ltd.
Aluminum. Cu 0.15, Mg 4.5-5.5, Si 0.6, Fe 0.7, Mn 1, Ti 2, bal Al.
M-temper: 36,000 TS; 17,000 YS; 18 El. For structural members; corrosion resistant.

ALMINAL W6
Renfrew Foundries Ltd.
Cu 0.15, Mg 4.5-5.5, Si 0.6, Fe 0.7, Mn 1, Ti 2, bal Al.
M-temper: 36,000 TS; 17,000 YS; 18 El. For structural members; corrosion resistant.

ALMINAL W7
Imperial Aluminium Co., Ltd.
Aluminum. Cu 0.15, Mg 6.5-7.5, Si 0.6, Fe 0.7, Mn 1, Ti 0.2, bal Al.
M-temper: 44,800 TS; 19,000 YS; 18 El. For structural members; corrosion resistant.

ALMINAL W7
Renfrew Foundries Ltd.
Cu 0.15, Mg 6.5-7.5, Si 0.6, Fe 0.7, Mn 1, Ti 0.2, bal Al.
M-temper: 44,800 TS; 19,000 YS; 18 El. For structural members; corrosion resistant.

ALMINAL W9
Imperial Aluminium Co., Ltd.
Aluminum. Cu 0.15, Mg 0.4-0.9, Si 0.3-0.7, Fe 0.6, Ti 0.2, bal Al.
M-temper: 15,700 TS; 10,000 YS; 15 El. W-temper: 19,500 TS; 11,000 YS; 18 El. WP-temper: 27,000 TS; 22,400 YS; 12 El. For structural members; age-hardened.

ALMINAL W9
Renfrew Foundries Ltd.
Cu 0.15, Mg 0.4-0.9, Si 0.3-0.7, Fe 0.6, Ti 0.2, bal Al.
M-temper: 15,700 TS; 10,000 YS; 15 El. W-temper: 19,500 TS; 11,000 YS; 18 El. WP-temper: 27,000 TS; 22,400 YS; 12 El. For structural members; age-hardened.

ALMN1 (821) 3110
J. & A. Erbsloh Aluminium
Aluminum. Si 0.2, Fe 0.6, Cu 0.05, Mn 1-1.4, Mg 0.3, Cr 0.05, Zn 0.1, Ti 0.03, 0.15 others, bal Al.
Nonhardenable alloy. Pressed: 95 N/mm^2 TS; 40 N/mm^2 YS; 30 Brin. Werkstoff Nr. 3.0515.08.

ALMNCU 3110
J. & A. Erbsloh Aluminium
Aluminum. Si 0.2, Fe 0.4-0.5, Cu 0.05-0.1, Mn 1.1-1.3, Mg 0.05, Cr 0.03, Zn 0.1, Ti 0.03, 0.25 others, bal Al.
Nonhardenable alloy. Pressed: 95 N/mm^2 TS; 40 N/mm^2 YS; 30 Brin. Werkstoff Nr. 3.0517.08.

ALMO NO. 1
Colt Industries
C, Mo, Cr, bal Fe.
Annealed: 110,600 TS; 88,600 YS; 26 El; 71.2 RA; 187 Brin. Heat treated: 167,500 TS; 141,500 YS; 17.5 El; 58.7 RA; 340 Brin. For crankshafts, axles, connecting rod. *Obsolete*

ALMO NO. 1 C.H.
Colt Industries
Low C, Mo, Cr, bal Fe.
Oil treated: 110,800 TS; 110,000 YS; 20.5 El; 72.9 RA; 212 Brin. For gears. *Obsolete*

ALMO NO. 2
Colt Industries
C, Mo, Cr, bal Fe.
Annealed: 105,000-138,800 TS; 84,700-130,000 YS; 24-17 El; 71.6-62.6 RA; 174-269 Brin. For general use, axles, crankshafts. *Obsolete*

ALMO NO. 2 C.H.
Colt Industries
Low C, Mo, Cr, bal Fe.
Case-hard: 105,000 TS; 95,000 YS; 22.5 El; 72.3 RA; 217 Brin. For general use, gears. *Obsolete*

ALMO NO. 3
Colt Industries
C, Mo, Cr, bal Fe.
Annealed: 94,500 TS; 79,500 YS; 30.5 El; 69.8 RA; 166 Brin. Hardened: 134,400 TS; 114,100 YS; 15 El; 57.8 RA; 277 Brin. For crankshafts, axles. *Obsolete*

ALMO NO. 3 C.H.
Colt Industries
Low C, Mo, Cr, bal Fe.
Rolled: 95,400 TS; 63,800 YS; 34 El; 82.5 RA. For gears. *Obsolete*

ALMOLD-20
AL Tech Specialty Steel Corp.
C 0.3, Mn 0.7, Si 0.5, Cr 1.7, Mo 0.4, bal Fe.
For zinc die casting dies and plastic molds. Type P20 tool steel, mold quality.

ALMOS
Compagnie des Mines
C 0.08-0.16, Cr 0.8-1.2, Mo 0.15-0.3, Mn 0.6-0.9, bal Fe.
Annealed: 78,000 TS; 50,000 YS; 25 El. Hardened: 170,000 TS; 114,000 YS; 8 El. For shafts, gears; case hardening steel.

ALNEON
Strasser Co.
Aluminum. Al 70-90, Cu 2-3, Zn 7-22, 0.4-10 other elements.
28,000-48,500 TS; 10,700-17,000 YS; 1-3 El; 100-150 Brin. For light alloy parts; age-hardening, high endurance limit.

ALNESIUM
American manufacture
Mg 3, bal Al.
For cases, containers, corrosion resistant.

ALNI
Harsco Corp. (Taylor-Wharton Div.)
Al 11-13, Ni 24-26, Co 0-0.5, bal Fe.
For magnets; Alnico III. *Obsolete*

ALNI
Russian manufacture
C 0-0.1, Ni 24, Al 13, Cu 3.5, bal Fe.
For permanent magnets; high permeability.

ALNIC
Manufacturer not listed.
Ni 25, Al 12, Fe 63.
For permanent magnets; high coercive force. *Obsolete*

ALNICO
Russian manufacture
C 0-0.1, Ni 17, Al 10, Co 12.5, Cu 6, bal Fe.
For permanent magnets; high permeability.

ALNICO 1
Electronic Memories & Magnetics Corp.
Al 12, Ni 21, Co 5, Cu 2, bal Fe.
Cast permanent magnet, heat treated. Peak energy product 1.4 x 10^6 max; residual induction 7.2 kilogauss; coercive force 470 oersted.

ALNICO 1
Stackpole Magnet Division
Al 12, Ni 21, Co 5, Cu 3, bal Fe.
Cast permanent magnet material. Cast: 7200 Residual Flux Density Br. (gauss) 470 Coercive Force, Hc, (oersteds). 1.40 Max. Energy Product, (BH) max (MGO) (MG.Oe) - Megagauss-oersteds.

ALNICO 1
Arnold Engineering Co.
Al 12, Ni 21, Co 5, Cu 3, bal Fe.
Cast permanent magnet material. Cast: 7200 G Residual Flux Density; 470 Oe Coercive Force; 1.40 MG ·Oe Maximum Energy Product.

ALNICO 1
Wallace Murray Corp.
Al 12, Ni 22, Co 5, Ti 0.35, bal Fe.
Maximum energy product, 1.38 x 10^6 max; residual flux density, 6800; coercive force 530. Cast, furnished heat treated.

ALNICO 160
Plansee Metallwerk Gesellschaft
Al, Ni, Co.
For permanent magnets; sintered.

ALNICO 2
Manufacturer not listed.
Al 10, Ni 17, Co 12.5, Cu 8, bal Fe.
Isotropic permanent magnet material. Cast: 3,100 TS; 450 Brin. Sintered: 6,500 TS; 430 Brin. For permanent magnets; cast or sintered. *Obsolete*

ALNICO 2
Arnold Engineering Co.
Al 10, Ni 19, Co 13, Cu 3, bal Fe.
Cast permanent magnet material. Cast: residual flux density 7500 gauss; coercive force 560 oersted; peak energy product 170 MGO max.

ALNICO 2
Wallace Murray Corp.
Al 10, Ni 17, Co 12.5, Cu 6, Ti 0.45, bal Fe.
Maximum energy product, 1.65 x 10^6 max; residual flux density, 7500; coercive force 550. Cast, furnished heat treated.

ALNICO 2
Electronic Memories & Magnetics Corp.
Al 10, Ni 19, Co 13, Cu 3, bal Fe.
Cast permanent magnet, heat treated. Peak energy product 1.7 x 10^6 max; residual induction 7.5 kilogauss; coercive force 560 oersted.

ALNICO 2 SINTERED
Electronic Memories & Magnetics Corp.
Al 10, Ni 19, Co 13, Cu 3, bal Fe.
Peak energy product 1.5 x 10^6 max; residual induction 7.1 kilogauss; coercive force 550 oersted. Permanent magnet.

ALNICO 2C
Wallace Murray Corp.
Al 10, Ni 21, Co 12.5, Cu 6, Ti 0.3, bal Fe.
Maximum energy product, 1.60 x 10^6 max; residual flux density, 7100; coercive force 580. Cast, furnished heat treated.

ALNICO 3
Manufacturer not listed.
Al 12, Ni 25, bal Fe.
Isotropic permanent magnet material. Cast: 1,200 TS; 450 Brin. For permanent magnets; hard, brittle. *Obsolete*

ALNICO 3
Arnold Engineering Co.
Al 12, Ni 25, Cu 3, bal Fe.
Cast permanent magnet material. Cast: residual flux density 7000 gauss; coercive force 480 oersted; peak energy product 1.35 MGO max.

ALNICO 3
Wallace Murray Corp.
Al 12, Ni 26, bal Fe.
Maximum energy products, 1.20 x 10^6 max; residual flux density, 6600; coercive force 450. Cast, furnished heat treated.

ALNICO 3
Electronic Memories & Magnetics Corp.
Al 12, NI 25, Cu 2, bal Fe.
Cast permanent magnet, heat treated. Peak energy product 1.35 x 10^6 max; residual induction 7.0 kilogauss; coercive force 480 oersted.

ALNICO 350
German manufacture
Al 7.8, Ni 15, Co 34, Cu 3.5, Ti 5, bal Fe.
For electrical and magnetic equipment; permanent magnet, high permeability.

ALNICO 4
Manufacturer not listed.
Al 12, Ni 28, Co 5, bal Fe.
Isotropic permanent magnet material. Cast: 9,100 TS; 450 Brin. For permanent magnets. *Obsolete*

ALNICO 4
Electronic Memories & Magnetics Corp.
Al 12, Ni 27, Co 5, Cu 2, bal Fe.
Cast permanent magnet, heat treated. Peak energy product 1.5 x 10^6 max; residual induction 5.6 kilogauss; coercive force 720 oersted.

ALNICO 4
Wallace Murray Corp.
Al 12, Ni 25, Co 5, Ti 0.4, bal Fe.
Maximum energy product, 1.40 x 10^6 max; residual flux density, 6300; coercive force 800. Cast, furnished heat treated.

ALNICO 4
Arnold Engineering Co.
Al 12, Ni 27, Co 5, bal Fe.
Cast permanent magnet material. Cast: residual flux density 5600 gauss; coercive force 720 oersted; peak energy product 1.35 MGO max.

ALNICO 4 SINTERED
Indiana General
Al 12, Ni 27, Co 5, Cu 2, bal Fe.
Pressed from powder and sintered. Residual induction, Br, 5.2 kilogauss; coercive force, Hc, 700.00 oersteds. Permanent magnet. *Obsolete*

ALNICO 400
Polish manufacture
Al 8, Ni 14, Co 24, Cu 3, Ti 0.5, bal Fe.
For electrical and magnetic equipment; permanent magnet, high permeability.

ALNICO 5
Manufacturer not listed.
Al 8, Ni 14, Co 24, Cu 3, bal Fe.
Anisotropic permanent magnet material. Cast: 5,400 TS; 500 Brin. Sintered: 450 Brin. Crystal oriented-cast; high energy product. *Obsolete*

ALNICO 5
Arnold Engineering Co.
Al 8, Ni 14, Co 24, Cu 3, bal Fe.
Cast permanent magnet material. Cast: residual flux density 12,800 gauss; coercive force 640 oersted; peak energy product 5.50 MGO max.

ALNICO 5
Wallace Murray Corp.
Al 8, Ni 14.5, Co 24, Cu 3, bal Fe.
Maximum energy product, 5.25 x 10^6 max; residual flux density, 12,700; coercive force 640. Cast, directional magnetic properties; furnished heat treated.

ALNICO 5
Electronic Memories & Magnetics Corp.
Al 8, Ni 14, Co 24, Cu 3, bal Fe.
Cast permanent magnet, cooled in magnetic field. Peak energy product 5.5 x 10^6 max; residual induction 12.5 kilogauss; coercive force 640 oersted.

ALNICO 5 CB
Thomas & Skinner Inc.
Al 8, Ni 14, Co 24, Cb, bal Fe.
For permanent magnets; high permeability. *Obsolete*

ALNICO 5 COL
Arnold Engineering Co.
Al 8, Ni 14, Co 24, Cu 3, bal Fe.
Cast permanent magnet material. Cast: residual flux density 13,500 gauss; coercive force 740 oersted; peak energy product 7.55 MGO max.

ALNICO 5 DG
Stackpole Magnet Division
Al 8, Ni 14, Co 24, Cu 3, bal Fe.
Permanent magnet. Residual flux: 13,300 Gauss. Coercive force: 670 Oersted. *Obsolete*

ALNICO 5 DG
Arnold Engineering Co.
Al 8, Ni 14, Co 24, Cu 3, bal Fe.
Cast permanent magnet material. Cast: residual flux density 13,300 gauss; coercive force 670 oersted; peak energy product 6.50 MGO max.

ALNICO 5 E
Wallace Murray Corp.
Al 8, Ni 15, Co 22.5, Cu 3.6, Ti 0.45, bal Fe.
Maximum energy product, 4.60 x 10^6 max; residual flux density, 11,800, coercive force 700. Cast, directional magnetic properties; furnished heat treated.

ALNICO 5 SINTERED
Electronic Memories & Magnetics Corp.
Al 8, Ni 14, Co 24, Cu 3, bal Fe.
Pressed from powder and sintered. Peak energy product 3.9 x 10^6 max; residual induction 10.9 kilogauss; coercive force 620 oersted. Cooled in magnetic field; permanent magnet.

ALNICO 5-7
Electronic Memories & Magnetics Corp.
Al 8, Ni 14, Co 24, Cu 3, bal Fe.
Cast permanent magnet, cooled in magnetic field. Peak energy product 7.50 x 10^6 max; residual induction 13.4 kilogauss; coercive force 730 oersted.

ALNICO 6
Manufacturer not listed.
Al 8, Ni 15, Co 24, Ti 1.2, bal Fe.
Anisotropic permanent magnet material. Cast or sintered. Cast: 2,300 TS; 560 Brin. High coercive force. *Obsolete*

ALNICO 6
Arnold Engineering Co.
Al 8, Ni 16, Co 24, Cu 3, Ti 1, bal Fe.
Cast permanent magnet material. Cast: residual flux density 10,500 gauss; coercive force 780 oersted; peak energy product 3.90 MGO max.

ALNICO 6
Wallace Murray Corp.
Al 8, Ni 15, Co 24, Cu 3.4, Ti 1.25, bal Fe.
Maximum energy product, 3.65 x 10^6 max; residual flux density, 10,400; coercive force 700. Cast, directional magnetic properties; furnished heat treated.

ALNICO 6
Electronic Memories & Magnetics Corp.
Al 8, Ni 16, Co 24, Cu 3, Ti 1, bal Fe.
Cast permanent magnet, heat treated. Peak energy product 3.5 x 10^6 max; residual induction 10.1 kilogauss; coercive force 750 oersted.

ALNICO 6 SINTERED
Electronic Memories & Magnetics Corp.
Al 8, Ni 16, Co 24, Cu 3, Ti 1, bal Fe.
Pressed from powder and sintered. Peak energy product 2.76 x 10^6 max; residual induction 8.8 kilogauss; coercive force 800 oersted. Cooled in magnetic field; permanent magnet.

ALNICO 7
Manufacturer not listed.
Al 8.5, Ni 18, Co 24, Cu 3.25, Ti 5, bal Fe.
Anisotropic permanent magnet material. Cast: 600 Brin. High coercive force. *Obsolete*

ALNICO 7
Arnold Engineering Co.
Al 8.5, Ni 18, Co 24, Cu 3.25, Ti 5, bal Fe.
Cast permanent magnet material. Cast: residual flux density 7700 gauss; coercive force 1050 oersted; peak energy product 2.85 MGO max.

ALNICO 7-5 ORIENTED
Indiana General
Al 7, Ni 14, Co 31, Cu 4, Ti 5, bal Fe.
Cast permanent magnet, cooled in magnetic field. Residual induction, Br, 8.5 kilogauss; coercive force, Hc, 1,040.00 oersteds. *Obsolete*

ALNICO 7-5 UNORIENTED
Indiana General
Al 7, Ni 14, Co 31, Cu 4, Ti 5, bal Fe.
Cast permanent magnet, heat treated. Residual induction, Br, 7.4 kilogauss; coercive force, Hc, 890.00 oersteds. *Obsolete*

ALNICO 7C
Thomas & Skinner Inc.
Al-Ni-Co, bal Fe.
For magnetic structures and devices; high coercive strength. *Obsolete*

ALNICO 8

Manufacturer not listed.
Ni, Co, Al, Cu, Ti, Fe.
Anisotropic permanent magnet material. Cast or sintered. High energy product; high coercive force. *Obsolete*

ALNICO 8

Arnold Engineering Co.
Co 35, Ni 15, Cu 4, Ti 5, Al 7, bal Fe.
Cast permanent magnet material. Cast: residual flux density 8200 gauss; coercive force 1650 oersted; peak energy product 5.3 MGO max.

ALNICO 8B

Electronic Memories & Magnetics Corp.
Ni 15, Co 35, Cu 3, Ti 5, Al 7, bal Fe.

ALNICO 8H

Electronic Memories & Magnetics Corp.
Cu 3, Ti 8, Al 8, Ni 14, Co 38, bal Fe.
Cast permanent magnet, cooled in magnetic field. Peak energy product 5.0×10^6 max; residual induction 7.0 kilogauss; coercive force 1900 oersted.

ALNICO 8H SINTERED

Electronic Memories & Magnetics Corp.
Al 7, Ni 14, Co 38, Cu 3, Ti 8, bal Fe.
Permanent magnet. Pressed from powder and sintered. Peak energy product 4.5×10^6 max; residual induction 6.5 kilogauss; coercive force 1800 oersted.

ALNICO 8HC

Thomas & Skinner Inc.
Cast Alnico magnet. Peak energy, 6.5×10^6 max; residual induction, 7700 Gauss; coercive force, 2200 Oersteds; incremental permeability, 2.8.

ALNICO 8HC

Stackpole Magnet Division
Al 8, Ni 14, Co 38, Cu 3, Ti 8, bal Fe.
Permanent magnet. Residual flux: 7200 Gauss. Coercive force: 1900 Oersted. *Obsolete*

ALNICO 8HE

Thomas & Skinner Inc.
Cast Alnico magnet. Peak energy, 6.75×10^6 max; residual induction, 8800 Gauss; coercive force, 1800 Oersteds; incremental permeability, 2.9.

ALNICO 8HE

Electronic Memories & Magnetics Corp.
Al 7, Ni 15, Co 35, Cu 4, Ti 5, bal Fe.
Cast permanent magnet, cooled in magnetic field. Peak energy product 6.0×10^6 max; residual induction 9.25 kilogauss; coercive force 1550 oersted.

ALNICO 8HE SINTERED

Electronic Memories & Magnetics Corp.
Al 7, Ni 15, Co 35, Cu 4, Ti 5, bal Fe.
Permanent magnet. Pressed from powder and sintered. Peak energy product 5.5×10^6 max; residual induction 8.6 kilogauss; coercive force 1500 oersted.

ALNICO 9

Manufacturer not listed.
Anisotropic permanent magnet material. Crystal oriented-cast only. Highest energy product alnico material; high coercive force. *Obsolete*

ALNICO 9

Arnold Engineering Co.
Co 35, Ni 15, Cu 4, Ti 5, Al 7, bal Fe.
Cast permanent magnet material. Cast: residual flux density 10,500 gauss; coercive force 1500 oersted; peak energy product 9.0 MGO max.

ALNICO 9

Electronic Memories & Magnetics Corp.
Co 35, Ni 15, Cu 4, Ti 5, Al 7, bal Fe.
Cast permanent magnet, cooled in magnetic field. Peak energy product 10.0×10^6 max; residual induction 10.5 kilogauss; coercive force 1600 oersted.

ALNICO 9NB

Thomas & Skinner Inc.
Cast Alnico magnet. Peak energy, 11.8×10^6 max; residual induction, 11,400 Gauss; coercive force, 1640 Oersteds; incremental permeability, 2.5.

ALNICO I

Colt Industries
C 0.1, Ni 22, Al 12, Co 5, bal Fe.
For permanent magnets; cast. *Obsolete*

ALNICO I

General Electric Co.
Al 12, Ni 20, Co 5, bal Fe.
Cast: 4,100 TS; 450 Brin. For permanent magnets; hard and brittle. *Obsolete*

ALNICO II

Colt Industries
C 0.1, Ni 18.5, Al 10, Co 13, bal Fe.
For permanent magnets; cast. *Obsolete*

ALNICO III

Colt Industries
Al 12, Ni 25, Si 0.04, bal Fe.
For permanent magnets; cast. *Obsolete*

ALNICO IV

Colt Industries
Al 8, Ni 15, Co 24, Cu 3, Ti 1, bal Fe.
For permanent magnets; cast. *Obsolete*

ALNICO V

Colt Industries
Al 8, Ni 14, Co 24, bal Fe.
For permanent magnets; cast. *Obsolete*

ALNICO VI

Colt Industries
Ni 15, Al 8, Co 22, bal Fe.
Cast: 20,000 TS. For permanent magnets; cast; Br 10,100, Hc 750. *Obsolete*

ALNICO VIII

Colt Industries
Al 7, Ni 15, Co 35, Fe 34, Cu 4, Ti 5.
Permanent magnets for motors, generators, focusing of electron beams, magnetic cores; high resistance to demagnetization. *Obsolete*

ALNICO VIII

General Electric Co.
Al 7, Ni 15, Co 35, Fe 34, Cu 4, Ti 5.
Permanent magnets for motors, generators, focusing of electron beams, magnetic cores; high resistance to demagnetization. *Obsolete*

ALNICO AL-2

Thomas & Skinner Inc.
Energy product, 1.8×10^6 max; residual induction, 7,500 Gauss; coercive force, 570 Oersteds.

ALNICO AL-5

Thomas & Skinner Inc.
Energy product, 5.5×10^6 max; residual induction, 12,700 Gauss; coercive force, 660 Oersteds.

ALNICO AL-5-7

Thomas & Skinner Inc.
Energy product, 7.5×10^6 max; residual induction, 13,500 Gauss; coercive force, 750 Oersteds.

ALNICO AL-5E

Thomas & Skinner Inc.
Energy product, 4.3×10^6 max; residual induction, 10,800 Gauss; coercive force, 725 Oersteds.

ALNICO AL-6

Thomas & Skinner Inc.
Energy product, 3.8×10^6 max; residual induction, 10,400 Gauss; coercive force, 810 Oersteds.

ALNICO AL-9

Thomas & Skinner Inc.
Energy product, 10.5×10^6 max; residual induction, 10,800 Gauss; coercive force, 1520 Oersteds.

ALNICUS GR. USM65

U.S. Magnet & Alloy Corp.
Al 8, Ni 14, Co 24, Cu 3, bal Fe.
Cast: 5500 TS; 500 Brin. For magnetic chucks, meters, starting devices, electrocardiographs; permanent magnets, oriented.

ALNICUS GR. USM75

U.S. Magnet & Alloy Corp.
Al 8, Ni 14, Co 24, Cu 3, bal Fe.
Cast: 5500 TS; 500 Brin. For magnetic chucks, meters, starting devices, electrocardiographs; permanent magnets, oriented.

ALNIFER

Texas Instruments Inc./Materials Control
Low C steel clad with Al on one side and Ni on other side. For electron tubes.

ALNILOY

NL Industries
Cu 4, Ni 4, Si 2, Fe 0-2, bal Al.
Sand cast: 30,000 TS; 15,000 YS; 2 El; 7 RA; 75 Brin. For ornaments. *Obsolete*

ALNILOY

Doehler-Jarvis Co.
Aluminum.
Aluminum alloy AlCu4Ni4Si2.

ALOI

Thomas Bolton Ltd.
For single plates in textile industry; non-gassing, non-distorting. *Obsolete*

ALOY-NUM

E.N. Egge Co.
Si 4, Cu 2, bal Al.
For pistons, cylinder heads. Age hardenable.

ALOYCO 18-8 MO

Walworth Co.
C 0.2, Cr 18-18.2, Ni 8, Mo 0.2, bal Fe.
For valves, fittings, corrosion resistant. *Obsolete*

ALOYCO 18-8 SCB

Walworth Co.
C 0-0.07, Cr 18-21, Ni 9-12, 1.0 Cb + Ta, bal Fe.
Cast: 75,000 TS; 35,000 YS; 40 El; 160 Brin. For valves, fittings, pumps. Corrosion resistant. ACI CF8C.

ALOYCO 18-8S

Walworth Co.
C 0-0.07, Cr 18-21, Ni 8-11, bal Fe.
Cast: 73,000 TS; 33,000 YS; 50 El; 160 Brin. For valves, fitting pumps. Corrosion resistant. ACI CF8.

ALOYCO 18-8S ELC

Walworth Co.
C 0-0.03, Cr 17-21, Ni 8-11, bal Fe.
Cast: 74,000 TS; 34,000 YS; 50 El, 160 Brin. For valves, fittings, pumps. Corrosion resistant. ACI CF3.

ALOYCO 18-8S MO
Walworth Co.
C 0-0.07, Cr 18-21, Ni 9-12, Mo 2-3, bal Fe.
Cast: 75,000 TS; 35,000 YS; 45 El; 160 Brin. For valves, fittings, pumps. Corrosion resistant. ACI CF8M.

ALOYCO 18-8S MO ELEC
Walworth Co.
C 0-0.03, Cr 17-21, Ni 9-13, Mo 2-3, bal Fe.
Cast: 75,000 psi TS; 34,000 YS; 45 El; 160 Brin. For valves, fittings, pumps. Corrosion resistant. ACI CF3M.

ALOYCO 20
Walworth Co.
C 0-0.07, Cr 19-22, Ni 27.5-30.5, Mo 2-3, Cu 3-4, bal Fe.
Cast: 67,000 TS; 28,000 YS; 45 El; 155 Brin. For valves, fittings, pumps. Corrosion resistant to sulfuric acid, seawater, nitric acid, organic acid. ACI CN7M.

ALOYCO 35
Walworth Co.
Cr 24-26, Ni 19-21, Mo 2.75-3.25, Cu 2.5-3, C 0-0.07, bal Fe.
Water quenched: 75,000 TS; 40,000 YS; 45 El; 45 RA; 160 Brin. For chemical equipment; to handle hot sulfuric acid solutions. *Obsolete*

ALOYCO 37
Walworth Co.
C 0-0.07, Ni 20, Cr 25, Mo 3, Cu 2.7, Si 3, Mn 0.7, bal Fe.
Annealed: 70,000 min TS; 40,000 min YS; 30 El; 45 RA; 145 Brin. For chemical plant equipment; stainless, austenitic castings. *Obsolete*

ALOYCO MONEL
Walworth Co.
C 0-0.35, Mn 0-1.5, Si 0-2, Cu 26-33, Fe 0-3.5, bal Ni.
Cast: 67,500 TS; 30,000 YS; 25 El; 160 Brin. For valves, fittings, pumps. ACI M35.

ALOYCO-25-20
Walworth Co.
C 0-0.2, Cr 22-26, Ni 19-22, bal Fe.
Cast: 70,000 TS; 32,000 YS; 30 El; 160 Brin. For valves, fittings, pumps. Corrosion resistant. ACI CK20.

ALOYCO-CA6NM
Walworth Co.
Co 0-0.06, Si 0-1, Ni 3.5-4.5, Mo 0.4-1, Cr 11.5-14, bal Fe.
Cast: 120,000 TS; 87,000 YS; 15 El; 260 Brin. For valves, fittings, pumps. Corrosion and cavitation resistant.

ALOYCO-INCONEL
Walworth Co.
C 0-0.4, Mn 0-1.5, Si 0-3, Cr 14-17, Fe 0-11, bal Ni.
Cast: 72,000 TS; 30,000 YS; 30 El; 160 Brin. For valves, fittings, pumps. ACI CY40.

ALOYCO-N-2
Walworth Co.
C 0-0.12, Mo 26-30, Cr 0-1, V 0.2-0.6, Fe 4-6, bal Ni.
Cast: 80,000 TS; 48,000 YS; 6 El; 170 Brin. For valves, fittings, pumps. Corrosion resistant. ACI Ni2M-1.

ALOYCO-N-3
Walworth Co.
C 0-0.12, Mo 16-18, Cr 15.5-17.95, W 3.75-5.25, V 0.2-0.4, Fe 4.5-7.5, bal Ni.
Cast: 75,000 TS; 48,000 YS; 4 El; 170 Brin. For valves, fittings, pumps. Corrosion resistant. ACI CW12-M-1.

ALOYCO-NICKEL
Walworth Co.
C 0-1, Mn 0-1.5, Si 0-2, bal Ni.
Cast: 55,000 TS; 20,000 YS; 10 El; 140 Brin. For valves, fittings, pumps. ACI CZ-100.

ALPACCA
English manufacture
Cu 65.9, Zn 19.25, Ni 14.6, Fe 0.5.
For base for plated table ware; corrosion resistant.

ALPAKKA (ALPACA)
English manufacture
Cu 60, Zn 19, Ni 15, Ag 2.
For base metal for silver plated tableware; resists corrosion.

ALPAX
Sterling International Technology Ltd.
Si 10-13, bal Al.
Sand cast: 23,500-26,800 TS; 6700-7800 YS; 5-8 El; 45-50 Brin; 4.5 Izod. For castings of thin sections, hydraulic parts. BS 1490 LM 6-M; BS 4L 33.

ALPAX
Koch Light Alloys Ltd.
Aluminum. Si 10-13, Mg 0.2-0.6, Mn 0.3-0.7, bal Al.
Sand cast: 24,000 TS; 9,000 YS; 6-8 El; 55 Brin. Chill cast: 30,500 TS; 12,500 YS; 10-12 El; 65 Brin. For light corrosion resistant castings; corrosion resistant.

ALPAX
Alais Forges et Camargue
Cu 4.5, Mg 0.5, Ti 0.4, bal Al.
36,000-43,000 TS; 29,000-36,000 YS; 1-2 El. For light alloy parts; age-hardenable.

ALPAX
Birmingham Aluminium Casting Co.
Aluminum. Si 10-13, Ti 0-0.2, Fe 0-0.6, Mn 0-0.5, bal Al.
Sand cast: 23,100 TS; 10,000 YS; 5-10 El; 50 RA; 60 Brin. Permanent mold: 28,000 TS; 12,000 YS; 8-14 El; 60 RA; 65 Brin. For water pumps, engine parts; corrosion resistant, good castability.

ALPAX "B" (ALPAX BETA)
Sterling International Technology Ltd.
Mg 0.2-0.6, Si 10-13, Mn 0.3-0.7, bal Al.
Sand cast: 24,600-29,000 TS; 14,500-16,000 YS; 1.5-3 El; 60-70 Brin; 1.5 Izod. General purpose sand or chill cast aluminum. BS 1490 LM 9-P.

ALPAX "G" (ALPAX GAMMA)
Sterling International Technology Ltd.
Mg 0.2-0.6, Si 10-13, Mn 0.3-0.7, bal Al.
Sand cast: 35,000-37,000 TS; 29,000-31,000 YS; 0-2 El; 90-100 Brin; 0.9 Izod. General purpose sand or chill cast aluminum. BS 1490 LM 9 WP; BS L 75.

ALPAX ALPHA
Birmingham Aluminium Casting Co.
Aluminum. Si 10-13, Mg 0.6, bal Al.
Sand cast: 22,000 TS; 2-4 El; 64 Brin. Permanent mold: 30,000 TS; 3-5 El; 70 Brin. For gear cases, instrument housings; corrosion resistant, good castability

ALPAX ALPHA
Koch Light Alloys Ltd.
Aluminum. Si 10-13, Mg 0.2-0.6, Mn 0.3-0.7, bal Al.
Sand cast: 22,500 TS; 9000 YS; 2-4 El; 64 Brin. Chill cast: 3-5 El; 70 Brin. For light corrosion resistant castings; corrosion resistant.

ALPAX BETA
Birmingham Aluminium Casting Co.
Aluminum. Si 10-13, Mg 0.2-0.6, Fe 0.6, Ti 0-0.2, bal Al.
Sand cast: 26,000 TS; 16,000 YS; 2-3 El; 75 Brin. Permanent mold: 34,000 TS; 18,000 YS; 2-4 El; 85 Brin. For gear boxes, housings, general castings; corrosion resistant, good castability.

ALPAX BETA
Koch Light Alloys Ltd.
Aluminum. Si 10-13, Mg 0.2-0.6, Mn 0.3-0.7, bal Al.
Sand cast: 24,000 TS; 16,000 YS; 2-3 El; 75 Brin. Chill cast: 34,000 TS; 20,000 YS; 2-4 El; 85 Brin. For light corrosion resistant castings; corrosion resistant.

ALPAX GAMMA
Birmingham Aluminium Casting Co.
Aluminum. Si 10-13, Mg 0.6, bal Al.
Aged: 36,000 TS; 30,000 YS 1-2 El; 95 Brin. For engine components, general castings; corrosion resistant, good castability.

ALPAX GAMMA
Koch Light Alloys Ltd.
Aluminum. Si 10-13, Mg 0.2-0.6, Mn 0.3-0.7, bal Al.
Sand cast: 38,000 TS; 32,000 YS; 1-2 El; 95 Brin. Chill cast: 36,000 TS; 1-3 El; 100 Brin. For light corrosion resistant castings; corrosion resistant.

ALPAX HAU COBALT
Koch Light Alloys Ltd.
Aluminum. Si 9.3, Co 0.45, Mg 0.2, bal Al.
Cast: 48,000 TS; 26,000 YS; 4 El. For light weight housings, casings, cover plates. Corrosion resistant.

ALPERM
Japanese manufacture
Al 14-16, bal Fe.
For magnetic laminations; high permeability.

ALPHA
Abex Corp.
Pb, Sb, bal Sn.
14,900 TS; 5.6 El; 6.1 RA; 35 Brin. For Babbitts, bearings; Babbitt metal. *Obsolete*

ALPHA
Allegheny Ludlum Steel
high C, bal Fe.
For drills.

ALPHA
Chase Brass & Copper Co.
Cu 64-68, Pb 0-0.8, Fe 0-0.07, bal Zn.
Annealed: 50,000 TS; 40 El. For condenser tubes, pipes; resists corrosion of natural water. *Obsolete*

ALPHA CHISEL
Midvale-Heppenstall Co.
C, W, bal Fe.
For chisels, tools. *Obsolete*

ALPHA CRUSHER
Abex Corp.
Sb, Pb, bal Sn.
For bearings; Babbitt metal. *Obsolete*

ALPHA GAMMA
Birmingham Aluminium Casting Co.
Aluminum. Si 10-13, Mg, bal Al.
For gears, bolts, springs, crankshafts; oil hardened, shock resistant.

ALPHA NO. 2
Midvale-Heppenstall Co.
C, alloy, bal Fe.
For cutting tools. *Obsolete*

ALPHA TOOL
Midvale-Heppenstall Co.
C, W, bal Fe.
For tools. *Obsolete*

ALPHA-TWO ALUMINIDE (24/11)
TIMET
Titanium. Al 13.8, Nb 21.5, O 0.07, bal Ti.
Developmental alloy. Beta annealed plus aged at 1200°F: 85 ksi TS; 55 ksi YS; 16 El. Ingot, billet, bar, sheet, plate, foil. For high temperature applications.

ALPHA-TWO ALUMINIDE (25/10/3/1)
TIMET
Titanium. Al 14, Nb 20, V 3.2, Mo 2, O 0.07, bal Ti.
Developmental alloy. Beta annealed plus aged at 1200°F:
115 ksi TS; 85 ksi YS; 10 El. Ingot, billet, bar, sheet, plate, foil.
High strength, more creep resistant version of the 24/11
alloy.

ALPINE AHD
Vereinigte Edelstahlwerke
C 0.5, Cr 1, V 0.1, bal Fe.
For gears, bolts, springs, crankshafts; oil hardened, shock
resistant. *Obsolete*

ALPINE AHK
Now VEW K455.

ALPINE AHKW
Vereinigte Edelstahlwerke
C 0.35, Cr 1, V 0.18, W 1.85, bal Fe.
For cold work tools, upsetters; oil hardened. *Obsolete*

ALPINE AKL
Now VEW K244.

ALPINE ANP3
Vereinigte Edelstahlwerke
C 0.5, Cr 1, Ni 3.25, bal Fe.
For gears, bolts, crankshafts, fasteners; oil hardened, shock
resistant. *Obsolete*

ALPINE BEC60
Vereinigte Edelstahlwerke
C 0.15, Cr 0.65, Mn 0.5, bal Fe.
For gears, cams, camshafts, fasteners; case hardened,
tough. *Obsolete*

ALPINE BEC80
Vereinigte Edelstahlwerke
C 0.2, Mn 1.25, Cr 1.15, bal Fe.
For gears, cams, camshafts, fasteners; case hardened,
tough. *Obsolete*

ALPINE BLH
Vereinigte Edelstahlwerke
C 0.2, Cr 1.15, Mn 1.25, bal Fe.
For gears, cams, camshafts, fasteners; case hardened,
tough. *Obsolete*

ALPINE BR55
Vereinigte Edelstahlwerke
C 0.56, Si 0.3, Mn 0.55, bal Fe.
Heat treated: 155,000 TS; 110,000 YS; 15 El; 45 RA; 350 Brin.
For tools, hammers, shafts, axes, gears, bolts; water
hardened. *Obsolete*

ALPINE ECR15
Now VEW E410.

ALPINE ECR20
Now VEW E400.

ALPINE ED
Now VEW E900.

ALPINE EFW
Now VEW E920.

ALPINE EJR35
Vereinigte Edelstahlwerke
C 0.13, Cr 0.7, Ni 3.5, bal Fe.
For gears, bolts, machine tool parts; case hardening steel,
shock resistant. *Obsolete*

ALPINE EJR45
Vereinigte Edelstahlwerke
C 0.13, Ni 4.5, Cr 1.1, bal Fe.
For gears, bolts, cams, pinions, camshafts; case hardening
steel, shock resistant. *Obsolete*

ALPINE EJRA
Vereinigte Edelstahlwerke
C 0.13, Cr 0.7, Ni 2.5, bal Fe.
For gears, pinions, cams, shafts, camshafts; case hardening
steel, shock resistant. *Obsolete*

ALPINE EL1
Vereinigte Edelstahlwerke
C 0.13, Ni 1.5, Cr 0.2, bal Fe.
For gears, fasteners, crankshafts; case hardening steel,
tough. *Obsolete*

ALPINE ERJ15
Vereinigte Edelstahlwerke
C 0.15, Cr 1.55, Ni 1.55, Mn 0.5, bal Fe.
For gears, bolts, camshafts, cams; case hardened, shock
resistant. *Obsolete*

ALPINE ERJ20
Vereinigte Edelstahlwerke
C 0.18, Cr 2, Ni 2, Mn 0.5, bal Fe.
For gears, bolts, camshafts; case hardened, shock resistant.
Obsolete

ALPINE ERM15
Now VEW E304.

ALPINE ERM20
Now VEW E300.

ALPINE EXTRA M
Vereinigte Edelstahlwerke
C 1, Mn 0.25, V 0.1, bal Fe.
For drills, taps, reamers, broaches; Type W2; water hardened.
Obsolete

ALPINE EXTRA MH
Now VEW K992.

ALPINE EXTRA SPEZIAL ZH II
Vereinigte Edelstahlwerke
C 0.85, Si 0-0.25, Mn 0-0.25, bal Fe.
Heat treated: 190,000 TS; 145,000 YS; 10 El; 30 RA; 400 Brin.
For springs, taps, cutters, drills; Type W1; water hardened.
Obsolete

ALPINE EXTRA ZAH
Now VEW K976.

ALPINE EXTRA ZH I
Now VEW K988.

ALPINE EXTRA ZH II
Now VEW K984.

ALPINE EZH
Vereinigte Edelstahlwerke
C 0.9-1, bal Fe.
For tools, lathe and planer cutters; water hardened. *Obsolete*

ALPINE FAM
Now VEW F180.

ALPINE FRS
Now VEW F204.

ALPINE HIR 25
Vereinigte Edelstahlwerke
C 0.28, Cr 0.7, Ni 2.5, bal Fe.
For gears, bolts, machine tool parts; oil hardened, shock
resistant. *Obsolete*

ALPINE HIR15 W/H
Vereinigte Edelstahlwerke
C 0.28-0.35, Si 0.35, Mn 0.6, Ni 1.5, Cr 0.5, bal Fe.
For gears, bolts, crankshafts, fasteners; oil hardened, shock
resistant. *Obsolete*

ALPINE HIR15 W/H
Vereinigte Edelstahlwerke
C 0.28-0.35, Si 0.35, Mn 0.6, Ni 1.5, Cr 0.5, bal Fe.
For gears, bolts, crankshafts, fasteners; oil hardened, shock
resistant. *Obsolete*

ALPINE HIRA 35 W/H
Vereinigte Edelstahlwerke
C 0.22-0.3, Si 0.35, Mn 0.6, Cr 0.7, Ni 3.5, bal Fe.
For gears, bolts, machine tool parts; oil hardened, shock
resistant. *Obsolete*

ALPINE HIRA 45
Now VEW V204.

ALPINE HIRA W/H
Vereinigte Edelstahlwerke
C 0.28-0.35, Cr 0.7, Ni 2.5, bal Fe.
For gears, bolts, machine tool parts; oil hardened, shock
resistant. *Obsolete*

ALPINE HMV15
Now VEW V930.

ALPINE HMV25
Vereinigte Edelstahlwerke
C 0.41, Si 0.25, Mn 0.65, Cr 1, bal Fe.
For gears, bolts, machine tool parts; oil hardened, tough.
Obsolete

ALPINE HMV35
Now VEW V762.

ALPINE HMV42
Now VEW V742.

ALPINE HMV50
Now VEW F550.

ALPINE HMV60
Vereinigte Edelstahlwerke
C 0.58, Cr 1, V 0.09, Mn 0.95, bal Fe.
For gears, bolts, crankshafts; oil hardened, shock resistant.
Obsolete

ALPINE HR10
Now VEW V500.

ALPINE HRIM15
Now VEW V155.

ALPINE HRIM20
Now VEW V145.

ALPINE HRM15
Now VEW V340.

ALPINE HRM35
Now VEW V330.

ALPINE HRM40
Now VEW V320.

ALPINE HRM45
Vereinigte Edelstahlwerke
C 0.42, Cr, Mo, bal Fe.
For gears, bolts, crankshafts, axles; oil hardened, tough.
Obsolete

ALPINE HRMV25
Vereinigte Edelstahlwerke
C 0.3, Cr 2.5, Mo 0.2, Mn 0.55, bal Fe.
For header dies, upsetters, sprockets; oil hardened, tough.
Obsolete

ALPINE KL3
Now VEW R100.

ALPINE MA3
Now VEW V960.

ALPINE MA4
Now VEW V945.

ALPINE MA5
Now VEW 935.

ALPINE MA6
Now VEW 920.

ALPINE MAC
Vereinigte Edelstahlwerke
C 0.41, Si 0.25, Mn 0.65, Cr 1, bal Fe.
For gears, bolts, crankshafts, machine tool parts; water hardened. *Obsolete*

ALPINE MN13
Now VEW K701.

ALPINE NALH
Vereinigte Edelstahlwerke
C 0.34, Cr 1.4, Al 1.1, Mn 0.6, bal Fe.
For oil refinery equipment; heat and creep resistant. *Obsolete*

ALPINE NALJ
Vereinigte Edelstahlwerke
C 0.33, Mn 0.5, Al 1.1, Cr 1.7, bal Fe.
For oil refinery equipment; heat and creep resistant. *Obsolete*

ALPINE NALM
Vereinigte Edelstahlwerke
C 0.32, Cr 1.1, Al 1.1, Mo 0.18, bal Fe.
For oil refinery equipment; heat and creep resistant. *Obsolete*

ALPINE NALW
Vereinigte Edelstahlwerke
C 0.27, Mn 0.6, Al 1.1, Cr 1.4, bal Fe.
For oil refinery equipment; heat and creep resistant. *Obsolete*

ALPINE NMV
Now VEW 304.

ALPINE PRG
Now VEW K204.

ALPINE PRIMA ZAH
Vereinigte Edelstahlwerke
C 0.7, Si 0-0.25, Mn 0-0.25, bal Fe.
Heat treated: 122,000-174,000 TS; 82,000-128,000 YS; 12-22 El; 37-52 RA; 241-352 Brin. For wheels, die blocks, girders, rails; Type W1; water hardened. *Obsolete*

ALPINE PRIMA-H
Vereinigte Edelstahlwerke
C 1.3, Si 0-0.25, Mn 0-0.25, bal Fe.
For blanking and forming dies; Type W1; water hardened. *Obsolete*

ALPINE RAV
Now VEW K510.

ALPINE SF
Now VEW F112.

ALPINE WAM
Now VEW W106.

ALPINE WAMV
Now VEW W105.

ALPINE WKL3
Now VEW K200.

ALPINE WMC
Now VEW W501.

ALPINE WMC
Vereinigte Edelstahlwerke
C 0.55, Cr 0.7, Mo 0.18, Ni 1.65, V 0.1, bal Fe.
For forging and heading dies; oil hardened, tough. *Obsolete*

ALPINE WMW
Vereinigte Edelstahlwerke
C 0.3, Cr 2.65, W 8.5, V 0.35, bal Fe.
For extrusion dies, rams, liners, punches; oil hardened, hot work steel. *Obsolete*

ALPINE WMWS
Vereinigte Edelstahlwerke
C 0.3, Co 2, Cr 2.4, V 0.25, W 8.5, bal Fe.
For extrusion dies, rams, liners, punches; oil hardened, non-deforming. *Obsolete*

ALPINE-EC3
Vereinigte Edelstahlwerke
C 0.13, Cr 0.4, bal Fe.
For gears, bolts, pinions, cams, camshafts. Case hardening, wear resistant. *Obsolete*

ALPLATE
Reynolds Metals Co.
Aluminum plated steel wire.

ALPLATIN
W. Seibel AG.
Aluminum.
Aluminum alloy AlZnMg.

ALPRO NO. 11
Belmont Metals Inc.
Deoxidizer for Cu.

ALPRO NO. 112
Belmont Metals Inc.
P 15, Cu 85.
Cu deoxidizer; for copper alloys.

ALPRO NO. 12
Belmont Metals Inc.
P 10, Cu 90.
Cu deoxidizer.

ALPRO NO. 126 A
Belmont Metals Inc.
Ni, bal Cu.
Hardener for Al bronzes.

ALPRO NO. 126 C
Belmont Metals Inc.
Ni, bal Cu.
Hardener for Al bronzes.

ALPRO NO. 126 D
Belmont Metals Inc.
Ni, bal Cu.
Hardener for Al bronzes.

ALPRO NO. 139
Belmont Metals Inc.
Fe 20, Al 80.
For adding Fe to Al alloys.

ALPRO NO. 140
Belmont Metals Inc.
Ni 20, Cu 40, Al 40.

ALPRO NO. 143
Belmont Metals Inc.
Ni 40, Cu 20, Al 40.
Hardener for Ni-Al alloys.

ALPRO NO. 146
Belmont Metals Inc.
Ni, Cr, bal Fe.
Gray iron hardener.

ALPRO NO. 15
Belmont Metals Inc.
Ni degasifying alloy.

ALPRO NO. 157
Belmont Metals Inc.
Ni 50, Fe 50.
Gray iron hardener.

ALPRO NO. 158
Belmont Metals Inc.
Ni 50, Mn 50.
For adding Mn to alloys.

ALPRO NO. 159
Belmont Metals Inc.
Ni 75, Cr 2.
Gray iron hardener.

ALPRO NO. 19 AX
Belmont Metals Inc.
Mn 30, Cu 70, C free.
For adding Mn to alloys.

ALPRO NO. 19 AZ
Belmont Metals Inc.
Mn 30, Cu 70, C free.
For adding Mn to alloys.

ALPRO NO. 19 BX
Belmont Metals Inc.
Copper. Mn 30, C 0-0.15, Fe 2.5, bal Cu.
For adding Mn to alloys.

ALPRO NO. 19 CX
Belmont Metals Inc.
Copper. Fe 0-4.5, Mn 30, C 0-0.25, bal Cu.
For adding Mn to alloys.

ALPRO NO. 220
Belmont Metals Inc.
Al 10, Cu 90.
For gears, cams, valves; resists abrasion, wear and shock.

ALPRO NO. 221
Belmont Metals Inc.
Al 10, Cu 89, Fe 1.
For ball bearings; resists abrasion wear and shock.

ALPRO NO. 223
Belmont Metals Inc.
Al 10, Cu 87, Fe 3.
For bearings.

ALPRO NO. 224
Belmont Metals Inc.
Al 10, Cu 86, Fe 4.
For bearings.

ALPRO NO. 300
Belmont Metals Inc.
Cu 8, Al 92.
For general light alloy castings.

ALPRO NO. 301
Belmont Metals Inc.
Al alloy.
For castings to be forged or die pressed.

ALPRO NO. 302
Belmont Metals Inc.
Cu 10, Al 90.
For die castings; non-hardenable.

ALPRO NO. 304
Belmont Metals Inc.
Si 2, Cu 8, bal Al.
For castings.

ALPRO NO. 306
Belmont Metals Inc.
Si 4, Cu 6, bal Al.
For castings.

ALPRO NO. 307
Belmont Metals Inc.
Al alloy.
For core boxes.

ALPRO NO. 308
Belmont Metals Inc.
Al alloy.
For cast cooking utensils.

ALPRO NO. 309
Belmont Metals Inc.
Al alloy.
For pistons.

ALPRO NO. 30A
Belmont Metals Inc.
Cu 50, Al 50.
Copper hardener for Al alloys.

ALPRO NO. 31
Belmont Metals Inc.
Al 75, Mn 25.
For adding Mn to alloys.

ALPRO NO. 310
Belmont Metals Inc.
Cu 7.5-8.5, 1.5 Fe + Si max, bal Al.
For castings; non-hardenable.

ALPRO NO. 311
Belmont Metals Inc.
Cu 9.5-10.5, 1.5 Fe + Si max, bal Al.
For die castings; non-hardenable.

ALPRO NO. 312
Belmont Metals Inc.
Si 2, Cu 8, Zn 2, bal Al.
For castings; non-hardenable.

ALPRO NO. 314
Belmont Metals Inc.
Si 4, Cu 6, bal Al.
For castings.

ALPRO NO. 315
Belmont Metals Inc.
Cu 7.5-8.5, Zn 1.75-2.25, 1.75 Fe + Si max, bal Al.
For castings; non-hardenable.

ALPRO NO. 316
Belmont Metals Inc.
Si 4, Cu 4, bal Al.
For castings.

ALPRO NO. 317
Belmont Metals Inc.
Si 2, Cu 6, bal Al.
For castings.

ALPRO NO. 320
Belmont Metals Inc.
Zn, bal Al.
For light alloy castings.

ALPRO NO. 330
Belmont Metals Inc.
Al alloy.
For match plates; low degree of top shrinkage.

ALPRO NO. 331
Belmont Metals Inc.
Si, bal Al.
For marine castings; resists sea-water corrosion.

ALPRO NO. 332
Belmont Metals Inc.
Si, bal Al.
For general utility castings; resists sea-water corrosion.

ALPRO NO. 333
Belmont Metals Inc.
Al alloy.
For high pressure castings; tough, dense, strong.

ALPRO NO. 335
Belmont Metals Inc.
Si 4.5-5.5, Fe 0-0.75, bal Al.
19,000-21,000 psi TS; 4-3 El; 37 Brin. For marine parts, castings; corrosion resistant.

ALPRO NO. 338
Belmont Metals Inc.
Si 11-13.5, Fe 0-0.8, bal Al.
25,000-27,500 psi TS; 7-6 El; 45 Brin. For die castings; corrosion resistant.

ALPRO NO. 34
Belmont Metals Inc.
Ni 20, Al 80.
Hardeners for Ni-Al alloys.

ALPRO NO. 35
Belmont Metals Inc.
Si 50, Al 50.
Si hardener for Al alloys.

ALPRO NO. 351
Belmont Metals Inc.
Si 5, bal Al.
For castings; sand or die cast.

ALPRO NO. 352
Belmont Metals Inc.
Si 12, bal Al.
For castings; die cast.

ALPRO NO. 360
Belmont Metals Inc.
Ni 2, Cu 6, bal Al.
For castings.

ALPRO NO. 361
Belmont Metals Inc.
Ni 2, Cu 6, Si 2, bal Al.
For castings.

ALPRO NO. 362
Belmont Metals Inc.
Ni 4, Cu 4, Al 92.
For pistons, light alloy parts.

ALPRO NO. 363
Belmont Metals Inc.
Ni 2, Cu 5, Si 2, Al 91.
For castings.

ALPRO NO. 364
Belmont Metals Inc.
Ni 2, Cu 4, Si 4, Al 90.
For castings; age-hardened.

ALPRO NO. 365
Belmont Metals Inc.
Ni 3, Cu 5, Si 2, Al 90.
For castings; age-hardened.

ALPRO NO. 370
Belmont Metals Inc.
Cu 4, Ni 2, Mg 1.5, Al 92.5.
At 65°F: 29,000 psi TS. At 482°F: 25,500 psi TS. For aeronautical parts; similar to "Y" alloy.

ALPRO NO. 371
Belmont Metals Inc.
Cu 2, Ni 1.5, Mg 1, Si 0.6, Al 94.9.
For pistons; similar to "Magnalite."

ALPRO NO. 380
Belmont Metals Inc.
Cu, Mn, bal Al.
For light alloy parts; modified duralumin type.

ALPRO NO. 385
Belmont Metals Inc.
Cu 6.5, Ni 1.5, Cr 0.1, Mg 0.5, Mn 0.5, Al 90.9.
For sand castings; modified Duralumin.

ALPRO NO. 39
Belmont Metals Inc.
Fe 10, Al 90.
For adding Fe in Al alloys.

ALPRO NO. 40
Belmont Metals Inc.
Ni 25, Cu 25, Al 50.
Hardener for Ni-Al alloys.

ALPRO NO. 41
Belmont Metals Inc.
Ni 25, Mn 25, Al 50.
Hardener for Ni-Al alloys.

ALPRO NO. 42
Belmont Metals Inc.
Ni 25, Sn 25, Al 50.
Hardener for Ni-Al alloys.

ALPRO NO. 44
Belmont Metals Inc.
Cu 25, Al 50, Mn 25.
For adding Mn to alloys.

ALPRO NO. 46
Belmont Metals Inc.
Fe 25, Al 50, Mn 25.
For adding Mn to alloys.

ALPRO NO. 50 B
Belmont Metals Inc.
Ni 50, Cu 50.
For addition of Ni to bronzes.

ALPRO NO. 55
Belmont Metals Inc.
Ni, bal Cu.
Deoxidizer of Cu alloys; red composition densifying alloy.

ALPROPAT
Belmont Metals Inc.
Cu, Si, bal Al.
Cast: 27,000 psi TS; 15,000 psi YS; 4 El; 60 Brin. For match plates and patterns; low shrinkage.

ALRAMAN
VMW
Aluminum.
Aluminum alloy AlMn.

ALRAY A 8
H.K. Porter Co., Inc.
Cr 20, Ni 80.
For electrical resistance; heat resistant.

ALRAY C

H.K. Porter Co., Inc.
Cr 15, Ni 62, bal Fe.
Annealed: 95,000-175,000 TS; 35,000 YS; 35-40 El; 60 RA.
For heat and corrosion resistant parts; corrosion and heat resistant. *Obsolete*

ALRAY D

H.K. Porter Co., Inc.
Cr 35, Ni 18, bal Fe.
Annealed: 75,000-160,000 TS; 35-40 El; 60 RA. For heat resistant parts; corrosion and heat resistant. *Obsolete*

ALRONZE GR.M

Olin Corp.
Al 9-10, Fe 3.5-4.5, bal Cu.
Annealed: 85,000-100,000 TS; 45,000-65,000 YS; 30-35 El.
Heat treated: 150,000-170,000 TS; 100,000-150,000 YS; 3 El.
For fuses, clips, springs. Heat treatable, high strength aluminum bronze; corrosion resistant. *Obsolete*

ALRONZE R 619

Olin Corp.
Cu 86, Al 9, Fe 4.
Annealed: 85,000-100,000 TS; 45,000-75,000 YS; 25-35 El.
Cold rolled: 97,000-145,000 TS; 110,000-124,000 YS; 2-30 El.
Very high strength after brazing; corrosion resistant.
Obsolete

ALRONZE STANDARD

Olin Corp.
Al 9-10, Fe 3.5-4.5, Zn 0-0.8, Pb 0-0.01, bal Cu.
Annealed: 85,000-100,000 TS; 45,000-65,000 YS; 30-35 El,
Rock B 80-90. Hard temper: 112,000-130,000 TS;
90,000-110,000 YS; 10 El, Rock B 96-101. For fuses, clips, springs, fasteners. High strength aluminum bronze, corrosion resistant. *Obsolete*

ALRONZE-619

Olin Corp.
Al 9.5, Fe 4, Cu 86.5.
Aged: 152,000 TS; 145,000 YS. For fuses, clips, springs, fasteners, terminals, diaphragms. High strength aluminum bronze. *Obsolete*

ALSEX

Allgemeine Elektrizitats-Gesellschaft
Si 0.5, Mg 0.5, bal Al.
Heat treated: 29,00-42,500 TS; 25,000-35,000 YS; 7-5 El; 90 Brin. For light alloy parts, forgings; conductivity 52% of copper.

ALSIA

Societe Alsia
Fe 0.7, Si 20, Cu 1, bal Al.
Cast: 20,000 TS; 18,000 YS; 1-2 El; 80-90 Brin. For pistons; same as "Alusil".

ALSICHROM-1 (CA30)

J.C. Soding & Halbach
C 0.05, Si 0.6, Mn 0.35, Cr 22, Co 0.5, Al 5.2, bal Fe
For heating elements for electric furnaces. Heat and oxidation resistant. *Obsolete*

ALSICHROM-2 (CAF)

J.C. Soding & Halbach
C 0.1, Si 0.8, Mn 0.3, Cr 21, Al 4.6, Co 0.5, bal Fe.
For heating elements for electric furnaces. Heat and oxidation resistant. *Obsolete*

ALSIFONT 5C

Soc. Alluminio Veneto per Azioni
Aluminum. Cu 4, Si 5.5, bal Al.
Heat treated: 28,000-32,000 TS; 20,000-23,000 YS; 1.5-4 El; 55-75 Brin. For light alloy parts; heat treatable.

ALSILIT 311

Metallwerk Olsberg GmbH
Aluminum.
Aluminum alloy AlSi6Cu3.

ALSILOY

NL Industries
Cu 0-0.6, Si 11.5-12.5, Fe 0-2, bal Al.
Die cast: 30,000-33,000 TS; 10,000-12,000 YS; 1.5-3 El; 70-75 Brin. For engineering parts, instrument parts and housings; good castability. *Obsolete*

ALSILOY

Doehler-Jarvis Co.
Aluminum.
Aluminum alloy AlSiFe(Cu).

ALSILOY 10

NL Industries
Si 10, Mg 0.5, bal Al.
Cast: 45,000 TS; 27,000 YS; 4.5 El; 75 Brin. For die castings; high fluidity. *Obsolete*

ALSILOY 3

NL Industries
Cu 4, Si 5, bal A;/.
Die cast: 30,000 TS; 14,000 YS; 1.5 El; 80 Brin. For general die castings. *Obsolete*

ALSILOY 5

NL Industries
Si 5, bal Al.
Cast: 30,000 TS; 14,000 YS; 7 El; 60 Brin. For die castings; high fluidity. *Obsolete*

ALSILOY 9

NL Industries
Cu 3.5, Si 9, bal Al.
Cast: 40,000 TS; 25,000 YS; 2.5 El; 80 Brin. For castings; good machinability. *Obsolete*

ALSILOY S-1

NL Industries
Si 11.5-12, Fe 0-2, Cu 0-0.6, bal Al.
Die cast: 42,000 TS; 19,000 YS; 3.5 El; 80 Brin. For die castings; corrosion resistant, good castability. *Obsolete*

ALSILOY S-5

NL Industries
Si 4-6, Cu 0-0.6, Fe 0-2, bal Al.
Die cast: 30,000 TS; 16,000 YS; 9 El; 70 Brin. For die castings; corrosion resistant, good castability. *Obsolete*

ALSIPLATE

Texas Instruments Inc./Materials Control
Ag clad to 2S aluminum alloy. For electrical contacts; Al alloys plated by lamination with silver.

ALSOCO M-1001

Aluminum Solder Corp. (ALSOCO)
Zn, Sn-Pb.
For aluminum solder; melting point: 728°F.

ALSOCO M-1002

Aluminum Solder Corp. (ALSOCO)
Zn, Sn-Pb.
For aluminum solder; melting point: 680°F.

ALSOCO M-1003

Aluminum Solder Corp. (ALSOCO)
For aluminum solder.

ALSOLDER 720

J.W. Harris Co., Inc.
Zinc/aluminum solder rod; for torch or other heat applications; melts at 700°F. For miscellaneous maintenance on aluminum base materials.

ALSTEEL

H. Boker & Co.
C 0.2-7, bal Fe.
For auto tools. *Obsolete*

ALT BLITZ

Fakirstahl Hoffmanns GmbH & Co. KG
C 0.79, Co 4.7, Cr 4.3, Mo 0.7, V 1.5, W 18, bal Fe.
For lathe and planer tools, hobs, reamers, taps; high speed steel.

ALTAM

NL Industries
Ti 68-88, Al 11, Fe 1.25, Si 1.18, bal TiO_2.
For making ductile titanium. *Obsolete*

ALTECH NITRALLOY 135 MODIFIED

Bethlehem Steel Corp.
C 0.4, Mn 0.5, Al 1, Cr 1.75, Mo 0.4, bal Fe.
Heat treated: 158,000 TS; 141,000 YS; 17 El; 56 RA; 320 Brin. For cylinder barrels, nitrided parts; nitriding steel.

ALTECH NITRALLOY 135 MODIFIED

AL Tech Specialty Steel Corp.
C 0.4, Mn 0.5, Al 1, Cr 1.75, Mo 0.4, bal Fe.
Heat treated: 158,000 TS; 141,000 YS; 17 El; 56 RA; 320 Brin. For cylinder barrels, nitrided parts; nitriding steel.

ALTEMP 19-19DL

AL Tech Specialty Steel Corp.
C 0.28-0.35, Cr 18-20, Ni 8-11, Mo 1-1.7, W 1-1.7, Cb 0.4, Ti 0-0.5, bal Fe.
Hot rolled: 118,000 TS; 69,000 YS; 58 El; 55 RA; 216 Brin.
Heat treated: 103,000 TS; 43,000 YS; 54 El; 57 RA; 189 Brin.
For jet engine and gas turbine components; austenitic, heat resistant. *Obsolete*

ALTEMP 625

Allegheny Ludlum Steel
Nickel. C 0-0.05, Cr 22, Mo 9, Mn 0.3, Si 0.25, Ti 0.3, Al 0.3, Fe 4, 3.5 Cb + Ta, bal Ni.
RT, annealed at 1920°F: 136,000 psi TS; 63,000 psi YS; 51.5 El. Austenitic superalloy; resistant to oxidation and corrosion. Temperature ranges cryogenic to 2000°F. Jet engine, aerospace, and chemical process applications. N06625

ALTEMP 718

Allegheny Ludlum Steel
Nickel. C 0.05, Cr 19, Mo 3, Ni 53, Mn 0.1, Si 0.1, Ti 0.9, Al 0.5, Co 0.9, 5.25 Cb + Ta, bal Fe.
At 1200°F: 168,000 psi TS; 148,000 psi YS; 19 El. Austenitic superalloy. Used in applications requiring high strength to 1400°F, oxidation resistant to 1800°F. N07718

ALTEMP A-286

Allegheny Ludlum Corp./Special Materials
Superalloy. C 0.04, Mn 0.2, P 0.02, S 0.01, Si 0.25, Cr 15, Ni 25.5, Mo 1.25, Ti 2.1, V 0.25, Al 0.25, B 0.006, bal Fe.
90,000 psi TS; 40,000 psi YS; 40 El. For high strength and corrosion resistance up to 1300°F. S66286

ALTEMP D979

Allegheny Ludlum Steel
Fe, Cr, Mo, W, Ti, V, bal Ni.
For high temperature applications; heat resistant. *Obsolete*

ALTEMP HX

Allegheny Ludlum Steel
Nickel. Cr 22, Mo 9, Co 1.5, C 0.1, Mn 0.5, Si 0.5, Ti 0.5, Fe 18, bal Ni.
Austenitic alloy with exceptional resistance to oxidation at elevated temperatures. N06002

ALTEMP L605

Allegheny Ludlum Steel
C 0.15, Mn 1.5, Cr 20, Ni 10, W 15, bal Co.
Annealed: 150,000 TS; 70,000 YS; 50 El; Rock C 23. At 1500 F: 50,000 TS; 35,000 YS; 17 El. Uses: jet engine and gas turbine parts, afterburners, exhaust cone assemblies, valves, springs, buckets. High temperature alloy. High mechanical properties to 1800 F. heat and oxidation resistant. *Obsolete*

ALTEMP M-252
Allegheny Ludlum Steel
C 0.15, Cr 19, Co 10, Mo 10, Ti 2.5, Al 0.8, Fe 0-5, bal Ni.
At 70 F: 193,000 TS. At 1000 F: 155,000 TS; 93,000 YS. At
1400 F: 122,000 TS; 80,000 YS. For jet engine and gas
turbine buckets; heat and oxidation resistant. *Obsolete*

ALTEMP S-590
Allegheny Ludlum Steel
C 0.38-0.47, Cr 20, Ni 20, Co 20, Mo 4, W 4, Co 4, bal Fe.
Rolled: 152,000 TS; 80,000 YS; 19 El; 25 RA. Heat treated:
140,000 TS; 42,000 YS; 20 El; 19 RA. At 1400 F: 61,500 TS;
50,000 YS; 26.6 El; 34.8 RA. For supercharger and jet engine
parts, turbine blades; age-hardenable, corrosion and heat
resistant. *Obsolete*

ALTEMP S-816
Allegheny Ludlum Steel
C 0.32-0.42, Cr 20, Ni 20, Co 42, W 4, Mo 4, Cb 4, Mn 1.2, Fe
0-5.
Heat treated: 140,000 TS; 67,000 YS; 31 El; 31 RA; 269 Brin.
At 1000 F: 121,000 TS; 46,000 YS; 27 El; 23 RA. At 1500 F:
73,000 TS; 40,500 YS; 23 El; 24 RA. For jet engine and
supercharger parts, turbine blades; age-hardenable, heat
resistant. *Obsolete*

ALTEMP V-36
Allegheny Ludlum Steel
C 0.25, Mn 0.8, Ni 20, Mo 4, W 2, Cb 2, Fe 2, Cr 25, bal Co.
For jet engine components; high heat resistant. *Obsolete*

ALTEN NO. 11
Alten Foundry & Machine Works
Si 1.7, Mn 0.5, TC 3.5, Cr 0.07, Mo 0.03, 2.7 graphite, bal Fe.
For castings, gears, and housings. Cast iron.

ALTEN NO. 15
Alten Foundry & Machine Works
Si 1.8, Mn 0.6, TC 3.5, Cr 1.4, Ni 0.16, Mo 0.6, V 0.08, Cu
0.04, 3.1 graphite, bal Fe.
For castings, gears, and housings. Cast iron.

ALTEN NO. 18
Alten Foundry & Machine Works
Si 1.8, Mn 0.5, TC 3.7, Cr 1, Ni 0.18, Mo 0.04, V 0.09, 3.0
graphite, bal Fe.
For castings, stokers, and grate bars. Cast iron.

ALTEN TYPE 30
Alten Foundry & Machine Works
C 3.3, Si 2.2, Mn 0.8, Ni 0-0.2, Cr 0-0.1, Mo 0-0.1, bal Fe.
Cast: 30,000 TS; 170-223 Brin. For valve components and
machinery castings. Cast iron. Transverse strength 2200 lb;
SAE 111.

ALTEN TYPE 35 HR
Alten Foundry & Machine Works
C 3.4, Si 2, Mn 0.8, Ni 1, Cr 0-0.1, Mo 0.4, bal Fe.
Cast: 35,000 TS; 190-210 Brin. For glass molds, stop cocks,
and heat resistant parts. Cast iron; heat resistant.

ALTEN TYPE 45 HR
Alten Foundry & Machine Works
C 3.4, Si 1.9, Mn 0.8, Cr 1.2, Mo 0.3, Ni 0-0.2, bal Fe.
Cast: 50,000 TS; 300-330 Brin. For furnace parts and heat
resistant parts. Cast iron; heat resistant.

ALTEN TYPE 45 WR
Alten Foundry & Machine Works
C 3.1, Si 2, Mn 0.7, 0.5 Ni min, 0.3 Cr min, 0.25 Mo min, bal
Fe.
Cast: 45,000 TS; 205-230 Brin. For valves, pressure castings,
and wear resistant parts. Cast iron; wear resistant; SAE 122.

ALTEN TYPE 50WR
Alten Foundry & Machine Works
TC 3.2, CC 0.7, Si 1.8, Mn 0.7, Ni 0.2, Cr 0.5, Mo 0.5, Cu 0.09,
bal Fe.
Cast: 50,000 TS; 215 Brin. For valves, wear plates, and
superheaters. Wear and corrosion resistant; C.I.

ALTIOR BRAND ALUMINUM SOLDER
Birkett, Billington & Newton Ltd.
Pb, bal Sn.
Al solder.

ALTIOR BRAND ANTIFRICTION METAL
Birkett, Billington & Newton Ltd.
Pb, Sb, bal Sn.
For bearings. White antifriction metal.

ALTMAG
Russian manufacture
Mg 5-8, Mn 0.5-1, Ti 0.1-0.5, bal Al.
For light alloy parts; age hardenable.

ALTO
Associated Steel Corp.
C, alloy, bal Fe.
Rolled: 151,000 TS; 98,000 YS; 51 El; 50 RA; 220 Brin. Heat
treated: 217,000 TS; 220,000 YS; 16 El; 49 RA; 555 Brin. For
axes, axles, drills, chisels, crowbars; non-tempering, shock
resistant, water hardened.

ALTOLOY A
Associated Steel Corp.
C, alloy, bal Fe.
For pneumatic tools, chisels; shock resistant; non-tempering,
water hardened.

ALTOLOY B
Associated Steel Corp.
C, alloy, bal Fe.
For pneumatic tools, chisels; tough, oil hardened.

ALU-SOL 45
Multicore
Lead. Pb 80.1, Sn 18, Ag 1.9.
Solder for aluminum joints. MP 352°F. FP 518°F.
Conductivity is 8.7% of copper. 3.8 kg/mm^2 TS.

ALUAL
American Steel Co.
Aluminum.
Aluminum 99.9.

ALUCABLE
Societe Electro-Cables
Si 0.55, Mg 0.4, bal Al.
For light alloy parts; non-hardenable.

ALUCHROM 1
VDM Nickel-Technologie AG
Cr 22, Al 5, bal Fe.
Annealed: 92,400-99,500 TS; 25-30 El. For heat resistant
parts; maximum operating temperature 1250°C. *Obsolete*

ALUCHROM II
VDM Nickel-Technologie AG
Cr 8, Al 6, bal Fe.
Annealed: 85,200-88,500 TS; 25-30 El. For heat resistant
parts; maximum operating temperature 1100°C. *Obsolete*

ALUCHROM O
VDM Nickel-Technologie AG
Cr 22-28, Al 5-5.5, bal Fe.
Annealed: 99,500-106,000 TS; 25-30 El. For heat resistant
parts; maximum operating temperature 1250°C.

ALUCHROM W
VDM Nickel-Technologie AG
Cr 14-16, Si 0-0.5, Al 3.5-5, C 0-0.08, Zr 0-0.3, Mn 0-0.6, bal
Fe.
Minimum values at 20°C: 600 N/mm^2 TS; 400 N/mm^2 YS;
18 El. For high temperature furnaces and kilns. Material No.
1.4725. K91670

ALUDUR
Giulini Werke A.G.
Aluminum. Mg 0.5, Si 0.7, Fe 0.45, bal Al.
Heat treated: 50,000 TS; 6 El; 90 Brin. For overhead
transmission lines; self aging alloy; high corrosion resistance.

ALUDUR 275
Aluminium-Werke Wutoschingen GmbH
Aluminum. Mg 1.5-3, Mn 0.5-1.5, Cr 0-0.3, bal Al.
For aircraft structural parts. *Obsolete*

ALUDUR 300
Aluminium-Werke Wutoschingen GmbH
Aluminum. Mg 2-4, Mn 0-0.4, Cr 0-0.4, bal Al.
For aircraft structural parts. *Obsolete*

ALUDUR 500
Aluminium-Werke Wutoschingen GmbH
Aluminum. Mg 4-5, Mn 0-0.8, Cr 0-0.3, bal Al.
Soft: 42,000 psi TS; 22,000 psi YS; 35 El; 65 Brin. Hard:
60,000 psi TS; 50,000 psi YS; 15 El; 100 Brin. For aircraft
structures, light alloy parts; corrosion resistant. *Obsolete*

ALUDUR 533
German manufacture
Si 0.3-1, Mn 0.3-0.8, Mg 0.5-1.3, bal Al.
Soft: 16,000-18,500 TS; 8,500-11,30 YS; 20-27 El; 30-40 Brin.
For structural fittings, hardware, aircraft components; age-
hardenable.

ALUDUR 570
German manufacture
Cu 2.5-5, Mg 0.2-1.8, Mn 0.3-1.5, Bal Al.
For light alloy parts; age hardenable.

ALUDUR 580
Aluminium-Werke Wutoschingen GmbH
Aluminum. Cu 2.5-5, Mg 0.2-1.8, Mn 0.3-1.5, bal Al.
For light alloy parts; age hardenable. *Obsolete*

ALUDUR 630/OM
Aluminium-Werke Wutoschingen GmbH
Aluminum. Cu 2.5-5, Mg 0.2-1.8, Mn 0.3-1.5, bal Al.
For light alloy parts; age hardenable. *Obsolete*

ALUDUR 700
Aluminium-Werke Wutoschingen GmbH
Aluminum. Mg 5.5-7.5, Mn 0-0.8, Cr 0-0.3, bal Al.
For light alloy parts; corrosion resistant. *Obsolete*

ALUDUR NO. 533D
Giulini Werke A.G.
Aluminum. Mg 0.8, Si 0.7, bal Al.
Heat treated: 45,000-50,000 TS; 8-10 El; 90-100 Brin. For
overhead transmission lines; corrosion resistant.

ALUDUR NO. 570D
Giulini Werke A.G.
Aluminum. Mn 0.8, Si 1, bal Al.
Heat treated: 60,000 TS; 17 El; 90-100 Brin. For overhead
transmission lines; corrosion resistant; self aging alloy.

ALUDUR-570
Manufacturer not listed.
Aluminum. Cu 3.5-5.5, Si 0.3-0.6, Mn 0.3-1, Mg 0.3-0.7, bal
Al.
Annealed: 26,000 TS; 10,000 YS; 22 El; 45 Brin. Hardened:
60,000 TS; 40,000 YS; 20 El; 105 Brin. For construction and
transportation equipment, fittings, screw machine products.
High strength, age hardenable.

ALUDUR-580
German manufacture
Cu 3.5-5.5, Si 0.3-0.6, Mn 0.3-1, Mg 0.3-0.7, bal Al.
Annealed: 26,000 TS; 10,000 YS; 22 El; 45 Brin. Hardened:
62,000 TS; 40,000 YS; 20 El; 105 Brin. For construction and
transportation equipment, fittings, hardware, screw machine
products, fasteners; high strength, age-hardenable.

ALUDUR-630
German manufacture
Cu 3.5-5.5, Si 0.3-0.6, Mn 0.3-1, Mg 0.3-0.7, bal Al.
Annealed: 26,000 TS; 10,000 YS; 22 El; 45 Brin. Age-
hardened: 60,000 TS; 40,000 YS; 20 El; 105 Brin. For fittings,
hardware, screw machined products, fasteners; age-
hardenable, high strength.

ALUFER
German manufacture
Al alloy.
For light alloy parts; non-hardenable.

ALUFLEX
Pechiney/E. & E. Kaye Ltd.
Mg 0.75, bal Al.
For electric conductors, electrical wires, and cable for braiding. High electrical conductivity.

ALUFONT "H"
Aluminium Industrie Aktiengesellschaft
Si 0.4, Fe 1.4, Mg 0.15, Cu 2, Zn 12, bal Al.
Sand cast: 43,000 TS; 36,000 YS; 0.5 El. Chill cast: 44,000 TS; 42,000 YS; 1.5 El. For large castings for general purposes; corrosion resistant.

ALUFONT "W"
Aluminium Industrie Aktiengesellschaft
Si 2, Fe 0.6, Mn 0.5, Cu 2, Zn 12, bal Al.
Sand cast: 35,000 TS; 20,000 YS; 3.5 El. Chill cast: 40,000 TS; 35,000 YS; 3.5 El. For large castings, light alloy parts.

ALUFONT 3
Aluminium Industrie Aktiengesellschaft
Ti 0.2-0.5, Cu 4-5, bal Al.
Sand cast: 41,000 TS; 28,000 YS; 6 El; 85 Brin. Chill cast: 53,000 TS; 37,000 YS; 6 El; 100 Brin. For automobile and aircraft construction; heat treatable.

ALUFONT II
Aluminium Industrie Aktiengesellschaft
Cu 4, Si 2, Mn 0.6, Mg 0.2, Ti 0.15, bal Al.
Heat treated: 33,000-55,000 TS; 27,000-45,000 YS. For castings where high strength, but no high corrosion resistance is required; high fluidity.

ALUFONT-47
Swiss Aluminium Ltd.
Aluminum. Si 0.15, Fe 0.3, Cu 4.2-4.9, Mn 0.01-0.05, Mg 0.15-0.3, Zn 0.07, Ti 0.15-0.25, bal Al.
Primary foundry alloy in ingot form. 300-420 N/mm^2 TS; 220-300 N/mm^2 YS; 5-18 El; 90-115 Brin.

ALUFONT-48
Swiss Aluminium Ltd.
Aluminum. Si 0.15, Fe 0.15, Cu 4.5-5, Mn 0.2-0.6, Mg 0.2-0.6, Zn 0.07, Ti 0.1-0.5, Ag 0.5, bal Al.
Primary foundry alloy in ingot form. 415-450 N/mm^2 TS; 345-390 N/mm^2 YS; 3-5 El; 105-120 Brin.

ALUFONT-52
Swiss Aluminium Ltd.
Aluminum. Si 0.15, Fe 0.15, Cu 4.5-5.2, Mn 0.01-0.05, Mg 0.03, Zn 0.07, Ti 0.15-0.3, bal Al.
Primary foundry alloy in ingot form. 300-400 N/mm^2 TS; 190-240 N/mm^2 YS; 8-23 El; 90-110 Brin.

ALUFONT-57
Swiss Aluminium Ltd.
Aluminum. Si 0.3, Fe 0.3, Cu 3.8-4.2, Mn 0.1, Mg 1.3-1.6, Zn 0.1, Ti 0.01-0.2, Ni 1.8-2.2, bal Al.
Primary foundry alloy in ingot form. 220-320 N/mm^2 TS; 180-220 N/mm^2 YS; 0.3-1.0 El; 90-125 Brin.

ALUFONT-II
German manufacture
C 3.7-4.3, Si 2, Mn 0.5-0.7, Mg 0.2, bal Al.
Cast: 21,000 TS; 14,000 YS; 2.5 El; 55 Brin. For manifold and valve bodies, machine frames, oil pans; good castability and machinability.

ALUFRAN
Pechiney/Soc. Vente de l'Aluminium
Tradename for several aluminum grades of ingots, billets, casting alloys.

ALUGIR
Ateliers de la Gironde
Cu 3, Mg 0.8, Ni 1.2, bal Al.
57,000-64,000 TS; 6-22 El. For light alloy parts; heat treatable.

ALUMA
Aluminium Belge, S.A.
Mn 1.5, bal Al.
14,500-18,500 TS; 7,000-9,000 YS; 30-40 El. For light alloy parts; corrosion resistant.

ALUMAG 15
Trefileries & Laminoirs du Havre
Mg 1.5, Mn 0.4, bal Al.
Annealed: 23,000 TS; 10,000 YS; 20 El; 32 Brin. Hardened: 40,000 TS; 35,000 YS; 55 Brin. For light alloy parts; corrosion resistant.

ALUMAG 25
Trefileries & Laminoirs du Havre
Mg 2.5, Mn 0.4, bal Al.
Soft: 43,000 TS; 14,000 YS; 22 El; 40 Brin. Hard: 60,000 TS; 39,000 YS; 67 Brin. For light alloy parts; corrosion resistant.

ALUMAG 35
Trefileries & Laminoirs du Havre
Mg 3.5, Mn 0.4, bal Al.
Soft: 36,000 TS; 17,000 YS; 25 El; 47 Brin. Hard: 55,000 TS; 46,000 YS; 78 Brin. For light alloy parts; corrosion resistant.

ALUMAG 50
Trefileries & Laminoirs du Havre
Mg 5, Mn 0.4, bal Al.
Soft: 43,000 TS; 22,000 YS; 25 El; 53 Brin. Hard; 64,000 YS; 54,000 YS; 95 Brin. For light alloy parts; corrosion resistant.

ALUMAG 65
Trefileries & Laminoirs du Havre
Mg 6.5, Mn 0.4, bal Al.
Soft: 49,000 TS; 27,000 YS; 27 El. Hard: 69,000 TS; 57,000 YS. For high temperature applications; high temperature properties.

ALUMAG GRADE 1
Cegedur Pechiney
Mg 4-5, bal Al.
For light alloy parts; high resistance to corrosion.

ALUMAG GRADE 2
Cegedur Pechiney
Mg 8-10, bal Al.
For light alloy parts; high resistance to corrosion.

ALUMAGNESE 10
Pechiney World Trade (USA) Inc.
Cu, Mg, Si, Fe, Mn, Ti, bal Al.
For light alloy parts; suitable for welding. *Obsolete*

ALUMAGNESE 20
Pechiney World Trade (USA) Inc.
Cu 0.15, Mg 1.75-2.75, Si 0.6, Fe 0.75, Mn 0.5, Ti 0.2, bal Al.
M-temper: 29,200 TS; 23 El. For light alloy parts; corrosion resistant. *Obsolete*

ALUMAGNESE 35
Pechiney World Trade (USA) Inc.
Cu 0.15, Mg 3-4, Fe 0.75, Si 0.6, Mn 1, Ti 0.2, bal Al.
M-temper: 36,000 TS; 18,000 YS; 23 El. For light alloy parts; corrosion resistant. *Obsolete*

ALUMAGNESE 50
Pechiney World Trade (USA) Inc.
Cu 0.15, Mg 4.5-5.5, Si 0.6, Fe 0.75, Mn 1, Ti 0.2, bal Al.
M-temper: 42,600 TS; 27,000 YS; 21 El. For light alloy parts; corrosion resistant. *Obsolete*

ALUMALUN
American Abrasive Metals Co.
Al, Si alloy with abrasive grains embedded in, wearing surface.
For floor plates, stair treads, car steps; wear resistant; anti-slip.

ALUMAN
Aluminium Industrie Aktiengesellschaft
Si 0-0.25, Fe 0-0.25, Cu 0-0.1, Mn 1.5, bal Al.
Soft: 15,000 TS; 7000 YS; 30 El; 30 Brin. Hard: 36,000 TS; 29,000 YS; 7 El; 50 Brin. Extruded: 17,000 TS; 10,000 YS; 24 El; 35 Brin. For paneling, roofing, containers, welded structures; corrosion resistant, non-hardenable.

ALUMAN (AW15)
Aluminium Walzwerke Singen GmbH
Aluminum. Mn 1.4-1.6, bal Al.
Soft: 17,000 TS; 7,000 YS; 35 El; 30 Brin. Hard: 32,000 TS; 30,000 YS; 4 El; 60 Brin. For heat exchangers, truck panels, tanks reflectors, ducts. Hardened by cold work only.

ALUMAN - 100
Aluminium Walzwerke Singen GmbH
Aluminum. Mn 1, bal Al.
Annealed: 90 MPa TS; 28 El. Hard: 200 MPa TS; 3 El. For containers.

ALUMAN-100
Swiss Aluminium Ltd.
Aluminum. Si 0.5, Fe 0.7, Cu 0.1, Mn 0.9-1.4, Mg 0.1, Cr 0.1, Zn 0.2, Ti 0.1, bal Al.
Extrusion alloy in billet form; not age hardenable. 95-140 N/mm^2 TS; 40-90 N/mm^2 YS; 15-17 El; 30 Brin. Also available as wrought alloy in rolling slab form.

ALUMAN-103
Swiss Aluminium Ltd.
Aluminum. Si 0.5, Fe 0.7, Cu 0.05-0.2, Mn 0.9-1.4, Mg 0.05, Cr 0.05, Zn 0.1, Ti 0.05, bal Al.
Extrusion alloy in billet form; not age hardenable. 95-130 N/mm^2 TS; 35-80 N/mm^2 YS; 18-20 El; 30 Brin. Also available as wrought alloy in rolling slab form.

ALUMAN-30
Lavorazione Leghe Leggere SpA
Mn 1-1.5, Cu 0-0.1, Fe 0-0.5, Si 0-0.5, Ti 0-0.05, bal Al.
Annealed: 11,400-18,500 TS; 5,700-10,000 YS; 25-45 El; 25-35 Brin. Hard: 17,000-35,600 TS; 24,200-35,600 YS; 3-8 El; 50-65 Brin. For cooking utensils, heat exchangers, storage tanks, plumbing fixtures.

ALUMAN-30
Alluminio SA
Mn 1-1.5, Cu 0-0.1, Fe 0-0.5, Si 0-0.5, Ti 0-0.05, bal Al.
Annealed: 11,400-18,500 TS; 5,700-10,000 YS; 25-45 El; 25-35 Brin. Hard: 17,000-35,600 TS; 24,200-35,600 YS; 3-8 El; 50-65 Brin. For cooking utensils, heat exchangers, storage tanks, plumbing fixtures.

ALUMANGAN
Versevorder Metallwerk
Aluminum.
Aluminum alloy AlMn.

ALUMANINGOT
Now SPECIALLOY 5813 ALUMANINGOT TM.

ALUMAR 4043, ETC. (15 ALLOYS)
Arcos Alloys
Bar aluminum welding wire; for welding similar aluminum alloys. Aluminum Association type 4043 and 15 other AA type composites. *Obsolete*

ALUMASOD
Continental Industries Corp.
Sn 76, Zn 23, Cd 1.
For Al solder.

ALUMAWELD

Johnson Mfg. Co.
Sn, Zn, Pb, and chemical catalyzer.
12,000 TS. For soldering of all metals, especially Al and white metals.

ALUMAWELD ALL METAL SOLDER

Johnson Mfg. Co.
Sn, Zn, Pb.
For soldering Al and Mg alloys; M.P. 550°F, no flux required for Mg.

ALUMAWELD H-3

Johnson Mfg. Co.
For solder for Al, Mg, foil capacitors; M.P. 510-650°F.

ALUMAWELD H-6

Johnson Mfg. Co.
For solder for Al and other metals; M.P. 775-785°F; hard and strong.

ALUMAWELD H-8

Johnson Mfg. Co.
For solder for Al and other metals; M.P. 715-740°F; good for friction soldering.

ALUMAWELD NO. 1

Johnson Mfg. Co.
For solder for Al and Mg alloys; M.P. 364-620°F.

ALUMAWELD NO. 2

Johnson Mfg. Co.
For solder for Al and Mg foil capacitors; M.P. 393-563°F.

ALUMAWELD NO. 22

Johnson Mfg. Co.
For solder for Al, Mg, foil capacitors; M.P. 393-635°F.

ALUMAWELD NO. 23

Johnson Mfg. Co.
For solder for Al, Mg, foil capacitors; M.P. 336-414°F; good wetting power.

ALUMAWELD NO. 7

Johnson Mfg. Co.
For solder for Al-bronze, Mg, and cast iron; M.P. 333-620°F.

ALUMAWELD SPECIAL

Johnson Mfg. Co.
Sn, Zn, Pb.
For soldering Al and Mg alloys; M.P. 550°F.

ALUMBRO

P.H. Muntz & Co., Ltd.
Cu 76, Zn 22, Al 2.
Drawn: 88,000 TS; 67,000 YS; 10 El; 175 Brin. Annealed: 54,000 TS; 50 El; 65 Brin. For condenser tubing to resist sea water corrosion, plates; high resistance to corrosion.

ALUMBRO

IMI Knoch Ltd.
Cu 76, Zn 22, Al 2.
Drawn: 88,000 TS; 67,000 YS; 10 El; 175 Brin. Annealed: 54,000 TS; 50 El; 65 Brin. For condenser tubing to resist sea water corrosion, plates; high resistance to corrosion.

ALUMEC

British Alcan Aluminium Ltd.
Aluminum.
Aluminum alloy AlZn4Mg3CuCr.

ALUMEL

Hoskins Mfg. Co.
Nickel. Ni 95, Mn 2, 2 Al (plus 8 minor constituents).
Ann: 85,000 TS. EMF: Negative. Thermocouple wire,-used withC Corrosion and heat resistant.

ALUMEND

Arcos Alloys
Cu 4, bal Al.
For aluminum welding rod. *Obsolete*

ALUMEND 1100

Arcos Alloys
Cu 0-0.2, 99.0 Al min, 1.0 Si + Fe max.
For coated aluminum welding rod. *Obsolete*

ALUMEND 4043

Arcos Alloys
Si 4.5-6, Fe 0-0.8, Cu 0-0.3, Ti 0-0.2, Zn 0-0.1, Mn 0-0.05, Mg 0-0.05, bal Al.
For coated aluminum welding rod. *Obsolete*

ALUMETAL

Braun-Steeples Co. Ltd.
Si 5, bal Al.
For Al solder. Corrosion resistant.

ALUMILIT A

Aluminiumwerke Rorschach AG
Aluminum.
Aluminum alloy Al99.7MgO.5.

ALUMINARK

Westinghouse Electric Corp.
Si 5, bal Al.
For aluminum welding electrodes; coated. *Obsolete*

ALUMINITE

American manufacture
Al 73, Zn 23, Cu 2.7, Fe 0.4, Si 0.2.
For light alloy castings; non-hardenable.

ALUMINIUM L-15

English manufacture
Mg 2-4, Mn 1.2-1.5, 0.2 Sb or Ti max, bal Al.
For light alloy castings; resists sea water corrosion.

ALUMINUM

Atomergic Chemetals Corp.
Al.
Purities, zone refined: 99.9999%, 99.999%, 99.99%; 99.9%.
Forms: ingot, rod, powder, wire, sheet, foil, single crystals.

ALUMINUM 359

Kaiser Aluminum & Chemical Corp.
Si 9, Mg 0.6, bal Al.
T61 temper: 48,000 TS; 37,000 YS; 6 El. T62 temper: 50,000 TS; 42,000 YS; 5 El. For aircraft and missile applications. Sand, permanent mold or plaster casting. Good strength, weldability and corrosion resistance. Heat treatable. *Obsolete*

ALUMINUM 1060

Now AA 1060.

ALUMINUM 108-Z

Aluminum Smelting & Refining Co., Inc.
Cu 3-4, Si 2.5-3.5, Zn 2-3, Fe 0-1, Ni 0-0.5, Ti 0-0.25, Mn 0-0.5, bal Al.
For valve bodies, manifolds, truck wheels, piano frames, and oil pans. Non-heat-treatable. *Obsolete*

ALUMINUM 1180

Reynolds Metals Co.
Si 0.09, Fe 0-0.09, Cu 0-0.01, Ti 0-0.02, 99.80% min Al.
For capacitor foil. *Obsolete*

ALUMINUM 1188

Reynolds Metals Co.
Si 0.06, Fe 0.06, Mn 0-0.01, 99.88% min Al.
For capacitor foil. *Obsolete*

ALUMINUM 1199

Reynolds Metals Co.
Si 0.006, Fe 0.006, Cu 0.006, 99.99% min Al.
For capacitor foil. *Obsolete*

ALUMINUM 1230

Reynolds Metals Co.
Cu 0-0.1, 0.07% Si + Fe, 99.30% min Al.
For cladding on Aluminum 2024. *Obsolete*

ALUMINUM 1235

Reynolds Metals Co.
Cu 0-0.05, 0.65% Si + Fe, 99.35% min Al.
For foil. *Obsolete*

ALUMINUM 2020

Aluminum Company of America
Li 1.1, Cu 4.5, Mn 0.5, Cd 0.2, bal Al.
T-6 temper: 86,000 TS; 79,000 YS; 6 El; 150 Brin. At 400 F: 61,000 TS; 57,000 YS; 7 El. For aircraft components; for high temperature service. *Obsolete*

ALUMINUM 2021

Aluminum Company of America
Cu 5.8-6.8, Mn 0.2-0.4, Ti 0.02-0.1, Zr 0.1-0.25, V 0.05-0.15, Cd 0.05-0.2, Sn 0.03-0.08, bal Al.
T8E31 temper: (At RT) 75,000 TS; 66,000 YS; 9 El. (At -320 F) 90,000 TS; 80,000 YS; 9 El. (At -423 F) 102,000 TS; 84,000 YS; 9 El. For cryogenic services, space vehicles, liquid-fueled low temperature tanks. Tough, weldable, good corrosion resistance. *Obsolete*

ALUMINUM 218SP

Aluminum Company of America
Mg 6.7, Si 0.2, Mn 0.1, bal Al.
Cast: 36,000 TS; 18,000 YS; 9 El. For dairy and food handling equipment, fittings for chemical and sewerage use, cooking utensils. Corrosion resistant. *Obsolete*

ALUMINUM 218SP

Reynolds Metals Co.
Mg 6.7, Si 0.2, Mn 0.1, bal Al.
Cast: 36,000 TS; 18,000 YS; 9 El. For dairy and food handling equipment, fittings for chemical and sewerage use, cooking utensils. Corrosion resistant. *Obsolete*

ALUMINUM 2219

Aluminum Company of America
Cu 5.8-6.8, Mn 0.2-0.4, Ti 0.02-0.1, Si 0-0.2, Fe 0-0.3, bal Al.
O-temper: 25,000 TS; 10,000 YS; 20 El. T62-temper: 61,000 TS; 42,000 YS; 11 El; 115 Brin. T87-temper: 70,000 TS; 58,000 YS; 10 El; 130 Brin. For aircraft and automotive engine parts, missiles, space vehicles. Heat treatable. High creep resistance. *Obsolete*

ALUMINUM 2219

Reynolds Metals Co.
Cu 5.8-6.8, Mn 0.2-0.4, Ti 0.02-0.1, Si 0-0.2, Fe 0-0.3, bal Al.
O-temper: 25,000 TS; 10,000 YS; 20 El. T62-temper: 61,000 TS; 42,000 YS; 11 El; 115 Brin. T87-temper: 70,000 TS; 58,000 YS; 10 El; 130 Brin. For aircraft and automotive engine parts, missiles, space vehicles. Heat treatable. High creep resistance. *Obsolete*

ALUMINUM 2364

Kaiser Aluminum & Chemical Corp.
Cu 3.5-5, Si 5.5-7.5, Fe 0-0.6, bal Al.
For general castings; corrosion resistant, cast. *Obsolete*

ALUMINUM 2393

Kaiser Aluminum & Chemical Corp.
Cu 3-4, Si 7-9, Fe 0.6-1, bal Al.
For pressure tight die castings; good castability, corrosion resistant. *Obsolete*

ALUMINUM 2618

Kaiser Aluminum & Chemical Corp.
Cu 2.5, Mg 1.5, Si 0.15-0.25, Ni 1.2, Fe 1, Ti 0.1, bal Al.
T61-temper: 64,000 TS; 54,000 YS; 11 El. For large engine components; for use up to 500 F, age-hardened. *Obsolete*

ALUMINUM 2EC

Kaiser Aluminum & Chemical Corp.
99.45% Al min.
T8-temper: 28,000 TS; 25,000 YS. For busways, conductors; 57% conductivity. *Obsolete*

ALUMINUM 3002
Aluminum Company of America
Si 0-0.08, Mn 0.1-0.25, Mg 0.05-0.2, bal Al.
O-temper: 14,000 TS; 6,000 YS; 35 El; 25 Brin. H25-temper:
22,000 TS; 20,000 YS; 14 El. For Reflectors. High reflectivity.
Obsolete

ALUMINUM 3005
Aluminum Company of America
Si 0.6, Fe 0.7, Cu 0.3, Mn 1-1.5, Mg 0.2-0.6, Cr 0.1, Zn 0.25,
bal Al.
O-temper: 20,000 TS; 8,000 YS; 25 El. H-14 temper: 26,000
TS; 25,000 YS; 5 El. For sheet metal structures, commercial
vehicles. Non-hardenable, corrosion resistant. *Obsolete*

ALUMINUM 3005
Revere Copper Products, Inc.
Mn 1-1.5, Mg 0.06-0.2, Ti 0.03, bal Al.
O-temper: 20,000 TS; 8,000 YS; 25 El. H14-temper: 26,000
TS; 25,000 YS; 5 El. For reflectors. High reflectivity. *Obsolete*

ALUMINUM 310
Aluminum Company of America
Si 10.5, Cr 1, Mg 0.3, Pb 1, Bi 0.5, bal Al.
Cast: 43,000 TS; 25,000 YS; 2.5 El; 80 Brin. For die castings,
housings, containers, machine tools. Free-cutting. Good
strength, weldability and corrosion resistance. *Obsolete*

ALUMINUM 3105
Aluminum Company of America
Mn 0.3-1.5, Mg 0.2-0.8, Si 0-0.6, Fe 0-0.7, Cu 0-0.4, Cr 0-0.25,
Zn 0-0.4, Ti 0-0.1, bal Al.
H25 temper: 26,000 TS; 24,000 YS; 8 El; 47 Brin. For building
products, boxes, covers, containers. Good corrosion
resistance and formability. *Obsolete*

ALUMINUM 3105
Reynolds Metals Co.
Mn 0.3-1.5, Mg 0.2-0.8, Si 0-0.6, Fe 0-0.7, Cu 0-0.4, Cr 0-0.25,
Zn 0-0.4, Ti 0-0.1, bal Al.
H25 temper: 26,000 TS; 24,000 YS; 8 El; 47 Brin. For building
products, boxes, covers, containers. Good corrosion
resistance and formability. *Obsolete*

ALUMINUM 333
English manufacture
Cu, Si, bal Al.
Cast: 34,000 TS; 19,000 YS; 2 El; 90 Brin. For castings,
pistons, sole plates; permanent molds.

ALUMINUM 344
Aluminum Company of America
Si 7, bal Al.
For cooking utensils; corrosion resistant. *Obsolete*

ALUMINUM 354
Now AA 354.0.

ALUMINUM 357
Aluminum Company of America
Si 6.5-7.5, Mg 0.45-0.6, Ti 0.1-0.2, Zn 0-0.03, Fe 0-0.15, bal
Al.
F-temper: 28,000 TS; 15,000 YS; 6 El. Sand cast (T6): 50,000
TS; 43,000 YS; 2 El. Permanent mold (T6): 52,000 TS; 43,000
YS; 5 El. For aircraft and missile structures, high velocity
blowers and impellers. Heat treatable, corrosion resistant.
Obsolete

ALUMINUM 357
Kaiser Aluminum & Chemical Corp.
Si 6.5-7.5, Mg 0.45-0.6, Ti 0.1-0.2, Zn 0-0.03, Fe 0-0.15, bal
Al.
F-temper: 28,000 TS; 15,000 YS; 6 El. Sand cast (T6): 50,000
TS; 43,000 YS; 2 El. Permanent mold (T6): 52,000 TS; 43,000
YS; 5 El. For aircraft and missile structures, high velocity
blowers and impellers. Heat treatable, corrosion resistant.
Obsolete

ALUMINUM 359
Aluminum Company of America
Si 9, Mg 0.6, bal Al.
T61 temper: 48,000 TS; 37,000 YS; 6 El. T62 temper: 50,000
TS; 42,000 YS; 5 El. For aircraft and missile applications.
Sand, permanent mold or plaster casting. Good strength,
weldability and corrosion resistance. Heat treatable.
Obsolete

ALUMINUM 360
Aluminum Company of America
Si 9.5, Mg 0.5, bal Al.
Die cast: 47,000 TS; 25,000 YS; 11 El. For general purpose
die castings, marine and outboard motor parts. High
corrosion resistance and good mechanical properties.
Obsolete

ALUMINUM 360
Kaiser Aluminum & Chemical Corp.
Si 9.5, Mg 0.5, bal Al.
Die cast: 47,000 TS; 25,000 YS; 11 El. For general purpose
die castings, marine and outboard motor parts. High
corrosion resistance and good mechanical properties.
Obsolete

ALUMINUM 363
Kaiser Aluminum & Chemical Corp.
Cu 2.5-3.5, Si 4.5-6, Mn 0.6, Mg 0.3, Zn 3-4.5, bal Al.
Sand cast: 30,000 TS; 20,000 YS; 3 El; 75 Brin. T6-temper:
46,000 TS; 28,000 YS; 5 El; 90 Brin. For pressure tight
castings; age-hardened, good castability. *Obsolete*

ALUMINUM 380B
Manufacturer not listed.
Cu 3.5, Si 9, Fe 1, Mn 0.5, Mg 0.1, Zn 2, Ni 0.5, bal Al.
Cast: 40,000 TS; 17,00 YS; 2.5 El. For general purpose die
castings. Corrosion resistant, high strength.

ALUMINUM 380Z
Manufacturer not listed.
Cu 3.5, Si 9, Fe 1, Mn 0.5, Mg 0.1, Zn 3, Ni 0.5, bal Al.
Cast: 40,000 TS; 17,000 YS; 2.5 El. For general purpose die
castings. Corrosion resistant, pressure tight.

ALUMINUM 384
Now AA 384.0.

ALUMINUM 385
Aluminum Company of America
Si, Sn, Pb, Cd, bal Al.
Die cast: 36,000 TS; 20,000 YS; 3 El. For connecting rods,
gear housings; good bearing surface, high fatigue properties.
Obsolete

ALUMINUM 390
Aluminum Company of America
Si 16-18, Cu 4-5, Mg 0.45-0.65, Fe 0.6-1.1, Ti 0-0.2, bal Al.
-F temper: 40,500 TS; 35,000 YS; 1 El; 120 Brin. -T5 temper:
43,000 TS; 38,500 YS; 1 El. For die cast engine blocks,
pistons, cylinders. Pressure-tight die casting, wear and
corrosion resistant. *Obsolete*

ALUMINUM 390
Reynolds Metals Co.
Si 16-18, Cu 4-5, Mg 0.45-0.65, Fe 0.6-1.1, Ti 0-0.2, bal Al.
-F temper: 40,500 TS; 35,000 YS; 1 El; 120 Brin. -T5 temper:
43,000 TS; 38,500 YS; 1 El. For die cast engine blocks,
pistons, cylinders. Pressure-tight die casting, wear and
corrosion resistant. *Obsolete*

ALUMINUM 392
Now AA 392.0.

ALUMINUM 3L-8
Mond Nickel Co. Ltd.
Si 11-13, bal Al.
Die cast: 24,500 TS; 15,500 YS; 1 El; 80 Brin. Sand cast:
20,000 TS; 1 El; 75 Brin. For housings, cases, general
engineering castings; corrosion resistant, good castability.

ALUMINUM 3L-8
British Aluminium Co., Ltd.
Si 11-13, bal Al.
Die cast: 24,500 TS; 15,500 YS; 1 El; 80 Brin. Sand cast:
20,000 TS; 1 El; 75 Brin. For housings, cases, general
engineering castings; corrosion resistant, good castability.

ALUMINUM 4245
Aluminum Company of America
Si 9.3-10.7, Fe 0-0.8, Cu 3.3-4.7, Mn 0-0.07, Cr 0-0.07, Zn
9.3-10.7, bal Al.
For brazing aluminum alloys, filler metal. MP 960-1040 F.
Obsolete

ALUMINUM 4543
Aluminum Company of America
Si 5-7, Fe 0-0.5, Mg 0.1-0.4, Ti 0-0.1, Cu 0-0.1, Cr 0-0.05, Zn
0-0.1, bal Al.
T61 temper: 28,000 TS; 17,000 YS; 15 El. For extruded
products, architectural trim and shapes. Age-hardenable,
corrosion resistant. *Obsolete*

ALUMINUM 4543
Reynolds Metals Co.
Si 5-7, Fe 0-0.5, Mg 0.1-0.4, Ti 0-0.1, Cu 0-0.1, Cr 0-0.05, Zn
0-0.1, bal Al.
T61 temper: 28,000 TS; 17,000 YS; 15 El. For extruded
products, architectural trim and shapes. Age-hardenable,
corrosion resistant. *Obsolete*

ALUMINUM 4643
Aluminum Company of America
Si 3.6-4.6, Fe 0.8, Cu 0-1, Mn 0.05, Mg 0.1-0.3, Cr 0.15, Zn
0.1, Ti 0.15, bal Al.
For weld filler metal in welding Aluminum 6061 T6 thick
plates. Heat treatable. *Obsolete*

ALUMINUM 4L-11
British Aluminium Co., Ltd.
Cu 7, Sn 0-2, Zn 0-1, bal Al.
Sand cast: 17,000-20,000 TS; 12,000-14,500 YS; 1.5-2.5 El;
60-70 Brin. For auto parts, general purpose castings; not
hardenable. *Obsolete*

ALUMINUM 5039
Kaiser Aluminum & Chemical Corp.
Si 0.13, Fe 0.17, Cu 0.06, Mn 0.25, Mg 3.75, Zn 3.1, Ti 0.08,
Cr 0.1, bal Al.
For weld filler alloy to join 7039 Aluminum Alloy. Self aging.
Obsolete

ALUMINUM 5040
Now AA 5040.

ALUMINUM 5056
Now AA 5056.

ALUMINUM 5080
Aluminum Company of America
Mn 0.5, Mg 4, Cr 0.2, Zn 2, Ti 0.1, bal Al.
For welding filler metal. For use with Aluminum 7006.
Obsolete

ALUMINUM 5083
Now AA 5083.

ALUMINUM 5086
Now AA 5086.

ALUMINUM 50S
Kaiser Aluminum & Chemical Corp.
Mg 1.5, bal Al.
S-D temper: 21,000 TS; 8,500 YS; 25 El. S-H 22 temper:
23,500 TS; 16,500 YS; 13 El. S-H 34 temper: 28,000 TS;
24,500 YS; 6 El. For kitchen utensils, deep drawn parts,
commercial appliances; good finishing characteristics.
Obsolete

ALUMINUM 5154
Now AA 5154.

ALUMINUM 5180
Aluminum Company of America
Zn 2, Mg 4, bal Al.
For filler metal in welding Aluminum 7106, for low temperature applications. High weld strength. *Obsolete*

ALUMINUM 5183
Kaiser Aluminum & Chemical Corp.
Si 0.4, Fe 0.4, Cu 0.1, Mn 0.3-1, Mg 4.5-5.2, Cr 0.05-0.25, Ti 0-0.15, Be 0-0.0005, bal Al.
For structural and marine applications, truck and trailer bodies, power shovels; good formability and weldability, welding electrodes.

ALUMINUM 5205
Aluminum Company of America
Si 0-0.15, Fe 0-0.7, Cu 0.03-0.1, Mn 0-0.1, Mg 0.1-0.6, Cr 0-0.1, bal Al.
For auto trim, instrument panels. Sheet alloy, excellent formability, antiglare sheet. *Obsolete*

ALUMINUM 5205
Reynolds Metals Co.
Si 0-0.15, Fe 0-0.7, Cu 0.03-0.1, Mn 0-0.1, Mg 0.1-0.6, Cr 0-0.1, bal Al.
For auto trim, instrument panels. Sheet alloy, excellent formability, antiglare sheet. *Obsolete*

ALUMINUM 5252
Now AA 5252.

ALUMINUM 5257
Aluminum Company of America
Si 0-0.08, Mg 0.2-0.6, Mn 0-0.03, bal Al.
H25 temper: 19,000 TS; 16,000 YS; 14 El; 32 Brin. H28 temper: 23,000 TS; 19,000 YS; 8 El; 43 Brin. For deeply drawn parts, headlamps, tail lamp bezels, anodized auto trim. Good formability. *Obsolete*

ALUMINUM 5357
Now AA 5357.

ALUMINUM 5557
Aluminum Company of America
Al alloy.
For decoration, bezels, appliance trim; high luster. *Obsolete*

ALUMINUM 55EC
Kaiser Aluminum & Chemical Corp.
99.45% Al min.
T8-temper: 29,000 TS; 25,000 YS. For busways, conductors; 55% conductivity. *Obsolete*

ALUMINUM 5652
Aluminum Company of America
Cu 0.04, Mn 0.01, Mg 2.2-2.8, Zn 0.1, Cr 0.15-0.35, 0.40% Fe + Si, bal Al.
For hydrogen peroxide equipment. Good corrosion resistance. *Obsolete*

ALUMINUM 5652
Reynolds Metals Co.
Cu 0.04, Mn 0.01, Mg 2.2-2.8, Zn 0.1, Cr 0.15-0.35, 0.40% Fe + Si, bal Al.
For hydrogen peroxide equipment. Good corrosion resistance. *Obsolete*

ALUMINUM 5657
Reynolds Aluminum
Aluminum. Mg 0.6-1, Si 0-0.08, Fe 0-0.1, Cu 0-0.1, Zn 0-0.05, bal Al.
H25 Temper: 23,000 TS; 20,000 YS; 12 El; 40 Brin. H38 Temper: 28,000 TS; 24,000 YS; 7 El; 50 Brin. For deep drawn parts and stampings, interior and exterior trim for automobiles and appliances. Bright, corrosion resistant.

ALUMINUM 5757
Aluminum Company of America
Cu 0.05-0.15, Mg 0.6-1, Cr 0.05-0.15, bal Al.
For automobile and appliance grills, trim, auto bumpers; bright finish, fabricating trim. *Obsolete*

ALUMINUM 6003
Reynolds Metals Co.
Si 0.35-1, Mg 0.8-1.5, Fe 0-0.6, Cr 0-0.35, bal Al.
For cladding on Alclad 2014 sheet and plate. Corrosion resistant. *Obsolete*

ALUMINUM 6011
Now AA 6011.

ALUMINUM 6051
Aluminum Company of America
Cu 0.05, Mg 0.6, Mn 0.01, Fe 0.4, Si 1, bal Al.
For light alloy parts; heat treatable. *Obsolete*

ALUMINUM 6051
Kaiser Aluminum & Chemical Corp.
Cu 0.05, Mg 0.6, Mn 0.01, Fe 0.4, Si 1, bal Al.
For light alloy parts; heat treatable. *Obsolete*

ALUMINUM 6063
Now AA 6063.

ALUMINUM 6066
Martin-Marietta Corp.
Cu 0.7-1.2, Si 0.9-1.8, Mg 0.8-1.4, Mn 0.6-1.1, Ti 0.2, Cr 0.4, bal Al.
O temper: 22,000 TS; 12,000 YS; 18 El; 43 Brin. T6 temper: 62,000 TS; 55,000 YS; 12 El; 120 Brin. For truck and trailer bodies, aircraft construction; age hardened, good forming and welding properties.

ALUMINUM 6070
Now AA 6070.

ALUMINUM 6071
Aluminum Company of America
Si 1.1-1.9, Fe 0.5, Cu 0.15-0.4, Mn 0.4-1, Mg 0.8-1.4, Cr 0.1, Zn 0.25, Ti 0.15, bal Al.
Rolled (T6): 50,000 TS; 48,000 YS; 6 El (min). For light weight structural weldments, commercial vehicles, process tanks. High strength sheet and plate. Heat treatable, corrosion resistant. *Obsolete*

ALUMINUM 6101
Now AA 6101.

ALUMINUM 613
Aluminum Company of America
Si 0-0.25, Fe 0-0.8, Mg 0.25-0.5, Cr 0-0.35, Zn 7-8, Ti 0-0.2, Cu 0.4-1, Mn 0-0.6, bal Al.
Sand cast and aged: 35,000 TS; 25,000 YS; 5 El; 74 Brin. For general castings, machinery parts, fittings, brackets, auto and airplane parts. Self-aging. No heat treatment required. *Obsolete*

ALUMINUM 613
Reynolds Metals Co.
Si 0-0.25, Fe 0-0.8, Mg 0.25-0.5, Cr 0-0.35, Zn 7-8, Ti 0-0.2, Cu 0.4-1, Mn 0-0.6, bal Al.
Sand cast and aged: 35,000 TS; 25,000 YS; 5 El; 74 Brin. For general castings, machinery parts, fittings, brackets, auto and airplane parts. Self-aging. No heat treatment required. *Obsolete*

ALUMINUM 613
Kaiser Aluminum & Chemical Corp.
Si 0-0.25, Fe 0-0.8, Mg 0.25-0.5, Cr 0-0.35, Zn 7-8, Ti 0-0.2, Cu 0.4-1, Mn 0-0.6, bal Al.
Sand cast and aged: 35,000 TS; 25,000 YS; 5 El; 74 Brin. For general castings, machinery parts, fittings, brackets, auto and airplane parts. Self-aging. No heat treatment required. *Obsolete*

ALUMINUM 6151
Now AA 6151.

ALUMINUM 6201
Now AA 6201.

ALUMINUM 6253
Reynolds Metals Co.
Mg 1-1.5, Cr 0.15-0.35, Fe 0-0.5, Zn 1.6-2.4, bal Al.
For cladding Aluminum 5056. *Obsolete*

ALUMINUM 6262
Now AA 6262.

ALUMINUM 6351
Now AA 6351.

ALUMINUM 6463
Now AA 6463.

ALUMINUM 6563
Aluminum Company of America
Si 0.2-0.45, Mg 0.35-0.7, Fe 0-0.15, Cu 0-0.15, Mn 0-0.05, bal Al.
T6-temper: 28,000 TS; 21,000 YS; 12 El. For trim applications; heat treatable extrusions, bright luster for anodizing. *Obsolete*

ALUMINUM 6951
Aluminum Company of America
Si 0.2-0.5, Cu 0.15-0.4, Fe 0-0.8, Mn 0-0.2, Fe 0-0.1, bal Al.
O-temper: 16,000 TS; 6,000 YS; 30 El; 28 Brin. T6-temper: 39,000 TS; 33,000 YS; 13 El; 82 Brin. For brazed assemblies. Heat treatable, clad with aluminum, 4343 or 4045 on one or both sides. *Obsolete*

ALUMINUM 7001
Martin-Marietta Corp.
Si 0.35, Fe 0.4, Cu 1.2-2.6, Mg 2.6-3.4, Cr 0.18-0.4, Zn 6.8-8, bal Al.
T6 temper: 98,000 TS; 88,000 YS; 10 El. For missile bodies, trailers, airframes, curtain walls; high strength.

ALUMINUM 7002
Aluminum Company of America
Mg 2.5, Zn 3.5, Cu 0.75, Cr 0.2, bal Al.
Heat treated: 67,000 TS; 57,000 YS; 12 El. For components of space equipment, armor plate for military vehicles and cryogenic uses. Heat treatable, weldable. *Obsolete*

ALUMINUM 7002
Reynolds Metals Co.
Mg 2.5, Zn 3.5, Cu 0.75, Cr 0.2, bal Al.
Heat treated: 67,000 TS; 57,000 YS; 12 El. For components of space equipment, armor plate for military vehicles and cryogenic uses. Heat treatable, weldable. *Obsolete*

ALUMINUM 7004
Now AA 7004.

ALUMINUM 7005
Now AA 7005.

ALUMINUM 7006
Aluminum Company of America
Si 0.1, Fe 0.17, Cu 0.04, Mn 0.22, Mg 2.24, Cr 0.12, Zn 4.1, Ti 0.01, bal Al.
T-6 temper: 65,200 TS; 58,600 YS; 10.5 El. T-63 temper: 62,000 TS; 57,000 YS; 15 El. For welded structures, flight weight and cryogenic applications, missiles. Excellent resistance to stress corrosion. Age-hardenable. Good weldability. *Obsolete*

ALUMINUM 7007
Aluminum Company of America
Mg 1.4-2.2, Cr 0.05-0.25, Zn 6-7, Ti 0.01-0.06, Zr 0.05-0.25, bal Al.
T6 temper: (At RT) 77,000 TS; 69,000 YS; 12 El. (At -320 F) 93,000 TS; 83,000 YS; 5 El. (At -423 F) 104,000 TS; 90,000 YS; 4 El. For cryogenic services. Tough, weldable, good corrosion resistance, heat treatable. *Obsolete*

ALUMINUM 7038
Kaiser Aluminum & Chemical Corp.
Zn 4-5, Mg 3-4, Si 0.2, Fe 0.35, Cu 0.1, Mn 0.1-0.35, Cr 0.5, Ti 0.1, bal Al.
For cryogenic equipment, armor plate and aerospace equipment. High strength and good welding qualities. *Obsolete*

ALUMINUM 7039
Now AA 7039.

ALUMINUM 7076
Now AA 7076.

ALUMINUM 7079
Now AA 7079.

ALUMINUM 7080
Aluminum Company of America
Cu 0.5-1.5, Mg 1.5-3, Zn 5-7, Mn 0.1-0.7, Ti 0-0.2, bal Al.
Forged T7 temper: 68,000 TS; 60,000 YS; 14 El. For large airframe parts, aircraft landing components. Heat treatable, resists stress corrosion. *Obsolete*

ALUMINUM 7106
Aluminum Company of America
Zn 4.2, Mg 2.2, Mn 0.2, Cr 0.1, bal Al.
At 75 F: 61,000 TS; 55,000 YS; 15 El. At -112 F: 64,000 TS; 60,000 YS; 15 El. For missile cases, space vehicle fuel and oxidizer tankage, large rocket motor cases. Heat treatable. High fatigue strength. Good weldability. Good cryogenic toughness. *Obsolete*

ALUMINUM 7175
Now AA 7175.

ALUMINUM 7178
Now AA 7178.

ALUMINUM 7179
Now AA 7179.

ALUMINUM 718
Aluminum Company of America
Cu 0-0.3, Fe 0-0.8, Si 11-13, Zn 0-0.2, bal Al.
For brazing alloy; M.P. 1070-1080 F. *Obsolete*

ALUMINUM 718
Kaiser Aluminum & Chemical Corp.
Cu 0-0.3, Fe 0-0.8, Si 11-13, Zn 0-0.2, bal Al.
For brazing alloy; M.P. 1070-1080 F. *Obsolete*

ALUMINUM 7275
Kaiser Aluminum & Chemical Corp.
Cu 1.2-2, Mg 2.1-2.9, Cr 0.18-0.4, Zn 5.1-6.1, Ti 0-0.1, bal Al.
For lightweight structures; age-hardenable. *Obsolete*

ALUMINUM 7277
Now AA 7277.

ALUMINUM 7279
Kaiser Aluminum & Chemical Corp.
Cu 0.4-0.8, Mn 0.1-0.3, Mg 2.9-3.7, Cr 0.1-0.25, Zn 3.8-4.8, bal Al.
For aircraft components; age-hardenable. *Obsolete*

ALUMINUM 8001
Now AA 8001.

ALUMINUM 8002
Aluminum Company of America
Fe 1.2-1.6, Ni 1.2-1.6, Cu 0-0.05, Si 0-0.25, bal Al.
For light alloy parts. *Obsolete*

ALUMINUM 8013
Kaiser Aluminum & Chemical Corp.
Cr 0.2-0.5, Ti 0-0.1, 0.25% max Si + Fe, bal Al.
For lightweight parts; corrosion resistant. *Obsolete*

ALUMINUM 8081
Alcan Canada Products Ltd.
Sn 18-22, Si 0-0.7, Mn 0-0.1, Fe 0-0.7, Cu 0.7-1.3, Ti 0-0.1, 0.00-0.15 others, bal Al.
H25 temper: 24,000 TS; 21,500 YS; 13 El. H112 temper: 28,000 TS; 25,000 YS; 10 El. For bearings. Steel backed automotive bearings, bearings for light duty diesel engines, gas turbines, farm tractors; improved corrosion resistance.

ALUMINUM 8212
Kaiser Aluminum & Chemical Corp.
Cu 0.1-0.4, Mn 0.15-0.5, Mg 0.05-0.35, Si 0-1, Fe 0-1, Zn 0-1, bal Al.
For lightweight parts. *Obsolete*

ALUMINUM 8280
Alcan Canada Products Ltd.
Cu 0.7-1.3, Ni 0.2-0.7, Si 1-2, Sn 5.5-7, bal Al.
Annealed: 17,000 TS; 70,000 YS; 28 El; 65 Rock H. H16 temper: 27,000 TS; 25,000 YS; 5 El; 92 Rock H. For bearings made from continuous coils for automotive engines. High load carrying capacity, excellent fatigue strength.

ALUMINUM A.W. 160
Aluminium Walzwerke Singen GmbH
Aluminum. Si 1, Mn 0.7, Mg 0.65, bal Al.
For light alloy parts; same as Anticorodal.

ALUMINUM A-43
American Smelting & Refining Co.
Si 5.3, bal Al.
Die cast: 33,000 psi TS; 14,000 psi YS; 9 El; 50 Brin; 19,000 psi shear strength; 17,000 fatigue strength. For marine castings, water jackets, meter housings, carburetor bodies. High fluidity and castability; corrosion resistant. Die casting alloy.

ALUMINUM A-57G
French manufacture
Si 6.5-7, bal Al.
Cast: 44,000 TS; 7 El; 100 Brin. Corrosion resistant.

ALUMINUM A357
Aluminum Company of America
Cu 0.2, Si 7.5, Mg 0.6, Zn 0.1, Ti 0.2, Fe 0.2, bal Al.
T6 sand cast: 46,000 TS; 36,000 YS; 3 El, 85 Brin. T6 permanent mold: 50,000 TS; 40,000 YS; 10 El; 100 Brin, 43,000 S.S. For cylinder heads, gear cases, oil pans, superchargers, transmission cases. High strength, pressure tight. *Obsolete*

ALUMINUM A357
Kaiser Aluminum & Chemical Corp.
Cu 0.2, Si 7.5, Mg 0.6, Zn 0.1, Ti 0.2, Fe 0.2, bal Al.
T6 sand cast: 46,000 TS; 36,000 YS; 3 El, 85 Brin. T6 permanent mold: 50,000 TS; 40,000 YS; 10 El; 100 Brin, 43,000 S.S. For cylinder heads, gear cases, oil pans, superchargers, transmission cases. High strength, pressure tight. *Obsolete*

ALUMINUM ALLOY B.E.S.A. NO. 361
English manufacture
Cu 6-8, Zn 0.1, Fe 0-1, Si 0-1, bal Al.
Chill cast: 20,160 YS; 3 El; 60 Brin. Sand cast: 18,000 TS; 1-2 El. For general castings, cylinder heads; good machinability.

ALUMINUM ALLOY B.E.S.A. NO. 362
English manufacture
Cu 11-13, Zn 0.1, Fe 0-1, Si 0-1, bal Al.
Chill cast: 20,160 TS; 80 Brin. For permanent mold castings, pistons; non-hardenable.

ALUMINUM ALLOY B.E.S.A. NO. 363
English manufacture
Cu 2.5-3, Zn 12.5-14.5, Fe 0-1, Si 0-1, bal Al.
Sand cast: 25,000-30,000 TS; 1 El; 70 Brin. For sand castings, crankcases; non-hardenable.

ALUMINUM ALLOY NO. 119
Aluminium Industrie Aktiengesellschaft
Si 0.4, Fe 1.9, Cu 4, bal Al.
28,000-32,000 TS; 11,000-20,000 YS; 16-22 El. For machine parts; wrought alloy. *Obsolete*

ALUMINUM ALLOY NO. 119 M
Aluminium Industrie Aktiengesellschaft
Si 0.4, Fe 1.9, Mg 0.5, Cu 4, bal Al.
Hard: 47,000-60,000 TS; 22,000-58,000 YS; 2-18 El. For machine parts; wrought alloy. *Obsolete*

ALUMINUM AS-10
Soc. Alluminio Veneto per Azioni
Cu 2.2, Si 10, Mg 1.4, Ni 1, bal Al.
Cast: 32,700-35,600 TS; 27,800-32,700 YS; 0.5-0.7 El; 100-120 Brin. For pistons; high temperature resistance.

ALUMINUM AS-10
Alluminio SA
Cu 2.2, Si 10, Mg 1.4, Ni 1, bal Al.
Cast: 32,700-35,600 TS; 27,800-32,700 YS; 0.5-0.7 El; 100-120 Brin. For pistons; high temperature resistance.

ALUMINUM AS-10
Industria Nazionale Alluminio, Mori
Cu 2.2, Si 10, Mg 1.4, Ni 1, bal Al.
Cast: 32,700-35,600 TS; 27,800-32,700 YS; 0.5-0.7 El; 100-120 Brin. For pistons; high temperature resistance.

ALUMINUM AS-5
Soc. Alluminio Veneto per Azioni
Cu 0.6-0.9, Si 4.5-5.5, Mg 0.5, bal Al.
Cast: 35,600-45,600 TS; 25,600-29,900 YS; 4-5 El; 90-100 Brin. For instrument housings, oil pans, crankcases; high fluidity and corrosion resistant.

ALUMINUM AS-5
Industria Nazionale Alluminio, Mori
Cu 0.6-0.9, Si 4.5-5.5, Mg 0.5, bal Al.
Cast: 35,600-45,600 TS; 25,600-29,900 YS; 4-5 El; 90-100 Brin. For instrument housings, oil pans, crankcases; high fluidity and corrosion resistant.

ALUMINUM AS-5
Alluminio SA
Cu 0.6-0.9, Si 4.5-5.5, Mg 0.5, bal Al.
Cast: 35,600-45,600 TS; 25,600-29,900 YS; 4-5 El; 90-100 Brin. For instrument housings, oil pans, crankcases; high fluidity and corrosion resistant.

ALUMINUM ASCR
Kaiser Aluminum & Chemical Corp.
99.9% Al.
For cables; high conductivity. *Obsolete*

ALUMINUM AW-15
Aluminium Walzwerke Singen GmbH
Aluminum. Mn 1.5, bal Al.
Hard worked: 37,000 TS; 3 El; 59 Brin. Annealed: 17,000 TS; 27 El; 32 Brin. For deep drawing, stamping and pressing; weldable, corrosion resistant.

ALUMINUM B-95,
Russian manufacture
Zn 5-7, Mg 1.8-2.8, Cu 1.4-2, Mn 0.2-0.6, Cr 0.1-0.25, bal Al.
For high strength castings; age-hardenable.

ALUMINUM B218
Now AA B535.

ALUMINUM BA355
British Aluminium Co., Ltd.
Cu 3.5-4.8, Mg 0.8, Si 0.9, Fe 1, Mn 1.2, Ti 0.3, bal Al.

ALUMINUM BA46

British Aluminium Co., Ltd.
Cu 0.1, Mg 0.2-0.6, Si 4.5-6, Mn 0.4, Ni 0.1, Fe 0.6, bal Al.
Heat treated: 33,600-39,000 TS; 31,400 YS; 70-75 Brin. For crankcases, gear cases, covers, housings; age-hardenable, good castability. *Obsolete*

ALUMINUM BRASS

Anaconda Co.
Al 1-6, Zn 24-42, Cu 55-71.
85,000-45,000 TS; 73,000-32,000 YS; 6-50 El; 8-33 RA; 114-193 Brin. For bushings, hardware. *Obsolete*

ALUMINUM BRASS

Chase Brass & Copper Co., Inc.
Copper. Cu 77.5, Zn 20.5, Al 2, As 0.055.
Annealed: 60,000 psi TS; 27,000 psi YS; 55 El. Drawn (35%): 85,000 psi TS; 65,000 YS; 15 El. Corrosion resistant; for condenser and heat exchanger tubes, ferrules.

ALUMINUM BRASS

Century Brass Products Inc.
Al 1.75-2.5, 75% min Cu, bal Zn.
Annealed: 60,000 TS; 50 El. Hard: 85,000 TS; 15 El. For condenser tubes; corrosion resistant. *Obsolete*

ALUMINUM BRONZE

American Manganese Bronze Co.
Copper. Cu 87.5, Al 4.1, Fe 1.7, Zn 6.6.
Hot rolled: 63,9000 TS; 35,000 YS; 6.5 El. Cold drawn: 98,000 TS 90,500 YS; 13.0 El; 14.0 RA. For bearings, gears; heavy duty.

ALUMINUM BRONZE

Janney Cylinder Co.
Cu 86-90, Fe 1-4, Al 8-12, Mn 1-4.
Cast: 60,000-90,000 TS; 4-20 El; 150-250 Brin. For centrifugal castings, valve seat rings; heat treatable.

ALUMINUM BRONZE

ACIPCO Steel Products Division
Cu 91-94, Fe 0-1.2, Al 6.6-7, Mn 0-1.
Sand cast: 71,000-112,000 TS; 25,300-50,000 YS; 21.7-3.0 El; For bearing metals, ball bearings, pin pivots, pump, propeller blades, bearings, gears; very close-grained castings; corrosion resistant. *Obsolete*

ALUMINUM BRONZE

Anaconda Co.
Al 7-9, Fe 2.5-4, Sn 0.5, bal Cu.
70,000 TS; 2.5 El. For bearings, gears; corrosion resistant. *Obsolete*

ALUMINUM BRONZE 135

Mackenzie's Sons Co. Inc.
Copper. Al 10, bal Cu.
For hardware; Al-bronze.

ALUMINUM BRONZE 136

Mackenzie's Sons Co. Inc.
Copper. Al 10, Fe 1, bal Cu.
For propellers; Al-bronze.

ALUMINUM BRONZE 5%

Anaconda Co.
Copper. Cu 95, Al 5.
Hard: 105,000 TS; 5 El. Soft: 52,000 TS; 70 El. For diaphragms to withstand pressure, condenser tubes. MP 1060°C.

ALUMINUM BRONZE 707

Olin Brass, Indianapolis
Copper. Al 7.15, Si 2, Cu 90.85.
Soft: 90,000 psi TS; 50,000 psi YS; 30 El; 165 Brin. For boat shafting, pump rods, and gears. Aluminum bronze; corrosion resistant.

ALUMINUM BRONZE 712

Olin Brass, Indianapolis
Copper. Cu 5, Si 1, Al 3.5, Cu 88.5.
Spring: 114,000 psi TS; 80,000 psi YS; 3 El; 97 Rock B. Soft: 65,000 psi TS; 24,000 psi YS; 55 El; 60 Rock B. For spring contacts, diaphragms, and bellows. Spring properties; tough; 12% conductivity.

ALUMINUM BRONZE 715

Olin Brass, Indianapolis
Copper. Al 2.9, Si 0.35, Cu 96.75.
Hard: 75,000 psi TS; 15 El. Soft: 40,000 psi TS; 55 El. For bolts, nuts, and screws. Aluminum bronze; corrosion resistant.

ALUMINUM BRONZE AZ

J.W. Harris Co., Inc.
Cu 88.2, Fe 1.5, Al 8.5-11.
Melting point: 1967°F approx. For gas tungsten arc and gas metal arc braze welding and building up on ferrous and non-ferrous alloys. AWS A5.7 ER CuAl-A2.

ALUMINUM BRONZE E-707

Accurate Brass Co.
Al 6.2-8, Si 1.6-2.2, bal Cu.
Forged: 85,000-90,000 TS; 30,000-40,000 YS; 20-25 El; 120-142 Brin. For marine hardware, propellers, gears, fasteners, bolts; high strength; corrosion resistant.

ALUMINUM BRONZE NO. 1

Knowsley Cast Metal Co., Ltd.
Al, Fe, bal C.
Cast: 71,700 TS; 20 El; 90-140 Brin. For high strength castings; corrosion resistant, good strength at high temperature.

ALUMINUM BRONZE NO. 2

Knowsley Cast Metal Co., Ltd.
Al, Fe, bal Cu.
Cast: 89,600 TS; 12 El; 156-187 Brin. For high strength castings; corrosion resistant, retains strength at high temperature.

ALUMINUM C-113

Now AA 2130.

ALUMINUM C-960

Aluminum Company of America
Si 16, bal Al.
For automotive engines, pistons. Cast: 45,000 TS; 42,000 YS; 1 El. High corrosion resistance. *Obsolete*

ALUMINUM C46

Industria Nazionale Alluminio, Mori
Si 0.8-1.2, Mg 0.3, Ti 0.2, Ni 1.5, Cu 9.5-10.5, bal Al.
Cast: 51,200-54,100 TS; 42,700-48,400 YS; 0.5-1 El; 125-140 Brin. For light alloy castings; general purpose.

ALUMINUM C58

Aluminum Company of America
Cu 0.15-0.4, Si 0.4-0.8, Mg 0.8-1.2, Cr 0.15-0.35, Fe 0-0.7, Mn 0-0.15, Zn 0-0.2, Ti 0-0.15, bal Al.
O-temper: 18,000 TS; 8,000 YS; 30 El; 30 Brin. T4-temper: 34,000 TS; 21,000 YS; 25 El; 65 Brin. T6-temper: 45,000 TS; 40,000 YS; 17 El; 90 Brin. Uses: porcelain enameled extruded parts. Heat treatable. Extrusion alloy. *Obsolete*

ALUMINUM C872

Aluminum Company of America
Si 0.2-0.6, Mg 0.45-0.9, Fe 0-0.35, Cu 0-0.1, Ti 0-0.1, bal Al.
T42 temper (min): 17,000 TS; 9,000 YS; 12 El. T5 temper (min): 22,000 TS; 16,000 YS; 8 El. T6 temper (min): 30,000 TS; 25,000 YS; 8-10 El. For trim, architectural, decorations. Extrusion alloy. *Obsolete*

ALUMINUM C914

Aluminum Company of America
Si 0.2-0.6, Mg 0.45-0.9, Fe 0-0.35, Cu 0-0.1, Mn 0-0.1, Zn 0-0.1, Cr 0-0.1, Ti 0-0.1, bal Al.
T5-temper: 27,000 TS; 21,000 YS; 12 El; 60 Brin. T6-temper: 35,000 TS; 31,000 YS; 10-12 El; 73 Brin. O-temper: 13,000 TS; 7,000 YS; 25 Brin. For constructional equipment and components. Heat treatable extrusion alloy. *Obsolete*

ALUMINUM C989

Aluminum Company of America
Si 0.2-0.6, Mg 0.45-0.9, Fe 0-0.35, Cu 0-0.1, Ti 0-0.1, bal Al.
T42 temper (min): 17,000 TS; 9,000 YS; 12 El. T5 temper (min): 22,000 TS; 16,000 YS; 8 El. T6 temper (min): 30,000 TS; 25,000 YS; 8-10 El. For trim, architectural, decorative applications. Extrusion alloy, heat treatable. *Obsolete*

ALUMINUM CD1S

Now AA 1160.

ALUMINUM CH-70

Aluminum Company of America
Cu 4.7, Ag 0.79, Mg 0.27, Mn 0.27, Ti 0.21, bal Al.
Heat treated: 60,000 TS; 50,000 YS; 5 El. For wing-spar castings, aerospace and undersea equipment. Cast alloy, heat treatable. *Obsolete*

ALUMINUM CHROMIUM

Foote Mineral Co.
10 Cr, 90 Al or 20 Cr, 80 Al.
For use in manufacture of non-ferrous alloys. *Obsolete*

ALUMINUM D132

Aluminum Company of America
Cu 2-4, Fe 0-1.2, Si 8.5-10.5, Mg 0.5-1.5, Ni 0.5-1.5, Ti 0.2, bal Al.
T5-temper: 36,000 TS; 28,000 YS; 1 El; 105 Brin. For pistons, cylinder heads; age-hardenable. *Obsolete*

ALUMINUM DTD 424

British Aluminium Co., Ltd.
Cu 3, Si 5, Mn 0.5, bal Al.
Sand cast: 20,200 TS; 12,300 YS; 2.5 El; 65 Brin. Permanent mold: 22,400 TS; ;12,300 YS; 2.5 El; 70 Brin. For general castings; corrosion resistant. *Obsolete*

ALUMINUM DTD-25

English manufacture
Si 5, bal Al.
For instrument housings, gear cases, general castings; corrosion resistant.

ALUMINUM DTD428

British Aluminium Co., Ltd.
Cu 7, Si 3, Zn 3, bal Al.
Sand cast: 19,100 TS; 12,200 YS; 1 El; 73 Brin. Permanent mold: 24,600 TS; 15,700 YS; 1 El; 80 Brin. For general castings; pressure tight. *Obsolete*

ALUMINUM ECH17

Aluminum Company of America
99.45% Al min.
17,000 TS; 15,000 YS. For bus bars, special parts; 61% electrical conductivity. *Obsolete*

ALUMINUM GENUINE

United American Metals Corp.
Pb, Sb, bal Sn.
For bearings; Babbitt metal. *Obsolete*

ALUMINUM GM-3889M

General Motors Corp./Central Foundry
Si 4, Cd 1, bal Al.
For bearings; requires steel backing.

ALUMINUM H10

English manufacture
Cu 0.07, Fe 0.2, Mn 0.58, Mn 0.61, Si 0.94, Ti, bal Al.

ALUMINUM HD-11
Bridgeport Brass Corp.
Al alloy.
For impact extrusions; ductile.

ALUMINUM HP356
Aluminum Company of America
Si 6.75-7.5, Mg 0.3, Cu 0-0.2, Fe 0-0.15, Mn 0-0.3, bal Al.
For aircraft fittings, pump parts, cylinder blocks; age-hardenable, good castability. *Obsolete*

ALUMINUM HP356
Kaiser Aluminum & Chemical Corp.
Si 6.75-7.5, Mg 0.3, Cu 0-0.2, Fe 0-0.15, Mn 0-0.3, bal Al.
For aircraft fittings, pump parts, cylinder blocks; age-hardenable, good castability. *Obsolete*

ALUMINUM HZM-100
Martin-Marietta Corp.
Zn 7.3, Mg 3, Cu 2, Cr 0.2, bal Al.
T6 temper: 100,000 TS; 92,000 YS; 9 El; 160 Brin. O temper: 40,000 TS; 26,000 YS; 10 El. For aircraft parts and structures; age hardenable.

ALUMINUM IRON BRONZE
Anaconda Co.
Cu 85-89, Al 6-9.5, Fe 3.5-7.5.
74,000 TS; 23 RA; 100 Brin. For propeller blades, pump parts, bearings. *Obsolete*

ALUMINUM IRON BRONZE "B"
Manufacturer not listed.
Cu 85.16, Al 9.43, Fe 4.74, Pb 0.38, P 0.09.
For strong corrosion resistant structural parts; corrosion resistant.

ALUMINUM IRON BRONZE "H"
Manufacturer not listed.
Cu 89.43, Al 6.97, Fe 3.41.
For strong corrosion resistant structural parts; corrosion resistant.

ALUMINUM IRON BRONZE "R"
Manufacturer not listed.
Cu 85.16, Al 6.6, Fe 7.52, Mn 0.5.
For strong corrosion resistant structural parts; corrosion resistant.

ALUMINUM K 183
Kaiser Aluminum & Chemical Corp.
Si 0.4, Fe 0.4, Cu 0.1, Mn 0.5-1, Mg 4-4.9, Cr 0.25, Zn 0.25, Ti 0.15, bal Al.
O-temper: 44,000 TS; 22,000 YS; 21 El; 73 Brin. H 113-temper: 46,000 TS; 33,000 YS; 16 El. For welded structures and girders; good weldability. *Obsolete*

ALUMINUM K 186
Kaiser Aluminum & Chemical Corp.
Mg 4, Mn 0.45, Cr 0.1, bal Al.
O-temper: 38,000 TS; 17,000 YS; 22 El. H 112-temper: 39,000 TS; 19,000 YS; 14 El. H 34-temper: 47,000 TS; 37,000 YS; 10 El. For pressure vessels, truck and trailer frames; good formability and weldability. *Obsolete*

ALUMINUM K-155
Kaiser Aluminum & Chemical Corp.
Mg 0.8, bal Al.
O-temper: 18,000 TS; 6,000 YS; 30 El; 28 Brin. H18 temper: 29,000 TS; 28,000 YS; 4 El; 41 Brin. H38-temper: 29,000 TS; 27,000 YS; 5 El; 51 Brin. For light reflectors, refrigerator shelves; now aluminum 5005. *Obsolete*

ALUMINUM KO-1
Manufacturer not listed.
Cu 4.5-5.2, Ag 0.4-1, Mg 0.18-0.35, Ti 0.15-0.35, Zn 0-0.4, bal Al.
Cast: 62,000-66,000 YS; 72,300-73,000 TS; 10-11 El. At 350°F: 60,000 TS; 57,000 YS; 11 El. At 500°F: 46,000 TS; 44,000 YS; 13 El. Heat treated: 65,000 TS; 55,000 YS; 3 El. For impellers, throttle levers, fins. Age hardenable.

ALUMINUM KS280
German manufacture
Si 21-22, Cu 1.5, Ni 1.5, Mn 0.6, Mg 0.5, Co 1.2, bal Al.
For bearings.

ALUMINUM KSS
Lavorazione Leghe Leggere SpA
Mg 1.7-2.3, Mn 1-1.4, bal Al.
For light alloy parts for marine service; corrosion resistant.

ALUMINUM L8N
Industria Nazionale Alluminio, Mori
Cu 11-12.5, Fe 0.7, Ni 0.5, bal Al.
Cast: 38,400-42,700 TS; 22,800-25,600 YS; 1-2 El; 110-125 Brin. For light alloy castings; general purpose.

ALUMINUM L8N
Alluminio SA
Cu 11-12.5, Fe 0.7, Ni 0.5, bal Al.
Cast: 38,400-42,700 TS; 22,800-25,600 YS; 1-2 El; 110-125 Brin. For light alloy castings; general purpose.

ALUMINUM L8T
Industria Nazionale Alluminio, Mori
Cu 11-12.5, Fe 1, Si 0.5, Ti 0.2, bal Al.
Cast: 38,400-42,700 TS; 22,800-25,600 YS; 1-1.5 El; 110-125 Brin. For light alloy castings; general purpose.

ALUMINUM L8T
Alluminio SA
Cu 11-12.5, Fe 1, Si 0.5, Ti 0.2, bal Al.
Cast: 38,400-42,700 TS; 22,800-25,600 YS; 1-1.5 El; 110-125 Brin. For light alloy castings; general purpose.

ALUMINUM LAC 10
British Aluminium Co., Ltd.
Cu 9.8, Mg 0.25, Fe 0.7, bal Al.
Sand cast: 18,000 TS; 13,500 YS; 75 Brin. WP-temper: 31,400 TS; 25,400 YS; 120 Brin. For general castings; age-hardened. *Obsolete*

ALUMINUM LAC 112A TYPE 2
British Aluminium Co., Ltd.
Cu 2.1, Si 10.3, Zn 0.9, bal Al.
Permanent mold: 25,300 TS; 12,000 YS; 2 El; 78 Brin. For pressure tight castings; corrosion resistant. *Obsolete*

ALUMINUM LAC 113B
British Aluminium Co., Ltd.
Cu 3.5, Zn 11, bal Al.
Sand cast: 24,600 TS; 13,500 YS; 1.5 El; 85 Brin. For general castings; good corrosion resistance. *Obsolete*

ALUMINUM LAC112A TYPE 1
British Aluminium Co., Ltd.
Cu 1.1, Si 10.3, bal Al.
Sand cast: 21,300 TS; 11,000 YS; 2 El; 70 Brin. For pressure tight castings; corrosion resistant. *Obsolete*

ALUMINUM LM11
English manufacture
Cu 4.2-5, Mg 0.15-0.35, Fe 0-0.5, Si 0-0.3, Ti 0-0.3, Zn 0-0.1, bal Al.
Chill cast: 40,000 TS; 9 El. For pressure die castings, cases, housings; heat treatable.

ALUMINUM MANGANESE
Foote Mineral Co.
Mn 30, Al 70.
For use in manufacture of non-ferrous alloys. *Obsolete*

ALUMINUM MOLYBDENUM
Foote Mineral Co.
Mo 25, Al 75.
For use in manufacture of non-ferrous alloys. *Obsolete*

ALUMINUM NO. 16
Stewart-Warner (Die Casting Div.)
Ni 4.5-5.5, Si 0-1, bal Al.
Die cast: 37,000 TS; 21,000 YS; 4.5 El; 80 Brin. For die castings, stove burner heads; heat resistant. *Obsolete*

ALUMINUM NO. 2EC
Aluminum Company of America
Si 0.3-0.6, Mg 0.4-0.8, Fe 0-0.3, Cu 0-0.05, Mn 0-0.01, Zn 0-0.05, bal Al.
Annealed: 17,000 TS; 7,500 YS; 35 El; 30 Brin. T6 temper: 29,000 TS; 25,000 YS. T61 temper: 15,000-20,000 TS; 8,000-15,000 YS. T62 temper: 27,000 TS; 22,000 YS. For electrical buses. High strength and electrical conductivity. Age hardenable. Corrosion resistant. *Obsolete*

ALUMINUM NO. 37
Stewart-Warner (Die Casting Div.)
Si 6.5-7.5, Mg 0.2-0.4, bal Al.
Cast: 40,000 TS; 27,000 YS; 5 El; 90 Brin. For permanent mold castings; dense; C.S. 24,000. *Obsolete*

ALUMINUM NO. 46
Stewart-Warner (Die Casting Div.)
Cu 3.5-4.5, Si 2.5-3.5, bal Al.
Cast: 28,000 TS; 16,000 YS; 2 El; 70 Brin. For permanent mold castings; O.S. 10,000. *Obsolete*

ALUMINUM OILITE
Chrysler Corp.
Al alloy.
For bearings; self-lubricating, porous.

ALUMINUM OILITE 201
Chrysler Corp.
Cu 4.5, Si 0.8, Mg 0.5, Al 94.
Sintered: 40,000 TS; 32,000 YS; 2.0 El; 90 RH; density, 2.5-2.6. Aluminum base powder metal.

ALUMINUM OILITE 601
Chrysler Corp.
Cu 0.25, Mg 1, Al 98.
Sintered: 35,000 TS; 30,000 YS; 2.0 El; 80 RH; density, 2.5-2.6. Aluminum base powder metal.

ALUMINUM P35
Lavorazione Leghe Leggere SpA
Mg 3.2-3.8, Mn 0.5, Si 0-0.3, Fe 0-0.4, bal Al.
For light alloy parts for marine service; corrosion resistant.

ALUMINUM S-12A
NL Industries
Si 11.76, Cu 0.53, Fe 0.81, Mg 0.01, C 0.19, Ni 0.04, Mn, bal Al.
As cast: 39,200 TS; 21,300 YS; 3.5 El. For die castings, instrument cases, housings. High corrosion resistant, pressure tight. *Obsolete*

ALUMINUM S457
Aluminum Company of America
Cu 0.05-0.2, Mn 0.15-0.45, Mg 0.8-1.2, bal Al.
For decorative trim, appliances; mirror-like finish. *Obsolete*

ALUMINUM S6063
Reynolds Aluminum
Aluminum. Si 0.5, Mg 0.8, Fe 0-0.35, bal Al.
T5: 27,000 TS; 21,000 YS; 12 El; 60 Brin. T42: 25,000 TS; 13,000 YS; 22 El. For architectural applications, portable irrigation systems, moldings and trim. Heat treatable extrusion alloy. *Obsolete*

ALUMINUM SC84A + MG
NL Industries
Si 8.8, Cu 4.4, Fe 0.82, Mg 0.27, Mn 0.25, Ni 0.01, bal Al.
Die cast: 48,600 TS; 25,000 YS; 3 El. At 400 F: 24,800 TS; 15,700 YS; 6 El. For die castings, pistons, sole plates, housings. Corrosion resistant, pressure tight. *Obsolete*

ALUMINUM SG-81
NL Industries
Si 8.6, Cu 0.42, Fe 0.75, Mg 0.62, Ni 0.46, Zn 0.07, Ti 0.16, Be 0.15, bal Al.
Die cast: 46,300 TS; 49,200 YS; 4.8 El. As aged: 54,500 TS; 46,100 YS; 2.5 El. At 500 F: 28,800 TS; 28,400 YS; 7 El. For die castings, missiles, housings, cases. Good high temperature properties, corrosion resistant. *Obsolete*

ALUMINUM SILICON BRONZE
Chase Brass & Copper Co.
Cu 91, Al 7, Si 2, Copper Alloy No. 637.
Light annealed: 92,000 TS; 53,000 YS; 35 El; 51,000 shear;
93 Rock B. For bolts, connectors, sleeves, gears, marine
hardware, studs, pump parts. ASTM B150 Alloy 1; CDA 637.
Obsolete

ALUMINUM SOLDER (A)
Manufacturer not listed.
Zn 30, Sn 65, Bi 5.
For Al solder.

ALUMINUM SOLDER (B)
Manufacturer not listed.
Zn 50, Cu 1.5, Sn 33, Sb 2, Pb 12.5.
For Al solder.

ALUMINUM SOLDER (C)
Manufacturer not listed.
Zn 57, Cd 43.
For Al solder.

ALUMINUM SOLDER (D)
English manufacture
Sn 67.5, Cu 1, Zn 16.5, bal Al.
7100 TS. For Al solder.

ALUMINUM SOLDER (E)
English manufacture
Sn 53, Zn 40, Al 7.
6700 TS. For Al solder.

ALUMINUM SOLDER RICHARDS
Manufacturer not listed.
Zn 25, Sn 71.5, Al 3.5.
For Al solder.

ALUMINUM SOLDER, FRISMUTH (A)
Manufacturer not listed.
Zn 47.5, Cu 5.5, Sn 31.5, Al 10.5, Ag 5.5.
For Al solder.

ALUMINUM SOLDER, FRISMUTH (B)
Manufacturer not listed.
Zn 47.4, Cu 5.3, Sn 36.8, Al 10.5.
For Al solder.

ALUMINUM SOLDER, FRISMUTH (C)
Manufacturer not listed.
Sn 67, Pb 27, Al 3.
For Al solder.

ALUMINUM SOLDER, GRIMM'S (A)
Manufacturer not listed.
Sn 69.1, Pb 8-28, Zn 1.44, Ag 0.72.
For Al solder.

ALUMINUM SOLDER, GRIMM'S (B)
Manufacturer not listed.
Sn 50, Pb 25, Zn 25.
For Al solder.

ALUMINUM SS6063
Reynolds Aluminum
Aluminum. Si 0.5, Mg 0.8, Fe 0-0.35, bal Al.
T5: 27,000 TS; 21,000 YS; 12 El; 60 Brin. For architectural
applications, portable irrigation systems, moldings and trim.
Heat treatable extrusion alloy. *Obsolete*

ALUMINUM STANDARD GRADES
Wabash Alloys, Inc.
Aluminum.
Casting alloys and secondary aluminum alloys.

ALUMINUM T6063
Reynolds Aluminum
Aluminum. Si 0.5, Mg 0.8, Fe 0-0.35, bal Al.
T5: 27,000 TS; 21,000 YS; 12 El; 60 Brin. For architectural
applications, portable irrigation systems, molding and trim.
Heat treatable extrusion alloy. *Obsolete*

ALUMINUM TITANIUM BRONZE
American manufacture
Cu 89-90, Al 9-10, Fe 1, Ti.
For propellers; strong, tough.

ALUMINUM UNION 123
Aluminium Union Ltd.
Aluminum. Si 5, bal Al.
Cast: 16,000 TS; 4 El; 40 Brin. For light alloy castings, aircraft;
corrosion resistant.

ALUMINUM UNION 125
Aluminium Union Ltd.
Aluminum. Si 4.5-5.5, Cu 1-1.5, Mg 0.4-0.6, bal Al.
Heat treated: 30,000 TS; 22,000 YS. For cylinder heads, valve
bodies; pressure proof castings.

ALUMINUM UNION 126
Aluminium Union Ltd.
Aluminum. Cu 1.4, Mg 0.5, Mn 0.75, Ni 0.75, Si 5, bal Al.
Heat treated: 26,000-32,000 TS; 1-3 El. For air cooled cylinder
heads; heat treatable.

ALUMINUM UNION 135
Aluminium Union Ltd.
Aluminum. Mg 0.3, Si 7, bal Al.
Heat treated: 26,000-32,000 TS; 6-12 El. For light alloy
castings; corrosion resistant.

ALUMINUM UNION 160
Aluminium Union Ltd.
Aluminum. Si 12, bal Al.
For light alloy castings; high fluidity.

ALUMINUM UNION 162
Power Products Inc.
Aluminum. Si 12-15, Ni 1.5-2.5, Mg 0.75-1.25, Cu 0.5-1, bal
Al.
For pistons; low thermal expansion.

ALUMINUM UNION 17S
Aluminium Union Ltd.
Aluminum. Cu 3.5-4.5, Mn 0.4-0.7, Mg 0.4-0.7, bal Al.
For light alloy parts, aircraft; heat treatable.

ALUMINUM UNION 218
Power Products Inc.
Aluminum. Cu 4, Mg 1.5, Ni 2, bal Al.
For pistons, cylinder heads; "Y" alloy.

ALUMINUM UNION 225
Power Products Inc.
Aluminum. Cu 4, bal Al.
Heat treated: 36,000 TS; 4 El. For aircraft light alloy castings;
age hardenable.

ALUMINUM UNION 22S
Aluminium Union Ltd.
Aluminum. Cu 3.5-4.7, Si 0.7-1.5, Mg 0.4-1, Mn 0.4-1.5, bal
Al.
Heat treated: 56,000 TS; 44,000 YS; 8 El. For light alloy parts,
aircraft; heat treatable.

ALUMINUM UNION 24S
Aluminium Union Ltd.
Aluminum. Cu 3.5-4.8, Mg 0.8-1.8, Mn 0.3-1.5, bal Al.
Heat treated: 56,000 TS; 35,000 YS; 15 El. For light alloy
parts, aircraft; heat treatable.

ALUMINUM UNION 27S
Aluminium Union Ltd.
Aluminum. Cu 4.5, Mn 0.8, Si 0.8, Sn 0.05, bal Al.
Heat treated: 58,000 TS; 30,000 YS. For forged aircraft parts;
heat treatable.

ALUMINUM UNION 350
Power Products Inc.
Aluminum. Mg 8.5-11.5, bal Al.
Cast: 36,000 TS; 22,000 YS; 14 El. Heat treated: 44,000 TS;
28,000 YS; 12 El. For light alloy castings; highest strength,
ductility and impact.

ALUMINUM UNION 35S
Aluminium Union Ltd.
Aluminum. Si 10-13, bal Al.
For light alloy paneling; non-heat treatable.

ALUMINUM UNION 51S
Aluminium Union Ltd.
Aluminum. Si 0.5-1.5, Mg 0.5-1.5, bal Al.
Heat treated: 40,000 TS; 31,000 YS; 8 El. For light alloy
architectural structure; heat treatable.

ALUMINUM UNION 57S
Aluminium Union Ltd.
Aluminum. Mg 1.5-3, Cr 0.1-0.35, bal Al.
Heat treated: 32,000 TS; 28,000 YS; 5 El. For light alloy
paneling, aircraft; heat treatable.

ALUMINUM UNION A51S
Aluminium Union Ltd.
Aluminum. Mg 0.6, Si 1, Cr 0.25, bal Al.
Heat treated: 40,000 TS; 15 El. For light alloy forgings; non-
hardenable.

ALUMINUM UNION NO. 14S
Aluminium Union Ltd.
Aluminum. Cu 4.4, Mg 0.35, Mn 0.75, Si 0.8, bal Al.
Heat treated: 62,000 TS; 18 El; 130 Brin. For forged aircraft,
light alloy parts; heat treatable.

ALUMINUM UNION NO. 3S
Aluminium Union Ltd.
Aluminum. Mn 0-1.5, bal Al.
Hard: 25,000 TS; 22,000 YS. Soft: 14,000 TS; 5,000 YS. For
light alloy body paneling and molding; corrosion resistant.

ALUMINUM V-95
Russian manufacture
Mg, Zn, Cu, bal Al.
For aircraft parts; age-hardenable.

ALUMINUM WELDING WIRE & ELECTRODES 1100
J.W. Harris Co., Inc.
Si 0-0.95, Fe 0-0.95, Cu 0.05-0.2, Mn 0.05, Zn 1, Be 0.0008,
99.0 Mg min, 0.05 others max.
AWS A5.10, Class ER 1100.

ALUMINUM WIRE & ELECTRODES 1188(D)
J.W. Harris Co., Inc.. Si 0.06, Fe 0.06, Cu 0.005, Mn 0.01, Mg
0.01, Zn 0.03, Ti 0.01, Be 0.0008, 0.05 others max, bal Al.
AWS A5.10, Class ER 1188.

ALUMINUM WIRE & ELECTRODES 2319(B)
J.W. Harris Co., Inc.. Si 0.2, Fe 0.3, Cu 5.8-6.8, Mn 0.2-0.4,
Mg 0.02, Zn 0.1, Ti 0.1-0.2, Be 0.0008, 0.05 others max, bal
Al.
AWS A5.10, Class ER 2319.

ALUMINUM WIRE & ELECTRODES 4043
J.W. Harris Co., Inc.
Si 4.5-6, Fe 0.8, Cu 0.3, Mn 0.05, Mg 0.05, Zn 0.1, Ti 0.2, Be
0.0008, 0.05 others max, bal Al.
AWS A5.10, Class ER4043.

ALUMINUM WIRE & ELECTRODES 5183
J.W. Harris Co., Inc.
Si 0.4, Fe 0.4, Cu 0.1, Mn 0.5-1, Mg 4.3-5.2, Cr 0.05-0.25, Zn
0.25, Ti 0.15, Be 0.0008, 0.05 others max, bal Al.
AWS A5.10, Class ER 5183.

ALUMINUM WIRE & ELECTRODES 5356

J.W. Harris Co., Inc.
Si 0.25, Fe 0.4, Cu 0.1, Mn 0.05-0.2, Mg 4.5-5.5, Cr 0.05-0.2, Zn 0.1, Ti 0.06-0.2, Be 0.0008, 0.05 others max, bal Al.
AWS A5.10, Class ER 5356.

ALUMINUM WIRE & ELECTRODES 5554

J.W. Harris Co., Inc.
Si 0.25, Fe 0.4, Cu 0.1, Mn 0.5-1, Mg 2.4-3, Cr 0.05-0.2, Zn 0.25, Ti 0.05-0.2, Be 0.0008, 0.05 others max, bal Al.
AWS A5.10, Class ER 5554.

ALUMINUM WIRE & ELECTRODES 5556

J.W. Harris Co., Inc.
Si 0.25, Fe 0.4, Cu 0.1, Mn 0.5-1, Mg 4.7-5.5, Cr 0.05-0.2, Zn 0.25, Ti 0.05-0.2, Be 0.0008, 0.05 others max, bal Al.
AWS A5.10, Class ER 5556.

ALUMINUM WIRE & ELECTRODES 5654

J.W. Harris Co., Inc.
Si 0-0.45, Fe 0-0.45, Cu 0.05, Mn 0.1, Mg 3.1-3.9, Cr 0.15-0.35, Zn 0.2, Ti 0.05-0.15, Be 0.0008, 0.05 others max, bal Al.
AWS A5.10, Class ER 5654.

ALUMINUM WIRE & ELECTRODES A356

J.W. Harris Co., Inc.
Si 6.5-7.5, Fe 0.2, Cu 0.2, Mn 0.1, Mg 0.2-0.4, Zn 0.1, Ti 0.2, Be 0.0008, bal Al.
AWS A5.10, Class R-A356.

ALUMINUM WIRE & ELECTRODES C 355

J.W. Harris Co., Inc.
Si 4.5-5.5, Fe 0.2, Cu 1-1.5, Mn 0.1, Mg 0.4-0.6, Zn 0.1, Ti 0.2, Be 0.0008, 0.05 others max, bal Al.
AWS A5.10, Class R-C 355.

ALUMINUM X2219

Aluminum Company of America
Cu 6, Mn 0.3, V 0.1, Zr 0.15, bal Al.
Heat treated: 38,000-62,000 TS; 43,000 YS; 16 El. For aircraft and auto engine components; age-hardenable, high temperature use. *Obsolete*

ALUMINUM X250

Aluminum Company of America
Mg 8, Cu 0.15, Mn 0.25, Zn 1.5, bal Al.
T4-temper: 50,000 TS; 28,000 YS; 16 El; 85 Brin. Aged: 58,100 TS; 33,500 YS; 18 El. For castings, aircraft engine cases and parts; age-hardenable, stress corrosion resistant. *Obsolete*

ALUMINUM X357

Kaiser Aluminum & Chemical Corp.
Si 6.5-7.5, Mg 0.45-0.6, Ti 0.1-0.2, bal Al.
Permanent mold: 50,000 TS; 43,000 YS; 5 El; 100 Brin. Sand cast: 52,000 TS; 43,000 YS; 2 El; 90 Brin. For aircraft and missile structures, impellers; age-hardenable. *Obsolete*

ALUMINUM XA140

Aluminum Company of America
Cu 8, Mg 6, Mn 0.5, Ni 0.5, B, Be, Ti, bal Al.
For components for aircraft and gas turbine engines; high strength at 400-600 F. *Obsolete*

ALUMINUM XB-805

English manufacture
Sn 6.5, Ni 1, Cu 1, bal Al.
For bearings; requires steel backing.

ALUMINUM XC65S

Aluminum Company of America
Al alloy.
For ship structures, trailer truck frames; high rigidity and strength. *Obsolete*

ALUMINUM XZM-100

Martin-Marietta Corp.
Al alloy.
Extruded: 92,000 TS; 11 El. For light alloy parts.

ALUMINUM Y TI

Industria Nazionale Alluminio, Mori
Cu 3.8-4.2, Mg 1.3-1.7, Ti 0.2, Ni 1.8-2.3, bal Fe.
For pistons, cylinder heads; age hardenable, heat resistant.

ALUMINUM Y TI

Soc. Alluminio Veneto per Azioni
Cu 3.8-4.2, Mg 1.3-1.7, Ti 0.2, Ni 1.8-2.3, bal Fe.
For pistons, cylinder heads; age hardenable, heat resistant.

ALUMINUM Y TI

E.N. Egge Co.
Cu 3.8-4.2, Mg 1.3-1.7, Ti 0.2, Ni 1.8-2.3, bal Fe.
For pistons, cylinder heads; age hardenable, heat resistant.

ALUMINUM-100

Vereinigte Deutsche Metallwerke AG
Mg, Zn, bal Al.
For light alloy parts; corrosion resistant. *Obsolete*

ALUMINUM-1145

Aluminum Company of America
99.45% min Al, 0.55% Si + Fe.
O-temper: 11,000 TS; 5,000 YS; 40 El; 20 Brin. H18 temper: 24,000 TS; 21,000 YS; 4 El; 44 Brin. For light sheet metal work, foil. Good corrosion resistance and weldability. *Obsolete*

ALUMINUM-1145

Reynolds Metals Co.
99.45% min Al, 0.55% Si + Fe.
O-temper: 11,000 TS; 5,000 YS; 40 El; 20 Brin. H18 temper: 24,000 TS; 21,000 YS; 4 El; 44 Brin. For light sheet metal work, foil. Good corrosion resistance and weldability. *Obsolete*

ALUMINUM-A108

Aluminum Company of America
Cu 4.5, Si 5.5, bal Al.
Cast: 28,000 TS; 16,000 YS; 2 El; 70 Brin. For ornamental grilles, reflectors, general purpose castings. Fair corrosion resistance. *Obsolete*

ALUMINUM-A108

Reynolds Metals Co.
Cu 4.5, Si 5.5, bal Al.
Cast: 28,000 TS; 16,000 YS; 2 El; 70 Brin. For ornamental grilles, reflectors, general purpose castings. Fair corrosion resistance. *Obsolete*

ALUMINUM-A13

Reynolds Metals Co.
Cu 1, Si 12, bal Al.

ALUMINUM-A13

Aluminum Company of America
Cu 1, Si 12, bal Al.

ALUMINUM-A132

Aluminum Company of America
Cu 0.8, Si 12, Mg 1.2, Ni 2.5, bal Al.
T551: 36,000 TS; 28,000 YS; 0.5 El; 105 Brin. T65: 47,000 TS; 43,000 YS; 0.5 El; 125 Brin. For permanent mold castings, automotive and diesel engine pistons, pulleys, sheaves. Age-hardenable. Good high temperature properties. *Obsolete*

ALUMINUM-A132

Kaiser Aluminum & Chemical Corp.
Cu 0.8, Si 12, Mg 1.2, Ni 2.5, bal Al.
T551: 36,000 TS; 28,000 YS; 0.5 El; 105 Brin. T65: 47,000 TS; 43,000 YS; 0.5 El; 125 Brin. For permanent mold castings, automotive and diesel engine pistons, pulleys, sheaves. Age-hardenable. Good high temperature properties. *Obsolete*

ALUMINUM-A140

Now AA A2400.

ALUMINUM-A214

Reynolds Metals Co.
Mg 4, Zn 1.8, bal Al.
Cast: 27,000 TS; 16,000 YS; 7 El; 60 Brin. For cooking utensils, architectural fittings. Fair resistance to hot cracking. Good corrosion resistance. *Obsolete*

ALUMINUM-A344

Now AA 440.0.

ALUMINUM-A356

Now AA A356.0.

ALUMINUM-A360

Aluminum Company of America
Si 9-10, Mg 0.45-0.6, Fe 0-0.6, bal Al.
Die cast: 41,000 TS; 23,000 YS; 5 El; 75 Brin. For general die castings, instrument housings and cases, outboard motor and marine castings. Corrosion resistant. Die cast. *Obsolete*

ALUMINUM-A360

Kaiser Aluminum & Chemical Corp.
Si 9-10, Mg 0.45-0.6, Fe 0-0.6, bal Al.
Die cast: 41,000 TS; 23,000 YS; 5 El; 75 Brin. For general die castings, instrument housings and cases, outboard motor and marine castings. Corrosion resistant. Die cast. *Obsolete*

ALUMINUM-A380

Aluminum Company of America
Si 7.5-9.5, Cu 3-4, Fe 0-0.6.
Die cast: 47,000 TS; 23,000 YS; 4 El; 80 Brin. For general purpose die castings, instrument housings and cases. Corrosion resistant. *Obsolete*

ALUMINUM-A380

Kaiser Aluminum & Chemical Corp.
Si 7.5-9.5, Cu 3-4, Fe 0-0.6.
Die cast: 47,000 TS; 23,000 YS; 4 El; 80 Brin. For general purpose die castings, instrument housings and cases. Corrosion resistant. *Obsolete*

ALUMINUM-A390

Aluminum Company of America
Si 16-18, Cu 4-5, Mg 0.45-0.65, Fe 0-0.5, Ti 0-0.2, bal Al.
Permanent mold T6 temper: 45,000 TS; 45,000 YS; 1 max El; 145 Brin. As cast: 29,000 TS; 29,000 YS; 1 max El; 110 Brin. For high strength castings, engine blocks, cylinders, pistons. Heat treatable, good resistance to heat cracking, excellent fluidity. *Obsolete*

ALUMINUM-A390

Reynolds Metals Co.
Si 16-18, Cu 4-5, Mg 0.45-0.65, Fe 0-0.5, Ti 0-0.2, bal Al.
Permanent mold T6 temper: 45,000 TS; 45,000 YS; 1 max El; 145 Brin. As cast: 29,000 TS; 29,000 YS; 1 max El; 110 Brin. For high strength castings, engine blocks, cylinders, pistons. Heat treatable, good resistance to heat cracking, excellent fluidity. *Obsolete*

ALUMINUM-A612

Now AA A7120.

ALUMINUM-A750

Aluminum Company of America
Si 2-3, Cu 0.7-1.3, Ni 0.3-0.7, Sn 5.5-7, bal Al.
Cast (T-5 temper): 20,000 TS; 11,000 YS; 5 El; 45 Brin. For bearings, large bearing for rolling mills. Resists galling and seizing. Less hot-short than aluminum 750. *Obsolete*

ALUMINUM-A750

Kaiser Aluminum & Chemical Corp.
Si 2-3, Cu 0.7-1.3, Ni 0.3-0.7, Sn 5.5-7, bal Al.
Cast (T-5 temper): 20,000 TS; 11,000 YS; 5 El; 45 Brin. For bearings, large bearing for rolling mills. Resists galling and seizing. Less hot-short than aluminum 750. *Obsolete*

ALUMINUM-AS132

Alluminio SA
Aluminum. Si 12, Cu 2, Mn 0.3, Fe 0.7, bal Al.
For crankcases, oil pump bodies, die castings for automobile components. Die cast alloy.

ALUMINUM-AS62

Alluminio SA

Aluminum. Si 6, Cu 2, Mg 0.3, bal Al.

Cast: 27,000 TS; 18,000 YS; 70 Brin. Heat treated: 36,000 TS 24,000 YS; 80 Brin. For automobile cylinder heads, crankcases, housings. Permanent molds, age hardening.

ALUMINUM-B133

VDM Nickel-Technologie AG

Mg, Zn, bal Al.

For light alloy parts; corrosion resistant. *Obsolete*

ALUMINUM-B750

Aluminum Company of America

Cu 1.7-2.3, Mg 0.6-0.9, Ni 0.9-1.5, Sn 5.5-7, Si 0-0.4, Fe 0-0.5, bal Al.

Cast (T-5 temper): 27,000-32,000 TS; 22,000-23,000 YS; 2-5 El; 65-70 Brin. For bearings and bushings carrying heavy loads. Resists galling and seizing. Higher compressive strength than aluminum 750. *Obsolete*

ALUMINUM-B750

Kaiser Aluminum & Chemical Corp.

Cu 1.7-2.3, Mg 0.6-0.9, Ni 0.9-1.5, Sn 5.5-7, Si 0-0.4, Fe 0-0.5, bal Al.

Cast (T-5 temper): 27,000-32,000 TS; 22,000-23,000 YS; 2-5 El; 65-70 Brin. For bearings and bushings carrying heavy loads. Resists galling and seizing. Higher compressive strength than aluminum 750. *Obsolete*

ALUMINUM-C355

Now AA C355.0.

ALUMINUM-C612

Aluminum Company of America

Cu 0.5, Zn 6.5, Fe 0-1.4, Mg 0.35, Si 0-0.35, Ti 0-0.25, bal Al. Cast and room aged: 35,000 TS; 18,000 YS; 8 El; 70 Brin. T5: 34,000 TS; 18,000 YS; 7 El. For general purpose castings. Brazed assemblies. Good brazing properties, high strength and ductility. *Obsolete*

ALUMINUM-C612

Reynolds Metals Co.

Cu 0.5, Zn 6.5, Fe 0-1.4, Mg 0.35, Si 0-0.35, Ti 0-0.25, bal Al. Cast and room aged: 35,000 TS; 18,000 YS; 8 El; 70 Brin. T5: 34,000 TS; 18,000 YS; 7 El. For general purpose castings. Brazed assemblies. Good brazing properties, high strength and ductility. *Obsolete*

ALUMINUM-CG86

NL Industries

Si 0.2, Cu 8.28, Fe 0.94, Mg 6.13, Mn 0.63, Ni 0.48, bal Al. As cast: 53,000 TS; 42,100 YS; 1.0 El. AT 400 F: 44,500 TS; 31,800 YS; 1.5 El. For die castings, pistons, housings. Good high temperature properties. *Obsolete*

ALUMINUM-CH91

Aluminum Company of America

Cu 6-8, Si 1-3, Zn 0-2.2, Fe 0-1.4, bal Al.

Die cast: 44,000 TS; 22,000 YS; 4 El; 28,000 shear strength. For commercial die castings, large pressure tight engine covers, grills. High strength. Similar to Aluminum 13.

ALUMINUM-D16T

Russian manufacture

Cu 4.4-4.52, Mg 1.6, Mn 0.7, Fe 0.38, Si 0.26, Ni 0.01, Zn 0.2, Ti 0.03, bal Al.

Heat treated: 70,000 TS; 56,000 YS; 130 Brin. For aircraft structures, fittings, hardware, fasteners; heat treatable, high strength.

ALUMINUM-D612

Aluminum Company of America

Si 0-0.15, Fe 0-0.4, Mg 0.5-0.65, Mn 0-0.1, Cr 0.4-0.6, Zn 5-6.5, Ti 0.15-0.25, Cu 0-0.2, bal Al.

Sand cast and aged: 35,000 TS; 25,000 YS; 9 El; 75 Brin. For general castings, transmission housings, manifolds, cylinder heads, ventilating fans. Self-aging. No heat treatment required. *Obsolete*

ALUMINUM-D612

Kaiser Aluminum & Chemical Corp.

Si 0-0.15, Fe 0-0.4, Mg 0.5-0.65, Mn 0-0.1, Cr 0.4-0.6, Zn 5-6.5, Ti 0.15-0.25, Cu 0-0.2, bal Al.

Sand cast and aged: 35,000 TS; 25,000 YS; 9 El; 75 Brin. For general castings, transmission housings, manifolds, cylinder heads, ventilating fans. Self-aging. No heat treatment required. *Obsolete*

ALUMINUM-D612

Reynolds Metals Co.

Si 0-0.15, Fe 0-0.4, Mg 0.5-0.65, Mn 0-0.1, Cr 0.4-0.6, Zn 5-6.5, Ti 0.15-0.25, Cu 0-0.2, bal Al.

Sand cast and aged: 35,000 TS; 25,000 YS; 9 El; 75 Brin. For general castings, transmission housings, manifolds, cylinder heads, ventilating fans. Self-aging. No heat treatment required. *Obsolete*

ALUMINUM-EC

Now AA 1350.

ALUMINUM-F132

Now AA F332.0.

ALUMINUM-F214

Aluminum Company of America

Mg 4, Si 0.5, bal Al.

Cast: 21,000 TS; 12,000 YS; 3 El; 50 Brin. For dairy and food handling equipment, cooking utensils, fittings for chemical and sewerage use. Corrosion resistant. Resists hot cracking. *Obsolete*

ALUMINUM-F214

Reynolds Metals Co.

Mg 4, Si 0.5, bal Al.

Cast: 21,000 TS; 12,000 YS; 3 El; 50 Brin. For dairy and food handling equipment, cooking utensils, fittings for chemical and sewerage use. Corrosion resistant. Resists hot cracking. *Obsolete*

ALUMINUM-G8A

NL Industries

Si 0.44, Cu 0.14, Fe 1.72, Mg 8.8, Mn 0.02, Ni 0.01, bal Al. As Cast: 47,900 TS; 28,500 YS; 4.2 El. For die castings, instrument cases, housings. Corrosion resistant.

ALUMINUM-G8A

Reynolds Metals Co.

Si 0.44, Cu 0.14, Fe 1.72, Mg 8.8, Mn 0.02, Ni 0.01, bal Al. As Cast: 47,900 TS; 28,500 YS; 4.2 El. For die castings, instrument cases, housings. Corrosion resistant.

ALUMINUM-G8A

Aluminum Company of America

Si 0.44, Cu 0.14, Fe 1.72, Mg 8.8, Mn 0.02, Ni 0.01, bal Al. As Cast: 47,900 TS; 28,500 YS; 4.2 El. For die castings, instrument cases, housings. Corrosion resistant.

ALUMINUM-HS30

English manufacture

Cu 0.1, Mg 0.4-1.5, Si 0.6-1.3, Fe 0.6, Zn 0.1, Mn 0.4-1, Cr 0.5, Ti 0.2, bal Al.

W.P. Temper: 46,000 TS; 42,800 YS; 11 El. At 320°F: 61,600 TS; 51,100 YS; 17.8 El. For aircraft and space vehicle structures and components; heat treatable, good weldability.

ALUMINUM-K

VDM Nickel-Technologie AG

Si 11.5-13.5, Cu 0.4-1.5, Mg 0.8-1.5, Ni 1-2, Mn 0-0.5, bal Al. At 75°F: 55,000 TS; 46,000 YS; 9 El; 120 Brin. At 700°F: 35,000 TS; 2000 YS; 90 El. For forged pistons; for elevated temperature use. *Obsolete*

ALUMINUM-L214

Aluminum Company of America

Cu 0.1, Si 0.75, Fe 0.75, Mn 0.5, Mg 3.3, Zn 0.1, Ni 0.1, bal Al.

Cast: 41,000 TS; 10 El. For hardware, support brackets. Fair resistance to hot cracking. Good corrosion resistance. *Obsolete*

ALUMINUM-L214

Reynolds Metals Co.

Cu 0.1, Si 0.75, Fe 0.75, Mn 0.5, Mg 3.3, Zn 0.1, Ni 0.1, bal Al.

Cast: 41,000 TS; 10 El. For hardware, support brackets. Fair resistance to hot cracking. Good corrosion resistance. *Obsolete*

ALUMINUM-MOLYBDENUM-TITANIUM-ZIRCONIUM

Reading Alloys, Inc.

Reactive/refractive. Al 30-32, C 0.1, Mo 21-24, Si 0.5, Ti 3-5, Zr 43-46.

Master alloy.

ALUMINUM-NICKEL NIAL-14TI

VDM Nickel-Technologie AG

Al 4-5, Ti 0.5, bal Ni.

For heating elements; heat and corrosion resistant. *Obsolete*

ALUMINUM-OY

English manufacture

Cu 4, bal Al.

For light alloy parts; age-hardened.

ALUMINUM-S12A

NL Industries

Si 11.76, Cu 0.53, Fe 0.81, Mg 0.01, C 0.19, Ni 0.04, bal Al. *Obsolete*

ALUMINUM-SC114A

NL Industries

Si 11.42, Cu 3.25, Fe 1.1, Mg 0.04, Mn 0.2, Ni 0.38, bal Al. As cast: 49,700 TS; 25,300 YS; 3.4 El. For die castings, instrument cases, housings, oil pans. Corrosion resistant, pressure tight. *Obsolete*

ALUMINUM-SC114A + MG

NL Industries

Si 12, Cu 3.94, Fe 0.97, Mg 0.51, Mn 0.3, Ni 0.01, bal Al. Die cast: 48,200 TS; 27,400 YS; 2.0 El. For die castings, instrument cases, housings, grills. Corrosion resistant. *Obsolete*

ALUMINUM-SC84A

NL Industries

Si 8.87, Cu 3.68, Fe 1.05, Mn 0.34, Ni 0.14, Mg 0.01, bal Al. As cast: 50,600 TS; 25,000 YS; 3.8 El. At 500 F: 19,000 TS; 14,600 YS; 11.2 El. For die castings, pistons, sole plates. Corrosion resistant, pressure tight. Good high temperature properties. *Obsolete*

ALUMINUM-SN122A

NL Industries

Si 12, Cu 0.79, Fe 1.12, Mg 1.23, Mn 0.07, Ni 2.5, bal Al. As cast: 46,600 TS; 34,700 YS; 0.9 El. At 300 F: 43,500 TS; 39,400 YS; 1.0 El. For die castings, auto and diesel engine pistons, pulleys, sheaves. Good high temperature strength. *Obsolete*

ALUMINUM-X149

Now AA 249.0.

ALUMINWELD

Lincoln Electric Co.

Si 4.5, Fe 0.8, Cu 0.3, Mg 0.5, Zn 0.1, Ti 0.2, bal Al. Aluminum alloy, arc welding electrodes. AWS Class Al-43.

ALUNEON

Manufacturer not listed.

Al alloy.

For light alloy parts.

ALUSAD

Metalltechnik Schmidt GmbH & Co.

Aluminum. Cu 6-6.5, Si 0.6-1.2, Fe 1.1-1.5, Zn 1-1.5, Mn 0.3-0.6, Pb 0-0.1, Mg 0-0.2, bal Al.

Aluminum shot.

ALUSIL

German manufacture
Cu 1-2, Si 20-21, Ni 0-0.7, Fe 0-0.7, bal Al.
For engine pistons, cylinder liners, pumps; low density and
low expansivity.

ALUSIN N. 280

German manufacture
Si 22, Cu 1, Ni 1, Co 1, Mg 1, Cr 0.5, bal Al.
Cast: 25,000-35,000 TS; 100-140 Brin. For pistons in engines,
pumps, cylinder liners; low density and low expansivity.

ALUVAC

Fonderie de Precision S.A.
Cu 4, bal Al.
For light alloy parts; heat treatable.

ALUWANGAN

Vesevorder Metallwerke GmbH
Mn 0.5-1.5, Cr 0-0.3, bal Al.
Soft: 16,000 TS; 6000 YS; 40 El. Hard: 29,000 TS; 27,000 YS;
10 El. For cooking utensils, heat exchangers, tanks, furniture;
good welding and forming properties.

ALUWE-1

Westfalische Kupfer und Messingwerke AG
Al alloy.
For light alloy parts.

ALUWE-22

Westfalische Kupfer und Messingwerke AG
Al alloy.
For light alloy parts.

ALUWE-4

Westfalische Kupfer und Messingwerke AG
Al alloy.
For light alloy parts.

ALUWE-53

Westfalische Kupfer und Messingwerke AG
Al alloy.
For light alloy parts.

ALUWE-55

Westfalische Kupfer und Messingwerke AG
Al alloy.
For light alloy parts.

ALUWE-57

Westfalische Kupfer und Messingwerke AG
Al alloy.
For light alloy parts.

ALUWE-6

Westfalische Kupfer und Messingwerke AG
Al alloy.
For light alloy parts.

ALUWE-8

Westfalische Kupfer und Messingwerke AG
Al alloy.
For light alloy parts.

ALVA 36

German manufacture
Pb 3, Sb 3, Cu 2, Mn, bal Al.
Wrought: 30-80 Brin. For bearings.

ALVA ALLOY HOLLOW DRILL STEEL

Colt Industries
C 0.8, Mn 0.3, Si 0.15, V 0.2, bal Fe.
For hollow drill rod for detachable bits; fatigue resistant.
Obsolete

ALVA DUPLEX FORGING

Colt Industries
C 0.4, Cr 1.2, V 0.2.
Heat treated: 130,000 TS; 112,000 YS; 20 El; 52 RA; 290 Brin.
For axles, pins, shafts. *Obsolete*

ALVA EXTRA

Crucible Materials Corp.
Tool material. C 1, V 0.25, bal Fe.
For blanking, threading, forming dies, taps; 1150°C max;
water hardening. W2.

ALVA SPECIAL

Colt Industries
C 1, V 0.25, bal Fe.
For blanking, threading, forming dies, taps; water hardening.
Obsolete

ALX CAST ALLOY

Allegheny Ludlum Steel
C 2, W 17, Cr 33, Co 40, B, bal Fe.
For cutting tool bits; fast cutting, centrifugally cast.

ALZEN

French manufacture
Al 66.6, Zn 33.4.
For strong light alloy parts; not hardenable.

ALZEN 305

French manufacture
Zn, Al, bal Cu.
For bearings; bronze.

ALZIN

Manufacturer not listed.
Zn 20, bal Al.
Cast: 18,000-28,000 TS; 2-3 El; 62-65 Brin. For light alloy
parts; non-hardenable.

ALZINC

French manufacture
Zn 20, Al 80.
21,000 TS; 2 El; 80 Brin. For light castings; Sibley casting
alloy.

ALZN4.5MG0.8 (711) 7080

J. & A. Erbsloh Aluminium
Aluminum. Si 0.1, Fe 0.3, Cu 0.05, Mn 0.05, Mg 0.7-0.9, Cr
0.15-0.2, Zn 4.5-4.9, Ti 0.2, 0.15 others, bal Al.
Hardenable alloy. Warm hardened: 320 N/mm^2 TS; 260
N/mm^2 YS; 95 Brin.

ALZN4.5MG1 (721) 7120

J. & A. Erbsloh Aluminium
Aluminum. Si 0.3, Fe 0.3, Cu 0.08, Mn 0.2-0.3, Mg 1.1-1.3, Cr
0.15-0.2, Zn 4.5-5, Ti 0.2, 0.15 others, bal Al.
Hardenable alloy. Warm hardened: 350 N/mm^2 TS; 290
N/mm^2 YS; 105 Brin. Werkstoff Nr. 3.4335.71.

AM

Walsingham Steel Co., Ltd.
C 1-1.2, Si 0-1, S 0-0.006, P 0-0.07, Mn 11-14, bal Fe.
High resistance to impact wear. Meets BS 1457; ASTM
A128-64.

AM 05

Otto Fuchs Metallwerke
Copper. Mg 0.5, bal Cu.
Extruded, wire, bar, tube for ornamental moldings.

AM 1

Acciaierie Valbruna s.p.a.
C 0.75-0.9, Cr 3.5-4.5, Mo 8-9.5, W 1-2, V 0.8-1.3, bal Fe.
Molybdenum high speed tool steel. W. Nr. 1.3346; AISI M1.
Obsolete

AM 10

German manufacture
Ce 10, bal Mg.
For pistons, supercharger impeller; ASTM-E10.

AM 10

Otto Fuchs Metallwerke
Aluminum. Mg 1, bal Al.
100-155 N/mm^2 TS; 40-135 N/mm^2 YS, 4-22 El. Good
formability, weldability, anodizing. Extrusions, wire, bar, tube
Type AlMg1. Similar to AA 5005.

AM 100A

Various foundries
Al 10, Mn 0.1-0, bal Mg.
F-Temper: 20,000-22,000 TS; 10,000-12,000 YS; 0-2 El. T4-
Temper: 34,000-40,000 TS; 10,000-13,000 YS; 6-10 El. T6-
Temper: 34,000-40,000 TS; 17,000-22,000 YS; 0-1 El.
Magnesium permanent mold casting with good pressure
tightness and strength and good weldability; for housings for
motors and power tools. ASTM B199-68; AMS 4483; QQ-
M-55; SAE 502.

AM 100A (MG INGOT)

Dow Chemical Co.
Magnesium. Al 9.4-10.6, Mn 0.13-0.5, Zn 0-0.2, Si 0-0.1, Cu
0-0.04, Ni 0-0.005, 0.30 others max, bal Mg.
ASTM B93.

AM 11

Otto Fuchs Metallwerke
Aluminum. Mg 1, bal Al.
Extrusions, wire, bar, tube for ornamental purposes.

AM 112

Aluminum Company of America
Cu 12, bal Mg.
For castings, pistons. *Obsolete*

AM 18

Otto Fuchs Metallwerke
Aluminum. Mg 1.9, bal Al.
145-205 N/mm^2 TS; 60-155 N/mm^2 YS; 4-17 El. Good
formability, weldability, corrosion resistance. Type AlMg2.
Similar to AA 5051.

AM 18 11 S

Bergische Stahl Industrie
C 0-0.07, S 0-1, Mn 0-2, Cr 16-18, Ni 10-12, N 0-0.1, bal Fe.
Nonmagnetic cast steel. SEW 1.3944; G-x 5 CrNi 18 11.

AM 18 15 MO

Bergische Stahl Industrie
C 0-0.07, Cr 18, Ni 15, Mo 0.3, bal Fe.
Nonmagnetic cast steel. SEW 1.3950; G-X 5 CrNi Mo 18 15.

AM 18.10

Bergische Stahl Industrie
C 0.12, Cr 17.5, Ni 10.5, bal Fe.
Solution annealed and aged: 64,000 TS min; 28,000 YS min;
20 El min. Non-magnetic steel casting. DIN G-X12 CrNi 1811;
W.-Nr. 1.3955.

AM 21

Otto Fuchs Metallwerke
Aluminum. Mg 2, Mn 0.8, bal Al.
175-255 N/mm^2 TS; 80-175 N/mm^2 YS; 4-17 El. Good
formability, weldability, corrosion resistance. Type AlMgMn.

AM 242

Aluminum Company of America
Al 8, Mn 0.2-0.4, bal Mg.
Sand cast: 23,000-25,000 TS; 12,000 YS; 2-5 El; 2.0 RA;
44-53 Brin. Heat treated: 29,000 TS; 10,000 YS; 5 El; 6 RA; 46
Brin. For sand and permanent mold castings; fair resistance
to corrosion. *Obsolete*

AM 244

Aluminum Company of America
Al 3.5-5, Zn 0-0.3, Si 0-0.3, 0.2% min Mn, bal Mg.
Cast: 24,000 TS; 9,000 YS; 6 El; 44 Brin. For fittings; sand
cast. *Obsolete*

AM 30

Otto Fuchs Metallwerke
Aluminum. Mg 2.9, Mn 0.3, bal Al.
175-255 N/mm^2 TS; 80-175 N/mm^2 YS; 4-17 El. Good
corrosion resistance, weldability and formability. Type AlMg3.

AM 350

Now CARPENTER PYROMET 350 and UNITEMP 350.

AM 350
Colt Industries
C 0.1, Mn 1, Si 0.4, Cr 16.5, Ni 4.25, Mo 2.75, N 0.1, bal Fe.
Precipitation hardening stainless steel with 10% Delta ferrite.
AMS 5546; AISI 633. S35000

AM 350
Allegheny Ludlum Steel
C 0.1, Mn 1, Si 0.4, Cr 16.5, Ni 4.25, Mo 2.75, N 0.1, bal Fe.
Precipitation hardening stainless steel with 10% Delta ferrite.
AMS 5546; AISI 633. S35000

AM 350
Cannon-Muskegon Corp.
C 0.1, Mn 1, Si 0.4, Cr 16.5, Ni 4.25, Mo 2.75, N 0.1, bal Fe.
Precipitation hardening stainless steel with 10% Delta ferrite.
AMS 5546; AISI 633. S35000

AM 350
Bishop Tube Co.
C 0.1, Mn 1, Si 0.4, Cr 16.5, Ni 4.25, Mo 2.75, N 0.1, bal Fe.
Precipitation hardening stainless steel with 10% Delta ferrite.
AMS 5546; AISI 633. S35000

AM 355
Cannon-Muskegon Corp.
C 0.15, Mn 1, Si 0.4, Cr 15.5, Ni 4.25, Mo 2.75, N 0.1, bal Fe.
Precipitation hardenable to 216,000 psi TS. Good strength,
wear and corrosion resistance. AMS 5359; AISI 634. S35500

AM 355
Allegheny Ludlum Steel
C 0.15, Mn 1, Si 0.4, Cr 15.5, Ni 4.25, Mo 2.75, N 0.1, bal Fe.
Precipitation hardenable to 216,000 psi TS. Good strength,
wear and corrosion resistance. AMS 5359; AISI 634. S35500

AM 355
Colt Industries
C 0.15, Mn 1, Si 0.4, Cr 15.5, Ni 4.25, Mo 2.75, N 0.1, bal Fe.
Precipitation hardenable to 216,000 psi TS. Good strength,
wear and corrosion resistance. AMS 5359; AISI 634. S35500

AM 36
Otto Fuchs Metallwerke
Aluminum. Mg 2.9, bal Al.
175-255 N/mm^2 TS; 80-175 N/mm^2 YS; 4-17 El. Modification
of AM 30 for decorative purposes.

AM 363
Allegheny Ludlum Steel
Stainless steel. C 0.04, Mn 0-0.5, Si 0-0.3, Cr 11.5, Ni 4.5, Ti
0.5, bal Fe.
Hot rolled: 138,000 TS; 131,000 YS; 11 El. Annealed: 124,000
TS; 105,000 YS; 8 El. Aged: 124,000 TS; 118,000 YS; 11.5 El.
For automobile mufflers, jet engine and gas turbine
components. Precipitation-hardenable. Ferromagnetic,
stainless, maraging.

AM 40
Otto Fuchs Metallwerke
Aluminum. Mg 4.5, Mn 0.8, Cr, bal Al.
255-275 N/mm^2 TS; 110-155 N/mm^2 YS; 12 El. Good
corrosion resistance, weldability and formability. Type
AlMg4.5Mn. Similar to AA 5083.

AM 503
Birmetals Ltd.
Magnesium. Mn 1.5, bal Mg.
Plate, sheet, strip, extrusions: 24,000-30,000 TS;
8,000-14,000 YS (est); 4-5 El; 35-55 HV. Low strength alloy
with very good welding characteristics; for fuel and oil tanks.

AM 537
German manufacture
Al, Zn, bal Mg.
For light alloy parts; heat treatable.

AM 54
Otto Fuchs Metallwerke
Aluminum. Mg 4.5, Mn 0.2, bal Al.
235-325 N/mm^2 TS; 110-235 N/mm^2 YS; 4-18 El. Good
corrosion resistance and weldability. Type AlMg5. Similar to
AA 5182.

AM 58
Otto Fuchs Metallwerke
Aluminum. Mg 5.2, Mn 0.5, bal Al.
235-325 N/mm^2 TS; 110-235 N/mm^2 YS; 4-18 El. Corrosion
resistant and weldability. For shipbuilding applications. Type
AlMg5; similar to AA 5056.

AM 6
German manufacture
Ce 6, Mn 2, bal Mg.
For light alloy parts for high temperature use; ASTM-EM62;
cast and wrought.

AM 60A (MG INGOT)
Dow Chemical Co.
Magnesium. Al 5.7-6.3, Mn 0.15, Zn 0-0.2, Si 0-0.2, Cu
0-0.25, Ni 0-0.01, Be 0.0004-0.001, 0.30 others max, bal Mg.
For use in die castings. ASTM B93.

AM 60B (MG INGOT)
Dow Chemical Co.
Magnesium. Al 5.7-6.3, Zn 0-0.2, Si 0-0.05, Cu 0-0.008, Ni
0-0.001, Be 0.0004-0.001, Fe 0-0.004, 0.27 Mn min, 0.01
others max, bal Mg.
For use in die castings. ASTM B93.

AM 61 S
Aluminum Company of America
Mn 0.9-1.5, Sn 6, bal Mg.
Extruded: 42,000 TS; 26,000 YS; 12-15 El; 54 Brin. Forged:
38,000 TS; 24,000 YS; 5-10 El; 45 Brin. For forgings and
extruded shapes; excellent corrosion resistance. *Obsolete*

AM 764
Aluminum Company of America
Mn 0.7-1, Sn 5, Zn 4, bal Mg.
Sand cast: 25,000 TS; 10,000 YS; 5-6 El; 45 Brin. Heat
treated: 36,000 TS; 24,000 YS; 3-5 El; 60 Brin. For sand
castings; aged alloy. *Obsolete*

AM 858
Bergische Stahl Industrie
C 0.27, Mo 8.5, Cr 7.5, Ni 5.5, bal Fe.
Solution annealed: 71,000 TS min; 31,000 YS min; 30 El min.
Non-magnetic steel casting. DIN G-X25 MnCrNi 886; W.-Nr.
1.3966.

AM 88S
Aluminum Company of America
Al 9.5-10.5, Zn 0-2, 0.1% min Mn, bal Mg.
For welding rod. *Obsolete*

AM C52S
Aluminum Company of America
Al 3, Zn 1, bal Mg.
Extruded: 35,000 TS; 60 Brin. For light alloy parts; sheets and
extrusions. *Obsolete*

AM CR
Walsingham Steel Co., Ltd.
C 1-1.2, Si 0-1, S 0-0.06, P 0-0.07, Cr 1.5-2, 11.0 Mn min, bal
Fe.
High resistance to impact wear. Meets ASTM A128-64 Gr.C.

AM-350
Allegheny Ludlum Steel
Stainless steel. C 0.08, Mn 0.8, Si 0.25, Cr 16.5, Ni 4.3, Mo
2.75, N 0.1, bal Fe.
Sheet, hard: 145,000 psi TS; 60,000 psi YS; 40 El; 20 Rock C.
For aircraft application.

AM-355 CAST
Allegheny Ludlum Steel
Stainless steel. C 0.1, Mn 0.8, Si 0.6, Cr 15, Ni 4.2, Mo 2.3, N
0.09, bal Fe.
Cast, RT: 215,000 psi TS; 165,000 psi YS; 15 El; 37.5 RA.
Corrosion resistant.

AM-355 WROUGHT
Allegheny Ludlum Steel
Stainless steel. C 0.13, Mn 0.95, Si 0.25, Cr 15.5, Ni 4.3, Mo
2.75, N 0.1, bal Fe.
Bar, RT: 216,000 psi TS; 182,000 psi YS; 19 El; 38.5 RA.
Corrosion resistant.

AM-362
Now ALMAR 362 and ALLEGHENY AM 362.

AM241
Aluminum Company of America
Al 7.5-8.5, Mn 0.2-0.4, Zn 0-0.3, bal Mg.
Sand cast: 23,000-25,000 TS; 11,000-12,000 YS; 3-5 El; 2 RA;
44-48 Brin. Heat treated: 29,000-36,000 TS; 10,000-12,000
YS; 6-10 El; 6-10 RA; 46-56 Brin. For sand and die castings;
heat treatable. *Obsolete*

AM246
Aluminum Company of America
Al 11.5-12.5, Mn 0.1-0.2, Zn 0-0.3, bal Mg.
Sand cast: 19,000 TS; 14,000 YS; 5 El; 65 Brin. T-51 temper:
19,000 TS; 16,000 YS; 70 Brin. For pistons, sand and mold
castings. *Obsolete*

AM53S
Aluminum Company of America
Al 4, Mn 0.2-0.4, bal Mg.
Annealed: 33,000 TS; 18,000 YS; 16 El; 26 RA; 56 Brin. Hard
rolled: 45,000 TS; 35,000 YS; 8 El; 8 RA; 62 Brin. For
extruded rods and shapes, rolled sheet; formerly "A.M.4.4".
Obsolete

AM555
Aluminum Company of America
Mg 5, bal Al.
41,000 TS; 18,000 YS; 33 El; 71 Brin. For rivers for fabrication
of Mg alloy parts. *Obsolete*

AM58S
Aluminum Company of America
Al 8.5, Mn 0.2-0.4, Zn 0.5, bal Mg.
Hot press forgings: 43,000-48,000 TS; 24,000-38,000 YS; 4-8
El; 70-76 Brin. For propeller blades, hot pressed forgings;
age-hardenable. *Obsolete*

AM67S
Aluminum Company of America
Mn 1, Sn 5, Zn 4, bal Mg.
Hot press forgings: 38,000-40,000 TS; 18,000-21,000 YS;
16-20 El; 55-61 Brin. Heat treated: 35,000-39,000 TS;
22,000-25,000 YS; 1.5-4.0 El; 62-65 Brin. For press forgings.
Obsolete

AM740
Aluminum Company of America
Mn 0.75-1.25, Sn 9.5-10.5, bal Mg.
Cast: 18,000-20,000 TS; 8,000-9,000 YS; 5 El; 42-45 Brin. For
sand castings; "A.M. 740". *Obsolete*

AMALLOY
American Alloys Corp.
Mg 7, Mn 0.1, Fe 0.15, Ti 0.14, Cu 0.01, bal Al.
As cast: 34,000-38,000 TS; 19,000 YS; 7-10 El; 70 Brin. For
dairy, agricultural, cookware, hardware, aircraft,
transportation and oil field equipment. Corrosion and impact
resistant. Sand, permanent mold and die castings.

AMALOY
Armstrong Bros. Tool Co.
C 0.6-0.7, Mn 0.7-0.9, Mo, bal Fe.
For tools, dies. Water hardened.

AMALOY-1
AMF Inc.
Pb 97.5, Sn 2.5.
For protective coating against corrosion; applied to ferrous and copper alloys.

AMALOY-2
AMF Inc.
Pb 90-97.5, Sn 2.5-10.
For protective coating against corrosion; applied to ferrous and copper alloys.

AMALOY-3
American manufacture
Ni, Cr, W.
For corrosion and heat resistant parts; corrosion and heat resistant.

AMANIMPHY
Creusot-Loire
Nickel alloy.
For motors, electrical equipment; non-magnetic. *Obsolete*

AMANOX 3805
Krupp Stahl AG
C 0.3-0.4, Si 0-1, Mn 19-20, P 0-0.06, S 0-0.03, Ni 0-1, N 0-0.1, bal Fe.
SEW 390/90; W. Nr. 1.3805.

AMANOX 3813
Krupp Stahl AG
C 0.3-0.5, Si 0-0.8, Mn 17-19, P 0-0.1, S 0-0.03, Cr 3-5, N 0.08-0.12, bal Fe.
SEL 90; W. Nr. 1.3813.

AMANOX 3816
Krupp Stahl AG
C 0-0.12, Si 0-1, Mn 17.5-20, P 0-0.06, S 0-0.03, Cr 17.5-20, Ni 0-1, N 0.4-0.7, bal Fe.
SEW 390/90; W. Nr. 1.3816.

AMANOX 3817 (1)
Krupp Stahl AG
C 0.3-0.5, Si 0-1, Mn 17-19, P 0-0.06, S 0-0.03, Cr 3-5, Ni 0-1, N 0.05-0.12, bal Fe.
SEW 390/90; W. Nr. 1.3817.

AMANOX 3817 (2)
Krupp Stahl AG
C 0.5-0.6, Si 0-1, Mn 17.5-20, P 0-0.06, S 0-0.03, Cr 4-5, Ni 0-1, N 0.05-0.12, bal Fe.
SEW 390/90; W. Nr. 1.3818.

AMANOX 3914
Krupp Stahl AG
C 0-0.03, Si 0-0.75, Mn 6-8, Cr 20-22, Mo 3-3.5, Ni 14-16, Nb 0.1-0.25, N 0.35-0.5, bal Fe.
DIN X2CrNiMnMoNNb211573; W. Nr. 1.3914. 480 N/mm^2 YS, 800-1050 N/mm^2 TS, 35 El.

AMANOX 3941
Krupp Stahl AG
C 0-0.05, Si 0-1, Mn 0-2, P 0-0.045, S 0-0.03, Cr 16.5-18.5, Ni 12-14, bal Fe.
SEL 90; W. Nr. 1.3941.

AMANOX 3945
Krupp Stahl AG
C 0-0.03, Si 0-1, Mn 0-2, Cr 17-19, Mo 0-0.75, Ni 10-12, N 0.1-0.2, bal Fe.
DIN X2CrNiN1811; W. Nr. 1.3945. 270 N/mm^2 YS, 550-750 N/mm^2 TS, 40 El.

AMANOX 3948
Krupp Stahl AG
C 0-0.05, Si 0-1, Mn 7-10, P 0-0.03, S 0-0.02, Cr 17.5-20, Mo 2.5-3.5, Ni 12-15, N 0.2-0.4, Nb 0.1-0.25, bal Fe.
SEL 90; W. Nr. 1.3948.

AMANOX 3949
Krupp Stahl AG
C 0-0.08, Si 0-1, Mn 17-19, P 0-0.06, S 0-0.03, Cr 12-14, Mo 0.2-0.8, Ni 1.5-3, N 0.1-0.35, Nb 0-0.25, bal Fe.
SEW 390/90; W. Nr. 1.3949.

AMANOX 3951
Krupp Stahl AG
C 0-0.03, Si 1-1.7, Mn 0-1.2, P 0-0.03, S 0-0.015, Cr 21.5-24, Mo 1.2-1.6, Ni 14-16, N 0.25-0.4, bal Fe.
SEL 90; W. Nr. 1.3951.

AMANOX 3952
Krupp Stahl AG
C 0-0.03, Si 0-1, Mn 0-2, Cr 16.5-18.5, Mo 2.5-3, Ni 13-15, N 0.1-0.3, bal Fe.
DIN X2CrNiMoN1814; W. Nr. 1.3952. 290 N/mm^2 YS, 550-800 N/mm^2 TS, 40 El.

AMANOX 3953
Krupp Stahl AG
C 0-0.03, Si 0-1, Mn 0-2, Cr 16.5-18.5, Mo 2.5-3, Ni 13.5-15.5, N 0-0.1, bal Fe.
DIN X2CrNiMo1815; W. Nr. 1.3953. 200 N/mm^2 YS, 500-700 N/mm^2 TS, 40 El.

AMANOX 3958
Krupp Stahl AG
C 0-0.07, Si 0-1, Mn 0-2, P 0-0.045, S 0-0.03, Cr 17-19, Ni 9-11, bal Fe.
SEL 90; W. Nr. 1.3958.

AMANOX 3962
Krupp Stahl AG
C 0.05-0.2, Si 0-0.6, Mn 5.5-6.5, Cr 10.5-12.5, Mo 0-0.75, Ni 9-11, N 0-0.1, bal Fe.
DIN X15CrNiMn1210, W. 1.3962. 230 N/mm^2 YS, 530-730 N/mm^2 TS, 40 El. Only produced as wire rod.

AMANOX 3964
Krupp Stahl AG
C 0-0.03, Si 0-1, Mn 4-6, Cr 20-21.5, Mo 3-3.5, Ni 15-17, Nb 0-0.25, N 0.2-0.35, bal Fe.
DIN X2CrNiMnMoNNb211653; W. Nr. 1.3964. 365 N/mm^2 YS, 700-950 N/mm^2 TS, 35 El.

AMANOX 3965
Krupp Stahl AG
C 0-0.1, Si 0-1, Mn 7.5-9.5, P 0-0.045, S 0-0.03, Cr 17-19, Ni 4.5-6.5, N 0.1-0.2, bal Fe.
SEL 90; W. Nr. 1.3965.

AMANOX 3968
Krupp Stahl AG
C 0-0.15, Si 0-1, Mn 17-19, P 0-0.08, S 0-0.03, Cr 11-13, Mo 0.3-0.8, Ni 1.5-2.5, bal Fe.
SEL 90; W. Nr. 1.3968.

AMANOX 3974
Krupp Stahl AG
C 0-0.03, Si 0-1, Mn 4.5-6.5, Cr 21-25, Mo 3-3.7, Ni 15-18, Nb 0-0.3, N 0.3-0.5, bal Fe.
DIN X2CrNiMnMoNbN231763; W. Nr. 1.3974 (as austenitic special stainless steel see code No. 1.4565). 420 N/mm^2 YS, 800-1000 N/mm^2 TS, 35 El.

AMANOX 3980
Krupp Stahl AG
C 0-0.08, Si 0-1, Mn 0-2, P 0-0.025, S 0-0.015, Cr 13.5-16, Mo 1-1.5, Ni 24-27, Al 0-0.35, B 0.003-0.01, N 0-0.05, Ti 1.9-2.3, V 0.1-0.5, bal Fe.
SEW 390/90; W. Nr. 1.3980.

AMAX AB-3
Climax Performance Materials Corp.
Mo 99.95.
Arc cast bar. 70,000-95,000 psi TS min; 60,000-80,000 psi YS min (0.2% offset); 15-30 El min.

AMAX ABL-3
Climax Performance Materials Corp.
Mo 99.97.
Arc cast bar, low carbon. 65,000-95,000 psi TS min; 55,000-80,000 psi YS min (0.2% offset); 15-35 El min.

AMAX ABT-3
Climax Performance Materials Corp.
Mo 99.25, Ti 0.4-0.55, Zr 0.06-0.12, C 0.01-0.03.
Arc cast bar. 90,000-115,000 psi TS min; 80,000-100,000 psi YS min (0.2% offset); 10-18 El min.

AMAX AFBT-2
Climax Performance Materials Corp.
Mo 99.25, Ti 0.4-0.55, Zr 0.06-0.12, C 0.01-0.03.
Arc cast forging billet.

AMAX AP-2
Climax Performance Materials Corp.
Mo 99.95.
Arc cast plate. 75,000-95,000 psi TS min; 60,000-80,000 psi YS min (0.2% offset); 1-10 El min.

AMAX AP-30W-1
Climax Performance Materials Corp.
Mo 66.9, W 27-33.
Arc cast plate. 80,000-100,000 psi TS min; 70,000-90,000 psi YS min (0.2% offset); 1 El min.

AMAX APL-2
Climax Performance Materials Corp.
Mo 99.97.
Arc cast plate, low carbon. 75,000-95,000 psi TS min; 60,000-80,000 psi YS min (0.2% offset); 1-10 El min.

AMAX APT-3
Climax Performance Materials Corp.
Mo 99.25, Ti 0.4-0.55, Zr 0.06-0.12, C 0.01-0.03.
Arc cast plate. 100,000-120,000 psi TS min; 85,000-100,000 psi YS min (0.2% offset); 8-10 El min.

AMAX AS-2
Climax Performance Materials Corp.
Mo 99.95.
Arc cast sheet. 100,000-115,000 psi TS min; 80,000-95,000 psi YS min (0.2% offset); 4-15 El min.

AMAX AS-30W-1
Climax Performance Materials Corp.
Mo 66.9, W 27-33.
Arc cast sheet. 110,000 psi TS min; 100,000 psi YS min (0.2% offset); 8-12 El min.

AMAX ASL-2
Climax Performance Materials Corp.
Mo 99.97.
Arc cast sheet, low carbon. 100,000-115,000 psi TS min; 80,000-95,000 psi YS min (0.2% offset); 4-15 El min.

AMAX AST-3
Climax Performance Materials Corp.
Mo 99.25, Ti 0.4-0.55, Zr 0.06-0.12, C 0.01-0.03.
Arc cast sheet. 120,000 psi TS min; 100,000 psi YS min (0.2% offset); 8-10 El min.

AMAX F-21-2
Climax Performance Materials Corp.
Mo 99.95.
Foil. 115,000 psi TS min; 95,000 psi YS min (0.2% offset); 1-3 El min.

AMAX LP
Climax Performance Materials Corp.
P 0.005-0.012, bal Cu.
Oxygen free. Electrical conductivity: 90% IACS min. Rod, tube, flat products; for plumbing, refrigerator units, steam lines, commutators. CDA 108.

AMAX METAL
American manufacture
Cu 81, Sn 11, Pb 7.4, P 0.3.
For bearings; heavy duty.

AMAX MMP-6
Climax Performance Materials Corp.
Mo 99.95.
Powder used in vacuum melted superalloys, powder
metallurgy, as a catalyst and coating; -100 mesh.

AMAX MMP-6
Climax Performance Materials Corp.
Mo 99.95.
Powder.

AMAX MP35N ALLOY
Climax Performance Materials Corp.
Mo 10, Ni 35, Co 35, Cr 20.
Annealed: 110,000 TS; 50,000 YS; 60 El; 7 Rock C. For
hardware items, fasteners, wire, prosthetic devices, chemical
and marine equipment, aircraft and aerospace components.
Corrosion resistant.

AMAX MZ COPPER
Climax Performance Materials Corp.
Mg 0.03, Zr 0.03, 99.95 Cu + Ag + Mg + Zr min.
Cold worked 40%: 365 MPa (53,000 psi) TS; 358 MPa (52,000
psi) YS; 3.5 El. Electrical conductivity: 93% IACS at 68°F. For
heat sinks, high temperature wire, lead frames,
semiconductor bases.

AMAX MZC COPPER
Climax Performance Materials Corp.
Mg 0.03-0.06, Zr 0.06-0.15, Cr 0.4-0.8, 99.95 Cu + Ag + Mg
+ Zr + Cr min.
Cold worked 40%, aged: 496 MPa (72,000 psi) TS; 455 MPa
(66,000 psi) YS; 10 El. Electrical conductivity: 80% IACS at
68°F. Electrical and electronic components, contacts, heat
sinks, molds for continuous casting.

AMAX OMP-6
Climax Performance Materials Corp.
Mo 99.8.
Powder.

AMAX OMPP-5
Climax Performance Materials Corp.
Mo 99.7.
Pellets sintered in hydrogen.

AMAX OMPP-6
Climax Performance Materials Corp.
Mo 99.8.
Pressed and sintered pellets.

AMAX OMPT-1
Climax Performance Materials Corp.
Mo 99.8.
Pressed and sintered pellets.

AMAX OMPZ-A
Climax Performance Materials Corp.
Mo 99.8.
Powder used in vacuum melted superalloys, powder
metallurgy, as a catalyst and coating, flame spray coating;
-200 mesh.

AMAX OMPZ-B
Climax Performance Materials Corp.
Mo 99.8.
Powder used in vacuum melted superalloys, powder
metallurgy, as a catalyst and coating, plasma spray coating;
-200 +325 mesh.

AMAX PB-2
Climax Performance Materials Corp.
Mo 99.95.
Powder metallurgy bar.

AMAX PP-22
Climax Performance Materials Corp.
Mo 99.95.
Powder metallurgy plate. 95,000-100,000 psi TS min;
85,000-90,000 psi YS (0.2% offset); 1-10 El.

AMAX PS-100
Climax Performance Materials Corp.
Mo 99.95.
Powder metallurgy sheet. 100,000-115,000 psi TS min;
80,000-95,000 psi YS min (0.2% offset); 5-18 El min.

AMAX PS-300
Climax Performance Materials Corp.
Mo 99.95.
Powder metallurgy sheet.

AMAX WF-2
Climax Performance Materials Corp.
Mo 99.95.
Foil.

AMAX WMP-3
Climax Performance Materials Corp.
Mo 99.5.
Powder.

AMAX WMPS-3
Climax Performance Materials Corp.
Mo 99.
Sintered corrugates produced in wafers.

AMAX WMPS-4
Climax Performance Materials Corp.
Mo 99.7.
Sintered, corrugates.

AMAX WP-1
Climax Performance Materials Corp.
Mo 99.95.
Plate produced from pressed and sintered powder metallurgy
sheet bar.

AMAX WS-2
Climax Performance Materials Corp.
Mo 99.95.
Sheet produced from pressed and sintered powder
metallurgy sheet bar.

AMAX XLP
Climax Performance Materials Corp.
P 0.001-0.005, bal Cu.
Oxygen free. Electrical conductivity: 98.16% IACS. Rod, tube,
flat products for electrical conductors and terminals,
waveguide tubing, thermostatic control tubing. CDA 103.

AMBEROID
Barker & Allen Ltd.
Ni 15, Zn 23, bal Cu.
Annealed: 50,000 TS; 20,000 YS; 60 El; 63 Brin. Hard:
114,000 TS; 110,000 YS; 2 El; 228 Brin. For costume jewelry,
hollow ware, optical equipment; nickel silver.

AMBO AST
Ambo-Stahl-Gesellschaft
C 0.45, Cr, Ni, W, bal Fe.
For machinery parts, tools and dies; oil hardening, tough.
Obsolete

AMBO BW O
Ambo-Stahl-Gesellschaft
C 0.85-0.95, Si 0.15-0.3, Mn 0.2-0.4, Cr 0.7-0.9, bal Fe.
Cold work took steel, for mandrels, punches. W.-Nr. 1.2036.

AMBO BW1
Ambo-Stahl-Gesellschaft
C 1.05, Cr 1, Mn 0.3, bal Fe.
For bearings, sleeves, liners; water hardened, wear resistant.

AMBO BWC
Ambo-Stahl-Gesellschaft
C 1, Cr 1.55, Mn 0.35, bal Fe.
For bearings, sleeves, liners; water hardened, wear resistant.

AMBO BWG
Ambo-Stahl-Gesellschaft
C 0.9, Cr 0.8, Mn 0.3, bal Fe.
For bearings, cutters, dies, sleeves; water hardened, wear
resistant. *Obsolete*

AMBO BWW
Ambo-Stahl-Gesellschaft
C 1.45, Cr 1.4, Mn 0.6, bal Fe.
For forming and blanking dies, bearings; water hardened,
wear resistant.

AMBO C15W3
Ambo-Stahl-Gesellschaft
Mn 0.37, C 0.15, Si 0.25, bal Fe.
Annealed: 70,000 TS; 40,000 YS; 25 El; 60 RA; 145 Brin. For
gears, pinions, camshafts, cams; case hardening steel.

AMBO C22W3
Ambo-Stahl-Gesellschaft
C 0.22, Si 0.25, Mn 0.45, bal Fe.
Annealed: 75,000 TS; 43,000 YS; 20 El; 55 RA; 150 Brin. For
gears, pinions, camshafts, cams; case hardening steel.

AMBO C35W3
Ambo-Stahl-Gesellschaft
C 0.35, Si 0.25, Mn 0.55, bal Fe.
Hot rolled: 85,000 TS; 54,000 YS; 30 El; 53 RA; 185 Brin. For
gears, pinions, shafts, bolts; water hardened.

AMBO C45W3
Ambo-Stahl-Gesellschaft
C 0.45, Si 0.25, Mn 0.65, bal Fe.
Hot rolled: 98,000 TS; 59,000 YS; 24 El; 45 RA; 212 Brin. For
gears, pinions, shafts, bolts, fasteners; water hardened.

AMBO C53W3
Ambo-Stahl-Gesellschaft
C 0.53, Si 0.38, Mn 0.55, bal Fe.
Normalized: 100,000 TS; 55,000 YS; 18 El; 26 RA; 200 Brin.
For gears, pinions, shafts, axles, bolts, fasteners; water
hardened.

AMBO C60W3
Ambo-Stahl-Gesellschaft
C 0.6, Si 0.25, Mn 0.65, bal Fe.
Heat treated: 160,000 TS; 115,000 YS; 12 El; 40 RA; 325 Brin.
For springs, rails, machine tool parts, axes; water hardened.

AMBO CMO
Ambo-Stahl-Gesellschaft
C 0.2, Mn 1.25, Cr 1.25, bal Fe.
For gears, cams, camshafts; case hardened, tough.

AMBO CMO4E
Ambo-Stahl-Gesellschaft
C 0.15, Cr 1, Mo 0.2, bal Fe.
For gears, bolts, machine tool parts; case hardened, tough.
Obsolete

AMBO CMO5E
Ambo-Stahl-Gesellschaft
C 0.2, Cr 1, Mo 0.2, Si 0.25, Mn 0.65, bal Fe.
For gears, bolts, machine tool parts; case hardened, tough.
Obsolete

AMBO CMV
Ambo-Stahl-Gesellschaft
C 0.25, Cr 1, Mo 0.2, bal Fe.
For gears, bolts, fasteners, machine tool parts; oil hardened,
tough. *Obsolete*

AMBO CMV25
Ambo-Stahl-Gesellschaft
C 0.33, Cr 1, Mo 0.2, Mn 0.65, bal Fe.
For gears, bolts, machine tool parts; oil hardened, tough.
Obsolete

AMBO CMV30
Ambo-Stahl-Gesellschaft
C 0.42, Cr 1, Mo 0.2, Mn 0.65, bal Fe.
For gears, studs, bolts, shafts; oil hardened, tough. *Obsolete*

AMBO CMV50
Ambo-Stahl-Gesellschaft
C 0.5, Cr 1, Mo 0.2, bal Fe.
For gears, bolts, studs, shafts; oil hardened, tough. *Obsolete*

AMBO CNM6
Ambo-Stahl-Gesellschaft
C 0.56, Ni, Cr, Mo, V, bal Fe.
For gears, bolts, crankshafts; oil hardened, shock resistant.

AMBO CNMO 6
Ambo-Stahl-Gesellschaft
C 0.3-0.38, Si 0-0.4, Mn 0.4-0.7, Cr 1.4-1.7, Mo 0.15-0.3, Ni 1.4-1.7, bal Fe.
Steel for cold extrusion. W.-Nr. 1.6582.

AMBO CNMO 8
Ambo-Stahl-Gesellschaft
C 0.26-0.33, Si 0.15-0.4, Mn 0.3-0.6, Cr 1.8-2.2, Mo 0.3-0.5, Ni 1.8-2.2, bal Fe.
Steel for cold extrusion. W.-Nr. 1.6580.

AMBO CNMV
Ambo-Stahl-Gesellschaft
C 0.55, Ni 1.65, Cr 0.9, Mo 0.18, V 0.1, bal Fe.
For gears, bolts, crankshafts; oil hardened, shock resistant.

AMBO CR 105
Ambo-Stahl-Gesellschaft
C 1-1.1, Si 0.15-0.3, Mn 0.25-0.4, Cr 0.4-0.6, bal Fe.
Water hardening tool steel.

AMBO CR13
Ambo-Stahl-Gesellschaft
C 1.65, Cr 11.5, V 0.1, bal Fe.
For blanking and forming dies, punches; air hardening, nondeforming.

AMBO CR13CO
Ambo-Stahl-Gesellschaft
C 1.65, Cr 11.5, Co, bal Fe.
For blanking and forming dies, punches; air hardened, nondeforming.

AMBO CR13MW
Ambo-Stahl-Gesellschaft
C 1.65, Cr 11.5, Mo, V, bal Fe.
For blanking and forming dies, punches; air hardened, nondeforming.

AMBO CRT
Ambo-Stahl-Gesellschaft
C 0.9, Mn 19, V 0.1, bal Fe.
For punches, dies, upsetters, crimpers; oil hardened, nondeforming.

AMBO CRV50
Ambo-Stahl-Gesellschaft
C 0.5, Mn 0.95, Cr 1.05, V 0.1, bal Fe.
For gears, springs, crankshafts, bolts, studs; oil hardened, shock resistant.

AMBO CRZ
Ambo-Stahl-Gesellschaft
Mn 0.3, C 2.1, Cr 11.5, bal Fe.
For blanking and forming dies, punches; oil hardened, nondeforming.

AMBO CRZW
Ambo-Stahl-Gesellschaft
C 2.1, Cr 11.5, W 0.7, bal Fe.
For blanking and forming dies, punches; oil hardened, nondeforming.

AMBO CSJ
Ambo-Stahl-Gesellschaft
C 0.61, Cr 1.18, V 0.1, Mn 0.7, Si 0.85, bal Fe.
For springs, valves, countershafts; oil hardened, shock resistant. *Obsolete*

AMBO DCM
Ambo-Stahl-Gesellschaft
C 0.4, Cr, Mo, Mn, bal Fe.
For gears, bolts, crankshafts, fasteners; oil hardened, tough.

AMBO DCN EXTRA
Ambo-Stahl-Gesellschaft
C 0.32-0.38, Si 0.15-0.3, Mn 0.4-0.6, Cr 1.2-1.5, Mo 0.2-0.4, Ni 3.8-4.3, bal Fe.
Hot work tool steel. W.-Nr. 1.2766.

AMBO DCN3
Ambo-Stahl-Gesellschaft
C 0.5, Cr 1.05, Ni 3.25, Mn 0.5, bal Fe.
For gears, bolts, crankshafts, studs; oil hardened, shock resistant.

AMBO DCN4
Ambo-Stahl-Gesellschaft
C 0.31, Cr 0.7, Ni 3.5, Mn 0.6, bal Fe.
For gears, bolts, machine tool parts; oil hardened, shock resistant.

AMBO DCN4E
Ambo-Stahl-Gesellschaft
C 0.4, Cr 0.7, Ni 3.5, Mn 0.6, bal Fe.
For gears, bolts, crankshafts, fasteners; oil hardened, shock resistant.

AMBO DCN5
Ambo-Stahl-Gesellschaft
C 0.35, Cr 1.1, Ni 4.5, Mn 0.6, bal Fe.
For gears, bolts, crankshafts; oil hardened, shock resistant.
Obsolete

AMBO DCN5E
Ambo-Stahl-Gesellschaft
C 0.4, Cr, Mn, Mo, bal Fe.
For gears, bolts, machine tool parts; oil hardened, tough.
Obsolete

AMBO DCNV
Ambo-Stahl-Gesellschaft
C 0.36, Mn 0.6, Cr 0.7, Ni 2.5, bal Fe.
For gears, bolts, crankshafts; oil hardened, shock resistant.
Obsolete

AMBO DCZ
Ambo-Stahl-Gesellschaft
C 0.4, Cr, Mn, Mo, bal Fe.
For gears, bolts, machine tool parts; oil hardened, tough.
Obsolete

AMBO DM2
Ambo-Stahl-Gesellschaft
C 0.15, Mn 0.37, Si 0.25, bal Fe.
Annealed: 70,000 TS; 40,000 YS; 25 El; 60 RA; 145 Brin. For gears, shafts, cams, camshafts; case hardening steel.
Obsolete

AMBO DM3
Ambo-Stahl-Gesellschaft
C 0.36, Mn 0.55, Si 0.25, bal Fe.
Hot rolled: 85,000 TS; 54,000 YS; 30 El; 53 RA; 185 Brin. For gears, pinions, shafts; water hardened. *Obsolete*

AMBO DM4
Ambo-Stahl-Gesellschaft
C 0.45, Si 0.25, Mn 0.65, bal Fe.
Hot rolled: 98,000 TS; 59,000 YS; 24 El; 45 RA; 212 Brin. For gears, pinions, shafts, axles, countershafts; water hardened.
Obsolete

AMBO DM5
Ambo-Stahl-Gesellschaft
C 0.6, Si 0.25, Mn 0.65, bal Fe.
Heat treated: 160,000 TS; 113,000 YS; 12 El; 40 RA; 321 Brin. For axles, guides, punches, crimpers; water hardened.
Obsolete

AMBO DM6
Ambo-Stahl-Gesellschaft
C 0.75, Si 0.38, Mn 0.7, bal Fe.
Heat treated: 175,000 TS; 130,000 YS; 12 El; 37 RA; 355 Brin. For springs, tools, dies, punches, crimpers; Type W1; water hardened. *Obsolete*

AMBO DR5
Ambo-Stahl-Gesellschaft
C 1.3, W 4.75, bal Fe.
For cutters, form tools; fast finishing steel, water hardened.

AMBO EM 1820
Ambo-Stahl-Gesellschaft
C 0.5-0.58, Mn 0.3-0.5, bal Fe.
Carbon tool steel. W.-Nr. 1.1820.

AMBO EM 2 E E
Ambo-Stahl-Gesellschaft
C 1.2-1.35, Si 0.1-0.25, Mn 0.1-0.25, bal Fe.
Water hardening tool steel. W.-Nr. 1560.

AMBO EM 2 EXTRA EXTRA
Ambo-Stahl-Gesellschaft
C 1.2-1.35, Si 0.1-0.25, Mn 0.1-0.25, bal Fe.
Carbon tool steel. W.-Nr 1.1560.

AMBO EM3 EXTRA
Ambo-Stahl-Gesellschaft
C 1.15, Si 0-0.25, Mn 0-0.25, bal Fe.
Annealed: 110,000 TS; 56,000 YS; 20 El; 40 RA; 210 Brin. For springs, cutters, reamers, taps, drills, hobs; water hardened; Type W1.

AMBO EM3 EXTRA EXTRA
Ambo-Stahl-Gesellschaft
C 1.1, Si 0-0.25, Mn 0-0.25, bal Fe.
Annealed: 110,000 TS; 56,000 YS; 20 El; 40 RA; 210 Brin. For springs, cutters, reamers, taps, drills, hobs; Type W1; water hardened.

AMBO EM3 PRIMA
Ambo-Stahl-Gesellschaft
C 0.75, Si 0.25-0.5, Mn 0.3-0.8, bal Fe.
Heat treated: 180,000 TS; 135,000 YS; 12 El; 35 RA; 375 Brin. For springs, rails, punches, upsetters; Type W1; water hardened.

AMBO EM3CV
Ambo-Stahl-Gesellschaft
C 1, Mn 0.07, Cr 1.1, bal Fe.
For bearings, liners, bushings; water hardened, wear resistant.

AMBO EM3W
Ambo-Stahl-Gesellschaft
C 1.2, V 0.1, Cr 0.2, W 1, bal Fe.
For bearings, cutters, tools, dies; water or oil hardened, wear resistant.

AMBO EM4 EXTRA
Ambo-Stahl-Gesellschaft
C 1, Si 0.25, Mn 0.25, bal Fe.
Annealed: 100,000 TS; 53,000 YS; 21 El; 42 RA; 200 Brin. For drills, taps, springs, reamers, hobs; Type W1; water hardened.

AMBO EM4 EXTRA EXTRA
Ambo-Stahl-Gesellschaft
C 1, Si 0.25, Mn 0.25, bal Fe.
Annealed: 100,000 TS; 53,000 YS; 21 El; 42 RA; 200 Brin. For springs, drills, reamers, taps; Type W1; water hardened.

AMBO EM4 PRIMA
Ambo-Stahl-Gesellschaft
C 0.6, Si 0.25-0.5, Mn 0.3-0.8, bal Fe.
Heat treated: 160,000 TS; 113,000 YS; 12 El; 40 RA; 325 Brin. For gears, bolts, crankshafts, rails, springs; water hardened.

AMBO EM5 EXTRA
Ambo-Stahl-Gesellschaft
C 0.85, Si 0.25, Mn 0.25, bal Fe.
Heat treated: 190,000 TS; 145,000 YS; 10 El; 30 RA; 400 Brin. For springs, tools, drills, dies, cutters; Type W1; water hardened.

AMBO EM5 EXTRA EXTRA
Ambo-Stahl-Gesellschaft
C 0.85, Si 0.25, Mn 0.25, bal Fe.
Heat treated: 190,000 TS; 145,000 YS; 10 El; 30 RA; 400 Brin. For springs, tools, drills, dies, cutters; Type W1; water hardened.

AMBO EM5 PRIMA
Ambo-Stahl-Gesellschaft
C 0.45, Si 0.25-0.5, Mn 0.3-0.8, bal Fe.
Hot rolled: 98,000 TS; 59,000 YS; 24 El; 45 RA; 215 Brin. For gears, pinions, bolts, fasteners, shafts; water hardened.

AMBO EM6 EXTRA
Ambo-Stahl-Gesellschaft
C 0.7, Si 0-0.25, Mn 0-0.25, bal Fe.
Heat treated: 175,000 TS; 128,000 YS; 12 El; 37 RA; 355 Brin. For springs, rails, hammers, machine tool parts; Type W1; water hardened.

AMBO EM6 EXTRA EXTRA
Ambo-Stahl-Gesellschaft
C 0.7, Si 0-0.25, Mn 0-0.25, bal Fe.
Heat treated: 175,000 TS; 128,000 YS; 12 El; 37 RA; 355 Brin. For springs, rails, hammers, cutters, punches; Type W1; water hardened.

AMBO EM6 PRIMA
Ambo-Stahl-Gesellschaft
C 0.35, Si 0.25-0.5, Mn 0.3-0.8, bal Fe.
Hot rolled: 85,000 TS; 54,000 YS; 30 El; 53 RA; 185 Brin. For gears, bolts, fasteners, machine tool parts; water hardened.

AMBO EMV
Ambo-Stahl-Gesellschaft
C 1, Si 0.2, Mn 0.25, V 0.1, bal Fe.
For drills, taps, reamers, broaches; Type W2; water hardened.

AMBO EN 15
Ambo-Stahl-Gesellschaft
C 0-0.18, Si 0.1-0.35, Mo 0.3-0.4, Cr 1.3-1.6, bal Fe.
For structural equipment to operate down to -100°C. W.-Nr. 1.5622.

AMBO FAH
Ambo-Stahl-Gesellschaft
C 0.55-0.62, Si 0.15-0.4, Mn 0.7-1.1, Cr 0.9-1.2, V 0.1-0.2, bal Fe.
For heavy sections to be flame or induction hardened. W.-Nr. 1.8161.

AMBO FAW
Ambo-Stahl-Gesellschaft
C 0.47-0.55, Si 0.15-0.4, Mn 0.7-1.1, Cr 0.9-1.2, V 0.1-0.2, bal Fe.
For heavy sections to be flame or induction hardened. W.-Nr. 1.8159.

AMBO FS 7
Ambo-Stahl-Gesellschaft
C 0.6-0.7, Si 1.5-1.8, Mn 0.7-1, bal Fe.
For springs. W.-Nr. 1.5028.

AMBO FSE
Ambo-Stahl-Gesellschaft
C 0.68-0.75, Si 1.5-1.8, Mn 0.6-0.8, bal Fe.
For springs. W.-Nr. 1.5029.

AMBO HHCA10
Ambo-Stahl-Gesellschaft
C 0-0.12, Si 1, Al 1, Cr 18, bal Fe.
For oil refinery equipment; creep and heat resistant. *Obsolete*

AMBO HHCA12
Ambo-Stahl-Gesellschaft
C 0-0.12, Si 1.5, Al 1.5, Cr 24, bal Fe.
For oil refinery equipment; creep and heat resistant. *Obsolete*

AMBO HHCA8
Ambo-Stahl-Gesellschaft
C 0-0.12, Al 0.8, Cr 6.5, bal Fe.
For oil refinery equipment; creep and heat resistant. *Obsolete*

AMBO HHCN100
Ambo-Stahl-Gesellschaft
C 0.15, Cr 19.5, Ni 9.5, bal Fe.
Annealed: 80,000 TS; 35,000 YS; 55 El; 75 RA; 150 Brin. For chemical plant equipment, tanks, mixers; Type 302; stainless, austenitic. *Obsolete*

AMBO HHCN120
Ambo-Stahl-Gesellschaft
C 0.15, Cr 24, Ni 19, bal Fe.
Annealed: 100,000 TS; 45,000 YS; 50 El; 65 RA; 185 Brin. For valves, pumps, turbine and jet parts; Type 310; corrosion and heat resistant. *Obsolete*

AMBO HHCN120S
Ambo-Stahl-Gesellschaft
C 0.2, Cr 25, Ni 4, Si 1.2, bal Fe.
Cast: 90,000 TS; 65,000 YS; 2 El; 212 Brin. For cylinder linings, bushings, valve seats and bodies; Type CC-50. *Obsolete*

AMBO KB 1
Ambo-Stahl-Gesellschaft
C 0.12-0.2, Si 0.1-0.35, Mn 0.4-0.8, Cr 0-0.3, Mo 0.25-0.3, bal Fe.
For high temperature equipment, to 530°C.

AMBO KB 3
Ambo-Stahl-Gesellschaft
C 0.13-0.2, Si 0.15-0.35, Mn 0.5-0.8, Cr 0.9-1.2, Mo 0.4-0.5, Ni 0-0.4, bal Fe.
For high temperature forgings, to 530°C. W.-Nr. 1.7337.

AMBO KB2
Ambo-Stahl-Gesellschaft
C 0.1-0.18, Si 0.1-0.35, Mn 0.4-0.7, Cr 0.8-1.15, Mo 0.45-0.65, bal Fe.
For high temperature piping to 530°C. W.-Nr. 1.7335.

AMBO MN 12
Ambo-Stahl-Gesellschaft
C 1.2, Mn 12.5, bal Fe.
For wear plates, dipper teeth, rail frogs; wear and abrasion resistant.

AMBO MN 2 S
Ambo-Stahl-Gesellschaft
C 0-0.24, Si 0-0.6, Mn 0-1.6, Cr 0.5-1.2, bal Fe.
Structural steel.

AMBO MS 1
Ambo-Stahl-Gesellschaft
C 0.58-0.65, Si 0.8-1, Mn 0.8-1.2, bal Fe.
Hot work tool steel. W.-Nr. 1.2826.

AMBO MS 125
Ambo-Stahl-Gesellschaft
C 1.2-1.3, Si 1.05-1.25, Mn 0.6-0.8, Cr 1.1-1.3, bal Fe.
Cold work tool steel, taps, threading tools. W.-Nr. 1.2109.

AMBO MS 2
Ambo-Stahl-Gesellschaft
C 0.42-0.5, Si 1.5-1.8, Mn 0.5-0.8, bal Fe.
For springs. W.-Nr. 1.0902.

AMBO MS 3
Ambo-Stahl-Gesellschaft
C 0.47-0.55, Si 1.5-1.8, Mn 0.5-0.8, bal Fe.
For springs. W.-Nr. 1.0903.

AMBO MS 4
Ambo-Stahl-Gesellschaft
Mn 0.9, C 0.53, Si 0.9, bal Fe.
For punches, chisels, pneumatic tools; oil hardened, tough.

AMBO MS 5
Ambo-Stahl-Gesellschaft
C 0.55, Si 1.7, Mn 0.7, bal Fe.
For springs, chisels, punches, pneumatic tools; oil hardened, shock resistant.

AMBO MS 6
Ambo-Stahl-Gesellschaft
C 0.6-0.68, Si 1.5-1.8, Mn 0.7-1, bal Fe.
For springs. W.-Nr. 1.0906.

AMBO MS 70
Ambo-Stahl-Gesellschaft
C 0.65-0.75, Si 1.5-1.8, Mn 0.6-0.8, bal Fe.
Cold work tool steel. W.-Nr 1.2823.

AMBO MS 90
Ambo-Stahl-Gesellschaft
C 0.85-0.95, Si 1.05-1.25, Mn 0.6-0.8, Cr 1.1-1.3, bal Fe.
Cold work tool steel as shear blades, etc. W.-Nr 1.2108.

AMBO MSIN
Ambo-Stahl-Gesellschaft
C 0.65, Si 1.7, Mn 0.7, bal Fe.
For springs, punches, pneumatic tools; oil hardened, shock resistant. *Obsolete*

AMBO MSJ
Ambo-Stahl-Gesellschaft
C 0.53, Si 0.9, Mn 0.9, bal Fe.
Annealed: 96,000 TS; 52,000 YS; 16 El; 23 RA; 170 Brin. For axles, gears, bolts, tie-rods, bushings; water hardened.

AMBO MSO
Ambo-Stahl-Gesellschaft
C 0.9, Mn 1.9, V 0.1, bal Fe.
For blanking and forming dies, thread rolling dies; oil hardened, nondeforming. *Obsolete*

AMBO NR 2 AF
Ambo-Stahl-Gesellschaft
C 0-0.12, Si 0-1, Mn 0-2, Cr 16-18, Ni 7-9, bal Fe.
Austenitic stainless steel, work hardens rapidly; for cold worked springs. W.-Nr. 1.4310. Similar to AISI 301.

AMBO NR 2 AZ
Ambo-Stahl-Gesellschaft
C 0-0.15, Si 0-1, Mn 0-2, Cr 17-19, Ni 8-10, S 0.15-0.35, bal Fe.
Free-machining austenitic stainless steel, for screw machined parts. W.-Nr. 1.4305. AISI 303.

AMBO NR 4 AW
Ambo-Stahl-Gesellschaft
C 0-0.03, Si 0-1, Mn 0-1, Cr 16.5-18.5, Ni 11-14, Mo 2-2.5, bal Fe.
Austenitic stainless steel; for chemical industry equipment. W.-Nr. 1.4404. AISI 316L.

AMBO NR2A
Ambo-Stahl-Gesellschaft
C 0-0.15, Cr 18, Ni 8.5, bal Fe.
Annealed: 80,000 TS; 35,000 YS; 55 El; 75 RA; 160 Brin. For chemical plant equipment, tanks, fermenters; Type 302; stainless, austenitic. *Obsolete*

AMBO NR2AE
Ambo-Stahl-Gesellschaft
C 0-0.12, Cr 18, Ni 9.5, Cb = 8 x C, bal Fe.
Annealed: 90,000 TS; 45,000 YS; 56 El; 65 RA; 160 Brin. Cold drawn: 100,000 TS; 65,000 YS; 45 El; 60 RA; 205 Brin. For welded chemical plant equipment, tanks, fixtures; Type 347; stainless, austenitic. *Obsolete*

AMBO NR2AS
Ambo-Stahl-Gesellschaft
C 0.07, Cr 18, Ni 9.5, bal Fe.
Annealed: 85,000 TS; 35,000 YS; 60 El; 70 RA; 150 Brin. Cold drawn: 180,000 TS; 125,000 YS; 10 El; 330 Brin. For welded chemical plant equipment, tanks, Type 304, stainless, austenitic. *Obsolete*

AMBO NR2ATE
Ambo-Stahl-Gesellschaft
C 0-0.12, Cr 18, Ni 9.5, Ti = 4 x C, bal Fe.
Annealed: 85,000 TS; 35,000 YS; 55 El; 65 RA; 150 Brin. Cold drawn: 95,000 TS; 60,000 YS; 40 El; 60 RA; 185 Brin. For welded chemical plant equipment, tanks; Type 321; stainless, austenitic. *Obsolete*

AMBO NR4AE
Ambo-Stahl-Gesellschaft
C 0-0.12, Cr 18, Mo 2, Ni 10.5, Ti = 4 x C, bal Fe.
Annealed: 85,000 TS; 35,000 YS; 50 El; 65 RA; 160 Brin. Cold drawn: 150,000 TS; 135,000 YS; 6 El; 300 Brin. For acid resistant equipment, welded; Type 316Ti; stainless, austenitic. *Obsolete*

AMBO NR4AS
Ambo-Stahl-Gesellschaft
C 0-0.07, Cr 18, Ni 10.5, Mo 2, bal Fe.
Annealed: 85,000 TS; 35,000 YS; 50 El; 65 RA; 160 Brin. Cold drawn: 150,000 TS; 135,000 YS; 6 El; 300 Brin. For acid resistant equipment; Type 316; stainless, austenitic. *Obsolete*

AMBO PD 1
Ambo-Stahl-Gesellschaft
C 0.22-0.3, Si 0.3-0.5, Mn 0.2-0.4, Cr 0.6-0.9, Mo 0.2-0.4, Ni 1.3-1.6, V 0.15-0.2, bal Fe.
Hot work tool steel. W.-Nr.1.2726.

AMBO PD4
Ambo-Stahl-Gesellschaft
C 0.28, Ni, Mo, bal Fe.
For gears, bolts, fasteners, shafts, oil hardened, tough.

AMBO PDZ
Ambo-Stahl-Gesellschaft
C 0.28, Ni, Cr, Mo, V, bal Fe.
For gears, bolts, shafts, crankshafts; oil hardened, shock resistant.

AMBO SP1V
Ambo-Stahl-Gesellschaft
C 0.38, Si, Cr, V, bal Fe.
For springs, gears, crankshafts; oil hardened, shock resistant.

AMBO SP2V
Ambo-Stahl-Gesellschaft
C 0.45, Si, Cr, V, bal Fe.
For springs, gears, crankshafts; oil hardened, shock resistant.

AMBO SPW3
Ambo-Stahl-Gesellschaft
C 1.42, W, V, bal Fe.
For blanking and forming dies, engravers' tools; water or oil hardened, wear resistant.

AMBO SPW4
Ambo-Stahl-Gesellschaft
C 0.45, Cr 1.4, Mo 0.7, V 0.3, Mn 0.7, bal Fe.
For forging and heading dies, upsetters, punches; oil hardened, tough.

AMBO SPW5
Ambo-Stahl-Gesellschaft
C 0.3, Cr 1.1, V 0.18, W 3.75, bal Fe.
For chisels, punches, pneumatic tools; oil hardened, shock resistant.

AMBO SPW6
Ambo-Stahl-Gesellschaft
C 0.3, Cr 2.35, V 0.6, W 4.25, Mn 0.3, bal Fe.
For pneumatic tools, extrusion rams and dies, upsetters; oil hardened, shock resistant.

AMBO SPWD
Ambo-Stahl-Gesellschaft
C 0.35, Si 0.9, Cr 1.05, V 0.18, W 1.85, bal Fe.
For pneumatic tools, punches, upsetters; oil hardened, shock resistant.

AMBO SPWH
Ambo-Stahl-Gesellschaft
C 0.55, Si 0.9, Cr 1.05, V 0.18, W 1.85, bal Fe.
For pneumatic tools, upsetters, rivet sets; oil hardened, shock resistant.

AMBO V111
Ambo-Stahl-Gesellschaft
C 0.95, Cr 4, V, W, Mo, bal Fe.
For lathe and planer tools, reamers, broaches; high speed steel.

AMBO V25
Ambo-Stahl-Gesellschaft
C 0.82, Cr 4.1, Mo 0.85, V 1.6, W 8.7, bal Fe.
For lathe and planner tools, reamers; high speed steel.

AMBO V35
Ambo-Stahl-Gesellschaft
C 0.86, Cr 4.1, Mo 0.85, V 2.5, W 12, bal Fe.
For lathe and planer tools, reamers, broaches; high speed steel.

AMBO V50
Ambo-Stahl-Gesellschaft
C 1.3, Cr 4.3, Mo 0.85, V 3.8, W 12, bal Fe.
For blanking and forming dies, engravers' tools; high speed steel.

AMBO V66
Ambo-Stahl-Gesellschaft
C 0.85, W, Mo, Cr, V, bal Fe.
For lathe and planer tools, reamers, taps, drills; high speed steel.

AMBO VANADIUM
Ambo-Stahl-Gesellschaft
C 0.74, Cr 4.1, V 1.1, W 18.5, bal Fe.
For lathe and planer tools, drills, taps, hobs; high speed steel.

AMBO VC 135
Ambo-Stahl-Gesellschaft
C 0.3-0.37, Si 0.15-0.4, Mn 0.6-0.9, Cr 0.9-1.2, bal Fe.
Heat treatable steel for axles, shafts. W.-Nr. 1.7033.

AMBO WCM1
Ambo-Stahl-Gesellschaft
C 1.05, Cr 1, Mn 0.9, W 1.15, bal Fe.
For cold work tools, cutters, form dies; oil hardened, tough.

AMBO WF 1000
Ambo-Stahl-Gesellschaft
C 0.08-0.15, Si 0.15-0.5, Mn 0.4-0.7, Cr 2-2.5, Mo 0.9-1.2, bal Fe.
For high temperature equipment to 530°C. W.-Nr. 1.7380.

AMBO WMA
Ambo-Stahl-Gesellschaft
C 0.45, Cr, Ni, W, bal Fe.
For upsetters, chisels, punches, pneumatic tools; oil hardened, tough. *Obsolete*

AMBO WMOV
Ambo-Stahl-Gesellschaft
C 0.45, Mo 0.45, Cr 1.35, V 0.8, W 0.45, bal Fe.
For forging and heading dies; oil hardened, tough.

AMBO WOF
Ambo-Stahl-Gesellschaft
C, Cr, Mo, bal Fe.
For machine tool parts; oil hardened, tough.

AMBRAC "B"
Anaconda Co.
Cu 65, Zn 5, Ni 30.
Hard: 120,000 TS; 2 El. Soft: 65,000 TS; 30 El. For laundry and paper machinery; corrosion resistant. *Obsolete*

AMBRAC 815
Anaconda Co.
Cu 64.5, Ni 20, Zn 7.5, Sn 3, Pb 5.
Cast: 40,000 TS; 25,000 YS; 15 El. For hardware, plumbers fixtures. *Obsolete*

AMBRAC 850
Anaconda Co.
Copper. Ni 20, Zn 4.4, Mn 0.6, bal Cu.
Soft: 50,000 TS; 18,000 YS; 35 El. Hard: 70,000 TS; 60,000 YS; 20 El. For plumbing, hardware. Corrosion resistant.

AMBRAC A
Anaconda Co.
Ni 20, Zn 5, Cu 75.
Cast: 44,400 TS; 14,800 YS; 19.4 El. Hot rolled: 50,000 TS; 16,000 YS; 29.0 El. Cold rolled: 80,000 TS; 70,000 YS; 13.0 El; 55.0 RA. Wire: 55,000-125,000 TS; 20,000 YS; 1.0 El. For laundry and paper machinery; resists corrosion. *Obsolete*

AMBRALLOY
A.M. Byers Co.
C 0.4, Mn 0.7, Ni 1.8, Cr 0.8, Mo 0.2, bal Fe.
Heat treated: 274,000 TS; 237,000 YS; 11 El; 35.4 RA. For structural members, cranes, derricks, bridges; high strength plate, heat treated at mill. *Obsolete*

AMBRALOY 606
Anaconda Co.
Copper. Al 5, Cu 95.
Soft: 55,000 TS; 22,000 YS; 65 El. Hard: 92,000 TS; 65,000 YS; 7 El. For condenser tubes, forged parts. Corrosion resistant.

AMBRALOY 612
Anaconda Co.
Copper. Al 8, Cu 92.
Soft: 65,000 TS; 25,000 YS; 65 El. Hard: 80,000 TS; 50,000 YS; 30 El. For screw machine parts, wire, rod. Corrosion resistant.

AMBRALOY 614
Anaconda Co.
Copper. Al 7, Fe 2.75, bal Cu.
Plate: 70,000 TS; 30,000 YS; 35 El. For condenser tubes, heat exchangers. High corrosion resistance. AMBRALOY 900.

AMBRALOY 630
Anaconda Co.
Copper. Cu 82, Al 9.5, Mn 1, Ni 5, Fe 2.5.
Soft: 90,000 TS; 12 El. Hard: 105,000 TS; 12 El. For forgings. Heat treatable.

AMBRALOY 929
Anaconda Co.
Al 10, Cu 90.
78,000-125,000 TS; 5-36 El. For rods; corrosion resistant. *Obsolete*

AMBRALOY-687
Anaconda Co.
Copper. Zn 22, Al 2, As 0.04, bal Cu.
Soft: 52,000 TS; 20,000 YS; 65 El. Hard: 85,000 TS; 60,000 YS; 10 El. For condenser tubes, ferrules. Corrosion resistant.

AMBRAZE 111
American Brazing Alloys Co.
Ni, Zn, bal Cu.
Cast: 80,000 TS. For brazing alloy; nickel silver. Melting point: 1700°F, corrosion resistant.

AMBRAZE 111FC
American Brazing Alloys Co.
Ni, Zn, bal Cu.
For brazing alloy; nickel silver.

AMBRAZE 131
American Brazing Alloys Co.
Ni, Zn, bal Cu.
For brazing alloy; nickel silver.

AMBRAZE 131 FC
American Brazing Alloys Co.
Ni, Zn, bal Cu.
For brazing alloy; nickel silver.

AMBRAZE 181
American Brazing Alloys Co.
P 5, Ag 15, bal Cu.
For brazing alloy; self-fluxing.

AMBRAZE 35
American Brazing Alloys Co.
Ag 35, Cu 26, Zn 21, Cd 18.
For brazing alloy for torch brazing; BAg-2.

AMBRAZE 430
American Brazing Alloys Co.
Ag alloy.
For silver solder for stainless steel; melting point: 425°F.

AMBRAZE 45
American Brazing Alloys Co.
Ag 45, Cu 15, Zn 16, Cd 24.
For brazing alloy for dissimilar metals; BAg-1; melting point: 1125°F-1145°F.

AMBRAZE 50
American Brazing Alloys Co.
Ag 50, Cu 15.5, Zn 16.5, Cd 18.
For general purpose brazing alloy for ferrous and non-ferrous alloys; BAg-1A; melting point: 1160°F-1175°F.

AMBRAZE-1001
American Brazing Alloys Co.
Ag 40, Cu 30, Zn 28, Ni 2.
For brazing alloys, filler metal. Corrosion resistant.

AMBRAZE-1010
American Brazing Alloys Co.
Ag 45, Cu 18, Zn 18, Cd 19.
For brazing, filler metal. Corrosion resistant.

AMBRAZE-1110
American Brazing Alloys Co.
Ag 41, Cu 18, Zn 15, Cd 27.
For brazing, filler metal. Corrosion resistant.

AMBRAZE-1201
American Brazing Alloys Co.
Ag 20, Cu 45, Zn 30, Cd 5.
For brazing, filler metal. Corrosion resistant.

AMBRAZE-1551
American Brazing Alloys Co.
Ag 56, Cu 22, Zn 17, Sn 5.
For brazing, filler metal. Corrosion resistant.

AMBRAZE-211
American Brazing Alloys Co.
P 7, bal Cu.
For brazing copper. Phosphor-copper, self-fluxing.

AMBRAZE-231
American Brazing Alloys Co.
P 7.3, Ag 2, bal Cu.
For brazing alloy. Self-fluxing.

AMBRAZE-311
American Brazing Alloys Co.
Si 12, Cu 4, Fe 1, bal Al.
For brazing aluminum alloys, filler metal. Corrosion resistant.

AMBRAZE-331
American Brazing Alloys Co.
Si 10, Cu 4, Fe 0.5, bal Al.
For brazing cast aluminum alloys.

AMBRAZE-411
American Brazing Alloys Co.
Cu 58, Sn 1, Fe 1, bal Zn.
For brazing bronzes, filler metal. Corrosion resistant.

AMBRAZE-411FC
American Brazing Alloys Co.
Cu 58, Sn 1, Fe 1, bal Zn.
For brazing bronzes. Flux coated, corrosion resistant.

AMBRAZE-611
American Brazing Alloys Co.
Al 9, Mn 0.15, Zn 2, bal Mg.
For brazing magnesium alloys, filler metal.

AMBRO
Manufacturer not listed.
Al 2, Cu 76, Zn 22.
For condenser tubes; corrosion resistant.

AMBRONZE 405
Anaconda Co.
Copper. Zn 4, Sn 1, P 0.03, bal Cu.
Hard: 60,000 TS; 50,000 YS; 6 El; 125 Brin. Soft: 40,000 TS; 15,000 YS; 40 El; 55 Brin. For electrical springs. Corrosion resistant.

AMBRONZE 413
Anaconda Co.
Copper. Zn 7, Sn 1, bal Cu.
Hard: 63,000 TS; 55,000 YS; 7 El; 137 Brin. Soft: 42,000 TS; 16,000 YS; 40 El; 57 Brin. For strips, tubes. Corrosion resistant.

AMBRONZE 430
Anaconda Co.
Copper. Zn 11, Pb 2, bal Cu.
Hard: 73,000 TS; 65,000 YS; 10 El; 160 Brin. Soft: 46,000 TS; 18,000 YS; 50 El; 59 Brin. For springs. Corrosion resistant.

AMBRONZE 4301
Anaconda Co.
Copper. Zn 10, Pb 2, bal Cu.
Soft: 20,000-46,000 TS; 55 El; 59 Brin. For condenser and heat exchanger tubes. Corrosion resistant to mine waters.

AMBRONZE-4222
Anaconda Co.
Copper. Zn 11, Sn 1, bal Cu.
Soft: 43,000 TS; 16,000 YS; 47 El. Hard: 72,000 TS; 60,000 YS; 8 El. For weather strip. Corrosion resistant.

AMC 74S
Aluminum Company of America
Al 2.5-3.5, Zn 2.5-3.5, Si 0-0.3, bal Mg.
Extruded: 42,000 TS; 30,000 YS; 19 El; 51 Brin. Aged: 44,000 TS; 34,000 YS; 13 El; 54 Brin. For pressings and extrusions; corrosion resistant. *Obsolete*

AMCARB D-10
American Carbide Corp.
WC 95.5, Co 4.5.
Sintered: 92 Rock A. 180,000 transverse strength. For extrusion press nibs, drawing dies, cutting tools. Sintered carbide, wear and abrasion resistant.

AMCARB D-15
American Carbide Corp.
Co 6, WC 94.
Sintered: 91.5 Rock A. 225,000 transverse strength. For drawing and extrusion dies, cutting tools. Sintered carbides. Wear and abrasion resistant.

AMCARB D-20
American Carbide Corp.
Co 6, WC 94.
Sintered: 90.8 Rock A, 275,000 transverse strength. For extrusion and drawing dies, cutting tools. Sintered carbide, wear and abrasion resistant.

AMCARB D-30
American Carbide Corp.
Co 9, WC 91.
Sintered: 89.5 Rock A. 325,000 transverse strength. For extrusion and drawing dies, header punches, cutting tools. Sintered carbide, wear and abrasion resistant.

AMCARB D-35
American Carbide Corp.
Co 9, WC 81, 10.0 TaC.
Sintered: 89.0 Rock A. 325,000 transverse strength. For extrusion and drawing dies, header punches, and inserts, cutters. Sintered carbide, wear and abrasion resistant.

AMCARB D-40
American Carbide Corp.
Co 13, WC 87.
Sintered: 88.5 Rock A. 400,000 transverse strength. For extrusion, drawing and header dies, cutters. Sintered carbide, wear and abrasion resistant.

AMCARB D-43
American Carbide Corp.
Co 13, WC 60, 27.0 TaC.
Sintered: 88 Rock A. 350,000 transverse strength. For extrusion and drawing dies. Wear and abrasion resistant. Sintered carbide.

AMCARB D-5
American Carbide Corp.
Co 3.5, WC 96.5.
Sintered: 92.5 Rock A. 170,000 transverse strength. For extrusion and drawing dies, cutting tools. Sintered carbide, wear and abrasion resistant.

AMCARB D-50
American Carbide Corp.
Co 15, WC 85.
Sintered: 87 Rock A. 425,000 transverse strength. For extrusion, drawing and header dies, cutters. Sintered carbide, wear and abrasion resistant.

AMCARB D-57
American Carbide Corp.
Co 15, WC 78, 7.0 TaC.
Sintered: 86.5 Rock A. 350,000 transverse strength. For extrusion, drawing and header dies. Sintered carbides, wear and abrasion resistant.

AMCARB D-60
American Carbide Corp.
Co 20, WC 75, 5.0 TaC.
Sintered: 85 Rock A. 390,000 transverse strength. For extrusion, drawing and header dies; light impact. Sintered carbides, wear and abrasion resistant.

AMCARB D-65
American Carbide Corp.
Co 20, WC 80.
Sintered: 85 Rock A. 400,000 transverse strength. For extrusion, drawing and header dies; light impact. Sintered carbides, wear and abrasion resistant.

AMCARB D-70
American Carbide Corp.
Co 25, WC 70, 5.0 TaC.
Sintered: 83 Rock A. 380,000 transverse strength. For extrusion, drawing and header dies; normal impact. Sintered carbides, wear and abrasion resistant.

AMCARB D-75
American Carbide Corp.
Co 25, WC 75.
Sintered: 83 Rock A. 390,000 transverse strength. For extrusion, drawing and header dies. Sintered carbides, wear and abrasion resistant.

AMCARB D-80
American Carbide Corp.
Co 27, WC 68, 5.0 TaC.
Sintered: 82 Rock A. 375,000 transverse strength. For extrusion, drawing and header dies; heavy impact. Sintered carbides, wear and abrasion resistant.

AMCARB D-85
American Carbide Corp.
Co 30, WC 65, 5.0 TaC.
Sintered: 81 Rock A. 360,000 transverse strength. For extrusion, drawing and header dies; very heavy impact. Sintered carbides, wear and abrasion resistant.

AMCARB T-38
American Carbide Corp.
WC 71, Co 9, 8.0 TiC, 12.0 TaC.
Sintered: 92.5 Rock A. For extrusion press nibs, drawing and header dies. Sintered carbide, wear and abrasion resistant.

AMCO COPPER
Climax Performance Materials Corp.
Electroytic tough pitch copper. For electrical applications and general usage in sheet, strip and rod applications.

AMCO OFHC
AMAX Corp.
Cu 99.99.
For electronic and electrical equipment, magnetrons, triodes; oxygen free high conductivity copper. *Obsolete*

AMCOAB D-55
American Carbide Corp.
Co 16, WC 57, 27.0 TaC.
Sintered: 86 Rock A. 360,000 transverse strength. For extrusion, drawing and header dies, punches. Sintered carbides, wear and abrasion resistant.

AMCOH
A. Milne & Co.
C 0.9, Mn 1.15, Cr 0.5, W 0.5, bal Fe.
Oil hardening; tool steel; AISI O1.

AMCOH EXTRA SPECIAL
A. Milne & Co.
C 0.7, W 18, Cr 4, V 1, bal Fe.
For tools, cutters, hobs; high speed steel.

AMCOH HOLLOW DIE
A. Milne & Co.
C 0.8, Cr 0.9, bal Fe.
For hollow dies; oil hardening.

AMCOH OIL HARDENING
A. Milne & Co.
C 0.9, Cr 0.5, W 0.5, Mn 1-1.25, bal Fe.
For tools, dies, punches; oil hardening.

AMCOLOY
A. Milne & Co.
C 0.75, Mn 0.75, Cr 0.9, Ni 1.75, Mo 0.35, bal Fe.
Oil hardening; for shafts, arbors, lathe centers; AISI L6.

AMCOLOY 70
A. Milne & Co.
C 0.75, Ni 1.25, Cr 0.8, Mo 0.25, V 0.15, bal Fe.
Heat treated: 121,000-208,000 TS; 90,000-180,000 PL; 29-45 Rock C. For brake dies, bushings, cams, shear blades, gears, set screws, forming and swaging dies, punches. Type L6; tough, oil hardening.

AMCOLOY 86-20
Ampco Pittsburgh Corp.
Ni 1.1-1.5, Be 2.25-3, bal Cu.
Sand cast: 160,000 TS; 100,000 YS; 2.5 El; 2.5 RA; 393 Brin. For welding dies, safety tools; age-hardenable. *Obsolete*

AMCR
Creusot-Loire
C 0.3, Mn 18, Si 0.6, Cr 10, Ni 1, bal Fe.
Water quenched: 700 N/mm^2, psi TS min. Nonmagnetic alloy. AFNOR Z 30 MCN 18-10.

AMCROM
Climax Performance Materials Corp.
Cr 0.4-1.2, bal Cu.
Aged: 80,000 TS. For electrical contacts, rocket nozzles, resistance welding tips, current carrying components. 80-90% electrical conductivity, age-hardenable, good resistance to softening at high temperatures.

AMDRY 721
Alloy Metals Inc.
C 0.15, Cr 6, B 1.25, Si 3.25, Fe 3.4, bal Ni.
Powder for flame spraying. 2050°F approximate fusing temperature.

AMDRY 722
Alloy Metals Inc.
C 0.05, B 1.25, Si 3.25, Fe 1, bal Ni.
Powder for flame spraying. 1925°F approximate fusing temperature.

AMDRY 723
Alloy Metals Inc.
C 0.05, B 1, Si 2.5, Fe 1, bal Ni.
Powder for flame spraying. 1950°F approximate fusing temperature.

AMDRY 724
Alloy Metals Inc.
C 0.05, B 1, Si 2.5, Fe 1, bal Ni.
Powder for flame spraying. 1950°F approximate fusing temperature.

AMDRY 754
Alloy Metals Inc.
C 0.45, Cr 10, B 1.25, Si 3.5, Fe 3.65, bal Ni.
Hard surfacing powder for flame spraying. 1980°F approximate fusing temperature.

AMDRY 755
Alloy Metals Inc.
C 0.55, Cr 12, B 2.4, Si 4, Fe 4.15, bal Ni.
Hard surfacing powder for flame spraying. 1930°F approximate fusing temperature.

AMDRY 756
Alloy Metals Inc.
C 0.6, Cr 13, B 2.75, Si 4.25, Fe 4.4, bal Ni.
Hard surfacing powder for flame spraying. 1910°F approximate fusing temperature.

AMDRY 761
Alloy Metals Inc.
C 0.65, Cr 14, B 3.15, Si 4.4, Fe 4.5, bal Ni.
Hard surfacing powder for flame spraying. 1890°F approximate fusing temperature.

AMDRY 769
Alloy Metals Inc.
C 0.65, Cr 14.5, B 3.15, Si 4.4, Fe 4.5, Cu 2.3, Mo 2.3, bal Ni.
Hard surfacing powder for flame spraying. 1900°F approximate fusing temperature.

AMDRY 850
Alloy Metals Inc.
C 0.7, Cr 19, B 3.25, Si 3.5, Fe 3.65, Ni 18, W 10, bal Co.
Hard facing powder for flame spraying. 2040°F approximate fusing temperature.

AMDRY ALLOYS
Now AMI ALLOYS.

AMER-LED A
U.S. Steel Corp.
C 0-0.15, Mn 0.75-1.25, Pb 0.15-0.35, S 0.25, bal Fe.
For screw machine products, screws, bolts; free-cutting. *Obsolete*

AMER-LED B
U.S. Steel Corp.
C 0-0.15, Mn 0.85-1.35, S 0.36, Pb 0.15-0.35, bal Fe.
For screw machine products, screws, bolts; free-cutting. *Obsolete*

AMERA-MAG
American Tank & Fabricating Co.
c 0.08-0.15, Mn 1.2, P 0.12, Al 0.05, Ni 0-0.1, Mo 0-0.1, bal Fe.
Rolled: 70,000 TS; 50,000 YS; 34 El; 53 RA; 147 Brin. For tanks, bodies; high tensile.

AMERCUT 1020
U.S. Steel Corp.
C 0.18-0.23, bal Fe.
65,000-85,000 TS; 50,000-70,000 YS; 20-30 El; 45-55 RA; 155-185 Brin. For screw products; SAE 1020. *Obsolete*

AMERCUT 4640
U.S. Steel Corp.
C 0.38-0.43, Mn 0.6-0.8, Ni 1.65-2, Mo 0.2-0.3, bal Fe.
Heat treated: 125,000 TS; 110,000 YS; 17 El; 50 RA. For engine studs, bolts, shafts; tough. *Obsolete*

AMERCUT B-1111
U.S. Steel Corp.
C 0.08-0.13, Mn 0.6-0.9, P 0.09-0.13, S 0.1-0.2, bal Fe.
Cold finish: 70,000-90,000 TS; 60,000-80,000 YS; 15-25 El; 40-50 RA; 160-190 Brin. For automatic screw machine operations; approximate machinability 125 ft/min. *Obsolete*

AMERCUT B-1112
U.S. Steel Corp.
C 0.08-0.13, Mn 0.6-0.9, P 0.09-0.13, S 0.16-0.23, bal Fe.
Cold finish: 80,0000-100,000 TS; 65,000-85,000 YS; 10-20 El; 40-50 RA; 170-200 Brin. For automatic screw machine operations; approximate machinability 180 ft/min. *Obsolete*

AMERCUT B-17
U.S. Steel Corp.
C 0.08-0.16, Mn 0.6-0.9, P 0.09-0.13, S 0.1-0.2, bal Fe.
Cold finish: 80,000-100,000 TS; 65,000-85,000 YS; 10-20 El; 40-50 RA; 170-200 Brin. For screw machine operations; approx machinability 150 ft/min. *Obsolete*

AMERCUT B-24
U.S. Steel Corp.
C 0.08-0.16, Mn 0.6-0.9, P 0.09-0.13, S 0.2-0.3, bal Fe.
Cold finish: 80,000-100,000 TS; 65,000-85,000 YS; 10-20 El; 40-50 RA; 170-200 Brin. For automatic screw machine operations; approx machinability 195 ft/min. *Obsolete*

AMERCUT B-28
U.S. Steel Corp.
C 0.08-0.16, Mn 0.6-0.9, P 0.09-0.13, S 0.2-0.3, bal Fe.
Cold finish: 80,0000-100,000 TS; 65,000-85,000 YS; 10-20 El; 40-50 RA; 170-200 Brin. For automatic screw machine operations; approximate machinability 215 ft/min. *Obsolete*

AMERCUT C-1115
U.S. Steel Corp.
C 0.13-0.18, Mn 0.7-0.1, P 0.13, bal Fe.
Rolled: 10-20 El. For screw machine parts, bolts, nuts, screws; free-cutting. *Obsolete*

AMERCUT C-1117
U.S. Steel Corp.
C 0.14-0.2, Mn 1-1.3, S 0.08-0.13, bal Fe.
Rolled: 15-25 El. For screw machine parts, bolts, nuts, screws; free-cutting. *Obsolete*

AMERCUT C-1118
U.S. Steel Corp.
C 0.14-0.2, Mn 1.3-1.6, S 0.08-0.13, bal Fe.
Rolled: 15-25 El. For screw machine parts, bolts, nuts, screws; free-cutting. *Obsolete*

AMERCUT C-1120
U.S. Steel Corp.
C 0.18-0.23, Mn 0.7-1, S 0.08-0.13, bal Fe.
Rolled: 15-25 El. For screw machine parts, bolts, nuts, screws; free-cutting. *Obsolete*

AMERCUT C-1132
U.S. Steel Corp.
C 0.27-0.34, Mn 1.35-1.65, S 0.08-0.13, bal Fe.
Rolled: 15-25 El. For screw machine parts, bolts, nuts, screws; free-cutting. *Obsolete*

AMERCUT C-1137
U.S. Steel Corp.
C 0.32-0.39, Mn 1.35-1.65, S 0.08-0.13, bal Fe.
For screw machine parts, cold headed parts; wire. *Obsolete*

AMERCUT C-1137
U.S. Steel Corp.
C 0.32-0.39, Mn 1.35-1.65, S 0.08-0.13, bal Fe.
Rolled: 15-25 El. For screw machine parts, bolts, gears, screws; free-cutting. *Obsolete*

AMERCUT C-1141
U.S. Steel Corp.
C 0.37-0.45, Mn 1.35-1.65, S 0.08-0.13, bal Fe.
Rolled: 15-25 El. For screw machine parts, bolts, gears, nuts, screws; free-cutting. *Obsolete*

AMERCUT X1020
U.S. Steel Corp.
C 0.15-0.25, bal Fe.
Cold drawn: 70,000-90,000 TS; 50,000-70,000 YS; 15-25 El; 50-55 RA; 160-190 Brin. For screw machine products; S.A.E. X1020. *Obsolete*

AMERHEAD 1330
U.S. Steel Corp.
C 0.28-0.33, Mn 1.6-1.9, P 0-0.04, S 0-0.04, bal Fe.
For screw machine parts, cold headed parts; wire. *Obsolete*

AMERHEAD 1335
U.S. Steel Corp.
C 0.33-0.38, Mn 1.6-1.9, P 0-0.04, S 0-0.04, bal Fe.
For screw machine parts, cold headed parts; wire. *Obsolete*

AMERHEAD 4037
U.S. Steel Corp.
C 0.35-0.4, Mn 0.75-1, Mo 0.2-0.3, bal Fe.
Drawn: 85,000-100,000 TS; 75,000-85,000 YS; 15-25 El; 50-65 RA. For bolts, shafts, screws; water hardened. *Obsolete*

AMERHEAD 4140
U.S. Steel Corp.
C 0.38-0.43, Mn 0.7-1, Cr 0.8-1.1, Mo 0.15-0.25, bal Fe.
For screw machine parts, cold headed parts; wire. *Obsolete*

AMERHEAD 4615
U.S. Steel Corp.
C 0.13-0.18, Ni 1.65-2, Mo 0.2-0.3, bal Fe.
For screw machine parts, cold headed parts; wire. *Obsolete*

AMERHEAD 8620
U.S. Steel Corp.
C 0.18-0.23, Mn 0.7-0.9, Cr 0.4-0.6, Ni 0.4-0.7, Mo 0.15-0.25, bal Fe.
Drawn: 78,000-93,000 TS; 68,000-80,000 YS; 15-25 El; 50-65 RA. For bolts, shafts; case hardened. *Obsolete*

AMERHEAD A-2317
U.S. Steel Corp.
C 0.15-0.2, Mn 0.4-0.6, Ni 3.25-3.75, bal Fe.
For screw machine parts, fasteners; wire. *Obsolete*

AMERHEAD A-2330
U.S. Steel Corp.
C 0.28-0.33, Ni 3.25-3.75, Mn 0.6-0.8, bal Fe.
For screw machine parts, fasteners; wire. *Obsolete*

AMERHEAD A-3115
U.S. Steel Corp.
C 0.13-0.18, Mn 0.4-0.6, Cr 0.55-0.75, Ni 1.1-1.4, bal Fe.
For screw machine parts, fasteners; wire. *Obsolete*

AMERHEAD A-3120
U.S. Steel Corp.
C 0.17-0.22, Mn 0.6-0.8, Cr 0.55-0.75, Ni 1.1-1.4, bal Fe.
For screw machine parts, fasteners; wire. *Obsolete*

AMERHEAD A-3135
U.S. Steel Corp.
C 0.33-0.38, Cr 0.5-0.7, Ni 1.1-1.4, bal Fe.
For screw machine parts fasteners; wire. *Obsolete*

AMERHEAD C-1018
U.S. Steel Corp.
C 0.15-0.2, Mn 0.6-0.9, P 0-0.04, S 0-0.05, bal Fe.
For screw machine parts, cold headed parts; wire. *Obsolete*

AMERHEAD NE9437
U.S. Steel Corp.
C 0.35-0.4, Mn 0.9-1.2, Cr 0.3-0.5, Ni 0.3-0.6, Mo 0.08-0.15, bal Fe.
Drawn: 90,000-105,000 TS; 75,000-85,000 YS; 12-25 El; 45-65 RA. For bolts, shafts; case hardened. *Obsolete*

AMERICAN 10
American Steel Foundries
C 0-0.25, Mn 0.5-0.7, Si 0.2-0.6, Mo 0.4-0.6, bal Fe.
Cast N-T: 65,000 TS; 35,000 YS; 20 El; 30 RA; 160-225 Brin. For high pressure and valve castings; high temperature steam resisting. *Obsolete*

AMERICAN ALLOY
American manufacture
Al 95, Cu 3, Mg 1, Mn 1.
For light alloy parts; heat treated.

AMERICAN BLACK HEART MALLEABLE IRON
Bethlehem Steel Corp.
C 2.8-3.5, Si 0.6-0.8, Mn, bal Fe.
20,000-46,000 TS; 0.2-15 El; 105-220 Brin. For plumbing, hardware, fittings, pipes; castings. *Obsolete*

AMERICAN BLACK HEART MALLEABLE IRON
Manufacturer not listed.
C 2.8-3.5, Si 0.6-0.8, Mn, bal Fe.
20,000-46,000 TS; 0.2-15 El; 105-220 Brin. For plumbing, hardware, fittings, pipes; castings. *Obsolete*

AMERICAN CA-14
American Steel Foundries
C 0-0.15, Mn 0-0.85, Si 0-1, Cr 11.5-15, bal Fe.
Cast: 80,000 TS; 50,000 YS; 20 El; 30 RA. For valve seats, chemical apparatus castings; corrosion resistant. *Obsolete*

AMERICAN CF-7
American Steel Foundries
C 0-0.07, Cr 18-21, Ni 8-11, bal Fe.
Cast W.Q.: 70,000 TS; 30,000 YS; 45 El; 50 RA; 150 Brin. For valves, fittings, pumps, castings; corrosion resisting to acetic acid. *Obsolete*

AMERICAN CF-7C
American Steel Foundries
C 0-0.07, Cr 18-21, Ni 9-12, Cb = 10 x C, bal Fe.
Water quenched, 2000 F: 71,000 TS; 35,000 YS; 45 El; 50 RA; 150 Brin. For stainless castings, valves, fittings; corrosion resistant, austenitic. *Obsolete*

AMERICAN CF-7MC
American Steel Foundries
C 0-0.07, Cr 18-20, Ni 8-10, Mo 2.5, Cb = 10 x C, bal Fe.
Water quenched, 2000 F: 72,000 TS; 35,000 YS; 45 El; 50 RA; 160 Brin. For stainless castings; corrosion and heat resistant, austenitic. *Obsolete*

AMERICAN CK-25
American Steel Foundries
C 0-0.25, Cr 23-27, Ni 19-22, bal Fe.
Cooled from 1900 F: 70,000 TS; 30,000 YS; 25 El; 25 RA; 170 Brin. For heat resistant castings; heat resistant. *Obsolete*

AMERICAN CN-7
American Steel Foundries
C 0-0.07, Cr 18-20, Ni 23-26, Mo, bal Fe.
Water quenched, 2000 F: 70,000 TS; 35,000 YS; 45 El; 45 RA; 150 Brin. For heat resisting castings; heat and corrosion resistant. *Obsolete*

AMERICAN CT-7M
American Steel Foundries
C 0-0.07, Cr 13-17, Ni 34-37, Mo 2.5, bal Fe.
Water quenched, 2000 F: 65,000 TS; 35,000 YS; 45 El; 45 RA; 150 Brin. For heat resisting castings, valves, fittings, pumps; heat and corrosion resistant to hot sulfuric and caustic solutions. *Obsolete*

AMERICAN HYLASTIC
American Steel Foundries
C 0.26-0.36, Mn 1.4-1.7, V 0.1, bal Fe.
Cast N-T: 90,000 TS; 60,000 YS; 22 El; 45 RA; 180 Brin. For railroad and structural castings; tempered. *Obsolete*

AMERICAN MARINE
United American Metals Corp.
Pb-Sn-Sb.
Cast: 8,750 TS; 8,000 YS; 1.0 El; 0.7 RA; 30 Brin. For Babbitts, bearings; Babbitt metal. *Obsolete*

AMERICAN MARINE GENUINE
United American Metals Corp.
For bearings for marine engines and locomotives; Babbitt; shock resistant. *Obsolete*

AMERICAN PISTON RING STANDARD IRON
Koppers Co. Inc.
CC 0.6-0.8, Mn 0.5-0.8, Si 1.2-3.2, 2.8-3.1% graphite C, bal Fe.
30,000 TS. For piston rings for autos and aero engines; castings. *Obsolete*

AMERICAN SILVER-1
American manufacture
Cu 49.4, Zn 20.7, Ni 24.2, Fe 1.3, Mn 3.8, Sn 0.5.
For ornaments, hardware; corrosion resistant.

AMERICAN SILVER-2
American manufacture
Cu 59, Zn 23, Ni 11, Pb 3, Al 1.5, 5 P + Sn.
For ornaments, hardware; corrosion resistant.

AMERICAN SILVER-3
American manufacture
Cu 49-58, Zn 21-24, Ni 15-24, 4 (Mn + Sn + Fe + Al + Pb) max.
For ornaments, hardware; corrosion resistant.

AMERICAN SUPER-TENS

U.S. Steel Corp.
C 0.72-0.93, Mn 0.4-1.1, Si 0.1-0.35, bal Fe.
Drawn: 250,000 min TS. For prestressed concrete. *Obsolete*

AMERICAN TYPE 12

American Steel Foundries
C 0.1-0.2, Mn 0.6-0.8, Si 0-0.6, Ni 2.5-3, bal Fe.
Cast: 70,000 TS; 40,000 YS; 26 El; 35 RA. For valves, fittings, pumps; high impact value at low temperature. *Obsolete*

AMERICAN TYPE 20

American Steel Foundries
C 0.2-0.25, Mn 0.65-0.85, Si 0.2-0.5, Cr 0.65-0.85, Mo 0.25-0.35, bal Fe.
Cast N-T: 100,000 TS; 70,000 YS; 22 El; 35 RA; 202 Brin. For valves, fittings, structural castings; in welded structures. *Obsolete*

AMERICAN TYPE 22

American Steel Foundries
C 0.15-0.3, Mn 0.65-0.85, Si 0-0.6, Cr 4-6.5, Mo 0.4-0.6, bal Fe.
Cast N-T: 100,000 TS; 65,000 YS; 18 El; 30 RA; 250-300 Brin. For oil refinery valves and fittings, pumps; high temperature steam and oil service. *Obsolete*

AMERICAN TYPE 23

American Steel Foundries
C 0.25-0.35, Mn 0.6-0.8, Si 0-0.6, Ni 1-1.5, Cr 0.5-0.8, bal Fe.
Cast: 99,000 TS; 57,000 YS; 22 El; 40 RA; 200 Brin. For gears, pinions, shafts, tough. *Obsolete*

AMERICAN TYPE 31

American Steel Foundries
C 0.2-0.3, Mn 0.7-1, Ni 2.5-3, Cr 1-1.5, Mo 0.2-0.4, bal Fe.
Cast N-T: 150,000 TS; 130,000 YS; 10 El; 20 RA; 300 Brin. For structural and aircraft castings; tough. *Obsolete*

AMERICAN VALVE SPRING

U.S. Steel Corp.
C 0.6-0.75, Mn 0.5-0.9, Si 0.12-0.3, bal Fe.
Heat treated: 195,000-300,000 TS; 48 El; 450 Brin. For valve springs; oil tempered wire. *Obsolete*

AMERSPRING

U.S. Steel Corp.
C 0.7-1, Mn 0.2-0.6, Si 0.12-0.3, bal Fe.
Hard drawn: 238,000-485,000 TS. For springs; music wire. *Obsolete*

AMERSTITCH

U.S. Steel Corp.
C 0.4-0.95, Mn 0.5-1.2, Si 0.1-0.3, bal Fe.
Drawn: 190,000-360,000 TS; for metal stitching wire. *Obsolete*

AMERSTRIP

U.S. Steel Corp.
C 0.05-1.35, bal Fe.
For springs; water hardening. *Obsolete*

AMERVAN

Manufacturer not listed.
Fe 63.5, V 35, Si 1.5.
For alloy steel making; vanadium additions.

AMI ALLOY 207

Alloy Metals Inc.
Fe 2.4, Cr 5.6, Si 3.2, B 2.8, bal Ni.
For honeycomb, wide gap, and general purpose high temperature brazing. 2000-2075°F brazing temperature. P50T9N.

AMI ALLOY 100

Alloy Metals Inc.
C 0.03, Cr 19, Si 10, bal Ni.
High temperature brazing alloy. 2075-2200°F brazing temperature. AMS 4782; B14Y3; B50TF81.

AMI ALLOY 101

Alloy Metals Inc.
Cr 11.5, Si 6, bal Ni.
For honeycomb, wide gap, and general purpose high temperature brazing. 2200-2250°F brazing temperature.

AMI ALLOY 102

Alloy Metals Inc.
Cr 15.2, Si 8, bal Ni.
For honeycomb, wide gap, and general purpose high temperature brazing. 2150-2200°F brazing temperature. B50T51A; B50T1403.

AMI ALLOY 103

Alloy Metals Inc.
Cr 17.1, Si 9.2, B 0.08, bal Ni.
For honeycomb, wide gap, and general purpose high temperature brazing. 2100-2150°F brazing temperature. B50T51B.

AMI ALLOY 104

Alloy Metals Inc.
Cr 11.4, Si 6.8, B 0.3, bal Ni.
For honeycomb, wide gap, and general purpose high temperature brazing. 2100-2150°F brazing temperature. P50T9K.

AMI ALLOY 105

Alloy Metals Inc.
Cr 13.3, Si 7.6, B 0.23, bal Ni.
For honey comb, wide gap, and general purpose high temperature brazing. 2100-2150°F brazing temperature. P50T9L.

AMI ALLOY 131

Alloy Metals Inc.
Cr 16, Ni 2, bal Fe.
Plasma spray powder. Similar to AISI 431 stainless.

AMI ALLOY 134

Alloy Metals Inc.
Cr 18, Ni 10, bal Fe.
Plasma spray powder. Similar to AISI 304 stainless.

AMI ALLOY 136

Alloy Metals Inc.
Cr 17, Mo 2.5, Ni 12, bal Fe.
Plasma spray powder. Similar to AISI 316 stainless.

AMI ALLOY 201

Alloy Metals Inc.
Fe 1.3, Cr 2.9, Si 2.8, B 1.9, bal Ni.
For honeycomb, wide gap, and general purpose high temperature brazing. 2000-2110°F brazing temperature. P50T9A.

AMI ALLOY 202

Alloy Metals Inc.
Fe 1, Cr 2.2, Si 2.6, B 1.6, bal Ni.
For honeycomb, wide gap, and general purpose high temperature brazing. 2000-2110°F brazing temperature. P50T9B.

AMI ALLOY 205

Alloy Metals Inc.
Fe 1.1, Cr 2.7, Si 2.8, B 1.9, bal Ni.
For honeycomb, wide gap, and general purpose high temperature brazing. 2000-2120°F brazing temperature. P50T9F.

AMI ALLOY 300

Alloy Metals Inc.
C 0.03, Cr 19.5, Si 9.5, Mn 9.5, bal Ni.
High temperature brazing alloy. 2025-2125°F brazing temperature. B50T50.

AMI ALLOY 301

Alloy Metals Inc.
C 3.9, Co 12, bal W.
Plasma spray and hard facing powder. PWA 1301.

AMI ALLOY 302

Alloy Metals Inc.
C 3.9, Co 12, bal W.
Plasma spray and hard facing powder. PWA 1302.

AMI ALLOY 304

Alloy Metals Inc.
C 12.5, Cr 87.
Plasma spray and hard facing powder. PWA 1304.

AMI ALLOY 305

Alloy Metals Inc.
Cr 5, Ni 20, 75.0 CrC.
Plasma spray and hard facing powder. PWA 1305.

AMI ALLOY 306

Alloy Metals Inc.
C 13, Cr 86.
Plasma spray and hard facing powder. PWA 1306; B50TF39.

AMI ALLOY 307

Alloy Metals Inc.
Cr 5, Ni 20, 75.0 CrC.
Plasma spray and hard facing powder. PWA 1307.

AMI ALLOY 308

Alloy Metals Inc.
Cr 3, Ni 12, 85.0 CrC.
Plasma spray and hard facing powder. PWA 1308.

AMI ALLOY 313

Alloy Metals Inc.
Mo 99.5.
Plasma spray and hard facing powder. PWA 1313.

AMI ALLOY 315

Alloy Metals Inc.
Cr 20, bal Ni.
Plasma spray powder. PWA 1315. B50TF40A.

AMI ALLOY 316

Alloy Metals Inc.
C 0.5, Cr 25.5, Ni 10.5, W 7.5, bal Co.
Plasma spray and hard facing powder. PWA 1316.

AMI ALLOY 317

Alloy Metals Inc.
Cr 20, bal Ni.
Plasma spray powder. PWA 1317; B50TF40B.

AMI ALLOY 318

Alloy Metals Inc.
C 0.5, Cr 25.5, Ni 10.5, W 7.5, bal Co.
Plasma spray and hard facing powder. PWA 1318.

AMI ALLOY 319

Alloy Metals Inc.
Cr 20, bal Ni.
Plasma spray powder. PWA 1319.

AMI ALLOY 336

Alloy Metals Inc.
Cr 5, Fe 5.5, Mo 24.5, bal Ni.
Plasma spray and hard facing powder. PWA 1336.

AMI ALLOY 338

Alloy Metals Inc.
Mo 99.5.
Plasma spray and hard facing powder. PWA 1338.

AMI ALLOY 358

Alloy Metals Inc.
Co 30, Y 0.9, bal Al.
Plasma spray powder. PWA 1358.

AMI ALLOY 400

Alloy Metals Inc.
C 0.4, Cr 19, Si 8, B 0.8, W 4, Ni 16.5, bal Co.
High temperature brazing alloy. 2150-2200°F brazing temperature. B50T56; PWA 713.

AMI ALLOY 500
Alloy Metals Inc.
Ni 36.5, Mn 5, bal Cu.
Plasma spray powder. B50TF72.

AMI ALLOY 716
Alloy Metals Inc.
C 0-0.03, Mn 37.5, Ni 9.5, bal Cu.
High temperature brazing alloy. 1750-1850°F brazing temperature.

AMI ALLOY 717
Alloy Metals Inc.
C 0.03, Mn 23.5, Ni 9, bal Cu.
High temperature brazing alloy. 1800-1850°F brazing temperature. B50TF80.

AMI ALLOY 74
Alloy Metals Inc.
C 0.75, Si 4.5, Cr 14.5, B 3.5, Fe 4.5, bal Ni.
Plasma spray and hard facing alloy powder. AMS 4775.

AMI ALLOY 75
Alloy Metals Inc.
C 0.9, Si 4, Cr 16.5, B 3.75, Fe 4, bal Ni.
Plasma spray and hard facing alloy powder. AMS 4775.

AMI ALLOY 750
Alloy Metals Inc.
C 0.9, Cr 16.5, Si 4, B 3.75, Fe 4, bal Ni.
High temperature brazing alloy. 1950-2200°F brazing temperature. AMS 4775.

AMI ALLOY 760
Alloy Metals Inc.
C 0.03, Cr 16.5, Si 4, B 3.75, Fe 4, bal Ni.
High temperature brazing alloy. 1975-2200°F brazing temperature. AMS 4776.

AMI ALLOY 770
Alloy Metals Inc.
C 0.03, Cr 7, Si 4.5, B 2.75, Fe 3, bal Ni.
High temperature brazing alloy. 1950-2150°F brazing temperature. AMS 4777.

AMI ALLOY 780
Alloy Metals Inc.
C 0.03, Si 4.5, B 3, Fe 1, bal Ni.
High temperature brazing alloy. 1925-2150°F brazing temperature. AMS 4778.

AMI ALLOY 780 B
Alloy Metals Inc.
Si 4.5, B 3, bal Ni.
Plasma spray powder. B50TF84.

AMI ALLOY 790
Alloy Metals Inc.
C 0.03, Si 3.5, B 2, Fe 1, bal Ni.
High temperature brazing alloy. 1975-2150°F brazing temperature. AMS 4779.

AMI ALLOY 912
Alloy Metals Inc.
Si 2.5, Cr 10, B 2.5, Fe 2.5, bal Ni.
Plasma spray and hard facing powder.

AMI ALLOY 914
Alloy Metals Inc.
C 0-0.03, Si 4.5, B 3, Co 20, bal Ni.
High temperature brazing alloy. 2050-2175°F brazing temperature. BTS1205.

AMI ALLOY 915
Alloy Metals Inc.
C 0-0.02, Cr 13, Si 4, B 2.75, Fe 4, bal Ni.
High temperature brazing alloy. 2050-2150°F brazing temperature. BMS7-141.

AMI ALLOY 916
Alloy Metals Inc.
C 0.6, Cr 13, Si 4, B 3, Fe 4, bal Ni.
Plasma spray and hard facing alloy. High temperature brazing alloy. 1950-2150°F brazing temperature. AMS 4775.

AMI ALLOY 930
Alloy Metals Inc.
C 0.01, Si 7, Cu 5, Mn 22.5, bal Ni.
High temperature brazing alloy. 1870-2000°F brazing temperature. BMS7-141.

AMI ALLOY 931
Alloy Metals Inc.
C 0.01, Si 8, Mn 17, bal Ni.
High temperature brazing alloy. 1900-2000°F brazing temperature.

AMI ALLOY 942
Alloy Metals Inc.
Ni 38, bal Cu.
Plasma spray powder. B50TF42.

AMI ALLOY 955
Alloy Metals Inc.
Al 18.5, Ni 81.5.
Plasma spray powder. PWA 1339.

AMI ALLOY 956
Alloy Metals Inc.
Al 5, Ni 95.
Plasma spray powder.

AMINO 3
Krupp Stahl AG
tool material.
See KRUPP 2083

AMNIC 10
Climax Performance Materials Corp.
Ni 10, bal Cu (others very low).
Annealed: 221 MPa (38,000 psi) TS; 131 MPa (19,000 psi) YS; 41 El. 50% cold worked: 476 MPa (69,000 psi) TS; 428 MPa (62,000 psi) YS; 4 El. High purity copper-nickel alloy for electronic and cryogenic applications.

AMNIC 30
Climax Performance Materials Corp.
Ni 30, bal Cu (others very low).
Annealed: 372 MPa (54,000 psi) TS; 152 MPa (22,000 psi) YS; 41 El. 50% cold worked: 552 MPa (80,000 psi) TS; 524 MPa (76,000 psi) YS; 4 El. High purity copper-nickel alloy for electronic and cryogenic applications.

AMOL
Atlas Metal & Alloys Co. Ltd.
C 0.8-0.9, bal Fe.
For tools, cutters; water hardening.

AMOLA
Now AISI 40XX STEELS.

AMOTUN
Atlantic Steel Corp.
C 0.85, Cr 4, Co 6, V 1.75, W 1.5, Mo 8, bal Fe.
For cutting tools; high speed steel.

AMOUTUN
Atlantic Steel Corp.
C 0.8, Cr 4, Mo 8, W 1.5, V 1, bal Fe.
Hardened: 64-66 Rock C. For lathe and planer cutters, drills, reamers, broaches, hobs, form cutters, taps. Type M1 high speed steel, high red-hardness.

AMPAL 601
Ampal, Inc.
Aluminum. Fe 0.16, Si 0.1, 99.7 Al min.
Atomized aluminum powder meeting MIL-A-512A, Type III, Grade F, Class 7 specifications. Used as a fuel material in explosives and as a reducing agent in chemical processing.

AMPAL 602
Ampal, Inc.
Aluminum. Fe 0.16, Si 0.1, 99.7 Al min.
Atomized aluminum powder for use in aluminothermic reactions and master alloy manufacture.

AMPAL 603
Ampal, Inc.
Aluminum. Fe 0.16, Si 0.1, 99.7 Al min.
Atomized aluminum powder for aluminothermic reactions and master alloy manufacture.

AMPAL 604
Ampal, Inc.
Aluminum. Fe 0.16, Si 0.1, 99.7 Al min.
40 mesh powder with somewhat improved flow properties than those offered by Ampal 601. Used in aluminothermic reactions and cored welding filler wire.

AMPAL 605
Ampal, Inc.
Aluminum. Fe 0.16, Si 0.1, 99.7 Al min.
Atomized aluminum powder for aluminothermic smelting and aluminum master alloys.

AMPAL 607
Ampal, Inc.
Aluminum. Fe 0.16, Si 0.1, 99.7 Al min.
Powder used in aluminothermic smelting and as a component in cored welding filler wire.

AMPAL 609
Ampal, Inc.
Aluminum. Fe 0.16, Si 0.1, 99.7 Al min.
Air-atomized granule used in aluminothermic reactions for smelting and thermite welding.

AMPAL 611
Ampal, Inc.
Aluminum. Fe 0.16, Si 0.1, 99.7 Al min.
Atomized aluminum powder meeting MIL-A-512-A Type III, Grade F, Class 6 specifications. Used in the chemical and ordnance industries.

AMPAL 615
Ampal, Inc.
Aluminum. Fe 0.16, Si 0.1, 99.7 Al min.
Atomized aluminum powder for explosives, pyrotechnics, thermit compounds, reducing agents, and plastic compound fillers.

AMPAL 625
Ampal, Inc.
Aluminum. Fe 0.16, Si 0.1, 99.7 Al min.
Free-flowing powder sized in the 100 to 325 mesh range. Used in chemical reducing reactions and as a fuel in combustion, refractory cutting systems.

AMPAL 637
Ampal, Inc.
Aluminum. Fe 0.16, Si 0.1, 99.7 Al min.
Ultra-fine, air-atomized powder used in the production of metallic pigments.

AMPAL 650
Ampal, Inc.
Aluminum. Fe 0.16, Si 0.1, 99.7 Al min.
Atomized aluminum powder (-100 mesh) for cored welding wires, aluminothermic smelting and chemical reductions.

AMPAL 660
Ampal, Inc.
Aluminum. Fe 0.16, Si 0.1, 99.7 Al min.
Moderately coarse aluminum powder for compaction in aluminum powder metallurgy and aluminized coatings.

AMPAL 670
Ampal, Inc.
Aluminum. Fe 0.16, Si 0.1, 99.7 Al min.
Atomized aluminum powder (12 to 325 mesh) for commercial explosives, aluminothermic reduction and chemical reducing reagent.

AMPCO
Wellman Dynamics Corp.
Sn, Zn, Pb, bal Cu.
For bearings; heavy duty. *Obsolete*

AMPCO 12
Ampco Metal
Copper. Al 8.9, Fe 2.9, Cu 88.
Cast: 60,000 TS; 30,000 YS; 40 El; 130 Brin. For bushings, adjusting nuts, fittings, inlet hoppers, machinery parts. Corrosion resistant. C95200

AMPCO 15
Ampco Metal
Copper. Al 9.3, Fe 3.1, Cu 87.1.
Wrought: 92,000 TS; 46,000 YS; 25 El; 25 RA; 174 Brin. For valve stems, washers, worm gears, cam rollers, bushings, shear pins, studs. Good bearing qualities, wear and corrosion resistant. C62300

AMPCO 16
Ampco Metal
Copper. Al 10.1, Fe 3.3, Cu 86.2.
Cast: 90,000 TS; 32,000 YS; 22 El; 155 Brin. For bushings, gears, worm wheels, machinery parts, trolley shoes, trolley wheels. Corrosion and shock resistant.

AMPCO 18
Ampco Metal
Copper. Al 10.5, Fe 3.5, Cu 85.5.
Cast: 77,000-90,000 TS; 37,000-42,000 YS; 10-14 El; 6-12 RA; 165-185 Brin. For acid equipment, gears, worm wheels, trolley shoes, piston rods, valve seats. Corrosion and shock resistant. Also UNS C62400. C95400

AMPCO 18-13
Ampco Pittsburgh Corp.
Al 10.6-11.2, Fe 3.4-4, bal Cu.
Rolled: 85,000-90,000 TS; 38,000-42,000 YS; 15-20 El; 18-22 RA; 149-159 Brin. For gears, worm wheels; high impact resistance. *Obsolete*

AMPCO 18-22
Ampco Metal
Copper. Al 10.5, Fe 3.5, Cu 85.5.
Heat treated: 90,000-100,000 TS; 45,000-55,000 YS; 3-7 El; 3-7 RA; 202-223 Brin. For bearings; heat treatable, wear resistant, impact resistant. C95400

AMPCO 18-23
Ampco Metal
Copper. Cu 85.5, Al 10.5, Fe 3.5.
Heat treated: 100,000 TS; 50,000 YS; 14 El; 14 RA; 179-207 Brin. For aircraft, propellers, motors, heavy duty worm gears; castings. C95400

AMPCO 18-33
Ampco Pittsburgh Corp.
Cu 84.6, Al 11.3, Fe 3.7, 0.4% special agents.
Heat treated: 90,000-100,000 TS; 45,000-55,000 YS; 8-12 El; 180-200 Brin. For aircraft parts. *Obsolete*

AMPCO 20
Ampco Metal
Copper. Al 11.3, Fe 3.8, Cu 84.4.
Cast: 83,000-90,000 TS; 40,000-43,000 YS; 4-6 El; 3.5 RA; 212-241 Brin. For cams, rollers, safety tools, welding jaws, worm gears, valve bodies. Corrosion and wear resistant.

AMPCO 20-13
Ampco Metal
Copper. Al 11.3, Fe 3.8, Cu 84.4.
Heat treated: 85,000-102,000 TS; 35,000-50,000 YS; 6-12 El; 192-207 Brin. For bearings, bushings, corrosion and wear resistant; non-seizing.

AMPCO 21
Ampco Metal
Copper. Al 13.1, Fe 4.5, Cu 82.
Cast: 75,000-95,000 TS; 55,000-60,000 YS; 1.5 El; 0.5 RA; 285-302 Brin. For valves, bushings, forming dies, slides, roller bushings, gears, worm gears. Corrosion and wear resistant.

AMPCO 21W
Ampco Metal
Copper. Cu 80.2, Al 12.8, Fe 4.5, 2.5 others max.
110 ksi TS; 60 ksi YS; 1.0 El; 286 Brin. For guide bushings, wear strips, die rings, inserts, forming rolls, wiping blocks.

AMPCO 22
Ampco Metal
Copper. Al 14.1, Fe 4.7, Cu 80.7.
Cast: 85,000 TS; 70,000 YS; 0.5 El; 321-341 Brin. For forming and drawing dies. Corrosion and wear resistant. High compressive strength.

AMPCO 22W
Ampco Metal
Copper. Cu 78.7, Al 13.8, Fe 5, 2.5 others max.
105 ksi TS; 62 ksi YS; 0.5 El; 332 Brin. For forming and drawing dies especially with stainless steel.

AMPCO 24
Ampco Metal
Copper base castings. 351-364 Brin. For draining and forming operations. *Obsolete*

AMPCO 25
Ampco Metal
Copper.
Copper base castings. 364-375 Brin. Tough, strong die material.

AMPCO 45
Ampco Metal
Copper. Cu 81, Al 10, Fe 2.5, Ni 5, Mn 0-1, 0.5 others max.
118 ksi TS; 75 ksi YS; 15 El; 228 Brin. For landing gear parts, valve guides, stems and seats, pump shafts and sleeves, faucet balls, plunger tips, gears, machine tool parts. High-strength wrought material for extra tough applications. C63000

AMPCO 483
Ampco Metal
Copper. Cu 81, Al 9, Fe 4, Ni 4.5, Mn 0-1, 0.5 others max.
92-105 ksi TS; 39-53 ksi YS; 22-24 El; 174-212 Brin. For marine propellers and other salt water applications. Corrosion and cavitation resistant. Also UNS 63200. C95800

AMPCO 495
Ampco Metal
Copper. Cu 75, Al 8, Fe 2.5, Ni 2, Mn 0-12.5, 0.5 others max.
96 ksi TS; 48 ksi YS; 28 El; 180 Brin. For pump and valve parts, hydraulic service, propellers. Heavy duty, corrosion resistant. C95700

AMPCO 8
Ampco Metal
Copper. Al 6.5, Fe 2.5, Cu 90.5.
Plate: 72,000 TS; 35,000 YS; 34 El; 130 Brin. Extruded: 82,000 TS; 55,000 YS; 35 El; 149 Brin. For wear plate service, slides, gibs, bushings, bearings, bolts, fittings. Alpha phase, corrosion resistant. High impact and fatigue strength. Also UNS C61300. C61400

AMPCO A-1
Ampco Metal
Copper. Al 8.5, Fe 3, Cu 87.5.
Cast: 80,000 TS; 29,000 YS; 30 El; 135 Brin. For bushings, bearings, gears, wear plates, connecting rods. Medium strength, light loads. C95200

AMPCO B-2
Ampco Metal
Copper. Al 10, Fe 1, Cu 88.5.
Bar: 85,000 TS; 42,000 YS; 23 El; 170 Brin. Cast: 77,000 TS; 29,000 YS; 30 El; 140 Brin. For bearings, liners, pump parts, pump impellers, bushings. Heat treatable, high strength. Aluminum bronze. For use with medium loads and average speeds. Corrosion resistant. Also UNS C61800. C95300

AMPCO BERYLLIUM COPPER
Ampco Pittsburgh Corp.
Cu 97.5, Be 2.5.
90,000-125,000 TS; 100,000-850,000 YS; 0-1 El; 350-420 Brin. For wire springs, forming dies, plastic dies, non-sparking tools, airplane propeller bushings; age-hardening. *Obsolete*

AMPCO C-3
Ampco Metal
Copper. Al 10.5, Fe 4, Cu 85, Ni 0-2.5, Mn 0-0.5.
Bar: 100,000 TS; 48,000 YS; 12 El; 192 Brin. Cast: 85,000 TS; 35,000 YS; 12 El; 183 Brin. For bearings, bushings, gears, valve guides and seats, pump and hydraulic valves. Aluminum bronze. Tough, corrosion and wear resistant. C95400

AMPCO C-3HT
Ampco Metal
Copper. Cu 85, Al 10.5, Fe 4, Ni 0-2.5, Mn 0-0.5, 0.5 others max.
95 ksi TS; 47 ksi YS; 10 El; 192 Brin. For bushings, bearings, gears, worms, worm wheels, valve guides and seats, pump parts, landing gear parts, shifter forks, wear plates, hydraulic valve parts, liners, steel mill slippers, pickling hooks. Impact and corrosion resistant. C95400

AMPCO D-4
Ampco Metal
Copper. Al 10, Fe 5, Ni 5, Cu 79.5.
Bar: 105,000 TS; 46,000 YS; 18 El; 195 Brin. Heat treated: 120,000 TS; 70,000 YS; 18 El; 240 Brin. For piston guides, valve seats, glands, pump fluid ends. High strength, tough Al-bronze, heat treatable. C95500

AMPCO D-4HT
Ampco Metal
Copper. Cu 79.5, Al 10, Fe 5, Ni 5, 0.5 others max.
115 ksi TS; 65 ksi YS; 14 El; 228 Brin. For aircraft components, valve and piston guides, valve seats, pump fluid ends, stuffing box nuts, glands. C95500

AMPCO M-4
Ampco Metal
Copper. Cu 77.6, Al 11, Fe 4.8, Ni 5.1, Mn 0-1, 0.5 others max.
130 ksi TS; 105 ksi YS (0.2% offset); 4 El; 269 Brin. For landing gear parts, oil field equipment, forming roll bearings, cryogenic valve components, truck bushings. High-duty, high-loading mechanical and corrosion service.

AMPCO TRODALOY
Ampco Pittsburgh Corp.
Cu alloy.
For welding electrodes. *Obsolete*

AMPCO TRODE 22
Ampco Pittsburgh Corp.
Al 11, Fe 2, bal Cu.
Cast: 65,000-75,000 TS; 294-342 Brin. For bronze welding rod; building up worn parts. *Obsolete*

AMPCO-TRODE 10
Ampco Metal
Copper. Al 9-11, Fe 0-1.5, bal Cu.
Cast: 60,000-77,000 TS; 25,000-35,000 YS; 25-30 El; 28-30 RA; 110-130 Brin. For bronze welding rod: aluminum bronze.

AMPCO-TRODE 12
Ampco Pittsburgh Corp.
Al 10, Fe 1, bal Cu.
Cast: 50,000-60,000 TS; 100-115 Brin. For bronze welding rod for cast iron; now Ampco-Trode 10. *Obsolete*

AMPCO-TRODE 150
Ampco Metal
Copper. Al 10-11, Fe 3-5, 0.50 others max, bal Cu.
Weld metal: 90 ksi TS; 40 ksi YS; 20 El; 166 Brin. For welding AMPCO 18 Alloy.

AMPCO-TRODE 16
Ampco Pittsburgh Corp.
Al 9.2, Fe 4.8, bal Cu.
Cast: 60,000-70,000 TS; 10-20 El; 147-160 Brin. For bronze welding rod; now Ampco-Trode 160. *Obsolete*

AMPCO-TRODE 160
Ampco Metal
Copper. Al 11-12, Fe 3-4.25, 0.6 others max, bal Cu.
Cast: 89,000 TS; 47,000 YS; 15 El; 17 RA; 177 Brin. For welding electrodes, bearing overlay; resists salt water and commercial acids.

AMPCO-TRODE 18
Ampco Pittsburgh Corp.
Al 10.3, Fe 5.8, bal Cu.
Cast: 70,000-80,000 TS; 8-14 El; 160-180 Brin. For bronze welding rod; building up worn parts; now Ampco-Trode 200. *Obsolete*

AMPCO-TRODE 20
Ampco Pittsburgh Corp.
Al 11.8, Fe 5.9, bal Cu.
Cast: 75,000-85,000 TS; 200-250 Brin. For bronze welding rod; building up worn parts; now Ampco-Trode 250. *Obsolete*

AMPCO-TRODE 200
Ampco Metal
Copper. Al 12-13, Fe 3-5, 0.6 others max, bal Cu.
Cast: 90,500 TS; 46,000 YS; 4 El; 5 RA; 212 Brin. For welding electrodes, surfacing, overlays; corrosion and acid resistant.

AMPCO-TRODE 21
Ampco Pittsburgh Corp.
Al 10, Fe 3, bal Cu.
Cast: 60,000-70,000 TS; 1.5-3.0 El; 275-310 Brin. For bronze welding rods; building up worn parts. *Obsolete*

AMPCO-TRODE 250
Ampco Metal
Copper. Al 13-14, Fe 3-5, 0.6 others max, bal Cu.
Cast: 77,000 TS; 54,000 YS; 1 El; 260 Brin. For welding electrodes, bushings, bearings, gears, slides; free from galling and scuffing.

AMPCO-TRODE 300
Ampco Metal
Copper. Al 14-15, Fe 3-5, 0.6 others max, bal Cu.
Cast: 80,000 TS; 69,000 YS; 0 El; 316 Brin. For welding electrodes, for overlaying, drawing and forming dies; wear resistant; good bearing qualities.

AMPCO-TRODE 40
Ampco Metal
Copper. Al 7-8, Fe 2-3, Ni 2-3, Mn 11-12, 0.6 others max, bal Cu.
Wrought: 95,500 TS; 56,000 YS; 27 El; 38 RA; 185 Brin. For manganese bronze welding rod. Wear resistant.

AMPCO-TRODE 46
Ampco Metal
Copper. Al 8.5-9.5, Ni 4-5.5, Fe 3-5, Mn 0.6-3.5, 0.50 others max, bal Cu.
Weld metal: 99 ksi TS; 58 ksi YS; 25 El; 187 Brin. For welding ship propellers, ship fittings, pump housings. AWS-A5.6 Class CuNiAl.

AMPCO-TRODE 7
Ampco Metal
Copper. Al 6-6.9, 0.50 others max, bal Cu.
Weld metal: 68 ksi TS; 28 ksi YS; 47 El; 125 Brin. For overlay deposits; not recommended for joining.

AMPCOLOY 10W-86
Ampco Pittsburgh Corp.
Cu-W.
160,000 TS; 260 Brin. For resistance welding electrode; heat treatable. *Obsolete*

AMPCOLOY 30
Ampco Pittsburgh Corp.
Al 9.1, Pb 9.1, bal Cu.
45,000-50,000 TS; 18,000-20,000 YS; 32-35 El; 25-30 RA; 50-60 Brin. For bearings; formerly "Atlas 10." *Obsolete*

AMPCOLOY 31
Ampco Pittsburgh Corp.
Pb 7-9, Sn 7-9, Ni 0-1, 0.65% max others, bal Cu.
Cast: 35,000 TS; 18,000 YS; 15 El; 10 RA; 55 Brin. For bearings and bushings in high speed and pressure services. Resists mild acids. *Obsolete*

AMPCOLOY 32
Ampco Pittsburgh Corp.
Cu 80, Sn 10, Pb 10.
28,000-35,000 TS; 14,000-18,000 YS; 8-12 El; 8-12 RA; 50-65 Brin. For spindle bearings; formerly "Atlas P." *Obsolete*

AMPCOLOY 34
Ampco Pittsburgh Corp.
Cu 71.5, Pb 23, Sn 5.5.
25,000-30,000 TS; 14,000-16,000 YS; 14-18 El; 12-14 RA; 50-60 Brin. For bearings, bushings; formerly "Atlas 20." *Obsolete*

AMPCOLOY 342
Ampco Pittsburgh Corp.
Sn 4.25-5.5, Pb 18-22, Zn 0-0.5, bal Cu.
For bearings, bushings; continuous cast rod. *Obsolete*

AMPCOLOY 35
Ampco Pittsburgh Corp.
Sn 6.25-7.5, Pb 6-8, Zn 2-4, 1.0% max others, bal Cu.
Cast: 45,000 TS; 20,000 YS; 20 El; 15 RA; 65 Brin. For lead screw nuts, bearings. Low strength bearing alloy. *Obsolete*

AMPCOLOY 38
Ampco Pittsburgh Corp.
Sn 6, Pb 20, bal Cu.
Cast: 30,000 TS; 12,000 YS; 12 El; F 66 Rock. For bearings, mine water pump parts, roll neck bearings. Corrosion resistant, good embeddability. *Obsolete*

AMPCOLOY 382
Ampco Pittsburgh Corp.
Cu 81-85, Sn 6.2-7.5, Zn 2-4, Pb 6-8.
Cast: 30,000 TS; 14,000 YS; 12 El. For bearings, bushings; continuous cast. *Obsolete*

AMPCOLOY 3W-86
Ampco Pittsburgh Corp.
Cu-W.
120,000 TS; 210 Brin. For resistance welding electrode; heat treatable. *Obsolete*

AMPCOLOY 3W1
Ampco Pittsburgh Corp.
64% W, bal Cu.
Bars: 67,000-135,000 TS; 130-165 Brin. For resistance welding electrodes in welding yellow brass and stainless steel. 43% electrical conductivity, corrosion and wear resistant. *Obsolete*

AMPCOLOY 3W10
Ampco Pittsburgh Corp.
W 70, bal Cu.
Bars: 90,000-160,000 TS; 205-208 Brin. For resistance welding electrodes, for flash welding. 38% electrical conductivity, wear and corrosion resistant. *Obsolete*

AMPCOLOY 3W20
Ampco Pittsburgh Corp.
W 74, bal Cu.
Bars: 95,000-170,000 TS; 228-250 Brin. For resistance welding electrodes, for heavy projection welding. 30% electrical conductivity, wear and corrosion resistance. *Obsolete*

AMPCOLOY 40
Ampco Pittsburgh Corp.
Al 10, Cu 90.
Cast: 65,000-70,000 TS; 24,000-28,000 YS; 25-30 El; 25-30 RA; 110-120 Brin. For bushings, pump valves; formerly "Atlas 90"; resists sulfuric acid. *Obsolete*

AMPCOLOY 405
Ampco Pittsburgh Corp.
Al 7, Si 2, bal Cu.
Extruded: 85,000-87,000 TS; 50,000-53,000 YS; 28-30 El; Rockwell B86-84. For valve stems, valve seats, gears, hardware; high strength, high ductility, wear resistant. *Obsolete*

AMPCOLOY 405
Ampco Pittsburgh Corp.
Al 7, Si 2, bal Cu.
1/2 Hard: 87,000 TS; 53,000 YS; 30 El; 170 Brin. For screw machine products, valve stems, gears, bushings, bearings; corrosion and wear resistant, extruded bars. *Obsolete*

AMPCOLOY 42
Ampco Pittsburgh Corp.
Cu 89, Al 10, Fe 1.
60,000-75,000 TS; 22,000-28,000 YS; 25-32 El; 25-30 RA; 110-135 Brin. For propellers, nuts; corrosion resistant. *Obsolete*

AMPCOLOY 44
Ampco Pittsburgh Corp.
Al 12, bal Cu.
Cast: 60,000-70,000 TS; 45,000-50,000 YS; 3-5 El; 200-229 Brin. For welding dies, electrode holders; non heat treatable. *Obsolete*

AMPCOLOY 45-HT
Ampco Metal
Al 9.7-10.9, Fe 2-3.5, Ni 4.5-5.5, Mn 0-1.5, 0.6 others max, bal Cu.
Wrought: 115,000 TS; 66,000 YS; 15 El; 15 RA; 223 Brin. For plunger tips, valve guides, shafts. Corrosion resistant. *Obsolete*

AMPCOLOY 46
Ampco Metal
Fe 2-6, Al 8-12, Ni 2-6, bal Cu.
eat treated: 90,000-96,000 TS; 43,000-55,000 YS; 3-12 El; 10 RA; 180-235 Brin. For valve bodies, gears, sleeves, liners. Heat treatable castings. *Obsolete*

AMPCOLOY 46-HT
Ampco Metal
Al 8-12, Fe 2-6, Ni 2-6, Mn 0-2, bal Cu.
Cast: 95,000 TS; 43,000 YS; 16 El; 16 RA; 234 Brin. For gears, shafts, valve bodies, shaft sleeves, liners; heat treatable; corrosion resistant. *Obsolete*

AMPCOLOY 49
Ampco Pittsburgh Corp.
Al 10, bal Cu.
For marine hardware; Al bronze, corrosion resistant.
Obsolete

AMPCOLOY 50
Ampco Metal
Cu 84, Sn 10, Ni 3.5, Pb 2.5.
Cast: 47,000 TS; 24,000 YS; 10 El; 20 RA; 80 Brin. For worm gears, bushings, nuts, gears, elevating screw units. *Obsolete*

AMPCOLOY 501
Ampco Pittsburgh Corp.
Mn 0.7, Si 1.5, bal Ni.
Cast: 50,000 TS; 25,000 YS; 25 El; 100 Brin. For valve and pump bodies, food and drug handling equipment. Corrosion and heat resistant. *Obsolete*

AMPCOLOY 502
Ampco Pittsburgh Corp.
Mn 0.3, bal Ni.
Forged: 75,000 TS; 35,000 YS; 30 El; 130 Brin. For valve trim, food and drug handling equipment. Corrosion and heat resistant. *Obsolete*

AMPCOLOY 521
Ampco Metal
Copper. Ni 30, Mn 1, Fe 0.5, Si 0.5, Cu 68.
Cast: 63,000 TS; 32,000 YS; 35 El; 120 Brin. For valve bodies and trim, valve stems for pipe fittings, propeller sleeves, pump impellers. Good corrosion resistance to 700°F. C96400

AMPCOLOY 522
Ampco Metal
Copper. Fe 0.5, Ni 30, Mn 0.5, Cu 68.5.
Forged: 55,000 TS; 23,000 YS; 40 El; 85 Brin. For valve bodies and trim, valve stems for pipe fittings, propeller sleeves, pump impellers. Corrosion resistant to 700°F. C71500

AMPCOLOY 525
Ampco Metal
Copper. Fe 1.5, Ni 10, Mn 1, Si 0.1, Cu 87.
Cast: 42,000 TS; 19,000 YS; 35 El; 100 Brin. For oil refinery equipment, propeller sleeves, pump impellers, marine hardware. Corrosion resistant to 600°F. C96200

AMPCOLOY 526
Ampco Metal
Copper. Fe 1.5, Ni 10, Cu 88.
Forged: 40,000 TS; 18,000 YS; 45 El; 35 Rock B. For oil refinery equipment, propeller sleeves, pump impellers, marine hardware. Corrosion resistant to 600°F. C70600

AMPCOLOY 53
Ampco Metal
Ni 5, Sn 5, Pb 2, bal Cu.
Obsolete

AMPCOLOY 531
Ampco Pittsburgh Corp.
Fe 8, Mn 1, Cr 15, bal Ni.
Rolled: 100,000 TS; 60,000 YS; 35 El; 180 Brin. For exhaust manifolds, furnace and heat treating equipment, jet engine parts, oil and gas burners. For subzero and high temperature service. *Obsolete*

AMPCOLOY 54
Ampco Metal
Copper. Sn 10.3, Ni 1.5, Cu 88.
Cast: 47,000 TS; 24,000 YS; 15 El; 12 RA; 93 Brin. For rims. C91600

AMPCOLOY 551
Ampco Metal
Cu 29, Al 3, Mn 1, bal Ni.
Rolled: 150,000 TS; 105,000 YS; 25 El; 280 Brin. For pump and valve parts, scrapers. High strength, wear and corrosion resistant. K-Monel. *Obsolete*

AMPCOLOY 552
Ampco Metal
Cu 30, Fe 1.4, Mn 1, bal Ni.
Forged: 80,000 TS; 45,000 YS; 35 El; 140 Brin. For propeller shafts, valve seats and stems, pump and compressor parts. Resists corrosion and elevated temperatures. Monel. *Obsolete*

AMPCOLOY 553
Ampco Metal
Cu 30, Fe 2, Mn 0.75, Si 4, bal Ni.
Cast: 125,000 TS; 100,000 YS; 2 El; 300 Brin. For pump wear rings, valve trim, steam nozzles. Non-magnetic, non-galling, corrosion resistant. Similar to S-Monel. *Obsolete*

AMPCOLOY 557
Ampco Metal
Cu 31, Fe 2, Mn 0.75, Si 3, bal Ni.
Cast: 100,000 TS; 65,000 YS; 8 El; 200 Brin. For pump wear rings, valve trim, steam nozzles. Corrosion resistant, non-magnetic. Similar to H-Monel. *Obsolete*

AMPCOLOY 558
Ampco Metal
Cu 33, Fe 1.2, Mn 0.8, Si 1.7, bal Ni.
Cast: 75,000 TS; 35,000 YS; 30 El; 130 Brin. For pump impellers, valve bodies and trim, bubble caps, food machinery, steam ejectors. Weldable; corrosion resistant. *Obsolete*

AMPCOLOY 559
Ampco Metal
Cu 33, Fe 1.2, Mn 0.8, Si 1.7, bal Ni.
Cast: 75,000 TS; 35,000 YS; 30 El; 130 Brin. For pump impellers, valve bodies, bubble caps, food machinery, steam ejectors. Corrosion resistant, strong and tough. *Obsolete*

AMPCOLOY 561
Ampco Pittsburgh Corp.
Cu 2, Fe 30, Mn 1, Mo 3, Ti 1, Cr 1, bal Ni.
Forged: 95,000 TS; 48,000 YS; 43 El; 156 Brin. For agitators, valves, pump fittings, heat exchangers. Resists sulfuric and phosphoric acids. Corrosion resistant (Ni-O-Nel). *Obsolete*

AMPCOLOY 570
Ampco Metal
Copper. Cu 71.8, Al 11.5, Fe 0.7, Ni 14, Co 0-1.5, 0.5 others max.
85 ksi TS; 53 ksi YS; 3 El; 202 Brin. For glass mold service. Good thermal conductivity and resistance to thermal fatigue cracking. C99300

AMPCOLOY 62
Ampco Metal
Copper. Zn 39.5, Mn 0.3, Al 1, Fe 1, Cu 58.
Cast: 65,000-75,000 TS; 28,000-38,000 YS; 20-35 El; 20-35 RA; 110-135 Brin. For castings, gears, cams; low tensile bronze. C86500

AMPCOLOY 64
Ampco Metal
Copper. Cu 64, Al 4, Fe 3, Mn 3.5, Zn 25.5.
Cast: 95,000 TS; 50,000 YS; 25 El; 192 Brin. For cast gears, gear cases, scrapers, housings. Medium strength bronze. C86200

AMPCOLOY 66
Ampco Metal
Copper. Al 6, Fe 3, Mn 3.5, Zn 25.5, Cu 62.
Cast: 110,000-120,000 TS; 60,000-70,000 YS; 10-15 El; 8-14 RA; 217-235 Brin. For castings, gears; high tensile bronze, nuts. C86300

AMPCOLOY 663HT
Ampco Metal
Copper. Cu 60.5, Mn 0-2.6, Pb 0-1, Si 0-1, 1.0 others max, bal Zn.
74 ksi TS; 43 ksi YS; 16 El; 149 Brin. Gear and bearing alloy.

AMPCOLOY 666
Ampco Metal
Copper. Pb 0.3, Mn 2.5, Cu 58.5, Si 0.7, Zn 38.
Extruded: 82,000-88,000 TS; 42,000-55,000 YS; 12-18 El; 170-187 Brin. For bushings, gears, bearings, cams, valve stems, connecting rods; high strength bearing bronze. C67400

AMPCOLOY 668
Ampco Metal
Copper. Cu 61, Pb 2.5, Mn 2.5, Si 1, Zn 33.
Extruded: 72,000-75,000 TS; 58,000-60,000 YS; 18-20 El; 144-162 Brin. For bushings, bearings, gears, cams, valve stems, lead screw nuts; free-cutting, leaded manganese bearing bronze. C67300

AMPCOLOY 71
Ampco Pittsburgh Corp.
Cu 86-90, Sn 5.5-6.5, Pb 1-2, Zn 3-5, Ni 0-1.
Cast: 40,000 TS; 20,000 YS; 30 El; 30 RA; 70 Brin. For pressure castings, gears, bearings, bushings; "Navy M" alloy. *Obsolete*

AMPCOLOY 711
Ampco Metal
Copper. Sn 11, P 0.2, Cu 88.3.
Cast: 55,000 TS; 30,000 YS; 16 El; 102 Brin. For gears, worm wheels. Gear bronze, tough. C90700

AMPCOLOY 712
Ampco Metal
Copper. Ni 1.5, Sn 11.5, P 0.2, Cu 86.8.
Cast: 60,000 TS; 32,000 YS; 16 El; 106 Brin. For gears, worm wheels. Gear bronze, tough. C91700

AMPCOLOY 715
Ampco Metal
Copper. Sn 12, P 0.2, Cu 87.8.
Cast: 60,000 TS; 32,000 YS; 16 El; 106 Brin. For gears, worm wheels. Gear bronze, tough, wear resistant. C90800

AMPCOLOY 72
Ampco Metal
Sn 7.5-9, Zn 3-5, bal Cu.
Cast: 46,000 TS; 24,000 YS; 26 El; 24 RA; 66 Brin. For castings; venturi meter bodies, slip rings. *Obsolete*

AMPCOLOY 74
Ampco Metal
Cu 84-86, Sn 4-6, Pb 4-6, Zn 4-6, Ni 0-1.5.
Cast: 38,000 TS; 20,000 YS; 26 El; 24 RA; 65 Brin. For valves, pipe elbows, hydraulic parts; "Ounce Metal;" for high pressure service. *Obsolete*

AMPCOLOY 742
Ampco Metal
Cu 86-89, Sn 9-11, Zn 1-3, Pb 0-0.3.
Cast: 40,000 TS; 18,000 YS; 14 El; R80F Brin. For gears, shafts; continuous cast rod. *Obsolete*

AMPCOLOY 79
Ampco Metal
Cu 86-89, Sn 9-11, Zn 1-3.
Cast: 46,000 TS; 22,000 YS; 24 El; 22 RA; 105 Brin. For gears, shafts, bearings, sleeves. Hard bronze, wear and corrosion resistant. *Obsolete*

AMPCOLOY 80
Ampco Pittsburgh Corp.
Be 2-2.75, Ni 0-1.25, bal Cu.
Heat treated: 90,000-125,000 TS; 85,000-100,000 YS; 0-4 El; 0-4 RA; 350-420 Brin. For castings. *Obsolete*

AMPCOLOY 83
Ampco Pittsburgh Corp.
Co 0.25, Be 1.8, bal Cu.
Wrought: 190,000 TS; 120,000 YS; 2.5 El. For resistance welding parts, shafts, bushings. Age-hardenable, high strength. *Obsolete*

AMPCOLOY 83-20
Ampco Pittsburgh Corp.
Be 1.7-2.05, Co 0.1-0.35, 0.60% max others, bal Cu.
Extruded: 190,000 TS; 120,000 YS; 2.5 El; 2.5 RA; 350 Brin.
For plunger tips. Age-hardenable, corrosion resistant.
Obsolete

AMPCOLOY 84
Ampco Pittsburgh Corp.
Be 2, 1.0% others, bal Cu.
For resistance welding parts, shafts, bushings, die inserts.
Age hardenable, wear resistant. *Obsolete*

AMPCOLOY 84-20
Ampco Pittsburgh Corp.
Be 1.5-2, Co 0-1.5, 2.5% max others, bal Cu.
Cast: 135,000 TS; 90,000 YS; 2 El; 2.5 RA; 351 Brin. For
plunger tips, die inserts, shafts, bushings. Age hardenable,
corrosion resistant. *Obsolete*

AMPCOLOY 86
Ampco Pittsburgh Corp.
Ni 1.1-1.5, Be 2.25-3, bal Cu.
Sand cast: 160,000 TS; 100,000 YS; 2.5 El; 2.5 RA; 393 Brin.
Centrifugal cast: 170,000 TS; 100,000 YS; 2.5 El; 2.5 RA; 393
Brin. For welding dies, non-sparking safety tools; see "Ampco
Beryllium Copper." *Obsolete*

AMPCOLOY 90
Ampco Metal
Copper. Cu 99.75, 0.25 others.
Cast: 25,000 TS; 9,000 YS; 40 El; 40 RA; 45 Brin. For
transformer secondaries; cable attachments, bus bars; high
conductivity. C81100

AMPCOLOY 900
Ampco Metal
Copper. Cu 99.98.
Extruded: 30,000 TS; 10,000 YS; 40 El; 40 Brin. For electric
motor components, conductors. Electrical conductivity 100%
min. OFHC copper. C10200

AMPCOLOY 901
Ampco Metal
Cu 99.9, 30 oz/ton Ag.
Cast: 26,000 TS; 8,000 YS; 40 El; 44 Brin. Forged: 31,000 TS;
10,000 YS; 45 El; 47 Brin. For high temperature (600°F)
electrical and acid resistant applications, heat exchangers,
computor segments. Electrical conductivity 95% min.
Obsolete

AMPCOLOY 91
Ampco Pittsburgh Corp.
Cu 97, Co 2.6, Be 0.4.
Heat treated: 90,000-115,000 TS; 45,000-55,000 YS; 10-20 El;
200-240 Brin. For welding electrodes; same as Ampco
Trodaloy 1. *Obsolete*

AMPCOLOY 910
Ampco Metal
Copper. Zr 0.15, Cu 99.8.
Forged: 55,000 TS; 47,000 YS; 14 El; 100 Brin. Extruded:
65,000 TS; 60,000 YS; 16 El; 119 Brin. For rotor wedges,
commutators, collector rings, switch gears, soldering tips,
welding electrodes. 90% electrical conductivity. C15000

AMPCOLOY 92
Ampco Metal
Copper. Al 11.5, Cu 88.2, 0.3 others.
Cast: 70,000 TS; 34,000 YS; 6 El; 150 Brin. For resistance
welding electrodes, hinges, bushings, shafts. 10-15%
electrical conductivity, corrosion resistant.

AMPCOLOY 92-H
Ampco Metal
Al, bal Cu.
Cast: 75,000 TS; 2 El; 240 Brin. For resistance welding
electrode; 20% conductivity. *Obsolete*

AMPCOLOY 92-S
Ampco Metal
Al, bal Cu.
Cast: 75,000 TS; 25 El; 137 Brin. For resistance welding
electrode; 20% conductivity. *Obsolete*

AMPCOLOY 94
Ampco Metal
Copper. Ni 1.3, Si 0.6, Cr 0.7, bal Cu.
Cast: 65,000 TS; 45,000 YS; 15 El; 146 Brin. For resistance
welding parts, shafts, bushings, plunger tips. Medium
conductivity, wear resistant.

AMPCOLOY 940
Ampco Metal
Copper. Cu 96.5, Ni 2.5, Si 0-0.6, Cr 0-0.4.
Wrought, cast: 100 ksi TS; 75 ksi YS; 13 El; 94 Brin. For
plunger tips, resistance welding holders and seam welder
wheels, flash, butt and projection welding dies, collets. Good
electrical conductivity. Also UNS C81540. C18000

AMPCOLOY 95
Ampco Pittsburgh Corp.
Be 0.1-1, Co 1.5-3, bal Cu.
Cast: 95,000 TS; 15 El; 396 Brin. Wrought: 105,000 TS; 15 El.
For welding electrodes; age-hardenable. *Obsolete*

AMPCOLOY 95-20
Ampco Pittsburgh Corp.
Co 1.3-3, Be 0.1-1, bal Cu.
Cast: 92,000 TS; 68,000 YS; 8 El; 9 RA; 217 Brin. For welding
electrodes, shafts, bushings; age-hardenable, good electrical
conductivity. *Obsolete*

AMPCOLOY 97
Ampco Metal
Copper. Cu 98.5, Cr 1.
Cast: 40,000-50,000 TS; 35,000-40,000 YS; 10-15 El; 140-150
Brin. For welding electrodes; same as Ampco Trodaloy 7.
Also UNS C81500. C18200

AMPCOLOY 97-20
Ampco Pittsburgh Corp.
Cr 1.5, 0.5% max others, bal Cu.
Cast: 51,000 TS; 40,000 YS; 17 El; 35 RA; 105 Brin. For
welding electrodes, bushings, inserts; age hardenable, high
electrical conductivity. *Obsolete*

AMPCOLOY 972
Ampco Metal
Copper. Cu 98.65, Cr 0-1.1, Zr 0-0.12.
78 ksi TS; 70 ksi YS; 13 El; 84 Brin. For high-temperature,
high-stress applications.

AMPCOLOY 98
Ampco Metal
Copper. Cu 98.5, Cr 0-1, 0.5 others max.
65 ksi TS; 61 ksi YS; 19 El; 70 Brin. For spot welding
aluminum and magnesium alloys, coated materials, brass
and bronze. C18200

AMPCOLOY 98
Ampco Pittsburgh Corp.
Cu 98.9, Cr 1, Si 0.1.
Cast: 40,000-50,000 TS; 18,000-25,000 YS; 20-25 El; 75-85
Brin. For resistance welding electrodes; formerly Atlas 98-21.
Obsolete

AMPCOLOY 99
Ampco Metal
Cu alloy.
Bar: 50,000-60,000 TS; 15,000-20,000 YS; 20-25 El; 100-115
Brin. For resistance welding electrode; for terneplate and
galvanized stock. *Obsolete*

AMPCOLOY A-1
Ampco Metal
Now AMPCO A-1.

AMPCOLOY A-2
Ampco Metal
Al 7-10, Fe 0.5-2, 0.6 others max, bal Cu.
Cast: 62,000 TS; 23,000 YS; 30 El; 30 RA; 101 Brin. For
pickling hooks, bushings, marine hardware. Corrosion
resistant. *Obsolete*

AMPCOLOY A-3
Ampco Pittsburgh Corp.
Cu 89, Al 10, Fe 1.
Cast: 70,000-75,000 TS; 25,000-32,000 YS; 30-40 El; 30-40
RA; 115-126 Brin. For gears, pumps, impellers, bearings;
corrosion resistant. *Obsolete*

AMPCOLOY A-323
Ampco Pittsburgh Corp.
Cu 89, Al 10, Fe 1.
Heat treated: 80,000-90,000 TS; 40,000-50,000 YS; 22-30 El;
24-32 RA; 143-156 Brin. For hardware; heat treatable;
formerly "Atlas 89A323." *Obsolete*

AMPCOLOY B-2
Ampco Metal
Now AMPCO B-2.

AMPCOLOY C-3
Ampco Metal
Now AMPCO C-3.

AMPCOLOY D-4
Ampco Metal
Now AMPCO D-4.

AMPCOLOY E-1
Ampco Metal
Al 9.5-11, Fe 0-1.5, bal Cu.
Cast: 70,000-80,000 TS; 40,000-45,000 YS; 12-20 El; 14-20
RA; 131-143 Brin. For castings, power shovels, worm wheels,
pistons, bushings, corrosion resistant. *Obsolete*

AMPCOLOY E-117
Ampco Pittsburgh Corp.
Cu 89, Al 10, Fe 1.
Cast: 90,000-100,000 TS; 70,000-80,000 YS; 2-6 El; 3-7 RA;
229-255 Brin. For propellers; formerly "Atlas 89E117."
Obsolete

AMPCOLOY E-123
Ampco Pittsburgh Corp.
Al 10-13, Fe 1.25, bal Cu.
Cast: 90,000-100,000 TS; 70,000-80,000 YS; 2-6 El; 3-7 RA;
229-255 Brin. For propellers; formerly "Atlas 89E117."
Obsolete

AMPCOLOY E-133
Ampco Pittsburgh Corp.
Cu 89, Al 10, Fe 1.
Cast: 80,000-100,000 TS; 45,000-55,000 YS; 12-15 El; 13-17
RA; 163-179 Brin. For hardware; formerly "Atlas 89E123".
Obsolete

AMPCOLOY E-2
Ampco Pittsburgh Corp.
Cu 89, Al 10, Fe 1.
Cast: 75,000-85,000 TS; 43,000-48,000 YS; 10-15 El; 14-20
RA; 160-180 Brin. For cam bearings, gears; formerly Atlas
89E2. *Obsolete*

AMPCOLOY E-3
Ampco Pittsburgh Corp.
Cu 89, Al 10, Fe 1.
Cast: 75,000-80,000 TS; 45,000-50,000 YS; 7-12 El; 9-14 RA;
202-229 Brin. For hardware; formerly Atlas 89E3. *Obsolete*

AMPCOLOY E-5
Ampco Metal
Al 9, Fe 3, bal Cu.
Bar: 90,000 TS; 48,000 YS; 18 El; 174 Brin. For valve seats,
guides, stems, gears, bearings. Wear and corrosion resistant
aluminum bronze. Tough. *Obsolete*

AMPCOLOY F-6
Ampco Metal
Al 10, Fe 3, bal Cu.
Cast: 85,000 TS; 31,000 YS; 20 El; 170 Brin. For bushings, bearings, gears, worm wheels, shifters. Resists squashing under load. *Obsolete*

AMPCOLOY M-100
Ampco Metal
Mo 100.
Sintered: 80,000 TS; 30 El; 90 Rock B. For facings on spot welder tips. High heat resistance. *Obsolete*

AMPCOLOY W-100
Ampco Metal
Bar: 50,000-200,000 TS; 76 Rock A. For resistance welding electrodes for Red Brass weldings. RWMA Class B-13. 30% electrical conductivity, corrosion and wear resistance. *Obsolete*

AMPHOS
Climax Performance Materials Corp.
P 0.03, bal Cu.
For condenser tubes, heat exchangers, piping; good hot and cold workability.

AMPHOS-40
Climax Performance Materials Corp.
Oxygen free copper with added phosphorus. For plating anodes.

AMS
Thyssen Edelstahlwerke AG
C 0.55, Cr 0.7, Mo 0.18, Ni 1.6, V 0.1, bal Fe.
For gears, shafts, crankshafts, bolts, studs; oil hardened, shock resistant. *Obsolete*

AMS
Acciaierie Valbruna s.p.a.
Stainless Steel. C 0-0.08, Si 0-1, Mn 0-2, P 0-0.045, S 0-0.03, Cr 19, Mo 3.5, Ni 13, bal Fe.
Austenitic. AISI 317; W. Nr. 1.4449.

AMS 5700
Now CARPENTER AMS 5700.

AMS EXTRA
Thyssen Edelstahlwerke AG
C 0.55, Cr 1.1, Ni 1.7, Mo 0.5, V 0.1, bal Fe.
Oil hardening steel for drop forging dies; tough, high red hardness. *Obsolete*

AMSCO 40
Abex Corp.
C 3.5, Cr 2, Mo 5, bal Fe.
Welded: 540 Brin. For hard facing electrode; abrasion resistant.

AMSCO 53
Abex Corp.
C 3, Cr 15, Mo, bal Fe.
Welded: 540 Brin. For hard facing electrode; abrasion and impact resistant.

AMSCO 60
Abex Corp. (AMSCO Div.)
C 2, Cr 4, bal Fe.
Welded: 500 Brin. For hard facing electrode; abrasion and impact resistant. *Obsolete*

AMSCO 77
Abex Corp.
C 2.5, Mn 1, Cr 19, Mo 1, bal Fe.
Weld hardness: 46 Rock C. Hardfacing electrode, abrasion and impact resistant.

AMSCO AIR HARDENING
Abex Corp. (AMSCO Div.)
C 1-1.5, Cr 2, Mo 1, bal Fe.
Cast: 450-550 Brin. For welding rod; abrasion and impact resistant to 400 F, air hardening. *Obsolete*

AMSCO AW-420
Abex Corp.
C 0.16, Mn 1.3, Si 0.75, Cr 12, bal Fe.
Weld hardness: 45 Rock C. Hardfacing wire for submerged melt welding; resistant to heat, abrasion, metal to metal wear and thermal shock.

AMSCO AW-72
Abex Corp.
C 0.27, Cr 5.8, Mo 0.8, W 1.5, bal Fe.
Weld hardness: 52 Rock C. Hardfacing wire for submerged melt welding of carbon and low-alloy steel parts; resistant to abrasion, impact, thermal shock and compressive loading.

AMSCO AW-79
Abex Corp.
C 0.15, Mn 2.2, Si 0.7, Cr 3.5, Mo 0.8, bal Fe.
Weld hardness: 44 Rock C. Hardfacing wire for submerged melt welding of carbon and low-alloy steel parts, resistant to impact, abrasion, and metal to metal wear.

AMSCO AW-84
Abex Corp.
C 0.13, Mn 2.2, Si 0.8, Mo 0.6, bal Fe.
Weld hardness: 30 Rock C. Build up wire for submerged melt welding of carbon and low-alloy steel parts; high compressive strength and impact resistant.

AMSCO AW-87
Abex Corp.
C 0.14, Mn 2.2, Si 0.8, Cr 2.6, Mo 0.6, bal Fe.
Weld hardness: 41 Rock C. Hardfacing and buildup wire for submerged melt welding of carbon and low-alloy steel parts; resistant to medium impact and abrasion.

AMSCO AW-CMS
Abex Corp.
C 0.3, Mn 15, Cr 15, Ni 1, bal Fe.
Weld hardness: 200 Brin, work hardens. Wire for submerged melt welding; impact and abrasion resistant.

AMSCO AW-NICROMANG
Abex Corp.
C 0.8, Mn 15, Cr 4, Ni 3.5, bal Fe.
As welded: 120,000 TS; 70,000 YS; 40 El; 35 RA. Weld hardness: 200 Brin, work hardens to 500 Brin. Manganese steel wire for submerged metal welding; impact and abrasion resistant.

AMSCO AW-THERMALLOY 4
Abex Corp. (AMSCO Div.)
C 0.8, Cr 20, Ni 8, Mo, bal Fe.
Welded: 250 Brin. For automatic submerged melt electrode; impact resistant at high temperature, work hardens. *Obsolete*

AMSCO AW-THERMALLOY 400
Abex Corp.
C 0.6, Mn 3, Cr 22, Ni 8, Mo 0.5, bal Fe.
Deposit hardness: 22 Rock C, work hardened. Hardfacing wire for submerged melt welding; resistant to abrasion and impact at high temperature.

AMSCO CHROMANAL
Abex Corp.
C 1-1.4, Mn 10-14, Si 0.2-1, Cr 1.3, bal Fe.
Heat treated: 95,000-145,000 TS; 53,000-65,000 YS; 27-63 El; 30-40 RA; 195-220 Brin. For crushers, hammer mills, liners, castings; work hardened up to 550 Brin.

AMSCO CHROME MOLY BALL MILL GRATE STEEL
Abex Corp.
C 0.5-0.7, Mn 0.5-1.5, Si 0.2-0.8, Cr 1.8-2.8, Mo 0.2-0.5, bal Fe.
Heat treated: 145,000-175,000 TS; 90,000-130,000 YS; 8-11 El; 12-20 RA. For ball mill grates; abrasion resistant.

AMSCO CHROME MOLY BALL MILL LINER STEEL
Abex Corp.
C 0.7-1, Mn 0.5-1.5, Si 0.2-0.8, Cr 1.8-2.8, Mo 0.2-0.5, bal Fe.
Heat treated: 150,000-185,000 TS; 95,000-140,000 YS; 4-8 El; 4-10 RA; 300-350 Brin. For ball mill liners; abrasion resistant.

AMSCO CHROMEFACE
Abex Corp. (AMSCO Div.)
C 4, Mn 6, Si 2, Cr 30, bal Fe.
Cast: 500-600 Brin. For hardfacing electrodes; abrasion resistant to 600 F, austenitic. *Obsolete*

AMSCO CM
Abex Corp.
C 0.5-1, Mn 0.5-1.5, Si 0.2-1, Cr 1.8-2.8, Mo 0.2-0.5, bal Fe.
Heat treated: 125,000-185,000 TS; 80,000-150,000 YS; 2-13 El; 2-25 RA; 300-400 Brin. For rod and ball mill grates liners, crushers; high abrasion and moderate impact resistant.

AMSCO CO-MANG
Abex Corp. (AMSCO Div.)
C 0.6, Mn 13, Mo 0.7, bal Fe.
Welded: 180 Brin. Worked: 500 Brin. For welding electrodes for austenitic Mn steel; coating of Mn compounds and other alloys, composite electrode. *Obsolete*

AMSCO CONOMANG
Abex Corp.
C 0.95, Mn 16, Cr 1.8, bal Fe.
As welded: 124,000 TS; 80,000 YS; 28 El; 25 RA. Weld hardness: 230 Brin, work hardens to 500 Brin. Manganese steel electrode for rebuilding manganese steel parts. Impact resistant.

AMSCO CS
Abex Corp.
C 0.2-0.35, Cr 0.7-1.5, Mn 0.6-1.5, Si 0.2-0.8, Ni 0.3-0.8, Mo 0.2-0.6, bal Fe.
Heat treated: 200,000-240,000 TS; 170,000-210,000 YS; 6-10 El; 10-30 RA; 300-500 Brin. For castings, dipper teeth; abrasion and wear resistant.

AMSCO DIEWELD
Abex Corp. (AMSCO Div.)
C 1, Cr 3.5, Mo 2.3, bal Fe.
500-600 Brin. For welding rod for forging dies, punches, hard facing; coated, air hardened. *Obsolete*

AMSCO F-1
Abex Corp.
Cr 15-18, Ni 34-37, bal Fe.
For electric furnace grids, hearth plates, rails, furnace chains; heat resistant, maximum operating temperature 2000°F.

AMSCO F-1-N
Abex Corp.
Cr 15-18, Ni 38-40, C, bal Fe.
For retorts, muffles, furnace castings, hearth plates; heat, corrosion and shock resistant.

AMSCO F-10
Abex Corp.
C 0.4, Cr 28, Ni 12, bal Fe.
For furnace parts, dampers, valves, beams; heat resistant.

AMSCO F-10H
Abex Corp.
C 0.4-0.6, Cr 28, Ni 12, bal Fe.
For furnace parts; heat resistant.

AMSCO F-10HH
Abex Corp.
C 0.6-0.85, Cr 28, Ni 12, bal Fe.
For furnace parts; heat resistant.

AMSCO F-12
Abex Corp.
Cr 27-30, Ni 7-10, C, bal Fe.
For heat treating furnaces, valves; creep resistant.

AMSCO F-13
Abex Corp.
Cr 25-28, Ni 34-37, C, bal Fe.
For furnace parts, retorts, muffles; resists heat and sulfur atmosphere to 2100°F.

AMSCO F-14
Abex Corp.
Cr 24-27, Ni 19-22, C, bal Fe.
Cast. For furnace parts; heat resistant.

AMSCO F-1H
Abex Corp.
C 0.4-0.6, Cr 16, Ni 35, bal Fe.
For furnace parts; heat resistant.

AMSCO F-1HH
Abex Corp.
C 0.6-0.85, Cr 16, Ni 35, bal Fe.
For furnace parts; heat resistant.

AMSCO F-2
Abex Corp.
Cr 21-24, Ni 3-6, bal Fe.
For furnace lead pots; heat resistant, maximum operating temperature 1800°F.

AMSCO F-3
Abex Corp.
Cr 26-29, Ni 0-3, bal Fe.
For rabble arms, rabble blades; corrosion and heat resisting.

AMSCO F-3H
Abex Corp.
C 0.4-0.6, Cr 28, Ni 2, bal Fe.
For furnace parts; heat resistant.

AMSCO F-3HH
Abex Corp.
C 0.6-0.85, Cr 28, Ni 2, bal Fe.
For furnace parts; heat resistant.

AMSCO F-4
Abex Corp.
Cr 6-9, Ni 18-21, bal Fe.
For high-strength castings; corrosion resistant.

AMSCO F-5
Abex Corp.
Cr 16-19, Ni 65-70, bal Fe.
For carburizing boxes. Oil burner parts; heat resistant, tough.

AMSCO F-6
Abex Corp.
Cr 12-15, Ni 58-64, C, bal Fe.
For furnace parts, heat treating boxes, retorts; corrosion and heat resistant.

AMSCO F-7H
Abex Corp.
C 0.5, Cr 25, Ni 12, bal Fe.
For furnace parts; heat resistant.

AMSCO F-8
Abex Corp.
Cr 17-22, Ni 7-10, bal Fe.
For pots in chemical plants, acid pumps; corrosion resistant.

AMSCO F-8N
Abex Corp. (AMSCO Div.)
C 0.12, Cr 8, Ni 0.5, Mo 0.5, bal Fe.
For welding rods. *Obsolete*

AMSCO HC
Abex Corp.
C, Cr, Si, Mn, bal Fe.
Heat treated: 60,000 TS; 7000 YS; 400-600 Brin. For castings, chute plates, liners; cast iron, abrasion and wear resistant.

AMSCO HC-250
Abex Corp.
C 2-3, Mn 0.5-1.8, Si 0.2-1, Cr 22-30, Mo 0.2-0.6, bal Fe.
Heat treated: 60,000-100,000 TS; 50,000-75,000 YS; 450-550 Brin. For pump parts, liners; erosion and abrasion resistant.

AMSCO HF-40
Abex Corp. (AMSCO Div.)
C 3.5, Mn 0.5, Cr 5, Mo 4, bal Fe.
Cast: 500-580 Brin. For hard facing electrode; martensitic cast iron, wear resistant. *Obsolete*

AMSCO HF-60
Abex Corp. (AMSCO Div.)
C 1.6, Mn 1, Cr 3, bal Fe.
Cast: 450-555 Brin. For hard facing electrode; abrasion resistant. *Obsolete*

AMSCO HF20
Abex Corp. (AMSCO Div.)
Steel core wire, coated with Cr, Mn, Mo, Si, V.
Welded: 600 Brin. For hard facing electrode; hot wear resistant. *Obsolete*

AMSCO MACHINE-FACE
Abex Corp. (AMSCO Div.)
C 0.4, Cr 1, Mo 0.2, bal Fe.
Welded: 293-331 Brin. For hard surfacing electrode; coated. *Obsolete*

AMSCO MANGANESE STEEL
Jelliff Corp.
Mn 10-14, C 1-1.4, Si 0.2-1, bal Fe.
Heat treated: 100,000-145,000 TS; 50,000-57,000 YS; 60-30 El; 40-30 RA; 155-200 Brin. For steam shovel teeth rolls, ball mill liners, wheels, gears and pinions; tough and shock resisting; resists abrasion.

AMSCO MANGANESE STEEL
Abex Corp.
Mn 10-14, C 1-1.4, Si 0.2-1, bal Fe.
Heat treated: 100,000-145,000 TS; 50,000-57,000 YS; 60-30 El; 40-30 RA; 155-200 Brin. For steam shovel teeth rolls, ball mill liners, wheels, gears and pinions; tough and shock resisting; resists abrasion.

AMSCO MO-MANG
Abex Corp. (AMSCO Div.)
C 0.8, Mn 14, Mo 0.2, bal Fe.
Welded: 200 Brin. For welding rod for building up worn parts; abrasion resistant, work hardens to 500 Brin, austenitic. *Obsolete*

AMSCO NI-HARD
Abex Corp. (AMSCO Div.)
C 3.5, Ni 4.5, Cr 1.5, bal Fe.
Welded: 600-700 Brin. For welding rod for hard facing; abrasive resistant to 600 F. *Obsolete*

AMSCO NICKEL MANGANESE
Abex Corp. (AMSCO Div.)
C 0.7, Mn 13-15, Si 0.9-1.2, Ni 2.5-3.5, bal Fe.
Welded: 200 Brin. Hardened: 550 Brin. For crusher rolls, dipper teeth, dredge parts; austenitic, high work hardenability. *Obsolete*

AMSCO NICROMANG
Abex Corp.
C 0.8, Mn 14.5, Cr 4, Ni 3.5, bal Fe.
As welded: 120,000 TS; 75,000 YS; 42 El; 35 RA. Weld hardness: 200 Brin, work hardened to 500 Brin. Manganese steel electrode for rebuilding manganese steel and joining manganese steel and other steels. Impact resistant.

AMSCO NO. 1
Abex Corp.
C 2.5, Cr 31, W 13, Co 50.
Weld hardness: 52 Rock C. Hardfacing rod and electrode; heat and corrosion resistant to 1200°F. For valves and valve seats, dies.

AMSCO NO. 12
Abex Corp.
C 1.5, Cr 30, W 9, Co 55.
Weld hardness: 45 Rock C. Hardfacing rod and electrode; heat and corrosion resistant to 1200°F. For buildup and repair valves and dies.

AMSCO NO. 217
Abex Corp. (AMSCO Div.)
C 3, W 15, Cr 8, bal Fe.
Cast: 650 Brin. For welding rod for hard surfacing; abrasion resistant to 1000 F. *Obsolete*

AMSCO NO. 459
Abex Corp. (AMSCO Div.)
C 3, Cr 4, Mo 4, Ni 1, bal Fe.
Cast: 600 Brin. For hard facing rod for dipper teeth; abrasion and impact resistant. *Obsolete*

AMSCO NO. 6
Abex Corp.
C 1.05, Cr 28, W 5, Cr 60.
Weld hardness: 41 Rock C. Hardfacing rod and electrode; heat and corrosion resistant to 1200°F. For buildup and repair valves and dies.

AMSCO RAILFACE
Abex Corp. (AMSCO Div.)
C 0.5, Cr 1.5, Mo 0.2, bal Fe.
Welded: 325-380 Brin. For welding electrode for rails; composite coated electrodes, abrasion and impact resistant. *Obsolete*

AMSCO RESISTWEAR
Abex Corp. (AMSCO Div.)
C 0.6, Cr 6, Mo 0.5, Mn, bal Fe.
Welded: 450-550 Brin. For welding electrode, hard surfacing; coated, composite electrode. *Obsolete*

AMSCO SA-33
Abex Corp. (AMSCO Div.)
C 3, Cr 12, Mo, V, bal Fe.
Welded: 550 Brin. For hard facing electrode; severe abrasion resistance. *Obsolete*

AMSCO SA-53
Abex Corp.
C 3, Cr 16, Mo 1, bal Fe.
Weld hardness: 52 Rock C. Hardfacing open arc wire, abrasion and impact resistant.

AMSCO SA-BUILD UP
Abex Corp. (AMSCO Div.)
C 0.5, Cr, V, bal Fe.
Welded: 260 Brin. For electrode for carbon and low alloy steel build up. *Obsolete*

AMSCO SA-CMO
Abex Corp.
C 0.3, Mn 16, Cr 16, Ni 1, bal Fe.
As welded: 124,000 TS; 82,000 YS; 38 El; 34 RA. Weld hardness: 230 Brin, work hardened. Wire for open arc rebuilding and joining of manganese, carbon and low-alloy steel parts. Impact resistant.

AMSCO SA-MANGANESE
Abex Corp.
C 0.9, Mn 17.5, V 0.5, bal Fe.
As welded: 128,000 TS; 85,000 YS; 25 El; 25 RA. Weld hardness: 235 Brin, work hardened to 500 Brin. Manganese steel wire for open arc rebuilding of manganese and other steel parts. Impact resistant with minimum flow.

AMSCO SA-NICROMANG
Abex Corp.
C 0.8, Mn 15, Cr 4, Ni 3.5, bal Fe.
As welded: 120,000 TS; 70,000 YS; 40 El; 35 RA. Weld hardness: 200 Brin, work hardened to 500 Brin. Manganese steel wire for open arc rebuilding and joining of manganese parts and to other steels. Impact resistant.

AMSCO SA-ROLL BUILD

Abex Corp.
C 0.9, Mn 17.5, V 0.5, bal Fe.
As welded: 230 Brin, work hardened up to 500 Brin.
Manganese steel wire for open arc automatic rebuilding of manganese steel parts. Impact resistant.

AMSCO SA-ROLL FACE

Abex Corp.
C 3, Cr 16, Mo 1, bal Fe.
Weld hardness: 52 Rock C. Hardfacing wire for open arc automatic applications; abrasion and impact resistant.

AMSCO SA-T40

Abex Corp.
C 0.6, Mn 3.5, Cr 22, Ni 9, Mo 0.5, bal Fe.
Weld hardness: 22 Rock C, work hardened. Hardfacing open arc wire, resistant to impact, heat, metal to metal wear and thermal shock.

AMSCO SA-TOUGHWEAR

Abex Corp.
C 2, Si 1.8, Cr 5, bal Fe.
Weld hardness: 48 Rock C. Hardfacing open arc wire; impact and abrasion resistant; for multiple layer buildup.

AMSCO SA-TUNGSITE

Abex Corp.
60% WC particles in a steel matrix. Hardfacing open arc wire, abrasion resistant.

AMSCO SUPER 20

Abex Corp.
C 6, Cr 22, Mo 7, W 5, bal Fe.
Weld hardness: 63 Rock C. Hardfacing electrode, resistant to abrasion at normal and elevated temperatures.

AMSCO SUPERCHROME

Abex Corp.
C 4.5, Si 2, Cr 3, bal Fe.
Weld hardness: 59 Rock C. Hardfacing electrode, abrasion resistant at normal and elevated temperatures.

AMSCO THERMALLOY 4

Now AMSCO THERMALLOY 400.

AMSCO THERMALLOY 400

Abex Corp.
C 0.6, Cr 22, Ni 8, Mo 0.5, bal Fe.
Weld hardness: 22 Rock C; work hardened. Hard facing electrode, resistant to heat, impact, metal to metal wear and thermal shock.

AMSCO TOOLFACE

Abex Corp. (AMSCO Div.)
C 0.8, Cr 4, Mo 9, W 1.5, V 1, bal Fe.
Welded: 575-675 Brin. For welding rod, composite cutting tools; abrasion and impact resistant to 1000 F. *Obsolete*

AMSCO TRACKRAIL

Abex Corp. (AMSCO Div.)
C 0.5, Cr, V, bal Fe.
Welded: 295 Brin. For automatic submerged melt electrode; build up for tractor track links. *Obsolete*

AMSCO TRACKWEAR

Abex Corp.
C 0.9, Mn 0-17.5, V 0.5, bal Fe.
As welded: 128,000 TS; 85,000 YS; 25 El; 25 RA. Weld hardness: 235 Brin, work hardened to 500 Brin. Manganese steel electrode for rebuilding manganese steel parts. Impact resistant with minimum flow.

AMSCO TUBE CHROMEFACE

Abex Corp.
C 4, Mn 1.5, Si 2, Cr 30, bal Fe.
Weld hardness: 60 Rock C. Hardfacing rod, resistant to sliding abrasion.

AMSCO TUBE TUNGSITE

Abex Corp.
60% WC particles in a steel or iron matrix. Hardfacing rod and electrode; abrasion resistant. Several particle sizes available.

AMSCO TUBE TUNGSITE 30 DOWN

Abex Corp.
50% WC particles in a steel matrix. Hardfacing rod; abrasion resistant.

AMSCO TUBE TUNGSITE SRB

Abex Corp.
38% WC particles in a steel matrix. Hardfacing rod; abrasion resistant with good impact resistance.

AMSCO TUNGCHROME

Abex Corp.
Tungsten carbide particles in a high chromium iron matrix. Hardfacing rod and electrode, abrasion resistant.

AMSCO TUNGROD

Abex Corp.
60% WC (smaller mesh size) particles in an iron or steel matrix. Hardfacing rod and electrode; abrasion resistant.

AMSCO V-MANG

Abex Corp. (AMSCO Div.)
C 0.8, Mn 14, Mo 1, bal Fe.
Welded: 175 Brin. For buildup electrode for Mn-steel parts; work hardens. *Obsolete*

AMSCO V-MANG

Abex Corp. (AMSCO Div.)
C 0.8, Mn 14, Mo 1, bal Fe.
Cast: 200 Brin. Heat treated: 550 Brin. For hard surfacing electrode; for build up on Mn steels, austenitic. *Obsolete*

AMSCO X-53

Abex Corp.
C 3.5, Cr 16, Mo 1, bal Fe.
Weld hardness: 52 Rock C. Hardfacing electrode, abrasion and impact resistant.

AMSCO-CMH

Abex Corp.
C, Cr, Mo, bal Fe.
Heat treated: 155,000 TS; 130,000 YS; 6 El; 7 RA; 300-400 Brin. For castings, ball and rod mill shell liners; air hardened, wear resistant.

AMSCO-CML

Abex Corp.
C, Cr, Mo, bal Fe.
Heat treated: 155,000 TS; 130,000 YS; 10 El; 15 RA; 275-375 Brin. For castings, trunnion liners, feeder scoops; air hardened, abrasion resistant.

AMSCO-M

Abex Corp.
C 1-1.4, Mn 10-14, Si 0.2-1, bal Fe.
Heat treated: 100,000-145,000 TS; 50,000-57,000 YS; 30-60 El; 30-40 RA; 195 Brin. For dipper buckets, crawler shoes, mill liners, screens; austenitic, work hardened, wear resistant.

AMSCO-MM

Abex Corp.
C 0.7-1.4, Mn 10-14, Si 0.2-1, Mo, bal Fe.
Heat treated: 90,000-160,000 TS; 45,000-75,000 YS; 30-65 El; 30-40 RA; 200 Brin. For mining and construction equipment; abrasion and impact resistant, tough, work hardened.

AMSCO-MMH

Abex Corp.
C, Mo, Mn, bal Fe.
Heat treated: 120,000 TS; 65,000 YS; 20 El; 18 RA. For castings, rolls, jaws, hammers, dipper teeth; wear resistant, austenitic.

AMSCO-MML

Abex Corp.
C, Mo, Mn, bal Fe.
Heat treated: 120,000 TS; 52,000 YS; 50 El; 40 RA. For castings, railroad frogs and crossings; austenitic, wear resistant.

AMSCO-MNI

Abex Corp.
C, Mn, Ni, bal Fe.
Heat treated: 110,000 TS; 48,000 YS; 55 El; 40 RA. For railroad frogs and crossings; austenitic, wear resistant.

AMSCO-MY

Abex Corp.
C 1-1.4, Cr 2, Mn 10-14, bal Fe.
Cast: 120,000 TS; 56,000 YS; 45 El; 30 RA; 200 Brin. Heat treated: 150,000 TS; 65,000 YS; 30 El; 30 RA. For castings, mill liners, screens, dipper teeth jaws, hammers; austenitic, wear resistant.

AMSIL

Climax Performance Materials Corp.
Ag 0.05, bal Cu.
Electrical conductivity: 101%. Wire: 48,000 TS; 40,000 YP; 15 El; 50 Rock B. For rotor and stator windings in high speed generators, commutators. High creep resistance, freedom from embrittlement.

AMSL

Acciaierie Valbruna s.p.a.
Stainless Steel. C 0-0.03, Si 0-1, Mn 0-2, P 0-0.045, S 0-0.03, Cr 19, Mo 3.5, Ni 13, bal Fe.
Austenitic. AISI 317L; W. Nr. 1.4438.

AMSULF

Climax Performance Materials Corp.
S 0.3, bal Cu.
Extruded: 32,000 TS; 47 El; 60 RA. Cold drawn: 57,000 TS; 6 El; 23 RA. For screw machine products, clamps, connectors, clips; high electrical conductivity, free-cutting.

AMTEL

Climax Performance Materials Corp.
Te 0.5, bal Cu.
Bar: 50,000 TS; 40,000 YP; 12 El. For screw machine products, hardware, fasteners, welding tips and nozzles. Free machining, immune to hydrogen embrittlement.

AMUTIT

Bohler Gesellschaft M.B.H.
Now BOHLER K465.

AMUTIT S, NOW VEW K460

Vereinigte Edelstahlwerke
C 1, Mn 1, Cr 0.5, W 0.6, V 0.1, bal Fe.
For cold forming, blanking and bending dies; non-deforming, oil hardening. AISI 01.

AMUTIT S, NOW VEW K460

Raven Steel & Tool Co.
C 1, Mn 1, Cr 0.5, W 0.6, V 0.1, bal Fe.
For cold forming, blanking and bending dies; non-deforming, oil hardening. AISI 01.

AMVILOY 1000

CMW Inc.
W, Ni, Cu.
Sintered: 112,000 TS; 270 Brin. For electrical upsets; 14% electrical conductivity. *Obsolete*

AMVILOY 2000

CMW Inc.
W, Ni, Cu.
Sintered: 110,000 TS; 290 Brin. For electrical upsets; 16% electrical conductivity. *Obsolete*

AMZIRC
Climax Performance Materials Corp.
Zr 0.7, bal Cu.
At 70°F: 72,000 TS; 62,000 YS; 12 El. At 750°F: 51,600 TS; 15 El. At 900°F: 39,400 TS; 19 El. For rectifier bases, rotor wedges, resistance welding wheels and tips; high conductivity, good high temperature strength.

AMZIRC-150
Anaconda Co.
Copper. Zr 0.17, Cu 99.83.
At RT: 71,000 TS; 62,000 YS. At 752°F: 52,000 TS; 45,000 YS; 8 El. At 932°F: 35,200 TS; 25,800 YS; 9 El. For resistance welding wheels and tips, commutators, contacts, switch blades. Annealed: 92% electrical conductivity. Precipitation hardening.

AN 40
Otto Fuchs Metallwerke
Aluminum. Cu 2.5, Mg 1.5, Fe, Ni, bal Al.
390-430 N/mm^2 TS; 305-375 N/mm^2 YS; 3-8 El. Heat treatable alloy with high temperature properties. Type AlCuM Similar to AA 2618.

AN 50
Otto Fuchs Metallwerke
Cu 2, Mg 1.5, Si 0.8, Fe, Ni, bal Fe.
355-390 N/mm^2 TS; 245-275 N/mm^2 YS; 4-6 El. Heat treatable alloy with high temperature properties. Type AlCuMgNiSi.

ANACONDA 1026
Anaconda Co.
Copper. Cu 81.5, Sn 4.25, Mn 0.15, bal Zn.
Die cast: 85,000 TS; 50,000 YS; 8 El. For hardware, fixtures. Corrosion resistant.

ANACONDA 205
Anaconda Co.
Cu 78.5, Zn 20, Pb 1.5.
Hard: 60,000 TS; 47,000 YS; 20 El; B70 Brin. Soft: 43,000 TS; 16,000 YS; 50 El; B10 Brin. For hardware, screw machine parts; free cutting. *Obsolete*

ANACONDA 211
Anaconda Co.
Cu 69, Zn 29.5, Pb 1.5.
Hard: 65,000 TS; 48,000 YS; 20 El; B75 Brin. Soft: 46,000 TS; 17,000 YS; 60 El; B15 Brin. For hardware, screw machine parts; free cutting. *Obsolete*

ANACONDA 24
Anaconda Co.
Cu 85, Zn 15.
Hard: 56,000 TS; 45,000 YS; 20 El; B60 Brin. Soft: 40,000 TS; 15,000 YS; 50 El; B5 Brin. For hardware; red brass. *Obsolete*

ANACONDA 271
Anaconda Co.
Copper. Cu 62, Zn 35, Pb 3.
Hard: 62,000 TS; 50,000 YS; 20 El; 77 Rock B. Soft: 47,000 TS; 32,000 YS; 60 El; 16 Rock B. For machined parts, hardware. Free cutting.

ANACONDA 293
Anaconda Co.
Copper. Zn 39, Pb 1, bal Cu.
Hard: 80,000 TS; 60,000 YS; 6 El; 142 Brin. Soft: 54,000 TS; 20,000 YS; 40 El; 79 Brin. For hardware, architectural uses. Leaded Muntz metal.

ANACONDA 32
Anaconda Co.
Cu 80, Zn 20.
Hard: 60,000 TS; 47,000 YS; 25 El; B70 Brin. Soft: 43,000 TS; 16,000 YS; 55 El; B10 Brin. For hardware; red brass. *Obsolete*

ANACONDA 605
Anaconda Co.
Copper. Zn 38.55, Sn 0.75, Pb 0.7, bal Cu.
Hard: 63,000 TS; 35,000 YS; 28 El; 102 Brin. Soft: 56,000 TS; 22,000 YS; 38 El; 83 Brin. For hardware, screw machine products. Leaded naval brass.

ANACONDA 61
Anaconda Co.
Copper. Cu 63, Zn 37.
Hard: 65,000 TS; 50,000 YS; 20 El; 75 Brin B. Soft: 46,000 TS; 17,000 YS; 60 El; 20 Brin B. For screw machine parts. Yellow brass.

ANACONDA 612
Anaconda Co.
Copper. Cu 60, Zn 29.5, Pb 2, Sn 0.75.
Hard: 63,000 TS; 35,000 YS; 25 El; B65 Brin. Soft: 56,000 TS; 22,000 YS; 35 El; B50 Brin. For screw machine parts. Free cutting.

ANACONDA 624
Anaconda Co.
Copper. Cu 60, Sn 1, Pb 1, Al 0.1, Si 0.15, bal Zn.
Die cast: 55,000 TS; 35,000 YS; 8 El. For hardware, fixtures. Free cutting.

ANACONDA 66
Anaconda Co.
Copper. Cu 60, Zn 40.
Soft: 54,000 TS; 20,000 YS; 40 El; 45 Brin B. For hardware, condenser tubes. High strength.

ANACONDA 67
Anaconda Co.
Cu 67, Zn 33.
For water pipes. *Obsolete*

ANACONDA 681
Anaconda Co.
Copper. Cu 57.8, Zn 40.27, Sn 0.95, Fe 0.85, Si 0.1, Mn 0.03. For oxyacetylene braze welding of steel, cast iron and copper alloys and for bearing surfaces. Welding rods. Low fuming bronze. 870°C MP. Corrosion resistant.

ANACONDA 719
Anaconda Co.
Ni 18, Zn 17, Cu 65.
Hard: 90,000 TS; 3 El. Soft: 58,000 TS; 40 El. For costume jewelry. *Obsolete*

ANACONDA 831
Anaconda Co.
Copper. Pb 1, Ni 1, bal Cu.
Bar: 85,000 TS; 75,000 YS; 5 El. For fasteners, screw machine products, electrical contacts, connectors. Free cutting, corrosion resistant.

ANACONDA 85
Anaconda Co.
Cu 85, Zn 15.
Soft: 40,000 TS; 15,000 YS; 50 El. Hard: 69,000 TS; 55,000 YS; 10 El. For water pipes, condenser tubes; for highly corrosive waters. *Obsolete*

ANACONDA 997
Anaconda Co.
Copper. Zn 40.3, Sn 0.95, Fe 0.85, bal Cu.
For welding rod. Low fuming.

ANACONDA BERYLLIUM COPPER
Anaconda Co.
Be 2-2.25, Ni 0.25-0.5, bal Cu.
Annealed: 50,000 TS; 65 El. Hard rolled: 112,000 TS. Heat treated: 196,000 TS; 3 El. For springs, high fatigue and strong bronze parts, fuse clips, switch blades, welding electrodes. *Obsolete*

ANACONDA COPPER 189
Anaconda Co.
Copper. Cu 98.8, Sn 0.75, Si 0.25, Mn 0.2.
For inert gas and oxyacetylene welding of copper for ductile and strong welds. Welding rods. 1075°C MP. Corrosion resistant.

ANACONDA COPPER 372
Anaconda Co.
Copper. Cu 98.8, Sn 0.75, Si 0.25, Mn 0.2.
For welding rod for oxyacetylene and inert gas arc welding of copper. 1967°F MP.

ANACONDA ELECTRO SHEET COPPER
Anaconda Co.
Cu.
For weather proofing, printed circuits; furnace brazing. *Obsolete*

ANACONDA PHOSPHOR BRONZE, GRADE "E"
Anaconda Co.
Cu 98.25, Sn 1.75.
Hard: 100,000 TS; 3 El. Soft: 50,000 TS; 33 El. For electrical parts; M.P. 1070 C. *Obsolete*

ANACONDA SPECIAL PHOSPHOR BRONZE
Anaconda Co.
Cu 88, Sn 4, Zn 4, Pb 4.
For corrosion, heat and abrasion resistance. *Obsolete*

ANACONDA SUPER NICKEL
Anaconda Co.
Cu 70, Ni 30.
For condenser tubes; for severe service. *Obsolete*

ANATOMICAL ALLOY
American manufacture
Bi 54, Sn 19, Pb 17, Hg 11, Cd.
For fusible alloy for anatomical impressions and casts; fusible.

ANCAST 410 MOD. (1)
Ancast, Inc.
Stainless steel. C 0-0.06, Mn 0-1, Si 0-1, Cr 11.5-14, Ni 3.5-4.5, Mo 0.4-1, P 0-0.04, S 0-0.03, bal Fe.
ASTM A-743 GR CA6NM.

ANCAST 410 MOD. (2)
Ancast, Inc.
Stainless steel. C 0-0.15, Mn 0-1, Si 0-0.65, Cr 11.5-14, Ni 0-1, Mo 0.15-1, P 0-0.04, S 0-0.04, bal Fe.
ASTM A-743 GR CA 15M.

ANCHOR ALLOY NO. 250
Anchor Alloys Inc.
Aluminum.
Aluminum solder; MP 390°F. For joining aluminum to aluminum.

ANCHOR ALLOY NO. 260
Anchor Alloys Inc.
Aluminum.
Aluminum solder; MP 650°F. For joining aluminum to aluminum or to dissimilar metals, including castings, die cast parts and aluminum foil.

ANCHOR ALLOY NO. 270
Anchor Alloys Inc.
Solder; MP 507°F. All purpose solder for all metals except magnesium; including aluminum, zinc die casting, copper, etc.

ANCHOR ALLOY NO. 304
Anchor Alloys Inc.
Solder, contains Ag, no Cd; MP 430°F; 14,000-20,000 TS.
For joining stainless steels to dissimilar metals.

ANCHOR ALLOY NO. 520
Anchor Alloys Inc.
Solder for repairing, sealing and filling cracks and blow holes in cast iron; MP 500°F.

ANCHOR ALLOY NO. 530
Anchor Alloys Inc.
Solder for repairing zinc die castings such as radiator grills, door handles; MP 700°F.

ANCHOR ALLOY NO. 95/5
Anchor Alloys Inc.
Refrigeration solder, contains Sn and Sb; used with copper sweat fittings; MP 450°F.

ANCHOR CHASER DIE
Teledyne Vasco
C 0.95-1.1, Alloy, bal Fe.
For tools, chasers. *Obsolete*

ANCHOR GALVANIZED BAR
Anchor Alloys Inc.
Galvanizing repair solder; MP 550°F. For repairing damaged or burned galvanized surface; no flux required.

ANCHOR REDIMIX SOLDER PASTE
Anchor Alloys Inc.
Solder paste for metals except aluminum; contains solder and flux; MP 370°F.

ANCHOR SPECIAL TAP
Teledyne Vasco
C 1.2, W 1.6, Cr 0.7, V 0.2, bal Fe.
For tools, dies. *Obsolete*

ANCHOR-TITE
Tungsten Widia Tool Corp
WC.
For tools, dies. Cemented WC.

ANCOLOY
Hoeganaes Corp.
Ni 1.75, Mo 0.5, C 0.02, Cu 1.5, bal Fe.
Sintered: 54,400 TS; 41,500 YS; 1-2 El. Heat treated: 115,000 TS; 104,000 YS; 0.3 El. For general machinery parts. Heat treatable. Prealloyed powder, high strength.

ANCOLOY SPONGE IRON POWDER
Hoeganaes Corp.
Fe 95, C 0.02, Cu 1.55, Ni 1.75, Mo 0.6, (0.45 H_2 loss).
Compressibility at 30 psi 6.2 g/cm³. For medium density P/M parts requiring heat treatment.

ANCOR 303 STAINLESS STEEL POWDER
Hoeganaes Corp.
C 0-0.03, Cr 17.5, Ni 12, Si 0.9, S 0.25, bal Fe.
Compressibility at 30 psi 6.05 g/cm³. For austenitic stainless P/M parts requiring machinability.

ANCOR 303L
Hoeganaes Corp.
Stainless steel.
Stainless steel high alloy powder. Free machining austenitic stainless.

ANCOR 304L
Hoeganaes Corp.
Stainless steel.
Stainless steel high alloy powder. Standard 18% austenitic stainless.

ANCOR 304L STAINLESS STEEL POWDER
Hoeganaes Corp.
C 0-0.03, Cr 18.5, Ni 10, Si 0.9, bal Fe.
Compressibility at 30 psi 6.1 g/cm³. Non-magnetic P/M parts.

ANCOR 316L
Hoeganaes Corp.
Stainless steel.
Stainless steel high alloy powder. Austenitic stainless.

ANCOR 316L STAINLESS STEEL POWDER
Hoeganaes Corp.
C 0-0.03, Cr 17.5, Ni 13, Si 0.9, Mo 2.2, bal Fe.
Compressibility at 30 psi 6.25 g/cm³. For corrosion resistant high strength P/M parts and filters.

ANCOR 410L
Hoeganaes Corp.
Stainless steel.
Stainless steel high alloy powder. Martensitic stainless.

ANCOR 410L STAINLESS STEEL POWDER
Hoeganaes Corp.
C 0-0.015, Cr 13.2, Si 0.09, bal Fe.
Compressibility at 30 psi 5.9 g/cm³. If blended to build up carbon content, the P/M parts can have good hardness and strength as well as corrosion resistance.

ANCOR 430L
Hoeganaes Corp.
Stainless steel.
Stainless steel high alloy powder. Ferritic stainless.

ANCOR 434L
Hoeganaes Corp.
Stainless steel.
Stainless steel high alloy powder. Ferritic stainless. Automotive applications.

ANCOR 50NI/50FE
Hoeganaes Corp.
Stainless steel.
Stainless steel high alloy powder. For soft electromagnetic applications.

ANCOR MH-100
Hoeganaes Corp.
Sponge iron powder. For P/M parts requiring low to medium densities.

ANCOR MH-100 SPONGE IRON POWDER
Hoeganaes Corp.
Fe 99, C 0.02, 0.045 H_2 (loss when heated).
Compressibility at 30 psi 6.3 g/cm³. For low and medium density P/M parts.

ANCOR MH-1024 M-1S SPONGE IRON
Hoeganaes Corp.
Fe 98, C 0.02, S 1, 0.45 H_2 (loss).
Compressibility at 30 psi 6.2 g/cm³. Improved machining P/M parts because of sulfur.

ANCOR MH-1024M
Hoeganaes Corp.
Sponge iron powder. For parts requiring machining after sintering.

ANCOR MH-818
Hoeganaes Corp.
Sponge iron powder. Used where very high green strength is required or when apparent density control is necessary.

ANCOR W-100 SPONGE IRON POWDER
Hoeganaes Corp.
Fe 97, C 0.08, S 0-0.02, 0.80 H_2 (loss).
100 mesh powder for flux core wire and small diameter stick electrodes.

ANCOR W-423A SPONGE IRON POWDER
Hoeganaes Corp.
Fe 97, C 0.08, S 0-0.02, 0.8 H_2 (loss).
40 mesh powder for low yield stick electrode coatings.

ANCOR W-428 SPONGE IRON POWDER
Hoeganaes Corp.
Fe 97, C 0.08, S 0-0.02, 0.8 H_2 (loss).
40 mesh powder for medium yield stick electrode coatings.

ANCORMET 101
Hoeganaes Corp.
Sponge iron powder. Provides fast carbon pick up during sintering and small dimensional change.

ANCORMET 101 SPONGE IRON POWDER
Hoeganaes Corp.
Fe 99, C 0.18, 0.40 H_2 (loss when heated).
Compressibility at 30 psi 6.2 g/cm³. For medium density, high strength P/M parts.

ANCORSPRAY 120
Hoeganaes Corp.
C 0.03, B 1.5, Si 2.5, Fe 1.5, bal Ni.
Self-fluxing atomized powder for spraying. Fusion temperature 2000°F; 13-15 Rock C. For soft coatings that may be hand finished.

ANCORSPRAY 125
Hoeganaes Corp.
C 0.06, B 1.5, Si 3.5, Fe 1.5, bal Ni.
Self-fluxing atomized powder for spraying. Fusion temperature 1975°F; 25-28 Rock C. For rebuilding damaged articles; may be refinished by machining. AMS 4779.

ANCORSPRAY 135
Hoeganaes Corp.
C 0.05, Cr 10.5, B 2, Si 3.25, Fe, bal Ni.
Self-fluxing atomized powder for spraying. Fusion temperature 1925°F; 33-38 Rock C. For build-up; either final coat or as an undercoat; soft, machinable.

ANCORSPRAY 140
Hoeganaes Corp.
C 0.3, Cr 7.5, B 1.35, Si 4, Fe 1.5, bal Ni.
Self-fluxing atomized powder for spraying. Fusion temperature 1925°F; 35-40 Rock C. For build-up; may be machined with carbide tools.

ANCORSPRAY 150
Hoeganaes Corp.
C 0.65, Cr 14, B 2.8, Si 3.8, Fe 4.2, bal Ni.
Fusion temperature 1900°F; 50-55 Rock C. For build-up; strong, hard, but still somewhat flexible. AMS 4775.

ANCORSPRAY 155
Hoeganaes Corp.
C 0.5, Cr 16, B 4, Si 4.25, Fe 4, Cu 2.5, Mo 2.5, W 2.5, bal Ni.
Fusion temperature 1900°F; 55-59 Rock C. For build-up of heavy coating, and for final coat. Hard, wear resistant.

ANCORSPRAY 160
Hoeganaes Corp.
C 0.9, Cr 16.5, B 3.25, Si 4.25, Fe 4.5, bal Ni.
Fusion temperature 1900°F; 60-64 Rock C. Hardest and most wear resistant grade; finished by grinding. AMS 4775.

ANCORSPRAY 250
Hoeganaes Corp.
C 0.1, Cr 18.5, B 3.2, Si 3.3, Fe 2, Ni 27, Mo 5.5, bal Co.
Fusion temperature 2050°F; 47-53 Rock C. For special build-up jobs; good hardness and wear resistance, corrosion resistant and resistant to thermal cracking.

ANCORSTEEL 1000
Hoeganaes Corp.
Atomized steel powder. Low carbon and low oxygen content with high compressibility.

ANCORSTEEL 1000 ATOMIZED IRON POWDER
Hoeganaes Corp.
Fe 99.2, C 0.015, 0.25 H_2 (loss).
Compressibility at 30 psi 6.7 g/cm³. For high density P/M parts.

ANCORSTEEL 1000B
Hoeganaes Corp.
Atomized steel powder. Provides greater compressibility than Ancorsteel 1000. For parts requiring high density. Used where good electromagnetic properties are required.

ANCORSTEEL 1000B ATOMIZED IRON POWDER
Hoeganaes Corp.
Fe 99.4, C 0.01, 0.12 H_2 (loss).
Compressibility at 30 psi 6.8 g/cm^3. For high density and electromagnetic P/M parts.

ANCORSTEEL 1000C
Hoeganaes Corp.
Atomized steel powder. For P/M parts requiring densities of about 7.0 g/cm^3.

ANCORSTEEL 1000G
Hoeganaes Corp.
Atomized steel powder. Increased green strength. Good compressibility.

ANCORSTEEL 1000M
Hoeganaes Corp.
S 0.45.
Atomized steel powder. Machinable. Properties similar to Ancorsteel 1000.

ANCORSTEEL 1015
Hoeganaes Corp.
Atomized steel powder. Low prealloyed carbon steel powder for fast sintering.

ANCORSTEEL 1015 ATOMIZED STEEL POWDER
Hoeganaes Corp.
Fe 99.1, C 0.15, 0.30 H_2 (loss).
Compressibility at 30 psi 6.5 g/cm^3. Fast sintering of high density P/M parts.

ANCORSTEEL 2000
Hoeganaes Corp.
Ni 0.45, Mo 0.6, Mn 0.3, bal Fe.
Atomized steel powder. For production of parts requiring heat treatment.

ANCORSTEEL 2000 ATOMIZED STEEL POWDER
Hoeganaes Corp.
Fe 98.2, Mo 0.6, Ni 0.45, Mn 0.3, C 0.2, 0.13 H_2 (loss).
Compressibility at 30 psi 6.6 g/cm^3. Low cost P/M parts requiring heat treatment.

ANCORSTEEL 45P
Hoeganaes Corp.
Atomized steel powder. High compressibility, high purity iron powder with a small amount of added phosphorus. Good ductility and tensile strength.

ANCORSTEEL 45P ATOMIZED STEEL POWDER
Hoeganaes Corp.
Fe 98.8, C 0.02, P 0.45, 0.13 H_2 (loss).
Compressibility at 30 psi 6.8 g/gm^3. Phosphorus speeds sintering and enhances properties; strength/ductility and electromagnetic.

ANCORSTEEL 4600V
Hoeganaes Corp.
Ni 1.8, Mo 0.5, Mn 0.25, bal Fe.
Atomized steel powder. For production of parts requiring higher hardenability than provided by Ancorsteel 2000.

ANCORSTEEL 4600V ATOMIZED STEEL POWDER
Hoeganaes Corp.
Fe 96.9, Ni 1.8, Mo 0.5, Mn 0.25, C 0.02, 0.13 H_2 (loss).
Compressibility at 30 psi 6.5 g/cm^3. Good hardenability parts requiring high impact strength.

ANDARD
Duke Steel Co. Inc.
C 1.1, bal Fe.
For tools, drills, water hardening.

ANFRILOY
Wellman Dynamics Corp.
Sn, Zn, Pb, bal Cu.
For bearings; heavy duty. *Obsolete*

ANGEL
Apex Steel Co.
C, alloy, bal Fe.
For high speed tools, high speed steel.

ANGSBURG
Manufacturer not listed.
Cu 71.9, Zn 27.6.
For tubes, fittings, brass.

ANHYSTER A
Creusot-Loire
Ni 35, bal Fe.
For magnets; high permeability. *Obsolete*

ANHYSTER B
Creusot-Loire
Ni 35, bal Fe.
For magnets; high permeability. *Obsolete*

ANHYSTER C
Creusot-Loire
Ni 45-50, bal Fe.
For magnets, relays, transformers; high permeability. *Obsolete*

ANHYSTER D
Creusot-Loire
Ni 45-50, bal Fe.
For magnets, relays, transformers; high permeability. *Obsolete*

ANHYSTER M
Creusot-Loire
Ni 77, Cr 1.6, Cu 5, bal Fe.
For relays, transformers, magnetic amplifiers; high permeability. *Obsolete*

ANHYSTER MAU MOLYBDENE
Creusot-Loire
Ni 77, Mo 3.9, Cu 4, bal Fe.
For relays, magnetic amplifiers; high permeability. *Obsolete*

ANIBAL
Creusot-Loire
Ni 42, Cr 2, C, bal Fe.
For clocks, balance wheels; corrosion resistant. *Obsolete*

ANKA "H"
Dunford Hadfields Ltd.
C 0.13-0.2, Cr 16-23, Ni 7-11, bal Fe.
For heat and corrosion resistant parts; stainless; heat and corrosion resistant. *Obsolete*

ANKA STEEL
Dunford Hadfields Ltd.
C 0.2, Si 0.6, Mn 0.6, Ni 6-10, Cr 16-20, bal Fe.
Annealed: 85,000 TS; 35,000 YS; 60 El; 70 RA; 150 Brin. Cold drawn: 125,000 TS; 95,000 YS; 25 El; 55 RA; 277 Brin. For chemical plant equipment, tanks, vessels, fittings; Type 301; stainless, austenitic. *Obsolete*

ANKO-2
Russian manufacture
Al 9, Cu 4, Ni 20, Co 15, bal Fe.
Heat treated magnetic alloy;.For electrical and magnetic equipment.

ANKO-3
Russian manufacture
Al 10, Ni 19, Co 18, Cu 3, bal Fe.
Permanent magnet alloy.

ANKO-4
Russian manufacture
Al 8, N: 14, Co 24, Cu 3, bal Fe.
Permanent magnet alloy. Similar to ALNICO 5.

ANNITE
Bisset Steel Co.
C 1.5, Cr 4, Co 1, Mo 1, bal Fe.
For dies, tools, punches, rotary shears, blanking dies; resists abrasion. *Obsolete*

ANNITE
NL Industries
C 1.5, Cr 4, Co 1, Mo 1, bal Fe.
For dies, tools, punches, rotary shears, blanking dies; resists abrasion. *Obsolete*

ANNITE NO. 1
Bethlehem Steel Corp.
C 2.25, Mn 0.3, Cr 13, V 0.2, Si 0.2, bal Fe.
For punches and shearing dies, drawing dies; oil hardened, non-deforming. *Obsolete*

ANNITE NO. 2
Bethlehem Steel Corp.
C 1.5, Cr 11.5, Mo 0.8, V 0.25, Si 0.25, bal Fe.
For punches and shearing dies, thread rolling dies; air hardened, non-deforming. *Obsolete*

ANODE
Lincoln Electric Co.
Non-ferrous alloy.
For welding electrode; general purpose welding. *Obsolete*

ANODE METAL-1
Manufacturer not listed.
Pb 94, Sb 6.
For batteries, plates; hard.

ANODE METAL-2
Manufacturer not listed.
Pb 98.8, Ag 1, As 0.2.
For lead storage battery parts; hard.

ANOREFRACT 50
Acieries et Forges d'Anor
C 0.25-0.35, Cr 24-26, Ni 12-14, bal Fe.
Heat resisting steel. AFNOR Z30 CN 25.12.

ANOREFRACT 50 NB
Acieries et Forges d'Anor
C 0.25-0.35, Cr 24-26, Ni 12-14, Nb 1.3-1.5, bal Fe.
Heat resisting steel.

ANORESIST 55T
Acieries et Forges d'Anor
C 0.4-0.6, Cr 17-19, Ni 37-39, bal Fe.
Heat resisting steel. For furnace parts. AFNOR Z50 NCS 38.78.

ANOREXACT 144
Acieries et Forges d'Anor
C 1.05-1.15, Si 0.15-0.3, Mn 0.2-0.4, Cr 1.1-1.3, W 1.2-1.4, V 0.15-0.25, bal Fe.
Low alloy tool steel. AFNOR 110 WCV 13.05.

ANOREXACT 344 B
Acieries et Forges d'Anor
C 0.52-0.62, Si 0.6-0.8, Mn 0.4-0.5, Cr 0.8-1, W 1.7-1.9, V 0.15-0.2, bal Fe.
Tool steel; shock resisting type. AFNOR 55 WCV 20.04.

ANOREXACT 44
Acieries et Forges d'Anor
C 0.75-0.85, Si 0.4-0.6, Mn 0.3-0.4, Cr 1-1.2, W 1.8-2.1, V 0.25-0.35, bal Fe.
Low alloy tool steel. AFNOR 80 WCV 20.04.

ANOREXACT 5B
Acieries et Forges d'Anor
C 0.65-0.7, Si 1.5-1.6, Mn 0.35-0.45, Ni 2.6-2.7, Cr 0.85-1, bal Fe.
Low alloy tool steel. AFNOR G7 NSC 11.

ANOREXACT SD 13
Acieries et Forges d'Anor
C 1.55-1.75, Si 0.25-0.4, Mn 0.2-0.4, Cr 11-12, Mo 0.5-0.7, V 0.1-0.5, W 0.4-0.6, bal Fe.
High chromium cold work tool steel. AFNOR Z 160 CDW 12.

ANOREXACT CNT7
Acieries et Forges d'Anor
C 0.43-0.47, Si 0.2-0.3, Mn 0.65-0.75, Cr 1.45-1.55, Ni 3.9-4, Mo 0.8-1, V 0.45-0.52, bal Fe.
Low alloy nickel steel. AFNOR 45 NCD 16.

ANOREXACT D11
Acieries et Forges d'Anor
C 0.85-0.95, Si 0.15-0.3, Mn 1.9-2.1, V 0.1-0.2, Cr 0.3-0.5, bal Fe.
Oil or water hardening cold work tool steel. AFNOR 90MV8.

ANOREXACT P5
Acieries et Forges d'Anor
C 0.2-0.25, Si 0.3-0.5, Mn 0.45-0.75, Cr 0.7-1, Ni 2.8-3.4, Mo 0.2-0.3, W 0.15-0.25, bal Fe.
Low alloy nickel steel. AFNOR 22 NC 11.

ANOREXACT SB 11
Acieries et Forges d'Anor
C 0.48-0.52, Cr 7.5-8.5, Si 0.9-1.1, Mn 0.45-0.55, Mo 1.4-1.6, V 0.45-0.55, bal Fe.
High alloy tool steel. AFNOR Z 50 CDV8.

ANOREXACT SB 55
Acieries et Forges d'Anor
C 0.95-1.05, S: 0.25-0.35, Mn 0.25-0.35, Cr 4.8-5.3, Mo 0.95-1.05, V 0.25-0.35, bal Fe.
Air hardening cold work tool steel. AFNOR Z 100 CDV 5.

ANOREXACT SB 7
Acieries et Forges d'Anor
C 0.55, Si 1, Mn 0.3, Cr 5, Mo 1.25, V 0.5, bal Fe.
Hot work tool steel. AFNOR Z 55 CDV 5. *Obsolete*

ANOREXACT SB6
Acieries et Forges d'Anor
C 0.3-0.4, Si 0.8-1, Mn 0.2-0.4, Cr 4.75-5.25, Mo 1.35-1.45, V 0.3-0.4, W 1.25-1.35, bal Fe.
Hot work tool steel. AFNOR Z 55 CDVW 5.

ANOREXACT SD 13 V
Acieries et Forges d'Anor
C 1.6, Si 0.3, Mn 0.3, Cr 12, Mo 0.8, V, bal Fe.
High chromium cold work tool steel. AFNOR Z 160 CDV 12-01. *Obsolete*

ANOREXACT SD 15
Acieries et Forges d'Anor
C 2-2.2, Si 0.2-0.3, Mn 0.2-0.3, Cr 12.5-13.5, bal Fe.
High chromium cold work tool steel. AFNOR Z 200 C 12.

ANOREXACT SD 3
Acieries et Forges d'Anor
C 0.35-0.45, Si 0.9-1, Mn 0.2-0.4, Cr 5-5.5, Mo 1.3-1.4, V 0.4-0.6, bal Fe.
Hot work tool steel. AFNOR Z 38 CDV 5.

ANOREXACT SD 9S
Acieries et Forges d'Anor
C 0.58-0.62, Si 1.5-1.6, Mn 0.9-1.1, Cr 0.9-1, Mo 0.2-0.3, bal Fe.
Tool steel; shock resisting type. AFNOR Y G0 SCD 6.

ANOREXACT W
Acieries et Forges d'Anor
C 0.35-0.45, Si 0.3-0.4, Mn 0.6-0.8, Cr 0.8-1.2, Mo 0.2-0.3, bal Fe.
Low alloy structural or tool steel. AFNOR Y 42 CD 4.

ANORINOX 100
Acieries et Forges d'Anor
C 0-0.08, Si 0.2-1, Mn 0.2-1, Cr 19-21, Ni 24-26, Mo 4-5, Cu 1.5-2, bal Fe.
Austenitic stainless steel. AFNOR Z 08 NCDV 25.20.05.

ANORINOX 17.4S
Acieries et Forges d'Anor
C 0-0.07, Si 0.4-0.8, Mn 0.4-0.8, Cr 15.5-17.5, Ni 4.1-4.4, Cu 3.1-3.4, Nb 0.25-0.3, bal Fe.
Precipitation hardening stainless steel. AFNOR Z06 CNV 16.04.

ANORINOX 35 A
Acieries et Forges d'Anor
C 0.28-0.34, Si 0.5-1, Mn 0.5-1, Cr 12.5-13.5, bal Fe.
Martensitic type stainless steel. AFNOR Z 30 C 13.

ANORINOX 70
Acieries et Forges d'Anor
C 0-0.08, Si 0-1.5, Mn 0-1.5, Cr 18-20, Ni 8-12, bal Fe.
Austenitic 18/8 type stainless steel. AFNOR Z 08 CN 18-08.

ANORINOX 80
Acieries et Forges d'Anor
C 0-0.08, Si 0-1.5, Mn 0-1.5, Cr 17-19, Ni 10-12, Mo 2-3, bal Fe.
Austenitic 18/8 plus Mo type stainless steel. AFNOR Z 08 CND 18-08; similar to AISI 316.

ANORINOX 90
Acieries et Forges d'Anor
C 0-0.08, Si 0-1.5, Mn 0-1.5, Cr 18-20, Ni 13-15, Mo 3-4, bal Fe.
Austenitic stainless steel. AFNOR Z 08 CND 18.13.03.

ANORINOX SR17
Acieries et Forges d'Anor
C 1-1.2, Si 1-1.2, Mn 0.4-0.6, Cr 16-17, Ni 1.5-1.7, Mo 0.8-1, bal Fe.
Martensitic stainless steel. AFNOR Z110 CND 17.

ANOXIN 1
Rochling Burbach GmbH
C 0-0.15, Cr 18, Ni 8.5, bal Fe.
Annealed: 80,000 TS; 35,000 YS; 55 El; 75 RA; 150 Brin. Cold drawn: 180,000 TS; 150,000 YS; 10 El; 250 Brin. For chemical plant equipment, tanks, mixers, agitators, filters; Type 302; stainless, austenitic. *Obsolete*

ANOXIN 2
Rochling Burbach GmbH
C 0-0.07, Cr 18, Ni 9.5, bal Fe.
Annealed: 85,000 TS; 35,000 YS; 60 El; 70 RA; 150 Brin. Cold drawn: 180,000 TS; 125,000 YS; 10 El; 330 Brin. For welded structures, chemical plant equipment, tanks; Type 304; stainless, austenitic. *Obsolete*

ANOXIN 20
Rochling Burbach GmbH
C 0-0.07, Cr 18, Ni 9.5, bal Fe.
Annealed: 85,000 TS; 35,000 YS; 60 El; 70 RA; 150 Brin. Cold drawn: 180,000 TS; 125,000 YS; 10 El; 330 Brin. For welded chemical plant equipment, tanks, mixers; Type 304; stainless, austenitic. *Obsolete*

ANOXIN 200
Rochling Burbach GmbH
C 0-0.03, Cr 18-20, Ni 8-12, bal Fe.
Annealed: 85,000 TS; 30,000 YS; 60 El; B 80 Rock. Cold drawn: 150,000 TS; 90,000 YS; 30 El; C 25 Rock. For welded structures, chemical plant equipment, vessels, tanks, trim. Type 304L stainless steel, austenitic, welding grade. *Obsolete*

ANOXIN 22
Rochling Burbach GmbH
C 0-0.12, Cr 18, Ni 9.5, Cb = 8 x C, bal Fe.
Annealed: 90,000 TS; 45,000 YS; 56 El; 65 RA; 160 Brin. Cold drawn: 100,000 TS; 65,000 YS; 40 El; 60 RA; 202 Brin. For welded chemical plant equipment, tanks, mixers; Type 347; stainless, austenitic. *Obsolete*

ANOXIN 2G
Rochling Burbach GmbH
C 0-0.12, Cr 18, Ni 9.5, Ti = 4 x C, bal Fe.
Annealed: 85,000 TS; 35,000 YS; 55 El; 65 RA; 150 Brin. Cold drawn: 95,000 TS; 60,000 YS; 40 El; 60 RA; 185 Brin. For welded structures, chemical plant equipment; Type 321; stainless, austenitic. *Obsolete*

ANOXIN 3
Rochling Burbach GmbH
C 0.08, Cr 18, Ni 8, bal Fe.
Annealed: 85,000 TS; 35,000 YS; 60 El; 70 RA; 150 Brin. Cold drawn: 180,000 TS; 125,000 YS; 10 El; 330 Brin. For welded structures, chemical plant equipment, tanks; Type 304; stainless, austenitic. *Obsolete*

ANOXIN 4
Rochling Burbach GmbH
C 0.05, Cr 18, Mo 2, Ni 10.5, Ti = 4 x C, bal Fe.
Annealed: 85,000 TS; 35,000 YS; 50 El; 60 RA; 160 Brin. For welded acid resistant chemical equipment, tanks; Type 316Ti, stainless, austenitic. *Obsolete*

ANOXIN 40
Rochling Burbach GmbH
C 0-0.07, Cr 18, Ni 10.5, Mo 2, bal Fe.
Annealed: 85,000 TS; 35,000 YS; 50 El; 65 RA; 160 Brin. Cold drawn: 150,000 TS; 135,000 YS; 6 El; 300 Brin. For acid resistant chemical plant equipment, tanks; Type 316; stainless, austenitic. *Obsolete*

ANOXIN 44
Rochling Burbach GmbH
C 0-0.12, Cr 18, Ni 10.5, Mo 2, Cb = 8 x C, bal Fe.
Annealed: 85,000 TS; 35,000 YS; 50 El; 65 RA; 160 Brin. For welded chemical plant equipment, tanks, mixers; Type 316Cb; stainless, austenitic. *Obsolete*

ANOXIN 4G
Rochling Burbach GmbH
C 0.15, Cr 18, Ni 8.5, Mo 2, bal Fe.
Annealed: 85,000 TS; 35,000 YS; 50 El; 65 RA; 160 Brin. Cold drawn: 150,000 TS; 135,000 YS; 6 El; 300 Brin. For acid resistant chemical plant equipment, tanks; Type 316; stainless, austenitic. *Obsolete*

ANOXIN 5
Rochling Burbach GmbH
C 0-0.07, Cr 17.5, Ni 17.5, Mo 2, Cu 2, Ti = 4 x C, bal Fe.
For valves, pumps, chemical plant equipment; stainless, austenitic. *Obsolete*

ANOXIN 5G
Rochling Burbach GmbH
C 0.1, Cr 18, Ni 9, Mo 2, Cu, bal Fe.
Annealed: 85,000 TS; 35,000 YS; 50 El; 65 RA; 160 Brin. For chemical plant equipment, tanks; Type 316; stainless, austenitic. *Obsolete*

ANOXIN 66
Rochling Burbach GmbH
C 0.05, Cr 18, Ni 9, Cu, bal Fe.
Annealed: 85,000 TS; 35,000 YS; 60 El; 70 RA; 150 Brin. For chemical plant equipment, pump parts, mixers; stainless, austenitic. *Obsolete*

ANOXIN 70
Rochling Burbach GmbH
C 0-0.08, Cr 16-18, Ni 10-14, Mo 2-3, bal Fe.
Annealed: 80,000 TS; 30,000 YS; 60 El; B 78 Rock. Cold rolled: 150,000 TS; 135,000 YS; 6 El; C 32 Rock. For chemical and textile plant equipment, agitators, kettles, acid tanks and vessels. Type 316 stainless steel, austenitic, acid resistant. *Obsolete*

ANOXIN 77
Rochling Burbach GmbH
C 0.1, Cr 17.5, Ni 13, Mo 2.5, Cb 0.8, bal Fe.
Annealed: 84,000 TS; 40,000 YS; 50 El; B 82 Rock. For welded structures and tanks, mixers, agitators, digesters. Welding grade, stabilized. Type 318 stainless steel; austenitic. *Obsolete*

ANOXIN 8
Rochling Burbach GmbH
C 0.1, Cr 18, Ni 9, Mo 2, Ti = 7 x C, bal Fe.
Annealed: 85,000 TS; 35,000 YS; 50 El; 65 RA; 160 Brin. For acid resistant chemical plant equipment, welded structures; Type 316Ti; stainless, austenitic. *Obsolete*

ANOXIN G
Rochling Burbach GmbH
C 0.15, Cr 18, Ni 8.5, bal Fe.
Annealed: 80,000 TS; 35,000 YS; 55 El; 75 RA; 150 Brin. Cold drawn: 180,000 TS; 150,000 YS; 10 El; 250 Brin. For chemical plant equipment, tanks, mixers, filters; Type 302; stainless, austenitic. *Obsolete*

ANOXIN ULTRA-C
Rochling Burbach GmbH
Ni 55, Cr 16, Mo 17, Fe 6, W 4.
For pumps, valves, pharmaceutical and chemical plant equipment. Heat and corrosion resistant. *Obsolete*

ANOXIN-1F
Rochling Burbach GmbH
C 0-0.15, Cr 16-18, Ni 6-8, bal Fe.
Annealed: 110,000 TS; 40,000 YS; 60 El; B 85 Rock. Hard: 185,000 TS; 140,000 YS; 8 El; C 41 Rock. For aircraft structural members, trailer bodies, auto wheel covers, household utensils. Type 301 stainless steel, austenitic. Work hardens rapidly. *Obsolete*

ANOXIN-2P
Forges et Acieries de Voelkingen
C 0-0.08, Cr 18-20, Ni 8-12, bal Fe.
Annealed: 85,000 TS; 35,000 YS; 60 El; 150 Brin. Cold drawn: 100,000-180,000 TS; 50,000-125,000 YS; 130-180 Brin. For architectural trim, chemical and pharmaceutical plant equipment. Type 304 stainless steel, austenitic. Good formability and weldability.

ANOXIN-4P
Forges et Acieries de Voelkingen
C 0-0.08, Cr 16-18, Ni 10-14, Mo 2-3, bal Fe.
Annealed: 80,000 TS; 30,000 YS; 60 El; 78 Rock B. Cold rolled: 150,000 TS; 135,000 YS; 6 El; 32 Rock C. For chemical and textile plant equipment, agitators, kettles, valve trim, acid tanks and vessels. Type 316 stainless steel, austenitic, acid resistant.

ANSCOL 45
Acme Steel Co.
C 0.16, Mn 0.7, 0.015 Cb min, bal Fe.
Plate: 60,000 TS; 45,000 YS; 30 El. For railroad and mine cars, bridges, booms, pressure vessels, derricks. Good fabricability and weldability. *Obsolete*

ANSCOL 50
Acme Steel Co.
C 0.18, Mn 0.7, Si 0.04, 0.015 Cb min, bal Fe.
Plate: 63,000 TS; 50,000 YS; 30 El; 140 Brin. For railroad and mine cars, bridges, booms, pressure vessels, derricks. Good fabricability and weldability. *Obsolete*

ANTACIRON
Antaciron, Inc.
Si 14.5, Fe 84.5.
Cast: 15,000-18,000 TS; 15,000-18,000 YS; 450-475 Brin. For equipment to resist corrosion acids carrying abrasive solids; finished by grinding.

ANTAG CHISEL
Midvale-Heppenstall Co.
C 0.7-0.8, bal Fe.
For chisels, tools. *Obsolete*

ANTI FRICTION BABBITT
Belmont Metals Inc.
Sn 5, Sb 15, bal Pb.
10,000 psi TS. For low speed bearings; melting point 522°F.

ANTICORODAL
Aluminium Walzwerke Singen GmbH
Aluminum. Mg 0.6-1.4, Si 0.6-1.6, Mn 0.6-1, bal Al.
Annealed: 21,000 TS; 8000 YS; 24 El. For window frames, fan blades, gutters, boats; good forming and welding properties.

ANTICORODAL - 112
Aluminium Walzwerke Singen GmbH
Aluminum. Mg 1, Si 1, Mn 0.7, bal Al.
Annealed: 110 MPa TS; 25 El. Solution heat treated: 310 MPa TS; 10 El. Bars, tubing for ship equipment.

ANTICORODAL 15
Aluminium Industrie Aktiengesellschaft
Si 0.5-1.5, Mn 0.2-1, Mg 0.5-1, Cd 0.5-1.5, bal Al.
Heat treated: 45,000-61,000 TS; 38,000-58,000 YS; 12-20 El; 90-120 Brin. For aircraft structures, highly stressed parts; corrosion resistant, for forgings and extrusions.

ANTICORODAL 15
United States Bronze Powders Inc.
Si 4.5, Mg 0.65, Mn 0.7, bal Al.
Heat treated: 36,000-46,000 TS; 26,000-36,000 YS; 1-2 El; 90-105 Brin. For light alloy parts; corrosion resistant.

ANTICORODAL 15
Alluminio SA
Si 4.2-5.2, Mg 0.6, Mn 0.7, Ti 0.15, bal Al.
Cast: 35,000-45,500 TS; 28,500-38,400 YS; 1-2 El; 90-105 Brin. For instrument housings, oil pans, crankcases; high fluidity and corrosion resistant.

ANTICORODAL 15
Soc. Alluminio Veneto per Azioni
Si 4.2-5.2, Mg 0.6, Mn 0.7, Ti 0.15, bal Al.
Cast: 35,000-45,500 TS; 28,500-38,400 YS; 1-2 El; 90-105 Brin. For instrument housings, oil pans, crankcases; high fluidity and corrosion resistant.

ANTICORODAL G
Soc. Alluminio Veneto per Azioni
Aluminum. Si 2, Mg 0.65, Mn 0.7, bal Al.
Heat treated: 36,000-42,000 TS; 29,000-39,000 YS; 1-3 El; 90-105 Brin. For light alloy parts.

ANTICORODAL G
Alluminio SA
Si 1.8-2.3, Mg 0.7, Mn 0.7, bal Al.
Cast: 35,600-42,700 TS; 31,300-39,900 YS; 1-3 El; 90-105 Brin. For light alloy castings; general purpose.

ANTICORODAL SPECIAL
Aluminium Industrie Aktiengesellschaft
Mg 2, Mn 1.5, bal Al.
Soft: 36,000 TS; 21,000 YS; 20 El. For punchings; same as Peraluman.

ANTICORODAL-045
Aluminium Walzwerke Singen GmbH
Aluminum. Mg 0.5, Si 0.5, bal Al.
Solution heat treated: 215 MPa TS; 12 El. Bars, rods, tubing for structural work.

ANTICORODAL-062
Swiss Aluminium Ltd.
Aluminum. Si 0.5-0.9, Fe 0.35, Cu 0.3, Mn 0.5, Mg 0.4-0.7, Cr 0.3, Zn 0.2, Ti 0.1, bal Al.
Extrusion age hardenable alloy in billet form. 270-330 N/mm^2 TS; 225-300 N/mm^2 YS; 6-8 El; 90 Brin.

ANTICORODAL-082
Swiss Aluminium Ltd.
Aluminum. Si 0.4-0.8, Fe 0.35, Cu 0.15-0.3, Mn 0.15, Mg 0.8-1.1, Cr 0.05-0.2, Zn 0.2, Ti 0.1, bal Al.
Extrusion age hardenable alloy in billet form. 310-360 N/mm^2 TS; 260-340 N/mm^2 YS; 8-10 El; 100 Brin.

ANTICORODAL-090
Swiss Aluminium Ltd.
Aluminum. Si 0.7-1.3, Fe 0.5, Cu 0.1, Mn 0.4-0.7, Mg 0.6-1, Cr 0.1, Zn 0.2, Ti 0.1, bal Al.
Extrusion age hardenable alloy in billet form. 310-360 N/mm^2 TS; 260-340 N/mm^2 YS; 8-10 El; 100 Brin.

ANTICORODAL-100
Swiss Aluminium Ltd.
Aluminum. Si 0.7-1.3, Fe 0.5, Cu 0.1, Mn 0.1-0.5, Mg 0.6-1, Cr 0.25, Zn 0.2, Ti 0.1, bal Al.
Extrusion age hardenable alloy in billet form. 310-370 N/mm^2 TS; 260-350 N/mm^2 YS; 8-10 El; 100 Brin.

ANTICORODAL-11
Alluminio SA
Aluminum. Si 0.6-1.2, Mg 0.5-0.85, Mn 0.25-0.7, Fe 0-0.45, bal Al.
TA16 Temper: 42,000-50,000 TS; 36,000-46,000 YS; 12-20 El; 90-120 Brin. For decorative parts, auto trim, metal furniture, marine structures. Corrosion resistant.

ANTICORODAL-110
Swiss Aluminium Ltd.
Aluminum. Si 0.7-1.3, Fe 0.5, Cu 0.1, Mn 0.4-0.8, Mg 0.6-1.2, Cr 0.1, Zn 0.2, Ti 0.1, bal Al.
Wrought alloy in rolling slab form.

ANTICORODAL-112
Swiss Aluminium Ltd.
Aluminum. Si 0.7-1.3, Fe 0.5, Cu 0.1, Mn 0.4-1, Mg 0.6-1.2, Cr 0.25, Zn 0.2, Ti 0.1, bal Al.
Extrusion age hardenable alloy in billet form. 310-370 N/mm^2 TS; 260-350 N/mm^2 YS; 8-10 El; 110 Brin.

ANTICORODAL-5 SI
Swiss Aluminium Ltd.
Si 4-6, Mg 0.4-1, Mn 0.5-1, bal Al.
Sand cast: 30,000 TS; 23,000 YS; 3.5 El; 75 Brin. Chilled cast: 40,000 TS; 37,000 YS; 2 El; 95 Brin. For architecture, ship and auto construction, chemical plants; heat treatable.

ANTICORODAL-5 SI
Aluminium Industries, AG
Si 4-6, Mg 0.4-1, Mn 0.5-1, bal Al.
Sand cast: 30,000 TS; 23,000 YS; 3.5 El; 75 Brin. Chilled cast: 40,000 TS; 37,000 YS; 2 El; 95 Brin. For architecture, ship and auto construction, chemical plants; heat treatable.

ANTICORODAL-5 SI
Aluminium Industrie Aktiengesellschaft
Si 4-6, Mg 0.4-1, Mn 0.5-1, bal Al.
Sand cast: 30,000 TS; 23,000 YS; 3.5 El; 75 Brin. Chilled cast: 40,000 TS; 37,000 YS; 2 El; 95 Brin. For architecture, ship and auto construction, chemical plants; heat treatable.

ANTICORODAL-50

Swiss Aluminium Ltd.
Aluminum. Si 5-6, Fe 0.15, Cu 0.01, Mn 0.1, Mg 0.4-0.8, Zn
0.1, Ti 0.01-0.2, bal Al.
Primary foundry alloy in ingot form. 200-270 N/mm^2 TS;
150-190 N/mm^2 YS; 2-10 El; 60-90 Brin.

ANTICORODAL-63

Lavorazione Leghe Leggere SpA
Si 0.2-0.6, Mg 0.45-0.85, Fe 0-0.35, Ti 0-0.1, bal Al.
Annealed: 11,400-17,000 TS; 5700-10,000 YS; 25-40 El;
25-35 Brin. TA16H20 Temper: 37,000-42,700 TS;
32,700-40,000 YS; 2-6 El; 80-90 Brin. For marine structures
and hardware. Corrosion resistant.

ANTICORODAL-63

Alluminio SA
Si 0.2-0.6, Mg 0.45-0.85, Fe 0-0.35, Ti 0-0.1, bal Al.
Annealed: 11,400-17,000 TS; 5700-10,000 YS; 25-40 El;
25-35 Brin. TA16H20 Temper: 37,000-42,700 TS;
32,700-40,000 YS; 2-6 El; 80-90 Brin. For marine structures
and hardware. Corrosion resistant.

ANTICORODAL-70

Swiss Aluminium Ltd.
Aluminum. Si 6.5-7.5, Fe 0.15, Cu 0.01, Mn 0.05, Mg
0.3-0.45, Zn 0.07, Ti 0.01-0.2, bal Al.
Primary foundry alloy in ingot form. 240-340 N/mm^2 TS;
220-280 N/mm^2 YS; 3-9 El; 80-110 Brin.

ANTICORODAL-71

Swiss Aluminium Ltd.
Aluminum. Si 6.8-7.5, Fe 0.15, Cu 0.03, Mn 0.015, Mg
0.3-0.4, Zn 0.07, Ti 0.01, bal Al.
Primary foundry alloy in ingot form. 220-250 N/mm^2 TS;
160-200 N/mm^2 YS; 2-6 El; 70-80 Brin.

ANTICORODAL-72

Swiss Aluminium Ltd.
Aluminum. Si 6.5-7.5, Fe 0.15, Cu 0.01, Mn 0.05, Mg 0.5-0.6,
Zn 0.07, Ti 0.01-0.2, bal Al.
Primary foundry alloy in ingot form. 250-350 N/mm^2 TS;
220-280 N/mm^2 YS; 1-6 El; 90-115 Brin.

ANTICORODAL-78 DV

Swiss Aluminium Ltd.
Aluminum. Si 6.5-7.5, Fe 0.15, Cu 0.01, Mn 0.05, Mg
0.3-0.45, Zn 0.07, Ti 0.01-0.2, bal Al.
Primary foundry alloy in ingot form. 140-320 N/mm^2 TS;
80-280 N/mm^2 YS; 2-6 El; 45-110 Brin.

ANTICORODAL-850

Swiss Aluminium Ltd.
Aluminum. Si 0.35-0.7, Fe 0.08, Cu 0.05-0.2, Mn 0.03, Mg
0.35-0.7, Cr 0.02, Zn 0.05, Ti 0.02, bal Al.
Extrusion alloy in billet form for electrical conductors.
240-310 N/mm^2 TS; 195-280 N/mm^2 YS; 12-14 El; 80 Brin.

ANTICORODAL-CASTING

Aluminium Industrie Aktiengesellschaft
Si 2.25, Fe 0.45, Mn 0.7, Mg 0.7, Ti 0.15, bal Al.
Chill cast: 20,000-28,000 TS; 16,000-22,000 YS; 1.5-5.0 El.
Heat treated: 28,000-40,000 TS; 22,000-30,000 YS; 1.0-5.0 El.
For window frames, chemical industry, architecture; high
corrosion resistance.

ANTICORODAL-WROUGHT

Lavorazione Leghe Leggere SpA
Si 0.5-1.5, Mn 0.2-1, Mg 0.5-1, Fe 0.3, bal Al.
Soft: 17,000 TS; 9000 YS; 30 El; 35 Brin. Hard: 55,000 TS;
45,000 YS; 15 El; 105 Brin. For architecture, window frames,
chemical plants; heat treatable.

ANTICORODAL-WROUGHT

Aluminium Industrie Aktiengesellschaft
Si 0.5-1.5, Mn 0.2-1, Mg 0.5-1, Fe 0.3, bal Al.
Soft: 17,000 TS; 9000 YS; 30 El; 35 Brin. Hard: 55,000 TS;
45,000 YS; 15 El; 105 Brin. For architecture, window frames,
chemical plants; heat treatable.

ANTIFRICTION METAL-A

Puget Sound Metal Works
Cu 8, Sn 83, Sb 9.
For special admiralty bearings for heavy loads; Babbitt.

ANTIFRICTION METAL-B

Puget Sound Metal Works
Cu 1.4, Sn 69, Sb 29.6.
For special admiralty bearings for under water; Babbitt.

ANTIKLOR

Manufacturer not listed.
C 0.5-0.7, Si 13-16, bal Fe.
For cast pumps and conduits; acid resistant, cast.

ANTIKORRO

Westa-Westdeutsche
C 0.9, Cr 18, Mo 1.1, V 1, bal Fe.
Annealed: 107,000 TS; 62,000 YS; 18 El; 35 RA; 220 Brin.
Heat treated: 280,000 TS; 270,000 YS; 3 El; 15 RA; 555 Brin.
For ball bearings, racers, liners, sleeves; stainless, wear
resistant.

ANTIKORRO AK1B

Westa-Westdeutsche
C 0-0.1, Cr 17.5, Ti = 7 x C, bal Fe.
Annealed: 80,000 TS; 50,000 YS; 25 El; 50 RA; 150 Brin. For
welded structures in chemical plants and oil refineries;
corrosion and heat resistant.

ANTIKORRO AK1W

Westa-Westdeutsche
C 0-0.12, Cr 13, Si 0.4, bal Fe.
Annealed: 75,000 TS; 40,000 YS; 35 El; 70 RA; 155 Brin. Cold
drawn: 100,000 TS; 85,000 YS; 17 El; 60 RA; 205 Brin. For
turbine blades, chemical plant equipment, cutlery; Type 410;
corrosion resistant.

ANTIKORRO AK2

Westa-Westdeutsche
C 0.22, Cr 17, Ni 1.5, bal Fe.
Annealed: 125,000 TS; 95,000 YS; 20 El; 55 RA; 260 Brin. For
oil refinery equipment; Type 431; corrosion resistant.

ANTIKORRO AK2S

Westa-Westdeutsche
C 0.2, Si 0.4, Cr 13, bal Fe.
Annealed: 95,000 TS; 50,000 YS; 25 El; 55 RA; 196 Brin. Cold
drawn: 105,000 TS; 85,000 YS; 17 El; 50 RA; 215 Brin. For
turbine blades, cutlery, gauges, instruments; Type 420;
corrosion resistant.

ANTIKORRO AK4

Westa-Westdeutsche
C 0.4, Si 0.4, Cr 13, bal Fe.
Annealed: 95,000 TS; 50,000 YS; 25 El; 55 RA; 200 Brin. For
cutlery, valves, surgical and dental instruments; Type 420;
corrosion resistant.

ANTIKORRO AK5

Westa-Westdeutsche
C 0.4, Si 0.4, Cr 13, bal Fe.
Annealed: 95,000 TS; 50,000 YS; 25 El; 55 RA; 200 Brin. For
cutlery, valves, surgical instruments; Type 420; stainless.

ANTIKORRO AKC

Westa-Westdeutsche
C 0.15, Cr 24, Ni 19, bal Fe.
Annealed: 100,000 TS; 45,000 YS; 50 El; 65 RA; 185 Brin. For
furnace parts, valves, pumps, turbine parts; Type 310;
stainless, austenitic.

ANTIKORRO AKC2

Westa-Westdeutsche
C 0.15, Cr 19.5, Ni 9.5, bal Fe.
Annealed: 80,000 TS; 35,000 YS; 55 El; 75 RA; 150 Brin. For
chemical plant equipment, tanks, vessels; Type 302;
stainless, austenitic.

ANTIKORRO AKH

Westa-Westdeutsche
C 0.85, Cr, V, bal Fe.
For cutting tools, bearings, dies, liners; water or oil hardened.

ANTIKORRO AKL

Westa-Westdeutsche
C 0-0.1, Cr 12.5, Ni 12, bal Fe.
For valves, oil refinery equipment; corrosion resistant.

ANTIKORRO AKSZ

Westa-Westdeutsche
C 0.05, Cr 18, Ni 10, Cu, bal Fe.
Annealed: 85,000 TS; 35,000 YS; 60 El; 70 RA; 150 Brin. For
chemical plant equipment; stainless, austenitic.

ANTIKORRO AKV

Westa-Westdeutsche
C 0-0.15, Cr 18, Ni 8.5, bal Fe.
Annealed: 80,000 TS; 35,000 YS; 55 El; 75 RA; 100 Brin. For
chemical plant equipment, tanks, mixers; Type 302; stainless,
austenitic.

ANTIKORRO AKV EXTRA

Westa-Westdeutsche
C 0-0.12, Cr 18, Mo 2, Ni 10.5, Ti = 4 x C, bal Fe.
Annealed: 85,000 TS; 35,000 YS; 50 El; 65 RA; 160 Brin. Cold
drawn: 150,000 TS; 135,000 YS; 6 El; 300 Brin. For welded
chemical plant equipment; Type 316 Ti; stainless, austenitic.

ANTIKORRO AKVM

Westa-Westdeutsche
C 0-0.15, Cr 18, Ni 8.5, bal Fe.
Annealed: 80,000 TS; 35,000 YS; 55 El; 75 RA; 160 Brin. For
chemical plant equipment, tanks; Type 302; stainless,
austenitic.

ANTIKORRO AKV3

Westa-Westdeutsche
C 0-0.12, Cr 18, Ni 9.5, Cb = 8 x C, bal Fe.
Annealed: 90,000 TS; 45,000 YS; 56 El; 65 RA; 160 Brin. Cold
drawn: 100,000 TS; 65,000 YS; 45 El; 60 RA; 205 Brin. For
welded chemical plant equipment, tanks, fermenters; Type
347; stainless, austenitic.

ANTIKORRO AKX SPEZIAL

Westa-Westdeutsche
C 0.2, Cr 25, Ni 4, bal Fe.
Cast: 90,000 TS; 65,000 YS; 2 El; 212 Brin. For furnace parts,
grids, conveyors, heat treating boxes; corrosion and heat
resistant.

ANTIKORRO AKX10

Westa-Westdeutsche
C 0-0.12, Cr 18, Al 1, bal Fe.
Annealed: 80,000 TS; 50,000 YS; 25 El; 50 RA; 150 Brin. For
oil refinery and chemical plant equipment; corrosion
resistant.

ANTIKORRO AKX12

Westa-Westdeutsche
C 0-0.12, Al 1.5, Cr 24, bal Fe.
Annealed: 90,000 TS; 50,000 YS; 30 El; 55 RA; 180 Brin. For
furnace parts, heat treating boxes; corrosion and heat
resistant.

ANTIMONIAL ADMIRALTY

Chase Brass & Copper Co.
Cu 71, Sn 1, Sb 0.04, Zn 27.96.
Annealed: 54,000 TS; 23,000 YS; 60 El. For condensers,
tubes; Admiralty Brass. *Obsolete*

ANTIMONIAL LEAD

NL Industries
Sb 0-25, Pb 75-100.
For storage battery plates, bullets, type metal, pipes.
Obsolete

ANTIMONY
Atomergic Chemetals Corp.
Sb.
Purities, zone refined: 99.9999%, 99.999%, 99.99%. Forms: ingot, shot, powder, granules, lump, single crystals, foil.

ANTINIT 1KB2
Vereinigte Edelstahlwerke
C 0-0.1, Cr 18, Ni 8.5, bal Fe.
Annealed: 80,000 TS; 35,000 YS; 55 El; 75 RA; 150 Brin. For chemical plant equipment, tanks; Type 302; stainless, austenitic. *Obsolete*

ANTINIT 1KB4
Vereinigte Edelstahlwerke
C 0-0.1, Cr 18, Ni 9.5, Mo 2, bal Fe.
Annealed: 85,000 TS; 35,000 YS; 50 El; 65 RA; 160 Brin. Cold drawn: 150,000 TS; 135,000 YS; 6 El; 300 Brin. For acid resistant chemical plant equipment, tanks, mixers, pumps; Type 316; stainless, austenitic. *Obsolete*

ANTINIT AS2
Now VEW A505.

ANTINIT AS2(H)
Vereinigte Edelstahlwerke
C 0-0.15, Cr 18, Ni 8.5, bal Fe.
Annealed: 80,000 TS; 35,000 YS; 55 El; 75 RA; 150 Brin. For chemical plant equipment, tanks, mixers, filters; Type 302; stainless, austenitic. *Obsolete*

ANTINIT AS2F
Now VEW A520.

ANTINIT AS2G
Now VEW A550G.

ANTINIT AS2H
Vereinigte Edelstahlwerke
C 0.2, Cr 19.5, Ni 7.5, bal Fe.
Annealed: 92,500 TS; 49,800 YS; 180 Brin. For rotating knives, from pulping engines; stainless, austenitic. *Obsolete*

ANTINIT AS2K
Vereinigte Edelstahlwerke
C 0.2, Mn 1.25, Cr 18, Ni 12, bal Fe.
Annealed: 78,300 TS; 35,600 YS; 160 Brin. For bone surgery nails and wire; corrosion resistant. *Obsolete*

ANTINIT AS2W
Now VEW A500.

ANTINIT AS2Z
Now VEW A506.

ANTINIT AS3W
Now VEW A122.

ANTINIT AS4K
Vereinigte Edelstahlwerke
C 0.2, Cr 18, Ni 12, Mo 2.1, bal Fe.
Annealed: 78,300 TS; 35,600 YS; 160 Brin. For parts for bone surgery; stainless, austenitic. *Obsolete*

ANTINIT AS4M
Now VEW A100.

ANTINIT AS4W
Now VEW A120.

ANTINIT AS5W
Now VEW A102.

ANTINIT ASA
Vereinigte Edelstahlwerke
C 0.1, Cr 12, Ni 12, bal Fe.
Water quenched: 78,000-100,000 TS; 31,000 YS; 56 El; 68 RA; 150 Brin. For tableware, cutlery; austenitic, stainless. *Obsolete*

ANTINIT EAS2
Now VEW A600.

ANTINIT EAS4
Now VEW A200.

ANTINIT EAS4M
Now VEW A205.

ANTINIT KW-10
Now VEW N100.

ANTINIT KW10 SPEZIAL
Vereinigte Edelstahlwerke
C 0.2, Cr 13, Mo 1.15, bal Fe.
Annealed: 95,000 TS; 50,000 YS; 25 El; 55 RA; 195 Brin. For turbine blades, cutlery, valves, pumps, knives; corrosion resistant; Type 420 Mo. *Obsolete*

ANTINIT KW100
Now VEW N690.

ANTINIT KW10A1
Now VEW N110.

ANTINIT KW10M
Now VEW N132.

ANTINIT KW10Z
Vereinigte Edelstahlwerke
C 0.15, Cr 12-14, Mo 0-0.6, S 0-0.15, bal Fe.
Annealed: 80,000 TS; 45,000 YS; 30 El; 165 Brin. Cold drawn: 100,000 TS; 85,000 YS; 13 El; 205 Brin. Heat treated: 160,000 TS; 130,000 YS; 15 El; 300 Brin. For screw machine products, fasteners, pump shafts, gears, valve seats. Type 416 stainless, magnetic. *Obsolete*

ANTINIT KW15
Now VEW N315.

ANTINIT KW15M
Vereinigte Edelstahlwerke
C 0.15, Cr 13.5, Mo 1.1, bal Fe.
Heat treated: 106,700 TS; 78,300 YS; 245 Brin. For steam turbine blades; used up to 1020 F. *Obsolete*

ANTINIT KW15Z
Now VEW N316.

ANTINIT KW20
Now VEW N320.

ANTINIT KW20M
Now VEW N330.

ANTINIT KW30
Now VEW N330.

ANTINIT KW35M
Now VEW N335.

ANTINIT KW40
Now VEW N540.

ANTINIT KW40M
Vereinigte Edelstahlwerke
C 0.4, Cr 14.5, Mo 0.6, bal Fe.
For surgical instruments, valves; corrosion resistant. *Obsolete*

ANTINIT KW5
Vereinigte Edelstahlwerke
C 0.1, Cr 14, bal Fe.
Annealed: 64,000-80,000 TS; 42,700-50,000 YS; 25 El; 50 RA; 150-160 Brin. For chemical plant equipment, shafts, fasteners; corrosion resistant; Type 430. *Obsolete*

ANTINIT KW50
Vereinigte Edelstahlwerke
C 0.5, Cr 14.5, bal Fe.
For surgical instruments, cutlery; corrosion resistant. *Obsolete*

ANTINIT KW50M
Vereinigte Edelstahlwerke
C 0.54, Cr 14.5, Mo 0.55, bal Fe.
For surgical instruments, cutlery; corrosion resistant. *Obsolete*

ANTINIT KW8
Now VEW N104.

ANTINIT KW80
Now VEW N685.

ANTINIT KWA
Now VEW N200.

ANTINIT KWB
Now VEW N350.

ANTINIT KWG
Vereinigte Edelstahlwerke
C 0.25, Cr 14, Ni 0.5, bal Fe.
Annealed: 100,000-120,000 TS; 64,000 YS; 20 El; 190 Brin. For armature and turbine parts; resists oxidation and erosion. *Obsolete*

ANTINIT KWMZ
Vereinigte Edelstahlwerke
C 0.18, Cr 17.5, Mo 1.8, S 0.2, bal Fe.
Annealed: 85,400 TS; 56,900 YS; 190 Brin. For bolts, nuts, screws, shafts; stainless, free-cutting. *Obsolete*

ANTINIT KWZ
Vereinigte Edelstahlwerke
C 0.15, Cr 17, S 0.2, bal Fe.
Annealed: 100,000-122,000 TS; 62,000 YS; 22 El; 60 RA; 235 Brin. For screw machine products; free-cutting, stainless. *Obsolete*

ANTINIT KWZA
Now VEW N310.

ANTINIT NG
Now VEW V622.

ANTINIT RAM
Now VEW A900.

ANTINIT RS 30 M
Vereinigte Edelstahlwerke
C 1, Cr 30, Mo 2, bal Fe.
Annealed: 64,000-85,000 TS; For corrosion resistant castings; for bleaching equipment. *Obsolete*

ANTINIT RS30
Vereinigte Edelstahlwerke
C 1.2, Cr 29, bal Fe.
For crushers, ball mills, grates, baffles; heat and wear resistant. *Obsolete*

ANTINIT SAS10
Now VEW A955.

ANTINIT SAS2
Now VEW A750.

ANTINIT SAS2 EXTRA
Vereinigte Edelstahlwerke
C 0-0.12, Cr 18, Ni 9.5, Cb = 8 x C, bal Fe.
Annealed: 90,000 TS; 45,000 YS; 56 El; 65 RA; 160 Brin. For welded structures, chemical plant equipment, tanks; Type 347; stainless, austenitic. *Obsolete*

ANTINIT SAS4
Now VEW A350.

ANTINIT SAS4M
Now VEW A354.

ANTINIT SAS5
Vereinigte Edelstahlwerke
C 0.08, Cr 17, Ni 10, Mo 2, bal Fe.
Annealed: 80,000 TS; 30,000 YS; 60 El; B 78 Rock. For chemical plant equipment, acid containers, vessels, valve trim, digesters. Stainless steel. Type 18-10 Mo, acid resistant. *Obsolete*

ANTINIT SAS5G
Vereinigte Edelstahlwerke
C 0-0.07, Cr 17, Ni 13, Mo 4.7, bal Fe.
Annealed: 95,000 TS; 50,000 YS; 50 El; 55 RA; 190 Brin. For acid resistant chemical plant equipment, mixers; Type 317; stainless, austenitic. *Obsolete*

ANTINIT SAS8
Now VEW A960.

ANTINIT SAST2
Now VEW A700.

ANTINIT SAST4
Now VEW A300.

ANTINIT SAST4M
Now VEW A305.

ANTINIT SKWA
Vereinigte Edelstahlwerke
C 0-0.1, Cr 17.5, Ti = 7 x C, bal Fe.
Annealed: 71,000-80,000 TS; 42,700-50,000 YS; 25 El; 50 RA; 160-180 Brin. For furnace parts, welded heat treating boxes; corrosion resistant. *Obsolete*

ANTINIT SKWL
Now VEW N238.

ANTINIT SKWN
Vereinigte Edelstahlwerke
C 0.06, Cr 17, Ni 8, Cb, bal Fe.
Annealed: 85,000 TS; 35,000 YS; 50 El; B 82 Rock. For welded structures, tanks, vessels, chemical plant equipment. Type 17-8 Cb stainless steel, stabilized, welding grade. *Obsolete*

ANTINIT-SIL18
Vereinigte Edelstahlwerke
C 0.15, Mn 7.5-10, Cr 17-19, Ni 4-6, bal Fe.
Annealed: 100,000 TS; 50,000 YS; 60 El; B 90 Rock. For sinks, cooking utensils, dairy and hospital equipment, chemical handling equipment, springs, tubular furniture. Type 202 stainless steel, austenitic, nonmagnetic. *Obsolete*

ANTIOX-1
Chiers-Chatillon
Cr 12-14, 0.15% min C, bal Fe.
Annealed: 95,000 TS; 45,000 YS; 25 El; 196 Brin. Oil hardened: 250,000 TS; 215,000 YS; 8 El; 512 Brin. For cutlery, surgical instruments, gears, shafts, gauges, needle valves. Type 420 stainless steel, hardenable. *Obsolete*

ANTIOX-2
Chiers-Chatillon
Cr 12-14, 0.15% min C, bal Fe.
Annealed: 95,000 TS; 45,000 YS; 25 El; 196 Brin. Heat treated: 250,000 TS; 215,000 YS; 8 El; 512 Brin. For cutlery, surgical instruments, gears, needle valves, gauges, shafts. Type 420 stainless steel, heat treatable. *Obsolete*

ANTIOX-3
Chiers-Chatillon
C 0-0.15, Cr 11.5-13.5, bal Fe.
Annealed: 75,000 TS; 45,000 YS; 30 El; B 82 Rock. Cold drawn: 95,000 TS; 80,000 YS; 15 El; B 92 Rock. For flat springs, knives, tableware. Type 410 stainless steel, hardenable. *Obsolete*

ANTITHERM F
Now VEW A700.

ANTITHERM FA
Now VEW H300.

ANTITHERM FB 30 S
Vereinigte Edelstahlwerke
C 0.2, Cr 25, Ni 3, Si 1, bal Fe.
Water-quenched: 92,000-113,000 TS; 56,000 YS. For SO 2 H 2 S resistance to 2200 F; heat resistant. *Obsolete*

ANTITHERM FB10
Now VEW H120.

ANTITHERM FB105
Vereinigte Edelstahlwerke
C 0.12, Si 2, Cr 18, bal Fe.
Annealed: 80,000 TS; 50,000 YS; 25 El; 50 RA; 150 Brin. For kitchen sinks, valves, pumps, furnace parts; Type 430; corrosion resistant. *Obsolete*

ANTITHERM FB10G
Now VEW H1220G.

ANTITHERM FB12
Now VEW H100.

ANTITHERM FB30SG
Vereinigte Edelstahlwerke
C 0.4, Cr 27, Ni 4, bal Fe.
Cast: 90,000 TS; 65,000 YS; 2 El; 212 Brin. For cylinders, liners, valve seats and bodies, furnace parts; heat and corrosion resistant. *Obsolete*

ANTITHERM FB7
Vereinigte Edelstahlwerke
C 0.12, Si 2, Cr 3, bal Fe.
Annealed: 71,200 TS; 50,000 YS; 165 Brin. For heat resistant parts; used up to 1300 F. *Obsolete*

ANTITHERM FB8
Now VEW H160.

ANTITHERM FB8G
Now VEW H160G.

ANTITHERM FB9
Now VEW H140.

ANTITHERM FBS105
Bohler Gesellschaft M.B.H.
C 0.06, Cr 17, Ni 10, bal Fe.

ANTITHERM FBS90
Bohler Gesellschaft M.B.H.
C 0-0.12, Si 2.3, Cr 6, bal Fe.

ANTITHERM FBS95
Bohler Gesellschaft M.B.H.
C 0-0.12, Si 2.2, Cr 13, bal Fe.

ANTITHERM FBS95
Vereinigte Edelstahlwerke
C 0-0.12, Si 2.2, Cr 13, bal Fe.
Annealed: 75,000 TS; 40,000 YS; 35 El; 70 RA; 155 Brin. For oil refinery and chemical plant equipment; corrosion and heat resistant. *Obsolete*

ANTITHERM FF
Now VEW H550.

ANTITHERM FFB
Now VEW H525.

ANTITHERM FFB 400
Now VEW H520.

ANTITHERM FFBG
Now VEW H537.

ANTITHERM FFG
Now VEW H551.

ANV-300
Russian manufacture
C 0-0.1, Cr 15.5, Ti 1.7, Al 5, W 8.5, B 0-0.1, Fe 0-5, bal Ni.
Cast nickel-base supervolly. For automotive turbine blades.

ANVIL BRASS
American manufacture
Cu 62.5, Zn 37.5.
For hardware, tubes, yellow brass.

ANVIL HEADING DIE
Manufacturer not listed.
C 0.6, Cr 0.8, Mo 0.2, bal Fe.
For tools, heading dies; oil hardened.

ANVILOY 1100
Rheinische Rohrenwerke
W 90, N 4, Mo 4, Fe 2.
125,000 psi TS; 30 Rock C. At 800°C: 49 RA. For high temperature tooling.

ANVILOY 1150
CMW Inc.
W 90, Ni 4, Mo 4, Fe 2.
140,000 psi TS; 34 Rock C. At 800°C: 75,000 psi TS. For die casting, hot extrusion dies and high strength applications.

ANVILOY 1200
CMW Inc.
W 90, Ni 4, Fe 2, Mo 4. *Obsolete*

ANVILOY WELD ROD
CMW Inc.
W, Ni, Fe.
For welding high density (tungsten base) materials.

AO 20
A. Milne & Co.
C 0.5, Mn 0.3, Si 0.75, Cr 1.15, V 0.2, W 2.5, bal Fe.
Shock resisting tool steel; Type S1.

AO 3900
General Motors Corp./Central Foundry Div
Si 16-18, Fe 0-0.9, Cu 4-5, Mn 0-0.35, Mg 0.45-0.65, Zn 0-0.9, Ti 0-0.2, bal Al.
Cast aluminum alloy 390.

AP201
Comalco Ltd.
Cu 4-5, Si 0-0.2, Fe 0-0.25, Mn 0-0.05, Mg 0-0.05, Ni 0-0.05, Zn 0-0.1, Ti 0-0.2, bal Al.
Sand cast, T4: 216 MPa TS min; 138 MPa YS; 8 El; 80 Brin. Permanent mold cast, T6: 324 MPa TS; 193 MPa YS; 10 El; 105 Brin. ADC H49-9; BS LM11.

AP301
Comalco Ltd.
Cu 5.5-6.5, Si 5.5-6.5, Fe 0-0.4, Mn 0-0.05, Mg 0.25-0.5, Ni 0-0.05, Zn 0-0.1, Ti 0-0.05, Sn 0-0.05, Cr 0-0.05, bal Al.
Permanent mold cast, T5: 241 MPa TS; 193 MPa YS; 4 El; 95 Br

AP303

Comalco Ltd.
Cu 2-4, Si 5-6.5, Fe 0-0.6, Mn 0-0.7, Mg 0-0.1, Ni 0-0.1, Zn 0-0.1, Ti 0-0.2, bal Al.
Sand cast, T6: 248 MPa TS; 165 MP YS; 2 El; 80 Brin. Permanent mold cast, T6: 310 MPa TS; 165 MPa YS; 2 El; 110 Brin. ADC H49-3; BS LM4.

AP309

Comalco Ltd.
Cu 1-1.5, Si 4.5-5.5, Fe 0.14-0.25, Mn 0-0.05, Mg 0.5-0.6, Zn 0-0.05, Ti 0-0.2, bal Al.
Sand cast, HT T6: 241 MPa TS; 172 MPa YS; 3 El; 80 Brin. Permanent mold cast, T62: 310 MPa TS; 276 MPa YS; 2 El; 105 Brin. ADC H49-13; BS LM16.

AP311

Comalco Ltd.
Cu 1-1.5, Si 4-6, Fe 0-0.15, Mn 0-0.05, Mg 0-0.05, Zn 0-0.1, Ti 0-0.2, bal Al.
Permanent mold cast, F1: 124 MPa TS; 48 MPa YS; 7 El; 40 Brin.

AP315

Comalco Ltd.
Cu 3-4.5, Si 10.5-12, Fe 0.6-1, Mn 0-0.1, Mg 0-0.1, Ni 0-0.1, Zn 0-0.1, Sn 0-0.1, bal Al.
Pressure cast, F1: 331 MPa TS; 165 MPa YS; 2.5 El. SAE 303.

AP33M

French manufacture
Cu 4-6, Mg 0.5, Ti 0.4, bal Al.
Heat treated: 47,000 TS. For light metal parts, gear cases; heat treatable.

AP403

Comalco Ltd.
Cu 0-0.1, Si 4.5-6, Fe 0-0.6, Mn 0-0.1, Mg 0-0.05, Zn 0-0.1, Ti 0-0.2, bal Al.
Sand cast, F1: 131 MPa TS; 55 MPa YS; 8 El; 40 Brin. Permanent mold cast, F1; 165 MPa TS; 62 MPa YS; 9 El; 45 Brin. ADC H49-14; BS LM18.

AP501

Comalco Ltd.
Cu 0-0.1, Si 0-0.3, Fe 0-0.3, Mn 0-0.5, Ti 0-0.2, Mg 3.6-4.5, Zn 0-0.1, bal Al.
Sand cast, F1: 172 MPa TS; 83 MPa YS; 9 El; 50 Brin. ADC H49-4; BS LM5.

AP601

Comalco Ltd.
Cu 0-0.05, Si 6.5-7.5, Fe 0.12-0.2, Mn 0-0.05, Mg 0.3-0.4, Zn 0-0.05, Ti 0-0.2, bal Al.
Sand cast, T6: 255 MPa TS; 166 MPa YS; 5 El; 70 Brin. Permanent mold cast, T61: 296 MPa TS; 207 MPa YS; 8 El; 105 Brin. ADC H49-18; BS LM25.

AP701

Comalco Ltd.
Cr 0-0.1, Si 0-0.25, Fe 0-0.4, Mn 0-0.1, Mg 0.5-0.75, Ni 0-0.1, Ti 0.15-0.25, Sn 0-0.05, Cr 0.4-0.6, Zn 4.8-5.7, bal Al.
Permanent mold cast, T1: 255 MPa TS; 179 MPa YS; 5 El; 80 Brin. ADC H49-17.

AP703

Comalco Ltd.
Cu 0-0.1, Si 0-0.15, Fe 0-0.15, Mn 0-0.1, Mg 0.65-0.85, Zn 6.5-7.5, Ti 0.1-0.25, Cr 0.06-0.15, bal Al.
Sand cast, T6: 345 MPa TS; 276 MPa YS; 9 El; 93 Brin.

APACHE

AL Tech Specialty Steel Corp.
C 0.7, Mn 2, Si 0.3, Cr 1, Mo 1.35, bal Fe.
Heat treated: 293,000 TS; 265,000 YS; 1 El; Usual 56-59 Rock C. For dies and punches, blanking and forming dies, gages, shear blades, bending tools, stripper plates, master hubs, air hardening. AISI A6 tool steel, wear resistant. Outstanding size stability in heat treatment.

APEX

Lehigh Steel Corp.
C 0.9, bal Fe.
For tools, drills, punches; drill rod, water hardened.

APEX "T" SERIES ZINC BASE ALLOY

Apex International Alloys, Inc.
Al, Cu, bal Zn.
For castings; die casting. *Obsolete*

APEX 110

Apex International Alloys, Inc.
Cu 10, bal Al.
SC-F temper: 23,000 TS; 16,000 YS; 1 El; 85 Brin. SC-T 61 temper: 36,000 TS; 32,000 YS; 0.5 El; 110 Brin. PM-F temper: 27,000 TS; 19,000 YS; 1.5 El; 90 Brin. For bearings; sand and permanent mold. *Obsolete*

APEX 12

Apex International Alloys, Inc.
Cu 6.5, Si 3, Zn 2, bal Al.
Sand cast: 17,000-23,000 TS; 14,000-15,000 YS; 3-1 El; 65-70 Brin. For general castings, sand and permanent mold castings; non-hardenable. *Obsolete*

APEX 12 PM

Apex International Alloys, Inc.
Cu 6.5, Si 3, Zn 2, bal Al.
Sand cast: 23,000 TS; 15,000 YS; 1.5 El; 70 Brin. Permanent mold: 29,000 TS; 22,000 YS; 1.0 El; 70 Brin. For sand and permanent mold castings; good machinability. *Obsolete*

APEX 125

Apex International Alloys, Inc.
Mg 9.7-10.5, Si 0-0.1, Ti 0-0.25, bal Al.
T4 temper: 57,000 TS; 29,000 YS; 30 El; 80 Brin. For sand castings, instrument cases, housing, oil pans. Age hardenable, corrosion resistant. *Obsolete*

APEX 13

Apex International Alloys, Inc.
Si 11-13, bal Al.
Die cast: 45,000 TS; 25,000 YS; 4 El; 70 Brin. For die castings, housings; corrosion resistant. *Obsolete*

APEX 28

Apex International Alloys, Inc.
Cu 1.7-2.3, Si 10, bal Al.
For light alloy castings; high fluidity. *Obsolete*

APEX 31

Apex International Alloys, Inc.
Cu 9-11, Fe 1.5, Si 3.5-4.5, Mg 0.15-0.35, Zn 0-1.5, Ti 0-0.25, Ni 0-1, bal Al.
For permanent mold castings. *Obsolete*

APEX 32

Apex International Alloys, Inc.
Cu 1, Si 12, Mg 1.1, bal Al.
PM-F temper: 27,000 TS; 19,000 YS; 1.0 El; 88 Brin. PM-T 65 temper: 47,000 TS; 43,000 YS; 0.5 El; 125 Brin. PM-T 551 temper: 36,000 TS; 28,000 YS; 0.5 El; 105 Brin. For pistons, high temperature uses; permanent mold, heat treatable. *Obsolete*

APEX 34

Apex International Alloys, Inc.
Cu 3, Si 9, Mg 1, Ni 1, bal Al.
PM-F temper: 35,000 TS; 26,000 YS; 1 El; 85 Brin. PM-T 5 temper: 36,000 TS; 28,000 YS; 1 El; 100 Brin. For pistons; permanent mold. *Obsolete*

APEX 35

Apex International Alloys, Inc.
Cu 10, Mg 0.3, bal Al.
SC-F temper: 26,000 TS; 21,000 YS; 0.5 El; 85 Brin. SC-T 65 temper: 40,000 TS; 30,000 YS; 0.5 El; 115 Brin. PM-T 65 temper: 48,000 TS; 36,000 YS 0.5 El; 140 Brin. For sand and permanent mold castings; wear resistant. *Obsolete*

APEX 36

Apex International Alloys, Inc.
Cu 0-0.6, Si 9-10, Mg 0.4-0.6, bal Al.
Cast: 46,000 TS; 27,000 YS; 3.5 El; 75 Brin. For light alloy castings; high fluidity. *Obsolete*

APEX 37

Apex International Alloys, Inc.
Cu 4.5, Si 4.5, bal Al.
SC-F temper: 25,000 TS; 15,000 YS; 2 El; 65 Brin. PM-F temper: 27,000 TS; 13,000 YS; 2.5 El; 70 Brin. For castings; sand and permanent mold. *Obsolete*

APEX 38

Apex International Alloys, Inc.
Cu 10, Si 4, Mg 0.25, bal Al.
PM-F temper: 32,000 TS; 24,000 YS; 1.5 El; 100 Brin. For sole plates, permanent mold castings; retains hardness at elevated temperatures. *Obsolete*

APEX 39

Apex International Alloys, Inc.
Cu 3.5, Si 9, bal Al.
DC-F temper: 47,000 TS; 27,000 YS; 3.5 El; 80 Brin. For die castings; good machinability. *Obsolete*

APEX 39-11

Apex International Alloys, Inc.
Cu 3-4, Si 10.5-11.5, Zn 0-1, Fe 0-1, Ni 0-0.5, bal Al.
For general die castings; TS 28,000; Charpy 2.2. Die cast: 44,000-46,000 TS; 23,000-27,000 YS; 2 El; 85 Brin. *Obsolete*

APEX 39A

Apex International Alloys, Inc.
Cu 3-4, Si 8.5-9.5, bal Al.
Cast: 44,000 TS; 26,000 YS; 3 El; 80 Brin. For light alloy castings; high fluidity. *Obsolete*

APEX 39B

Apex International Alloys, Inc.
Cu 3.5-4.5, Si 8.9-9.5, Zn 1.5-2.5, Fe 0-1.2, Mg 0-0.1, bal Al.
Die cast: 43,000 TS; 22,000 YS; 3 El; 80 Brin. For general die castings; corrosion resistant. *Obsolete*

APEX 39P

Apex International Alloys, Inc.
Cu 4, Si 8, bal Al.
SC-F temper: 26,000 TS; 15,000 YS; 2 El; 65 Brin. PM-F temper: 32,000 TS; 20,000 YS; 2.5 El; 75 Brin. PM-T6 temper: 40,000 TS; 27,000 YS; 3.5 El; 95 Brin. For sand and permanent mold castings; for thin sections. *Obsolete*

APEX 400

Apex International Alloys, Inc.
Cu 0-0.1, Mg 3.6-4.2, bal Al.
24,000-26,000 TS; 10,000 YS; 8-6 El; 50-60 Brin. For light alloy parts, cooking utensils, marine castings; similar to "Permbrite." *Obsolete*

APEX 405

Apex International Alloys, Inc.
Zn 5, Mg 0.4, Mn 0.2, bal Al.
For die castings. *Obsolete*

APEX 41

Apex International Alloys, Inc.
Cu 1.5, Si 5, Mg 0.5, bal Al.
SC-F temper: 25,000 TS; 12,000 YS; 2.5 El; 65 Brin. SC-T 6 temper: 35,000 TS; 25,000 YS; 2.5 El; 80 Brin. PM-T 6 temper: 43,000 TS; 27,000 YS; 4 El; 90 Brin. For sand and permanent mold castings; heat treatable. *Obsolete*

APEX 41-10

Apex International Alloys, Inc.
Cu 1-1.5, Si 4.5-5.5, Mg 0.45-0.6, Zn 0.1, Mn 0.1, bal Al.
Sand cast: 27,000 TS; 18,000 YS; 2.5 El; 65 Brin. T6 temper: 42,000 TS; 30,000 YS; 4.0 El; 90 Brin. Good pressure tightness, good resistance to hot cracking. SAE 335; AMS 4215; ASTM B26-68 SC51B; ASTM B108-68 SC51B. *Obsolete*

APEX 417

Apex International Alloys, Inc.
Cu 0-0.1, Si 0-0.15, Mg 6-7.5, Ti 0.2, bal Al.
Cast: 40,000 TS; 21,000 YS; 13 El; 70 Brin. For agricultural and aircraft castings, marine parts; corrosion and impact resistant. *Obsolete*

APEX 418

Apex International Alloys, Inc.
Mg 8, bal Al.
DC-F temper: 43,000 TS; 27,000 YS; 8 El; 80 Brin. For die castings; corrosion resistant. *Obsolete*

APEX 41A

Apex International Alloys, Inc.
Cu 1-1.5, si 4.5-5.5, Mg 0.4-0.6, Cr 0-0.25, bal Al.
T4-Temper: 25,000 TS; 14,000 YS; 3.5 El; 75 Brin. T6-Temper: 35,000 TS; 25,000 YS; 2.5 El; 80 Brin. For light alloy castings; age-hardenable. *Obsolete*

APEX 420

Apex International Alloys, Inc.
Mg 10, bal Al.
SC-T4 temper: 46,000 TS; 25,000 YS; 14 El; 75 Brin. For sand castings; high strength, corrosion resistant. *Obsolete*

APEX 43

Apex International Alloys, Inc.
Si 5, bal Al.
SC-F temper: 21,000 TS; 10,000 YS; 5 El; 45 Brin. PM-F temper: 25,000 TS; 9000 YS; 5 El; 45 Brin. Dc-F temper: 30,000 TS; 14,000 YS; 7 El; 60 Brin. For castings; pressure tight, corrosion resistant. *Obsolete*

APEX 43A

Apex International Alloys, Inc.
Cu 0-0.1, Si 4.5-6, Zn 0-0.3, bal Al.
Sand cast: 19,000 TS; 9000 YS; 7 El; 40 Brin. Permanent mold: 24,000 TS; 9000 YS; 9 El; 45 Brin. For castings; corrosion resistant. *Obsolete*

APEX 43C

Apex International Alloys, Inc.
Cu 0-0.3, Si 4.5-6, Zn 0-0.3, bal Al.
Sand cast: 19,000 TS; 9000 YS; 7 El; 40 Brin. Permanent mold: 24,000 TS: 9000 YS; 5 El; 65 Brin. For castings, high fluidity. *Obsolete*

APEX 440

Apex International Alloys, Inc.
Cu 0-0.2, Mg 0.5-0.65, Zn 5.2-6, Cr 0.4-0.6, Ti 0.15-0.25, bal Al.
Cast: 28,000 TS; 16,000 YS; 12 El; 55 Brin. Aged: 36,000 TS; 25,000 YS; 6 El; 75 Brin. For sand castings, housings, oil pans, instrument cases, gear cases. Self aging. *Obsolete*

APEX 45 ALLOY TERNALLOY

Apex International Alloys, Inc.
Cr 0.3, Mg 1.5-2.5, Zn 3.5-4.5, bal Al.
For light alloy parts. *Obsolete*

APEX 46 A

Apex International Alloys, Inc.
Si 7, Mg 0.3, bal Al.
SC-F temper: 23,000 TS; 12,000 YS; 3.5 El; 55 Brin. SC-T 6 temper: 33,000 TS; 24,000 YS; 4 El; 70 Brin. PM-T 6 temper: 40,000 TS; 27,000 YS; 5 El; 90 Brin. For sand and permanent mold castings; pressure tight. *Obsolete*

APEX 46-10

Apex International Alloys, Inc.
Cu 0.1, Si 6.5-7.5, Mg 0.28-0.4, Zn 0.1, Mn 0.1, bal Al.
Sand cast: 24,000 TS; 14,000 YS; 4 El; 55 Brin. T6 temper: 40,000 TS; 30,000 YS; 5 El; 85 Brin. Good pressure tightness, good resistance to hot cracking. SAE 336; ASTM B26-68 SG70B; ASTM B108-68 SG70B. *Obsolete*

APEX 461

Apex International Alloys, Inc.
Mg 0.2, Zn 5.5, Cu 1.4, bal Al.
For general purpose castings; room temperature aging. *Obsolete*

APEX 470

Apex International Alloys, Inc.
Zn 7.5, Cu 0.7, Mg 0.4, bal Al.
For general purpose castings; room temperature aging. *Obsolete*

APEX 50

Apex International Alloys, Inc.
Cu 3.7-4.7, Si 2.5-3.5, bal Al.
SC-F temper: 24,000 TS; 14,000 YS; 2.5 El; 60 Brin. SC-T6-temper: 36,000 TS; 26,000 YS; 1.5 El; 85 Brin. PM-T6 temper: 40,000 TS; 26,000 YS; 4 El; 85 Brin. For light castings: sand and permanent mold. *Obsolete*

APEX 50-50 COPPER ALLOY

Apex International Alloys, Inc.
Al 50, Cu 50.
For rich hardener; for alloying. *Obsolete*

APEX 51

Apex International Alloys, Inc.
Cu 4, Si 6.9, bal Al.
SC-F temper: 23,000 TS; 13,000 YS; 2 El; 55 Brin. PM-F temper: 28,000 TS; 16,000 YS; 2 El; 70 Brin. PM-T 6 temper: 45,000 TS; 28,000 YS; 4 El; 105 Brin. For castings; sand and permanent mold. *Obsolete*

APEX 52

Apex International Alloys, Inc.
Cu 4.4, bal Al.
SC-F temper: 22,000 TS; 10,000 YS; 3 El; 50 Brin. SC-T temper: 36,000 TS; 24,000 YS; 5 El; 75 Brin. SC-T6Z temper: 40,000 TS; 30,000 YS; 2 El; 95 Brin. For sand castings; age-hardenable. *Obsolete*

APEX 52 A

Apex International Alloys, Inc.
Cu 4-4.8, Si 0-1.5, Ti 0-0.2, bal Al.
T4-temper: 32,000 TS; 16,000 YS; 8.5 El; 60 Brin. T6-temper: 36,000 TS; 24,000 YS; 5 El; 75 Brin. For castings; age hardenable. *Obsolete*

APEX 53

Apex International Alloys, Inc.
Cu 4.5, Si 2.5, bal Al.
PM-F temper: 30,000 TS; 14,000 YS; 3.5 El; 70 Brin. PM-T6 temper: 40,000 TS; 26,000 YS; 5 El; 90 Brin. For permanent mold castings; age-hardenable. *Obsolete*

APEX 56

Apex International Alloys, Inc.
Cu 4, Si 5, bal Al.
DC-F temper: 38,000 TS; 22,000 YS; 3 El; 70 Brin. For die castings; good machinability. *Obsolete*

APEX 56A

Apex International Alloys, Inc.
Cu 3-4, Si 4.5-5.5, bal Al.
For castings; age hardenable. *Obsolete*

APEX 59

Apex International Alloys, Inc.
Cu 4, Ni 2, Mg 1.5, bal Al.
F-temper: 28,000 TS; 24,000 YS; 1 El; 80 Brin. T6-temper: 54,000 TS; 51,000 YS; 0.8 El; 120 Brin. For high temperature castings; heat treatable. *Obsolete*

APEX 59A

Apex International Alloys, Inc.
Cu 3.5-4.5, Si 0-0.6, Mg 1.3-1.8, Ni 1.7-2.3, Ti 0-0.2, bal Al.
T4-Temper: 37,000 TS; 30,000 YS; 1 El; 95 Brin. T6-temper: 38,000 TS; 100 Brin. For castings; age-hardenable. *Obsolete*

APEX 60

Apex International Alloys, Inc.
Cu 3-4, Si 5.5-6.5, Mg 0.15, Zn 1, Mn 0.5, Ni 0.3, bal Al.
Sand cast, T6: 32,000-38,000 TS; 20,000-28,000 YS; 2.5-3.0 El; 90 Brin. Permanent mold cast, T61: 40,000-50,000 TS; 24,000-42,000 YS; 2 El; 100 Brin. ASTM B 108-68 SC64D; ASTM B 26-68 SC64D. *Obsolete*

APEX 60-40 COPPER ALLOY

Apex International Alloys, Inc.
Al 60, Cu 40.
For rich hardeners for Zn base die cast alloys. *Obsolete*

APEX 62

Apex International Alloys, Inc.
Zn 11, Cu 2.7, Mg 0.3, bal Al.
Sand cast: 29,000 TS; 17,000 YS; 2.5 El; 70 Brin. Permanent mold: 35,000 TS; 30,000 YS; 1.0; 105 Brin. For castings, light alloy parts. *Obsolete*

APEX 630

Apex International Alloys, Inc.
Al 6, Zn 3, bal Mg.
SC-F temper: 29,000 TS; 14,000 YS; 6 El; 50 Brin. SC-T4 temper: 40,000 TS; 14,000 YS; 12 El; 55 Brin. SC-T6 temper: 40,000 TS; 19,000 YS; 5 El; 70 Brin. For sand castings; heat treatable. *Obsolete*

APEX 66

Apex International Alloys, Inc.
Cu 5.5-5.7, Si 5-6, Mg 0.25-0.6, Zn 0.8, Mn 0.8, bal Al.
Permanent mold cast; T5: 32,000-35,000 TS; 25,000-30,000 YS; 95 Brin. Good fluidity, resistance to hot cracking. SAE 300; ASTM B108-68 CS66A. *Obsolete*

APEX 70

Apex International Alloys, Inc.
Cu 3-4, Si 5.5-6.5, Mg 0.1, Zn 1, Mn 0.5, Ni 0.35, bal Al.
Sand cast; T6: 36,000 TS; 24,000 YS; 2.0 El; 80 Brin; 22,000 shear strength. Permanent mold cast; T6: 40,000 TS; 27,000 YS; 3.0 El; 99 Brin. Good, strong, general purpose aluminum casting. SAE 326, 329; ASTM B108-68 SC64D. ASTM B26-68 SC64D. *Obsolete*

APEX 700

Apex International Alloys, Inc.
Cu 2.5, Ni 1.2, Si 7.5, bal Al.
PM-T5 temper: 20,000 TS; 85,000 YS; 10 El; 45 Brin. For bearings; permanent mold. *Obsolete*

APEX 700A

Apex International Alloys, Inc.
Sn 6.2, Cu 1, Ni 1, bal Al.
T5-temper: 20,000 TS; 8500 YS; 10 El; 55 Brin. For bearings; age hardened. *Obsolete*

APEX 75

Apex International Alloys, Inc.
Cu 7, Si 5.5, Mg 0.5, bal Al.
PM-T 6 temper: 35,000 TS; 27,000 YS; 90 Brin. PM-T 51 temper: 40,000 TS; 31,000 YS; 1.0 El; 100 Brin. For pistons; permanent mold. *Obsolete*

APEX 75-25 SILICON HARDENER

Apex International Alloys, Inc.
Al 75, Si 25.
For rich hardener for alloying. *Obsolete*

APEX 750

Apex International Alloys, Inc.
Cu 0.7-1.3, Si 0-0.7, Mn 0-0.1, Ni 0.7-1.3, Sn 5.5-7, bal Al.
Sand or permanent mold cast. T5 cond: 23,000 TS; 11,000 YS; 12 El; 45 Brin; 15,000 shearing strength. AA No. 850.0 T5. *Obsolete*

APEX 77

Apex International Alloys, Inc.
Cu 7, Si 2.5, bal Al.
SC-F temper: 23,000 TS; 15,000 YS; 1.5 El; 70 Brin. For patterns; sand cast. *Obsolete*

APEX 78

Apex International Alloys, Inc.
Cu 4, Si 8, bal Al.
For match plates; plaster mold casting. *Obsolete*

APEX 80/20 COPPER HARDENER

Apex International Alloys, Inc.
Cu 20, bal Al.
For alloying hardener. *Obsolete*

APEX 910

Apex International Alloys, Inc.
Al 9, Zn 0.6, bal Mg.
DC-F temper: 33,000 TS; 22,000 YS; 3 El; 60 Brin. For die castings. *Obsolete*

APEX 920

Apex International Alloys, Inc.
Al 9, Zn 2, bal Mg.
SC-F temper: 24,000 TS; 14,000 YS; 2 El; 65 Brin. SC-T4 temper: 40,000 TS; 16,000 YS; 10 El; 65 Brin. SC-T6 temper: 40,000 TS; 23,000 YS; 2 El; 85 Brin. For castings; heat treatable. *Obsolete*

APEX 95-5 ALUMINUM SILICON ALLOY

Apex International Alloys, Inc.
Si 4.5-6, Zn 0-0.2, 1.0 Fe + Cu, bal Al.
17,000-21,000 TS; 9,000 YS; 4-2 El; 40 Brin. For thin, intricate dense castings; easy to cast. *Obsolete*

APEX 97 DC

Apex International Alloys, Inc.
Mg 1, Al 97-0.
For alloying to zinc; shot. *Obsolete*

APEX 97 DCZ

Apex International Alloys, Inc.
Zinc hardener. ASTM ZG71A. *Obsolete*

APEX A 750

Apex International Alloys, Inc.
Cu 0.7-1.3, Si 2-3, Mn 0-0.1, Ni 0.3-0.7, Sn 5.5-7, bal Al.
Sand or permanent mold cast. T5 cond: 20,000 TS; 11,000 YS; 5.0 El; 45 Brin; 14,000 shearing strength. AA No. 850.OT5. *Obsolete*

APEX ALLOY NO. 2

Apex International Alloys, Inc.
Al 3.9-4.3, Cu 2.5-2.9, Mg 0.02-0.05, Fe 0-0.075, Pb 0-0.004, Cd 0-0.003, bal Zn.
Die cast: 52,000 TS; 93,000 compressive strength; 7 El; 100 Brin; 46,000 shearing strength. General purpose die casting, motor frames, household hardware, novelties. *Obsolete*

APEX ALLOY NO. 3

Apex International Alloys, Inc.
Al 3.9-4.3, Cu 0.1, Mg 0.025-0.05, Fe 0-0.075, Pb 0-0.004, Sn 0-0.003, bal Zn.
Die cast: 41,000 TS; 60,000 compressive strength; 10 El; 76 Brin; 31,000 shearing strength. For housings, cases, general die castings, motor frames, household hardware. ASTM: B240-64, AG40A. *Obsolete*

APEX ALLOY NO. 5

Apex International Alloys, Inc.
Al 3.9-4.3, Cu 0.75-1.25, Mg 0.03-0.06, Fe 0-0.075, Pb 0-0.004, Cd 0-0.003, bal Zn.
Die cast: 47,000 TS; 87,000 compressive strength; 7 El; 88 Brin; 38,000 shearing strength. For housings, cases, general die castings, motor frames, radio parts, hardware. ASTM: B240-64, AC41. *Obsolete*

APEX ALLOY NO. 7

Apex International Alloys, Inc.
Al 3.9-4.3, Cu 0-0.1, Mg 0.1-0.2, Fe 0-0.075, Pb 0-0.004, Cd 0-0.003, Ni 0.005-0.02, bal Zn.
Die cast: 41,000 TS; 87,000 compressive strength; 14 El; 76 Brin; 31,000 shearing strength. General die casting, automotive. SAE 903. *Obsolete*

APEX AS-1

Apex International Alloys, Inc.
95 Al min.
For deoxidizing steel. *Obsolete*

APEX AS-2

Apex International Alloys, Inc.
92 Al min.
For deoxidizing steel. *Obsolete*

APEX AS-3

Apex International Alloys, Inc.
90 Al min.
For deoxidizing steel. *Obsolete*

APEX AS-4

Apex International Alloys, Inc.
85 Al min.
For deoxidizing steel. *Obsolete*

APEX B750

Apex International Alloys, Inc.
Cu 1.7-2.3, Si 0-0.4, Mg 0.6-0.9, Mn 0-0.1, Ni 0.9-1.5, Sn 5.5-7, bal Al.
Sand or permanent mold cast. T5 cond: 32,000 TS; 23,000 YS; 5.0 El; 70 Brin; 21,000 shearing strength, 11,000 Limit. AA No. B850.OT5. *Obsolete*

APEX BRONZE

Michigan Smelting & Refining Co.
Cu 86, Al 9.5, bal Fe.
For marine parts, hardware; Sillman Bronze.

APEX D 400

Apex International Alloys, Inc.
Mg 4, Si 1.8, bal Al.
DC-F temper: 41,000 TS; 20,000 YS; 10 El. For die casting; corrosion resistant. *Obsolete*

APEX DRILL RODS

Lehigh Steel Corp.
C 1, Mn, bal Fe.
Type AISI W1 water hardening tool steel.

APEX NO. 405 ZINC BASE ALLOY

Apex International Alloys, Inc.
Al 4.1, Mg 0.03, bal Zn.
Cast: 36,000 YS; 4 El; 62 Brin. For die castings, housing; Izod 18. *Obsolete*

APEX NO. 4100 ZINC BASE ALLOY

Apex International Alloys, Inc.
Al 4.1, Cu 1.25, bal Zn.
Cast: 39,000 TS; 10 El; 71 Brin. For die castings, housing; Izod 18. *Obsolete*

APEX NO. 4102 ZINC BASE ALLOY

Apex International Alloys, Inc.
Al 4.1, Cu 1, bal Zn.
Cast: 41,000 TS; 4 El; 73 Brin. For die castings, housing; Izod 17. *Obsolete*

APEX NO. 6

Apex International Alloys, Inc.
Al 4.1, Cu 1, bal Zn.
DC-F temper: 43,500 TS; 7.3 El; 90 Brin. For die castings. *Obsolete*

APEX NO. 7 ZINC BASE ALLOY

Apex International Alloys, Inc.
Al 4.1, Cu 2.7, bal Zn.
Cast: 47,000 TS; 8 El; 83 Brin. For die castings housings; Izod 15. *Obsolete*

APEX NO. 810

Apex International Alloys, Inc.
Al 7.5, Zn 0.7, bal Mg.
For light alloy castings; heat treatable. *Obsolete*

APEX P 400

Apex International Alloys, Inc.
Mg 4, Zn 1.8, bal Al.
PM-F temper: 27,000 TS; 16,000 YS; 7 El; 60 Brin. For castings; permanent mold. *Obsolete*

APEX PATTERN ALUMINUM ALLOY

Apex International Alloys, Inc.
Cu 4-6, Si 2-3.5, 1.5 Zn + Fe, bal Al.
16,000-22,000 TS; 16,000 YS; 3-1 El; 68 Brin. For matchplate and patterns; age hardenable. *Obsolete*

APEX PISTON ALLOY

Apex International Alloys, Inc.
Cu 9.25-10.75, Zn 0-0.2, Mg 0.15-0.35, Fe 1-1.25, bal Al.
Chill cast: 34,000 TS; 1 El; 95-125 Brin. For pistons; heat treated. *Obsolete*

APEX PRESSURE ALUMINUM ALLOY

Apex International Alloys, Inc.
Cu 3-5, Si 2-4, 1.2 Fe + Zn, bal Al.
17,000-21,000 TS; 17,000 YS; 2 El; 60 Brin. For castings requiring extreme density; age hardenable. *Obsolete*

APEX Z-33

Apex International Alloys, Inc.
Si 4, Cu 3, Zn 3, bal Al.
F-temper: 24,000 TS; 13,000 YS; 3.5 El; 55 Brin. T6-temper: 30,000 TS; 16,000 YS; 3.0 El; 65 Brin. For sand and permanent mold castings: good castability. *Obsolete*

APEX Z-39

Apex International Alloys, Inc.
Cu 3-4, Si 8.5-9.5, Zn 2-3, Fe 0-1, bal Al.
Die cast: 46,000 TS; 25,000 YS; 3 El; 80 Brin. For general die castings; shearing strength 29,000. *Obsolete*

APEX Z-39-11

Apex International Alloys, Inc.
Cu 3-4, Si 10.5-11.5, Zn 2-3, Fe 0-1, bal Al.
For general die castings; shearing strength 28,000. Die cast: 46,000 TS; 27,000 YS; 3 El; 80 Brin. *Obsolete*

APEX Z-50

Apex International Alloys, Inc.
Cu 3.5-4.5, Si 2.5-3.5, Zn 2.5, bal Al.
F-temper: 24,000 TS; 14,000 YS; 2.5 El; 60 Brin. T4-temper: 30,000 TS; 18,000 YS; 4 El; 65 Brin. T6-temper: 36,000 TS; 26,000 YS; 1.5 El; 85 Brin. For sand and permanent mold castings; heat treatable. *Obsolete*

APEX Z-51

Apex International Alloys, Inc.
Cu 4-5, Si 5-6, Zn 1.5-2.5, Fe 0-1, Mn 0-0.6, bal Al.
Cast: 23,000 TS; 13,000 YS; 2 El; 55 Brin. Aged: 24,000 TS; 16,000 YS; 1 El; 75 Brin. For gear cases, housings, machine tool castings; age hardenable. *Obsolete*

APEX Z28

Apex International Alloys, Inc.
Cu 1.7-2.3, Si 9.5-10.5, Mn 0.5, Mg 0.1, Zn 2-3, Ni 0.5, bal Al.
Aluminum die casting. As cast; 44,000 TS; 23,000 YS; 4.0 El; 75 Brin. 26,000 shearing strength. For housings, motor frames, radio parts, hardware. *Obsolete*

APFI

Acciaierie Valbruna s.p.a.
Stainless Steel. C 0-0.25, Si 0-1, Mn 0-2, Cr 25, Ni 20.5, P 0.045, S 0.03, bal Fe.
Austenitic. W. Nr 1.4845; AISI 310.

APFI-S

Acciaierie Valbruna s.p.a.
Stainless Steel. C 0-0.07, Si 0-1, Mn 0-2, Cr 28, Ni 20.5, P 0.045, S 0.03, bal Fe.
Austenitic. W. Nr. 1.4335; AISI 310S.

APFI-SI

Acciaierie Valbruna s.p.a.
Stainless Steel. C 0-0.2, Si 0-2, Mn 2, P 0-0.045, S 0-0.03, Cr 25, Ni 20, bal Fe.
Heat resisting steel. AISI 314; W. Nr. 1.4841.

APFR

Acciaierie Valbruna s.p.a.
Stainless Steel. C 0-0.2, Si 0-1, Mn 0-2, Cr 23, Ni 13, P 0.04, S 0.03, bal Fe.
Austenitic. AISI 309.

APFR-S

Acciaierie Valbruna s.p.a.
Stainless Steel. C 0-0.08, Si 0-1, Mn 0-2, Cr 22, Ni 13.5, P 0.04, S 0.03, bal Fe.
Austenitic. W. Nr. 1.4833; AISI 309S.

APFR-SI

Acciaierie Valbruna s.p.a.
Stainless Steel. C 0.11, Si 0-2, Mn 1.75, P 0-0.045, S 0-0.03, Cr 20.5, bal Fe.
Heat resisting steel. W. Nr. 1.4828.

APHIT

Manufacturer not listed.
Cu 70, Ni 20, Zn 5.5, Cd 4.5.
For condenser tubes, chemical equipment; corrosion resistant.

APHTIT

English manufacture
Cu 70-75, Ni 20-21, Zn 2.4-5.5, Cd 1.8-4.5.
For corrosion resistant parts.

APIS

Saarstahl AG
C 1.65, Cr 12, Co, bal Fe.
For blanking and forming dies, punches, cutters; air hardened, non-deforming.

APM

Acciaierie Valbruna s.p.a.
Stainless Steel. C 0-0.08, Si 0-1, Mn 0-2, Cr 17, Ni 11, Mo 2.25, P 0.045, S 0.03, bal Fe.
Austenitic. W. Nr. 1.4401; AISI 316.

APM SUP.

Acciaierie Valbruna s.p.a.
Stainless Steel. C 0-0.06, Si 0-0.75, Mn 0-2, Cr 17, Ni 12.5, Mo 2.75, P 0.03, S 0.03, bal Fe.
Austenitic. W. Nr. 1.4436; AISI 316 spec.

APM-M-276

Aluminum Company of America
Al, Al $_2O_2$.
Sintered: 51,000 TS; 34,000 YS; 7 El; 92 Brin. For fasteners, fan blades, bolts; sintered, high strength at 1000 F. *Obsolete*

APM-M-276

Aluminum Company of America
Al, aluminum oxide.
Sintered: 51,000 TS; 34,000 YS; 7 El; 92 Brin. For fasteners, fan blades, bolts; sintered, high strength at 1000 F. *Obsolete*

APM-M470

Aluminum Company of America
Aluminum oxide + Al.
For high temperature applications; heat resistant to 900 F. *Obsolete*

APMC

Acciaierie Valbruna s.p.a.
Stainless Steel. C 0-0.08, Si 0-1, Cr 17, Ni 11.75, Mo 2.25, Mn 0-2, P 0.04, S 0.03, Nb + Ta = 10 x C, bal Fe.
Stabilized austenitic. W. Nr. 1.4580; AISI 318.

APML

Acciaierie Valbruna s.p.a.
Stainless Steel. C 0.03, Si 0-1, Mn 0-2, Cr 17, Ni 12, Mo 2.25, P 0.04, S 0.03, bal Fe.
Austenitic. W. Nr. 1.4404; AISI 316L.

APML TEK

Acciaierie Valbruna s.p.a.
Stainless Steel. C 0-0.03, Si 0-0.75, Mn 0-2, Cr 17.5, Ni 13.75, Mo 2.75, P 0.04, S 0.03, bal Fe.
Austenitic. W. Nr. 1.4435; AISI 316 L spec.

APMLN

Acciaierie Valbruna s.p.a.
C 0-0.03, Si 0-1, Mn 0-2, P 0-0.045, S 0-0.03, Cr 17, Mo 2.75, Ni 13, N 0.17, bal Fe.
Austenitic. AISI 316LN; W. Nr. 1.4429.

APMT

Acciaierie Valbruna s.p.a.
Stainless Steel. C 0-0.08, Si 0-1, Mn 0-2, Cr 17.25, Ni 12, Mo 2.25, P 0.045, S 0.03, Ti = 5 x C min, bal Fe.
Stabilized austenitic. W. Nr. 1.4571; AISI 316 Ti.

APMZ

Acciaierie Valbruna s.p.a.
Stainless Steel. C 0-0.08, Si 0-1, Mn 0-2, P 0-0.2, S 0-0.25, Cr 17, Mo 2, Ni 12, bal Fe.
Austenitic. AISI 316S.

APOLLO

Cytemp Specialty Steel Div.
C 0.45, Cr 2, V 0.2, bal Fe.
For die casting dies; flying shears; oil hardening. *Obsolete*

APOLLO

Time Steel Service Inc.
C 2.5, Cr 12, V 4, Mo 0.8, bal Fe.
High carbon, high chromium tool and die steel; air or oil hardened. AISI D7.

APOLLO CHROMSTEEL

Apollo Metals Inc.
Ni-Cr plated steel. For heat resistant parts; heat resistant to 800°F.

APOLLO CROM

English manufacture
Cr plated sheet Zn.
For corrosion resistant parts; stainless.

APOLLOY

Apollo Steel Co.
C 0.08, Cu 0.25, bal Fe.
For building construction; corrosion resisting to weather.

APPLAUSE

J.M. Ney Co.
Pd 55, Ag 35.
White gold color alloy for porcelain to metal restorations subject to stress. Melting range: 2140-2280°F; casting temperature: 2500°F; 200 Brin; 240 HV.

APS 10, ETC.

Now POMPEY APS 10, ETC.

APS-10 M 4

Societe Nouvelle des Acieries de Pompey
C 0.12, Cr 2, Mo 0.3, Al 0.3, bal Fe.
Tube: 100,000 psi TS min; 80,000 psi YS min. For gas well piping. Resists H_2S corrosion and stress cracking.

APW 431

Engelhard Corp.
Now ENGELHARD ALLOY NO. 356.

APW NO. 129

Engelhard Corp.
Now ENGELHARD ALLOY NO. 8409.

APW NO. 133

Engelhard Corp.
Pd 80, Ag 14.5, Al 5.5.
For high temperature brazing; 1925-2050°F MP; good oxidation resistance. *Obsolete*

APW NO. 18

Engelhard Minerals & Chemicals Corp.
Ag 45, Cu 30, Zn 25.
For silver solder; M.P. 1250-1370 F. *Obsolete*

APW NO. 200

Engelhard Corp.
Now ENGELHARD ALLOY NO. 3004.

APW NO. 201

Engelhard Minerals & Chemicals Corp.
Ag 20, Zn 30, Cd 5, Cu 45.
For silver solders; brazing. *Obsolete*

APW NO. 238

Engelhard Corp.
Now ENGELHARD ALLOY NO. 4606.

APW NO. 241

Engelhard Corp.
Now ENGELHARD ALLOY NO. 7666.

APW NO. 242

Engelhard Corp.
Now ENGELHARD ALLOY NO. 5769.

APW NO. 243

Engelhard Corp.
Now ENGELHARD ALLOY NO. 8292.

APW NO. 253

Engelhard Corp.
Au 20, Cu 80.
For brazing filler metal; 1850-1880°F MP. *Obsolete*

APW NO. 255

Engelhard Corp.
Now ENGELHARD ALLOY NO. 3202.

APW NO. 259

Engelhard Corp.
Au 94, Cu 6.
For brazing filler metal; 1742-1796°F MP. *Obsolete*

APW NO. 260

Engelhard Corp.
Now ENGELHARD ALLOY NO. 10355.

APW NO. 261

Engelhard Corp.
Now ENGELHARD ALLOY NO. 5496.

APW NO. 265

Engelhard Corp.
Au 72, Ni 22, Cr 6.
For brazing filler metal; 1785-1835°F MP. *Obsolete*

APW NO. 440

Engelhard Corp.
Now ENGELHARD ALLOY NO. 8169.

APW NO. 441

Engelhard Corp.
Now ENGELHARD ALLOY NO. 8177.

APW-301

Engelhard Minerals & Chemicals Corp.
Ag 72, Cu 28.
For silver solder; M.P. 1430 F. *Obsolete*

APW-4
Engelhard Minerals & Chemicals Corp.
Ag 50, Cu 15.5, Zn 16.5, Cd 18.
For silver solder; M.P. 1160-1175 F. *Obsolete*

APW-A11
Engelhard Minerals & Chemicals Corp.
Ag 20, Cu 45, Zn 35.
For silver brazing; MP 1430-1500 F. *Obsolete*

APW-A14
Engelhard Minerals & Chemicals Corp.
Ag 40, Cu 36, Zn 24.
For silver brazing; MP 1330-1445 F. *Obsolete*

APW-A25
Engelhard Minerals & Chemicals Corp.
Ag 50, Cu 34, Zn 16.
For silver brazing; MP 1272-1425 F. *Obsolete*

APW-A28
Engelhard Minerals & Chemicals Corp.
Ag 50, Cu 28, Zn 22.
For silver brazing; MP 1250-1340 F. *Obsolete*

APW-A33
Engelhard Minerals & Chemicals Corp.
Ag 60, Cu 25, Zn 15.
For silver brazing; MP 1260-1325 F. *Obsolete*

APW-A4
Engelhard Minerals & Chemicals Corp.
Ag 10, Cu 52, Zn 38.
For silver brazing; MP 1510-1600 F. *Obsolete*

APW-A49
Engelhard Minerals & Chemicals Corp.
Ag 80, Cu 16, Zn 4.
For silver brazing; MP 1360-1490 F. *Obsolete*

APW-A850
Engelhard Minerals & Chemicals Corp.
Ag 85, Mn 15.
For silver brazing; MP 1760-1778 F. *Obsolete*

APW-B-211
Engelhard Minerals & Chemicals Corp.
Ag 40, Cu 18, Zn 15, Cd 27.
For silver solder brazing; M.P. 1120-1205 F. *Obsolete*

APW-B-217
Engelhard Minerals & Chemicals Corp.
Ag 45, Cu 17.6, Zn 19.4, Cd 18.
For silver brazing; M.P. 1140-1190 F; obsolete. *Obsolete*

APW-C-250
Engelhard Minerals & Chemicals Corp.
Ag 40, Cu 30, Zn 28, Ni 2.
For silver solder for brazing carbide tips; M.P. 1220-1435 F; obsolete. *Obsolete*

APW-EASY
Engelhard Minerals & Chemicals Corp.
Ag 65, Cu 20, Zn 15.
For silver brazing; MP 1280-1325 F. *Obsolete*

APW-G-355
Engelhard Corp.
Now ENGELHARD ALLOY NO. 3558.

APW-HARD
Engelhard Minerals & Chemicals Corp.
Ag 75, Cu 22, Zn 3.
For silver brazing; MP 1365-1450 F. *Obsolete*

APW-L-451
Engelhard Minerals & Chemicals Corp.
Ag alloy.
For silver solder, brazing; M.P. 1144 F, all purpose alloy. *Obsolete*

APW-MEDIUM
Engelhard Minerals & Chemicals Corp.
Ag 70, Cu 20, Zn 10.
For silver brazing; MP 1335-1390 F. *Obsolete*

AQUATOUGH 100
British Steel Corp.
C 0.95, Mn 0.3, bal Fe.
Heat treated: 184,000 TS; 118,000 YS; 10 El; 30 RA; 375 Brin. For arbors, axes, forming dies, springs; water hardened; Type W1. *Obsolete*

AQUATOUGH 70
British Steel Corp.
C 0.7, Mn 0.3, bal Fe.
Water hardening tool steel; for chisels, caulking tools, cold heading dies, rivet snaps. *Obsolete*

AQUILA
Westa-Westdeutsche
C 0.15, Cr, Mo, bal Fe.
For cams, gears, camshafts; case hardening steel, tough.

AR
United States Steel Corp.
Si 0.25, C 0.4, Mn 1.75.
As rolled: 225 Brin. Mild forming. Resistant to sliding abrasion. For chutes, conveyer troughs.

AR 235 (2)
Bethlehem Steel Corp.
C 0.35-0.5, Mn 1.4-2, bal Fe.
Rolled: 110,000 TS; 70,000 YS; 16 El; 220 Brin. Heat treated: 155,000 TS; 130,000 YS; 10 El; 330 Brin. For shovels, crushers, scraper blades, hoppers. Wear resistant. Was Bethlehem Abrasion Resisting.

AR 360
Algoma Steel Corp. Ltd.
HSLA steel. C 0.25, Mn 1.5, P 0.035, S 0.035, Si 0.5, Cb 0.06, Mo 0-0.55, bal Fe.
1105-1310 MPa (160-190 ksi) TS; 1035-1240 MPa (150-180 ksi) YS; 15 El min; 360-415 Brin. Quenched and tempered steel plate for abrasion resistance. For use in truck and hopper liners, wear plates, chutes, buckets, crushers, excavators and other heavy equipment. Corrosion resistant.

AR 425
Algoma Steel Corp. Ltd.
HSLA steel. C 0.25, Mn 1.5, P 0.035, S 0.035, Si 0.5, Cb 0.06, Mo 0-0.55, B 0.003, bal Fe.
1310 MPa (190 ksi) TS; 1240 MPa (180 ksi) YS; 12 El min; 425 Brin. Quenched and tempered steel plate for abrasion resistance. For use in truck and hopper liners, wear plates, chutes, buckets, crushers, excavators and other heavy equipment. Corrosion resistant.

AR-213
American manufacture
C 0.12, Cr 20.3, W 4.5, Al 4.8, Ta 2.3, Fe 0.6, Zr 0.34, bal Co.
Age-hardened: 47 Rock C. For gas turbine components. Age-hardenable, corrosion and heat resistant. Sulphidation and oxidation resistant.

AR-235 (1)
Bethlehem Steel Corp.
C 0.35-0.5, Mn 1.4-2, Si 0.15-0.3, bal Fe.
As rolled: 235 Brin min. Abrasion resistant steel plate, 3/16 to 15 in. thick.

AR-300
United States Steel Corp.
C 0.3, Mn 1.5, Si 0.3, B 0.001, bal Fe.
Water quenched and tempered: 315 Brin. Mild forming and welding. Resistant to sliding and light impact abrasion. For chutes, hoppers.

AR-350
United States Steel Corp.
C 0.3, Mn 1.5, Si 0.3, B 0.001, bal Fe.
Water quenched and tempered: 360 Brin. Mild forming and welding. Resistant to sliding and light impact abrasion. For frames, conveyer troughs.

AR-360
Joseph T. Ryerson & Son Inc.
C 0.31, Mn 1.4, Si 0.24, Cr 0.5, Mo 0.11, B 0.004, bal Fe.
Heat treated: 360-400 Brin. Wear resistant plate.

AR-360
United States Steel Corp.
C 0.3, Mn 1.4, Si 0.25, Cr 0.55, Mo 0.11, B 0.001, bal Fe.
Water quenched and tempered: 400 Brin. Mild forming and welding. Resistant to sliding and moderate impact abrasion. For deck plates, hoppers, dump-bodies.

ARA
Universal Cyclops
C, alloy, bal Fe.
For tools, drills, cutters. *Obsolete*

ARAPAHO
AL Tech Specialty Steel Corp.
C 0.5, Mn 0.8, Si 0.25, Cr 3.25, Mo 1.45, bal Fe.
General purpose air hardening shock steel for blanking and forming dies and plastic molds requiring exceptional toughness. AISI S-7.

ARBOGA
English manufacture
Tungsten carbide.
For tools, hard cutting tools; sintered alloy.

ARC 1628
Creusot-Loire
C 0.03, Ni 65, Fe 6, Mo 26, V 0.4.
Corrosion resistant alloy. For valves, pumps for hot acids and other chemicals. AFNOR ND 27 Fe V; W.Nr. 2.4600; similar to HASTELLOY B.

ARC 2266 CB
Creusot-Loire
C 0-0.1, Cr 18, Ni 13, Mo, Cb, bal Fe.
For chemical plant equipment; Type 318; stainless, austenitic. *Obsolete*

ARC 2266 TRS
Creusot-Loire
C 0-0.06, Cr 17, Ni 12, Mo 3, bal Fe.
Annealed: 85,000 TS; 35,000 YS; 50 El; 65 RA; 160 Brin. For chemical plant equipment; Type 316L; stainless, austenitic. *Obsolete*

ARC 2702 CB
Creusot-Loire
C 0-0.1, Cr 18, Ni 10, Cb, bal Fe.
Annealed: 90,000 TS; 45,000 YS; 56 El; 65 RA; 160 Brin. For welded, chemical plant equipment; Type 347; stainless, austenitic. *Obsolete*

ARC 2702 M
Creusot-Loire
C 0-0.03, Cr 18, Ni 10, Ti, bal Fe.
Annealed: 85,000 TS; 35,000 YS; 55 El; 65 RA; 150 Brin. For welded, chemical plant equipment; Type 321; stainless, austenitic. *Obsolete*

ARC 2702 SP
Creusot-Loire
C 0-0.06, Cr 18, Ni 10, Ti, bal Fe.
Annealed: 85,000 TS; 35,000 YS; 55 El; 65 RA; 150 Brin. For welded, chemical plant equipment; Type 321; stainless, austenitic. *Obsolete*

ARC 2702 TRS

Creusot-Loire
C 0-0.04, Cr 18, Ni 10, bal Fe.
Annealed: 85,000 TS; 35,000 YS; 65 El; 75 RA; 145 Brin. For chemical plant equipment; Type 304L; stainless, austenitic. *Obsolete*

ARC 2702B

Creusot-Loire
C 0-0.08, Cr 17-19, Ni 8-11, Ti = 5 x C, bal Fe.
Normalized: 93,000 TS; 36,000 YS; 45 El; 60 RA; 165 Brin.
Annealed: 87,000 TS; 33,000 YS; 57 El; 73 RA; 155 Brin. For chemical plant equipment; Type 321; stainless, austenitic. *Obsolete*

ARC 6015

Creusot-Loire
C 0.04, Ni 59, Cr 15.5, Mo 17, W 4.5, V 0.35, Fe 5.
Corrosion resistant alloy; for use with sulfuric and phosphoric acids, chlorine. AFNOR NC 17 DWY; similar to HASTELLOY C.

ARC 7915

Creusot-Loire
C 0.1-0.25, Cr 15, Ni 78, bal Fe.
For chemical plant equipment; corrosion and heat resistant. *Obsolete*

ARC M1628

Now ARC 1628.

ARC WEL

Northeast Metals Co.
Zn, bal Cu.
For welding rod.

ARC ZCR

Creusot-Loire
C 0-0.1, Cr 24, Ni 5, Mo, bal Fe.
For furnace parts and equipment; heat resistant.

ARC-6015

Creusot-Loire
Ni 60, Cr 15, Mo 15, Fe 5, W 5.
For pumps, valves, chemical and plastic plant equipment. Heat and corrosion resistant. *Obsolete*

ARC.2233 A2

Creusot-Loire
C 0-0.03, Mn 1.8, Ni 14, Cr 17, Mo 2.5, bal Fe.
Annealed: 80,000 TS; 30,000 YS; 40 El; B 78 Rock. Cold rolled: 130,000 TS; 75,000 YS; 30 El; C 27 Rock. For chemical and pharmaceutical plant equipment, tanks, digesters, vessels. Stainless, austenitic, acid resistant. AISI Type 316L. *Obsolete*

ARC.2233 A8

Creusot-Loire
C 0-0.08, Mn 1.8, Ni 14, Cr 17, Mo 2.5, bal Fe.
Annealed: 85,000 TS; 35,000 YS; 50 El; B 82 Rock. Cold rolled: 140,000 TS; 120,000 YS; 10 El; C 28 Rock. For chemical and pharmaceutical plant equipment, valve trim, tanks, digesters, evaporators. Stainless, austenitic, acid resistant. AISI Type 316. *Obsolete*

ARC.2233 B6

Creusot-Loire
C 0-0.06, Ni 11.5, Cr 19, Mo 2.2, Ti 0.5, bal Fe.
Annealed: 85,000 TS; 35,000 YS; 50 El; B 82 Rock. Cold rolled: 140,000 TS; 120,000 YS; 10 El; C 28 Rock. For welded chemical and plastic plant equipment, digesters, tanks. Stabilized for welding, stainless, austenitic. AISI Type 316 Ti. *Obsolete*

ARC.2266 A2

Creusot-Loire
C 0-0.03, Mn 1.2, Ni 13, Cr 19, Mo 3, bal Fe.
Annealed: 85,000 TS; 40,000 YS; 60 El; 70 RA; 135 Brin. Cold rolled: 150,000 TS; 125,000 YS; 15 El; 300 Brin. For chemical and plastic plant equipment, digesters, mixers, agitators, valve trim, kettles, tanks. Acid resistant, austenitic. AISI Type 317 L, weldable. *Obsolete*

ARC.2266 A6

Creusot-Loire
C 0-0.06, Mn 1.2, Ni 13, Cr 19, Mo 3, bal Fe.
Annealed: 85,000 TS; 40,000 YS; 60 El; 135 Brin. Cold rolled: 150,000 TS; 125,000 YS; 15 El; 300 Brin. For chemical and plastic plant equipment, digesters, agitators, valve trim, tanks, vessels. Austenitic, acid resistant. AISI Type 317. Non-hardenable. *Obsolete*

ARC.2266 B6

Creusot-Loire
C 0-0.06, Ni 12, Cr 18.5, Mo 3, Ti 0.4, bal Fe.
Annealed: 85,000 TS; 35,000 YS; 50 El; 160 Brin. Cold rolled: 125,000 TS; 85,000 YS; 30 El; 240 Brin. For welded structures, chemical and plastic plant equipment, tanks, agitators. Stabilized welding grade, austenitic, stainless. AISI Type 316 Ti. *Obsolete*

ARC.2702 A2

Creusot-Loire
C 0-0.03, Mn 1.8, Ni 11, Cr 19, bal Fe.
Annealed: 85,000 TS; 35,000 YS; 60 El; 150 Brin. Cold drawn: 140,000 TS; 90,000 YS; 30 El; 250 Brin. For architectural trim, kitchen equipment, chemical and pharmaceutical plant equipment. Stainless, austenitic, good weldability. AISI Type 304L. *Obsolete*

ARC.2702 B2

Creusot-Loire
C 0-0.03, Ni 11, Cr 18, Ti 0.2, bal Fe.
Annealed: 85,000 TS; 35,000 YS; 55 El; 150 Brin. Cold drawn: 95,000 TS; 60,000 YS; 40 El; 185 Brin. For welded structures, oil refinery and chemical plant equipment, exhaust systems. Stabilized stainless steel, welding grade, austenitic. AISI Type 321. *Obsolete*

ARC.2702 B6

Creusot-Loire
C 0-0.06, Ni 10, Cr 18, Ti 0.4, bal Fe.
Annealed: 85,000 TS; 35,000 YS; 55 El; 150 Brin. Cold drawn: 95,000 TS; 60,000 YS; 40 El; 185 Brin. For welded structures, chemical and oil refinery equipment, tanks, exhaust systems. Stabilized stainless steel, austenitic, nonmagnetic. AISI Type 321. *Obsolete*

ARC.2702 B8

Creusot-Loire
C 0-0.08, Ni 10, Cr 18, Ti = 5 x C, bal Fe.
Annealed: 85,000 TS; 35,000 YS; 55 El; 150 Brin. Cold drawn: 95,000 TS; 60,000 YS; 40 El; 185 Brin. For welding structures, chemical and oil refinery equipment, tanks, vessels, shafts. Stabilized stainless steel, austenitic, nonmagnetic. AISI Type 321. *Obsolete*

ARCALOY 308

Chemetron Corp.
Weld metal: 0.07 max C, 19.0 Cr, 9.5 Ni, 1.6 Mn, 0.50 Si, bal Fe. As welded: 85,000-95,000 psi TS; 40-50 El. Covered electrode for welding 18-8 type austenitic stainless steels. AWS E308-15,16; ASME F-5, A-8.

ARCALOY 308ELC

Chemetron Corp.
Weld metal: 0.04 max C, 19.0 Cr, 9.5 Ni, 1.0 Mn, 0.30 Si, bal Fe. As welded: 80,000-90,000 psi TS; 40-50 El. For welding low carbon austenitic stainless steels. AWS Class E308-15,16; ASME F-5, A-8

ARCALOY 309

Chemetron Corp.
Weld metal: 0.10 max C, 23.0 Cr, 13.0 Ni, 1.6 Mn, 0.50 Si, bal Fe. As welded: 85,000-95,000 psi TS; 35-45 El. Covered electrode for welding type 309 stainless steel. AWS E309-15,16; ASME F5, A-8.

ARCALOY 309 CB

Chemetron Corp.
Weld metal: 0.10 max C, 23.0 Cr, 13.0 Ni, 0.80 Cb, 1.6 Mn, 0.60 Si, bal Fe. As welded: 85,000-95,000 psi TS; 30-40 El. Covered electrode to inhibit carbide precipitation when welding type 309 and other stainless steels. AWS E309Cb-15,16; ASM F-5, A-8.

ARCALOY 309 MO

Chemetron Corp.
Weld metal: 0.10 max C, 23.0 Cr, 13.0 Ni, 2.2 Mo, 1.7 Mn, 0.50 Si, bal Fe. As welded: 85,000-95,000 psi TS; 35-45 El. Covered electrode, often used for welding type 316 clad steels. AWS E309Mo-15; ASME F-5, A-8.

ARCALOY 310

Chemetron Corp.
Weld metal: 0.20 max C, 26.0 Cr, 21.0 Ni, 1.8 Mn, 0.40 Si, bal Fe. As welded: 85,000-95,000 psi TS; 35-45 El. Covered electrode for welding type 310 stainless steel. AWS E310-15,16; ASME F-5, A-9.

ARCALOY 310CB

Chemetron Corp.
Weld metal: 0.12 max C, 26.0 Cr, 21.0 Ni, 0.80 Cb, 1.8 Mn, 0.40 Si, bal Fe. As welded: 85,000-95,000 psi TS; 30-40 El. Covered electrode for welding types 310,321 and 347 stainless steels. AWS E310Cb-15,16; ASM F-5, A-9.

ARCALOY 310MO

Chemetron Corp.
Weld metal: 0.12 max C, 26.0 Cr, 21.0 Ni, 2.0 Mo, 1.8 Mn, 0.40 Si bal Fe. As welded: 85,000-95,000 psi TS; 35-45 El. Covered electrode for welding type 316 clad steels; also for lining digesters in the paper industry. AWS E310Mo-15,16; ASME F-5, A-9.

ARCALOY 312

Chemetron Corp.
Weld metal: 0.15 max C, 29.0 Cr, 9.5 Ni, 1.9 Mn, 0.50 Si, bal Fe. As welded: 110,000-120,000 psi TS; 22-25 El. Covered electrode for welds requiring high strength and stainless properties. AWS E312-15,16; ASME F-5.

ARCALOY 316

Chemetron Corp.
Weld Metal: 0.07 max C, 18.0 Cr, 13.0 Ni, 2.25 Mo, 1.7 Mn, 0.40 Si, bal Fe As welded: 85,000-95,000 psi TS; 35-45 El. Covered electrode for welding type 316 steel. AWS E316-15,16; ASME F-5,A-8.

ARCALOY 316CB

Chemetron Corp.
Weld metal: 0.07 max C, 18.0 Cr, 12.0 Ni, 2.25 Mo, 0.80 Cb, 1.6 Mn, 0.60 Si, bal Fe. As welded: 85,000-95,000 psi TS; 30-40 El. Covered electrode for welding types 316 and 316 ELC stainless.

ARCALOY 316ELC

Chemetron Corp.
Weld metal: 0.04 max C, 18.0 Cr, 13.0 Ni, 2.25 Mo, 1.0 Mn, 0.30 Si, bal Fe. As welded: 80,000-90,000 psi TS; 35-45 El. Covered electrode for welding types 316 and 316 ELC stainless steels. AWS E316L-15,16; ASME F-5,A-8.

ARCALOY 317

Chemetron Corp.
Weld metal: 0.07 max C, 19.0 Cr, 13.0 Ni, 3.5 Mo, 1.7 Mn, 0.50 Si, bal Fe. As welded: 85,000-95,000 psi TS; 35-45 El. Covered electrode for welding 317 stainless. AWS E317-15,16; ASME F-5, A-8.

ARCALOY 317ELC
Chemetron Corp.
Weld metal: 0.04 max C, 19.0 Cr, 13.0 Ni, 3.5 Mo, 1.7 Mn, 0.50 Si, bal Fe. As welded: 80,000-90,000 psi TS; 35-45 El. Covered electrode for welding type 317 stainless. ASME F-5, A-8.

ARCALOY 318
Chemetron Corp.
Weld metal: 0.07 max C, 18.0 Cr, 12.0 Ni, 2.25 Mo, 0.80 Cb, 1.6 Mn, 0.60 Si, bal Fe. As welded: 85,000-95,000 psi TS; 30-40 El. Covered electrode for welding types 316, 316 ELC stainless steels. AWS E318-15,16; ASME F-5, A-8.

ARCALOY 320
Chemetron Corp.
Weld metal: 0.07 max C, 20.0 Cr, 29.0 Ni, 3.0 Cu, 2.0 Mo, 0.5 Cb, 1.5 Mn, 0.40 Si, bal Fe. As welded: 75,000-85,000 psi TS; 35-40 El. Covered electrode for welding Carpenter 320 and Durimet 20. AWS E320-15.

ARCALOY 330
Chemetron Corp.
Weld metal: 0.25 max C, 15.0 Cr, 35.0 Ni, 1.6 Mn, 0.30 Si, bal Fe. As welded: 75,000-85,000 psi TS; 25-35 El. Covered electrode for welding type 330 stainless steel and castings of similar alloy. AWS E330-15,16; ASME F-5.

ARCALOY 330HC
Chemetron Corp.
Weld metal: 0.80 C, 15.0 Cr, 35.0 Ni, 2.0 Mn, 0.80 Si, bal Fe. As welded: 75,000-85,000 psi TS; 25-35 El. Covered electrode for welding high carbon grades of castings. Similar to ACI HT.

ARCALOY 347
Chemetron Corp.
Weld metal: 0.07 max C, 19.0 Cr, 9.5 Ni, 0.80 Cb, 1.6 Mn, 0.60 Si, bal Fe. As welded: 85,000-95,000 psi TS; 35-45 El. Covered electrode for welding types 347 and 321 stainless steels. AWS E347-15,16; ASME F-5,A-8.

ARCALOY 410
Chemetron Corp.
Weld metal: 0.10 max C, 12.5 Cr, 0.60 Mn, 0.40 Si, bal Fe. Welded annealed: 80,000-90,000 psi TS; 30-35 El. Covered electrode for welding type 410 stainless. AWS E410-15,16; ASME F4, A-6.

ARCALOY 430
Chemetron Corp.
Weld metal: 0.10 max C, 16.0 Cr, 0.60 Si, 0.60 Mn, bal Fe. Welded annealed: 75,000-80,000 psi TS; 30-35 El. Covered electrode for welding type 430 stainless. AWS E430-15,16; ASME F-5, A-7.

ARCALOY 502
Chemetron Corp.
Weld metal: 0.05 max C, 5.10 Cr, 0.56 Mo, 0.55 Mn, 0.40 Si, bal Fe. Welded annealed: 79,000 psi TS; 22-35 El. Covered electrode for welding type 502 steel. AWS E502-15; ASME F-4, A-4.

ARCALOY 8N12
Chemetron Corp.
Weld metal: 0.03 C, 14.0 Cr, 5.5 Mn, 9.0 Fe, 0.75 Si, 1.9 Cb, 0.60 Ti, bal Ni. As welded: 96,000 psi TS; 62,000 psi YS; 44 El. Covered electrode for welding many dissimilar alloys: Monel, Inconel, carbon steels, cupronickels. AWS Class ENiCrFe-3.

ARCALOY 9N10
Chemetron Corp.
Weld metal: 0.03 C, 29.0 Cu, 0.9 Si, 0.74 Fe, 0.80 Ti, 64.3 Ni, 3.7 Mn, 0.25 Al. As Welded: 79,500 psi TS; 51,000 psi YS; 33 El. Covered electrode for welding Monel and similar alloys. AWS Class ENiCu-2.

ARCALOY ER 308
Chemetron Corp.
Weld Metal: 0.07 max C, 20.5 Cr, 9.75 Ni, 1.75 Mn, 0.40 Si bal Fe. Bare wire or rod for welding 18-8 stainless steel.

ARCALOY ER 308ELC
Chemetron Corp.
Weld metal: 0.03 max C, 20.5 Cr, 9.75 Ni, 1.75 Mn, 0.40 Si. Bare wire or rod for welding low carbon 18-8 stainless steels.

ARCALOY ER 309
Chemetron Corp.
Weld metal: 0.12 C, 24.25 Cr, 13.50 Ni, 1.75 Mn, 0.40 Si, bal Fe. Bare wire or rod for welding type 309 stainless steel.

ARCALOY ER 310
Chemetron Corp.
Weld metal: 0.08-0.15 C, 26.5 Cr, 21.5 Ni, 1.75 Mn, 0.40 Si, bal Fe. Bare wire or rod for welding type 310 and many dissimilar steels.

ARCALOY ER 312
Chemetron Corp.
Weld metal: 0.08-0.15 C, 30.0 Cr, 8.75 Ni, 1.75 Mn, 0.40 Si, bal Fe. Bare wire or rod for welding stainless steels to mild steels and for welding high strength steels.

ARCALOY ER 316ELC
Chemetron Corp.
Weld metal: 0.03 max C, 19.0 Cr, 13.25 Ni, 2.2 Mo, 1.75 Mn, 0.40 Si, bal Fe. Bare wire or rod for welding type 316elc.

ARCALOY ER 347
Chemetron Corp.
Weld metal: 0.07 max C, 20.0 Cr, 9.75 Mi, 0.80 Cb+Ta, 1.75 Mn, 0.40 Si, bal Fe. Bare wire or rod for welding types 347 and 321 stainless steels.

ARCAST
C.E. Philips & Co.
C 3, Si 2, bal Fe.
For welding electrode; for cast iron.

ARCAST-67
Arwood Corp.
Zn 5-9, Mg 0.96-1.4, Ti 0.1-0.4, Cr 0.025-0.15, bal Al. Cast: 67,000 TS; 57,000 YS; 7 El. For general purpose castings, casings, housing, hardware. Good castability, heat treatable. *Obsolete*

ARCHANGEL
Apex Steel Co.
C, alloy, bal Fe.
For high speed tools; high speed steel.

ARCHITECTURAL BRONZE C-94
Delta Metal (BW) Ltd.
Mn, Zn, bal Cu.
Extruded: 86,000 TS; 38,000 YS; 25 El; 145 Brin. For casements, hand rails, show cases, panels, architectural; corrosion resistant. *Obsolete*

ARCHITECTURAL BRONZE-385
Anaconda Co.
Copper. Cu 56, Zn 41.5, Pb 2.5.
Soft rod: 60,000 TS; 20,000 YS; 25 El; 65 Rock B. For ornamental hardware, architectural applications, hinges, lock bodies, hardware. Free cutting, corrosion resistant.

ARCHITECTURAL BRONZE-385
Chase Brass & Copper Co., Inc.
Copper. Cu 58.5, Zn 38.25, Pb 3.25.
Extruded: 60,000 psi TS; 20,000 psi YS; 30 El. For architectural parts and fittings.

ARCM 098
Creusot-Loire
C 0-0.1, Mn 0-1.5, Si 0-1.5, Ni 18-21, Cr 24-27, Mo 1.5-2.5, bal Fe.
Cast, annealed: 135-165 Brin. Austenitic stainless; corrosion resistant to nitric acid and mixed sulfo-nitrics. AFNOR Z 8 CND25.19.2M.

ARCM 1628
Creusot-Loire
C 0-0.12, Cr 0-1, Mo 26-30, V 0.2-0.6, Fe 4-6, bal Ni.
Cast, annealed: 500 N/mm^2 psi TS min. Good resistance to acids. ACE N-12M (Hastelloy B).

ARCM 2233 C10
Creusot-Loire
C 0-0.1, Mn 0-1.5, Si 0-1.5, Ni 12-15, Cr 17-19, Mo 2-2.5, Nb = 10 x C, bal Fe.
Cast, annealed: 490 N/mm^2 psi TS min. Austenitic stainless, weldable, for pumps, valves, faucets; improved resistance to corrosion.

ARCM 2233 C8
Creusot-Loire
C 0-0.08, Mn 0-1.5, Si 0-1.5, Ni 10.5-12.5, Cr 17-19.5, Mo 2-2.5, Nb = 8 x C, bal Fe.
Cast, annealed: 490 N/mm^2 psi TS min. Austenitic stainless, weldable, for pumps, valves, faucets; improved resistance to corrosion. AFNOR Z4CNDNb18.12M; W.Nr. 1.4581.

ARCM 2266 A10
Creusot-Loire
C 0-0.1, Mn 0-1.5, Si 0-1.5, Ni 9-11, Cr 17-20, Mo 3-3.5, bal Fe.
Cast, annealed: 490 N/mm^2 psi TS min. Austenitic stainless for pumps, valves, faucets; improved resistance to corrosion. AFNOR Z8CND 18.10.3M.

ARCM 2266 A3
Creusot-Loire
C 0-0.03, Mn 0-1.5, Si 0-1.5, Ni 9-13, Cr 17-21, Mo 2-3, bal Fe.
Cast, annealed: 490 N/mm^2 psi TS min. Austenitic stainless, weldable, for pumps, valves, faucets; improved resistance to corrosion. AFNOR Z3CND20.10M; ACI CF-3M.

ARCM 2266 A5
Creusot-Loire
C 0-0.05, Mn 0-1.5, Si 0-1.5, Ni 12-15, Cr 17.5-20.5, Mo 3-3.5, bal Fe.
Cast, annealed: 490 N/mm^2 psi TS min. Austenitic stainless for pumps, valves, faucets; improved resistance to corrosion. AFNOR Z4CND19.13M; similar to AISI 317.

ARCM 2266 CU
Creusot-Loire
C 0-0.06, Mn 0-1, Si 0-1, Ni 7-9, Cr 23.5-25.5, Mo 3-4, Cu 1-2, bal Fe.
Cast, annealed: 600 N/mm^2 psi TS min. Austenitic stainless; resistant to stress corrosion. AFNOR Z4CNUD 25.8M.

ARCM 2702 A12
Creusot-Loire
C 0-0.12, Mn 0-1.5, Si 0-2, Ni 8-10, Cr 17-19.5, bal Fe.
Cast, annealed: 490 N/mm^2 psi TS min. Austenitic stainless for faucets, valves, pump parts; weldable. AFNOR Z10CN18.9; similar to ACI CF20.

ARCM 2702 A3
Creusot-Loire
C 0-0.03, Mn 0-1.5, Si 0-2, Ni 8-12, Cr 18-21, bal Fe.
Cast, annealed: 490 N/mm^2 psi TS min. Austenitic stainless steel; weldable for faucets, valves, pumps. AFNOR Z3 CN19.9M; ACI CF-3.

ARCM 2702 A8
Creusot-Loire
C 0.08, Mn 0-1.5, Si 0-2, Ni 8-11, Cr 18-21, bal Fe.
Cast, annealed: 490 N/mm^2 psi TS min. Austenitic stainless for pumps, valves, parts requiring corrosion resistance. AFNOR Z6CN19.9N; ACI-CF8.

ARCM 2702 C8
Nyby, Granges, AB
C 0-0.08, Mn 0-1.5, Si 0-2, Ni 9-12, Cr 18-21, Nb = 8 x C, bal Fe.
Cast, annealed: 490 N/mm^2 TS min. Stabilized austenitic stainless, preferred for welded pumps, valves, faucets. AFNOR Z4CNNb19.10M; ACI CF-8C.

ARCM 30 MO

Creusot-Loire
C 0.9-1.2, Mn 0-1.5, Cr 28-30, Mo 1.5-2.5, 1.5 Si max.
Full annealed: 270-340 Brin. Stainless casting; extreme
resistance to both corrosion and abrasion. AFNOR Z100
CD29.2M.

ARCM 6015

Creusot-Loire
C 0-0.12, Cr 15.5-17.5, Mo 16-18, V 0.2-0.4, Fe 4-7, W
3.75-5.25, bal Ni.
Cast, annealed: 500 N/mm^2 psi TS min. Good resistance to
acids and hyperchlorites. ACI CW 12 M (Hastelloy C).

ARCM 8510

Creusot-Loire
C 0-0.1, Si 8-10, Cu 2-4, Fe 0-2, bal Ni.
Cast stainless alloy. Resists sulfuric acid in all concentrations;
also organic acids (Hastelloy D).

ARCM 990

Creusot-Loire
C 0.5-0.7, Mn 0-1, Si 0-1, Ni 4-6, Cr 21.5-23.5, Cu 2-3, bal Fe.
Quenched, tempered: 240-280 Brin. Stainless; for both
abrasion and corrosion resistance. AFNOR Z60 CNW 22.5 M.

ARCO SSS 100B, A514 A517 GRADE L

Armco
C 0-0.2, Mn 0.4-0.7, Si 0.2-0.35, Cr 1.15-1.65, Mo 0.25-0.4,
Cu 0.2-0.4, B 0.0015-0.005, 0.04-0.10 Ti or V, bal Fe.
2 in. max thickness: 115,000-135,000 TS; 100,000 YS; 16-18
El. Quenched and tempered steel for buildings, bridges,
ordnance, ships, earth moving equipment, crane booms.
ASTM A514, A517. *Obsolete*

ARCOLOY

American Radiator Co.
Cu 95, Si, P.
55,000 TS. For range boilers, storage tanks; corrosion
resistant. *Obsolete*

ARCOS 16-8-2

Arcos Alloys
Stainless steel. C 0.07, Mn 1.7, Si 0.4, Cr 15.5, Ni 8.5, Mo 1.5,
bal Fe.
For welding electrodes; stainless steel type 16-8-2.

ARCOS 16-8-2/15

Arcos Alloys
Stainless steel. C 0.05, Mn 2.2, Si 0.3, Cr 16, Ni 8.5, Mo 1.5,
bal Fe.
For welding electrodes; stainless steel for high temperature
service; free from sigma-phase embrittlement.

ARCOS 307

Arcos Alloys
C 0.07-0.17, Mn 3.3-4.75, Mo 0-0.25, Cr 18-20.5, Ni 9-10.5,
bal Fe.
Welded: 84,000-96,000 TS; 56,000-62,000 YS; 37-46 El; 35-49
RA. For welding electrodes; for armor, coated.

ARCOS 308

Arcos Alloys
Stainless steel. C 0.07, Cr 19, Ni 9, bal Fe.
For welding rod; stainless.

ARCOS 308, ETC.

Arcos Alloys
Stainless steel.
One of 36 bare stainless steel wires made to 24 AISI standard
stainless steel grades plus 12 modified grades. For welding
or buildup on stainless steel.

ARCOS 308HC

Arcos Alloys
Stainless steel. C 0.14, Mn 1.7, Si 0.4, Cr 20, Ni 9.7, bal Fe.
For welding electrodes; specially coated for welding
austenitic Cr-Ni stainless steel.

ARCOS 308L

Arcos Alloys
Stainless steel. C 0.035, Mn 1.8, Si 0.4, Cr 20, Ni 9.8, bal Fe.
For welding electrodes; stainless steel type 308L.

ARCOS 309

Arcos Alloys
Stainless steel. C 0.1, Cr 25, Ni 12, bal Fe.
For weld metal; stainless.

ARCOS 309 CB

Arcos Alloys
C 0.1, Cr 25, Ni 12, Cb 0.9-1.1, bal Fe.
For welding rod; corrosion and heat resistant.

ARCOS 309 MO

Arcos Alloys
Stainless steel. C 0.07, Mn 1.8, Si 0.4, Cr 23, Ni 12.8, Mo 2.5,
bal Fe.
For welding electrodes; Mo-bearing stainless steel for root
pass welding of 316, 316L, 319 clad steel.

ARCOS 310

Arcos Alloys
Stainless steel. Cr 25, Ni 20, C, bal Fe.
For welding electrodes; E-310-16; stainless.

ARCOS 310 CB

Arcos Alloys
C 0.1, Mn 2, Si 0.5, Cr 26, Ni 21, 0.8 Cb + Ta, bal Fe.
For welding electrodes; Cb-stabilized for root pass welds in
type 347 clad plate.

ARCOS 310 HC

Arcos Alloys
Mn 1.8, Si 0.4, Cr 26.5, Ni 21, 0.22 C (also 0.3 C and 0.4 C),
bal Fe.
For welding electrodes; welding ACI types CK-20, HK, HL, HN
castings, supplied in three carbon ranges.

ARCOS 310 MO

Arcos Alloys
Stainless steel. C 0.1, Mn 1.8, Si 0.4, Cr 25.5, Ni 21.5, Mo 2.3,
bal Fe.
For welding electrodes; Mo-bearing stainless steel.

ARCOS 310HC

Arcos Alloys
Si 0.4, Cr 26.5, Ni 21, Mn 1.8, 0.22 C (also 0.30 C, or 0.40 C),
bal Fe.
For welding electrodes; welding ACI types CK-20, HK, HL, HN
castings; supplied in three carbon ranges.

ARCOS 312

Arcos Alloys
C 0.12, Mn 1.8, Si 0.5, Cr 29, Ni 9.2, bal Fe.
For welding electrodes; for joining dissimilar metals.

ARCOS 316

Arcos Alloys
Stainless steel. C 0.06, Mn 1.8, Si 0.5, Cr 18.5, Ni 12.5, bal
Fe.
For welding electrodes; stainless steel type 316.

ARCOS 316L

Arcos Alloys
Stainless steel. C 0.035, Mn 1.8, Si 0.4, Cr 18.5, Ni 12.5, Mo
2.2, bal Fe.
For welding electrodes; stainless steel type 316L.

ARCOS 317L

Arcos Alloys
Stainless steel. C 0.35, Mn 1.8, Si 0.4, Cr 19.5, Ni 13, Mo 3.5,
bal Fe.
For welding electrodes; stainless steel type 317L.

ARCOS 318

Arcos Alloys
Stainless steel. C 0.06, Mn 1.8, Si 0.5, Cr 18.5, Ni 12.5, Mo
2.2, bal Fe.
For welding electrodes; stainless steel type 318.

ARCOS 320

Arcos Alloys
Stainless steel. C 0.06, Mn 1.7, Si 0.4, Cr 20.5, Ni 34, Mo 2.5,
Cu 3.5, 0.6 Cb + Ta, bal Fe.
For welding electrodes; stainless steel type 320 (Cb-3 plate).

ARCOS 320LR

Arcos Alloys
Stainless steel. C 0.015, Mn 1.75, Si 0.07, Cr 20.5, Ni 33.5, Cu
3.25, Mo 2.5, bal Fe.
For welding electrodes; stainless steel type 320LR (Cb-3
plate) low residual.

ARCOS 330

Arcos Alloys
Stainless steel. C 0.2, Mn 1.7, Si 0.5, Cr 15.5, Ni 35, bal Fe.
For welding electrodes; stainless steel type 330.

ARCOS 330HC

Arcos Alloys
Mn 1.8, Si 0.6, Cr 16, Ni 34, 0.30 C (also 0.40 C), bal Fe.
For welding electrodes; welding ACI types HT and HU
castings; supplied in two carbon ranges.

ARCOS 347

Arcos Alloys
Stainless steel. C 0.06, Mn 1.7, Si 0.5, Cr 19.5, Ni 10, bal Fe.
For welding electrodes; stainless steel type 347.

ARCOS 349

Arcos Alloys
Stainless steel. Cr 19, Ni 9, C, W, Mo, bal Fe.
For welding electrodes; stainless, austenitic.

ARCOS 410

Arcos Alloys
Stainless steel. C 0.08, Mn 0.9, Si 0.4, Cr 12.7, Ni 0.2, Mo 0.4,
bal Fe.
For welding electrodes; martensitic stainless steel, type 410.

ARCOS 410NI

Arcos Alloys
C 0.04, Mn 0.6, Si 0.4, Cr 12, Ni 4, Mo 0.8, bal Fe.
For welding electrodes; welding CA6NM casting.

ARCOS 4NIA

Arcos Alloys
C 0.04, Mn 3, Si 0.6, Cr 15, Ni 69, Mo 1.3, Cb 1.6, Fe 8.9.
For welding electrodes; Inconel, class ENiCrFe-2.

ARCOS 502

Arcos Alloys
Stainless steel. C 0.05, Mn 0.6, Si 0.11, Cr 4.82, Mo 0.53, bal
Fe.

ARCOS 502-15/16

Arcos Alloys
C 0.05, Mn 0.6, Si 0.11, P 0.013, S 0.008, Cr 4.82, Mo 0.53,
Co 0.04, bal Fe.
Heat treated: 88.0 ksi TS; 49.2 ksi YS; 22 El; 56.5 RA. Coated
electrodes used for shielded metal arc welding of 5Cr-1/2Mo
alloyed steel.

ARCOS 505

Arcos Alloys
C 0.2, Cr 8-10, Mo 1.5, bal Fe.
For welding electrodes; for high temperatures, corrosion
resistant.

ARCOS 505-15/16

Arcos Alloys
C 0.06, Mn 0.75, Si 0.32, P 0.017, S 0.018, Cr 8.57, Mo 1.02,
Cu 0.03, bal Fe.
Heat treated: 69.6 ksi TS; 35.3 ksi YS; 36 El; 74.6 RA. Coated
electrodes used for shielded metal arc welding of 9Cr-1Mo
alloyed steel.

ARCOS 630

Arcos Alloys
Stainless steel. C 0.05, Mn 0.5, Si 0.5, Cr 16.4, Ni 4.7, Cu 3.4,
bal Fe.

ARCOS 700
Arcos Alloys
Stainless steel. C 0.05, Mn 2.5, Si 0.4, Cr 20, Ni 25.3, Mo 5, bal Fe.
For welding electrodes; stainless steel Jessop 700 plate.

ARCOS 700-15
Arcos Alloys
C 0.05, Mn 0.5-2.5, Si 0.42, S 0.02, P 0.02, Cr 20, Ni 25.3, Mo 5, bal Fe.
As welded: 80 ksi TS; 35 ksi YS; 30 El; 40 RA. Coated electrodes used for shielded metal arc welding of Jessop 700.

ARCOS 803
Arcos Alloys
Cu 70, Ni 30.
For welding electrodes; for Cu-Ni alloys.

ARCOS 813
Arcos Alloys
C 0.03, Mn 0.8, Si 0.1, Ni 30.5, Fe 0.6, Ti 0.4, Al 0.2, bal Cu.
For bare cupro-nickel welding wire, class RCuNi.

ARCOS 816
Arcos Alloys
C 0.09, Mn 0.8, Si 0.08, Ni 65.5, Ti 2.5, Al 1, bal Cu.
For bare Monel welding wires; class ER-NiCu-7.

ARCOS 8N12
Arcos Alloys
Ni 69.8, C 0.04, Mn 5.62, Si 0.83, S 0.004, P 0.004, Ti 0.55, Fe 5.5, Cr 15, Co 0.04, 2.10 Cb + Ta.
As welded: 97.9 ksi TS; 63.7 ksi YS; 48 El; 61 RA. Coated electrodes used for shielded metal arc welding of Inconel grades 600 and 601.

ARCOS 9N10
Arcos Alloys
C 0.03, Mn 3.7, Si 0.8, Ni 64, Cu 28, Fe 0.3, Ti 0.8, Al 0.4.
For welding electrodes; Monel, meeting classes ENiCu-1 (3N10), ENiCu-2 (4N10), ENiCu-4 (8N10), also 9N10.

ARCOS UNIA (ENICRFE-2)
Arcos Alloys
Ni 75, Cr 14, bal Fe.
For welding electrodes; "Inconel," heat and corrosion resistant.

ARCOSARC 100
Arcos Alloys
Mn 1.5.
0.5 Mo steel. 100,000 psi TS. For flux-cored welding wire. *Obsolete*

ARCOSARC 110T
Arcos Alloys
Ni 2, Cr 1.
0.5 Mo steel. 110,000 psi TS. For flux-cored welding wire. *Obsolete*

ARCOSARC 1CM
Arcos Alloys
Cr 1.25.
0.5 Mo steel. For flux-cored welding wire. *Obsolete*

ARCOSARC 2CM
Arcos Alloys
Cr 2.25.
1.0 Mo steel. For flux-cored welding wire. *Obsolete*

ARCOSARC 410 NI
Arcos Alloys
C 0.04, Mn 0.6, Si 0.4, Cr 12, Ni 4, Mo 0.8, bal Fe.
For flux-cored welding wire; welding CA6NM castings. *Obsolete*

ARCOSARC 5M
Arcos Alloys
Cr 5.
0.5 Mo steel. For flux-cored welding wire. *Obsolete*

ARCOSARC 70
Arcos Alloys
Mild steel, 70,000 psi TS. For flux-cored welding wire; class E70T-2. *Obsolete*

ARCOSARC 70X
Arcos Alloys
Mild steel, 70,000 psi TS. For flux-cored welding wire, multi-pass deposition without slag removal; class E70T-2. *Obsolete*

ARCOSARC 72
Arcos Alloys
Mild steel, 70,000 psi high impact strength. For flux-cored welding wire; class E70T-1. *Obsolete*

ARCOSARC 80
Arcos Alloys
1.0 Ni steel. 80,000 psi high impact strength. For flux-cored welding wire. *Obsolete*

ARCOSARC CI
Arcos Alloys
Cast iron. For flux-cored welding wire. *Obsolete*

ARCOSARC SS-1
Arcos Alloys
Stainless steel.
Type 308 stainless steel. For flux-cored welding wire. *Obsolete*

ARCOSARC SS-2
Arcos Alloys
Stainless steel.
Type 309 stainless steel. For flux-cored welding wire. *Obsolete*

ARCOSARC SS-3
Arcos Alloys
Stainless steel.
Type 316 stainless steel. For flux-cored welding wire. *Obsolete*

ARCTIC
Now CRANE ARCTIC.

ARCTIC D
British Steel plc
C 0.08-0.14, Mn 0-1.5, Si 0-0.5, Nb 0.1, N 0-0.009, bal Fe.
High yield stress/ultimate tensile strength ratio; weldable; with high notch tough properties at low temperatures. For liquid gas ship steel. *Obsolete*

ARCTIC GRADE X70
Climax Performance Materials Corp.
C 0.05, Mn 1.6, Si 0.05, Mo 0.25, Cb 0.06, bal Fe.
For arctic pipe lines.

ARCUM 2266 A8
Creusot-Loire
C 0-0.08, Mn 0-1.5, Si 0-1.5, Ni 9-12, Cr 18-21, Mo 2-3, bal Fe.
Cast, annealed: 490 N/mm^2 psi TS min. Austenitic stainless for pumps, valves, faucets; improved resistance to corrosion. AFNOR Z5CND20.10M; similar to ACI CF-8M.

ARDAL METAL
Ardal Ltd.
Cu 2, Fe 1.7, Ni 0.6, bal Al.
31,000 TS; 17,000 YS; 15 El; 60 Brin. For bearings, pistons, wire, tubes. Cannot be hardened by heat treatment.

ARDENT
Osborn Steels Ltd.
C 1.5, W 4, Cr 0.5, bal Fe.
For upsetters, forging dies; high abrasion resistance, cold work steel. *Obsolete*

ARDHO NO. 2
Spencer Clark Metal Industries Ltd.
C 0.47, Mn 0.47, Si 0.9, Ni 1.1, Cr 0.6, bal Fe.
Nickel-chromium tool steel, for smiths' tools, cold chisels, punches.

ARDOLOY 1A
Herbert Cutanit Ltd.
Sintered carbide tool material. Cutting tool for semi-finishing and finishing cast iron and non-ferrous materials; suitable for milling. For turning and reaming all aluminum alloys. *Obsolete*

ARDOLOY 2A
Herbert Cutanit Ltd.
Sintered carbide tool material. Shock-resistant cutting tool for planing cast iron; general wear-resistant application. ISO K30. *Obsolete*

ARDOLOY AD
Herbert Cutanit Ltd.
Sintered carbide tool material. General-purpose cutting tool for machining cast iron and non-ferrous materials at medium feeds and speeds. ISO M10, M20. *Obsolete*

ARDOLOY AF
Herbert Cutanit Ltd.
Sintered carbide tool material. Hard grade of straight tungsten carbide; particularly for finishing cast iron, non-ferrous, plastics and abrasive non-metallics. ISO K05. *Obsolete*

ARDOLOY AK
Herbert Cutanit Ltd.
Sintered carbide tool material. Tough cutting tool for heavy roughing cuts at slow speeds; good on interrupted cuts. ISO P40, P50. *Obsolete*

ARDOLOY CN10
Herbert Cutanit Ltd.
Sintered carbide tool material. Hard wear-resistant grade of straight tungsten carbide; for finish turning and boring of cast iron and non-ferrous metal. ISO K10. *Obsolete*

ARDOLOY CN20(K)
Herbert Cutanit Ltd.
Sintered carbide tool material. General-purpose cutting tool for use on cast iron, non-ferrous metal and stainless steel. ISO K20. *Obsolete*

ARDOLOY CN30(M)
Herbert Cutanit Ltd.
Sintered carbide tool material. For planing cast iron; tough, shock resistant. ISO K30. *Obsolete*

ARDOLOY CN01(B)
Herbert Cutanit Ltd.
Sintered carbide tool material. Hard wear-resistant straight tungsten grade due to low cobalt content and fine-grain structure; for finishing and boring cast iron and non-ferrous metal. ISO K01. *Obsolete*

ARDOLOY CR 20
Herbert Cutanit Ltd.
Sintered carbide tool material. For general purpose, semi-finishing of steels; resistant to wear and cratering. ISO P20. *Obsolete*

ARDOLOY CR10
Herbert Cutanit Ltd.
Sintered carbide tool material. For medium and high speed cutting speeds and finishing steel; resistant to wear and cratering. ISO P10. *Obsolete*

ARDOLOY CR25(T)
Herbert Cutanit Ltd.
Sintered carbide tool material. General purpose cutting grade for all types of steel at medium speeds and feeds. ISO P20. *Obsolete*

ARDOLOY CR30

Herbert Cutanit Ltd.
Sintered carbide tool material. General purpose grade for roughing and semi-finishing all types of steel at medium feeds and speeds. ISO P30. *Obsolete*

ARDOLOY CR40(R)

Herbert Cutanit Ltd.
Sintered carbide tool material. Tough grade of cutting tool for roughing cuts at slow speeds and heavy loads. ISO P40. *Obsolete*

ARDOLOY ICC

Herbert Cutanit Ltd.
Sintered carbide tool material. Grade of cutting tool for semi-finishing and finishing of cast iron and non-ferrous material; used on all aluminum alloys. ISO K10. *Obsolete*

ARDOLOY S200

Herbert Cutanit Ltd.
Sintered carbide tool material. General purpose cutting tool particularly for lathe turning most metals. Resistant to wear and cratering. ISO P20. *Obsolete*

ARDOLOY S3

Herbert Cutanit Ltd.
Sintered carbide tool material. General purpose cutting tool for all types of steel at medium speeds and feeds. ISO P30. *Obsolete*

ARDOLOY S48

Herbert Cutanit Ltd.
Sintered carbide tool material. Tough cutting tool for roughing operations, particularly on steel. ISO P40. *Obsolete*

ARDOLOY SK2

Herbert Cutanit Ltd.
Sintered carbide tool material. Titanium-carbide grade for finish turning and boring steel; resistant to flank and crater wear. ISO PO1. *Obsolete*

ARDORIT 13A1

Hoffman & Co. KG
C 0-0.12, Si 1.2, Al 1, Cr 13, bal Fe.
Annealed: 75,000 TS; 40,000 YS; 30 El; 70 RA; 160 Brin. For oil refinery equipment; Type 410 Al; stainless, creep resistant.

ARDORIT 18A1

Hoffman & Co. KG
C 0-0.12, Si 1, Al 1, Cr 18, bal Fe.
Annealed: 80,000 TS; 50,000 YS; 22 El; 48 RA; 150 Brin. For oil refinery equipment; Type 430 Al; heat and creep resistant.

ARDORIT 20/10

Hoffman & Co. KG
C 0.15, Cr 19.5, Ni 9.5, bal Fe.
Annealed: 80,000 TS; 35,000 YS; 55 El; 75 RA; 150 Brin. Cold drawn: 180,000 TS; 150,000 YS; 10 El; 250 Brin. For chemical plant equipment, tanks, mixers, agitators; Type 302; stainless, austenitic.

ARDORIT 24/20

Hoffman & Co. KG
C 0.15, Si 2, Cr 24, Ni 19, bal Fe.
Annealed: 100,000 TS; 45,000 YS; 50 El; 65 RA; 185 Brin. For furnace parts, pumps, valves, turbine parts; Type 310; stainless, austenitic.

ARDORIT 25N

Hoffman & Co. KG
C 0.2, Si 1.2, Cr 25, Ni 4, bal Fe.
Aged: 115,000 TS; 80,000 YS; 15 El. Cast: 70,000 TS; 65,000 YS; 2 El; 190 Brin. For furnace parts, heat treat boxes, baffles; heat and corrosion resistant.

ARDORIT 6

Hoffman & Co. KG
C 0-0.12, Si 0.8, Al 0.8, Cr 6.5, bal Fe.
Bar: 70,000 TS; 30,000 YS; 28 El; 65 RA; 160 Brin. For oil refinery equipment; creep and heat resistant.

AREMITE

Robbins & Meyers Inc.
synthetic cast iron.
30,000-50,000 TS; 165 Brin. For pipes, fittings, hardware.

ARGAL

Gerhardi & Co.
Mg 2-4, Mn 0-0.4, Cr 0-0.3, bal Al.
Soft: 28,000 TS; 13,000 YS; 30 El; 47 Brin. Hard: 40,000 TS; 35,000 YS; 10 El; 73 Brin. For aircraft tanks and fittings, marine parts, fuel lines; resists seawater corrosion.

ARGALIUM

Rieger Gebr.
Aluminum. Cr 0.3, Ti 0.2, Mg 1.5-3, Mn 0.6-1.3, Si 0-1.3, bal Al.
Soft: 26,000 TS; 10,000 YS; 20 El; 45 Brin. Hard: 41,000 TS; 36,000 YS; 5 El; 77 Brin. For roofing, hydraulic tubing, architectural trim; good forming and welding properties.

ARGELITE

English manufacture
Al 90, Cu 6, Si 2, Bi 2.
For light alloy parts; non-hardenable.

ARGENT

Plouff Metallographic Institute
Ag 2, bal Sn.
6775 TS; 3.7 El. Al solder. Sc-14.

ARGENT FRANCAIS

French manufacture
Cu-Zn-Ni.
For jewelry, ornaments; Moussets Silver.

ARGENTAL

Manufacturer not listed.
Sn 10, Co 5, bal Cu.
For jewelry; corrosion resistant.

ARGENTAL

English manufacture
Al 60-75, Ag 15-18, Zn 7-20, Cu 3-5.
For jewelry; corrosion resistant.

ARGENTALIUM

English manufacture
Ag 5, Mg 0.1-1, bal Al.
For light alloy parts; non-hardenable.

ARGENTAN

Plouff Metallographic Institute
Al 2, Ag 0.5, bal Cu.
Rolled: 23,660 TS; 2.3 El. For contacts.

ARGENTAN

English manufacture
Cu 48.35, Zn 34.05, Ni 17.6.
For table cutlery, ornaments; high resistance to corrosion.

ARGENTAN

American manufacture
Cu 50-90, Ni 3-40, Al 0-10, Zn 0-40.
For cutlery; corrosion resistant.

ARGENTAN (NEUSILBER)

German manufacture
Ni 26, Cu 56, Zn 18.
For chemical equipment construction; substitute for Ag, a nickel silver.

ARGENTAN RUSSIAN (CAST)

Russian manufacture
Cu 57.52, Zn 18.94, Ni 20.35, Fe 3.15.
For ornaments, hardware; corrosion resistant.

ARGENTAN SHEET

Russian manufacture
Cu 40-65, Zn 17-32, Ni 15-30.
For ornaments, jewelry, hardware; corrosion resistant.

ARGENTAN, BERLIN

German manufacture
Cu 55.5, Zn 29.1, Ni 15.5.
For electrical resistances; corrosion resistant.

ARGENTAN, CHINESE

Chinese manufacture
Cu 40.4, Zn 25.4, Ni 31.6, Fe 2.6.
For ornaments, jewelry, hardware; corrosion resistant.

ARGENTAN, ENGLISH

English manufacture
Cu 63.36, Zn 17, Ni 19.13, Fe 0.38.
For domestic ware; corrosion resistant.

ARGENTAN, FRENCH

French manufacture
Cu 50.32, Zn 30.94, Ni 18.4.
For electrical resistances; corrosion resistant.

ARGENTAN, RUSSIAN

Russian manufacture
Cu 63.88, Zn 17.58, Ni 17.58, Fe 0.33, Pb 0.32.
For ornaments, architectural purposes; corrosion resistant.

ARGENTAN, SOLDER

Russian manufacture
Cu 35, Zn 57, Ni 8.
For solder; corrosion resistant.

ARGENTAN, VIENNA

Austrian manufacture
Cu 55.6, Zn 21.8, Ni 22.2, Fe 0.38.
For ornaments, jewelry, hardware; corrosion resistant.

ARGENTIN

English manufacture
Sn 85, Sb 14.5, Cu 0.5.
For bearings; antifriction.

ARGENTINE METAL

American manufacture
Sn 85.5, Sb 14.5.
For statuettes and small ornaments; expands on cooling.

ARGESTE 120

Ergst, Stahlwerke
C 0-0.1, Cr 11.5-13.5, Ni 12-14, bal Fe.
Austenitic, good corrosion resistance. For cutleries, jewelry, stamped articles. W. Nr. 4307. *Obsolete*

ARGESTE 13 A1

Ergst, Stahlwerke
C 0-0.08, Cr 12-14, Mn 0-1, Si 0-1, Al 0.1-0.3, bal Fe.
Annealed: 50-65 Kg/mm 2 TS; 30 Kg/mm 2 YS; 20 El. Magn type stainless. W. Nr. 4002. *Obsolete*

ARGESTE 13 PX

Ergst, Stahlwerke
C 0-0.1, Cr 13.5-15.5, bal Fe.
Welding wire. W. Nr. 4009. *Obsolete*

ARGESTE 135 PX

Ergst, Stahlwerke
C 0-0.06, Cr 17-19, Mo 4-5, Ni 12.5-15, bal Fe.
Welding wire. W. Nr. 4447. *Obsolete*

ARGESTE 17 LT

Ergst, Stahlwerke
C 0-0.1, Cr 16-18, Mo 1.5-2, Ti = 7 x C, bal Fe.
Annealed: 50-65 Kg/mm 2 TS; 30 Kg/mm 2 YS; 20 El. Magnet non-hardenable stainless, weldable. For parts and welded assemblies of higher chemical stability. W. Nr. 4523. *Obsolete*

ARGESTE 17 NB

Ergst, Stahlwerke

C 0-0.1, Cr 16-18.5, Nb = 12 x C, bal Fe.
Annealed: 45-60 Kg/mm 2 TS; 30 Kg/mm 2 YS; 20 El.
Magnet non-hardenable stainless, weldable. For welded
assemblies of stainless. W. Nr. 4511. *Obsolete*

ARGESTE 17 PX

Ergst, Stahlwerke

C 0-0.1, Cr 16.5-18.5, bal Fe.
Welding wire. W. Nr. 4015. *Obsolete*

ARGESTE 17 UZ

Ergst, Stahlwerke

C 0.1, Cr 15.5-17.5, Mo 0.25, S 0.15-0.25, bal Fe (plus special
additions).
Magnetic, non-hardenable stainless, free machining. For
special machined parts requirements. *Obsolete*

ARGESTE 180 K

Ergst, Stahlwerke

C 0-0.03, Cr 15.5-16.5, Ni 17.5-18.8, bal Fe.
Austenitic, good ductility, corrosion resistant. For cold
headed stainless parts. W. Nr. 4321. *Obsolete*

ARGESTE 182 RX

Ergst, Stahlwerke

C 0-0.06, Cr 17.5-20, Mo 2-2.5, Ni 20-22, Cu 1.8-2.2, Nb = 12
x C min, bal Fe.
Welding wire. W. Nr. 4507. *Obsolete*

ARGESTE 3956 AM

Ergst, Stahlwerke

C 0-0.12, Cr 16.5-18.5, Ni 11-13, bal Fe.
Austenitic, ductile, corrosion resistant, less tendency to work
harden. For cold formed or stamped stainless parts. Werkstoff
Nr. 3956; AISI Type 305 stainless. Formerly 80 AM.

ARGESTE 3962 AM

Ergst, Stahlwerke

C 0.05-0.2, Cr 10.5-12.5, Ni 9-11, Mn 5.5-6.5, bal Fe.
Austenitic, corrosion resistant. Formerly 80 AB. Werkstoff Nr.
3962.

ARGESTE 4001 IW

Ergst, Stahlwerke

C 0-0.08, Cr 13-15, Si 0-1, Mn 0-1, bal Fe.
Annealed: 50-65 kg/mm^2 TS; 30 kg/mm^2 YS; 20 El.
Magnetic type stainless. For structural parts, building fittings.
Werkstoff Nr. 4001. Formerly 13W.

ARGESTE 4006 IH

Ergst, Stahlwerke

C 0.08-0.12, Cr 12-14, Mn 0-1, Si 0-1, bal Fe.
Martensitic type stainless, hardenable. For structural parts in
water and steam. Werkstoff Nr. 4006, AISI type stainless.
Formerly 13H.

ARGESTE 4016 IM

Mannesmannrohren-Werke AG

C 0-0.1, Cr 15.5-17.5, bal Fe.
Annealed: 45-60 kg/mm^2 TS; 30 kg/mm^2 YS; 20 El. Ferritic,
non-hardenable stainless. For structural parts, armatures,
building fittings. W. Nr. 4016, AISI Type 430. Formerly 17.
Obsolete

ARGESTE 4021 YB

Ergst, Stahlwerke

C 0.17-0.22, Cr 12-14, Mn 0-1, Si 0-1, bal Fe.
Magnetic, hardenable, corrosion resistant. For structural parts
requiring higher strength. Werkstoff Nr. 4021. Formerly K 20.

ARGESTE 4024 YA

Ergst, Stahlwerke

C 0.12-1.17, Cr 12-14, bal Fe.
Magnetic, hardenable, corrosion resistant. For structural parts
and turbine blades. Werkstoff Nr. 4024. Formerly K 15.

ARGESTE 4034 YD

Ergst, Stahlwerke

C 0.4-0.5, Cr 12-14, bal Fe.
Magnetic, hardenable, corrosion resistant. For high
temperature parts, arbors, spindles, bolts, shafts. Formerly K
40. Werkstoff Nr. 4034.

ARGESTE 4057 YN

Ergst, Stahlwerke

C 0.17-0.25, Cr 16-18, Ni 1-2.5, bal Fe.
Magnetic, hardenable, corrosion resistant. Heat treated parts
have good strength to 500°C. Werkstoff Nr. 4057; AISI Type
431. Formerly K 20N.

ARGESTE 4104 IU

Ergst, Stahlwerke

C 0.1-0.17, Cr 15.5-17.5, Mo 0.2-0.3, Si 0.15-0.25, bal Fe.
Magnetic, free machining, nonhardenable stainless. For
stainless parts made on automatic screw machines; bolts,
studs, nuts. Formerly 17 U. Werkstoff Nr. 4104.

ARGESTE 4112 YL

Ergst, Stahlwerke

C 0.85-0.95, Cr 17-19, Mo 1-1.3, bal Fe.
Magnetic, hardenable, corrosion resistant. For cutlery,
kitchen knives, harden shafts operating at elevated
temperature. Formerly K 90L. Werkstoff Nr. 4112; AISI Type
440B stainless.

ARGESTE 4113 IL

Ergst, Stahlwerke

C 0-0.1, Cr 15.5-17.5, Mo 0.9-1.3, bal Fe.
Annealed: 45-65 kg/mm^2 TS; 30 kg/mm^2 YS; 30 El.
Magnetic, stainless. For parts for motor cars requiring
corrosion resistance. Werkstoff Nr. 4113. Formerly 17L.

ARGESTE 4120 YL

Ergst, Stahlwerke

C 0.17-0.22, Cr 12-14, Mo 1-1.3, bal Fe.
Magnetic, hardenable, corrosion resistant. For high quality
structural parts as turbine blades. Werkstoff Nr. 4120.
Formerly K 20L.

ARGESTE 4122 YL

Ergst, Stahlwerke

C 0.33-0.43, Cr 15.5-17.5, Mo 1-1.3, bal Fe.
Magnetic, hardenable, corrosion resistant. For arbors,
spindles, bolts operating at elevated temperatures. Formerly
K 35L. Werkstoff Nr. 4122.

ARGESTE 4301 PA

Ergst, Stahlwerke

Cr 17-20, Ni 9-11.5, C 0-0.07, bal Fe.
Austenitic, hardenable by cold work only, corrosion resistant,
weldable. For food processing equipment, dairy equipment.
Werkstoff Nr. O1; AISI Type 304 stainless. Formerly 80 P.

ARGESTE 4305 UA

Ergst, Stahlwerke

C 0-0.15, Cr 17-19, Ni 8-10, S 0.1-0.2, bal Fe.
Annealed: 50-70 kg/mm^2 TS; 22 kg/mm^2 YS; 50 El.
Austenitic, free machining, corrosion resistant. For stainless
bolts, nuts, studs, shafts made on automatic screw machines.
Formerly 80 U. Werkstoff Nr. 4305; AISI Type 303 stainless.

ARGESTE 4306 LA

Ergst, Stahlwerke

C 0-0.03, Cr 17-20, Ni 10-12.5, bal Fe.
Austenitic, weldable, corrosion resistant. For welded
assemblies for food, dairy and chemical industries. Formerly
80 LC. Werkstoff Nr. 4306; AISI Type 304L.

ARGESTE 4310 FA

Ergst, Stahlwerke

C 0-0.15, Cr 16-18, Ni 7-9, bal Fe.
Austenitic, corrosion resistant; hardens readily by cold work.
For cold worked wire and strip for springs for operation to
350°C. Formerly 80 FH. Werkstoff Nr. 4310; AISI Type 301
stainless.

ARGESTE 4310 IT

Ergst, Stahlwerke

C 0-0.1, Cr 16-18, Ti = 7 x C, bal Fe.
Annealed: 45-60 kg/mm^2 TS; 30 kg/mm^2 YS; 20 El.
Magnetic, nonhardenable stainless, weldable. For welded
assemblies of stainless. Werkstoff Nr. 4510. Formerly 17T.

ARGESTE 4401 PA

Ergst, Stahlwerke

C 0-0.07, Cr 16.5-18.5, Mo 2-2.5, Ni 10.5-13.5, bal Fe.
Austenitic, resistant to corrosion. For parts and apparatus for
textile industry. Werkstoff Nr. 4401; AISI Type 316 stainless.
Formerly 82 P.

ARGESTE 4404 LA

Ergst, Stahlwerke

C 0-0.03, Cr 16.5-18.5, Mo 2-2.5, Ni 11-14, bal Fe.
Austenitic, resistant to corrosion. For parts and apparatus for
chemical and textile industry. Formerly 82 LC. Werkstoff Nr.
4404; AISI Type 316L stainless.

ARGESTE 4435 LA

Ergst, Stahlwerke

C 0-0.03, Cr 16.5-18.5, Mo 2.5-3, Ni 12.5-15, bal Fe.
Austenitic, weldable, resistant to corrosion. For welded
assemblies requiring good corrosion resistance. Formerly 83
LC. Werkstoff Nr. 4435.

ARGESTE 4436 PA

Ergst, Stahlwerke

C 0-0.07, Cr 16.5-18.5, Mo 2.5-3, Ni 12-14.5, bal Fe.
Austenitic, weldable, resistant to corrosion. For welded
assemblies requiring corrosion resistance. Werkstoff Nr.
4436. Formerly 83 P.

ARGESTE 4449 PA

Ergst, Stahlwerke

C 0-0.07, Cr 16-18, Mo 4-5, Ni 12.5-14.5, bal Fe.
Austenitic, good corrosion resistance. For parts and
apparatus with high pitting stability. Werkstoff Nr. 4449.
Formerly 135 P.

ARGESTE 4505 BA

Ergst, Stahlwerke

C 0-0.07, Cr 16.5-18.5, Mo 2-2.5, Ni 19-21, Cu, Nb, bal Fe.
Austenitic, ductile, corrosion resistant. For parts in chemical
industry in contact with sulfuric acid. Formerly 182 Nb.
Werkstoff Nr. 4505.

ARGESTE 4506 TA

Ergst, Stahlwerke

C 0-0.07, Cr 16.5-18.5, Mo 2-2.5, Ni 19-21, Cu, Ti, bal Fe.
Austenitic, ductile, corrosion resistant. For parts in chemical
industry in contact with sulfuric acid. Formerly 182 RT.
Werkstoff Nr. 4506.

ARGESTE 4541 TA

Ergst, Stahlwerke

C 0-0.1, Cr 17-19, Ni 9-11.5, Ti = 5 x C, bal Fe.
Austenitic, weldable, corrosion resistant. For welded
assemblies for food, chemicals; and for operation at elevated
temperatures. Formerly 80 T. Werkstoff Nr. 4541; AISI Type
321 stainless.

ARGESTE 4550 BA

Ergst, Stahlwerke

C 0-0.1, Cr 17-19, Ni 9-11.5, Nb = C, bal Fe.
Austenitic, weldable, corrosion resistant. For welded
assemblies for food, chemicals, and for operation at elevated
temperature. Formerly 80 Nb. Werkstoff Nr. 4550; AISI Type
347 stainless.

ARGESTE 4568 GA

Ergst, Stahlwerke

C 0-0.09, Cr 16-18, Ni 6.5-7.75, plus special additions, bal Fe.
Austenitic, corrosion resistant. Formerly 80 SG. Werkstoff Nr.
4568.

ARGESTE 4571 TB
Ergst, Stahlwerke
C 0-0.1, Cr 16.5-18.5, Mo 2-2.5, Ni 10.5-13.5, Ti = 5 x C min, bal Fe.
Austenitic, weldable, resistant to corrosion. For welded assemblies for chemical, textile, and cellulose industries. Formerly 82 T. Werkstoff Nr. 4571.

ARGESTE 4580 BA
Ergst, Stahlwerke
C 0-0.1, Cr 16.5-18.5, Mo 2-2.5, Ni 10.5-13.5, Nb = 8 x C min, bal Fe.
Austenitic, weldable, resistant to corrosion. For welded assemblies in chemical industry. Werkstoff Nr. 4580. Formerly 82 Nb.

ARGESTE 58 M
Ergst, Stahlwerke
C 0-0.1, Cr 17-19, Ni 4.5-6.5, Mn 7.5-9.5, bal Fe.
Austenitic, corrosion resistant. For household articles, washing machines, rinsing tubs. W. Nr. 4371; similar to AISI Type 202. *Obsolete*

ARGESTE 80 K
Ergst, Stahlwerke
C 0-0.12, Cr 17-19, Mo 8-10, bal Fe.
Annealed: 50-70 Kg/mm 2 TS; 22 Kg/mm 2 YS; 50 El.
Austenitic, hardenable only by cold work, corrosion resistant. For dairy equipment, food handling, dental instruments. W. Nr. 4300; AISI Type 302 stainless. *Obsolete*

ARGESTE 80 MN
Ergst, Stahlwerke
C 0-0.2, Cr 17-20, Ni 7.5-9.5, Mn 5.5-7.5, bal Fe.
Welding wire. W. Nr. 4370. *Obsolete*

ARGESTE 80 PX
Ergst, Stahlwerke
C 0-0.08, Cr 18-20, Ni 8.5-10.5, bal Fe.
Welding wire. W. Nr. 4302. *Obsolete*

ARGESTE 80 PXLC
Ergst, Stahlwerke
C 0-0.025, Cr 18-20, Ni 9-11.
Welding wire. W. Nr. 4316; AISI Type 304L stainless. *Obsolete*

ARGESTE 80 UZ
Ergst, Stahlwerke
C 0-0.15, Cr 17-19, Ni 8-10, S 0.1-0.2, bal Fe (plus special additions).
Austenitic, free machining, corrosion resistant. For stainless screw machine parts, as fasteners. *Obsolete*

ARGESTE 80 X
Ergst, Stahlwerke
C 0-0.07, Cr 18-20, Ni 8-10, Nb = 12 x C min, bal Fe.
Welding wire. W. Nr. 4551. *Obsolete*

ARGESTE 82 PX
Ergst, Stahlwerke
C 0-0.06, Cr 18-20, Mo 2-2.5, Ni 9-11, bal Fe.
Welding wire. W. Nr. 4402. *Obsolete*

ARGESTE 82 PXLC
Ergst, Stahlwerke
C 0-0.025, Cr 17-19, Mo 2-2.5, Ni 9.5-11.5, bal Fe.
Welding wire. W. Nr. 4428. *Obsolete*

ARGESTE 82 SG
Ergst, Stahlwerke
C 0-0.08, Cr 14-16, Mo 2-2.5, Ni 6.5-7.75, bal Fe (plus special additions).
Austenitic, corrosion resistant. W. Nr. 4532. *Obsolete*

ARGESTE 82 X
Ergst, Stahlwerke
C 0-0.07, Cr 18-20, Mo 2-2.5, Ni 9.5-11.5, Nb = 12 x C min, bal Fe.
Welding wire. W. Nr. 4575. *Obsolete*

ARGESTE 83 NB
Ergst, Stahlwerke
C 0-0.1, Cr 16.5-18.5, Mo 2.5-3, Ni 12-14.5, Nb = 8 x C min, bal Fe.
Austenitic, weldable, very good resistance to corrosion. For parts and assemblies for textile, fuel and rubber industries. W. Nr. 4583. *Obsolete*

ARGESTE 83 PX
Ergst, Stahlwerke
C 0-0.06, Cr 18-20, Mo 2.5-3, Ni 10-12, bal Fe.
Welding wire. W. Nr. 4403. *Obsolete*

ARGESTE 83 PXLC
Ergst, Stahlwerke
C 0-0.025, Cr 17-19, Mo 2.5-3, Ni 10-13, bal Fe.
Welding wire. W. Nr. 4430. *Obsolete*

ARGESTE 83 T
Ergst, Stahlwerke
C 0-0.1, Cr 16.5-18.5, Mo 2.5-3, Ni 12-14, Ti = 5 x C min, bal Fe.
Austenitic, weldable, very good resistance to corrosion. For parts and assemblies in chemical, textile and cellulose industries. W. Nr. 4573. *Obsolete*

ARGESTE 83 X
Ergst, Stahlwerke
C 0-0.07, Cr 18-20, Mo 2.5-3, Ni 10-13, Nb = 12 x C min, bal Fe.
Welding wire. W. Nr. 4576. *Obsolete*

ARGESTE K 100
Ergst, Stahlwerke
C 1-1.1, Cr 16-17, Mo 0.4-0.6, bal Fe.
Magnetic, hardenable to file hardness, corrosion resistant. For stainless ball bearings, races, hardened shafts, safety razor blades. W. Nr. 4125; AISI Type 440C stainless. *Obsolete*

ARGESTE K 15L
Ergst, Stahlwerke
C 0.12-0.17, Cr 12-14, Mo 1-1.3, bal Fe.
Magnetic, hardenable, corrosion resistant. For steam turbine blades and similarly stressed parts. W. Nr. 4119. *Obsolete*

ARGESTE K 50
Ergst, Stahlwerke
C 0.5-0.6, Cr 13-15, Mo 0.5-0.6, bal Fe.
Hardenable, magnetic, corrosion resistant. For cutlery, table knives, instrument parts, requiring hard, stainless surfaces. W. Nr. 4110. *Obsolete*

ARGILITE
French manufacture
Al 90, Cu 6, Si 2, Bi 2.
For automotive engine parts.

ARGO HIGH SPEED
English manufacture
C 0.82, Mn 0.17, Cr 1.74, W 4.87, V 0.1, bal Fe.
For tools, cutters, dies, reamers, gages, punches; high speed steel.

ARGO-BOND
Johnson Matthey plc
Ag 23, Cu 35, Zn 27, Cd 15.
Cadmium bearing silver brazing alloy; 610-735°C MP; DIN 8513.

ARGO-BRAZE 25
Johnson Matthey plc
Ag 25, Cu 38, Zn 33, Mn 2, Ni 2.
Silver brazing alloys for tungsten carbide; 710-810°C MP; DIN 8513.

ARGO-BRAZE 27
Johnson Matthey plc
Ag 27, Cu 38, Zn 20, Mn 9.5, Ni 5.5.
Silver brazing alloys for tungsten carbide; 680-830°C MP; DIN 8513 L-Ag27.

ARGO-BRAZE 40
Johnson Matthey plc
Ag 40, Cu 30, Zn 28, Ni 2.
Silver brazing alloys for tungsten carbide; 670-780°C MP; DIN 8513.

ARGO-BRAZE 49H
Johnson Matthey plc
Ag 49, Cu 16, Zn 23, Mn 7.5, Ni 4.5.
Silver brazing alloys for tungsten carbide; 680-705°C MP; DIN 8513 L-Ag49.

ARGO-BRAZE 50
Johnson Matthey plc
Ag 50, Cu, Zn, Cd, Ni, Mn.
Silver brazing alloy; 639-668°C MP.

ARGO-BRAZE 56
Johnson Matthey plc
Ag 56, Cu, Zn, Ni.
Silver brazing alloy; 600-711°C MP.

ARGO-FLO
Johnson Matthey plc
Ag 38, Cd, Zn.
Silver brazing alloy; 605-651°C MP.

ARGO-SWIFT
Johnson Matthey plc
Ag 30, Cu 28, Zn 21, Cd 21.
Cadmium bearing silver brazing alloy; 607-685°C MP; DIN 8513 L-Ag30Cd.

ARGOFIL
IMI Rod & Wire
Mn 0.25, Si 0.25, bal Cu.
For filter rod for argon arc and inert gas shielded metal arc welding of copper. M.P. 1083°C.

ARGOID METAL
English manufacture
Cu 52.6, Zn 25.8, Ni 21.6.
For ornamental parts; Nickel Silver.

ARGOZOIL
American manufacture
Cu 54, Zn 20-28, Ni 14, Sn 2, Pb 2-10.
For ornamental parts; corrosion resistant.

ARGUS
Nassau Smelting & Refining Co.
Sn, Pb, Sb.
For bearings. *Obsolete*

ARGUZOID
German manufacture
Cu 48-56, Ni 13-21, Zn 23-31, Sn 0-4, Pb 0-4.
For cutlery, ornaments, utensils; German Silver.

ARGYROID (ARGIROIDE)
English manufacture
Cu, Ni, Zn.
For resistances, ornaments; Nickel Silver.

ARGYROLITH
English manufacture
Cu 50-70, Ni 10-20, Zn 5-30.
For resistances, ornaments; Nickel Silver.

ARGYROPHAN
English manufacture
Cu, Ni, Zn.
For resistances, ornaments; Nickel Silver.

ARISTOCRAT
Lumen Bearing Co.
Sn 85, Sb 7.5, Cu 7.5.
For Babbitt, bearings. *Obsolete*

ARISTOLOY
Engelhard Corp.
Ag 67-70, Sn 25-29, Cu 3-5, Zn 0-1.
For dental amalgams and dies; U.S.P. 1963085.

ARISTOLOY 302
Copperweld Steel Co.
Cr 17-19, Ni 8-10, C 0.08-0.2, Mn 0-2, Si 0-1, bal Fe.
For stainless parts; Type 302; resists scaling to 1450 F.
Obsolete

ARISTOLOY 303
Copperweld Steel Co.
Cr 17-19, Ni 8-10, C 0-0.15, S or Se or Zn, bal Fe.
For stainless parts; Type 303; free machining. *Obsolete*

ARISTOLOY 304
Copperweld Steel Co.
Cr 18-20, Ni 8-11, C 0-0.08, Mn 0-2, Si 0-1, bal Fe.
For stainless parts; Type 304; heat and corrosion resistant.
Obsolete

ARISTOLOY 305
Copperweld Steel Co.
Cr 17-19, Ni 10-13, C 0-0.12, Mn 0-2, Si 0-2, bal Fe.
For stainless parts; Type 305; corrosion resistant. *Obsolete*

ARISTOLOY 306
Copperweld Steel Co.
Cr 18-20, Ni 9-11, C 0-0.08, Mn 0-2, bal Fe.
For stainless parts for welding; Type 306; heat and corrosion resistant. *Obsolete*

ARISTOLOY 307
Copperweld Steel Co.
Cr 20-22, Ni 10-12, C 0.08-0.2, Mn 0-1.5, bal Fe.
For stainless parts; Type 307; corrosion resistant. *Obsolete*

ARISTOLOY 308
Copperweld Steel Co.
Cr 19-21, Ni 10-12, C 0-0.08, Mn 1.5, bal Fe.
For stainless parts; Type 308; heat and corrosion resistant.
Obsolete

ARISTOLOY 309
Copperweld Steel Co.
Cr 22-24, Ni 12-15, C 0-0.2, Mn 0-2, bal Fe.
For stainless parts; resists scaling to 1800 F. *Obsolete*

ARISTOLOY 310
Copperweld Steel Co.
C 0-0.25, Cr 24-26, Ni 19-22, bal Fe.
For heat and corrosion resisting parts; resists scaling to 2000 F. *Obsolete*

ARISTOLOY 316
Copperweld Steel Co.
Cr 16-18, Ni 10-14, C 0-0.1, Mn 0-2, Mo 2-3, bal Fe.
For stainless parts; heat and corrosion resistant. *Obsolete*

ARISTOLOY 317
Copperweld Steel Co.
Cr 18-20, Ni 11-14, C 0-0.1, Mn 0-2, Mo 3-4, bal Fe.
For stainless parts; heat and corrosion resistant. *Obsolete*

ARISTOLOY 321
Copperweld Steel Co.
C 0-0.08, Mn 0-2, Si 0-1, Cr 17-19, Ni 8-11, Ti = 5 x C, bal Fe.
For heat and corrosion resisting parts; heat and corrosion resisting. *Obsolete*

ARISTOLOY 347
Copperweld Steel Co.
Cr 17-19, Ni 9-12, C 0-0.1, Mn 0-2, Si 0-1, Cb bal Fe.
For stainless parts; heat and corrosion resistant. *Obsolete*

ARISTOLOY 4-6 CR-MO
Copperweld Steel Co.
Cr 4-6, C 0-0.15, Mo 0.4-0.6, bal Fe.
For oil tubes, stills, oil refineries; heat and creep resistant.
Obsolete

ARISTOLOY 4027 LEADED
Copperweld Steel Co.
C 0.25-0.3, Mn 0.7-0.9, Mo 0.2-0.3, Pb 0.15-0.35, bal Fe.
Annealed: 80,000 TS; 65,000 YS; 25 El; 220 Brin. Hardened: 250,000 TS; 210,000 YS; 8 El; 477 Brin. For gears, shafts for truck and automobile transmissions. Free-cutting, water or oil hardening. *Obsolete*

ARISTOLOY 410
Copperweld Steel Co.
Cr 11-13.5, C 0-0.15, Mn 0-1, Si 0-1, bal Fe.
For corrosion resistant parts; resists scaling to 1250 F.
Obsolete

ARISTOLOY 414
Copperweld Steel Co.
Cr 11-13.5, Ni 1.2-2.5, C 0-0.15, bal Fe.
For corrosion resistant parts; resists scaling to 1250 F.
Obsolete

ARISTOLOY 4140
Copperweld Steel Co.
C 0.4, Cr 0.8-1.1, Mo 0.15-0.25, bal Fe.
Rolled: 130,000 TS; 95,000 YS; 15 El; 40 RA; 270 Brin.
Annealed: 90,000 TS; 65,000 YS; 27 El; 55 RA; 185 Brin. For gears, shafts, axles, fittings, bolts; oil hardened. *Obsolete*

ARISTOLOY 416
Copperweld Steel Co.
Cr 12-14, C 0-0.15, Mo 0-0.6, S 0-0.07, Se 0-0.07, bal Fe.
For corrosion resistant parts; free machining. *Obsolete*

ARISTOLOY 420
Copperweld Steel Co.
C 0-0.15, Mn 0-0.7, Si 0.5, Cr 12-14, bal Fe.
For corrosion resisting parts; air or oil hardening. *Obsolete*

ARISTOLOY 430
Copperweld Steel Co.
Cr 14-18, C 0-0.12, Mn 0-1, Si 0-1, bal Fe.
For corrosion resistant parts; stainless. *Obsolete*

ARISTOLOY 446
Copperweld Steel Co.
C 0-0.35, Cr 23-27, bal Fe.
For heat and corrosion resisting parts; resists scaling to 2000 F. *Obsolete*

ARISTOLOY 7-9 CR-MO
Copperweld Steel Co.
Cr 7-9, C 0-0.15, Mo, bal Fe.
For oil tubes, stills, oil refineries; creep and heat resistant.
Obsolete

ARK
Jessop-Saville Ltd.
C 0.7, Cr 3.7, Mo 0.6, W 14, V 1, bal Fe.
For tools, cutters, punches; high speed steel.

ARK HIGH SPEED
Jessop-Saville Ltd.
C 0.84, Mn 1.53, W, Cr, bal Fe.
For tools, high speed cutters; high speed steel.

ARK SUPERIOR
Jessop-Saville Ltd.
C 0.76, Cr 4.25, Mo 0.6, V 1.4, W 18, bal Fe.
For drills, broaches, reamers, hobs, gear cutters; high speed steel.

ARK SUPERIOR EXTRA
Jessop-Saville Ltd.
C 0.78, Cr 4.25, Mo 0.6, V 1.4, W 22, bal Fe.
For form cutters, tools for chilled roll turning; high speed steel.

ARK SUPERLATIVE
Jessop-Saville Ltd.
C 0.8, Cr 4.8, Mo 0.6, V 1.6, W 18.5, Co 5.7, bal Fe.
For tools, cutters, milling cutters; high speed steel.

ARK SUPREME
Jessop-Saville Ltd.
C 0.8, W 22, Cr 5, Co 17, V 1.7, Mo 0.6, bal Fe.
For tools, cutters, planing and boring tools; high speed steel.

ARK TRIUMPH
Saville & Co. Ltd., J.J.
C 1.2, Cr 4.35, V 4.5, W 14, bal Fe.
For form tools, broaches, reamers; high speed steel.

ARK TRIUMPH
Jessop-Saville Ltd.
C 1.2, Cr 4.35, V 4.5, W 14, bal Fe.
For form tools, broaches, reamers; high speed steel.

ARKIT
French manufacture
Co 38, Cr 30, W 16, Ni 10, Mo 4, C 2.5.
For high speed cutting tips of lathe tools; high heat resistance.

ARKO
American manufacture
Cu 80, Zn 20.
For tubes, fittings, hardware; drawn and spun.

ARKTURUS
Hover, Gebruder, Edelstahlwerk
C 0.55, Cr 1.05, V 0.18, W 1.85, bal Fe.
For cold work tools, headers, upsetters; oil hardened, tough.

ARM 12
Russian manufacture
C 0.2, Cr 11.5, Ni 0-0.5, Mo 0.5, V 0.25, W 2.5, Co 2.2, B 0.01, bal Fe.
For high temperature applications, furnace parts; heat resistant to 600°C., corrosion resistant.

ARM 4
Russian manufacture
C 0.08-0.14, Mn 0.3-1, Si 0-0.6, Cr 14-16, Ni 14-16, Mo 1.8-2.2, W 0.8-1.2, Co 2.5-3, Ti 0.3, bal Fe.
Annealed: 90,000 TS; 40,000 YS; 40 El; 82 Rock B. For high temperature applications; welded structures, agitators; heat and corrosion resistant.

ARM 6
Russian manufacture
Cr 16, Ni 14, Mo 2.5, W 6, Cb 1, V 0.4, C, bal Fe.
Annealed: 85,000 TS; 35,000 YS; 50 El; 80 Rock B. For high temperature applications, chemical plant equipment, fasteners, mixers; heat and corrosion resistant.

ARMACAST
Duraloy Blaw-Knox/Union Steel Casting
C 0.25-0.3, Cr 2.65-3.15, Mo 0.45-0.55, bal Fe.
Cast: 110,000 TS; 90,000 YS; 20 El; 45 RA; 200-225 Brin. For road construction and railway equipment castings; tough.

ARMALOY
Armstrong Bros. Tool Co.
C, alloy, bal Fe.
For cutting tools, bits, blades, ratchets, wrenches. Cast alloy.

ARMASTEEL
General Motors Corp./Central Foundry Div
C 2.45-2.75, Si 1.25-1.55, S 0.025-0.075, Mn 0.3-0.6, bal Fe.
Cast: 55,000-108,000 psi TS; 42,000-95,000 psi YS; 8-1.5 El; 143-285 Brin. For gears, rocker arms, universal joints, and camshafts; pearlitic malleable iron.

ARMASTEEL GM 84 M

General Motors Corp./Central Foundry Div
C 2.45-2.75, Si 1.25-1.55, Mn 0.3-0.6, S 0.025-0.075, P
0-0.05, bal Fe.
Pearlitic malleable cast iron. 100,000 psi TS; 80,000 psi YS; 2
El; 241-269 Brin. For parts requiring good strength;
connecting rods. SAE Grade 70002; ASTM Grade 80002.

ARMASTEEL GM 85 M

General Motors Corp./Central Foundry Div
C 2.45-2.75, Si 1.25-1.55, Mn 0.3-0.6, S 0.025-0.075, P
0-0.05, bal Fe.
Pearlitic malleable cast iron. 80,000 psi TS; 60,000 psi YS; 3
El; 197-241 Brin. For additional local rehardening by flame or
induction; plant carriers. SAE & ASTM Grade 60003.

ARMASTEEL GM 85 M MODIFIED

General Motors Corp./Central Foundry Div
C 2.45-2.75, Si 1.25-1.55, Mn 0.3-0.6, S 0.025-0.075, P
0-0.05, bal Fe.
Pearlitic malleable cast iron. 90,000 psi TS; 60,000 psi YS; 3
El; 217-269 Brin. For automotive crankshafts and
transmission gears. SAE & AST Grade 60003. *Obsolete*

ARMASTEEL GM 86 M

Gimo-Osterby Bruks, A.B.
C 2.55, Si 1.4, Mn 0.45, S 0.12, P 0.05, bal Fe.
Pearlitic malleable cast iron. 70,000 TS; 48,000 YS; 4 El;
163-207 Brin. For lightly stressed automotive type parts; e.g.,
certain compressor crankshafts. SAE Grade 48005; ASTM
Grade 48004.

ARMASTEEL GM 88 M

General Motors Corp./Central Foundry Div
C 2.45-2.75, Si 1.25-1.55, Mn 0.3-0.6, S 0.025-0.075, P
0-0.05, bal Fe.
Pearlitic malleable cast iron. 105,000 psi TS; 85,000 psi YS; 2
El; 269-302 Brin; 250,000 psi compressive strength (ultimate).
For parts requiring good strength and wear resistance; gears.
ASTM Grade 80002.

ARMATURE ELECTRIC

Follansbee Steel Co.
C 0.04, Si 0.5, bal Fe.
Annealed: 45,000 TS; 25,000 YS; 25 El. For rotating
equipment and field poles; high permeability. *Obsolete*

ARMCO 12 SR

Armco
Stainless steel. Cr 12, C 0-0.02, Si 0.5, Al 1.2, Ti 0.3, Cb 0.6,
bal Fe.
Sheet, strip, pipe and tubing. More oxidation resistant and
more creep resistant than Type 409. Ferritic, non-hardenable.

ARMCO 13

Armco Steel Corp.
C 0-0.12, Cr 10, bal Fe.
For cutlery; corrosion resistant. *Obsolete*

ARMCO 15 TYPE 425

Armco Steel Corp.
C 0-0.12, Cr 14-16, bal Fe.
For corrosion resistant parts. *Obsolete*

ARMCO 15-5 PH

Armco
Stainless steel. C 0.07, Cr 14-15.5, Ni 3.5-5.5, Cu 2.5-4.5, bal
Fe.
Martensitic; precipitation hardenable. Condition H 900:
204,000 TS; 182,000 YS; 14 El. For springs, strong
stampings; better transverse ductility and toughness than
17-7 PH.

ARMCO 16-6

Armco Steel Corp.
C 0.08-0.15, Cr 0-16, Ni 0-6, Mn 0-4, bal Fe.
Annealed; 110,000 TS, 40,000 YS; 60 El; 70 RA; Rockwell
B85. For chemical plant equipment; Type 301. *Obsolete*

ARMCO 16-6S

Armco Steel Corp.
C 0.08-0.15, Cr 0-16, Ni 0-6, S, bal Fe.
Annealed: 105,000 TS; 40,000 YS, 55 El; 67 RA; 165 Brin. For
chemical plant equipment; Type 301S; stainless. *Obsolete*

ARMCO 17-10P

Armco
C 0-0.15, Cr 16-18, Ni 9.5-12, P 0.2-0.4, bal Fe.
Austenitic stainless; precipitation hardenable. Solution
treated: 89,000 TS; 38,000 YS; 70 El; 143 Brin. Double aged:
144,000 TS; 99,000 YS; 20 El; 31 Rock C; 70,000 psi
endurance limit. *Obsolete*

ARMCO 17-4 PH

Armco
Stainless steel. C 0.05, Cr 17, Ni 4, Cu 4, bal Fe.
Age-hardened: 210,000 TS; 200,000 YS; 6-15 El; 450 Brin.
For gears, cams, chains, valves, pump parts; age-hardenable.

ARMCO 17-7 PH

Armco
Stainless steel. C 0.09, Cr 16-18, Ni 6.5-7.75, Al 0.75-1.25, bal
Fe.
Semi-austenitic; precipitation hardened. Annealed: 130,000
TS; 40,000 YS; 35 El. Hardened: 200,000-265,000 TS;
185,000-260,000 YS; 2-9 El (wire can be stronger). For
springs, instrument stampings such as gears, levers, cams,
fasteners.

ARMCO 18 CR-CB HP-10

Armco
Stainless steel. Cr 17.6-18.25, Ni 0-0.4, C 0-0.02, Ti 0.25-0.35,
Cb 0.55-0.65, bal Fe.
Sheet, strip. Good oxidation and corrosion resistance plus
good formability. Ferritic, non-hardenable.

ARMCO 18 SR

Armco
Stainless steel. C 0.05, Mn 0.5, Si 1, Cr 18, Ni 0.5, Al 2, Ti 0.4,
bal Fe.
Ferritic sheet and strip, weldable. RT: 80,000-90,000 TS;
60,000-70,000 YS; 25-30 El; 90 Rock B. 1000°F: 48,700 TS;
33,900 YS; 30 El. For industrial ovens, heaters, furnace tubes,
annealing boxes, baffle plates, pyrometer tubes; resists
scaling at high temperature.

ARMCO 18-2 MN

Now ARMCO NITRONIC 32.

ARMCO 18-20 SMO

Armco Steel Corp.
C 0-0.1, Mn 0-1.5, Cr 16-18, Ni 0-14, Mo 2-3, bal Fe.
85,000 TS; 40,000 YS; 60 El; 135 Brin. For stainless parts;
chemical plant equipment; stainless, austenitic. *Obsolete*

ARMCO 18-9 LW

Armco
Stainless steel. C 0-0.1, Mn 0-2, Si 0-1, Cr 17-19, Ni 8-10, Cu
3-4, bal Fe.
Annealed: 78,000 TS. Cold drawn: 152,000 TS. For cold
formed parts, fasteners, ferrules; good formability, austenitic.

ARMCO 19-9

Armco Steel Corp.
C 0-0.2, Cr 19, Ni 9, bal Fe.
For heat and corrosion resistant, parts; stainless, heat and
corrosion resistant. *Obsolete*

ARMCO 19-9 TYPE 305

Armco Steel Corp.
C 0.08-0.2, Cr 18-20, Ni 8-10, bal Fe.
Annealed: 85,000 TS; 35,000 YS; 55 El; 70 RA; 150 Brin. For
chemical plant equipment; Type 305; stainless. *Obsolete*

ARMCO 19-9 TYPE 306

Armco Steel Corp.
C 0-0.11, Cr 18-20, Ni 8-10, bal Fe.
Annealed: 85,000 TS; 35,000 YS; 55 El; 70 RA; 150 Brin. For
chemical plant equipment; Type 306; stainless. *Obsolete*

ARMCO 20-10 MN TYPE 307

Armco
C 0.08-0.2, Cr 19-22, Ni 9-12, bal Fe.
Annealed: 85,000 TS; 35,000 YS; 55 El; 70 RA; 150 Brin. For
chemical plant equipment; Type 304; stainless. *Obsolete*

ARMCO 20-10 TYPE 307

Armco Steel Corp.
C 0.08-0.2, Cr 19-22, Ni 9-12, bal Fe.
Annealed: 85,000 TS; 35,000 YS; 55 El; 70 RA; 150 Brin. For
chemical plant equipment; Type 307; stainless. *Obsolete*

ARMCO 20-45-5

Cannon-Muskegon Corp.
C 0.05, Mn 5, Si 0.4, Cr 20, Ni 45, Mo 2.25, Cb 0.15, bal Fe.
For heat exchanger and condenser tubing.

ARMCO 20-45-5

Armco Steel Corp.
C 0.05, Mn 5, Si 0.4, Cr 20, Ni 45, Mo 2.25, Cb 0.15, bal Fe.
For heat exchanger and condenser tubing.

ARMCO 21-6-9

Now ARMCO NITRONIC 40.

ARMCO 22-13-5

Now ARMCO NITRONIC 50.

ARMCO 22-4-9

Armco
Stainless steel. C 0.45-0.6, Mn 7-10, Cr 20-23, Ni 3-5, N 0.4, C
0-0.12, bal Fe.
Heat treated: 162,000 TS; 102,000 YS; 9.0 El; 9.0 RA; 344
Brin. At 900°F: 114,000 TS; 58,000 YS; 20.0 El; 21.0 RA; 236
Brin. At 1600°F: 38,000 TS; 25,000 YS; 27.0 El; 39.0 RA; 177
Brin. For high temperature steam valves and gas turbines;
age hardenable, corrosion, wear and heat resistant.

ARMCO 311 DQ

Armco
Stainless steel. Cr 17.25, Ni 4.5, C 0-0.04, Cu 2.4, Mn 2.5, N
0.15, bal Fe.
Rod and wire, sheet, strip, pip and tubing. Higher strength
than Types 301 and 304 with comparable ductility and
formability. Austenitic, non-hardenable by heat treatment.

ARMCO 4-79 NI

Armco Steel Corp.
Mo 4, Ni 79, bal Fe.
For electrical equipment; high permeability. *Obsolete*

ARMCO 400

Armco
Stainless steel. C 0-0.05, Mn 0-1, Si 0-1, Cr 12-13, Ni 0-0.5, Al
0-0.5, bal Fe.
Ferritic, non-hardenable, weldable. RT: 61,500 TS; 29,500 YS;
35 El. 1200°F: 15,000 TS; 10,000 YS; 63 El. For office
furniture, dryer drums, air conditioning panels, sink strainers;
resists scaling up to 1300°F.

ARMCO 409 HP-10

Armco
Stainless steel. C 0-0.01, Cr 10.75-11.25, Ti 0.2-0.3, bal Fe.
Ferritic, nonhardenable. Sheet: 68,000 TS; 40,000 YS. Low
cost sheet; formable.

ARMCO 410 CB

Armco
Stainless steel. C 0-0.15, Cr 11.5-13.5, Cb 0-0.25, bal Fe.
Martensitic, magnetic, hardenable. Hardened, tempered,
500°F: 195,000 TS; 161,000 YS; 16 El; 43 Rock C; Charpy 65
ft·lb. Tempered, 1100°F: 137,000 TS; 121,000 YS; 19 El; 277
Brin; 73,000 psi endurance limit.

ARMCO 420 F

Armco
C 0.15-0.45, Cr 12-14, Se 0.18-0.35, bal Fe.
Annealed: 95,000 TS; 50,000 YS; 20 El; 50 RA; 195 Brin. For
cutlery, ball bearings, dental and surgical instruments; Type
420 F; free-cutting, corrosion resistant.

ARMCO 439 HP-10
Stainless steel. Cr 17.1-17.5, Ni 0-0.4, C 0-0.02, Ti 0.4-0.5, bal Fe.
Sheet, strip. Good oxidation and corrosion resistance plus good formability. Ferritic, non-hardenable.

ARMCO 48 NI
Armco Steel Corp.
Nickel-Iron alloy.
High permeability, low coercive force, low hysteresis loss. For instrument transformers, generators, and motors; soft magnetic alloy. *Obsolete*

ARMCO 48 NI-R
Armco Steel Corp.
Nickle-Iron alloy.
High permeability, low coearcive force, low hysteresis loss. For instrument transformers, generators and motors; soft magnetic alloy. *Obsolete*

ARMCO 9% NICKEL A353
Armco
C 0-0.13, Mn 0-0.9, Si 0.15-0.3, Ni 8.5-9.5, bal Fe.
2 in. max thickness: 100,000-120,000 TS; 75,000 YS; 20 El. Heat treated nickel alloy steel, weldable; for storage and process vessels for liquefied gases to -320°F. Double normalized and tempered. *Obsolete*

ARMCO 9% NICKEL A553 GRADE A
Armco
C 0-0.13, Mn 0-0.9, Si 0.15-0.3, Ni 8.5-9.5, bal Fe.
2 in. max thickness: 100,000-120,000 TS; 100,000 YS; 20 El. Quenched and tempered: nickel alloy steel, weldable; for storage and process vessels, for liquefied gases to -320°F. *Obsolete*

ARMCO A-286
Armco
C 0.05, Mn 1.45, Si 0.5, Cr 14.75, Ni 25.25, Mo 1.3, Ti 2.15, Al 0.15, V 0.3, B 0.005, bal Fe.
Austenitic, precipitation hardenable, stainless. Annealed: 90,000 TS; 35,000 YS; 45 El; 83 Rock B. Solution treated and aged: 150,000 TS; 100,000 YS; 25 El; 34 Rock C. Very good high temperature properties such as creep strength, fatigue, scaling, impact up to 1300°F; for high temperature fasteners, etc. AMS 5731, 5732, etc.

ARMCO ALUMINIZED STEEL TYPES 1 AND 2
Armco
Sheet steel coated with Al. Rolled: 45,000 TS; 30,000 YS; 35 El; 55 Rock B. For structural uses, mufflers; corrosion resistant.

ARMCO ARMATURE
Armco Steel Corp.
Si 0.5, bal Fe.
For D.C. motors, armatures; high magnetic permeability. *Obsolete*

ARMCO BETA 3
Armco Steel Corp.
C 11.5, Zr 6, Sn 4.5, bal Ti.
Wrought: 100,000-110,000 TS; 90,000 YS; 15 El. Beta alloy: good cold formability; age hardenable to 135,000 TS. AMS 4977. *Obsolete*

ARMCO C 42, ETC.
Now ARMCO HIGH STRENGTH C 42, ETC.

ARMCO CRYONIC 5
Armco
C 0-0.13, Mn 0.3-0.6, Ni 4.75-5.25, Mo 0.2-0.35, Al 0.05-0.12, Ni 0-0.02, bal Fe.
For cryogenic tanks. *Obsolete*

ARMCO CT N
C 0.18, Mn 1.15-1.6, Si 0.15-0.3, Cb 0.02-0.05, bal Fe.
Normalized: 70,000-90,000 TS; 50,000 YS min (up to 2 in.). HSLA steel; for structural purposes. ASTM A633 Gr. C. *Obsolete*

ARMCO CT QT
Manufacturer not listed.
Quenched and tempered condition of Armco CT: YS 60,000 psi min.

ARMCO DI MAX M 15
Armco
Cold rolled, fully processed silicon sheet steel; non-oriented; low core loss. For high efficiency distribution transformers and large power transformers.

ARMCO DI MAX M 19
Armco
Cold rolled, fully processed silicon sheet steel; non-oriented; low core loss. For 2, 4 or many pole large rotating machine laminations.

ARMCO DI MAX M 22
Armco
Cold rolled fully and semi-processed silicon sheet steel; non-oriented; low core loss. For transformers, radio chokes, large rotating machines.

ARMCO DI MAX M 27
Armco
Cold rolled fully and semi-processed silicon sheet steel; non-oriented; low core loss. For transformers, radio chokes, domestic appliance motors.

ARMCO DI MAX M 36
Armco
Cold rolled fully and semi-processed silicon sheet steel; non-oriented; low core loss. For transformers, magnetos, motor laminations.

ARMCO DI MAX M 43
Armco
Cold rolled fully and semi-processed silicon sheet steel; non-oriented; low core loss. For generators and motor laminations.

ARMCO DI MAX M 45
Armco
Cold rolled fully and semi-processed silicon sheet steel; non-oriented; low core loss. For generators, motor laminations.

ARMCO DI MAX M 47
Armco
Cold rolled semi-processed silicon sheet steel; non-oriented; low core loss. For motor laminations.

ARMCO DQ
Armco
Steel.
Hot-rolled: 50 ksi (345 MPa) TS; 35 ksi (241 MPa) YS; 38 El; 50 Rock B. Cold-rolled: 44 ksi (303 MPa) TS; 27 ksi (186 MPa) YS; 42 El; 42 Rock B. Drawability.

ARMCO DQSK
Armco
Steel.
Hot-rolled: 50 ksi (345 MPa) TS; 35 ksi (241 MPa) YS; 40 El; 50 Rock B. Cold-rolled: 43 ksi (296 MPa) TS; 24 ksi (165 MPa) YS; 43 El; 42 Rock B. Drawability.

ARMCO ELECTROMAGNET IRON
Armco Steel Corp.
C 0-0.03, bal Fe.
Rolled: 44,000 TS; 23,000 YS; 37 El; 70 RA; 83 Brin. Cold drawn: 73,000 TS; 69,000 YS; 12 El; 63 RA; 142 Brin. For cores of D.C. electrical and magnetic equipment; high permeability. *Obsolete*

ARMCO ENAMELING IRON
Armco
C 0-0.1, bal Fe.
For enameling purposes; sheet. *Obsolete*

ARMCO FIELD GRADE
Armco Steel Corp.
Si 0.25, bal Fe.
For intermittent duty fractional H. P. motors. *Obsolete*

ARMCO GALVANIZED PAINTGRIP
Armco Steel Corp.
Galvanized steel.
For structures. *Obsolete*

ARMCO H T 50 Y
Armco Steel Corp.
C 0-0.12, bal Fe.
Annealed: 75,000 TS; 55,000 YS; 20 El. For corrosion resistant parts; corrosion and abrasion resistant. *Obsolete*

ARMCO HIGH STRENGTH
Armco
See ARMCO GAINEX and FORMABLE series.

ARMCO HIGH STRENGTH 55Y
Armco Steel Corp.
C 0-0.12, Alloy, bal Fe.
Annealed: 75,000 TS; 53,000 YS; 20 El. For structures. *Obsolete*

ARMCO HIGH TENSILE
Armco Steel Corp.
C 0-0.12, Cu 0.35, Ni 0.5, Mo 0.05, bal Fe.
Rolled: 65,000 TS; 50,000 YS; 28 El; 60 RA. For trucks, railroad cars, automobiles. *Obsolete*

ARMCO I-F
Armco
Steel.
Hot-rolled: 50 ksi (345 MPa) TS; 30 ksi (207 MPa) YS; 43 El; 54 Rock B. Cold-rolled: 45 ksi (310 MPa) TS; 22 ksi (152 MPa) YS; 44 El; 44 Rock B. Drawability.

ARMCO INGOT IRON
Armco Steel Corp.
C 0.012, Mn 0.017, P 0.005, S 0.025, bal Fe.
Treated: 38,000-45,000 TS; 20,000-25,000 YS; 40 El; 50 RA; 85 Brin. Annealed: 38,000 TS; 19,000 YS; 40 El; 50 RA; 85 Brin. For pipes, roofing, tanks, culverts, stoves, metal furniture; ductile. *Obsolete*

ARMCO INTERMEDIATE TRANSFORMER
Armco Steel Corp.
High Si, bal Fe.
For electrical purposes, transformers; for magnetic cores. *Obsolete*

ARMCO LO-TEMP, CLASS 1
Armco
C 0-0.2, Mn 0.7-1.35, Si 0.15-0.5, Cr 0-0.25, N 0-0.25, Mo 0.08, Cu 0-0.35, bal Fe.
1-1/4 in. max thickness: 70,000-90,000 TS; 50,000 YS; 18 El (8 in. gage). Weldable, fine-grain, good toughness to -75°F. For off-shore platforms, ships, process and pressure tanks, storage tanks. ASTM A537, Cl. 1. *Obsolete*

ARMCO LONG TERNES
Armco
Pb-Sn coated steel. Electronic chassis and fuel tanks. ASTM Standard A440.

ARMCO LTM
Armco
C 0-0.14, Mn 0.09-1.35, Si 0.15-0.3, bal Fe.
3/4 in. max thickness: 62,000-82,000 TS; 42,000 YS; 20 El (8 in. gage). For pressure vessels; fine grain; weldable. Good toughness to -75°F. for pipes, tanks, refrigerated containers. *Obsolete*

ARMCO LTM-N
Manufacturer not listed.
Normalized condition of ARMCO LTM: 42,000 psi min YS. ASME SA 662. ARMCO LTM QT. Quenched and tempered condition of ARMCO LTM: 50,000 psi min YS. ASTM A 678 Gr A (Structured).

ARMCO NI-COP
Armco
C 0-0.07, Mn 0.4-0.7, Ni 0.7-1, Cr 0.6-0.9, Cu 1-1.3, Mo 0.15-0.25, Cb 0.02, bal Fe.
General structural applications, usually when low ambient temperatures are encountered. *Obsolete*

ARMCO NITRONIC 32
Armco
Stainless steel. C 0-0.15, Si 0-1, Mn 11-14, P 0-0.06, S 0-0.03, Cr 16.5-19, Ni 0.5-2.5, N 0.2-0.45, bal Fe.
Annealed: 120,000 psi (827 MPa) TS; 65,000 psi (448 MPa) YS; 55 El; 70 RA; 9 Rock B. Non-magnetic, austenitic stainless for pole line hardware, springs, cold-headed parts.

ARMCO NITRONIC 33
Armco
Stainless steel. C 0-0.08, Si 0-1, Mn 11.5-14.5, P 0-0.06, S 0-0.03, Cr 17-19, Ni 2.25-3.75, N 0.2-0.4, bal Fe.
Annealed: 120,000 psi (827 MPa) TS; 65,000 psi (448 MPa) YS; 55 El; 70 RA; 96 Rock B. Austenitic stainless for heat exchangers, chemical equipment. ASTM A242 XM-29.

ARMCO NITRONIC 40
Armco
C 0-0.08, Si 0-1, Mn 8-10, P 0-0.06, S 0-0.03, Cr 19-21, Ni 5.5-5.7, N 0.15-0.4, bal Fe.
Annealed: 112,000 psi (772 MPa) TS; 68,000 psi (469 MPa) YS; 44 El; 96 Rock B. High tensile: 145,000 psi (1000 MPa) TS; 130,000 psi (896 MPa) YS; 20 El; 34 Rock C. Aircraft exhaust systems; cryogenic applications. Austenitic stainless. ASM 5656; 5595.

ARMCO NITRONIC 50
Armco
C 0-0.06, Si 0-1, Mn 4-6, P 0-0.04, S 0-0.03, Cr 20.5-23.5, Ni 11.5-13.5, Mo 1.5-3, N 0.2-0.4, Cb 0.1-0.3, V 0.1-0.3, bal Fe.
Annealed: 120,000 psi (862 MPa) TS; 65,000 psi (448 MPa) YS; 45 El; 65 RA; 23 Rock C. Austenitic stainless for petrochemical, pulp and paper, food processing industries. ASM 5861, 5764; ASTM A240 XM-19.

ARMCO NITRONIC 60
Armco
C 0-0.1, Si 3.5-4.5, Mn 7-9, Cr 16-18, Ni 8-9, N 0.08-0.18, bal Fe.
Austenitic stainless steel with good galling resistance.

ARMCO ORIENTED M-2
Armco
Iron-silicon electrical steel sheet with oriented properties; low core loss; thickness 7 mils.

ARMCO ORIENTED M-3
Armco
Iron-silicon electrical steel sheet with oriented properties; low core loss; thickness 9 mils.

ARMCO ORIENTED M-4
Armco
Iron-silicon electrical steel sheet with oriented properties; low core loss; thickness 11 mils.

ARMCO ORIENTED M-5
Armco
Iron-silicon electrical steel sheet with oriented properties; low core loss; thickness 11, 12 mils.

ARMCO ORIENTED M-6
Armco
Iron-silicon electrical steel sheet with oriented properties; low core loss; thickness 12, 14 mils.

ARMCO ORIENTED M-7
Armco Steel Corp.
Iron silicon electrical steel sheet with oriented properties; low core loss; thickness 14 mils. *Obsolete*

ARMCO ORIENTED M-8
Armco Steel Corp.
Iron silicon electrical steel sheet with oriented properties; low core loss. *Obsolete*

ARMCO ORIENTED T
Armco Steel Corp.
Thin iron silicon sheets, electrical steel, with directional magnetic properties; good permeability; for transformers, yoke cores. *Obsolete*

ARMCO ORIENTED T-S
Armco Steel Corp.
Thin iron-silicon sheets, electrical steel, with directional magnetic properties; designed for 400 cycle service at inductions above 16 kilogausses. *Obsolete*

ARMCO ORIENTED TG
Armco Steel Corp.
Thin iron-silicon sheets, electrical steel, with directional magnetic properties; designed for 400 Hz and higher frequency service. *Obsolete*

ARMCO PH 12 9 MO
Armco
C 0.035, Mn 0.002, Si 0.03, Cr 11.85, Ni 8.65, Mo 1.5, Al 1.65, N 0.005, bal Fe.
Precipitation hardening steel; experimental. *Obsolete*

ARMCO PH 13-8 MO
Armco
Stainless steel. C 0-0.05, Cr 12.25-13.25, Ni 7.5-8.5, Mo 2-2.5, Al 0.9-1.35, bal Fe.
Martensitic; precipitation hardenable. Condition H 950: 225,000 TS; 205,000 YS; 12 El. For springs, stampings, fasteners; better transverse properties than 17-7 PH.

ARMCO PH 14-8 MO
Armco
C 0.05, Cr 13.5-15.5, Ni 7.5-9.5, Mo 2-3, Al 0.75-1.25, bal Fe.
Semi-austenitic; precipitation hardened. Condition SRH 900: 230,000 TS; 215,000 YS; 6 El. For springs, stampings, watch and instrument parts, fasteners; stainless. *Obsolete*

ARMCO PH 15-7 MO
Armco
Stainless steel. C 0-0.09, Cr 14-16, Ni 6.5-7.75, Mo 2-3, Al 0.75-1.5, bal Fe.
Semi-austenitic; precipitation hardened. Annealed: 130,000 TS; 55,000 YS; 35 El. Hardened: 220,000-265,000 TS; 210,000-260,000 YS; 2-7 El; 45-50 Rock C. For springs, watch and instrument parts such as gears, cams, levers, fasteners.

ARMCO QTC
Armco
C 0-0.2, Mn 1.15-1.5, Si 0.2-0.5, Cr 0-0.2, Ni 0-0.15, Mo 0-0.08, 0.2 Cu min, 0.0015 B min, bal Fe.
3/4 in. max thickness: 100,000-120,000 TS; 80,000 YS; 18 El. Quenched and tempered, fine-grain, weldable; for off-shore platforms, ships, railroad cars. *Obsolete*

ARMCO R A TYPE 434A
Armco Steel Corp.
C 0-0.12, Cr 16-18, Si 1, Cu 1, bal Fe.
For corrosion resistant parts. *Obsolete*

ARMCO RADIO 1
Armco Steel Corp.
Si, bal Fe.
For small transformers. *Obsolete*

ARMCO RADIO 2
Armco Steel Corp.
Si, bal Fe.
For small transformers. *Obsolete*

ARMCO RADIO 3
Armco Steel Corp.
Si, bal Fe.
For small transformers. *Obsolete*

ARMCO RADIO 4
Armco Steel Corp.
Si, bal Fe.
For chokes, electrical purposes. *Obsolete*

ARMCO SPECIAL ELECTRIC
Armco Steel Corp.
Si 2.6-2.9, bal Fe.
For A. C. motors and generators, electrical purposes; for magnetic cores. *Obsolete*

ARMCO SSS 100 A514 A517 GRADE E
Armco
C 0-0.2, Mn 0.4-0.7, Si 0.2-0.35, Cr 1.4-2, Mo 0.4-0.6, Cu 0.2-0.4, B 0.0015-0.005, 0.04-0.1 Ti or V, bal Fe.
2-1/2 in. max thickness: 115,000-135,000 TS; 100,000 YS; 16-18 El. Quenched and tempered steel for bridges, buildings, crane booms, ships, earth moving equipment. ASTM A514, A517. *Obsolete*

ARMCO SSS 100A A 514 A517 GRADE D
Armco
C 0-0.2, Mn 0.4-0.7, Si 0.2-0.35, Cr 0.85-1.2, Ni 0.15-0.25, Cu 0.2-0.4, B 0.0015-0.005, 0.04-0.1 Ti or V, bal Fe.
1-1/4 in. max thickness: 115,000-135,000 TS; 100,000 YS; 16-18 El. Quenched and tempered steel for bridges, buildings, crane booms, ships, earth moving equipment. ASTM A514, A517. *Obsolete*

ARMCO SUPER-LO-TEMP, CLASS 2
Armco
C 0-0.2, Mn 0.7-1.35, Si 0.15-0.5, Cr 0-0.25, Ni 0-0.25, Mo 0-0.08, Cu 0-0.35, bal Fe.
1-1/4 in. max thickness: 80,000-100,000 TS; 60,000 YS; 22 El. Quenched and tempered, fine-grain, excellent low temperature properties. For refrigerated storage tanks. ASTM A537. *Obsolete*

ARMCO TI 40
Armco Steel Corp.
Essentially pure titanium.
Wrought: 50,000 TS; 40,000 YS; 20 El. For aircraft ducting, corrosion resisting parts in chemical industry. AMS 4902 (sheet). *Obsolete*

ARMCO TI 55
Armco Steel Corp.
Commercially pure titanium.
Wrought: 65,000 TS; 55,000 YS; 18 El. For anodizing racks, welding wire, aircraft industry, chemical industry. AMS 4900 (sheet). *Obsolete*

ARMCO TI 70
Armco Steel Corp.
Commercially pure titanium.
Wrought: 80,000 TS; 70,0000 YS; 15 El. For racks in chemical industry, anodizing racks, aircraft industry. AMS 4921 (bar). *Obsolete*

ARMCO TI-5AL-2 1/2 SN
Armco Steel Corp.
Al 4-6, Sn 2-3, bal Ti.
Wrought: 115,000 TS; 110,000 YS; 10 El. Forgeable, weldable, good creep resistance to 900 F; alpha alloy. For aircraft parts and low temperature equipment. AMS 4926, 4966. *Obsolete*

ARMCO TI-6AL-2SN-4ZR-6MO
Armco Steel Corp.
Al 6, Sn 2, Zr 4, Mo 2, bal Ti.
Heat treated: 130,000 TS; 120,000 YS; 10 El. Alpha-beta alloy; good high temperature creep strength; for jet engine components. AMS 4975. *Obsolete*

ARMCO TI-6AL-4V
Armco Steel Corp.

Al 5.5-6.5, V 3.5-4.5, bal Ti.

Wrought: 130,000 TS; 120,000 YS; 10 El. Alpha-beta alloy, weldable, heat treatable, good properties to 800 F; for compressor blades and wheels, spacers, cryogenic applications, rocket cases, marine equipment. AMS 4928; MIL-T-9047, Type III Comp A. *Obsolete*

ARMCO TI-6AL-6V-2SN
Armco Steel Corp.

Al 6, V 6, Sn 2, bal Ti.

Wrought: 140,000-150,000 TS; 130,000-140,000 YS; 10 El. Alpha-beta alloy; for rocket engine cases, ordnance and aircraft components. AMS 4978; MIL-T-9047. *Obsolete*

ARMCO TI-8AL-1MO-1V
Armco Steel Corp.

Al 7.5-8.5, Mo 0.75-1.25, V 0.75-1.25, bal Ti.

Wrought: 130,000 TS; 120,000 YS; 10 El. Alpha-beta alloy, good creep properties at elevated temperature; for jet engine and aircraft parts. AMS 4972. *Obsolete*

ARMCO TRAN COR H 2 (0 012)
Armco

Iron silicon electrical steel sheet with oriented properties; very low core loss and high permeability. For power and distribution transformers.

ARMCO TRAN COR H 3 (0 014)
Armco

Iron silicon electrical steel sheet with oriented properties; very low core loss and high permeability. For power and distribution transformers.

ARMCO TRAN-COR 58
Armco Steel Corp.

Si 4-4.4, bal Fe.

For distribution transformers, electrical purposes. *Obsolete*

ARMCO TRAN-COR 65
Armco Steel Corp.

Si 4-4.4, bal Fe.

For electrical purposes, power and distribution transformers; low core loss. *Obsolete*

ARMCO TRAN-COR 72
Armco Steel Corp.

Si 3-3.4, bal Fe.

For electrical purposes, large generators, transformers; for magnetic cores. *Obsolete*

ARMCO TRAN-COR A-6
Armco Steel Corp.

Hot rolled silicon sheet steel, nonoriented, low core loss transformer grade; for large magnetic amplifiers, saturable reactors. *Obsolete*

ARMCO TRAN-COR M-14
Armco Steel Corp.

Hot rolled silicon sheet steel, nonoriented, low core loss transformer grade; for distribution transformers and large power transformers. *Obsolete*

ARMCO TRAN-COR T
Armco Steel Corp.

Thin iron-silicon sheets, electrical steel, nonoriented; good permeability; for transformers, chokes, converters, generators. *Obsolete*

ARMCO TRAN-COR X
Armco Steel Corp.

Si, bal Fe.

For high frequency transformers. *Obsolete*

ARMCO TRAN-COR XX
Armco Steel Corp.

Si, bal Fe.

For high frequency transformers; 0.80 watt/lb maximum core loss. *Obsolete*

ARMCO TRAN-COR XXX
Armco Steel Corp.

Si, bal Fe.

For high frequency transformers; 0.73 watt/lb maximum core loss. *Obsolete*

ARMCO TRANCOR 52
Armco Steel Corp.

Si 4.7, bal Fe.

For transformers. *Obsolete*

ARMCO TYPE 301
Armco

Stainless steel. C 0.1-0.15, Cr 16-18, Ni 6-8, bal Fe.

Annealed: 90,000 TS; 45,000 YS; 55 El; 65 RA; 143 Brin. For chemical plant equipment; Type 301; austenitic.

ARMCO TYPE 302
Armco

Stainless steel. C 0.08-0.15, Cr 17-19, Ni 8-10, Si 2-3, bal Fe.

For heat resistant parts; heat and corrosion resistant.

ARMCO TYPE 303
Armco

Stainless steel. C 0-0.15, Cr 17-19, Ni 8-10, S 0-0.07, Zr 0-0.6, P, Se, Mo, bal Fe.

Cold drawn: 100,000 TS; 60,000 YS; 40 El; 53 RA; 212 Brin. Annealed: 90,000 TS; 35,000 YS; 50 El; 55 RA; 160 Brin. For screw machine products, hardware, bolts, nuts, screws; Type 303; free-cutting.

ARMCO TYPE 304
Armco

Stainless steel. C 0-0.08, Cr 18-20, Ni 8-10.5, bal Fe.

Austenitic; non-hardenable by heat treatment. Annealed: 85,000 TS; 35,000 YS; 60 El; 150 Brin. Good grade of 18-8 type stainless steel; for food, beverage, meat handling equipment, AISI 304.

ARMCO TYPE 304 L
Armco

Stainless steel. C 0-0.03, Cr 18-20, Ni 8-10, bal Fe.

Austenitic; non-hardenable by heat treatment. Annealed: 80,000 TS; 30,000 YS; 60 El; 140 Brin. For welded assemblies without subsequent heat treatment; for food, beverage, meat processing equipment. AISI 304L.

ARMCO TYPE 308
Armco

Stainless steel. C 0-0.08, Cr 19-21, Ni 10-12, bal Fe.

Annealed: 85,000 TS; 35,000 YS; 50 El; 65 RA; 165 Brin. For chemical and plastic plant equipment; austenitic.

ARMCO TYPE 308L
Armco

Stainless steel. Cr 19-21, Ni 10-12, C 0-0.03, bal Fe.

Rod and wire. Extra-low carbon variation of Type 308 eliminates harmful carbide precipitation during welding. Austenitic, non-hardenable by heat treatment.

ARMCO TYPE 309
Armco

Stainless steel. C 0-0.2, Cr 22-24, Ni 12-15, bal Fe.

Annealed: 90,000 TS; 40,000 YS; 45 El; 55 RA; 150 Brin. For furnace parts, heat treating boxes; Type 309; heat and corrosion resistant.

ARMCO TYPE 309S
Armco

Stainless steel. Cr 22-24, Ni 12-15, C 0-0.08, bal Fe.

Bar, rod and wire, billets. Similar to Type 309 but carbon lowered to minimize carbide precipitation and improve weldability. Welding wire. Austenitic, non-hardenable by heat treatment.

ARMCO TYPE 310S
Armco

Stainless steel. C 0-0.08, Cr 24-26, Ni 19-22, bal Fe.

Annealed: 95,000 TS; 40,000 YS; 45 El; 65 RA; 185 Brin. Cold drawn: 125,000 TS; 90,000 YS; 20 El; 60 RA; 220 Brin. For furnace equipment, chemical plant equipment; austenitic.

ARMCO TYPE 316
Armco

Stainless steel. C 0-0.1, Cr 16-18, Ni 10-14, Mo 2-3, bal Fe.

Annealed: 80,000 TS; 30,000 YS; 60 El; 70 RA; 150 Brin. Cold drawn: 90,000 TS; 60,000 YS; 45 El; 60 RA; 180 Brin. For chemical and textile plant equipment; Type 316; austenitic.

ARMCO TYPE 316 L
Armco

Stainless steel. C 0-0.03, Cr 16-18, Ni 10-14, Mo 2-3, bal Fe.

Annealed: 80,000 TS; 30,000 YS; 60 El; 70 RA; 140 Brin. Cold drawn: 90,000 TS; 60,000 YS; 45 El; 60 RA; 190 Brin. For chemical and textile plant equipment, welded structures; Type 316L; austenitic.

ARMCO TYPE 317
Armco

Stainless steel. Cr 18-20, Ni 11-15, C 0-0.08, Mo 3-4, bal Fe.

Pipe and tubing. Higher alloy content improves basic advantages of Type 316. Austenitic, non-hardenable by heat treatment.

ARMCO TYPE 317L
Armco

Stainless steel. Cr 18-20, Ni 11-15, C 0-0.03, Mo 3-4, bal Fe.

Bar, rod and wire, billets, pipe and tubing. Extra-low carbon version of Type 317 for welded structures. Austenitic, non-hardenable by heat treatment.

ARMCO TYPE 321
Armco

Stainless steel. C 0-0.08, Mn 0-2, Si 0-0.75, Cr 17-19, Ni 8-11, Ti = 5 x C, bal Fe.

Annealed: 90,000 TS; 35,000 YS; 55 El; 60 RA; 160 Brin. For high temperature service and welded corrosion resistant; stabilized, austenitic.

ARMCO TYPE 347
Armco

Stainless steel. Cr 17-19, Ni 9-13, C 0-0.08, Cb + Ta = 10 x C min, bal Fe.

Bar, rod and wire, billets, pipe and tubing. Characteristics similar to Type 321; stabilized by Cb and Ta. Austenitic, non-hardenable by heat treatment.

ARMCO TYPE 348
Armco

Stainless steel. C 0-0.1, Mn 0-1.5, Cr 17-19, Ni 9-12, Cb = 10 x C, bal Fe.

Annealed: 85,000 TS; 40,000 YS; 45 El; 65 RA; 140 Brin. Cold drawn: 110,000 TS; 65,000 YS; 40 El; 60 RA; 220 Brin. For welded structures, chemical plant equipment; austenitic.

ARMCO TYPE 403
Armco

Stainless steel. Cr 11.5-13, C 0-0.15, Si 0-0.5, bal Fe.

Bar, billets. Special high-quality variation of Type 410 for highly stressed parts. Martensitic, hardenable.

ARMCO TYPE 410 SE
Armco Steel Corp.

C 0-0.15, Cr 12-14, 0.15% min Se, bal Fe.

For corrosion resistant parts; Type 416 Se; corrosion resistant. *Obsolete*

ARMCO TYPE 410S
Armco

Stainless steel. C 0-0.15, Cr 11.5-13.5, bal Fe.

Annealed: 75,000 TS; 41,000 YS; 35 El; 70 RA; 155 Brin. Cold drawn: 110,000 TS; 85,000 YS; 23 El; 65 RA; 230 Brin. For pumps, valves, pressure vessels, motor shafts; corrosion resistant.

ARMCO TYPE 416
Armco

Stainless steel. C 0-0.15, Cr 12-14, P 0-0.07, Zr 0-0.6, S, Se, Mo, bal Fe.

Cold drawn: 100,000 TS; 85,000 YS; 13 El; 40 RA; 205 Brin. Annealed: 75,000 TS; 40,000 YS; 30 El; 60 RA; 105 Brin. For screw machine products, screws, shafts, gears; Type 416; corrosion resistant, free-cutting.

ARMCO TYPE 420
Armco
Cr 12-14, 0.15 C min, bal Fe.
Annealed: 95,000 TS; 50,000 YS; 25 El; 55 RA; 195 Brin. Cold drawn: 105,000 TS; 85,000 YS; 17 El; 50 RA; 215 Brin. For cutlery, ball bearings, surgical instruments; Type 420; corrosion resistant.

ARMCO TYPE 430
Armco
C 0-0.12, Cr 14-18, bal Fe.
Annealed: 75,000 TS; 50,000 YS; 30 El; 65 RA; 155 Brin. Cold drawn: 85,000 TS; 70,000 YS; 20 El; 60 RA; 185 Brin. For trim, hardware, fixtures; industrial equipment; Type 430; corrosion and heat resistant.

ARMCO TYPE 431
Armco
C 0-0.2, Cr 15-17, Ni 1.5-2.5, bal Fe.
Annealed: 125,000 TS; 95,000 YS; 20 El; 55 RA; 250 Brin. Cold drawn: 130,000 TS; 110,000 YS; 15 El; 35 RA; 270 Brin. For aircraft and marine parts, bolts; type 431; corrosion resistant.

ARMCO UNIVIT
Armco
C 0-0.008, Mn 0.2, bal Fe.
Annealed: 40,000 TS; 25,000 YS; 45 El; 35 Rock B. Decarburized enameling steel. For enameling drawn parts, appliances and plumbing.

ARMCO VNT
Armco
C 0-0.22, Mn 1.15-1.5, Si 0.15-0.5, V 0.04-0.11, N 0.01-0.03, bal Fe.
80,000-100,000 TS; 60,000 YS; 23 El. For pressure vessels, truck bodies, heavy machinery, construction, farm equipment. Weldable, good shock resistance to -75°F. ASTM A 633. *Obsolete*

ARMCO VNT N
Armco
C 0-0.22, Mn 1.15-1.5, Si 0.15-0.5, V 0.04-0.11, N 0.01-0.03, Cu 0-0.35, Cr 0-0.25, Ni 0-0.25, Mo 0-0.08, bal Fe.
Normalized: 80,000-100,000 TS; 60,000 YS min; 23 El (2 in.). For pressure vessels, truck bodies, farm equipment. Weldable; nil-ductility transition temperature -50°F. ASTM A633 Gr. E (Struc.). *Obsolete*

ARMCO X
Armco Steel Corp.
Fe alloy.
For transformers; oriented. *Obsolete*

ARMCO ZGPG ZINCGRIP-PAINTGRIP
Armco Steel Corp.
Galvanized sheet steel phosphated. For structures, sheet metal work; mill bonderized. *Obsolete*

ARMCO ZINCROMETAL
Armco
Zn-rich painted steel. Auto body components.

ARMELEC
Empire Sheet & Tin Plate Co.
Si 0.5, bal Fe.
For armatures, motors; good permeability.

ARMIDE
Armstrong Bros. Tool Co.
Carbide alloy TaC.
250,000 TS. For tipped cutters and tools. Hard, tough and wear resisting.

ARMIDE (350)
Armstrong Bros. Tool Co.
Sintered carbide tool material.
For cutting tools; light roughing and general finishing cuts on steel.

ARMIDE (370)
Armstrong Bros. Tool Co.
Sintered carbide tool material.
Cutting tools for heavy roughing cuts on steel.

ARMIDE (78)
Armstrong Bros. Tool Co.
Sintered carbide tool material.
For cutting tools, finishing and light roughing cuts on steel.

ARMIDE (78B)
Armstrong Bros. Tool Co.
Sintered carbide tool material.
For cutting tools, general purpose machining of steel.

ARMIDE (883)
Armstrong Bros. Tool Co.
Sintered carbide tool material.
For cutting tools, general purpose cutting of cast iron and non-ferrous metals.

ARMIDE GRAY
Armstrong Bros. Tool Co.
Carbide.
For tipped tools; for machining cast iron and brass.

ARMIDE RED
Armstrong Bros. Tool Co.
Carbide.
For tipped tools; for machining steel.

ARMIN 90
Stahlwerke Sudwestfalen
C 1.3, Cr 4.3, Mo 0.85, V 3.8, W 12, bal Fe.
For engravers' tools, blanking and forming dies; high speed steel. *Obsolete*

ARMORLOY
Now MCKAY ARMORLOY.

ARMORLOY A-5
Teledyne McKay
C 0-0.14, Mn 4-4.5, Cr 19, Ni 9, bal Fe.
Welded: 92,000 TS; 65,000 YS; 42 El. For welding electrodes for armor; shielded arc, work hardenable. *Obsolete*

ARMORLOY A-8
Teledyne McKay
Stainless steel. C 0.08, Mn 3.6, Si 0.5, Cr 20.2, Ni 9.4, Mo 1.15, bal Fe.
Covered welding electrode. 90,000 psi TS; 67,000 psi YS; 44 El. AWS E307-15. For armor plate and dissimilar metals.

ARMORLOY A-9
Teledyne McKay
Stainless steel. C 0.1, Mn 1.65, Si 0.45, Cr 20, Ni 9.5, Mo 2.1, bal Fe.
Covered welding electrode. 100,000 psi TS; 73,000 psi YS; 37 El.

ARMORLOY C
Teledyne McKay
Now MCKAY 310, WIRE

ARMSTEEL 88M
General Motors Corp./Central Foundry Div
C 2.45-2.75, Si 1.25-1.55, Mn 0.3-0.6, S 0.025-0.075, P 0-0.05, bal Fe.
Cast: 105,000 psi TS; 85,000 psi YS; 2 El; 302 Brin. For transmission gears and shafts, and universal joints; pearlitic malleable iron.

ARMSTRONG
American manufacture
Cr 12, Si 5, C 0.5, bal Fe.
For heat resisting steel parts; heat resisting stainless steel.

ARMSTRONG HIGH SPEED
Armstrong Bros. Tool Co.
C 0.7, W 18, Cr 4, V 1, bal Fe.
For tools, cutters. High speed steel.

ARMSTRONG METAL
English manufacture
C 0.1, Mn 4-6, Cr 17, Ni 8, Cu 3, bal Fe.
Annealed: 70,000 TS; 52 El; 71 RA; 126 Brin. For heat resisting parts; corrosion resistant.

ARMSTRONG SELF-HARDENING
Armstrong Bros. Tool Co.
C, alloy, bal Fe.
For tool shanks. Oil hardening.

ARMSTRONG-1
American manufacture
Cr 3-50, Si 0.5-8, C, bal Fe.
For heat and corrosion resisting parts; heat resisting.

ARMSTRONG-2
American manufacture
Cr 21.6, Si 3.4, C 2.35, Mn 2.25, bal Fe.
For heat and corrosion resisting parts; heat resistant cast iron.

ARMSTRONG-3
American manufacture
Cr 12, Si 5, C 0.45, bal Fe.
For heat and corrosion resisting parts; heat and corrosion resistant.

ARNE
Uddeholm Corp.
C 0.9, Mn 1.2, Cr 0.5, W 0.5, V 0.1, bal Fe.
For dies; for cold work. AISI O1.

ARNGRIM 2
Uddeholm Corp.
C 0.68, Cr 0.25, W 6, bal Fe.
For permanent magnets. *Obsolete*

ARNGRIM 3
Uddeholm Corp.
C 0.72, Cr 0.65, W 6, bal Fe.
For permanent magnets. *Obsolete*

ARNOX-I
Arnold Engineering Co.
Barium ferrite + iron oxide.
Coercive force 1800 oersted; residual induction 2200 gauss; peak energy product 1,000,000 max. For magnets in door latches, holding assemblies, DC motors, focusing devices. Permanent magnet, nonconducting. Nonoriented.

ARNOX-III
Arnold Engineering Co.
Oxides.
Residual induction 1500 gauss; coercive force 1100 oersted; peak energy product 400,000 max. For permanent magnets in toys and novelties, timing motors. Nonoriented. Molded type magnet. High coercive force and low residual induction and energy product.

ARNOX-V
Arnold Engineering Co.
Barium ferrite.
Residual induction 3950 gauss; coercive force 2200 oersted; peak energy product 3,500,000 max. For permanent magnets in loudspeakers, motors, generators, magnetic separation applications. Ceramic magnet. High coercive force. Magnetized in direction of orientation.

ARNOX-VA
Arnold Engineering Co.
Barium ferrite.
Maximum energy product 3,500,000. Remanence, residual induction 3850 gauss; coercive force 2000 oersted. For permanent nonconducting magnets. Hard brittle ceramic, high coercive force; low flux density.

ARO
Time Steel Service Inc.
C 0.5, Mn 0.7, Si 0.7, V 0.2, Mo 0.45, bal Fe.
Oil or water hardened shock resistant tool steel. AISI S2.

AROSTIT C15
Hoffman Elektrogusstahlwerk, Alb.
C 0.25, Cr 14.5, Ni 0-1, bal Fe.
Annealed: 95,000 TS; 50,000 YS; 25 El; 55 RA; 195 Brin. Cold
drawn: 105,000 TS; 85,000 YS; 17 El; 50 RA; 215 Brin. For
cutlery, valves, surgical instruments; Type 420; corrosion
resistant.

AROSTIT C13
Hoffman Elektrogusstahlwerk, Alb.
C 0-0.12, Si 0.4, Cr 13, bal Fe.
Annealed: 75,000 TS; 40,000 YS; 35 El; 70 RA; 180 Brin. For
turbine blades, chemical plant equipment; Type 410;
corrosion resistant.

AROSTIT C13S
Hoffman Elektrogusstahlwerk, Alb.
C 0-0.12, Cr 16.5, Mo 0.25, S 0.2, bal Fe.
Annealed: 70,000 TS; 40,000 YS; 30 El; 55 RA; 150 Brin. Cold
drawn: 130,000 TS; 120,000 YS; 2 El; 270 Brin. For chemical
plant equipment, bolts, oil refinery parts; Type 430F;
corrosion and heat resistant.

AROSTIT C18
Hoffman Elektrogusstahlwerk, Alb.
C 0.25, Cr 17, Ni 0-1, bal Fe.
Annealed: 75,000 TS; 42,000 YS; 28 El; 53 RA; 160 Brin. For
chemical plant equipment, bolts, shafts; Type 430; corrosion
and heat resistant.

AROSTIT C20
Hoffman Elektrogusstahlwerk, Alb.
C 0.2, Cr 13, bal Fe.
Annealed: 95,000 TS; 50,000 YS; 25 El; 55 RA; 195 Brin. Cold
drawn: 105,000 TS; 85,000 YS; 17 El; 50 RA; 215 Brin. For
turbine blades, cutlery, surgical instruments; Type 420;
corrosion resistant.

AROSTIT C20N
Hoffman Elektrogusstahlwerk, Alb.
C 0.2, Cr 13, Mo 1.15, bal Fe.
Annealed: 100,000 TS; 55,000 YS; 20 El; 50 RA; 205 Brin. For
turbine blades, cutlery, valves; corrosion resistant.

AROSTIT C22K
Hoffman Elektrogusstahlwerk, Alb.
C 0.22, Cr 17, Ni 1.5, bal Fe.
Annealed: 125,000 TS; 95,000 YS; 20 El; 55 RA; 260 Brin.
Cold drawn: 130,000 TS; 110,000 YS; 15 El; 35 RA; 270 Brin.
For pump shafts, marine hardware, valve trim; Type 431;
corrosion and heat resistant.

AROSTIT C22L
Hoffman Elektrogusstahlwerk, Alb.
C 0.35, Cr 16.5, Mo 1.15, bal Fe.
Annealed: 130,000 TS; 100,000 YS; 18 El; 52 RA; 275 Brin.
For pump shafts, marine hardware, valve trim; corrosion
resistant.

AROSTIT C30
Hoffman Elektrogusstahlwerk, Alb.
C 0.4, Cr 29, Si 1.3, bal Fe.
For furnace parts and equipment, heat treat boxes; heat
resistant, ferritic.

AROSTIT C30H
Hoffman Elektrogusstahlwerk, Alb.
C 1.2, Si 1.3, Cr 29, bal Fe.
For wear plates, rolls, crushers; corrosion and heat resistant.

AROSTIT C30N
Hoffman Elektrogusstahlwerk, Alb.
C 1.3, Si 1.3, Cr 29, Mo 2, bal Fe.
For wear plates, rolls, crushers; corrosion and heat resistant.

AROSTIT C40
Hoffman Elektrogusstahlwerk, Alb.
C 0.4, Cr 13, bal Fe.
Annealed: 100,000 TS; 55,000 YS; 22 El; 52 RA; 200 Brin.
Cold drawn: 110,000 TS; 90,000 YS; 15 El; 45 RA; 225 Brin.
For turbine blades, gages, valve trim, cutlery; Type 420;
corrosion resistant.

AROSTIT C90N
Hoffman Elektrogusstahlwerk, Alb.
C 0.9, Cr 18, Mo 1.5, V 1, bal Fe.
Annealed: 107,000 TS; 62,000 YS; 18 El; 35 RA; 220 Brin. For
cutlery, valve parts, bearings, instruments; Type 440B;
corrosion resistant, hardenable.

AROSTIT CK12
Hoffman Elektrogusstahlwerk, Alb.
C 0-0.07, Cr 17, Mo 4.7, Ni 13, bal Fe.
Annealed: 90,000 TS; 40,000 YS; 45 El; 65 RA; 170 Brin. For
acid resistant equipment, mixers, agitators; Type 317;
stainless, austenitic.

AROSTIT CK4
Hoffman Elektrogusstahlwerk, Alb.
C 0.4, Si 1.3, Cr 27, Ni 4, bal Fe.
For furnace equipment, heat treat boxes; heat resistant.

AROSTIT CK8
Hoffman Elektrogusstahlwerk, Alb.
C 0.15, Si 1.5, Cr 18, Ni 8.5, bal Fe.
Annealed: 80,000 TS; 35,000 YS; 50 El; 65 RA; 160 Brin. For
chemical plant equipment, tanks, mixers; Type 302; stainless,
austenitic.

AROSTIT CK8 EXTRA
Hoffman Elektrogusstahlwerk, Alb.
C 0-0.12, Cr 18, Ni 9.5, Ti = 4 x C, bal Fe.
Annealed: 85,000 TS; 35,000 YS; 55 El; 65 RA; 150 Brin. For
welded structures, chemical plant equipment; Type 321;
stainless, austenitic.

AROSTIT CK8SI
Hoffman Elektrogusstahlwerk, Alb.
C 0-0.1, Cr 18, Ni 8.5, bal Fe.
Annealed: 85,000 TS; 35,000 YS; 60 El; 70 RA; 140 Brin. For
welded structures, chemical plant equipment; Type 302;
stainless, austenitic.

AROSTIT CK9N
Hoffman Elektrogusstahlwerk, Alb.
C 0.15, Si 2, Mo 2, Ni 9.5, Cr 18, bal Fe.
Annealed: 80,000 TS; 30,000 YS; 60 El; 80 RA; 135 Brin. Cold
drawn: 150,000 TS; 135,000 YS; 6 El; 300 Brin. For acid
resistant equipment, mixers, filters, agitators; Type 316;
stainless, austenitic.

AROSTIT CK9N EXTRA
Hoffman Elektrogusstahlwerk, Alb.
C 0-0.12, Cr 18, Ni 10.5, Mo 2, Ti = 4 x C, bal Fe.
Annealed: 85,000 TS; 35,000 YS; 50 El; 75 RA; 160 Brin. Cold
drawn: 150,000 TS; 135,000 YS; 6 El; 30 RA. For welded
structures, chemical plant equipment; Type 316 Ti; Stainless
austenitic.

ARPOCALLOY
Arpocalloy Co.
C, Ni, Mo, bal Fe.
For tool shanks; cast iron.

ARQ 280, ARQ 320, ARQ 360
British Steel plc
C 0-0.22, Si 0.15-0.35, Mn 0-1.6, P 0-0.03, S 0-0.02, Mo 0-0.2,
Nb 0-0.06, Ti 0-0.05, B 0-0.003, bal Fe.
Quenched and tempered: 750 (ARQ 280) to 1150 (ARQ 360)
N/mm^2 TS; 280 (ARQ 280) to 360 (ARQ 360) Brin. For
dumper trucks, mining machinery, and earth moving
equipment. Structural plate; high abrasion resistance.

ARRESTITE
Republic Steel Corp.
C 0.9, Mn 1.5, Cr 0.2, bal Fe.
For general tools; non-deforming. *Obsolete*

ARROW
Latrobe Steel Co.
C 0.3, Cr 1, Mo 0.6, V 0.2, bal Fe.
For tools, dies; water hardening. *Obsolete*

ARROW (LATROBE)
Latrobe Steel Co.
C 0.26, Cr 0.9, V 0.2, bal Fe.
For tools, arbors, shafts, gears. *Obsolete*

ARROW HIGH SPEED
John A. Crowley Inc.
C 0.7, W 18, Cr 4, V 2, bal Fe.
For tools, dies, broaches, drills; high speed steel.

ARROW NO. 16
Manufacturer not listed.
C 0.95, Mn 1.2, Cr 0.5, W 0.5, V 0.1, bal Fe.
For tools and dies; non-deforming oil hardening.

ARROW OIL HARDENING
Boyd-Wagner Co.
C 0.6, W 1.5, Mn 0.9, Cr 3, V 1, bal Fe.
88,800 TS; 55,100 YS; 30 El; 189 Brin. For tools, dies,
perforated dies, taps, intricate shapes; nonshrinking.

ARROW OIL HARDENING
Manufacturer not listed.
C 0.7, W 2, bal Fe.
For tools, dies; oil hardened.

ARROW OIL HARDENING SWEDISH STEEL
Boyd-Wagner Co.
C 0.95, Mn 1.25, Cr 0.5, W 0.5, V 0.1, bal Fe.
For thread gages, taps, master tools, dies; nonshrinking.

ARROW OIL HARDENING SWEDISH STEEL
Boyd-Wagner Co.
C 0.95, Mn 1.25, Cr 0.5, W 0.5, V 0.1, bal Fe.
For thread gages, taps, master tools, dies; nonshrinking.

ARROW SPECIAL
Manufacturer not listed.
C 1.1, bal Fe.
For tools, dies.

ARS
Wallace Murray Corp.
C 2.35, Cr 12, V 4, Mo 1, bal Fe.
Air or oil hardening, cold work tool steel, chromium type; AISI
D7.

ARSENIC
Atomergic Chemetals Corp.
As (crystalline and amorphous).
Purities: 99.9999%, 99.999%, 99.99%, commercial grade.
Forms: lump, powder.

ARSENIC BRONZE
American manufacture
Cu 80, Sn 10, Pb 9.2, As 0.8.
30,000 TS; 60 Brin. For bearings; heavy duty.

ARSENICAL ADMIRALTY 30
Olin Brass, Indianapolis
Copper. Cu 71, Sn 1, As 0.03, Zn 28.
Annealed: 53,000 psi TS; 65 El. Drawn: 85,000 psi TS; 10 El.
For condenser and heat exchanger; embrittlement-free.

ARSENICAL ADMIRALTY-443
Anaconda Co.
Copper. Cu 71, Zn 27.96, Sn 1, As 0.04.
Soft: 48,000 TS; 18,000 YS; 65 El; 25 Rock B. For condenser
tubes, heat exchangers. Resists dezincification; corrosion
resistant.

ARSENICAL ALUMINUM BRASS 54
Olin Brass, Indianapolis
Copper. Cu 76, Al 2, As 0.03, Zn 22.
Annealed: 60,000 psi TS; 27,000 psi YS; 55 El. For condensers and heat exchangers; free of embrittlement.

ARSENICAL ALUMINUM BRONZE 53
Olin Brass, Indianapolis
Copper. Al 5.5, As 0.25, Cu 94.25.
Annealed: 60,000 psi TS; 22,000 psi YS; 60 El. For condenser tubes; resists seawater corrosion.

ARSENICAL COPPER
Chase Brass & Copper Co.
Cu 99.7, As 0.2, P 0.02.
Annealed: 34,000 TS; 10,000 YS; 45 El. Drawn: 55,000 TS; 50,000 YS; 8 El. For condenser, heat exchanger, boiler tubes; corrosion resisting. *Obsolete*

ART CASTERS BRASS
Belmont Metals Inc.
Pb 0.1-4, Si 0.1-4, Zn 30-35, bal Cu.
63,000 psi TS. For jewelry and decorative castings; melting point 1625-1650°F.

ART DIE
Columbia Tool Steel Co.
C 0.95, Cr 0.3, V 0.02, Mn 0.25, bal Fe.
Annealed: 170 Brin. For jewelers' dies; deep hardening. *Obsolete*

ARTISAN
J.F. Jelenko & Co.
Precious metal. Au 69, Pd 18.5, Ag 9.
96,000 psi TS; 75,000 psi YS; 6 El; 190 Brin. Dental alloy for fusing porcelain to metal.

ARZADE
Arcade Malleable Iron Co.
C 3, Si 1.5, bal Fe.
For castings; malleable cast iron.

ARZITE
Arcade Malleable Iron Co.
C 3, Si, Mn, bal Fe.
For gears, fittings, housings; malleable iron.

ARZITE METAL
Arcade Malleable Iron Co.
C 3.2, Si 1.5, Ni 1.5, Cr 0.8, bal Fe.
For castings; corrosion resistant malleable iron.

ARZON METAL
Arcade Malleable Iron Co.
Si 0.9, Mn 0.85, C 2.3, S 0.09, P 0.16, bal Fe.
Cast: 90,000 TS; 80,000 YS; 5 El. For malleable iron castings, fittings, plumbing; close grain; white fracture.

AS 05
Otto Fuchs Metallwerke
Aluminum. Mg 0.5, Si 0.5, bal Al.
130-245 N/mm^2 TS; 70-195 N/mm^2 YS; 10-15 El. Heat treatable, corrosion resistant, weldable and good anodizing characteristics. Type AlMgSi0.5. Similar to AA 6063.

AS 08
Otto Fuchs Metallwerke
Aluminum. Mg 0.8, Si 0.8, bal Al.
195-275 N/mm^2 TS; 100-195 N/mm^2 YS; 12-16 El. Heat treatable, corrosion resistant, weldable with good anodizing characteristics. Type AlMgSi0.8.

AS 10
Soc. Alluminio Veneto per Azioni
Aluminum. Cu 2-2.5, Si 9.5-10.5, Mg 1.4, Ni 0.8-1.2, bal Al.
Cast: 36,000-50,000 TS; 34,000-42,000 YS; 0.3-0.5 El; 95-125 Brin. For light alloy parts; corrosion resistant.

AS 10
Otto Fuchs Metallwerke
Aluminum. Mg 0.9, Si 1, Mn 0.6, Cr, bal Al.
195-313 N/mm^2 TS; 100-225 N/mm^2 YS; 6-18 El. Heat treatable, corrosion resistant, weldable with good anodizing characteristics. Type AlMgSi1.

AS 15
Otto Fuchs Metallwerke
Aluminum. Mg 0.8, Si 1, Mn 0.5, bal Al.
195-315 N/mm^2 TS; 100-255 N/mm^2 YS; 6-18 El. Heat treatable, corrosion resistant, weldable with good anodizing characteristics. Type AlMgSi1.

AS 17
Otto Fuchs Metallwerke
Aluminum. Mg 0.8, Si 1, Mn 0.6, Cr, Be, bal Al.
195-315 N/mm^2 TS; 100-255 N/mm^2 YS; 6-18 El. Special mining quality. Type AlMgSi1Be.

AS 20
Otto Fuchs Metallwerke
Aluminum. Cu 0.25, Mg 0.9, Si 0.7, Cr, bal Al.
193-315 N/mm^2 TS; 100-255 N/mm^2 YS; 6-18 El. Heat treatable, corrosion resistant, weldable with good anodizing characteristics. Type AlMgSi1Cu. Similar to AA 6061.

AS 303
Comalco Ltd.
Cu 2-4, Si 4-6, Fe 0-0.08, Mn 0-0.07, Zn 0-0.5, Ti 0-0.2, Sn 0-0.05, Mg 0-0.15, Ni 0-0.3, bal Al.
Sand cast, T6: 248 MPa TS; 138 MPa YS; 1 El; 100 Brin. ADC H49-3; BS LM4.

AS 305
Comalco Ltd.
Cu 2-4, Si 8.5-10.5, Fe 0-0.9, Mn 0-0.5, Mg 0.6-1.5, Ni 0-0.5, Zn 0-1, Ti 0-0.25, bal Al.
Permanent mold cast, T6: 248 MPa TS; 193 MPa YS; 1 El; 105 Brin.

AS 307
Comalco Ltd.
Cu 0.7-2.5, Si 9-11.5, Fe 0-1, Mn 0-0.5, Mg 0-0.3, Ni 0-1, Zn 0-1.2, Ti 0-0.2, Sn 0-0.2, bal Al.
Pressure cast, F1: 248 MPa TS; 3 El; 80 Brin. ADC H49-2; BS LM2.

AS 313
Comalco Ltd.
Cu 3-4, Si 7.5-9.5, Fe 0-1.3, Mn 0-0.5, Mg 0-0.3, Ni 0-0.5, Sn 0-0.2, Zn 0-3, Ti 0-0.2, bal Al.
Pressure cast, F1: 269 MPa TS; 159 MPa YS; 2 El, 85 Brin. ADC H49-16; BS LM24.

AS 315
Comalco Ltd.
Cu 3-4.5, Si 10.5-12, Fe 0-1.3, Mn 0-0.5, Mg 0-0.1, Ni 0-0.5, Zn 0-1, Ti 0-0.35, Sn 0-0.35, bal Al.
Pressure cast, F1; 331 MPa YS; 165 MPa YS; 2.5 El. SAE 303.

AS 317
Comalco Ltd.
Cu 1.5-2.5, Si 8-8, Fe 0-0.08, Mn 0.2-0.8, Mg 0-0.3, Ni 0-0.5, Zn 0-1, Ti 0-0.2, Sn 0-0.1, bal Al.
Sand cast, HT-T6: 280 MPa TS min; 250 MPa YS; 0.5 El min; 110 Brin. BS LM27.

AS 401
Comalco Ltd.
Cu 0-0.6, Si 11-13, Fe 0-1, Mn 0-0.5, Zn 0-0.4, Ti 0-0.2, Sn 0-0.15, Mg 0-0.15, Ni 0-0.5, bal Al.
Pressure cast, F1: 262 MPa TS; 131 MPa YS; 2 El, 85 Brin. BS LM20.

AS 41 A (INGOT)
Dow Chemical Co.
Magnesium. Al 3.7-4.8, Mn 0.22-0.48, Zn 0-0.1, Si 0.6-1.4, Cu 0-0.04, Ni 0-0.002, 0.30 others max, bal Mg.
Mg ingots to be used for die castings.

AS 41A (DIE CASTING)
Various foundries
Al 3.5-5, Mn 0.2-0.5, Zn 0-0.12, Si 0.5-1.5, Cu 0-0.06, Ni 0-0.03, 0.30 others max, bal Mg.
As cast: 32,000 psi TS; 22,000 psi YS; 4 El.

AS 5
Soc. Alluminio Veneto per Azioni
Aluminum. Cu 1-3, Si 5, Mg 0.5, bal Al.
Heat treated: 50,000-57,000 TS; 40,000-46,000 YS; 2-5 El; 110-140 Brin. For light alloy parts; age hardenable.

AS 55
Manufacturer not listed.
W 5, Zr 1, Cr 0.05, C 0.06, Y, bal Cb.
For jet engines, missile components.

AS 55
General Electric Co.
W 5, Zr 1, Cr 0.05, C 0.06, Y, bal Cb.
For jet engines, missile components.

AS 601
Comalco Ltd.
Cu 0.25, Si 6.5-7.5, Fe 0-0.5, Mn 0-0.35, Mg 0.3-0.5, Zn 0-0.35, Ti 0-0.25, bal Al.
Sandcast, T6: 228 MPa TS; 152 MPa YS; 4 El; 75 Brin. ADC H49-18. BS LM25.

AS 607
Comalco Ltd.
Cu 0-0.1, Si 10-13, Fe 0-0.6, Mn 0.3-0.7, Mg 0.2-0.6, Ni 0-0.1, Zn 0-0.1, Ti 0-0.2, Sn 0-0.05, bal Al.
Sand cast, T6: 240 MPa TS min; 230 MPa YS; 0.1 El; 95 Brin. BS LM9.

AS 9
Otto Fuchs Metallwerke
Aluminum. Mg 0.4, Si 10, bal Al.
For welding rod.

AS NO. 46 GREEN LABEL
Ackerlind Steel Co., Inc.
C 0.9, Mn 1.1, Cr 0.5, W 0.5, V 0.2, bal Fe.
For gages, spindles, punches, forming and trimming dies; Type O1; non-deforming, oil hardened. *Obsolete*

AS41XB (INGOT)
Dow Chemical Co.
Magnesium. Al 3.7-4.8, Mn 0.35-0.6, Zn 0-0.1, Si 0.6-1.4, Cu 0-0.015, Ni 0-0.001, Fe 0-0.0035, Be 0.0005-0.0015, 0.01 others max, bal Mg.
Mg ingots to be used for die castings.

ASARCO ZDC NO. 7
American Smelting & Refining Co.
Al 3.5-4.3, Cu 0-0.25, Mg 0.005-0.02, Ni 0.005-0.02, Cd 0-0.003, bal Zn.
Die cast: 41,000 psi TS; 10 El; 82 Brin. For hardware, housings, instrument cases, die castings.

ASARCOLO 117
American Smelting & Refining Co.
Pb 23, Cd 5, Bi 45, Sn 8, In 19.
Cast: 5400 psi TS; 12 Brin. For hermetic seals, heat transfer medium, fusible cores; 117°F MP.

ASARCOLO 158
American Smelting & Refining Co.
Pb 27, Cd 10, Bi 50, Sn 13.
6000 psi TS; 9 Brin. For foundry cores, pipe bending, anchoring. 158°F MP; fusible alloy.

ASARCOLO 158-162
American Smelting & Refining Co.
Bi 50, Pb 25, Cd 12.5, Sn 12.5.
For fire protection devices; fusible alloy; 158-162°F MP.

ASARCOLO 158-190
American Smelting & Refining Co.
Bi 42.5, Pb 37.7, Sn 11.3, Cd 8.5.
5500 psi TS; 9 Brin. For foundry patterns, coating core boxe spotting fixtures. 158-190°F MP; fusible alloy.

ASARCOLO 205
American Smelting & Refining Co.
Bi 52, Pb 32, Sn 16.
For fire protection devices; fusible alloy; 205°F MP.

ASARCOLO 205-215
American Smelting & Refining Co.
Bi 50, Pb 31, Sn 19.
For fire protection devices; fusible alloy; 205-215°F MP.

ASARCOLO 217-440
American Smelting & Refining Co.
Pb 28.5, Sn 14.5, Sb 9, Bi 48.
13,000 psi TS; 19 El. For anchoring punches in stamping dies, making chuck jaws; 217-440°F MP; fusible alloy.

ASARCOLO 243-260
American Smelting & Refining Co.
Indium alloy.
For soldering and sealing; 243-260°F MP; fusible alloy.

ASARCOLO 255
American Smelting & Refining Co.
Pb 45, Bi 55.
6400 psi TS; 10 Brin. For foundry molds dies, proof casting cavities; 255°F MP; fusible alloy.

ASARCOLO 255-300
American Smelting & Refining Co.
Bi 50, Pb 50.
For fire protection devices; fusible alloy; 255-300°F MP.

ASARCOLO 281
American Smelting & Refining Co.
Bi 58, Sn 42.
8000 psi TS; 22 Brin. For molds and patterns; 281°F MP; fusible alloy.

ASARCOLO 281-338
American Smelting & Refining Co.
Bi 40, Sn 60.
8000 psi TS; 22 Brin. For soldering and sealing; 281-338°F MP; fusible alloy.

ASARCOLO 362
American Smelting & Refining Co.
Pb 38, Sn 62.
For fire protection devices; fusible alloy; 362°F MP.

ASARCOLOY NO. 7
American Smelting & Refining Co.
Ni 0.75-3, bal Cd.
Cast: 164,000 TS; 117,000 YS; 19 El; 43 RA; 33 Brin. For bearings; for severe service.

ASARCON 100
American Smelting & Refining Co.
Sn 10, Zn 2, Cu 88.
Cast: 51,000 TS; 28,000 YS; 18 El; 92 Brin. For gears, bearings, liners, sleeves; SAE 62; hard bronze.

ASARCON 1010
American Smelting & Refining Co.
Sn 10, Pb 10, Cu 80.
Cast: 41,000 TS; 26,000 YS; 10 El; 80 Brin. For bearings, bushings, liners; SAE 64 leaded bronze.

ASARCON 1010B
American Smelting & Refining Co.
Sn 10, Pb 10, bal Cu.
Cast: 45,000 psi TS; 29,000 psi YS; 5 El; 93 Brin. For gears, bearings, bushings, shafts; continuous cast.

ASARCON 102
American Smelting & Refining Co.
Sn 10, Pb 2, Cu 88.
Cast: 49,000 TS; 25,000 YS; 18 El; 86 Brin. For bushings, bearings, sleeves; SAE 63; leaded gun metal.

ASARCON 102N
American Smelting & Refining Co.
Cu 84, Sn 10, Pb 2.5, Ni 3.5, bal Zn.
Cast: 53,000 TS; 31,000 YS; 15 El; 92 Brin. For hardware, marine parts, liners; free-cutting, leaded bronze.

ASARCON 110
American Smelting & Refining Co.
Sn 11, Cu 89.
Cast: 51,000 TS; 29,000 YS; 18 El; 100 Brin. For gears, worms, shafts; Al-bronze; continuous cast.

ASARCON 210
American Smelting & Refining Co.
Sn 2.5, Pb 10, Zn 7.5, Cu 80.
Cast: 34,000 TS; 18,000 YS; 22 El; 62 Brin. For bearings, bushings, liners; leaded bronze.

ASARCON 230
American Smelting & Refining Co.
Sn 1.5-2.5, Pb 27-35, Ni 0.25-0.75, bal Cu.
Cast: 16,300 psi TS; 9000 psi YS; 10 El; 30 Brin. For self-lubricating bearings, bushings and seals, jet engine fuel pumps. Leaded bronze; free-machining.

ASARCON 3
American Smelting & Refining Co.
Al 4.15, Mg 0.5, bal Zn.
Die cast: 40,000 psi TS; 63,000 psi YS; 8 El; 80 Brin. For d castings, hardware, door knobs, instrument cases, coin chutes, ornaments. High fluidity and castability.

ASARCON 310 N
American Smelting & Refining Co.
Cu 77, Sn 3, Pb 10, Zn 8, Ni 2.
Cast: 48,000 TS; 28,000 YS; 8 El; 80 Brin. For bearings. Continuous cast.

ASARCON 50N
American Smelting & Refining Co.
Cu 88, Sn 5, Zn 2, Ni 5.
H: 80,000 TS; 55,000 YS; 10 El; 180 Brin.

ASARCON 520
American Smelting & Refining Co.
Sn 5, Pb 20, Cu 75.
Cast: 28,700 TS; 22,800 YS; 8 El; 57 Brin. For bearings, bushings, liners; leaded bronze.

ASARCON 55
American Smelting & Refining Co.
Sn 5, Pb 5, Zn 5, Cu 85.
Cast: 45,000 TS; 21,400 YS; 28 El; 72 Brin. For water pump impellers, bushings, fittings; SAE 40; leaded red brass.

ASARCON 59
American Smelting & Refining Co.
Sn 5, Pb 9, Zn 1, bal Cu.
Cast: 38,000 TS; 21,000 YS; 20 El; 66 Brin. For bearings, bushings, sleeves; SAE 66; leaded bronze.

ASARCON 61
American Smelting & Refining Co.
Sn 6, Pb 1.5, Zn 4.5, Cu 88.
Cast: 45,500 TS; 23,000 YS; 35 El; 76 Brin. For gears, bearings, bushings; SAE 622; continuous cast.

ASARCON 616
American Smelting & Refining Co.
Cu 78, Sn 6, Pb 16.
34,200 TS; 23,000 YS; 12 El; 62 Brin.

ASARCON 7
American Smelting & Refining Co.
Al 3.5-4.3, Cu 0-0.25, Mg 0.01-0.02, Ni 0.005-0.02, Cd 0-0.003, bal Zn.
Die cast: 41,000 psi TS; 14 El; 76 Brin. For die castings, hardware, instrument cases, ornaments, coin chutes. High fluidity and castability.

ASARCON 77
American Smelting & Refining Co.
Sn 7, Pb 7, Zn 3, Cu 83.
Cast: 40,000 TS; 27,000 YS; 16 El; 72 Brin. For bearings, spring bushings, thrust washers; SAE 660; leaded bronze.

ASARCON 773
American Smelting & Refining Co.
Sn 7, Pb 7, Zn 3, Cu 83.
Cast: 40,000 TS; 27,000 YS; 16 El; 72 Brin. For bearings, bushings, liners; SAE 660; leaded bronze.

ASARCON 80
American Smelting & Refining Co.
Sn 8.
Cast: 49,000 TS; 23,000 YS; 18 El; 77 Brin. For gears, bearings, sleeves; SAE 620; hard bronze.

ASC 1
Atlantic Steel Casting Co.
C 0.25-0.3, Mn 0.7, Si 0.4, bal Fe.
Annealed: 70,000 TS; 40,000 YS; 24 El; 150 Brin. For castings for construction, railroad.

ASC 11
Atlantic Steel Casting Co.
C 0.4, Mn 1.15, Si 0.4, Cr 1.15, Mo 0.4, bal Fe.
Physical properties depend on heat treatment. High strength castings. Similar to wrought AISI 4140.

ASC 15
Atlantic Steel Casting Co.
C 0.4, Mn 1.15, Si 0.4, Cr 1.15, Ni 1.75, Mo 0.4, bal Fe.
Physical properties depend on heat treatment. High strength casting dies. Similar to wrought AISI 4340.

ASC 2
Atlantic Steel Casting Co.
C 0.4-0.5, Mn 0.7, Si 0.4, bal Fe.
Annealed: 80,000 TS; 40,000 YS; 18 El; 170 Brin. For castings for dies. Flame hardenable, wear resistant.

ASC 3
Atlantic Steel Casting Co.
C 0.6-0.8, Mn 0.7, Si 0.4, bal Fe.
Normalized and tempered: 110,000 TS; 60,000 YS; 10 El; 190 Brin. For castings for dies. Flame hardenable, wear resistant.

ASC 5
Atlantic Steel Casting Co.
C 0.2, Mn 0.7, Si 0.5, Cr 1.5, Mo 1.1, bal Fe.
Normalized and tempered: 70,000 TS; 40,000 YS; 20 El; 160 Brin. For turbine casings and supports, high temperature steel castings.

ASC 6
Atlantic Steel Casting Co.
C 0.18, Mn 0.6, Cr 2.5, Mo 1.1, bal Fe.
Normalized and tempered: 70,000 TS; 40,000 YS; 20 El; 160 Brin. For turbine casings and supports, high temperature steel castings.

ASC 7
Atlantic Steel Casting Co.
C 0.25, Mn 0.6, Si 0.5, Mo 0.6, bal Fe.
Normalized and tempered: 70,000 TS; 45,000 YS; 22 El; 160 Brin. For turbine casings and supports, high temperature steel castings.

ASC 8
Atlantic Steel Casting Co.
C 0.3, Mn 1.3, Si 0.5, Mo 0.4, bal Fe.
Normalized and tempered: 90,000 TS; 60,000 YS; 180 Brin.
Quenched: 200,000 TS; 150,000 YS; 400 Brin. High strength
casting dies. Weldable, water hardenable.

ASC-B10
Commerce Pattern Co.
Copper. Be 0.4, Co 2.6, bal Cu.
Annealed: 50,000 TS; 25,000 YS; 30 El; 10 RA; 240 Brin. For
current carrying springs and switch parts. Heat treatable;
corrosion and wear resistant.

ASC-B165
Commerce Pattern Co.
Copper. Be 1.7, Co 0.2-0.35, bal Cu.
Annealed: 60,000 TS; 28,000 YS; 60 El; 79 Brin. Aged:
200,000 TS; 185,000 YS; 1 El; 410 Brin. For electrical
contacts, springs, slips, and diaphragms. Heat treatable;
corrosion and wear resistant.

ASC-B25
Commerce Pattern Co.
Copper. Be 1.9, Co 0.3, bal Cu.
Annealed: 75,000 TS; 30,000 YS; 50 El; 70 RA; 380 Brin.
Aged: 205,000 TS; 170,000 YS; 1 El; 3 RA; 380 Brin. For
electrical contacts, springs, and diaphragms. Age
hardenable; corrosion resistant.

ASCO
Pasminco Europe (Mazak) Ltd.
C 1.2-1.7, bal Fe.
For general tools; water hardening.

ASCOLOY NO. 66
Allegheny Ludlum Steel
C 0.12, Si 0.43, Mn 0.4, Cr 18.5, bal Fe.
Hot rolled: 65,000-85,000 TS; 45,000-55,000 YS; 25-35 El; 60
RA; 150-160 Brin. For furnace parts; changed to "Allegheny";
corrosion and heat resistant. *Obsolete*

ASGM
Pechiney/SMG
Si 0.7-1.3, Fe 0-0.5, Mn 0.4-1, Mg 0.6-1.2, bal Al.
Wrought aluminum alloy, AA 6082. For production of high
pressure gas containers.

ASH-21
Allen-Sherman-Hoff Pump Co.
C 3-3.6, Ni 4-4.75, Cr 1.4-3.5, Si 0.4-0.7, bal Fe.
Sand cast: 45,000 TS; 600 Brin. Permanent mold: 55,000 TS;
675 Brin. For cams, dies, rollers, bearing races; white cast
iron, corrosion resistant, hard.

ASHBERRY METAL-A
English manufacture
Sn 80, Zn 1, Sb 14, Cu 2, Ni 2, Al.
For tableware and utensils; same as "Brittania."

ASHBERRY METAL-B
English manufacture
Sn 78-80, Zn 0-2.8, Sb 14-19, Cu 0-3.
For utensils, bearings; Babbitt.

ASHBERRY METAL-C
English manufacture
Sn 80, Sb 14, Cu 2, Zn 1, Ni 3.
For utensils, bearings; Babbitt.

ASHBERRY METAL-D
English manufacture
Sn 79, Sb 15, Cu 3, Zn 2, Ni 1.
For utensils, bearings; Babbitt.

ASM-122
Acieries de Sambra & Meuse
C 0-0.05, Cr 13, Mo 0.5, bal Fe.
Annealed: 95,000 TS; 50,000 YS; 25 El; 92 Rock B. For
cutlery, surgical instruments, gages, needle valves, bearings.
Corrosion resistant, hardenable, heat treatable.

ASM-123
Acieries de Sambra & Meuse
C 0-0.3, Cr 13, Mo 0.25, Ni 1.25, bal Fe.
Annealed: 98,000 TS; 52,000 YS; 22 El; 95 Rock B. For
bearings, cutlery, gages, needle valves, surgical instruments.
Corrosion resistant, hardenable.

ASP 11
Allmetal Screw Products Co., Inc.
C 0-0.04, Si 0-0.8, Mn 0-1.5, Cu 0-0.3, Ni 5.3-6.9, Cr 23.5-25,
Mo 1.45-1.95, Nb 0.3-0.5, B 0.001-0.003, bal Fe.
Water quenched: 49.6 kg/mm^2 YS; 68.4 kg/mm^2 TS; 36.6 El;
70.5 RA; 229 Brin; 34 ft·lb IS (Charpy). Magnetic, ferritic-
austenitic stainless with excellent corrosion resistance. For
hardware for chemical equipment.

ASP 23 H.S.S
Ackerlind Steel Co., Inc.
C 1.27, Cr 4.2, Mo 5, W 6.4, V 3.1, bal Fe.
High speed steel for cutting tools, broaches, and cold work
applications. High tensile strength; high toughness; excellent
grindability.

ASP 30 H.S.S
Ackerlind Steel Co., Inc.
C 1.27, Cr 4.2, Mo 5, W 6.4, V 3.1, Co 8.5, bal Fe.
High speed steel; cutting tools for hard to machine materials.
High red hardness and temper resistance; good grindability.

ASP 60 H.S.S
Ackerlind Steel Co., Inc.
C 2.3, Cr 4, Mo 7, W 6.5, Co 10.5, V 6.5, bal Fe.
High-carbon high-alloy high-speed steel; 69 Rock C. For
cutting tools. Good grindability.

ASR 1
French manufacture
C 0.36, Cr 18, Ni 9, W 10, Ti 2, Co 12, bal Fe.
For gas turbine components; heat and creep resistant.

ASR 2
French manufacture
Cr 18, Ni 8, C, 10 W + Co, bal Fe.
For gas turbines; heat creep and oxidation resistant.

ASTAR 811-C
Westinghouse Electric Corp.
W 8, Re 1, Hf 1, C 0.025, bal Ta.
For space power systems utilizing liquid alkali metals as
coolants. High creep resistance, good ductility and
weldability.

ASTRA
George Cook & Co., Ltd.
C 0.55, W 2, Si 0.5, Cr 1, bal Fe.
For cold punches, shears, piercers; oil hardened, tough.

ASTRALLOY GR. 1
Lukens Steel
C 0.23-0.27, Mn 0.7-1, P 0-0.02, S 0-0.02, Ni 3.25-3.75, Cr
1.25-1.75, Mo 0.2-0.3, bal Fe.
Normalized: 241,000 TS; 157,000 YS; 39 RA; 477 Brin.
Quenched: 250,000 TS; 190,000 YS; 39 RA. For paper and
mining industry equipment, coal mine chutes and hoppers.
High toughness and ductility. Good weldability. Abrasion
resistant.

ASTRALLOY GR. 2
Bethlehem Steel Corp.
C 0.29-0.33, Mn 0.7-1, Ni 3.25-3.75, Cr 1.25-1.75, Mo 0.2-0.3,
0.02% S and I, bal Fe.
Normalized: 250,000 TS; 160,000 YS; 35 RA; 500 Brin. For
paper and mining industry equipment, coal mine chutes and
hoppers. High toughness and ductility. Good weldability,
abrasion resistant. *Obsolete*

ASTRALLOY GR. 2
Lukens Steel
C 0.29-0.33, Mn 0.7-1, Ni 3.25-3.75, Cr 1.25-1.75, Mo 0.2-0.3,
0.02% S and I, bal Fe.
Normalized: 250,000 TS; 160,000 YS; 35 RA; 500 Brin. For
paper and mining industry equipment, coal mine chutes and
hoppers. High toughness and ductility. Good weldability,
abrasion resistant. *Obsolete*

ASTRO GRADE 1
Astro
Titanium. C 0-0.08, Fe 0-0.2, N 0-0.03, O 0-0.18, bal Ti. Sheet:
0.015 H; bar: 0.0125 H; billet: 0.0100 H.
35,000 psi TS (ultimate); 25,000 psi YS (0.2% offset); 24 El (in
2 in., sheet >0.025 in. thick); 30 RA (bar); density 0.163
lb/in.^{3}; ASTM B265 Gr 1, ASTM B337 Gr 1., ASTM B348 Gr 1,
ASTM B338 Gr 1, ASTM B381 Gr 1, ASTM F67 Gr 1, ASTM
F467 Gr 1, ASTM F468 Gr 1. Used in the chemical and
marine industries. Corrosion resistant, with maximum ease of
formability.

ASTRO GRADE 12
Astro
Titanium. C 0-0.08, N 0-0.03, Fe 0-0.3, H 0-0.015, O 0-0.25,
Mo 0.2-0.4, Ni 0.6-0.9, bal Ti.
70,000 psi TS (ultimate); 50,000 psi YS (0.2% offset); 18 El (in
2 in., sheet >0.025 in. thick); 25 RA (bar); density 0.163
lb/in.^{3}; ASTM B265 Gr 12, ASTM B348 Gr 12, ASTM 338 Gr
12, ASTM B337 Gr 12, ASTM B381 Gr 12, ASTM F467 Gr 12,
ASTM F468 Gr 12. For corrosion resistance in the chemical
industry where media is mildly reducing or varies between
oxidizing and reducing.

ASTRO GRADE 2
Astro
Titanium. C 0-0.1, Fe 0-0.3, N 0-0.03, O 0-0.25, bal Ti. Sheet:
0.015 H; bar: 0.0125 H; billet: 0.0100 H.
50,000 psi TS (ultimate); 40,000 psi YS (0.2% offset); 20 El (in
2 in., sheet >0.025 in. thick); 30 RA (bar); 25-40 ft·lb IS
(Charpy V-notch, at room temperature); density 0.163 lb/in.^{3};
AMS 4902, ASTM B265 Gr 2, ASTM B337 Gr 2, ASTM B338
Gr 2, ASTM B 348 Gr 2, ASTM B381 Gr 2, ASTM F67 Gr 2,
ASTM F467 Gr 2, ASTM F468 Gr 2. Used in the chemical and
marine industries. Corrosion resistant, with higher strength
and ease of formability.

ASTRO GRADE 3
Astro
Titanium. C 0-0.1, Fe 0-0.3, N 0-0.05, O 0-0.35, bal Ti. Sheet:
0.015 H; bar: 0.0125 H; billet: 0.0100 H.
65,000 psi TS (ultimate); 55,000 psi YS (0.2% offset); 18 El (in
2 in., sheet >0.025 in. thick); 30 RA (bar); 20-40 ft·lb IS
(Charpy V-notch, at room temperature); density 0.163 lb/in.^{3};
AMS 4900, ASTM B265 Gr 3, ASTM B337 Gr 3, ASTM B348
Gr 3, ASTM B338 Gr 3, ASTM B381 Gr 3, ASTM F67 Gr 3.
Used in the chemical and marine industries. Corrosion
resistant.

ASTRO GRADE 4
Astro
Titanium. C 0-0.1, Fe 0-0.5, N 0-0.05, O 0-0.4, bal Ti. Sheet:
0.015 H; bar: 0.0125 H; billet: 0.0100 H.
80,000 psi TS (ultimate); 70,000 psi YS (0.2% offset); 15 El (in
2 in., sheet >0.025 in. thick); 25 RA (bar); density 0.163
lb/in.^{3}; AMS 4901, 4921, ASTM B348 Gr 4, ASTM B381 Gr 4,
ASTM F67 Gr 4, ASTM F467 Gr 4, ASTM F468 Gr 4. Used in
the chemical and marine industries. Corrosion resistant

ASTRO GRADE 7
Astro
Titanium. C 0-0.1, Fe 0-0.3, N 0-0.03, Pd 0.12-0.25, O 0-0.25,
bal Ti. Sheet: 0.15 H; bar: 0.0125 H.
50,000 psi TS (ultimate); 40,000 psi YS (0.2% offset); 20 El (in
2 in., sheet >0.025 in. thick); 25 RA (bar); density 0.163
lb/in.^{3}; ASTM B265 Gr 7, ASTM B348 Gr 7, ASTM B337 Gr 7,
ASTM B338 Gr 7, ASTM B381 Gr 7, ASTM F467 Gr 7, ASTM
F468 Gr 7. For corrosion resistance in the chemical industry,
where media is mildly reducing or varies between oxidizing
and reducing.

ASTROLOY
New CARPENTER ASTROLOY.

ASTROLOY
Cannon-Muskegon Corp.
C 0.05, Cr 15, Mo 5, Ti 3.5, Al 4, Zr 0.06, Co 17, bal Ni.
At room temperature: 190,000 TS; 130,000 YS; 15 El. For turbine wheels and blades, jet engine components, nozzles, compressor discs. High heat and corrosion resistance. High creep and stress-rupture strength.

ASTROLOY
Teledyne Allvac
C 0.05, Cr 15, Mo 5, Ti 3.5, Al 4, Zr 0.06, bal Ni.
At room temperature: 190,000 TS; 130,000 YS; 15 El. For turbine wheels and blades, jet engine components, nozzles, compressor discs. Heat and corrosion resistant. Creep and stress-rupture strength.

AT 250
Now LESCALLOY MARVAC 250.

AT NICKEL
Now NICKEL 205.

AT-3
Russian manufacture
Al 3.2, Cr 0.84, Fe 0.4, Si 0.34, B 0.01, bal Ti.
At 70°F: 126,000 TS; 122,000 YS; 16 El; 52 RA. At 750°F: 84,000 TS; 80,000 YS; 16 El; 67 RA. At 1110°F: 62,000 TS; 60,000 YS; 21 El; 86 RA. For aircraft structures, high temperature fasteners; Pseudoalpha alloy; corrosion resistant.

AT-4
Russian manufacture
Al 4.4, Cr 0.79, Fe 0.6, Si 0.4, B 0.01, bal Ti.
At 70°F: 147,000 TS; 146,000 YS; 12 El; 41 RA. At 750°F: 100,000 TS; 92,000 YS; 15 El; 55 RA. At 1110°F: 58,000 TS; 55,000 YS; 28 El; 84 RA. For high temperature fasteners, aircraft structures; Pseuodoalpha alloy; corrosion resistant.

AT-6
Russian manufacture
Al 5.8, Cr 0.64, Fe 0.38, Si 0.32, B 0.01, bal Ti.
At 70°F: 162,000 TS; 157,000 YS; 13 El; 36 RA. At 750°F: 102,000 TS; 91,000 YS; 12 El; 58 RA. At 1110°F: 90,000 TS; 84,000 YS; 20 El; 78 RA. For high temperature fasteners, aircraft structures; corrosion resistant.

AT-8
Russian manufacture
Al 6.5, Cr 0.73, Fe 0.39, Si 0.4, B 0.01, bal Ti.
At 750°F: 162,000 YS; 156,000 YS; 15 El; 38 RA. At 750°F: 119,000 TS; 109,000 YS; 10 El; 47 RA. At 1110°F: 82,000 TS; 74,000 YS; 16 El; 67 RA. For airplane structures, high temperature fasteners; corrosion resistant.

AT-NICKEL
Inco Alloys International
99.0 Ni min.
For chemical plant equipment. Welding grade. Good resistance to alkaline salts and organic acids. *Obsolete*

ATERITE
Barber Asphalt Co.
Ni 35-44, Cu 36-55, Fe 5-20, Zn 0-5.
For hardware, valves, fixtures. *Obsolete*

ATERITE
Barber Asphalt Co.
Ni 10, Cu 65, Fe 2, Zn 23.
Rolled: 53,000-163,000 TS; 40,000-110,000 YS; 1-59 El; 17-62 RA. Cast: 55,000-110,000 TS; 35,000-110,000 YS; 0.6-45 El; 10-36 RA. For hardware, valves, plumbing fixtures, cocks, chemical equipment; corrosion resistant. *Obsolete*

ATERITE
Barber Asphalt Co.
Cu 62.4, Zn 18.9, Ni 12.6, Pb 2, Fe 2.9.
For hardware, chemical equipment. *Obsolete*

ATG 33
Creusot-Loire
C 0.06, Ni 45.5, Cr 25.5, Co 3.25, Mo 3.25, W 3.25, bal Fe.
High temperature alloy; for radiant tubes, furnace parts; heat treat fixtures. AFNOR Z 6 NCKDW 45. Similar to RA 333.

ATG C 1
Creusot-Loire
C 0.04, Ni 52, Cr 19, Mo 3, Ti 0.8, Al 0.5, Nb 5.25, bal Fe.
Good high temperature properties. For rocket motors, pump bodies, jet engines. AFNOR NC19FeNb; INCONEL 718.

ATG E
Creusot-Loire
C 0.1, Cr 22, Co 1.5, Mo 9, W 0.6, Fe 18.5, bal Ni.
Good strength at elevated temperature. For aircraft and jet engine parts; weldable. AFNOR NC22FeD; HASTELLOY X.

ATG E 2
Creusot-Loire
C 0.07, Cr 21.5, Mo 9, Nb 3.65, Fe 2, bal Ni.
Good strength at elevated temperature. For fuel nozzles, after burners. AFNOR NC22FeDNb; INCONEL 625.

ATG F
Creusot-Loire
C 0.05, Cr 15, Ti 2.5, Al 0.7, Fe 7, bal Ni.
Good strength at elevated temperature. For gas turbine parts, furnace equipment. AFNOR NC15TNbA; INCONEL X750.

ATG H
Creusot-Loire
C 0.1, Ni 10, Cr 20, W 15, Fe 0-3, bal Co.
Good strength at elevated temperature. For turbine blades and discs. AFNOR KC20WN; HAYNES ALLOY NO. 25.

ATG M 2
Creusot-Loire
C 0.18, Cr 10, Co 15, Mo 3, Ti 4.7, Al 5.5, V 1, bal Ni.
Good strength at elevated temperature. For jet engine parts, turbine blades. AFNOR NK15CAT; IN 100.

ATG R
Creusot-Loire
C 0.06, Cr 19.5, Co 0-5, Ti 0.4, Fe 0-5, bal Ni.
Good strength at elevated temperatures. For combustion chambers, turbines. AFNOR NC20T; similar to NIMONIC 75.

ATG S 3
Creusot-Loire
Ti 2.5, Al 1.5, Fe 0-1, C 0.07, Cr 19, 0.07 C 19 Cr, bal Ni.
Good strength at elevated temperature. For gas turbine blades. AFNOR NC20TA; NIMONIC 80A.

ATG S 4
Creusot-Loire
C 0.07, Cr 19, Co 19, Ti 2.5, Al 1.5, Fe 0-1, bal Ni.
Good strength at elevated temperatures. For jet engine parts. AFNOR NC20KTA; NIMONIC 90.

ATG S 8
Creusot-Loire
C 0.12, Cr 15, Co 27, Mo 3, Ti 2.1, Al 3, Fe 0-4.
Good strength at elevated temperature. For gas turbine parts. AFNOR NK27CADT; INCONEL 700.

ATG S 9
Creusot-Loire
C 0.12, Cr 13, Co 1, Mo 4.5, Ti 0.7, Fe 0-2, Al 6, Nb 2, bal Ni.
Good strength at elevated temperature. For jet engine components. AFNOR NC13AD; INCONEL 713 C.

ATG W 1
Creusot-Loire
C 0.06, Cr 20, Co 13, Mo 4, Ti 3, Al 1.25, Fe 0-2, bal Ni.
High strength at elevated temperature. For jet engine components. AFNOR NC20K14; WASPALOY.

ATG W 2
Creusot-Loire
C 0.1, Cr 18, Co 18, Mo 4, Ti 3, Al 3, Fe 0-4, bal Ni.
High strength at elevated temperatures. For engine components. AFNOR NC20KDTA; UDIMET 500.

ATG W 3
Creusot-Loire
C 0.01, Cr 15, Co 18, Mo 5, Ti 3, Al 4, Fe 0-4, bal Ni.
High strength at elevated temperatures. For combustion chambers, turbine blades. AFNOR NK18CDAT. Similar to UDIMET 700.

ATG W 4
Creusot-Loire
C 0.07, Cr 18, Co 15, Mo 3, W 1.5, Ti 5, Al 2.5, bal Ni.
High strength at elevated temperatures. For land-based gas turbine blades. AFNOR NCK18TDA; UDIMET 710.

ATG W O
Creusot-Loire
C 0.06, Cr 20, Co 20, Mo 5.9, Ti 2.15, Al 0.45, bal Ni.
High strength at elevated temperatures. For gas turbine components. AFNOR NCK20D; WIGGIN C-263.

ATG X
Creusot-Loire
C 0.12, Ni 20, Cr 21, Co 20, Mo 3, W 2.5, Nb 1, bal Fe.
High temperature alloy; for gas turbine blades, jet engine parts. AFNOR Z12CNKDW 20. Similar to N-155.

ATG XX
Creusot-Loire
C 0.4, Ni 20, Cr 20, Co 20, Mo 4, W 4, Nb 4, bal Fe.
High temperature alloy. For jet engines, turbine wheels and buckets. AFNOR Z42CKNDW 20. Similar to S590.

ATG-M
Creusot-Loire
Ni 35, Cr 20, bal Fe.
For gas engine and turbine discs; heat resistant. *Obsolete*

ATGS
French manufacture
C 0.1, Ni 75, Cr 20, Ti 2, Al 2.
For gas turbine components; heat and oxidation resistant. *Obsolete*

ATHA CHAMPION
Colt Industries
C 0.9, Cr 3.7, V 1.9, W 13.7, bal Fe.
For tools, dies; non-deforming. *Obsolete*

ATHA CHROME ROLL
Colt Industries
C 0.9, Cr 1.35, bal Fe.
For rolls for tire rims, drawing dies; oil hardening. *Obsolete*

ATHA NO. 2500
Colt Industries
Ni 22.9, Cr 5.42, Si 1.65, Cu 0.78, Mn 0.8, C 0.24, bal Fe.
98,000 TS; 32.3 El; 58.6 RA. For furnace parts; resists high temperatures. *Obsolete*

ATHA NO. 2600
Colt Industries
Ni 22, Cr 8, Si 1.75, Cu 1, bal Fe.
For furnace parts, heat treating boxes; heat resistant. *Obsolete*

ATHA PNEU
Crucible Materials Corp.
Tool material. C 0.5, W 2.75, Cr 1.25, V 0.25, bal Fe.
For pneumatic chisels, rivet busters; hot work steel. AISI S1.

ATHA RIM ROLL
Colt Industries
C 1, Cr 0.5, bal Fe.
For forming rolls; water hardening. *Obsolete*

ATHOS
English manufacture
Ni 22, Cr 8, Si 1.75, Cu 1, Mn 0.7, C 0-0.5, bal Fe.
For stainless parts; heat and corrosion resistant.

ATLAN
Atlantic Steel Corp.
C 0.9, Cr 0.5, Mn 1.15, W 0.5, bal Fe.
For tools, cutters, drawing dies; non-deforming. AISI O1.

ATLAN H.C.C. DIE
Atlantic Steel Corp.
C 1.5, Cr 12, V 1, Mo 0.8, bal Fe.
For tools, blanking dies; non-deforming. AISI D2.

ATLAN NON-SHRINK DIE
Atlantic Steel Corp.
C 0.9, Mn 1.5, bal Fe.
For tools dies; non-deforming.

ATLANTALOY 24
Atlantic Casting & Engineering Co.
Cu 86.75, Al 8.5, Fe 3.5, Mn 1.25.
Cast: 66,200 TS; 28 El. For castings; plaster mold. *Obsolete*

ATLANTALOY NO 22
Atlantic Casting & Engineering Co.
Cu 91, Al 7, Si 2.
Cast: 61,400 TS; 44.5 El; 95 Brin. For castings; plaster mold.
Obsolete

ATLANTALOY NO. 10 YELLOW BRASS
Atlantic Casting & Engineering Co.
Cu 58, Sn 1, Pb 1, Al 0.3, bal Zn.
Cast: 50,000 psi TS; 20,000 psi YS; 25 El; 70 Brin. For
hardware castings, brush boxes; free cutting. CDA 857.

ATLANTALOY NO. 12
Atlantic Casting & Engineering Co.
Cu 56.7, Zn 40, Al 1.2, Fe 0.75, Mn 0.9.
Cast: 81,000 TS; 22 El; 111 Brin. For castings, instrument
parts; plaster mold. *Obsolete*

ATLANTALOY NO. 13B SILICON BRONZE
Atlantic Casting & Engineering Co.
Cu 82.5, Si 4, Zn 13.5.
Cast: 68,000 psi TS; 30,000 psi YS; 17 El; 120 Brin. Hardware
castings, levers. CDA 875.

ATLANTALOY NO. 20
Atlantic Casting & Engineering Co.
Cu 82.5, Al 11.2, Fe 5, Mn 1.2.
Cast: 74,500 TS; 41,500 YS; 7 El; 160 Brin. For castings,
instrument parts; plaster mold, corrosion resistant. *Obsolete*

ATLANTALOY NO. 20C BERYLLIUM COPPER
Atlantic Casting & Engineering Co.
Be 2, Co 0.5, Si 0.25, bal Cu.
As cast: 133 Brin. Hardened and aged: 160,000 psi TS;
150,000 psi YS; 1 El; 400 Brin. For brush holders, electrical
contacts. AMS 4890.

ATLANTALOY NO. 21
Atlantic Casting & Engineering Co.
Cu 89, Al 10, Fe 1.
Cast: 79,300 TS; 34,900 YS; 22 El; 156 Brin. For castings;
plaster mold. *Obsolete*

ATLANTALOY NO. 23
Atlantic Casting & Engineering Co.
Cu 79, Al 11, Fe 5, Ni 5.
Cast: 77,600 TS; 55,500 TS; 5.5 El; 190 Brin. For castings;
plaster mold. *Obsolete*

ATLANTALOY NO. 25 ALUMINUM BRONZE
Atlantic Casting & Engineering Co.
Cu 81.5, Ni 4, Fe 4, Al 10.5.
As cast: 98,000 psi TS; 45,000 psi YS; 14 El; 190 Brin. Heat
treated: 117,000 psi TS; 70,000 psi YS; 240 Brin. For gears,
levers, non-sparking tools. CDA 955.

ATLANTALOY NO. 31 MANGANESE BRONZE
Atlantic Casting & Engineering Co.
Cu 64, Fe 3, Al 6, Mn 3.5, bal Zn.
Cast: 115,000 psi TS; 68,000 psi YS; 18 El; 210 Brin.
Manganese bronze casting for gears, marine hardware. CDA
863; AMS 4862.

ATLANTALOY NO. 40
Atlantic Casting & Engineering Co.
Cu 0.8, Mg 0.4, Zn 8, bal Al.
Cast: 38,000 psi TS; 25,000 psi YS; 4 El; 70 Brin. For light
castings, housings.

ATLANTALOY NO. 42
Atlantic Casting & Engineering Co.
Cu 91, Al 3.25, Fe 1.25, Si 4.5.
Cast: 65,800 TS; 42,800 YS; 10 El; 121 Brin. For castings;
plaster mold. *Obsolete*

ATLANTALOY NO. 60
Atlantic Casting & Engineering Co.
Cu 1.3, Al 93.2, Mg 0.5, Si 5.
Cast: 16,000 TS; 15,000 YS; 1.0 El. For castings; plaster
mold. *Obsolete*

ATLANTALOY NO. 70C BERYLLIUM COPPER
Atlantic Casting & Engineering Co.
Be 0.05, Cr 0.8, bal Cu.
Cast, heat treated: 53,000 psi TS; 36,000 psi YS; 11 El.
Electrical contacts; good electrical conductivity.

ATLANTALOY NO. 8A MANGANESE BRONZE
Atlantic Casting & Engineering Co.
Cu 58, Sn 0.5, Fe 1, Al 1, Zn 39.5.
Cast: 71,000 psi TS; 28,000 psi YS; 30 El; 130 Brin. For
housings, hardware, geared impellers. CDA 865.

ATLANTIC
Atlantic Steel Corp.
C 0.5, W 18, Cr 4, V 1, Mo 0.15, bal Fe.
For tools, cutters; high speed steel. *Obsolete*

ATLANTIC "V"
Atlantic Steel Corp.
C 0.8, Cr 4.25, V 2, W 18.5, Mo 0.65, bal Fe.
For cutting tools; high steel.

ATLANTIC 33C
Atlantic Steel Corp.
C 0.33, Cr 0.76, Co 0.5, Mo 0.75, Cu 0.75, bal Fe.
For rivet sets, pneumatic tools, blacksmith tools;
nontempering, water hardening. *Obsolete*

ATLANTIC 44
Atlantic Steel Corp.
C 0.44, Cr 0.75, Mn 0.4, Mo 0.75, Si 0.6, Cu 0.7, bal Fe.
For dies, shear blades, pneumatic tools; tough, shock
resistant. *Obsolete*

ATLANTIC C
Atlantic Steel Corp.
C 0.65, Cr 4, Co 4, W 18, Mo 1, bal Fe.
For shaping tools, millers, cutters; high speed steel.

ATLANTIC DIE
Atlas Steels Ltd.
C 0.7, Mn 0.4, Cr 1, Ni 1.65, bal Fe.
For blanking and cold working dies; oil hardened. *Obsolete*

ATLANTIC DIE
Atlantic Steel Corp.
C 0.7, Cr 0.75, Ni 1.5, Mo 0.25, bal Fe.
Special purpose die steel. AISI L6.

ATLANTIC H.S.
Atlantic Steel Corp.
C 0.75, W 18, Cr 4, V 1, bal Fe.
High speed steel. AISI T1.

ATLANTIC N.T.
Atlantic Steel Corp.
C 0.4, Mn 0.4, Si 0.65, Cr 1, Ni 0.5, Mo 0.75, Cu 0.5, bal Fe.
Shock resistant tool steel.

ATLANTIC NO. 33
Atlantic Steel Corp.
C 0.33, Cr 0.75, Mo 0.75, Cu 0.75, bal Fe.
Heat treated: 135,000 TS; 116,000 YS; 18 El; 54 RA; 277 Brin.
For chisels, dies, blacksmith tools, hot work tools; water
hardened, non-tempering.

ATLANTIC STANDARD
Atlantic Steel Corp.
C 0.8-1.2, Si 0.25, Mn 0.25, bal Fe.
Water hardened: 166,000-216,000 TS; 110,000-150,000 YS;
11-15 El; 32-37 RA; 330-600 Brin. For taps, drills, reamers,
punches, stamps, knurls, mandrels. Type W1; water
hardened.

ATLAS "Q"
Atlas Specialty Steels
C 1.2, Cr 0.5, bal Fe.
For tools, taps, reamers; water hardening. *Obsolete*

ATLAS 12-12
Atlas Specialty Steels
C 0-0.12, Cr 12-14, Ni 12-14, bal Fe.
Annealed: 80,000 TS; 35,000 YS; 50 El; 60 RA; 135 Brin. For
cutlery, valves; low rate of work hardening. *Obsolete*

ATLAS 20
Atlas Specialty Steels
C 0-0.08, Mn 2, Cr 19-21, Mo 2-3, Ni 27-30, Cu 3-4, bal Fe.
For chemical and paper industries; high resistance to H_2SO_4.
Obsolete

ATLAS 20
Ampco Pittsburgh Corp.
Cu 75, Pb 20, Sn 5.
Cast: 25,000-30,000 TS; 14,000-16,000 YS; 14-18 El; 12-14
RA; 50-60 Brin. For bearings for use under poor lubrication.
Obsolete

ATLAS 301
Atlas Specialty Steels
C 0-0.15, Cr 16-18, Ni 6-8, bal Fe.
Annealed: 110,000 TS; 40,000 YS; 60 El; 70 RA; 165 Brin. For
doctor blades and springs, auto and furniture trim; high
strength and wear resistant stainless.

ATLAS 302
Atlas Specialty Steels
C 0.08-0.2, Cr 17-19, bal Fe.
Annealed: 80,000 TS; 35,000 YS; 50 El; 70 RA; 150 Brin. For
chemical plant equipment; austenitic stainless, Type 302.

ATLAS 303
Atlas Specialty Steels
C 0-0.15, Cr 17-19, Ni 8-10, P 0.09-0.17, 0.07 Se min, bal Fe.
Annealed: 90,000 TS; 45,000 YS; 40 El; 60 RA; 170 Brin. For
screw machine products, shafts, fasteners; free-cutting
stainless, Type 303.

ATLAS 303 MX
Atlas Specialty Steels
C 0-0.15, Cr 18, Ni 9, Mn 1.5, 0.15 Se min, bal Fe.
Free-machining austenitic stainless for automatic screw
machines.

ATLAS 304
Atlas Specialty Steels
C 0-0.08, Mn 0-2, Cr 18-20, Ni 8-10, bal Fe.
Annealed: 85,000 TS; 35,000 YS; 60 El; 70 RA; 160 Brin. Cold drawn: 180,000 TS; 125,000 YS; 10 El; 330 Brin. For architectural molding and trim, kitchen equipment, chemical plant equipment; austenitic stainless, Type 304.

ATLAS 304 XL
Atlas Specialty Steels
C 0.03, Mn 2, P 0.045, S 0.03, Si 1, Cr 18-20, Ni 8-10.5, N 0.1, bal Fe.
37,000 YS; 60 El. High temperature strength, weldable, good machinability; corrosion resistant.

ATLAS 305
Atlas Specialty Steels
C 0-0.12, Cr 16-19, Ni 8-11, bal Fe.
Annealed: 85,000 TS; 38,000 YS; 50 El; 60 RA; 140 Brin. For spinning, cold heading and drawing operations; low rate of work hardening, stainless.

ATLAS 308
Atlas Specialty Steels
C 0-0.08, Cr 19-21, Ni 10-12, bal Fe.
Annealed: 85,000 TS; 35,000 YS; 50 El; 60 RA; 150 Brin. For welding rods; heat and corrosion resistant.

ATLAS 310
Atlas Specialty Steels
C 0-0.25, Mn 0-2, Cr 24-26, Ni 19-22, bal Fe.
Annealed: 100,000 TS; 45,000 YS; 50 El; 65 RA; 185 Brin. For furnace parts and equipment, heat treat boxes, baffles; austenitic, heat resistant, Type 310.

ATLAS 316
Atlas Specialty Steels
C 0-0.1, Cr 16-18, Ni 10-14, Mo 2-3, bal Fe.
Annealed: 80,000 TS; 30,000 YS; 60 El; 80 RA; 150 Brin. Cold drawn: 150,000 TS; 135,000 YS; 6 El; 300 Brin. For chemical plant equipment, tanks, evaporators, valve trim; austenitic stainless, Type 316.

ATLAS 316 XL
Atlas Specialty Steels
C 0.03, Mn 2, P 0.045, S 0.03, Si 1, Cr 16-18, Ni 10-14, Mo 2-3, N 0.1-1, bal Fe.
42,000 YS; 55 El. High temperature strength, weldable, good machinability; corrosion resistant.

ATLAS 317
Atlas Specialty Steels
C 0-0.1, Cr 18-20, Ni 11-14, Mo 3-4, bal Fe.
Annealed: 90,000 TS; 40,000 YS; 45 El; 160 Brin. For chemical plant equipment, tanks; high corrosion resistance.

ATLAS 321
Atlas Specialty Steels
C 0-0.08, Cr 17-19, Ni 8-11, Ti = 5 x C, bal Fe.
Annealed: 85,000 TS; 33,000 YS; 58 El; 75 RA; 180 Brin. Cold drawn: 95,000 TS; 60,000 YS; 40 El; 60 RA; 190 Brin. For welded structures, chemical plant equipment, tanks; stabilized stainless, Type 321.

ATLAS 330
Atlas Specialty Steels
C 0-0.25, Cr 14-16, Ni 33-36, bal Fe.
Annealed: 80,000 TS; 40,000 YS; 30 El; 30 RA; 185 Brin. At 1800°F: 147,000 TS; 43 El; 40 RA. For furnace parts and equipment, cracking units, fixtures; heat resistant, austenitic, Type 330.

ATLAS 331
Atlas Specialty Steels
C 0-0.08, Cr 19-21, Ni 30-34, bal Fe.
Annealed: 95,000 TS; 45,000 YS; 45 El; 60 RA; 160 Brin. For carburizing boxes, heater tubing, nitriding fixtures; good resistance to thermal shock.

ATLAS 347
Atlas Specialty Steels
C 0-0.08, Cr 17-19, Ni 9-12, Cb = 10 x C, bal Fe.
Annealed: 90,000 TS; 35,000 YS; 50 El; 65 RA; 160 Brin. Cold drawn: 100,000 TS; 65,000 YS; 40 El; 60 RA; 212 Brin. For welded structures, chemical plant equipment; stabilized stainless, Type 347.

ATLAS 4-79
Atlas Specialty Steels
C 0-0.06, Ni 78-80, Co 0.85, Mn 0.3-0.8, bal Fe.
For communication and telephone equipment; high permeability. *Obsolete*

ATLAS 40
Atlas Specialty Steels
C 0.24, Cr 1.2, Mo 0.26, Ni 3.5, Al 1.2, bal Fe.
For plastic molds, die casting dies; precipitation hardening. *Obsolete*

ATLAS 403
Atlas Specialty Steels
C 0-0.15, Cr 11.5-13.5, bal Fe.
Annealed: 75,000 TS; 40,000 YS; 34 El; 72 RA; 155 Brin. Heat treated: 215,000 TS; 140,000 YS; 14 El; 54 RA; 415 Brin. For dental and surgical instruments, valves, cutlery; corrosion resistant, hardenable, Type 403.

ATLAS 405
Atlas Specialty Steels
C 0-0.08, Cr 11.5-13.5, Al 0.1-0.3, bal Fe.
Sub-annealed: 65,000 TS; 40,000 YS; 20 El; 50 RA; 165 Brin. For mufflers, oil refinery equipment, annealing boxes; creep resistant.

ATLAS 410
Atlas Specialty Steels
C 0-0.15, Cr 11.5-13.5, bal Fe.
Annealed: 75,000 TS; 35 El; 70 RA; 155 Brin. Cold drawn: 100,000 TS; 85,000 YS; 60 El; 205 Brin. For flat springs, tableware, valve parts, turbine blades; corrosion resistant, hardenable, Type 410.

ATLAS 416
Atlas Specialty Steels
C 0-0.15, Cr 12-14, S 0.18-0.35, bal Fe.
Annealed: 75,000 TS; 40,000 YS; 30 El; 60 RA; 155 Brin. Heat treated: 110,000 TS; 85,000 YS; 18 El; 55 RA; 230 Brin. For screw machine products, shafts, valve trim; corrosion resistant, free-cutting, Type 416.

ATLAS 420
Atlas Specialty Steels
Cr 12-14, 0.15 C min, bal Fe.
Annealed: 95,000 TS; 50,000 YS; 25 El; 55 RA; 196 Brin. Cold drawn: 105,000 TS; 85,000 YS; 17 El; 50 RA; 215 Brin. For cutlery, surgical instruments, valve trim; corrosion resistant, hardenable, Type 420.

ATLAS 420F
Atlas Specialty Steels
Cr 12-14, S 0.18-0.35, 0.15 C min, bal Fe.
Annealed: 95,000 TS; 50,000 YS; 20 El; 50 RA; 196 Brin. Cold drawn: 105,000 TS; 85,000 YS; 15 El; 45 RA; 215 Brin. For cutlery, surgical instruments, screw machine products; corrosion resistant, free-cutting, Type 420 F.

ATLAS 430
Atlas Specialty Steels
C 0-0.12, Cr 14-18, bal Fe.
Annealed: 70,000 TS; 40,000 YS; 30 El; 55 RA; 160 Brin. Cold drawn: 130,000 TS; 120,000 YS; 2 El; 190 Brin. For oil refinery and chemical plant equipment, hardware, bolts; corrosion resistant, Type 430.

ATLAS 431
Atlas Specialty Steels
C 0-0.2, Cr 15-17, Ni 1.2-2.5, bal Fe.
Annealed: 95,000 TS; 50,000 YS; 20 El; 50 RA; 195 Brin. For oil refinery and chemical plant equipment; corrosion resistant, Type 431.

ATLAS 440A
Atlas Specialty Steels
C 0.6-0.75, Cr 16-18, Mo 0-0.75, bal Fe.
Annealed: 95,000 TS; 55,000 YS; 20 El; 240 Brin. Heat treated: 275,000 TS; 240,000 YS; 2 El; 555 Brin. For bearings, cutlery, surgical instruments, pivots; corrosion resistant, hardenable, Type 440A.

ATLAS 440B
Atlas Specialty Steels
C 0.75-0.95, Cr 16-18, Mo 0-0.75, bal Fe.
Annealed: 107,000 TS; 62,000 YS; 18 El; 35 RA; 220 Brin. Cold drawn: 120,000 TS; 95,000 YS; 9 El; 20 RA; 250 Brin. For bearings, valves, surgical instruments, cutlery; corrosion resistant, hardenable, Type 440B.

ATLAS 440C
Atlas Specialty Steels
C 0.95-1.2, Cr 16-18, Mo 0-0.75, bal Fe.
Annealed: 100,000 TS; 60,000 YS; 18 El; 200 Brin. Heat treated: 280,000 TS; 250,000 YS; 2 El; 575 Brin. For bearings, cutlery, valves, pivots, surgical instruments; corrosion resistant, hardenable, Type 440C.

ATLAS 442
Atlas Specialty Steels
C 0-0.2, Cr 23-27, bal Fe.
Sub-annealed: 80,000 TS; 50,000 YS; 20 El; 50 RA; 165 Brin. For oil burner furnace and boiler parts; high corrosion and oxidation resistance.

ATLAS 446
Atlas Specialty Steels
C 0-0.35, Cr 23-27, bal Fe.
Annealed: 75,000 TS; 45,000 YS; 35 El; 65 RA; 160 Brin. Cold drawn: 175,000 TS; 155,000 YS; 2 El; 25 RA; 250 Brin. For oil burner parts, heat treat boxes, furnace equipment; corrosion resistant, Type 446.

ATLAS 45% NI ALLOY
Atlas Specialty Steels
C 0-0.06, Mn 0.4-1, Ni 43.5-46.5, Co 0-0.6, bal Fe.
Half hard: 140,000 TS; 128,000 YS; 1.5 El; 331 Brin. For telephone equipment; high permeability and high induction. *Obsolete*

ATLAS 501
Atlas Specialty Steels
C 0-0.1, Mn 0-1, Cr 5, bal Fe.
Annealed: 70,000 TS; 30,000 YS; 28 El; 65 RA; 160 Brin. Heat treated: 175,000 TS; 135,000 YS; 15 El; 50 RA; 370 Brin. For oil refinery equipment, valve trim, furnace parts; good to 1200°F, creep resistant.

ATLAS 502
Atlas Specialty Steels
C 0-0.1, Mn 0-1, Cr 4-6, Mo 0.5, bal Fe.
Annealed: 75,000 TS; 35,000 YS; 26 El; 62 RA; 170 Brin. Heat treated: 175,000 TS; 135,000 YS; 15 El; 50 RA; 370 Brin. For oil refinery equipment, valve trim, furnace parts; good to 1200°F, creep resistant.

ATLAS 78 1/2 NI ALLOY
Atlas Specialty Steels
C 0-0.06, Ni 77.5-79.5, Co 0-0.6, bal Fe.
Annealed: 86,000 TS; 35,000 YS; 41 El; 130 Brin. For communication circuits; high permeability. *Obsolete*

ATLAS 89-A-207
Ampco Pittsburgh Corp.
Cu 94, Al 5, Fe 1.
Treated: 40,000-50,000 TS; 18,000-22,000 YS; 40-60 El; 40-60 RA; 60-80 Brin. For machine parts requiring great ductility. *Obsolete*

ATLAS 89-A-3
Ampco Pittsburgh Corp.
Cu 89.2, Al 9.8, Fe 1.
Cast: 70,000-75,000 TS; 25,000-32,000 YS; 30-40 El; 30-40 RA; 115-126 Brin. For bearings, gears, machine parts. *Obsolete*

ATLAS 89-A-303
Ampco Pittsburgh Corp.
Cu 89.2, Al 9.8, Fe 1.
Normalized: 68,000-80,000 TS; 28,000-40,000 YS; 22-30 El; 25-30 RA; 115-131 Brin. For bearings, gears, machine parts. *Obsolete*

ATLAS 89-A-318
Ampco Pittsburgh Corp.
Cu 89.2, Al 9.8, Fe 1.
Single treatment: 80,000-95,000 TS; 45,000-55,000 YS; 12-18 El; 14-20 RA; 156-170 Brin. For welding jaws, gears, bearings, wearplates, shifter forks. *Obsolete*

ATLAS 89-A-323
Ampco Pittsburgh Corp.
Cu 89.2, Al 9.8, Fe 1.
Double treatment: 80,000-90,000 TS; 40,000-50,000 YS; 22-30 El; 24-32 RA; 143-156 Brin. For welding jaws, gears, bearings, wearplates, shifter forks. *Obsolete*

ATLAS 89-E-103
Ampco Pittsburgh Corp.
Cu 88.7, Al 10.3, Fe 1.
Normalized: 80,000-85,000 TS; 38,000-45,000 YS; 17-23 El; 17-23 RA; 131-143 Brin. For heavy duty bronze service. *Obsolete*

ATLAS 89-E-106
Ampco Pittsburgh Corp.
Cu 88.7, Al 10.3, Fe 1.
Normalized: 75,000-80,000 TS; 30,000-35,000 YS; 12-18 El; 14-20 RA; 131-149 Brin. For heavy duty bronze service. *Obsolete*

ATLAS 89-E-108
Ampco Pittsburgh Corp.
Cu 88.7, Al 10.3, Fe 1.
Normalized: 75,000-85,000 TS; 30,000-40,000 YS; 10-18 El; 12-20 RA; 131-141 Brin. For heavy duty bronze service. *Obsolete*

ATLAS 89-E-117
Ampco Pittsburgh Corp.
Cu 88.7, Al 10.3, Fe 1.
Single treatment: 90,000-100,000 TS; 70,000-80,000 YS; 2-6 El; 3-7 RA; 229-255 Brin. For machine parts. *Obsolete*

ATLAS 89-E-133
Ampco Pittsburgh Corp.
Cu 88.7, Al 10.3, Fe 1.
Triple treatment: 80,000-100,000 TS; 45,000-55,000 YS; 12-15 El; 13-17 RA; 163-179 Brin. For heavy duty gear, bearing, welding jaws. *Obsolete*

ATLAS 89-E-135
Ampco Pittsburgh Corp.
Cu 88.7, Al 10.3, Fe 1.
Triple treatment: 80,000-90,000 TS; 30,000-40,000 YS; 14-22 El; 16-24 RA; 149-166 Brin. For heavy duty gear, bearings, welding jaws. *Obsolete*

ATLAS 89-E-203
Ampco Pittsburgh Corp.
Cu 88.2, Al 10.8, Fe 1.
Corrective treatment: 80,000-85,000 TS; 45,000-50,000 YS; 14-20 El; 16-22 RA; 160-180 Brin. For machine parts. *Obsolete*

ATLAS 89E-123
Ampco Pittsburgh Corp.
Cu 88.7, Al 10, Fe 1, 0.30% special elements.
80,000-95,000 TS; 42,000-55,000 YS; 12-15 El; 13-17 RA; 160-180 Brin. For gears, bearings; heat treatable. *Obsolete*

ATLAS 90
Ampco Pittsburgh Corp.
Cu 90, Al 10.
Cast: 70,000 TS; 28,000 YS; 25 El; 25 RA; 120 Brin. For gears; corrosion resistant. *Obsolete*

ATLAS 93
AL Tech Specialty Steel Corp.
C 0.55, Mn 0.55, Si 0.2, Cr 0.7, Mo 0.4, bal Fe.
Shock resisting tool steel.

ATLAS A
AL Tech Specialty Steel Corp.
C 0.3, Cr 3.5, W 9, V 0.45, bal Fe.
W-C hot work tool steel for heavy duty tooling requirements; punches, mandrels, dies. AISI H-21.

ATLAS ACX
Atlas Specialty Steels
C 0.83, W 18.5, Cr 4, V 1.7, Mo 1, Co 10.5, bal Fe.
For hogging cutters, tools; high speed steel. *Obsolete*

ATLAS AHT-28
Atlas Steels Ltd.
C 0.3, Mn 0.5, Cr 1.4, Ni 4, Mo 2, bal Fe.
Heat treated: 241,000 TS; 204,000 YS; 12.5 El; 477 Brin. For gears, rolls, camshafts, transmission components, carbide bit holders; oil or air hardenable.

ATLAS AHT-28
Rio Algom Corp.
C 0.3, Mn 0.5, Cr 1.4, Ni 4, Mo 2, bal Fe.
Heat treated: 241,000 TS; 204,000 YS; 12.5 El; 477 Brin. For gears, rolls, camshafts, transmission components, carbide bit holders; oil or air hardenable.

ATLAS ALLOY 909
Atlas Foundry & Machine Co.
Stainless steel. C 0.03, Mn 1, Si 1, P 0.04, S 0.03, Mo 0.4-1, Cr 11.5-14, Ni 3.5-4.5, bal Fe.
Low carbon alloy similar to martensitic stainless steel Grade CA-6NM for use in hydrogen sulfide environments. 90 TS min; 65 YS min; 18 El min; 30 RA min; 23 Rock C max.

ATLAS ALLOY 948
Atlas Foundry & Machine Co.
Stainless steel. C 0-0.03, Mn 1.5, Si 1.5, P 0.04, S 0-0.01, Mo 2-3, Cu 3-4, Cr 19-22, Ni 27.5-30.5, bal Fe.
Low carbon alloy similar to cast austenitic stainless steel CN-7M for use in sulfuric acid and other reducing chemicals. 62 TS min; 25 YS min; 35 El min. Meets ASTM A351, A743, and A744.

ATLAS ALLOY 958
Atlas Foundry & Machine Co.
Stainless steel. C 0-0.03, Mn 1.5, Si 1, N 0.1-0.3, Mo 4-5, Cr 24-26, Ni 6-8, bal Fe.
Chloride corrosion resistant duplex stainless steel for pumps and valves in seawater and other aggressive environments. 100 TS min; 75 YS min; 18 El min.

ATLAS ALPHA
Atlas Specialty Steels
C 0.8, bal Fe.
For punches, dies, blacksmith tools; water hardened. *Obsolete*

ATLAS ALPHA-8
Atlas Specialty Steels
C 0.8, Mn 0.3, bal Fe.
For hammer dies, picks, set screws, wedges; water hardening, shock resistant. *Obsolete*

ATLAS B
Allegheny Ludlum Steel
C 0.4, Cr 3, W 11, V 0.5, bal Fe.
For hot work tools, dies, punches; hot work steel. *Obsolete*

ATLAS BRAKE DIE
Atlas Specialty Steels
C 0.5, Mn 1.1, Cr 0.66, Mo 0.15, bal Fe.
Heat treated: 130,000 TS; 118,000 YS; 20 El; 262 Brin. For brake press dies, gears, drive shafts; high toughness and wear resistance.

ATLAS CW26
Atlas Specialty Steels
C 2.4, Mn 0.4, V 5, Mo 1.1-8.15, bal Fe.
For drawing and stamping dies; high abrasion and wear resistance. *Obsolete*

ATLAS D319
Atlas Specialty Steels
C 0-0.07, Mn 2, Cr 17.5-19.5, Mo 2.25-3, Ni 13-15, bal Fe.
For pulp and paper industry; high corrosion resistance. *Obsolete*

ATLAS DIE CASTING STEEL
Atlas Specialty Steels
C 0.4, Mn 0.7, Cr 0.6, Mo 0.15, Ni, bal Fe.
For die casting dies; oil hardened. *Obsolete*

ATLAS DOUBLE EXTRA
Atlas Steels Ltd.
C 0.9-1.1, bal Fe.
For drills, punches, reamers, hobs; Type W 1. *Obsolete*

ATLAS EXTRA
Atlas Specialty Steels
C 1.05, Mn 1.25, bal Fe.
For tools, shear blades, cutters; oil hardening. *Obsolete*

ATLAS FNS
Atlas Specialty Steels
C 1.5, Cr 12, Mo 0.8, bal Fe.
For tools, lamination dies; hot work steel.

ATLAS HOBBING IRON
Atlas Specialty Steels
C 0.05, Mn 0.2, Si 0.15, bal Fe.
For plastic mold dies; hubbed cavity. *Obsolete*

ATLAS HOLLOW DRILL
Atlas Steels Ltd.
C 0.75-0.85, bal Fe.
For hollow drills. *Obsolete*

ATLAS HOT DIE
Farrelloy Co.
C 0.5, Cr 4, W 18, V 1, bal Fe.
For hot dies; punches, shears; hot work steel.

ATLAS HOT DIE
Allegheny Ludlum Steel
C 0.5, Cr 4, W 18, V 1, bal Fe.
For hot dies; punches, shears; hot work steel.

ATLAS HW 24
Atlas Specialty Steels
C 0.35, Mn 0.55, Si 1.3, Cr 3.5, Mo 4.25, V 0.85, bal Fe.
Hot work tool steel. *Obsolete*

ATLAS HW 7
Atlas Specialty Steels
C 0.45, Si 1, Cr 5, W 3.75, Mo 1, V 0.5, Co 0.5, bal Fe.
At 1000°F: 217,000 TS; 8.5 El; 40 RA; 477 Brin. For hot punches, shears, forging dies, cold punches; high hot strength and shock resistance. *Obsolete*

ATLAS IRON NO. 1
ACIPCO Steel Products Division
Si 2-2.15, 2.9-3.0% graphitic carbon, 0.5-0.6% combined carbon, bal Fe.
200-250 Brin. For dies; high test iron Tr.S.-6900. *Obsolete*

ATLAS IRON NO. 2
ACIPCO Steel Products Division
Si 1.85-1.95, 2.75% graphitic carbon, 0.50% combined carbon, bal Fe.
200 Brin. For valves; high test iron Tr.S.-6100. *Obsolete*

ATLAS IRON NO. 3

ACIPCO Steel Products Division
Si 1.5-1.75, 2.7% graphitic carbon, 0.60% combined carbon, bal Fe.
350 Brin. For abrasion iron hammers; high test iron
Tr.S.-5000. *Obsolete*

ATLAS M 3

Atlas Specialty Steels
C 1.03, Mn 0.25, Si 0.3, Cr 4, Mo 6, W 6.25, V 2.5, bal Fe.
High speed steel. *Obsolete*

ATLAS M 34

Atlas Specialty Steels
C 0.9, W 1.45, Cr 3.75, Mo 8.7, Co 8.25, V 2.05, bal Fe.
For tool bits, reamers, form tools; high red hardness.
Obsolete

ATLAS M 4

Atlas Specialty Steels
C 1.25, W 6, Cr 4.25, Mo 4.75, bal Fe.
For reamers, tool bits, form rolls, drill; high abrasion and wear resistance. *Obsolete*

ATLAS NN

Atlas Specialty Steels
C 2.2, Cr 12, bal Fe.
For tools dies, blanking dies; non-deforming.

ATLAS NO. 50A

Allegheny Ludlum Steel
C 0.3, Cr 3.5, W 9, V 0.4, bal Fe.
For hot work tools, punches; hot work tool steel.

ATLAS NO. 50B

Allegheny Ludlum Steel
C 0.4, Cr 3, W 11, V 0.5, bal Fe.
For hot work tools, punches; hot work tool steel.

ATLAS NO. 57

Atlas Pattern & Model Works
Cu 82, Sn 11, Zn 3.5, Pb 0.5, 3.0 Ni-Sn hardener.
For gears, bearings; wear resistant.

ATLAS NO. 83A

Atlas Pattern & Model Works
Cu 85, Sn 2, Zn 5, Pb 8.
For gears, worms, nuts, bearings.

ATLAS NO. 89 A-2

Ampco Pittsburgh Corp.
Cu 94, Al 5, Fe 1.
Cast: 40,000-50,000 TS; 20,000-150,000 YS; 30-40 El; 35-50 RA; 60-80 Brin. For bushings and machine parts. *Obsolete*

ATLAS NO. 89 E-1

Ampco Pittsburgh Corp.
Cu 88.7, Al 10.3, Fe 1.
Cast: 70,000-80,000 TS; 40,000-45,000 YS; 12-18 El; 15-22 RA; 131-143 Brin. For welding jaws, chains, gears, bearings, cams, wear plates. *Obsolete*

ATLAS NO. 89 E-2

Ampco Pittsburgh Corp.
Cu 88.2, Al 10.8, Fe 1.
Cast: 73,000-83,000 TS; 43,000-48,000 YS; 10-16 El; 16-20 RA; 160-180 Brin. For welding blocks, wear surfaces, chains. *Obsolete*

ATLAS NO. 89 E-3

Ampco Pittsburgh Corp.
Cu 87.5, Al 11.5, Fe 1.
Cast: 75,000-80,000 TS; 45,000-50,000 YS; 7-12 El; 9-14 RA; 202-229 Brin. For welding blocks, wearing surfaces, chains. *Obsolete*

ATLAS NO. 89 E-4

Ampco Pittsburgh Corp.
Cu 86.6, Al 12.4, Fe 1.
Cast: 65,000-75,000 TS; 55,000-65,000 YS; 3-9 El; 5-10 RA; 255-302 Brin. For welding blocks, chains, wearing surfaces. *Obsolete*

ATLAS NO. 89 E-5

Ampco Pittsburgh Corp.
Cu 86, Al 13.8, Fe 1.
Cast: 60,000-70,000 TS; 2-4 El; 3-6 RA; 302-332 Brin. For welding blocks, wearing surfaces, chains. *Obsolete*

ATLAS NON-MAG

Atlas Specialty Steels
C 0.4, Mn 11.5-12.5, Ni 7-8.5, bal Fe.
For transformer parts, retainer rings; low magnetic permeability. *Obsolete*

ATLAS REFINED

Atlas Steels Ltd.
C 0.95, bal Fe.
For tools, cutters, dies; water hardening. *Obsolete*

ATLAS REFINED-10

Atlas Specialty Steels
C 1, Mn 0.25, bal Fe.
For stamping and blanking dies, arbors, lathe centers, drills; water hardening. *Obsolete*

ATLAS REFINED-8

Atlas Specialty Steels
C 0.8, Mn 0.3, bal Fe.
For flaring tools, vise jaws, sledges, drills; water hardening. *Obsolete*

ATLAS ROLL

Rio Algom Corp.
C 1.05, Mn 0.3, S 0.015, P 0.02, Cr 1.45, bal Fe.
Annealed: 95,000 TS; 54,000 YS; 26 El; 55 RA; 192 Brin.
Hardened: 189,000 TS; 131,000 YS; 14 El, 35 RA; 331 Brin.
For rolls, feed and pinch rolls, bearings, flaring tools. Water hardening; wear and fatigue resistant.

ATLAS SPECIAL ALLOY 10

Atlas Specialty Steels
C 1.05, bal Fe.
For tools, striking dies; water hardening. *Obsolete*

ATLAS SPECIAL ALLOY 8

Atlas Specialty Steels
C 0.8, Mn 0.2, V 0.2, bal Fe.
For impact tools, rivet sets; water hardened, shock resistant. *Obsolete*

ATLAS STAINLESS NO. 20

Atlas Specialty Steels
C 0.07, Ni 29, Cr 20, Mo 2-3, Cu 4, bal Fe.
Cast: 65,000-75,000 TS; 28,000-38,000 YS; 35-50 El; 40-50 RA; 120-150 Brin. For chemical plant equipment; resists mixed acids, austenitic. *Obsolete*

ATLAS SUPERIOR

Rio Algom Corp.
C 0.4, Mn 1.1, S 0.08, P 0.02, Mo 0.15, bal Fe.
Rolled: 108,000 TS; 84,000 YS; 18 El; 46 RA; 229 Brin. For arbors, bolts, camshafts, collets, compressor shafts, guide bars, mandrels. Free machining, water hardening.

ATLAS SUPERIOR

Atlas Brass Foundry
Sn, bal Cu.
For springs, bearings; corrosion resistant.

ATLAS TRIPLE EXTRA

Atlas Specialty Steels
C 1.3, W 3.5, bal Fe.
For fast finishing tools, drills; water hardening. *Obsolete*

ATLAS X-10

Atlas Specialty Steels
C 1.05, Mn 0.25, bal Fe.
For shear blades, dies and cutting tools; water hardening.

ATLAS X-12

Atlas Specialty Steels
C 1.2, Mn 0.25, bal Fe.
For drills, taps, reamers, files, cold forming dies; water hardening. *Obsolete*

ATLAS XLO

Atlas Specialty Steels
C 0.55, Mn 0.7, V 0.8, Cr 2, Mo 0.55, bal Fe.
For drop forging die blocks; oil hardening.

ATLAS XX-95

Atlas Specialty Steels
C 0.95, bal Fe.
For cold beading dies; water hardened. *Obsolete*

ATLAS XXX

Atlas Specialty Steels
C 1.35, W 3.75, bal Fe.
For tools, forming rolls; maximum wear resistance. *Obsolete*

ATLAS XXX

Atlas Steels Ltd.
C 0.7, W 18, Cr 4, V 1, bal Fe.
For tools, dies, punches; high speed steel. *Obsolete*

ATLAS-CM

Atlas Steels Ltd.
C 0.4, Mn 1.2, Mo 0.15, Cr 0.6, bal Fe.
Normalized: 100,000 TS; 65,000 YS; 24 El; 53 RA; 217 Brin.
For machinery parts, bolts axles, jigs, fasteners, gears, shafts.
Shock resistant, tough, water hardening.

ATLAS-CM

Rio Algom Corp.
C 0.4, Mn 1.2, Mo 0.15, Cr 0.6, bal Fe.
Normalized: 100,000 TS; 65,000 YS; 24 El; 53 RA; 217 Brin.
For machinery parts, bolts axles, jigs, fasteners, gears, shafts.
Shock resistant, tough, water hardening.

ATLAS-KK

Atlas Specialty Steels
C 1.1, Cr 1.7, Mo 0.4, V 1.4, bal Fe.
For bearings, machinery parts; oil.hardening. *Obsolete*

ATLAS-SPS

Atlas Steels Ltd.
C 0.4, Mn 0.75, Cr 0.6, Ni 1.25, Mo 0.15, bal Fe.
Annealed: 105,000 TS; 70,000 YS; 25 El; 212 Brin. Hardened: 145,000 TS; 132,000 YS; 20 El; 293 Brin. For heavy duty shafts, gears, axles, spindles, tool holders, feed screws. Oil hardening, shock resistance.

ATLAS-SPS

Rio Algom Corp.
C 0.4, Mn 0.75, Cr 0.6, Ni 1.25, Mo 0.15, bal Fe.
Annealed: 105,000 TS; 70,000 YS; 25 El; 212 Brin. Hardened: 145,000 TS; 132,000 YS; 20 El; 293 Brin. For heavy duty shafts, gears, axles, spindles, tool holders, feed screws. Oil hardening, shock resistance.

ATLOY D.D.

Atlantic Steel Corp.
C, alloy, bal Fe.
Heat treated: 310,000 TS. For bushings, gears, clutches, gages, spindles; tough, non-deforming.

ATLOY FM

Atlantic Steel Corp.
C, B, Pb, alloy, bal Fe.
For gears, arbors, worms, collets, motor shafts; free-cutting, fatigue and wear resistant.

ATLOY H.T.
Atlantic Steel Corp.
C, Cr, Mn, bal Fe.
Heat treated: 112,000-260,000 TS; 95,000-230,000 YS; 9.5-21.5 El; 26-62.5 RA; 241-514 Brin. For arbors, axles, bolts, cams, gears, jaws; fatigue and wear resistant.

ATLOY Z
Atlantic Steel Corp.
C, B, alloy, bal Fe.
Heat treated: 280,000 TS. For gears, axles, connecting rods, propeller shafts; high strength.

ATMODIE
Columbia Tool Steel Co.
Tool material. C 1.5, Cr 12, Mo 0.8, V 0.85, bal Fe.
For blanking and forming dies, punches, shear blades; tough and wear resistant; air hardening; AISI Type D2.

ATMODIE 4
Columbia Tool Steel Co.
Tool material. C 2.22, Mn 0.4, Si 0.45, Mo 0.8, Cr 11.65, V 0.2, bal Fe.
Air hardening cold work tool steel. For coining dies, slither's, forming mandrels. AISI D4.

ATMODIE 5
Columbia Tool Steel Co.
Tool material. C 1.55, Mn 0.4, Si 0.4, Mo 0.8, Cr 11.9, V 0.5, Co 3.25, Ni 0.4, bal Fe.
High hardenability, high wear resistance; air hardening cold work tool steel. For draw dies, trim dies, extrusion mandrels. AISI D5.

ATMODIE SMOOTHCUT
Columbia Tool Steel Co.
Tool material. C 1.5, Cr 12, Mo 0.9, V 1, bal Fe.
For drawing and blanking dies, punches, mandrels, gages; air or oil hardening; free machining, wear and corrosion resistant. AISI D2.

ATMOS
Ludlow Steel Corp.
C, alloy, bal Fe.
For dies, punches; air hardened.

ATOM ARC 10018
Chemetron Corp.
Mn 1.2, Si 0.54, Ni 1.73, Mo 0.33, weld metal: 0.05 C, bal Fe.
As welded: 103,000 psi TS; 96,000 psi YS; 24 El. Iron powder, low hydrogen electrode for welding where joints must equal 100,000 psi tensile strength. AWS Class E 10018-M.

ATOM ARC 12018
Chemetron Corp.
Mn 1.9, Si 0.25, Cr 0.85, Ni 2, Mo 0.5, weld metal: 0.05 C, bal Fe.
As welded: 132,000 psi TS; 120,000 psi YS; 20 El. Iron powder, low hydrogen electrode for welding low-alloy, high-tensile steels requiring weld strengths of 120,000 psi minimum tensile, AWS Class E 12018-M.

ATOM ARC 4130
Chemetron Corp.
Mn 1.25, Si 0.4, Cr 0.5, Ni 1.28, Mo 0.2, weld metal: 0.18 C, bal Fe.
Welded, Quench + Temper 1100°F: 138,000 psi TS; 121,000 psi YS; El. All-position, iron-powder, low-hydrogen electrode to weld AISI 8630, 4130 and similar alloys.

ATOM ARC 4140
Chemetron Corp.
Mn 0.95, Si 0.62, Cr 0.8, Mo 0.33, weld metal: 0.38 C, bal Fe.
All-position, iron-powder, low-hydrogen electrode to weld AISI 4140 and similar steels, including castings. Properties after heat treatment are similar to AISI 4140 base metal.

ATOM ARC 4340
Chemetron Corp.
C 38, Mn 0.94, Si 0.7, Cr 0.8, Mo 0.4, Ni 1.3, weld metal: 0, bal Fe.
All-position, iron-powder, low-hydrogen electrode to weld AISI 4340 an similar steels, including castings. Properties after heat treatment are similar to AISI 4340 base metal.

ATOM ARC 502
Chemetron Corp.
Mn 0.62, Si 0.6, Cr 5.72, Mo 0.62, weld metal: 0.05 C, bal Fe.
Welded, stress annealed 1025°F: 96,000 psi TS; 78,000 psi YS; 22 Iron powder, low hydrogen electrode for welding 4-6 Cr steels as AISI 501,502. AWS Class E501-16(-18).

ATOM ARC 7018
Chemetron Corp.
Mn 1.1, Si 0.5, weld metal: 0.06 C, bal Fe.
As welded: 75,000 psi TS; 68,000 psi YS; 34 El. Iron-powder, low hydrogen electrode for welding wide variety of carbon steels. AWS class E7018.

ATOM ARC 7018 MO
Chemetron Corp.
Mn 0.75, Si 0.56, Mo 0.53, weld metal: 0.05 C, bal Fe.
As welded: 79,000 psi TS; 68,000 psi YS; 31 El. Iron-powder, low hydrogen electrode for welding low alloy, high tensile steels of 50,000 psi YS, and also the 0.5 Mo steels. AWS class E7018-A1.

ATOM ARC 8018
Chemetron Corp.
Mn 1.06, Ni 1.04, weld metal: 0.05 C, bal Fe.
As welded: 84,000 psi TS; 73500 psi YS; 30 El. Iron-powder, low hydrogen electrode for welding high strength steels in the 70,000-80,000 psi tensile strength range and low temperature operation. AWS Class E 8018-C3.

ATOM ARC 8018 C1
Chemetron Corp.
Mn 1.06, Si 0.31, Ni 2.37, weld metal: 0.04 C, bal Fe.
As welded: 88,500 psi TS; 73,800 psi YS; 28 El. Iron-powder, low hydrogen electrode for welding 2 1/3% Ni steels for low temperature operation. AWS Class E8018-C1.

ATOM ARC 8018 CM
Chemetron Corp.
Mn 0.68, Si 0.6, Cr 1.24, Mo 0.4, weld metal: 0.05 C, bal Fe.
As welded: 92,000 psi TS; 83,000 psi YS; 27 El. Iron-powder, low hydrogen electrode for welding low Cr-Mo steels in power piping and boiler work. AWS Class E 8018-B2.

ATOM ARC 8018 N
Chemetron Corp.
Mn 0.84, Si 0.37, Ni 0.3, weld metal: 0.05 C, bal Fe.
As welded: 94,000 psi TS; 83,000 psi YS; 25 El. Iron-powder, low hydrogen electrode for welding 2-4% Ni steels; for low temperature equipment; and for welds subject to impact. AWS Class E 8018-C2.

ATOM ARC 8018 NM
Chemetron Corp.
Si 0.4, Ni 1, Mo 0.5, weld metal: 0.06 C 1.10 Mn, bal Fe.
As welded: 93,000 psi TS; 82,000 psi YS; 25 El. Iron-powder, low hydrogen electrode for welding quenched and tempered Mn-Ni-Mo steels, especially for pressure vessels. AWS Class E 8018-G.

ATOM ARC 9018
Chemetron Corp.
Mn 1.11, Si 0.32, Ni 1.72, Mo 0.28, weld metal: 0.05 C, bal Fe.
As welded: 94,400 psi TS; 84,600 psi YS; 27 El. Iron-powder, low hydrogen electrode for welding HY-80 and other high tensile, quenched and tempered steels. AWS Class E 9018-M.

ATOM ARC 9018CM
Chemetron Corp.
Mn 0.75, Si 0.6, Cr 2.2, Mo 1.05, weld metal: 0.05 C, bal Fe.
As welded: 96,000 psi TS; 83,000 psi YS; 25 El. Iron-powder, low hydrogen electrode for welding 2% Cr steels. AWS Class E 9018-B3.

ATOM ARC 9018HT
Chemetron Corp.
Mn 0.8, Si 0.65, Cr 2.3, Mo 1, weld metal: 0.14 C, bal Fe.
Welded, normalized + tempered 1275°F: 99,000 psi TS; 82,000 psi YS; 24 El. For welding Cr-Mo steel castings in all-position, with iron powder, low, hydrogen electrode.

ATOM ARC "T"
Chemetron Corp.
Mn 1.53, Si 0.27, Cr 0.31, Ni 1.88, Mo 0.42, weld metal: 0.06 C, bal Fe.
As welded: 115,000 psi TS; 103,000 psi YS; 22 El. Iron powder, low hydrogen electrode for high strength welds usable also at low temperature. AWS Class E 11018-M.

ATOMINPHY B2
Creusot-Loire
C 0-0.1, Mn 0-2, Si 0-1.5, Ni 13-15, Cr 17-20, Co 0-0.2, Be 1.5-2, bal Fe.
Stainless casting: 220-250 Brin. Protective screens, nuclear energy. AFNOR Z8CNB19.14M.

ATR ALLOY
Metropolitan-Vickers Electrical Co. Ltd.
Cu 0.5, Mo 0.5, Sn 1.5, Fe 0.12, Cr 0.1, Ni 0.05, bal Zr.
For nuclear reactors. Resists CO_2 corrosion.

ATR ALLOY
Associated Electrical Industries Ltd.
Cu 0.5, Mo 0.5, Sn 1.5, Fe 0.12, Cr 0.1, Ni 0.05, bal Zr.
For nuclear reactors. Resists CO_2 corrosion.

ATRAX A 6, ETC.
Now A 6.

ATRIX 1
Rochling Burbach GmbH
C 0.19, Si 0.5, Mn 1.15, bal Fe.
For gears, shafts, machine tool parts; case hardening steel. *Obsolete*

ATRIX 100
Saarstahl AG
C 0.17, Si 0.3, Mn 1.1, bal Fe.
Heat treated: 410-590 N/mm^2 TS. Weldable, forgeable, for boiler flanges. W.-Nr. 0481. *Obsolete*

ATRIX 101
Saarstahl AG
No 0-2.05808e+028.

ATRIX 10A
Rochling Burbach GmbH
C 0.24, Cr 1.15, Mo 0.25, bal Fe.
For gears, bolts, cams, camshafts; case hardening or water hardening steel. *Obsolete*

ATRIX 10E
Rochling Burbach GmbH
C 0.16, Cr 1.05, Mo 0.25, bal Fe.
For gears, bolts, camshafts, cams; case hardening steel, tough. *Obsolete*

ATRIX 10S
Rochling Burbach GmbH
C 0.34, Cr 1.55, Ni 1.55, Mo 0.2, bal Fe.
For gears, bolts, crankshafts, fasteners; oil hardened, shock resistant. *Obsolete*

ATRIX 10SS
Rochling Burbach GmbH
C 0.28, Mn 0.55, Mo 0.2, Ni, Cr, bal Fe.
For gears, bolts, crankshafts, fasteners; oil hardened, shock resistant. *Obsolete*

ATRIX 110
Saarstahl AG
Now SAARSTAHL 20 MNMONI 55.

ATRIX 12
Rochling Burbach GmbH
C 0.24, Cr 1.15, Mo 0.25, Mn 0.55, bal Fe.
For gears, bolts, crankshafts, fasteners; oil or water hardened, tough. *Obsolete*

ATRIX 120
Saarstahl AG
Now SAARSTAHL 15 MO 3.

ATRIX 122
Saarstahl AG
Now SAARSTAHL 14 MOV 63.

ATRIX 140 CU
Saarstahl AG
Now SAARSTAHL 15 NICUMONB 5.

ATRIX 15A
Rochling Burbach GmbH
C 0.24, Cr 1.25, Mo 0.4, Mn 0.55, bal Fe.
For gears, bolts, fasteners, pinions; oil or water hardened, tough. *Obsolete*

ATRIX 15E
Rochling Burbach GmbH
C 0.16, Cr 1.05, Mo 0.45, Mn 0.65, bal Fe.
For gears, cams, camshafts, machine tool parts; case hardening steel. *Obsolete*

ATRIX 15S
Rochling Burbach GmbH
C 0.28, Cr 1.15, Ni 1.15, Mo 0.45, bal Fe.
For gears, bolts, fasteners, oil refinery equipment; oil hardened, shock resistant. *Obsolete*

ATRIX 15SS
Rochling Burbach GmbH
C 0.28, Cr 1.25, Mo 0.35, Mn 0.45, Ni 1.15, bal Fe.
For gears, bolts, fasteners, shafts; oil hardened, shock resistant. *Obsolete*

ATRIX 20
Rochling Burbach GmbH
C 0.24, Cr 1.35, Mo 0.55, V 0.2, bal Fe.
For gears, bolts, machine tool parts, shafts; oil hardened, shock resistant. *Obsolete*

ATRIX 200
Saarstahl AG
Now SAARSTAHL 13 CRMO 44.

ATRIX 203
Saarstahl AG
Now SAARSTAHL 24 CRMO 5.

ATRIX 204
Saarstahl AG
Now SAARSTAHL 10 CRMO G10.

ATRIX 231
Saarstahl AG
Now SAARSTAHL 21 CRMOV 511.

ATRIX 232
Saarstahl AG
Mn 0.45, Cr 1.35, Mo 0.55, Si 0.25, C 0.24, V 0.2, bal Fe.
Heat treated: 685-835 N/mm^2 TS. Resistant to elevated temperatures up to 530°C. W.-Nr.7733. *Obsolete*

ATRIX 234
Saarstahl AG
Now SAARSTAHL 21 CRMOV 57.

ATRIX 300
Saarstahl AG
Now SAARSTAHL 22 NIMOCR 37.

ATRIX 321
Saarstahl AG
Now SAARSTAHL 23 CRNIMO 747.

ATRIX 380
Saarstahl AG
Now SAARSTAHL 20 CRMONIV 47.

ATRIX 381
Saarstahl AG
Now SAARSTAHL 28 CRMONIV 49.

ATRIX 382
Saarstahl AG
Now SAARSTAHL 30 CRMONIV 511.

ATRIX 3R
Rochling Burbach GmbH
C 0.12, Si 1.15, Cr 2.5, Mo 0.45, bal Fe.
For gears, cams, camshafts, pinions; case hardening steel, tough. *Obsolete*

ATRIX 4R
Rochling Burbach GmbH
C 0.15, Mo 0.3, Mn 0.6, Si 0.25, bal Fe.
Annealed: 70,000 TS; 40,000 YS; 25 El; 60 RA; 145 Brin. For gears, shafts, camshafts, bolts; case hardening steel, tough. *Obsolete*

ATRIX 5R
Rochling Burbach GmbH
C 0.13, Cr 0.85, Mo 0.45, bal Fe.
For gears, cams, camshafts; case hardening steel, tough. *Obsolete*

ATRIX N 3
Saarstahl AG
Now SAARSTAHL 24 CRMO 10.

ATRIX N 9
Saarstahl AG
Now SAARSTAHL 20 CRMOV 135.

ATRIX N10
Rochling Burbach GmbH
C 0.21, Cr 3, Mo 0.4, V 0.8, W 0.4, bal Fe.
For oil refinery equipment, heat exchangers; heat and creep resistant. *Obsolete*

ATS
Thyssen Edelstahlwerke AG
C 0.05, Mn 1.27, Cr 16.5, Ni 12.3, Cb 1.1, bal Fe.
For turbine blades and nozzles. Austenitic, corrosion and heat resistant. *Obsolete*

ATS-103
Thyssen Edelstahlwerke AG
C 0.38-0.47, Mn 1-2, Cr 19-21, Ni 19-21, Co 19-21, Mo 3.5-4.5, W 3.5-4.5, Cb 3.5-4.5, bal Fe.
Rolled: 152,000 TS; 80,000 YS; 19 El; 25 RA; 260 Brin. Aged: 140,000 TS; 42,000 YS; 20 El; 19 RA; 300 Brin. For supercharger and jet engine components, turbine blades and wheels. Corrosion and heat resistant. *Obsolete*

ATS-113
Thyssen Edelstahlwerke AG
C 0.4, Mn 1.2, Cr 20, Ni 20, Co 43, W 4, Fe 3, Cb 4.
Heat treated: 140,000 TS; 67,000 YS; 31 El; 31 RA; 269-352 Brin. For turbine blades and rotors; high temperature hardware, jet engine parts. High heat and corrosion resistant to 1500 F. *Obsolete*

ATS-360
Thyssen Edelstahlwerke AG
Cr 13-16, Co 18-22, Mo 4-5, C 0-0.2, Al 4.6, Ti 1.2, bal Ni.
At 20 C: 144,000 TS; 116,000 YS; 7 El; 7 RA. At 500 C: 132,000 TS; 105,000 YS; 8 El; 12 RA. For gas turbine and jet engine components, fasteners. Similar to Nimonic 105. High creep and oxidation resistance. *Obsolete*

ATS-390
Thyssen Edelstahlwerke AG
Cr 14-16, Co 13-17, Mo 3.6, Al 5, Ti 4, C 0.12-0.2, B 0.02, Zr 0.2, bal Ni.
Heat treated: 180,000 TS; 125,000 YS; 27 El; 28 RA; 370-440 Vickers. At 800 C: 147,000 TS; 109,000 YS; 19 El; 19 RA. For gas turbine and jet engine components, missiles, space equipment. Similar to Nimonic 115. High creep and oxidation resistance. *Obsolete*

ATS2
Thyssen Edelstahlwerke AG
C 0.08, Mn 1.3, Cr 17, Mo 2.2, Ni 13.5, 0.9% Cb or Ta, bal Fe.
For valves, jet engine components; high temperature resistant. *Obsolete*

ATS6
Thyssen Edelstahlwerke AG
C 0.08, Mn 1.3, Cr 17, Mo 1.4, Ni 13, V 0.7, Cb 0.9, N 2 , bal Fe.
For jet engine and missile components; high heat resistant. *Obsolete*

ATSCO
Atlantic Steel Corp.
C, alloy, bal Fe.
For tools, water hardened. AISI W1.

ATSCO AR
Atlantic Steel Corp.
C 0.9, Mn 0.7, Mo 0.3, Cr 1.1, bal Fe.
For shovel teeth, bucket lips; water hardened, abrasion resistant.

ATSCO EXTRA
Atlantic Steel Corp.
C 0.8-1.1, Si 0.25, Mn 0.25, bal Fe.
Water hardened: 166,000-216,000 TS; 110,000-150,000 YS; 11-15 El; 32-37 RA; 330-600 Brin. For taps, drills, reamers, punches, stamps, knurls, mandrels. Type W1; water hardened.

ATSCO SPECIAL
Atlantic Steel Corp.
C 0.5-1, Mn 0.5-1, Mo 0.3, bal Fe.
For rock drills, concrete wreckers, demolition tools; water hardened.

ATSIL
Atlantic Steel Corp.
C 0.5, Mn 0.6, W 0.5, Mo 0.3, Si 1.3, bal Fe.
For shear blades, rivet sets, cutting tools; shock resistant. AISI S5.

ATSINA
Chemalloy Electronics Corp.
C 0.7, Cr 0.5, W 5, bal Fe.
For magnets, electrical machinery. *Obsolete*

ATV-3
Creusot-Loire
Ni 35, Cr 11, W 3, bal Fe.
For turbine and gas turbine parts; heat resistant. *Obsolete*

ATV-R
Creusot-Loire
Ni 30, Cr 15, Mo, Al, bal Fe.
For turbine and gas engine parts; heat resistant. *Obsolete*

ATV-S
Creusot-Loire
Ni 35, Cr 11, C 0.15, Al 1, Ti 1, bal Fe.
For turbine and gas engine parts; heat resistant. *Obsolete*

ATVS 2

Creusot-Loire
C 0.04, Ni 26, Cr 13.5, Co 1, Mo 2.75, Ti 1.8, Al 0.3, bal Fe.
Good strength at elevated temperatures. For turbine discs, extrusion dies, high temperature bolts. AFNOR Z3NCT25; DISCALOY.

ATVS 7

Creusot-Loire
C 0.1, Ni 30, Cr 18, Co 20, Ti 2, Al 0.8, bal Fe.
Good strength at elevated temperature. AFNOR Z10NKC30.

ATVS 7 MO

Creusot-Loire
Cr 18, Co 20, Mo 3, Ti 2.75, C 0.06, Ni 37, bal Fe.
Good strength at elevated temperature. For turbine blades and parts. AFNOR Z6NKCDT38; REFRACTOLOY 26.

ATVS MO

Creusot-Loire
C 0-0.05, Cr 15, Ni 26, Mo 1.2, Ti 2, V 0.3, N 0.2, bal Fe.
Wrought, annealed, aged: 930 N/mm² psi TS min.
Austenitic, non-magnetic stainless, good strength; resistant to high temperature oxidation. AFNOR Z6NCTDV25.15.2; AISI 660.

ATW 432 ATOMIZED IRON POWDER

Hoeganaes Corp.
Fe 97, C 0.08, S 0-0.03, 0.80 H_2 (loss).
40 mesh powder for high yield stick electrode coatings.

AU177

Acciaierie Valbruna s.p.a.
Stainless Steel. C 0-0.15, Si 0-1, Mn 0-2, Cr 17, Ni 7, bal Fe.
Austenitic; work hardens rapidly. W. Nr.1.4310; AISI 301.

AU188

Acciaierie Valbruna s.p.a.
Stainless Steel. C 0-0.12, Si 0-1, Mn 0-1.25, Cr 18, Ni 8.75, P 0.04, S 0.03, bal Fe.
Austenitic. W. Nr. 1.4300; AISI 302.

AU188Z

Acciaierie Valbruna s.p.a.
Stainless Steel. C 0-0.12, Si 0-1, Mn 0-1.5, Cr 18, S 0.25, Ni 9, Mo 0-0.6, P 0.045, bal Fe.
Free machining austenitic. W. Nr. 1.4305; AISI 303.

AU4SG

Pechiney/SMG
Si 0.5-1.2, Fe 0-0.7, Cu 3.5-5, Mn 0.4-1.2, Mg 0.2-0.8, bal Al.
Wrought, heat treatable aluminum alloy; AA2014. For production of high quality weapon parts.

AUBERT & DUVAL 56 A

Aubert & Duval
Alloy steel. C 0.28, Cr 1.5, Mo 0.7, V 0.3, bal Fe.
Annealed: 950 MPa TS; 17 El; 170 Brin. For aircraft construction.

AUBERT & DUVAL 56 C

Aubert & Duval
Alloy steel. C 0.1, Cr 10.5, Co 6, Mo 0.8, Nb 0.5, V 0.3, bal Fe.
For aircraft applications, turbine compressors and parts requiring good creep properties between 400 and 600 °C.

AUBERT & DUVAL 56 T 5

Aubert & Duval
C 0.2, Cr 11, Mo 0.7, Nb 0.4, V 0.18, bal Fe.
Annealed: 95,000 TS; 40,000 YS; 25 El; 92 Rock B.
Hardened: 240,000 TS; 205,000 YS; 8 El; 50 Rock C. For valves, bearings, cutlery, surgical instruments, gears. Corrosion resistant, hardenable.

AUBERT & DUVAL 56 T G

Aubert & Duval
C 0.2, Cr 1.5, Mo 0.6, V 0.2, bal Fe.
For high temperature equipment. AFNOR 20CDV6.

AUBERT & DUVAL 67 T

Aubert & Duval
Alloy steel. C 0.08, Cr 18, Ni 12, Ti = 5 x C, bal Fe.
Water quenched: 520 MPa TS; 52 El. For petroleum industries, forgings, superheated headers, or tubes, bolts for superheated steam plants.

AUBERT & DUVAL 67 T.N

Aubert & Duval
Alloy steel. C 0.08, Cr 18, Ni 12, Nb = 10 x C, bal Fe.
Water quenched: 550 MPa TS; 47 El. For petroleum industries, forgings, superheated headers, or tubes, bolts for superheated steam plants.

AUBERT & DUVAL 819 A

Aubert & Duval
Alloy steel. C 0.38, Ni 4, Cr 1.75, Mo 0.5, bal Fe.
Annealed: 1150-1900 MPa TS; 10-16 El; 245 Brin. For aircraft components, notably undercarriage parts.

AUBERT & DUVAL 819B

Aubert & Duval
C 0.35, Ni 3.8, Cr 1.7, Mo 0.3, bal Fe.
For gears, shafts, machinery parts; oil hardened, shock resistant.

AUBERT & DUVAL 820

Aubert & Duval
Alloy steel. C 0.4, Ni 4.5, Cr 1.5, Mo 0.5, bal Fe.
Annealed: 1200-2000 MPa TS; 10-14 El; 285 Brin. Large section parts.

AUBERT & DUVAL 836 J

Aubert & Duval
Alloy steel. C 0.3, Cr 2, Ni 2, Mo 0.4, bal Fe.
Annealed: 1100 MPa TS; 17 El; 225 Brin. For shafts, pinions, gears.

AUBERT & DUVAL 897 D.MO

Aubert & Duval
Alloy steel. C 0.3, Ni 3, Cr 0.75, Mo 0.2, bal Fe.
Annealed: 1000-1750 MPa TS; 11-18 El; 220 Brin. For shafts, spindles, gears.

AUBERT & DUVAL 897D

Aubert & Duval
C 0.3, Ni 3, Cr 0.7, bal Fe.
For crankshafts, axles, connecting rods; AISI 3335; oil hardened, shock resistant.

AUBERT & DUVAL A.D 3 M

Aubert & Duval
Alloy steel. C 0.1, Ni 3, Cr 0.7, Mo 0.2, bal Fe.
Annealed: 900-1100 MPa TS; 12-16 El; 180 Brin. For shafts, spindles, pins, gears.

AUBERT & DUVAL A.P.X 4

Aubert & Duval
Alloy steel. C 0.06, Cr 16, Ni 4, Mo 1, bal Fe.
Annealed: 1000-1200 MPa TS; 16-18 El; 269 Brin. For welded fabrications exposed to sea water; hydraulic pumps and turbines, shafts, pins, bolts.

AUBERT & DUVAL A.P.X.

Aubert & Duval
C 0.16, Ni 2, Cr 17, bal Fe.
Annealed: 107,000 TS; 62,000 YS; 18 El; 35 RA; 220 Brin. For cutlery, valves, surgical instruments, ball bearings; Type 431; corrosion resistant.

AUBERT & DUVAL A.P.Z. 2

Aubert & Duval
C 0.8, Ni 1.5, Cr 20, Si 1.8, bal Fe.
Heat treated: 143,000 TS. For exhaust valves; corrosion and heat resistant.

AUBERT & DUVAL AD3

Aubert & Duval
C 0.08, Ni 3.1, Cr 0.65, bal Fe.
For shafts, cams, camshafts, machinery parts; AISI 3310; case hardened shock resistant.

AUBERT & DUVAL B X 1

Aubert & Duval
C 0.14, Ni 1.5, Cr 0.9, Mo 0.15, bal Fe.
Carburizing steel; core hardenable to 135 kg/mm² TS; 105 kg/mm² YS; 13 El. Similar to AFNOR 16NCD6. *Obsolete*

AUBERT & DUVAL B X 3

Aubert & Duval
C 0.1, Mn 0.75, Ni 1.4, Cr 1, Mo 0.2, bal Fe.
Alloy carburizing steel. AFNOR 10NCD6. *Obsolete*

AUBERT & DUVAL B X O

Aubert & Duval
C 0.19, Ni 1.5, Cr 0.9, Mo 0.2, bal Fe.
Carburizing steel; core hardenable to 140 kg/mm² TS; 110 kg/mm² YS; 12 El. AFNOR 18NCD6; similar to AISI 4320. *Obsolete*

AUBERT & DUVAL B.M. 3

Aubert & Duval
C 0.12, Cr 2.25, Mo 1, bal Fe.
Steel for high temperature operation. AFNOR 12CD9 10.

AUBERT & DUVAL B.M.V. 4

Aubert & Duval
Tool material. C 0.4, Cr 5, Mo 1.3, V 0.45, bal Fe.
Deep hardening steel for highly stressed heavy equipment; or hot work tool steel. Similar to AFNOR 40CDV20; AISI H11 tool steel.

AUBERT & DUVAL BM 6

Aubert & Duval
C 0.15, Si 0.3, Mn 0.45, Cr 4-6, Mo 0.5, bal Fe.
For stainless requirements or high temperature operation. AFNOR Z15CD5.05. *Obsolete*

AUBERT & DUVAL C.M.S.MO

Aubert & Duval
Alloy steel. C 0.35, Ni 1.3, Cr 1, Mo 0.2, bal Fe.
Annealed: 950-1150 MPa TS; 13-17 El; 215 Brin. For shafts, spindles, gears.

AUBERT & DUVAL C.N.S.

Aubert & Duval
C 0.35, Ni 1.2, Cr 0.85, bal Fe.
For crankshafts, axles, bolts, gears; AISI 3135; oil hardened, shock resistant.

AUBERT & DUVAL CNS3

Aubert & Duval
C 0.5-0.6, Ni 0.8-1.3, Cr 0.5-1, bal Fe.
For die blocks; hot work steel, oil hardened. *Obsolete*

AUBERT & DUVAL E.A 1

Aubert & Duval
Alloy steel. C 20, Ti 2, Al 1.25, Co 0-2, Fe 0-5, bal Ni.
Solution treated: 1000 kg/mm² TS min; 20 El min. For racing car and other engines functioning at high temperatures.

AUBERT & DUVAL F 65

Aubert & Duval
C 0.35, Cr 1, Mo 0.2, bal Fe.
For axles, gears, machinery parts; AISI 4137; oil hardened.

AUBERT & DUVAL F 66 J

Aubert & Duval
C 0.32, Cr 1, Mo 0.2, bal Fe.
For gears, shafts, machinery parts; AISI 4130; oil hardened.

AUBERT & DUVAL F 66 S

Aubert & Duval
C 0.25, Cr 1, Mo 0.2, bal Fe.
For gears, shafts, machinery parts; AISI 4125; tough.

AUBERT & DUVAL F.A.D.H.

Aubert & Duval
C 0.16, Ni 3.2, Cr 1, Mo 0.25, bal Fe.
Deep hardening carburizing steel; core hardenable to 140 kg/mm² TS; 110 kg/mm² YS; 13 El. For highly stressed heavy sections in aircraft and automotive industry.

AUBERT & DUVAL F.A.D.S.
Aubert & Duval
C 0.15, Ni 4.25, Cr 1.2, Mo 0.2, bal Fe.
Deep hardening carburizing steel; core hardenable to 145 kg/mm^2 TS; 120 kg/mm^2 YS; 12 El. For highly stressed heavy sections in aircraft and automotive industry. AFNOR 16NCD17.

AUBERT & DUVAL F.D.M.A.
Aubert & Duval
C 0.3, Ni 3.5, Cr 1.2, Mo 0.45, bal Fe.
For crankshafts, gears, connecting rod; oil hardened, shock resistant.

AUBERT & DUVAL FAD
Aubert & Duval
C 0-0.17, Ni 3-3.5, Cr 0.8-1.3, bal Fe.
For gears, cams, crankshafts; AISI 9315; case hardened, shock resistant. *Obsolete*

AUBERT & DUVAL G.H 4
Aubert & Duval
Alloy steel. C 0.4, Cr 3, Mo 1, V 0.2, bal Fe.
Annealed: 1400-2000 MPa TS; 10-13 El; 220 Brin. For piston pins; highly stressed aircraft parts.

AUBERT & DUVAL G.K 5 S
Aubert & Duval
Alloy steel. C 0.25, Cr 3.25, Mo 0.6, bal Fe.
Annealed: 700-1250 MPa TS; 16-27 El; 190 Brin. For components gears with sharp edges or angles that must withstand impact or friction.

AUBERT & DUVAL G.K.H.
Aubert & Duval
C 0.32, Cr 3, Mo 1, V 0.2, bal Fe.
For elevated temperature operation. AFNOR 32 CDV 12.

AUBERT & DUVAL GK3
Aubert & Duval
C 0.3, Cr 3, Mo 0.4, bal Fe.
For gears, shafts, axles, machinery parts; nitriding steel.

AUBERT & DUVAL GK5
Aubert & Duval
C 0.15-0.25, Cr 2.7-3.3, Mo 0.2-0.5, Mn 0.8, bal Fe.
For gears, shafts, machinery parts; nitriding steel. *Obsolete*

AUBERT & DUVAL JD 19
Aubert & Duval
C 0.2, Mn 0.8, Ni 0.55, Cr 0.5, Mo 0.2, Cu 0-0.35, bal Fe.
Carburizing steel. AFNOR 20NCD2. *Obsolete*

AUBERT & DUVAL L.K 3
Aubert & Duval
C 0.4, Cr 1.8, Al 1, Mo 0.25, bal Fe.
For gears, axles, machinery parts; nitriding steel.

AUBERT & DUVAL L.K 5
Aubert & Duval
C 0.3, Cr 1.8, Mo 0.25, Al 1, bal Fe.
For gears, shafts, machine tool parts; nitriding steel.

AUBERT & DUVAL M.E.P.
Aubert & Duval
C 0.45, Cr 1.2, W 2, Mo 0.3, V 0.25, bal Fe.
For impact tools, rivet sets, chisels, dies; cold work steel, oil hardened.

AUBERT & DUVAL M.M.R. 0
Aubert & Duval
Tool material. C 0.17, Ni 1.6, Cr 0.85, bal Fe.
For shafts, cams, gears, axles, machine tool parts; AISI 3120; case hardened, shock resist.

AUBERT & DUVAL M.M.R. 1
Aubert & Duval
C 0.15, Ni 1.6, Cr 0.85, bal Fe.
For gears, cams, camshafts; AISI 3115; case hardened.

AUBERT & DUVAL M.M.R. 3
Aubert & Duval
C 0.12, Ni 1.6, Cr 0.85, bal Fe.
For gears, cams, camshafts; AISI 3115; case hardened.

AUBERT & DUVAL M.O.C 2 J
Aubert & Duval
Alloy steel. C 0.45, Cr 1.1, Mo 0.3, bal Fe.
Annealed: 1000 MPa TS; 16 El; 230 Brin. For heavily stressed machine parts.

AUBERT & DUVAL M.O.C 3 J
Aubert & Duval
Alloy steel. C 0.4, Cr 1, Mo 0.2, bal Fe.
Annealed: 1000 MPa TS; 18 El; 230 Brin. For shafts and gears.

AUBERT & DUVAL M.U.V.
Aubert & Duval
C 1, V 0.15, bal Fe.
For punches, dies, threading taps, drawing dies; water hardened; Type W2.

AUBERT & DUVAL MES
Aubert & Duval
C 0.3-0.4, Cr 2.5-3, W 2.5-3, bal Fe.
For rivet sets, extrusion rams and dies; hot work steel, oil or air hardened. *Obsolete*

AUBERT & DUVAL MOC2
Aubert & Duval
C 0.42, Cr 1.6, Mo 0.85, bal Fe.
For gears, axles, shafts, bolts, spindles; AISI 4145; oil hardened.

AUBERT & DUVAL N.C 40 M
Aubert & Duval
C 0.4, Ni 1.8, Cr 0.8, Mo 0.25, bal Fe.
Deep hardening structural steel; for axles, gears, torsion bars. AFNOR 30 NCD 10; AISI 4340.

AUBERT & DUVAL N.C.A.V.
Aubert & Duval
C 0.12, Ni 3.3, Cr 0.75, bal Fe.
For cams, gears, camshafts, machinery parts; AISI 3310; case hardened, shock resistant.

AUBERT & DUVAL N.C.A.V. 4
Aubert & Duval
C 0.16, Ni 3.3, Cr 0.75, bal Fe.
For camshafts, cams, gears, machinery parts; AISI 3316; case hardened, shock resistant.

AUBERT & DUVAL N.C.A.V. 5
Aubert & Duval
C 0.18, Ni 3.4, Cr 0.85, bal Fe.
For camshafts, cams, gears, machinery parts; AISI 3320; case hardened, shock resistant.

AUBERT & DUVAL NC36
Aubert & Duval
C 0.35-0.45, Cr 0.5-1, Ni 3-3.5, Mn 0.8, bal Fe.
For gears, shafts, machinery parts; AISI 3340; oil hardened, shock resistant. *Obsolete*

AUBERT & DUVAL NCAV2
Aubert & Duval
C 0.14, Ni 3.3, Cr 0.75, bal Fe.
For shafts, gears, camshafts, machinery parts; AISI 3316; case hardened, shock resistant.

AUBERT & DUVAL P.E.R. 2 U
Aubert & Duval
Nickel. C 0.09, Cr 20, Co 19, Ti 2.5, Al 1.35, Fe 0-1, bal Ni.
High temperature alloy for blades and wheels for jet or turbo-prop engines. AFNOR NCK 20 TA.

AUBERT & DUVAL R 2 C
Aubert & Duval
C 1, Cr 2, bal Fe.
For stamping tools, dies, jigs, templates; oil hardened.

AUBERT & DUVAL R.A.D.
Aubert & Duval
C 1, Cr 1.5, bal Fe.
For roller bearings, thrust bearings; AISI 52100; water hardened, wear resistant.

AUBERT & DUVAL R12S
Aubert & Duval
C 0.6-0.7, W 0.4-0.8, V 0.15-0.4, bal Fe.
For punches, dies, chisels, impact tools; water hardened, tough. *Obsolete*

AUBERT & DUVAL RA 6
Aubert & Duval
Alloy steel. C 0.85, Mo 5.5, W 5.5, Cr 4.5, V 2, bal Fe.
Annealed: 875 Vickers; 240 Brin. For wire drawing dies.

AUBERT & DUVAL S.C.V.
Aubert & Duval
C 0.15, Cr 1.25, Mo 0.9, V 0.25, bal Fe.
For elevated temperature operation. AFNOR 15 CDV 6.

AUBERT & DUVAL S.M.1
Aubert & Duval
C 0.6, Ni 2.7, Cr 0.8, Mo 0.35, V 0.15, bal Fe.
For die blocks; hot work steel, air hardened.

AUBERT & DUVAL S.M.H
Aubert & Duval
Alloy steel. C 1, Cr 5, Mo 1, V 0.3, bal Fe.
Annealed: 675-740 Vickers; 240 Brin. Cold work tools. For cutting and stamping dies, drawing dies, rolling mill bearings.

AUBERT & DUVAL S.M.V.
Aubert & Duval
C 0.35, Cr 5, Mo 1.3, W 1.5, V 0.4, bal Fe.
For forging dies, punches, mandrels, hot shear blades; air hardened, hot work steel.

AUBERT & DUVAL S.O.S. 3
Aubert & Duval
C 0.4, Cr 8.5, Si 3.25, bal Fe.
For motor car inlet valves, also for super-charged diesel engines. Similar to EN 52; AFNOR Z 45 CS 09 03.

AUBERT & DUVAL SMV 3
Aubert & Duval
C 0.4, Cr 5, Mo 1.3, V 0.4, bal Fe.
Hot work tool steel. AFNOR Z 38 CDWV 5; similar to AISI H12.

AUBERT & DUVAL SY 3
Aubert & Duval
Alloy steel. C 1.6, Cr 12, Mo 0.8, V 0.8, Co 0.8, bal Fe.
Annealed: 800 Vickers; 240 Brin. For cutting or cropping tools.

AUBERT & DUVAL T.A 2
Aubert & Duval
Alloy steel. C 0.35, Ni 3.7, Cr 1.8, Mo 0.3, bal Fe.
Annealed: 400-550 Vickers; 245 Brin. For forging dies.

AUBERT & DUVAL T.A 3
Aubert & Duval
Tool material. C 0.4, Ni 4.4, Mo 0.2, Cr 1.6, bal Fe.
Hot work tool steel. AFNOR 32 NDC 18.12.

AUBERT & DUVAL V 300
Aubert & Duval
C 0.45, Si 1.6, Cr 0.6, Mo 0.25, Mn 0.6, bal Fe.
For suspension springs, torsion bars; AISI 9245; tough.

AUBERT & DUVAL V.C.E
Aubert & Duval
Alloy steel. C 0.5, Cr 1, V 0.15, bal Fe.
Annealed: 1200-2150 MPa TS; 8-12 El; 210 Brin. For machine parts and springs subject to exceptional stress and strain.

AUBERT & DUVAL VCD
Aubert & Duval
C 0.3-0.4, Cr 1-1.5, V 0.2-0.4, bal Fe.
For punches, rivet sets, dies, impact tools; water hardened, shock resistant. *Obsolete*

AUBERT & DUVAL X 13
Aubert & Duval
C 0.3, Cr 13, bal Fe.
For valves, cutlery, turbine blades, surgical instruments; Type 420; corrosion resistant.

AUBERT & DUVAL X 13 B.C
Aubert & Duval
Alloy steel. C 0.06, Cr 13, bal Fe.
Annealed: 580-700 MPa TS; 21-26 El; 145 Brin. For petroleum, food and chemical industries. Nuclear installations.

AUBERT & DUVAL X 13 D
Aubert & Duval
Stainless steel. C 0.2, Cr 13, bal Fe.
Martensitic; hardenable to 140 kg/mm^2 TS; 115 kg/mm^2 YS; 15 El. For steam turbine blades, valves, cocks, fittings. AFNOR Z 20 C 13; G.B 25.62, En 56 C; similar to AISI 420.

AUBERT & DUVAL X 13 E
Aubert & Duval
Stainless steel. C 0.13, Cr 13, Ni 0.5, bal Fe.
Martensitic; hardenable to 135 kg/mm^2 TS. For compressor blades and wheels; pump and valve parts. AFNOR Z 12 C 13; G.B. En 56 B.

AUBERT & DUVAL X 13 M
Aubert & Duval
Alloy steel. C 0.1, Cr 13, bal Fe.
Annealed: 650-850 MPa TS; 19-28 El; 165 Brin. For steam turbine blades, pump pistons and rotors, marine engine pistons; cocks, taps and fittings for petroleum industry; equipment for the chemical industry.

AUBERT & DUVAL X 13 T 1
Aubert & Duval
Alloy steel. C 0.4, Cr 13, bal Fe.
Annealed: 550-680 Vickers; 210 Brin. For cutlery, piston rods, pump parts for synthetic fibers.

AUBERT & DUVAL X 13 V.D
Aubert & Duval
Alloy steel. C 0.12, Cr 11.5, Ni 2.5, Mo 1.6, V 0.3, bal Fe.
Annealed: 1050-1350 MPa TS; 15-17 El; 255 Brin. For machine and engine parts.

AUBERT & DUVAL X 13 W 3
Aubert & Duval
Alloy steel. C 0.18, Cr 13, W 3, Ni 2, bal Fe.
Annealed: 1100-1500 MPa TS; 14-16 El; 290 Brin. For compressor blades; fasteners for high temperatures.

AUBERT & DUVAL X 15 D 2
Aubert & Duval
Stainless steel. C 0.08, Cr 15, Ni 7, Mo 2.5, Al 1, bal Fe.
Precipitation hardening stainless. AFNOR Z 8 CND 15 07.

AUBERT & DUVAL X 15 U 5
Aubert & Duval
Alloy steel. C 0-0.07, Cr 15, Ni 5, Cu 3, bal Fe.
Annealed and solution treated: 970-1370 MPa TS; 13-16 El; 331 Brin. For forgings and machine parts.

AUBERT & DUVAL X 16 D 3
Aubert & Duval
C 0.12, Cr 16, Ni 4.8, Mo 3, bal Fe.
HT: 150 kg/mm^2 TS; 115 kg/mm^2 YS; 16 El. Good strength and corrosion resistance. AFNOR Z 15 CND 16 04 03; USA AM 355.

AUBERT & DUVAL X 17 T
Aubert & Duval
Ni 13, Cr 17, C 0.1, W 3.5, Ti 0.5, bal Fe.
High temperature alloy; used up to 750°C. For jet engine and gas turbine parts.

AUBERT & DUVAL X 17 U 4
Aubert & Duval
Stainless steel. C 0-0.07, Cr 16.5, Ni 4, Cu 4, bal Fe.
Precipitation hardening stainless. Z 8 CNU 17 04; similar to 17-4 PH.

AUBERT & DUVAL X 18
Teledyne Vasco
C 0-0.12, Ni 8-11, Cr 17-20, bal Fe.
Annealed: 80,000 TS; 35,000 YS; 55 El; 75 RA; 150 Brin. For chemical plant equipment, tanks, mixers; Type 302; stainless, austenitic. *Obsolete*

AUBERT & DUVAL X 18 B.C.
Aubert & Duval
Stainless steel. C 0.02, Ni 11, Cr 19, bal Fe.
Annealed: 85,000 TS; 35,000 YS; 60 El; 70 RA; 150 Brin. For chemical plant equipment, tanks, mixers; Type 304; austenitic.

AUBERT & DUVAL X 18 J
Aubert & Duval
Stainless steel. C 0.06, Cr 19, Ni 10, bal Fe.
Austenitic. AFNOR Z 6 CN 18 09; AISI 304.

AUBERT & DUVAL X 18 M
Aubert & Duval
Alloy steel. C 0.06, Cr 17, Ni 12, Mo 2.5, bal Fe.
Water quenched: 580 MPa TS; 55 El. For chemical, wine, paper, petroleum, and electrical industries; shipbuilding, aircraft construction, railway rolling stock, household utensils, building trade hardware; bolts, nuts and screws.

AUBERT & DUVAL X 18 M.B.C.
Aubert & Duval
Stainless steel. C 0.02, Ni 12, Cr 17, Mo 2.5, bal Fe.
Annealed: 85,000 TS; 35,000 YS; 50 El; 65 RA; 160 Brin. For acid resistant chemical plant equipment; Type 316; austenitic.

AUBERT & DUVAL X 18 M.NB
Aubert & Duval
Stainless steel. C 0.08, Cr 18, Ni 13, Mo 2.5, Nb = 10 x C, bal Fe.
Stabilized austenitic; for welded chemical equipment. AFNOR Z6 CNDNb 17-12; modified AISI 316.

AUBERT & DUVAL X 18 M.P
Aubert & Duval
Stainless steel. C 0.08, Cr 17, Ni 12, Mo 2.5, Ti = 5 x C, bal Fe.
Stabilized austenitic; for welded chemical equipment. AFNOR Z 8 CNDT 17-12; modified AISI 316.

AUBERT & DUVAL X 18 NB
Aubert & Duval
Stainless steel. C 0.08, Cr 18, Ni 10.5, Nb = 10 x C, bal Fe.
Stabilized austenitic. AFNOR Z 6 CNNb 18-11; AISI 347.

AUBERT & DUVAL X 18 P
Aubert & Duval
Stainless steel. C 0-0.12, Ni 9-12, Cr 17-20, Ti = 5 x C, bal Fe.
Annealed: 85,000 TS; 35,000 YS; 55 El; 65 RA; 150 Brin. For welded chemical plant equipment, mixers, tanks; Type 321; stainless, austenitic. *Obsolete*

AUBERT & DUVAL X 18 P.A
Aubert & Duval
Alloy steel. C 0.08, Cr 18, Ni 11, Ti > 5 x C, bal Fe.
Water quenched: 600 MPa TS; 55 El. For chemical, wine, paper, petroleum, and electrical industries; shipbuilding, aircraft construction, railway rolling stock, household utensils, building trade hardware; bolts, nuts and screws.

AUBERT & DUVAL X 18 U
Aubert & Duval
Alloy steel. C 0-0.08, Cr 18, Ni 9, S 0.14, bal Fe.
Water quenched: 600 MPa TS; 55 El. For parts mass produced on automatic lathes.

AUBERT & DUVAL X 19 T 4
Aubert & Duval
Alloy steel. C 0.45, Si 2.5, Cr 18, Ni 9, W 1, bal Fe.
Austenitic, nitrided steel. Water quenched: 900 kg/mm^2 TS; 30 El. For motorcar, lorry and diesel engine valves.

AUBERT & DUVAL X 20 DU
Aubert & Duval
Alloy steel. C 0-0.07, Cr 20, Ni 9, Mo 2.5, Cu 1.5, bal Fe.
Quenched: 680-800 MPa TS; 30-45 El. For chemical industries, petroleum refineries, food industries, papermaking.

AUBERT & DUVAL X 20 T
Aubert & Duval
C 0.25, Ni 10, Cr 22, W 2.1, Si 1.2, bal Fe.
Heat treated: 114,000 TS; 65,000 YS; 40 El. For valves for diesel engines; austenitic, corrosion and heat resistant.

AUBERT & DUVAL X 20 T 2
Aubert & Duval
Alloy steel. C 0.25, Si 1, Cr 21, Ni 12, W 2.7, bal Fe.
Oil or water quenched: 700 MPa TS; 40 El. For aviation industries, gas turbine casings, rings, guideblades; land and marine gas turbines; nozzles, stationary and rotating blades.

AUBERT & DUVAL X 203
Aubert & Duval
Alloy steel. C 0.12, Cr 20, Ni 20, Co 20, Mo 3.5, W 3, Nb 1, bal Fe.
Heat treated: 850 MPa TS; 45 El. For gas turbine components; combustion chamber parts, blades; bolts for elevated temperatures.

AUBERT & DUVAL X 21 R B
Aubert & Duval
Alloy steel. C 0.5, Si 0.25, Mn 9, Cr 21, Ni 4, Nb 0.4, 0.4 N$_2$, bal Fe.
Solution treated: 1000 kg/mm^2 TS; 15 El. For motorcar, lorry and diesel engine valves.

AUBERT & DUVAL X 25
Aubert & Duval
Stainless steel. C 0-0.15, Ni 18-20, Cr 24-26, bal Fe.
Annealed: 100,000 TS; 45,000 YS; 50 El; 65 RA; 185 Brin. For furnace parts, valve parts, jet engine parts; Type 310; stainless, austenitic. *Obsolete*

AUBERT & DUVAL X.D.B. STEEL
Aubert & Duval
Stainless steel. C 1, Cr 17, Mo 0-0.5, bal Fe.
Martensitic stainless steel; hardenable to 700 HV; for valve seats and guide rings. Similar specs: AFNOR Z 100 C 17; AISI 440 C. *Obsolete*

AUBERT & DUVAL X.D.B.D
Aubert & Duval
Alloy steel. C 1, Cr 17, Mo 0.5, bal Fe.
Annealed: 640-700 Vickers; 230 Brin. For ball or roller bearings exposed to corrosive agents or high temperatures (up to 500°C); valve seats or guiding rings in contact with steam.

AUBERT & DUVAL X.M 114
Aubert & Duval
C 0.5, Si 0-0.25, Mn 9, Cr 21, Ni 4, bal Fe.
Solution treated and aged: 105 kg/mm^2 TS; 65 kg/mm^2 YS; 11 El; 300 Brin. For exhaust valves. AFNOR Z 50 CMN 22.

AUBERT & DUVAL X.N 26 T.W
Aubert & Duval
Alloy steel. C 0.05, Ni 26, Cr 15, Ti 2, Mo 1.25, V 0.25, bal Fe.
Solution treated: 1000 MPa TS; 25 El. For discs and blades for gas or steam turbines.

AUBERT & DUVAL-56R

Aubert & Duval

C 0.08, Cr 12, Mo 1, V 0.25, W 1, bal Fe.

Annealed: 75,000 TS; 35,000 YS; 30 El; 70 RA; 82 Rock B.
Cold drawn: 95,000 TS; 80,000 YS; 15 El; 60 RA; 92 Rock B.
For springs, table flatware, knives, oil refinery and chemical
plant equipment. Corrosion resistant. *Obsolete*

AUBURN CAST

Hammond & Irving Inc.

C 0.7-1, bal Fe.

For tools, drills, taps; water hardened.

AUBURN EXTRA

Hammond & Irving Inc.

C 0.8-1.2, bal Fe.

For tools, drills, taps; water hardened.

AUBURN HOT DIE

Hammond & Irving Inc.

C 0.5, W 4, bal Fe.

For tools, hot shears; hot work steel.

AUBURN PERFECTION

Hammond & Irving Inc.

C, alloy, bal Fe.

For tools, punches; water hardened.

AUBURN SPECIAL

Hammond & Irving Inc.

C 0.7-1.4, bal Fe.

For tools, drills, taps; water hardened.

AUBURN STANDARD

Hammond & Irving Inc.

C 0.7-0.9, bal Fe.

For tools, drills, taps; water hardened.

AUBURN TOOL MAKER

Hammond & Irving Inc.

C 0.9, Mn 1.2, bal Fe.

For tools, dies, punches; non-deforming.

AUDEN WIRE

Johnson Matthey plc

Au 75, bal Pt.

Wire for dental purposes.

AUDIO 101

Manufacturer not listed.

Si 2.25-2.75, bal Fe.

For laminations for electrical equipment; high permeability.

AUDIO 117

Manufacturer not listed.

Si 1-1.5, bal Fe.

For laminations for electrical equipment; high permeability.

AUDIO 130

Manufacturer not listed.

Si 0.5-0.75, bal Fe.

For laminations for electrical equipment; high permeability.

AUDIO 145

Manufacturer not listed.

Si 0.25-0.5, bal Fe.

For laminations for electrical equipment; high permeability.

AUDIO 165

Manufacturer not listed.

Si 0-0.25, bal Fe.

For laminations for electrical equipment; high permeability.

AUDIO 52

Manufacturer not listed.

Si 4.5-4.75, bal Fe.

For laminations for electrical equipment; high permeability.

AUDIO 58

Manufacturer not listed.

Si 3.5-3.75, bal Fe.

For laminations for electrical equipment; high permeability.

AUDIO 65

Manufacturer not listed.

Si 3.25-3.75, bal Fe.

For laminations for electrical equipment; high permeability.

AUDIO 72

Manufacturer not listed.

Si 3.25-3.75, bal Fe.

For laminations for electrical equipment; high permeability.

AUDIO 82

Manufacturer not listed.

Si 2.25-2.75, bal Fe.

For laminations for electrical equipment; high permeability.

AUDIOLOY

Colt Industries

Ni 48, Fe 52.

For magnets; high permeability. *Obsolete*

AUER METAL-A

German manufacture

Fe 35, Ce 65.

For gas and cigarette lighters; pyrophoric.

AUER METAL-B

German manufacture

Fe 53, Mn 30, Sb 10, 7 Misch metal.

For gas and cigarette lighters; pyrophoric, sparking alloy.

AUER METAL-C

German manufacture

Fe 35, Ce 35, 29 Misch metal.

For gas and cigarette lighters; pyrophoric.

AUER METAL-D

German manufacture

Fe 35, Ce 35, La 24, Yb 3, Er 2.

For gas and cigarette lighters; pyrophoric.

AUFRILOY

Wellman Dynamics Corp.

Pb, Sn, bal Cu.

For bearings, bushings; heavy duty. *Obsolete*

AUGER

Ziv Steel & Wire Co.

C, bal Fe.

For machine tool parts; water hardened.

AUGER

Bethlehem Steel Corp.

C 0.8, Mn 0-0.4, Si 0-0.4, bal Fe.

For augers, mining tools; for soft rock, clay, coal. *Obsolete*

AUGER SECTION A

Ziv Steel & Wire Co.

C 0.75, bal Fe.

For tools, mining drills; water hardened.

AUGER SECTION B

Ziv Steel & Wire Co.

C 0.75, bal Fe.

For tools, mining drills; water hardened.

AUMAT

Johnson Matthey plc

Au 83.3, bal Pt.

Wire for dental purposes.

AUR O MET 245

Aurora Industries

Cu 88, Sn 6, Pb 1.5, Zn 4.5.

40,000 psi TS; 20,000 psi YS; 30 El; 14.3% IACS electrical
conductivity. CDA 922; SAE 622; QQ-C-390 Alloy D4.
Obsolete

AUR-O-MET 115

Aurora Industries

Sn 5, Pb 5, Zn 5, bal Cu.

Cast: 32,000-38,000 TS; 15,000-17,000 YS; 20-40 El; B 24
Rock. For water pump impellers, fittings for gasoline and oil
lines, bushings, miscellaneous castings. Red brass, free-
cutting, good castability. SAE 40. *Obsolete*

AUR-O-MET 11B

Aurora Industries

Al 10.5, Fe 1, bal Cu.

Sand cast: 75,000-85,000 psi TS; 32,000-42,000 psi YS; 20-35
El; 71-81 RA. Heat treatable aluminum bronze, corrosion
resistant. For gears, worm wheels, valve seats, valve guides,
shafts. ASTM B148-52-9B; SAE 68 B; CDA 953.

AUR-O-MET 123

Aurora Industries

Sn 3, Pb 7, Zn 9, bal Cu.

Sand cast: 29,000-39,000 psi TS; 13,000-17,000 psi YS; 18-30
El; 22 Rock B (approximate). Semi-red brass for general
purpose plumbing and machine parts; ductile and corrosion
resistant; free machining. ASTM B145-52 5A; CDA 844.
Obsolete

AUR-O-MET 141

Aurora Industries

Cu 59, Zn 39.5, Si 1.5.

Cast: 70,000-78,000 TS; 35,000-40,000 YS; 15-20 El; B 78-82
Rock. For fittings, hardware, permanent mold and shell mold
castings. Yellow brass, corrosion resistant. *Obsolete*

AUR-O-MET 145

Aurora Industries

Zn 14, Si 5, bal Cu.

Die cast: 70,000-92,000 psi TS; 40,000-60,000 psi YS; 15-20
El; 20 RA; 128-160 Brin. For gears, water pumps, impellers,
marine parts; corrosion and wear resistant. CDA 875.

AUR-O-MET 150

Aurora Industries

Cu 60, Zn 39.5, Al 0.5.

Cast: 40,000-45,000 psi TS; 11,000-14,000 psi YS; 23-28 El;
40-50 Rock B. For fittings, hardware, permanent and shell
mold castings. Yellow brass, corrosion resistant. CDA 858.

AUR-O-MET 305

Aurora Industries

Sn 10, Pb 10, P 0-0.25, bal Cu.

Cast: 27,000-39,000 TS; 15,000-22,000 YS; 26-34 El; B 27
Rock. For wrist pins, piston pins, bushings, bearings. Good
antifriction qualities. High leaded tin bronze; ASTM B144-52
(3A). *Obsolete*

AUR-O-MET 315

Aurora Industries

Sn 7, Pb 7, Zn 2, bal Cu.

Cast: 30,000-38,000 TS; 17,000-21,000 YS; 12-20 El; B 20
Rock. For bearings, spring bushings, thrust washers. High
leaded tin bronze; SAE 660. *Obsolete*

AUR-O-MET 403

Aurora Industries

Sn 1, Pb 3, Zn 29, bal Cu.

Cast: 30,000-38,000 TS; 11,000-15,000 YS; 20-35 El; B 10
Rock. For commercial bronze castings, radiator parts, fittings,
battery terminals. Good machinability. Yellow brass leaded;
SAE 41. *Obsolete*

AUR-O-MET 420
Aurora Industries
Sn 0.75, Pb 0.75, Zn 37, Fe 1.25, Al 0.75, Mn 0.5, bal Cu.
Cast: 60,000-78,000 psi TS; 20,000-40,000 psi YS; 15-30 El;
40-51 Rock B. For propellers, marine hardware, shafts, gears,
fittings. Good corrosion resistance. Manganese bronze, high
strength. ASTM-B147-52 7A; CDA 864.

AUR-O-MET 421
Aurora Industries
Zn 39.25, Fe 1.25, Al 1.25, Mn 0.25, bal Cu.
Cast: 65,000-80,000 psi TS; 25,000-40,000 psi YS; 20-40 El;
54-68 Rock B. For gear shift forks, spiders, fittings, brackets,
landing gears. Good corrosion resistance, high strength.
Manganese bronze, SAE 43; CDA 865.

AUR-O-MET 424
Aurora Industries
Zn 26, Fe 3, Al 5.5, Mn 3.5, bal Cu.
Cast: 110,000-120,000 TS; 70,000-85,000 YS; 15-18 El; B
89-97 Rock. For propellers, impellers, gears, fittings, marine
hardware. Good corrosion and wear resistance. Manganese
bronze, high strength; SAE 430-B. *Obsolete*

AUR-O-MET 56
Aurora Industries
Al 10.5, Ni 4.75, Fe 4.25, bal Cu.
Sand cast: 90,000-100,000 psi TS; 50,000-60,000 psi YS; 5-10
El; 89-93 Rock B. For gears, machine and structural parts,
impellers; corrosion and wear resistant. ASTM B148-52-9D;
CDA 955.

AUR-O-MET X-10
Aurora Industries
Al, Fe, bal Cu.
Die cast: 115,000 TS; 105,000 YS; 1.5 El; 269 Brin. For gears,
hardware; corrosion resistant. *Obsolete*

AUR-O-MET X-14
Aurora Industries
Al, Fe, bal Cu.
Die cast: 104,000 TS; 46,000 YS; 8.8 El; 187 Brin. For gears,
hardware; Al-bronze. *Obsolete*

AURIGA MARK I
Sowers Mfg. Co.
C 0.25, Mn 0.7, bal Fe.
Annealed: 67,200-78,500 TS; 34,000 YS; 20 El. For structural
castings. Water hardening.

AURIGA MARK III
David Brown Foundries Co.
C 0.4, Mn 0.7, bal Fe.
Annealed: 83,000-101,000 TS; 42,000 YS; 15 El. For gears
and wear resistant castings; water hardened.

AURIGA MARK IX
David Brown Foundries Co.
C 0.3, Cr, Mo, bal Fe.
Hardened: 102,000-141,000 TS; 70,000-101,000 YS; 12-15 El.
For high strength castings; oil hardened.

AURIGA MARK VIII
David Brown Foundries Co.
C 0.2, Mo, bal Fe.
Normalized: 67,200 TS; 40,300 YS; 20 El. For high
temperature castings to 1020°F; heat resistant.

AURIGA MARK VIII A
David Brown Foundries Co.
C 0.15, Cr, Mo, bal Fe.
Normalized: 70,000 TS; 44,800 YS; 15 El. For castings; high
creep strength to 1020°F.

AURIGA MARK X
Sowers Mfg. Co.
Mn 11-14, C, bal Fe.
For wear resistant castings. Mn steel.

AURIGA MARK XV
David Brown Foundries Co.
C 0.3, Cr 3, Mo, bal Fe.
Hardened: 102,000-141,000 TS; 78,000-101,000 YS; 12-15 El.
For high strength castings; oil hardened.

AURIGA MARK XVI
David Brown Foundries Co.
C 0.3, Cr, Ni, Mo, bal Fe.
Hardened: 102,000-141,600 TS; 78,000-100,000 YS; 12-15 El.
For high strength castings; oil hardened.

AURIGA MARK XVII
David Brown Foundries Co.
C 0.15, Cr 18, Ni 8, W 1.5, bal Fe.
Annealed. For heat resistant castings; corrosion and heat
resistant.

AURIGA MARK XXI
David Brown Foundries Co.
C 0.2, Cr 5, Mo, bal Fe.
Hardened: 90,000 TS; 60,000 YS; 18 El. For castings for high
temperature service; heat resistant.

AURIGA MARK XXII
David Brown Foundries Co.
C 0.2, Cr 9, Mo 1.2, bal Fe.
Hardened: 90,000 TS; 60,000 YS; 18 El. For castings for high
temperature service; heat resistant.

AURIGA MARK XXIII
David Brown Foundries Co.
C 0.2, Cr 13, bal Fe.
Hardened: 78,500 TS; 49,600 YS; 20 El. For castings;
corrosion resistant.

AURIGA MARK XXIV B
David Brown Foundries Co.
C 0.15, Cr 18, Ni 8, Cb, bal Fe.
Annealed: 68,500 TS; 30,300 YS; 20 El. For castings;
stainless, austenitic.

AURIGA V
David Brown Foundries Co.
C 0.55, Si 0.4, Mn 0.7, bal Fe.
Steel casting.

AUSCO 80
Auto Specialties Mfg. Co.
C 1.3-1.4, Si 0.7, Mn 0.9, Cr 0-0.1, Cu 0-0.1, Ni 0.5, Mo 0.15,
bal Fe.
Heat treated: 110,000 TS; 80,000 YS; 7 El; 270 Brin. For
crankshafts; castings.

AUSMAN 12
Forjas Alavesas S.A.
C 1.2, Mn 12.5, Si 0.5, bal Fe.
Austenitic manganese steel for resistance to abrasion; for
excavator bucket teeth, wear plates. IHA F-642; DIN x 120 MN
12.

AUSTALON
Howmet Corp.
C 0.08, Cr 18, Ni 8, Mo 2, bal Fe.
For orthodontic and denture applications; stainless,
austenitic.

AUSTENITE
T. Inman & Co. Ltd.
C 0.65, Cr 4, W 14, V 0.5, bal Fe.
Tungsten high speed tool steel. For drills, turning tools,
cutters.

AUSTINOX
Societe Nouvelle des Acieries de Pompey
C 0.06, Cr 18, Ni 10, bal Fe.
Annealed: 85,000 psi TS; 35,000 psi YS; 60 El; 70 RA; 150
Brin. For chemical plant equipment, tanks, vessels, filters;
Type 304; stainless, austenitic.

AUSTINOX B
Societe Nouvelle des Acieries de Pompey
C 0.06, Cr 18, Ni 12, Mo 2-3, bal Fe.
Annealed: 85,000 psi TS; 35,000 psi YS; 50 El; 65 RA; 160
Brin. For acid resistant chemical plant equipment; Type 316;
stainless, austenitic.

AUSTINOX F
Societe Nouvelle des Acieries de Pompey
C 0.1, Cr 18, Ni 9, S 0.2, Mo, bal Fe.
Free machining austenitic stainless steel. AISI 303.

AUSTINOX S
Societe Nouvelle des Acieries de Pompey
C 0.08, Cr 18, Ni 10, Ti = 5 x C, bal Fe.
Annealed: 85,000 psi TS; 35,000 psi YS; 55 El; 65 RA; 150
Brin. For welded chemical plant equipment, mixers, tanks;
Type 321; stainless, austenitic.

AUSTINOX SB
Societe Nouvelle des Acieries de Pompey
C 0.08, Cr 18, Ni 12, Mo 2-3, Ti = 5 x C, bal Fe.
Annealed: 85,000 psi TS; 35,000 psi YS; 50 El; 65 RA; 160
Brin. For welded acid resistant chemical plant equipment;
Type 316 Ti; stainless, austenitic.

AUSTRIAN (GERSDORF)
Austrian manufacture
Cu 50-60, Zn 20-25, Ni 20-25.
For costume jewelry; German silver.

AUSTRIAN ALLOY (SPANDAU)
Austrian manufacture
Cu 4-6, Al 2-3.5, bal Zn.
For die castings; Spandau Alloy.

AUSTRIAN JOURNAL BOX
Austrian manufacture
Cu 92.5, Zn 7.5.
For bearing purposes; gilding metal.

AUTO
Westa-Westdeutsche
C 0.41, Cr 1.1, Mn 0.7, Si 0.25, bal Fe.
For gears, bolts, shafts, axles, bolts; oil or water hardened,
shock resistant.

AUTO DIE
Crucible Specialty Metals
C, alloy, bal Fe.
For tools; oil hardening. *Obsolete*

AUTO DIE
Colt Industries
C, alloy, bal Fe.
For tools; oil hardening. *Obsolete*

AUTO EXTRA N.C.
Vereinigte Edelstahlwerke
C 0.3, Cr 0.75, Ni 3.5, bal Fe.
Heat treated: 150,000 TS; 106,000 YS; 14 El; 50 RA; 310 Brin.
For crankshafts, axles, valves, connecting rods, gears; oil
hardened. *Obsolete*

AUTO EXTRA NC W/H
Rochling Burbach GmbH
C 0.22-0.3, Cr 0.8, Ni 3.5, bal Fe.
Heat treated: 135,000-142,000 TS; 92,500-113,800 YS. For
gears, bolts, studs, shafts, fasteners; oil hardened, shock
resistant. *Obsolete*

AUTO EXTRA PA
Now VEW V204.

AUTO FIRST QUALITY N.C.
Vereinigte Edelstahlwerke
C 0.3, Cr 0.5, Ni 1.5, bal Fe.
For crankshafts, axles, connecting rods, gears; water
hardened. *Obsolete*

AUTO LITE A1-5
Electric Auto-Lite Co.
Si 4.5-6, Fe 0-2, bal Al.
Cast: 30,000 TS; 5 El. For die castings; high fluidity.

AUTO LITE A1-6
Electric Auto-Lite Co.
Si 11-13, Fe 0-2, bal Al.
Cast: 37,000 TS; 1.8 El. For die castings; high fluidity.

AUTO NC W/H
Rochling Burbach GmbH
C 0.28-0.35, Cr 0.7, Ni 2.5, bal Fe.
For gears, bolts, shafts, axles, crankshafts; oil hardened, shock resistant. *Obsolete*

AUTO NC W/H
Rochling Burbach GmbH
C 0.28-0.35, Cr 0.7, Ni 2.5, bal Fe.
For gears, bolts, shafts, axles, crankshafts; oil hardened, shock resistant. *Obsolete*

AUTO NO. 1
Colt Industries
C 0.5, Ni 1.8, Cr 1, bal Fe.
For punches, shock tools; tough. *Obsolete*

AUTO PRIMA NC W/H
Vereinigte Edelstahlwerke
C 0.28-0.35, Cr 0.5, Ni 1.5, bal Fe.
Heat treated: 120,000-128,000 TS; 71,200-85,000 YS. For gears, bolts, shafts, axles, crankshafts; oil hardened, shock resistant. *Obsolete*

AUTO SUMUS
Stahlwerke R. & H. Plate
C 0.74, Cr 4.1, V 1.1, W 18.5, bal Fe.
For lathe and planer tools, drills, reams, taps; high speed steel. *Obsolete*

AUTO-LITE A1-9
Electric Auto-Lite Co.
Si 7.5-9.5, Cu 3-4, Fe 0-2, bal Al.
Cast: 43,000 TS; 2 El. For die castings; high fluidity.

AUTO-LITE Z-3
Electric Auto-Lite Co.
Al 3.5-4.2, Mg 0.03-0.08, bal Zn.
Cast: 43,000 TS; 10 El; 82 Brin. For die castings.

AUTOCHROM CV4
Westa-Westdeutsche
C 0.5, Mn 0.85, Cr 1, V 0.09, bal Fe.
For gears, bolts, springs, crankshafts; oil hardened, shock resistant.

AUTOCRAT
United American Metals Corp.
alloy, bal Cu.
For bearings. *Obsolete*

AUTOCRAT BUSHING BRONZE
United American Metals Corp.
P, Sn, bal Cu.
34,000 TS; 10 El; 68 Brin. For bushings, bearings; bearing bronze. *Obsolete*

AUTOGRIP
Pechiney/Eurotungstene
WC, W, others.
Antiskid and wear-resistant alloy for anti-skid studs; automotive snow tires and ice chains with studded straps.

AUTOMANG
Rankin Mfg. Co.
20% total Mn, Cr, Mo, Ni, Cu, Si, bal Fe.
Self-shielded, flux-cored wire for joining and build-up of high manganese and other steels. As deposited: 20-24 Rock C. Work hardened: 50-54 Rock C.

AUTOMANG 2
Rankin Mfg. Co.
31% total Mn, Cr, Si, Mo, Ni, bal Fe.
Cr-Mn austenitic manganese welding wire, for semi-automatic (DC reverse) build-up on high manganese and other steels. As deposited: 15-20 Rock C. Work hardened: 47-57 Rock C.

AUTOMANG 3
Rankin Mfg. Co.
TC 36, Mn, Cr, Si, bal Fe.
Cr-Mn austenitic manganese welding wire for semi-automatic (DC reverse) build-up and joining of high manganese and other steels. As deposited: 15-20 Rock C. Work hardened: 47-57 Rock C.

AUTOMATIC AW-79
Abex Corp.
C 0.3, Mn 2, Cr 4, Mo 1, bal Fe.
Welded: 450 Brin. For tractor rolls and idlers, steel car wheels; hard-facing electrode.

AUTOMOTIVE DIE
CCS Braeburn Alloy Steel
C, alloy, bal Fe.
For dies, tools. *Obsolete*

AUTOPAN
VDM Nickel-Technologie AG
Mg 0.6-1, Si 0.6-1.2, Cr 0-0.3, Pb 0.5-2.5, Sn, Cd, Bi, bal Al.
Annealed: 21,000 TS; 8000 YS; 20 El. For screw machine products, fasteners; free-cutting. *Obsolete*

AUTOSPEZIAL 2
Stahlwerke R. & H. Plate
C 0.45, W, Cr, V, bal Fe.
For rams, punches, shears, crimpers, upsetters; hot work steel, oil hardened. *Obsolete*

AUTOTHERMIC
English manufacture
steel coated with Al.
For fire walls; heat resistant.

AUTOVALVE STEEL
Carpenter Technology Corp.
Ni, C, bal Fe.
For valves, spark plugs; low coefficient of resistivity; corrosion resistant. *Obsolete*

AV 1
Crucible Materials Corp.
Tool material. C 0.5, Mn 7.5, Si 0.17, Cr 21.5, Ni 3, W 2, Cb 1, V 0.3, N 0.5, bal Fe.
For diesel engine valves.

AVC
American manufacture
Mo 70, W 30.
For rocket nozzle components subject to molten zinc.

AVESTA 153 MA
Avesta AB
C 0-0.06, Cr 18.5, Ni 9.5, Si, N, Ce, bal Fe.
245 N/mm^2 YS (at 50°C). Hot rolled plate (10-30 mm thick): 190 Brin. Cold rolled sheet (2.5-5 mm thick): 190 Brin. For plate, sheet, and welded consumables. Austenitic.

AVESTA 17-11-2
Avesta AB
C 0-0.05, Cr 17, Ni 11, Mo 2.2, bal Fe.
196 N/mm^2 YS (at 50°C). Hot rolled plate (10-30 mm thick): 160 Brin. Cold rolled sheet (2.5-5 mm thick): 160 Brin. For plate, sheet, bar, pipe, and welded consumables. Austenitic.

AVESTA 17-11-2L
Avesta AB
C 0-0.03, Cr 17, Ni 11.5, Mo 2.2, bal Fe.
187 N/mm^2 YS (at 50°C). Hot rolled plate (10-30 mm thick): 160 Brin. Cold rolled sheet (2.5-5 mm thick): 160 Brin. For plate, sheet, bar, pipe, and welded consumables. Austenitic.

AVESTA 17-11-2LN
Avesta AB
C 0-0.03, Cr 17.5, Ni 11, Mo 2.2, N, bal Fe.
250 N/mm^2 YS (at 50°C). Hot rolled plate (10-30 mm thick): 190 Brin. Cold rolled sheet (2.5-5 mm thick): 190 Brin. For plate, sheet, bar, pipe, and welded consumables. Austenitic.

AVESTA 17-11-2TI
Avesta AB
C 0-0.08, Cr 17, Ni 11, Mo 2.2, Ti, bal Fe.
197 N/mm^2 YS (at 50°C). Hot rolled plate (10-30 mm thick): 155 Brin. Cold rolled sheet (2.5-5 mm thick): 155 Brin. For plate, sheet, bar, pipe, and welded consumables. Austenitic.

AVESTA 17-12-2.5
Avesta AB
C 0-0.05, Cr 17, Ni 11, Mo 2.7, bal Fe.
196 N/mm^2 YS (at 50°C). Hot rolled plate (10-30 mm thick): 160 Brin. Cold rolled sheet (2.5-5 mm thick): 160 Brin. For plate, sheet, bar, pipe, and welded consumables. Austenitic.

AVESTA 17-12-2.5L
Avesta AB
C 0-0.03, Cr 17.5, Ni 13, Mo 2.7, bal Fe.
187 N/mm^2 YS (at 50°C). Hot rolled plate (10-30 mm thick): 160 Brin. Cold rolled sheet (2.5-5 mm thick): 160 Brin. For plate, sheet, bar, pipe, and welded consumables. Austenitic.

AVESTA 17-12-2.5LN
Avesta AB
C 0-0.03, Cr 17.5, Ni 11.5, Mo 2.7, N, bal Fe.
255 N/mm^2 YS (at 50°C). Hot rolled plate (10-30 mm thick): 190 Brin. For plate, pipe, and welded consumables. Austenitic.

AVESTA 17-14-4LN
Avesta AB
C 0-0.03, Cr 17, Ni 13, Mo 4.2, N, bal Fe.
260 N/mm^2 YS (at 50°C). Hot rolled plate (10-30 mm thick): 180 Brin. For plate, sheet, pipe, and welded consumables. Austenitic.

AVESTA 17-5N
Avesta AB
C 0-0.15, Cr 17, Ni 5, Mn, N, bal Fe.
Sheet; austenitic.

AVESTA 17-7
Avesta AB
C 0-0.15, Cr 16.5, Ni 7, bal Fe.
Sheet; austenitic.

AVESTA 18-10L
Avesta AB
C 0-0.03, Cr 18.5, Ni 9.5, bal Fe.
168 N/mm^2 YS (at 50°C). Hot rolled plate (10-30 mm thick): 160 Brin. Cold rolled sheet (2.5-5 mm thick): 155 Brin. For plate, sheet, bar, pipe and welded consumables. Austenitic.

AVESTA 18-10TI
Avesta AB
C 0-0.08, Cr 17.5, Ni 9.5, Ti, bal Fe.
191 N/mm^2 YS (at 50°C). Hot rolled plate (10-30 mm thick): 155 Brin. Cold rolled sheet (2.5-5 mm thick): 160 Brin. For plate, sheet, bar, pipe, and welded consumables. Austenitic.

AVESTA 18-12
Avesta AB
C 0-0.06, Cr 18.5, Ni 11.5, bal Fe.
Sheet; austenitic.

AVESTA 18-13-3L
Avesta AB
C 0-0.03, Cr 18.5, Ni 13.5, Mo 3.2, bal Fe.
196 N/mm^2 YS (at 50°C). Hot rolled plate (10-30 mm thick): 160 Brin. For plate, sheet, bar, pipe, and welded consumables. Austenitic.

AVESTA 18-8N
Avesta AB
C 0-0.05, Cr 18.5, Ni 8.5, N, bal Fe.
240 N/mm^2 YS (at 50°C). Hot rolled plate (10-30 mm thick): 190 Brin. For plate, pipe, and welded consumables. Austenitic.

AVESTA 18-9
Avesta AB
C 0-0.05, Cr 18.5, Ni 9, bal Fe.
186 N/mm^2 YS (at 50°C). Hot rolled plate (10-30 mm thick): 170 Brin. Cold rolled sheet (2.5-5 mm thick): 165 Brin. For plate, sheet, bar, pipe, and welded consumables. Austenitic.

AVESTA 18-9LN
Avesta AB
C 0-0.03, Cr 18.5, Ni 9.5, N, bal Fe.
228 N/mm^2 YS (at 50°C). Hot rolled plate (10-30 mm thick): 190 Brin. For plate, pipe, and welded consumables. Austenitic.

AVESTA 20-12SI
Avesta AB
C 0-0.2, Cr 20, Ni 12, Si, bal Fe.
For plate, sheet, and pipe. Austenitic.

AVESTA 2205
Avesta AB
C 0-0.03, Cr 22, Ni 5.5, Mo 3, N, bal Fe.
400 N/mm^2 YS (at 50°C). Hot rolled plate (10-30 mm thick): 240 Brin. Cold rolled sheet (2.5-5 mm thick): 240 Brin. For plate, sheet, bar, pipe, and welded consumables. Ferritic-austenitic.

AVESTA 23-13
Avesta AB
C 0-0.08, Cr 22.5, Ni 12.5, bal Fe.
For plate, sheet, and welded consumables. Austenitic.

AVESTA 248 SV
Avesta AB
C 0-0.05, Cr 16, Ni 5, Mo 1, bal Fe.
615 N/mm^2 YS (at 50°C). Hot rolled plate (10-30 mm thick): 290 Brin. For plate, bar, and welded consumables. Martensitic.

AVESTA 249 MV
Avesta AB
C 0.05, Cr 17, bal Fe.
Annealed: 74,000 TS; 52,000 YS; 30 El; 150 Brin. For pressed and welded articles subject to slight corrosion attack, oil refinery equipment, and oil burners. Corrosion resistant; type 430; ferritic. *Obsolete*

AVESTA 25-20
Avesta AB
C 0-0.08, Cr 25, Ni 20, bal Fe.
186 N/mm^2 YS (at 50°C). Hot rolled plate (10-30 mm thick): 150 Brin. For plate, sheet, bar, pipe, and welded consumables. Austenitic.

AVESTA 25-5-1L
Avesta AB
C 0-0.03, Cr 25, Ni 5, Mo 1.5, bal Fe.
365 N/mm^2 YS (at 50°C). Hot rolled plate (10-30 mm thick): 190 Brin. For plate, bar, pipe, and welded consumables. Ferritic-austenitic.

AVESTA 252 M
Avesta AB
C 0.07, Cr 23, Ni 14.5, bal Fe.
Annealed: 100,000 TS; 50,000 YS; 40 El; 170 Brin. For furnace parts, heat treating boxes, and oil burners. Type 309S. Corrosion and heat resistant; austenitic. *Obsolete*

AVESTA 252 S
Avesta AB
C 0-0.03, Cr 21.5, Ni 15, Mo 2.7, N 0.2, bal Fe.
Annealed: 106,000 TS; 50,000 YS; 50 El; 190 Brin. For pipes, fittings, and valves. For salt water applications. Corrosion resistant; austenitic. *Obsolete*

AVESTA 253 MA
Avesta AB
C 0-0.1, Cr 21, Ni 11, Si, N, Ce, bal Fe.
280 N/mm^2 YS (at 50°C). Hot rolled plate (10-30 mm thick): 190 Brin. Cold rolled sheet (2.5-5 mm thick): 190 Brin. For plate, sheet, bar, and welded consumables. Austenitic.

AVESTA 254
Avesta AB
C 0.1, Cr 19.5, Ni 20.5, bal Fe.
Annealed: 85,000 TS; 36,000 YS; 54 El; 160 Brin. For furnace parts. Low sensitivity to sigma phase formation. Corrosion and heat resistant; austenitic. *Obsolete*

AVESTA 254 EM
Avesta AB
Cr 25, Ni 21, C 0-0.08, bal Fe.
Annealed: 93,000 TS; 43,000 YS; 50 El; 170 Brin. For annealing boxes, tanks, and recuperators. Type 310 S; corrosion and heat resistant; austenitic. *Obsolete*

AVESTA 254 SLX
Avesta AB
C 0-0.02, Cr 20, Ni 25, Mo 4.5, Cu 1.5, bal Fe.
Annealed: 85,000 TS; 36,000 YS; 50 El; 155 Brin. For pumps, chemical plant equipment. Resists sulfuric acid. Corrosion resistant; austenitic. *Obsolete*

AVESTA 254 SMO
Avesta AB
C 0-0.02, Cr 20, Ni 18, Mo 6.2, Cu, N, bal Fe.
270 N/mm^2 YS (at 50°C). Hot rolled plate (10-30 mm thick): 180 Brin. Cold rolled sheet (2.5-5 mm thick): 180 Brin. For plate, sheet, bar, pipe, and welded consumables. Austenitic.

AVESTA 393
Avesta AB
C 0.12, Cr 13.5, bal Fe.
Annealed: 78,000 TS; 50,000 YS; 30 El; 190 Brin. Heat treated: 185,000 TS; 157,000 YS; 17 El; 390 Brin. For press and caul plates in hardboard, chipboard and plastic laminating industries. Type 410; hardenable; corrosion resistant. *Obsolete*

AVESTA 393 M
Avesta AB
C 0-0.07, Cr 14, bal Fe.
Annealed: 74,000 TS; 47,000 YS; 26 El; 150 Brin. For table flatware, wood pulp conveyor screws, steam and water turbine parts and recuperators. Non-hardenable; corrosion resistant. *Obsolete*

AVESTA 3RE60
Avesta AB
C 0-0.03, Cr 18.5, Ni 5, Mo 2.7, bal Fe.
For plate, pipe, and welded consumables. Ferritic-austenitic.

AVESTA 453 E
Avesta AB
C 0.09, Cr 26, Ni 4.5, bal Fe.
Annealed: 100,000 TS; 70,000 YS; 25 El; 210 Brin. For furnace plates and lead and salt bath crucibles. Ferritic-austenitic; heat resistant. *Obsolete*

AVESTA 453 S
Avesta AB
C 0.08, Cr 26, Ni 5, Mo 1.5, bal Fe.
Annealed: 100,000 TS, 70,000 YS; 20 El; 220 Brin. For valves, pumps, shafts and equipment in the cellulose, dyeing, chemical and dairy industries. Corrosion resistant; ferritic-austenitic; Type 329. *Obsolete*

AVESTA 453 SG
Avesta AB
C 0-0.04, Cr 22, Ni 6, Mo 1.7, bal Fe.
Annealed: 90,000 TS; 56,000 YS; 20 El; 220 Brin. Cast material, for ship propellers. Corrosion resistant; ferritic-austenitic. *Obsolete*

AVESTA 664 M
Avesta AB
C 0-0.07, Cr 17, Ni 5.5, Mn 6.5, bal Fe.
Annealed: 100,000 TS; 50,000 YS; 70 El; 190 Brin. For sinks and food processing equipment. Type 202; corrosion resistant; austenitic. *Obsolete*

AVESTA 664 MV
Avesta AB
C 0-0.05, Cr 18, Ni 5, Mn 8, bal Fe.
Annealed: 106,000 TS; 54,000 YS; 60 El; 190 Brin. For nitric acid and brewery equipment. Type 204; corrosion resistant; austenitic. *Obsolete*

AVESTA 724L
Avesta AB
C 0-0.03, Cr 17.5, Ni 13.5, Mo 2.7, bal Fe.
For plate, sheet, pipe, and welded consumables. Austenitic.

AVESTA 739 G
Avesta AB
C 0.12, Cr 13, Ni 1, bal Fe.
Annealed: 100,000 TS; 64,000 YS; 20 El; 210 Brin. For castings, water turbine parts, vanes, and bottom plates. Corrosion resistant. *Obsolete*

AVESTA 739 S
Avesta AB
C 0.14, Cr 13.5, Mo 1, bal Fe.
Annealed: 93,000 TS; 64,000 YS; 18 El; 190 Brin. Hardened: 192,000 TS; 128,000 YS; 10 El; 385 Brin. For paper mill knives and ship propellers. Corrosion resistant; hardenable. *Obsolete*

AVESTA 739 SG
Avesta AB
C 0.12, Cr 13, Ni 1, Mo 1, bal Fe.
Annealed: 92,000 TS; 58,000 YS; 16 El; 50 RA; 190 Brin. For castings and ship propellers. Corrosion resistant. *Obsolete*

AVESTA 831
Avesta AB
C 0.2, Si 1, Cr 25, Ni 0.8, bal Fe.
Annealed: 83,000 TS; 57,000 YS; 30 El; 175 Brin. For furnace equipment. Type 446; heat resistant; non-hardenable. *Obsolete*

AVESTA 832 H
Avesta AB
C 0.11, Cr 17, Ni 8, bal Fe.
Annealed: 107,000 TS; 50,000 YS; 65 El; 200 Brin. For press and caul plates in hardboard, chipboard and plastic laminating industries and chain type bottle conveyors. Type 302; corrosion resistant; austenitic. *Obsolete*

AVESTA 832 M
Avesta AB
C 0-0.05, Cr 18, Ni 0, bal Fe.
Annealed: 85,000 TS; 40,000 YS; 70 El; 155 Brin. Cold rolled strip. For household articles and in the architectural industry. For equipment in the food-processing industry. Type 304, corrosion resistant; austenitic. *Obsolete*

AVESTA 832 MV
Avesta AB
C 0-0.05, Cr 18, Ni 9, bal Fe.
Annealed: 83,000 TS; 37,000 YS; 62 El; 150 Brin. Hot rolled. For welded structures, acid vessels and other equipment in the food processing and chemical industries. Type 304; corrosion resistant; austenitic. *Obsolete*

AVESTA 832 MVN
Avesta AB
C 0-0.05, Cr 18.5, Ni 8.5, N 0.2, bal Fe.
Annealed: 90,000 TS; 48,000 YS; 50 El; 160 Brin. High proof stress stainless steel. Cryogenic applications. Corrosion resistant; austenitic. *Obsolete*

AVESTA 832 MVNB
Avesta AB
C 0-0.08, Cr 17.5, Ni 9.5, Cb, bal Fe.
Annealed: 86,000 TS; 40,000 YS; 56 El; 160 Brin. For nitric acid industries, aircraft internal combustion and jet engine parts and furnace equipment. Type 347; corrosion resistant; austenitic. *Obsolete*

AVESTA 832 MVR
Avesta AB
C 0-0.03, Cr 18.5, Ni 10, bal Fe.
Annealed: 77,000 TS; 36,000 YS; 60 El; 150 Brin. For nitric acid equipment and nuclear industries. Type 304 L; corrosion resistant; austenitic. *Obsolete*

AVESTA 832 MVRN
Avesta AB
C 0-0.03, Cr 18.5, Ni 10.5, N 0.2, bal Fe.
Annealed: 88,000 TS; 40,000 YS; 50 El; 150 Brin. High proof stress stainless steel. For equipment in chemical industries. Corrosion resistant; austenitic. *Obsolete*

AVESTA 832 MVT
Avesta AB
C 0-0.08, Cr 17.5, Ni 10, Ti, bal Fe.
Annealed: 86,000 TS; 40,000 YS; 55 El; 160 Brin. For nitric acid industries, aircraft internal combustion and jet engine parts, and furnace equipment. Type 321; corrosion resistant; austenitic. *Obsolete*

AVESTA 832 SF
Avesta AB
C 0-0.05, Cr 17, Ni 11, Mo 2.3, bal Fe.
Annealed: 85,000 TS; 42,000 YS; 58 El; 150 Brin. Equipment in the chemical, cellulose and food processing industries. Type 316; corrosion resistant; austenitic. *Obsolete*

AVESTA 832 SFR
Avesta AB
C 0-0.03, Cr 17, Ni 11.5, Mo 2.3, bal Fe.
Annealed: 83,000 TS; 42,000 YS; 58 El; 150 Brin. For equipment in the chemical and food processing industries. Type 316 L; corrosion resistant; austenitic. *Obsolete*

AVESTA 832 SFT
Avesta AB
C 0-0.08, Cr 17, Ni 12, Mo 2.3, Ti, bal Fe.
Annealed: 85,000 TS; 42,000 YS; 55 El; 160 Brin. For equipment in the chemical industries. Type 316 T; corrosion resistant; austenitic. *Obsolete*

AVESTA 832 SI
Avesta AB
C 0-0.02, Si 4, Cr 17.5, Ni 14.5, bal Fe.
Annealed: 96,000 TS; 42,000 YS; 55 El; 160 Brin. Vessels, pipes, and valves. Especially good resistance against highly concentrated nitric acid. Austenitic. *Obsolete*

AVESTA 832 SK
Avesta AB
C 0-0.05, Cr 17, Ni 11.5, Mo 2.7, bal Fe.
Annealed: 85,000 TS; 42,000 YS; 58 El; 150 Brin. For the chemical and cellulose industries. Type 316; corrosion resistant; austenitic. *Obsolete*

AVESTA 832 SKER (832 SKR-5)
Avesta AB
C 0-0.03, Cr 17.5, Ni 13.5, Mo 2.7, bal Fe.
Annealed: 85,000 TS; 40,000 YS; 58 El; 150 Brin. Low ferrite grade mainly for the urea and acetic acid industries. Type 316; corrosion resistant; austenitic. *Obsolete*

AVESTA 832 SKNB
Avesta AB
C 0-0.06, Cr 17, Ni 13, Mo 2.7, Cb, bal Fe.
Annealed: 85,000 TS; 42,000 YS; 55 El; 160 Brin. For equipment in the chemical industries. Type 316 Cb; corrosion resistant; austenitic. *Obsolete*

AVESTA 832 SKR
Avesta AB
C 0-0.03, Cr 17.5, Ni 13, Mo 2.7, bal Fe.
Annealed: 85,000 TS; 42,000 YS; 57 El; 150 Brin. For equipment in the chemical industries. Type 316 L; corrosion resistant; austenitic. *Obsolete*

AVESTA 832 SKRN
Avesta AB
C 0-0.03, Cr 17.5, Ni 13, Mo 2.7, N 0.2, bal Fe.
Annealed: 90,000 TS; 50,000 YS; 50 El; 160 Brin. High proof stress stainless steel. For equipment in the chemical and cellulos industries. For chemical tankers. Corrosion resistant; austenitic. *Obsolete*

AVESTA 832 SKT
Avesta AB
C 0-0.08, Cr 17, Ni 13, Mo 2.7, bal Fe.
Annealed: 85,000 TS; 42,000 YS; 55 El; 160 Brin. For equipment in the chemical industries. Type 316 Ti; corrosion resistant; austenitic. *Obsolete*

AVESTA 832 SL
Avesta AB
C 0-0.05, Cr 16.5, Ni 15, Mo 4.3, bal Fe.
Annealed: 88,000 TS; 46,000 YS; 51 El; 150 Brin. For equipment in the chemical industries. Highly resistant to acids; austenitic. *Obsolete*

AVESTA 832 SN
Avesta AB
C 0-0.5, Cr 18.5, Ni 14.5, Mo 3.3, bal Fe.
Annealed: 85,000 TS; 43,000 YS; 50 El; 150 Brin. For sulfite digesters and equipment in the chemical industries. Type 317; corrosion resistant; austenitic. *Obsolete*

AVESTA 832 SNR
Avesta AB
C 0-0.03, Cr 18.5, Ni 14.5, Mo 3.3, bal Fe.
Annealed: 85,000 TS; 43,000 YS; 50 El; 150 Brin. For sulfite digesters and equipment in the chemical industries. Type 317 L; corrosion resistant. *Obsolete*

AVESTA 832 SV
Avesta AB
C 0-0.05, Cr 17.5, Ni 9.5, Mo 1.5, bal Fe.
Annealed: 84,000 TS; 40,000 YS; 61 El; 150 Brin. For equipment in the cellulose and food processing industry and window frames. Corrosion resistant; austenitic. *Obsolete*

AVESTA 832T/NB
Avesta AB
C 0.1, Cr 18, Ni 9.5, Cb, bal Fe.
Annealed: 90,000 TS; 45,000 YS; 56 El; 65 RA; 150 Brin. For welded chemical plant equipment, tanks, and mixers. Type 347; corrosion resistant. *Obsolete*

AVESTA 904L
Avesta AB
C 0-0.02, Cr 20, Ni 25, Mo 4.5, Cu, bal Fe.
190 N/mm^2 YS (at 50°C). Hot rolled plate (10-30 mm thick): 155 Brin. Cold rolled sheet (2.5-5 mm thick): 155 Brin. For plate, sheet, bar, pipe, and welded consumables. Austenitic.

AVESTA ELI-T 18-2
Avesta AB
C 0-0.015, Cr 18, Mo 2.2, Ti, bal Fe.
310 N/mm^2 YS (at 50°C). For sheet, pipe, and welded consumables. Ferritic.

AVESTA SKR-4
Avesta AB
C 0-0.03, Cr 17.5, Ni 11, Mo 2.9, N, bal Fe.
250 N/mm^2 YS (at 50°C). Hot rolled plate (10-30 mm thick): 185 Brin. For plate, sheet, pipe, and welded consumables. Austenitic.

AVESTA SNR-4
Avesta AB
C 0-0.03, Cr 18.5, Ni 13.5, Mo 3.6, N, bal Fe.
260 N/mm^2 YS (at 50°C). Hot rolled plate (10-30 mm thick): 185 Brin. For plate, pipe, and welded consumables. Austenitic.

AVIAL
Bidault-Elion SA
Si 0.5, Cu 2.5, Mg 0.6, Ni 1, Cr 0.7, bal Al.
Heat treated: 57,000 TS; 26 El. For light alloy parts; hardenable.

AVIALITE-915
Anaconda Co.
Copper. Cu 89.25, Al 9.25, Sn 0.4, Ni 0.5, Fe 0.6.
Soft: 80,000 TS; 40,000 YS; 22 El. Hard: 95,000 TS; 55,000 YS; 16 El. For valve seats, spark plug bushings in aircraft engines. Same as "Millard metal."

AVIOL
Russian manufacture
Si 0.7, Mg 0.6, bal Al.
Annealed: 17,000 TS; 7000 YS; 27 El; 40 Brin. For aircraft parts.

AVIONAL
Lavorazione Leghe Leggere SpA
Cu 3-4, Mg 0.6, Mn 0.6, Si 0-0.3, Fe 0-0.3, bal Al.
For aircraft parts; age hardenable.

AVIONAL
Aluminium Walzwerke Singen GmbH
Cu 2.5-5, Mg 0.2-1.8, Mn 0.3-1.5, bal Al.
Annealed: 27,000 TS; 11,000 YS; 22 El; 47 Brin. Heat Treated: 72,000 TS; 57,000 YS; 130 Brin. For aircraft structures and fittings; age-hardenable.

AVIONAL
Alluminio SA
Cu 2.5-5, Mg 0.2-1.8, Mn 0.3-1.5, bal Al.
Annealed: 27,000 TS; 11,000 YS; 22 El; 47 Brin. Heat Treated: 72,000 TS; 57,000 YS; 130 Brin. For aircraft structures and fittings; age-hardenable.

AVIONAL 102
Aluminium Walzwerke Singen GmbH
Aluminum. Cu 4, Mg 1.5, bal Al.
Solution heat treated: 380 MPa TS; 10 El. Bars, forgings, for vehicle and machine construction.

AVIONAL 411
Aluminium Industrie Aktiengesellschaft
Si 1, Cu 4.8, Mg 0.5, Mn 0.8, bal Al.
Heat treated: 64,000-74,000 TS; 54,000-63,000 YS; 8-12 El; 115-130 Brin. For light alloy parts; hardenable.

AVIONAL D
Lavorazione Leghe Leggere SpA
Cu 3.5-5, Mg 0.2-1.5, Mn 0.2-1.5, Si 0-1, bal Al.
Soft: 25,000 TS; 11,000 YS; 22 El; 50 Brin. Heat treated: 58,000-87,000 TS; 38,000 YS; 21-2 El; 105 Brin. For aviation and automobile parts; corrosion resisting; heat treatable.

AVIONAL D
Aluminium Industrie Aktiengesellschaft
Cu 3.5-5, Mg 0.2-1.5, Mn 0.2-1.5, Si 0-1, bal Al.
Soft: 25,000 TS; 11,000 YS; 22 El; 50 Brin. Heat treated: 58,000-87,000 TS; 38,000 YS; 21-2 El; 105 Brin. For aviation and automobile parts; corrosion resisting; heat treatable.

AVIONAL D TI
Aluminium Industrie Aktiengesellschaft
Si 0.2, Cu 3-8, Mn 0.5, Mg 0.5, Fe 0.2, Ti 0.16, bal Al.
For aircraft parts; age hardenable.

AVIONAL M
Aluminium Industrie Aktiengesellschaft
Si 0.7, Cu 4.2, Mg 0.65, Mn 0.7, bal Al.
Heat treated: 57,000-65,000 TS; 40,000-44,000 YS; 16-20 El; 105-120 Brin. For aircraft parts; age hardened.

AVIONAL S
Aluminium Industrie Aktiengesellschaft
Cu 3.5-5, Mg 0.2-1.5, Mn 0.2-1.5, Fe 0.2, Si 0.1-1, bal Al.
Extruded: 71,000 TS; 54,000 YS; 15 El; 125 Brin. For aircraft
parts, automotive construction; heat treatable.

AVIONAL SK
Aluminium Industrie Aktiengesellschaft
Cu 3.5-5, Mn 0.5-0.8, Mg 0.5-0.8, bal Al.
Sheet: 65,000-72,000 TS; 48,000-56,000 YS; 14-18 El. For
light alloy parts; age hardenable.

AVIONAL Z
Aluminium Industrie Aktiengesellschaft
Si 0.4, Fe 0.3, Mn 0.75, Mg 0.8, Cu 4.5, bal Al.
Wrought: 62,700-71,500 TS; 37,000-46,000 YS; 12-15 El. For
aviation and automobile parts; similar to "Duralumin."

AVIONAL-102
Swiss Aluminium Ltd.
Aluminum. Si 0.2-0.8, Fe 0.5, Cu 3.5-4.5, Mn 0.4-0.9, Mg
0.5-0.8, Cr 0.1, Zn 0.25, Ti 0.1, bal Al.
Extrusion age hardenable alloy in billet form. 380-470
N/mm^2 TS; 250-360 N/mm^2 YS; 8-10 El; 110 Brin.

AVIONAL-14
Aluminium Industrie Aktiengesellschaft
Cu 4.4, Si 0.8, Mg 0.4, Mn 0.8, bal Al.
Annealed: 27,000-30,000 TS; 11,400-17,800 YS; 12-18 El;
45-55 Brin. TA-Temper: 67,500-75,400 TS; 58,300-64,000 YS;
6-9 El; 125-140 Brin. For aircraft structures, general
engineering components. Age hardenable; high strength.

AVIONAL-14
Lavorazione Leghe Leggere SpA
Cu 4.4, Si 0.8, Mg 0.4, Mn 0.8, bal Al.
Annealed: 27,000-30,000 TS; 11,400-17,800 YS; 12-18 El;
45-55 Brin. TA Temper: 67,500-75,400 TS; 58,300-64,000 YS;
6-9 El; 125-140 Brin. For aircraft structures, general
engineering components. Age hardenable; high strength.

AVIONAL-14
Alluminio SA
Cu 4.4, Si 0.8, Mg 0.4, Mn 0.8, bal Al.
Annealed: 27,000-30,000 TS; 11,400-17,800 YS; 12-18 El;
45-55 Brin. TA-Temper: 67,500-75,400 TS; 58,300-64,000 YS;
6-9 El; 125-140 Brin. For aircraft structures, general
engineering components. Age hardenable; high strength.

AVIONAL-152
Swiss Aluminium Ltd.
Aluminum. Si 0.5, Fe 0.5, Cu 3.8-4.9, Mn 0.3-0.9, Mg 1.2-1.8,
Cr 0.1, Zn 0.25, Ti 0.15, bal Al.
Extrusion age hardenable alloy in billet form. 440-560
N/mm^2 TS; 315-450 N/mm^2 YS; 8-10 El; 125 Brin.

AVIONAL-20
Aluminium Industrie Aktiengesellschaft
Cu, Mg, bal Al.
For clad sheet for orthopedics; age hardened, malleable as
heat treated.

AVIONAL-21
Lavorazione Leghe Leggere SpA
Cu 2.5, Si 0.3, Mg 0.3, bal Al.
Annealed: 14,200-21,400 TS; 5700-10,000 YS; 25-40 El;
35-45 Brin. TN-Temper: 38,400-50,000 TS; 21,500-28,400 YS;
16-23 El; 80-90 Brin. For aircraft components, marine and
transportation equipment.

AVIONAL-21
Alluminio SA
Cu 2.5, Si 0.3, Mg 0.3, bal Al.
Annealed: 14,200-21,400 TS; 5700-10,000 YS; 25-40 El;
35-45 Brin. TN Temper: 38,400-50,000 TS; 21,500-28,400 YS;
16-23 El; 80-90 Brin. For aircraft components, marine and
transportation equipment.

AVIONAL-22
Aluminium Industrie Aktiengesellschaft
Cu 3.5-5, Si 0-1, Mn 0.2-1.5, Mg 0.2-1.5, bal Al.
Annealed: 22,750 TS; 8500 YS; 30 El; 45 Brin. Heat Treated:
62,500 TS; 42,500 YS; 20 El; 120 Brin. For aircraft and bus
construction, machine tool parts; age hardened, high fatigue
strength.

AVIONAL-22
Alluminio SA
Cu 3.5-5, Si 0-1, Mn 0.2-1.5, Mg 0.2-1.5, bal Al.
Annealed: 22,750 TS; 8500 YS; 30 El; 45 Brin. Heat Treated:
62,500 TS; 42,500 YS; 20 El; 120 Brin. For aircraft and bus
construction, machine tool parts; age hardened, high fatigue
strength.

AVIONAL-24
Aluminium Industrie Aktiengesellschaft
Cu 3.5-5, Si 0.1, Mn 0.2-1.5, Mg 0.2-1.5, bal Al.
Annealed: 28,500 TS; 12,750 YS; 22 El; 55 Brin. Heat Treated:
74,000 TS; 51,000 YS; 15 El; 150 Brin. For aircraft and bus
construction, machine tool parts; age-hardened, high fatigue
strength.

AVIONAL-24
Alluminio SA
Cu 3.5-5, Si 0.1, Mn 0.2-1.5, Mg 0.2-1.5, bal Al.
Annealed: 28,500 TS; 12,750 YS; 22 El; 55 Brin. Heat Treated:
74,000 TS; 51,000 YS; 15 El; 150 Brin. For aircraft and bus
construction, machine tool parts; age-hardened, high fatigue
strength.

AVIONAL-25
Aluminium Industrie Aktiengesellschaft
Cu, Mg, bal Al.
Annealed: 22,750 TS; 8500 YS; 30 El; 50 Brin. Heat treated:
74,000 TS; 68,250 YS; 10 El; 150 Brin. For aircraft and bus
body construction; age hardened, corrosion resistant.

AVIONAL-660
Swiss Aluminium Ltd.
Aluminum. Si 0.5-1.2, Fe 0.7, Cu 3.9-5, Mn 0.4-1.2, Mg
0.2-0.8, Cr 0.1, Zn 0.25, Ti 0.15, bal Al.
Wrought alloy in rolling slab form.

AVIONAL-662
Swiss Aluminium Ltd.
Aluminum. Si 0.5-1.2, Fe 0.7, Cu 3.9-5, Mn 0.4-1.2, Mg
0.2-0.8, Cr 0.1, Zn 0.25, Ti 0.15, bal Al.
Extrusion age hardenable alloy in billet form. 450-510
N/mm^2 TS; 400-480 N/mm^2 YS; 6-7 El; 135 Brin.

AVIONALPLAT
Aluminium Industrie Aktiengesellschaft
Al alloy.
For light alloy parts. Avional coated with Al.

AVONAL-23
Aluminium Industrie Aktiengesellschaft
Cu 3.5-5, Mg 0.2-1.5, Mn 0.2-1.5, Si 0-1, bal Al.
Annealed: 27,000 TS; 11,500 YS; 20 El; 50 Brin. Heat treated:
71,000 TS; 48,250 YS; 20 El; 125 Brin. For aircraft
construction, machine tool parts; age hardened, high fatigue
strength.

AVONMOUTH
British Metal Corp., Ltd.
Al alloy.
For light alloy parts.

AVRIL 40/60 SOLDER
G.A. Avril Co.
Lead. Sn 40, 0.50 others, bal Pb.
Low melting point alloy for soldering.

AVRIL 50/50 SOLDER
G.A. Avril Co.
Lead. Sn 50, 0.50 others, bal Pb.
Low melting point alloy for soldering.

AVRIL 60/40 SOLDER
G.A. Avril Co.
Lead. Sn 60, 0.50 others, bal Pb.
Low melting point alloy for soldering.

AVRIL 63/37 SOLDER
G.A. Avril Co.
Lead. Sn 63, 0.50 others, bal Pb.
Low melting point alloy for soldering.

AVRIL 95/5 SOLDER
G.A. Avril Co.
Lead. Sn 95, Sb 5.
Low melting point alloy for soldering.

AVRIL C8345
G.A. Avril Co.
Copper. Cu 87.5, Sn 2.5, Pb 2, Zn 7, Ni 1.
Ingot: 37,000 TS; 15,000 YS; 34 El.

AVRIL C836
G.A. Avril Co.
Copper. Cu 85, Sn 5, Pb 5, Zn 5.
Ingot: 37,000 TS; 17,000 YS; 30 El.

AVRIL C838
G.A. Avril Co.
Copper. Cu 83, Sn 4, Pb 6, Zn 7.
Ingot: 35,000 TS; 16,000 YS; 25 El.

AVRIL C844
G.A. Avril Co.
Copper. Cu 81, Sn 3, Pb 7, Zn 9.
Ingot: 34,000 TS; 15,000 YS; 26 El.

AVRIL C854
G.A. Avril Co.
Copper. Cu 67, Sn 1, Pb 3, Zn 29.
Ingot: 34,000 TS; 12,000 YS; 35 El.

AVRIL C858
G.A. Avril Co.
Copper. Cu 58, Sn 1, Pb 1, Zn 40.
Ingot: 55,000 TS; 30,000 YS; 15 El.

AVRIL C863
G.A. Avril Co.
Copper. Cu 63, Zn 25, Fe 3, Al 6, Mn 3.
Ingot: 119,000 TS; 83,000 YS; 18 El.

AVRIL C865
G.A. Avril Co.
Copper. Cu 58, Sn 0.5, Zn 39.5, Fe 1, Al 1.
Ingot: 71,000 TS; 28,000 YS; 30 El.

AVRIL C873
G.A. Avril Co.
Copper. Cu 95, Si 4, Mn 1.
Ingot: 45,000 TS; 18,000 YS; 20 El.

AVRIL C875
G.A. Avril Co.
Copper. Cu 82, Zn 14, Si 4.
Ingot: 67,000 TS; 30,000 YS; 21 El.

AVRIL C876
G.A. Avril Co.
Copper. Cu 90, Zn 5.5, Si 4.5.
Ingot: 66,000 TS; 32,000 YS; 20 El.

AVRIL C903
G.A. Avril Co.
Copper. Cu 88, Sn 8, Zn 4.
Ingot: 45,000 TS; 21,000 YS; 30 El.

AVRIL C905
G.A. Avril Co.
Copper. Cu 88, Sn 10, Zn 2.
Ingot: 45,000 TS; 22,000 YS; 25 El.

AVRIL C907
G.A. Avril Co.
Copper. Cu 89, Sn 11.
Ingot: 44,000 TS; 22,000 YS; 20 El.

AVRIL C916
G.A. Avril Co.
Copper. Cu 88, Sn 10.5, Ni 1.5.
Ingot: 44,000 TS; 22,000 YS; 16 El.

AVRIL C922
G.A. Avril Co.
Copper. Cu 88, Sn 6, Pb 1.5, Zn 4.5.
Ingot: 40,000 TS; 20,000 YS; 30 El.

AVRIL C923
G.A. Avril Co.
Copper. Cu 87, Sn 8, Pb 0.5, Zn 4.
Ingot: 40,000 TS; 20,000 YS; 25 El.

AVRIL C925
G.A. Avril Co.
Copper. Cu 87, Sn 11, Pb 1, Ni 1.
Ingot: 44,000 TS; 20,000 YS; 20 El.

AVRIL C926
G.A. Avril Co.
Copper. Cu 87, Sn 10, Pb 1, Zn 2.
Ingot: 44,000 TS; 20,000 YS; 30 El.

AVRIL C927
G.A. Avril Co.
Copper. Cu 88, Sn 10, Pb 2.
Ingot: 42,000 TS; 21,000 YS; 20 El.

AVRIL C929
G.A. Avril Co.
Copper. Cu 84, Sn 10, Pb 2.5, Ni 3.5.
Ingot: 45,000 TS; 25,000 YS; 20 El.

AVRIL C932
G.A. Avril Co.
Copper. Cu 83, Sn 7, Pb 7, Zn 3.
Ingot: 35,000 TS; 18,000 YS; 20 El.

AVRIL C937
G.A. Avril Co.
Copper. Cu 80, Sn 10, Pb 10.
Ingot: 35,000 TS; 18,000 YS; 20 El.

AVRIL C938
G.A. Avril Co.
Copper. Cu 78, Sn 7, Pb 15.
Ingot: 30,000 TS; 16,000 YS; 18 El.

AVRIL C943
G.A. Avril Co.
Copper. Cu 70, Sn 5, Pb 25.
Ingot: 27,000 TS; 13,000 YS; 10 El.

AVRIL C952
G.A. Avril Co.
Copper. Cu 88, Fe 3, Al 9.
Ingot: 80,000 TS; 27,000 YS; 35 El.

AVRIL C953
G.A. Avril Co.
Copper. Cu 89, Fe 1, Al 10.
Ingot: 75,000 TS; 27,000 YS; 25 El.

AVRIL C954
G.A. Avril Co.
Copper. Cu 85, Fe 4, Al 11.
Ingot: 85,000 TS; 35,000 YS; 18 El.

AVRIL C955
G.A. Avril Co.
Copper. Cu 81, Fe 4, Al 11, Ni 4.
Ingot: 100,000 TS; 44,000 YS; 12 El.

AVRIL C976
G.A. Avril Co.
Copper. Cu 64, Sn 4, Pb 4, Zn 20, Ni 8.
Ingot: 40,000 TS; 24,000 YS; 20 El.

AVRIL DREADNAUGHT BABBITT
G.A. Avril Co.
Tin. Sn 89, Sb 7.5, Cu 3.5.
Bearing material. 24.50 Brin; 6100 psi YP.

AVRIL OUTLASTA BABBITT
G.A. Avril Co.
Tin. Sn 10, Sb 15, Cu 0-0.5, bal Pb.
Bearing material. 22.50 Brin; 3550 psi YP.

AVRIL SUPER NICKEL BABBITT
G.A. Avril Co.
Tin. Sn 84, Sb 8, Cu 8.
Bearing material. 27.00 Brin; 6600 psi YP.

AVROCAN M7-12
Atlas Specialty Steels
C 0.4, Si 1, Cr 5, Mo 1.4, V 0.5, bal Fe.
Heat treated: 225,000-305,000 TS; 195,000-255,000 YS; 8-15 El; 30-50 RA; 45-56 Rock C. For extrusion dies, forging dies and inserts, hot punches, header and gripper dies; tough, hot work tool steel, Type H11. *Obsolete*

AVS
Acciaierie Valbruna s.p.a.
Stainless Steel. C 0.8, Si 2.25, Mn 0.6, Cr 20, Ni 1.35, P 0.03, S 0.03, bal Fe.
For valves. W. Nr. 1.4747.

AVS METAL
Ste de Produits Metallurgiques
C 0.2-0.4, bal Fe.
For structural parts; water hardened.

AVW
Acciaierie Valbruna s.p.a.
Stainless Steel. C 0.45, Si 2.5, Mn 1.15, Cr 18, Ni 9, W 1, P 0.035, S 0.03, bal Fe.
For valves. W. Nr. 1.4873.

AW DYNALLOY 50
Alan Wood Steel Co.
C 0.13, Mn 0.9, Cu 0.35, Ni 0.45, Mo 0.08, bal Fe.
Plate: 50,000 min YS; 70,000 min TS; 22 min El. For railroad gondola and hopper cars, automobile and truck bodies and frames, lift trucks, conveyors, bridges, buildings. Good forming and welding properties ASTM A242. High strength, low alloy steel. *Obsolete*

AW V-STEEL
Alan Wood Steel Co.
C 0-0.22, Mn 0-1.25, 0.02% V min, bal Fe.
V-45 Plate: 65,000 min TS; 45,000 min YS. V-50 Plate: 70,000 min TS; 50,000 min YS. V-55 Plate: 70,000 min TS; 55,000 min YS. V-60 Plate: 75,000 min TS; 60,000 min YS. V-65 Plate: 80,000 min TS; 65,000 min YS. For railroad and mine cars, truck and bus bodies. High strength, low alloy steel. Tough Readily workable and weldable. *Obsolete*

AW-TEN
Alan Wood Steel Co.
C 0.18, Mn 0.75, Si 0.04, Cb 0.02, Cu 0.25, bal Fe.
Hot Rolled: 70,000 min TS; 50,000 min YS; 18 min El. For mine cars, bridges, railroad cars, booms, agriculture equipment. Good formability and weldability. High strength, low alloy structural steel. *Obsolete*

AWARNITE (NATURAL)
English manufacture
Ni 75, Fe 25.
For heat and corrosion resistant parts.

AWCO ALLOYS
British Alcan Wire Ltd.
Aluminum.
Now BAW ALLOYS.

AWCO-07
Aluminium Wire & Cable Co., Ltd.
Mg 7, Mn 0.25, bal Al.
For welding rod. *Obsolete*

AWCO-21
Aluminium Wire & Cable Co., Ltd.
Mg 2.25, Mn 0.4, bal Al.
Annealed: 26,000 TS; 11,200 YS; 48 Brin. Hard: 45,000 TS; 40,000 YS; 85 Brin. For wire; corrosion resistant. *Obsolete*

AWCO-24
Aluminium Wire & Cable Co., Ltd.
Mg 0.5, Si 0.5, bal Al.
S.T.-temper: 27,000 TS; 13,500 YS; 50 Brin. S.T.A.-temper: 33,600 TS; 27,000 YS; 70 Brin. Drawn: 47,000 TS; 40,000 YS; 90 Brin. For wire; heat treatable. *Obsolete*

AWCO-25
Aluminium Wire & Cable Co., Ltd.
Mg 0.7, Si 1, bal Al.
S.T.-temper: 33,600 TS; 20,000 YS; 60 Brin. S.T.A.-temper: 45,000 TS; 40,000 YS; 95 Brin. For wire; heat treatable. *Obsolete*

AWCO-27
Aluminium Wire & Cable Co., Ltd.
Mg 3.5, Mn 0.5, bal Al.
Annealed: 35,000 TS; 16,200 YS; 55 Brin. Hard: 51,000 TS; 43,500 YS; 100 Brin. For wire; corrosion resistant. *Obsolete*

AWCO-28
Aluminium Wire & Cable Co., Ltd.
Mg 5, Mn 0.35, bal Al.
Annealed: 40,000 TS; 18,000 YS; 65 Brin. Hard: 60,000 TS; 49,000 YS; 115 Brin. For weaving wire, rivets; corrosion resistant. *Obsolete*

AWCO-301
Aluminium Wire & Cable Co., Ltd.
Cu 4.1, Mg 0.6, Si 0.45, Mn 0.5, bal Al.
S.T.A.-temper: 61,500 TS; 36,000 YS; 105 Brin. For wire, rivets; heat treatable. *Obsolete*

AWCO-303
Aluminium Wire & Cable Co., Ltd.
Cu 4.5, Mg 0.5, Si 0.75, Mn 0.75, bal Al.
S.T.-temper: 62,500 TS; 38,000 YS; 115 Brin. S.T.A.-temper: 69,000 TS; 58,000 YS; 135 Brin. For wire, bolts; heat treatable. *Obsolete*

AWCO-304
Aluminium Wire & Cable Co., Ltd.
Cu 2, Mg 0.4, bal Al.
S.T.A.-temper: 43,500 TS; 22,200 YS; 80 Brin. For wire, rivets. *Obsolete*

AWCO-31
Aluminium Wire & Cable Co., Ltd.
Cu 5, bal Al.
For welding rod. *Obsolete*

AWCO-35
Aluminium Wire & Cable Co., Ltd.
Cu 3, Mg 0.6, Sb 0.6, Sn 0.15, bal Al.
Drawn: 53,000 TS; 49,200 YS; 110 Brin. For machined parts; free-cutting, heat treatable. *Obsolete*

AWCO-40
Aluminium Wire & Cable Co., Ltd.
Si 10-13, bal Al.
For welding rod. *Obsolete*

AWCO-45
Aluminium Wire & Cable Co., Ltd.
Si 4.5-6, bal Al.
For welding rod. *Obsolete*

AWCO-60

Aluminium Wire & Cable Co., Ltd.
Mn 1.25, bal Al.
Annealed: 15,700 TS; 7,800 YS; 30 Brin. Hard: 33,600 TS; 25,800 YS; 60 Brin. For wire; corrosion resistant. *Obsolete*

AWCO-EP

Aluminium Wire & Cable Co., Ltd.
99.5% Al min.
23,000-28,000 TS. For conductors, cables; high conductivity. *Obsolete*

AWCO-SP

Aluminium Wire & Cable Co., Ltd.
99.99% Al min.
Annealed: 8,500 TS. Hard: 16,000 TS. For wire; high conductivity. *Obsolete*

AWCO-SP12

Aluminium Wire & Cable Co., Ltd.
Mg 1.25, bal Al.
Rolled: 29,200 TS. For costume jewelry. *Obsolete*

AWCO-SP16

Aluminium Wire & Cable Co., Ltd.
Cu 0.25, Mg 0.7, Si 0.3, bal Al.
Rolled: 45,000 TS. For costume jewelry, pens, bracelets; good for anodizing. *Obsolete*

AWX-45

Alan Wood Steel Co.
C 0.14, Mn 0.5, Si 0.04, Cb 0.02, bal Fe.
Rolled: 63,000 TS; 45,000 min YS; 25 El; B 78 Rock. For trucks, plow frames, pressure vessels. Good weldability and formability. *Obsolete*

AWX-50

Alan Wood Steel Co.
C 0.14, Mn 0.5, Si 0.04, Cb 0.02, bal Fe.
Plate: 50,000 min YS; 67,000 TS; 22 El; B 81 Rock. For trucks, plow frames, pressure vessels. Good formability and weldability. High strength, low alloy steel. *Obsolete*

AWX-55

Alan Wood Steel Co.
C 0.14, Mn 0.5, Si 0.04, Cb 0.02, bal Fe.
Plate: 55,000 min YS; 71,000 TS; 20 El; B 83 Rock. For trucks, plow frames, pressure vessels. Good formability and weldabiltiy. High strength, low alloy steel. *Obsolete*

AXALOY

Timken-Detroit Axle Co.
C 0.4, bal Fe.
For truck axles; water hardened.

AXITE

Axelson Mfg. Co.
Co, Cr-W.
For valve seats, hard facing electrodes; wear resistant.

AXLE STEEL

Sanderson Kayser Ltd.
C 0.35-0.45, Mn 1.2, Ni 1, Si 0-0.3, bal Fe.
Heat treated: 89,000-112,000 TS; 22 min El; 46 min RA; 174-223 Brin. For axles, gears, shafts; oil or water hardened. *Obsolete*

AXLO

British Steel Corp.
C, alloy, bal Fe.
Heat treated: 124,000-234,000 TS; 104,000-218,000 YS; 13-21 El; 39-57 RA; 302-477 Brin. For crankshaft, gears, die blocks; shock resistant. *Obsolete*

AXLOY 20

Axelson Mfg. Co.
C 3.6, Si 2.5, Mn 0.6, bal Fe.
Cast: 20,000 TS, 183 Brin. For glass molds; soft cast iron.

AXLOY 30

Axelson Mfg. Co.
C 3.4, Si 1.8, Mn 0.7, Ni 0.25-1, Cr 0.4, Mo 0.4, bal Fe.
Normalized: 30,000-35,000 TS; 180 Brin. For machinery castings, gears; cast iron, wear resistant.

AXLOY 30 CN

Axelson Mfg. Co.
C 3.2, Si 2.5, Ni 1.5, bal Fe.
For hydraulic cylinders, valves; cast iron.

AXLOY 30 CNM

Axelson Mfg. Co.
C 3.3, Ni 0.25-1, Cr 0.25-1, Mo 0.25-1, bal Fe.
Cast: 35,000 TS; 207 Brin. For liners, cams, cylinder blocks, gears; cast iron.

AXLOY 30 CRN

Axelson Mfg. Co.
C 3.2, Ni 1.5, Cr 0.8, Si 2.5, bal Fe.
Cast: 30,0000 TS. For oil field valves, pump liners; cast iron, corrosion resistant.

AXLOY 35

Axelson Mfg. Co.
C 3.2, Si 1.9, Mn 0.85, bal Fe.
Normalized: 35,000-45,000 TS; 220 Brin. For machinery castings, gears; cast iron, wear resistant.

AXLOY 35 CN

Axelson Mfg. Co.
C 3, Si 2, Ni 2, Cr 1, bal Fe.
Cast: 35,000 TS. For brake drums, gears, pump liners; abrasion and wear resistant.

AXLOY 50

Axelson Mfg. Co.
C 3, Si 2.2, Mn 1, bal Fe.
Normalized: 50,000-60,000 TS; 240 Brin. For gears, machinery castings; cast iron.

AXLOY 50 CN

Axelson Mfg. Co.
C 3.2, Si 2.2, Ni 2, Cr 1, bal Fe.
Cast: 50,000 TS. For hydraulic cylinders, valves, steam chests; cast iron.

AXLOY 50 CNM

Axelson Mfg. Co.
C 3.2, Si 2, Ni 2, Cr 1, bal Fe.
Cast: 50,000 TS. For cylinder heads, pressure gates; cast iron.

AXLOY CNM

Axelson Mfg. Co.
C 3, Si 2, Ni 1, bal Fe.
Cast: 35,000 TS. For dies, pistons, liners, cams; wear resistant.

AZ

British Metal Corp., Ltd.
Al alloy.
For light alloy parts.

AZ 100 A

Now AM 100 A.

AZ 14

Otto Fuchs Metallwerke
Aluminum. Mg 1.2, Mn 0.1, Zn 4.7, Cr, Zr, bal Al.
315-390 N/mm^2 TS; 275-295 N/mm^2 YS; 6-10 El. High strength heat treatable alloy, corrosion resistant with good weldability and formability. Type AlZnMg1. Similar to AA 7005.

AZ 24

Otto Fuchs Metallwerke
Aluminum. Cu 0.1, Mg 2, Si 0.2, Mn 0.25, Zn 4, Zr, bal Al.
370-385 N/mm^2 TS; 310-330 N/mm^2 YS; 3-8 El. Heat treatable alloy. Type AlZnMg2.

AZ 31C

Dow Chemical Co.
Magnesium. Al 2.4-3.6, Zn 0.5-1.5, bal Mg.
Extruded: 37,000 psi TS; 28,000 psi YS; 12 El; 49 Brin. O-temper: 37,000 psi TS; 22,000 psi YS; 21 El; 56 Brin. For truck bodies, dock boards; good formability.

AZ 40

Otto Fuchs Metallwerke
Aluminum. Cu 0.7, Mg 3.2, Mn 0.1, Zn 4.5, Cr, bal Al.
460-500 N/mm^2 TS; 385-430 N/mm^2 YS; 5-8 El. High strength heat treatable alloy, special extrusion quality. Type AlSnMgCu0.5. Similar to AA 7079.

AZ 54

Otto Fuchs Metallwerke
Aluminum. Cu 0.7, Mg 3.4, Mn 0.1, Zn 4.6, Cr, bal Al.
High strength heat treatable alloy. Type AlZnMgCu0.5. Similar to AA 7079.

AZ 63

Otto Fuchs Metallwerke
Aluminum. Cu 1.4, Mg 2.3, Zn 5.7, Cr, bal Al.
Special extrusion quality, hardenable to: 420-530 N/mm^2 TS; 355-46 N/mm^2 YS; 5-8 El. Type AlZnMgCu1.5. Similar to AA 7075.

AZ 64

Otto Fuchs Metallwerke
Aluminum. Cu 1.4, Mg 2.5, Zn 5.7, Cr, bal Al.
420-530 N/mm^2 TS; 355-460 N/mm^2 YS; 5-8 El. High strength heat treatable alloy. Type AlZnMgCu1.5. Similar to AA 7075.

AZ 67

Otto Fuchs Metallwerke
Aluminum. Cu 1, Mg 2.3, Zn 5.7, Cr, Be, bal Al.
420-530 N/mm^2 TS; 355-460 N/mm^2 YS; 5-8 El. Special mining quality. Type AlZnMgCu1Be.

AZ 74

Otto Fuchs Metallwerke
Aluminum. Cu 1, Mg 2.4, Zn 5.8, Cr, Ag, bal Al.
450-550 N/mm^2 TS; 390-490 N/mm^2 YS; 5-9 El. High strength alloy with resistance to stress-corrosion cracking. Type AlZnMgCuAl.

AZ 75

Otto Fuchs Metallwerke
Aluminum. Cu 1, Mg 2.5, Zn 6, Cr, Zr, Ag, bal Al.
450-550 N/mm^2 TS; 390-490 N/mm^2 YS; 5-10 El. High strength alloy, mainly for die forgings having low level of internal stresses. Type AlZnMgCuAgZr.

AZ 79

Otto Fuchs Metallwerke
Aluminum. Cu 1, Mg 2.5, Zn 6, Cr, Ag, bal Al.
450-550 N/mm^2 TS; 390-490 N/mm^2 YS; 5-10 El. Improved hardenability and fracture toughness. Type AlZnMgCuAg.

AZ 92 A (WELD ROD)

Dow Chemical Co.
Magnesium. Al 8.3-9.7, Zn 1.7-2.3, 0.15 Mn min, 0.30 others max, bal Mg.
Melting point: 1110°F. For welding rod and electrodes.

AZ 92A

J.W. Harris Co., Inc.
Magnesium base rod. For use on magnesium base materials.

AZ.31

Birmetals Ltd.
Zn 1, Mn 0.3, Al 3, bal Mg
Plate, sheet, strip, extrusions. 28,000-34,000 TS; 14,000-20,000 proof stress; 8-12 El; up to 20,000 fatigue strength. Good general purpose alloy with medium strength and good corrosion resistance; weldable. *Obsolete*

AZ31A
Dow Chemical Co.
Al 2.5-3.5, Zn 0.6-1.4, 0.20% Mn min, bal Mg.
Heat treated: 32,000-40,000 TS; 16,000-29,000 YS; 4-12 El.
For aircraft structural parts; age hardenable. *Obsolete*

AZ31B
Dow Chemical Co.
Magnesium. Al 2.5-3.5, Zn 0.7-1.3, bal Mg.
O-temper: 32,000-37,000 psi TS; 18,000-22,000 psi YS; 21-12 El; 46 Brin. F-temper: 32,000-38,000 psi TS; 20,000-29,000 psi YS; 7-18 El; 49 Brin. H24-temper: 37,000-42,000 psi TS; 24,000-32,000 YS; 6-19 El; 73 Brin. Sheet, extrusions; formable, weldable; for use to 300°F. ASTM B107; QQ-M-31b.

AZ5G
French manufacture
Zn 5, Mg 0.5, Ti 0.2, Cr 0.3, bal Al.
Aged: 31,300-35,600 TS; 18,500-22,800 YS; 5-9 El; 70 Brin.
For aircraft components, high strength castings; age-hardenable, shock resistant.

AZ61A
Dow Chemical Co.
Magnesium. Al 5.8-7.2, Zn 0.4-1.5, bal Mg.
F-temper: 38,000-46,000 psi TS; 21,000-33,000 psi YS; 7-17 El; 60 Brin. General purpose alloy for use up to 300°F. Used as welding rod and wire, extruded bar and plate and tube, and forgings. ASTM B107-69, B91-68, A5.19-69.

AZ63A
Various foundries
Al 6, Zn 3, Mn 0.15-0, bal Mg.
F-temper: 24,000-29,000 TS; 10,000-14,000 YS; 4-6 El. T6-temper: 34,000-40,000 TS; 16,000-19,000 YS; 3-5 El. Sand and permanent mold castings; age hardenable, good corrosion resistance, pressure tight. For airplane wheel and brake castings, gear housings. ASTM B93-66; AMS 4420, 4422; QQ-M-55b; SAE 50.

AZ63A (MG INGOT)
Dow Chemical Co.
Magnesium. Al 5.5-6.5, Mn 0.18-0.32, Zn 2.7-3.3, Cu 0-0.04, Ni 0-0.005, Si 0-0.1, 0.3 others max, bal Mg.
Ingots for remelting. ASTM B93.

AZ80A
Dow Chemical Co.
Magnesium. Al 7.8-9.2, Zn 0.2-0.8, bal Mg.
T5-temper: 45,000-55,000 psi TS; 30,000-40,000 psi YS; 4-8 El; 82 Brin. Extruded rod and forgings; good strength. ASTM B107, B91; AMS 4360.

AZ81A
Various foundries
Al 7.6, Zn 0.7, Mn 0.13-0, bal Mg.
T4-temper: 34,000-40,000 TS; 10,000-12,000 YS; 7-15 El. Sand and permanent mold castings; age hardenable, good corrosion resistance. For aircraft equipment where good elongation and toughness are required. ASTM B80-69; QQ-M-56; SAE 505.

AZ8GU
French manufacture
Zn 7-8.5, Mg 1.7-3, Cu 1-2, Cr 0.2, Mn 0.4, bal Al.
Hardened: 95,000 TS; 85,000 YS; 8 El. For light alloy parts; age-hardenable.

AZ8GU
Pechiney/SMG
Si 0-0.25, Fe 0-0.35, Cu 1.2-1.9, Mg 0.2-2.9, Cr 0.1-0.22, Zn 7.2-8.2, bal Al.
Wrought, heat treatable aluminum alloy; AA7049. For production of high quality weapon parts.

AZ91A
Various foundries
Al 9, Zn 0.7, Mn 0.13-0, bal Mg.
F-temper: 34,000 TS; 23,000 YS; 3 El. Magnesium die cast alloy with good pressure tightness and corrosion resistance. For housings and small precision parts. ASTM B94-57; QQ-M-38; AMS 4490; SAE 501.

AZ91A (MG INGOT)
Dow Chemical Co.
Magnesium. Al 8.5-9.5, Be 0.0003-0.001, Mn 0-0.15, Zn 0.45-0.9, Cu 0-0.08, Ni 0-0.08, Si 0-0.2, 0.30 others max, bal Mg.
Ingots for remelting. ASTM B93-66. *Obsolete*

AZ91B
Manufacturer not listed.
Similar to AZ91A.

AZ91B (MG INGOT)
Dow Chemical Co.
Magnesium. Al 8.5-9.5, Be 0.0004-0.001, Zn 0.5-0.9, Cu 0-0.25, Ni 0-0.01, Si 0-0.2, 0.15 Mn min, 0.30 others max, bal Mg.
Ingots for remelting for die casting and permanent mold casting use only. ASTM B93.

AZ91C
Fansteel/Wellman Dynamics
Magnesium. Al 8.1-9.3, Zn 0.4-1, 0.13 Mn min, bal Mg.
Magnesium casting alloy. Cast: 23,000-34,000 psi TS; 11,000-16,000 psi YS; 2-7 El; 52-66 Brin.

AZ91C
Various foundries
Al 8.7, Zn 0.7, 0.13 Mn min, bal Mg.
F-temper: 18,000-24,000 TS; 10,000-14,000 YS; 0.2 El. T6-temper: 34,000-40,000 TS; 16,/00-19,000 YS; 3-5 El. Sand and permanent mold castings, good pressure tightness; age hardenable. For light weight housings for motors and various accessories. ASTM B80-69; QQ-M-56; QQ-M-55; SAE 504.

AZ91C (MG INGOT)
Dow Chemical Co.
Magnesium. Al 8.3-9.2, Mn 0.15-0.32, Zn 0.5-0.9, Cu 0-0.04, Ni 0-0.005, Si 0-0.1, 0.30 others max, bal Mg.
Ingot for remelting. ASTM B93.

AZ91D (MG INGOT)
Dow Chemical Co.
Magnesium. Al 8.5-9.5, Be 0.0004-0.001, Zn 0.5-0.9, Cu 0-0.01, Ni 0-0.001, Si 0-0.02, Mn 0.17-0.32, Fe 0-0.004, 0.01 others max, bal Mg.
Ingots for remelting for die casting and permanent mold casting use only. ASTM B93.

AZ91E
Fansteel/Wellman Dynamics
Magnesium. Al 8.1-9.3, Zn 0.4-1, Ni 0-0.001, 0.17-0.35 Mn min, Fe:Mn ratio shall not exceed 0.032,, bal Mg.
Magnesium casting alloy. Cast: 23,000-34,000 psi TS; 11,000-16,000 psi YS; 2-7 El; 52-66 Brin.

AZ91E (MG INGOT)
Dow Chemical Co.
Magnesium. Al 8.3-9.2, Mn 0.17-0.25, Zn 0.5-0.9, Cu 0-0.01, Ni 0-0.001, Si 0-0.1, Fe 0-0.0004, 0.01 others max, bal Mg.
Ingot for remelting. ASTM B93.

AZ92A
Fansteel/Wellman Dynamics
Magnesium. Al 8.3-9.7, Zn 1.6-2.4, 0.10 Mn min, bal Mg.
Magnesium casting alloy. Cast: 23,000-34,000 psi TS; 11,000-18,000 psi YS; 1-6 El; 65-84 Brin.

AZ92A (MG INGOT)
Dow Chemical Co.
Magnesium. Al 8.5-9.5, Mn 0.13-0.32, Zn 1.7-2.3, Cu 0-0.04, Ni 0-0.005, Si 0-0.1, 0.30 others max, bal Mg.
Ingot for remelting. ASTM B93.

AZA
American Alloys Corp.
Cu 0.45-0.85, Fe 0.5, Si 0.15, Zn 7.1-7.8, Mg 0.2-0.5, Ti 0.1-0.2, Mn 0.3, Cr 0.2, Be 0.0003, Ni 0.05, bal Al.
Aged: 38,00 TS; 26,000 YS; 5 El; 77 Brin. For hardware, aircraft and oil field equipment, dairy and agriculture equipment. Pressure tight castings.

AZM
Birmetals Ltd.
Magnesium. Zn 1, Mn 0.3, Al 6, bal Mg.
Extrusions and forgings: 30,000-36,000 TS; 20,000-22,000 proof stress; 7-10 El; 55-70 HV. Medium strength alloy, machines readily, difficult to cold form and to weld; may need protection against corrosion.

AZOWIT 25 S
Thyssen Edelstahlwerke AG
C 0.3-0.37, Mn 0.6-0.9, Al 0.8-1.1, Cr 1-1.3, bal Fe.
Oil hardenable; to be case hardened by nitriding; good core hardness and strength. DIN 34 Cr Al S5. *Obsolete*

AZOWIT 30 M
Thyssen Edelstahlwerke AG
C 0.2-0.34, Cr 2.3-2.7, Mo 0.15-0.25, V 0.1-0.2, bal Fe.
Oil hardenable; to be case hardened by nitriding to surface hardness of 47-56 Rock C. DIN 31 Cr Mo V 9. *Obsolete*

AZOWIT 30 ML
Thyssen Edelstahlwerke AG
C 0.3, Si 0.25, Mn 0.5, Cr 2.5, Mo 0.2, V 0.15, bal Fe.
Alloy steel, case hardenable by nitriding. DIN 31 Cr Mo V 9; Werkstoff Nr. 1.8514. *Obsolete*

AZOWIT 32 M
Thyssen Edelstahlwerke AG
C 0.28-0.35, Cr 2.8-3.3, Mo 0.3-0.5, bal Fe.
Oil hardenable; to be case hardened by nitriding to 48-57 Rc surface hardness. DIN 32 Cr Mo 12. *Obsolete*

AZOWIT 35
Thyssen Edelstahlwerke AG
C 0.3-0.37, Mn 0.6-0.9, Al 0.8-1.1, Cr 1.2-1.5, bal Fe.
Oil hardenable; to be case hardened by nitriding; good core hardness and strength. DIN 34 Cr Al 6. *Obsolete*

AZOWIT 35 M
Thyssen Edelstahlwerke AG
C 0.3-0.37, Mn 0.6-0.9, Al 0.8-1.1, Cr 1-1.3, Mo 0.15-0.25, bal Fe.
Oil hardenable; to be case hardened by nitriding; good core hardness and strength. DIN 34 Cr Al Mo 5. *Obsolete*

AZOWIT 35 ML
Thyssen Edelstahlwerke AG
C 0.32, Si 0.25, Mn 0.6, Cr 1.1, Mo 0.2, Al 1.1, bal Fe.
Alloy steel, case hardenable by nitriding. DIN 34 Cr Al Mo 5; Werkstoff Nr. 1.8544. *Obsolete*

AZOWIT 38 M
Thyssen Edelstahlwerke AG
C 0.3-0.37, Al 0.8-1.1, Cr 1.5-1.8, Mo 0.15-0.25, Ni 0.9-1.1, bal Fe.
Oil hardenable; to be case hardened by nitriding; good core hardness and strength. DIN 34 Cr Al Ni 7. *Obsolete*

B & B

Manufacturer not listed.
Co 30, Cr 15, Mo 5, Ti 2.5, Al 3, B 0.5, bal Ni.
For jet engine components, buckets; high heat resistance.

B & K AZ-31A

Brooks & Perkins Inc.
Al 2.5-3.5, Zn 0.6-1.4, 0.20% min Mn, bal Mg.
32,000-40,000 TS; 16,000-29,000 YS; 2-12 El. For light alloy
parts; readily formed. *Obsolete*

B & W 12-14 CR

Babcock & Wilcox Co.
C 0-0.12, Cr 12-14, bal Fe.
For corrosion resistant parts; corrosion resistant. *Obsolete*

B & W ALLOY NO. 1000

Babcock & Wilcox Co.
Cu 2.35, Mn 1.5, Si 2, Cr 18, W 9, bal Fe.
Cast and treated: 600-680 Brin. For abrasion resisting parts,
ceramic and cold drawn dies; not shock resisting. *Obsolete*

B & W NO. 445

Babcock & Wilcox Co.
C 0.3, Cr 8-10, Mo 1.5, bal Fe.
Cast: 90,000 TS; 180 Brin. For oil refinery and boiler parts;
heat and corrosion resistant. *Obsolete*

B & W NO. 5202

Babcock & Wilcox Co.
C 1, Mn 2, Cr 1.5, Mo 0.5, Si 0.8, bal Fe.
Cast: 150,000 TS; 500-650 Brin. For grinding and pulverizing
equipment; wear resistant. *Obsolete*

B & W NO. 602

Babcock & Wilcox Co.
C 0-0.07, Cr 17-19, Ni 7.5-10.5, bal Fe.
For stainless parts; stainless. *Obsolete*

B & W NO. 603

Babcock & Wilcox Co.
C 0-0.07, Cr 16-23, Ni 7-11, Ti, bal Fe.
Cast: 70,000-85,000 TS; 130-160 Brin. For stainless parts;
stainless. *Obsolete*

B & W NO. 604

Babcock & Wilcox Co.
C 0.08, Cr 18, Ni 8, bal Fe.
Cast: 70,000-85,000 TS; 130-160 Brin. For stainless parts;
stainless, austenitic. *Obsolete*

B & W NO. 640S

Babcock & Wilcox Co.
C 0.08, Cr 19, Ni 9, Mo 3, bal Fe.
Cast: 88,000 TS; 170 Brin. For paper mill plugs, wire guides;
wear and heat resistant. *Obsolete*

B & W NO. 690

Babcock & Wilcox Co.
C 0-0.18, Cr 24-30, Ni 7-11, Mo, bal Fe.
For heat and corrosion resistant parts; heat and corrosion
resistant. *Obsolete*

B & W NO. 692

Babcock & Wilcox Co.
C 0-0.1, Cr 24-30, Ni 7-11, Mo, bal Fe.
For heat and corrosion resistant parts; heat and corrosion
resistant. *Obsolete*

B & W NO. 951

Babcock & Wilcox Co.
C 1.5, Cr 20, Ni 2, bal Fe.
Cast: 80,000 TS. For heat and wear resistant parts; heat and
wear resistant. *Obsolete*

B 15 LC2

Creusot-Loire
C 1.6, Cr 12, Mo 0.9, V 0.9, bal Fe.
Cold work tool steel; air hardening. For stamping dies, thread
rollers, shears. AFNOR Z160CDV 12; AISI D2.

B 15 M

Creusot-Loire
C 2, Cr 12, Mo 0.6, bal Fe.
Cold work tool steel; air hardening. For stamping and
trimming dies, shears. AFNOR Z190CD12; similar to AISI D3.

B 1900

Cannon-Muskegon Corp.
C 0.1, Cr 8, Co 10, Mo 6, Ti 1, Al 6, B 0.015, Zr 0.1, Ta 4, bal
Ni.
At room temperature: 141,000 TS; 120,000 YS; 8 El. At
1400°F: 138,000 TS; 117,000 YS; 4 El. For high temperature
applications, jet engines, turbines. Heat and corrosion
resistant. Casting alloy. Good thermal fatigue and creep
resistance.

B ALLOY

American manufacture
Zn 25, Cu 3, bal Al.
For light alloy parts; non-hardenable.

B E NO 4 ALLOY

American manufacture
Cu 4, Mg 0.25, Fe 0-0.5, Si 0-0.1, bal Al.
For boxes, covers, face-plates; U.S.N. Bureau of Engineering
Alloy.

B ELECTROMAL

Belle City Malleable Co.
C 2.3-2.4, Si 1.2-1.3, Cu 1, Ni 4, bal Fe.
Cast: 60,000-70,000 TS; 45,000 YS; 15 El; 143-170 Brin. For
castings; malleable iron. *Obsolete*

B K SPECIAL

Paul Bergsøe & Søn
Sn 90.3, Sb 6.5, Cu 3, Ni 0.2.
Cast: 12,800 TS; 25 Brin. MP: 440-610°F. Shock resistant. For
engine bearings.

B MONEL

International Nickel Inc.
Ni 67, Cu 30.
For stainless parts; corrosion resistant. *Obsolete*

B MONEL

Wilbur B. Driver Co.
Ni 67, Cu 30.
For stainless parts; corrosion resistant. *Obsolete*

B NO. 4

Cytemp Specialty Steel Div.
C 0.7, W 18, Cr 4, V 1, bal Fe.
For drills, reamers, cutters, hobs, taps, broaches; high speed
steel, oil hardened. *Obsolete*

B NO. 9

Cytemp Specialty Steel Div.
W 1.5, Cr 4, V 1, C 0.7, Mo 9.5, bal Fe.
For drills, reamers, hobs, broaches, taps, cutters; high speed
steel, oil hardened. *Obsolete*

B&W 0117

Babcock & Wilcox Co.
C 0-0.22, Mn 0.85-1.25, 0.2 Cu min, 0.02 V min, bal Fe.
Hot finished or normalized: 70,000 TS; 50,000 YS; 25 El.
Shock resistant; good resistance to atmospheric corrosion.

B&W 1118FM

Babcock & Wilcox Co.
C 0.14-0.2, Mn 1.3-1.6, S 0.08-0.13, P 0-0.04, bal Fe.
Hot finished: 70,000-75,000 TS; 40,000-45,000 YS; 75-85
Rock B. Cold drawn: 90,000-105,000 TS; 75,000-90,000 YS;
92-100 Rock B. For machining applications.

B&W 52100

Babcock & Wilcox Co.
C 0.95-1.1, Mn 0.25-0.45, Cr 1.3-1.6, bal Fe.
Annealed: 90,000 TS; 51,000 YS; 31 El; 180 Brin. Normalized:
161,000 TS; 106,000 YS; 7 El; 373 Brin. Hardened, oil
quenched and tempered: TS in excess of 300,000 psi is
obtainable. For bearings, bushings, tools.

B&W CARBON MOLY (T/P-1)

Babcock & Wilcox Co.
C 0.1-0.2, Mn 0.3-0.8, Si 0.1-0.5, Mo 0.44-0.65, bal Fe.
Annealed: 55,000 TS min; 30,000 YS min; 30 El min; 80 Rock
B max. For service conditions requiring higher creep strength
than carbon steel with no increase in corrosion or oxidation
resistance. Oxidation resistant to 1050°F in air.

B&W CARBON STEEL

Babcock & Wilcox Co.
C 0.06-0.18, Mn 0.27-0.63, bal Fe.
Annealed: 47,000 TS min; 26,000 YS min; 35 El min; 77 Rock
B max. For tubular heat exchangers, condensers, and similar
heat transfer apparatus.

B&W CROLOY 1-1/4 (T/P-11)

Babcock & Wilcox Co.
C 0-0.15, Mn 0.3-0.6, Si 0.5-1, Cr 1-1.5, Mo 0.44-0.65, bal Fe.
Annealed: 60,000 TS min; 30,000 YS min; 30 El min; 85 Rock
B max. Good creep strength properties; more corrosion
resistant than Cr free grades. Oxidation resistant to 1100°F in
air.

B&W CROLOY 1/2 (T/P-2)

Babcock & Wilcox Co.
C 0.1-0.2, Mn 0.3-0.61, Si 0.1-0.3, Cr 0.5-0.81, Mo 0.44-0.65,
bal Fe.
Annealed: 60,000 TS min; 30,000 YS min; 30 El min; 85 Rock
B max. Superior to CARBON MOLY for creep strength and
graphitization; for high temperature steam piping. Oxidation
resistant to 1075°F in air.

B&W CROLOY 12 (T/P 410)

Babcock & Wilcox Co.
C 0-0.15, Mn 0-1, Si 0-0.75, Cr 11.5-13.5, bal Fe.
Annealed: 60,000 TS min; 30,000 YS min; 20 El min; 95 Rock
B max. Normalized and tempered: 106,000-180,000 TS;
85,000-175,000 YS; 15-25 El; 21-39 Rock C. For use where
mechanical properties plus corrosion resistance are
important. Oxidation resistant to 1300°F in air.

B&W CROLOY 12-3W

Babcock & Wilcox Co.
Stainless steel. C 0.15-0.2, Mn 0-0.5, Cr 12-14, Ni 1.8-2.2, W
2.5-3.5, bal Fe.
Martensitic, stainless that can be heat treated to high
hardness.

B&W CROLOY 12A1 (T/P 405)

Babcock & Wilcox Co.
C 0-0.08, Mn 0-1, Si 0-0.75, Cr 11.5-13.5, Al 0.1-0.3, bal Fe.
Annealed: 60,000 TS min; 30,000 YS min; 20 El min; 95 Rock
B max. Non-hardenable by heat treating; weldable. Good
oxidation resistance.

B&W CROLOY 16-1

Babcock & Wilcox Co.
C 0-0.03, Mn 0-1, Cr 14-16, Ni 1-1.5, bal Fe.
Annealed: 60,000 TS min; 30,000 YS min; 20 El min; 95 Rock
B max. In tubular operations requiring increased resistance
to chloride stress corrosion cracking.

B&W CROLOY 16-6 PH

Babcock & Wilcox Co.
C 0.025-0.045, Mn 0.7-0.9, Cr 15-16, Ni 7-8, Al 0.25-0.45, Ti
0.3-0.5, bal Fe.
Solution annealed: 134,000 TS; 110,000 YS; 16 El; 28 Rock
C. Solution annealed and aged: 190,000 TS; 185,000 YS; 15
El; 40 Rock C. Good combination of hardness, strength and
corrosion resistance.

B&W CROLOY 18 (T/P 430)

Babcock & Wilcox Co.
C 0-0.12, Mn 0-1, Cr 16-18, Ni 0-0.5, bal Fe.
Annealed: 60,000 TS min; 35,000 YS min; 20 El min; 90 Rock
B max. Ferritic, magnetic, non-hardenable steel with good
corrosion resistance; good for nitration work and nitric acid
manufacture.

B&W CROLOY 2 (T/P-3B)
Babcock & Wilcox Co.
C 0-0.15, Mn 0.3-0.6, Si 0-0.5, Cr 1.65-2.35, Mo 0.44-0.65, bal Fe.
Annealed: 60,000 TS min; 30,000 YS min; 30 El min; 85 Rock B max. For resisting both oxidation and corrosion, with good high temperature strength. Oxidation resistant to 1150°F in air.

B&W CROLOY 2 AI
Babcock & Wilcox Co.
C 0-0.15, Mn 0-0.5, Si 1-1.4, Cr 1.75-2.25, Mo 0.45-0.65, bal Fe.
Annealed: 60,000 TS min; 30,000 YS min; 30 El min; 85 Rock B max. Good corrosion resistance to high temperature (vapor phase) acids. Slightly greater resistance to oxidation.

B&W CROLOY 2-1/4 (T/P 22)
Babcock & Wilcox Co.
C 0-0.15, Mn 0.3-0.6, Si 0-0.5, Cr 1.9-2.6, Mo 0.87-1.13, bal Fe.
Annealed: 60,000 TS min; 30,000 YS min; 30 El min; 85 Rock B max. High creep strength for polymerization and high pressure cracking. Oxidation resistant to 1175°F in air.

B&W CROLOY 27 (T/P 446)
Babcock & Wilcox Co.
C 0-0.2, Mn 0-1.5, Cr 23-30, Ni 0-0.5, 0.10-0.25 N_2, bal Fe.
Annealed: 70,000 TS min; 40,000 YS min; 18 El min; 95 Rock B max. Good resistance to oxidation at 1500-2100°F, and resists attack by slag flue dust.

B&W CROLOY 299
Babcock & Wilcox Co.
Stainless steel. C 0.12-0.25, Mn 14-15.5, Cr 16.5-18.5, Ni 1.15-1.75, 0.32-0.40 N_2, bal Fe.
Annealed: 115,000 TS; 68,000 YS; 72 El; 18 Rock C. Cold worked: 173,000 TS; 160,000 YS; 26 El; 39 Rock C. Austenitic, stainless steel; work hardened to high strength without becoming magnetic.

B&W CROLOY 304
Babcock & Wilcox Co.
C 0-0.08, Mn 0-2, Cr 18-20, Ni 8-11, bal Fe.
Annealed: 75,000 TS min; 30,000 YS min; 28 El min; 90 Rock B max. Resistant to corrosion and heat (available also as 304H, 304L, 321, 347, 348; 1/8, 1/4, 1/2 hard).

B&W CROLOY 310
Babcock & Wilcox Co.
C 0-0.15, Mn 0-2, Cr 24-26, Ni 19-22, bal Fe.
Annealed: 75,000 TS min; 35,000 YS min; 25 El min; 90 Rock B max. For extreme resistance to oxidation and corrosion, for high pressure, high-temperature applications.

B&W CROLOY 316
Babcock & Wilcox Co.
C 0-0.08, Mn 0-2, Cr 16-18, Ni 11-14, Mo 2-3, bal Fe.
Annealed: 75,000 TS min; 30,000 YS min; 28 El min; 90 Rock B max. Superior to Type 304 in resistance to creep and corrosion (also available as 316H and 316L).

B&W CROLOY 3M (T/P 21)
Babcock & Wilcox Co.
C 0-0.15, Mn 0.3-0.6, Si 0-0.5, Cr 2.65-3.35, Mo 0.8-1.06, bal Fe.
Annealed: 60,000 TS min; 30,000 YS min; 30 El min; 85 Rock B max. Better creep properties and corrosion and oxidation resistance than CROLOY 2. Oxidation resistant to 1175°F in air.

B&W CROLOY 5 (T/P 5)
Babcock & Wilcox Co.
C 0-0.15, Mn 0.3-0.6, Si 0-0.5, Cr 4-6, Mo 0.45-0.65, bal Fe.
Annealed: 60,000 TS min; 30,000 YS min; 30 El min; 85 Rock B max. Normalized: 160,000 TS; 120,000 YS; 16 El; 34 Rock C. Good corrosion resistance, creep strength. Oxidation resistant to 1200°F in air.

B&W CROLOY 5 SI (T/P 5B)
Babcock & Wilcox Co.
C 0.15, Mn 0.3-0.6, Si 1-2, Cr 4-6, Mo 0.45-0.65, bal Fe.
Annealed: 60,000 TS min; 30,000 YS min; 30 El min; 89 Rock B max. For operating conditions where oxidation is a primary requirement; resistant to scaling by oxidation. Oxidation resistant to 1300°F in air.

B&W CROLOY 600
Babcock & Wilcox Co.
C 0-0.15, Mn 0-1, Cr 14-17, Fe 6-10, 72.0 Ni min.
Annealed: 80,000 TS min; 35,000 YS min; 30 El min; 92 Rock B max. Good high temperature strength and useful oxidation resistance up to 2100°F.

B&W CROLOY 7 (T/P 7)
Babcock & Wilcox Co.
C 0-0.15, Mn 0.3-0.6, Si 0.5-1, Cr 6-8, Mo 0.45-0.65, bal Fe.
Annealed: 60,000 TS min; 30,000 YS min; 30 El min; 89 Rock B max. Corrosion resistance is intermediate between CROLOYS 5 and 9. Oxidation resistant to 1250°F in air.

B&W CROLOY 800
Babcock & Wilcox Co.
C 0-0.1, Mn 0-1.5, Cr 19-23, Ni 30-35, Al 0.15-0.6, Ti 0.15-0.6, bal Fe.
Annealed: 75,000 TS min; 30,000 YS min; 30 El min; 95 Rock B max. Resistant to elevated temperature oxidation and carburization.

B&W CROLOY 9M (T/P 9)
Babcock & Wilcox Co.
C 0-0.15, Mn 0.3-0.6, Si 0.25-1, Cr 8-10, Mo 0.9-1.1, bal Fe.
Annealed: 60,000 TS min; 30,000 YS min; 30 El min; 89 Rock B max. Normalized: 185,000 TS; 150,000 YS; 17 El; 42 Rock C. For severe operating conditions where high corrosion and oxidation resistance are essential. Oxidation resistant to 1300°F in air.

B&W IRON
Babcock & Wilcox Co.
C 0-0.05, Mn 0.25-0.4, Cu 0-0.1, bal Fe.
Hot finished: 45,000-50,000 TS; 30,000-41,000 YS; 49-70 El; 53-57 Rock B. Cold drawn: 68,000-86,000 TS; 60,000-80,000 YS; 25-42 El; 76-88 Rock B. For electric motor and generator housings; good magnetic characteristics.

B&W NICOLOY 3 1/2
Babcock & Wilcox Co.
C 0.06-0.12, Mn 0.31-0.64, Si 0.18-0.37, Ni 3.18-3.82, bal Fe.
Normalized: 65,000 TS min; 35,000 YS min; 30 El min; 90 Rock B max. Combination of high strength and resistance to brittle fracture at temperatures down to -150°F.

B&W NICOLOY 9
Babcock & Wilcox Co.
C 0-0.13, Mn 0-0.9, Si 0.13-0.32, Ni 8.4-9.6, bal Fe.
Double normalized and tempered: 100,000 TS min; 75,000 YS min; 22 El min. Combination of high strength and resistance to brittle fracture at temperatures down to -320°F.

B&W NO. 1101
Babcock & Wilcox
C 0.4, Cr 26, Ni 20, bal Fe.
Cast: 190 Brin. For heat resisting castings, furnace parts; heat resistant to 2000°F.

B&W NO. 1440
Babcock & Wilcox
C 0.45, Cr 1.5, Mo 7.5, bal Fe.
Annealed: 250 Brin. For tube mill piercer points; wear resistant.

B&W NO. 1501
Babcock & Wilcox
C 0.6, Cr 27, Ni 9.5, bal Fe.
Cast: 200 Brin. For oil burner parts; heat and corrosion resistant.

B&W NO. 441
Babcock & Wilcox
C 0.2, Cr 5, Mo 0.55, bal Fe.
Annealed: 250 Brin. max. For oil still headers, castings, pumps, pipes; moderate corrosion resistance.

B&W NO. 5150
Babcock & Wilcox
C 1, Cr 1.4, bal Fe.
Heat treated: 300-500 Brin. For grinding units, wear plates; abrasion resistant.

B&W NO. 600
Babcock & Wilcox
C 0.2, Cr 19.5, Ni 9.5, bal Fe.
Cast: 130-160 Brin. For corrosion resisting parts; chemical industries; austenitic alloy; high corrosion resistance.

B&W NO. 642
Babcock & Wilcox
C 0.08, Cr 19.5, Ni 10.5, Mo 2.5, bal Fe.
Cast: 140-165 Brin. For corrosion resisting parts; paper mill sulfite; resists mineral acids; austenitic; corrosion resistant.

B&W NO. 661
Babcock & Wilcox
C 0.35, Cr 26, Ni 12.5, Si 2, bal Fe.
Cast: 75,000 TS; 190 Brin. For oil still tube supports, furnace parts; heat and corrosion resistant to 1800°F.

B&W NO. 800
Babcock & Wilcox
C 1.25, Cr 25, Ni 11, bal Fe.
Heat treated: 300 Brin. For heat resisting parts, wire guides; heat resistant.

B&W NO. 850
Babcock & Wilcox
C 1.55, Cr 19, Ni 7, bal Fe.
Heat treated: 320-430 Brin. For rolling mill plugs, wire guides; wear and heat resistant. *Obsolete*

B&W STROLOY 1
Babcock & Wilcox Co.
C 0.14-0.2, Mn 0.6-1, Si 0.15-0.35, Cr 0.4-0.65, Ni 0.7-1, Mo 0.4-0.6, V 0.03-0.08, B 0.001, bal Fe.
Quenched and tempered: 115,000-145,000 TS; 100,000 YS min; 15 El min (up to 1.5 in. thick). Good strength, toughness and weldability; for structural applications.

B&W STROLOY 2A
Babcock & Wilcox Co.
C 0.15-0.21, Mn 0.7-1, Si 0.2-0.35, Cr 0.8-1.1, Ni 0.4-0.7, Mo 0.2-0.3, B 0.002-0.004, bal Fe.
Quenched and tempered: 110,000-145,000 TS; 100,000 YS min; 15 El min (up to 0.750 in. thick). Good strength, toughness and weldability; for structural applications.

B&W STROLOY 5C
Babcock & Wilcox Co.
C 0.15-0.21, Mn 0.7-1, Si 0.2-0.35, Cr 0.75-1.1, Mo 0.15-0.25, B 0.001-0.005, bal Fe.
Quenched and tempered: 110,000-145,000 TS; 100,000 YS min; 15 El min (up to 0.375 in. thick). Good strength, toughness and weldability; for structural applications.

B-10 HIGH SPEED STEEL
Cytemp Specialty Steel Div.
C 0.8, Cr 4.5, W 18.5, V 2, Co 9, bal Fe.
For lathe and form tools, reamers, broaches; high speed steel for heavy cuts. *Obsolete*

B-110
United States Bronze Powders Inc.
Cu 85, Zn 15.
Bronze powder, 60 mesh. For fabrication of compacted-sintered mechanical P/M articles.

B-120VCA
Now CRUCIBLE B 120 VAC.

B-129

United States Bronze Powders Inc.
Cu 78.5, Pb 1.5, bal Zn.
Bronze powder, 60 mesh. For fabrication of compacted-sintered mechanical P/M articles.

B-155

United States Bronze Powders Inc.
Cu 70, Pb 1.5, bal Zn.
Bronze powder, 60 mesh. For fabrication of compacted-sintered mechanical P/M articles.

B-161

United States Bronze Powders Inc.
Cu 89, Pb 1.5, bal Zn.
Bronze powder, 60 mesh. For fabrication of compacted-sintered mechanical P/M articles.

B-174

United States Bronze Powders Inc.
Cu 63.5, Pb 1.5, bal Zn.
Bronze powder, 60 mesh. For fabrication of compacted-sintered mechanical P/M articles.

B-1900

Now ALLOY B-1900.

B-1910

Cannon-Muskegon Corp.
C 0.1, Cr 10, Co 10, Mo 3, Ti 1, Al 6, B 0.015, Zr 0.1, Ta 7, bal Ni.
High temperature alloy; for jet engine blades.

B-24 BRONZE

Koppers Co. Inc.
Cu 77-81, Sn 19.5-20.5.
230 Brin. For air brake rings; corrosion resistant. *Obsolete*

D-0C

Koppers Co. Inc.
Cu 86-90, Sn 6-8, Pb 2-4, Ni 1-2, P 0.2-0.7.
Cast: 35,000-45,000 TS; 75 Brin. For locomotive main cylinder packing and segmental rings; wear resisting. *Obsolete*

B-4

Atrax Cemented Carbide
Sintered carbide. 350,000 transverse strength; 14.0-14.25 g/cm^3 density; 87.5-88.5 RA. Industry code: C-12-13-14.

B-412

United States Bronze Powders Inc.
Bronze powder, 60 mesh. For fabrication of compacted-sintered mechanical P/M articles.

B-47 HOT WORK

AL Tech Specialty Steel Corp.
C 0.4, Mn 0.35, Cr 4.2, W 4.2, Co 4.2, V 2.2, Mo 0.4, bal Fe.
At 800°F: 178,000 TS; 9 El; 36 RA. At 1200°F: 110,000 TS; 16.5 El; 37 RA. W-Cr-Co hotwork tool steel for heavy duty applications; forging die inserts, hotwork punches, brass extrusion tooling. AISI H-19.

B-50

Champion Steel Co.
C 0.5, Mn 0.85, Cr 1.1, V 0.15, Mo 0.25, bal Fe.
Oil hardening tool steel for shafts, arbors, lathe centers.

B-66

Now WESTINGHOUSE B-66.

B-76

Now HEPPENSTALL B76.

B-ARROW 1733

S.K.F. Industries Inc.
C, alloy, bal Fe.
For tools. *Obsolete*

B-CARBON VANADIUM

Chemalloy Electronics Corp.
C 0.8, V 0.2, bal Fe.
For tools, cutters. *Obsolete*

B-D CU

Anaconda Co.
B 0.01, bal Cu.
Hard: 48,000 TS; 40,000 YS; 6 El; B 50 Rock. Soft: 33,000 TS; 10,000 YS; 45 El; F 45 Rock. For electrical and electronic equipment, magnetrons, vacuum switchgear. High resistance to grain growth and resistance to thermal stress cracking. *Obsolete*

B-ELITE

Thyssen Edelstahlwerke AG
C 0.1, Mn 0.65, Si 0.12, bal Fe.
For welding electrodes; general purpose. *Obsolete*

B-ELITE HS

Thyssen Edelstahlwerke AG
C 0.2, Mn 1.5, Si 0.2, Al 0.15, Ti 0.25, bal Fe.
For welding electrodes; general purpose. *Obsolete*

B-ELITE KVA

Vereinigte Edelstahlwerke
C 0.2, Mn 1, Zr 0.4, bal Fe.
Welded: 80,000 TS; 20-25 El. For welding electrodes; cored, high ductility. *Obsolete*

B-ELITE KVA

Thyssen Edelstahlwerke AG
C 0.18, Mn 1.1, Si 0.6, Al 0.1, Zr 0.3, bal Fe.
For welding electrodes; general purpose. *Obsolete*

B-F H.T. LOW CARBON

Bonney-Floyd Co.
C 0.3, Mn 0.6-0.8, bal Fe.
Heat treated: 80,000-128,000 TS; 53,000-98,000 YS; 10-30 El; 18-65 RA; 165-250 Brin. For general structural parts; cast metal; IZ 14-18. *Obsolete*

B-F HIGH SPEED

Now REPUBLIC T1.

B-F N.C.M.

Bonney-Floyd Co.
C 0.25-0.35, Mo 0.15-0.25, Ni 1.75-2.25, Cr 0.75-1, bal Fe.
Heat treated: 100,000-220,000 TS; 80,000-182,000 YS; 2-22 El; 14-55 RA; 220-460 Brin. For dipper teeth, gears, rollers, wheels; resists abrasion and shock. *Obsolete*

B-F TOOL

Republic Steel Corp.
C 0.9-1.2, bal Fe.
For tools, drills, taps; water hardening. *Obsolete*

B-F UNIQUE NO. 20

Republic Steel Corp.
C, Cr, V, bal Fe.
For flying shears; tough. *Obsolete*

B-METAL (BAHN-METALL)

Maywood Chemical Works
Li 0.03-0.05, Ca 0.68-0.76, Na 0.62-0.72, K 0.02-0.04, Al 0.2, bal Pb.
For locomotive bearings and journal bearings in railroad cars; retains high hardness at high temperature.

B-METAL (BAHN-METALL)

Hans-Heinrich Hutte GmbH
Li 0.03-0.05, Ca 0.68-0.76, Na 0.62-0.72, K 0.02-0.04, Al 0.2, bal Pb.
For locomotive bearings and journal bearings in railroad cars; retains high hardness at high temperature.

B-METAL (BAHN-METALL)

Schaefer und Schael, A.G.
Li 0.03-0.05, Ca 0.68-0.76, Na 0.62-0.72, K 0.02-0.04, Al 0.2, bal Pb.
For locomotive bearings and journal bearings in railroad cars; retains high hardness at high temperature.

B-QUALITY

Sanderson Kayser Ltd.
high C, W, Cr, bal Fe.
For die blank, cold drawing, dies; oil hardening. *Obsolete*

B-W STANDARDIZED

Brown-Wales Co.
C 0.15-0.25, Mn 0.3-0.6, bal Fe.
Rolled: 55,000-65,000 TS; 35,000-40,000 YS; 30 El; 55 RA. Heat treated: 65,000-75,000 TS; 40,000-45,000 YS; 20 El; 60 RA; 158 Brin. For general use, boiler plates.

B-XX

Brighton Electric Steel Casting
C 0.3-0.4, bal Fe.
For gears, shafts. Water hardening.

B-Y HOT WORK

Universal Cyclops
C, W, Cr, V, bal Fe.
For hot work tools; oil hardened. *Obsolete*

B. & W. 16-3-3

Babcock & Wilcox Co.
C 0-0.15, Cr 15.5-17, Ni 12.5-14.5, Mo 2-3.25, bal Fe.
Water quenched: 95,000 TS; 38,000 YS; 55 El; 160 Brin. For acetic acid, sulfite liquor corrosion resistant parts, corrosion resistant. *Obsolete*

B. & W. 441

Babcock & Wilcox Co.
C 0.3, Cr 4-6, Mo 0.6, bal Fe.
Annealed: 100,000 TS; 50,000 YS; 25 El; 50 RA; 200 Brin. For oil still headers, castings, pumps, pipe; moderate corrosion resistance. *Obsolete*

B. & W. 9 (CROLOY 9)

Babcock & Wilcox Co.
C 0-0.15, Mn 0-0.5, Si 0-0.5, Cr 8-10, Mo 1.25-1.75, bal Fe.
Annealed: 78,000 TS; 40,000 YS; 30 El; 165 Brin. For oil refining still and exchanger tubes; for service up to 1250 F. *Obsolete*

B. & W. 900

Babcock & Wilcox Co.
C 0.25, Cr 17-19, bal Fe.
Cast: 95,000 TS; 45,000 YS; 10 El; 15 HA; 200 Brin. For castings and pump parts, flanges, nozzles; corrosion resistant; nitric acid, sea water and mine water resistant. *Obsolete*

B. & W. ALLOY NO. 1100

Babcock & Wilcox Co.
C 0.25, Mn 0.5, Si 1.85, Cr 25, Ni 20, bal Fe.
Cast: 70,000-85,000 TS; 40,000 YS; 10-75 El; 160-180 Brin. For furnace rolls, grates, retorts, skid rails, chains; Rockwell "B" 83-89. *Obsolete*

B. A./"Y" ALLOY

British Aluminium Co., Ltd.
Cu 4, Ni 2, Mg 1.5, Si 0-0.7, Fe 0-0.75, bal Al.
Hot rolled: 39,400 TS; 26,900 YS; 20 El; 30 RA. Cold rolled: 54,000 TS; 34,000 YS; 23 El; 33 RA. Heat treated: 35,000-50,000 TS; 30,000-36,000 YS; 1.5-6.5 El; 90-110 Brin. Chill cast: 27,000 TS; 22,000 YS; 1.0-3.3 El; 85-105 Brin. For pistons in gasoline engines; reciprocating parts in aero engines, air cooled cylinder heads; similar to Duralumin, high strength at elevated temperatures. *Obsolete*

B.50 QUALITY
Delta Metal (BW) Ltd.
Cu 60, Pb 1, bal Zn.
Extruded: 55,000 TS; 30 El; 30 RA; 110 Brin. Drawn: 64,000 TS; 26 El; 25 RA; 115 Brin. For screw machine products, hardware; yellow brass. *Obsolete*

B.A. 2014, ETC.
Now AA 2014, ETC.

B.A. 212
British Aluminium Co., Ltd.
Mg 0.6, bal Al.
Alloy sheet and strip for bright anodizing.

B.A. 226
British Aluminium Co., Ltd.
Mg 1, bal Al.
Alloy aluminum tubing.

B.A. 24
Now AA 6063.

B.A. 25
Now AA 6082.

B.A. 28
Now AA 5056 A.

B.A. 40M
British Aluminium Co., Ltd.
Si 10-13, bal Al.
Cast: 24,000-33,000 TS; 4-5 El; 70-75 Brin. For light alloy parts; die castings. *Obsolete*

B.A. 60
Now AA 3103.

B.A. 705
British Aluminium Co., Ltd.
Cu 0.5, Mg 2.5, Mn 0.5, Zn 5.7, bal Al.
Aluminium alloy; solution treated and aged.

B.A.-23
British Aluminium Co., Ltd.
Mg 1-1.5, Si 0.5-0.75, bal Al.
Sand cast: 14,000 TS; 7,500 YS; 3 El; 50 Brin. Hardened: 36,000 TS; 26,000 YS; 12 El; 85 Brin. For electrical fittings, architectural purposes; sand or die cast, age-hardenable. *Obsolete*

B.A.-34
British Aluminium Co., Ltd.
Cu 9-11, Fe 1-1.3, Mg 0.2-0.3, bal Al.
Cast: 24,000 TS; 22,000 YS; 1.0 El; 105 Brin. Hardened: 46,000 TS; 36,000 YS; 0.5 El; 140 Brin. For pistons; age hardenable. *Obsolete*

B.A.-37
British Aluminium Co., Ltd.
Cu 7.5-8.5, Si 1.5-2.5, bal Al.
Cast: 20,000-26,000 TS; 1.5-3.0 El; 75 Brin. For light alloy parts; sand and die castings. *Obsolete*

B.A./40 J
British Aluminium Co., Ltd.
Si 8, bal Al.
Sand cast: 20,000-22,000 TS; 9,500-10,000 YS; 6-11 El; 47 Brin. For medium strength castings; corrosion resistant. *Obsolete*

B.A./40D ALLOY
British Aluminium Co., Ltd.
Si 10-13, bal Al.
Sand cast: 23,500-24,000 TS; 11,000-13,500 TS; 5-8 El; 50-55 Brin. Chill cast: 29,000-31,300 TS; 12,000-14,500 YS; 8-15 El; 55-60 Brin. For light alloy parts, wrought forms, corrosion resistant castings; resists sea water corrosion; modified alloy. *Obsolete*

B.A./50 ALLOY
British Aluminium Co., Ltd.
Ni, bal Al.
Chill cast: 20,000 TS; 13 El. For light alloy parts; hard, ductile. *Obsolete*

B.A./60A ALLOY
British Aluminium Co., Ltd.
Mn 1.25, bal Al.
Hard: 29,000 TS; 25,000 YS; 3-5 El; 28 Brin. Soft: 14,000 TS; 5,000 YS; 40 El; 55 Brin. For light alloy parts, tubes, sections; corrosion resistant. *Obsolete*

B.A.C. BRIGHTWAY
Henry Wiggin & Co. Ltd.
Ni 80, Cr 20.
For fusion coating of poppet valves; resists heat and corrosion. *Obsolete*

B.B
Jessop-Saville Ltd.
C 0.6, W 9.5, Cr 2.5, V 0.1, bal Fe.
For tools, dies; hot die steel.

B.B. HOT DIE
Jessop-Saville Ltd.
C 0.4, W 10, Cr 3.5, V 0.5, bal Fe.
For hot dies, punches; hot die steel.

B.B.D.C. STANDARD ALLOY
English manufacture
Cu 88.5, Pb 0.25, Ni 1, Sn 10, P 0.25.
For gears, bearings; tough.

B.C. HOTWORK
Teledyne Firth Sterling
C 0.6, W 18, Cr 4, V 1, bal Fe.
For cutters, tools, dies; high speed steel. *Obsolete*

B.C. NO. 10
British Steel Corp.
C 1, Mn 0.2, bal Fe.
For tools; general. *Obsolete*

B.C. NO. 12
British Steel Corp.
C 1.2, Mn 0.2, bal Fe.
For tools; general. *Obsolete*

B.C. NO. 8
British Steel Corp.
C 0.8, Mn 0.2, bal Fe.
For tools; general. *Obsolete*

B.H.T.A. METAL
Manufacturer not listed.
Cu 2, Ni 1.25-1.75, Fe 1.75-2, Mg 0.8, Si 0-0.5, bal Al.
Heat treated: 60,000 TS; 47,000 YS; 10 El. For light alloy parts, forgings; Duralumin type, age-hardening.

B.K.L.
British Steel Corp.
C 1.5, Cr 12, bal Fe.
Hardened: 212-601 Brin. For dies for plastic molds; corrosion resistant. *Obsolete*

B.M.S
Hidalgo Steel Co. Inc.
C 0.7, Cr 5, W 1.5, bal Fe.
For tools, chisels, hard punches, shear blades; air hardening.

B.N.F. COPPER-NICKEL-IRON ALLOY
BNF Metals Technology Centre.
Ni 5-10, Fe 1-2, Mn 0.3-0.8, bal Cu.
For marine parts and hardware. Resists sea water corrosion. Included in BS 2870. *Obsolete*

B.N.F. LEAD ALLOY NO. 1
BNF Metals Technology Centre.
Cd 0.25, Sb 0.5, bal Pb.
For lead sheathing of electric cables. Included in BS 801. *Obsolete*

B.N.F. LEAD ALLOY NO. 2
Vacuumschmelze GmbH
Cd 0.25, Sn 1.5, bal Pb.
For water pipes, lead sheathing of electric cables; corrosion resistant. *Obsolete*

B.N.F. LEAD ALLOY NO. 3
BNF Metals Technology Centre.
Cd 0.15, Sn 0.4, bal Pb.
For sheathing of ship cables. Included in BS 801. *Obsolete*

B.O.H.
H. Boker & Co.
C, alloy, bal Fe.
For tools, dies; oil hardened. *Obsolete*

B.R.S.
British Steel Corp.
C 1, Cr 1.2, bal Fe.
For ball races. *Obsolete*

B.S. SEEWASSER
Karl Schmidt Co.
Aluminum. Mg 7.5, Mn 0.3, bal Al.
Heat treated: 30,000 TS; 20,600 YS; 4 El; 10 RA; 80 Brin. Rolled: 64,000 TS; 45,000 YS; 8 El; 50 RA; 120 Brin. For furniture, interior light fixtures, wire, castings; resists sea water corrosion.

B.S.2
Acieries de Champagnole
C 0.57, Cr 0.3, Mn 0.85, Si 1.9, V 0.2, Mo 0.3, bal Fe.
Oil or water hardening tool steel. For stamps, cold working dies, pneumatic tools, cold cutting tools for thick metal, and shock resisting tools. AFNOR: 60 SMD 08.03; AISI 35.

B.S.C. "A"
British Steel Corp.
C 0.45-0.5, Ni 0-0.5, Mn 0.7-0.9, bal Fe.
Normalized: 89,600-112,000 TS; 44,800 YS; 20 El; 35 RA; 163-212 Brin. Oil treated: 100,800-134,000 TS; 62,700 YS; 18 El; 40 RA; 197-285 Brin. For gun and other mechanism parts, keys, shafts, gears, cylinders; suitable for surface hardening. *Obsolete*

B.S.C. "A-31"
British Steel Corp.
C 0.5-0.6, Mn 0.6-0.8, bal Fe.
Normalized: 100,800 TS; 18 El; 40 RA; 197-241 Brin. Oil treated: 123,200 TS; 65,000-85,000 YS; 15 El; 35 RA; 235-285 Brin. For cylinders, gears, cams, mechanism parts; suitable for surface hardening, oil hardening, shock resistant. *Obsolete*

B.S.C. "B.C.T.B."
British Steel Corp.
C 0.55-0.65, Cr 0.5-0.8, Mn 0.5-0.8, bal Fe.
Normalized: 144,000 TS; 67,200 YS; 15 El; 277 Brin. Oil treated: 126,000-160,000 TS; 90,000-112,000 YS; 18-23 El; 341 Brin. For mandrels, gear wheels, dies, axles, pinions, cylinder liners. *Obsolete*

B.S.C. "C.H.-2N."
British Steel Corp.
C 0.1-0.15, Ni 2-2.5, bal Fe.
Normalized: 56,000 TS; 30,200 YS; 30 El; 55 RA; 103 Brin. For general case-hardened objects; IZ 50. *Obsolete*

B.S.C. "C.H.-3N."
British Steel Corp.
C 0.1-0.15, Ni 2.75-3.5, Cr 0-0.3, bal Fe.
Normalized: 67,200 TS; 36,000 YS; 28 El; 55 RA; 137 Brin. Hardened: 104,400 TS; 67,200 YS; 18 El; 45 RA. For general case-hardened objects; gears, cams, cam shafts, fasteners; IZ 40. *Obsolete*

B.S.C. "C.H.-5 N."
British Steel Corp.
C 0.09-0.16, Ni 4.6-5.5, Cr 0-0.2, bal Fe.
Normalized: 71,700 TS; 50,000 YS; 28 El; 55 RA. Hardened: 190,000 TS; 112,000 YS; 13 El; 40 RA. For gear wheels; tough core. *Obsolete*

B.S.C. "C.H.M.S."
British Steel Corp.
C 0.1-0.18, Mn 0.6-0.9, bal Fe.
Normalized: 58,300 TS; 36,500 YS; 28 El; 50 RA; 111 Brin. Hardened: 69,400 TS; 50,000 YS; 20 El; 50 RA. For wheels, cam shafts, pins, levers, spindles; IZ 50. *Obsolete*

B.S.C. "C.H.N.C."
British Steel Corp.
C 0.09-0.16, Ni 3-3.75, Cr 0.8-1.1, bal Fe.
Hardened: 100,800-207,000 TS; 78,500-123,200 YS; 12-20 El; 35-55 RA; 192-277 Brin. For heavily loaded gears, shafts; IZ 18-45. *Obsolete*

B.S.C. "C.H.N.M."
British Steel Corp.
C 0.15-0.2, Ni 1.75-2.1, Mo 0.2-0.35, bal Fe.
Normalized: 74,000-85,200 TS; 52,000 YS; 25 El; 50 RA; 143-166 Brin. Hardened: 112,000-123,200 TS; 89,600 YS; 17 El; 40 RA; 228-255 Brin. For ball and roller races; IZ 50. *Obsolete*

B.S.C. "C.O.M.O."
British Steel Corp.
C 0.07-0.15, Cu 0.2-0.35, Mo 0.2, bal Fe.
Normalized: 58,000-78,400 TS; 33,000 YS; 25 El; 50 RA; 111-152 Brin. For boiler and superheater tubes, boiler drums; high creep limit. *Obsolete*

B.S.C. "H.T.C."
British Steel Corp.
C 0.37-0.43, Mn 0.6-0.9, Ni 0-0.5, bal Fe.
Normalized: 78,400-100,800 TS; 40,200 YS; 23 El; 40 RA; 140-197 Brin. Oil treated: 89,600-112,000 TS; 53,000 YS; 22 El; 45 RA; 170-241 Brin. For crankshafts, shafts, spindles, aero engine cylinders, axles, connecting rods; suitable for surface hardening. *Obsolete*

B.S.C. "H.T.C.N."
British Steel Corp.
C 0.35-0.45, Mn 0.6-1.2, Ni 0.5-1, bal Fe.
Oil treated: 94,000-112,000 TS; 22 El; 45 RA; 174-223 Brin. For crankshafts, spindles, aero engine cylinders, connecting rods; IZ 35. *Obsolete*

B.S.C. "N.C."
British Steel Corp.
C 0.18-0.25, Ni 3-3.75, Cr 0.4-0.8, bal Fe.
Oil treated: 112,000-134,000 TS; 89,600 YS; 19 El; 55 RA; 217-285 Brin. For bolts, studs, shafts, axles, crankshafts; for highly stressed parts; IZ 40. *Obsolete*

B.S.C. "NILEX."
British Steel Corp.
Ni 36, C, bal Fe.
Rolled: 89,000-112,000 TS; 40 El; 55 RA. For use where minimum coefficient of expansion is required. *Obsolete*

B.S.C. "S.H.N.C."
British Steel Corp.
C 0.25-0.32, Ni 3.75-4.5, Cr 1-1.5, bal Fe.
Air hardened: 224,000-280,000 TS; 168,000 YS; 12 El; 25 RA; 444-555 Brin. Hardened and tempered: 134,400-156,800 TS; 98,500 YS; 18 El; 50 RA; 240-311 Brin. For highly stressed shafts, gears, tubes, turnbuckles; air hardening steel; IZ 15-35. *Obsolete*

B.S.C. "SI-CR"
British Steel Corp.
C 0.4-0.5, Si 3.5-4.25, Mn 0.4-0.6, Cr 7.5-8.5, bal Fe.
Oil treated: 123,200-150,000 TS; 100,800 YS; 18 El; 45 RA; 255-320 Brin. For exhaust valves for high duty, non-scaling steel, notch-brittle when cold. *Obsolete*

B.S.C. "SI-MN"
British Steel Corp.
C 0.5-0.6, Si 1.5-2, Mn 0.6-1, bal Fe.
179,000-224,000 TS; 145,600 YS; 7 El; 364-460 Brin. For spring plates, laminated automobile springs; oil hardening. *Obsolete*

B.S.C. "V.A. 65"
British Steel Corp.
C 0.22-0.28, Ni 2.75-3.5, Cr 1-1.4, Mo 0.15-0.25, bal Fe.
Oil treated: 145,600-157,000 TS; 125,400 YS; 17 El; 40 RA; 293-321 Brin. For highly stressed aero engine connecting rods; also known as "Non-Brit;" IZ 35. *Obsolete*

B.S.C. "V.C.M."
British Steel Corp.
C 0.3-0.35, Ni 0.5-0.7, Cr 0.7-1.5, Mo 1-1.5, bal Fe.
Oil treated: 123,000-145,600 TS; 112,000 YS; 18 El; 50 RA; 248-293 Brin. For studs, bolts, used in high temperature and high pressure steam generating plants; high creep limit; IZ 40. *Obsolete*

B.S.C. "V.H.R.D."
British Steel Corp.
C 0.7-1, Si 0.15-0.2, Mn 0.3-0.5, bal Fe.
For mandrels, drills, dies, wear resisting parts; suitable for surface hardening. *Obsolete*

B.S.C. "V.N.C.A."
British Steel Corp.
C 0.25-0.35, Ni 3-3.75, Cr 0.5-0.8, bal Fe.
Oil treated: 123,000-145,600 TS; 100,800 YS; 18 El; 50 RA; 240-341 Brin. For aero and automotive engine crankshafts, propellers, shafts, gear shafts, pressure vessels; great toughness; IZ 40. *Obsolete*

B.S.C. "V.N.C.G."
British Steel Corp.
C 0.28-0.32, Ni 3-3.75, Cr 0.55-0.7, bal Fe.
Oil treated: 224,000-280,000 TS; 179,200 YS; 10 El; 25 RA; 429-555 Brin. For shock resisting gears; IZ 10. *Obsolete*

B.S.C. 3 1/2 NS
British Steel Corp.
C 0.35-0.45, Ni 3.25-3.75, Cr 0.3, bal Fe.
Oil treated: 123,000-145,600 TS; 18 El; 50 RA; 241-293 Brin. For crankshafts, axles, connecting rods, forgings; IZ 35. *Obsolete*

B.S.C. 3 NS
British Steel Corp.
C 0.25-0.37, Ni 2.75-3.75, Cr 0.3, bal Fe.
Normalized: 76,000-112,000 TS; 44,800 YS; 24 El; 45 RA; 152-229 Brin. Oil treated: 100,800-132,200 TS; 71,000 YS; 22 El; 50 RA; 201-277 Brin. For crankshafts, axles, connecting rods, forgings; IZ 35. *Obsolete*

B.S.C. 5 CC
British Steel Corp.
C 1, Cr 5, V 0.45, Mo 1.1, Mn 0.6, bal Fe.
For brick mold liners, thread rolling dies, cold work steel, air hardenable. *Obsolete*

B.S.C. A.W.
British Steel Corp.
C 0.7, Cr 3.75, W 14, V 0.65, bal Fe.
For drills, cutters, reamers, blanking and piercing dies; high speed steel. *Obsolete*

B.S.C. C 12
British Steel Corp.
C 1.8, Cr 12, V 0.5, Mo 1, bal Fe.
For dies, wortle plates, drawing dies; non-deforming. *Obsolete*

B.S.C. C.D.
British Steel Corp.
C 2, Cr 12, bal Fe.
For dies, cold press cutting dies; non-deforming. *Obsolete*

B.S.C. C.V.M.
British Steel Corp.
C 0.35, Cr 5, V 0.3, Mo 1.25, Si 0.9, bal Fe.
For hot forging and die casting dies; oil hardening. *Obsolete*

B.S.C. CVM2
British Steel Corp.
C 0.36, W 1.4, Cr 5, V 0.3, Mo 1.35, Si 0.9, bal Fe.
For blanking and forging dies, punches, shears; hot work steel; air or oil hardened. *Obsolete*

B.S.C. CVM3
British Steel Corp.
C 0.4, Cr 5, V 1.1, Mo 1.35, Si 0.9, bal Fe.
For blanking and forging dies, punches, shears; hot work steel, oil hardened. *Obsolete*

B.S.C. H.S.M.
British Steel Corp.
C 0.3, Cr 3, W 10, V 0.3, bal Fe.
For hot forging dies; hot work steel. *Obsolete*

B.S.C. M.I.C.
British Steel Corp.
C 0.9, Mn 1.6, Cr 0.35, bal Fe.
For tools, dies, taps, gauges; non-deforming. *Obsolete*

B.S.C. M.I.C.4.
British Steel Corp.
C 0.95, W 0.5, Cr 0.5, V 0.2, Mn 1.25, bal Fe.
For paper and metal slitters, wood knives, shears; cold work steel, oil hardened. *Obsolete*

B.S.C. M.I.C.8.
British Steel Corp.
C 1.4, Mo 0.25, Si 1, Mn 0.3-1, bal Fe.
For gages, jigs, drawing and forming dies; graphitic steel, oil hardened. *Obsolete*

B.S.C. MED. "C."
British Steel Corp.
C 0.25-0.35, Mn 0.4-0.8, Ni 0-0.5, bal Fe.
Normalized: 58,200-78,400 TS; 36,000 YS; 25 El; 50 RA; 109-163 Brin. *Obsolete*

B.S.C. SUPER "C.H.N.C."
British Steel Corp.
C 0.14-0.18, Ni 4-4.5, Cr 1-1.3, Mo 0-0.3, bal Fe.
Hardened: 190,000 TS; 145,600 YS; 12 El; 35 RA. For exceedingly high strength core; IZ 25. *Obsolete*

B.S.C. SUPER "S.H.N.C."
British Steel Corp.
C 0.25-0.32, Ni 3.75-4.5, Cr 1-1.5, Mo 0.15-0.25, bal Fe.
Air hardened: 224,000-280,000 TS; 179,000 YS; 12 El; 25 RA; 450-555 Brin. Hardened and tempered: 134,400-168,000 TS; 100,800 YS; 18 El; 50 RA; 250-311 Brin. For forgings, tubes, shafts; air hardening steel; IZ 15-35. *Obsolete*

B.S.C. SUPER "V.N.C.A."
British Steel Corp.
C 0.25-0.35, Ni 3-3.75, Cr 0.5-0.8, Mo 0.15-0.25, bal Fe.
Oil treated: 132,000-156,000 TS; 112,000 YS; 18 El; 45 RA; 269-341 Brin. For aero and automotive engine crankshafts, propellers, shafts, gear shafts, pressure vessels; great toughness; IZ 40. *Obsolete*

B.S.C. SUPER C 12
British Steel Corp.
C 1.6, Cr 12, V 0.5, bal Fe.
For dies, press dies; non-deforming. *Obsolete*

B.S.C. SUPERTOUGH "A"
British Steel Corp.
C 0.1-0.15, Mn 1.4-1.8, bal Fe.
Normalized: 76,000-98,600 TS; 50,000 YS; 25 El; 55 RA; 143-207 Brin. Oil hardened: 76,000-98,600 TS; 56,000 YS; 30 El; 60 RA; 174-217 Brin. Case hardened: 89,600-100,800 TS; 44,800 YS; 28 El; 50 RA; 143-174 Brin. For gear wheels, cam shafts, pins, levers, spindles, railway work; tough core; IZ 50-80. *Obsolete*

B.S.C. SUPERTOUGH "B"
British Steel Corp.
C 0.25-0.3, Mn 1.4-1.8, bal Fe.
Oil hardened: 98,500-123,200 TS; 71,700 YS; 20 El; 50 RA; 196-255 Brin. For crankshafts, connecting rods, railway axles, armature spindles, tough core; IZ 50. *Obsolete*

B.S.C. SUPERTOUGH "C"
British Steel Corp.
C 0.25-0.3, Mn 1.4-1.8, Mo 0.2-0.3, bal Fe.
Oil hardened: 112,000-134,000 TS; 89,600 YS; 20 El; 50 RA; 223-286 Brin. For crankshafts, connecting rods, railway axles, armature spindles; IZ 50; tough. *Obsolete*

B.S.C. SUPERTOUGH "C-20"
British Steel Corp.
C 0.15-0.2, Mn 1.5-1.8, Ni 0.5-0.7, Mo 0.25-0.3, bal Fe.
Oil hardened: 89,600-112,000 TS; 67,200 YS; 20 El; 50 RA; 183-235 Brin. For general use, axles, railway parts; IZ 70; good workability. *Obsolete*

B.S.C. V.A.P.
British Steel Corp.
C 0.78, Cr 4, W 18, V 1.3, bal Fe.
For twist drills, broaches, cutters; high speed steel. *Obsolete*

B.S.SEEWASSER
Vereinigte Leichtmetallwerke G.m.b.H.
Aluminum. Mg 5-10, bal Al.
Soft: 47,000 TS; 22,000 YS; 25 El. Hard: 78,000 TS; 65,000 YS; 3 El. For light alloy parts, ship and seaplane parts; corrosion resistant to sea water.

B.T.G. STEEL
Manufacturer not listed.
Ni 60, Fe 25, Cr 12, Mn 2, C 0.5, W 3.
Sand cast: 50,000-70,000 TS; 40,000-70,000 YS; 2-10 El; 130-180 Brin. Rolled: 90,000-109,000 TS; 50,000- 70,000 YS; 25-45 El; 50-70 RA; 180-200 Brin. For heating elements; heat and corrosion resistant.

B/43
Eagle & Globe Steel Ltd.
Steel. C 0.38-0.43, S 0-0.05, P 0-0.05, Mn 0.7-0.9, bal Fe.
Cold rolled strip steel. For seat belt fittings, transmission chain. UK BS1449 CS.40; SAE 1040.

B/55
Eagle & Globe Steel Ltd.
Steel. C 0.5-0.55, S 0-0.05, P 0-0.05, Mn 0.7-0.9, bal Fe.
Cold rolled strip steel: for forming applications, hardenable grade. Hardened and tempered strip steel: for drop hammer belting. UK BS1449 CS.50; SAE 1050; DIN 1.7222 C.53.

B/70
Eagle & Globe Steel Ltd.
Steel. C 0.65-0.7, S 0-0.05, P 0-0.05, Mn 0.65-0.85, bal Fe.
Cold rolled strip steel: for general presswork spring fasteners. Hardened and tempered strip steel: for general springs. UK BS1449 CS.70; SAE 1070; DIN 1.7222 C.67.

B/82
Eagle & Globe Steel Ltd.
Crinoline steel. C 0.75-0.82, S 0-0.05, P 0-0.05, Mn 0.65-0.85, bal Fe.
Cold rolled strip steel: for springs, decorators' cutlery, automotive clutch plates, circlips, ice skate blades. Hardened and tempered strip steel: for springs, knives, low duty saws, decorators' cutlery. UK BS1449 CS.80; SAE 1080; DIN 1.7222 C.75.

B1080
Comalco Ltd.
99.80 Al min.
O-temper: 62 MPa TS; 28 MPa YS; 45 El. H18-temper: 138 MPa TS; 131 MPa YS; 6 El. Chemical and process plant and equipment.

B1200
Comalco Ltd.
99.0 Al min.
O-temper: 90 MPa TS; 34 MPa YS; 35 El. H18-temper: 165 MPa TS; 152 MPa YS; 5 El; 44 Brin. Spinnings, holloware and general sheet metal work.

B1914
Sorcery Metals
Cr 10, Co 10, Mo 3, Al 5.5, Ti 5.3, B 0.1, bal Ni (C, Zr as low as possible).
Cast, annealed (2100°F, 1 h, air cooled; annealed 1650°F, 10 h, air cooled) at RT: 145,000 psi TS; 112 ksi YS; 12 El. At 1400°F: 150,000 psi TS; 131 ksi YS; 20 El. Good high temperature creep strength. For turbine blades, and wheels.

B1925
Sorcery Metals
Cr 12, Co 8.5, Mo 1.8, W 4.5, Ta 4, Al 3.5, Ti 4, B 0.1, bal Ni (C, Zr as low as possible).
Cast, annealed: 160-174 ksi TS; 140-145 ksi YS; 4-8 El (at room temperature, and up to 1400°F). Good high temperature creep strength. For turbine blades, and wheels.

B1964
Sorcery Metals
Cr 8.8, Co 10, Mo 1, W 8.5, Ta 2.5, Al 3.5, Ti 5.3, Zr 0.02, C 0.02, B 0.11, Zr 0.02, bal Ni.
Cast, annealed (1975°F, 4 h, air cooled; 1650°F, 10 h, air cooled) at RT: 176,000 psi TS; 144 ksi YS; 7 El. At 1400°F: 168,000 psi TS; 135 ksi YS; 7 El. Good high temperature creep strength. For turbine blades, and wheels.

B1981
Sorcery Metals
Cr 16, Co 8.5, Mo 1.8, W 2.6, Ta 1.8, Cb 0.9, Al 3.4, Ti 3.4, B 0.1, bal Ni (C, Zr as low as possible).
Cast, annealed (2050°F, 2 h, air cooled; + 1550°F, 24 h, air cooled) at RT: 160,000 psi TS; 140 ksi YS; 4.5 El. At 1400°F: 163,000 psi TS; 128 ksi YS; 12 El. Good high temperature creep strength.

B201
Fansteel/Wellman Dynamics
Aluminum. Cu 4.5-5, Si 0.05, Fe 0.05, Mn 0.2-0.5, Zn 0.05, Mg 0.2-0.3, Ti 0.15-0.35, Be 0.05, Ag 0.4-0.8, bal Al.
Aluminum casting alloy (integrally attached coupon). Cast: 62,000 psi TS; 55,000 psi YS; 5 El.

B2014
Comalco Ltd.
Cu 4.4, Si 0.8, Mn 0.8, Mg 0.6, bal Al.
T4-temper: 448 MPa TS; 290 MPa YS; 20 El. Aircraft structures, forgings, heavy duty structural applications.

B24 Q-TEMP
United States Steel Corp.
C 0.2-0.25, Mn 0.8-1.1, Si 0.15-0.3, B 0.0005, bal Fe.
Heat treated: 120-200 ksi TS; 92-180 ksi YS; 375 Brin. For high strength parts.

B33
Now WESTINGHOUSE B-33.

B5083
Comalco Ltd.
Mg 4.5, Mn 0.7, Cr 0.15, bal Al.
O-temper: 290 MPa TS; 148 MPa YS; 22 El; 70 Vickers. H321-temper: 331 MPa TS; 228 MPa YS; 16 El; 82 Brin. Cryogenics, marine aircraft, drilling rigs.

B6063
Comalco Ltd.
Mg 0.7, Si 0.4, bal Al.
O-temper: 90 MPa TS; 48 MPa YS; 30 El; 30 Vickers. T6 Temper: 241 MPa TS; 214 MPa YS; 12 El; 73 Brin. Furniture, general purpose extrusions.

B6101
Comalco Ltd.
Mg 0.6, Si 0.5, bal Al.
T5 temper: 207 MPa TS; 179 MPa YS; 12 El. Electrical conductors.

B6351
Comalco Ltd.
Mg 0.6, Si 1, Mn 0.6, bal Al.
T4-temper: 241 MPa TS; 165 MPa YS; 20 El; 70 Vickers. T6-temper: 331 MPa TS; 310 MPa YS; 11 El; 103 Vickers. Transport applications.

B80
Sintered Products Ltd.
Bronze, sintered. 6.8-7.2 g/cm³ density; 14 kg/mm² TS; 5 El min; 18-26 porosity. Medium duty bronze. Meets BSS A110; MPIE BT-0010-S.

B80/0 1-138
Sintered Products Ltd.
Bronze, sintered. 5.8-8.0 g/cm³ density; 9.5-31.5 kg/mm² TS. For bearings, piston rings, structural parts.

B80/1
Sintered Products Ltd.
Bronze-graphite, sintered. 6.8-7.2 g/cm³ density; 14 kg/mm² TS; 6 El min; 16-22 porosity. For bearings. Meets SAE 842; BSS A113; MPIE BT-0010-S.

B85
Sintered Products Ltd.
Bronze, sintered. 8.0-8.4 g/cm³ density; 35 kg/mm² TS; 8 El min; 5-9 porosity. Heavy duty bronze. Meets MPIE BT-0100-W.

B88
Now COLUMBIUM XB-88, WESTINGHOUSE XB-88.

B9-HIGH SPEED
Cytemp Specialty Steel Div.
C 8.84, Cr 4.2, W 18.5, V 2, bal Fe.
For tools, taps, reamers; high speed steel. *Obsolete*

B99
Manufacturer not listed.
W 22, Hf 2, C 0.07, bal Cb.
High temperature, refractory alloy.

BA 46
British Aluminium Co., Ltd.
Mg 0.5, Si 5, Mn 0.35, bal Al.
P-temper: 21,300 TS; 1 El. WP-temper: 33,600 TS; 2 El. For general engineering castings; corrosion resistant, age-hardenable. *Obsolete*

BA. 21
Now AA 5251.

BA. 27
Now AA 5154A.

BA. 33
British Aluminium Co., Ltd.
Cu 3.5-4.5, Ni 1.8-2.3, Mg 1.2-1.7, bal Al.
Cast: 22,500 TS; 21,500 YS; 1 El; 90 Brin. Heat treated: 31,500 TS; 30,000 YS; 1 El; 105 Brin. For engine parts, pistons, cylinder heads; age-hardenable. *Obsolete*

BA. 703
British Aluminium Co., Ltd.
Cu 1, Mg 2.6, Mn 0.25, Zn 5.7, Cr 0.1, bal Al.
WP-temper: 78,000 TS; 67,000 YS; 7 El. For light alloy parts; age hardened.

BA. LM8

British Aluminium Co., Ltd.
Mg 0.3-0.8, Si 3.5-6, Mn 0-0.5, bal Al.
Sand cast: 17,920 TS min; 11,200 PS; 2 El min. WP-sand cast: 33,600 TS min; 31,400 PS; 5 El min. WP-chill cast: 40,320 TS min; 31,400 PS; 2 El min. For cylinder heads, propeller gear boxes, gear housings, crankcases. Pressure-tight leak-proof castings.

BA.301

British Aluminium Co., Ltd.
Cu 3.5-5, Mg 0.4-1.2, Mn 0.4-1.2, Si 0-0.7, bal Al.
Solution treated: 54,000 TS; 31,500 YS; 15 El.; 105 Brin. For aircraft, road and rail transport, structures; high strength, heat treatable. *Obsolete*

BA.303

British Aluminium Co., Ltd.
Cu 3.5-4.8, Mg 0-0.6, Mn 0-1.2, Si 0-1.5, bal Al.
Solution treated: 54,500 TS; 31,500 YS; 15 El; 115 Brin. Aged: 62,500 TS; 51,500 YS; 8 El; 135 Brin. For aircraft, road and rail transport, structures; heat treatable, high strength. *Obsolete*

BA.306

British Aluminium Co., Ltd.
Cu 1.7, Mg 0.7, Si 1, Mn 0.6, bal Al.
W-temper: 44,800 TS; 27,000 YS; 15 El. WP-temper: 58,000 TS; 49,000 YS; 8 El. For light alloy parts; age-hardened. *Obsolete*

BA.32

British Aluminium Co., Ltd.
Cu 4-5, bal Al.
Cast: 31,500 TS; 18,000 YS; 7 El; 80 Brin. Hardened: 47,000 TS; 47,000 YS; 1 El; 135 Brin. For aircraft and transport castings; age-hardenable. *Obsolete*

BA.352

British Aluminium Co., Ltd.
Cu 3.5-5, Mg 0.4-1.2, Mn 0.4-1.2, Si 0-0.7, bal Al.
Solution treated: 54,000 TS; 31,500 YS; 15 El. For aircraft, road and rail transport structures; heat treatable, high strength. *Obsolete*

BA.353

British Aluminium Co., Ltd.
Cu 3.5-4.8, Mg 0-0.6, Si 0-1.5, Mn 0-1.2, bal Al.
Solution treated: 54,000 TS; 31,500 YS; 15 El. Aged: 58,000 TS; 45,000 YS; 8 El. For aircraft, road and rail transport, structures; heat treatable, high strength. *Obsolete*

BA.40

British Aluminium Co., Ltd.
Si 10-13, bal Al.
Sand cast: 23,500 TS; 9,000 YS; 5 El; 55 Brin. Chill cast: 27,000 TS; 10,000 YS; 7 El; 60 Brin. For marine and general engineering castings; good corrosion resistance. *Obsolete*

BA.41

British Aluminium Co., Ltd.
Si 10-13, Mg 0-0.6, Mn 0-0.6, bal Al.
Aged: 24,500 TS; 15,500 YS; 1.5 El; 75 Brin. Hardened: 34,500 TS; 34,500 YS; 1.0 El; 105 Brin. For marine and general engineering castings; age-hardenable. *Obsolete*

BA.45

British Aluminium Co., Ltd.
Si 4.5-5.5, bal Al.
Sand cast: 18,000 TS; 9,000 YS; 4 El; 40 Brin. Chill cast: 22,500 TS; 10,000 YS; 6 El; 50 Brin. For hollowware, castings; good corrosion resistance. *Obsolete*

BA.701

British Aluminium Co., Ltd.
Cu 1.3, Mg 2.9, Mn 0.4, Zn 5.7, Cr 0.1, bal Al.
WP-temper: 81,000 TS; 69,400 YS; 9 El. For light alloy parts; age-hardened. *Obsolete*

BA.704

British Aluminium Co., Ltd.
Cu 1, Mg 0.25, Zn 6, Cr 0.1, bal Al.
WP-temper: 85,000 TS; 74,000 YS; 5 El. For light alloy parts; age-hardened. *Obsolete*

BA.751

British Aluminium Co., Ltd.
Zn 4.5-6.5, Mg 2-3.5, Cu 0-1.5, Mn 0.3-1, bal Al.
Age hardened: 67,000-71,500 TS; 58,000-60,500 YS; 8 El. For aircraft structures; age-hardenable. *Obsolete*

BA35

British Aluminium Co., Ltd.
Cu 3, Mg 0.5, Sb 0.6, Sn 0.2, bal Al.
Extruded: 36,000 TS; 15,500 YS; 10 El; 80 Brin. For machinery parts, fasteners; free cutting. *Obsolete*

BA42

British Aluminium Co., Ltd.
Si 12-14, Ni 1-3, Mg 0.8-1.3, Cu 0.7-1.2, bal Al.
Sand cast: 24,000 TS; 0.5 El; 130 Brin. Permanent mold: 36,000 TS; 0.5 El; 140 Brin. For pistons, cylinder heads; low expansion, corrosion resistant. *Obsolete*

BABBIT

Now BABBITT, GRAPHO BABBITT and NBD ARMATURE BABBITT, ETC.

BABBIT

Manufacturer not listed.
Lead base antifriction metals.

BACHITE

Bachite Development Corp.
C 0.2, Cr 18, Ni 8, bal Fe.
For stainless parts, chemical equipment; austenitic.

BACK CASE METAL

English manufacture
Cu 62, Zn 20, Ni 18.
For heat and corrosion resistant parts.

BADALL

Badell Co. Inc.
C 0.25, S 0.028, Si 0.65, Mo 0.15, Mn 0.85, Cr 0-0.6, B 0-0.0025, bal Fe.
Plate: 245,000 TS; 210,000 YS; 11 El; 44 RA; 500 Brin. Plate: 190,000 TS; 175,000 YS; 16 El; 55 RA; 360-400 Brin. For severe service applications, pressure vessels, wear plates. Heat treated, wear resistant plates.

BADGER

Atlas Specialty Steels
C 1.2, W 1.5, Cr 0.4, V 0.2, bal Fe.
For tools, cutters, broaches; hot work tools; non-deforming. *Obsolete*

BADGER

Latrobe Steel Co,
Tool material. C 0.94, Si 0.3, Mn 1.2, W 0.5, Cr 0.5, bal Fe.
Heat treated: 250,000 TS; 225,000 YS; 8 El; 16 RA; 50 Rock C. For blanking dies, punches, reamers, threading dies, nondeforming; oil hardening tool steel; AISI O1.

BADGER STEEL

Farrelloy Co.
C 0.9-1.2, bal Fe.
For taps, reamers, threading dies.

BAHN ALUMINIUM

Allgemeine Elektrizitats-Gesellschaft
Cu 8, bal Al.
Extruded: 29,000 TS; 13,000 YS; 20 El; 50 Brin. Cold drawn: 36,000 TS; 17,000 YS; 10 El; 70 Brin. For light alloy parts, electric pantographs.

BAILY'S METAL

English manufacture
Cu 82.1, Sn 12.8, Zn 5.1.
For bearings, corrosion resistant castings.

BAIN ALLOY

U.S. Steel Corp.
C 0.1-0.2, Mn 0-0.6, Si 0.4-0.75, Cr 1.5-2, Mo 0.6-0.8, bal Fe.
Normalized and drawn: 70,000 TS; 50,000 YS; 30 El; 140-160 Brin. For oil cracking still tubes, hot oil piping, steam pipes; corrosion resistant. *Obsolete*

BAIN BOLT STEEL

U.S. Steel Corp.
C 0.35-0.45, Mn 0.4-0.6, Si 0.45-0.75, Cr 1.5-2, Mo 0.6-0.8, bal Fe.
Drawn: 120,000 TS; 20 El; 50 RA; 290 Brin. For stud bolts; for high temperature and high pressure flange joints. *Obsolete*

BAIN FLANGE STEEL

U.S. Steel Corp.
C 0.15-0.25, Mn 0.4-0.6, Si 0.45-0.75, Cr 1.5-2, Mo 0.6-0.8, bal Fe.
Normalized: 70,000 TS; 30 El; 65 RA; 160 Brin. For flange forgings; for high temperature and high pressure piping. *Obsolete*

BAIN STEEL

Manufacturer not listed.
C 0.15-0.25, Mn 0.4-0.65, Si 0.4-0.75, Cr 1.5-2, Mo 0.6-0.8, bal Fe.
Cast: 75,000 TS; 45,000 YS; 20 El; 45 RA. For high temperature castings and tubing; heat resistant. *Obsolete*

BAIN TUBE STEEL

U.S. Steel Corp.
C 0.1-0.2, Mn 0.4-0.6, Si 0.45-0.75, Cr 1.5-2, Mo 0.6-0.8, bal Fe.
Normalized: 65,000 TS; 30 El; 65 RA; 160 Brin. For seamless tubes; for high temperature and high pressure service. *Obsolete*

BAKADIE

CCS Braeburn Alloy Steel
Low C, Ni, Mo, bal Fe.
Annealed: 130 Brin. For bakelite molds, dies. *Obsolete*

BAKEDIE NO. 2

CCS Braeburn Alloy Steel
C 0.2, Cr 18, Ni 8, bal Fe.
For colored stainless steel parts; stainless. *Obsolete*

BAKER 1729

Engelhard Corp.
Now ENGELHARD ALLOY NO. 7294.

BAKER 1780

Engelhard Corp.
Now ENGELHARD ALLOY NO. 5264.

BAKER ALLOY 1757

Engelhard Corp.
Now ENGELHARD ALLOY NO. 7989.

BAKER ALLOY NO. 1534

Engelhard Corp.
Now ENGELHARD ALLOY NO. 5348.

BAKER ALLOY NO. 1765

Engelhard Corp.
Now ENGELHARD ALLOY NO. 6031.

BAKER ALLOY NO. 934

Engelhard Corp.
Pt alloy.
For electrical resistances; resistance wire. *Obsolete*

BAKER CONTACT ALLOY NO. 846

Engelhard Corp.
Now ENGELHARD ALLOY NO. 8466.

BAKER NO. 175L

Engelhard Corp.
Now ENGELHARD ALLOY NO. 2352.

BAL-CUT 1035
Bliss & Laughlin Steel Co.
Pb 0.15-0.25, C 0.3-0.4, bal Fe.
For gears, shafts; free machining. *Obsolete*

BAL-CUT STEELS
Bliss & Laughlin Steel Co.
C 0.15, Pb 0.15-0.25, bal Fe.
Cold finished: 110,000 TS; 90,000 YS; 12 El; 35 RA; 202 Brin.
For machinery parts; free machining. *Obsolete*

BAL-CUT X-1314
Bliss & Laughlin Steel Co.
C 0.1-0.2, Mn 1-1.3, Pb 0.15-0.25, bal Fe.
For gears, shafts; free machining. *Obsolete*

BAL-CUT X-1335
Bliss & Laughlin Steel Co.
Pb 0.15-0.25, C 0.3-0.4, Mn 1.35-1.65, bal Fe.
For gears, shafts; free machining. *Obsolete*

BALANCED LINE, REGULAR SOLDER
J.M. Ney Co.
Gold color solder, color matched to BALANCED LINE casting
gold alloys. Melting range: 1465-1585°F.

BALCO
Gilby-Fodor S.A.
Fe 29-31, bal Ni.
Annealed: 80,000 TS; 55,000 YS; 25 El. For thermometer
bulbs, and ballast tubes. Heat resistant up to 590°C.

BALCO
Wilbur B. Driver Co.
Nickel. Fe 30, bal Ni.
Rolled: 70,000 TS. For voltage resistors. Magnetic, high
permeability.

BALCO ALLOY 1100.01
BALCO
Aluminum. Si 0-0.25, Fe 0.45-0.6, Cu 0.1-0.2, Mn 0-0.05, Mg
0-0.015, Ti 0.005-0.03, bal Al.
Rolling ingots.

BALCO ALLOY 1145.02
BALCO
Aluminum. Si 0-0.12, Fe 0.3-0.4, Cu 0-0.05, Mn 0-0.02, Mg
0-0.005, Zn 0-0.03, Pb 0-0.01, Li 0-0.0003, bal Al.
Rolling ingots.

BALCO ALLOY 1200.01
BALCO
Aluminum. Si 0-0.25, Fe 0.45-0.6, Cu 0-0.03, Mn 0-0.05, Mg
0-0.015, Ti 0.005-0.03, bal Al.
Rolling ingots.

BALCO ALLOY 1350.0
BALCO
Aluminum. Si 0-0.1, Fe 0-0.4, Cu 0-0.05, Mn 0-0.01, Mg
0-0.02, Ti 0-0.02, Cr 0-0.03, 0.05 Ti + V max, bal Al.
Extrusion ingots.

BALCO ALLOY 3003.01
BALCO
Aluminum. Si 0.15-0.25, Fe 0.5-0.7, Cu 0.09-0.2, Mn 1-1.2,
Mg 0-0.015, Ti 0.005-0.03, bal Al.
Rolling ingots.

BALCO ALLOY 3004.00
BALCO
Aluminum. Si 0.13-0.17, Fe 0.4-0.45, Cu 0.13-0.17, Mn 1-1.1,
Mg 0.9-1.05, Cr 0-0.02, Zn 0-0.05, Ti 0.01-0.022, bal Al.
Rolling ingots.

BALCO ALLOY 400
Carpenter Technology Corp.
Ni 70, Fe 30.
For use in thermometer bulbs and other applications that
require a high temperature coefficient of resistance. ASTM
B-267.

BALCO ALLOY 5005.02
BALCO
Aluminum. Si 0-0.1, Fe 0-0.2, Cu 0-0.1, Mg 0.5-0.7, Ti 0-0.03,
bal Al.
Rolling ingots.

BALCO ALLOY 5052.00
BALCO
Aluminum. Si 0-0.24, Fe 0-0.29, Cu 0-0.08, Mn 0-0.08, Mg
2.5-2.77, Cr 0.16-0.34, Zn 0-0.08, Ti 0.02-0.04, bal Al.
Rolling ingots.

BALCO ALLOY 5182.00
BALCO
Aluminum. Si 0-0.18, Fe 0.15-0.35, Cu 0.035-0.065, Mn
0.3-0.4, Mg 4.31-4.74, Cr 0-0.04, Zn 0-0.18, Ti 0-0.03, bal Al.
Rolling ingots.

BALCO ALLOY 6005.00
BALCO
Aluminum. Si 0.65-0.9, Fe 0.15-0.25, Cu 0-0.1, Mn 0-0.1, Mg
0.45-0.6, Zn 0-0.1, Ti 0-0.05, Cr 0-0.05, bal Al.
Extrusion ingots.

BALCO ALLOY 6063.20
BALCO
Aluminum. Si 0.34-0.42, Fe 0.16-0.22, Cu 0-0.03, Mn 0-0.03,
Mg 0.45-0.55, Zn 0-0.025, Ti 0-0.02, Cr 0-0.03, bal Al.
Extrusion ingots.

BALCO ALLOY 6063.30
BALCO
Aluminum. Si 0.39-0.47, Fe 0.16-0.22, Cu 0-0.03, Mn 0-0.03,
Mg 0.47-0.57, Zn 0-0.025, Ti 0-0.02, Cr 0-0.03, bal Al.
Extrusion ingots.

BALCO ALLOY 6351.01
BALCO
Aluminum. Si 0.9-1.2, Fe 0.15-0.25, Cu 0-0.03, Mn 0.45-0.65,
Mg 0.55-0.7, Zn 0-0.03, Ti 0.005-0.02, bal Al.
Extrusion ingots.

BALCO ALLOY 8079.00
BALCO
Aluminum. Si 0.05-0.15, Fe 0.75-1, Cu 0-0.05, Mn 0-0.02, Mg
0-0.02, Cr 0-0.02, Zn 0-0.02, Ti 0-0.03, bal Al.
Rolling ingots.

BALCO ALLOY 99.6 EC
BALCO
Aluminum. Si 0-0.1, Fe 0-0.3, Cu 0-0.01, Mn 0-0.01, Mg
0-0.01, Cr 0-0.005, Zn 0-0.02, Ti 0-0.005, V 0-0.005, B
0-0.015, bal Al.
Standard ingots.

BALCO ALLOY 99.7
BALCO
Aluminum. Si 0-0.1, Fe 0-0.2, Cu 0-0.03, Mn 0-0.03, Mg
0-0.03, Cr 0-0.03, Zn 0-0.03, Ti 0-0.03, V 0-0.03, B 0-0.01, bal
Al.
Standard ingots.

BALCO ALLOY 99.7 EC
BALCO
Aluminum. Si 0-0.1, Fe 0-0.2, Cu 0-0.01, Mn 0-0.01, Mg
0-0.01, Cr 0-0.005, Zn 0-0.02, Ti 0-0.005, V 0-0.005, B
0-0.015, bal Al.
Standard ingots.

BALCO ALLOY 99.8
BALCO
Aluminum. Si 0-0.08, Fe 0-0.12, Cu 0-0.02, Mn 0-0.02, Mg
0-0.02, Cr 0-0.02, Zn 0-0.02, Ti 0-0.02, V 0-0.02, B 0-0.01, bal
Al.
Standard ingots.

BALDER
Uddeholm Corp.
C 0.05, Si 0.1, Mn 0.1, bal Fe.
For electrical equipment; soft magnet iron. *Obsolete*

BALDWIN AH
Baldwin Steel Co.
C 1.1, Mn 0.75, Cr 5.5, V 0.3, Mo 1.2, bal Fe.
For dies, gages, forming rolls, master hobs, cutters; non-
deforming, air hardened.

BALDWIN NO. 1
Baldwin Steel Co.
C 0.75, Cr 0.91, Mo 0.4, Mn 0.81, Ni 1.78, bal Fe.
For dies, punches, shear blades, knives, rolls; oil hardened,
shock resistant.

BALDWIN NO. 711
Baldwin Steel Co.
C 0.35, Cr 0.86, Mn 0.81, Mo 0.35, Si 0.48, Cu 0.34.
Heat treated: 267,000-275,000 TS; 209,000-221,000 YS;
13-15 El; 37-51 RA; 560-580 Brin. For chisels, punches, hard
tools, furnace bars, dies, shears; non-tempering, water
hardened.

BALDWINS P. 25
British Steel Corp.
C 0-0.35, Cr 25, bal Fe.
Annealed: 85,000 TS; 50,000 YS; 30 El; 55 RA; 180 Brin. For
furnace parts and equipment, heat treating boxes; Type 446;
heat resistant. *Obsolete*

BALDWINS P.25-12
British Steel Corp.
C 0-0.2, Cr 24, Ni 12, bal Fe.
Annealed: 90,000 TS; 40,000 YS; 50 El; 65 RA; 170 Brin. For
furnace parts, pumps, oil burners; Type 309; heat resistant.
Obsolete

BALDWINS P.25-22
British Steel Corp.
C 0-0.25, Cr 25, Ni 20, Si 1.5-3, bal Fe.
For furnace parts; Type 314; heat resistant. *Obsolete*

BALDWINS P.E.C.
British Steel Corp.
C 0-0.08, Cr 18, Ni 9, Cb, bal Fe.
Annealed: 90,000 TS; 45,000 YS; 50 El; 65 RA; 160 Brin. For
welded chemical plant equipment, tanks; Type 347; stainless,
austenitic. *Obsolete*

BALDWINS P.E.H.
British Steel Corp.
C 0.17-0.25, Cr 18, Ni 8, bal Fe.
Annealed: 85,000 TS; 40,000 YS; 50 El; 70 RA; 170 Brin. For
chemical plant equipment, tanks, mixers; Type 302; stainless,
austenitic. *Obsolete*

BALDWINS P.E.L.
British Steel Corp.
C 0.1, Cr 18, Ni 8, bal Fe.
Annealed: 80,000 TS; 35,000 YS; 55 El; 75 RA; 150 Brin. For
chemical plant equipment, tanks, mixers, filters; Type 302;
stainless, austenitic. *Obsolete*

BALDWINS P.E.T.
British Steel Corp.
C 0-0.08, Cr 18, Ni 8, Ti, bal Fe.
Annealed: 85,000 TS; 35,000 YS; 55 El; 65 RA; 150 Brin. For
welded chemical plant equipment, tanks, mixers; Type 321;
stainless, austenitic. *Obsolete*

BALDWINS P.K.H.
British Steel Corp.
C 0.26-0.37, Cr 13, bal Fe.
Annealed: 100,000 TS; 55,000 YS; 22 El; 52 RA; 210 Brin. For
valves, cutlery, surgical and dental instruments; Type 420;
corrosion resistant. *Obsolete*

BALDWINS P.K.I.
British Steel Corp.
C 0.07-0.12, Cr 17, bal Fe.
Annealed: 80,000 TS; 50,000 YS; 25 El; 50 RA; 160 Brin. For
oil refinery equipment, oil burners and heaters, bolts; Type
430; corrosion resistant. *Obsolete*

BALDWINS P.K.L.
British Steel Corp.
C 0.13-0.18, Cr 13, bal Fe.
Annealed: 75,000 TS; 40,000 YS; 35 El; 70 RA; 155 Brin. For turbine blades, valves, cutlery, surgical instruments; Type 410; corrosion resistant. *Obsolete*

BALDWINS P.K.M.
British Steel Corp.
C 0.19-0.25, Cr 13, bal Fe.
Annealed: 95,000 TS; 50,000 YS; 25 El; 55 RA; 195 Brin. For valves, cutlery, surgical instruments; Type 420; corrosion resistant. *Obsolete*

BALDWINS P.K.N.
British Steel Corp.
C 0-0.25, Cr 17, Ni 2, bal Fe.
Annealed: 125,000 TS; 95,000 YS; 20 El; 55 RA; 260 Brin. For pumps, marine hardware, valves; Type 431; corrosion and heat resistant. *Obsolete*

BALDWINS P.M.H.
British Steel Corp.
C 0-0.12, Cr 18, Ni 10, Mo 2.5, bal Fe.
Annealed: 85,000 TS; 35,000 YS; 50 El; 65 RA; 160 Brin. For acid resistant chemical plant equipment, mixers; Type 316; stainless, austenitic. *Obsolete*

BALDWINS P.M.L.
British Steel Corp.
C 0.07-0.12, Cr 18, Ni 9, Mo 1.5, bal Fe.
Annealed: 85,000 TS; 35,000 YS; 50 El; 65 RA; 160 Brin. For acid resistant chemical plant equipment; Type 316; stainless, austenitic. *Obsolete*

BALDWINS P.T.L.
British Steel Corp.
C 0.08-0.16, Cr 12, Ni 10, bal Fe.
For valves, pump parts; corrosion resistant. *Obsolete*

BALFORS ULTRA-CAPITAL + 1
English manufacture
C, alloy, bal Fe.
For cutters, tools; high speed steel.

BALFOSTEEL
Ekstrand & Tholand Co.
C 0.7-1.2, bal Fe.
For punches, dies, reamers, saws. Water hardening.

BALFOUR 00
Darwins Alloy Castings
C 0.6, Cr 1, W 2, V 0.3, bal Fe.
For dies, punches, trimmers; oil hardened, non-deforming. *Obsolete*

BALFOUR 227
Darwins Alloy Castings
C 0.28, W 9, Cr 3, V 0.3, bal Fe.
For bolt headers, extrusion dies and liners; hot-work steel, tough. *Obsolete*

BALFOUR 293
Darwins Alloy Castings
C 0.36, Cr 3, W 9, V 0.3, bal Fe.
Annealed: 230 Brin. For extrusion and hot piercing dies, rams, liners; hot-work steel, oil hardened. *Obsolete*

BALFOUR 351
Darwins Alloy Castings
C 0.5, Cr 1, W 2, V 0.3, bal Fe.
For punches, forging dies, shear blades; hot-work steel, oil hardened. *Obsolete*

BALFOUR 4.T.S.S.
Darwins Alloy Castings
C 1.25, Cr 1.2, W 4.5, V 0.25, bal Fe.
For reamers, rolls, punches, drawing dies; oil hardened, abrasion resistant. *Obsolete*

BALFOUR A.B.75
Darwins Alloy Castings
C 0.6, Si 2, Mn 0.85, Cr 0.25, bal Fe.
For pneumatic chisels, shear blades, punches; oil hardened, tough. *Obsolete*

BALFOUR A.G.S.
Darwins Alloy Castings
C 1.05, Cr 1.3, Mn 0.35, Si 0.25, bal Fe.
Annealed: 210 Brin. For forming rolls, cams, jigs, trimming dies; oil hardened, tough. *Obsolete*

BALFOUR BLUE LABEL
Darwins Alloy Castings
C 0.7-0.8, Mn 0.3, bal Fe.
For drills, taps, punches, crimpers; Type W1; water hardened. *Obsolete*

BALFOUR DARWINS AR1
E.A. Balfour Steel
Corrosion resistant alloy useful in handling sulfuric acid.

BALFOUR DARWINS AR2
E.A. Balfour Steel
Nickel-chromium-copper-molybdenum alloy for resistance to hot concentrated sulfuric acid.

BALFOUR DARWINS AR3
E.A. Balfour Steel
Nickel-chromium-copper-iron alloy for resistance to sulfuric and phosphoric acids; used in pickeling processes and in drainage disposal plants.

BALFOUR DARWINS AR4A
E.A. Balfour Steel
Nickel-molybdenum-iron alloy for use with hydrochloric and sulfuric acids.

BALFOUR DARWINS AR5B
E.A. Balfour Steel
Nickel-molybdenum-iron alloy; higher molybdenum than AR4A for greater corrosion resistance.

BALFOUR DARWINS AR6C
E.A. Balfour Steel
Nickel-molybdenum-chromium-iron alloy for handling oxidizing acids as nitric acid, and mixed acids and salts containing free chlorine.

BALFOUR DARWINS AR7
E.A. Balfour Steel
Duplex, age hardenable, corrosion resistant stainless steel with good strength, weldability, resistance to corrosion and galling; used particularly in wood pulp processing.

BALFOUR DSW
Darwins Alloy Castings
C 1.3, W 4.5, bal Fe.
For dies, cutters; water or oil hardened. *Obsolete*

BALFOUR E.X.D.I.
Darwins Alloy Castings
C 0.33, Cr 1.5, W 5.5, Ni 3.75, bal Fe.
For extrusion dies, upsetting dies, mandrels; hot-work steel, oil hardened. *Obsolete*

BALFOUR NSS3
Darwins Alloy Castings
C 0.95, Mn 2, Si 0.3, bal Fe.
For punches, reamers, taps, dies, cams, pawls; oil hardened, non-deforming. *Obsolete*

BALFOUR P.R.N.2
Darwins Alloy Castings
C 1.05, Mn 1.1, Cr 1.5, bal Fe.
For blanking and forming dies, master hobs; oil hardened, abrasion resistant. *Obsolete*

BALFOUR R. 9030
Darwins Alloy Castings
C 0.3, Cr 0.75, Mo 0.5, Ni 2.75, bal Fe.
For plastic mold dies, piercing punches; hot-work steel, oil hardened. *Obsolete*

BALFOUR R9030
Darwins Alloy Castings
C 0.95, Mn 1.25, Cr 0.5, W 0.5, bal Fe.
For tools, dies, punches, forming dies, crimpers, upsetters; oil hardened, non-deforming. *Obsolete*

BALFOUR RSD
Darwins Alloy Castings
C 0.95, Mn 1.25, W 0.5, Cr 0.5, bal Fe.
For tools, dies, punches, headers, upsetters; non-deforming, oil hardened. *Obsolete*

BALFOUR S.C. 25
Darwins Alloy Castings
C 1.5, V 0.25, Mo 0.75, Cr 12, bal Fe.
For blanking and forming dies, shear blades; air hardened, non-deforming. *Obsolete*

BALFOUR S.C. 26
Darwins Alloy Castings
C 1.8, Cr 12, Mn 0.3, Si 0.2, bal Fe.
For rolling dies, shear blades, rim rolls; air hardened, non-deforming. *Obsolete*

BALFOUR S.D. 20
Darwins Alloy Castings
C 0.35, Mn 0.6, Cr 1.2, Ni 1.75, bal Fe.
Annealed: 200 Brin. For plastic mold dies, punches, extrusion dies; oil hardened, shock resistant. *Obsolete*

BALFOUR S.I.W.
Darwins Alloy Castings
C 0.65, Si 1.1, Mn 0.7, Cr 0.2, W 1.3, bal Fe.
Annealed: 220 Brin. For pneumatic chisels, rivet snaps, shear blades; oil hardened, fatigue resistant. *Obsolete*

BALFOUR SC13
Darwins Alloy Castings
C 2.15, Cr 11.5, Mo 0.75, bal Fe.
For blanking and forming dies, engravers' rolls, hobs; non-deforming, air hardened. *Obsolete*

BALFOUR SCP
Darwins Alloy Castings
C 0.9, V 0.15, bal Fe.
For chisels, drills, taps, punches, reamers, hobs; Type W2; water hardened. *Obsolete*

BALFOUR TIO. H
Darwins Alloy Castings
C 0.9, Mn 1.2, Cr 0.4, W 0.5, bal Fe.
For blanking and forming dies, cams, pawls; oil hardened, non-deforming. *Obsolete*

BALFOUR TSS
Darwins Alloy Castings
C 1.3, W 4.5, bal Fe.
For dies, cutters; water or oil hardened. *Obsolete*

BALFOURS SHOE DIE
Adams & Osgood Steel Co.
C 0.5, Cr 0.6, Mo 0.3, bal Fe.
For tools, leather cutters; oil hardened.

BALL BEARING STAINLESS STEEL
Firth-Vickers Stainless Steels Ltd.
C 0.4, Si 0.25, Mn 0.3, Cr 11.5, bal Fe.
Annealed: 110,000 TS; 72,000 YS; 20 El; 240 Brin. For ball bearings; corrosion resistant.

BALLAST
Gilby-Fodor S.A.
Ni 99.8.
Soft: 60,000 TS; 18,000 YS; 50 El. For cathodes and filaments for electronic tubes. Heat resistant.

BALLAST NICKEL
Wilbur B. Driver Co.
Nickel. Ni 99.6.
Annealed: 56,000 TS. For ballast wire for voltage control electrical equipment; resistance 52 ohm/mil-ft.

BALTIC C.D.V. 1
Joseph Beardshaw & Son Ltd.
C 2.25, Cr 11.5, Mo 0.5, V 0.3, bal Fe.
For blanking and extrusion dies; oil hardened, non-deforming.

BALTIC C.D.V. 2
Joseph Beardshaw & Son Ltd.
C 1.6, Cr 12, Mo 0.8, V 0.5, bal Fe.
For blanking and extruding dies, gages, shear blades; air hardened, non-deforming.

BALTIC C.L. 15
Joseph Beardshaw & Son Ltd.
C 0.38-0.42, Ni 3-3.5, Mn 0-0.6, bal Fe.
For chisels, snaps, tools, punches; tough and shock resistant.

BALTIC C.L. 222
Joseph Beardshaw & Son Ltd.
C 0.35, Cr 0.8, Mo 0.7, W 0.5, bal Fe.
For punches, snaps, shear blades, pneumatic tools; non-tempering, water hardened.

BALTIC C.L. 224
Joseph Beardshaw & Son Ltd.
C 0.55, Ni 1.5, Cr 0.75, Mo 0.3, bal Fe.
For drop forging dies; tough and water resistant.

BALTIC C.L. 225
Joseph Beardshaw & Son Ltd.
C 0.5, Mn 0.5, Ni 1.25, Cr 0.5, Mo 0.25, bal Fe.
For drop forging dies; tough and wear resistant.

BALTIC C.L. 400
Joseph Beardshaw & Son Ltd.
C 0.9, Cr 3.75, Mo 0.5, V 0.5, bal Fe.
For forging and extrusion dies, hot shears and punches; hot die steel, air or oil hardened.

BALTIC C.L. 40T
Joseph Beardshaw & Son Ltd.
C 0.55, Si 2, Mn 0.9, Cr 0.3, V 0.2, bal Fe.
For punches, shear blades, concrete breakers; tough, shock resistant.

BALTIC C.L. 444
Joseph Beardshaw & Son Ltd.
C 0.35, Cr 5, Si 0.9, Mo 1.5, W 1.25, V 0.3, bal Fe.
For forging and extrusion dies, hot shears and punches; shock resistant, hot die steel.

BALTIC C.L. 444W
Joseph Beardshaw & Son Ltd.
C 0.35, Si 0.9, Cr 5, W 5, V 0.3, bal Fe.
For hot heading and gripping dies; oil hardened, hot die steel.

BALTIC C.L. 45
Joseph Beardshaw & Son Ltd.
C 0.5, Cr 1.2, Mn 0.7, V 0.2, bal Fe.
For chisels, pneumatic tools, pistons; tough, shock resistant.

BALTIC C.L. 60
Joseph Beardshaw & Son Ltd.
C 0.6, Cr 0.6, Mn 0.7, bal Fe.
For vice grips, general tools; water hardened.

BALTIC C.L. 666
Joseph Beardshaw & Son Ltd.
C 0.6-0.7, Mn 0.4, W 6, Cr 0.6, bal Fe.
For coal cutters, picks; air or oil hardened.

BALTIC L.C.H.D.
Joseph Beardshaw & Son Ltd.
C 0.3, W 9-10, Cr 2.75, V 0.4, bal Fe.
For dies, inserts and liners; oil hardened, hot die steel.

BALTIC P.C.S.K.
Joseph Beardshaw & Son Ltd.
C 0.45, W 2, Cr 1.5, V 0.25, bal Fe.
For chisels, snaps, pneumatic tools; tough and shock resistant.

BALTOC
Crucible Specialty Metals
C 0.7, W 18, Cr 4, V 1, Co 5, bal Fe.
For hogging tools for heavy cuts; high speed steel. *Obsolete*

BALTOC
Colt Industries
C 0.7, W 18, Cr 4, V 1, Co 5, bal Fe.
For hogging tools for heavy cuts; high speed steel. *Obsolete*

BAND FILE
Colt Industries
C 1.25, Cr 0.5, bal Fe.
For files; water or oil hardening. *Obsolete*

BARAL
Calloy Ltd.
Ba 0-50, bal Al.
For getters in electrical discharge devices and vacuum tubes; good stability in air.

BARBERITE
Barber Asphalt Co.
Cu 88, Ni 5, Sn 5, Si 1.5, Mn 0.5, C 0.04, Fe 0.5.
Cast: 56,800-60,000 TS; 41,500-48,000 YS; 6.0-6.6 El; 6.6-6.8 RA; 77 Brin. For marine parts, ornaments, fixtures; corrosion resisting casting. *Obsolete*

BARE-BRITE A632
Airco Vacuum Metals
Bare wire for inert gas welding of T1 steel, HY 80 and similar grades. MIL-E-19822A, Type B88.

BARE-BRITE AX-110
Airco Vacuum Metals
Low alloy bare wire for welding high strength low alloy steel. 120, 000 psi TS in weld area. MIL-E-23765/2 Type 120S-1.

BARE-BRITE AX-140
Airco Vacuum Metals
C 0.1, Mn 1.8, Si 0.35, Ni 2.25, Cr 1, Mo 0.55, bal Fe.
Low alloy bare wire for welding high strength low alloy steel. 140,000 psi TS in weld area. MIL-E-24355 Type 140S.

BARE-BRITE AX-90
Airco Vacuum Metals
Low alloy bare wire for welding high strength low alloy steel. 100, 000 psi TS in weld area. For ships, military vessels and equipment, and earth moving equipment. MIL-E-23765/2 Type 100 S-1.

BARIO
English manufacture
Cr 21.4, Ni 57.4, Fe 1, W 15.4, C 0.3.
For resistor elements; stainless and corrosion resisting.

BARIO SOFT
Manufacturer not listed.
Co 60, Cr 20, W 20.
For tools, heat and corrosion resistant parts; corrosion and heat resistant.

BARIO, SHEET
Manufacturer not listed.
Ni 90, Cr 4.3, W 1.2, Si 0.3, trace Co, Cu, Fe.
For tools, high corrosion resistant parts; stainless and corrosion resistant.

BARIO-HARD
Manufacturer not listed.
Co 30, Cr 30, W 25, Mn 10, Ti 5.
For tools, high temperature applications; corrosion and heat resistant.

BARIUM
Alcan Metal Powders Division
99.9, 99.5 to 99.7, 98.0 Ba.
Forms: sticks, rod, billets, powder, lump, wire.

BARIUM 13 CR
Barium Stainless Steel Corp.
C 0.12, Cr 12-15, bal Fe.
Annealed: 65,000-75,000 TS; 40,000-45,000 YS; 30 El; 60 RA.
For stainless parts; corrosion resistant.

BARIUM 17 CR
Barium Stainless Steel Corp.
C 0.1, Cr 16-23, bal Fe.
Annealed: 70,000-80,000 TS; 40,000-45,000 YS; 28 El; 55 RA.
For stainless trim; corrosion resistant.

BARIUM 18-8
Barium Stainless Steel Corp.
C 0.05-0.2, Cr 16-23, Ni 7-11, bal Fe.
Annealed: 80,000-89,000 TS; 35,000-45,000 YS; 50-60 El; 55-65 RA. For stainless utensils; heat and corrosion resistant.

BARIUM DIE STEEL
Barium Stainless Steel Corp.
C 0.4-0.6, Ni 1-2, Cr 0.5-1, Mo 0.2-0.3, bal Fe.
For forging dies; in three grades of varying hardness.

BARMAG
Calloy Ltd.
Ba 35, bal Mg.
For getters in electrical discharge devices and vacuum tubes.

BARNITE
English manufacture
Cu 5.5, Mg, Cr, Si, Ti, bal Al.
For light alloy parts; age-hardenable.

BAROS
Creusot-Loire
Ni 90, Cr 10.
For pen points, balance weights; heat resisting. *Obsolete*

BARR ALLOY NO. 00C
McCallum-Hatch Bronze Co. Inc.
Cu 84, Sn 16.
Cast: 25,000-35,000 TS; 20,000-28,000 YS; 1-2 El; 80-100 Brin. For bearings for turn tables and movable bridges; compression of 0.1%-18000.

BARR ALLOY NO. 1
McCallum-Hatch Bronze Co. Inc.
Cu 88, Sn 8-10, Zn 2-4.
Cast: 40,000-50,000 TS; 18,000-22,000 YS; 20-40 El; 20-30 RA; 65-74 Brin. For steam and hydraulic castings, air valves, small gears; U.S.N. "G" Bronze; "Gun Metal"; "SAE No. 62."

BARR ALLOY NO. 11
McCallum-Hatch Bronze Co. Inc.
Cu 89, Al 10, Fe 1.
Cast: 65,000-80,000 TS; 23,000-27,000 YS; 20-27 El; 20-27 RA; 93-100 Brin. Heat treated: 75,000-95,000 TS; 55,000-65,000 YS; 3-15 El; 140-200 Brin. For gears, feed nits, bearings; resists wear, repeated shock and corrosion.

BARR ALLOY NO. 14
McCallum-Hatch Bronze Co. Inc.
Cu 90, Sn 6.5, Pb 1.5, Zn 2.
Cast: 34,000-40,000 TS; 16,000-19,000 YS; 23-35 El; 23-53 RA; 50-60 Brin. For plain bearings, backs of Babbitt-lined shells; low coefficient of friction and expansion.

BARR ALLOY NO. 15
McCallum-Hatch Bronze Co. Inc.
Cu 88.5, Sn 11, Pb 0.25, P 0.25.
Sand cast: 33,000-40,000 TS; 21,000-24,000 YS; 10-15 El; 8-15 RA; 7-80 Brin. Chill cast: 50,000 TS; 30,000 YS; 6 El; 100 Brin. For worm gears mating with hardened worms; "SAE-65"; compression of 0.1% 17000.

BARR ALLOY NO. 20
McCallum-Hatch Bronze Co. Inc.
Cu 65, Zn 23, Fe 2, Mn 3, Al 7.
Cast: 90,000-100,000 TS; 50,000-60,000 YS; 20-5 El; 180-200 Brin. For general castings; compression of 0.1% 65000.

BARR ALLOY NO. 25 B
McCallum-Hatch Bronze Co. Inc.
Cu 69, Sn 5, Pb 25, Ni 1.
20,000-24,000 TS; 14,000-16,000 YS; 16-12 El. For general castings, bearings; acid resisting.

BARR ALLOY NO. 4
McCallum-Hatch Bronze Co. Inc.
Cu 80, Sn 10, Pb 10.
Cast: 30,000-50,000 TS; 19,000-21,000 YS; 20-10 El; 20-10 RA; 55-65 Brin. For high speed bearings; for high pressures.

BARR ALLOY NO. 48
McCallum-Hatch Bronze Co. Inc.
Cu 84, Sn 10, Pb 2.5, Ni 3.5.
Sand cast: 40,000-50,000 TS; 25,000-28,000 YS; 25-15 El; 20-10 RA; 80-93 Brin. For gears; compression of 0.1% 20000-24000.

BARR ALLOY NO. 5
McCallum-Hatch Bronze Co. Inc.
Cu 85, Sn 5, Pb 5, Zn 5.
Cast: 30,000-38,000 TS; 15,000-19,000 YS; 15-20 El; 15-20 RA; 50-59 Brin. For bearings, pumps, steam valves; carburetors; "Ounce Metal."

BARR ALLOY NO. 9
McCallum-Hatch Bronze Co. Inc.
Cu 58, Zn 40, Fe 1.5, Al 0.5-1.5, Mn 0-0.25.
Grade A: 70,000-80,000 TS; 30,000-34,000 YS; 25-40 El; 100 Brin. Grade B: 80,000-90,000 TS; 40,000-45,000 YS; 15-25 El; 120 Brin. For propeller blades and hubs, valve stems, engine framing; compression of 0.1%-20000.

BARRONIA METAL
Barronia Metals, Ltd.
Cu 83, Pb 0.5, Sn 4, bal Zn.
Cast: 42,400 TS; 20,400 YS; 38 El; 33 RA; 72 Brin. Drawn: 84,000 TS; 74,000 YS; 18 El; 46 RA; 171 Brin. For condenser tubes, heat exchangers, evaporators, fittings; for superheated steam work.

BARROW B.H. BRAND PIG IRON
Barrow Haematite Steel Co., Ltd.
Si 1.5-4, S 0-0.025, P 0-0.03, Mn 0.25-1, 4.0 TC approx, bal Fe.
For metallurgical applications. For making steel and cast iron.

BARROW B.H.R. BRAND (CUPOLA GRADE)
Barrow Haematite Steel Co., Ltd.
TC 3.25-3.45, Si 0.5-1.1, S 0.12-0.18, P 0.045, Mn 0.25, bal Fe.
For metallurgical applications; for malleable foundries; various grades.

BARROW B.X. BRAND PIG IRON
Barrow Haematite Steel Co., Ltd.
Si 1.5-4, S 0-0.02, P 0-0.025, Mn 0.25-1.5, 4.0 TC approx, bal Fe.
For metallurgical applications; for making steel and cast iron.

BARROW B.X.X. BRAND PIG IRON
Barrow Haematite Steel Co., Ltd.
Si 1.5-3, S 0-0.02, P 0-0.02, Mn 0.25-1.5, 4.0 TC approx, bal Fe.
For metallurgical applications; for making steel and cast iron.

BARROW B.X.X. BRAND PIG IRON "SWEDISH"
Barrow Haematite Steel Co., Ltd.
Si 1-2.5, S 0-0.012, P 0-0.02, Mn 0-0.25, 4.0 TC approx, bal Fe.
For metallurgical applications; for making steel and cast iron.

BARROW BHS BESSEMER HAEMATITE PIG IRON
Barrow Haematite Steel Co., Ltd.
Si 1.5-4.5, S 0-0.03, P 0-0.35, Mn 0.25-1.2, 4.0 TC approx, bal Fe.
For metallurgical applications. For making steel.

BARROW BHS FOUNDRY HAEMATITE PIG IRON
Barrow Haematite Steel Co., Ltd.
Si 2-3.5, S 0-0.035, P 0-0.03, Mn 0.6-1, 3.85 TC approx, bal Fe.
For metallurgical applications; for general foundry work.

BARROW BRAND FOUNDRY HAEMATITE PIG IRON
Barrow Haematite Steel Co., Ltd.
Si 1.5-3, S 0-0.02, P 0.03, Mn 0.9-1.25, 4.0 TC approx, bal Fe.
For metallurgical applications, for heat resisting castings; for making alloy steels.

BARROW S.P. BRAND FOUNDRY PIG IRON
Barrow Haematite Steel Co., Ltd.
TC 3.5-3.75, Si 1.5-4, S 0.03-0.2, Mn 1-1.5, P 0.1-0.3, bal Fe.
For foundry iron for cylinders, valves, pumps, steam chests. For making steel and cast iron.

BARTO
Now CARPENTER BARTO.

BARWORTH B.C.M.
Barworth Flockton Ltd.
C 1, Mn 1.5, Cr 0.5, Si 0.4, bal Fe.
For press tools, gages, taps, dies; nondeforming, oil hardened.

BARWORTH B.M.S.
Barworth Flockton Ltd.
C 0.3-0.35, Ni 3, Cr 1, Mo 0.4, bal Fe.
For plastic mold dies; oil hardened.

BARWORTH B.S.C.C.
Barworth Flockton Ltd.
C 1.5, Mn 1, Cr 1, W 0.5, bal Fe.
For press tools, taps, dies; oil hardened, wear resistant.

BARWORTH B.S.W. 10CO
Barworth Flockton Ltd.
C 0.8, W 22, Cr 4.5, V 1.5, Mo 1, Co 10, bal Fe.
For cutting tools, reamers, hobs, broaches; high speed steel.

BARWORTH B.S.W. 14
Barworth Flockton Ltd.
C 0.65, W 14, Cr 3.75, V 0.5, bal Fe.
For drills, turning tools, hot punches, die cores; high speed steel.

BARWORTH B.S.W. 16
Barworth Flockton Ltd.
C 0.75, W 16, Cr 4.5, V 1, Mo 1.5, bal Fe.
For lathe and planer tools, drills, punches, taps; high speed steel.

BARWORTH B.S.W. 18
Barworth Flockton Ltd.
C 0.75, W 18, Cr 4.5, V 1, Mo 0.5, bal Fe.
For lathe and planer tools, hot punches; high speed steel.

BARWORTH B.S.W. 18V2
Barworth Flockton Ltd.
C 0.75-0.8, W 18, Cr 4.5, V 2, Mo 0.5, bal Fe.
For lathe and planer tools, drills, reamers, taps; high speed steel.

BARWORTH B.S.W. 22
Barworth Flockton Ltd.
C 0.76, W 22, Cr 4, V 1.25, bal Fe.
For drills, turning tools, cutters, hot punches; high speed steel.

BARWORTH B.S.W. 5CO
Barworth Flockton Ltd.
C 0.8, W 20, Cr 4.5, V 1.5, Mo 1, Co 5, bal Fe.
For cutters, drills, reamers, hobs, broaches; high speed steel.

BARWORTH B.S.W. 6/6/2
Barworth Flockton Ltd.
C 0.8-0.87, Cr 4, W 6.5, V 2, Mo 5, bal Fe.
For lathe and planer tools, reamers, drills, hobs; high speed steel.

BARWORTH C.D.S.H.
Barworth Flockton Ltd.
C 2, Cr 13, bal Fe.
For forging and blanking dies, punches; oil or air hardened, abrasion resistant.

BARWORTH C.R.V.
Barworth Flockton Ltd.
C 0.45-0.5, Cr 2, V 0.2, bal Fe.
For die casting dies; oil hardened, for short runs.

BARWORTH C.T.S.
Barworth Flockton Ltd.
C 0.5-0.55, Cr 1, W 2, Si 1, Mn 0.6, bal Fe.
For hot stamping and pressing dies, pneumatic tools; cold work steel.

BARWORTH H.D.3
Barworth Flockton Ltd.
C 0.35, Cr 3.5, W 3, V 0.3, bal Fe.
For hot and cold stamping dies, extrusion dies; hot work steel, oil hardened.

BARWORTH J.H.D.
Barworth Flockton Ltd.
C 0.3, Cr 3, W 10, V 0.2, bal Fe.
For stamping, pressing and extrusion dies; hot work steel, oil hardened.

BARWORTH J.H.D.N.
Barworth Flockton Ltd.
C 0.3, Cr 3, W 10, V 0.2, Ni 2, bal Fe.
For stamping, pressing and extrusion dies; hot work steel, oil hardened.

BARWORTH N.T. TYPE A
Barworth Flockton Ltd.
C 0.4, Mn 0.3, Cr 0.6, bal Fe.
For chisels, shock tools; water hardened, tough.

BARWORTH N.T. TYPE B
Barworth Flockton Ltd.
C 0.4, Mn 0.7, Cr 1, bal Fe.
For chisels, pneumatic tools; oil hardened, shock resistant.

BARWORTH S.C.R.
Barworth Flockton Ltd.
C 1, Mn 1, Cr 0.8, bal Fe.
For press tools, taps, dies; oil hardened, wear resistant.

BARWORTH T.C.
Barworth Flockton Ltd.
C 0.4, Cr 5, W 2, Mo 1, bal Fe.
For die casting dies; oil hardened, for long runs.

BATALBRA
Birmingham Battery & Metal Co., Ltd.
Cu 76, Zn 22, Al 2, As 0.03.
Inhibited brass for condenser and sea water pipeline, chemical plant equipment, heat exchangers.

BATH METAL-1
English manufacture
Cu 83, Zn 17.
For plumbing, pipe; red brass.

BATH METAL-2
English manufacture
Cu 55, Zn 45.
For bath fixtures; yellow brass.

BATNAVAL
Birmingham Battery & Metal Co., Ltd.
Cu 62, Zn 37, Sn 1.
For heat exchangers, marine condensers, power station condensers.

BATNICKON
Birmingham Battery & Metal Co., Ltd.
Ni 5, Fe 1.2, Mn 0.5, bal Cu.
For ship pipeline.

BATS 2Z
SMC (Shieldalloy Metallurgical Corp.)
B 2.5, Al 11, Ti 36, Si 8, Zr 6, bal Fe.
Master alloy for ladle additions.

BATS 50
SMC (Shieldalloy Metallurgical Corp.)
B 0.5, Al 13, Ti 20, Si 7, Zr 5, Mn 8, bal Fe.
Master alloy for ladle additions.

BATS L
SMC (Shieldalloy Metallurgical Corp.)
B 2, Al 17, Ti 47, Si 4.5, Zr 3.5, Mn 6, bal Fe.
Master alloy for ladle additions.

BATS TT
SMC (Shieldalloy Metallurgical Corp.)
B 1.5, Al 11, Ti 58, Zr 3.5, Mn 3, bal Fe.
Master alloy for ladle additions.

BATTERIUM
Batterium Metal & Vislok, Ltd.
Cu 89, Al 9, Ni 1, 1.0 other metal.
78,000-101,000 TS; 35-48 El; 158-168 Brin. For chemical apparatus, valves, plug cocks, plate terminals; non-corrodible, acid resistant.

BATTERY COPPER
American manufacture
Cu 94, Zn 6.
For batteries, gilding metal.

BATTERY PLATES
American manufacture
Pb 94, Sb 6.
For storage battery plates; hard.

BATURNAL
Birmingham Battery & Metal Co., Ltd.
Pb 2, Zn 35, bal Cu.
Free-machining brass, particularly for tubing.

BAUDRINS METAL
English manufacture
Cu 72, Zn 7.1, Ni 17, Fe 2.5, Co 1.8, Al 0-0.5.
For ornamental and corrosion resistant parts.

BAUDRINS METAL NO. 1
Manufacturer not listed.
Cu 72, Ni 16.6, Sn 2.5, Zn 7.1, Co 1.8.
For ornamental and corrosion resistant parts; corrosion resistant.

BAUDRINS METAL NO. 2
Manufacturer not listed.
Al 0.5, Cu 75, Fe 1.5, Ni 16, Sn 2.75, Zn 2.25, Co 2.
For ornamental and corrosion resistant parts; corrosion resistant.

BAUSH A-5 CASTING METAL
Bausch Machine Tool Co.
Cu 3, Ni 2, bal Al.
Cast: 24,000 TS; 0.5 El; 100 Brin. Heat treated: 30,000 TS. For heat treating castings; highly resistant to corrosion. *Obsolete*

BAUSH DURALUMIN GRADE A
Bausch Machine Tool Co.
Cu 3, Mg 0.3, Fe, Si, bal Al.
Annealed: 22,000-28,000 TS; 10-14 El; 54-60 Brin. Heat treated: 50,000-55,000 TS; 25,000-30,000 YS; 15-18 El; 83-93 Brin. For light alloy parts for airplanes, dirigibles and automobiles; age-hardening. *Obsolete*

BAUSH DURALUMIN GRADE B
Bausch Machine Tool Co.
Cu 4, Mn 0.6, Mg 0.5, Cr 0.1, Fe, Si, bal Al.
Annealed: 25,000 TS; 14 El; 60 Brin. Heat treated: 55,000-62,000 TS; 30,000-36,000 YS; 18-25 El; 93-100 Brin. For light alloy parts for automobiles, airplanes, dirigibles; age-hardening; sheet grade. *Obsolete*

BAW 1050A
British Alcan Wire Ltd.
Aluminum. Al 99.5, Si 0.25, Fe 0.4, Cu 0.05, Mn 0.05, Mg 0.05, Zn 0.07, Ti 0.05, Be 0.0008.
Soft: 75 N/mm^2 TS. Hard: 140-195 N/mm^2 TS. General mechanical wire, welding wire, rivet stock.

BAW 1080A
British Alcan Wire Ltd.
Aluminum. Al 99.8, Si 0.15, Fe 0.15, Cu 0.03, Mn 0.02, Mg 0.02, Zn 0.06, Ti 0.02, Be 0.0008.
Soft: 70 N/mm^2 TS. Hard: 130-160 N/mm^2 TS. General mechanical wire, welding wire.

BAW 1350
British Alcan Wire Ltd.
Aluminum. Al 99.5.
Soft: 75 N/mm^2 TS. Hard: 160-200 N/mm^2 TS. For electrical conductors.

BAW 2011
British Alcan Wire Ltd.
Aluminum. Si 0.4, Fe 0.7, Cu 5-6, Zn 0.3, bal Al.
Aluminum wire.

BAW 2014A
British Alcan Wire Ltd.
Aluminum. Si 0.5-0.9, Fe 0.5, Cu 3.9-5, Mn 0.4-1.2, Mg 0.2-0.8, Cr 0.1, Zn 0.25, Ti 0.15, bal Al.
Aluminum wire.

BAW 2017
British Alcan Wire Ltd.
Aluminum. Si 0.2-0.8, Fe 0.7, Cu 3.5-4.5, Mn 0.4-1, Mg 0.4-0.8, Cr 0.1, Zn 0.25, Ti 0.15, bal Al.
Aluminum wire.

BAW 2024
British Alcan Wire Ltd.
Aluminum. Si 0.5, Fe 0.5, Cu 3.8-4.9, Mn 0.3-0.9, Mg 1.2-1.8, Cr 0.1, Zn 0.25, Ti 0.15, bal Al.
Aluminum wire.

BAW 2117
British Alcan Wire Ltd.
Aluminum. Si 0.8, Fe 0.7, Cu 2.2-3, Mn 0.2, Mg 0.2-0.5, Cr 0.1, Zn 0.25, bal Al.
Aluminum wire.

BAW 3103
British Alcan Wire Ltd.
Aluminum. Si 0.5, Fe 0.7, Cu 0.1, Mn 0.9-1.15, Mg 0.3, Cr 0.1, Zn 0.2, bal Al.
Soft: 115 N/mm^2 TS. Hard: 205-245 N/mm^2 TS. General mechanical wire, welding wire.

BAW 4043A
British Alcan Wire Ltd.
Aluminum. Si 4.5-6, Fe 0.8, Cu 0.3, Mn 0.05, Mg 0.05, Zn 0.1, Ti 0.2, Be 0.0008, bal Al.
Welding and brazing wire.

BAW 4047A
British Alcan Wire Ltd.
Aluminum. Si 11-13, Fe 0.8, Cu 0.3, Mn 0.15, Mg 0.1, Zn 0.2, Be 0.0008, bal Al.
Brazing wire.

BAW 4145 A
British Alcan Wire Ltd.
Aluminum. Si 10, Cu 4, bal Al.
Brazing wire. *Obsolete*

BAW 5052
British Alcan Wire Ltd.
Aluminum. Si 0.25, Fe 0.4, Cu 0.1, Mn 0.1, Mg 2.2-2.8, Cr 0.15-0.35, Zn 0.1, bal Al.
Aluminum wire.

BAW 5056A
British Alcan Wire Ltd.
Aluminum. Si 0.4, Fe 0.5, Cu 0.1, Mn 0.1-0.6, Mg 4.5-5.6, Cr 0.2, Zn 0.2, Ti 0.2, Be 0.0008, bal Al.
Soft: 300 N/mm^2 TS. Hard: 400-450 N/mm^2 TS. General mechanical wire, rivet and bolt stock, welding wire, zip fastening wire.

BAW 5154A
British Alcan Wire Ltd.
Aluminum. Si 0.5, Fe 0.5, Cu 0.1, Mn 0.1-0.5, Mg 3.1-3.9, Cr 0.25, Zn 0.2, Ti 0.2, Be 0.0008, bal Al.
Soft: 250 N/mm^2 TS. Hard: 355 N/mm^2 TS. General mechanical wire, rivet stock, welding wire.

BAW 5183
British Alcan Wire Ltd.
Aluminum. Si 0.4, Fe 0.4, Cu 0.1, Mn 0.5-1, Mg 4.3-5.2, Cr 0.05-0.25, Zn 0.25, Ti 0.15, Be 0.0008, bal Al.
Aluminum wire.

BAW 5251
British Alcan Wire Ltd.
Aluminum. Si 0.4, Fe 0.5, Cu 0.15, Mn 0.1-0.5, Mg 1.7-2.4, Cr 0.15, Zn 0.15, Ti 0.15, bal Al.
Soft: 200 N/mm^2 TS. Hard: 280-310 N/mm^2 TS. General mechanical wire.

BAW 5356
British Alcan Wire Ltd.
Aluminum. Si 0.25, Fe 0.4, Cu 0.1, Mn 0.05-0.2, Mg 4.5-5.5, Cr 0.05-0.2, Zn 0.1, Ti 0.06-0.2, Be 0.0008, bal Al.
Aluminum wire.

BAW 5554
British Alcan Wire Ltd.
Aluminum. Si 0.25, Fe 0.4, Cu 0.1, Mn 0.5-1, Mg 2.4-3, Cr 0.05-0.2, Zn 0.25, Ti 0.05-0.2, Be 0.0008, bal Al.
Aluminum wire.

BAW 5556A
British Alcan Wire Ltd.
Aluminum. Si 0.25, Fe 0.4, Cu 0.1, Mn 0.6-1, Mg 5-5.5, Cr 0.05-0.2, Zn 0.2, Ti 0.05-0.2, Be 0.0008, bal Al.
Welding wire.

BAW 5754
British Alcan Wire Ltd.
Aluminum. Si 0.25, Fe 0.4, Cu 0.05, Mn 0.6, Mg 2.6-3.4, Cr 0.3, Zn 0.2, Ti 0.15, bal Al.
Aluminum wire.

BAW 6061
British Alcan Wire Ltd.
Aluminum. Si 0.4-0.8, Fe 0.7, Cu 0.15-0.4, Mn 0.15, Mg 0.8-1.2, Cr 0.04-0.35, Zn 0.25, Ti 0.15, bal Al.
Aluminum wire.

BAW 6063
British Alcan Wire Ltd.
Aluminum. Si 0.2-0.6, Fe 0.35, Cu 0.1, Mn 0.1, Mg 0.45-0.9, Cr 0.1, Zn 0.1, Ti 0.1, bal Al.
Aluminum wire.

BAW 6082
British Alcan Wire Ltd.
Aluminum. Si 0.7-1.3, Fe 0.5, Cu 0.1, Mn 0.4-1, Mg 0.6-1.2, Cr 0.25, Zn 0.2, Ti 0.1, bal Al.
Aluminum wire.

BAW 6101A
British Alcan Wire Ltd.
Aluminum. Si 0.5, Mg 0.6, bal Al.
Solution treated, cold worked and precipitation treated: 300 N/mm^2 TS. Electrical conductors.

BAW 7050
British Alcan Wire Ltd.
Aluminum. Si 0.12, Fe 0.15, Cu 2-2.6, Mn 0.1, Mg 1.9-2.6, Cr 0.04, Zn 5.7-6.7, Ti 0.06, bal Al.
Aluminum wire.

BAW 7075
British Alcan Wire Ltd.
Aluminum. Si 0.4, Fe 0.5, Cu 1.2-2, Mn 0.3, Mg 2.1-2.9, Cr 0.18-0.28, Zn 5.1-6.1, Ti 0.2, bal Al.
Aluminum wire.

BAW 99.98
British Alcan Wire Ltd.
Aluminum. Si 0.01, Fe 0.006, Cu 0.003, Zn 0.01, Ti 0.003, bal Al.
Aluminum wire.

BAXTRON-DBA
DuPont de Nemours & Co., E.I.
WC + ceramic + metal.
For cutting tool inserts; milling cutters, wear components.
Ceramic-carbide composite. *Obsolete*

BAXTRON-DBW
DuPont de Nemours & Co., E.I.
WC + oxide + metal.
Sintered: 539,000 Tr.R.; 91.8 RA. For cutting tools, dies, wear components, punches. High strength and hardness at elevated temperatures. Composite. *Obsolete*

BAZAR
Barker & Allen Ltd.
Ni 10, Zn, bal Cu.
For domestic utensils, ornaments; nickel silver.

BBB WELDING ELECTRODE
Becker Bros. Carbon Corp.
C 0.2, bal Fe.
For welding electrode.

BC-35
Forjas Alavesas S.A.
C 0.37, Mn 0.75, Si 0.25, Cr 1, bal Fe.
Chromium structural steel. AFNOR 38C4; DIN 37Cr4; AISI 5135.

BC-40
Forjas Alavesas S.A.
C 0.41, Mn 0.76, Si 0.25, Cr 1, bal Fe.
Chromium structural steel. AFNOR 45C4; AISI 5140; BS EN 18-D.

BC-NO. 10V
British Steel Corp.
C 1, V 0.22, Mn 0.2, bal Fe.
For coining and embossing dies, engraving tools; Type W2; water hardened. *Obsolete*

BCNI
Friedr. Lohmann GmbH
C 0.55, Cr 1.05, Ni 3.25, bal Fe.
For cold heading dies, hobbing punches, cutlery dies. Oil hardened, shock resistant.

BCV-42
Lukens Steel
C 0-0.25, Mn 0-1.5, Si 0.15-0.3, Cr 0.4-0.65, Cu 0.25-0.4, V 0.02-0.1, bal Fe.
Plate, normalized: 63,000 psi TS; 42,000 psi YS; 19 El. For structural purposes, particularly transmission tower uprights.

BCV-46
Lukens Steel
C 0-0.25, Mn 0-1.5, Si 0.15-0.3, Cr 0.4-0.65, Cu 0.25-0.4, V 0.02-0.1, bal Fe.
Plate, normalized: 67,000 psi TS; 46,000 psi YS; 19 El. For structural purposes, particularly transmission tower uprights.

BCV-50
Lukens Steel
C 0-0.25, Mn 0-1.5, Si 0.15-0.3, Cr 0.4-0.65, Cu 0.25-0.4, V 0.02-0.1, bal Fe.
Plate, normalized: 70,000 psi TS; 50,000 psi YS; 19 El. For structural purposes, particularly transmission tower uprights.

BCV-55
Lukens Steel
C 0-0.25, Mn 0-1.5, Si 0.15-0.3, Cr 0.4-0.65, Cu 0.25-0.4, V 0.02-0.1, bal Fe.
Plate, normalized: 70,000 psi TS; 55,000 psi YS; 19 El. For structural purposes, particularly transmission tower uprights.

BCV-60
Lukens Steel
C 0-0.25, Mn 0-1.5, Si 0.15-0.3, Cr 0.4-0.65, Cu 0.25-0.4, V 0.02-0.1, bal Fe.
Plate, quenched and tempered: 70,000 psi TS; 60,000 psi YS; 19 El. For structural purposes, particularly transmission tower uprights.

BCV-70
Lukens Steel
C 0-0.25, Mn 0-1.5, Si 0.15-0.3, Cr 0.4-0.65, Cu 0.25-0.4, V 0.02-0.1, bal Fe.
Plate, quenched and tempered: 80,000 psi TS; 70,000 psi YS; 16 El. For structural purposes, particularly transmission tower uprights.

BCV-X-50
Lukens Steel
C 0-0.24, Mn 1-1.6, Si 0.15-0.5, Cu 0-0.35, Ni 0-0.25, Cr 0-0.25, Mo 0-0.08, bal Fe.
Plate: 70,000-90,000 psi TS; 50,000 psi YS min; 20 El min. For structural purposes; transmission lines requiring improved low temperature impact.

BCV-X-52
Lukens Steel
C 0-0.24, Mn 1-1.6, Si 0.15-0.5, Cu 0-0.35, Ni 0-0.25, Cr 0-0.25, Mo 0-0.08, bal Fe.
Plate: 73,000-93,000 psi TS; 52,000 psi YS min; 20 El min. For structural purposes; transmission lines requiring improved low temperature impact.

BCV-X-56
Lukens Steel
C 0-0.24, Mn 1-1.6, Si 0.15-0.5, Cu 0-0.35, Ni 0-0.25, Cr 0-0.25, Mo 0-0.08, bal Fe.
Plate: 75,000-95,000 psi TS; 56,000 psi YS min; 20 El min. For structural purposes; transmission lines requiring improved low temperature impact.

BCV-X-60
Lukens Steel
C 0-0.24, Mn 1-1.6, Si 0.15-0.5, Cu 0-0.35, Ni 0-0.25, Cr 0-0.25, Mo 0-0.08, bal Fe.
Plate: 80,000-100,000 psi TS; 60,000 psi YS min; 22 El min. For structural purposes; transmission lines requiring improved low temperature impact.

BD. 30
Peninsular Steel Co.
C 0.51, Cr 0.95, Mn 0.87, Mo 0.2, bal Fe (typical).
Prehardened to 248-293 Brin. For dies for mechanically and hand operated press brakes for forming operations.

BE 12
Le Bronze Industriel
Sn 11.5, Zn 1, 0.10 P min, bal Cu.
Phosphor bronze, cast or wrought; 80-95 Brin. AFNOR: UE 12 Z1; UE 12P; ASTM B103-74.

BE 20
Le Bronze Industriel
Sn 18, bal Cu.
Bronze; wrought; 170 Brin.

BE 5
Le Bronze Industriel
Sn 5, bal Cu.
Tin bronze, wrought; 70-100 Brin.

BEACON
Edgcomb Metals Co.
C 0.8-1, bal Fe.

BEARCAT
Latrobe Steel Co.
C 0.5, Si 0.25, Mn 0.75, Cr 3.25, Mo 1.4.
Shock resisting die steel. For cold and hot tooling applications; blades, swaging dies, gripper dies, chisels and punches. AISI S7.

BEARCAT
Bethlehem Steel Corp.
C 0.5, Mn 0.7, Si 0.3, Cr 3.25, Mo 1.4, bal Fe.
Heat treated: 145,000-342,000 TS; 130,000-205,000 YS; 4-20 El; 7-60 RA; 310-590 Brin. For rivet sets, punches, chisels, hobs; shock resistant, air or oil hardened. AISI S7. *Obsolete*

BEARDSHAW H.M. 1
Joseph Beardshaw & Son Ltd.
C 0.95, Mn 1.6, Mo 0.25, V 0.25, bal Fe.
For punches, gages, shear blades, reamers; oil hardened, non-deforming.

BEARING METAL NO. 600
English manufacture
Cu 59.7, Zn 35, Sn 0.2, Pb 0.33, Mn 2.6, Al 1.38, Si 0.7.
For bearings, bushings; tough.

BEARING STANDARD
Bethlehem Steel Corp.
C 1, Mn 0.35, Si 0.23, Cr 1.35, bal Fe.
For tools, bearings; water hardened. *Obsolete*

BEARINGOY
Studebaker Chemical Co.
Pb, Cu, Sb, bal Sn.
For machine bearings; long life under poor lubrication.

BEARITE
A.W. Cadman Mfg. Co.
Pb 80.1, Cu 0.37, Bi 0.13, Sb 16.75, 2.65 hardener.
At 70°F: 8750 TS; 1.5 El; 29.1 Brin. At 212°F: 24.5 Brin. For bearings not subject to vibration or pounding; high compressive value, low coefficient friction.

BEARIUM 82
Bearium Metals Corp.
Cu 70, Pb 28, bal Sn.
Cast: 18,000 TS; 8,000 YS; 14 El; 35 Brin. For bearings, bushings. *Obsolete*

BEARIUM B-10

Bearium Metals Corp.
Pb 20, Cu 70, bal Sn.
Cast: 25,500 TS; 12,000 YS; 10 El; 55 Brin. For bearings, bushings. Non-scoring; severe applications.

BEARIUM B-11

Bearium Metals Corp.
Pb 17.5, Cu 70, bal Sn.
Cast: 30,000 TS; 8 El; 195 Brin. For bearings, bushings. Severe applications.

BEARIUM B-4

Bearium Metals Corp.
Cu 70, Pb 26, bal Sn.
Cast: 21,650 TS; 9750 YS; 16 El; 40 Brin. For bearings, bushings. Poor lubrication.

BEARIUM B-6

Bearium Metals Corp.
Pb 24, Cu 70, bal Sn.
Cast: 22,500 TS; 14 El; 45 Brin. For bearings for severe service and large installations. *Obsolete*

BEARIUM B-8

Bearium Metals Corp.
Cu 70, Pb 22, Sn 8.
Cast: 24,500 TS; 11,500 YS; 12 El; 50 Brin. For bearings, bushings, thrust washers. For heavy duty requirements.

BEAUCALLOY

Beaumont Birch Co.
C 1.4, Ni, Cr, bal Fe.
Hardened: 175,000 TS; 150,000 YS; 400 Brin. For conveyor chains; casting.

BEAUTYWELD

Eutectic Corp.
Electrode for AC-DC metallic arc welding all low carbon steel sheet, forms, plate; 75,000 psi TS.

BEAVER

Atlas Specialty Steels
C 0.68, Si 1, Cr 8.25, V 1, Mo 1.4, Ni 1.5, bal Fe.
For slitter knives, blanking dies, cold work tools; high wear resistance.

BEAVER

Teledyne Vasco
C 0.7, W 18, Cr 4, V 1, bal Fe.
For tools, cutters, drills; high speed steel. *Obsolete*

BEAVER

Farrelloy Co. (American Solder&Flux)
C 0.7, W 18, Cr 4, bal Fe.
For high speed tools. *Obsolete*

BEAVER BABBITT

United American Metals Corp.
Sb, Pb, Sn.
For machinery bearings; Babbitt. *Obsolete*

BEAVER D-2

Fagersta Bruks Aktiebolag
C 1.5, Cr 12, V 0.4, Mo 1, bal Fe.
Air Hardened: 288,000 TS; 215,000 YS; 1 El; C 58 Rock. For blanking and drawing dies, punches, gages, broaches, hobs, shear blades. Precision ground flat stock, AISI Type D2, wear and abrasion resistant. Air harden. *Obsolete*

BEAVER PRECISION GROUND FLATS & SQUARES

Teledyne Pittsburgh Tool Steel
C 1.55, Si 0.38, Mn 0.25, Cr 12, Mo 0.8, V 0.8, bal Fe.
AISI D2. Cold work tool and die steel.

BECKET ALLOY

English manufacture
Cr 25-30, Si 3, C 1.5-3, bal Fe.
For stainless castings; corrosion resistant.

BEDCO ALLOY

Bedford Tool & Forge Co.
C 0.55, Mn 1, Si 2, Mo 0.4, V 0.35, bal Fe.
Heat treated: 350,000 TS; 295,000 YS; 6 El; 12 RA; 600 Brin. For chisels, punches, pneumatic and beading tools; shock resistant, oil hardened. AISI S5.

BEDCO M-2

Bedford Tool & Forge Co.
C 0.65, Cr 4, V 2, W 6.5, Mo 5, bal Fe.
Hardened: 58-60 Rock C. For hot extrusion dies, punches, shear blades, forging mandrels. High speed steel for hot working; shock resisting.

BEDCO M-2 HIGH SPEED

Bedford Tool & Forge Co.
C 0.82, Mo 5, W 6.5, Cr 4, V 2, bal Fe.
Oil hardened: 63-65 Rock C. For lathe and planer cutters, drills, taps, chasers, drawing dies, punches hobs. Type M2 high speed steel, high red-hardness.

BEDCO T1

Bedford Tool & Forge Co.
C 0.75, Mn 0.3, Si 1.3, Cr 4, V 1.15, W 18, bal Fe.
Standard grade of tungsten high speed steel for cutting tools; AISI T1.

BEDEL "DIAMONT BEDEL"

Acieries S.A. Bedel
W 4, Cr 4, V, bal Fe.
For cutting tools; oil hardened.

BEDEL 1 COURONNES 0

Acieries S.A. Bedel
C, bal Fe.
For wood tools used without hardening; crucible steel.

BEDEL 1 COURONNES 1

Acieries S.A. Bedel
C, bal Fe.
For saws, lathe tools; crucible steel.

BEDEL 1 COURONNES 2

Acieries S.A. Bedel
C, bal Fe.
For cutlery, taps, chisels, mandrels, razors, metal saws; crucible steel.

BEDEL 1 COURONNES 3

Acieries S.A. Bedel
C, bal Fe.
For drills, borers, mining bars, chisels, engraving tools, stone tools, needle dies; crucible steel.

BEDEL 1 COURONNES 4

Acieries S.A. Bedel
C, bal Fe.
For rams, shear blades, dies, hammers; crucible steel.

BEDEL 1 COURONNES 5

Acieries S.A. Bedel
C, bal Fe.
For surgical instruments, forge tools; crucible steel.

BEDEL 1 TREFLE 5

Acieries S.A. Bedel
Ni 4.5, C, Cr, Mo, bal Fe.
Heat treated: 157,000 TS; 135,000 YS; 15 El. For valves operating at high temperatures; heat resistant.

BEDEL 2 COURONNES 0

Acieries S.A. Bedel
C, bal Fe.
For wood tools used without hardening; crucible steel.

BEDEL 2 COURONNES 1

Acieries S.A. Bedel
C, bal Fe.
For saws for hard wood, lathe tools; crucible steel.

BEDEL 2 COURONNES 2

Acieries S.A. Bedel
C, bal Fe.
For general cutlery, chisels, mandrels, razors, metal saws; crucible steel.

BEDEL 2 COURONNES 3

Acieries S.A. Bedel
C, bal Fe.
For drills, borers, mining bars, chisels, punches, stone tools; crucible steel.

BEDEL 2 COURONNES 4

Acieries S.A. Bedel
C, bal Fe.
For shear blades, cutting and stamping dies, hammers; crucible steel.

BEDEL 2 COURONNES 5

Acieries S.A. Bedel
C, bal Fe.
For surgical instruments, forge tools; crucible steel.

BEDEL 3 COURONNES 0

Acieries S.A. Bedel
C, bal Fe.
For wood tools used without hardening; crucible steel.

BEDEL 3 COURONNES 1

Acieries S.A. Bedel
C 0.9-1.1, bal Fe.
For saws for hard wood, lathe tools; crucible steel.

BEDEL 3 COURONNES 2

Acieries S.A. Bedel
C 1-1.1, bal Fe.
For general cutlery, taps, chisels, mandrels, razors, metal saws; crucible steel.

BEDEL 3 COURONNES 3

Acieries S.A. Bedel
C 1.2-1.3, bal Fe.
For drills, borers, chisels, engraving tools, needle dies, stone tools; crucible steel.

BEDEL 3 COURONNES 4

Acieries S.A. Bedel
C 0.9, bal Fe.
For hot and cold shear blades, hammers, cutting and stamping dies; crucible steel.

BEDEL 3 COURONNES 5

Acieries S.A. Bedel
C, bal Fe.
For forge tools, surgical instruments; crucible steel.

BEDEL 405

Acieries S.A. Bedel
C 1.6, Cr 4.5, W 12, Mo 0.4, V 0.5, Co 5, bal Fe.
For special tools, engravers tools; textile needles. High carbon, high-speed steel, wear and abrasion resistant.

BEDEL BCN

Acieries S.A. Bedel
C 0.4, Ni 2-3, Cr, bal Fe.
Heat treated: 135,000-242,000 TS; 121,000-213,000 YS; 10-15 El. For general engineering construction; oil hardening.

BEDEL BCN 3 S 2

Acieries S.A. Bedel
Ni 3, High C, Cr, bal Fe.
Heat treated: 270,000 TS; 242,000 YS; 6 El. For general engineering construction; oil hardening.

BEDEL BCN 4 S

Acieries S.A. Bedel
Ni 4, C, Cr, bal Fe.
Air hardened: 260,000-270,000 TS; 220,000-235,000 YS; 6-8 El. For general engineering construction; air hardening.

BEDEL BK 3 B

Acieries S.A. Bedel
Ni 3.5, low C, Cr, bal Fe.
Annealed: 78,000 TS; 58,000 YS; 30 El. Heat treated: 135,000 TS; 128,000 YS; 11 El. For crankshafts, general case hardened parts for severe service; case hardening steel.

BEDEL BN 23

Acieries S.A. Bedel
Ni 25, C, Cr, bal Fe.
Annealed: 100,000 TS; 45 El. For chemical plant equipment; non-magnetic; corrosion and acid resistant.

BEDEL BN 33

Acieries S.A. Bedel
Ni 33, C, Cr, bal Fe.
Annealed: 115,000 TS; 30 El. For chemical plant equipment; high acid and corrosion resistant.

BEDEL BNAV 0

Acieries S.A. Bedel
Ni 3, C, Cr, Mo, bal Fe.
Heat treated: 228,000 TS; 199,000 YS; 9 El. For general engineering construction; oil hardening.

BEDEL BNAV 1

Acieries S.A. Bedel
Ni 4, C, Cr, Mo, bal Fe.
Heat treated: 315,000 TS; 260,000 YS; 6 El. For general engineering construction; air hardening.

BEDEL BNAV 2

Acieries S.A. Bedel
Ni 4, medium C, Cr, Mo, bal Fe.
Heat treated: 242,000 TS; 228,000 YS; 9 El. For aircraft motor crankshafts; oil hardening.

BEDEL BNAV 3

Acieries S.A. Bedel
Ni 4, low C, Cr, Mo, bal Fe.
Annealed: 107,000 TS. Heat treated: 143,000 TS; 135,000 YS; 14 El. For crankshafts, general case hardened parts for severe service; case hardening steel.

BEDEL C.3

Acieries S.A. Bedel
Ni 2.5, low C, bal Fe.
Annealed: 85,000 TS; 78,000 YS; 20 El. For aviation and marine case hardened parts; case hardening steel.

BEDEL C.5

Acieries S.A. Bedel
Ni 5, low C, bal Fe.
Annealed: 90,000 TS; 78,000 YS; 20 El. For aviation and marine case hardened parts; case hardening steel.

BEDEL CHROME 4

Acieries S.A. Bedel
Cr 2-3, C, bal Fe.
For cold stamping dies and punches; non-deforming.

BEDEL CKN

Acieries S.A. Bedel
Ni 3, low C, Cr, bal Fe.
Annealed: 85,000 TS. Heat treated: 164,000 TS; 143,000 YS; 9 El. For crankshafts, general case hardened parts for severe service; case hardening steel.

BEDEL DOUBLE TREFLE

Acieries S.A. Bedel
W 14, C, Cr, bal Fe.
For general usage, medium speed cutters; high speed steel.

BEDEL DOUBLE TREFLE E

Acieries S.A. Bedel
W 15-16, C, bal Fe.
For tools, dies; oil hardening.

BEDEL INO 11

Acieries S.A. Bedel
Ni 6, Cr 25, C, bal Fe.
For rolled or forged parts, pump and furnace parts, valves; heat and corrosion resistant; non-hardenable.

BEDEL INO 12

Acieries S.A. Bedel
Cr 20, Ni 8, C, Mo, bal Fe.
For valves operating up to 900°C; austenitic, heat resistant.

BEDEL INO 13

Acieries S.A. Bedel
Ni 8, Cr 20, C, bal Fe.
For carburizing boxes, recuperator tubes, salt bath pots; heat and corrosion resistant; non-hardenable.

BEDEL INO 14

Acieries S.A. Bedel
Ni 6, Cr 25, C, bal Fe.
For carburizing boxes, recuperator tubes, salt bath pots; heat and corrosion resistant; non-hardenable.

BEDEL INO 15

Acieries S.A. Bedel
Ni 6, Cr 25, C, bal Fe.
For carburizing boxes, recuperator tubes, salt bath pots; heat and corrosion resistant; non-hardenable.

BEDEL INO 16

Acieries S.A. Bedel
Ni 6, Cr 25, C, Si, bal Fe.
For carburizing boxes, recuperator tubes, salt bath pots; heat and corrosion resistant; non-hardenable.

BEDEL INO 2

Acieries S.A. Bedel
Cr 13, medium C, bal Fe.
Heat treated: 170,000 TS; 157,000 YS; 5 El. For marine valves; corrosion resistant.

BEDEL INO 20

Acieries S.A. Bedel
Cr 25, Ni 20, C, bal Fe.
For pump, furnace and valve parts, carburizing boxes; heat and corrosion resistant; non-hardenable.

BEDEL INO 21

Acieries S.A. Bedel
C, high Ni, high Cr, bal Fe.
For pump, furnace and valve parts, carburizing boxes; heat and corrosion resistant; non-hardenable.

BEDEL INO 22

Acieries S.A. Bedel
Cr 25, Ni 20, C, bal Fe.
For pump, furnace and valve parts, carburizing boxes; heat and corrosion resistant; non-hardenable.

BEDEL INO 23

Acieries S.A. Bedel
Cr 20, Ni 20, C, Si, bal Fe.
For pump, furnace and valve parts, carburizing boxes; heat and corrosion resistant; non-hardenable.

BEDEL INO 3

Acieries S.A. Bedel
Cr 13, C, V, bal Fe.
Heat treated: 135,000 TS; 110,000 YS; 10 El. For automobile valves; corrosion resistant.

BEDEL INO 31

Acieries S.A. Bedel
Cr 14, Ni 15, C, W, Mo, bal Fe.
For valves operating up to 900°C; austenitic, heat resistant.

BEDEL INO 32

Acieries S.A. Bedel
Cr 15, Ni 20, C, Mo, bal Fe.
For valves operating up to 900°C; austenitic, heat resistant.

BEDEL INO 33

Acieries S.A. Bedel
Ni 25, Cr 10, C, W, bal Fe.
For valves operating up to 900°C; austenitic, heat resistant.

BEDEL INO 40

Acieries S.A. Bedel
Ni 35, Cr 15, C, bal Fe.
For turbine blading, marine parts; maximum heat and corrosion resistant.

BEDEL INO 41

Acieries S.A. Bedel
Ni 40, Cr 10, C, bal Fe.
For marine parts; maximum heat and corrosion resistant.

BEDEL INO 50

Acieries S.A. Bedel
Cr 15, Ni 50, C, bal Fe.
For carburizing boxes, furnace parts; maximum heat and corrosion resistant.

BEDEL INO 51

Acieries S.A. Bedel
Ni 50, Cr 15, C, bal Fe.
For carburizing boxes, furnace parts; maximum heat and corrosion resistant.

BEDEL INO 6

Acieries S.A. Bedel
Cr 17, C, V, bal Fe.
For cutlery, surgical instruments; high heat and corrosion resistant.

BEDEL M.G.B.A

Acieries S.A. Bedel
C, Co, bal Fe.
For magnets; magnet steel.

BEDEL M.G.B.C

Acieries S.A. Bedel
Co 35, W 8, C, bal Fe.
For magnets; magnet steel.

BEDEL M.G.B.S

Acieries S.A. Bedel
W 5-6, C, Cr, bal Fe.
For magnets; magnet steel.

BEDEL NO. 600 C

Acieries S.A. Bedel
Ni 1.5, low C, Cr, Mo, V, bal Fe.
Heat treated: 150,000-177,000 TS; 143,000-157,000 YS; 12 El. For crankshafts, general case hardened parts for severe service; case hardening steel.

BEDEL NO. 644

Acieries S.A. Bedel
C, W, Cr, V, bal Fe.
For high speed cutters; high speed steel.

BEDEL QUATRE TREFLES

Acieries S.A. Bedel
C, W, Cr, V, Co, bal Fe.
For high speed cutters; high speed steel for extra hard pieces at high speed.

BEDEL QUATRE TREFLES C

Acieries S.A. Bedel
Co 10, C, Mo, V, bal Fe.
For tools, dies; oil hardening.

BEDEL R.B. 6

Acieries S.A. Bedel
Cr 2.5, Ni 0.5, C, bal Fe.
For bearings, dies, punches; not shock resistant.

BEDEL R.B. 7

Acieries S.A. Bedel
Cr 1.5, high C, bal Fe.
For bearings, dies, punches; not shock resistant.

BEDEL SOC

Acieries S.A. Bedel
Cr 13, C, Mo, Co, bal Fe.
Heat treated: 145,000 TS; 130,000 YS; 6 El. For valves operating at temperatures up to 800°C; very oxidation resistant.

BEDEL SOC 2

Acieries S.A. Bedel
Cr 20, C, high Co, bal Fe.
Heat treated: 135,000 TS; 115,000 YS; 7.5 El. For valves operating at temperatures up to 800°C; very oxidation resistant.

BEDEL SPECIAL 000

Acieries S.A. Bedel
Cr 13, C, bal Fe.
For dies, punches, rolling mill rolls; non-deforming.

BEDEL SPECIAL F

Acieries S.A. Bedel
W 4, C, Cr, Mo, bal Fe.
For wire drawing dies, cold drawn wire dies; non-deforming.

BEDEL SPECIAL FA

Acieries S.A. Bedel
Cr 13, C, Ni, Mo, V, bal Fe.
For dies and punches; non-deforming.

BEDEL SPECIAL FB

Acieries S.A. Bedel
Cr 13, C, Mo, V, bal Fe.
For dies and punches; non-deforming.

BEDEL SPECIAL FK

Acieries S.A. Bedel
Cr 13, C, Mo, V, bal Fe.
For dies and punches; non-deforming.

BEDEL SPECIAL FL

Acieries S.A. Bedel
Cr 13, C, bal Fe.
For wire drawing, dies for cold drawn hard steel; non-deforming.

BEDEL SPECIAL OZ

Acieries S.A. Bedel
W 10, C, Cr, V, bal Fe.
For hot forging tools, stamping and punching, hot shear blades; hot work steel.

BEDEL SPECIAL X

Acieries S.A. Bedel
C, Cr, V, bal Fe.
For calipers, gages, dies, punches; non-deforming.

BEDEL SPECIAL Y.2

Acieries S.A. Bedel
Ni 3, Cr 0.7, C, bal Fe.
For pneumatic or hydraulic rivet sets; tough.

BEDEL SPECIAL Y.5

Acieries S.A. Bedel
W 2, V 1, C, bal Fe.
For pneumatic chisels, punches, shear blades; tough.

BEDEL SPECIAL Z3

Acieries S.A. Bedel
W 10, C, Cr, Ni, V, bal Fe.
For tools, dies; oil hardening.

BEDEL SPECIAL Z4

Acieries S.A. Bedel
W 10, C, Cr, V, Co, bal Fe.
For tools, dies; oil hardening.

BEDEL TRIPLE TREFLE

Acieries S.A. Bedel
W 18, C, Cr, bal Fe.
For heavy roughing and milling cutters; high speed steel.

BEDEL TRIPLE TREFLE E.S

Acieries S.A. Bedel
Co 6, C, W, Cr, V, bal Fe.
For high speed cutters; super high speed steel.

BEDEL TRIPLE TREFLE ESD

Acieries S.A. Bedel
Co 5, C, Mo, V, bal Fe.
For tools, dies; oil hardening.

BEDEL TRIPLE TREFLE F.R

Acieries S.A. Bedel
C, high W, high Cr, bal Fe.
For taps, forming tools, bores, finishing tools; high speed steel.

BEDEL TRIPLE TREFLE-E

Acieries S.A. Bedel
C 0.75, W 18, Cr 5, Mo 1, V 1, bal Fe.
For tools, cutters, drills, reamers, broaches, lathe and planer tools. High-speed steel, high red hardness.

BEDEL UN TREFLE 1

Acieries S.A. Bedel
W 2-3, C, Cr, V, bal Fe.
For hot and cold forging punches, dies; hot work steel.

BEDEL UN TREFLE 2

Acieries S.A. Bedel
Ni 4, C, Cr, Mo, bal Fe.
For hot stamping dies and punches, pipe mandrel, shear blades; shock resistant.

BEDEL UN TREFLE 3

Acieries S.A. Bedel
C, Cr, Ni, bal Fe.
For pneumatic chisels, hot slicing; shock resistant.

BEDEL UN TREFLE COURONNE

Acieries S.A. Bedel
C, Cr, bal Fe.
For hot punches and dies for screw machines; hot work steel.

BELAIS WHITE GOLD-1

Engelhard Corp.
Au 75-85, Ni 8-18, Zn 4-14.
For jewelry, ornaments; corrosion resistant. *Obsolete*

BELAIS WHITE GOLD-2

Engelhard Corp.
Au 75-85, Ni 10-18, Zn 2-9, 0.05-0.5 Pt or 0.5-2.0 Mn.
For jewelry, ornaments; corrosion resistant. *Obsolete*

BELECTRIC NO. 0

Belle City Malleable Co.
C 3, Si 2, bal Fe.
Cast: 26,000-33,000 TS; 170-190 Brin. For machinery castings, pumps; cast iron. *Obsolete*

BELECTRIC NO. 1

Belle City Malleable Co.
C 3.25, Si 2, bal Fe.
Cast: 35,000-42,000 TS; 200-220 Brin. For gears, pinions, brake drums, brake shoes, pumps; cast iron. *Obsolete*

BELECTRIC NO. 2

Belle City Malleable Co.
C 2.9, Si 2, bal Fe.
Cast: 42,000-50,000 TS; 220-240 Brin. For gears, pinions, brake drums, pumps; cast iron. *Obsolete*

BELECTRIC NO. 3

Belle City Malleable Co.
C 3, Si 2, Alloys, bal Fe.
Cast: 50,000-65,000 TS; 230-300 Brin. For gears, pinions, brake shoes, pumps; cast iron. *Obsolete*

BELGIAN ALUMINUM PISTON ALLOY

Belgian manufacture
Al 90.5, Cu 7, Zn 2.5.
For pistons; non-hardenable.

BELL BRAND

Patriarche & Bell
C 0.7-1, bal Fe.
For tools; water hardened.

BELL BRASS

English manufacture
Cu 64, Zn 35, Sn 0.85.
For bells.

BELL BRONZE

A. Cohn Ltd.
Cu 80, Sn 20.
For bells.

BELL METAL, JAPANESE KARAKANE-1

Japanese manufacture
Cu 61-72, Sn 14-25, Pb 0.14, Zn 0-9.4, Fe 0-3.
For bells; corrosion resistant.

BELL METAL, JAPANESE KARAKANE-2

Japanese manufacture
Cu 71.42, Sn 14.2, Pb 14.3.
For bells; free-cutting.

BELL METAL, JAPANESE KARAKANE-3

Japanese manufacture
Cu 70, Sn 19, Zn 3, Pb 8.
For bells; free-cutting.

BELL METAL, JAPANESE KARAKANE-4

Japanese manufacture
Cu 65.95, Sn 17.25, Zn 3.45, Pb 10.35.
For bells; free-cutting.

BELL METAL, JAPANESE KARAKANE-5

Japanese manufacture
Cu 64, Sn 24, Zn 9, Fe 3.
For bells; corrosion resistant.

BELL METAL, JAPANESE KARAKANE-6

Japanese manufacture
Cu 61, Sn 18, Zn 6, Fe 3, Pb 12.
For bells; free-cutting.

BELL METAL, MUSICAL

American manufacture
Cu 84, Sn 16.
For bells; bronze.

BELL METAL-1

English manufacture
Al 83, Mn 10, Cd 7.
For bells, chimes, whistles, ornaments.

BELL METAL-2

English manufacture
Cu 85-60, Sn 15-40.
For bells, chimes; hard bronze.

BELL SPECIAL

Patriarche & Bell
C 1.2, V 0.2, bal Fe.
For tools, fixtures, jigs; water hardened.

BELLER NO. 4

Chiers-Chatillon
C 0.75, bal Fe.
For tools, taps, springs, punches; water hardened.

BELMALLOY
Belle City Malleable Co.
C 2.15-2.35, Si 1, Mn 1, bal Fe.
Cast: 70,000-80,000 TS; 45,000-50,000 YS; 5-10 El; 179-207 Brin. For tractor parts, axle parts; pearlitic malleable cast iron. *Obsolete*

BELMONT
Time Steel Service Inc.
C 0.9, Mn 1.5, Si 0.25, Mo 0.3, bal Fe.
Oil hardened tool steel. AISI O2.

BELMONT NO. 1015
Belmont Metals Inc.
Sn 10, Sb 15, bal Pb.
10,500 psi TS. Babbitt for low speed bearings; melting point 514°F.

BELMONT NO. 2405
Belmont Metals Inc.
Bi 40, Sn 60.
8000 psi TS. Low melting alloy for sprinklers, anchoring, encapsulating, patterns; melting point 281-338°F.

BELMONT NO. 2431
Belmont Metals Inc.
Pb 37.7, Sn 11.3, Cd 8.5, bal Bi.
5400 psi TS. Low melting alloy for anchoring, tube bending, molds, chucks; melting point 160-190°F.

BELMONT NO. 2451
Belmont Metals Inc.
Pb 22.6, Sn 8.3, Cd 5.3, In 19, bal Bi.
5400 psi TS. Low melting alloy for soldering, lens grinding, chucking, sprinklers; melting point 117°F.

BELMONT NO. 2481
Belmont Metals Inc.
Sn 14.5, Pb 28.5, bal Bi.
13,000 psi TS. Low melting alloy for soldering, anchoring, tube bending, chucking; melting point 218-440°F.

BELMONT NO. 2491
Belmont Metals Inc.
Sn 12, Pb 18, In 21, bal Bi.
6300 psi TS. Low melting alloy for soldering, lens grinding, anchoring; melting point 136°F.

BELMONT NO. 2505
Belmont Metals Inc.
Sn 13.3, Pb 26.7, Cd 10, bal Pb.
5990 psi TS. Low melting alloy for tube bending, radiation shielding, anchoring, chucking, molds; melting point 158°F.

BELMONT NO. 2562
Belmont Metals Inc.
Sn 44.5, bal Bi.
6400 psi TS. For anchoring, tube bending, sprinklers, soldering, chucking; melting point 255°F.

BELMONT NO. 2581
Belmont Metals Inc.
Sn 42, bal Bi.
8000 psi TS. For anchoring, chucking, soldering, molds; melting point 281°F.

BELMONT NO. 4
Belmont Metals Inc.
Sn 1, Sb 10, bal Pb.
8500 psi TS. Babbitt for low speed bearings; melting point 498°F.

BELMONT NO. 40
Belmont Metals Inc.
Sn 40, Zn 60.
For aluminum solder; melting point 700°F.

BELMONT NO. 60
Belmont Metals Inc.
Sn 60, Zn 40.
For aluminum solder; melting point 650°F.

BELMONT NO. 70
Belmont Metals Inc.
Sn 70, Zn 30.
For soldering aluminum; melting point 620°F.

BELT LACE
American manufacture
Cu 62, Zn 38.
For belt lacing, light alloy parts; high impact strength.

BEMAL-1
Yorkshire Imperial Metals Ltd.
Cu 70.11, Pb 0.32, Zn 29.43.
For condenser tubes. *Obsolete*

BEMAL-2
Yorkshire Imperial Metals Ltd.
Cu 70, Zn 29, P 0.45, Fe 0.15.
For condenser tubes. *Obsolete*

BEMIT
German manufacture
Al 96.91, Cu 2, Mn 0.45, W 0.27.
For light alloy welded, drawn or stamped parts; also called "Benit"; German alloy.

BEN2
Le Bronze Industriel
Sn 11, Ni 2, bal Cu.
Bronze. As cast: 80 Brin.

BEN5
Le Bronze Industriel
Sn 11, Ni 5, bal Cu.
Nickel bronze. As cast: 80 Brin. ASTM B584.

BENDALLOY
Cerro Metal Products Co.
Now CERROBEND.

BENDIX A2 FLAT GROUND STOCK
Bendix Steel Co.
C 1, Mn 0.5, Cr 5, V 0.3, Mo 1.25, bal Fe.
Air hardening tool steel. AISI A2.

BENDIX COMMERCIAL DRILL ROD
Bendix Steel Co.
C 1, Mn 0.3, Si 0.2, bal Fe.
Water hardening tool steel. AISI W1.

BENDIX D2
Bendix Steel Co.
C 1.5, Mn 0.4, Si 0.3, Cr 12, V 0.9, Mo 0.8, bal Fe.
High carbon, high chromium tool steel. AISI D2.

BENDIX D2 FLAT GROUND STOCK
Bendix Steel Co.
C 1.55, Mn 0.25, Si 0.38, Cr 12, V 0.8, Mo 0.8, bal Fe.
High carbon, high chromium tool steel. AISI D2.

BENDIX O1
Bendix Steel Co.
C 0.9, Mn 1.2, Si 0.35, Cr 0.5, V 0.2, W 0.5, bal Fe.
Oil hardening tool steel. AISI O1.

BENDIX O1 DRILL ROD
Bendix Steel Co.
C 0.9, Mn 1.2, Cr 0.5, V 0.2, W 0.5, bal Fe.
Oil hardening tool steel. AISI O1.

BENDIX O1 FLAT GROUND STOCK
Bendix Steel Co.
C 0.9, Mn 1.2, Si 0.35, Cr 0.5, V 0.2, W 0.5, bal Fe.
Oil hardening tool steel. AISI O1.

BENECKE
Alexander Benecke, Inc.
C 0.7-1, alloy, bal Fe.
For tools, dies; oil hardening.

BENEDICT METAL
Riverside Metals Corp.
Cu 75-80, Ni 20-25, 0.75% max Fe + Mn.
Hard: 88,300 TS; 83,000 YS; 6 El; 35 RA; 148 Brin. Annealed: 50,000 TS; 30,000 YS; 46 El; 63 RA; 76 Brin. For condensers, feed water heaters, coinage; corrosion resistant. *Obsolete*

BENEDICT NICKEL
Anaconda Co.
Cu 79, Ni 20, Fe 0.36.
For sheets, tubes, extruded parts; corrosion resistant. *Obsolete*

BENEDICT NICKEL-812
Anaconda Co.
Cu 55, Ni 12.5, Zn 20, Sn 2, Pb 10.5.
Cast: 30,000 TS; 20,000 YS; 15 El. For hardware, plumbers fixtures. *Obsolete*

BENEDICT PLATE
English manufacture
Cu 57, Zn 28, Ni 15.
For white metal for flat work; corrosion resistant.

BENSON
Hewitt Metals Corp.
Sn 10, Sb 15, bal Pb.
For bearings and bushings. Babbitt. SAE 14.

BENUM
George Cook & Co., Ltd.
C 0.3, Mn 0.5, Cr 1.3, Si 0.3, Mo 0.3, Ni 4.2, bal Fe.
For molds, dies; air hardened, tough.

BENZ AVIATEK
English manufacture
Al 80, Cu 6, Zn 12, Fe 1.5.
For pistons; non-hardenable.

BERA AUTO A
Paul Bergsoe & Son
Sn 88.7, Sb 7.8, Cu 3.5.
Cast: 12,800 TS; 26 Brin. MP: 440-620°F. For engine bearings. Shock resistant.

BERA AUTO SPECIAL
Paul Bergsoe & Son
Sn 90.3, Sb 6.5, Cu 3, Ni 0.2.
Cast: 12,800 TS; 25 Brin. MP: 440-610°F. For engine bearings. Shock resistant.

BERA COMMON A
Paul Bergsoe & Son
Sn 55, Sb 10, Cu 2.5, Pb 32.5.
Cast: 11,400 TS; 22 Brin. MP: 355-620°F. For refrigerator and generator bearings. Good castability.

BERA COMMON B
Paul Bergsoe & Son
Sn 24.5, Sb 13, Cu 0.5, Pb 62.
Cast: 12,800 TS; 26 Brin. MP: 355-535°F. For refrigerator and generator bearings. Good castability.

BERA DIESEL
Paul Bergsoe & Son
Sn 83.5, Sb 8, Cu 6.5, Pb 2.
Cast: 17,100 TS; 31 Brin. MP: 355-710°F. For diesel engine bearings. Wear resistant.

BERA MARINE
Paul Bergsoe & Son
Sn 85, Sb 10, Cu 5.
Cast: 19,900 TS; 28 Brin. MP: 440-630°F. For engine bearings. Shock resistant.

BERA STEAM A
Paul Bergsoe & Son
Sn 80, Sb 11.5, Cu 5.5, Pb 3.
Cast: 17,800 TS; 34 Brin. MP: 355-675°F. For steam turbine and generator bearings. Wear resistant.

BERA STEAM B
Paul Bergsoe & Son
Sn 74, Sb 9, Cu 4, Pb 13.
Cast: 14,200 TS; 28 Brin. MP: 355-625°F. For steam turbine
bearings. Wear resistant.

BERACO
Paul Bergsoe & Son
Sn 10, Sb 13.5, Cu 0.5, Pb 76.
Cast: 12,900 TS; 30 Brin. MP: 470-705°F. For transmission
bearings.

BERALITE 35
Cooper Metallurgical Corp.
Be 35, bal Al.
Drawn: 62,800 TS; 58,400 YS; 1.5 El; 94 Brin. Annealed:
41,400 TS; 23,100 YS; 14.5 El; 68 Brin. For instruments;
corrosion resistant.

BERALOY 1
Wilbur B. Driver Co.
Co 1.8-2.4, Be 0.3-0.5, bal Cu.
Rolled: 110,000 TS; 95,000 YS; 10 El; 200 Brin. For springs;
50% electrical conductivity. *Obsolete*

BERALOY A
Gilby-Fodor S.A.
Be 2, Co 0.25, bal Cu.
Heat treated: 180,000 TS; 95,000 YS; 1 El. For springs.
Corrosion resistant; hardenable.

BERALOY B
Wilbur B. Driver Co.
Be 1.9-2.2, Ni 0.25-0.5, bal Cu.
Annealed: 66,000 TS; 26,000 YS; 60 El. Heat treated: 200,000
TS; 179,000 YS; 2 El. For diaphragms, springs; age-
hardenable. *Obsolete*

BERALOY C
Wilbur B. Driver Co.
Be 0.4-0.5, Co 2.5-2.7, bal Cu.
Annealed: 30,000 TS. Heat treated: 110,000 TS; 10 El. For
springs; age-hardenable. *Obsolete*

BERALOY D
Wilbur B. Driver Co.
Be 1.6-1.8, Co 0.25-0.5, bal Cu.
For springs, diaphragms; age hardened, corrosion resistant.
Obsolete

BERALOY NO. 25
Wilbur B. Driver Co.
Cu 97.75, Be 2, Co 0.25.
Annealed: 66,000 TS. Heat treated: 175,000 TS. For springs,
electric contacts, diaphragm valves, bearings; age-
hardenable. *Obsolete*

BERDO ALLOY NO. 1
Barronia Metals, Ltd.
Cu, Sn, Pb, Ni.
30,000 TS; 20,000 YS; 1.5 El; 2.6 RA; 99 Brin. For bearings;
heavy service; CYP-38,000.

BERDO ALLOY NO. 2
Barronia Metals, Ltd.
Cu, Sn, Pb, Ni.
75 Brin. For bearings; general service; CYP-28,000.

BERDO ALLOY NO. 3
Barronia Metals, Ltd.
Cu, Sn, Pb, Ni.
44 Brin. For bearings; light service; CYP-14,000.

BERDO NO. 6
Barronia Metals, Ltd.
Cu-Sn-Ni-Pb-P.
For mill bearings, rock drill parts, gears; wear resistant.

BERDO NO. 7
Barronia Metals, Ltd.
Cu-Sn-Ni-Pb-P.
For mill bearings, rock drill parts, gears; wear resistant.

BERGAL
Bergmann Elektrizitawerke
Al 94.1, Cu 4, Mg 0.8, Mn 0.7, Si 0.4.
For light alloy parts; age hardenable.

BERGISCHE 115 W 2 C
Bergische Stahl Industrie
C 1.05-1.15, Si 0.15-0.3, Mn 0.3-0.4, Cr 1.1-1.3, W 1.8-2, V
0.15-0.25, bal Fe.
Cold work tool steel. W.-Nr. 1.2521.

BERGISCHE 134 MO
Bergische Stahl Industrie
C 0-0.07, Si 0-1, Mn 0-1.5, Cr 12-13.5, Ni 3.5-5.5, Mo 0-0.7,
bal Fe.
Stainless steel casting. W.-Nr. 1.4313.

BERGISCHE 151 M
Bergische Stahl Industrie
C 1.05-1.15, Si 0-1, Mn 0-1, Cr 14-16, Mo 0.4-0.6, V 0.1-0.15,
bal Fe.
Martensitic stainless steel. W.-Nr. 1.4111.

BERGISCHE 165 CC
Bergische Stahl Industrie
C 1.55-1.75, Si 0.25-0.4, Mn 0.2-0.4, Co 1.2-1.4, Mo 0.5-0.6,
bal Fe.
Cold work tool steel. W.-Nr. 1.2880.

BERGISCHE 165 CCM
Bergische Stahl Industrie
C 1.65, Si 0.4, Mn 0.35, Co 1.5, Cr 13.5, Mo 1.2, bal Fe.
Cold work tool steel. W.-Nr. 1.2885.

BERGISCHE 188 E
Bergische Stahl Industrie
C 0-0.07, Si 0-1, Mn 0-2, Cr 16.5-18.5, N 10.5-13.5, Mo 2-2.5,
bal Fe.
Austenitic stainless steel. W.-Nr. 1.4401. Similar to AISI 316.

BERGISCHE 188 ES
Bergische Stahl Industrie
C 0-0.1, Si 0-1, Mn 0-2, Cr 16.5-18.5, N 10.5-13.5, Mo 2-2.5,
Nb .= 8 x C min, bal Fe.
Austenitic stainless steel. W.-Nr. 1.4580.

BERGISCHE 188 EST
Bergische Stahl Industrie
C 0-0.1, Si 0-1, Mn 0-2, Cr 16.5-18.5, Ni 10.5-13.5, Mo 2-2.5,
Ti = 5 x C min, bal Fe.
Austenitic stainless steel. W.-Nr. 1.4571.

BERGISCHE 189 M
Bergische Stahl Industrie
C 0.85-0.95, Si 0-1, Mn 0-1, Cr 17-19, Mo 1-1.3, V 0.07-0.12,
bal Fe.
Martensitic stainless steel. W.-Nr. 1.4112. Similar to AISI
440C.

BERGISCHE 20 VM
Bergische Stahl Industrie
C 0.2, Mn 1.3, bal Fe.
Low alloy cast iron. SEW 1.1133; GS-20 Mn 5.

BERGISCHE 25 VM
Bergische Stahl Industrie
C 0.25, Mn 1, bal Fe.
Low alloy cast iron. SEW 1.1136; GS-24 Mn 4.

BERGISCHE 30 VM
Bergische Stahl Industrie
C 0.3, Mn 1.3, bal Fe.
Low alloy cast iron. SEW 1.1165; GS-30 Mn 5.

BERGISCHE 35 SWC
Bergische Stahl Industrie
C 0.3-0.4, Si 0.8-1.1, Mn 0.2-0.4, Cr 0.9-1.2, W 1.8-2.1, V
0.15-0.2, bal Fe.
Hot work tool steel. W.-Nr. 1.2541.

BERGISCHE 40 VMS
Bergische Stahl Industrie
C 0.38, Mn 1, Si 0.5, bal Fe.
Low alloy steel casting. GS 38 Mn Si 4.

BERGISCHE 45 SWC
Bergische Stahl Industrie
C 0.4-0.5, Si 0.8-1.1, Mn 0.2-0.4, Cr 0.9-1.2, W 1.8-2.1, V
0.15-0.2, bal Fe.
Hot work tool steel. W.-Nr. 1.2542.

BERGISCHE 45 VM
Bergische Stahl Industrie
C 0.45, Mn 1, bal Fe.
Low alloy steel casting. GS-46 Mn 4.

BERGISCHE 50 CMV
Bergische Stahl Industrie
C 0.47-0.55, Si 0.15-0.35, Mn 0.8-1.1, Cr 0.9-1.2, V 0.07-0.12,
bal Fe.
Cold work tool steel. W.-Nr. 1.2241.

BERGISCHE 60 SWC
Bergische Stahl Industrie
C 0.55-0.65, Si 0.5-0.7, Mn 0.2-0.4, Cr 0.9-1.2, W 1.8-2.1, V
0.15-0.2, bal Fe.
Hot work tool steel. W.-Nr. 1.2550.

BERGISCHE 70 WM
Bergische Stahl Industrie
C 0.68-0.78, Si 0.2-0.4, Mn 0.4-0.6, Cr 0.4-0.6, Mo 0.25-0.4, W
0.4-0.7, V 0.15-0.3, bal Fe.
Cold work tool steel. W.-Nr. 1.2604.

BERGISCHE 702 CN
Bergische Stahl Industrie
C 0.45-0.55, Si 0.15-0.35, Mn 0.4-0.6, Cr 0.9-1.2, Ni 3-3.5, bal
Fe.
Cold work tool steel. W.-Nr. 1.2721.

BERGISCHE 702 W
Bergische Stahl Industrie
C 0.4-0.5, Si 0.15-0.3, Mn 0.3-0.5, Cr 1.2-1.5, Mo 0.15-0.35, Ni
3.8-4.3, 0.50 W (optional), bal Fe.
Cold work tool steel. W.-Nr. 1.2767.

BERGISCHE 719 CNM
Bergische Stahl Industrie
C 0.16-0.22, Si 0.15-0.3, Mn 0.3-0.5, Cr 1.1-1.4, Mo 0.15-0.25,
Ni 3.8-4.3, 0.50 W (optional), bal Fe.
Cold work tool steel. W.-Nr. 1.2764.

BERGISCHE 85 WCV
Bergische Stahl Industrie
C 0.75-0.85, Si 0.15-0.4, Mn 0.3-0.5, Cr 0.4-0.6, W 0.6-0.8, bal
Fe.
Cold work tool steel. W.-Nr. 1.2511.

BERGISCHE-CMVW11
Bergische Stahl Industrie
C 0.2, Cr 12.5, Mo 1.1, Ni 0.6, V 0.3, W 0.5, bal Fe.

BERGISCHE-CMW11
Bergische Stahl Industrie
C 0.2, Cr 12.5, Mo 1.1, Ni 0.6, V 0.3, bal Fe.

BERGIT 1
Bergische Stahl Industrie
C 0.1, Ni, Cr, Mo, Cu, bal Fe.

BERGIT A
Bergische Stahl Industrie
C 0.1, Ni, Mo, bal Fe.

BERGSOE F

Paul Bergsoe & Son
Sn 88.7, Sb 7.8, Cu 3.5.
Cast: 17,000 TS; 25 Brin. MP: 440-620°F. Shock resistant. For engine bearings.

BERGSOE G

Paul Bergsoe & Son
Sn 88, Sb 8, Cu 4.
Cast: 18,500 TS; 27 Brin. MP: 440-625°F. For engine bearings. Shock resistant.

BERGSOE M

Paul Bergsoe & Son
Sn 6, Sb 16, Pb 79.
Cast: 29 Brin. MP: 470-520°F. For transmission bearings.

BERGSOE P

Paul Bergsoe & Son
Sb 16, Pb 84.
Cast: 19 Brin. MP: 475-520°F. For transmission bearings.

BERGSOE WM

Paul Bergsoe & Son
Sn 5, Sb 12, Pb 83.
Cast: 25 Brin. MP: 470-490°F. For transmission bearings.

BERGSOL MN. CU

Paul Bergsoe & Son
Sn 8, Sb 14, Cu 0.5, Pb 77.5.
Cast: 29 Brin. MP: 470-715°F. For transmission bearings.

BERGSTAHL AM16-13N

Bergische Stahl Industrie
C 0-0.05, Cr 18, Ni 8, Cb 0.7, bal Fe.
Annealed: 90,000 TS; 45,000 YS; 56 El; 65 RA; 160 Brin. For chemical plant equipment; stainless, austenitic; Type 347. *Obsolete*

BERGSTAHL AM17

Bergische Stahl Industrie
C 0.4, Mn, bal Fe.
For gears, bolts, shafts, machine tool parts; water hardened. *Obsolete*

BERGSTAHL AM17-13E

Bergische Stahl Industrie
C 0-0.1, Cr 17, Ni 13, Mo 2, bal Fe.
Annealed: 85,000 TS; 35,000 YS; 50 El; 65 RA; 160 Brin. Cold drawn: 150,000 TS; 135,000 YS; 6 El; 300 Brin. For acid resistant chemical plant equipment; Type 316; stainless, austenitic. *Obsolete*

BERGSTAHL AM10-0

Bergische Stahl Industrie
C 0-0.18, Cr 18, Ni 8, bal Fe.
Annealed: 80,000 TS; 35,000 YS; 55 El; 75 RA; 150 Brin. Cold drawn: 180,000 TS; 150,000 YS; 10 El; 250 Brin. For chemical plant equipment, tanks, mixers, filters; Type 302; stainless, austenitic. *Obsolete*

BERGSTAHL AM858

Bergische Stahl Industrie
C 0-0.1, Cr, Ni, Mn, bal Fe.
Annealed: 80,000 TS; 35,000 YS; 55 El; 75 RA; 150 Brin. For chemical plant equipment; stainless. *Obsolete*

BERGSTAHL AMG10

Bergische Stahl Industrie
C 0.3, Ni, Mn, Si, bal Fe.
For machine tool parts; oil hardened. *Obsolete*

BERGSTAHL C1H

Bergische Stahl Industrie
C 0.8, Cr, bal Fe.
For bearings, liners, sleeves, bushings; water hardened. *Obsolete*

BERGSTAHL C2H

Bergische Stahl Industrie
C 0.65, Cr, Ni, bal Fe.
For springs, bolts, axes, hammers; oil hardened, tough. *Obsolete*

BERGSTAHL CM1

Bergische Stahl Industrie
C 0.25, Cr 1.2, Mo 0.4, bal Fe.
For gears, bolts, shafts, fasteners; oil hardened, tough. *Obsolete*

BERGSTAHL CM1B

Bergische Stahl Industrie
C 0.35, Cr 1.2, Mo 0.25, bal Fe.
For gears, bolts, shafts, machine tool parts; oil hardened, shock resistant. *Obsolete*

BERGSTAHL CM1H

Bergische Stahl Industrie
C 0.45, Cr 1.3, Mo 0.4, bal Fe.
For gears, bolts, machine tool parts; oil hardened, shock resistant. *Obsolete*

BERGSTAHL CM3

Bergische Stahl Industrie
C 0.15, Cr, Mo, bal Fe.
For gears, bolts, machine tool parts; case hardening steel. *Obsolete*

BERGSTAHL CMVW2

Bergische Stahl Industrie
C 0.22, Cr 0.75, Mo 0.25, V 0.3, bal Fe.
For gears, bolts, machine tool parts; case hardening steel. *Obsolete*

BERGSTAHL CMW2

Bergische Stahl Industrie
C 0.22, Cr 1.2, Mo 0.45, bal Fe.
For machine tool parts, gears, fasteners; case hardening steel. *Obsolete*

BERGSTAHL CMW3

Bergische Stahl Industrie
C 0.1, Cr, Mo, bal Fe.
For machine tool parts, gears, fasteners; case hardening steel. *Obsolete*

BERGSTAHL CMW4

Bergische Stahl Industrie
C 0.22, Cr, Mo, V, bal Fe.
For machine tool parts, gears, fasteners; case hardening steel. *Obsolete*

BERGSTAHL CN10

Bergische Stahl Industrie
C 0.4, Cr 22, Ni 9.5, bal Fe.
Cast: 85,000 TS; 45,000 YS; 35 El; 165 Brin. For heat treat boxes, burner tips, conveyors, chains; Type HF; corrosion and heat resistant. *Obsolete*

BERGSTAHL CN15

Bergische Stahl Industrie
C 0.4, Cr 28, Ni 14, bal Fe.
Cast: 75,000 TS; 47,000 YS; 17 El; 25 RA; 200 Brin. For heat treat boxes, furnace parts, conveyor belts, hearth plates; Type HH; corrosion and heat resistant. *Obsolete*

BERGSTAHL CN20

Bergische Stahl Industrie
C 0.4, Cr 25, Ni 19, bal Fe.
Cast: 75,000 TS; 50,000 YS; 17 El; 170 Brin. For furnace parts, retorts, stack dampers; Type HK; corrosion and heat resistant. *Obsolete*

BERGSTAHL CN35

Bergische Stahl Industrie
C 0.5, Cr 25, Ni 30, bal Fe.
For retorts, heat treat boxes, dampers; corrosion and heat resistant. *Obsolete*

BERGSTAHL CN38

Bergische Stahl Industrie
C 0.25, Cr 17, Ni 38, bal Fe.
Cast: 70,000 TS; 40,000 YS; 10 El; 12 RA; 170 Brin. For heat treat boxes, salt pots, furnace parts; Type HT; corrosion and heat resistant. *Obsolete*

BERGSTAHL CN5

Bergische Stahl Industrie
C 0.4, Si 1.3, Cr 27, Ni 4, bal Fe.
Cast: 90,000 TS; 65,000 YS; 2 El; 212 Brin. For cylinder liners, bushings, valve seats; Type CC-50; heat resistant. *Obsolete*

BERGSTAHL CN7

Bergische Stahl Industrie
C 0.3, Si 1.3, Cr 22, Ni 7, bal Fe.
For furnace parts, heat treat boxes, conveyors; heat and corrosion resistant. *Obsolete*

BERGSTAHL CN8

Bergische Stahl Industrie
C 0.4, Cr 22, Ni 9.5, bal Fe.
Cast: 85,000 TS; 45,000 YS; 35 El; 165 Brin. For heat treat boxes, burner tips, conveyors, chains; Type HF; corrosion and heat resistant. *Obsolete*

BERGSTAHL CV1

Bergische Stahl Industrie
C 0.24, Cr 1, V 0.2, Mn 0.6, bal Fe.
For gears, bolts, machine tool parts; oil hardened, shock resistant. *Obsolete*

BERGSTAHL CV1H

Bergische Stahl Industrie
C 0.45, Cr 1, V 0.2, Mn 0.6, bal Fe.
For gears, bolts, springs, crankshafts; oil hardened, shock resistant. *Obsolete*

BERGSTAHL CV2

Bergische Stahl Industrie
C 0.35, Cr 1.2, V 0.2, Mn 0.6, bal Fe.
For gears, crankshafts, axles, bolts; oil hardened, shock resistant. *Obsolete*

BERGSTAHL E12

Bergische Stahl Industrie
C 1.25, Cr 1, bal Fe.
For bearings, cutters, bushings, liners; water hardened, wear resistant. *Obsolete*

BERGSTAHL E16

Bergische Stahl Industrie
C 0.8, Cr, bal Fe.
For bearings, liners, bushings; water hardened, wear resistant. *Obsolete*

BERGSTAHL E2

Bergische Stahl Industrie
C 2, Cr 11.5, bal Fe.
For blanking and forming dies, punches; oil hardened, nondeforming. *Obsolete*

BERGSTAHL E28

Bergische Stahl Industrie
C 2.7, Cr, bal Fe.
For liners, nozzles; abrasion resistant. *Obsolete*

BERGSTAHL MNA1

Bergische Stahl Industrie
C 1.2, Mn, bal Fe.
For dies, cutters, tools; oil hardened. *Obsolete*

BERGSTAHL MNA2

Bergische Stahl Industrie
C 1.25, Mn, bal Fe.
For dies, cutters, tools; oil hardened. *Obsolete*

BERGSTAHL MVW2

Bergische Stahl Industrie
C 0.2, Mo, V, bal Fe.
For gears, bolts, fasteners, shafts; case hardened. *Obsolete*

BERGSTAHL MVW4
Bergische Stahl Industrie
C 0.2, Mo, V, bal Fe.
For gears, bolts, fasteners, shafts; case hardened. *Obsolete*

BERGSTAHL MW2
Bergische Stahl Industrie
C 0.22, Mn 0.65, Mo 0.4, Cr 0-0.3, bal Fe.
For gears, bolts, machine tool parts; water hardened, tough. *Obsolete*

BERGSTAHL P6M
Bergische Stahl Industrie
C 0.12, Cr, Mo, bal Fe.
For gears, bolts, camshafts, cams; case hardened. *Obsolete*

BERGSTAHL P7M
Bergische Stahl Industrie
C 0.12, Cr, Mo, bal Fe.
For gears, bolts, camshafts, cams; case hardened. *Obsolete*

BERGSTAHL SP120
Bergische Stahl Industrie
C 1.2, Si 1.3, Cr 29, bal Fe.
For wear plates, crushers, rolls; abrasion and heat resistant. *Obsolete*

BERGSTAHL VCN170
Bergische Stahl Industrie
C 0.4, Ni, Cr, bal Fe.
For gears, bolts, crankshafts; oil hardened, tough. *Obsolete*

BERGSTAHL VMS
Bergische Stahl Industrie
C 0.45, Mn, Si, bal Fe.
Hot rolled: 98,000 TS; 59,000 YS; 24 El; 54 RA; 212 Brin. For gears, axles, bolts, shafts; water hardened. *Obsolete*

BERGSTROM TYPE S
Bergstrom Alloys Corp.
C, Mn, Ni, bal Fe.
300 Brin. For hard surfacing electrodes; work hardening rod.

BERKSHIRE
Carpenter Technology Corp.
C 1.2, W 1.34, bal Fe.
For taps, cutters, engravers' tools. Fast finishing steel, water hardened.

BERKSHIRE, HIGH MANGANESE
Carpenter Technology Corp.
C 1.2, Mn 1, Si 0.4, Cr 0.5, V 0.2, W 1.4, bal Fe.
Water or oil hardening high carbon tool steel for shafts, arbors, drill bushings, lathe centers.

BERLIN NO. 1
German manufacture
Cu 56, Zn 29, Ni 16.
For electrical resistances, ornaments; corrosion resistant.

BERLIN NO. 2
German manufacture
Cu 48, Zn 24, Ni 24, Fe 3.6.
For ornamental parts; spinning, drawing, stamping.

BERMAX
Federal-Mogul Corp.
Sb 9-11, Cu 0-0.5, bal Pb.
Cast: 11,000 TS; 4.5 El; 20 Brin. For bearings; anti-friction, nonmagnetic.

BERRY METAL
Berry Metal Co.
Pb 10, Ni 8, Sb 2, bal Cu.
Cast: 20,000 TS; 13,000 YS; 6 El. For bearings; wear resistant.

BERRYDUR
Vacuumschmelze GmbH
Cr 15, Ni 60, Fe 16, Mn 2, Mo 7, Be 0.5.
For heat and corrosion resisting parts; heat and corrosion resistant parts; heat and corrosion resistant. *Obsolete*

BERRYDUR CONTRACID
Vacuumschmelze GmbH
Be, bal Cu.
600 Brin. For springs. *Obsolete*

BERRYDUR-CU
Vacuumschmelze GmbH
Be 2.5, bal Cu.
360 Brin. For springs; nonmagnetic. *Obsolete*

BERSCH METAL
German manufacture
Al 93, Ni 7.
For bearings.

BERTHIER'S ALLOY-1
English manufacture
Cu 68, Ni 32.
For evaporators, stills; corrosion resistant.

BERTHIER'S ALLOY-2
English manufacture
Cu 72, Zn 25, Pb 2, Sn 1.2.
For screw machine products; free-cutting.

BERUDA ALLOY
English manufacture
Cu-Zn.
For strong corrosion resistant parts; high tensile brass.

BERYL-TRODE
Ampco Pittsburgh Corp.
Be, bal Cu.
For Be-Cu electrode for welding Be-Cu alloys; coated. *Obsolete*

BERYLCO 10
NGK Metals Corp.
Copper. Be 0.4, Co 2.6, bal Cu.
Annealed: 50,000 TS; 25,000 YS; 30 El; 50 RA; 70 Brin. Heat treated: 125,000 TS; 105,000 YS; 8 El; 10 RA; 240 Brin. For springs, resistance welding electrodes; age-hardenable. C17500

BERYLCO 10 C
NGK Metals Corp.
Be 0.6, Co 2.6, bal Cu.
Cast: 60,000 TS; 30,000 YS; 20 El; 35 RA; 90 Brin. Annealed: 40,000 TS; 20,000 YS; 30 El; 50 RA; 70 Brin. Hardened: 90,000 TS; 75,000 YS; 10 El; 15 RA; 185 Brin. For resistance welding parts, switches, contacts; age-hardenable, corrosion resistant.

BERYLCO 14
NGK Metals Corp.
Copper. Be 0.2-0.6, Ni 1.4-2.2, bal Cu.
After age hardening: 55,000 max TS; 45,000 max YS; 20-35 El. C17510

BERYLCO 165
NGK Metals Corp.
Be 1.6-1.8, Co 0.2-0.35, bal Cu.
Annealed: 70,000 TS; 30,000 YS; 40 El; 45 RA; 95 Brin. Hardened: 185,000 TS; 155,000 YS; 2 El; 3 RA; 355 Brin. For springs, terminal plugs, jewelry, clips; heat treatable,/corrosion resistant.

BERYLCO 185
NGK Metals Corp.
Be 1.5-2, Fe 0-0.25, 0.5 max Ni or Co, bal Cu.
Annealed: 68,000 TS; 49 El. Heat treated: 195,000 TS; 3 El. For springs; age-hardenable.

BERYLCO 200C
NGK Metals Corp.
Be 2.05-2.25, Co 0.35-0.65, bal Ni.
For bearings, gears, cams, plungers, drawing dies; age-hardenable, corrosion and wear resistant.

BERYLCO 20C
NGK Metals Corp.
Be 2-2.25, Co 0.35-0.6, bal Cu.
Heat treated: 110,000-160,000 TS; 95,000-140,000 YS; 8-2 El; 10-3 RA; 250-440 Brin. For plastic molds, safety tools, cams, pump parts; wear and impact resistant, age-hardenable.

BERYLCO 20CR
NGK Metals Corp.
Be 2-2.25, Co 0.35-0.65, bal Cu.
Cast: 70,000 TS; 40,000 YS; 15 El; 138 Brin. Heat treated: 155,000 TS; 115,000 YS; 0 El; 380 Brin. For plastic molds, safety tools, gears, bushings, bearings; age hardenable.

BERYLCO 225C
Kawecki Berylco Industries Inc.
Be, bal Cu.
Heat treated: 110,000-165,000 TS; 90,000-140,000 YS; 1-3 El; 255-350 Brin. For safety tools with cutting edge, cams, pump parts; corrosion resistant, age-hardenable. *Obsolete*

BERYLCO 25
NGK Metals Corp.
Copper. Be 2, Co 0.3, bal Cu.
Heat treated: 200,000 TS; 170,000 YS; 2 El; 3 RA; 400 Brin. For springs, diaphragms; age-hardenable. C17200

BERYLCO 250-C
NGK Metals Corp.
Be 2.6-2.8, Co 0.35-0.65, bal Cu.
For bearings, gears, cams, valves, drawing dies; age-hardenable, corrosion and wear resistant.

BERYLCO 25S
NGK Metals Corp.
Be 2, Co 0.3, bal Cu.
Annealed: 165,000 TS; 135,000 YS; 5 El; 8 RA; 350 Brin. 4 No. Hard: 190,000 TS; 160,000 YS; 2 El; 3 RA; 400 Brin. For springs, diaphragms, clips, bellows; corrosion and heat resistant.

BERYLCO 275CR
NGK Metals Corp.
Be 2.6-2.8, Co 0.35-0.65, bal Cu.
Heat treated: 125,000-175,000 TS; 110,000-145,000 YS; 5-2 El; 8-3 RA; 270-415 Brin. For plastic mold dies, safety tools; age-hardenable, corrosion resistant.

BERYLCO 40
NGK Metals Corp.
Be 2, Co 0.25, bal Cu.
Rolled: 98,000 TS; 2 El. Heat treated: 180,000 TS; 95,000 YS; 1 El. For springs; age-hardenable.

BERYLCO 50
NGK Metals Corp.
Be 0.3-0.55, Co 1.4-1.7, Ag 1, bal Cu.
Heat treated: 110,000-95,000 TS; 90,000-65,000 YS; 5-10 El; 5-10 RA 195-240 Brin. For springs, resistance welding electrodes and dies; age-hardenable.

BERYLCO 70
NGK Metals Corp.
Be 0.1, Cr 0.5, bal Cu.
Heat treated: 80,000 TS; 60,000 YS; 15 El; 25 RA; 150 Brin. For springs, current carrying parts; age-hardenable.

BERYLCO ALLOYS
Haynes International, Inc.
Copper.
High performance spring applications; 220 ksi TS max.

BERYLCO CR-1

NGK Metals Corp.
Be 2.75, C 0-0.5, bal Ni.
Heat treated: 125,000-195,000 TS; 200,000-180,000 YS; 2-0 El; 300-520 Brin. Cast: 115,000 TS; 60,000 YS; 5 El; 240 Brin. For aircraft fuel pumps, impellers, core drill bits; age hardened, corrosion resistant castings.

BERYLCO CR-2

NGK Metals Corp.
Be 2.75, C 0.75-1.1, bal Ni.
For aircraft fuel pumps, impellers, core drill bits; age hardened, corrosion and wear resistant.

BERYLCO HPA

NGK Metals Corp.
Be 98, Fe 0.18, Al 0.15, Mg 0.08, Mn 0.02, Cu 0.02, Si 0.1, Zn 0.05, Ni 0.06, 1.80 BeO.
Vacuum hot pressed: 40,000 TS; 30,000 YS; 1 El; 80 Rock B. Hot Extruded and Annealed: 70,000 TS; 45,000 YS; 10 El; 90 Rock B. For nuclear space applications.

BERYLCO NICKEL 440

NGK Metals Corp.
Be 1.95, Ti 0.5, bal Ni.
Annealed: 105,000 TS; 45,000 YS; 40 El; 70 Rock B. Hard Drawn: 170,000 TS; 165,000 YS; 2 El; 35 Rock C. HT-Temper: 270,000 TS; 230,000 YS; 8 El; 51 Rock C. For heat resistant springs, and switches, diaphragms, bellows, retainer clips, electrical shunts, contact springs. High strength and corrosion resistant. Age-hardenable.

BERYLCO-717

NGK Metals Corp.
Be 0.5, Fe 0.7, Ni 30, bal Cu.
Bar: 110,000-140,000 TS; 10-15 El. Age-hardened: 145,000 min TS; 125,000 min YS; 10 min El; 29 Rock C. For marine hardware, piping, high velocity heat exchangers, hydrophone parts. High sea water corrosion resistant. Age-hardenable.

BERYLCO-717 C

NGK Metals Corp.
Be 0.5, Fe 1, Mn 0.95, C 0.06, Ni 29, bal Cu.
Cast: 76,000 TS; 44,700 YS; 14 El; 15.3 RA; 82 Rock B. Heat Treated: 121,500 TS; 84,400 YS; 14.5 El; 20 Rock C. For marine hardware, heat exchangers. Age hardenable. High sea water corrosion resistant.

BERYLDUR

NGK Metals Corp.
Be 0.8-1.2, Ni 0.2-0, 4.0 others, bal Cu.
Annealed: 55,000-65,000 TS; 18,000-30,000 YS; 40-60 El. Aged: 125,000-140,000 TS; 105,000-125,000 YS; 12-20 El; 230-300 Brin. For springs, clips, connectors; age-hardenable, high strength, corrosion resistant.

BERYLLIUM

Atomergic Chemetals Corp.
Be.
Purities: distilled 99.99+%, grade AA 99.96+%, grade A 99.87%, nuclear grade 99.5+%. Forms: flake, powder, plate, sheet, foil, wire, rod, single crystals.

BERYLLIUM BRONZE 1.0% BE

Vacuumschmelze
Be 1, bal Cu.
Untempered: 67,000 TS; 62,000 YS; 7.5 El. For corrosion resistant parts; corrosion resistant.

BERYLLIUM BRONZE 1.0% BE

Siemans & Halske AG.
Be 1, bal Cu.
Untempered: 67,000 TS; 62,000 YS; 7.5 El. For corrosion resistant parts; corrosion resistant.

BERYLLIUM BRONZE 1.5% BE

Vacuumschmelze
Be 1.5, bal Cu.
Untempered: 68,000 TS; 45 El. Tempered: 94,000 TS; 8 El. For corrosion resistant parts; corrosion resistant.

BERYLLIUM BRONZE 1.5% BE

Siemans & Halske AG.
Be 1.5, bal Cu.
Untempered: 68,000 TS; 45 El. Tempered: 94,000 TS; 8 El. For corrosion resistant parts; corrosion resistant.

BERYLLIUM BRONZE 2.0% BE

Vacuumschmelze
Be 2, bal Cu.
Untempered: 72,000 TS; 15,500 YS; 40 El. Tempered: 121,000 TS; 100,000 YS; 1 El. For corrosion resistant parts; corrosion resistant.

BERYLLIUM BRONZE 2.0% BE

Siemans & Halske AG.
Be 2, bal Cu.
Untempered: 72,000 TS; 15,500 YS; 40 El. Tempered: 121,000 TS; 100,000 YS; 1 El. For corrosion resistant parts; corrosion resistant.

BERYLLIUM BRONZE 2.5% BE

Siemens-Schuckert AG.
Cu 97.5, Be 2.5.
Soft: 62,000 TS; 20,000 YS; 52 El; 66 RA; 98 Brin. Hard: 176,000 TS; 160,000 YS; 0.6 El; 396 Brin. For springs and parts subjected to frictional wear; wear resistant.

BERYLLIUM BRONZE 2.5% BE

Vacuumschmelze
Cu 97.5, Be 2.5.
Soft: 62,000 TS; 20,000 YS; 52 El; 66 RA; 98 Brin. Hard: 176,000 TS; 160,000 YS; 0.6 El; 396 Brin. For springs and parts subjected to frictional wear; wear resistant.

BERYLLIUM COPPER ALLOY 10

Criterion Metals Inc.
Copper. Be 0.4-0.75, Co 2.4-2.7, rem Cu.
Thin gauge sheet, various tempers: 20-60 ksi YS min; 35-70 ksi TS min. C17500

BERYLLIUM COPPER ALLOY 165

Criterion Metals Inc.
Copper. Be 1.6-1.79, 0.20 Co and Ni min, 0.60 Co, Ni and Fe max, rem Cu.
Thin gauge sheet, various tempers: 28-155 ksi YS min; 60-175 ksi TS min. C17000

BERYLLIUM COPPER ALLOY 25

Criterion Metals Inc.
Copper. Be 1.8-2, 0.20 Co and Ni min, 0.60 Co, Ni and Fe max, rem Cu.
Thin gauge sheet, various tempers: 28-155 ksi YS min; 60-175 ksi TS min. C17200

BERYLLIUM COPPER NO. 175

Ohio Brass Co.
Be 2.15, Ni 0.35, bal Cu.
Heat treated: 160,000-185,000 TS; 140,000-160,000 YS; 2-5 El; 360-450 Brin. For springs, clips, diaphragms; age hardenable, high fatigue strength.

BERYLLIUM COPPER-175

Anaconda Co.
Cu 97.5, Be 2.15, Ni 0.35.
Soft: 70,000 TS; 30,000 YS; 45 El. Hard: 190,000 TS; 97,000 YS; 2 El. For springs, diaphragms. *Obsolete*

BERYLLIUM I-400

Brush Wellman
Be 92, Al 0-0.2, C 0-0.5, Fe 0-0.3, Mg 0-0.1, Si 0-0.15, 4.5 BeO.
Unnotched: 129,000 TS; 86,400 YS; 10.6 El. Notched: 127,100 TS. For super-critical gyro applications. For aerospace structures, pressure vessels, jet engine turbine wheels and blades. High stiffness-weight ratio.

BERYLLIUM MALLEABLE

Manufacturer not listed.
Be 99.8, Ti 0.2.
For X-ray windows, camera shutters; very malleable.

BERYLLIUM MANGANESE BRONZE 1.0% BE

Vacuumschmelze
Be 1, Mn 10, bal Cu.
Untempered: 74,000 TS; 38,000 YS; 22 El. Tempered: 146,000 TS; 127,700 YS; 2.3 El. For corrosion resistant parts; corrosion resistant.

BERYLLIUM MANGANESE BRONZE 1.0% BE

Siemans & Halske AG.
Be 1, Mn 10, bal Cu.
Untempered: 74,000 TS; 38,000 YS; 22 El. Tempered: 146,000 TS; 127,700 YS; 2.3 El. For corrosion resistant parts; corrosion resistant.

BERYLLIUM MANGANESE BRONZE 1.5% BE

Vacuumschmelze
Be 1.5, Mn 3, bal Cu.
Untempered: 78,000 TS; 38,000 YS; 25 El. Tempered: 150,000 TS; 135,000 YS; 4.5 El. For corrosion resistant parts; corrosion resistant.

BERYLLIUM MANGANESE BRONZE 1.5% BE

Siemans & Halske AG.
Be 1.5, Mn 3, bal Cu.
Untempered: 78,000 TS; 38,000 YS; 25 El. Tempered: 150,000 TS; 135,000 YS; 4.5 El. For corrosion resistant parts; corrosion resistant.

BERYLLIUM S-200-C

Brush Wellman
Be 98, Al 0-0.16, C 0-0.15, Fe 0-0.18, Mg 0-0.08, Si 0-0.08, 2.0 BeO.
Hot pressed: 44,800 TS; 33,800 YS; 1.6 El. For aerospace structures, pressure vessels, jet engine components. High stiffness-weight ratio.

BERYLLIUM-ALUMINUM

NGK Metals Corp.
Be 5, Mg 1, bal Al.
For master alloy for remelting; pigs.

BERYLLIUM-COBALT

NGK Metals Corp.
Be 50, bal Co.
For master alloy for remelting.

BERYLLIUM-COPPER

NGK Metals Corp.
Be 4, bal Cu.
Used to introduce Be in Cu alloys; master alloy.

BERYLLIUM-IRON

NGK Metals Corp.
Be 50, bal Fe.
Used to introduce Be in Cu-Fe alloys; master alloy.

BERYLLIUM-MAGNESIUM-ALUMINUM

NGK Metals Corp.
Be 5, Mg 5, bal Al.
For master alloy for remelting.

BERYLLIUM-NICKEL

NGK Metals Corp.
Be 50, bal Ni.
Used to introduce Be in Cu-Ni alloys; master alloy.

BERYLLIUM-SILVER

NGK Metals Corp.
Be 50, bal Ag.
For master alloy for remelting.

BERYVAC 170

Vacuumschmelze GmbH
Be 1.0-1.8, 0.2 Ni and/or Co max, bal Cu.
For springs, plugs, connectors; heat treatable; corrosion resistant. *Obsolete*

BERYVAC 200

Vacuumschmelze GmbH
Be 1.8-2.05, 0.2 Ni and/or Co max, bal Cu.
For springs, diaphragms; can be age hardened. *Obsolete*

BERYVAC 520

Vacuumschmelze GmbH
Be 2, bal Ni.
Springs, diaphragms, plug connections; can be age hardened. 520 Vickers; 1500 N/mm^2 YS; 1800 N/mm^2 TS.

BERYVAC 60

Vacuumschmelze GmbH
Be 0.4-0.6, Co 2.5, bal Cu.
For springs, resistance welding electrodes; can be age hardened. *Obsolete*

BERYVAC M 25

Vacuumschmelze GmbH
Be 1.8-2.05, 0.2 Ni and/or Co max, Pb, bal Cu.
Free cutting grade (round bar only). Can be age hardened. *Obsolete*

BESCOLOY

Brighton Electric Steel Casting
C, bal Fe.
For piercer points.

BESPLATE (1)

McGean-Rohco, Inc.
Ni.
For anodes.

BESPLATE (2)

McGean-Rohco, Inc.
C 0.2-0.6, Si 0.2-0.8, bal Ni.
Nickel anodes for electroplating.

BEST

Boyd-Wagner Co.
C 1, Mn 0.3, bal Fe.
78,000 TS; 41,600 YS; 28 El; 167 Brin. For tools, taps, reamers, dies; "Soderfors Best;" nondeforming.

BEST

Bethlehem Steel Corp.
C 0.7-1.1, V 0.2, bal Fe.
For tools, punches, dies, broaches, reamers, shear blades; water hardened. *Obsolete*

BEST

Manufacturer not listed.
C 0.7-1.1, bal Fe.
For general tools, taps, drills; water hardening.

BEST

Thyssen Edelstahlwerke AG
C 1, V 0.1, Mn 0.25, bal Fe.
For cutters, dies, tools, drills; Type W2; water hardened. *Obsolete*

BEST (DISSTON)

Disston Inc.
C 0.7-1.2, bal Fe.
For tools, drills, taps; water hardened. *Obsolete*

BEST 2

Henckels Zwillingwerke, G.A.
C 1, Si 0-0.25, Mn 0-0.25, bal Fe.
Annealed: 100,000 TS; 53,000 YS; 21 El; 42 RA; 200 Brin. For springs, drills, taps, reamers, broaches; Type W1; water hardened.

BEST 3K

Henckels Zwillingwerke, G.A.
C 0.85, Si 0-0.25, Mn 0-0.25, bal Fe.
Heat treated: 190,000 TS; 145,000 YS; 10 El; 30 RA; 400 Brin. For springs, drills, taps, punches, tools; Type W1; water hardened.

BEST 4W

Henckels Zwillingwerke, G.A.
C 0.7, Si 0-0.25, Mn 0-0.25, bal Fe.
Heat treated: 174,000 TS; 128,000 YS; 12 El; 37 RA; 355 Brin. For springs, rails, hammers, crimpers; Type W1; water hardened.

BEST BRONZE

English manufacture
Cu 90, Zn 10.
For jewelry trade as base for fire enameling, primers, bullet shells; gilding brass.

BEST TYPE METAL

English manufacture
Pb 50, Sn 25, Sb 25.
For type metal.

BETA CHISEL STEEL

Midvale-Heppenstall Co.
C, alloy, bal Fe.
For chisels, shear blades, scarfing, tools, rivet busters; shock resistant. *Obsolete*

BETA III

Now CRUCIBLE BETA III.

BETH-CU-LOY

Bethlehem Steel Corp.
C 0.2, 0.20 Cu min, bal Fe.
For sheet metal construction, roofing, siding; resists atmospheric corrosion.

BETH-LED

Bethlehem Steel Corp.
C, Mn, S, Pb, bal Fe.
For screw machine products; free-cutting.

BETHADUR 301

Bethlehem Steel Corp.
C 0.12, Cr 17, Ni 7, bal Fe.
For stainless parts, cooking utensils; stainless, austenitic. *Obsolete*

BETHADUR 301X

Bethlehem Steel Corp.
C 0.1-0.2, Cr 16-18, Ni 7-8.5, bal Fe.
For stainless parts, chemical plant equipment; stainless. *Obsolete*

BETHADUR 304

Bethlehem Steel Corp.
C 0.08, Cr 19, Ni 10, bal Fe.
For welded construction, chemical equipment; non-magnetic, austenitic, corrosion resistant. *Obsolete*

BETHADUR 305

Bethlehem Steel Corp.
C 0.1, Cr 18, Ni 12, bal Fe.
For welded construction followed by corrective treatment; non-magnetic, austenitic, corrosion resistant. *Obsolete*

BETHADUR 306

Bethlehem Steel Corp.
C 0-0.11, Cr 18-20, Ni 8-10, bal Fe.
For welded construction; non-magnetic, austenitic, corrosion resistant. *Obsolete*

BETHADUR 307

Bethlehem Steel Corp.
C 0.08-0.2, Cr 19-22, Ni 9-12, bal Fe.
For welded construction followed by corrective treatment; non-magnetic, austenitic, corrosion resistant. *Obsolete*

BETHADUR 308

Bethlehem Steel Corp.
C 0.1, Cr 20, Ni 11, bal Fe.
For welded construction, valves; non-magnetic, austenitic, corrosion resistant. *Obsolete*

BETHADUR 309

Bethlehem Steel Corp.
C 0.12, Cr 23, Ni 13.5, bal Fe.
For furnace parts, petroleum stills; heat, stainless and corrosion resistant. *Obsolete*

BETHADUR 310

Bethlehem Steel Corp.
C 0.15, Mn 1, Si 1, Ni 20.5, Cr 25, bal Fe.
For furnace parts, heat exchangers; heat resistant. *Obsolete*

BETHADUR 316

Bethlehem Steel Corp.
C 0.08, Cr 17, Ni 12, Mo 2.5, bal Fe.
For chemical and textile equipment; corrosion and heat resistant. *Obsolete*

BETHADUR 317

Bethlehem Steel Corp.
C 0.08, Mn 1, Si 0.75, Ni 12.5, Cr 19, Mo 3.5, bal Fe.
For high temperature parts; corrosion and heat resistant. *Obsolete*

BETHADUR 320

Bethlehem Steel Corp.
C 0-0.2, Cr 17-19, Ni 7-9.5, Ti = 4% x C, bal Fe.
For stainless parts, welded chemical plant equipment; stainless. *Obsolete*

BETHADUR 321

Bethlehem Steel Corp.
C 0.08, Cr 20, Ni 11, Ti = 5 x C, bal Fe.
For stainless parts, welded parts; stainless, stabilized. *Obsolete*

BETHADUR 345

Bethlehem Steel Corp.
C 0-0.15, Cr 17-19, Ni 8-12, Cb = 6-10% x 3, bal Fe.
For stainless parts, welded tanks; stainless. *Obsolete*

BETHADUR 346

Bethlehem Steel Corp.
C 0-0.15, Cr 18-20, Ni 8-12, Cb = 6-10% x C, bal Fe.
For stainless and heat resistant parts; stainless and heat resistant. *Obsolete*

BETHADUR 347

Bethlehem Steel Corp.
C 0.08, Cr 18, Ni 10.5, Cb = 10 x C, bal Fe.
For stainless and heat resistant parts; stainless and heat resistant, stabilized. *Obsolete*

BETHADUR 348

Bethlehem Steel Corp.
C 0-0.15, Cr 18-20, Ni 8-12, Cb = 10 x C, bal Fe.
For heat and corrosion resistant parts; stainless. *Obsolete*

BETHADUR 403

Bethlehem Steel Corp.
C 0.1, Cr 11.5-13, bal Fe.
For steam turbine blades; corrosion resistant. *Obsolete*

BETHADUR 405

Bethlehem Steel Corp.
C 0.08, Mn 0.4, Si 0.5, Cr 12.5, Al 0.2, bal Fe.
For turbine blades, annealing boxes; corrosion and heat resistant. *Obsolete*

BETHADUR 410

Bethlehem Steel Corp.
C 0.1, Cr 12, Mo 0.3, bal Fe.
Heat treated: 100,000 TS; 70,000 YS; 20 El; 60 RA; 200-240 Brin. For turbine blades, valves, cutlery, bolts; corrosion resisting, hardenable. *Obsolete*

BETHADUR 414

Bethlehem Steel Corp.
C 0.1, Mn 0.4, Si 0.5, Ni 1.75, Cr 12.25, bal Fe.
For valve parts, scraper knives, spring parts; corrosion and heat resistant. *Obsolete*

BETHADUR 420

Bethlehem Steel Corp.
C 0.3, Cr 13, bal Fe.
Heat treated: 250,000 TS; 214,000 YS; 10 El; 22 RA; 514 Brin.
For surgery and dental tools, table cutlery; corrosion
resistant, hardenable.

BETHADUR 425

Bethlehem Steel Corp.
C 0-0.12, Cr 16-18, bal Fe.
For corrosion resistant parts; corrosion resistant. *Obsolete*

BETHADUR 431

Bethlehem Steel Corp.
C 0.15, Ni 1.7, Cr 16, bal Fe.
For starter cylinders, valves; corrosion resisting. *Obsolete*

BETHADUR 440 C

Bethlehem Steel Corp.
C 1.1, Mn 0.4, Si 0.05, Cr 17, bal Fe.
For needle valves, ball bearings, oil pumps; hardenable,
corrosion resistant. *Obsolete*

BETHADUR 440-B

Bethlehem Steel Corp.
C 0.85, Mn 0.7, Si 0.4, Cr 17, bal Fe.
For cutlery, surgical instruments; hardenable, corrosion
resistant. *Obsolete*

BETHADUR 501

Bethlehem Steel Corp.
C 0.15, Cr 5, bal Fe.
For chemical engineering equipment, refinery parts;
stainless, heat resistant. *Obsolete*

BETHADUR 501C

Bethlehem Steel Corp.
C 0.11-0.15, Cr 4-6, bal Fe.
For oil refinery equipment; corrosion resistant. *Obsolete*

BETHADUR 501D

Bethlehem Steel Corp.
C 0-0.1, Cr 4-6, bal Fe.
For oil refinery equipment; corrosion resistant. *Obsolete*

BETHADUR 502

Bethlehem Steel Corp.
C 0.08, Cr 4-6, bal Fe.
For oil refinery equipment; corrosion resistant. *Obsolete*

BETHADUR NO. 3

Bethlehem Steel Corp.
C 0.12, Ni 19, Cr 9, bal Fe.
Annealed: 76,500 TS; 54,000 YS; 42 El; 50 RA; 128 Brin. For
chemical plants; not used for atmospheric corrosion;
corrosion resistant to sulfuric acid and HCl. *Obsolete*

BETHADUR NO. 302

Bethlehem Steel Corp.
C 0.12, Cr 17-19, Ni 9-10, bal Fe.
Water quench: 95,000 TS; 57,000 YS; 42 El; 75 RA; 152 Brin.
For ship fittings, food containers, ornamental trim; non-
magnetic, austenitic, corrosion resistant. *Obsolete*

BETHADUR NO. 430

Bethlehem Steel Corp.
C 0.1, Cr 16, bal Fe.
Annealed: 75,000 TS; 58,000 YS; 28 El; 62 RA; 170 Brin. For
exterior trim, ornaments, cold rivets, tanks, retorts, nitric acid
equipment; ductile, corrosion resistant. *Obsolete*

BETHADUR NO. 440

Bethlehem Steel Corp.
C 1.15, Cr 17, Mn 0.3, bal Fe.
Annealed: 118,000 TS; 78,000 YS; 16 El; 19 RA; 241 Brin. For
ball bearings, cutlery, hard stainless parts; corrosion and
abrasion resistant. *Obsolete*

BETHADUR NO. 440

Bethlehem Steel Corp.
C 0.65, Cr 15-18, bal Fe.
Oil treated: 258,000 TS; 195,000 YS; 3 El; 6 RA; 495 Brin. For
balls for ball bearings, ball races, cutlery, valves, instruments,
hardened machine parts; corrosion, heat and abrasion
resistant. *Obsolete*

BETHADUR NO. 440-A

Bethlehem Steel Corp.
C 0.65, Cr 15-18, bal Fe.
Oil treated: 258,000 TS; 195,000 YS; 3 El; 6 RA; 495 Brin. For
balls for ball bearings, ball races, cutlery, valves, instruments,
hardened machine parts; corrosion, heat and abrasion
resistant. *Obsolete*

BETHADUR NO. 442

Bethlehem Steel Corp.
C 0.35, Cr 18-23, bal Fe.
For nitric acid industry; corrosion and heat resistant.
Obsolete

BETHADUR NO. 486

Bethlehem Steel Corp.
Cr 25, C 0.25, bal Fe.
Water quench: 95,000 TS; 50,000 YS; 15 El; 20 RA; 187 Brin.
For protection tubes, conveyor chains, furnace parts, bolts,
nuts, racks; corrosion and heat resistant to S-atmosphere.
Obsolete

BETHALLOY

Bethlehem Steel Corp.
C 0.75, Mn 0.75, Si 0.3, Cr 0.9, Mo 0.37, Ni 1.8, bal Fe.
For tools, broaches, taps, slitting cutters, reamers; oil
hardened, tough.

BETHALON 416

Bethlehem Steel Corp.
C 0-0.15, Cr 12-14, Mo 0-0.6, 0.07% min S, bal Fe.
Annealed: 75,000 TS; 40,000 YS; 30 El; 60 RA; 155 Brin. Heat
treated: 200,000 TS; 150,000 YS; 12 El; 40 RA; 415 Brin. For
screw machine products, valve trim, pump shafts; Type 416,
stainless, hardenable. *Obsolete*

BETHALON 430F

Bethlehem Steel Corp.
C 0-0.12, Cr 16-18, Mo, S, Se, bal Fe.
For automatic and screw machine parts; not to be welded;
similar to Bethadur No. 3 plus machinability. *Obsolete*

BETHALON D

Bethlehem Steel Corp.
C 0.1, S 0.3-0.5, Cr 21, bal Fe.
For automatic and screw machine parts; not to be welded.
Obsolete

BETHANIZED PRODUCTS

Bethlehem Steel Corp.
Zn coated steel. For wire for severe duty and weather
resistance; can be worked and deformed. *Obsolete*

BETHCO MACHINE SCREW

Bethlehem Steel Corp.
C 0.3, bal Fe.
For machine screws; special cold heading steel, special
process. *Obsolete*

BETHCO WOOD SCREW

Bethlehem Steel Corp.
C 0.25, bal Fe.
For wood screws; special cold-heading steel, special process.
Obsolete

BETHLEHEM 33A

Bethlehem Steel Corp.
C 0.16, Mn 1.3, S, bal Fe.
Water treated: 73,000-119,000 TS; 48,000-89,000 YS; 19-35
El; 55-63 RA; 156-241 Brin. For carburized parts, machinery
parts; free machining. *Obsolete*

BETHLEHEM 33B

Bethlehem Steel Corp.
C 0.3, Mn 1.3, S, bal Fe.
Water treated: 82,000-143,000 TS; 52,000-118,000 YS; 15-19
El; 48-54 RA; 162-293 Brin. For machinery parts; free
machining. *Obsolete*

BETHLEHEM 33C

Bethlehem Steel Corp.
C 0.4, Mn 1.3, S 0.08, bal Fe.
Annealed: 90,000 TS; 62,000 YS; 26 El; 47 RA; 183 Brin. Heat
treated: 142,000-178,000 TS; 102,000-161,000 YS; 12-13 El;
36-39 RA; 285-375 Brin. For screw machine products, bolts,
nuts, hardware; free-cutting, water hardened. *Obsolete*

BETHLEHEM 445

Bethlehem Steel Corp.
Cr 3.75, C 0.85, Mn 0.3, Ni 0-0.15, Si 0.25, bal Fe.
For spike dies, forming dies, heading tools; for forming and
shaping of hot metals; high toughness and hardness.
Obsolete

BETHLEHEM 5% CHROMIUM AIR HARDENING

Bethlehem Steel Corp.
C 1, Cr 5.25, Mo 1.1, V 0.25, bal Fe.
For dies; wear resisting, air hardening. *Obsolete*

BETHLEHEM 5% CR AIR HARDENING

Bethlehem Steel Corp.
C 1, Cr 5.25, Mo 1.1, V 0.25, bal Fe.
For dies; air hardened, low distortion. *Obsolete*

BETHLEHEM 6 NI

Bethlehem Steel Corp.
Ni 3.4-3.6, C 0.3, Mn 1.05, Cr 0.25, S 0.2, bal Fe.
For tools, gears, shafts; oil hardening. *Obsolete*

BETHLEHEM 6-6

Bethlehem Steel Corp.
C 0.7, Mo 6, W 6, bal Fe.
For tools, cutters; high speed steel. *Obsolete*

BETHLEHEM 66 HS

Now M-2 HIGH SPEED STEEL.

BETHLEHEM 67 TAP

Bethlehem Steel Corp.
C 1.2, Cr 0.65, Mn 0.25, V 0.2, W 1.4, bal Fe.
For dies, cutting tools, taps, drills; oil hardening. *Obsolete*

BETHLEHEM 88-80

Bethlehem Steel Corp.
Cr 2.5, C 0.8, Mo, bal Fe.
For castings, ball mill liners, crushers, rolls, screen plates;
resists severe abrasion. *Obsolete*

BETHLEHEM O1 DRILL ROD

Bethlehem Steel Corp.
C 0.9, Mn 1.2, Cr 0.5, V 0.2, W 0.5, bal Fe.
For pin headers, special punches, precision gages, dies,
drills, plugs. AISI Type O1; oil-hardened tool steel. *Obsolete*

BETHLEHEM AH-5

Bethlehem Steel Corp.
C 1, Mn 0.6, Cr 5.25, Mo 1.1, V 0.25, bal Fe.
For cutters, tools, dies, punches; air hardened, tough, wear
resistant. AISI A2. *Obsolete*

BETHLEHEM AIR DIE

Bethlehem Steel Corp.
C 0.95, Mn 2, Cr 2.2, Mo 1.1, bal Fe.
For dies; air hardened. *Obsolete*

BETHLEHEM ALLOY HOLLOW DRILL STEEL

Bethlehem Steel Corp.
C 0.28, Mn 0.95, Cr 0.65, Ni 2.25, Mo 0.3, bal Fe.
For hollow drills and forged on bits; shock and wear resistant.
Obsolete

BETHLEHEM AUGER DRILL
Bethlehem Steel Corp.
C 0.75, Mn 0.2, Si 15, bal Fe.
For auger drills; water hardening. *Obsolete*

BETHLEHEM B-14
Bethlehem Steel Corp.
C, Cr, Mo, V, bal Fe.
For high temperature applications, 900-1100 F; heat resistant. *Obsolete*

BETHLEHEM B-7
Bethlehem Steel Corp.
C, Cr, Mo, bal Fe.
For high temperature applications, 900-1100 F; heat resistant. *Obsolete*

BETHLEHEM B-7A
Bethlehem Steel Corp.
C, Cr, Mo, bal Fe.
For high temperature applications, 900-1100 F; heat resistant. *Obsolete*

BETHLEHEM BA-H
Bethlehem Steel Corp.
C 1, Mn 2, Si 0.23, Cr 2.2, Mo 1, bal Fe.
For tools, dies, punches; air hardening. *Obsolete*

BETHLEHEM BFS
Bethlehem Steel Corp.
C 1.3, W 4, bal Fe.
For tools, finishing tools; keen cutting edge. *Obsolete*

BETHLEHEM BROACHING STEEL
Bethlehem Steel Corp.
C 0.8, Si 0.2, Mn 0.25, bal Fe.
For stone working tools, drills; Type W1; water hardened. *Obsolete*

BETHLEHEM CARBON
Bethlehem Steel Corp.
C 1, Mn 0.25, Si 0.2, bal Fe.
For cutting tools; water hardening. *Obsolete*

BETHLEHEM CR-MO-V-HIGH V
Bethlehem Steel Corp.
C 0.38, Cr 5.2, Mo 1.2, V 1, Si 1, Mn 0.4, bal Fe.
Heat treated: 220,000 TS; 185,000 YS; 12 El; 38 RA; 440 Brin.
For die casting dies, shear blades, header dies; hot work steel, shock resistant. *Obsolete*

BETHLEHEM CR-MO-W
Now CROMO W.

BETHLEHEM EXTRA SPECIAL
Bethlehem Steel Corp.
W 14, Cr 4, V 2, C, bal Fe.
For tools, cutters; high speed steel. *Obsolete*

BETHLEHEM H.V.
Bethlehem Steel Corp.
W 18, Cr 4.5, V 2, Mo 0.75, C, bal Fe.
For heavy duty cutting tools; high speed steel. *Obsolete*

BETHLEHEM HM
Now M-1 HIGH SPEED STEEL.

BETHLEHEM M-10
Now M-10 HIGH SPEED STEEL.

BETHLEHEM MOKUT
Bethlehem Steel Corp.
C, Mo, W, Cr, bal Fe.
For tools, cutters; high speed steel. *Obsolete*

BETHLEHEM MOLY
Bethlehem Steel Corp.
C, Cr, Mo, bal Fe.
For bearings, injector parts. *Obsolete*

BETHLEHEM NO. 1 PERMANENT MAGNET
Bethlehem Steel Corp.
W 6, C 0.65, bal Fe.
For magnets; magnet steel. *Obsolete*

BETHLEHEM NO. 235
Bethlehem Steel Corp.
C 0.35-0.5, Mn 1.4-2, bal Fe.
235 Brin. For shovels, crushers, scraper blades, hoppers; wear resistant. *Obsolete*

BETHLEHEM NO. 300
Bethlehem Steel Corp.
C 0.7, Mn 1, Si 0.6, bal Fe.
300 Brin. For shovels, hoppers, crushers, scraper blades, abrasion resistant. *Obsolete*

BETHLEHEM NO. 6 NICKEL STEEL
Bethlehem Steel Corp.
Ni 34-36, C 0.05, bal Fe.
For heating elements, instruments; low coefficient of expansion. *Obsolete*

BETHLEHEM NO. 7
Bethlehem Steel Corp.
Ni 40, bal Fe.
For electrical equipment; low coefficient of expansion. *Obsolete*

BETHLEHEM PIVOT STEEL
Bethlehem Steel Corp.
C 1.2, Cr 1.65, Mo 0.4, bal Fe.
For pivot bearings; wear resisting. *Obsolete*

BETHLEHEM SHANK STEEL
Bethlehem Steel Corp.
C 0.55, Mn 0.9, Si 2, V 0.25, bal Fe.
For tool shanks; tough, shock resisting. *Obsolete*

BETHLEHEM SILICO-MANGANESE SPRING STEEL
Bethlehem Steel Corp.
C 0.7, Mn 0.9, Si 0.2, bal Fe.
Heat treated: 210,000-235,000 TS; 180,000-195,000 YS; 9-12 El; 30-35 RA; 415 Brin. For springs; oil hardened.

BETHLEHEM SILVERY MAYARI IRON
Bethlehem Steel Corp.
C 2-2.7, Mn 2.5-3.5, Si 8-12, Ni 1, Cr 2, bal Fe.
50,000 TS. For pig iron for high strength iron or semi-steel castings with good machinability; alloy pig iron; *Obsolete*

BETHLEHEM SOLID DRILL
Bethlehem Steel Corp.
C 0.8, bal Fe.
For drills, bits, blacksmith tools; shock resistant.

BETHLEHEM SPECIAL GEAR STEEL
Bethlehem Steel Corp.
C 0.4, Ni 1.5, Cr 0.9, Mo 0.2, bal Fe.
For gears, pinions, transmission units; grain-size control. *Obsolete*

BETHLEHEM SPECIAL HS
Now T-1 HIGH SPEED STEEL.

BETHLEHEM STAINLESS TYPE "A"
Bethlehem Steel Corp.
Mn 0.3, C 0.35, Cr 14, bal Fe.
Hot rolled: 95,000 TS; 65,000 YS; 30 El; 55 RA; 179 Brin. For stainless steel parts. *Obsolete*

BETHLEHEM STANDARD BEARINGS
Bethlehem Steel Corp.
C 1, Cr 1.35, bal Fe.
For bearings, rollers, master gauges; bearing steel.

BETHLEHEM STONE DRESSING
Bethlehem Steel Corp.
C 1.15, Mn 0.25, Si 0.2, bal Fe.
For tools for stone dressing; water hardened. *Obsolete*

BETHLEHEM SUPERIOR HOLLOW DRILL
Bethlehem Steel Corp.
C 0.8, Low P and S, bal Fe.
For hollow drills and bits; resistance to abrasion, shock and vibration. *Obsolete*

BETHLEHEM T1 HIGH SPEED (B.S.H.S.)
Bethlehem Steel Corp.
C 0.73, Cr 4, V 1.1, W 18, bal Fe.
General purpose high-speed cutting tool, all types of cutting operations. AISI Type T1; high-speed tool steel. *Obsolete*

BETHLEHEM TOUGH
Bethlehem Steel Corp.
C 0.3, Cr 1.3, V 0.2, bal Fe.
For tools, punches, cold cutters, shear blades, flogging chisels; tough and hard. *Obsolete*

BETHLEHEM X
Bethlehem Steel Corp.
C 0.7-1.1, bal Fe.
For general tools, quarrying tools, crow bars, soft rock drills, chisels; water hardened.

BETHLEHEM XLC
Bethlehem Steel Corp.
C 0.8-1.2, bal Fe.
For dies, cold cutters, chisels, punches, shear blades; regular quality carbon tool steel. *Obsolete*

BETHLEHEM XXX
Bethlehem Steel Corp.
C 0.9-1.35, bal Fe.
For finishing tools and cutters; special quality. *Obsolete*

BETHLEHEM XXX SPECIAL
Bethlehem Steel Corp.
C 0.7, W 18, Cr 4, V 1, bal Fe.
For tools, finishing tools, cutters; high speed steel. *Obsolete*

BETHNAMEL
Bethlehem Steel Corp.
C 0.1, bal Fe.
Rolled: 26,000 TS; 43,000 YS; 38 El; 77 Brin. For porcelain enameled articles; good drawability. *Obsolete*

BF 302 STAINLESS
Bonney-Floyd Co.
C 0.08-0.2, Cr 17-19, Ni 8-10, bal Fe.
For castings; stainless. *Obsolete*

BF 316 STAINLESS
Bonney-Floyd Co.
C 0.2, Cr 18, Ni 8, Mo 3, bal Fe.
For stainless parts; stainless, acid resistant. *Obsolete*

BF 4-6 CRMO
Bonney-Floyd Co.
C 0.3, Cr 4-6, Mo, bal Fe.
For castings; corrosion resistant. *Obsolete*

BF 410 STAINLESS
Bonney-Floyd Co.
C 0-0.15, Mn 0-2, Si 0-1.5, Cr 11.5-13.5, bal Fe.
For castings; corrosion resistant. *Obsolete*

BF 954
Bradley & Foster Ltd.
C 0.3, Cr 13, bal Fe.
For knives, crusher jaws, guide plates, and blasting nozzles. Wear resistant.

BF-CRMO
Bonney-Floyd Co.
C 0.3, Cr 0.8, Mo 0.2, bal Fe.
For castings; water hardening. *Obsolete*

BF-MM
Bonney-Floyd Co.
C 0.25-0.35, Mo 0.2-0.25, Mn 1-1.25, Si 0.3, bal Fe.
Heat treated: 90,000-210,000 TS; 65,000-183,000 YS; 7-25 El;
25-55 RA; 196-480 Brin. For power shovels, tractor tread
steel, mine car wheels; resists shock and wear; IZ-3-56.
Obsolete

BFD
Wallace Murray Corp.
C 1.2, Mn 0.35, Si 0.25, Cr 0.6, V 0.2, W 1.5, bal Fe.
AISI Type O7; oil hardening tool steel.

BFS
Bethlehem Steel Corp.
O 1.3, Mn 0.28, W 3.5, bal Fe.
Water or oil hardenable tool steel, high hardness and wear
resistance as heat treated; for lathe centers, drill bushings.
Obsolete

BG42
Now LESCALLOY BG42.

BHS
Thyssen Edelstahlwerke AG
C 0.25, Mn 1.5, Si 0.2, Al 0.15, Ti 0.25, bal Fe.
For welding electrodes. *Obsolete*

BHT 80
Nippon Steel USA Inc.
C 0.06, Si 0.67, Mn 2.52, P 0.015, S 0.007, Cu 0.3, Ni 0.3, Nb
0.04, bal Fe.
4.5 mm thick plate, quenched and tempered: 122,000 psi TS;
105,000 psi YS; 20 El. High strength steel, good formability
for welded structures as earth moving equipment and crane
booms.

BICALOY
British Insulated Callender's Cables
Cu alloy.
For resistance welding electrodes. Wear and deformation
resistant. *Obsolete*

BICOP OXYGEN FREE COPPER
Manufacturer not listed.
Cu.
For electrical motors, generators; high conductivity.

BIDDERY, HEINIES
Manufacturer not listed.
Zn 84.3, Cu 11.4, Pb 2.9, Sn 1.4.
For bearings; Babbitt.

BIDERY (BIDDERY)
American manufacture
Zn 90.2, Cu 6.3, Pb 2.6, Sn 0.8.
For buttons, ornaments; free-cutting.

BIDERY BUTTONS "A"
Manufacturer not listed.
Cu 48.5, Zn 33.3, Sn 6.06, Pb 12.15.
For buttons, ornamental parts; free-cutting.

BIDERY BUTTONS "B"
Manufacturer not listed.
Cu 48.5, Zn 33.32, Sn 6.06, Pb 12.15.
For buttons; free-cutting.

BIERMAN TUNGSTEN BRONZE
German manufacture
Cu 95, Sn 3.4, W 1.6.
For strong corrosion resistant parts.

BIG J
Joseph Jackman & Co. Ltd.
C 0.7, W 18, Cr 4, V 1, bal Fe.
For high speed tools. High speed steel.

BIH
Thyssen Edelstahlwerke AG
C 0.3, Mn 1.1, Si 0.2, Cr 1, Ti 0.2, bal Fe.
For welding electrodes. *Obsolete*

BILAME A
Creusot-Loire
Nickel alloy bimetal.
For thermostats, fire detectors; bimetal. *Obsolete*

BILAME AS
Creusot-Loire
Nickel alloy bimetal.
For thermostats, fire detectors; bimetal. *Obsolete*

BILAME BC
Creusot-Loire
Nickel alloy bimetal.
For thermostats, fire detectors; bimetal. *Obsolete*

BILGEN BRONZE
German manufacture
Cu 97, Sn 1.9, Fe 0.5, Pb 0.2.
For electrical bronze; corrosion resistant.

BINAL
Alloy Technology International, Inc.
Sintered cast boron carbide. Made to customer's
requirement. *Obsolete*

BINDING BRASS
American manufacture
Cu 63.25, Zn 35, Pb 1.75.
For automatic screw machine products; free-cutting.

BINNEY HEAT RESISTING ALLOY
Binney Castings Co.
C, Cr, Ni, bal Fe.
Cast. For furnace parts; resists heat to 1400 F. *Obsolete*

BINNEY NO. 71
Binney Castings Co.
C, alloy, bal Fe.
For heat treating furnaces and equipment; heat resistant.
Obsolete

BINNEY NO. 73
Binney Castings Co.
C, alloy, bal Fe.
For heat treating furnaces and equipment; heat resistant.
Obsolete

BIOSIL
Degussa AG
Precious metal
Nonprecious metal alloy for dentistry and dental engineering.

BIRDSBORO 50
Birdsboro Corp.
C, alloy, bal Fe.
For rolls; water hardening. *Obsolete*

BIRDSBORO CA15
Birdsboro Corp.
C 0-0.15, Cr 12.5, bal Fe.
Cast: 90,000-110,000 TS; 65,000-80,000 YS; 18-25 El. Mildly
corrosion resistant. *Obsolete*

BIRDSBORO DA
Birdsboro Corp.
C 0.2-0.3, Cr 0.6, Mo 0.4, Mn 1.5, Ni 0-0.7, bal Fe.
Normalized: 110,000 TS; 85,000 YS; 17 El; 35 RA; 225 Brin.
For rolls; wear resistant. *Obsolete*

BIRDSBORO HY 80
Birdsboro Corp.
C 0-0.2, Cr 1.5, Ni 2.85, Mo 0.45, Mn 0.65, bal Fe.
80,000-95,000 YS; 20-30 El. High yield, good notch
toughness. *Obsolete*

BIRDSBORO METAL
Birdsboro Corp.
C 3, Si 1.5, Mn 0.8, Alloy, bal Fe.
For rolls; alloy cast iron. *Obsolete*

BIRDSBORO NO. 26
Birdsboro Corp.
C, Mo, Cu, bal Fe.
For structures. *Obsolete*

BIRDSBORO NO. 30
Birdsboro Corp.
C 0.3, Mo 0.2, Cu 1, bal Fe.
85,000 TS; 55,000 YS; 22 El; 40 RA; 180 Brin. For
engineering construction. Corrosion and fatigue resistant.
Obsolete

BIRDSBORO-20 CN
Birdsboro Corp.
C 0-0.25, Mn 0.7, bal Fe.
Cast: 60,000-75,000 TS; 30,000-45,000 YP; 115-140 Brin;
22-30 El. For general and structural applications, gears,
housings. Excellent weldability. AISI 1020. *Obsolete*

BIRDSBORO-20A6
Birdsboro Corp.
C 0-0.25, Mn 0.9, Cr 0.5, Ni 0.5, Mo 0.5, bal Fe.
Cast: 80,000-120,000 TS; 50,000-90,000 YP; 150-210 Brin;
14-20 El. For general and structural applications, gears,
shafts, housings. Good weldability. AISI 8620 modified.
Obsolete

BIRDSBORO-22 CMN
Birdsboro Corp.
C 0-0.25, Mn 1, bal Fe.
Cast: 70,000-80,000 TS; 40,000-55,000 YP; 22-30 El; 30-50
RA. For general and structural applications, gears, housings.
ASTM A216, WCC. *Obsolete*

BIRDSBORO-25 CN
Birdsboro Corp.
C 0-0.3, Mn 0.7, bal Fe.
Cast: 65,000-80,000 TS; 35,000-55,000 YP; 125-160 Brin;
24-30 El. For general and structural applications, gears,
housings. Excellent weldability. AISI 1025. *Obsolete*

BIRDSBORO-25A6
Birdsboro Corp.
C 0-0.3, Mn 0.9, Cr 0.5, Ni 0.5, Mo 0.5, bal Fe.
Cast: 105,000-145,000 TS; 85,000-120,000 YP; 14-17 El;
30-35 RA. For general and structural applications, gears,
housings, shafting. AISI 8625 modified. *Obsolete*

BIRDSBORO-30 CN
Birdsboro Corp.
C 0-0.35, Mn 0.7, bal Fe.
Cast: 65,000-80,000 TS; 35,000-55,000 YP; 125-160 Brin;
22-30 El; 30-50 RA. For general and structural applications,
gears, housings. Excellent weldability. AISI 1030. *Obsolete*

BIRDSBORO-30A1
Birdsboro Corp.
C 0-0.35, Mn 0.9, Mo 0.2, Cu 1, bal Fe.
Cast: 80,000-105,000 TS; 50,000-90,000 YS; 150-210 Brin;
15-30 El; 30-40 RA. For general and structural applications,
gears, housings. Good weldability. *Obsolete*

BIRDSBORO-30A2
Birdsboro Corp.
C 0-0.35, Mn 0.8, Cr 0.95, Mo 0.2, bal Fe.
Cast: 90,000-120,000 TS; 65,000-100,000 YP; 14-20 El; 30-40
RA; 180-260 Brin. For general and structural castings, shafts,
gears, housings. AISI 413. Shock resistant. *Obsolete*

BIRDSBORO-30A3
Birdsboro Corp.
C 0-0.35, Mn 0.75, Cr 0.85, Ni 2, Mo 0.3, bal Fe.
Heat treated: 120,000-145,000 TS; 95,000-120,000 YP; 14-20 El; 30-40 RA; 200-250 Brin. For general and structural castings, gears, shafting, housing. AISI 4330. Shock resistant. *Obsolete*

BIRDSBORO-30A4
Birdsboro Corp.
C 0-0.35, Mn 1.5, Cr 0.5, Ni 0.5, Mo 0.5, bal Fe.
Cast: 120,000-150,000 TS; 95,000-125,000 YP; 9-14 El; 22-30 RA; 240-310 Brin. For general and structural castings, shafts, housings, gears. Tough, shock resistant. *Obsolete*

BIRDSBORO-30A6
Birdsboro Corp.
C 0-0.35, Mn 0.9, Cr 0.5, Ni 0.5, Mo 0.5, bal Fe.
Cast: 120,000-145,000 TS; 95,000-120,000 YP; 14-20 El; 30-40 RA. For general and structural applications, gears, shafts, housings. AISI 8630 modified. *Obsolete*

BIRDSBORO-40 CN
Birdsboro Corp.
C 0-0.5, Mn 0.8, bal Fe.
Cast: 80,000-95,000 TS; 40,000-60,000 YP; 140-180 Brin; 30-40 RA; 18-28 El. For general and structural applications, gears, housings. Weldable with caution. AISI 1040. *Obsolete*

BIRDSBORO-40A2
Birdsboro Corp.
C 0-0.45, Mn 0.8, Cr 0.95, Mo 0.2, bal Fe.
Cast: 90,000-165,000 TS; 60,000-140,000 YP; 9-20 El; 22-40 RA; 240-310 Brin. For general and structural castings, gears, shafts, housings. AISI 414. Tough. *Obsolete*

BIRDSBORO-40A3
Birdsboro Corp.
C 0.45, Mn 0.75, Cr 0.85, Ni 2, Mo 0.3, bal Fe.
Heat treated: 180,000-210,000 TS; 150,000-180,000 YP; 20-25 RA; 330-400 Brin. For general and structural castings, shafts, gears, housings. AISI 4340. Shock resistant. *Obsolete*

BIRDSBORO-40A5
Birdsboro Corp.
C 0-0.45, Mn 0.95, Cr 0.5, Ni 0.5, Mo 0.5, bal Fe.
Cast: 120,000-150,000 TS; 95,000-125,000 YP; 9-14 El; 22-30 RA; 290-350 Brin. For general and structural castings, shafts, gears, housings. High strength, shock resistant. AISI 8640 modified. *Obsolete*

BIRDSBORO-45 CN
Birdsboro Corp.
C 0.5, Mn 0.8, bal Fe.
Cast: 80,000-95,000 TS; 40,000-60,000 YP; 140-180 Brin; 18-28 El; 30-40 RA. For general and structural applications, gears, housings. Weldable with caution. *Obsolete*

BIRDSBORO-45A2
Birdsboro Corp.
C 0-0.5, Mn 0.8, Cr 0.95, Mo 0.2, bal Fe.
Cast: 120,000-165,000 TS; 95,000-140,000 YP; 14-9 El; 22-30 RA; 300 Brin. For general and structural castings; gears, shafts, housings. AISI 4145. Tough. *Obsolete*

BIRDSBORO-45A7
Birdsboro Corp.
C 0-0.5, Mn 0.8, Cr 2, Mo 0.5, bal Fe.
Heat treated: 175,000-210,000 TS; 145,000-180,000 YP; 6-12 El; 12-30 RA; 290-350 Brin. For general and structural castings, gears, shafts, housings. Tough, shock resistant. *Obsolete*

BIRDSBORO-50 CN
Birdsboro Corp.
C 0-0.55, Mn 0.8, bal Fe.
Cast: 80,000-95,000 TS; 50,000-65,000 YP; 22-30 El; 25-40 RA. For general and structural applications, gears, housings. AISI 1050. *Obsolete*

BIRDSBORO-CMO
Birdsboro Corp.
C 0-0.25, Mn 1, Mo 0.25, bal Fe.
Cast: 65,000-80,000 TS; 35,000-50,000 YP; 24-30 RA; 35-50 RA. For general and structural applications, gears, shafts, housings. AISI 4024 modified. Tough. *Obsolete*

BIRDSBORO-CNI
Birdsboro Corp.
C 0-0.45, Mn 0.8, Ni 3.5, bal Fe.
Cast: 120,000-145,000 TS; 95,000-120,000 YP; 14-20 El; 0-40 RA. For general and structural castings, shafts, gears, housings. AISI 2340. Tough, shock resistant. *Obsolete*

BIRDSBORO-CRMO1
Birdsboro Corp.
C 0-0.25, Cr 0.5, Mo 0.5, bal Fe.
Cast: 70,000-80,000 TS; 45,000-55,000 YS; 22-30 El; 35-50 RA. For general and structural castings, shafts, gears, housings. ASTM A356 Alloy 5. *Obsolete*

BIRDSBORO-CRMO2
Birdsboro Corp.
C 0-0.2, Mn 0.65, Cr 1.25, Mo 0.5, bal Fe.
Cast: 70,000-80,000 TS; 45,000-55,000 YP; 22-30 El; 35 RA. For general and structural castings, shafts, gears, housings. ASTM A356 Alloy A. *Obsolete*

BIRDSBORO-CRMO2 + V
Birdsboro Corp.
C 0-0.18, Mn 0.55, Cr 1.25, Mo 0.5, V 0.2, bal Fe.
Cast: 70,000-80,000 TS; 40,000-55,000 YP; 20-30 El; 35-50 RA. For general and structural castings, shafting, gears, housings. Shock resistant. *Obsolete*

BIRDSBORO-CRMO3
Birdsboro Corp.
C 0-0.18, Mn 0.55, Cr 2.3, Mo 1, bal Fe.
Cast: 70,000-105,000 TS; 40,000-85,000 YP; 17-30 El; 30-45 RA. For general and structural castings, shafts, housings. Tough, shock resistant. ASTM A217, Alloy WC9. *Obsolete*

BIRDSBORO-CRMO4
Birdsboro Corp.
C 0-0.2, Mn 0.55, Cr 5.25, Mo 0.55, bal Fe.
Cast: 90,000-120,000 TS; 60,000-95,000 YP; 18-25 El; 35-45 RA. For general and structural castings, gears, shafts, housings. ASTM A217, Alloy C5. *Obsolete*

BIRDSBORO-LC2
Birdsboro Corp.
C 0-0.25, Mn 0.65, Ni 2.5, bal Fe.
Cast: 80,000 TS; 50,000 YP; 24 El; 35 RA. For general and structural applications, gears, shafts, housings. Shock resistant. *Obsolete*

BIRDSBORO-LC3
Birdsboro Corp.
C 0-0.15, Mn 0.65, Ni 3.5, bal Fe.
Cast: 60,000-80,000 TS; 40,000-55,000 YP; 24-30 El; 35-50 RA. For general and structural castings, gears, housings, shafts. ASTM A352, LC3; AISI 4812 modified. Tough, shock resistant. *Obsolete*

BIRDSBORO-MM
Birdsboro Corp.
C 0-0.32, Mn 1.5, Mo 0.5, bal Fe.
Cast: 80,000-120,000 TS; 50,000-90,000 YS; 14-22 El; 150-210 Brin; 30-35 RA. For general and structural applications, gears, housings. Good weldability. ASTM A148. *Obsolete*

BIRMABRIGHT
Birmingham Aluminium Casting Co.
Aluminum. Mg 3-6, Mn 0.25-0.75, bal Al.
Sand cast: 20,000-22,000 TS; 11,000-15,000 YS; 3.0-5.5 El; 54-60 Brin. Drawn: 34,000-43,000 TS; 20,000-27,000 YS; 15.0-25.0 El; 70-80 Brin. For tubes, rivets, bolts, nuts, screws, architectural metal work; corrosion resistant to salt water.

BIRMABRIGHT
Sterling International Technology Ltd.
Mg 3-6, Mn 0.3-0.7, bal Al.
Sand cast: 20,000-22,400 TS; 20,000-22,400 YS; 3-6 El; 55-60 Brin; 5.8 Izod. Good corrosion resistance; takes good finish. For decorative parts. BS 1490 LM 5-M.

BIRMABRIGHT 5454
Birmetals Ltd.
Aluminum. Mn 0.7, Mg 2.8, bal Al.
Sheet and strip, condition O: 215-285 MPa TS; 80 min MPa YS; 8-12 El. AA 5454.

BIRMABRIGHT B.B. 7
Birmetals Ltd.
Mg 7, bal Al.
40,000 TS; 18,000 YS; 15 El. For light weight construction; corrosion resistant; difficult to fabricate. *Obsolete*

BIRMABRIGHT BB 1-X
Birmetals Ltd.
Mg 1, bal Al.
Soft: 15,000 TS; 7,000 YS; 30 El. 1/4 H: 18,000 TS; 15,000 YS; 7 El. 3/4 H: 24,000 TS; 21,000 YS; 4 El. For marine parts; resists sea water corrosion. *Obsolete*

BIRMABRIGHT BB 17
Birmetals Ltd.
Aluminum. Mg 0.6, bal Al.
Sheet and strip, annealed.: 110 MPa TS; 20 El.

BIRMABRIGHT BB 5
Birmetals Ltd.
Aluminum. Mn 0.3, Mg 5, bal Al.
Sheet and strip, condition O: 278 MPa TS; 147 MPa YS; 25 El. AA 5056A.

BIRMABRIGHT BB 7
Birmetals Ltd.
Mg 7, bal Al.
Soft: 44,000 TS; 20,000 YS; 20 El. Extruded: 43,000 TS; 21,000 YS; 20 El. For marine construction, fixtures; resists sea water corrosion. *Obsolete*

BIRMABRIGHT BB1
Birmetals Ltd.
Aluminum. Mn 0.5, Mg 1, bal Al.
Drawn tube, condition O: 162 MPa TS; 70 MPa YS; 25 El.

BIRMABRIGHT BB2
Birmetals Ltd.
Aluminum. Mn 0.25, Mg 2, bal Al.
Sheet and strip, condition H6: 230 MPa TS; 190 MPa YS; 8 El. AA 5251.

BIRMABRIGHT BB3
Birmetals Ltd.
Aluminum. Mn 0.25, Mg 3.5, bal Al.
Sheet and strip, condition H4; 290 MPa TS; 260 MPa YS; 7 El. AA 5154A.

BIRMABRIGHT BB4
Birmetals Ltd.
Aluminum. Mn 0.7, Mg 4.7, bal Al.
Sheet and strip, condition H4; 370 MPa TS; 290 MPa YS; 8 El. AA 5083.

BIRMAL
Birmingham Aluminium Casting Co.
Aluminum. Mg 2-6, Mn 0.25-0.75, bal Al.
Cast: 22,000 TS; 15,000 YS; 3.0 El; 58 Brin. For light alloy parts; heat treatable.

BIRMAL L4
Birmingham Aluminium Casting Co.
Aluminum. Cu 2-4, Mg 0.5-1.5, Si 8.5-10.5, Ni 0.5-1.5, Ti 0.2, bal Al.
Permanent mold: 32,500 TS; 0 El; 120 Brin. For pistons; age-hardenable.

BIRMAL MB7
Birmingham Aluminium Casting Co.
Aluminum. Cu 0.7-1.3, Mg 0.75-1.2, Si 0.35-0.85, Ni 1.6, Sn 7, Ti 0.2, bal Al.
Cast. For bearings; age-hardenable.

BIRMAL P83
Birmingham Aluminium Casting Co.
Aluminum. Cu 3-4, Si 7.5-9.5, Fe 0-1.3, Zn 0-1, bal Al.
Die cast: 40,400 TS; 15,700 YS; 3 El; 85 Brin. For general purpose die castings; SAE 306.

BIRMALITE
Birmingham Aluminium Casting Co.
Aluminum. Cu 9-11, Mg 0.1-0.5, Fe 1, bal Al.
Cast: 19,000 TS; 17,000 YS; 0.5 El; 80 Brin. Heat treated: 40,000 TS; 0.5 El; 150 Brin. For pistons; heat treatable.

BIRMASIL
Birmingham Aluminium Casting Co.
Si 11-14, bal Fe.
For light alloy castings; die cast alloy.

BIRMASIL SPECIAL
Birmingham Aluminium Casting Co.
Aluminum. Si 10-13, Ni 2.5-3.5, Ti 0.2, Mg 0.6, bal Al.
Chill cast: 36,000-42,000 TS; 16,000-22,000 YS; 3-6 El; 70-80 Brin. Sand cast: 28,000-29,000 TS; 14,000-16,000 YS; 2-4 El; 52-60 Brin. For automotive engine water cooled cylinder blocks, cylinder heads, fire brick molds, motor car fittings; British Patent 342, 152; good corrosion resistance.

BIRMETAL B.B. 016
Birmetals Ltd.
Aluminum. Mg 1, Si 0.6, Cu 0.25, bal Al.

BIRMETAL B.B. 019
Birmetals Ltd.
Aluminum. Mn 1.25, Si 0.8, Cu 0.15, Ti 0.2, bal Al.

BIRMETAL BBZ 36
Birmetals Ltd.
Aluminum. Cu 0.3, Mg 3.5, Mn 0.3, Zn 6.5, Cr 0.3, bal Al.

BIRMETAL BMB 055
Birmetals Ltd.
Aluminum. Mg 0.5, Si 0.4, bal Al.
Extruded, annealed: 110 MPa TS; 75 MPa YS; 30 El. AA 6063.

BIRMETAL BMB 065
Birmetals Ltd.
Same as BMB 055 but more suitable for bright anodizing.

BIRMETAL BMB 071
Birmetals Ltd.
Aluminum. Mn 0.06, Mg 0.7, Si 1, bal Al.
Sheet and strip: 120 MPa TS; 20 El; 35 HV. Annealed. AA 6082.

BIRMETAL BMB 1306
Birmetals Ltd.
Aluminum. Cu 0.75, Mg 2.75, Mn 0.5, Zn 5.5, bal Al.

BIRMETAL BMB 2024
Birmetals Ltd.
Aluminum. Cu 4.4, Mn 0.6, Mg 1.5, bal Al.
Clad sheet and strip: 430 MPa TS; 285 MPa YS; 18 El (TB conditioned). AA 2024.

BIRMETAL BMB 2308
Birmetals Ltd.
Aluminum. Cu 1.75, Mg 2, Mn 0.2, Zn 6.5, Cr 0.15, bal Al.

BIRMETAL BMB 240
Birmetals Ltd.
Aluminum. Cu 2, Mg 0.5, bal Al.
Sheet and strip, TB condition: 290 MPa TS; 140 MPa YS; 24 El. AA 2117.

BIRMETAL BMB 473
Birmetals Ltd.
Aluminum. Cu 4, Mg 0.7, Mn 0.7, Si 0.3, bal Al.

BIRMETAL BMB 478
Birmetals Ltd.
Aluminum. Cu 4.3, Mn 0.8, Mg 0.7, Si 0.8, bal Al.
Sheet and strip: 440 MPa TS; 310 MPa YS; 18 El; 120 HV (TB condition). AA 2014.

BIRMETAL BMB 551
Birmetals Ltd.
Aluminum. Cu 4.5, Mg 0.5, Mn 0.7, Si 1, Fe 0.75, bal Al.

BIRMETAL BMB 761
Birmetals Ltd.
Cu 1, Mg 1, Si 1, bal Al.
W(T4): 38,100 TS; 22,400 YS; 15 El. WP(T6): 56,000 TS; 44,800 YS; 6 El. For structures, marine trim; age hardenable. *Obsolete*

BIRMETAL BMB Z12
Birmetals Ltd.
Cu 1.6, Mg 2.5, Cr 0.16, Zn 5.8.
Extruded, condition TF: 600 MPa TS; 560 MPa YS; 8 El; 190 HV. AA 7075.

BIRMETAL-230
Birmetals Ltd.
Cu 2.5, Mg 0.3, bal Al.
Heat treated: 43,000 TS; 24,000 YS; 27 El; 70 Brin. For rivets. Heat treatable, corrosion resistance. *Obsolete*

BIRMETAL-477
Birmetals Ltd.
Cu 4.4, Mg 0.7, Si 0.3, Mn 0.6, bal Al.
Heat treated: 70,000 TS; 60,000 YS; 13 El; 135 Brin. For rivets, fittings, hardware. High strength, age-hardenable, good fatigue strength. *Obsolete*

BIRMETAL-AM503
Birmetals Ltd.
Mn 1.5, bal Mg.
Plate: 29,000-40,000 TS; 11,000-22,000 YS; 5-14 El. Extrusion: 44,000 TS; 29,000 YS; 4 El; 55 Vickers. For fuel tanks, oil tanks, welded structures, bearing housings, valve and pump bodies, trailer bodies. Good weldability, sho *Obsolete*

BIRMETAL-AZ31
Birmetals Ltd.
Zn 1, Mn 0.3, Al 3, bal Mg.
Extrusion: 38,000 TS; 29,000 YS; 15 El; 49 Brin. H24 sheet: 42,000 TS; 32,000 YS; 15 El. For general purpose component parts, truck cabs, shipping containers. Good corrosion resistance, good bendability. *Obsolete*

BIRMETAL-AZM
Birmetals Ltd.
Magnesium. Mn 0.3, Al 6, bal Mg.

BIRMETAL-ZW1
Birmetals Ltd.
Magnesium. Zn 1.2, Zr 0.4-0.8, bal Mg.

BIRMETAL-ZW3
Birmetals Ltd.
Magnesium. Zn 3, Zr 0.4-0.8, bal Mg.

BIRMETAL-ZW6
Birmetals Ltd.
Zn 5.5, Zr 0.4-0.8, bal Mg.
Extruded: 49,000 TS; 38,000 YS; 10 El. Heat treated: 51,000 TS; 42,000 YS; 8 El. For floor beams, aircraft structural members. High strength, good corrosion resistant. Shock resistant. *Obsolete*

BIRMID 112
Birmingham Aluminium Casting Co.
Aluminum. Cu 0.7-2.5, Si 9-11.5, Ti 0-0.2, bal Al.
Cast: 36,000 TS; 13,500 YS; 1.5 El; 70 Brin. For instrument housings; die castings.

BIRMID 21
Birmingham Aluminium Casting Co.
Aluminum. Cu 2-4, Mn 0.3-0.7, Zn 0-0.2, Ti 0-0.2, bal Al.
Sand cast: 21,300 TS; 9000 YS; 1.7 El; 60 Brin. Permanent mold: 27,000 TS; 10,000 YS; 2.0 El; 70 Brin. For moderately stressed parts; not heat treatable.

BIRMID 298/304
Birmingham Aluminium Casting Co.
Aluminum. Cu 4-5, Ti 0.05-0.3, Si 0-0.25, bal Al.
Sand cast: 40,400 TS; 4 El. Permanent mold: 44,800 TS; 9 El. Heat treated: 42,600 TS; 27,000 YS; 5 El; 100 Brin. For high strength, shock resistant castings; age-hardenable.

BIRMID 300
Birmingham Aluminium Casting Co.
Aluminum. Mg 9.5-11, Ti 0.2, Si 0.25, Mn 0.1, bal Al.
Sand cast: 42,600 TS; 22,400 YS; 10 El; 76 Brin. Permanent mold; 49,300 TS; 24,600 YS; 18 El; 80 Brin. For high strength castings; age-hardenable.

BIRMID 428
Birmingham Aluminium Casting Co.
Aluminum. Cu 6-8, Si 2-4, Zn 2-4, Mn 0-0.6, Fe 0-1, bal Al.
Cast: 24,600 TS; 14,600 YS; 1 El; 80 Brin. For housings, gear cases; light stressed parts.

BIRMID D7
Birmingham Aluminium Casting Co.
Aluminum. Cu 2-4, Si 4-6, Mn 0.3-0.7, Ni 0.25, Zn 0.3, Ti 0.2, bal Al.
Sand cast: 20,200 TS; 2 El; 70 Brin. Permanent mold: 22,400 TS; 2 El; 75 Brin. For heavy duty castings; age-hardenable, good castability.

BIRMID D8
Birmingham Aluminium Casting Co.
Aluminum. Cu 2-4, Si 4-6, Mn 0.3-0.7, Ni 0.35, Zn 0.3, Ti 0.2, bal Al.
Sand cast: 20,200 TS; 89,600 YS; 2 El; 60 Brin. Heat treated: 47,000 TS; 40,000 YS; 1 El; 110 Brin. For gears, cases, hardware, instrument housings; age-hardenable, good castability.

BIRMIDAL
Birmingham Aluminium Casting Co.
Aluminum. Mg 0.5, Si 3.5-6, Ti 0-0.2, bal Al.
Cast: 25,000 TS; 16,000 YS; 2 10 El; 60 Brin. Heat treated. 44,000 TS; 32,000 YS; 1-5 El; 95 Brin. For gear cases, instrument housings; age-hardenable.

BIRMIDAL
Sterling International Technology Ltd.
Mg 0.3-0.8, Si 3.5-6, bal Al.
For grades of castings of same compositions, 18,000-43,000 TS. General purpose castings; properties vary with varying heat treatments. BS 1490 LM 8-M, 8-P, 8-W, 8-WP.

BIRMIDIUM "Y"
Birmingham Aluminium Casting Co.
Aluminum. Cu 3.5-4.5, Ni 1.8-2.3, Mg 1.2-1.7, bal Al.
Sand cast: 25,000 TS; 20,000 YS; 0.5 El; 85 Brin. Aged: 32,000 TS; 29,000 YS; 1.0 El; 100 Brin. For pistons, cylinder heads, crankcases, oil pans; age-hardenable, heat resistant.

BIRMIDIUM "Y" ALLOY
Birmingham Aluminium Casting Co.
Aluminum. Cu 3.5-4.5, Ni 1.8-2.0, Mg 1.2-1.7, bal Al.
Heat treated: 35,000-42,000 TS; 2-3 El. For automobile pistons, piston heads, crank cases; corrosion resistant.

BIRMINGHAM
English manufacture
Cu 50-62, Zn 20-32, Ni 12-30.
For German silver; corrosion resistant.

BIRMINGHAM 21

Birmingham Aluminium Casting Co.
Aluminum. Cu 2-4, Si 4-6, Zn 0-2, Ti 0-0.2, Mg 0-0.15, bal Al.
Sand cast: 21,300 TS; 9000 YS; 1.7 El; 60 Brin. Permanent mold: 27,000 TS; 11,000 YS; 2 El; 70 Brin. For cast parts subject to moderate stress; sand and permanent mold castings.

BIRMINGHAM 298/304

Birmingham Aluminium Casting Co.
Aluminum. Cu 4-5, Mg 0-0.1, Si 0-0.25, Ti 0.05-0.3, bal Al.
T4-temper: 34,800 TS; 14,000 YS; 12 El; 65 Brin. T6-temper: 42,000 TS; 27,000 YS; 5 El; 100 Brin. For shock resistant castings; sand and permanent mold cast, age-hardenable.

BIRMINGHAM 300

Birmingham Aluminium Casting Co.
Aluminum. Cu 0-0.1, Mg 9.5-11, Ti 0-0.2, Si 0-0.25, bal Al.
Sand cast: 43,000 TS; 22,400 YS; 10 El; 76 Brin. Permanent mold: 49,000 TS; 25,000 YS; 18 El; 80 Brin. For highly stressed castings; solution heat treated only.

BIRMINGHAM 428

Birmingham Aluminium Casting Co.
Aluminum. Cu 6-8, Si 2-4, Zn 2-4, Ti 0-0.2, bal Al.
Permanent mold: 25,000 TS; 12,500 YS; 1 El; 80 Brin. For castings for light stress; permanent mold casting.

BIRMINGHAM L.4

Birmingham Aluminium Casting Co.
Aluminum. Cu 2-4, Fe 0-1.2, Mg 0.5-1.5, Ti 0-0.2, Si 8.5-10.5, bal Al.
Permanent mold: 32,000 TS; 0 El; 90-120 Brin. For pistons; permanent mold castings.

BIRMINGHAM P.83

Birmingham Aluminium Casting Co.
Aluminum. Cu 3-4, Si 7.5-9.5, Fe 0-1.3, Mg 0-0.1, Zn 0-1, Sn 0-0.3, bal Al.
Die cast: 40,400 TS; 16,000 YS; 3 El; 85 Brin. For instruments, casings; die castings.

BIRSO

Birkett, Billington & Newton Ltd.
C, bal Fe.
For gears, housings, castings.

BISBO ARMA

General Motors Corp./Central Foundry Div
Si 1.4, Mn 0.4, Bi 0.025, TC 2.5, CC 0.7, bal Fe.
Heat treated: 70,000-100,000 psi TS; 48,000-80,000 psi YS; 2-4 El; 163-269 Brin. For crankshafts, rocker arms, universal joint yokes, and gears. Pearlitic malleable iron. *Obsolete*

BISCO

Bisset Steel Co.
C 0.4-0.45, Mn 0.85, Cr 1, Mo 0.2, bal Fe.
Heat treated: 285-341 Brin. Bars 3/4 round to 6 in. round; for shafts.

BISCO AIRPRO

Bisset Steel Co.
C 1, bal Fe.
For broaches, trimming dies; water hardening. *Obsolete*

BISCO ANNITE NO. 2

Bisset Steel Co.
C 1.5, Cr 11.5, V 0.25, Mo 0.8, bal Fe.
For punches, dies, gauges; nondeforming. *Obsolete*

BISCO BEST

Bisset Steel Co.
C 0.7-1.2, bal Fe.
For taps, chisels, cutters; water hardening. *Obsolete*

BISCO TOOL STEEL TUBING

Bisset Steel Co.
C 1, Mn 1, Cr 1.5, Mo 0.2, bal Fe.
For blanking and cold forming dies; oil hardening. *Obsolete*

BISHILITE NO. 1

Mitsubishi Metals America Corp.
Cr 30, W 12, C 2.5, Ni 0-3, Fe 0-3, bal Co.
Bare or covered electrode for hard facing. As welded: 46-54 Rock C. Excellent metal-to-metal wear. High pressure pump seal rings, crashers and cutters.

BISHILITE NO. 21

Mitsubishi Metals America Corp.
Cr 27, Mo 5, C 0.25, Ni 2.5, Fe 0-2, bal Co.
Bare or covered electrode for hard-facing applications. As welded: 20-33 Rock C. For high temperature and high pressure valve and hot forging dies.

BISHILITE NO. 6

Mitsubishi Metals America Corp.
Cr 28, W 4, C 1, Ni 0-3, Fe 0-3, bal Co.
Bare or uncovered electrode for hard facing. As welded: 37-44 Rock C. For metal-to-metal wear resistance and high temperature applications, valves for marine and power generators, and exhaust valves for engines.

BISHOP 29-17

Bishop Tube Co.
Ni 29, Co 17, Fe 53.
1/2 hard welded or seamless tubing: 85,000-100,000 psi TS; 60,000-80,000 psi YS; 10-15 El. Coefficient of thermal expansion: 6 x 10^6 in./in.·°C, 20-500°C. For glass to metal seals.

BISHOP 42

Bishop Tube Co.
Ni 42, Fe 57.
1/2 hard seamless tubing: 90,000-110,000 psi TS; 45,000-65,000 psi YS; coefficient of thermal expansion: 5.3x10^6 in./in.·°C, 20-500°C. For glass to metal seals.

BISHOP 52

Bishop Tube Co.
Ni 52, Fe 47.
1/2 hard welded or seamless tubing: 85,000-100,000 psi TS; 45,000-65,000 psi YS; 10-25 El. Coefficient of thermal expansion: 9.8x10^6 in./in.·°C, 20-500°C. For glass to metal seals.

BISMO 18-4-1

Bethlehem Steel Corp.
C 0.7-0.9, Cr 4, W 1.5, Mo 8.5, V 1.5, bal Fe.
For tools, cutters, reamers; high speed steel. *Obsolete*

BISMO M-2

Bisset Steel Co.
C 0.82, Cr 4, W 6.5, Mo 5, V 1.9, bal Fe.
For cutting tools; high speed steel. *Obsolete*

BISMO M-3

Bisset Steel Co.
C 1, Cr 4, W 6.2, Mo 6.2, V 2.4, bal Fe.
For lathe and planer tools, reamers, broaches, hobs; high speed steel, oil hardened. *Obsolete*

BISMO M-3

Bethlehem Steel Corp.
C 1, Cr 4, W 6.2, Mo 6.2, V 2.4, bal Fe.
For lathe and planer tools, reamers, broaches, hobs; high speed steel, oil hardened. *Obsolete*

BISMUTH

Atomergic Chemetals Corp.
Bi.
Purities, zone refined: 99.9999%, 99.9995%, 99.999%, 99.99%. Forms: rod, shot, needle, powder, ingot, foil, single crystals.

BISMUTH BRASS-1

American manufacture
Cu 52, Ni 30, Zn 12, Pb 5, Bi 1.
For ornaments, hardware; corrosion resistant.

BISMUTH BRASS-2

American manufacture
Cu 47, Ni 31, Zn 21, Bi 1, Sn 1.
For ornaments, hardware; corrosion resistant.

BISMUTH BRONZE

American manufacture
Cu 45-53, Ni 10-33, Zn 20-22, Sn 15-16, Bi 1, Al 0-0.1.
For ornaments, hardware; corrosion resistant.

BISON

Time Steel Service Inc.
C 0.5, Si 0.75, Cr 1.15, V 0.2, W 2.5, bal Fe.
Oil hardened shock resistant tool steel. AISI S1.

BIT & JAR STEEL

Midvale-Heppenstall Co.
C, Mn, bal Fe.
For tools, oil well tools. *Obsolete*

BKE

Russian manufacture
Cd 3-4, bal Cu.
For electrical contacts; high conductivity.

BKS

Thyssen Edelstahlwerke AG
C 0.85, Si 0.1-0.4, Mn 0.5-0.7, bal Fe.
Heat treated: 190,000 TS; 145,000 YS; 10 El; 30 RA; 400 Brin. For springs, tools, cutters, punches; Type W1; water hardened. *Obsolete*

BKS EXTRA

Thyssen Edelstahlwerke AG
C 0.8, Cr 1.1, V 0.1, bal Fe.
For bearings, liners, sleeves, cutters; water hardened, wear resistant. *Obsolete*

BKS SPECIAL

Thyssen Edelstahlwerke AG
C 0.85, Si 0.1-0.4, Mn 0.5-0.7, bal Fe.
Heat treated: 190,000 TS; 145,000 YS; 10 El; 30 RA; 400 Brin. For tools, dies, springs, cutters; Type W1; water hardened. *Obsolete*

BKS-3

Thyssen Edelstahlwerke AG
C 0.85, Cr 1, bal Fe.
For bearings, cutters, liners; water hardened, wear resistant. *Obsolete*

BLA-CALOY

Black-Clawson Co.
C 3.3, Mn 0.7, Si 2, Mg 0.05, bal Fe.
For gears, shafts, cams, housings; ductile cast iron.

BLACK BEAUTY

Empire Sheet & Tin Plate Co.
C 0.2, bal Fe.
For construction steel; black oxide sheet.

BLACK DEVIL

Champion Rivet Co.
C 0.09, Mn 0.3, bal Fe.
For welding rod; E 6020.

BLACK DEVIL E B

Champion Rivet Co.
C 0.07, Mn 0.39, Si 0.21, bal Fe.
Welded: 66,000 TS; 56,000 YS; 26 El; 45 RA. For welding electrodes; arc, E-6030.

BLACK DEVIL NO. 75

Champion Rivet Co.
C 0.09, Mn 0.37, Si 0.2, Mo 0.56, bal Fe.
Welded: 70,000-74,000 TS; 57,000-62,000 YS; 27-34 El; 50-65 RA. For welding electrodes; E-7020.

BLACK DIAMOND
Crucible Materials Corp.
Tool material. C 1, bal Fe.
For tools, cutters, drillers, dies, hammers; water hardening.
AISI W1.

BLACK HEART FERRITIC 22/14/14
Ley's Malleable Castings Co. Ltd.
C 2.3-2.6, Si 1.3-1.55, Mn 0.4-0.57, S 0.15-0.25, P 0-0.08, bal Fe.
Ferritic malleable iron castings. 49,300 TS; 31,400 YS; 14 El; 146 Brin. For automotive, truck and tractor industries.

BLACK LABEL
Peninsular Steel Co.
C 1.1, Cr 0.5, V 0.2, bal Fe.
For tools and dies; water hardening.

BLACK LABEL CAST (EXTRA)
Jessop-Saville Ltd.
C, bal Fe.
For tools and dies; water hardening.

BLACK STREAK ALNICO NO. 4
Wallace Murray Corp.
Al 12, Ni 28, Co 5, bal Fe.
For cast permanent magnets; high coercive force.

BLACKALLOY
Now BLACKALLOY TX90.

BLACKALLOY 525
Blackalloy Company of America, Inc.
Cobalt. Co 43-46, Cr 26-29, W 16-18, bal Fe.
Cast: 63-65 Rock C. For cutting and boring tools. Hard; abrasion and heat resistant. Nonmagnetic.

BLACKALLOY 700
Blackalloy Company of America. Inc.
Co 9-11, 89-91 WC.
91.4 RA. High impact and shock resistant; high strength, fine grained carbide.

BLACKALLOY TX-90
Blackalloy Company of America, Inc.
Cobalt. Co 40-42, Cr 26-29, W 19-21, bal Fe.
Cast: 66-67 Rock C. For tools, cutters, form tools, reamers, drills. Centrifugally cast. High heat and abrasion resistance. Nonmagnetic.

BLACKOR
Blackor Co.
W 87, C 9.
For oilwell core bits, cutters, tools; W_2C hard facing compound.

BLACKOR
Lincoln Electric Co.
W 87, C 9.
For hard facing electrode; abrasion resistant. *Obsolete*

BLACKSKIN ADMIRALTY
Phelps Dodge Industries
Cu 70, Zn 29, 1.0 Sn.

BLANCO B SPEZIAL
Robert-Leyer-Pritzkow & Co.
C 0-0.1, Cr 18, Ni 8.5, bal Fe.
Annealed: 80,000 TS; 35,000 YS; 55 El; 75 RA; 150 Brin. Cold drawn: 180,000 TS; 150,000 YS; 10 El; 250 Brin. For chemical plant equipment, tanks, mixers, vessels; Type 302; stainless, austenitic.

BLANCO B SUPER
Robert-Leyer-Pritzkow & Co.
C 0-0.1, Cr 18, Mo 2, Ni 9.5, bal Fe.
Annealed: 85,000 TS; 35,000 YS; 50 El; 65 RA; 160 Brin. Cold drawn: 150,000 TS; 135,000 YS; 6 El; 300 Brin. For acid resistant chemical plant equipment, tanks; Type 316; stainless, austenitic.

BLANCO BA
Robert-Leyer-Pritzkow & Co.
C 0.1, Cr 18, Ni 8.5, bal Fe.
Annealed: 80,000 TS; 35,000 YS; 55 El; 75 RA; 150 Brin. Cold drawn: 180,000 TS; 150,000 YS; 10 El; 250 Brin. For chemical plant equipment, tanks, fermenters; Type 302; stainless, austenitic.

BLANCO BA88
Robert-Leyer-Pritzkow & Co.
C 0.12, Cr 18, Ni 8, Mo 0.25, S 0.2, bal Fe.
Annealed: 80,000 TS; 35,000 YS; 40 El; 60 RA; 150 Brin. For screw machine products, fasteners; Type 303; stainless, free-cutting.

BLANCO CM2
Robert-Leyer-Pritzkow & Co.
C 0.35, Cr 16.5, Mo 1.15, bal Fe.
Annealed: 90,000 TS; 55,000 YS; 20 El; 45 RA; 180 Brin. For oil refinery equipment, furnace parts; corrosion and heat resistant.

BLANCO CN12
Robert-Leyer-Pritzkow & Co.
C 0-0.1, Cr 12.5, Ni 12, bal Fe.
For valves, pumps; corrosion and heat resistant.

BLANCO CNM
Robert-Leyer-Pritzkow & Co.
C 0-0.07, Cr 17, Ni 13, Mo 4.7, bal Fe.
Annealed: 90,000 TS; 40,000 YS; 45 El; 60 RA; 180 Brin. For acid resistant chemical plant equipment; Type 317; stainless, austenitic.

BLANCO CT
Robert-Leyer-Pritzkow & Co.
C 0-0.12, Cr 18, Ni 19.5, Ti = 4 x C, bal Fe.
Annealed: 85,000 TS; 35,000 YS; 55 El; 65 RA; 150 Brin. For welded structures, chemical plant equipment; Type 321; stainless, austenitic.

BLANCO CT2N
Robert-Leyer-Pritzkow & Co.
C 0-0.12, Cr 18, Ni 9.5, Cb = 8 x C, bal Fe.
Annealed: 90,000 TS; 45,000 YS; 56 El; 65 RA; 160 Brin. Cold drawn: 100,000 TS; 65,000 YS; 40 El; 60 RA; 202 Brin. For welded structures, chemical plant equipment; Type 347; stainless, austenitic.

BLANCO CU
Robert-Leyer-Pritzkow & Co.
C 0-0.07, Cr 17.5, Mo 2, Ni 17.5, Cu 2, Ti = 7 x C, bal Fe.
For valves, pumps, chemical plant equipment; corrosion and heat resistant.

BLANCO F
Vereinigte Edelstahlwerke
C 0.08, Cr 17, bal Fe.
Annealed: 80,000 TS; 50,000 YS; 25 El; 50 RA; 150 Brin. Cold drawn: 130,000 TS; 120,000 YS; 2 El; 200 Brin. For gears, shafts, furnace parts, heat treating boxes; corrosion and heat resistant; Type 430. *Obsolete*

BLANCO G
Robert-Leyer-Pritzkow & Co.
C 0.2, Cr 13, bal Fe.
Annealed: 95,000 TS; 50,000 YS; 25 El; 55 RA; 196 Brin. Cold drawn: 105,000 TS; 85,000 YS; 17 El; 50 RA; 215 Brin. For turbine blades, cutlery, valves, instruments; Type 420; stainless, hardenable.

BLANCO H
Now VEW N540.

BLANCO K(H)
Robert-Leyer-Pritzkow & Co.
C 0.4, Cr 13, bal Fe.
Annealed: 100,000 TS; 55,000 YS; 20 El; 50 RA; 200 Brin. For valves, cutlery, surgical and dental instruments; Type 420; stainless, hardenable.

BLANCO L1
Robert-Leyer-Pritzkow & Co.
C 0-0.12, Cr 13, bal Fe.
Annealed: 75,000 TS; 40,000 YS; 35 El; 70 RA; 155 Brin. Cold drawn: 100,000 TS; 85,000 YS; 17 El; 60 RA; 205 Brin. For turbine blades, surgical instruments; Type 410; stainless.

BLANCO L2
Robert-Leyer-Pritzkow & Co.
C 0-0.12, Cr 13, bal Fe.
Annealed: 75,000 TS; 40,000 YS; 35 El; 70 RA; 155 Brin. Cold drawn: 100,000 TS; 85,000 YS; 17 El; 60 RA; 205 Brin. For turbine blades, cutlery, surgical and dental instruments; Type 410; stainless.

BLANCO L3
Robert-Leyer-Pritzkow & Co.
C 0-0.12, Cr 13, Si 0.4, bal Fe.
Annealed: 75,000 TS; 40,000 YS; 35 El; 70 RA; 155 Brin. Cold drawn: 100,000 TS; 85,000 YS; 17 El; 60 RA; 205 Brin. For cutlery, turbine blades, instruments; Type 410; stainless.

BLANCO M
Now VEW N320.

BLANCO M1
Robert-Leyer-Pritzkow & Co.
C 0.2, Si 0.4, Cr 13, bal Fe.
Annealed: 95,000 TS; 50,000 YS; 25 El; 55 RA; 196 Brin. Cold drawn: 105,000 TS; 85,000 YS; 17 El; 50 RA; 215 Brin. For cutlery, valves, turbine blades, knives; Type 420; stainless, hardenable.

BLANCO M2
Robert-Leyer-Pritzkow & Co.
C 0.22, Si 0.4, Cr 17, Ni 1.5, bal Fe.
Annealed: 125,000 TS; 95,000 YS; 20 El; 55 RA; 260 Brin. Cold drawn: 130,000 TS; 110,000 YS; 15 El; 35 RA; 270 Brin. For pumps, marine hardware, valve trim, shafts; Type 431; stainless

BLANCO M3
Robert-Leyer-Pritzkow & Co.
C 0.4, Si 0.4, Cr 13, bal Fe.
Annealed: 95,000 TS; 50,000 YS; 25 El; 55 RA; 195 Brin. Cold drawn: 105,000 TS; 85,000 YS; 17 El; 50 RA; 215 Brin. For valves, pumps, turbine blades, cutlery, knives; Type 420; stainless, hardenable.

BLANCO M3E
Robert-Leyer-Pritzkow & Co.
C 0.9, Si 0.4, Cr 18, Mo 1.15, V 1, bal Fe.
For valves, bearings, instrument pivots, rollers; corrosion and wear resistant, hardenable.

BLANCO M4
Robert-Leyer-Pritzkow & Co.
C 0.12, Cr 16.5, Mo 0.25, S 0.2, bal Fe.
Annealed: 80,000 TS; 50,000 YS; 25 El; 50 RA; 160 Brin. For gears, shafts, screw machine products; Type 430F; stainless, free-cutting.

BLANCO RCM
Robert-Leyer-Pritzkow & Co.
C 0.2, Cr 13, Mo 1.15, bal Fe.
Annealed: 95,000 TS; 50,000 YS; 25 El; 55 RA; 196 Brin. For valves, cutlery, valve trim, turbine blades; Type 420 Mo; stainless, hardenable.

BLANCO SPEZIAL
Robert-Leyer-Pritzkow & Co.
C 0.07, Cr 18, Ni 9.5, bal Fe.
Annealed: 85,000 TS; 35,000 YS; 60 El; 70 RA; 150 Brin. Cold drawn: 180,000 TS; 125,000 YS; 10 El; 330 Brin. For welded structures, chemical plant equipment, tanks; Type 304; stainless, austenitic.

BLANCO SPEZIAL
Now VEW A550.

BLANCO SUPER

Robert-Leyer-Pritzkow & Co.
C 0-0.07, Cr 18, Mo 2, Ni 10.5, bal Fe.
Annealed: 85,000 TS; 35,000 YS; 50 El; 65 RA; 160 Brin. Cold drawn: 150,000 TS; 135,000 YS; 6 El; 300 Brin. For acid resistant chemical plant equipment, tanks; Type 316; stainless, austenitic.

BLANCO SUPER EXTRA

Robert-Leyer-Pritzkow & Co.
C 0-0.12, Cr 18, Mo 2, Ni 10.5, Ti = 4 x C, bal Fe.
Annealed: 85,000 TS; 35,000 YS; 50 El; 65 RA; 160 Brin. Cold drawn: 150,000 TS; 135,000 YS; 6 El; 300 Brin. For welded structures, chemical plant equipment, tanks. Type 316 Ti; stainless, austenitic.

BLANCO SUPER EZN

Robert-Leyer-Pritzkow & Co.
C 0-0.12, Cr 18, Mo 2, Ni 10.5, Cb = 8 x C, bal Fe.
Annealed: 85,000 TS; 35,000 YS; 50 El; 65 RA; 160 Brin. For welded structures, chemical plant equipment, tanks; Type 316 Cb; stainless, austenitic.

BLANCO T70

Robert-Leyer-Pritzkow & Co.
C 0-0.1, Cr 17.5, Ti = 7 x C, bal Fe.
Annealed: 80,000 TS; 50,000 YS; 25 El; 50 RA; 150 Brin. For heat treating boxes, furnace parts, rabble arms; Type 430 Ti; stainless, ferritic.

BLANCO W

Robert-Leyer-Pritzkow & Co.
C 0-0.12, Si 0.4, Cr 13, bal Fe.
Annealed: 75,000 TS; 40,000 YS; 35 El; 70 RA; 155 Brin. Cold drawn: 100,000 TS; 85,000 YS; 17 El; 60 RA; 205 Brin. For valve trim, turbine blades, cutlery; Type 410; stainless.

BLANKO-BLECH

Robert-Leyer-Pritzkow & Co.
Cu 80, Ni 20.
For turbine blades, condenser tubes; corrosion resistant.

BLATT (LEAF) GOLD

English manufacture
Cu 77, Zn 23.
For gold leaf substitute, signs; corrosion resistant.

BLATT (LEAF) SILVER

English manufacture
Sn 91, Zn 8.3, Pb 0.4, Fe 0.2.
For bearings, tin foil for wrappers; Babbitt.

BLAUPUNKT

SWB Stahlformguss Gesellschaft mbH
C 0.82, Cr 4.1, Mo 0.85, V 1.6, W 8.7, bal Fe.
For lathe and planer tools, drills, reamers, taps; high speed steel. *Obsolete*

BLAW-KNOX C-1

Duraloy Blaw-Knox
C 0.08-0.2, Mn 0.5-1.1, Si 0.2-0.4, bal Fe.
Cast: 55,000 TS; 27,000 YP; 25 El; 45 RA; 135-155 Brin. For general machinery castings, housings, shafts, case-hardened parts. *Obsolete*

BLAW-KNOX C-1-0

Duraloy Blaw-Knox
C 0-0.3, Mn 0-0.7, Si 0.35-0.45, bal Fe.
Cast: 60,000 TS; 30,000 YS; 30 El; 45 RA; 135-155 Brin. For general machinery castings, gears, housings, casings. Water hardening. *Obsolete*

BLAW-KNOX C-1-1

Duraloy Blaw-Knox
C 0.2-0.25, Mn 0.6-0.8, Si 0.35-0.45, bal Fe.
Cast: 60,000 TS; 30,000 YP; 30 El; 45 RA; 135-155 Brin. For general machinery castings, gears, housings, casings. Water hardening. *Obsolete*

BLAW-KNOX C-1-2

Duraloy Blaw-Knox
C 0.2-0.3, Mn 0.6-0.8, Si 0.3-0.45, bal Fe.
Cast: 65,000 TS; 33,000 YP; 26 El; 40 RA; 140-155 Brin. For general purpose castings, gears, shafts, housings. Water hardening. *Obsolete*

BLAW-KNOX C-1-3

Duraloy Blaw-Knox
C 0.3-0.35, Mn 0.65-0.85, Si 0.3-0.45, bal Fe.
Cast: 70,000 TS; 35,000 YS; 24 El; 36 RA; 145-160 Brin. For general machinery castings, gears, shafts, casings. Water hardening. *Obsolete*

BLAW-KNOX C-1-4

Duraloy Blaw-Knox
C 0.35-0.4, Mn 0.7-0.9, Si 0.3-0.45, bal Fe.
Cast: 75,000 TS; 40,000 YP; 22 El; 33 RA; 150-165 Brin. For general machinery castings, housings, gears, shafts. Water hardening. *Obsolete*

BLAW-KNOX C-1-5

Duraloy Blaw-Knox
C 0.4-0.45, Mn 0.7-0.9, Si 0.3-0.45, bal Fe.
Cast: 80,000 TS; 42,000 YP; 20 El; 30 RA; 155-170 Brin. For general machinery castings, housings, shafts, gears. Water hardening. *Obsolete*

BLAW-KNOX C-1-6

Duraloy Blaw-Knox
C 0.5-0.55, Mn 0.7-0.9, Si 0.3-0.45, bal Fe.
Cast: 85,000 TS; 45,000 YP; 18 El; 25 RA; 160-175 Brin. For gears, shafts, housings, general purpose castings. Water hardening. *Obsolete*

BLAW-KNOX C-2

Duraloy Blaw-Knox
C 0.27-0.32, Mn 0.9-1.3, Si 0.3-0.45, Ni 0.3-0.4, Mo 0.1-0.18, bal Fe.
Cast: 85,000 TS; 62,000 YP; 22 El; 50 RA; 170-300 Brin. For gears, countershafts, axles, housings, castings. Oil hardening, shock resistant. *Obsolete*

BLAW-KNOX C-2-A

Duraloy Blaw-Knox
C 0.32-0.38, Mn 1.05-1.25, Si 0.3-0.45, Ni 0.3-0.4, Mo 0.1-0.18, bal Fe.
Cast: 90,000 TS; 65,000 YP; 20 El; 45 RA; 185-310 Brin. For gears, housings, shafts. Oil hardening. *Obsolete*

BLAW-KNOX C-3-A

Duraloy Blaw-Knox
C 0.35-0.4, Mn 1.15-1.35, Si 0.3-0.45, Mo 0.2, bal Fe.
Cast: 95,000 TS; 65,000 YP; 20 El; 45 RA; 190-320 Brin. For gears, axles, shafts, housings. Oil or water hardening. *Obsolete*

BLAW-KNOX C-4

Duraloy Blaw-Knox
C 0.3-0.4, Mn 1-1.2, Si 0.3-0.45, Mo 0.2, V 0.1, bal Fe.
Cast: 80,000 TS; 50,000 YP; 22 El; 40 RA; 150-170 Brin. For gears, housings, axles, shafts. Water or oil hardening. *Obsolete*

BLAW-KNOX C-5

Duraloy Blaw-Knox
C 0.25-0.35, Mn 0.8-1, Si 0.3-0.45, Ni 0.7-0.8, Mo 0.35-0.4, bal Fe.
Cast: 90,000 TS; 65,000 YP; 24 El; 50 RA; 180-330 Brin. For gears, shafts, axles, bolts. Water or oil hardening. *Obsolete*

BLAW-KNOX C-6

Duraloy Blaw-Knox
C 0.4-0.5, Mn 0.5, Si 0.4, Cr 2.75-3.25, Mo 0.3-0.4, bal Fe.
Cast: 110,000 TS; 90,000 YP; 16 El; 40 RA; 225-400 Brin. For gears, shafts, axles. Oil hardening, tough. *Obsolete*

BLAW-KNOX C-6-M

Duraloy Blaw-Knox
C 0.25-0.3, Mn 0.5-0.6, Si 0.3-0.45, Cr 2.75-3.25, Mo 0.2-0.3, bal Fe.
Cast: 100,000 TS; 75,000 YP; 19 El; 50 RA; 225-350 Brin. For gears, shafts, housings. Oil hardening, shock resistant. *Obsolete*

BLAW-KNOX C-7

Duraloy Blaw-Knox/Union Steel Casting
C 0.3-0.4, Mn 0.8-1, Si 0.3-0.45, Cr 0.8, Ni 1.5, Mo 0.4, bal Fe.
Cast: 125,000 TS; 100,000 YP; 15 El; 35 RA; 220-365 Brin. For gears, shafts, housings. Shock resistant, tough.

BLAW-KNOX C-8

Duraloy Blaw-Knox/Union Steel Casting
C 0.25-0.35, Mn 0.7-0.9, Si 0.3-0.45, Cr 0.5, Ni 0.5, Mo 0.4, bal Fe.
Cast: 115,000 TS; 90,000 YP; 22 El; 50 RA; 200-325 Brin. For gears, shafts, housings. Shock resistant. Water hardening.

BLEIZINNBRONZE 10

Oederlin & Co. Ltd.
Copper. Zn, Sn, bal Cu.
Cast: 26,000 TS; 15 El; 70 Brin. For hardware; bronze.

BLEIZINNBRONZE 7

Oederlin & Co. Ltd.
Zn, Sn, bal Cu.
Cast: 25,000 TS; 8 El; 100 Brin. For hardware; bronze.

BLEIZINNBRONZE 8

Oederlin & Co. Ltd.
Copper. Zn, Sn, bal Cu.
Cast: 22,000 TS; 8 El; 60 Brin. For hardware; bronze.

BLENDALLOY 22-1000 HIGH PURITY NICKEL

Spang Specialty Metals
C 0-0.005, Fe 0-0.005, bal Ni.
Powder metallurgy product in sheet, strip, rod, wire and billet form suitable for plater bars, tube cathodes, fluorescent lamp components, and resistance thermometers.

BLENDALLOY 22-9604

Spang Specialty Metals
C 0-0.005, Fe 0-0.005, W 3.8-4.2, bal Ni.
Powder metallurgy product in sheet, strip, rod, wire, and billet form. Suitable for specialized electronic and magnetic applications.

BLENDALLOY 22-9800

Spang Specialty Metals
C 0-0.005, Fe 0-0.005, Ni 98, Ti 0.5, Mg 0.25.
For springs requiring high electrical conductivity and parts requiring good thermal conductivity. High purity. *Obsolete*

BLENDALLOY 25-4200 ("42-ALLOY")

Spang Specialty Metals
C 0-0.005, Ni 42, Fe 58.
For electronic, magnetic and electrical applications. Glass sealing. High purity. *Obsolete*

BLENDALLOY 25-4206

Spang Specialty Metals
C 0-0.005, Ni 43, Fe 48, Ti 2.5, Cr 5, Al 0.5.
For springs, mechanical filters, tuning forks. High purity. *Obsolete*

BLENDALLOY 25-4400

Spang Specialty Metals
Ni 44, 0.02 others max, bal Fe.
Free from non-metallic inclusions. For mechanical filters. *Obsolete*

BLENDALLOY 25-4601 FM

Spang Specialty Metals
C 0-0.005, Ni 46, Fe 53, Mn 0.5, Se 0.1.
For glass sealing applications. High purity. *Obsolete*

BLENDALLOY 25-4803
Spang Specialty Metals
C 0-0.005, Ni 48, Fe 49, Mo 3.
High magnetic permeability with minimum energy loss. For communication and electronic equipment. High purity. *Obsolete*

BLENDALLOY 25-5000 (ORTHONOL)
Spang Specialty Metals
C 0-0.005, Ni 50, Fe 50.
Soft magnetic alloy with very high squareness and high core gain. For saturable reactors, magnetic amplifiers, switching devices, and power inverter-converter applications. High purity. *Obsolete*

BLENDALLOY 25-5025 (ORTHONAL)
Spang Specialty Metals
Ni 47.8-48.2, Mn 0.15-0.3, 0.02 others max, bal Fe.
Powder metallurgy product in foil form. Soft magnetic strip with an extremely square hysteresis loop. Magnetic properties vary with thickness.

BLENDALLOY 25-5200 ("52" ALLOY)
Spang Specialty Metals
C 0-0.005, Ni 52, bal Fe.
Precise thermal expansion, uniformly fine-grained microstructure, excellent electrical and thermal conductivity. For glass sealing in dry reed switches and mercury wetted relays. High purity. *Obsolete*

BLENDALLOY 25-7904 (PERMALLOY)
Spang Specialty Metals
C 0-0.005, Ni 80, Fe 15, Mo 5.
Soft magnetic alloy with high initial permeability. For use in transformers, reed relays, amplifiers, vacuum tubes, microphones, loud speakers, cables. High purity. *Obsolete*

BLENDALLOY 25-8000
Spang Specialty Metals
C 0-0.005, Ni 80, bal Fe.
For electronic and magnetic applications. High purity. *Obsolete*

BLENDALLOY 25-8004
Spang Specialty Metals
Ni 80-80.4, Mo 4.1-4.4, C 0-0.003, 0.02 others max, bal Fe.
Soft magnetic powder metallurgy product in foil form with square hysteresis loop and low coercive force.

BLENDALLOY 25-8004 (PERMALLOY)
Spang Specialty Metals
C 0-0.005, Ni 80, Fe 15, Mo 5.
Soft magnetic alloy with high squareness and high core gain, low coercive force. For preamplifiers and modifiers, converters. High purity. *Obsolete*

BLENDALLOY 25-8300
Spang Specialty Metals
C 0-0.005, Ni 83, bal Fe.
For electronic and magnetic applications. High purity. *Obsolete*

BLENDALLOY 26-5446
Spang Specialty Metals
Cu 46, C 0.002, bal Ni.
For electronic tube and glass-to-metal sealing applications. *Obsolete*

BLENDALLOY 32-9010
Spang Specialty Metals
C 0-0.005, Fe 0-0.005, Co 90, Ni 10.
For electronic, magnetic, magnetostrictive, and plating applications. *Obsolete*

BLENDALLOY 33-1000, GRADE 1
Spang Specialty Metals
Co 99.5, 10 ppm Cu, 900 ppm Fe, 800 ppm Ni, 150 ppm S.
For use in electroplating, alloying. *Obsolete*

BLENDALLOY 33-9505
Spang Specialty Metals
Fe 5, 10 ppm Cu, 1000 ppm Ni, 300 ppm S, bal Co.
Ductile bar and strip. *Obsolete*

BLITZ FEE
Fakirstahl Hoffmanns GmbH & Co. KG
C 0.86, Cr 4.3, Mo 0.85, V 2.1, W 12, bal Fe.
For lathe and planer tools, drills, taps, hobs; high speed steel.

BLOCK BRASS
American manufacture
Cu 66.5, Zn 32, Pb 1.5.
For free machining brass parts.

BLOCK'S ALLOY
American manufacture
Co 54, Ni 45, Si 0.9.
For tools; corrosion and heat resistant.

BLOMBIT
Allgemeine Elektrizitats-Gesellschaft
Ag 2-7, Cu 93-98.
Forged: 72,000-78,000 TS; 5-8 El; 120-140 Brin. For electrodes and tips for spot welding; 83-90% conductivity.

BLUE ANCHOR DRILL ROD
Teledyne Vasco
C 1.2-1.35, bal Fe.
Cold drawn: 85,000 TS; 60,000 YS; 25 El; 50 Brin. For tools, taps, drills, reamers; water hardening. *Obsolete*

BLUE CHIP
Teledyne Firth Sterling
Steel. C 0.7, Mn 0.25, Cr 4, W 18, V 1, bal Fe.
Annealed: 105,100 TS; 75,000 YS; 14 El; 17 RA; 230 Brin. For tools, cutters, reamers, drills, punches, dies, shears, taps; high speed steel. *Obsolete*

BLUE CHIP SUPERIOR
Teledyne Firth Sterling
C 0.67, Mn 0.22, Cr 4.12, W 15.91, V 0.71, bal Fe.
For tools, high speed cutter; high speed steel. *Obsolete*

BLUE DEVIL 100
Champion Rivet Co.
C 0.5, Mo 0.5, bal Fe.
103,000 TS; 100,000 YS; 19 El; 47 RA. For welding rod; flux coated. *Obsolete*

BLUE DEVIL 85
Champion Rivet Co.
C 0.2, Mo 0.2, bal Fe.
85,000 TS; 75,000 YS; 24 El; 52 RA. For welding rod; flux coated. *Obsolete*

BLUE DEVIL YOLOY
Champion Rivet Co.
C 0.2, Cu 0.5, bal Fe.
80,000 TS; 75,000 YS; 25 El; 50 RA. For welding rod, flux coated.

BLUE DIAMOND
Northfield Iron Co.
C, alloy, bal Fe.
For road machinery cutting edges. Wear resistant.

BLUE DOT SPECIAL
Allied Steel & Tractor Products Inc.
C, alloy, bal Fe.
For maintenance and repair operations; oil hardening.

BLUE GOLD
American manufacture
Au 75, Fe 25.
For ornaments, jewelry; corrosion resistant.

BLUE LABEL
Peninsular Steel Co.
C 1-1.1, V 0.15-0.25, Mn, Si, bal Fe.
For cold swaging dies, forming rolls, knives, punches, collets, pipe cutters, drill bushings. AISI W2 water hardening tool steel.

BLUE LABEL
Ackerlind Steel Co., Inc.
C 1, Mn 0.3, Si 0.25, bal Fe.
Type W1 water hardening tool steel.

BLUE LABEL
Edgar T. Ward's Sons Co.
C 0.8-1.3, bal Fe.
For tools. *Obsolete*

BLUE LABEL
Wallace Murray Corp.
C 0.8-1.1, bal Fe.
For general tools; "Heller's Special Tool."

BLUE LABEL
Adams & Osgood Steel Co.
C, bal Fe.
For tools, drills, taps; water hardened.

BLUE LABEL EXTRA
Wallace Murray Corp.
C 0.6-1.4, Si 0.25, Mn 0.25, bal Fe.
For drills, taps, cutters, punches; Type W1; water hardening.

BLUE LABEL SF-2
Peninsular Steel Co.
Same as blue label but surface finished on two sides.

BLUE RIBAND VICTORY NO. 7
H. Russell & Co. Ltd.
C 0.7, W 18, Cr 4, V 1, Co 5, bal Fe.
For cutters, tools; high speed steel.

BLUE SEAL
Sanderson Kayser Ltd.
C 0.9-1, bal Fe.
For tools, gauges, springs; water hardened. *Obsolete*

BLUE STREAK 18-4-1
Diehl Steel Co.
C 0.72, Cr 4, W 18, V 1, bal Fe.
For cutting tools; high speed steel.

BLUE STREAK COBALT
Diehl Steel Co.
C 1.25, Cr 4, V 3, W 9, Mo 3, Co 9, bal Fe.
High speed steel, extra wear and red hardness properties; tungsten-cobalt type.

BLUE STREAK MOLY
Diehl Steel Co.
C 0.8, Cr 4, W 5.75, Mo 4.5, V 1.6, bal Fe.
For cutting tools; high speed steel.

BLUE TIP NAVAL BRASS
Mueller Brass Co.
Cu 60, Sn 0.75, bal Zn.
Hard: 68,000 TS; 45,000 YS; 35 El. Annealed: 63,000 TS; 40,000 YS; 40 El. For bolts, nuts, marine hardware, valve stems; corrosion resistant. *Obsolete*

BLUEDAC
Champion Rivet Co.
C 0.08, Mn 0.57, Si 0.21, bal Fe.
Cast: 63,000 TS; 55,000 YS; 28 El; 55 RA. Welded: 73,000 TS; 64,000 YS; 23 El; 35 RA. For welding electrodes; arc welding all positions; E-6011.

BM
Thyssen Edelstahlwerke AG
C 1.1, Mn 13, Si 0.25, bal Fe.
For welding electrodes. *Obsolete*

BM 2

Acciaierie Valbruna s.p.a.
C 0.75-0.9, Cr 3.5-4.5, Mo 4.5-5.5, W 5.5-7, V 1.5-2.2, bal Fe.
Mo-W high speed tool steel. W. Nr. 1.3343; AISI M2.
Obsolete

BM 78

Sintered Products Ltd.
Cu 9, C 1, bal Fe.
5.9-6.2 g/cm^3 density; 34.5 kg/mm^2 TS; 1 El; 5-200 Vickers
min. Sintered; machinable.

BMS

Thyssen Edelstahlwerke AG
C 1.1, Cr 1.1, W 1.1, bal Fe.
For metal saws; water hardened. *Obsolete*

BMS EXTRA

Thyssen Edelstahlwerke AG
C, W, bal Fe.
For metal saws; water hardened. *Obsolete*

BMSC

Thyssen Edelstahlwerke AG
C 1.15, Cr 0.65, V 0.1, bal Fe.
For bearings, sleeves, liners; oil or water hardened. *Obsolete*

BMSW

Thyssen Edelstahlwerke AG
C 1.15, W, Cr, bal Fe.
For cutters, bearings; water hardened. *Obsolete*

BNC STEEL

English manufacture
C, Cr, Ni, bal Fe.
Hardened: 194,000 TS; 15 El; 50 RA; 400 Brin. For gears,
pinions; case hardening.

BND

Billiton International Metals B.V.
C 0.7, Cr, bal Fe.
For tools, gripper dies; tough.

BO-STAN

Alloy Technology International, Inc.
Stainless steel. B 1.2.
Type 304 stainless steel. Annealed: 104,500 TS; 7.5 El; 5.5
RA; 220 Brin. Cold worked: 420 Brin. For pressurized water
reactors and controls. Sintered, corrosion and heat resistant.
Obsolete

BOBIERRE'S METAL

French manufacture
Cu 58-66, Zn 34-42.
For bolts, nuts, sheathing; yellow brass.

BOCHLET FCHD

Berted Foundry Co.
C 0-0.14, Mn 17-19, Cr 10-13, Si 0.5-2, Ti 0.6, bal Fe.
For turbine blades, jet engine components; austenitic, heat
resistant.

BOCHUM 85C7

SWB Stahlformguss Gesellschaft mbH
C 0.85, Cr 1.75, Mn 0.35, bal Fe.
For bearings, liners, forming dies; water hardened, wear
resistant. *Obsolete*

BOCHUM BC13V

SWB Stahlformguss Gesellschaft mbH
C 1.15, Cr 0.65, bal Fe.
For blanking and forming dies; water hardened, wear
resistant. *Obsolete*

BOCHUM BCMH

SWB Stahlformguss Gesellschaft mbH
C 1, Mn 1, Si 0.37, bal Fe.
Annealed: 100,000 TS; 53,000 YS; 21 El; 42 RA; 200 Brin. For
drills, taps, reamers, cutters; Type W1; water hardened.
Obsolete

BOCHUM BCMW

SWB Stahlformguss Gesellschaft mbH
C 0.9, Si 0.37, Mn 1, bal Fe.
Heat treated: 190,000 TS; 145,000 YS; 10 El; 30 RA; 400 Brin.
For drills, taps, reamers, broaches, hobs; Type W1; water
hardened. *Obsolete*

BOCHUM BCR

SWB Stahlformguss Gesellschaft mbH
C 0.85, Cr, bal Fe.
For bearings, liners, sleeves; water hardened, wear resistant.
Obsolete

BOCHUM BCVH

SWB Stahlformguss Gesellschaft mbH
C 0.9, Cr, V, bal Fe.
For bearings, liners, sleeves; water hardened, wear resistant.
Obsolete

BOCHUM BCVW

SWB Stahlformguss Gesellschaft mbH
C 0.8, Cr, V, bal Fe.
For bearings, liners, sleeves; water hardened, wear resistant.
Obsolete

BOCHUM BFK6

SWB Stahlformguss Gesellschaft mbH
C 0.6, Si 0.25-0.5, Mn 0.3-0.8, bal Fe.
Heat treated: 160,000 TS; 113,000 YS; 12 El; 40 RA; 325 Brin.
For gears, rails, punches, hammers; water hardened.
Obsolete

BOCHUM BFK7

SWB Stahlformguss Gesellschaft mbH
C 0.75, Si 0.25-0.5, Mn 0.3-0.8, bal Fe.
Heat treated: 185,000 TS; 140,000 YS; 15 El; 40 RA; 400 Brin.
For springs, tools, punches, crimpers; water hardened.
Obsolete

BOCHUM BFK8

SWB Stahlformguss Gesellschaft mbH
C 0.9, Si 0.25-0.5, Mn 0.3-0.8, bal Fe.
Heat treated: 190,000 TS; 145,000 YS; 10 El; 30 RA; 400 Brin.
For springs, tools, cutters, drills; Type W1; water hardened.
Obsolete

BOCHUM BGK9

SWB Stahlformguss Gesellschaft mbH
C 0.85, Si 0.1-0.4, Mn 0.5-0.7, bal Fe.
For springs, tools, drills, taps, reamers; water hardened; Type
W1. *Obsolete*

BOCHUM BSW10

SWB Stahlformguss Gesellschaft mbH
C 1.2, W, Cr, bal Fe.
For cutters, bearings; water hardened, wear resistant.
Obsolete

BOCHUM BSW18

SWB Stahlformguss Gesellschaft mbH
C 1.2, W, Cr, bal Fe.
For cutters, bearings; water hardened, wear resistant.
Obsolete

BOCHUM C58G

SWB Stahlformguss Gesellschaft mbH
C 0.25, Cr 14.5, Ni 0-0.1, bal Fe.
For cutlery, valves, surgical and dental instruments; stainless,
hardenable. *Obsolete*

BOCHUM C68G

SWB Stahlformguss Gesellschaft mbH
C 0.25, Cr 17, Ni 0-1.8, bal Fe.
Annealed: 125,000 TS; 95,000 YS; 20 El; 55 RA; 260 Brin. For
pumps, marine hardware, cutlery; corrosion and heat
resistant. *Obsolete*

BOCHUM CSG

SWB Stahlformguss Gesellschaft mbH
C 0.6, Cr, V, bal Fe.
For springs, gears, crankshafts; oil hardened, shock resistant.
Obsolete

BOCHUM D10

SWB Stahlformguss Gesellschaft mbH
C 1, Si 0-0.25, Mn 0-0.25, bal Fe.
Annealed: 100,000 TS; 53,000 YS; 20 El; 42 RA; 200 Brin. For
springs, cutters, hobs, drills, taps; Type W1; water hardened.
Obsolete

BOCHUM D11

SWB Stahlformguss Gesellschaft mbH
C 1.15, Si 0-0.25, Mn 0-0.25, bal Fe.
Annealed: 110,000 TS; 56,000 YS; 18 El; 40 RA; 210 Brin. for
springs, cutters, taps, drills, hobs; Type W1; water hardened.
Obsolete

BOCHUM D13

SWB Stahlformguss Gesellschaft mbH
C 1.3, Si 0-0.25, Mn 0-0.25, bal Fe.
For engravers' tools, blanking and forming dies; Type W1;
water hardened. *Obsolete*

BOCHUM D7

SWB Stahlformguss Gesellschaft mbH
C 0.7, Si 0-0.25, Mn 0-0.25, bal Fe.
Heat treated: 175,000 TS; 128,000 YS; 12 El; 37 RA; 355 Brin.
For rails, axes, hammers, crimpers; water hardened; Type
W1. *Obsolete*

BOCHUM D8

SWB Stahlformguss Gesellschaft mbH
C 0.85, Si 0-0.25, Mn 0-0.25, bal Fe.
Heat treated: 190,000 TS; 145,000 YS; 10 El; 30 RA; 400 Brin.
For drills, taps, springs, reamers; Type W1; water hardened.
Obsolete

BOCHUM EK10

SWB Stahlformguss Gesellschaft mbH
C 0.1, Si 0.25, Mn 0.37, bal Fe.
Cold drawn: 72,000 TS; 60,000 YS; 22 El; 58 RA; 145 Brin.
For gears, shafts, fasteners; case hardening steel. *Obsolete*

BOCHUM EK15

SWB Stahlformguss Gesellschaft mbH
C 0.15, Si 0.25, Mn 0.37, bal Fe.
Annealed: 70,000 TS; 40,000 YS; 25 El; 60 RA; 145 Brin. For
gears, shafts, fasteners; case hardening steel. *Obsolete*

BOCHUM EU

SWB Stahlformguss Gesellschaft mbH
C 0.15, Si 0.25-0.5, Mn 0.3-0.8, bal Fe.
Annealed: 70,000 TS; 40,000 YS; 25 El; 60 RA; 145 Brin. For
machine tool parts, gears, fasteners; case hardened.
Obsolete

BOCHUM K55 SPEZIAL

SWB Stahlformguss Gesellschaft mbH
C 0.55, Si 0.1-0.4, Mn 0.5-0.7, bal Fe.
Annealed: 100,000 TS; 55,000 YS; 15 El; 20 RA; 180 Brin. For
axles, gears, bolts, shafts; water hardened. *Obsolete*

BOCHUM K65

SWB Stahlformguss Gesellschaft mbH
C 0.6, Si 0.1-0.4, Mn 0.5-0.7, bal Fe.
Heat treated: 115,000-160,000 TS; 77,000-113,000 YS; 12-23
El; 230-320 Brin. For wheels, die blocks, girders, clutch discs;
water hardened. *Obsolete*

BOCHUM KPN

SWB Stahlformguss Gesellschaft mbH
C 0.5, Cr 1.05, Ni 3.25, Mn 0.5, bal Fe.
For gears, bolts, crankshafts, axles; oil hardened, shock
resistant. *Obsolete*

BOCHUM M100

SWB Stahlformguss Gesellschaft mbH
C 0.22, Cr 1.25, Mn 1.3, bal Fe.
For gears, bolts, camshafts, cams; water or oil hardened.
Obsolete

BOCHUM M80

SWB Stahlformguss Gesellschaft mbH
C 0.16, Cr 0.95, Mn 1.15, Si 0.25, bal Fe.
For gears, bolts, camshafts, cams; case hardened, tough.
Obsolete

BOCHUM M90

SWB Stahlformguss Gesellschaft mbH
C 0.2, Mn 1.25, Cr 1.15, bal Fe.
For gears, bolts, camshafts, cams; case hardened, tough.
Obsolete

BOCHUM MMB3

SWB Stahlformguss Gesellschaft mbH
C, alloy, bal Fe.
For machine tool parts; oil hardened. *Obsolete*

BOCHUM MN18G

SWB Stahlformguss Gesellschaft mbH
C 1.3, Mn, Cr, bal Fe.
For cutters, wear plates; oil or water hardened. *Obsolete*

BOCHUM MNH(G)

SWB Stahlformguss Gesellschaft mbH
C 1.2, Mn 12.5, Si 0.4, bal Fe.
For tracks, frogs, dipper teeth shovels; wear and abrasion
resistant. *Obsolete*

BOCHUM MNHW

SWB Stahlformguss Gesellschaft mbH
C 1.2, Si 0.4, Mn 12.5, bal Fe.
For wear plates, dipper teeth, shovels; wear and abrasion
resistant. *Obsolete*

BOCHUM PKS EXTRA

Stahlwerke Sudwestfalen
C 1, V 0.1, Mn 0.25, bal Fe.
Heat treated: 200,000 TS; 130,000 YS; 8 El; 25 RA; 400 Brin.
For reamers, drills, drawing and stamping dies; oil or water
hardened, wear resistant. *Obsolete*

BOCHUM PKS EXTRA SPEZIAL

SWB Stahlformguss Gesellschaft mbH
C 1, Cr, V, bal Fe.
Heat treated: 200,000 TS; 130,000 YS; 8 El; 25 RA; 400 Brin.
For reamers, drills, drawing dies; oil or water hardened.
Obsolete

BOCHUM PKS10H

Stahlwerke Sudwestfalen
C 1.1, Si 0-0.25, Mn 0-0.25, bal Fe.
Heat treated: 200,000 TS; 130,000 YS; 8 El; 25 RA; 400 Brin.
For springs, taps, drills, reamers, cutters, hobs; Type W1;
water hardened. *Obsolete*

BOCHUM QKS10W

Stahlwerke Sudwestfalen
C 1, Si 0-0.25, Mn 0-0.25, bal Fe.
Heat treated: 200,000 TS; 130,000 YS; 8 El; 25 RA; 400 Brin.
For springs, taps, drills, reamers, cutters, hobs; Type W1;
water hardened. *Obsolete*

BOCHUM RR10.13

SWB Stahlformguss Gesellschaft mbH
C 0-0.12, Si 0.4, Cr 13, bal Fe.
Annealed: 75,000 TS; 40,000 YS; 35 El; 70 RA; 155 Brin. For
turbine blades, cutlery, valves, surgical instruments; Type
410; stainless. *Obsolete*

BOCHUM RR10.13MO

SWB Stahlformguss Gesellschaft mbH
C 0-0.12, Cr 13, Mo 0.2, bal Fe.
Annealed: 75,000 TS; 40,000 YS; 35 El; 70 RA; 155 Brin. For
turbine blades, cutlery, valves, surgical instruments; Type
410; stainless. *Obsolete*

BOCHUM RR10.13T

SWB Stahlformguss Gesellschaft mbH
C 0-0.12, Cr 13, Si 0.4, bal Fe.
Annealed: 75,000 TS; 40,000 YS; 35 El; 70 RA; 155 Brin. For
turbine blades, cutlery, valves, knives; Type 410; stainless.
Obsolete

BOCHUM RR10.17MOS

SWB Stahlformguss Gesellschaft mbH
C 0.12, Cr 16.5, Mo 0.25, S 0.2, bal Fe.
Annealed: 80,000 TS; 50,000 YS; 20 El; 45 RA; 150 Brin. For
screw machine products; Type 430F; stainless, free-cutting.
Obsolete

BOCHUM RR10.17S

SWB Stahlformguss Gesellschaft mbH
C 0.1, Ni 20, Cr 25, Mo, Si, Cu, bal Fe.
For chemical plant equipment, tanks; stainless, austenitic.
Obsolete

BOCHUM RR15.13

SWB Stahlformguss Gesellschaft mbH
C 0-0.15, Si 0.4, Cr 13, bal Fe.
Annealed: 75,000 TS; 40,000 YS; 35 El; 70 RA; 155 Brin. For
turbine blades, cutlery, valves, surgical instruments; Type
410; stainless. *Obsolete*

BOCHUM RR20.13

SWB Stahlformguss Gesellschaft mbH
C 0.2, Si 0.4, Cr 13, bal Fe.
Annealed: 95,000 TS; 50,000 YS; 25 El; 55 RA; 195 Brin. For
turbine blades, cutlery, valves; Type 420; stainless. *Obsolete*

BOCHUM RR20.13MO

SWB Stahlformguss Gesellschaft mbH
C 0.2, Cr 13, Mo 1.15, bal Fe.
Annealed: 95,000 TS; 50,000 YS; 25 El; 55 RA; 195 Brin. For
turbine blades, valves, cutlery; Type 420 Mo; stainless.
Obsolete

BOCHUM RR20.13NI

SWB Stahlformguss Gesellschaft mbH
C 0.2, Cr 13, Ni, bal Fe.
For valves, cutlery, surgical and dental instruments; corrosion
resistant. *Obsolete*

BOCHUM RR22.17NI

SWB Stahlformguss Gesellschaft mbH
C 0.22, Cr 17, Ni 1.5, Si 0.4, bal Fe.
Annealed: 125,000 TS; 95,000 YS; 20 El; 55 RA; 260 Brin. For
pumps, marine hardware, valves; Type 431; heat and
corrosion resistant. *Obsolete*

BOCHUM RR35.17MO

SWB Stahlformguss Gesellschaft mbH
C 0.35, Cr 16.5, Mo 1.15, bal Fe.
For chemical plant equipment; corrosion resistant. *Obsolete*

BOCHUM RR40.13

SWB Stahlformguss Gesellschaft mbH
C 0.4, Si 0.4, Cr 13, bal Fe.
Annealed: 100,000 TS; 55,000 YS; 20 El; 50 RA; 200 Brin. For
valves, cutlery, pump parts; Type 420; stainless. *Obsolete*

BOCHUM RR45.13MO

SWB Stahlformguss Gesellschaft mbH
C, alloy, bal Fe.
For machine tool parts; oil hardened, tough. *Obsolete*

BOCHUM RR8.17

SWB Stahlformguss Gesellschaft mbH
C 0.08, Cr 17, Si 0.4, bal Fe.
Annealed: 80,000 TS; 50,000 YS; 25 El; 50 RA; 150 Brin. For
oil refinery equipment, sinks, soot blowers; Type 430;
stainless. *Obsolete*

BOCHUM RR8.17E

SWB Stahlformguss Gesellschaft mbH
C 0-0.1, Cr 17.5, Ti = 7 x C, bal Fe.
Annealed: 80,000 TS; 50,000 YS; 25 El; 50 RA; 150 Brin. For
welded oil refinery equipment; Type 430 Ti; stainless.
Obsolete

BOCHUM RR90.17MOV

SWB Stahlformguss Gesellschaft mbH
C 0.9, Cr 18, Mo 1.15, V 1, bal Fe.
For bearings, cutlery, valves; oil hardened, wear and
corrosion resistant. *Obsolete*

BOCHUM U35

SWB Stahlformguss Gesellschaft mbH
C 0.35, Si 0.4, Mn 0.6, bal Fe.
Hot rolled: 85,000 TS; 54,000 YS; 30 El; 53 RA; 185 Brin. For
gears, bolts, axles, shafts, fasteners; water hardened.
Obsolete

BOCHUM U45

SWB Stahlformguss Gesellschaft mbH
C 0.45, Si 0.4, Mn 0.6, bal Fe.
Hot rolled: 98,000 TS; 59,000 YS; 24 El; 45 RA; 212 Brin. For
axles, gears, bolts, crankshafts; water hardened. *Obsolete*

BOCHUM U60

SWB Stahlformguss Gesellschaft mbH
C 0.6, Si 0.4, Mn 0.6, bal Fe.
Heat treated: 160,000 TS; 113,000 YS; 12 El; 40 RA; 320 Brin.
For wheels, die blocks, girders, springs, rails; water hardened.
Obsolete

BOCHUM U75

SWB Stahlformguss Gesellschaft mbH
C 0.75, Si 0.4, Mn 0.6, bal Fe.
Heat treated: 180,000 TS; 135,000 YS; 12 El; 36 RA; 375 Brin.
For springs, tools, hammers, rails, girders, Type W1; water
hardened. *Obsolete*

BOCHUM U90

SWB Stahlformguss Gesellschaft mbH
C 0.9, Si 0.4, Mn 0.6, bal Fe.
Heat treated: 190,000 TS; 145,000 YS; 10 El; 30 RA; 400 Brin.
For springs, taps, reamers, drills; Type W1; water hardened.
Obsolete

BOCHUM VK22

SWB Stahlformguss Gesellschaft mbH
C 0.22, Si 0.25, Mn 0.45, bal Fe.
Annealed: 78,000 TS; 40,000 YS; 22 El; 58 RA; 140 Brin. For
screws, bolts, gears; water hardened. *Obsolete*

BOCHUM VK35

SWB Stahlformguss Gesellschaft mbH
C 0.35, Si 0.25, Mn 0.45, bal Fe.
Hot rolled: 85,000 TS; 54,000 YS; 30 El; 53 RA; 185 Brin. For
gears, shafts, axles, bolts; water hardened. *Obsolete*

BOCHUM VK45

SWB Stahlformguss Gesellschaft mbH
C 0.45, Si 0.25, Mn 0.65, bal Fe.
Hot rolled: 98,000 TS; 58,000 YS; 24 El; 45 RA; 212 Brin. For
gears, bolts, axles, shafts, crankpins; water hardened.
Obsolete

BOCHUM VK53

SWB Stahlformguss Gesellschaft mbH
C 0.53, Si 0.25, Mn 0.65, bal Fe.
Annealed: 96,000 TS; 52,000 YS; 16 El; 23 RA; 170 Brin. For
gears, axles, bolts, shafts; water hardened. *Obsolete*

BOCHUM VK60

SWB Stahlformguss Gesellschaft mbH
C 0.61, Si 0.25, Mn 0.65, bal Fe.
Heat treated: 160,000 TS; 115,000 YS; 12 El; 40 RA; 325 Brin.
For wheels, die blocks, rails, girders; water hardened.
Obsolete

BOCHUM W3G

SWB Stahlformguss Gesellschaft mbH
C 0.9, Cr, bal Fe.
For bearings, liners, cutters; oil or water hardened, wear resistant. *Obsolete*

BOCHUMER BC SPEZIAL

Bochumer Verein
C 1.45, Mn 0.6, Cr 1.4, bal Fe.
For bearings, liners, sleeves; water hardened, abrasion resistant. *Obsolete*

BOCHUMER BC1

Bochumer Verein
C 1, Cr 1.55, bal Fe.
For bearings, sleeves, liners, blanking dies; water hardened, wear resistant. *Obsolete*

BOCHUMER BC2

Bochumer Verein
C 0.9, Cr 0.8, Mn 0.3, bal Fe.
For springs, tools, punches, dies, hammers; Type W1; water hardened. *Obsolete*

BOCHUMER BCE SPEZIAL

Bochumer Verein
C 1.45, Cr 1.4, Mn 0.6, bal Fe.
For bearings, liners, sleeves; water hardened, abrasion resistant. *Obsolete*

BOCHUMER BCL200

Bochumer Verein
C 2.1, Cr 11.5, Mn 0.3, bal Fe.
For blanking and forming dies, punches; oil or air hardened, nondeforming. *Obsolete*

BOCHUMER BCOL155

Bochumer Verein
C 1.65, Cr 11.5, Mn 0.3, bal Fe.
For blanking and forming dies, punches; air hardened, nondeforming. *Obsolete*

BOCHUMER BCOV40

Bochumer Verein
C 0.45, Cr 1.4, Mo 0.7, V 0.3, Mn 0.7, bal Fe.
For gears, bolts, fasteners, machine tool parts; oil hardened, tough. *Obsolete*

BOCHUMER BCS120

Bochumer Verein
C 1.25, Si 1.15, Cr 1.2, Mn 0.7, bal Fe.
For bearings, sleeves, liners; water or oil hardened, wear resistant. *Obsolete*

BOCHUMER BCSV/G

Bochumer Verein
C 0.61, Cr 1.18, V 0.1, Mn 0.7, bal Fe.
For springs, gears, crankshafts; oil hardened, shock resistant. *Obsolete*

BOCHUMER BCSV/K

Bochumer Verein
C 0.61, Cr 1.18, V 0.1, Mn 0.75, bal Fe.
For springs, gears, crankshafts; oil hardened, shock resistant. *Obsolete*

BOCHUMER BCV115

Bochumer Verein
C 1.15, Cr 0.65, V 0.1, Mn 0.3, bal Fe.
For cutters, dies, forming dies; water hardened, wear resistant. *Obsolete*

BOCHUMER BCVOW

Bochumer Verein
C 0.45, Mo 0.45, Cr 1.35, V 0.8, W 0.45, bal Fe.
For cold work tools, upsetters, headers; oil hardened, tough. *Obsolete*

BOCHUMER BCWL200

Bochumer Verein
C 2.1, Cr 11.5, W 0.7, bal Fe.
For blanking and forming dies, punches; oil hardened, nondeforming. *Obsolete*

BOCHUMER BFS5 1/2

Bochumer Verein
C 0.46, Si 1.7, Mn 0.65, bal Fe.
For springs, punches, crimpers; oil hardened, tough. *Obsolete*

BOCHUMER BFS5 1/2+

Bochumer Verein
C 0.51, Si 1.7, Mn 0.65, bal Fe.
For springs, punches, crimpers; oil hardened, tough. *Obsolete*

BOCHUMER BFS6

Bochumer Verein
C 0.6, Si, Mn, bal Fe.
Heat treated: 160,000 TS; 115,000 YS; 12 El; 40 RA; 325 Brin.
For springs, punches, crimpers; oil hardened, tough. *Obsolete*

BOCHUMER BMS1

Bochumer Verein
C 0.53, Si 0.9, Mn 0.9, bal Fe.
Heat treated: 160,000 TS; 113,000 YS; 12 El; 40 RA; 320 Brin.
For gears, springs, shafts; water or oil hardened. *Obsolete*

BOCHUMER BMV85

Bochumer Verein
C 0.9, Mn 1.9, V 0.1, bal Fe.
For punches, forming dies, crimpers, upsetters; oil hardened, shock resistant. *Obsolete*

BOCHUMER BNCO/G

Bochumer Verein
C 0.35, Mo 0.25, Cr 1.35, Ni 3.9, bal Fe.
For gears, bolts, crankshafts, studs; oil hardened, shock resistant. *Obsolete*

BOCHUMER BNCO/K

Bochumer Verein
C 0.19, Cr 1.25, Mo 0.2, Ni 3.75, bal Fe.
For gears, bolts, cams, studs, camshafts; case hardening steel, shock resistant. *Obsolete*

BOCHUMER BVA3

Bochumer Verein
C 0.2, Mo 0.55, Mn 1, Si 0.25, bal Fe.
For gears, bolts, cams, studs, camshafts; case hardening steel, tough. *Obsolete*

BOCHUMER BVT125 EXTRA

Bochumer Verein
C 1.3, W 4.75, Cr 0-0.2, bal Fe.
For cutters, form cutters, engravers' tools; water hardened. *Obsolete*

BOCHUMER BVT130

Bochumer Verein
C 0.2, Cr 1.25, Mo 0.22, Mn 0.55, bal Fe.
For gears, bolts, machine tool parts; case hardening steel, tough. *Obsolete*

BOCHUMER BVT130V

Bochumer Verein
C 0.22, Cr 12, Mo 1, Ni 0.4, V 0.3, bal Fe.
Annealed: 95,000 TS; 40,000 YS; 25 El; 55 RA; B 92 Rock.
Heat Treated: 240,000 TS; 20,000 YS; 10 El; 25 RA; C 50 Rock. For cutlery, surgical instruments, knives, shafts, gears, scissors. Corrosion resistant, hardenable. *Obsolete*

BOCHUMER BVT130VSO

Bochumer Verein
C 0.2, Cr 12, Mo 1, V 0.3, W 0.5, bal Fe.
Annealed: 95,000 TS; 40,000 YS; 25 El; 55 RA; B 92 Rock.
Heat Treated: 240,000 TS; 200,000 YS; 10 El; 25 RA; C 50 Rock. For gears, shafts, cutlery, hardware, surgical instruments. Corrosion resistant, hardenable. *Obsolete*

BOCHUMER BVT40

Bochumer Verein
C 0.16, Cr 1.05, Mo 0.45, Mn 0.65, bal Fe.
For gears, pinions, machine tool parts; case hardening steel, tough. *Obsolete*

BOCHUMER BVT40N

Bochumer Verein
C 0.16, Cr, Mo, V, bal Fe.
For gears, machine tool parts; case hardening steel, tough. *Obsolete*

BOCHUMER BVT50

Bochumer Verein
C 0.24, Cr 1.25, Mo 0.45, bal Fe.
For gears, pinions, bolts, fasteners, shafts; oil hardened, tough. *Obsolete*

BOCHUMER BVT50N

Bochumer Verein
C 0.22, Cr, Mo, V, bal Fe.
For bolts, cams, camshafts, fasteners; case hardening steel, tough. *Obsolete*

BOCHUMER BVT60

Bochumer Verein
C 0.24, Cr 1.35, Mo 0.55, V 0.2, bal Fe.
For bolts, machine tool parts, shafts, studs; oil hardened, tough. *Obsolete*

BOCHUMER BVT60N

Bochumer Verein
C 0.22, Cr 1.15, Mo 0.25, V 0.2, bal Fe.
For bolts, machine tool parts, shafts, studs; oil hardened, tough. *Obsolete*

BOCHUMER BVT90

Bochumer Verein
C 0.18, Cr, Mo, V, bal Fe.
For gears, bolts, machine tool parts; case hardening steel, tough. *Obsolete*

BOCHUMER BWC100

Bochumer Verein
C 1, Mn 0.9, Cr 1, W 1.15, bal Fe.
For cutters, dies, shears, upsetters; oil hardened, tough. *Obsolete*

BOCHUMER BWCO SUPRA

Bochumer Verein
C 0.65, Cr 3.75, Mo 0.85, V 0.7, W 8.5, bal Fe.
For lathe and planer tools, reamers, taps, drills; high speed steel. *Obsolete*

BOCHUMER BWCV25 EXTRA

Bochumer Verein
C 0.3, Cr, V, W, bal Fe.
For upsetters, crimpers, header dies; oil hardened, tough. *Obsolete*

BOCHUMER BWCV25 SPEZIAL

Bochumer Verein
C 0.3, Si 1, Cr 1.1, V 0.18, W 3.75, bal Fe.
For upsetters, riveters, header dies; oil hardened, tough. *Obsolete*

BOCHUMER BWCV25 SUPRA

Bochumer Verein
C 0.3, Cr 2.65, V 0.35, W 8.5, bal Fe.
For extrusion press dies and rams, upsetters; hot work steel, oil hardened. *Obsolete*

BOCHUMER BWCV30 SPEZIAL
Bochumer Verein
C 0.35, Cr 1.05, V 0.18, W 1.85, bal Fe.
For header dies, upsetters, crimpers; oil hardened, tough.
Obsolete

BOCHUMER BWCV40 SPEZIAL
Bochumer Verein
C 0.45, Cr 1.05, V 0.2, W 1.85, bal Fe.
For header dies, upsetters, crimpers; oil hardened, tough.
Obsolete

BOCHUMER BWCV50 SPEZIAL
Bochumer Verein
V 0.18, W 1.85, 0.55C 1.05 Cr, bal Fe.
For header dies, upsetters, crimpers; oil hardened, tough.
Obsolete

BOCHUMER BWV115
Bochumer Verein
C 1.2, Cr 0.2, V 0.1, W 1, bal Fe.
For cutters, bearings, crimpers; water hardened, wear resistant. *Obsolete*

BOCHUMER CBV
Bochumer Verein
C 0.15, Cr 0.65, Mn 0.5, Si 0.25, bal Fe.
For gears, bolts, machine tool parts; case hardened.
Obsolete

BOCHUMER CBV/H
Bochumer Verein
C 0.36, Mn 0.65, Cr 1.55, bal Fe.
For gears, bolts, fasteners; oil hardened, tough. *Obsolete*

BOCHUMER CBV/V
Bochumer Verein
C 0.22, Cr, V, bal Fe.
For gears, bolts, machine tool parts; oil hardened, tough.
Obsolete

BOCHUMER CBV1
Bochumer Verein
C 0.33, Cr 1, Mn 0.65, bal Fe.
For gears, bolts, fasteners; oil hardened, tough. *Obsolete*

BOCHUMER CBV2
Bochumer Verein
C 0.37, Mn 0.65, Cr 1, bal Fe.
For gears, bolts, fasteners; oil hardened, tough. *Obsolete*

BOCHUMER CBVO/D
Bochumer Verein
C 0.3, Mn 0.55, Cr 1.5, Mo 0.18, bal Fe.
For gears, bolts, machine tool parts; oil hardened, tough.
Obsolete

BOCHUMER CBVO/G
Bochumer Verein
C 0.4, Cr, Mn Mo, V, bal Fe.
For gears, bolts, machine tool parts; oil hardened, tough.
Obsolete

BOCHUMER CBVO/H
Bochumer Verein
C 0.5, Mn 0.65, Cr 1, Mo 0.2, bal Fe.
For gears, bolts, machine tool parts; oil hardened, tough.
Obsolete

BOCHUMER CBVO/M
Bochumer Verein
C 0.25, Mn 0.65, Cr 1, Mo 0.2, bal Fe.
For gears, bolts, machine tool parts; oil hardened, tough.
Obsolete

BOCHUMER CBVO/MD
Bochumer Verein
C 0.21, Mn 0.65, Cr 0.85, Mo 0.25, bal Fe.
For gears, bolts, machine tool parts; oil hardened, tough.
Obsolete

BOCHUMER CBVO/TM
Bochumer Verein
C 0.24, Mn 0.55, Cr 1.15, Mo 0.25, bal Fe.
For gears, bolts, machine tool parts; oil hardened, tough.
Obsolete

BOCHUMER CBVO/V
Bochumer Verein
C 0.3, Mn 0.55, Cr 2.5, Mo 0.2, V 0.15, bal Fe.
For gears, bolts, machine tool parts; oil hardened, tough.
Obsolete

BOCHUMER CBVO/W
Bochumer Verein
C 0.16, Mn 0.65, Cr 1.05, Mo 0.2, bal Fe.
For gears, cams, camshafts; case hardened. *Obsolete*

BOCHUMER CBVO/Z
Bochumer Verein
C 0.42, Mn 0.65, Cr 1.05, Mo 0.2, bal Fe.
For gears, bolts, crankshafts, fasteners; oil hardened, tough.
Obsolete

BOCHUMER CBVZ
Bochumer Verein
C 0.41, Cr 1, Mn 0.65, Si 0.25, bal Fe.
For gears, bolts, machine tool parts; oil hardened, tough.
Obsolete

BOCHUMER CMBV/H
Bochumer Verein
C 0.2, Mn 1.25, Cr 1.15, bal Fe.
For gears, cams, camshafts, fasteners; case hardened, tough. *Obsolete*

BOCHUMER CMBV/W
Bochumer Verein
C 0.16, Mn 1.15, Cr 0.95, bal Fe.
For gears, cams, camshafts, fasteners; case hardened, tough. *Obsolete*

BOCHUMER CNBV1W
Bochumer Verein
C 0.15, Cr 1.55, Ni 1.55, bal Fe.
For gears, bolts, machine tool parts; case hardened, tough.
Obsolete

BOCHUMER CNBV2W
Bochumer Verein
C 0.18, Cr 2, Ni 2, Mn 0.5, bal Fe.
For gears, bolts, machine tool parts; case hardened, tough.
Obsolete

BOCHUMER CNBVO/M
Bochumer Verein
C 0.25, Mn 0.5, Cr 1, Mo 0.18, Ni 1.5, bal Fe.
For gears, bolts, fasteners; oil hardened, shock resistant.
Obsolete

BOCHUMER CRVO
Bochumer Verein
C 0.33, Mn 0.65, Cr 1, Mo 0.2, bal Fe.
For gears, bolts, machine tool parts; oil hardened, tough.
Obsolete

BOCHUMER CRW
Bochumer Verein
C 0.85, Cr 1.75, Mn 0.35, bal Fe.
For bearings, bushings, liners, sleeves; water or oil hardened, wear resistant. *Obsolete*

BOCHUMER CV23
Bochumer Verein
C 0.22, Mn 0.65, Cr 1.1, V 0.2, bal Fe.
For gears, pinions, bolts, shafts; oil hardened, tough.
Obsolete

BOCHUMER CV40/H
Bochumer Verein
C 0.42, Cr, V, bal Fe.
For gears, bolts, springs, crankshafts; oil hardened, tough.
Obsolete

BOCHUMER CV48
Bochumer Verein
C 0.5, Cr 1, V 0.09, bal Fe.
For gears, bolts, springs, studs; oil hardened, shock resistant.
Obsolete

BOCHUMER CV58
Bochumer Verein
C 0.5, Mn 0.85, Cr 1, V 0.1, bal Fe.
For gears, bolts, springs, crankshafts; oil hardened, shock resistant. *Obsolete*

BOCHUMER EMC
SWB Stahlformguss Gesellschaft mbH
C 0.2, Mn 1.25, Cr 1.15, bal Fe.
For camshafts, cams, gears; case hardened. *Obsolete*

BOCHUMER ES5
Bochumer Verein
C 0.85, Si 0.25, Mn 0.25, bal Fe.
Heat treated; 188,000 TS; 143,000 YS; 12 El; 35 RA; 390 Brin.
For springs, drills, taps, reamers; Type W1; water hardened.
Obsolete

BOCHUMER ES6
Bochumer Verein
C 0.7, Si 0.25, Mn 0.25, bal Fe.
Heat treated: 175,000 TS; 128,000 YS; 12 El; 37 RA; 355 Brin.
For springs, rails, punches, hammers; Type W1; water hardened. *Obsolete*

BOCHUMER EVL
Bochumer Verein
C, Cr, Ni, bal Fe.
For machine tool parts; oil hardened, shock resistant.
Obsolete

BOCHUMER F EXTRA
Bochumer Verein
C, alloy, bal Fe.
For machine tool parts; oil hardened. *Obsolete*

BOCHUMER FCK
Bochumer Verein
C, Cr, bal Fe.
For machine tool parts; oil hardened. *Obsolete*

BOCHUMER FRS
Bochumer Verein
C 0.67, Si 1.3, Cr 0.5, Mn 0.5, bal Fe.
For punches, crimpers, upsetters; oil or water hardened.
Obsolete

BOCHUMER GKS
Bochumer Verein
C 0.15, Mn 1.1, Si, bal Fe.
For machine tool parts; case hardened. *Obsolete*

BOCHUMER HMFB EXTRA
Bochumer Verein
C 0.55, Ni, Cr, Mo, V, bal Fe.
For forging dies, punches, upsetters, shears; oil hardened, shock resistant. *Obsolete*

BOCHUMER HMFB SPEZIAL
Bochumer Verein
C 0.55, Cr 0.7, Mo 0.18, Ni 1.65, V 0.1, bal Fe.
For forging and header dies, punches; oil hardened, shock resistant. *Obsolete*

BOCHUMER HMS 35
Bochumer Verein
C 0.1, Si 0.25, Mn 0.37, bal Fe.
Annealed: 64,000 TS; 48,000 YS; 28 El; 65 RA; 135 Brin. For plastic mold dies, rivets, screws, fan blades; case hardened. *Obsolete*

BOCHUMER HMS 40
Bochumer Verein
C 0.15, Si 0.25, Mn 0.37, bal Fe.
Annealed: 70,000 TS; 55,000 YS; 25 El; 60 RA; 145 Brin. For screws, bolts, fan blades, bushings, gears; case hardened. *Obsolete*

BOCHUMER HMS 45
Bochumer Verein
C 0.22, Si 0.25, Mn 0.45, bal Fe.
Annealed: 73,000 TS; 61,000 YS; 22 El; 58 RA; 150 Brin. For gears, bolts, fan blades, camshafts; water hardened. *Obsolete*

BOCHUMER HMS 55
Bochumer Verein
C 0.35, Si 0.25, Mn 0.55, bal Fe.
Hot rolled: 85,000 TS; 54,000 YS; 30 El; 53 RA; 185 Brin. For gears, shafts, axles, bolts, screws; water hardened. *Obsolete*

BOCHUMER HMS 65
Bochumer Verein
C 0.45, Si 0.25, Mn 0.65, bal Fe.
Hot rolled: 98,000 TS; 59,000 YS; 24 El; 45 Ra; 212 Brin. For axles, gears, bolts, tie rods, bushings; water hardened. *Obsolete*

BOCHUMER HMS 80
Bochumer Verein
C 0.6, Si 0.25, Mn 0.65, bal Fe.
Heat treated: 160,000-115,000 TS; 113,000-77,000 YS; 12-23 El; 40-54 RA; 321-229 Brin. For wheels, die blocks, girders, shafts; water hardened. *Obsolete*

BOCHUMER HMS 85
Bochumer Verein
C 0.6, Si 0.25, Mn 0.65, bal Fe.
Heat treated: 115,000-160,000 TS; 77,000-113,000 YS; 23-12 El; 54-40 RA; 229-331 Brin. For wheels, die blocks, rails, girders; water hardened. *Obsolete*

BOCHUMER HSS
Bochumer Verein
C, alloy, bal Fe.
For machine tool parts, gears; oil hardened, tough. *Obsolete*

BOCHUMER HSS EXTRA
Bochumer Verein
C, alloy bal Fe.
For machine tool parts, gears; oil hardened, tough. *Obsolete*

BOCHUMER HSS SPEZIAL
Bochumer Verein
C, alloy, bal Fe.
For machine tool parts, gears; oil hardened, tough. *Obsolete*

BOCHUMER KMC 17E
Bochumer Verein
C 0.2, Cr 1.15, Mn 1.25, bal Fe.
For camshafts, cams, gears; case hardened. *Obsolete*

BOCHUMER KSS/CV
Bochumer Verein
C, Cr, V, bal Fe.
For saws, saw blades; oil hardened. *Obsolete*

BOCHUMER LFS
Bochumer Verein
C, Mn, Si, bal Fe.
For springs; oil hardened. *Obsolete*

BOCHUMER LKLS
Bochumer Verein
C, Si, Mn, bal Fe.
For machine tool parts. *Obsolete*

BOCHUMER LKRS
Bochumer Verein
C, Cr, bal Fe.
For machine tool parts. *Obsolete*

BOCHUMER M14
Bochumer Verein
C 0.17, Si 0.3, Mn 1.05, bal Fe.
For gears, cams, camshafts; case hardened. *Obsolete*

BOCHUMER M17
Bochumer Verein
C 0.19, Si 0.5, Mn 1.15, bal Fe.
For gears, cams, camshafts; case hardened. *Obsolete*

BOCHUMER MBV
Bochumer Verein
C 0.4, Si 0.37, Mn 0.95, bal Fe.
Hot rolled: 90,000 TS; 58,000 YS; 17 El; 50 RA; 200 Brin. For gears, shafts, machine tool parts; water hardened. *Obsolete*

BOCHUMER MBV/M
Bochumer Verein
C 0.3, Si 0.25, Mn 1.35, bal Fe.
For gears, machine tool parts, shafts; water hardened. *Obsolete*

BOCHUMER MCV 24
Bochumer Verein
C 0.27, Mn, Cr, V, bal Fe.
For gears, bolts, machine tool parts; oil hardened, shock resistant. *Obsolete*

BOCHUMER MINIMUM R
Bochumer Verein
C, alloy, bal Fe.
For machine tool parts; oil hardened. *Obsolete*

BOCHUMER MNFS
Bochumer Verein
C 0.46, Mn, Si, bal Fe.
Hot rolled: 98,000 TS; 60,000 YS; 24 El; 45 RA; 215 Brin. For gears, shafts, machine tool parts; water hardened. *Obsolete*

BOCHUMER MS
Bochumer Verein
C 0.55, Cr 1.05, V 0.18, W 1.85, bal Fe.
For header dies, upsetters, dies, crimpers; oil hardened, tough. *Obsolete*

BOCHUMER MSBV
Bochumer Verein
C 0.37, Si 1.25, Mn 1.25, bal Fe.
For gears, shafts, fasteners; oil or water hardened. *Obsolete*

BOCHUMER MSS
Bochumer Verein
C, alloy, bal Fe.
For machine tool parts; oil hardened. *Obsolete*

BOCHUMER MSS EXTRA
Bochumer Verein
W 2, C, bal Fe.
For saws; oil hardened. *Obsolete*

BOCHUMER MSS SPEZIAL
Bochumer Verein
W 1, C, bal Fe.
For saws; oil hardened. *Obsolete*

BOCHUMER MSS/CV
Bochumer Verein
C, Cr, V, bal Fe.
For saws; oil hardened. *Obsolete*

BOCHUMER MV38
Bochumer Verein
C 0.42, Si 0.25, Mn 1.75, V 0.1, bal Fe.
For gears, bolts, crankshafts, dies; oil hardened, tough. *Obsolete*

BOCHUMER NFS
Bochumer Verein
C, alloy, bal Fe.
For files; oil hardened. *Obsolete*

BOCHUMER NZF S/H
Bochumer Verein
C 0.65, Si 1.7, Mn 0.7, bal Fe.
For springs, upsetters, punches; oil hardened, tough. *Obsolete*

BOCHUMER NZF S/M
Bochumer Verein
C 0.51, Si 1.7, Mn 0.65, bal Fe.
For springs, upsetters, punches; oil hardened, tough. *Obsolete*

BOCHUMER NZF S/O
Bochumer Verein
C 0.55, Si 1.7, Mn 0.7, bal Fe.
For springs, punches, upsetters; oil hardened, tough. *Obsolete*

BOCHUMER NZF S/W
Bochumer Verein
C 0.46, Si 1.7, Mn 0.65, bal Fe.
For springs, punches, rivet sets; oil hardened, tough. *Obsolete*

BOCHUMER P400
Bochumer Verein
C, alloy, bal Fe.
For machine tool parts; oil hardened. *Obsolete*

BOCHUMER P600
Bochumer Verein
C, alloy, bal Fe.
For machine tool parts; oil hardened. *Obsolete*

BOCHUMER PFS
Bochumer Verein
C, alloy, bal Fe.
For machine tool parts; oil hardened. *Obsolete*

BOCHUMER PFS/1
Bochumer Verein
C, alloy, bal Fe.
For machine tool parts; oil hardened. *Obsolete*

BOCHUMER PKL/V-KPV
Bochumer Verein
C 1, Mn 0.25, V 0.1, bal Fe.
For header dies, forming rolls; Type W2; water hardened. *Obsolete*

BOCHUMER PNC45
Bochumer Verein
C 0.45, Cr 1.05, Ni 3.25, bal Fe.
For gears, bolts, crankshafts, forging dies; oil hardened, shock resistant *Obsolete*

BOCHUMER RKS
Bochumer Verein
C, alloy, bal Fe.
For machine tool parts; oil hardened, tough. *Obsolete*

BOCHUMER RKS SPEZIAL
Bochumer Verein
C, alloy, bal Fe.
For machine tool parts; oil hardened, tough. *Obsolete*

BOCHUMER SAFS
Bochumer Verein
C 0.7, Si 1.7, Mn 0.7, bal Fe.
Heat treated: 340,000 TS; 280,000 YS; 5 El; 20 RA; 600 Brin.
For springs; oil hardened, shock resistant. *Obsolete*

BOCHUMER SC 60
Bochumer Verein
C 0.67, Si 1.3, Cr 0.5, Mn 0.5, bal Fe.
Heat treated: 320,000 TS; 250,000 YS; 8 El; 25 RA; 550 Brin.
For springs; oil hardened, shock resistant. *Obsolete*

BOCHUMER SFS60
Bochumer Verein
C 0.6, bal Fe.
Heat treated: 160,000 TS; 113,000 YS; 12 El; 40 RA; 320 Brin.
For wheels, die blocks, rails, springs; water hardened.
Obsolete

BOCHUMER SFS65
Bochumer Verein
C 0.69, bal Fe.
Heat treated: 175,000 TS; 128,000 YS; 12 El; 37 RA; 350 Brin.
For springs, clutch discs, girders, rails; water hardened.
Obsolete

BOCHUMER SFS70
Bochumer Verein
C 0.72, bal Fe.
Heat treated: 175,000 TS; 128,000 YS; 12 El; 37 RA; 350 Brin.
For springs, clutch discs, girders, rails; Type W1; water
hardened. *Obsolete*

BOCHUMER SMFS/O
Bochumer Verein
C 0.6, Si, Mn, bal Fe.
Heat treated: 160,000 TS; 113,000 YS; 12 El; 40 RA; 320 Brin.
For wheels, die blocks, rails, springs; water hardened.
Obsolete

BOCHUMER SPEZIAL REZISTANCE STAHL
Bochumer Verein
C 0.3, Si 1, Cr 1.1, V 0.18, W 3.75, bal Fe.
For extrusion rams, dies; hot work steel, oil hardened.
Obsolete

BOCHUMER SPEZIAL W
Bochumer Verein
C 0.3, Cr 2.65, V 0.35, W 8.5, bal Fe.
For extrusion rams, dies; hot work steel, oil hardened.
Obsolete

BOCHUMER SSCV
Bochumer Verein
C, V, bal Fe.
For machine tool parts; water hardened. *Obsolete*

BOCHUMER SSS
Bochumer Verein
C, bal Fe.
For machine tool parts; water hardened. *Obsolete*

BOCHUMER SSS1
Bochumer Verein
C 0.7, Si 1.7, Mn 0.7, bal Fe.
For springs, punches, upsetters, chisels; oil hardened, shock
resistant. *Obsolete*

BOCHUMER SSS1 EXTRA
Bochumer Verein
C, W, V, bal Fe.
For heading dies, cutters, liners; oil or water hardened.
Obsolete

BOCHUMER SSSK
Bochumer Verein
C 0.5, Cr 1, V 0.09, Mn 0.85, bal Fe.
For springs, gears, fasteners, crankshafts; oil hardened,
shock resistant. *Obsolete*

BOCHUMER TSE
Bochumer Verein
C, Cr, Mn, bal Fe.
For machine tool parts; water hardened *Obsolete*

BOCHUMER TSV
Bochumer Verein
C, bal Fe.
For machine tool parts; water hardened. *Obsolete*

BOCHUMER UFS
Bochumer Verein
C, bal Fe.
For machine tool parts; water hardened. *Obsolete*

BOCHUMER UGS
Bochumer Verein
C, bal Fe.
For machine tool parts; water hardened. *Obsolete*

BOCHUMER UMS45
Bochumer Verein
C 0.22, Mn 0.45, Si 0.25, bal Fe.
Annealed: 73,000 TS; 41,000 YS; 22 El; 58 RA; 140 Brin. For
screws, bolts, gears, shafts, rivets; case hardened. *Obsolete*

BOCHUMER WFS SPEZIAL
Bochumer Verein
C 0.5, Cr, Mo, V, bal Fe.
For header and forming dies, punches; oil hardened, tough.
Obsolete

BOCHUMER WFS SUPRA
Bochumer Verein
C 0.3, Cr 2.35, V 0.6, W 4.25, bal Fe.
For extrusion dies, rams, liners, punches; oil hardened,
tough. *Obsolete*

BOCHUMER WK45
Bochumer Verein
C 0.45, Si 0.25-0.5, Mn 0.3-0.8, bal Fe.
Hot rolled: 98,000 TS; 59,000 YS; 24 El; 45 RA; 212 Brin. For
axles, gears, bolts, fasteners, crankshafts; water hardened.
Obsolete

BOCHUMER WK60
Bochumer Verein
C 0.6, Si 0.25-0.5, Mn 0.3-0.8, bal Fe.
Heat treated: 160,000 TS; 113,000 YS; 12 El; 40 RA; 320 Brin.
For wheels, die blocks, springs, girders, rails; water hardened.
Obsolete

BOCHUMER ZFS
Bochumer Verein
C 0.7, Si 1.7, Mn 0.7, bal Fe.
For springs, punches, upsetters; oil hardened, shock
resistant. *Obsolete*

BOCHUMER ZMCV
Bochumer Verein
C 1.4, Mn 0.3, Cr 0.3, V 0.1, bal Fe.
For engravers tools, cutters, bearings; water hardened, wear
resistant. *Obsolete*

BODVAR 1
Uddeholm Corp.
C 1, Cr 3.6, W 0.5, Co 2, bal Fe.
For permanent magnets. *Obsolete*

BODVAR 2
Uddeholm Corp.
C 0.95, Cr 5.5, W 0.5, Co 3.25, bal Fe.
For permanent magnets. *Obsolete*

BODYRITE SOLDER
American Smelting & Refining Co.
Pb, bal Sn.
For solder; for auto bodies.

BOFORS 2R107
Bofors AB
C 0.55, Cr 14, bal Fe.
Annealed: 110,000 TS; 72,000 YS; 20 El; 240 Brin. Hardened:
156,000 TS; 128,000 YS; 10 El; 325 Brin. For edge tools,
instruments, springs, intake valves, cutlery, surgical
instruments. AISI Type 420, hardenable, corrosion resistant.
Obsolete

BOFORS 2R27
Bofors AB
C 0-0.1, Cr 13.5, Ni 0-0.3, bal Fe.
Annealed: 71,100 TS; 42,700 YS; 160 Brin. For corrosion
resistant parts; ferritic, corrosion resistant. *Obsolete*

BOFORS 2R29
Bofors AB
C 0.09, Cr 17, bal Fe.
Annealed: 80,000 TS; 50,000 YS; 25 El; 60 Brin. For
household articles, meat hooks, furnace grates, furnace
parts. Type 430 stainless, non-hardenable. *Obsolete*

BOFORS 2R37
Bofors AB
C 0.3, Cr 13, bal Fe.
Annealed: 75,000 TS; 48,000 YS; 26 El; 150 Brin. Hardened:
250,000 TS; 215,000 YS; 8 El; 52 Rock C. For chemical and
oil refinery equipment, surgical instruments, marine
hardware. Corrosion resistant steel, Type 420, stainless,
hardenable. *Obsolete*

BOFORS 2R47
Bofors AB
C 0.2, Cr 13, Ni 0-0.3, bal Fe.
Hardened: 106,700 TS; 92,500 YS; 240 Brin. For corrosion
resistant parts; hardenable martensitic. *Obsolete*

BOFORS 2R57
Bofors AB
C 0.3, Cr 13, Ni 0.4, bal Fe.
Annealed: 78,000 TS; 50,000 YS; 25 El; 155 Brin. Hardened:
145,000 TS; 117,000 YS; 25 El; 280 Brin. For chemical and oil
refinery equipment, surgical instruments, gears, shafts.
Corrosion and heat resistant. Martensitic. *Obsolete*

BOFORS 2R77
Bofors AB
C 0.35, Cr 13.5, Ni 0-0.3, bal Fe.
Hardened: 135,000 TS; 107,000 YS; 300 Brin. For corrosion
resistant parts; hardenable, martensitic. *Obsolete*

BOFORS 2RA27
Bofors AB
C 0.06, Cr 11.5-14.5, Al 0.1-0.3, bal Fe.
Annealed: 70,000 TS; 35,000 YS; 30 El; 60 RA; 160 Brin. Cold
Drawn: 85,000 TS; 70,000 YS; 20 El; 60 RA; 185 Brin. For
heat treating equipment, oil refinery equipment. Type 405
corrosion resistant steel. *Obsolete*

BOFORS 2RL2
Bofors AB
C 0.1, Cr 13, S 0.2, bal Fe.
Annealed: 72,000 TS; 35,000 YS; 30 El; 155 Brin. Hardened:
100,000 TS; 78,000 YS; 25 El; 210 Brin. For screw, bolts,
fasteners, shafts, screw machine products. Type 416
stainless, free-cutting. *Obsolete*

BOFORS 2RM2
Bofors AB
C 0.08, Cr 13, Ni 6, bal Fe.
Heat treated casting: 106,000-142,000 TS; 74,000 YS min; 10
El min; 230-290 Brin. For water turbine castings, turbine
blades and wheels, ship propellers, feed screws for the
cellulose industry. Hardenable, stainless cast alloy. *Obsolete*

BOFORS 2RO-189
Bofors AB
C 0.85, Cr 17, Ni 0-0.3, Mo 0.6, bal Fe.
Hardened: 590 Brin. For corrosion resistant edge tools;
martensitic, corrosion resistant, hardenable. *Obsolete*

BOFORS 2RO26

Bofors AB
C 0.1, Cr 12, Ni 0.5, Mo 0.5, bal Fe.
Annealed: 78,000 TS; 50,000 YS; 27 El; 165 Brin. Hardened: 107,000 TS; 85,000 YS; 22 El; 220 Brin. For tableware, knives, turbine parts, machinery components. Resistant to stress corrosion cracking. Martensitic, hardenable. *Obsolete*

BOFORS 2RO27

Bofors AB
C 0.1, Cr 13.5, Mo 1.2, bal Fe.
Annealed: 78,000 TS; 50,000 YS; 25 El; 155 Brin. Hardened: 135,000 TS; 107,000 YS; 25 El; 280 Brin. For chemical and oil refinery equipment. Martensitic, corrosion and heat resistant. *Obsolete*

BOFORS 2RO46

Bofors AB
C 0.2, Cr 12, Mo 1.2, bal Fe.
Hardened: 142,000 TS; 121,000 YS; 17 El; 300 Brin. For steam turbine blading, compressor parts, gas turbine components, propeller shafts. Corrosion and creep resistant, hardenable. *Obsolete*

BOFORS A286

Bofors AB
C 0.06, Cr 15, Ni 25, Mo 1.3, V 0.3, Ti 2, bal Fe.
Rolled: 150,000 TS; 105,000 YS; 22 El; 290 Brin. For gas turbine components, jet engine parts, afterburners. High heat and creep resistance, austenitic. *Obsolete*

BOFORS ARO-75

Bofors AB
C 0.3, Cr 1, Mo 0.25, Al 1.1, bal Fe.
Heat treated: 121,000 TS; 99,200 YS; 245-290 Brin. For fuel pump pistons, valve parts; nitriding steel. *Obsolete*

BOFORS B10

Bofors AB
C 0.45, bal Fe.
For machine parts, gears; construction steel. *Obsolete*

BOFORS B12

Bofors AB
C 0.6, bal Fe.
For machinery parts; construction steel. *Obsolete*

BOFORS B14

Bofors AB
C 0.7, Si 0.2, Mn 0.6, bal Fe.
Stress relieved: 107,000-121,000 TS; 57,000 YS; 13 El; 220-250 Brin. Heat treated: 175,000 TS; 130,000 YS; 12 El; 350 Brin. For machinery components, gears, shafts, control rods, axes, hammers, springs. AISI 1070 steel. Wear resistant, water hardening. *Obsolete*

BOFORS B15 V

Bofors AB
C 0.75, bal Fe.
For tools, fixtures; water hardening. *Obsolete*

BOFORS B15T

Bofors AB
C 0.75, Si 0.2, Mn 0.6, bal Fe.
Stress relieved: 110,000 TS; 60,000 YS; 12 El; 240 Brin. Heat treated: 180,000 TS; 135,000 YS; 10 El; 370 Brin. For rock drills, chisels, reamers, shafts, axes, hammers, springs, hand tools, pliers. Water hardening, AISI 1075 steel. Wear resistant. *Obsolete*

BOFORS B20V

Bofors AB
C 1, Si 0.2, Mn 0.3, bal Fe.
For tools, drills, reamers; water hardened; Type W1. *Obsolete*

BOFORS B24V

Bofors AB
C 1.2, Si 0.2, Mn 0.3, bal Fe.
For reamers, drills, taps, hobs, cutters; water hardened; Type W1. *Obsolete*

BOFORS B28V

Bofors AB
C 1.4, bal Fe.
For knives, trimming tools; water hardening. *Obsolete*

BOFORS B4

Bofors AB
C 0.25, bal Fe.
For machine parts; construction steel. *Obsolete*

BOFORS B4V

Bofors AB
C 0.15, N 0.65, bal Fe.
Heat treated: 71,000-142,000 TS; 36,000-42,000 YS; 150-300 Brin. For gears, shafts; case hardening. *Obsolete*

BOFORS B7

Bofors AB
C 0.35, bal Fe.
For machinery parts, construction steel. *Obsolete*

BOFORS CR83

Bofors AB
C 0.4, Cr 0.8, Ni 1.25, bal Fe.
114,500 TS; 85,100 YS; 240 Brin. For machine parts, shafts, crankshafts; water or oil hardened. *Obsolete*

BOFORS CRO861

Bofors AB
C 0.35, Mn 0.7, Cr 1.4, Ni 1.4, Mo 0.2, bal Fe.
Hardened: 156,000-178,000 TS; 128,000 YS; 12 El; 45 RA; 330-370 Brin. For shafts, gears, connecting rods, bolts, crankshafts. AISI 4337, oil hardening, tough, shock resistant. *Obsolete*

BOFORS DR34

Bofors AB
C 0.15, Mn 0.7, Cr 0.8, Ni 1.5, bal Fe.
Hardened: 100,000-206,000 TS; 57,000-100,000 YS; 10-14 El; 210-430 Brin. For gears, bolts, camshafts, shafting worms, chains. Case hardening, tough. *Obsolete*

BOFORS DR44

Bofors AB
C 0.2, Mn 0.7, Cr 0.8, Ni 1.5, bal Fe.
Hardened: 114,000-228,000 TS; 71,000-114,000 YS; 10-12 El; 240-480 Brin. For gears, bolts, camshafts, worms. Case hardening, tough. *Obsolete*

BOFORS DRO-1133

Heppenstall Co.
Tool material. C 0.55, Cr 0.6, Ni 1.6, Mo 0.3, bal Fe.
For drop forging dies; oil hardened.

BOFORS DRS16

Bofors AB
C 0.8, Cr 21, Ni 1.5, Si 2, bal Fe.
Hardened: 137,000 TS; 111,000 YS; 15 El; 290 Brin. For outlet valves for internal combustion engines. Martensitic, heat and corrosion resistant. Resists leaded fuels. *Obsolete*

BOFORS FR-86

Bofors AB
C 0.35, Cr 1.2, Ni 2.6, bal Fe.
Rolled: 128,000 TS; 107,000 YS; 270-300 Brin. For crankshafts, shafts; oil hardened. *Obsolete*

BOFORS HR 44

Bofors AB
C 0.2, Ni 3, Cr 0.7, bal Fe.
Heat treated: 142,200-213,000 TS; 92,500-142,200 YS; 300-430 Brin. For gears, cams, shafts; case hardened. *Obsolete*

BOFORS HR-19

Bofors AB
C 0.55, Cr 1.5, Ni 3, bal Fe.
For cold and hot work dies, bakelite dies; oil hardening. *Obsolete*

BOFORS HR33

Bofors AB
C 0.15, Cr 0.75, Ni 3, bal Fe.
Heat treated: 99,500-185,000 TS; 71,100-113,800 YS; 210-390 Brin. For gears, shafts; case hardened. *Obsolete*

BOFORS HRD-1243

Bofors AB
C 0.55, Cr 1, Ni 3, Mo 0.3, bal Fe.
For cold and hot work dies, forging dies; oil hardening. *Obsolete*

BOFORS HRP 1152

Bofors AB
C 0.5, Mn 0.85, Cr 1.05, V 0.15, bal Fe.
Annealed: 105,000 TS; 75,000 YP; 27 El; 16 Rock C. Hardened: 300,000 TS; 263,000 YP; 4 El; 570 Brin. For gears, springs, shafts, machinery parts, axles, clutches, small tools. Oil hardened, shock resistant. *Obsolete*

BOFORS IR34

Bofors AB
C 0.13, Mn 0.4, Cr 0.7, Ni 3.5, bal Fe.
Hardened: 120,000-192,000 TS; 78,000-114,000 YS; 10-12 El; 250-400 Brin. For gears, worms, camshafts, bolts, fasteners. AISI 3310 steel, case hardening, tough and shock resistant. *Obsolete*

BOFORS IR74

Bofors AB
C 0.3, Mn 0.6, Cr 0.75, Ni 3.5, bal Fe.
For gears, pinions, shafts; oil hardening, tough. *Obsolete*

BOFORS KR35

Bofors AB
C 0.12, Cr 1.25, Ni 4.5, bal Fe.
For camshafts, crankshafts, gears; case hardening, tough. *Obsolete*

BOFORS KR75

Bofors AB
C 0.3, Cr 1.3, Ni 4.3, bal Fe.
Heat treated: 156,500 TS; 128,000 YS; 330-370 Brin. For shafts, gears, wear parts; air hardening. *Obsolete*

BOFORS N82

Bofors AB
C 0.4, Mn 1.25, bal Fe.
99,500 TS; 71,000 YS; 240 Brin. For machinery parts; water or oil hardened. *Obsolete*

BOFORS N91

Bofors AB
C 0.45, Mn 0.8, bal Fe.
Normalized: 85,000-107,000 TS; 46,000 YS; 18 El; 180-230 Brin. For induction and flame hardened machinery components, gears, pinions, shafting. AISI 1045 steel. Water hardening. *Obsolete*

BOFORS NON-SHRINKING STEEL RT-1733

S.K.F. Industries Inc.
C 0.85-0.95, Mn 1-1.25, Cr 0.5-0.6, W 0.5-0.6, bal Fe.
For blanking dies, broaches, cutters, tools, drills, plugs, gages; non-shrinking tool steel.

BOFORS P02

Bofors AB
C 85, Cr 4.3, W 6.4, Mo 5, V 2, bal Fe.
For cutters, twist drills; high speed steel. *Obsolete*

BOFORS P03

Bofors AB
C 0.85, W 7, Cr 4, Mo 3-5, V 2, bal Fe.
Heat treated: 64-67 Rock C. For lathe and planer tools, milling cutters, drills, broaches, reamers. High speed steel. High red hardness, wear and abrasion resistant. *Obsolete*

BOFORS P10

Bofors AB
C 0.75, Cr 4, W 18, V 1.2, bal Fe.
For dies, shears, cutters; high speed steel. *Obsolete*

BOFORS P121
Bofors AB
C 0.6-1.05, V 0.1, bal Fe.
Annealed: 120,000 TS; 70,000 YS; 15 El; 220 Brin. Hardened: 185,000 TS; 135,000 YS; 13 El; 400 Brin. For axes, cutting tools, drills, springs, punches. Water hardening, wear resistant. Type W2. *Obsolete*

BOFORS P15
Bofors AB
C 0.8, W 18, Cr 4, V 1.5, Co 2, Mo 1, bal Fe.
For cutting tools; high speed steel. *Obsolete*

BOFORS P171
Bofors AB
C 0.8, V 0.1, bal Fe.
For pneumatic tools; water hardened. *Obsolete*

BOFORS P181
Bofors AB
C 0.9, V 0.1, bal Fe.
For tools, dies, punches; water hardening. *Obsolete*

BOFORS P211
Bofors AB
C 1, V 0.1, bal Fe.
For cold working dies, pneumatic pistons; water hardening. *Obsolete*

BOFORS Q10
Bofors AB
C 0.8, Cr 4, W 18, V 1.7, Co 10.5, Mo 1, bal Fe.
For cutters, millers, tool bits, hobs; high speed steel. *Obsolete*

BOFORS Q5
Bofors AB
C 0.8, Cr 4, W 18, V 1.7, Co 5.5, Mo 1, bal Fe.
For cutters, taps, hobs, reamers; high speed steel. *Obsolete*

BOFORS QRO45
Bofors AB
C 0.3, Cr 2.8, Mo 2.8, V 0.5, Co 2.8, bal Fe.
Hardened: 45-55 Rock C. For die casting and forging dies, extrusion dies, mandrels, punches. Oil or air hardening. Hot work steel. Develops high red hardness and toughness to 1300°F. *Obsolete*

BOFORS QX2
Bofors AB
C 1, W 7, Cr 4, Mo 3-5, V 2, Co 9.5, bal Fe.
Hardened: 64-67 Rock C. For lathe and planer tools, milling cutters, drills, broaches, reamers, shears. High speed steel. High red-hardness. Wear and abrasion resistant. *Obsolete*

BOFORS R10-214
Bofors AB
C 0-0.1, Cr 27, Ni 5, Mo 1.5, bal Fe.
Rolled: 85,330 TS; 64,000 YS; 210 Brin. For sulfite-cellulose industries; high corrosion resistance. *Obsolete*

BOFORS R10214
Bofors AB
C 0.2, Cr 23-28, Ni 2.5-5, Mo 1-2, bal Fe.
Normalized: 103,000 TS; 78,000 YS; 18 El; 45 RA; 235 Brin. Annealed: 95,000 TS; 41,000 YS; 29 El; 60 RA; 225 Brin. For valves, pumps, furnace parts; Type 327; heat resistant. *Obsolete*

BOFORS R309
Bofors AB
C 1.5, Cr 2, bal Fe.
For files, trimmer tools; oil hardened. *Obsolete*

BOFORS RCK3
Bofors AB
C 0.15, Cr 25, Ni 13, N 0.2, bal Fe.
Annealed: 94,000 TS; 50,000 YS; 54 El; 170 Brin. For feeding plates in clinker coolers in cement industry. Austenitic, stainless, high oxidation resistance. Type 309S. *Obsolete*

BOFORS RCK4
Bofors AB
C 0.2, Cr 21, Ni 12, N 0.2, bal Fe.
Annealed: 117,000 TS; 67,000 YS; 41 El; 240 Brin. For outlet valve heads in internal combustion engines. Austenitic, corrosion resistant, oxidation resistant. *Obsolete*

BOFORS RCT
Bofors AB
C 0.4, Ni 12, Cr 14.5, Si 2.3, bal Fe.
Hardened: 106,700 TS; 64,000 YS; 240 Brin. For engine valves; heat resistant. *Obsolete*

BOFORS RCT3
Bofors AB
C 0.2, Cr 21, Ni 13, Si 1.2, W 3, bal Fe.
Annealed: 104,000 TS; 47,000 YS; 27 El; 220 Brin. For stator blades, turbolators, rings in jet engines, gas and steam turbines, diesel engine valves. Austenitic, high heat resistance. *Obsolete*

BOFORS RD-653
Bofors AB
C 0.25, Cr 1.1, Mo 0.2, bal Fe.
Rolled: 128,000 TS; 99,500 YS; 270-310 Brin. For shafts, cams, gears; oil hardened. *Obsolete*

BOFORS RE39
Bofors AB
C 0.2, Cr 17.5, Ni 2.1, bal Fe.
Heat treated: 121,000-135,000 TS; 92,500-112,000 YS; 15-20 El; 45-60 RA; 260-288 Brin. For pumps, spindles, propeller shafts; Type 431; corrosion resistant. *Obsolete*

BOFORS REB-210
Bofors AB
C 0.15, Cr 22.5, Ni 24, bal Fe.
71,000 TS; 28,500 YS; 180 Brin. For heat resistant parts; resists S and SO_2. *Obsolete*

BOFORS RES210
Bofors AB
C 0.2, Cr 22-24, Ni 12-15, Mn 0-2, Si 0-1, bal Fe.
Annealed: 85,000-95,000 TS; 40,000-50,000 YS; 45-55 El; 150-185 Brin. For heat treating boxes, furnace parts, chemical plant equipment; Type 309; austenitic, heat resistant. *Obsolete*

BOFORS RIM 200
Bofors AB
C 0.04, Cr 18, Ni 9, bal Fe.
Annealed: 85,000 TS; 35,000 YS; 55 El; 80 Rock B. For architectural trim, kitchen equipment, chemical and textile plant equipment. Type 304 stainless, austenitic. *Obsolete*

BOFORS RIM 213
Bofors AB
C 0.07, Cr 18, Ni 11, Mo 2.3, Ti 0.4, bal Fe.
Annealed: 85,000 TS; 38,000 YS; 50 El; 165 Brin. For chemical plant equipment and structural parts operating at 900-1150°F. Stabilized, austenitic, stainless, weldable grade. *Obsolete*

BOFORS RIM 290
Bofors AB
C 0-0.03, Cr 18, Ni 10, bal Fe.
Annealed: 76,000 TS; 32,000 YS; 60 El; 140 Brin. For welded structures, chemical and textile plant equipment, food and drug equipment. Type 304L stainless, austenitic steel. *Obsolete*

BOFORS RIM-210
Bofors AB
C 0-0.07, Cr 19, Ni 10, Mo 1.5, bal Fe.
Annealed: 78,400 TS; 28,500 YS; 165 Brin. For sulfite, cellulose industries; high corrosion resistance. *Obsolete*

BOFORS RIM-215
Bofors AB
C 0-0.08, Cr 18, Ni 11, Mo 2.7, bal Fe.
Annealed: 82,500 TS; 165 Brin. For chemical industries; corrosion resistant; austenitic. *Obsolete*

BOFORS RIM-29
Bofors AB
C 0-0.08, Cr 18, Ni 18, bal Fe.
Annealed: 82,500 TS; 28,500 YS; 60 El; 70 RA; 165 Brin. Cold drawn: 125,000 TS; 95,000 YS; 25 El; 55 RA; 277 Brin. For chemical and oil refinery equipment, tanks, vessels; Type 302; stainless, austenitic. *Obsolete*

BOFORS RIM-291
Bofors AB
C 0-0.08, Cr 18-20, Ni 8-11, Mn 0-2, bal Fe.
Annealed: 90,000 TS; 45,000 YS; 60 El; 135 Brin. Cold drawn: 180,000 TS; 150,000 YS; 10 El; 330 Brin. For chemical plant equipment, welded structures; Type 304; stainless, austenitic. *Obsolete*

BOFORS RIM-294
Bofors AB
C 0-0.08, Cr 18, Ni 10, Mn 0.8, Ti 0.5, bal Fe.
Annealed: 82,500 YS; 165 Brin. For chemical industries; heat and corrosion resistant. *Obsolete*

BOFORS RIM-92
Bofors AB
C 0-0.08, Cr 18-20, Ni 8-11, Mn 0-2, bal Fe.
Annealed: 90,000 TS; 45,000 YS; 60 El; 135 Brin. Cold drawn: 180,000 TS; 150,000 YS; 10 El; 330 Brin. For chemical plant equipment, welded structures; Type 304; stainless, austenitic. *Obsolete*

BOFORS RIM21
Bofors AB
C 0.08, Cr 18, Ni 11, Mo 1.5, bal Fe.
Annealed: 85,000 TS; 38,000 YS; 60 El; 180 Brin. For acid resisting shafts, bolts, chemical plant equipment. Austenitic, stainless, non-hardening. *Obsolete*

BOFORS RIM295
Bofors AB
C 0-0.08, Cr 17-19, Ni 9-12, Cb = 10 x C, bal Fe.
Annealed: 85,000-95,000 TS; 35,000-45,000 YS; 55-50 El; 175 Brin. For welded structures, chemical plant equipment; Type 347; corrosion and heat resistant. *Obsolete*

BOFORS RK 214
Bofors AB
C 0-0.1, Cr 28, Ni 4.5, bal Fe.
Annealed: 85,300 TS; 64,000 YS; 210 Brin. For furnace parts, nitric acid contact; heat and corrosion resistant. *Obsolete*

BOFORS RLH 2
Bofors AB
C 0.1, Cr 18, Ni 8, S 0.25, bal Fe.
Annealed: 85,000 TS; 28,000 YS; 60 El; 160 Brin. For screws, bolts, fasteners, shafts, screw machine products. Type 303 stainless, free-cutting, austenitic, non-hardenable. *Obsolete*

BOFORS RNK 29
Bofors AB
C 0.09, Cr 18, Ni 5, Mn 8.5, N 0.2, bal Fe.
Annealed: 104,000 TS; 56,000 YS; 59 El; 200 Brin. For shafts, valve parts, bolts, fittings, foodstuff and chemical equipment; Type 202 stainless, austenitic. *Obsolete*

BOFORS RO-346
Bofors AB
C 0.15, Cr 0.8, Mo 0.6, bal Fe.
For camshafts, crankshafts, gears; case hardened, tough. *Obsolete*

BOFORS RO-7155
Bofors AB
C 0.3, Cr 2.7, Mo 0.5, bal Fe.
For shafts, gears; oil hardened. *Obsolete*

BOFORS RO-752
Bofors AB
C 0.35, Cr 1.05, Mo 0.2, bal Fe.
For shafts, gears; tough. *Obsolete*

BOFORS RO-952
Bofors AB
C 0.4, Cr 1.1, Mo 0.25, bal Fe.
For shafts, gears; oil hardened, tough. *Obsolete*

BOFORS RO211
Bofors AB
C 0.12, Cr 2.5, Mo 1.1, bal Fe.
Normalized: 88,000 TS; 70,000 YS; 26 El; 190 Brin. For steam and gas turbine parts operating up to 1100°F. Martensitic, corrosion resistant. *Obsolete*

BOFORS RO4154
Bofors AB
C 0.25, Cr 3, Mo 0.5, bal Fe.
Hardened: 124,000 TS; 102,000 YS; 20 El; 260 Brin. For construction elements subjected to stresses at temperatures up to 1100°F. Martensitic. Good impact strength and resistance to temper brittleness. *Obsolete*

BOFORS ROP-63
Bofors AB
C 0.3, Mn 0.3, Cr 2.5, V 0.2, Ni 0.5, Mo 0.3, bal Fe.
Heat treated: 142,200 TS; 121,000 YS; 295-330 Brin. For gears, shafts, nitrided parts; nitriding steel. *Obsolete*

BOFORS ROP10
Bofors AB
C 0.08, Cr 3, Mo 0.8, V 0.2, bal Fe.
For hobbed molds for pressing and injection molding of plastics, hobbed dies for zinc die castings. Case-hardening, shock resistant. *Obsolete*

BOFORS ROP18
Bofors AB
C 0.2, Cr 1.3, Mo 0.5, V 0.2, bal Fe.
For hobbed hot work dies, die casting and drop forging dies. Hot work steel, oil or air hardening. *Obsolete*

BOFORS ROP19
Bofors AB
C 0.4, Cr 5.3, Mo 1.4, V 1, Si 1, bal Fe.
Annealed: 98,000 TS; 74,000 YS; 28 El; 210 Brin. Heat treated: 290,000 TS; 228,000 YS; 3 El; 55 Rock C. For die casting and hot pressing dies, extrusion dies, mandrels, punches. Type H13 hot work steel. Tough and red hard. *Obsolete*

BOFORS ROP21
Bofors AB
C 1, Cr 5.3, V 0.2, Mo 1.1, bal Fe.
For cold work dies, shears, press tools; air hardened. *Obsolete*

BOFORS ROP43
Bofors AB
C 0.23, Cr 12, Mo 1.2, Ni 0.6, V 0.3, bal Fe.
Annealed: 95,000 TS; 40,000 YS; 25 El; 92 Rock B. Hardened: 240,000 TS; 205,000 YS; 9 El; 48 Rock C. For cutlery, bearings, valves, surgical instruments. Corrosion resistant, hardenable. *Obsolete*

BOFORS ROP46
Bofors AB
C 0.2, Cr 12, Mo 0.5, V 0.35, bal Fe.
Annealed: 90,000 TS; 38,000 YS; 26 El; 90 Rock B. Hardened: 235,000 TS; 200,000 YS; 9 El; 50 Rock C. For valves, cutlery, bearings, surgical instruments. Corrosion resistant, hardenable. *Obsolete*

BOFORS ROP5462
Bofors AB
C 0.24, Cr 1.3, Mo 0.5, V 0.2, bal Fe.
Hardened: 121,000 TS; 107,000 YS; 20 El; 260 Brin. For steam turbine rings, bolts, screws, nuts at temperatures to 1025°F. Martensitic, high temperature steel. *Obsolete*

BOFORS ROP57
Bofors AB
C 1.5, Cr 12, V 0.2, Mo 0.8, bal Fe.
For dies, stamping and shearing tools; oil or air hardened. *Obsolete*

BOFORS ROP9653
Bofors AB
C 0.45, Cr 1.5, Mo 0.7, V 0.3, bal Fe.
For die casting and hot pressing dies, extrusion dies and mandrels, punches, forging dies. Hot work, oil hardening steel. *Obsolete*

BOFORS ROPT
Bofors AB
C 0.2, Cr 3, Mo 0.6, W 0.5, V 0.8, bal Fe.
Hardened: 128,000 TS; 107,000 YS; 17 El; 275 Brin. For turbine discs for jet engines, gas and steam turbines, shafts, rotors, bolts. Martensitic, high creep resistance. *Obsolete*

BOFORS RP1152
Now BOFORS HRP 1152.

BOFORS RR212
Bofors AB
C 0.35, Cr 23-27, bal Fe.
Annealed: 90,000 TS; 60,000 YS; 20 El; 45 RA; 180 Brin. For furnace parts, heat treating boxes; Type 446; heat resistant. *Obsolete*

BOFORS RR27
Bofors AB
C 0-8, Cr 11.5-13, Al 0.1-0.3, bal Fe.
Annealed: 71,000 TS; 42,600 YS; 22 El; 70 RA; 150 Brin. Heat treated: 175,000 TS; 145,000 YS; 21 El; 64 RA; 352 Brin. For oil refinery and chemical plant equipment; Type 405; corrosion resistant. *Obsolete*

BOFORS RR412
Bofors AB
C 0.35, Cr 23-27, bal Fe.
Annealed: 90,000 TS; 60,000 YS; 20 El; 45 RA; 180 Brin. For furnace parts, heat treating boxes; Type 446; heat resistant. *Obsolete*

BOFORS RR77
Bofors AB
C 0-0.15, Cr 12-14, 0.15 C min, bal Fe.
Annealed: 88,000 TS; 40,000 YS; 32 El; 68 RA; 170 Brin. For cutlery, valve trim, turbine blades; Type 420; stainless, hardenable. *Obsolete*

BOFORS RT 225
Bofors AB
C 1.3, W 5, bal Fe.
For impact extrusion dies; water hardened. *Obsolete*

BOFORS RT-1733
Bofors AB
C 0.9, Cr 0.6, W 0.6, Mn 1.2, bal Fe.
For cutting, threading tools; oil hardening. *Obsolete*

BOFORS RT27
Bofors AB
C 0.3, Cr 1, W 5.5, Ni 3.5, bal Fe.
For extrusion dies, mandrels; oil hardened, hot work steel. *Obsolete*

BOFORS RT45
Bofors AB
C 0.3, Cr 3, W 10, Ni 2, V 0.3, bal Fe.
For die casting dies, hot work dies; hot work steel, oil hardened. *Obsolete*

BOFORS RT46
Bofors AB
C 0.3, Cr 12, W 12, V 0.5, bal Fe.
For die casting and pressing dies, extrusion dies and mandrels, punches. Hot work steel; Type H23. Develops red hardness and toughness up to 1300°F. *Obsolete*

BOFORS RT60
Bofors AB
C 2, Cr 13, W 1, bal Fe.
For stamping and shearing tools, dies; oil hardening. *Obsolete*

BOFORS RTO 712
Bofors AB
C 0.4, Cr 1.2, W 2.5, Mo 0.3, V 0.15, bal Fe.
For chisels, pneumatic tools; tough. *Obsolete*

BOFORS RTO-912
Bofors AB
C 0.45, Cr 1, W 2.5, Mo 0.25, V 0.15, bal Fe.
For chisels, shear blades; hot work steel. *Obsolete*

BOFORS S-145
Bofors AB
C 0.55, Si 1.75, Mn 0.75, bal Fe.
Hardened: 185,000 TS; 164,000 YS; 380-440 Brin. For springs; tough, oil hardened. *Obsolete*

BOFORS SIR
Bofors AB
C 0.2, Cr 25, bal Fe.
Annealed: 71,000 TS; 43,000 YS; 170 Brin. For furnace parts, heat treating boxes, quenching baskets. Corrosion and heat resistant to 2000°F. *Obsolete*

BOFORS SRO
Bofors AB
C 0.45, Cr 9, Mo 0.3, Si 2.8, bal Fe.
Hardened: 144,000 TS; 121,000 YS; 30 El; 320 Brin. For valves, oil refinery equipment. Martensitic, heat and corrosion resistant. *Obsolete*

BOFORS SRO2
Bofors AB
C 0.45, Cr 13, Ni 1, Si 2.8, bal Fe.
Heat treated: 142,000 TS; 121,000 YS; 16 El; 300 Brin. Uses: Intake valves, outlet valves. Martensitic steel, corrosion and oxidation resistant. *Obsolete*

BOHLER
Vereinigte Edelstahlwerke
C 0.015, Mn 1, Si 1.2, Cr 17.5, Ni 16, Mo 1.75, Cb 2, V 0.12, bal Fe.
For gas engine parts; high heat resistance. *Obsolete*

BOHLER 2M
Now VEW F180.

BOHLER 2NCMO
Vereinigte Edelstahlwerke
C 0.3, Cr 0.7, Ni 1.9, Mo 0.25, bal Fe.
Heat treated: 121,000 TS; 92,500 YS. For turbine and motor parts; oil hardened, shock resistant. *Obsolete*

BOHLER 3 NM
Vereinigte Edelstahlwerke
C 0.32, Ni 3, bal Fe.
Hardened: 113,000 TS; 71,000 YS; 20 El; 50 RA; 240 Brin. For bolts, shafts, gears, connecting rods; tough. *Obsolete*

BOHLER 3NIMO
Vereinigte Edelstahlwerke
C 0.15, Ni 3.5, Mo 0.25, bal Fe.
For gears, cams, camshafts; case hardened. *Obsolete*

BOHLER 3WKZ
Vereinigte Edelstahlwerke
W 9.5, Cr 2.5, V 0.1, C, bal Fe.
For tools, dies, hot work dies; hot work steel. *Obsolete*

BOHLER 5 NW
Now VEW P602.

BOHLER 701
Vereinigte Edelstahlwerke
C 0.45, Si 3, Cr 9, bal Fe.
Heat treated: 135,200 TS; 106,700 YS. For inlet and exhaust valves; corrosion and heat resistant. *Obsolete*

BOHLER 751
Now VEW K980.

BOHLER 851
Vereinigte Edelstahlwerke
C 0.85, bal Fe.
Hardened: Rock C 65. Annealed: 85,000-100,000 TS; 175-204 Brin. For mines, quarries and road construction drills, hard rock drills, punches. Water hardening, wear resistant. *Obsolete*

BOHLER A100
Bohler Gesellschaft M.B.H.
Stainless steel. C 0-0.07, Cr 17, Ni 12, Mo 2.7, bal Fe.
Austenitic stainless steel for chemical equipment; improved corrosion resistance. W. Nr. 1.4436; AISI 316.

BOHLER A101
Bohler Gesellschaft M.B.H.
Stainless steel. C 0.05, Si 0.5, Mn 1.8, Cr 17.8, Mo 2.7, Ni 12, S 0.2, bal Fe.
Austenitic stainless steel for textile, paper, cellulose and rayon industries. Quenched: 500-700 N/mm^2 TS; 30-40 El; 205 N/mm^2 0.2% proof stress.

BOHLER A102
Bohler Gesellschaft M.B.H.
Stainless steel. C 0-0.07, Cr 17, Ni 13, Mo 4.3, bal Fe.
Austenitic stainless steel; good corrosion resistance; for chemical equipment. W. Nr. 1.4449; similar to AISI 317.

BOHLER A122
Bohler Gesellschaft M.B.H.
Stainless steel. C 0-0.06, Cr 18, Ni 8.5, Mo 1.6, bal Fe.
Austenitic stainless steel; for fittings for chemical plant equipment. Similar to W. Nr. 1.4420. *Obsolete*

BOHLER A128
Bohler Gesellschaft M.B.H.
Stainless steel. C 0-0.12, Si 0-2, Mn 0-1.5, Cr 18.5, Mo 2.25, Ni 10, bal Fe.
Austenitic stainless casting; for pumps, stirring gear, filters. W. Nr. 1.4410. *Obsolete*

BOHLER A205
Bohler Gesellschaft M.B.H.
Stainless steel. C 0-0.03, Cr 18, Ni 13.5, Mo 2.8, bal Fe.
Austenitic stainless steel, weldable, for chemical plant equipment. W. Nr. 1.4435; AISI 316L.

BOHLER A220
Bohler Gesellschaft M.B.H.
Stainless steel. C 0-0.03, Cr 17.5, Mo 2.7, Ni 14.5, bal Fe.
Austenitic stainless steel, weldable, for chemical plant equipment. W. Nr. 1.4435; AISI 316L.

BOHLER A305
Bohler Gesellschaft M.B.H.
Stainless steel. C 0-0.1, Cr 17.5, Ni 13.5, Mo 3, Ti = 5 x C, bal Fe.
Stabilized austenitic stainless steel, weldable, for chemical plant equipment. *Obsolete*

BOHLER A350
Bohler Gesellschaft M.B.H.
Stainless steel. C 0-0.04, Cr 17, Ni 14, Mo 2.2, Nb > 10 x C < 0.80, bal Fe.
Stabilized austenitic stainless steel; weldable; for chemical plant equipment. W. Nr. 1.4580.

BOHLER A400
Bohler Gesellschaft M.B.H.
Stainless steel. C 0-0.03, Cr 17, Ni 13.5, Mo 4.3, N 0.15, bal Fe.
Austenitic stainless steel. W. Nr. 1.4439.

BOHLER A405
Bohler Gesellschaft M.B.H.
Stainless steel. C 0-0.02, Cr 25, Ni 22, Mo 2.2, N 0.12, bal Fe.
Austenitic stainless steel; good high temperature properties. W. Nr. 1.4465.

BOHLER A410
Bohler Gesellschaft M.B.H.
Stainless steel. C 0-0.03, Cr 17.5, Mo 2.7, Ni 13.5, N 0.18, bal Fe.
Austenitic stainless steel; weldable; for parts and equipment in chemical and textile industries. W. Nr. 1.4429.

BOHLER A505
Bohler Gesellschaft M.B.H.
Stainless steel. C 0-0.12, Cr 18, Ni 8.5, bal Fe.
Austenitic stainless steel. W. Nr. 1.4304; AISI 302.

BOHLER A506
Bohler Gesellschaft M.B.H.
Stainless steel. C 0-0.08, Si 0.5, Mn 1.8, Cr 17.3, Ni 8.3, S 0.25, bal Fe.
Free machining austenitic stainless steel for screws, bolts, and nuts mass produced. Quenched: 500-700 N/mm^2 TS; 35 El; 195 N/mm^2 0.2% proof stress. W. Nr. 1.4305.

BOHLER A511
Bohler Gesellschaft M.B.H.
Stainless steel. C 0-0.07, Cr 19, Ni 11, bal Fe.
Austenitic stainless steel for chemical industry. W. Nr. 1.4303; AISI 305. *Obsolete*

BOHLER A520
Bohler Gesellschaft M.B.H.
Stainless steel. C 0-0.12, Cr 17.5, Ni 7.5, bal Fe.
Austenitic stainless steel for springs, trailer bodies, and wheelcovers; work hardens. W. Nr. 1.4310; AISI 301.

BOHLER A522
Bohler Gesellschaft M.B.H.
Stainless steel. C 0-0.09, Cr 12.5, Ni 12.5, bal Fe.
Austenitic stainless; for valves, pump parts. W. Nr. 1.4307.

BOHLER A610
Bohler Gesellschaft M.B.H.
Stainless steel. C 0-0.012, Cr 17.5, Ni 15, bal Fe.
Free machining austenitic stainless steel. W. Nr. 1.4361.

BOHLER A905
Bohler Gesellschaft M.B.H.
Stainless steel. C 0-0.03, Si 0.4, Mn 5.8, Cr 26, Mo 2.2, Ni 3.8, W 0.5, N 0.33, bal Fe.
Stainless steel for seawater delivery and natural gas production. Quenched: 750 N/mm^2 TS; 25-30 El; 590 N/mm^2 0.2% proof stress. W. Nr. 1.4462.

BOHLER A955
Bohler Gesellschaft M.B.H.
Stainless steel. C 0-0.05, Cr 17.5, Ni 22.5, Mo 3.2, Cu 1.8, Nb = 8 x C, bal Fe.
Austenitic stainless for components in chemical, petroleum, and dye plants. W. Nr. 1.4586.

BOHLER A960
Bohler Gesellschaft M.B.H.
Stainless steel. C 0-0.07, Cr 18, Ni 20, Mo 2, Cu 2, Nb = 8 x C, bal Fe.
Welded austenitic stainless steel for parts and equipment in chemical plants. W.Nr 1.4505. *Obsolete*

BOHLER A976
Bohler Gesellschaft M.B.H.
Stainless steel. C 0-0.06, Si 0.55, Mn 1.3, Cr 17.8, Ni 13, B 1, bal Fe.
Austenitic stainless steel for components for radioactive waste disposal in nuclear engineering. Quenched: 600 N/mm^2 TS; 300 N/mm^2 0.2% proof stress.

BOHLER AC0
Vereinigte Edelstahlwerke
C 0.27, Al 1.1, Cr 1.4, bal Fe.
For oil refinery equipment, fasteners; heat and creep resistant. *Obsolete*

BOHLER AC1
Vereinigte Edelstahlwerke
C 0.34, Al 1.1, Cr 1.4, bal Fe.
For oil refinery equipment, bolts; heat and creep resistant. *Obsolete*

BOHLER AC3
Vereinigte Edelstahlwerke
C 0.3-0, Cr, V, bal Fe.
For gears, shafts, crankshafts, machine tool parts; oil or water hardened. *Obsolete*

BOHLER ACE
Now VEW V810.

BOHLER ACN
Now VEW V820.

BOHLER ACV
Now VEW V350.

BOHLER AS-2
Now VEW A505.

BOHLER AS-4
Vereinigte Edelstahlwerke
C 0-0.1, Cr 16-18, Ni 10-14, Mo 0-2, Cb, bal Fe.
Annealed: 85,000-95,000 TS; 35,000-45,000 YS; 50-60 El; 60-75 RA; 150-190 Brin. For valve seats, cones and spindles, piston rods; Type 316 Cb; austenitic, stainless. *Obsolete*

BOHLER AS-8
Vereinigte Edelstahlwerke
C 0-0.1, Cr 16-18, Ni 10-14, Mo 2-3, bal Fe.
Annealed: 85,000-95,000 TS; 35,000-45,000 YS; 50-60 El; 60-75 RA; 150-190 Brin. For valve seats, kitchen utensils, chemical plant equipment; Type 316; stainless, austenitic. *Obsolete*

BOHLER AS2G
Now VEW A505G.

BOHLER AUTO MS
Vereinigte Edelstahlwerke
C 0.5, Mn, Si, bal Fe.
For crankshafts, connecting rods, axles, cranks, levers; oil hardened. *Obsolete*

BOHLER AUTO SPECIAL PA
Vereinigte Edelstahlwerke
C 0.35, Cr 1.25, Ni 4.45, bal Fe.
For crankshafts, gear shafts, axles, steering gear parts; oil hardened. *Obsolete*

BOHLER AWP
Vereinigte Edelstahlwerke
C 0.45, Si 1.4, Mn 0.8, Cr 13, Ni 13, W 1.4, bal Fe.
For hot extension dies, die casting dies; hot work steel. *Obsolete*

BOHLER AZH
Vereinigte Edelstahlwerke
C 0.85, Mn 0.3, Si 0.25, bal Fe.
For tools, drills, taps; water hardening. *Obsolete*

BOHLER B-ELITE
Vereinigte Edelstahlwerke
C 0.2, bal Fe.
For welding wire and electrodes; for electric arc welding. *Obsolete*

BOHLER B-ELITE 18
Vereinigte Edelstahlwerke
C, bal Fe.
For welding wire and electrodes; for electric arc welding.
Obsolete

BOHLER B-ELITE U
Vereinigte Edelstahlwerke
C 0.1, bal Fe.
For welding wire and electrodes; electric arc welding.
Obsolete

BOHLER B-SPECIAL
Vereinigte Edelstahlwerke
C, bal Fe.
For welding wire; for gas welding. *Obsolete*

BOHLER B304
Bohler Gesellschaft M.B.H.
Tool material. C 0.73, Cr 0.5, Mo 0.35, V 0.25, W 0.6, bal Fe.
Cold work tool steel for paper shears, wood cutting saws. W.
Nr. 1.2604.

BOHLER B400
Bohler Gesellschaft M.B.H.
Tool material. C 0.8, Si 0.35, Mn 0.4, Cr 0.55, V 0.2, bal Fe.
Cold work tool steel for paper shears, saws. W. Nr. 1.2235.

BOHLER B406
Bohler Gesellschaft M.B.H.
Tool material. C 0.84, Cr 0.55, V 0.25, bal Fe.
Cold work tool steel; for cutters, saws, bearings.

BOHLER B4125A
Bohler Gesellschaft M.B.H.
Tool material. C 0.8, Cr 0.5, Ni 0.35, V 0.2, bal Fe.
Cold work tool steel for cutters, saws, bearings.

BOHLER B53A
Bohler Gesellschaft M.B.H.
Tool material. C 0.75, Cr 0.3, Ni 2.35, bal Fe.
Cold work tool steel for cutters, saws, bearings.

BOHLER BATS
Vereinigte Edelstahlwerke
C, bal Fe.
For bats for threshing machines. *Obsolete*

BOHLER BH
Vereinigte Edelstahlwerke
C, bal Fe.
For welding wire and electrodes; for electric arc welding.
Obsolete

BOHLER BH-EXTRA
Vereinigte Edelstahlwerke
C, bal Fe.
For welding wire and electrodes; for electric arc welding.
Obsolete

BOHLER BHC
Vereinigte Edelstahlwerke
C 0.7, Mn 2, Cr 1, Ti 0.2, bal Fe.
Welded: 220,000 TS. For hard facing rod; wear resistant.
Obsolete

BOHLER BHS
Vereinigte Edelstahlwerke
C 0.22, Mn 1.6, Ti 0.3, bal Fe.
Welded: 87,000 TS. For welding wire; wear resistant.
Obsolete

BOHLER BHW
Vereinigte Edelstahlwerke
C, bal Fe.
For welding wire and electrodes; for electric arc welding.
Obsolete

BOHLER BLADE STEEL
Vereinigte Edelstahlwerke
C, bal Fe.
For blades. *Obsolete*

BOHLER BLITZ
Now VEW V960.

BOHLER BM
Vereinigte Edelstahlwerke
C, bal Fe.
For welding wire and electrodes; for arc welding. *Obsolete*

BOHLER BOREAS
Vereinigte Edelstahlwerke
C, bal Fe.
For drawing plates of steel wire and hard bronze wire; for arc
welding. *Obsolete*

BOHLER BROWNIE
Vereinigte Edelstahlwerke
C, bal Fe.
For mining drills; for work on medium hard rocks. *Obsolete*

BOHLER BROWNIE EXTRA
Vereinigte Edelstahlwerke
C, bal Fe.
For mining drills; for work on hardest rocks. *Obsolete*

BOHLER BS2
Vereinigte Edelstahlwerke
C 0.55, Si 0.1-0.4, Mn 0.5-0.7, bal Fe.
Heat treated: 155,000 TS; 110,000 YS; 15 El; 45 RA; 320 Brin.
For gears, bolts, shafts, axes, fasteners; water hardened.
Obsolete

BOHLER BTH
Now VEW K960.

BOHLER BW
Vereinigte Edelstahlwerke
C, bal Fe.
For welding electrodes; for electric arc welding. *Obsolete*

BOHLER BW XII
Vereinigte Edelstahlwerke
C 0.1, Mn 1, Ni 0.5, bal Fe.
Welded: 67,000 TS; 22-26 El. For welding wire; gas welding.
Obsolete

BOHLER BW-IX
Vereinigte Edelstahlwerke
C, bal Fe.
For welding wire; for gas welding. *Obsolete*

BOHLER BW-VII
Vereinigte Edelstahlwerke
C, bal Fe.
For welding wire; for gas welding. *Obsolete*

BOHLER C/2
Vereinigte Edelstahlwerke
C 0.85, Cr 0.5, bal Fe.
For wood saws, band saws; water hardened. *Obsolete*

BOHLER CC
Now VEW S300.

BOHLER CC SPECIAL
Vereinigte Edelstahlwerke
C 0.65, Cr 4.1, W 18.5, V 1.35, Co 18.5, Mo 0.8, bal Fe.
For super high speed cutters; high speed steel. *Obsolete*

BOHLER CC55 SPEZIAL
Now VEW S307.

BOHLER CC55N
Now VEW S308.

BOHLER CM EXTRA
Vereinigte Edelstahlwerke
C 0.58, Cr 1, Mn 0.95, V 0.09, bal Fe.
For springs, gears, crankshafts, bolts; oil hardened, shock
resistant. *Obsolete*

BOHLER CMO
Now VEW V340.

BOHLER CNME
Vereinigte Edelstahlwerke
C 0.56, Ni, Cr, Mo, V, bal Fe.
For gears, bolts, crankshafts; oil hardened, shock resistant.
Obsolete

BOHLER CNW
Vereinigte Edelstahlwerke
C 0.45, Cr 1.35, Ni 1.65, bal Fe.
For tools, dies, gears, shafts; oil hardening. *Obsolete*

BOHLER COH
Vereinigte Edelstahlwerke
C, Co, bal Fe.
For magnets; magnetic steel. *Obsolete*

BOHLER COK
Vereinigte Edelstahlwerke
C 0.95, Mn 0.3, Cr 8.5, Co 15.5, Mo 1, bal Fe.
For magnets, regulators for furnace installations; magnet
steel. *Obsolete*

BOHLER COM
Vereinigte Edelstahlwerke
C, Co, bal Fe.
For half ring and horseshoe magnets; magnet steel.
Obsolete

BOHLER CRUCIBLE QUALITY
Vereinigte Edelstahlwerke
C 0.7-1.2, bal Fe.
For tools, dies; crucible quality steel. *Obsolete*

BOHLER CRV
Now VEW F550.

BOHLER CS
Vereinigte Edelstahlwerke
C 0.67, Si 1.3, Mn 0.5, Cr 0.5, bal Fe.
For high temperature springs; heat resistant. *Obsolete*

BOHLER CSF
Now VEW K243.

BOHLER CSI
Vereinigte Edelstahlwerke
C 0.7, Si 1.3, Cr 0.5, bal Fe.
For tools, dies, punches; shock resistant. *Obsolete*

BOHLER CV
Vereinigte Edelstahlwerke
C 1.2, Cr 0.8, V 0.1, bal Fe.
For drills, taps; water or oil hardened. *Obsolete*

BOHLER D SPECIAL
Vereinigte Edelstahlwerke
C 1.2-1.3, Mn 0.6-0.8, Si 0.25-1.05, Cr 1.1-1.3, bal Fe.
For tools, dies; oil hardening. *Obsolete*

BOHLER D220
Bohler Gesellschaft M.B.H.
C 0.2, Cr 1.3, Mo 1.1, V 0.3, bal Fe.
Oil hardening; for bolts and nuts resistant to elevated
temperatures to 530°C. W. Nr. 1.8070.

BOHLER D230
Bohler Gesellschaft M.B.H.
C 0.22, Cr 1.3, Mo 0.75, V 0.25, bal Fe.
Oil hardening; for parts resistant to elevated temperature. W.
Nr. 1.7709.

BOHLER D240
Bohler Gesellschaft M.B.H.
C 0.24, Cr 1.4, Mo 0.55, V 0.2, bal Fe.
Oil hardening; for bolts and nuts resistant to elevated temperatures to 530°C. W. Nr. 1.7733.

BOHLER D310
Bohler Gesellschaft M.B.H.
C 0.13, Cr 5.3, Mo 0.55, bal Fe.
Air or oil hardening; for tubes for petroleum distilling, and for hydrogenation plants. W. Nr. 1.7362.

BOHLER D320
Bohler Gesellschaft M.B.H.
C 0.1, Cr 2.3, Mo 1, bal Fe.
Oil hardening; for steam boiler and super heater tubes up to 530°C. W. Nr. 1.7380.

BOHLER D330
Bohler Gesellschaft M.B.H.
C 0.13, Cr 1, Mo 0.55, bal Fe.
Oil hardening; for steam boilers and superheater tubes to 530°C. W. Nr. 1.7335.

BOHLER D500
Bohler Gesellschaft M.B.H.
C 0.15, Mo 0.3, bal Fe.
Oil hardening; for flanges resistant to elevated temperature to 530°F. W. Nr. 1.5415.

BOHLER D502
Bohler Gesellschaft M.B.H.
C 0.22, Mn 0.6, Cr 0-0.3, Mo 0.4, bal Fe.
For thick walled, high pressure tubes, or small forgings, for elevated temperature operation. W. Nr. 1.5419. *Obsolete*

BOHLER DC 7
Vereinigte Edelstahlwerke
C 0.2, Si 0.3, Mn 1, Cr 0.5, bal Fe.
Hardened: 78,000-85,000 TS; 42,000 YS; 28 El; 142-176 Brin.
For bolts, valves, fittings; case hardening. *Obsolete*

BOHLER DCM 10
Vereinigte Edelstahlwerke
C 0.15, Si 0.3, Mn 0.7, Cr 1.1, Mo 0.2, bal Fe.
Hardened: 85,000-105,000 TS; 57,000 YS; 24 El; 60 RA; 176-223 Brin. For bolts for high temperature; case hardening *Obsolete*

BOHLER DCM 12
Vereinigte Edelstahlwerke
C 0.25, Mn 0.6, Cr 1.2, Mo 0.2, bal Fe.
Hardened: 92,000-113,000 TS; 71,000 YS; 22 El; 60 RA; 192-235 Brin. For high temperature bolting; good creep resistance. *Obsolete*

BOHLER DCM 15
Vereinigte Edelstahlwerke
C 0.25, Mn 0.5, Cr 1.3, Mo 0.4, bal Fe.
Hardened: 106,000-120,000 TS; 85,000 YS; 22 El; 60 RA; 223-253 Brin. For bolting for high temperature; good creep resistance. *Obsolete*

BOHLER DCM195
Vereinigte Edelstahlwerke
C 0.1, Cr 5.5, Mo 0.5, bal Fe.
Heat treated: 85,400 TS; 56,900 YS. For power engines; used up to 1120 F. *Obsolete*

BOHLER DCM42
Now VEW V340.

BOHLER DCM54
Vereinigte Edelstahlwerke
C 0.25, Cr 1.3, Mo 0.45, bal Fe.
Heat treated: 113,800 TS; 85,400 YS. For power engines; used up to 970 F. *Obsolete*

BOHLER DCM910
Now VEW D320.

BOHLER DCMV 20
Vereinigte Edelstahlwerke
C 0.18, Cr 1, Si 0.4, Mn 0.6, Mo 0.45, V 0.2, bal Fe.
Hardened: 106,000-120,000 TS; 85,000 YS; 21 El; 50 RA; 223-248 Brin. For high temperature bolting; good creep resistance. *Obsolete*

BOHLER DCMV 30
Vereinigte Edelstahlwerke
C 0.18, Cr 1.2, Mo 0.5, V 0.3, bal Fe.
Hardened: 113,000-128,000 TS; 85,000 YS; 19 El; 50 RA; 235-262 Brin. For high temperature bolting; good creep resistance. *Obsolete*

BOHLER DCMV511
Now VEW D220.

BOHLER DCMV55
Now VEW D240.

BOHLER DCMV7
Vereinigte Edelstahlwerke
C 0.1, Cr 1.8, Mo 0.3, Si 1.1, V 0.3, bal Fe.
Heat treated: 71,200 TS; 42,700 YS. For oil plant pipes and vessels; used up to 1022 F. *Obsolete*

BOHLER DCMVW12
Vereinigte Edelstahlwerke
C 0.22, Cr 3, Mo 0.5, V 0.8, W, bal Fe.
Heat treated: 114,000 TS; 78,300 YS. For engine parts; used up to 986 F. *Obsolete*

BOHLER DMO3
Now VEW D500.

BOHLER DMO6
Vereinigte Edelstahlwerke
C 0.15, Mo 0.55, Mn 0.45, bal Fe.
Annealed: 64,000 TS; 41,300 YS. For steam boiler fittings; used up to 970 F. *Obsolete*

BOHLER DMV83
Now VEW D404.

BOHLER E110
Bohler Gesellschaft M.B.H.
C 0.18, Si 0.3, Mn 0.5, Cr 1.7, Mo 0.3, Ni 1.5, bal Fe.
Deep hardening carburizing steel; high core strength. For plate wheels, driving pinions and highly stressed gears and cog wheels. W. Nr. 1.6587.

BOHLER E115
Bohler Gesellschaft M.B.H.
C 0.2, Cr 0.5, Ni 0.6, Mo 0.2, bal Fe.
Low alloy carburizing steel. For gears, pinions, arbors, bushings. W. Nr. 1.6523; AISI 8620. *Obsolete*

BOHLER E154
Bohler Gesellschaft M.B.H.
C 0.2, Ni 0.5, Cr 0.5, Mo 0.2, bal Fe.
Carburizing steel. *Obsolete*

BOHLER E2
Vereinigte Edelstahlwerke
C 0.15, Si 0.25, Mn 0.37, bal Fe.
Annealed: 70,000 TS; 40,000 YS; 25 El; 60 RA; 145 Brin. For gears, bolts, machine tool parts; case hardened. *Obsolete*

BOHLER E200
Bohler Gesellschaft M.B.H.
C 0.14, Si 0.3, Mn 0.45, Cr 0.7, Ni 3.5, bal Fe.
Alloy carburizing steel; deep hardening. For cams, camshafts, gears, universal joints. W. Nr. 1.5752.

BOHLER E204
Bohler Gesellschaft M.B.H.
C 0.13, Cr 1.1, Ni 4.5, bal Fe.
Alloy carburizing steel; deep hardening. For gears, cams, high stressed gear wheels. W. Nr. 1.5860.

BOHLER E220
Bohler Gesellschaft M.B.H.
C 0.18, Si 0.3, Mn 0.5, Ni 2, Cr 2, bal Fe.
Alloy carburizing steel; deep hardening, high core strength. For gears, shafts, heavy bolts, highly stressed parts. W. Nr. 1.5920.

BOHLER E224
Bohler Gesellschaft M.B.H.
C 0.18, Mn 0.5, Cr 2, Ni 2, bal Fe.
Alloy carburizing steel; high core strength. For gears, cams, cog wheels. W. Nr. 1.5920.

BOHLER E230
Bohler Gesellschaft M.B.H.
C 0.17, Si 0.3, Mn 0.5, Cr 1.5, Ni 1.6, bal Fe.
Alloy carburizing steel. For gears, cams, camshafts, chain wheels. W. Nr. 1.5919.

BOHLER E234
Bohler Gesellschaft M.B.H.
C 0.16, Ni 1.3, Cr 0.65, bal Fe.
Alloy carburizing steel. For cams, camshafts, pinions, bearings. W. Nr. 1.5713.

BOHLER E300
Bohler Gesellschaft M.B.H.
C 0.2, Cr 1.2, Mo 0.25, bal Fe.
Alloy carburizing steel. For cams, pinions, gears, bearings. W. Nr. 1.7264.

BOHLER E320
Bohler Gesellschaft M.B.H.
C 0.2, Mn 0.75, Cr 0.4, Mo 0.45, bal Fe.
Carburizing steel; for small gears, cams, arbors, bushings. W. Nr. 1.7321.

BOHLER E321
Bohler Gesellschaft M.B.H.
C 0.2, Mn 0.75, P 0.035, S 0.02-0.035, Cr 0.4, Mo 0.45, bal Fe.
Carburizing steel. W. Nr. 1.7323. *Obsolete*

BOHLER E400
Bohler Gesellschaft M.B.H.
C 0.2, Mn 1.25, Cr 1.15, bal Fe.
Alloy carburizing steel. For cams, pinions, bearings. W. Nr. 1.7147.

BOHLER E401
Bohler Gesellschaft M.B.H.
C 0.2, Si 0.25, Mn 1.3, Cr 1.2, S 0.03, bal Fe.
Alloy carburizing steel. For bearings, gears, cams. W. Nr. 1.7149.

BOHLER E406
Bohler Gesellschaft M.B.H.
C 0.18, Mn 1.15, Cr 1, bal Fe.
Alloy carburizing steel. For cams, bearings, pinions. W. Nr. 1.7168. *Obsolete*

BOHLER E410
Bohler Gesellschaft M.B.H.
C 0.17, Si 0.3, Mn 1.2, Cr 0.9, bal Fe.
Low alloy carburizing steel. For bearings, cams, pinions, arbors. W. Nr. 1.7131.

BOHLER E411
Bohler Gesellschaft M.B.H.
C 0.17, Si 0.25, Mn 1.2, Cr 0.95, S 0.03, bal Fe.
Low alloy carburizing steel. For pinions, shafts, bearings, cams. W. Nr. 1.7139.

BOHLER E416
Bohler Gesellschaft M.B.H.
C 0.16, Mn 1.1, Cr 1.05, B, bal Fe.
Low alloy carburizing steel. For gears, shafts, universal joints. W. Nr. 1.7160. *Obsolete*

BOHLER E502
Bohler Gesellschaft M.B.H.
C 0.2, Mn 0.9, Cr 1, V 0.1, bal Fe.
Low alloy carburizing steel. For pinions, gears, cams, bearings. W. Nr. 1.7510. *Obsolete*

BOHLER E525
Bohler Gesellschaft M.B.H.
C 0.15, Mn 0.5, Cr 0.65, bal Fe.
Low alloy carburizing steel. For piston pins, roller bearings, cams. W. Nr. 1.7015.

BOHLER E900
Bohler Gesellschaft M.B.H.
C 0.1, Si 0.25, Mn 0.4, bal Fe.
Carburizing steel. For light loaded cams, small machine parts. W. Nr. 1.1121; AISI 1010.

BOHLER E920
Bohler Gesellschaft M.B.H.
C 0.15, Si 0.25, Mn 0.4, bal Fe.
Carburizing steel. For light loaded cams, small machine parts. W. Nr. 1.1141; AISI 1015.

BOHLER EB100
Vereinigte Edelstahlwerke
C 0.2, Mn 1.25, Cr 1.15, bal Fe.
For gears, cams, camshafts; case hardened. *Obsolete*

BOHLER EB30
Vereinigte Edelstahlwerke
C 0.13, Cr 4, bal Fe.
For oil refinery equipment; creep and heat resistant. *Obsolete*

BOHLER EB60
Now VEW E525.

BOHLER EB80
Vereinigte Edelstahlwerke
C 0.16, Mn 1.05, Cr 0.95, bal Fe.
Heat treated: 142,500 TS; 86,000 YS. For gears, cams, camshafts; case hardened. *Obsolete*

BOHLER EB95
Vereinigte Edelstahlwerke
C 0.2, Mn 1.25, Cr 1.15, bal Fe.
Heat treated: 171,000 TS; 100,000 YS. For gears, cams, camshafts; case hardened. *Obsolete*

BOHLER EBK
Vereinigte Edelstahlwerke
C 0.16, bal Fe.
For machine tool parts; case hardened. *Obsolete*

BOHLER EBKW
Vereinigte Edelstahlwerke
C 0.1, bal Fe.
Heat treated: 85,400 TS. For machine tool parts; case hardened. *Obsolete*

BOHLER EBM
Now VEW M100.

BOHLER ECL 100
Now VEW E300.

BOHLER ECL 80
Now VEW E304.

BOHLER ECN 100
Vereinigte Edelstahlwerke
C 0.15, Cr 1, Ni 0.8, bal Fe.
For gears, pinions, crankshafts; carburizing steel, tough. *Obsolete*

BOHLER ECN 200
Now VEW E220.

BOHLER ECN150
Now VEW E230.

BOHLER EL
Now VEW M150.

BOHLER EMC
Vereinigte Edelstahlwerke
C 0.4, Cr 1, bal Fe.
Heat treated: 241,800 TS; 185,000 YS. For structural parts, gears, shafts; oil hardened. *Obsolete*

BOHLER ENA
Vereinigte Edelstahlwerke
C 0-0.16, Ni 1.5, Cr 0.2, bal Fe.
For gears, cams, camshafts; case hardening, tough. *Obsolete*

BOHLER EPB EXTRA
Vereinigte Edelstahlwerke
C 0.14, Cr 1, Ni 4.3, bal Fe.
For plastic mold dies; case hardened. *Obsolete*

BOHLER EPB EXTRA M
Vereinigte Edelstahlwerke
C 0.19, Ni 4, Mo 0.2, Cr 1.25, bal Fe.
Heat treated: 200,000 TS. For gears, bolts, camshafts, cams; case hardened, shock resistant. *Obsolete*

BOHLER EPB EXTRA W
Now VEW M130.

BOHLER EPB PRIMA
Vereinigte Edelstahlwerke
C 0.17, Cr 1.5, Ni 1.5, bal Fe.
For plastic mold dies; case hardened. *Obsolete*

BOHLER EPB SPEZIAL
Now VEW M120.

BOHLER ES
Vereinigte Edelstahlwerke
C 0-0.17, Mn, Si, bal Fe.
For gears, cams, camshafts; case hardening steel. *Obsolete*

BOHLER ES PRIMA
Now VEW E216.

BOHLER ES SPECIAL
Now VEW E200.

BOHLER ES2
Vereinigte Edelstahlwerke
C 0.15, Si 0.25, Mn 0.37, bal Fe.
Annealed: 70,000 TS; 40,000 YS; 25 El; 60 RA; 145 Brin. For gears, bolts, machine tool parts; case hardened. *Obsolete*

BOHLER ESC
Now VEW K240.

BOHLER ESK
Vereinigte Edelstahlwerke
C 0.15, Cr 0.8, bal Fe.
For gears, shafts; case hardening. *Obsolete*

BOHLER EWH
Now VEW V920.

BOHLER EXTRA FM
Now VEW K505.

BOHLER EXTRA H
Vereinigte Edelstahlwerke
C 1.1, V 0.1, bal Fe.
For nail machine dies and cutters, cutting dies, coining dies; water hardening. *Obsolete*

BOHLER EXTRA HARD
Vereinigte Edelstahlwerke
C 1.4, Cr 0.5, V 1.1, bal Fe.
For cutting tools, razors, surgical instruments, chisels, drawing mandrels; oil hardening. *Obsolete*

BOHLER EXTRA K
Now VEW K310.

BOHLER EXTRA K 5
Vereinigte Edelstahlwerke
C 1, Cr 5, Mo 1, V, bal Fe.
For forming and blanking tools; air hardened, tough. *Obsolete*

BOHLER EXTRA MG
Vereinigte Edelstahlwerke
C 1.05, Mn 1, Cr 0.6, bal Fe.
Water hardened: 220,000 TS, 155,000 YP; 11 El; 600 Brin. Oil hardened: 188,000 TS; 120,000 YP; 10 El; 400 Brin. For taps, threading dies, chasers, reamers, broaches, gauges, knives, bearings, plastic mold dies. Tough, oil harden or water harden. *Obsolete*

BOHLER EXTRA MITTEL-HART
Vereinigte Edelstahlwerke
C 1.25, bal Fe.
For milling cutters, file cutters, stone picks, knives, cams; water hardening. *Obsolete*

BOHLER EXTRA RAPID 300
Vereinigte Edelstahlwerke
C 1.3, Cr 4.3, Mo 0.85, V 3.8, W 12, bal Fe.
For forming and blanking dies, engravers' tools; high speed steel. *Obsolete*

BOHLER EXTRA S
Now VEW K761.

BOHLER EXTRA SC
Now VEW K240.

BOHLER EXTRA SOFT
Vereinigte Edelstahlwerke
C 0.7, bal Fe.
For hammers, axes, tools, shafts, rails; Type W1; water hardened. *Obsolete*

BOHLER EXTRA TOUGH
Vereinigte Edelstahlwerke
C 1.12, Si 0.25, Mn 0-0.23, bal Fe.
For taps, threading dies, reamers, mint dies; water hardening; AISI W1. *Obsolete*

BOHLER EXTRA WEICH
Vereinigte Edelstahlwerke
C 0.7, Si 0-0.25, Mn 0-0.25, bal Fe.
Heat treated: 174,000 TS; 128,000 YS; 12 El; 37 RA; 355 Brin. For springs, rails, punches, hammers, axes; water hardened; Type W1. *Obsolete*

BOHLER EXTRA ZAH
Now VEW K980.

BOHLER EXTRA ZAHNHART
Now VEW K990.

BOHLER EXTRA ZH100
Now VEW K990.

BOHLER F100
Bohler Gesellschaft M.B.H.
C 0.5, Si 1.7, Mn 0.7, bal Fe.
Silicon-manganese spring steel; oil hardening. Laminated springs for rail vehicles, bumper springs, shock resisting tools. W. Nr. 1.0903; similar to AISI 9255.

BOHLER F105
Bohler Gesellschaft M.B.H.
C 0.65, Si 1.7, Mn 0.9, bal Fe.
Silicon-manganese spring steel; oil hardening. Laminated or helical springs for automotive. W. Nr. 1.0906; similar to AISI 9260.

BOHLER F108
Bohler Gesellschaft M.B.H.
C 0.6, S 2, Mn 0.9, bal Fe.
Silicon-manganese spring steel; oil hardening. For laminated or helical springs for vehicles. W. Nr. 1.0909; similar to AISI 9260.

BOHLER F110
Bohler Gesellschaft M.B.H.
C 0.55, Si 1.7, Mn 0.9, bal Fe.
Silicon-manganese spring steel; oil hardening. Laminated or helical springs for automotive. W. Nr. 1.0904; similar to AISI 9255.

BOHLER F114
Bohler Gesellschaft M.B.H.
C 0.48, Si 1.65, Mn 0.7, bal Fe.
For laminated or elliptical springs. W. Nr. 1.0902.

BOHLER F120
Vereinigte Edelstahlwerke
C 1.2, bal Fe.
Hardened: Rock C 66. Annealed: 85,000-100,000 TS; 175-204 Brin. For machinist files, drills, cutters, reamers. Water hardening, wear resistant. *Obsolete*

BOHLER F124
Bohler Gesellschaft M.B.H.
C 0.6, Si 0.8, Mn 1.1, bal Fe.
Silicon-manganese spring steel; oil hardening.

BOHLER F128
Bohler Gesellschaft M.B.H.
C 0.6, Si 1.65, Mn 0.9, Cr 0.3, bal Fe.
Silicon-manganese spring steel; oil hardening. Laminated springs, plate and spiral springs. W. Nr. 1.0961.

BOHLER F130
Now VEW K995.

BOHLER F145
Vereinigte Edelstahlwerke
C 1.45, bal Fe.
Hardened: Rock C 66. Annealed: 85,000-100,000 TS; 175-204 Brin. For precision and circular files, knurling and rotary files. Water hardening, wear resistant. *Obsolete*

BOHLER F180
Bohler Gesellschaft M.B.H.
C 0.5, Mn 1.8, bal Fe.
Water or oil hardening spring steel. W. Nr. 1.0913.

BOHLER F200
Bohler Gesellschaft M.B.H.
C 0.67, Si 1.3, Mn 0.5, Cr 0.5, bal Fe.
Oil hardening spring steel. For valve springs, helical springs. W. Nr. 1.7103.

BOHLER F300
Bohler Gesellschaft M.B.H.
C 0.55, Mn 0.85, Cr 0.75, bal Fe.
Helical springs, torsion bar springs; oil hardening. W. Nr. 1.7176; AISI 5155.

BOHLER F500
Bohler Gesellschaft M.B.H.
C 0.52, Mn 0.9, Cr 1.05, Mo 0.2, V 0.1, bal Fe.
Oil hardening spring steel. W. Nr. 1.7701; AISI 4150. *Obsolete*

BOHLER F550
Bohler Gesellschaft M.B.H.
C 0.5, Mn 1, Cr 1.05, V 0.13, bal Fe.
Chrome-vanadium spring steel. For coiled springs; oil hardening. W. Nr. 1.8159; AISI 6150.

BOHLER FB10
Now VEW H120.

BOHLER FB30G
Vereinigte Edelstahlwerke
C 0.6, Si 1.5, Cr 29, bal Fe.
For oil refinery equipment; creep and heat resistant. *Obsolete*

BOHLER FB30S
Vereinigte Edelstahlwerke
C 0.2, Si 1.2, Cr 25, Ni 4, bal Fe.
Cast: 90,000 TS; 65,000 YS; 2 El; 212 Brin. For cylinder liners, bushings, valve seats and bodies; corrosion and heat resistant. *Obsolete*

BOHLER FB8
Now VEW H160.

BOHLER FB8G
Now VEW H160G.

BOHLER FC
Now VEW K205.

BOHLER FF
Vereinigte Edelstahlwerke
C 0.15, Si 2, Cr 19.5, Ni 9.5, bal Fe.
Annealed: 80,000 TS; 35,000 YS; 55 El; 75 RA; 150 Brin. For chemical plant equipment, tanks, mixers; Type 302; stainless, austenitic. *Obsolete*

BOHLER FF
Vereinigte Edelstahlwerke
C 0-0.2, Cr 22-24, Ni 12-15, Mn 0-2, Si 0-3.5, bal Fe.
Annealed: 85,000-95,000 TS; 40,000-50,000 YS; 45-55 El; 150-185 Brin. For heat treat boxes, furnace parts, refinery equipment; Type 309; austenitic, heat resistant. *Obsolete*

BOHLER FFB
Vereinigte Edelstahlwerke
C 0-0.25, Cr 24-26, Ni 19-22, bal Fe.
Annealed: 95,000 TS; 45,000 YS; 50 El; 65 RA; 180 Brin. At 1200 F: 57,000 TS; 22,000 YS; 32 El; 45 RA. For valves, pumps, engine components, furnace parts; Type 310; austenitic, heat resistant. *Obsolete*

BOHLER FFBG
Vereinigte Edelstahlwerke
C 0.4, Si 2, Cr 26, Ni 14, bal Fe.
Annealed: 90,000 TS; 45,000 YS; 45 El; 170 Brin. For furnace parts, valves, pumps; Type HI; corrosion and heat resistant. *Obsolete*

BOHLER FFG
Vereinigte Edelstahlwerke
C 0.4, Si 2, Cr 22, Ni 9.5, bal Fe.
Cast: 85,000 TS; 45,000 YS; 35 El; 165 Brin. For furnace parts, heat treating boxes, conveyors; Type HF; corrosion and heat resistant. *Obsolete*

BOHLER FIRST QUALITY
Vereinigte Edelstahlwerke
C 0.7-1.3, bal Fe.
For general purpose tools and dies; water hardening. *Obsolete*

BOHLER FM
Now VEW K505.

BOHLER FM EXTRA
Vereinigte Edelstahlwerke
C 1.5, Cr 1.4, V 0.1, bal Fe.
For tools, dies, bearings; wear resistant. *Obsolete*

BOHLER FOX A 7
Vereinigte Edelstahlwerke
C 0.12, Mn 7, Cr 19, Ni 0.5, bal Fe.
Welded: 95,000 TS; 48 El. For stainless welding electrodes; austenitic, corrosion resistant. *Obsolete*

BOHLER FOX DCMS
Vereinigte Edelstahlwerke
C 0.15, Mn 0.6, Cr 0.8, Mo 0.4, bal Fe.
Welded: 86,000 TS; 27 El. For welding electrodes; for high temperature service. *Obsolete*

BOHLER FOX DMO
Vereinigte Edelstahlwerke
C 0.1, Mn 0.6, Mo 0.4, bal Fe.
Welded: 80,000 TS; 33 El. For welding wire; Mo coated. *Obsolete*

BOHLER FOX DUR 600
Vereinigte Edelstahlwerke
C 0.4, Si 3, Cr 10, bal Fe.
Welded: 30,000 TS. For welding electrodes, hard facing; coated, abrasion resistant. *Obsolete*

BOHLER FOX FB 30 S
Vereinigte Edelstahlwerke
C 0.1, Mn 1, Cr 25, Ni 3, 0.25% nitrogen gas, bal Fe.
For welding electrodes; coated, heat resistant. *Obsolete*

BOHLER FOX GFW
Vereinigte Edelstahlwerke
Cu-Ni.
For welding electrodes; Monel metal. *Obsolete*

BOHLER FOX MSU
Vereinigte Edelstahlwerke
C 0.1, Mn 0.6, bal Fe.
Welded: 76,000 TS; 32 El. For welding wire; coated. *Obsolete*

BOHLER FOX SAS 4
Vereinigte Edelstahlwerke
C 0.08, Cr 18, Ni 10, Mo 2.2, Nb 0.8, bal Fe.
Welded. For welding electrodes; coated, stainless. *Obsolete*

BOHLER FOX SAS 8
Vereinigte Edelstahlwerke
C 0.08, Cr 17, Ni 15, Mo 2, Cu 2, Nb 0.8, bal Fe.
Welded. For welding electrodes; coated, stainless. *Obsolete*

BOHLER FOX SPE
Vereinigte Edelstahlwerke
C 0.1, Mn 0.6, bal Fe.
Welded: 68,000 TS; 34 El. For welding wire; coated. *Obsolete*

BOHLER FOX UMZ
Vereinigte Edelstahlwerke
C 0.1, Mn 0.6, bal Fe.
Welded: 68,000 TS; 34 El. For welding wire; coated. *Obsolete*

BOHLER FOX WKZ 50
Vereinigte Edelstahlwerke
C 0.3, Cr 2.3, W 4.5, V 8.6, bal Fe.
Welded: 210,000 TS. For welding electrodes hard facing; coated. *Obsolete*

BOHLER FSB
Vereinigte Edelstahlwerke
C 0-0.25, Cr 24-26, Ni 19-22, bal Fe.
Annealed: 95,000 TS; 45,000 YS; 50 El; 65 RA; 180 Brin. At 1200 F: 57,000 TS; 22,000 YS; 32 El; 45 RA. For valves, pumps, furnace parts and equipment; Type 310; austenitic, heat resistant. *Obsolete*

BOHLER G55
Vereinigte Edelstahlwerke
C 0.63, Cu 0.2, bal Fe.
Heat treated: 121,000 TS; 78,300 YS. For rifle barrels; water hardened. *Obsolete*

BOHLER GA
Vereinigte Edelstahlwerke
C, bal Fe.
For welding wire or electrodes; for gas or electric welding.
Obsolete

BOHLER GCMO
Vereinigte Edelstahlwerke
C 0.55, Cr 1.1, Ni 0.4, Mo 0.45, V 0.1, bal Fe.
For drop forging dies; hot work steel, oil hardened. *Obsolete*

BOHLER GCN4
Vereinigte Edelstahlwerke
C 0.35, Cr 1.3, Ni 4, Mo 0.25, bal Fe.
For drop forging dies; hot work steel, oil hardened. *Obsolete*

BOHLER GEODURIT SH
Vereinigte Edelstahlwerke
carbides.
For hard facing with carbides; tubes filled with carbides.
Obsolete

BOHLER GMC
Vereinigte Edelstahlwerke
C 0.4, Cr, Mn, Mo, bal Fe.
For bolts, gears, machine tool parts; oil hardened, tough.
Obsolete

BOHLER GMCA
Vereinigte Edelstahlwerke
C 0.4, Mn 1.35, Cr 1.85, Mo 0.2, bal Fe.
For tools, dies, punches; non-deforming. *Obsolete*

BOHLER GMME
Vereinigte Edelstahlwerke
C 0.56, Ni, Cr, Mo, V, bal Fe.
For gears, bolts, crankshafts; oil hardened, tough. *Obsolete*

BOHLER GMNE
Vereinigte Edelstahlwerke
C 0.55, Cr 0.7, Mo 0.5, Ni 1.7, bal Fe.
For tools, dies, gears, shafts; shock resistant. *Obsolete*

BOHLER GNM
Now VEW W501.

BOHLER GNME
Now VEW W500.

BOHLER GSF
Vereinigte Edelstahlwerke
C 0.67, Si 1.3, Mn 0.5, Cr 0.5, bal Fe.
Heat treated: 175,000 TS; 150,000 YS; 350 Brin. For springs,
tools, punches. Water hardening, shock resistant. *Obsolete*

BOHLER GSI
Vereinigte Edelstahlwerke
C 0.5, Cr 0.7, Mn 0.9, bal Fe.
For drop forging dies for soft materials; water hardening.
Obsolete

BOHLER H VII/VI
Vereinigte Edelstahlwerke
C 0.22, Si 0.25, Mn 0.45, bal Fe.
Annealed: 73,000 TS; 41,000 YS; 22 El; 58 RA; 140 Brin. For
gears, bolts, machine tool parts; water hardened. *Obsolete*

BOHLER H100
Bohler Gesellschaft M.B.H.
C 0-0.1, Si 1.2, Mn 0.5, Cr 23.8, Al 1.5, bal Fe.
Stainless and temperature resisting steel. For heat treatment
processes. Annealed: 520-720 N/mm² TS; 280 N/mm² 0.2%
proof stress; 7-10 El. W. Nr. 1.4762.

BOHLER H102
Bohler Gesellschaft M.B.H.
Stainless steel. C 0.06, Mn 0.8, Cr 27, N 0.12, bal Fe.
Stainless and oxidation resisting steel for elevated
temperature operations. AISI 446.

BOHLER H103
Bohler Gesellschaft M.B.H.
Stainless steel. C 0.45, Si 2, Cr 23, bal Fe.
Stainless cast steel for elevated temperature operations. W.
Nr. 1.4745. *Obsolete*

BOHLER H120
Bohler Gesellschaft M.B.H.
Stainless steel. C 0-0.12, Cr 18, Si 1.2, Al 1, bal Fe.
Stainless and temperature resisting steel for furnace parts, oil
refinery equipment. W. Nr. 1.4742.

BOHLER H140
Bohler Gesellschaft M.B.H.
Stainless steel. C 0-0.12, Si 1.1, Cr 12.5, Al 1, bal Fe.
Stainless and temperature resisting steel for furnace parts, oil
refining equipment. W. Nr. 1.4724.

BOHLER H160
Bohler Gesellschaft M.B.H.
C 0-0.12, Si 0.75, Mn 0.4, Cr 6.8, Al 0.75, bal Fe.
Heat resisting steel for furnace and boiler construction. W. Nr.
1.4713. Annealed: 420-620 N/mm² TS; 220 N/mm² 0.2%
proof stress; 15-20 El.

BOHLER H161
Bohler Gesellschaft M.B.H.
C 0.3, Si 1.75, Mn 0-1, Cr 6-8, bal Fe.
Cast steel for elevated temperature operation. W. Nr. 1.4710.
Obsolete

BOHLER H300
Bohler Gesellschaft M.B.H.
C 0.13, Si 1, Cr 25, Ni 4, bal Fe.
Heat resisting steel for furnace parts. W. Nr. 1.4821.

BOHLER H301
Bohler Gesellschaft M.B.H.
Stainless steel. C 0.4, Si 1.5, Mn 0-1.5, Cr 27, Ni 4.5, bal Fe.
Stainless casting for high temperature parts. W. Nr. 1.4823.
Obsolete

BOHLER H304
Bohler Gesellschaft M.B.H.
C 0.2, Si 0.35, Mn 1.2, Cr 25, Ni 4, bal Fe.
Heat resisting alloy for furnace parts and heat treating
equipment. Quenched: 600-850 N/mm² TS; 400 N/mm²
0.2% proof stress; 12-16 El. W. Nr. 1.4876; similar to AISI 330.

BOHLER H500
Bohler Gesellschaft M.B.H.
C 0.07, Si 0.4, Cr 21, Ni 32, Ti 0.3, Al 0.3, bal Fe.
Heat resisting alloy for furnace parts and heat treating
equipment. W. Nr. 1.4876; similar to AISI 330.

BOHLER H520
Bohler Gesellschaft M.B.H.
C 0-0.12, Si 1.3, Mn 0.65, Cr 15.8, Ni 35, bal Fe.
Heat resisting alloy for furnace parts and heat treating
equipment. Quenched: 550-800 N/mm² TS; 230 N/mm²
0.2% proof stress; 22-30 El. W. Nr. 1.4864.

BOHLER H522
Bohler Gesellschaft M.B.H.
C 0-0.08, Si 0.5, Cr 25, S 20, bal Fe.
For gas turbines, furnace equipment. W. Nr. 1.4845; similar to
AISI 310.

BOHLER H525
Bohler Gesellschaft M.B.H.
C 0.08, Si 1.7, Mn 1.2, Cr 24.8, Ni 19.8, bal Fe.
For furnace and heat treating equipment. Quenched:
550-800 N/mm² TS; 230 N/mm² 0.2% proof stress; 22-30 El.
W. Nr. 1.4841; similar to AISI 314.

BOHLER H532
Bohler Gesellschaft M.B.H.
Stainless steel. C 0-0.08, Si 1.7, Mn 1.2, Cr 25.3, Ni 19.8, bal
Fe.
Stainless steel for high temperature operations. Quenched:
550-800 N/mm² TS; 230 N/mm² 0.2% proof stress; 22-30 El.
AISI 310.

BOHLER H550
Bohler Gesellschaft M.B.H.
C 0.09, Si 1.7, Mn 1.2, Cr 19.5, Ni 11.5, bal Fe.
Heat resisting steel for furnace parts, oil refinery equipment.
Quenched: 550-750 N/mm² TS; 230 N/mm² 0.2% proof
stress; 22-30 El. W. Nr. 1.4828; AISI 309, 305.

BOHLER H566
Bohler Gesellschaft M.B.H.
Stainless steel. C 0.08, Mn 1.4, Si 0.3, Cr 23, Ni 14, bal Fe.
Stainless steel for high temperature operations. AISI 309S.

BOHLER H700
Bohler Gesellschaft M.B.H.
C 0.46, Si 2.9, Mn 0.4, Cr 9, bal Fe.
For exhaust valves in automotive engines. Annealed:
900-1100 N/mm² TS; 700 N/mm² 0.2% proof stress; 14 El;
40 RA; 266-325 Brin. W. Nr. 1.4718; similar to SAE HNV 3.

BOHLER H710
Bohler Gesellschaft M.B.H.
C 0.38, Si 2.2, Cr 9.5, Mo 0.85, bal Fe.
For exhaust valves in automotive engines. W. Nr. 1.4731.

BOHLER H730
Bohler Gesellschaft M.B.H.
C 0.8, Si 2, Cr 19.5, Ni 1.4, bal Fe.
For exhaust valves in automotive engines. W. Nr. 1.4747;
similar to SAE HNV6.

BOHLER H734
Bohler Gesellschaft M.B.H.
C 0.85, Si 0.35, Mn 1.3, Cr 17.5, Mo 2.4, V 0.5, bal Fe.
For exhaust valves in automotive engines. Annealed:
1000-120 N/mm² TS; 800 N/mm² 0.2% proof stress; 7 El; 12
RA; 296-355 Brin.

BOHLER H800
Bohler Gesellschaft M.B.H.
C 0.43, Si 2.5, Mn 1.2, Cr 18, Ni 9, W 1, bal Fe.
For exhaust valves in automotive engines. 800-1000 N/mm²
TS; 380 N/mm² 0.2% proof stress; 25 El; 35 RA. W. Nr.
1.4873; similar to SAE EV 5.

BOHLER H850
Bohler Gesellschaft M.B.H.
C 0.52, Si 0.13, Mn 8.8, Cr 20.8, Ni 3.6, S 0.06, N 0.45, bal Fe.
For exhaust valves in heavy duty engines. 950-1200 N/mm²
TS; 580 N/mm² 0.2% proof stress; 32 Rock C; 8 El; 10 RA. W.
Nr. 1.4871; similar to SAE EV 8.

BOHLER H851
Bohler Gesellschaft M.B.H.
C 0.61, Si 0-0.25, Mn 10.5, Cr 21, Mo 0.9, V 0.9, Nb 1.1, N
0.5, bal Fe.
For exhaust valves in heavy duty engines. 1000-1250 N/mm²
TS; 800 N/mm² 0.2% proof stress; 35 Rock C; 8 El; 10 RA.

BOHLER H852
Bohler Gesellschaft M.B.H.
C 0.55, Si 0-0.25, Mn 8.5, Cr 20.8, Ni 2.3, N 0.35, bal Fe.
For exhaust valves in heavy duty engines. 900-1150 N/mm²
TS; 550 N/mm² 0.2% proof stress; 28 Rock C; 8 El; 10 RA.

BOHLER H860
Bohler Gesellschaft M.B.H.
C 0.7, Si 0.65, Mn 6.1, Cr 21.3, Ni 1.6, N 0.25, bal Fe.
For exhaust valves in heavy duty engines. 880-1130 N/mm²
TS; 540 N/mm² 0.2% proof stress; 28 Rock C; 8 El; 10 RA.

BOHLER HARD
Vereinigte Edelstahlwerke
C 0.45, Si 0.25, Mn 0.56, bal Fe.
For gears, bolts, machine tool parts; water hardened.
Obsolete

BOHLER HARD CORE
Vereinigte Edelstahlwerke
C, bal Fe.
For core insertions. *Obsolete*

BOHLER HART
Vereinigte Edelstahlwerke
C 1.4, Mn 0.3, Si 0.25, bal Fe.
For tools, cutters, hobs; keen cutting edge. *Obsolete*

BOHLER HH
Vereinigte Edelstahlwerke
C 0.61, Si 0.25, Mn 0.65, bal Fe.
Heat treated: 160,000 TS; 113,000 YS; 12 El; 40 RA; 325 Brin.
For hammers, axes, punches; water hardened. *Obsolete*

BOHLER HM
Now VEW V930.

BOHLER HMA
Now VEW V940.

BOHLER HSB
Vereinigte Edelstahlwerke
C 0.8, Cr 0.5, Ni 0.4, bal Fe.
For saw blades; water or oil hardened. *Obsolete*

BOHLER IMP
Vereinigte Edelstahlwerke
C 0.65, Cr 3.75, Mo 0.85, V 0.07, W 8.5, bal Fe.
For lathe and planer tools, reamers, drills, hobs; high speed
steel. *Obsolete*

BOHLER IN 1 A
Vereinigte Edelstahlwerke
C 0.25, Si 0.3, Mn 0.7, Cr 3, Mo 0.3, bal Fe.
Hardened: 92,000-108,000 TS; 64,000 YS; 18 El; 192-226
Brin. For chemical processing equipment; creep resistant.
Obsolete

BOHLER IN 10
Vereinigte Edelstahlwerke
C 0.22, Si 0.3, Mn 0.4, Cr 0.3, Mo 0.4, W 0.4, V 0.8, bal Fe.
Hardened: 120,000-142,000 TS; 106,000 YS; 18 El; 55 RA;
253-300 Brin. For chemical processing equipment; creep
resistant. *Obsolete*

BOHLER IN 5
Now VEW D328.

BOHLER IN 8
Vereinigte Edelstahlwerke
C 0.17, Si 0.3, Mn 0.4, Cr 2.5, Mo 0.5, W 0.5, V 0.05, bal Fe.
Hardened: 92,000-117,000 TS; 64,000 YS; 22 El; 55 RA;
192-239 Brin. For chemical processing equipment; creep
resistant. *Obsolete*

BOHLER IN 9
Now VEW D204.

BOHLER IN5A
Vereinigte Edelstahlwerke
C 0.1, Cr 2.7, V 0.2, bal Fe.
For high pressure vessels. *Obsolete*

BOHLER INC5
Now VEW D310.

BOHLER INOIL
Vereinigte Edelstahlwerke
C 0.1, Cr 5, Mo 0.5, bal Fe.
Hardened: 71,000-92,000 TS; 65,000 YS; 22 El:, 50 RA;
145-192 Brin. For chemical processing equipment; creep
resistant. *Obsolete*

BOHLER INVAR STEEL
Vereinigte Edelstahlwerke
C, alloy, bal Fe.
For component parts of barometers and optical instruments;
low coefficient of expansion. *Obsolete*

BOHLER K
Vereinigte Edelstahlwerke
C 2.25, Cr 13, bal Fe.
For tools, dies; non-deforming. *Obsolete*

BOHLER K-100
Vereinigte Edelstahlwerke
C 1.05, Cr 0.4, Si 0.2, Mn 0.3, bal Fe.
For ball bearings, races; water hardened. *Obsolete*

BOHLER K-100 S
Vereinigte Edelstahlwerke
C 1.1, Cr 1.2, bal Fe.
For gears, cams, rams, slides, spindles; water or oil
hardening. *Obsolete*

BOHLER K-100/1
Vereinigte Edelstahlwerke
C 1.2, Cr 0.8, bal Fe.
For mint and medal dies, press dies, rolls; water hardening.
Obsolete

BOHLER K-3
Vereinigte Edelstahlwerke
C, Cr, bal Fe.
For magnets; magnet steel. *Obsolete*

BOHLER K-3 S
Vereinigte Edelstahlwerke
C, alloy, bal Fe.
For plug and ring gages, rolls for rubbers; oil hardening.
Obsolete

BOHLER K100
Bohler Gesellschaft M.B.H.
C 2, Si 0.2, Mn 0.3, Cr 11.5, bal Fe.
Cold work tool steel; air or oil hardening. For punches,
blanking and forming dies, trimming dies. W. Nr. 1.2080;
similar to AISI D3.

BOHLER K100 W/S
Vereinigte Edelstahlwerke
C 0.9, Si 0.25, Mn 0.3, Cr 0.8, bal Fe.
For bearings, bushings, cutters; water hardened, wear
resistant. *Obsolete*

BOHLER K102
Bohler Gesellschaft M.B.H.
Tool material. C 2.9, Cr 12, bal Fe.
Cold work tool steel; air or oil hardening. For trimming and
coining dies, thread rollers. W. Nr. 1.2086.

BOHLER K103
Bohler Gesellschaft M.B.H.
Tool material. C 2, Cr 12, Mo 0.5, V 0.5, W 1, bal Fe.
Cold work tool steel; air or oil hardening. For punching and
trimming dies, coining dies, broaches, thread rolling dies. W.
Nr. 1.2000, modified AISI D3.

BOHLER K105
Bohler Gesellschaft M.B.H.
Tool material. C 1.6, Si 0.35, Mn 0.3, Cr 11.5, Mo 0.6, W 0.5,
V 0.2, bal Fe.
Cold work tool steel; air or oil hardening. For broaches,
forming dies, punches. W. Nr. 1.2601.

BOHLER K107
Bohler Gesellschaft M.B.H.
Tool material. C 2.1, Si 0.35, Mn 0.35, Cr 11.5, W 0.7, bal Fe.
Cold work tool steel; air or oil hardening. Heavy duty
punching and trimming dies, reamers, broaches. W. Nr.
1.2436; similar to AISI D3.

BOHLER K110
Bohler Gesellschaft M.B.H.
Tool material. C 1.55, Si 0.3, Mn 0.3, Cr 11.5, V 1, Mo 0.7, bal
Fe.
Cold work tool steel; air or oil hardening. For punching and
forming dies, trimming dies, thread rolling dies. W. Nr.
1.2379; AISI D2.

BOHLER K116
Bohler Gesellschaft M.B.H.
Tool material. C 1.65, Cr 11.5, V 0.1, bal Fe.
Cold work tool steel; air or oil hardening. For punching,
forming and trimming dies, thread rolling dies. W. Nr. 1.2201.

BOHLER K150
Now VEW K200.

BOHLER K190 ISOMATRIX PM
Bohler Gesellschaft M.B.H.
Tool material. C 2.3, Si 0.4, Mn 0.4, Cr 12.5, V 4, Mo 1.1, bal
Fe.
Cold work tool steel; air or oil hardening. For punching and
forming dies, trimming dies, thread rolling dies.

BOHLER K200
Bohler Gesellschaft M.B.H.
Tool material. C 1, Si 0.3, Mn 0.35, Cr 1.5, bal Fe.
Cold work tool steel; water or oil hardening. For reamers,
lathe centers, drills, bearings. W. Nr. 1.2067.

BOHLER K201
Bohler Gesellschaft M.B.H.
Tool material. C 0.63, Si 1.1, Cr 0.6, Mn 1.1, bal Fe.
Cold work tool steel; water or oil hardening. For bushings,
liners, hand tools. W. Nr. 1.2064.

BOHLER K205
Bohler Gesellschaft M.B.H.
Tool material. C 1.4, Cr 0.6, bal Fe.
Cold work tool steel; water or oil hardening. For files, needle
files, precision tools. W. Nr. 1.2008.

BOHLER K240
Bohler Gesellschaft M.B.H.
Tool material. C 0.9, Mn 0.7, Si 1.2, Cr 1.2, bal Fe.
Cold work tool steel; oil hardening. For punches, shear
blades, engravers' tools, bushings. W. Nr. 1.2108.

BOHLER K243
Bohler Gesellschaft M.B.H.
Tool material. C 0.63, Si 1.7, Cr 0.9, bal Fe.
Cold work tool steel, shock resisting type. For chisels, rivet
sets, staking tools.

BOHLER K244
Bohler Gesellschaft M.B.H.
Tool material. C 0.63, Si 1.5, Mn 0.6, Cr 0.7, bal Fe.
Cold work tool steel; shock resisting type. For pneumatic
chisels, rivet sets, staking tools.

BOHLER K245
Bohler Gesellschaft M.B.H.
Tool material. C 0.63, Si 1.1, Mn 1.1, Cr 0.6, bal Fe.
Cold work tool steel; shock resisting type. For chisels, staking
tools, rivet sets. W. Nr. 1.2101.

BOHLER K300
Bohler Gesellschaft M.B.H.
Tool material. C 0.53, Si 0.9, Mn 0.5, Cr 8.3, Mo 1.2, W 1.2,
bal Fe.
Cold work tool steel; oil hardening. Shear blades for metal
sheet. W. Nr. 1.2631.

BOHLER K301
Bohler Gesellschaft M.B.H.
Tool material. C 0.4, Cr 8, Mo 1.6, W 1.35, V 0.5, bal Fe.
Cold work tool steel; oil hardening. Special tools. W. Nr.
1.2631.

BOHLER K305
Bohler Gesellschaft M.B.H.
Tool material. C 0.98, Si 0.3, Mn 0.5, Cr 5.1, Mo 1, V 0.15, bal Fe.
Cold work tool steel; air hardening. For forming, blanking, embossing. W. Nr. 1.2363; AISI A2.

BOHLER K306
Bohler Gesellschaft M.B.H.
Tool material. C 0.51, Si 0.95, Mn 0.3, Cr 5, Mo 1.4, V 1.4, bal Fe.
Cold work tool steel; air hardening. W. Nr. 1.2345.

BOHLER K310
Bohler Gesellschaft M.B.H.
Tool material. C 0.87, Si 0.3, Mn 0.4, Cr 1.8, Mo 0.3, V 1.4, bal Fe.
Cold work tool steel. For cold rolls, cams, press discs.

BOHLER K311
Bohler Gesellschaft M.B.H.
Tool material. C 0.45, Mn 0.9, Cr 1.8, Mo 0.25, bal Fe.
Cold work tool steel. For chisels, center punches. W. Nr. 1.2328. *Obsolete*

BOHLER K400
Bohler Gesellschaft M.B.H.
Tool material. C 1.4, Cr 0.25, W 3.3, V 0.25, bal Fe.
Cold work tool steel; oil or water hardening. For engravers' tools, fast finishing tools. W. Nr. 1.2562.

BOHLER K405
Bohler Gesellschaft M.B.H.
Tool material. C 1.2, Cr 0.2, W 1, V 0.1, bal Fe.
Cold work tool steel; water hardening. For fast finishing tools for short runs, rotary files, counter bores. W. Nr. 1.2516.

BOHLER K450
Bohler Gesellschaft M.B.H.
Tool material. C 0.48, Si 0.9, Mn 0.3, Cr 1, W 2, V 0.18, bal Fe.
Cold work tool steel; oil hardening. For sheet shears, pneumatic chisels; shock resisting type. W. Nr. 1.2542; similar to AISI S1.

BOHLER K451
Bohler Gesellschaft M.B.H.
Tool material. C 0.8, W 2.6, Mo 0.4, Cr 0.7, bal Fe.
Cold work tool steel. For punches, header dies, upsetters.

BOHLER K455
Bohler Gesellschaft M.B.H.
Tool material. C 0.63, Si 0.6, Mn 0.3, Cr 1.1, W 2, V 0.18, bal Fe.
Hot or cold work tool steel; oil hardening. For header dies, upsetters, chisels. W. Nr. 1.2550.

BOHLER K457
Bohler Gesellschaft M.B.H.
Tool material. C 1.2, Cr 1.6, W 1.5, V 0.15, bal Fe.
Cold work tool steel; oil hardening. Cutting tools for leather, plastic and other nonmetallic materials. W. Nr. 1.2519.

BOHLER K458
Bohler Gesellschaft M.B.H.
Tool material. C 1.1, Cr 1.2, W 1.3, V 0.2, bal Fe.
Cold work tool steel; oil hardening. Cutting tools for wood, leather, plastics. W. Nr. 1.2519.

BOHLER K460
Bohler Gesellschaft M.B.H.
Tool material. C 0.95, Si 0.3, Mn 1.1, Cr 0.5, W 0.55, V 0.12, bal Fe.
Cold work tool steel; oil hardening. For reamers, hand taps, pipe threading tools. W. Nr. 1.2510; AISI O1.

BOHLER K465
Bohler Gesellschaft M.B.H.
Tool material. C 1.05, Si 0.25, Mn 0.95, Cr 1, W 1.1, bal Fe.
Cold work tool steel; oil hardening. For cold forming and cold heading dies, punches. W. Nr. 1.2419.

BOHLER K466
Bohler Gesellschaft M.B.H.
Tool material. C 0.42, Cr 1.05, W 1.9, bal Fe.
Cold work tool steel; oil hardening. For cold header dies, punches for thin sheet, pneumatic tools. W. Nr. 1.2542.

BOHLER K467
Bohler Gesellschaft M.B.H.
Tool material. C 1.2, W 1, bal Fe.
Cold work tool steel; water hardening. For center drills, counterbores, reamers. W. Nr. 1.2414. *Obsolete*

BOHLER K505
Bohler Gesellschaft M.B.H.
Tool material. C 1.45, Si 0.25, Mn 0.6, Cr 1.4, V 0.12, bal Fe.
Abrasion resistant cold work steel; oil hardening. For blanking and drawing dies, punches, bushings. W. Nr. 1.2063.

BOHLER K506
Bohler Gesellschaft M.B.H.
Tool material. C 0.48, Mn 1, Cr 1.2, V 0.3, bal Fe.
Cold work tool steel; oil hardening. For cold heading and forming. Similar to W. Nr. 1.2241. *Obsolete*

BOHLER K508
Bohler Gesellschaft M.B.H.
Tool material. C 0.5, Mn 0.9, Cr 1.05, V 0.1, bal Fe.
Cold work tool steel; oil hardening. For carbide tool shanks, screwdrivers, hatchets, various hand tools. W. Nr. 1.2241; similar to AISI 6150.

BOHLER K510
Bohler Gesellschaft M.B.H.
Tool material. C 1.18, Si 0.25, Mn 0.3, Cr 0.7, V 0.1, bal Fe.
Cold work tool steel; oil or water hardening. For drills, reamers, countersinks, scraping tools, punches. W. Nr. 1.2210.

BOHLER K511
Bohler Gesellschaft M.B.H.
Tool material. C 0.32, Mn 0.5, Cr 0.55, V 0.1, bal Fe.
Cold work tool steel; water hardening. For hand tools such as screwdrivers and wrenches. W. Nr. 1.2208. *Obsolete*

BOHLER K600
Bohler Gesellschaft M.B.H.
Tool material. C 0.45, Ni 4, Cr 1.3, Mo 0.25, bal Fe.
Cold work tool steel; oil or air hardening. For embossing dies, forming dies, shear blades. W. Nr. 1.2767.

BOHLER K605
Bohler Gesellschaft M.B.H.
Tool material. C 0.45, Si 0.25, Mn 0.4, Cr 1.3, Mo 0.25, Ni 4, bal Fe.
Cold work tool steel; air or oil hardening. For cold heading dies, shear blades. W. Nr. 1.2721.

BOHLER K618
Bohler Gesellschaft M.B.H.
Tool material. C 0.7, Mn 1.3, Si 1, Cr 1, Ni 1, bal Fe.
Heat treated: 227,000 TS; 350 Brin. For machine tool parts; oil hardened. *Obsolete*

BOHLER K630
Bohler Gesellschaft M.B.H.
Tool material. C 0.85, Si 0.25, Mn 0.35, Ni 0.75, V 0.13, bal Fe.
Cold work tool steel. For cold heading tools.

BOHLER K700
Bohler Gesellschaft M.B.H.
C 1.23, Si 0.4, Mn 12.5, bal Fe.
Austenitic manganese steel. For wear and abrasion resisting parts. Quenched: 800-1000 N/mm^2 TS; 350 N/mm^2 0.2% proof stress; 35 El min; 35 RA min; 200 Brin. W. Nr. 1.3401.

BOHLER K701
Bohler Gesellschaft M.B.H.
C 1.2, Mn 12.5, bal Fe.
Austenitic manganese steel. For wear and abrasion resisting parts. W. Nr. 1.3401. *Obsolete*

BOHLER K720
Bohler Gesellschaft M.B.H.
Tool material. C 0.9, Si 0.25, Mn 2, Cr 0.35, V 0.13, bal Fe.
Cold work tool steel; oil hardening. For punches, trimming dies, small shears. W. Nr. 1.2842; similar to AISI O2.

BOHLER K722
Bohler Gesellschaft M.B.H.
Tool material. C 0.6, Si 0.9, Mn 1, Cr 0.3, bal Fe.
Cold work tool steel; oil hardening. Stamping or heading dies, rivet sets. W. Nr. 1.2826.

BOHLER K724
Bohler Gesellschaft M.B.H.
C 0.5, Mn 1.8, bal Fe.
Oil or water hardenable. W. Nr. 1.0913. *Obsolete*

BOHLER K746
Bohler Gesellschaft M.B.H.
Tool material. C 1.43, Mn 0.65, Si 1.1, Mo 0.25, bal Fe.
Cold work tool steel; water hardening. For drills, reamers, punches. W. Nr. 1.2833; AISI W1.

BOHLER K760
Bohler Gesellschaft M.B.H.
Tool material. C 1, Mn 0.2, Si 0.2, V 0.13, bal Fe.
Cold work tool steel; water hardening. For drills, reamers, punches. W. Nr. 1.2833; AISI W1.

BOHLER K765
Bohler Gesellschaft M.B.H.
Tool material. C 1.45, V 3.25, bal Fe.
Cold work tool steel; water hardening. For deep drawing tools and dies, cold heading dies. W. Nr. 1.2838.

BOHLER K935
Bohler Gesellschaft M.B.H.
C 0.35, Mn 0.65, bal Fe.
Carbon tool steel; water hardening. For screwdrivers, wrenches.

BOHLER K945
Bohler Gesellschaft M.B.H.
C 0.44, Si 0.3, Mn 0.7, bal Fe.
Carbon tool steel; water hardening. For hand tools, axes, hammers, screwdrivers. W. Nr. 1.1730.

BOHLER K950
Bohler Gesellschaft M.B.H.
Tool material. C 0.5, Si 0.25, Mn 0.65, bal Fe.
Cold work tool steel; water hardening.

BOHLER K960
Bohler Gesellschaft M.B.H.
Tool material. C 0.6, Si 0.3, Mn 0.7, bal Fe.
Cold work tool steel; water hardening. For tool shanks, punches, dies. W. Nr. 1.1740.

BOHLER K970
Bohler Gesellschaft M.B.H.
Tool material. C 0.7, Si 0.25, Mn 0.75, bal Fe.
Cold work tool steel; water hardening. For hand saws, knives. W. Nr. 1.1744; AISI W1.

BOHLER K980
Bohler Gesellschaft M.B.H.
Tool material. C 0.8, Si 0.2, Mn 0.2, bal Fe.
Hot or cold work tool steel; water hardening. Cold: reamers, taps, drills. Hot: forming or punching dies. W. Nr. 1.1525; AISI W1.

BOHLER K985
Bohler Gesellschaft M.B.H.
Tool material. C 0.86, Si 0.3-0.5, Mn 0.65, bal Fe.
Carbon tool steel; water hardening. For taps, reamers, drills, woodworking tools. W. Nr. 1.1830; AISI W1.

BOHLER K990
Bohler Gesellschaft M.B.H.
Tool material. C 1, Si 0.2, Mn 0.2, bal Fe.
Carbon tool steel; water hardening. For drills, reamers, taps, punches. W. Nr. 1.1545 or 1.1645; AISI W1.

BOHLER K991
Bohler Gesellschaft M.B.H.
C 1.05, bal Fe.
Carbon tool steel; water hardening. For drills, reamers, taps, punches. AISI W1. *Obsolete*

BOHLER K995
Bohler Gesellschaft M.B.H.
Tool material. C 1.3, Si 0.2, Mn 0.35, bal Fe.
Cold work tool steel; water hardening. For files. W. Nr. 1.1663.

BOHLER K996
Bohler Gesellschaft M.B.H.
Tool material. C 1.25, Si 0.2, Mn 0.3, Cr 0.3, bal Fe.
Cold work tool steel; water hardening. For drawing dies, mandrels, reamers, punches. W. Nr. 1.2002.

BOHLER KHMU
Vereinigte Edelstahlwerke
C 0.1, Cu 0.4, bal Fe.
For body structures, acid containers; rust resistant. *Obsolete*

BOHLER KHS
Now VEW F126.

BOHLER KHSW
Now VEW F100.

BOHLER KK
Vereinigte Edelstahlwerke
C 1, Si 0.2, Cr 1.1, Mn 0.7, bal Fe.
For ball bearings, races; water hardened. *Obsolete*

BOHLER KL
Now VEW K463.

BOHLER KMC
Vereinigte Edelstahlwerke
C 0.33, Cr 1, Mn 0.65, Si 0.25, bal Fe.
For gears, bolts, machine tool parts; water hardened.
Obsolete

BOHLER KMCW
Now VEW V510.

BOHLER KNIFE STEEL
Vereinigte Edelstahlwerke
C 0.7, bal Fe.
For knives, tools; water hardening. *Obsolete*

BOHLER KP
Vereinigte Edelstahlwerke
C 1.3, W 4.75, bal Fe.
For high duty cold drawing and pressing dies, bending punches, drawing mandrels; oil hardening. *Obsolete*

BOHLER KPV
Vereinigte Edelstahlwerke
C 1.45, V 1.2, W 1, bal Fe.
For tools, dies, bearings; water hardening. *Obsolete*

BOHLER KR SPECIAL
Raven Steel & Tool Co.
C 2.1, Mo 0.8, Cr 12, V, bal Fe.
For tools, dies, plug gages, drawing dies; non-deforming.
Obsolete

BOHLER KR SPECIAL
Vereinigte Edelstahlwerke
C 2.1, Mo 0.8, Cr 12, V, bal Fe.
For tools, dies, plug gages, drawing dies; non-deforming.
Obsolete

BOHLER KW-60
Vereinigte Edelstahlwerke
C 0.4, Cr 13, bal Fe.
For knives, scissors, surgical instruments, gages, calipers, valve seats, piston rods; rust and abrasion resisting. *Obsolete*

BOHLER KW20
Now VEW N320.

BOHLER KW30
Now VEW N530.

BOHLER KW40
Now VEW N540.

BOHLER KW5
Vereinigte Edelstahlwerke
C 0-0.15, Cr 11.5-13, bal Fe.
Heat treated: 120,000-135,000 TS; 110,000-117,000 YS; 15-16 El; 58-63 RA; 220-240 Brin. For steam turbine blades, valves, cutlery; Type 410; corrosion resistant. *Obsolete*

BOHLER KW60-1
Vereinigte Edelstahlwerke
Cr 12-14, 0.15% C min, bal Fe.
Annealed: 95,000 TS; 50,000 YS; 25 El; 55 RA; 195 Brin. Cold drawn: 105,000 TS; 85,000 YS; 17 El; 50 RA; 215 Brin. For cutlery, valves, turbine blades, gages, gears; Type 420; stainless, hardenable. *Obsolete*

BOHLER KWA
Now VEW N200.

BOHLER L125
Bohler Gesellschaft M.B.H.
C 0.4, Cr 20, Ni 20, Mo 4, W 4, Nb 3.8, Fe 0-5, bal Co.
High temperature alloy; for components of gas turbines. W. Nr. 2.4989.

BOHLER L208
Bohler Gesellschaft M.B.H.
C 2.4, Cr 31, W 18, Co 45, bal Fe.
For hard facing electrodes; heat and abrasion resistant.
Obsolete

BOHLER L216
Bohler Gesellschaft M.B.H.
Cobalt. C 2, Cr 32, W 14, Co 50.
For corrosion resistant castings. *Obsolete*

BOHLER L219
Bohler Gesellschaft M.B.H.
Cobalt. C 1.2, Cr 25, W 4, Co 65, bal Fe.
Hard facing electrodes; for dies, exhaust valves.

BOHLER L300
Bohler Gesellschaft M.B.H.
Nickel. C 0.12, Cr 20, Ti 0.4, Fe 0-5, Mn 0.6, bal Ni.
High temperature alloy. W. Nr. 2.4951.

BOHLER L318
Bohler Gesellschaft M.B.H.
Nickel. Ni 60, Mo 17, Cr 16, bal Fe.
For boilers, tanks, chemical plant equipment; heat and corrosion resistant. W. Nr. 2.4811. *Obsolete*

BOHLER L535
Bohler Gesellschaft M.B.H.
Mo 4, Sn 2, Al 4, bal Ti.
For high temperature operations. Annealed: 1080-1200 N/mm² TS; 930-1070 N/mm² 0.2% proof stress; 12-14 El.

BOHLER L610
Bohler Gesellschaft M.B.H.
Aluminum. Si 0.6, Cr 0.2, Cu 0.3, Mg 1, bal Al.
Forgings for aircraft, aerospace, and automotive industries.
Solution annealed and artificially aged: 260 N/mm² TS min; 245 N/mm² 0.2% proof stress min; 4-0 El min; 80 Brin.
A96061

BOHLER L620
Bohler Gesellschaft M.B.H.
Aluminum. Si 0.2, Ni 1.1, Cu 2.3, Mg 1.5, Fe 1.2, bal Al.
Forgings for aircraft, aerospace, and automotive industries.
Solution annealed and artificially aged: 390-410 N/mm² TS min; 330 N/mm² 0.2% proof stress min; 3-7 El min; 125 Brin.

BOHLER L625
Bohler Gesellschaft M.B.H.
Aluminum. Si 0.9, Mn 0.8, Cu 4.5, Mg 0.5, bal Al.
Forgings for aircraft, aerospace, and automotive industries.
Solution annealed and artificially aged: 430-450 N/mm² TS min; 380 N/mm² 0.2% proof stress min; 3-7 El min; 125 Brin.

BOHLER L630
Bohler Gesellschaft M.B.H.
Aluminum. Cr 0.3, Cu 1.6, Mg 2.5, Zn 5.6, bal Al.
Forgings for aircraft, aerospace, and automotive industries.
Solution annealed and artificially aged: 425-455 N/mm² TS min; 365 N/mm² 0.2% proof stress min; 3-7 El min; 130 Brin.
A97075

BOHLER L635
Bohler Gesellschaft M.B.H.
Aluminum. Mn 0.3, Cr 0.2, Cu 0.6, Mg 3.3, Zn 4.3, bal Al.
Forgings for aircraft, aerospace, and automotive industries.
Solution annealed and artificially aged: 470-500 N/mm² TS min; 400-430 N/mm² 0.2% proof stress min; 3-6 El min; 130 Brin.

BOHLER LIGHTNING
Vereinigte Edelstahlwerke
C 0.5, Cr 1.6, Mo, V, bal Fe.
Heat treated: 156,000 TS; 142,000 YS; 350 Brin. For rifle barrels; tough. *Obsolete*

BOHLER M100
Bohler Gesellschaft M.B.H.
C 0.2, Si 0.3, Mn 1.2, Cr 1.1, bal Fe.
For molds. Good core strength. 205 Brin max. W. Nr. 1.2162.

BOHLER M112
Bohler Gesellschaft M.B.H.
C 0.2, Mn 1, Cr 1.2, Mo, bal Fe.
To be carburized and hardened for plastic molding dies or structural purposes. Good core strength. W. Nr. 1.2160.
Obsolete

BOHLER M120
Bohler Gesellschaft M.B.H.
C 0.14, Si 0.3, Mn 0.45, Cr 0.7, Ni 3.5, bal Fe.
For plastic molds. 220 Brin. W. Nr. 1.2735.

BOHLER M130
Bohler Gesellschaft M.B.H.
C 0.19, Si 0.25, Mn 0.3, Cr 1.3, Mo 0.2, Ni 4.1, bal Fe.
For plastic molds. Good core strength. 250 Brin. W. Nr. 1.2764.

BOHLER M150
Bohler Gesellschaft M.B.H.
C 0-0.07, Cr 3.8, Mo 0.5, bal Fe.
To be carburized and hardened for plastic mold dies, or for structural purposes. Designed to be hobbed before heat treating. W. Nr. 1.2341.

BOHLER M152
Bohler Gesellschaft M.B.H.
C 0.06, Cr 4, Mo 0.5, bal Fe.
To be hobbed, carburized and hardened for plastic mold dies or die casting dies. W. Nr. 1.2341. *Obsolete*

BOHLER M200
Bohler Gesellschaft M.B.H.
C 0.4, Si 0.4, Mn 1.5, Cr 1.9, Mo 0.2, S 0.07, bal Fe.
Plastic mold or die casting die steel; oil hardenable. 230 Brin. W. Nr. 1.2312.

BOHLER M201 ECOPLUS
Bohler Gesellschaft M.B.H.
C 0.4, Si 0.3, Mn 1.5, Cr 2, Mo 0.2, bal Fe.
Plastic mold or die casting die steel; oil hardenable. 230 Brin.
W. Nr. 1.2311.

BOHLER M210
Bohler Gesellschaft M.B.H.
C 0.4, Si 0.4, Mn 1.5, Cr 1.9, Mo 0.2, bal Fe.
Plastic mold or die casting die steel; oil hardenable. W. Nr.
1.2312; similar to AISI P20. *Obsolete*

BOHLER M252
Bohler Gesellschaft M.B.H.
C 0.35, Si 0.35, Mn 0.6, Cr 1.3, Ni 4.5, bal Fe.
Plastic mold or die casting die steel; oil hardening. *Obsolete*

BOHLER M300
Bohler Gesellschaft M.B.H.
C 0.38, Si 0.4, Mn 0.65, Cr 16, Mo 1, Ni 0.8, bal Fe.
Stainless plastic mold or die cast die steel. Air or oil
hardening. W. Nr. 1.2316.

BOHLER M310
Bohler Gesellschaft M.B.H.
C 0.46, Si 0.4, Mn 0.4, Cr 13, bal Fe.
Stainless plastic mold or die casting die steel; air or oil
hardening. 225 Brin. W. Nr. 1.2083; similar to AISI 420.

BOHLER M312
Bohler Gesellschaft M.B.H.
C 0.48, Cr 13.5, Mo 0.35, Ni 0.25, Mn 0.4, bal Fe.
Stainless plastic mold or die casting die steel; air or oil
hardening. W. Nr. 1.2083; similar to AISI 420.

BOHLER M390 ISOMATRIX PM
Bohler Gesellschaft M.B.H.
C 1.9, Si 0.3, Cr 20, Mo 1, V 4, Mn 0.3, W 0.6, bal Fe.
Stainless plastic mold or die casting die steel; wear and
corrosion resistant. Annealed: 280 Brin approx.

BOHLER M751
Vereinigte Edelstahlwerke
C 0.75, bal Fe.
Hardened: Rock C 64. Annealed: 85,000-100,000 TS; 175-204
Brin. For soft to hard rock drills, punches, cutters. Water
hardening, wear resistant. *Obsolete*

BOHLER M851
Vereinigte Edelstahlwerke
C 0.85, bal Fe.
Hardened: Rock C 65. Annealed: 85,000-100,000 TS; 175-204
Brin. For hard rock drills, cutters. Water hardening, wear
resistant. *Obsolete*

BOHLER MAGNET STEEL
Vereinigte Edelstahlwerke
W 6.3, C 0.7, bal Fe.
For magnets; water hardening. *Obsolete*

BOHLER MARTIN STEEL
Vereinigte Edelstahlwerke
C 0.05-1.1, bal Fe.
For general use, machinery parts; all grades of carbon steel.
Obsolete

BOHLER ME
Vereinigte Edelstahlwerke
C 0.7, Si 0-0.25, Mn 0.34, W 5.8, bal Fe.
For magnets; magnet steel. *Obsolete*

BOHLER ME-6
Vereinigte Edelstahlwerke
C, W, bal Fe.
For meter magnets; magnet steel. *Obsolete*

BOHLER MIDDLE HARD 100
Vereinigte Edelstahlwerke
C 1, W 1, bal Fe.
Annealed: 84,000 TS; 60,000 YS; 26 El; 185 Brin. Heat
treated: 280,000 TS; 535 Brin. For heading and flanging
tools, bending and blanking dies, center bits, plugs, jaws,
cold heading dies. Water hardening cold work tool steel.
Obsolete

BOHLER MIDDLE HARD 115
Vereinigte Edelstahlwerke
C 1.2, W 1, bal Fe.
Annealed: 84,000 TS; 60,000 YS; 26 El; 185 Brin. Hardened:
210,000 TS; 185,000 YS; 2 El; 400 Brin. For beading and
flanging tools, bending and blanking dies, center bits, plugs,
jaws, cold heading dies. Water or oil hardening cold work
steel. *Obsolete*

BOHLER MITTEL-HART
Vereinigte Edelstahlwerke
C 1.15, Si 0-0.25, Mn 0.23, bal Fe.
For turning tools, milling cutters, crown hammers, knives,
mint dies; water hardening. *Obsolete*

BOHLER ML
Vereinigte Edelstahlwerke
C, alloy, bal Fe.
For rifle barrels; oil hardening. *Obsolete*

BOHLER MO RAPID EXTRA 1200
Now VEW S700.

BOHLER MO RAPID EXTRA 3
Now VEW S610.

BOHLER MO RAPID EXTRA 500
Now VEW S705.

BOHLER MO RAPID EXTRA 800
Now VEW S500.

BOHLER MO RAPID EXTRA 9
Now VEW S401.

BOHLER MO RAPID EXTRA V30
Now VEW S607.

BOHLER MOLETTE
Vereinigte Edelstahlwerke
C, bal Fe.
For female and embossed relief molettes; case hardening
steel. *Obsolete*

BOHLER MPA
Vereinigte Edelstahlwerke
C 0.25, Cr 1, Ni 4.2, Mo, bal Fe.
Heat treated: 170,000 TS; 135,000 YS; 15 El; 50 RA; 360 Brin.
For shafts, gears, bolts; tough. *Obsolete*

BOHLER MPD
Now VEW W322.

BOHLER MPD EXTRA
Vereinigte Edelstahlwerke
C 0.3, Cr 1.8, Ni 1.3, W 0.9, Mo 3.2, Co 3.2, bal Fe.
For die casting tools, spreaders; oil hardened. *Obsolete*

BOHLER MS STEEL
Vereinigte Edelstahlwerke
C 0.45-0.85, bal Fe.
For general purpose tools; water hardening. *Obsolete*

BOHLER MS45
Vereinigte Edelstahlwerke
C 0.45, Si 0.3, Mn 0.6, bal Fe.
Hot rolled: 98,000 TS; 59,000 YS; 24 El; 45 RA; 212 Brin. For
gears, bolts, shafts, machine tool parts; water hardened.
Obsolete

BOHLER MS60
Vereinigte Edelstahlwerke
C 0.6, Si 0.25-0.5, Mn 0.3-0.8, bal Fe.
Heat treated: 160,000 TS; 113,000 YS; 12 El; 40 RA; 325 Brin.
For hammers, axes, springs, punches; water hardened.
Obsolete

BOHLER MS70
Vereinigte Edelstahlwerke
C 0.75, Si 0.25-0.5, Mn 0.3-0.8, bal Fe.
Heat treated: 185,000 TS; 140,000 YS; 14 El; 38 RA; 375 Brin.
For springs, tools, punches, hammers, axes; water hardened;
Type W1. *Obsolete*

BOHLER MS85
Vereinigte Edelstahlwerke
C 0.85, Si 0.25-0.5, Mn 0.3-0.8, bal Fe.
Heat treated: 190,000 TS; 145,000 YS; 10 El; 30 RA; 400 Brin.
For springs, tools, cutters, drills, taps; Type W1; water
hardened. *Obsolete*

BOHLER MS90
Vereinigte Edelstahlwerke
C 0.9, Si 0.25-0.5, Mn 0.3-0.8, bal Fe.
Heat treated: 195,000 TS; 150,000 YS; 10 El; 30 RA; 420 Brin.
For springs, drills, reamers, broaches; Type W1; water
hardened. *Obsolete*

BOHLER MSI
Vereinigte Edelstahlwerke
C 0.6, Si 0.95, Mn 0.95, bal Fe.
For tools, dies, punches, springs; oil hardening. *Obsolete*

BOHLER MST
Now VEW K720.

BOHLER MST
U.N. Alloy Steel Corp.
C 0.9, Mn 2, V 0.2, bal Fe.
For stay bolt traps, broaches, reamers, gages, cutting dies;
non-deforming; AISI 02. *Obsolete*

BOHLER MY EXTRA
Now VEW K450.

BOHLER MY EXTRA W
Vereinigte Edelstahlwerke
C 0.35, Si 0.9, Mn 0.3, Cr 1.05, V 0.18, W 1.85, bal Fe.
For forging and header dies, punches; oil hardened, tough.
Obsolete

BOHLER MYA
Now VEW K247.

BOHLER MYAH
Now VEW K507.

BOHLER MYD
Vereinigte Edelstahlwerke
C 0.35, Si 1.5, Cr 1.35, V 0.1, bal Fe.
For tools, dies, punches; oil hardening. *Obsolete*

BOHLER N104
Bohler Gesellschaft M.B.H.
C 0.06, Cr 12.5, bal Fe.
Corrosion resistant steel. For structural parts in water and
steam. W. Nr. 1.4000.

BOHLER N106
Bohler Gesellschaft M.B.H.
C 0-0.08, Cr 13-15, bal Fe.
Corrosion resisting steel. For tableware, building fittings. W.
Nr. 1.4001. *Obsolete*

BOHLER N108
Bohler Gesellschaft M.B.H.
C 0.07, Cr 12.5, Al 0.2, bal Fe.
Ferritic type corrosion resisting steel. Weldable; for annealing
boxes, furnace parts, tableware. W. Nr. 1.4002; AISI 405.

BOHLER N205
Bohler Gesellschaft M.B.H.
C 0-0.04, Cr 16.5, Ti > 7 x C, bal Fe.
Stabilized ferritic corrosion resisting steel. For welded oil refinery and dairy equipment. W. Nr. 1.4510; modified AISI 430.

BOHLER N238
Bohler Gesellschaft M.B.H.
Stainless steel. C 0-0.04, Cr 17, Mo 1.75, Ti > 7 x C, bal Fe. Ferritic type stainless, weldable. For corrosion resistant equipment. W. Nr. 1.4523.

BOHLER N242
Bohler Gesellschaft M.B.H.
C 0.06, Cr 19.5, Mo 2.2, S 0.15, bal Fe.
For welded assemblies with good corrosion and heat resisting properties.

BOHLER N244
Bohler Gesellschaft M.B.H.
C 0-0.03, Si 0.5, Mn 1, Cr 16.5, Mo 0.25, S 0.25, bal Fe.
For welded assemblies with good corrosion and heat resisting properties. Annealed: 400-600 N/mm^2 TS; 30 El.

BOHLER N310
Bohler Gesellschaft M.B.H.
Stainless steel. C 0.14, Si 0.4, Mn 1.4, Cr 16, Mo 0.25, S 0.25, bal Fe.
Free machining, ferritic type stainless. For stainless bolts, and fasteners. Annealed: 540-740 N/mm^2 TS; 16 El. Hardened and tempered: 450 N/mm^2 0.2% proof stress; 640-840 N/mm^2 TS; 11 El. W. Nr. 1.4104; similar to AISI 430 F.

BOHLER N315
Bohler Gesellschaft M.B.H.
Stainless steel. C 0.14, Cr 12.5, bal Fe.
Martensitic stainless steel for turbine blades, cutlery, tableware. W. Nr. 1.4024; AISI 410-420.

BOHLER N316
Bohler Gesellschaft M.B.H.
Stainless steel. C 0.13, Cr 12.5, S 0.23, bal Fe.
Martensitic stainless for hardenable threaded parts. W. Nr. 1.4005; AISI 416.

BOHLER N320
Bohler Gesellschaft M.B.H.
Stainless steel. C 0.2, Si 0.4, Mn 0.4, Cr 12.5, bal Fe.
Martensitic stainless for cutlery, surgical instruments, dental tools. Hardened and tempered: 450-550 N/mm^2 0.2% proof stress; 650-950 N/mm^2 TS; 12-15 El. W. Nr. 1.4021; AISI 420.

BOHLER N324
Bohler Gesellschaft M.B.H.
Stainless steel. C 0.2, Si 0.4, Cr 13, bal Fe.
Martensitic stainless for valves, cutlery, surgical and dental tools. AISI 420. *Obsolete*

BOHLER N330
Bohler Gesellschaft M.B.H.
Stainless steel. C 0.2, Cr 13, Mo 1.15, bal Fe.
Martensitic stainless steel. For turbine blades, valve cones. W. Nr. 1.4120. *Obsolete*

BOHLER N335
Bohler Gesellschaft M.B.H.
Stainless steel. C 0.38, Si 0.4, Mn 0.65, Cr 16, Mo 1, Ni 0.8, bal Fe.
Martensitic stainless steel. For cutlery, high temperature valves and fittings, arbors, spindles, bolts. As tempered: 550 N/mm^2 0.2% proof stress; 750-950 N/mm^2 TS; 12 El W. Nr. 1.4122.

BOHLER N350
Bohler Gesellschaft M.B.H.
Stainless steel. C 0.19, Si 0.25, Mn 0.4, Cr 15.9, Ni 1.6, bal Fe.
Martensitic stainless steel. For marine hardware. As tempered: 550 N/mm^2 0.2% proof stress; 750-950 N/mm^2 TS; 12-14 El. W. Nr. 1.4057; similar to AISI 431.

BOHLER N351
Bohler Gesellschaft M.B.H.
Stainless steel. C 0.2, Cr 16.8, Ni 1.8, bal Fe.
Stainless and heat resisting steel. For marine equipment. W. Nr. 1.4057; similar to AISI 431.

BOHLER N352
Bohler Gesellschaft M.B.H.
Stainless steel. C 0.17, Cr 15.5, Ni 2.2, bal Fe.
Martensitic stainless steel; good hardenability. AISI 431.

BOHLER N359
Bohler Gesellschaft M.B.H.
C 0.1, Cr 13, Ni 1, bal Fe.
Corrosion resistant steel casting; for pump parts, valves, rotors. W. Nr. 1.4008; ACI CA-15.

BOHLER N400
Bohler Gesellschaft M.B.H.
C 0-0.029, Cr 13, Ni 4.1, bal Fe.
Corrosion resistant casting. For hydraulic turbine equipment. W. Nr. 1.4313.

BOHLER N530
Bohler Gesellschaft M.B.H.
Stainless steel. C 0.33, Cr 12.5, bal Fe.
Martensitic stainless steel. For valves, springs, cutlery, surgical and dental instruments. W. Nr. 1.4028; AISI 420.

BOHLER N540
Bohler Gesellschaft M.B.H.
Stainless steel. C 0.46, Si 0.4, Mn 0.4, Cr 13, bal Fe.
Hardenable martensitic stainless. For cutlery, springs, surgical instruments. Annealed: 800 N/mm^2 TS max. W. Nr. 1.4034; similar to AISI 420.

BOHLER N555
Bohler Gesellschaft M.B.H.
Stainless steel. C 0.6, Cr 14, Mo 0.55, bal Fe.
Hardenable martensitic stainless steel. For cutlery, shears, cutting and forming tools. W. Nr. 1.4110.

BOHLER N685
Bohler Gesellschaft M.B.H.
Stainless steel. C 0.9, Si 0.45, Mn 0.4, Cr 17.5, Mo 1.1, V 0.1, bal Fe.
Hardenable martensitic stainless steel. For ball or roller bearings, cutlery, shears, surgical equipment. Annealed: 265 Brin max. W. Nr. 1.4112; similar to AISI 440B.

BOHLER N688
Bohler Gesellschaft M.B.H.
Stainless steel. C 1, Cr 16, Mo 0.8, Co 2, bal Fe.
Hardenable martensitic stainless steel. For ball or roller bearings, cutlery, shears, surgical and dental equipment. W. Nr. 1.4535; similar to AISI 440C. *Obsolete*

BOHLER N690
Bohler Gesellschaft M.B.H.
Stainless steel. C 1.07, Si 0.4, Mn 0.4, Cr 17, Mo 1.1, Co 1.5, V 0.1, bal Fe.
High carbon martensitic stainless. For ball or roller bearings, valves, pump parts. Annealed: 285 Brin max. W. Nr. 4528; similar to AISI 440C.

BOHLER N692
Bohler Gesellschaft M.B.H.
Stainless steel. C 0.9, Cr 16.5, Mo 0.5, V 0.25, Co 1.4, bal Fe.
Martensitic stainless steel. For cutlery, knife blades. W. Nr. 1.4535; similar to AISI 440B. *Obsolete*

BOHLER N702
Bohler Gesellschaft M.B.H.
C 0.04, Cr 16.5, Ni 4, Cu 4, Nb 0.4, bal Fe.
Corrosion resistant steel. W. Nr. 1.4542; similar to 17-4 PH.

BOHLER NAB
Vereinigte Edelstahlwerke
C 0.75, Mn 0.9, bal Fe.
For dies and parts of textile machines; water hardened. *Obsolete*

BOHLER NBS
Now VEW K805.

BOHLER NBSN
Now VEW W500.

BOHLER NEEDLE DIE STEEL
Vereinigte Edelstahlwerke
C, alloy, bal Fe.
For dies. *Obsolete*

BOHLER NH
Vereinigte Edelstahlwerke
C 0.35, Si 0.25, Mn 0.55, bal Fe.
Hot rolled: 85,000 TS; 54,000 YS; 30 El; 52 RA; 185 Brin. For gears, bolts, shafts, machine tool parts; water hardened. *Obsolete*

BOHLER NI
Vereinigte Edelstahlwerke
C 0.25, Ni 1.55, bal Fe.
For shafts, axles, connecting rods, cranks, levers; oil hardening. *Obsolete*

BOHLER NIP29
Vereinigte Edelstahlwerke
Ni 29, bal Fe.
For heat conductors; controlled expansion. *Obsolete*

BOHLER NIP36
Vereinigte Edelstahlwerke
Ni 36, bal Fe.
For heat conductors; controlled expansion. *Obsolete*

BOHLER NIP50
Vereinigte Edelstahlwerke
Ni 49, bal Fe.
For electrical motors, laminations; high permeability. *Obsolete*

BOHLER NMH
Now VEW V130.

BOHLER NO. 16
Vereinigte Edelstahlwerke
C 0.9, V 0.1-0.2, bal Fe.
For tools, drills, chasers; water hardened. *Obsolete*

BOHLER NO. 2 SI
Now VEW F105.

BOHLER NO. 3 NW
Vereinigte Edelstahlwerke
C 0.14, Ni 3, bal Fe.
Heat treated: 100,000 TS; 64,000 YS; 22 El; 55 RA; 235 Brin. For gears, cams, camshafts; case hardening steel. *Obsolete*

BOHLER NO. 36 N
Vereinigte Edelstahlwerke
C, alloy, bal Fe.
For component parts of barometers, optical instruments; low coefficient of expansion. *Obsolete*

BOHLER NO. 4
Vereinigte Edelstahlwerke
C 0.5, Cr 1.5, W 2.25, V 0.25, bal Fe.
For hot work tools and dies; hot work steel. *Obsolete*

BOHLER NO. 5 NM
Vereinigte Edelstahlwerke
C 0.25, Ni 5, bal Fe.
For auto parts, crankshafts. *Obsolete*

BOHLER NO. 711
Vereinigte Edelstahlwerke
Fe 50, 50% other metals.
For hard facing work, well drill bits, cutting tools; does not require heat treatment for hardening. *Obsolete*

BOHLER NO. 751
Vereinigte Edelstahlwerke
C 0.7, bal Fe.
Annealed: 85,000-100,000 TS; 175-204 Brin. Heat treated: 174,000 TS; 128,000 YS; 12 El; 355 Brin. For soft to hard rock drills, hammers, punches. Water hardening, wear resistant. *Obsolete*

BOHLER NO. 90
Vereinigte Edelstahlwerke
C 0.9, Si 0-0.2, Mn 0.3, bal Fe.
For cold heading and forming dies for heavy duty; water hardening. *Obsolete*

BOHLER NW
Vereinigte Edelstahlwerke
C 0.4, Cr 1.25, Ni 4.4, bal Fe.
For drop forging dies, hot press swages, clipping tools, heavy anvil blocks; air hardening. *Obsolete*

BOHLER NWM
Now VEW K800.

BOHLER OFH70
Vereinigte Edelstahlwerke
C 0.56, Si 0.3, Mn 0.55, bal Fe.
Heat treated: 160,000 TS; 112,000 YS; 12 El; 40 RA; 320 Brin. For wheels, die blocks, rails, girders, axles; water hardened. *Obsolete*

BOHLER P505
Bohler Gesellschaft M.B.H.
Stainless steel. C 0-0.03, Cr 23, Ni 15, Mo 1.4, N 0.3, bal Fe.
Stainless steel. W. Nr. 1.3951.

BOHLER P530
Bohler Gesellschaft M.B.H.
Stainless steel. C 0-0.07, Mn 19, Cr 13.5, Ni 2.2, N 0.24, Mo 0.5, bal Fe.
Nonmagnetizable stainless steel for welded construction. W. Nr. 1.3949.

BOHLER P550
Bohler Gesellschaft M.B.H.
C 0.45, Mn 18, Cr 4.4, bal Fe.
Nonmagnetizable steel; for electrical equipment. W. Nr. 1.3813.

BOHLER P600
Bohler Gesellschaft M.B.H.
C 0.05, Si 0.2, Mn 0.5, Ni 9, bal Fe.
Steel tough at subzero temperatures. For tanks and containers for cryogenic operations. W. Nr. 1.5662.

BOHLER P602
Bohler Gesellschaft M.B.H.
C 0.12, Ni 5, bal Fe.
Oil hardenable steel for low temperature operations. *Obsolete*

BOHLER P758
Bohler Gesellschaft M.B.H.
C 0.05, Co 24, Ni 14, Al 8, Cu 3, Ti, bal Fe.
For permanent magnets. W. Nr. 1.3761. *Obsolete*

BOHLER P760
Bohler Gesellschaft M.B.H.
C 0.05, Co 24, Ni 21, Co 8, Cu 3, bal Fe.
Permanent magnet. W. Nr. 1.3760; ALNICO 400 or 500. *Obsolete*

BOHLER P764
Bohler Gesellschaft M.B.H.
C 0.05, Co 32, Ni 15, Al 7, Cu 4.5, Ti 5, bal Fe.
Permanent magnet. W. Nr. 1.3758; ALNICO 350. *Obsolete*

BOHLER P804
Bohler Gesellschaft M.B.H.
C 0-0.15, Ni 42, bal Fe.
Low thermal expansion; for instruments. W. Nr. 1.3917.

BOHLER P906
Bohler Gesellschaft M.B.H.
C 0-0.08, Si 0-0.15, Mn 0-0.5, bal Fe.
Soft magnetic metal; for magnet cores, solenoids. W. Nr. 1.1009. *Obsolete*

BOHLER PA2
Vereinigte Edelstahlwerke
C 0.35, Cr 1.3, Ni 4.5, bal Fe.
For gears, bolts, forging and heading dies; oil hardened, shock resistant. *Obsolete*

BOHLER PANTHER
Vereinigte Edelstahlwerke
C, bal Fe.
For general use; water hardening. *Obsolete*

BOHLER PAZ
Vereinigte Edelstahlwerke
C 0.35, Cr 1.3, Ni 4.5, bal Fe.
For gears, bolts, crankshafts, dies, tools; oil hardened, shock resistant. *Obsolete*

BOHLER PNA
Vereinigte Edelstahlwerke
C 0.08, Mn 1-1.25, bal Fe.
Welded: 55,000 TS; 45,000 YS; 33-40 El; 60-80 RA. For welding rod. *Obsolete*

BOHLER PPA
Now VEW E204.

BOHLER PRIMA H
Vereinigte Edelstahlwerke
C 1.3, Si 0.2, Mn 0.2, bal Fe.
For engravers' tools, forming and blanking dies; Type W1; water hardened. *Obsolete*

BOHLER PRIMA HART
Vereinigte Edelstahlwerke
C 1.3, bal Fe.
For tools, cutters, dies, drills; water hardened. *Obsolete*

BOHLER PRIMA MITTELHART 100
Now VEW K990.

BOHLER PRIMA MITTELHART 115
Vereinigte Edelstahlwerke
C 1.15, bal Fe.
Annealed: 110,000 TS; 56,000 YS; 19 El; 40 RA; 210 Brin. For tools, cutters, drills, taps, reamers; water hardened. *Obsolete*

BOHLER PRIMA WEICH
Now VEW K971.

BOHLER PRIMA ZAH
Now VEW K980.

BOHLER PT 15
Vereinigte Edelstahlwerke
C 0.13, Cr 0.2, Ni 1.5, bal Fe.
For gears, bolts, camshafts, cams, fasteners; case hardened. *Obsolete*

BOHLER PV35
Vereinigte Edelstahlwerke
C 1.5, Cr, Si, bal Fe.
For cutters, bearings, bushings; water or oil hardened. *Obsolete*

BOHLER R100
Bohler Gesellschaft M.B.H.
Cr 1, Mn 0.35, Cr 1.55, bal Fe.
Oil or water hardening steel for ball or roller bearings, bushings. W. Nr. 1.3505; AISI 52100.

BOHLER R110
Bohler Gesellschaft M.B.H.
C 1, Si 0.6, Mn 1.1, Cr 1.55, bal Fe.
Oil or water hardening steel for ball or roller bearings, bushings; for larger sections than VEW R100. W. Nr. 1.3520; similar to AISI 52100. *Obsolete*

BOHLER RAPID STEEL
Vereinigte Edelstahlwerke
W 14, Cr 3.9, C 0.7, bal Fe.
For cutters, tools; high speed steel. *Obsolete*

BOHLER REMANENCELESS STEEL
Vereinigte Edelstahlwerke
C, bal Fe.
For electrical equipment; high permeability. *Obsolete*

BOHLER S200
Bohler Gesellschaft M.B.H.
C 0.75, Cr 4.1, W 18, V 1.1, bal Fe.
High speed steel; for lathe and planer tools. W. Nr. 1.3355; S18-0-1; AISI T1.

BOHLER S201
Bohler Gesellschaft M.B.H.
C 0.8, Cr 4.3, W 17.5, V 1.8, Mo, bal Fe.
For lathe tools, drills, reamers, broaches. W. Nr. 1.3357; AISI T2.

BOHLER S203
Bohler Gesellschaft M.B.H.
C 0.86, C 4.1, W 12, V 2.5, V 0.85, Mo 0.85, bal Fe.
For lathe and planer tools, broaches, taps, milling cutters. *Obsolete*

BOHLER S205
Bohler Gesellschaft M.B.H.
C 0.82, Cr 4.3, W 8.7, V 1.5, Mo 0.85, bal Fe.
High speed steel; for drills, milling cutters, taps, lathe tools.

BOHLER S300
Bohler Gesellschaft M.B.H.
C 0.76, Cr 4.2, W 18, V 1.5, Mo 0.6, Co 9.5, bal Fe.
Cobalt-tungsten high speed steel for lathe tools, milling cutters, threading tools. W. Nr. 1.3256; S18-1-2-10; similar to AISI T5.

BOHLER S302
Bohler Gesellschaft M.B.H.
C 0.6, Cr 4, W 17.5, V 1.1, Co 17.5, bal Fe.
Cobalt-tungsten high speed steel. For lathe and planer tools, milling cutters, hot shears. Good red hardness.

BOHLER S305
Bohler Gesellschaft M.B.H.
C 0.81, Cr 4.3, W 18, V 1.5, Mo 0.65, Co 4.8, bal Fe.
For lathe and planer tools, milling cutters, threading tools. Cobalt-tungsten high speed steel; good red hardness. Annealed: 300 Brin. AISI T4; W. Nr. 1.3255.

BOHLER S307
Bohler Gesellschaft M.B.H.
C 1.5, Cr 4.8, W 12.5, V 5, Co 5, bal Fe.
High carbon high speed steel. For special lathe and threading tools. Good wear resistance. Similar to AISI T15.

BOHLER S308
Bohler Gesellschaft M.B.H.
C 1.35, Cr 4.3, W 12, V 3.7, Co 4.8, Mo 0.8, bal Fe.
High carbon high speed steel. For lathe tools, form tools, threading tools. Wear resistant; good red hardness. Annealed: 300 Brin. W. Nr. 1.3202.

BOHLER S400
Bohler Gesellschaft M.B.H.
C 1.02, Cr 3.8, Mo 8.7, V 2, W 1.8, bal Fe.
High speed steel; for milling cutters. Annealed: 280 Brin. W. Nr. 1.3348; AISI M7.

BOHLER S401
Bohler Gesellschaft M.B.H.
C 0.83, Cr 3.8, W 1.8, V 1.2, Mo 8.7, bal Fe.
Molybdenum high speed steel. For drills, reamers, form
cutters, lathe tools. Annealed: 280 Brin. W. Nr. 1.3346; AISI
M1.

BOHLER S404
Bohler Gesellschaft M.B.H.
C 0.89, Cr 4.1, W 1.2, V 1.9, Mo 4.5, bal Fe.
Molybdenum high speed steel. For drills, reamers, form
cutters, lathe tools. Annealed: 280 Brin.

BOHLER S500
Bohler Gesellschaft M.B.H.
C 1.1, Cr 3.9, W 1.4, Mo 9.2, V 1.2, Co 7.8, bal Fe.
Cobalt-molybdenum high speed steel. For twist drills,
reamers, threading cutters, lathe tools, milling cutters. For
difficult machining; good red hardness. Annealed: 280 Brin.
W. Nr. 1.3247; AISI M42.

BOHLER S600
Bohler Gesellschaft M.B.H.
C 0.9, Cr 4.1, W 6.4, Mo 5, V 1.8, bal Fe.
Molybdenum-tungsten high speed steel. For drills, reamers,
milling cutters, lathe tools. Annealed: 280 Brin. W. Nr. 1.3343;
AISI M2.

BOHLER S604
Bohler Gesellschaft M.B.H.
C 1, Cr 4.3, W 6.3, V 1.8, Mo 5, bal Fe.
For twist drills, taps, broaches, lathe tools, milling cutters. W.
Nr. 1.3342; AISI M2.

BOHLER S607
Bohler Gesellschaft M.B.H.
C 1.21, Cr 4.1, W 6.4, V 2.9, Mo 5, bal Fe.
High speed steel. For lathe tools, thread cutters, milling
cutters, broaches. Annealed: 280 Brin. W. Nr. 1.3344; AISI M3
Class 2.

BOHLER S610
Bohler Gesellschaft M.B.H.
C 0.99, Cr 4, W 2.9, V 2.4, Mo 2.7, bal Fe.
For twist drills, milling cutters, broaches. Annealed: 280 Brin.
W. Nr. 1.3333.

BOHLER S620
Bohler Gesellschaft M.B.H.
C 1.12, Cr 4.3, W 6.4, V 1.9, Mo 5, Al 1.1, bal Fe.
For twist drills, milling cutters, broaches. Annealed: 280 Brin.

BOHLER S690 ISOMATRIX PM
Bohler Gesellschaft M.B.H.
C 1.33, Cr 4.3, W 5.9, V 4.1, Mo 4.9, bal Fe.
For twist drills, milling cutters, broaches. Annealed: 280 Brin.

BOHLER S700
Bohler Gesellschaft M.B.H.
C 1.26, Cr 4, W 9.3, V 3.2, Mo 3.6, Co 10, bal Fe.
High speed steel. For lathe tools, threading cutters.
Annealed: 300 Brin. W. Nr 1.3207.

BOHLER S705
Bohler Gesellschaft M.B.H.
C 0.92, Cr 4.1, W 6.4, Mo 5, V 1.9, Co 4.8, bal Fe.
High speed steel. For lathe and planer tools, milling cutters.
Annealed: 280 Brin. W. Nr. 1.3243; AISI M41.

BOHLER SAS2
Now VEW A750.

BOHLER SAS4
Now VEW A350.

BOHLER SAS4MN
Vereinigte Edelstahlwerke
C 0.07, Cr 17.5, Ni 11.5, Mo 2.7, Cb, bal Fe.
Annealed: 85,000 TS; 35,000 YS; 50 El; 65 RA; 160 Brin. For
welded acid resistant chemical plant equipment; Type 318;
stainless, austenitic. *Obsolete*

BOHLER SAS8
Now VEW A960.

BOHLER SC EXTRA
Vereinigte Edelstahlwerke
C 1.25, Si 1.15, Cr 1.2, bal Fe.
For tools, dies; oil hardening. *Obsolete*

BOHLER SCV
Vereinigte Edelstahlwerke
C 0.6, Mn 1, Cr 1.2, V 0.1, bal Fe.
For tools, dies, springs; oil hardening. *Obsolete*

BOHLER SHF
Vereinigte Edelstahlwerke
C 1.1, Cr 1.2, N 0.4, V, bal Fe.
For cold work tools; water hardened. *Obsolete*

BOHLER SIC20
Now VEW H730.

BOHLER SK3
Vereinigte Edelstahlwerke
C 1.05, Cr 1, Mn 0.3, bal Fe.
For bearings, cutters, liners, sleeves; water hardened, wear
resistant. *Obsolete*

BOHLER SKVL
Vereinigte Edelstahlwerke
C 0.08, Cr 17.5, Mo 1.2, bal Fe.
Annealed: 80,000 TS; 50,000 YS; 25 El; 50 RA; 150 Brin. For
oil refinery equipment, chemical plant equipment; corrosion
resistant. *Obsolete*

BOHLER SMF
Vereinigte Edelstahlwerke
C 0.55, Si 1.7, Mn 0.8, bal Fe.
Annealed: 115,000 TS; 78,000 YS; 22 El; 223 Brin. Oil
hardened: 295,000 TS; 275,000 YS; 2 El; 575 Brin. For
springs, lock washers, chisels, collets, hand tools. Water
hardening, tough, shock resistant. *Obsolete*

BOHLER SPECIAL
Vereinigte Edelstahlwerke
C 2.4, Cr 12, V 4, Mo, bal Fe.
For blanking and shaping dies; high wear resistance, non-
deforming. *Obsolete*

BOHLER SPECIAL EXTRA MG
Vereinigte Edelstahlwerke
C 1.05, Mn 1, Cr 0.6, bal Fe.
Annealed: 100,000 TS; 54,000 YP; 20 El; 197 Brin. Water
hardened: 200,000 TS; 140,000 YP; 11 El; 400 Brin. For taps,
threading dies, chasers, reamers, broaches, knives, beading
and bending tools, gauges. Tough, oil hardening. *Obsolete*

BOHLER SPECIAL K
Now VEW K100.

BOHLER SPECIAL K5
Now VEW K305.

BOHLER SPECIAL K8
Now VEW K300.

BOHLER SPECIAL KMV
Now VEW K110.

BOHLER SPECIAL KN
Now VEW K116.

BOHLER SPECIAL KNL
Now VEW K105.

BOHLER SPECIAL KR
Now VEW K107.

BOHLER SPECIAL KRM
Vereinigte Edelstahlwerke
C 2.1, Cr 12, Mo, V, bal Fe.
For blanking and forming dies; oil hardened, non-deforming.
Obsolete

BOHLER SPECIAL KV
Vereinigte Edelstahlwerke
C, Cr, V, bal Fe.
For wire drawing plates; oil hardening. *Obsolete*

BOHLER SPECIAL VERY HARD
Now VEW K400.

BOHLER SPECIAL W-43
Vereinigte Edelstahlwerke
C, alloy, bal Fe.
For rifle barrels; oil hardening. *Obsolete*

BOHLER SPEZIAL EXTRA HART
Vereinigte Edelstahlwerke
C 1.4, Cr 0.3, V 0.1, Mn 0.3, bal Fe.
For blanking and forming dies, engravers tools; oil or water
hardened, wear resistant. *Obsolete*

BOHLER SPEZIAL KR
Now VEW K107.

BOHLER SPEZIAL ZAH
Vereinigte Edelstahlwerke
C 1.42, W, V, bal Fe.
For bearings, cutters, blanking dies; oil hardened, wear
resistant. *Obsolete*

BOHLER SPI
Vereinigte Edelstahlwerke
C 0.65, Si 1.7, bal Fe.
For chisels, upsetters, punches; oil hardened, shock
resistant. *Obsolete*

BOHLER SPN
Now VEW F180.

BOHLER SPU
Vereinigte Edelstahlwerke
C 1.15, Si 0-0.25, Mn 0-0.25, bal Fe.
For springs, taps, reamers, drills; Type W1; water hardened.
Obsolete

BOHLER SPV
Vereinigte Edelstahlwerke
C 0.9, V 0.1, Mn 1.9, Si 0.25, bal Fe.
For forming and blanking dies, punches; oil hardened, non-
deforming. *Obsolete*

BOHLER SSC
Now VEW K510.

BOHLER SSW
Vereinigte Edelstahlwerke
C 1.2, W 1, bal Fe.
For tools, dies; oil hardening. *Obsolete*

BOHLER SSWV
Now VEW K405.

BOHLER SUPER RAPID
Vereinigte Edelstahlwerke
W 15, Cr 3.7, C 0.8, V 0.2, Mo 0.2, bal Fe.
For high speed cutters, tools; high speed steel. *Obsolete*

BOHLER SUPER RAPID EXTRA
Now VEW S200.

BOHLER SUPER RAPID EXTRA 214
Vereinigte Edelstahlwerke
C 0.75, Cr 4, W 18, Mo 1, V 1.55, Co 2.5, bal Fe.
For high speed cutters, taps, drills, reamers; high speed steel.
Obsolete

BOHLER SUPER RAPID EXTRA 500
Now VEW S305.

BOHLER SUPER RAPID EXTRA HV
Vereinigte Edelstahlwerke
C 0.9, W 14.5, V 2, Mo 1, Co 4.5, bal Fe.
For milling cutters; abrasive resistant, high speed steel.
Obsolete

BOHLER SUPER RAPID EXTRA MO
Now VEW S600.

BOHLER SVM
Vereinigte Edelstahlwerke
C 0.6, Si 2, Mn 0.85, Cr 0.25, Mo 0.25, V 0.2, bal Fe.
Oil hardened: 340,000 TS; 283,000 YP; 5 El; 601 Brin.
Annealed: 107,000 TS; 64,000 YP; 27 El; 212 Brin. For shear
blades, pneumatic tools, punches, chisels. Type S-5 tool
steel. Shock resistant. *Obsolete*

BOHLER SVM
U.N. Alloy Steel Corp.
C 0.6, Si 2, Mn 0.85, Cr 0.25, Mo 0.25, V 0.2, bal Fe.
Oil hardened: 340,000 TS; 283,000 YP; 5 El; 601 Brin.
Annealed: 107,000 TS; 64,000 YP; 27 El; 212 Brin. For shear
blades, pneumatic tools, punches, chisels. Type S-5 tool
steel. Shock resistant. *Obsolete*

BOHLER SW
Vereinigte Edelstahlwerke
C 1, Si 0.2, W 0.9, Mn 0.3, bal Fe.
For tools, dies; oil or water hardening. *Obsolete*

BOHLER T240
Bohler Gesellschaft M.B.H.
C 0.13, Cr 16, Ni 13.5, W 2.75, Ti 0.5, bal Fe.
Austenitic temperature resisting alloy. For parts for steam and
gas turbine engines. W. Nr. 1.4962.

BOHLER T245
Bohler Gesellschaft M.B.H.
C 0-0.1, Cr 16.5, Ni 16.5, W 3, Nb/Ta = 10 x C, bal Fe.
Austenitic temperature resisting alloy. Blades for steam
turbine, turbo blades, rotor wheels. W. Nr. 1.4945. *Obsolete*

BOHLER T250
Bohler Gesellschaft M.B.H.
C 0.06, Cr 16.5, Ni 13.5, Mo 1.3, V 0.7, Nb/Ta = 10 x C, bal
Fe.
Austenitic temperature resisting alloy. Parts for steam turbine
plants. W. Nr. 1.4988.

BOHLER T255
Bohler Gesellschaft M.B.H.
C 0.06, Cr 16.5, Ni 16.5, Mo 1.9, Nb/Ta = 10 x C, bal Fe.
Austenitic temperature resisting alloy. Parts for steam and
gas turbine engines. W. Nr. 1.4981.

BOHLER T270
Bohler Gesellschaft M.B.H.
C 0-0.08, Mn 0-2, Cr 18, Ni 11, bal Fe.
Austenitic temperature resisting alloy. For elevated
temperature pipelines. W. Nr. 1.4948. *Obsolete*

BOHLER T275
Bohler Gesellschaft M.B.H.
C 0.06, Si 0.4, Mn 1.1, Cr 16, Ni 12.5, Nb = 10 x C, bal Fe.
Austenitic temperature resisting alloy. Parts for steam and
gas turbines. Quenched: 245 N/mm^2 0.2% proof stress;
510-690 N/mm^2 TS; 22 El. W. Nr. 1.4961.

BOHLER T502
Bohler Gesellschaft M.B.H.
C 0.2, Cr 11.5, Mo 1, Ni 0.7, W 0.5, V 0.3, bal Fe.
Heat resisting alloy; hardenable; for elevated temperature
hardware. W. Nr. 1.4935; similar to AISI 616.

BOHLER T550
Bohler Gesellschaft M.B.H.
C 0.23, Si 0.35, Mn 0.7, Cr 11.8, Mo 1.1, Ni 0.75, V 0.3, bal
Fe.
Martensitic temperature resisting alloy. For engine
components; heat resistant to 550°C. Hardened and
tempered: 600-700 N/mm^2 0.2% proof stress; 800-1050
N/mm^2 TS; 11-14 El. W. Nr. 1.4922.

BOHLER T552
Bohler Gesellschaft M.B.H.
C 0.12, Cr 11.7, Mo 1.7, Ni 2.75, V 0.3, N 0.035, bal Fe.
Heat resisting alloy; hardenable; for elevated temperature
hardware. W. Nr. 1.4939.

BOHLER T558
Bohler Gesellschaft M.B.H.
C 0.2, Cr 12, Mo 1, Ni 0.65, V 0.3, bal Fe.
Corrosion resistance at elevated temperatures; hardenable;
piping hot gas or liquids. W. Nr. 1.4922.

BOHLER T560
Bohler Gesellschaft M.B.H.
C 0.17, Cr 11, Mo 0.6, Ni 0.5, V 0.3, Nb 0.28, B, bal Fe.
Corrosion resistance at elevated temperatures; hardenable;
weldable. W. Nr. 1.4913.

BOHLER T602
Bohler Gesellschaft M.B.H.
C 0.2, Cr 12.5, Mo 1.1, Ni 0.5, bal Fe.
Temperature resisting alloy; hardenable. Parts for thermal
power plants. W. Nr. 1.4921.

BOHLER T651
Bohler Gesellschaft M.B.H.
C 0.21, Si 0.35, Mn 0.5, Cr 13.3, bal Fe.
Temperature resisting alloy; hardenable. Parts for thermal
power plants. Hardened and tempered: 450-550 N/mm^2
0.2% proof stress; 650-950 N/mm^2 TS; 8-15 El. W. Nr.
1.4921.

BOHLER THM
Now VEW K451.

BOHLER TOUGH
Vereinigte Edelstahlwerke
C 0.85, Si 0-0.25, Mn 0.3, bal Fe.
For cold chisels, hammers, shear blades, punches, press
dies, knives; water hardening. *Obsolete*

BOHLER TW
Vereinigte Edelstahlwerke
C 1.15, W 0.9, bal Fe.
For twist drills, taps, files; water hardening. *Obsolete*

BOHLER TWR
Now VEW K459.

BOHLER TWV
Vereinigte Edelstahlwerke
C 1.2, W 1.2, V 1, bal Fe.
For circular knives, drawing rings, cartridge shell dies,
piercing punches; water hardening. *Obsolete*

BOHLER TWVW
Now VEW K403.

BOHLER TWW
Vereinigte Edelstahlwerke
C 1.25, Cr 0.85, W 1.6, bal Fe.
For milling cutters, saw blades; oil hardening. *Obsolete*

BOHLER UF100
Vereinigte Edelstahlwerke
C 1, Si 0.2, Mn 0.4, bal Fe.
Water hardened: 220,000 TS; 155,000 YP; 10 El; 600 Brin.
For springs, cutters, drills. Water hardening, wear resistant.
Obsolete

BOHLER UM 1
C 0.33, Mn 17.5, Cr 3, bal Fe.
Water quenched: 135,000 TS; 60,000 YS; 46 El; 48 RA. For
nonmagnetic parts; austenitic. *Obsolete*

BOHLER UM2
Vereinigte Edelstahlwerke
C 0.65, Mn 9, Cr 3, Ni 7.3, bal Fe.
Water quenched: 126,000 TS; 80,000 YS; 54 El; 44 RA. For
non-magnetic structural parts; non-magnetic, austenitic.
Obsolete

BOHLER UM2M
Vereinigte Edelstahlwerke
C 0.2, Mn 8, Cr 8, Ni 6, bal Fe.
Annealed: 85,400 TS; 34,200 YS. For electrical equipment;
non-magnetic. *Obsolete*

BOHLER UM8
Now VEW P550.

BOHLER UMB
Vereinigte Edelstahlwerke
C 0.2, Mn 6, Cr 11, Ni 10, bal Fe.
Cold drawn: 213,400 TS; 170,700 YS. For electrical
equipment; non-magnetic wires. *Obsolete*

BOHLER US
Vereinigte Edelstahlwerke
C, alloy, bal Fe.
For die casting dies; oil hardening. *Obsolete*

BOHLER US 25
Vereinigte Edelstahlwerke
C 0.3, Cr 2.35, V 0.25, bal Fe.
For tools, dies, mandrels; oil hardening. *Obsolete*

BOHLER US SPEZIAL
Now VEW W326.

BOHLER US ULTRA
Now VEW W300.

BOHLER US ULTRA 2
Now VEW W302.

BOHLER US ULTRA 4
Now VEW W304.

BOHLER USK
Vereinigte Edelstahlwerke
C 0.45, Cr 1.6, Mo 0.6, V, bal Fe.
For cold working and stamping dies; oil hardened. *Obsolete*

BOHLER V110
Bohler Gesellschaft M.B.H.
Alloy steel. C 0.32, Cr 1.1, Mo 0.25, Ni 3.4, bal Fe.
Alloy steel for heavy automotive equipment; deep hardening;
oil hardening. W. Nr. 1.6746.

BOHLER V130
Bohler Gesellschaft M.B.H.
C 0.4, Cr 0.8, Ni 1.4, Mo 0.4, V, bal Fe.
Alloy structural steel; for automotive forgings and couplings;
oil hardening. W. Nr. 1.6565; similar to AISI 4340.

BOHLER V145
Bohler Gesellschaft M.B.H.
Alloy steel. C 0.3, Si 0.3, Mn 0.5, Ni 2, Cr 2, Mo 0.35, bal Fe.
Alloy steel for automotive shafts, crankshafts, gears; oil
hardening. Hardened and tempered: 700-1050 N/mm^2 YS
min; 900-1450 N/mm^2 TS; 9-12 El. W. Nr. 1.6580.

BOHLER V155
Bohler Gesellschaft M.B.H.
Alloy steel. C 0.34, Si 0.3, Mn 0.5, Ni 1.5, Cr 1.5, Mo 0.2, bal Fe.
Alloy steel for automotive parts such as shafts, connecting rods, gears; oil hardening. Hardened and tempered: 600-1000 N/mm² YS min; 800-1400 N/mm² TS; 9-13 El. W. Nr. 1.6582.

BOHLER V157
Bohler Gesellschaft M.B.H.
Alloy steel. C 0.4, Cr 1, Ni 2, Mo 0.25, bal Fe.
Alloy steel for automotive parts such as axles, gears, crankshafts, spline couplings. Oil hardening; similar to AISI 4340. *Obsolete*

BOHLER V165
Bohler Gesellschaft M.B.H.
Alloy steel. C 0.36, Ni 1, Cr 1, Mo 0.2, bal Fe.
Alloy steel for automotive parts such as shafts, axles, spline couplings; oil hardening. W. Nr. 1.6511.

BOHLER V174
Bohler Gesellschaft M.B.H.
C 0.36, Ni 0.85, C 0.8, Mo 0.2, bal Fe.
For structural parts; oil hardening. W. Nr. 1.6506; similar to AISI 8637.

BOHLER V204
Bohler Gesellschaft M.B.H.
C 0.35, Mn 0.5, Cr 1.2, Ni 4.5, bal Fe.
Alloy structural steel; deep hardening; for heavy shafts on earth moving equipment. W. Nr. 1.5864.

BOHLER V214
Bohler Gesellschaft M.B.H.
C 0.31, Cr 0.7, Ni 3.5, bal Fe.
Alloy structural steel; deep hardening, for diesel crankshafts, drive couplings. W. Nr. 1.5755.

BOHLER V228
Bohler Gesellschaft M.B.H.
C 0.4, Mn 0.8, Cr 0.65, Ni 1.25, bal Fe.
Alloy structural steel, for automotive and machine parts; oil hardening. W. Nr. 1.5711; similar to SAE 3140.

BOHLER V304
Bohler Gesellschaft M.B.H.
Alloy steel. C 0.31, Cr 3.1, Mo 0.4, bal Fe.
Alloy steel, deep hardening; for shafts, axles, large machine parts; may also be nitrided. W. Nr. 1.8515.

BOHLER V310
Bohler Gesellschaft M.B.H.
Alloy steel. C 0.5, Cr 1, Mo 0.2, V, bal Fe.
Alloy steel for shafts, arbors, bushings, axles; oil hardening. W. Nr. 1.7228; AISI 4150.

BOHLER V320
Bohler Gesellschaft M.B.H.
Alloy steel. C 0.41, Si 0.3, Mn 0.7, Cr 1.1, Mo 0.2, bal Fe.
Alloy steel for cog wheels, connecting rods, spline couplings. Hardened and tempered: 500-900 N/mm² YS; 750-1300 N/mm² TS; 10-14 El. W. Nr. 1.7223; AISI 4142.

BOHLER V330
Bohler Gesellschaft M.B.H.
Alloy steel. C 0.34, Si 0.3, Mn 0.7, Cr 1.1, Mo 0.2, bal Fe.
Alloy steel for bolts, shafts, arbors; weldable; oil hardening. Hardened and tempered: 450-800 N/mm² YS; 700-1200 N/mm² TS; 11-15 El. W. Nr. 7220; AISI 4130, 4135. G41350

BOHLER V340
Bohler Gesellschaft M.B.H.
C 0.26, Si 0.3, Mn 0.7, Cr 1.1, Mo 0.25, bal Fe.
For shafts, bolts, lever arms. Hardened and tempered: 400-700 N/mm² YS; 650-1100 N/MM² TS; 12-16 El. W. Nr. 1.7218; similar to AISI 4130.

BOHLER V350
Bohler Gesellschaft M.B.H.
C 0.3, Si 0.3, Mn 0.6, Cr 2.5, Mo 0.2, V 0.15, bal Fe.
For bolts, crankshafts, die casting dies. Hardened and tempered: 700-1050 N/mm² YS; 900-1450 N/mm² TS; 9-12 El. W. Nr. 1.7707.

BOHLER V354
Bohler Gesellschaft M.B.H.
C 0.16, Si 0-0.2, Mn 0.95, Cr 1.4, Mo 0.9, V 0.25, bal Fe.
For bolts, crankshafts, die casting dies. Hardened and tempered: 635-930 N/mm² YS; 830-1250 N/mm² TS; 10-12 El. W. Nr. 1.7734.

BOHLER V444D
Vereinigte Edelstahlwerke
C 0.45, Ni, Cr, W, bal Fe.
For heading and forging dies; oil hardened, tough. *Obsolete*

BOHLER V500
Bohler Gesellschaft M.B.H.
C 0.41, Mn 0.7, Cr 1, bal Fe.
For bolts, shafts, machine tool parts; oil hardening. W. Nr. 1.7035; similar to AISI 5140.

BOHLER V510
Bohler Gesellschaft M.B.H.
C 0.33, Mn 0.7, Cr 1, bal Fe.
For bolts, shafts, machine parts; oil or water hardening. W. Nr. 1.7033; AISI 5132. *Obsolete*

BOHLER V520
Bohler Gesellschaft M.B.H.
C 0.46, Mn 0.75, Cr 0.6, bal Fe.
For shafts, bolts, machine parts; oil or water hardening. W. Nr. 1.7006.

BOHLER V560
Bohler Gesellschaft M.B.H.
C 0.65, Mn 0.5, Cr 0.85, V 0.2, bal Fe.
Oil hardening steel for shafts. W. Nr. 1.8161; similar to AISI 5160.

BOHLER V6 N
Now VEW K630.

BOHLER V622
Bohler Gesellschaft M.B.H.
C 0.13, Ni 8.3, bal Fe.
Corrosion resisting steel; for rifle barrels. W. Nr. 1.5662.

BOHLER V734
Bohler Gesellschaft M.B.H.
C 0.15, Mn 2, bal Fe.
Water or oil hardening steel; may be carburized before hardening. W. Nr. 1.5074. *Obsolete*

BOHLER V742
Bohler Gesellschaft M.B.H.
C 0.42, Si 0.25, Mn 1.75, bal Fe.
For axles, chain wheels, bolts, shafts. W. Nr. 1.5223; AISI 1340. *Obsolete*

BOHLER V762
Bohler Gesellschaft M.B.H.
C 0.37, Si 1.25, Mn 1.25, bal Fe.
For crankshafts, axles, shock resisting tools and parts; oil or water hardening. W. Nr. 1.5122.

BOHLER V800
Bohler Gesellschaft M.B.H.
C 0.42, Cr 1.65, Al 1.1, Mo 0.35, bal Fe.
Nitriding steel. W. Nr. 1.8509.

BOHLER V810
Bohler Gesellschaft M.B.H.
C 0.32, Cr 1.1, Al 1, Mo 0.2, bal Fe.
Nitriding steel. W. Nr. 1.8507.

BOHLER V820
Bohler Gesellschaft M.B.H.
C 0.34, Si 0.3, Mn 0.5, Cr 1.7, Mo 0.2, Ni 1, Al 0.95, bal Fe.
Nitriding steel for large cross section parts. Hardened and tempered: 600-850 N/mm² YS; 800-1050 N/mm² TS; 12-13 El. W. Nr. 1.8550.

BOHLER V918
Bohler Gesellschaft M.B.H.
Carbon steel. C 0.18, Si 0.6, Mn 0.6, bal Fe.
Carbon steel; water hardening; for small fasteners. W. Nr. 1.0443. *Obsolete*

BOHLER V920
Bohler Gesellschaft M.B.H.
C 0.22, Si 0.25, Mn 0.45, bal Fe.
For small fasteners; water hardening. W. Nr 1.1151; AISI 1023.

BOHLER V922
Bohler Gesellschaft M.B.H.
C 0.2, Si 0.45, Mn 1.45, bal Fe.
For high pressure vessels and tubes for low temperature operation down to -100°C. W. Nr. 1.1169. *Obsolete*

BOHLER V923
Bohler Gesellschaft M.B.H.
Carbon steel. C 0.3, Si 0.4, Mn 0.4, bal Fe.
Carbon steel; water hardening; for small fasteners. W. Nr. 1.0551. *Obsolete*

BOHLER V930
Bohler Gesellschaft M.B.H.
C 0.3, Si 0.25, Mn 1.35, bal Fe.
For shafts, bolts, fasteners; water or oil hardening. W. Nr. 1.1165.

BOHLER V935
Bohler Gesellschaft M.B.H.
C 0.35, Si 0.25, Mn 0.65, bal Fe.
For bolts, shafts, fasteners; water hardening. W. Nr. 1.1181; AISI 1035.

BOHLER V936
Bohler Gesellschaft M.B.H.
Carbon steel. C 0.34, Si 0.8, Mn 0.8, bal Fe.
Carbon steel; water hardening; for small fasteners. W. Nr. 1.0553. *Obsolete*

BOHLER V940
Bohler Gesellschaft M.B.H.
C 0.4, Si 0.37, Mn 0.95, bal Fe.
For bolts, shafts, machine parts; water hardening. W. Nr. 1.1157.

BOHLER V943
Bohler Gesellschaft M.B.H.
Carbon steel. C 0.4, Mn 0.65, bal Fe.
Carbon steel; for shafts, axles, bolts. W. Nr. 1.1186; AISI 1040.

BOHLER V945
Bohler Gesellschaft M.B.H.
C 0.45, Si 0.25, Mn 0.65, bal Fe.
For bolts, shafts, gears; water hardening. W. Nr. 1.1191; AISI 1042.

BOHLER V946
Bohler Gesellschaft M.B.H.
C 0.45, Si 0.35, Mn 0.68, bal Fe.
Designed particularly for shafts, axles, to be flame or induction hardened on the surface or localized area. W. Nr. 1.1193; AISI 1045. *Obsolete*

BOHLER V953
Bohler Gesellschaft M.B.H.
C 0.56, Si 0.3, Mn 0.55, bal Fe.
For gears, shafts, worms, camshafts to be induction or flame hardened. W. Nr. 1.1213; similar to AISI 1055. *Obsolete*

BOHLER V955
Bohler Gesellschaft M.B.H.
C 0.56, Si 0.25, Mn 0.7, bal Fe.
For shafts, connectors, automotive parts; water hardening. W. Nr. 1.1203; AISI 1045. *Obsolete*

BOHLER V960
Bohler Gesellschaft M.B.H.
C 0.61, Si 0.25, Mn 0.75, bal Fe.
For axles, shafts, pins; water hardening. Hardened and tempered: 450 N/mm^2 YS; 690-1000 N/mm^2 TS; 11-14 El. W. Nr. 1.1221; AISI 1060.

BOHLER V969
Bohler Gesellschaft M.B.H.
C 0.69, Si 0.25, Mn 0.7, bal Fe.
For shafts, pins, hand tools, springs. W. Nr. 1.1231; AISI 1078.

BOHLER VB 135
Now VEW V510.

BOHLER VB150
Vereinigte Edelstahlwerke
C 0.5, Cr 1.05, V 0.1, Mn 0.95, bal Fe.
For springs, gears, bolts, crankshafts; oil hardened, shock resistant. *Obsolete*

BOHLER VB200
Now VEW V554.

BOHLER VBS 135
Vereinigte Edelstahlwerke
C 0.4, Mn 1.2, Si 1.2, bal Fe.
Hardened: 128,000-145,000 TS; 97,000 YS; 14 El; 45 RA; 270-295 Brin. For bolts, gears, shafts, connecting rods; tough, wear resistant. *Obsolete*

BOHLER VBV140
Vereinigte Edelstahlwerke
C 0.5, Mn 1.75, V 0.1, bal Fe.
Hot rolled: 97,000 TS; 59,000 YS; 40 El; 200 Brin. Heat treated: 142,300 TS; 106,700 YS. For valve lifters, crankshafts; oil hardened, tough. *Obsolete*

BOHLER VC2
Now VEW B400.

BOHLER VCL 140
Now VEW V320.

BOHLER VCL125
Now VEW V340.

BOHLER VCL135
Now VEW V330.

BOHLER VCL150
Now VEW V310.

BOHLER VCL230
Now VEW V350.

BOHLER VCL240
Vereinigte Edelstahlwerke
C 0.5, Cr 1, Mo 0.2, Mn 0.6, bal Fe.
For gears, bolts, crankshafts; oil hardened, tough. *Obsolete*

BOHLER VCN 100
Now VEW V165.

BOHLER VCN 200
Now VEW V145.

BOHLER VCN150
Now VEW V155.

BOHLER VCN400W
Vereinigte Edelstahlwerke
C 0.25, Cr 1, Ni 4, Mo, bal Fe.
For crankshafts, gears; heavy forgings, oil hardened. *Obsolete*

BOHLER VSI
Vereinigte Edelstahlwerke
C 0.6, Si 1.7, bal Fe.
Heat treated: 113,800 TS; 85,400 YS. For valves. *Obsolete*

BOHLER VSK
Vereinigte Edelstahlwerke
C 0.45, Si 1.5, Cr 15, Ni 13, W 3, bal Fe.
Annealed: 106,700 TS; 71,200 YS. For exhaust valves; corrosion and heat resistant. *Obsolete*

BOHLER VSW
Now VEW H800.

BOHLER W 150
Vereinigte Edelstahlwerke
C 1, Cr 1.5, bal Fe.
For balls, races, rollers, bearings; water hardening bearing steel. *Obsolete*

BOHLER W100
Bohler Gesellschaft M.B.H.
Tool material. C 0.29, Si 0.25, Mn 0.3, Cr 2.7, W 8.5, V 0.35, bal Fe.
Hot work tool steel; oil hardening. For pressure casting molds, hot extrusion dies. Hardened and tempered: 38-52 Rock C; 800-1350 N/mm^2 TS; 600-1100 N/mm^2 0.2% proof stress. W. Nr. 1.2581; similar to AISI H21.

BOHLER W100
Vereinigte Edelstahlwerke
C 1, Cr 0.9, bal Fe.
For balls, rollers, races, bearings; bearing steel. *Obsolete*

BOHLER W103
Bohler Gesellschaft M.B.H.
Tool material. C 0.23, W 8.5, Cr 2.4, Ni 1.65, V 0.13, bal Fe.
Hot work tool steel; oil hardening. For pressure casting molds, hot extrusion dies. W. Nr. 1.2759. *Obsolete*

BOHLER W105
Bohler Gesellschaft M.B.H.
Tool material. C 0.32, Si 0.25, Mn 0.3, Cr 2.4, W 4.3, V 0.6, bal Fe.
Hot work tool steel for pressure casting molds, cores, dies for nonferrous metals. Hardened and tempered: 36-52 Rock C; 750-1350 N/mm^2 TS; 600-1100 N/mm^2 0.2% proof stress. W. Nr. 1.2567.

BOHLER W106
Bohler Gesellschaft M.B.H.
Tool material. C 0.3, Cr 1, W 3.75, V 0.18, bal Fe.
Hot work tool steel; oil hardening. For extrusion dies, rams, pressing mandrels for nonferrous metals.

BOHLER W108
Bohler Gesellschaft M.B.H.
Tool material. C 0.4, Cr 4.3, W 4.3, Co 4.3, Mo 0.4, V 1.9, bal Fe.
Hot work tool steel; oil hardening. For hot extruding dies, mandrels, pressure casting molds for brass. W. Nr. 1.2678; AISI H19.

BOHLER W300
Bohler Gesellschaft M.B.H.
Tool material. C 0.36, Si 1.1, Mn 0.4, Cr 5, Mo 1.3, V 0.4, bal Fe.
Hot work tool steel; oil hardening. Pressure casting molds for light metal, extrusion press tools, forging dies. Hardened and tempered: 30-54 Rock C; 600-1300 N/mm^2 TS; 400-1100 N/mm^2 0.2% proof stress. W. Nr. 1.2343; AISI H11.

BOHLER W301
Bohler Gesellschaft M.B.H.
C 0.38, Cr 5, Mo 1.3, V 0.5, bal Fe.
Hot work die steel; oil hardening. Pressure casting molds, forging dies. W. Nr. 1.7783; similar to AISI H11.

BOHLER W302
Bohler Gesellschaft M.B.H.
Tool material. C 0.39, Si 1, Mn 0.4, Cr 5.1, Mo 1.3, V 1, bal Fe.
Hot work tool steel; oil hardening. For extrusion press rams and liners, die casting dies, forging dies. Hardened and tempered: 32-55 Rock C; 600-1300 N/mm^2 TS; 400-1100 N/mm^2 0.2% proof stress. W. Nr. 1.2344; similar to AISI H13.

BOHLER W303
Bohler Gesellschaft M.B.H.
Tool material. C 0.39, Si 0.25, Mn 0.3, Cr 5, Mo 2.9, V 0.55, bal Fe.
Hot work tool steel; oil hardening. For extrusion press rams and liners, die casting dies, forging dies. Hardened and tempered: 35-54 Rock C; 700-1350 N/mm^2 TS; 580-1150 N/mm^2 0.2% proof stress. W. Nr. 1.2344; similar to AISI H13.

BOHLER W304
Bohler Gesellschaft M.B.H.
Tool material. C 0.37, Cr 5.2, W 1.3, Mo 1.4, V 0.3, bal Fe.
Hot work tool steel; oil hardening. For extrusion rams and liners, die casting dies, forging dies. W. Nr. 1.2606; similar to AISI H12.

BOHLER W320
Bohler Gesellschaft M.B.H.
Tool material. C 0.31, Si 0.3, Mn 0.35, Cr 2.9, Mo 2.8, V 0.5, bal Fe.
Hot work tool steel; oil hardening. For die casting dies, heading dies, rams, molds. 36-52 Rock C; 700-1350 N/mm^2 TS; 580-1100 N/mm^2 0.2% proof stress. W. Nr. 1.2365.

BOHLER W321
Bohler Gesellschaft M.B.H.
Tool material. C 0.39, Si 0.3, Mn 0.35, Cr 2.9, Mo 2.8, V 0.5, Co 2.9, bal Fe.
Hot work tool steel; oil hardening. For die casting dies, extrusion dies, forging dies. Hardened and tempered: 36-53 Rock C; 730-1350 N/mm^2 TS; 600-1120 N/mm^2 0.2% proof stress. W. Nr. 1.2889.

BOHLER W322
Bohler Gesellschaft M.B.H.
Tool material. C 0.3, Cr 1.9, Ni 2.6, W 0.9, Mo 2.6, Nb 0.25, bal Fe.
Hot work tool steel; oil hardening. For die casting dies, hot forming dies.

BOHLER W326
Bohler Gesellschaft M.B.H.
Tool material. C 0.45, Si 0.3, Mn 0.7, Cr 1.4, Mo 0.75, V 0.3, bal Fe.
Hot work tool steel, oil hardening. For pressing punches, heading dies, shears. Hardened and tempered: 32-51 Rock C; 500-1250 N/mm^2 TS; 400-1100 N/mm^2 0.2% proof stress. W. Nr. 1.2323.

BOHLER W327
Bohler Gesellschaft M.B.H.
Tool material. C 0.45, Cr 1.5, V 0.8, Mo 0.5, W 0.5, bal Fe.
Hot work tool steel; for light upsetting tools, shear knives, pressing punches. W. Nr. 1.2603.

BOHLER W329
Bohler Gesellschaft M.B.H.
Tool material. C 0.21, Si 0.3, Mn 0.3, Cr 2.4, Mo 0.45, bal Fe.
Hot work tool steel; oil hardening. For pressure casting; may be case hardened. Annealed: 200 Brin. W. Nr. 1.2313.

BOHLER W335

Bohler Gesellschaft M.B.H.
Tool material. C 0.38, Si 0.65, Mn 1.7, Cr 2.6, Mo 2.6, V 0.75, Nb 0.12, bal Fe.
Hot work tool steel; oil hardening. For pressure casting; may be case hardened. Annealed: 230 Brin. Hardened and tempered: 40-53 Rock C; 750-950 N/mm^2 TS; 640-810 N/mm^2 0.2% proof stress. W. Nr. 1.2313.

BOHLER W50

Vereinigte Edelstahlwerke
C 1.05, Cr 0.5, bal Fe.
For balls, rollers, bearings; bearing steel. *Obsolete*

BOHLER W500

Bohler Gesellschaft M.B.H.
Tool material. C 0.55, Si 0.25, Mn 0.75, Cr 1.1, Mo 0.5, Ni 1.7, V 0.1, bal Fe.
Hot work tool steel; oil hardening. For forging and upsetting dies. Hardened and tempered: 36-50 Rock C; 600-1200 N/mm^2 TS; 350-1000 N/mm^2 0.2% proof stress. W. Nr. 1.2714.

BOHLER W501

Bohler Gesellschaft M.B.H.
Tool material. C 0.55, Cr 0.7, Ni 1.65, Mo 0.3, V 0.1, bal Fe.
Hot work tool steel; oil hardening. Forging dies, extrusion dies. W. Nr. 1.2713.

BOHLER W502

Bohler Gesellschaft M.B.H.
Tool material. C 0.35, Cr 1.4, Ni 4, Mo 0.25, bal Fe.
Hot work tool steel; oil hardening. For pressing dies, roll rings. W. Nr. 1.2766.

BOHLER W600

Bohler Gesellschaft M.B.H.
Tool material. C 0.21, Mo 3.3, Ni 3.2, bal Fe.
Hot work tool steel. W. Nr. 1.2777.

BOHLER W701

Bohler Gesellschaft M.B.H.
Tool material. C 0.53, Si 0.35, Mn 0.45, Cr 4.3, Ni 12, Mo 0.5, W 12, Co 1.5, V 1, bal Fe.
Hot work tool steel; air or oil hardening. For extrusion presses. W. Nr. 1.2758. *Obsolete*

BOHLER W703

Bohler Gesellschaft M.B.H.
Tool material. C 0.45, Si 1.1, Mn 0.9, Cr 9.5, Mo 1.4, Ni 8.5, W 1.7, Co 1.2, V 1, bal Fe.
Hot work tool steel; air or oil hardening. For extrusion presses. Hardened and tempered: 40-43 Rock C; 550-1200 N/mm^2 TS; 450-950 N/mm^2 0.2% proof stress. W. Nr. 1.2758.

BOHLER W705

Bohler Gesellschaft M.B.H.
Tool material. C 0.15, Cr 10, Mo 5, Co 10, V 0.5, bal Fe.
Hot work tool steel; may be case hardened. W. Nr. 1.2886.

BOHLER W720

Bohler Gesellschaft M.B.H.
C 0-0.03, Si 0-0.1, Mn 0-0.1, Ni 18.5, Co 9, Mo 5.3, Ti 0.6, Al 0.1, bal Fe.
Cold work tools; for stressed components in aerospace industry. Solution annealed: 980-1130 N/mm^2 TS; 650 N/mm^2 0.2% proof stress. W. Nr. 1.6358.

BOHLER W725

Bohler Gesellschaft M.B.H.
C 0-0.03, Ni 18, Co 12.3, Mo 4, Ti 1.65, bal Fe.
To be hubbed before heat treating. For plastic mold or die cast dies. W. Nr. 1.6356.

BOHLER WACE

Vereinigte Edelstahlwerke
C 0.33, Cr 1.1, Al 1.1, O 0, Fe 0, Mo, bal Fe.
For gears, cams, camshafts; nitriding steel. *Obsolete*

BOHLER WACV

Vereinigte Edelstahlwerke
C 0.3, Cr 2.5, Mo 0.2, V 0.15, Mn 0.55, bal Fe.
For die casting and plastic mold dies; oil hardened, tough. *Obsolete*

BOHLER WB

Vereinigte Edelstahlwerke
C 0.5, Mn 0.9, bal Fe.
For drop forging dies; water hardening. *Obsolete*

BOHLER WD15

Vereinigte Edelstahlwerke
C 0.3, Cr 0.8, Ni 4, Mo 1.5, bal Fe.
For mandrels, piercers; hot work steel. *Obsolete*

BOHLER WD17

Vereinigte Edelstahlwerke
C 0.25, Cr 0.8, Ni 4.5, W 4.5, Mo 0.4, bal Fe.
For mandrels, piercers; hot work steel. *Obsolete*

BOHLER WD3

Vereinigte Edelstahlwerke
C 0.45, Si 1, Mn 1, V, bal Fe.
For piercing dies and tools; oil hardened, tough. *Obsolete*

BOHLER WD6

Vereinigte Edelstahlwerke
C 0.28, Cr 0.7, Ni 1.7, V, bal Fe.
For piercing dies and tools; oil hardened, tough. *Obsolete*

BOHLER WEICH

Vereinigte Edelstahlwerke
C 0.65, Mn 0.3, Si 0.25, bal Fe.
For tools, punches, springs; water hardening. *Obsolete*

BOHLER WFO

Vereinigte Edelstahlwerke
C 0.6, Si 0.25, Mn 0.65, bal Fe.
Hot rolled: 115,000 TS; 70,000 YS; 16 El; 240 Brin. Oil hardened: 160,000 TS; 113,000 YS; 13 El; 320 Brin. For springs, hand tools, hammers, wrenches. Oil or water hardening. *Obsolete*

BOHLER WH

Vereinigte Edelstahlwerke
C 0.22, Si 0.25, Mn 0.45, bal Fe.
Annealed: 73,000 TS; 41,000 YS; 22 El; 58 RA; 140 Brin. For rivets, nails, gears, bolts, fasteners; case hardened, water hardened. *Obsolete*

BOHLER WKD

Vereinigte Edelstahlwerke
C 0.5, Cr 4, W 18, V 1, bal Fe.
For piercing and punching tools; high speed steel. *Obsolete*

BOHLER WKV

Now VEW W327.

BOHLER WKW2

Vereinigte Edelstahlwerke
C 0.2, Cr 14, bal Fe.
For stainless parts, dies; corrosion resistant. *Obsolete*

BOHLER WKW2M

Vereinigte Edelstahlwerke
Cr 13, Mo 1.2, C, bal Fe.
For plastic mold dies; corrosion resistant. *Obsolete*

BOHLER WKW4

Now VEW M310.

BOHLER WKW6

Vereinigte Edelstahlwerke
C 0.4, Cr 13.5, bal Fe.
For tools, dies, cutlery, tableware; hardenable, corrosion resistant. *Obsolete*

BOHLER WKW8

Vereinigte Edelstahlwerke
Cr 16, Mo 1, C, bal Fe.
For plastic mold dies; corrosion resistant. *Obsolete*

BOHLER WKZ

Now VEW W100.

BOHLER WKZ

Vereinigte Edelstahlwerke
W 9, Cr 3, C, V, bal Fe.
For hot work tools, extrusion press tools; oil hardened, hot work steel. *Obsolete*

BOHLER WKZ100

Vereinigte Edelstahlwerke
C 0.55, Cr 4, W 10, V 1, Mo 0.8, bal Fe.
For hot working mandrels, punches; hot work tool steel. *Obsolete*

BOHLER WKZ50

Now VEW W105.

BOHLER WM

Now VEW W106.

BOHLER WM2

Vereinigte Edelstahlwerke
C 0.35, Si 0.9, W 1.85, Cr 0.05, V 0.2, bal Fe.
For tools, dies, hot work tools; hot work steel. *Obsolete*

BOHLER WMD

Now VEW W320.

BOHLER WON

Vereinigte Edelstahlwerke
C 0.8, Mn 0.3, V 0.3, bal Fe.
For stamping tools, mint and medal dies, cold beater tools, special dies; water hardening. *Obsolete*

BOHLER WPD

Vereinigte Edelstahlwerke
C 0.25, Cr 1.8, W 0.9, Mo 2.8, Ni 2.5, bal Fe.
For hot press tools, press jaws, hot rolling mandrels, hollow punches; hot work steel. *Obsolete*

BOHLER WPN

Now VEW W103.

BOHLER WPZ

Vereinigte Edelstahlwerke
C 0.3, Cr 2.25, W 9, V 0.3, Co 2.2, bal Fe.
For injection and compression molding and extrusion dies, spline bushings; hot work steel. *Obsolete*

BOHLER WV

Vereinigte Edelstahlwerke
C 1.2, Cr 0.2, W 1, V 0.1, bal Fe.
For drills, taps; water or oil hardened. *Obsolete*

BOHLER Z-II

Vereinigte Edelstahlwerke
C, bal Fe.
For wire drawing cast steel plates. *Obsolete*

BOHLER ZCS

Vereinigte Edelstahlwerke
C 0.5, Si 0.9, Cr 1.15, bal Fe.
For crankshafts, gears, connecting rods; oil hardening. *Obsolete*

BOHLER ZE

Vereinigte Edelstahlwerke
C, bal Fe.
For wire drawing cast steel plates. *Obsolete*

BOHLER ZK
Vereinigte Edelstahlwerke
C 1.4-1.5, Si 0.25-1.15, Cr 1.3-1.5, V 0.08-0.15, bal Fe.
For tools, dies, pneumatic chisels, rivet sets; tough, shock resistant. *Obsolete*

BOHLER ZM
Vereinigte Edelstahlwerke
C, Mn, bal Fe.
For leaf springs. *Obsolete*

BOHLER ZNM
Vereinigte Edelstahlwerke
C 0.5, Cr 0.7, Ni 1.7, Mo, bal Fe.
Heat treated: 243,000 TS; 213,000 YS. For gears; air or oil hardening. *Obsolete*

BOHLER ZNM4
Vereinigte Edelstahlwerke
C 0.4, Cr 1.3, Ni 4, Mo, bal Fe.
Heat treated: 228,000 TS; 213,000 YS. For gears, bolts, crankshafts; oil hardened, shock resistant. *Obsolete*

BOHLER ZRH
Vereinigte Edelstahlwerke
C 0.53, Si 1.6, bal Fe.
Heat treated: 113,800 TS; 78,300 YS. For gears, bolts, shafts; oil hardened, tough. *Obsolete*

BOHLER ZRW
Vereinigte Edelstahlwerke
C 0.5, Si 1.6, bal Fe.
Heat treated: 106,700 TS; 71,200 YS. For gears, bolts, crankshafts; oil hardened, shock resistant. *Obsolete*

BOHLER ZSV
Vereinigte Edelstahlwerke
C 0.42, Si 1.2, Mn 1, Cr 1.2, V, bal Fe.
For adjustable spanners; oil hardened, tough. *Obsolete*

BOHLER-FOX SFW
Vereinigte Edelstahlwerke
Ni-Cu.
For welding electrodes for cast iron; Monel metal, coated. *Obsolete*

BOHLERIT
Vereinigte Edelstahlwerke
WC + Co, TiC + WC + Co.
For cutting tools, dies; sintered carbides. *Obsolete*

BOHN ALLOY 70
Bohn Aluminium & Brass Corp.
Sn 0.75-1.75, Pb 2-3.5, Cu 70-73, bal Zn.
Cast: 35,000 TS; 12,000 YS; 25 El; 45 Brin. For marine hardware; leaded yellow brass. *Obsolete*

BOHN ALLOY 80
Bohn Aluminium & Brass Corp.
Cu 78-82, Sn 9.25-10.75, Pb 8.5-10.7, Ni 0-0.7, Zn 0-0.7.
Cast: 25,000 TS; 16,000 YS; 8 El; 60 Brin. For bearings; high leaded bronze; SAE 64. *Obsolete*

BOHN ALLOY 88C
Bohn Aluminium & Brass Corp.
Cu 85-89, Sn 7.7-9, Pb 0.4-0.9, Zn 3-5, Ni 0-0.75.
Cast: 36,000 TS; 18,000 YS; 70 Brin. For bearings; leaded tin bronze; SAE 621. *Obsolete*

BOHN ALLOY XXX
Bohn Aluminium & Brass Corp.
Cu 63-67, 13 hardener, bal Zn.
Cast: 110,000 TS; 75,000 YS; 12 El; 20 Brin. For ingots; Mn bronze. *Obsolete*

BOHN ALLOY XXX
Ste Metallurgique de Knutange
Cu 63-67, 13 hardener, bal Zn.
Cast: 110,000 TS; 75,000 YS; 12 El; 20 Brin. For ingots; Mn bronze. *Obsolete*

BOHN ALLOY XXA
Bohn Aluminium & Brass Corp.
Cu 60-68, Al 3-7, Mn 2-5, Fe 2-4, Sn 0-1.5, Pb 0-0.2, bal Zn.
Cast: 95,000 TS; 45,000 YS; 20 El; 148 Brin. For ship propellers, marine hardware; manganese bronze, corrosion resistant. *Obsolete*

BOHN NO. 10-90
Bohn Aluminium & Brass Corp.
Sn 10, Pb 90, Sb 0-0.5.
For solder; MP 310 F. *Obsolete*

BOHN NO. 100
Bohn Aluminium & Brass Corp.
Cu 100.
20,000-25,000 TS; 20-30 El; 80-85 Brin. For conductors, electrical equipment; conductivity copper. *Obsolete*

BOHN NO. 100C
Bohn Aluminium & Brass Corp.
Cu 98, Zn 2. *Obsolete*

BOHN NO. 15/85
Bohn Aluminium & Brass Corp.
Sn 14-15, Sb 0-0.4, bal Pb.
For solder; M.P. 440-550 F. *Obsolete*

BOHN NO. 20-80
Bohn Aluminium & Brass Corp.
Sn 20, Pb 80, Sb 0-0.5.
For solder; MP 361-525 F. *Obsolete*

BOHN NO. 25-75
Bohn Aluminium & Brass Corp.
Sn 25, Pb 75, Sb 0-0.5.
For solder; MP 361-571 F. *Obsolete*

BOHN NO. 30-70
Bohn Aluminium & Brass Corp.
Sn 30, Pb 70, Sb 0-0.5.
For solder; MP 361-494 F. *Obsolete*

BOHN NO. 35-65
Bohn Aluminium & Brass Corp.
Sn 35, Pb 65, Sb 0-0.25.
For solder. *Obsolete*

BOHN NO. 40-60
Bohn Aluminium & Brass Corp.
Sn 40, Pb 60, Sb 0-0.25.
For solder; MP 361-460 F. *Obsolete*

BOHN NO. 50-50
Bohn Aluminium & Brass Corp.
Sn 50, Pb 50, Sb 0-0.25.
For solder; MP 361-420 F. *Obsolete*

BOHN NO. 60-40
Bohn Aluminium & Brass Corp.
Sn 60, Pb 40, Sb 0-0.25.
For solder; MP 357.8 F. *Obsolete*

BOHN NO. 7
Bohn Aluminium & Brass Corp.
Sn 3.25, Pb 87.75, Sb 9, Cu 0-0.5.
14,600 TS; 3,400 YS; 19 Brin. For bearings, bushings; anti-friction. *Obsolete*

BOHN NO. 97.5/2.5
Bohn Aluminium & Brass Corp.
Ag 2.4-2.6, bal Pb.
For solder; M.P. 579 F. *Obsolete*

BOHN NO. R-45
Bohn Aluminium & Brass Corp.
Cu 45, Ni 10, Mn 0-0.25, Sn 0-0.15, bal Zn.
Wrought: 63,000 TS; 26,000 YS; 5 El; 114 Brin. For hardware; wrought alloy, corrosion resistant. *Obsolete*

BOHN NO. R-47
Bohn Aluminium & Brass Corp.
Cu 47, Ni 8, Mn 0.25-0.5, Sn 0-0.15, bal Zn.
Wrought: 65,000-70,000 TS; 32,000-35,000 YS; 30-40 El; 100-125 Brin. For hardware; wrought alloy, corrosion resistant. *Obsolete*

BOHN NO.45/55
Bohn Aluminium & Brass Corp.
Sn 44-45, Sb 0-0.4, bal Pb.
For solder; M.P. 361-414 F. *Obsolete*

BOHN NO.95/5
Bohn Aluminium & Brass Corp.
Pb 94-95, Ag 5-6, bal Pb.
For solder; M.P. 580-700 F. *Obsolete*

BOHNALITE 25S
Bohn Aluminium & Brass Corp.
Cu 3.9-5, Si 0.5-1.2, Mn 0.4-1.2, bal Al.
T6-temper: 55,000 TS; 30,000 YS; 16 El; 100 Brin. For bearing blocks, connecting rods; forgings, heat treatable. *Obsolete*

BOHNALITE 62S
Bohn Extruded Product Div.
Mg 0.8-1.2, Si 0.4-0.8, Cu 0.15-0.4, Ti 0-0.15, bal Al.
Annealed: 17,500 TS; 6500 YS; 30 El; 28 Brin. T4-temper: 35,000 TS; 21,000 YS; 25 El; 65 Brin. T6-temper: 45,000 TS; 40,000 YS; 17 El; 95 Brin. For doors, window frames, trim; age hardenable.

BOHNALITE A
Bohn Extruded Product Div.
Sn 5-7, Cu 0.75-1.25, Si 1.5-2.2, Ni 0.75-1.25, bal Al.
Cast. For bearings; similar to Alcoa X750.

BOHNALITE E
Bohn Aluminium & Brass Corp.
Cu 7-8.5, Si 0-0.5, Fe 0-1, bal Al.
18,000-23,000 TS; 12,000 YS; 1.5-3.0 El; 55-65 Brin. For crankcases, oil pans, hub caps, camshaft housing; S.A.E. No. 30. *Obsolete*

BOHNALITE I
Bohn Aluminium & Brass Corp.
Cu 11-13, Si 1, Fe 0-1.2, bal Al.
21,000-25,000 TS; 16,000 YS; 0-2 El; 70-75 Brin. For castings; S.A.E. No. 32. *Obsolete*

BOHNALITE J-2
Bohn Aluminium & Brass Corp.
Cu 9-11, Si 3.5-4.5, Mg 0.15-0.35, bal Al.
HTA: 36,000 TS; 25,000 YS; 100 Brin. For leak proof castings; permanent mold. *Obsolete*

BOHNALITE K
Bohn Aluminium & Brass Corp.
Cu 6-8, Si 5-6, Mg 0.2-0.6, bal Al.
Aged: 27,000 TS; 19,000 YS; 90 El. For automotive pistons; permanent mold. *Obsolete*

BOHNALITE L-2
Bohn Aluminium & Brass Corp.
Si 9-12, Cu 1-3, Ni 0.5-2.5, Mg 0.5-1.5, Ti 0-0.2, bal Al.
T551-temper: 31,000 min TS; 90-120 Brin. For pistons; age hardenable. *Obsolete*

BOHNALITE L-3
Bohn Aluminium & Brass Corp.
Si 8-10, Cu 3-4.5, Ti 0-0.2, Fe 0-1, Mn 0-0.8, bal Al.
F-temper: 34,000 TS; 19,000 YS; 90 Brin. T5-temper: 34,000 TS; 25,000 YS; 100 Brin. T6-temper: 42,000 TS; 30,000 YS; 105 Brin. For hydraulic pistons, covers; age hardenable. *Obsolete*

BOHNALITE L-4
Bohn Aluminium & Brass Corp.
Si 8.5-10.5, Cu 2-4, Mg 0.5-1.5, Ni 0.5-1.5, bal Al.
Aged: 31,000 TS; 85 Brin. For automotive pistons; permanent mold. *Obsolete*

BOHNALITE L-6
Bohn Aluminium & Brass Corp.
Si 11-13, Cu 0.5-2.75, Mg 0.7-1.3, Fe 0-1.3, Ti 0-0.15, bal Al.
Cast: 31,000 min TS; 90-120 Brin. For pistons; age
hardenable. *Obsolete*

BOHNALITE M-10
Bohn Aluminium & Brass Corp.
Mg 9.5-10.6, bal Al.
Solution treated: 42,000 TS; 20,000 YS; 12 El; 50 Brin.
Obsolete

BOHNALITE M-4
Bohn Aluminium & Brass Corp.
Mg 3.25-4.5, bal Al.
Cast: 22,000 TS; 12,000 YS; 6 El; 46 Brin. For dairy
equipment, cooking utensils; corrosion resistant. *Obsolete*

BOHNALITE M-4A
Bohn Aluminium & Brass Corp.
Mg 3.5-4.5, Zn 1.5-2.5, bal Al.
Cast: 22,000 TS; 14,000 YS; 2.5 El; 60 Brin. For food
equipment; permanent mold. *Obsolete*

BOHNALITE O
Bohn Aluminium & Brass Corp.
Cu 2.25-3.25, Zn 12.5-14.5, bal Al.
22,000-29,000 TS; 19,000 YS; 1-4 El; 60-70 Brin. For crank
cases, oil pans, transmission cases; S.A.E. No. 31. *Obsolete*

BOHNALITE O-2
Bohn Aluminium & Brass Corp.
Cu 2.25-3.25, Fe 1.5, Zn 9-11.5, bal Al.
25,000-32,000 TS; 20,000 YS; 2-6 El; 65-80 Brin. For crank
cases; S.A.E. No. 31A. *Obsolete*

BOHNALITE R-1
Bohn Aluminium & Brass Corp.
Cu 1.25-1.75, Ni 0.75-1, Mg 0.08-0.2, Fe 0.05-1.4.
Normalized: 20,000 TS; 2 El; 50 Brin. For sand castings;
general purpose. *Obsolete*

BOHNALITE S
Bohn Aluminium & Brass Corp.
99% min Al.
Wrought: 1,000-15,000 TS; 4000 YS; 30-40 El; 20-30 Brin. For
light structures; S.A.E. No. 25. *Obsolete*

BOHNALITE S-17
Bohn Aluminium & Brass Corp.
Cu 3.5-4.5, Mg 0.2-0.8, Mn 0.4-1, bal AL.
Heat treated: 58,000 TS; 35,000 YS; 15 El; 100 Brin. For
tubing, airplane and dirigible parts; S.A.E. No. 26. *Obsolete*

BOHNALITE S-25
Bohn Aluminium & Brass Corp.
Cu 4-4.8, Si 0.5-1.1, Mn 0.5-1.1, bal Al.
Heat treated: 58,000 TS; 35,000 YS; 15 El; 100 Brin. For
connecting rods, aircraft propellers, hardware, fittings;
S.A.E.No. 27. *Obsolete*

BOHNALITE S-3
Bohn Aluminium & Brass Corp.
Cu 0-0.2, Mn 1-1.5, bal Al.
Wrought: 15,000-19,000 TS; 6000 YS; 25-40 El; 28-50 Brin.
For automobile body panels, gas tanks, panelling; S.A.E. No.
29. *Obsolete*

BOHNALITE S-43
Bohn Aluminium & Brass Corp.
Cu 0-0.2, Si 4.5-6, bal Al.
Wrought: 19,000 TS; 10,000 YS; 18-30 El; 35-40 Brin. For
automobile body parts; S.A.E. No. 35. *Obsolete*

BOHNALITE S-51
Bohn Aluminium & Brass Corp.
Si 0.6-1.2, Mg 0.5-0.9, bal Al.
Heat treated: 45,000 TS; 30,000 YS; 14 El; 95 Brin. For
airplane engine crankcases, automobile hardware; S.A.E. 28.
Obsolete

BOHNALITE S-53
Bohn Aluminium & Brass Corp.
Cu 0-0.05, Si 0.75-1, Mg 1.1-1.4, Cr 0.25, bal Al.
Wrought: 17,000-25,000 TS; 10,000-16,000 YS; 18-30 El; 40
Brin. For light alloy parts. *Obsolete*

BOHNALITE U
Bohn Aluminium & Brass Corp.
Cu 0-0.3, Si 12.5-13, Fe 0-0.8, bal Al.
24,000-31,000 TS; 11,000 YS; 15 El; 45-65 Brin. For corrosion
resisting tight castings; permanent mold. *Obsolete*

BOHNALITE W-5
Bohn Aluminium & Brass Corp.
Cu 4-5, Si 5-6, bal Al.
Cast: 22,000 TS; 1 El; 60 Brin. For pressure tight castings;
sand casting. *Obsolete*

BOHNALITE W-6
Bohn Aluminium & Brass Corp.
Cu 2.5-4.5, Si 5-8, Fe 0-1.2, Mg 0.2, bal Al.
Cast: 22,000 TS; 1 El; 65 Brin. Aged: 24,000 TS; 1 El; 65 Brin.
T6-temper: 32,000 TS; 1 El; 85 Brin. For sand castings; heat
treatable. *Obsolete*

BOHNALITE X-1 S
Bohn Aluminium & Brass Corp.
Al 4, Mn 0.3, bal Mg.
Extruded: 40,000 TS; 26,000 YS; 15 El; 48 Brin. For extruded
shapes and sheets; C.U.S. 58,000, S.S. 20,000. *Obsolete*

BOHNALITE X-10
Bohn Permanent Mold Div.
Al 9-11, Mn 0.1, bal Mg.
Cast: 22,000 TS; 12,000 YS; 1-3 El; 53 Brin. Heat treated:
36,000 TS; 17,000 YS; 1-4 El; 65 Brin. For castings;
permanent mold, sand or die cast.

BOHNALITE X 11 S
Bohn Aluminium & Brass Corp.
Mn 1, Sn 6, bal Mg.
Forged: 35,000 TS; 19,000 YS; 5 El; 45 Brin. For hammer
forgings; End. limit 9000. *Obsolete*

BOHNALITE X-2
Bohn Aluminium & Brass Corp.
Al 10, Mn 0.1, bal Mg.
Sand cast: 29,000 TS; 20,000 YS; 1 El; 63 Brin. For light alloy
castings; high strength. *Obsolete*

BOHNALITE X-3
Bohn Aluminium & Brass Corp.
Al 6, Mn 0.3, bal Mg.
Sand cast: 28,000-31,000 TS; 7000-9000 YS; 0-12 El; 48
Brin. For high strength light alloy casting; heat treatable.
Obsolete

BOHNALITE X-3 S
Bohn Aluminium & Brass Corp.
Al 6, Mn 0.3, bal Mg.
Extruded: 42,000 TS; 30,000 YS; 14 El; 50 Brin. For light alloy
parts; C.U.S. 61,000. *Obsolete*

BOHNALITE X-4
Bohn Aluminium & Brass Corp.
Al 2, Mn 0.2, Cu 4, Cd 2, bal Mg.
Sand cast: 24,000 TS; 8,000 YS; 6 El; 40 Brin. For light alloy
pistons; C.U.S. 39,000. *Obsolete*

BOHNALITE X-5
Bohn Aluminium & Brass Corp.
Al 8, Mn 0-0.2, bal. Mg.
Heat treated: 31,000-35,000 TS; 9000-10,000 YS; 0-12 El; 50
Brin. For high strength castings; age hardened. *Obsolete*

BOHNALITE X-6
Bohn Aluminium & Brass Corp.
Mn 1.5, bal Mg.
Wrought: 38,000-42,000 TS; 22,000-25,000 YS; 5-8 El; 40
Brin. For light alloy parts; max corrosion resistance.
Obsolete

BOHNALITE X-7
Bohn Aluminium & Brass Corp.
Al 12, Mn 0.1, bal Mg.
Sand cast and heat treated: 29,000 TS; 21,000 YS; 0.5 El; 75
Brin. For pistons and castings; high hardness. *Obsolete*

BOHNALITE Y-2
Bohn Aluminium & Brass Corp.
Cu 3-4, Si 0.5-0.75, Fe 0.75-1.75, Mg 0.5-0.75, Ni 0.5-0.75,
bal Al.
Heat treated: 32,000-50,000 TS; 24,000-28,000 YS; 0.5-5.0
El; 80-115 Brin. For castings, cylinder heads; sand casting
alloy, age hardened. *Obsolete*

BOHNALLOY 422
Bohn Bearing Div.
Pb 20-24, Sn 3.5-5, Ni 0-0.3, Zn 0-0.3, Fe 0-0.15, Sb 0-0.1, P
0-0.03, S 0-0.02, Si 0-0.005, bal Cu.
High load capacity; resistant to oil corrosion. Similar to
PMB24.

BOHNALLOY MB201
Bohn Bearing Div.
Sn 11-23, Cu 0.7-1.3, Si 0-0.7, Fe 0-0.7, Mn 0-0.1, Ti 0-0.1,
bal Al.
Not subject to acid attack; moderate loads. For engine
bearings and bushings.

BOHNALLOY MB6
Bohn Bearing Div.
Si 3.5-4.5, Bi 0.6-1.5, Fe 0-0.35, Cu 0.05-0.2, Mn 0-0.03, Mg
0.05-0.2, Zn 0-0.04, Ti 0-0.02, Cd 0-0.05, bal Al.
Bearings and bushings. AA4013.

BOHNALLOY MB7
Bohn Bearing Div.
Sn 6.5-7.5, Ni 1.5-1.8, Cu 0.7-1.3, Mg 0.75-1.25, Si 0.5-1.25,
Fe 0.6, Ti 0.1, Mn 0.25, Zn 0.15, bal Al.
Solid permanent mold cast bearings and bushings. Similar to
SAE 770.

BOHNALLOY MB8
Bohn Bearing Div.
Sn 5.5-7, Ni 0.2-0.7, Cu 0.7-1.3, Si 1-2, Fe 0.7, Mn 0.1, Ti 0.1,
bal Al.
Bearings and bushings. SAE 780; AA8280.

BOHNALLOY NO. 15
Bohn Bearing Div.
Sn 9.2-10.7, Sb 14-16, Cu 0.5, As 0.6, Bi 0.1, Cd 0.05, 0.005
Zn, 0.005 Al, bal Pb.
High lead babbitt. Main and rod bearings and bushings,
moderate loads. SAE 14, ASTM B23 Alloy 7. *Obsolete*

BOHNALLOY NO. 18
Bohn Bearing Div.
Sn 0.9-1.3, Sb 14-15.5, Cu 0.5, As 0.8-1.2, Bi 0.1, Zn 0.005, Al
0.005, Cd 0.02, bal Pb.
High lead babbitt. Main and rod bearings and bushings,
moderate loads. SAE 15, ASTM B23 Alloy 15. *Obsolete*

BOHNALLOY NO. 3
Bohn Bearing Div.
Sb 7-8, Pb 0.5, Cu 3-4, Fe 0.08, As 0.1, Bi 0.08, Zn 0.005, Al
0.005, bal Sn.
Tin base babbitt. Main and rod bearings and bushings; high
corrosion resistance. SAE 12; ASTM B23 Alloy No. 2.
Obsolete

BOHNALLOY NO. 9
Bohn Bearing Div.
Sn 5-7, Sb 9-11, Cu 0.5, As 0.25, Bi 0.1, Zn 0.005, Al 0.005,
Cd 0.05, bal Pb.
High lead babbitt. Main and rod bearings and bushings,
moderate loads. SAE 13, ASTM B23 Alloy No. 13. *Obsolete*

BOHNALLOY PMB15

Bohn Bearing Div.

Pb 12.5-16.5, Sn 3-6, Zn 0-0.75, Ni 0-0.5, Fe 0-0.15, Sb 0-0.15, S 0-0.015, Al 0-0.005, Si 0-0.005, P 0-0.03, bal Cu. Intermediate to heavy loads with oscillating or rotating shafts; requires hard shaft transmission, rocker arm or gear bushings.

BOHNALLOY PMB20

Bohn Bearing Div.

Cu 77-81, Sn 9-11, Pb 8-11, Zn 0-0.75, Sb 0-0.2, Ni 0-0.5, Fe 0-0.15, P 0-0.03.
Bushings and wear plates. Hard shaft desirable. SAE 797.

BOHNALLOY PMB20A

Bohn Bearing Div.

Cu 77-81, Sn 9-11, Pb 8-11, Zn 0-0.75, Sb 0-0.5, Ni 0-0.5, Fe 0-0.15, P 0-0.03.
Bushings. Hard shaft desirable. Softer than PMB20. SAE 797.

BOHNALLOY PMB21

Bohn Bearing Div.

Cu 70-75, Sn 3-4, Pb 21-25, Zn 0-0.3, Sb 0-0.5, Ni 0-0.5, Fe 0-0.15, P 0-0.03.
Rocker arm, transmission and camshaft bushings; intermediate load applications. SAE 794 or 799.

BOHNALLOY PMB22

Bohn Bearing Div.

Cu 83-89, Sn 3.5-4.5, Pb 7-9, Zn 0-4, Fe 0-0.15, P 0-0.03, Sb 0-0.5, Ni 0-0.5.
Transmission and chassis bushings, medium to high load applications. SAE 798.

BOHNALLOY PMB224

Bohn Bearing Div.

Pb 22-26, Sn 2-2.75, Ni 0-0.3, Zn 0-0.1, Fe 0-0.15, Sb 0-0.1, P 0-0.03, S 0-0.02, Si 0-0.005, Al 0-0.005, bal Cu.
Engine bearings and bushings, high load capacity, good resistance to erosion. SAE 49.

BOHNALLOY PMB24

Bohn Bearing Div.

Pb 23-27, Sn 0.6-1, Fe 0-0.15, Ni 0-0.1, Sb 0-0.1, Zn 0-0.1, P 0-0.03, bal Cu.
Heavy duty rod and main bearings, hard or soft shaft. SAE 49.

BOHNALLOY PMB26

Bohn Bearing Div.

Pb 23-27.5, Sn 1-2, Fe 0-0.07, P 0-0.06, Ni 0-0.07, Zn 0-0.1, Sb 0-0.3, S 0-0.02, Al 0-0.005, Si 0-0.005, bal Cu.
High load capacity; resistant to corrosion. For engine bearings and bushings.

BOHNALLOY PMB30

Bohn Bearing Div.

Pb 29-33, Sn 0-0.5, Fe 0-0.25, Ni 0-0.1, Sb 0-0.1, Zn 0-0.1, P 0-0.03, bal Cu.
Rod and main bearings, medium to heavy loads, hard or soft shaft without overplate. SAE 48.

BOHNALLOY PMB42

Bohn Bearing Div.

Cu 54-58, Pb 40-44, Sn 1.5-3.5, Fe 0-0.25, Sb 0-0.25, Ni 0-0.1, Zn 0-0.1, P 0-0.03.
Intermediate bearing material usually without overlay. May be used on soft shaft. Similar to SAE 484.

BOHNALLOY R-56

Bohn Extruded Product Div.

Cu 56-59, Al 0.7-1.2, Fe 0.8-1.3, Mn 0-0.5, bal Zn.
Hard: 80,000 TS; 50,000 YS; 12 El; 85 Rock B. Soft: 70,000 TS; 25,000 YS; 20 El; 70 Rock B. For wrought architectural shapes, marine hardware; wrought alloy Mn bronze.

BOHNALLOY R55

Bohn Extruded Product Div.

Cu 56-58, Pb 2.25-3.5, Fe 0-0.35, bal Zn.
Soft: 60,000 psi TS; 20,000 psi YS; 30 El; 55 Rock B. For wrought architectural parts, casements, lock bodies, extrusions. Architectural bronze.

BOHNALLOY R58NE

Bohn Extruded Product Div.

Cu 58-62, Pb 1.6-2.5, Fe 0-0.3, bal Zn.
Soft: 58,000 psi TS; 23,000 psi YS; 40 El; 40 Rock B. For forgings and good machinability. Forging brass.

BOHNALLOY R60L

Bohn Extruded Product Div.

Cu 59-62, Pb 0.4-1, Sn 0.5-1, Fe 0-0.1, bal Zn.
Half hard: 62,000 TS; 40,000 YS; 18 El; 65 Rock B. Soft: 54,000 TS; 22,000 YS; 32 El; 50 Rock B. Low leaded naval brass; for marine parts and screw machine products.

BOHNALLOY R60N

Bohn Extruded Product Div.

Cu 59-62, Sn 0.5-1, Pb 0-0.2, Fe 0-0.1, bal Zn.
Half hard 62,000 TS; 40,000 YS; 22 El; 65 Rock B. Soft: 54,000 TS; 22,000 YS; 40 El; 50 Rock B. For water pump parts, shafting, bushings and turnbuckle bands. Naval brass, uninhibited.

BOHNALLOY R61

Bohn Extruded Product Div.

Cu 60-63, Pb 2.5-3.7, Fe 0-0.25, bal Zn.
Half hard: 60,000 TS; 45,000 YS; 18 El; 75 Rock B. Soft: 48,000 TS; 18,000 YS; 45 El; 20 Rock B. For high speed automatic screw machine parts. Free cutting brass.

BOHNALLOY R62

Bohn Extruded Product Div.

Pb 2-3, Fe 0-0.17, Cu 59-64.5, bal Zn. Rod: 61.0 Cu min.
Soft: 52,000 TS; 25,000 YS; 25 El; 55 Rock B. For moderate thread rolling, slight cold working. Extra high leaded brass.

BOHNALLOY R63

Bohn Extruded Product Div.

Pb 1.3-2.3, Fe 0-0.1, Cu 59-64.5, bal Zn. Rod: 61.0 Cu min.
Extruded: 55,000 TS; 40,000 YS; 30 El; 65 Rock B. For thread rolling, staking and bending. High leaded brass.

BOHNALLOY R64

Bohn Extruded Product Div.

Cu 63-66, Pb 0.5-1.5, Fe 0-0.15, bal Zn.
Extruded: 55,000 TS; 40,000 YS; 30 El; 65 Rock B. For heading, upsetting, roll threading and riveting. Low leaded brass.

BOHNALLOY R67

Bohn Extruded Product Div.

Cu 58-63, Pb 0.4-3, Fe 0-0.5, Sn 0-0.3, Ni 0-0.25, Al 0-0.25, Mn 2-3.5, Si 0.5-1.5, bal Zn.
Half hard: 87,000 TS; 65,000 YS; 15 El; 85 Rock B. For bearings and bushings. Free cutting bearing bronze.

BOHNALLOY R71

Bohn Extruded Product Div.

Cu 57-60, Pb 0-0.35, Fe 0-0.25, Al 0.5-2, Mn 2-3.5, Si 0.5-1.5, bal Zn.
Half hard: 89,000 TS; 60,000 YS; 15 El; 85 Rock B. Soft: 68,000 TS; 40,000 YS; 20 El; 75 Rock B. For bushings, connecting rods, marine hardware and gears. Forgeable bearing alloy.

BOHNALLOY R72N

Bohn Extruded Product Div.

Cu 55-58, Pb 0.25-1.5, Fe 0-0.35, Ni 1.5-2.5, Al 0-0.25, Mn 2-3.5, Si 0.5-1.5, bal Zn.
Half hard 1 in. rod: 75,000 TS; 55,000 YS; 19 El; 81 Rock B. For gear blanks and hydraulic cylinder barrels. Free cutting bearing bronze.

BOHNOLLOY 53

Bohn Aluminium & Brass Corp.

Cu 53, Pb 45, Ni 2.
Cast: 6,000-8,000 TS; 4,000-5,000 YS; 5-7 El; 25-30 Brin. For castings; castings alloy. *Obsolete*

BOHNOLLOY 56

Bohn Aluminium & Brass Corp.

Cu 56-59, Zn 41, Al 0.75-1.5, Fe 0.75-1.75, Mn 0.5.
Sand cast: 70,000-80,000 TS; 30,000-40,000 YS; 20-35 El; 109 Brin. For brackets, spiders, gear shifter forks, airplane tail skids; Mn bronze. *Obsolete*

BOHNOLLOY 57

Bohn Aluminium & Brass Corp.

Cu 57, Zn 34, Al 2-4, Fe 1.2-2.5, Mn 2.5-4.
Sand cast: 95,000-105,000 TS; 45,000-55,000 YS; 25-35 El; 148 Brin. For marine propellers, valve stems; corrosion resistant. *Obsolete*

BOHNOLLOY 58

Bohn Aluminium & Brass Corp.

Cu 45-47, Ni 9-11, bal Zn.
Soft: 65,000 TS; 40,000 YS; 5 El; 95 Rock B. For hardware; wrought alloy, white brass; architectural applications; weldable. *Obsolete*

BOHNOLLOY 58

Bohn Aluminium & Brass Corp.

Cu 55-64, Ni 18, Fe 0-0.35, bal Zn.
Cast: 30,000 TS; 20 YS. For trimmings, control brackets, levers, fittings, plumbing; S.A.E. No. 42. *Obsolete*

BOHNOLLOY 59

Bohn Aluminium & Brass Corp.

Cu 61-68, Al 5-7, Fe 2-4, Mn 2.5-5, bal Zn.
Cast: 110,000 TS; 60,000 YS; 12 El; 190 Brin. For sand castings; high strength. *Obsolete*

BOHNOLLOY 68

Bohn Aluminium & Brass Corp.

Cu 65-67, Sn 0.5-1.5, Pb 1.5-3, bal Zn.
Cast: 28,000-33,000 TS; 10,000-12,000 YS; 22-30 El; 45-52 Brin. For castings, radiator parts, battery terminals, fittings; yellow brass. *Obsolete*

BOHNOLLOY 70D

Bohn Aluminium & Brass Corp.

Cu 75, Sn 5, Pb 20.
Cast: 19,000-23,000 TS; 12,000-15,000 YS; 12-16 El; 37-45 Brin. For castings, bearings; casting alloy. *Obsolete*

BOHNOLLOY 75

Bohn Aluminium & Brass Corp.

Cu 75, Sn 6, Pb 16, bal Zn.
Cast: 18,000-25,000 TS; 11,000-13,000 YS; 11-17 El; 45-55 Brin. For castings, journals; journal brass. *Obsolete*

BOHNOLLOY 79

Bohn Aluminium & Brass Corp.

Cu 78-82, Sn 2.5-3.5, Pb 6.25-7.75, Ni 0-0.5, Fe 0-0.35, 7.5% min Zn.
Cast: 22,000-29,000 TS; 12,000 YS; 10-18 El; 50-55 Brin. For castings, valves, fittings; semi-red casting brass. *Obsolete*

BOHNOLLOY 80C

Bohn Aluminium & Brass Corp.

Cu 79, Sn 9, Pb 11, Zn 1.
Sand cast: 29,000-35,000 TS; 18,000-21,000 YS; 6-10 El; 63-75 Brin. For castings. *Obsolete*

BOHNOLLOY 82

Bohn Aluminium & Brass Corp.

Cu 84-86, Sn 13-15, P 0-1.
Sand cast: 30,000-32,000 TS; 25,000-27,000 YS; 1.0-1.5 El; 83-93 Brin. For castings, thrust bearing washers; sand casting. *Obsolete*

BOHNOLLOY 82D

Bohn Aluminium & Brass Corp.
Cu 79, Sn 17, Zn 3, Ni 1.
Sand cast: 30,000-35,000 TS; 23,000-28,000 YS; 1-4 El;
115-130 Brin. For castings; casting alloy. *Obsolete*

BOHNOLLOY 83

Bohn Aluminium & Brass Corp.
Cu 81-85, Sn 6.25-7.5, Pb 6-8, Zn 2-4.
Cast: 30,000 TS; 14,000 YS; 12 El; 50 Brin. For bearings,
bushings; heavy duty. *Obsolete*

BOHNOLLOY 84

Bohn Aluminium & Brass Corp.
Cu 84-86, Sn 5-6, Pb 9-10, Max Zn.
Sand cast: 25,000-31,000 TS; 13,000-16,000 YS; 10-15 El;
44-52 Brin. For bronze backed bearings, connecting rod
bearings; bearing back bronze. *Obsolete*

BOHNOLLOY 85

Bohn Aluminium & Brass Corp.
Cu 85, Sn 5, Pb 5, Zn 5.
Cast: 26,000-32,000 TS; 15,000-18,000 YS; 15-20 El; 49-55
Brin. For castings, water pump impellers, bushings, fittings
Obsolete

BOHNOLLOY 86

Bohn Aluminium & Brass Corp.
Cu 86.5, Sn 11, Pb 1.25, P 0.25, Ni 1.
and cast: 35,000-42,000 TS; 14,000-18,000 YS; 2-8 El;
0-95 Brin. For castings, gears; casting alloy. *Obsolete*

BOHNOLLOY 86D

Bohn Aluminium & Brass Corp.
Cu 84, Sn 10, Pb 2.5, Ni 3.5.
Sand cast: 38,000-45,000 TS; 12,000-16,000 YS; 15 El;
75-90 Brin. For castings; casting alloy. *Obsolete*

BOHNOLLOY NO. 152

Bohn Aluminium & Brass Corp.
99.75% Ag min.
For bearings; corrosion resistant. *Obsolete*

BOHNOLLOY NO. 17 B

Bohn Aluminium & Brass Corp.
Sn 7-8, Sb 10-12, Cu 0.3-0.75, As 0.3-0.6, bal Pb.
Cast. For bearings; Babbitt. *Obsolete*

BOHNOLLOY NO. 2

Bohn Aluminium & Brass Corp.
Cu 2.25-3.75, Sb 7-8.5, Pb 0-0.2, bal Sn.
For bearings; anti-friction. *Obsolete*

BOHNOLLOY NO. 4

Bohn Aluminium & Brass Corp.
Cu 4-5, Sb 4-5, bal Sn.
Cast: 12,900 TS; 4400 YS; 20-35 El; 17 Brin. For bearings,
aircraft engine linings; S.A.E. No. 10. *Obsolete*

BOHNOLLOY NO. 45

Bohn Aluminium & Brass Corp.
Cu 52-57, Pb 38-42, Ag 4.5-5.5.
For bearings; heavy duty. *Obsolete*

BOHNOLLOY NO. 5

Bohn Aluminium & Brass Corp.
Cu 5-6.5, Sb 6-7.5, bal Sn.
Cast: 10,000-13,000 TS; 25-35 El; 15-25 Brin. For bearings,
for lining connecting rod and shaft bearings; S.A.E. No. 11.
Obsolete

BOHNOLLOY NO. 6

Bohn Aluminium & Brass Corp.
Sn 85, Cu 7.5, Sb 7.5, Pb 0-0.6.
Cast: 12,000-15,000 TS; 30-40 El; 15-28 Brin. For bearings,
bushings; anti-friction. *Obsolete*

BOHNOLLOY NO. 66

Bohn Aluminium & Brass Corp.
Cu 64-70, Pb 30-35, Ag 0-1.
For bearings; heavy duty. *Obsolete*

BOHNOLLOY NO. 70 A

Bohn Aluminium & Brass Corp.
Cu 67-74, Pb 26-31, Ag 0.75-1.5.
For bearings, heavy duty. *Obsolete*

BOHNOLLOY NO. 70 C

Bohn Aluminium & Brass Corp.
Cu 67-74, Ag 0-1, bal Pb.
For bearings; heavy duty. *Obsolete*

BOHNOLLOY NO. 74

Bohn Aluminium & Brass Corp.
Cu 68-75, Pb 23-27, Sn 2-4.
For bearings; heavy duty. *Obsolete*

BOHNOLLOY NO. 80 B

Bohn Aluminium & Brass Corp.
Cu 77-81, Pb 8-11, Sn 9-11.
For bearings; heavy duty. *Obsolete*

BOHNOLLOY R 90 AA

Bohn Aluminium & Brass Corp.
Al 9.5-10.75, Fe 2.5-3.75, bal Cu.
Wrought: 80,000 TS; 36,000 YS; 20 El; 137 Brin. For hardware
bolts; Al bronze, heat treatable. *Obsolete*

BOHNOLLOY R-53

Bohn Aluminium & Brass Corp.
Cu 53-56, Al 1.2-2, Fe 0.7-1.3, Mn 1-2, Ni 1-2, bal Zn.
Extruded: 78,000 TS; 35,500 YS; 15 El; 145 Brin. For
extrusions; Mn bronze. *Obsolete*

BOHNOLLOY R-54A

Bohn Aluminium & Brass Corp.
Cu 57-60, Sn 0.5-1.5, Fe 0.8-2, Mn 0.2-0.5, Al 0.1-0.25, Pb
0-0.2, bal Zn.
Extruded: 55,000 TS; 22,000 YS; 25 El; 100 Brin. For
extruded shapes, hardware, pump rods, valve stems;
corrosion resistant. *Obsolete*

BOHNOLLOY R-58E

Bohn Aluminium & Brass Corp.
Cu 58.5-62, Pb 1.5-2.5, bal Zn.
Forged: 58,000 TS; 23,000 YS; 40 El; 75 Brin. For forgings;
forging brass. *Obsolete*

BOHNOLLOY R-59B

Bohn Aluminium & Brass Corp.
Cu 58.5-61.5, Sn 0.1, Pb 1.5-2.5, Fe 0.3, Al 0.5, Ni 0.5, bal Zn.
Wrought: 55,000 TS; 25,000 YS; 25-40 El; 80-90 Brin. For
brass forgings. *Obsolete*

BOHNOLLOY R-59C

Bohn Aluminium & Brass Corp.
Cu 57.5-60, Pb 2.25-3.25, Fe 0-0.15, Sn 0-0.2, bal Zn.
Wrought: 55,000 TS; 25,000 YS; 20 El; 100 Brin. For brass
fittings; forging brass. *Obsolete*

BOHNOLLOY R-59T

Bohn Aluminium & Brass Corp.
Cu 50-01, Pb 0.5-1, bal Zn.
Wrought: 45,000 TS; 22,000 YS; 25-35 El; 85-100 Brin. For
propeller shafts, tubing, piston rods, shafts, bushings,
bearings; S.A.E. 76. *Obsolete*

BOHNOLLOY R-90

Bohn Aluminium & Brass Corp.
Sn 0-0.2, Fe 0-1, Al 10, bal Cu.
Wrought: 70,000-80,000 TS; 30,000-40,000 YS; 6-15 El. For
worm wheels, gears, valve seats and guides, forgings; Al
bronze. *Obsolete*

BOHNOLLOY R-90A

Bohn Aluminium & Brass Corp.
Fe 3.5, Al 10, bal Cu.
Wrought: 85,000 TS; 38,000 YS; 25 El; 135 Brin. Forged:
90,000 TS; 55,000 YS; 3 El; 165 Brin. For valve seats; Al
bronze. *Obsolete*

BOHNOLLOY R-90H

Bohn Aluminium & Brass Corp.
Cu 84.5, Fe 3.5, Al 12.
Wrought: 95,000 TS; 60,000 YS; 1-4 El; 180-210 Brin. For
gears, shafts; wrought alloy, heat treatable. *Obsolete*

BOHNOLLOY R-90M

Bohn Aluminium & Brass Corp.
Cu 85.5, Fe 3.5, Al 11.
Wrought: 90,000 TS; 50,000 YS; 1-8 El; 165-185 Brin. For
gears, shafts; wrought alloy, heat treatable. *Obsolete*

BOHNOLLOY R-90S

Bohn Aluminium & Brass Corp.
Fe 3.5, Sn 0-0.5, Al 8.5, bal Cu.
Wrought: 85,000 TS; 32,000 YS; 15-25 El; 120-145 Brin. For
worm wheels, gears, valve parts, forgings; Al bronze.
Obsolete

BOHNOLLOY R-90V

Bohn Aluminium & Brass Corp.
Cu 79, Fe 5, Al 11, Ni 5.
Wrought: 105,000-125,000 TS; 70,000-80,000 YS; 3-5 El;
235-254 Brin. For valve seats; wrought alloy. *Obsolete*

BOHNOLLOY R54B

Bohn Aluminium & Brass Corp.
Cu 57-60, Pb 0.5-0.7, Sn 1, Fe 1, Mn 0.35, bal Zn.
1/2 H-temper: 78,000 TS; 55,000 YS; 15 El; 150 Brin. Soft:
68,000 TS; 35,000 YS; 25 El; 116 Brin. For pump rods, valve
stems and bodies, shafts; manganese bronze. *Obsolete*

BOHNOLLOY R54C

Bohn Aluminium & Brass Corp.
Cu 57-60, Pb 1-1.2, Sn 1, Fe 1, Mn 0.35, bal Zn.
1/2 H-temper: 78,000 TS; 55,000 YS; 15 El; 150 Brin. Soft:
68,000 TS; 35,000 YS; 25 El; 116 Brin. For pump rods, valve
stems and bodies, shafts; manganese bronze, free-cutting.
Obsolete

BOHNOLLOY R60M

Bohn Aluminium & Brass Corp.
Cu 60, Pb 2, Sn 0.75, bal Zn.
1/2 H-temper: 65,000 TS; 48,000 YS; 15 El; 150 Brin. Soft:
54,000 TS; 24,000 YS; 30 El; 83 Brin. For bearings, hardware;
light loads, free-cutting. *Obsolete*

BOHNOLLOY R64R

Bohn Aluminium & Brass Corp.
Cu 63, Pb 1.75, bal Zn.
For swaged, riveted and rolled parts; free-cutting. *Obsolete*

BOHNOLLOY R65

Bohn Aluminium & Brass Corp.
Cu 62.5, Pb 1.25, bal Zn.
For spun and swaged parts; free-cutting and good workability
Obsolete

BOHNOLLOY R66

Bohn Aluminium & Brass Corp.
Cu 61, Pb 3.5, bal Zn.
For screw machine products, fasteners; free-cutting.
Obsolete

BOHNOLLOY R68

Bohn Aluminium & Brass Corp.
Cu 60, Pb 0.75, Si 0.8, Mn 2.5, bal Zn.
1/2 H-temper: 74,000 TS; 40,000 YS; 12 El; 150 Brin. For
bearings, bushings; free-cutting. *Obsolete*

BOHNOLLOY R69

Bohn Aluminium & Brass Corp.
Cu 60, Si 0.8, Mn 2.5, bal Zn.
Hard: 78,000 TS; 60,000 YS; 9 El; 165 Brin. Soft: 60,000 TS;
30,000 YS; 30 El; 80 Brin. For bearings, bushings; good
ductility. *Obsolete*

BOHNOLLOY R93

Bohn Aluminium & Brass Corp.
Cu 91, Si 12, Al 7.
Soft: 85,000 TS; 35,000 YS; 25 El; 160 Brin. For pole line hardware, nuts, bolts; corrosion resistant. *Obsolete*

BOHNOLLOY W 90 VI

Bohn Aluminium & Brass Corp.
Al 9-11, Fe 2-3.5, Ni 4.5-5.5, Mn 0.5-1.5, bal Cu.
Wrought: 95,000 TS; 35,000 YS; 15 El; 190 Brin. For hardware; Al bronze, heat treatable. *Obsolete*

BOHNOLLOY-12

Bohn Aluminium & Brass Corp.
Cu 87-91.5, Sn 3.5-4.5, Pb 3.5-4.5, Zn 1.5-4, Fe 0-0.1.
For general purpose bearings, transmission bushings, thrust washers, piston pin bushings. Moderate to high loads, bearing bronze. *Obsolete*

BOKEBIT

H. Boker & Co.
C 0.8, Cr 4.24, W 18.8, Mo 8, V 2, Co 9, bal Fe.
For tools, cutters; high speed steel. *Obsolete*

BOKER POWER CHISEL

H. Boker & Co.
C 0.53, W 2, Cr 1.3, V 0.25, Mn 0.3, bal Fe.
For tools, chisels, crimpers, upsetters; oil hardened, shock resistant. *Obsolete*

BOKER SPECIAL NO. 847

H. Boker & Co.
C 0.8, Cr 4, Mo 8.5, W 1.5, V 1.1, bal Fe.
For cutters, reamers, hobs, taps, drills, broaches; Type M1; high speed steel. *Obsolete*

BOKER SUPER COBALT

H. Boker & Co.
C 0.8, Cr 4.2, W 19, Mo 0.8, V 2, Co 12, bal Fe.
For tools, cutters; high speed steel. *Obsolete*

BOLONEY

Oman Non-Friction Metal Co.
Sn, Pb, bal Cu.
For bearings; heavy duty.

BOLSTER SILVER

English manufacture
Cu 65.5, Zn 16, Ni 18, P 0.5.
For ornaments, hardware; spun and drawn.

BOLTOMET 103

Thomas Bolton Ltd.
Copper. Cu 99.99.
Oxygen free H.C. copper, electronic grade. B.S. No. C 110.

BOLTOMET 104

Thomas Bolton Ltd.
Copper. Cu 99.95.
Oxygen free H.C. copper. B.S. No. C 103.

BOLTOMET 105

Thomas Bolton Ltd.
Copper. Cu 99.9.
Electrolytic tough pitch H.C. copper. B.S. No. C 101.

BOLTOMET 107

Thomas Bolton Ltd.
Copper. Cu 99.9.
Fire refined tough pitch H.C. copper. B.S. No. C 102.

BOLTOMET 110

Thomas Bolton Ltd.
Copper. Cu 99.9, Ag 0.04.
Tough pitch H.C. copper/silver. B.S. No. C 101.

BOLTOMET 111

Thomas Bolton Ltd.
Copper. Cu 99.9, Ag 0.025.
Tough pitch H.C. copper/silver. B.S. No. C 101.

BOLTOMET 112

Thomas Bolton Ltd.
Copper. Ag 0.09, 99.95 Cu min.
Oxygen free H.C. copper strip.

BOLTOMET 113

Thomas Bolton Ltd.
Copper. Ag 0.07, 99.90 Cu min.
Tough pitch H.C. copper. For commutator bars.

BOLTOMET 114

Thomas Bolton Ltd.
Copper. Ag 0.09, Cu 99.9.
Tough pitch H.C. copper/silver. B.S. No. C 101.

BOLTOMET 115

Thomas Bolton Ltd.
Copper. Ag 0.15, 99.90 Cu min.
Tough pitch H.C. copper.

BOLTOMET 117

Thomas Bolton Ltd.
Copper. Ag 0.09, 99.90 Cu min.
Low oxygen H.C. copper strip. *Obsolete*

BOLTOMET 123

Thomas Bolton Ltd.
Copper. P 0.03, Cu 99.85.
Phosphor deoxidized non-arsenical copper. B.S. No. C 106.

BOLTOMET 160

Thomas Bolton Ltd.
Copper. P 0.06, Cu 99.85.
High phosphorus deoxidized non-arsenical copper wire. *Obsolete*

BOLTOMET 162

Thomas Bolton Ltd.
Copper. Cu 99.85, P 0.04.
Phosphorus deoxidized copper. Nominal resistivity: 0.023 micro-ohm-meter.

BOLTOMET 210

Thomas Bolton Ltd.
Copper. Cd 0.9, P 0.01, bal Cu.
1% cadmium copper. B.S. No. C 108.

BOLTOMET 302

Thomas Bolton Ltd.
Copper. Sn 0.4, P 0.05, bal Cu.
0.4% tin bronze rotor bars.

BOLTOMET 304

Thomas Bolton Ltd.
Copper. Sn 1, P 0.02, bal Cu.
1% tin bronze; nominal resistivity: 0.035 micro-ohm-meter.

BOLTOMET 305

Thomas Bolton Ltd.
Copper. Sn 1, P 0.08, bal Cu.
1% tin bronze strip. *Obsolete*

BOLTOMET 307

Thomas Bolton Ltd.
Copper. Sn 1.25, P 0.025, bal Cu.
1.25% tin bronze; nominal resistivity: 0.041 micro-ohm-meter.

BOLTOMET 309

Thomas Bolton Ltd.
Copper. Sn 2.5, P 0.045, bal Cu.
2.5% tin bronze; nominal resistivity: 0.061 micro-ohm-meter.

BOLTOMET 317

Thomas Bolton Ltd.
Copper. Sn 5, P 0.13, bal Cu.
5% tin bronze. B.S. No. PB 102.

BOLTOMET 320

Thomas Bolton Ltd.
Copper. Sn 6.15, P 0.24, bal Cu.
6% tin bronze wire. B.S. No. PB 103. *Obsolete*

BOLTOMET 338

Thomas Bolton Ltd.
Copper. Sn 7.7, P 0.3, bal Cu.
8% tin bronze strip and wire. *Obsolete*

BOLTOMET 510

Thomas Bolton Ltd.
Copper. Zn 10, Cu 90.
90/10 brass strip. B.S. No. CZ 101. *Obsolete*

BOLTOMET 514

Thomas Bolton Ltd.
Copper. Zn 15, Cu 85.
85/15 brass strip. B.S. No. CZ 102. *Obsolete*

BOLTOMET 516

Thomas Bolton Ltd.
Copper. Zn 20, Cu 80.
80/20 brass strip and wire. B.S. No. CZ 103. *Obsolete*

BOLTOMET 518

Thomas Bolton Ltd.
Copper. Zn 30, Cu 70.
70/30 brass strip. B.S. No. CZ 106. *Obsolete*

BOLTOMET 520

Thomas Bolton Ltd.
Copper. Zn 35, Cu 65.
65/35 brass wire. B.S. No. CZ 107. *Obsolete*

BOLTOMET 522

Thomas Bolton Ltd.
Copper. Zn 37, Cu 63.
63/37 brass forged and machine parts. B.S. No. CZ 108.

BOLTOMET 565

Thomas Bolton Ltd.
Copper. Zn 36, Cu 64.
64136 brass forged and machined parts. B.S. No. CZ 108.

BOLTOMET 611

Thomas Bolton Ltd.
Copper. Cu 59, Pb 1.8, bal Zn.
Leaded brass forged and machined parts. B.S. No. CZ 120/122.

BOLTOMET 613

Thomas Bolton Ltd.
Copper. Cu 58, Pb 2.5, bal Zn.
Leaded brass rod; free machining. B.S. No. CZ 121.

BOLTOMET 710

Thomas Bolton Ltd.
Copper. Cu 62.5, Sn 1.2, bal Zn.
Naval brass. Nominal resistivity: 0.071 micro-ohm-meter. B.S. No. CZ 112.

BOLTOMET 807

Thomas Bolton Ltd.
Copper. Al 10, Fe 5, Ni 5, bal Cu.
Aluminum bronze rod. B.S. No. CA 104; DTD 197.

BOLTOMET 814

Thomas Bolton Ltd.
Copper. Cr 0.7, bal Cu.
Copper-chromium. *Obsolete*

BOLTOMET 818

Thomas Bolton Ltd.
Copper. Cr 0.7, Mg 0.05, bal Cu.
Copper-chromium-magnesium. B.S. No. CC 101.

BOLTOMET 819

Thomas Bolton Ltd.
Copper. Cr 0.75, Mg 0.045, Zr 0.15, bal Cu.
Copper-chromium-zirconium.

BOLTOMET 851

Thomas Bolton Ltd.
Copper. Al 6.2, Si 2.2, Fe 0.6, Mn 0.5, bal Cu.
Aluminum-silicon bronze.

BOLTOMET 876

Thomas Bolton Ltd.
Copper. Ni 1.85, Si 0.65, Cr 0.3, Fe 0.15, bal Cu.
Copper-nickel-silicon.

BOLTOMET 912

Thomas Bolton Ltd.
Copper. Ni 31, Mn 1, Fe 0.8, bal Cu.
Cupro-nickel 70/30 in forged and machined parts. B.S. No.
CN 107.

BOLTOMET 917

Thomas Bolton Ltd.
Copper. S 0.4, P 0.005, bal Cu.
Sulfurized copper. B.S. No. C 111.

BOLTOMET 965

Thomas Bolton Ltd.
Copper. Cu 98.25, Si 1.75.
Copper-silicon rotor bars.

BOLTOMET 968

Thomas Bolton Ltd.
Copper. Cu 97.25, Si 2.75.
Copper-silicon rotorbars.

BOLTON 1% TIN BRONZE

Thomas Bolton Ltd.
Sn 1, P, bal Cu.
For telephone and telegraph wire; high electrical
conductivity. *Obsolete*

BOLTON 1.5% TIN PHOSPHOR BRONZE

Thomas Bolton Ltd.
Sn 1.5, P 0.15, bal Cu.
Annealed: 39,200 TS; 11,000 YS; 62 El; 70 Brin. Hard: 61,500
TS; 53,500 YS; 22 El; 140 Brin. For flexible tubing; corrosion
resistant. *Obsolete*

BOLTON 2.5% TIN PHOSPHOR BRONZE

Thomas Bolton Ltd.
Sn 2.5, P 0.035, bal Cu.
Annealed: 44,000 TS; 13,500 YS; 57 El; 80 Brin. Hard: 68,000
TS; 59,600 YS; 21 El; 157 Brin. For rotor bars for motors; high
electrical conductivity. *Obsolete*

BOLTON NO. 10 PHOSPHOR BRONZE

Thomas Bolton Ltd.
Sn 5, P 0.15, bal Cu.
Annealed: 49,500 TS; 20,200 YS; 64 El; 85 Brin. Hard: 88,000
TS; 60,500 YS; 20 El; 210 Brin. For spring contacts, worm
gears, bolts, fuse clips; corrosion resistant, high strength.
Obsolete

BOLTON NO. 11 PHOSPHOR BRONZE

Thomas Bolton Ltd.
Sn 6, P 0.15, bal Cu.
Annealed: 52,000 TS; 17,500 YS; 68 El; 82 Brin. Hard:
117,000 TS; 102,000 YS; 3.5 El; 241 Brin. For springs,
electrical contacts; strong and corrosion resistant. *Obsolete*

BOLTON NO. 12 PHOSPHOR BRONZE

Thomas Bolton Ltd.
Sn 3.5, P 0.06, bal Cu.
Annealed: 44,500 TS; 14,500 YS; 62 El; 79 Brin. Hard: 79,500
TS; 68,500 YS; 18 El; 180 Brin. For rotor bars for motors;
corrosion resistant. *Obsolete*

BOLTON NO. 14 PHOSPHOR BRONZE

Thomas Bolton Ltd.
Sn 10, P 0.035, bal Cu.
Annealed: 62,500 TS; 25,000 YS; 67 El; 105 Brin. Hard:
106,300 TS; 78,000 YS; 21 El; 235 Brin. For springs, bellows;
wear and corrosion resistant. *Obsolete*

BOLTON NO. 15 PHOSPHOR BRONZE

Thomas Bolton Ltd.
Sn 8, P 0.15, bal Cu.
Annealed: 54,300 TS; 22,000 YS; 77 El; 93 Brin. Hard: 97,500
TS; 76,000 YS; 20 El; 232 Brin. For diaphragms, bellows; high
strength, corrosion resistant. *Obsolete*

BOLTON NO. 16 PHOSPHOR BRONZE

Thomas Bolton Ltd.
Sn 7, P 0.25, bal Cu.
Wire: 123,000 TS. For wire cloth for paper-making equipment;
corrosion resistant, high strength. *Obsolete*

BOLTONS "SPECIAL" BEARING METAL

Thomas Bolton Ltd.
For bearings, bushing. *Obsolete*

BONDED CARBIDE

CCS Braeburn Alloy Steel
C 0.7, Mn 0.2, Cr 4.5, V 1.5, W 18, Mo 0.7, Co 10, bal Fe.
For tools, dies, cutters, drills, taps, reamers. High speed steel.
Obsolete

BONDUR

Vereinigte Leichtmetallwerke G.m.b.H.
Aluminum. Cu 3.5-5.5, Si 0.3-0.5, Mn 0.25-1, Mg 0.2-0.7, bal
Al.
Annealed: 31,000 TS; 15 El. Heat treated: 65,000 TS; 43,000
YS; 16 El; 125 Brin. For light alloy parts, aircraft parts; age-
hardening. Duralumin type alloy.

BONDUR

VAW Vereinigte Aluminium-Werke AG
Aluminum. Cu 3.5-5.5, Si 0.4, Mn 0.3-1, Mg 0.2-0.7, bal Al.
Heat treated: 74,000-80,000 TS; 57,000-63,000 YS; 8-12 El;
130-150 Brin. For structural components, transportation
equipment, fittings, fasteners. Age hardenable, high strength.

BONDURPLATE

Vereinigte Leichtmetallwerke G.m.b.H.
Aluminum. Cu 2.5-5, Mg 0.2-1.8, Mn 0.3-1.5, Si 0-1.2, bal Al.
Clad with 0-1.5 Mn, 0-1.0 Mg, 0-1.2 Si, bal Al. Aged:
55,000-60,000 TS; 35,000-45,000 YS; 20-15 El. For aircraft
construction, wings, fuselage; age-hardenable clad alloy.

BONDWICH

Texas Instruments Inc./Materials Control

Shim material with silver brazing alloy clad on both sides. For
sandwich brazing of carbide-tipped cutting tools; shim
absorbs stresses during brazing and cutting. *Obsolete*

BONGRIP

Pechiney/Eurotungstene
WC plus alloy.
For anti-skid tire studs and studded straps; for snow and ice
chains.

BONNEY-FLOYD 4-6 CR-MO

Bonney-Floyd Co.
C 0.21-0.35, Cr 4-6, Mo, Cu, bal Fe.
For heat and corrosion resistant parts; heat and corrosion
resistant. *Obsolete*

BONNEY-FLOYD NIRRESIST

Bonney-Floyd Co.
Ni 16-23, Cu 5-7, 1.1% C min, Cr, bal Fe.
For heat and corrosion resistant parts; heat and corrosion
resistant. *Obsolete*

BONNEY-FLOYD STAINLESS

Bonney-Floyd Co.
C 0.36-0.5, Ni 31-39, Cr 16-23, Si, bal Fe.
For heat and corrosion resistant parts; heat and corrosion
resistant. *Obsolete*

BONNEY-FLOYD STAINLESS "N"

Bonney-Floyd Co.
C 0.13-0.5, Cr 16-23, Ni 7-11, Mo, bal Fe.
For stainless, heat and corrosion resistant parts; stainless
heat and corrosion resistant. *Obsolete*

BONNEY-FLOYD STAINLESS "O"

Bonney-Floyd Co.
C 0.36-0.5, Cr 12-15, Mo, Cu, bal Fe.
For corrosion resistant parts; corrosion resistant. *Obsolete*

BONNEY-FLOYD STAINLESS "S"

Bonney-Floyd Co.
C 0.13-1.1, Cr 24-30, Ni 12-15, bal Fe.
For heat and corrosion resistant parts; heat and corrosion
resistant. *Obsolete*

BOOTH 20S

Alcan-Booth Industries, Ltd.
Cu 56-62, Fe 0.75, Mn 1.25, Pb 0.5, Sn 0.5, bal Zn.
70,000 TS; 30,000 YS; 28 El; 120 Brin. For extrusions;
corrosion resistant. *Obsolete*

BOOTH 20SA

Alcan-Booth Industries, Ltd.
Cu 57-58, Fe 0.75, Mn 1-1.2, bal Zn.
70,000 TS; 30,000 YS; 25 El; 120 Brin. For extrusions;
corrosion resistant. *Obsolete*

BOOTH 40D

British Aluminium Co., Ltd.
Si 5-14, bal Al.
Hard: 31,000 TS; 5 El. Soft: 20,000 TS; 30 El. For marine
Obsolete

BOOTH 76A

Alcan-Booth Industries, Ltd.
Al 1.8-2.5, Cu 76-78, bal Zn.
60,000 TS; 20,000 YS; 30 El; 70 Brin. For condenser tubes;
corrosion resistant. *Obsolete*

BOOTH A1

Alcan-Booth Industries, Ltd.
Cu 55-60, Pb 2-3.5, bal Zn.
Annealed: 60,000 TS; 30,000 YS; 30 El; 100 Brin. Cold drawn:
75,000 TS; 60,000 YS; 20 El; 170 Brin. For screws, nuts, bolts;
free cutting. *Obsolete*

BOOTH AD

Alcan-Booth Industries, Ltd.
Cu 70-72, Sn 1, bal Zn.
54,000 TS; 18,000 YS; 45 El; 65 Brin. For tubes; corrosion
resistant. *Obsolete*

BOOTH ADA

Alcan-Booth Industries, Ltd.
Cu 70-72, As 0.01-0.05, Sn 1, bal Zn.
50,000 TS; 18,000 YS; 40 El; 75 Brin. For condenser tubes;
embrittlement free. *Obsolete*

BOOTH ALM

Alcan-Booth Industries, Ltd.
Mn 1.2, bal Al.
Rolled: 25,800 TS; 3 El. For cooking utensils; good corrosion
resistance. *Obsolete*

BOOTH D1

Alcan-Booth Industries, Ltd.
Cu 56.5-60, Pb 1-2.5, bal Zn.
Annealed: 60,000 TS; 25,000 YS; 40 El; 35 RA; 100 Brin. Cold
drawn: 70,000 TS; 50,000 YS; 25 El; 170 Brin. For screws,
bolts, nuts; free cutting. *Obsolete*

BOOTH B2C
Alcan-Booth Industries, Ltd.
Cu 58-61, Pb 0.75, bal Zn.
60,000 TS; 24,000 YS; 35 El; 100 Brin. For extrusions; free-cutting. *Obsolete*

BOOTH B76
Alcan-Booth Industries, Ltd.
Cu 61-63, Pb 1.5-2, bal Zn.
Annealed: 55,000 TS; 20,000 YS; 40 El; 80 Brin. Cold drawn: 66,000 TS; 35,000 YS; 25 El; 150 Brin. For hardware; free cutting. *Obsolete*

BOOTH CAP COPPER
Alcan-Booth Industries, Ltd.
Cu 96-97.5, bal Zn.
Soft: 35,000 TS; 50 El; 35 Brin. Hard: 50,000 TS; 10 El; 80 Brin. For small pressings; red brass. *Obsolete*

BOOTH J5K
Alcan-Booth Industries, Ltd.
Cu 56-62, Fe 0.75, Mn 0.75, Pb 0.5, Sn 1.2-2, bal Zn.
75,000 TS; 30,000 YS; 25 El; 120 Brin. For extrusions; free-cutting. *Obsolete*

BOOTH M.G. 3
Alcan-Booth Industries, Ltd.
Cu 0.15, Mg 3-4, Si 0.6, Fe 0.7, Mn 1, Cr 0.5, Ti 0.2, bal Al.
O-temper: 31,400 TS; 13,500 YS; 18 El. 1/4 H-temper: 38,100 TS; 26,500 YS: 8 El. For structural members; corrosion resistant. *Obsolete*

BOOTH MG1
Now ALCAN GB-B57S.

BOOTH MG2
Now ALCANGB-M57S.

BOOTH MG3
Now ALCAN GB-54S.

BOOTH MG3
Alcan-Booth Industries, Ltd.
Mg 3, bal Al.
O-temper: 31,400 TS; 13,500 YS; 18 El. 1/4 H-temper: 38,100 TS; 24,700 YS; 8 El. For light weight structural parts; corrosion resistant, not heat treatable. *Obsolete*

BOOTH MG3C
Now ALCAN GB-B53S.

BOOTH MG5
Alcan-Booth Industries, Ltd.
Fe 0-0.6, Mg 4.5-5.5, Mn 0.15, bal Al.
Annealed: 38,000 TS; 18,000 YS; 18 El; 40 RA. 1/4H-temper: 43,000 TS; 30,000 YS; 8 El; 60 RA. For rivets, marine parts; not hardenable, corrosion resistant. *Obsolete*

BOOTH MG5S
Now ALCAN GB-D54S.

BOOTH MG7
Alcan-Booth Industries, Ltd.
Mg 7, Fe 0.6, Cu 0.15, Si 0.6, Ti 0.2, bal Al.
Annealed: 50,000 TS; 24,000 YS; 26 El; 70 RA. Hard: 70,000 TS; 62,000 YS; 4 El; 115 RA. For rivets, light alloy parts; weldable and ductile. *Obsolete*

BOOTH MV3
Alcan-Booth Industries, Ltd.
Cu 59-61, Pb 2-3, bal Zn.
Annealed: 60,000 TS; 20,000 YS; 30 El; 100 Brin. Cold drawn: 70,000 TS; 50,000 YS; 2 El; 160 Brin. For screws, bolts, nuts; free cuttting. *Obsolete*

BOOTH T2
Alcan-Booth Industries, Ltd.
Al 2-3, Cu 54-56, Fe 0.5, Mn 1, Pb 1.5-2, bal Zn.
80,000 TS; 40,000 YS; 15 El; 150 Brin. For extrusions; corrosion resistant. *Obsolete*

BOOTH YDA
Alcan-Booth Industries, Ltd.
Cu 57-59, Fe 0.5, Mn 1-1.5, Ni 0.5, Sn 1, bal Zn.
70,000 TS; 30,000 YS; 25 El; 120 Brin. For extrusions; corrosion resistant. *Obsolete*

BOOTH YDB
Alcan-Booth Industries, Ltd.
Cu 50-61, Pb 0.75-1, bal Zn.
Annealed: 55,000 TS; 20,000 YS; 40 El; 80 Brin. Cold drawn: 70,000 TS; 40,000 YS; 25 El; 150 Brin. For hardware; free cutting. *Obsolete*

BORA
Thyssen Edelstahlwerke AG
C 2, Cr 12, W 0.7, bal Fe.
For high duty blanking and forming dies; oil hardened, non-deforming. *Obsolete*

BORA 12
Thyssen Edelstahlwerke AG
C 2, Cr 12, bal Fe.
For shear blades, blanking and forming dies; oil hardened, nondeforming. *Obsolete*

BORA 318
Thyssen Edelstahlwerke AG
C 2.1, Cr 13, Mo 0.4, W 0.7, Co 1, bal Fe.
For cutting tools, blanking and forming dies; oil hardened, nondeforming. *Obsolete*

BORA 5
Thyssen Edelstahlwerke AG
C 1, Cr, Mo, V, bal Fe.
For cutters, dies; oil hardened. *Obsolete*

BORA SPECIAL
Thyssen Edelstahlwerke AG
C 1.6, Cr 12, Mo 0.7, V 0.4, W 0.5, bal Fe.
For punches, blanking and forming dies; air hardened, nondeforming. *Obsolete*

BORAL
Brooks & Perkins Inc.
Composite of boron carbide and aluminum. High neutron absorbing properties. For shielding and spent fuel storage containers.

BORAWIRE
Johnson Matthey plc
Au 61, bal Pt.
Wire for dental purposes.

BORCHER ALLOY
English manufacture
Ni 24, Cr 32.5, Ag 0.5, Mo 1.8, bal Fe.
For chemical apparatus, crucibles, pyrometer tubes; heat and corrosion resistant.

BORCHER ALLOY 1
American manufacture
Cr 30, Co 35, Ni 35.
For chemical apparatus; heat and corrosion resistant.

BORCHER ALLOY 2
American manufacture
Cr 36, Fe 60, Mo 4.
For heat treating and annealing pots; heat and corrosion resistant.

BORCHER ALLOY 3
American manufacture
Cr 65, Fe 35.
For pyrometer tubes, crucibles; heat and corrosion resistant.

BORCHER ALLOY 4
American manufacture
Co 34, Ni 34, Cr 30, Ag 2.
For chemical apparatus; heat and corrosion resistant.

BORCHER ALLOY 5
American manufacture
Co 35, Ni 35, Cr 30, Mo 0.5-5.
For heat resisting parts, annealing pots; heat and corrosion resistant.

BORCHER'S "A"
Manufacturer not listed.
Ni 65-68, Cr 30, Au 0.5-5, Ag 0.25-1.5.
For resistances, heat and corrosion resistant parts; heat and corrosion resistant.

BORCHER'S "B"
Manufacturer not listed.
Ni 52-68, Cr 30, Ag 0.15-1.5, bal Au.
For heat and corrosion resistant parts; heat and corrosion resistant.

BORCOLOY GR. 5
General Aircraft Equipment Co.
Co 12, B, bal Fe.
For tools, cutters. *Obsolete*

BORCOLOY GR. 6
General Aircraft Equipment Co.
Co 20, B, bal Fe.
For tools, cutters. *Obsolete*

BORCOLOY GR. 7
General Aircraft Equipment Co.
B, Co, bal Fe.
For tools, cutters. *Obsolete*

BORE 2
Uddeholm Corp.
C 1.1, Cr 0.3, W 1, V 0.1, bal Fe.
For twist drills, finishing cutters; water hardened. *Obsolete*

BORE 2 LEDLOY
Uddeholm Corp.
C 1.1, Cr 0.3, W 1, V 0.1, Pb 0.2, bal Fe.
For twist drills, punches; water hardened. *Obsolete*

BORITE
Swedish American Steel Corp.
C 1-1.2, bal Fe.
For tools, drills; water hardening.

BORITE HOLLOW
Swedish American Steel Corp.
C 0.7, Si 0.2, Mn 0.2, bal Fe.
Heat treated: 175,000 TS; 128,000 YS; 12 El; 37 RA; 360 Brin. For drills, tools, springs, hammers; Type W1; water hardened.

BORITE SOLID DRILL
Swedish American Steel Corp.
C 0.75, Si 0.2, Mn 0.2, bal Fe.
For moil points; blacksmith tools; shock resistant.

BORIUM
Stoody Company
WC + W_2C.
For welding rod for hard facing, hard inserts; hard and abrasion resistant.

BOROBEST 2D
Bochumer Verein
C 0.2, Cr 13, bal Fe.
Annealed: 95,000 TS; 50,000 YS; 25 El; 55 RA; 195 Brin. For cutlery, turbine blades, surgical instruments; Type 420; stainless, hardenable. *Obsolete*

BOROBEST 4K
Bochumer Verein
C 0.4, Cr 13, bal Fe.
Annealed: 95,000 TS; 50,000 YS; 20 El; 50 RA; 200 Brin. For cutlery, valves, surgical instruments; Type 420; stainless, hardenable. *Obsolete*

BOROD
Stoody Company
62 WC + 38 steel.
For welding rod for hard facing; hard and abrasion resistant.

BOROFIL
Now IMI 161.

BOROLITE
Borolite Corp.
Boron carbide.
For cutting tools; sintered carbide.

BORON
Atomergic Chemetals Corp.
B (crystalline and amorphous).
Standard purities: 99.9999%, 99.999%, 99.9%, 99.5%, 95-97%, 85-92% commercial grade. Forms: zone refined rod, lump, powder, single crystals, Isotopes B_{10}, B_{11} (various enrichments).

BORON DEOXIDIZED COPPER 109
Anaconda Co.
B 0.01, bal Cu.
Hard: 48,000 TS; 40,000 YS; 6 El; B 50 Rock. Soft: 33,000 TS; 10,000 YS; 45 El; F 45 Rock. For electrical and electronic parts, magnetrons, synchrotons, vacuum switchgear. Resists oxide penetration and thermal stress cracking. Resists grain growth. *Obsolete*

BORON DEOXIDIZED COPPER 1170
Anaconda Co.
Copper. B 0.01, bal Cu.
Hard: 48,000 TS; 40,000 YS; 6 El; 50 Rock B. Soft: 33,000 TS; 10,000 YS; 45 El; 45 Rock F. For electrical and electronic parts, magnetrons, synchrotrons, vacuum switchgear. Resists oxide penetration and thermal stress cracking; resists grain growth.

BORON-T
British Steel Corp.
C 0.4, Mn 1.6, B 0.003, 0.8% Ni + Cr + Mo max, bal Fe.
Oil harden: 116,000-160,000 TS; 95,000-152,000 YS; 18-27 El; 48-57 RA; 24-70 IZ. For fasteners, crankshafts, axles, shafts, gears, pinions. Oil hardening, impact resistant. *Obsolete*

BOROSIL
SIMETCO
B 3-4, Si 38-42, bal Fe.
For boron additions to steel. *Obsolete*

BOROTAL Z7
German manufacture
Cu 3-4, Fe 0-2, Pb 0-3, Zn 0-3, 0.10 graphite, bal Al.
For bearings.

BOROTEC 10009
Eutectic Corp.
Powder for spray coating that resists metal-to-metal friction.
55-62 Rock C.

BOROTO B.K
German manufacture
Br 31, graphitic white metal.
For bearings; for high stresses.

BOROTO B.L
German manufacture
graphitic white metal.
35 Brin. For bearings; high friction and heavy torque.

BOROTO B.N
German manufacture
graphitic white metal.
10 Brin. For bearings; for high speeds.

BOROTO B.R
German manufacture
graphitic white metal.
27 Brin. For bearings; high friction and speed

BORTAM
NL Industries
Ti 16-18, Al 13-15, Mn 22-24, Si 20-25, C 0-1, B 1.5-2.
For steel working additions; adds deep hardening properties. *Obsolete*

BOSCH (AL-20)
Robert Bosch Metallwerk GmbH
Al 88, Zn 11, Mg 0.5.
Annealed: 46,000 TS; 36,000 YS; 18 El; 25 RA. Pressed: 50,000 TS; 18 El; 20 RA; 100 Brin. For high quality parts for fine mechanism; good hot-working properties.

BOSCH (AL-34)
Robert Bosch Metallwerk GmbH
Al 89, Cu 11.
Chill cast: 21,000 TS; 1-2 El; 80 Brin. Sand cast: 20,000 TS; 1-2 El; 75 Brin. For automobile pistons, sand and permanent mold parts; good machinability.

BOSCH (AL-36)
Robert Bosch Metallwerk GmbH
Al 89, Cu 6.5, Ni 3.5.
Annealed: 19,000 TS; 1-2 El. Die cast: 27,000 TS; 2 El; 80 Brin. For die castings; good fluidity and machinability.

BOSCH (AL-5)
Robert Bosch Metallwerk GmbH
Cu 6, Si 3, bal Al.
Sand cast: 21,200 TS; 1-2 El. Chill cast: 24,400 TS; 1-2 El; 80 Brin. Die cast: 28,000 TS; 1-2 El. For Al castings; very good machinability.

BOSCH (AL-7)
Robert Bosch Metallwerk GmbH
Al 89, modifying agent + bal Zn.
Annealed: 36,000 TS; 20 El. Soft drawn: 42,500 TS; 12 El; 30 RA; 85 Brin. For parts for machinery and apparatus, pressed, forged and drawn parts; aged at room temperature.

BOSCH (AL-9)
Robert Bosch Metallwerk GmbH
Cu 4, Si 2, Ni 1.5, bal Al.
Die cast: 28,000 TS; 2 El; 80 Brin. For ordinary die cast parts; good fluidity.

BOSCH (AN-4)
Robert Bosch Metallwerk GmbH
Zn 4, Cu 3, modifying agent + bal Al.
Die cast: 50,100 TS; 2 El; 110 Brin. For die castings; high tensile and bending strength.

BOSCH (BR-14)
Robert Bosch Metallwerk GmbH
Cu 86, Sn 10, Zn 2, Pb 2.
Sand cast: 36,000 TS; 6 El; 80 Brin. For bearings, phosphor bronze castings; corrosion and high wear resistant, tough, hard.

BOSCH (BR-32)
Robert Bosch Metallwerk GmbH
Cu 76, Al, Ni, Fe, Mn.
Pressed: 120,000 TS; 72,000 YS; 0.5 El; 20 RA; 250 Brin. For gears, forged parts; special bronze; wear resistant.

BOSCH (BR-6)
Robert Bosch Metallwerk GmbH
Cu 66, Sn 6, bal Zn.
Pressed: 50,000 TS; 0.5 El; 140 Brin. For bearings subject to heavy loads and wear; hard, dense bronze.

BOSCH (CU-11)
Robert Bosch Metallwerk GmbH
90 plus Cu.
Annealed: 35,000 TS; 40 El. Soft drawn: 38,000 TS; 30 El. For bars, shapes, tubes; corrosion resistant. *Obsolete*

BOSCH (CU-28)
Robert Bosch Metallwerk GmbH
Cu 99.9.
Annealed: 40,300 TS; 40 El. Pressed: 33,000 TS; 40 El; 60 RA; 50 Brin. For bar, shape, tube, pressed parts; electrolytic copper; corrosion resistant.

BOSCH (ME-1)
Robert Bosch Metallwerk GmbH
Cu 58, Pb 1.8, bal Zn.
Annealed: 67,000 TS; 27 El; 30 RA; 115 Brin. Pressed: 62,500 TS; 18 El; 20 RA; 105 Brin. For hot pressed parts for machines, apparatus and equipment; easily workable.

BOSCH (ME-16)
Robert Bosch Metallwerk GmbH
Cu 61.5, Pb 2, bal Zn.
Pressed: 50,000 TS; 35 El; 40 RA; 80 Brin. Soft drawn: 58,000 TS; 30 El; 40 RA; 90 Brin. For rod, bar, and shapes handled by automatic machinery; cold workable.

BOSCH (ME-17)
Robert Bosch Metallwerk GmbH
Cu 63, bal Zn.
Pressed: 46,000 TS; 40 El min; 55 RA; 75 Brin. Soft drawn: 55,000 TS; 35 El min; 50 RA; 85 Brin. For cold formed rods and shapes, rivets; good cold forming.

BOSCH (ME-2)
Robert Bosch Metallwerk GmbH
Cu 58, Pb 2.2, bal Zn.
Annealed: 67,000 TS; 30 El; 35 RA; 110 Brin. Soft drawn: 72,000 TS; 20 El; 40 RA; 130 Brin. For automatic parts, spring boxes, ornamental and hot pressed parts; easily workable.

BOSCH (ME-25) (BOSCH METAL)
Robert Bosch Metallwerk GmbH
Cu 57, Mn, Al, Ni, bal Zn.
Pressed: 88,000 TS; 15 El; 20 RA; 135 Brin. Soft drawn: 94,000 TS; 12 El; 15 RA; 155 Brin. For bushings, bars, shapes; resists sea water corrosion and wear.

BOSCH (ME-3)
Robert Bosch Metallwerk GmbH
Cu 59, Pb 2.5, bal Zn.
Pressed: 65,000 TS; 30 El; 38 RA. Soft drawn: 72,000 TS; 20 El; 25 RA; 120 Brin. For screws and turned parts for fine work, clock manufacture; somewhat cold workable.

BOSCH (ME-4)
Robert Bosch Metallwerk GmbH
Cu 59, Pb 1, bal Zn.
Pressed: 65,000 TS; 30 El; 45 RA. Soft drawn: 72,000 TS; 20 El; 25 RA; 100 Brin. For wire, hot pressed and forged parts; can be cold drawn.

BOSCH (ME-5) RIVET BRASS
Robert Bosch Metallwerk GmbH
Cu 62, Pb 0.5, bal Zn.
Pressed: 58,000 TS; 40 El; 60 RA. Soft drawn: 65,000 TS; 30 El. For clock manufacture and fine mechanical work, rivets, tubes; cold workable.

BOSCH (ME-6) (BOSCH SPECIAL)
Robert Bosch Metallwerk GmbH
Cu 57, Ni, Mn, bal Zn.
Pressed: 72,000 TS; 30 El; 40 RA. Soft drawn: 78,500 TS; 20 El. For piston rods, screws, valve spindles, condenser shells; corrosion resistant to sea water.

BOSCH AL 1
Robert Bosch Metallwerk GmbH
Cu 2.5-5, Mg 0.2-1.0, Mn 0.3-1.5, bal Al.
For aircraft structures; age hardenable.

BOSCH AL 12
Robert Bosch Metallwerk GmbH
Al 99.5.
Annealed: 13,000 TS; 5000 YS; 45 El; 23 Brin. Hard: 24,000 TS; 22,000 YS; 15 El; 44 Brin. For structures, machine tool parts; corrosion resistant.

BOSCH AL 24
Robert Bosch Metallwerk GmbH
Mg 0.6-1.4, Si 0.6-1.6, Mn 0.6-1, Cr 0-0.3, bal Al.
Annealed: 21,000 TS; 8000 YS; 24 El; 36 Brin. Hard: 32,000 TS; 29,000 YS; 6 El. For structural members.

BOSCH AL 25
Robert Bosch Metallwerk GmbH
Mg 4-5.5, Mn 0-0.8, Cr 0-0.3, bal Al.
Annealed: 42,000 TS; 22,000 YS; 35 El; 65 Brin. For light alloy parts, marine structures; resists gasoline corrosion.

BOSCH AL 27
Robert Bosch Metallwerk GmbH
Cu 2.5-5, Mg 0.2-1.8, Mn 0.3-1.5, Pb, Sn, Cd, Bi, bal Al.
For screw machine products; free-cutting.

BOSCH AL 29
Robert Bosch Metallwerk GmbH
Mg 6-10, Mn 0.2-0.7, Fe 0-1.5, bal Al.
For gasoline meters, aircraft parts; corrosion resistant.

BOSCH AL 39
Robert Bosch Metallwerk GmbH
Mn 1-1.5, Cr 0-0.3, bal Al.
Annealed: 16,000 TS; 6000 YS; 40 El; 28 Brin. Hard: 29,000 TS; 27,000 YS; 10 El; 55 Brin. For light alloy tanks, formed parts; good formability and corrosion resistance.

BOSCH AL 42
Robert Bosch Metallwerk GmbH
Si 5-6.5, Cu 2-3, Mn 0.2-0.6, Fe 0-1.5, bal Al.
For light alloy parts; high corrosion resistance.

BOSCH AM-14
Robert Bosch Metallwerk GmbH
Cu 90, Al 10.
Pressed: 85,000 TS; 25 El; 150 Brin. For parts exposed to H_2SO_4 and HNO_3; acid resistant.

BOSCH AM-158
Robert Bosch Metallwerk GmbH
Mn 1.5, bal Al.
Pressed: 17,800 TS; 20 El; 50 RA; 35 Brin. Drawn: 21,300 TS; 6 El; 30 RA; 50 Brin. For rod, tube, pressings; corrosion resistant.

BOSCH ME-15
Robert Bosch Metallwerk GmbH
Cu 78, Pb 4, Sn 3, bal Zn.
Sand cast: 32,000 TS; 15 El; 70 Brin. For housings, fittings; free-cutting.

BOSCH NS-1
Robert Bosch Metallwerk GmbH
Cu, Ni, Fe, Mn.
Pressed: 85,000 TS; 30 El; 30 RA; 150 Brin. Hard: 100,000 TS; 15 El; 20 RA; 180 Brin. For plates of nickel color; corrosion resistant.

BOSCH PB-7
Robert Bosch Metallwerk GmbH
Pb 74, Sn 10, Sb 15, Cu 1.
Cast: 8000 TS; 30 Brin. For bearing shells; anti-friction.

BOUND BROOK
GKN Powder Met Inc.
Sn, P, graphite, bal Cu.
For bearings, bushings; sintered. *Obsolete*

BOURBOUNES
English manufacture
Sn 51, Al 49, Fe 0.3, Cu 0.3.

BOURNE FULLER AIR HARDENING
Republic Steel Corp.
C 0.5, W 3.5, bal Fe.
For bull-dozing dies, gripper dies, piercers, extrusion punches; for hot work. *Obsolete*

BOURNE FULLER AIR HARDENING
Republic Steel Corp.
C 0.5, W 3.5, bal Fe.
For bull-dozing dies, gripper dies, piercers, extrusion punches; for hot work. *Obsolete*

BOURNE-FULLER "H-C"
Republic Steel Corp.
C, Cr, bal Fe.
For special die work. *Obsolete*

BOW AND ARROW
Osborn Steels Ltd.
C 0.9, bal Fe.
For tools, dies; water hardened. *Obsolete*

BOW WIRE
English manufacture
Cu 93, Zn 2, Sn 5.
For corrosion resistant wire.

BOWCO
Boyd-Wagner Co.
C, W, Mn, Cr, V, Mo, Co, bal Fe.
For reamers, punches, dies, stamping and cutting tools; high speed steel.

BOWCO 7720
Boyd-Wagner Co.
C, Cr, Mn, Mo, bal Fe.
For blanking dies; oil hardening.

BOWCO COBALT
Boyd-Wagner Co.
C, W, Co, Mn, Cr, V, Mo, bal Fe.
For tools for trimming and heavy cutting on hard and abrasive material; high speed steel.

BOWCO FAST FINISHING
Boyd-Wagner Co.
C 1.3, Cr 1.8, W 4, V 0.2, bal Fe.
For dies, cutting tools; oil hardening.

BOWCO NONSHRINK
Boyd-Wagner Co.
C 0.9, Mn 1.2, Cr 0.5, W 0.5, Si 0.35, bal Fe.
For tools, dies; oil hardening, nondeforming.

BOWCO OIL HARDENING
Boyd-Wagner Co.
C 0.9, Mn 1.3, Cr 0.5, W 0.5, V 0.1, bal Fe.
For thread gages, taps, dies; nonshrinking.

BOWCO ONE STAR
Boyd-Wagner Co.
C 0.7, W 14, Cr 4, V 2, bal Fe.
For tools, cutters for intermittent cutting; high speed steel.

BOWCO WATER HARDENING TUBING
Boyd-Wagner Co.
C 1-1.1, bal Fe.
For forming and blanking dies; water hardened.

BOWER 315
Republic Steel Corp.
C 0.1-0.15, Mn 0.5, Si 0.3, Ni 2.8, Cr 1.5, Mo 5, bal Fe.
For bearings for service up to 600°F, case hardened.

BOWER 315
Federal-Mogul Corp.
C 0.1-0.15, Mn 0.5, Si 0.3, Ni 2.8, Cr 1.5, Mo 5, bal Fe.
For bearings for service up to 600°F, case hardened.

BOWSTEEL FC
Bowsteel Distributors Corp.
C 1, V 0.25, bal Fe.
For tools, cutters, taps, reamers; Type W2; water hardened.

BOWSTEEL FCCR
Bowsteel Distributors Corp.
C 1.5, Cr 12, Mo 1, Co 3, bal Fe.
For blanking and forming tools and dies; Type D5; air hardened, non-deforming.

BOWSTEEL FND
Bowsteel Distributors Corp.
C 0.9, Mn 1, Cr 0.5, W 0.5, bal Fe.
For dies, punches, broaches, taps, hobs; Type O1; oil hardened, non-deforming.

BOYD
Time Steel Service Inc.
C 0.75, Mn 0.75, Cr 0.9, Ni 1.75, Mo 0.35, bal Fe.
Tool steel, miscellaneous applications as dies, wood working tools, shears. AISI L6.

BOYD-WAGNER B-W POINT 5
Boyd-Wagner Co.
C 0.02, Mn 0.18, S 0.012, P 0.004, trace Si, bal Fe.
For plastic molds, white metal, die casting dies; water hardening.

BOYD-WAGNER CHROME MAGNET
Boyd-Wagner Co.
C, Cr, bal Fe.
For magnets; magnet steel.

BOYD-WAGNER EZ9W
Boyd-Wagner Co.
C 0.35, W 5, bal Fe.
For punches, shear tools, gripper dies; hot work steel, oil hardened.

BOYD-WAGNER NO. 4 SWEDISH
Boyd-Wagner Co.
C 1.05, Cr 0.5, V 0.1, bal Fe.
For wood and machine screw cold header dies; water hardening.

BOYD-WAGNER NO. 41 SWEDISH
Boyd-Wagner Co.
C 0.7, Mn 0.6, bal Fe.
For chipping chisels; water hardening.

BOYD-WAGNER NO. 9 SWEDISH
Boyd-Wagner Co.
C 0.9, Cr 0.5, bal Fe.
For large cold header dies; water hardening.

BOYD-WAGNER TUNGSTEN MAGNET
Boyd-Wagner Co.
C, W, bal Fe.
For magnets; magnet steel.

BP 24
Saarstahl AG
Now SAARSTAHL 1.5752.

BP 24 EXTRA
Saarstahl AG
Now SAARSTAHL 1.2764.

BP 28
Saarstahl AG
Now SAARSTAHL 1.2162.

BP 41
Saarstahl AG
C 0.23, Mn 0.45, Cr 13.5, Ni 0.9, bal Fe.
Corrosion resistant steel for use with corrosive plastics. *Obsolete*

BP 42
Saarstahl AG
Now SAARSTAHL 1.2316.

BP313

Comalco Ltd.
Cu 3-4, Si 7.5-9.5, Fe 0-0.6, Mn 0-0.1, Mg 0-0.1, Ni 0-0.1, Zn 0-0.1, bal Al.
Pressure cast: F1: 324 MPa TS; 159 MPa YS; 4 El; 80 Brin.
ADC H49-16; BS LM24.

BP401

Comalco Ltd.
Cu 0-0.1, Si 11-13, Fe 0-0.4, Mn 0-0.1, Mg 0-0.05, Ni 0-0.05, Zn 0-0.1, Ti 0-0.2, bal Al.
Permanent mold cast, F1: 207 MPa TS; 90 MPa YS; 9 El; 60 Brin. ADC H49-5; BS LM6.

BP601

Comalco Ltd.
Cu 0-0.05, Si 6.5-7.5, Fe 0-0.11, Mn 0-0.05, Mg 0.3-0.4, Zn 0-0.05, Ti 0-0.2, bal Al.
Permanent mold cast, T6: 276 MPa TS; 165 MPa YS; 17 El; 85 Brin. BS LM25.

BP605

Comalco Ltd.
Cu 0-0.1, Si 9-10, Fe 0-0.6, Mn 0-0.05, Mg 0.45-0.6, Zn 0-0.05, bal Al.
Permanent mold cast: 303 MPa TS; 248 MPa YS; 5 El; 100 Brin. ADC H49-23.

BPS

W. Ossenberg & Cie Edelstahlwerke
C 0.21, Mn 1.2, Si 0.3, Cr 1, bal Fe.
For plastic molds, measuring instruments. W.-Nr. 1.2162.

BPS 2

W. Ossenberg & Cie Edelstahlwerke
C 0.4, Mn 0.3, Si 0.4, Cr 13, bal Fe.
Dies and molds for corrosive plastics. W.-Nr. 1.2083.

BPS 3

W. Ossenberg & Cie Edelstahlwerke
C 0.21, Mn 1.2, Si 0.3, Cr 1, bal Fe.
For plastic molds, measuring instruments. W.-Nr. 1.2082.

BR-2 DIE STEEL

Latrobe Steel Co.
C 2.5, Si 0.3, Mn 0.75, Cr 5.25, Mo 1.1, V 4.5, bal Fe.
Hardened: 55-66 Rock C. For dies, brick and tile mold liners and sleeves requiring extreme abrasion resistance; air hardenable, good temperature resistance. *Obsolete*

BR-3

Now LATROBE BR3.

BR-3 & BR-4

Now LATROBE BR-3 and BR-4.

BR-4 FM

Now LATROBE BR-4FM.

BR-NICKEL 99.6

Vereinigte Deutsche Nickel-Werke AG
Nickel. Cu 0.02, Fe 0.05, C 0.03, Mn 0.15, Si 0.03, 99.7 Ni + Co.
Strip, wire, bar, forged and turned parts for electrical, automotive, optical, fastener and welding industries, controls and apparatus, drawing and stamping.

BRAE-CAST

CCS Braeburn Alloy Steel
C 2, Mo 1.5, W 20, Cr 35, Co 45, B 0.8, Fe 2.
Cast: 650 Brin. For cutters, tools; cast. *Obsolete*

BRAEBURN COBALT

CCS Braeburn Alloy Steel
C 0.65-0.73, Cr 3.75-4.25, V 0.85-0.1, W 17.5-18.5, Co 4-4.5, Mo 0.75-1, bal Fe.
For tools and cutters for lathe and planer work; high speed steel. *Obsolete*

BRAEBURN EXTRA

CCS Braeburn Alloy Steel
C 0.7-1, Si 0.25, Mn 0.25, bal Fe.
For drills, reamers, cutters; water hardened, Type W1.
Obsolete

BRAEBURN HIGH SPEED

CCS Braeburn Alloy Steel
C 0.76, Mn 0.2, Cr 2.85, W 18.43, V 0.87, bal Fe.
For tools, high speed cutters; high speed steel. *Obsolete*

BRAEBURN M-10

CCS Braeburn Alloy Steel
C 0.88, Cr 4, Mo 8, V 2, bal Fe.
For milling cutters, end mills, form tools, taps; high speed steel. *Obsolete*

BRAEBURN M-33

CCS Braeburn Alloy Steel
C 0.9, W 1.5, Mo 9.5, Cr 4, V 1.15, Co 8, bal Fe.
Cobalt-molybdenum high speed steel; for extra cutting ability. AISI M-33. *Obsolete*

BRAEBURN M-7

CCS Braeburn Alloy Steel
C 1, Cr 3.75, V 2.05, W 1.75, Mo 8.75, bal Fe.
For broaches, reamers, end mills, counterbores; high speed steel. *Obsolete*

BRAEBURN S.O.D.

CCS Braeburn Alloy Steel
C 0.95, Mn 1.65, V 0.15, bal Fe.
For dies, tools, punches; non-deforming. *Obsolete*

BRAEBURN SPECIAL

CCS Braeburn Alloy Steel
C 0.7-1.4, Mn 0.25, Si 0.25, bal Fe.
For shears, reamers, drills, punches, hobs, broaches; water hardened; Type W1. *Obsolete*

BRAEBURN STAINLESS

CCS Braeburn Alloy Steel
C 0.35, Cr 13, bal Fe.
For corrosion resisting parts; chemical equipment; corrosion resistant. *Obsolete*

BRAEBURN STANDARD

CCS Braeburn Alloy Steel
C 0.6-1.4, Si 0.25, Mn 0.3, bal Fe.
For drills, taps, shear knives, reamers, broaches; Type W1; water hardened. *Obsolete*

BRAEBURN T-15

CCS Braeburn Alloy Steel
C 1.5, W 13, Cr 4.5, V 4.75, Co 5, bal Fe.
Hardened: C 65-67 Rock. For lathe tools, form cutters, drills, broaches, boring tools, punches, milling cutters. Type T-15 high-speed steel. High red-hardness. Abrasion resistant.
Obsolete

BRAEBURN W 5

CCS Braeburn Alloy Steel
C 1.1, Cr 0.5, bal Fe.
Water hardening tool steel; for extra wear properties. AISI W-5. *Obsolete*

BRAEBURN-BDM

CCS Braeburn Alloy Steel
C 0.3-0.35, Mn 0.8, Si 0.5, Cr 1.7, Mo 0.4, bal Fe.
Hardened: C 36-42 Rock. For extrusion and die casting dies, machined plastic molds, holder blocks, compression molding dies. Oil hardening, tough. *Obsolete*

BRAECUT

CCS Braeburn Alloy Steel
C 1.15, Cr 4.25, V 2.25, W 5.25, Co 12, Mo 6.25, bal Fe.
For cutting tools for high temperature alloys. High speed steel. *Obsolete*

BRAEFOUR

CCS Braeburn Alloy Steel
C 1.25, Cr 4.5, V 4, W 5.5, Mo 4.5, bal Fe.
Hardened: 64-66 Rock C. For cutting, finishing and form tools, broaches, end mills, hobs, gauges, lathe and planer tools, reamers. Type M-4 high speed steel. Excellent abrasion resistance and high red-hardness. *Obsolete*

BRAEMAX

CCS Braeburn Alloy Steel
C 1.1, W 1.5, Mo 9.5, Cr 3.75, V 1.15, Co 8, bal Fe.
Hardened: 66-70 Rock C. For broaches, form cutters, hobs, end mills, milling cutters, gear shaper cutters, taps. High speed steel. High red-hardness and abrasion resistance.
Obsolete

BRAEMOW

CCS Braeburn Alloy Steel
C 0.8, Mo 5, W 5, Cr 4, V 1, bal Fe.
For tools, cutters, taps, reamers, broaches; high speed steel.
Obsolete

BRAEMOW M-2

CCS Braeburn Alloy Steel
C 0.8, Mn 0.2, Cr 4.2, V 2, W 6.5, Mo 5, bal Fe.
For tools, cutters. High speed steel. *Obsolete*

BRAEMOW SPECIAL

CCS Braeburn Alloy Steel
C 0.65, Mo 5, W 6.5, Cr 4.2, V 1.9, bal Fe.
Heat treated: 550-640 Brin. For planing and shaping tools; high speed steel. *Obsolete*

BRAETUF

CCS Braeburn Alloy Steel
C 1.15, Cr 4, V 2, W 5, Mo 6.75, Co 7.25, bal Fe.
Hardened: C 69 Rock. For cutting tools for machining tough superalloys, cold extrusion punches, broaches, reamers, milling cutters, drills. Modified Type M44; high-speed steel.
Obsolete

BRAETWIST

CCS Braeburn Alloy Steel
C 0.9, Mn 0.3, Si 0.3, Mo 9.5, W 1.6, Cr 3.75, V 1.15, Co 8, bal Fe.
Cobalt-molybdenum high speed steel for drills, taps, reamers, broaches, milling cutters and end mills. AISI M-33.
Obsolete

BRAEVAC 718

CCS Braeburn Alloy Steel
C 0.06, Mn 0.05, Cr 18-19, Ni 53, Mo 3, Ti 1, Al 0.5, B 0.004, 5.2% Cb + Ta, bal Fe.
Heat treated: 200,000 TS; 177,000 YS; 22 El. At 1200 F: 167,000 TS; 148,000 YS; 20 El. For aircraft and aerospace structures, high temperature fasteners. High strength at elevated temperatures, precipitation hardening. *Obsolete*

BRAEVAN

CCS Braeburn Alloy Steel
C 1.02, Mo 5.7, W 6.05, Cr 4, V 2.5, bal Fe.
Heat treated: 630-660 Brin. For broaches, reamers, form tools, drills. High speed steel. *Obsolete*

BRAEVAN 2

CCS Braeburn Alloy Steel
C 1.15, Mo 5.5, W 5.6, Cr 4, V 3.3, bal Fe.
For broaches, reamers, end mills, form tools. High speed steel. *Obsolete*

BRAEVAN-M-3

Bisset Steel Co.
C 1, W 6.2, Mo 5.6, V 2.5, Cr 4, bal Fe.
For lathe and planer tools, reamers, broaches; high speed steel.

BRAGE 1

Uddeholm Corp.
C 1.05, Cr 0.5, V 0.1, bal Fe.
For threading tools; water hardened. *Obsolete*

BRAGE 2
Uddeholm Corp.
W 1, V 0.1, C 1.1, Cr 0.25, bal Fe.
For threading tools, twist drills; water hardened. *Obsolete*

BRAKE DIE
Peninsular Steel Co.
C 0.51, Cr 0.95, Mn 0.87, Mo 0.2, bal Fe.
Hardened: 140,000 TS; 293 Brin. For press brake dies; oil hardening.

BRAKE DIE
Bethlehem Steel Corp.
C 0.51, Cr 0.95, Mn 0.87, Mo 0.2, bal Fe.
For dies, press brakes; preheat treated. *Obsolete*

BRAKE DIE
CCS Braeburn Alloy Steel
C, alloy, bal Fe.
For dies, tools, fixtures. *Obsolete*

BRAND 110
Flockton, Tompkin & Co., Ltd.
C 0.55-0.65, bal Fe.
For drop forging die blocks; water hardened.

BRAND 113
Flockton, Tompkin & Co., Ltd.
C 0.5, Ni 1, bal Fe.
For drop forging die blocks; water hardened.

BRAND 123
Flockton, Tompkin & Co., Ltd.
C, Ni, Cr, bal Fe.
For cold chisels and stakes; oil hardened, shock resistant.

BRAND 124
Flockton, Tompkin & Co., Ltd.
C 0.5, Ni 3, Cr 0.75, bal Fe.
For drop forging dies; oil hardened, shock resistant.

BRAND 125
Flockton, Tompkin & Co., Ltd.
C 0.5, Ni 1.5, Cr 1, Mo 0.25, bal Fe.
For drop forging dies; oil hardened, shock resistant.

BRAND 131
Flockton, Tompkin & Co., Ltd.
C, Ni, Cr, bal Fe.
For cold chisels and stakes; oil hardened, shock resistant.

BRAND 133
Flockton, Tompkin & Co., Ltd.
C 0.15, bal Fe.
Annealed: 70,000 TS; 40,000 YS; 25 El; 60 RA; 143 Brin. For gears, pinions, shafts; case hardened.

BRAND 134
Flockton, Tompkin & Co., Ltd.
C 0.15, Ni 3, bal Fe.
For gears, pinions, camshafts, cams; case hardened, tough.

BRAND 135
Flockton, Tompkin & Co., Ltd.
C 0.15, Ni 2, Mo, bal Fe.
For gears, pinions, camshafts, shafts; case hardened, tough.

BRAND 138
Flockton, Tompkin & Co., Ltd.
C 0.15, Ni 5, bal Fe.
For gears, pinions, camshafts, cams; case hardened, tough.

BRAND 140
Flockton, Tompkin & Co., Ltd.
C 0.15, Ni, Cr, bal Fe.
For gears, pinions, camshafts, fasteners; case hardened, tough.

BRAND F.T. 1000
Flockton, Tompkin & Co., Ltd.
C 0.2-0.3, Cr 18-20, Ni 0.4, Mo 1.2, bal Fe.
For furnace parts, glass molds; heat and scale resistant to 1150°C.

BRAND F.T. 20-30
Flockton, Tompkin & Co., Ltd.
C 0.12, Cr 25, Ni 20, bal Fe.
For furnace parts, glass molds; heat and scale resistant to 1250°F.

BRASS COMMON HIGH
Anaconda Co.
Cu 65, Zn 35.
Annealed: 45,000 TS; 20,000 YS; 50 El; Hard: 70,000 TS; 65,000 YS; 3 El. For lamp fixtures, automobile parts, ornaments; deep drawing. *Obsolete*

BRASS DEEP DRAWING
Anaconda Co.
Cu 70, Zn 30.
Wire: 50,000-125,000 TS; 2-50 El. For seamless tubes, cartridges, primers, shot shells; deep drawing. *Obsolete*

BRASS E-133
Accurate Brass Co.
Cu 58.5-60.5, Pb 1.5-2, bal Zn.
Forged: 55,000-60,000 TS; 20,000-25,000 YS; 30-40 El; 80-100 Brin. For hardware, machinery parts; leaded brass, free-cutting.

BRASS LK80-3L
Russian manufacture
Cu 79.28, Pb 0.28, Zn 17.35, Fe 0.24, Si 2.8.
Cast: 63,000 TS; 24,000 YS; 36 El; 44 RA. For hardware, plumbing; red brass.

BRASS LMTS58-2
Russian manufacture
Cu 57.2, Zn 40.15, Fe 0.08, Mn 1.54, P 0.01.
Rolled: 72,000 TS; 48,000 YS; 26 El; 60 RA. For marine parts; strong and corrosion resistant.

BRASS LS59-1
Russian manufacture
Cu 55.32, Pb 1.42, Zn 40.25.
Rolled: 75,000 TS; 55,000 YS; 18 El; 31 RA.

BRASS STEEL
American Nickeloid Co.
brass coated steel.
For fabricated parts; easily formed, stamped, drawn.

BRASS TIN
American Nickeloid Co.
Brass coated tin.
For fabricated parts; easily formed, stamped, drawn. *Obsolete*

BRASS, DRAWING
Anaconda Co.
Copper. Cu 67-70, Zn 30-33.
Sand cast: 40,000 TS; 23,000 YS; 35 El; 35 RA; 45 Brin. Hard rolled: 67,200 TS; 67,000 YS; 15 El; 50 RA; 145 Brin. For seamless tubes. Deep drawing.

BRASS, ESCUTCHEON
American manufacture
Cu 64.5, Zn 35.07, Pb 0.43.
For escutcheon pins; yellow brass.

BRASS, HIGH
Anaconda Co.
Copper. Cu 65, Zn 35.
Cold rolled: 47,000-75,000 TS; 20,000-60,000 YS; 5-60 El; 5-75 RA; 45-180 Brin. Used for drawing, forming and spinning parts. High strength.

BRASS, KRUPP NICKEL
Krupp Stahl AG
copper. Cu 48.5, Zn 24.3, Ni 24.3, Fe 2.9.
For ornamental and corrosion resistant parts. *Obsolete*

BRASS, LANCASHIRE
English manufacture
Cu 73, Zn 25, Pb 2.
For parts to be brazed or soldered; free-cutting.

BRASS, LEADED HIGH
Anaconda Co.
Copper. Cu 65, Pb 0.5-1.5, bal Zn.
For cupped, formed or drawn parts; does not foul cutting tools.

BRASS, LEADED LOW
American manufacture
Cu 78, Zn 20, Pb 1.7, traces Fe.
Hard: 80,000 TS; 5 El. Soft: 40,000 TS; 35 El. For rivets, pins, wire; free-cutting.

BRASS, LEADED SCREEN WIRE
American manufacture
Cu 69, Zn 30, Pb 1, traces Fe.
For screens, hardware; high ductility.

BRASS, MANGANESE
American manufacture
Cu 54-85, Zn 2-40, Mn 1-25, Fe 0-2.4, Ni 0-2.5, Al, Pb.
For propellers, marine parts, bolts, fittings; corrosion resistant.

BRASS, MANGANESE NICKEL
American manufacture
Cu 53-66, Zn 5-40, Ni 2-18, Mn 1.5-20, Al, Fe, Sn, Pb.
For condenser tubes, pump parts, corrosion resisting fittings; corrosion resistant.

BRASS, PIN WIRE
American manufacture
Cu 61, Zn 39.
Wire: 51,000 TS; 20 El. For brass pins, condenser tubes, water pipes; corrosion resistant.

BRASSOID
American Nickeloid Co.
Brass coated Zn.
For fabricated parts; easily formed, stamped, drawn. *Obsolete*

BRASTIL
NL Industries
Si 4-5, Cu 81, bal Zn.
Cast: 90,000-95,000 TS; 70,000 YS; 10-7 El; 160-180 Brin. For bearings, general castings; corrosion resistance. *Obsolete*

BRAZE 051
Handy & Harmon
Ag 5, Cu 58, Zn 37.
MP: 1545°F; FP: 1615°F. Brazing nichrome resistance elements, or simultaneous brazing and heat treating of steel.

BRAZE 053
Handy & Harmon
Low melting. Ag 5, Cd 95.
MP: 640°F; FP: 740°F. High temperature solder for medium strength joints.

BRAZE 056 (TEC Z)
Handy & Harmon
Low melting. Ag 4.5-5.5, Zn 15.6-17.6, Cd 77.9-78.9, 0.15 others max.
MP: 480°F; FP: 600°F. High temperature solder for medium strength joints.

BRAZE 071
Handy & Harmon
Copper. Ag 7, Cu 85, Sn 8.
MP: 1225°F; FP: 1805°F. Used when heat treatment follows brazing.

BRAZE 090

Handy & Harmon
Copper. Ag 9, Cu 53, Zn 38.
MP: 1410°F; FP: 1565°F. For brazing copper based alloys such as band instruments.

BRAZE 200

Handy & Harmon
Copper. Ag 20, Cu 45, Zn 30, Cd 5.
MP: 1140°F; FP: 1500°F. For brazing ferrous and non-ferrous alloys, good color match for yellow brass.

BRAZE 202

Handy & Harmon
Copper. Ag 20, Cu 45, Zn 35.
MP: 1315°F. For brazing steel.

BRAZE 250

Handy & Harmon
Copper. Ag 25, Cu 52.5, Zn 22.5.
MP· 1250°F; FP: 1575°F. For brazing ferrous and non ferrous metals that are not damaged by 1600°F temperature.

BRAZE 251

Handy & Harmon
Ag 25, Cu 57.5, Zn 17.5.
For brazing alloy; M.P. 1255-1625 F. *Obsolete*

BRAZE 252

Handy & Harmon
Ag 25, Cu 38, Zn 33, Mn 2, Ni 2.
MP: 1305°F; FP: 1475°F. Filler metal for tungsten carbide, stainless steel and steel.

BRAZE 255

Handy & Harmon
Ag 25, Cu 40, Zn 33, Sn 2.
MP: 1270°F; FP: 1435°F. Filler metal for ferrous and nonferrous joints not requiring high ductility or impact strength.

BRAZE 285

Handy & Harmon
Zinc. Ag 28.5, Cu 32.3, Zn 34.2, Mn 5.
MP: 1305°F; FP: 1385°F. Economical narrow melt range filler metal for ferrous and non-ferrous alloys.

BRAZE 300

Handy & Harmon
Copper. Ag 30, Cu 38, Zn 32.
MP: 1250°F; FP: 1410°F. For brazing steel and non-ferrous alloys melting above 1450°F, as nickel-silver knife handles, electrical equipment. AWS BAg-20.

BRAZE 351

Handy & Harmon
Ag 35, Cu 32, Zn 33.
MP: 1265°F; FP: 1390°F. Intermediate temperature filler metal for use with ferrous and nonferrous materials.

BRAZE 380

Handy & Harmon
Ag 38, Cu 32, Zn 28, Sn 2.
MP: 1200°F; FP: 1330°F. Free flowing, cadmium-free filler metal used with ferrous and nonferrous base metals.

BRAZE 400

Handy & Harmon
Precious metal. Ag 40, Cu 36, Zn 24.
MP: 1235°F; FP: 1415°F. For brazing copper base alloys, Monel, mild steel; can braze wide joints.

BRAZE 401

Handy & Harmon
Precious metal. Ag 40, Cu 30, Zn 30.
MP: 1245°F; FP: 1340°F. For brazing copper alloys, mild steel, nickel and Monel, wide gap joints.

BRAZE 402

Handy & Harmon
Ag 40, Cu 30, Zn 28, Sn 2.
MP: 1200°F; FP: 1310°F. Free-flowing medium temperature filler metal for ferrous and nonferrous alloys.

BRAZE 403

Handy & Harmon
Precious metal. Ag 40, Cu 30, Zn 28, Ni 2.
MP: 1220°F; FP: 1435°F. For brazing tungsten carbide tool tips, stainless food handling equipment. AWS BAg-4.

BRAZE 404

Handy & Harmon
Precious metal. Ag 40, Cu 30, Zn 25, Ni 5.
MP: 1220°F; FP: 1580°F. For brazing tungsten carbide tips; stainless steel.

BRAZE 440

Handy & Harmon
Ag 44, Cu 27, Zn 13, Cd 15, P 1.
MP: 1100°F; FP: 1220°F. Low melting filler metal for brazing electrical contacts and molybdenum or copper-tungsten electrodes.

BRAZE 450

Handy & Harmon
Precious metal. Ag 45, Cu 30, Zn 25.
MP: 1225°F; FP: 1370°F. For brazing ships piping, band instruments, aircraft engine oil coolers, brass lamps. AWS BAg-5.

BRAZE 495

Handy & Harmon
Precious metal. Ag 49, Cu 16, Zn 23, Mn 7.5, Ni 4.5.
MP: 1160°F; FP: 1300°F. For low temperature brazing of tungsten carbides and stainless steel.

BRAZE 501 (ETX)

Handy & Harmon
Precious metal. Ag 50, Cu 34, Zn 16.
MP: 1250°F; FP: 1425°F. For brazing steam turbine blading and heavily galvanized or tinned steel. AWS BAg-6.

BRAZE 502, 503 (VTG)

Handy & Harmon
Ag 50, Cu 50.
MP: 1435°F; FP: 1600°F. For applications similar to Brazes 720 and 721, except where better gap filling is needed.

BRAZE 505

Handy & Harmon
Precious metal. Ag 50, Cu 20, Zn 28, Ni 2.
MP: 1220°F; FP: 1305°F. For brazing 300 series stainless food handling equipment.

BRAZE 506

Handy & Harmon
Ag 50, Cu 22, Zn 20, Cd 7, Sn 1.
MP: 1165°F; FP: 1240°F. Low melting filler metal for brazing silver and iron alloys; also for joining synthetic diamond tools.

BRAZE 541

Handy & Harmon
Precious metal. Ag 54, Cu 40, Zn 5, Ni 1.
MP: 1340°F; FP: 1575°F. For atmosphere furnace brazing steel and stainless; for applications up to 700°F. AMS 4772B; AWS BAg-13.

BRAZE 559

Handy & Harmon
Precious metal. Ag 56, Cu 42, Ni 2.
MP: 1420°F; FP: 1640°F. For furnace brazing of steels and high temperature alloys where zinc fumes are undesirable. AMS-4765; AWS BAg-13a.

BRAZE 560

Handy & Harmon
Precious metal. Ag 56, Cu 22, Zn 17, Sn 5.
MP: 1145°F; FP: 1205°F. For brazing food handling equipment requiring low melting cadmium free alloy. AWS BAg-7.

BRAZE 580

Handy & Harmon
Precious metal. Ag 57.5, Cu 32.5, Sn 7, Mn 3.
MP: 1120°F; FP: 1345°F. For brazing tungsten and chrome carbides, and vacuum brazing of high manganese stainless steel.

BRAZE 600

Handy & Harmon
Precious metal. Ag 60, Cu 25, Zn 15.
MP: 1245°F; FP: 1325°F. For brazing Monel and other nickel base alloys, silverware.

BRAZE 603

Handy & Harmon
Precious metal.
Now BRAZE 603, 604 (VTG).

BRAZE 603, 604 (VTG)

Handy & Harmon
Precious metal. Ag 60, Cu 30, Sn 10.
MP: 1115°F; FP: 1325°F. For brazing marine heat exchangers, ferrous and non-ferrous alloys, vacuum tube seals. AMS 4773A; AWS BAg-18.

BRAZE 630

Handy & Harmon
Precious metal. Ag 63, Cu 28.5, Sn 6, Ni 2.5.
MP: 1275°F; FP: 1475°F. For brazing 400 series stainless, alloy steels, food handling equipment. Good for combined furnace brazing and hardening for low alloy steel. AMS 4774A.

BRAZE 650

Handy & Harmon
Precious metal. Ag 65, Cu 20, Zn 15.
MP: 1240°F; FP: 1325°F. For brazing silverware, iron and nickel alloys. ASME BAg-9.

BRAZE 655

Handy & Harmon
Precious metal. Ag 65, Cu 28, Mn 5, Ni 2.
MP: 1385°F; FP: 1560°F. For jet engine components, brazing alloy.

BRAZE 700

Handy & Harmon
Precious metal. Ag 70, Cu 20, Zn 10.
MP: 1275°F; FP: 1360°F. For brazing silverware when subsequent joints are made with Braze 650. ASME BAg-10.

BRAZE 715, 716 (VTG)

Handy & Harmon
Ag 71.5, Cu 28, Ni 0.5.
MP: 1435°F; FP: 1465°F. Filler metal and high conductivity, similar to Braze 720, but suitable for both ferrous and nonferrous alloys.

BRAZE 720 (BT)

Handy & Harmon
Precious metal.
Now BRAZE 720, 721 (VTG).

BRAZE 720, 721 (VTG)

Handy & Harmon
Precious metal. Ag 72, Cu 28.
MP and FP: 1435°F. Eutectic alloy for brazing electric components requiring highest electrical and thermal conductivity; low volatile materials and impurities; for furnace brazing. AWS BAg-8.

BRAZE 750
Handy & Harmon
Precious metal. Ag 75, Cu 22, Zn 3.
MP: 1365°F; FP: 1450°F. For brazing silverware, for step brazing or subsequent enameling; iron and nickel based alloys. ASME BAg-11.

BRAZE 752
Handy & Harmon
Ag 75, Zn 25.
For brazing alloy; M.P. 1300-1330 F. *Obsolete*

BRAZE 800
Handy & Harmon
Precious metal. Ag 80, Cu 16, Zn 4.
MP: 1340°F; FP: 1490°F. For brazing silver, iron and nickel based alloys.

BRAZE 852
Handy & Harmon
Precious metal. Ag 85, Mn 15.
MP: 1760°F; FP: 1780°F. For brazing stainless steel; Stellite and Inconel.

BRAZE 999
Handy & Harmon
Precious metal. Ag 99.9.
MP: 1761°F; FP: 1061°C. For metallizing ceramics to be used as conductors.

BRAZE ATT
Now BRAZE 200 (ATT).

BRAZE BT
Now BRAZE 720 (BT).

BRAZE DE
Now BRAZE 450 DE.

BRAZE DT
Now BRAZE 400 DT.

BRAZE EASY
Now BRAZE 650 (EASY).

BRAZE ETX
Now BRAZE 501 ETX.

BRAZE HARD
Now BRAZE 750 (HARD).

BRAZE IT
Now BRAZE 800 (IT).

BRAZE MEDIUM
Now BRAZE 700 (MEDIUM).

BRAZE NE
Now BRAZE 250 NE.

BRAZE RT
Now BRAZE 600 (RT).

BRAZE SS
Now BRAZE 403 (SS).

BRAZE TL
Now BRAZE 090 (TL).

BRAZINAL
A.E. Ullman & Associates
Aluminum. Si, bal Al.
Cast: 45,000 TS; 80 Brin. Brazing alloy for aluminum.

BRAZING BRASS
Anaconda Co.
Copper. Cu 75-80, Zn 20-25.
For brazing.

BRAZING METAL
Anaconda Co.
Copper. Cu 84-86, Zn 14-16.
Cast: 50,000-95,000 TS; 7 El; 180 Brin. For brazed joints on steel parts. Tough and ductile.

BRAZING METAL "F"
English manufacture
Cu 85, Zn 15.
For brazing metal, water pipe, architectural purposes; low melting point.

BREARLEY B
Dunford Hadfields Ltd.
C 0-0.15, Cr 14-16, bal Fe.
Heat treated: 68,000-80,000 TS; 20 El; 130-153 Brin. For pressings, cooking utensils; Type 430; corrosion resistant. *Obsolete*

BREARLEY B.B.H.
Dunford Hadfields Ltd.
C 0.35-0.4, Cr 11-12.5, bal Fe.
Heat treated: 212,000-280,000 TS; 450-580 Brin. For ball bearings and engine valves; Type 420; stainless steels. *Obsolete*

BREARLEY C
Dunford Hadfields Ltd.
C 0-0.15, Cr 16-18, bal Fe.
Annealed: 60,000-80,000 TS; 50,000 YS; 30 El; 50 RA; 130-155 Brin. For boiler and furnace fittings, kitchen sinks; Type 430; corrosion resistant. *Obsolete*

BREARLY "K"
Dunford Hadfields Ltd.
C 0.25, Cr 16-18, bal Fe.
67,000-90,000 TS. For chemical plant, particularly for HNO 3 ; corrosion resistant. *Obsolete*

BREARLY A
Dunford Hadfields Ltd.
C 0-0.12, Cr 12-14, bal Fe.
Heat treated: 80,000-100,000 TS; 25 El; 152-207 Brin. For turbine blades and shrouds, tanks, cooking utensils; Type 405; corrosion resistant. *Obsolete*

BREARLY PATENT
American manufacture
Cr 9-16, C 0-0.7, bal Fe.
For cutlery, surgical instruments; stainless and corrosion resisting.

BREDA CM10
Breda Co.
C 0.09-0.12, Cr 13, Si 0-0.5, bal Fe.
Annealed: 75,000 TS; 40,000 YS; 35 El; 70 RA; 155 Brin. For valves, cutlery, surgical and dental instruments; Type 403; corrosion resistant.

BREDA CM13
Breda Co.
C 0.26-0.37, Cr 13, bal Fe.
Annealed: 95,000 TS; 50,000 YS; 25 El; 55 RA; 195 Brin. For valves, cutlery, surgical and dental instruments; Type 420; corrosion resistant.

BREDA CMC
Breda Co.
C 0.38-0.45, Cr 13, bal Fe.
Annealed: 100,000 TS; 55,000 YS; 23 El; 52 RA; 210 Brin. For valves, cutlery, surgical and dental instruments; Type 420; corrosion resistant.

BREDA LYS
Breda Co.
C 0.07-0.12, Cr 17, bal Fe.
Annealed: 80,000 TS; 50,000 YS; 25 El; 50 RA; 150 Brin. For oil refining equipment, oil burners and heaters; Type 430; corrosion resistant.

BREDA NK1
Breda Co.
C 0-0.25, Cr 17, Ni 2, bal Fe.
Annealed: 125,000 TS; 95,000 YS; 20 El; 55 RA; 260 Brin. For pumps, marine hardware, valves; Type 431; corrosion and heat resistant.

BREDA RAS
Breda Co.
C 0.4, Cr 10, Mo 1, Si 2.5, bal Fe.
For oil refinery equipment; heat resistant.

BREDA RF302
Breda Co.
C 0.35, Cr 25, bal Fe.
Annealed: 85,000 TS; 50,000 YS; 30 El; 55 RA; 180 Brin. For furnace parts and equipment; Type 446; heat resistant.

BREDA-NK2
Breda Co.
C 0-0.2, Cr 24, Ni 12, bal Fe.
Annealed: 90,000 TS; 40,000 YS; 50 El; 65 RA; 170 Brin. For furnace parts, pumps, oil burners, heat treat boxes; Type 309; stainless, austenitic.

BREDA-NK3
Breda Co.
C 0-0.25, Cr 25, Ni 20, Si 0-1.5, bal Fe.
Annealed: 100,000 TS; 45,000 YS; 50 El; 65 RA; 185 Brin. For furnace parts, pumps, valves, turbine and jet parts; Type 310; stainless, austenitic.

BRESCIANA AK
Bresciana Metallurgica
C 0.09-0.12, Cr 13, Si 0-0.5, bal Fe.
Annealed: 75,000 TS; 40,000 YS; 35 El; 70 RA; 155 Brin. For turbine blades, valves, cutlery knives; Type 403 and 410; corrosion resistant.

BRESCIANA ATK
Bresciana Metallurgica
C 0.07-0.12, Cr 17, bal Fe.
Annealed 80,000 TS; 50,000 YS; 25 El; 50 RA; 150 Brin. For oil refinery equipment, oil burners and heaters; Type 430; corrosion resistant.

BRESCIANA AU
Bresciana Metallurgica
Stainless steel. C 0.17-0.25, Cr 18, Ni 8, bal Fe.
Annealed: 80,000 TS; 35,000 YS; 55 El; 75 RA; 150 Brin. For chemical plant equipment, tanks, mixers; Type 302; austenitic.

BRESCIANA AUS
Bresciana Metallurgica
Stainless steel. C 0.11-0.16, Cr 18, Ni 8, bal Fe.
Annealed: 80,000 TS; 35,000 YS; 55 El; 75 RA; 150 Brin. For chemical plant equipment, tanks, mixers; Type 301 and 302; austenitic.

BRESCIANA AUT
Bresciana Metallurgica
Stainless steel. C 0.09-0.15, Cr 18, Ni 9, bal Fe.
Annealed: 85,000 TS; 35,000 YS; 55 El; 65 RA; 150 Brin. For welded chemical plant equipment, tanks; Type 321; austenitic.

BRESCIANA CNV
Bresciana Metallurgica
C 0-0.2, Cr 24, Ni 12, bal Fe.
Annealed: 90,000 TS; 40,000 YS; 50 El; 65 RA; 170 Brin. For furnace parts, heat treating boxes, pumps, oil burners; Type 309; corrosion and heat resistant.

BRESCIANA INC
Bresciana Metallurgica
C 0.26-0.37, Cr 13, bal Fe.
Annealed: 100,000 TS; 55,000 YS; 22 El; 52 RA; 200 Brin. For valves, cutlery, surgical instruments; Type 420; corrosion resistant.

BRESCIANA KK
Bresciana Metallurgica
C 0-0.25, Cr 25, bal Fe.
Annealed: 85,000 TS; 50,000 YS; 30 El; 55 RA; 180 Brin. For furnace parts and equipment; Type 446; heat resistant.

BRESCIANA MIC
Bresciana Metallurgica
C 0.38-0.45, Cr 13, bal Fe.
Annealed: 110,000 TS; 60,000 YS; 20 El; 50 RA; 220 Brin. For valves, cutlery, surgical instruments; Type 420; corrosion resistant.

BRESCIANA VV
Bresciana Metallurgica
C 0-0.25, Cr 25, Ni 20, Si 0-1.5, bal Fe.
Annealed: 100,000 TS; 45,000 YS; 50 El; 65 RA; 185 Brin. For furnace parts, valves, pumps, turbine components; Type 310; corrosion and heat resistant.

BRICROME
British Piston Ring Co.
TC 1.75, Si 1.5, Mn 0.8, Cr 30, bal Fe.
Cast: 72,000 TS; 310 Brin. For piston rings, cylinder liners; wear resistant.

BRIDGE BRONZE A
English manufacture
Cu 80, Sn 20, P 0.1.
3 El; 1.8 RA; 80 Brin. For tubes; P-Bronze.

BRIDGE BRONZE B
English manufacture
Cu 85, Sn 15, P 0.1.
50 Brin. For springs, electrical parts; P-Bronze.

BRIDGE BRONZE C
English manufacture
Cu 80, Sn 10, P 0.7-1, Pb 10.
15 El; 74 Brin. For bearings; heavy duty P-Bronze.

BRIDGE BRONZE D
English manufacture
Cu 88, Sn 10, P 0.3, Zn 2.
For bearings, gears, worm wheels; P-Bronze.

BRIDGEPORT BRONZE
Olin Brass, Indianapolis
Copper. Cu 60, Sn 0.75, bal Zn.
Drawn: 75,000 psi TS; 15 El. For condenser tubes, water pipes, nuts, and bolts; corrosion resistant.

BRIDGEPORT FORGING ROD
Olin Brass, Indianapolis
Copper. Cu 59.5, Pb 1.75, Sn 0.2, Zn 38.75.
Annealed: 52,000 psi TS; 20,000 psi YS; 45 El. For forgings and hardware; free cutting.

BRIDGEPORT JEWELRY BRONZE NO. 92
Olin Brass, Indianapolis
Copper. Cu 89, Sn 1.9, Zn 9.1.
Hard: 90,000 psi TS; 72,000 psi YS; 3 El; 90 Rock B. For costume jewelry and electrical contacts; resembles red gold.

BRIDGEPORT MANGANESE BRONZE 19
Olin Brass, Indianapolis
Cu 58.5, Fe 1, Sn 0.75, Mn 0.3, bal Zn.
Annealed: 60,000 psi TS; 15 El. Drawn: 85,000 psi TS; 35 El. For welding rod, forgings, and brazing; corrosion resistant. *Obsolete*

BRIDGEPORT NO. 1
Olin Brass, Indianapolis
Copper. Cu 66, Zn 34.
Hard: 74,000 psi TS; 60,000 psi YS; 8 El. Soft: 47,000 psi TS; 15,000 psi YS; 62 El. For hardware and general drawing and forming; yellow brass.

BRIDGEPORT NO. 100
Olin Brass, Indianapolis
Copper. Ag, bal Cu.
Soft: 32,000 psi TS; 10,000 psi YS; 45 El; 40 Rock F. For radio, television, radar and computer parts; bus conductors. Silver bearing copper.

BRIDGEPORT NO. 101
Olin Brass, Indianapolis
Copper. Ag, bal Cu.
Soft: 32,000 psi TS; 10,000 psi YS; 45 El; 40 Rock F. For radio, television, radar and computer parts; wave guides and bus conductors. Does not soften as readily as pure copper during soldering. Silver bearing copper.

BRIDGEPORT NO. 102
Olin Brass, Indianapolis
Copper. Cu 99.9-100.
Hard: 50,000 psi TS; 45,000 psi YS; 4 El; 100 Brin. Soft: 34,000 psi TS; 11,000 psi YS; 45 El; 45 Brin. For electrical uses and hollowware; tough pitch copper.

BRIDGEPORT NO. 103
Olin Brass, Indianapolis
Copper. 99.92 Cu min.
Hard: 53,000 psi TS; 51,000 psi YS; 3 El; 56 Rock B. Soft: 32,000 psi TS; 8000 psi YS; 43 El; 33 Rock F. For electronic and radar parts, bus conductors, wave guides, transistor and rectifier bases, and heat sinks. OFHC copper; 100% electrical conductivity.

BRIDGEPORT NO. 104
Olin Brass, Indianapolis
Copper. bal Cu.
For electronic parts, vacuum tube parts; oxygen free copper.

BRIDGEPORT NO. 105
Olin Brass, Indianapolis
Copper. Cu 99.9-99.95, P 0.02.
Hard: 55,000 psi TS; 50,000 psi YS; 8 El; 107 Brin. Soft: 32,000 psi TS; 10,000 psi YS; 45 El; 40 Brin. For deep drawing and water tubes; deoxidized copper. *Obsolete*

BRIDGEPORT NO. 106
Olin Brass, Indianapolis
99.4 Cu min.
Hard: 55,000 psi TS; 50,000 psi YS; 8 El; 60 Rock B. Soft: 32,000 psi TS; 10,000 psi YS; 45 El; 40 Rock F. For tubular bus bars. Deoxidized copper. *Obsolete*

BRIDGEPORT NO. 108
Olin Brass, Indianapolis
Copper. P 0.02, As 0.3, 99.4 Cu min.
Hard: 40,000 psi TS; 32,000 psi YS; 25 El; 77 Rock F. For condensers, heat exchangers. Inhibited. Arsenical deoxidized copper.

BRIDGEPORT NO. 110
Olin Brass, Indianapolis
Copper. P 0.02, Cu 99.98.
Hard: 50,000 psi TS; 45,000 psi YS; 14 El; 40 Rock B. Soft: 32,000 psi TS; 10,000 psi YS; 45 El; 40 Rock F. For heat sinks. 85% electrical conductivity; corrosion resistant.

BRIDGEPORT NO. 112
Olin Brass, Indianapolis
Copper. Cu 99.38, P 0.02, Te 0.6.
1/2 H-temper: 44,000 psi TS; 42,000 psi YS; 20 El; 78 Brin. For screw machine products; high conductivity, free-cutting.

BRIDGEPORT NO. 120
Olin Brass, Indianapolis
Cu 99.7, S 0.3.
1/2 H-temper: 44,000 psi TS; 42,000 psi YS; 20 El; 78 Brin. For screw machine products, fasteners; high conductivity, free-cutting. *Obsolete*

BRIDGEPORT NO. 1232
Olin Brass, Indianapolis
Copper. Si 3, Cu 97.
Welded: 55,000 psi TS; 55 El. For welding rod; bronze.

BRIDGEPORT NO. 133
Olin Brass, Indianapolis
Copper. Cu 59, Pb 1.75, Zn 39.25.
Hard: 72,000 psi TS; 50,000 psi YS; 15 El; 144 Brin. Soft: 52,000 psi TS; 20,000 psi YS; 45 El; 72 Brin. For hardware, machine tool parts; BPT forging rod.

BRIDGEPORT NO. 134
Olin Brass, Indianapolis
Copper. Pb 2.25, Zn 38, Cu 59.75.
Rolled: 75,000 psi TS; 52,000 psi YS; 22 El; 80 Rock B. For hot forged and machined parts, hardware, fasteners, and bolts. 27% electrical conductivity. Forging brass; corrosion resistant; free-cutting.

BRIDGEPORT NO. 14
Olin Brass, Indianapolis
Copper. Zn 40.25, Cu 59.75.
Hard: 72,000 psi TS; 50,000 psi YS; 25 El; 78 Rock B. Soft: 54,000 psi TS; 21,000 psi YS; 50 El; 80 Rock F. For architectural trim, fasteners, condenser plates, and heat exchangers. Good hot working qualities; high corrosion resistance.

BRIDGEPORT NO. 141
Olin Brass, Indianapolis
Copper. Cu 62.25, Zn 37.75.
Annealed: 54,000 psi TS; 21,000 psi YS; 45 El; 80 Rock F. Half hard: 70,000 psi TS; 50,000 psi YS; 10 El; 75 Rock B. For architectural trim, fasteners, hardware, heat exchangers, brazing rod, and condenser tubes. Muntz metal; good corrosion resistance.

BRIDGEPORT NO. 142
Olin Brass, Indianapolis
Copper. Cu 62, Zn 38.
Hard: 56,000 psi TS; 23,000 psi YS; 50 El; 82 Rock F. For condensers, heat exchangers. Free from dezincification. Arsenical muntz metal; inhibited.

BRIDGEPORT NO. 1426
Olin Brass, Indianapolis
Copper. Cu 95.5, Al 3.5, Si 1.
For tubing; corrosion resistant.

BRIDGEPORT NO. 1552
Olin Brass, Indianapolis
Copper. Sn 1.9, Zn 6.1, Cu 92.
For fuse clips, springs, diaphragms; good electrical conductivity.

BRIDGEPORT NO. 16
Olin Brass, Indianapolis
Copper. Cu 65, bal Zn.
For wood and machine screws; yellow brass.

BRIDGEPORT NO. 18
Olin Brass, Indianapolis
Copper. Cu 67, Pb 0.5, Zn 32.5.
Hard: 75,000 psi TS; 60,000 psi YS; 7 El. For garden sprayers; free-cutting.

BRIDGEPORT NO. 19
Olin Brass, Indianapolis
Copper. Mn 0.3, Sn 0.7, Fe 1, Cu 58.5, Zn 39.5.
Hard: 83,000 psi TS; 55,000 psi YS; 25 El; 165 Brin. Soft: 72,000 psi TS; 30,000 psi YS; 45 El; 100 Brin. For bolts, valve parts, and tie rods; manganese bronze.

BRIDGEPORT NO. 192
Olin Brass, Indianapolis
Copper. Cu 58, Sn 0.9, Fe 0.6, Zn 39.8.
For welding rod, brazing; non-fuming.

BRIDGEPORT NO. 2
Olin Brass, Indianapolis
Copper. Pb 1.8, Cu 63, Zn 35.2.
Hard: 55,000 psi TS; 42,000 psi YS; 30 El; 75 Rock B. Soft: 45,000 psi TS; 15,000 psi YS; 55 El; 65 Rock F. For screw machine parts requiring some cold working such as roll threading and knurling; binding posts and toggle switch parts. High leaded brass; free-cutting.

BRIDGEPORT NO. 24
Olin Brass, Indianapolis
Copper. Cu 60, Sn 0.75, Zn 39.25.
Hard: 75,000 psi TS; 53,000 psi YS; 20 El; 156 Brin. Soft: 57,000 psi TS; 25,000 psi YS; 47 El; 100 Brin. For nuts, bolts, marine hardware; naval brass.

BRIDGEPORT NO. 25
Olin Brass, Indianapolis
Copper. Cu 90, Zn 10.
Soft: 38,000 psi TS; 12,000 psi YS; 45 El. For threaded fasteners; resists season cracking.

BRIDGEPORT NO. 26
Olin Brass, Indianapolis
Copper. Cu 95, Zn 5.
Hard: 56,000 psi TS; 50,000 psi YS; 5 El; 114 Brin. Soft: 35,000 psi TS; 11,000 psi YS; 45 El; 50 Brin. For jewelry; gilding metal.

BRIDGEPORT NO. 28
Olin Brass, Indianapolis
Copper. Cu 60, Pb 0.6, Sn 0.65, Zn 38.75.
Hard: 75,000 psi TS; 53,000 psi YS; 20 El; 156 Brin. Soft: 57,000 psi TS; 25,000 psi YS; 47 El; 100 Brin. For marine hardware; leaded naval brass.

BRIDGEPORT NO. 285
Olin Brass, Indianapolis
Copper. Cu 82, Si 4.25, Pb 0.15, Zn 13.6.
Annealed: 95,000 psi TS; 50,000 psi YS; 30 El; 185 Brin. For valve stems and pump parts; corrosion and wear resistant.

BRIDGEPORT NO. 29
Olin Brass, Indianapolis
Copper. Cu 60, Sn 0.65, Pb 1.75, Zn 37.6.
Hard: 75,000 psi TS; 53,000 psi YS; 15 El; 162 Brin. For hardware; leaded naval brass.

BRIDGEPORT NO. 3
Olin Brass, Indianapolis
Copper. Cu 65.5, Zn 33.9, Pb 0.6.
Hard: 74,000 psi TS; 60,000 psi YS; 8 El. Soft: 49,000 psi TS; 17,000 psi YS; 57 El. For hardware, bolts, and studs. Low leaded brass.

BRIDGEPORT NO. 30
Olin Brass, Indianapolis
Copper. Cu 71, As 0.03, Sn 1, Zn 27.97.
Soft: 53,000 psi TS; 22,000 psi YS; 65 El; 68 Brin. For condenser tubes; arsenical admiralty metal.

BRIDGEPORT NO. 32
Olin Brass, Indianapolis
Copper. Cu 98.53, P 0.07, Sn 1.4.
Hard: 65,000 psi TS; 50,000 psi YS; 8 El; 137 Brin. Soft: 40,000 psi TS; 14,000 psi YS; 48 El; 56 Brin. For flexible hose; phosphor-bronze; corrosion resistant.

BRIDGEPORT NO. 34
Olin Brass, Indianapolis
Cu 89.85, P 0.15, Sn 10.
For springs; phosphor-bronze, corrosion resistant. *Obsolete*

BRIDGEPORT NO. 35
Olin Brass, Indianapolis
Sn 8, P 0.1, Cu 91.9.
Hard: 93,000 psi TS; 72,000 psi YS; 10 El; 93 Rock B. Soft: 60,000 psi TS; 24,000 psi YS; 63 El; 82 Rock F. For springs, fuse clips, contacts, meter parts, diaphragms, bellows, snap switches, cutter pins, and terminals. Phosphor bronze; 13% electrical conductivity. *Obsolete*

BRIDGEPORT NO. 36
Olin Brass, Indianapolis
Copper. Sn 5.5, P 0.15, Cu 94.35.
Annealed: 47,000 psi TS; 19,000 psi YS; 64 El; 26 Rock B. Hard: 81,000 psi TS; 75,000 psi YS; 10 El; 87 Rock B. Spring: 100,000 psi TS; 80,000 psi YS; 4 El; 95 Rock B. For springs, fuse clips, contacts, relay parts, snap switches, terminals, diaphragms, bellows, and cotter pins. Phosphor-bronze; 18% electrical conductivity.

BRIDGEPORT NO. 37
Olin Brass, Indianapolis
Copper. Cu 70, Zn 30.
Hard: 76,000 psi TS; 63,000 psi YS; 8 El. Soft: 47,000 psi TS; 15,000 psi YS; 62 El. For cartridge cases and deep drawn parts; cartridge brass.

BRIDGEPORT NO. 41
Olin Brass, Indianapolis
Copper. Pb 0.75, Cu 60, Zn 39.25.
Annealed: 54,000 psi TS; 21,000 psi YS; 50 El; 80 Rock F. 1/4 hard: 72,000 psi TS; 50,000 psi YS; 25 El; 78 Rock B. For screw machine products, bolts, fasteners, hardware, and valve stems. Low leaded brass; free cutting; corrosion resistant.

BRIDGEPORT NO. 42
Olin Brass, Indianapolis
Copper. Zn 38.65, Pb 1.1, Cu 60.25.
Hard: 80,000 psi TS; 60,000 psi YS; 6 El; 85 Rock B. Soft: 54,000 psi TS; 20,000 psi YS; 80 Rock F. For screw machine products, bolts, fasteners, hardware. Corrosion resistant. Free-cutting muntz metal.

BRIDGEPORT NO. 45
Olin Brass, Indianapolis
Copper. Cu 61, Zn 38.35, Sn 0.65.
Hard: 75,000 psi TS; 50,000 psi YS; 20 El; 80 Rock B. Soft: 54,000 psi TS; 25,000 psi YS; 50 El; 55 Rock B. For bolts, fasteners, propellers, valve stems, rivets, marine hardware. Good resistance to salt water corrosion. Cold heading Naval Brass. Corrosion resistant.

BRIDGEPORT NO. 5
Olin Brass, Indianapolis
Copper. Cu 80, Zn 20.
Hard: 74,000 psi TS; 59,000 psi YS; 7 El. Soft: 44,000 psi TS; 14,000 psi YS; 50 El. For tubing, hardware, and diaphragms. Low brass.

BRIDGEPORT NO. 511
Olin Brass, Indianapolis
Copper. Cu 88.85, Ni 10, Fe 1.15.
Hard: 50,000 psi TS; 45,000 psi YS; 18 El. Soft: 45,000 psi TS; 15,000 psi YS; 40 El. For condensers and heat exchangers; cupro-nickel; corrosion resistant.

BRIDGEPORT NO. 520
Olin Brass, Indianapolis
Copper. Cu 78.85, Ni 20, Fe 0.4, Mn 0.75.
Soft: 49,000 psi TS; 14,000 psi YS; 40 El. For condensers and heat exchangers; cupro-nickel; corrosion resistant.

BRIDGEPORT NO. 521
Olin Brass, Indianapolis
Copper. Ni 20, Cu 80.
Hard: 103,000 psi TS; 101,000 psi YS; 2 El. Spring: 85,000 psi TS; 81,300 psi YS; 2.5 El. For wave guides, radar equipment, relay springs. Resists stress corrosion cracking.

BRIDGEPORT NO. 53
Olin Brass, Indianapolis
Copper. Al 5, As 0.25, Cu 94.75.
Soft: 60,000 psi TS; 22,000 psi YS; 60 El. For propellers, hardware; aluminum bronze.

BRIDGEPORT NO. 531
Olin Brass, Indianapolis
Copper. Cu 67.75, Ni 31, Fe 0.5, Mn 0.75.
Soft: 60,000 psi TS; 25,000 psi YS; 45 El; 75 Brin. For condensers and heat exchangers; cupro-nickel; corrosion resistant.

BRIDGEPORT NO. 54
Olin Brass, Indianapolis
Copper. Zn 20.87, Al 2.1, As 0.03, Cu 77.
Soft: 60,000 psi TS; 27,000 psi YS; 55 El. For hardware; corrosion resistant; aluminum bronze.

BRIDGEPORT NO. 548
Olin Brass, Indianapolis
Cu 48, Ni 9.5, Cu 42.5.
For nickel silver welding rod; corrosion resistant. *Obsolete*

BRIDGEPORT NO. 555
Olin Brass, Indianapolis
Copper. Cu 55, Ni 18, Zn 27.
Drawn: 100,000 psi TS; 3 El. Spring: 115,000 psi TS; 90,000 psi YS; 2.5 El; 99 Rock B Brin. For telephone switch parts, springs; corrosion resistant.

BRIDGEPORT NO. 558
Olin Brass, Indianapolis
Copper. Ni 12, Cu 56.5, Zn 31.5.
Annealed: 55,000 psi TS; 22,000 psi YS; 50 El; 40 Rock B. Hard: 100,000 psi TS; 90,000 psi YS; 5 El; 93 Rock B. For spring parts and contacts for telephone boards, radios, controls, springs, resistance wire, and diaphragms. Good workability and weldability. Corrosion resistant nickel silver.

BRIDGEPORT NO. 565
Olin Brass, Indianapolis
Copper. Cu 65, Ni 18, Zn 17.
Hard: 85,000 psi TS; 75,000 psi YS; 3 El; 172 Brin. Soft: 60,000 psi TS; 30,000 psi YS; 35 El; 64 Brin. For hollowware, zippers, tableware, optical goods; nickel silver, corrosion resistant.

BRIDGEPORT NO. 566
Olin Brass, Indianapolis
Copper. Cu 65, Ni 12, Mn 0.15, Zn 22.85.
For jewelry, hardware; nickel silver, corrosion resistant.

BRIDGEPORT NO. 567
Olin Brass, Indianapolis
Copper. Cu 65.25, Ni 9.75, Mn 0.15, Zn 24.85.
Hard: 86,000 psi TS; 75,000 psi YS; 4 El; 180 Brin. Soft: 60,000 psi TS; 28,000 psi YS; 36 El; 100 Brin. For hardware, hollowware, and jewelry; nickel silver, corrosion resistant.

BRIDGEPORT NO. 6
Olin Brass, Indianapolis
Copper. Zn 35.35, Pb 3.4, Cu 61.25.
Hard: 58,000 psi TS; 45,000 psi YS; 25 El; 144 Brin. Soft: 49,000 psi TS; 20,000 psi YS; 40 El; 61 Brin. For screw machine products; free-cutting brass.

BRIDGEPORT NO. 606
Olin Brass, Indianapolis
Copper. Cu 96.05, Si 3, Mn 0.95.
Hard: 94,000 psi TS; 58,000 psi YS; 8 El. Soft: 60,000 psi TS; 25,000 psi YS; 60 El. For pole line hardware and stampings; silicon bronze; corrosion resistant.

BRIDGEPORT NO. 609
Olin Brass, Indianapolis
Copper. Cu 98, Si 2.
Annealed: 45,000 psi TS; 15,000 psi YS; 50 El. Drawn: 80,000 psi TS; 50,000 psi YS; 12 El. For transmission lines, marine hardware, bolts, and welding rod; corrosion resistant.

BRIDGEPORT NO. 62
Olin Brass, Indianapolis
Copper. Cu 61.5, Zn 36.5, Pb 2.
Hard: 74,000 psi TS; 60,000 psi YS; 7 El; 150 Brin. Soft: 49,000 psi TS; 17,000 psi YS; 52 El. For hardware, bolts, nuts; free-cutting.

BRIDGEPORT NO. 63

Olin Brass, Indianapolis
Copper. Cu 65.5, Zn 33.4, Pb 1.1.
Hard: 74,000 psi TS; 60,000 psi YS; 7 El. Soft: 49,000 psi TS;
17,000 psi TS; 54 El. For hardware, bolts, and nuts; free-
cutting.

BRIDGEPORT NO. 632

Olin Brass, Indianapolis
Copper. Fe 0.1, Si 2.95, Cu 96.95.
Hard: 94,000 psi TS; 58,000 psi YS; 8 El; 93 Rock B. Soft:
60,000 psi TS; 25,000 psi YS; 60 El; 85 Rock F. For bolts,
fasteners, springs, marine fittings, and hardware. High fatigue
strength. High silicon bronze; corrosion resistant.

BRIDGEPORT NO. 635

Olin Brass, Indianapolis
Cu 97.5, Ni 1.9, Si 0.6.
Annealed: 40,000 psi TS; 12,000 psi YS; 50 El; 56 Brin. Aged:
103,000 TS; 97,000 psi YS; 17 El; 216 Brin. For cold
headed bolts, fasteners, switch gear, springs, and contacts.
Age-hardenable; corrosion resistant. *Obsolete*

BRIDGEPORT NO. 64

Olin Brass, Indianapolis
Copper. Cu 67, Zn 31.25, Pb 1.75.
Hard: 75,000 psi TS; 60,000 psi YS; 7 El; 150 Brin. Soft:
52,000 psi TS; 20,000 psi YS; 50 El; 70 Brin. For screw
machine products; leaded brass.

BRIDGEPORT NO. 69

Olin Brass, Indianapolis
Copper. Cu 70, Zn 30.
Hard: 65,000 psi TS; 25 El. Soft: 50,000 psi TS; 60 El. For
hollow rivets, and fasteners; good formability, ductile.

BRIDGEPORT NO. 707

Olin Brass, Indianapolis
Copper. Al 7, Si 2, Cu 91.
Soft: 90,000 psi TS; 50,000 psi YS; 30 El; 85 Rock B. For
pump rods, gears, valve stems, pole-line hardware, and oil
burner nozzles. High strength and corrosion resistance;
Duronze III.

BRIDGEPORT NO. 708

Olin Brass, Indianapolis
Copper. Cu 91.5, Si 1.75, Al 6.75.
Annealed: 85,000 psi TS; 40,000 psi YS; 30 El; 165 Brin. For
valve stems, bolts, and fasteners. Corrosion and wear
resistant; hot forgeable.

BRIDGEPORT NO. 712

Olin Brass, Indianapolis
Copper. Si 1, Al 3.5, Cu 95.5, bal Cu.
Aluminum silicon bronze. For bolts, cold headed nuts, and
pole line hardware. CDA C63600.

BRIDGEPORT NO. 715

Olin Brass, Indianapolis
Copper. Cu 96.75, Al 2.9, Si 0.35.
Hard: 75,000 psi TS; 15 El. Soft: 40,000 psi TS; 55 El. For
bolts, nuts, and screws. Aluminum bronze.

BRIDGEPORT NO. 77

Olin Brass, Indianapolis
Zn 30, Hg 0.05, Cu 69.95.
Hard: 78,000 psi TS; 64,000 psi YS; 8 El; 110 Brin. Soft:
47,000 psi TS; 15,000 psi YS; 65 El; 58 Brin. For heat
exchangers and condensers; resists bio-fouling and
dezincification. *Obsolete*

BRIDGEPORT NO. 819

Olin Brass, Indianapolis
Zn 4, Sn 4, Pb 4, Cu 88.
For bearings, bushings, and thrust washers; free machining
bronze. CDA C83600. *Obsolete*

BRIDGEPORT NO. 820

Olin Brass, Indianapolis
Copper. Cu 95.35, P 0.15, Sn 4.5.
Hard: 81,000 psi TS; 75,000 psi YS; 10 El; 172 Brin. Soft:
50,000 psi TS; 21,000 psi YS; 52 El; 71 Brin. For springs,
clutch discs, and diaphragms; phosphor bronze.

BRIDGEPORT NO. 828

Olin Brass, Indianapolis
Copper. Cu 92, Sn 1.9, Zn 6.1.
Hard: 72,000 psi TS; 8 El. 1/2 H-temper: 58,000 psi TS; 20 El.
Spring: 90,000 psi TS; 2 El; 172 Brin. For electrical switches,
springs, and contacts; substitute for phosphor bronze; 37%
electrical conductivity.

BRIDGEPORT NO. 835

Olin Brass, Indianapolis
Cu 98.85, Sn 0.75, Si 0.25, Mn 0.15.
For copper welding rods. *Obsolete*

BRIDGEPORT NO. 840

Olin Brass, Indianapolis
Copper. Cu 98.6, Sn 1.4.
Hard: 65,000 psi TS; 10 El. Soft: 40,000 psi TS; 40 El. For
pole line hardware, and marine parts; corrosion resistant.

BRIDGEPORT NO. 85

Olin Brass, Indianapolis
Copper. Cu 85, Zn 15.
Hard: 70,000 psi TS; 57,000 psi YS; 5 El. Soft: 40,000 psi TS;
12,000 psi TS; 47 El. For flexible hose and deep drawn parts;
red brass.

BRIDGEPORT NO. 87

Olin Brass, Indianapolis
Copper. Cu 88, Zn 12.
For costume jewelry; resembles 14K gold.

BRIDGEPORT NO. 89

Olin Brass, Indianapolis
Copper. Cu 89.5, Pb 2, Zn 8.5.
Hard: 52,000 psi TS; 45,000 psi YS; 18 El; 58 Rock B. Soft:
37,000 psi TS; 12,000 psi YS; 45 El; 55 Rock F. For hardware;
free-cutting.

BRIDGEPORT NO. 90

Olin Brass, Indianapolis
Cu 88.5, Pb 2, Ni 1, P 0.07, Zn 8.43.
Hard: 70,000 psi TS; 60,000 psi YS; 12 El; 144 Brin. For screw
machine products and hardware; free-cutting, corrosion
resistant. *Obsolete*

BRIDGEPORT NO. 92

Olin Brass, Indianapolis
Copper. Cu 89, Zn 9.1, Sn 1.9.
Hard: 90,000 psi TS; 72,000 psi YS; 3 El; 90 Rock B. For
springs, spring contacts, and slide contacts. Substitute for
phosphor bronze. Corrosion resistant.

BRIDGEPORT NO. 980

Olin Brass, Indianapolis
Cu 99, Cd 1.
Hard: 55,000 psi TS; 45,000 psi YS; 6 El; 65 Rock B. Soft:
35,000 psi TS; 15,000 psi YS; 60 El; 35 Rock F. For electrical
applications; high electrical conductivity. *Obsolete*

BRIDGEPORT NO. 985

Olin Brass, Indianapolis
Cu 99.1, Cd 0.9.
Hard: 55,000 psi TS; 48,000 psi YS; 6 El; 116 Brin. Soft:
37,000 psi TS; 12,000 psi YS; 50 El; 45 Brin. For trolley wire
and marine hardware; cadmium copper. *Obsolete*

BRIDGEPORT NO. 992

Olin Brass, Indianapolis
Copper. Zr 0.13, Cu 99.8.
Annealed: 30,000 psi TS; 10,000 psi YS; 45 El; 40 Brin. Aged:
60,000 psi TS; 55,000 psi YS; 15 El; 116 Brin. For resistance
welding electrodes and grid wires. Heat treatable; high
conductivity.

BRIDGIT SOLDER

J.W. Harris Co., Inc.
Ag 0-2, Cu 3, Ni 0-1, Sb 5, bal Sn.
Melting range: 460°F solidus; 630°F liquidus. ASTM B32-89.
Lead-free, nickel/silver bearing solder for use in potable
water systems.

BRIGHT ALLOY

Hanson-Van Winkle-Munning Co.
For anodes; electroplating.

BRIGHT CAP GILDING

English manufacture
Cu 90, Zn 9.9, Pb 0.4-0.1.
For gilding, jewelry; corrosion resistant.

BRIGHT E-3

Now CRUCIBLE BRIGHT E-3.

BRIGHT EXTRUDED BRONZE

Chase Brass & Copper Co.
Cu 58.5, Zn 38.7, Pb 2.5.
As extruded: 56,000 TS; 20,000 YS; 28 El; 33,000 shear; 56
Rock B. For architectural, ornamental hardware, hinges.
Obsolete

BRIGHTRAY A

Henry Wiggin & Co. Ltd.
Ni 80, Cr 20.
118,000 TS. For electrical resistances; can be used up to
1150 C. *Obsolete*

BRIGHTRAY ALLOY C

Inco Alloys International
Nickel. Ni 80, Cr 19, Fe 0-0.5, Si 1.5.
Cold drawn: 106,000 TS; 47 El; 63 RA; 196 Brin. For electrical
resistances, heating elements. Heat resistant to 1150°C.

BRIGHTRAY ALLOY F

Inco Alloys International
Ni 37, Cr 18, Si 2, bal Fe.
For electrical resistances, heating elements, electric furnace
elements. Heat resistant to 1000°C.

BRIGHTRAY ALLOY S

Inco Alloys International
Nickel. Ni 80, Cr 20.
Annealed: 107,000 TS; 50,000 YS; 47 El; 63 RA; 187 Brin. For
heavy industrial electrical and resistance heating elements;
rod and strip.

BRIGHTRAY C

Inco Alloys International
Nickel.
Now BRIGHTRAY ALLOY C.

BRIGHTRAY F

Inco Alloys International
Now BRIGHTRAY ALLOY F.

BRIGHTRAY H

Henry Wiggin & Co. Ltd.
Ni 80, Cr 20.
Annealed: 95,000 TS; 30 El; 55 RA; 185 Brin. Cold drawn:
65,000 TS; 1 El; 1 RA; 260 Brin. For resistors, heating
elements; heat resistant to 1250 C. *Obsolete*

BRIGHTRAY N

Henry Wiggin & Co. Ltd.
Ni, Cr.
For electrical resistances, heating elements; heat resistant.
Obsolete

BRIGHTRAY S

Inco Alloys International
Nickel.
Now BRIGHTRAY ALLOY S.

BRIGHTWAY ALLOY 35

Inco Alloys International
C 0.05, Ni 35, Fe 42, Cr 20, Si 2.
Electrical resitivity: 102 micro ohm/cm at 20°C. Electrical resistance alloy for heating elements up to 1050°C. ASTM B344.

BRIGHTWAY ALLOY B

Inco Alloys International
Cr 15-17, Ni 58-60, Fe 20.
Annealed: 99,000 TS; 41,000 YS; 44 El; 66 RA; 190 Brin. For heating elements, electrical resistances; heat resistant to 950°C.

BRIGHTWAY B

Inco Alloys International
Now BRIGHTRAY ALLOY B.

BRIGHTWAY H

Henry Wiggin & Co. Ltd.
Ni 80, Cr 20.
Annealed: 95,000 TS; 30 El; 55 RA; 185 Brin. For electrical resistances, heating elements; operating range 1100-1250 C. *Obsolete*

BRILLALUMAG 3

Trefileries & Laminoirs du Havre
Mg 3, bal Al.
Annealed: 39,000 TS; 14,000 YS; 30 El; 35 Brin. Hardened: 53,000 TS; 50,000 YS; 4 El; 65 Brin. For decorative, light alloy parts; takes high polish.

BRILLALUMAG 5

Trefileries & Laminoirs du Havre
Mg 5, bal Al.
Annealed: 47,000 TS; 31,000 YS; 28 El; 48 Brin. Hardened: 69,000 TS; 50,000 YS; 3 El; 90 Brin. For decorative, light alloy parts; takes high polish.

BRILLIANT

Swedish American Steel Corp.
C 0.7, Cr 4, V 1, W 18, bal Fe.
For twist drills, reamers, taps, milling cutters; high speed steel.

BRILLIANT AXL

Fagersta Bruks Aktiebolag
C 0.7, Cr 4.5, W 18.5, Co 0-0.6, V 1.2, bal Fe.
For twist drills, lathe and planer tools, cutters; high speed steel. *Obsolete*

BRILLIANT MM

Swedish American Steel Corp.
C 0.7, Mo 5, W 6, Cr 4, V 4, bal Fe.
For cutters, dies; Type M4; high speed steel.

BRILLIANT WKE

Fagersta Bruks Aktiebolag
C 0.8, Cr 4.5, Mo 1.2, W 18.5, Co 5.5, V 1.6, bal Fe.
For milling cutters, hobs, drills, taps; high speed steel. *Obsolete*

BRILLIANT WKE EXTRA

Fagersta Bruks Aktiebolag
C 0.8, Cr 4.5, Mo 1, W 18.5, Co 10, V 1.6, bal Fe.
For milling cutters, planer and shaper tools; high speed steel. *Obsolete*

BRILLIANT WW

Swedish American Steel Corp.
C 0.7, W 18, Cr 4, V 1, bal Fe.
For drills, taps, cutters, reamers; Type T1; high speed steel.

BRILLUM

Alcoa of Great Britain
Cu 1.5, Ni 2, bal Al.
For pistons; light alloy.

BRILYBDENUM

British Piston Ring Co.
C 3.2, Si 2, Ni 0.4, Cr 0.2, bal Fe.
For piston rings; heat resistant.

BRIMCO

Bridgeport Rolling Mills Co.
Cu 87-88, Sn 2.36-3.25, P 0.016-0.06, Mn 0.21-0.23, bal Zn.
For contacts; corrosion resistant.

BRIMCO BRONZE

Bridgeport Rolling Mills Co.
Sn 2.5, Zn 10, bal Cu.
Hard: 77,000 TS; 70,000 YS; 8 El; Rock B 83. Spring: 94,000 TS; 81,000 YS; 4 El; Rock B 90. For fuse clips, fasteners, lock washers, springs, electrical components. Good formability and strength. *Obsolete*

BRIMCOLLOY-100

Bridgeport Rolling Mills Co.
Cu 86, Sn 0.5, Zn 13.5.
Half hard: 54,000 TS; 48,500 YS; 18 El; B 70 Rock. Hard: 67,500 TS; 63,000 YS; 5 El; B 78 Rock. Spring: 80,500 TS; 76,000 YS; 3 El; B 87 Rock. For electrical contacts, jewelry chain relay components. Corrosion resistant. Readily fabricated. *Obsolete*

BRIMCOLLOY-200

Bridgeport Rolling Mills Co.
Cu 86, Sn 1, Zn 13.
Hard: 77,500 TS; 65,200 YS; 5 El; 82 Rock B. Spring: 84,000 TS; 76,000 YS; 2 El; 88 Rock B. For electrical contacts, springs, fuse clips, heat exchangers, condensers. Corrosion resistant, readily fabricated.

BRIMCOLLOY-300

Bridgeport Rolling Mills Co.
Cu 86, Sn 2, Zn 12.
Hard: 75,000 TS; 67,000 YS; 8 El; B 82 Rock. Spring: 88,000 TS; 75,000 YS; 3 El; B 89 Rock. For electrical contacts, springs, fuse clips, retention devices, heat exchangers. Corrosion resistant, readily fabricated. *Obsolete*

BRIMCOLLOY-400

Bridgeport Rolling Mills Co.

Cu 87, Sn 2, Pb 2, bal Zn.
Half hard: 58,000 TS; 49,000 YS; 14 El; B 70 Rock. Hard: 76,000 TS; 70,000 YS; 10 El; B 80 Rock. For meter and gauge components, bearings, thrust washers, valve parts. Corrosion resistant, self-lubricating, free-cutting. *Obsolete*

BRIMOL

British Piston Ring Co.
C 2.8, Si 2, Ni 14, Cr 3, Cu 7, bal Fe.
For cylinder liners, rings; austenitic.

BRINALLOY (NS-8)

Lunkenheimer Co.
Ni 70, Cr 16, Si 10.
Cast: 120,000 TS; 600 Brin. For valve seats and discs; resists wear, galling and corrosion.

BRINALLOY (NS-9)

Lunkenheimer Co.
Ni 73, Cr 15, Si 8.
Cast: 332 Brin. For valve seats and discs.

BRISTAHL

Nihon Jyokiko Seikosho Goshi
C 0.2, Cr 18, Ni 8, bal Fe.
For stainless steel parts; stainless.

BRISTOL

English manufacture
Cu 60.8-75.7, Zn 24.3-39.2.
For hardware, clocks.

BRISTOL

European manufacture
C 0.3, Cr 0.9, Ni 0.65, Mo 1, Al 0.6, bal Fe.
Heat treated: 150,000-200,000 TS; 135,000-175,000 YS; 18-16 El; 55-50 RA; 300-400 Brin. For nitrided parts, gears, cams; nitriding steel.

BRISTOL ALLOY BUTTONS

English manufacture
Cu 57-61, Zn 36-37, Sn 2.7-5.3.
For buttons, fixtures; corrosion resistant.

BRISTOL BRASS

English manufacture
Cu 76-61, Zn 24-39.
For condenser tubes, pipes, ornamental purposes; same as "Princes Metal."

BRITEST

British Piston Ring Co.
TC 3.1, Si 2, Mn 0.8, Cr 1, bal Fe.
Cast: 58,000 TS; 280 Brin. For piston rings, cylinder liners; wear resistant.

BRITISH ALUMINUM NO. 12

British Aluminium Co., Ltd.
Cu, bal Al.
15,000 TS; 2 El. For light alloy parts; non-porous castings *Obsolete*

BRITISH ALUMINUM NO. 4

British Aluminium Co., Ltd.
Cu, bal Al.
Sand cast: 14,500 TS; 4 El. For light alloy parts; shock resistant, very ductile. *Obsolete*

BRITISH ALUMINUM NO. 6

British Aluminium Co., Ltd.
Cu, Zn, bal Al.
29,000 TS; 4 El. For light alloy parts; excellent machinability. *Obsolete*

BRITISH ALUMINUM NO. 6A

British Aluminium Co., Ltd.
Al.
Al.
36,000 TS; 2 El. For light alloy parts; takes high polish. *Obsolete*

BRITISH ALUMINUM PISTON ALLOY

Metal Castings Ltd.
Al 85, Cu 14, Mn 1.
For pistons, castings; non-hardenable.

BRITISH ALUMINUM PISTON ALLOY

Metal Castings Ltd.
Al 94.5, Ni 5.5.
For pistons, castings; non-hardenable.

BRITISH ALUMINUM PISTON ALLOY

Metal Castings Ltd.
Al 88, Cu 12.
For pistons, castings; non-hardenable.

BRITISH I STEEL

Manufacturer not listed.
C 0.45, Mn 0.62, Cr 2.8, Ni 0.43, Mo 0.9, V 0.2, bal Fe.
At 70°F: 167,000 TS; 3.7 El; 7.0 RA. At 1000°F: 88,000 TS; 9.2 El; 17.5 RA. At 1200°F: 48,000 TS; 25 El; 55.4 RA. For ordnance mortar tubes. Good fatigue resistance.

BRITISH NAVY ANTIFRICTION METAL

Puget Sound Metal Works
Cu 5, Sn 85, Sb 10.
For admiralty lining, plastic bearings; Babbitt.

BRITOR

Manufacturer not listed.
W, C.
For welding; welding to steel.

BRITTANIA METAL-1

A. Johnson & Co.
Sn 80-94, Zn 1.5-5, Cu 0.9, Sb 4-16.2, Pb 1-8.5, Bi 2-8.
10,000 TS; 11-19 El; 10-25 RA. For bearings, table ware;
heavy duty. *Obsolete*

BRITTANIA METAL-2

A. Johnson & Co.
Sn 91, Cu 1.5, Sb 5-7.
80,000 TS; 50 El; 8 Brin. For bearings, pewter, table ware;
corrosion resistant. *Obsolete*

BRITTANIA METAL-3

A. Johnson & Co.
Sn 85-91, Zn 0-3, Cu 0.2-1, Sb 9-11.
For bearings, table ware. *Obsolete*

BRITTANIA, ENGLISH

English manufacture
Sn 90-85, Zn 0-3, Cu 1.3, Sb 5-10.
For bearings, table ware; corrosion resistant.

BRITTANIA, GERMAN

German manufacture
Sn 70-94, Zn 0-5, Cu 1.8-5, Sb 3.7-5, Pb 0-9.
For bearings; Babbitt.

BRIX

English manufacture
Ni 60-75, Cr 15-20, Cu 5, Si 4, Ti 3, Al 2, W 1-4, B 1.
For heating elements, heat and corrosion resisting parts.

BRM

Agawam Tool Co.
C 0.8, Cr 4.15, V 2.1, W 19, V 0.6, bal Fe.
For cutting tools; oil hardening.

BRM

Ziv Steel & Wire Co.
C 0.75-0.85, Cr 4.2, V 2.1, W 19, Mo 0.6, bal Fe.
For tools and cutters; oil hardening, high speed steel.

BROACHING

Bethlehem Steel Corp.
C 0.8, Mn 0.4, Si 0.3, bal Fe.
For cutting hard stone, granite, quartz. AISI Type W1 water
hardening tool steel. *Obsolete*

BROCKHOUSE

Brockhouse Casting Co.
C 0.2, Cr 18, Ni 8, bal Fe.
For case hardening boxes, steam boiler parts; heat resisting.

BROLUNICK

English manufacture
Cu 82, Al 7, Ni 5.5, Fe 4, Mn 2.
For corrosion resisting parts; Al-Bronze; tough.

BROLUNICK BRONZE

French manufacture
Al 7, Ni 5.5, Fe 4, Mn 2, bal Cu.
For gears, housings, propellers, marine hardware; tough,
corrosion resistant.

BROMET

Australian manufacture
W, C.
For dies; sintered.

BRONCO

Texas Instruments Inc./Materials Control
Copper clad with P bronze. For springs; high current carrying
capacity. *Obsolete*

BRONWITE

American Smelting & Refining Co.
Mn 20, Zn 20, Al 1, bal Cu.
Die cast: 75,000 psi TS; 40,000 psi YS; 20 El. Sand cast:
60,000-70,000 psi TS; 30,000-35,000 psi YS; 25-40 El. For
general die castings, hardware, fixtures. For die casting
copper alloys. Die castable bronze, corrosion resistant.

BRONZ-ROD NO. 61

Marquette Corp.
Cu 97, Si 2.85, Fe 0.15.
Welded: 55,000 TS; 22,000 YS; 8 El; 85 Rock F. For brazing
rod for steels and cast iron; high tensile brazing.

BRONZALUN

American Abrasive Metals Co.
Copper. Cu 85, Al 15, Bronze with abrasive grains cast in the
metal.
For floor plates, stair treads, car steps and door saddles; wear
resistant and anti-slip.

BRONZARK

Westinghouse Electric Corp.
Sn, bal Cu.
For welding electrodes; for cast iron or copper. *Obsolete*

BRONZE AU-CADMIUM

Gilby-Fodor S.A.
Cd 1, bal Cu.
For heating elements. Useful operating temperature up to
350°C.

BRONZE CLAD COPPER

Texas Instruments Inc./Materials Control
CDA 524 bronze clad to one or both sides of CDA 102
copper. CDA 524 bronze is used as braze filler metal and
CDA 102 copper as heat exchanger structural member.

BRONZE DEVIL

Champion Rivet Co.
Sn 8.5, P 0.25, bal Cu.
For P-bronze welding rods; arc welding.

BRONZE FILTER POWDER

Alcan Metal Powders Division
Cu 90, Sn 10.
Powder for compressing and sintering to make bronze filters;
10 to 325 mesh grades.

BRONZE OTSSN3-7-5-1

Russian manufacture
Cu 85.08, Pb 3.58, Zn 7.35, P 0.01, Sn 3.4, Ni 0.3.
Cast: 27,000 TS; 13,000 YS; 17 El; 26 RA. For hardware,
plumbing; free-cutting.

BRONZE WABBLER

American manufacture
Sn, bal Cu.
For rolls; tough and hard.

BRONZE, PHOSPHOR GRADE "A"

Anaconda Co.
Copper. Cu 96, Sn 3.75, P 0.25.
Cold rolled: 90,000 TS; 45,000 YS; 10 El. Wire: 150,000 TS;
0.5 El. For springs, electrical switches, diaphragms. High
strength.

BRONZE, PHOSPHOR GRADE "C"

Anaconda Co.
Copper. Cu 92, Sn 8.
Plate: 50,000 TS; 20,000 YS; 73 El; 60 Brin. Wire: 150,000 TS.
For springs, switches, fittings. Tough.

BRONZE, PHOSPHOR GRADE "D"

Anaconda Co.
Copper. Cu 89.5, Sn 10.5.
Cold rolled: 92,000 TS; 83,000 YS; 29 El; 218 Brin. For worm
wheels, pumps. Tough.

BRONZE, STEAM FITTING

American manufacture
Cu 88, Sn 8, Zn 2, Pb 2.
For steam fittings; pressure tight.

BRONZE, STEAM VALVE

American manufacture
Cu 88, Sn 10, Zn 2.
For steam valves.

BRONZE, STEEL STAHL

American manufacture
Cu 52-59, Zn 36-43, Fe 1, Mn 2.5-3, Al 1.
For propellers, marine parts, hardware; same as "Uchatius
Bronze."

BRONZE, WEATHERSTRIP

American manufacture
Cu 89, Zn 9.5, Sn 1.5.
For weather strips.

BRONZE, WHITE

American manufacture
Cu 54, Zn 42, Ni 4, 0.3 Fe + Al.
For ornamental and architectural parts; corrosion resistant.

BRONZE, WIRE

American manufacture
Cu 98.75, Sn 1.2, Pb 0.05.
For electrical purposes, roofing, gutters.

BRONZEND

Arcos Alloys
Mn, bal Cu.
For bronze welding rod. *Obsolete*

BRONZEND E

Arcos Alloys
Si 3, Mn 1.5, bal Cu.
Welded: 65,000 TS; 40,000 YS; 20 El; 20 RA; 100 Brin. For
bronze welding electrodes. *Obsolete*

BRONZEND P

Arcos Alloys
Sn 5, Cu 95, P 0.3.
Welded: 55,000 TS; 35,000 YS; 20 El; 25 RA; 85 Brin. For
welding electrodes. *Obsolete*

BRONZOCHROM 10185

Eutectic Corp.
Nickel base alloy powder for overlays on ferrous, and nickel
base alloys.

BRONZOCHROM NO. 185

Eutectic Corp.
Zn, Ni, bal Cu.
180-235 Brin. For superfacing on steel and cast iron; MP
1000-1200 F. *Obsolete*

BRONZOCHROM NO. 10G

Eutectic Corp.
Zn, bal Cu.
240-290 Brin. For brazing rod on iron and copper; MP
1000-1200 F. *Obsolete*

BRONZOCHROM NO. 187

Eutectic Corp.
Zn, bal Cu.
300 Brin. For brazing rod on copper; MP 1000-1200 F.
Obsolete

BRONZSTOX

A.W. Cadman Mfg. Co.
Sn, Pb, bal Cu.
For bearings, bushings; heavy loads. *Obsolete*

BROTERNAL
IMI Knoch Ltd.
Cu 92, Mn 1, Al 7.
Drawn: 112,000 TS; 85,000 YS; 6 El; 200 Brin. For periscope tubes, paper mill and dye work equipment; Al bronze, corrosion resistant.

BROWN & SHARPE PLAIN
Brown & Sharpe Mfg. Co.
C 0.9, Mn 1, Cr 0.5, W 0.5, bal Fe.
Oil hardening flat ground stock. AISI O1.

BROWN & SHARPE WATER HARDENING
Brown & Sharpe Mfg.Co.
C 1.08-1.18, Mn 0.2-0.3, Cr 0.4-0.5, bal Fe.
Annealed: 97,000 TS; 56,000 YS; 27 El; 45 RA; 170-200 Brin. For tools, dies, fixtures, gages, templates; ground flat stock. *Obsolete*

BROWN BAILEY BB4K-TI
Dunford Hadfields Ltd.
C 0-0.1, Cr 16-18, Ni 10-14, Mo 1.7-2.7, Cb = 10 x C, bal Fe.
Annealed: 85,000-95,000 TS; 35,000-45,000 YS; 50-60 El; 60-75 RA; 150-190 Brin. For acid resistant chemical plant equipment; Type 316 Cb; stainless, austenitic. *Obsolete*

BROWN BAILEYS BB 2 K
Dunford Hadfields Ltd.
C 0.2, Cr 18-22, Ni 7-12, other alloys, bal Fe.
89,600 TS. For chemical plant equipment resisting sulfurous, acetic and sulfuric acids; resists severe chemical corrosion. *Obsolete*

BROWN BAILEYS QS
Dunford Hadfields Ltd.
C 0.4, Cr 6-8, Si 1.
For valves for internal combustion engines; heat and corrosion resistant. *Obsolete*

BROWN BAILEYS QSS
Dunford Hadfields Ltd.
C 0.4, Cr 6-8, Si 3, bal Fe.
For valves for internal combustion engines; heat and corrosion resistant. *Obsolete*

BROWN BAYLEY BBH
Dunford Hadfields Ltd.
Cr 12-14, 0.15% C min, bal Fe.
Annealed: 88,000 TS; 40,000 YS; 32 El; 68 RA; 170 Brin. Heat treated: 256,000 TS; 190,000 YS; 6 El; 10 RA; 540 Brin. For cutlery, valve trim, springs, turbine blades; Type 420; stainless, hardenable. *Obsolete*

BROWN LABEL
Peninsular Steel Co.
C 0.5, Mn 0.2, Cr 1.15, W 2.5, V 0.2, Si 0.75, bal Fe.
For coining and swaging dies, chisels; oil hardening, tough; AISI S1, shock resistant tool steel.

BROWN LABEL
Wallace Murray Corp.
C 0.85-1, Cr 3-4, bal Fe.
For tools, hot work tools and dies, hot shear blades, bull dies; "Heller's Hot Die Steel."

BROWN-BAYLEYS NO. 33
Dunford Hadfields Ltd.
Medium C, Cr, Si, bal Fe.
For internal combustion engine valves; heat and corrosion resistant. *Obsolete*

BROWNIE EXTRA
Raven Steel & Tool Co.
C 0.8, bal Fe.
For drills, taps, reamers, cutters; Type W1; water hardened.

BRT
Now BETHLEHEM BRT.

BRUSH 10
Brush Wellman
Be 0.4-0.7, Co 2.4-2.7, bal Cu.
HT temper: 110,000-135,000 TS; 95-102 Rock B; 50% IACS. High electrical conductivity and good formability. For springs, switches, instrument parts, clutch rings, resistance welding electrodes. C17500

BRUSH 10-C
Brush Wellman
Be 0.45-0.65, Co 2.4-2.6, bal Cu.
Heat treated: 90,000-120,000 TS; 70,000-90,000 YS; 5-12 El; 5-18 RA; 195-262 Brin. For slip rings, contact arms, switch parts; age-hardenable, good conductivity.

BRUSH 10-C
Brush Wellman Corp.
Be 0.45-0.65, Co 2.4-2.6, bal Cu.
Heat treated: 90,000-120,000 TS; 70,000-90,000 YS; 5-12 El; 5-18 RA; 195-262 Brin. For slip rings, contact arms, switch parts; age-hardenable, good conductivity.

BRUSH 10C
Brush Wellman
Be 0.45-0.8, Co 2.4-2.7, bal Cu.
Solution annealed and aged: 90,000-110,000 TS; 92-100 Rock B; 50% IACS. Age hardenable, good conductivity. For slip rings, contact arms, switch parts.

BRUSH 10X
Brush Wellman
Be 0.4-0.7, Co 2.35-2.7, bal Cu.
HT temper: 110,000-130,000 TS; 95-102 Rock B. High strength up to 800°F. Corrosion resistant, heat treatable. For wind tunnel components, springs, switch gears, resistance welding electrodes. C17500

BRUSH 125
Brush Wellman
Be 1.8-2, 0.12-0.18 Co + Ni, bal Cu.
Magnetic mass susceptibility -0.5 x 10^{-1}. Magnetic permeability 0.999997. Heat treated: 200,000 TS; 175,000 YS; 2-5 El, 42 Rock C. For strain members in nonmagnetic cables, wire forms, springs. High strength, nonmagnetic, heat treatable, corrosion resistant. C17200 *Obsolete*

BRUSH 165
Brush Wellman
Be 1.6-1.79, 0.20 Co + Ni min, 0.60 Co + Ni + Fe max, bal Cu.
HT temper: 180,000-210,000 TS; 38-44 Rock C. Corrosion and wear resistant. For corrosion and wear resistant parts. C17000

BRUSH 165C
Brush Wellman
Be 1.6-1.85, Co 0.2-0.65, bal Cu.
Solution annealed and aged: 145,000-155,000 TS; 34-39 Rock C. Corrosion resistant, heat treatable, tough. For plastic tooling and pressure containers. C82400

BRUSH 165CT
Brush Wellman
Be 1.6-1.85, Co 0.2-0.65, Ti 0.02-0.12, bal Cu.
Solution annealed and aged: 145,000-155,000 TS; 34-39 Rock C.

BRUSH 17
Brush Wellman
Ni 31, Fe 0.7, Mn 0-1, Be 0.3-0.7, Si 0-0.15, bal Cu.
Cast: 76,000 TS; 45,000 YS; 14 El; 82 Rock B. Age hardened: 120,000 TS; 85,000 YS; 14 El; 20 Rock C. For marine hardware, sonar cases, heat exchangers, desalinization equipment. Corrosion resistant, age hardenable. *Obsolete*

BRUSH 174
Brush Wellman
Copper. Be 0.15-0.5, Co 0.35-0.6, bal Cu.
HT temper: 110,000-130,000 TS; 95 Rock B; 27 Rock C; 50% IACS. High strength, high conductivity, superior resistance to stress relaxation. For automotive, computer, appliance, electrical/electronic applications. C17410

BRUSH 18
Brush Wellman
Ni 30, Fe 0.62, Mn 0-1.5, Be 0.4-0.6, Si 0.7-0.9, bal Cu.
Cast: 100,000 TS; 70,000 YS; 12 El; 94 Rock B. Age hardened: 120,000 TS; 80,000 YS; 13 El; 30 Rock C. For marine hardware, desalinization equipment, heat exchangers, valve bodies, sonar cases. Corrosion resistant, age hardenable. *Obsolete*

BRUSH 190
Brush Wellman
Be 1.8-2, 0.20 Co + Ni min, 0.60 Co + Ni + Fe, bal Cu.
HM temper: 135,000-150,000 TS; 28-35 Rock C. Corrosion resistant, heat treatable. For springs, shafts, fasteners. C17200

BRUSH 20-C
Brush Wellman
Be 1.9-2.15, Co 0.35-0.65, bal Cu.
Heat treated: 150,000-175,000 TS; 125,000-160,000 YS; 1-4 El; 1-5 RA; 352-426 Brin. For bearings, gears, valve and pump parts; age-hardenable, wear and corrosion resistant.

BRUSH 200C
Brush Wellman
Be 2, up to 0.4 additives, bal Ni.
Annealed: 130,000 TS; 57,000 YS; 30 El; 97 Rock B. Aged: 240,000 TS; 190,000 YS; 5 El; 54 Rock C. For safety tools, plastic molds and cores, fuel pump impellers, mechanical seals. Good structural stability at elevated temperatures. *Obsolete*

BRUSH 20C
Brush Wellman
Be 1.9-2.15, Co 0.35-0.7, Si 0.2-0.35, bal Cu.
Solution annealed and aged: 150,000-175,000 TS; 38-43 Rock C. Age hardenable, wear and corrosion resistant. For bearings, gears, valve and pump parts.

BRUSH 20CT
Brush Wellman
Be 1.9-2.25, Co 0.35-0.7, Si 0.2-0.35, Ti 0.02-0.12, bal Cu.
Solution annealed and aged: 150,000-175,000 TS; 38-43 Rock C.

BRUSH 21C
Brush Wellman
Be 1.9-2.15, Co 1-1.2, Si 0.2-0.35, bal Cu.
Solution annealed and aged: 150,000-175,000 TS; 38-43 Rock C.

BRUSH 220C
Brush Wellman
Be 2-2.3, C 0-0.4, bal Ni.
Corrosion and wear resistant. For corrosion and wear resistant parts.

BRUSH 220CC
Brush Wellman
Be 2-2.3, Cr 0-0.8, C 0-0.4, bal Ni.
For corrosion and wear resistant parts. Corrosion and wear resistant. *Obsolete*

BRUSH 221C
Brush Wellman
Be 2-2.3, C 0.5-1, bal Ni.
For corrosion and wear resistant parts. Corrosion and wear resistant. *Obsolete*

BRUSH 221CC

Brush Wellman
Be 2-2.3, Cr 0-0.8, C 0.5-1, bal Ni.
For corrosion and wear resistant parts. Corrosion and wear resistant. *Obsolete*

BRUSH 240-C

Brush Wellman Corp.
Be 2.35-2.55, Ni 1-1.2, bal Cu.
Heat treated: 155,000-195,000 TS; 130,000-180,000 YS; 1-7 El; 1-8 RA; 360-435 Brin. For bearings, gears, valve and pump parts; age-hardenable, high strength. *Obsolete*

BRUSH 240C

Brush Wellman
Be 2.4-2.6, Ni 1.2-1.3, bal Cu.
Heat treated: 76,000-193,000 TS; 68,000-85,000 YS; 0-34 El; 2-33 RA; 135-387 Brin. For springs, diaphragms, instrument parts. Age hardenable. *Obsolete*

BRUSH 245C

Brush Wellman
Be 2.25-2.55, Co 0.35-0.65, Si 0.2-0.35, bal Cu.
Solution annealed and aged: 165,000-180,000 TS; 40-45 Rock C. Heat treatable, corrosion resistant, impact resistant. For plastic tooling, pressure containers. C82600

BRUSH 245CT

Brush Wellman
Be 2.25-2.55, Co 0.35-0.65, Si 0.2-0.35, Ti 0.02-0.12, bal Cu.
Solution annealed and aged: 165,000-180,000 TS; 40-45 Rock C.

BRUSH 25

Brush Wellman
Be 1.8-2, 0.20 Co + Ni min, 0.60 Co + Ni + Fe max, bal Cu.
HT temper: 190,000-220,000 TS; 38-45 Rock C. Age hardenable, high conductivity. For resistance welding dies and jaws. C17200

BRUSH 250C

Brush Wellman
Be 2.5-2.75, Co 0.35-0.65, bal Cu.
Cast: 140,000-165,000 TS; 110,000-130,000 YS; 0-2 El; 0-2 RA; 42-48 Rock C. For plastic molds, deep drawing dies, zinc die casting dies. High wear resistance. *Obsolete*

BRUSH 260-C

Brush Wellman
Be 2.55-2.8, C 0-0.4, bal Ni.
Cast: 115,000-125,000 TS; 60,000-70,000 YS; 7-12 El; 5-10 RA; 250-283 Brin. Heat treated: 200,000-220,000 TS; 190,000-210,000 YS; 0-2 El; 0-1 RA; 500-560 Brin. For aircraft parts, molds for plastics; age-hardenable, heat resistant to 800°F.

BRUSH 260CC

Brush Wellman
Be 2.55-2.8, Cr 0-0.8, C 0-0.4, bal Ni.
For springs, bellows. Age hardenable. *Obsolete*

BRUSH 261CC

Brush Wellman
Be 2.55-2.8, Cr 0-0.8, C 0.5-1, bal Ni.
For springs, bellows. Age hardenable. *Obsolete*

BRUSH 275-C

Brush Wellman
Be 2.5-2.75, Co 0.35-0.65, bal Cu.
Heat treated: 140,000-165,000 TS; 110,000-130,000 YS; 0-2 El; 0-2 RA; 393-460 Brin. For plastic molds, deep drawing dies; age-hardenable, wear resistant.

BRUSH 275C

Brush Wellman
Copper. Be 2.5-2.85, Co 0.35-0.7, Si 0.2-0.35, bal Cu.
Solution annealed and aged: 180,000-195,000 TS; 43-47 Rock C. Age hardenable, wear resistant. For springs, diaphragms, instrument parts. C82800

BRUSH 275CT

Brush Wellman
Be 2.5-2.85, Co 0.35-0.7, Si 0.2-0.35, Ti 0.02-0.12, bal Cu.
Solution annealed and aged: 180,000-195,000 TS; 43-47 Rock C.

BRUSH 290

Brush Wellman
Be 1.8-2, 0.20 Co + Ni, 0.60 Co + Ni + Fe max, bal Cu.
TM04 temper: 140,000 TS min; 28-38 Rock C.

BRUSH 3

Brush Wellman
Copper. Be 0.2-0.6, Ni 1.4-2.2, bal Cu.
HT temper: 110,000-135,000 TS; 95-102 Rock B; 50% IACS. High electrical and thermal conductivity, good formability. For resistance welding electrodes, molds, heat sinks, plunger tips for die casting, heavy duty connectors, current carrying springs, switch and instrument parts. C17510

BRUSH 35

Brush Wellman
Be 0.23-0.32, Ni 1.4-1.5, bal Cu.
Cast: 54,000 TS; 21,000 YS; 15 El; 14 RA; 110 Brin. Heat treated: 84,000 TS; 64,000 YS; 4 El; 7 RA; 189 Brin. For diaphragms, springs, instrument parts. Age hardenable. *Obsolete*

BRUSH 35-C

Brush Wellman
Be 0.25-0.5, Ni 1.4-1.6, bal Cu.
Heat treated: 70,000-90,000 TS; 50,000-70,000 YS; 5-17 El; 5-25 RA; 180-210 Brin. For resistance welding dies and jaws; age-hardenable, high conductivity.

BRUSH 360

Brush Wellman
Be 1.85-2.05, Ti 0.4-0.6, bal Ni.
HT temper: 270,000 TS min; 83-90 Rock 15-N. For military aircraft connectors, fatigue and temperature resistant springs, switches, relays, burn-in connectors.

BRUSH 3C

Brush Wellman
Be 0.35-0.8, Ni 1-2, bal Cu.
Solution annealed and aged: 90,000-100,000 TS; 92-100 Rock B; 50% IACS.

BRUSH 50

Brush Wellman
Be 0.25-0.5, Co 1.4-1.7, Ag 0.9-1.1, bal Cu.
Annealed: 35,000-55,000 TS; 25,000 YS; 20-35 El; 35 Rock B. HT temper: 120,000 TS; 110,000 YS; 8-20 El; 95-102 Rock B. For resistance welding applications, electrodes. Heat treatable, corrosion resistant. *Obsolete*

BRUSH 50C

Brush Wellman
Be 0.4-0.65, Co 1.4-1.7, Ag 1-1.15, bal Cu.
Cast: 45,000-60,000 TS; 15,000-35,000 YS; 50-65 Rock B. Heat treated: 100,000 TS; 75,000 YS; 3-15 El; 92-100 Rock B. For resistance welding electrodes, dies and holders. Heat treatable, corrosion resistant. RWMA Class 3. *Obsolete*

BRUSH 55C

Brush Wellman
Be 0.45-0.65, Co 2.4-2.6, bal Cu.
Cast: 90,000-120,000 TS; 70,000-90,000 YS; 5-12 El; 5-18 RA; 90-103 Rock B. For resistance welding electrodes, circuit breaker and switch parts, slip rings, contact arms, welder bearings. Good conductivity, resists temperatures to 700°F. RWMA Class 3. *Obsolete*

BRUSH BERYLLIUM ALUMINUM 1.0% BE

Brush Wellman
Be 0.9-1.05, Mg 0.5-1.5, bal Al.
For master alloy for remelting.

BRUSH BERYLLIUM ALUMINUM 1.5% BE

Brush Wellman
Be 1.15-1.5, bal Al.
For master alloy for remelting.

BRUSH BERYLLIUM ALUMINUM 2.6% BE

Brush Wellman
Be 2.5-2.7, bal Al.
For master alloy for remelting.

BRUSH BERYLLIUM ALUMINUM 5.0% BE

Brush Wellman
Be 4-6, bal Al.
For master alloy for remelting.

BRUSH BERYLLIUM COPPER

Brush Wellman
Be 3.5-4.5, bal Cu.
For master alloy for remelting.

BRUSH BERYLLIUM MAGNESIUM

Brush Wellman
Be as required, bal Mg.
For master alloy for remelting.

BRUSH BERYLLIUM NICKEL (1)

Brush Wellman
Be 5-7, bal Ni.
For master alloy for remelting.

BRUSH BERYLLIUM NICKEL (2)

Brush Wellman
Be 1.85-2.05, Ti 0.4-0.6, bal Ni.
HT temper: 270,000 TS min; 83-90 Rock 15-N. For military and aircraft connectors, fatigue and temperature resistant springs, switches, relays, burn-in connectors.

BRUSH BERYLLIUM NO. 6

Brush Wellman
Same as BRUSH 240C. *Obsolete*

BRUSH CHROMIUM COPPER

Brush Wellman
Cr 0.4-1.2, Fe 0-0.15, Si 0-0.1, Zr 0-0.7, bal Cu.
HT temper: 50,000-65,000 TS; 60-75 Rock B. Heat treatable, corrosion resistant, high strength, high electrical conductivity. For resistance welding electrodes, seam welding wheels, electrical switchgear, electrode holder jaws, cable connectors, electrical/electronic components.

BRUSH M220C

Brush Wellman
Be 1.8-2.3, C 0.3-0.5, bal Ni.
High strength, high thermal conductivity, good wear and corrosion resistance. For molds, plungers, forming tools and plungers in the glass industry.

BRUSH M25

Brush Wellman
Be 1.8-2, Pb 0.2-0.6, 0.20 Co + Ni min, 0.60 Co + Ni + Fe max, bal Cu.
HT temper: 190,000-220,000 TS; 38-45 Rock C. Free cutting, heat treatable, corrosion resistant. For gears, shafts, screw machine products, fasteners. C17300

BRUSH M260C

Brush Wellman
Be 2.55-2.8, C 0-0.4, bal Ni.
Cast: 115,000-125,000 TS; 60,000-70,000 YS; 7-12 El; 5-10 RA; 24-30 Rock C. Aged: 210,000-230,000 TS; 200,000-210,000 YS; 1-2 El; 0-1 RA; 50-54 Rock C. For turbine wheels, glass molds, valve bodies, fuel injection tips. Age hardenable, corrosion resistant, temperature resistant. *Obsolete*

BRUSH MOLDMAX

Brush Wellman
Be 1.6-2, 0.20 Co + Ni min, 0.60 Co + Ni + Fe max, bal Cu.
High hardness: 36-42 Rock C. Low hardness: 26-32 Rock C.
High strength, wear and corrosion resistance. For pinch-offs,
neck rings and handle inserts for blow molds; injection molds
and mold inserts; injection nozzles and manifolds for hot
runner systems.

BRUSH PROTHERM

Brush Wellman
Be 0.2-0.7, 1.40-2.20 Ni + Co, bal Cu.
HT temper: 90-102 Rock B. High conductivity, wear and
corrosion resistance. For reaction injection molding;
expandable polystyrene foam processing; injection blow
molds for corrosive plastics; nozzle tips, edge gates and
manifolds for hot runner molds.

BRUSH QMV

Brush Wellman
Be, bal Cu.
For corrosion and wear resistant parts. Age hardenable.
Obsolete

BRUSH WIRE

American manufacture
Cu 64.25, Zn 35, Sn 0.75.
For wire.

BRUSH ZRII-5% BE

Brush Wellman
Be 5, bal Zr.
For brazing Zircalloy alloys. High temperature brazing alloy.
Obsolete

BRUSH-261 C

Brush Wellman
Be 2.55-2.8, C 0.5-1, bal Ni.
For aircraft parts, molds for plastics; age-hardenable, heat
resistant to 800°F.

BRYAN

Time Steel Service Inc.
C 1.1, Mn 0.3, Si 0.5, Cr 0.25, bal Fe.
Water hardened carbon tool steel, good hardness and wear;
AISI W4.

BRYIRON

Fillmore Foundry Inc.
TC 2.96, Mn 0.73, Si 1.2, Ni 1.5, bal Fe.
51,000 TS; 221 Brin. For liners, pistons; transverse strength
4160.

BRYMILL BRM-1

British Rolling Mills Ltd.
C 0.1-0.18, Mn 0.5-1.1, bal Fe.
Annealed: 60,000 TS; 36,000 YS; 38 El; 70 Rock B. For gears,
fasteners, bolts, machine tool parts. Case hardened.

BRYMILL BRM-2

British Rolling Mills Ltd.
C 0.15, bal Fe.
Annealed: 47,000 TS; 50 El. For deep pressings; deep
stamping steel. *Obsolete*

BRYMILL BRM-4

British Rolling Mills Ltd.
C 0.2-0.25, bal Fe.
Drawn: 82,900 TS; 17 El; 45 RA. For shafting, general
engineering applications; case hardening. *Obsolete*

BRYMILL BRM-5

British Rolling Mills Ltd.
C 0.25-0.35, bal Fe.
Drawn: 89,600 TS; 15 El; 40 RA. For bolts, studs, jack
spindles; water hardened.

BRYMILL BRM-6

British Rolling Mills Ltd.
C 0.35-0.45, bal Fe.
Drawn: 103,000 TS; 10 El; 35 RA. For axles, shafts, racks,
clutch plates, chain links; water hardened.

BRYMILL CH-1

British Rolling Mills Ltd.
C 0.2, bal Fe.
Hardened: 78,000 TS; 30 El; 55 RA. For case hardened parts;
case hardening steel. *Obsolete*

BRYMILL CH-10

British Rolling Mills Ltd.
C 0.25, bal Fe.
Hardened: 98,500 TS; 30 El; 60 RA. For case hardened parts;
case hardening steel. *Obsolete*

BRYMILL CH-2

British Rolling Mills Ltd.
C 0.12-0.18, Mn 0.9-1.5, S 0.08-0.15, bal Fe.
Heat treated: 98,500 TS; 20 El; 50 RA. For screw machine
products, gears, bolts, fasteners; case hardened, free-cutting.

BRYMILL CH-8

British Rolling Mills Ltd.
C 0.08-0.55, Ni 2.75-3.5, bal Fe.
Heat treated: 107,500 TS; 30 El; 50 RA. For gears, bolts,
crankshafts, camshafts; case hardened or through hardened,
tough.

BRYMILL DH-11

British Rolling Mills Ltd.
C 0.4, Ni 1, bal Fe.
For shafts, axles, gears, machinery parts; case hardened.

BRYMILL DH12

British Rolling Mills Ltd.
C 0.3-0.4, Mn 1.3-1.7, bal Fe.
For gears, bolts, machine tool parts; oil hardened.

BRYMILL DH13

British Rolling Mills Ltd.
C 0.25-0.35, Mn 0.35-0.75, Ni 2.75-3.5, bal Fe.
For gears, bolts, crankshafts; oil hardened, shock resistant.

BRYMILL X

British Rolling Mills Ltd.
C 0.1, bal Fe.
Annealed: 45,000 TS; 55 RA. For stampings and pressings;
deep stamping steel. *Obsolete*

BS 1

Thyssen Edelstahlwerke AG
C, bal Fe.
For rock drills; water hardening. *Obsolete*

BS 4

Thyssen Edelstahlwerke AG
C, bal Fe.
For rock drills; water hardening. *Obsolete*

BS-SEEWASSER 05

Vereinigte Leichtmetallwerke G.m.b.H.
Aluminum. Mg 4-5.5, Mn 0-0.8, Cr 0-0.3, bal Al.
Soft: 42,000 TS; 22,000 YS; 35 El; 65 Brin. Hard: 60,000 TS;
50,000 YS; 10 El; 105 Brin. For aircraft and marine parts;
good corrosion resistance.

BS-SEEWASSER 07

Vereinigte Leichtmetallwerke G.m.b.H.
Aluminum. Mg 5.5-7.5, Mn 0-0.8, Cr 0-0.3, bal Al.
For aircraft and marine parts; good corrosion resistance.

BS-SEEWASSER 63/03

Vereinigte Leichtmetallwerke G.m.b.H.
Aluminum. Mg 2-4, Mn 0-0.4, bal Al.
Soft: 28,000 TS; 13,000 YS; 30 El; 47 Brin. Hard: 40,000 TS;
35,000 YS; 10 El; 73 Brin. For aircraft tanks, and fittings, fuel
lines, marine parts; resists sea water corrosion.

BS401

Comalco Ltd.
Cu 0-0.14, Si 11-13, Fe 0-0.6, Mn 0-0.5, Zn 0-0.14, Ti 0-0.2,
Sn 0-0.05, Mg 0-0.1, Ni 0-0.1, bal Al.
Permanent mold; cast; 207 MPa TS; 62 MPa YS; 9 El; 60 Brin.
ADC H49-5. BS LM 6.

BSC-HC-110

British Steel plc
862 N/mm^2 TS; 758-965 N/mm^2 YS. Seamless tube casing
steel for high external pressure. Oil industry applications.

BSC-HC-95

British Steel plc
758 N/mm^2 TS; 655-965 N/mm^2 YS. Seamless tube casing
steel for high external pressure. Oil industry applications.

BSC-SR-80

British Steel plc
655 N/mm^2 TS; 552-655 N/mm^2 YS; 22 Rock C max.
Seamless tube casing steel for use in sulfide stress-cracking
environment. Oil industry applications.

BSC-SR-95

British Steel plc
C 0-0.35, Si 0-0.4, Mn 0.4-1, P 0-0.02, S 0-0.015, Cr 0.8-1.4,
Mo 0.15-0.75, bal Fe.
724 N/mm^2 TS; 655-758 N/mm^2 YS. Seamless tube casing
steel for use in sulfide stress-cracking environment. Oil
industry applications.

BSC-V-150

British Steel plc
C 0-0.35, Si 0-0.4, Mn 0.5-1.6, P 0-0.04, S 0-0.04, Cr 0-1.5,
Mo 0.1-0.8, V 0-0.12, bal Fe.
1103 N/mm^2 TS; 104-1241 N/mm^2 YS. Seamless tube
casing steel. Oil industry applications.

BSC-WHC-95

British Steel plc
758 N/mm^2 TS; 655-965 N/mm^2 YS. Welded tube casing
steel for high external pressure. Oil industry applications.

BSC-XT-130

British Steel plc
C 0-0.35, Si 0-0.4, Mn 0.5-0.9, P 0-0.03, S 0-0.02, Cr 0.5-1.3,
Mo 0.1-0.75, V 0-0.12, bal Fe.
965 N/mm^2 TS; 896-1069 N/mm^2 YS. Seamless tube casing
steel. Oil industry applications; high strength; tough.

BSC-XT-140

British Steel plc
C 0-0.35, Si 0.2-0.4, Mn 0.5-0.9, P 0-0.03, S 0-0.02, Cr 1-1.5,
Mo 0.1-0.75, V 0-0.12, bal Fe.
1034 N/mm^2 TS; 965-1138 N/mm^2 YS. Seamless tube
casing steel. High strength; tough. For oil industry
applications.

BSC-XT-155

British Steel plc
C 0-0.35, Si 0-0.4, Mn 0.5-0.9, P 0-0.03, S 0-0.02, Cr 1-1.5,
Mo 0.1-0.75, V 0-0.12, bal Fe.
1138 N/mm^2 TS; 1069-1241 N/mm^2 YS. Seamless tube
casing steel. High strength; tough. For oil industry
applications.

BSEM 558

Sterling Metals Ltd.
Same as RR.350. *Obsolete*

BSF

Thyssen Edelstahlwerke AG
C 0.33, Cr 1, Mo 0.2, bal Fe.
For gears, bolts, machine tool parts; water hardened, tough.
Obsolete

BSF DMOC

Thyssen Edelstahlwerke AG
C, Cr, Mo, bal Fe.
15-18 El. For machinery parts; oil hardened. *Obsolete*

BSI 30 VMS
Bergische Stahl Industrie
C 0.3, Mn 1.25, bal Fe.
Heat treated: 85,000-128,000 TS; 57,000-71,000 YS; 12-16 El. Low alloy steel casting. Werkstoff Nr. 1.5066. *Obsolete*

BSI AM 12
Bergische Stahl Industrie
C 1.2, Mn 13, bal Fe.
Solution annealed and aged: 85,000 min TS; 42,500 min YS; 25 min El. Non-magnetic steel casting. Werkstoff Nr. 1.3802; DIN G-X120 Mn 12. *Obsolete*

BSI AM 16.13 S
Bergische Stahl Industrie
C 0.05, Cr 16.5, Ni 12.5, bal Fe.
Solution annealed and aged: 64,000 min TS; 28,000 min YS; 20 min El. Non magnetic steel casting. DIN G-X5CrNi1613. *Obsolete*

BSI AM 16.13 SS
Bergische Stahl Industrie
C 0.07, Cr 16.5, Ni 12.5, Nb, bal Fe.
Solution annealed and aged: 64,000 min TS; 28,000 min YS; 20 min El. Non-magnetic steel casting. DIN G-X10CrNiNb1613; Werkstoff Nr. 1.3959. *Obsolete*

BSI AM 16.16 ES
Bergische Stahl Industrie
C 0.05, Cr 16.5, Ni 16, Mo 2.25, bal Fe.
Solution annealed and aged: 64,000 min TS; 28,000 min YS; 20 min El. Non-magnetic steel casting. DIN G-X5CrNiMo1616. *Obsolete*

BSI AM 16.16 ESS
Bergische Stahl Industrie
C 0.07, Cr 16.5, Ni 16, Mo 2.25, Nb, bal Fe.
Solution annealed and aged: 64,000 min TS; 28,000 min YS; 20 min El. Non-magnetic steel casting. DIN G-X7CrNiMoNb1816. *Obsolete*

BSI AM G 10
Bergische Stahl Industrie
C 3.1, Ni 10.5, Mn 5.75, Si 2, bal Fe.
Non-magnetic stainless steel casting. DIN G-X300 NiMnSi106. *Obsolete*

BSI CM 1B
Bergische Stahl Industrie
C 0.35, Cr 1.3, Mo 0.45, bal Fe.
Heat treated: 78,000-150,000 TS; 42,500-100,000 YS; 9-20 El. Low alloy steel casting. *Obsolete*

BSI GS-38D
Bergische Stahl Industrie
C 0.15, bal Fe.
Annealed: 54,000 min TS; 28,000 min YS; 25 min El. Carbon dynamo steel casting. *Obsolete*

BSI GS-45D
Bergische Stahl Industrie
C 0.2, bal Fe.
Annealed: 64,000 min TS; 35,000 min YS; 22 min El. Carbon dynamo steel casting. *Obsolete*

BSI GS-C 25
Bergische Stahl Industrie
C 0.22, bal Fe.
Annealed: 64,000 TS; 35,000 min YS; 22 min El. Carbon steel casting; useful to -40 C. Werkstoff Nr. 1.0416. *Obsolete*

BSI NI 2
Bergische Stahl Industrie
C 0.1, Ni 1.55, bal Fe.
Heat treated: 64,000-85,000 TS; 38,000 min YS; 20 min El. Low alloy steel casting, useful to -50 C. Werkstoff Nr. 1.5621. *Obsolete*

BSI NI 3
Bergische Stahl Industrie
C 0.1, Ni 3.55, bal Fe.
Heat treated: 71,000-92,000 TS; 51,000 min YS; 20 min El. Low alloy steel casting; good low temperature properties to -100 C. Werkstoff Nr. 1.5638; ASTM A352 Gr.LC3. *Obsolete*

BSI SHF
Bergische Stahl Industrie
C 0.38, Ni 2.1, Cr 1.1, Mo 0.45, bal Fe.
Heat treated: 255,000 min TS; 212,000 min YS; 5 min El. Low alloy steel casting. *Obsolete*

BSI V12
Bergische Stahl Industrie
C 0.29, Cr 1.35, Mo 0.4, V 0.1, bal Fe.
Heat treated: 128,000 TS min; 100,000 YS min; 10 El min. Low alloy steel casting (formerly PZA) W.- Nr. 1.7725.

BSI V22
Bergische Stahl Industrie
C 0.35, Cr 2.45, Mo 0.4, V 0.1, bal Fe.
Heat treated: 100,000-120,000 TS; 78,000-92,000 YS; 11-12 El. Low alloy steel casting (formerly PZB) W.- Nr. 1.7755.

BSO SPEZIAL
Thyssen Edelstahlwerke AG
C 0.41, Cr 1, Mn 0.65, Si 0.25, bal Fe.
For gears, bolts, machine tool parts; water hardened. *Obsolete*

BSW
Thyssen Edelstahlwerke AG
C 0.25, Cr 1, Mo 0.2, bal Fe.
For gears, bolts, fasteners, shafts; oil hardened, tough. *Obsolete*

BSW, DMOCN
Thyssen Edelstahlwerke AG
C, Cr, Mo, bal Fe.
18-23 El. For machinery parts. *Obsolete*

BT-3
Forjas Alavesas S.A.
C 0.15, Mn 0.5, Si 0.25, Ni 3.5, bal Fe.
For low temperature operation, down to -100°C.

BT-4
Russian manufacture
Al 4, Mn 1.5, bal Ti.
Titanium alloy.

BT-9
Forjas Alavesas S.A.
C 0.1, Mn 0.7, Si 0.25, Ni 9, bal Fe.
For low temperature operation, down to -200°C.

BT-O
United States Bronze Powders Inc.
C 0.1, Mn 0.9, Si 0-0.1, bal Fe.
For low temperature operation down to -70°C. Similar to ASTM A333 Gr 1.

BTF ALLOY
Bedford Tool & Forge Co.
C 0.55, Mn 1, Mo 0.35, Cr 0.3, Si 2, bal Fe.
For punches, rivet sets, pneumatic tools; oil hardened, shock resistant.

BTG
Creusot-Loire
Ni 60, Cr 12, W 2, bal Fe.
For ammonia synthesis tubes; heat and NH_3 resistant.

BTH ALLOY NO. 12
British Thomson Houston Co. Ltd.
Ni Cr-Fe.
For seals to lead glass; metal to glass seal.

BTR
Bethlehem Steel Corp.
C 0.7, Mn 1.2, Cr 0.5, V 0.2, W 0.5, bal Fe.
For tools, master tools and dies; taps, hobs, reamers, broaches; non-deforming; oil hardened. *Obsolete*

BU-NITE
English manufacture
Ni-Al, bal Cu.
For pistons; retain strength up to 370°C.

BUCKEYE BRONZE
Manufacturer not listed.
Sn 41.6, Cu 0.6, Pb 15.6, Zn 37.8, Al 4.3.
For bearings; Babbitt.

BUCKEYE M1
Time Steel Service Inc.
C 0.78, Cr 3.9, V 1.05, W 1.6, Mo 8.5, bal Fe.
High speed tool steel, molybdenum type. AISI M1.

BUCKEYE M10
Time Steel Service Inc.
C 0.86, Cr 4, V 2, Mo 8.5, bal Fe.
High speed tool steel, molybdenum type. AISI M10.

BUCKEYE M2
Time Steel Service Inc.
C 0.8, Cr 4, V 1.75, W 6, Mo 5, bal Fe.
High speed tool steel, Mo-W type. AISI M2.

BUCKEYE M3 CLASS 1
Time Steel Service Inc.
C 1, Cr 4, V 2.5, W 6.25, Mo 6.25, bal Fe.
High speed tool steel, Mo-W type. AISI M3 Class 1.

BUCKEYE M3 CLASS 2
Time Steel Service Inc.
C 1.15, Cr 4, V 3, W 6, Mo 5.5, bal Fe.
High speed tool steel, Mo-W type. AISI M3 Class 2.

BUCKEYE M4
Time Steel Service Inc.
C 1.3, Cr 4, V 4, W 5.5, Mo 4.5, bal Fe.
High speed steel for cutting tools, high hardness and good wear; AISI M4.

BUCKEYE M42
Time Steel Service Inc.
C 1.1, Cr 3.75, V 1.15, W 1.7, Mo 9.5, Co 8, bal Fe.
High speed tool steel, Mo-Co type. AISI M42.

BUCKEYE M43
Time Steel Service Inc.
C 1.2, Cr 3.75, V 1.6, W 2.7, Mo 8, Co 8, bal Fe.
High speed tool steel, Mo-Co type. AISI M43.

BUCKEYE M7
Time Steel Service Inc.
C 1, Cr 3.75, V 2, W 1.75, Mo 8.75, bal Fe.
High speed tool steel, molybdenum type. AISI M7.

BUCKEYE T1
Time Steel Service Inc.
C 0.75, Cr 4, V 1, W 18, bal Fe.
High speed tool steel, tungsten type. AISI T1.

BUCKEYE T15
Time Steel Service Inc.
C 1.5, Cr 4.75, V 5, W 12.5, Co 5, bal Fe.
High speed tool steel W-Co-V type. AISI T15.

BUCKEYE T2
Time Steel Service Inc.
C 0.83, Cr 4.25, V 2.1, W 18.5, Mo 0.85, bal Fe.
High speed tool steel, tungsten type. AISI T2.

BUCKEYE T4
Time Steel Service Inc.
C 0.75, Cr 4.5, V 1, W 19, Co 5, bal Fe.
High speed tool steel, tungsten-cobalt type. AISI T4.

BUCKEYE T5
Time Steel Service Inc.
C 0.8, Cr 4, V 2, W 18, Co 8, bal Fe.
High speed tool steel, tungsten-cobalt type. AISI T5.

BUCKLE BRASS-1
English manufacture
Cu 90, Zn 9, Pb 1.
For bullets, shells, tubes, cartridges; free-cutting.

BUCKLE BRASS-2
English manufacture
Cu 65, Zn 34, Pb 1.
For bullets; free-cutting.

BUDERUS RCC
Now RCC.

BUDERUS RCW2
Rochling Burbach GmbH
C 0.28, W 9.5, Cr 2.5, V 0.1, bal Fe.
For hot dies; hot work steel. *Obsolete*

BUDERUS RSZ
Rochling Burbach GmbH
C 0.7, W 18, Cr 4, V 1, bal Fe.
For tools, dies, cutters; high speed steel. *Obsolete*

BUDERUS RSZ ESPECIAL
Rochling Burbach GmbH
W 19, Cr 4, V 2.2, C, bal Fe.
For tools, dies, cutters; high speed steel. *Obsolete*

BUDERUS RT11
Now RB11.

BUDERUS RT12
Rochling Burbach GmbH
C 0.8, bal Fe.
For tools, punches; water hardened. *Obsolete*

BUDERUS RT6
Now RT6.

BUDERUS RT9
Rochling Burbach GmbH
C 0.9-1, bal Fe.
For tools, drills, springs; water hardened. *Obsolete*

BUDERUS RTC20
Rochling Burbach GmbH
C 0.6, Cr 1, V 0.2, Mn 0.5, bal Fe.
For tools, crimpers, pliers; oil hardened. *Obsolete*

BUDERUS RVS
Rochling Burbach GmbH
C 0.95, Mn 1.2, W 0.5, Cr 0.5, bal Fe.
For tools, dies; non-deforming. *Obsolete*

BUDERUS T76
Rochling Burbach GmbH
C 1.3, W 4.5, bal Fe.
For tools, dies, cutters; fast finishing steel. *Obsolete*

BUDERUS TRW2
Rochling Burbach GmbH
C 0.5, W 2.25, Cr 1.5, V 0.25, bal Fe.
For tools, dies, hot work dies; hot work steel. *Obsolete*

BUFF ALOY
Buffalo Wire Works Co. Inc.
High C, bal Fe.
For wire cloth; abrasion resisting.

BUFLOKAST
Duraloy
C 3.2, Si 2.4, Mn 0.7, bal Fe.
For drums, dryers, kettles, chemical engineering equipment.
Cast iron for chromium plating.

BUILDTEC 10225
Eutectic Corp.
Nickel base alloy powder for spraying on cast iron.

BUILDTEC 25685
Eutectic Corp.
Alloy powder for metal spraying one-step machinable coating
with good wear resistance.

BULL ALLOY
English manufacture
Sb 18.3, Fe 0.6, Pb 80.9.
For bearings; antifriction.

BULL'S METAL
Bull's Metal & Marine Ltd.
Al 0.5-1.5, Mn 0.5-1, Sn 0.5-1, Fe 0.8-1.2, Cu 57-60, bal Zn.
Cast: 65,000 TS; 30,000 YS; 25 El; 30 RA; 420 Brin. For
marine hardware, propellers; corrosion resistant.

BULL'S WHITE METAL B
Bull's Metal & Marine Ltd.
Sn 80, Sb 10, Pb 6, Cu 4.
For bearings, linings for marine; white metal, wear resistant.

BULLDOG
Pennsylvania Steel Corp.
C 0.5, Cr 1.2, V 0.25, W 2.5, bal Fe.
For chisels, upsetters, crimpers, punches; Type S1; shock
resistant.

BULLET BRASS
American manufacture
Cu 90, Zn 9-10, Pb 0-1.
For bullets, shells, cartridges.

BUNDYWELD STEEL TUBING
Bundy Corp.
C 0.2, bal Fe.
For tubing, gasoline and oil lines, refrigerator coils; hydrogen
welded, copper coated, rolled steel tubing.

BUNTING 100
NL Industries
Cu 87.5-90.5, Fe 0-1, C 0-1.75, Sn 9.5-10.5.
Sintered: density 5.8-6.2. For bearings. SAE 840; ASTM B438
Grade 1; Type 1. *Obsolete*

BUNTING 101
NL Industries
Cu 87.5-90.5, Fe 0-1, C 0-1.75, Sn 9.5-10.5.
Sintered: 6.4-6.8 density. For bearings. SAE 841; ASTM B 438
Grade 1; Type 2. *Obsolete*

BUNTING 102
NL Industries
Cu 87.5-90.5, Fe 0-1, C 0-1.75, Sn 9.5-10.5.
Sintered: density 6.4-6.8; 13,500-16,000 TS; 11,000-15,000
YS (Comp). For mechanical components. SAE 841; ASTM
B255 Type 1. *Obsolete*

BUNTING 104
NL Industries
Cu 82.6-88.5, Fe 0-1, Pb 2-4, C 0-1.75, Sn 9.5-10.5.
Sintered: 6.5-6.9 density. For bearings. SAE 843; ASTM B438
Grade 2; Type 1. *Obsolete*

BUNTING 105
NL Industries
Cu 87.5-90.5, Sn 9.5-10.5, Fe 1, C 0-1.75.
Sintered: 6.8-7.2 density; 16,000-20,000 TS; 15,000-20,000
YS (comp). For mechanical components. SAE 842; ASTM
B255 Type 2. *Obsolete*

BUNTING 108
NL Industries
Cu 77-80, Sn 0-0.1, Pb 1-2, Fe 0-0.25, Ni 0-0.1, bal Zn.
Sintered: 7.2-7.6 density; 7.0 porosity, 9-10 El; 24,000 TS.
For mechanical components. MPIF CZP-0218-T; SAE 890;
ASTM B282 Type 1. *Obsolete*

BUNTING 109
NL Industries
Cu 77-80, Sn 0-0.1, Pb 1-2, Fe 0-0.25, Ni 0-0.1, bal Zn.
Sintered: 7.6-8.0 density; 10.0-13.0 El; 28,000 psi TS; 14,000
psi CS. For mechanical components. MPIF CZP-0218-U; SAE
891; ASTM B282; Type 2. *Obsolete*

BUNTING 200
NL Industries
C 0.25-0.6, 95.9% Fe min, 3.0% others max.
Sintered: density 5.7-6.1; porosity 18. For bearings. SAE 851;
ASTM B439 Gr2. *Obsolete*

BUNTING 201
NL Industries
C 0-0.25, 96.25 Fe min, 0.0-3.0 others.
Sintered: 5.7-6.1 density; 18.0 porosity. For bearings. SAE
850; ASTM B439 Grade 1. *Obsolete*

BUNTING 202
NL Industries
Cu 7-11, C 0-0.3, 86.5% Fe min.
Sintered: density 5.8-6.2; 29,500-34,000 TS; 28,500-30,000
YS (Comp). For mechanical components. SAE 862 is similar.
Obsolete

BUNTING 203
NL Industries
Cu 18-22, bal Fe.
Sintered: 5.8-6.2 density. For bearings. SAE 863; ASTM B439
Grade 4. *Obsolete*

BUNTING 204
NL Industries
Cu 7-11, bal Fe.
Sintered: 5.8-6.2 density; 18.0 porosity. For bearings. SAE
862; ASTM B439 Grade 3. *Obsolete*

BUNTING 205-1
NL Industries
Cu 1-2.5, C 0.6-1, 94.5 Fe min.
Sintered: 6.0 density max; 20 Rock B. Hardened: 40,000 psi
TS; 28 Rock C. For mechanical components. MPIF FC-0208-
N; SAE 864-A; ASTM B426 Grade 1; Type 1. *Obsolete*

BUNTING 205-10
NL Industries
Cu 6-11, C 0.6-1, 86.0 Fe min.
Sintered: 6.0-6.4 density; 60 Rock B. Hardened: 56,000 psi
TS; 33 Rock C. For mechanical components. MPIF FC-0808-
P; ASTM B426 Grade 3; Type 2; SAE 866-B. *Obsolete*

BUNTING 205-11
NL Industries
Cu 6-11, C 0.6-1, 86.0 Fe min.
Sintered: 6.4-6.8 density. For mechanical components. MPIF
FC-0808-R; ASTM B426 Grade 3; Type 3. *Obsolete*

BUNTING 205-12
NL Industries
Cu 6-11, c 0.6-1, 86.0 Fe min.
Sintered: 6.8 density min. For mechanical components. MPIF
FC-0808-S; ASTM B426 Grade 3; Type 4. *Obsolete*

BUNTING 205-2
NL Industries
Cu 1-2.5, C 0.6-1, 94.5 Fe min.
Sintered: 6.0-6.4 density; 50 Rock B. Hardened: 49,000 psi
TS; 36 Rock C. For mechanical components. MPIF FC-0208-
P; SAE 864-B; ASTM B426 Grade 1; Type 2. *Obsolete*

BUNTING 205-3
NL Industries
Cu 1-2.5, C 0.6-1, 94.5 Fe min.
Sintered: 6.4-6.8 density; 65 Rock B. Hardened: 89,000 psi TS; 40 Rock C. For mechanical components. MPIF FC-0208-R; ASTM B426 Grade 1; Type 3. *Obsolete*

BUNTING 205-4
NL Industries
Cu 1-2.5, C 0.6-1, 94.5 Fe min.
Sintered: 6.8 density min; 80 Rock B. Hardened: 124,000 psi TS; 44 Rock C. For mechanical components. MPIF FC-0208-S; ASTM B426 Grade 1; Type 4. *Obsolete*

BUNTING 205-5
NL Industries
Cu 2.5-6, C 0.6-1, 91.0 Fe min.
Sintered: 6.0 density max; 30 Rock B. Hardened: 45,000 psi TS; 28 Rock C. For mechanical components. MPIF FC-0508-N; ASTM B426 Grade 2; Type 1; SAE 865-A. *Obsolete*

BUNTING 205-6
NL Industries
Cu 2.5-6, C 0.6-1, 91.0 Fe min.
Sintered: 6.0-6.4 density; 60 Rock B. Hardened: 54,000 psi TS; 40 Rock C. For mechanical components. MPIF FC-0508-P; ASTM B426 Grade 2; Type 2; SAE 865-B. *Obsolete*

BUNTING 205-7
NL Industries
Cu 2.5-6, C 0.6-1, 91.0 Fe min.
Sintered: 6.4-6.8 density; 75 Rock B. Hardened: 92,000 psi TS; 47 Rock C. For mechanical components. MPIF FC-0508-R; ASTM B426 Grade 2; Type 3. *Obsolete*

BUNTING 205-8
NL Industries
Cu 2.5-6, C 0.6-1, 91.0 Fe min.
Sintered: 6.8 density min. For mechanical components. MPIF FC-0508-S; ASTM B426 Grade 2; Type 4. *Obsolete*

BUNTING 205-9
NL Industries
Cu 6-11, C 0.6-1, 86.0 Fe min.
Sintered: 6.0 density max; 30 Rock B. Hardened: 45,000 psi TS; 26 Rock C. For mechanical components. MPIF FC-0808-N; ASTM B426 Grade 3; Type 1; SAE 866-A. *Obsolete*

BUNTING 206-1
NL Industries
Cu 0-2.5, Ni 1-3, C 0.6-0.9, 91.6 Fe min.
Sintered: 6.4-6.8 density; 62 Rock B. Hardened: 100,000 psi TS; 34 Rock C. For mechanical components. MPIF FN-0208-R; ASTM B484 Grade 1; Type 1; Class C. *Obsolete*

BUNTING 206-2
NL Industries
Cu 0-2, Ni 3-5.5, C 0.6-0.9, 89.6 Fe min.
Sintered: 6.8-7.2 density; 79 Rock B. Hardened: 135,000 psi TS; 45 Rock C. For mechanical components. MPIF FN-0208-S; ASTM B484 Grade 1; Type 2; Class G. *Obsolete*

BUNTING 207-IA
NL Industries
C 0-0.25, 97.75 Fe min.
Sintered: 5.7-6.1 density: 16,000 psi TS; 2.0 El. For mechanical components. MPIF F-0000-N; ASTM B310 Type 1; Class A; SAE 850. *Obsolete*

BUNTING 207-IB
NL Industries
C 0.26-0.6, 97.40 Fe min.
Sintered: 5.7-6.1 density; 20,000 psi TS; 1.5 El. Hardened: 30,000 psi TS. For mechanical components. MPIF F-0005-N; ASTM B310 Type 1; Class B; SAE 851. *Obsolete*

BUNTING 207-IC
NL Industries
C 0.61-1, 97.0 Fe min.
Sintered: 5.7-6.1 density; 18.0 porosity; 35 Rock B. Hardened: 40,000 psi TS. For mechanical components. MPIF F-0008-N; ASTM B310 Type 1; Class C; SAE 852. *Obsolete*

BUNTING 207-IIA
NL Industries
C 0-0.25, 97.75 Fe min.
Sintered: 6.1-6.5 density; 20,000 psi TS; 3.0 El. For mechanical components. MPIF F-0000-P; ASTM B310 Type 2; Class A; SAE 853. *Obsolete*

BUNTING 207-IIB
NL Industries
C 0.26-0.6, 97.40 Fe min.
Sintered: 6.1-6.5 density; 26,000 psi TS; 5.0 El. Hardened: 40,000 psi TS. For mechanical components. MPIF F-0005-P; ASTM B310 Type 2; Class B. *Obsolete*

BUNTING 207-IIC
NL Industries
C 0.61-1, 97.0 Fe min.
Sintered: 6.1-6.5 density; 34,000 psi TS; 50 Rock B. Hardened: 50,000 psi TS. For mechanical components. MPIF F-0008-P; ASTM B310 Type 2; Class C; SAE 855. *Obsolete*

BUNTING 207-IIIA
NL Industries
C 0-0.25, 97.75 Fe min.
Sintered: 6.5-6.9 density; 26,000 psi TS; 5.0 El. For mechanical components. MPIF F-0000-R; ASTM B310 Type 3; Class A. *Obsolete*

BUNTING 207-IIIB
NL Industries
C 0.26-0.6, 97.40 Fe min.
Sintered: 6.5-6.9 density; 34,000 psi TS; 0.0 El. Hardened: 50,000 psi TS. For mechanical components. MPIF F-0005-R; ASTM B310 Type 3; Class B. *Obsolete*

BUNTING 207-IIIC
NL Industries
C 0.61-1, 97.0 Fe min.
Sintered: 6.5-6.9 density; 44,000 psi TS; 1.0 El. Hardened: 64,000 psi TS. For mechanical components. MPIF F-0008-R; ASTM B310 Type 3; Class C. *Obsolete*

BUNTING 207-IVA
NL Industries
C 0-0.25, 97.75 Fe min.
Sintered: 6.9-7.3 density; 30,000 psi TS; 60 Rock F. For mechanical components. MPIF F-0000-S; ASTM B310 Type 4; Class A. *Obsolete*

BUNTING 207-IVB
NL Industries
C 0.26-0.6, 97.40 Fe min.
Sintered: 6.9-7.3 density; 50,000 psi TS; 4.0 El. Hardened: 70,000 psi TS. For mechanical components. MPIF F-0005-S; ASTM B310 Type 4; Class B. *Obsolete*

BUNTING 207-IVC
NL Industries
C 0.61-1, 97.0 Fe min.
Sintered: 6.9-7.3 density; 60,000 psi TS; 2.0 El. Hardened: 80,000 psi TS. For mechanical components. MPIF F-0008-S; ASTM B310 Type 4; Class C. *Obsolete*

BUNTING 208
NL Industries
Cu 15-25, Fe 69-84, C 0.61-1.
Sintered: 7.1 density min. 85,000 TS; 120,000 YS (comp). For mechanical components. SAE 872; ASTM B303 Class C. *Obsolete*

BUNTING 954
NL Industries
Cu 85, Fe 4, Al 11.
Cast: 102,000 TS; 45,000 YS; 18 El; 170 Brin. Bearings for heavy loads. SAE 954; AMS 4870B. *Obsolete*

BUNTING NO. 143
NL Industries
Cu 85, Sn 14, Pb 1, Zn 0.25, Fe 0.2, Sb 0.2, P 0.1.
Cast: 40,000 TS; 15,000 YS; 10 El; 85 Brin. For bearings; C.S. 24,000. *Obsolete*

BUNTING NO. 158
NL Industries
Cu 70, Sn 5, Pb 25, Fe 0.2, Sb 0.2, Ni 0.25, P 0.03.
Cast: 20,000 TS; 10,000 YS; 8 El; 38 Brin. For bearings; C.S. 13,000. *Obsolete*

BUNTING NO. 162
NL Industries
Cu 70, Sn 9, Pb 21, Fe 0.2, Sb 0.2, Ni 0.25, P 0.03.
Cast: 30,000 TS; 15,000 YS; 15 El; 55 Brin. For bearings; C.S. 20,000. *Obsolete*

BUNTING NO. 170
NL Industries
Cu 0.25, Sn 10, Fe 0.08, Sb 15, Ni 0.2, P 0.08, bal Pb.
Cast. For bearings; Babbitt. *Obsolete*

BUNTING NO. 183
NL Industries
Cu 63-67, Sn 3.5-4.5, Pb 3-5, Ni 19-21, bal Zn.
Cast: 30,000 TS; 17,000 YS; 8 El; 95 Brin. For bearings to resist food acids; corrosion resistant. *Obsolete*

BUNTING NO. 188
NL Industries
Sn 8-10, Pb 1-1.5, P 0.1, Ni 5, bal Cu.
For high strength castings; age hardenable. *Obsolete*

BUNTING NO. 327
NL Industries
Sn 10, Pb 2.5, Zn 1, P 0.1, Fe 0.2, Sb 0.2, bal Cu.
Cast: 40,000 TS; 18,000 YS; 20 El; 65 Brin. For bearings; CS 22,000. *Obsolete*

BUNTING NO. 78
NL Industries
Cu 66-70, Sn 3.5-4.5, Pb 26-30.
Cast: 18,000 TS; 8,000 YS; 5 El; 36 Brin. For bearings; plastic. *Obsolete*

BUNTING NO. 905
NL Industries
Cu 88, Sn 10, Zn 2, Fe 0.06, Sb 0.2, P 0.1.
Cast: 40,000 TS; 18,000 YS; 20 El; 67 Brin. For bearings for aircraft; CS 23,000. *Obsolete*

BUNTING NO. 907
NL Industries
Cu 89, Sn 11, P 0.2.
Cast: 40,000 TS; 18,000 YS; 20 El; 67 Brin. For gears; CS 23,000. *Obsolete*

BUNTING NO. 923
NL Industries
Cu 88, Sn 8, Pb 1, Zn 3, Fe 0.2, Sb 0.2, P 0.1.
Cast: 35,000 TS; 17,000 YS; 15 El; 65 Brin. For bearings; CS 17,000. *Obsolete*

BUNTING NO. 925
NL Industries
Cu 86.5, Sn 11, Pb 1.2, Fe 0.3, Sb 0.2, Ni 1, P 0.25.
Cast: 40,000 TS; 18,000 YS; 12 El; 75 Brin. For bearings; CS 23,000. *Obsolete*

BUNTING NO. 932
NL Industries
Cu 83, Sn 7, Pb 7, Zn 3, P 0.07, Fe 0.2, Sb 0.2.
Cast: 34,000 TS; 16,000 YS; 20 El; 58 Brin. For bearings; CS 20,000. *Obsolete*

BUNTING NO. 935

NL Industries
Cu 85, Sn 5, Pb 9, Zn 1, Fe 0.2, Sb 0.2, P 0.1.
Cast: 28,000 TS; 14,000 YS; 15 El; 50 Brin. For bearings; CS 17,000; low friction. *Obsolete*

BUNTING NO. 937

NL Industries
Cu 78.5-81.5, Sn 9-11, Pb 9-11.
Cast: 35,000 TS; 17,000 YS; 20 El; 60 Brin. For bearings for machinery and automobiles; CS 21,000. *Obsolete*

BUNTING NO. 938

NL Industries
Sn 8, Pb 15, Zn 1, Fe 0.2, Sb 0.2, P 0.05, bal Cu.
Cast: 30,000 TS; 14,000 YS; 15 El; 52 Brin. For bearings; CS 18,000. *Obsolete*

BUNTING NO. 941

NL Industries
Sn 4.5, Pb 20.5, Zn 1, Fe 0.2, Sb 0.2, P 0.05, bal Cu.
Cast: 22,000 TS; 11,000 YS; 12 El; 40 Brin. For bearings for electric motors; CS 14,000. *Obsolete*

BURDEN BEST

Burden Iron Co.
C 0.05, Si 0.1, 2.5% SiO 2.
For plates, pipe; wrought iron. *Obsolete*

BURDOX 9

Burdox Inc.
Cu 56-60, Sn 0-1, Mn 0.25-1, bal Zn.
For welding rod; melting point: 1400-1600°F.

BURDOX 91

Burdox Inc.
Cu 57-61, Sn 0.25-1, Pb 0-0.05, Al 0-0.01, bal Zn.
For welding rod; melting point: 1630-1650°F.

BURDOX 92

Burdox Inc.
Cu 58-60, Sn 0.6-0.9, Fe 0.35-0.5, Mn 0.15-0.3, Ni 0.25-0.4, bal Zn.
For welding rod.

BURGESS ALUMINUM SOLDER

English manufacture
Sn 76, Zn 21, Al 3.
For aluminum solder.

BURGESS-PARR NO. 85 ALLOY

Stainless Foundry & Engineering Inc.
Ni, Cr, Co, Mo.
For chemical apparatus; acid resistant. *Obsolete*

BURLOY A

Canada Electric Steel Castings Ltd.
C 0-0.12, Cr 16-23, Ni 7-11, bal Fe.
For stainless and corrosion resistant parts; stainless, heat and corrosion resistant.

BURLOY B

Canada Electric Steel Castings Ltd.
C 0.21-0.35, Cr 12-15, bal Fe.
For corrosion resistant parts; corrosion resistant.

BURLOY C

Canada Electric Steel Castings Ltd.
C 0-0.12, Cr 24-30, Ni 12-15, bal Fe.
For heat and corrosion resistant parts; heat and corrosion resistant.

BURLOY C-1

Canada Electric Steel Castings Ltd.
C 0.36-0.5, Cr 16-23, Ni 7-11, Si, bal Fe.
For heat and corrosion resistant parts; heat and corrosion resistant.

BURNDY NO. 101

Burndy Corp.
Cu.
Cast: 20,000 TS; 9,000 YS; 50 El; 35 Brin. For castings.

BURNDY NO. 102 AND 103

Burndy Corp.
Cast alloys from BURNDY NO. 101.

BURNDY NO. 111

Burndy Corp.
Zn 37, Fe 1.5, Al 1.5, Mn 0.25, bal Cu.
Cast: 70,000-80,000 TS; 30,000-34,000 YS; 25-40 El; 100-105 Brin. For castings; high strength. *Obsolete*

BURNDY NO. 113

Burndy Corp.
Cu 92.9, Sn 1.7, Pb 2.2, Zn 3.2.
Cast: 22,600 TS; 9,200 YS; 18 El. For castings; free-cutting. *Obsolete*

BURNDY NO. 113LM

Burndy Corp.
Li 1, bal Cu.
For electrical convectors; high conductivity. *Obsolete*

BURNDY NO. 115

Burndy Corp.
Burndy cast alloy. 30,000 psi TS min; 14,200 psi YS min; 20 El.

BURNDY NO. 202

Burndy Corp.
Cu 85, Sn 5, Pb 5, Zn 5.
Cast: 26,000-37,000 TS; 8,000-13,000 YS; 15-20 El; 45-59 Brin. For castings, bearings; pressure tight. *Obsolete*

BURNDY NO. 206

Burndy Corp.
Cu 80, Sn 10, Pb 10, Trace P.
Cast: 22,000-23,000 TS; 9,000-12,700 YS; 6-9 El; 46-65 Brin. For castings, bearings; pressure tight. *Obsolete*

BURNDY NO. 215

Burndy Corp.
Leaded commercial bronze, half hard. 50,000 psi TS min; 30,000 psi YS min; 14 El; 80% machinability.

BURNDY NO. 302

Burndy Corp.
Cu 89, Fe 1, Al 10.
Cast: 60,000-80,000 TS; 9,000-13,000 YS; 15-27 El; 93-100 Brin. For castings; Al bronze. *Obsolete*

BURNDY NO. 308

Burndy Corp.
Cu 88, Sn 10, Zn 2.
Cast: 30,000-45,000 TS; 10,000-16,000 YS; 15-30 El; 57-74 Brin. For castings; Gun Metal. *Obsolete*

BURNDY NO. 309

Burndy Corp.
Cu 86, Sn 9.5, Pb 2.5, Zn 2.
Cast: 30,000-40,000 TS; 17,000-19,000 YS; 15-25 El; 52-72 Brin. For castings; modified Gun Metal. *Obsolete*

BURNDY NO. 314

Burndy Corp.
Cu 96, Mn 1, Si 3.
Cast: 50,000 TS; 30,000 YS; 24 El; 103 Brin. For castings; "Everdur." *Obsolete*

BURNDY NO. 316

Burndy Corp.
Cu 88, Fe 4, Al 8.
Cast: 75,000 TS; 15-20 El; 100 Brin. For castings; Al bronze. *Obsolete*

BURNDY NO. 329

Burndy Corp.
Everdur 1014, soft. 88,000 psi TS; 44,000 psi YS; 25 El; 60% machinability.

BURNDY NO. 330

Burndy Corp.
Aluminum bronze. 55,000 psi TS min; 21,000 psi YS min; 6 El.

BURNDY NO. 335

Manufacturer not listed.
Spec. Aluminum-Nickel bronze. 90,000 psi min TS; 40,000 psi min YS; 6 El; 35% machinability.

BURNDY NO. 339

Burndy Corp.
Aluminium bronze. 85,000 psi TS min; 28,000 psi YS min; 10 El; 35% machinability.

BURNDY NO. 60

Burndy Corp.
Ductile cast iron. 60,000 psi TS min; 45,000 psi YS min; 10 El min.

BURNDY NO. 701

Burndy Corp.
Everdur 1015, soft. 40,000 psi TS; 15,000 psi YS; 50 El; 30% machinability.

BURNDY NO. 715

Burndy Corp.
DURIUM VI, hard. 75,000 psi TS; 15 El; 30% machinability.

BURNDY NO. 801

Burndy Corp.
Phosphor bronze, hard. 75,000 psi TS; 63,000 psi YS; 15 El; 80% machinability.

BURNDY NO. 802

Burndy Corp.
Phosphor bronze, hard. 94,000 psi TS; 13 El; 20% machinability.

BURNDY NO. 851

Burndy Corp.
High brass, half hard. 57,000 psi TS; 25,000 psi YS; 18 El; 100% machinability.

BURNDY NO. 86

Burndy Corp.
Gilding metal, hard. 57,000 psi TS min; 54,000 psi YS, typical; 6 El.

BURNDY NO. 87

Burndy Corp.
Gilding metal, hard. 50,000 psi TS min; 46,000 psi YS, typical; 5 El.

BURR-A

English manufacture
Cu 62, Zn 38.
For pipes, tubes, hardware.

BURR-B

English manufacture
Cu 90, Zn 10.
For window screen wire, radiators.

BURYS C. 16. B.

Burys & Co. Ltd.
C 0.5, Si 1, Cr 1.5, W 2, V 0.2, bal Fe.
For pneumatic tool pistons, chisels, riveters; tough and impact resistant.

BURYS H.18.S.

Burys & Co. Ltd.
C 1, Cr 1.5, bal Fe.
For ball and roller bearings, forming rolls; oil or water hardened.

BUSH BRAND BEARING METAL
Thomas Bolton Ltd.
P, alloy, bal Cu.
For bearings; fatigue and wear resistant. *Obsolete*

BUSH HAMMER
Colt Industries
C 1, V 0.1, bal Fe.
For tools, cutters. *Obsolete*

BUSHINGS
English manufacture
Cu 85-86.2, Sn 10.2-11, Zn 3.6-4.
For bushings, bearings; high strength.

BUSTER ALLOY
Columbia Tool Steel Co.
Tool material. C 0.5, W 2.2, Cr 1.25, Mn 0.25, Si 1, V 0.25, bal Fe.
Heat treated: 283,000 TS; 246,000 YS; 10 El; 53 Rock C.
Annealed: 190 Brin. For dies, chisels, shear blades, punches, forging dies. Type S1 hot work tool steel; shock resistant.

BUSTER ALLOY 50
Columbia Tool Steel Co.
Tool material. C 0.5, Si 0.8, W 2.25, Cr 1.35, V 0.25, bal Fe.
Oil hardened, shock resistant tool steel. For rivet busters, pneumatic chisels, punches. AISI S1.

BUSTER ALLOY 60
Columbia Tool Steel Co.
Tool material. C 0.58, Si 0.8, W 2.25, Cr 1.35, V 0.25, bal Fe.
Oil hardened, shock resistant tool steel. For rivet sets, chisels, heading dies. AISI S1.

BUSTER BRAND
Columbia Tool Steel Co.
C, Cr, W, bal Fe.
For tools, chisels, shear blades; shock resisting. *Obsolete*

BUTTON
American manufacture
Zn 80, Cu 20.
For die castings, ornaments; red brass.

BUTTON ALLOY-1
English manufacture
Cu 50-60, Zn 30-45, Sn 0-10.
For buttons, ornaments; corrosion resistant.

BUTTON ALLOY-2
English manufacture
Cu 43, Zn 57.
For buttons, ornaments; low strength.

BUTTON BRASS
English manufacture
Cu 90, Zn 10, Sn 0.5.
For buttons, ornaments.

BVR 30
Krupp Stahl AG
alloy steel.
See Krupp 34CUNIMO6

BVT 130V
Krupp Stahl AG
stainless steel.
See CRONIDUR 4922

BVT 30
Krupp Stahl AG
alloy steel. C 0.22, Mn 0.55, Cr 0-0.3, Mo 0.35, bal Fe.
For thick-walled high pressure tubes to operate up to 530°C.
W. Nr. 1.5419. *Obsolete*

BVT 35
Krupp Stahl AG
alloy steel.
See KRUPP 13CRMO44

BVT 40
Krupp Stahl AG
alloy steel.
See KRUPP 16CRMO44

BVT 50
Krupp Stahl AG
alloy steel. C 0.22, Cr 1.05, Mo 0.45, Ni 0-0.6, bal Fe.
Forgings for steam turbines. *Obsolete*

BVT 60
Krupp Stahl AG
alloy steel.
See KRUPP 24CRMOV55

BVT 90
German manufacture
C 0.3, Cr 13, Mo 2, V 1, bal Fe.
For turbine rotors; high heat resistant.

BVT-130
Bochumer Verein
C 0.2, Cr 12, Mo 1.2, bal Fe.
Annealed: 82,000 TS; 38,000 YS; B 82 Rock. For chemical plant and oil refinery equipment, gears, shafts, marine hardware. Corrosion resistant, hardenable. *Obsolete*

BW XII
Thyssen Edelstahlwerke AG
C 0.1, Mn 1.3, Si 0.1, Ni 0.4, bal Fe.
For welding electrodes; general purpose, gas. *Obsolete*

BX-3
Now JESSOP BX-3.

BYLOY GR. W2
A.M. Byers Co.
C 0-0.3, Mn 1.1-1.6, 0.10% Mo min, 0.10% Cu min, 0.0005% B min, bal Fe.
Rolled: 105,000 TS; 80,000 YS; 21 El; 48 RA; 210 Brin. Heat treated: 145,000 TS; 135,000 YS; 15 El; 55 RA; 340 Brin. For shovels, ore chutes, dippers, graders, drag buckets; abrasion and wear resistant. *Obsolete*

BYRWILL "B"
Byrwill Co.
Zn 91-93, Pb 2-3, Cu 6-7.
28,000 TS; 67 Brin. For bearings; antifriction

C 1023
Union Carbide Corp.
C 0.15, Cr 15.5, Co 9.7, Mo 8.4, Ti 3.6, Al 4.2, bal Ni.
For nozzle vanes.

C 105W2
Westig (U.K.) Ltd.
Carbon steel. C 1.05, Si 0.2, Mn 0.35, P 0-0.03, S 0-0.03, bal Fe.
Carbon tool steel equivalent to JIS SK3.

C 18
Bergische Stahl Industrie
C 0.15-0.23, Si 0-1, Mn 0-1, Cr 16-18, Ni 1.5-2.5, bal Fe.
Martensitic stainless steel. W.-Nr. 1.4057. Similar to AISI 431.

C 18 8 S
Bergische Stahl Industrie
C 0-0.07, Si 0-2, Mn 0-2, Cr 17.5-20, Ni 9-11, Mo 0-0.7, bal Fe.
Austenitic stainless cast steel for low temperature service. SEW 6902; G-X 6 CrNi 18 10.

C 18 8 SS
Bergische Stahl Industrie
C 0-0.1, Si 0-1, Mn 0-2, Cr 17-19, Ni 10-12, Mo 0-0.5, Nb 1, bal Fe.
Austenitic stainless cast steel for low temperature service. SEW 1.6905; G-X 7 CrNiNb 18 10.

C 18 E
Bergische Stahl Industrie
C 0.33-0.43, Si 0-1, Mn 0-1, Cr 15.5-17.5, Mo 1-1.3, Ni 0-1, bal Fe.
Martensitic stainless steel. W.-Nr. 1.4122.

C 18 W
Bergische Stahl Industrie
C 0-0.1, Si 0-1, Mn 0-1, Cr 15.5-17.5, bal Fe.
Ferritic stainless steel. W.-Nr. 1.4016.

C 40
German manufacture
60 Al_2O_3, 40 mixed carbides.
For hard cutters to machine steel and cast iron; Sintered Alloy.

C 46
Italian manufacture
Cu 10, Si 1, Mg 0.25, Ti 0.15, Ni 1.5, bal Al.
For castings; heat resistant.

C C M
Now SIMONDS CCM.

C N M
Johnson Matthey
C, Cr, Ni, bal Fe.
For alloy steel castings; wear resistant. *Obsolete*

C PHOSPHOR BRONZE
Criterion Metals, Inc.
Copper. Cu 91.82, Zn 0-0.2, Sn 8, P 0.18, Fe 0-0.1, Pb 0-0.05.
Thin gauge sheet, various tempers: 56-110 ksi TS min; 23-105 ksi YS min. ASTM B-103. C52100

C Q 12
TRW Inc.
WC 90.7, Co 9, 0.3 TaC.
340,000 transverse strength; 91.0 Rock A; density 14.60.
Sintered carbide cutting tool; high shock resistance; for heavy cuts or rough work including slow speed and heavy feed conditions.

C Q 13
TRW Inc.
WC 85, Co 15.
400,000 transverse strength; 88.0 Rock A; density 14.0.
Sintered carbide tool material.

C Q 14
TRW Inc.
WC 87, Co 13.
400,000 transverse strength; 89.3 Rock A; density 14.20.
Sintered carbide tool material.

C Q 2
TRW Inc.
WC 93.7, Co 6, 0.3 TaC.
290,000 transverse strength. 92.0 Rock A; density 14.95.
Sintered carbide cutting tool; very good shock and wear resistance; for general machining.

C Q 22
TRW Inc.
WC 89.5, Co 10, 0.5 TaC.
400,000 transverse strength; 91.2 Rock A; density 14.51.
Sintered carbide cutting tool; high load strength, for slow speed operations under heavy load, as in high temperature alloys.

C Q 23
TRW Inc.
WC 93.5, Co 6, 0.5 TaC.
270,000 transverse strength; 92.5 Rock A; density 14.95.
Sintered carbide cutting tool; exceptional abrasion resistance and good shock resistance; for machining high temperature alloys.

C Q 3
TRW Inc.
WC 95.5, Co 4, 0.5 TaC.
250,000 transverse strength; 92.3 Rock A; density 15.10.
Sintered carbide cutting tool; excellent abrasion resistance, for high speed finishing cuts.

C Q 4
TRW Inc.
WC 96.5, Co 3, 0.5 TaC.
240,000 transverse strength; 92.7 Rock A; density 15.15.
Sintered carbide cutting tool; very high wear resistance, for high speed finishing.

C-1 H-19
Teledyne Firth Sterling
Tool material. 92.0 WC, 8.0 binder.
For general wear applications. 92.0 Rock A.

C-10 CE-301
Teledyne Firth Sterling
Tool material.
For cold extrusion punches. 89.0 Rock A.

C-10 H-91
Teledyne Firth Sterling
Tool material. WC 88.5, 11.0 binder, 0.5 TiC + TaC + NbC.
For wire bar and tube drawing dies, powdered metal compacting dies, seal rings, cold heading and impact extrusions. 89.7 Rock A.

C-103
Now COLUMBIUM C-103.

C-11 CE-304
Teledyne Firth Sterling
Tool material.
For cold extrusion punches. 88.0 Rock A.

C-11 H-05
Teledyne Firth Sterling
Tool material. WC 85, 15.0 binder.
For wire bar and tube drawing dies, powdered metal compacting dies, seal rings, cold heading and impact extrusions. 90.3 Rock A.

C-11 H-81
Teledyne Firth Sterling
Tool material. WC 86.5, 13.0 binder, 0.5 TiC + TaC + NbC.
For wire bar and tube drawing dies, powdered metal compacting dies, seal rings, cold heading and impact extrusions. 88.6 Rock A.

C-11 R-72
Teledyne Firth Sterling
Tool material. WC 90, 10.0 binder.
For rod mill rolls. 89.0 Rock A.

C-12 H-71
Teledyne Firth Sterling
Tool material. WC 83.5, 16.0 binder, 0.5 TiC + TaC + NbC.
For wire bar and tube drawing dies, powdered metal compacting dies, seal rings, cold heading and impact extrusions. 87.3 Rock A.

C-12 R-61
Teledyne Firth Sterling
Tool material. WC 85, 15.0 binder.
For rod mill rolls. 86.0 Rock A.

C-13 ND-20
Teledyne Firth Sterling
Tool material. WC 75, 20.0 binder, 5.0 TiC + TaC + NbC.
For heading dies. 86.0 Rock A.

C-13 R-52
Teledyne Firth Sterling
Tool material. WC 82, 18.0 binder.
For rod mill rolls. 84.0 Rock A.

C-13 RMA-3
Teledyne Firth Sterling
Tool material.
For rod mill rolls.

C-13 RMA-5
Teledyne Firth Sterling
Tool material. WC 80, 20.0 binder.
For rod mill rolls. 82.5 Rock A.

C-14 ND-25
Teledyne Firth Sterling
Tool material. WC 70, 25.0 binder, 5.0 TiC + TaC + NbC.
For heading dies. 83.0 Rock A.

C-14 ND-30
Teledyne Firth Sterling
Tool material. WC 70, 30.0 binder.
For heading dies. 81.0 Rock A.

C-207
General Electric Co.
W 7.5, Zr 0.8, Ti 0.2, C 0.1, bal Cr.
For gas turbine engines. High stress-rupture strength above 1800 F. High resistance to thermal shock and fatigue, excellent oxidation resistance. *Obsolete*

C-242
Rolls-Royce Mfg. Co.
Nickel. C 0.3, Co 10, Cr 20, Mo 10, Fe 0-1, bal Ni.
For high temperature applications, jet engine and gas turbine components. Heat and oxidation resistant.

C-276
Now HASTELLOY ALLOY C-276 and NICKELVAC H-C276.

C-4 HOT DIE STEEL
Delsteel Inc.
C 0.95, Cr 4, bal Fe.
For hot work dies, gripper dies, bolt and rivet header dies; hot work steel.

C-501
F.A.C.A.
Mg 2.1, Mn 1.5, bal Al.
For light alloy parts; similar to "Pearlman."

C-502
F.A.C.A.
Si 11-14, bal Al.
For light alloy parts; similar to "Silumin."

C-503

F.A.C.A.
Cu 4, Ti 0.4, bal Al.
Heat treated: 40,000-43,000 TS; 26,000-27,000 YS; 4-5 El;
80-90 Brin. For light alloy parts; age hardenable.

C-55

Heppenstall Co.
C 0.5, Cr 0.75, Ni 1.25, Mo 0.3, bal Fe.
For dies; oil hardening. *Obsolete*

C-67

Swiss manufacture
Sn 5, Al 7, bal Cu.
Non-staining, gold color. For watch cases.

C-74 ARMOR

Duraloy
C 0.25-0.3, Mn 1-1.6, Si 0.3-0.4, Cr 0.5-1, Ni 0.75-1.45, Mo
0.4-0.6, bal Fe.
Cast: 120,000 TS; 95,000 YP; 18 El; 40 RA; 225-270 Brin. For
gears, countershafts, axles, crankshafts. Oil hardening,
tough, shock resistant.

C-BRITE 29-4

Airco Vacuum Metals
Cr 29, Mo 4, bal Fe.

C-CAST ALLOYS

Cominco Metals
Lead.
See COMINCO ALLOYS.

C-W

Great Western Steel Co. ·
C 0.9, Mn 1.1, Cr 0.5, W 0.5, Si 0.35, bal Fe.
For drills, milling cutters, dies; oil hardened.

C-XB VALVE STEEL

Now CARPENTER C-XB VALVE STEEL.

C.B.F. CHROMIUM BRONZE

Stone Manganese - J. Stone & Co. Ltd.
Sn 1.5-3.5, Cr 0.5-2, Fe 0.5-3, Mn 0-0.5, bal Cu.
47,000 TS; 21,000 YS; 30-40 El; 90-130 Brin. For engine valve
guides, gears, rock drill twist nuts; heavy duty.

C.C. CHISEL

Crucible Specialty Metals
C 0.47, Cr 0.7, W 1.1, bal Fe.
For punches, pneumatic chisels; shock and fatigue resistant.
Obsolete

C.C. CHISEL

Colt Industries
C 0.47, Cr 0.7, W 1.1, bal Fe.
For punches, pneumatic chisels; shock and fatigue resistant.
Obsolete

C.C.A.

Colt Industries
C 1.5, Cr 12, Mo 0.8, V 0.25, bal Fe.
For blanking dies, punches, plug gauges, wire drawing dies;
non-deforming. *Obsolete*

C.C.S. DIE

Colt Industries
C, Cr, W, Si, bal Fe.
For tools, dies. *Obsolete*

C.C.S. DIE CASTING

Hawkridge Bros Co.
C 0.35, Si 1.15, Cr 5.25, Co 0.4, W 4.25, bal Fe.
For Al or brass die casting dies; oil or air hardened. *Obsolete*

C.D.C. MANGANESE ALLOY NO. 762

Chicago Development Corp.
Mn 62-65, bal Cu.
Quenched and aged: 78,000 TS; 38,000 YS; 35 El; 70-75
Rock B. High electrical resistivity and damping capacity.

C.D.C. MANGANESE ALLOY NO. 780

Chicago Development Corp.
Mn 80, Cu 20.
Quenched: 68,000 TS; 24,000 YS; 35 El. Cold rolled: 130,000
TS; 115,000 YS; 9 El. High electrical resistivity and high
damping properties.

C.D.C. NO. 595

Chicago Development Corp.
Cu 5, Sn 95.
Wire; for electronic uses. *Obsolete*

C.D.C. NO. 720

Chicago Development Corp.
Cu 60, Ni 20, Mn 20.
Hardenable by cold work and/or heat treatment. Annealed:
98,000 TS; 80,000 YS; 30 El; 140 Vickers. Hardened: 200,000-
220,000 TS; 475-515 Vickers. For springs, diaphragms.

C.D.C. NO. 772

Chicago Development Corp.
Mn 72, Cu 18, Ni 10.
Wrought: 115,000 TS; 95,000 YS; 6.5 El; 220 Vickers.
Electrical resitivity: 1050 ohm/circular mil-ft. Low temperature
coefficient of resistance. For low temperature resistance
applications, rheostats, circuit breaker parts, miscllaneous
electrical appliance controls.

C.D.W

Acieries de Champagnole
C 0.9, Cr 4.25, W 1.5, Mo 4.25, V 1.3, bal Fe.
Air or oil hardens to 62-65 Rock C. For drills, reamers,
broaches, and threading tools. AFNOR: Z 90 CDWV 04.04.

C.E.S. NO. 2

Industrial Steels Inc.
C 0.36-0.5, Ni 16-23, Cr 7-11, bal Fe.
For heat and corrosion resistant parts; heat and corrosion
resistant.

C.F.S.

Latrobe Steel Co.
Tool material. C 0.9-1.3, Cr 0.5, bal Fe.
For dies, draw dies; water hardened. AISI W5. *Obsolete*

C.H.Q. DIE STEEL

Teledyne Firth Sterling
Steel. C, bal Fe.
For cold header dies; cold work steel. *Obsolete*

C.H.W.

Latrobe Steel Co.
Tool material. C 0.5, Cr 2.75, V 0.5, W 15, bal Fe.
For dies, punches, shear blades, trimmers; hot work steel.
Obsolete

C.I.G. NO. 400

Compressed Industrial Gases Co.
Sn 10, bal Cu.
For bronze welding rod; fluxed.

C.I.G. NO. 401

Compressed Industrial Gases Co.
Mn, Sn, bal Cu.
For bronze welding rod; fluxed.

C.L.W.

Latrobe Steel Co.
Tool material.
Now CLW.

C.M.A. BEARING METAL

English manufacture
Pb, alkali earth metals.
For bearings.

C.M.A.1

Stone Manganese - J. Stone & Co. Ltd.
Mn 10-15, Al 7-8, Fe 2.5-3.5, bal Cu.
Cast: 94,000-105,000 TS; 40,000-49,000 YS; 20-35 El;
165-200 Brin. For propellers, marine hardware. Aluminum
bronze.

C.M.P

Cold Metal Products Company, Inc.
C 0.2, Ni 0.7, bal Fe.
For machinery parts; precision cold rolled strip steel.
Obsolete

C.M.S.

Miller Steel Co.
C 0.1-0.2, Mn 1.2-1.5, S 0.075-0.15, bal Fe.
Rolled: 80,000 TS; 75,000 YS; 22 El; 49 RA; 192 Brin. For
machinery parts; free machining.

C.M.V

Sanderson Kayser Ltd.
C 0.38, Si 1.05, Mn 0.35, Cr 5.25, Mo 1.35, V 1, bal Fe.
Hot work tool steel. B.S. 4659 Type BH13. AISI H13.

C.N. DIE

H. Boker & Co.
C, Cr, bal Fe.
For tools. *Obsolete*

C.R.U.

S.K.F. Industries Inc.
C, bal Fe.
For tools. *Obsolete*

C.R.U. COMPOSITE

S.K.F. Industries Inc.
C, Cr, bal Fe.
For tools. *Obsolete*

C.R.U. EXCELSIOR

S.K.F. Industries Inc.
C, alloy, bal Fe.
For high speed cutting tools. *Obsolete*

C.R.U. EXCELSIOR EXTRA

S.K.F. Industries Inc.
C, alloy, bal Fe.
For high speed cutting tools. *Obsolete*

C.R.U. EXCELSIOR EXTRA XII

S.K.F. Industries Inc.
C, alloy, bal Fe.
For high speed cutting tools. *Obsolete*

C.R.U. EXCELSIOR XXII

S.K.F. Industries Inc.
C, alloy, bal Fe.
For high speed cutting tools. *Obsolete*

C.S.M.

Crucible Specialty Metals
C 0.7-1.4, Mn 0.75, Cr 0.75, bal Fe.
For wrenches, valve parts, studs, bolts, chisels, springs,
shafts, gears; tough, fatigue resistant. *Obsolete*

C.S.M.

Colt Industries
C 0.7-1.4, Mn 0.75, Cr 0.75, bal Fe.
For wrenches, valve parts, studs, bolts, chisels, springs,
shafts, gears; tough, fatigue resistant. *Obsolete*

C.S.M. STEEL

Evans Steel Co.
C 0.15, Cr 0.3-0.7, bal Fe.
Oil treated: 110,000 TS; 25 El; 68 RA; 219 Brin. For springs,
gears, axles, shafts; tough.

C.T. METAL
American Smelting & Refining Co.
Pb alloy.
For slush castings, statues; low MP.

C.T.C. FERRY
Henry Wiggin & Co. Ltd.
Ni 45, Cu 55.
Annealed: 73,000 TS; 34,000 YS; 47 El; 77 RA; 155 Brin. For instrument resistances; heat and corrosion resistant. *Obsolete*

C.T.U.
Eagle & Globe Steel Ltd.
Tool material. C 0.36, Si 1, Cr 5, Mo 1.3, W 1.4, V 0.4, bal Fe. Hot work steel similar to ADIC. AS1239 H12A; AISI H12; Werkstoff 1.2606.

C.T.V. HOT WORK
Jessop-Saville Ltd.
C 0.4, Cr 3.5, W 2.4, V 0.5, bal Fe.
For bolt and extrusion dies; hot work steel; oil hardening.

C.V.M.
Great Western Steel Co.
C 1, Mn 0.7, Cr 5, Mo 1.1, V 0.25, bal Fe.
For dies, blanking, trimming and forming dies; air hardening.

C.V.S.
British Steel Corp.
C 0.4, V 0.2, Cr 1.5, bal Fe.
Annealed: 110,000 TS; 90,000 YS; 20 El; 50 RA; 277 Brin. Heat treated: 130,000 TS; 120,000 YS; 17 El; 50 RA; 302 Brin. For locomotive crank pins; shock resistant. *Obsolete*

C.V.S.S.
British Steel Corp.
C 0.5, V 0.2, Cr 1.5, bal Fe.
Heat treated: 100,000 ±10,000 TO; 170,000-190,000 YS, 8-12 El; 20-30 RA; 430-437 Brin. For springs; shock resistant. *Obsolete*

C.W. CHISEL
Cytemp Specialty Steel Div.
C 0.7, bal Fe.
For tools, chisels; oil hardening. *Obsolete*

C.W. OIL
Hoyland Steel Co.
C 0.85-0.95, Mn 1-1.2, Cr 0.4-0.6, W 0.4-0.6, bal Fe.
For tools, dies; non-deforming.

C.W.3
Acieries de Champagnole
C 1.25, W 3, Mn 0-0.4, Si 0.3, bal Fe.
Water hardening tool steel. For dies for cartridge cases, threading brass and stamping light alloys and thin steel sheet. AFNOR: 125 W30; AISI F2.

C.Y. ALLOY
Fellsain Wycliffe Foundries, Ltd.
C 3.2, Si 1.8, Cr 1, bal Fe.
Cast: 60,000-70,000 TS; 0.5 El; 250 Brin. For castings, gears; abrasion resistant, cast iron. *Obsolete*

C.Y.W. CHOICE DIE
Teledyne Firth Sterling
C 1, Cr 3.5, Mn 0.5, bal Fe.
Hardened: 425-625 Brin. For dies, mandrels, nut piercers, swaging and piercing dies; hot work steel. *Obsolete*

C120AV
Now CRUCIBLE C 120AV.

C129
Now COLUMBIUM C-129.

C196
Anaconda Co.
Copper. Cu 98.7, Fe 1, P 0.3.
Annealed: 45,000 psi TS; 21,000 psi YS; 36 El. Hard: 62,000 psi TS; 59,000 psi YS; 5 El. Electrical conductivity: 74-76% IACS. Good strength, good conductivity; retains fine grain to 850°C. For current conducting springs.

C20CB3
Now CARPENTER STAINLESS NO. 20 CB-3.

C263
Now WIGGIN ALLOY 263.

C355
Fansteel/Wellman Dynamics
Aluminum. Cu 1-1.5, Si 4.5-5.5, Fe 0.2, Mn 0.1, Zn 0.1, Mg 0.4-0.6, Ti 0.2, bal Al.
Aluminum casting alloy. Cast: 41,000-50,000 psi TS; 31,000-40,000 psi YS; 2-3 El.

C355.0
Bunting Bearings Corp.
Aluminum. Cu 1-1.5, Si 4.5-5.5, Fe 0-0.2, Mn 0-0.1, Zn 0-0.1, Mg 0.4-0.6, Ti 0-0.2, bal Al.
MIL-A-21180D specification alloy C355.0.

C5152
Comalco Ltd.
Mg 2, bal Al.
O-temper: 186 MPa TS; 76 MPa YS; 24 El; 48 Vickers. H38-temper: 283 MPa TS; 241 MPa YS; 5 El; 92 Vickers. Sheet metal work, hydraulic tube.

C5154
Comalco Ltd.
Mg 3.5, Cr 0.25, bal Al.
O-temper: 241 MPa TS; 117 MPa YS; 27 El; 59 Vickers. H32-temper: 269 MPa TS; 207 MPa YS; 15 El; 92 Vickers. Storage tanks, welded structures.

C62
Keystone Carbon Co.
High graphite bronze. Sintered: 11,000 psi TS; 16% porosity; 6.0-6.4 density; 25 Rock H. For bearings requiring quiet motor operation.

C63 QUALITY
Delta Metal (BW) Ltd.
Ni 10, Cu, bal Zn.
Rolled: 80,000 TS; 44,000 YS; 22 El; 12 RA; 145 Brin. Drawn: 90,000 TS; 48,000 YS; 20 El; 10 RA; 150 Brin. For hot stampings, hardware, architectural trim; silver bronze. *Obsolete*

C64 5.8-6.2
Keystone Carbon Co.
Sn 9.5-10.5.
Sintered: 11,000 psi TS; 25% porosity; 25 Rock H; 5.8-6.2 density. For light loaded bearings. ASTM B438-83A Gr. 1, Class B, Type I; SAE 840.

C64 6.4-6.8
Keystone Carbon Co.
Sn 9.5-10.5.
Sintered: 13,000 psi TS; 18% porosity, 40 Rock H; 6.4-6.8 density. For bearings. ASTM B438-83A Gr. 1, Class B, Type II; SAE 841.

C64, 6.81-7.2 DENSITY
Keystone Carbon Co.
Sintered Cu-Sn bronze. 6.81-7.2 density; 20,000 psi TS; 12% porosity; 60 Rock H. Self lubricating for moving parts. ASTM B255-83A, Type II.

C65
Keystone Carbon Co.
Staking grade phosphor bronze. Sintered: 25,000 psi TS; 10.0 El; 14% porosity; 7.0-7.4 density; 80 Rock H. High density bearing; good ductility; can be staked.

C71
Keystone Carbon Co.
Leaded bronze. Sintered: 12,000 psi TS; 18% porosity; 1.5% El; 6.5-6.9 density; 30 Rock H. For bearings difficult to keep oiled. ASTM B438-83A Gr. 2, Class B, Type I; SAE 843.

C80
Sintered Products Ltd.
Pure copper sintered. 8-11 porosity. Good electrical properties.

C90 QUALITY
Delta Metal (BW) Ltd.
Cu 55-60, Pb 0-0.2, Mn 2-3, bal Zn.
Extruded: 70,000 TS; 32,000 YS; 30 El; 30 RA; 120 Brin. Drawn: 76,000 TS; 38,000 YS; 20 El; 25 RA; 130 Brin. For high pressure air service tubes; Parson's manganese bronze. *Obsolete*

C92 QUALITY
Delta Metal (BW) Ltd.
Cu 55-60, Pb 0-0.2, Mn 2-3, bal Zn.
Rolled: 76,000 TS; 38,000 YS; 25 El; 30 RA; 135 Brin. For disc and slotted valves; Parson's manganese bronze. *Obsolete*

CA DOUBLE DIAMOND
Colt Industries
C 1, Cr 1.5, Mo 0.35, bal Fe.
For rock drills; hollow drill steel. *Obsolete*

CA NICKEL
Henry Wiggin & Co. Ltd.
Co 4, bal Ni.
For magnetostrictive transducers; high electromechanical coupling coefficient. *Obsolete*

CA-10, ETC.
Now CARMET CA-10, ETC.

CA6NM
American manufacture
C 0-0.06, Mn 0-1, Si 0-0.65, Cr 11.5-14, Ni 3-5, Mo 0-1, bal Fe.
Corrosion resistant casting alloy. ASTM A296-77 CA-6NM.

CABLE WRAP
Texas Instruments Inc./Materials Control
Copper clad 430 stainless steel copper. Volume ratio 16/68/16. Used to armor buried telephone cables and wire providing corrosion, electrical, and gopher protection.

CABOT ALLOY NO. 625
Haynes International, Inc.
Nickel. Co 0-1, Cr 20-23, Mo 8-10, Fe 0-5, Si 0-0.5, Mn 0-0.5, C 0-0.1, Al 0-0.4, Ti 0-0.4, 3.15-4.15 Cb + Ta, bal Ni.
High temperature and corrosive applications.

CABOT ALLOY NO. 800
Haynes International, Inc.
Ni 30-35, Cr 19-23, Al 0.15-0.6, Ti 0.15-0.6, Si 0-1, Mn 0-1.5, Co 0-2, Cu 0-0.75, C 0-0.1, bal Fe.
Fe-Ni-Cr wrought alloy with good resistance to oxidation and carburization at elevated temperatures. Petroleum industry. UNS NO8800; ASTM B408 and B409.

CABOT ALLOY NO. 800 H
Haynes International, Inc.
Ni 30-35, Cr 19-23, Al 0.15-0.6, Ti 0.15-0.6, Si 0-1, Mn 0-1.5, Co 0-2, Cu 0-0.75, C 0.05-0.1, bal Fe.
Solution heat-treated version of No. 800 for improved elevated temperature properties. UNS NO8800; ASTM B408 and B409.

CABOT ALLOY NO. 825
Haynes International, Inc.
Nickel. Ni 38-46, Co 0-2, Cr 19.5-23.5, Mo 2.5-3.5, Cu 1.5-3, Fe 22, Si 0-0.5, Mn 0-1, C 0-0.05, Al 0-0.2, Ti 0.6-1.2, S 0 0.03.
Resistance to acids and seawater. Plate, sheet, strip, billet, bar, wire, pipe, tubing for chemical process equipment, propeller shafts, tank trucks.

CABOT ALLOY NO. R-41

Haynes International, Inc.
Nickel. Cr 18-20, Fe 0-5, C 0.05-0.12, Si 0-0.5, Co 10-12, Mn 0-0.1, Ti 3-3.3, Mo 9-10.5, Al 1.4-1.6, bal Ni.
High temperature strength for aircraft components.

CADMET

Castings Development Co.
WC + Co.
For cutting tool; sintered carbides.

CADMIUM

Atomergic Chemetals Corp.
Cd.
Purities, zone refined: 99.9999%, 99.9995%, 99.999%, 99.9+% (nuclear grade). Forms: bar, rod, sheet, foil, shot, powder, wire, single crystals.

CADMIUM AMALGAM

English manufacture
Cd, bal Hg.
For modeling purposes, fusible plugs; softens when moderately heated, soft as wax.

CADMIUM COPPER

Chase Brass & Copper Co., Inc.
Copper. Cu 99, Cd 1.
Annealed: 36,000 psi TS; 12,000 psi YS; 57 El. Hard drawn: 73,000 psi TS; 69,000 psi YS; 9 El. High conductivity; for trolley wire, connectors, switch gear.

CADMIUM COPPER 985

Olin Brass, Indianapolis
Cu 99.1, Cd 0.9.
Hard: 55,000 psi TS; 48,000 psi YS; 6 El; 116 Brin. Soft: 37,000 psi TS; 12,000 psi YS; 50 El; 45 Brin. For trolley wire and marine hardware; corrosion resistant. *Obsolete*

CADMIUM COPPER, 0.1% CADMIUM

Chase Brass & Copper Co., Inc.
Cu 99.9, Cd 0.1.
Annealed: 32,000 TS; 45 El. Rolled hard: 51,000 TS; 6 El. High conductivity, resistance to softening. Cooling fins for radiators and air conditioners.

CADMIUM COPPER, 0.2% CADMIUM

Chase Brass & Copper Co., Inc.
Cu 99.8, Cd 0.2.
Annealed: 32,000 TS; 45 El. Rolled hard: 51,000 TS; 6 El. Good conductivity, resistance to softening. For electrical contacts, terminals, springs.

CADMIUM SILVER

English manufacture
Cd, bal Ag.
For corrosion resistant parts.

CAF 303

Carondelet Foundry Co.
C 0-0.16, Mn 0-1.5, Si 0-2, Cr 18-21, Ni 9-12, Mo 0-1.5, Se 0.2-0.35, bal Fe.
Free-machining, corrosion resistant steel casting. ACI CF-16F; ASTM A-296.

CAF 304

Carondelet Foundry Co.
C 0-0.08, Mn 0-1.5, Cr 18-21, Ni 8-11, bal Fe.
Corrosion resistant steel casting. ACI CF-8; ASTM A-296, A-351 CF-8.

CAF 310

Carondelet Foundry Co.
C 0-0.2, Mn 0-2, Si 0-2, Cr 23-27, Ni 19-22, bal Fe.
Corrosion resisting steel casting; for high temperature operation. ACI CK-20; ASTM A-351 CK-20.

CAF 312

Carondelet Foundry Co.
C 0-0.3, Mn 0-1.5, Si 0-2, Cr 26-30, Ni 8-11, bal Fe.
Corrosion resistant steel casting. ACI CE-30; ASTM A-296.

CAF 317

Carondelet Foundry Co.
C 0-0.08, Mn 0-1.5, Si 0-1.5, Cr 18-21, Ni 9-13, Mo 3-4, bal Fe.
Corrosion resistant steel casting. ACI CG-8M; ASTM A-296.

CAF 318

Carondelet Foundry Co.
C 0-0.08, Mn 0-1.5, Si 0-2, Cr 18-21, Ni 9-12, Mo 2-3, Cb 0.7, bal Fe.
Corrosion resistant steel casting; weldable.

CAF 320

Carondelet Foundry Co.
C 0-0.08, Mn 0-1.5, Si 0-1, Cr 18-21, N 9-12, Mo 2-3, bal Fe.
Corrosion resistant steel casting. ACI CF-8M; ASTM A-351 CF8M.

CAF 327

Carondelet Foundry Co.
C 0-0.5, Mn 0-1.5, Si 0-2, Cr 26-30, Ni 4-7, Mo 0-0.5, bal Fe.
Corrosion resistant steel casting. ACI-HD.

CAF 329H

Carondelet Foundry Co.
C 0-0.2, Mn 0-1.5, Si 0-2, Cr 22-26, Ni 12-15, bal Fe.
Corrosion resisting steel casting. ACI CF-20; AMS 5358.

CAF 379

Carondelet Foundry Co.
C 0.2-0.5, Mn 0-2, Si 0-2, Cr 26-30, Ni 8-11, Mo 0-0.5, bal Fe.
Corrosion resistant steel casting. ACI HE; ASTM A-297.

CAF 404

Carondelet Foundry Co.
C 0-0.12, Mn 0-1, Si 0-2, Cr 11-13, Ni 0-1, Cu 1.5-2, bal Fe.
Cast stainless steel.

CAF 420

Carondelet Foundry Co.
C 0.2-0.4, Mn 0-1, Si 0-1.5, Cr 11.5-14, Ni 0-1, Mo 0-0.5, bal Fe.
Cast stainless steel. ACI CA-40; ASTM A-296.

CAF 509

Carondelet Foundry Co.
C 0-0.5, Mn 0-2, Si 0-2, Cr 24-28, Ni 11-14, Mo 0-0.5, bal Fe.
Corrosion resistant steel casting. ACI HH.

CAF 602

Carondelet Foundry Co.
C 0.2-0.4, Mn 0-2, Si 0-2, Cr 18-23, Ni 8-12, Mo 0-0.5, bal Fe.
Corrosion resistant steel casting. ACI HF; ASTM A-297.

CAF 620

Carondelet Foundry Co.
C 0-0.07, Mn 0-1.5, Si 0-1.5, Cr 19-22, Ni 27.5-30.5, Mo 2-3, Cu 3-4, bal Fe.
Corrosion resistant steel casting. ACI CN-7M; ASTM A-351 CN-7M.

CAF 720

Carondelet Foundry Co.
C 0-0.07, Mn 0-1.5, Si 0-1.5, Cr 19-22, Ni 27.5-30.5, Mo 2-3, Cu 3-4, 0.6-1.0 Cb + Ta, bal Fe.
Corrosion resistant steel casting.

CAF 78

Carondelet Foundry Co.
C 0-0.5, Mn 0-1, Si 0-2, Cr 26-30, Ni 0-4, Mo 0-0.5, bal Fe.
Corrosion resistant steel casting. ACI HC.

CAF 820

Carondelet Foundry Co.
C 0-0.07, Mn 0-1.5, Si 0-1.5, Cr 19-22, Ni 35-38, Mo 2-3, Cu 3-4, Cb 1, bal Fe.
Corrosion resisting steel casting.

CALCIUM

Atomergic Chemetals Corp.
Ca.
Purities: distilled 99.9%, 99.5%, 99%, 98%. Forms: granule, ingot, extruded form, sheet, powder.

CALDUR

Uddeholm Corp.
C 0.25, Cr 12, W 7, Co 4, bal Fe.
Tempered at 600°F: 50.5 Rock C. Tempered at 900°F: 54 Rock C. Tempered at 1200°F: 45 Rock C. For die casting dies for brass and aluminum, forging and extrusion dies. Hot work tool steel. Resists heat checking. Good hardness and toughness. *Obsolete*

CALIDO (ELALCO)

Abex Corp.
Ni 64, Cr 8, Mn 3, Fe 25.
For electrical resistances; heat resistant. *Obsolete*

CALIDO (ELALCO)

Driver Harris Co.
Ni 60, Cr 16, Fe 24.
100,000 TS; 56,000 YS; 28 El; 59 RA. For dipping baskets, electrical resistance materials; resists oxidation up to 1000 C. *Obsolete*

CALITE "C"

Calorizing Co.
Ni 40, Cr 5.5, Fe 60, Al 4.1.
For temperature resisting alloys; heat resisting up to 1800 F. *Obsolete*

CALITE "S"

Calorizing Co.
C 0.04, Cr 17, Al 0.08, bal Fe.
75,000 TS; 50,000 YS; 35 El; 55 RA. For stainless parts, bolts, rivets; resists oxidation up to 1650 F. *Obsolete*

CALITE 25-10

Calorizing Co.
C 0.25-0.45, Cr 23-28, Ni 10-14, Si 0.25-1, Mn 0.25-1, bal Fe.
Cast: 81,000 TS; 45,000 YS; 16 El; 180 Brin. Annealed: 87,000 TS; 51,000 YS; 5 El; 180 Brin. For furnace parts, beams, tube supports; heat resistant to 1950°F.

CALITE A

Calorizing Co.
C 0.4, Cr 15-18, Ni 35-37, Si 1-1.7, Mn 1.5, bal Fe.
At 1600°F: 18,000 TS; 26 El. Cast: 68,000 TS; 36,200 YS; 8 El; 156 Brin. For carburizing boxes, hearth plates, retorts; corrosion and heat resistant to 2000°F.

CALITE A-15

Calorizing Co.
C 0-0.15, Cr 15-18, Ni 35-37, Si 1-1.75, Mn 1-1.75, bal Fe.
Cast: 71,000 TS; 34,000 YS; 19 El; 112 Brin. Annealed: 90,650 TS; 44,200 YS; 13 El. For retorts, fan blades; heat resistant to 2000°F.

CALITE B 28

Calorizing Co.
C 0.35, Cr 25-28, Ni 10-12, Mo 1, N 0.1.
Cast: 85,000 TS; 57,000 YS; 24 El; 10 RA; 187 Brin. For furnace parts, beams, tube supports; to sustain loads at high temperatures.

CALITE B 28 N

Calorizing Co.
C 0.3-0.35, Cr 24-26, Ni 10-12, Si 0.75-1.75, Mn 0.75-1.75, 0.1 N_2, bal Fe.
Cast: 84,100 TS; 47,400 YS; 10 El; 217 Brin. Annealed: 86,000 TS; 46,500 YS; 7 El. For furnace parts, hangers, supports; heat resistant.

CALITE B 29

Calorizing Co.
C 0.29-0.34, Cr 28-30, Ni 15-17, Si 1-1.75, Mn 0.7-1.7, Mo 0.9-1.1, 0.1 N_2, bal Fe.
Cast: 64,600 TS; 45,000 YS; 8 El. Annealed: 85,500 TS; 50,650 YS; 6 El. For furnace parts; heat resistant to 2100°F.

CALITE B-18
Calorizing Co.
C 0.29-0.36, Cr 18-20, Ni 8-10, Mo 1, bal Fe.
Cast: 80,000 TS; 35,000 YS; 9 El; 3 RA; 170 Brin. Annealed: 105,000 TS; 48,100 YS; 17 El. For beams supporting loads at elevated temperatures, oil still tube supports; heat resistant up to 1650°F; resists sulfur; formerly Calite B.

CALITE B-18 LC
Calorizing Co.
C 0-0.08, Cr 18-20, Ni 8-10, Si 0.75-1.75, Mn 1-1.75, Mo 0.9-1.1, 0.1 N₂, bal Fe.
Cast: 156 Brin. For welded stainless equipment; heat resistant to 1650°F.

CALITE B-18 MO
Calorizing Co.
C 0-0.08, Cr 18-20, Ni 12.5-14, Mo 2.9-3.1, bal Fe.
Cast: 78,600 TS; 34,100 YS; 44 El; 143 Brin. Annealed: 77,200 TS; 33,300 YS; 37 El; 149 Brin. For paper and pulp equipment; corrosion resistant.

CALITE B29
Calorizing Co.
C 0.3-0.35, Cr 28-30, Ni 15-17, Si 1.5, Mn 1.5, Mo 1, N 0.1, bal Fe.
Cast: 65,000 TS; 45,000 YS; 10 El; 4 RA; 170 Brin. Aged: 90,000 TS; 50,000 YS; 4 El; 4 RA; 190 Brin. For furnace parts; heat resistant to 2050 F. *Obsolete*

CALITE BL
Calorizing Co.
C 0.35, Cr 21, Ni 9, bal Fe.
Cast: 75,000 TS; 30 El; 23 RA. For tube supports; heat resistant to 1800 F. *Obsolete*

CALITE BL-28
Calorizing Co.
C 0.35, Cr 26, Ni 10, bal Fe.
Cast: 75,000 TS; 25 El; 14 RA. For furnace parts, tube supports; heat resistant to 2000 F. *Obsolete*

CALITE D
Calorizing Co.
C, alloy, bal Fe.
For carburizing boxes; heat resistant; cast alloy. *Obsolete*

CALITE E
Calorizing Co.
Ni 7.8, Cr 17.28, Fe 73.62, Mn 0.45, C 0.14, Al 0.18, Cu 0.14.
Rolled, annealed: 85,000 TS; 50 El; 60 RA; 165 Brin. For stainless parts, bolts, nuts, forged and rolled parts; resists oxidation up to 1800°F.

CALITE E 28
Calorizing Co.
C 0-0.2, Cr 22-26, Ni 12-14, bal Fe.
Rolled: 95,000 TS; 40 El; 50 RA; 156 Brin. For malleable bolts, nuts, rivets, welding wire, furnace parts; heat resistant to 1950°F.

CALITE F
Calorizing Co.
C 0.36-0.5, Ni 24-30, Cr 16-23, bal Fe.
Heat and corrosion resistant parts. *Obsolete*

CALITE N
Calorizing Co.
C 0.55, Cr 17-20, Ni 65-68, Si 1-1.7, Mn 1-1.7, bal Fe.
At 100°F: 50,000 TS; 2 El; 1 RA; 190 Brin. At 2100°F: 8000 TS; 12 El; 13 RA. For furnace parts, heat treatment equipment, carburizing boxes; resists heat up to 2100°F; poor S resistance.

CALITE N-2
Calorizing Co.
C 0.35, Cr 14, Ni 60, bal Fe.
Cast. For carburizing boxes, furnace parts; heat resistant to 2000 F. *Obsolete*

CALITE NCT-3
Calorizing Co.
Cr 25, Ni 18, C, bal Fe.
Cast: 78,000 TS; 10 El; 7 RA. For parts to resist heat and corrosion, furnace parts; heat resistant *Obsolete*

CALITE NIROSTA KA2
Calorizing Co.
C 0.2, Si 0.74, Mn 0.27, Cr 18.5, Ni 10.5, bal Fe.
65,000-75,000 TS; 28,000-32,000 YS; 55-60 El; 60-65 RA; 155 Brin. For stainless parts, valves, pipe fittings; scaling temperature 1650 F. *Obsolete*

CALITE R
Calorizing Co.
C, alloy, bal Fe.
For apparatus to resist corrosion of boiling sulfuric acid; *Obsolete*

CALITE S-28
Calorizing Co.
C, 28% min Cr, bal Fe.
For corrosion resistant applications; corrosion resistant. *Obsolete*

CALITE-NIROSTA KA4
Calorizing Co.
C 0.16, Cr 18, Ni 8, Mo 2, bal Fe.
For equipment and apparatus in paper and pulp industries. *Obsolete*

CALIX
Uddeholm Corp.
C 0.4, Cr 1.3, W 4, V 0.25, bal Fe.
For upsetters, punches, extrusion dies and mandrels; hot work steel, oil hardened. *Obsolete*

CALLIFLEX
GTE Sylvania
W alloy.
For thermostats. *Obsolete*

CALLINITE
GTE Sylvania
W alloy.
For facing alloy; high conductivity. *Obsolete*

CALLINITE ST-1
GTE Sylvania
Ag-W.
For contact facing; high heat and electrical conductivity. *Obsolete*

CALLINITE ST-2
GTE Sylvania
Ag-W.
For contact facing, contacts for switches, relays; high heat and electrical conductivity. *Obsolete*

CALLINITE ST-3
GTE Sylvania
Ag-W.
For contact facing, contacts; high conductivity. *Obsolete*

CALLINITE TC-1
GTE Sylvania
W-Cu.
For contact facing, for welding electrodes; high electrical conductivity. *Obsolete*

CALLINITE TC-2
GTE Sylvania
W-Cu.
For contact facing for welding electrodes. *Obsolete*

CALLINITE TC-3
GTE Sylvania
W-Cu.
For contact facing for arcing tips and circuit breakers; high hardness. *Obsolete*

CALLOY
Commonwealth Aircraft Corp.
C 1.1, Mn 5.5, Cr 1.5, Mo 1.5, Si 1.1, Co 1.5, bal Fe.
Annealed: 120,000 psi TS (825 MPa). Hardened: 58-64 Rock C. Tough and abrasion resistant.

CALLOY (CALAL)
Calloy Ltd.
Ca 8-26, bal Al.
For deoxidation of steel.

CALLOY CADMIUM BARIUM
Calloy Ltd.
Cd, Ba.
For alloys for bearings.

CALLOY CADMIUM CALCIUM
Calloy Ltd.
Cd, Ca.
For alloys for bearings.

CALLOY CADMIUM STRONTIUM
Calloy Ltd.
Cd, Sr.
For alloys for bearings.

CALLOY LEAD BARIUM
Calloy Ltd.
Ba, bal Pb.
For alloys for bearings.

CALLOY LEAD CALCIUM
Calloy Ltd.
Ca, bal Pb.
For alloys for bearings.

CALLOY LEAD STRONTIUM
Calloy Ltd.
Sr, bal Pb.
For alloys for bearings.

CALLOY STRONTIUM ALUMINUM
Calloy Ltd.
Sr 0-50, bal Al.
For getters in electrical discharge devices and vacuum tubes; good stability in air.

CALMALLOY NO. 1
General Electric Co.
Cu 30, Ni 68, Fe 2.
For electrical equipment. Magnetically soft, high permeability. *Obsolete*

CALMALLOY NO. 2
General Electric Co.
Ni 88, Cu 10, Fe 2.
For electrical and magnetic equipment. Magnetically soft alloy. High magnetic permeability. *Obsolete*

CALMAR 18-8 CB
Warman Steel Casting Co.
Cr 17-20, Ni 7-10, Cb 0.9-1.25, C, bal Fe.
Annealed: 80,000-85,000 TS; 40,000-45,000 YS; 40-45 El. For castings; abrasion and corrosion resistant.

CALMAR 19-9
Warman Steel Casting Co.
Stainless Steel. C 0-0.15, Cr 18-21, Ni 8-10, bal Fe.
Heat treated: 85,000 TS; 40,000 YS; 64 El. For stainless castings; stainless.

CALMAR 19-9M
Warman Steel Casting Co.
Stainless Steel. C 0-0.15, Cr 18-21, Ni 8-10, bal Fe.
Heat treated: 90,000 TS; 45,000 YS; 50 El. For stainless castings; stainless.

CALMAR 19-9S
Warman Steel Casting Co.
Stainless Steel. C 0-0.07, Cr 18-21, Ni 8-10, bal Fe.
Heat treated: 75,000 TS; 37,000 YS; 55 El. For stainless castings; stainless.

CALMAX
Uddeholm Corp.
C 0.28, Cr 12, W 7, Co 9, V 0.4, bal Fe.
Hardened: 54-55 Rock C. For cores in die casting dies, extrusion dies, forging dies, mandrels. Hot work tool steel, high hot hardness, air or oil hardened.

CALMET
Caloriz Corp. of Great Britain Ltd.
Ni-Cr.
For thermocouples, resistance wire; heat resistant to 1920°F.

CALMOLLOY
GTE Sylvania
W alloy.
For grid wire. *Obsolete*

CALOMIC
Telcon Metals Ltd.
Ni 60, Cr 16, bal Fe.
Electrical resistant alloy for up to 900°C; for heating element.

CALOR 10
Haeckerstahl GmbH
C 0.3, Cr 2.6, V 0.35, W 8.5, bal Fe.
For extrusion rams and liners, punches; dies; hot work steel, oil hardened.

CALOR 10 CO
Haeckerstahl GmbH
C 0.3, Co 2, Cr 2.4, V 0.25, W 8.5, bal Fe.
For extrusion rams and liners, dies, punches; hot work steel, oil hardened.

CALOR 304
Haeckerstahl GmbH
C 0.45, Cr 4, Mo 0.7, V 0.3, bal Fe.
For gears, punches, crimpers, crankshafts; oil hardened, tough.

CALOR 5
Haeckerstahl GmbH
C 0.3, W, Cr, V, bal Fe.
For upsetters, riveters, punches; hot work steel, oil hardened.

CALOR LR
Haeckerstahl GmbH
C 0.3, Cr 1.1, V 0.18, W 3.75, bal Fe.
For cold water tools, upsetters, punches; oil hardened, tough.

CALORITE-1
Manufacturer not listed
Ni 65, Cr 12, Mn 8, Fe 15.
For resistance wire in electrical heating apparatus; high heat resistance. *Obsolete*

CALORITE-2
Manufacturer not listed
Ni 65, Cr 12, Fe 23.
For resistance wire in electrical heating apparatus; high heat resistance. *Obsolete*

CALORIZED 1% MO STEEL
Calorizing Co.
C 0.15, Mo 1, bal Fe.
For tubes, bolts, gas pyrolysis, polymerization equipment; 35% Al- 1/32" case. *Obsolete*

CALORIZED 4-6 CR MO
Calorizing Co.
C 0.15, Mo 0.5, Cr 4-6, bal Fe.
For parts to resist erosion and hydrogen sulfide; 35% Al-1/32 in. case.

CALORIZED DM STEEL
Calorizing Co.
C 0.12, Cr 1.25, Mo 0.6, Si 1, Mn 0.8, bal Fe.
For tubes, bolts, cracking tubes, super heaters; 35% Al- 1/32" case. *Obsolete*

CALORIZED STEEL
Calorizing Co.
Steel coated with Al-rich surface alloy.
For general usage, machinery parts, tubes, plates, pots, refinery tubes; improves resistance to oxidation and scaling to 1500 F. *Obsolete*

CALOXO 25 12
Warman Steel Casting Co.
Stainless Steel. C 0-0.25, Cr 24-27, Ni 11-14, bal Fe.
Annealed: 75,000-85,000 TS; 40,000-45,000 YS; 30-35 El. For stainless castings; stainless.

CALOXO 26
Warman Steel Casting Co.
Cr 25-30, 0.25 C min, bal Fe.
For corrosion resisting parts; corrosion resistant.

CALOXO 26-4
Warman Steel Casting Co.
Cr 25-30, Ni 2.5-4, 0.25 C min, bal Fe.
For heat and corrosion resisting parts; corrosion and heat resistant.

CALOXO 28
Warman Steel Casting Co.
C 0-0.25, Cr 26-30, bal Fe.
For corrosion and heat resisting castings; corrosion and heat resistant.

CALOXO 28-10
Warman Steel Casting Co.
Cr 27-30, Ni 8-12, C 0-0.25, bal Fe.
For heat and corrosion resisting parts; heat and corrosion resistant.

CALOXO 28-10 M
Warman Steel Casting Co.
Cr 27-30, Ni 8-12, Mo 3-4.5, 0.25 C min, bal Fe.
For heat and corrosion resisting parts; heat and corrosion resistant.

CALOXO 35-15
Warman Steel Casting Co.
C 0-0.35, Cr 15-18, Ni 35-38, bal Fe.
For heat and corrosion resisting castings; heat and corrosion resistant.

CALSIFER 75
Cyprus Foote Mineral Co.
Si 74-79, Al 0.75-1.25, Ca 0.5-1, bal Fe.
Ferrosilicon for inoculation of gray and ductile iron; also a source of silicon in iron and steel. *Obsolete*

CALSUN BRONZE 951
Anaconda Co.
Copper. Cu 95.5, Sn 2, Al 2.5.
Soft: 52,000 TS; 40 RA. Hard: 135,000 TS; 1 RA. For electrical conductors. Corrosion resistant.

CALUMET
Colt Industries
C 0.3, bal Fe.
Rolled: 65,000-75,000 TS; 40,000-50,000 YS. For machine tool parts, gears, shafts; easy to machine, water hardened. *Obsolete*

CALUMET
Gulf Steel Corp.
C 1, Mn 2, Cr 0.9, Mo 0.9, bal Fe.
For punches, blanking and forming dies; air hardened.

CALUMET BOX AND PIN
Colt Industries
C, alloy, bal Fe.
For mining boxes and pins; water hardening. *Obsolete*

CALUMET BR
Calumet Steel Castings Corp.
C 0.3-0.4, Mn 0.9-1.1, Si 0.25-0.45, Mo 0.2-0.3, bal Fe.
Normalized: 95,000-115,000 TS; 60,000-95,000 YS; 18-30 El; 30-50 RA; 150-200 Brin. For tool bit bodies, plows, bucket teeth sprockets; castings. *Obsolete*

CALUMET E
Calumet Steel Castings Corp.
C 0.36-0.43, Mn 1.2, Mo 0.25, bal Fe.
Cast: 83,000 TS; 59,000 YS; 23 El; 37 RA; 194 Brin. Heat treated: 125,000 TS; 114,000 YS; 17 El; 40 RA; 286 Brin. For guides for rolling mills. Wear resistant; tough.

CALUMET G
Vereinigte Leichtmetallwerke G.m.b.H.
C 0.32-0.39, Ni 0.7-1.3, Mo 0.6, bal Fe.
Cast: 100,000 TS; 65,000 YS; 20 El; 35 RA; 200 Brin. Heat treated: 200,000 TS; 173,000 YS; 6 El; 10 RA; 600 Brin. For gears, connecting rods, cams, dies; tough, wear resistant.

CALUMET GH
Calumet Steel Castings Corp.
C 0.4-0.5, Mn 0.5-0.8, Cr 0.45-0.75, Ni 1-1.5, bal Fe.
Heat treated: 110,000-180,000 TS; 80,000-120,000 YS; 10-20 El; 20-35 RA; 200-350 Brin. For gears, cams, rollers, structures; castings. *Obsolete*

CALUMET GR
Calumet Steel Castings Corp.
C 0.3-0.4, Mn 0.5-0.8, Cr 0.45-0.75, Ni 1-1.5, bal Fe.
Heat treated: 80,000-110,000 TS; 50,000-90,000 YS; 20-30 El; 38-50 RA; 160-230 Brin. For crankshafts, axles, cams, links; castings, abrasion resistant. *Obsolete*

CALUMET H
Calumet Steel Castings Corp.
C 0.25-0.35, Mn 1.6, Si 0.45, bal Fe.
Cast: 77,000 TS; 48,000 YS; 30 El; 56 RA; 168 Brin. Heat treated: 127,000 TS; 115,000 YS; 17 El; 47 RA; 286 Brin. For sprockets, gears, and housings. Abrasion resistant.

CALUMET M
Calumet Steel Castings Corp.
C 0.32-0.39, Mn 0.7, Cr 1, Mo 0.2, bal Fe.
Cast: 96,000 TS; 63,000 YS; 22 El; 42 RA; 178 Brin. Heat treated: 126,000 TS; 106,000 YS; 12 El; 29 RA; 330 Brin. For gears, shafts, sprockets, and housings. Abrasion resistant.

CALUMETAL B
Calumet Steel Castings Corp.
C 0.26-0.33, Mn 1.2, Mo 0.25, bal Fe.
Heat treated: 130,000 TS; 93,000 YS; 16 El; 16 RA; 260 Brin. Cast: 83,000 TS; 59,000 YS; 23 El; 37 RA; 194 Brin. For guides for rolling mills. Wear resistant; tough.

CALUMETAL C-5
Calumet Steel Castings Corp.
C 0-0.2, Cr 4-6.5, Mo 0.45-0.65, bal Fe.
Annealed: 90,000 TS; 60,000 YS; 18 El; 35 RA; 230 Brin. For castings for high temperature service up to 1400°F. Corrosion and heat resistant.

CALUMETAL L
Calumet Steel Castings Corp.
C 0.38-0.43, Mn 0.7, Si 0.4, Ni 1.5-2, Cr 0.7, Mo 0.3, bal Fe.
Normalized: 100,000-130,000 TS; 70,000-95,000 YS; 10-20 El; 25-40 RA; 210-250 Brin. Heat treated: 120,000-250,000 TS; 100,000-225,000 YS; 3-18 El; 7-40 RA; 250-500 Brin. For machine tool parts, gears, shafts, and castings. Shock resistant; oil hardened.

CALUMETAL MC
Calumet Steel Castings Corp.
C 0-0.2, Mn 0.7, Si 0.4, Cr 0.8-1.1, Mo 0.15-0.25, bal Fe.
Heat treated: 100,000-200,000 TS; 90,000-180,000 YS; 6-18 El; 35-60 RA; 200-390 Brin. For gears, pinions, shafts, and castings. Tough, carburizing steel.

CALUMETAL NE
Calumet Steel Castings Corp.
C 0.28-0.33, Mn 0.7, Si 0.4, Ni 0.45-0.8, Cr 0.4-0.7, Mo 0.2-0.3, bal Fe.
Normalized: 90,000-115,000 TS; 65,000-90,000 YS; 22-28 El; 40-55 RA; 190-230 Brin. Heat treated: 110,000-215,000 TS; 90,000-170,000 YS; 8-25 El; 30-55 RA; 200-450 Brin. For gears, pinions, shafts, machinery parts, castings; tough, water hardened. *Obsolete*

CALUMETAL WC6
Calumet Steel Castings Corp.
C 0-0.2, Mn 0.7, Si 0.4, Cr 1-1.5, Mo 0.4-0.6, bal Fe.
Annealed: 70,000 TS; 40,000 YS; 20 El; 35 RA; 200 Brin. For pressure tight castings for high temperature uses. Tough; oil hardened.

CALUMETAL WC9
Calumet Steel Castings Corp.
C 0-0.18, Mn 0.6, Si 0.4, Cr 2-2.75, Mo 0.9-1.1, bal Fe.
Annealed: 90,000 TS; 40,000 YS; 20 El; 35 RA; 200 Brin. For pressure tight castings to operate up to 1000°F.

CALUMETAL WCI
Calumet Steel Castings Corp.
C 0-25, Mn 0.7, Si 0-0.6, Mo 0.45-0.65, bal Fe.
Annealed: 65,000 TS; 35,000 YS; 24 El; 35 RA; 200 Brin. For castings, gears, bolts, housings. For high temperature service.

CALUMETAL X-1
Calumet Steel Castings Corp.
Stainless steel. C 0-0.08, Mn 0-1.5, Si 0-1.5, Ni 8-11, Cr 18-21, bal Fe.
Annealed: 85,000 TS; 35,000 YS; 55 El; 60 RA; 120 Brin. For food and chemical plant equipment, and castings. Stainless; austenitic.

CALUMETAL X-15
Calumet Steel Castings Corp.
C 0-0.35, Ni 33-37, Cr 14-17, bal Fe.
Cast: 65,000 TS; 40,000 YS; 6 El; 10 RA; 170 Brin. At 1600°F: 18,500 TS; 14,000 YS; 22 El; 53 RA; 55 Brin. For conveyors, furnace parts, and heat treating boxes. Resists oxidation and carburization.

CALUMETAL X-2
Calumet Steel Castings Corp.
Stainless steel. C 0-0.08, Mn 0-1.5, Si 0-1.5, Ni 8-11, Cr 18-21, S 0-0.7, bal Fe.
Annealed: 85,000 TS; 35,000 YS; 40 El; 50 RA; 120 Brin. For screw machine products. Stainless; austenitic; free-cutting.

CALUMETAL X-4
Calumet Steel Castings Corp.
Stainless steel. C 0-0.08, Ni 8-11, Cr 18-21, Mo 2-3, bal Fe.
Annealed: 85,000 TS; 40,000 YS; 50 El; 55 RA; 160 Brin. For castings to handle hot chlorides and acids. Austenitic; stainless; Type 316.

CALUMETAL X-5
Calumet Steel Castings Corp.
Stainless steel. C 0-0.3, Mn 0-1.5, Cr 11-13, Ni 23-27, bal Fe.
Cast: 90,000 TS; 40,000 YS; 20 El; 30 RA; 170 Brin. For furnace parts and heat treating boxes. Austenitic; stainless.

CALUMETAL X-8
Calumet Steel Castings Corp.
Stainless steel. C 0-0.15, Cr 11.5-13.5, bal Fe.
Heat treated: 97,000-150,000 TS; 70,000-100,000 YS; 10-20 El; 40-55 RA; 190-350 Brin. For chemical plant equipment and stainless castings. Corrosion resistant; ACI-CK20.

CALYPSO 25 A
G.O. Carlson Inc.
Fe 0.2, Si 0.15, Cu 4.5-5.5, Zn 1.7-2.3, Mg 0.2-0.35, Mn 0.05, Ni 0.05, Ti 0.15-0.25, bal Al.
Cast, solution treated and aged: 450 MPa TS; 390 MPa YS; 7 El. High strength, for gun breeches. *Obsolete*

CALYPSO 41 R
G.O. Carlson Inc.
Fe 0.15-0.3, Si 10.2-11.8, Cu 0.05, Zn 0.1, Mg 0.05, Mn 0.1, Ni 0.05, Ti 0.05, bal Al.
Permanent mold, as cast: 170 MPa TS; 75 MPa YS; 17 El. For high voltage electric line equipment. *Obsolete*

CALYPSO 43 P
G.O. Carlson Inc.
Fe 0.45-0.61, Si 12.5-14, Cu 0.03, Zn 0.05, Mg 0.03, Mn 0.1, Ni 0.03, Ti 0.035, Co 0.1, Cr 0.03, bal Al.
Permanent mold, as cast: 150 MPa TS; 80 MPa YS; 4 El. For thin section permanent mold castings, as tool boxes, lawn mower parts, letter boxes. *Obsolete*

CALYPSO 61 S
G.O. Carlson Inc.
Fe 0.15, Si 10-12, Cu 0.1, Zn 0.1, Mg 0.1-0.25, Mn 0.3, Ni 0.1, Pb 0.05, Sn 0.05, Ti 0.2, bal Al.
Permanent mold, as cast: 185 MPa TS; 80 MPa YS; 15 El. Low pressure wheel castings. *Obsolete*

CALYPSO 67 B
G.O. Carlson Inc.
Fe 0.15, Si 6.5-7.5, Cu 0.05, Zn 0.1, Mg 0.25-0.4, Mn 0.1, Ni 0.05, Ti 0.08-0.16, bal Al.
Permanent mold cast, solution treated and aged: 290 MPa TS; 200 MPa YS; 18 El. Suspension arms for automotive. *Obsolete*

CALYPSO 67 B1
G.O. Carlson Inc.
Fe 0.15, Si 6.5-7.5, Cu 0.05, Zn 0.1, Mg 0.45-0.6, Mn 0.1, Ni 0.05, Ti 0.08-0.16, bal Al.
Permanent mold, solution treated and aged: 340 MPa TS; 285 MPa YS; 10 El. For high quality aeronautical components. *Obsolete*

CALYPSO 67 N
G.O. Carlson Inc.
Fe 0.15, Si 6.5-7.5, Cu 0.05, Zn 0.1, Mg 0.25-0.4, Mn 0.1, Ni 0.05, Ti 0.08-0.16, bal Al.
Permanent mold cast, solution treated and aged: 290 MPa TS; 200 MPa YS; 18 El. For disc brake fittings, pivots and suspension arms. *Obsolete*

CALYPSO 67 N1
G.O. Carlson Inc.
Fe 0.15, Si 6.5-7.5, Cu 0.05, Zn 0.1, Mg 0.45-0.6, Mn 0.1, Ni 0.05, Ti 0.08-0.16, bal Al.
Permanent mold cast, solution treated and aged: 340 MPa TS; 285 MPa YS; 10 El. High strength castings. *Obsolete*

CALYPSO 67 R
G.O. Carlson Inc.
Fe 0.15, Si 6.5-7.5, Cu 0.05, Zn 0.1, Mg 0.25-0.4, Mn 0.05, Ni 0.05, Ti 0.08-0.16, Cr 0.05, bal Al.
Permanent mold cast: 195 MPa TS; 95 MPa YS; 15 El. For automotive wheels. *Obsolete*

CALYPSO 67 R1
G.O. Carlson Inc.
Fe 0.15, Si 6.5-7.5, Cu 0.05, Zn 0.1, Mg 0.45-0.65, Mn 0.05, Ni 0.05, Ti 0.08-0.16, Cr 0.05, bal Al.
Permanent mold, solution treated and aged: 350 MPa TS; 295 MPa YS; 10 El. For mechanical parts and electronic assemblies. *Obsolete*

CALYPSO 73 A
G.O. Carlson Inc.
Fe 0.3, Si 3.5-4.5, Cu 0.05, Zn 12-14, Mg 0.15-0.25, Mn 0.05, Ti 0.1, bal Al.
Permanent mold, as cast: 310 MPa TS; 230 MPa YS; 5 El. High strength as cast. One use: spirit levels. *Obsolete*

CALYPSO 82 N
G.O. Carlson Inc.
Fe 0.6, Si 10.5-12, Cu 0.8-1.2, Zn 0.05, Mg 1-1.4, Mn 0.1, Ni 0.8-1.2, Ti 0.1, bal Al.
Permanent mold cast; stress relieved: 240 MPa TS; 220 MPa YS. For petroleum engine pistons. *Obsolete*

CALYPSO 82 P
G.O. Carlson Inc.
Fe 0.3, Si 11.8-13.2, Cu 0.8-1.2, Zn 0.05, Mg 1.1-1.6, Mn 0.1, Ni 0.8-1.2, Ti 0.1, bal Al.
Permanent mold cast, stress relieved: 240 MPa TS; 220 MPa YS. Engine pistons and cylinder heads for air-cooled automotive engines. *Obsolete*

CALYPSO 85 R
G.O. Carlson Inc.
Fe 0.15, Si 4.5-5.5, Cu 2.5-3.5, Zn 0.2, Mg 0.25-0.4, Mn 0.1, Ni 0.05, Pb 0.05, Sn 0.05, Ti 0.05-0.1, Cr 0.05, bal Al.
Permanent mold cast, solution treated and aged: 400 MPa TS; 300 MPa YS; 2 El. Good strength, machinability, decorative anodizing; for textile beams, automotive. *Obsolete*

CALYPSO 92 A
G.O. Carlson Inc.
Fe 1.2-1.6, Si 0.15, Zn 0.1, Mg 0.05, Mn 0.05, Ti 0.05, Co 1.4-1.8, bal Al.
Die cast: 115 MPa TS; 45 MPa YS; 17 El. Good strength at elevated temperatures. *Obsolete*

CAMBRILOY 1
Cambridge Wire Cloth Co.
C 0-0.1, Si 0.5-1, Mo 0-0.05, 1.00 Cr min, bal Fe.
For woven wire conveyor belts, wire cloth and slings.

CAMBRILOY 3
Cambridge Wire Cloth Co.
C 0-0.15, Cr 2.75-3.25, Si 1-1.4, Mo 0.45-0.65, bal Fe.
For woven wire conveyor belts, wire cloth and slings.

CAMBRILOY 35-19
Cambridge Wire Cloth Co.
C 0-0.15, Cr 20, Ni 35, Si 2, bal Fe.
For woven wire conveyor belts, wire cloth and slings.

CAMBRILOY 35-19 CB
Cambridge Wire Cloth Co.
C 0-0.15, Cr 20, Ni 35, Si 2, Cb 1.2, bal Fe.
For woven wire conveyor belts, wire cloth and slings; high temperature operation.

CAMBRILOY 5
Cambridge Wire Cloth Co.
C 0-0.1, Cr 5, Si 0-1, bal Fe.
For woven wire conveyor belts, wire cloth and slings. *Obsolete*

CAMBRILOY 80-20
Cambridge Wire Cloth Co.
C 0-0.15, Si 1.2, Fe 1, Cr 20, bal Ni.
For woven wire conveyor belts, wire cloth and slings; high temperature operation.

CAMBRILOY 80-20 CB
Cambridge Wire Cloth Co.
C 0-0.15, Si 1.2, Cb 1.2, Cr 20, bal Ni.
For woven wire conveyor belts, wire cloth and slings.

CAMBRILOY A
Cambridge Wire Cloth Co.
Cr 80, Ni 20.
For furnace parts, conveyor belts; heat resistant to 2050 F in S-free atmosphere. *Obsolete*

CAMBRILOY AL
Cambridge Wire Cloth Co.
Aluminum coated carbon steel for woven wire conveyor belts, wire cloth and slings. *Obsolete*

CAMELIA METAL
English manufacture
Cu 70, Pb 15, Zn 10, Sn 4.2, Fe 0.5.
For hardware; free-cutting.

CAMEO
J.F. Jelenko & Co.
Precious metal. Ag 52.5, Pd 27, Ag 16.
100,000 psi TS; 80,000 psi YS; 10 El; 200 Brin. Dental alloy for fusing porcelain to metal.

CAMITE
Cleveland Automatic Machinery Co.
C, W, bal Fe.
For cutting tool tips; similar to "Carboloy." *Obsolete*

CAMLOY
Cameron & Son Ltd.
Ni 25-35, Cr 10-20, bal Fe.
90,000-112,000 TS; 35-40 El. For marine parts, valves and tubing for superheated steam, turbine blading, fittings, tanks; stainless and heat resisting. *Obsolete*

CAMVAC 200
Now NICKEL 200.

CAMVAC 400
Now NICKEL 400.

CAMVAC 600
Now INCONEL 600.

CAMVAC 800
Now INCOLOY 800.

CAMVAC B
Now HASTELLOY B.

CAN METALS
German manufacture
Pb 95, Ca 1.75, Cu 1.35, Sr 1, Ba 1.
For bearings.

CANNON
Darwin & Milner Inc.
Tool material. C 0.7, W 16, Cr 3.5, V 1, bal Fe.
For high speed tools, planing and lathe tools, twist drills, punches; high speed steel.

CANNON SPECIAL
Darwin & Milner Inc.
Tool material. C 0.7-0.75, Cr 4, W 18-20, Mo 0.5, V 2-2.25, bal Fe.
For tools, cutters.

CANNON VANADIUM
Darwin & Milner Inc.
Tool material. C 1, Cr 4, W 18-20, Mo 0.5-0.8, V 3.5, bal Fe.
For tools, cutters; high speed steel.

CANNON-MUSKEGON NO. 12
Cannon-Muskegon Corp.
Cr 29, W 8, C 1.5, Si 1, Fe 0-3, bal Co.
For hardfacing bushings, saw teeth, pressure bars, entry guides. Wear and abrasion resistant.

CANNON-MUSKEGON NO. 40
Cannon-Muskegon Corp.
Cr 15, Fe 4.5, B 3.25, Si 4.25, C 0.75, bal Ni.
For hardfacing gear teeth, plump plungers, valve plugs. Wear, oxidation and abrasion resistant.

CANNON-MUSKEGON NO. 41
Cannon-Muskegon Corp.
Cr 13, Fe 4, B 3, Si 4, C 0.6, bal Ni.
For hardfacing and welding, oil expeller screws, scraper blades. Heat and wear resistant.

CANNON-MUSKEGON NO. 6
Cannon-Muskegon Corp.
Cr 28, W 4, C 1, bal Co.
For hard facing bushings, pressure valves, tong bits, hot trimming dies. Heat and abrasion resistant.

CANNONITE
Textron Inc.
C 2.7, Si 1.5, bal Fe.
50,000 TS; 0 El; 250 Brin. For diesel engine and refrigeration parts, crankshafts, pistons, cylinders brake drums; high test cast iron; transverse strength 8300.

CANON
Creusot-Loire
C, alloy, bal Fe.
For machine tool parts; oil hardened. *Obsolete*

CANON SUPERIOR
Creusot-Loire
C, alloy, bal Fe.
For machine tool parts; oil hardened. *Obsolete*

CANTRAL A
Societe des AFC
Si 12, Cu 2.5, Mg 1.25, Mn 1.25, Ni 2.5, Ti 0.25, bal Al.
Cast: 24,000 psi TS; 1 El; 110 Brin. Heat treated: 33,000 psi TS; 1 El; 140 Brin. For light alloy parts; heat treatable.

CANZLER BRASS
Perry Equipment Corp.
Zn 30, bal Cu.
For welding wire; brass. *Obsolete*

CANZLER NO. 1
Perry Equipment Corp.
Sn 10, bal Cu.
For welding wire; bronze. *Obsolete*

CANZLER NO. 4
Perry Equipment Corp.
Cu.
For welding wire; copper. *Obsolete*

CAP GILDING
American manufacture
Cu 90, Zn 10.
For ornamental window screens; commercial bronze.

CAP M4 POWDER
Cytemp Specialty Steel Div.
C 1.3, Mo 4.5, W 5.5, Cr 4.5, V 4, bal Fe.
Hardened: 64-66 Rock C. For cutters, broaches, reamers, milling cutters, lathe and planer tools, taps, counterbore tools. Good abrasion resistance, tough. AISI Type M4. High speed steel.

CAP T-15
Cytemp Specialty Steel Div.
C 1.5, W 12.5, Cr 4.75, V 5, Co 5, bal Fe.
Oil hardened: 64-66 Rock C. For heavy duty cutters, form tools, lathe and planer tools, broaches, milling cutters, blanking dies, punches. High red-hardness and abrasion resistance. AISI T15. High speed steel.

CAPALOY
Oscap Mfg. Co.
Pt alloy.
190 Brin. For laboratory ware. *Obsolete*

CAPI 205-WI
Certified Alloy Products Inc.
C 0.4, Cr 21, W 11, Cb 2, Fe 2, bal Co.
As cast: 120,000 TS; 90,000 YS; 5 El; 10 RA; 341 Brin. High temperature strength, oxidation resistance. For jet engine vanes. *Obsolete*

CAPI 224
Certified Alloy Products Inc.
C 0.25, Cr 28, Mo 6, bal Co.
121,000 TS; 79,000 YS; 12 El; 14 RA; 269 Brin. High yield strength; resists body fluids. For hip joints, dental prosthetics. *Obsolete*

CAPI 309X
Certified Alloy Products Inc.
C 0.1, Cr 22, Mo 9, W 1, Co 1, Fe 18, bal Ni.
As cast: 85,000 TS; 48,000 YS; 15 El; 20 RA; 179 Brin. Good strength and oxidation resistance to 2200°F; for furnace hardware burner parts, retorts, vanes. *Obsolete*

CAPI 320
Certified Alloy Products Inc.
C 0.05, Cr 21, Mo 9, Cb 4, bal Ni.
As cast: 80,000 TS; 45,000 YS; 30 El; 30 RA; 170 Brin. Easily fabricated, corrosion resistant. For burner hardware, cryogenic and chemical equipment. *Obsolete*

CAPI 462
Certified Alloy Products Inc.
C 0-0.1, Mn 0-0.3, Si 1, Fe 0-1, Ti 0-0.5, Al 0-0.25, Cr 48-52, bal Ni.
Cast: 80,000 min TS; 50,000 min YS; 5 min El; 30 max Rock C. For parts to resist corrosion at elevated temperatures.

CAPI 462
Centrifugal Products Inc.
C 0-0.1, Mn 0-0.3, Si 1, Fe 0-1, Ti 0-0.5, Al 0-0.25, Cr 48-52, bal Ni.
Cast: 80,000 min TS; 50,000 min YS; 5 min El; 30 max Rock C. For parts to resist corrosion at elevated temperatures.

CAPI 864 CB
Certified Alloy Products Inc.
C 0.05, Cr 19, Ni 10.5, Cb 1, bal Fe.
Solution annealed: 78,000 TS; 34,000 YS; 54 El; 57 RA. Columbium stabilized stainless; for general corrosion resistant and heat resistant parts. *Obsolete*

CAPI 960
Certified Alloy Products Inc.
C 0.35-0.45, Mn 0-1.5, Si 0-1.5, Cr 23-25, Ni 21-23, Mo 0-0.5, Cb 1.25-2, bal Fe (100 ppm max Pb).
Cast alloy similar to HK40 but stabilized with Cb to resist thermal cycling in petrochemical environments.

CAPI 960
Centrifugal Products Inc.
C 0.35-0.45, Mn 0-1.5, Si 0-1.5, Cr 23-25, Ni 21-23, Mo 0-0.5, Cb 1.25-2, bal Fe (100 ppm max Pb).
Cast alloy similar to HK40 but stabilized with Cb to resist thermal cycling in petrochemical environments.

CAPITAL 305
Eagle & Globe Steel Ltd.
Tool material. C 0.8, Cr 3.85, Mo 8.7, W 1.65, V 1.15, bal Fe.
High speed steel for general purpose use. Similar to Ultra Capital. AS1239 M1A; AISI M1.

CAPITAL 398
Eagle & Globe Steel Ltd.
Tool material. C 0.85, Cr 4, Mo 5, W 6.5, V 1.9, Co 5, bal Fe.
High speed steel. Gives long life in general machine shop work and heavy duty cutting. AS1239 M35A; AISI M35.

CAPITAL 405
Eagle & Globe Steel Ltd.
Tool material. C 1.1, Cr 3.75, Mo 9, W 1.5, V 1.2, Co 8, bal Fe.
High speed steel. Gives long life in general machine shop work and heavy duty cutting. High performance capabilities. AS1239 M42A: AISI M42; Werkstoff 1.3247.

CAPITAL 562 HIGH SPEED STEEL
Eagle & Globe Steel Ltd.
Tool material. C 0.83, Cr 4, Mo 5, W 6.4, V 1.9, bal Fe.
Annealed: 250 Brin. Standard general purpose high speed
steel. Excellent toughness and a good combination of red
hardness and wear resistance. AS1239 M2A-8; AISI M2;
BS4659 BM2; Werkstoff 1.3343.

CAPITAL-562
Darwins Alloy Castings
C 0.8, Cr 4.2, W 6, V 2, Mo 5, bal Fe.
For lathe and planer tools, milling cutters; high-speed steel.
Obsolete

CAPITO-VK5M
Capito & Klein
C 0.2, Cr 13, Mo 1.2, bal Fe.
Annealed: 95,000 TS; 40,000 YS; 25 El; 55 RA; 92 Rock B.
Heat treated: 240,000 TS; 200,000 YS; 8 El; 25 RA; 50 Rock
C. For gears, shafts, surgical instruments, knives, scissors,
cutlery. Corrosion resistant, hardenable.

CAPIVAC 326
Certified Alloy Products Inc.
C 0.05, Cr 19, Mo 3, Cb 5, Fe 18, Al 0.5, Ti 1, bal Ni.
Aged: 150,000 TS; 115,000 YS; 20 El; 20 RA; 363 Brin. For
structural parts up to 1400°F; cryogenic gear. *Obsolete*

CAPIVAC 371
Certified Alloy Products Inc.
C 0.1, Cr 13, Mo 5, Cb 2, Al 6, Ti 0.8, bal Ni.
As cast: 120,000 TS; 110,000 YS; 5 El; 8 RA; 331 Brin. High
temperature strength; for gas turbine blades, wheels.
Obsolete

CAPIVAC IV
Certified Alloy Products Inc.
C 0.55-0.65, Cr 23-26, Ni 9-11, Fe 0-1.5, Ta 3.5-4.5, W 6.5-7.5,
Al 0.35-0.6, Ti 0.2-0.5, Cu 0-0.5, bal Co.
Cast: 105,000 TS; 70,000 min YS; 3.0 min El. Stress rupture:
23 hr min at 9000 psi at 2000°F. For high temperature
operation.

CAPIVAC IV
Centrifugal Products Inc.
C 0.55-0.65, Cr 23-26, Ni 9-11, Fe 0-1.5, Ta 3.5-4.5, W 6.5-7.5,
Al 0.35-0.6, Ti 0.2-0.5, Cu 0-0.5, bal Co.
Cast: 105,000 TS; 70,000 min YS; 3.0 min El. Stress rupture:
23 hr min at 9000 psi at 2000°F. For high temperature
operation.

CAPSULE METAL
American manufacture
Pb 92, Sn 8.
For bearing.

CAR-VAN
Allied Steel & Tractor Products Inc.
C 0.75-1.1, V 0.2, bal Fe.
For tools, dies, shear blades, punches; water hardening.

CAR-VAN SPECIAL
Allied Steel & Tractor Products Inc.
C 0.8-1.35, V 0.2, bal Fe.
For tools, dies, drills, cutters; water hardening.

CARAS
George Cook & Co., Ltd.
alloy, bal Fe.
For molds; case hardened, oil hardened.

CARB-X
American manufacture
C, bal Fe.
Used in manufacturing cast iron molds; ferro-alloy,
exothermic.

CARBDI
Eagle & Globe Steel Ltd.
Tool material. C 0.2, Mn 1, Cr 1.2, Mo 0.2, Si 0.3, bal Fe.
High hardenability for case hardened molds. For
compression transfer or injection molds for abrasive plastics.
AS1239 P100A.

CARBIDE G1
German manufacture
Co 6, 94 WC.
For cutting tools; sintered carbide.

CARBIDE H1
German manufacture
Co 6, 94 WC.
For cutting tools; sintered carbide.

CARBIDE H2
German manufacture
Co 7, 1.50 TiC, 91.5 WC.
For cutting tools; sintered carbide

CARBIDE L2
German manufacture
Co 8, 14.0-15.0 TiC, 77-78 WC.
For cutting tools; sintered carbide.

CARBIDE LO
German manufacture
Co 7, 5.0 TiC, 88 WC.
For cutting tools; sintered carbide.

CARBIDE T15K6
Russian manufacture
Co 6, 15.0 TiC, 79.0 WC.
For cutting tools; sintered carbide.

CARBIDE T5K10
Russian manufacture
Co 9, 6.0 TiC, 85.0 WC.
For cutting tools; sintered carbide.

CARBIDE WK3
Russian manufacture
Co 3, WC.
For cutting tools; sintered carbide.

CARBIDE WK6
Russian manufacture
Co 6, 94.0 WC.
For cutting tools; sintered carbide.

CARBIDE WK8
Russian manufacture
Co 8, 92.0 WC.
For cutting tools; sintered carbide.

CARBIDIE CD-10N
Carbidie
Co 10, Sintered carbide.
Nickel binder tungsten carbide used for low shock, corrosive
environments. Seal ring faces, pump bearings, pump
plungers, nozzles, and food processing wear parts. 89.5-91.5
Rock A.

CARBIDIE CD-10N (NON-MAG)
Carbidie
Co 10, Sintered carbide.
Nickel binder tungsten carbide used in the manufacture of
ferrite magnets. 89.5-91.5 Rock A.

CARBIDIE CD-12N
Carbidie
Co 12, Sintered carbide.
Nickel binder tungsten carbide used in the can tooling
industry for carbide tooling where lubricants attack the cobalt
binder grades. 88.0-89.5 Rock A.

CARBIDIE CD-18
Carbidie
Co 12, Sintered carbide.
Special wear resistant die grade for extra abrasive stamping
applications; gall resistant. Stamping and lamination tooling,
drawing and forming; can tooling. 89-90 Rock A.

CARBIDIE CD-20
Carbidie
Co 3, Sintered carbide.
Extreme wear applications where no shock is involved;
highest compression strength. Glass and plastic cut-off
tooling; gages, nozzles, valve seats. 91.8-92.8 Rock A.

CARBIDIE CD-24X
Carbidie
Co 5, Sintered carbide.
Wear applications in no-shock or light shock tooling, such as
ceramic compacting; will hold a highly polished, wear
resistant finish. Valve balls and seats, bearings and seals;
corrosion resistant uses, pyrochemical liners and plungers,
gages, centerless grinder blades. 91.5-92.5 Rock A.

CARBIDIE CD-30
Carbidie
Co 6, Sintered carbide.
Wear and die applications. Light shock metal powder tooling,
drawing, blanking, boring bars and tooling, seal rings, drills.
90.5-92.0 Rock A.

CARBIDIE CD-337
Carbidie
Co 11, Sintered carbide.
High strength, wear resistance. For cold forming, structural-
wear-impact applications; swaging hammers, mandrels,
rotary percussion applications, cold forming, back extrusion
punches. 88.0-89.0 Rock A.

CARBIDIE CD-35
Carbidie
Co 9, Sintered carbide.
Wear applications with good wear and light shock. Draw dies,
valve lifter discs, compacting dies, stamping dies. 89.8-90.8
Rock A.

CARBIDIE CD-355
Aiken Industries Inc./Carbide Div.
Sintered carbide.
Shock and wear resistant. Rod mill rolls, guide inserts.
Obsolete

CARBIDIE CD-35F
Carbidie
Co 9, Sintered carbide.
Wear and gall resistant; will hold high polish. Aluminum,
stainless and brass draw applications, wire straightening,
bearings and seals, valve ball and seats. 90.2-91.2 Rock A.

CARBIDIE CD-36 1/4
Carbidie
Co 10, Sintered carbide.
Wear and die grade; moderate shock resistance. Normal
metal powder tooling; blanking and drawing; slitter knives.
89.5-90.5 Rock A.

CARBIDIE CD-38 3/4
Carbidie
Co 12, Sintered carbide.
Forming and blanking grade; wear resistant; will resist light
shock. Structural parts, wire bar and tube drawing. 88.8-89.8
Rock A.

CARBIDIE CD-40
Carbidie
Co 13, Sintered carbide.
Die and blanking grade, non-ferrous material. Wear resistant
blanking, forming, drawing; general purpose metal powder
tooling; slitter knives for paper and steel, can die rings.
88.5-89.5 Rock A.

CARBIDIE CD-45
Carbidie
Co 14, Sintered carbide.
Wear resistant ferrous blanking grade. Very light impact applications, forming dies, drawing, general can tooling. 88.0-90.0 Rock A.

CARBIDIE CD-50
Carbidie
Co 15, Sintered carbide.
General all purpose die, stamping, drawing. Will take moderate shock with good die resistance, crush rolls, coining dies. 87.5-88.5 Rock A.

CARBIDIE CD-53
Carbidie
Co 16, Sintered carbide.
Shock resistant, medium impact-shock applications, spring tooling, heavy forming dies, slitters. 87.0-88.0 Rock A.

CARBIDIE CD-55
Carbidie
Co 17, Sintered carbide.
Tube gripper jaws, pulverizing hammers, header dies. 86.0-87.0 Rock A.

CARBIDIE CD-60
Carbidie
Co 20, Sintered carbide.
Moderate shock, cold heading grade. Nail grippers, and tooling, spring tooling, crush hammers. 83.0-84.5 Rock A.

CARBIDIE CD-630
Carbidie
Co 6, Sintered carbide.
Submicron material for use as circuit board drills, stamping dies for mylar, end mill blanks, lead wire trimmers. High hardness, low shock. 92.0-93.5 Rock A.

CARBIDIE CD-636
Carbidie
Co 10, Sintered carbide.
Submicron material for use in the most abrasive stamping and wear applications. High strength and hardness, and medium shock resistance. 90.5-91.5 Rock A.

CARBIDIE CD-650
Carbidie
Co 15, Sintered carbide.
Submicron material for use in highly abrasive and severe stamping and wear applications. Extreme strength, hardness, and shock resistance. 89.0-90.5 Rock A.

CARBIDIE CD-6N
Carbidie
Co 6, Sintered carbide.
Nickel binder tungsten carbide used for low shock, corrosive environments. Seal ring faces, nozzles, parts in food processing wear parts. 90.5-92.0 Rock A.

CARBIDIE CD-70
Carbidie
Co 25, Sintered carbide.
Severe shock, cold heading grade. Heavy bolt making and sizing. Nail tooling. 81.5-83.0 Rock A.

CARBIUM
A. Gayer Co.
Cu 4-5, bal Al.
For light alloy parts; heat treatable.

CARBO
English manufacture
WC, Co.
For cutting tools; cemented.

CARBO TOOL
Bisset Steel Co.
C 1, bal Fe.
For tools, dies, drills, taps; water hardening. *Obsolete*

CARBOBRONZE
English manufacture
Cu 92, Sn 8, P 0.3.
For tubing; resists acid and alkalis.

CARBOLOY 1078A
General Electric Co.
WC-TiC-Co.
Sintered: Rockwell A92. For cutters, tools, dies; sintered carbide. *Obsolete*

CARBOLOY 115
General Electric Co.
Co 11.5, WC 88.5.
Sintered cemented carbide. 400,000 transverse strength; 88.5 Rock A. For percussive mining bits and large size impact punches; good shock and wear resistance.

CARBOLOY 120
General Electric Co.
Co 12, WC 88.
Sintered cemented carbide. 410,000 transverse strength; 86.5 Rock A. For rotary and percussive mining tools and small impact extrusion punches; impact and wear resistant.

CARBOLOY 190
General Electric Co.
Co 25, WC 75.
Sintered cemented carbide. 450,000 transverse strength; 84.0 Rock A. For heavy impact die and punch applications; high toughness.

CARBOLOY 210
General Electric Co.
Co 4, WC 28, 2.0 TaC, 64.0 TiC, 2 Cr_3C_2.
Sintered cemented carbide. 100,000 transverse strength; 94.5 Rock A. For finish machining of steel and cast iron; high wear resistance at elevated temperatures.

CARBOLOY 231
General Electric Co.
Co 10, WC 90.
Sintered cemented carbide. 400,000 transverse strength; 87.8 Rock A. For mining applications; shock resistant.

CARBOLOY 241
General Electric Co.
Co 10, WC 90.
Sintered cemented carbide. 400,000 transverse strength; 88.4 Rock A. For mining applications; shock and wear resistant.

CARBOLOY 248
General Electric Co.
Co 11, WC 89.
Sintered cemented carbide. 450,000 transverse strength; 89.5 Rock A. For rock drilling and interrupted metal cutting; wear resistance relative to shock resistance.

CARBOLOY 258
General Electric Co.
Co 13, WC 87.
Sintered cemented carbide. 470,000 transverse strength; 88.5 Rock A. For mining and for cold extrusion; shock resistant.

CARBOLOY 268
General Electric Co.
Co 16, WC 84.
Sintered cemented carbide. 490,000 transverse strength; 87.0 Rock A. For mining and for impact extrusion; high shock resistance.

CARBOLOY 320
General Electric Co.
Co 6, WC 64, 4.5 TaC, 25.5 TiC.
Sintered cemented carbide. 200,000 transverse strength; 93.0 Rock A. For high-speed finishing of steel and high tensile cast irons; wear resistant.

CARBOLOY 350
General Electric Co.
Co 4.5, WC 71, 12.0 TaC, 12.5 TiC.
Sintered cemented carbide. 200,000 transverse strength; 92.4 Rock A. For light-roughing and semi-finishing of steel, ferrous castings, stainless steel and some high-temperature alloys; wear resistant; tough.

CARBOLOY 370
General Electric Co.
Co 8.5, WC 72, 11.5 TaC, 8.0 TiC.
Sintered cemented carbide. 250,000 transverse strength; 91.2 Rock A. For heavy duty roughing of steels, ferrous castings, stainless steel and some high-temperature alloys; shock and wear resistant.

CARBOLOY 390
General Electric Co.
Sintered complex tungsten carbide, uncoated. For heavy duty and extremely heavy duty roughing of carbon, alloy and chromium stainless steels at relatively low speeds.

CARBOLOY 44A
General Electric Co.
Co 6, WC 94.
Sintered cemented carbide. 320,000 transverse strength; 91.0 Rock A. For semi-roughing of cast iron, non-ferrous and non-metallics; medium high shock resistance.

CARBOLOY 514
General Electric Co.
Sintered TiC coated cemented carbide. For finishing and precision finishing of steel and cast iron.

CARBOLOY 516
General Electric Co.
Sintered TiC coated cemented carbide. For finishing and precision finishing of steel and cast iron.

CARBOLOY 518
General Electric Co.
Sintered complex tungsten carbide, TiC coated. For light to heavy roughing of alloy steels, tool steels and stainless steel.

CARBOLOY 519
General Electric Co.
Sintered complex tungsten carbide, TiC coated. For machining.

CARBOLOY 523
General Electric Co.
Sintered TiC coated cemented carbide. For general purpose and finish machining of cast iron and non-ferrous materials.

CARBOLOY 545
General Electric Co.
Sintered complex tungsten carbide, Al_2O_3 coated. For high-speed finishing cast iron and steel.

CARBOLOY 55A
General Electric Co.
Co 13, WC 87.
Sintered cemented carbide. 390,000 transverse strength; 88.2 Rock A. For heavy duty roughing of cast iron, non-ferrous and non-metallics; variety of die applications; high shock resistance.

CARBOLOY 55B
General Electric Co.
Co 16, WC 84.
Sintered cemented carbide. 420,000 transverse strength; 86.8 Rock A. For medium and large dies; cold work, piercing, blanking and drawing; high shock resistance.

CARBOLOY 570
General Electric Co.
Sintered complex tungsten carbide, Al_2O_3 coated. For machining.

CARBOLOY 608
General Electric Co.
Ni 15, W 2, Sintered cemented carbide; 83 Cr 3 C 2 .
100,000 TrS; A 89.5 Rock. For component applications that require high resistance to oxidation and corrosion. *Obsolete*

CARBOLOY 616
General Electric Co.
Ni 16, Cr 2, WC 82.
Sintered cemented carbide. 370,000 transverse strength; 87.8 Rock A. For components rather than cutting tool.

CARBOLOY 779
General Electric Co.
Co 9, WC 91.
Sintered cemented carbide. 340,000 transverse strength; 89.5 Rock A. For medium-size wire drawing dies, bar and tube dies, shape dies, cupping and compression dies; mining tool and rock-drilling applications.

CARBOLOY 77A
General Electric Co.
WC + TaC + Co.
Sintered: 200,000 TS. For cutters for light cuts on steel. *Obsolete*

CARBOLOY 77B
General Electric Co.
Co 16, WC 57, 27.0 TaC.
Sintered cemented carbide. 390,000 transverse strength; 85.0 Rock A. For cutting hot flash, extrusion of aluminum wire bar and tube.

CARBOLOY 78
General Electric Co.
Co 8, WC 76, 4.0 TaC, 12.0 TiC.
Sintered cemented carbide. 250,000 transverse strength; 92.0 Rock A. For light-roughing and finishing of steels; high abrasion resistance.

CARBOLOY 78B
General Electric Co.
Co 9, WC 79, 4.0 TaC, 8.0 TiC.
Sintered cemented carbide. 260,000 transverse strength; 91.2 Rock A. For medium roughing of steel; die applications.

CARBOLOY 820
General Electric Co.
Co 10, WC 90.
Sintered cemented carbide. 450,000 transverse strength; 51.0 Rock A. For finish and semi-finish machining.

CARBOLOY 831A
General Electric Co.
WC + Co, Ti.
Sintered: 125,000 TS. For cutters for boring hard steel. *Obsolete*

CARBOLOY 860
General Electric Co.
Co 5, WC 91, 4.0 TaC.
Sintered cemented carbide. 270,000 transverse strength; 92.0 Rock A. For machining high tensile cast iron; good wear resistance.

CARBOLOY 883
General Electric Co.
Co 6, WC 94.
Sintered cemented carbide. 290,000 transverse strength; 92.0 Rock A. For general purpose machining of non-steel work materials; also used for small compacting dies, burnishing rings and nozzles; high wear resistance.

CARBOLOY 895
General Electric Co.
Co 6, WC 94.
Sintered cemented carbide. 250,000 transverse strength; 92.9 Rock A. For light machining of cast iron and non-ferrous materials.

CARBOLOY 90
General Electric Co.
Co 10, WC 90.
Sintered cemented carbide. 350,000 transverse strength; 89.0 Rock A. For rock drilling; wear resistant.

CARBOLOY 905
General Electric Co.
Co 4, WC 92, 4.0 TaC.
Sintered cemented carbide. 240,000 transverse strength; 92.2 Rock A. For light finishing of cast iron, non-ferrous and non-metallic materials.

CARBOLOY 906
General Electric Co.
WC, Co.
Sintered: 250,000 TS. For cutters for hogging cuts on cast iron. *Obsolete*

CARBOLOY 907
General Electric Co.
Co 6, WC 74, 20.0 TaC.
Sintered cemented carbide. 270,000 transverse strength; 91.3 Rock A. For finish machining of fine-grained cast iron, malleable iron, aluminum and magnesium alloys; crater and abrasive resistant.

CARBOLOY 999
General Electric Co.
Co 3, WC 97.
Sintered cemented carbide. 230,000 transverse strength; 92.9 Rock A. For machining cast iron, non-ferrous metals, and non-metallics; also for fine wire dies and small nozzles; abrasion resistant.

CARBOLOY HM-1 HEVIMET
General Electric Co.
W 90, Ni 7.5, Cu 2.5.
Sintered alloy. 300,000 transverse strength; 25 Rock C. For weights, counter balances, radiation shielding, vibration damping; high-density tungsten alloy.

CARBOLOY HM-3 HEVIMET
General Electric Co.
W 90, Ni 7.5, Cu 2.5.
Sintered alloy. For governor counter weights, radiation shielding, vibration damping; high-density tungsten alloy.

CARBOLOY O-30
General Electric Co.
Sintered cemented oxide. 90,000 transverse strength; 94.0 Rock A. For high-speed machining of cast iron and steel; high heat and wear resistance.

CARBOMANG
Detroit Alloy Steel Co.
C 0.9-1, Mn 1-1.25, Cr 0.45-0.6, W 0.4-0.6, bal Fe.
For tools and dies, machine parts; castings; oil hardenable; cast to shape.

CARBON
Driver Harris Co.
C, Ni.
For radio tubes; heat resistant. *Obsolete*

CARBON
U.S. Steel Corp.
C 0.1-0.2, Mn 0.3-0.6, Mo 0.45-0.65, Si 0.1-0.2, bal Fe.
Hot rolled: 60,000 TS; 35,000 YS; 40 El; 150 Brin. For oil cracking still tubes, hot oil piping, high temperature steam piping; superior creep strength to plain C-steel. *Obsolete*

CARBON + GRAPHITE
Atomergic Chemetals Corp.
C.
Purities: 99.9999%, 99.999%, 99.99%. Forms: rod, powder, electrodes, single crystals.

CARBON 1/2 MO
U.S. Steel Corp.
C 0.2-0.3, Mn 0-0.75, Si 0-0.3, Mo 0.45-0.65, bal Fe.
Normalized and drawn: 70,000 TS; 50,000 YS; 30 El; 170 Brin. For high temperature high pressure steam piping; high tensile and creep strength. *Obsolete*

CARBON BRONZE
English manufacture
Pb 15, Sn 10, bal Cu.
Cast: 28,000 TS; 6 El. For bearings; plastic bronze.

CARBON CHISEL
Bisset Steel Co.
C 0.7-1, bal Fe.
For tools, chisels; water hardening. *Obsolete*

CARBON COLD HEADER
Latrobe Steel Co.
Tool material. C 0.9, bal Fe.
For cold heading dies, water hardening. *Obsolete*

CARBON COLD HEADER
Bethlehem Steel Corp.
C 0.95, Mn 0.4, Si 0.4, bal Fe.
For header dies, gripper dies, swaging and forming dies.

CARBON COLD HEADER NO. V
Latrobe Steel Co.
C 0.93, Mn 0.25, Cr 0.1, bal Fe.
For cold forming dies; water hardening. *Obsolete*

CARBON COLD HEADER WITH MO
Latrobe Steel Co.
C 0.9, Mn 0.25, Mo 0.3, Cr 0-1, bal Fe.
For large cold header dies; water hardening. *Obsolete*

CARBON COLD HEADER WITH V
Latrobe Steel Co
C 0.9, Mn 0.25, V 0.2, Cr 0-0.1, bal Fe.
For cold header dies and hammers; water hardening. *Obsolete*

CARBON DRILL ROD
Cytemp Specialty Steel Div.
C 1.2, bal Fe.
For tools; water hardened. *Obsolete*

CARBON MOLY STEEL
Bonney-Floyd Co.
C 0.25-0.45, Cr 0.8-1.1, Mo 0.15-0.25, bal Fe.
Heat treated: 197-401 Brin. For castings, gears, shafts; abrasion and wear resisting, heat treated. *Obsolete*

CARBON SHOE DIE
Colt Industries
C 0.75, bal Fe.
For shoe dies; water hardening. *Obsolete*

CARBON STEEL
Wallace Murray Corp.
C 0.7-1.4, bal Fe.
For tools, drills, taps; water hardening.

CARBON TOOL DOUBLE EXTRA
Midvale-Heppenstall Co.
C, Cr, bal Fe.
For tools, dies; water hardened. *Obsolete*

CARBON VANADIUM STEEL
Flockton, Tompkin & Co., Ltd.
C 0.8-1.3, V, bal Fe.
For shear blade, punches, chisels, heading and swaging dies; water hardened; Type W2.

CARBON-FORD
Republic Steel Corp.
C 0.15-0.45, Mn 0.3-0.85, Si 0-0.09, bal Fe.
For machinery parts. *Obsolete*

CARBON-MOLYBDENUM STEEL

Lukens Steel
C 0.18-0.28, Mo 0.4-0.6, Mn 0.5-0.9, bal Fe.
65,000-87,000 TS; 145 Brin. *Obsolete*

CARBON-VANADIUM

Midvale-Heppenstall Co.
C, V, bal Fe.
For tools, cold heading; cold work steel. *Obsolete*

CARBON-VANADIUM DRILL ROD

Teledyne Allvac
C 1, V 0.2, bal Fe.
For drill rod, punches; water hardened.

CARBONDALE SILVER

American manufacture
Cu 66, Ni 18, Zn 16.
Hard: 96,000 TS; 2 El; 158 Brin. Soft: 58,000 TS; 33 El; 77
Brin. For spinning and drawing, flatware, spoons, forks,
knives, etc., to be plated; German silver.

CARBONIZED NICKEL

Harrison Alloys Inc.
Nickel. Ni 99.5.
Strip for electron tube plate.

CARBORTAM

NL Industries
Ti 16-17, Si 2.5-3, C 6.5-7.5, B 1-1.25, bal Fe.
For steel making conditions; adds deep hardening
properties. *Obsolete*

CARBORUNDUM ZIRCONIUM GR. 21

Carborundum Co.
N 0.007, Hf 0.02, 0.17 Fe + Cr, 99.5 Zr min.
Annealed: 65,000 TS; 38,000 YS; 19 El; 31 RA; 76 Rock B.
For atomic reactor components, fuel sheathing, reactor
structures. Thermal neutron absorption 0.18-0.20 barns.
Obsolete

CARBORUNDUM ZIRCONIUM GR. 32

Carborundum Co.
Sn 1.5, Fe 0.14, Cr 0.1, Ni 0.05, bal Zr.
Annealed: 70,000 TS; 43,000 YS; 28 El; 150 Brin. At 1000°F:
18,000 TS; 10,000 YS; 45 El. Cold reduced 60%: 110,000 TS;
102,000 YS; 35 RA. For nuclear reactors, fuel sheathing,
boiling and pressurized water reactors. Resists irradiated
water corrosion. *Obsolete*

CARBOTAM

SMC (Shieldalloy Metallurgical Corp.)
B 2, Al 3, Ti 15-20, Si 4, bal Fe.
Master alloy for ladle additions.

CARBURITE

American manufacture
C 47-48, Fe 28, S 0.3, P 0.2.
For recarburizer in steel; for steel making.

CARBURIZED TUNGSTEN CARBIDE

Kennametal Inc.

Matrix hardener for hot-pressed, impregnated-diamond
products; wear modifier in hot-pressed matrices.

CARDINAL RAPID

Thyssen Edelstahlwerke AG
C 0.7, W 18, Cr 4, V 1, bal Fe.
For high speed cutters; high speed steel.

CARECO

American manufacture
Pb-Sn-Sb.
For bearings; white bearing alloy.

CAREND

Arcos Alloys
C 0.03, bal Fe.
For welding electrodes for sheet metal; Armco iron type.
Obsolete

CARILLOY CARBON MOLYBDENUM

U.S. Steel Corp.
C 0.38-0.45, Mn 0.65-0.95, P 0-0.04, S 0-0.04, Si 0.2-0.35, Mo
0.2-0.3, bal Fe.
110,050 TS; 77,430 YS; 13.2 El; 48.1 RA. For auto forgings;
test on 1/2 in. rounds. *Obsolete*

CARILLOY CARBON-MANGANESE

U.S. Steel Corp.
Mn 1.5-2, Si 0.15-0.3, C, bal Fe.
Quenched and tempered: 100,000-120,000 TS; 75,000 Elastic
Limit; 20 El; 50 RA. For forgings; Class L over 2 to 7 in. ASTM,
A-18. *Obsolete*

CARILLOY CARBON-MANGANESE

U.S. Steel Corp.
P 0-0.04, S 0-0.05, 1.4% Mn min, C, bal Fe.
85,000 TS; 48,000 YS; 16 El. For chain links. *Obsolete*

CARILLOY CARBON-MANGANESE

U.S. Steel Corp.
C 0-0.55, Mn 1.1-1.5, P 0-0.05, S 0-0.05, bal Fe.
Quenched and tempered: 105,000 TS; 70,000 YS; 22 El; 40
RA. For forgings; water hardened. *Obsolete*

CARILLOY CARBON-MANGANESE

U.S. Steel Corp.
C 0.35-0.5, Mn 1.5-2, P 0-0.05, S 0-0.055, Si 0.15-0.3, bal Fe.
100,000-125,000 TS; 200-250 Brin. For abrasion resisting
parts; add 0.2% Cu min for corrosion resisting properties.
Obsolete

CARILLOY CARBON-MANGANESE

U.S. Steel Corp.
C 0.35-0.5, Mn 1.5-2, P 0-0.05, S 0-0.055, Si 0.15-0.3, bal Fe.
Quenched and tempered: 150,000 TS; 321 Brin. For abrasion
resisting parts; add 0.2% Cu min for corrosion resisting
properties. *Obsolete*

CARILLOY CARBON-MANGANESE

U.S. Steel Corp.
P 0-0.04, S 0-0.05, 1.4% Mn min, C, bal Fe.
Quenched and tempered: 105,000 TS; 60,000 YS; 16 El.
Obsolete

CARILLOY CARBON-MANGANESE

U.S. Steel Corp.
Mn 1.5-2, Si 0.15-0.3, C, bal Fe.
Quenched and tempered: 100,000-120,000 TS; 75,000 Elastic
Limit; 18 El; 45 RA. For forgings; Class L over 7 to 10 in.
ASTM, A-18. *Obsolete*

CARILLOY CARBON-MANGANESE-COPPER

U.S. Steel Corp.
C 0.2-0.35, Mn 1-1.3, Cu 0.4-0.5, bal Fe.
For still tubes; corrosion resisting properties. *Obsolete*

CARILLOY CARBON-MOLYBDENUM

U.S. Steel Corp.
C 0.2-0.25, Mn 0.65-0.95, P 0-0.04, S 0-0.04, Si 0.2-0.35, Mo
0.2-0.3, bal Fe.
78,950 TS; 53,340 YS; 21 El; 59.1 RA; 156 Brin. For auto
forgings; test on 13/16 in. rounds. *Obsolete*

CARILLOY CARBON-MOLYBDENUM

U.S. Steel Corp.
C 0.6-0.7, Mn 0.65-0.95, P 0-0.04, S 0-0.04, Si 0.2-0.35, Mo
0.15-0.3, bal Fe.
143,700 TS; 100,100 YS; 8.5 El; 33.3 RA. For springs; test on
1-3/4 x # 7. *Obsolete*

CARILLOY CARBON-MOLYBDENUM

U.S. Steel Corp.
C 0.15-0.25, Mn 0.3-0.6, Si 0.15-0.3, Mo 0.9-1.1, P 0-0.05, S
0-0.05, bal Fe.
For turbine diaphragms; carburizing grade. *Obsolete*

CARILLOY CARBON-MOLYBDENUM

U.S. Steel Corp.
C 0.5-0.6, Mn 0.65-0.95, P 0-0.04, S 0-0.04, Si 0.2-0.35, Mo
0.02-0.3, bal Fe.
For auto forgings; water hardened. *Obsolete*

CARILLOY CARBON-MOLYBDENUM

U.S. Steel Corp.
C 0.38-0.45, Mn 0.65-0.95, P 0-0.04, S 0-0.04, Si 0.2-0.35, Mo
0.2-0.3, bal Fe.
110,050 TS; 77,430 YS; 13.2 El; 48.1 RA. For auto forgings;
test on 1/2 in. rounds. *Obsolete*

CARILLOY CARBON-MOLYBDENUM

U.S. Steel Corp.
C 0-0.25, P 0-0.04, S 0-0.04, Mo 0.5-0.6, bal Fe.
Annealed: 60,000 TS; 30,000 YS; 20 El; 40 RA. For gears,
shafts, axles; tough. *Obsolete*

CARILLOY CARBON-VANADIUM

U.S. Steel Corp.
C 0.4-0.55, Mn 0.65-0.95, Ni 0-0.25, Cr 0-0.15, 0.15% Si min,
0.15% N min, bal Fe.
Normalized and tempered: 90,000 TS; 60,000 YS; 22 El; 44
RA. For gears, shafts, axles; Class B 5 to 9 in. diameter ABR-
M-104-34. *Obsolete*

CARILLOY CARBON-VANADIUM

U.S. Steel Corp.
C 0.4-0.55, Mn 0.65-0.95, Ni 0-0.25, Cr 0-0.15, 0.15% Si min,
0.15% V min, bal Fe.
Normalized and tempered: 90,000 TS; 58,000 YS; 21 El; 42
RA. For gears, shafts, axles; Class B 9 to 13 in. diameter
AAR-M-104-34. *Obsolete*

CARILLOY CARBON-VANADIUM

U.S. Steel Corp.
C 0.1-0.15, Mn 0.3-0.6, P 0-0.04, S 0-0.04, Si 0-0.1, V
0.15-0.25, bal Fe.
For carburized parts; cold rolling. *Obsolete*

CARILLOY CARBON-VANADIUM

U.S. Steel Corp.
C 0.35-0.45, Mn, Si, 0.15% V min, bal Fe.
For forgings, gears, shafts; water hardened. *Obsolete*

CARILLOY CARBON-VANADIUM

U.S. Steel Corp.
C 0.4-0.55, Mn 0.65-0.95, Ni 0-0.25, Cr 0-0.15, 0.15% Si min,
0.15% V min, bal Fe.
Normalized and tempered: 90,000 TS; 60,000 YS; 24 El; 48
RA. For gears, shafts, axles; Class B under 5 in. diameter
AAR-M-104-34. *Obsolete*

CARILLOY CARBON-VANADIUM

U.S. Steel Corp.
C 0.45-0.55, Mn 0.4-0.6, 0.18% V min, bal Fe.
For gears, shafts; water hardened. *Obsolete*

CARILLOY CHROMIUM-MANGANESE-SILICON

U.S. Steel Corp.
C 0-0.17, Mn 1.05-1.4, P 0-0.35, S 0-0.04, Si 0.6-0.9, Cr
0.3-0.6, bal Fe.
75,000-90,000 TS; 45,000-54,000 YS. For boiler plate; firebox
quality, ASME, S-28, Grade A. *Obsolete*

CARILLOY CHROMIUM-MANGANESE-SILICON

U.S. Steel Corp.
C 0-0.25, Mn 1.05-1.4, P 0-0.035, S 0-0.04, Si 0.6-0.9, Cr
0.3-0.6, bal Fe.
85,000-100,000 TS; 46,750-55,000 YS. For boiler plate;
firebox quality; ASME, S-28, Grade B. *Obsolete*

CARILLOY CHROMIUM-MOLYBDENUM

U.S. Steel Corp.
C 0.7-0.8, Mn 0.5-0.8, Si 0.15-0.25, Cr 1-1.3, Mo 0.4-0.6, bal
Fe.
For forgings, springs, mandrels, tools; oil hardened.
Obsolete

CARILLOY CHROMIUM-VANADIUM

U.S. Steel Corp.

C 0.68-0.8, Mn 0.25-0.4, Cr 0.4-0.6, V 0.15-0.2, 0.15% Si min, bal Fe.

For springs; strip. *Obsolete*

CARILLOY CR-MN-SI

U.S. Steel Corp.

C 0-0.25, Mn 1.05-1.4, P 0-0.035, S 0-0.04, Si 0.6-0.9, Cr 0.3-0.6, bal Fe.

85,000-100,000 TS; 46,750-55,000 YS. For boiler plate; firebox quality; ASME, S-28, Grade B. *Obsolete*

CARILLOY FC

United States Steel Corp.

C 0.5, Mn 1.1, Si 0.25, Cr 0.75, Mo 0.2, S 0.08, bal Fe.

Heat treated bar and plate: 130 ksi TS; 110 ksi YS; 20 El. Free machining; high strength steel for gears, shafts, structural plates.

CARILLOY HIGH CHROMIUM NICKEL

U.S. Steel Corp.

C 0.6-0.8, Mn 0.2-0.4, P 0-0.04, S 0-0.04, Si 0.1-0.2, Ni 2.25-2.75, Cr 4-4.75, bal Fe.

Annealed. For shear blades; tough. *Obsolete*

CARILLOY HIGH NICKEL

U.S. Steel Corp.

Mn 0.4-0.8, P 0-0.04, S 0-0.05, 3% Ni min, C, bal Fe.

Annealed (minimum): 80,000 TS; 50,000 YS; 22 El; 40 RA. For forgings; Class H under 12 in. diameter. ASTM, A-18. *Obsolete*

CARILLOY HIGH NICKEL

U.S. Steel Corp.

Mn 0.4-0.8, P 0-0.04, S 0-0.05, 3% Ni min, C, bal Fe.

Annealed (minimum): 80,000 TS; 50,000 YS; 21 El; 38 RA. For forgings; A-18, Class H over 12 in. to 20 in. diameter ASTM. *Obsolete*

CARILLOY HIGH NICKEL

U.S. Steel Corp.

Mn 0.4-0.8, P 0-0.04, S 0-0.05, 3% Ni min, C, bal Fe.

Quenched and tempered: 100,000 TS; 70,000 YS; 20 El; 41 RA. For forgings; Class I up to 4 in. diameter ASTM, A-18. *Obsolete*

CARILLOY HIGH NICKEL

U.S. Steel Corp.

Mn 0.4-0.8, P 0-0.04, S 0-0.05, 3% Ni min, C, bal Fe.

Quenched and tempered: 85,000 TS; 55,000 YS; 20 El; 41 RA. For forgings; Class I not over 20 in. diameter ASTM, A-18. *Obsolete*

CARILLOY HIGH NICKEL

U.S. Steel Corp.

Mn 0.4-0.8, P 0-0.04, S 0-0.05, 3% Ni min, C, bal Fe.

Quenched and tempered: 90,000 TS; 60,000 YS; 20 El; 41 RA. For forgings; Class I over 7 to 10 in. diameter ASTM, A-18. *Obsolete*

CARILLOY HIGH NICKEL

U.S. Steel Corp.

Mn 0.4-0.8, P 0-0.04, S 0-0.05, 3% Ni min, C, bal Fe.

Quenched and tempered: 100,000 TS; 70,000 YS; 20 El; 41 RA. For forgings; Class I up to 4 in. diameter ASTM, A-18. *Obsolete*

CARILLOY HIGH NICKEL

U.S. Steel Corp.

Mn 0.4-0.8, P 0-0.04, S 0-0.05, 3% Ni min, C, bal Fe.

Quenched and tempered: 100,000 TS; 65,000 YS; 20 El; 41 RA. For forgings; Class I over 4 to 7 in. diameter ASTM, A-18. *Obsolete*

CARILLOY HIGH-CARBON LOW-CHROMIUM

U.S. Steel Corp.

C 1.2-1.4, Mn 0-0.4, P 0-0.04, S 0-0.05, Si 0.18-0.3, Cr 0.2-0.4, bal Fe.

For razor blades; cold rolled heat treated. *Obsolete*

CARILLOY HIGH-CARBON LOW-CHROMIUM

U.S. Steel Corp.

C 1-1.15, Mn 0.25-0.4, P 0-0.025, S 0-0.025, Si 0.15-0.3, Cr 0.1-0.2, bal Fe.

For flat wire; cold rolled heat treated. *Obsolete*

CARILLOY HIGH-CHROMIUM-MOLYBDENUM

U.S. Steel Corp.

C 0-0.15, Mn 0-0.5, P 0-0.03, S 0-0.03, Si 0-0.5, Cr 4-6, Mo 0.45-0.65, Al 0-0.05, bal Fe.

Drawn: 167,410 TS; 107,770 YS; 11.5 El; 27.2 RA. For still tubes; high temperature work. *Obsolete*

CARILLOY LOW NICKEL CHROMIUM

U.S. Steel Corp.

C 0.35-0.45, Mn 0.5-0.8, P 0-0.04, S 0-0.045, Ni 1-1.5, Cr 0.45-0.75, bal Fe.

Normalized, quenched and tempered: 95,000 TS; 60,000 YS; 20 El; 50 RA. For motor shafts; Grade A over 4 to 7 in. diameter, G.E. Co. B5C9S1. *Obsolete*

CARILLOY LOW NICKEL CHROMIUM

U.S. Steel Corp.

C 0.35-0.45, Mn 0.5-0.8, P 0-0.04, S 0-0.045, Ni 1-1.5, Cr 0.45-0.75, bal Fe.

Normalized, quenched and tempered: 100,000 TS; 65,000 YS; 20 El; 50 RA. For motor shafts; Grade A over 2 to 4 in. diameter, G.E. Co. B5C9S1. *Obsolete*

CARILLOY LOW NICKEL CHROMIUM

U.S. Steel Corp.

P 0-0.04, S 0-0.05, Si 0.15-0.3, Ni 1-1.5, Cr 0.45-0.75, C, Mn, bal Fe.

Quenched and tempered: 100,000-120,000 TS; 75,000 YS; 18 El; 45 RA. For forgings; Class L over 7 to 10 in. diameter ASTM, A-18. *Obsolete*

CARILLOY LOW NICKEL CHROMIUM

U.S. Steel Corp.

P 0-0.04, S 0-0.05, Si 0.15-0.3, Ni 1-1.5, Cr 0.45-0.75, C, Mn, bal Fe.

Quenched and tempered: 100,000-120,000 TS; 75,000 YS; 20 El; 50 RA. For forgings; Class L over 2 to 7 in. diameter ASTM, A-18. *Obsolete*

CARILLOY LOW NICKEL CHROMIUM

U.S. Steel Corp.

C 0.35-0.45, Mn 0.5-0.8, P 0-0.04, S 0-0.045, Ni 1-1.5, Cr 0.45-0.75, bal Fe.

Normalized, quenched and tempered: 105,000 TS; 70,000 YS; 20 El; 50 RA. For motor shafts; Grade A 2 in. and under in diameter, G.E. Co. B5-C9S1. *Obsolete*

CARILLOY MANGANESE-MOLYBDENUM

U.S. Steel Corp.

C 0.1-0.2, Mn 0.9-1.2, P 0-0.04, S 0-0.05, Si 0.15-0.3, Mo 0.35-0.5, bal Fe.

77,300 TS; 55,820 YS; 21.7 El; 58.7 RA. For structural parts; test on 3/4 in. plate. *Obsolete*

CARILLOY MANGANESE-MOLYBDENUM

U.S. Steel Corp.

C 0.1-0.2, Mn 1.3-1.6, P 0-0.04, S 0-0.05, Si 0.3-0.5, Mo 0.35-0.5, bal Fe.

97,750 TS; 64,200 YS; 21.5 El; 38.6 RA. For structural parts; test on 1/2 in. plate. *Obsolete*

CARILLOY MANGANESE-MOLYBDENUM

U.S. Steel Corp.

C 0.5-0.6, Mn 0.9-1.2, P 0-0.04, S 0-0.04, Si 0.15-0.3, Mo 0.45-0.75, bal Fe.

For seamless tube; good weldability. *Obsolete*

CARILLOY MANGANESE-MOLYBDENUM

U.S. Steel Corp.

C 0.23-0.3, Mn 1.3-1.6, P 0-0.04, S 0-0.05, Si 0-0.25, 0.15% Mo min, bal Fe.

For seamless drill pipe; tough. *Obsolete*

CARILLOY MANGANESE-MOLYBDENUM

U.S. Steel Corp.

C 0.2-0.3, Mn 0.8-1.1, P 0-0.04, S 0-0.05, Si 0.1-0.2, Mo 0.2-0.3, bal Fe.

84,840 TS; 49,550 YS; 21.5 El; 41.2 RA. For structural parts; test on 6 x 6 x 13/16 in. angle. *Obsolete*

CARILLOY MANGANESE-VANADIUM

U.S. Steel Corp.

C 0.75-0.85, Mn 1.4-1.75, Si 0.1-0.25, V 0.15-0.3, bal Fe.

For shear blades; tough. *Obsolete*

CARILLOY MEDIUM NICKEL

British Standards Institution

C 0.2-0.3, Mn 0.65-0.95, P 0-0.045, S 0-0.05, Si 0.15-0.35, Ni 2.5-3, Cr 0-0.15, bal Fe.

Normalized and tempered: 80,000 TS; 55,000 YS; 28 El; 60 RA. For forgings; Class C under 8 in. diameter AAR-M-104-34. *Obsolete*

CARILLOY MEDIUM NICKEL

U.S. Steel Corp.

C 0.2-0.3, Mn 0.65-0.95, P 0-0.045, S 0-0.05, Si 0.15-0.35, Ni 2.5-3, Cr 0-0.15, bal Fe.

Normalized and tempered: 80,000 TS; 55,000 YS; 28 El; 55 RA. For forgings; Class C 8 in. and over in diameter AAR-M-104-34. *Obsolete*

CARILLOY NI-CR-MO

U.S. Steel Corp.

C 0.55-0.65, Mn 0.6-0.8, Si 0.2-0.3, Ni 1.5-2, Cr 0.8-1.1, Mo 0.3-0.4, bal Fe.

Annealed: 202-241 Brin. For gears, axles, crankshafts; tough, oil hardened. *Obsolete*

CARILLOY NI-CR-MO

U.S. Steel Corp.

C 0.5-0.6, Mn 0.5-0.8, Si 0.2-0.3, Ni 1-1.5, Cr 0.5-0.8, 0.20 % Mo min, bal Fe.

For shear blades; tough, oil hardened. *Obsolete*

CARILLOY NI-CR-MO

U.S. Steel Corp.

C 0.35-0.45, Mn 0.5-0.8, Si 0-0.35, Ni 1.5-2, Cr 0.5-0.8, V 0-0.05, Mo 0.3-0.4, bal Fe.

For forgings, gears, shafts; modified spec., G.E. Co. B5A14S3. *Obsolete*

CARILLOY NI-CR-MO-V

U.S. Steel Corp.

C 0-0.4, Mn 0-0.75, Si 0-0.3, Ni 0-3.25, Cr 0-1.25, V 0-0.25, Mo 0.3-0.5, bal Fe.

Normalized and annealed: 90,000 TS; 15 El; 24 RA. For turbine rotor; radial body tests, G.E. Co. B31A1S1. *Obsolete*

CARILLOY NI-V-MO

U.S. Steel Corp.

C 0.3-0.4, Mn 0.5-0.8, Si 0-0.3, Ni 2.5-3.25, Mo 0.3-0.5, 0.15% V min, bal Fe.

Normalized and annealed: 90,000 TS; 15 El; 24 RA. For gears, shafts, axles, crankshafts; shock resistant. *Obsolete*

CARILLOY NICKEL STEEL

U.S. Steel Corp.

C 0.08-0.18, Mn 0.3-0.6, P 0-0.04, S 0-0.045, Si 0-0.05, Ni 0.2-0.5, bal Fe.

For seamless tubes; good weldability. *Obsolete*

CARILLOY NICKEL-CHROMIUM-MOLYBDENUM

U.S. Steel Corp.

C 0.5-0.6, Mn 0.5-0.8, Si 0.2-0.3, Ni 1-1.5, Cr 0.5-0.8, 0.20 % Mo min, bal Fe.

For shear blades; tough, oil hardened. *Obsolete*

CARILLOY NICKEL-CHROMIUM-MOLYBDENUM

U.S. Steel Corp.
C 0.55-0.65, Mn 0.6-0.8, Si 0.2-0.3, Ni 1.5-2, Cr 0.8-1.1, Mo 0.3-0.4, bal Fe.
Annealed: 202-241 Brin. For gears, axles, crankshafts; tough, oil hardened. *Obsolete*

CARILLOY NICKEL-CHROMIUM-MOLYBDENUM

U.S. Steel Corp.
C 0.35-0.45, Mn 0.5-0.8, Si 0-0.35, Ni 1.5-2, Cr 0.5-0.8, V 0-0.05, Mo 0.3-0.4, bal Fe.
For forgings, gears, shafts; modified spec., G.E. Co. B5A14S3. *Obsolete*

CARILLOY NICKEL-VANADIUM-MOLYBDENUM

U.S. Steel Corp.
C 0.3-0.4, Mn 0.5-0.8, Si 0-0.3, Ni 2.5-3.25, Mo 0.3-0.5, 0.15% V min, bal Fe.
Normalized and annealed: 90,000 TS; 15 El; 24 RA. For gears, shafts, axles, crankshafts; shock resistant. *Obsolete*

CARILLOY SILICON-VANADIUM

U.S. Steel Corp.
C 0.68-0.78, Mn 0.3-0.5, P 0-0.025, S 0-0.025, Si 0.5-0.7, 0.15% min V, bal Fe.
Annealed. For shear blades; water hardening. *Obsolete*

CARILLOY STRUCTURAL MANGANESE

U.S. Steel Corp.
C 0-0.35, Mn 0-1.8, Si 0.1-0.3, bal Fe.
Rolled: 82,000-100,000 TS; 47,000 YS; 30 El. For rivets. *Obsolete*

CARILLOY STRUCTURAL MANGANESE

U.S. Steel Corp.
C 0-0.4, Mn 0-1.8, Si 0.1-0.3, bal Fe.
Rolled: 90,000 TS; 55,000 YS; 30 El. For structural purposes. *Obsolete*

CARILLOY STRUCTURAL MANGANESE

U.S. Steel Corp.
C 0-0.35, Mn 0-1.8, Si 0.1-0.3, bal Fe.
Rolled: 82,000-100,000 TS; 47,000 YS; 30 El. For rivets. *Obsolete*

CARILLOY STRUCTURAL NICKEL

U.S. Steel Corp.
C 0-0.3, Mn 0-0.6, P 0-0.03, S 0-0.045, 3.25% Ni min, bal Fe.
70,000-80,000 TS; 45,000 YS; 40 El. For rivets; ASTM, A-8. *Obsolete*

CARILLOY STRUCTURAL NICKEL

U.S. Steel Corp.
C 0-0.45, Mn 0-0.7, P 0-0.04, S 0-0.05, 3.25% Ni min, bal Fe.
95,000-110,000 TS; 55,000 YS; 25 El. For eye bars; ASTM, A-8. *Obsolete*

CARILLOY STRUCTURAL NICKEL

U.S. Steel Corp.
C 0-0.45, Mn 0-0.7, P 0-0.04, S 0-0.05, 3.25% Ni min, bal Fe.
Annealed: 90,000-105,000 TS; 52,000 YS; 20 El; 35 RA. For bridge pins; ASTM, A-8. *Obsolete*

CARILLOY STRUCTURAL NICKEL

U.S. Steel Corp.
C 0-0.45, Mn 0-0.7, P 0-0.04, S 0-0.05, 3.25% Ni min, bal Fe.
85,000-100,000 TS; 50,000 YS; 25 RA. For structural parts; add 0.2% Cu min for corrosion resisting properties. *Obsolete*

CARILLOY STRUCTURAL STEEL

U.S. Steel Corp.
C 0-0.45, Mn 0-0.7, P 0-0.04, S 0-0.05, Ni 0-3.25, bal Fe.
90,000-105,000 TS; 52,000 YS; 20 El; 35 RA. For eye bars; ASTM, A-8. *Obsolete*

CARILLOY STRUCTURAL STEEL

U.S. Steel Corp.
C 0-0.45, Mn 0-0.7, P 0-0.04, S 0-0.05, 3.25% Ni min, bal Fe.
95,000-110,000 TS; 55,000 YS; 16 El; 25 RA. For bridge rollers; ASTM, A-8. *Obsolete*

CARIRON

Fillmore Foundry Inc.
TC 2.96, Mn 0.75, Si 1.2, Ni 1.55, bal Fe.
50,600 TS; 221 Brin. For liners, pistons; transverse strength 4160.

CARLOY

Benedict-Miller, Inc.
C, Mo, bal Fe.
For tools; water hardened. *Obsolete*

CARLSON 22-4-9 STAINLESS STEEL

G.O. Carlson Inc.
Stainless steel. C 0.45-0.6, Mn 7-10, P 0-0.045, S 0-0.12, Si 0-1, Cr 20-23, Ni 3-5, N 0.3-0.5, bal Fe.
Hot rolled and aged (plate), RT: 296-1124 MPa TS; 179-676 MP YS (0.2%); 10-50 El (in 2 in.); 13-63 RA; 105-321 Brin. For die inserts, punches, high temperature steam valves, gas turbine parts, mechanical wear-resistant parts.

CARLSON 309S CB

G.O. Carlson Inc.
Stainless steel. C 0-0.08, Mn 0-2, Si 0-1, Cr 22-24, Ni 12-15, Cb = 8 x C, bal Fe.
Austenitic stainless steel.

CARLSON 600

G.O. Carlson Inc.
C 0.08, Mn 0.5, S 0.008, Si 0.25, Cr 15.5, Fe 8, Cu 0.25, Ni 76.
Nickel base superalloy. ACI CY-40; INCONEL 600.

CARLSON 825

G.O. Carlson Inc.
C 0-0.03, Mn 0.5, Si 0.25, Cr 21.5, Ni 42, Mo 3, Ti 0.9, Al 0.1, Cu 2.25, Fe 30.
Corrosion and heat resistant alloy. Reference: INCOLOY 825.

CARLSON ALLOY 330

G.O. Carlson Inc.
C 0-0.1, Mn 0-2, S 0-0.03, P 0-0.03, Si 0.75-1.5, Ni 34-37, Cu 0-1, Cr 17-20, Pb 0-0.005, Sn 0-0.025, bal Fe.
Nickel-iron-chromium-silicon austenitic alloy for furnace containers, heat treating equipment, gas turbine components, heat exchangers, furnace fans, petrochemical furnace parts, salt pots - neutral, cyanide, hot pressing platens. 70.0 ksi TS; 30.0 ksi YS; 30 El. N08330

CARLSON ALLOY C 20 PLUS

G.O. Carlson Inc.
C 0-0.07, Mn 0-2, S 0-0.035, P 0-0.045, Si 0-1, Cu 3-4, Cr 19-21, Mo 2-3, Ni 32-38, Cb + Ta = 8 x C min (up to 1.00), bal Fe.
Oxidation and corrosion resistant nickel-molybdenum-chromium alloy for chemical process equipment, pulp and paper, air pollution control, ore processing, waste treatment and disposal, steel pickling, marine equipment, nuclear reactors, aerospace and gas turbines. 80.0 ksi TS; 35.0 ksi YS; 30 El. N08020

CARLSON ALLOY C 625

G.O. Carlson Inc.
C 0-0.1, Mn 0-0.5, S 0-0.015, P 0-0.015, Si 0-0.5, Co 0-1, Cr 20-23, Mo 8-10, Fe 0-5, Al 0-0.4, Ti 0-0.4, 3.15-4.15 Cb = Ta, 58.0 Ni min.
Nickel-iron-chromium-silicon austenitic alloy for furnace containers, heat treating equipment, gas turbine components, heat exchangers, furnace fans, petrochemical furnace parts, salt pots - neutral, cyanide, hot pressing platens. 70.0 ksi TS; 30.0 ksi YS; 30 El. As rolled: 75.0 ksi TS; 40.0 ksi YS; 25 El. N06625

CARLSON ALLOY GOC 276

G.O. Carlson Inc.
C 0-0.01, Mn 0-1, S 0-0.03, P 0-0.04, Si 0-0.08, Co 0-2.5, Cr 14.5-16.5, Mo 15-17, Fe 4-7, W 3-4.5, V 0.35, bal Ni.
Corrosion resistant nickel-molybdenum-chromium alloy for chemical process equipment, pulp and paper, air pollution control, ore processing, and waste treatment and disposal. 100.0 ksi TS; 41.0 ksi YS; 40 El. N10276

CARLSON C 200

G.O. Carlson Inc.
Nickel. C 0-0.15, Mn 0-0.35, S 0-0.01, Si 0-0.35, Cu 0-0.25, Fe 0-0.4, 99.0 Ni min.
Wrought nickel for electronic parts, food processing equipment, caustic manufacturing and storage, aerospace and missile parts, fluorine electrolysis and synthetic fiber manufacture. Annealed: 55.0 ksi TS; 15.0 ksi YS; 40 El.

CARLSON C 201

G.O. Carlson Inc.
Nickel. C 0-0.02, Mn 0-0.35, S 0-0.01, Si 0-0.35, Cu 0-0.25, Fe 0-0.4, 99.0 Ni min.
Wrought nickel for electronic parts, food processing equipment, caustic manufacturing and storage, aerospace and missile parts, fluorine electrolysis and synthetic fiber manufacture. Annealed: 55.0 ksi TS; 15.0 ksi YS; 40 El.

CARLSON C 400

G.O. Carlson Inc.
Nickel. C 0-0.3, Mn 0-2, S 0-0.024, Si 0-0.5, Ni 63-70, Cu 28-34, Fe 0-2.5.
Nickel-copper alloy for pump parts and shafts, valves, marine fixtures, crude petroleum, stills, gasoline and fresh water storage tanks, chlorinated solvents and process vessels. Annealed: 70.0 ksi TS; 28.0 ksi YS; 35 El. As rolled: 75.0 ksi TS; 40.0 ksi YS; 25 El.

CARLSON C 600

G.O. Carlson Inc.
Nickel. C 0-0.15, Mn 0-1, S 0-0.015, Si 0-0.5, Cr 14-17, Cu 0-0.5, Ti 0.1-0.5, Fe 6-10, 72.0 Ni min.
Nickel-chromium alloy for electronic components, nuclear parts, chemical and food processing equipment, heat exchangers, evaporators, cracking units, heat treating and carburizing atmospheres. Annealed: 80.0 ksi TS; 35.0 ksi YS; 30 El.

CARLSON C 600 MIL

G.O. Carlson Inc.
Nickel. C 0.03-0.15, Mn 0-1, S 0-0.015, Si 0-0.5, Cr 14-17, Cu 0-0.5, Ti 0.1-0.5, Al 0.05-0.35, Co 0.1, Fe 6-10, 72.0 Ni min.
Nickel-chromium alloy for electronic components, nuclear parts, chemical and food processing equipment, heat exchangers, evaporators, cracking units, heat treating and carburizing atmospheres. Annealed: 80.0 ksi TS; 35.0 ksi YS; 30 El.

CARLSON C 625

G.O. Carlson Inc.
Nickel. C 0-0.1, Mn 0-0.5, S 0-0.015, Si 0-0.5, Cr 20-23, Ti 0-0.4, Al 0-0.4, Co 0-1, Mo 8-10, Fe 0-5, 58.0 Ni min, 3.15-4.15 Cb + Ta.
Nickel-chromium alloy for marine atmosphere and sea water applications, pollution control equipment (scrubbers), and high temperature process equipment. Annealed: 120.0 ksi TS; 60.0 ksi YS; 30 El.

CARLSON C 800

G.O. Carlson Inc.
Nickel. C 0-0.1, Mn 0-1.5, S 0-0.015, Si 0-1, Cr 19-23, Ni 30-35, Cu 0-0.75, Ti 0.15-0.6, Al 0.15-0.6, bal Fe.
Nickel-chromium alloy for heat exchangers and process equipment, carburizing fixtures and retorts, furnace components, desulfurizers, flare tips, ammonia effluent coolers and methane reformers. Annealed: 75.0 ksi TS; 30.0 ksi YS; 30 El.

CARLSON C 800 H

G.O. Carlson Inc.
Nickel. C 0.05-0.1, Mn 0-1.5, S 0-0.015, Si 0-1, Cr 19-23, Ni 30-35, Cu 0-0.75, Ti 0.15-0.6, Al 0.15-0.6, bal Fe.
Nickel-chromium alloy for heat exchangers and process equipment, carburizing fixtures and retorts, furnace components, desulfurizers, flare tips, ammonia effluent coolers and methane reformers. Solution treated: 70.0 ksi TS; 25.0 ksi YS; 30 El.

CARLSON C 825 ESR

G.O. Carlson Inc.
Nickel. C 0-0.05, Mn 0-1, S 0-0.03, Si 0-0.5, Cr 19.5-23.5, Ni 38-46, Cu 1.5-3, Ti 0.6-1.2, Al 0-0.2, Fe 22, 2.50-3.50 Cb + Ta.
Nickel-chromium alloy for phosphoric acid evaporators, pickling tank heaters, chemical processing equipment, propeller shafts, spent nuclear fuel components. Annealed: 85.0 ksi TS; 35.0 ksi YS; 30 El.

CARLTON

Time Steel Service Inc.
C 0.95, Mn 2, Cr 2.2, Mo 1.1, bal Fe (contains lead).
Free machining grade of air hardened tool steel. AISI A4.

CARMELIA BRONZE

Manufacturer not listed
Pb, Zn, bal Cu.
For machinery bearings; heavy duty.

CARMET

Allegheny Ludlum Steel
WC.
For tools; cemented carbides.

CARMET ACA-1

Allegheny Ludlum Steel
WC.
For roughing tools for steel; sintered. *Obsolete*

CARMET C.A. 51

Allegheny Ludlum Steel
WC.
90 RA. For high speed planer tools, heavy cuts on steel; sintered; transverse strength 225,000.

CARMET CA-1

Allegheny Ludlum Steel
WC + Co.
For tipped tools; for steel.

CARMET CA-10

Allegheny Ludlum Steel
WC + Co.
For cutting tool bits, dies; sintered, for heavy cuts; transverse strength 320,000.

CARMET CA-10

Carmet Materials
Sintered carbide. Cold work dies for blanking dies.

CARMET CA-11

Allegheny Ludlum Steel
WC.
86 RA. For cold header dies; sintered; transverse strength 350,000.

CARMET CA-11

Carmet Materials
Sintered carbide. Cold header dies; cold work dies for blanking.

CARMET CA-12

Allegheny Ludlum Steel
WC.
89 RA. For cold forming dies; sintered; transverse strength 300,000.

CARMET CA-12

Carmet Materials
Sintered carbide. For cold forming dies; rough machining non-ferrous metals.

CARMET CA-2

Allegheny Ludlum Steel
WC + Co.
For tipped tools; for finishing cuts.

CARMET CA-20

Allegheny Ludlum Steel
WC.
84 RA. For cold header dies; sintered; transverse strength 375,000.

CARMET CA-21

Allegheny Ludlum Steel
WC.
86 RA. For blanking and cold work dies; sintered; transverse strength 275,000.

CARMET CA-22

Allegheny Ludlum Steel
WC.
For blanking and cold work dies; sintered. *Obsolete*

CARMET CA-3

Allegheny Ludlum Steel
WC + Co.
For tipped tools; for roughing cuts.

CARMET CA-3

Carmet Materials
Sintered carbide. For rough cutting operations on cast iron, non-ferrous metals and non-metallics. For gages.

CARMET CA-310

Carmet Materials
Sintered carbide. For cutting tools.

CARMET CA-315

Carmet Materials
Sintered carbide. For cutting tools.

CARMET CA-4

Allegheny Ludlum Steel
WC + Co.
For tipped tools; for soft metals.

CARMET CA-4

Carmet Materials
Sintered carbide. Cutting tools for finish cutting cast iron, non-ferrous metals and non-metallics. Also for gages.

CARMET CA-425

Carmet Materials
Sintered carbide. For cold header dies.

CARMET CA-5

Allegheny Ludlum Steel
WC + Co.
For cutting tool bits for slow speeds and heavy cuts on steel; sintered; transverse strength 200,000.

CARMET CA-51

Carmet Materials
Sintered carbide. For cutting tools, heavy roughing cuts on cast iron and on steel.

CARMET CA-6

Allegheny Ludlum Steel
WC + Co.
For cutting tool bits, steel precision boring; sintered, for light precision cuts; transverse strength 150,000.

CARMET CA-604

Allegheny Ludlum Steel
WC, TaC, TiC.
For cutting tools for high speed machines; sintered carbides.

CARMET CA-606

Allegheny Ludlum Steel
WC.
For cutting tools, dies; sintered carbides.

CARMET CA-606

Carmet Materials
Sintered carbide. For cutting tools for finish cutting of steel.

CARMET CA-608

Allegheny Ludlum Steel
TiC.
For cutting tools; sintered carbides.

CARMET CA-609

Allegheny Ludlum Steel
WC.
For cutting tools, dies; sintered carbides.

CARMET CA-610

Allegheny Ludlum Steel
TiC.
For cutting tools; sintered carbides.

CARMET CA-610

Carmet Materials
Sintered carbide. For cutting tools, for roughing and general purpose cutting of steel.

CARMET CA-7

Allegheny Ludlum Steel
WC + Co.
For cutting tool bits for hard cast iron; sintered, wear resistant; transverse strength 190,000.

CARMET CA-7

Carmet Materials
Sintered carbide. Cutting tools for finish cutting of non ferrous metals and non-metallics.

CARMET CA-704

Carmet Materials
Sintered carbide. For cutting tools, precision finishing and boring on steel.

CARMET CA-711

Carmet Materials
Sintered carbide. For cutting tools, finish cutting of steel.

CARMET CA-720

Carmet Materials
Sintered carbide. For cutting tools, for roughing and general purpose cutting of steel.

CARMET CA-740

Carmet Materials
Sintered carbide. For cutting tools, heavy roughing interrupted cuts.

CARMET CA-8

Allegheny Ludlum Steel
WC + Co.
For cutting tool bits for light precision cuts on cast iron; sintered, high resistance to wear; transverse strength 190,000.

CARMET CA-8

Carmet Materials
Sintered carbide. For cutting tools for finish cutting of non-ferrous metals and non-metallics.

CARMET CA-815

Allegheny Ludlum Steel
WC.
For hot extrusion dies, precision gages; sintered carbides.

CARMET CA-815

Carmet Materials
Sintered chromium carbide. Used for gages.

CARMET CA-9
Allegheny Ludlum Steel
WC + Co.
For cutting tool bits for soft plastics, gage blanks; sintered; transverse strength 220,000.

CARMET CA-9443
Carmet Materials
Coated sintered CA-443.

CARMET CA-9721
Carmet Materials
Coated sintered CA-721.

CARMET CA-9740
Carmet Materials
Coated sintered CA-740.

CARMET CA-B
Carmet Materials
Al_2O_3 + TiC hot-pressed ceramic. For rough and finished machining of hard cast irons and hardened steel.

CARMET CA-W
Carmet Materials
Al_2O_3 base sintered ceramic. For machining cast irons, heat-treated steel.

CARNOLIA
Charles Carr Ltd.
Pb-Sn-Sb.
For antifriction alloy for bearings; Babbitt.

CARO
Manufacturer not listed
Cu 91.6, Sn 8.5, P 0.19.
For tubes; corrosion resistant.

CARO BRONZE
Wrought Bearing Metals Inc.
Cu 91.2, Sn 8.5, P 0.3.
Annealed: 63,000 TS; 60 El. Hard: 99,000 TS; 12 El. For bearings and wearing parts; wrought.

CAROBRONZE
Carobronze Ltd.
Copper. Cu 91.2, Sn 8.5, P 0.3.
60,000-95,000 TS; 20-50 El; 90-190 Brin. For solid cold drawn tubes, bushes for highly stressed bearings; resists corrosion.

CARPALOY NO. 2
Carpenter Technology Corp.
C 0.3, Cr 13, bal Fe.
Hardened: 260,000 TS; 11 El; 32 RA; 512 Brin. For tools, cutlery, scissors, check valves, surgical instruments; superseded by "Carpenter Stainless Steel No. 2." Obsolete

CARPALOY NO. 3
Carpenter Technology Corp.
C 0.19, Cr 20.94, Ni 0.17, Cu 0.92, bal Fe.
Superseded by "Carpenter Stainless Steel No. 3". 87,000 TS; 40,000 YS; 23 El; 52 RA. For heat and corrosion resisting parts. Obsolete

CARPENTAL
Alumetal (Montecatini Edison)
Aluminum.
Aluminum alloy AlZnMg (Cu-free).

CARPENTER "D.D."
Carpenter Technology Corp.
C, Cr, Ni, bal Fe.
Heat treated: 165,000-195,000 TS; 145,000-165,000 YS; 16-18 El; 52-60 RA; 302-332 Brin. For axles; for extra high duty service. Obsolete

CARPENTER "J.Y." STEEL NO. 656
Carpenter Technology Corp.
C 0.6, Cr 1, V 0.2, bal Fe.
For roller paths in turrets of battleships, roller bearings.
Obsolete

CARPENTER 15-15LC STAINLESS
Carpenter Technology Corp.
Stainless steel. C 0.03, Mn 15.3, Si 0.4, P 0.025, S 0.003, Cr 16.3, Ni 1.1, Mo 1.1, Cu 0.56, N 0.4, bal Fe.
Annealed, RT: 427 MPa YS; 633 MPa TS; 50 El; 60 RA; 210 Brin. Austenitic, nitrogen strengthened stainless steel for oil and gas industry applications such as nonmagnetic drill collars, stabilizers and MWD housings.

CARPENTER 19 ALLOY
Carpenter Technology Corp.
At 20°C: 25,000 psi YS; 60,000 psi TS; 35 El. Nickel molybdenum/nickel thermocouple used in hydrogen, dissociated ammonia and vacuum furnaces to 2500°F. Resists green rot.

CARPENTER 20 ALLOY
Carpenter Technology Corp.
At 20°C: 75,000 psi YS; 130,000 psi TS; 35 El. Nickel molybdenum/nickel thermocouple used in hydrogen, dissociated ammonia and vacuum furnaces to 2500°F. Resists green rot.

CARPENTER 20CB-3
Now CARPENTER STAINLESS 20 CB-3.

CARPENTER 21-12 VALVE STEEL
Carpenter Technology Corp.
C 0.2, Mn 1.25, Si 0.8, Cr 21, Ni 11.5, bal Fe.
For diesel and gasoline engine valves. SAE EV-3.

CARPENTER 21-12N VALVE STEEL
Carpenter Technology Corp.
C 0.2, Mn 1.25, Si 0.5, Cr 21, Ni 11.5, N 0.2, bal Fe.
For diesel and gasoline engine valves. SAE EV-4.

CARPENTER 21-55N VALVE STEEL
Carpenter Technology Corp.
C 0.2, Mn 5, Si 0.5, Cr 21, Ni 4.5, N 0.3, bal Fe.
Austenitic. For automotive exhaust valves. SAE EV-7.

CARPENTER 22-3 ALLOY
Carpenter Technology Corp.
C 0.1, Mn 0.5, Si 0.25, Cr 3.1, Ni 22, bal Fe.
As annealed: 483 MPa TS; 276 MPa YS; 35 El; 74 Rock B. Austenitic, nonmagnetic ferrous alloy with a high coefficient of thermal expansion. For electronics applications.

CARPENTER 302HQ-FM STAINLESS
Carpenter Technology Corp.
Stainless steel. C 0-0.06, Mn 0-2, P 0-0.04, S 0-0.14, Si 0-1, Cr 16-19, Ni 9-11, Cu 1.3-2.4, bal Fe.
25.4-mm round bar, annealed, RT: 221 MPa YS; 510 MPa TS; 57 El; 74 RA. Modification of Carpenter Stainless Custom Flo 302 HQ designed for cold headed parts that need drilling, slotting, broaching, etc. Also for automatic screw machine operations where thread rolling or cold form tapping is critical.

CARPENTER 430F SOLENOID QUALITY
Carpenter Technology Corp.
C 0.06, Mn 0.5, P 0.03, S 0.3, Si 0.5, Cr 17.5, Mo 0.3, bal Fe.
Corrosion resistant, soft, magnetic steel for solenoids. Properties vary with annealing treatment.

CARPENTER 430FR SOLENOID QUALITY
Carpenter Technology Corp.
C 0.06, Mn 0.5, Si 1.25, P 0.02, S 0.3, Cr 17.5, Mo 0.3, bal Fe.
Corrosion resistant, magnetic, soft metal for solenoid valves.

CARPENTER 434 HS
Carpenter Technology Corp.
C 0.12, Mn 1, Si 1, Cr 16-18, Mo 0.75-1.25, bal Fe.
Closely controlled chemical balance of Type 434; for dependable response to heat treatment. AISI 434. Obsolete

CARPENTER 49
Carpenter Technology Corp.
Ni 49, bal Fe.
For low expansion alloy; low coefficient of expansion.
Obsolete

CARPENTER 5-F
Now CARPENTER STAINLESS NO. 5F.

CARPENTER 636 ALLOY (TYPE 422)
Carpenter Technology Corp.
C 0.2-0.25, Mn 0-1, Si 0-1, Cr 12-14, Ni 0.5-1, Mo 0.75-1.25, W 0.75-1.25, V 0.2-0.5, bal Fe.
Air or oil hardenable corrosion resistant steel. Hardened, tempered at 1200°F, RT: 149,000 TS; 125,000 YS. At 1100°F: 76,000 TS; 72,000 YS. For buckets and blades in steam turbines, high temperature bolting, compressor parts. AISI 422.

CARPENTER 709 TYPE 2
Carpenter Technology Corp.
C 0.45, Mn 0.55, Si 0.25, Cr 1, Mo 0.5, V 0.3, bal Fe.
Heat treated: 142,000-260,000 TS; 130,000-230,000 YS; 10-20 El; 41-60 RA; 340-510 Brin. For steam turbine valves, pressure vessels. Operating temperature up to 1200°F.

CARPENTER 80-20
Carpenter Technology Corp.
C 0.15-0.3, Mn 0.6-1, Fe 0-1, Si 0.3, Cr 19-21, bal Ni.
Annealed: 121,000 TS; 60,000 YS; 35 El; 50 RA; 205 Brin. For aircraft valves, aircraft engine exhaust systems. Resists corrosion of leaded gasoline exhausts.

CARPENTER A-8
Carpenter Technology Corp.
C 0.55, Mn 0.3, Si 0.9, Cr 5, Mo 1.25, W 1.2, bal Fe.
Medium carbon, air hardening tool steel used for slitting cutters, shear blades, forming dies and blanking dies.

CARPENTER ALLOY 182-FM
Now CARPENTER PROJECT 70 182-FM.

CARPENTER AMS 5616
Carpenter Technology Corp.
C 0.15-0.2, Mn 0.5, Si 0.5, Cr 12-14, Ni 1.8-2.2, Mo 0-0.5, W 2.5-3.5, bal Fe.
Air or oil hardenable corrosion resistant steel. Heat treated: about 217,000 TS max. At 1100°F: 135,000 TS; 120,000 YS; 18 El. For steam turbine buckets and blades, gas turbine parts, high temperature bolts. AMS 5616; Greek Ascoloy.

CARPENTER ASTROLOY
Carpenter Technology Corp.
C 0.06, Cr 15, Co 15, Ti 3.5, Al 4.4, Mo 5.25, B 0.03, Zr 0-0.06, bal Ni.
Precipitation hardening nickel base superalloy. For turbine discs in aircraft gas turbine engines. Obsolete

CARPENTER B
Carpenter Technology Corp.
Mo 26-30, Fe 4-6, C 0-0.12, bal Ni.
Rolled: 140,000 TS; 65,000 YS; 45 El; 45 RA; 235 Brin. For tubing; resists hydrochloric acid, phosphoric acid and up to 60% sulfuric acid. Obsolete

CARPENTER BARTO
Carpenter Technology Corp.
C 0.5, Mn 0.5, Si 0.25, Cr 1, Ni 1.75, bal Fe.
Heat treated: 310,000 TS; 24,000 YS; 5 El; 56 Rock C. Tough, shock resistant, oil hardenable alloy steel for expander punches, feeder rolls, clutch parts, vise jaws and chuck jaws.

CARPENTER C
Carpenter Technology Corp.
Fe 6, Mn 1, Si 1, Cr 15, Mo 17, C 0.15, W 5, bal Ni.
Rolled: 130,000 TS; 65,000 YS; 25 El; 210 Brin. For tubing; resists phosphoric anhydrides, sulfites, and free chlorides.
Obsolete

CARPENTER C-XB VALVE STEEL

Carpenter Technology Corp.
C 0.8, Mn 0.4, Si 2.25, Cr 20, Ni 1.5, bal Fe.
Hardenable silicon-chrome alloy. At 1200°F: 39,500 TS; 29,500 YS; 33.5 El. For exhaust valves, valve seat inserts, intake valves in internal combustion engines.

CARPENTER CHROME MAGNET

Carpenter Technology Corp.
Cr 3.5, C 0.95, bal Fe.
For magnetos, meters, electrical apparatus, permanent magnets; permanent magnet steel. *Obsolete*

CARPENTER CHROME STEEL NO. 12 S

Carpenter Technology Corp.
C 1, Cr 1.2, bal Fe.
For balls. *Obsolete*

CARPENTER CHROME-NICKEL STAINLESS WITH B

Carpenter Technology Corp.
Now NEUTROSORB.

CARPENTER CONSUMET CORE IRON

Carpenter Technology Corp.
C 0-0.06, Mn 0-0.2, Si 0-0.25, V 0-0.1, bal Fe.
For solenoid switches, armatures. High magnetic permeability.

CARPENTER CONSUMET M-50 BEARING STEEL

Now CARPENTER CONSUMET M-50 HIGH SPEED STEEL.

CARPENTER CONSUMET M50 HIGH SPEED STEEL

Carpenter Technology Corp.
C 0.8, Mn 0.25, Si 0.25, Cr 4, Ni 0.1, V 1, Mo 4.5, bal Fe.
High speed steel used mainly for bearings in aircraft and gas turbine engines operating at temperatures up to 800°F. AISI M50.

CARPENTER CRB-7 ALLOY

Carpenter Technology Corp.
Alloy steel. C 1.1, Mn 0.4, Si 0.3, Cr 14, Mo 2, V 1, Cb 0.25, bal Fe.
Annealed: 751 MPa YS; 875 MPa TS; 10.9 El; 18.9 RA; 97 Rock B. Corrosion and wear resistant secondary hardening high temperature bearing steel with high heat treated hardness and high hardness at elevated temperatures. Similar to AISI Type 440 stainless steel.

CARPENTER D-6A

Carpenter Technology Corp.
Now CARPENTER D6-AC.

CARPENTER D6-AC

Carpenter Technology Corp.
C 0.48, Mn 0.75, Si 0.25, Cr 1.1, Ni 0.55, Mo 1, V 0.1, bal Fe.
Tough alloy tool steel.

CARPENTER DOUBLE VACUUM MELTED (VIN-VAR)

Now CARPENTER CONSUMET M-50 HIGH SPEED STEEL.

CARPENTER ELECTRICAL IRON

Carpenter Technology Corp.
C 0-0.02, Mn 0.12, Si 0.12, P 0.01, S 0.01, Cr 0.1, Ni 0.08, Al 0.03, bal Fe.
Soft magnetic iron for solenoids, magnetic pole pieces, magnetic circuit core members.

CARPENTER FIBRALLOY 460

Carpenter Technology Corp.
C 0.02, Mn 0.04, Si 0.04, Cr 11.25, Ni 9, Cu 2.25, Al 0.8, Cb 0.45, bal Fe.
Martensitic, precipitation hardening stainless steel. 45 Rock C. Tough, ductile; retains hardness and oxidation resistance at elevated temperature. Designed for use in the synthetic fiber textile industry. *Obsolete*

CARPENTER FREE CUT INVAR 36

Carpenter Technology Corp.
C 0-0.12, Mn 0.35, Se 0.2, 36 Ni + Co, bal Fe.
For low expansion alloy. Free cutting.

CARPENTER GLASS SEALING 27

Carpenter Technology Corp.
C 0-0.15, Cr 28, bal Fe.
Annealed: 85,000 TS; 55,000 YS; 25 El; 185 Brin. For glass sealing, metal to glass.

CARPENTER GLASS SEALING 42

Carpenter Technology Corp.
C 0-0.1, Mn 0.5, Si 0.25, Ni 41.5, bal Fe.
Cold drawn: 120,000 TS; 3 El; 240 Brin. Annealed: 82,000 TS; 30 El; 140 Brin. For glass-to-metal seals; for hard and soft glass.

CARPENTER GLASS SEALING 42-6

Carpenter Technology Corp.
C 0-0.1, Mn 0.5, Si 0.25, Ni 42.5, Cr 5.75, bal Fe.
Strip and wire with thermal expansion matching characteristics of 0120 glass. For sealing into glass.

CARPENTER GLASS SEALING 45-6

Carpenter Technology Corp.
C 0-0.1, Mn 0.3, Si 0.3, Ni 45, Cr 6, bal Fe.
Annealed: 80,000 TS; 40,000 YS; 30 El; 80 Rock B. For metal-to-glass sealing (0120 and 9010 glass). Controlled coefficient of expansion. Vacuum melted, low gas content.

CARPENTER GLASS SEALING 46

Carpenter Technology Corp.
Ni 46, bal Fe.
For metal-to-glass seals. Controlled expansion.

CARPENTER GLASS SEALING 52

Carpenter Technology Corp.
C 0-0.1, Mn 0.5, Si 0.25, Ni 51, bal Fe.
Annealed: 80,000 TS; 40,000 YS; 35 El; 83 Rock B. For glass-to-metal seals; for soft glass and ceramics.

CARPENTER GLASS SEALING 52 PHOTOETCH

Carpenter Technology Corp.
C 0.1, Mn 0.5, Si 0.25, Ni 50.5, bal Fe.
Strip designed to produce sharp, square edges in photo-etching and to match expansion of soft glasses and some ceramics.

CARPENTER GREEN LABEL DRILL ROD

Carpenter Technology Corp.
C 1.2, Mn 0.2, Si 0.2, bal Fe.
AISI Type W1 water hardening tool steel.

CARPENTER H-46

Carpenter Technology Corp.
C 0.15-0.2, Cr 10-14, Ni 0.3-0.6, Mo 0.5-0.8, V 0.2-0.4, Cb 0.2-0.6, N 0.06-0.1, bal Fe.
At 70°F: 151,000 TS, 124,000 YS, 20 El, 56 RA, 302 Brin. At 1200°F: 60,500 TS; 56,200 YS; 30 El; 76 RA. For jet aircraft engine compressor blades and rotor discs. Good strength and ductility to 1200°F.

CARPENTER H-9 DOUBLE HEADER DIE STEEL

Carpenter Technology Corp.
C 0.9, Mn 0.4, Si 0.4, bal Fe.
Water hardening tool steel for solid and gripper cold heading dies, inserts, coining dies, knurls. AISI W1. G10900

CARPENTER HAMPDEN

Carpenter Technology Corp.
C 2.1, Mn 0.35, Si 0.3, Cr 12, Ni 0.5, bal Fe.
Oil hardening high carbon, high chromium die steel with extra wear resistance. For slitting cutters, lamination dies, cold rolls, blanking and forming dies. AISI D3.

CARPENTER HI MAG PERM

Erie Steel Co.
C 0.03-0.05, P 0.005-0.009, Si 0.01-0.02, Cr 0.03-0.07, Al 0.006-0.01, Mo 0.04-0.07, bal Fe.
Annealed: 40,000 TS; 20,000 YS; 40 El; 78 RA; 69 Brin. For electrical applications, magnetic control devices, magnetic clutches and chucks.

CARPENTER HI-SHOCK 60 (1)

Carpenter Technology Corp.
C 0.68, Mn 0.5, Si 0.3, Cr 1, Ni 0.5, Mo 1, Cu 2.5, V 0.15, bal Fe.
Air hardened: 226,000-363,000 TS; 206,000-316,000 YS; 1.6-10.2 El; 3.1-30.9 RA; 500-580 Brin. For hobs, punches, mandrels, tools and dies. Air hardened, shock resistant.

CARPENTER HI-SHOCK 60 (2)

Carpenter Technology Corp.
C 0.68, Mn 0.35, Si 0.3, Cr 1, Ni 0.5, Mo 1, Cu 2.5, V 0.15, bal Fe.
Air or oil hardening tool steel for large blanking and forming dies, trimming dies, shear blades, mandrels, forming tools.

CARPENTER HIGH NICKEL

Carpenter Technology Corp.
Ni 23-30, bal Fe.
73,300 TS; 39,600 YS; 68.3 El. For wire drawing dies; tough. *Obsolete*

CARPENTER HIGH PERMEABILITY 49 (1)

Carpenter Technology Corp.
C 0.02, Mn 0.5, Si 0.35, Ni 48, bal Fe.
Three grades with varying permeability depending on treatment and size or shape. For rotors, transformers, solenoid cores and magnetic shields.

CARPENTER HIGH PERMEABILITY 49 (2)

Carpenter Technology Corp.
C 0.05, Mn 0.05, Si 0.35, Ni 48, bal Fe.
Annealed: 70,000 TS; 22,000 YS; 45 El. Flux density: 15,000 Gausses. Low hysteresis loss. Available in three grades for magnet cores, solenoids, laminations, transformers.

CARPENTER HIPERCO 27

Carpenter Technology Corp.
C 0.015, Mn 0.25, Si 0.25, Cr 0.6, Co 27, bal Fe.
High magnetic saturation alloy for use in magnetic flux carrying members, magnetic pole caps, and laminations for aircraft motors and generators.

CARPENTER HIPERCO 50

Carpenter Technology Corp.
C 0.01, Mn 0.05, Si 0.05, Co 48.75, V 1.9, bal Fe.
High magnetic saturation alloy for use as magnetic pole caps, lamination for aircraft motors and generators, transformer laminations, and tape toroids.

CARPENTER HIPERCO 50FM

Carpenter Technology Corp.
C 0.04, Mn 0.8, Si 0.4, Co 49, V 1.9, Se 0.2, bal Fe.
Annealed residual induction: 22,000 Gausses. For sonar applications, ultrasonic transducers.

CARPENTER HIPERNOM

Carpenter Technology Corp.
Mo 4.2, Ni 80, Si 0.35, C 0.02, Mn 0.5, bal Fe.
Soft magnetic alloy capable of high permeability. For shielding applications.

CARPENTER HY-RA 49

Carpenter Technology Corp.
Ni 49, bal Fe.
For electronic devices, magnetic amplifiers. High permeability.

CARPENTER HY-RA 80 (1)

Carpenter Technology Corp.
C 0.015, Mn 0.5, Si 0.15, Ni 79, Mo 4.2, bal Fe.
For electrical equipment, electronic devices. Square hysteresis loop properties, high permeability.

CARPENTER HY-RA 80 (2)
Carpenter Technology Corp.
Ni 79, C 0.05, Mo 4, Mn 0.5, Si 0.15, bal Fe.
For electronic devices, magnetic amplifiers. High
permeability.

CARPENTER HYMU 80
Carpenter Technology Corp.
C 0.15, Mn 0.5, Si 0.35, Ni 80, Mo 4.2, bal Fe.
Unoriented alloy with high initial permeability and maximum
permeability with minimum hysteresis loss. For transformer
cores, tape wound toroids, and laminations.

CARPENTER HYMU 80 MARK II
Carpenter Technology Corp.
C 0.015, Mn 0.5, Si 0.3, Ni 80, Mo 4.6, bal Fe.
Unoriented high initial permeability alloy; maximum
permeability with minimum hysteresis loss; slightly better
than HyMu 80. For transformer lamps, tape toroids, and
magnetic pick-up head laminations.

CARPENTER HYMU 800 (1)
Carpenter Technology Corp.
C 0.03, Mn 0.5, Si 0.35, Ni 80, Mo 4, bal Fe.
Vacuum melted soft magnetic alloy for tape wound toroids
and laminations. Minimum permeability of 60,000 Gausses at
40 Gausses.

CARPENTER HYMU 800 (2)
Carpenter Technology Corp.
C 0.01, Mn 0.5, Si 0.15, Ni 80, Mo 5, bal Fe.
Soft magnetic material. For laminations or toroids.

CARPENTER HYMU 800 PHOTOCHEM QUALITY
Carpenter Technology Corp.
C 0.01, Mn 0.5, Si 0.15, Ni 80, Mo 5, bal Fe.
Vacuum melted soft magnetic sheet for use with
photochemical techniques for lamination or toroids.
Minimum permeability of 60,000 Gausses at 40 Gausses.

CARPENTER INVAR 36
Carpenter Technology Corp.
C 0-0.12, Mn 0.35, 36 Ni + Co, bal Fe.
Rolled: 90,000 TS; 70,000 YS; 20 El; 60 RA; 90 Brin B. For
precision instruments, radio parts, bi-metal. Low coefficient of
expansion.

CARPENTER INVAROD
Carpenter Technology Corp.
C 0-0.1, Mn 2.5, Si 0-0.2, Ni 36, Ti 0.75, bal Fe.
Welding alloy for joining Invar 36 to itself or other metals
without preheating or postheating.

CARPENTER JASON STEEL NO. 12-324
Carpenter Technology Corp.
C 1, bal Fe.
For balls. *Obsolete*

CARPENTER JY ROLL STEEL
Carpenter Technology Corp.
C 0.85, Mn 0.3, Si 0.3, Cr 2, V 0.2, bal Fe.
High carbon, low alloy steel used for cold rolls.

CARPENTER KOVAR
Carpenter Technology Corp.
C 0-0.02, Mn 0.3, Si 0.2, Ni 29, Co 17, bal Fe.
Vacuum melted, low expansion alloy for making hermetic
seals. Coefficient of expansion: 5.06 x 10^{-6} in./in.·^{o}F,
77-1292°F. Formerly NICOSEAL.

CARPENTER KOVAR PHOTOETCH QUALITY
Carpenter Technology Corp.
C 0-0.02, Mn 0.3, Si 0.2, Ni 29, Co 17, bal Fe.
Designed to secure square edges in photoetching of "flatpak"
lead preforms for integrated circuits. Corrosion resistant;
controlled thermal expansion. Formerly NICOSEAL.

CARPENTER KR TYPE B
Carpenter Technology Corp.
C 1, Cr 4, bal Fe.
For tools, dies; air hardening. *Obsolete*

CARPENTER L-605
Carpenter Technology Corp.
C 0.12, Mn 1.55, Cr 20, Ni 12, W 15, bal Co.
For jet engine components, afterburners. Heat and corrosion
resistant.

CARPENTER LAPELLOY
Carpenter Technology Corp.
C 0.25-0.35, Mn 0.95-1.25, Si 0-0.5, Cr 11-12, Ni 0-0.5, Mo
2.5-3, V 0.2-0.3, bal Fe.
Air or oil hardenable corrosion resistant steel. Hardened, at
1200°F: 60,000 TS; 50,000 YS. For steam turbine buckets
and blades, compressor blades, valve stems.

CARPENTER LAPELLOY C
Carpenter Technology Corp.
C 0.2-0.25, Mn 0.65-1, Cr 11-12, Ni 0-0.5, Mo 2.5-3, Cu
1.75-2.25, Ni 0.06-0.1, bal Fe.
Heat treated: 135,000-203,000 TS; 105,000-170,000 YS;
17-18 El; 47-55 RA. For compressor wheels, turbine shafts,
compressor buckets, blades, bolts; for high stressed parts
operating to 1200°F.

CARPENTER LOW EXPANSION 39
Carpenter Technology Corp.
Ni 39, bal Fe.
For instruments. Low expansion alloy.

CARPENTER LOW EXPANSION 42
Carpenter Technology Corp.
Ni 42, bal Fe.
Annealed: 82,000 TS; 30 El; 140 Brin. Cold drawn: 120,000
TS; 3 El; 240 Brin. For low expansion and high temperature
applications. Coefficient of expansion: 3 x 10^{-6}in./in·^{o}F at
70-650°F.

CARPENTER LOW EXPANSION 43-PH
Carpenter Technology Corp.
C 0.03, Mn 0.5, Si 0.5, Cr 5.25, Ni 42.5, Al 0.5, Ti 2.5, bal Fe.
Solution treated: 90 ksi TS; 35 ksi YS; 40 El; 75 Rock B. Age
hardenable, ferromagnetic, austenitic alloy with a constant
modulus of elasticity. For springs, diaphragms and supports
in timing and measuring devices, continuous weighing
systems, flow meters, gyro suspensions, motor speed
switches, bourdon tubes, tuning forks.

CARPENTER MANIFLEX-FM
Carpenter Technology Corp.
Stainless steel. C 0.2, Mn 1.25, Si 0.8, P 0.02, S 0.2, Cr 21, Ni
11.5, bal Fe.
Free machining austenitic stainless with excellent high
temperature strength and hardness. For shafts and bushings
in manifold exhaust heat control valves and emission control
devices.

CARPENTER MARINALOY 17
Carpenter Technology Corp.
Stainless steel. C 0.07, Mn 1, Si 1, Cr 15.5-17.5, Ni 3-5, Cu
3-5, 0.15-0.45 Cb + Ta, bal Fe.
Martensitic, age hardening stainless steel with high strength
and excellent corrosion resistance. 135,000 TS min. For boat
shafting.

CARPENTER MARINALOY HN
Carpenter Technology Corp.
Stainless steel. C 0.08, Mn 2, Si 1, Cr 18-20, Ni 8-10.5, N
0.16-0.3, bal Fe.
Nitrogen strengthened, austenitic stainless steel with unique
combination of strength, toughness and corrosion resistance.
For boat shafting.

CARPENTER MEL-TROL K-W
Carpenter Technology Corp.
C 1.3, Mn 0.3, Si 0.3, W 3.5, bal Fe.
Water hardening tool steel with extra wear and abrasion
resistance; for burnishing tools, tube drawing dies, wire
drawing dies, reamers. AISI F2. *Obsolete*

CARPENTER MEL-TROL SPEED STAR
Now SPEED STAR.

CARPENTER MEL-TROL STAR-ZENITH
Now STAR ZENITH.

CARPENTER MIDAS STEEL NO. 9-961
Carpenter Technology Corp.
C 1.2, Cr 1, bal Fe.
For ball races. *Obsolete*

CARPENTER MIRROMOLD
Carpenter Technology Corp.
C 0.1, Mn 0.2, V 0.1, bal Fe.
Low carbon steel for easy cold hubbing and subsequent case
hardening for plastic molds. AISI P1.

CARPENTER NIAL ALLOY
Carpenter Technology Corp.
Ni 95, bal Al.
At 20°C: 28,000 psi YS; 83,000 psi TS; 45 El. For
thermocouples, extension wires.

CARPENTER NICKEL 200 ALLOY
Carpenter Technology Corp.
C 0-0.07, Mn 0-0.35, Si 0-0.25, Fe 0-0.25, Cu 0-0.15, S 0-0.01,
99.5 Ni + Co.
60-120 ksi TS; 20-110 YS; 5-45 El. Commercially pure nickel
with good corrosion resistance and relatively low electrical
resistivity. For food handling equipment, caustic solutions,
general corrosion resistant parts and structures, in
electrical/electronic components, in sonic devices. ASTM
B-160.

CARPENTER NICKEL 201 ALLOY
Carpenter Technology Corp.
C 0-0.02, Mn 0-0.35, Si 0-0.25, Fe 0-0.25, Cu 0-0.15, S 0-0.01,
99.5 Ni + Co.
60-120 ksi TS; 20-110 YS; 5-45 El. Low carbon modification of
Carpenter Nickel 200 Alloy for electronic industry
applications, cold forming operations and corrosion resistant
equipment.

CARPENTER NICKEL 205 ALLOY
Carpenter Technology Corp.
C 0.07, Mn 0-0.35, Si 0-0.15, Fe 0-0.2, Cu 0-0.15, Mg
0.01-0.08, Ti 0.01-0.05, S 0-0.008, 99.5 Ni + Co.
65-135 ksi TS; 25-120 ksi YS; 5-45 El. Wrought nickel alloy for
use in electronic components such as vacuum tube
components, pins, terminals, support wires, lead wires,
shields, tubes, etc. ASTM F-9.

CARPENTER NICKEL 206 ALLOY
Carpenter Technology Corp.
C 0-0.002, Mn 0-0.35, Si 0-0.15, Fe 0-0.2, Cu 0-0.15, Mg
0.01-0.08, Ti 0.01-0.05, S 0-0.008, 99.5 Ni + Co.
65-135 ksi TS; 25-120 ksi YS; 5-45 El. Low carbon
modification of Carpenter Nickel 205 Alloy for applications
requiring minimum evolution of gas in components (support
rods and wire) where high temperature assembly (glass
sealing) is performed.

CARPENTER NICKEL 211 ALLOY
Carpenter Technology Corp.
Mn 4.25-5.25, C 0-0.2, Si 0-0.15, Fe 0-0.75, Cu 0-0.25, S
0-0.015, 93.7 Ni + Co.
77-86 ksi TS; 35-36 ksi YS; 38-40 El. Hard, strong, sulfur-
resistant alloy with good electrical resistivity. For light bulb
fuses, vacuum tube grids and in assemblies where sulfur is
present in heating flames. ASTM F-290.

CARPENTER NICROSIL
Carpenter Technology Corp.
Cr 14.4, Si 1.4, Mg 0.1, bal Ni.
At 20°C: 60,000 psi YS; 110,000 psi TS; 30 El. For thermocouples, extension wires.

CARPENTER NIMARK 200
Carpenter Technology Corp.
C 0.008, Ni 18.5, Mo 4.25, Co 7.5, Ti 0.2, Al 0.1, bal Fe.
Low carbon, high nickel maraging steel.

CARPENTER NIMARK 300
Carpenter Technology Corp.
C 0-0.03, Ni 18-19, Mo 4.7-5.1, Co 8-9.5, Ti 0.5-0.8, Al 0.05-0.15, Zr 0-0.03, B 0-0.005, Ca 0-0.05, bal Fe.
Maraging type alloy. Hardened: 294,000 TS; 290,000 YS; 11 El; 52 Rock C. Weldable. For high strength parts and assemblies.

CARPENTER NIMARK II
Carpenter Technology Corp.
C 0-0.03, Ni 19.5-20.5, Ti 1.2-1.6, Al 0.15-0.35, Cb 0.4-0.6, Zr 0-0.05, B 0-0.003, Ca 0-0.05, bal Fe.
Maraging type alloy; as hardened: 265,000 TS; 255,000 YS; 12 El; 51 Rock C. Weldable; for high strength parts and assemblies. *Obsolete*

CARPENTER NIMARK I
Now NIMARK 250.

CARPENTER NISIL
Carpenter Technology Corp.
Si 4.2, Mg 0.1, bal Ni.
At 20°C: 55,000 psi YS; 95,000 psi TS; 35 El. For thermocouples, extension wires.

CARPENTER NITREX I & II AMPERSAND
Now NITREX I & II.

CARPENTER NO. 1-JR
Carpenter Technology Corp.
C 0-0.15, Cr 12-14, Al 3.25-4.5, bal Fe.
At 70°F: 86,000 TS; 25 El; 57 RA. At 1400°F: 13,500 TS; 77 El; 93 RA. For magnetic cores, resistors. Corrosion and oxidation resistant.

CARPENTER NO. 11 SPECIAL
Carpenter Technology Corp.
C 1.05, Mn 0.2, Si 0.2, bal Fe.
Water hardening tool steel for drills, taps, reamers, punches, blanking dies, bushings. AISI W1. G10910

CARPENTER NO. 158
Carpenter Technology Corp.
C 0.1, Mn 0.5, Si 0.3, Cr 1.5, Ni 3.5, bal Fe.
Deep hardening carburizing steel for heavy duty gears, power tool cams, clutch levers.

CARPENTER NO. 2 SAMSON
Carpenter Technology Corp.
C 0.2, Cr, Ni, bal Fe.
For machine tool parts; case hardened, tough. *Obsolete*

CARPENTER NO. 2-317
Carpenter Technology Corp.
C 0.15, Cr, Ni, bal Fe.
For case-hardened parts, gears, crankshafts; case-hardening steel. *Obsolete*

CARPENTER NO. 2-408
Carpenter Technology Corp.
C 0.15, Cr, Ni, bal Fe.
For case-hardened pinions, shafts and parts for automotive and machine tool construction; case-hardening. *Obsolete*

CARPENTER NO. 2-720
Carpenter Technology Corp.
C 0.18, Cr 0.9, V 0.2, bal Fe.
For case-hardened parts; case-hardening steel. *Obsolete*

CARPENTER NO. 200
Carpenter Technology Corp.
C, alloy, bal Fe.
For machine tool parts. *Obsolete*

CARPENTER NO. 3-314
Carpenter Technology Corp.
Cr 1.5, C 0.35, Ni 3.5, bal Fe.
Heat treated: 285,000 TS; 250,000 YS; 6 El; 18 RA; 490 Brin.
Heat treated: 120,000 TS; 100,000 YS; 23 El; 62 RA; 235 Brin.
For extrusion rams, press parts, heavy shafts; oil hardening. *Obsolete*

CARPENTER NO. 3-317
Carpenter Technology Corp.
C, Cr, Ni, bal Fe.
Hardened: 195,000 TS; 170,000 YS; 10 El; 39 RA; 430 Brin.
Toughened: 125,000 TS; 95,000 YS; 35 El; 65 RA; 235 Brin.
For case-hardened parts, gears, pinions, crane wheels; case-hardening steel. *Obsolete*

CARPENTER NO. 3-427
Carpenter Technology Corp.
Cr 1, C 0.3, Mo 0.2, bal Fe.
Heat treated: 230,000 TS; 210,000 YS; 10 El; 42 RA; 460 Brin.
Heat treated: 140,000 TS; 125,000 YS; 20 El; 61 RA; 260 Brin.
For gears, spline shafts, bolts, nuts, tubing; tough, shock and wear resisting. *Obsolete*

CARPENTER NO. 3-547
Carpenter Technology Corp.
Ni 3.5, C 0.35, bal Fe.
Heat treated: 210,000 TS; 180,000 YS; 12 El; 42 RA; 400 Brin.
Heat treated: 110,000 TS; 85,000 YS; 24 El; 65 RA; 215 Brin.
For shafts, gears; shock resistant. *Obsolete*

CARPENTER NO. 3-720
Carpenter Technology Corp.
C 0.3, Cr 0.9, V 0.2, bal Fe.
For studs, bolts; tough. *Obsolete*

CARPENTER NO. 30
Carpenter Technology Corp.
Ni 30, bal Fe.
For electrical parts; non-magnetic. *Obsolete*

CARPENTER NO. 37-7FM
Carpenter Technology Corp.
Ni 37, bal Fe.
For instruments; low coefficient of expansion. *Obsolete*

CARPENTER NO. 4
Carpenter Technology Corp.
Cr 0.6, C 0.4, Ni 1.25, bal Fe.
Heat treated: 110,000-240,000 TS; 90,000-210,000 YS; 8-22 El; 25-60 RA; 230-440 Brin. For tempered set screws, shanks of high speed tools, machine tool parts; oil hardening. *Obsolete*

CARPENTER NO. 4-317
Carpenter Technology Corp.
Cr 1, C 0.4, Ni 1.7, bal Fe.
Heat treated: 125,000 TS; 100,000 YS; 24 El; 62 RA; 255 Brin.
Heat treated: 165,000 TS; 140,000 YS; 18 El; 52 RA; 302 Brin.
For crankshaft, transmission shaft, propellers, axles, etc.; tough as treated. *Obsolete*

CARPENTER NO. 4-408
Carpenter Technology Corp.
Cr 0.75, C 0.4, Ni 3, bal Fe.
Heat treated: 260,000 TS; 220,000 YS; 8 El; 20 RA; 440 Brin.
Heat treated: 120,000 TS; 100,000 YS; 23 El; 63 RA; 240 Brin.
For shafts, gears; easy machine at high hardness. *Obsolete*

CARPENTER NO. 404
Carpenter Technology Corp.
C 0-0.05, Cr 11-12, Ni 1.25-2, N 0-0.03, Mn 0-1, Si 0-0.5, bal Fe.
Heat treated: 108,000-163,000 TS; 95,000-125,000 YS; 10-23 El; 40-70 RA; 180-350 Brin. For steam turbine buckets, blades and bucket covers; for highly stressed parts at temperatures up to 1050°F. *Obsolete*

CARPENTER NO. 408 PUNCH STEEL
Carpenter Technology Corp.
C 0.5, Mn 0.5, Si 0.25, Cr 0.75, Ni 3, bal Fe.
Heat treated: 275,000 TS; 250,000 YS; 8 El; 56 Rock C. For punches, clutch parts, vise jaws, chuck jaws, feeder rolls. Tough, wear resistant.

CARPENTER NO. 426
Carpenter Technology Corp.
Ni 42, Cr 6, bal Fe.
Annealed: 73,000 TS; 30,000 YS. Cold drawn: 140,000 TS; 135,000 YS. For glass to metal seals; controlled expansion. *Obsolete*

CARPENTER NO. 436
Carpenter Technology Corp.
C 0.15, Ni 1.75, Mo 0.25, bal Fe.
For case-hardened parts; free machining. *Obsolete*

CARPENTER NO. 478
Carpenter Technology Corp.
C 0.1-0.65, bal Fe.
Treated: 65,000-87,000 TS; 35,000-60,000 YS; 23-29 El; 54-56 RA; 146-217 Brin. For general automobile parts. *Obsolete*

CARPENTER NO. 481
Carpenter Technology Corp.
C 0.55, Mn 0.8, Si 1.9, Cr 0.25, Mo 0.4, bal Fe.
Oil hardening, shock resistant steel for pneumatic tools, chipping chisels, rivet sets, shear blades, collets.

CARPENTER NO. 484-FM
Carpenter Technology Corp.
C 1, Mn 0.8, Si 0.3, Cr 5.25, Mo 1.1, V 0.2, Plus alloy sulfides, bal Fe.
Free machining air hardening tool steel; for large blanking dies, thread roller dies, coining dies, gages. AISI A2. *Obsolete*

CARPENTER NO. 492
Carpenter Technology Corp.
C, bal Fe.
Untreated: 80,000 TS; 20 El; 38 RA; 156 Brin. Treated: 95,000 TS; 22 El; 47 RA; 228 Brin. For steel parts that must be welded, axles. *Obsolete*

CARPENTER NO. 5-317
Carpenter Technology Corp.
C 0.5, Mn 0.5, Si 0.25, Cr 1, Ni 1.75, bal Fe.
Nickel chrome alloy steel. Oil hardenable to above 55 Rock C. For shafts, pinions, vise jaws.

CARPENTER NO. 5-427
Carpenter Technology Corp.
C 0.5, Cr 1, Mo 0.2, bal Fe.
For gears, splines, shafts; shock resistant. *Obsolete*

CARPENTER NO. 5-720 TYPE D
Carpenter Technology Corp.
Cr 0.9, C 0.5, V 0.2, bal Fe.
Heat treated: 280,000 TS; 240,000 YS; 9 El; 32 RA; 540 Brin.
Heat treated: 140,000 TS; 125,000 YS; 16 El; 52 RA; 300 Brin.
For high duty leaf springs, coil springs, gears, shafts; high resistance to shock and fatigue. *Obsolete*

CARPENTER NO. 500
Carpenter Technology Corp.
Ni 5, Low C, bal Fe.
For turbine blades, case-hardened gears and pinions; wear resistant. *Obsolete*

CARPENTER NO. 500 EXTRA SPECIAL
Carpenter Technology Corp.
C 0.12, Ni 5, bal Fe.
Hardened: 175,000 TS; 150,000 YS; 11 El; 40 RA; 332 Brin.
For bevel gears and pinions for transmission; case-hardening steel. *Obsolete*

CARPENTER NO. 610-FM
Carpenter Technology Corp.
C 1.5, Mn 0.5, Si 0.3, Cr 12, Mo 0.8, V 0.9, alloy sulfides, bal Fe.
Air hardening, high carbon, high chromium tool steel. Free machining grade for blanking and forming dies, drawing dies, edging rolls, extrusion dies. AISI D2.

CARPENTER NO. 872
Carpenter Technology Corp.
C 0.15, Mn 0.8, Cr 0.5, Ni 0.5, Mo 0.22, bal Fe.
For case hardened gears, shafts; carburized. *Obsolete*

CARPENTER NO. 874
Carpenter Technology Corp.
C 0.45, Mn 0.9, Cr 0.5, Ni 0.5, Mo 0.22, bal Fe.
Heat treated: 115,000-225,000 TS; 105,000-200,000 YS; 10-24 El; 37-65 RA; 235-444 Brin. For gears, shafts, bolts; water hardened. *Obsolete*

CARPENTER NO. 883
Carpenter Technology Corp.
C 0.37, Mn 0.35, Si 1, Cr 5.25, V 1, Mo 1.3, bal Fe.
Chrome hot work tool steel, air or oil hardenable for forging dies, die casting dies, extrusion dies. AISI H 13.

CARPENTER PERMANENT MAGNET TYPE A
Carpenter Technology Corp.
C 0.9, Cr 2, bal Fe.
For magnetic units; permanent magnets. *Obsolete*

CARPENTER PERMANENT MAGNET TYPE B
Carpenter Technology Corp.
C 1, Cr 3.75, bal Fe.
For magnetic units; permanent magnets. *Obsolete*

CARPENTER PERMANENT MAGNET TYPE C
Carpenter Technology Corp.
C 0.9, Cr 6, bal Fe.
For magnetic units; permanent magnets. *Obsolete*

CARPENTER PERMANENT MAGNET TYPE D
Carpenter Technology Corp.
C 1.05, Cr 4.1, Mo 0.35, bal Fe.
For magnetic units; permanent magnets. *Obsolete*

CARPENTER PERMANENT MAGNET TYPE E
Carpenter Technology Corp.
C 0.6, Cr 0.2, W 5.5, bal Fe.
For magnetic units; permanent magnet. *Obsolete*

CARPENTER PERMANENT MAGNET TYPE F
Carpenter Technology Corp.
C 0.8, Cr 2.75, W 8.75, Mo 0.2, Co 17.5, bal Fe.
For magnetic units; permanent magnets. *Obsolete*

CARPENTER PERMANENT MAGNET TYPE G
Carpenter Technology Corp.
C 0.9, Cr 5.75, W 3.75, Co 36, bal Fe.
For magnetic units; permanent magnets. *Obsolete*

CARPENTER PYROMET 31
Carpenter Technology Corp.
C 0.04, Cr 22.7, Ni 55.5, Mo 2, Ti 2.3, Cb 0.85, B 0.005, bal Fe.
Precipitation hardenable superalloy. At 1500°F: 100,000-115,000 psi (687-798 MPa) TS; 93,000 psi (640 MPa) YS; 16-21 El. For truck and locomotive diesel valves, gas and oil well equipment.

CARPENTER PYROMET 41
Carpenter Technology Corp.
C 0.6-0.12, Cr 18-20, Mo 9-10.5, Co 10-12, Ti 3-3.3, Al 1.4-1.6, B 0.003-0.01, Fe 0-5, bal Ni.
Aged: 206,000 TS; 154,000 YS; 14 El. For jet engine and aircraft components, high temperature bolts. Age hardenable, useful to 1800°F.

CARPENTER PYROMET 538
Carpenter Technology Corp.
Stainless steel. C 0.03, Mn 8-10, Si 0-1, Cr 18-21.5, Ni 5.5-7.5, N 0.15-0.4, bal Fe.
Strengthened austenitic stainless. RT: 112,000 TS; 65,000 YS; 42 El. At 1000°F: 71,000 TS; 29,000 YS; 35 El. For steam and autoclave parts, airframe and aircraft engine, chemical processing equipment.

CARPENTER PYROMET 751
Carpenter Technology Corp.
C 0.04, Mn 0.7, Si 0.3, Cr 15, S 0.007, Cb 1, Fe 6.75, Ti 2.5, Al 1.2, Cu 0.05, bal Ni.
Solution treated and aged: 1207 MPa TS; 758 MPa YS; 20 El. Age hardenable alloy with good high temperature stress rupture properties. Formerly X-751.

CARPENTER PYROMET 800
Carpenter Technology Corp.
C 0.1, Mn 1.5, Si 1, Cr 19-23, Ni 30-35, Al 0.15-0.6, Ti 0.15-0.6, Cu 0-0.75, bal Fe.
Nonmagnetic Ni-Cr-Fe alloy. At 1000°F: 72,000 TS; 31,700 YS; 38.5 El. At 1500°F: 24,800 TS; 14,200 YS; 91 El. Weldable, corrosion resistant. For high temperature piping.

CARPENTER PYROMET 80A
Carpenter Technology Corp.
C 0.06, Cr 20, S 0.007, Ti 2.35, Al 1.25, Cu 0.05, Co 1, bal Ni.
Age hardened: 145,000 TS; 90,000 YS; 39 El. At 1200°F: 115,000 TS; 80,000 YS; 21 El. For high temperature applications. Heat treatable. High creep and fatigue resistance. Oxidation and corrosion resistance.

CARPENTER PYROMET 860
Carpenter Technology Corp.
C 0-0.1, Mn 0-1, Si 0-1, Cr 12-16, Ni 40-45, Mo 5-7, Co 3.5-4.5, Ti 2.75-3.75, Al 0.75-1.5, B 0.0008-0.012, bal Fe.
Austenitic Fe-Ni base precipitation hardening alloy. Hardened, RT: 180,000 TS; 115,000 YS; 21 El. At 1200°F: 159,000 TS; 125,000 YS; 19 El. At 1500°F: 106,000 TS; 103,000 YS; 16 El. For corrosion and scale resisting parts at high temperatures, turbine engine parts, steam turbine bolting.

CARPENTER PYROMET 88
Carpenter Technology Corp.
C 0.03, Mn 2.2, Cr 16.4, Fe 6.7, Ti 3.05, bal Ni.
High strength, high temperature nickel base alloy. Age hardenable; useful to 1500°F. Good corrosion resistance.

CARPENTER PYROMET 901
Carpenter Technology Corp.
C 0-0.1, Mn 0-1, Cr 11-14, Ni 40-45, Mo 5-7, Ti 2.35-3.1, Cu 0-0.5, Al 0-0.35, B 0.01-0.02, bal Fe.
At 70°F: 175,000 TS; 125,000 YS; 15 El; 19 RA. At 1000°F: 156,000 TS; 113,000 YS; 17 El; 29 RA. At 1500°F: 81,000 TS; 79,000 YS; 13 El; 20 RA. For aircraft, gas turbines, rotors, compressor discs, hubs, shafts. Precipitation hardened, heat and corrosion resistant to 1400°F.

CARPENTER PYROMET 95
Carpenter Technology Corp.
C 0.15, Mn 0-0.15, Si 0-0.2, Cr 14, Co 8, Ti 2.5, Al 3.5, W 3.5, Cb 3.5, bal Ni.
Precipitation hardening nickel base superalloy for turbine and compressor discs, shafts and seals in aircraft gas turbine engines.

CARPENTER PYROMET CTX-1
Carpenter Technology Corp.
C 0.03, Mn 0-0.2, Si 0-0.2, Cr 0-0.2, Mo 0-0.2, Cu 0-0.5, Ni 37.7, Ti 1.75, Al 1, B 0.0075, Co 16, 3.0 Cb + Ta, bal Fe.
Precipitation hardening alloy with good stress rupture properties to 1200°F. Low coefficient of expansion. For ordnance hardware, turbine blades, springs, gauge blocks, die casting dies.

CARPENTER PYROMET M-252
Carpenter Technology Corp.
C 0.1-0.2, Cr 18-20, Mo 9-10.5, Co 9-11, Al 0.75-1.25, Ti 2.25-2.75, Zr 0.02-0.15, B 0.001-0.01, Fe 0-5, bal Ni.
Heat treated: 175,000 TS; 110,000 YS; 25 El. At 1500°F: 91,000 TS; 84,000 YS; 24 El. For jet engine and gas turbine buckets. Resists heat to 1600°F. Precipitation hardening.

CARPENTER PYROMET N-155
Carpenter Technology Corp.
C 0.08-0.16, Mn 1-2, Si 0-1, Cr 20-22.5, Ni 19-21, Co 18.5-21, Mo 2.5-3.5, W 2-3, Cu 0-0.5, N 0.1-0.2, 0.75-1.25 Cb + Ta, bal Fe.
Hardened, RT: 118,000 TS; 59,000 YS; 40 El. At 1800°F: 19,000 TS (131 MPa). For aircraft tail cones and tail pipes, exhaust manifolds, combustion chambers, afterburners.

CARPENTER PYROMET X-15
Carpenter Technology Corp.
C 0-0.03, Cr 15, Co 20, Mo 2.9, Ni 0-0.2, bal Fe.
Precipitation hardenable martensitic alloy. Hardened, RT: 235,000 TS; 215,000 YS; 17 El. At 1000°F: 190,000 TS; 170,000 YS; 18 El. For highly stressed parts operating at temperatures up to 1050°F. *Obsolete*

CARPENTER PYROMET X-23
Carpenter Technology Corp.
C 0-0.02, Cr 10, Co 10, Mo 5.5, Ni 7, bal Fe.
Low carbon, martensitic alloy with high strength from cryogenic temperatures to 1000°F (538°C).

CARPENTER PYROMET X-751
Now CARPENTER PYROMET 75.

CARPENTER PYROTOOL 15
Carpenter Technology Corp.
C 0.03, Cr 15, Co 20, Mo 2.9, Ni 0.2, bal Fe.
Low carbon, martensitic alloy with good strength up to 1050°F. Solution treatable. Hardened, at 1100°F: 145,000 TS; 125,000 YS. For arbors, cams, collets, dies, fixtures. *Obsolete*

CARPENTER PYROTOOL EX
Carpenter Technology Corp.
C 0.05, Mn 0.2, Si 0.2, Cr 14, Ni 42.5, Mo 6, Co 4, Ti 3, Al 1.2, bal Fe.
Austenitic, precipitation hardenable nickel base alloy for high strength in the 1000-1500°F range. Hardened, at 1500°F: 106,000 TS; 103,000 YS. For high temperature tooling, extrusion dies, forging dies, rams, liners.

CARPENTER PYROTOOL M
Carpenter Technology Corp.
C 0.12, Mn 0.2, Si 0.2, Cr 19, Mo 10, Co 10, Al 1, Ti 2.5, Fe 2, bal Ni.
Austenitic, precipitation hardenable nickel base alloy for operation up to 1500°F. Hardened, at 1500°F: 91,000 TS; 84,000 YS. For high temperature tooling, extrusion dies, forging dies, mandrels, dummy blocks.

CARPENTER PYROTOOL V

Carpenter Technology Corp.
C 0.04, Mn 0.25, Si 0.25, Cr 14.5, Ni 27, Mo 1.25, Ti 3, V 0.2, bal Fe.
Austenitic, precipitation hardenable iron-base alloy. Hardened, at 1500°F: 60,000 TS; 49,000 YS. For high temperature tooling, liners, extrusion dies, forging dies, dummy blocks.

CARPENTER S-7

Carpenter Technology Corp.
C 0.5, Mn 0.7, Si 0.3, Cr 3.25, Mo 1.4, bal Fe.
Air hardening tool steel with high impact and shock resistance. For blanking dies, rivet sets, pneumatic tools. AISI S7.

CARPENTER SAMSON NO. 1

Carpenter Technology Corp.
C 0.1, Cr 0.6, Ni 1.25, bal Fe.
For tools and dies of intricate shape; case-hardening steel. *Obsolete*

CARPENTER SAMSON NO. 2

Carpenter Technology Corp.
C 0.15, Cr 0.6, Ni 1.25, bal Fe.
For general case-hardened parts, gears, studs, bolts, roller bearings, set screw; case-hardening steel. *Obsolete*

CARPENTER SAMSON NO. 3

Carpenter Technology Corp.
C 0.3, Cr 0.6, Ni 1.25, bal Fe.
Untreated: 85,000 TS; 55,000 YS; 24 El; 58 RA; 166 Brin. Heat treated: 120,000 TS; 95,000 YS; 20 El; 56 RA; 241 Brin. For hot rolled axles, shafts, bolts, set screws; tough. *Obsolete*

CARPENTER SAMSON NO. 4A

Carpenter Technology Corp.
C 0.4, Cr 0.6, Ni 1.25, bal Fe.
Heat treated: 115,000-132,000 TS; 90,000-108,000 YS; 18-21 El; 50-57 RA; 269-286 Brin. For axles, steering knuckles, crank shafts. *Obsolete*

CARPENTER SAMSON NO. 4B

Carpenter Technology Corp.
C 0.4, Cr 0.6, Ni 1.25, bal Fe.
Heat treated: 115,000-140,000 TS; 95,000-115,000 YS; 16-19 El; 45-52 RA; 269-302 Brin. For cold drawn parts. *Obsolete*

CARPENTER SAMSON NO. 4C

Carpenter Technology Corp.
C 0.4, Cr 0.6, Ni 1.25, bal Fe.
Heat treated: 90,000-115,000 TS; 80,000-91,000 YS; 16-19 El; 45-54 RA; 207-241 Brin. For cold drawn parts. *Obsolete*

CARPENTER SAMSON NO. 5

Carpenter Technology Corp.
C 0.5, Cr 0.6, Ni 1.25, Mn 0.6, bal Fe.
For machinery parts, gears, clutch parts; tough, wear resistant. *Obsolete*

CARPENTER SEAT RING DIE STEEL

Carpenter Technology Corp.
C 0.65, Cr 3, Mo 3, V 1, W 3.5, bal Fe.
Oil or air hardenable modified high speed tool steel designed for hot forming of automotive valves. Good wear resistance at elevated temperatures.

CARPENTER SIL NO. 1

Carpenter Technology Corp.
C 0.45, Mn 0.5, Si 3.25, Cr 8.5, bal Fe.
At 77°F: 133,000 TS; 82,500 YS; 22.5 El; 49.0 RA; 269 Brin. At 1000°F: 87,500 TS; 53,000 YS; 40.5 El; 71.0 RA; 125 Brin. For intake valves, exhaust valve stems. Hardenable, heat resistant.

CARPENTER SILICON CORE IRON A

Carpenter Technology Corp.
C 0.05, Mn 0.15, Si 1, bal Fe.
Saturation 21,000 Gausses; residual induction 6500 Gausses; Coercive force 0.90 Oersted; maximum permeability 4500. For motor armatures, pole pieces, solenoid switches, relays.

CARPENTER SILICON CORE IRON A-FM

Carpenter Technology Corp.
C 0-0.05, Mn 0-0.15, Si 0-1, P 0.18, bal Fe.
Free machining silicon core iron. DC U max 4500 at H = 1.36 Oersted. Coercive force from 10,000 Gausses: 0.70-0.80 Oersted. Residual induction from 10,000 Gausses: 6000 Gausses. Saturation at H = 200: 21,000 Gausses. For machined magnet cores.

CARPENTER SILICON CORE IRON B

Carpenter Technology Corp.
C 0.05, Mn 0.15, Si 2.5, bal Fe.
Saturation 20,600 Gausses. Residual induction from 10,000 Gausses: 6000 Gausses. Coercive force from 10,000 Gausses: 0.70 Oersted. Maximum permeability 7000. For solenoid switches, armatures, pole pieces.

CARPENTER SILICON CORE IRON B-FM

Carpenter Technology Corp.
C 0.05, Mn 0.4, Si 2.5, P 0.12, bal Fe.
Free machining grade of Carpenter Silicon Core Iron b. Properties and uses similar.

CARPENTER SILICON CORE IRON C

Carpenter Technology Corp.
C 0.03, Mn 0.15, Si 4, bal Fe.
Saturation 20,000 Gausses; residual induction 4000 Gausses; coercive force 0.60 Oersted; maximum permeability 4000. For motor armatures, pole pieces, solenoid switches, relays.

CARPENTER SOLAR

Carpenter Technology Corp.
C 0.5, Mn 0.4, Si 1, Mo 0.5, bal Fe.
Water hardening tool steel with more than normal toughness. For chisels, pneumatic tools, rivet busters, screw drivers. AISI S2.

CARPENTER SPECIAL

Carpenter Technology Corp.
C 1.1, Mn 0.2, bal Fe.
For tools, taps, reamers, drills, jigs, fixtures, dies. *Obsolete*

CARPENTER STAINLESS 18-18 PLUS

Carpenter Technology Corp.
Stainless steel. C 0.15, Mn 17-19, Si 0-1, Cr 17-19, Mo 0.75-1.25, Cu 0.75-1.25, N 0.4-0.6, bal Fe.
Nickel-free high strength austenitic stainless. Annealed, RT: 120 ksi (827 MPa) TS; 69 ksi (476 MPa) YS; 65 El; 75 RA. Good high temperature strength, good corrosion resistance.

CARPENTER STAINLESS 18CR-2NI-12MN

Carpenter Technology Corp.
Stainless steel. C 0-0.15, Mn 11-14, Si 0-1, Cr 16.5-19, Ni 0.5-2.5, N 0.2-0.45, bal Fe.
Annealed: 122,000 psi (841 MPa) TS; 68,000 psi (469 MPa) YS; 58 El; 76 RA. Can be cold worked (40%) to 206,000 psi TS. Austenitic. For springs, fasteners, pump shafts.

CARPENTER STAINLESS 304 + B

Carpenter Technology Corp.
C 0-0.08, Cr 18-20, Ni 11-14, Mn 0-2, 2% min B, bal Fe.
Annealed: 100,000 TS; 50,000 YS; 10 El; 10 RA; 210-240 Brin. For control rods and shielding for nuclear reactors; corrosion resistant. *Obsolete*

CARPENTER STAINLESS CUSTOM FLO 316HQ

Carpenter Technology Corp.
Stainless steel. C 0-0.03, Mn 0-2, P 0-0.03, S 0-0.015, Si 0-1, Cr 16-18.25, Ni 10-14, Mo 2-3, Cu 3-4, bal Fe.
Low carbon, copper modified Type 316L for cold headed nuts, bolts, screws, fittings, etc. Used in paper pulp handling equipment, process equipment for producing photographic chemicals, inks, rayon, rubber, textile bleaches and dyestuffs, high temperature equipment.

CARPENTER STAINLESS NO. 2-B

Carpenter Technology Corp.
C 1, Cr 17, bal Fe.
For cutlery, scissors, surgical instruments, ball bearings. *Obsolete*

CARPENTER STAINLESS NO. 404

Carpenter Technology Corp.
Stainless steel. C 0-0.05, Mn 0-1, Si 0.5, Cr 11-12.5, Ni 1.25-2, N 0.03, bal Fe.
Hardenable to 163,000 TS; weldable, tough, corrosion resistant. For stressed parts to 1050°F.

CARPENTER STAINLESS NO. 6-20

Carpenter Technology Corp.
C 0-0.12, Cr 18-23, bal Fe.
For stainless parts; stainless, heat resistant. *Obsolete*

CARPENTER STAINLESS NO. 7 (TYPE 329)

Carpenter Technology Corp.
C 0-0.15, Cr 25-30, Ni 3.5-6, Mo 1-2, bal Fe.
Precipitation hardenable, chromium stainless. Annealed: 105,000 TS; 80,000 YS; 230 Brin. Hardened: 32-50 Rock C. For valves, valve fittings, pump parts. *Obsolete*

CARPENTER STAINLESS NO. D-1

Carpenter Technology Corp.
C 0.1, Cr 14-16, bal Fe.
Annealed: 70,000 TS; 50,000 YS; 30 El. For pyrometer protection tubes, strip heaters; resists heat to 1200 F. *Obsolete*

CARPENTER STAINLESS NO. N-1

Carpenter Technology Corp.
C 0.1, Cr 13, Ni 2, bal Fe.
Hard rolled: 220,000 TS. For mild springs, cutlery, spatulas, rules, tapes; stainless, magnetic. *Obsolete*

CARPENTER STAINLESS NO. N-1 (TYPE 414)

Carpenter Technology Corp.
Stainless steel. C 0-0.15, Mn 0-1, Si 0-1, Cr 11.5-13.5, Ni 1.25-2.5, bal Fe.
Air or oil hardening corrosion resistant steel. As quenched: 210,000 TS; 155,000 YS; 15 El. For high strength bolts, nuts, studs and other hardware operating up to about 900°F. AISI 414.

CARPENTER STAINLESS TYPE 316N

Carpenter Technology Corp.
Stainless steel. C 0.08, Mn 2, Si 1, Cr 16-18, Ni 10-14, Mo 2-3, N 0.1-0.16, bal Fe.
Nitrogen strengthened Type 316 to increase strength with minimum effect on ductility and corrosion resistance. AISI 316.

CARPENTER STAINLESS TYPE 330

Carpenter Technology Corp.
Stainless steel. C 0.08, Mn 2, Si 0.75-1.5, Cr 17-20, Ni 34-37, bal Fe.
Austenitic, nonhardenable, heat and corrosion resistant alloy; weldable and machinable. For high temperature applications. AISI 330.

CARPENTER STAINLESS TYPE 347F SE

Carpenter Technology Corp.
Stainless steel. C 0.08, Mn 2, Si 1, Cr 17-19, Ni 9-13, Se 0.15, Cb + Ta = 10 x C, bal Fe.
Improved machinability, good high temperature scale resistance. Not recommended for welding.

CARPENTER STAINLESS TYPE 443

Carpenter Technology Corp.
Stainless steel. C 0-0.2, Cr 18-23, Cu 0.9-1.25, bal Fe.
Heat resistant, nonhardenable, corrosion resistant steel. Annealed: 90,000 TS; 50,000 YS; 22 El. At 1300°F: 26,000 TS; 11,000 YS; 41 El. For furnace parts, miscellaneous hardware, magnetic cores.

CARPENTER STAINLESS TYPE 446

Carpenter Technology Corp.
Stainless steel. C 0-0.15, Cr 24-30, Ni 0-0.5, bal Fe.
Chromium corrosion resistant steel.

CARPENTER STAR MAX FM

Carpenter Technology Corp.
Free machining grade of Carpenter Star Max.

CARPENTER STAR-ZENITH LOW CARBON
Carpenter Technology Corp.
C 0.5, W 18, Cr 4, V 1, bal Fe.
For cutting tools, drills, punches. High speed steel.

CARPENTER SUPER SAMSON
Carpenter Technology Corp.
C 0.1, Mn 0.3, Si 0.2, Cr 5, Mo 0.9, V 0.25, bal Fe.
Low carbon, chromium air hardening steel to be case hardened for plastic molds or for die casting dies. AISI P4.

CARPENTER SUPERSTAR
Carpenter Technology Corp.
C 1.08, Si 0.25, Mn 0.25, Cr 3.75, Mo 9.5, W 1.5, V 1.15, Co 8, bal Fe.
High carbon molybdenum-cobalt high speed steel. For broaches, taps, endmills, gear cutters. AISI Type M42.

CARPENTER T-K
Carpenter Technology Corp.
C 0.35, Mn 0.3, Si 0.3, Cr 3.5, W 9, V 0.4, bal Fe.
Air or oil hardenable hot work tool steel. For hot shear blades, hot gripper dies, die casting dies, hot punches, hot extrusion dies. AISI H21.

CARPENTER TEMPERATURE COMPENSATOR 30 (1)
Carpenter Technology Corp.
Type 1. 32% nickel-iron alloy with magnetic permeability that decreases at a controlled rate with increased temperature to compensate for variations in ambient temperature.

CARPENTER TEMPERATURE COMPENSATOR 30 (2)
Carpenter Technology Corp.
Types 2, 3, 4 and 5. 30% nickel-iron alloys with magnetic permeabilities that decrease at controlled rates with increases in temperature to compensate for temperature variations by using as shunts in watt-hour meters, speedometers, tachometers, voltage regulators.

CARPENTER TEMPERATURE COMPENSATOR 32
Carpenter Technology Corp.
C 0.12, Mn 0.6, Si 0.25, Ni 32.5, bal Fe.
Type 1. As annealed: 483 MPa TS; 276 MPa YS; 35 El; 75 Rock B. Nickel-iron alloy with magnetic permeability that decreases at a controlled rate with increased temperature. For use in electrical circuits, voltage regulators, as shunt material in tachometers and speedometers.

CARPENTER TGS
Carpenter Technology Corp.
C 0.2, Mn 1.3, Si 0.2, bal Fe.
Oil hardening carburizing grade steel for case hardened gears, guides, cams, and gauges.

CARPENTER TITANIUM 6-4
Carpenter Technology Corp.
Al 6, V 4, bal Ti.
Heat treatable, weldable, strong titanium alloy. Annealed: 130,000 TS; 120,000 YS; 10 El. Heat treated: 160,000 TS; 150,000 YS; 8 El. Widely used titanium alloy for good strength to weight applications; corrosion resistant. AMS 4967, 4928. *Obsolete*

CARPENTER TITANIUM 6-6-2
Carpenter Technology Corp.
Al 5.5, V 5.5, Sn 2, Fe 0.5, Cu 0.5, bal Ti.
Corrosion resistant high strength alloy. Room temperature: 180,000 TS; 170,000 YS; 6 El. 600 F: 150,000 TS; 132,000 YS; 15 El. AMS 4971. *Obsolete*

CARPENTER TYPE D
Carpenter Technology Corp.
C 0.5, Cr 0.9, V 0.2, bal Fe.
Heat treated: 218,000 TS; 195,000 YS; 10 El; 35 RA. For automobile springs. *Obsolete*

CARPENTER V 57
Carpenter Technology Corp.
C 0.05, Cr 14, Ni 26.5, Mo 1.5, Ti 3, Al 0.25, B 0.007, bal Fe.
Rolled: 175,000 TS; 125,000 YS; 21 El; 35 RA; 331 Brin. For jet engine blades and buckets, turbine wheels, torque rings. Vacuum melted superalloy with high heat resistance up to 1300°F. AISI No. 663.

CARPENTER V S M
Carpenter Technology Corp.
C 0.7, Si 1.1, Cr 3, Mo 5.25, Mn 0.5, bal Fe.
Heat treated: 550-640 Brin. For valve seat inserts and stems, guides, spindles. Resists heat checking and wear. *Obsolete*

CARPENTER VACUMET KOVAR (NICOSEAL)
Carpenter Technology Corp.
C 0-0.02, Mn 0.3, Si 0.2, Ni 29, Co 17, bal Fe.
Vacuum melted low-expansion alloy for making hermetic seals with the harder Pyrex glasses and ceramic materials. At 70-392°F: 2.89 x 10^{-6} in./in.·°F. *Obsolete*

CARPENTER VACUMET KOVAR (NICOSEAL) PHOTO
Carpenter Technology Corp.
C 0-0.02, Mn 0.3, Si 0.2, Ni 29, Co 17, bal Fe.
Photoetching quality strip grade of Carpenter Vacumet Nicoseal. *Obsolete*

CARPENTER VACUMET NICOSEAL
Now KOVAR.

CARPENTER VACUMET NICOSEAL PHOTO-ETCH
Now KOVAR PHOTO-ETCH QUALITY.

CARPENTER VACUUM MELTED 52100
Carpenter Technology Corp.
C 1, Mn 0.3, Si 0.25, Cr 1.4, bal Fe.
Vacuum melted, high carbon, chromium-bearing steel for ball and roller bearings. G12516

CARPENTER VEGA-FM
Carpenter Technology Corp.
C 0.7, Mn 2.25, Si 0.3, Cr 1, Mo 1.35, alloy sulfides, bal Fe.
Air hardening, nondeforming die steel, free machining grade. For blanking, forming and trimming dies, mandrels, punches, precision tools. AISI A6.

CARPENTER WEAR RESISTANT HYMU 800A
Carpenter Technology Corp.
C 0-0.005, Mn 0.01, Si 0-0.01, Ni 80, Mo 5, 0.3 Al_2O_3, bal Fe.
Soft magnetic alloy with good abrasive wear resistance for use in tape recorder head applications. Magnetic saturation 7000 Gausses.

CARR'S QUALITY "O" BRAND
Richard W. Carr & Co., Ltd.
C 0.35, Mn 1.5, bal Fe.
Carbon-manganese steel. Similar to AISI 1335.

CARR'S QUALITY 06S
Richard W. Carr & Co., Ltd.
C 0.95, V 0.25, bal Fe.
Carbon-vanadium tool steel. AISI W2.

CARR'S QUALITY 08S
Richard W. Carr & Co., Ltd.
C 1.45, Cr 12.5, Mo 1, Co 3, bal Fe.
High chrome-cobalt cold work tool steel. AISI D5.

CARR'S QUALITY 09B
Richard W. Carr & Co., Ltd.
C 0.95, Mn 1.25, W 0.5, Cr 0.5, V 0.3, bal Fe.
Tungsten-chrome general purpose oil-hardening tool steel. AISI O1.

CARR'S QUALITY 12S
Richard W. Carr & Co., Ltd.
C 0.3, Cr 3.25, W 9.5, V 0.4, bal Fe.
Tungsten-chrome hot work tool steel. AISI H21.

CARR'S QUALITY 14S
Richard W. Carr & Co., Ltd.
C 2.2, Cr 12, Mo 0.8, V 0.4, bal Fe.
High carbon-chrome hot work tool steel. AISI D4.

CARR'S QUALITY 23S
Richard W. Carr & Co., Ltd.
C 2.2, Cr 12, bal Fe.
High carbon-high chrome cold work tool steel. AISI D3.

CARR'S QUALITY 24S
Richard W. Carr & Co., Ltd.
C 0.24, Ni 2.5, Cr 3, W 8.5, bal Fe.
Tungsten-chrome-nickel hot work tool steel.

CARR'S QUALITY 28F
Richard W. Carr & Co., Ltd.
C 0.1, Mn 0.3, Cr 5, Mo 1, V 0.3, bal Fe.
Chrome carburizing steel.

CARR'S QUALITY 32S
Richard W. Carr & Co., Ltd.
C 1, Cr 5, Mo 1, V 0.3, bal Fe.
Chrome air hardening cold work tool steel. AISI A2.

CARR'S QUALITY 53S
Richard W. Carr & Co., Ltd.
C 0.4, Si 1, Cr 5.25, Mo 1.4, V 1, bal Fe.
Chrome hot work tool steel. AISI H13.

CARR'S QUALITY 58S
Richard W. Carr & Co., Ltd.
C 0.35, Si 1, Cr 5.25, Mo 1.5, W 1.25, V 0.3, bal Fe.
Special chrome hot work tool steel. AISI H12.

CARR'S QUALITY 65S
Richard W. Carr & Co., Ltd.
C 0.4, Si 1, Cr 5.25, Mo 1.4, V 0.4, bal Fe.
Chrome hot work tool steel; AISI H11.

CARR'S QUALITY 67S
Richard W. Carr & Co., Ltd.
C 1.35, W 3.75, C 0.75, bal Fe.
Carbon-tungsten tool steel. AISI F3.

CARR'S QUALITY 69S
Richard W. Carr & Co., Ltd.
C 1.5, Cr 12, Mo 0.8, V 0.9, bal Fe.
High carbon-chrome cold work tool steel. AISI D2.

CARR'S QUALITY 74S
Richard W. Carr & Co., Ltd.
C 0.3, Cr 2.3, Mo 3, W 4.25, V 0.6, bal Fe.
Tungsten-chrome hot work tool steel.

CARR'S QUALITY 82S
Richard W. Carr & Co., Ltd.
C 0.33, Si 1, Cr 3, Mo 2.8, W 1, V 0.9, Co 3, bal Fe.
Chrome-molybdenum-cobalt hot work tool steel.

CARR'S QUALITY BCC
Richard W. Carr & Co., Ltd.
C 0.5, Si 1, W 2.25, Cr 1.2, V 0.25, bal Fe.
Heavy duty general purpose tool steel. AISI S1.

CARR'S QUALITY BCD 37
Richard W. Carr & Co., Ltd.
C 0.37, W 2, Cr 1.6, V 0.3, bal Fe.
Nontempering chisel steel.

CARR'S QUALITY BLUE LABEL
Richard W. Carr & Co., Ltd.
C 0.7, Mn 0.3, bal Fe.
Water hardening plain carbon tool steel. AISI W1.

CARR'S QUALITY P 1000
Richard W. Carr & Co., Ltd.
C 0.1, Mn 0.4, Ni 0.4, Cr 13, bal Fe.
Martensitic chrome stainless. AISI 410.

CARR'S QUALITY P 1001
Richard W. Carr & Co., Ltd.
C 0.16, Mn 0.4, Ni 0.4, Cr 13, bal Fe.
Martensitic chrome stainless steel.

CARR'S QUALITY P 1002
Richard W. Carr & Co., Ltd.
C 0.21, Mn 0.4, Ni 0.4, C 13, bal Fe.
Martensitic chrome stainless. AISI 420.

CARR'S QUALITY P 1003
Richard W. Carr & Co., Ltd.
C 0.3, Mn 0.4, Ni 0.4, Cr 13, bal Fe.
Martensitic chrome stainless steel. Similar to AISI 420.

CARR'S QUALITY P 1008
Richard W. Carr & Co., Ltd.
C 0.35, Mn 0.4, Ni 1, Cr 13, bal Fe.
Stainless mold steel.

CARR'S QUALITY P 1009
Richard W. Carr & Co., Ltd.
C 0.18, Mn 0.6, Ni 2.1, Cr 17, bal Fe.
Martensitic stainless steel. AISI 431.

CARR'S QUALITY P 1010
Richard W. Carr & Co., Ltd.
C 0-0.16, Cr 18, Ni 8, bal Fe.
Austenitic stainless steel. AISI 302.

CARR'S QUALITY P 1011
Richard W. Carr & Co., Ltd.
C 0-0.15, Cr 18, 8.0 Ni + Ti, bal Fe.
Stabilized austenitic stainless steel. AISI 321.

CARR'S QUALITY P 1012
Richard W. Carr & Co., Ltd.
C 0-0.15, Cr 18.5, 10.0 Ni + Ti, bal Fe.
Stabilized austenitic stainless steel. AISI 321.

CARR'S QUALITY P 1013
Richard W. Carr & Co., Ltd.
C 0-0.15, Cr 13, Ni 13, bal Fe.
Austenitic stainless steel.

CARR'S QUALITY P 1014
Richard W. Carr & Co., Ltd.
C 0-0.08, Cr 19, Ni 9.5, bal Fe.
Austenitic stainless steel. AISI 304.

CARR'S QUALITY P 1015
Richard W. Carr & Co., Ltd.
C 0-0.15, Cr 18, Ni 8, Nb, bal Fe.
Stabilized austenitic stainless. AISI 347.

CARR'S QUALITY P 1016
Richard W. Carr & Co., Ltd.
C 0-0.15, Cr 18.5, Ni 10, Nb, bal Fe.
Stabilized austenitic stainless. AISI 347.

CARR'S QUALITY P 1017
Richard W. Carr & Co., Ltd.
Ni 10, Mo 2, C 0-0.12, Cr 18.5, Ti or Nb, bal Fe.
Stabilized austenitic stainless.

CARR'S QUALITY P 1018
Richard W. Carr & Co., Ltd.
C 0-0.12, Cr 18.5, Ni 10, Mo 3, Ti or Nb, bal Fe.
Stabilized austenitic stainless steel. AISI 316.

CARR'S QUALITY P 151
Richard W. Carr & Co., Ltd.
C 0.15, Mn 0.8, bal Fe.
Low carbon steel for case hardening.

CARR'S QUALITY P 153
Richard W. Carr & Co., Ltd.
C 0.17, Mn 0.45, Ni 2, Mo 0.25, bal Fe.
Nickel-molybdenum steel for case-hardening intermediate size sections.

CARR'S QUALITY P 155
Richard W. Carr & Co., Ltd.
C 0.15, Mn 0.45, Ni 3.4, Cr 1, bal Fe.
Nickel-chrome carburizing steel; for heavy sections.

CARR'S QUALITY P 158
Richard W. Carr & Co., Ltd.
C 0.15, Mn 0.4, Ni 4.25, Cr 1.2, Mo 0.3, bal Fe.
Ni-Cr-Mo carburizing steel; for very heavy sections.

CARR'S QUALITY P 20
Richard W. Carr & Co., Ltd.
C 0.32, Mn 0.8, Cr 1.6, Mo 0.4, bal Fe.
Prehardened chrome-molybdenum plastic mold steel. AISI P20.

CARR'S QUALITY P 256
Richard W. Carr & Co., Ltd.
C 0.55, Mn 0.7, bal Fe.
Bolster steel.

CARR'S QUALITY P 280
Richard W. Carr & Co., Ltd.
C 0.6, Mn 0.6, Cr 0.6, bal Fe.
Carbon-chrome steel.

CARR'S QUALITY P 552
Richard W. Carr & Co., Ltd.
C 0.3, Mn 0.5, Ni 4.25, Cr 1.25, Mo 0.3, bal Fe.
Nickel-chrome-molybdenum structural steel; deep hardening; plastic molds.

CARR'S QUALITY P 553
Richard W. Carr & Co., Ltd.
C 0.4, Mn 0.6, Ni 1.5, Cr 1.1, Mo 0.3, bal Fe.
Nickel-chrome-molybdenum steel. Similar to AISI 4340.

CARR'S QUALITY P 558
Richard W. Carr & Co., Ltd.
C 0.3, Mn 0.6, Ni 2.5, Cr 0.7, Mo 0.5, bal Fe.
Nickel-chrome-molybdenum structural steel; deep hardening; tough.

CARR'S QUALITY P 564
Richard W. Carr & Co., Ltd.
C 0.3, Mn 0.6, Ni 3, Cr 0.75, bal Fe.
Nickel-chrome structural steel.

CARR'S QUALITY P 602
Richard W. Carr & Co., Ltd.
C 0.4, Mn 0.65, Cr 1.1, Mo 0.3, bal Fe.
Chrome-molybdenum hot work tool steel.

CARR'S QUALITY P 609
Richard W. Carr & Co., Ltd.
C 0.4, Mn 0.8, Cr 1, bal Fe.
Chrome hot work tool steel.

CARR'S QUALITY P 618
Richard W. Carr & Co., Ltd.
C 0.4, Si 0.3, Mn 0.5, Ni 0.3, Cr 3, Mo 1, V 0.25, bal Fe.
Chrome-molybdenum-vanadium hot work tool steel.

CARR'S QUALITY P 704
Richard W. Carr & Co., Ltd.
C 0.95, Mn 1.65, Cr 0.2, bal Fe.
Manganese oil-hardening general purpose tool steel. AISI O2.

CARR'S QUALITY P 720
Richard W. Carr & Co., Ltd.
C 1, Mn 0.4, Cr 1.4, bal Fe.
Carbon-chrome oil-hardening tool steel. AISI L1.

CARR'S QUALITY RED LABEL
Richard W. Carr & Co., Ltd.
C 0.9, Mn 0.3, bal Fe.
Water hardening plain carbon tool steel. AISI W1.

CARR'S QUALITY ST BRAND
Richard W. Carr & Co., Ltd.
C 0.7, Mn 0.3, bal Fe.
Water hardening plain carbon tool steel.

CARR'S QUALITY YELLOW LABEL
Richard W. Carr & Co., Ltd.
C 1.15, Mn 0.3, bal Fe.
Water hardening plain carbon tool steel. AISI W1.

CARRIAGE WHEEL BEARING
English manufacture
Cu 84, Sn 16.
For bearings; corrosion eresistant.

CARTER WHITE GOLD
English manufacture
Au 83.3, Ni 16.7.
For jewelry, ornaments; corrosion resistant.

CARTOS
Heppenstall Co.
Tool material. C 0.93, Mn 0.4, Si 0.6, V 0.09, bal Fe.
Annealed: 250 Brin. Heat treated: 620 Brin. For shear knives, trimmers. Oil hardened; wear resistant.

CARTOS 2V72
Heppenstall Co.
Tool material. C 0.8, Mn 0.4, Si 0.6, V 0.1, bal Fe.
Water hardened tool steel; AISI W2.

CARTOS 2V90
Heppenstall Co.
Tool material. C 0.9, Mn 0.4, Si 0.6, V 0.1, bal Fe.
Water hardened tool steel; AISI W2.

CARTRIDGE BRASS
Chase Brass & Copper Co.
Cu 70, Zn 30.
Annealed: 47,000 TS; 15,000 YS; 62 El. Rolled: 76,000 TS; 64,000 YS; 8 El. For cartridge cases, primer cups, springs; deep drawing.

CARTRIDGE BRASS 260
Olin Corp.
Cu 70, Zn 30.
Annealed: 45,000-65,000 TS; 10,000-34,000 YS; 35-67 El. All purpose deep drawing brass. For all deep drawn brass parts. Cold rolled: 49,000-110,000 TS; 20,000-100,000 YS; 1-56 El. For less drastic forming, spring clips.

CARTRIDGE BRASS 260
Anaconda Co.
Cu 70, Zn 30.
Annealed: 45,000-65,000 TS; 10,000-34,000 YS; 35-67 El. All purpose deep drawing brass. For all deep drawn brass parts. Cold rolled: 49,000-110,000 TS; 20,000-100,000 YS; 1-56 El. For less drastic forming, spring clips.

CARTRIDGE BRASS 42
Anaconda Co.
Copper. Cu 70, Zn 30.
1/4 H temper: 54,000 TS; 40,000 YS; 43 El; 63 Brin. 1/2 H temper: 62,000 TS; 52,000 YS; 23 El; 120 Brin. H temper: 76,000 TS; 63,000 YS; 8 El; 156 Brin. For cartridges, primers, shot shells, tanks, fasteners. Deep drawing, high ductility.

CARTRIDGE GILDING
American manufacture
Cu 93, Zn 7.
For cartridge shells, ornaments, base for fire enameling; gilding metal.

CARTUN

Delsteel Inc.
C 1.35, W 2.75, bal Fe.
For fast finishing tools, cutters; water or oil hardened.

CARVAN

Bethlehem Steel Corp.
C 0.8, bal Fe.
For cutting tools; water hardening. *Obsolete*

CASAR

Fakirstahl Hoffmanns GmbH & Co. KG
C 1.3, Mo 0.85, V 3.8, W 12, Cr 4.3, bal Fe.
For lathe and planer tools, reamers, broaches, taps; high speed steel.

CASAR C

Fakirstahl Hoffmanns GmbH & Co. KG
C 1.35, Cr 4.2, Mo, W, V, bal Fe.
For blanking and forming dies, engravers' tools; high speed steel.

CASCADE

Latrobe Steel Co.
Tool material. C 0.2, Cr 0.25, Ni 4.1, V 0.2, Si 0.3, Mn 0.3, Al 1.2, bal Fe.
207-350 Brin. For plastic mold zinc die casting dies; can be nitrided.

CASE DIE

Wallace Murray Corp.
C 0.45, W 1, Mo 0.2, Cr 0.9, bal Fe.
Shock resisting tool steel; AISI S3.

CASINO

Manufacturer not listed
C 0.7, W 18, Cr 4, V 1, bal Fe.
For shapers, cutters; high speed steel.

CASONA

Osborn Steels Ltd.
C 0.1, Si 0-0.25, Mn 0-0.25, bal Fe.
Annealed: 64,000 TS; 40,000 YS; 28 El; 65 RA; 132 Brin. For deep hobbing dies; case hardened. *Obsolete*

CAST 14-14

Manufacturer not listed
C 1, Mn 0.8, Si 3, Cr 14.5, Ni 14.5, bal Fe.
Exhaust valve steel. SAE EV 10.

CAST COMPOSITE

Edgar Allen Balfour Ltd.
C, bal Fe.
For tools, blanking dies; water hardening.

CAST COMPOSITE

A. Milne & Co.
C 1-1.05, bal Fe.
For tools, drills, taps; water hardening.

CAST-I-NICKEL

Champion Rivet Co.
Ni.
For cast iron welding rod; flux coated, machinable.

CAST-I-STEEL

Champion Rivet Co.
C 0.1-0.14, Mn 0.4-0.6, Si 0-0.02, bal Fe.
For non-machinable welds on cast iron; coated arc electrode.

CASTALOY

Detroit Alloy Steel Co.
C 1.5-1.6, Cr 12-14, Mo 0.7-0.8, Mn 0.45-0.55, bal Fe.
For dies; easily machined; 26-30 Rock C.

CASTALOY

Fisher Scientific Co.
Al 4.1, Mg 0.04, bal Zn.
Die cast: 40,300 TS; 5 El; 74 Brin. For chemical apparatus, clamps, holders. SAE903.

CASTDIE

Columbia Tool Steel Co.
C 0.35, Cr 5.25, Mo 1.35, V 0.5, Si 0.95, bal Fe.
For die casting dies; oil hardened. *Obsolete*

CASTEC

Eutectic Corp.
Now CASTEC 3055. *Obsolete*

CASTEC 3055

Eutectic Corp.
53,000 psi TS. For AC-DC all-position welding of cast iron; machinable deposits.

CASTEEL 15

Calumet Steel Castings Corp.
C 0.13-0.18, Mn 0.6, Si 0.4, bal Fe.
Normalized: 57,000-65,000 TS; 32,000-37,000 YS; 30-35 El; 50-55 RA; 110-125 Brin. For gears, pinions, and shafts. Case hardening.

CASTEEL 22

Calumet Steel Castings Corp.
C 0.18-0.25, Mn 0.6, Si 0.4, bal Fe.
Normalized: 65,000-75,000 TS; 38,000-45,000 YS; 25-30 El; 45-50 RA; 130-145 Brin. For gears, pinions, housings, and castings. Case hardening.

CASTEEL 28

Calumet Steel Castings Corp.
C 0.26-0.32, Mn 0.7, Si 0.4, bal Fe.
Normalized: 75,000-85,000 TS; 45,000-55,000 YS; 20-25 El; 36-45 RA; 140-160 Brin. For structural and machinery parts, and castings. Water hardened.

CASTEEL 40

Calumet Steel Castings Corp.
C 0.37-0.43, Mn 0.7, Si 0.4, bal Fe.
Normalized: 85,000-95,000 TS; 55,000-65,000 YS; 15-20 El; 25-35 RA; 165-180 Brin. Heat treated: 90,000-110,000 TS; 62,000-70,000 YS; 18-23 El; 40-50 RA; 180-220 Brin. For gears, pinions, shafts, housings, and castings. Water hardened.

CASTINGWELD

Westinghouse Electric Corp.
C, bal Fe.
For welding electrodes; nonmachinable, welds on cast iron. *Obsolete*

CASTOR 1939

Uddeholm Corp.
C 0.86, Co 2.8, Cr 4.3, Mo 0.85, V 2, W 12, bal Fe.
For lathe and planer tools, reamers, broaches; high speed steel. *Obsolete*

CASTOR 3

Uddeholm Corp.
C 0.72, Cr 4.5, W 18, V 1.2, bal Fe.
For tools, cutters; high speed steel. *Obsolete*

CASTOR 32

Uddeholm Corp.
C 0.8, Cr 4, Mo 5, W 6.5, V 1.9, bal Fe.
For lathe and planer tools, hobs, reamers; high speed steel. *Obsolete*

CASTOR 7

Uddeholm Corp.
C 0.8, Cr 4.5, Mo 1.25, W 18.5, Co 2.5, V 1.6, bal Fe.
For tools, cutters; high speed steel. *Obsolete*

CASTOR 8

Uddeholm Corp.
C 0.8, Cr 4.5, Mo 1.25, W 18.5, Co 5.5, V 1.6, bal Fe.
For tools, cutters; high speed steel. *Obsolete*

CASTOR 9

Uddeholm Corp.
C 0.8, Cr 4.5, Mo 1, W 18.5, Co 10.5, V 1.6, bal Fe.
For tools, cutters; high speed steel. *Obsolete*

CAT 497 STARRETT PRECISION GROUND DIE

Now STARRETT NO. 497.

CAT BRAND 14.0 W

Flockton, Tompkin & Co., Ltd.
C 0.65, W 14.5, Cr 3.75, V 0.5, bal Fe.
For drills, punches, taps, piercing and blanking dies; high speed steel.

CAT BRAND S.H.X.

Flockton, Tompkin & Co., Ltd.
C 0.65, W 15.5, Cr 3.9, V 0.5, bal Fe.
For slitting saws; high speed steel.

CATALOY

Manufacturer not listed
Pb, Cu, Sn.
Catalyst for non-ferrous metals. *Obsolete*

CATALOY NO. 1

Manufacturer not listed
Copper. Pb 20, Cu 72, Sn 4, bal others.
26,750 TS; 18 El; 25 RA; 50 Brin. For bearings; heavy duty. *Obsolete*

CATALOY NO. 2

Manufacturer not listed
Copper. Pb 25, Cu 68, Sn 3, bal others.
21,350 TS; 13 El; 25 RA; 50 Brin. For bearings; heavy duty. *Obsolete*

CATALOY NO. 3

Manufacturer not listed
Copper. Pb 30, Cu 63, Sn 3, bal others.
Cast: 19,550 TS; 11.7 El; 19.5 RA; 39 Brin. For bearings; heavy duty. *Obsolete*

CATALOY NO. 4

Manufacturer not listed
Copper. Pb 25, Cu 70, Sn 1, bal others.
Cast: 13,700 TS; 10 El; 9.7 RA; 34 Brin. For bearings; heavy duty. *Obsolete*

CATALOY NO. 5

Manufacturer not listed
Copper. Pb 40, Cu 52, bal others.
Cast: 9,910 TS; 5.5 El; 9.8 RA; 27 Brin. For bearings; heavy duty. *Obsolete*

CATALOY NO. 6

Manufacturer not listed
Copper. Pb 40, Cu 51, Sn 1, bal others.
Cast: 10,500 TS; 5.5 El; 19.5 RA; 30 Brin. For bearings; heavy duty. *Obsolete*

CATALOY NO. 7

Manufacturer not listed
Lead. Pb 50, Cu 42, bal others.
Cast: 6,400 TS; 10 El; 8.6 RA; 13 Brin. For bearings; heavy duty. *Obsolete*

CATALOY NO. 8S

Manufacturer not listed
Copper. Pb 13, Cu 80, Sn 5, bal others.
Cast: 30,700 TS; 18.8 El; 24.9 RA; 50 Brin. For bearings; heavy duty. *Obsolete*

CATALOY NO. 9N

Manufacturer not listed
Copper. Pb 11, Cu 80, Sn 5, bal others.
Cast: 31,000 TS; 14.1 El; 8.2 RA; 53 Brin. For bearings; heavy duty. *Obsolete*

CATAWBA

Jessop Steel Co.
C, Si, Mn, bal Fe.
For tools, punches, upsetters; shock resistant. *Obsolete*

CATHALOY A-30
Superior Tube Company
C 0.03-0.1, Cu 0-0.05, Fe 0-0.1, Al 0.03-0.08, Mg 0.01-0.06, bal Ni.
Tempered: 75,000-120,000 TS; 15,000-110,000 YS; 4-50 El; 125-260 Brin. For electronic valves; cathode nickel. *Obsolete*

CATHALOY A-31
Superior Tube Company
W 3.7-4.2, Cu 0-0.1, Mg 0.01-0.06, Si 0.02-0.06, bal Ni.
Tempered: 85,000-140,000 TS; 15,000-125,000 YS; 4-50 El; 150-350 Brin. For electronic valves; cathode nickel.

CATHALOY A-33
Superior Tube Company
W 1.75-2.25, Mn 0.05-0.1, C 0.05-0.1, Zr 0.04-0.8, bal Ni + Co.
Tubing for electron tubes; all purpose cathode. Active.

CATHALOY A-34
Superior Tube Company
W 3.75-4.25, Mg 0.01-0.06, Si, C, low Mn, bal Ni + Co.
Tubing for electron tubes subject to shock, vibration or use above rated heater voltage. Active.

CATHALOY A32
Superior Tube Company
W 2-2.5, Al 0.05, Cu 0.02, C 0.03-0.1, Mg 0.01-0.06, bal Ni.
85,000-140,000 TS; 15,000-125,000 YS; 4-50 El; 150-350 Brin. For cathodes in electron tubes; high emission and shock resistant. *Obsolete*

CATHALOY P-50
Superior Tube Company
Cu 0-0.04, Fe 0-0.05, C 0-0.05, Ti 0-0.01, bal Ni.
Tempered: 75,000-120,000 TS; 15,000-110,000 YS; 4-50 El; 125-250 Brin. For cathodes in electron tubes; ASTM Alloy 22.

CATHALOY P-51
Superior Tube Company
W 3.75-4.25, Cu 0-0.04, Fe 0-0.05, bal Ni.
Tempered: 85,000-140,000 TS; 65,000-125,000 YS; 4-50 El; 150-350 Brin. For electronic tubes, cathodes; passive cathode material.

CATHALOY P-52
Superior Tube Company
Cb 0.4-0.7, Mo 1.5-2.5, bal Ni + Co.
Tubing as nonemitting alloy for disc cathode shanks in electron tubes. Passive.

CATHALOY P-53
Superior Tube Company
Pure Ni + Co.
Tubing for low sublimation disc cathode shanks. Passive.

CATHALOY-1
Superior Tube Company
Al 0.1, bal Ni.
For television cathode tubes; high emission. *Obsolete*

CATHALOY-2
Superior Tube Company
W 4, bal Ni.
For television cathode tubes; high emission. *Obsolete*

CATHERMALITE
Jessop Steel Co.
C, alloy, bal Fe.
For annealing and carburizing boxes, flue dampers, furnace parts, acid containers; heat and acid resisting, casting. *Obsolete*

CATHODE NICKEL
Wilbur B. Driver Co.
Nickel. Ni 96-99.
Selected additives. Annealed: 60,000 TS. For cathodes in electron tubes; several types.

CAV1
Monarch Steel Corp.
C 0.6, Si 0.9, Mn 0.9, Cr 1.35, V 0.15, bal Fe.
For tools, dies. Shock resistant.

CAV2
Thyssen Edelstahlwerke AG
C 0.5, Si 1.5, Mn 0.75, Cr 1.2, V 0.15, bal Fe.
For tools, dies; shock resistant.

CAV3
Thyssen Edelstahlwerke AG
C 0.35, Si 1.5, Mn 0.7, Cr 1.1, V 0.15, W 1.5-7, bal Fe.
For tools, dies; shock resistant.

CAZIN
Manufacturer not listed
Cd 82.6, Zn 17.4.
For solder for steel; M.P. 263°F.

CB 10W-2.5ZR
Now COLUMBIUM 10W -2.5ZR.

CB 10W-2.5ZR
Now WAH CHANG 10W -2.5ZR.

CB 12
Now FANSTEEL 80.

CB 45
Now STELCO CB 45-60.

CB 7 CU
Ancast, Inc.
Stainless steel. C 0-0.07, Mg 0-0.7, Si 0-1, Cr 14-15.5, Ni 4.5-5.5, P 0-0.035, S 0-0.03, Cu 2.5-3.2, Cb 0.15-0.35, N 0-0.05, bal Fe.
ACI CB-7 Cu-2.

CB 74
Union Carbide Corp.
Zr 5, W 10, bal Cb.
At 70°F: 88,000 TS; 70,000 YS; 26% El. At 2200°F: 39,000 TS; 33,000 YS; 25% El. (Recrystallized 1 hr. a 2700°F). For space vehicles, nuclear reactors. Good combination of density, strength and oxidation resistance at high temperatures.

CB = 33TA-1ZR
Now FANSTEEL 82.

CB CUPRON ALLOY
Carpenter Technology Corp.
Cu 55, Ni 45.
Annealed: 414 MPa TS; 25 El. Cold worked: 793 MPa TS; 25 El. For use in rheostats and controls up to 540°C and in fine sizes in resistors and electrical instruments.

CB-1
American manufacture
W 30, Zr 1, C 0.06, N 0.032, bal Cb.
Similar to XB-88.

CB-10HF-1TI-0.5ZR
Teledyne Wah Chang Albany
Hf 9-11, Ti 0.7-1.3, Zr 0-0.7, bal Cb.
Cold rolled: 105,000 TS; 96,500 YS; 5 El. Recrystallized: 59,000 TS; 45,400 YS; 26 El. For high temperature applications, space vehicles. Good TIG weldability.

CB-10W-10HF
Now COLUMBIUM C-129.

CB-10W-1ZR-0.1C
Now COLUMBIUM D-43.

CB-10W-5ZR
Cabot Corporation
W 10, Zr 5, bal Cb.
Rolled: 130,000 TS; 128,000 YS; 2 El. At 2400 F: 89,000 TS; 74,000 YS; 28 El. Space vehicles, missiles, nuclear reactors. High temperature alloy. Combines high strength with excellent ductility at high temperature. *Obsolete*

CB-15W-5MO-1ZR
Now COLUMBIUM 15W-5MO-IZR.

CB-1ZR
Now WAH CHANG WC-1ZR.

CB-20W-10TI-6MO
Now COLUMBIUM D-41.

CB-28 TA-10W-1ZR
Now FANSTEEL FS-85.

CB-28W-2HF-0.07C
Now COLUMBIUM XB-88.

CB-751
Now FANSTEEL 80.

CB-752
Now COLUMBIUM CB-752.

CB-752
Cabot Corporation
W 10, Zr 5, bal Cb.
Rolled: 130,000 TS; 128,000 YS; 2 El. At 2400 F: 89,000 TS; 74,000 YS; 28 El. For space vehicles, nuclear reactors. High temperature alloy, combines high strength with excellent ductility at high temperature. *Obsolete*

CB-753
Cabot Corporation
V 5, Zr 1.25, bal Cb.
At 75 F: 85,000 TS. At 2400 F: 18,000 min TS. For nuclear industry, aerospace equipment. Low nuclear cross-section. Good fracture ductility. *Obsolete*

CB-TZM
American manufacture
Cb 1.5, Ti 0.5, Zr 0.3, C 0.03, bal Mo.

CB/V 42
Stelco Steel
HSLA steel. C 0.21, Mn 1.35, Si 0.3, 0.01 V min, 0.005 Cb min, bal Fe.
Wrought: 60,000 psi TS; 42,000 psi YS; 24 El. Meets ASTM A572.

CB/V 45
Algoma Steel Corp. Ltd.
HSLA steel. C 0-0.22, Mn 0-1.35, Cb 0-0.005, V 0-0.01, bal Fe. 45,000 YS min.

CB/V 45
Stelco Steel
HSLA steel. C 0-0.22, Mn 0-1.35, Cb 0-0.005, V 0-0.01, bal Fe. 45,000 psi YS min.

CB/V 50
Algoma Steel Corp. Ltd.
HSLA steel. C 0-0.23, Mn 0-1.35, Cb 0-0.005, V 0-0.01, bal Fe. 50,000 psi YS min.

CB/V 50
Stelco Steel
HSLA steel. C 0-0.23, Mn 0-1.35, Cb 0-0.005, V 0-0.01, bal Fe. 50,000 psi YS min.

CB/V 55

Stelco Steel
HSLA steel. C 0.25, Mn 1.35, Si 0.3, 0.01 V min, 5.00 Cb min, bal Fe.
Wrought: 70,000 psi TS; 55,000 psi YS; 20 El. Meets ASTM A572.

CB/V 60

Algoma Steel Corp. Ltd.
HSLA steel. C 0-0.26, Mn 0-1.35, Cb 0-0.005, V 0-0.01, bal Fe.
60,000 psi YS min.

CB/V 60

Stelco Steel
HSLA steel. C 0-0.26, Mn 0-1.35, Cb 0-0.005, V 0-0.01, bal Fe.
60,000 psi YS min.

CB/V 65

Stelco Steel
HSLA steel. C 0.26, Mn 1.35, Si 0.3, 0.005 Cb min, 0.01 V min, bal Fe.
Wrought: 80,000 psi TS; 65,000 psi YS; 15 El. Meets ASTM A572.

CB132

Now COLUMBIUM CB-132M.

CBS 600

Now TIMKEN CBS 600.

CBS-1000M VIM-VAR

Latrobe Steel Co.
C 0.13, Mn 0.55, Si 0.5, Cr 1.05, Ni 3, Mo 4.5, V 0.4, Al 0.06.
Carburizing bearing and gear steel. For service temperatures up to 800°F.

CBS-50 NIL VIM-VAR

Latrobe Steel Co.
C 0.13, Mn 0.25, Si 0.2, Cr 4.2, Ni 3.4, Mo 4.25, V 1.2.
Nickel added variant of AISI M50; for use in aircraft engine bearings and gears to about 600°F.

CBS-600 VIM-VAR

Latrobe Steel Co.
C 0.19, Mn 0.6, Si 1.1, Cr 1.45, Mo 1, Al 0.06.
Carburizing bearing and gear steel. For service temperatures up to 600°F.

CBX CUPRON ALLOY

Carpenter Technology Corp.
Cu 53, Ni 44, Mn 3.
Annealed: 414 MPa TS; 25 El. Cold worked: 896 MPa TS. For use in vitreous enameled, glass, ceramic coated or bare resistors, rheostats and shunts in power controls up to 281°C.

CC 4000

Creusot-Loire
C 0.08-0.15, Mn 0-1, Si 0-1, Ni 0.5-1.5.
Stainless casting; annealed: 170-290 Brin. For pumps and faucets handling cold dilute organic acids in food industries. AFNOR Z12CN13M; W.Nr. 1.4008.

CC 4027

Creusot-Loire
C 0.18-0.25, Mn 0-1, Si 0-1, Cr 12.5-14.5, bal Fe.
Stainless casting; annealed 200-250 Brin. Pumps and faucets for food industries. AFNOR Z20C13M; similar to ACI CA-40.

CC/1 BIS

Creusot-Loire
C 0-0.15, Mn 0-1, Si 0-1, Cr 11.5-14, bal Fe.
Stainless casting; annealed: 180-220 Brin. For petroleum and food industries. AFNOR Z12C13M; ACI CA 15.

CC/2

Creusot-Loire
C 0.25-0.35, Si 0-1, Cr 13-15, Mn 1, bal Fe.
Stainless casting; annealed: 250-000 Brin. Faucets and pumps for paper industries. AFNOR Z30C13M; similar to ACI CA-40.

CC/2C

Creusot-Loire
C 0.35-0.45, Mn 0-1, Si 0-1, Cr 13-15, bal Fe.
Stainless casting; annealed: 260-300 Brin. For pump parts in paper industries. AFNOR Z38C13M; W.Nr. 1.4034.

CC/4 BIS

Creusot-Loire
C 0.15-0.25, Mn 0-1, Si 0-1, Ni 1.5-3, Cr 15-18, bal Fe.
Stainless casting; annealed: 275-310 Brin. For parts subject to seawater or to dilute organic acids. AFNOR Z 20 Cn 17.2 M; similar to AISI 431.

CCM 5

Bergische Stahl Industrie
C 0.9-1.05, Si 0.2-0.4, Mn 0.4-0.7, Cr 4.8-5.5, Mo 0.9-1.2, V 0.1-0.3, bal Fe.
Cold work tool steel. W.-Nr. 1.2363.

CCN

Bergische Stahl Industrie
C 1.9-2.2, Si 0.2-0.4, Mn 0.2-0.4, Cr 11-12, bal Fe.
Cold work tool steel. W.-Nr. 1.2080.

CCNW

Bergische Stahl Industrie
C 2-2.25, Si 0.25-0.4, Mn 0.2-0.4, Cr 11-12, W 0.6-0.8, bal Fe.
Cold work tool steel. W.-Nr. 1.2436.

CCR

Henry A. Kries & Sons Co.
C 2.25, Cr 13, bal Fe.
For tools, dies; nondeforming.

CCS

Delta Enfield Metals Ltd.
Now ERM CCS (M3).

CCS/Z

Delta Enfield Metals Ltd.
Now ERM CCS/Z (M328).

CCS/Z (M328)

Delta Enfield Metals Ltd.
Cr 0.5-1.2, Zr 0.03-0.2, Si 0.01-0.1, bal Cu.
Bar: 54,000-77,000 TS; 15-30 El; 110-165 Brin. Forged: 52,000-67,000 TS; 17-32 El; 105-135 Brin; 78-85% electrical conductivity. For resistance welding equipment, rotor bars. BS 4577, ISO 5182 and RWMA alloy A/2/2.

CCZ

George Cook & Co., Ltd.
C 0.35, Cr 3, Mo 2.75, V 0.6, Co 3, bal Fe.
Hot work tool steel; for die-casting dies and hot extrusion dies.

CD 18

Now CARBIDIE CD 18.

CD 4M CU

Ancast, Inc.
Stainless steel. C 0-0.04, Mn 0-1, Si 0-1, Cr 24.5-26.5, Ni 4.75-6, Mo 1.75-2.25, P 0-0.04, S 0-0.04, Cu 2.75-3.25, bal Fe.
ASTM A-351 GR CD4MCu, ASTM A-743 GR CD4MCu, ASTM A-744 GR CD4MCu.

CD-4M CU

Now COOPER ALLOY CD4M CU and OHIOLOY CD4M CU.

CD20 AND CD25

Herbert Cutanit Ltd.
Sintered carbide tool material. For general purpose wire drawing dies and for powder metal pressing dies. *Obsolete*

CD30

Herbert Cutanit Ltd.
Sintered carbide tool material. Hardest cutanit grade recommended for lamination die work. *Obsolete*

CD40

Herbert Cutanit Ltd.
Sintered carbide tool material. General purpose insert material for bar and tube drawing dies, and for press tool dies and punches. *Obsolete*

CD4MCU

Cooper Alloy Corp.
Stainless steel. C 0.04, Cr 25-27, Ni 4.75-6, Mo 1.75-2.25, Cu 2.75-3.25, bal Fe.
Annealed: 112,700 psi TS; 81,000 psi YS; 28 El; 248 Brin. Aged: 141,000 psi TS; 100,000 psi YS; 25 El; 302 Brin. For pumps, valves, impellers, and high pressure components. Precipitation hardening; corrosion and heat resistant.

CD50

Herbert Cutanit Ltd.
Sintered carbide tool material. For swaging dies, dies for cold heading, bar and tube dies, notching dies. *Obsolete*

CD55 F

Herbert Cutanit Ltd.
Sintered carbide tool material. For lamination dies and punches. *Obsolete*

CD55, CD60 AND CD65

Herbert Cutanit Ltd.
Sintered carbide tool materials. For cold heading and extrusion dies in nut, bolt and fastener manufacture. CD65 is the toughest; CD55 the most wear resistant. *Obsolete*

CDC ALLOY 50

Chicago Development Corp.
Al 5, Mn 50, Fe 45.
For electrical resistors; low expansion. *Obsolete*

CDC MANGANESE ALLOY NO. 730

Chicago Development Corp.
Cu 40, Ni 30, Mn 30.
Soft: 85,000 TS; 30 El; 130 Brin. Rolled: 145,000 TS; 2 El; 280 Brin. Hard: 270,000 TS; 580 Brin. For watch and clock parts; age hardenable, corrosion resistant. *Obsolete*

CDC MANGANESE NO. 715

Chicago Development Corp.
Mn, Ni, bal Cu.
For springs; corrosion resistant.

CDC VAN-AD

Chicago Development Corp.
V 75, bal Ti.
For addition agent to titanium alloys; for alloying. *Obsolete*

CDO1

Herbert Cutanit Ltd.
Sintered carbide tool material. For die insert requiring extreme wear resistance, as wire drawing die or pressing abrasive material. *Obsolete*

CDV

Spencer Clark Metal Industries Ltd.
C 0.39, Si 1, Mo 1.5, Cr 5, V 1, bal Fe.
Chromium-vanadium hot work steel, for forging dies for nonferrous metals.

CEC 57L

Eutectic Corp.
309-16 stainless steel; low carbon content. For AC-DC welding of stainless steels; good oxidation resistance at elevated temperatures; 85,000 psi TS.

CEC IMPACT

Columbia Tool Steel Co.
Tool material. C 0.58, Mn 0.85, Si 1.95, Cr 0.3, V 0.25, bal Fe.
Oil or water hardened, shock resistant tool steel. For pneumatic tools, chisels, rivet sets. AISI S4.

CEC SMOOTHCUT

Columbia Tool Steel Co.
C 0.55, Mn 0.8, Si 2, Mo 0.4, bal Fe.
For punches, crimpers; Type S5; oil hardened. *Obsolete*

CECO

English manufacture
Cu 62.5, Pb 32, Sn 4.6, Ni 0.9.
18,000 TS; 3 El; 0.35 RA; 52 Brin. For bearings, bushings; heavy duty.

CECOLLOY

Chambersburg Engineering Co.
C 0.3, Mn 0.9, Si 1.3, Ni 0.6, Mo 0.5, bal Fe.
56,000 TS; 255 Brin. For steam cylinder lines, valves; cast iron. *Obsolete*

CECOLLOY A

Chambersburg Engineering Co.
TC 3, Mo 0.5, Ni 0.6, bal Fe.
40,000-60,000 TS. For anvils, frames for forges, forming dies, steam cylinder liners, large valves. Series of synthetic alloys; made in air furnaces.

CECOLLOY B

Chambersburg Engineering Co.
TC 2.8, Mo 0.5, Cr 0.35, bal Fe.
For liners, frames, castings, valves. Shock resistant.

CECOLLOY C

Chambersburg Engineering Co.
TC 3, Mo 0.5, Ni 1.5, bal Fe.
For liners, frames, castings, valves; shock resistant. *Obsolete*

CEKAS

American manufacture
Cr 11.2, Ni 59.7, Mn 2, bal Fe.
For heat and corrosion resisting parts; stainless and corrosion resistant.

CEKAS

C. Kuhbier & Sohn
C 0.16, Cr 1.05, Mo 0.25, Mn 0.65, bal Fe.
For gears, bolts, machine tool parts; case hardened, tough.

CEKAS EXTRA 2

C. Kuhbier & Sohn
C 0-0.1, Si 0-1.3, Mn 0-1, Co 0-1.2, Cr 19-22, Al 0-5.5, bal Fe.
For high temperature equipment. W.-Nr. 1.4767.

CEKAS EXTRA 3

C. Kuhbier & Sohn
C 0-0.1, Si 0-1, Mn 0-0.6, Cr 22-25, Al 0-0.6, Ti 0-1, bal Fe.
For high temperature equipment. W.-Nr. 1.4765.

CEKAS M151

C. Kuhbier & Sohn
C 0.15, Cr 13, Mo 1, bal Fe.
Annealed: 75,000 TS; 40,000 YS; 30 El; 82 Rock B. Heat treated: 135,000 TS; 105,000 YS; 10 El; 30 Rock C. For flat springs, hardware, fittings, tableware, chemical plant and oil refining equipment. Corrosion resistant. Hardenable.

CEKAS M152

C. Kuhbier & Sohn
C 0.2, Cr 13, Mo 1, bal Fe.
Annealed: 95,000 TS; 50,000 YS; 25 El; 92 Rock B. Oil hardened: 250,000 TS; 215,000 YS; 8 El; 50 Rock C. For surgical instruments, cutlery, pivots, ball bearings, gears. Corrosion resistant, hardenable.

CELEBRATED 101

Flockton, Tompkin & Co., Ltd.
C 0.75, W 18, Cr 4.2, V 1.5, bal Fe.
For swaging dies, hobs, lathe and planer tools; high speed steel.

CELERO

Disston Inc.
C 1.35, Cr 0.25, W 2.75, bal Fe.
For finishing tools, plug gages; oil hardening. *Obsolete*

CELFOR

Sanderson Kayser Ltd.
C 0.2, bal Fe.
For gears, pinions, shafts; case hardening steel. *Obsolete*

CELLINI

Bethlehem Steel Corp.
C 0.8, Mn 0.9, Cr 0.5, bal Fe.
For tools, dies. *Obsolete*

CELLO VANADIUM

McInnes Steel Co.
C 0.9, Mn 1.2, Cr 0.5, V 0.15, W 0.5, bal Fe.
For dies, taps, reamers, hobs, broaches; oil hardened, non-deforming. *Obsolete*

CELSIT

Vereinigte Edelstahlwerke
C, Co, W, Cr, bal Fe.
For cutting tools; does not require heat treatment for hardening. *Obsolete*

CELSIT K

Vereinigte Edelstahlwerke
C 2, Cr 34, W 16, Co 44.
Annealed: 80,000 TS; 50,000 YS; 25 El; 45 RA; 187 Brin. For cutting tools; heat, wear and corrosion resistant. *Obsolete*

CELSIT N

Now VEW L208.

CELSIT P

Now VEW L216.

CELSIT SEO

Vereinigte Edelstahlwerke
C 3.5, Mn 1.2, Cr 31, bal Fe.
Welded. For hard facing electrodes; wear resistant. *Obsolete*

CELSIT V

Now VEW L219.

CELSIT V

Vereinigte Edelstahlwerke
C 1.2, Cr 25, W 4, Co 65, bal Fe.
Welded: Rockwell C43. For hard facing electrodes; for exhaust valves and dies. *Obsolete*

CELSIT V-300

Thyssen Edelstahlwerke AG
C 1.7, Cr 28, W 6, Co 33, bal Fe.
For hard facing electrode; wear resistant. *Obsolete*

CELSIT V-400

Thyssen Edelstahlwerke AG
C 1.6, Cr 26, W 4.5, Mo 4.5, Co 20, bal Fe.
For hard facing electrode; wear resistant. *Obsolete*

CELSIT VEO

Thyssen Edelstahlwerke AG
C 1.9, Cr 30, bal Fe.
For hard facing electrode; wear resistant. *Obsolete*

CELTO

Keasby & Matteson Co.
C 0.2, bal Fe.
For welding electrodes; shielded arc.

CEMENTED TUNGSTEN CARBIDE H8

Teledyne Firth Sterling
C 5.5, W 86.5, Co 8.
For hard cutting tools, tool tips.

CENCO

Central Brass & Aluminum Foundry Co.
Cu, Sb, Sn.
For bearings, bushings. *Obsolete*

CENTANIN

Isabellenhutte
Cu 67, Ni 5, Mn 27.
Annealed: 72,000-78,000 TS. Temperature coefficient resistance per °C: +/-0.00002. For electrical equipment and instruments. Resistance alloy. Maximum working temperature 300°C. *Obsolete*

CENTAUR

Jessop-Saville Ltd.
C, bal Fe.
For tools, drills, taps; water hardened.

CENTRA STEEL

General Motors Corp.
C 1.7, Si 2.25, Mg 0.4, S 0.1, P 0.05, B 0.01, bal Fe.
For crankshafts, gears, agricultural equipment; castings, good wear resistance. *Obsolete*

CENTRAL PURE IRON

Republic Steel Corp.
C 0.14, Mn, S, P, Si, Cu, bal Fe.
Rolled: 38,000-42,000 TS; 22,000-25,000 YS; 38-42 El; 55-60 RA. For culverts. *Obsolete*

CENTRAL V

Societe des AFC
Si 12, Mg 1.25, Mn 2.5, Ti 0.25, bal Al.
Cast: 23,000 psi TS; 1 El; 100 Brin. Heat treated: 31,000 psi TS; 1 El; 130 Brin. For light alloy parts; heat treatable.

CENTRALLOY

Central Iron & Steel Co.
C 0.15, Ni 0.8, Cr 0.2, bal Fe.
For bus and railway bodies; high strength.

CENTRARD

Sheepbridge Engineering Ltd
C 2.7, Cr 1.5-1.75, Al 1.5-1.75, bal Fe.
Annealed: 45,000-56,000 TS; 250 Brin. For cylinder liners, general parts to resist severe abrasion; nitrogen hardened alloy cast iron.

CENTRARD

Sheepbridge Alloy Castings Ltd.
C 2.7, Cr 1.5-1.75, Al 1.5-1.75, bal Fe.
Annealed: 45,000-56,000 TS; 250 Brin. For cylinder liners, general parts to resist severe abrasion; nitrogen hardened alloy cast iron.

CENTREX HT

Ludlow Steel Corp.
C, alloy, bal Fe.
Pre-heat-treated to 150,000 TS; 115,000 YS; 24 El; 302 Brin. For gears, bolts, shafts; abrasion resistant.

CENTREX P & G

Ludlow Steel Corp.
C, alloy, bal Fe.
Pre-heat-treated to 155,000 TS; 135,000 YS; 24 El; 302 Brin. For gears, bolts, shafts; shock and fatigue resistant.

CENTREX PBD

Ludlow Steel Corp.
C, alloy, bal Fe.
Heat treated: 248-293 Brin. For brake dies. Resists severe wear and high impact forces.

CENTREX PRESS BRAKE

Now CENTREX-PBD.

CENTREX RHINO-CHEK

Ludlow Steel Corp.
C, alloy, bal Fe.
Heat treated: 370 Brin; 175,000 TS; 150,000 YS. For fan blades; lining chutes. Wear and corrosion resistant.

CENTREX RHINO-TUF

Ludlow Steel Corp.
C, alloy, bal Fe.
Heat treated: 175,000 TS; 150,000 YS; 360-400 Brin (can be furnished harder). For anchors, coal chutes, conveyers, furnace liners, grader blades, mixers. Special heat treated wear and corrosion resistant plate.

CENTRICAST MARK 10-HI-TEN

Sheepbridge Engineering Ltd.
C 3.3, Si 2.1, P 0.1, Ni 1.6, Al 0.75, bal Fe.
Cast: 63,000-90,000 TS; 10-3 El. For rollers, shafts; nitriding cast iron.

CENTRICAST MARK 10-HI-TEN

Sheepbridge Alloy Castings Ltd.
C 3.3, Si 2.1, P 0.1, Ni 1.6, Al 0.75, bal Fe.
Cast: 63,000-90,000 TS; 10-3 El. For rollers, shafts; nitriding cast iron.

CENTRICAST MARK 11

Sheepbridge Engineering Ltd.
C 3.2, Si 2.3, Mn 0.7, P 0.1, Cr 0.2, Ni 0.6, Mo 0.15, bal Fe.
Cast: 52,000 TS; 207-255 Brin. For cylinder liners; cast iron, wear resistant.

CENTRICAST MARK 11

Sheepbridge Alloy Castings Ltd.
C 3.2, Si 2.3, Mn 0.7, P 0.1, Cr 0.2, Ni 0.6, Mo 0.15, bal Fe.
Cast: 52,000 TS; 207-255 Brin. For cylinder liners; cast iron, wear resistant.

CENTRICAST MARK 12-4K6

Sheepbridge Engineering Ltd.
C 3.4, Si 2.3, P 0.6, Cr 0.3, bal Fe.
Cast: 36,000 TS; 229-293 Brin. For automobile piston rings; cast iron, wear resistant.

CENTRICAST MARK 12-4K6

Sheepbridge Alloy Castings Ltd.
C 3.4, Si 2.3, P 0.6, Cr 0.3, bal Fe.
Cast: 36,000 TS; 229-293 Brin. For automobile piston rings; cast iron, wear resistant.

CENTRICAST MARK 14

Sheepbridge Engineering Ltd.
C 3.2, Si 2.4, P 0.6, Cr 0.4, Mo 0.6, bal Fe.
Cast: 45,000 TS; 255-293 Brin. For aircraft piston rings; cast iron, wear resistant.

CENTRICAST MARK 14

Sheepbridge Alloy Castings Ltd.
C 3.2, Si 2.4, P 0.6, Cr 0.4, Mo 0.6, bal Fe.
Cast: 45,000 TS; 255-293 Brin. For aircraft piston rings; cast iron, wear resistant.

CENTRICAST MARK 15

Sheepbridge Engineering Ltd.
C 2.9, Si 2.4, P 0.4, Cr 0.9, Mo 0.9, bal Fe.
Cast: 58,000 TS; 269-302 Brin. For aircraft compression and piston rings; cast iron, wear resistant.

CENTRICAST MARK 15

Sheepbridge Alloy Castings Ltd.
C 2.9, Si 2.4, P 0.4, Cr 0.9, Mo 0.9, bal Fe.
Cast: 58,000 TS; 269-302 Brin. For aircraft compression and piston rings; cast iron, wear resistant.

CENTRICAST MARK 16

Sheepbridge Engineering Ltd.
C 1.9, Si 1.9, P 0.06, Cr 16, bal Fe.
Cast: 78,000 TS; 269-321 Brin. For piston rings, cylinder liners; heat and abrasion resistant.

CENTRICAST MARK 16

Sheepbridge Alloy Castings Ltd.
C 1.9, Si 1.9, P 0.06, Cr 16, bal Fe.
Cast: 78,000 TS; 269-321 Brin. For piston rings, cylinder liners; heat and abrasion resistant.

CENTRICAST MARK 17

Sheepbridge Engineering Ltd.
C 1.2, Si 2, Cr 33, bal Fe.
Cast: 78,000 TS; 285-341 Brin. For liners and pump cylinders; heat and abrasion resistant.

CENTRICAST MARK 17

Sheepbridge Alloy Castings Ltd.
C 1.2, Si 2, Cr 33, bal Fe.
Cast: 78,000 TS; 285-341 Brin. For liners and pump cylinders; heat and abrasion resistant.

CENTRICAST MARK 18

Sheepbridge Engineering Ltd.
C 2.7, Si 2, Ni 14, Cr 1.75, Cu 7, bal Fe.
Cast: 31,500 TS; 175-235 Brin. For cylinder liners; austenitic, cast iron, corrosion resistant.

CENTRICAST MARK 18

Sheepbridge Alloy Castings Ltd.
C 2.7, Si 2, Ni 14, Cr 1.75, Cu 7, bal Fe.
Cast: 31,500 TS; 175-235 Brin. For cylinder liners; austenitic, cast iron, corrosion resistant.

CENTRICAST MARK 2

Sheepbridge Engineering Ltd.
C 3.3, Si 2.5, P 0.6, Cr 0.3, bal Fe.
Cast: 36,000 TS; 241-302 Brin. For cylinder liners, general castings; cast iron.

CENTRICAST MARK 2

Sheepbridge Alloy Castings Ltd.
C 3.3, Si 2.5, P 0.6, Cr 0.3, bal Fe.
Cast: 36,000 TS; 241-302 Brin. For cylinder liners, general castings; cast iron.

CENTRICAST MARK 20

Sheepbridge Engineering Ltd.
C 3, Si 2.5, P 0.1, Cr 1.3, bal Fe.
Cast: 58,000 TS; 269-311 Brin. For valve seats; heat resistant, cast iron.

CENTRICAST MARK 20

Sheepbridge Alloy Castings Ltd.
C 3, Si 2.5, P 0.1, Cr 1.3, bal Fe.
Cast: 58,000 TS; 269-311 Brin. For valve seats; heat resistant, cast iron.

CENTRICAST MARK 21

Sheepbridge Engineering Ltd.
C 2.8, Si 2.7, P 0.2, Cr 1.5, Al 1, bal Fe.
Cast: 54,000 TS; 269-311 Brin. For cylinder liners; nitriding cast iron.

CENTRICAST MARK 21

Sheepbridge Alloy Castings Ltd.
C 2.8, Si 2.7, P 0.2, Cr 1.5, Al 1, bal Fe.
Cast: 54,000 TS; 269-311 Brin. For cylinder liners; nitriding cast iron.

CENTRICAST MARK 23

Sheepbridge Engineering Ltd.
TC 3.3, Si 2.1, Mn 0.3, P 0.1, Ni 1.6, bal Fe.
Normalized: 90,000 TS; 260 Brin. Annealed: 60,000 TS; 223 Brin. For rollers, cylinder liners; spheroidal cast iron.

CENTRICAST MARK 23

Sheepbridge Alloy Castings Ltd.
TC 3.3, Si 2.1, Mn 0.3, P 0.1, Ni 1.6, bal Fe.
Normalized: 90,000 TS; 260 Brin. Annealed: 60,000 TS; 223 Brin. For rollers, cylinder liners; spheroidal cast iron.

CENTRICAST MARK 24-LODED

Sheepbridge Alloy Castings Ltd.
C 3.2, Si 2.9, Mn 0.8, P 0.7, Cr 0.7, bal Fe.
Cast iron, harder and more wear resistant than ordinary gray iron. As cast: 241-302 Brin. Liners for internal combustion engines.

CENTRICAST MARK 27

Dewramet Ltd.
C 3.3, Si 2.2, Mn 0.8, P 0.1, Cr 0.15, V 0.25, bal Fe.
As cast: 212-293 Brin. For cylinder liners. *Obsolete*

CENTRICAST MARK 28

Sheepbridge Alloy Castings Ltd.
C 3.2, Si 1.8, Mn 0.7, P 0.25, Cu 0.8, V 0.4, Ti 0.04, bal Fe.
As cast: 210-262 Brin. For large diesel engine liners, particularly in the marine industry.

CENTRICAST MARK 29

Sheepbridge Alloy Castings Ltd.
C 3.2, Si 2.5, Mn 0.8, P 0.5, Cr 0.5, bal Fe.
As cast: 229-302 Brin. General purpose cast iron for motor vehicle industry.

CENTRICAST MARK 3

Sheepbridge Engineering Ltd.
C 3.2, Si 1.9, P 0.2, Ni 1.4, Cr 0.4, bal Fe.
Cast: 40,000 TS; 217-285 Brin. For gears, cylinder liners; heat treated castings.

CENTRICAST MARK 3

Sheepbridge Alloy Castings Ltd.
C 3.2, Si 1.9, P 0.2, Ni 1.4, Cr 0.4, bal Fe.
Cast: 40,000 TS; 217-285 Brin. For gears, cylinder liners; heat treated castings.

CENTRICAST MARK 4

Sheepbridge Engineering Ltd.
C 3.2, Si 2.1, Mn 0.8, P 0.2, Ni 2, bal Fe.
Cast: 38,000 TS; 195-255 Brin. For gears, cylinder liners; alloy cast iron.

CENTRICAST MARK 4

Sheepbridge Alloy Castings Ltd.
C 3.2, Si 2.1, Mn 0.8, P 0.2, Ni 2, bal Fe.
Cast: 38,000 TS; 195-255 Brin. For gears, cylinder liners; alloy cast iron.

CENTRICAST MARK 5

Sheepbridge Engineering Ltd.
C 3.3, Si 1.8, P 0.2, bal Fe.
Cast: 36,000 TS; 179-241 Brin. For piston rings, liners; cast iron.

CENTRICAST MARK 5

Sheepbridge Alloy Castings Ltd.
C 3.3, Si 1.8, P 0.2, bal Fe.
Cast: 36,000 TS; 179-241 Brin. For piston rings, liners; cast iron.

CENTRICAST MARK 9

Sheepbridge Engineering Ltd.
C 3.1, Si 2, P 0.4, Cr 0.7, bal Fe.
Cast: 58,000 TS; 269-302 Brin. For piston rings, automobile castings; cast iron.

CENTRICAST MARK 9

Sheepbridge Alloy Castings Ltd.
C 3.1, Si 2, P 0.4, Cr 0.7, bal Fe.
Cast: 58,000 TS; 269-302 Brin. For piston rings, automobile castings; cast iron.

CEOT

Saarstahl AG
Now SAARSTAHL 1.2323X2.

CERALLOY 400

Ronson Metals Corp.
Th 80, bal Ce + Al.
For vacuum tube getter; continuous getter.

CERALLOY 420

Ronson Metals Corp.
Fe 20, Mg 2, bal Mischmetal.
For vacuum tube getter, anode in voltage regular gas tubes; flash getter.

CERALLOY 75

Ronson Metals Corp.
Fe 25, bal Mischmetal.
For additives for ductile cast iron; desulfurizer and graphite nodularizer.

CERALLOY 75M2

Ronson Metals Corp.
Fe 23, Mg 2, bal Mischmetal.
For additives for ductile cast iron; desulfurizer and graphite nodularizer.

CERALLOY 80A

Ronson Metals Corp.
Al 20, bal Mischmetal.
For steel deoxidizer and desulfurizer, scavenger.

CERALLOY 90M

Ronson Metals Corp.
Mg 10, bal Mischmetal.
For additives for ductile cast iron; desulfurizer and graphite nodularizer.

CERALLOY MISCHMETAL 100X

Ronson Metals Corp.
99.9 rare earth metals, typically, 50 Ce + 27 La + 16 Nd + 5 Pr, bal other rare earth metals.
Additive for non-ferrous alloys.

CERALLOY MISCHMETAL 95 M

Ronson Metals Corp.
Mg 5, bal mixed rare earth metals.
Ignition alloy.

CERALLOY MISCHMETAL FG

Ronson Metals Corp.
Fe 4, bal mixed rare earth metals.
Additive for iron and steel; desulfurizer and graphite nodularizer.

CERALLOY MISCHMETAL M

Ronson Metals Corp.
Mg 2.5, bal mixed rare earth metals.
Additive for magnesium alloys.

CERALUMIN "ASM"

Stone Manganese - J. Stone & Co. Ltd.
Aluminum. Cu 1-2, Ni 0.25-1, Fe 0-0.5, Si 0.7-1.3, Mg 0.4-0.8, Cb 0-0.03, bal Al.
Sand cast: 38,000 TS; 4-7 El; 95 Brin. Chill cast: 47,000 TS; 14-21 El; 95 Brin. For castings: good corrosion resistance.

CERALUMIN 21

Stone Manganese - J. Stone & Co. Ltd.
Aluminum. Cu 3.5-4.5, Co 0.5-1, Mg 1.2-2.5, bal Al.
Cast: 54,000-58,000 TS; 20-22 El. For sand and die castings for elevated temperature service. High strength and hardness at elevated temperatures.

CERALUMIN 22

Stone Manganese - J. Stone & Co. Ltd.
Aluminum. Si, Cu, Ce, bal Al.
For light alloy parts; forgings.

CERALUMIN A

Stone Manganese - J. Stone & Co. Ltd.
Aluminum. Si 1.25, Cu 2.5, Mg 0.8, Fe 1.2, Ni 1.5, Ce 0.15, bal Al.
Heat treated: 51,000-60,000 TS; 45,000-54,000 YS; 1 El; 130-140 Brin. For light alloy parts; heat treatable.

CERALUMIN B

Stone Manganese - J. Stone & Co. Ltd.
Aluminum. Cu 1.5, Ni 1.5, Mg 0.2, Fe 0.7, Si 1.5, Ce 0.05-0.3, bal Al.
Sand cast: 26,000 TS; 20,000 YS; 2-3 El; 80 Brin. Permanent mold: 31,000 TS; 20,000 YS; 4-9 El; 82 Brin. For pistons, cylinder heads, gear boxes, impellers; good castability.

CERALUMIN C

Stone Manganese - J. Stone & Co. Ltd.
Aluminum. Cu 2-3, Ni 1-2, Fe 1-1.5, Si 1-1.4, Ce 0.05-0.2, Mg 0.5-1, Cb 0.05-0.3, bal Al.
Aged: 63,000 TS; 55,000 YS; 9 El; 145 Brin. Permanent mold: 59,000 TS; 49,000 YS; 1-4 El; 135 Brin. For castings for high temperature use; age hardenable.

CERALUMIN D

Stone Manganese - J. Stone & Co. Ltd.
Aluminum. Cu 2-3, Ni 1-2, Mg 0.5-1, Si 1.2, Fe 1.2, Ce 0.05-0.2, bal Al.
Heat treated: 28,000-40,000 TS; 25,000 YS; 1-5 El; 90-100 Brin. For cylinder heads, general castings; age hardenable.

CERALUMIN F

Stone Manganese - J. Stone & Co. Ltd.
Aluminum. Cu 1.3-30, Ni 1.5-2, Ce 0.05-0.3, Mg 0.5-1.25, Cr 0.03, Fe 1, Si 0.9, bal Al.
For light alloy castings; heat treatable.

CERAMAGNET A

Stackpole Magnet Division
Permanent magnet. Residual flux density: 1900-2300 Gauss. Coercive force: 1580-1860 Oersted.

CERAMAGNET A20 SEGMENTS

Stackpole Magnet Division
Permanent magnet. Residual flux density: 2850 Gauss. Coercive force: 2400 Oersted.

CERAMAGNET A20 SLEEVES

Stackpole Magnet Division
Permanent magnet. Residual flux density: 2575 Gauss. Coercive force: 2130 Oersted.

CERAMISEAL

Wilbur B. Driver Co.
Fe 48, Ni 27, bal Co.
For ceramic-to-metal seals; high temperature applications; M.P. 2590 F. *Obsolete*

CERAMVAR

Wilbur B. Driver Co.
Superalloy. Fe 48, Ni 27, Co 25.
For ceramic to metal seals. Expansion characteristics match aluminum.

CERIUM

Atomergic Chemetals Corp.
Ce, metallothermic and electrolytic.
Purities: 99.9%, 99.6%. Forms: ingot, turnings, powder, wire, sheet, rod, lump, foil.

CERIUM METAL

Ronson Metals Corp.
Comparatively pure Ce; for alloying.

CERMETI-10

Dynamet Technology, Inc.
Titanium. 10% TiC, 90% Ti-6Al-4V.
Titanium alloy matrix containing TiC particulate. Various grades indicate percentage of TiC ("-10" indicates 10%, "-20" indicates 20%, etc.). Increase in TiC% indicates increase in modulus at less ductility and improved elevated temperature strength. At 10%: 125 ksi TS; 120 ksi YS; 1-2 El; E = 20 x 10^6 psi.

CERRO 9500-1

Cerro Metal Products Co.
Bi 95, Sn 5.
For solder. Soft.

CERROBASE

Cerro Metal Products Co.
Bi 5-58, Pb 1-51.5, Sn 27-60, Cd 0-10.
6400 psi TS; 60-70 El (2 in. slow loading); 10.2 Brin. Series of fusible alloys of varying composition.

CERROBASE

Cerro Metal Products Co.
Bi 55.5, Pb 44.5.
Cast: 6200 TS; 4000 YS; 64 El; 10 Brin. For molds, tube filler, proof casting forging dies; M.P. 123.5°C; formerly "Basaloy."

CERROBEND

Cerro Metal Products Co.
Bi 50, Pb 26.7, Sn 13.3, Cd 10.
Cast: 5990 TS; 200 El; 9.2 Brin. For filler for bending tube, low melting solder and sealing glass to metal. 158°F MP. Formerly "Bendalloy."

CERROCAST

Cerro Metal Products Co.
Bi 40, Sn 60.
Cast: 8000 TS; 200 El; 22 Brin. For wax pattern molds; 281-338°F MP.

CERRODENT

Cerro Metal Products Co.
Bi 38.14, Pb 26.42, Sn 31.67, Cd 2.64, Cu 0.06, Sb 1.07.
For dental models. 75-118°C MP.

CERROLOW 105

Cerro Metal Products Co.
Sn 7.97, Bi 42.91, Pb 21.7, Cd 5.09, In 18.33, Hg 4.
For fusible alloy. 100-110°F MP. *Obsolete*

CERROLOW 117

Cerro Metal Products Co.
Bi 44.7, Pb 22.6, Sn 8.3, Cd 5.3, In 19.1.
Cast: 5400 TS; 1.5 El; 9-12 Brin. For low melting alloys, solders, and fuses. Eutectic alloy; 117°F MP.

CERROLOW 117 B

Cerro Metal Products Co.
Sn 11.3, Bi 44.7, Pb 22.6, Cd 5.3, In 16.1.
For fusible alloy. 117-120°F MP.

CERROLOW 136

Cerro Metal Products Co.
Bi 49, Pb 18, Sn 12, In 21.
Cast: 6300 TS; 50 El; 14 Brin. For low melting alloys, solder, and fuses. 136°F MP; eutectic alloy.

CERROLOW 136 B

Cerro Metal Products Co.
Sn 15, Bi 49, Pb 18, In 18.
For fusible alloy. 136-156°F MP.

CERROLOW 140

Cerro Metal Products Co.
Sn 12.6, Bi 47.5, Pb 25.4, Cd 9.5, In 5.
For fusible alloy. 134-149°F MP.

CERROLOW 147

Cerro Metal Products Co.
Sn 12.77, Bi 48, Pb 25.63, Cd 9.6, In 4.
Cast: 4950 TS; 13.5 El; 11 Brin. For thermal safety controls. 142-149°F MP.

CERROLOW 174

Cerro Metal Products Co.
Bi 57, Sn 17, In 26.
For safety devices, fusible elements, and solder. 174°F MP.

CERROMATRIX

Cerro Metal Products Co.
Bi 48, Sn 14.5, Pb 28.5, Sb 9.
Cast: 13,000 TS; 1 El; 19 Brin. For dental alloys and fusible alloy for die mounting. Expands on cooling; 218-440 F MP.

CERROSAFE

Cerro Metal Products Co.
Bi 42.5, Pb 37.7, Sn 11.3, Cd 8.5.
Cast: 5400 TS; 200 El; 9 Brin. For toy casting sets. 158-194°F MP. Formerly "Saffalloy".

CERROSEAL 35
Cerro Metal Products Co.
Sn 50, In 50.
For soldering glass to metal. 250-260°F MP; useful in vacuum and low vapor pressure.

CERROTRU
Cerro Metal Products Co.
Bi 58, Sn 42.
Cast: 8000 TS; 200 El; 22 Brin. For castings, molds, and patterns. Zero volume change from molten to solid state; 280°F MP.

CERTANIUM 250
Certanium Alloys & Research Co.
For hard surfacing. Resists high stress abrasion. Hard and tough electrode for AC and DC welding. Composite of an austenitic matrix plus TiC and CrC and silicides.

CERTIFIED OXYGEN FREE
Criterion Metals, Inc.
Copper. Cu 99.99.
Thin gauge sheet, various tempers: 23-52 ksi TS min; 5-51 ksi YS min. C10100

CESCO DIAMOND
Crucible Electric Steel Co.
C 0.5, Mn 0.85-1, Si 2.2, Mo 0.5, V 0.25-0.35, bal Fe.
Heat treated: 231,600-320,500 TS; 185,700-259,400 YS; 2-5 El; 5-27 RA; 659 Brin. For hollow and solid drills, concrete busters, hammer pistons, springs; shock resistant.

CESCO DIAMOND SPECIAL
Crucible Electric Steel Co.
C 0.5-0.65, Si 2, Mn 1, Mo 0.6-0.7, V 0.34-0.36, bal Fe.
Heat treated: 350,000 TS; 290,000 YS; 6 El; 11 RA; 615 Brin.
For chisels, concrete busters, drills, caulking and beading tools; shock and abrasion resistant.

CESCO SPECIAL
Crucible Electric Steel Co.
C 0.55, Mn 1, Mo 0.4, V 0.35, Si 2, bal Fe.
For tools; shock resistant.

CESIUM
Atomergic Chemetals Corp.
Cs.
Purities: 99.99%, 99.9+%, 99.5%. Packaging: glass ampules, steel containers (under vacuum, inert gas, oil).

CETAL
Metallgesellschaft Reuterweg
Si 6.5, Cu 3, Zn 10, bal Al.
Cast: 26,000-31,000 TS; 0.5-3.0 El; 65-90 Brin. For light alloy parts; non-hardenable.

CETO
N.V. Philips Co.
Th 80, Al 15, 5.0 Mischmetal.
For getters. Absorbs gases.

CF-43
American manufacture
C 0.5, Cr 25, Ni 10, W 7.5, Fe 1.5, bal Co.
For marine gas turbines, jet engine components. Similar to X-40.

CF10MC
Cooper Alloy Corp.
Stainless steel. C 0.1, Si 1.5, Mn 1.5, Cr 15-18, Ni 13-16, Mo 1.75-2.25, Cb = 10 x C min (1.20 max), bal Fe.
Cast: 483 MPa TS; 207 MPa YS; 20 El. Heat resistant alloy; ASTM A-351 CF10MC.

CG 27
Now CRUCIBLE CG27.

CG ALLOY
Cyprus Foote Mineral Co.
Mg 4-5, Ti 8.5-10.5, Ce 0.2-0.35, Ca 4-5.5, Al 1-1.5, Si 48-52, bal Fe.
Ferroalloy for production of compacted graphite iron. *Obsolete*

CH750
Eagle & Globe Steel Ltd.
Alloy steel. C 0.2, Mn 1.6, bal Fe.
For automotive components, gears and splines, and small case hardened components requiring core tensile strength of approximately 900 MPa TS combined with impact toughness.

CHACE NO. 1050
W.M. Chace Co.
Ni 36, Cr 3, bal Fe/100 Cu/22 Ni, bal Fe.
Thermostatic bimetal; low electrical resistivity.

CHACE NO. 2400
W.M. Chace Co.
Ni 36, Cr 3, bal Fe/22 Ni, bal Fe.
Thermostatic bimetal; range 0-300°F.

CHACE NO. 2500
W.M. Chace Co.
Ni 50, Cr 3, bal Fe/25 Ni, bal Fe.
Thermostatic bimetal; maximum temperature 1000°F.

CHACE NO. 3700
W.M. Chace Co.
Ni 40, Cr 3, bal Fe/22 Ni, bal Fe.
For thermostatic bimetal; range 100-550°F.

CHACE NO. 4700
W.M. Chace Co.
Ni 38, Cr 7, bal Fe/19 Ni, 7.0 Cr, bal Fe.
For thermostatic bimetal; corrosion resistant.

CHACE NO. 6125
W.M. Chace Co.
Ni 36, Cr 3, bal Fe/100 Ni/22 Ni, bal Fe.
For thermostatic bimetal; range 0-300°F.

CHACE NO. 6150
W.M. Chace Co.
Ni 36, Cr 3, bal Fe/100 Ni/22 Ni, bal Fe.
Thermostatic bimetal; range 0-300°F.

CHACE NO. 6650
W.M. Chace Co.
Ni 36, bal Fe/18 Cu, bal Mn.
For thermostatic bimetal; high deflection.

CHACE NO. 6850
W.M. Chace Co.
Ni 36, bal Fe/18 Cu, bal Mn.
For thermostatic bimetal; high electrical resistivity.

CHACE NO. 772
W.M. Chace Co.
Cu 18, Ni 10, bal Mn.
Treatment (50% Reduction): 105,000 TS; 90,000 YS; 8.5 El; 220 Brin. High expansion, high resistivity. Maximum temperature: 600°F.

CHAIN BRONZE
American manufacture
Sn 4.9, P 0.1, bal Cu.
Rolled: 28,000-69,000 TS; 15,000-40,000 YS; 30-6 El. For chains, springs, diaphragms; high strength, corrosion resistant.

CHAIN IRON
English manufacture
Mn 0-0.1, Si, bal Fe.
For chains for cranes, slings, hoists, steam shovels, marine uses; wrought iron.

CHAMAX
Acieries de Champagnole
C 0.82, Cr 4.5, W 18, Mo 0.8, V 1.75, Co 5, bal Fe.
Air or oil hardens to 63-66 Rock C. For lathe tools, form tools, gear cutters, counter bores, and planing tools. AFNOR: Z 85 WK 18.05. AISI T4. Germany: S 18.1.2.5 (E 18 Co 5); W. Nr. 1.3255.

CHAMAX 00
Acieries de Champagnole
Cr 0.95-4.5, W 18, Mo 0.7, V 1.75, Co 5, bal Fe.
Air or oil hardens to 62-67.5 Rock C. For lathe tools, milling cutters, and gear cutters. AFNOR: Z 95 WK 18.05.

CHAMAX 11
Acieries de Champagnole
C 0.82, Cr 4.5, W 18, Mo 0.8, V 1.75, Co 11, bal Fe.
Air or oil hardens to 63-67 Rock C. For lathe tools and drills for special jobs requiring high speeds, high temperature, special alloys. AFNOR: Z 85 WK 18.10. AISI T5 or T6. Germanay: S 18.1.2.10 (E 18 Co 10); W. Nr. 1.3265.

CHAMAX. CV
Acieries de Champagnole
C 1.55, Cr 4.3, W 12.5, Mo 0.8, V 5, Co 5, bal Fe.
Air or oil hardens to 62-68 Rock C. For tools for machining stainless steels and punches for silicon sheet. AFNOR: Z 150 WKV 12.05.05. AISI T15. Germany: S 12.1.1.5 (EV 4 Co); W. Nr. 1.3202.

CHAMAX. CV. 11
Acieries de Champagnole
C 1.65, Cr 4.4, W 12, Mo 0.8, V 5, Co 10, bal Fe.
Air or oil hardens to 62-69 Rock C. For lathe tools for finish cuts at high speeds and machining special materials.
AFNOR: Z 160 WKV 12.10.05.

CHAMET BRONZE B
Chase Brass & Copper Co.
Cu 62.25, Zn 37.1, Sn 0.65.
Drawn: 62,000 TS; 35,000 YS; 35 El. For rivets, screws, bolts; corrosion resisting. *Obsolete*

CHAMET BRONZE, TYPE A
Chase Brass & Copper Co.
Cu 60, Zn 39.25, Sn 0.75.
Annealed: 63,000 TS; 30,000 YS; 40 El. Hard: 75,000 TS; 53,000 YS; 20 El. For welding rod, bolts, screws; resists sea water. *Obsolete*

CHAMET LEADED BRONZE
American manufacture
Zn 38.5, Pb 0.75, Sn 0.75, bal Cu.
For fittings, pipes, hardware; free-cutting, high strength.

CHAMFERTRODE
Eutectic Corp.

For chamfering and grooving electrodes; (AC-DC) on ferrous or non-ferrous.

CHAMPAGNOLE C.8.V.M.
Acieries de Champagnole
C 0.8, Cr 0.5, Mn 0.4, Si 0.4, V 0.25, Mo 0.25, bal Fe.
Water hardening tool steel; for engraving tools, stamps, anvils, shock resisting hand and pneumatic tools. AFNOR: 80 CDV 02; AISI W 2. *Obsolete*

CHAMPAGNOLE C10
Acieries de Champagnole
C 1.1, bal Fe.
For tools, taps; water hardened. *Obsolete*

CHAMPAGNOLE C12
Acieries de Champagnole
C 1.2, bal Fe.
For tools, dies, drills; water hardened. *Obsolete*

CHAMPAGNOLE C7
Acieries de Champagnole
C 0.75, bal Fe.
For tools, springs, punches; water hardened. *Obsolete*

CHAMPAGNOLE C8
Acieries de Champagnole
C 0.75, bal Fe.
For tools, springs, punches; water hardened. *Obsolete*

CHAMPAGNOLE C9
Acieries de Champagnole
C 0.95, bal Fe.
For tools, springs, cutters, drills; water hardened. *Obsolete*

CHAMPAGNOLE CR.3
Acieries de Champagnole
C 1.35, Cr 4, Mn 0.5, Si 0.25, Mo 0.4, bal Fe.
Oil hardening tool steel; for files, shears, plane blades, lock laminations. AFNOR: 130 C 16. *Obsolete*

CHAMPAGNOLE CRED.
Acieries de Champagnole
C 2.25, Cr 13, bal Fe.
For tools, dies, punches; non-deforming. *Obsolete*

CHAMPAGNOLE T.E.D.C. STEEL
Acieries de Champagnole
C 0.8, Cr 4.2, W 12.5, Mo 1.8, V 1.8, Co 1.2, bal Fe.
Air or oil harden to 63-66 Rc. For twist drills, reamers, taps, cutters. AFNOR: Z 80 WDVK 12.02.02. *Obsolete*

CHAMPAGNOLE T1
Acieries de Champagnole
C 0.5, Cr 1.5, W 2.2, V 0.25, bal Fe.
For hot work tools and dies; hot work steel. *Obsolete*

CHAMPAGNOLE T5
Acieries de Champagnole
C 1.3, Cr 4.5, bal Fe.
For tools, dies; air or oil hardened. *Obsolete*

CHAMPALOY
Crucible Materials Corp.
Tool material. C 0.75, Ni 1.5, Cr 0.75, Mo 0.3, bal Fe.
Heat treated: 208,000 TS; 180,000 YS; 450 Brin. For tools, collets, jigs, shear blades, rolls; oil hardening. AISI L6.

CHAMPALOY NO. 2
Colt Industries
C 0.65, Cr 0.75, Mo 0.25, bal Fe.
For tools; oil hardening. *Obsolete*

CHAMPION 2
Champion Rivet Co.
C 0-0.1, Cr 2, Mo 0.5, bal Fe.
Welded: 70,000-75,000 TS; 35-30 El. For welding rods for low alloy steels; subject to air hardening.

CHAMPION 255
Champion Steel Co.
C 0.6, Mn 0.8, Si 2, Cr 0.2, Mo 0.2, V 0.2, bal Fe.
Annealed: 107,000 TS; 64,000 YP; 27 El; 212 Brin. Oil hardened: 145,000-340,000 TS; 127,000-283,000 YP; 5-24 El; 20-44 RA; 293-611 Brin. For shear blades, pneumatic tools, punches, caulking tools. Type S5 shock resisting tool steel. Oil or water hardening.

CHAMPION 308
Champion Rivet Co.
C 0-0.07, Cr 18.5-20, Ni 10.5-11.25, bal Fe.
Welded: 85,000-95,000 TS; 50-40 El. For welding rods for stainless steel; stainless, austenitic; Type 308.

CHAMPION 309
Champion Rivet Co.
C 0-0.1, Cr 22.5-24, Ni 13-14, bal Fe.
Welded: 85,000-95,000 TS; 35-45 El. For welding rods for stainless steel; resists oxidation to 2000°F; Type 309.

CHAMPION 309 CB
Champion Rivet Co.
C 0-0.1, Cr 22.5-24, Ni 13-14, Cb 0.8-1, bal Fe.
Welded: 85,000-95,000 TS; 30-40 El. For welding rods for stainless steel; prevents carbide precipitation; austenitic.

CHAMPION 310
Champion Rivet Co.
C 0.09-0.17, Cr 25-26.5, Ni 20-21, bal Fe.
Welded: 85,000-95,000 TS; 35-45 El. For welding rods for 5% Cr steel; will not air harden; austenitic.

CHAMPION 310 CB
Champion Rivet Co.
C 0-0.1, Cr 25-26.5, Ni 20-21, Cb 0.8-1, bal Fe.
Welded: 85,000-95,000 TS; 30-40 El. For welding rods for stainless steel; resists carbide precipitation; stabilized.

CHAMPION 310 MO
Champion Rivet Co.
C 0-0.1, Cr 25-26.5, Ni 20-21, Mo 2-2.5, bal Fe.
Welded: 85,000-95,000 TS; 35-45 El. For welding rods for stainless steel; acid resistant, stainless, austenitic.

CHAMPION 316
Champion Rivet Co.
C 0-0.07, Cr 17.5-19, Ni 12.5-13.5, Mo 2-2.5, bal Fe.
Welded: 85,000-95,000 TS; 40-50 El. For welding rods for stainless steel; acid resistant, austenitic.

CHAMPION 316 CB
Champion Rivet Co.
C 0-0.07, Cr 17.5-19, Ni 12.5-13.5, Mo 2-2.5, Cb 0.5-0.75, bal Fe.
Welded: 85,000-95,000 TS; 35-45 El. For welding rods for stainless steel; stabilized, stainless.

CHAMPION 317
Champion Rivet Co.
C 0-0.07, Cr 17.5-19, Ni 12.5-13.5, Mo 3.5-4, bal Fe.
Welded: 85,000-95,000 TS; 40-50 El. For welding rods for stainless steel; stainless, austenitic.

CHAMPION 330
Champion Rivet Co.
C 0-0.15, Cr 14.5-16, Ni 34-35, bal Fe.
Welded: 65,000-70,000 TS; 25-30 El. For welding rods for furnace construction; heat resistant.

CHAMPION 347
Champion Rivet Co.
C 0-0.07, Cr 18.5-20, Ni 10.5-11.25, Cb 0.75-1, bal Fe.
Welded: 85,000-95,000 TS; 35-45 El. For welding rods for stainless steel; stabilizer, stainless; Type 347.

CHAMPION 410
Champion Rivet Co.
C 0-0.1, Cr 12-13.5, Mo 0.4-0.65, bal Fe.
Welded: 85,000-95,000 TS; 30-35 El. For welding rods for oil refinery vessels; corrosion and heat resistant.

CHAMPION 430
Champion Rivet Co.
C 0-0.1, Cr 15.25-16.75, bal Fe.
Welded: 75,000-80,000 TS; 30-35 El. For welding rods to resist HNO$_3$; corrosion resistant.

CHAMPION 442
Champion Rivet Co.
C 0-0.1, Cr 17.5-19, bal Fe.
For welding rods for high temperature applications; corrosion and heat resistant.

CHAMPION 446
Champion Rivet Co.
C 0-0.13, Cr 27-28.5, bal Fe.
For welding rods for high temperature applications; resists oxidation at high temperature.

CHAMPION 502
Champion Rivet Co.
C 0-0.1, Cr 4.5-6, Mo 0.4-0.65, bal Fe.
For welding rods for oil refinery equipment; corrosion resistant.

CHAMPION 9
Champion Rivet Co.
C 0-0.1, Cr 9, Mo 0.5, bal Fe.
Welded: 90,000-95,000 TS. For welding rods for low Cr alloys; high creep resistance.

CHAMPION C-30
Champion Rivet Co.
C 0.22-0.26, Mn 0.45, Mo 0.4-0.6, V 0.1-0.15, bal Fe.
Welded: 215 Brin. For welding electrodes; for worn surfaces, wear resistant.

CHAMPION DOUBLE SPECIAL
American manufacture
C 1.22, Mn 0.25, W 7.53, bal Fe.
For tools, dies; oil hardening.

CHAMPION EXTRA
Crucible Materials Corp.
Tool material. C 1.2, W 1.25, Cr 0.3, bal Fe.
For cutters, broaches, drills; oil hardening.

CHAMPION GRAY DEVIL NO. 2
Champion Rivet Co.
C 0.09, Mn 0.35, bal Fe.
Welded: 78,000 TS; 67,000 YS; 24 El; 45 RA. For welding rod; flux coated, E 6012.

CHAMPION HW1
Champion Steel Co.
C 0.35, Cr 5, V 0.4, W 1.5, Mo 1.5, bal Fe.
Hot work tool steel, air or oil hardening, for forging and hot forming dies. AISI H12.

CHAMPION HW21
Champion Steel Co.
C 0.35, Cr 3.5, W 9, bal Fe.
Hot work tool steel, air or oil hardening, tungsten type; for hot forming operations. AISI H21.

CHAMPION HW3
Champion Steel Co.
C 0.35, Si 1, Cr 5, V 1, Mo 1.5, bal Fe.
Annealed: 98,000 TS; 74,000 YS; 28 El; 210 Brin. Hardened: 135,000-290,000 TS; 100,000-228,000 YS; 3-16 El; 7-48 RA; 27-55 Rock C. For forging and heading dies, compression tools, die casting dies, piercing and forming punches, bolt and swaging dies. Type H13 hot work steel, red-tough, shock and impact resistant.

CHAMPION NO. 48-1
Champion Rivet Co.
C, alloy, bal Fe.
550 Brin. For hard surfacing electrodes; for high abrasion, little shock or impact. *Obsolete*

CHAMPION NO. 58-12
Champion Rivet Co.
C, alloy, bal Fe.
500 Brin. For hard surfacing electrodes; for high abrasion and impact. *Obsolete*

CHAMPION NO. 59
Champion Rivet Co.
C 3, Cr 4, Mo 4, bal Fe.
550 Brin. For hard surfacing electrodes; for high carbon and manganese steel, abrasion resistant.

CHAMPION NO. 64-6
Champion Rivet Co.
C, alloy, bal Fe.
420 Brin. For hard surfacing electrodes; for valve seats and hot punches. *Obsolete*

CHAMPION NON-CHANGEABLE
Colt Industries
C 0.9, Mn 1.35, Cr 0.5, V 0.09, bal Fe.
For tools, stamping dies, taps, reamers, holes, gauges; nondeforming, oil hardening. *Obsolete*

CHAMPION S1
Champion Steel Co.
C 0.5, Cr 1.5, W 2.5, bal Fe.
Oil hardening tool steel, shock resisting type. AISI Type S1 shock resisting tool steel.

CHAMPION TOOL
Colt Industries
C 1.2, W 1.25, Cr 0.3, bal Fe.
For reamers, taps, threading dies, punches, blanking dies; oil hardening. *Obsolete*

CHAMPION TOOL
American manufacture
C 0.77, Mn 0.18, bal Fe.
For tools, punches; water hardening.

CHAMPLAIN
North American Steel Corp.
C 0.9, W 0.5, Cr 0.5, Mn 1.2, V 0.2, bal Fe.
Heat treated: 170,000-280,000 TS; 155,000-272,000 YS; 2-4 El; 2-18 RA; 335-535 Brin. For dies, knives, guides, shear blades, vise jaws, forming rolls. Oil hardened, non-deforming, tough.

CHANNELER
U.S. Steel Corp.
C 1, Mn 0.5, bal Fe.
For stone cutting tools, augers, paving drills; water hardening. *Obsolete*

CHANNELLER
Bethlehem Steel Corp.
C 0.8, Mn 0.4, Si 0.3, bal Fe.
For cutting soft stone, marble, limestone. Type W1 water hardening tool steel. *Obsolete*

CHAR-PAC
United States Steel Corp.
C 0.17, Mn 1.25, Si 0.35, Cu 0.25, Ni 0.15, Cr 0.12, Mo 0.04, bal Fe.
Heat treated: 60,000 min YS; 80,000-100,000 TS; 15 min Charpy. Normalized: 70,000-95,000 TS; 50,000 min YS. For storage tanks, penstocks, bridges and buildings, pressure vessels. Heat treated carbon steel plate.

CHARDON-S
Compagnie Ateliers et Forges de la Loire
C 0.6-1.1, bal Fe.
Annealed: 100,000 TS; 52,000 YS; 20 El; 200 Brin. Water hardened: 185,000-220,000 TS; 125,000-155,000 YS; 32-37 RA; 11-13 El; 330-600 Brin. For tools, cutters, hammers, chisels, punches. Water hardening, Type W1 tool steel.

CHARLES LEONARD
Jessop-Saville Ltd.
C 0.7-1, Mn 0.3, Si 0.3, bal Fe.
For tools, cutters, drills; water hardened; Type W1.

CHARPY ALLOY
English manufacture
Cu 5.6, Sn 83.3, Sb 11.1.
For bearings; Babbitt.

CHARPY PHOSPHOR BRONZE
American manufacture
Sn 12.2-13.4, P 0.4, bal Cu.
For bearings, gears, bushings; heavy duty.

CHASE 149 BRONZE
Chase Brass & Copper Co.
Cu 88.5, Ni 1.1, P 0.22, Te 0.5, Zn 9.7.
Rolled: 60,000-70,000 TS; 40,000-50,000 YS; 6-15 El. For screw machine parts, forgings. *Obsolete*

CHASE 192 HC
Chase Brass & Copper Co., Inc.
Cu 98.97, Fe 1, P 0.03.
Annealed, 1/4 hard: 50,000 TS; 37,000 YS; 25 El. Rolled hard: 65,000 TS; 60,000 YS; 7 El. Good conductivity, resistance to softening. For electrical contacts, terminals, springs, lead frames, cable wrap.

CHASE 50 RE-50 MO
Chase Brass & Copper Co.
Re 47.5-48.5, Mo 51.5-52.5.
Recrystallized: 150,000 TS; 123,000 YS; 16-20 El; 350 VHN. Wrought: 240,000 TS; 210,000 YS; 600 VHN. For electron tube components, high temperature furnace parts, nuclear and space propulsion and power generation systems, crucibles, thermocouple protection tubes. High strength and ductility to 4200 F. *Obsolete*

CHASE 690
Chase Brass & Copper Co., Inc.
Cu 73.3, Zn 22.7, Al 3.4, Ni 0.6.
Annealed: 82,000 TS; 52,000 YS; 35 El. Rolled hard: 113,000 TS; 101,000 YS; 4 El. For springs, bellows, diaphragms, fuse clips, switch parts.

CHASE 75W-25RE
Chase Brass & Copper Co.
W 75, Re 25.
Cold drawn: 400,000 TS; 2 El. Annealed: 245,000 TS; 15 El. For thermocouples, electron tube components; good for 5000 F, ductile and strong. *Obsolete*

CHASE ALLOY NO. 58
Chase Brass & Copper Co.
Cu 96.68, Ni 1.1, P 0.22.
Rolled: 88,000-110,000 TS; 80,000-100,000 YS. For springs, electric conductors; now Phosnic Bronze. *Obsolete*

CHASE BRIGHT EXTRUDED BRONZE
Chase Brass & Copper Co.
Cu 58.5, Zn 38.7, Pb 2.5, Al 0.3.
Extruded: 62,000 TS; 24,000 YS; 20 El. For architectural sections; free-cutting. *Obsolete*

CHASE CUPRO-NICKEL 80-20
Chase Brass & Copper Co.
Cu 80, Ni 20.
Rolled: 45,000 YS; 35 El. For condenser tubes; corrosion resistant to dilute hydrochloric acid, sulfates, sulfites, alkalis. *Obsolete*

CHASE HIGH STRENGTH TANK BRASS
Chase Brass & Copper Co.
Zn 30, bal Cu.
Rolled: 53,000 TS; 20,500 YS; 52 El; 68 Brin. For tanks, radiators; high yield strength, good drawability. *Obsolete*

CHASE NICKEL ALUMINUM BRONZE
Chase Brass & Copper Co.
Cu 92, Ni 4, Al 4.
Annealed: 70,000 TS; 30,000 YS; 48 El. Hard: 120,000 TS; 90,000 YS; 3 El. For condenser tubing, pipe and corrosion resistant purposes; corrosion resistant. *Obsolete*

CHASE NICKEL SILVER 20%
Chase Brass & Copper Co.
Cu 75, Zn 5, Ni 20.
Rolled: 50,000 TS; 35 El. For condenser tubes; corrosion resistant to brine, alkalis, dilute hydrochloric acid. *Obsolete*

CHASE TELNIC BRONZE
Chase Brass & Copper Co.
Cu 98.3, Ni 1.1, P 0.22, Te 0.5, bal Zn.
Drawn hard: 72,000 TS; 62,000 YS; 27 El; 50 RA. Heat treated: 58,000 TS; 38,000 YS. For forgings and screw machine parts; heat treatable. *Obsolete*

CHASE VALVE STEM BRONZE
Chase Brass & Copper Co.
Cu 57, Pb 3.5, Zn 39.5.
Rolled: 70,000 TS; 40,000 YS; 15 El. For valve stems. *Obsolete*

CHATEAUGAY LOW PHOSPHORUS PIG IRON
Chateaugay Ore & Iron Co.
C 4, Si 0.75-4.05, Mn 0.1-2, 0.035 S and P max, bal Fe.
For casting rolls, gears, cylinders, steel and iron; produced from New York State ores.

CHATILLON 3100
Chiers-Chatillon
C 0-0.12, Cr 14-18, Mo, bal Fe.
Annealed: 65,000-90,000 TS; 40,000-55,000 YS; 20-35 El; 40-60 RA; 130-160 Brin. For hardware, fittings, oil burners, fasteners, storage tanks. Type 430 Mo stainless steel. *Obsolete*

CHATILLON 3308
Chiers-Chatillon
C 0-0.08, Cr 17-19, Ni 9-12, Ti, bal Fe.
Annealed: 85,000 TS; 35,000 YS; 55 El. Cold drawn: 95,000 TS; 60,000 YS; 40 El; B 90 Rock. For exhaust systems, oil refinery equipment; welded structures, superheaters. Type 321 stainless steel, stabilized, welding quality. *Obsolete*

CHATILLON 3333
Chiers-Chatillon
C 0-0.15, Cr 17-19, Ni 8-10, bal Fe.
Annealed: 90,000 TS; 40,000 YS; 50 El; B 85 Rock. Cold rolled: 150,000 TS; 110,000 YS; 30 El; B 100 Rock. For chemical and pharmaceutical plant equipment, valve trim, food handling equipment. Type 302 stainless steel, austenitic. *Obsolete*

CHATILLON 3400
Chiers-Chatillon
C 0-0.08, Cr 16-18, Ni 10-14, Mo 2-4, bal Fe.
Annealed: 85,000 TS; 35,000 YS; 50 El; B 80 Rock. Cold rolled: 150,000 TS; 135,000 YS; 6 El; C 30 Rock. For chemical plant equipment, agitators, digesters, valve trim, evaporators. Type 18-8 Mo stainless steel, austenitic, acid resistant. *Obsolete*

CHATILLON 3408
Chiers-Chatillon
C 0.08, Cr 17, Ni 12, Mo, Ti, bal Fe.
Annealed: 85,000 TS; 35,000 YS; 50 El; B 82 Rock. Cold rolled: 130,000 TS; 90,000 YS; 30 El; B 100 Rock. For chemical plant equipment, agitators, welded structures. Type 18-8 Mo Ti stainless steel, austenitic, welding grade. *Obsolete*

CHATILLON 5250
Chiers-Chatillon
C 0.95, Mn 1.2, Cr 0.5, W 0.5, bal Fe.
For tools, dies, cutters; non-deforming. *Obsolete*

CHATILLON 5650
Chiers-Chatillon
W 9.5, Cr 2.5, V 0.1, C, bal Fe.
For hot work tools and dies; hot work steel. *Obsolete*

CHATILLON 5654
Chiers-Chatillon
W 9.5, Cr 2.5, V 0.1, C, bal Fe.
For hot work tools and dies; hot work steel. *Obsolete*

CHATILLON 5655
Chiers-Chatillon
W 9.5, Cr 2.5, V 0.1, C, bal Fe.
For hot work tools and dies; hot work steel. *Obsolete*

CHATILLON 5755
Chiers-Chatillon
W 19, Cr 4, V 2.2, C, bal Fe.
For tools, dies, cutters; high speed steel. *Obsolete*

CHATILLON 5815
Chiers-Chatillon
W 18, Co 5, Cr 4, V 1.3, C, bal Fe.
For tools, dies, cutters; high speed steel. *Obsolete*

CHATILLON 5820
Chiers-Chatillon
W 18, Co 9, Cr 4, V 2, C, bal Fe.
For tools, dies, cutters; high speed steel. *Obsolete*

CHATILLON 5830
Chiers-Chatillon
C 0.5, Cr 1.5, W 2.2, V 0.2, bal Fe.
For tools, dies; hot work steel. *Obsolete*

CHATILLON 709
St. Jacques, Societe des Usines
C 0-0.21, Cr 0.85-1.2, Mo 0.9-1.1, V 0.15-0.3, bal Fe.
Heat treated: 105,000-178,000 TS; 92,400-106,000 YS; 8-17
El. For steam turbine rotors and stators, high temperature
bolts; creep resistant, structural steel.

CHATILLON MN SI
Chiers-Chatillon
C 0.5, Mn 0.7, Si 1.8-2, bal Fe.
For tools, dies, punches; tough. *Obsolete*

CHATILLON MF NO. 1
Chiers-Chatillon
C 2.25, Cr 13, bal Fe.
For tools, dies; non-deforming. *Obsolete*

CHATILLON TSM NO. 2
Chiers-Chatillon
C 1.3, Cr 4.5, bal Fe.
For tools, dies; air hardened. *Obsolete*

CHATILLON W18
Chiers-Chatillon
W 18, Cr 4, V 1, C, bal Fe.
For tools, dies, cutters; high speed steel. *Obsolete*

CHD
Spencer Clark Metal Industries Ltd.
C 0.33, Mn 0.35, Si 1, Cr 5, W 1.5, V 0.3, Mo 1.5, bal Fe.
Chromium-tungsten hot work steel, for die casting dies for
light alloys, extrusion dies, punches.

CHECKNO
Bisset Steel Co.
C 0.4, Cr 3.5, V 1, W 8, bal Fe.
For tools, hot working dies, punches; hot work steel.
Obsolete

CHECKNO NO. 1
Bethlehem Steel Corp.
C 0.45, Mn 0.25, Cr 3.5, W 13.25, Si 0.2, bal Fe.
For hot work dies; hot work steel, oil hardened. *Obsolete*

CHECKNO NO. 1
Bisset Steel Co.
C 0.45, W 13.25, Cr 3.5, bal Fe.
For extrusion dies and liners; oil hardened, hot work steel.
Obsolete

CHECKNO NO. 2
Bethlehem Steel Corp.
C 0.35, Mn 0.25, Cr 3.5, W 11, V 0.45, Si 0.2, bal Fe.
For hot work dies; hot work steel, oil hardened. *Obsolete*

CHECKNO NO. 2
Bisset Steel Co.
C 0.35, Cr 3.5, W 11, V 0.45, bal Fe.
For extrusion dies and liners, punches; oil hardened, hot
work steel. *Obsolete*

CHECKNO NO. 3
Bethlehem Steel Corp.
C 0.28, Mn 0.3, Si 0.3, Cr 3.25, W 8.5, V 0.3, bal Fe.
For hot work dies, hot work steel, oil hardened. *Obsolete*

CHECKNO NO. 3
Bisset Steel Co.
C 0.28, Cr 3.25, W 8.5, V 0.3, bal Fe.
For extrusion dies and liners, punches; oil hardened, hot
work steel. *Obsolete*

CHECKWAITE 8
AL Tech Specialty Steel Corp.
C 0.1, Mn 1-2, P 0-0.04, S 0-0.03, Si 0-1, Cr 18.5-21.5, Ni
23-26, Mo 1.75-2.75, bal Fe.
For balance weights. Vacuum consumable electrode melted.
Stainless, austenitic, non-magnetic.

CHECO
American manufacture
Ag 89, Sn 10, Pt 1.
For dentures.

CHEMALLOY
Chemalloy Electronics Corp.
Low melting. Zn, Pb, plus additives.
Homogenized alloy for fluxless soldering of aluminum,
copper base alloys, galvanized iron. MP: 500°F (higher
melting grades are available).

CHEMALLOY A28C
Abex Corp.
Cr 18-20, Ni 8-10, C, Cb, bal Fe.
Cast: 80,000 TS; 35,000 YS; 45 El; 55 RA; 160 Brin. For valve
parts, pumps, anchors, condensers; for welded sections,
corrosion resistant. *Obsolete*

CHEMALLOY A28F
Abex Corp.
Cr 18-20, Ni 8-10, Se 0.25, C, bal Fe.
Cast: 75,000 TS; 35,000 YS; 50 El; 60 RA; 150 Brin. For
valves, pumps, condensers, anchors, conveyors; corrosion
resistant; free machining. *Obsolete*

CHEMALLOY A28M
Abex Corp.
Cr 18-20, Ni 8-10, Mo 2-3, C, bal Fe.
Cast: 75,000 TS; 30,000 YS; 45 El; 60 RA; 170 Brin. For
valves, pumps, piping, retorts, condensers; resists acid
pitting, corrosion resistant. *Obsolete*

CHEMALLOY A32
Abex Corp.
Cr 20-22, Ni 10-12, C, bal Fe.
Cast, at 2000 F, and water quenched: 75,000 TS; 34,000 YS;
45 El; 50 RA; 145 Brin. For valves, pumps, anchors,
condensers; acid resistant. *Obsolete*

CHEMALLOY A32M
Abex Corp.
Cr 20-22, Ni 10-12, Mo 2-3, C, bal Fe.
Cast, at 2000 F, and water quenched: 80,000 TS; 38,000 YS;
40 El; 50 RA; 150 Brin. For valves, pumps, anchors,
conveyors, condensers; acid resistant. *Obsolete*

CHEMALLOY A38
Abex Corp.
Cr 23-26, Ni 12-14, C, bal Fe.
Cast, at 2050 F, and water quenched: 80,000 TS; 45,000 YS;
35 El; 45 RA; 165 Brin. For pumps, valves, filters, fittings;
corrosion resistant. *Obsolete*

CHEMALLOY A38M
Abex Corp.
Cr 23-26, Ni 12-14, Mo 2-3, C, bal Fe.
Cast: 85,000 TS; 40,000 YS; 40 El; 40 RA; 165 Brin. For
pumps, valves, filters, vessels, fittings, condensers; acid
resistant. *Obsolete*

CHEMALLOY A15
Abex Corp.
Cr 23-26, Ni 19-21, C, bal Fe.
Cast: 70,000 TS; 40,000 YS; 30 El; 35 RA; 170 Brin. For
pumps, valves, filters, fittings, condensers, marine service;
corrosion resistant in reducing acids. *Obsolete*

CHEMALLOY A45N
Abex Corp.
Cr 19-22, Ni 23-26, C, bal Fe.
Cast, at 2050 F, and water quenched: 70,000 TS; 38,000 YS;
25 El; 35 RA; 170 Brin. For pumps, valves, filters, fittings,
condensers; corrosion resistant in reducing acids. *Obsolete*

CHEMALLOY A50N
Abex Corp.
Cr 15-17, Ni 34-37, Mo 0.7-0.9, C, bal Fe.
Cast, at 2050 F, and water quenched: 70,000 TS; 35,000 YS;
25 El; 35 RA; 170 Brin. For pumps, valves, filters, rakes,
fittings, acid pipes; corrosion resistant to reducing acids.
Obsolete

CHEMALLOY A52N
Abex Corp.
Cr 20-23, Ni 26-29, Mo 2-3, Cu 3-4, Si 1.5-2.5, C, bal Fe.
Cast: 70,000 TS; 35,000 YS; 40 El; 45 RA; 145 Brin. For
pumps, valves, fittings; corrosion resistant in non-oxidizing
media. *Obsolete*

CHEMALLOY F12
Abex Corp.
Cr 11-14, C, bal Fe.
Cast: 95,000 TS; 75,000 YS; 25 El; 45 RA; 225 Brin. For valve
parts, gears, pumps, impellers, dies; heat resistant. *Obsolete*

CHEMALLOY F12F
Abex Corp.
Cr 11-14, Se 0.25, C, bal Fe.
Cast. For valve parts, gears, pumps, impellers; heat resistant;
more machinable than F12 grade. *Obsolete*

CHEMALLOY F20
Abex Corp.
Cr 18-21, C, bal Fe.
Cast: 90,000 TS; 55,000 YS; 20 El; 25 RA; 190 Brin. For oil
refineries and steam plants, valves, conveyors; heat and
corrosion resistant. *Obsolete*

CHEMALLOY F32F
Abex Corp.
Cr 26-29, Ni 3-6, Mo 1-1.5, C, bal Fe.
Cast. For pumps, condensers, marine parts; heat and
corrosion resistant. *Obsolete*

CHEMALLOY F6
Abex Corp.
Cr 4.5-6, Mo 0.5, bal Fe.
Cast: 110,000 TS; 75,000 YS; 20 El; 40 RA; 230 Brin. For dies,
molds, valve trim, steam turbines; heat resistant to 1250 F.
Obsolete

CHEMALLOY F8
Abex Corp.
Cr 7-9, Mo 0.5-1, C, bal Fe.
Cast: 110,000 TS; 75,000 YS; 20 El; 40 RA; 200-360 Brin. For
dies, molds, valve trim, steam turbine parts, fittings; heat
resistant; high strength and ductility to 1300 F. *Obsolete*

CHEMALLOY H1
Abex Corp.
Mo 21, Fe 21, bal Ni.
Cast: 75,000 TS; 45,000 YS; 10 El; 175 Brin. For pumps,
valves, condensers, cylinder liners; corrosion resistant to
halogen ions as HCl, HBr, HF. *Obsolete*

CHEMALLOY H2
Abex Corp.
Mo 29, Fe 6, bal Ni.
Cast: 80,000 TS; 55,000 YS; 8 El; 210 Brin. For pumps,
valves, condensers, fittings; resists boiling HCl and
phosphorous acid. *Obsolete*

CHEMALLOY H3
Abex Corp.
Cr 14, Mo 16, W 4, Fe 6, bal Ni.
Cast: 75,000 TS; 45,000 YS; 12 El; 195 Brin. For valves,
pumps, condensers, fittings, cylinder liners; resists free
chlorine, hypochlorites, sulphurous acid. *Obsolete*

CHEMALLOY H4

Abex Corp.
Si 10, Cu 3, bal Ni.
Cast: 40,000 TS; 40,000 YS; 475 Brin. For valves, pumps, fittings, centrifuge rotars; hard and brittle. *Obsolete*

CHEMALLOY N1

Abex Corp.
Ni 15, Cr 6, Cu 2, bal Fe.
Cast: 30,000 TS; 12,000 YS; 150 Brin. For pumps, valves, chains, filters, fittings, pump seals; corrosion resistant to alkalis; Ni-Resist Type 1. *Obsolete*

CHEMALLOY N2

Abex Corp.
Ni 20, Cr 2, bal Fe.
Cast: 30,000 TS; 12,000 YS; 150 Brin. For pumps, valves, chains, filters; corrosion resistant to alkalis; Ni-Resist Type 2. *Obsolete*

CHEMALLOY N3

Abex Corp.
Ni 30, Cr 3, bal Fe.
Cast: 30,000 TS; 15,000 YS; 135 Brin. For pumps, valves, chains, dryers, filters; corrosion and heat resistant; Ni-Resist Type 3. *Obsolete*

CHEMALLOY N4

Abex Corp.
Ni 30, Cr 5, Si 5, bal Fe.
Cast: 30,000 TS; 15,000 YS; 165 Brin. For valves, pumps, fittings, filters, rolls; stain resistant. *Obsolete*

CHEMALLOY N5

Abex Corp.
Ni 35, bal Fe.
Cast: 20,000 TS; 10,000 YS; 125 Brin. For salt rolls, molds, chains; "Minovar" minimum thermal expansion. *Obsolete*

CHEMBRITE A608

Airco Vacuum Metals
C 0.1, Mn 1.9, Si 0.68, Ni 0.06, Mo 0.5, bal Fe.
Low alloy bare for machine welding. 100,000 psi TS. AWS E 70 S-1B.

CHEMBRITE A675

Airco Vacuum Metals
C 0.12, Mn 1.19, Si 0.59, bal Fe.
Bare mild steel welding wire for CO_2 and inert gas shielded arc welding. AWS E 70 S-3.

CHEMBRITE A681

Airco Vacuum Metals
C 0.11, Mn 1.64, Si 0.86, bal Fe.
Mild steel bare wire for CO_2 and inert gas welding of steel. AWS E 70 S-6.

CHEMICAL LEAD

NL Industries
Pb 99.93, Cu 0.07, Ag 0.005.
Cast: 1750 TS; 1000 YS; 40 El; 90 RA; 5.5 Brin. Rolled: 2000 TS; 1000 YS; 54 El; 5.5 Brin. For tank linings, valves, pipes, fuses; ME 2,000,000. *Obsolete*

CHENITE

English manufacture
Ti, V, Cr, Hf.
For cutting tools, dies, air hardening.

CHEROKEE

AL Tech Specialty Steel Corp.
C 1.5, Cr 12, Ni 0.5, Mo 0.7, Co 3, bal Fe.
Usual heat treated hardness: 59-61 Rock C. High performance cold work punch and die steel capable of long runs. AISI D-5.

CHEVIS

George Cook & Co., Ltd.
C, alloy, bal Fe.
For punches, dies for sheet metal blanking and forming; water hardened. *Obsolete*

CHEVRE NO. 2 1/2

Chiers-Chatillon
C 1.1, bal Fe.
For tools, taps, drills; water hardened. *Obsolete*

CHEVRE NO.3

Chiers-Chatillon
C 0.95, bal Fe.
For tools, dies, springs; water hardened. *Obsolete*

CHEYENNE

AL Tech Specialty Steel Corp.
C 0.8, Cr 4, W 6, Mo 5, V 2, bal Fe.
General purpose high speed steel for a wide variety of cutting tools and difficult blanking and forming punch and die requirements. AISI M-2.

CHICKASAU

AL Tech Specialty Steel Corp.
C 0.6, Mn 0.85, Si 2, Mo 0.45, V 0.25, bal Fe.
Oil hardening shock resisting steel capable of high toughness with moderate wear resistance. For punches, dies, shear blades, stamps, and machinery parts. AISI S-5.

CHINA BRASS

Chinese manufacture
Cu 56.6, Zn 27-37, Sn 0.2-1, Pb 0-0.8.
For sheet, hardware, fittings; good workability.

CHINESE BRONZE

Chinese manufacture
Pb 10-20, Zn 1-14, Sn 1-13, bal Cu.
For bearings; heavy duty.

CHINESE GERMAN SILVER

Chinese manufacture
Cu 40.4-41, Zn 25.4-26.5, Ni 30.8-31.6, Fe 2.6-2.7.
For electrical resistances, ornaments; corrosion resistant.

CHINESE SILVER

Chinese manufacture
Cu 58, Zn 17.5, Ni 11.5, Ag 2.
For ornaments, electrical parts; nickel silver.

CHINESE SPECULUM-1

English manufacture
Cu 81, Sn 11, Sb 8.5.
For mirrors, optical grading.

CHINESE SPECULUM-2

English manufacture
Cu 81, Pb 9, Sb 8, Sn 2.
For mirrors; (Elsner's).

CHIPPAWAY

Jessop Steel Co.
C 0.6-1.4, bal Fe.
For pneumatic hammers, chisels, vise jaws, rock drills, punches; water hardening. *Obsolete*

CHIPPER

Uddeholm Corp.
C 0.5, Si 1, Mn 0.5, Cr 8, Mo 1.5, V 0.5, bal Fe.
Cold work tool steel. 300,000 psi TS; 250,000 psi YP; 6 El; 15 RA; 58 Rock C. High alloy steel; for knives, blades, shears and dies.

CHIPPER KNIFE

Disston Inc.
C 0.7, Cr 1.15, V 0.25, Mo 0.7, bal Fe.
For shears, dies, chipper knives; oil hardening. *Obsolete*

CHIPPEWA

Atlas Specialty Steels
C 0.45, Mn 0.6, Si 0.25, V 0.4, Cr 3, Mo 0.25, bal Fe.
For mining bit shanks; hollow shank steel. *Obsolete*

CHISEL 3581

Darwin & Milner Inc.
Tool material. C 0.55, Mn 0.7, V 0.2, Mo 0.35, Si 1.6, bal Fe.
For pneumatic tools, chisels; oil hardened; shock resistant.

CHITONAL

Lavorazione Leghe Leggere SpA
Si 1.5, Cu 5, bal Al.
For light alloy parts, ship parts, bi-metal, coated "Avional."

CHITONAL-14

Lavorazione Leghe Leggere SpA
Cu 3.9-5, Si 0.5-1.2, Mg 0.2-0.8, Mn 0.4-1.2, bal Al.
Annealed: 27,000-30,000 TS; 11,400-14,200 YS; 16-20 El.
TA-Temper: 66,800-75,400 TS; 58,300-66,800 YS; 6-12 El. For general engineering components. Age hardenable, high strength.

CHITONAL-14

Alluminio SA
Cu 3.9-5, Si 0.5-1.2, Mg 0.2-0.8, Mn 0.4-1.2, bal Al.
Annealed: 27,000-30,000 TS; 11,400-14,200 YS; 16-20 El.
TA-Temper: 66,800-75,400 TS; 58,300-66,800 YS; 6-12 El. For general engineering components. Age hardenable, high strength.

CHITONAL-24

Lavorazione Leghe Leggere SpA
Cu 3.8-4.9, Mg 1.2-1.8, Mn 0.3-0.9, bal Al.
Annealed: 27,000 TS; 14,000 YS; 14-22 El; TH06N-Temper: 74,000 TS; 62,000 YS; 10 El. For aircraft structures, general engineering components. Age hardenable, high strength.

CHITONAL-24

Alluminio SA
Cu 3.8-4.9, Mg 1.2-1.8, Mn 0.3-0.9, bal Al.
Annealed: 27,000 TS; 14,000 YS; 14-22 El; TH06N-Temper: 74,000 TS; 62,000 YS; 10 El. For aircraft structures, general engineering components. Age hardenable, high strength.

CHIZ-ALAIR

Manufacturer not listed
C 0.5, Mn 0.7, Si 0.25, Cr 3.25, Mo 1.4, bal Fe.
Air hardening; for tools, dies. AISI S7 shock resisting tool steel.

CHIZ-ALLOY

Manufacturer not listed
C 0.5, Cr 1.15, W 2.5, V 0.2, Si 0.75, bal Fe.
For hand and pneumatic chisels and punches; shock resisting; oil hardening octagons and hexagons; AISI S1 tool steel.

CHLORIMET 2

Duriron Co. Inc.
Nickel. Ni 60, Mo 30-33, Fe 3, Si 0-1, Mn 0-1.
Cast, heat treated: 76,000 psi TS; 40,000 psi YS; 20 El; 200 Brin. For corrosion resistant pumps and valves; resistant to HCl, boiling H_2SO_4, boiling H_3PO_4, and brine.

CHLORIMET 3

Duriron Co. Inc.
Nickel. Ni 58, Cr 17-20, Mo 17-20, Fe 0-3, Si 0-1, Mn 0-1, C 0-0.03.
Cast, heat treated: 72,000 psi TS; 40,000 psi YS; 25 El; 200 Brin. For corrosion resistant pumps and valves; resistant to HCl, hot H_2SO_4, hot H_3PO_4, wet chlorine, ferric chloride, and bleach.

CHOYCE-77

Champion Steel Co.
C 0.9, Mn 1.2, V 0.2, W 0.5, Cr 0.5, bal Fe.
Annealed: 84,000 TS; 60,000 YS; 26 El; 185 Brin. Heat treated: 145,000-280,000 TS; 125,000-272,000 YS; 2-8 El; 2-31 RA; 290-535 Brin. For blanking and bending dies, master tools, knurling tools, punches, cutters, master dies, gauges. Type O1, oil hardening; cold work steel, shock resisting.

CHRISTOFLE METAL
English manufacture
silver plated German Silver (2.0 Ag).
For cutlery, hardware; corrosion resistant.

CHRO-MOW
Crucible Materials Corp.
Tool material. C 0.35, Si 1.05, Cr 5, W 1.25, V 0.35, Mo 1.35, bal Fe.
Air hardens to 51-53 Rock. For hot work punches, shell piercing tools, mandrels for Al and extrusion. Type H12. Hot work tool steel.

CHROGO "U-42"
Manufacturer not listed
Au 40, Cu 45, Ni 14, Cr 1, trace Pt.
For dental alloy; corrosion resistant.

CHROM SPECIAL
Now VEW K100.

CHROM SPECIAL EXTRA
Now VEW K107.

CHROM SPECIAL SUPRA
Now VEW K103.

CHROM-SUPER SERVICE
Central Engineering & Supply Co.
Cr, Sb, Pb, bal Sn.
Cast. For Babbitt; antifriction.

CHROM-TEC 19222
Eutectic Corp.
Alloy powder for metal spraying. Corrosion resistant dense coating; machinable. *Obsolete*

CHROM-X
Chromium Mining & Smelting Corp. Ltd.
ferrochrome, C, Si.
For chromium addition to steel; for alloy steel.

CHROMADOR STEEL
British Steel plc
C 0-0.3, Mn 0.7-1, Cr 0.7-1, Cu 0.25-0.5, bal Fe.
Rolled: 83,000-98,500 TS; 52,000 YS; 17 El; 40 RA; 160-200 Brin. For structural work, tanks, and rivets. High tensile steel; resists corrosion. *Obsolete*

CHROMAL
Swedish manufacture
Cr 2-4, Ni, Mn, bal Al.
Rolled: 55,000-60,000 TS. For airplane parts, propellers, cooking utensils, milk and oil separators; non-hardenable.

CHROMAL STEEL
English manufacture
Cr 0.8, Mn 0.8, Mo 0.8, C 0.3, bal Fe.
For gears, pinions, shafts; water hardening.

CHROMALLOY
General Electric Co.
C 0.2, Cr 1, Mo 1, V 0.1, Si 0.75, Mn 0.5, bal Fe.
Heat treated: 138,000 TS; 117,000 YS; 7 El; 45 RA. For high temperature welded structures; good strength and weldability. *Obsolete*

CHROMALLOY
Universal Cyclops
C 0.2, Cr 1, Mo 1, V 0.1, Si 0.75, Mn 0.5, bal Fe.
Heat treated: 138,000 TS; 117,000 YS; 7 El; 45 RA. For high temperature welded structures; good strength and weldability. *Obsolete*

CHROMALOID
American Nickeloid Co.
Cr, coated zinc.
For plumbing fixtures; hardware; bonded Zn base. *Obsolete*

CHROMALOY V
A.C. Scott & Co. Ltd.
Ni 80, Cr 20.
Annealed: 95,000 TS; 25-35 El; 185 Brin. For heating and resistance elements. Heat resistant to 1150°F.

CHROMAN "AO"
Vacuumschmelze GmbH
Ni 88, Cr 11, Mn 1.
For heating elements; acid resisting. *Obsolete*

CHROMAN "B"
Vacuumschmelze GmbH
Fe 20, Ni 64, Cr 15, Mn 1.
At 20 C: 87,000 TS; 48 El. At 500 C: 73,000 TS; 46 El. For heating elements; acid and heat resistant. *Obsolete*

CHROMAN "BO"
Vacuumschmelze GmbH
Ni 84, Cr 15, Mn 1.
For heating elements; acid resisting. *Obsolete*

CHROMAN "C"
Vacuumschmelze GmbH
Fe 10, Ni 69, Cr 20, Mn 1.
For heating elements; acid resisting. *Obsolete*

CHROMAN "CO"
Vacuumschmelze GmbH
Ni 79, Cr 20, Mn 1.
For heating elements; acid resisting. *Obsolete*

CHROMAN "D"
Vacuumschmelze GmbH
Fe 9, Ni 65, Cr 25, Mn 1.
For heating elements; acid resisting. *Obsolete*

CHROMAN "E"
Vacuumschmelze GmbH
Fe 16, Ni 50, Cr 33, Mn 1.
For heating elements; acid resisting. *Obsolete*

CHROMAN ALLOY
Vacuumschmelze GmbH
Ni, Cr, Mn, bal Fe.
For heating elements; acid resisting. *Obsolete*

CHROMAN B2 MO
Vacuumschmelze GmbH
Ni 61.4, Cr 18.5, Fe 14.5, Mn 3, Mo 2, Si 0.6.
For jet engine parts; heat and oxidation resistant. *Obsolete*

CHROMANG
Arcos Alloys
C 0.1, Mn 4, Cr 19, Ni 9, bal Fe.
Welded: 96,000 TS; 62,000 YS; 37 El; 35 RA. For welding electrodes for armor; stainless. *Obsolete*

CHROMAX
Harrison Alloys Inc.
Superalloy.
Now NAI-NICR 35.

CHROMAX
Empire Steel Castings Co.
Cr 1.25-1.5, C 0.3-1, Mo 0.35-0.45, bal Fe.
Used for resistance to abrasion. *Obsolete*

CHROMAX (CAST)
Driver Harris Co.
Ni 35, Cr 15, Fe 50.
Cast: 60,000 TS; 3.5 El; 2 RA; 168 Brin. For furnace parts, carburizing boxes, enameling fixtures; resists abrasion, impact, strain, heat and corrosion. *Obsolete*

CHROMCARB N-6006
Eutectic Corp.
Electrode for hard but tough overlays on steel; AC-DC. 57-60 Rock C. For crusher jaws, clam shell bucket lips, and dipper teeth.

CHROMDIE
Faitout Iron & Steel Co.
C 1.5, Cr 12, V 0.4, Mo 1, bal Fe.
Hardened: 278,000 TS; 214,000 YS; 1 El; 567 Brin. For blanking and drawing dies; wire drawing and stamping dies, broaches, hobs, punches, gauges, thread rolling dies. Type D2 air hardening, nondeforming tool steel, tough.

CHROME
George Cook & Co., Ltd.
C 1.4, Mn 0.35, Cr 0.6, Si 0.2, bal Fe.
For reamer blades, impact extrusion dies; water hardened, wear resistant. *Obsolete*

CHROME "B-15"
Hidalgo Steel Co. Inc.
Cr 15, C, bal Fe.
For tools, cold and threading dies; punch stamps; non-deforming.

CHROME 3 1/2%
Indiana General
Cr 3.5, C 1, bal Fe.
For permanent magnets; Br-9500; Hc-60. *Obsolete*

CHROME 6%
Indiana General
Cr 6, C 1, bal Fe.
For permanent magnets; Br-8500; Hc-65. *Obsolete*

CHROME 9A
Creusot-Loire
C 1, Cr 5, Mo 1.2, V 0.4, bal Fe.
Cold work tool steel; air hardening. AFNOR Z 100 CCV 5; AISI A2.

CHROME ALLOY PRODUCTS KA-2
Magnolia Anti-Friction Metal Co.
C 0.13, Cr 18.5, Ni 9, bal Fe.
For stainless parts; corrosion and heat resistant.

CHROME ALLOY PRODUCTS KA-2
Chicago Steel Foundry
C 0.13, Cr 18.5, Ni 9, bal Fe.
For stainless parts; corrosion and heat resistant.

CHROME ALLOY PRODUCTS KA-2 MO
Magnolia Anti-Friction Metal Co.
C 0.14, Cr 20, Ni 9, Mo 3, bal Fe.
For stainless parts; corrosion and heat resistant.

CHROME ALLOY PRODUCTS KA-2 MO
Chicago Steel Foundry
C 0.14, Cr 20, Ni 9, Mo 3, bal Fe.
For stainless parts; corrosion and heat resistant.

CHROME ALLOY PRODUCTS KA-2S
Magnolia Anti-Friction Metal Co.
C 0.06, Cr 20, Ni 9, bal Fe.
For stainless parts; corrosion and heat resistant.

CHROME ALLOY PRODUCTS KA-2S
Chicago Steel Foundry
C 0.06, Cr 20, Ni 9, bal Fe.
For stainless parts; corrosion and heat resistant.

CHROME ALLOY PRODUCTS KNC-3
Magnolia Anti-Friction Metal Co.
C 0.17, Cr 25, Ni 19, Si 1.75, bal Fe.
For furnace parts; corrosion and heat resistant.

CHROME ALLOY PRODUCTS KNC-3
Chicago Steel Foundry
C 0.17, Cr 25, Ni 19, Si 1.75, bal Fe.
For furnace parts; corrosion and heat resistant.

CHROME ALLOY PRODUCTS STAINLESS NO. 7
Magnolia Anti-Friction Metal Co.
C 0.25, Cr 20.5, Cu 1, bal Fe.
For stainless parts; stainless.

CHROME ALLOY PRODUCTS STAINLESS NO. 7
Chicago Steel Foundry
C 0.25, Cr 20.5, Cu 1, bal Fe.
For stainless parts; stainless.

CHROME B
Hidalgo Steel Co. Inc.
C, Cr, bal Fe.
For cutting dies; non-deforming.

CHROME BALL STEEL
Crucible Specialty Metals
C 1, Cr 1.2, bal Fe.
Hardened: 650 Brin. For balls, ball bearings; water hardening. *Obsolete*

CHROME BEARING STEEL
Crucible Specialty Metals
C 1, Cr 1.2, bal Fe.
Hardened: 650 Brin. For ball bearing races; water hardening. *Obsolete*

CHROME BRASS
American Nickeloid Co.
Cr coated brass.
For fabricated parts; easily formed, stamped, drawn.

CHROME CAST
Atlantic Steel Casting Co.
C 0-1.1, Cr 16-23, bal Fe.
For heat, wear and corrosion resistant parts; heat, wear and corrosion resistant.

CHROME DIE
Associated Steel Corp.
C 1.5, Cr 12, Mo 1, V 1, bal Fe.
Cold-work tool steel; used with manganese rod in "two-tone" arc method. AISI D2.

CHROME DIE
Allegheny Ludlum Steel
C 0.5, Cr 5, bal Fe.
For hot dies, punches, shears; hot work steel.

CHROME DIE
Farrelloy Co.
C 0.5, Cr 5, bal Fe.
For hot dies, punches, shears; hot work steel.

CHROME DRILL ROD
Colt Industries
C 1.1, Cr 1.1, bal Fe.
For drills, taps; water or oil hardening. *Obsolete*

CHROME FIRMINY NO. 0
Creusot-Loire
C 1.15, Cr 3.25, bal Fe.
For blanking, drawing and forming dies; oil hardened, wear resistant. *Obsolete*

CHROME HOT DIE
Wallace Murray Corp.
C 0.8-1, Cr 3-4, bal Fe.
For bolt and rivet dies, forging mandrels, piercing dies; hot work steel, air hardening. *Obsolete*

CHROME HOT WORK
A. Milne & Co.
C 0.9, Cr 3.75, bal Fe.
For tools, dies, mandrels; hot work steel.

CHROME IRON
English manufacture
Si 13-14, Cr, bal Fe.
For pumps, valves, corrosion resisting vessels; similar to "Duriron."

CHROME MAGNET
Allegheny Ludlum Steel
C 0.9-1, Si 0.15-0.25, Cr 3.5-4, bal Fe.
For permanent magnet; magnet steel. *Obsolete*

CHROME MAGNET
Teledyne Vasco
C 0.9-1, Si 0.15-0.25, Cr 3.5-4, bal Fe.
For permanent magnet; magnet steel. *Obsolete*

CHROME MAGNET M-154
Crucible Specialty Metals
C 0.9, Cr 3.6, bal Fe.
For permanent magnets for speakers, meters, etc.; high magnetic values. *Obsolete*

CHROME MAGNET M-31
Crucible Specialty Metals
C 0.85, Cr 2.7, bal Fe.
For permanent magnets; oil hardening. *Obsolete*

CHROME MANGANESE STEEL
Lukens Steel
C 0.4, Cr 0.5, Mn 0.9, bal Fe.
Rolled: 100,000 TS; 55,000 YS; 25 El; 200 Brin. For machinery parts; water hardening. *Obsolete*

CHROME MOLY ROLL
Crucible Specialty Metals
C 1, Cr 1.2, Mo 0.3, bal Fe.
For straightening and reducing rolls; deep hardening, high toughness. *Obsolete*

CHROME MOLY ROLL
Colt Industries
C 1, Cr 1.2, Mo 0.3, bal Fe.
For straightening and reducing rolls; deep hardening, high toughness. *Obsolete*

CHROME NICKEL
VDM Nickel-Technologie AG
Cr 20, Ni 77, Mn 2.
Annealed: 121,000 TS; 23 El. Hard: 195,000 TS; 1 El. For heat resistances, electrical heating elements, rheostats; maximum operating temperature 1150°C. *Obsolete*

CHROME NICKEL IRON
VDM Nickel-Technologie AG
Cr 20, Ni 70, Fe 8, Mn 2.
Annealed: 114,000 TS; 24 El. Hard drawn: 182,000 TS; 1 El. For electrical heating elements, rheostats, electrical measuring instruments; maximum operating temperature 1150°C. *Obsolete*

CHROME NICKEL SILVER
American Nickeloid Co.
Ni, Zn, bal Cu.
For automatic fabrication, sheet and strip; nickel silver sheet bonded with chromium, rust proof. *Obsolete*

CHROME NICKEL SKATE BLADE
Colt Industries
C 0.75, Ni 1.25, Cr 0.6, bal Fe.
For ice skate blades; oil hardening. *Obsolete*

CHROME ROLL
Midvale-Heppenstall Co.
C 0.9, Cr 1, bal Fe.
For tools, cutters, dies; water hardening.

CHROME SILICON SPRING
U.S. Steel Corp.
C 0.5-0.6, Mn 0.5-0.8, Si 1.2-1.6, Cr 0.5-0.8, bal Fe.

CHROME STEEL
American Nickeloid Co.
Cr bonded to steel.
For floor plates, reflectors, trim and molding, toasters; resists heat to 750°F; Cr bonded to steel.

CHROME TIN
American Nickeloid Co.
Cr coated Sn.
Obsolete

CHROME VANADIUM CHISEL
Colt Industries
C 0.7, Cr 0.35, V 0.2, bal Fe.
For screw drivers, cold chisels, punches; oil hardening. *Obsolete*

CHROME VANADIUM D
CCS Braeburn Alloy Steel
C, Cr, V, bal Fe.
For tools, drills, taps. *Obsolete*

CHROME VANADIUM DIE CASTING
H.K. Porter Co. (Vulcan-Kidd Div.)
C 0.5, Cr 1.5, V 0.2, bal Fe.
For tools and dies. *Obsolete*

CHROME VANADIUM G
CCS Braeburn Alloy Steel
C, Cr, V, bal Fe.
For tools. *Obsolete*

CHROME VANADIUM H
CCS Braeburn Alloy Steel
C, Cr, V, bal Fe.
For tools. *Obsolete*

CHROME VANADIUM K
CCS Braeburn Alloy Steel
C, Cr, V, bal Fe.
For tools. *Obsolete*

CHROME VANADIUM SCREW DRIVER
Colt Industries
C 0.75, Cr 0.25, V 0.25, bal Fe.
For screw drivers; water or oil hardening. *Obsolete*

CHROME VANADIUM SPRING WIRE
Crucible Specialty Metals
C 0.5, Cr 1, V 0.2, bal Fe.
For aircraft exhaust valve springs, all coil and flat springs; high fatigue resistance. *Obsolete*

CHROME VANADIUM SPRING WIRE
Colt Industries
C 0.5, Cr 1, V 0.2, bal Fe.
For aircraft exhaust valve springs, all coil and flat springs; high fatigue resistance. *Obsolete*

CHROME VANADIUM TOOL
CCS Braeburn Alloy Steel
C 0.4-0.8, Cr 0.7, V 0.2, bal Fe.
For tools. *Obsolete*

CHROME-COPPER
American Nickeloid Co.
Cr coated copper.
For stampings, display cases, refrigerators, kitchen equipment; pure Cu base bonded with Ni then Cr on surface.

CHROME-COPPER-NICKEL STEEL
Lukens Steel
Cr 0.65-0.85, Ni 0.75, C 0.12, Cu 0.45-0.65, bal Fe.
Rolled: 65,000 TS; 35,000 YS; 30 El; 50 RA; 135 Brin. For machinery parts; low temperature resistance. *Obsolete*

CHROME-MOLY
Detroit Alloy Steel Co.
C 2.75-3, Si 2-2.6, Mn 0.8-1, Cr 0.4, Mo 0.4-0.6, bal Fe.
Cast: 35,000-45,000 TS; 230-270 Brin. For heavy dies; tough, close grained.

CHROME-NICKEL STEEL ARMOR PLATE
English manufacture
Ni 4, Cr 2, C 0.33, Mn 0.32, Si 0.06, S 0.03, P 0.14, bal.
For armor plate; oil hardening.

CHROME-VANADIUM DIE
Midvale-Heppenstall Co.
C, Cr, V, bal Fe.
For hot dies. *Obsolete*

CHROMEL 1
Hoskins Mfg. Co.
Nickel. Ni 37, Cr 21, Si 2, bal Fe.
Drawn wire: 115,000 psi TS. For heating elements; corrosiona

CHROMEL 70/30
Hoskins Mfg. Co.
Nickel. Ni 70, Cr 30.
Drawn wire: 128,000 psi TS. For heating elements to 2150°F; corrosion and heat resistant.

CHROMEL A
Hoskins Mfg. Co.
Nickel. Cr 20, Fe 0.5, Si 1, bal Ni.
21 gauge wire: 120,000 TS. Elec. resistance: 650 ohms/circular mil ft. For restance wire for heating elements up to 2150°F; corrosion and heat resistant. Toasters, electric furnaces, electric ranges.

CHROMEL AA
Hoskins Mfg. Co.
Nickel. Cr 20, Fe 8.3, Si 2, bal Ni.
21 gauge wire: 130,000 TS. Elec. resistance: 700 ohms/circular mil ft. Resistance wire for furnaces operating in controlled atmosphere to 2250°F; more resistant to reducing atmospheres.

CHROMEL B
Hoskins Mfg. Co.
Ni 85, Cr 15.
147 Brin. For heating elements; corrosion and heat resistant. *Obsolete*

CHROMEL C
Hoskins Mfg. Co.
Nickel. Cr 16, Fe 25.5, Si 1.3, bal Ni.
21 gauge wire: 110,000 TS. Elec. resistance: 675 ohms/circular mil ft. Resistance wire for heating elements up to 1850°F; corrosion and heat resistant. Flat irons, dryers, waffle and sandwich grills.

CHROMEL D
Hoskins Mfg. Co.
Ni 35, Cr 18, Fe 47.
Cold drawn: 85,000 TS; 30 El. For heating elements; corrosion and heat resistant to 2000 F. *Obsolete*

CHROMEL D
Hoskins Mfg. Co.
Superalloy. Cr 18.5, Fe 44, Si 1.5, bal Ni.
21 gauge wire: 105,000 TS. Elec. resistance: 600 ohms/circular mil ft. Resistance wire for heating elements to 1800°F. Industrial furnaces operating to 1800°F. Good hot strength.

CHROMEL P
Hoskins Mfg. Co.
Nickel. Ni 90, Cr 10, (plus 9 minor constituents).
Ann: 95,000 TS. EMF: positive. Thermocouple wire-matched with Alumel for temperature measurement to 2400°F; corrosion and heat resistant.

CHROMEL R
Hoskins Mfg. Co.
Ni 80, Cr 20.
Wire: 95,000-175,000 TS; 50,000 YS; 35 El. Annealed: 95,000 TS; 25-35 El; 55 RA; B 85-90 Rock. For precision wound resistors, potentiometers, heating elements. High heat and oxidation resistant. *Obsolete*

CHROMENAR 308, ETC.
Arcos Alloys
Stainless steel.
Now ARCOS 308, etc.

CHROMEND 1 MA
Arcos Alloys
C 0.5, Mn 0.75, Si 0.6, Cr 0.8-1.1, Mo 0.45-0.65, bal Fe.
Welded: 85,000-100,000 TS; 68,000-88,000 YS; 20-23 El; 50-65 RA. For welding rod; repairing 1% Cr steel castings and pipe. *Obsolete*

CHROMEND 14/75
Arcos Alloys
Now ARCOS UNIA (ENiCrFe-2).

CHROMEND 15/85
Arcos Alloys
Cr 15, Ni 85.
For welding electrodes; heat resisting. *Obsolete*

CHROMEND 16-8-2
Arcos Alloys
Now ARCOS 16-8-2/15.

CHROMEND 19/9 MN
Arcos Alloys
C 0.07-0.17, Mn 3.3-4.75, Mo 0-0.25, Cr 18-20.5, Ni 9-10.5, bal Fe.
Welded: 84,000-96,000 TS; 56,000-62,000 YS; 37-46 El; 35-49 RA. For welding electrodes; for armor, coated. *Obsolete*

CHROMEND 19/9 WMO
Arcos Alloys
Cr 19, Ni 9, C, W, Mo, bal Fe.
For welding electrodes; stainless, austenitic. *Obsolete*

CHROMEND 25/12 CB
Arcos Alloys
C 0.1, Cr 25, Ni 12, Cb 0.9-1.1, bal Fe.
For welding rod; corrosion and heat resistant. *Obsolete*

CHROMEND 25/3 MO
Arcos Alloys
Cr 25, Ni 3, C, Mo, bal Fe.
For welding electrodes; stainless. *Obsolete*

CHROMEND 28/3 MO
Arcos Alloys
Cr 28, Ni 3, Mo 1, C, bal Fe.
For welding rod; heat resistant. *Obsolete*

CHROMEND 2M
Arcos Alloys
C 0.1, Cr 2, Mo 0.9, bal Fe.
Welded: 150,000 TS; 120,000 YS; 4 El; 10 RA; 325 Brin. For welding rod; for low Cr steel. *Obsolete*

CHROMEND 2MA
Arcos Alloys
C 0.03, Mn 0.7, Si 0.6, Cr 2-2.5, Mo 0.9-1.2, bal Fe.
Welded: 113,000 TS; 99,000 YS; 18 El; 48 RA. Stress relieved (1350°F): 92,000 TS; 85,000 YS; 25 El; 72 RA. For welding rod, type E9018-B3L; welding 2-1/4 Cr, 1 Mo steel. *Obsolete*

CHROMEND 307
Arcos Alloys
Now ARCOS 307.

CHROMEND 308
Arcos Alloys
Now ARCOS 308.

CHROMEND 308HC
Arcos Alloys
Now ARCOS 308HC.

CHROMEND 308L
Arcos Alloys
Now ARCOS 308L.

CHROMEND 309
Arcos Alloys
Now ARCOS 309.

CHROMEND 309 CB
Arcos Alloys
Now ARCOS 309 CB.

CHROMEND 309MO
Arcos Alloys
Now ARCOS 309MO.

CHROMEND 310
Arcos Alloys
Now ARCOS 310.

CHROMEND 310 CB
Arcos Alloys
Now ARCOS 310 CB.

CHROMEND 310 MO
Arcos Alloys
Now ARCOS 310 Mo.

CHROMEND 310HC
Arcos Alloys
Now ARCOS 310HC.

CHROMEND 312
Arcos Alloys
Cr 29, Ni 9, C, bal Fe.
For welding rod; heat resistant. *Obsolete*

CHROMEND 316
Arcos Alloys
Now ARCOS 316.

CHROMEND 316L
Arcos Alloys
Now ARCOS 316L.

CHROMEND 317
Arcos Alloys
Now ARCOS 317L.

CHROMEND 318
Arcos Alloys
Now ARCOS 318.

CHROMEND 320
Arcos Alloys
Now ARCOS 320.

CHROMEND 330
Arcos Alloys
Now ARCOS 330.

CHROMEND 330HC
Arcos Alloys
Now ARCOS 330HC.

CHROMEND 347
Arcos Alloys
Now ARCOS 347.

CHROMEND 349
Arcos Alloys
Now ARCOS 349.

CHROMEND 410

Arcos Alloys
Now ARCOS 410.

CHROMEND 410NI

Arcos Alloys
Now ARCOS 410NI.

CHROMEND 430

Arcos Alloys
C 0.1, Cr 16, bal Fe.
For welding rod; corrosion resistant. *Obsolete*

CHROMEND 442

Arcos Alloys
C 0.1, Cr 18, bal Fe.
For welding rod; corrosion resistant. *Obsolete*

CHROMEND 446

Arcos Alloys
Cr 28, C, bal Fe.
For welding rod; heat resistant. *Obsolete*

CHROMEND 505

Arcos Alloys
Now ARCOS 505.

CHROMEND 5M

Arcos Alloys
C 0.1, Cr 4-6, Mo 0.5, bal Fe.
For welding rod. *Obsolete*

CHROMEND 8/18

Arcos Alloys
Cr 8, Ni 18, C, bal Fe.
For welding rod; corrosion resistant. *Obsolete*

CHROMEND CMV

Arcos Alloys
C 0.3, Cr 1.25, Mo 0.5, V 0.25, bal Fe.
Timken 17/22A(S) steel. Heat treated: 140,000-240,000 psi
TS. For welding electrodes. *Obsolete*

CHROMEND HN

Arcos Alloys
C 0.1, Cr 17, Ni 25, bal Fe.
For welding rod; heat and corrosion resistant. *Obsolete*

CHROMEND KS

Arcos Alloys
C 0.1, Si 2.25, Cr 18, Ni 8, bal Fe.
For welding rod; stainless. *Obsolete*

CHROMEND W

Arcos Alloys
Cr 18, Ni 8, C, bal Fe.
For welding electrodes; stainless, austenitic. *Obsolete*

CHROMESCO 1

Vallourec S.A.
C 0.14, Cr 0.5, Mo 0.5, bal Fe.
Good creep rupture properties up to 550°C. For boilers,
steam pipes, superheaters.

CHROMESCO 3

Vallourec S.A.
C 0-0.15, Cr 1.9-2.6, Mo 0.8-1.2, bal Fe.
Annealed: 60,000 TS; 30,000 YS; 30 El; 163 Brin. For oil
refinery tubes and equipment; ASTM-A-200.

CHROMET

German manufacture
Si 10, bal Al.
For bearings.

CHROMETOUGH

British Steel plc
C 0.55, Mn 0.55, Cr 0.5, V 0.2, bal Fe.
Oil hardening general purpose tool steel. For boring bars,
chisels, and rivet snaps. *Obsolete*

CHROMEWEAR

Teledyne Vasco
C 2.3, Mn 0.7, Si 0.4, Cr 5.25, V 4.75, W 1.1, Mo 1.1, bal Fe.
Heat treatable to high hardness and wear resistance; for tube
manufacturing rolls, brick mold liners. Type A7; air hardening
tool steel.

CHROMEWEAR 300

Teledyne Vasco
C 2.7, Si 0.4, Mn 0.7, Cr 8.25, V 4.5, Mo 1.12, bal Fe.
Wear resistant tool steel; air hardenable to 66 Rock C. For
deep drawing dies, brick mold liners. Liners for sand slingers,
extrusion punches for ceramics.

CHROMEWEAR 400

Teledyne Vasco
C 2.2-3.5, Cr 2.5-5, V 6-10, Mo 1.5-3, Si 0-1, S 0-0.03, P
0-0.03, Mn 0-1, W 0-1, Co 0-5.
62-63 Rock C. For brick mold liners, sandslinger liners, ore
handling equipment and other wear applications.

CHROMEWELD 4-6

Lincoln Electric Co.
Cr 4-6, C, Mo, bal Fe.
For welding electrodes; for 4-6 Cr steel. *Obsolete*

CHROMEX 2

Russian manufacture
C 1.2-1.7, Cr 32-35, Si 1.2-2, bal Fe.
For furnace equipment; heat resistant, cast iron.

CHROMEX 3

Russian manufacture
C 1.8-2.8, Cr 32, Si 1.5-2.5, bal Fe.
For furnace equipment; heat resistant, cast iron.

CHROMIC-A

Riverside Metals Corp.
Cr 20, Ni 80.
Annealed: 100,000 TS; 40,000 YS; 35 El. For resistors,
heating elements; high electrical resistance, for service to
2150 F. *Obsolete*

CHROMIC-C

Riverside Metals Corp.
Cr 18, Ni 60, bal Fe.
Annealed: 95,000 TS; 35,000 YS; 35 El. For resistors, heating
elements; high electrical resistance, for service up to 1750 F.
Obsolete

CHROMIDIUM 2

Midland Motor Cylinder Co.
C 3.1-3.5, Si 2-2.4, Mn 0.75-1, Cr 0.2-0.4, bal Fe.
Cast: 37,000 TS; 200-240 Brin. For brake drums, brake discs;
72,000 transverse strength; cast iron.

CHROMIDIUM-1

Midland Motor Cylinder Co.
TC 3.2, CC 0.7, Si 2.2, Cr 0.25-0.45, Mn 0.8, bal Fe.
Cast: 35,000-45,000 TS; 35,000-45,000 YS; 0 El; 0 RA;
200-250 Brin. For automobile engine blocks, cylinders, brake
drums; wear resistant, cast iron.

CHROMIN

German manufacture
For anodes for Cr-plating.

CHROMIN "D"

Wilbur B. Driver Co.
Ni 30.1, Fe 63.9, Mn 0.64, C 0.2, Cr 4-7.
Annealed: 83,000 TS. For low temperature heater works,
motor starters, fans, irons, toasters; also known as Comet and
No. 193 Alloy. *Obsolete*

CHROMIUM

Atomergic Chemetals Corp.
Cr.
Purities: 99.999%, 99.99%, 99.9%, 99.6 + % commercial
grade. Forms: crystal bar (ductile), degassed and regular
powders, pellet, foil, rod (hot pressed), tubules, flake, lump,
composites, single crystals, discs (arc melted).

CHROMIUM BORIDE B401

American manufacture
CrB.
Sintered. For high heat applications; operating temperature
range above 2000°F.

CHROMIUM BORON

Foote Mineral Co.
B 20, Cr 80.
For use in manufacture of non-ferrous alloys. *Obsolete*

CHROMIUM COPPER-182

Anaconda Co.
Copper. Cu 99.14, Cr 0.85, Si 0.01.
Hard rod: 70,000 TS; 60,000 YS; 20 El; 82 Rock B. Soft rod:
35,000 TS; 15,000 YS; 40 El. For cable connectors, electronic
devices, resistance welding electrode tips, grid supports,
circuit breakers. Age hardenable, corrosion resistant.
Hardened electrical conductivity: 80% IACS.

CHROMIUM PERMALLOY

Bell Telephone Laboratories
Cr 3.8, Ni 78.5, Fe 17.5.
For high frequency apparatus; high permeability. *Obsolete*

CHROMIUM-COPPER 999

Anaconda Co.
Copper. Cu 99.05, Cr 0.85, Si 0.1.
Soft: 35,000 TS; 15,000 YS; 40 El; 877 Brin. Hard: 72,000 TS;
61,000 YS; 20 El. For electrical apparatus. High conductivity.

CHROMNICKEL 1

VDM Nickel-Technologie AG
Ni 78, Cr 20, Mn 2.
For furnace parts, electric heating elements; heat and
corrosion resistant. *Obsolete*

CHROMNICKEL II

VDM Nickel-Technologie AG
Ni 70, Cr 20, Mn 2, Fe 8.
For electrical heating elements; heat and acid resistant.
Obsolete

CHROMO-LOY

Grand Northern Products Ltd.
C 3, Si 2, Mn 0.43, Cr 21.6, Mo 4.2, Ni 3.4, bal Fe.
Coated, hardfacing electrode. As arc welded: 70,000 psi TS
approx; 56 Rock C. For overlay or buildup on roll crushers,
shovel teeth, bucket lips and oil field tool joints; tough and
wear resistant. *Obsolete*

CHROMO-N

Bethlehem Steel Corp.
C 0.23, Si 1.25, Ni 0.75, Cr 10, Mo 1.2, V 1, bal Fe.
For extrusion mandrels, hot-work tools and dies; hot-work
steel. *Obsolete*

CHROMODUR-22

Krupp Stahl AG
stainless steel.
See CRONIDUR 4923 *Obsolete*

CHROMODUR-33

Krupp Stahl AG
stainless steel. C 0.2, Cr 12, Mo 1, Ni 0.4, V 0.3, W 0.5, bal Fe.
Annealed: 95,000 TS; 40,000 YS; 25 El; 92 Rock B. Heat
treated: 240,000 TS; 205,000 YS; 10 El; 48 Rock C. For
cutlery, hardware, knives, oil refinery and chemical plant
equipment. Corrosion resistant, hardenable. *Obsolete*

CHROMOLD-VM

Teledyne Vasco
C 0.06-0.12, Si 0.15-0.25, Mn 0.1-0.4, S 0-0.01, P 0-0.01, Cr
2.15-2.45, bal Fe.
Heat treated: 110,000-120,000 TS; 70,000-90,000 YS; 20-25
Rock C in core and 64 Rock C in case. Plastic molds, cavities.
Vacuum melted. Case hardening steel. Easy hubbing for
mold cavities.

CHROMSOL
Union Carbide Corp.
Cr 62, Mn 5, Si 1.5, C 5.25, bal Fe.
For chromium additions to steel melts. Steel alloying agent.

CHROMTEC 10680
Eutectic Corp.
Nickel base alloy powder for steel, stainless, and nickel alloys.

CHRONIFER F 14
Now WESTIG 4006.

CHRONIFER F 17
Now WESTIG 4016.

CHRONIFER SPEZIAL
Westig GmbH
C 0-0.12, Si 1, Mn 2, Cr 18, Ni 9, bal Fe.
Non hardenable, austenitic stainless steel. Werkstoff Nr. 1.4300; AISI-SAE 302. *Obsolete*

CHRONIFER SPEZIAL 4 SUPRA
Now WESTIG 4401.

CHRONIFER SPEZIAL D
Now WESTIG 4305.

CHRONIFER SPEZIAL EXTRA
Now WESTIG 4541.

CHRONIFER SPEZIAL SUPRA
Now WESTIG 4301.

CHRONIFER V-13
Now WESTIG 4021.

CHRONIKA 1565
Main Metal Ltd.
Nickel. Cr 0.2, Ni 57, Cr 15, bal Fe.
For high temperature applications. X15NiCr5715. Heat resistant.

CHRONIKA 2035
Main Metal Ltd.
C 0.15, Ni 30, Cr 21, bal Fe.
For high temperature applications. X15NiCr3021. Heat resistant.

CHRONIKA 2080
Main Metal Ltd.
Nickel. C 0.15, Ni 77, Cr 18, bal Fe.
For high temperature applications. X15NiCr7718. Heat resistant.

CHRONIKA 2520
Main Metal Ltd.
C 0.15, Si 2, Cr 24, Ni 19, bal Fe.
For furnace parts and heat treat boxes. X15CrNiSi2419. Corrosion and heat resistant.

CHRONIN
English manufacture
Ni 83.7, Cr 14.7.
For resistance alloy.

CHRONIN "85"
Vereinigte Deutsche Nickel-Werke AG
Cr 13, Ni 85, Fe, Mn, Al impurities, bal Cu.
Annealed: 100,800 TS; 30 El. For electric resistances, resistors, thermocouple element against Ni; heat resistance to 1200°C. *Obsolete*

CHRONIN 100
Vereinigte Deutsche Nickel-Werke AG
Cr 18, Ni 80, Fe, Mn, Al, bal Cu.
Annealed: 117,000 TS; 30 El. For electric resistance, resistors, thermocouple element; heat resistant to 1200°C. *Obsolete*

CHRONIN 110
Vereinigte Deutsche Nickel-Werke AG
Nickel.
Now IGS-CHRONIN 110.

CHRONIN 110TI
Vereinigte Deutsche Nickel-Werke AG
Nickel. Fe 3, C 0.1, Cr 20, Ti 0.4, bal Ni + Co.
For apparatus and welding applications.

CHRONIT 14
J.C. Soding & Halbach
C 0.2, Cr 13, bal Fe.
Annealed: 95,000 TS; 50,000 YS; 25 El; 92 Rock B.
Hardened: 250,000 TS; 215,000 YS; 8 El; 52 Rock C. For cutlery, surgical instruments, rules, gears, valves, springs, pivots. Corrosion resistant. Hardenable.

CHRONIT 14 M
J.C. Soding & Halbach
C 0.18, Cr 13, Mo 0.6, Ni 0.7, bal Fe.
Annealed: 100,000 TS; 55,000 YS; 20 El; 95 Rock B.
Hardened: 240,000 TS; 210,000 YS; 8 El; 50 Rock C. For cutlery, surgical instruments, gears, springs, valve parts. Corrosion resistant. Hardenable.

CHRONIT 14 MS
J.C. Soding & Halbach
C 0.18, Cr 13, Mo 0.6, Ni 0.7, others, bal Fe.
Annealed: 100,000 TS; 55,000 YS; 20 El; 95 Rock B.
Hardened: 240,000 TS; 210,000 YS; 8 El; 50 Rock C. For cutlery, surgical instruments, gears, springs, valve parts. Corrosion resistant. Hardenable.

CHRONIT 14 N
J.C. Soding & Halbach
C 0.17, Cr 13, Ni 1.1, bal Fe.
Annealed: 95,000 TS; 52,000 YS; 22 El; 82 Rock B.
Hardened: 220,000 TS; 180,000 YS; 12 El; 48 Rock B. For cutlery, surgical instruments, gears, springs, valve trim. Corrosion resistant. Hardenable.

CHRONIT 14 P
J.C. Soding & Halbach
C 0.1, Cr 12.3, Mo 0.6, Ni 0.5, bal Fe.
Annealed: 85,000 TS; 45,000 YS; 30 El; 85 Rock B. For chemical and oil refinery equipment, valve trim. Corrosion resistant.

CHRONIT 1417
J.C. Soding & Halbach
Stainless steel. C 0-0.07, Cr 17, Mo 4.5, Ni 13.5, Cb = 8 x C, bal Fe.
Annealed: 90,000 TS; 45,000 YS; 35 El; 160 Brin. For chemical and oil refinery equipment, digesters, acid tanks, agitators. Austenitic, non-hardenable. Welding grade, stainless, non-magnetic.

CHRONIT 1618
J.C. Soding & Halbach
C 0-0.08, Cr 17.5, Mo 2.25, Ni 20, Cu 2, Nb = 8 x C, bal Fe.
Cast, quenched: 65,000-92,000 TS; 25,000 YS; 15 El; 130-180 Brin. Acid resistant chrome-nickel austenitic cast steel. DIN G-X8CrNiMoCu1818.

CHRONIT 165 M
J.C. Soding & Halbach
C 0-0.07, Cr 16, Mo 2, Ni 6, others, bal Fe.
Annealed: 85,000 TS; 35,000 YS; 50 El; 80 Rock B.
Precipitation hardenable to about 400 Brin. For chemical and pharmaceutical plant equipment digesters, valve trim, acid containers, tanks. Corrosion resistant.

CHRONIT 18 M
J.C. Soding & Halbach
C 0.2, Mo 16.5, Ni 1.2, bal Fe.
Annealed: 80,000 TS; 45,000 YS; 26 El; 140 Brin. For hardware, burner parts, oil refinery equipment, fasteners, furnace parts, storage tanks. Corrosion resistant.

CHRONIT 18 N
J.C. Soding & Halbach
C 0.2, Cr 16.5, Ni 1.2, bal Fe.
Annealed: 82,000 TS; 43,000 YS; 28 El; 140 Brin. For hardware, burner parts, oil refinery equipment, furnace parts. Corrosion resistant.

CHRONIT 1810 MN
J.C. Soding & Halbach
C 0.1, Cr 17.5, Ni 10, Mn 10, bal Fe.
Annealed: 100,000 TS; 50,000 YS; 60 El; B 90 Rock. For cooking utensils, dairy and chemical equipment, springs, sinks. Stainless, non-magnetic, austenitic. *Obsolete*

CHRONIT 1811 N
J.C. Soding & Halbach
C 0.05, Cr 17, Ni 10.5, bal Fe.
Annealed: 80,000 TS; 30,000 YS; 50 El; B 80 Rock. Cold drawn: 125,000 TS; 85,000 YS; 20 El; B 98 Rock. For chemical plant equipment, agitators, evaporators, digesters, kettles. Stainless, non-magnetic, austenitic. *Obsolete*

CHRONIT 2025
J.C. Soding & Halbach
C 0-0.08, Cr 20, Mo 2.75, Ni 25, Cu 2, Cb = 8 x C, bal Fe.
Annealed: 95,000 TS; 45,000 YS; 45 El; 90 Rock B. For furnace parts, heat treating boxes, carburizing boxes, valves, pumps, jet engine parts. Non-hardenable. Welding grade, heat resistant, austenitic.

CHRONIT 216 U
J.C. Soding & Halbach
C 0.05, Cr 15.5, Ni 10, Cb 0.5, Cu 2.2, bal Fe.
Annealed: 75,000 TS; 30,000 YS; B 80 Rock. For chemical plant equipment, welded structures. Welding grade, stabilized. Stainless, non-magnetic, austenitic. *Obsolete*

CHRONIT 218
J.C. Soding & Halbach
Stainless steel. C 0-0.15, Cr 18, Ni 9, bal Fe.
Annealed: 90,000 TS; 40,000 YS; 50 El; 85 Rock B. Cold rolled: 150,000 TS; 100,000 YS; 25 El; 100 Rock B. For chemical and pharmaceutical equipment, valve trim, fasteners, molding, acid tanks, digesters. Stainless, austenitic, non-magnetic.

CHRONIT 218 E
J.C. Soding & Halbach
Stainless steel. C 0-0.1, Cr 18.5, Ni 10, Cb = 8 x C, bal Fe.
Annealed: 85,000 TS; 35,000 YS; 50 El; 82 Rock B. Cold rolled: 130,000 TS; 90,000 YS; 30 El; 100 Rock B. For welded structures, chemical plant equipment, vessels, agitators. Austenitic, stabilized. Welding grade, stainless, non-magnetic.

CHRONIT 218 S
J.C. Soding & Halbach
Stainless steel. C 0-0.07, Cr 18.5, Ni 10, bal Fe.
Annealed: 85,000 TS; 35,000 YS; 50 El; 82 Rock B. For chemical plant equipment, tanks, evaporators, agitators, kettles. Stainless, austenitic, non-magnetic.

CHRONIT 274 R
J.C. Soding & Halbach
Stainless steel. C 0.04, Cr 27, Ni 4, bal Fe.
Cast, annealed: 95,000-120,000 TS; 4 El; 210-260 Brin. Ferrite-austenite carbide structure. Stainless and acid-resistant cast steel. DIN G-XCrNi274.

CHRONIT 274 RM
J.C. Soding & Halbach
Stainless steel. C 0.4, Cr 27, Mo 2.25, Ni 4, bal Fe.
Cast, quenched: 210-260 Brin; 95,000-120,000 TS; 4 El. Ferrite-austenite carbide; stainless and acid-resistant cast steel.

CHRONIT 30
J.C. Soding & Halbach
Stainless steel. C 1, Cr 28, bal Fe.
Cast, annealed: 260-330 Brin. Ferrite-carbide structure; stainless and acid resistant cast steel. DIN G X1200i28.

CHRONIT 30 W

J.C. Soding & Halbach
Stainless steel. C 0.7, Cr 28, bal Fe.
Cast, annealed: 210-280 Brin. Ferrite-carbide structure; stainless and acid-resistant cast steel. DIN G-X70Cr29.

CHRONIT 30 C

J.C. Soding & Halbach
Stainless steel. C 1.5, Cr 28, bal Fe.
Cast, annealed: 260-330 Brin. Ferrite-carbide structure; stainless and acid-resistant cast steel.

CHRONIT 30 M

J.C. Soding & Halbach
Stainless steel. C 1, Cr 28, Mo 2.25, Cu 1.25, bal Fe.
Cast, annealed: 260-330 Brin. Ferrite-carbide structure: stainless and acid-resistant cast steel. DIN G-X120CrMo 292.

CHRONIT 30 MW

J.C. Soding & Halbach
Stainless steel. C 0.7, Cr 28, Mo 2.25, bal Fe.
Cast, annealed: 219-280 Brin. Ferrite-carbide structure; stainless and acid-resistant cast steel. DIN G-XCrMo292.

CHRONIT 418

J.C. Soding & Halbach
Stainless steel. C 0-0.15, Cr 18, Mo 2.25, Ni 10, bal Fe.
Annealed: 90,000 TS; 40,000 YS; 40 El; 85 Rock B. For chemical plant and oil refinery equipment digesters, valve trim, evaporators. Stainless, austenitic, non-magnetic.

CHRONIT 418 E

J.C. Soding & Halbach
Stainless steel. C 0-0.1, Cr 17.5, Mo 2.25, Ni 11.5, Cb = 9 x C, bal Fe.
Annealed: 90,000 TS; 40,000 YS; 40 El; 85 Rock B. For welded structures, tanks, vessels, chemical plant and oil refinery equipment austenitic. Welding grade, stainless, non-magnetic.

CHRONIT 418 S

J.C. Soding & Halbach
Stainless steel. C 0-0.07, Cr 18, Mo 2.25, Ni 11, bal Fe.
Annealed: 85,000 TS; 35,000 YS; 50 El; 82 Rock B. For chemical plant equipment, kettles, agitators, evaporators. Stainless, austenitic, non-magnetic.

CHRONIT CNM100

Now VEW K618.

CHRONIT CR150H

Vereinigte Edelstahlwerke
C 1, Cr 1.4, bal Fe.
Heat treated: 213,400 TS; 400 Brin. For machine tool parts; oil hardened, wear resistant. *Obsolete*

CHRONIT CR150W

Vereinigte Edelstahlwerke
C 0.55, Cr 1.6, bal Fe.
Heat treated: 199,200 TS; 380 Brin. For wear resistant machine tool parts; oil hardened, wear resistant. *Obsolete*

CHRONIT CSFG

Vereinigte Edelstahlwerke
C 0.6, Mn 0.6, Si 1.7, Cr 0.8, bal Fe.
Heat treated: 184,900 TS; 380 Brin. For wear resistant machine tool parts; oil hardened, wear resistant. *Obsolete*

CHRONIT SPEZIAL KG

Vereinigte Edelstahlwerke
C 2, Cr 12, bal Fe.
For blanking and forming dies; oil hardened, non-deforming. *Obsolete*

CHRONIT VHS

Now VEW F126.

CHRONIT VM

Now VEW K724.

CHRONIT VMG

Bohler Gesellschaft M.B.H.
Cobalt. C 1.3, Cr 25, W 4, Co 67, Fe 0-2.
Cast: 42-45 Rock C. For hardfacing. Wear and corrosion resistant.

CHRONIT VMH

Bohler Gesellschaft M.B.H.
C 0.9, Mn 1.8, bal Fe.
Heat treated: 171,000 TS; 350 Brin. For wear resistant parts, punches; oil hardened.

CHRONIT ZII

Bohler Gesellschaft M.B.H.
C 2, Cr 2, bal Fe.
Heat treated: 185,000 TS; 370 Brin. For wear resistant parts; oil hardened.

CHRONITE 275 RM

J.C. Soding & Halbach
Stainless steel. C 0.1, Cr 27, Mo 1.5, Ni 5, bal Fe.
Cast, quenched: 85,000-110,000 TS; 50,000 YS; 8 El; approx 240 Brin. Ferrite-austenite structure; stainless and acid-resistant cast steel. DIN G-X10CrNiMo275.

CHRONITE 418 ES

J.C. Soding & Halbach
Stainless steel. C 0-0.1, Cr 17.5, Mo 2.25, Ni 11.5, Cb = 9 x C, bal Fe.
Annealed: 85,000 TS; 35,000 YS; 50 El; 82 Rock B. For acid tanks, evaporators, kettles, chemical plant equipment. Austenitic, stabilized. Welding grade, stainless, non-magnetic.

CHRONITHERM 20

J.C. Soding & Halbach
C 0.1, Si 2, Mn 0.5, Cr 24, Ni 19, bal Fe.
Annealed: 95,000 TS; 43,000 YS; 48 El; B 90 Rock. For heating elements. Heat and oxidation resistant, austenitic. *Obsolete*

CHRONITHERM 20/SPEZ.

J.C. Soding & Halbach
C 0.1, Si 2.6, Mn 0.5, Cr 24, Ni 19, bal Fe.
Annealed: 95,000 TS; 43,000 YS; 46 El; B 90 Rock. For heating elements. Heat and oxidation resistant, austenitic. *Obsolete*

CHRONITHERM 30

J.C. Soding & Halbach
C 0.07, Si 2.3, Mn 1, Cr 19.5, Ni 33, bal Fe.
For heating elements. Heat and oxidation resistant. *Obsolete*

CHRONITHERM 30/SPEZ.

J.C. Soding & Halbach
C 0.07, Si 2.3, Mn 1, Cr 19.5, Ni 33, bal Fe.
For heating elements. Heat and oxidation resistant. *Obsolete*

CHRONITHERM 60

J.C. Soding & Halbach
C 0.1, Si 0.8, Mn 1, Cr 17.5, Ni 62, bal Fe.
For heating elements. Heat and oxidation resistant. *Obsolete*

CHRONITHERM 60/SPEZ.

J.C. Soding & Halbach
C 0.1, Si 0.8, Mn 1, Cr 17.5, Ni 62, bal Fe.
For heating elements. Heat and oxidation resistant. *Obsolete*

CHRONITHERM 80

J.C. Soding & Halbach
C 0.06, Si 1.6, Mn 0.2, Cr 19, Ni 77, bal Fe.
For heating elements. Heat and oxidation resistant. *Obsolete*

CHRONITHERM 80/SPEZ.

J.C. Soding & Halbach
C 0.06, Si 1.6, Mn 0.2, Cr 19, Ni 77, bal Fe.
For heating elements. Heat and oxidation resistant. *Obsolete*

CHRONOS

Now VEW K700.

CHRYSIODE

Manufacturer not listed
Ag 92, Al 8.

CHRYSITE

English manufacture
Cu 63, Zn 37, Pb 0.24.
For dental alloy.

CHRYSOKALK-1

English manufacture
Cu 95, Zn 4.5, Pb 0.5.
For cheap jewelry; tarnishes easily.

CHRYSOKALK-2

English manufacture
Cu 59, Zn 40, Pb 1.
For tubes.

CHRYSOKALK-3

English manufacture
Cu 91, Zn 7.9, Pb 1-6.
For cheap jewelry.

CHRYSORIN

English manufacture
Cu 72-63, Zn 28-37.
40,000-56,000 TS; 20,000-35,000 YS; 78-35 El; 78-63 RA. For tubes, pipes, plumbing; high strength.

CHUGAL

Russian manufacture
C 2.5-3.2, Si 1.6-2.3, Al 5.5-20, bal Fe.
For furnace equipment; cast iron.

CICRON-1

Forjas Alavesas S.A.
C 0.65, Si 0-1.2, Cr 0.8, Mo 0.3, bal Fe.
Hot work tool steel.

CICRON-2

Forjas Alavesas S.A.
C 0.6, Si 0-1.2, Cr 0.8, Mo 0.3, bal Fe.
Hot work tool steel.

CIMET (MALLEABLE)

Driver Harris Co.
Cr 22-26, Ni 12-14, bal Fe.
For mine water pumps, corrosion resisting parts; corrosion and heat resistant. *Obsolete*

CIMET, CAST

Driver Harris Co.
Cr 27-29, Ni 11-13, bal Fe.
For pumps, screens, valves, furnace parts, mining machinery; corrosion resistant to acid mine waters and many chemicals. *Obsolete*

CINDAL

D. & J. Tullis Ltd.
Zn 0.8, Mg 0.3, Cr 0.3, bal Al.
27,000-58,000 TS; 5-20 El. For light alloy parts; heat treatable.

CINDAL ALLOY "J-551"

Cindal Aluminium Ltd.
Aluminum. Si 5, bal Al.
22,000-26,000 TS; 11,000-15,000 YS; 15-17 El. Used to resist acetic acid, edible oils, acids, beer, cider, fruit juices; stainless and non-corrosive in sea water.

CINDALL "J-12"

Cindal Aluminium Ltd.
Aluminum. Al alloy.
15,500-21,200 TS; 11,000-13,000 YS; 9-11 El. For resistance to industrial waters and alkaline solutions, radiators, water meters; nonhardenable.

CINDALL "L-316"
Cindal Aluminium Ltd.
Aluminum. Al alloy.
29,000 TS; 16,000 YS; 2-4 El; 68 Brin. For exhaust manifolds, automobile parts; sand and die castings.

CINDALL 50 "A"
Cindal Aluminium Ltd.
Aluminum. Al alloy.
36,000 TS; 22,500 YS; 1-1.5 El; 140 Brin. Domestic castings, light gear wheels, light pulleys, bearings and brushings; to replace cast iron where lightness and corrosion resistance are required.

CINDALL E-11 "A"
Cindal Aluminium Ltd.
Aluminum. Al alloy.
31,500 TS; 14,000 YS; 5-8 El; 66 Brin. For water pipes, carburetor parts; sand or chill castings subjected to atmospheric, sea water, or other corrosive influences; boron modified.

CINDALL E-11 "B"
Cindal Aluminium Ltd.
Aluminum. Al alloy.
36,000 TS; 18,500 YS; 4-6 El; 72 Brin. For propellers, water cooled cylinders, cylinder heads, pump impellers, stern tubes, water pump bodies; sand castings, boron modified.

CINDALL ALLOY "J-51"
Cindal Aluminium Ltd.
Aluminum. Al alloy.
20,000-24,000 TS; 9,000-11,000 YS; 24-28 El. For parts to resist sea water, fruit acids, beer and cider vats; non-corrosive to sea water.

CINIDUR
Manufacturer not listed
C 0.25, Cr 19, Ni 24, Mo 2, W 1, Ti 2.25, Al 1, bal Fe.
For oil refinery equipment, furnace parts; heat resistant.

CINSEAL
ITT Components Group Europe
Ni 29, Fe 53, Co 17.5, Mn 0.5.
For telecommunications, hermetic seals; metal-to-glass seal.

CIRCLE "C"
Hoover Ball & Bearing Co.
C 0.77, Co 9, W 18.5, Cr 4.5, V 2, Mo 1, bal Fe.
For tools, cutters, drills, tools for milling, slotting, forming; super high speed steel; tough, heavy duty. *Obsolete*

CIRCLE C
Teledyne Vasco
C 0.8, W 18.5, Mo 0.95, Cr 4.75, V 2.05, Co 8.9.
66 Rock C. Tungsten type high speed tool steel; Izod: 20 feet.

CIRCLE L 1
Lebanon Steel Foundry
Mn 1-1.5, C 0.25, V or Mo, bal Fe.
Air hardened: 85,000-100,000 TS; 55,000-65,000 YS; 22-30 El; 40-55 RA; 160-200 Brin. For general castings; wear and corrosion resistant. *Obsolete*

CIRCLE L 10
Now LEBANON C5.

CIRCLE L 11
Now LEBANON 442.

CIRCLE L 11
Lebanon Steel Foundry
C 0.75, Cr 18, 0.75 others, bal Fe.
525 Brin. For sand pumps, corrosion resistant castings; hard stainless steel. *Obsolete*

CIRCLE L 12
Now LEBANON CA15.

CIRCLE L 12M
Now LEBANON CA15M.

CIRCLE L 13
Now LEBANON CA40.

CIRCLE L 130
Now LEBANON CHW.

CIRCLE L 15
Now LEBANON CC50.

CIRCLE L 16
Lebanon Steel Foundry
Cr 27-30, C 2.25, bal Fe.
Hardened: 180,000 TS; 450-600 Brin. For sand pumps, chemical pumps. *Obsolete*

CIRCLE L 17
Now LEBANON 17-4.

CIRCLE L 18
Lebanon Steel Foundry
C 0-0.07, Cr 37.5, Al 7.5, bal Fe.
For cast resistors; heat resistant. *Obsolete*

CIRCLE L 19
Now LEBANON LC2.

CIRCLE L 2
Lebanon Steel Foundry
Mn 1.4, C 0.3, Cr 0.75, Mo 0.3, bal Fe.
Air hardened: 100,000-120,000 TS; 70,000-90,000 YS; 17-24 El; 30-55 RA; 190-250 Brin. For crankshafts, valves, fittings; wear and corrosion resistant. *Obsolete*

CIRCLE L 205A
Now LEBANON 8630.

CIRCLE L 205AL
Now LEBANON 8630-1.

CIRCLE L 205B
Now LEBANON 8630-2.

CIRCLE L 205C
Now LEBANON 8630-3.

CIRCLE L 205D
Now LEBANON 8630-4.

CIRCLE L 206
Now LEBANON 8613.

CIRCLE L 209
Now LEBANON WC6.

CIRCLE L 219
Now LEBANON LC3.

CIRCLE L 22
Now LEBANON CF8.

CIRCLE L 22 AGXM
Now LEBANON CG8M.

CIRCLE L 22 XML
Now LEBANON CF3M.

CIRCLE L 22L
Now LEBANON CF3.

CIRCLE L 22M
Now LEBANON CF8C.

CIRCLE L 22XM
Now LEBANON CF8M.

CIRCLE L 23
Now LEBANON CF20.

CIRCLE L 25M
Lebanon Steel Foundry
C 0-0.2, Cr 21, Ni 10, Se 0.25, bal Fe.
Water quenched: 80,000 TS; 40,000 YS; 40 El; 40 RA; 160 Brin. For free machining stainless parts; free machining, stainless. *Obsolete*

CIRCLE L 3
Now LEBANON 4140.

CIRCLE L 30H
Now LEBANON HH.

CIRCLE L 31
Now LEBANON CE30.

CIRCLE L 31H
Now LEBANON HE.

CIRCLE L 32
Now LEBANON HT.

CIRCLE L 33
Now LEBANON 33.

CIRCLE L 34
Now LEBANON CN7M.

CIRCLE L 4
Lebanon Steel Foundry
Cr 1.25-2, C 0.5-0.8, Mo 0.5-1, bal Fe.
Cast and heat treated: 125,000-275,000 TS; 90,000-225,000 YS; 3-12 El; 3-20 RA; 250-600 Brin. For cams, mixer blades rolls, pressing dies; wear and abrasion resistant. *Obsolete*

CIRCLE L 40
Lebanon Steel Foundry
C 0.5, Cr 28, Ni 8, bal Fe.
For heat resisting parts; heat resistant. *Obsolete*

CIRCLE L 41
Now LEBANON HX.

CIRCLE L 42
Lebanon Steel Foundry
C 0.4, Cr 12, Ni 60, bal Fe.
Cast: 78,000 TS; 33,000 YS; 15 El; 160 Brin. For heat resisting parts; heat resistant. *Obsolete*

CIRCLE L 43
Lebanon Steel Foundry
C 0.4, Cr 19, Ni 38, bal Fe.
Cast: 72,000 TS; 47,000 YS; 15 El; 170 Brin. For heat resisting parts; heat resistant. *Obsolete*

CIRCLE L 431
Now LEBANON 431.

CIRCLE L 44
Lebanon Steel Foundry
C 0.4, Cr 10, Ni 30, bal Fe.
Cast: 68,000 TS; 33,000 YS; 15 El; 140 Brin. For heat resisting; heat resistant. *Obsolete*

CIRCLE L 45
Lebanon Steel Foundry
C 0.4, Cr 20, Ni 25, bal Fe.
Cast. For heat resisting parts; heat resistant. *Obsolete*

CIRCLE L 46
Now LEBANON CK20.

CIRCLE L 46H
Now LEBANON HK.

CIRCLE L 47

Lebanon Steel Foundry
C 0.4, Cr 30, Ni 20, bal Fe.
For heat resisting parts; heat resistant. *Obsolete*

CIRCLE L 48

Lebanon Steel Foundry
C 0.4, Cr 30, Ni 30, bal Fe.
For heat resisting castings; maximum load carrying ability at high temperature. *Obsolete*

CIRCLE L 5

Now LEBANON 4330.

CIRCLE L 5

Lebanon Steel Foundry
Ni 2, C 0.3, Cr 0.75, Mo 0.3, bal Fe.
Air hardened: 100,000-120,000 TS; 65,000-85,000 YS; 18-24 El; 30-50 RA; 190-250 Brin. For valves, fittings and pressure parts for high temperatures; for high stressed parts. *Obsolete*

CIRCLE L 8

Now LEBANON WC8.

CIRCLE L 9

Now LEBANON WC1.

CIRCLE L 91

Now LEBANON C-12.

CIRCLE L NO. 25

Lebanon Steel Foundry
C 0-0.2, Cr 20-23, Ni 9-11, bal Fe.
Cast: 70,000-85,000 TS; 35,000-45,000 YS; 30-40 El; 35-45 RA; 130-170 Brin. For heat and corrosion resisting parts; heat and corrosion resistant. *Obsolete*

CIRCLE L-106

Lebanon Steel Foundry
C 0.17, Mn 1.3, Mo 0.25, bal Fe.
Cast: 125,000 TS; 90,000 YS; 10 El; 22 RA; 260 Brin. For castings. *Obsolete*

CIRCLE L-119

Lebanon Steel Foundry
C 0.2, Mn 0.65, Ni 1.75, bal Fe.
Cast: 78,000 TS; 52,000 YS; 28 El; 50 RA; 165 Brin. For castings. *Obsolete*

CIRCLE L-430

Lebanon Steel Foundry
C 0.35, Si 1.25, Cr 24.5, Ni 11, bal Fe.
Cast: 90,000 TS; 50,000 YS; 15 El; 190 Brin. For chemical equipment; stainless. *Obsolete*

CIRCLE L-431

Lebanon Steel Foundry
C 0.35, Si 1.25, Cr 28.5, Ni 9.5, bal Fe.
Cast: 85,000 TS; 45,000 YS; 15 El; 170 Brin. For chemical equipment; stainless. *Obsolete*

CIRCLE L22AG

Lebanon Steel Foundry
C 0.07, Cr 18, Ni 8, Ag 0.25-0.3, bal Fe.
For stainless parts, chemical plant equipment; stainless against sea water. *Obsolete*

CIRCLE L22AGXM

Lebanon Steel Foundry
C 0.07, Cr 18, Ni 8, Mo 3, Ag 0.25, bal Fe.
For stainless parts, chemical plant equipment; resists sea water corrosion. *Obsolete*

CIRCLE L6

Lebanon Steel Foundry
C 0.15, Ni 1.75, Mo 0.25, bal Fe.
Cast: 103-250 TS; 67,000 YS; 20 El; 40 RA; 207 Brin. For carburized parts requiring tough core and hard surface; carburizing steel castings. *Obsolete*

CIRCLE LA

Now LEBANON 1040.

CIRCLE LB

Now LEBANON WCA.

CIRCLE LB20

Now LEBANON WCB.

CIRCLE LCD4MCU

Now LEBANON CD.

CIRCLE LHB

Now LEBANON HAB.

CIRCLE LHC

Now LEBANON HAC.

CIRCLE LIN

Now LEBANON INC.

CIRCLE LM

Now LEBANON ME.

CIRCLE M

Teledyne Firth Sterling
Steel. C 0.85, W 6, Mo 5, Cr 4, V 2, Co 8, bal Fe.
For drills, hobs, reamers, taps, broaches, cutters; high speed steel, Type T5. *Obsolete*

CIRCLEF L 109

Now LEBANON WC6A.

CIRCULAR SAW BIT

Colt Industries
C 0.75, Mo 0.3, bal Fe.
For saw teeth; water hardening. *Obsolete*

CIRCULAR SAW PLATE 55

Colt Industries
C 0.67, Ni 0.75, Cr 0.6, Mo 0.2, bal Fe.
For saws; oil hardening. *Obsolete*

CIRCULAR SAW PLATE B

Colt Industries
C 0.8, Ni 2.5, Cr 0.2, Mo 0.15, bal Fe.
For saws; oil hardening. *Obsolete*

CIRCULAR SAW PLATE C

Colt Industries
C 0.82, Ni 0.7, Cr 0.4, bal Fe.
For saws; water or oil hardening. *Obsolete*

CITROEN

English manufacture
Cu 12, Mg 0.1, Mn 0.5, Fe 0.5, Si 0.5, bal Al.
For pistons; cast.

CK 15

Thyssen Edelstahlwerke AG
C 0.15, Si 0.25, Mn 0.37, bal Fe.
Annealed: 70,000 TS; 40,000 YS; 25 El; 60 RA; 145 Brin. For gears, shafts, machine tool parts; case hardened. *Obsolete*

CK 22

Thyssen Edelstahlwerke AG
C 0.22, Si 0.25, Mn 0.45, bal Fe.
Annealed: 73,000 TS; 41,000 YS; 22 El; 58 RA; 140 Brin. For fan blades, bushings, gears, camshafts; water hardened, case hardened. *Obsolete*

CK 35

Thyssen Edelstahlwerke AG
C 0.35, Si 0.25, Mn 0.55, bal Fe.
Hot rolled: 85,000 TS; 54,000 YS; 30 El; 53 RA; 185 Brin. For gears, shafts, axles, bolts; water hardened. *Obsolete*

CK 45

Thyssen Edelstahlwerke AG
C 0.45, Si 0.25, Mn 0.65, bal Fe.
Hot rolled: 98,000 TS; 59,000 YS; 24 El; 45 RA; 212 Brin. For gears, bolts, axles, shafts, machine tool parts; water hardened. *Obsolete*

CK 60

Thyssen Edelstahlwerke AG
C 0.61, Si 0.25, Mn 0.65, bal Fe.
Heat treated: 160,000 TS; 113,000 YS; 12 El; 40 RA; 320 Brin. For wheels, die blocks, springs, rails, girders; water hardened. *Obsolete*

CK1065 (K65A)

Eagle & Globe Steel Ltd.
Carbon steel. C 0.6-0.7, S 0-0.03, P 0-0.03, Mn 0-0.9, Si 0-0.35, bal Fe.
Cold rolled annealed steel strip. For nuts, mower blades, automotive components, seat recliners, springs. SAE 1065.

CLAD R-303

Reynolds Metals Co.
Zn 6.5-7.5, Cu 1, Mg 2, Fe 0.3, Si 0.2, Cr 0.2, Ni 0.1, bal Al.
Annealed: 27,000 TS; 11,500 YS; 19 El. Heat treated: 75,000 TS; 69,000 YS; 9 El. For light alloy structures; clad. *Obsolete*

CLAMERS ALLOY

Manufacturer not listed
Ni 5-25, Co 5-25, bal Fe.
For electrical machinery; heat and corrosion resistant.

CLARITE

Columbia Tool Steel Co.
Tool material. C 0.7, W 18, Cr 4, V 1.1, Mn 0.25, Mo 0.6, bal Fe.
Heat treated: 64 Rock C; 400,000 bend strength; 300,000 torsion strength. For lathe and planer tools, drills, punches, dies, cutters, springs. Type T1; high speed steel; high red hardness.

CLARITE HW

Columbia Tool Steel Co.
C 0.55, W 18, bal Fe.
For hot punches, chisels, upsetters; hot work steel, oil hardened. Cast: 45,000-55,000 TS; 60-70 El. For bronze castings; corrosion resistant. *Obsolete*

CLARITE HW 60

Columbia Tool Steel Co.
Tool material. C 0.63, Mn 0.3, Si 0.3, Cr 4, W 18, V 1.1, bal Fe.
Hot work tool steel.

CLARITE HW26

Columbia Tool Steel Co.
Tool material. C 0.53, W 18, Cr 4, V 1, bal Fe.
For punches, piercing tools, shear blades, forming rolls. AISI Type H26; hot work tool steel.

CLARITE HW50

Columbia Tool Steel Co.
C 0.53, W 18, Cr 4, V 1, bal Fe.
For piercing punches, hot shell punches, hot blanking dies. Hot work steel Type H 26. *Obsolete*

CLARITE M

Columbia Tool Steel Co.
C 0.8, W 18, V 1, bal Fe.
For hot work tools and dies; hot work steel. *Obsolete*

CLARITE-L

Columbia Tool Steel Co.
C 0.6, W 18, V 1, Cr 4, bal Fe.
For hot work tools and dies, punches, hot work steel. *Obsolete*

CLARK'S ALLOY

English manufacture
Cu 75, Zn 7.2, Ni 14, Co 1.9.
For chemical equipment; corrosion resistant.

CLARK'S PATENT

English manufacture
Cu 75, Ni 14, Zn 7.2, Sn 1.9, Co 1.9.
For corrosion resisting parts; corrosion resistant.

CLARUS METAL

A.B. Gabriel & Co., Ltd.
Cu 1.5, Si 4, bal Al.
For light alloy parts; will not oxidize; 60% stronger than Al.

CLASS "P"

Edgar Allen Balfour Ltd.
C 0.7-0.9, bal Fe.
For tools, punches, dies, engraved dies, trimming dies; water hardening.

CLAY-LOY

Colorado Fuel & Iron Co.
C 0.22, Mn 1.25, Si 0.35, Cu 0.5, V 0.2, bal Fe.
Rolled: 70,000 TS; 50,000 YS. For railroad and bus bodies; high strength, low alloy construction steel.

CLEBRIUM-1

English manufacture
Ni 2, Cr 13.1, Mn 0.75, Mo 3.6, Si 1.5, C 2.6, bal Fe.
For heat and corrosion resisting cast iron castings; high heat resistant.

CLEBRIUM-2

English manufacture
Ni 4.6, Cr 18.3, Mn 2.8, C 2, Cu 2, bal Fe.
For heat and corrosion resisting cast iron castings.

CLEREMONT DRILL ROD

Crucible Materials Corp.
Tool material. C 1.05, bal Fe.
For tools, pivots; drill rod.

CLETALOY 10-BC

Chase Brass & Copper Co.
Cu alloy.
For pressure projection welding; density 13.75; sintered. *Obsolete*

CLETALOY 10-CC

Chase Brass & Copper Co.
Cu alloy.
For electrodes; density 12.0; sintered. *Obsolete*

CLETALOY 10-CSS

Chase Brass & Copper Co.
Cu alloy.
For breaker contacts and facing; 45-50% electrical conductivity, density 13.5; sintered. *Obsolete*

CLETALOY 10-CT

Chase Brass & Copper Co.
Cu alloy.
For projection weld die facing; density 13.75; sintered. *Obsolete*

CLETALOY 10-CTA

Chase Brass & Copper Co.
Cu alloy.
For electrodes; density 13.75; sintered. *Obsolete*

CLETALOY 20-BC

Chase Brass & Copper Co.
Cu alloy.
For pressure projection welding; density 14.15; sintered. *Obsolete*

CLETALOY 20-CC

Chase Brass & Copper Co.
Cu alloy.
For electrodes; density 12.4; sintered. *Obsolete*

CLETALOY 20-CT

Chase Brass & Copper Co.
Cu alloy.
For projection weld die facing; density 14.15; sintered. *Obsolete*

CLETALOY 20-CTA

Chase Brass & Copper Co.
Cu alloy.
For electrodes; density 14.15; sintered. *Obsolete*

CLETALOY 30-BC

Chase Brass & Copper Co.
Cu alloy.
For pressure projection welding; density 14.35; sintered. *Obsolete*

CLETALOY 30-CC

Chase Brass & Copper Co.
Cu alloy.
For electrodes; density 12.7; sintered. *Obsolete*

CLETALOY 30-CT

Chase Brass & Copper Co.
Cu alloy.
For projection weld die facing; density 14.35; sintered. *Obsolete*

CLETALOY 30-CTA

Chase Brass & Copper Co.
Cu alloy.
For electrodes; density 14.35; sintered. *Obsolete*

CLETALOY CT-65

Chase Brass & Copper Co.
Cu-W.
For spot welding electrode for stainless steel. *Obsolete*

CLETALOY CT-86

Chase Brass & Copper Co.
Cu-W.
For spot welding electrode for nonferrous metals. *Obsolete*

CLETALOY CT-A

Chase Brass & Copper Co.
Cu-W.
For spot welding electrode. *Obsolete*

CLETALOY L-N-14

Chase Brass & Copper Co.
Cu-W.
For spot welding. *Obsolete*

CLETALOY TA

Chase Brass & Copper Co.
Ag alloy.
For circuit breaker contacts; density 12.8; sintered. *Obsolete*

CLETALOY TS

Chase Brass & Copper Co.
Ag alloy.
For circuit breaker facing; density 13.2; sintered. *Obsolete*

CLEVITE 100

Now CLEVITE F-100.

CLEVITE 112

Now CLEVITE F-112.

CLEVITE 153

Now CLEVITE F-153.

CLEVITE 250

Now CLEVITE F-250.

CLEVITE 500

Gould Inc.
Pb 44-58, Sn 0.5-1.5, bal Cu.
Sintercast with steel backing. For bearings.

CLEVITE 77

Now CLEVITE F-77.

CLEVITE F-1

Gould Inc.
Cu 3-3.5, Sb 7.25-7.75, Te 0.1-0.14, bal Sn.
For engine bearings; steel backed Babbitt.

CLEVITE F-100

Gould Inc.
Pb 9-11, Sn 9-11, bal Cu.
For transmission bushings and washers; sintered on steel back. Formerly CLEVITE 100.

CLEVITE F-112

Gould Inc.
Pb 21-27, Sn 1.75-2.75, bal Cu.
For bearings; with steel backing. Formerly CLEVITE 112.
Overlay: 8-12 Sn, 2-3 Cu, bal Pb.

CLEVITE F-153

Gould Inc.
Sn 3.5-4.5, Cd 0.75-1.4, bal Al.
For bearings, aluminum clad to steel. Formerly CLEVITE 153.
Overlay: 8-12 Sn, 2-3 Cu, bal Pb.

CLEVITE F-154

Gould Inc.
Si 3.5-4.5, Cd 0.75-1.4, Cu 0.05-0.15, Mg 0.1-0.2, bal Al.
Overlay: 87.5 Pb, 10 Sn, 2.5 Cu. Trimetal/steel backed. For heavy-duty automotive, truck and diesel engine bearings.

CLEVITE F-17

Gould Inc.
Pb 8-12, Sn 8-12, Sb 7-8, bal Cu.
For heavy duty bearings; with steel backing. Formerly F-17 TRIMETAL. Overlay: 3-5 Cu, bal Sn.

CLEVITE F-23

Gould Inc.
Sn 0.9-1.25, Sb 14.75-15.5, As 0.8-1.1, Cu 0-0.6, bal Pb.
For bearings. SAE 15. Formerly F-23.

CLEVITE F-250

Gould Inc.
Pb 21-25, Sn 3-4, bal Cu.
For transmission bushings and washers; sintered on steel back. Formerly CLEVITE 250.

CLEVITE F-4

Gould Inc.
Pb 7-9, Sn 3.5-4.5, Zn 4, bal Cu.
For wrist pin and transmission bearings; with steel backing. Formerly CLEVITE NO. 8.

CLEVITE F-5

Gould Inc.
Pb 9-11, Sn 9-11, Zn 0-0.5, bal Cu.
For wrist pin and transmission bearings; with steel backing. Formerly CLEVITE NO. 10.

CLEVITE F-66

Gould Inc.
Si 3.5-4.5, Cu 0.5-1, Pb 7.5-9.5, Sn 1.25-1.75, bal Al.
Bimetal/steel backed. For intermediate range plain bearings, bushings and washers.

CLEVITE F-7

Gould Inc.
Pb 21-25, Sn 3-4, Zn 3, bal Cu.
For steering knuckle bushings; steel backing. Formerly CLEVITE NO. 25.

CLEVITE F-77

Gould Inc.
Pb 22-26, Sn 0.15-0.5, bal Cu.
For bearings; with steel backing. Formerly CLEVITE 77.
Overlay: 8-12 Sn, 2-3 Cu, bal Pb.

CLEVITE NO. 10
Now CLEVITE F-5.

CLEVITE NO. 25
Now CLEVITE F-7.

CLEVITE NO. 8
Now CLEVITE F-4.

CLEVITE S-56
Gould Inc.
Cu 80-86, Sn 3.5-5, Pb 3.5-4.5, Zn 4.
For bushings, bearings; SAE 791. Formerly NO. 444 ALLOY.

CLICHIER METAL
English manufacture
Pb 33, Sn 48, Bi 9, Sb 11.
For bearings, fuses; antifriction.

CLICHIER METAL
English manufacture
Pb 5, Sn 80, Bi 15.
For bearings, fuses; antifriction.

CLICHIER METAL
English manufacture
Pb 50, Sn 36, Cd 14.
For bearings, fuses; antifriction.

CLICKER DIE
Disston Inc.
C 0.75, Mo 0.25, bal Fe.
For tools, knives; water hardening. *Obsolete*

CLICKER DIE WELDING ROD
Colt Industries
C 0.45, Ni 1.4, Cr 0.55, bal Fe.
For welding rods; tough. *Obsolete*

CLICKFLEX
Vacuumschmelze GmbH
See VACOFLEX.

CLICKING DIE
Disston Inc.
C 0.55, Mo 0.2, bal Fe.
For dies; water hardening. *Obsolete*

CLICKING DIE
Disston Inc.
C 0.4, Cr 0.7, bal Fe.
For clicking dies; water hardening. *Obsolete*

CLIMAX
Driver Harris Co.
Ni 24, Fe 73, Mn 2.6.
Rolled: 76,000 TS. For electrical parts; heat and corrosion resistant. *Obsolete*

CLIMAX
Colt Industries
Ni 25, Mn 1, Fe 74.
75,000 TS. For heat and corrosion resistant parts. *Obsolete*

CLIMAX 12
Climax Performance Materials Corp.
C 3-3.5, Mn 0.5-0.8, Si 0.5-0.8, Cr 11-14, Mo 0.5-1, Cu 0-1, bal Fe.
Hardened: 60-67 Rock C. For grinding balls, classifier wear shoes, shot blast wheel blades. Martensitic casting.

CLIMAX 12-2 HCS
Climax Performance Materials Corp.
C 1.3-1.45, Mn 12-14, Si 0.3-0.75, Mo 1.8-2.1, P 0-0.05, Cr 0-0.5, bal Fe.
Alloy casting, austenitic as quenched, work hardenable, high abrasion resistance and good toughness. For crusher liners, scraper blades, heavy-duty grizzly screens.

CLIMAX 12-2 LCS
Climax Performance Materials Corp.
C 1.15-1.3, Mn 12-14, Si 0.3-0.75, Mo 1.8-2.1, P 0-0.05, Cr 0-0.5, bal Fe.
Alloy casting, austenitic as quenched, work hardenable, high abrasion resistance and good toughness. For crusher liners, scraper blades, heavy-duty grizzly screens.

CLIMAX 15-2-1
Climax Performance Materials Corp.
C 2.8-3.5, Mn 0.6-0.9, Si 0.4-0.8, Cr 14-16, Mo 1.9-2.2, Cu 0.8-1.2, bal Fe.
Casting, hardened: 60-67 Rock C. For rod and ball mill liners, tires and grinding rings for roller mill pulverizers.

CLIMAX 15-3 HC
Climax Performance Materials Corp.
C 3.2-3.6, Mn 0.7-1, Si 0.3-0.8, Cr 14-16, Mo 2.5-3, bal Fe.
High carbon grade martensitic white cast iron; hard and abrasion resistant. For shot blast impeller blades and liners, and sand pump impellers, jaw plates in small jaw crushers, garbage disposal wearing parts.

CLIMAX 15-3 LC
Climax Performance Materials Corp.
C 2.4-2.8, Mn 0.5-0.8, Si 0.3-0.8, Cr 14-16, Mo 2.4-2.8, bal Fe.
Lower carbon grade martensitic white iron; hard and abrasion resistant. For heavier section castings for rod and ball mill liners, tires for roll mill pulverizers and heavy pulverizer hammers.

CLIMAX 15-3 MC
Climax Performance Materials Corp.
C 2.8-3.2, Mn 0.6-0.9, Si 0.3-0.8, Cr 14-16, Mo 2.5-3, bal Fe.
Medium carbon grade martensitic white cast iron; hard and abrasion resistant. For die bushings in clay product molds, impact pulverizer blow bars, chute liners.

CLIMAX 15-3 XHC
Climax Performance Materials Corp.
C 3.6-4.3, Mn 0.7-1, Si 0.3-0.8, Cr 14-16, Mo 2.5-3, bal Fe.
Extra high carbon grade martensitic white cast iron; hard, brittle, abrasion resistant. For unstressed liners and parts handling abrasive slurries without impact, and fluidized solids up to about 1200°F.

CLIMAX 18-2 (18 CR-2 MO)
Climax Performance Materials Corp.
C 0-0.04, Cr 18-20, Mo 1.75-2.25, Mn 0-1, Si 0-1, Cu + Ni, Ti = 5 x C + N (0.25 min), Cb = 9 x C + N min,, bal Fe.
60% cold rolled: 829 N/mm^2 TS; 823 N/mm^2 YS; 5.5 El. Ferritic, stainless steel with very good resistance to corrosion, stress-corrosion cracking, and good formability. Weldable.

CLIMAX 20-2-1
Climax Performance Materials Corp.
C 2.6-2.9, Mn 0.6-0.9, Si 0.4-0.9, Cr 18-21, Mo 1.4-2, Cu 0.8-1.2, bal Fe.
Annealed: 38-43 Rock C. Hardened: 60-67 Rock C. Alloy steel casting, martensitic as quenched. For rod and ball mill liners, sand and dredge-pump parts, clay working machine parts, pulverizer impactor and blow bars.

CLIMAX 321
Climax Performance Materials Corp.
C 3.3-3.6, Cr 1.75-2.25, Mo 0.7-1.1, Ni 2.75-3.25, bal Fe.
For ball mill parts, grinding balls, pug mill knives; martensitic, cast iron, abrasion resistant.

CLIMAX 6-1
Climax Performance Materials Corp.
C 1.2-1.35, Mn 5.5-6.75, Si 0.4-0.7, Mo 0.9-1.1, P 0-0.05, Cr 0-0.5, bal Fe.
Austenitic in water quenched condition. Good abrasion resistance casting but less tough than Hatfield type manganese steel. For ball mill liners, ball mill discharge grates, grizzly screens, drag chain and crusher liners, scoop lips.

CLIMAX 6-2-1
Climax Performance Materials Corp.
C 1.05-1.2, Mn 5.25-6.5, Si 0.3-0.7, Cr 1.5-2, Mo 0.9-1.1, P 0-0.05, bal Fe.
Lean alloy casting, austenitic after quench. Good abrasion resistance, work hardenable. For ball mill liners and discharge grates, scoop lips, drag chain and crusher liners subject to rapid wear and moderate impact.

CLIMAX ALLOY 42
Now CLIMAX 15-3.

CLIMAX CR13MO
Climax Performance Materials Corp.
C 0.25-0.4, Mn 0.4-0.6, Si 0.4-0.8, Cr 13-14, Mo 0.6-0.75, bal Fe.
Annealed: 180-220 Brin. Hardened: 500-550 Brin. For grinding mill liners, sand and dredge pump castings. Abrasion and corrosion resistant.

CLIMAX FERROMOLYBDENUM
Climax Performance Materials Corp.
Mo 58-64, Si 0-1, C 0-0.1, bal Fe.
For metallurgical applications in steel and cast iron; Mo additions.

CLIMAX MACHINERY
Colt Industries
C 0.35, Mn, bal Fe.
For shafts, gears, pinions. *Obsolete*

CLIMELT LOW-CARBON MOLYBDENUM
Climax Performance Materials Corp.
C 0-0.005, 99.97 Mo min.
Vacuum arc melted. For space power generator parts, grinding quills, chemical handling equipment, machined components.

CLIMELT MO-30W
Climax Performance Materials Corp.
W 30, bal Mo.
At 72°F: 121,500 TS; 106,900 YS; 26 El; 40 RA; 198 Brin. At 1800°F: 65,700 TS; 25 EL; 77 RA; 80 Brin. For high temperature applications; high heat resistance; has good resistance to attack by molten zinc and certain other liquid metals.

CLIMELT MOLYBDENUM
Climax Performance Materials Corp.
C 0-0.03, 99.94 Mo min.
Vacuum arc melted. For rocket nozzle parts, electrical and electronic parts, spot welding tips, brazing contacts, boring bars, furnace heating elements and supports.

CLIMELT TZM
Climax Performance Materials Corp.
C 0.01-0.04, Ti 0.4-0.55, Zr 0.06-0.09, bal Mo.
At 72°F: 144,000 TS; 129,000 YS; 21 El; 46 RA; 172 Brin. At 1600°F: 88,000 TS; 48,000 YS; 21 El; 76 RA; 82 Brin. For high temperature applications, heat engines, heat exchangers; heat and corrosion resistant, high strength and hardness at elevated temperatures.

CLINCHING SCREW WIRE
English manufacture
Cu 69, Zn 29.5, Pb 1.5.
For brass screws.

CLIPPER
Manufacturer not listed
C 0.73, W 18, Cr 4, V 1.1, bal Fe.
For cutting tools. AISI T1. High speed steel.

CLOCK BRASS-243
Anaconda Co.
Copper. Zn 37, Pb 2, Cu 62.
Hard: 75,000 TS; 60,000 YS; 10 El. Soft: 45,000 TS; 17,000 YS; 50 El. For clock gears and frame, meter parts. Free milling.

CLOMO

Uddeholm Corp.
C 0.97, Cr 1.15, Mo 0.32, bal Fe.
For hollow mine drills; oil hardened. *Obsolete*

CLOVERLEAF

E.A. Williams & Sons
Sn, Pb, bal Cu.
For bearings, bushings.

CLUNISE

English manufacture
Cu 40, Ni 32, Zn 25, Fe 2.6.
For decorative parts.

CLW

Latrobe Steel Co.
Tool material. C 0.33, Cr 3.3, V 0.45, W 9.15, Si 0.45, Mn 0.25, bal Fe.
For extrusion and trimming dies, punches, shears; hot work steel. AISI H22.

CLW NO. 1

Latrobe Steel Co.
C 0.3, W 9, Cr 3.3, V 0.5, bal Fe.
For hot headers, punches, extrusion rams and dies; oil hardened, hot work steel.

CLY-DIE

Osborn Steels Ltd.
C 0.1-0.3, Cr 13, bal Fe.
Annealed: 95,000 TS; 50,000 YS; 25 El; 55 RA; 195 Brin. For plastic mold dies; corrosion resistant. *Obsolete*

CLYDALL 12 SPECIAL

Osborn Steels Ltd.
C 0.78, Cr 4, Co 12, V 1.35, W 22, bal Fe.
Annealed: 240 Brin. For roll turning tools, tool bits, cutters; high speed steel. *Obsolete*

CLYDALL 5 SPECIAL

Osborn Steels Ltd.
C 0.78, Cr 4, Co 5.5, V 1.35, W 18.5, bal Fe.
Annealed: 240 Brin. For broaches, drills, lathe and planer tools; high speed steel. *Obsolete*

CLYDE ALLOY

Steel Co., Ltd.
C 0.3-0.6, bal Fe.
For machinery parts; water hardening.

CLYDMO

Osborn Steels Ltd.
C 0.83, Cr 4, Mo 5, V 1.8, W 6.25, bal Fe.
Annealed: 240 Brin. For lathe and planer tools, reamers, hobs, drills; high speed steel. *Obsolete*

CM

A. Milne & Co.
C 0.38, Si 1, Cr 5, V 0.45, Mo 1.25, bal Fe.
Air or oil hardening; hot work tool steel; for forging dies, and hot forming tools. AISI H11.

CM 1

Bergische Stahl Industrie
C 0.25, Cr 1, Mo 0.25, bal Fe.
Heat treated; 85,000-170,000 TS; 64,000-120,000 YS; 5-16 El. Low alloy steel casting. Also useful to -110°C. W.-Nr. 1.7218.

CM 1 H

Bergische Stahl Industrie
C 0.42, Cr 1, Mo 0.2, bal Fe.
Heat treated: 106,000-184,000 TS; 78,000-142,000 YS; 4-12 El. Low alloy steel casting. W.-Nr. 1.7225.

CM 1 K

Bergische Stahl Industrie
C 0.34, Cr 1, Mo 0.25, bal Fe.
Heat treated: 78,000-150,000 TS; 42,500-106,000 YS; 9-20 El. Low alloy steel casting. W.-Nr. 1.7220.

CM 2 H

Bergische Stahl Industrie
C 0.5, Mn 0.7, Cr 1, Mo 0.25, bal Fe.
Low alloy cast steel. GS-50 CrMo 4.

CM 3

Acciaierie Valbruna s.p.a.
C 1-1.1, Cr 3.5-4.5, Mo 4.5-5.5, W 5.5-7, V 2.2-2.6, bal Fe.
Mo-W high speed tool steel. AISI M3 Class 1. *Obsolete*

CM 44

Walsingham Steel Co., Ltd.
C 0.4-0.48, Si 0.8-1, Mn 0.5-0.7, Cr 2.8-3.1, Mo 0.3-0.4, bal Fe.
350-500 Brin.

CM 469

Cannon-Muskegon Corp.
C 0.03, Cr 60, Mo 25, Fe 14.
For jet engine parts. High heat and oxidation resistant.

CM 60

United Engineering Steels Ltd.
C 0.13-0.18, Si 0.15-0.35, Mn 0.8-1.1, 0.80 Cr + Mo + Ni max, 0.0005 B min, bal Fe.
Low alloy carburizing grade for small parts.

CM 70

United Engineering Steels Ltd.
C 0.13-0.19, Si 0.15-0.35, Mn 1.1-1.4, 0.80 Cr + Mo + Ni max, 0.0005 B min, bal Fe.
Similar to CM 60, but higher strength.

CM 718

Cannon-Muskegon Corp.
C 0.08, Cr 20, Ni 52, Co 0.4, Cb 5.25, Mo 3, Ti 0.83, Al 0.55, B 0.004, Cu 0.02, bal Fe.
Annealed: 120,200 TS; 53,400 YS; 0.45 El. Aged: 191,000 TS; 172,000 YS; 22 El; 45 Rock C. For aircraft structural parts operating at 800-1400°F. High oxidation and corrosion resistance.

CM 80

United Engineering Steels Ltd.
C 0.17-0.23, Si 0.15-0.35, Mn 1.2-1.5, 0.80 Cr + Mo + Ni max, 0.0005 B min, bal Fe.
Similar to CM 70, but higher strength.

CM 90

United Engineering Steels Ltd.
C 0.2-0.25, Si 0.15-0.35, Mn 1.3-1.6, 0.80 Cr + Mo + Ni max, 0.0005 B min, bal Fe.
Similar to CM 80, but higher strength.

CM L-605

Cannon-Muskegon Corp.
C 0.12, Mn 1.65, Cr 19.8, Ni 9.9, W 15.2, Fe 1.6, bal Co.
At room temperature: 144,300 TS; 65,400 YS; 55 El. At 1400°F: 73,000 TS; 49,000 YS; 20 El; 25 Rock C. For jet engine afterburners, exhaust cone assemblies, nozzle diaphragm valves, high temperature springs, turbine buckets. Good oxidation and corrosion resistance.

CM N-155

Cannon-Muskegon Corp.
C 0.1, Mn 1.5, Cr 20.75, Ni 19.85, Co 19.5, Mo 2.95, W 2.35, Cb 1.15, Cu 0.2, bal Fe.
At room temperature: 112,000 TS; 57,000 YS; 41 El, 77 Rock B. At 1500°F: 39,000 TS; 25,000 YS; 35 El. For turbine rotors and blading, afterburner rings, rocket chambers. High oxidation and corrosion resistance.

CM-I

Atrax Cemented Carbide
Sintered carbide. 325,000 transverse strength; 14.3-14.5 g/cm³ density; 88.3-89.3 RA. Industry code: C-16.

CM-2

Atrax Cemented Carbide
Sintered carbide. 325,000 transverse strength; 14.4-14.6 g/cm³ density; 89.0-90.0 RA. Industry code: C-16.

CM-3

Atrax Cemented Carbide
Sintered carbide. 350,000 transverse strength; 14.1-14.35 g/cm³ density; 88.3-89.3 RA. Industry code: C-16.

CM-R41

Cannon-Muskegon Corp.
C 0.09, Cr 19, Co 11, Mo 10, Ti 3, Al 1.5, bal Ni.
At 70°F: 206,000 TS; 154,000 YS; 14 El. At 1500°F: 126,000 TS; 118,000 YS; 14 El. For jet engine components and high speed airframes. Afterburners, turbine castings, combustion liners, fasteners. Precipitation hardening. High temperature alloy. Vacuum melted.

CM-WASPALLOY

Cannon-Muskegon Corp.
C 0.06, Mn 0.08, S 0.01, P 0.004, Si 0.1, Cr 20.5, Co 14.2, Mo 4.2, Ti 3, Al 1.5, Zr 0.03, B 0.003, Fe 0.05, Cu 0.05, bal Ni.
Annealed: 121,400 TS; 69,700 YS; 54 El; 90 Rock B. Aged: 186,000 TS; 138,000 YS; 26 El; 39 Rock C. For jet engine turbine buckets and discs, high temperature bolts, missile systems. Precipitation hardened. High temperature strength.

CM1KM

Bergische Stahl Industrie
C 0.35, Mn 0.8, Cr 1, Mo 1, bal Fe.
Low alloy cast steel. GS-34 CrMoMn 44.

CM25

German manufacture
C 0.3-0.4, V 0.2, Cr, bal Fe.
For turbine blades for jet engines; water hardening.

CMA

Acciaierie Valbruna s.p.a.
C 0.33, Si 0.35, Mn 0.65, P 0-0.03, S 0-0.035, Cr 1.65, Mo 0.35, Al 1, bal Fe.
Nitriding steel. W. Nr. 1.8507.

CMA-DE

Acciaierie Valbruna s.p.a.
C 0.33, Si 0.35, Mn 0.55, P 0-0.03, S 0-0.035, Cr 1.65, Mo 0.2, Ni 1.5, Al 1, bal Fe.
Nitriding steel. W. Nr. 1.8550.

CMC

Thyssen Edelstahlwerke AG
C 0.45, Cr 1.4, Mo 0.7, V 0.3, bal Fe.
For gears, bolts, crankshafts, upsetters; oil hardened, tough. *Obsolete*

CMCW

Bergische Stahl Industrie
C 2-2.25, Si 0.2-0.4, Mn 0.2-0.4, Co 0.8-1.1, Cr 11.5-12.5, Mo 0.3-0.5, W 0.6-0.8, bal Fe.
Cold work tool steel. W.-Nr. 1.2884.

CMH

Acciaierie Valbruna s.p.a.
C 0.27-0.34, Mn 0.4-0.7, Cr 2.7-3.3, Mo 0.3-0.4, bal Fe.
Cr-Mo structural steel. 30 CrMo 12. *Obsolete*

CML

Acciaierie Valbruna s.p.a.
C 0.41, Si 0.35, Mn 0.65, P 0-0.03, S 0-0.035, Cr 1.65, Mo 0.35, Al 1, bal Fe.
Nitriding steel. W. Nr. 1.8509.

CMLR

Acciaierie Valbruna s.p.a.
C 0.42, Si 0.35, Mn 0.5, P 0-0.03, S 0-0.035, Cr 1.65, Mo 0.35, Al 0.4, bal Fe.
Nitriding steel.

CMN

Acciaierie Valbruna s.p.a.
C 0.31, Si 0.35, Mn 0.55, P 0-0.03, S 0-0.035, Cr 3, Mo 0.35, Ni 0-0.3, bal Fe.
Nitriding steel. W. Nr. 1.8515.

CMP(3)
Creusot-Loire
C 0.95, Mn 1.25, Cr 0.5, W 0.5, bal Fe.
For tools, cutters; oil hardened, non-deforming. *Obsolete*

CMS
Thyssen Edelstahlwerke AG
C 0.45, Mn 1.3, Cr 2, Mo 0.3, bal Fe.
For upsetting tools, liners and containers in extrusion presses, forging die blocks; shock resistant. *Obsolete*

CMSZ
Union Carbide Corp.
Cr 40-56, Mn 4-6, Si 13-21, Zr 1.25-1.75, C 3-5, bal Fe.
For ladle additions of Cr to cast iron; for harder and stronger iron.

CMSZ-4 MIXTURE
Union Carbide Corp.
Cr 45-49, Mn 4-6, Si 18-21, Zr 1.25-1.75, C 3-4.5, bal Fe.
For ladle additions of Cr to cast iron; no increase in chill.

CMSZ-5 MIXTURE
Union Carbide Corp.
Cr 50-56, Mn 4-6, Si 13.5-16, Zr 0.75-1.25, bal Fe.
For ladle additions of Cr to cast iron; no increase in chill.

CMV
Acciaierie Valbruna s.p.a.
C 0.31, Si 0.35, Mn 0.5, P 0-0.03, S 0-0.035, Cr 2.6, Mo 0.35, V 0.15, bal Fe.
Nitriding steel. W. Nr. 1.8519.

CMVA
Walsingham Steel Co., Ltd.
C 0.24-0.28, Si 0.3-0.5, Mn 0.7-0.8, Ni 0-0.5, Cr 0.95-1.15, Mo 0.45-0.55, V 0.03-0.07, bal Fe.
38 tsi YS min; 47 tsi TS min; 17 El.

CMVB
Walsingham Steel Co., Ltd.
C 0.3-0.4, Si 0.4-0.6, Mn 0.5-0.7, Cr 2.5-3, Mo 0.7-1, Y 0.1-0.2, bal Fe.
300 Brin max.

CMW
Sanderson Kayser Ltd.
C 0.35, Cr 5, W 1.3, Mo 1.7, Si 1, bal Fe.
For hot work tools, die casting dies; hot work steel. AISI H12.

CMW
A. Milne & Co.
C 0.35, Cr 5, W 1.3, Mo 1.7, Si 1, bal Fe.
For hot work tools, die casting dies; hot work steel. AISI H12.

CMW 100
CMW Inc.
Cu, Co, Be.
Rolled: 110,000 TS; 10 El; 100 Rock B (heat treated). Cast: 95,000 TS; 6 El; 95 Rock B (heat treated). Electrical conductivity; 48% IACS. For resistance welding electrodes, conducting arms, springs. RWMA Class 3.

CMW 1000
CMW Inc.
Ni 3.5, Cu 1.5, bal W.
110,000 TS; 85,000 YS; 3 El, 27 Rock C. Sintered density: 18.00 g/cm^3. For weights, radiation shielding.

CMW 1000
CMW Inc.
Ni 6, Cu 4.
110,000 TS; 80,000 YS; 4 El, 25 Rock C. Sintered density: 17.00 g/cm^3. For weights, radiation shielding and boring bars.

CMW 150
CMW Inc.
Cu, Ag, Be.
110,000 TS; 95 Rock B (heat treated). Electrical conductivity: 52% IACS. For resistance welding dies and fixtures. *Obsolete*

CMW 28
CMW Inc.
Cu, Zr.
Forgings: 54,000 TS, 48,000 YS; 90% IACS electrical conductivity; 65 Rock B (heat treated). Rod: 70,000 TS, 63,000 YS; 90% IACS electrical conductivity; 75 Rock B (heat treated). For resistance welding electrodes.

CMW 3
CMW Inc.
Cu, Cr.
Electrical conductivity: 80% IACS (heat treated). Sand cast: 50,000 TS, 46,000 YS, 20 El; 110 Brin. Bar: 75,000 TS; 70,000 YS; 15 El; 150 Brin. For resistance welding electrodes and high electrical conductivity, high strength applications (RWMA Class 2).

CMW 3000
CMW Inc.
Ni 7, Fe 3, bal W.
130,000 TS; 85,000 YS; 18 El; 25 Rock C. Sintered density: 17.00 g/cm^3. For armor piercing projectiles, high strength weights.

CMW 3950
CMW Inc.
Ni 3.5, Fe 1.5, bal Fe.
130,000 TS; 90,000 YS; 15 El; 28 Rock C. Sintered density: 18.00 g/cm^3. For armor piercing projectiles, high strength weights.

CMW 3970
CMW Inc.
Ni 2.1, Fe 0.9, bal W.
130,000 TS; 90,000 YS; 10 El; 30 Rock C. Sintered density: 18.50 g/cm^3. For armor piercing projectiles, high strength weights.

CMW 44 STRIP
CMW Inc.
Cu, Cr, Cd.
Electrical conductivity: 78% IACS. 70,000 psi TS; 75 Rock B (heat treated). For high electrical conductivity; high strength applications. *Obsolete*

CMW 53
CMW Inc.
Cu, Ni, Si, Mn.
Cast: 60,000 TS, 180 Brin (heat treated). Electrical conductivity: 45% IACS. For flash welding dies, bearings and bushings.

CMW 53 B
CMW Inc.
Cu, Ni, Be.
Cast: 80,000 TS, 90 Rock B (heat treated). Electrical conductivity: 45% IACS. For flash welding dies, bearings, bushings.

CMW 73
CMW Inc.
Cu, Be, Co.
Cast: 110,000 TS; 38 Rock C (heat treated). Wrought: 170,000 TS; 38 Rock C (heat treated). Electrical conductivity: 20% IACS. Good abrasion resistance for current carrying shafts, collets, and bearings. RWMA Class 4 welding electrodes.

CMW 80 MO-20 CU
CMW Inc.
Cu 20, Mo 80.
Electrical conductivity: 32% IACS. Hardness: 90 Rock B. Density: 9.85 g/cm^3. 90,000 psi YS; 100,000 psi TS; 3.6 x 10^{-6} in./in.·°F thermal expansion; 84 Btu/h/ft·°F thermal conductivity; 160,000 psi flexural strength; 35 x 10^6 psi modulus of elasticity. Custom material engineered for specific applications.

CMW 8CC
CMW Inc.
Cu, Cd, Si.
Cast: 50,000 TS; 2 El; 50% IACS electrical conductivity. For electrical contacts. *Obsolete*

CMW D1058
CMW Inc.
C 10, bal Ag.
Electrical conductivity: 35% IACS. Hardness: 3 Rock F. Density: 6.4 g/cm^3. For electrical contacts, resistant to sliding wear.

CMW D154F
CMW Inc.
13.3 CdO, bal Ag.
Annealed: 40,000 TS; 70 Rock F. Worked: 47,000 TS; 90 Rock F. Electrical conductivity: 75% IACS. For electrical contacts; resistant to welding and high surge currents.

CMW D155F
CMW Inc.
17.0 CdO, bal Ag.
Annealed: 40,000 TS; 70 Rock F. Worked: 50,000 TS; 90 Rock F. Electrical conductivity: 70% IACS. For electrical contacts; resistant to welding and high surge currents.

CMW D158F
CMW Inc.
C 1, bal Ag.
Annealed: 23,000 TS; 36 Rock F. Worked: 35,000 TS; 68 Rock F. Electrical conductivity: 99% IACS. For electrical contacts; reduced sticking; low friction.

CMW D355F
CMW Inc.
20.0 CdO, bal Ag.
Annealed: 40,000 TS; 70 Rock F. Worked: 51,000 TS; 90 Rock F. Electrical conductivity: 68% IACS. For electrical contacts; resistant to welding and high surge currents.

CMW D50
CMW Inc.
Ni 15, bal Ag.
Electrical conductivity: 65% IACS. 50 Rock F annealed hardness. Density: 10.0 g/cm^3. For electrical contacts; resistant to mechanical wear.

CMW D505F
CMW Inc.
Ni 5, bal Ag.
Annealed: 24,000 TS; 32 Rock F. Worked: 84 Rock F. Electrical conductivity: 95% IACS. For electrical contacts.

CMW D50F
CMW Inc.
Ni 15, bal Ag.
27,000 TS; 40 Rock F (annealed hardness). 89 Rock F (worked); Electrical conductivity: 80% IACS. For electrical contacts; resistant to mechanical wear.

CMW D510F
CMW Inc.
Ni 10, bal Ag.
Annealed: 25,000 TS; 35 Rock F. Worked: 89 Rock F. Electrical conductivity: 87% IACS. For electrical contacts.

CMW D54 (1)
CMW Inc.
10.0 CdO, bal Ag.
Electrical conductivity: 75% IACS. 40 Rock F (annealed hardness). Density: 9.80 g/cm^3. For electrical contacts; especially resistant to welding and high surge currents.

CMW D54 (2)
CMW Inc.
Ag, CdO.
Sintered: 18,000 TS. For electrical contacts. 80% conductivity.

CMW D54F

CMW Inc.
10.0 CdO, bal Ag.
Annealed: 39,000 TS; 71 Rock F. Worked: 46,000 TS; 90 Rock F. Electrical conductivity: 82% IACS. For electrical contacts; resistant to welding and high surge currents.

CMW D54X

CMW Inc.
10.0 CdO, bal Ag.
Annealed: 27,000 TS, 45 Rock F. Electrical conductivity: 75% IACS. For electrical contacts; resistant to welding and high surge currents.

CMW D55

CMW Inc.
15.0 CdO, bal Ag.
Electrical conductivity: 55% IACS. 25 Rock F annealed hardness. Density: 9.6 g/cm^3. For electrical contacts; resistant to welding and high surge currents.

CMW D55F

CMW Inc.
15.0 CdO, bal Ag.
Annealed: 40,000 TS; 70 Rock F. Worked: 48,000 TS; 90 Rock F. Electrical conductivity: 72% IACS. For electrical contacts; resistant to welding and to high surge currents.

CMW D55X

CMW Inc.
15.0 CdO, bal Ag.
Annealed: 30,000 TS; 50 Rock F. Electrical conductivity: 65% IACS. For electrical contacts, resistant to welding and to high surge current.

CMW D56

CMW Inc.
Ni 30, bal Ag.
Electrical conductivity: 55% IACS. 45 Rock F (annealed). Density: 9.7 g/cm^3. For electrical contacts; resistant to mechanical wear.

CMW D57F

CMW Inc.
Fe 10, bal Ag.
Annealed: 31,000 TS; 48 Rock F. Worked: 39,000 TS; 81 Rock F. Electrical conductivity: 90% IACS. For electrical contacts; AC resistance loads.

CMW D58

CMW Inc.
C 5, bal Ag.
Density: 8.65 g/cm^3. Electrical conductivity: 55%; 25 Rock F. For electrical contacts; resistant to sliding wear.

CMW D581F

CMW Inc.
C 0.5, bal Ag.
Annealed: 24,000 TS; 44 Rock F. Worked: 36,000 TS; 72 Rock F. Electrical conductivity: 102% IACS. For electrical contacts for reduced sticking, low friction.

CMW D582F

CMW Inc.
C 0.25, bal Ag.
Annealed: 25,000 TS; 45 Rock F. Worked: 37,000 TS; 73 Rock F. Electrical conductivity: 103% IACS. For electrical contacts for reduced sticking, low voltage DC.

CMW D583F

CMW Inc.
C 0.75, bal Ag.
Annealed: 24,000 TS; 39 Rock F. Worked: 35,000 TS; 70 Rock F. Electrical conductivity: 100% IACS. For electrical contacts for reduced sticking, low friction.

CMW D63X

CMW Inc.
0.41 MgO, 0.25 NiO, bal Ag.
Heat treated: 70,000 TS; 97 Rock F. Electrical conductivity: 70% IACS. For electrical contacts; high hardness; resists annealing.

CMW D64F

CMW Inc.
0.3 CaO, bal Ag.
Annealed: 24,000 TS; 14 Rock F. Worked: 40,000 TS; 82 Rock F. Electrical conductivity: 101% IACS. For electrical contacts, voltage regulators; low voltage DC.

CMW K-TUNGSTEN

CMW Inc.
W.
Wrought: 300,000 TS. For contacts for ignition system; 32% conductivity. *Obsolete*

CMXA

Acciaierie Valbruna s.p.a.
Stainless Steel. C 0.65, Si 0-1, Mn 0-1, Cr 17, Mo 0-0.75, P 0.04, S 0.03, bal Fe.
Martensitic. AISI 440A.

CMXB

Acciaierie Valbruna s.p.a.
Stainless Steel. C 0.85, Si 0-1, Mn 0-1, Cr 17, Mo 0-0.75, P 0.04, S 0.03, bal Fe.
Martensitic. AISI 440B.

CMXBM

Acciaierie Valbruna s.p.a.
Stainless Steel. C 0.9, Si 0-1, Mn 0-1, P 0-0.04, S 0-0.03, Cr 18, Mo 1.15, Ni 0-0.3, V 0.1, bal Fe.
Martensitic. W. Nr. 1.4112.

CMXC

Acciaierie Valbruna s.p.a.
Stainless Steel. C 1.1, Si 0-1, Mn 0-1, Cr 17, Mo 0.55, P 0.04, S 0.03, bal Fe.
Martensitic. W. Nr. 1.4125. AISI 440C.

CN-7MCB3

Stainless Foundry & Engineering Inc.
C 0.05, Cr 30, Ni 33, Mo 2.5, Cb 0.5, bal Fe.
Cast 70,000 TS; 32,000 YS; 30 El; 163 Brin. For chemical plant equipment, valves, agitators, pumps and tanks; resists mixed acids.

CN01

Herbert Cutanit Ltd.
Sintered carbide tool material. Hardness: over 1750 VHN. For finish-machining cast iron and non-ferrous at high speeds, light cuts. ISO K01. *Obsolete*

CN10

Herbert Cutanit Ltd.
Sintered carbide tool material. Hardness: over 1750 VHN. For high-speed machining of nodular iron and semi-steel. ISO K10 and M10. *Obsolete*

CN15

Herbert Cutanit Ltd.
Sintered carbide tool material. Hardness: 1625-1725 VHN. For finish machining with light feeds on cast iron, non-ferrous and non-metallics. ISO K10. *Obsolete*

CN20

Bergische Stahl Industrie
C 0-0.2, Si 1.8-2.3, Mn 0-2, Cr 24-26, Ni 19-21, bal Fe.
Heat resisting steel. W.-Nr. 1.4841; similar to AISI 310.

CN20

Herbert Cutanit Ltd.
Sintered carbide tool material. Hardness: 1550-1650 VHN. For general purpose machining cast iron, non-ferrous and non-metallics. ISO K20. *Obsolete*

CN25

Herbert Cutanit Ltd.
Sintered carbide tool material. Hardness: 1475-1575 VHN. For machining cast iron, non-ferrous and stainless under vibration and some shock; tough. ISO K20-K30. *Obsolete*

CN30

Herbert Cutanit Ltd.
Sintered carbide tool material. Hardness: 1425-1500 VHN. For machining cast iron and non-ferrous on old machines or at slow speeds; also for interrupted cuts. ISO K30. *Obsolete*

CN40

Herbert Cutanit Ltd.
Sintered carbide tool material. Hardness: 1275-1375 VHN. For heavy duty planing and interrupted cuts of cast iron; tough, shock resistant. ISO K40. *Obsolete*

CNC 1

Creusot-Loire
C 0.16, Mn 0.85, Ni 0.95, Cr 0.95, bal Fe.
Carburizing steel; oil hardening. AFNOR 16NC6.

CNDB 1

Creusot-Loire
C 0.38, Mn 0.65, Ni 0.85, Cr 0.85, Mo 0.2, bal Fe.
Oil hardenable to 980-1130 N/mm^2 psi TS. For structural parts.

CNDB1-SV75

Creusot-Loire
C 0.38, Ni 0.85, Cr 0.85, Mo 0.2, bal Fe.
Bars, treated: 980-1130 N/mm^2 WTS; 835 N/mm^2 min YS; 12 El; 293 Brin min. For structural purposes.

CNDC 1

Creusot-Loire
C 0.18, Mn 0.75, Ni 1.35, Cr 0.85, Mo 0.2, bal Fe.
Alloy carburizing steel; deep hardening. AFNOR 18 NCD 6.

CNK2

Thyssen Edelstahlwerke AG
C 0.3, Cr 18, Ni 8, W 1.5, bal Fe.
For cutlery, plastic mold dies, valves; stainless, austenitic. *Obsolete*

CNS NO. 1 (AISI D2)

Jessop Steel Co.
C 1.55, Cr 12.5, V 8.8, Mo 0.8, bal Fe.
For blanking and forming dies, punches; air hardened, nondeforming. AISI D2.

CNS-2

Jessop Steel Co.
C 2.15, Cr 12, Co 0.25, V 0.18, bal Fe.
For blanking and forming dies, punches; oil hardened, nondeforming. AISI D3.

CNS-TH

Thyssen Edelstahlwerke AG
C 0.25-0.3, Cr 1-1.3, W 4.5, Mo 0.3-0.5, bal Fe.
For drop forging dies; oil or water hardening. *Obsolete*

CNS2H

Thyssen Edelstahlwerke AG
C, Cr, Ni, W, bal Fe.
For mandrels, stamps, bakelite dies; oil hardened. *Obsolete*

CO 12

Now HAYNES STELLITE ALLOY NO. 12.

CO 19

Now HAYNES STELLITE ALLOY NO. 19.

CO 21

Now HAYNES STELLITE ALLOY NO. 21.

CO 31

Now HAYNES STELLITE ALLOY NO. 31.

CO 6
Now HAYNES STELLITE ALLOY NO. 6B.

CO ALLOY 21
Ancast, Inc.
Reactive and refractory. C 0.2-0.3, Mn 0-1, Si 0-1, Cr 25-29, Ni 1.75-3.75, Mo 5-6, Fe 0-3, P 0-0.04, S 0-0.04, B 0-0.007, bal Co.
ASTM A-567 GR1.

CO ALLOY 25
Ancast, Inc.
Reactive and refractory. C 0.05-0.15, Mn 1-2, Si 0-1, Cr 19-21, Ni 9-11, W 14-16, Fe 0-3, bal Co.
H.S. 25 (L605).

CO ALLOY 31
Ancast, Inc.
Reactive and refractory. C 0.45-0.55, Mn 0-1, Si 0-1, Cr 24.5-26.5, Ni 9.5-11.5, Mo 0-0.5, W 7-8, Fe 0-2, P 0-0.04, S 0-0.04, bal Co.
AMS 5382G.

CO ALLOY 6
Ancast, Inc.
Reactive and refractory. C 0.9-1.4, Mn 0-1, Si 0-1.5, Cr 27-31, Ni 0-3, Mo 0-1.5, W 3.5-5.5, Fe 0-3, P 0-0.04, S 0-0.04, bal Co.
AMS 5387A.

CO ALLOY X-40
Ancast, Inc.
Reactive and refractory. C 0.45-0.55, Mn 0-1, Si 0-1, Cr 24.5-26.5, Ni 9.5-11.5, W 7-8, Fe 0-2, P 0-0.04, S 0-0.04, B 0.005-0.015, bal Co.
ASTM A-567 GR2.

CO ALLOY X-45
Ancast, Inc.
Reactive and refractory. C 0.2-0.3, Mn 0.4-1, Si 0.75-1, Cr 24.5-26.5, Ni 9.5-11.5, W 7-8, Fe 0-2, P 0-0.04, S 0-0.04, B 0.005-0.015, bal Co.
ASTM A-567 GR 13.

CO J
Now HAYNES STELLITE STAR J METAL.

CO MAJOR
Jessop Steel Co.
C 0.7, W 18, Cr 4, V 1, Co 5, bal Fe.
For tools, cutters, lathe and planer tools; high speed steel. *Obsolete*

CO-6
Latrobe Steel Co.
Tool material. C 0.89, W 5.75, Cr 4.1, V 1.9, Mo 5.25, Co 8, bal Fe.
For tools, cutters; high speed steel.

CO-CO
Teledyne Vasco
C 0.7, W 18, Cr 4, V 1, Co 5, bal Fe.
For high speed cutters and turning tools; high speed steel. *Obsolete*

CO-ELINVAR
Creusot-Loire
Ni 36, Cr 12, Co, bal Fe.
For chronometers, hair springs; controlled expansion. *Obsolete*

CO-ELINVAR
American manufacture
Co 57-60, Fe 25-35, Ni 8-15.
For instruments; constant modulus.

CO-ELINVAR
Wallace Murray Corp.
Ni 36, Cr 12, Co, bal Fe.
For chronometers, hair springs; controlled expansion. *Obsolete*

CO-NETIC A-152
Polymer Corp. Ltd.
Fe, bal Ni.
Sheet: 21,000 YP; 43-47 Rock B. For magnetic shielding.

CO-NETIC AA
Magnetic Shield Corp.
Nickel.
Magnetic shielding alloy with high magnetic permeability. Perfection annealed: 64,000 TS; 18,500 YS; 27 El. Stress annealed: 85,000 TS; 33,000 YS; 32 El.

CO-NETIC AA
Polymer Corp. Ltd.
Fe 20, bal Ni.
Sheet: 57,000 TS; 28,500 YP; 47-53 Rock B. For electrostatic shielding. High permeability.

CO-NETIC B
Magnetic Shield Corp.
Nickel.
Magnetic shielding alloy with high magnetic permeability. Stress annealed: 80,000 TS; 27,000 YS; 32 El.

CO-NETIC B
Polymer Corp. Ltd.
Fe, bal Ni.
Sheet: 58,000 TS; 19,000 YP. For magnetic shielding.

CO-NETIC S3-6
Magnetic Shield Corp.
Nickel.
Magnetic shielding alloy with high magnetic permeability. Stress annealed: 42,000 TS; 27,000 YS; 38 El.

COANAILIUM
R.W. Coan Ltd.
Aluminum. Al alloy.
For marine parts; corrosion resistant.

COAST CM-119
Coast Metals Inc.
For hard facing and hot friction guides Resists molten Cu.

COAST METAL
Coast Metals Inc.
Fe, Cr, Ni, Mo, C.
For hard facing rod; wear and corrosion resistant. *Obsolete*

COAST METALS NP
Coast Metals Inc.
Ni 50, Fe 30, Si 12, P 4, Mo 4.
For brazing stainless steel atomic fuel elements for service in 565°F. pressurized water. Corrosion resistant.

COAST NO. 1
Coast Metals Inc.
C 4, Cr 16, Ni 6, bal Fe.
Cast: 420-550 Brin. For hard facing welding rod. Austenitic; wear and heat resistant.

COAST NO. 10
Coast Metals Inc.
C 4, Cr 16, Ni 2, Mo 8, bal Fe.
Cast: 480-600 Brin. For hard facing welding rod. Non-magnetic; hot abrasion resistant.

COAST NO. 100X (1)
Coast Metals Inc.
C 1, Cr 3, Ni 3, Mo 7, bal Fe.
For hard facing electrodes. Arc welding; can be forged.

COAST NO. 100X (2)
Coast Metals Inc.
C 0.7, W 18, Cr 4, V 1, bal Fe.
Cast: 450-530 Brin. For hard facing welding rod. High speed steel.

COAST NO. 101
Coast Metals Inc.
C 4, Cr 16, Ni 6, bal Fe.
Cast: 420-550 Brin. For hard facing welding rod. Austenitic; wear and heat resistant.

COAST NO. 104
Coast Metals Inc.
C 4, Cr 16, Ni 6, Si 5, bal Fe.
Cast: 530-600 Brin. For hard facing welding rod. Austenitic; wear and heat resistant.

COAST NO. 106
Coast Metals Inc.
Co-Cr alloy.
For hard facing welding rod. *Obsolete*

COAST NO. 107
Coast Metals Inc.
C 3, Cr 25, Ni 12, Mo 8, bal Fe.
Cast: 300-430 Brin. For hard facing welding rod. Heat and abrasion resistant.

COAST NO. 108
Coast Metals Inc.
C 3, Cr 16, Ni 6, Co 20, bal Fe.
Cast: 400-530 Brin. For hard facing welding rod. Heat and abrasion resistant.

COAST NO. 109
Coast Metals Inc.
C 1, Cr 29, bal Fe.
Cast: 450-550 Brin. For hard facing welding rod; heat resistant. *Obsolete*

COAST NO. 11
Coast Metals Inc.
C, Cr, Ni, bal Fe.
Cast: 450-560 Brin. For hard facing electrode; corrosion resistant. *Obsolete*

COAST NO. 110
Coast Metals Inc.
C 4, Cr 16, Ni 2, Mo 8, bal Fe.
Cast: 480-600 Brin. For hard facing welding rod. Non-magnetic; hot abrasion resistant.

COAST NO. 111
Coast Metals Inc.
C, Cr, Ni, bal Fe.
Cast: 450-560 Brin. For hard facing electrode. Corrosion resistant. *Obsolete*

COAST NO. 112
Coast Metals Inc.
C 4, Cr 16, Ni 6, Mg 8, bal Fe.
Cast: 520-590 Brin. For hard facing electrode. Resists heat and dry abrasion.

COAST NO. 115
Coast Metals Inc.
C 4, Cr 16, Ni 4, Mo 6.5, Co 20, bal Fe.
For hard facing rod for arc welding. Corrosion and wear resistant.

COAST NO. 117
Coast Metals Inc.
C 3, Cr 40, Mo 8, Co 30, Si 1, bal Fe.
Welded: 530-600 Brin. For arc welding rod for chemical equipment. Corrosion and abrasion resistant.

COAST NO. 118
Coast Metals Inc.
C 3, Cr 25, Ni 15, Mo 8, Co 30, Si 1, bal Fe.
Cast: 400-470 Brin. For hard facing welding rod. Heat and abrasion resistant.

COAST NO. 119
Coast Metals Inc.
C 1, Cr 35, Cu 2, bal Fe.
Welded: 300-400 Brin. For arc welding rod for rolling mills. Heat and abrasion resistant.

COAST NO. 140
Coast Metals Inc.
C 3, Cr 30, Ni 40, W 14, bal Fe.
For arc welding rod for valves. Corrosion, heat and wear resistant.

COAST NO. 15
Coast Metals Inc.
C 4, Cr 16, Ni 4, Mo 6.5, Co 20, bal Fe.
Welded: 530-560 Brin. For hard facing rod for acetylene welding. Corrosion and wear resistant; tough; hard.

COAST NO. 1600-N
Coast Metals Inc.
Mn Cu-Ni.
Wire, powder or foil for brazing and welding 300 and 400 stainless steels, AM 350, H11, and SAE 4130 steels.

COAST NO. 17
Coast Metals Inc.
C 3, Cr 40, Mo 8, Co 30, Si 1, bal Fe.
Welded: 530-600 Brin. For acetylene welding rod for chemical equipment. Corrosion and abrasion resistant.

COAST NO. 18
Coast Metals Inc.
C 3, Cr 25, Ni 15, Mo 8, Co 30, Si 1, bal Fe.
Cast: 400-470 Brin. For hard-facing electrode. Heat and abrasion resistant.

COAST NO. 19
Coast Metals Inc.
C 1, Cr 35, Cu 2, bal Fe.
Welded: 300-400 Brin. For acetylene welding rod for rolling mills. Heat and abrasion resistant.

COAST NO. 190
Coast Metals Inc.
C 3, Cr 5, Ni 3, bal Fe.
For hard surfacing electrodes. Resists abrasion and impact; arc welded.

COAST NO. 4
Coast Metals Inc.
C 4, Cr 16, Ni 6, Si 5, bal Fe.
Cast: 600-630 Brin. For hard facing welding rod. Austenitic; wear and heat resistant.

COAST NO. 40
Coast Metals Inc.
C 3, Cr 30, Ni 40, W 14, bal Fe.
Welded: 380-420 Brin. For acetylene welding rod for valves. Corrosion, heat and wear resistant.

COAST NO. 50
Coast Metals Inc.
Ni 93.25, Si 3.5, B 2.25, 1.0 others.
Cast: 390-440 Brin. High temperature nickel brazing alloy; MP 1825°F. For thin walled joints, and wide or close tolerance joints. Oxidation resistant.

COAST NO. 50B
Coast Metals Inc.
Si 2.5, B 1.5, bal Ni.
Metal powder for repair brazing of castings and overlays on dies and molds. Machinable.

COAST NO. 50C
Coast Metals Inc.
Si 3, B 1.8, bal Ni.
Brazing alloy for casting repairs and overlays on dies and molds. Machinable; metal powders.

COAST NO. 52
Coast Metals Inc.
Ni 91.25, Si 4.5, B 4.5, 1.0 others.
Cast: 570-620 Brin. For brazing alloy; MP 1825°F. Oxidation resistant; high strength joints.

COAST NO. 52 SPECIAL
Coast Metals Inc.
Si 4.5, C 0.15, Co 20, B 2.9, bal Ni.
For brazing high temperature alloys. Brazing temperature 2150-2175°F.

COAST NO. 53
Coast Metals Inc.
Ni 4.5, B 2.9, Cr 7, Fe 3, 0.5 others.
Cast: 570-620 Brin. For brazing alloy; MP 1825°F. Oxidation resistant.

COAST NO. 54
Coast Metals Inc.
C 0.35, Cr 9, Si 2.5, Fe 2.1, B 1.66, bal Ni.
Metal powder for repair or build-up on valves and valve seats. Average hardness 38 Rock C; machinable.

COAST NO. 55
Coast Metals Inc.
C 0.35, Cr 10, Si 4.35, Fe 4, B 2.1, bal Ni.
Metal powder for build-up or repair on pump sleeves and wear rings. Average hardness 45 Rock C.

COAST NO. 56
Coast Metals Inc.
Ni 72.5, Si 4, B 3.75, Cr 16, Fe 4, 1.0 others.
Cast: 570-620 Brin. For brazing alloy; MP 1880°F. Oxidation resistant.

COAST NO. 6
Coast Metals Inc.
Co-Cr alloy.
For hard facing welding rod; wear resistant. *Obsolete*

COAST NO. 60
Coast Metals Inc.
Ni-Cr-Si.
Alloy powder for nuclear brazing.

COAST NO. 62
Coast Metals Inc.
Mn-Ni-Co.
Alloy powder for brazing wide gaps and for high temperature strengths with good corrosion resistance, on such alloys as Rene 41, 7-15 Mo, A-286, tungsten. AMS 4780.

COAST NO. 7
Coast Metals Inc.
C 3, Cr 25, Ni 12, Mo 8, bal Fe.
Cast: 300-450 Brin. For hard facing welding rod. Heat and abrasion resistant.

COAST NO. 8
Coast Metals Inc.
C 3, Cr 16, Ni 6, Co 20, bal Fe.
Cast: 400-530 Brin. For hard facing welding rod. Heat and abrasion resistant.

COAST NO. 80A
Coast Metals Inc.
Mo, Cr, W, bal Ni.
For hard facing electrode. Corrosion and abrasion resistant.

COAST NO. 81
Coast Metals Inc.
Mo 30, Fe 5, B 2, bal Ni.
Metal powder for repair or build-up of chemical and petro-chemical plant equipment. Excellent corrosion resistance.

COAST NO. 9
Coast Metals Inc.
C 1, Cr 29, bal Fe.
Cast: 450-550 Brin. For hard facing welding rod. Heat resistant. *Obsolete*

COAST NO. 90
Coast Metals Inc.
C 3, Cr 5, Ni 3, bal Fe.
For hard surfacing electrode. Resists abrasion and impact; acetylene welded.

COAST NO. 91A
Coast Metals Inc.
C, Cr, bal Fe.
For hard facing electrode. Abrasion and impact resistant.

COAST NO. 92A
Coast Metals Inc.
C, Cr, Si, Mn, bal Fe.
For hard facing electrode. Impact and abrasion resistant.

COAST NO. 96A
Coast Metals Inc.
C, Cr, Mn, Si, bal Fe.
For hard facing electrode. Work hardens; wear resistant.

COAST NO. 98A
Coast Metals Inc.
Cr, Mn, Si, bal Fe.
For hard facing electrode. Corrosion and impact resistant.

COAST NO. X (1)
Coast Metals Inc.
C 0.7, W 18, Cr 4, V 1, bal Fe.
Cast: 450-530 Brin. For hard facing welding rod. High speed steel.

COAST NO. X (2)
Coast Metals Inc.
C 1, Cr 3, Ni 3, Mo 7, bal Fe.
For acetylene welding rod for hot shears. Forgeable; tough.

COATING METAL 560
NL Industries
Pb 95, Sn 2.5, Zn 2.5.
For coated iron for roofing, pipes. *Obsolete*

COBAFLUX
Bethlehem Steel Corp.
5 Cr, 4 W, 37 Co, or 4 Cr, 9 W, 19 Co, bal Fe.
For magnets, magnetos, scientific instruments. *Obsolete*

COBAFLUX "B"
Bethlehem Steel Corp.
C 0.9, Cr 4.5, W 9, Co 19, bal Fe.
For cutting tools. *Obsolete*

COBAFLUX MAGNET A
Bethlehem Steel Corp.
C 0.7, Cr 5, W 4, Co 37, bal Fe.
For permanent magnets. *Obsolete*

COBAL 1
Uddeholm Corp.
C 1, Cr 6.5, Mo 1, Co 7.5, bal Fe.
For permanent magnets. *Obsolete*

COBAL 3
Uddeholm Corp.
C 0.95, Mn 0.7, Cr 5.7, W 4.5, Co 15.5, bal Fe.
For permanent magnets. *Obsolete*

COBAL 4
Uddeholm Corp.
C 0.85, Mn 0.7, Cr 5.5, W 4.5, Co 35, bal Fe.
For permanent magnets. *Obsolete*

COBALT
Hidalgo Steel Co. Inc.
C 0.7, W 18, Cr 4, V 1, Co 5, bal Fe.
For tools, cutters; high speed steel.

COBALT
Latrobe Steel Co.
Tool material. C 0.7, Co 5, W 18, Cr 4, V 1, bal Fe.
For tools for heavy hogging cuts; high speed steel.

COBALT
Atomergic Chemetals Corp.
Co.
Purities: 99.999%, 99.99%, 99.9%, 99.7% commercial grade.
Forms: sponge, powder, rod, wire, sheet, foil, platelets, cakes, single crystals.

COBALT
CCS Braeburn Alloy Steel
C 0.74, W 18, Cr 4, V 1, Co 5, Mo 0.8, bal Fe.
For lathe and planer tools, reamers, hobs, broaches, drills.
High speed steel, oil hardened. *Obsolete*

COBALT 1
Arcos Alloys
C 2-3, Si 0.4-2, Cr 26-35, W 11-14, Ni 0-3, Mo 0-1, bal Co.
For bare (1B) and coated (1C) hardfacing electrodes; grades RCoCr-C and ECoCr-C. *Obsolete*

COBALT 12
Arcos Alloys
C 1.2-1.7, Cr 26-32, W 7-9.5, Ni 0-3, Mo 0-1, bal Co.
For bare (12B) and coated (12C) hardfacing electrodes; grades RCoCr-B and ECoCr-B. *Obsolete*

COBALT 125
Thyssen Edelstahlwerke AG
C 1, Mn 1.1, Cr 8.5, Mo 1.25, Co 6.25, bal Fe.
For permanent magnets. *Obsolete*

COBALT 160
Thyssen Edelstahlwerke AG
C 1, Cr 8.5, Mo 1.5, Co 10.5, bal Fe.
For permanent magnets. *Obsolete*

COBALT 17%
Indiana General
C 0.75, W 6, bal Fe.
For permanent magnets; Br 9500, Hc 150. *Obsolete*

COBALT 2
Arcos Alloys
C 2-3, Si 0-1, Cr 29-35, W 16-19, Ni 0-2.5, bal Co.
For bare hardfacing electrodes; abrasion resistant at high temperatures. *Obsolete*

COBALT 200
Thyssen Edelstahlwerke AG
C 1, Cr 8.5, Mo 1.5, Co 15.5, bal Fe.
For permanent magnets. *Obsolete*

COBALT 3%
Electronic Memories & Magnetics Corp.
Co 3.25, Cr 4, C 1, bal Fe.
Peak energy product 0.38×10^6 max; residual induction 9700 gauss; coercive force 80 oersted. Permanent magnet.

COBALT 300
Thyssen Edelstahlwerke AG
C 0.9, Cr 4.5, Mo 0.3, Co 30, W 4.5, bal Fe.
For permanent magnets. *Obsolete*

COBALT 36%
Indiana General
Cr 4, W 5, C 0.85, Co 36, bal Fe.
For permanent magnets; Br 8500, Hc 240. *Obsolete*

COBALT 6
Arcos Alloys
C 0.9-1.4, Si 0.4-2, Cr 26-32, W 3-6, Ni 0-3, Mo 0-1, bal Co.
For bare (6B) and coated (6C) hardfacing electrodes, grades RCoCr-A and ECoCr-A. *Obsolete*

COBALT 7
Arcos Alloys
C 0.2-0.4, Cr 25-29, Ni 2-4, Mo 5-6, bal Co.
For bare (7B) and coated (7C) hardfacing electrodes; impact resistant, crack resistant. *Obsolete*

COBALT I
Thyssen Edelstahlwerke AG
C 0.7, W 18, Cr 4, V 1.6, Mo 0.6, Co 9.5, bal Fe.
For lathe and planer tools, hobs, drills, broaches; high speed steel, oil hardened. *Obsolete*

COBALT II
Thyssen Edelstahlwerke AG
C 0.8, Cr 4.2, Mo 0.65, V 1.6, W 18.5, Co 5, bal Fe.
For lathe and planer tools, reamers, broaches, taps; high speed steel, oil hardened. *Obsolete*

COBALT II
Thyssen Edelstahlwerke AG
C 0.7, W 18, Cr 4, Co 5, V 1, bal Fe.
For lathe and planer tools, drills, hobs, reamers; high speed steel, oil hardened. *Obsolete*

COBALT III
Thyssen Edelstahlwerke AG
C 0.8, Cr 4.25, Mo 0.6, V 2.25, W 12.5, Co 2.75, bal Fe.
For tools, dies, cutters; high speed steel. *Obsolete*

COBALT III
Thyssen Edelstahlwerke AG
C 0.7, W 18, Cr 4, V 1, Co 5, bal Fe.
For lathe and planer tools, drills, hobs, reamers; high speed steel, oil hardened. *Obsolete*

COBALT III, NX
Thyssen Edelstahlwerke AG
C 0.8, Cr 4.25, Mo 0.6, V 2, W 10, Co 2.5, bal Fe.
For tools, dies, cutters; high speed steel. *Obsolete*

COBALT ALUMINUM BRONZE
Ampco Metal
Al 10.5, Fe 3.5, Co 2.5, bal Cu.
Cast: 90,000 TS; 45,000 YS; 15 El; 20 RA; 180 Brin. Forged: 120,000 TS; 80,000 YS; 10 El; 12 RA; 241 Brin. For pump bodies, shafting, heavy loaded bearings, impellers, steel mill slippers and screwdown nuts. Wear and corrosion resistant. *Obsolete*

COBALT ASCOLOY
American manufacture
C 0.2, Cr 12.25, Co 5, W 3, V 0.25, bal Fe.
High temperature alloy; heat and corrosion resistant.

COBALT CHROME
Latrobe Steel Co.
Tool material. C 1.5, Cr 12.25, Co 3.1, Mo 0.85, Si 0.5, Mn 0.5, bal Fe.
For broaches, burnishing tools, valves, punches; high abrasion resistance, non-deforming.

COBALT CHROME FM
Latrobe Steel Co.
C 1.35, Cr 12.5, Ni 0.3, Co 3, Mo 0.8, S, bal Fe.
For burnishing tools, valves, punches, broaches; high abrasion resistance, non-deforming. *Obsolete*

COBALT HIGH SPEED
McInnes Steel Co.
C 0.7, Cr 4, W 18, Co 5, bal Fe.
For high speed cutting tools, dies; high speed steel. *Obsolete*

COBALT I
Thyssen Edelstahlwerke AG
C 0.75, Cr 4.2, Mo 0.85, V 1.6, W 18, Co 9.5, bal Fe.
For lathe and planer tools, slotting tools, form cutters; high speed steel, heavy duty. *Obsolete*

COBALT MAGNET
Wallace Murray Corp.
C, alloy, bal Fe.
For magnets.

COBALT MAGNET 37%
Latrobe Steel Co.
C 0.9, W 2.5, Cr 4, Co 37, bal Fe.
For electrical equipment, permanent magnets; magnet steel. *Obsolete*

COBALT MAGNET STEEL (17%)
Colt Industries
C, Cr, W, Co, bal Fe.
For permanent magnets. *Obsolete*

COBALT MAGNET STEEL (35%)
Crucible Specialty Metals
C, Cr, W, Co, bal Fe.
For permanent magnets. *Obsolete*

COBALT MAJOR
Jessop Steel Co.
C 0.7, W 19, Cr 4, V 2, Co 8, bal Fe.
Annealed: 228-241 Brin. For lathe and planer tools, reamers, broaches, hobs; high speed steel. *Obsolete*

COBALT SPECIAL
Thyssen Edelstahlwerke AG
C 0.65, Cr 4.2, Mo 0.65, V 1.7, W 18, Co 15.5, bal Fe.
For lathe and planer tools, form cutters; high speed steel, heavy duty. *Obsolete*

COBALT STEEL
CCS Braeburn Alloy Steel
Mn 0.15-0.3, C 0.73, Cr 3.75-4.25, V 0.85-1, W 17.5-18.5, Co 4-4.5, Mo 0.75-1, bal Fe.
For tools, high speed cutting tools; high speed steel. *Obsolete*

COBALTCROM
Darwin & Milner Inc.
Tool material. Co 3.7, Cr 13.6, C, bal Fe.
For dies, broaches, cutters, valves; nonscaling.

COBALTCROM KXK STAINLESS
Darwin & Milner Inc.
Tool material. C 1, Cr 18, Co 1.1, V 0.2, Mo 1.1, bal Fe.
For dies, press tools, pump valves; corrosion and abrasion resistant.

COBALTCROM PRK-HT
Darwin & Milner Inc.
C, Co, Cr, bal Fe.
For welding rod; coated.

COBALTCROM STAINLESS
Darwin & Milner Inc.
Tool material. Co 3.3, Cr 18, Mo 0.75, C 1.1, V 0.25, Mn 0.3, bal Fe.
For special cutlery and corrosion resistant parts; air hardened steel; nondeforming.

COBALTLOY
W.A. Zelnicker
C 1.5, Mo 0.9, Cr 12, V 0.5, Co 3.25, bal Fe.
For hot forming dies; abrasion resistant.

COBAMET X5
Kennametal Inc.
metallurgical-grade Co.
93-98 Rock B. Hot-press matrix powder used in the manufacture of impregnated-diamond and cubic boron nitride products for sawing, drilling, grinding, and honing tools.

COBAMET XB
Kennametal Inc.
metallurgical-grade Co.
93-98 Rock B. Hot-press matrix powder used in the manufacture of impregnated-diamond and cubic boron nitride products for sawing, drilling, grinding, and honing tools.

COBANIC
Gilby-Fodor S.A.
Ni 54.5, Co 44.5, Fe 1.
Annealed: 80,000 TS; 60,000 YS; 25 El. For cathodes and filaments in electronic tubes. Heat resistant.

COBANIC
Wilbur B. Driver Co.
Nickel. Ni 55, Co 45, C 0.1, bal Fe.
Annealed: 65,000-80,000 TS; 38 El. For vacuum tube filament wire; heat and corrosion resistant.

COBAR 6
Arcos Alloys
C 1, Cr 30, W 4.5, Ni 2, Fe 2, bal Co.
Weld metal. Cored wire hardfacing electrode. *Obsolete*

COBEND 1
Arcos Alloys
C 2-3, Cr 26-33, W 11-14, Ni 0-3, Mo 0-1, bal Co.
Welded: 45-48 Rock C. For hardfacing electrodes; Grade ECoCr-C; resists metal to metal wear.

COBEND 12
Arcos Alloys
C 1.2-1.7, Cr 26-32, W 7-9.5, Ni 0-3, Mo 0-1, bal Co.
Welded: 35-39 Rock C. For hardfacing electrodes; grade ECoCr-B.

COBEND 6
Arcos Alloys
C 0.9-1.4, Cr 26-32, W 3-6, Ni 0-3, Mo 0-1, bal Co.
Welded: 37-40 Rock C; work hardened: 47-51 Rock C. For hardfacing electrodes; grade ECoCr-A; resists abrasion, corrosion, galling, oxidation, erosion in 1200-1800°F range.

COBEND 7
Arcos Alloys
C 0.2-0.4, Cr 25-29, Ni 2-4, Mo 5-6, bal Co.
Welded: 34-37 Rock C; work hardened: 40-47 Rock C. For hardfacing electrodes; resistant to cracking, erosion, hot abrasion.

COBEND 7W
Arcos Alloys
C 0.25-0.45, Cr 24-26, W 4.5-6, Ni 2-4, Mo 4.5-6, bal Co.
Welded: 35-37 Rock C. For hardfacing electrodes; for cutting edge of hot forming tools. *Obsolete*

COBENIUM
Wilbur B. Driver Co.
Co 40, Cr 20, Ni 15, Mo 7, Mn 2, Be 0.04, C 0.15, bal Fe.
Cold worked: 250,000-300,000 TS; 240,000-290,000 YS; 1 El; 60 RA; 372 Brin. For springs, high temperature applications; heat treatable, corrosion and heat resistant. *Obsolete*

COBITE
Columbia Tool Steel Co.
Tool material. C 0.81, W 18.4, Cr 4.25, V 2.05, Co 8.75, bal Fe.
W-Co high speed tool steel; very high red hardness. For lathe tools, boring tools, cutoff tools. AISI T5.

COBRA
Ludlow Steel Corp.
C 0.7, W 18, Cr 4, V 1, bal Fe.
Hardened: 64-66 Rock C. For cutters, lathe and planer tools, reamers, hobs, form cutters. High-speed steel, Type T1, high red hardness.

COCHROME
Cochrane Corp.
C 0.3, Cr 1.2, Mo 0.3, bal Fe.
For valve and valve fittings, castings for 250 lb steam pressure. High strength.

COCHROME
English manufacture
Co 60, Fe 24, Cr 12, Mn 2.
For heating elements, filaments; high heat resistance.

COCK BRONZE
English manufacture
Sn 8-10, Zn 2-6, bal Cu.
For cocks, fittings, hardware; corrosion resistant.

COCUMAN
GTE Products Corp./Wesgo Div.
Co 10, Cu 58.5, Mn 31.5.
Brazing alloy available in foil, flexibraze, wire, powder, extrudable paste and preform. PWA. Liquidus 1830°F. Solidus 1645°F.

CODE-ARC
Now AIRCO CODE-ARC.

CODE-ROD NO. 120
Marquette Corp.
C 0.08-0.13, Mn 0.45-0.65, Si 0.15-0.25, S 0.03, P 0.03, bal Fe.
Welded: 70,000-75,000 TS; 60,000-65,000 YS; 22-27 El; 50-55 RA. For all-purpose welding rod; AWS-E6012.

COE BRONZE
Anaconda Co.
Copper. Cu 89.5, Sn 10.5.
Hard sheet: 115,000 TS; 95,000 YS; 5 El; 190 Brin. Soft sheet: 60,000 TS; 40,000 YS; 65 El; 74 Brin. For gears. Tough.

COFLEX
Engelhard Corp.
Bimetal. For thermostatic bimetals; corrosion resistant. *Obsolete*

COG BRONZE
English manufacture
Sn 11, Zn 4, bal Cu.
For cogs, worms; tough, corrosion resistant.

COG WHEEL BRAND NO. II
Phosphor Bronze Co. Ltd.
11.0 Sn phosphor bronze.
Valves, bearings, bushes, pumps and connection rods.

COG WHEEL BRAND NO. VII
Phosphor Bronze Co. Ltd.
12.0 Sn phosphor bronze.
Worm and gear wheels, piston rings, wear plates, gib keys.

COG WHEEL BRAND NO. VIII
Phosphor Bronze Co. Ltd.
14.0 Sn phosphor bronze.
Heavy duty; for high speed bearings.

COG WHEEL BRAND NO. XI
Phosphor Bronze Co. Ltd.
10.0 Pb phosphor bronze.
For rolling mills, crushing equipment.

COGNE CX
Cogne Co.
C 0.38-0.45, Cr 13, bal Fe.
Annealed: 100,000 TS; 55,000 YS; 20 El; 50 RA; 210 Brin. For bearings, valves, cutlery, surgical instruments; Type 420; corrosion resistant.

COGNE FEOX
Cogne Co.
C 0.07-0.12, Cr 17, bal Fe.
Annealed: 80,000 TS; 50,000 YS; 25 El; 50 RA; 150 Brin. For oil refinery equipment, oil burners and heaters, dairy and food equipment; Type 430; corrosion resistant.

COGNE FNOX
Cogne Co.
C 0-0.25, Cr 17, Ni 2, bal Fe.
Annealed: 125,000 TS; 95,000 YS; 20 El; 55 RA; 260 Brin. For pumps, marine hardware, valves; Type 431; corrosion and heat resistant.

COGNE IOX1
Cogne Co.
C 0.13-0.18, Cr 13, bal Fe.
Annealed: 85,000 TS; 45,000 YS; 25 El; 55 RA; 200 Brin. For valves, cutlery, surgical instruments, turbine blades; Type 410 and 420; corrosion resistant.

COGNE IOX3
Cogne Co.
C 0.26-0.37, Cr 13, bal Fe.
Annealed: 85,000 TS; 50,000 YS; 25 El; 55 RA; 195 Brin. For valves, cutlery, turbine blades, surgical instruments; Type 420; corrosion resistant.

COGNE IOXA
Cogne Co.
Stainless steel. C 0.06, Cr 13, Al 0.2, bal Fe.
Annealed: 70,000 TS; 42,000 YS; 25 El; 160 Brin. Cold drawn: 85,000 TS; 70,000 YS; 20 El; 185 Brin. For annealing boxes, quenching racks, oil refinery equipment; not hardenable. Type 405, magnetic.

COGNE IOXO
Cogne Co.
C 0.09-0.12, Cr 13, C 0-0.5, bal Fe.
Annealed: 75,000 TS; 40,000 YS; 35 El; 70 RA; 155 Brin. For turbine blades, valves, cutlery, surgical instruments; Type 403; corrosion resistant.

COGNE IOXOO
Cogne Co.
C 0-0.08, Cr 13, bal Fe.
Annealed: 75,000 TS; 40,000 YS; 35 El; 70 RA; 155 Brin. For turbine blades, valves, cutlery, surgical instruments; Type 403 and 410; corrosion resistant.

COGNE KXO2
Cogne Co.
C 0-0.15, Cr 13, Ni 2, bal Fe.
Annealed: 80,000 TS; 40,000 YS; 32 El; 68 RA; 160 Brin. For oil refinery equipment; Type 414; corrosion resistant.

COGNE LIOX
Cogne Co.
C 0.08-0.16, Cr 12, Ni 12, bal Fe.
For valves, pump parts; corrosion resistant.

COGNE NIOX
Cogne Co.
Stainless steel. C 0.1, Cr 18, Ni 8, bal Fe.
Annealed: 80,000 TS; 35,000 YS; 55 El; 75 RA; 155 Brin. For chemical plant equipment, mixers, tanks, filters; Type 301 and 302; austenitic.

COGNE NIOX-C
Cogne Co.
Stainless steel. C 0.06, Cr 18, Ni 12, Cb 0.5, bal Fe.
Annealed: 90,000 TS; 40,000 YS; 55 El; 82 Rock B. Cold drawn: 100,000 TS; 65,000 YS; 40 El; 95 Rock B. For exhaust manifolds, steam pipes, radiant superheaters, welded structures. Type 347, stabilized, austenitic.

COGNE NIOX-D
Cogne Co.
Stainless steel. C 0.11-0.16, Cr 18, Ni 8, bal Fe.
Annealed: 80,000 TS; 35,000 YS; 55 El; 75 RA; 155 Brin. For chemical plant equipment, tanks, mixers, filters; Type 302; austenitic.

COGNE NIOX-L
Cogne Co.
Stainless steel. C 0-0.03, Cr 18-20, Ni 8-12, bal Fe.
Annealed: 77,000 TS; 30,000 YS; 60 El; 110 Brin. Cold drawn: 100,000-180,000 TS; 50,000-125,000 YS; 10-50 El; 180-330 Brin. For kitchen utensils, architectural trim, welded components, chemical and textile plant equipment. Type 304L.

COGNE NIOX-M
Cogne Co.
Stainless steel. C 0-0.1, Cr 19, Ni 12, Mo 3.5, bal Fe.
Annealed: 90,000 TS; 40,000 YS; 48 El; 62 RA; 170 Brin. For acid resistant chemical plant equipment; Type 317; austenitic.

COGNE NIOX-MT
Cogne Co.
Stainless steel. C 0-0.12, Cr 18, Ni 8, Mo 2.5, Ti, bal Fe.
Annealed: 85,000 TS; 35,000 YS; 50 El; 65 RA; 160 Brin. For welded acid resistant chemical plant equipment; Type 316 Ti; austenitic.

COGNE NIOX-S
Cogne Co.
Stainless steel. C 0-0.08, Cr 18-20, Ni 8-12, bal Fe.
Annealed: 85,000 TS; 35,000 YS; 60 El; 150 Brin. Cold drawn: 100,000-180,000 TS; 50,000-125,000 YS; 10-50 El; 180-330 Brin. For kitchen equipment, architectural trim, welded components, chemical and textile plant equipment. Type 304.

COGNE NIOX-T
Cogne Co.
Stainless steel. C 0-0.08, Cr 18, Ni 8, Ti, bal Fe.
Annealed: 85,000 TS; 35,000 YS; 55 El; 65 RA; 150 Brin. For welded chemical plant equipment, tanks, mixers; Type 321; austenitic.

COGNE RIOX
Cogne Co.
Stainless steel. C 0-0.25, Cr 25, Ni 20, Si 0-1.5, bal Fe.
Annealed: 100,000 TS; 45,000 YS; 50 El; 65 RA; 185 Brin. For furnace parts, valves, pumps, turbine and jet parts; Type 310 and 314, austenitic.

COGNE SIOX
Cogne Co.
C 0.04, Cr 10, Mo 1, Si 2.5, bal Fe.
For oil refinery equipment; creep and heat resistant.

COGNE VIOX
Cogne Co.
C 0-0.2, Cr 24, Ni 12, bal Fe.
Annealed: 90,000 TS; 40,000 YS; 50 El; 65 RA; 170 Brin. For furnace parts, heat treat boxes, pumps, oil burners; Type 309; corrosion and heat resistant.

COGNE-IOXKM
Cogne Co.
C 0.11, Cr 12.75, Mo 0.5, bal Fe.
Annealed: 75,000 TS; 35,000 YS; 30 El; 92 Rock B. For springs, table flatware, oil refinery equipment. Corrosion resistant, hardenable.

COIN BRONZE
Now IMI 341.

COIN SILVER
Fansteel Metals
Precious metal. Ag 90, Cu 10.
Annealed: 68 Rock-15T. For electrical contacts. Obsolete

COINAGE BRONZE
American manufacture
Sn 3-4, Zn 1-2, bal Cu.
For coins; corrosion resistant.

COL-GRAPH
Columbia Tool Steel Co.
Tool material. C 1.45, Mn 0.8, Si 1.2, Cr 0.2, Mo 0.25, bal Fe.
For blanking dies, gages, machine parts, taps, wear plates.
AISI Type O6; oil hardened tool steel.

COLALLOY
Colonial Alloys Co.
Mg 2, Mn 1, Si 1, bal Al.
Heat treated: 12,000-61,000 TS; 20-110 Brin. For chemical and food handling equipment; corrosion resistant.

COLCLAD 13CR AL
British Steel plc
Layer of stainless. AISI 405 steel on mild steel backing plate. Obsolete

COLCLAD 18/10/2 ELC
British Steel plc
Layer of stainless. AISI 316L steel on mild steel backing plate. Obsolete

COLCLAD 18/10/2 TI
British Steel plc
Layer of titanium stabilized AISI 316 stainless steel on mild steel backing plate. Obsolete

COLCLAD 18/8 ELC
British Steel plc
Layer of stainless. AISI 304L steel on mild steel backing plate. Obsolete

COLCLAD 18/8 TI
British Steel plc
Layer of stainless. AISI 321 steel on mild steel backing plate. Obsolete

COLCLAD INCONEL
British Steel plc
Layer of Inconel on mild steel backing plate. Obsolete

COLCLAD L NICKEL
British Steel plc
Layer of nickel on mild steel backing plate. Obsolete

COLCLAD MONEL
British Steel plc
Layer of Monel on mild steel backing plate. Obsolete

COLD HEADER NO. 4
Boyd-Wagner Co.
C 1, Mn 0.3, Si 0.2, bal Fe.
For cold heading dies; water hardened.

COLD HEADER VANADIUM
Jessop Steel Co.
C 1, Mn 0.25, Si 0.2, V 0.2, bal Fe.
Water hardens to 60 Rock C. Can be tempered to desired hardness from 200-800°F. For taps, dies, punches and pneumatic tools. AISI Type W2 water hardening tool steel.

COLD HEADING 4
Cerro Metal Products Co.
Cu 66.5, bal Zn.
60,000 TS; 35,000 YS; 42 El; 75 RA; 112 Brin. For cold upsetting work, rivets, and thread rolled products. For cold deformation; CA 270.

COLDIE
CCS Braeburn Alloy Steel
C 0.9, V 0.2, Si 0.25, Mn 0.25, bal Fe.
For blanking and forming dies, rivet sets. Water hardened; Type W2. Obsolete

COLESCO
Manufacturer not listed
Pb 77, Sn 8, Sb 14, Cu 1.
For bearings.

COLFORM "E.T.D." 1527
LaSalle Steel Co.
C 0.22-0.29, Mn 1.2-1.5, Si 0.15-0.3, bal Fe.
Cold finished: 120,000 TS min; 92,000 YS min; 14 El; 25-34 Rock C min. For cold forming or cold heading without subsequent heat treatment for strong parts. Obsolete

COLFORM "E.T.D." 1541 A
LaSalle Steel Co.
C 0.4-0.45, Mn 1.3-1.65, Si 0.15-0.3, bal Fe.
Cold finished: 150,000 TS min; 130,000 YS min; 120,000 proof strength min; 32 Rock C min. For cold forming or cold heading without subsequent heat treatment for strong parts. Obsolete

COLFORM "E.T.D." 1541 B
LaSalle Steel Co.
C 0.36-0.44, Mn 1.35-1.65, Si 0.15-0.3, bal Fe.
Cold finished: 140,000 TS min; 125,000 YS min; 10 El; 28 Rock C min. For cold forming or cold heading without subsequent heat treatment for strong parts. Obsolete

COLLET BRASS
American manufacture
Cu 61, Zn 37, Pb 2.5.
For hot forged parts, collets; free-cutting.

COLMO 1000
British Steel plc
C 0-0.13, Si 0.1-0.3, Mn 0.4-0.7, Cr 0.25-0.5, Mo 0.5-0.7, V 0.22-0.3, bal Fe.
For boiler and pressure vessels; good creep resistance up to 575°C. Obsolete

COLMO 900 GRADE 29
British Steel plc
C 0.18-0.25, Si 0.1-0.3, Mn 0.5-0.9, Mo 0.45-0.65, bal Fe.
For boiler and pressure vessels; good creep resistance to 480°C. Obsolete

COLMO 900 GRADE 31
British Steel plc
C 0.2-0.27, Si 0.1-0.3, Mn 0.5-0.9, Mo 0.45-0.65, bal Fe.
For boiler and pressure vessels; good creep resistance to 480°C. Obsolete

COLMO 950
British Steel plc
C 0-0.17, Si 0.1-0.3, Mn 0.4-0.7, Cr 0.7-1.1, Mo 0.45-0.65, bal Fe.
For boiler and pressure vessels; good creep resistance to 520°C. Obsolete

COLMONOY NO. 1
Wall Colmonoy Corp.
C 0.9, Cr 11, B 2.5, Si 1.75, Mn 0.4, bal Fe.
Coated rods. 58-63 Rock C. For hard facing alloy; abrasion resistant.

COLMONOY NO. 10
Wall Colmonoy Corp.
C, B, Ni, bal Fe.
Cast: 600 Brin. For welding rod; corrosion and wear resistant. Obsolete

COLMONOY NO. 15
Wall Colmonoy Corp.
Cu 93, P 7.
78-82 Rock B; 1460°F MP (approx). Powder for build-up or braze welding of copper and copper alloys.

COLMONOY NO. 20
Wall Colmonoy Corp.
C 0.25, Cr 5, B 1, Si 3, Fe 3.5, bal Ni.
Cast: 15-20 Rock C. For glass mold repairs; corrosion and heat resistant. Obsolete

COLMONOY NO. 21
Wall Colmonoy Corp.
C 0.25, Cr 5, Si 3.25, B 1.25, Fe 1, bal Ni.
As cast: 26-31 Rock C; 2050°F MP (approx). For medium hard facing for glass container molds; heat, impact and corrosion resistant. Also available in rod and fuseweld powder.

COLMONOY NO. 22
Wall Colmonoy Corp.
C 0.1, Si 3.15, Fe 0.75, B 1.25, bal Ni.
Cast: 28-33 Rock C; 1925°F MP (approx). Atomized powder for machinable build-up on shafts and worn or undersize parts; good impact resistance. Available in rod and fuseweld powder.

COLMONOY NO. 221
Wall Colmonoy Corp.
C 2.75, Cr 16, Si 1, Mo 1, Mn 1.5, bal Fe.
Welded. 430-480 Brin. For hard facing alloy; abrasion resistant. *Obsolete*

COLMONOY NO. 23A
Wall Colmonoy Corp.
C 0.1, Fe 1, Si 2.3, B 1.25, bal Ni.
Cast: 14-19 Rock C; 1950°F MP (approx). Atomized powder for machinable build-up on shafts and worn or undersize parts; good impact resistance. Available in rod and fuseweld powder.

COLMONOY NO. 24
Wall Colmonoy Corp.
C 0.1, Si 2.3, Fe 1, B 1.25, bal Ni.
Cast: 14-19 Rock C; 1950°F MP (approx). Atomized powder for machinable build-up on shafts and worn or undersize parts; good impact resistance. Also available in rod and fusewold powder.

COLMONOY NO. 3
Wall Colmonoy Corp.
C, W, Cr, B, bal Fe.
Cast: 650 Brin. For casting, welding rod; contains borides. *Obsolete*

COLMONOY NO. 4
Wall Colmonoy Corp.
C 0.45, Cr 10, B 2, Si 2.25, Fe 2.5, bal Ni.
Cast: 35-40 Rock C. For hard facing alloy; corrosion and heat resistant. Also available in rod and powder.

COLMONOY NO. 42SA
Wall Colmonoy Corp.
Same as Colmonoy No. 43, but as sprayweld powder.

COLMONOY NO. 43
Wall Colmonoy Corp.
C 0.45, Cr 10, Fe 2.75, Si 2.75, B 2, bal Ni.
35-40 Rock C; 2025°F MP (approx). For building up and surface finishing shafts and wear surfaces; machinable with carbide tools; impact resistance better than Colmonoy No. 53. For application with fusewelder.

COLMONOY NO. 4SA
Wall Colmonoy Corp.
Now COLMONOY NO. 42SA.

COLMONOY NO. 5
Wall Colmonoy Corp.
C 0.65, Cr 11.5, B 2.5, Si 3.75, Fe 4.25, bal Ni.
Cast: 45-50 Rock C. For hard facing alloy; corrosion and heat resistant. Also available in rod and powder.

COLMONOY NO. 52SA
Wall Colmonoy Corp.
Same as Colmonoy No. 53, but as sprayweld powder.

COLMONOY NO. 53
Wall Colmonoy Corp.
C 0.65, Cr 11.5, Fe 4.25, Si 3.75, B 2.5, bal Ni.
45-50 Rock C. 1950°F MP (approx). For hard facing shafts, wear surface; good red hardness, corrosion resistance, and improved shock resistance. For application with fusewelder.

COLMONOY NO. 56
Wall Colmonoy Corp.
C 0.7, Cr 12.5, B 2.75, Si 4, Fe 4.5, bal Ni.
50-55 Rock C. For hard facing alloy; abrasion, galling and corrosion resistance. Available in rod and powder.

COLMONOY NO. 5SA
Wall Colmonoy Corp.
Now COLMONOY NO. 52SA.

COLMONOY NO. 6
Wall Colmonoy Corp.
C 0.75, Cr 13.5, B 3, Si 4.25, Fe 4.75, bal Ni.
Cast: 56-61 Rock C. For hard facing alloy; corrosion and heat resistant. Also available in rod and powder.

COLMONOY NO. 62
Wall Colmonoy Corp.
C 0.75, Cr 14, Si 4.5, Fe 4.75, B 3, bal Ni.
58-63 Rock C. 1875°F MP (approx). For hard surfacing shafts, sleeves, valve trim; good resistance to abrasion, corrosion and galling. For application with spraywelder.

COLMONOY NO. 62SA
Wall Colmonoy Corp.
Same as Colmonoy No. 62, but as atomized powder.

COLMONOY NO. 63
Wall Colmonoy Corp.
C 0.7, Cr 14, Fe 3.75, Si 3.75, B 3.5, bal Ni.
58-63 Rock C. 1875°F MP (approx). For hard surfacing by fusewelder torch on shafts, sleeves, valve trim; good resistance to abrasion, corrosion and galling. For application with fusewelder.

COLMONOY NO. 69
Wall Colmonoy Corp.
C 0.7, Cr 14, Si 4, B 3, Cu 1.75, Mo 1.75, Fe 4.35, bal Ni.
Cast: 58-63 Rock C; 1900°F MP (approx). For hard surfacing, usually by spraywelder; for shafts, sleeves, valve trim; good resistance to heat, abrasion, corrosion and galling.

COLMONOY NO. 7
Wall Colmonoy Corp.
C, B, bal Fe.
Cast: 650 Brin. For welding rod; wear and corrosion resistant. *Obsolete*

COLMONOY NO. 70
Wall Colmonoy Corp.
C 0.4-0.7, Cr 10-13, B 1.7-3.2, W 15-17, Si 2.5-4, Fe 2.5-4, bal Ni.
As cast: 50-55 Rock C. For hard facing; wear, heat and corrosion resistant. Available in rod and powder.

COLMONOY NO. 705
Wall Colmonoy Corp.
Colmonoy No. 63 with tungsten carbide particles added.
Cast: 58-63 Rock C; 1900°F MP (approx). For hard facing surfaces requiring extreme abrasion resistance as buffing fixtures. For application with fusewelder.

COLMONOY NO. 72
Wall Colmonoy Corp.
C 0.75, Cr 10, W 12, Si 0.6, Fe 3.5, B 2.7, bal Ni.
Cast: 58-63 Rock C; 1940°F MP (approx). For hard surfacing; good resistance to fretting, corrosion and abrasion. Available in rod and powder.

COLMONOY NO. 730
Wall Colmonoy Corp.
C 2.3, Si 2.34, Cr 8.45, Fe 2.5, B 1.75, Co 2.5, W 35.1, bal Ni.
Tough nickel-chrome-tungsten matrix alloy used to hold fine, hard tungsten carbide particles. Used on pump plungers and sleeves under fine-particulate abrasive conditions. Minimizes packing wear. Finished by grinding. 58-63 Rock C; 1940°F MP (approx). Crushed and atomized powder applied by spraywelder.

COLMONOY NO. 75
Wall Colmonoy Corp.
C 0.7, Cr 13.5, B 3, Si 4.2, Fe 4.75, WC, bal Ni.
54-59 Rock C. For hard facing; resists galling and abrasion. For application with spraywelder.

COLMONOY NO. 750
Wall Colmonoy Corp.
C 3.1, Si 1.8, Cr 6.5, Fe 1.9, B 1.35, Co 6.25, W 43.5, bal Ni.
Tough nickel-chrome-tungsten matrix alloy used to hold hard tungsten carbide particles. Used for the most severe abrasive conditions. Finished by grinding. 58-63 Rock C; 1940°F MP (approx). Crushed and atomized powder applied by spraywelder.

COLMONOY NO. 8
Wall Colmonoy Corp.
C 0.95, Cr 26, B 3.3, Si 4, Fe 1, bal Ni.
Weld: 53-58 Rock C. For hard facing by spraywelder. Resists oxidation and corrosion. High abrasion resistance.

COLMONOY NO. 805
Wall Colmonoy Corp.
Same as Colmonoy No. 6 with chromium boride added. For application with fusewelder.

COLMONOY NO. 83PTA
Wall Colmonoy Corp.
C 2, Si 1.4, Cr 20, Fe 1.4, B 1, W 34, bal Ni.
Tough nickel-chrome-tungsten-boron matrix alloy containing chromium carbides with the addition of extremely hard tungsten carbide particles for good abrasive wear protection. Good edge retention. Specifically for PTA application. 49-56 Rock C; 2250°F MP (approx). Crushed and atomized powder.

COLMONOY NO. 845
Wall Colmonoy Corp.
C 2.4, Cr 12.7, B 0.8, Si 1.2, Fe 1.1, W 51.2, bal Ni.
Nickel-chromium tungsten alloy. Resistance to extreme abrasion. Good corrosion resistance to a broad range of media. 57-62 Rock C; 2250°F MP (approx); 100,000 psi TS. Crushed and atomized powder for application with fusewelder torch.

COLMONOY NO. 9
Wall Colmonoy Corp.
Cr 7-11, B 1-2, Mo 12-16, bal Fe.
Cast: 670 Brin. For welding alloy, hard facing; for tipping tools. *Obsolete*

COLMONOY NO. C-290
Wall Colmonoy Corp.
C 0.45, Si 2.5, Ni 37, B 1.5, Cr 13.25, bal Fe.
For hard facing by spraywelder, building up shaft bearing surfaces. *Obsolete*

COLMONOY NO. C-395
Wall Colmonoy Corp.
C 0.5, Si 2.5, Cr 13.25, Ni 37, B 1.5, bal Fe.
Weld metal: 35-45 Rock C; 2700°F MP (approx). Metallizing powder for spray welder for build-up to resist metal-to-metal wear in rotary type engines.

COLMONOY NO. HC240
Wall Colmonoy Corp.
C 5, Mn 1.5, Si 1.5, Cr 30, bal Fe.
Weld metal: 60 Rock C (approx). Semiautomatic tube wire for build-up of high abrasion resistant and hot wear coatings. *Obsolete*

COLMONOY NO. HC240
Wall Colmonoy Corp.
C 4.3, Cr 24, Si 3, Mn 2.9, bal Fe.
Cast: C 60 Rock (approximately). Wire for semiautomatic welder for buildup of hard, abrasion resistant surfaces having good red-hardness. *Obsolete*

COLMONOY OVERLAY
Wall Colmonoy Corp.
Ni, B, bal Fe.
For hard facing electrode; paste form. *Obsolete*

COLMONOY SPECIAL NO. 1
Wall Colmonoy Corp.
C 1, Cr 13, Si 3, B 3, Mn 0.7, bal Fe.
Coated rods. As cast: 60-65 Rock C; 2450°F MP (approx). For abrasion resistant coatings as coal chutes, dredge cutters.

COLMONOY SPECIAL NO. 4
Wall Colmonoy Corp.
Ni 80-83, Cr 13-16, B 2.5-3.5.
40 Rc. For hard facing electrodes; resists galling and corrosion. *Obsolete*

COLMONOY SWEAT-ON PASTE
Wall Colmonoy Corp.
Cr 82, B 18.
70 Rock C. For welding hard facing alloy paste; wear and abrasion resisting.

COLMONOY WCR 200
Wall Colmonoy Corp.
Cr, B, W, Fe.
For welding alloy; tipping tools. *Obsolete*

COLMONOY WCR 400
Wall Colmonoy Corp.
Cr, B, W, Fe.
For welding alloy; tipping tools. *Obsolete*

COLOMO
Uddeholm Corp.
C 0.95, Cr 1.1, Mo 0.33, bal Fe.
For tools, cutters, mining drills; oil hardened, tough. *Obsolete*

COLONIAL 795 F
Teledyne Vasco
C 1, Cr 17.5, Ni 1, Si 0.15, bal Fe.
For corrosion resistant parts; corrosion resistant. *Obsolete*

COLONIAL GRIPPER
Teledyne Vasco
C 0.5, Cr 2, bal Fe.
For tools, hot work dies and tools; hot work steel. *Obsolete*

COLONIAL HIGH SERVICE
English manufacture
Cu 0.9, Mn 0.2, V 0.14, bal Fe.
For tools, drills, taps; water hardening.

COLONIAL HIGH SPEED
Teledyne Vasco
C 0.52, Mn 0.28, Cr 3.8, W 18.25, V 0.24, bal Fe.
For high speed cutting tools; high speed steel.

COLONIAL NO. 100
Colonial Metals Co.
Sn 4.5-5.7, Pb 2.75-4, Zn 1-9.5, bal Cu.
For bearings, pump impellers; free-cutting. *Obsolete*

COLONIAL NO. 101
Colonial Metals Co.
Sn 4.5-5.75, Pb 2-3, Zn 9.5-16, bal Cu.
For bushings, bearings, pump impellers; free-cutting. *Obsolete*

COLONIAL NO. 101 (CDA 842)
Colonial Metals Co.
Cu 80, Sn 5, Pb 2.5, Zn 12.5.
For pipe fittings, elbows, bushings.

COLONIAL NO. 110
Colonial Metals Co.
Sn 4.5-5.5, Pb 4.25-7, Zn 2-7, bal Cu.
For pump impellers, bushings, bearings; free-cutting, red brass. *Obsolete*

COLONIAL NO. 115
Colonial Metals Co.
Sn 4-5.5, Pb 4.25-7, Zn 4-10, bal Cu.
Cast: 33,000-46,000 TS; 17,000-24,000 YS; 15-35 El; 12-32 RA; 55-65 Brin. For hardware, plumbing; red brass, free-cutting. *Obsolete*

COLONIAL NO. 115 (CDA 836)
Colonial Metals Co.
Cu 85, Sn 5, Pb 5, Zn 5.
For plumbing, hardware, impellers, valves.

COLONIAL NO. 120
Colonial Metals Co.
Sn 3-4.5, Pb 6-10, Zn 5-10, bal Cu.
Cast: 30,000-38,000 TS; 12,000-17,000 YS; 15-27 El; 12-25 RA; 50-60 Brin. For hardware, plumbing; red brass, free-cutting. *Obsolete*

COLONIAL NO. 120 (CDA 838)
Colonial Metals Co.
Cu 83, Sn 4, Pb 6, Zn 7.
For low pressure valves, plumbing, hardware.

COLONIAL NO. 123
Colonial Metals Co.
Sn 2-4, Pb 5-8, Zn 6-18, bal Cu.
Cast: 29,000-39,000 TS; 13,000-17,000 YS; 18-30 El; 15-27 RA; 50-60 Brin. For hardware, plumbing; plumbers brass, free-cutting. *Obsolete*

COLONIAL NO. 123 (CDA 844)
Colonial Metals Co.
Cu 81, Sn 3, Pb 7, Zn 9.
For low pressure valves, plumbing, hardware.

COLONIAL NO. 125
Colonial Metals Co.
Sn 2-4, Pb 5-8, Zn 6-18, bal Cu.
Cast: 30,000-40,000 TS; 12,000-16,000 YS; 20-35 El; 15-30 RA; 50-60 Brin. For hardware, plumbing; plumbers brass, free-cutting. *Obsolete*

COLONIAL NO. 125 (CDA 845)
Colonial Metals Co.
Cu 78, Sn 3, Pb 7, Zn 12.
For low pressure valves, plumbing, air and gas fittings.

COLONIAL NO. 130
Colonial Metals Co.
Sn 1-3, Pb 4-10, Zn 8-20, bal Cu.
For hardware, plumbing; free-cutting. *Obsolete*

COLONIAL NO. 130 (CDA 848)
Colonial Metals Co.
Cu 76, Sn 3, Pb 6, Zn 15.
For low pressure valves, plumbing, air and gas fittings.

COLONIAL NO. 131
Colonial Metals Co.
Sn 1-2, Pb 1-2, Zn 2-8, bal Cu.
For hardware, plumbing; free-cutting. *Obsolete*

COLONIAL NO. 131 (CDA 833)
Colonial Metals Co.
Cu 93, Sn 1.5, Pb 1.5, Zn 4.
For terminal ends for electrical cables.

COLONIAL NO. 132
Colonial Metals Co.
Cu 87, Sn 2, Pb 2, Zn 9.
For electrical fittings.

COLONIAL NO. 14
Teledyne Vasco
C 0.6-1.2, bal Fe.
For tools; water hardening.

COLONIAL NO. 193
Colonial Metals Co.
Sn 21-23, 0.50% max others, bal Cu.
For bearings. *Obsolete*

COLONIAL NO. 194
Colonial Metals Co.
Sn 18-20, bal Cu.
For bearings. *Obsolete*

COLONIAL NO. 194 (CDA 913)
Colonial Metals Co.
Cu 81, Sn 19.
For bearings, bridge plates, bells.

COLONIAL NO. 195
Colonial Metals Co.
Sn 16-18, bal Cu.
For bearings. *Obsolete*

COLONIAL NO. 196
Colonial Metals Co.
Sn 14-16, bal Cu.
For bearings.

COLONIAL NO. 197
Colonial Metals Co.
Sn 13-15, Zn 0-1.5, Pb 0.2, bal Cu.
For bearings. *Obsolete*

COLONIAL NO. 197 (CDA 910)
Colonial Metals Co.
Cu 85, Sn 14, Zn 1.
For bearings, piston rings.

COLONIAL NO. 198
Colonial Metals Co.
Sn 12.5-14.5, Zn 2.5-4.5, Pb 0-1, bal Cu.
For bearings. *Obsolete*

COLONIAL NO. 199
Colonial Metals Co.
Sn 12-14, bal Cu.
For bearings.

COLONIAL NO. 200
Colonial Metals Co.
Sn 10.75-12.25, Pb 0.75-1.25, Zn 0-0.1, bal Cu.
For bearings; free-cutting. *Obsolete*

COLONIAL NO. 200 (CDA 925)
Colonial Metals Co.
Cu 87, Sn 11, Pb 1, Ni 1.
For gears, automotive rings.

COLONIAL NO. 201
Colonial Metals Co.
Sn 10.75-12, Pb 0.8-1.2, Zn 1-4, bal Cu.
For bearings. *Obsolete*

COLONIAL NO. 205. (CDA 907)
Colonial Metals Co.
Sn 9.75-12, bal Cu.
For bearings, bushings, gears.

COLONIAL NO. 206
Colonial Metals Co.
Sn 9-11, Pb 1-2.5, bal Cu.
For bearings. *Obsolete*

COLONIAL NO. 206 (CDA 927)
Colonial Metals Co.
Cu 88, Sn 10, Pb 2.
For bearings, impellers, steam fittings.

COLONIAL NO. 206A
Colonial Metals Co.
Same as No. 206 with Ni added. For bearings, better machining than 210.

COLONIAL NO. 210
Colonial Metals Co.
Sn 9-10.7, Zn 1-7, Pb 0-0.2, bal Cu.
For bearings. *Obsolete*

COLONIAL NO. 210 (CDA 905)
Colonial Metals Co.
Cu 88, Sn 10, Zn 2.
For bearings, gears, steam fittings.

COLONIAL NO. 215
Colonial Metals Co.
Sn 9-10.7, Zn 1-7, Pb 0.8-1.2, bal Cu.
For bearings. *Obsolete*

COLONIAL NO. 215 (CDA 926)
Colonial Metals Co.
Cu 87, Sn 10, Pb 1, Zn 2.
For bearings, gears, steam fittings.

COLONIAL NO. 220
Colonial Metals Co.
Sn 9-10.7, Pb 1.2-2.7, Zn 1-5, bal Cu.
For bearings. *Obsolete*

COLONIAL NO. 221
Colonial Metals Co.
Sn 9-10.7, Pb 2.7-4.2, Zn 1-5, bal Cu.
For bearings. *Obsolete*

COLONIAL NO. 225
Colonial Metals Co.
Sn 7.5-9.7, Zn 1-7, Pb 0.2-0.4, bal Cu.
For bearings. *Obsolete*

COLONIAL NO. 225 (CDA 903)
Colonial Metals Co.
Cu 88, Sn 8, Zn 4.
For bearings, gears, steam fittings.

COLONIAL NO. 230
Colonial Metals Co.
Sn 7.5-9, Pb 0.8-1.2, Zn 1-7, bal Cu.
Cast: 33,000-43,000 TS; 16,000-24,000 YS; 18-30 El; 15-30 RA; 60-75 Brin. For bearings; Gun Metal. *Obsolete*

COLONIAL NO. 230 (CDA 923)
Colonial Metals Co.
Cu 87, Sn 8, Zn 4.
For valves, high pressure steam castings.

COLONIAL NO. 235
Colonial Metals Co.
Cu 87, Sn 8, Zn 4, Pb.
For bearings.

COLONIAL NO. 240
Colonial Metals Co.
Sn 7.5-9, Pb 2.7-4.2, Zn 1-6, bal Cu.
For bearings. *Obsolete*

COLONIAL NO. 241
Colonial Metals Co.
Sn 7.5-9, Pb 4.2-6.2, Zn 1-6, bal Cu.
For bearings. *Obsolete*

COLONIAL NO. 242. (CDA 902)
Colonial Metals Co.
Sn 6-8, Zn 0-0.5, bal Cu.
For bearings.

COLONIAL NO. 245
Colonial Metals Co.
Sn 5.7-7.5, Pb 1-2, Zn 1-7, bal Cu.
Cast: 34,000-42,000 TS; 16,000-21,000 YS; 20-35 El; 16-30 RA; 60-70 Brin. For valves, pump parts; free-cutting. *Obsolete*

COLONIAL NO. 245 (CDA 922)
Colonial Metals Co.
Cu 88, Sn 6, Pb 1.5, Zn 4.5.
For valves, fittings, for use up to 550°F.

COLONIAL NO. 250
Colonial Metals Co.
Sn 5.7-7.5, Pb 2-3.2, Zn 2-6, bal Cu.
For bearings. *Obsolete*

COLONIAL NO. 251
Colonial Metals Co.
Sn 5.7-7.5, Pb 3.2-5.2, Zn 1-6, bal Cu.
For bearings. *Obsolete*

COLONIAL NO. 253
Colonial Metals Co.
Sn 4.5-5.7, Pb 1-2, Zn 1-9, bal Cu.
For bearings. *Obsolete*

COLONIAL NO. 255
Colonial Metals Co.
Cu 90, Sn 5, Pb 2.5, Zn 2.5.
For plaques.

COLONIAL NO. 256
Colonial Metals Co.
Sn 3-4.5, Pb 1.7-3, Zn 1-9, bal Cu.
For bearings. *Obsolete*

COLONIAL NO. 257
Colonial Metals Co.
Sn 3-4, Pb 0.5-1, Cu 85-90, bal Zn.
For bearings. *Obsolete*

COLONIAL NO. 295
Colonial Metals Co.
Sn 15-17, Pb 4-6, Zn 0.25-1, bal Cu.
For bearings. *Obsolete*

COLONIAL NO. 295 (CDA 928)
Colonial Metals Co.
Cu 76, Sn 16, Pb 5.
For piston rings.

COLONIAL NO. 296
Colonial Metals Co.
Sn 12-14, Pb 14-16, Zn 0-0.5, Ni 0.5-1, bal Cu.
For bearings. *Obsolete*

COLONIAL NO. 296 (CDA 940)
Colonial Metals Co.
Cu 72, Sn 13, Pb 15.
For bearings.

COLONIAL NO. 296.5
Colonial Metals Co.
Sn 10.5-12.5, Pb 7-10, bal Cu.
For bearings. *Obsolete*

COLONIAL NO. 297
Colonial Metals Co.
Sn 9.5-11.5, Pb 4.2-6, 0.75-1.5% others, bal Cu.
For bearings. *Obsolete*

COLONIAL NO. 298
Colonial Metals Co.
Sn 9-11, Pb 4.2-6.2, Zn 1-5, bal Cu.
For bearings. *Obsolete*

COLONIAL NO. 299
Colonial Metals Co.
Sn 9-11, Pb 9-11, Zn 0.15-0.75, bal Cu.
For bearings. *Obsolete*

COLONIAL NO. 3
Teledyne Vasco
C 0.33, W 10, Cr 3.5, V 0.5, bal Fe.
For dies, tools, extrusion dies; hot work steel. *Obsolete*

COLONIAL NO. 300
Colonial Metals Co.
Sn 9-10.7, Pb 8.5-11.7, Sb 0-0.25, bal Cu.
For bearings. *Obsolete*

COLONIAL NO. 301
Colonial Alloys Co.
C 0.1, Cr 18, Ni 8, bal Fe.
Annealed: 72,000 TS; 57,000 YS; 32 El; 50 RA; 110 Brin. For chemical plant equipment; Type 301; stainless, austenitic. *Obsolete*

COLONIAL NO. 305
Colonial Metals Co.
Sn 9-10.7, Pb 8.5-21.7, 0.75-1.5% others, bal Cu.
Cast: 27,000-37,000 TS; 15,000-22,000 YS; 6-12 El; 5-11 RA; 55-70 Brin. For bearings; bearing bronze. *Obsolete*

COLONIAL NO. 305 (CDA 937)
Colonial Metals Co.
Cu 80, Sn 10, Pb 10.
For high speed and heavy pressure bearings.

COLONIAL NO. 310
Colonial Metals Co.
Sn 7.5-9, Pb 7.2-8.7, 0.75-1.5% others, bal Cu.
For bearings. *Obsolete*

COLONIAL NO. 310 (CDA 934)
Colonial Metals Co.
Cu 84, Sn 8, Pb 8.
For bearings and bushings.

COLONIAL NO. 311
Colonial Metals Co.
Cu 84, Sn 8, Pb 8.
For bearings.

COLONIAL NO. 312
Colonial Metals Co.
Sn 7.5-9, Pb 8.7-11, Zn 0-3, 0.75-1.5% others, bal Cu.
For bearings. *Obsolete*

COLONIAL NO. 312 (CDA 944)
Colonial Metals Co.
Cu 81, Sn 8, Pb 11.
For general utility bushings and bearings.

COLONIAL NO. 313
Colonial Metals Co.
Sn 7.5-9, Pb 12-16, Ni 0.25-0.5, bal Cu.
For bearings. *Obsolete*

COLONIAL NO. 314
Colonial Metals Co.
Sn 7.5-9, Pb 12-16, 0.25-0.75% others, bal Cu.
For bearings. *Obsolete*

COLONIAL NO. 315
Colonial Metals Co.
Sn 5.5-7.5, Pb 6-10.5, Zn 0-4, 0.75-1.5% others, bal Cu.
Cast: 30,000-38,000 TS; 17,000-21,000 YS; 12-20 El; 10-22 RA; 55-65 Brin. For bearings, bushings; bearing bronze. *Obsolete*

COLONIAL NO. 319
Colonial Metals Co.
Sn 5.7-7.5, Pb 12-16, 0.75-2.0% others, bal Cu.
Cast: 25,000-30,000 TS; 14,000-20,000 YS; 10-18 El; 8-15 RA; 50-60 Brin. For bearings, bushings; bearing bronze. *Obsolete*

COLONIAL NO. 319 (CDA 938)
Colonial Metals Co.
Cu 78, Sn 7, Pb 15.
For bearings, mine water impellers.

COLONIAL NO. 320
Colonial Metals Co.
Sn 5.7-7.5, Pb 16.7-29.7, Zn 0-3, bal Cu.
For bearings. *Obsolete*

COLONIAL NO. 321
Colonial Metals Co.
Sn 5.7-7.5, Pb 16-29, 0.75-1.5% others, bal Cu.
For bearings. *Obsolete*

COLONIAL NO. 321 (CDA 945)
Colonial Metals Co.
Cu 73, Sn 7, Pb 20.
For high speed, low load bearings.

COLONIAL NO. 322
Colonial Metals Co.
Sn 4-5.7, Pb 21-26, 1.5% max others, bal Cu.
Cast: 23,000-30,000 TS; 11,000-15,000 YS; 7-16 El; 5-12 RA; 42-55 Brin. For bearings, bushings; bearing bronze. *Obsolete*

COLONIAL NO. 322 (CDA 943)
Colonial Metals Co.
Cu 70, Sn 5, Pb 25.
For high speed bearings for light loads.

COLONIAL NO. 323
Colonial Metals Co.
Sn 4-6, Pb 21.7-26, Zn 0-0.5, bal Cu.
For bearings. *Obsolete*

COLONIAL NO. 324
Colonial Metals Co.
Sn 4-5.7, Pb 26-32, 0.75-1.5% impurities, bal Cu.
For bearings. *Obsolete*

COLONIAL NO. 325
Colonial Metals Co.
Sn 4.5-5.7, Pb 11.7-21.7, Zn 0-3, bal Cu.
For bearings. *Obsolete*

COLONIAL NO. 325 (CDA 941)
Colonial Metals Co.
Cu 70, Sn 5.5, Pb 18.5, Zn 3.
For high speed, low load bearings.

COLONIAL NO. 326
Colonial Metals Co.
Sn 3.5-5.5, Pb 8-10, Zn 0-4, bal Cu.
For bearings. *Obsolete*

COLONIAL NO. 326 (CDA 935)
Colonial Metals Co.
Cu 85, Sn 5, Pb 9.
For small bearings and bushings.

COLONIAL NO. 351 (CDA 932)
Colonial Metals Co.
Cu 83, Sn 7, Pb 7, Zn 3.
For general utility bearings and bushings.

COLONIAL NO. 36
Teledyne Vasco
C, Cr, V, bal Fe.
For hot work tools. *Obsolete*

COLONIAL NO. 4
Teledyne Vasco
C 1.3, W 3.5, bal Fe.
For tools, reamers, plug gages, forming dies. *Obsolete*

COLONIAL NO. 400
Colonial Metals Co.
Sn 0.5-1.7, Pb 2-4, Cu 70-74, bal Zn.
Cast: 35,000-40,000 TS; 12,000-14,000 YS; 25-40 El; 20-40 RA; 40-55 Brin. For plumbing, hardware; leaded yellow brass, free-cutting. *Obsolete*

COLONIAL NO. 400 (CDA 852)
Colonial Metals Co.
Cu 72, Sn 1, Pb 3, Zn 24.
For plumbing fittings, ornamental, and hardware.

COLONIAL NO. 403
Colonial Metals Co.
Cu 66-70, Sn 0.5-1.7, Pb 2-4, bal Zn.
Cast: 30,000-38,000 TS; 11,000-15,000 YS; 20-35 El; 15-30 RA; 40-60 Brin. For plumbing, hardware; yellow brass, free-cutting. *Obsolete*

COLONIAL NO. 403 (CDA 854)
Colonial Metals Co.
Cu 67, Sn 1, Pb 3, Zn 29.
For general purpose yellow brass.

COLONIAL NO. 405
Colonial Metals Co.
Cu 0-66, Sn 1, Pb 3, bal Zn.
Yellow brass; for plumbing, hardware.

COLONIAL NO. 405.1
Colonial Metals Co.
Cu 0-65, Sn 0.5-1.5, Pb 0-0.5, bal Zn.
For hardware, screw machine products; yellow brass. *Obsolete*

COLONIAL NO. 405.1 (CDA 858)
Colonial Metals Co.
Cu 58, Sn 1, Pb 1, Zn 40.
For general purpose yellow brass die castings.

COLONIAL NO. 405.2
Colonial Metals Co.
Cu 0-65, Sn 0.5-1.5, Pb 0.5-1, bal Zn.
For hardware, screw machine products; free-cutting, yellow brass. *Obsolete*

COLONIAL NO. 405.2 (CDA 857)
Colonial Metals Co.
Cu 63, Sn 1, Pb 1, Zn 35.
For ornamental and hardware fittings.

COLONIAL NO. 406
Colonial Metals Co.
Cu 0-66, Sn 1, Pb 1, bal Zn.
For hardware and fittings.

COLONIAL NO. 407
Colonial Metals Co.
Cu 66-72, Sn 0-0.5, Pb 0-0.5, bal Zn.
For hardware, screw machine products; yellow brass. *Obsolete*

COLONIAL NO. 407 (CDA 853)
Colonial Metals Co.
Cu 70, Zn 30.
For yellow brass hardware.

COLONIAL NO. 407.5
Colonial Metals Co.
Cu 90-95, Sn 0-0.2, Pb 0-0.5, bal Zn.
For clips, fasteners, springs; good corrosion resistance. *Obsolete*

COLONIAL NO. 407.5 (CDA 834)
Colonial Metals Co.
Cu 90, Zn 10.
For moderate conductivity, rotating bends.

COLONIAL NO. 408
Colonial Metals Co.
Cu 83-86, Sn 0-0.2, Pb 0-0.5, bal Zn.
For clips, fuse parts, hardware; good corrosion resistance. *Obsolete*

COLONIAL NO. 408
Colonial Metals Co.
Cu 85, Zn 15.
For brazing applications.

COLONIAL NO. 409
Colonial Metals Co.
Cu 0-61, Pb 1.5-2, bal Zn.
For screw machine products, hardware; free-cutting, yellow brass. *Obsolete*

COLONIAL NO. 410
Teledyne Vasco
Cr 13.5, Ni 1, C 0-0.12, S 0-0.025, bal Fe.
Heat treated: 128,000 TS; 113,000 YS; 21 El; 55 RA; 269 Brin. For valve parts, golf club heads; corrosion resisting. *Obsolete*

COLONIAL NO. 410
Teledyne Vasco
Cr 13.5, Ni 1, C 0-0.12, S 0-0.025, bal Fe.
Heat treated: 128,000 TS; 113,000 YS; 21 El; 55 RA; 269 Brin. For valve parts, golf club heads; corrosion resisting. *Obsolete*

COLONIAL NO. 410 (CDA 973)
Colonial Metals Co.
Cu 56, Sn 2, Pb 10, Ni 12, Zn 20.
For ornamental and hardware castings.

COLONIAL NO. 411
Colonial Metals Co.
Sn 0-5, Pb 0-10, Ni 14-18, bal Cu + Zn.
Cast: 35,000-45,000 TS; 17,000-24,000 YS; 15-30 El; 15-30 RA; 65-80 Brin. For bearings; nickel silver, leaded. *Obsolete*

COLONIAL NO. 411 (CDA 974)
Colonial Metals Co.
Cu 59, Sn 3, Pb 5, Ni 17, Zn 16.
For valves, hardware, fittings.

COLONIAL NO. 412
Colonial Metals Co.
Sn 0-5, Pb 0-10, Ni 18-22, bal Cu + Zn.
Cast: 40,000-60,000 TS; 17,000-30,000 YS; 15-25 El; 11-22 RA; 76-120 Brin. For bearings; nickel silver, leaded. *Obsolete*

COLONIAL NO. 412 (CDA 976)
Colonial Metals Co.
Cu 64, Sn 4, Pb 4, Ni 20, Zn 8.
For marine hardware, sanitary fittings.

COLONIAL NO. 413
Colonial Metals Co.
Sn 0-5, Pb 0-10, Ni 22-27, bal Cu + Zn.
For valves and sanitary fittings.

COLONIAL NO. 413B (CDA 978)
Colonial Metals Co.
Cu 66, Sn 5, Pb 2, Ni 25, Zn 2.
For valves, sanitary fittings, musical instruments.

COLONIAL NO. 414
Colonial Metals Co.
Sn 0-5, Pb 0-10, Ni 27-33, bal Cu + Zn.
For bearings; nickel silver, leaded. *Obsolete*

COLONIAL NO. 415
Colonial Metals Co.
Sn 0-1, Pb 0-0.1, Al 8-13, bal Cu + Zn.
Cast: 70,000-87,000 TS; 25,000-30,000 YS; 22-38 El; 20-36 RA; 110-140 Brin. For hardware, propellers, gears; Al Bronze. *Obsolete*

COLONIAL NO. 415A (CDA 952)
Colonial Metals Co.
Cu 88, Fe 3, Al 9.
For general aluminum bronze castings.

COLONIAL NO. 415B (CDA 953)
Colonial Metals Co.
Cu 89, Fe 1, Al 10.
For marine equipment, nuts, gears.

COLONIAL NO. 415C (CDA 954)
Colonial Metals Co.
Cu 85, Fe 4, Al 11.
For gears, bushings, valve seats.

COLONIAL NO. 415D (CDA 955)
Colonial Metals Co.
Cu 81, Ni 4, Fe 4, Al 11.
For corrosion resistant parts, ship propellers.

COLONIAL NO. 420
Colonial Metals Co.
Zn, Mn, bal Cu.
Cast: 60,000-65,000 TS. For hardware, propellers, gears; Mn Bronze. *Obsolete*

COLONIAL NO. 420 (CDA 864)
Colonial Metals Co.
60,000 psi tensile strength Mn bronze. Free machining manganese bronze.

COLONIAL NO. 421
Colonial Metals Co.
Cu 56-69, Fe 0.75-2, Pb 0-0.3, Ni 0-0.5, bal Zn.
Cast: 65,000-68,000 TS; 28,000-40,000 YS; 20-35 El; 20-40 RA; 115-150 Brin. For propellers, gears, hardware; Bronze. *Obsolete*

COLONIAL NO. 421 (CDA 865)
Colonial Metals Co.
65,000 psi tensile strength Mn bronze. For valve stems, gears, machinery.

COLONIAL NO. 422
Colonial Metals Co.
Zn, Mn, bal Cu.
Cast: 80,000-90,000 TS. For hardware, gears, propellers; Mn Bronze. *Obsolete*

COLONIAL NO. 422 (CDA 867)
Colonial Metals Co.
80,000 psi tensile strength Mn bronze. For high strength free machining parts.

COLONIAL NO. 423
Colonial Metals Co.
Zn, Mn, bal Cu.
Cast: 90,000-100,000 TS. For hardware, gears, propellers; Mn Bronze. *Obsolete*

COLONIAL NO. 423 (CDA 861)
Colonial Metals Co.
90,000 psi tensile strength Mn bronze. For marine castings, bushings, and bearings.

COLONIAL NO. 424
Colonial Metals Co.
Cu 60-68, Fe 2-4, Ni 0-1, Pb 0-0.1, bal Zn.
Cast: 100,000-120,000 TS; 65,000-90,000 YS; 12-18 El; 5-18 RA; 190-235 Brin. For hardware, gears, propellers; Mn Bronze. *Obsolete*

COLONIAL NO. 424 (CDA 863)
Colonial Metals Co.
110,000 psi tensile strength Mn bronze. For high strength manganese bronze.

COLONIAL NO. 500
Colonial Metals Co.
Sn 0-2, Si 2-5.5, Fe 0-2.5, Mn 0-1.5, 90% min Cu.
For bolts, gears, fasteners; Si Bronze. *Obsolete*

COLONIAL NO. 500A (CDA 875)
Colonial Metals Co.
Silicon bronze 12A. For bearings, bells, impellers, pumps.

COLONIAL NO. 500B (CDA 874)
Colonial Metals Co.
Cu 83, Zn 14, Si 3.
Silicon brass 13A. For small boat propellers, valve stems.

COLONIAL NO. 500C (CDA 875)
Colonial Metals Co.
Cu 82, Zn 14, Si 4.
Silicon brass 13B. For small boat propellers, valve stems.

COLONIAL NO. 500D (CDA 876)
Colonial Metals Co.
Cu 90, Zn 5.5, Si 4.5.
Silicon bronze 13C. For valve stems.

COLONIAL NO. 500E (CDA 878)
Colonial Metals Co.
Cu 82, Zn 14, Si 4.
Silicon brass ZS144A. For thin wall die castings, brush holders.

COLONIAL NO. 500F (CDA 879)
Colonial Metals Co.
Cu 65, Zn 34, Si 1.
Silicon brass ZS331A. For general purpose die casting. ASTM 176-70 Z5-331A.

COLONIAL NO. 6
Teledyne Vasco
See NON-SHRINKABLE.

COLONIAL NO. 6
Teledyne Vasco
Cr 0.5, V 0.15, W 0.5, 0.95 C, 1.3 Mn, bal Fe.
Hardened: 140,000 TS; 120,000 YS; 8 El; 30 RA; 302 Brin. For blanking dies, chasers, gages, hobs, master tools; non-shrinking, oil hardening.

COLONIAL NO. 600A (CDA 957)
Colonial Metals Co.
Cu 75, Ni 2, Fe 3, Al 8, Mn 12.
For propellers, impellers, pumps.

COLONIAL NO. 600B (CDA 958)
Colonial Metals Co.
Cu 81, Ni 5, Fe 4, Al 9, Mn 1.
For propeller hubs, blades for salt water.

COLONIAL NO. 7
Teledyne Vasco
C 0.6-1.4, V 0.18, bal Fe.
For tools, chisels, pneumatic riveters, cutting tools, thread cutting dies; tough, shock and vibration resistant.

COLONIAL STAINLESS IRON, C N C
Teledyne Vasco
Cr 9, C 0.1, Ni 2.75-3.25, Mn 0.1-0.2, Si 2.75-3.25, Cu 1.75-2.25, bal Fe.
Rolled: 153,200 TS; 7 El; 1.2 RA; 429 Brin. Annealed: 137,600 TS; 24 El; 54 RA; 285 Brin. For stainless parts; subject to grain growth. *Obsolete*

COLORADO METAL
American manufacture
Cu 57, Ni 25, Zn 18.
For electrical resistances, corrosion resisting parts; corrosion resistant.

COLORADO WATER HARDENING
Sanderson Kayser Ltd.
C 0.9-1, bal Fe.
For tools, drills, taps; water hardened. *Obsolete*

COLORCOAT
British Steel plc
Mild steel; prepainted; strip.

COLORFARM AP
British Steel plc
Mild steel; prepainted; galvanized. For animal housing.

COLOSSO
Hidalgo Steel Co. Inc.
Cr 0.3, Ni 0.5, W 0.3, V 0.1, Mo 0.6, bal Fe.
For tools, flogging and pneumatic tools, rivet sets; no drawing of temper.

COLSPRAY 120
Wall Colmonoy Corp.
Low carbon steel powder, for use primarily in making extra-thick build-ups. Good machinability.

COLSPRAY 200
Wall Colmonoy Corp.
Machinable nickel-base alloy that forms corrosion resistant bearing surfaces. 80-85 Rock B; 2-4% porosity; 0.250 in. thickness max. For shafts, journals, seal areas, pump sleeves, bedways.

COLSPRAY 300
Wall Colmonoy Corp.
Machinable stainless steel powder. Good metal-to-metal wear resistance and compressive strength. Can be machined or ground to very fine surface. 85 Rock B; 6% porosity max; 0.150 in. thickness max. For motor shafts, journals, machine ways.

COLSPRAY 395
Wall Colmonoy Corp.
High chromium, high nickel composite powder, with wear-resistant chromium borides. Similar to Colspray 200, but produces harder, denser coating. Good coating for resisting metal-to-metal wear, dry or lubricated. 45 Rock C 6% porosity max; 0.070-0.090 in. thickness. For impeller fans and grinder shafts, spindles.

COLSPRAY 400
Wall Colmonoy Corp.
Wear-resistant, martensitic stainless steel alloy, high in chromium; good resistance to abrasive wear and shock. Machinable. 30 Rock C; 3% porosity max; 0.080-0.090 thickness. For pistons, pump parts, wear rings, cylinder liners.

COLSPRAY 500
Wall Colmonoy Corp.
General-purpose aluminum-bronze alloy for soft bearing surfaces. High tensile strength; low coefficient of friction. Machines to smooth finish. 70 Rock B hardness; 5% porosity max; 0.125-0.150 in. thickness. For propeller shafts, pistons, conveyor shafts.

COLSPRAY 955
Wall Colmonoy Corp.
Self-bonding nickel-aluminum powder used to make initial bond coats on any clean, oxide-free metal surface except copper. Thickness: 0.004-0.008 in. Thermal spray: 7000-8000 psi bond strength. Plasma spray: 9000-10,000 bond strength.

COLT HOT
A. Finkl & Sons Co.
C, Cr, Ni, Mo, bal Fe.
For dies, tools. *Obsolete*

COLTUF 28
British Steel Corp.
C 0-0.16, Mn 1.2, Si 0-1.25, bal Fe.
Rolled: 63,000-72,000 TS; 38,000 YS; 25 El. For ship construction; high ductility, no noted embrittlement. *Obsolete*

COLUMAX
Permanent Magnet Assoc.
Co 25, Cu 3, Cb 0.8, Al 7.8, Ni 13.5, bal Fe.
Heat treatable, for permanent magnets.

COLUMAX-5
Thomas & Skinner, Inc.
Al 8, Ni 14, Co 24, Cu 3, bal Fe.
Peak energy product, 7,500,000; residual induction, 13,500
Gauss; coercive force, 750 Oersteds; 5000 TS; 8000 Tr.M.R.,
50 Rock C. For permanent magnets in electrical and
magnetic equipment and instruments. Similar to Alnico V-7,
cast magnet, hardenable. Hard and brittle.

COLUMBIA 16
Columbia Tool Steel Co.
Tool material. C 0.98, Mn 0.35, Si 1.38, Cr 4.25, Mo 2.5, V 1.1,
W 0.4, bal Fe.
Cold work tool steel; for punches, forming rolls, thread rolling
dies.

COLUMBIA A 8
Columbia Tool Steel Co.
Tool material. C 0.56, Cr 5, W 1.25, Mo 1.25, Si 1, bal Fe.
For shear blades, forming rolls, heavy blanking dies,
punches. Type A8; air hardened tool steel.

COLUMBIA CCB
Columbia Tool Steel Co.
C 0.41, Mo 0.8, Cr 1.5, Ni 4.25, bal Fe.
For forging dies, dummy blocks, backer blocks, hot shear
blades. Hot work tool steel. *Obsolete*

COLUMBIA DOUBLE SPECIAL
Columbia Tool Steel Co.
C 1.3, W 3.2, Cr 0.55, Si 0.4, Mn 0.25, bal Fe.
Annealed: 200 Brin. For machining chilled rolls, drawing and
master dies; fast finishing. *Obsolete*

COLUMBIA ELECTREX
Columbia Tool Steel Co.
C 1, Si 0.25, Mn 0.35, bal Fe.
For drills, taps, punches, hobs, reamers; Type W1; water
hardened. *Obsolete*

COLUMBIA EXTRA
Columbia Tool Steel Co.
Tool material. C 1.06, Mn 0.25, Si 0.25, bal Fe.
Water hardened tool steel. For threading dies, shear blades,
hand stamps. AISI W1-2.

COLUMBIA EXTRA HEADERDIE
Columbia Tool Steel Co.
Tool material. C 0.95, Mn 0.35, Si 0.25, bal Fe.
Water hardened tool steel. For stamping dies, punches,
heading dies. AISI W1-2H.

COLUMBIA EXTRA VANADIUM
Columbia Tool Steel Co.
C 1, V 0.2, bal Fe.
For tools, cutters, dies; tough. *Obsolete*

COLUMBIA SPECIAL
Columbia Tool Steel Co.
Tool material. C 1.06, Mn 0.3, Si 0.25, Cr 0.2, V 0.05, bal Fe.
Water hardened tool steel. For taps, reamers, paper knives,
gages. AISI W1-1.

COLUMBIA SPECIAL WIRE DRAWING
Columbia Tool Steel Co.
C 2.3, W 10.5, Cr 1.85, bal Fe.
For wire drawing ties; used unhardened. *Obsolete*

COLUMBIA SPRING
Columbia Tool Steel Co.
C 0.95, Mn 0.35, Si 0.25, bal Fe.
For springs, clips, tools; water hardened. *Obsolete*

COLUMBIA STANDARD
Columbia Tool Steel Co.
Tool material. C 1.06, Mn 0.3, Si 0.25, bal Fe.
Water hardened tool steel. For blacksmith tools, dowel pins,
drift pins. AISI TYPE W1-3.

COLUMBIA STANDARD DIE BLOCK
Columbia Tool Steel Co.
C 0.8, Si 0.3, Mn 0.3, bal Fe.
For die blocks, general tools; Type W1; water hardened.
Obsolete

COLUMBIA VANADIUM STANDARD
Columbia Tool Steel Co.
C 0.95, V 0.2, Mn 0.3, Si 0.25, bal Fe.
For drills, hobs, reamers, cutters, broaches; Type W 2; water
hardening. *Obsolete*

COLUMBIUM (NIOBIUM)
Atomergic Chemetals Corp.
Cb (Nb).
Purities, zone refined: 99.99+%, dendritic 99.99+%, nuclear
grade 99.8%, 99.5%. Forms: powder, crystal bar, sintered bar,
wire, sheet, ingot, single crystals.

COLUMBIUM 10W-5 ZR
Teledyne Wah Chang Albany
W 10, Zr 5, bal Cb.
Annealed: 102,000 TS; 85,000 YS; 12 El; 9 RA. For space
vehicles and nuclear reactors. High strength at elevated
temperatures. Insensitive to notch stress concentration.
Obsolete

COLUMBIUM B-88
Westinghouse Electric Corp.
W 28, Hf 2, C 0.067, bal Cb.
For gas turbine buckets. Heat resistant. High rupture
strength.

COLUMBIUM CB 132 M
Universal Cyclops
C 0.001, Zr 1.5, W 15, Mo 5, Ta 20, Cb 58.5.
Extruded: 131,800 TS; 122,100 YS; 4 El at 2400°F: 56,700
TS; 49,200 YS; 37 El (extruded). Heat treated: 89,200 TS; 0.5
El. At 2400°F: 58,300 TS; 49,900 YS; 20 El; (heat treated). For
jet engine and gas turbine blades. High strength, high
temperature alloy.

COLUMBIUM CB 132 M
Teledyne Wah Chang Albany
C 0.001, Zr 1.5, W 15, Mo 5, Ta 20, Cb 58.5.
Extruded: 131,800 TS; 122,100 YS; 4 El at 2400°F: 56,700
TS; 49,200 YS; 37 El (extruded). Heat treated: 89,200 TS; 0.5
El. At 2400°F: 58,300 TS; 49,900 YS; 20 El; (heat treated). For
jet engine and gas turbine blades. High strength, high
temperature alloy.

COLUMBIUM CB-752
Union Carbide Corp.
W 9-11, Zr 2-3, bal Cb.
Annealed sheet: 78,000 TS; 58,000 YS; 20 min El. At 2200°F:
28,000 TS; 22,000 YS; 25 min El. For high temperature
applications. High heat and corrosion resistance.

COLUMBIUM CB-752
Teledyne Wah Chang Albany
W 9-11, Zr 2-3, bal Cb.
Annealed sheet: 78,000 TS; 58,000 YS; 20 min El. At 2200°F:
28,000 TS; 22,000 YS; 25 min El. For high temperature
applications. High heat and corrosion resistance.

COLUMBIUM D-10
Du Pont Co.
C 0.01, 0.0175 O_2, 0.0020 H_2, 0.01 N_2, bal C.
Recrystallized sheet: 30,000 TS min; 15,000 YS min; 15 El
min. Recrystallized bar: 25,000 TS min; 12,000 YS min. 25 El
min. For high temperature applications. High heat resistance.

COLUMBIUM D-11
Manufacturer not listed
Zr 0.75-1.25, bal Cb.
Bar (recrystallized) 30,000 min TS; 16,000 min YS; 25 min El.
Extruded: 40,000 TS; 20,000 YS; 25 El. At 1800°F: 27,000 TS;
16,000 YS; 23 El. Sheet (recrystallized): 35,000 min TS;
20,000 min YS; 12 min El. For space vehicles and nuclear
reactors. High heat resistance.

COLUMBIUM D-14
Du Pont Co.
Zr 4.4-5.6, bal Cb.
At 78°F: 68,000 TS; 56,000 YS; 15 El. At 2000°F: 32,000 TS;
22,000 YS; 40 El. Stress relieved sheet: 55,000 TS min;
45,000 YS min; 12 El min. For hot surfaces on space re-entry
vehicles. Moderate strength alloy for elevated temperature
service.

COLUMBIUM D-36
Du Pont Co.
Ti 9-11, Zr 4-6, bal Cb.
Sheet, recrystallized: 80,000 TS; 71,000 YS; 20 El. At 800°F:
65,000 TS; 45,000 YS; 18 El. At 2000°F: 22,000 TS; 20,000
YS; 50 El. Bar, recrystallized: 65,000 TS min; 58,000 YS min;
20 El min. For space re-entry vehicles. High heat resistance.

COLUMBIUM D-40
Du Pont Co.
W 15, Zr 1, C 0.06, Mo, bal Cb.
Cold rolled: 125,000 TS; 85,000 YS; 25 El. At 2200°F: 50,000
TS; 30,000 YS; 22 El. For space missiles and rocket
components. Heat resistant. Retains usable strength to
2500°F.

COLUMBIUM D-41
Du Pont Co.
W 19-21, Ti 9-11, Mo 5.5-6.5, bal Cb.
Extruded: 130,000 TS; 127,000 YS; 5.3 El. At 2000°F: 56,000
TS; 53,000 YS; 26 El. At 2400°F: 34,000 TS; 29,000 YS; 33 El.
For space vehicles and nuclear reactors. High heat
resistance.

COLUMBIUM D-43
Du Pont Co.
W 10, Zr 1, C 0.1, bal Cb.
Sheet: 92,000 TS; 78,000 YS; 22 El. At 2000°F: 50,000 TS;
45,000 YS; 15 El. For missiles and space craft. High creep
and rupture strength.

COLUMBIUM SU-16
Stauffer Chemical Corp.
W 11, Mo 3, Hf 2, C 0.08, bal Cb.
At 2000 F, stress relieved bars have a stress rupture life of
1,000 hours with a 28,000 stress. At 2200 F, the 1,000 hour
rupture stress is 14,800. For applications requiring high
strength at 2000-2500 F; aircraft gas turbine blades, space
vehicles, military equipment. High temperature strength and
excellent room temperature ductility. *Obsolete*

COLUMBIUM SU-31
Stauffer Chemical Corp.
W 17, Hf 3.5, C 0.12, bal Cb.
For gas turbine blades. High heat resistance and creep
strength. *Obsolete*

COLUMBIUM-15W-5MO-1ZR
Teledyne Wah Chang Albany
W 15, Mo 5, Zr 1, bal Cb.
Annealed: 115,000 TS; 95,000 YS; 5 El, 8 RA. At 2200°F:
43,500 TS; 39,000 YS; 22 El; 40 RA. For missile and space
vehicle components, high temperature fasteners. High heat,
oxidation and corrosion resistant. *Obsolete*

COLUMBUS K 0109
Schmidt & Clemens Edelstahlwerke
C 0.8, Co 4.7, Cr 4.3, Mo 0.7, V 1.5, W 18, bal Fe.
For lathe and planer tools, taps, hobs, drills; high speed
steel, oil hardened. *Obsolete*

COLUMBUS K011

Schmidt & Clemens Edelstahlwerke
C 0.76, Co 10, Cr 4.2, Mo 0.8, V 1.8, W 18, bal Fe.
For lathe and planer tools, milling cutters, drills; high speed steel, oil hardened. *Obsolete*

COLUMBUS K03

Schmidt & Clemens Edelstahlwerke
C 0.86, Co 2.8, Cr 4.3, Mo 0.85, V 2, W 12, bal Fe.
For lathe and planer tools, hobs, taps, reamers, drills; high speed steel, oil hardened. *Obsolete*

COLUMBUS K055

Schmidt & Clemens Edelstahlwerke
C 0.79, Co 4.75, Cr 4.3, Mo 0.75, V 1.5, W 18, bal Fe.
For lathe and planer tools, drills, taps, hobs; high speed steel, oil hardened. *Obsolete*

COLUMBUS MO

Schmidt & Clemens Edelstahlwerke
C 0.82, Cr 4.1, Mo 0.85, V 1.6, W 8.7, bal Fe.
For lathe and planer tools, drills, taps, hobs; high speed steel, oil hardened. *Obsolete*

COLUMBUS SS10

Hover, Gebruder, Edelstahlwerk
C 0.82, Cr 4, Mo 0.8, V 1.6, W 8.7, bal Fe.
For lathe and planer tools, drills, reamers, hobs; high speed steel, oil hardened. *Obsolete*

COLUMBUS SS11

Schmidt & Clemens Edelstahlwerke
C 0.86, Cr 4, Mo 0.8, V 2.5, W 12, bal Fe.
For lathe and planer tools, reamers, broaches, hobs; high speed steel, oil hardened. *Obsolete*

COLUMBUS SS13

Schmidt & Clemens Edelstahlwerke
C 1.3, Cr 4.3, Mo 0.85, V 3.8, W 12, bal Fe.
For special milling cutters; high speed steel, oil hardened. *Obsolete*

COLUMBUS SS19

Schmidt & Clemens Edelstahlwerke
C 0.74, Cr 4.1, V 1, W 18.5, bal Fe.
For lathe and planer tools, drills, reamers, taps; high speed steel, oil hardened. *Obsolete*

COLVILLE STAINLESS IRON

British Steel Corp.
Cr 16-17, Mn 0.4, C 0.1, bal Fe.
For heat and corrosion resisting parts; stainless and corrosion resistant. *Obsolete*

COMALLOY

English manufacture
Mo 17, Co 12, bal Fe.
For permanent magnets, electrical and magnetic equipment; precipitation hardened, high permeability.

COMANCHE

Pyramid Steel Company
C 0.9, Mn 1.2, Cr 0.5, V 0.2, W 0.5, bal Fe.
Oil hardened tool steel; AISI O1.

COMBARLOY

Thomas Bolton Ltd.
Ag, bal Cu.
For cummutator bars; high conductivity. *Obsolete*

COMET

Carpenter Technology Corp.
C 1.05, Mn 0.25, bal Fe.
For taps, reamers, dies for blanking, trimming and forming. Water hardening. "Carpenter No. 11."

COMET

Driver Harris Co.
Fe 65, Cr 5, Ni 30.
85,000 TS; 47,000 YS; 37 El; 69 RA. For dipping baskets, electric resistance wire, rheostats; maximum operating temperature 700 C; heat resisting. *Obsolete*

COMINCO ALLOY NO. 412

Cominco Metals
Lead. Pb 78-82, Sb 17-21, As 0-1.
Bluish-white, silvery gray metal. MP 320°C.

COMINCO ALLOY NO. 413

Cominco Metals
Lead. Pb 72-81, Sb 17-21, As 2-7.
Bluish-white, silvery gray metal. MP 350°C.

COMINCO ALLOY NO. 414

Cominco Metals
Lead. Pb 86-89, Sb 0-2, As 10-12.
Bluish-white, silvery gray metal. MP 410°C.

COMINCO BISMUTH METAL

Cominco Metals
99.99 Bi min.
Silver-white metal with bright luster. MP 271°C. BP 1564°C.

COMINCO CADMIUM METAL

Cominco Metals
99.97 Cd min.
Malleable silver-white, lustrous metal. MP 321°C. BP 765°C.

COMINCO INDIUM METAL

Cominco Metals
99.97 In min.
Soft, lustrous, silvery white metal with a bluish tinge. MP 157°C. BP 2080°C.

COMINCO LEAD

Cominco Metals
Lead. 99.0 Pb min.
Malleable, bluish-white, silvery gray metal. MP 328°C. BP 1740°C.

COMINCO NO. 2570

Cominco Trail
Zn alloy.
Forged: 60,000-72,000 TS; 45,000-60,000 YS; 14 El; 120 Brin. For industrial applications, hardware, fire fighting equipment, automotive parts, plumbing fixtures, solenoid valves.

COMINCO NO. 2573

Cominco Trail
Zn alloy.
Forged: 62,000-75,000 TS; 45,000-60,000 YS; 13 El; 125 Brin. For industrial applications, hardware, fire fighting equipment, automotive parts, plumbing fixtures, solenoid valves.

COMINCO NO. 3130

Cominco Trail
Zn alloy.
Forged: 35,000 TS; 23,000 YS; 30 El; 65 Brin. For structural and pressure components, plumbing fixtures, solenoid valves. Good creep resistance.

COMINCO NO. 3330

Cominco Trail
Zn alloy.
Forged: 48,000 TS; 35,000 YS; 20 El; 90 Brin. For structural and pressure components, plumbing fixtures, valves, regulators. Good creep resistance.

COMINCO SILVER METAL

Cominco Metals
Silver. 99.9 Ag min.
Soft, ductile, malleable, lustrous, white metal. MP 962°C. BP 2163°C.

COMINCO ZINC METAL

Cominco Metals
Zinc. 99.0 Zn min.
Zinc metal; meets ASTM B6. MP 419°C. BP 907°C.

COMMANDO

Atlas Specialty Steels
C 1.2, Mn 0.25, Si 0.2, bal Fe.
For dies; drill rod. *Obsolete*

COMMANDO

Wallace Murray Corp.
C 0.45, Cr 1.4, Si 0.4, W 2, bal Fe.
For hot punches, tools; hot work steel; shock resistant. AISI S1.

COMMELL SPECIAL

English manufacture
C 1.2, Mn 0.21, W 2.92, bal Fe.
For tools, dies, cutters; fast finishing tool steel.

COMMERCE H

Commerce Pattern Co.
Copper. Cu alloy.
Cast: 65,000-75,000 TS; 12,000-16,000 YS; 2-10 El; 116-165 Brin. For resistance welding electrodes; 10-15% electrical conductivity.

COMMERCIAL BRASS

Anaconda Co.
Copper. Cu 65, Zn 35.
Hard: 76,000 TS; 4 El; 153 Brin. Soft: 45,000 TS, 60 El; 52 Brin. For fixtures, radiators, ornaments. Must be annealed.

COMMERCIAL BRONZE

Chase Brass & Copper Co.
Cu 90, Zn 10.
Sheets: 32,000 TS; 10,000 YS; 40 El; 50 Brin. Wire: 100,000 TS; 50,000 YS; 1 El. For window screen wire, automobile radiators, hardware, ornaments.

COMMERCIAL BRONZE 220

Anaconda Co.
Copper. Zn 10, bal Cu.
Sheet: 32,000 TS; 10,000 YS; 40 El; 50 Brin. Wire: 100,000 TS; 50,000 YS; 1 El. For window screen wire, auto radiators, hardware. Corrosion resistant.

COMMERCIAL BRONZE 226

Anaconda Co.
Copper. Zn 12.5, bal Cu.
Hard: 65,000 TS; 52,000 YS; 7 El; 125 Brin. Soft: 39,000 TS; 13,000 YS; 45 El; 72 Brin. For costume jewelry, slide fasteners; base for gold plate.

COMMERCIAL BRONZE 25

Olin Brass, Indianapolis
Copper. Cu 90, Zn 10.
Hard: 61,000 psi TS; 54,000 psi YS; 5 El; 125 Brin. Soft: 38,000 psi TS, 12,000 psi YS, 45 El, 53 Brin. For weather stripping; corrosion resistant.

COMMERCIAL CARBON DRILL ROD

Latrobe Steel Co.
C 0.95-1.05, Mn 0.3-0.5, Si 0.15-0.3, bal Fe.
AISI Type W1 water hardening tool steel.

COMMERCIAL CASTINGS B. C

English manufacture
Cu 62, Zn 30, Sn 6, Pb 2.
For name plates, oil cups, instrument cases.

COMMERCIAL WATER HARDENING DRILL ROD

Bethlehem Steel Corp.
C 1, Mn 0.3, Si 0.2, bal Fe.
For automatic screw machine parts, dowel pins, mandrels, punches, small tools, dies. AISI Type W1 water-hardened tool steel.

COMMON FORMULA

English manufacture
Cu 55, Zn 25, Ni 20.
For electrical resistance; German silver.

COMMON TOMBAC

French manufacture
Zn 28, Cu 72.
Rolled: 42,000 TS; 25,000 YS; 35 El; 30 RA; 45 Brin. For cartridge cases, condenser tubes, brazing; corrosion resistant; ductile.

COMMON TYPE METAL "A"

American manufacture
Pb 60, Sn 10, Sb 30.
For type metal.

COMMON TYPE METAL "B"

American manufacture
Pb 55.5, Sn 40, Sb 4.5.
For type metal.

COMMONWEALTH

General Steel Industries
C 0.3, Ni 2, bal Fe.
80,000 TS. For trunk frames; tough.

COMO

CCS Braeburn Alloy Steel
C 0.7, Mo 9, W 1.5, Cr 4, V 1, Co 5, bal Fe.
For lathe and planer tools, drills, hobs, reamers. High speed steel. AISI M-30. *Obsolete*

COMO

Darwin & Milner Inc.
C 0.76, Cr 4, V 1, W 1.5, Mo 8.75, bal Fe.
For high speed tools; high speed steel. *Obsolete*

COMOL-17

Colt Industries
Co 12, Mo 17, bal Fe.
Bar: 1,100,000 (BdHd) maximum; 10,000 Br; 230 Hc; 6900 Bo. Bar: 125,000 TS; 50,000 Tr.S; C 60 Rock. For electrical and magnetic equipment. Permanent magnet. High permeability. *Obsolete*

COMOL-20

American Magnesium Co.
Co 12, Mo 20, bal Fe.
Cast: 1,250,000 (BdHd) maximum; 1,250,000; residual flux density 8,550. For electrical and magnetic equipment. Permanent magnet. High permeability. *Obsolete*

COMOL-20

Manufacturer not listed
Co 12, Mo 20, bal Fe.
Cast: 1,250,000 (BdHd) maximum; residual flux density 8,550. For electrical and magnetic equipment. Permanent magnet. High permeability. *Obsolete*

COMPAX

Uddeholm Corp.
C 0.5, Si 0.3, Mn 0.7, Cr 3.2, Mo 1.4, bal Fe.
Chromium-molybdenum alloyed steel. 54-58 Rock C. Suitable for plastic molding, including injection, compression and transfer molds, and for shear blades and cropping tools. AISI S7.

COMPENSATOR ALLOY

English manufacture
Fe-Ni.
For compensating shunts for electrical equipment; temperature sensitive, magnetic.

COMPLEX ENGLISH METAL

English manufacture
Sn 87, Sb 6, Ni 2, Cu 2, W 1.5, Zn 1, Bi 0.5.
For bearings.

COMPO

GKN Powder Met Inc.
Cu 88.5, Sn 10, 1.5% graphite.
Sintered: up to 20,000 TS; 5-10 El; 20-90 Brin. For self-lubricating porous bronze bushings; porosity and oil content is 10 to 40% by volume. *Obsolete*

COMPO 59-E

GKN Powder Met Inc.
Cu 36.5, Sn 4, Fe 58.5, Graphite.
Sintered: 12,500 TS; 5.9 density; 28% oil by volume. For light loaded, low cost bearings, for self-aligning bearings in motors. *Obsolete*

COMPO 60-Y

GKN Powder Met Inc.
Cu 89, Sn 9.75, 1.25% graphite.
Sintered: 10,000 TS; 6.0 density; 30% oil by volume. High oil content plus MoS 2 for additional lubrication for very light loads, as cameras, instruments, light duty appliances. *Obsolete*

COMPO 61-A

GKN Powder Met Inc.
Cu 87, Sn 9.5, 3.5% graphite.
Sintered: 6,500 TS; 6.1 density; 20% oil by volume. For bearings used in light duty applications where noise elimination is important. *Obsolete*

COMPO 61-A, ETC.

Now GKN 61-A, ETC.

COMPO 62-E

GKN Powder Met Inc.
Cu 36.5, Sn 4, Fe 58.5, 1% graphite.
Sintered: 15,000 TS; 6.2 density; 23% oil by volume. Good load bearing qualities; for fan motors, tractors and agricultural equipment. *Obsolete*

COMPO 63-H

GKN Powder Met Inc.
Cu 89, Sn 9.75, 1.25% graphite.
Sintered: 12,000 TS; 6.3 density; 27% oil content by volume. For light loaded, self-aligning bearings in fractional horse power motors. *Obsolete*

COMPO 66-H

GKN Powder Met Inc.
Cu 89, Sn 9.5, 1.25% graphite.
Sintered: 14,700 TS; 6.6 density; 23% oil by volume. Standard bearing material, good load bearing qualities. ASTM B438-67, Gr. 1, Type II; SAE 841; PMPMA BT-0010-R. *Obsolete*

COMPO 66-Q

GKN Powder Met Inc.
Cu 90, Sn 10.
Sintered: 16,000 TS; 6.6 density; 25% oil by volume. For bearings and parts requiring light machining, boring or staking; good load bearing qualities. ASTM B438-67, Gr. 1, Type II; SAE 841; PMPMA BT-0010-R. *Obsolete*

COMPO 66-R

GKN Powder Met Inc.
Cu 87, Sn 9.5, 3.5% graphite.
Sintered: 15,000 TS; 6.6 density; 18% oil by volume. For bearings having oscillating and reciprocating loads; for bearings die cast into place. *Obsolete*

COMPO 70-H

GKN Powder Met Inc.
Cu 89, Sn 9.75, 1.25% graphite.
Sintered: 18,000 TS; 7.0 density; 15% oil by volume. For heavy duty bearings as in agricultural or construction equipment, and in industrial work. ASTM B255-61, Type II; SAE 842; PMPMA BT-0010-S. *Obsolete*

COMPO 70-Q

GKN Powder Met Inc.
Cu 90, Sn 10.
Sintered: 19,500 TS; 7.0 density; 20% oil by volume. For bearings requiring heavy machining, severe staking, spinning, and where ductility is important; good for shock and high loading. ASTM B255-61, Type II; SAE 842; PMPMA BT-0010-S. *Obsolete*

COMPO 70-R

GKN Powder Met Inc.
Cu 87, Sn 9.5, 3.5% graphite.
Sintered: 18,000 TS; 7.0 density; 5% oil by volume. For bearings for oscillating and reciprocating loads; for die casting thin walled bearings into place. *Obsolete*

COMPOSITE

H. Boker & Co.
C, bal Fe.
For tools, dies; water hardening. *Obsolete*

COMPOSITE DIE

Jessop-Saville Ltd.
C, bal Fe.
For tools, dies; water hardening.

COMPOSITION BRASS

Manufacturer not listed
Cu 84-85, Zn 4-6, Sn 4-6, Pb 4-6.
26,000 TS; 12,000 YS; 15 El. For bearings, screws, hardware (ounce metal).

COMPOSITION NO. 1

Janney Cylinder Co.
Sn 6, Pb 3, Zn 4, bal Cu.
34,000 TS; 23,000 YS; 20 El; 75 Brin. For centrifugal castings; general purpose.

COMSOL

Johnson Matthey plc
Ag, Sn, Pb.
Cast: 56,000 TS; 40 El. For soft solder; 296°C MP; retains strength at high temperature.

CON-PAC

U.S. Steel Corp.
C 0.17, Mn 1.35, Si 0.4, Cu 0.3, bal Fe.
Heat treated: 80,000 minimum YS; 100,000 minimum TS. For storage tanks, penstocks, bridges and buildings, railroad cars, trucks. Heat treated carbon steel plates. *Obsolete*

CON-PAC 100

United States Steel Corp.
C 0.18, Mn 1.25, Si 0.3, Ni 0.15, Cr 0.15, Mo 0.04, B 0.001, bal Fe.
Water quenched and tempered: 120 ksi TS; 100 ksi YS; 20 El. For heavy duty equipment, trucks, mobile machinery. ASTM A678.

CON-PAC 80

United States Steel Corp.
C 0.18, Mn 1.25, Si 0.3, Ni 0.15, Cr 0.15, Mo 0.04, B 0.001, bal Fe.
Water quenched and tempered: 100 ksi TS; 80 ksi YS; 20 El. For heavy duty equipment, trucks, mobile machinery. ASTM A678.

CON-PAC 90

United States Steel Corp.
C 0.18, Mn 1.25, Si 0.3, Ni 0.15, Cr 0.15, Mo 0.04, B 0.001, bal Fe.
Water quenched and tempered: 110 ksi TS; 90 ksi YS; 20 El. For heavy duty equipment, trucks, mobile machinery. ASTM A678.

CON-PAC M

United States Steel Corp.
C 0.18, Mn 1.3, Si 0.4, bal Fe.
Water quenched and tempered: 90 ksi TS; 75 ksi YS; 22 El. Readily formed and welded. Construction equipment, off-shore platforms, ship hulls. Tough. ASTM A678.

CONDENSER FOIL
English manufacture
Pb 90, Sn 9, Sb 1.
For accumulators; accumulator metal.

CONDOR SPECIAL
Farrelloy Co.
C 0.7-1.4, Mn, Si, bal Fe.
For drill rods, tools; see Pompton Special.

CONDUCTAL
Pechiney/Soc. Vente de l'Aluminium
Tradename of several aluminum alloys. For busbars and insulated conductors.

CONDULOY
Brush Wellman
Same as BRUSH 35. *Obsolete*

CONEL
Waltham Precision Instruments
Ni 38.5, Mn 3.3, Si 3.3, Cr 4.6, bal Fe.
Cold worked and annealed: 185,000 psi TS; 145,000 psi YS; 4 El. For springs; low thermoelastic coefficient.

CONFLEX 216
Metal & Controls and General Plate
10% copper layer on AISI 1065 steel. Heat treated: 215,000 TS; 185,000 YS; 5 El. For current carrying springs. Magnetic, 16% conductivity.

CONFLEX 316
Metal & Controls and General Plate
Cu 10, 90 steel.
Carbon spring steel clad with Cu on both sides. Annealed: 74,000 TS; 65,000 YS; 25 El. Rolled: 130,000 TS; 110,000 YS; 2 El. Heat treated: 213,000 TS; 185,000 YS; 5 El. For current carrying springs. 16% conductivity. High elasticity, low contact resistance.

CONFLEX 326
Metal & Controls and General Plate
Cu 20, 80 steel.
Carbon spring steel clad with Cu on both sides. Annealed: 70,000 TS; 61,000 YS; 25 El. Rolled: 120,000 TS; 110,000 YS; 2 El. Heat treated: 204,000 TS; 177,000 YS; 6 El. For current carrying springs. High elasticity, low contact resistance.

CONFLEX 335S
Metal & Controls and General Plate
Core of age hardened 17-7 PH double clad with copper. Aged: 155,000 TS; 140,000 YS; 2.5 El. For current carrying springs. 30-40% conductivity.

CONFLEX 545
Metal & Controls and General Plate
AISI 6150 steel double clad on a copper base. Heat treated: 145,000 TS; 133,000 YS; 4-5 El. For flat or cantilever type springs. Two thin copper layers clad on the outer surfaces to provide resistance to corrosion. Five layers bonded together. Not brazed together. Copper is 45% of cross-section.

CONFLEX 720
Metal & Controls and General Plate
Mn 20, Ni 20, bal Cu.
Hardened: 102,650-175,600 TS; 49,800-149,300 YS; 0.5-15 El; 200-425 Vickers. For springs, diaphragms, clips, watch cases, pencaps. Nonmagnetic, corrosion resistant. Age-hardenable.

CONGO
CCS Braeburn Alloy Steel
C 0.8, Cr 4, V 1.5, W 4, Mo 5, Co 12, bal Fe.
For form tools, cut-off knives, reamer blades. High speed steel, oil hardened. *Obsolete*

CONGO HOT WORK
CCS Braeburn Alloy Steel
C 0.1, V 0.5, Cr 3.5, W 4, Mo 5, Co 25, bal Fe.
For brass die, casting dies, master hobs; age hardenable, air quenched. *Obsolete*

CONICO
Manufacturer not listed
Cu 50, Co 29, bal Ni.
For magnets for electrical equipment; magnetically soft.

CONLY
Constrictor Ltd.
Cu 4-5, bal Al.
For bicycles; tubes.

CONPERNIK
Westinghouse Electric Corp.
Fe 50, Ni 50, trace Mn.
Annealed: 55,000 psi TS; 20,000 psi YS; 33 El. For laminations for transformers, choke coils; high magnetic permeability. *Obsolete*

CONQUEROR
Lehigh Steel Corp.
C 0.9, Mn 0.4, Si 0.3, bal Fe.
For stone drills and tools, punches, chisels; tough and wear resistant.

CONQUEROR
Joseph Beardshaw & Son Ltd.
C 0.7-0.85, bal Fe.
Heat treated: 188,000 TS; 143,000 YS; 12 El; 35 RA; 385 Brin. For press tools, chisels, springs; Type W1; water hardened.

CONQUEROR 14%
Joseph Beardshaw & Son Ltd.
C 0.65, W 14, Cr 4, V 0.75, bal Fe.
For shear blades, punches, drills, taps; high speed steel.

CONQUEROR AA
Joseph Beardshaw & Son Ltd.
C 0.7, W 17, Cr 4, V 1, Mo 0.5, bal Fe.
For cutters, taps, reamers, hobs, drills; high speed steel.

CONQUEROR HOLLOW DRILL
Lehigh Steel Corp.
C 0.75, Mn 0.3, Si 0.2, bal Fe.
For hollow stone drills, pneumatic tools; abrasion and shock resistant.

CONQUEROR L.C.
Joseph Beardshaw & Son Ltd.
C 0.45, W 14, Cr 2.7, V 0.4, bal Fe.
For forging and extrusion dies, mandrels; oil hardened, hot die steel.

CONQUEROR O.H.D.
Joseph Beardshaw & Son Ltd.
C 1, Mn 0.25, Cr 1.6, W 0.5, bal Fe.
For cutters, gages, drawing dies, rolls, reamers; cold work steel, oil hardened.

CONSERVALOY
Allegheny Ludlum Steel
C 0.6, Cr 22, Mn 8.5, N 0.35, bal Fe.
For valves; corrosion and heat resistant. *Obsolete*

CONSIL 852
Handy & Harmon
Precious metal. Ag 85, 15.0 CdO.
For electric contacts. Internally oxidized.

CONSIL 866
Handy & Harmon
Precious metal. Ag 86.6, Cd 13.4.
For electric contacts.

CONSIL 880
Handy & Harmon
Precious metal. Ag 88, Cd 12.
For electric contacts.

CONSIL 900
Handy & Harmon
Precious metal. Ag 90, 10.0 CdO.
For electric contacts. Internally oxidized.

CONSIL 910
Handy & Harmon
Precious metal. Ag 91, Cd 9.
For electric contacts.

CONSIL 983
Handy & Harmon
Mg 1.7, Ag 98.3.
For secondary emitter for television camera tubes. *Obsolete*

CONSIL 995
Handy & Harmon
Precious metal. Mg 0-0.25, Ni 0-0.25, 99.4 Ag min.
Annealed: 28,000-34,000 TS; 30-40 El; 60-65 Rock 15 T. Cold worked: 55,000 TS; 6-8 El; 50-80 Rock 30 T. Hardened: 68,000 TS; 56,000 YS; 5-15 El; 68 Rock 30 T. For electrical contacts, cable connectors, spring clips, relay springs. Air hardened. High electrical conductivity.

CONSTAHL
Nihon Jyokiko Seikosho Goshi
C 0.1, Cr 19, Ni 9, bal Fe.
For stainless parts; stainless.

CONSTANT
Midvale-Heppenstall Co.
C 0.9, Mn 1.2, W 0.5, Cr 0.5, bal Fe.
For tools, cutters, hobs, reamers, dies, taps, chasers; nondeforming, oil hardening. *Obsolete*

CONSTANT
Styria-Stahl Steirische Gusstahlwerke AG
C 0.95, Mn 1.25, Cr 0.5, W 0.5, bal Fe.
For tools, dies; non-deforming. *Obsolete*

CONSTANT SPECIAL
Now VEW K720.

CONSTANTAN
Wilbur B. Driver Co.
Cu 45-60, Ni 40-55, Mn 0-1.4.
Annealed: 60,000-70,000 TS; 20,000-30,000 YS; 40-60 El; 50-70 RA; 100-120 Brin. Rolled: 140,000 TS; 125,000 YS; 1 El; 5 RA; 300 Brin. For electrical resistances, thermocouples; high heat resistance.

CONSTANTAN
Driver Harris Co.
Cu 45-60, Ni 40-55, Mn 0-1.4.
Annealed: 60,000-70,000 TS; 20,000-30,000 YS; 40-60 El; 50-70 RA; 100-120 Brin. Rolled: 140,000 TS; 125,000 YS; 1 El; 5 RA; 300 Brin. For electrical resistances, thermocouples; high heat resistance.

CONSTANTAN
Hoskins Mfg. Co.
Copper. Fe 0.2, Mn 0.7, Ni 46, Cu 54.
For resistance alloy, rheostats, thermocouples; heat resistant.

CONSTANTIN
Driver Harris Co.
Ni 44-46, Cu 54, Fe 0-0.4, Mn 0-1.34.
For resistance alloy, rheostats, thermocouples, heat resistant. *Obsolete*

CONSTRUCTAL 1
Vereinigte Leichtmetallwerke G.m.b.H.
Aluminum. Cu, Zn, Mg, bal Al.
For light alloy parts.

CONSTRUCTAL 2
Vereinigte Leichtmetallwerke G.m.b.H.
Aluminum. Cu 1.2, Mg 0.9, Si 0.6, Ti 0.5, bal Al.
Aged: 54,000-60,000 TS; 18-25 El; 95-115 Brin. For aircraft and heavy duty forgings; age-hardenable.

CONSTRUCTAL 20/42
Vereinigte Leichtmetallwerke G.m.b.H.
Aluminum. Zn 4.5, Mg 2.5, Si 0.4, Mn 0.6, Cu 0.1, bal Al.
Aged: 65,000-71,000 TS; 51,000-60,000 YS; 10-12 El. For light alloy parts; high strength, age hardenable.

CONSTRUCTAL 20/53
Vereinigte Leichtmetallwerke G.m.b.H.
Aluminum. Zn 5, Mg 3, Si 0.4, Cr 0-0.3, Cu 0.1, bal Al.
Aged: 71,000-77,000 TS; 64,000-68,000 YS; 8 El. For light alloy parts; high strength, age-hardenable.

CONSTRUCTAL 8
Vereinigte Leichtmetallwerke G.m.b.H.
Aluminum. Zn 7, Mg 2.5, Si 0.2, Mn 1, bal Al.
Heat treated: 67,000-74,000 TS; 15-18 El; 120-140 Brin. For light alloy structural parts; heat treatable.

CONSTRUCTAL 87
Vereinigte Leichtmetallwerke G.m.b.H.
Aluminum. Fe 0.28, Mn 1.24, Zn 6.87, Mg 1.62, Si 0.75, bal Al.
For light alloy structural parts; heat treatable.

CONSUMET
Now CARPENTER CONSUMET.

CONSUMET 882
Carpenter Technology Corp.
C 0.35, Cr 5, Mo 1.5, V 0.4, bal Fe.
Heat treated: 300,000 TS; 250,000 YS; 6 El; 55 Rock C. For die casting dies, extrusion and forging dies, hot punches, aircraft structural fittings. Vacuum melted Type H11 tool steel.

CONTACT BRONZE
Criterion Metals, Inc.
Copper. Cu 88, Zn 10, Sn 2, Fe 0-0.05, Pb 0-0.05.
Thin gauge sheet, various tempers: 46-82 Rock B; 41-92 ksi TS min; 13-87 ksi YS min (0.2% offset). C42500

CONTACT BRONZE
Olin Brass, Indianapolis
Copper. Zn 8.95, Sn 1.9, P 0.15, Cu 89.
Annealed: 48,000 psi TS; 54 El; 67 Brin. Hard: 78,000 psi TS; 7.5 El; 165 Brin. Spring: 97,500 psi TS; 3 El; 190 Brin. For electrical contacts, switches, and springs; good spring properties.

CONTACT BRONZE 92
Olin Brass, Indianapolis
Copper. Cu 89, P 0.15, Sn 1.9, bal Zn.
Hard: 90,000 psi TS; 72,000 psi YS; 3 El; 185 Brin. For electrical contacts; corrosion resistant.

CONTEX
Degussa AG
Anti-fluxing agent for the jewelry industry.

CONTRACID
Vacuumschmelze GmbH
Be 0.6, Ni 60, Cr 15, Mo 7, bal Fe.
For springs; corrosion resistant. *Obsolete*

CONTRACID B 10 W
Vacuumschmelze GmbH
Fe 12, Ni 61, Cr 15, W 10, Mn 2.
At 25 C: 125,000 TS; 112,000 YS; 22 El. At 500 C: 109,000 TS; 100,000 YS; 18 El. For chemical apparatus, surgical instruments; heat and corrosion resistant. *Obsolete*

CONTRACID B 2.5 M
Vacuumschmelze GmbH
Fe 19.5, Ni 61, Cr 15, Mo 2.5, Mn 2.
For chemical apparatus, surgical instruments; heat and corrosion resistant. *Obsolete*

CONTRACID B 4 M
Vacuumschmelze GmbH
Fe 18, Ni 61, Cr 15, Mo 4, Mn 2.
For chemical apparatus, surgical instruments; heat and corrosion resistant. *Obsolete*

CONTRACID B 6 W
Vacuumschmelze GmbH
Fe 16, Ni 61, Cr 15, W 6, Mn 2.
For chemical apparatus, surgical instruments; heat and corrosion resistant. *Obsolete*

CONTRACID B 7 M
Vacuumschmelze GmbH
Fe 16, Ni 60, Cr 15, Mo 7, Mn 2.
At 20 C: 136,000 TS; 120,000 YS; 23 El. At 500 C: 107,000 TS; 91,000 YS; 28 El. For chemical apparatus, surgical instruments; heat and corrosion resistant. *Obsolete*

CONTRACID B W M C
Vacuumschmelze GmbH
Fe 14, Ni 58, Co 3, Cr 15, Mo 3, W 5, Mn 2.
For chemical apparatus, surgical instruments; heat and corrosion resistant. *Obsolete*

CONTRACID B7MO
Vacuumschmelze GmbH
Ni 60, Cr 15, Fe 16, Mn 2, Mo 7.
For jet engine parts; heat and corrosion resistant. *Obsolete*

CONTRACID-B2M
Vacuumschmelze GmbH
Ni 60, Cr 18, Mo 2, bal Fe.
For clinical apparatus, surgical instruments, pumps, valves. Corrosion and acid resistant. *Obsolete*

CONTRALOY
Michiana Products Corp.
Copper. Cr 28, Ni 15, C, bal Fe.
For heat and corrosion resistant parts; now Michiana No. 63.

COO 75
Creusot-Loire
C 0.48, Mn 0.4, Si 0.25, bal Fe.
For plastic molds or die-casting molds for aluminum or zinc. AFNOR $Y_2 48$.

COOK A.L.Z.
George Cook & Co., Ltd.
C 0.4, Cr 5, W 0.1, Si 1, Mo 1.5, V 0.6, bal Fe.
For punches, pneumatic tools; oil hardened, tough.

COOK B.K.V.
George Cook & Co., Ltd.
C 1.9, Mn 0.35, Cr 13.5, Si 0.6, bal Fe.
For punches, plastic mold dies, form tools; air or oil hardened, nondeforming.

COOK C.K.K.
George Cook & Co., Ltd.
C 1, Cr 1.4, Mn 0.3, Si 0.25, bal Fe.
For bearings, bushings, blanking and forming dies; water hardened, wear and abrasion resistant.

COOK C.M.C.
George Cook & Co., Ltd.
C 0.9, Mn 1.3, Cr 0.95, Si 0.3, bal Fe.
For blanking and forming tools, taps, punches; oil hardened, nondeforming.

COOK C.R.P.
George Cook & Co., Ltd.
C 0.9, Mn 1.3, Cr 1, Si 0.25, bal Fe.
For press tools, gauges, master hobs, rollers; oil hardened, nondeforming.

COOK CCV
George Cook & Co., Ltd.
C 0.5, Mn 0.75, Cr 1, V 0.15, bal Fe.
Oil hardening tool steel. For chuck jaws, hammer pistons, axes. Similar to AISI 6150.

COOK CRP O1
George Cook & Co., Ltd.
C 0.95, Mn 1.25, Cr 0.5, V 0.2, W 0.5, bal Fe.
Oil hardening tool and die steel. For jigs, fixtures, slitting cutters, blanking tools. BS 4659 B01; AISI O1.

COOK E.T.A.
George Cook & Co., Ltd.
C 0.9, Mn 0.35, Si 0.2, V 0.1, bal Fe.
For drawing, pressing and forming dies; cold work steel, water hardened.

COOK ETH. (RED LABEL)
George Cook & Co., Ltd.
C 1.14, Mn 0.4, V 0.15, bal Fe.
Water hardening tool steel. For mandrels, knurling tools.

COOK H.D.Z.
George Cook & Co., Ltd.
C 0.4, Cr 3, W 9, V 3, bal Fe.
For extrusion dies and rams, punches; hot work steel, oil hardened. *Obsolete*

COOK I.L.O.
George Cook & Co., Ltd.
C, alloy, bal Fe.
For drawing, forming and coining dies; water hardened. *Obsolete*

COOK M.K.Z.
George Cook & Co., Ltd.
C 0.38, Cr 3, W 9, Mo 0.4, V 0.2, bal Fe.
For hot forging and extrusion dies and tools; hot work steel, oil hardened.

COOK M.Y.
George Cook & Co., Ltd.
C 0.58, Mn 0.4, Cr 1, Si 0.95, Mo 0.4, bal Fe.
For chisels, caulking and riveting tools; oil hardened, shock resistant. *Obsolete*

COOK SILVER LABEL
George Cook & Co., Ltd.
C 1.4, Mn 0.4, Cr 0.6, bal Fe.
Water hardening tool steel. Punches and dies, lathe and planer tools.

COOK W.Z.
George Cook & Co., Ltd.
C 0.4, Cr 3, W 11, V 0.2, Mo 0.4, Mn 0.35, bal Fe.
For extrusion dies, die casting dies, tools; hot work steel, oil hardened. *Obsolete*

COOK'S ALLOY
English manufacture
Sb 56-69, Zn 32-44.

COOPER
English manufacture
Au 40-50, Pd 50-60.
For dental alloy; corrosion resistant.

COOPER 13-4
Cooper Alloy Corp.
Stainless steel. C 0.06, Si 1, Mn 1, Cr 11.5-14, Ni 3.5-4.5, Mo 0.4-1, bal Fe.
Cast: 758 MPa TS; 552 MPa YS; 15 El; 35 RA. Corrosion resistant casting. ACI CA6NM.

COOPER 14H
Cooper Alloy Corp.
Stainless steel. C 0.25-0.35, Cr 11-14, Mn 0-1, Ni 0-0.75, bal Fe.
Annealed: 99,000 TS; 55,000 YS; 27 El; 48 RA; 190 Brin. Hardened: 250,000 TS; 225,000 YS; 0 El; 0 RA; 550 Brin. For grinding parts and castings. Wear and abrasion resistant.

COOPER 14I
Cooper Alloy Corp.
Stainless steel. C 0.12-0.2, Cr 11-14, Mn 0-1, Ni 0-0.75, bal Fe.
Cast: 115,000-120,000 TS; 90,000-100,000 YS; 25-20 El; 60-55 RA; 200-240 Brin. For valve trim, fittings, and pump parts. Corrosion resistant.

COOPER 14S
Cooper Alloy Corp.
Stainless steel. C 0.1, Cr 12.5, Ni 0.5, Si 0.75, bal Fe.
Heat treated: 197,000-116,000 TS; 139,000-96,000 YS; 16-23 El; 42-60 RA; 372-230 Brin. For valves, valve trim, and pump parts. Corrosion resistant.

COOPER 14SM
Cooper Alloy Corp.
Stainless steel. C 0.15, Si 0.65, Mn 1, Cr 11.5-14, Ni 1, Mo 0.15-1, bal Fe.
Cast: 621 MPa TS; 448 MPa YS; 18 El; 30 RA. Corrosion resistant casting. ACI CA-15M.

COOPER 15B
Cooper Alloy Corp.
Tool material. C 1.4-1.6, Si 0.3-0.6, Mn 0.2-0.4, Cr 11-12.5, Ni 0.25, Mo 0.7-1, V 0.75-1.25, bal Fe.
Abrasion resistant cast alloy.

COOPER 16
Cooper Alloy Corp.
Stainless steel. C 0.12-0.3, Cr 16-20, bal Fe.
Annealed: 70,000-100,000 TS; 50,000-75,000 YS; 5-12 El; 5-15 RA; 200-225 Brin. For valve trim for high pressure steam, and pump valves. Stainless steel; formerly "Sweetaloy No. 16."

COOPER 16A
Cooper Alloy Corp.
Stainless steel. C 0.12-0.2, Si 1, Mn 1, Cr 15-17, Ni 1.25-2.5, bal Fe.
Abrasion resistant cast alloy. AISI 431.

COOPER 17
Cooper Alloy Corp.
Stainless steel. C 0-0.2, Cr 18-20, Ni 8-10, bal Fe.
Annealed: 75,000-85,000 TS; 40,000-50,000 YS; 40-50 El; 40-55 RA; 180-160 Brin. For oil refineries, canneries, dairies, pump valves, cocks, and fittings. Best general acid resisting base iron; "Sweetaloy No. 17."

COOPER 17 LL
Cooper Alloy Corp.
Stainless steel. Cr 18-20, Ni 8-10, Se 0.2-0.3, C 0-0.2, Si 0-2, C 0-0.2, Si 0-2, bal Fe.
Water quenched.: 75,000-85,000 TS; 40,000-50,000 YS; 40-55 El; 40-55 RA; 130-160 Brin. For fittings, valves, and pumps. Corrosion resistant; free-cutting.

COOPER 17 SLL
Cooper Alloy Corp.
Stainless steel. Cr 18-20, Ni 8-10, Se 0.2-0.3, C 0-0.1, Si 0-2, bal Fe.
Water quenched: 70,000-80,000 TS; 35,000-45,000 YS; 45-60 El; 60-45 RA; 130-160 Brin. For fittings, valves, and pumps. Corrosion resistant; free-cutting.

COOPER 17-4 PH
Cooper Alloy Corp.
Stainless steel. C 0.05, Cr 16.5, Ni 4.5, Si 0.65, Cu 2.8, bal Fe.
Heat treated: 165,000 TS; 144,000 YS; 19 El; 49 RA; 401 Brin. For high strength, corrosion resistant parts and castings. Heat treatable; stainless.

COOPER 17-S
Cooper Alloy Corp.
Stainless steel. C 0.1, Cr 18, Ni 8-10, bal Fe.
Cast: 84,500 TS; 45,000 YS; 50 El; 50 RA; 165 Brin. For stainless parts, valves, pumps, and fittings. Stainless; corrosion resistant.

COOPER 17ELC
Cooper Alloy Corp.
Stainless steel. C 0.025, Cr 20, Ni 8.5, Si 1.25, bal Fe.
Annealed: 85,000 TS; 43,000 YS; 54 El; 59 RA; 145 Brin. For chemical plant equipment and stainless castings. Stainless; austenitic.

COOPER 17GM
Cooper Alloy Corp.
Stainless steel. C 0.06-0.08, Cr 17-19, Ni 12-14, Mo 0-0.5, bal Fe.
Annealed: 78,000 TS; 38,000 YS; 55 El; 140 Brin. For chemical plant equipment, agitators, mixers, vessels, valve bodies, and pumps. Corrosion resistant; austenitic.

COOPER 17SCB
Cooper Alloy Corp.
Stainless steel. C 0.8, Cr 18-20, Ni 8-10, Cb = 10 x C, bal Fe.
Cast: 80,000 TS; 40,000 YS; 47 El; 50 RA; 150 Brin.
Annealed: 90,000 TS; 55,000 YS; 55 El; 55 RA; 140 Brin. For stainless steel castings, pumps, oil refining. Stainless; resists acids; stabilized tough welding.

COOPER 17SELC
Cooper Alloy Corp.
Stainless steel. C 0-0.03, Cr 17-21, Ni 8-12, bal Fe.
Annealed: 77,000 TS; 37,000 YS; 55 El; 140 Brin. For autoclaves, filter press plates, pasteurizers, mixing kettles, pumps, pump sleeves, and spray nozzles. ACI-CF3; AISI 304; austenitic; stainless.

COOPER 17SM
Cooper Alloy Corp.
Stainless steel. C 0-0.08, Cr 19, Ni 10, Si 1.25, Mo 2.5, bal Fe.
Water quenched: 90,000 TS; 55,000 YS; 50 El; 55 RA; 155 Brin. For paper mill equipment. Corrosion resistant; austenitic.

COOPER 17SM ELC
Cooper Alloy Corp.
Stainless steel. C 0-0.3, C 17-21, Ni 9-13, Mo 2-3, bal Fe.
Annealed: 80,000 TS; 42,000 YS; 50 El; 155-170 Brin. For chemical plant equipment, acid mixers, filter presses, acid pumps, and spray nozzles. ACI-CF3M; AISI 316 L. Corrosion resistant; austenitic.

COOPER 17XM
Cooper Alloy Corp.
Stainless steel. C 0-0.08, Cr 18-21, Ni 9-13, Mo 3-4, bal Fe.
Annealed: 80,000 TS; 42,000 YS; 50 El; 155-170 Brin. For chemical plant equipment, acid mixers and pumps, valves, filter presses, and spray nozzles. ACI-CG8M; AISI 317. Corrosion resistant; austenitic.

COOPER 17XM ELC
Cooper Alloy Corp.
Stainless steel. C 0.03, Si 1.5, Mn 1.5, Cr 17-21, Ni 9-13, Mo 3-4, bal Fe.
Corrosion resistant casting. ACI CG3M.

COOPER 19A
Cooper Alloy Corp.
Stainless steel. C 0-0.35, Cr 26-30, Ni 2-3, bal Fe.
Cast: 75,000-100,000 TS; 50,000-65,000 YS; 15-25 El; 15-25 RA; 190-210 Brin. For furnace parts and stainless castings. Heat and corrosion resistant.

COOPER 19BH
Cooper Alloy Corp.
Cast iron. C 2.5-2.8, Si 0.3-0.7, Mn 1, Cr 27-30, Ni 1.5-2.5, bal Fe.
Abrasion resistant cast alloy.

COOPER 20
Cooper Alloy Corp.
Stainless steel. C 0.3-0.5, Ni 34-37, Cr 13-18, bal Fe.
Cast: 60,000-70,000 TS; 35,000-40,000 YS; 6-12 El; 6-12 RA; 165-185 Brin. For furnace parts, enameling racks, tube supports, and retorts. Resists heat and flue gases at high temperature; "Sweetaloy No. 20."

COOPER 21
Cooper Alloy Corp.
Nickel. C 0.3-0.5, Ni 62-68, Cr 12-18, bal Fe.
Cast: 60,000-75,000 TS; 30,000-50,000 YS; 4-10 El; 4-10 RA; 170-185 Brin. For carburizing boxes, pots, and retorts. Heat resisting alloy; "Sweetaloy No. 21."

COOPER 22
Cooper Alloy Corp.
Stainless steel. C 0.3-0.4, Cr 24-28, Ni 11-14, bal Fe.
Cast: 80,000-100,000 TS; 40,000-55,000 YS; 20-30 El; 20-30 RA; 165-190 Brin. For sulfite pulp mills, digester fittings, pump valves, and furnace parts. Resists sulfurous acids and ammonia; "Sweetaloy No. 22."

COOPER 22 P
Cooper Alloy Corp.
Stainless steel. Cr 27-30, Ni 8-11, C 0-0.3, Si 0-2, Mn 0-1.5, bal Fe.
Cast: 80,000-100,000 TS; 40,000-55,000 YS; 10-30 El; 10-30 RA; 170-210 Brin. For furnace parts and sulfite fittings. Heat and corrosion resistant; austenitic.

COOPER 22 S
Cooper Alloy Corp.
Stainless steel. Mn 0-1.5, Cr 24-28, Ni 11-14, C 0-0.2, Si 0-2, bal Fe.
Water quenched: 80,000-90,000 TS; 35,000-55,000 YS; 25-45 El; 25-55 RA; 135-190 Brin. For fittings and valves. Corrosion and heat resistant to hot HNO_3.

COOPER 23
Cooper Alloy Corp.
C 0.35-1, Cr 15-19, Ni 64-68, Si 0-2.5, Mn 0-2, Mo 0-0.5, bal Fe.
Cast: 65,000 TS; 36,000 YS; 9 El; 176 Brin. Aged: 73,000 TS; 44,000 YS; 9 El; 185 Brin. For autoclaves, carburizing boxes, cyanide pots, roller hearths, and salt pots. ACI-HX; heat and corrosion resistant.

COOPER 24
Cooper Alloy Corp.
C 0.35-0.75, Cr 17-21, Ni 37-41, Si 0-2.5, Mn 0-2, Mo 0-0.5, bal Fe.
As cast: 70,000 TS; 40,000 YS; 9 El; 170 Brin. For carburizing retorts, cyanide pots, dipping baskets, muffles, and lead pots. ACI-HU; heat and corrosion resistant.

COOPER 26
Cooper Alloy Corp.
Stainless steel. C 0.2-0.6, Cr 28-32, Ni 18-22, Si 0-2, Mn 0-2, Mo 0-0.5, bal Fe.
Cast: 82,000 TS; 52,000 YS; 19 El; 192 Brin. For enameling furnace parts, furnace skids, stack dampers, and radiant tubes. ACI-HL; heat and corrosion resistant.

COOPER 27
Cooper Alloy Corp.
Stainless steel. C 0.2-0.5, Cr 26-30, Ni 14-18, Si 0-2, Mn 0-2, Mo 0-0.5, bal Fe.
Cast: 80,000 TS; 45,000 YS; 12 El; 180 Brin. For billet skids, brazing fixtures, conveyor rollers, furnace rails, lead pots, and retorts. ACI-HI; heat and corrosion resistant.

COOPER 28
Cooper Alloy Corp.
Stainless steel. C 0.2-0.4, Cr 19-23, Ni 9-12, Si 0-2, Mn 0-2, Mo 0-0.5, bal Fe.
Cast: 85,000 TS; 45,000 YS; 35 El; 165 Brin. For arc furnace electrodes, annealing boxes, trays and baskets, burner tips, conveyor belts, and chains. ACI-HF; AISI 302 B; corrosion resistant; austenitic.

COOPER 29
Cooper Alloy Corp.
Stainless steel. C 0.2-0.5, Cr 26-30, Ni 8-11, Si 0-2, Mn 0-2, Mo 0-0.5, bal Fe.
Cast: 95,000 TS; 45,000 YS; 20 El; 200 Brin. For billet skids, furnace chains and conveyors, oil burner parts, rabble arms and blades, and recuperators. ACI-HE; heat and corrosion resistant; austenitic.

COOPER 30
Cooper Alloy Corp.
Stainless steel. C 0-0.5, Cr 26-30, Ni 4-7, Si 0-2, Mn 0-1.5, Mo 0-0.5, bal Fe.
Cast: 85,000 TS; 48,000 YS; 16 El; 190 Brin. For brazing furnace parts, furnace blowers, rabble shoes, salt pots, recuperators, and gas burners. ACI-HD; heat and corrosion resistant.

COOPER 50
Cooper Alloy Corp.
C 0.12, Si 1, Mn 1, Cr 15.5-17.5, Mo 16-18, V 0.2-0.4, W 3.75-5.25, Fe 4.5-7.5, bal Ni.
Cast: 496 MPa TS; 317 MPa YS; 4 El. Corrosion resistant casting. ACI CW-12M-1; ASTM A-296/A-494.

COOPER 51
Cooper Alloy Corp.
Nickel. C 0.12, Si 1, Mn 1, Cr 1, Mo 26-30, V 0.2-0.6, Fe 4-6, bal Ni.
Cast: 503 MPa TS; 317 MPa YS; 6.0 El. Corrosion resistant casting. ACI N-12M-1; ASTM A-296/A-494.

COOPER 52
Cooper Alloy Corp.
Nickel. C 0.12, Si 7.5-8.5, Mn 0.5-1.25, Cr 1, Cu 2-4, Fe 2, bal Ni.
Corrosion resistant casting.

COOPER 60
Cooper Alloy Corp.
Nickel. C 0.1, Si 1, Mn 0.3, Cr 58-62, Fe 1, bal Ni.
Cast: 760 MPa TS; 590 MPa YS. Corrosion resistant casting; ASTM A-560.

COOPER ALLOY 14
Cooper Alloy Corp.
C 0.12-0.35, Cr 11-14, bal Fe.
Cast: 75,000 TS; 50,000 YS; 20 El; 30 RA; 180 Brin. For corrosion resistant castings; valve trim. *Obsolete*

COOPER ALLOY 14A
Cooper Alloy Corp.
C 0.1-0.15, Cr 11.5-14, Ni 1.25-2.5, Mo 0-0.5, bal Fe.
Annealed: 115,000 TS; 90,000 YS; 235 Brin. For furnace parts, chemical and textile plant equipment. AISI 414; corrosion resistant. *Obsolete*

COOPER ALLOY 15
Cooper Alloy Corp.
C 1.5-1.7, Cr 12-14, Mo 1.1-1.5, Ni 0-0.75, Co 3-3.5, V 0.5-0.75, bal Fe.
Annealed: 235 Brin. Hardened: 600 Brin. For grinding parts, castings; abrasion and wear resistant. *Obsolete*

COOPER ALLOY 15A
Cooper Alloy Corp.
C 1.5, Cr 12-14, V 1, Mo 1, bal Fe.
For valves, pump bodies; corrosion and heat resistant. *Obsolete*

COOPER ALLOY 16 C3
Cooper Alloy Corp.
C 0-0.3, Cr 20-24, Ni 0-0.75, Cu 0.9-1.2, bal Fe.
Annealed: 70,000-95,000 TS; 35,000-70,000 YS; 2-7 El; 2-7 RA; 185-235 Brin. For castings for copper plating solutions; corrosion resistant. *Obsolete*

COOPER ALLOY 16 H
Cooper Alloy Corp.
C 0.95-1.1, Cr 16-18, Ni 0-0.75, Mn 0-1, bal Fe.
Cast: 240 Brin. Hardened: 620 Brin. For wear resistant castings; corrosion resistant. *Obsolete*

COOPER ALLOY 16C
Cooper Alloy Corp.
C 0.3, Cr 20-22, Cu 1, bal Fe.
Cast: 70,000 TS; 35,000 YS; 3 El; 2 RA; 235 Brin. For rabble blades on ore roasting furnaces; heat resistant. *Obsolete*

COOPER ALLOY 17 MO
Cooper Alloy Corp.
C 0-0.2, Cr 18-20, Ni 8-10, Mo 3, bal Fe.
Annealed: 85,000 TS; 45,000 YS; 35 El; 35 RA; 155 Brin. For stainless, heat and corrosion resisting parts, pumps; corrosion and heat resistant. *Obsolete*

COOPER ALLOY 17 SE
Cooper Alloy Corp.
Cr 18-20, Ni 8-10, Se 0.3, C, bal Fe.
Cast: 85,000 TS; 45,000 YS; 50 El; 50 RA; 165 Brin. For stainless parts; stainless, corrosion resistant; free machining. *Obsolete*

COOPER ALLOY 17-S-MO
Cooper Alloy Corp.
C 0-0.1, Cr 18-20, Ni 8-12, Mo 2-3, bal Fe.
Cast: 85,000 TS; 45,000 YS; 50 El; 50 RA; 165 Brin. For stainless, heat and corrosion resisting parts, valves; corrosion resistant, stainless. *Obsolete*

COOPER ALLOY 17C
Cooper Alloy Corp.
C 0.07-1, Cr 18, Ni 8, bal Fe.
Cast: 84,000 TS; 42,000 YS; 47 El; 50 RA; 165 Brin. For oil refineries, pumps, dyeing and bleach equipment; corrosion and heat resistant. *Obsolete*

COOPER ALLOY 17G
Cooper Alloy Corp.
C 0.15-0.3, Cr 17-20, Ni 8-11, Mo 0-0.5, bal Fe.
Annealed: 80,000 TS; 38,000 YS; 46 El; 170 Brin. For chemical plant equipment, agitators, mixers, filter presses, valve bodies, pump components. MIL-S-17509 Cl.I; corrosion resistant, austenitic. *Obsolete*

COOPER ALLOY 17M-ELC
Cooper Alloy Corp.
C 0.025, Cr 19, Ni 12, Si 1.25, Mo 2.5, bal Fe.
Annealed: 84,000 TS; 43,000 YS; 46 El; 59 RA. For chemical plant equipment and castings; acid resistant, austenitic. *Obsolete*

COOPER ALLOY 18
Cooper Alloy Corp.
C 0-0.35, Ni 20-22, Cr 8-10, bal Fe.
Annealed: 65,000-79,000 TS; 35,000-46,500 YS; 20-30 El; 30-35 RA; 150-180 Brin. For cracking coil plugs, in oil refineries, thermocouple wells, electric resistance grids; erosion, heat and corrosion resistant; "Sweetaloy No. 18." *Obsolete*

COOPER ALLOY 18-8
Cooper Alloy Corp.
C 0.1, Cr 18, Ni 8, bal Fe.
For stainless parts; austenitic, stainless. *Obsolete*

COOPER ALLOY 19
Cooper Alloy Corp.
C 0-0.35, Cr 27-30, Ni 0.5-3, bal Fe.
Cast: 50,000-90,000 TS; 35,000-60,000 YS; 1-3 El; 1-5 RA; 170-205 Brin. For equipment to resist S gases at high temperature, corrosive acids, molten metals; excellent abrasion resistance; "Sweetaloy No. 19." *Obsolete*

COOPER ALLOY 19 AM
Cooper Alloy Corp.
C 0-0.1, Cr 28, Ni 3, Si 0.75, Mo 1.5, bal Fe.
Cast: 90,000 TS; 75,000 YS; 10 El; 20 RA; 200 Brin. For paper mill equipment; corrosion resistant. *Obsolete*

COOPER ALLOY 19 AMO
Cooper Alloy Corp.
C 0-0.35, Cr 27-30, Ni 2-4, Mo 1-1.5, bal Fe.
Annealed: 90,000 TS; 75,000 YS; 10 El; 20 RA; 200 Brin. For soot blower parts; acid mine water pumps; heat and corrosion resistant. *Obsolete*

COOPER ALLOY 19AH
Cooper Alloy Corp.
C 1, Cr 28, Ni 3, Si 0.75, bal Fe.
Cast: 350 Brin. For roasting furnace blades; corrosion, heat and abrasion resistant. *Obsolete*

COOPER ALLOY 19B
Cooper Alloy Corp.
C 0-0.1, Cr 26-30, Ni 3-4, Si 0-1, Mn 0-0.1, bal Fe.
Cast: 90,000 TS; 65,000 YS; 18 El; 185 Brin. For chemical and food processing equipment, fittings, gears, impellers, jet engine components. ACI Type CC-10, heat and abrasion resistant. *Obsolete*

COOPER ALLOY 20-10
Cooper Alloy Corp.
C 0-0.12, Cr 20.5, Ni 10.5, Si 1.25, bal Fe.
Water quenched: 80,000 TS; 50,000 YS; 55 El; 55 RA; 150 Brin. For stainless parts; stainless, austenitic. *Obsolete*

COOPER ALLOY 21 B
Cooper Alloy Corp.
C 0-0.07, Si 1-1.5, Mo 2-4, Cr 18-20, Ni 22-25, Cu 1-2, bal Fe.
Water quenched: 65,000-75,000 TS; 30,000-40,000 YS; 25-45 El; 25-50 RA; 130-160 Brin. For fittings, valves; corrosion resistant, austenitic. *Obsolete*

COOPER ALLOY 21 D
Cooper Alloy Corp.
C 0-0.07, Si 3-3.5, Mo 2.5-3, Cr 18-20, Mn 0.4-0.7, Ni 24-25, Cu 1.5-2, bal Fe.
Water quenched: 65,000-75,000 TS; 30,000-40,000 YS; 25-40 El; 25-40 RA; 130-160 Brin. For fittings, valves; corrosion resistant, austenitic. *Obsolete*

COOPER ALLOY 21 E
Cooper Alloy Corp.
C 0-0.15, Si 0-1.5, Cr 20, Ni 80, Mn 0-1.
Cast: 70,000-95,000 TS; 30,000-45,000 YS; 10-30 El; 10-30 RA; 160-190 Brin. For heat resistant castings; Nichrome type alloy. *Obsolete*

COOPER ALLOY 21A
Cooper Alloy Corp.
C 0-0.2, Cr 18-20, Ni 24-26, Mo 2-4, Si 1.5, bal Fe.
Annealed: 65,000 TS; 30,000 YS; 25 El; 25 RA; 130 Brin. For heat and corrosion resisting parts; corrosion resistant to hot H 2 SO 4 . *Obsolete*

COOPER ALLOY 21C
Cooper Alloy Corp.
C 0.07-0.1, Cr 15-20, Ni 20-25, Mo 3, Si 1.5, bal Fe.
Cast. For acid resistant castings; stainless. *Obsolete*

COOPER ALLOY 22 SMO
Cooper Alloy Corp.
C 0-0.1, Si 0-2, Cr 24-28, Ni 11-14, Mo 2.5-3.5, bal Fe.
Water quenched: 85,000-95,000 TS; 45,000-70,000 YS; 15-25 El; 15-25 RA; 160-190 Brin. For fittings, valves; pumps; corrosion resistant to hot sulfite liquors. *Obsolete*

COOPER ALLOY 22 W
Cooper Alloy Corp.
C 0.15-0.3, Si 0.75-2, Cr 20-25, Ni 10-14, W 2.5-3.5, bal Fe.
Normalize: 93,000 TS; 52,000 YS; 31 El; 30 RA; 185 Brin. For highly stressed castings; heat and creep resistant, austenitic. *Obsolete*

COOPER ALLOY 22M
Cooper Alloy Corp.
C 0.2-0.5, Cr 24-28, Ni 10-12, Mo 3, bal Fe.
For valves, pumps, fittings; resists sulfite liquors. *Obsolete*

COOPER ALLOY 22PM
Cooper Alloy Corp.
C 0.2-0.5, Cr 29, Ni 9, Mo 3, bal Fe.
Cast: 90,000 TS; 60,000 YS; 30 El; 180 Brin. For pumps, valve fittings; resists sulfite liquors. *Obsolete*

COOPER ALLOY 22S CB
Cooper Alloy Corp.
C 0-0.1, Cr 25, Ni 12, Si 1, Cb 0-1, bal Fe.
Water quenched: 90,000 TS; 55,000 YS; 25 El; 25 RA; 175 Brin. For corrosion and heat resistant parts; resists intergranular corrosion. *Obsolete*

COOPER ALLOY 25
Cooper Alloy Corp.
C 0.2-0.5, Cr 19-23, Ni 23-27, Si 0-2, Mn 0-2, Mo 0-0.5, bal Fe.
Cast: 68,000 TS; 38,000 YS; 17 El; 160 Brin. At 1400 F: 42,000 TS; 11 El. For brazing fixtures, torch nozzles, trays, furnace parts, radiant tubes. ACI-HN, heat and corrosion resistant. *Obsolete*

COOPER ALLOY 25-20
Cooper Alloy Corp.
C 0.2-0.3, Cr 24-26, Ni 19-20, bal Fe.
Cast: 65,000 TS; 45,000 YS; 22 El; 25 RA; 180 Brin. For furnace parts; heat resistant over 1600 F. *Obsolete*

COOPER ALLOY 25-20S
Cooper Alloy Corp.
C 0-0.2, Cr 23-27, Ni 19-22, Si 0-2, Mn 0-1.5, bal Fe.
Annealed: 76,000 TS; 38,000 YS; 37 El; 144 Brin. For furnace parts, salt pots, heat treating boxes, digesters, pumps. Heat and corrosion resistant, austenitic. ACI Type CK-20. *Obsolete*

COOPER ALLOY 31
Cooper Alloy Corp.
C 0-0.2, Cr 8-10, Si 0-1, Mn 0.35-0.65, Mo 0.9-1.2, bal Fe.
Annealed: 95,000 TS; 65,000 YS; 23 El; 180 Brin. Normalized: 107,000 TS; 81,000 YS; 21 El; 220 Brin. At 1100 F: 44,000 TS; 32,000 YS; 36 El; 58 RA. For furnace rollers, refinery fittings, trunnions, fan blades. ACI-HA castings, heat and corrosion resistant. *Obsolete*

COOPER ALLOY 531
Cooper Alloy Corp.
C 0-0.1, Cr 10-14, Ni 27-30, Mo 3-4, Cu 3-4, Sb 0.4-0.6, Si 0.5-0.8, bal Fe.
Cast: 50,000-70,000 TS; 40,000-50,000 YS. For castings for oil refineries and chemical plants; corrosion and heat resistant. *Obsolete*

COOPER ALLOY 58
Cooper Alloy Corp.
C 0-0.2, Cr 23, Ni 58, Si 0.5, Mo 6, Cu 7, W 1.5, bal Fe.
Cast: 65,000 TS; 55,000 YS; 5 El; 5 RA; 175 Brin. For sulfuric acid equipment; "Illium G"; acid resistant. *Obsolete*

COOPER ALLOY 60
Cooper Alloy Corp.
C 0-0.05, Cr 58-62, Si 0-1, Mn 0-0.3, 1.5% max others, bal Ni.
Cast: 137,500 TS; 90,000 YS; 1.5 El; 1.0 RA; 375 Brin. For furnace parts, where fuel oil containing vanadium pentoxide is burned. Heat resistant. *Obsolete*

COOPER ALLOY P.H.20
Cooper Alloy Corp.
C 0-0.07, Cr 21, Ni 29, Mo 3, Cu 3, Cb 0-1, N 0.15, bal Fe.
For corrosion resistant and non-galling castings; austenitic, stainless, age-hardenable. *Obsolete*

COOPER ALLOY S-21 W
Cooper Alloy Corp.
C 0-0.4, Cr 15, Ni 25, W 3.5, bal Fe.
For heat, erosion and corrosion resistant parts; heat and corrosion resistant. *Obsolete*

COOPER ALLOY S-23
Cooper Alloy Corp.
C 0-0.2, Cr 2, Ni 40, bal Fe.
For temperature control application; low thermal expansion. *Obsolete*

COOPER ALLOY: INCONEL
Cooper Alloy Corp.
Stainless steel. C 0.4, Si 3, Mn 1.5, Cr 14-17, Fe 11, bal Ni.
Cast: 483 MPa TS; 193 MPa YS; 30 El. Corrosion resistant casting. ACI CY-40; ASTM A-296 CY-40.

COOPER ALLOY: NICKEL
Cooper Alloy Corp.
Nickel. C 1, Si 2, Mn 1.5, Fe 3, Cu 1.25, bal Ni.
Cast: 345 MPa TS; 124 MPa YS; 10 El. Corrosion resistant casting. ACI CZ-100; ASTM A-296 CZ100.

COOPER PH 55A
Cooper Alloy Corp.
Stainless steel. C 0-0.08, Si 3-3.75, Cr 20, Ni 9, Mo 4, bal Fe.
For corrosion resistant castings. Austenitic; stainless; high-strength.

COOPER V2B
Cooper Alloy Corp.
C 0-0.07, Cr 19, Ni 10, Cu 2, Mo 3.5, Si 3, Be 0.15, bal Fe.
Rolled: 140,000 TS; 109,000 YS; 13 El; 16 RA; 302 Brin. Cast: 109,000 TS; 88,400 YS; 16 El; 22 RA; 269 Brin. Heat treated: 152,000 TS; 122,000 YS; 3 El; 2 RA; 365 Brin. For valve discs, plug cocks, impellers, sleeves; stainless, non-galling, hardenable. *Obsolete*

COOPER'S SPECULUM
English manufacture
Cu 57.8, Sn 27.3, Zn 3.6, As 1.2, Pt 10.
For mirrors, reflectors.

COOPERITE
English manufacture
Ni 80, W 14, Zr 6.
For cutting tools; extreme hardness.

COP R LOY
Wheeling-Pittsburgh Steel Corp.
Ingot iron or low C steel containing small % of Cu.
Rolled: 50,000-80,000 TS; 25,000-40,000 YS; 25-40 El; 30-50 RA; 100-150 Brin. For structural parts, boiler tubes, roofing, wire fencing, pipe; corrosion resistant. *Obsolete*

COP-SIL-LOY (1)
Cop-Sil-Loy Inc.
Cu-Ag-Pb.
For reducing wear on friction surfaces. Low coefficient of friction.

COP-SIL-LOY (2)
Cop-Sil-Loy Inc.
Pb 47-53, bal Cu.
For special parts; sintered.

COPALOY
Michigan Smelting & Refining Co.
Sn, Sb, Pb.
For bearings; antifriction.

COPAN
American manufacture
Sn 80-90, Sb 10-15, Cu 2-5, Pb 0.2.
For bearings; Babbitt.

COPASTAR C-1
Mitsubishi Metal America Corp.
Copper. Al 7.5, Fe 3.5, Ni 2.5, Mn 0.5, bal Cu.
Special aluminum bronze. Discoloration resistance alloy; for hardware and material for building and housing industry, tableware, fixtures, musical instruments.

COPEL
Hoskins Mfg. Co.
Copper. Cu 55, Ni 45.
60,000 TS; 40 El. For electrical instruments, rheostats, hot resistances; resistance alloy up to 800°F.

COPELMET-D
Metro Cutanit Ltd.
W + Cu.
For electrical contacts, welding electrodes. *Obsolete*

COPELMET-P
Metro Cutanit Ltd.
W + Cu.
For electrical contacts, resistance welding electrodes. *Obsolete*

COPELMET-PW
Metro Cutanit Ltd.
W + Cu.
For electrical contacts, resistance welding electrodes. *Obsolete*

COPELMET-PW4
Metro Cutanit Ltd.
WC + Cu.
Sintered carbide. For facing of electrical contacts, spot, projection and butt welding electrodes, hot riveting dies. *Obsolete*

COPERNICK
Western Electric Co.
Fe 50, Ni 50.
For transformer cores; resistance alloy.

COPPCO 110
Copperweld Steel Co.
C 1.1, Mn 0.3, Si 0.25, bal Fe.
For cutters, knives, drills, hobs, reamers; Type WI; water hardened. *Obsolete*

COPPCO 120
Copperweld Steel Co.
C 1.2, Mn 0.3, Si 0.5, Cr 0.2, bal Fe.
For knives, drills, reamers, taps; Type WI; water hardened. *Obsolete*

COPPCO 200
Copperweld Steel Co.
C 0.9, Mn 1.15, Cr 0.5, W 0.5, bal Fe.
For tools, dies, taps, gauges, master tools; non-deforming, oil hardening. *Obsolete*

COPPCO 75
Copperweld Steel Co.
C 0.75, Mn 0.3, Si 0.25, bal Fe.
For cutters, tools, shears; Type WI; water hardened. *Obsolete*

COPPCO ACE
Copperweld Steel Co.
C 0.2-0.7, Ni 1.5, Cr 0.55, bal Fe.
For tools, cutters, dies, punches; oil hardening. *Obsolete*

COPPCO ACE A TEMPER
Copperweld Steel Co.
C 0.7, Cr 0.55, Ni 1.5, bal Fe.
For punches, collets, forming dies; tough. *Obsolete*

COPPCO ACE B TEMPER
Copperweld Steel Co.
C 0.5, Cr 0.55, Ni 1.5, bal Fe.
For collets, shear blades, pawls; tough. *Obsolete*

COPPCO ACE C TEMPER
Copperweld Steel Co.
C 0.2, Mn 0.6, Cr 0.55, Ni 1.5, bal Fe.
For carburized tools and parts; tough. *Obsolete*

COPPCO CRO-TUNG
Copperweld Steel Co.
C 0.5, Cr 1.4, W 2.5, bal Fe.
For punches, chisels, tools; tough. *Obsolete*

COPPCO EXTRA
Copperweld Steel Co.
C 0.95, Mn 0.25, bal Fe.
For cold heading dies, pneumatic piston tools; tough.
Obsolete

COPPCO FAST FINISHING
Copperweld Steel Co.
C 1.3, W 3.75, bal Fe.
For fast finishing tools, cutters; oil hardened. *Obsolete*

COPPCO HOT WORK NO. 1
Copperweld Steel Co.
C 0.35, Si 0.9, Cr 5, Mo 1.5, W 1.1, bal Fe.
For hot work tools and dies, punches; hot work steel.
Obsolete

COPPCO SHOCK
Copperweld Steel Co.
C 0.55, Mn 0.8, Si 1.5, Cr 0.2, Mo 0.35, bal Fe. For chisels, pneumatic tools; tough, resists shock. *Obsolete*

COPPCO SPECIAL
Copperweld Steel Co.
C 0.98, bal Fe.
For burnishing tools, dies, reamers, cutters; water hardening.
Obsolete

COPPCO STANDARD
Copperweld Steel Co.
C 0.9, Mn 1.2, bal Fe.
For broaches, gauges, milling cutters, reamers; non-deforming. *Obsolete*

COPPCO UNIVERSAL
Copperweld Steel Co.
C 1, Mn 0.25, Si 0.25, bal Fe.
For pneumatic tools, punches, dies; water hardening.
Obsolete

COPPER
Atomergic Chemetals Corp.
Cu.
Purities, zone refined: 99.9999%, 99.9997%, 99.999%, 99.99+% (OFHC). Forms: rod, powder, wire, sheet, foil, ingot, shot, single crystals.

COPPER ALLOY Z30A
American manufacture
Pb 0-1.5, 57.0 Cu min, 30 Zn min.
Die cast: 55,000 TS; 30,000 YS; 15 El; 10 Brin. For die castings, hardware; yellow brass, good machinability.

COPPER ALLOY ZS144A
Manufacturer not listed
Cu 80-83, Si 3.75-4.25, bal Zn.
Die Cast: 85,000 YS; 50,000 YS; 25 El; 70 Charpy; 85-90 Rock B. For die castings.

COPPER ALLOY ZS331A
Manufacturer not listed
Cu 63-67, Si 0.75-1.25, bal Zn.
Die Cast: 70,000 TS; 35,000 YS; 25 El; 50 Charpy; 68-72 Rock B. For die castings.

COPPER ALNICO
Harsco Corp. (Taylor-Wharton Div.)
Al 9-11, Ni 16-18, Co 12-13, Cu 5.5-6.5, bal Fe.
For magnets; Alnico II. *Obsolete*

COPPER ARC
C.E. Phillips & Co.
C 0.2, Cu 0.5, bal Fe.
For weldng electrodes; for cast iron.

COPPER BEARING LOW-METALLOID
U.S. Steel Corp.
C, 0.2% Cu min, bal Fe.
For culverts, building, general fabrication; hot rolled, annealed and galvanized sheets. *Obsolete*

COPPER BEARING STEEL
U.S. Steel Corp.
C 0.08-0.18, Mn 0.3-0.6, Si 0.1-0.2, Cu 0.2-0.35, bal Fe.
For pipes, boiler tubes. *Obsolete*

COPPER CLAD 304 SS
Texas Instruments Inc./Materials Control
CDA 194 copper clad type 304 stainless steel. Volume ratio 20/80. For automotive airbag switches.

COPPER CLAD 304 SS
Texas Instruments Inc./Materials Control
CDA copper clad to type 304 stainless steel. For button cell battery anode caps. Ratio of copper to type 304 stainless steel can be specified from 5-50%.

COPPER CLAD 304 SS CLAD NICKEL
Texas Instruments Inc./Materials Control
Type 304 stainless steel clad with CDA 101 copper on one side and 201 nickel on the other. For button cell battery anode caps.

COPPER CLAD 304L SS
Texas Instruments Inc./Materials Control
CDA 122 copper clad to one or both sides of type 304L stainless steel. CDA 122 copper is used as braze filler metal and type 304L stainless steel as structural member of brazed heat exchanger unit.

COPPER CLAD 430 SS
Texas Instruments Inc./Materials Control
Type 430 stainless steel clad with CDA 122 copper on one or both sides. CDA 122 copper is used as braze filler metal and type 409 stainless steel as structural member.

COPPER CLAD AL & CU CLAD AL CLAD CU
Texas Instruments Inc./Materials Control
Copper clad aluminum in various ratios for applications requiring solderability, electrical and thermal conductivity and light weight. Tempers A, B, C, and D provided for applications from deep drawing to stamping.

COPPER IRON
Criterion Metals, Inc.
Copper. Zn 0-0.05, P 0.01-0.04, Fe 0.8-1.2, Pb 0-0.025, 98.7 Cu min.
Thin gauge sheet, various tempers: 40-74 ksi TS min. C19200

COPPER M3
Russian manufacture
Cu 99.86, Pb 0.01, Zn 0.03, Fe 0.01, Sn 0.02.
Rolled: 32,000 TS; 43 El; 69 RA.

COPPER NICKEL 30%
Chase Brass & Copper Co., Inc.
Copper. Cu 69.5, Ni 30, Fe 0.5.
Annealed: 54,000 psi TS; 20,00 psi YS; 55 El. Drawn (35%): 85,000 psi TS; 74,000 psi YS; 5 El. For condenser and heat exchanger tubes, ferrules.

COPPER NICKEL, 10%
Chase Brass & Copper Co., Inc.
Copper. Cu 88.6, Ni 10, Fe 1.4.
Annealed: 47,000 psi TS; 18,000 psi YS; 48 El. Drawn (35%): 70,000 psi TS; 65,000 psi YS; 10 El. For condenser and heat exchanger tubes.

COPPER STEEL
American Nickeloid Co.
Cu coated steel.
For fabricated parts; easily stamped, formed, drawn.

COPPER STEEL
U.S. Steel Corp.
C, 0.2% Cu min, bal Fe.
For building and general fabrication; bars, plates, shapes.
Obsolete

COPPER STEEL
American manufacture
C 0.37, Mn 0.17, Cu 4, Si 0.22, bal Fe.
Rolled: 138,000 TS; 116,000 YS; 11 El; 23 RA; 302 Brin. Annealed: 97,000 TS; 80,500 YS; 16 El; 42 RA; 212 Brin. For structural parts; corrosion resistant.

COPPER STEEL
U.S. Steel Corp.
C, 0.2% Cu min, bal Fe.
For woven wire fencing and barbed wire; wire products.
Obsolete

COPPER STEEL
Manufacturer not listed
C 0.37, Mn 0.17, Cu 4, Si 0.22, bal Fe.
Rolled: 138,000 TS; 116,000 YS; 11 El; 23 RA; 302 Brin. Annealed: 97,000 TS; 80,500 YS; 16 El; 42 RA; 212 Brin. For structural parts; corrosion resistant. *Obsolete*

COPPER STEEL
American Nickeloid Co.
Copper coated steel.
For fabricated parts; easily stamped, formed, drawn.
Obsolete

COPPER TIN
American Nickeloid Co.
Cu coated Sn.
For fabricated parts; easily stamped, formed, drawn.
Obsolete

COPPER TITANIUM ALLOY
Climax Performance Materials Corp.
Ti 4.3, bal Cu.
Solution heat treated, 40% cold worked, aged: 1076 MPa (156,000 psi) TS; 965 MPa (140,000 psi) YS; 5 El. Electrical conductivity: 10% IACS at 68°F. For non-sparking tools, springs, electrical contacts, diaphragms.

COPPER'S GOLD
English manufacture
Cu 67-81, Pt 19-30, Zn 0-4.
For jewelry, ornaments; corrosion resistant.

COPPER'S MIRROR
English manufacture
Cu 58, Sn 28, Pt 9.5, Zn 3.5, As 1.5.
For reflectors, mirrors; corrosion resistnt.

COPPER'S PEN METAL-1
English manufacture
Cu 50, Au 25, Ag 25.
For pen points; corrosion resistant.

COPPER'S PEN METAL-2
English manufacture
Cu 12, Ag 38, Pt 50.
For pen points; corrosion resistant.

COPPER, DEOXIDIZED, DHP
Chase Brass & Copper Co., Inc.
P 0.02, 99.9 Cu min.
Annealed: 32,000 TS; 10,000 YS; 45 El. Rolled hard: 50,000 TS; 45,000 YS; 6 El. Deoxidized, high phosphorus. Readily brazed and welded. For heat exchangers.

COPPER, DEOXIDIZED, DLP
Chase Brass & Copper Co., Inc.
P 0.007, 99.9 Cu min.
Annealed: 32,000 TS; 10,000 YS; 45 El. Rolled hard: 50,000 TS; 45,000 YS; 6 El. Deoxidized, low phosphorus. For welded or brazed components.

COPPER, DEOXIDIZED, XLP
Chase Brass & Copper Co., Inc.
P 0.003, 99.95 Cu min.
Annealed: 32,000 TS; 10,000 YS; 45 El. Rolled hard: 50,000 TS; 45,000 YS; 6 El. Extra low phosphorus. For high conductivity components.

COPPER, ELECTROLYTIC TOUGH PITCH
Chase Brass & Copper Co.
O 0.04, 99.9 Cu min.
Annealed: 32,000 TS; 10,000 YS; 45 El. Rolled hard: 50,000 TS; 45,000 YS; 6 El. Electrical conductors, roofing, switches.

COPPER, OXYGEN FREE HIGH CONDUCTIVITY
Chase Brass & Copper Co.
99.99 Cu min.
Annealed: 32,000 TS; 10,000 YS; 45 El. Rolled hard: 50,000 TS; 45,000 YS; 6 El. Electrical conductors. Certified or uncertified.

COPPER, PHOSPHORUS
Chase Brass & Copper Co., Inc.
Copper. 99.0 Cu min, 0.003, 0.007 or 0.02 P.
Annealed: 32,000 psi TS; 10,000 psi YS; 45 El. Hard rolled: 50,000 psi TS; 45,000 psi YS; 6 El. For electrical conductor heat exchangers, kettles, tanks.

COPPER, SILVER BEARING
Chase Brass & Copper Co., Inc.
Copper. Ag 0.027-0.085, 99.9 Cu min.
Annealed: 33,000 psi TS; 10,000 psi YS; 50 El. Hard drawn: 55,000 psi TS; 50,000 psi YS; 10 El. For motor commutators, switches; high softening temperature.

COPPER-BERYLLIUM 50
Kawecki Berylco Industries Inc.
Be 0.38, Co 1.55, Ag 1, bal Cu.
Annealed: 45,000 TS; 27 El; 67 Brin. Cold rolled: 70,000 TS; 13 El; 132 Brin. Aged: 120,000 TS; 14 El; 234 Brin. For springs, contacts, diaphragms; corrosion and wear resistant.

COPPER-BERYLLIUM 50
Brush Wellman
Be 0.38, Co 1.55, Ag 1, bal Cu.
Annealed: 45,000 TS; 27 El; 67 Brin. Cold rolled: 70,000 TS; 13 El; 132 Brin. Aged: 120,000 TS; 14 El; 234 Brin. For springs, contacts, diaphragms; corrosion and wear resistant.

COPPER-FLO
Johnson Matthey plc
Cu 92.6, P 7.4.
Phosphorus bearing silver brazing alloy; 714-810°C MP; DIN 8513 L-CuP7.

COPPER-FLO NO. 3
Johnson Matthey plc
Cu 93.8, P 6.2.
Phosphorus bearing silver brazing alloy; 714-890°C MP; DIN 8513 L-CuP6.

COPPER-NICKEL (4%)
Henry Wiggin & Co. Ltd.
Cu 96, Ni 4.
For locomotive fire box stays. *Obsolete*

COPPER-NICKEL 10%
Henry Wiggin & Co. Ltd.
Ni 10, Cu 90.
Annealed: 40,000 TS; 40 El. For dynamo rings. *Obsolete*

COPPER-NICKEL 10%
Vereinigte Deutsche Metallwerke AG
Ni 10, Cu 90.
Annealed: 40,000 TS; 40 El. For dynamo rings. *Obsolete*

COPPER-NICKEL 15%
Henry Wiggin & Co. Ltd.
Ni 15, Cu 85.
Annealed: 42,500 TS; 40 El; For condenser tubes, coinage, medallions, bullet jackets, chemical works. *Obsolete*

COPPER-NICKEL 15%
Vereinigte Deutsche Metallwerke AG
Ni 15, Cu 85.
Annealed: 42,500 TS; 40 El. For condenser tubes, coinage, medallions, bullet jackets, chemical works. *Obsolete*

COPPER-NICKEL 20%
Henry Wiggin & Co. Ltd.
Ni 20, Cu 80.
Annealed: 45,000 TS; 17,000 YS; 38 El; 65 Brin. For condenser tubes. *Obsolete*

COPPER-NICKEL 40%
VDM Nickel-Technologie AG
Ni 40-45, Cu 55-60.
Annealed: 70,000 TS; 38 El. For electrical resistors, thermocouple elements; heat resistant. *Obsolete*

COPPER-NICKEL 401
Now MONEL ALLOY 401.

COPPER-NICKEL-TIN
Waterbury Rolling Mills Inc.
Ni 8.5-10.5, Sn 2, bal Cu.
UNS C 725 00.

COPPER-NICKEL-TIN ALLOY 725
Criterion Metals, Inc.
Copper. Cu 91, Zn 0-0.05, Sn 2, Mn 0-0.2, Fe 0-0.6, Ni 9, Pb 0-0.05.
Thin gauge sheet, various tempers: 45-90 ksi TS min; 18-89 ksi YS min. C72500

COPPER-NICKEL-TITANIUM ALLOY
Climax Performance Materials Corp.
Ni 5, Ti 2.5, bal Cu.
Solution, heat treated, 90% cold work, aged: 745 MPa (108,000 psi) TS; 634 MPa (92,000 psi) YS; 10 El. Electrical conductivity: 53% IACS at 68°F. For electronic applications at elevated temperatures, current carrying springs, heat sinks, switch parts, radar components for high temperature service.

COPPER-SILICON ALUMINUM 8-2
Birmingham Aluminium Casting Co.
Aluminum. Cu 8, Si 2.5, Fe 0.8, bal Al.
Cast: 17,000 TS; 1 El; 70 Brin. For gear boxes, radiator tanks; non-hardenable.

COPPER-SILICON STEEL
Ford Motor Co.
Cast: 107,000 TS; 92,000 YS; 1.7 El; 2.2 RA; 269 Brin. For cast crankshafts. *Obsolete*

COPPER/INVAR/COPPER (12.5/75/12.5)
Texas Instruments Inc./Materials Control
12.5 Cu, 75 Invar, 12.5 Cu (nominal).
Active constraining plans for printed circuit boards. Low thermal expansion.

COPPER/INVAR/COPPER (20/60/20)
Texas Instruments Inc./Materials Control
20 Cu, 60 Invar, 20 Cu (nominal).
Constraining core for printed circuit boards. Low thermal expansion; heat sink applications.

COPPER/INVAR/COPPER (5/90/5)
Texas Instruments Inc./Materials Control
5 Cu, 90 Invar, 5 Cu (nominal).
CTE match. Low thermal expansion.

COPPERCLAD
GTE Products Corp.
42 Ni steel core with pure Cu sleeve. For sealing in vacuum tubes.

COPPERIOR
Becker Stahlwerk AG
C 0.1, Cu, bal Fe.
For roofing, culverts; corrosion resistant.

COPPEROID
Youngstown Steel
0.2 C min, bal Fe.
For sheet for construction purposes, automobile furniture; resists rust and atmospheric corrosion.

COPPEROID
American manufacture
Cu alloy.
For bearings; copper bearing alloy.

COPPEROID
American Nickeloid Co.
Cu coated Zn.
For fabricated parts; easily stamped, formed, drawn. *Obsolete*

COPPERWELD WIRE
British Insulated Callender's Cables
Cu on steel core. *Obsolete*

COPPRO NITRIDING G
Copperweld Steel Co.
C 0.3-0.4, Cr 0.9-1.4, Mo 0.15-0.25, Al 0.85-1.2, bal Fe.
For nitrided parts; nitriding steel. *Obsolete*

COPPRO NITRIDING G-MODIFIED
Copperweld Steel Co.
C 0.38-0.45, Cr 1.4-1.8, Mo 0.3-0.45, Al 0.85-1.2, bal Fe.
For nitrided aircraft parts; nitriding steel. *Obsolete*

COPPRO NITRIDING H
Copperweld Steel Co.
C 0.2-0.3, Cr 0.9-1.4, Mo 0.15-0.25, Al 0.85-1.2, bal Fe.
For nitriding parts; nitriding steel. *Obsolete*

COPR-TRODE
Ampco Metal
Copper. Sn 0-1, Mn 0-0.5, Si 0-0.5, P 0-0.15, 98.0 min Cu, 0.50 others max.
Weld metal: 29 ksi TS; 8 ksi YS; 29 El; 54 Brin. For welding deoxidized copper castings with gas metal-arc and gas-tungsten arc.

COPREX
Wakefield Corp.
Cu alloy.
For bearings; sintered. *Obsolete*

COR-TEN "A"
Spartan Redheugh Ltd.
C 0-0.12, Mn 0.2-0.5, P 0.07-0.15, S 0-0.05, Si 0.25-0.75, Cu 0.25-0.55, Cr 0.3-1.25, Ni 0-0.65, bal Fe.
Plate: 3290 kg/cm^2 YP min; 4690 kg/cm^2 TS min; 21 El. High strength low alloy steel that resists atmospheric corrosion five to eight times better than structural carbon steel.

COR-TEN "B"
Spartan Redheugh Ltd.
C 0.1-0.19, Mn 0.9-1.25, P 0-0.04, S 0-0.05, Si 0.15-0.3, Cu 0.25-0.4, Cr 0.4-0.65, V 0.02-0.1, bal Fe.
Plate: 3500 kg/cm^2 YP min; 4900 kg/cm^2 TS min; 21 El. High strength low alloy steel that resists atmospheric corrosion four times better than carbon steel.

COR-TEN A
British Steel plc
C 0-0.12, Si 0.25-0.75, Mn 0.2-0.5, P 0.07-0.15, S 0-0.05, Cr 0.3-1.25, Ni 0-0.65, Cu 0.25-0.55, bal Fe.
High-strength low-alloy weldable steel. Corrosion resistant.

COR-TEN A
British Steel Corp.
C 0.1, Mn 0.4, P 0.12, S 0-0.05, Si 0.5, Cu 0.4, Ti 0.02-0, Cr 1, Ni 0-0.65, bal Fe.
Plate: 70,000 TS; 50,000 YS; 19 El. For bridges, booms, derricks, mine cars, bus and truck bodies. High strength, low alloy steel. ASTM A 242.

COR-TEN A
United States Steel Corp.
C 0.1, Mn 0.4, P 0.12, S 0-0.05, Si 0.5, Cu 0.4, Ti 0.02-0, Cr 1, Ni 0-0.65, bal Fe.
Plate: 70,000 TS; 50,000 YS; 19 El. For bridges, booms, derricks, mine cars, bus and truck bodies. High strength, low alloy steel. ASTM A 242.

COR-TEN B

Alan Wood Steel Co.
C 0.1-0.19, Mn 0.9-1.25, Cu 0.25-0.4, Cr 0.4-0.65, V 0.02-0.1, bal Fe.
Plate: 70,000 TS; 50,000 YS; 19 El. For structures, derricks, booms, mine cars, truck and bus bodies, bridges. High strength low alloy steel. *Obsolete*

COR-TEN B

Inland Steel Co.
C 0.1-0.19, Mn 0.9-1.25, Cu 0.25-0.4, Cr 0.4-0.65, V 0.02-0.1, Ni 0-0.65, bal Fe.
Plate: 70,000 TS; 50,000 YS; 19 El. For structures, derricks, booms, mine cars, truck and bus bodies, bridges. High strength, low alloy steel. ASTM A 588.

COR-TEN B

British Steel Corp.
C 0.1-0.19, Mn 0.9-1.25, Cu 0.25-0.4, Cr 0.4-0.65, V 0.02-0.1, Ni 0-0.65, bal Fe.
Plate: 70,000 TS; 50,000 YS; 19 El. For structures, derricks, booms, mine cars, truck and bus bodies, bridges. High strength, low alloy steel. ASTM A 588.

COR-TEN B

United States Steel Corp.
C 0.1-0.19, Mn 0.9-1.25, Cu 0.25-0.4, Cr 0.4-0.65, V 0.02-0.1, Ni 0-0.65, bal Fe.
Plate: 70,000 TS; 50,000 YS; 19 El. For structures, derricks, booms, mine cars, truck and bus bodies, bridges. High strength, low alloy steel. ASTM A 588.

COR-TEN B

Inland Steel Co.
C 0.1-0.19, Mn 0.9-1.25, Cu 0.25-0.4, Cr 0.4-0.65, V 0.02-0.1, bal Fe.
Plate: 70,000 TS; 50,000 YS; 19 El. For structures, derricks, booms, mine cars, truck and bus bodies, bridges. High strength low alloy steel. *Obsolete*

COR-TEN B-QT

United States Steel Corp.
C 0.16, Mn 1.15, Si 0.25, Ni 0-0.65, Cr 0.5, V 0.04, Cu 0.25-0.4, bal Fe.
Water quenched and tempered: 90 ksi TS; 70 ksi YS; 19 El. High strength, corrosion resistant steel for bridges, towers, buildings. ASTM A588.

COR-TEN B1

British Steel plc
C 0.1-0.19, Si 0.15-0.65, Mn 0.9-1.25, P 0-0.04, S 0-0.05, Cr 0.5-0.65, Cu 0.25-0.4, Al 0.01-0.06, V 0.02-0.1, bal Fe.
High-strength low-alloy weldable steel. Corrosion resistant.

COR-TEN C

United States Steel Corp.
C 0.12-0.19, Mn 0.9-1.35, Ni 0-0.65, Cu 0.25-0.4, Cr 0.4-0.7, V 0.04-0.1, bal Fe.
Plate: 80,000 TS; 60,000 YS; 16 El. For structures, bridges, bus and truck bodies, booms, mine cars, derricks. High strength-low alloy steel.

COR-TEN C

Inland Steel Co.
C 0.12-0.19, Mn 0.9-1.35, Ni 0-0.65, Cu 0.25-0.4, Cr 0.4-0.7, V 0.04-0.1, bal Fe.
Plate: 80,000 TS; 60,000 YS; 16 El. For structures, bridges, bus and truck bodies, booms, mine cars, derricks. High strength-low alloy steel.

CORBIN

English manufacture
Al 87.5, Cu 12.5.
For pistons; non-hardenable.

CORE IRON

Carpenter Technology Corp.
C 0.06, V 1, bal Fe.
For magnetic and electrical equipment. High permeability.

CORINTH

Allegheny Ludlum Steel
C 0.9, Mn 0.3, bal Fe.
For tools, cutters; water hardening. AISI W1.

CORINTH

AL Tech Specialty Steel Corp.
C 0.9, Mn 0.3, bal Fe.
For tools, cutters; water hardening. AISI W1.

CORMET-A

Corning Glass Co.
Nickel.
Sintered: 8,200 TS; 8,100 YS. For non-contacting conveyor for extra sensitive materials; sintered porous Ni, maximum operating temperature 575 F. *Obsolete*

CORMIN BRONZE

Manufacturer not listed
Cu 44, Ni 37, Sn 11, Pb 8.
For hardware; corrosion resistant.

CORNISH BRONZE

English manufacture
Sn 9.6, Pb 16.5, P 0.8, bal Cu.
For bearings; heavy duty.

CORNIX-1

Rochling Burbach GmbH
C 0.2, Mn 0.4, Si 0.3, Cr 13, Mo 1.15, bal Fe.
Annealed: 95,000 TS; 50,000 YS; 25 El; B 92 Rock. For acid containers, mixers, chemical plant and oil refinery equipment. Corrosion resistant, hardenable. *Obsolete*

CORNIX-10

Rochling Burbach GmbH
C 0.1, Cr 16, Ni 13, Cb, bal Fe.
Annealed: 85,000 TS; 35,000 YS; 50 El; B 80 Rock. For acid containers, mixers, digesters, welded tanks and structures. Type 16-13 Cb stainless steel, austenitic, corrosion resistant. *Obsolete*

CORNIX-2

Stahlwerke Sudwestfalen
C 0.2, Mn 0.5, Cr 12, Mo 1.15, Ni 0.5, V 0.3, bal Fe.
Annealed: 95,000 TS; 40,000 YS; 25 El; 50 RA; B 92 Rock.
Heat treated: 240,000 TS; 205,000 YS; 9 El; 26 RA; C 50 Rock. For surgical instruments, hardware, oil refinery and chemical plant equipment. Corrosion resistant, hardenable. *Obsolete*

CORNIX-40

Rochling Burbach GmbH
C 0.08, Cr 16, Ni 14, Mo, Cb, bal Fe.
Annealed: 90,000 TS; 40,000 YS; 40 El; B 85 Rock. For acid containers, mixers, digesters, welded tanks and structures. Type 16-14 MoCb stainless steel, acid resistant, austenitic. *Obsolete*

CORODENT

Johnson Matthey plc
Au 71, bal Pt.
For dental purposes.

COROMANT C1

Sandvik Steel Co.
WC, Co.
For cutting tools; sintered carbide. *Obsolete*

COROMANT C3

Sandvik Steel Co.
WC, Co.
For finishing cutting tools; sintered carbide. *Obsolete*

COROMANT C5

Sandvik Steel Co.
TaC, WC, Co.
For cutting tools for roughing; sintered carbide. *Obsolete*

COROMANT C6

Sandvik Steel Co.
WC, TiC, Co.
For general purpose cutting tools; sintered carbide. *Obsolete*

COROMANT C7

Sandvik Steel Co.
WC, TiC, Co.
For cutting tools, dies; sintered carbide. *Obsolete*

COROMANT C8

Sandvik Steel Co.
WC, TiC.
For finishing cutters and tools; sintered carbide. *Obsolete*

COROMANT F0A

Sandvik Steel Co.
Sintered carbide tool material. For finish cuts and precision machining of steel; crater resistant grade. *Obsolete*

COROMANT H05

Sandvik Steel Co.
Sintered carbide tool material. Cutting tool for precision finishing of cast iron, hard irons and steels, and heavy metals. *Obsolete*

COROMANT H13

Sandvik Steel Co.
Sintered carbide stool material. Cutting tools for general purpose machining of cast iron, steels, hard and tough metals. *Obsolete*

COROMANT H1P

Sandvik Steel Co.
Sintered carbide tool material. For cutting tools for machining cast iron; Good wear resistance and toughness.

COROMANT H20

Sandvik Steel Co.
Sintered carbide tool material. Cutting tools for general purpose rough machining of cast iron and non-ferrous metals. *Obsolete*

COROMANT R1P

Sandvik Steel Co.
Sintered carbide tool material. Cutting tools for finish cuts on nickel base, high strength, heat resistant alloys. *Obsolete*

COROMANT S1P

Sandvik Steel Co.
Sintered carbide tool material. For light roughing and finishing cuts on steel; high wear resistance.

COROMANT S2

Sandvik Steel Co.
Sintered carbide tool material. Cutting tools for general purpose machining of steel; tough. *Obsolete*

COROMANT S4

Sandvik Steel Co.
Sintered carbide tool material. Cutting tools for rough machining of steel, good toughness and wear resistance. *Obsolete*

COROMANT S6

Sandvik Steel Co.
WC, TaC, Co.
For cutting tools for roughing; sintered carbide.

COROMANT SH

Sandvik Steel Co.
WC, TaC, Co.
For cutting tools for general machining; sintered carbide. *Obsolete*

CORONA EXTRA M

Stahlwerke Kabel, C.
C 0.85, W 9, Mo 1, Cr 4, bal Fe.
For milling cutters, lathe and planer tools, hobs; high speed steel.

CORONA EXTRA V
Stahlwerke Kabel, C.
C 0.86, Cr 4.1, Mo 0.9, V 2.5, W 12, bal Fe.
For lathe and planer tools, drills, hobs; high speed steel.

CORONA KOBALT 10
Stahlwerke Kabel, C.
C 0.76, Co 10, Cr 4.2, Mo 0.8, V 1.8, W 18, bal Fe.
For lathe and planer tools, milling cutters, hobs; high speed steel.

CORONA KOBALT 3
Stahlwerke Kabel, C.
C 0.86, Co 2.8, Cr 4.3, Mo 0.85, V 2, W 12, bal Fe.
For lathe and planer tools, drills, hobs, reamers; high speed steel.

CORONA KOBALT 5
Stahlwerke Kabel, C.
C 1.35, W, Cr, V, Co, bal Fe.
For fast finishing cutters; high-speed steel.

CORONA KOBALT 5W
Stahlwerke Kabel, C.
C 0.8, Co 4.7, Cr 4.3, Mo 0.75, V 1.5, W 18, bal Fe.
For lathe and planer tools, reamers, broaches, taps, hobs, drills; high speed steel.

CORONA PRIMA M
Stahlwerke Kabel, C.
C 0.95, W, Mo, Cr, V, bal Fe.
For cutters, drills, reamers, broaches; high speed steel.

CORONA PRIMA V
Stahlwerke Kabel, C.
C 0.82, Cr 4.1, Mo 0.8, V 1.6, W 8.7, bal Fe.
For reamers, drills, broaches, cutters; high speed steel.

CORONA PRIMA W
Stahlwerke Kabel, C.
C 0.74, Cr 4, V 1, W 18.5, bal Fe.
For lathe and planer tools, reamers, broaches; high speed steel.

CORONA VIERBRENZ V
Stahlwerke Kabel, C.
C 1.3, Cr 4.3, Mo 0.8, V 3.8, W 12, bal Fe.
For cutters, fast finishing tools; high speed steel.

CORONEL
English manufacture
Ni 65-70, Cu 30-35.
Cast: 64,000 TS; 32,000 YS; 34 EL; 32 RA. Rolled: 78,000 TS; 48,000 YS; 42 El; 35 RA. For valves, pumps, turbine blades; similar to "Monel Metal."

CORONZE
Olin Corp.
Al 2.8, Si 1.8, Co 0.4, bal Cu.
Annealed: 82,000 TS; 54,000 YS; 38 El. Spring temper: 125,000 TS; 111,000 YS; 4 El. For heat exchangers, glass to metal seals, springs, electrical contacts. Aluminum bronze, corrosion and oxidation resistant. High strength. *Obsolete*

CORORESIST 19300
Eutectic Corp.
Low carbon stainless alloy powder for metal spraying.

COROSOLOY
United States Steel Corp.
C, alloy, bal Fe.
For hard surfacing electrodes; corrosion resisting. *Obsolete*

COROVAC AF
Vacuumschmelze GmbH
Soft magnetic powder composite materials based on powders of various relatively high density iron based grades with low coercivity; similar to COROVAC EF.

COROVAC EF
Vacuumschmelze GmbH
Soft magnetic powder composite materials based on powders of various relatively high density iron based grades with 50 to 175 microns permeability.

COROVAC EM
Vacuumschmelze GmbH
Soft magnetic powder composite materials based on powders of various relatively high density iron based grades; low permeability for large pot cores up to 230 mm diameter especially power chokes.

COROVAC NP
Vacuumschmelze GmbH
Soft magnetic powder composite materials based on powders of various relatively high density iron based grades; low permeability and high Q-factor for HF applications.

CORRESIST 13
Pose-Marre Edelstahlwerk G.m.b.H.
C 0-0.12, Si 0.4, Cr 13, bal Fe.
Annealed: 75,000 TS; 40,000 YS; 35 El; 70 RA; 155 Brin. For turbine blades, valves, pumps, surgical instruments; Type 410; corrosion resistant. *Obsolete*

CORRESIST 13H
Pose-Marre Edelstahlwerk G.m.b.H.
C 0.25, Cr 14.5, Ni 0-1, bal Fe.
Annealed: 95,000 TS; 50,000 YS; 25 El; 55 RA; 200 Brin. For turbine blades, valves, pumps, surgical and dental instruments; corrosion resistant, hardenable. *Obsolete*

CORRESIST 17
Pose-Marre Edelstahlwerk G.m.b.H.
C 0.08, Si 0.4, Cr 17, bal Fe.
Annealed: 80,000 TS; 50,000 YS; 25 El; 50 RA; 150 Brin. Cold drawn: 130,000 TS; 120,000 YS; 2 El; 185 Brin. For oil refinery and dairy equipment, sinks, oil burners; Type 430; corrosion and heat resistant. *Obsolete*

CORRESIST 17/12 NB
Pose-Marre Edelstahlwerk G.m.b.H.
Stainless steel. C 0-0.07, Cr 17, Ni 12, Nb = 8 x C, bal Fe.
As quenched: 50-75 kp/mm^2 TS; 23 kp/mm^2 YS min; 40 El min. Rolled and wrought stainless steel. For chemical and petrochemical industries. *Obsolete*

CORRESIST 17H
Pose-Marre Edelstahlwerk G.m.b.H.
C 0.1-0.25, Cr 17, Ni 2, bal Fe.
Heat treated: 113,800-135,100 TS; 85,300 minimum YS; 14 minimum El. Ferritic-pearlitic chromium stainless. For food production industries, for pumps, valves, tubes in chemical industries. *Obsolete*

CORRESIST 17M
Pose-Marre Edelstahlwerk G.m.b.H.
C 0-0.1, Mo 1.8, Cr 17, Ti = 7 x C, bal Fe.
Annealed: 90,000 TS; 55,000 YS; 22 El; 48 RA; 170 Brin. For welded oil refinery and chemical plant equipment; corrosion resistant. *Obsolete*

CORRESIST 18/10 MONB
Pose-Marre Edelstahlwerk G.m.b.H.
Stainless steel. C 0-0.07, Cr 18, Ni 10, Mo 2.2, Nb = 8 x C, bal Fe.
As quenched: 50-75 kp/mm^2 TS; 23 kp/mm^2 YS min; 40 El min. Rolled and wrought stainless steel. For chemical and petrochemical industries. *Obsolete*

CORRESIST 18/10 MOS
Pose-Marre Edelstahlwerk G.m.b.H.
Stainless steel. C 0-0.07, Cr 18, Ni 10, Mo 2.2, bal Fe.
As quenched: 50-70 kp/mm^2 TS; 21 kp/mm YS min; 45 El min. Rolled and wrought stainless steel. For chemical and petrochemical industries. *Obsolete*

CORRESIST 18/10 MOTI
Pose-Marre Edelstahlwerk G.m.b.H.
Stainless steel. C 0-0.07, Cr 18, Ni 10, Mo 2.2, Ti = 5 x C, bal Fe.
As quenched: 50-75 kp/mm^2 TS; 23 kp/mm^2 YS; 40 El min. Rolled and wrought stainless steel. For chemical and petrochemical industries. *Obsolete*

CORRESIST 18/18
Pose-Marre Edelstahlwerk G.m.b.H.
Stainless steel. C 0-0.13, Cr 18, Ni 18, bal Fe.
As quenched: 50-70 kp/mm^2 TS; 21 kp/mm^2 YS min; 40 El min. Rolled and wrought stainless steel. For chemical and petrochemical industries. *Obsolete*

CORRESIST 18/20 MOCU
Pose-Marre Edelstahlwerk G.m.b.H.
C 0-0.07, Cr 18, Ni 20, Mo 2.2, Cu 2, Nb = 8 x C, bal Fe.
Quenched: 71,100-106,700 TS; 32,700 minimum YS; 40 minimum El. Austenitic, stainless; very resistant to attack by H 2 SO 4 . *Obsolete*

CORRESIST 18/8
Pose-Marre Edelstahlwerk G.m.b.H.
Stainless steel. C 0-0.13, Cr 18, Ni 18, bal Fe.
As quenched: 50-70 kp/mm^2 TS; 22 kp/mm^2 YS min; 50 El min. Rolled and wrought stainless steel. For chemical and petrochemical industries. *Obsolete*

CORRESIST 18/8 M
Pose-Marre Edelstahlwerk G.m.b.H.
C 0-0.12, Cr 18, Mo 2, Ni 10.5, Cb = 8 x C, bal Fe.
Annealed: 85,000 TS; 35,000 YS; 50 El; 65 RA; 160 Brin. For welded acid resistant equipment, tanks, mixers; Type 316 Cb; stainless, austenitic. *Obsolete*

CORRESIST 18/8 MW
Pose-Marre Edelstahlwerk G.m.b.H.
C 0-0.07, Cr 18, Mo 2, Ni 10.5, bal Fe.
Annealed: 85,000 TS; 35,000 YS; 50 El; 65 RA; 160 Brin. For acid resistant chemical plant equipment, tanks; Type 316; stainless, austenitic. *Obsolete*

CORRESIST 18/8 N
Pose-Marre Edelstahlwerk G.m.b.H.
C 0-0.15, Cr 18, Ni 8.5, bal Fe.
Annealed: 80,000 TS; 35,000 YS; 55 El; 75 RA; 150 Brin. Cold drawn: 180,000 TS; 150,000 YS; 10 El; 250 Brin. For chemical plant equipment, tanks, mixers, filters; Type 302; stainless, austenitic. *Obsolete*

CORRESIST 18/8 NB
Pose-Marre Edelstahlwerk G.m.b.H.
Stainless steel. C 0-0.07, Cr 18, Ni 10, Nb = 8 x C, bal Fe.
As quenched: 50-75 kp/mm^2 TS; 21 kp/mm^2 YS min; 40 El min. Rolled and wrought stainless steel. For chemical and petrochemical industries. *Obsolete*

CORRESIST 18/8 S
Pose-Marre Edelstahlwerk G.m.b.H.
Stainless steel. C 0-0.07, Cr 18, Ni 10, bal Fe.
As quenched: 50-70 kp/mm^2 TS; 19 kp/mm^2 YS min; 50 El min. Rolled and wrought stainless steel. For chemical and petrochemical industries. *Obsolete*

CORRESIST 18/8 TI
Pose-Marre Edelstahlwerk G.m.b.H.
Stainless steel. C 0-0.07, Cr 18, Ni 10, Ti = 5 x C, bal Fe.
As quenched: 50-75 kp/mm^2 TS; 21 kp/mm^2 YS; 40 El min. Rolled and wrought stainless steel. For chemical and petrochemical industries. *Obsolete*

CORRESIST 18/8 W
Pose-Marre Edelstahlwerk G.m.b.H.
C 0-0.07, Cr 18, Ni 9.5, bal Fe.
Annealed: 85,000 TS; 35,000 YS; 60 El; 70 RA; 150 Brin. Cold drawn: 180,000 TS; 125,000 YS; 10 El; 330 Brin. For welded chemical plant equipment, tanks; Type 304; stainless, austenitic. *Obsolete*

CORRESIST 25/25 MOTI

Pose-Marre Edelstahlwerk G.m.b.H.
Stainless steel. C 0-0.06, Cr 25, Ni 25, Mo 2.2, Ti = 5 x C, bal Fe.
As quenched: 50-70 kp/mm^2 TS; 23 kp/mm^2 YS min; 35 El min. Rolled and wrought stainless steel. For chemical and petrochemical industries. *Obsolete*

CORRESIST 28

Pose-Marre Edelstahlwerk G.m.b.H.
C 0.08, Cr 28, bal Fe.
Annealed: 85,000 TS; 50,000 YS; 30 El; 55 RA; 180 Brin. For furnace parts, heat treating boxes, retorts; heat resistant. *Obsolete*

CORRESIST 28/5

Pose-Marre Edelstahlwerk G.m.b.H.
C 0.4, Si 1.3, Cr 27, Ni 4, bal Fe.
Cast: 90,000 TS; 65,000 YS; 2 El; 212 Brin. Heat treated: 97,000 TS; 65,000 YS; 18 El; 210 Brin. For cylinder liners, bushings, valve seats and bodies; Type CC-50; corrosion and heat resistant. *Obsolete*

CORRESIST 28H

Pose-Marre Edelstahlwerk G.m.b.H.
C 1.2, Si 1.3, Cr 29, bal Fe.
For crushers, rollers, skids, grates; heat and wear resistant. *Obsolete*

CORRESIST 28MH

Pose-Marre Edelstahlwerk G.m.b.H.
C 1.2, Si 1.3, Cr 29, Mo 2, bal Fe.
For crushers, rollers, skids, grates; heat and wear resistant. *Obsolete*

CORRESIST G 14

Pose-Marre Edelstahlwerk G.m.b.H.
Stainless steel. C 0.2, Cr 13.5, bal Fe.
Tempered: 590-790 N/mm^2 TS; 440 N/mm^2 YS min; 12 El min. Cast stainless steel. For chemical and petrochemical industries.

CORRESIST G 17

Pose-Marre Edelstahlwerk G.m.b.H.
Stainless steel. C 0.22, Cr 17, Ni 1.5, bal Fe.
Tempered: 780-980 N/mm^2 TS; 590 N/mm^2 YS min; 4 El min. Cast stainless steel. For chemical and petrochemical industries.

CORRESIST G 18/10 MONB

Pose-Marre Edelstahlwerk G.m.b.H.
Stainless steel. C 0-0.06, Cr 19, Ni 12, Mo 2.2, Nb = 8 x C, bal Fe.
As quenched: 450-650 N/mm^2 TS; 210 N/mm^2 YS min; 30 El min. Cast stainless steel. For chemical and petrochemical industries.

CORRESIST G 18/10 MOS

Pose-Marre Edelstahlwerk G.m.b.H.
Stainless steel. C 0-0.07, Cr 19, Ni 11, Mo 2.2, bal Fe.
As quenched: 450-650 N/mm^2 TS; 210 N/mm^2 YS min; 30 El min. Cast stainless steel. For chemical and petrochemical industries.

CORRESIST G 18/18

Pose-Marre Edelstahlwerk G.m.b.H.
Stainless steel. C 0-0.1, Cr 18, Ni 18, bal Fe.
As quenched: 450-650 N/mm^2 TS; 180 N/mm^2 YS min; 20 El min. Cast stainless steel. For chemical and petrochemical industries.

CORRESIST G 18/8 ELC

Pose-Marre Edelstahlwerk G.m.b.H.
Stainless steel. C 0-0.03, Cr 18, Ni 10, bal Fe.
As quenched: 450-650 N/mm^2 TS; 175 N/mm^2 YS min; 30 El min. Cast stainless steel. For chemical and petrochemical industries.

CORRESIST G 18/8 NB

Pose-Marre Edelstahlwerk G.m.b.H.
Stainless steel. C 0-0.06, Cr 19, Ni 10, Nb = 8 x C, bal Fe.
As quenched: 450-650 N/mm^2 TS; 200 N/mm^2 YS min; 30 El min. Cast stainless steel. For chemical and petrochemical industries.

CORRESIST G 18/8 S

Pose-Marre Edelstahlwerk G.m.b.H.
Stainless steel. C 0-0.07, Cr 19, Ni 10, bal Fe.
As quenched: 450-650 N/mm^2 TS; 200 N/mm^2 YS min; 20 El min. Cast stainless steel. For chemical and petrochemical industries.

CORRESIST G 26/6

Pose-Marre Edelstahlwerk G.m.b.H.
Stainless steel. C 0-0.08, Cr 26, Ni 7, bal Fe.
As quenched: 590-790 N/mm^2 TS; 420 N/mm^2 YS min; 20 El min. Cast stainless steel. For chemical and petrochemical industries.

CORRESIST G 28

Pose-Marre Edelstahlwerk G.m.b.H.
Stainless steel. C 0.6, Cr 28, bal Fe.
Cast stainless steel. For chemical and petrochemical industries.

CORRESIST G 28 H

Pose-Marre Edelstahlwerk G.m.b.H.
Stainless steel. C 1.3, Cr 28, bal Fe.
Cast, stainless steel. For chemical and petrochemical industries.

CORRESIST G 28 HMO

Pose-Marre Edelstahlwerk G.m.b.H.
Stainless steel. C 1.3, Cr 28, Mo 2.2, bal Fe.
Cast, stainless steel. For chemical and petrochemical industries.

CORRESIST G 28 MO

Pose-Marre Edelstahlwerk G.m.b.H.
Stainless steel. C 0.5, Cr 28, Mo 2.2, bal Fe.
Cast stainless steel. For chemical and petrochemical industries.

CORRESIST G 28/5

Pose-Marre Edelstahlwerk G.m.b.H.
Stainless steel. C 0.4, Cr 28, Ni 5, bal Fe.
As quenched: 550-750 N/mm^2 TS; 3 El min. Cast stainless steel. For chemical and petrochemical industries.

CORRESIST G 28/5 MO

Pose-Marre Edelstahlwerk G.m.b.H.
Stainless steel. C 0.4, Cr 28, Ni 5, Mo 2.2, bal Fe.
As quenched: 550-750 N/mm^2 TS; 3 El min. Cast stainless steel. For chemical and petrochemical industries.

CORRESIST G14

Pose-Marre Edelstahlwerk G.m.b.H.
C 0.25, Cr 14.5, Ni 0-1, bal Fe.
Annealed: 95,000 TS; 50,000 YS; 25 El; 55 RA; 195 Brin. For cutlery, valves, turbine blades, surgical instruments; Type 420; corrosion resistant. *Obsolete*

CORRESIST G17

Pose-Marre Edelstahlwerk G.m.b.H.
C 0.25, Cr 17, Ni 0-1.8, bal Fe.
Annealed: 125,000 TS; 95,000 YS; 20 El; 55 RA; 260 Brin. For pumps, valves, marine hardware; Type 431; corrosion resistant. *Obsolete*

CORRESIST G18/10 MONB

Pose-Marre Edelstahlwerk G.m.b.H.
C 0-0.1, Cr 18, Ni 11, Mo 2.2, Nb = 8 x C, bal Fe.
Quenched: 64,000-92,400 TS; 29,900 minimum YS; 20 minimum El. Austenitic, stainless, weldable casting. For parts and welded assemblies for use in chemical industry and food plants. *Obsolete*

CORRESIST G18/10 MOS

Pose-Marre Edelstahlwerk G.m.b.H.
C 0-0.07, Cr 18, Ni 11, Mo 2.2, bal Fe.
Quenched: 64,000-92,400 TS; 29,900 minimum YS; 20 minimum El. Austenitic, stainless; high corrosion resistance; for parts used in chemical plants. Casting. *Obsolete*

CORRESIST G18/20 MOCU

Pose-Marre Edelstahlwerk G.m.b.H.
C 0-0.07, Cr 18, Ni 20, Mo 2.2, Cu 2, Nb = 8 x C, bal Fe.
Quenched: 64,000-92,400 TS; 25,600 minimum YS; 15 min El. Austenitic, stainless, weldable; excellent corrosion resistance; casting. For pickling hooks for H 2 SO 4 . *Obsolete*

CORRESIST G18/8

Pose-Marre Edelstahlwerk G.m.b.H.
C 0.15, Si 1.5, Cr 18, Ni 8.5, bal Fe.
Annealed: 80,000 TS; 35,000 YS; 55 El; 75 RA; 150 Brin. For chemical plant equipment, tanks, mixers, filters; Type 302; stainless, austenitic. *Obsolete*

CORRESIST G18/8 M

Pose-Marre Edelstahlwerk G.m.b.H.
C 0.15, Cr 18, Mo 2, Ni 9.5, bal Fe.
Annealed: 85,000 TS; 35,000 YS; 50 El; 65 RA; 160 Brin. For acid resistant chemical plant equipment, tanks; Type 316; stainless, austenitic. *Obsolete*

CORRESIST G18/8 NB

Pose-Marre Edelstahlwerk G.m.b.H.
C 0-0.08, Cr 19, Ni 9, Nb = 8 x C, bal Fe.
Quenched: 64,000-92,400 TS; 28,400 minimum YS; 20 minimum El. Austenitic, stainless, weldable casting. For welded assemblies in food, beverage and dairy industries. *Obsolete*

CORRESIST G18/8 S

Pose-Marre Edelstahlwerk G.m.b.H.
C 0-0.07, Cr 19, Ni 9, bal Fe.
Quenched: 64,000-92,400 TS; 28,400 minimum YS; 20 minimum El. Austenitic, stainless casting. For pumps, shafts, armatures in the chemical industry. *Obsolete*

CORRESIST G27/4

Pose-Marre Edelstahlwerk G.m.b.H.
C 0.4, Cr 27, Ni 4, bal Fe.
Cast: 90,000 TS; 65,000 YS; 2 El; 212 Brin. Heat treated: 97,000 TS; 65,000 YS; 18 El; 210 Brin. For cylinder liners, bushings, valves; Type CC-50; corrosion and heat resistant. *Obsolete*

CORRESIST G28

Pose-Marre Edelstahlwerk G.m.b.H.
C 0.4, Cr 29, bal Fe.
Cast: 90,000 TS; 65,000 YS; 2 El; 212 Brin. Heat treated: 97,000 TS; 65,000 YS; 18 El; 210 Brin. For valve bodies and seats, furnace parts, liners; heat resistant; Type CC-50. *Obsolete*

CORRESIST G28 MO

Pose-Marre Edelstahlwerk G.m.b.H.
C 0.3-0.6, Cr 28, Mo 2.2, bal Fe.
Annealed: 210-280 Brin. Ferritic, magnetic, stainless casting. For chemical plant, food industries, high temperature parts. *Obsolete*

CORRESIST G28H

Pose-Marre Edelstahlwerk G.m.b.H.
C 1.2, Si 1.3, Cr 29, bal Fe.
For crushers, rollers, skids, grates, rabbles; heat and wear resistant. *Obsolete*

CORRESIST G28HM

Pose-Marre Edelstahlwerk G.m.b.H.
C 1.2, Si 1.3, Cr 29, Mo 2, bal Fe.
For crushers, rollers, grates, rabble arms; heat and wear resistant. *Obsolete*

CORRESIST-13HMO

Pose-Marre Edelstahlwerk G.m.b.H.

C 0.2, Cr 13, Mo 1, bal Fe.

Annealed: 95,000 TS; 50,000 YS; 25 El; 150 Brin. Hardened: 180,000 TS; 150,000 YS; 16 El; C 42 Rock. For chemical and oil refinery equipment, valve trim, gauges, surgical instruments. Corrosion resistant, hardenable, martensitic. *Obsolete*

CORRIX METAL

German manufacture

Cu 88.1, Al 8.7, Fe 3.2.

Cast: 78,000-90,000 TS; 30-20 El; 30 RA. For worm wheels, gears, marine and mining applications; corrosion resistant to acids and alkalis.

CORROCHROM-13G

J.C. Soding & Halbach

C 0.2, Cr 13, bal Fe.

Annealed: 95,000 TS; 50,000 YS; 25 El; 150 Brin. Hardened: 240,000 TS; 170,000 YS; 12 El; C 47 Rock. For chemical and oil refinery equipment, surgical instruments, knives, gauges, valve trim. Corrosion resistant. Heat treatable. *Obsolete*

CORROCHROM-13M

J.C. Soding & Halbach

C 0.4, Cr 13, bal Fe.

Annealed: 95,000 TS; 50,000 YS; 25 El; 150 Brin. Hardened: 250,000 TS; 192,000 YS; 10 El; C 49 Rock. For surgical instruments, knives, cutlery, scissors, shears. Corrosion resistant. Heat treatable. *Obsolete*

CORROCHROM-13T

J.C. Soding & Halbach

C 0.2, Cr 13, Mo 1.2, bal Fe.

Annealed: 95,000 TS; 50,000 YS; 25 El; 150 Brin. Hardened: 180,000 TS; 150,000 YS; 16 El; C 42 Rock. For chemical and oil refining equipment, surgical instruments, valves, pump shafts. Corrosion resistant. Heat treatable. *Obsolete*

CORROCHROM-13W

J.C. Soding & Halbach

C 0.12, Cr 13, bal Fe.

Annealed: 85,000 TS; 40,000 YS; 30 El; B 85 Rock. Hardened: 150,000 TS; 110,000 YS; 22 El; C 38 Rock. For chemical and oil refinery equipment, valve trim. Corrosion resistant, hardenable. *Obsolete*

CORROCHROM-170

J.C. Soding & Halbach

C 0.35, Cr 16.5, Mo 1.2, bal Fe.

Annealed: 125,000 TS; 95,000 YS; 20 El; C 24 Rock. Cold drawn: 130,000 TS; 110,000 YS; 15 El; C 26 Rock. For pump shafts, valve trim, hardware, surgical instruments. Corrosion resistant. Non-hardenable. *Obsolete*

CORROCHROM-17A

J.C. Soding & Halbach

C 0.13, Cr 16.5, Mo 0.25, S 0.2, bal Fe.

Annealed: 85,000 TS; 45,000 YS; 25 El; 140 Brin. For furnace and heater parts, oil burners, hardware. Non-hardenable. Corrosion resistant. Free machining. *Obsolete*

CORROCHROM-17F

J.C. Soding & Halbach

C 0.1, Cr 17.5, bal Fe.

Annealed: 80,000 TS; 45,000 YS; 28 El; 140 Brin. Cold rolled: 130,000 TS; 120,000 YS; 12 El; 190 Brin. For automotive and architectural trim, hardware, oil burner parts, tanks, furnace parts. Corrosion resistant, non-hardenable. *Obsolete*

CORROCHROM-17N

J.C. Soding & Halbach

C 0.2, Cr 17, Ni 1.4, bal Fe.

Annealed: 125,000 TS; 95,000 YS; 20 El; 260 Brin. Cold drawn: 130,000 TS; 110,000 YS; 15 El; 280 Brin. For pump shafts, hardware, surgical instruments, valve trim. Corrosion resistant. Type 431. *Obsolete*

CORROCHROM-17T

J.C. Soding & Halbach

C 0.1, Cr 17.5, Ti = 7 x C, bal Fe.

Annealed: 88,000 TS; 48,000 YS; 20 El; 145 Brin. For welded structures, tanks, hardware, trim, oil burners. Non-hardenable. Corrosion resistant. Welding grade. *Obsolete*

CORROCHROM-18E

J.C. Soding & Halbach

C 0.9, Cr 18, Mo 1.2, V 0.1, bal Fe.

Heat treated: 280,000 TS; 265,000 YS; 3 El; C 55 Rock. For dental and surgical instruments, bearings, pump and valve parts, bushings, nozzles. Corrosion and wear resistant. Heat treatable. *Obsolete*

CORROCHROM-18T

J.C. Soding & Halbach

C 0.1, Cr 17, Mo 1.7, Ti = 7 x C, bal Fe.

Annealed: 85,000 TS; 45,000 YS; 22 El; 140 Brin. For welded structures, hardware, furnace and heater parts. Non-hardenable. Corrosion resistant. Weldable grade. *Obsolete*

CORROCHRONI-1310K

J.C. Soding & Halbach

C 0.07, Cr 17, Mo 4.5, Ni 13.5, bal Fe.

Annealed: 80,000 TS; 35,000 YS; 70 El; B 70 Rock. Cold rolled: 150,000 TS; 125,000 YS; 15 El; C 30 Rock. For textile and dye works equipment, paper and pulp mill equipment, chemical plant mixers and agitators. Corrosion and heat resistant. Austenitic. Type 317 stainless steel. *Obsolete*

CORROCHRONI-1810A

J.C. Soding & Halbach

C 0.1, Cr 17.5, Mo 2.3, Ni 11.5, Ti = 5 x C, bal Fe.

Annealed: 90,000 TS; 40,000 YS; 50 El; B 83 Rock. At 1000 F: 55,000 TS; 34,000 YS; 36 El; 69 RA. For welded structures, chemical and pharmaceutical plant equipment, acid tanks, digesters, vessels. Austenitic, non-hardenable. Welding grade, acid resistant, non-magnetic. *Obsolete*

CORROCHRONI-1810M

J.C. Soding & Halbach

C 0.1, Cr 17.5, Mo 2.3, Ni 11.5, Cb = 8 x C, bal Fe.

Annealed: 90,000 TS; 40,000 YS; 50 El; B 82 Rock. At 1200 F: 45,000 TS; 30,000 YS; 32 El; C 66 Rock. For welded structures, chemical and pharmaceutical plant equipment, digesters, acid tanks, mixers. Austenitic, stabilized, non-hardenable. Welding grade, stainless, non-magnetic. *Obsolete*

CORROCHRONI-1810U

J.C. Soding & Halbach

C 0.07, Cr 17.5, Mo 2.3, Ni 11.5, bal Fe.

Annealed: 85,000 TS; 35,000 YS; 60 El; B 80 Rock. Cold drawn: 100,000 TS; 65,000 YS; 40 El; B 95 Rock. For chemical and pharmaceutical plant equipment, acid containers, mixers, tanks, agitators. Acid resistant, austenitic, non-magnetic. *Obsolete*

CORROCHRONI-1812

J.C. Soding & Halbach

C 0.07, Cr 17.5, Mo 2.8, Ni 13, bal Fe.

Annealed: 80,000 TS; 30,000 YS; 60 El; B 78 Rock. Cold rolled: 125,000 TS; 90,000 YS; 30 El; C 28 Rock. For chemical and plastic plant equipment, digesters, tanks, agitators, valve trim. Type 316, non-hardenable. Austenitic, stainless, non-magnetic. *Obsolete*

CORROCHRONI-1812C

J.C. Soding & Halbach

C 0.1, Cr 17.5, Mo 2.8, Ni 13, Cb = 8 x C, bal Fe.

Annealed: 80,000 TS; 30,000 YS; 60 El; B 78 Rock. Cold rolled: 125,000 TS; 90,000 YS; 30 El; C 28 Rock. For chemical and plastic plant equipment, digesters, tanks, agitators, valve trim, welded structures. Non-hardenable, austenitic. Welding grade, stainless, non-magnetic. *Obsolete*

CORROCHRONI-187F

J.C. Soding & Halbach

C 0.15, Cr 17, Ni 7.5, bal Fe.

Annealed: 90,000 TS; 40,000 YS; 50 El; B 85 Rock. Half hard: 125,000 TS; 75,000 YS; 12 El; C 25 Rock. For chemical plant and oil refinery equipment, fasteners, valve and pump parts. Austenitic, stainless, non-magnetic. *Obsolete*

CORROCHRONI-188C

J.C. Soding & Halbach

C 0.12, Cr 18, Ni 9, bal Fe.

Annealed: 90,000 TS; 40,000 YS; 50 El; B 85 Rock. Cold rolled: 140,000 TS; 120,000 YS; 25 El; B 100 Rock. For chemical plant and oil refinery equipment, hardware, fasteners, pump parts. Austenitic, stainless, non-magnetic. *Obsolete*

CORROCHRONI-188T

J.C. Soding & Halbach

C 0.15, Cr 18, S 0.2, Ni 9, bal Fe.

Annealed: 90,000 TS; 35,000 YS; 50 El; 160 Brin. Cold drawn: 125,000 TS; 95,000 YS; 20 El; 277 Brin. For screw machine products, rivets, fasteners, gears, shafts, fittings. Free-machining, stainless, non-magnetic. *Obsolete*

CORROCHRONI-189E

J.C. Soding & Halbach

C 0.1, Cr 18, Ni 10, Ti = 5 x C, bal Fe.

Annealed: 85,000 TS; 35,000 YS; 55 El; 150 Brin. Cold drawn: 95,000 TS; 60,000 YS; 40 El; 180 Brin. For exhaust systems, welded structures, engine manifolds, refinery equipment. Austenitic, stabilized. Welding grade, stainless, non-magnetic. *Obsolete*

CORROCHRONI-189N

J.C. Soding & Halbach

C 0.1, Cr 18, Ni 10, Cb = 8 x C, bal Fe.

Annealed: 85,000 TS; 35,000 YS; 60 El; B 80 Rock. Cold drawn: 100,000 TS; 65,000 YS; 40 El; B 95 Rock. For aircraft collector rings, exhaust manifolds, welded structures, superheaters. Austenitic, non-hardenable. Welding grade, stainless, non-magnetic. *Obsolete*

CORROCHRONI-189S

J.C. Soding & Halbach

C 0.07, Cr 18, Ni 10, bal Fe.

Annealed: 85,000 TS; 35,000 YS; 60 El; B 80 Rock. For chemical plant and pharmaceutical equipment, tanks, agitators, mixers. Non-hardenable. Stainless, austenitic, non-magnetic. *Obsolete*

CORRODUR 13 4 MO

Bergische Stahl Industrie

C 0.07, Si 0-1, Mn 0-1.5, Cr 12-13.5, Mo 0-0.7, Ni 3.5-5, bal Fe.

Stainless steel casting. W.-Nr. 1.4313; DIN G-X 5 CrNi 13 4.

CORRODUR 14

Bergische Stahl Industrie

C 0.25, Cr 14, Ni 0-1, bal Fe.

Annealed: 95,000 TS; 50,000 YS; 25 El; 55 RA; 195 Brin. For turbine blades, cutlery, valves; Type 420; corrosion resistant. *Obsolete*

CORRODUR 16-13 N

Bergische Stahl Industrie

C 0.1, Cr 16, Ni 16, Cb = 8 x C, bal Fe.

For valves, pumps; corrosion and heat resistant. *Obsolete*

CORRODUR 17-13 E3S

Bergische Stahl Industrie

C 0-0.1, Cr 17, Ni 13, Mo 3, bal Fe.

Annealed: 90,000 TS; 40,000 YS; 42 El; 60 RA; 170 Brin. For acid resistant chemical plant equipment; Type 317; stainless, austenitic. *Obsolete*

CORRODUR 17-13 E5S
Bergische Stahl Industrie
C 0.1, Cr 17, Ni 13, Mo 5, bal Fe.
Annealed: 90,000 TS; 40,000 YS; 42 El; 60 RA; 170 Brin. For acid resistant chemical plant equipment; Type 317; stainless, austenitic. *Obsolete*

CORRODUR 17.13 ESS
Bergische Stahl Industrie
C 0.05, Cr 17, Ni 13.5, Mo 4.75, bal Fe.
Solution annealed and aged: 69,000 TS min; 35,000 YS min; 20 El min. Corrosion resistant casting. W.-Nr. 1.4448; DIN G-X 6 CrNiMo 1713.

CORRODUR 18
Bergische Stahl Industrie
C 0.25, Cr 17, Ni 0-1.8, bal Fe.
For furnace parts, retorts, heat treating boxes; Type CB-30; corrosion and heat resistant.

CORRODUR 18-8
Bergische Stahl Industrie
C 0-0.15, Cr 18, Ni 8.5, bal Fe.
Annealed: 80,000 TS; 35,000 YS; 55 El; 75 RA; 150 Brin. Cold drawn: 180,000 TS; 150,000 YS; 10 El; 250 Brin. For chemical plant equipment, tanks, agitators, filters; Type 302; stainless, austenitic. *Obsolete*

CORRODUR 18-8 E
Bergische Stahl Industrie
C 0.15, Cr 18, Mo 2, Ni 9.5, bal Fe.
Annealed: 85,000 TS; 35,000 YS; 50 El; 65 RA; 160 Brin. Cold drawn: 150,000 TS; 135,000 YS; 6 El; 300 Brin. For acid resistant, chemical plant equipment, tanks; Type 316; stainless, austenitic. *Obsolete*

CORRODUR 18-8 ES
Bergische Stahl Industrie
C 0-0.07, Cr 18, Mo 2, Ni 10.5, bal Fe.
Annealed: 85,000 TS; 35,000 YS; 50 El; 65 RA; 160 Brin. Cold drawn: 150,000 TS; 135,000 YS; 6 El; 300 Brin. For acid resistant chemical plant equipment, tanks; Type 316; stainless, austenitic. *Obsolete*

CORRODUR 18-8 ESS
Bergische Stahl Industrie
C 0-0.12, Cr 18, Mo 2, Ni 10.5, Cb = 8 x C, bal Fe.
Annealed: 85,000 TS; 35,000 YS; 50 El; 65 RA; 160 Brin. For welded acid resistant chemical plant equipment; Type 316 Cb; stainless, austenitic. *Obsolete*

CORRODUR 18-8 S
Bergische Stahl Industrie
C 0-0.07, Cr 18, Ni 9.5, bal Fe.
Annealed: 85,000 TS; 35,000 YS; 60 El; 70 RA; 150 Brin. Cold drawn: 180,000 TS; 125,000 YS; 10 El; 330 Brin. For welded chemical plant equipment, tanks, vessels; Type 304; stainless, austenitic. *Obsolete*

CORRODUR 18-8 SS
Bergische Stahl Industrie
C 0-0.07, Cr 18, Ni 9.5, bal Fe.
Annealed: 85,000 TS; 35,000 YS; 60 El; 70 RA; 150 Brin. Cold drawn: 180,000 TS; 125,000 YS; 10 El; 330 Brin. For welded chemical plant equipment, tanks, mixers; Type 304; stainless, austenitic. *Obsolete*

CORRODUR 18.8
Bergische Stahl Industrie
C 0.1, Cr 18, Ni 8.5, bal Fe.
Solution annealed and aged: 64,000-85,000 TS; 30,000 YS min; 25 El min. Corrosion resistant casting. W.-Nr. 1.4312; DIN G-X 10 CrNi 188.

CORRODUR 18.8 E
Bergische Stahl Industrie
C 0.1, Cr 18, Ni 9.5, Mo 2.25, bal Fe.
Solution annealed and aged: 71,000-92,000 TS; 30,000 YS min; 25 El min. Corrosion resistant casting. W.-Nr. 1.4410; DIN G-X CrNiMo 189.

CORRODUR 18.8 E 3 S
Bergische Stahl Industrie
C 0-0.07, Si 0-2, Mn 0-2, Cr 16.5-18.5, Mo 2.5-3, Ni 11.5-13.5, bal Fe.
Stainless steel casting. W.-Nr. 1.4437; DIN G-X 6 CrNiMo 18 12.

CORRODUR 18.8 E 3 SS
Bergische Stahl Industrie
C 0-0.07, Si 0-2, Mn 0-2, Cr 16.5-18.5, Mo 2.5-3, Ni 11.5-13.5, Nb, bal Fe.
Stainless steel casting. DIN G-X 7 CrNiMoNb 18 12.

CORRODUR 18.8 ES
Bergische Stahl Industrie
C 0.07, Cr 17.5, Ni 10.5, Mo 2.25, bal Fe.
Solution annealed and aged: 64,000-85,000 TS; 30,000 YS min; 25 El min. Corrosion resistant casting. W.-Nr. 1.4408; DIN G-X 6 CrNiMo 1810.

CORRODUR 18.8 ESS
Bergische Stahl Industrie
C 0.08, Cr 17.5, Ni 11, Mo 2.25, Nb, bal Fe.
Solution annealed and aged: 64,000-85,000 TS; 30,000 YS min; 20 El min. Corrosion resistant casting. W.-Nr. 1.4581; DIN G-X7CrNiMoNb 1810.

CORRODUR 20 25 ESS
Bergische Stahl Industrie
C 0.05, Ni 25, Cr 20, Mo 3.25, Cu 1.35, bal Fe.
Solution annealed and aged: 64,000-78,000 TS; 28,000 YS min; 15 El min. Corrosion resistant casting. W.-Nr. 1.4500; DIN G-X7NiCrMoCu2520.

CORRODUR 20-25 E
Bergische Stahl Industrie
C 0.12, Ni 25, Cr 20, Cu 2, Mo 2, bal Fe.
Annealed: 100,000 TS; 50,000 YS; 45 El; 60 RA; 195 Brin. For furnace parts, valves, pumps, turbines; corrosion and heat resistant. *Obsolete*

CORRODUR 24-7 E
Bergische Stahl Industrie
C 0.27, Cr 24, Ni 7, Mo 2, bal Fe.
For furnace parts, heat treat boxes, retorts; corrosion and heat resistant, austenitic. *Obsolete*

CORRODUR 28 4
Bergische Stahl Industrie
C 0.35, Cr 27, Ni 4, bal Fe.
As cast: 64,000-85,000 TS; 42,500 YS min; 4 El min. Corrosion resistant casting. W.-Nr. 1.4340; DIN G-X40CrNi274.

CORRODUR 28 4 MO
Bergische Stahl Industrie
C 0-0.1, Si 0-1, Mn 0-2, Cr 26-28, Mo 1.3-2, Ni 4-5, bal Fe.
Corrosion resistant cast steel. W.-Nr. 1.4460; DIN G-X 8 CrNiMo 27 5.

CORRODUR 28-4
Bergische Stahl Industrie
C 0.4, Cr 27, Ni 4, bal Fe.
Cast: 90,000 TS; 65,000 YS; 2 El; 212 Brin. Heat treated: 97,000 TS; 65,000 YS; 18 El; 210 Brin. For cylinder liners, bushings, valve seats and bodies; Type CC-50; heat resistant. *Obsolete*

CORRODUR 30
Bergische Stahl Industrie
C 0.6, Cr 29, bal Fe.
As cast: 57,000-64,000 TS. Corrosion resistant casting. W.-Nr. 1.4085; DIN G-X70Cr29.

CORRODUR 30 E
Bergische Stahl Industrie
C 0.65, Cr 28, Mo 2, bal Fe.
As cast: 57,000-64,000 TS. Corrosion resistant casting. W.-Nr. 1.4136; DIN G-X70 CrMo292.

CORRODUR 30 EW
Bergische Stahl Industrie
C 0.15, Cr 28, Mo 2, bal Fe.
Stainless steel casting. DIN G-X 15 CrMo 29.

CORRODUR 30 W
Bergische Stahl Industrie
C 0.15, Cr 28, bal Fe.
Stainless steel casting. DIN G-X 15 Cr 29.

CORRODUR 30E
Bergische Stahl Industrie
C 1.2, Si 1.3, Cr 29, Mo 2, bal Fe.
For crushers, rollers, grates, rabble arms; heat and wear resistant. *Obsolete*

CORRODUR 30EH
Bergische Stahl Industrie
C 1.9, Cr 29, Mo 2, Cu 2, bal Fe.
For crushers, rollers, grates, rabble arms; heat and wear resistant. *Obsolete*

CORRODUR 30H
Bergische Stahl Industrie
C 1.8, Si 1.3, Cr 29, bal Fe.
For crushers, rollers, grates, rabble arms; heat and wear resistant. *Obsolete*

CORRODUR BERGIT B
Bergische Stahl Industrie
Mo 28.5, bal Ni.
Solution annealed and aged: 78-92 ksi TS; 57 ksi YS min; 8 El min. Corrosion resistant nickel base coating. W.-Nr. 2.4810, DIN G-X 8 NiMo 6530.

CORRODUR BERGIT C
Bergische Stahl Industrie
Mo 19, Cr 17, bal Ni.
Solution annealed and aged: 71-85 ksi TS; 57 ksi YS min; 5 El min. Corrosion resistant nickel base coating. W.-Nr. 2.4472; DIN G-X 8 NiMoCr 6018.

CORRODUR C 14 N
Bergische Stahl Industrie
C 0.19, Cr 13, Ni 0.5, bal Fe.
Annealed: 78,000-92,000 TS; 50,000 YS min; 10 El min. Heat treated: 85,000-115,000 TS; 64,000 YS min; 8 El min. Corrosion resistant casting. Werkstoff Nr. 1.4021; DIN G-X 20 Cr13. *Obsolete*

CORRODUR C 14 W
Bergische Stahl Industrie
C 0.12, Cr 12.5, Ni 0.5, bal Fe.
Annealed: 71,000-92,000 TS; 42,500 YS min; 2 El min. Heat treated: 85,000-112,000 TS; 57,000 YS min; 4 El min. Corrosion resistant casting. W.-Nr. 1.4008; DIN G-X 12 Cr14. ASTM A296 Gr.CA 15.

CORRODUR C 17.13 E3S
Bergische Stahl Industrie
C 0.05, Cr 16.5, Ni 13.5, Mo 2.5, bal Fe.
Solution annealed and aged: 69,000 TS min; 30,000 YS min; 25 El min. Corrosion resistant casting. DIN G-X5CrNiMo 1713. *Obsolete*

CORRODUR C 18
Bergische Stahl Industrie
C 0.22, Cr 17, Ni 1.25, bal Fe.
Annealed: 85,000-106,000 TS; 57,000 YS min; 2 El min. Heat treated: 106,000-128,000 TS; 85,000 YS min; 4 El min. Corrosion resistant casting. Werkstoff Nr. 1.4059; DIN G-X25CrNi17. *Obsolete*

CORRODUR C 18 E
Bergische Stahl Industrie
C 0.37, Cr 16.5, Mo 1.15, bal Fe.
Annealed: 85,000-106,000 TS; 64,000 YS min; 2 El min. Heat treated: 106,000-128,000 TS; 85,000 YS min; 4 El min. Corrosion resistant casting. Werkstoff Nr. 1.4122; DIN G-X35CrMo17. *Obsolete*

CORRODUR C 18.8 ESSD

Bergische Stahl Industrie
C 0.05, Cr 17, Ni 11.5, Mo 2.25, Mn 1.5, Nb, bal Fe.
Solution annealed and aged: 69,000 TS min; 30,000 YS min; 20 El min. Corrosion resistant casting. DIN G-X5CrNiMoNb1712. *Obsolete*

CORRODUR C 18.8 O

Bergische Stahl Industrie
C 0.1, Cr 18, Ni 8.5, Mo 0-0.4, bal Fe.
Corrosion resistant casting. *Obsolete*

CORRODUR C 18.8 S

Bergische Stahl Industrie
C 0.07, Cr 18, Ni 9.5, bal Fe.
Solution annealed and aged: 70,000 TS min; 30,000 YS min; 25 El min. Corrosion resistant casting. Werkstoff Nr. 1.4308; DIN G-X6CrNi 189. ASTM A296 Gr. CF 8. *Obsolete*

CORRODUR C 18.8 SS

Bergische Stahl Industrie
C 0.8, Cr 18, Ni 10, Nb, bal Fe.
Solution annealed and aged: 70,000 TS min; 30,000 YS min; 20 El min. Corrosion resistant casting. Werkstoff Nr. 1.4552; DIN G-X7CrNiNb189. ASTM A296 Gr. CF 8C. *Obsolete*

CORRODUR C 18.8 SSO

Bergische Stahl Industrie
C 0.08, Cr 18, Ni 10, Mo 0-0.4, Nb, bal Fe.
Corrosion resistant casting. *Obsolete*

CORRODUR C 22.7

Bergische Stahl Industrie
C 0.16, Cr 20.5, Ni 4.25, bal Fe.
Solution annealed and aged: 85,000-100,000 TS; 54,000 YS min; 25 El min. Corrosion resistant casting. DIN G-X20CrNi214. *Obsolete*

CORRODUR C 24.7 E

Bergische Stahl Industrie
C 0.27, Cr 23, Ni 7, Mo 2.25, Cu 1.35, bal Fe.
As cast: 85,000-100,000 TS; 57,000 YS min; 4 El min.
Solution annealed and aged: 85,000-100,000 TS; 42,500 YS min; 6 El min. Corrosion resistant casting. DIN G-X30CrNiMo247. *Obsolete*

CORRODUR C 30 EH

Bergische Stahl Industrie
C 1.9, Cr 29, Mo 2, Cu 1.35, bal Fe.
Hardenable to 500 Brin max. Corrosion resistant casting. DIN G-X190CrMo292. *Obsolete*

CORRODUR C 30 H

Bergische Stahl Industrie
C 1.8, Cr 29, bal Fe.
Hardenable to 500 Brin max. Corrosion resistant casting. DIN G-X180Cr29. *Obsolete*

CORRODUR C14

Bergische Stahl Industrie
C 0.22, Cr 14.5, bal Fe.
Annealed: 78,000-92,000 TS; 50,000 YS min; 3 El min. Heat treated: 85,000-115,000 TS; 64,000 YS min; 12 El min. Corrosion resistant casting. W.-Nr. 1.4027; DIN G-X 22 Cr 14.

CORRODUR N 15

Bergische Stahl Industrie
C 2.9, Si 2.4, Ni 14.5, Cu 6, Cr 2, bal Fe.
Austenitic nickel-copper cast steel. DIN G-X300 NiCuCr156. *Obsolete*

CORRODUR SP 120

Bergische Stahl Industrie
C 1.2, Cr 29, bal Fe.
As cast: 57,000-64,000 TS. Corrosion resistant casting. Werkstoff Nr. 1.4086; DIN G-X120Cr29. *Obsolete*

CORROFEST 13/1

Gusstahl-Handels GmbH
C 0.1, Cr 13, bal Fe.
Annealed: 75,000 TS; 40,000 YS; 35 El; 70 RA; 155 Brin. Cold drawn: 100,000 TS; 85,000 YS; 17 El; 60 RA; 205 Brin. For furnace parts and heat treat boxes. X10Cr13; heat resistant.

CORROFEST 13/2

Gusstahl-Handels GmbH
C 0.2, Si 0.4, Cr 13, bal Fe.
Annealed: 95,000 TS; 50,000 YS; 25 El; 50 RA; 195 Brin. Heat treated: 250,000 TS; 215,000 YS; 8 El; 25 RA; 500 Brin. For cutlery, knives, valves, and surgical instruments. X20Cr13; corrosion resistant, hardenable.

CORROFEST 13/2 MO

Gusstahl-Handels GmbH
C 0.2, Cr 13, Mo 1.15, bal Fe.
Annealed: 100,000 TS; 55,000 YS; 22 El; 48 RA; 200 Brin. For cutlery, knives, and chemical plant equipment. X20CrMo13; corrosion resistant; hardenable.

CORROFEST 13/4

Gusstahl-Handels GmbH
C 0.4, Si 0.4, Cr 13, bal Fe.

CORROFEST 17/1

Gusstahl-Handels GmbH
C 0-0.1, Si 0.4, Cr 17.5, bal Fe.
Annealed: 80,000 TS; 45,000 YS; 30 El; 55 RA; 170 Brin. For furnace parts and heat treat boxes. X8Cr17; corrosion and heat resistant.

CORROFEST 17/1 MO

Gusstahl-Handels GmbH
C 0-0.1, Si 0.4, Cr 17.5, Mo 1, Ti = 7 x C, bal Fe.
Annealed: 80,000 TS; 45,000 YS; 30 El; 55 RA; 170 Brin. For chemical plant equipment. X8CrMoTi17; corrosion and heat resistant.

CORROFEST 17/2

Gusstahl-Handels GmbH
C 0.22, Si 0.4, Cr 17, Ni 1.5, bal Fe.
Annealed: 80,000 TS; 45,000 YS; 30 El; 55 RA; 170 Brin. For stainless machine tool parts. X22CrNi17; corrosion resistant; heat treatable.

CORROFEST 17/9

Gusstahl-Handels GmbH
C 0.9, Si 0.4, Cr 18, Mo 1.15, V 1, bal Fe.
For bearings, bushings, and pivots. X90CrMoV18; corrosion resistant; hardenable.

CORROFEST 18/8

Gusstahl-Handels GmbH
C 0-0.15, Si 0.4, Cr 18, Ni 8, bal Fe.
Annealed: 80,000 TS; 35,000 YS; 55 El; 65 RA; 150 Brin. Cold drawn: 180,000 TS; 125,000 YS; 10 El; 330 Brin. For chemical plant equipment, tanks, vessels, and piping. Type 302; stainless; austenitic.

CORROFESTAL

Deutsche Messingwerke
Mg 0.5-2, Si 0.3-1.5, Mn 0-1.5, bal Al.
For light alloy parts; corrosion resistant, hardenable.

CORROFOND M-10

Montecatini Settore Alluminio
Mg 10, Mn 0.4, bal Al.
Heat treated: 50,000-56,000 TS; 21,000-26,000 YS; 6-13 El; 75-85 Brin. For marine and aircraft parts; heat treatable, resists sea water corrosion.

CORROFOND M3

Montecatini Settore Alluminio
Mg 3, Mn 0.3, bal Al.
Cast: 21,000-29,000 TS; 8,000-12,000 YS; 6-10 El; 45-55 Brin. For light alloy parts, corrosion resistant.

CORROFOND M5

Montecatini Settore Alluminio
Mg 5, Mn 0.4, bal Al.
Heat treated: 28,000-33,000 TS; 13,000-17,000 YS; 8-12 El; 60-80 Brin. For light alloy parts; corrosion resistant.

CORROFOND M7

Montecatini Settore Alluminio
Mg 7, Mn 0.4, bal Al.
Heat treated: 37,000-42,000 TS; 19,000-22,000 YS; 5-11 El; 70-80 Brin. For marine hardware; resists sea water corrosion, age-hardenable.

CORROFOND S2

Montecatini Settore Alluminio
Si 2, Mg 0.65, Mn 0.7, bal Al.
Heat treated: 36,000-42,000 TS; 29,000-39,000 YS; 1-3 El; 90-105 Brin. For light alloy parts.

CORROFOND S4

Montecatini Settore Alluminio
Si 4.2, Mn 0.5, Ti 0.15, bal Al.
Cast: 19,000-26,000 TS; 7000-13,000 YS; 0-15 El; 45-60 Brin. For light alloy parts; corrosion resistant.

CORROFOND S45

Montecatini Settore Alluminio
Si 4.5, Mg 0.65, Mn 0.7, bal Al.
Heat treated: 36,000-46,000 TS; 26,000-36,000 YS; 1-2 El; 90-105 Brin. For light alloy parts; corrosion resistant, age-hardened.

CORRONEL 210

Henry Wiggin & Co. Ltd.
Mo 28, Fe 6, bal Ni.
For chemical plant equipment, petroleum equipment; high corrosion resistance to hydrochloric acid and acid chlorides. *Obsolete*

CORRONEL B

Gilby-Fodor S.A.
Ni 66, Mo 28, Fe 6.
Rolled: 134,400 TS; 62,000 YS; 40 El; 40 RA; 250 Brin. For heat exchangers, pump parts, and reaction vessels. Corrosion resistant.

CORRONEL B

Henry Wiggin & Co. Ltd.
Ni 66, Mo 28, Fe 6.
For chemical plant equipment; acid resistant, resists hydrochloric and phosphoric acids. *Obsolete*

CORRONEL-220

Henry Wiggin & Co. Ltd.
Mo 28, Fe 0-3, V 1.6-2.25, bal Ni.
Rolled: 157,000 TS; 82,000 YS; 37 El; 38 RA. For chemical and petroleum plant equipment, pickling plant. Acid resistant. *Obsolete*

CORRONEL-230

Henry Wiggin & Co. Ltd.
C 0-0.08, Si 0-0.6, Cu 0-1, Fe 0-5, Mn 0-1, Ti 0-1, Al 0-0.5, Cr 35-37, bal Ni.
Annealed: 105,700 TS; 46,000 YS; 43 El; 69 RA. For chemical and petrochemical industries, refinery equipment, mixers, agitators. High corrosion resistance. It resists non-oxidizing acid liquors. *Obsolete*

CORRONIL

Henry Wiggin & Co. Ltd.
Ni 70, Cu 26, Mn 4.
Hot rolled: 81,000 TS; 36,000 YS; 46 El; 66 RA; 150 Brin. For corrosion resisting vessels, electric resistance purposes, chemical equipment; corrosion and heat resistant. *Obsolete*

CORRONIUM

English manufacture
Cu 80, Zn 15, Sn 5.
4,000 TS; 58 El; 57 RA; 80 Brin. Cast or wrought.

CORRONON 184E

Stahlwerk Stahlschmidt GmbH & Co.
C 0-0.12, Cr 18, Mo 2, Ni 10.5, bal Fe.
Annealed: 80,000 TS; 30,000 YS; 60 El; 80 RA; 135 Brin. Cold drawn: 150,000 TS; 135,000 YS; 6 El; 300 Brin. For chemical plant equipment, mixers, agitators; filters; Type 316; stainless, austenitic.

CORRONON 184S

Stahlwerk Stahlschmidt GmbH & Co.
C 0-0.07, Cr 18, Mo 2, Ni 10.5, bal Fe.
Annealed: 80,000 TS; 30,000 YS; 60 El; 80 RA; 135 Brin. Cold drawn: 150,000 TS; 135,000 YS; 6 El; 300 Brin. For chemical plant equipment, mixers, agitators, filters; Type 316L stainless steel; stainless, austenitic.

CORRONON 188

Stahlwerk Stahlschmidt GmbH & Co.
C 0.12, Cr 18, Ni 8, bal Fe.
Annealed: 80,000 TS; 35,000 YS; 55 51; 65 RA; 150 Brin. Cold drawn: 180,000 TS; 135,000 YS; 10 El; 330 Brin. For chemical plant equipment, tanks, vessels, mixers; X12CrNi18-8; austenitic, stainless.

CORRONON 188E

Stahlwerk Stahlschmidt GmbH & Co.
C 0-0.12, Cr 18, Ni 9.5, Ti = 4 x C, bal Fe.
Annealed: 85,000 TS; 35,000 YS; 55 El; 65 RA; 150 Brin. Cold drawn: 95,000 TS; 60,000 YS; 40 El; 60 RA; 185 Brin. For welded structures, chemical plant equipment; Type 321; austenitic, stainless.

CORRONON 188S

Stahlwerk Stahlschmidt GmbH & Co.
C 0-0.07, Cr 18, Ni 9.5, bal Fe.
Annealed: 80,000 TS; 35,000 YS; 55 El; 65 RA; 150 Brin. Cold drawn: 180,000 TS; 135,000 YS; 10 El; 330 Brin. For kitchen equipment, architectural trim; Type 304; stainless, austenitic.

CORRONON 20/10

Stahlwerk Stahlschmidt GmbH & Co.
C 0.15, Si 2, Cr 19.5, Ni 9.5, bal Fe.
Annealed: 85,000 TS; 40,000 YS; 50 El; 60 RA; 180 Brin. For chemical plant equipment; X15CrNiSi199; corrosion and heat resistant.

CORRONON 23/20

Stahlwerk Stahlschmidt GmbH & Co.
C 0.15, Si 2, Cr 24, Ni 19, bal Fe.
For furnace parts and equipment, heat treat boxes; corrosion and heat resistant, austenitic.

CORRONON-1

Stahlwerk Stahlschmidt GmbH & Co.
C 0-0.12, Si 0.4, Cr 13, bal Fe.
Annealed: 90,000 TS; 45,000 YS; 28 El; 55 RA; 180 Brin. For turbine blades, cutlery, knives; X10Cr13; corrosion resistant.

CORRONON-10

Stahlwerk Stahlschmidt GmbH & Co.
C 0-0.12, Si 1, Al 1, Cr 18, bal Fe.
Annealed: 80,000 TS; 45,000 YS; 30 El; 55 RA; 170 Brin. For oil refinery and chemical plant equipment; X10CrA118; corrosion and heat resistant.

CORRONON-11

Stahlwerk Stahlschmidt GmbH & Co.
C 0.2, Si 1.2, Cr 25, Ni 4, bal Fe.
For furnace parts and equipment, heat treat boxes; X20CrNiSi254; heat resistant.

CORRONON-12

Stahlwerk Stahlschmidt GmbH & Co.
C 0-0.12, Si 1.5, Al 1.5, Cr 24, bal Fe.
For furnace parts and equipment, heat treat boxes; X10CrA124; heat resistant.

CORRONON-16

Stahlwerk Stahlschmidt GmbH & Co.
C 0-0.1, Si 0.4, Cr 17.5, bal Fe.
Annealed: 80,000 TS; 45,000 YS; 30 El; 55 RA; 170 Brin. For chemical plant and oil refinery equipment; X8Cr17; corrosion and heat resistant.

CORRONON-16N

Stahlwerk Stahlschmidt GmbH & Co.
C 0.22, Ni 17, bal Fe.
For chemical plant equipment; X22Ni17; corrosion resistant.

CORRONON-16S

Stahlwerk Stahlschmidt GmbH & Co.
C 0-0.1, Cr 17.5, Ti = 7 x C, bal Fe.
Annealed: 80,000 TS; 45,000 YS; 30 El; 55 RA; 170 Brin. For welded structures; X8CrTi17; heat resistant, stabilized.

CORRONON-18

Stahlwerk Stahlschmidt GmbH & Co.
C 0-0.1, Cr 17.5, Mo 1, Ti = 7 x C, bal Fe.
Annealed: 85,000 TS; 48,000 YS; 28 El; 52 RA; 180 Brin. For chemical plant equipment; X8CrMoTi17; heat and corrosion resistant.

CORRONON-18H

Stahlwerk Stahlschmidt GmbH & Co.
C 0.9, Cr 18, Mo 1.15, V 1, bal Fe.
For bearings, linings; X90CrMoV18.

CORRONON-2

Stahlwerk Stahlschmidt GmbH & Co.
C 0.2, Cr 13, bal Fe.
Annealed: 95,000 TS; 50,000 YS; 25 El; 50 RA; 195 Brin. Heat treated: 250,000 TS; 215,000 YS; 8 El; 25 RA; 500 Brin. For cutlery, knives, turbine blades; Type 320; corrosion resistant.

CORRONON-8

Stahlwerk Stahlschmidt GmbH & Co.
C 0-0.12, Si 0.8, Al 0.8, Cr 5-6, bal Fe.
For oil refinery equipment; X10CrA17; creep resistant.

CORRONON-9

Stahlwerk Stahlschmidt GmbH & Co.
C 0.2, Cr 13, Al 1, bal Fe.
Annealed: 95,000 TS; 50,000 YS; 25 El; 50 RA; 195 Brin. Heat treated: 250,000 TS; 215,000 YS; 8 El; 25 RA; 500 Brin. For cutlery, knives, turbine blades, surgical instruments; X20CrA113; corrosion resistant.

CORROSALLOY 13

Sheepbridge Alloy Castings Ltd.
C 0-0.25, Cr 11.5-13.5, Ni 0-1, bal Fe.
Cast: 101,000 TS; 15 El; 160-260 Brin. For chemical plant equipment; Britain BS 1630 B; corrosion resistant.

CORROSALLOY 18

Sheepbridge Alloy Castings Ltd.
C 0-0.25, Ni 1-1.3, Cr 16-20, bal Fe.
Cast: 123,200 TS; 10 El; 250 Brin. For chemical plant and oil refinery equipment; Britain S 80; corrosion and heat resistant, hardenable.

CORROSALLOY DU

Sheepbridge Alloy Castings Ltd.
C 0-0.1, Ni 28-30, Cr 19-21, Mo 2-3, Cu 3.5-4.5, bal Fe.
Cast: 62,800 TS; 30 El; 160 Brin. For chemical plant equipment; corrosion resistant to H_2SO_4.

CORROSALLOY HT

Sheepbridge Alloy Castings Ltd.
Si 0-1, Mn 0-1, Cr 15-17.5, Ni 3-5, Cu 3-5, C 0.07, bal Fe.
Precipitation hardening stainless steel casting. Hardenable to 300-350 Brin.

CORROSALLOY M

Sheepbridge Alloy Castings Ltd.
C 0-0.12, Ni 10-12, Cr 18-20, Mo 2.5-3.5, Cb = 10 x C, bal Fe.
Cast: 78,500 TS; 25 El; 180-220 Brin. For acid resistant chemical plant equipment, welded parts; Britain BS 163 B; stainless, austenitic.

CORROSALLOY NDP

Sheepbridge Alloy Castings Ltd.
C 0-0.12, Ni 8-10, Cr 18-20, Cb = 10 x C, bal Fe.
Cast: 78,500 TS; 30 El; 170-210 Brin. For welded chemical plant equipment, mixers; Britain BS 1631 B; stainless, austenitic.

CORROSALLOY S

Sheepbridge Alloy Castings Ltd.
C 0-0.15, Ni 8-10, Cr 18-20, bal Fe.
Cast: 78,500 TS; 30 El; 170-210 Brin. For chemical plant and oil refinery equipment; austenitic, stainless; Britain 1631 A.

CORROSALLOY S

Sheepbridge Alloy Castings Ltd.
C 0-0.08, Si 0-2, Mn 0-1.5, Ni 8-11, Cr 18-21, bal Fe.
Austenitic stainless casting for corrosion resisting applications. ASTM A351 CF8.

CORROSIL

Centrifugal Products Inc.
C 0.7, Si 14, bal Fe.
As cast: 514 Brin. Corrosion and wear resistant iron, corrosion control anodes.

CORROSIRON

Pacific Foundry Co.
C 0.8-1, Si 13.5-14.5, bal Fe.
Cast: 16,000 TS; 16,000 YS; 0 El; 0 RA; 350 Brin. For acid pans, drains, pumps, valves, cocks; corrosion resistant, not forgeable.

CORROSIRON

Bethlehem Foundry & Machine Co.
C 0.8-1, Si 13.5-14.5, bal Fe.
Cast: 16,000 TS; 16,000 YS; 0 El; 0 RA; 350 Brin. For acid pans, drains, pumps, valves, cocks; corrosion resistant, not forgeable.

CORROSIST B

Sheepbridge Alloy Castings Ltd.
C 0-0.15, Mo 28-32, Fe 0-6, bal Ni.
Cast: 78,500 TS; 6 El; 250-300 Brin. For pumps, valves, turbine parts, chemical plant equipment; resists HCl.

CORROSIST C

Sheepbridge Alloy Castings Ltd.
C 0-0.15, Mo 17-20, W 3-4, Fe 0-8, Cr 14-16, bal Ni.
Cast: 74,000 TS; 10 El; 250-300 Brin. For drains, pumps, chemical plant equipment; resists HCl and wet Cl_2.

CORROSIST D

Sheepbridge Alloy Castings Ltd.
C 0-0.15, Cu 2-3.5, Fe 0-5, Si 9-10.5, bal Fe.
Cast: 49,300 TS; 550-600 Brin. For valves, pumps, drains, chemical plant equipment; resists boiling H_2SO_4, hard and brittle.

CORROSIST IL

Sheepbridge Alloy Castings Ltd.
C 0-0.15, Cr 22-24, Mo 3.5-4.5, Cu 7-8, Fe 0-7, bal Ni.
Cast: 56,000 TS; 12 El; 190-220 Brin. For chemical plant equipment; resists H_2SO_4.

CORSAIR

Now ELECTRITE CORSAIR.

CORSAIR

Latrobe Steel Co.
Tool material. C 1.02, W 6.1, Cr 4, V 2.4, Mo 6, bal Fe.
For broaches, form tools; high speed steel.

CORSONITE B

British Steel Corp.
Cu-Co-Si.
For Cu alloys. *Obsolete*

CORSONITE C

British Steel Corp.
Cu-Fe-Si.
For Cu alloys. *Obsolete*

CORSONITE D
British Steel Corp.
Cu-Cr-Si.
For Cu alloys. *Obsolete*

CORSONITE E
British Steel Corp.
Cu-Cr.
For Cu alloys. *Obsolete*

CORSONITE F
British Steel Corp.
Cu-Co.
For Cu alloys. *Obsolete*

CORTEN
Now REPUBLIC CORTEN USS CORTEN.

CORVIC BRONZE
Chase Brass & Copper Co.
Cu 98.5, Sn 1.5.
For chemical engineering equipment; corrosion resistant.
Obsolete

COS-AR-COR
COSIPA-Companhia Siderurgica Paulista
Alloy steel. C 0-0.16, Mn 0-1.2, P 0-0.03, S 0-0.015, Si 0-0.5, Cu 0.2-0.5, Cr 0.4-0.7, Nb + V + Ti 0.15 max, bal Fe.
Weldable; resistant to atmospheric corrosion. 375 N/mm^2 min YS; 490-630 N/mm^2 TS.

COSHOCTON STAINLESS 320 (IN 840)
Coshocton Stainless
Stainless steel. Cr 19.5, Ni 19.5, C 0-0.08, bal Fe.
Heating element tubing for elevated temperature service.

COSHOCTON STAINLESS 332 (IN 800)
Coshocton Stainless
Stainless steel. Cr 20.5, Ni 31.5, C 0-0.1, bal Fe.
Heating element tubing for elevated temperature service.

COSHOCTON STAINLESS 434/AL (1)
Coshocton Stainless
Stainless steel. 60% stainless/40% aluminum.
For automotive trim. Improved corrosion resistance.

COSHOCTON STAINLESS 434/AL (2)
Coshocton Stainless
Stainless steel. 40% stainless/60% aluminum.
For automotive trim. Improved corrosion resistance.

COSINT 1000
Hoeganaes Corp.
Co 20, Cr 16, W 5.5, Mo 2.5, Al 5.5, Ti 2.5, B 0.02, Zr 0.3, bal Ni.
For high temperature applications; prealloyed powder, sintered, heat resistant.

COSMOLOY-F
American manufacture
C 0-0.04, Cr 15, Mo 3.8, W 2.2, Ti 3.4, Al 4.7, B 0.08, Zr 0.07, bal Ni.
For high temperature applications; heat and corrosion resistant.

COSMOS
Lumen Bearing Co.
Pb 76, Sn 9, Sb 15.
Cast: 11,000 TS; 1 El; 28 Brin. For shaft bearings; Pb base Babbitt. *Obsolete*

COTHIAS ALLOY
English manufacture
Al 93, Cu 6.5, Zn 0.5.
For light weight castings.

COUGAR
Pyramid Steel Company
C 1.55, Mn 0.04, Cr 11.5-12, V 0.9, Mo 0.8, bal Fe.
Chromium, cold work tool steel. AISI D2.

COURSIER
Compagnie Ateliers et Forges de la Loire
C 0.6-0.7, bal Fe.
Hardened: 58-60 Rock C. For cutting tools, drills, knives, hammers, chisels. Water hardening.

COUSSINAL A
French manufacture
Sn 3, Pb 2, Sb 3, bal Al.
For bearings; anti-friction.

COUSSINAL C
French manufacture
Sn 4, Cu 4, Mg 3, Mn 1, bal Al.
For bearings; antifriction.

COVAN
CCS Braeburn Alloy Steel
C 0.9, Cr 4, V 2, Mo 8, Co 8, bal Fe.
High speed steel for cutting tools, Mo-Co type, good red hardness. AISI M34. *Obsolete*

COVAN 55
Atlantic Steel Co.
HSLA steel. C 0-0.23, Mn 0-1.35, P 0-0.03, S 0-0.04, Si 0.15-0.3, Co 0.005-0.05, V 0.01-0.05, bal Fe.
High strength, low alloy steel for structural components in metal building systems. 55,000 psi YS min; 65,000 psi TS min, 18 El min. Meets ASTM A572-Grade 50.

COVANDUR
Carpenter Technology Corp.
C, alloy, bal Fe.
For machine tool parts; oil hardened. *Obsolete*

COWLES
English manufacture
Al 1.25-11, bal Cu.
For strong corrosion resistant parts.

COWLES "A-1"
American manufacture
Cu 80, Al 10.
For hardware, gears; Al-bronze.

COWLES "A-2"
American manufacture
Cu 80, Al 10.
For hardware, gears; Al-bronze.

COWLES "B"
American manufacture
Cu 92.5, Al 7.5.
For hardware, gears; Al-bronze.

COWLES "C-1"
American manufacture
Cu 94.5, Al 5.5.
For hardware, gears; Al-bronze.

COWLES "C-2"
American manufacture
Cu 95, Al 5.
For hardware, gears; Al-bronze.

COWLES "C-3"
American manufacture
Cu 95, Al 5.
For hardware, gears; Al-bronze.

COWLES "D"
American manufacture
Cu 97.5, Al 2.5.
For hardware, gears; Al-bronze.

COWLES "E"
American manufacture
Cu 98.75, Al 1.25.
For hardware; corrosion resistant.

COWLES ALUMINUM BRONZE
English manufacture
Cu 88.4, Al 9.74, Fe 0.43, Si 1.36.
For gears, trolley wheels, worm wheels; tough, wear resistant.

COWLES HIGH MANGANESE BRASS
American manufacture
Cu 67-80, Mn 15-18, Zn 5-13, Al 0.1, Si.
For strong corrosion resistant parts; corrosion resistant.

COWLES SPECIAL "A"
American manufacture
Cu 80, Al 11.
For hardware, gears; Al-bronze.

CP 401
Comalco Ltd.
Cu 0-0.1, Si 12-13, Fe 0-0.4, Mn 0-0.05, Mg 0-0.05, Ni 0-0.05, Zn 0-0.1, bal Al.
Permanent mold cast, F1: 207 MPa TS; 90 MPa YS; 9 El; 60 Brin. ADC1 H40-5; DS LM6.

CP 601
Comalco Ltd.
Cu 0.05, Si 6.5-7.5, Fe 0.12-0.2, Mg 0.3-0.4, Zn 0-0.05, Ti 0-0.05, Mn 0-0.05, bal Al.
Sand cast, T6: 255 MPa TS; 186 MPa YS; 5 El; 70 Brin.
Permanent mold cast, T6: 276 MPa TS; 186 MPa YS; 10 El; 100 Brin. BS LM25.

CP30, CP50
British Steel plc
Cr 1.5, Ni 0.75, Mo 0.5, bal Fe.
Armor plate steel. High hardness; tough.

CPI
Osborn Steels Ltd.
C 0.5, W 2.25, Cr 1.5, V 0.25, bal Fe.

CPM 10V
Crucible Materials Corp.
C 2.45, Mn 0.5, Si 0.9, Cr 5.25, V 9.75, Mo 1.3, S 0.07.
For tooling, rollforming rolls, wear parts. Tool steel. AISI A11.

CPM 9V
Crucible Materials Corp.
C 1.78, Mn 0.5, Si 0.9, Cr 5.25, V 9, Mo 1.3.
For knives, blades, tooling, rolls. Tool steel.

CPM REX 20
Crucible Materials Corp.
C 1.3, Mn 0.35, Si 0.25, Cr 3.75, V 2, W 6.25, Mo 10.5, S 0.07.
High speed steel. For use in machining operations requiring heavy cuts, high speeds and feeds, and for difficult to machine materials that are high in hardness and abrasion resistance. AISI M62.

CPM REX 25
Crucible Materials Corp.
C 1.8, Mn 0.3, Si 0.35, Cr 4, V 5, W 12.5, Mo 6.5, S 0.07.
High speed steel. Used in machining operations requiring heavy cuts, high speeds and feeds, and for difficult to machine materials high in hardness and abrasion resistance.

CPM REX 45 (HS)
Crucible Materials Corp.
C 1.3, Mn 0.4-0.7, Si 0.5, Cr 4.05, V 3.05, W 6.25, Mo 5, Co 8.25, S 0-0.03.
High speed steel. For tools, punches and dies, broaches, reamers, drills. Good wear resistance and tough.

CPM REX M2
Crucible Materials Corp.
Now REX M2.

CPM REX M2HS
Crucible Materials Corp.
C 1, Mg 0.8, S 0.27, Si 0.3, Cr 4.15, V 1.95, W 6.4, Mo 5.
High carbon; high speed steel. For tools, cutters, saws, drills, knives. AISI M2.

CPM REX M2S

Crucible Materials Corp.
Tool material. C 1, Cr 4.15, W 6.4, Mo 5, V 1.95, S 0.15, bal Fe.
Free machining Mo-W high speed steel. AISI M2.

CPM REX M35 HC HS

Crucible Materials Corp.
C 0.98, Cr 4.15, Mn 0.8, V 2, Si 0.3, W 6, S 0.27, Mo 5, Co 5.
For drills, tools, cutters. High speed steel. AISI M35.

CPM REX M35-2

Crucible Materials Corp.
Tool material. C 1.2, Cr 4, V 3, W 6.25, Mo 6.25, S 0.15, bal Fe.
Free machining high speed steel. AISI M3 C12.

CPM REX M35S

Crucible Materials Corp.
Tool material. C 0.85, Cr 4.15, V 2, W 6, Mo 5, Co 5, S 0.1, bal Fe.
High speed steel, free machining grade. For form tools, deep hole drills, hobs. AISI M35.

CPM REX M3HCHS

Crucible Materials Corp.
C 1.3, Mn 0.8, S 0.27, Si 0.55, Cr 4, V 3, W 6, Mo 5.
High speed steel. For tools, cutters, hobs, and broaches. AISI Type M3.

CPM REX M4

Crucible Materials Corp.
C 1.35, Mn 0.3, Si 0.3, Cr 4, V 4, W 5.75, Mo 4.5.
High speed steel. High resistance to cratering and wear in cold work punches, die inserts, and cutting applications involving high speeds and light cuts.

CPM REX M42

Crucible Materials Corp.
Now REX M42.

CPM REX T15

Crucible Materials Corp.
High speed steel. C 1.55, Mn 0.3, Si 0.3, Cr 4, V 5, W 12.25, Co 5.
High speed steel designed for use in machining operations requiring heavy cuts, high speeds and feeds, and for difficult to machine materials that are high in hardness and abrasion resistance. For tools, cutters, drills, hobs, taps.

CPM REX T15S

Now REX T15S.

CPM T440V

Crucible Materials Corp.
C 2.15, Mn 0.4, Si 0.4, Cr 17, V 5.5, Mo 0.4.
Stainless tool steel for food, plastics and chemical processing equipment, knives, bearings, bushings, cams, gears, and valves.

CPO

LaSalle Steel Co.
Cold finished, chrome-plated carbon steel bar; as per order.

CPT 64AV

Gould Inc.
C 0.02, Fe 0.3, Al 6, V 4, bal Ti.
Sintered: 119,000 TS; 107,000 YS; 23 Rock C. Heat treated: 146,000 TS; 133,000 YS; 27 Rock C. For aircraft bearing housings. Heat treatable. Corrosion resistant. Sintered powders.

CQ-12

TRW Inc.
Sintered carbide tool material. For roughing cuts on cast iron, steel and non-ferrous metals and for wear parts. *Obsolete*

CQ-13

TRW Inc.
Sintered carbide tool material. For wear parts; moderate shock. *Obsolete*

CQ-14

TRW Inc.
Sintered carbide tool material. For wear parts, will stand heavy shock. *Obsolete*

CQ-15

TRW Inc.
Sintered carbide tool material. For wear parts and light impact. *Obsolete*

CQ-16

TRW Inc.
Sintered carbide tool material. For wear parts and heavy impact applications. *Obsolete*

CQ-2

TRW Inc.
Sintered carbide tool material. General purpose machining of cast iron, non-ferrous and non-metallic materials. *Obsolete*

CQ-23

TRW Inc.
Sintered carbide tool material. For cutting and forming plastics, glass and highly abrasive materials and laminates. *Obsolete*

CQ-3

TRW Inc.
Sintered carbide tool material. For light roughing cuts and some finishing cuts on cast iron and non-ferrous metals. *Obsolete*

CQ-4

TRW Inc.
Sintered carbide tool material. Finish cutting and precision boring of cast iron and non-ferrous metals. *Obsolete*

CR 05

Herbert Cutanit Ltd.
Sintered carbide tool material. Hardness: above 1650 VHN. For continuous finishing cuts on ferritic steel and steel castings. ISO PO1. *Obsolete*

CR 10

Herbert Cutanit Ltd.
Sintered carbide tool material. Hardness: 1525-1625 VHN. For finishing and light to medium roughing on ferritic steel and steel castings. ISO P10. *Obsolete*

CR 12

Acieries de Champagnole
C 2, Cr 12, Mn 0.3, Si 0.3, bal Fe.
Air, oil or salt, hardenable to 65 Rock C. For cold stamping and shearing metal sheet, thread rolling, sheet punches, wood working cutters and tools. AFNOR Z 200 C 12. Similar to AISI D3. German: W. Nr. 1.2080.

CR 13 CO 3

TradeARBED Inc.
C 1.5, Cr 12-13, Mo 1, Co 3, bal Fe.
Cold work tool steel, punching and coining dies. AISI D5. *Obsolete*

CR 15

Herbert Cutanit Ltd.
Sintered carbide tool material. Hardness: 1525-1625 VHN. For general-purpose machining of ferritic steel with light and medium cuts. ISO P10 - P20. *Obsolete*

CR 2

Sumitomo Metal America Inc.
C 0-0.2, Si 0-0.75, Mn 0-1.4, Cu 0.2-0.7, Ni 0-0.65, Cr 0.2-1.2, Mo 0-0.2, 0.15 Cb + V max, bal Fe.
50 mm thick plate max, sheet and coil: 59,000-104,000 TS; 31,000-63,000 psi YP min. Atmospheric corrosion resisting steel for structures, bridges. Available in four strength grades: CR2-41, CR2-50, CR2-53, CR2-60.

CR 20

Herbert Cutanit Ltd.
Sintered carbide tool material. Hardness: 1475-1575 VHN. For medium to heavy cuts at high speeds on ferritic steel. ISO P20. *Obsolete*

CR 25

Herbert Cutanit Ltd.
Sintered carbide tool material. Hardness: 1450-1550 VHN. For general-purpose machining with medium feeds and medium speeds, including some interrupted cuts. ISO P20, P30. *Obsolete*

CR 30

Herbert Cutanit Ltd.
Sintered carbide tool material. Hardness: 1425-1525 VHN. For heavy interrupted cuts at medium to low speeds, using throwaway tips. ISO P30. *Obsolete*

CR 4 B

Sumitomo Metal America Inc.
C 0-0.15, Si 0-0.55, Mn 0-1.5, Cr 0.8-1.5, 0.2 Cu min, 0.15 Cb + V max, bal Fe.
50 mm thick plate max, sheet or coil: 59,000-89,000 TS; 33,000-43,000 YP min. Seawater resisting steel. Available in two strength grades: CR4B-41, CR4B-50.

CR 4 T

Sumitomo Metal America Inc.
C 0-0.08, Si 0.35-0.75, Mn 0-1.2, Cu 0-0.4, Cr 1.8-2.3, Mo 0.1-0.3, bal Fe.
6-100 mm thick plate: 71,000-89,000 TS; 31,000 YP min. Seawater resisting steel.

CR 40

Herbert Cutanit Ltd.
Sintered carbide tool material. Hardness: 1400-1500 VHN. For heavy interrupted turning and planing of steel where high metal removal is required. ISO P30, P40 and M30. *Obsolete*

CR 40

Walsingham Steel Co., Ltd.
C 0.36-0.43, Si 0-0.5, Mn 0.6-0.75, Cr 0.5-0.7, bal Fe.
Carbon chromium steel; abrasion resistant.

CR 5 MO

TradeARBED Inc.
C 1, Mo 1, Cr 5, bal Fe.
Air hardening type cold work tool steel. AISI A2. *Obsolete*

CR 50

Herbert Cutanit Ltd.
Sintered carbide tool material. Hardness: 1175-1275 VHN. For heavy cutting at low speeds, interrupted cutting, and under bad conditions; tough. ISO P50, M40. *Obsolete*

CR 50

Walsingham Steel Co., Ltd.
C 0.45-0.55, Si 0-0.75, Mn 0.5-1, Cr 0.8-1.2, bal Fe.
Carbon chromium steel; abrasion resistant. Meets BS 1956 Gr. A, B; ASTM A148-65 Gr. 105-85.

CR 6 MO

TradeARBED Inc.
Stainless steel. Mn 0-1, Si 0-1, Cr 4-6, Mo 0.4-0.65, 0.10 C min, bal Fe.
Martensitic stainless steel. AISI 501. *Obsolete*

CR Z

TradeARBED Inc.
C 2.25, Cr 12, bal Fe.
Cold work tool steel, punching and coining dies. AISI D3. *Obsolete*

CR Z S

TradeARBED Inc.
High chromium cold work tool steel. *Obsolete*

CR-MO BEARING
Bethlehem Steel Corp.
C 1.05, Cr 1.35, Mo 0.25, bal Fe.
For bearing races, master gauges; wear resisting. *Obsolete*

CR-MO-W
Peninsular Steel Co.
C 0.33, Si 1.05, W 1.55, Mo 1.65, Cr 5, bal Fe.
For hot work dies; air hardening.

CR-NI ROTOR FM
Carpenter Technology Corp.
C 0.55, Mn 0.6, Si 0.3, Cr 1.5, Ni 4, bal Fe.
Air hardening tool steel. *Obsolete*

CR-VICALLOY
Japanese manufacture
Co 52, 12 (Cr + V), bal Fe.
For electrical and magnetic equipment. Permanent magnet.

CR. N.O
Acieries de Champagnole
C 0.38, Cr 1.6, Ni 3.7, Mo 0.3, Mn 0.4, Si 0.3, bal Fe.
Deep hardening alloy steel, air or oil hardening. For cold forming tools, cold-heading and rivet sets. AFNOR: 35 NCD 15.

CR. T
Acieries de Champagnole
C 0.5, Cr 0.7, W 2, V 0.15, Mo 0.35, Mn 0.35, Si 0.5, bal Fe.
Oil or water hardening tool steel. For pneumatic tools, riveting hammers, and chisels. AFNOR: 50 WCS 20.03; AISI S1.

CR.5
Acieries de Champagnole
C 1, Cr 5, V 0.5, Mo 1.1, Mn 0.6, Si 0.3, bal Fe.
Air or oil hardening tool steel. For stamping and drawing dies, shears, and cold forming tube dies. AFNOR: Z 100 CDV 05; AISI A2.

CR.5.T.M
Acieries de Champagnole
C 0.35, Cr 5, W 1.4, V 0.4, Mo 1.4, Mn 0.25, Si 1, bal Fe.
Air or oil hardening tool and die steel. For hot work forming and forging. AFNOR: Z 35 CDWVS 05; AISI H12.

CR.5.T.M.S
Acieries de Champagnole
C 0.5, Cr 4.7, W 1.1, V 0.35, Mo 1.2, Mn 0.3, Si 0.8, bal Fe.
Air or oil hardening tool steel. For hot and cold working shear blades, dies, ejectors, bending tools, and forming tools. AFNOR: Z 50 CDWVS 05.

CR.5.V.4
Acieries de Champagnole
C 1.15, Cr 5, V 3.9, Mo 1.1, Mn 0.4, Si 1, bal Fe.
Air or oil hardening tool steel. For stamping and drawing dies requiring good abrasive wear; for crushers and grinders in stone and gravel work. AFNOR: Z 110 CVDS 05.04.

CR.5.V.6
Acieries de Champagnole
C 1.42, Cr 6.25, V 5.9, Mo 1, Mn 0.3, Si 1, bal Fe.
Air or oil hardening tool and die steel. For stamping silicon laminations and abrasive materials. AFNOR: Z 135 CVDS 06.06.

CR.5.V.M
Acieries de Champagnole
Cr 0.35-5.5, V 1, Mo 1.1, Mn 0.3, Si 1, bal Fe.
Air or oil hardening tool and die steel. For hot work stamping copper and copper alloys and molds for plastics. AFNOR: Z 35 CDVS 05; AISI H13.

CR.8.V.3
Acieries de Champagnole
C 0.83, Cr 8, V 2.5, Mo 1.5, Mn 0.3, Si 1, bal Fe.
Air hardening tool and die steel. For cutting and shaping tools, and shears. AFNOR: Z 85 CVDS 08.03.

CR.E
Acieries de Champagnole
C 2, Cr 12.5, W 1, V 0.3, Mo 0.3, Mn 0.5, Si 0.4, bal Fe.
Air or oil hardening tool steel. For Zendzimir rolls, wood working tools, and brick making tools. AFNOR: Z 200 CW 13; AISI D3.

CR.E.15
Acieries de Champagnole
C 1.55, Cr 12, V 0.8, Mo 0.8, Mn 0.25, Si 0.3, Co 0.8, bal Fe.
Air or oil hardening tool and die steel. For stamping dies for silicon laminations, swaging dies, and gages.

CR.E.4.V
Acieries de Champagnole
C 2.3, Cr 12.5, V 4.1, Mo 1.1, Mn 0.7, Si 0.4, bal Fe.
Oil or air hardening tool and die steel. For punches, shears, cutting tools for abrasive materials, and work in sand and gravel. AFNOR: Z 230 CVD 13.04; AISI D7.

CR1A
Sumitomo Metal America Inc.
C 0-0.13, Si 0.2-0.8, Mn 0-1.4, Cu 0.25-0.35, Cr 1-1.5, bal Fe.
50 mm thick plate max, sheet and coil: 59,000-89,000 TS; 31,000-49,000 YP min. Sulfuric acid resisting steel for chimneys, chemical equipment, etc. Available in two strength grades: Cr1A-41, Cr1A-50.

CR2R-H
Sumitomo Metal America Inc.
C 0-0.12, Si 0.25-0.75, Mn 0.2-0.5, p 0.07-0.15, Cu 0.25-0.55, Ni 0-0.45, Cr 0.3-1, bal Fe.
19 mm thick plate max, sheet or coil: 70,000 TS max; 50,000 YP max. Atmospheric corrosion resisting steel for vehicles.

CR4A
Sumitomo Metal America Inc.
C 0-0.55, Mn 0-1.2, P 0.07-0.15, Ni 0-0.65, Cr 0.3-0.8, 0.2 Cu min, 0.15 Cb + V max, bal Fe.
19 mm thick plate max, sheet or coil: 59,000-89,000 TS; 33,000-47,000 YP min. Seawater resisting steel. Available in two strength grades: Cr4A-41, Cr4A-50.

CR550
British Steel plc
C 0-0.18, Si 0-0.5, Mn 0-1.6, P 0-0.025, S 0-0.008, Nb 0-0.06, Al 0-0.06, V 0.12, bal Fe.
High-strength weldable structural steel.

CR690
British Steel plc
C 0-0.2, Si 0-0.5, Mn 0-1.5, P 0-0.035, S 0-0.015, V 0-0.2, bal Fe.
High-strength weldable structural steel.

CRAIG GOLD
Manufacturer not listed
Cu 80, Ni 10, Zn 10.
For jewelry, ornaments.

CRALCLAD
Japanese manufacture
Al alloy.
For light alloy parts; clad duralumin.

CRALFER
English manufacture
C 3, Mn, Si, Al, bal Fe.
For machinery castings; cast iron, growth resistant.

CRAMP ALLOY NO. 100 (P.M.G. METAL)
Baldwin-Lima-Hamilton Corp.
Cu 78, Zn 2, 20% special alloy.
45,000-58,000 TS; 22,000-28,000 YS; 18-30 El; 90-100 Brin.
For marine parts, tail shaft liners, stem bushings, valves, valve parts, pump casings, gears, pressure vessels, bearings. *Obsolete*

CRAMP ALLOY NO. 101 (PARSONS TURBADIUM)
Baldwin-Lima-Hamilton Corp.
Copper. Cu 47-52, Zn 40-46, Ni 1-2.5.
Cast: 65,000-75,000 TS; 33,000-40,000 YS; 15-18 El; 90 Brin.
For pump impellers, turbine runners, buckets for water wheels; resists erosive action of water.

CRAMP ALLOY NO. 114
Baldwin-Lima-Hamilton Corp.
Nickel. Ni 67, Cu 29, Fe 1.5, Mn 0.9, Si 1.25.
Cast: 65,000-80,000 TS; 32,500-45,000 YS; 25-40 El; 25-40 RA; 125-150 Brin. For impellers, valve seats; acid resistant; "Monel Metal."

CRAMP ALLOY NO. 115
Baldwin-Lima-Hamilton Corp.
Copper. Sn 5, Ni 5, Zn 2, bal Cu.
Heat treated: 50,000-85,000 TS; 22,000-55,000 YS; 10-40 El; 95-150 Brin. For gears, slippers, bearings, valve bodies; hardenable, heavy duty.

CRAMP ALLOY NO. 118
Baldwin-Lima-Hamilton Corp.
Nickel. Ni 63, Cu 30, Fe 2, Mn 0.9, Si 3-4.
Cast: 100,000-130,000 TS; 75,000-95,000 YS; 1-4 El; 250-325 Brin. For valve seats; S-Monel; acid resistant.

CRAMP ALLOY NO. 119
Baldwin-Lima-Hamilton Corp.
Cu 29.5, Fe 1.5, Mn 1, Si 2.3, bal Ni.
Cast: 90,000-115,000 TS; 40,000-60,000 YS; 15-25 El; 190-250 Brin. For steam nozzles, chemical plant equipment; corrosion resistant. *Obsolete*

CRAMP ALLOY NO. 12
Baldwin-Lima-Hamilton Corp.
Copper. Cu 87.75, Sn 6.25, Zn 4, Pb 1.5, Ni 0.5.
Cast: 32,000-38,000 TS; 16,000-18,000 YS; 20-25 El; 40-50 Brin. For valves, fittings, cocks, pressure castings; similar to U.S. Navy Composition "M."

CRAMP ALLOY NO. 13
Baldwin-Lima-Hamilton Corp.
Copper. Cu 83, Sn 13.5, Zn 3.5.
Cast: 23,000-32,000 TS; 13,000-15,000 YS; 2-4 El; 55-65 Brin.
For bearings for heavy duty as for bridges, trunnions, machines, center disks; corrosion resistant to weak sulfurous acid.

CRAMP ALLOY NO. 151
Baldwin-Lima-Hamilton Corp.
Copper. Cu 78-82, Al 8-11, Fe 3-6, Ni 3-6.
Heat treated: 80,000-115,000 TS; 35,000-70,000 YS; 5-20 El; 172-300 Brin. For ship propellers; heat treatable, nongalling, corrosion resistant.

CRAMP ALLOY NO. 174
Baldwin-Lima-Hamilton Corp.
Copper. Be 0.2-0.35, Ni 1.4-1.5, bal Cu.
Cast: 50,000 TS; 20,000 YS; 16 El; 100 Brin. Heat treated: 80,000 TS; 55,000 YS; 17 El; 210 Brin. For welding wheels; heat treatable, 40% electrical conductivity.

CRAMP ALLOY NO. 175
Baldwin-Lima-Hamilton Corp.
Copper. Cr 0.3-0.9, Fe 0-0.5, Si 0-0.5, bal Cu.
Cast: 50,000 TS; 40,000 YS; 15 El; 90 Brin. For welding wheels; heat treatable, 75% electrical conductivity.

CRAMP ALLOY NO. 176
Baldwin-Lima-Hamilton Corp.
Aluminum. Cu 4-5, Fe 1, Si 0-1.5, Ti 0-0.25, bal Al.
Cast: 32,000 TS; 16,000 YS; 8 El; 60 Brin. For cylinder heads, gear cases; age hardenable.

CRAMP ALLOY NO. 177
Baldwin-Lima-Hamilton Corp.
Aluminum. Zn 4-5, Mn 0.4-0.65, Cr 0.3-0.6, bal Al.
Cast: 31,000 TS; 7 El. For gear cases, general castings; age hardens at room temperature.

CRAMP ALLOY NO. 20
Baldwin-Lima-Hamilton Corp.
Copper. Cu 85, Sn 5, Zn 5, Pb 5.
Cast: 30,000 TS; 25 El. For bearings for general service; similar to "Ounce" metal.

CRAMP ALLOY NO. 250
Baldwin-Lima-Hamilton Corp.
C 3, Si 1, Mn 0.8, Ni 1.2, bal Fe.
Cast: 40,000-50,000 TS; 40,000-50,000 YS; 0 El; 0 RA; 175-225 Brin. For locomotive cylinders and parts; alloy cast iron. *Obsolete*

CRAMP ALLOY NO. 252
Baldwin-Lima-Hamilton Corp.
C 2.75-3.25, Si 1-2, Mn 0.8, bal Fe.
Cast: 30,000-40,000 TS; 30,000-40,000 YS; 0 El; 0 RA; 170-230 Brin. For locomotive cylinders, bushings; cast iron. *Obsolete*

CRAMP ALLOY NO. 257
Baldwin-Lima-Hamilton Corp.
TC 2.5-3, Si 1-2, Mn 0.75-1, bal Fe.
Cast: 40,000-50,000 TS; 175-225 Brin. For castings; cast iron. *Obsolete*

CRAMP ALLOY NO. 261
Baldwin-Lima-Hamilton Corp.
TC 3.3, Si 2, Mn 0.9, Ni 2, Mo 0.5, bal Fe.
Cast: 40,000 TS; 180-220 Brin. For engine cylinder heads, combustion chambers; cast iron. *Obsolete*

CRAMP ALLOY NO. 263
Baldwin-Lima-Hamilton Corp.
TC 3-3.4, Si 1.2-1.7, Mn 0.6-0.9, Ni 2.75-3.25, Mo 0.2-0.3, bal Fe.
Cast: 40,000-50,000 TS; 200-275 Brin. For auto die blocks, die inserts, bushings; cast iron. *Obsolete*

CRAMP ALLOY NO. 265
Baldwin-Lima-Hamilton Corp.
C 2.5-3.25, Si 1-2, Mn 0.75-1.25, Ni 12-15, Cr 1-3, Cu 5-7, bal Fe.
Cast: 20,000-35,000 TS; 125-200 Brin. For furnace equipment, high temperature applications; alloy cast iron, corrosion resistant. *Obsolete*

CRAMP ALLOY NO. 268
Baldwin-Lima-Hamilton Corp.
TC 2.8-3.3, Si 1-2, Mn 1, Mo 0.5-1, Cr 0-0.5, bal Fe.
For diesel engine liners, engine frames, bearing caps; cast iron. *Obsolete*

CRAMP ALLOY NO. 271
Baldwin-Lima-Hamilton Corp.
TC 2.5-3, Si 1-1.5, Ni 25-35, Cr 0-6, Mn 1, bal Fe.
Cast: 20,000-40,000 TS; 130-200 Brin. For cylinder liners, piston inserts; austenitic cast iron. *Obsolete*

CRAMP ALLOY NO. 272
Baldwin-Lima-Hamilton Corp.
TC 2.75-3.25, Si 1-2, Mn 0.75-1, bal Fe.
Cast: 35,000-45,000 TS; 170-250 Brin. For cylinders and valve bushings; cast iron. *Obsolete*

CRAMP ALLOY NO. 273
Baldwin-Lima-Hamilton Corp.
TC 2.8-3.2, Si 1-2, Mn 0.7-1, Ni 1-2, Mo 0.65, Cr 0-0.5, bal Fe.
Cast: 40,000-50,000 TS. For diesel engine cylinder heads; cast iron. *Obsolete*

CRAMP ALLOY NO. 274
Baldwin-Lima-Hamilton Corp.
TC 2.5-3, Si 1-2, Mn 0-1, Mo 0.5-1, Cr 0-0.5, Ni 1-2, bal Fe.
Cast: 45,000-60,000 TS. For diesel engine parts; cast iron. *Obsolete*

CRAMP ALLOY NO. 275
Baldwin-Lima-Hamilton Corp.
TC 2.6-3.8, Si 0.15-1.5, Ni 3-6, Cr 0.5-2.5, Mn 0.15-1.5, bal Fe.
Cast: 50,000-80,000 TS; 500-600 Brin. For pump plungers, rolls for metal rolling, grinding plates; wear and abrasion resistant; "Ni-Hard."

CRAMP ALLOY NO. 275
Baldwin-Lima-Hamilton Corp.
TC 2.6-3.8, Si 0.15-1.5, Ni 3-6, Cr 0.5-2.5, Mn 0.15-1.5, bal Fe.
Cast: 50,000-80,000 TS; 500-600 Brin. For pump plungers, rolls for metal rolling, grinding plates; wear and abrasion resistant; "Ni-Hard."

CRAMP ALLOY NO. 276
Baldwin-Lima-Hamilton Corp.
Cr 20-30, C 1-3, Mn 0.5-1, Si 0.5-2, Ni 0-3, bal Fe.
Cast: 50,000-75,000 TS; 200-500 Brin. For pug mill knives, liners, sand blast nozzles, pump impellers. *Obsolete*

CRAMP ALLOY NO. 278
Baldwin-Lima-Hamilton Corp.
TC 2.5-3, Si 1.5-2.5, Ni 1.5-4, Mo 0.5-1.2, Mn 1, Cr 0-0.15, bal Fe.
Cast: 50,000 TS; 200-400 Brin. For hydraulic cylinders, crankshafts, gears, rolls. *Obsolete*

CRAMP ALLOY NO. 45 (PARSONS MN BRONZE)
Baldwin-Lima-Hamilton Corp.
Copper. Cu 55-60, Zn 38-42, Pb 0-0.2, Al 0-1.5, Mn 0-3.5, Sn 0-1.5, Fe 0.4-2.
Cast: 65,000-75,000 TS; 33,000-38,000 YS; 20-30 El; 100 Brin. For ship propellers, heavy duty bushings, bearings, spur and bevel gears, valve stems, valve bodies; corrosion resistant, tough.

CRAMP ALLOY NO. 49
Baldwin-Lima-Hamilton Corp.
Copper. Cu 60-70, Zn 20-30, bal special alloy.
Also known as SUPER STRENGTH BRONZE. Cast: 90,000-115,000 TS; 45,000-68,000 YS; 10-25 El; 185-240 Brin. For housing nuts, trunnion bearings, expansion plates, worms, worm wheels, spur and bevel gears, valve stems; resists high compressive loads and shock.

CRAMP ALLOY NO. 49A
Baldwin-Lima-Hamilton Corp.
Copper. Cu 60-70, Zn 20-30, Fe, Al, bal Mn.
Cast: 120,000 TS; 90,000 YS; 10 El; 10 RA; 240 Brin. For bushings, worms, bearings, gears, valve bodies; Mn bronze, tough.

CRAMP ALLOY NO. 5
Baldwin-Lima-Hamilton Corp.
Copper. Cu 79.3, Sn 10, Pb 10, P 0.7.
Cast: 28,000-32,000 TS; 18,000-19,000 YS; 4-8 El; 70 Brin. For bearings for medium speeds and light loads, cross-head and mill bearings; SAE 64.

CRAMP ALLOY NO. 50
Baldwin-Lima-Hamilton Corp.
Copper. Cu 60-70, Zn 20-30, Fe, Al, bal Mn.
Cast: 90,000 TS; 45,000 YS; 25 El; 20 RA; 175 Brin. Forged: 93,000 TS; 50,000 YS; 20 El; 20 RA; 195 Brin. For bushings, worms, bearings, gears, valve bodies; Mn bronze, tough.

CRAMP ALLOY NO. 50A
Baldwin-Lima-Hamilton Corp.
Copper. Cu 60-70, Zn 20-30, Fe, Al, bal Mn.
Cast: 100,000 TS; 50,000 YS; 20 El; 18 RA; 150 Brin. Forged: 105,000 TS; 60,000 YS; 18 El; 18 RA; 210 Brin. For bushings, worms, bearings, gears, valve bodies; Mn bronze, tough.

CRAMP ALLOY NO. 54 (PARSONS WHITE BRASS)
Baldwin-Lima-Hamilton Corp.
Tin. Sn 61-78, Cu 4.5-5, Sb 6-11, Pb 3.5-13.
Cast: 12,250 TS; 10,750 YS; 2.7 El; 32-38 Brin. For bearings for heavy duty service, main and cross-head bearings; Babbitt.

CRAMP ALLOY NO. 69 (HYDRAULIC METAL)
Baldwin-Lima-Hamilton Corp.
Copper. Cu 86.5, Sn 11, Zn 2, Pb 0.5.
35,000-40,000 TS; 16,500-18,500 YS; 8-10 El; 60-65 Brin. For valve bodies, sleeves, pump parts, cylinder bodies, bearings; for high hydraulic pressures.

CRAMP ALLOY NO. 7
Baldwin-Lima-Hamilton Corp.
Copper. Cu 75, Pb 18, Sn 7.
Cast: 17,000-32,000 TS; 8,000-12,000 YS; 7-10 El; 45-50 Brin. For bearings of low compressive load and where lubrication is poor; acid corrosion resistant.

CRAMP ALLOY NO. 71
Baldwin-Lima-Hamilton Corp.
Copper. Cu 87.5, Sn 11, Ni 1.5.
Also known as CRAMP'S NICKEL BRONZE. Cast: 45,000-50,000 TS; 25,000-27,000 YS; 15-20 El; 85-95 Brin. For worms, worm gears, bevel and spur gears, high speed bearings, bushings wearing parts; for high contact speeds and heavy loads.

CRAMP ALLOY NO. 73
Baldwin-Lima-Hamilton Corp.
Copper. Sn 10, Pb 10, P 0.7, bal Cu.
Cast: 28,000-32,000 TS; 18,000-19,000 YS; 4-8 El; 70 Brin. For bearings, bushings, shaft sleeves; SAE 64; bearing bronze.

CRAMP ALLOY NO. 78
Baldwin-Lima-Hamilton Corp.
Copper. Cu 85-89, Al 8-10, Fe 1-2.
Cast: 65,000-95,000 TS; 47,000-57,000 YS; 4-20 El; 130-190 Brin. For acid resistant containers and parts; resistant to acid corrosion at high temperature.

CRAMP ALLOY NO. 79
Baldwin-Lima-Hamilton Corp.
Sn 8-11, Pb 3-6, bal Cu.
Cast: 26,500-32,000 TS; 17,500-19,500 YS; 5-8 El; 40-50 Brin. For bearings, bushings, sleeves, liners; for high speeds. *Obsolete*

CRAMP ALLOY NO. 83
Baldwin-Lima-Hamilton Corp.
Copper. Cu 87-89, Al 9-11, Fe 3-4.
Heat treated: 75,000-85,000 TS; 35,000-40,000 YS; 8-20 El; 165-185 Brin. For bearings, worm gears, spur gears; resists acid corrosion.

CRAMP ALLOY NO. 84
Baldwin-Lima-Hamilton Corp.
Copper. Cu 78-82, Al 9-11, Fe 3-6, Ni 3-6.
Heat treated: 90,000-115,000 TS; 40,000-70,000 YS; 5-20 El; 5-15 RA; 172-300 Brin. For structural parts; acid resisting.

CRAMP ALLOY NO. 91
Baldwin-Lima-Hamilton Corp.
Copper. Cu 87.5, Sn 8, Zn 4, Ni 0.5.
Cast: 35,000-45,000 TS; 17,500-21,000 YS; 20-30 El; 60-70 Brin. For shaft liners, bushings, pump parts, valve fittings, flanges, connections; similar to "Composition G."

CRAMP ALLOY NO. 97 (CRAMP COPPER ALLOY)
Baldwin-Lima-Hamilton Corp.
Copper. Cu 99-99.8.
17,000-22,000 TS; 5,000-7,000 YS; 30-40 El. For electrical installations for high conductivity; 75-85% electrical conductivity.

CRAMP ALLOY NO. 99
Horace T. Potts Co.
Copper. Si 3.5, Fe 1.5, Zn 4, bal Cu.
Cast: 45,000-55,000 TS; 20,000-28,000 YS; 15-23 El; 120-130 Brin. For gears, bushings, cross-head bearings; silicon bronze.

CRAMP'S ELFUR IRON NO. 11
Baldwin-Lima-Hamilton Corp.
C 3-3.5, Si 1.75-2.25, P 0.25-0.35, Mn 0.65-0.75, Ni 1-2, bal Fe.
Cast: 30,000-35,000 TS; 175-185 Brin. For diesel engine piston skirts; Tr.S. 2000-2400; Tr. D. 0.22-0.24. *Obsolete*

CRAMP'S ELFUR IRON NO. 4
Baldwin-Lima-Hamilton Corp.
C 3.15-3.5, Si 1.25-1.75, P 0.2-0.3, Mn 0.6-0.8, bal Fe.
Cast: 30,000-35,000 TS. For piston, piston bodies, casings; Tr.S. 2300-2800; Tr.D. 0.22-0.24. *Obsolete*

CRAMP'S ELFUR IRON NO. 5
Baldwin-Lima-Hamilton Corp.
C 3.25-3.75, Si 2-2.5, P 0.4-0.6, Mn 0.6-0.8, bal Fe.
Cast: 27,500-35,500 TS; 145-180 Brin. For parts for diesel engines, bodies, bearings, levers; Tr.S. 2000-2500; Tr.D. 0.20-0.22. *Obsolete*

CRAMP'S ELFUR IRON NO. 6
Baldwin-Lima-Hamilton Corp.
C 3.25-3.75, Si 1.9-2.4, P 0.25-0.35, Mn 0.5-0.65, Ni 0.2-0.5, bal Fe.
Cast: 30,000-35,000 TS; 175-200 Brin. For locomotive casings, piston valves, piston rings, packing rings; Tr.S. 2100-2600; Tr.D. 0.22-0.24. *Obsolete*

CRAMP'S ELFUR IRON NO. 7
Baldwin-Lima-Hamilton Corp.
C 3-3.5, Si 1.2-1.4, P 0.25-0.35, Mn 0.6-0.8, Ni 0.75-1.25, bal Fe.
Cast: 32,000-37,000 TS; 180-270 Brin. For locomotive cylinder bushings, hydraulic cylinders, crank cases; Tr.S. 2200-2700; Tr.D. 0.22-0.24. *Obsolete*

CRAMP'S ELFUR IRON NO. 8 (NI-RESIST)
Baldwin-Lima-Hamilton Corp.
C 2.5-3.25, Si 1-2, P 0.04-0.3, Mn 0.75-1.25, Ni 12-15, Cr 1-3, Cu 5-7, bal Fe.
Cast: 20,000-35,000 TS; 125-200 Brin. For heat resistant and corrosion resistant parts; corrosion resistant; Tr.S. 2000-2500. *Obsolete*

CRAMP'S ELFUR IRON NO. 9
Baldwin-Lima-Hamilton Corp.
C 3.25-3.75, Si 1.75-2.25, P 0.2-0.3, Mn 0.6-0.8, Ni 2-3, Cr 0.2-0.4, bal Fe.
Cast: 30,000-37,000 TS; 150-190 Brin. For oil pump cylinder bushings, liners, sleeves; Tr.S. 2000-2500; Tr.D. 0.23-0.25. *Obsolete*

CRAMP'S SUPER-STRENGTH BRONZE, GRADE "A"
Baldwin-Lima-Hamilton Corp.
Copper. Cu 57, Zn 28, bal hardeners of Mn, Fe, Al.
Cast: 115,000 TS; 68,000 YS; 10 El; 10 RA; 240 Brin.
Wrought: 118,000 TS; 70,000 YS; 8 El; 8 RA. For busings, bearings, gears, worms, die castings, shafts, valve stems, tube sheets; Mn bronze; C.Y.P. 65000.

CRAMP'S SUPER-STRENGTH BRONZE, GRADE "B"
Baldwin-Lima-Hamilton Corp.
Copper. Cu 57, Zn 28, bal hardeners of Mn, Fe, Al.
Cast: 110,000 TS; 63,000 YS; 15 El; 15 RA; 225 Brin.
Wrought: 115,000 TS; 65,000 YS; 15 El; 15 RA. For bushings, bearings, gears, worms, die castings, shafts, valve stems, tube sheets, pump impellers; Mn bronze; C.Y.P. 60000.

CRAMP'S SUPER-STRENGTH BRONZE, GRADE "C"
Baldwin-Lima-Hamilton Corp.
Copper. Cu 57, Zn 28, bal hardeners of Mn, Fe, Al.
Cast: 100,00 TS; 55,000 YS; 20 El; 18 RA; 200 Brin. Wrought: 105,000 TS; 60,000 YS; 18 El; 18 RA. For bushings, bearings, gears, worms, die castings, pump impellers, valve stems; Mn bronze; C.Y.P. 50000.

CRAMP'S SUPER-STRENGTH BRONZE, GRADE "D"
Baldwin-Lima-Hamilton Corp.
Copper. Cu 57, Zn 28, bal hardeners of Mn, Fe, Al.
Cast: 90,000 TS; 45,000 YS; 25 El; 20 RA; 185 Brin. Wrought: 93,000 TS; 50,000 YS; 20 El; 20 RA. For bushings, bearings, valve stems, shafts, gears, worms, die castings; Mn bronze; C.Y.P. 40000.

CRANE
Manufacturer not listed
C 0.7-1.2, bal Fe.
78,000 TS; 41,000 YS; 150 Brin. For tools, cold chisels; water-hardening.

CRANE "ARCTIC"
Crane Co.
C 0-0.15, Mn 0.5-0.8, Ni 3-4, bal Fe.
Cast: 483-655 MPa TS; 276 MPa YS; 24 El. Valves, fittings for low temperature service. ASTM A352, Grade LC3.

CRANE 18-8 MO
Crane Co.
C 0.1, Cr 18-20, Ni 9-10, Mo 2-4, bal Fe.
Cast: 85,000 TS; 48,000 YS; 53 El; 150 Brin. For valves, fittings; heat and abrasion resistant. *Obsolete*

CRANE 5 CR-MO
Crane Co.
C 0.3, Cr 4-6, Mo 0.55, bal Fe.
Cast: 110,000 TS; 80,000 YS; 18 El; 245 Brin. For valves, fittings, piping; corrosion resistant. *Obsolete*

CRANE CARBON CAST STEEL
Crane Co.
C 0.25, Mn 0.65, Si 0.4, bal Fe.
Cast: 76,000 TS; 45,000 YS; 30 El; 50 RA; 155 Brin. For valves, fittings. *Obsolete*

CRANE CAST MANGANESE BRONZE
Crane Co.
Zn, Mn, Pb, bal Cu.
60,000-80,000 TS. For valve stems, yoke sleeves and bushings; wear resistant. *Obsolete*

CRANE EXCELLOY (CAST)
Crane Co.
C 0-0.15, Mn 0-1, Si 0-1, Ni 0-1, Cr 11.5-14, Mo 0.15-1, bal Fe.
Cast: 621-793 MPa TS; 448 MPa YS; 18 El. Corrosion resistant cast alloy for valves, fittings. ASTM A487, Grade CA15M.

CRANE EXELLOY (WROUGHT)
Crane Co.
C 0-0.12, Mn 0-1, Si 0-1, Ni 0-0.5, Cr 11.5-13.5, bal Fe.
Forged or rolled: 690 MPa TS; 552 MPa YS; 15 El. Corrosion resistant alloy for valves, fittings. ASTM A182, Grade F6.

CRANE FERROSTELL
Crane Co.
P 0-0.75, S 0-0.15, C, Si, bal Fe.
Higher test gray iron. Cast: 214 MPa TS; 3300 transverse strength. ASTM A 126, Class B.

CRANE HARDENED STAINLESS STEEL
Crane Co.
Cr 16-18, C 0.6-0.8, bal Fe.
For discs, seats, valves; resists wear and galling. *Obsolete*

CRANE HIGH TENSILE
Crane Co.
P 0-0.75, S 0-0.15, C, Si, bal Fe.
High test cast iron. Cast: 283 MPa TS; 4000 transverse strength. ASTM A125, Class C.

CRANE NO. 1
Torrey S. Crane Co.
Pb 75, Sn 12.5, Sb 12.5.
25.2 Brin. For bearings; Babbitt.

CRANE NO. 2
Crane Co.
C 0.3, Ni 2, Cr 0.75, Mo 0.25, bal Fe.
Cast: 100,000 TS. For castings, valve fittings; for service at 750-1100°F, high pressure service.

CRANE NO. 2
Torrey S. Crane Co.
Pb 75, Sn 10, Sb 15.
25.6 Brin. For bearings; Babbitt.

CRANE NO. 3
Crane Co.
C 0-0.25, Mn 0.5-0.8, Si 0-0.6, Ni 2-3, bal Fe.
Cast: 483-655 MPa TS; 276 MPa YS; 24 El. Cast nickel steel for low temperature operations. ASTM A352, Grade LC2.

CRANE NO. 3
Torrey S. Crane Co.
Pb 80, Sn 5, Sb 15.
25.4 Brin. For bearings; Babbitt.

CRANE NO. 3 NICKEL ALLOY
Crane Co.
C 0.3, Ni 2-2.5, bal Fe.
Cast: 75,000 TS; 55,000 YS; 32 El; 55 RA. For valves, fittings; low temperature applications. *Obsolete*

CRANE NO. 4 CARBON-MOLYBDENUM
Crane Co.
C 0.15-0.3, Mo 0.4-0.6, Mn 0.5-0.7, Si 0.2-0.6, bal Fe.
Cast: 80,000 TS; 50,000 YS; 26 El; 50 RA; 165 Brin. For valves and fittings; good creep resistance. *Obsolete*

CRANE NO. 44 NICKEL ALLOY
Crane Co.
Ni, Sn, bal Cu.
For valve seats; wear resistant, resists dilute acids and alkalis. *Obsolete*

CRANE NO. 49 (NICKEL ALLOY)
Crane Co.
Cu 30.5-34, Sn 10.5-13, Ni 52-56, Mn 0.3-0.75.
60,000 psi TS; 414 MPa TS. For valves, fittings; corrosion resistant.

CRANE NO. 5
Crane Co.
C 0-0.2, Mn 0.4-0.7, Cr 4-6.5, Mo 0.45-0.65, bal Fe.
Cast: 621-793 MPa TS; 414 MPa YS; 18 El. For valves, fittings. ASTM A217, Grade C5.

CRANE NO. 5 CHROME-MOLYBDENUM
Crane Co.
C 0.15-0.3, Cr 4-6.5, Mo 0.4-0.65, Mn 0.45-1, bal Fe.
Cast: 105,000 TS; 78,000 YS; 22 El; 52 RA; 212 Brin. For valves, fittings; for temperature up to 1100 F. *Obsolete*

CRANE NO. 6
Torrey S. Crane Co.
Sn 85, Pb 0.25, Sb 10.75, Cu 4.
30.8 Brin. For bearings; Babbitt.

CRANE NO. 7
Torrey S. Crane Co.
Sn 81.5, Pb 4, Sb 10.5, Cu 4.
34.6 Brin. For bearings; Babbitt.

CRANE NO. 7 CHROME-MOLYBDENUM
Crane Co.
C 0-0.18, Cr 1-1.5, Mo 0.4-0.6, Mn 0.75, bal Fe.
Cast: 78,000 TS; 52,000 YS; 28 El; 60 RA; 185 Brin. For valves, fittings; for temperature up to 1000°F. ASTM A217, Grade WC6.

CRANE NO. 8
Torrey S. Crane Co.
Sn 40, Pb 48.5, Sb 10.5, Cu 1.
23.1 Brin. For bearings; Babbitt.

CRANE NO. 9 CHROME-MOLYBDENUM

Crane Co.
Cr 2-3, Mo 0.8-1.1, Mn 0.7, C 0-0.18, bal Fe.
Cast: 85,000 TS; 40,000 YS; 20 El; 55 RA; 180 Brin. For valves, fittings; high temperature steam service.

CRANE SPECIAL BRASS

Crane Co.
Sn, Pb, bal Cu.
38,000 TS; 17,000 YS; 10 El. For valves, fittings; for temperatures up to 550 F. *Obsolete*

CRANE SPECIAL BRONZE

Crane Co.
Sn 5.5-6.5, Pb 1-2, Ni 0-1, Zn 3-5, Cu 86-90.
Cast: 342 MPa TS; 110 MPa YS; 22 El. Steam or valve bronze castings. ASTM B61, Alloy 922. Formerly CRANE SPECIAL BRASS.

CRANE STEAM BRASS

Crane Co.
Sn 5, Pb 5, Zn 5, bal Fe.
Cast: 37,000 TS; 16,000 YS; 26 El. For valves, fittings; for temperature up to 450 F. *Obsolete*

CRANE STEAM BRONZE

Crane Co.
Sn 4-6, Pb 4-6, Zn 4-6, Ni 0-1, Cu 84-86.
Cast: 207 MPa TS; 97 MPa YS; 20 El. For valves, fittings. ASTM B62, Alloy 836. Formerly CRANE STEAM BRASS.

CRANE-1

Uddeholm Corp.
C, alloy, bal Fe.
For tool machinery parts; oil hardened. *Obsolete*

CRANE-2

Uddeholm Corp.
C 0.5, Cr 1.05, Ni 3.25, bal Fe.
For gears, bolts, crankshafts; oil hardened, shock resistant. *Obsolete*

CRANELOY 20

Crane Co.
C 0-0.07, Ni 29, Cr 20, Mo 2-3, Cu 4, bal Fe.
Cast: 65,000-75,000 TS; 28,000-38,000 YS; 35-50 El; 40-50 RA; 120-150 Brin. For chemical plant equipment, mixers; resists mixed acids, austenitic. *Obsolete*

CRANELOY 20 STAINLESS (FORGED-WROUGHT)

Crane Co.
C 0.07, Mn 0-2, Si 0-1, Ni 32-38, Cr 19-21, Mo 2-3, bal Fe.
Wrought: 500 MPa TS; 240 MPa YS; 30 El. Austenitic stainless for chemical equipment. Similar to ASTM B473.

CRANELOY 20 STAINLESS STEEL

Crane Co.
C 0-0.07, Mn 0-1.5, Si 0-1.5, Ni 27.5-30.5, Cr 19-22, Mo 2-3, Cu 3-4, bal Fe.
Cast: 431 MPa TS; 172 MPa YS; 35 El. Austenitic stainless for chemical equipment. ASTM A361, Grade CN7M.

CRASCO ALLOY

Detroit Steel Co.
C, W, Cr bal Fe.
For tools, chisels; shock resistant.

CRASCO BLACK LABEL

Detroit Steel Co.
C 1.5, Mn 0.4, Cr 11-13, Mo 0.8, V 0.2, bal Fe.
For tools, dies, hot working dies and long run dies; air hardening, non-shrinking.

CRASCO BROACH STEEL

Detroit Steel Co.
C, alloy, bal Fe.
For broaches, punches, dies; resists heavy pressures.

CRASCO DRILL ROD

Detroit Steel Co.
c 1.05, Mn 0.25, Si 0.25, bal Fe.
For pivots, bearings; drill rod.

CRASCO FINISHING

Detroit Steel Co.
C 1.4, V 0.2, Cr 4, Si 0.5, Mn 0.2, bal Fe.
For cutters, engravers' tools; fast finishing tool steel.

CRASCO GREEN LABEL

Detroit Steel Co.
C 1, bal Fe.
For tools, dies; water hardening.

CRASCO HIGH SPEED

Detroit Steel Co.
C 0.7, Cr 4, W 18, V 1.1, bal Fe.
For tools, cutters, drills, reamers; high speed steel.

CRASCO HIGH SPEED MOLY NO. 2

Detroit Steel Co.
C 0.83, Cr 4.25, W 6.4, Mo 5, V 1.8, bal Fe.
For tools, cutters, drills, broaches; high speed steel.

CRASCO NO. 7

Detroit Steel Co.
C 5, Cr 0.65, Mn 0.4, V 0.15, 1, bal Fe.
For tools, dies; water or oil hardening.

CRASCO RED LABEL

Detroit Steel Co.
C 0.9, Mn 1.2, Cr 0.5, W 0.5, bal Fe.
For tools dies; cutters, punches; oil hardening, non-shrinking.

CRASCO SHANK STEEL

Detroit Steel Co.
C 0.95, bal Fe.
For tools, tool shanks; water hardening.

CRASCO SPECIAL VANADIUM

Detroit Steel Co.
C 0.9-1.05, V 0.15-0.25, bal Fe.
For tools, dies; water hardening.

CRASCO WHITE LABEL

Detroit Steel Co.
C 1, bal Fe.
For tools, dies; water hardeing.

CRASCO YELLOW LABEL

Detroit Steel Co.
C 0.45-0.5, Mn 0.25, Cr 0.95, V 0.2, bal Fe.
For gears, shafts; oil hardening.

CRASFLOY

Continental Foundry & Machine Co.
C, Ni, Cr, bal Fe.
For rolls; hard alloy.

CRESCENT

Spencer Clark Metal Industries Ltd.

Carbon tool steel available in six tempers and carbon ranges. No. 1: 1.20-1.40 C; No. 2: 1.05-1.15 C; No. 3: 0.90-1.00 C; No. 4: 0.80-0.90 C; No. 5: 0.70-0.80 C; No. 6: 0.60-0.70 C.

CRESCENT DOUBLE SPECIAL

Colt Industries
C 1.3, W 3.6, bal Fe.
For finishing tools, drawing dies; oil hardening. *Obsolete*

CRESCENT EXTRA

Colt Industries
C 0.7-1, bal Fe.
For punches, dies, cutters, general tools; water hardening. *Obsolete*

CRESCENT HOT WORK NO. 2

Colt Industries
C 0.95, Cr 3.75, bal Fe.
For tools, gripper and header dies, bloom shears; hot work steel. *Obsolete*

CRESCENT RIM ROLL

Colt Industries
C 1, Cr 0.5, bal Fe.
For shaping rolls, drawing dies; oil hardening. *Obsolete*

CRESCENT SPECIAL

Colt Industries
C 1, bal Fe.
For tools, dies; water hardening. *Obsolete*

CRESCENT SPECIAL CARBON TOOL

Colt Industries
C 0.9, Mn 0.35, Si 0.2, bal Fe.
For blanking and coining dies, reamers, taps, striking dies; water hardened. *Obsolete*

CRESCENT SPECIAL WIRE DIE

Colt Industries
C 2.3, W 0.5, Mo 0.3, bal Fe.
For wire drawing dies; oil hardening. *Obsolete*

CRESCENT STEEL

Colt Industries
W 6.7, C 2.1, Mn 2.7, Si 0.1, bal Fe.
For tools, dies. *Obsolete*

CRESCENT TOOL

Colt Industries
C 0.7-1, bal Fe.
For punches, dies, cutters, general tools; water hardening. *Obsolete*

CREST A.H.

St. Lawrence Steel Co.
C 1, Mn 0.7, Cr 5.5, V 0.3, Mo 1.15, bal Fe.
Air hardened tool steel for blanking and trimming dies, shear blades, rolling dies, broaches. AISI A2.

CREST O.H.

St. Lawrence Steel Co.
C 0.9, Mn 1.15, Cr 0.65, V 0.25, W 0.5, bal Fe.
Oil hardened tool steel for blanking and bending dies, punches, knurling tools, gages. AISI O1.

CRESTALOY

Crescent Tool Co.
C 0.5, Cr 1.5, V 0.2, bal Fe.
For tools, wrenches; oil hardened.

CRESTON 8

Allied Steel & Tractor Products Inc.
C 0.9-1.4, bal Fe.
For tools, dies, cutters; water hardening.

CRESTON NO. 5

Allied Steel & Tractor Products Inc.
C 0.75-1.1, bal Fe.
For cold battering tools; water hardening.

CRESTON NO. 6

Allied Steel & Tractor Products Inc.
C 0.8-1.2, bal Fe.
For tools, dies; water hardening.

CRESTON NO. 7

Allied Steel & Tractor Products Inc.
C 0.9-1.3, bal Fe.
For tools, dies, cutters; water hardening.

CREUSABRO 32
Creusot-Loire
C 0.2, Mn 1.2, Cr 1.3, Mo 0.25, bal Fe.
Normalized: 142,000 psi TS; 106,000 psi YS; 10 El. For cement plant equipment, ventilators, screens, coal and mine equipment. Wear resistant.

CREUSABRO 360
Creusot-Loire
C 0.2, Mn 1.6, Cr 1.3, Mo 0.2, Cu, B, bal Fe.
360 Brin. Weldable. For mine construction, public works.

CREUSABRO 41
Creusot-Loire
C 0.35, Mn 0.8, Ni 1.4, Cr 2.75, Mo 0.3, Si 0-0.4, bal Fe.
Heat treated: 235,000 TS; 192,000 YS; 4.5 El. For gears, wear parts, highly stressed components. Abrasion resistant. *Obsolete*

CREUSABRO 43
Creusot-Loire
C 0.4, Mn 0.4, Ni 3.75, Cr 1.7, Si 0-0.4, bal Fe.
Heat treated: 242,000 TS; 192,000 YS; 4.5 El. For gears, gear rings, cams, punches, stamping dies. Abrasion resistant. *Obsolete*

CREUSABRO M
Creusot-Loire
C 1.2, Mn 13, Si 0.6, bal Fe.
Quenched: 128,000 psi TS; 50,000 psi YS; 35 El. For shafts, scraper buckets, crushers, grinders, safes, helmets. Abrasion and wear resistant.

CREUSABRO ML
Creusot-Loire
C 0.45, Mn 0.6, Si 2, bal Fe.
Heat treated: 214,000 psi TS; 115,000 psi YS; 5 El. For conveyors, wear plates. Abrasion and wear resistant.

CREUSABRO MLD
Creusot-Loire
C 0.45, Mn 0.6, 0.6% Ni-Cr-Mo max, bal Fe.
Heat treated: 221,000 TS; 171,000 YS; 5 El. For mining chisels, wedges, springs. Abrasion resistant. *Obsolete*

CREUSANBRO 400
Creusot-Loire
C 0.28, Mn 0.75, Ni 1.4, Cr 1.4, Mo 0.5, bal Fe.
Treated: 380 Brin. Weldable. For strong welded structures.

CREUSELSO 22
Creusot-Loire
C 0-0.15, S 0-0.3, Mn 0.5, Si 0.1, bal Fe.
Normalized: 50,000-62,000 TS; 28,000 YS min. For engineering and mechanical structures requiring resistance to brittle fracture. Good weldability. *Obsolete*

CREUSELSO 26
Creusot-Loire
C 0-0.2, Si 0.2, Mn 0.8, bal Fe.
Normalized: 59,000-72,000 TS; 34,000 YS min. For engineering and mechanical structures requiring resistance to brittle fractures. Good weldability. *Obsolete*

CREUSELSO 31
Creusot-Loire
C 0-0.2, Si 0.3, Mn 1, bal Fe.
Normalized: 65,000-78,000 TS; 40,000 YS min. For engineering and mechanical structures requiring resistance to brittle fracture. Good weldability. *Obsolete*

CREUSELSO 34SS
Creusot-Loire
C 0-0.17, Mn 1.3, Si 0-0.35, Ni 0-0.5, bal Fe.
Normalized: 490-610 N/mm² psi TS; 335 N/mm² psi YS min; 24 El. Containers for liquefied gases (propane, etc.).

CREUSELSO 38
Creusot-Loire
C 0-0.2, Mn 1.2, Si 0.4, 0.70 Ni + Cr + Mo + Cu max, bal Fe.
Normalized: 490-640 N/mm² WTS. Weldable; for pressure vessels. AFNOR E375.

CREUSELSO 38 AK
Creusot-Loire
C 0-0.2, Mn 0-1.5, Si 0-0.55, bal Fe.
Normalized: 510-640 N/mm² psi TS; 335 N/mm² psi YS min; 18 El min. Parts to be used at low temperature. AFNOR E 375 FP.

CREUSELSO 38W
Creusot-Loire
C 0.2, Si 0-0.55, Mn 0-1.5, 0.70 Ni + Cr + Mo + Cu max, bal Fe.
Normalized: 510-640 N/mm² psi TS; 335 N/mm² psi YS min; 18 El min. Weldable, for pressure vessels. AFNOR E 375 C.

CREUSELSO 42
Creusot-Loire
C 0-0.22, Mn 1.4, Si 0.4, 0.70 Ni + Cr + Mo + Cu max, bal Fe.
Normalized: 540-650 N/mm² WTS. Weldable; for pressure vessels, construction. AFNOR E 420.

CREUSELSO 42 FP
Creusot-Loire
C 0-0.22, Mn 1.6, Si 0-0.55, bal Fe.
Normalized: 540-650 N/mm² psi TS; 370 N/mm² psi YS min; 17 El min. Parts to be used at low temperature. AFNOR E 420 FP.

CREUSELSO 42C
Creusot-Loire
C 0.22, Si 0-0.55, Mn 0-1.6, 0.70 Ni + Cr + Mo + Cu max, bal Fe.
Normalized: 540-650 N/mm² psi TS; 370 N/mm² psi YS min; 17 El min. Weldable; for pressure vessels. AFNOR E 420 C.

CREUSELSO 47
Creusot-Loire
C 0-0.18, Mn 1.5, Si 0.35, Ni 0.4, V 0.07, 0.70 Cr + Mo + Cu max, bal Fe.
Normalized: 550-710 N/mm² psi TS. Weldable; for marine platforms, mine structures. AFNOR E 460.

CREUSELSO 47 FP
Creusot-Loire
C 0-0.18, Mn 1.5, Si 0.35, Ni 0.4, V 0.07, bal Fe.
Normalized: 560-710 N/mm² psi TS; 440 N/mm² psi YS min; 24 El. Parts for cryogenic operation. AFNOR E 460 FP.

CREUSELSO 47C
Creusot-Loire
C 0-0.18, Mn 1.5, Si 0.35, Ni 0.4, V 0.07, 0.70 Cr + Mo + Cu max, bal Fe.
Normalized: 440 N/mm² psi YS min. Weldable; for pressure vessels; structures. AFNOR E 460 C.

CREUSELSO 50
Creusot-Loire
C 0.2, Mn 1.4, Si 0.4, Mo 0.25, bal Fe.
Rolled: 72,000 TS; 45,000 YS; 30 El; 150 Brin. Heat treated: 100,000 TS; 62,000 YS; 20 El; 200 Brin. For penstocks, pressure vessels, structural steelwork, bridges, derricks. Good weldability. Resists brittle fracture. *Obsolete*

CREUSEM E
Creusot-Loire
C 0-0.08, Mn 0-0.375, Si 0.225, bal Fe.
Normalized: 310 N/mm² psi TS min; 155 N/mm² psi YS min; 35 El min. Soft magnetic material; for relays.

CREUSEM G
Creusot-Loire
C 0-0.05, Mn 0-0.3, Si 0.2, bal Fe.
Normalized: 290 N/mm² psi TS min; 145 N/mm² psi YS min. Soft magnetic material; for relays.

CREUSOT 0.5 FO
Creusot-Loire
C 0-0.18, Cr 0.5, Mo 0.5, bal Fe.
Normalized and tempered: 265 N/mm² psi YS min. Plate for pressure vessels; weldable. AFNOR 15 CD 2-05. ASTM A 387 Gr. 2.

CREUSOT 0.5DF03
Creusot-Loire
C 0.4, Cr 0.5, Ni 0.6, Mo 0.2, bal Fe.
Oil hardened: 142,000-200,000 TS; 120,000 YS; 9 El. For gears, pinions, axles, crankshafts, fasteners, bolts. AISI 8640 steel. Tough and shock resistant. *Obsolete*

CREUSOT 0.5DF06
Creusot-Loire
C 0.2, Cr 0.5, Ni 0.5, Mo 0.2, bal Fe.
Oil hardened: 142,000-200,000 TS; 120,000 YS; 7 El. For gears, pinions, bolts, fasteners, crankshafts, engine components, AISI 8620. Case hardening, tough, shock resistant. *Obsolete*

CREUSOT 0.5F08
Creusot-Loire
C 0.15, Cr 0.55, Mo 0.55, bal Fe.
Heat treated: 65,000 TS; 40,000 YS; 22 El. For boilers, steam manifolds, reactors, superheater pipes, turbine casings. ASTM A356 Gr. 5. *Obsolete*

CREUSOT 1.1 FO
Creusot-Loire
C 0-0.18, Cr 1, Mo 0.5, bal Fe.
Normalized and tempered: 295 N/mm² psi YS min. Plate for pressure vessels; weldable. AFNOR 15 CD 4-05. ASTM A387 Gr. 12.

CREUSOT 1.1FO8
Creusot-Loire
C 0.13, Cr 1, Mo 0.55, bal Fe.
Heat treated: 65,000 TS; 45,000 YS; 22 El. For boilers, steam manifolds, turbine casings, reactors, superheater pipes. ASTM A 387 Gr. B. *Obsolete*

CREUSOT 1.2 MOV
Creusot-Loire
C 0.15, Mo 0.5, Mn 1.2, V 0.07, bal Fe.
Normalized and tempered: 345 N/mm² psi YS min. Plate for pressure vessels; weldable. AFNOR 15 MDV 4-05.

CREUSOT 1.2DF3
Creusot-Loire
C 0.4, Cr 0.6, Ni 1.2, bal Fe.
Oil hardened: 114,000-157,000 psi TS; 100,000 psi YS; 9 El. For gears, clutches, connecting flanges, bolts, fasteners, shafts. AISI 3140 steel, tough, shock resistant.

CREUSOT 1.2DF6
Creusot-Loire
C 0.2, Cr 0.6, Ni 1.2, bal Fe.
Heat treated: 107,000-142,000 psi TS; 92,000 psi YS; 10 El. For bolts, levers, shafts, pins, fasteners, gears. Case-hardening. AISI 3120 steel, tough, shock resistant.

CREUSOT 1.3FOV
Creusot-Loire
C 0.24, Cr 1.25, Mo 0.6, V 0.25, bal Fe.
Heated treated: 100,000 psi TS; 82,000 psi YS; 16 El; 25 Rock C. For rotors, hot bolts, hardware, gears, shafts. Tough, shock resistant.

CREUSOT 1.4DF06
Creusot-Loire
C 0.14-0.2, Ni 1.2-1.6, Cr 0.8-1.2, Mo 0.1-0.3, bal Fe.
Heat treated: 229,000 psi TS; 130,000 psi YS; 6 El. For gears, bolts, camshafts; case hardened, oil quenched.

CREUSOT 1.4DF07/8
Creusot-Loire
C 0.1-0.16, Ni 1.2-1.6, Cr 0.8-1.2, Mo 0.1-0.3, bal Fe.
Heat treated: 221,000 psi TS; 128,000 psi YS; 7 El. For gears, bolts, camshafts, cams; case hardened, oil quenched.

CREUSOT 1.4DF3
Creusot-Loire
C 0.3-0.38, Ni 1.2-1.6, Cr 0.8-1.2, bal Fe.
Heat treated: 295,000 psi TS; 200,000 psi YS; 4 El. For gears, bolts, machine tool parts; oil hardened, tough.

CREUSOT 1.4DF5
Creusot-Loire
C 0.22-0.3, Ni 1.2-1.6, Cr 0.8-1.15, bal Fe.
Heat treated: 250,000 psi TS; 166,000 psi YS; 6 El. For gears, bolts, machine tool parts; oil hardened, tough.

CREUSOT 1.4DF6
Creusot-Loire
C 0.16-0.22, Ni 1.2-1.6, Cr 0.85-1.2, bal Fe.
Heat treated: 222,000 psi TS; 143,00 psi YS; 7 El. For gears, bolts, machine tool parts; case hardened, oil quenched.

CREUSOT 1.4DF7
Creusot-Loire
C 0.11-0.18, Ni 1.2-1.6, Cr 0.85-1.2, bal Fe.
Heat treated: 213,000 psi TS; 128,000 psi YS; 7 El. For gears, bolts, machine tool parts; case hardened, oil quenched.

CREUSOT 1.4DF8
Creusot-Loire
C 0.07-0.12, Mn 0.6-0.9, Ni 1.2-1.6, Cr 0.85-1.2, bal Fe.
Heat treated: 192,000 psi TS; 121,000 psi YS; 8 El. For gears, bolts, machine tool parts; case hardened, oil quenched.

CREUSOT 1.5D06S
Creusot-Loire
C 0-0.2, Cr 0.6, Ni 1.5, Mo 0.25, bal Fe.
For gears, bolts, shafts, bridge members; resists low temperature.

CREUSOT 1.5DF02
Creusot-Loire
C 0.28, Cr 1.4, Ni 1.5, Mo 0.38, bal Fe.
Oil hardened: 200,000 psi TS; 170,000 psi YS; 8 El. For aircraft engine parts, propeller shafts, gears, linings, fasteners. Tough, shock resisting.

CREUSOT 1.5DF0V1S
Creusot-Loire
C 0-0.22, Cr 1.3, Ni 1.5, Mo 0.27, V 0.17, bal Fe.
For welded structures, bridge members; heat resistant to 550°C.

CREUSOT 1.5DFO
Creusot-Loire
C 0.55, Ni 1.5, Cr 1, Mo 0.3, bal Fe.
Annealed: 262 Brin; 110,000 TS. Hardened: 208,000 TS; C 45 Rock. For hot stamping dies, punches, brake dies, bushings, set screws. Type L 6 hot work tool steel, oil harden, shock resistant. *Obsolete*

CREUSOT 1.5DFOV4S
Creusot-Loire
C 0-0.15, Cr 1.3, Ni 1.5, Mo 0.27, V 0.17, bal Fe.
For welded structures, bridge members; heat resistant to 550 C. *Obsolete*

CREUSOT 1.7DFOV
Creusot-Loire
C 0.55, Ni 1.5, Cr 1, Mo 0.3, V 0.1, bal Fe.
For hot stamping dies, punches. Oil hardening hot-work tool steel.

CREUSOT 1.8DFO 2/3
Creusot-Loire
C 0.4, Cr 0.8, Ni 1.8, Mo 0.25, bal Fe.
Oil hardened: 156,000-214,000 psi TS; 128,000 psi YS; 9 El. For gears, shafts, crankshafts, fasteners. AISI 4340 steel. Tough and shock resistant.

CREUSOT 1.8DFO 6/7
Creusot-Loire
C 0.2, Cr 0.8, Mo 0.25, Ni 1.8, bal Fe.
Oil hardened: 156,000-214,000 psi TS; 128,000 psi YS; 8 El. For gears, pinions, cams, crankshafts, cam shafts, pawls. AISI 4320 steel. Case hardening, tough and shock resistant.

CREUSOT 12FFB
Creusot-Loire
C 1.6, Cr 12, Mo 0.7, V 0.3, bal Fe.
Annealed: 248 Brin, 120,000 TS. Air hardened: 278,000 TS; 214,000 YS; C 56 Rock. For shear blades, dies, wear plates, wood tools. Type D 2 cold work tool steel, oil or air hardening. Non-deforming. *Obsolete*

CREUSOT 12FFO
Creusot-Loire
C 1.9, Cr 12, Mo 1, bal Fe.
Annealed: 248 Brin. Hardened: C 64-65 Rock; 525,000 C.S. For shear blades, dies, wood tools, blanking and lamination dies, rolls, knurls, thread rolling dies. Wear and abrasion resistant. Type D 4 cold work tool steel, oil or air hardening; non-deforming. *Obsolete*

CREUSOT 13FF
Creusot-Loire
C 2, Cr 13, bal Fe.
Annealed: 248 Brin. Hardened: C 64-66 Rock. For punches, cutting off tools, flanging and rolling dies, ceramic molds, shear blades for silicon steels, blanking and forming dies. Type D 3 cold work tool steel, oil or air hardening, wear resistant. *Obsolete*

CREUSOT 13FFCO
Creusot-Loire
C 2, Cr 13.5, Mo 1, V 0.5, Co 3, bal Fe.
Annealed: 277 Brin. Hardened: C 64-66 Rock. For wood tools, shear blades, dies, blanking and forming dies, punches, ceramic molds. Cold working tool steel, oil or air harden. Non-deforming, wear resistant. *Obsolete*

CREUSOT 14FF
Creusot-Loire
C 1.5, Cr 14, Mo 1, Co 2, bal Fe.
Annealed: 277 Brin; 120,000 TS. Hardened: C 63 Rock; 200,000 TS; 1 El. For punches, cut-off blades, milling cutters, wood tools, shear blades, rolls, mandrels. Type D5 cold work steel, air or oil harden. *Obsolete*

CREUSOT 1F3
Creusot-Loire
C 0.4-0.48, Cr 0.75-1.2, Mn 0.6-0.9, Si 0-0.4, bal Fe.
Heat treated: 193,000 TS; 128,000 YS; 7 El. For gears, bolts, shafts, axles; oil hardened. *Obsolete*

CREUSOT 1F3/4
Creusot-Loire
C 0.34-0.42, Cr 0.75-1.2, Mn 0.6-0.9, Si 0-0.4, bal Fe.
Heat treated: 295,000 TS; 192,000 YS; 3 El. For gears, bolts, crankshafts; oil hardened. *Obsolete*

CREUSOT 1F4/5
Creusot-Loire
C 0.3-0.37, Cr 0.8-1.2, Mn 0.6-0.9, Si 0-0.4, bal Fe.
Heat treated: 280,000 TS; 175,000 YS; 4 El. For gears, bolts, crankshafts; oil hardened. *Obsolete*

CREUSOT 1F5/6
Creusot-Loire
C 0.25-0.32, Cr 0.8-1.2, Mn 0.6-0.9, Si 0-0.4, bal Fe.
Heat treated: 235,000 TS; 150,000 YS; 5 El. For gears, bolts, crankshafts; oil hardened. *Obsolete*

CREUSOT 1FO3
Creusot-Loire
C 0.37-0.45, Mn 0.6-0.9, Cr 0.8-1.2, Mo 0.15-0.35, bal Fe.
Heat treated: 200,000 TS; 144,000 YS; 8 El. For gears, bolts, machine tool parts; oil hardened, tough. *Obsolete*

CREUSOT 1FO4
Creusot-Loire
C 0.31-0.39, Cr 0.85-1.2, Mo 0.15-0.35, Mn 0.6-0.9, bal Fe.
Annealed: 90,000 TS; 58,000 YS; 25 El; B 90 Rock. Heat treated: 185,000 TS; 168,000 YS; 12 El; 370 Brin. For gears, countershafts, bolts, fasteners, shafts, axles. Oil hardening, shock resistant. *Obsolete*

CREUSOT 1FO5/6
Creusot-Loire
C 0.2-0.3, Mn 0.6-0.9, Cr 0.8-1.2, Mo 0.15-0.35, bal Fe.
Heat treated: 242,000 TS; 150,000 YS; 5 El. For gears, bolts, crankshafts, axles; oil hardened, tough. *Obsolete*

CREUSOT 1FO7
Creusot-Loire
C 0.15-0.22, Cr 0.85-1.2, Mo 0.15-0.35, bal Fe.
Heat treated: 228,000 TS; 143,000 YS; 6 El. For gears, bolts, camshafts, cams; case hardened, oil quenched. *Obsolete*

CREUSOT 1FO8
Creusot-Loire
C 0.08-0.15, Cr 0.8-1.2, Mo 0.15-0.35, bal Fe.
Heat treated: 186,000 TS; 114,000 YS; 7 El. For gears, bolts, camshafts, cams; case hardened, oil quenched. *Obsolete*

CREUSOT 1FV2
Creusot-Loire
C 0.45-0.55, Cr 1.2-1.6, Mn 0-0.4, V 0.2, Si 0-0.4, bal Fe.
Heat treated: 213,000 TS; 143,000 YS; 7 El. For gears, bolts, springs; oil hardened, shock resistant. *Obsolete*

CREUSOT 2.2 FO
Creusot-Loire
C 0-0.15, Cr 2.25, Mo 1, bal Fe.
Normalized and tempered; 295 N/mm^2 psi YS min; 510-610 N/mm^2 WTS; 21 El. Plate for pressure vessels, weldable. AFNOR 10 CD 9-10. ASTM 387 Gr. 22.

CREUSOT 2D10
Creusot-Loire
C 0.07-0.14, Ni 1.8-2.3, Mn 0-0.5, bal Fe.
Rolled: 100,000 TS; 50,000 YS; 16 El. For gears, bolts, camshafts, cams; case hardened, oil quenched. *Obsolete*

CREUSOT 2D11
Creusot-Loire
C 0-0.1, Ni 1.8-2.3, Mn 0-0.5, bal Fe.
Annealed: 65,000 TS; 40,000 YS; 36 El; 140 Brin. For cams, camshafts, gears, pinions, fasteners. Case hardening, tough, shock resistant. *Obsolete*

CREUSOT 2D8
Creusot-Loire
C 0.15-0.23, Ni 1.8-2.3, Mn 0-0.6, bal Fe.
Heat treated: 114,000 TS; 65,000 YS; 17 El. For gears, bolts, camshafts, cams; case hardened, oil quenched. *Obsolete*

CREUSOT 2D9
Creusot-Loire
C 0.1-0.15, Ni 1.8-2.3, bal Fe.
Annealed: 70,000 TS; 43,000 YS; 32 El; 150 Brin. For cams, fasteners, bolts, camshafts, gears. Case hardening, tough, shock resistant. *Obsolete*

CREUSOT 2DF6S
Creusot-Loire
C 0-0.22, Cr 0.6, Ni 2, bal Fe.
For gears, bolts, shafts, bridge members; resists low temperature. *Obsolete*

CREUSOT 3.5 DFO1
Creusot-Loire
C 0.3-0.4, Ni 3.2-3.8, Cr 1.2-1.7, Mo 0.2-0.5, bal Fe.
Heat treated: 250,000 psi TS; 208,000 psi YS; 5 El. For gears, bolts, machine tool parts; oil hardened, tough.

CREUSOT 3.5D8
Creusot-Loire
C 0.2, Ni 3.5, Mn 0-0.9, Si 0-0.35, bal Fe.
Heat treated: 100,000-128,000 psi TS; 78,000 psi YS; 14 El. For cryogenic vessels, tanks, equipment. Shock resistant at cryogenic temperatures. ASTM A 203D.

CREUSOT 3.5DF 3/4
Creusot-Loire
C 0.35, Cr 1.6, Ni 3.5, bal Fe.
Oil hardened: 150,000-178,000 TS; 113,000 YS. For highly loaded gears, pinions, rollers, swivels, clutches. AISI 3335. Tough, shock resistant. *Obsolete*

CREUSOT 3.5DF 7/8
Creusot-Loire
C 0.1, Cr 1.6, Ni 3.5, bal Fe.
Oil hardened: 130,000-170,000 TS; 106,000 YS; 11 El. For shafts, gears, pinions, universal joints. AISI 3310 steel, tough, shock resistant. Case hardening. *Obsolete*

CREUSOT 3D8
Creusot-Loire
C 0.14-0.22, Ni 2.7-3.3, Mn 0-0.6, bal Fe.
Heat treated: 128,000 TS; 79,000 YS; 14 El. For gears, bolts, camshafts, cams; case hardened, oil quenched. *Obsolete*

CREUSOT 3DF4/5
Creusot-Loire
C 0.26-0.33, Ni 2.7-3.3, Cr 0.6-0.9, bal Fe.
Heat treated: 150,000 psi TS; 100,000 psi YS; 13 El. For gears, bolts, machine tool parts; oil hardened, tough.

CREUSOT 3DF7/8
Creusot-Loire
C 0.11-0.16, Ni 2.7-3.3, Cr 0.6-0.9, Mn 0.3-0.6, bal Fe.
Heat treated: 207,000 psi TS; 120,000 psi YS; 9 El. For gears, bolts, machine tool parts; case hardened, tough

CREUSOT 3DF8
Creusot-Loire
C 0.08-0.14, Ni 2.7-3, Cr 0.6-0.9, Mn 0.3-0.6, bal Fe.
Heat treated: 166,000 psi TS; 100,000 psi YS; 12 El. For gears, bolts, camshafts, cams; case hardened, oil quenched.

CREUSOT 3DFO1
Creusot-Loire
C 0.3-0.38, Cr 2.5-3, Ni 1.2-1.6, Mo 0.2-0.4, bal Fe.
Heat treated: 235,000 psi TS; 192,000 psi YS; 4 El. For gears, bolts, machine tool parts; oil hardened, tough.

CREUSOT 3DFO2
Creusot-Loire
C 0.25-0.35, Ni 2.7-3.3, Cr 0.6-1, Mo 0.2-0.5, bal Fe.
Heat treated: 177,000 TS; 127,000 YS; 10 El. For gears, bolts, machine tool parts; oil hardened, tough. *Obsolete*

CREUSOT 3DFO5
Creusot-Loire
C 0.12-0.19, Ni 3-3.6, Cr 0.7-1.1, Mo 0.15-0.35, bal Fe.
Heat treated: 173,000 TS; 136,000 YS; 8 El. For gears, bolts, camshafts, cams; case hardened, oil quenched. *Obsolete*

CREUSOT 3DFO8
Creusot-Loire
C 0.08-0.13, Cr 0.6-0.9, Ni 2.7-3.3, Mo 0.2-0.4, bal Fe.
Heat treated: 144,000 TS; 114,000 YS; 10 El. For gears, bolts, camshafts, cams; case hardened, oil quenched. *Obsolete*

CREUSOT 3FO
Creusot-Loire
C 0.25-0.35, Cr 2.5-3.5, Mo 0.3-0.6, bal Fe.
Heat treated: 178,000 TS; 128,000 YS; 9 El. For gears, bolts, crankshafts; oil hardened, tough. *Obsolete*

CREUSOT 3FP
Creusot-Loire
C 0.3, W 3, Cr 3.2, Mo, V, bal Fe.
Annealed: 215 Brin. For nail dies, riveting tools, punches. Hot work tool steel, tough. *Obsolete*

CREUSOT 4.5DF1
Creusot-Loire
C 0.37-0.45, Ni 4-4.6, Cr 1.5-2, bal Fe.
Heat treated: 250,000 TS; 200,000 YS; 5 El. For gears, crankshafts; oil hardened, tough. *Obsolete*

CREUSOT 4.5DFO1
Creusot-Loire
C 0.35-0.45, Ni 4.1-4.9, Cr 1.4-1.8, Mo 0.35-0.6, bal Fe.
Heat treated: 260,000 TS; 207,000 YS; 4 El. For gears, bolts, machine tool parts; oil hardened, tough. *Obsolete*

CREUSOT 4DF01
Creusot-Loire
C 0.3-0.4, Ni 3.5-4.5, Cr 1.5-2, Mo 0.35-0.6, bal Fe.
Heat treated: 270,000 psi TS; 207,000 psi YS; 4 El. For gears, bolts, machine tool parts; oil hardened, tough.

CREUSOT 4DF2
Creusot-Loire
C 0.3-0.38, Ni 3.5-4, Cr 1.5-1.8, bal Fe.
Heat treated: 242,000 psi TS; 193,000 psi YS; 5 El. For gears, bolts, crankshafts; oil hardened, tough.

CREUSOT 4DF6
Creusot-Loire
C 0.08-0.15, Ni 3.4-3.9, Cr 0.7-1.1, bal Fe.
Heat treated: 228,000 psi TS; 135,000 psi YS; 8 El. For gears, bolts, camshafts, cams; case hardened, oil quenched.

CREUSOT 4DFO2
Creusot-Loire
C 0.28-0.36, Ni 3.5-4.5, Cr 1.1-1.5, Mo 0.35-0.6, bal Fe.
Heat treated: 242,000 psi TS; 200,000 psi YS; 4 El. For gears, bolts, machine tool parts; oil hardened, tough. *Obsolete*

CREUSOT 4DFO2/3
Creusot-Loire
C 0.2-0.3, Ni 3.8-4.5, Cr 1-1.4, Mo 0.35-0.6, bal Fe.
Heat treated: 192,000 TS; 135,000 YS; 9 El. For gears, bolts, machine tool parts; oil hardened, tough. *Obsolete*

CREUSOT 4DO2
Creusot-Loire
C 0.15-0.25, Ni 3.7-4.4, Mo 0.7-1.2, bal Fe.
Heat treated: 207,000 TS; 178,000 YS; 6 El. For gears, bolts, camshafts; case hardened, oil quenched. *Obsolete*

CREUSOT 4DO6
Creusot-Loire
C 0.09-0.15, Ni 3.7-4.4, Mo 0.7-1.2, bal Fe.
Heat treated: 172,000 psi TS; 135,000 psi YS; 8 El. For gears, bolts, camshafts, cams; case hardened, oil quenched.

CREUSOT 4PFOV
Creusot-Loire
C 0.25, W 4.2, Cr 1.2, Mo 0.4, V 0.25, bal Fe.
Annealed: 238 Brin. For piercing punches, mandrels, press dies. Hot work tool steel. *Obsolete*

CREUSOT 5 FO
Creusot-Loire
C 0.1, Cr 5, Mo 0.5, bal Fe.
Normalized and tempered: 370 N/mm^2 psi YS min; 590 N/mm^2 psi TS min; 15 El min. Plate for pressure vessels; weldable. AFNOR Z 10 CD 5. ASTM A 387.

CREUSOT 5D6
Creusot-Loire
C 0.13-0.2, Mn 0.4-0.7, Ni 4.7-5.4, bal Fe.
Heat treated: 213,000 psi TS; 127,000 psi YS; 6 El. For gears, bolts, machine tool parts; case hardened, oil quenched.

CREUSOT 5D6/7
Creusot-Loire
C 0.1-0.16, Mn 0-0.6, Ni 4.7-5.4, bal Fe.
Heat treated: 200,000 TS; 125,000 YS; 8 El. For gears, bolts, machine tool parts; case hardened, oil quenched. *Obsolete*

CREUSOT 5D8
Creusot-Loire
C 0.08-0.14, Mn 0-0.6, Ni 4.7-5.4, bal Fe.
Heat treated: 178,000 TS; 112,000 YS; 9 El. For gears, bolts, machine tool parts; case hardened, oil quenched. *Obsolete*

CREUSOT 5FAO
Creusot-Loire
C 0-0.17, Cr 5, Mo 0.5, Al 1, bal Fe.
Heat treated: 114,000 TS; 92,000 YS; 10 El; 240 Brin. Annealed: 70,000 TS: 30,000 YS; 28 El; 160 Brin. For heat exchanger tubes, fasteners, oil refinery equipment, valve trim, fittings. Heat and corrosion resistant. Resists scaling to 1100 F. *Obsolete*

CREUSOT 5FOP
Creusot-Loire
C 0.35, Cr 5, W 1.2, Mo 1.2, V 0.5, bal Fe.
Annealed: 238 Brin; 110,000 TS; 70,000 YS. Hardened: 216,000 TS; 185,000 YS; C 58 Rock. For forging dies, punches, shear blades, drills, screwing dies, bolt heading dies. Type H 12 hot work tool steel, tough, air or oil harden, shock resistant. *Obsolete*

CREUSOT 5FOV
Creusot-Loire
C 0.35, Cr 5, Mo 1.5, V 0.4, bal Fe.
Annealed: 235 Brin; 102,000 TS; 66,000 YS. Hardened: 300,000 TS; 250,000 YS; 6 El; C 55 Rock. For forging dies, punches, screwing dies, shear blades, gripper and header dies, high temperature bolts. Type H 11 hot work tool steel, tough, shock resistant. *Obsolete*

CREUSOT 9D6
Creusot-Loire
C 0-0.08, Mn 0-0.8, Si 0-0.35, Ni 8.5-10, bal Fe.
Normalized: 93,000 psi TS; 71,000 psi YS. Sheet: 70,000 psi TS min; 60,000 psi YS min; 18 El min. For liquid gas industry, storage or transportation tanks, weldable cryogenic steel, good low temperature ductility.

CREUSOT A 0
Creusot-Loire
C 0.72, Mn 0.65, Si 0.45, bal Fe.
Annealed: 125,000-135,000 TS; 321 Brin. Heat treated: 185,000 TS; 136,000 YS; 13 El; C 39 Rock. For bands, springs, crusher components, hand tools, punches, hammers, shovels, harrows. AISI 1070 steel, water harden. *Obsolete*

CREUSOT A 0/1H
Creusot-Loire
C 0.67-0.75, Si 0-0.4, Mn 0.6-0.85, bal Fe.
Heat treated: 174,000 TS; 128,000 YS; 12 El; 37 RA; 355 Brin. For springs, rails, die blocks, clutch discs; water hardened. *Obsolete*

CREUSOT A 00
Creusot-Loire
C 0.8-0.93, Mn 0.7-1, Si 0.4, bal Fe.
Water hardened: 345,000-400,000 TS. Annealed: 100,000 TS; 55,000 YS; 20 El; 198 Brin. For crusher bars, axes, tools, dies, springs, bearings. AISI 1085 steel, water hardening. *Obsolete*

CREUSOT A 0B
Creusot-Loire
C 0.65-0.8, Si 0-0.45, Mn 0.5-0.8, bal Fe.
Heat treated: 175,000 TS; 130,000 YS; 12 El; 36 RA; 360 Brin. For die blocks, rails, tie rods, springs; water hardened. *Obsolete*

CREUSOT A 10
Creusot-Loire
C 0-0.16, S 0.04, P 0.04, bal Fe.
Annealed: 55,000 TS; 32,000 YS; 26 El. For boiler drums, steam manifolds, pressure vessels. ASTM A 285 Gr. A. *Obsolete*

CREUSOT A 11
Creusot-Loire
C 0.1, Mn 0.45, Si 0-0.3, bal Fe.
Water hardened: 64,000-100,000 TS; 43,000 YS. Annealed: 55,000 TS; 29,000 YS; 30 El; 110 Brin. For boiler plates, chains, rivets, tubes. AISI 1010 steel. *Obsolete*

CREUSOT A 2
Creusot-Loire
C 0.6, Mn 0.65, Si 0.45, bal Fe.
Annealed: 110,000-130,000 TS; 269 Brin. Oil hardened: 170,000 TS; 145,000 YS; 7 El. For springs, thrust bearings, cams, agricultural machinery, hammers. AISI 1060 steel. *Obsolete*

CREUSOT A 3/4
Creusot-Loire
C 0.5, Mn 0.65, Si 0.45, bal Fe.
Annealed: 110,000 TS; 55,000 YS; 13 El; 210 Brin. Oil hardened: 120,000-150,000 TS; 90,000 YS. For drive clutches, gears, cams, shafts, axles, hammers. AISI 1050 steel. *Obsolete*

CREUSOT A 5
Creusot-Loire
C 0.4, Mn 0.65, Si 0.45, bal Fe.
Annealed: 98,000 TS; 50,000 YS; 36 El; 180 Brin. Water hardened: 128,000-157,000 TS; 100,000 YS; 9 El. For pins, axles, hammers, fasteners, gears, bolts, shafts. AISI 1040 steel, water harden. *Obsolete*

CREUSOT A 6/7
Creusot-Loire
C 0.3, Mn 0.65, Si 0.45, bal Fe.
Water hardened: 100,000-135,000 TS; 71,000 YS. Annealed: 84,000 TS; 42,000 YS; 18 El; 163 Brin. For shafts, crankshafts, gears, stub axles, levers, cylinders. AISI 1030 steel, water harden. *Obsolete*

CREUSOT A 9
Creusot-Loire
C 0.2, Mn 0.65, Si 0.45, bal Fe.
Water hardened: 79,000-114,000 TS; 52,000 YP. Annealed: 54,000 TS; 29,000 YS; 30 El; 140 Brin. For slides, ship components, transmission shafts, fasteners, gears, pinions, cams. AISI 1020 steel, case harden. *Obsolete*

CREUSOT A/0H
Creusot-Loire
C 0.75-0.85, Si 0-0.4, Mn 0.5-0.75, bal Fe.
Heat treated: 190,000 TS; 145,000 YS; 12 El; 35 RA; 390 Brin. For springs, rails, die blocks, tools, hammers; water hardened. *Obsolete*

CREUSOT A10B
Creusot-Loire
C 0.1-0.18, Si 0-0.3, Mn 0.3-0.6, bal Fe.
Annealed: 70,000 TS; 40,000 YS; 25 El; 60 RA; 140 Brin. For bolts, screws, nails, cams, bushings; case hardened. *Obsolete*

CREUSOT A10HC
Creusot-Loire
C 0.09-0.16, Si 0-0.35, Mn 0.6-0.9, bal Fe.
Annealed: 70,000 TS; 40,000 YS; 25 El; 60 RA; 145 Brin. For screws, bolts, gears, cams, bushings; case hardened, water quenched. *Obsolete*

CREUSOT A11B
Creusot-Loire
C 0.06-0.15, Si 0-0.3, Mn 0.3-0.6, bal Fe.
Annealed: 64,000 TS; 40,000 YS; 28 El; 65 RA; 130 Brin. For rivets, nails, bolts, nuts, fan blades. *Obsolete*

CREUSOT A11C
Creusot-Loire
C 0.06-0.12, Si 0-0.2, Mn 0.3-0.6, bal Fe.
Annealed: 64,000 TS; 40,000 YS; 28 El; 65 RA; 130 Brin. For nails, rivets, bolts, cams; case hardened, water quenched. *Obsolete*

CREUSOT A11HC
Creusot-Loire
C 0.06-0.12, Si 0-0.15, Mn 0.4-0.65, bal Fe.
Annealed: 64,000 TS; 40,000 YS; 28 El; 65 RA; 130 Brin. For nails, rivets, bolts, nuts, cams; case hardened, water quenched. *Obsolete*

CREUSOT A1B
Creusot-Loire
C 0.6-0.7, Si 0-0.45, Mn 0.5-0.8, bal Fe.
Heat treated: 170,000 TS; 116,000 YS; 10 El; 38 RA; 350 Brin. For die blocks, rails, tie rods, springs; water hardened. *Obsolete*

CREUSOT A1H
Creusot-Loire
C 0.6-0.68, Si 0-0.4, Mn 0.65-0.9, bal Fe.
Heat treated: 165,000 TS; 115,000 YS; 12 El; 40 RA; 330 Brin. For rails, springs, die blocks; water hardened. *Obsolete*

CREUSOT A2B
Creusot-Loire
C 0.55-0.65, Si 0-0.45, Mn 0.3-0.8, bal Fe.
Heat treated: 160,000 TS; 113,000 YS; 12 El; 40 RA; 325 Brin. For die blocks, rails, tie rods, springs; water hardened. *Obsolete*

CREUSOT A3/4B
Creusot-Loire
C 0.45-0.55, Si 0-0.45, Mn 0.5-0.8, bal Fe.
Annealed: 96,000 TS; 52,000 YS; 16 El; 23 RA; 170 Brin. For axles, gears, crankshafts, bolts; water hardened. *Obsolete*

CREUSOT A3B
Creusot-Loire
C 0.5-0.6, Si 0-0.45, Mn 0.3-0.8, bal Fe.
Heat treated: 150,000 TS; 110,000 YS; 15 El; 42 RA; 310 Brin. For die blocks, clutch discs, springs; water hardened. *Obsolete*

CREUSOT A3H
Creusot-Loire
C 0.48-0.58, Si 0-0.4, Mn 0.5-0.8, bal Fe.
Annealed: 96,000 TS; 52,000 YS; 16 El; 23 RA; 170 Brin. For axles, gears, bolts, shafts, crankshafts; water hardened. *Obsolete*

CREUSOT A4/5B
Creusot-Loire
C 0.4-0.5, Si 0-0.45, Mn 0.5-0.8, bal Fe.
Hot rolled: 98,000 TS; 59,000 YS; 24 El; 45 RA; 212 Brin. For axles, gears, bolts, shafts; water hardened. *Obsolete*

CREUSOT A4/5H
Creusot-Loire
C 0.4-0.48, Si 0-0.4, Mn 0.5-0.8, bal Fe.
Hot rolled: 98,000 TS; 59,000 YS; 24 El; 45 RA; 212 Brin. For axles, gears, bolts, shafts; water hardened. *Obsolete*

CREUSOT A4H
Creusot-Loire
C 0.44-0.52, Si 0-0.4, Mn 0.5-0.8, bal Fe.
Hot rolled: 98,000 TS; 59,000 YS; 24 El; 45 RA; 212 Brin. For axles, gears, bolts, shafts; water hardened. *Obsolete*

CREUSOT A5/6H
Creusot-Loire
C 0.34-0.4, Si 0-0.4, Mn 0.5-0.8, bal Fe.
Hot rolled: 88,000 TS; 56,000 YS; 28 El; 52 RA; 190 Brin. For gears, bolts, crankshafts; water hardened. *Obsolete*

CREUSOT A5H
Creusot-Loire
C 0.38-0.45, Si 0-0.4, Mn 0.5-0.8, bal Fe.
Hot rolled: 91,000 TS; 58,000 YS; 27 El; 50 RA; 200 Brin. For gears, bolts, crankshafts; water hardened. *Obsolete*

CREUSOT A6/7B
Creusot-Loire
C 0.25-0.35, Si 0-0.45, Mn 0.5-0.8, bal Fe.
Hot rolled: 80,000 TS; 50,000 YS; 30 El; 56 RA; 165 Brin. For gears, shafts, bolts, screws; water hardened. *Obsolete*

CREUSOT A6B
Creusot-Loire
C 0.3-0.4, Si 0-0.45, Mn 0.5-0.8, bal Fe.
Hot rolled: 85,000 TS; 54,000 YS; 30 El; 53 RA; 185 Brin. For gears, shafts, axles, bolts; water hardened. *Obsolete*

CREUSOT A6H
Creusot-Loire
C 0.3-0.36, Si 0-0.4, Mn 0.5-0.8, bal Fe.
Hot rolled: 85,000 TS; 54,000 YS; 30 El; 53 RA; 185 Brin. For gears, bolts, shafts, fasteners; water hardened. *Obsolete*

CREUSOT A7B
Creusot-Loire
C 0.22-0.32, Si 0-0.45, Mn 0.5-0.8, bal Fe.
Hot rolled: 70,000 TS; 45,000 YS; 31 El; 58 RA; 145 Brin. For gears, bolts, armature shafts, brackets; water hardened. *Obsolete*

CREUSOT A8B
Creusot-Loire
C 0.2-0.3, Si 0-0.45, Mn 0.5-0.8, bal Fe.
Hot rolled: 70,000 TS; 45,000 YS; 31 El; 58 RA; 145 Brin. For gears, bolts, armature shafts, brackets; water hardened. *Obsolete*

CREUSOT A9B
Creusot-Loire
C 0.15-0.25, Si 0-0.45, Mn 0.5-0.8, bal Fe.
Annealed: 73,000 TS; 42,000 YS; 22 El; 58 RA; 145 Brin. For bolts, cams, bushings, rivets, gears; case hardened, water quenched. *Obsolete*

CREUSOT A9H
Creusot-Loire
C 0.15-0.22, Si 0-0.35, Mn 0.5-0.8, bal Fe.
Annealed: 73,000 TS; 41,000 YS; 22 El; 58 RA; 140 Brin. For fan blades, bushings, gears, bolts; case hardened, water quenched. *Obsolete*

CREUSOT AM10/11S
Creusot-Loire
C 0-0.16, Si 0.2, 0.40% min Mn, 0.04% max S and P, bal Fe.
For case hardened and structural parts, bridges; good weldability. *Obsolete*

CREUSOT AM11S
Creusot-Loire
C 0.12, Si 0.2, 0.40% min Mn, 0.04% max S and P, bal Fe.
For case hardened parts, structural parts, bridges; good weldability. *Obsolete*

CREUSOT AM6/7S
Creusot-Loire
C 0-0.25, Si 0-0.35, 0.60% min Mn, 0.04% max S and P, bal Fe.
For pressure tanks, bridges, hoists, machinery parts. *Obsolete*

CREUSOT AM8/9S
Creusot-Loire
C 0-0.22, Si 0-0.35, 0.60% min Mn, 0.04% max S and P, bal Fe.
For pressure tanks, conduits, naval construction; good weldability. *Obsolete*

CREUSOT AM9/10S
Creusot-Loire
N 0-0.18, C 0-0.25, 0.40% min Mn, 0.04% max S and P, bal Fe.
For case hardened and structural parts, bridges; good weldability. *Obsolete*

CREUSOT B2/3
Creusot-Loire
C 0.47-0.55, B 0.001, Mn 0.65-1.1, Si 0.3, bal Fe.
Heat treated: 213,000-320,000 TS; 177,000 YS; 6 El; 625 Brin.
For reactor equipment; water hardened. *Obsolete*

CREUSOT B3/4
Creusot-Loire
C 0.42-0.49, B 0.001, Mn 0.65-1.1, Si 0.3, bal Fe.
Heat treated: 165,000-290,000 TS; 322-578 Brin. For reactor
equipment; water hardened. *Obsolete*

CREUSOT B4/5
Creusot-Loire
C 0.37-0.44, B 0.001, Mn 0.65-1.1, Si 0.3, bal Fe.
Heat treated: 157,000-280,000 TS; 311-534 Brin. For reactor
equipment; water hardened. *Obsolete*

CREUSOT B5/6
Creusot-Loire
C 0.33-0.4, B 0.001, Mn 0.65-1.1, Si 0.3, bal Fe.
Heat treated: 128,000-242,000 TS; 256-477 Brin. For reactor
equipment; water hardened. *Obsolete*

CREUSOT B6
Creusot-Loire
C 0.3-0.37, B 0.001, Mn 0.65-1.1, Si 0.3, bal Fe.
Heat treated: 114,000-208,000 TS; 223-410 Brin. For reactor
equipment; water hardened. *Obsolete*

CREUSOT BD5/6
Creusot-Loire
C 0.32-0.39, Cr 0.2-0.5, Ni 0.2-0.4, B 0.001, Mo 0.08-0.15, bal
Fe.
Heat treated: 178,000-260,000 TS; 150,000-228,000 YS; 5-9
El. For reactor equipment; oil hardened. *Obsolete*

CREUSOT BD8/9
Creusot-Loire
C 0.17-0.23, Cr 0.2-0.5, Ni 0.2-0.4, Mo 0.08-0.15, bal Fe.
Heat treated: 144,000-186,000 TS; 135,000-156,000 YS; 6-9
El. For bolts, gears, machine tool parts; case hardened, oil
quenched. *Obsolete*

CREUSOT BD9/10
Creusot-Loire
C 0.12-0.18, Cr 0.2-0.5, Ni 0.2-0.4, Mo 0.08-0.15, B 0.001, bal
Fe.
Heat treated: 178,000 TS; 150,000 YS; 7 El. For reactor
equipment; oil hardened, case hardened. *Obsolete*

CREUSOT BF1/2
Creusot-Loire
C 0.55-0.65, Cr 0.3-0.7, B 0.001, Mn 0.65-1, bal Fe.
Heat treated: 330,000 TS; 650 Brin. For reactor equipment; oil
hardened. *Obsolete*

CREUSOT BF4/5
Creusot-Loire
C 0.37-0.45, Mn 0.65-1.1, Cr 0.3-0.7, B 0.001, bal Fe.
Heat treated: 193,000-222,000 TS; 157,000-185,000 YS; 7-8
El. For reactor equipment; oil hardened. *Obsolete*

CREUSOT C 1/2
Creusot-Loire
C 1.15, bal Fe.
Annealed: 205 Brin; 100,000 TS; 55,000 YS. Hardened: Rock
C 65; 220,000 TS; 155,000 YS. For milling cutters, taps,
reamers, drills, form tools, gauges. Type W1 water hardening
tool steel. Wear and abrasion resistant. *Obsolete*

CREUSOT C 4/5
Creusot-Loire
C 0.8, bal Fe.
Hardened: Rock C 64. Annealed: 192 Brin., 115,000 TS,
65,000 YS. Tempered: 185,000 TS; 135,000 YS; 13 El; C 39
Rock. For cut off and flanging tools, drills, springs, hand
tools, drills, hammers. Type W1 water hardening tool steel,
wear and abrasion resistant. *Obsolete*

CREUSOT CF
Creusot-Loire
C 1, Cr 1.4, bal Fe.
Annealed: 215 Brin.; 110,000 TS; 88,000 YS. Heat treated:
240,000 TS; 230,000 YS; 444 Brin. For taps, reamers,
stamping dies, bearings, punches, bushings. Type 52100
steel, water or oil harden. *Obsolete*

CREUSOT CMP
Creusot-Loire
C 1, Mn 1, Cr 0.5, W 0.5, bal Fe.
Annealed: 235 Brin.; 85,000 TS; 70,000 YS. Heat treated:
250,000 TS; 225,000 YS; 8 El; C 50 Rock. For cold stamping
dies, cutting tools, punches, rivet sets, upsetters. Type O1
cold working tool steel. *Obsolete*

CREUSOT CMV
Creusot-Loire
C 0.9, Mn 2, V 0.1, bal Fe.
Annealed: 235 Brin.; 110,000 TS. Hardened: C 63-65 Rock.
For cutting tools, flanging tools, cold headers, upsetters,
punches, molds, blanking and forming dies. Type O2 cold
work tool steel, deep hardening, shock resistant. *Obsolete*

CREUSOT DF3
Creusot-Loire
C 0.3-0.38, Ni 2.5-3, Cr 0.6-0.9, Mn 0.35-0.65, bal Fe.
Heat treated: 290,000 TS; 200,000 YS; 4 El. For gears, bolts,
machine tool parts; oil hardened, tough. *Obsolete*

CREUSOT DF4/5
Creusot-Loire
C 0.25-0.33, Ni 2.5-3, Cr 0.6-0.9, Mn 0.3-0.6, bal Fe.
Heat treated: 157,000 TS; 107,000 YS; 12 El. For gears, bolts,
machine tool parts; oil hardened, tough. *Obsolete*

CREUSOT DF5
Creusot-Loire
C 0.22-0.3, Ni 2.5-3, Cr 0.6-0.9, Mn 0.3-0.6, bal Fe.
Heat treated: 150,000 TS; 101,000 YS; 12 El. For gears, bolts,
machine tool parts; oil hardened, tough. *Obsolete*

CREUSOT DF6
Creusot-Loire
C 0.17-0.25, Ni 2.5-3, Cr 0.6-0.9, Mn 0.35-0.65, bal Fe.
Heat treated: 165,000 TS; 107,000 YS; 10 El. For gears, bolts,
camshafts, cams; case hardened, oil quenched. *Obsolete*

CREUSOT DF7
Creusot-Loire
C 0.11-0.18, Ni 2.5-3, Cr 0.6-0.9, bal Fe.
Heat treated: 213,000 TS; 128,000 YS; 7 El. For gears, bolts,
camshafts, cams; case hardened, oil quenched. *Obsolete*

CREUSOT DF7/8
Creusot-Loire
C 0.1-0.16, Ni 2.5-3, Cr 0.6-0.9, bal Fe.
Heat treated: 200,000 TS; 115,000 YS; 8 El. For gears, bolts,
camshafts, cams; case hardened, oil quenched. *Obsolete*

CREUSOT DF8
Creusot-Loire
C 0.08-0.14, Ni 2.5-3, Cr 0.6-0.9, bal Fe.
Heat treated: 157,000 TS; 94,000 YS; 11 El. For gears, bolts,
camshafts, cams; case hardened, oil quenched. *Obsolete*

CREUSOT DFO3
Creusot-Loire
C 0.23-0.33, Ni 2.4-3, Cr 0.5-0.8, Mo 0.2-0.5, bal Fe.
Heat treated: 235,000 TS; 200,000 YS; 5 El. For gears, bolts,
machine tool parts; oil hardened, tough. *Obsolete*

CREUSOT DMF
Creusot-Loire
C 0.6-0.8, Mn 7-9, Ni 7-9, Cr 3-4.5, O 0.0000 0.0000, Fe
0.0000 0.0000, bal Fe.
Annealed: 114,000 TS; 43,000 YS; 40 El. For corrosion
resistant parts; stainless, austenitic. *Obsolete*

CREUSOT DOF1
Creusot-Loire
C 0.35-0.45, Ni 4-5, Cr 0.3-0.6, Mo 0.7-1.2, bal Fe.
Heat treated: 270,000 TS; 208,000 YS; 4 El. For gears, bolts,
machine tool parts; oil hardened, tough. *Obsolete*

CREUSOT FO4
Creusot-Loire
C 0.31-0.39, Cr 0.85-1.2, Mo 0.15-0.35, Mn 0.6-0.9, bal Fe.
Heat treated: 285,000 TS; 192,000 YS; 5 El. For gears, bolts,
machine tool parts; oil hardened, tough. *Obsolete*

CREUSOT FO6/7S
Creusot-Loire
C 0-0.25, Mo 0.25, Cr 1, bal Fe.
For construction members, heat exchangers. *Obsolete*

CREUSOT FO6S
Creusot-Loire
C 0-0.28, Cr 1, Mo 0.25, Mn 0.6, bal Fe.
For construction members, heat exchangers. *Obsolete*

CREUSOT FO7S
Creusot-Loire
C 0-0.22, Mn 0.7, Cr 0.45, Mo 0.45, bal Fe.
For pressure tanks, gas cylinders. *Obsolete*

CREUSOT FO8S
Creusot-Loire
C 0-0.2, Cr 0.45, Mo 0.45, Mn 0-0.35, bal Fe.
For pressure tanks, gas cylinders. *Obsolete*

CREUSOT FOV
Creusot-Loire
C 0.25, Cr 1, Mo 1, V 0.25, bal Fe.
Heat treated: 100,000 TS; 82,000 YS; 16 El. For rotors, discs,
hot bolts. ASTM A356 Gr. 9. Tough, shock resistant. *Obsolete*

CREUSOT FP
Creusot-Loire
C 0.45, W 2, Cr 1, bal Fe.
Annealed: 205 Brin. Hardened: C 55 Rock; 300,000 TS;
260,000 YS. For shear and cut-off blades, punches, dies,
chisels, riveting tools, cold battering tools. Type S1 cold work
tool steel, oil hardening. *Obsolete*

CREUSOT FR
Creusot-Loire
C 0.9-1.1, Cr 1.25-1.65, Mn 0-0.45, Si 0-0.4, bal Fe.
For bearings, liners, bushings; water hardened. *Obsolete*

CREUSOT LMO
Creusot-Loire
C 1.4, Si 1, Mn 1, Mo 0.25, bal Fe.
Annealed: 217 Brin. For cutting tools, slides, dies. Oil
hardening, abrasion resistant. *Obsolete*

CREUSOT M
Creusot-Loire
C 1.2, Mn 12, Si 0.6, bal Fe.
Water quenched: 880 N/mm^2 psi TS min. Non-magnetic
alloy. AFNOR Z 120 M 12.

CREUSOT MF6/7S
Creusot-Loire
C 0-0.22, Si 0-0.35, Mn 1.3, bal Fe.
For pressure tanks, bridges, hoists, machinery parts.
Obsolete

CREUSOT ML
Creusot-Loire
C 0.4-0.5, Si 1.6-2.1, Mn 0.4-0.8, bal Fe.
Heat treated: 242,000 TS; 163,000 YS; 5 El. For punches,
pneumatic tools, springs; water or oil hardened, shock
resistant. *Obsolete*

CREUSOT MLD
Creusot-Loire
C 0.4-0.5, Si 1.6-2.1, Mn 0.4-0.8, bal Fe.
Heat treated: 242,000 TS; 170,000 YS; 5 El. For punches, springs, pneumatic tools; oil hardened, shock resistant. *Obsolete*

CREUSOT MO 6/7
Creusot-Loire
C 0-0.2, Mn 0.8, Mo 0.52, bal Fe.
Normalized: 300 N/mm^2 psi YS min. Weldable; plate for pressure vessels. AFNOR 18 MD 4.05. ASTM A 204 Gr. C.

CREUSOT MO7
Creusot-Loire
C 0.15, Mo 0.55, Mn 1, bal Fe.
Heat treated: 70,000 TS; 40,000 YS; 20 El. For boilers, steam manifolds, turbine casings, reactors, nuclear plants, superheater pipes. ASTM A356 Gr. 2. *Obsolete*

CREUSOT MV6/7S
Creusot-Loire
C 0-0.22, Si 0-0.35, Mn 1.4, bal Fe.
For pressure tanks, naval construction, machinery parts. *Obsolete*

CREUSOT P1
Creusot-Loire
C 1.15, W 2, bal Fe.
Annealed: 248 Brin. Hardened: C 67 Rock. For dies, drawing dies, drills, reamers, broaches, twist drills, burnishing tools. Type F2 and F3 water hardening tool steel. *Obsolete*

CREUSOT P2
Creusot-Loire
C 1.15, W 5, Cr 0.6, bal Fe.
Annealed: 240 Brin. Hardened: Rock C 47. For dies, drawing dies, drills, chisels, reamers, twist drills. Type F2 water hardening tool steel. *Obsolete*

CREUSOT S2V
Creusot-Loire
C 0.75, Cr 4, W 13, V 0.7, Mo 0.5, bal Fe.
Annealed: 248 Brin. For milling cutters, twist drills, lathe tools, reamers. High speed steel. *Obsolete*

CREUSOT S3
Creusot-Loire
C 0.8, Cr 4, W 18, V 1, bal Fe.
Annealed: 240 Brin.; 100,000 TS; 75,000 YS; 14 El; 17 RA. For milling cutters, twist drills, reamers, lathe tools, planing tools. Type T1 high speed steel, good red-hardness. *Obsolete*

CREUSOT S3V
Creusot-Loire
C 0.8, Cr 4.5, W 18, V 1.8, Mo 0.7, bal Fe.
Annealed: 255 Brin. Hardened: C 63-65 Rock. For milling cutters, twist drills, broaches, reamers, taps, lathe and planer tools. Type T2 high speed steel, high red-hardness. *Obsolete*

CREUSOT SISA
Creusot-Loire
C 0.35, W 10, Cr 3, V 0.4, bal Fe.
Annealed: 217 Brin. Hardened: 243,000 TS; 215,000 YP; 12 El; 37 RA. For nail dies, punches, shear blades, die casting dies, mandrels, extrusion dies and rams, trimmer dies. Type H21 hot work tool steel, high red-hardness. *Obsolete*

CREUSOT SO3
Creusot-Loire
C 0.8, Cr 4.5, W 6.5, Mo 5, V 2, bal Fe.
Annealed: 228 Brin. Hardened: C 63-65 Rock. For milling cutters, twist drills, reamers, broaches, taps, chasers, saws, lathe tools. Type M2 high speed steel, high red-hardness. *Obsolete*

CREUSOT SS10
Creusot-Loire
C 0.8, Cr 5, W 18, V 1.7, Mo 1, Co 10, bal Fe.
Annealed: 302 Brin. Hardened: C 65 Rock. For milling cutters, drills, broaches, reamers, taps, lathe and planer tools, shapers. Type T5 high speed steel, high red-hardness. *Obsolete*

CREUSOT SS14V
Creusot-Loire
C 1.7, Cr 4.5, W 14, V 4, Mo 0.5, Co 14, bal Fe.
Annealed: 340 Brin. Hardened: C 65-67 Rock. For lathe and planer tools, broaches, chasers, taps, milling cutters. High speed steel; high hot-hardness. *Obsolete*

CREUSOT SS15
Creusot-Loire
C 0.75, Cr 5, W 18, V 1.7, Mo 0.5, Co 15, bal Fe.
Annealed: 320 Brin. Hardened: C 65 Rock. For broaches, lathe and planer tools, drills, milling cutters, hobs, reamers. Type T6 high speed steel, high red-hardness. *Obsolete*

CREUSOT SS5
Creusot-Loire
C 0.8, Cr 4.5, W 18, V 1.5, Mo 1, Co 5, bal Fe.
Annealed: 268 Brin. Hardened: C 65 Rock. For piercers, lathe and planing tools, cutters, reamers, broaches, taps, chasers, cut-off tools, drills. Type T4 high speed steel, high red-hardness. *Obsolete*

CREUSOT SS5V
Creusot-Loire
C 1.5, Cr 4.5, W 14, V 4, Mo 0.5, Co 6, bal Fe.
Annealed: 302 Brin. Hardened: C 64-66 Rock. For hard dies, lathe and planing tools, reamers, broaches, drills, blanking dies, form tools. Type T15 high speed steel, high red-hardness. *Obsolete*

CREUSOT SV4
Creusot-Loire
C 1.3, Cr 4.5, W 12, V 4.5, Mo 0.8, bal Fe.
Annealed: 277 Brin. For finishing tools, taps, reamers, twist drills. High speed steel. *Obsolete*

CRH 4-17
Tokoshu Seiko Co. Ltd.
C 0.1, Cr 13, Mo 0.45, Ni 0-0.6, bal Fe.
Annealed: 75,000 TS; 40,000 YS; 35 El; 70 RA; 155 Brin. Cold drawn: 100,000 TS; 85,000 YS; 17 El; 60 RA; 205 Brin. For oil refinery and chemical plant equipment, knives, hardware, table flatware. Corrosion resistant.

CRILLEY METALS
Metal Sales Corp.
Hg, Sn, bal Cu.
23,000 TS. For bearings; compressive strength 92,000.

CRIPMORE NO. 2
Bethlehem Steel Corp.
C 0.35, Mn 0.35, Cr 5, W 1.35, Mo 1.75, Si 1, bal Fe.
Hot work dies, hot work steel. *Obsolete*

CRISIL
Johnston Steel & Wire Co. Inc.
C 0.55, Cr 0.5-0.8, Si 1.2-1.6, Mn 0.5-0.8, bal Fe.
Wire: 280,000-335,000 TS; 45 RA min. For mechanical springs, operating under high stress and moderately high temperatures. Hard drawn steel wire.

CRISTITE 1
Commercial Alloys Co.
W 17, Cr 10, C 3.5, Mo 2.5, bal Fe.
For facing dredgers, dipper or shovel teeth and bucket lips, agricultural implements, knives, tools, drills, chisels; resists abrasion and high temperatures.

CRISTITE 2
Commercial Alloys Co.
W 17, Cr 12, C 3, Mo 2, Ti 0.75, Hf 0.25, bal Fe.
For hard facing electrodes; wear resistant.

CRISTITE 3
Commercial Alloys Co.
Cr 16, C 3, Ti 0.8, Hf 0.28, Ni 6, bal Fe.
For hard facing electrodes; wear resistant.

CRITNIC
American manufacture
Ni 97, C 0.2, Ti.
For heat and corrosion resistant parts; corrosion and heat resistant.

CRM 4
Chrysler Corp.
C 0-0.02, Mn 0-0.4, Si 0-0.1, Cr 0-0.5, Ni 0-0.5, Cu 0-0.2, Al 5.5-6.5, bal Fe.
Low carbon low strength; wrought ferritic alloy; good oxidation resistance to 1200°F.

CRM 6D
Chrysler Corp.
C 1-1.1, Cr 20-22, Ni 5, Mn 5, W 1, Mo 1, Cb 1, Si 0.55, B 0.002-0.008, N 0-0.05, bal Fe.
Cast, austenitic; good strength above 1500°F. For high temperature rotating parts as turbine wheels.

CRM 6D
Cannon-Muskegon Corp.
C 1-1.1, Cr 20-22, Ni 5, Mn 5, W 1, Mo 1, Cb 1, Si 0.55, B 0.002-0.008, N 0-0.05, bal Fe.
Cast, austenitic; good strength above 1500°F. For high temperature rotating parts as turbine wheels.

CRM SPECIAL
Thyssen Edelstahlwerke AG
C 1, Mn 1.1, W 1, Cr 1, bal Fe.
For tools, dies; oil hardening. *Obsolete*

CRM-15D
Chrysler Corp.
C 1-1.1, Cr 20-22, Ni 5, Mn 5, W 2, Mo 2, Cb 2, Si 0.55, B 0.002-0.008, N 0.15-0.25, bal Fe.
Cast, austenitic; good strength above 1500°F. For high temperature roatating parts as turbine wheels.

CRM-15D
Cannon-Muskegon Corp.
C 1-1.1, Cr 20-22, Ni 5, Mn 5, W 2, Mo 2, Cb 2, Si 0.55, B 0.002-0.008, N 0.15-0.25, bal Fe.
Cast, austenitic; good strength above 1500°F. For high temperature roatating parts as turbine wheels.

CRM-17D
Chrysler Corp.
C 0.65-0.75, Cr 18-20, Ni 5, Mn 5, W 1, Mo 1, Cb 2, Si 0.55, B 0.002-0.008, N 0.15-0.25, bal Fe.
Cast, austenitic; for operation at 1200-1400°F as nozzle blades and nozzle support struts.

CRM-17D
Cannon-Muskegon Corp.
C 0.65-0.75, Cr 18-20, Ni 5, Mn 5, W 1, Mo 1, Cb 2, Si 0.55, B 0.002-0.008, N 0.15-0.25, bal Fe.
Cast, austenitic; for operation at 1200-1400°F as nozzle blades and nozzle support struts.

CRM-18D
Chrysler Corp.
C 0.7-0.8, Cr 22.5-24.5, Ni 5, Co 5, Mn 5, W 1, Mo 1, Cb 2, Si 0.55, B 0.002-0.008, N 0.25-0.35, bal Fe.
Cast, austenitic; for operation up to 1800°F such as first stage nozzle and burner vortex.

CRM-18D
Cannon-Muskegon Corp.
C 0.7-0.8, Cr 22.5-24.5, Ni 5, Co 5, Mn 5, W 1, Mo 1, Cb 2, Si 0.55, B 0.002-0.008, N 0.25-0.35, bal Fe.
Cast, austenitic; for operation up to 1800°F such as first stage nozzle and burner vortex.

CRMN 55
Kawasaki Steel Corp.
C 0.18, Mn 1.28, P 0.016, S 0.009, Si 0.29, Cr 0.7, bal Fe.
88,000 psi TS; 61,200 psi YS; 29 El. Cr-Mn low-alloy steel for abrasion resistant applications.

CRMN 60
Kawasaki Steel Corp.
C 0.3, Mn 1.27, P 0.021, S 0.014, Si 0.35, Cr 0.67, bal Fe.
111,000 psi TS; 64,000 psi YS; 20 El. Cr-Mn low-alloy steel for abrasion resistant applications.

CRMO 18
Walsingham Steel Co., Ltd.
C 0.16-0.2, Si 0.3-0.6, Mn 0.5-0.8, Ni 0-0.4, Cr 1-1.5, Mo 0.45-0.65, Cu 0-0.4, bal Fe.
Meets BS 1504-621.

CRMO 23
Walsingham Steel Co., Ltd.
C 0.2-0.25, Si 0.3-0.75, Mn 0.3-0.7, Ni 0-0.4, Cr 2.5-3.5, Mo 0.35-0.6, Cu 0-0.4, bal Fe.
For service at elevated temperatures, hardenable. Meets BS 1461.

CRMO 60
Walsingham Steel Co., Ltd.
C 0.55-0.65, Si 0-0.75, Mn 0.5-1, Cr 0.8-1.5, Mo 0.2-0.4, bal Fe.
Hardenable; high abrasion resistant steel. Meets BS 1956 Gr. C.

CRMO A
Walsingham Steel Co., Ltd.
C 0.5-0.6, Si 0.3-0.6, Mn 0.5-0.8, Ni 0.4-0.8, Cr 2.3-3, Mo 0.3-0.5, Cu 0-0.4, bal Fe.
Hardenable; abrasion resistant.

CRMO B
Walsingham Steel Co., Ltd.
C 0.55-0.65, Si 0.3-0.6, Mn 0.5-0.8, Ni 0.4-0.8, Cr 2.3-3, Mo 0.3-0.5, Cu 0-0.3, bal Fe.
Hardenable; abrasion resistant.

CRMO D
Walsingham Steel Co., Ltd.
C 0-0.18, Si 0.3-0.5, Mn 0.45-0.65, Ni 0-0.4, Cr 2-2.75, Mo 0.9-1.2, Cu 0-0.4, bal Fe.
Meets ASTM A217-65 Gr.WC9.

CRMO E
Walsingham Steel Co., Ltd.
C 0.15-0.22, Si 0.35-0.6, Mn 0.6-0.9, Cr 0.6-0.8, Mo 0.2-0.3, bal Fe.

CRMO F
Walsingham Steel Co., Ltd.
C 0.28-0.33, Si 0.3-0.5, Mn 0.6-0.8, Cr 2-2.5, Mo 0.2-0.3, bal Fe.
Oil hardenable, tough steel.

CRMO G
Walsingham Steel Co., Ltd.
C 0.35-0.45, Si 0.3-0.7, Mn 0.6-0.85, Cr 0.8-1.1, Mo 0.25-0.35, bal Fe.
Similar to AISI 4140.

CRO-13 MO
Pont-St.-Martin
C 0.15, Cr 13, Mo 0-0.6, bal Fe.
Annealed: 75,000 TS; 40,000 YS; 35 El; 70 RA; 155 Brin. Cold drawn: 100,000 TS; 85,000 YS; 17 El; 60 RA; 205 Brin. For oil refinery and chemical plant equipment, knives, hardware, table flatware. Corrosion resistant.

CRO-MO-LOY
Atlas Specialty Steels
C 1, Mn 1, Cr 5, V 0.25, Mo 1, Co 1, bal Fe.
For tools, dies, shear blades; air hardening, non-deforming.

CRO-MOL C-10
Continental Foundry & Machine Co.
C, Cr, V, bal Fe.
For dies; rams, saw blocks; casting.

CRO-SIL NO. 10
Colt Industries
C 0.08-0.2, Cr 10, Si 3, bal Fe.
For corrosion resistant parts; corrosion resistant. *Obsolete*

CRO-SIL NO. 14
Colt Industries
C 0.08-0.2, Cr 10, Si 3, bal Fe.
For corrosion resistant parts; corrosion resistant. *Obsolete*

CROBALITE NO. 1
Crobalt Inc.
Co 48, Cr 30, W 14.
Welded: 550 Brin. For hard facing electrode. Corrosion and abrasion resistant.

CROBALITE NO. 12
Crobalt Inc.
Co 52, Cr 30, W 10.
480 Brin. For hard facing electrode. Corrosion and abrasion resistant.

CROBALITE NO. 2
Crobalt Inc.
Co 40, Cr 33, W 18, Carbides.
620 Brin. For tools, cutters, and dies. Cast to shape; wear resistant.

CROBALITE NO. 3
Crobalt Inc.
Co 40, Cr 33, W 20, Carbides.
630 Brin. For tools, cutters, and dies. Cast to shape; wear resistant.

CROBALITE NO. 6
Crobalt Inc.
Co 57, Cr 30, W 5.
For hard facing electrode. Corrosion and abrasion resistant.

CROBALT
Michigan Tool Co.
Co 40-50, Cr 25-30, W 14-20.
Cast: 75,000 TS. For cutters, tools; wear resistant.

CROBALT NO. 1
Crobalt Inc.
Co 48, Cr 30, W 14, high C, bal Fe.
Cast: 75,000 TS; 0 El; 0 RA; 580-620 Brin. For cutting tool bits and milling cutters. High red-hardness.

CROBALT NO. 2
Crobalt Inc.
Co 40, Cr 33, W 18, bal Fe.
Cast: 61-62 Rock C. For cutters, tools, and punches. Cast alloy; wear resistant.

CROBALT NO. 3
Crobalt Inc.
Co 40, Cr 33, W 20, bal Fe.
Cast: 63-64 Rock C. For cutters, tools, and punches. Cast alloy; wear resistant.

CROCAR
Teledyne Vasco
Cr 12, C 2.15, V 0.75, Co 0.5, Si 0.3, bal Fe.
103,000 TS; 51,000 YS; 14 RA; 212 Brin. For thread rolling, threading, dies, extrusion dies, gages; heat and abrasion resisting.

CROCAR FM
Teledyne Vasco
C 2.15, Cr 12, V 0.75, Co 0.5, bal Fe.
Annealed: 103,000 TS; 51,000 YS; 14 El; 212 Brin. For thread rolling dies, extrusion dies, gages; oil or air hardened, free-cutting. *Obsolete*

CROCEM-18
Forjas Alavesas S.A.
C 0.16, Mn 1.1, Si 0.25, Cr 1, bal Fe.
Chromium carburizing steel. DIN 16 Mn Cr 5; AFNOR 16 MC5; UNI 16MC5.

CRODI
Atlas Specialty Steels
C 0.35, Mn 0.5, W 1.2, Cr 5, V 0.3, Mo 1.4, bal Fe.
For punches, dies, shear blades, extension dies; hot work steel. *Obsolete*

CRODON
Chromium Corp. of America
Cr, Fe.
For anodes in electrolytic copper cells; electroplating alloy.

CRODUR
SWB Stahlformguss Gesellschaft mbH
C 2.1, Cr 11.5, bal Fe.
For blanking and forming dies, punches; oil or air hardened, nondeforming. *Obsolete*

CRODUR SPEZIAL
SWB Stahlformguss Gesellschaft mbH
C 2.1, Cr 11.5, W 0.7, bal Fe.
For blanking and forming dies, punches; oil or air hardened, nondeforming. *Obsolete*

CRODUR V
SWB Stahlformguss Gesellschaft mbH
C 2.2, Cr 11.5, V 0.1, Mo 0.2, bal Fe.
For blanking and forming dies, punches; oil hardened, nondeforming. *Obsolete*

CRODUR ZAH
SWB Stahlformguss Gesellschaft mbH
C 1.65, Cr 11.5, V 0.1, bal Fe.
For blanking and forming dies, punches; air hardened, nondeforming. *Obsolete*

CROFER 106
VDM Nickel-Technologie AG
C 0-0.12, Cr 6-7, Si 0.6-0.9, Al 0.6-0.9, bal Fe.
For oil refinery equipment; heat and creep resistant. *Obsolete*

CROFER 113
VDM Nickel-Technologie AG
C 0-0.12, Cr 12-14, Si 1-1.3, Al 0.8-1.1, bal Fe.
For oil refinery equipment; heat and creep resistant. *Obsolete*

CROFER 118
Garford Engineering Co.
C 0-0.12, Cr 18, V 0.95, bal Fe.
For oil refinery equipment, furnace parts; corrosion and heat resistant. *Obsolete*

CROFER 119
VDM Nickel-Technologie AG
C 0-0.12, Cr 17-19, Si 0.8-1.1, Al 0.8-1.1, bal Fe.
For oil refinery equipment; heat and creep resistant. *Obsolete*

CROFER 124
VDM Nickel-Technologie AG
C 0-0.12, Cr 23-25, Al 1.3-1.6, bal Fe.
For oil refinery equipment; corrosion and heat resistant. *Obsolete*

CROFER 1300
VDM Nickel-Technologie AG
C 0-0.12, Cr 12-14, bal Fe.
Annealed: 75,000 TS; 40,000 YS; 35 El; 70 RA; 155 Brin. For valves, cutlery, turbine blades; corrosion resistant. *Obsolete*

CROFER 1300AL
VDM Nickel-Technologie AG
C 0.06, Cr 13, Al 0.2, bal Fe.
Annealed: 75,000 TS; 45,000 YS; 20 El; 180 Brin. For heat treating boxes, furnace parts, oil refinery equipment. Type 405; stainless steel, ferritic, corrosion and heat resistant. *Obsolete*

CROFER 1700
VDM Nickel-Technologie AG
C 0-0.1, Cr 15.5-17.5, bal Fe.
Annealed: 80,000 TS; 50,000 YS; 25 El; 50 RA; 150 Brin. For oil and chemical plant equipment. Type 430; stainless. *Obsolete*

CROFER 1700NB
VDM Nickel-Technologie AG
C 0.08, Cr 17, Ni 8, Cb 0.5, bal Fe.
Annealed: 85,000 TS; 40,000 YS; 50 El; 82 Rock B. For chemical plant equipment, welded tanks and structures. Type 17-8Cb. Welding grade, austenitic, stainless. *Obsolete*

CROFER 1700TI
VDM Nickel-Technologie AG
C 0-0.1, Cr 16-18, Ti = 7 x C, bal Fe.
Annealed: 80,000 TS; 50,000 YS; 25 El; 50 RA; 150 Brin. For welded oil and chemical plant equipment. Type 430Ti; stainless. *Obsolete*

CROFER 1702TI
VDM Nickel-Technologie AG
C 0-0.1, Cr 16-18, Mo 1.5-2, Ti = 7 x C, bal Fe.
Annealed: 125,000 TS; 95,000 YS; 20 El; 55 RA; 260 Brin. For pumps, marine hardware, valves. Type 431Ti; stainless. *Obsolete*

CROFER 1919
VDM Nickel-Technologie AG
C 0.15, Cr 19.5, Ni 9.5, Si 2, bal Fe.
Annealed: 80,00 TS; 35,000 YS; 55 El; 75 RA; 150 Brin. For chemical plant equipment, tanks, mixers, filters. Type 302; stainless, austenitic. *Obsolete*

CROFER 218
VDM Nickel-Technologie AG
C 0-0.12, Cr 18, Si 2, bal Fe.
Annealed: 80,000 TS; 50,000 YS; 25 El; 50 RA; 150 Brin. For oil refinery and dairy equipment, bolts. Type 430; corrosion resistant. *Obsolete*

CROFER 230
VDM Nickel-Technologie AG
C 0.12, Cr 29, bal Fe.
Annealed: 85,000 TS; 50,000 YS; 30 El; 55 RA; 180 Brin. For furnace parts, heat treat boxes; heat resistant. *Obsolete*

CROFER 2419
VDM Nickel-Technologie AG
C 0.15, Si 2, Cr 24, Ni 19, bal Fe.
Annealed: 100,000 TS; 45,000 YS; 50 El; 65 RA; 185 Brin. For valves, pumps, turbine and jet parts. Type 310; stainless, austenitic. *Obsolete*

CROFER 2420
VDM Nickel-Technologie AG
C 0.15, Cr 24, Ni 19, Mn 1.2, bal Fe.
Annealed: 100,000 TS; 45,000 YS; 50 El; 65 RA; 185 Brin. For valves, pumps, turbine and jet parts. Type 310; stainless, austenitic. *Obsolete*

CROFER 2504
VDM Nickel-Technologie AG
C 0.2, Cr 25, Ni 4, bal Fe.
Cast: 90,000 TS; 65,000 YS; 2 El; 212 Brin. For cylinder liners, valve seats and bodies. Type CC-50; corrosion and heat resistant. *Obsolete*

CROFORM
English manufacture
Co 60, Cr 30, Mo 5, bal Fe.
Cast: 100,000-119,000 TS; 50,000-95,000 YS; 5-4 El; 390 Brin. For dental alloy; corrosion resistant.

CROLOY
Disston Inc.
C 1.5, Cr 12-14, V 1, Mo 0.8, bal Fe.
For dies, gauges, broaches, wearing plates; non-deforming. *Obsolete*

CROLOY 1
Babcock & Wilcox Co.
C 0-0.15, Cr 0.8-1.2, Mo 0.45-0.6, bal Fe.
Tubes: 60,000 TS; 30,000 YS; 30 El; 170 Brin. For high temperature service; resists high temperature and pressure. *Obsolete*

CROLOY 1-1/4
Now B & W CROLOY 1-1/4.

CROLOY 1-3/4
Babcock & Wilcox Co.
C 0-0.15, Cr 1.45-2, Mo 0.6-0.8, bal Fe.
Tubes: 60,000 TS; 25,000 YS; 30 El; 170 Brin. For oil refinery equipment, heat exchangers; corrosion resistant to hot oils. *Obsolete*

CROLOY 1/2
Now B & W CROLOY 1/2.

CROLOY 12
Now B & W CROLOY 12.

CROLOY 12 AL
Now B & W CROLOY 12 AL.

CROLOY 15-15N
Babcock & Wilcox Co.
C 0-0.15, Cr 16, Ni 15, Mo 1.5, W 1.4, Cb 1, bal Fe.
Rolled: 75,000 TS; 30,000 YS; 35 El; 200 Brin. For pipes for high temperature and pressure service; corrosion resistant, high temperature use to 1500 F. *Obsolete*

CROLOY 16-1
Now B & W CROLOY 16-1.

CROLOY 16-13-3
Babcock & Wilcox Co.
C 0-0.08, Cr 16-18, Ni 11-14, Mo 2-3, Mn 0-2, Si 0-0.75, bal Fe.
At 70 F: 82,500 TS; 42,300 YS; 55 El; 72 RA; 200 Brin. At 1200 F: 53,750 TS; 35,750 YS; 42 El; 55 RA. For heat exchangers, steam piping, high temperature applications; corrosion resistant, austenitic. *Obsolete*

CROLOY 16-13-8
Babcock & Wilcox Co.
C 0.13, Mn 1.5, Si 0.6, Cr 15.5-17, Mo 2.5-3.25, Ni 12.5-14.5, bal Fe.
For tubing; for high temperature use. *Obsolete*

CROLOY 16-6 PH
Now B & W CROLOY 16-6 PH.

CROLOY 16-8-2
Babcock & Wilcox Co.
C 0.1, Cr 15, Ni 8, Mo 1.5, bal Fe.
For welding rods; corrosion and heat resistant, non-magnetic. *Obsolete*

CROLOY 18
Now B & W CROLOY 18.

CROLOY 18-13-3
Now B & W CROLOY 316.

CROLOY 18-8
Now B & W CROLOY 304.

CROLOY 18-8 STABILIZED
Babcock & Wilcox Co.
C 0.07, Mn 0.6, Si 0.75, Cr 17-19, Ni 10-12, Ti or Cb, bal Fe.
Annealed: 75,000 TS; 30,000 YS; 35 El; 200 Brin. For tubing; for high temperature use. *Obsolete*

CROLOY 18-8S
Babcock & Wilcox Co.
C 0.07, Mn 0.6, Si 0.75, Cr 17-19, Ni 8-10.5, bal Fe.
Annealed: 75,000 TS; 30,000 YS; 35 El; 200 Brin. For tubing; for high temperature use. *Obsolete*

CROLOY 19-9DL
Babcock & Wilcox Co.
C 0.28-0.55, Mn 0.7-1.5, Cr 18-20, Ni 8-11, Mo 1-1.7, W 1-1.7, Cb 0.4, bal Fe.
H.R.: 118,000 TS; 69,000 YS; 58 El; 55 RA; 216 Brin. For jet engine and gas turbine components; corrosion and heat resistant. *Obsolete*

CROLOY 2
Now B & W CROLOY 2.

CROLOY 2 SI
Babcock & Wilcox Co.
C 0.2, Si 2, Cr 2, bal Fe.
Tube: 60,000 TS; 25,000 YP. For heat exchangers in sulfuric acid plants. Good strength at high temperatures. Resists SO 2 corrosion. *Obsolete*

CROLOY 2-1/4
Now B & W CROLOY 2-1/4.

CROLOY 25-12
Babcock & Wilcox Co.
C 0-0.15, Mn 0-2, Si 0-0.75, Cr 22-24, Ni 12-15, bal Fe.
At R T: 95,000 TS; 48 El; 50,000 YS; B 85 Rock. At 1500 F: 26,000 TS; 23 El. For hot petroleum equipment; furnace parts; chemical plant equipment. Excellent corrosion and oxidation resistance. Good creep properties. *Obsolete*

CROLOY 25-12
Babcock & Wilcox Co.
C 0-0.15, Mn 0-2, Si 0-0.75, Cr 22-24, Ni 12-15, bal Fe.
At R T: 95,000 TS; 48 El; 50,000 YS; B 85 Rock. At 1500 F: 26,000 TS; 23 El. For hot petroleum equipment; furnace parts; chemical plant equipment. Excellent corrosion and oxidation resistance. Good creep properties. *Obsolete*

CROLOY 25-20
Now B & W CROLOY 310.

CROLOY 27
Now B & W CROLOY 27.

CROLOY 3M
Now B & W CROLOY 3M.

CROLOY 5
Now B & W CROLOY 5.

CROLOY 5 SI
Now B & W CROLOY 5 SI.

CROLOY 5 TI
Babcock & Wilcox Co.
C 0.15, Mn 0.5, Cr 4-6, Si 0.5, Mo 0.45-0.65, Ti, bal Fe.
Annealed: 60,000 TS; 30,000 YS; 30 El; 179 Brin. For tubing for oil refineries; for high temperature use. *Obsolete*

CROLOY 5CB
Babcock & Wilcox Co.
C 0-0.12, Cr 4-6, Mo 0.45-0.65, Cb = 10 x C, bal Fe.
Tubes: 60,000 TS; 25,000 YS; 30 El; 170 Brin. For oil refinery equipment, heat exchangers; corrosion resistant to hot oils. *Obsolete*

CROLOY 5M
Babcock & Wilcox Co.
C 0.15-0.2, Cr 4-6, Mo 0.45-0.65, bal Fe.
Annealed: 60,000 TS; 25,000 YS; 30 El; 163 Brin. For high temperature tubes; resists oxidation to 1200 F. *Obsolete*

CROLOY 7
Now B & W CROLOY 7.

CROLOY 8M
Babcock & Wilcox Co.
C 0-0.15, Si 1, Cr 7-9, Mo 0.9-1.1, bal Fe.
Tubes: 60,000 TS; 25,000 YS; 30 El; 170 Brin. For oil refinery equipment, heat exchangers; corrosion resistant to hot oils. *Obsolete*

CROLOY 9M
Now B & W CROLOY 9M.

CROMA
Lehigh Steel Corp.
C 0.35, Cr 1, Mn 0.8, bal Fe.
For machinery parts. Formerly LEHIGH CROMA.

CROMADUR
Krupp Stahl AG
stainless steel. C 0-0.15, Mn 18, Cr 12.5, V 1, Ni 0.2, bal Fe.
For blading for gas turbine on jet engine; heat resistant. *Obsolete*

CROMAL
Russian manufacture
C 1.8, Cr 20-22, Si 2.2, Al 3-4, bal Fe.
For furnace equipment; corrosion resistant, cast iron.

CROMALOY
A.C. Scott & Co. Ltd.
Ni 61, Cr 16, Mn 4, Fe 20.
107,500 TS. For electric heating devices. Heat resistant to 950°C.

CROMALOY
Midvale-Heppenstall Co.
Si 1.5-3, low C, 21% min Cr, bal Fe.
For corrosion and heat resisting parts; corrosion and heat resistant. *Obsolete*

CROMALOY IV
A.C. Scott & Co. Ltd.
Ni 80, Cr 20.
134,500 TS. For heating appliances for high temperature work. Heat resistant up to 1150°C.

CROMANSIL
Union Carbide Corp.
C 0.2, Cr 0.4-0.6, Mn 1.1-1.4, Si 0.7-0.8, bal Fe.
Rolled: 90,000 TS; 60,000 YS; 28 El; 62 RA; 200 BHN. For boilers and pressure vessels; heat treated after welding.

CROMANSIL
Union Carbide Corp.
Mn 1.1-1.4, Si 0.7-0.8, C 0.25, Cr 0.4-0.6, bal Fe.
Rolled: 115,000 TS; 70,000 YS; 25 El; 60 RA; 220 BHN. For seamless tubing, boilers, pressure vessels; good workability.

CROMANSIL
Union Carbide Corp.
C 0.1, Cr 0.4-0.6, Mn 1.1-1.4, Si 0.7-0.8, bal Fe.
Rolled: 65,000 TS; 45,000 YS; 45 El; 70 RA; 130 BHN. For staybolts, boilers, pressure vessels, ship plates, tanks; ductile.

CROMANSIL
Union Carbide Corp.
C 0.3, Cr 0.4-0.6, Mn 1.1-1.4, Si 0.7-0.8, bal Fe.
Rolled: 140,000 TS; 90,000 YS; 20 El; 50 RA; 265 BHN. Heat treated: 172,000 TS; 131,000 YS; 11 El; 32 RA; 364 BHN. For heat treated parts, gears, building, bridges; Iz-11.

CROMANSIL
Union Carbide Corp.
C 0.15, Cr 0.4-0.6, Mn 1.1-1.4, Si 0.7-0.8, bal Fe.
Rolled: 140,000 TS; 101,000 YS; 38 El; 65 RA; 150 BHN. For staybolts, ship plates, tanks, pressure vessels.

CROMANSIL
Union Carbide Corp.
C 0.25, Cr 0.4-0.6, Mn 1.1-1.4, Si 0.7-0.8, bal Fe.
Rolled: 115,000 TS; 70,000 YS; 25 El; 80 RA; 220 Brin. For seamless tubing, boilers, pressure vessels; good workability. *Obsolete*

CROMAR
SWB Stahlformguss Gesellschaft mbH
C 1, Cr, Mo, V, bal Fe.
For tools, dies, cutters; oil hardened. *Obsolete*

CROMAR W
SWB Stahlformguss Gesellschaft mbH
Cr 1, Mo, W, V, bal Fe.
For engravers' tools, reamers, broaches; high speed steel. *Obsolete*

CROMAX F
Edgar Allen Balfour Ltd.
C 0.5-0.55, Si 0.2-0.35, Mn 0.5-0.8, Cr 0.7-0.9, bal Fe.
For gears, housings, shafts; castings.

CROMAX F13
Creusot-Loire
C 1.2-1.5, Mn 0-1.5, Si 0-1.5, Cr 28-30, bal Fe.
Full annealed: 260-330 Brin (casting). Stainless steel; for pumps requiring corrosion resistance and extreme abrasion resistance. AFNOR Z 130 C 29 M.

CROMAX F6 MO
Creusot-Loire
C 0.5-0.7, Mn 0-1.5, Si 0-1.5, Cr 28-30, Mo 1.5-2.5, bal Fe.
Full annealed: 220-270 Brin. Stainless steel; for pumps requiring resistance to both corrosion and abrasion. AFNOR Z 60 CD 29.2 M.

CROMAX H
Edgar Allen Balfour Ltd.
C 0.7-0.75, Si 0.2-0.35, Mn 0.5-0.8, Cr 1.8-2, bal Fe.
For machinery parts, wear plates; castings.

CROMAZ-H
Creusot-Loire
C, alloy, bal Fe.
For furnace parts, heat treating boxes; high heat resistance to 1750 F. *Obsolete*

CROMAZ-N
Creusot-Loire
C, alloy, bal Fe.
For furnace parts, heat treating boxes; high heat resistance to 1750 F. *Obsolete*

CROME
Bethlehem Steel Corp.
C 0.5, Mn 0.5, Cr 1, Mo 0.35, bal Fe.
For impact tools; shock resistant. *Obsolete*

CROMETEX NO. 1
Grayborn Steel Co.
C 0.38-0.43, Si 0.3, Cr 0.8, Mo 0.25, Ni 1.8, Mn 0.7, bal Fe.
For arbors, axles, gears, piston rods; shock resistant.

CROMEX-1
Russian manufacture
C 0.5-1, Cr 25-32, Si, bal Fe.
For furnace parts and equipment; heat resistant.

CROMIC A
George W. Prentiss & Co.
Ni 80, Cr 20.
Annealed: 119,000 TS; 24 El. For resistance wire; heat and corrosion resistant to 2100 F. *Obsolete*

CROMIC C
George W. Prentiss & Co.
Ni 65, Cr 15, bal Fe.
Annealed: 100,000 TS; 23 El. For resistance wire; heat and corrosion resistant to 1650 F. *Obsolete*

CROMIC D
George W. Prentiss & Co.
C 0.13-0.2, Cr 20, Ni 24-30, bal Fe.
For resistance wire; heat and corrosion resistant to 1400 F. *Obsolete*

CROMIMPHY 1
Creusot-Loire
C 0-0.15, Cr 11.5-13.5, bal Fe.
Heat treated: 120,000-135,000 TS; 110,000-117,000 YS; 15-16 El; 58-63 RA; 220-240 Brin. For cutlery, valves, turbine blades; type 410; corrosion resistant. *Obsolete*

CROMIMPHY 1 BIS
Creusot-Loire
C 0.15, Cr 13, bal Fe.
Annealed: 75,000 TS; 40,000 YS; 35 El; 70 RA; 155 Brin. For turbine blades, valves, cutlery; Type 410; corrosion resistant. *Obsolete*

CROMIMPHY 1 BIS MO
Creusot-Loire
C 0.17-0.2, Cr 13, Mo 0.6, Ni 0-0.5, bal Fe.
Annealed: 90,000 TS; 50,000 YS; 25 El; 50 RA. Heat treated: 240,000 TS; 210,000 YS; 9 El; C 50 Rock. For oil refinery and chemical plant equipment, surgical instruments, cutlery. Corrosion resistant, hardenable. *Obsolete*

CROMIMPHY 2
Creusot-Loire
C 0.1, Si 0.25, Mn 0.4, Cr 13, bal Fe.
Annealed: 75,000 TS; 40,000 YS; 30 El; R 82 Rock. Heat treated: 135,000 TS; 105,000 YS; 10 El; C 30 Rock. For cutlery, knives, springs, shafts, table flatware. AISI Type 410 corrosion and heat resistant. *Obsolete*

CROMIMPHY 2100
Creusot-Loire
C, alloy, bal Fe.
For oil refinery equipment; oxidation resistant. *Obsolete*

CROMIMPHY 2200
Creusot-Loire
C, alloy, bal Fe.
For oil refinery equipment; oxidation resistant. *Obsolete*

CROMIMPHY 2C
Creusot-Loire
C 0.35, Cr 13, bal Fe.
Annealed: 95,000 TS; 50,000 YS; 25 El; 55 RA; 196 Brin. For turbine blades, valves, cutlery; Type 420; corrosion resistant. *Obsolete*

CROMIMPHY 33
Creusot-Loire
C 0.3, Cr 12, Mo 0.3, Ni 1, bal Fe.
Annealed: 95,000 TS; 50,000 YS; 25 El; 55 RA. Heat treated: 250,000 TS; 215,000 YS; 8 El; C 50 Rock. For oil refinery and chemical plant equipment, cutlery, surgical instruments, shafts. Corrosion resistant, hardenable. *Obsolete*

CROMIMPHY 4
Creusot-Loire
C 0-0.12, Cr 14-18, bal Fe.
Annealed: 70,000 TS; 40,000 YS; 30 El; 55 RA; 150 Brin. Cold drawn: 130,000 TS; 120,000 YS; 2 El; 185 Brin. For oil refinery equipment, bolts, kitchen sinks; Type 430; stainless, ferritic. *Obsolete*

CROMIMPHY 4 BIS
Creusot-Loire
C 0.15, Cr 17, Ni 1.8, bal Fe.
Annealed: 125,000 TS; 95,000 YS; 20 El; 55 RA; 260 Brin. For pumps, marine hardware, valves; Type 431; corrosion and heat resistant. *Obsolete*

CROMIMPHY A1007 MO
Creusot-Loire
C 0-0.12, Cr 12.5-13.5, Al 0.5, Mo 0.7, bal Fe.
Annealed: 75,000 TS; 40,000 YS; 35 El; 70 RA; 155 Brin. Cold drawn: 100,000 TS; 85,000 YS; 17 El; 60 RA; 205 Brin. For oil refinery and chemical plant equipment, hardware, knives. Corrosion resistant. *Obsolete*

CROMIMPHY A10M
Creusot-Loire
C 0-0.15, Cr 13, Mo 0.5-5, bal Fe.
Annealed: 75,000 TS; 40,000 YS; 35 El; 70 RA; 155 Brin. Cold drawn: 100,000 TS; 85,000 YS; 17 El; 60 RA; 205 Brin. For oil refinery and chemical plant equipment; knives, hardware, flatware. Corrosion resistant. *Obsolete*

CROMIMPHY A15
Creusot-Loire
Cr 12-14, 0.15% C min, bal Fe.
Annealed: 88,000 TS; 40,000 YS; 32 El; 68 RA; 170 Brin. For cutlery, valve trim, turbine blades; Type 420; stainless, hardenable. *Obsolete*

CROMIMPHY A15MO
Creusot-Loire
C 0.18, Mn 0.4, Si 0.1, Cr 13, Mo 0.5, bal Fe.
Annealed: 90,000 TS; 38,000 YS; 26 El; 55 RA; B 90 Rock. Heat treated: 250,000 TS; 190,000 YS; 10 El; 32 RA; C 49 Rock. For cutlery, knives, bearings, hardware, surgical instruments. Corrosion resistant, hardenable. *Obsolete*

CROMIMPHY A18MO
Creusot-Loire
C 0-0.25, Cr 13, Mo 0.5, bal Fe.
Annealed: 95,000 TS; 40,000 YS; 25 El; 55 RA; B 92 Rock. Heat treated: 250,000 TS; 210,000 YS; 8 El; 25 RA; C 50 Rock. For surgical instruments, bearings, shafts, gears, knives, cutlery. Corrosion resistant, hardenable. *Obsolete*

CROMIMPHY A20MO
Creusot-Loire
C 0.22, Mn 0.4, Si 0.2, Cr 12.5, Mo 1, bal Fe.
Annealed: 95,000 TS; 40,000 YS; 25 El; 55 RA; B 92 Rock. Heat treated: 250,000 TS; 215,000 YS; 8 El; 25 RA; C 50 Rock. For surgical instruments, cutlery, hardware, bearings, knives. Corrosion resistant, hardenable. *Obsolete*

CROMIMPHY A2100
Creusot-Loire
C 0.1, Cr 12, Mo 0.6, V 0.25, bal Fe.
Annealed: 75,000 TS; 35,000 YS; 30 El; B 90 Rock. For oil refinery and chemical plant equipment, springs, table flatware. Corrosion resistant. *Obsolete*

CROMIMPHY A2200
Creusot-Loire
C 0.14-0.2, Cr 10-13, Mo 0.6-1.2, V 0.3, bal Fe.
Annealed: 80,000 TS; 40,000 YS; 25 El; B 95 Rock. Hardened: 220,000 TS; 185,000 YS; 12 El; C 45 Rock. For surgical instruments, bearings, valves, cutlery. Corrosion resistant, hardenable. *Obsolete*

CROMIMPHY A3100
Creusot-Loire
C 0.06-0.1, Cr 11-13, Mo 0.8-1.2, Ni 0-0.5, V 0.2-0.3, W 0.8-1.2, bal Fe.
Annealed: 75,000 TS; 35,000 YS; 30 El; 70 RA; B 82 Rock. Cold drawn: 95,000 TS; 80,000 YS; 15 El; 60 RA; B 92 Rock. For springs, table flatware, knives, chemical plant equipment. Corrosion resistant. *Obsolete*

CROMIMPHY A3200
Creusot-Loire
C 0.2-0.25, Cr 11-12.5, Mo 0.9-1.2, V 0.2-0.3, W 0.9-1.2, bal Fe.
Annealed: 90,000 TS; 40,000 YS; 25 El; B 92 Rock. Hardened: 245,000 TS; 210,000 YS; 8 El; C 50 Rock. For cutlery, surgical instruments, gears, shafts, hardware. Corrosion resistant, hardenable. *Obsolete*

CROMIMPHY A5
Creusot-Loire
C 0.05, Si 0.25, Mn 0.4, Cr 13, bal Fe.
Annealed: 75,000 TS; 40,000 YS; 35 El; 155 Brin. For furnace parts, soot blowers, ladles, for non-ferrous molten metals, stirring rods. Corrosion resistant. *Obsolete*

CROMIMPHY A8
Creusot-Loire
C 0-0.08, Cr 11.5-13, Al 0.1-0.3, bal Fe.
Annealed: 80,000 TS; 40,000 YS; 23 El; 70 RA; 150 Brin. Heat treated: 175,000 TS; 145,000 YS; 21 El; 64 RA; 352 Brin. For oil refinery and chemical plant equipment; Type 405; corrosion resistant. *Obsolete*

CROMIMPHY A8MO
Creusot-Loire
C 0.08, Mn 0.4, Si 0.5, Cr 13, Mo 0.5, bal Fe.
Annealed: 75,000 TS; 40,000 YS; 35 El; 70 RA; 150 Brin. Cold drawn: 100,000 TS; 85,000 YS; 17 El; 60 RA; 205 Brin. For oil refinery and chemical plant equipment; knives, hardware, table flatware. Corrosion resistant. *Obsolete*

CROMIMPHY NO. 1
Creusot-Loire
C 0.09-0.12, Cr 13-18, bal Fe.
Heat treated: 92,000-114,000 TS; 71,000-91,000 YS; 13-25 El. For cooking utensils, food handling equipment, turbine blades; corrosion resistant; Type 403. *Obsolete*

CROMIMPHY NO. 1
Creusot-Loire
C 0.09-0.12, Cr 13-18, bal Fe.
Heat treated: 92,000-114,000 TS; 71,000-91,000 YS; 13-25 El. For cooking utensils, food handling equipment, turbine blades; corrosion resistant; Type 403. *Obsolete*

CROMIMPHY NO. 2
Creusot-Loire
C 0.19-0.27, Cr 12-14, bal Fe.
Heat treated: 107,000-143,000 TS; 85,000-123,000 YS; 10-22 El; 150-300 Brin. For surgical instruments, pump valves, pistons; Type 420; corrosion resistant. *Obsolete*

CROMIMPHY NO. 4
Creusot-Loire
C 0.07-0.12, Cr 16-18, bal Fe.
For chemical plant equipment; Type 430; corrosion resistant. *Obsolete*

CROMIMPHY NO. 6
Creusot-Loire
C 0-0.1, Cr 30, bal Fe.
For furnace parts, rabble arms, pots; heat and corrosion resistant. *Obsolete*

CROMIN-1
Gilby-Fodor S.A.
Ni 9-11, Cr 19-21, bal Fe.
For heating elements. Useful operating temperature up to 650°F.

CROMINO 0
Ambo-Stahl-Gesellschaft
C 0-0.12, Cr 13, bal Fe.
Annealed: 75,000 TS; 40,000 YS; 35 El; 70 RA; 155 Brin. For turbine blades, valves, cutlery, surgical instruments; Type 410; stainless.

CROMINO 1
Ambo-Stahl-Gesellschaft
C 0.2, Si 0.4, Cr 13, bal Fe.
Annealed: 95,000 TS; 50,000 YS; 25 El; 55 RA; 195 Brin. For turbine blades, valves, cutlery, surgical instruments; Type 420; stainless, hardenable.

CROMINO 1 M
Ambo-Stahl-Gesellschaft
C 0.17-0.22, Si 0-1, Mn 0-1, Cr 12-14, Mo 0.9-1.3, Ni 0-1, bal Fe.
Martensitic stainless steel; for valve cones, turbine blades. W.-Nr. 1.4120.

CROMINO 100
Ambo-Stahl-Gesellschaft
C 0.85-0.95, Si 0-1, Mn 0-1, Cr 15.5-17.5, Mo 0.4-0.6, V 0.2-0.3, Co 1.2-1.8, bal Fe.
Martensitic stainless; for cutlery. W.-Nr. 1.4535.

CROMINO 15
Ambo-Stahl-Gesellschaft
C 0.12-0.17, Si 0-1, Mn 0-1, Cr 12-14, bal Fe.
Martensitic stainless steel, for structural parts. W.-Nr. 1.4024; similar to AISI 410.

CROMINO 15 M
Ambo-Stahl-Gesellschaft
C 0.12-0.17, Si 0-1, Mn 0-1, Cr 12-14, Mo 1-1.3, bal Fe.
Martensitic stainless steel, for steam turbine blades. W.-Nr. 1.4119.

CROMINO 17
Ambo-Stahl-Gesellschaft
C 0.1-0.17, Si 0-1, Mn 0-1.5, Cr 15.5-17.5, Mo 0.2-0.3, S 0.15-0.35, bal Fe.
Ferritic type stainless steel; free cutting for machined parts. W.-Nr. 1.4104. AISI 430 F.

CROMINO 18
Ambo-Stahl-Gesellschaft
C 0.08, Si 0.4, Cr 17, bal Fe.
Annealed: 80,000 TS; 50,000 YS; 25 El; 50 RA; 150 Brin. Cold drawn: 130,000 TS; 120,000 YS; 2 El; 185 Brin. For oil refinery and food processing equipment, sinks; Type 430; stainless, ferritic.

CROMINO 2
Ambo-Stahl-Gesellschaft
C 0.4, Si 0.4, Cr 13, bal Fe.
Annealed: 95,000 TS; 50,000 YS; 25 El; 55 RA; 195 Brin. For cutlery, valves, surgical and dental instruments; Type 420; stainless, hardenable. *Obsolete*

CROMINO 8
Ambo-Stahl-Gesellschaft
C 0.85-0.95, Si 0-1, Mn 0-1, Cr 17-19, Mo 0.9-1.3, V 0.07-0.12, bal Fe.
Martensitic stainless steel; for cutlery. W.-Nr. 1.4112; similar to AISI 440 C.

CROMINO 8M
Ambo-Stahl-Gesellschaft
C 0.9, Cr 18, Mo 0.2, bal Fe.
Annealed: 110,000 TS; 65,000 YS; 18 El; 35 RA; 220 Brin. Heat treated: 280,000 TS; 270,000 YS; 3 El; 15 RA; 555 Brin. For cutlery, valves, ball bearings, surgical instruments; corrosion and wear resistant, hardenable.

CROMINO 8MS
Ambo-Stahl-Gesellschaft
C 0.12, Cr 16.5, Mo 0.25, S 0.2, bal Fe.
Annealed: 80,000 TS; 50,000 YS; 25 El; 50 RA; 150 Brin. For screw machine products; Type 430F; stainless, free-cutting.

CROMINO X
Ambo-Stahl-Gesellschaft
C 0-0.08, Si 0-1, Mn 0-1, Cr 13-15, bal Fe.
Ferritic type stainless steel; for structural parts and tableware. W.Nr. 1.4001. Similar to AISI 405.

CROMINO-I
TradeARBED Inc.
Stainless steel. C 0-0.15, Mn 0-1, Si 0-1, Cr 11.5-13.5, bal Fe.
Martensitic stainless steel. AISI 410. *Obsolete*

CROMINO-IV

TradeARBED Inc.
Stainless steel. C 0-0.2, Mn 0-1, Si 0-1, Cr 15-17, Ni 1.25-2.5, bal Fe.
Martensitic stainless steel. AISI 431. *Obsolete*

CROMNICKEL

VDM Nickel-Technologie AG
Cr 20, Ni 77, Mn 2.
For heating resistances up to 1100°C. *Obsolete*

CROMO

Bisset Steel Co.
C 0.9, Mn 1-2, Cr 0.5, W 0.5, bal Fe.
For tools, chisels, bull-dozers; shock tool steel. *Obsolete*

CROMO

Uddeholm Corp.
C 0.4, Mn 0.65, Cr 1.1, Mo 0.25, bal Fe.
For gears, shafts, bolts; oil hardened.

CROMO HIGH V

Peninsular Steel Co.
C 0.4, Cr 5, Mo 1.2, V 1.5, bal Fe.
Hot work tool steel; for die casting dies, shear blades; oil hardened; AISI H13.

CROMO HIGH V

Bethlehem Steel Corp.
C 0.4, Cr 5, Mo 1.2, V 1.5, bal Fe.
Hot work tool steel; for die casting dies, shear blades; oil hardened; AISI H13.

CROMO N

Peninsular Steel Co.
C 0.23, Mn 0.6, Si 1.25, Ni 0.75, Cr 10, Mo 1.2, V 1, N 0.1, bal Fe.
For die casting dies, extrusion mandrels, forging and extrusion tooling. Resists heat checking and cracking. Air hardening hot work tool steel.

CROMO N

Bethlehem Steel Corp.
C 0.25, Mn 1, Si 1, Ni 1, Cr 11, Mo 1, V 0.5, W 0.95, N 0.1, bal Fe.
For hot-work applications, extrusion mandrels, forging die inserts, pierce punches, die casting dies. *Obsolete*

CROMO V

Bethlehem Steel Corp.
C 0.35, Cr 5, V 0.4, Mo 1.5, bal Fe.
Heat treated: 220,000 TS; 186,000 YS; 13 El; 40 RA. At 1000°F: 145,000 TS; 105,000 YS; 19 El; 65 RA. For extrusion dies, gripper and header dies, aluminum and magnesium forging dies. Type H11; hot-work steel. Resists heat checking. *Obsolete*

CROMO W

Bethlehem Steel Corp.
C 0.35, Si 1.05, W 1.55, Mo 1.65, Cr 5.15, bal Fe.
Heat treated: 216,000 TS; 185,000 YS; 14 El; 52 RA; 56 Rock C. For trimmer and hot-forging dies, die-casting dies, hot shear blades, punches. Type H12; hot-work steel, oil hardened. *Obsolete*

CROMO W-55

Bethlehem Steel Corp.
C 0.55, Si 0.85-1.1, W 1-1.5, Cr 4.8-5.1, Mo 1-1.5, bal Fe.
Annealed: 103,000 TS; 65,000 YS; 24 El; 18 Rock C. Heat treated: 290,000 TS; 235,000 YS; 9 El; 22 RA; 54 Rock C. For blanking and beading dies, cold-forming dies, pneumatic tools. Type A8; hot-work steel, air hardened. Tough, shock resistant. *Obsolete*

CROMO WV

Bethlehem Steel Corp.
Cr 5, V 0.3, C, bal Fe.
For dies for hot extruding aluminum; hot work steel, resists heat checking. *Obsolete*

CROMO-CO

Bethlehem Steel Corp.
Stainless steel. C 0.27, Mn 0.6, Si 1.25, Cr 11, Ni 0.75, Co 10, Mo 1.2, W 0.45, V 1, N 0.1, bal Fe.
Martensitic, stainless steel. *Obsolete*

CROMOCO

Teledyne Firth Sterling
C 1.6, Cr 12, Co 1, Mo 1, bal Fe.
For dies, coining and blanking dies; air hardening. *Obsolete*

CROMODI

Pyramid Steel Company
C, alloy, bal Fe.
Heat treated: 280 Brin. For brake dies; preheat treated, non-deforming.

CROMODIE

Firth Brown Ltd.
C 0.35, Cr 5, Mg 1.5, V 1, Si 1, Mn 0.3, bal Fe.
For punches, crimpers, dies; oil hardened, tough. *Obsolete*

CROMODIE-W

Firth Brown Ltd.
C 0.31, Si 0.9, Mn 0.3, Cr 5, Mo 1.75, V 0.25, W 1.1, bal Fe.
For punches, crimpers, dies; oil hardened, tough. *Obsolete*

CROMOL C-9

Continental Foundry & Machine Co.
C, Cr, Mo, V, bal Fe.
Air hardened: 115,000-181,500 TS; 90,000-127,000 YS; 14-20 El; 46-63 RA; 217-375 Brin. For rams, saw blocks, dies; alloy steel casting.

CROMONITE

Continental Foundry & Machine Co.
C, alloy, bal Fe.
For chill rolls; wear resistant.

CROMOTEX NO. 2

Grayborn Steel Co.
C 0.43-0.48, Mn 0.8, Cr 0.8-1.1, Mo 0.2, Ni 1.8, bal Fe.
For arbors, axles, gears, piston rods; shock resistant.

CROMOTUNG

Ziv Steel & Wire Co.
C, alloy, bal Fe.
For taps, drills, reamers, dies, punches; oil hardened, tough.

CROMOVAN

Teledyne Firth Sterling
Steel. C 1.4-1.7, Cr 12-14, Mo 0.5-1, V 1-1.5, bal Fe.
Annealed: 110,000 TS; 210 Brin. Hardened: 278,000 TS; 214,000 YS; 1 El; 0.2 RA; 570 Brin. For dies for trimming, blanking extrusion, swaging, pressing, thread rolling, etc., taps, punchers, reamers, broaches, shear blades; heat and wear resistant. "Chromovan Triple Die Steel"; nondeforming. *Obsolete*

CROMOVAN F.M.

Teledyne Firth Sterling
Steel. C 1.55, Cr 12, V 1, Mo 1, S 0.12, bal Fe.
For blanking and drawing dies, punchers, gages; air hardened, nondeforming. *Obsolete*

CROMVA

Bethlehem Steel Corp.
C 0.35-0.5, Mn 0.6, Cr 1, V 0.2, bal Fe.
For hot work dies; hot work steel. *Obsolete*

CROMVA

Bisset Steel Co.
C 0.35, Cr 1, V 0.2, bal Fe.
For punches, dies, upsetters, crimpers; water or oil hardened. *Obsolete*

CROMVA

Bisset Steel Co.
C 0.5, Mn 0.6, Cr 1, V 0.2, bal Fe.
For tools, die casting dies; oil hardening. *Obsolete*

CRONI 11

Forjas Alavesas S.A.
C 0.14, Mn 0.8, Si 0.25, Ni 1, Cr 1, bal Fe.
Ni-Cr carburizing steel. AFNOR 16NC6; DIN 15 Cr Ni 6; BS EN-352.

CRONI 19

Forjas Alavesas S.A.
C 0.19, Mn 0.9, Si 0.25, Ni 1.1, Cr 0.95, bal Fe.
Ni-Cr carburizing steel. AFNOR 20 NC6; BS EN 352; WNI 19CN5.

CRONI-13

Forjas Alavesas S.A.
C 0.13, Mn 0.5, Si 0.25, Ni 2.9, Cr 0.7, bal Fe.
Ni-Cr carburizing steel; deep hardening. AFNOR 14NC11; UNI 15NC11; BS EN-36A.

CRONIDUR 4903

Krupp Stahl AG
C 0.08-0.12, Si 0.2-0.5, Mn 0.3-0.6, P 0-0.02, S 0-0.01, Cr 8-9.5, Mo 0.85-1.05, Ni 0-0.4, Al 0-0.04, N 0.03-0.07, Nb 0.06-0.1, V 0.18-0.25, bal Fe.
SEL 90; W. Nr. 1.4903.

CRONIDUR 4909

Krupp Stahl AG
C 0.015-0.03, Si 0-0.5, Mn 1.6-2, P 0-0.035, S 0-0.02, Cr 17-18, Mo 2.3-2.7, Ni 12-12.5, B 0-0.002, Co 0-0.25, Cu 0-0.05, N 0.06-0.08, bal Fe.
SEL 90; W. Nr. 1.4909.

CRONIDUR 4910

Krupp Stahl AG
C 0-0.04, Cr 16-18, Mo 2-2.8, Ni 12-14, N 0.1-0.18, B 0.0015-0.005, bal Fe.
Austenitic. DIN X3rNiMoN1713; W. Nr. 1.4910. 260 N/mm² YS, 550-750 N/mm² TS, 35 El.

CRONIDUR 4913

Krupp Stahl AG
Alloy Steel. C 0.16-0.22, Cr 10-11.5, Mo 0.5-1, Ni 0.3-0.8, N 0.05-0.1, V 0.1-0.3, B 0-0.005, Nb 0.3-0.6, bal Fe.
Martensitic. DIN X19CrMoVNbN111; W. Nr. 1.4913. 780 N/mm² YS, 900-1050 N/mm² TS, 10 El.

CRONIDUR 4919

Krupp Stahl AG
C 0.04-0.08, Si 0-0.75, Mn 0-2, P 0-0.045, S 0-0.03, Cr 16-18, Mo 2-2.5, Ni 12-14, bal Fe.
SEL 90; W. Nr. 1.4919.

CRONIDUR 4921

Krupp Stahl AG
C 0.15-0.23, Si 0.1-0.5, Mn 0.3-0.8, P 0-0.045, S 0-0.03, Cr 11-12.5, Mo 0.8-1.2, Ni 0-0.8, bal Fe.
SEW 670/69; W. Nr. 1.4921.

CRONIDUR 4922

Krupp Stahl AG
Alloy Steel. C 0.17-0.23, Cr 10-12.5, Mo 0.8-1.2, Ni 0.3-0.8, N 0.03-0.07, V 0.25-0.35, bal Fe.
Martensitic. DIN X20CrMoV121; W. Nr. 1.4922. 500 N/mm² YS, 700-850 N/mm² TS, 16 El.

CRONIDUR 4923

Krupp Stahl AG
Alloy Steel. C 0.18-0.24, Cr 11-12.5, Mo 0.8-1.2, Ni 0.3-0.8, N 0.03-0.07, V 0.25-0.35, bal Fe.
Martensitic. DIN X22CrMoV121; W. Nr. 1.4923. 600-700 N/mm² YS, 800-1050 N/mm² TS, 11-14 El.

CRONIDUR 4926

Krupp Stahl AG
Alloy Steel. C 0.2-0.26, Cr 11-12.5, Mo 0.8-1.2, Ni 0.3-0.8, N 0.03-0.07, V 0.25-0.35, bal Fe.
Martensitic. DIN X21CrMoV121; W. Nr.1.4926. 600 N/mm² YS, 750-900 N/mm² TS, 16 El.

CRONIDUR 4935

Krupp Stahl AG

C 0.17-0.25, Si 0.1-0.5, Mn 0.3-0.8, P 0-0.045, S 0-0.03, Cr 11-12.5, Mo 0.8-1.2, Ni 0.3-0.8, V 0.25-0.35, W 0.4-0.6, bal Fe. SEL 90; W. Nr. 1.4935.

CRONIDUR 4938

Krupp Stahl AG

Alloy Steel. C 0.08-0.15, Cr 11-12.5, Mo 1.5-2, Ni 2-3, N 0.02-0.04, V 0.25-0.4, bal Fe.
Martensitic. DIN X11CrNiMo12 and X12CRNiMo12, W.Nrs. 1.4938, 1.4933, 1.4939. For W. Nr. 1.4938, max N may equal 0.0500. 785 N/mm^2 YS, 930-1130 N/mm^2 TS, 14 El.

CRONIDUR 4941

Krupp Stahl AG

C 0.04-0.1, Cr 17-18.5, Mo 0.3-0.6, Ni 9.5-11.5, Ti 0-0.8, 5 X C less than or equal to Ti,bal Fe.
Austenitic. DIN X8CrNiTi1810; W. Nr. 1.4941. 195 N/mm^2 YS, 490-640 N/mm^2 TS, 35 El.

CRONIDUR 4948

Krupp Stahl AG

C 0.04-0.08, Si 0-0.75, Mn 0-2, P 0-0.045, S 0-0.03, Cr 17-19, Mo 0-0.5, Ni 10-12, bal Fe.
SEL 90; W. Nr. 1.4948.

CRONIDUR 4949

Krupp Stahl AG

C 0-0.04, Cr 17-19, Mo 0.2-0.5, Ni 9.5-11.5, N 0.1-0.18, bal Fe.
Austenitic. DIN X3CrNiN1811; W. Nr. 1.4949. 240 N/mm^2 YS, 500-700 N/mm^2 TS, 35 El.

CRONIDUR 4958

Krupp Stahl AG

C 0.03-0.08, Cr 19-22, Al 0.2-0.5, Ni 30-32.5, N 0-0.03, Ti 0.2-0.5, P 0-0.015, bal Fe.
Austenitic. DIN x5NiCrAlTi3120; W. Nr. 1.4958. 170-200 N/mm^2 YS, 500-750 N/mm^2 TS, 35 El.

CRONIDUR 4959

Krupp Stahl AG

C 0.05-0.1, Cr 19-22, Al 0.25-0.65, Ni 30-34, N 0-0.03, Ti 0.25-0.65, P 0-0.015, bal Fe.
Austenitic. DIN X8NiCrAlTi3221; W. Nr. 1.4959. 170 N/mm^2 YS, 500-750 N/mm^2 TS, 35 El.

CRONIDUR 4961

Krupp Stahl AG

C 0.04-0.1, Cr 15-17, Ni 12-14, Nb 0-1.2, 10 X C less than or equal to Nb, bal Fe.
Austenitic. DIN X3CrNiNb1613; W. Nr. 1.4961. 205 N/mm^2 YS, 510-690 N/mm^2 TS, 35 El.

CRONIDUR 4962

Krupp Stahl AG

C 0-0.15, Si 0-0.5, Mn 0-1, P 0-0.045, S 0-0.03, Cr 15-17, Ni 12.5-14.5, Ti 0.4-0.6, W 2.5-3, bal Fe.
SEL 90; W. Nr. 1.4962.

CRONIDUR 4967

Krupp Stahl AG

stainless steel. C 0.1, Cr 20, Ni 10, W 15, Fe 0-3, bal Co. For gas turbine and after burners. *Obsolete*

CRONIDUR 4971

Krupp Stahl AG

C 0.08-0.16, Si 0-1, Mn 0-2, P 0-0.045, S 0-0.03, Cr 20-22.5, Mo 2.5-3.5, Ni 19-21, Co 18.5-21, N 0.1-0.2, Nb 0.75-1.25, W 2-3, bal Fe.
SEL 90; W. Nr. 1.4971.

CRONIDUR 4980

Krupp Stahl AG

C 0-0.08, Si 0-1, Mn 1-2, P 0-0.03, S 0-0.03, Cr 13.5-16, Mo 1-1.5, Ni 24-27, Al 0-0.35, B 0.003-0.01, Ti 1.9-2.3, V 0.1-0.5, bal Fe.
SEL 90; W. Nr. 1.4980.

CRONIDUR 4981

Krupp Stahl AG

C 0.04-0.1, Cr 15.5-17.5, Mo 1.6-2, Ni 15.5-17.5, Nb 0-1.2, 10 X C less than or equal to Nb, bal Fe.
Austenitic. DIN X8CrNiMoNb1616; W. Nr. 1.4981. 215 N/mm^2 YS, 530-690 N/mm^2 TS, 35 El.

CRONIDUR 4986

Krupp Stahl AG

C 0.04-0.1, Si 0.3-0.6, Mn 0-1.5, P 0-0.045, S 0-0.03, Cr 15.5-17.5, Mo 1.6-2, Ni 15.5-17.5, B 0.05-0.1, Nb 0-1.2, bal Fe.
DIN 17240/76; W. Nr. 1.4986.

CRONIDUR 4988

Krupp Stahl AG

C 0.04-0.1, Cr 15.5-17.5, Mo 1.1-1.5, Ni 12.5-14.5, N 0.06-0.14, Nb 0-1.2, V 0.6-0.8, 10 X C less than or equal to Nb, bal Fe.
Austenitic. DIN X8CrNiMoVNb1613; W. Nr. 1.4988. 255 N/mm^2 YS, 540-740 N/mm^2 TS, 30 El.

CRONIFER 1212

VDM Nickel-Technologie AG

C 0-0.1, Cr 11.5-13.5, Ni 12-14, bal Fe.
For valves, pumps; corrosion resistant. *Obsolete*

CRONIFER 1613NB

VDM Nickel-Technologie AG

C 0-0.1, Cr 15-17, Ni 12-14, Nb = 10 x C + 0.4, bal Fe.
For chemical plant equipment, stainless. *Obsolete*

CRONIFER 1613NBN

VDM Nickel-Technologie AG

C 0-0.1, Cr 15.5-17.5, Ni 12.5-14.5, Mo 1.1-1.5, V 0.6-0.8, N 0.1, Nb = 10 x C + 0.4, bal Fe.
For chemical plant equipment; stainless. *Obsolete*

CRONIFER 1616NB

VDM Nickel-Technologie AG

C 0-0.1, Cr 15.5-17.5, Ni 15.5-17.5, Mo 1.6-2, Nb = 10 x C + 0.4, bal Fe.
For chemical plant equipment; stainless. *Obsolete*

CRONIFER 1704

VDM Nickel-Technologie AG

C 0-0.15, Cr 16-18, Ni 3.5-5.5, Mn 5.5-7.5, N 0.17-0.25, bal Fe.
For chemical plant equipment; stainless, austenitic. *Obsolete*

CRONIFER 1707

VDM Nickel-Technologie AG

C 0-0.15, Cr 16-18, Ni 7-8, bal Fe.
Annealed: 80,000 TS; 35,000 YS; 55 El; 75 RA; 150 Brin. For chemical plant equipment. Type 301; stainless, austenitic. *Obsolete*

CRONIFER 1713LCN-ALLOY 317LN

VDM Nickel-Technologie AG

Ni 13-14, Cr 17-18.5, Mo 4-4.5, C 0-0.03, N 0.12-0.2, bal Fe.
Minimum values at 20°C: 580 N/mm^2 TS; 285 N/mm^2 YS; 35 El. For condensers in fossil fuel and nuclear power stations. Material No. 1.4439. S31726

CRONIFER 1805

VDM Nickel-Technologie AG

C 0-0.15, Cr 17-19, Ni 4-6, Mn 7.5-10, N 0.17-0.25, bal Fe.
For chemical plant equipment; stainless, austenitic. *Obsolete*

CRONIFER 1808

VDM Nickel-Technologie AG

C 0-0.15, Si 0.4, Cr 18, Ni 8, bal Fe.
Quenched: 78,100-106,000 TS; 50 El; 130-180 Brin. For textile and food industries; stainless, austenitic. *Obsolete*

CRONIFER 1809

VDM Nickel-Technologie AG

C 0-0.07, Si 0.4, Cr 18, Ni 9, bal Fe.
Quenched: 78,100-99,500 TS; 50 El; 135-180 Brin. For textile and food equipment; stainless, austenitic. *Obsolete*

CRONIFER 1809 NB

VDM Nickel-Technologie AG

C 0-0.12, Cr 18, Ni 9.5, Cb = 8 x C, bal Fe.
Annealed: 90,000 TS; 45,000 YS; 56 El; 65 RA; 160 Brin. Cold drawn: 100,000 TS; 65,000 YS; 40 El; 60 RA; 205 Brin. For welded chemical plant equipment, tanks, mixers. Type 347; stainless, austenitic. *Obsolete*

CRONIFER 1809 TI

VDM Nickel-Technologie AG

C 0-0.12, Cr 18, Ni 9.5, Ti = 4 x C, bal Fe.
Annealed: 85,000 TS; 35,000 YS; 55 El; 65 RA; 150 Brin. Cold drawn: 95,000 TS; 60,000 YS; 40 El; 60 RA; 185 Brin. For welded chemical plant equipment, tanks, filters. Type 321; stainless, austenitic. *Obsolete*

CRONIFER 1809NC

VDM Nickel-Technologie AG

C 0.03, Cr 17-19, Ni 9-11, bal Fe.
Annealed: 85,000 TS; 35,000 YS; 65 El; 75 RA; 140 Brin. For chemical plant equipment. Type 304L; stainless, austenitic. *Obsolete*

CRONIFER 1810

VDM Nickel-Technologie AG

C 0-0.07, Si 0.4, Cr 18, Mo 2, Ni 10, bal Fe.
Quenched: 78,100-99,500 TS; 45 El; 140-180 Brin. For chemical equipment; stainless, austenitic. *Obsolete*

CRONIFER 1810 TI

VDM Nickel-Technologie AG

C 0-0.12, Cr 18, Mo 2, Ni 10.5, Ti = 4 x C, bal Fe.
Annealed: 85,000 TS; 40,000 YS; 50 El; 65 RA; 170 Brin. For welded acid resistant chemical plant equipment. Type 316Ti; stainless, austenitic. *Obsolete*

CRONIFER 1810NB

VDM Nickel-Technologie AG

C 0-0.1, Cr 16.5-18.5, Ni 10.5-12.5, Mo 2-2.5, Cb = 8 x C, bal Fe.
Annealed: 85,000 TS; 35,000 YS; 50 El; 65 RA; 160 Brin. For welded chemical plant equipment. Type 316Cb; stainless, austenitic. *Obsolete*

CRONIFER 1812

VDM Nickel-Technologie AG

C 0-0.07, Cr 16.5-18.5, Ni 11.5-13.5, Mo 2.5-3, bal Fe.
Annealed: 85,000 TS; 35,000 YS; 50 El; 65 RA; 160 Brin. For acid resistant equipment. Type 316; stainless, austenitic. *Obsolete*

CRONIFER 1812LCN-ALLOY 316LN

VDM Nickel-Technologie AG

Ni 11.5-14, Cr 16.5-18, Mo 2.5-3, C 0-0.03, N 0.14-0.22, bal Fe.
Minimum values at 20°C: 580 N/mm^2 TS; 295 N/mm^2 YS; 35 El. For condensers, strippers and reactor in urea fertilizer plants. Material No. 1.4429. S31653

CRONIFER 1812NB

VDM Nickel-Technologie AG

C 0-0.1, Cr 16.5-18.5, Ni 11.5-13.5, Mo 2.5-3, Nb = 8 x C, bal Fe.
Annealed: 85,000 TS; 35,000 YS; 50 El; 65 RA; 160 Brin. For acid resistant equipment. Type 316Cb; stainless, austenitic. *Obsolete*

CRONIFER 1812TI

VDM Nickel-Technologie AG

C 0-0.1, Cr 16.5-18.5, Ni 11.5-13.5, Mo 2.5-3, Ti = 5 x C, bal Fe.
Annealed: 85,000 TS; 35,000 YS; 50 El; 65 RA; 160 Brin. For welded chemical plant equipment. Type 316Ti; stainless, austenitic. *Obsolete*

CRONIFER 1813

VDM Nickel-Technologie AG

C 0-0.07, Cr 16.5-18.5, Ni 11.5-13.5, bal Fe.
Annealed: 85,000 TS; 35,000 YS; 60 El; 70 RA; 150 Brin. For chemical plant equipment. Type 304; stainless, austenitic. *Obsolete*

CRONIFER 1815LCSI-ALLOY 1815
VDM Nickel-Technologie AG
Ni 14.5-15.5, Cr 17-18.5, Mo 0-0.2, Cu 0-0.5, C 0-0.01, Si 3.7-4.3, N 0-0.08, bal Fe.
Minimum values at 20°C: 500 N/mm² TS; 225 N/mm² YS; 30 El. For production facilities for highly concentrated nitric acid (95-99%). Material No. 1.4361. S01815

CRONIFER 1818NB
VDM Nickel-Technologie AG
C 0-0.07, Cr 16.5-18.5, Ni 16.5-18.5, Mo 2-2.5, Cu 1.8-2.2, Nb = 8 x C, bal Fe.
For acid resistant equipment; stainless, austenitic. *Obsolete*

CRONIFER 1818TI
VDM Nickel-Technologie AG
C 0-0.07, Cr 16.5-18.5, Ni 16.5-18.5, Mo 2-2.5, Cu 1.8-2.2, Ti = 5 x C, bal Fe.
Annealed: 85,000 TS; 35,000 YS; 50 El; 65 RA; 160 Brin. For acid resistant equipment. Type 316Ti; stainless, austenitic. *Obsolete*

CRONIFER 1910 NB
VDM Nickel-Technologie AG
C 0-0.12, Cr 18, Mo 2, Ni 10.5, Cb = 8 x C, bal Fe.
Annealed: 85,000 TS; 40,000 YS; 50 El; 65 RA; 170 Brin. For welded acid resistant chemical plant equipment. Type 316Cb; stainless, austenitic. *Obsolete*

CRONIFER 1925HMO
VDM Nickel-Technologie AG
Ni 24.5-25.5, Cr 20-21, Mo 6-6.8, Cu 0.8-1, C 0-0.02, N 0.15-0.2, bal Fe.
Minimum values at 20°C: 650 N/mm² TS; 300 N/mm² YS; 40 El. For chloride phosphoric and sulfuric acid, salt concentration and crystallization, pulp and paper industry. Material No. 1.4529. N08925

CRONIFER 1925LC-ALLOY 904L
VDM Nickel-Technologie AG
Ni 24.5-25.5, Cr 20-21, Mo 4.5-5, Cu 1.2-1.8, C 0-0.02, N 0.05-0.1, bal Fe.
Minimum values at 20°C: 520 N/mm² TS; 220 N/mm² YS; 40 El. For plants for mineral acids, tanks for mineral acids. Material No. 1.4539. N08925

CRONIFER 2012
VDM Nickel-Technologie AG
C 0-0.2, Cr 19-21, Ni 11-13, Si 1.8-2.3, bal Fe.
For high temperature applications; heat resistant. *Obsolete*

CRONIFER 2205LCN-ALLOY 318LN
VDM Nickel-Technologie AG
Ni 4.5-6.5, Cr 21-23, Mo 2.5-3.5, C 0-0.03, N 0.08-0.2, bal Fe.
Minimum values at 20°C: 680 N/mm² TS; 480 N/mm² YS; 25 El. For chemical and petrochemical processing, marine service. For temperatures up to 280°C (540°F). Material No. 1.4462. S31803

CRONIFER 2328
VDM Nickel-Technologie AG
Ni 26-28, Cr 22-24, Mo 2.5-3, Cu 2.5-3.5, C 0-0.03, Ti 0.4-0.7, bal Fe.
Minimum values at 20°C: 500 N/mm² TS; 210 N/mm² YS; 35 El. For chemical processes; production, storage transportation of acids, leaching of sulfide ores. Material No. 1.4503.

CRONIFER 2418MON-ALLOY 24
VDM Nickel-Technologie AG
Ni 17-18, Cr 24-25, Mo 4-4.5, C 0-0.03, Mn 6-6.5, N 0.4-0.5, Nb 0-0.3, bal Fe.
Minimum values at 20°C: 800 N/mm² TS, 420 N/mm² YS, 30 El. For seawater piping, condensers, transport systems for slurries, filtration processes, crude oil transportation systems. Material No. 1.4565.

CRONIFER 2419MON
VDM Nickel-Technologie AG
Ni 17-18, Cr 24-25, Mo 4-4.5, Cu 0.5-1, C 0-0.03, Mn 6-6.5, N 0.4-0.5, Nb 0.1-0.3, bal Fe.
Minimum values at 20°C: 850 N/mm² TS; 470 N/mm² YS; 35 El. For seawater piping, condensers, transport systems for slurries, filtration processes, crude oil transportation systems. Material No. 1.4566.

CRONIFER 2504
VDM Nickel-Technologie AG
C 0.15-0.25, Cr 24-26, Ni 3.5-4.5, Si 0.8-1.3, bal Fe.
For furnace parts, salt pots; heat resistant. *Obsolete*

CRONIFER 2520
VDM Nickel-Technologie AG
C 0-0.2, Cr 24-26, Ni 17-21, Si 1.8-2.3, bal Fe.
For high temperature applications; heat resistant. *Obsolete*

CRONIFER 2520-ALLOY 314
VDM Nickel-Technologie AG
Ni 19-21, Cr 24-25, Si 1.5-2, C 0-0.15, 0.08 rare earths max, bal Fe.
Minimum values at 20°C: 550 N/mm² TS; 230 N/mm² YS; 30 El. For furnace components, petrochemical service. Material No. 1.4841. S31400

CRONIFER 2520NV
VDM Nickel-Technologie AG
C 0-0.2, Cr 23-25, Ni 19-21, Si 1-1.5, bal Fe.
For furnace parts, heat treat boxes; austenitic, corrosion and heat resistant. *Obsolete*

CRONIFER 2520NV-ALLOY 310S
VDM Nickel-Technologie AG
Ni 19-22, Cr 24-26, Si 0.2-0.7, C 0-0.08, 0.08 rare earths max, bal Fe.
Minimum values at 20°C: 500 N/mm² TS; 210 N/mm² YS; 35 El. Material No. 1.4845. S31008

CRONIFER 2521LC-(ALLOY 310L)
VDM Nickel-Technologie AG
Ni 20-21, Cr 24-25, Mo 0-0.1, C 0-0.02, Si 0-0.15, bal Fe.
Minimum values at 20°C: 470 N/mm² TS; 180 N/mm² YS; 40 El. For nitrate fertilizer, nitric acid production, nuclear fuel reprocessing. Material No. 1.4335.

CRONIFER 2525LCN
VDM Nickel-Technologie AG
Ni 24-25, Cr 24-25.5, Mo 2-2.5, C 0-0.02, N 0.1-0.16, bal Fe.
Minimum values at 20°C: 540 N/mm² TS; 255 N/mm² YS; 30 El. For highly strength components in urea plants. Material No. 1.4465. N08310

CRONIFER 2525TI
VDM Nickel-Technologie AG
Ni 24-25.5, Cr 24-25, Mo 2-2.5, C 0-0.03, Ti 0-0.6, 10.0 Ti/C min, bal Fe.
Minimum values at 20°C: 490 N/mm² TS; 205 N/mm² YS; 30 El. For equipment handling mineral acids. Material No. 1.4577.

CRONIFER 2803MO
VDM Nickel-Technologie AG
Ni 3-4, Cr 28-29, Mo 1.8-2.5, C 0-0.015, N 0-0.02, 0.03 C + N max, 12.0 Nb/C + N min, bal Fe.
Minimum values at 20°C: 600 N/mm² TS; 500 N/mm² YS; 16 El. For seawater cooled condenser tubes, equipment for seawater desalination and for handling hot sulfuric acid. Material No. 1.4575. S44660

CRONIFER II EXTRA
VDM Nickel-Technologie AG
Ni 65, Cr 15, bal Fe.
Annealed: 92,400-106,600 TS; 25-30 El; 45-55 RA; 130 Brin. For heat resistant parts; maximum operating temperature 1125°C. *Obsolete*

CRONIFER III
VDM Nickel-Technologie AG
Ni 30, Cr 20, bal Fe.
Annealed: 92,400-99,500 TS; 30-32 El; 50 RA; 150 Brin. For heat resistant parts; maximum operating temperature 1100°C. *Obsolete*

CRONIFER III EXTRA
VDM Nickel-Technologie AG
Ni 30, Cr 20, bal Fe.
Annealed: 92,400-99,500 TS; 30-32 El; 50 RA; 150 Brin. For heat resistant parts; maximum operating temperature 1150°C. *Obsolete*

CRONIFER IV
VDM Nickel-Technologie AG
Ni 20, Cr 24, C, bal Fe.
Rolled: 85,000-108,000 TS; 43,000 YS. For furnace parts, salt pots, heat treating equipment; heat resistant to 1050°C in continuous operation. *Obsolete*

CRONIFER IV-EXTRA
VDM Nickel-Technologie AG
Ni 20, Cr 24, C, bal Fe.
Rolled: 85,000-108,000 TS; 43,000 YS. For furnace parts, salt pots, heat treating equipment; heat resistant to 1100°C in continuous operation. *Obsolete*

CRONIFER S 1925
VDM Nickel-Technologie AG
Cu 1.2-1.8, Ni 24.5-25.5, Cr 20-21, Mo 4.5-5, C 0-0.02, N 0.05-0.1, bal Fe.
Filler metal. Minimum values at RT: 550 N/mm² TS; 250 N/mm² YS; 25 El. Material No. 1.4519.

CRONIMO
Colt Industries
C 0.5, Ni 3.25, Cr 1.1, Mo 0.25, bal Fe.
For dies; oil hardening. *Obsolete*

CRONIRO
GTE Products Corp./Wesgo Div.
Au 72, Ni 22, Cr 6.
Brazing alloy available in foil, flexibraze, wire, powder, extrudable paste and preform. Liquidus 1832°F. Solidus 1787°F.

CRONIRO
Western Gold & Platinum Co.
Au 72, Ni 22, Cr 6.
Melting point: 1787-1832°F. For high temperature brazing.

CRONIT
English manufacture
Ni 60, Cr 40.
For resistance alloys.

CRONITE
Cronite Foundry Co. Ltd.
Nickel. Ni 63.5-87, Cr 13.5-16, Mn 1, Si 0.4, Al 0.8, bal Fe
At 20°C: 67,000-78,000 TS; 0.5-1.5 El; 200-250 Brin. At 1000°C: 25,000 TS. For carburizing boxes, grids, fire doors, furnace parts; maximum operating temperature 1000°C; heat resistant.

CRONITE 428
Cronite Foundry Co. Ltd.
Stainless steel. Ni 12, Cr 25, bal Fe.
For high temperature uses; corrosion and heat resistant.

CRONITE ORD
Cronite Foundry Co. Ltd.
Nickel. Ni 55, Cr 18, bal Fe.
For high temperature uses; corrosion and heat resistant.

CRONITE SR
Cronite Foundry Co. Ltd.
Stainless steel. Ni 35, Cr 20, bal Fe.
For high temperature uses; corrosion and heat resistant.

CRONITE W.X.2
Cronite Foundry Co. Ltd.
Ni, Cr, W, Fe.
70,000-120,000 TS. For drastic temperature conditions.

CRONITUNG
Crucible Specialty Metals
C, alloy, bal Fe. *Obsolete*

CRONIX 70-NICR 70/30
VDM Nickel-Technologie AG
Cr 29-31, Fe 0-1, Si 1-1.5, Al 0-0.2, C 0-0.07, Mn 0-1, 0.01-0.04 rare earths, bal Ni.
Minimum values at 20°C: 650 N/mm^2 TS; 280 N/mm^2 YS; 25 El. For electric furnace elements and components. Material No. 2.4658. N06008

CRONIX 80-NICR 80/20
VDM Nickel-Technologie AG
Cr 19-21, Fe 0-1, Si 1-1.5, Al 0-0.2, C 0-0.08, Mn 0-1, 0.01-0.04 rare earths, bal Ni.
Minimum values at 20°C: 650 N/mm^2 TS; 280 N/mm^2 YS; 30 El. For electric furnace elements and components, industrial furnace elements. Material No. 2.4869. N06003

CROTERITE IV, ETC.
Now DELTA CROTERITE IV, ETC.

CROTORITE
Delta Metal (BW) Ltd.
Cu 89.5, Pb 0.06, Fe 0.56, Ni 6.32, Al 3.08, Mn 0.62.
50,000-80,000 TS; 16-32 RA. For paper mill machinery; corrosion resistant. *Obsolete*

CROUSE FUSIBLE ALLOY
English manufacture
Bi 45, Sn 25, Pb 25, Cd 5.
For fusible alloy, fire extinguishers; melting point 88°C.

CROVA
SWB Stahlformguss Gesellschaft mbH
C 1.15, Cr 0.65, V 0.1, bal Fe.
For cutters, drills, bearings; water hardened, wear resistant. *Obsolete*

CROVA 115
SWB Stahlformguss Gesellschaft mbH
C 1.2, Cr 0.65, V 0.1, bal Fe.
For bearings, cutters, forming dies; water hardened, wear resistant. *Obsolete*

CROVA 31
SWB Stahlformguss Gesellschaft mbH
C 0.31, Cr 0.65, V 0.1, bal Fe.
For gears, bolts, fasteners, crankshafts; oil hardened, tough. *Obsolete*

CROVA 50
SWB Stahlformguss Gesellschaft mbH
C 0.5, Mn 0.95, Cr 1.05, V 0.1, bal Fe.
For gears, bolts, springs, crankshafts; oil hardened, shock resistant. *Obsolete*

CROVAC
Vacuumschmelze GmbH
Cr, Co, Fe.
Deformable permanent magnet material for moving magnet systems, rotational frequency pick-ups, polarized relays and switching magnets. CROVAC 10/130: 12 kJ/m^3 maximum energy product. CROVAC 15/150: 15 kJ/m^3 maximum energy product. CROVAC 10/380: 32 kJ/m^3 maximum energy product. CROVAC 15/400: 34 kJ/m^3 maximum energy product.

CROVAN
Atlas Specialty Steels
C 0.35, Si 1, Cr 5, Mo 1.4, V 0.9, bal Fe.
At 1000°F: 186,200 TS; 149,100 YS; 18.6 El; 48.2 RA; 477 Brin. For die casting dies, extrusion tools, plastic molds; resists thermal shock, non-deforming. *Obsolete*

CROVANI
Haeckerstahl GmbH
C 0.58, Cr 1.05, V 0.1, bal Fe.
For springs, gears, bolts, studs; oil hardened, shock resistant.

CROW
Allegheny Ludlum Steel
C 1.2, Cr 0.5, bal Fe.
For drills, reamers, cutlery dies; gages, tools, drawing dies; water hardening. *Obsolete*

CROWN
Latrobe Steel Co.
Tool material. C 0.5, Cr 0.9, V 0.2, bal Fe.
For tools shafting; high fatigue resistance. AISI L2. *Obsolete*

CROWN
Jessop-Saville Ltd.
C 0.44, Cr 1.3, V 0.15, W 2.3, bal Fe.
For hot or cold trimming dies, punches, chipping chisels, pneumatic tools, mandrels, hot heading dies. Oil hardening, shock resistant.

CROWN
A. Cohn Ltd.
Cu 87-88, Sn 8-10, Zn 2-5.
Cast: 32,000 TS; 16,000 YS; 21 El. For bearings, pumps, engine casting; heavy duty.

CROWN "W"
Manufacturer not listed
C 0.8, W 2.5, bal Fe.
For tools, dies; oil hardened.

CROWN W 20
Boyd-Wagner Co.
C, W, bal Fe.
For tools. *Obsolete*

CRUCAST CA-15
Colt Industries
C 0.11, Cr 12.5, bal Fe.
Annealed: 95,000 TS; 65,000 YS; 25 El; 35 RA; 180 Brin. Heat treated: 120,000-190,000 TS; 105,000-160,000 YS; 5-18 El; 12-45 RA; 250-382 Brin. For chemical plant equipment, castings; corrosion resistant, hardenable. *Obsolete*

CRUCAST CB30
Colt Industries
C 0.17, Cr 20, Ni 1, bal Fe.
Annealed: 75,000 TS; 50,000 YS; 4 El; 4 RA; 170 Brin. For chemical plant equipment, castings; corrosion resistant. *Obsolete*

CRUCAST CF-8
Colt Industries
C 0.06, Cr 19, Ni 10, bal Fe.
Annealed: 71,000 TS; 40,000 YS; 50 El; 50 RA; 140 Brin. For chemical plant equipment, fittings, valve bodies; stainless, austenitic. *Obsolete*

CRUCAST CF-8C
Colt Industries
C 0.06, Cr 19, Ni 10, Cb 1, bal Fe.
Annealed: 73,000 TS; 42,000 YS; 50 El; 50 RA; 150 Brin. For welded equipment, valve bodies, mixers, filters; austenitic, stainless. *Obsolete*

CRUCAST CF-8M
Colt Industries
C 0.06, Cr 19, Ni 10, Mo 2.4, bal Fe.
Annealed: 75,000 TS; 45,000 YS; 50 El; 50 RA; 160 Brin. For chemical plant equipment, fittings, valve bodies, filters; stainless, austenitic, acid resistant. *Obsolete*

CRUCAST CF16FA
Colt Industries
C 0.12, Cr 19, Ni 10, S 0.3, Mo 0.6, bal Fe.
Annealed: 71,000 TS; 40,000 YS; 45 El; 45 RA; 140 Brin. For chemical plant equipment, castings; stainless, austenitic. *Obsolete*

CRUCAST CF20
Colt Industries
C 0.17, Cr 19, Ni 10, bal Fe.
Annealed: 73,000 TS; 42,000 YS; 50 El; 50 RA; 150 Brin. For valve bodies, pump bodies; corrosion resistant, austenitic. *Obsolete*

CRUCAST HH
Colt Industries
C 0.4, Cr 25, Ni 13, bal Fe.
Cast: 80,000 TS; 40,000 YS; 18 El; 23 RA; 170 Brin. For chain links, heat treating fixtures, furnace parts; austenitic, corrosion and heat resistant. *Obsolete*

CRUCAST HK
Colt Industries
C 0.4, Cr 26, Ni 20, bal Fe.
Cast: 75,000 TS; 40,000 YS; 16 El; 17 RA; 165 Brin. For high temperature castings, furnace equipment; heat resistant to 2100 F; austenitic. *Obsolete*

CRUCAST HT
Colt Industries
C 0.5, Cr 15, Ni 35, bal Fe.
Cast: 66,000 TS; 37,000 YS; 7 El; 8 RA; 160 Brin. For quenching fixtures, carburizing boxes; austenitic, corrosion and heat resistant. *Obsolete*

CRUCIA
Republic Steel Corp.
C 0.8-0.95, Mn 0.3-0.5, Cr 0.2-0.4, bal Fe.
For tools, dies, springs, cutters. *Obsolete*

CRUCIA STEEL
Republic Steel Corp.
C 0.75-1.15, Mn 0.3-0.5, Si 0.1-0.5, Cr 0.15-3.25, bal Fe.
For tools, dies, cutters, reamers, taps. *Obsolete*

CRUCIBLE 120 ZA
Crucible Materials Corp.
Titanium. Zr 12, Al 4.5, bal Ti.
For high temperature applications; heat and creep resistant.

CRUCIBLE 13% MN
Colt Industries
C 1.2, Mn 13, bal Fe.
For railroad frogs and tracks, wear plates; work hardens, wear and abrasion resistant. *Obsolete*

CRUCIBLE 1383
Colt Industries
C 0.08, Cr 11-13, Ni 14-16, bal Fe.
Annealed: 75,000 TS; 40,000 YS; 55 El; 72 RA; 116 Brin. Cold worked: 80,000 TS; 60,000 YS; 50 El; 70 RA; 140 Brin. For corrosion resistant parts; stainless. *Obsolete*

CRUCIBLE 17 CR-4 NI
Colt Industries
C 0-0.07, Si 0-1, Cr 15.5-17.5, Ni 3-5, Cu 3-5, Mn 0-1, 0.15-0.45% Cb + Ta, bal Fe.
Precipitation hardened 900 F: 200,000 TS; 185,000 YS; 14 El; 420 Brin. (values vary with treatment). Precipitation hardening stainless; for valves, pump parts, turbine blades, shafts, fasteners, gears, instrument parts. AMS 5643, ASTM A 564-66. *Obsolete*

CRUCIBLE 201
Colt Industries
C 0-0.15, Mn 5.5-7.5, Cr 16-18, Ni 3.5-5.5, N 0-0.25, bal Fe.
Annealed: 115,000 TS; 55,000 YS; 55 El; 185 Brin. 1/2 H-temper: 150,000 TS; 105,000 YS; 25 El. For chemical plant equipment; stainless, austenitic. *Obsolete*

CRUCIBLE 202
Colt Industries
C 0-0.15, Cr 17-19, Mn 7.5-10, Ni 4-6, bal Fe.
Annealed: 100,000 TS; 50,000 YS; 60 El; 185 Brin. For chemical plant equipment; stainless, austenitic. *Obsolete*

CRUCIBLE 218
Now HALCOMB 218.

CRUCIBLE 223

Crucible Materials Corp.
Stainless steel. C 0.08, Mn 12, Si 15.5, Ni 0-0.5, Mo 0-0.5, Cu 1, N 0.25, bal Fe.
Austenitic manganese steel. For earth moving equipment.

CRUCIBLE 25-25

Crucible Materials Corp.
Tool material. C 0.05, Cr 25, Ni 25, bal Fe.
Annealed: 90,000 TS; 40,000 YS; 40 El. For heating elements, thermal reactors.

CRUCIBLE 26-1

Crucible Materials Corp.
Stainless steel. C 0.03, Cr 26, Mo 1, bal Fe.
High temperature alloy.

CRUCIBLE 26-1

Crucible Materials Corp.
Tool material. C 0-0.04, Mn 0-0.75, Si 0-0.75, Ni 0-0.3, Cr 25, Mo 0.75, Cb 0-0.2, N 0-0.04, Ti 0.2-1, bal Fe.

CRUCIBLE 3003

Crucible Materials Corp.
Stainless steel. C 0.6, Mn 4.5, Ni 15, Cr 12.5, Al 3, bal Fe.
For corrosion resistant parts; age hardenable, high strength, stainless.

CRUCIBLE 301

Crucible Materials Corp.
Stainless steel. C 0-0.2, Ni 7, Cr 17, bal Fe.
Annealed: 110,000 TS; 40,000 YS; 60 El; 165 Brin. Rolled: 185,000 TS; 140,000 YS; 8 El; 410 Brin. For aircraft structural members, trailer bodies, diaphragms, household utensils; Type 301; stainless, austenitic.

CRUCIBLE 302

Crucible Materials Corp.
Stainless steel. C 0-0.2, Cr 18, Ni 9, bal Fe.
Annealed: 90,000 TS; 40,000 YS; 60 El; 70 RA; 165 Brin. For aircraft structural members, chemical plant equipment; Type 302; stainless, austenitic.

CRUCIBLE 302B

Colt Industries
C 0-0.2, Ni 9, Cr 18, Si 2.5, bal Fe.
Annealed: 90,000 TS; 40,000 YS; 50 El; 65 RA; 160 Brin. For heat treat fixtures, tube supports, high temperature applications; Type 302B; stainless, austenitic. *Obsolete*

CRUCIBLE 303

Crucible Materials Corp.
Stainless steel. C 0-0.2, S 0.3, Ni 9, Cr 18, bal Fe.
Annealed: 85,000 TS; 40,000 YS; 45 El; 60 RA; 170 Brin. For screw machine products, bolts, screws, fasteners; Type 303; free-cutting, stainless.

CRUCIBLE 303 PLUS

Crucible Materials Corp.
Stainless steel. C 0-0.15, Mn 0-2, Si 0-1, P 0-0.2, Cr 17-19, Ni 8-10, Mo 0-0.6, 0.15 S min, bal Fe.
Annealed: 90,000 TS; 35,000 YS; 50 El; 170 Brin. Good machinability, austenitic, corrosion resistant. For stainless hardware to be made on automatic screw machines, fasteners, bolts, studs. AMS 5640.

CRUCIBLE 303 PLUS X

Crucible Materials Corp.
Stainless steel. C 0-0.15, Mn 2.5-4.5, Si 0-1, P 0-0.2, Cr 17-19, Ni 7-10, Mo 0-0.6, 0.15 S min, bal Fe.
Annealed: 90,000 TS; 35,000 YS; 50 El; 170 Brin. Austenitic, good machinability, corrosion resistant. For stainless screw machine parts, studs, bolts, shafts, fasteners, chemical plants. AMS 5640.

CRUCIBLE 304

Crucible Materials Corp.
Stainless steel. C 0-0.08, Ni 8.5, Cr 19, bal Fe.
Annealed: 85,000 TS; 35,000 YS; 60 El; 70 RA; 160 Brin. For architectural molding and trim; Type 304; stainless, austenitic.

CRUCIBLE 304 PLUS

Crucible Materials Corp.
Stainless steel. C 0.08, Mn 0-2, Si 0-1, Cr 10-18, Ni 8-12, bal Fe.
Annealed: 85,000 TS; 35,000 YS; 60 El; 170 Brin. Austenitic, nonmagnetic, corrosion resistant; ductile. For shafts, bar and fountain equipment, food and dairy equipment, valves, marine equipment. AMS 5639.

CRUCIBLE 304L

Crucible Materials Corp.
Stainless steel. C 0-0.03, Ni 9.5, Cr 19, bal Fe.
Annealed: 85,000 TS; 35,000 YS; 60 El; 70 RA; 160 Brin. For welded structures, chemical plant equipment; Type 304L; austenitic, stainless.

CRUCIBLE 305

Crucible Materials Corp.
Stainless steel. C 0-0.12, Ni 11.5, Cr 18, bal Fe.
Annealed: 85,000 TS; 35,000 YS; 55 El; 70 RA; 180 Brin. For spun parts, cold heading, special drawing; Type 305; austenitic, stainless.

CRUCIBLE 307

Crucible Materials Corp.
Stainless steel. C 0-0.08, Mn 4, Ni 9.5, Cr 20.5, bal Fe.
For welding electrodes; austenitic, stainless.

CRUCIBLE 308

Crucible Materials Corp.
Stainless steel. C 0-0.08, Ni 11, Cr 20, bal Fe.
For welding electrodes; austenitic, stainless.

CRUCIBLE 309

Crucible Materials Corp.
Stainless steel. C 0-0.2, Ni 13.5, Cr 23, bal Fe.
Annealed: 90,000 TS; 40,000 YS; 50 El; 65 RA; 160 Brin. For furnace parts, heat treat boxes, boiler baffles; Type 309; austenitic corrosion and heat resistant.

CRUCIBLE 309B

Colt Industries
C 0-0.2, Si 2-3, Cr 22-24, Ni 12-15, bal Fe.
For furnace parts, and equipment, heat treating boxes; heat resistant, stainless, austenitic. *Obsolete*

CRUCIBLE 309S

Crucible Materials Corp.
Stainless steel. C 0-0.08, Ni 13.5, Cr 23, bal Fe.
Annealed: 90,000 TS; 40,000 YS; 50 El; 65 RA; 150 Brin. For furnace parts, heat treat boxes, baffles, tube supports; Type 309S; austenitic, corrosion and heat resistant.

CRUCIBLE 310

Crucible Materials Corp.
Stainless steel. C 0-0.25, Ni 20.5, Cr 25, bal Fe.
Annealed: 100,000 TS; 45,000 YS; 50 El; 65 RA; 180 Brin. For furnace parts and equipment, heat treat boxes, pots; Type 370; austenitic, heat resistant.

CRUCIBLE 310CB

Crucible Materials Corp.
Stainless steel. C 0-0.08, Ni 20.5, Cr 25, Cb 1, bal Fe.
Annealed: 100,000 TS; 45,000 YS; 50 El; 65 RA; 185 Brin. For welded parts, valves, pumps, furnace parts; austenitic, stainless, heat resistant.

CRUCIBLE 310S

Crucible Materials Corp.
Stainless steel. C 0-0.08, Ni 20.5, Cr 25, bal Fe.
Annealed: 100,000 TS; 45,000 YS; 50 El; 65 RA; 180 Brin. For furnace parts and equipment, heat treat boxes; Type 310S; austenitic heat resistant.

CRUCIBLE 311

Colt Industries
C 0-0.25, Si 2.5, Ni 25, Cr 19, bal Fe.
For heat and scale resistant parts; austenitic, stainless. *Obsolete*

CRUCIBLE 312

Colt Industries
C 0.1, Ni 9, Cr 30, bal Fe.
For welding electrodes; ferritic-austenitic, stainless. *Obsolete*

CRUCIBLE 314

Colt Industries
C 0-0.25, Si 2.25, Ni 20.5, Cr 24.5, bal Fe.
Annealed: 100,000 TS; 50,000 YS; 45 El; 60 RA; 180 Brin. For heat and scale resistant parts, furnace parts; austenitic, stainless, heat resistant. *Obsolete*

CRUCIBLE 316

Crucible Materials Corp.
Stainless steel. C 0-0.1, Ni 12, Cr 17, Mo 2.5, bal Fe.
Annealed: 80,000 TS; 30,000 YS; 60 El; 80 RA; 140 Brin. Cold drawn: 150,000 TS; 135,000 YS; 6 El; 300 Brin. For chemical plant equipment, agitators, digesters, kettles; Type 316; stainless, austenitic.

CRUCIBLE 316 PLUS

Crucible Materials Corp.
Stainless steel. C 0-0.08, Mn 0-2, Si 0-1, Cr 16-18, Ni 10-14, Mo 2-3, bal Fe.
Annealed: 85,000 TS; 35,000 YS; 55 El; 170 Brin. Austenitic, very good corrosion resistance, nonmagnetic, weldable. For parts for food and beverage equipment, pulp handling, chemical plants. AMS 5648 AISI Type 316 stainless.

CRUCIBLE 316L

Crucible Materials Corp.
Stainless steel. C 0-0.03, Ni 12, Cr 17, Mo 2.5, bal Fe.
Annealed: 80,000 TS; 30,000 YS; 60 El; 80 RA; 140 Brin. Cold drawn: 150,000 TS; 135,000 YS; 6 El; 300 Brin. For chemical plant equipment, agitators, kettles, digesters; Type 316L; stainless, austenitic.

CRUCIBLE 317

Colt Industries
C 0-0.1, Ni 12.5, Cr 19, Mo 3.5, bal Fe.
Annealed: 85,000-90,000 TS; 40,000-45,000 YS; 45-50 El; 160-170 Brin. For chemical plant equipment, evaporators, kettles, digesters; Type 317; austenitic, stainless. *Obsolete*

CRUCIBLE 318

Colt Industries
C 0-0.08, Ni 14, Cr 18, Mo 2.5, Cb 1, bal Fe.
For welded parts; stabilized, austenitic, stainless. *Obsolete*

CRUCIBLE 319L

Crucible Materials Corp.
Stainless steel. C 0.025, Cr 18, Ni 13.35, Mo 2.5, bal Fe.
Annealed: 84,000 TS; 39,000 YS; 52 El. For corrosion resistant parts in process industries.

CRUCIBLE 321

Crucible Materials Corp.
Stainless steel. C 0-0.08, Ni 9.5, Cr 18, Ti 0.5, bal Fe.
Annealed: 85,000 TS; 35,000 YS; 55 El; 65 RA; 165 Brin. Cold drawn: 95,000 TS; 60,000 YS; 40 El; 60 RA; 185 Brin. For chemical plant equipment, welded construction; stainless; Type 321; austenitic, stabilized.

CRUCIBLE 325

Colt Industries
C 0.4, Si 1.5, Ni 22, Cr 8.5, Cu 1.2, bal Fe.
Annealed: 80,000-110,000 TS; 40,000-75,000 YS; 30-55 El; 45-65 RA; 130-205 Brin. For crude oil and refinery equipment, heat treat boxes; Type 325; heat resistant, austenitic. *Obsolete*

CRUCIBLE 329

Colt Industries
C 0.15, Ni 4, Cr 26.5, Mo 1.25, bal Fe.
Annealed: 95,000 TS; 50,000 YS; 45 El; 60 RA; 160 Brin. For chemical plant equipment; Type 329; heat resistant. *Obsolete*

CRUCIBLE 330
Colt Industries
C 0-0.25, Ni 34.5, Cr 15, bal Fe.
Annealed: 80,000 TS; 40,000 YS; 30 El; 30 RA; 160 Brin. For furnace parts and equipment, fixtures, pyrometer tubes; Type 330; stainless, austenitic, heat resistant. *Obsolete*

CRUCIBLE 3311
Crucible Materials Corp.
Stainless steel. C 0.15, Ni 23, Cr 21.5, Al 3.25, bal Fe.
For corrosion resistant parts; age hardenable, high strength, stainless.

CRUCIBLE 3329
Crucible Materials Corp.
Stainless steel. C 0.4, Ni 6.25, Cr 24, Mo 4.25, bal Fe.
For corrosion resistance and strength; hardenable by sigma phase, stainless.

CRUCIBLE 347
Crucible Materials Corp.
Stainless steel. C 0-0.08, Ni 10.5, Cr 18, Cb 1, bal Fe.
Annealed: 90,000 TS; 40,000 YS; 50 El; 70 RA; 180 Brin. For welded structures, chemical plant equipment; Type 347; stainless, stabilized.

CRUCIBLE 347-F-SE
Crucible Materials Corp.
Stainless steel. C 0-0.08, Cr 17-19, Ni 9-13, Cb = 10 x C, P, Se, bal Fe.
For chemical plant equipment; free-cutting, stainless, austenitic.

CRUCIBLE 35% NICKEL-IRON
Crucible Materials Corp.
Nickel. Ni 35, bal Fe.
Annealed: 70,000 TS; 24,000 YS; 36 El; 68 RA; 143 Brin. Cold drawn: 90,000 TS; 70,000 YS; 20 El; 60 RA; 185 Brin. For instruments, geodetic parts, thermostats; controlled thermal expansion low.

CRUCIBLE 403
Crucible Materials Corp.
Stainless steel. C 0-0.15, Cr 12, bal Fe.
Annealed: 75,000 TS; 40,000 YS; 35 El; 70 RA; 140 Brin. Cold drawn: 100,000 TS; 85,000 YS; 60 El; 205 Brin. For flat springs, hardware, cutlery, tableware; Type 403; corrosion resistant.

CRUCIBLE 405
Colt Industries
C 0-0.08, Cr 12.5, Al 0.2, bal Fe.
Annealed: 70,000 TS; 40,000 YS; 30 El; 60 RA; 150 Brin. Cold drawn: 85,000 TS; 70,000 YS; 20 El; 60 RA; 185 Brin. For welded parts, annealing boxes, quenching racks; non-hardenable, ferritic, stainless. *Obsolete*

CRUCIBLE 406
Colt Industries
C 0.1, Cr 13, Al 4, bal Fe.
Annealed: 86,000 TS; 68,000 YS; 25 El; 57 RA; 160 Brin. For electrical resistance parts; non-hardenable, ferritic, stainless. *Obsolete*

CRUCIBLE 410
Crucible Materials Corp.
Stainless steel. C 0-0.15, Cr 12, bal Fe.
Annealed: 75,000 TS; 40,000 YS; 35 El; 70 RA; 140 Brin. Cold drawn: 100,000 TS; 85,000 YS; 60 El; 205 Brin. For flat springs, hardware, cutlery, tableware; Type 410; corrosion resistant. AMS 5613.

CRUCIBLE 414
Crucible Materials Corp.
Stainless steel. C 0-0.15, Ni 2, Cr 12, bal Fe.
Annealed: 115,000-120,000 TS; 90,000-105,000 YS; 15-20 El; 60 RA; 235-255 Brin. For corrosion resistant parts, hardenable, martensitic, stainless.

CRUCIBLE 416
Crucible Materials Corp.
Stainless steel. C 0-0.15, S 0.3, Cr 13, bal Fe.
Annealed: 75,000 TS; 40,000 YS; 30 El; 60 RA; 155 Brin. Heat treated: 110,000 TS; 85,000 YS; 18 El; 55 RA; 230 Brin. For screw machine products, gears, shafts; Type 416; free-cutting, stainless.

CRUCIBLE 416 PLUS
Crucible Materials Corp.
Stainless steel. C 0-0.15, Mn 0-1.25, Si 0-1, P 0-0.06, Cr 12-14, Mo 0-0.6, 0.15 S min, bal Fe.
Annealed: 75,000 TS; 40,000 YS; 30 El; 155 Brin. Free machining, magnetic, hardenable, corrosion resistant. For corrosion resistant hardware, screw machine parts, bolts, nuts, shafts. AISI 416; AMS 5610.

CRUCIBLE 416 PLUS X
Crucible Materials Corp.
Stainless steel. C 0-0.15, Mn 1.5-2.5, Cr 13, Si 0-1, P 0-0.06, Mo 0-0.6, 0.15 S min, bal Fe.
Annealed: 75,000 TS; 40,000 YS; 30 El; 155 Brin. Free machining, magnetic, hardenable, corrosion resistant. Parts made on automatic screw machines, bolts, studs, nuts, shafts. AISI 416. AMS 5610.

CRUCIBLE 42% NICKEL-IRON
Crucible Materials Corp.
Nickel. Ni 42, C 0-0.08, bal Fe.
Annealed: 80,000 TS; 30,000 YS; 30 El. For metal to glass seals, thermostats, temperature controls; low coefficient of expansion (controlled).

CRUCIBLE 420
Crucible Materials Corp.
Stainless steel. C 0.35, Cr 13, bal Fe.
Annealed: 95,000 TS; 50,000 YS; 25 El; 55 RA; 190 Brin. Cold drawn: 105,000 TS; 85,000 YS; 17 El; 50 RA; 215 Brin. For cutlery, dental and surgical instruments, valve trim; Type 420; stainless, hardenable.

CRUCIBLE 420 DENSIFIED
Crucible Materials Corp.
Stainless steel. C 0.3-0.4, Mn 0-1, Si 0-1, Cr 12-14, bal Fe.
Air hardenable to 250,000 TS. For plastic molds, glass molds, injection molds.

CRUCIBLE 420-F-SE
Crucible Materials Corp.
Stainless steel. Cr 12-14, 0.15 C min, P, Se, bal Fe.
For valves, cutlery, knives, free-cutting, corrosion resistant.

CRUCIBLE 420F
Crucible Materials Corp.
Stainless steel. C 0.35, S 0.3, Cr 13, bal Fe.
Annealed: 95,000 TS; 50,000 YS; 25 El; 55 RA; 190 Brin. Cold drawn: 105,000 TS; 85,000 YS; 17 El; 50 RA; 215 Brin. For cutlery, dental and surgical instruments, valve trim; Type 420F; free-cutting, stainless.

CRUCIBLE 420S
Crucible Materials Corp.
Stainless steel. C 0.2, Cr 13, bal Fe.
Annealed: 95,000 TS; 50,000 YS; 25 El; 55 RA; 196 Brin. For corrosion resistant parts, cutlery, surgical instruments; hardenable, free-cutting, stainless.

CRUCIBLE 422
Crucible Materials Corp.
Stainless steel. C 0.23, Mn 0.9, Si 0.14, Ni 0.7, Cr 13.2, V 0.25, W 1, Mo 1, bal Fe.
At 80°F: 149,000 TS; 125,000 YS; 18 El; 52 RA. At 1000°F: 96,000 TS; 82,000 YS; 25 El; 67 RA. For compressor blades, valves, furnace parts; stainless, ferritic.

CRUCIBLE 422M
Colt Industries
C 0.28, Mn 0.9, Si 0.2, Ni 0.2, Cr 12, V 0.5, W 1.7, Mo 2.2, bal Fe.
Rolled: 177,000 TS; 140,000 YS; 16 El; 46 RA; 350 Brin. Heat treated: 253,000 TS; 180,000 YS; 11 El; 36 RA. For jet engine parts, blades, disks, valves, bearings; stainless, hardenable. *Obsolete*

CRUCIBLE 430
Crucible Materials Corp.
Stainless steel. C 0-0.12, Cr 16, bal Fe.
Annealed: 70,000 TS; 40,000 YS; 30 El; 55 RA; 140 Brin. Cold drawn: 130,000 TS; 120,000 YS; 2 El. For kitchen sinks, chemical and dairy equipment, bolts; Type 430; stainless, ferritic.

CRUCIBLE 430-F-SE
Crucible Materials Corp.
Stainless steel. C 0.12, Cr 14-18, 0.15 Se min, bal Fe.
For stainless parts, shafts, valves; free-cutting, corrosion resistant.

CRUCIBLE 430F
Crucible Materials Corp.
Stainless steel. C 0-0.12, S 0.3, Cr 16, bal Fe.
Annealed: 70,000 TS; 40,000 YS; 30 El; 55 RA; 140 Brin. Cold drawn: 130,000 TS; 120,000 YS; 2 El. For screw machine products, bolts, fasteners, gears; Type 430F; stainless, free-cutting.

CRUCIBLE 431
Crucible Materials Corp.
Stainless steel. C 0-0.2, Ni 2, Cr 16, bal Fe.
Annealed: 125,000 TS; 95,000 YS; 20 El; 55 RA; 260 Brin. For corrosion resistant parts, valves, marine hardware; hardenable, martensitic, stainless.

CRUCIBLE 440-F-SE
Crucible Materials Corp.
Stainless steel. C 0.95-1.2, Cr 16.5-18.5, P, Se, bal Fe.
For bearings, valves, pivots; stainless, free-cutting, hardenable.

CRUCIBLE 440A
Crucible Materials Corp.
Stainless steel. C 0.65, Cr 17, bal Fe.
Annealed: 95,000 TS; 55,000 YS; 20 El; 250 Brin. Heat treated: 275,000 TS; 240,000 YS; 2 El; 555 Brin. For cutlery, valves, dental and surgical instruments, bearings; Type 440A; stainless, hardenable.

CRUCIBLE 440B
Crucible Materials Corp.
Stainless steel. C 0.85, Cr 17, bal Fe.
Annealed: 107,000 TS; 62,000 YS; 18 El; 35 RA; 220 Brin. Heat treated: 280,000 TS; 270,000 YS; 3 El; 15 RA; 555 Brin. For cutlery, valves, dental and surgical instruments, bearings: Type 440B; stainless, hardenable.

CRUCIBLE 440BM
Crucible Materials Corp.
Stainless steel. C 0.95, Cr 18, Mo 0.5, bal Fe.
Annealed: 107,000 TS; 62,000 YS; 18 El; 35 RA; 220 Brin. For corrosion and wear resistant parts, cutlery, bearings; hardenable, martensitic, stainless.

CRUCIBLE 440BMF
Crucible Materials Corp.
Stainless steel. C 0.95, Si 0.3, Cr 18, Mo 0.5, bal Fe.
Annealed: 107,000 TS; 62,000 YS; 18 El; 35 RA; 220 Brin. For corrosion and wear resistant parts; free-cutting, hardenable, stainless.

CRUCIBLE 440C
Crucible Materials Corp.
Stainless steel. C 1.05, Cr 17, bal Fe.
Annealed: 110,000 TS; 65,000 YS; 15 El; 30 RA; 225 Brin. Heat treated: 290,000 TS; 275,000 YS; 2 El; 12 RA; 575 Brin. For cutlery, valves, bearings, pivots; Type 440C; stainless, hardenable.

CRUCIBLE 442

Crucible Materials Corp.
Stainless steel. C 0-0.25, Cr 20.5, bal Fe.
For corrosion and heat resistant parts; non-hardenable, ferritic, stainless.

CRUCIBLE 446

Crucible Materials Corp.
Stainless steel. C 0-0.2, Mn 0-1.5, Cr 23-27, bal Fe.
Annealed: 80,000 TS; 50,000 YS; 25 El; 86 Rock B. Magnetic, corrosion resistant, non-hardenable, weldable. For heat treat equipment, radio tube parts, rotary driers and retorts, tank cars, combustion chambers. Type 446.

CRUCIBLE 501

Crucible Materials Corp.
Alloy steel. C 0.2, Cr 4-6, bal Fe.
Annealed: 70,000 TS; 30,000 YS; 28 El; 65 RA; 160 Brin. Heat treated: 175,000 TS; 135,000 YS; 15 El; 50 RA; 370 Brin. For furnace parts, valve trim, oil refinery equipment; Type 501; creep resistant.

CRUCIBLE 502

Crucible Materials Corp.
Alloy steel. C 0-0.1, Cr 4-6, Mo 0.5, bal Fe.
Annealed: 75,000 TS; 35,000 YS; 26 El; 60 RA; 160 Brin. For furnace parts, valve trim, oil refinery equipment; Type 502; creep resistant.

CRUCIBLE 52 CB BEARING STEEL

Crucible Materials Corp.
Tool material. C 0.75-0.9, Mn 0.3-0.45, Si 0.6-0.9, Cr 0.8-1.1, Ni 0.25, Mo 0.5-0.65, bal Fe.
Oil hardenable to 62-66 Rock. For bearings, bearing races, hardened shafts.

CRUCIBLE 52 CB VAR BEARING STEEL

Colt Industries
C 0.85, Mn 0.35, Si 0.75, Cr 1, Ni 0.1, Mo 0.6, bal Fe.
Oil hardenable to 62-66 Rc. For bearings, bearing races, hardened shafts. *Obsolete*

CRUCIBLE 56

Colt Industries
C 0.4, Cr 3.3, Mn 0.6, Mo 2.75, V 0.4, Si 1, bal Fe.
At 70 F: 300,000 TS; 240,000 YS. At 1000 F: 230,000 TS; 200,000 YS. For high temperature applications, shears, punches; heat resistant to 1000 F, hot work steel. *Obsolete*

CRUCIBLE A 55

Crucible Materials Corp.
Titanium. C 0-0.2, H 0-0.015, N 0-0.08, bal Ti.
Annealed: 75,000 TS; 65,000 YS; 25 El; 50 RA. For non-structural aircraft parts; commercial titanium.

CRUCIBLE A 70

Crucible Materials Corp.
Titanium. C 0.05-0.15, N 0-0.07, bal Ti.
Annealed: 90,000 TS; 80,000 YS; 20 El; 40 RA. For aircraft parts; commercial titanium.

CRUCIBLE A-110 AT

Colt Industries
Al 5, Si 2.5, bal Ti.
Cold drawn: 125,000 TS; 120,000 YS; 18 El; 40 RA; 350 Brin. At 600 F: 88,000 TS; 67,000 YS; 17 El; 45 RA. At 1200 F: 39,000 TS; 25,000 YS; 38 El; 59 RA. For compressor blades, jet engine components; high fatigue and creep resistance. *Obsolete*

CRUCIBLE A-40

Crucible Materials Corp.
Titanium. C 0.1, N 0.07, H 0.01, bal Ti.
Rolled: 65,000 TS; 50,000 YS; 28 El; 50 RA. For high temperature applications; for maximum ductility and formability.

CRUCIBLE A286

Now A-286.

CRUCIBLE AFC 77

Crucible Materials Corp.
Stainless steel. C 0.15, Cr 14.5, Co 13, Mo 5, V 0.4, bal Fe.
High strength, high temperature alloy. AMS 5748.

CRUCIBLE AM 350

Now AM 350.

CRUCIBLE B-120 VCA

Colt Industries
V 13.5, Cr 11, Al 3, bal Ti.
Rolled: 125,000 TS; 120,000 YS; 10 El. Aged: 200,000 TS; 180,000 YS; 5 El; 5-20 RA. For missile and aircraft components, honeycomb structures; age-hardened, good formability. *Obsolete*

CRUCIBLE B110 MO

Colt Industries
Mo 30, bal Ti.
For high temperature applications. *Obsolete*

CRUCIBLE BETA III

Crucible Materials Corp.
Titanium. Mo 11.5, Zr 6, Sn 4.5, bal Ti.
Heat treated: 205,000 TS; 191,000 YS; 7 El; 29 RA. Annealed: 122,000 TS; 107,000 YS; 20 El; 32 RA. For aircraft and structural fasteners. Heat treatable beta alloy.

CRUCIBLE BRIGHT E3

Crucible Materials Corp.
Stainless steel. C 0-0.08, Cr 12, Ni 0-0.5, Ti = 5 x C, bal Fe.
Corrosion resistant steel. For mufflers, sink hardware, plumbing hardware.

CRUCIBLE C-110 M

Colt Industries
Mn 8, bal Ti.
Annealed: 130,000 TS; 120,000 YS; 16 El; 32 RA; 350 Brin.
For jet engine components; for structural airframes. *Obsolete*

CRUCIBLE C-120 AV

Crucible Materials Corp.
Titanium. Al 6, V 4, bal Ti.
Annealed: 140,000 TS; 130,000 YS; 15 El; 40 RA. Heat treated: 190,000 TS; 180,000 YS; 10 El; 30 RA. For jet engine components, ordnance equipment; high temperature strength, creep resistant.

CRUCIBLE C-120AM

Colt Industries
Al 4, Mn 4, bal Ti.
Rolled: 130,000 TS; 120,000 YS; 10 El. For aircraft and jet engine components; heat resistant, low density. *Obsolete*

CRUCIBLE C-130 AM

Colt Industries
Mn 4, Al 4, C 0-0.2, bal Ti.
Rolled: 150,000 TS; 140,000 YS; 18 El; 40 RA; 350 Brin. For jet engine components, airframe forgings; alpha-beta alloy. *Obsolete*

CRUCIBLE C-135 AMO

Colt Industries
Al 6.3-7.3, Mo 3.5-4.5, C 0-0.15, H 0-0.0125, bal Ti.
Rolled: 150,000-170,000 TS; 145,000-155,000 YS; 13-18 El; 30-45 RA. For jet engine components, compressors, missile parts; alpha-beta type, high temperature stability. *Obsolete*

CRUCIBLE C105 VA

Colt Industries
V 14-17, Al 2.25-3.25, bal Ti.
Solution-annealed: 105,000 TS; 30,000 YS; 16 El; 30 RA. Aged: 190,000 TS; 182,000 YS; 5 El; 17 RA. For high temperature applications, aircraft and missile components; heat treatable, good formability and ductility. *Obsolete*

CRUCIBLE C115 AMOV

Colt Industries
Al 4, Mo 3, V 1, bal Fe.
Rolled: 115,000 YS. For high temperature applications; serviceable to 1100 F, heat treatable. *Obsolete*

CRUCIBLE C130AMO

Colt Industries
Al 6.5, Mo 3.75, bal Ti.
At room temperature: 160,000 TS; 150,000 YS; 15 El; 40 RA. At 1000 F: 100,000 TS; 90,000 YS. For jet engine discs and blades, fasteners; age-hardenable, alpha-beta alloy. *Obsolete*

CRUCIBLE CCS

Colt Industries
C 0.4, Si 1.15, Mn 0.3, Cr 5.25, W 4.25, bal Fe.
Hardened: C 55-58 Rock. For die casting dies, forging and extrusion tools, dummy blocks. Good high temperature properties. Type H14 hot work steel, air harden. *Obsolete*

CRUCIBLE CG27

Crucible Materials Corp.
Stainless steel. C 0.05, Mo 5.75, Cr 13, Ni 38, Al 1.6, Ti 2.5, Cb 0.7, B 0.01, bal Fe.
Austenitic, forgeable high temperature alloy. Turbine wheel forgings for use up to 1500°F.

CRUCIBLE CMC

Colt Industries
C, Cr, Mn, bal Fe.
For jet engine and missile components; stainless, resists oxidation to 2100 F. *Obsolete*

CRUCIBLE COLLET

Colt Industries
C 0.6, Ni 0.4, Cr 0.8, Mn 0.9, Mo 0.2, bal Fe.
For spring collets; oil hardened, shock resistant. *Obsolete*

CRUCIBLE COPPER BOND

Colt Industries
Copper welded to steel core.
For nails, staples, line wire. *Obsolete*

CRUCIBLE CPM REX 76 HIGH SPEED STEEL

Crucible Materials Corp.
Tool material. C 1.5, Cr 3.75, V 3.1, W 10, Mo 5.25, Co 9, bal Fe.
Super high speed steel, hardenable to 70 Rock. High abrasion resistance and superior red hardness. For difficult machining operations as lathe tools, milling cutters, end mills.

CRUCIBLE CSA39

Colt Industries
Ni 27, Cr 18, Mo 9, W 3, bal Fe.
For high temperature applications. *Obsolete*

CRUCIBLE CSM 420

Crucible Materials Corp.
Stainless steel. C 0.3-0.4, Mn 0-1, Si 0-1, Cr 12-14, bal Fe.
Martensitic stainless steel, hardenable to about 500 Brin. For plastic molds, glass molds, transfer molds. AISI Type 420.

CRUCIBLE CSM 6 DENSIFIED

Crucible Materials Corp.
Tool material. C 0.7, Mn 2, Si 0.3, Cr 1, Mo 1.35, bal Fe.
Air hardening tool steel; for injection molds, compression molds, transfer molds, lens molds. AISI A6.

CRUCIBLE CSM NO. 2

Colt Industries
C 0.3, Cr 1.65, Mo 0.45, bal Fe.
For plastic mold and die casting dies; oil hardened. *Obsolete*

CRUCIBLE CVM

Colt Industries
C 0.5, Cr 1, V 0.2, Mn 0.3, bal Fe.
For plastic mold and die casting dies; oil hardened. *Obsolete*

CRUCIBLE D-319

Colt Industries
C 0.07-0, Cr 17.5-19.5, Ni 11-15, Mn 0.05-9, bal Fe.
For chemical plant equipment; stainless, austenitic. *Obsolete*

CRUCIBLE D-6
Colt Industries
C 0.46, Ni 0.55, Cr 1, Mo 1, V 0.08, Mn 0-0.75, bal Fe.
Heat treated: 150,000-299,000 TS; 140,700-211,000 YS;
8.5-18.4 El; 19.0-60.8 RA; 360-570 Brin. For aircraft landing
gears and structures, rocket motor cases, pressure vessels;
ultra high strength steels. *Obsolete*

CRUCIBLE DIAMOND BRAND
Colt Industries
C 1.65, Cr 18, V 1, bal Fe.
For wear resistant parts, dies; nondeforming, stainless.
Obsolete

CRUCIBLE EZ
Crucible Materials Corp.
Stainless steel. C 0-0.08, Cr 10.5-11.75, Ni 0-0.5, Ti = 5 x C
min, bal Fe.
Corrosion resisting steel; nonhardenable.

CRUCIBLE HNM
Crucible Materials Corp.
Stainless steel. C 0.3, Mn 3.5, Si 0.5, Ni 9.5, Cr 18.5, P 0.25,
bal Fe.
At 80°F: 168,000 TS; 124,000 YS; 19.5 El; 31.5 RA; 380 Brin.
At 1200°F: 89,000 TS; 80,000 YS; 19 El; 38.5 RA. At 1800°F:
49,000 TS; 46,000 YS; 4.0 El; 16.5 RA. For aircraft and jet
engine components, structural members; age hardened,
stainless.

CRUCIBLE HNM
Colt Industries
C 0.3, Mn 3.5, Si 0.5, Ni 9.5, Cr 18.5, P 0.25, bal Fe.
At 80 F: 168,000 TS; 124,000 YS; 19.5 El; 31.5 RA; 380 Brin.
At 1200 F: 89,000 TS; 80,000 YS; 19.0 El; 38.5 RA.]At 1800 F:
49,000 TS; 46,000 YS; 4.0 El; 16.5 RA. For aircraft and jet
engine components, structural members; age hardened,
stainless. *Obsolete*

CRUCIBLE HNM
Colt Industries
C 0.3, Ni 0.5, Mn 3.5, Cr 18, bal Fe.
Annealed: 116,000 TS; 56,000 YS; 57 El; 192 Brin. Heat
treated: 168,000 TS; 124,000 YS; 19 El; 352 Brin. For
corrosion resistant parts; hardenable, corrosion resistant.
Obsolete

CRUCIBLE HOLDER BLOCK STEEL
Crucible Materials Corp.
Tool material. C 0.5, Mn 1.25, S 0.08, Cr 0.65, Mo 0.18, bal
Fe.
Hardenable to 300,000 TS max. For backers for forging dies,
brake dies, frames for plastic molds.

CRUCIBLE HRB
Colt Industries
C 0.12, Cr 2.25, Mo 0.9, Ti 0.15, V 0.5, B 0.025, Mn 0.4, bal
Fe.
Heat treated: 160,000 TS; 151,000 YS; 17 El; 60 RA. For high
temperature bolts, compressor blades and wheel; high
strength and good ductility to 1100 F. *Obsolete*

CRUCIBLE HTX
Colt Industries
C 0.45, Cr 21, Ni 8.5, Mn 8.5, Mo 1.5, P 0.23, bal Fe.
For corrosion resistant parts; stainless. *Obsolete*

CRUCIBLE JAIL BAR STEEL
Colt Industries
C, Mn, Mo, bal Fe.
For jail bars; tough. *Obsolete*

CRUCIBLE LAPELLOY
Crucible Materials Corp.
Stainless steel. C 0.3, Mn 1, Cr 12, Ni 0.3, Mo 2.75, V 0.25,
bal Fe.
For high temperature bolts and struts.

CRUCIBLE M-252
Colt Industries
C 0.15, Mn 0.5, Cr 19, Ni 55, Co 10, Mo 10, Ti 2.5, Al 1, Fe 2.
At 70 F: 175,000 TS; 98,000 YS; 25 El. At 1600 F: 71,200 TS;
70,500 YS; 39.5 El; 55.5 RA. For jet engine and gas turbine
parts; oxidation resistant to 1800 F. *Obsolete*

CRUCIBLE M-308
Colt Industries
C 0.04, Cr 13, Ni 33, Mo 3, W 6.4, Ti 2, Al 0.3, Zr 0.2, bal Fe.
For jet engine and gas turbine components; high oxidation
resistance. *Obsolete*

CRUCIBLE M50 VAR
Crucible Materials Corp.
Alloy steel. C 0.8, Cr 4, V 1, Mo 4.25, bal Fe.
Hardenable to 64 Rock. Used largely for bearings operating
at temperatures up to 800°F.

CRUCIBLE NCR124MOD
Colt Industries
C 0.55, Mn 5, Ni 9.5, bal Fe.
Cold drawn: 120,000-125,000 TS; 90,000-100,000 YS; 16 El;
30 RA; 248-331 Brin. For aircraft parts, bolts, studs; corrosion
and heat resistant. *Obsolete*

CRUCIBLE NITRARD NO. 1
Chemalloy Electronics Corp.
C 1.1, Cr 12, Mo, V, bal Fe.
For dies; non-deforming. *Obsolete*

CRUCIBLE NITRIDING 135 MOD
Crucible Materials Corp.
Alloy steel. C 0.35, Mn 0.8, S 0.1, Cr 1.15, Mo 0.2, V 1.1, bal
Fe.
For gears, shafts, camshafts, cams; nitriding steel.

CRUCIBLE ORTHOPEDIC STEEL
Colt Industries
C, alloy, bal Fe.
For orthopedic braces. *Obsolete*

CRUCIBLE PHV
Colt Industries
C 0.27, Ni 2.8, Cr 1.15, V 0.4, Mo 0.25, Al 1.15, bal Fe.
Heat treated: 170,000 TS; 159,000 YS; 16 El; 365 Brin. For
dies; oil hardened, tough. *Obsolete*

CRUCIBLE RENE 41
Crucible Materials Corp.
Superalloy. C 0.12, Cr 19, Co 11.3, Mo 10, Ti 3, Al 1.5, bal Ni.
At 70°F: 206,000 TS; 154,000 YS; 14 El. At 1500°F: 126,000
TS; 118,000 YS; 14 El. For jet engine components, after
burner parts; severely stressed high temperature applications.

CRUCIBLE S7
Crucible Materials Corp.
Tool material. C 0.5, Cr 3.25, Si 0.25, Mn 0.7, Mo 1.5, bal Fe.
Shock resisting tool steel. For punches, shears, chisels, rivet
sets, plastic molds.

CRUCIBLE SA22
Crucible Materials Corp.
Alloy steel. C 0.22, Mn 1.15, Si 1, Ni 1.15, Cr 0.5, Mo 0.4, bal
Fe.
Heat treated: 156,000 TS; 129,000 YS; 15 El; 51 RA; 340 Brin.
For structures, aircraft sheet components; good weldability,
resists softening or tempering.

CRUCIBLE SCB
Crucible Materials Corp.
Stainless steel. C 0-0.08, Ni 13.5, Cr 23, Cb 1, bal Fe.
For welded parts; stabilized austenitic stainless steel.

CRUCIBLE SELF-TEM
Colt Industries
C 0.35, Mn 0.7, Cr 0.85, W 0.5, Mo 0.4, bal Fe.
Heat treated: 137,000 TS; 120,000 YS; 20 El; 65 RA; 300 Brin.
For shafting, bolts, machinery parts; oil hardened, shock
resistant. *Obsolete*

CRUCIBLE STAINLESS IRON
Colt Industries
C 0.2, Cr 12-24, bal Fe.
For stainless articles; stainless and corrosion resistant.
Obsolete

CRUCIBLE STAINLESS IRON NO. 12
Colt Industries
C 0-0.12, Cr 12-15, bal Fe.
For stainless parts. *Obsolete*

CRUCIBLE STAINLESS IRON NO. 18
Colt Industries
Cr 17-19, C, bal Fe.
For stainless parts. *Obsolete*

CRUCIBLE STAINLESS IRON NO. 24
Colt Industries
Cr 24-27, C, bal Fe.
For stainless parts. *Obsolete*

CRUCIBLE STAINLESS STEEL
Colt Industries
Cr 11-30, C, bal Fe.
For stainless articles; corrosion resistant. *Obsolete*

CRUCIBLE STAINLESS STEEL NO. 12
Colt Industries
Cr 11.5-13, C, bal Fe.
For stainless parts; corrosion resistant. *Obsolete*

CRUCIBLE STAINLESS STEEL NO. 18
Colt Industries
Cr 17-19, C, bal Fe.
For stainless parts; corrosion resistant. *Obsolete*

CRUCIBLE STAINLESS STEEL NO. 24
Colt Industries
Cr 23-30, C, bal Fe.
For stainless parts; corrosion resistant. *Obsolete*

CRUCIBLE T303 PLUS X
Crucible Materials Corp.
Stainless steel. C 0-0.15, Mn 2.5-4.5, Cr 17-19, Ni 7-10, Mo
0-0.6, bal Fe.
Austenitic stainless steel; modified. AISI 302.

CRUCIBLE TXCR
Colt Industries
C 0.4, Mn 4.25, Si 0.8, Cr 2.4, Mo 1.3, Ni 3.75, bal Fe.
Heat treated: 136,000 TS; 4.7 El; 4.0 RA; 380 Brin. For
automotive exhaust valves; heat resistant. *Obsolete*

CRUCIBLE U.H.S. 260
Colt Industries
C, alloy, bal Fe.
For aircraft landing gears; high strength. *Obsolete*

CRUCIBLE UHS 260
Colt Industries
C 0.35, Mn 1.25, Si 1.5, Cr 1.25, Mo 0.35, V 0.2, bal Fe.
Heat treated: 270,000 TS; 230,000 YS; 11 El; 40 RA; 520 Brin.
For landing gears, bomb release hooks, axles; shock
resistant, high strength. *Obsolete*

CRUCIBLE VERSASTEEL
Colt Industries
C 1, Mn 0.3, Si 2, Cr 4.25, V 1.15, Mo 2.5, W 0.3, bal Fe.
Heat treated: 232,000-360,000 TS; 204,000-307,000 YS; 2-15
El; 2.5-3.0 RA; C 47-61 Rock. For thread chasers, drills, taps,
reamers, dies, punches, thread rolling and swaging dies,
slitter knives, cold shears. Tough and wear resistant.
Obsolete

CRUCIBLE WASPALOY
Colt Industries
C 0.07, Cr 19, Ni 56, Co 14, Mo 4.3, Ti 3, Al 1.3, Fe 1.
At 70 F: 188,000 TS; 115,000 YS; 28 El; 25.5 RA. At 1400 F:
117,000 TS; 99,000 YS; 28 El; 41 RA. For jet engine turbine
buckets and discs, high temperature bolts; high oxidation
resistance. *Obsolete*

CRUCIBLE WF11

Crucible Materials Corp.
Stainless steel. C 0.15, Cr 19-21, Ni 9-11, bal Fe.
Heat treated: 160,000 TS; 85,000 YS; 55 El. For aircraft turbine blades, after burners; corrosion and heat resistant.

CRUCIBLE-F

Accurate Brass Co.
C 0.4, Ni 1.5, bal Fe.
Cast: 80,000 TS; 45,000 YS; 180 Brin. For crankshafts, gears, machinery castings; oil hardening, shock resistant.

CRUCIN

Vereinigte Deutsche Nickel-Werke AG
Ni 44, Cu 56.
For kitchen utensils, kettles; corrosion and wear resistant. *Obsolete*

CRUICBLE AM 355

Now AM 355

CRUSADER

Latrobe Steel Co.
Tool material. C 1.02, Cr 4, W 6.1, V 2.4, Mo 5, bal Fe.
Hardened: 65-68 Rock C. For form cutters, roll turning cutters, lathe and planer tools. Good machinability rating. Type M-3, Type 2 high-speed steel, high red-hardness and edge toughness.

CRUSADER XL

Latrobe Steel Co.
Tool material.
Now CRUSADER.

CRUSCA "V.J." EXTRA

Colt Industries
C, W, bal Fe.
For tools, twist drills, cutting and blanking dies. *Obsolete*

CRUSCA "V.J." SPECIAL

Colt Industries
C, W, bal Fe.
For tools, twist drills, cutters, cutting and blanking dies. *Obsolete*

CRUSCA 12B

Colt Industries
C 0.1, Ni 3.5, Mn 0.5, Cr 1.6, bal Fe.
Heat treated; 165,000 TS; 135,000 YS; 16 El; 50 RA; 380 Brin. For machine cut plastic molds; carburizing steel. *Obsolete*

CRUSCA COLD HUBBING

Colt Industries
C 0.07, Mn 0.15, bal Fe.
Heat treated: 60,000 TS; 40,000 YS; 30 El; 65 RA; 120 Brin. For cold hubbed dies for plastic molds; case hardening steel, water hardened. *Obsolete*

CRUSCA FLAME HARDENING

Colt Industries
C 0.7, Mn 0.8, Cr 0.3, bal Fe.
For flame hardened parts; water hardening. *Obsolete*

CRUSCA NEW PROCESS HOLLOW DRILL

Colt Industries
C 0.8, bal Fe.
For tools, hollow drills; water hardening. *Obsolete*

CRUSCA SOLID DRILL

Colt Industries
C 0.8, bal Fe.
For solid drills, tools; water hardening. *Obsolete*

CRUSCO STEEL

Crucible Steel Castings Co.
C 0.7, W 18, Cr 4, V 1, Co 5, bal Fe.
For piercing points, rolling mill plugs, dies, forming rolls. High abrasion resistance at high temperature. *Obsolete*

CRUSHER BEARING BRONZE

Janney Cylinder Co.
Sn, Pb, bal Cu.
For bearings; heavy loads. *Obsolete*

CRWXI

Thyssen Edelstahlwerke AG
C 0.9, Cr 1.7, bal Fe.
For rolls for steel mills; oil hardened. *Obsolete*

CRYOGENIC TENELON

United States Steel Corp.
C 0.1, Mn 15.1, Si 0.7, Cr 17.5, Ni 5.5, N 0.38, bal Fe.
High strength at low temperatures.

CRYOPERM

Vacuumschmelze GmbH
Ni 77.
Soft magnetic alloy for cryogenic engineering.

CRYSTALLOY

Transformer Steels Ltd.
Si 3, bal Fe.
For transformers, lamination for motors and generators. High magnetic permeability.

CS

Thyssen Edelstahlwerke AG
C 0.8, Cr 1, Si 0.8, bal Fe.
For shears, knives, cutters; oil or water hardened. *Obsolete*

CS 2700

R. Lavin & Sons, Inc.
Aluminum casting alloy with 0.6 Cu maximum. Sand cast: 24,000 TS; 16,000 YS; 2 El; 57 Brin. T6: 37,000 TS; 27,000 YS; 2 El; 80 Brin. Improved corrosion resistance. *Obsolete*

CS 2700 D

R. Lavin & Sons, Inc.
Die cast aluminum alloy. CS 2700. As cast: 41,000 TS; 8.5 El. *Obsolete*

CS1020

Eagle & Globe Steel Ltd.
Carbon steel. C 0.15-0.25, Mn 0.3-0.9, P 0-0.06, S 0-0.06, bal Fe.
430 MPa TS; 13 El. General purpose cold finished mild steel for machined parts; case hardened and weldable. BSS970 EN Series EN3B.

CS1030

Eagle & Globe Steel Ltd.
Carbon steel. C 0.25-0.35, Mn 0.3-0.9, P 0-0.06, S 0-0.06, bal Fe.
555 MPa TS; 9 El. General purpose medium high tensile steel for shafts and medium stressed parts; weldable and can be heat treated. BSS970 EN Series EN6A.

CS5

Teledyne Firth Sterling
Tool material. Al_2O_3/TiC.
Hot pressed ceramic for machining cast irons and high hardness materials for higher speeds.

CSA

Colt Industries
C 0.2-0.3, Mn 5, Cr 18-20, Ni 6, Mo 1.5, W 1.5, Cb 0.4, bal Fe.
For discs and wheels for gas turbines; high heat resistance. *Obsolete*

CSA

Thyssen Edelstahlwerke AG
C 0.6, Si 1.6, Mn 0.6, Cr 0.8, bal Fe.
For tools, springs, chisels; shock resistant. *Obsolete*

CSA-G40.8

Algoma Steel Corp. Ltd.
C 0.15-0.22, Mn 0.8, N 0-0.008, Si 0-0.35, bal Fe.
Rolled: 65,000-85,000 TS; 30,000-40,000 YS; 20 El. For structures, bridges, buildings; high yield strength and good weldability. *Obsolete*

CSF-10

Cyprus Foote Mineral Co.
Ce 9-11, Si 36-40, Al 0-0.5, Ca 0-0.2, 10.5-15.0 total rare earths, bal Fe.
Ferroalloy for inoculant for gray iron and nodularizing supplement for ductile iron. *Obsolete*

CSM 2

Slater Steels Corp.
Tool material. C 0.3, Cr 1.65, Si 0.65, Mo 0.4, Mn 0.75, bal Fe.
Standard tool steel for the production of machine cut plastic molds and zinc die casting dies.

CSM 2

Eagle & Globe Steel Ltd.
Tool material. C 0.3, Si 0.5, Mn 0.8, Cr 1.65, Mo 0.4, bal Fe.
Heat treated: 300 Brin. For machine cut molds and dies, casting dies for zinc and lead. AS1239 P102A; AISI P20.

CSM 21

Crucible Materials Corp.
Stainless steel. C 0.02, Cr 15, Ni 4.75, Cu 3.5.
150,000-200,000 TS psi (RT); 110,000-185,000 YS psi (RT); 10-14 El (in 2 in., RT). For molds; corrosion resistant holder blocks and machinery parts; resistance to a wide variety of corrosive media. While corrosion resistance is good in all conditions, aging is recommended for best performance in severe environments. Nitriding reduces corrosion resistance.

CSM-2

Crucible Specialty Steel
C 0.3, Mn 0.75, Si 0.5, Cr 1.65, Mo 0.4, bal Fe.
For Zn and Sn die casting dies; oil or water hardened.

CSM-2

Colt Industries
C 0.3, Mn 0.75, Si 0.5, Cr 1.65, Mo 0.4, bal Fe.
For Zn and Sn die casting dies; oil or water hardened.

CSN 13-123

Polish manufacture
C 0.23, Mn 1.1, V 0.2, bal Fe.
For gears, shafts, hardware; cast case hardening and water hardening.

CSNG

Teledyne Firth Sterling
Tool material.
Silicon nitride for machining gray cast irons at high speeds.

CSP

Thyssen Edelstahlwerke AG
C 1.2, Cr 0.5, bal Fe.
For tools, drills, hobs; water hardening. *Obsolete*

CSV1

Thyssen Edelstahlwerke AG
C 0.61, Mn 0.75, Cr 1.2, V 0.1, bal Fe.
For springs, punches; oil hardened, shock resistant. *Obsolete*

CU-BE 250

Telcon Metals Ltd.
Be 2, Co 0.25, bal Cu.
Heat treated: 200,000 TS; 180,000 YS; 1-3 El. For wire and strip springs, clips, connectors. Non-magnetic, non-sparking. Age hardenable, corrosion resistant.

CU-BE 275

Telcon Metals Ltd.
Be 2.3-2.8, Co 0.3-0.6, bal Cu.
Beryllium copper alloy.

CU-BE 50

Telcon Metals Ltd.
Be 0.5, Co 2.5, bal Cu.
Annealed: 50,000 TS; 25,000 YS; 30 El; 30 Rock B. Heat treated: 120,000 TS; 105,000 YS; 10 El; 100 Rock B. For wire and strip springs, clips, connectors. Age hardenable; corrosion resistant. Non-magnetic and non-sparking.

CU-LEAD-ITE NO. 1
Manufacturer not listed
Copper. Cu 70, Pb 20, Sn 10.
Cast: 31,000 TS; 11 El; 12 RA; 68 Brin. For nuts, parts for pneumatic tools subjected to severe pounding and abuse; tough, hard casting. *Obsolete*

CU-LEAD-ITE NO. 2
Manufacturer not listed
Copper. Cu 68, Pb 25, Sn 7.
Cast: 25,000 TS; 16 El; 15 RA; 48 Brin. For severe service such as bearings in rolling mills, wire mills, diesel engines, ball mills and gyratory crushers; heavy duty. *Obsolete*

CU-LEAD-ITE NO. 3
Manufacturer not listed
Copper. Cu 65, Pb 30, Sn 5.
Cast: 23,200 TS; 17 El; 18 RA; 43 Brin. For all round bearing service in high speed engines, locomotives, mills and railroads; heavy duty. *Obsolete*

CU-LEAD-ITE NO. 4
Manufacturer not listed
Copper. Cu 60, Pb 38, Sn 2.
Cast: 18,500 TS; 14 El; 15 RA; 38 Brin. For bearings for all small machinery, deep well pumps, conveyors, connecting rods, automobiles and trucks; heavy duty. *Obsolete*

CU-LEAD-ITE NO. 5
Manufacturer not listed
Cu 50, Pb 50.
Cast: 6000 TS; 20 El; 21 RA; 14 Brin. For railroad locomotives, piston packers, metallic packing, bibs, etc.; withstands superheated steam. *Obsolete*

CU-LEAD-ITE NO. 6
Manufacturer not listed
Copper. Cu 66, Pb 34.
Cast: 8910 TS; 15 El; 16 RA; 28 Brin. For small high speed machinery, pumps, electric drills, loose pulleys, small tools; heavy duty. *Obsolete*

CU30
Herbert Cutanit Ltd.
Sintered carbide tool material. For mining and quarrying. National Coal Board grade H (England). *Obsolete*

CU35
Herbert Cutanit Ltd.
Sintered carbide tool material. For mining and quarrying. National Coal Board grade M. *Obsolete*

CU40
Herbert Cutanit Ltd.
Sintered carbide tool material. For mining and quarrying. National Coal Board grade T. *Obsolete*

CU50
Herbert Cutanit Ltd.
Sintered carbide tool material. For mining and quarrying. National Coal Board grade XT. *Obsolete*

CUBE INJECTION MOLD
Telcon Metals Ltd.
Be 2.7, Co 0.5, bal Cu.
Heat treated: 195,000 TS; 170,000 YS; 2 El. For plastic injection molds and zinc die casting dies. Age-hardenable. Corrosion resistant. *Obsolete*

CUBE-ALLOY
Handy & Harmon
BeO, bal Cu.
At 68 F: 80,000 TS; 63,000 YS; 12 El; 50 RA. At 1600 F: 23,000 TS; 22,000 YS; 2 El; 1 RA. For motor windings, springs, magnet wire, cables, welding electrode tips. Beryllia-strengthened copper. High resistance to creep. *Obsolete*

CUBEX
Westinghouse Electric Corp.
Si 3.2, bal Fe.
For electrical and magnetic equipment, motors, transformers; cores. Doubly grain oriented. *Obsolete*

CUBOND 14L
SCM Metal Products Inc.
One of several copper brazing pastes for use in brazing ferrous components in a reducing atmosphere furnace.

CUBRAZ
Wall Colmonoy Corp.
99.0 Cu min.
Brazing temperature: 2000-2100°F. Copper powder mixed in gel-type binder for air powered applicators. For brazing iron or steel assemblies. AWS BCu-1a; AMS 4740A.

CUFENIUM
American manufacture
Cu 60-72, Ni 20.5-22, bal Fe.
For tableware; nickel silver.

CUFENLOY-30
Phelps Dodge Industries
Ni 29.1, Fe 0.5, Mn 0.35, bal Cu.
At 75°F: 77,000 TS; 61,000 YS; 5 El. At 700°F: 62,000 TS; 19,000 YS; 62 El. At 1050°F: 43,000 TS; 16,000 YS; 48 El. For heat exchanger tubes; high strength, ductility and stress corrosion resistant.

CUFENLOY-40
Phelps Dodge Industries
Ni 41, Fe 2, Mn 1, bal Cu.
Rolled: 85,000 TS. For high-pressure feed water heaters; high strength, high temperature alloy. *Obsolete*

CUFERCO
Westinghouse Electric Corp.
Cu 96, Fe 2, Co 2.
For spot welding tips; hardenable; electric conductivity equal to 70% Cu. *Obsolete*

CUIVRALU/CUIVRAL
Cegedur Pechiney
Aluminum.
Copper plated aluminum.

CUIVRE POLI
French manufacture
Cu 70, Zn 30.
For cartridges, shell cases, condenser tubes; maximum ductility.

CUIVRELECT
French manufacture
Cu alloy.
For welding rods.

CULVER NCS
Jonas & Culver Ltd.
C 0.5, W 2, Cr 1.65, V 0.25, bal Fe.
Annealed: 70,000 TS; 30 El; 64 RA; 165 Brin. Heat treated: 245,000 TS; 5 El; 18 RA; 525 Brin. For punches, dies, chisels, pneumatic tools; shock resistant, oil hardened.

CUMANITE
Abex Corp. (AMSCO Div.)
C, Mn, bal Fe.
For machine tool parts; wear resistant. *Obsolete*

CUMBERLAND
Brown-Wales Co.
C 0.18-0.23, bal Fe.
Rolled: 60,000-70,000 TS; 35,000-45,000 YS; 25-35 El; 50-60 RA. For shafts, gears; case hardening.

CUMLOY
Now WEST NO. 9, 10, ETC.

CUNIC
George W. Prentiss & Co.
Ni 45, Cu 55.
Annealed: 62,000 TS; 25 El. For rheostats, shunts, thermocouples; low temperature resistance.

CUNICO
General Electric Co.
Cu 50, Ni 21, Co 29.
Rolled: 85,000 TS; 210 Brin. For permanent magnets; high coercive force. *Obsolete*

CUNICO
Indiana General
Cu 50, Ni 21, Co 29.
Rolled: 85,000 TS; 210 Brin. For permanent magnets; high coercive force. *Obsolete*

CUNICO 1
Electronic Memories & Magnetics Corp.
Cu 50, Ni 21, Co 29.
Rolled: 85,000 TS; 210 Brin. Annealed: residual induction 3400 gauss; coercive force 710 oersted; Bo 2000. For magnetic and electrical equipment. Permanent magnet, high coercive force.

CUNIFE
Hoskins Mfg. Co.
Copper. Cu 60, Ni 20, Fe 20.
Ductile permanent magnet alloy. Speedmeters, instruments, electronic equipment, and control systems.

CUNIFE 1
Electronic Memories & Magnetics Corp.
Cu 60, Ni 20, Fe 20.
Permanent magnet, cold reduced. Peak energy product 1.4 x 10^6 max; residual induction 5.5 kilogauss; coercive force 530 oersted. Most ductile grade of permanent magnet.

CUNIFE I
General Electric Co.
Cu 60, Ni 20, Fe 20.
Rolled: 120,000 TS; 125 Brin. For permanent magnets; high coercive force. *Obsolete*

CUNIFER 10-ALLOY CUNI 90/10
VDM Nickel-Technologie AG
Ni 9-11, Fe 1-1.8, Mn 0.5-1, bal Cu.
Minimum values at 20°C: 300 N/mm^2 TS; 100 N/mm^2 YS; 30 El. For seawater piping systems, condensers and heat exchangers in seawater desalination equipment, sheathing of platform legs. Material No. 2.0872. C70600

CUNIFER 30-ALLOY CUNI 70/30
VDM Nickel-Technologie AG
Ni 30-32, Fe 0.4-1, Mn 0.5-1.5, bal Cu.
Minimum values at 20°C: 350 N/mm^2 TS; 120 N/mm^2 YS; 35 El. For seawater piping systems, condensers and heat exchangers in seawater desalination equipment. Material No. 2.0882. C71500

CUNIFER B 7030-FM 67
VDM Nickel-Technologie AG
Ni 29-32, Fe 0.4-1.4, Mn 0.5-1, Ti 0.2-0.5, Si 0-0.1, C 0-0.05, bal Cu.
Submerged-arc and electroslag overlay welding on carbon steel or pressure vessel steel for use in the chemical and petrochemical industries, marine applications, intermediate layer of Nicorros B 6530 or Nickel B 9604 necessary. Material No. 2.0837. C71581

CUNIFER S 7030-FM 67
VDM Nickel-Technologie AG
Ni 29-32, Fe 0.4-1.4, Mn 0.5-1.5, Ti 0.2-0.5, C 0-0.05, bal Cu.
Filler metal. Minimum values at RT: 350 N/mm^2 TS; 180 N/mm^2 YS; 25 El. Material No. 2.0837. C71581

CUNIFER S 9010
VDM Nickel-Technologie AG
Ni 9-11, Fe 0.5-1.5, Mn 0.3-1, Ti 0.2-0.5, C 0-0.05, bal Cu.
Filler metal. Minimum values at RT: 300 N/mm^2 TS; 180 N/mm^2 YS; 20 El. Material No. 2.0873.

CUNILOY
American manufacture
Cu 25, Mn 3.8, Pb 1, bal Ni.
For pump rods, valve parts; corrosion resistant.

CUNIP
Handy & Harmon
Copper. Ni 1.1, P 0.2-0.3, bal Cu.
Wire or strip: 57,000-98,000 TS; 3-5 El. For electron tube components, cathode supports, tuning fingers, spring clips. Heat treatable. High strength and electrical conductivity.

CUNISIL 647
Anaconda Co.
Copper. Si 0.6, Ni 1.9, bal Cu.
Heat treated: 100,000 TS; 85,000 YS; 15 El; 95 Rock B. For electrical equipment, electrical hardware, machined mechanical fasteners. Precipitation hardening; corrosion resistant. 35% electrical conductivity.

CUNISIL 837
Anaconda Co.
Copper. Cu 97.5, Ni 1.9, Si 0.6.
Precipitation hardened: 90,000 TS; 70,000 YS; 8 El; 90 Rock B. For electrical apparatus. Corrosion resistant.

CUPA
Vereinigte Silberhammerwerke Hetzel
Cu plated Al.

CUPAL
Ambolt Machine Tool Co.
Aluminum. Cu clad Al sheet.
For panels for railroad cars; corrosion resisting.

CUPAL
American manufacture
Al 80, Cu 20.
For copper-clad aluminum; 10 Cu-80 Al-10 Cu.

CUPALLOY
Russian manufacture
Cr 0.5, bal Cu.
For electrical equipment, motors; high conductivity.

CUPALOY
A.W. Cadman Mfg.Co.
Cr 0.5, Ag 0.1, bal Cu.
Rolled: 70,000 TS; 65,000 YS; 15 El; 50 RA; 150 Brin. Cast: 55,000 TS; 35,000 YS; 30 El; 60 RA; 125 Brin. For commutators, slip rings, terminal studs; age-hardable, high electrical conductivity. *Obsolete*

CUPALOY
Westinghouse Electric Corp.
Cr 0.5, Ag 0.1, bal Cu.
Rolled: 70,000 TS; 65,000 YS; 15 El; 50 RA; 150 Brin. Cast: 55,000 TS; 35,000 YS; 30 El; 60 RA; 125 Brin. For commutators, slip rings, terminal studs; age-hardable, high electrical conductivity. *Obsolete*

CUPLAT
Western Gold & Platinum Co.
Pt 40, Cu 60.
For brazing cathode structures. Melting point: 1185-1216°C; corrosion resistant, low vapor pressure.

CUPRALINOX 100
Le Bronze Industriel
Al 10, bal Cu.
Aluminum bronze, cast or wrought: 140-180 Brin. AFNOR: UA10.

CUPRALINOX 115
Le Bronze Industriel
Al 13.5, Fe 4.5, Mn 3.5, bal Cu.
Aluminum bronze, cast or wrought: 290-350 Brin. Forming dies for stainless steel.

CUPRALINOX C
Le Bronze Industriel
Al 8.5, bal Cu.
Aluminum bronze, wrought. Annealed: 90-140 Brin. Hardened: 180-230 Brin. AFNOR: UA9.

CUPRALINOX CN
Le Bronze Industriel
Al 6, Ni 2, bal Fe.
Aluminum bronze, plate. AFNOR: UN6 N2.

CUPRALINOX NC2
Le Bronze Industriel
Al 9, Ni 5, Fe 2.5, Mn 0.5, bal Cu.
Aluminum-nickel bronze, wrought or cast: 152 Brin min. AFNOR: UA9 NFe; ASTM B171; 628.

CUPRALINOX NC4
Le Bronze Industriel
Al 10, Ni 5, Fe 4, bal Cu.
Aluminum-nickel bronze, wrought or cast: 170-210 Brin. AFNOR: UA10N; SAE 701C.

CUPRALINOX NCK
Le Bronze Industriel
Bronze, wrought, non-magnetic grade.

CUPRALINOX NCL
Le Bronze Industriel
Al 12, Ni 6, Fe 5, bal Cu.
Aluminum-nickel-iron bronze, wrought: 240-300 Brin. DIN 17665 CuAl11Ni.

CUPRALINOX NCS
Le Bronze Industriel
Al 11, Ni 5, Fe 5, bal Cu.
Aluminum-nickel-iron bronze, wrought: 185-235 Brin. AFNOR: UA11N.

CUPRALINOX NCVB
Le Bronze Industriel
Bronze, cast; for glass molds.

CUPRALINOX TM
Le Bronze Industriel
Al 9, Ni 2, Fe 2, Mn 1.5, bal Cu.
Aluminum bronze, cast or wrought: 115-180 Brin. AFNOR: UA9 NFe.

CUPRALINOX VE
Le Bronze Industriel
Al 8.5, Mn 5, bal Cu.
Aluminum-manganese bronze, wrought, 1/4 hard: 100-150 Brin; 1/2 hard: 150-195 Brin; hard: 190-230 Brin.

CUPRALINOX VN3
Le Bronze Industriel
Al 8.5, Ni 2, Fe 2, Mn 6, bal Cu.
Aluminum-manganese bronze, wrought: 152 Brin min. AFNOR: UA9 NFe.

CUPRALINOX VNC
Le Bronze Industriel
Al 9.5, Fe 2.5, Mn 7, bal Cu.
Aluminum-manganese bronze, wrought: 170-210 Brin. AFNOR: UA10 M.

CUPRALIUM
German manufacture
Cu 7-8, bal Al.
For light alloy parts; non-hardenable.

CUPRALIUM 12
German manufacture
Cu 11.5-12.5, bal Al.
For light alloy castings; non-hardenable.

CUPRALUM
Knapp Mills Inc.
Copper. Lead clad copper.
For chemical plant equipment. Acid resistant.

CUPRALUMIN 12
Manufacturer not listed
Cu 10, bal Al.
Cast: 21,000-26,000 TS; 20,000-24,000 YS; 0.5-1.5 El; 80-90 Brin. For light alloy parts; non-hardenable.

CUPRALUMIN 8
Manufacturer not listed
Fe 0.5, Si 1.5, Cu; bal Al.
For light alloy parts; similar to Alcoa 12.

CUPRANIUM
English manufacture
Ni, Zn, bal Cu.
For corrosion resistant parts.

CUPRO-ALUMINUM
American manufacture
Al 10, Fe, bal Cu.
For worm wheels, gears; aluminum bronze.

CUPRO-NICKEL 10%
Chase Brass & Copper Co.
Cu 89.25, Ni 10, Fe 0.75.
Annealed: 42,000 TS; 14,000 YS; 45 El. For condenser tubes, heat exchanger tubes; corrosion resistant. *Obsolete*

CUPRO-NICKEL 10% 706
Anaconda Co.
Copper. Cu 88.35, Ni 10, Fe 1.25, Mn 0.4.
Hard tube: 60,000 TS; 57,000 YS; 15 El; 68 Rock B. Soft tube: 44,000 TS; 22,000 YS; 46 El; 25 Rock B. For condenser tubes, heat exchangers, marine equipment, oil refinery condensers and evaporators. Resists general corrosion and stress corrosion cracking.

CUPRO-NICKEL 10% 755
Anaconda Co.
Copper. Ni 10, Mn 0.4, Fe 1.25, bal Cu.
Soft: 44,000 TS; 22,000 YS; 46 El. Hard: 60,000 TS; 57,000 YS; 15 El. For condenser tubes. Resists seawater corrosion.

CUPRO-NICKEL 10%-510
Olin Brass, Indianapolis
Ni 10, Fe 0.85, Cu 89.15.
Annealed: 46,000 psi TS; 42 El. Drawn: 75,000 psi TS; 10 El. For condensers and heat exchangers; corrosion resistant. *Obsolete*

CUPRO-NICKEL 10%-511
Olin Brass, Indianapolis
Copper. Cu 88.85, Ni 10, Fe 1.15.
Hard: 50,000 psi TS; 45,000 psi YS; 18 El. Soft: 45,000 psi TS; 15,000 psi YS; 40 El. For condensers and heat exchangers; corrosion resistant.

CUPRO-NICKEL 20%
Chase Brass & Copper Co.
Cu 78.9, Ni 20, Fe 0.35.
Annealed: 48,000 TS; 16,000 YS; 45 El. For condenser tubes, heat exchanger tubes; corrosion resistant. *Obsolete*

CUPRO-NICKEL 20% 710
Anaconda Co.
Copper. Cu 78.75, Ni 20, Fe 0.75, Mn 0.5.
Annealed: 50,000 TS; 22,000 YS; 45 El. For feedwater heaters. Corrosion resistant.

CUPRO-NICKEL 20%-520

Olin Brass, Indianapolis
Copper. Cu 78.85, Ni 20, Fe 0.4, Mn 0.75.
Soft: 49,000 psi TS; 14,000 psi YS; 40 El. For condensers and heat exchangers; corrosion resistant.

CUPRO-NICKEL 30% 702

Anaconda Co.
Copper. Cu 68.9, Ni 30, Mn 0.6, Fe 0.5.
Hard: 77,000 TS; 70,000 YS; 5 El; 162 Brin. Soft: 55,000 TS; 22,000 YS; 40 El; 71 Brin. For condenser tubes. Corrosion resistant.

CUPRO-NICKEL 30% 707

Anaconda Co.
Copper. Cu 64.15, Ni 30, Fe 5.25, Mn 0.6.
Annealed: 74,000 TS; 36,000 YS; 30 El. For heat exchanger tubes. High strength, corrosion resistant.

CUPRO-NICKEL 30% 715

Anaconda Co.
Copper. Cu 68.9, Ni 30, Mn 0.6, Fe 0.5.
Hard tube: 70,000 TS; 60,000 YS; 10 El; 80 Rock B. Soft tube: 55,000 TS; 22,000 YS; 45 El; 35 Rock B. For vessel condensers, saltwater tubes, heat exchangers, cold headed fasteners. Tough and corrosion resistant.

CUPRO-NICKEL 30% 716

Anaconda Co.
Copper. Cu 64.15, Ni 30, Fe 5.25, Mn 0.6.
Soft tube: 74,000 TS; 36,000 YS; 30 El. Drawn: 112,000 TS; 90,000 YS; 12 El. For feedwater heater tubes, heat exchanger tubes, high pressure air and hydraulic lines. Corrosion resistant, high strength. Electrical conductivity: 4.6.

CUPRO-NICKEL 30%-531

Olin Brass, Indianapolis
Copper. Cu 67.75, Ni 31, Mn 0.5, Fe 0.75.
Soft: 60,000 psi TS; 25,000 psi YS; 45 El; 75 Brin. For condensers and heat exchangers; corrosion resistant.

CUPRO-NICKEL 826

Anaconda Co.
Copper. Ni 30, Mn 0.6, Si 0.15, bal Cu.
For welding rod for steel and copper-nickel.

CUPRO-NICKEL 9% 725

Anaconda Co.
Copper. Cu 88.78, Ni 9, Sn 2, Mn 0.22.
Sheet, hard: 80,000 TS; 76,000 YS; 3 El; 90 Rock B. Soft: 52,000 TS; 22,000 YS; 40 El; 42 Rock B. For connectors used in telephone, computer and other electrical and electronic systems, tableware, boat hardware.

CUPRO-NICKEL BULLET JACKETS

American manufacture
Ni 15, Cu 85.
For bullet jackets; ductile.

CUPRO-NICKEL COMMERCIAL

American manufacture
Cu 60-98, Ni 2-40.
For hardware; corrosion resistant.

CUPRO-NICKEL DRIVING BANDS

American manufacture
Cu 95-97.5, Ni 2.5-5.
For driving bands; corrosion resistant.

CUPRO-NICKEL ELECTRODES

Chemetron Corp.
Si 0.07, Ni 31, Mn 1.3, Fe 0.55, Cu 68, Weld metal: 0.03 C.
As welded: 52,000 psi TS; 35,000 psi YS; 40 El. For welding 70-30, 90-10 and similar alloys.

CUPRO-NICKEL LOCOMOTIVE TUBES

American manufacture
Cu 97, Ni 3.
For locomotive tubes; corrosion resistant.

CUPRO-NICKEL NO. 300 ALLOY

Manufacturer not listed
Cu 89, Ni 11.
For corrosion resistant parts; corrosion resistant.

CUPRO-SILICON

American manufacture
Si 55, bal Cu.
For hardener for copper alloys.

CUPROCHROME

Wilbur B. Driver Co.
Copper. Cr 1, bal Cu.
Annealed: 40,000 TS; 27,000 YS; 25 El. For electron tubes; high conductivity.

CUPRODIE

Now FINKL CUPRODIE.

CUPROMAGNESIUM

American manufacture
Cu 90, Mg 10.
For cast iron inoculant; graphite spheroidizer.

CUPROMANGANESE

American manufacture
Cu 90, Mn 10.
For staybolts, heat resisting parts; heat resistant.

CUPROMANGANESE TUBES

American manufacture
Cu 96, Mn 4.
For tubes; corrosion resistant.

CUPRON

Gilby-Fodor S.A.
Ni 45, Cu 55.
Annealed: 85,000 TS; 50,000 YS; 50 El. For strain gauges and rheostats. Low temperature coefficient of resistance.

CUPRON

Wilbur B. Driver Co.
Copper. Cu 55, Ni 45.
Annealed: 62,000 TS. For rheostats, voltmeters, shunts, resistances; similar to "Advance," "Ideal," "Constantan" and "la-la."

CUPRON THERMOCOUPLE ALLOY

Carpenter Technology Corp.
Cu 55, Ni 45.
At 20°C: 50,000 psi YS; 85,000 psi TS; 45 El. For thermocouples, extension wires.

CUPRONAR 900

Arcos Alloys
Sn 1.
For welding wire; deoxidized copper, AWS RCu, ECu.
Obsolete

CUPRONAR 910

Arcos Alloys
Si 2.8-4, Fe 0.5, 94.0 Cu min.
For welding wire; silicon bronze, AWS RCuSi-A, ECuSi.
Obsolete

CUPRONAR 920A

Arcos Alloys
Sn 4-6, P 0.1-0.35, 93.5 Cu min.
For welding wire; phosphorus bronze, AWS RCuSn-A, ECuSn-A. *Obsolete*

CUPRONAR 920C

Arcos Alloys
Sn 7-9, P 0.05-0.35, bal Cu.
For welding wire; phosphorus bronze, AWS ECuSn-C.
Obsolete

CUPRONAR 950

Arcos Alloys
Al 6-9, Si 0.1, bal Cu.
For welding wires; aluminum bronze, AWS ECuAl-Al.
Obsolete

CUPROR

American manufacture
Cu 94, Al 5.8.
For pump rods, valve stems; corrosion resistant.

CUPROSIL NS5

Le Bronze Industriel
Ni 2, Si 0.5, bal Cu.
Cast or wrought; hardness 160-200 Brin. AFNOR UN35.

CUPROSIL SI45Z

Le Bronze Industriel
Si 3, Fe 2, Zn 3, bal Cu.
Silicon bronze, wrought: 90-190 Brin. AFNOR US3 2Fe; AMS 4616B.

CUPROTEC 10180

Eutectic Corp.
Copper-base alloy powder for brazing thin walled copper alloy sections. *Obsolete*

CUPROTHAL 180

Kanthal Corp.
Ni 22, bal Cu.
Annealed: 50,000-100,000 TS; 40 max El. For heating and resistance elements; max operating temperature 1000°F, nonmagnetic.

CUPROTHAL 180

Kanthal A.B.
Ni 22, bal Cu.
Annealed: 50,000-100,000 TS; 40 max El. For heating and resistance elements; max operating temperature 1000°F, nonmagnetic.

CUPROTHAL 294

Kanthal Corp.
Ni 45, bal Cu.
Rolled: 60,000-100,000 TS; 30 El. For resistors; low electrical resistance, nonmagnetic maximum temperature 1000°F.

CUPROTHAL 294

Kanthal A.B.
Ni 45, bal Cu.
Rolled: 60,000-100,000 TS; 30 El. For resistors; low electrical resistance, nonmagnetic maximum temperature 1000°F.

CUPROTHAL 30

Kanthal Corp.
Ni 2, bal Cu.
Annealed: 30,000-60,000 TS; 60 max El. For heating and resistance elements; max operating temperature 600°F, nonmagnetic.

CUPROTHAL 30

Kanthal A.B.
Ni 2, bal Cu.
Annealed: 30,000-60,000 TS; 60 max El. For heating and resistance elements; max operating temperature 600°F, nonmagnetic.

CUPROTHAL 60

Kanthal Corp.
Ni 6, bal Cu.
Annealed: 35,000-70,000 TS; 55 max El. For heating and resistance elements; max operating temperature 600°F, nonmagnetic.

CUPROTHAL 60

Kanthal A.B.
Ni 6, bal Cu.
Annealed: 35,000-70,000 TS; 55 max El. For heating and resistance elements; max operating temperature 600°F, nonmagnetic.

CUPROTHAL 90
Kanthal Corp.
Ni 11, bal Cu.
Annealed: 35,000-75,000 TS; 50 max El. For heating and resistance elements; max operating temperature 750°F, nonmagnetic.

CUPROTHAL 90
Kanthal A.B.
Ni 11, bal Cu.
Annealed: 35,000-75,000 TS; 50 max El. For heating and resistance elements; max operating temperature 750°F, nonmagnetic.

CUPROTHERM
Now WIELAND K80.

CUPROVAC E
Crucible Materials Corp.
Cu.
For electrical equipment; gas free, high purity copper. *Obsolete*

CUPTEN-G
NKK Corp.
C 0-0.12, Si 0-0.6, Mn 0-0.6, P 0.06-0.12, S 0-0.04, Cu 0.2-0.6, Cr 0.4-1.2, Mo 0-0.35, V 0-0.1, bal Fe.
Hot rolled, 1.4-13.0 mm thick: 355 N/mm^2 YS. Cold rolled, 0.5-2.6 mm thick: 325 N/mm^2 YS. Weather resistant steel.

CUPTEN-R
NKK Corp.
C 0-0.12, Si 0.25-0.75, Mn 0.2-0.5, P 0.07-0.15, S 0-0.04, Cu 0.25-0.55, Ni 0-0.45, Cr 0.3-1, bal Fe.
Hot rolled, 1.4-13.0 mm thick: 355 N/mm^2 YS. Cold rolled, 0.5-2.6 mm thick: 315 N/mm^2 YS. Weather resistant steel.

CUROLOY
Birdsboro Corp.
C 3, Si 1.2, Mn 0.9, Alloy, bal Fe.
For rolls; alloy cast iron. *Obsolete*

CURTISOL
Curtis-Wright Corp.
Silver alloy.
For solder for titanium alloys; M.P. 1300-1400°F.

CURTISS
English manufacture
Al 95.2, Cu 2.5, Mg 1.5.
For pistons; heat treatable.

CUSIL
GTE Products Corp./Wesgo Div.
Ag 72, Cu 28.
Brazing alloy available in foil, flexibraze, wire, powder, extrudable paste and preform. Liquidus 1436°F. Solidus 1436°F.

CUSIL
Western Gold & Platinum Co.
Ag 72, Cu 28.
Melting point: 1436°F. For high temperature brazing. Eutectic alloy. Excellent flow; high vapor pressure.

CUSIL-ABA
GTE Products Corp./Wesgo Div.
Ag 63, Ti 1.75, bal Cu.
Brazing alloy used in the manufacture of protective devices such as surge arresters. Liquidus 1500°F. Solidus 1435°F. 39,300 psi YS; 50,200 psi TS; 20 El; 110 Knoop.

CUSILOY
American manufacture
Cu 95, Si 1-3, Sn 1-1.5, Fe 0.7-1.
Soft: 50,000-60,000 TS; 15,000-20,000 YS; 50 El. For wire; corrosion resistant.

CUSILTIN 10
GTE Products Corp./Wesgo Div.
Ag 60, Cu 30, Sn 10.
Brazing alloy available in foil, flexibraze, wire, powder, extrudable paste and preform. Liquidus 1324°F. Solidus 1116°F. Meets AMS-4773.

CUSILTIN 5
GTE Products Corp./Wesgo Div.
Ag 68, Cu 27, Sn 5.
Brazing alloy available in foil, flexibraze, wire, powder, extrudable paste and preform. Liquidus 1400°F. Solidus 1369°F.

CUSTOM 455
Carpenter Technology Corp.
C 0.03, Cr 11.75, Ni 8.5, Ti 1.2, Cb 0.3, Cu 2.25, bal Fe.
Hardened: 332,000 TS; 244,000 YS; 13 El; 50 Rock C. For high temperature bolts, springs, valves. Maraging, precipitation hardening, tough.

CUSTOM 630, ETC.
Now CARPENTER CUSTOM 630.

CUT TRODE
Eutectic Corp.
C, Cu, bal Fe.
For cutting electrodes; exothermic coating. *Obsolete*

CUTANIT
Saville & Co. Ltd., J.J.
Mo, Ti, WC.
For taps, tools, cutters; cemented carbides.

CUTANIT
Jessop-Saville Ltd.
Mo, Ti, WC.
For taps, tools, cutters; cemented carbides.

CUTIN
GTE Products Corp./Wesgo Div.
Cu 85, Sn 15.
Brazing alloy available in foil, flexibraze, wire, powder, extrudable paste and preform. Liquidus 1760°F. Solidus 1468°F.

CUTLERY
Dunford Hadfields Ltd.
C 0.3-0.4, Cr 12-14, bal Fe.
123,000-157,000 TS; 500 Brin. For cutting tools, high tensile engineering requirements; corrosion resistant. *Obsolete*

CUTTER ALLOY
Columbia Tool Steel Co.
C 1.05, Mn 0.25, Cr 0.25, W 0.25, bal Fe.
Annealed: 110,000 TS. For cutting tools for hard materials; water hardening. *Obsolete*

CUTTRODE 1
Eutectic Corp.
Electrode for AC-DC to cut, pierce, clean castings, remove flash and risers; all metals.

CUYO
Hoover Ball & Bearing Co.
C 0.2, Cr 18, Ni 8, bal Fe.
Annealed: 75,000 TS. For stainless parts, furnace parts; stainless and heat resisting. *Obsolete*

CUZINAL
Olin Brass, Indianapolis
Copper, Cu 77, Al 2.1, As 0.03, Zn 20.87.
Annealed: 50,000 psi TS; 65 El. Drawn: 85,000 psi TS; 10 El. For condenser and heat exchangers; embrittlement free.

CV
Now DARWIN CV

CV 110
Thyssen Edelstahlwerke AG
C 0.3, Si 0.3, Mn 0.4, Cr 1.4, Mo 1.05, bal Fe.
For gears, shafts; water hardening. *Obsolete*

CV 120
Thyssen Edelstahlwerke AG
C 0.3, Si 0.3, Mn 0.4, Cr 1.4, Mo 0.55, V 0.4, bal Fe.
For gears, shafts; water hardening. *Obsolete*

CV 30
Thyssen Edelstahlwerke AG
C 0.3, Si 0.3, Mn 0.2, Cr 1, Mo 0.2, bal Fe.
For gears, shafts; water hardening. *Obsolete*

CV 60
Thyssen Edelstahlwerke AG
C 0.3, Si 0.3, Mn 0.4, Cr 1.35, Mo 0.55, bal Fe.
For gears, shafts; water hardening. *Obsolete*

CV 70
Thyssen Edelstahlwerke AG
C 0.15, Si 0.3, Mn 0.4, Cr 1, Mo 0.3, V 0.25, bal Fe.
For gears, shafts; case hardened. *Obsolete*

CV1 EXTRA
Thyssen Edelstahlwerke AG
C 1.2, Cr 0.7, V 0.1, bal Fe.
For cutting tools, bearings; wear resistant; keen cutting edge. *Obsolete*

CVFS
Thyssen Edelstahlwerke AG
C, Cr, Mn, bal Fe.
For springs; water hardening. *Obsolete*

CW 11
W. Ossenberg & Cie Edelstahlwerke
C 1.2, Mn 0.25, Si 0.2, W 1, V 0.1, bal Fe.
Cold work tool steel for center drills, twist drills, milling cutters. W.-Nr. 1.2516.

CW01
Herbert Cutanit Ltd.
Sintered carbide material. For wear parts; wear and shock resistant. *Obsolete*

CW25
Herbert Cutanit Ltd.
Sintered carbide material. For wear-resisting applications under good conditions with little shock. *Obsolete*

CW30
Herbert Cutanit Ltd.
Sintered carbide material. For wear-resisting applications where some shock is encountered. *Obsolete*

CW330
Contact Technologies, Inc.
Ag 50.
Silver-tungsten carbide. Density: 12.0 g/cm^3; specific resistivity: 0.043 micro-ohm-m; hardness: 30T70; strength: 90,000 psi. Electrical contacts for circuit breaker applications.

CW335
Contact Technologies, Inc.
Ag 35.
Silver-tungsten. Density: 14.3 g/cm^3; specific resistivity: 0.037 micro-ohm-m; hardness: 30T77; strength: 110,000 psi. Electrical contacts for circuit breaker applications.

CW350
Contact Technologies, Inc.
Ag 50.
Silver-tungsten. Density: 13.2 g/cm^3; specific resistivity: 0.031 micro-ohm-m; hardness: 30T60; strength: 100,000 psi. Electrical contacts for circuit breaker applications.

CW366

Contact Technologies, Inc.
99.0 W min.
Density: 18.0 g/cm^3; specific resistivity: 0.067 micro-ohm-m; hardness: 30N55; strength: 85,000 psi; coefficient of expansion: 4.5 x 10^{-6}/$^\circ$C. Heat sinks for semiconductor applications.

CW375

Contact Technologies, Inc.
Ag 50, W 30, WC 20.
Silver-tungsten-tunsten carbide. Density: 12.6 g/cm^3; specific resistivity: 0.034 micro-ohm-m; hardness: 30T87; strength: 100,000 psi. Electrical contacts for circuit breaker applications.

CW378

Contact Technologies, Inc.
Ag 60.
Silver-tungsten carbide. Density: 11.6 g/cm^3; specific resistivity: 0.034 micro-ohm-m; hardness: 30T55; strength: 90,000 psi. Electrical contacts for circuit breaker applications.

CW50

Herbert Cutanit Ltd.
Sintered carbide material. For wear-resisting applications where heavy shock is encountered. *Obsolete*

CWM, S 431

Thyssen Edelstahlwerke AG
C, Cr, bal Fe.
Water treated: 78,500-100,800 TS; 50,000-72,000 YS; 22-27 El. *Obsolete*

CX3

TRW Inc.
WC 59, Co 14, 27.0 TaC.
360,000 transverse strength; 86.7 Rock A; density 13.91. Sintered carbide tool material.

CY-12

TRW Inc.
Sintered carbide tool material.
For roughing cuts on carbon and low alloy steel. *Obsolete*

CY-14

TRW Inc.
Sintered carbide tool material.
For light roughing cuts and some finishing cuts on carbon and alloy steels. *Obsolete*

CY-16

TRW Inc.
Sintered carbide tool material.
For heavy roughing cuts on carbon and alloy steels. *Obsolete*

CY-17

TRW Inc.
Sintered carbide tool material.
For heavy roughing cuts at heavy feeds on carbon and alloy steels. *Obsolete*

CY-2

TRW Inc.
Sintered carbide tool material.
For general purpose and light roughing cuts on steel. *Obsolete*

CY-31

TRW Inc.
Sintered carbide tool material.
For finish cutting and precision boring of carbon and alloy steels. *Obsolete*

CY-4

TRW Inc.
Sintered carbide tool material.

CY-5

TRW Inc.
Sintered carbide tool material.
For general purpose machining of carbon and low alloy steels. *Obsolete*

CY/A

Follsain-Wycliffe Foundries, Ltd.
C 2.4-3, Si 0.5-1.5, Mn 0.2-0.8, Cr 0-2, P 0-0.15, bal Fe.
White cast iron; 250 Brin min. Abrasion resistant.

CY/C

Follsain-Wycliffe Foundries, Ltd.
C 2.4-3.2, Si 0-1, Mn 0.5-1.5, Cr 22-28, Mo 0-1.5, Ni 0-1, Cu 0-1.2, P 0-0.1, bal Fe.
High chrome abrasion resistant casting; 450 Brin min.

CY/H

Follsain-Wycliffe Foundries, Ltd.
C 2.4-3.4, Si 0.5-1.5, Mn 0.2-0.8, Cr 0-2, P 0-0.15, bal Fe.
White cast iron; 400 Brin min. Abrasion resistant.

CY12

TRW Inc.
WC 83.5, Co 13, 3.5 TiC.
300,000 transverse strength; 90 Rock A; density 13.30.
Sintered carbide cutting tool; excellent shock resistance; for heavy feed, low speed machining of rough and irregular steel.

CY14

TRW Inc.
WC 75, Co 6, 10.0 TaC, 9.0 TiC.
250,000 transverse strength; 92.6 Rock A; density 12.60.
Sintered carbide cutting tool; excellent wear resistance and resistance to cratering; for general purpose and finishing cuts.

CY16

TRW Inc.
WC 72, Co 8.5, 11.5 TaC, 8.0 TiC.
275,000 transverse strength; 91.3 Rock A; density 12.6.
Sintered carbide cutting tool; high resistance to cratering and to high temperatures; for general purpose machining.

CY17

TRW Inc.
WC 71, Co 9.5, 11.5 TaC, 8.0 TiC.
300,000 transverse strength; 91.0 Rock A; density 12.50.
Sintered carbide cutting tool; very high shock and good wear resistance; for milling or operations on rough work.

CY2

TRW Inc.
WC 76, Co 9, 15.0 TiC.
200,000 transverse strength; 92.0 Rock A; density 11.20.
Sintered carbide cutting tool; excellent wear resistance; for light and general purpose machining.

CY31

TRW Inc.
WC 76, Co 4, 12.0 TaC, 8.0 TiC.
180,000 transverse strength; 93.5 Rock A; density 12.90.
Sintered carbide cutting tool; highest abrasion resistance; for high speed precision boring and turning.

CY5

TRW Inc.
WC 82, Co 10, 8.0 TiC.
260,000 transverse strength; 91.0 Rock A; density 12.45.
Sintered carbide cutting tool; for general purpose and semi-roughing.

CYCLO

Cytemp Specialty Steel Div.
C 0.7, Mn 1.2, bal Fe.
For tools, dies, punches; oil hardening. *Obsolete*

CYCLOID GEAR STEEL

Carpenter Technology Corp.
Si 1.5, C 0.4, Mn 0.9, bal Fe.
Tempered: 235,000 TS; 185,000 YS; 7.2 El; 26 RA; 444 Brin.
Annealed: 85,000 TS; 57,000 YS; 24 El; 58 RA; 166 Brin. For gears, pinions, shafts; silico-manganese tempering steel. *Obsolete*

CYCLOPE

TradeARBED Inc.
C 0.35, W 9, Cr 3.5, bal Fe.
Hot work tool steel, for dies. AISI H21. *Obsolete*

CYCLOPS (S5)

Cytemp Specialty Steel Div.
C 0.55, Si 2, Mn 0.9, Mo 0.4, bal Fe.
Heat treated: 338,000 TS; 281,000 YS; 5 El; 600 Brin. Heat treated: 220,000 TS; 207,000 YS; 11 El; 455 Brin. For shear blades, punches, pneumatic tools, rivet sets, chisels. Shock and impact resistant. Type S5.

CYCLOPS 14 MV

Now UNITEMP 14MV.

CYCLOPS 17A

Now UNILOY 325.

CYCLOPS 2570

Cytemp Specialty Steel Div.
C 0.47, Mn 0.3, Si 1, Cr 8.5, Mo 1.15, V 1.15, bal Fe.
Shear blades, forging dies. *Obsolete*

CYCLOPS 42 (M42)

Cytemp Specialty Steel Div.
C 1.1, Cr 3.75, V 1.2, W 1.5, Mo 9.5, Co 8, bal Fe.
Heat treated: 63-67 Rock C. For finish machining, boring, precision turning, at high speeds; good temperature and wear resistance. AISI Type M42. High speed tool steel.

CYCLOPS 67

Now CYCLOPS S 5.

CYCLOPS L6

Cytemp Specialty Steel Div.
C 0.75, Mn 0.4, Si 0.25, Cr 1, Ni 1.5, bal Fe.
Heat treated: 190,000-305,000 TS; 178,000-280,000 YS; 5-12 El; 17-38 RA; 43-61 Rock C. For arbors, blanking dies, clutch parts, forming dies, brake dies, punches pinions, shear blades, spindles, swages. Tough and wear resistant. AISI Type L6. Oil hardening tool steel. *Obsolete*

CYCLOPS N-9

Now CYCLOPS L6.

CYCLOPS B 6X

Now THERMOLD H 26.

CYCLOPS B-10

Cytemp Specialty Steel Div.
C 0.8, Mo 1, W 18.5, Cr 4.5, V 2, Co 9, bal Fe.
For cutters, lathe and planer tools, form tools, checking tools, cut-off tools. High abrasion resistance and red-hardness. AISI T5 high speed steel. *Obsolete*

CYCLOPS B-42

Universal Cyclops
C, alloy, bal Fe.
For tools, dies. *Obsolete*

CYCLOPS B-44

Now THERMOLD H22.

CYCLOPS B-6

Cytemp Specialty Steel Div.
C 0.75, W 18, V 1, bal Fe.
For lathe tools, taps, reamers, twist drills, milling cutters, dies, lathe centers; high speed steel; wear and abrasion resistant. *Obsolete*

CYCLOPS B-9
Cytemp Specialty Steel Div.
C 0.84, Mn 0.25, Si 0.3, Cr 4.5, W 18.5, V 2.25, Mo 0.5, bal Fe.
For drills, taps, broaches, form tools, milling cutters; high speed steel. *Obsolete*

CYCLOPS B44J
Now THERMOLD H21.

CYCLOPS EXTRA
Now CYCLOPS W1.

CYCLOPS K
Cytemp Specialty Steel Div.
C, Cr, W, bal Fe.
For hot work dies, punches; hot work steel. *Obsolete*

CYCLOPS K-L DIE STEEL
Manufacturer not listed
C 0.35, Cr 6, W 6, bal Fe.
For coining dies, punches, hot heading dies; oil hardened.

CYCLOPS K-M DIE STEEL
Manufacturer not listed
C 0.45, Cr 6, W 6, bal Fe.
For blanking, forming and gripper dies, hot piercers; water hardened.

CYCLOPS L1
Cytemp Specialty Steel Div.
C 1.05, Mn 0.35, Si 0.3, Cr 1.4, bal Fe.
Water hardened: 237,000 TS; 226,000 YP; 444 Brin. Cold drawn: 107,000 TS; 87,000 YP; 17 El; 55 RA; 229 Brin. For gauges, knurls, knife edges, taps, dies, arbors, rolls. AISI Type L1. Oil hardening. Wear resistant. *Obsolete*

CYCLOPS L2
Cytemp Specialty Steel Div.
C 0.5, Mn 0.7, Si 0.25, Cr 1, V 0.2, bal Fe.
Annealed: 103,000 TS; 74,000 YS; 27 El; 52 RA; 201 Brin. Hardened: 298,000 TS; 263,000 YP; 1 El; 5 RA; 610 Brin. For gears, forgings, arbors, crankpins, chuck jaws, die rings, gun barrels, jack screws, rivet sets, shear blades. Tough, shock resisting. AISI Type L2. Oil hardening. Fatigue resistant. *Obsolete*

CYCLOPS M-T
Manufacturer not listed
C 0.75, V 1, W 1.5, Mo 7, bal Fe.
For tools; see Motung.

CYCLOPS M4
Cytemp Specialty Steel Div.
Now CAP M4 POWDER.

CYCLOPS R.B.C.
Universal Cyclops
C, alloy, bal Fe.
For tools. *Obsolete*

CYCLOPS S1
Cytemp Specialty Steel Div.
C 0.5, Mn 0.35, Si 0.3, Cr 1.5, W 2.25, V 0.25, Mo 0.3, bal Fe.
Hardened: 260,000-280,000 TS; 48-55 Rock C. For bolt header dies, chipping and caulking tools, concrete drills, pneumatic tools, shear blades, track tools. Type S1. Tough and fatigue resistant. *Obsolete*

CYCLOPS S2
Cytemp Specialty Steel Div.
C 0.5, Mn 0.45, Si 1.1, V 0.2, Mo 0.5, bal Fe.
Water hardened: 235,000-323,000 TS; 229,000-300,000 YS; 4-10 El; 12 RA; 47-58 Rock C. For hand and pneumatic tools, chisels, stamps, spindles, pipe cutters, rivet sets; flaring tools. Tough and shock resistant. Type S2. Tough and shock resistant. *Obsolete*

CYCLOPS SCK
Cytemp Specialty Steel Div.
C 0.7, Mn 0.35, Si 1, Cr 8.5, V 1, Mo 1.4, Ni 1.5, bal Fe.
Heat treated: 55-61 Rock C. For cold shears, trimmers, woodworking chippers, punches; air hardening; resists tempering, good wear resistance. *Obsolete*

CYCLOPS SPECIAL COLD HEADER
Now CYCLOPS W2.

CYCLOPS T 15
Cytemp Specialty Steel Div.
Now CAP T-15.

CYCLOPS W1
Cytemp Specialty Steel Div.
C 0.8-1.25, Mn 0.25, Si 0.25, bal Fe.
Annealed: 95,000 TS; 50,000 YS; 22 El; 195 Brin. Water hardened: 200,000-215,000 TS; 138,000-152,000 YS; 11-12 El; 400-600 Brin. For punches, beading tools, cold heading dies, axles, drills, reamers, files, woodworking tools. Wear and abrasion resistant. Water hardening high carbon steel. *Obsolete*

CYCLOPS W2
Cytemp Specialty Steel Div.
C 0.8-1.25, Mn 0.25, Si 0.25, V 0.25, bal Fe.
Annealed: 100,000 TS; 55,000 YS; 21 El; 200 Brin. Water hardened: 216,000 TS; 152,000 YS; 11 El; 32 RA; 600 Brin. For punches, beading tools, cold heading dies, axles, drills, reamers, woodworking tools, files, die rings. Wear and abrasion resistant. Water hardening high carbon steel. *Obsolete*

CYKLOP
Ambo-Stahl-Gesellschaft
C 0.3, Cr 2.65, V 0.35, W 8.5, bal Fe.
For extrusion press rams and liners, punches; hot work steel, oil hardened.

CYKLOP CO
Ambo-Stahl-Gesellschaft
C 0.3, Co 2, Cr 2.4, V 0.25, W 8.5, bal Fe.
For extrusion press rams and liners, punches; hot work steel.

CYKLOP EXTRA
Ambo-Stahl-Gesellschaft
C 0.65, Cr 3.75, Mo 0.85, V 0.7, W 8.5, bal Fe.
For lathe and planer tools, reamers, drills, hobs; high speed steel.

CYLINDER IRON
General Motors Corp./Central Foundry
C 3-3.25, Mn 0.4-0.75, Si 2-2.25, bal Fe.
For engine cylinders, pistons, piston rings; cast iron.

CYMBAL METAL
American manufacture
Cu 78, Zn 22.
For architectural and ornamental parts; red brass.

CYPRUS BRONZE
Manufacturer not listed
Cu 65, Pb 30, Sn 5.
17,000 TS; 6-8 El; 41 Brin. For bearings, hardware; heavy.

D & D SPECIAL O.H.

Denman & Davis Co.
C 0.7, Cr 0.8, Mo 0.2, bal Fe.
For tools and dies. Oil hardening. *Obsolete*

D & J ANTIFRICTION METAL-1

English manufacture
Sb 10, Zn 80-85, Sn 5-8.
For bearings, bushings; Babbitt.

D & J ANTIFRICTION METAL-2

English manufacture
Cu 1.6, Sb 0.4, Zn 52, Sn 46.
For bearings, bushings; Babbitt.

D & M-305

Darwin & Milner Inc.
Carbon steel. C, bal Fe.
For arc welding rods; coated.

D & M-7-0

Darwin & Milner Inc.
Alloy steel. C, alloy, bal Fe.
For arc welding electrodes; coated, self hardened.

D & M-7-G

Darwin & Milner Inc.
Alloy steel. C, alloy, bal Fe.
For arc welding electrodes; coated, hard surfacing; air hardened.

D & M-92

Darwin & Milner Inc.
Alloy steel. C, alloy, bal Fe.
For welding rods; shock resistant.

D & M-LT

Darwin & Milner Inc.
Alloy steel. C, alloy, bal Fe.
For arc welding rods; coated.

D & M-U2

Darwin & Milner Inc.
Alloy steel. C, alloy, bal Fe.
For welding rods; self hardening.

D 14

Now COLUMBIUM D-14.

D 22 S

Thyssen Edelstahlwerke AG
C, Cr, Ni, Mo, bal Fe.
13-15 El. For machinery parts; oil hardened. *Obsolete*

D 9 MO

Disston Inc.
C 1.15, Mo 0.2, bal Fe.
For taps, knives; water hardening. *Obsolete*

D PHOSPHOR BRONZE

Criterion Metals, Inc.
Copper. Cu 89.82, Zn 0-0.2, Sn 10, P 0.18, Fe 0-0.1, Pb 0-0.05.
Thin gauge sheet, various tempers: 58-122 ksi TS min; ASTM B-103. C52400

D-1 DISSTON

Disston Inc.
C 0.8, bal Fe.
For tools; general purpose. *Obsolete*

D-10

Keystone Carbon Co.
Brass and graphite.
For oil-less bearing; powder metallurgy, operating temperature from 69-800 F. *Obsolete*

D-11

D.A.B. Industries Inc.
Sn 87, Cu 6.25, Sb 6.75.
Tin based Babbitt; half shell bearings, bushings and thrust washers. Soft, excellent embeddability.

D-12

D.A.B. Industries Inc.
Sn 89, Cu 3.5, Sb 7.5.
Tin based Babbitt; half shell bearings, bushings and thrust washers. Good corrosion resistance.

D-13

D.A.B. Industries Inc.
Pb 84, Sn 6, Sb 10.
Lead based Babbitt; half shell bearings, bushings and thrust washers. For lightly loaded pump bearings.

D-14

D.A.B. Industries Inc.
Pb 75, Sn 10, Sb 15.
Lead based Babbitt; half shell bearings, bushings and thrust washers. Good score resistance.

D-15

D.A.B. Industries Inc.
Pb 83, Sn 1, Sb 15, As 1.
Lead based Babbitt; half shell bearings, bushings and thrust washers.

D-2

Now CLEVITE F-1.

D-2

Gould Inc.
Cu 3-3.5, Sb 7.25-7.75, Te 0.1-0.14, bal Sn.
For engine bearings; steel backed Babbitt.

D-2-DISSTON

Disston Inc.
C, bal Fe.
For tools; water hardening. *Obsolete*

D-29 CHISEL STEEL

Disston Inc.
C 0.5, Si 1.4, Mo 0.4, bal Fe.
For chisels, shear blades. *Obsolete*

D-29 KNIFE

Disston Inc.
C 0.65, Si 1.4, V 0.1, Mo 0.4, bal Fe.
For chipper knives; tough. *Obsolete*

D-406, ETC.

Now DIAMOND D-406, ETC.

D-48

D.A.B. Industries Inc.
Cu 70, Pb 29, Sn 1.
Copper lead alloy for half shell bearings for connecting rod and main bearings. Moderately hard.

D-50

D.A.B. Industries Inc.
When plated with 8Sn-92Pb as half shell bearing can be used for extremely heavy duty main and connecting rod bearings. OMPOSITION 9.9 Ag min.

D-51

D.A.B. Industries Inc.
Al 95, Si 4, Cd 1.
Aluminum based alloy for half shell bearings for intermediate loads.

D-5116

VILLARES
Now N7131.

D-5119

VILLARES
Now N7147.

D-52A

D.A.B. Industries Inc.
Cu 52, Pb 44, Sn 4.
Copper lead alloy for half shell bearings for intermediate loaded main and connecting rod bearings.

D-56

D.A.B. Industries Inc.
Al 95, Si 4, Cd 1.
Aluminum based alloy for half shell bearings. Should be over plated with 8Sn-92Pb for highly loaded main and connecting rod bearings.

D-57

D.A.B. Industries Inc.
Cu 80, Pb 10, Sn 10.
Copper-lead alloy for bushings and thrust washers. Maximum shock and load capacity. For steering knuckles, wear plates.

D-58

D.A.B. Industries Inc.
Cu 88, Pb 8, Sn 4.
Copper-lead alloy for bushings and thrust washers. General purpose loading.

D-59

D.A.B. Industries Inc.
Cu 73.5, Pb 23, Sn 3.5.
Copper-lead alloy for bushings and thrust washers. For intermediate loads on oscillating and rotating shafts.

D-6 AC

Republic Steel Corp.
C 0.46, Mn 0.76, Si 0.22, Cr 1.1, Ni 0.5, Mo 1, V 0.08, bal Fe.
Quenched and tempered: 195 ksi min TS; 180 ksi min YS; 8 min El; 25 min RA. For rocket motor case rings.

D-6 AC

Ladish Co., Inc. (developer)
C 0.46, Mn 0.76, Si 0.22, Cr 1.1, Ni 0.5, Mo 1, V 0.08, bal Fe.
Quenched and tempered: 195 ksi min TS; 180 ksi min YS; 8 min El; 25 min RA. For rocket motor case rings.

D-6 AC

Titanium Metals Corp.
C 0.46, Mn 0.76, Si 0.22, Cr 1.1, Ni 0.5, Mo 1, V 0.08, bal Fe.
Quenched and tempered: 195 ksi min TS; 180 ksi min YS; 8 min El; 25 min RA. For rocket motor case rings.

D-61

D.A.B. Industries Inc.
Al 91, Sn 6.5, Cu 1, Si 1.5.
Aluminum based alloy for half shell bearings. Hard, usually plated; for highly loaded main and connecting rod engine bearings.

D-6158

VILLARES
Now N8161.

D-62

D.A.B. Industries Inc.
Al 90.5, Sn 6.5, Cu 1, Si 1.5.
Aluminum based alloy for bushings and thrust washers. For highly loaded transmission and motor bushings.

D-63

D.A.B. Industries Inc.
Al 79, Sn 20, Cu 1.
Aluminum based alloy for bushings and thrust washers. Intermediate to heavy loaded transmission and camshaft bushings.

D-C-33

Manufacturer not listed
C 0.33, Si 1.05, W 1.55, Mo 1.65, Cr 5.15, bal Fe.
Air hardening; for tools, dies. AISI H12 hot work steel.

D-C-33-VA
Manufacturer not listed
C 0.33, Si 1.05, W 1.55, Mo 1.65, Cr 5.15, V, bal Fe.
Air hardening tool & die steel. AISI H 12 Hot work steel.

D-DIE
Specialty Steel Co. of America
C 1.55, Cr 11.5, V 0.9, Mo 0.8, bal Fe.
Air or oil hardened cold work tool steel, high carbon-high chromium type. For punching and trimming dies, gages, broaches. AISI D2.

D-H 60-OHM ALLOY
Harrison Alloys Inc.
Copper.
Now HAI-60 ALLOY.

D-H 90 OHM ALLOY
Harrison Alloys Inc.
Copper.
Now HAI-90 ALLOY.

D-H A-NICKEL
Harrison Alloys Inc.
Nickel.
Now HAI-NI 200.

D-H D-NICKEL
Harrison Alloys Inc.
Nickel.
Now HAI-NI 211.

D-H DURANICKEL
Driver Harris Co.
Nickel. Ni 93.5, Fe 0.35, Mn 0.3, Si 0.5, C 0.15, Al 4.5.
Wrought: 90,000-250,000 TS; 30,000-150,000 YS; 2-50 El; 15-65 RA; 140-380 Brin. For electrical equipment; Z-Nickel; Type A. *Obsolete*

D-H MANGANESE ALLOY NO. 720
Driver Harris Co.
Cu 60, Ni 20, Mn 20.
For springs, diaphragms; age hardening. *Obsolete*

D-H NI-HARD
Darbyshire Steel Co. Inc.
Ni 4.5, C 3.2, Cr 1.5, Si 1, bal Fe.
For ball and rod mill liners, pump casings, impellers; heat resistant castings.

D-H NO. 11
Harrison Alloys Inc.
Copper.
Now HAI-11 ALLOY.

D-H NO. 11
Driver Harris Co.
Cu-Ni.
For thermocouples, extension leads; 30 ohms/c.m.f. *Obsolete*

D-H NO. 111
Driver Harris Co.
Cu, bal Ni.
Rolled: 30,000-60,000 TS. For corrosion resistant parts; corrosion resistant. *Obsolete*

D-H NO. 129
Driver Harris Co.
Ni (pure).
For control of electron emission in the cathode of electronic tubes. Trace element control. *Obsolete*

D-H NO. 133
Driver Harris Co.
Nickel. Si 3, bal Ni.
Obsolete

D-H NO. 14
Driver Harris Co.
Ni 42, Cr 5, bal Fe.
Cold drawn: 140,000 TS; 135,000 YS. For high temperature applications; heat resistant. *Obsolete*

D-H NO. 142
Harrison Alloys Inc.
Nickel.
Now HAI-42 ALLOY.

D-H NO. 146
Harrison Alloys Inc.
Nickel.
Now HAI-46 ALLOY.

D-H NO. 146
Driver Harris Co.
Ni 46, bal Fe.
For electrical equipment; high permeability. *Obsolete*

D-H NO. 152
Driver Harris Co.
Nickel. Ni 51, bal Fe.
Rolled: 70,000-150,000 TS. For thermostats, glass seals, high frequency transformers; controlled expansion. *Obsolete*

D-H NO. 245
Driver Harris Co.
C 0.05, Cr 20, Al 5, Si 1.5, bal Ni.
Wire: 130,000-180,000 TS; 70,000-130,000 YS; 10-40 El; 195-250 Brin. For heating elements; for operating temperature to 2300 F. *Obsolete*

D-H NO. 30
Harrison Alloys Inc.
Copper.
Now HAI-30 ALLOY.

D-H NO. 33
Driver Harris Co.
C 0.05, Si 3, bal Ni.
For thermocouples; stable in reducing atmosphere. *Obsolete*

D-H NO. 399
Driver Harris Co.
Si 0.15-0.25, bal Ni.
For cathode nickel; heat resistant; active type. *Obsolete*

D-H NO. 446
Driver Harris Co.
Cr 28, bal Fe.
For glass-to-metal seal; heat resistant. *Obsolete*

D-H NO. 499
Driver Harris Co.
100% Ni plus trace elements control for control of electron, emission in the cathode of electron tubes.
For electron tube cathodes. Passive type. *Obsolete*

D-H NO. 52
Harrison Alloys Inc.
Nickel.
Now HAI-52 ALLOY.

D-H NO. 599
Driver Harris Co.
Si 0.15-0.25, bal Ni.
For cathode nickel; active type. *Obsolete*

D-H NO. 799
Driver Harris Co.
Si 0.12-0.2, bal Ni.
For cathode nickel; active type. *Obsolete*

D-H NO. 899
Driver Harris Co.
Si 0.02-0.07, bal Ni.
For cathode nickel; normal type. *Obsolete*

D-H NO. 95
Driver Harris Co.
Ni 10, bal Cu.
For resistances; 90 ohms/c.m.f. *Obsolete*

D-H NO. 99
Harrison Alloys Inc.
Nickel.
Now HAI-NI 99.

D-H NO. 999
Driver Harris Co.
Ni.
For cathode nickel; passive type. *Obsolete*

D-H NO. R. 63
Harrison Alloys Inc.
Nickel.
Now HAI-63 ALLOY.

D-H PERMANICKEL
Driver Harris Co.
Nickel. Mn 0.2-0.6, Fe 0-0.6, C 0-0.4, Si 0-0.35, bal Ni.
Age-hardened: 180,000-230,000 TS; 130,000-175,000 YS; 2-10 El. For springs; Z-Nickel; Type B. *Obsolete*

D-H PURE NICKEL
Driver Harris Co.
Nickel. Ni 99.8.
For passive and active electron tube cathodes; wire or ribbon. *Obsolete*

D-H SPECIAL ALLOY STEEL
Darbyshire Steel Co. Inc.
C, alloy, bal Fe.
For tools, dies. *Obsolete*

D-H T1 (TYPE KP)
Harrison Alloys Inc.
Nickel.
Now HAI-KP.

D-H TOOL STEEL
Darbyshire Steel Co. Inc.
C, alloy, bal Fe.
For tools, dies. *Obsolete*

D-H Z-NICKEL
Driver Harris Co.
Nickel. C 0-0.3, Al 4-4.75, bal Ni.
For high-temperature applications; two types, see Duranickel or Permanickel. *Obsolete*

D-H-E-NICKEL
Driver Harris Co.
Nickel. Mn 2, bal Ni.
For lead wire, lamp bulbs, grid wire, radio tubes. *Obsolete*

D-H-S BRONZE NO. 1
Koppers Co. Inc.
Cu 62-66, Al 4.5-7, Fe 2-3.5, Mn 3-4, bal Zn.
120,000 TS; 95,000 YS; 12 El; 12 RA; 240 Brin. For gears, bearings, valve stems; resists acid, wear and shock.

D-H-S BRONZE NO. 2
Koppers Co. Inc.
Cu 62-66, Al 4.5-7, Fe 2-3.5, Mn 3-4, bal Zn.
115,000-120,000 TS; 90,000-95,000 YS; 12-16 El; 12-19 RA; 235-275 Brin. For worm gears, rolling mill housing nuts; resists acid, wear and shock.

D-H-S BRONZE NO. 3
Koppers Co. Inc.
Cu 62-66, Al 4.5-7, Fe 2-3.5, Mn 3-4, bal Zn.
105,000 TS; 60,000 YS; 17 El; 17 RA; 200 Brin. For gears, bearings; resists acid, wear and shock.

D-H-S BRONZE NO. 4

Cu 62-66, Al 4.5-7, Fe 2-3.5, Mn 3-4, bal Zn.
90,000 TS; 45,000 YS; 25 El; 25 RA; 185 Brin. For gears, bearings; resists acid, wear and shock.

D-H180 OHM ALLOY

Driver Harris Co.
Ni 23, bal Cu.
For electrical resistances; 180 ohms/c.m.f. *Obsolete*

D-H180-OHM ALLOY

Harrison Alloys Inc.
Copper.
Now HAI-180 ALLOY.

D-M-E NO. 1

D-M-E Company
C 0.3, Mn 0.75, P 0-0.04, S 0-0.05, Si 0.25, bal Fe.
Annealed: 231 Brin; 70,000 TS; 45,000 YS; 25 EL; 55 RA. Medium carbon (SAE 1030 type), silicon-killed forging quality steel. For plastic molds and die cast dies, mold plates and holder blocks.

D-M-E NO. 1

Powder Alloys Corp.
C 0.28-0.34, Mn 0.6-0.9, bal Fe.
Plate: 165-185 Brin. For plastic molds. AISI C1030, water hardening.

D-M-E NO. 2

D-M-E Company
C 0.3, Mn 1.2, P 0.03, S 0.04, Si 0.28, Mo 0.2, Cr 0.75, bal Fe.
Plate, heat treated: 269-321 Brin; 150,000 TS; 135,000 YS; 18 El; 45 RA. AISI 4130 type steel. For plastic molds and die casting dies, mold plates, holder blocks.

D-M-E NO. 2

Detroit Mold Engineering Co.
C 0.42-0.49, Cr 0.8-1.1, Mn 0.6-0.9, 0.15 V min, bal Fe.
Hardened: 107,500 TS; 88,000 YS; 15.5 El; 33.8 RA; 225-300 Brin. For die casting mold bases and molds, cavity plates; for Zn die casting.

D-M-E NO. 2H

D-M-E Company
Now D-M-E NO. 2.

D-M-E NO. 2H

Powder Alloys Corp.
C 0.55, Mn 1.15, Mo 0.2, Cr 0.75, S 0-0.08, P 0-0.035, Si 0.25, bal Fe.
Plate: 252-302 Brin; 150,000 TS; 130,000 YS; 20 El; 60 RA. For plastic molds and die casting dies, holder blocks. Free-cutting, water or oil hardening.

D-M-E NO. 3

D-M-E Company
C 0.3, Mn 0.85, P 0.015, S 0.003, Si 0.3, Cr 1.1, V 0.08, Mo 0.55, bal Fe.
Plate, heat treated: 277-331 Brin; 155,000 TS; 140,000 YS; 18 El; 45 RA. P-20 AISI 4130 (modified) type steel. For cavity plates and insert blocks in plastic molds and die cast dies. High hardness, good machinability and polishability.

D-M-E NO. 3

Powder Alloys Corp.
C 0.36, Mn 0.85, Si 0.25, Cr 1, V 0.2, Mo 0.5, bal Fe.
Plate: 262-311 Brin. Heat treated: 206,000 TS; 175,000 YP; 12 El; 418 Brin. For cavity plates and inserts in die casting dies; zinc die casting dies, aluminum die casting dies.

D-M-E NO. 5

D-M-E Company
C 0.4, Mn 0.4, Si 1.1, Mo 1, Cr 5, V 1.1, bal Fe.
Annealed: 200 Brin. Heat treated: 235,000 TS; 180,000 YS; 14 El; 50 RA; 509 Brin. Thermal shock resistant hot work die steel, AISI-SAE H-13 type. For cavity inserts and plates for die cast dies and plastic molds.

D-M-E NO. 5

Powder Alloys Corp.
C 0.35, Cr 5, Mo 1.5, V 1, bal Fe.
Annealed: 209 Brin. Heat treated: 270,000 TS; 240,000 YS; 10 El; 30 RA; 540 Brin. For heavy duty compression tools, die casting dies, bolt dies, aluminum extrusion dies, plastic molds. Vacuum degassed forgings, Type H13 tool steel, ground flat stock.

D-M-E NO. 6

D-M-E Company
Stainless steel. C 0.3-0.4, Mn 0-1, Si 0-1, Cr 13, bal Fe.
Annealed: 200-240 Brin; 14-23 Rock C. Heat treated: 250,000 TS; 215,000 YS; 8 El; 25 RA; 512 Brin. T-420 type steel. Readily machinable.

D-M-E NO. 7

D-M-E Company
Stainless steel. C 0.3, Mn 1.1, S 0.15, Si 0.7, Cr 14, bal Fe.
Heat treated: 180,000 TS; 165,000 YS; 11 El; 25 RA; 305-342 Brin; 33-37 Rock C. AISI 420-F (modified) type holder block stainless steel.

D-Z-L MARINE NICKEL GENUINE

Magnolia Metal Corp.
Sn 88, Cu, bal Sb.
Cast: 17,0000 TS; 9510 YS; 29.6 Brin. For bearings subject to shock and heat; pouring temperature 850-925°F.

D.70

British Steel Corp.
C 0-0.03, Ni 4-4.5, Cr 11.5-12.5, Mo 4-4.5, Co 14-15, Ti, Al, Nb, B, Zr, bal Fe.
For jet engine and gas turbine components, aircraft and aerospace equipment, missiles. Stainless maraging steel. Low work hardening. Precipitation hardening. *Obsolete*

D.C. 33

Acieries de Champagnole
C 0.32, Cr 3, V 0.5, Mo 2.6, Mn 0.3, Si 0.3, bal Fe.
Air or oil hardening tool and die steel. For hot work forging dies, shears and saws. AFNOR: 32 CDV.12.30; AISI H10.

D.G. ALLOY

International Development Corp.
Cu 88, Sn 7.5, Ni 2.5, Zn 2.
25-15 El; 70-80 Brin. For shafts, staybolts, bearings, stuffing boxes; corrosion resistant.

D.G. CYCLONE

British Steel Corp.
C 0.75, Cr 4.5, W 18, V 1.25, Co 5.7, bal Fe.
For tools, cutters; high speed steel. *Obsolete*

D.N.S.

British Steel Corp.
C 0.25-0.35, Mo 0.5-0.7, bal Fe.
Heat treated: 85,000-120,000 TS; 60,000-95,000 YS; 31-40 El; 70-80 RA. For nuts, gears, shafts, bolts, fasteners; water hardened. *Obsolete*

D.N.V.

Now SIMONDS DNV.

D.O.H.

Spencer Clark Metal Industries Ltd.
C 0.9, Si 0.25, Mn 1.7, bal Fe.
Oil hardening, cold work tool steel, for bushes, collets, lathe centers.

D.T.D. 49B

British manufacture
C 0.35-0.45, Si 1-1.75, Mn 0.5-1, Cr 12.5-14.5, Ni 12.5-14.5, W 2-3, bal Fe.
For aircraft valves; austenitic.

D.Y.O.

Carpenter Technology Corp.
C 0.9, Mn 0.3, Si 0.25, Cr 4, W 14.5, V 0.5, bal Fe.
Heat treated: 205,000 TS. For tools, forging dies. Red-hard, hot work steel. Similar to AISI H25. *Obsolete*

D10

Keystone Carbon Co.
Graphite bronze. Sintered: 10,000 psi TS; 1.0 El; 30 Rock H; 6.5-6.9 density. Oiless bearing for light loads, slow speeds.

D1045

Comalco Ltd.
99.45 Al min.
O-temper: 83 MPa TS; 28 MPa YS; 40 El. H19-temper: 186 MPa TS; 165 MPa YS. Electrical conductors.

D1150

Comalco Ltd.
99.35 Al min.
O-temper: 69 MPa TS; 28 MPa YS; 35 El. H18-temper: 145 MPa TS; 138 MPa YS; 5 El; 50 Vickers. Sheet metal components requiring decorative finishing.

D12 ALLOY

Birmingham Aluminium Casting Co.
Aluminum. Cu 6.2-7.2, Mg 0.15-0.4, Si 5-6, Fe 0-1.25, bal Al.
Heat treated: 40,000-49,000 TS; 34,000 YS; 0 El; 130 Brin. For pistons, cylinder heads; age-hardenable.

D2-SA TOOL

Disston Inc.
C, bal Fe.
For tools,; general purpose. *Obsolete*

D29

Teledyne Firth Sterling
C, alloy, bal Fe.
For tools, dies; oil hardening. *Obsolete*

D3005

Comalco Ltd.
Mn 1.2, Mg 0.35, bal Al.
O-temper: 131 MPa TS; 62 MPa YS; 25 El. H18-temper: 241 MPa TS; 228 MPa YS; 4 El. High strength foil, roofing sheet.

D319L

G.O. Carlson Inc.
Stainless steel. C 0-0.03, Mn 0-2, Si 0-1, Cr 17.5-19.5, Ni 11-15, Mo 2.25-3, bal Fe.
Austenitic stainless steel.

D421

Spencer Clark Metal Industries Ltd.
C 0.44, Mn 0.4, Si 0.7, Cr 1.3, W 2.3, V 0.3, bal Fe.
Tungsten shock resisting steel, for hand and pneumatic chisels.

D5050

Comalco Ltd.
Mg 1.4, bal Al.
O-temper: 145 MPa TS; 55 MPa YS; 24 El; 36 Brin. H38-temper: 221 MPa TS; 200 MPa YS; 6 El; 63 Brin. Coiled tubes, refrigerator trim.

D6-CO

Disston Inc.
C 0.8, Cr 4, Co 6, V 1.75, W 18, bal Fe.
For tools, cutters; high speed steel. *Obsolete*

D6201

Comalco Ltd.
Mg 0.7, Si 0.6, bal Al.
T8 Temper: 310 MPa TS; 290 MPa YS; 4 El. Electrical conductors.

D6463

Comalco Ltd.
Mg 0.7, Si 0.4, bal Al.
T1-temper: 152 MPa TS; 90 MPa YS; 20 El. T6-temper: 241 MPa TS; 214 MPa YS; 12 El; 74 Brin. Extrusions for trim requiring decorative finishing.

D69

Keystone Carbon Co.
Diluted bronze 75 25. Sintered: 17,000 psi TS; 18% porosity; 6.2-6.6 density; 50 Rock H. For oscillatory-type bearings.

D6C
Now UNITEMP D6C.

D7 ALLOY
Birmingham Aluminium Casting Co.
Aluminum. Cu 2.2-2.8, Si 4.5-5.5, Mn 0.3-0.7, bal Al.
Cast: 26,000-36,000 TS; 8,000-12,000 YS; 10-5 El; 60-65 Brin.
For heavy duty castings; age-hardenable.

D7 QUALITY
Delta Metal (BW) Ltd.
Cu, bal Zn.
Extruded: 55,000 TS; 50 El; 45 RA. For rivets; cold working qualities. *Obsolete*

D8 (D8W, D8WP)
Sterling International Technology Ltd.
Cu 2-4, Si 4-6, Mn 0.3-0.7, bal Al.
Sand cast, solution treated (D8W): 22,400-26,800 TS; 12,300 YS; 2-3 El; 1-3 Izod. Chill cast, solution treated and aged (D8WP): 45,000-47,000 TS; 40,000 YS; 110 Brin; 0.6 Izod. Castings for good strength, hydraulic pressure. BS L79.

D8 ALLOY
Birmingham Aluminium Casting Co.
Aluminum. Cu 2-4, Si 3-6, Mn 0.3-0.7, Ni 0-0.35, bal Al.
Cast: 22,000-35,000 TS; 10,000-30,000 YS; 3-2 El; 60-100 Brin. For hydraulic units, gear cases; hardenable.

D8011
Comalco Ltd.
Fe 0.8, Si 0.7, bal Al.
Sheet for bottle closures.

D9 ALLOY
Birmingham Aluminium Casting Co.
Aluminum. Cu 3.2-4, Si 4.5-5.5, Mn 0.3-0.7, bal Al.
36,000-44,000 TS; 15,000-28,000 YS; 8-4 El; 75-105 Brin. For light alloy parts; hardenable.

D9-VA
Disston Inc.
C 1.15, Mn 0.6, V 0.2, Si 0.6, bal Fe.
For taps, twist drills, reamers, tools; water hardening. *Obsolete*

D979
Now UDIMET D-979.

DA-105
NL Industries
Cu 10, Si 5, Mg 0.3, bal Al.
Cast: 42,000 TS; 35,000 YS; 120 Brin. For castings; non-hardenable. *Obsolete*

DA-47
D.A.B. Industries Inc.
Cu 52, Pb 44, Sn 4.
Copper-lead alloy for half shell bearings for intermediate loads on connecting rod and main bearings.

DA-49
D.A.B. Industries Inc.
Cu 75, Pb 24, Sn 1.
Copper lead alloy for half shell bearings for highly loaded main and connecting rod bearings.

DA-8
NL Industries
Mg 8, bal Al.
Cast: 38,000 TS; 24,000 YS; 2.5 El; 65 Brin. For castings; high impact strength. *Obsolete*

DA-95
D.A.B. Industries Inc.
Cu 73.5, Pb 23, Sn 3.5.
Copper lead alloy for half shell bearings for heavy duty main and connecting rod bearings.

DA-99
NL Industries
Al 99.
For cables; high conductivity. *Obsolete*

DACAR
Jessop Steel Co.
C 0.7-1.2, bal Fe.
For machinery parts; water hardening. *Obsolete*

DAIDO ST-1
Daido Steel Co. Ltd.
C 0.13-0.18, Mn 1.35-1.85, Si 0.3-0.6, Cu 0-0.35, bal Fe.
Rolled: 78,000 TS min; 51,000 YP min; 18 El min. For buildings, bridges, agricultural equipment, structural members, case hardened parts. Constructional steel, shock resistant.

DAIDO ST-2
Daido Steel Co. Ltd.
C 0.15-0.2, Mn 1.2-1.5, Si 0.3-0.65, Cu 0-0.35, bal Fe.
Rolled: 78,000 TS min; 51,000 YP min; 18 El min. For buildings, bridges, structural members, agricultural equipment. Constructional steel, tough.

DAIMLER BEARING METAL
English manufacture
Cu 76, Sn 3, Zn 20, Pb 1.
For bearings, bushings; corrosion resistant.

DAIRY METAL
Amesbury Brass & Foundry Co.
Copper. Cu 68, Sn 1.5, Ni 28, Pb 0.5, Al 2.
For stainless filters; corrosion resisting.

DAIRYWHITE
American manufacture
Zn, Ni, bal Cu.
For dairy equipment; nickel silver, corrosion resistant.

DALTON FUSIBLE ALLOY
English manufacture
Bi 60, Sn 15, Pb 25.
For fusible alloy, fire extinguishers; melting point 92°C.

DAMAR
English manufacture
Cu 76, Pb 13, Sn 11.
For bearings; heavy duty.

DAMASCITE
Chrylser Corp.
Fe.
For molded iron parts, machine pads; strength of carbon steel. *Obsolete*

DAMASCUS
Abex Corp.
Sn, Pb, bal Cu.
For bearings. *Obsolete*

DAMASCUS
Latrobe Steel Co.
Tool material. C 0.55, Cr 0.25, Mn 0.9, V 0.2, Si 1.95, bal Fe.
For chisels, stamps, cold cutters, punches, shears; oil hardening. *Obsolete*

DAMASCUS
Damascus Steel Casting Co.
Sn, Pb, bal Cu.
For bearings.

DAMASCUS BRONZE
English manufacture
Pb 13, Sn 10, bal Cu.
For bearings, ornaments; plastic bronze.

DAMPING
American manufacture
Mn 87, Cu 13.
For alloy for meters; low temperature resistance coefficient.

DAMSTADT BELL METAL
German manufacture
Cu 72.5-74, Sn 21.7-21.1, Pb 2.12, Fe 0.05-0.19, Ni 2-2.6.
For bells; corrosion resistant.

DANA AUTO SPECIAL
Paul Bergsoe & Son
Sb 6.5, Cu 3, bal Sn.
Cast: 69 El; 24-25 Brin. For engine bearings; shock resistant, Babbitt metal.

DANA COMMON
Paul Bergsoe & Son
Sn 10, Sb 13.5, Cu 0.5, bal Pb.
Cast: 42 El; 28-30 Brin. For transmission bearings; Babbitt.

DANA DIESEL
Paul Bergsoe & Son
Pb 2, Cu 6.5, Sb 8, bal Sn.
Cast: 22 El; 30-31 Brin. For diesel engine bearings; wear resistant, Babbitt metal.

DANA STEAM
Paul Bergsoe & Son
Sb 11.5, Cu 5.5, Pb 3, bal Sn.
Cast: 19 El; 33-35 Brin. For steam turbine and generator bearings; wear resistant, Babbitt metal.

DANALLOY-I
Inland Electronics Products Corp.
Au, Ni-Mg, bal Ag.
Rolled: 82,000 TS; 78,000 YS. For circuit board retainers, micro-circuit back-up plates, relay contacts. High heat and electrical conductivity. Non-magnetic. Heat treatable.

DANALLOY-II
Inland Electronics Products Corp.
Au, Ni-Mg, bal Ag.
Rolled: 66,000 TS; 60,000 YS. For circuit board retainers, micro-circuit back-up plates, relay contacts. High heat and electrical conductivity. Non-magnetic. Heat treatable.

DANDELION METAL
English manufacture
Pb 72, Sb 18, Sn 10.
For heavy duty machine bearings and locomotive cross-head linings; high strength Babbitt.

DANISH MINT
English manufacture
Cu 92, Al 6, Ni 2.
For coinage; corrosion resistant.

DANNEMORA
Ryer Inc. Ltd.
C 0.7, W 18, Cr 4, V 1, bal Fe.
For coining dies. High speed steel.

DANNEMORA AD 95
Great Western Steel Co.
C, Cr, W, V, bal Fe.
For coining dies; oil hardening.

DANNEMORA BEST
Manufacturer not listed
C 0.9-1.1, bal Fe.
For drills, reamers, punches, broaches; Type W1; water hardened.

DANNEMORA DB59
Achorn Steel Co.
C 0.7, W 18, Cr 4, V 1, Co 5, bal Fe.
For cutters, hobs, high speed steel.

DANNEMORA EXTRA BEST
Adams & Osgood Steel Co.
C 0.7-1.2, bal Fe.
For tools; water hardened.

DANNEMORA NO. 0
Adams & Osgood Steel Co.
C, W, bal Fe.
For tools, dies; oil hardened.

DANNEMORA SELF-HARDENING
Adams & Osgood Steel Co.
C, W, bal Fe.
For tools, cutters, dies; oil hardened.

DANNEMORA STANDARD
Manufacturer not listed
C 0.9-1.1, bal Fe.
For tools, drills, taps; water hardened.

DANNEMORA VERY BEST
Manufacturer not listed
C 1-1.2, bal Fe.
For tools, cutters, drills; water hardened.

DARCET FUSIBLE ALLOY
English manufacture
Bi 50, Sn 25, Pb 25.
For boiler safety plugs, fire extinguishers; melting point 93°C.

DARGRAPH
Darwin & Milner Inc.
Tool material. C 1.45, Si 1, Mo 0.25, bal Fe.
Annealed: 84,500 TS; 49,500 YS; 25 El; 40 RA; 197 Brin. Heat treated: 164,000-218,000 TS; 136,000-177,000 YS; 8.5-13 El; 2-14 RA; 302-388 Brin. For wear plates, cams, cutters, dies, punches; oil hardened; wear resistant; graphite steel.

DARK RED GOLD
English manufacture
Au 50, Cu 50.
For ornaments; corrosion resistant.

DART
Latrobe Steel Co.
Tool material. C 0.42, Si 1, Mn 0.55, Cr 3.3, Mo 2.4, V 0.35, bal Fe.
Hardened: 53-59 Rock C. Hot work die steel with extra high carbon for higher hardness and wear resistance; forging dies, hot forming and press dies. AISI Type H 10 hot work tool steel.

DARWIN
Darwin & Milner Inc.
C, alloy, bal Fe.
For high speed tools, hacksaw blades, tools, cutters; super high speed steel.

DARWIN
Balfour Darwins Ltd.
C, alloy, bal Fe.
For high speed tools, hacksaw blades, tools, cutters; super high speed steel.

DARWIN "ALNI"
Darwins Alloy Castings
Fe 59, Cu 4, Ni 24, Al 13.
For permanent magnets; magnetic steel. *Obsolete*

DARWIN "BKM"
Darwin & Milner Inc.
C, Ni, Cr, bal Fe.
For Bakelite molds; oil hardening.

DARWIN "BKM"
Balfour Darwins Ltd.
C, Ni, Cr, bal Fe.
For Bakelite molds; oil hardening

DARWIN "DCI"
Darwin & Milner Inc.
C, alloy, bal Fe.
For electrical equipment, magnetic parts; magnetic alloy.

DARWIN "DCI"
Balfour Darwins Ltd.
C, alloy, bal Fe.
For electrical equipment, magnetic parts; magnetic alloy.

DARWIN "TTS"
Darwin & Milner Inc.
C, W, bal Fe.
For tools; oil hardening.

DARWIN "TTS"
Balfour Darwins Ltd.
C, W, bal Fe.
For tools; oil hardening.

DARWIN "W" BRAND
Darwin & Milner Inc.
Tool material. C 0.45, Cr 0.85-0.95, W 1, bal Fe.
For tools, dies, punches; oil hardened.

DARWIN 19
Darwin & Milner Inc.
Tool material. C 0.7, W 18, Cr 4, V 2, Co 8, bal Fe.
For tools, cutters, reamers; high speed steel.

DARWIN 3581
Darwin & Milner Inc.
Tool material. C, alloy, bal Fe.
For chisels; tough and shock resistant.

DARWIN 505 SPECIAL
Darwin & Milner Inc.
Tool material. C 0.8, Cr 4, V 2, W 18, Mo 0.8, Co 9, bal Fe.
High speed steel, cutting tool, tungsten-cobalt type; good red hardness; AISI T5.

DARWIN 5V
Darwin & Milner Inc.
Tool material. C 1.5, Cr 5, V 5, W 12.5, Co 5, bal Fe.
High speed tool steel; Cr-W-V-Co type. For cutting tools for tough metal; good red hardness and good wear properties; AISI T15.

DARWIN 93
Darwin & Milner Inc.
Tool material. C 0.3, Cr 3.5, V 0.3, W 10, Si 0.4, bal Fe.
For dies; hot work steel.

DARWIN ACD
Darwin & Milner Inc.
Tool material. C, bal Fe.
For dies; water hardened.

DARWIN BEST WARRANTED
Darwin & Milner Inc.
High C, bal Fe.
For tools; water hardening.

DARWIN BEST WARRANTED
Balfour Darwins Ltd.
High C, bal Fe.
For tools; water hardening.

DARWIN BRAKE DIE STEEL
Darwin & Milner Inc.
Tool material. C, Cr, Mo, bal Fe.
For brake dies; oil hardened.

DARWIN BRAND "H"
Darwin & Milner Inc.
Tool material. C 0.95, Cr 0.4, V 0.2, Mn 1, bal Fe.
For dies, blanking, forming and trimming dies, taps, broaches; oil hardened; nondeforming; resistant to wear

DARWIN BRAND L-35
Darwin & Milner Inc.
Tool material. C, alloy, bal Fe.
For heading dies.

DARWIN C C
Darwin & Milner Inc.
Tool material. C 0.55-0.65, Cr 0.3-0.4, W 0.6-0.75, Mo 0.55-0.65, Si 0.95-1.1, bal Fe.
For tools, punches; oil hardened.

DARWIN C V
Darwin & Milner Inc.
Tool material. C 1.1-1.15, Cr 0.2-0.4, V 0.25-0.35, bal Fe.
For tools, taps, drills; water hardened.

DARWIN C.L. NO. 1
Darwin & Milner Inc.
Alloy steel. C, alloy, bal Fe.
For arc welding electrodes; hard surfacing; coated. Tough and abrasion resistant.

DARWIN COBALT MAGNET
Darwin & Milner Inc.
Alloy steel. Co 6, C, bal Fe.
For permanent magnets.

DARWIN COBALT MAGNET
Darwin & Milner Inc.
Alloy steel. Co 35, C, bal Fe.
For permanent magnets.

DARWIN COBALT MAGNET
Darwin & Milner Inc.
Alloy steel. Co 3, C, bal Fe.
For permanent magnets.

DARWIN COBALT MAGNET
Darwin & Milner Inc.
Alloy steel. Co 9, C, bal Fe.
For permanent magnets.

DARWIN COBALT MAGNET
Darwin & Milner Inc.
Alloy steel. Co 15, C, bal Fe.
For permanent magnets.

DARWIN D93
Darwin & Milner Inc.
C, alloy, bal Fe.
For hot working dies and tools; hot die steel.

DARWIN D93
Balfour Darwins Ltd.
C, alloy, bal Fe.
For hot working dies and tools; hot die steel.

DARWIN DUREX
Darwin & Milner Inc.
W 18, Cr 4, V 1, high C, bal Fe.
For tools, cutters, reamers, dies, gauges, punches; high speed steel.

DARWIN DUREX
Balfour Darwins Ltd.
W 18, Cr 4, V 1, high C, bal Fe.
For tools, cutters, reamers, dies, gauges, punches; high speed steel.

DARWIN EE
Darwin & Milner Inc.
Tool material. C 1.35, Cr 0.15, W 4, V 0.05, bal Fe.
For cutters, shears, tools; water hardened.

DARWIN EXTRA QUALITY
Darwin & Milner Inc.
C 1, bal Fe.
For drills, taps, tools, cutters, reamers; Type W1; water hardened.

DARWIN EXTRA QUALITY
Balfour Darwins Ltd.
C 1, bal Fe.
For drills, taps, tools, cutters, reamers; Type W1; water hardened.

DARWIN EXTRA SPECIAL
Darwin & Milner Inc.
C 1.1, bal Fe.
For drills, taps, hobs, reamers; Type W1, water hardened.

DARWIN EXTRA SPECIAL
Balfour Darwins Ltd.
C 1.1, bal Fe.
For drills, taps, hobs, reamers; Type W1, water hardened.

DARWIN EXTRA TOUGH
Darwin & Milner Inc.
Tool material. C, alloy, bal Fe.
For punches, chisels, pneumatic tools; Type S6; oil hardened; shock resistant.

DARWIN FLAME HRD
Darwin & Milner Inc.
Alloy steel. C 0.5, Mn 1.2, Si 0.5, Cr 1.4, V 0.1, Mo 0.4, bal Fe.
Air hardened steel for shafts, arbors.

DARWIN H.A.W.
Darwin & Milner Inc.
Tool material. C 0.35, Cr 5, V 0.4, Mo 1.5, bal Fe.
For shears, punches, hot work tools; Type H11; hot work steel.

DARWIN H.W.S.
Darwin & Milner Inc.
Tool material. C 0.35, Cr 5, V 0.4, W 1.5, Mo 1.5, bal Fe.
For dies, punches, hot work tools; Type H12; hot work steel.

DARWIN I W I
Darwin & Milner Inc.
Tool material. C 0.4, Cr 3, W 9-10, V 0.2-0.3, bal Fe.
For tools, dies, punches; hot work steel.

DARWIN LOW AIR
Darwin & Milner Inc.
Tool material. C 0.75, Mn 2, Si 0.3, Cr 1, Mo 1.35, bal Fe.
Air hardened tool steel; AISI A6.

DARWIN M3
Darwin & Milner Inc.
Tool material. C 1, Cr 4, V 2.7, W 6, Mo 5, bal Fe.
For reamers, drills, taps, lathe and planer tools; Type M3; high speed steel.

DARWIN MT-6
Darwin & Milner Inc.
Tool material. C 0.85, W 6, Cr 4, V 1.5, Mo 6, bal Fe.
For cutters, tools; high speed steel.

DARWIN N-32
Darwin & Milner Inc.
Tool material. C 0.38, Ni 3-4, Mn 0.6, bal Fe.
For chisels; oil hardened.

DARWIN NO. 1 AIR HARDENING
Darwin & Milner Inc.
Tool material. C 1.45-1.6, Cr 11-12, V 0.2-0.35, Mo 0.7-0.8, bal Fe.
For tools, dies; nondeforming.

DARWIN NO. 1 FM
Darwin & Milner Inc.
Tool material. C 1.5, Cr 12, V 0.3, Mo 0.8, S 0.2, bal Fe.
For dies, tools, punches, crimpers; air or oil hardened; good machinability; resists galling.

DARWIN NO. 1366
Darwin & Milner Inc.
C 0.7, Cr 4.5, W 20, V 2, Co 12, bal Fe.
For lathe and planer tools, form cutters; high speed steel, oil hardened.

DARWIN NO. 1366
Balfour Darwins Ltd.
C 0.7, Cr 4.5, W 20, V 2, Co 12, bal Fe.
For lathe and planer tools, form cutters; high speed steel, oil hardened.

DARWIN NO. 505
Darwin & Milner Inc.
Tool material. C 0.65-0.75, Cr 3.75-4.25, W 17-18, V 1.5-1.75, Co 7-8, Mo 0.7-1, bal Fe.
For tools, cutters; high speed steel.

DARWIN OHT
Darwin & Milner Inc.
C 0.9, Mn 1, Cr 0.5, W 0.5, bal Fe.
For crimpers, punches, jaws, dies, cutters; Type OI; non-deforming, oil hardened.

DARWIN OHT
Balfour Darwins Ltd.
C 0.9, Mn 1, Cr 0.5, W 0.5, bal Fe.
For crimpers, punches, jaws, dies, cutters; Type OI; non-deforming, oil hardened.

DARWIN P-20
Darwin & Milner Inc.
Tool material. C 0.3, Cr 0.75, Mo 0.25, bal Fe.
Oil or water hardened tool steel; for molds; AISI P20.

DARWIN PRK-33
Darwin & Milner Inc.
Cast iron. C 3.7, Cr 13.5, bal Fe.
For welding electrodes; air hardened.

DARWIN SPECIAL
Darwin & Milner Inc.
Tool material. C 1.2, bal Fe.
Annealed: 100,000 TS; 53,000 YS; 21 El; 42 RA; 200 Brin. For drills, taps, hobs, reamers, cutters; Type W1; water hardened.

DARWIN SPECIAL CARBON
Darwin & Milner Inc.
Tool material. C 0.9, bal Fe.
Heat treated: 190,000 TS; 145,000 YS; 10 El; 30 RA; 400 Brin.
For drills, taps, hobs, reamers, lathe cutters; Type W1; water hardened.

DARWIN STANDARD
Darwin & Milner Inc.
Tool material. C 0.8-1.1, Si 0.25, Mn 0.25, bal Fe.
Water hardened: 166,000-216,000 TS; 110,000-150,000 YS; 11-15 El; 32-37 RA; 330-600 Brin. For taps, drills, reamers, punches, stamps, knurls, mandrels. Type W1; water hardened.

DARWIN TEMPER TOUGH
Darwin & Milner Inc.
Tool material. C 0.7, Cr 0.75, Ni 1.5, Mo 0.25, bal Fe.
For brake dies, bushings, jigs, cams, chucks; Type L6; oil hardened.

DARWIN WARRANTED
Darwin & Milner Inc.
high C, bal Fe.
For tools; water hardening.

DARWIN WARRANTED
Balfour Darwins Ltd.
high C, bal Fe.
For tools; water hardening.

DARWIN'S DUROR
Darwins Alloy Castings
C, alloy, bal Fe.
For files; water hardened. *Obsolete*

DARWIN'S T.T.S.
Darwins Alloy Castings
W 1.5, C, bal Fe.
For surgical instruments, taps, reamers, drills, cutters, hacksaws; water hardened. *Obsolete*

DARWIN'S TRIPLE LIFE
Darwins Alloy Castings
C, alloy, bal Fe.
For files; water hardened. *Obsolete*

DARWIN-1
Darwin & Milner Inc.
Tool material. C 1.5, Cr 12, Mo 1, bal Fe.
For drawing and forming dies, punches; Type D2; air hardened; nondeforming.

DARWIN-SSC
Darwin & Milner Inc.
Tool material. C, bal Fe.
For machine tool parts; water hardened.

DARWINS 1031
Darwins Alloy Castings
Ni, bal Fe.
For thermostats; controlled expansion. *Obsolete*

DARWINS 1366
Darwins Alloy Castings
C 0.73, W 20, Cr 4.5, V 2, Co 12, bal Fe.
Heat treated: 64-66 Rock C. For boring and turning tools, broaches, milling cutters. High-speed steel; Type T6, high red-hardness, abrasion resistant. *Obsolete*

DARWINS 168
Darwins Alloy Castings
Stainless steel. C 0.75, Cr 17, bal Fe.
For knives, bearings, cutlery, valves; Type 440A; stainless. *Obsolete*

DARWINS 18/8
Darwins Alloy Castings
Stainless steel. C 0-0.14, Ni 7.5-9, Cr 17.5-19, bal Fe.
Annealed: 78,000 TS; 27,000 YS; 30 El; 130 Brin. Cold drawn: 180,000 TS; 150,000 YS; 10 El; 250 Brin. For food, chemical and brewing process equipment; Type 302; stainless, austenitic. *Obsolete*

DARWINS 18/8 (316)
Darwins Alloy Castings
Stainless steel. C 0-0.14, Cr 17.5-19, Ni 10-12, Mo 2-3, bal Fe.
Bar: 78,000 TS; 27,000 YS; 30 El. For equipment handling sulfite liquors; Type 316; stainless, austenitic. *Obsolete*

DARWINS 18/8 (316A)
Darwins Alloy Castings
Stainless steel. C 0-0.08, Cr 17.5-19, Ni 8.5-10.5, Mo 1.25-2.25, Ti 0.25-0.4, bal Fe.
Bar: 78,000 TS; 27,000 YS; 30 El. For acid resistant chemical plant equipment; Type 316Ti; stainless, austenitic. *Obsolete*

DARWINS 18/8 CB
Darwins Alloy Castings
Stainless steel. C 0-0.1, Cr 17.5-19, Ni 10-12, Cb = 10 x C, bal Fe.
Bar: 78,000 TS; 27,000 YS; 30 El. For welded chemical plant equipment; Type 347; austenitic, stainless. *Obsolete*

DARWINS 18/8 FZ
Darwins Alloy Castings
Stainless steel. C 0-0.14, Ni 7.9-9, Cr 17.5-19, S 0.2, Mo 0-0.5, bal Fe.
Bar: 78,000 TS; 27,000 YS; 30 El. For stainless screw machine products; Type 303; free-cutting, stainless. *Obsolete*

DARWINS 18/8 LC
Darwins Alloy Castings
Stainless steel. C 0-0.07, Ni 8-10, Cr 17.5-19, bal Fe.
Annealed: 78,000 TS; 27,000 YS; 30 El. Cold drawn: 180,000 TS; 125,000 YS; 10 El; 330 Brin. For welded chemical plant and dyeing equipment; Type 304; stainless, austenitic. *Obsolete*

DARWINS 18/8 MO
Darwins Alloy Castings
Stainless steel. C 0-0.14, Cr 17.5-19, Ni 8-10, Mo 3-3.5, bal Fe.
Bar: 78,000 TS; 27,000 YS; 30 El. For acid resistant equipment; Type 317; stainless, austenitic. *Obsolete*

DARWINS 26
Darwins Alloy Castings
Ni, bal Fe.
For glass to metal seals; controlled expansion. *Obsolete*

DARWINS 426
Darwins Alloy Castings
Ni, bal Fe.
For glass to metal seals; controlled expansion. *Obsolete*

DARWINS 55
Darwins Alloy Castings
C 0-0.2, Cr 23, Ni 55, Mo 4, W 2, Fe 0-10, Cu 5.
For chemical plant and pickling equipment; resists hot or cold H_2SO_4. *Obsolete*

DARWINS 6/5/2
Darwins Alloy Castings
C 0.85, W 6, Cr 4, V 2, Mo 5, bal Fe.
Heat treated: 63-67 Rock C. For woodworking machine tools, taps, slitting saws, reamers, boring and planing tools, drawing dies. High-speed steel; Type M2. *Obsolete*

DARWINS 654A
Darwins Alloy Castings
C 0-0.1, Ni 60, Mo 20, bal Fe.
For chemical plant equipment; resists HCl and H_2SO_4. *Obsolete*

DARWINS 655B
Darwins Alloy Castings
C 0-0.1, Ni 65, Mo 27, bal Fe.
For chemical plant equipment; resists HCl and H_2SO_4. *Obsolete*

DARWINS 656C
Darwins Alloy Castings
C 0-0.1, Cr 14, Ni 58, Mo 17, W 5, bal Fe.
For chemical plant and high temperature equipment; resists HNO_3 and HCl. *Obsolete*

DARWINS COBALT FAST WORK
Darwins Alloy Castings
C, Co, Mo, bal Fe.
For press tools, cutters; oil hardened. *Obsolete*

DARWINS D.C.C.M.
Darwins Alloy Castings
C, alloy, bal Fe.
For punches, crimpers, upsetters; cold-work steel. *Obsolete*

DARWINS DSC
Darwins Alloy Castings
Stainless steel. C 0.3-0.4, Cr 13-14, Ni 0-1, bal Fe.
Heat treated: 100,000 TS; 20 El; 152-255 Brin. For cutlery, knives, valves; hardenable, stainless; EN56D. *Obsolete*

DARWINS EXTRA
Darwins Alloy Castings
C, alloy, bal Fe.
For files; water hardened. *Obsolete*

DARWINS F
Darwins Alloy Castings
Ni, bal Fe.
For glass to metal seals; controlled expansion. *Obsolete*

DARWINS H.W.1
Darwins Alloy Castings
C, alloy, bal Fe.
For pressure casting dies; hot-work steel, resists thermal shock. *Obsolete*

DARWINS H.W.2
Darwins Alloy Castings
C, alloy, bal Fe.
For pressure casting dies; hot-work steel, abrasion resistant. *Obsolete*

DARWINS H.W.3
Darwins Alloy Castings
C, alloy, bal Fe.
For extrusion and drawing dies, forging tools; hot-work steel, oil hardened. *Obsolete*

DARWINS H.W.4
Darwins Alloy Castings
C, alloy, bal Fe.
For die casting dies; hot-work steel, resists thermal shock. *Obsolete*

DARWINS H.W.5
Darwins Alloy Castings
C, alloy, bal Fe.
For hot piercing and forming punches, bolt dies, forging tools; hot-work steel, oil hardened. *Obsolete*

DARWINS HS22
Darwins Alloy Castings
C 0.75, W 22, Cr 5, V 2, bal Fe.
For roll turning tools, lathe and planer cutters, hacksaws; high-speed steel. *Obsolete*

DARWINS K-ALLOY
Darwins Alloy Castings
C 0-0.15, Cr 23, Ni 23, Mo 2, Cu 4, bal Fe.
For chemical plant equipment; resists H_2SO_4. *Obsolete*

DARWINS N.1932
Darwins Alloy Castings
C, alloy, bal Fe.
For chisels, punches, upsetters; shock resistant, oil hardened. *Obsolete*

DARWINS S61
Darwins Alloy Castings
Stainless steel. C 0-0.12, Cr 12-14, Ni 0-1, bal Fe.
Heat treated: 100,000 TS; 20 El; 155-255 Brin. For pump rods, valves, cutlery, surgical instruments; Type 410; stainless. *Obsolete*

DARWINS S61F
Darwins Alloy Castings
Stainless steel. C 0-0.12, Cr 12-14, S 0.2-0.3, Ni 0-1, bal Fe.
Heat treated: 100,000 TS; 20 El; 155-207 Brin. For pump rods, valves, cutlery, surgical instruments; Type 410F; stainless, free-cutting. *Obsolete*

DARWINS S62
Darwins Alloy Castings
Stainless steel. C 0.18-0.25, Cr 12-14, Ni 0-1, bal Fe.
Heat treated: 100,000 TS; 20 El; 155-255 Brin. For surgical instruments, chemical plant equipment; EN56C-D; stainless hardenable. *Obsolete*

DARWINS S62F
Darwins Alloy Castings
Stainless steel. C 0.18-0.25, Cr 12-14, Ni 0-1, S 0.2-0.3, bal Fe.
Heat treated: 100,000 TS; 20 El; 152-255 Brin. For screw machine products, bolts, screws; Type 420F; stainless, free-cutting. *Obsolete*

DARWINS S80
Darwins Alloy Castings
Stainless steel. C 0.12-0.18, Cr 16-18, Ni 1.6-2, bal Fe.
Heat treated: 123,000 TS; 15 El; 248 Brin. For pumps, fittings, valves; hardenable, stainless; EN57. *Obsolete*

DARWINS S80F
Darwins Alloy Castings
Stainless steel. C 0.15-0.25, Cr 16-18, Ni 1-3, Mo 0-0.5, S 0.25, bal Fe.
Heat treated: 123,000 TS; 15 El; 248 Brin. For pumps, fittings, valves; hardenable, stainless, free-cutting. *Obsolete*

DARWINS VANADIA
Darwins Alloy Castings
C, W, alloy, bal Fe.
For hacksaw blades; water hardened. *Obsolete*

DARWINS WDP
Darwins Alloy Castings
Stainless steel. C 0-0.07, Ni 8-10, Cr 17.5-19, Ti = 5 x C, bal Fe.
Bar: 78,000 TS; 27,000 YS; 30 El. For welded structures and chemical plant equipment; Type 321; austenitic, stainless. *Obsolete*

DAUPHINOX A3
Forges et Acieries de Bonpertuis
C 0-0.03, Cr 18, Ni 10, bal Fe.
Austenitic stainless steel, low carbon. For brewery, dairy, cheese industry equipment. AFNOR Z 03 CN 18-10. AISI 304L; W. Nr. 4306.

DAUPHINOX A3I
Forges et Acieries de Bonpertuis
C 0-0.07, Cr 18, Ni 10, bal Fe.
Austenitic stainless steel. Navy, railroad, aeronautics equipment. AFNOR Z 06 CN 18-10. AISI 304; W.-Nr. 4301.

DAUPHINOX A3II
Forges et Acieries de Bonpertuis
C 0-0.12, Cr 18, Ni 10, bal Fe.
Austenitic stainless steel. Equipment for paper mills. AFNOR Z 10 CN 18-10. AISI 302; W.-Nr. 4300.

DAUPHINOX A3M2
Forges et Acieries de Bonpertuis
C 0-0.03, Cr 18, Ni 13, Mo 2.5, bal Fe.
Low carbon type 316 stainless. For chemical industry equipment: organic, acetic, phosphoric acids; fiber industry. AFNOR Z 03 CND 18-13. AISI 316L; W.-Nr. 4404.

DAUPHINOX A3ML
Forges et Acieries de Bonpertuis
C 0-0.1, Cr 18, Ni 12, Mo 2.5, T = 4 x C, bal Fe.
Stabilized Type 316 stainless steel. For welded assemblies for food and beverage equipment; also navy and aeronautics. AFNOR Z 08 CNDT 18-12. AISI 316 T; W.-Nr. 4571.

DAUPHINOX A3MO
Forges et Acieries de Bonpertuis
C 0-0.1, Cr 18, Ni 12, Mo 2.5, bal Fe.
Austenitic stainless steel; extra corrosion resistance. Chemical, photography, cellulose industry equipment. AFNOR Z 08 CND 18-12. AISI 316; W.-Nr. 4401.

DAUPHINOX A3T
Forges et Acieries de Bonpertuis
C 0-0.12, Cr 18, Ni 10, Ti = 4 x C, bal Fe.
Titanium stabilized austenitic steel. For welded stainless structures. AFNOR Z 10 CNT 18-10. AISI 321; W.-Nr. 4541.

DAUPHINOX D1
Forges et Acieries de Bonpertuis
C 0.08, Cr 13, bal Fe.
Semi-ferritic type stainless; generally not hardenable. Fittings, slide calipers, knife handles. AFNOR Z 08 C 13. W.-Nr.-4000.

DAUPHINOX D2
Forges et Acieries de Bonpertuis
C 0.08, Cr 16, bal Fe.
Ferritic stainless steel; not hardenable. Corrosion resistant fittings. AFNOR Z 08 C 16. AISI 430; W.-Nr. 4016.

DAUPHINOX D2N
Forges et Acieries de Bonpertuis
C 0.15, Cr 16, Ni 2, bal Fe.
Martensitic stainless steel, hardenable to 29-46 Rock C. For parts used in sea water and superheated steam. AFNOR Z 15 CN 16.2. AISI 431; W.-Nr. 4057.

DAUPHINOX D2S
Forges et Acieries de Bonpertuis
C 0.08, Cr 16, S 0.2, bal Fe.
Free machining ferritic stainless steel; not hardenable. For screws, bolts, nuts, threaded fittings. AFNOR Z 08 CF 16. AISI 430F; W.-Nr. 4104.

DAUPHINOX D3
Forges et Acieries de Bonpertuis
C 0.08, Cr 18, bal Fe.
Ferritic stainless steel; not hardenable. Fittings; good corrosion resistance. AFNOR Z 08 C 18. Similar to AISI 430 and W.-Nr.- 4016.

DAUPHINOX T1
Forges et Acieries de Bonpertuis
C 0.4, Cr 13.5, bal Fe.
Martensitic stainless steel; hardenable to 53-56 Rock C. Special cutlery steel; can be highly polished. AFNOR Z 40 C 14. AISI 420; W.-Nr. 4034.

DAUPHINOX T10MC
Forges et Acieries de Bonpertuis
C 1, Cr 15, Mo 0.8, Co 1, bal Fe.
Martensitic stainless steel; hardenable to 56-59 Rock C. For cutlery, surgical instruments. AFNOR Z 100 CDK 15. Similar to AISI 440 B or C. W.-Nr. 4535.

DAUPHINOX T1MO
Forges et Acieries de Bonpertuis
C 0.5, Cr 13.5, Mo 0.8, bal Fe.
Martensitic stainless steel; hardenable to 55-56 Rock C. For cutlery, surgical instruments. AFNOR Z 50 CD 14. W.-Nr. 4110.

DAUPHINOX T1ST
Forges et Acieries de Bonpertuis
C 0.7, Cr 17, bal Fe.
Martensitic stainless steel; hardenable to 55-58 Rock C. For cutlery, surgical instruments. AFNOR Z 70 C 17. AISI 440A.

DAUPHINOX T5
Forges et Acieries de Bonpertuis
C 0.5, Cr 14, bal Fe.
Martensitic stainless steel; hardenable to 54-57 Rock C. For surgical instruments. AFNOR Z 50 Cr 14.

DAUPHINOX T5MO
Forges et Acieries de Bonpertuis
C 0.5, Cr 14, Mo 0.25, bal Fe.
Martensitic stainless steel; hardenable to 52-55 Rock C. For cutlery, surgical instruments. AFNOR Z 50 CD 14.

DAUPHINOX T7MO
Forges et Acieries de Bonpertuis
C 0.7, Cr 14, Mo 0.8, bal Fe.
Martensitic stainless steel; hardenable to 53-57 Rock C. For cutlery, surgical instruments. AFNOR Z 70 CD 14.

DAUPHINOX TP
Forges et Acieries de Bonpertuis
C 0.2, Cr 13.5, bal Fe.
Martensitic stainless steel; hardenable to 46-49 Rock C. For surgical instruments, knife springs, turbine blades and wheels. AFNOR Z 20 C 13. AISI 420; W-Nr. 4021.

DAUPHINOX TP1
Forges et Acieries de Bonpertuis
C 0.3, Cr 13.5, bal Fe.
Martensitic stainless steel; hardenable to 48-52 Rock C. For cutlery, scissors, surgical instruments. AFNOR Z 30 C 13. AISI 420.

DAUPHINOX TPMO
Forges et Acieries de Bonpertuis
C 0.2, Cr 13.5, Mo 0.8, bal Fe.
Martensitic stainless steel; hardenable to 46-49 Rock C. Paper stock beater and refiner blades, shears. AFNOR Z 20 CD 14. AISI 420; W.-Nr. 4120.

DAUPHINOX TPO
Forges et Acieries de Bonpertuis
C 0.15, Cr 13, bal Fe.
Martensitic stainless steel, hardenable to 42-45 Rock C. For plastic molds, turbine blades and wheels. AFNOR Z-15 C 13. AISI 416-410; W.-Nr. 4024.

DAVIGNON
English manufacture
Au 58, Cu 37, Al 5.
For jewelry, ornaments; corrosion resistant.

DAVIS METAL
Chapman Valve Mfg. Co.
Copper. Mn 1.5, Ni 29, Fe 2, Cu 67, C 0.5, Si.
Cast: 60,000 TS; 35,000 YS; 18 El; 120 Brin. For valves and fittings, turbine blades, throttle valves; corrosion and heat resisting.

DAVIS METAL
Chapman Valve Mfg. Co.
Copper. C 0.2, Mn 0.3-1, Ni 25, Fe 6, Cu 67, Pb 0.8.
For valves, fittings.

DAWSON
Time Steel Service Inc.
C 0.65, Mn 0.9, Si 2, bal Fe.
Shock resistant tool steel, water hardened, AISI S4.

DAWSON'S BRONZE
English manufacture
Cu 83.9, Sn 15.9, Pb 0.1, As 0.05.
Cast: 30,000 TS; 146 Brin. For journal bearings; very fluid.

DBK HOM-3
Duraloy
Superalloy.
Now DURALOY MO-RE 3.

DBK HT-CB
Duraloy
Superalloy.
Now DURALOY MO-RE 15.

DBL
Allegheny Ludlum Steel
C 0.8, Si 0.5-0.8, Cr 4, W 5.5, V 1.5, Mo 3.5-4, bal Fe.
For tools, cutters; high speed steel. *Obsolete*

DBL 3
Now LUDLUM DBL 3.

DBL-2
Allegheny Ludlum Steel
C 0.8, Mn 0.3, Si 0.3, Cr 4, V 2, W 6, Mo 5, bal Fe.
For tools, cutters; high speed steel. *Obsolete*

DC
Now FINKL DC.

DC-66
Now KLOSTER DC-66.

DCM ALLOY
Howmet Corp.
C 0-0.08, Cr 15, Ti 3.5, B 0.08, Al 4.6, Mo 4.5-6, Fe 4-6, bal Ni.
Heat treated: 140,000 TS; 116,000 YS; 5 El; 9 RA. For turbine blades, jet engine components; age hardenable, high stress rupture strength.

DCM ALLOY
General Electric Co.
C 0-0.08, Cr 15, Ti 3.5, B 0.08, Al 4.6, Mo 4.5-6, Fe 4-6, bal Ni.
Heat treated: 140,000 TS; 116,000 YS; 5 El; 9 RA. For turbine blades, jet engine components; age hardenable, high stress rupture strength.

DCM ALLOY
Stellite Division
C 0-0.08, Cr 15, Ti 3.5, B 0.08, Al 4.6, Mo 4.5-6, Fe 4-6, bal Ni.
Heat treated: 140,000 TS; 116,000 YS; 5 El; 9 RA. For turbine blades, jet engine components; age hardenable, high stress rupture strength.

DCM ALLOY
Special Metals Corp.
C 0-0.08, Cr 15, Ti 3.5, B 0.08, Al 4.6, Mo 4.5-6, Fe 4-6, bal Ni.
Heat treated: 140,000 TS; 116,000 YS; 5 El; 9 RA. For turbine blades, jet engine components; age hardenable, high stress rupture strength.

DCN 2W
Thyssen Edelstahlwerke AG
C, Cr, Ni, bal Fe.
19-22 El. For machinery parts. *Obsolete*

DCN 3W
Thyssen Edelstahlwerke AG
C, Cr, Ni, bal Fe.
20 El. For machinery parts. *Obsolete*

DCN 60
Thyssen Edelstahlwerke AG
C, Ni, bal Fe.
For electrical resistance wire. *Obsolete*

DCN 80
Thyssen Edelstahlwerke AG
C, Ni, bal Fe.
For electrical resistance wire. *Obsolete*

DCN EXTRA
Thyssen Edelstahlwerke AG
C, Cr, Ni, W, Mo, bal Fe.
13-16 El. For machinery parts. *Obsolete*

DCN SPEZIAL
Thyssen Edelstahlwerke AG
C, Cr, Ni, W, Mo, V, bal Fe.
12-15 El. For machinery parts. *Obsolete*

DCNA
Thyssen Edelstahlwerke AG
C, Cr, Ni, bal Fe.
7-8 El. For machinery parts; oil hardened. *Obsolete*

DCNO
Thyssen Edelstahlwerke AG
C, Cr, Ni, Mo, bal Fe.
8-10 El. For machinery parts; oil hardened. *Obsolete*

DCNR
Thyssen Edelstahlwerke AG
C 0.38, Ni 4, Cr 1.3, Mo 0.25, V 0.1, bal Fe.
For forging dies, hot work rollers; oil or air hardened, tough. *Obsolete*

DE LUXE-10
Federal Foundries & Steel Co. Ltd.
C 1.05, Mn 0.25, Si 0.2, bal Fe.
For dies and forming tools.

DE LUXE-12
Federal Foundries & Steel Co. Ltd.
C 1.2, Mn 0.25, Si 0.2, bal Fe.
For cutters and nonferrous metals.

DE LUXE-9
Federal Foundries & Steel Co. Ltd.
C 0.9, Mn 0.25, Si 0.2, bal Fe.
For header dies and reamers.

DECOBRA
English manufacture
Cu 74.4, Zn 5.4, Ni 19, Fe 0.8, Mn 0.4.
For ornamental parts, hardware; corrosion resistant.

DECORAL
Aluminium Laufen AG
Aluminum.
Aluminum alloy AlZn5.5Mg1.

DEEFIVE
Handy & Harmon
Au 60-85, Cu, Pt, Zn, bal Ag.
Cast: 112,000 TS; 1 El; 237 Brin. For dentures, dental inlays; cast, hard. *Obsolete*

DEEFOUR
Handy & Harmon
Au 60-85, Cu, Pt, Zn, bal Ag.
Cast: 113,000 TS; 1 El; 249 Brin. For dentures; cast, hard. *Obsolete*

DEEFOURTEEN
Handy & Harmon
Au 60-85, Cu, Pt, Zn, bal Ag.
Cast: 113,000 TS; 1 El; 243 Brin. For dental inlays, dentures; cast, hard. *Obsolete*

DEELITE
Handy & Harmon
Au 60-85, Cu, Pt, Zn, bal Ag.
Cast: 97,000 TS; 2 El; 189 Brin. For dentures, dental inlays; cast, hard. *Obsolete*

DEEONE
Handy & Harmon
Au 60-85, Cu, Pt, Zn, bal Ag.
Cast: 32,000 TS; 20 El; 65 Brin. For dental inlays; soft. *Obsolete*

DEEP HARDENING BERKSHIRE
Carpenter Technology Corp.
C 1.23, Mn 1, Cr 0.5, V 0.2, bal Fe.
Cold work tool steel for thread chasing dies.

DEEPEP
Handy & Harmon
Au 60-85, Cu, Pt, Zn, bal Ag.
Wrought: 154,000 TS; 3 El; 270 Brin. For dentures; wire. *Obsolete*

DEEPEX
Murex Ltd.
C 0.15, Mn 0.35, bal Fe.
Steel welding rod.

DEESEVEN
Handy & Harmon
Au 60-85, Cu, Pt, Zn, bal Ag.
Cast: 95,000 TS; 1 El; 200 Brin. For dental inlays; cast, hard. *Obsolete*

DEESIX
Handy & Harmon
Au 60-85, Cu, Pt, Zn, bal Ag.
Cast: 86,000 TS; 2 El; 189 Brin. For dentures, dental inlays; cast, hard. *Obsolete*

DEETHREE
Handy & Harmon
Au 60-85, Cu, Pt, Zn, bal Ag.
Cast: 106,000 TS; 1 El; 219 Brin. For dental inlays, dentures; castings, hard. *Obsolete*

DEETWO
Handy & Harmon
Au 60-85, Cu, Pt, Zn, bal Ag.
Cast: 53,000 TS; 18 El; 105 Brin. For dental inlays. *Obsolete*

DEFIHEAT
Armco Steel Corp.
C 0.3, Mn 0.5, Si 0.75, Cr 25-30, bal Fe.
Annealed: 75,000-100,000 TS; 55,000-80,000 YS; 23-35 El; 45-65 RA; 160-200 Brin. For stainless articles, furnace parts; corrosion and heat resistant. *Obsolete*

DEFIRUST
Armco Steel Corp.
C 0-0.12, Mn 0.4, Si 0.5, Cr 12-15, bal Fe.
Annealed: 75,000 TS; 44,000 YS; 37 El; 78 RA; 152 Brin. Heat treated: 110,000-180,000 TS; 90,000-168,000 YS; 10-22 El; 50-65 RA; 229-352 Brin. For stainless articles, turbine blades, valves; corrosion resisting, difficult weldability. *Obsolete*

DEFIRUST N
Armco Steel Corp.
C 0.12, Cr 10-13.5, Ni 0-2, bal Fe.
For corrosion resistant parts; corrosion resistant. *Obsolete*

DEFIRUST NO. 410
Armco Steel Corp.
C 0-0.12, Cr 10-13.5, bal Fe.
For turbine blades; corrosion resistant. *Obsolete*

DEFIRUST NO. 416
Armco Steel Corp.
C 0-0.12, Si 0.5, Cr 12-14, bal Fe.
For corrosion resistant parts; free machining. *Obsolete*

DEFIRUST SPECIAL
Armco Steel Corp.
C 0-0.12, Mn 0.4, Si 0.5, Cr 16-18, bal Fe.
Annealed: 75,000 TS; 42,000 YS; 37 El; 70 RA; 156 Brin. Cold drawn: 105,000 TS; 65,000-85,000 YS; 15-23 El; 54-65 RA; 170-210 Brin. For stainless articles, nitric acid equipment, ornamental trim; non-hardenable, corrosion resistant. *Obsolete*

DEFIRUST, TURBINE
Armco Steel Corp.
Cr 11.5-13, C, bal Fe.
Tempered: 105,000 TS; 77,000 YS; 23 El; 68 RA; 212 Brin. For turbine blades, valves, structural and machine parts; corrosion resistant. *Obsolete*

DEFISTAIN
Armco Steel Corp.
C 0-0.2, Mn 0.5, Si 0.5, Cr 17-19, Ni 7-9, bal Fe.
Annealed: 85,000 TS; 40,000 YS; 61 El; 75 RA; 160 Brin. Cold worked: 85,000-160,000 TS; 34,000-150,000 YS; 18-61 El; 52-75 RA; 135-260 Brin. For stainless articles, chemical apparatus, valves, trim; corrosion resistant; non-magnetic. *Obsolete*

DEFISTAIN, MACHINING
Armco Steel Corp.
Cr 18-20, Ni 8-10, S 0.3, C, bal Fe.
Annealed: 90,000 TS; 35,000 YS; 53 El; 64 RA; 150 Brin. Cold drawn: 100,000 TS; 85,000 YS; 35 El; 50 RA; 215 Brin. For valve and pump parts; free machining, stainless. *Obsolete*

DEFISTAIN, SPECIAL
Armco Steel Corp.
C 0-0.07, Cr 19, Ni 9, bal Fe.
Annealed: 90,000 TS; 35,000 YS; 65 El; 70 RA; 142 Brin. For welded structures, especially if annealing after welding is impracticable; stainless, corrosion resistant. *Obsolete*

DEGUBOND 4
Degussa AG
Precious metal.
Precious metal alloy for dentistry and dental engineering.

DEGUBOND U
Degussa AG
Precious metal.
Precious metal alloy for dentistry and dental engineering.

DEGUDENT
Degussa AG
Precious metal.
Precious metal alloy for dentistry and dental engineering.

DEGULFIT
Degussa AG
Soldering alloy; meets DIN 1707.

DEGULOR
Degussa AG
Precious metal.
Precious metal alloy for dentistry and dental engineering.

DEGUSSA ALLOY
English manufacture
Cu 66, Zn 34, traces Fe.
For hardware, lamp fixtures, ornamental parts; yellow brass.

DEKORSIL
Vereinigte Metall. Ranshofen-Berndorf
Aluminum.
Aluminum alloy AlSi5.

DELAIR
Delsteel Inc.
C 1, Cr 5, Mo 1, V 0.4, bal Fe.
Annealed: 105,000 TS; 52,000 YS; 26 El; 18 Rock C. Heat treated: 253,000 TS; 200,000 YS; 3 El; 53 Rock C. For blanking and trimming dies, cutters, engravers tools, shear blades, broaches, thread rolling dies. Type A2; air hardened tool steel. Tough, wear resistant.

DELAWARE EXTRA
Delsteel Inc.
C 0.6-1.5, bal Fe.
For tools and parts, drills; water hardened.

DELAWARE H.S.
Delsteel Inc.
C 0.82, Mo 5, W 6.5, Cr 4, V 2, bal Fe.
Heat treated: 63-67 Rock C. For lathe and planer tools, drills, taps, chasers, reamers, drawing dies. Type M2 high-speed steel. High toughness and wear resistant.

DELAWARE S.T.
Delsteel Inc.
C, Mn, Si, V, Mo, bal Fe.
Alloy tool steel; shock resistant.

DELAWARE STANDARD
Delsteel Inc.
C 0.7-1.2, bal Fe.
For tools and parts, taps; water hardened.

DELCAR
Deloro Stellite Ltd.
Tube carbide rods and electrodes (3 grades) for hard facing. Cannot be machined or ground. For ore crusher rolls, grab bucket teeth, coal plow picks.

DELCAR
Delsteel Inc.
C 0.6-1.4, bal Fe.
For drills, taps, springs, hobs, reamers, punches; Type W1; water hardened.

DELCONDEX 10
Alcan-Booth Industries, Ltd.
Ni 10, Mn 0.8, Fe 0.8, bal Cu.
Cold drawn: 62,700 TS; 44,800 YS; 14 El; 145 Brin. For condenser tubes, heat exchangers and fittings; resists impingement corrosion. *Obsolete*

DELCONDEX 12
Alcan-Booth Industries, Ltd.
Ni 10.5, Fe 1.4, Mn 0.8, bal Cu.
For condenser tubes; corrosion resistant. *Obsolete*

DELCONDEX 20
Alcan-Booth Industries, Ltd.
Ni 20, Fe 0.5, Mn 0.7, bal Cu.
For condenser tubes, boiler feed water heaters; excellent for brackish water. *Obsolete*

DELCONDEX 30
Alcan-Booth Industries, Ltd.
Ni 31, Fe 0.8, Mn 0.8, bal Cu.
Annealed: 60,500 TS; 20,200 YS; 42 El; 95 Brin. Cold drawn: 82,900 TS; 67,200 YS; 14 El; 172 Brin. For marine condenser tubes; resists sea water corrosion. *Obsolete*

DELCONDEX 31
Alcan-Booth Industries, Ltd.
Ni 30.5, Fe 0.6, Mn 0.7, bal Cu.
For marine condenser tubes; corrosion resistant. *Obsolete*

DELCONDEX 32
Alcan-Booth Industries, Ltd.
Ni 31, Fe 2, Mn 2, bal Cu.
For condenser tubes for marine construction; resists abrasion and sea water impingement. *Obsolete*

DELCONION
Alcan-Booth Industries, Ltd.
Ni 5.5, Fe 1.2, Mn 0.6, bal Cu.
Cold drawn: 74,000 TS; 67,000 YS; 3 El; 146 Brin. For sea water pipes and condensers; good formability. *Obsolete*

DELCROME 300 SERIES
Deloro Stellite Ltd.
Iron.
Standard iron base alloy.

DELCROME 410
Deloro Stellite Ltd.
Iron.
Standard iron base alloy.

DELCROME 450
Deloro Stellite Ltd.
C 0.2, Cr 11, Ni 2, Mo 1, bal Fe.
Cast: 45 Rock C. For submerged arc welding; good general properties.

DELCROME 50V
Deloro Stellite Ltd.
Iron. Cr 27, C 2.7, Si 1, Mn 1, V 1, bal Fe.
Standard iron base alloy.

DELCROME 50V
Deloro Stellite Ltd.
Cr 27, C 2.75, V 0.75, bal Fe.
Cast: 51-54 Rock C. High wear resistance, heat treatable for machining. Hardfacing electrode for cold abrasion application.

DELCROME 550
Deloro Stellite Ltd.
Cr 15, C 0.5, Mo 0.4, bal Fe.
Wire for hard surfacing, multilayer deposits on steel rolls, wear plates operating at temperature below 500°C. 51 Rock C.

DELCROME 600
Deloro Stellite Ltd.
Cr 27, C 3, Mo 0.7, bal Fe.
Wire for hard surfacing; maximum resistance to wear and abrasion. 54 Rock C.

DELCROME 94GV
Deloro Stellite Ltd.
Iron. Cr 28, C 3, Si 1.4, Mn 1.2, bal Fe.
Standard iron base alloy.

DELCROME R
Deloro Stellite Ltd.
Cr 30, W 3, Mn 6, Fe 61.
Cast: 53 Rock C. For hardfacing, especially on wearing parts of earth moving equipment.

DELCROME-C
Deloro Stellite Ltd.
Cr 21, C 3.75, bal Fe.
Cast: 54 Rock C. For tappet tips and rocker pads in internal combustion engines. Wear resistant. Resists mineral abrasion.

DELETOTS ALLOY
Manufacturer not listed
Cu 80, Zn 18, Mn 2.
For brass solder, cartridge cases; extremely ductile.

DELFER
Deloro Stellite Ltd.
Co 16.5, Cr 13.5, W 5.5, Mo 9, C 2.5, bal Fe.
Cast: 100,000 TS; 600 Brin. For hardfacing electrodes for abrasion resisting castings. Heat and wear resistant. *Obsolete*

DELFER B
Deloro Stellite Ltd.
Co 6, Cr 18, C 3.2, Mo 16, V 2, bal Fe.
Hardfacing electrode; for extreme conditions of abrasive wear by hard gritty particles; 62 Rock C.

DELHI GRADE A
Associated Steel Corp.
Cr 16-18, C 0-0.1, Si 0-0.5, Mn 0.5, bal Fe.
Rolled: 110,000 TS; 10 El; 25 RA; 200 Brin. Annealed: 70,000 TS; 45,000 YS; 30 El; 60 RA; 160 Brin. For ornamental work, tanks; maximum operating temperature 1600°F; austenitic, corrosion resistant.

DELHI HARD
Associated Steel Corp.
Cr 17.5, C 1, Ni 0.12, Si 1.13, bal Fe.
Annealed: 90,000 TS; 40,000 YS; 25 El; 50 RA; 195 Brin. Heat treated: 225,000 TS; 185,000 YS; 9 El; 25 RA; 420 Brin. For structures; see Silcrome 17.

DELHI IRON
Associated Steel Corp.
Cr 16.5-18, C 0.1-0.11, Si 0.75-1, bal Fe.
For replacing galvanized iron, for roofing, automobile body sheets; see Silcrome 17.

DELHI IRON GRADE "S"
Associated Steel Corp.
Cr 11.5-14, C 0-0.12, Si 0.5, Mn 0.5, bal Fe.
Annealed: 45,000 TS; 30,000 YS; 30 El; 60 RA; 130 Brin. Heat treated: 130,000 TS; 100,000 YS; 21 El; 68 RA; 235 Brin. For turbine blades, pump rods, valves, cutlery, machine parts; see Silcrome 12.

DELHI SPECIAL
Associated Steel Corp.
C 0.14, Mn 0.46, Si 0.64, Cr 17, bal Fe.
At 70°F: 65,000 TS; 40,000 YS; 32 El; 63 RA. At 1500°F: 10,000 TS; 8000 YS; 90 El; 98 RA. For furnace linings, conveyors, furnace parts; heat and corrosion resistant.

DELHI TOUGH IRON
Associated Steel Corp.
Cr 17, Si 1.25, Mn 0.3, C 0.07, bal Fe.
For corrosion resistant parts; corrosion resistant; see Silcrome 17.

DELLOY
Delloy Metals
Co, Cr-W.
For hard facing welding electrodes; heat and corrosion resistant.

DELLOY DC-7
Delloy Metals
WC + Co.
For cutters for iron and nonferrous; sintered carbide.

DELLOY DS-8
Delloy Metals
WC + Co.
For cutters for steel; sintered carbide.

DELLOY G-P
Delloy Metals
C 0.7, W 18, Cr 4, V 1, bal Fe.
For cutters, tools; high speed steel.

DELLOY NO. 4
Delloy Metals
Co, Cr-W.
For cutting tools.

DELLOY NO. 6
Delloy Metals
Co, Cr-W.
For cutting tools.

DELLOY NO. 7
Delloy Metals
Co, Cr-W.
For cutting tools.

DELLOY NO. 9
Delloy Metals
Co, Cr-W.
For cutting tools.

DELORO 1300K
Deloro Stellite, Inc.
C 0.05-0.12, Mn 0.3-1, Si 0.5-1, Cr 27-29, Co 48-52, bal Fe.
Cast: 78,000 TS; 46,000 YS; 8 El; 250 Brin. For furnace parts, grates, quenching baskets, salt bath electrodes. Good oxidation resistance at 2200 F and excellent thermal shock resistance. *Obsolete*

DELORO 15
Deloro Stellite Ltd.
Nickel. C 0.05, Si 2, B 1, Fe 1, Cu 20, bal Ni.
Standard nickel base alloy.

DELORO 22
Deloro Stellite Ltd.
Nickel. C 0.05, Si 2.5, B 1.4, Fe 0.8, bal Ni.
Standard nickel base alloy.

DELORO 25
Deloro Stellite Ltd.
Nickel. C 0.05, Si 3.5, B 1.5, Fe 0.8, bal Ni.
Standard nickel base alloy.

DELORO 30
Deloro Stellite Ltd.
Nickel. Cr 2.3, C 0.12, Si 2.7, B 1.4, Fe 1, bal Ni.
Standard nickel base alloy.

DELORO 35
Deloro Stellite Ltd.
Nickel. Cr 5, C 0.15, Si 3.2, B 1.5, Fe 2, bal Ni.
Standard nickel base alloy.

DELORO 38
Deloro Stellite Ltd.
Nickel. C 0.07, Si 3, B 2.1, Fe 0.8, bal Ni.
Standard nickel base alloy.

DELORO 40
Deloro Stellite Ltd.
Nickel. Cr 7.5, C 0.25, Si 3.5, B 1.7, Fe 2.5, bal Ni.
Standard nickel base alloy.

DELORO 40 G
Deloro Stellite Ltd.
Cr 7.5, Ni 82, Fe 5, Si 4, B 1.2.
Melt range 1810-2160°F. As cast: 386 N/mm^2 TS; 255 N/mm^2 YS; 1 El; 34 Rock C.

DELORO 45
Deloro Stellite Ltd.
Nickel. Cr 9, C 0.35, Si 3.6, B 1.8, Fe 2.7, bal Ni.
Standard nickel base alloy.

DELORO 50
Deloro Stellite Ltd.
Nickel. Cr 11, C 0.45, Si 3.9, B 2.3, Fe 2.8, bal Ni.
Standard nickel base alloy.

DELORO 55
Deloro Stellite Ltd.
Nickel. Cr 17, C 0.7, Mo 2.4, Si 4.5, B 3.4, Fe 3.5, Cu 2.4, bal Ni.
Standard nickel base alloy.

DELORO 60
Deloro Stellite Ltd.
Nickel. Cr 15, C 0.75, Si 4.4, B 3.2, Fe 3.5, bal Ni.
Standard nickel base alloy.

DELORO 625
Deloro Stellite Ltd.
Nickel. Cr 22, C 0.1, Mo 9, Fe 5, bal Ni.
Standard nickel base alloy.

DELORO ALLOY 40 G
Deloro Stellite, Inc.
nickel. C 0.3, Cr 7.5, Si 4, Fe 5, B 1.2, bal Ni.
As cast: 741 N/mm^2 TS; 432 N/mm^2 YS; 10 El; 29-38 Rock C. Hard facing alloys; good corrosion resistance; machinable.

DELORO ALLOY 45
Deloro Stellite Ltd.
Cr 7.5, Fe 1.5, Si 4, B 1.5, bal Ni.
Cast: 35-42 Rock C. Hardfacing electrode for abrasion and corrosion resistance.

DELORO ALLOY 45
Deloro Stellite, Inc.
Cr 7.5, Fe 1.5, Si 4, B 1.5, bal Ni.
Cast: 35-42 Rock C. Hardfacing electrode for abrasion and corrosion resistance.

DELORO ALLOY 50
Deloro Stellite Ltd.
Cr 10, Mo 4, Si 4, B 1.5, bal Ni.
Cast: 49-52 Rock C. Hardfacing electrode for abrasion and corrosion resistance.

DELORO ALLOY 50
Deloro Stellite, Inc.
Cr 10, Mo 4, Si 4, B 1.5, bal Ni.
Cast: 49-52 Rock C. Hardfacing electrode for abrasion and corrosion resistance.

DELORO ALLOY 60
Deloro Stellite Ltd.
Cr 15, Fe 4.5, Si 4.5, B 3, bal Ni.
Cast: 59-62 Rock C. Hardfacing electrode for abrasion and corrosion resistance.

DELORO ALLOY 60
Deloro Stellite, Inc.
Cr 15, Fe 4.5, Si 4.5, B 3, bal Ni.
Cast: 59-62 Rock C. Hardfacing electrode for abrasion and corrosion resistance.

DELORO ALLOY B
Deloro Stellite Ltd.
C 0.1, Mo 29, Fe 5, bal Ni.
Cast: 80,000 TS; 22-28 Rock C. High corrosion and oxidation resistance for chemical industry.

DELORO ALLOY B
Deloro Stellite, Inc.
C 0.1, Mo 29, Fe 5, bal Ni.
Cast: 80,000 TS; 22-28 Rock C. High corrosion and oxidation resistance for chemical industry.

DELORO ALLOY C
Deloro Stellite Ltd.
Cr 17, W 5, C 0.1, Mo 17, Fe 6, bal Ni.
Cast: 80,000 TS; 23-31 Rock C. For hardfacing drop forging dies, readily machinable.

DELORO ALLOY C
Deloro Stellite, Inc.
Cr 17, W 5, C 0.1, Mo 17, Fe 6, bal Ni.
Cast: 80,000 TS; 23-31 Rock C. For hardfacing drop forging dies, readily machinable.

DELORO ALLOY RT1
Deloro Stellite Ltd.
Cr 15, W 4, C 0.4, Ni 53, Mo 15, Fe 5, Nb 7.
Cast: 300 Brin (approximately). For hardfacing and build-up; corrosion resistant and resistant to abrasive wear and impact at high temperatures. *Obsolete*

DELORO NISTELLE C
Deloro Stellite Ltd.
Nickel. Cr 16.5, C 0.1, W 4.5, Mo 17, Si 1, Fe 6, Mn 1, V 0.3, bal Ni.
Standard nickel base alloy.

DELORO NISTELLE C 4 C
Deloro Stellite Ltd.
Nickel. Co 1, Cr 16, C 0.01, Mo 16, Fe 2, Mn 0.5, bal Ni.
Standard nickel base alloy.

DELORO PW22
Deloro Stellite Ltd.
Cr 1.2, Ni 95, Si 2.5, B 1.3.
Cast: 200-250 Brin (approximately). For hardfacing and build-up; good corrosion resistance; machinable.

DELORO SF40
Deloro Stellite Ltd.
Cr 7.5, Fe 1.5, Si 4, B 1.5, bal Ni.
Cast: 39 Rock C. For hardfacing; machinable. Abrasion and corrosion resistant.

DELORO SF50
Deloro Stellite Ltd.
Cr 10, Fe 4, Si 4, B 1.5, bal Ni.
Cast: 51 Rock C. For hardfacing. Wear and corrosion resistant.

DELORO SF56
Deloro Stellite Ltd.
Cr 16, Cu 2, W 2, C 0.5, Mo 2, Fe 4, Si 4, B 4, bal Ni.
Hardfacing alloy, powder form; deposited 60 Rock C.

DELORO SF60
Deloro Stellite Ltd.
Cr 16, Fe 4.5, Si 4.5, B 3, bal Ni.
Cast: 60 Rock C. For hardfacing. Abrasion and corrosion resistant.

DELORO STELLITE 1
Deloro Stellite, Inc.
Reactive & refractory.
Now STELLITE ALLOY 1.

DELORO STELLITE 100
Deloro Stellite, Inc.
Reactive & refractory.
Now STELLITE ALLOY 100.

DELORO STELLITE 12
Deloro Stellite, Inc.
Reactive & refractory.
Now STELLITE ALLOY 12.

DELORO STELLITE 20
Deloro Stellite, Inc.
Reactive & refractory.
Now STELLITE ALLOY 20.

DELORO STELLITE 250
Deloro Stellite, Inc.
Reactive & refractory.
Now STELLITE ALLOY 250.

DELORO STELLITE 3
Deloro Stellite, Inc.
Reactive & refractory.
Now STELLITE ALLOY 3.

DELORO STELLITE 4
Deloro Stellite, Inc.
Reactive & refractory.
Now STELLITE ALLOY 4.

DELORO STELLITE 6
Deloro Stellite, Inc.
Reactive & refractory.
Now STELLITE ALLOY 6.

DELORO STELLITE 7
Deloro Stellite, Inc.
Reactive & refractory.
Now STELLITE ALLOY 7.

DELORO STELLITE 8
Deloro Stellite, Inc.
Reactive & refractory.
Now STELLITE ALLOY 8.

DELORO STELLITE X-40
Deloro Stellite, Inc.
Reactive & refractory.
Now DELORO ALLOY X-40.

DELORO X-40
Deloro Stellite, Inc.
Reactive & refractory. Cr 25, W 7, C 0.3, Ni 10, bal Co.
Cast: 96,000 psi TS; 30-35 Rock C. Corrosion resistant; high temperature strength; resistant to thermal shock.

DELSTEEL ALLOY
Delsteel Inc.
C, Mn, S, V, Mo, bal Fe.
Alloy tool steel, shock resistant type. AISI S5.

DELSTEEL ALLOY S.T.
Delsteel Inc.
C 0.35-0.42, V 0.2, Mn 0.4, Mo 0.5, S 0.018, bal Fe.
Water quenched: 280,000 TS; 240,000 YS; 9.5 El; 29 RA; 570 Brin. For chisels, rivet sets, shear blades; tough. *Obsolete*

DELTA ALUMINUM BRONZE CA1
Delta Metal (BW) Ltd.
Al 10, Fe 5, Ni 5, Mn 0.25, bal Cu.
Light drawn: 110,000 TS; 67,000 YS; 15 El; 220 Brin. High strength aluminum bronze, corrosion resistant.

DELTA ALUMINUM BRONZE CA11
Delta Metal (BW) Ltd.
Al 6.2, Fe 0.9, Mn 0.3, Si 2.2, bal Cu.
Light drawn: 80,000 TS; 40,000 YS; 35 El. 25 ft·lb IS (Izod); for parts subject to shock.

DELTA ALUMINUM BRONZE CA12
Delta Metal (BW) Ltd.
Al 6.2, Fe 0.7, Mn 0.2, Si 2.2, bal Cu.
Light drawn: 80,000 TS; 45,000 YS; 35 El. For parts subject to shock; 30 ft·lb IS (Izod); non-magnetic.

DELTA ALUMINUM BRONZE CA14
Delta Metal (BW) Ltd.
Al 10, Fe 3, Ni 1.5, Mn 3, Sn 0.9, bal Cu.
Light drawn: 170 Brin. Wear resistant; for parts subject to stress corrosion.

DELTA ALUMINUM BRONZE CA15
Delta Metal (BW) Ltd.
Al 3, Ni 6.75, Mn 0.5, bal Cu.
Rolled: 90,000 TS; 50,000 YS; 20 El; 180 Brin. For paper mill beater bars, etc.

DELTA ALUMINUM BRONZE CA16
Delta Metal (BW) Ltd.
Al 11, Fe 5.5, Ni 4.5, Mn 1.5, bal Cu.
For high strength stamped parts.

DELTA ALUMINUM BRONZE CA2
Delta Metal (BW) Ltd.
Al 10.5, Fe 4.5, Ni 4.5, Mn 0.5, bal Cu.
Light drawn: 115,000 TS; 45,000 YS; 20 El; 160 Brin. Strong, more ductile aluminum bronze, corrosion resistant.

DELTA ALUMINUM BRONZE CA3
Delta Metal (BW) Ltd.
Al 9.2, Fe 4.2, Ni 4.2, Mn 0.25, bal Cu.
Light drawn: 100,000 TS; 47,000 YS; 25 El; 170 Brin. For bronze parts subject to shock; 20 ft. lbs. Izod.

DELTA ALUMINUM BRONZE CA4
Delta Metal (BW) Ltd.
Al 11.5, Fe 5.5, Ni 5.5, Mn 1.5, bal Cu.
Light drawn: 130,000 TS; 60,000 YS; 5 El; 230 Brin. For high strength hot-stampings.

DELTA ALUMINUM BRONZE CA5
Delta Metal (BW) Ltd.
Al 9.6, Fe 2.25, Ni 1.5, Mn 0.3, bal Cu.
Light drawn: 90,000 TS; 47,000 YS; 15 El; 170 Brin. Wear resistant; for valve guides on internal combustion engines.

DELTA ALUMINUM BRONZE CA6
Delta Metal (BW) Ltd.
Al 8.6, Fe 2, Mn 0.2, bal Cu.
Light drawn: 90,000 TS; 43,000 YS; 25 El; 160 Brin. High impact strength; for pump spindles, valve stems, etc.

DELTA ALUMINUM BRONZE CA7
Delta Metal (BW) Ltd.
Al 9.6, Fe 2, Mn 0.25, bal Cu.
Light drawn: 90,000 TS; 43,000 YS; 25 El; 160 Brin. For corrosion resistant bolts and hardware.

DELTA ALUMINUM BRONZE CA8
Delta Metal (BW) Ltd.
Al 9.4, Mn 0.25, bal Cu.
Light drawn: 78,000 TS; 37,000 YS; 25 El; 160 Brin. Resistant to sulphuric acid.

DELTA ALUMINUM BRONZE CA9
Delta Metal (BW) Ltd.
Al 10, Mn 0.25, Pb 1.5, bal Cu.
Light drawn: 78,000 TS; 36,000 YS; 10 El; 155 Brin. For machined parts.

DELTA ANTIFRICTION METAL
Ste des Acieries de Micheville
Sn, Sb, bal Cu.
For lining of bearings, bolts, nuts, castings. *Obsolete*

DELTA BRONZE II
Seymour Products Co.
Cu 55-65, Zn 30-44.9, Fe 0.1-5.
Cast: 76,000 TS; 10 El. Extruded: 84,000 TS; 14 El. For tubes, ornamental and sanitary fittings, automobile parts; corrosion resistant.

DELTA BRONZE III
Seymour Products Co.
Cu 40-98, Zn 1.8-45, Fe 0.1-5, Sn 0.1-10.
Cast: 49,000 TS; 30 El. Extruded: 72,000 TS; 29 El. For solid drawn tubes for hydraulic purposes, condensers, gears, valves, pump parts; corrosion resistant.

DELTA BRONZE NO. IX
Delta Metal (BW) Ltd.
Cu alloy.
For bearings, antifriction lining; low frictional resistance *Obsolete*

DELTA BRONZE NO. IXA
Delta Metal (BW) Ltd.
Cu alloy.
For bearings, antifriction lining; low frictional resistance *Obsolete*

DELTA BRONZE NO. VII
Delta Metal (BW) Ltd.
Cu alloy.
For valves, spindles, internal combustion engine parts; heat and corrosion resistant. *Obsolete*

DELTA BRONZE NO. VIII
Delta Metal (BW) Ltd.
Cu alloy.
For marine parts, bearings having no casings; shock and wear resistant. *Obsolete*

DELTA C101 WIRE
Delta Extruded Metals Co., Ltd.
Copper. Cu 99.9, bal others.
Electrolytic tough pitch HC copper suitable for the production of complex cold formed products where good ductility or electrical characteristics are a requirement. 250-340 N/mm^2 min TS.

DELTA C103 WIRE
Delta Extruded Metals Co., Ltd.
Copper. Cu 99.95, bal others.
Oxygen free HC copper used for cold formed products where resistance to hydrogen embrittlement is an additional requirement. 250-370 N/mm^2 min TS.

DELTA CA1
Delta (Manganese Bronze) Ltd.
Copper. Cu 79.75, Al 10, Fe 5, Ni 5, Mn 0.25.
High-strength aluminum bronze possessing excellent corrosion resistance and high-temperature mechanical properties. It is suitable for machining and hot stamping. Forms available: rod, bar, section, rolled products. Extruded: 700-770 N/mm^2 TS; 15-25 El. Rolled: 700-770 N/mm^2 TS; 15-25 El. Drawn: 700-850 N/mm^2 TS; 15-20 El.

DELTA CA10
Delta (Manganese Bronze) Ltd.
Copper. Cu 91.25, Al 6.75, Si 2.
Silicon aluminum bronze, which is highly corrosion resistant with good mechanical and machining properties. It is also suitable for hot stamping. Also known as SA4. Forms available: rod, bar, hollow rod, rolled products, coil. Extruded: 500-550 N/mm^2; 33-40 El. Drawn: 550-620 N/mm^2 TS; 30-35 El. Rolled: 500-550 N/mm^2 TS; 33-40 El.

DELTA CA12
Delta (Manganese Bronze) Ltd.
Copper. Cu 90.7, Al 6.2, Fe 0.7, Mn 0.2, Si 2.2.
By restricting iron content, this alloy has low magnetic permeability and will extrude as hollow rod. Forms available: rod, bar, hollow rod, rolled products. Extruded: 525-585 N/mm^2 TS; 35-40 El. Rolled: 525-600 N/mm^2 TS; 30-40 El. Drawn: 550-620 N/mm^2 TS; 35-40 El.

DELTA CA13
Delta (Manganese Bronze) Ltd.
Copper. Cu 93, Al 7.
Simple all alpha-phase aluminum bronze with good ductility and exceptional impact strength. The alloy has good corrosion resistance and is commonly used in cold forming operations. Forms available: rod, coil, bar. Extruded: 390-420 N/mm^2 TS; 50-70 El. Drawn: 430-500 N/mm^2 TS; 30-40 El.

DELTA CA16
Delta (Manganese Bronze) Ltd.
Copper. Cu 77.5, Al 11, Fe 5.5, Ni 4.5, Mn 1.5.
Extremely hard alloy suitable for the hot stamping of high-strength parts. Forms available: rod, bar. Extruded: 710-800 N/mm^2 TS; 5-10 El. Drawn: 770-850 N/mm^2 TS; 5-10 El.

DELTA CA18
Delta (Manganese Bronze) Ltd.
Copper. Cu 82, Al 10, Fe 3, Ni 5.
High-strength aluminum bronze possessing excellent corrosion resistance and high-temperature mechanical properties. Forms available: rod, bar, rolled products. Drawn: 610-690 N/mm^2 TS; 20-25 El. Rolled: 610-690 N/mm^2 TS; 20-30 El.

DELTA CA19
Delta (Manganese Bronze) Ltd.
Copper. Cu 86, Al 10, Fe 2, Mn 2.
Aluminum bronze containing manganese and iron, meeting German specification DIN 17665 Cu Al 10 Fe. Medium-strength alloy with good ductility. Forms available: rod, bar. Drawn: 680-740 N/mm^2 TS; 20-25 El. Rolled: 650-730 N/mm^2 TS; 15-22 El.

DELTA CA20
Delta (Manganese Bronze) Ltd.
Copper. Cu 83, Al 9, Fe 1, Ni 2, Mn 5.
High-strength complex aluminum bronze possessing excellent corrosion resistance. Forms available: rod. Drawn: 660-780 N/mm^2 TS; 15-25 El.

DELTA CA21
Delta (Manganese Bronze) Ltd.
Copper. Cu 82.5, Al 10, Fe 3, Ni 4, Mn 0.5.
High-strength aluminum bronze possessing excellent corrosion resistance and high-temperature mechanical properties, meets French specifications. Forms available: rod, bar, rolled products. Drawn: 690-770 N/mm^2 TS; 18-25 El. Rolled: 670-720 N/mm^2 TS; 20-25 El.

DELTA CA22
Delta (Manganese Bronze) Ltd.
Copper. Cu 84.9, Al 9.5, Fe 1.75, Ni 3.25, Mn 0.6.
Medium- to high-strength aluminum bronze with good ductility, meets French specifications. Drawn: 540-700 N/mm^2 TS; 20-25 El. Rolled: 510-570 N/mm^2 TS; 20-25 El.

DELTA CA23
Delta (Manganese Bronze) Ltd.
Copper. Cu 81.3, Al 9.2, Fe 3.7, Ni 4.5, Mn 1.3.
As extruded: 640-720 N/mm^2 TS; 20-30 El. Rolled: 650-750 N/mm^2 TS; 20-30 El. Forged: 650-750 N/mm^2 TS; 20-30 El.

DELTA CA3
Delta (Manganese Bronze) Ltd.
Copper. Cu 82.15, Al 9.2, Fe 4.2, Ni 4.2, Mn 0.25.
Lower percentages of alloying elements, more ductile alloy used for its good impact strength. Extruded: 620-700 N/mm^2 TS; 25-30 El. Rolled: 620-700 N/mm^2 TS; 25-30 El. Drawn: 620-700 N/mm^2 TS; 25-30 El.

DELTA CA6
Delta (Manganese Bronze) Ltd.
Copper. Cu 89.2, Al 8.6, Fe 2, Mn 0.2.
Absence of nickel gives this alloy greater ductility and a high impact strength. Forms available: rod, bar, hollow rod, section, rolled products. Rolled: 540-590 N/mm^2 TS; 28-35 El. Drawn: 590-650 N/mm^2 TS; 25-30 El.

DELTA CA7
Delta (Manganese Bronze) Ltd.
Copper. Cu 88.15, Al 9.6, Fe 2, Mn 0.25.
Higher aluminum content makes this alloy a little tougher than Delta CA6. It is commonly used for high-strength corrosion-resistant bolts and other fasteners. Forms available: rod, bar, hollow rod, rolled products. Rolled: 540-590 N/mm^2 TS; 25-30 El. Drawn: 590-650 N/mm^2 TS; 20-25 El.

DELTA CA8

Delta (Manganese Bronze) Ltd.
Copper. Cu 90.35, Al 9.4, Mn 0.25.
Medium-strength aluminum bronze with good ductility and a very high impact strength. It has reasonable machinability and good corrosion resistance. Forms available: rod, bar, hollow rod, rolled products. Heat treated: 520-600 N/mm^2 TS; 35-40 El. Rolled: 480-560 N/mm^2 TS; 25-30 El. Drawn: 500-590 N/mm^2 TS; 25-30 El.

DELTA CF1 WIRE

Delta Extruded Metals Co., Ltd.
Copper. Cu 62.5, bal Zn.
Cold forming brass extensively used in the fasteners industry. Suitable for heading, thread rolling, bending, re-drawing and weaving. 340-770 N/mm^2 min TS.

DELTA CF22 GENERAL PURPOSE ROD AND BAR

Delta Extruded Metals Co., Ltd.
Copper. Cu 60.5, bal Zn.
General purpose 60/40 brass (Muntz Metal) suitable for limited cold forging and more severe bending operations. Good hot work properties. 340 N/mm^2 min TS; 25 min El.

DELTA CF23 GENERAL PURPOSE ROD AND BAR

Delta Extruded Metals Co., Ltd.
Copper. Cu 60.5, Pb 0.5, bal Zn.
Leaded 60/40 brass (Muntz Metal) having reasonable machinability combined with good cold working properties (e.g., bending.) 350-380 N/mm^2 min TS; 25-28 min El.

DELTA CF23 HIGH SPEED MACHINING ROD

Delta Extruded Metals Co., Ltd.
Copper. Cu 60.5, Pb 0.5, bal Zn.
Leaded 60-40 brass (Muntz metal) having reasonable machinability combined with good cold working properties, e.g., bending. 350-380 N/mm^2 min TS; 25-28 min El.

DELTA CF24 FORGING ROD

Delta Extruded Metals Co., Ltd.
Copper. Cu 61.5, Pb 1, bal Zn.
Leaded hot forging brass formulated to provide good cold working properties. Extensively used for the production of components that are subjected to cold swaging/thread rolling.

DELTA CF4 WIRE

Delta Extruded Metals Co., Ltd.
Copper. Cu 63.5, bal Zn.
Cold forming brass having slightly greater ductility than Delta CF1. Used where cold forming operations are more demanding. 340-770 N/mm^2 min TS.

DELTA CF7 WIRE

Delta Extruded Metals Co., Ltd.
Copper. Cu 70, bal Zn.
Superior cold forming brass suitable for the more complex cold heading/extrusion operations. 340-770 N/mm^2 min TS.

DELTA CROTERITE IV

Delta Metal (BW) Ltd.
Ni, Al, Mn, bal Cu.
Annealed: 50,000 TS; 18,000 YS; 48 El; 95 Brin. Hardened: 100,000 TS; 20 El. For paper mill equipment, pump rods, piston rods, turbine blading; tough, corrosion resistant.

DELTA CROTERITE V

Delta Metal (BW) Ltd.
Al, bal Cu.
Cast: 65,000 TS; 25,000 YS; 35 El; 25 RA; 100 Brin. Hard rolled: 90,000 TS; 40,000 YS; 30 El; 50 RA; 140 Brin. For pump rods, piston rods, valves, impellers, exhaust manifolds, worm wheels, gears; resists high temperatures and corrosion.

DELTA CROTERITE Z

Delta Metal (BW) Ltd.
Fe 2, Cu, Al, Mn, bal Zn.
Extruded: 84,000 TS; 38,000 YS; 28 El; 25 RA; 140 Brin. Drawn: 84,000 TS; 42,000 YS; 25 El; 25 RA; 150 Brin. For hardware; corrosion resistant.

DELTA ENFIELD 3 A

Delta Enfield Metals Ltd.
Now DELTA ENFIELD ERM 3A.

DELTA ENFIELD ALW

Delta Enfield Metals Ltd.
Te 0.6-1, Ni 0.06-1, bal Cu.
Bar: 37,000-61,000 TS; 14-20 El; 80-100 Brin. Forged: 37,000-43,000 TS; 14-20 El; 80-95 Brin. For spot and seam welding of aluminum alloys. High conductivity (85-90%) resistance welding electrode. BS 4577 and ISO 5182 alloy A/1/2. *Obsolete*

DELTA ENFIELD ERM 3A

Delta Enfield Metals Ltd.
Ni 0.8-1.2, P 0.16-0.25, S 0.1-0.15, bal Cu.
Bar: 60,000-72,000 TS; 25-34 El; 130-160 Brin. Forged: 55,000-72,000 TS; 18-32 El; 120-155 Brin. For electrical switchgear and resistance welding equipment, electrical contacts. 50-60% electrical conductivity. BS 4577 ISO 5182. RWMA alloy A/4/1. C19000

DELTA HIGH TENSILE BRASS HT1

Delta Metal (BW) Ltd.
Cu 57, Pb 1, Sn 0.75, Fe 0.75, Mn 1.5, bal Zn.
Light drawn: 80,000 TS; 40,000 YS; 22 El; 135 Brin. General purpose high tensile brass; for pump rods, valve spindles, etc.

DELTA HIGH TENSILE BRASS HT10

Delta Metal (BW) Ltd.
Cu 57, Pb 0.75, Sn 0.3, Fe 1, Al 2.5, Mn 1.5, bal Zn.
For hot stampings; easily machined.

DELTA HIGH TENSILE BRASS HT11

Delta Metal (BW) Ltd.
Cu 58, Pb 0.75, Fe 0.75, Al 1.5, Mn 1, bal Zn.
Light drawn: 78,000 TS; 40,000 YS; 15 El. For hot stampings; easily machined.

DELTA HIGH TENSILE BRASS HT12

Delta Metal (BW) Ltd.
Cu 57.5, Pb 0.85, Fe 0.65, Al 0.3, Mn 1.75, bal Zn.
Light drawn: 160 Brin. For conveyer belt segments; hard wearing alloy.

DELTA HIGH TENSILE BRASS HT13

Delta Metal (BW) Ltd.
Cu 58, Pb 2.25, Sn 1, Fe 0.35, Al 0.35, Mn 1.5, bal Zn.
Light drawn: 72,000 TS; 32,000 YS; 20 El; 115 Brin. For valve spindles, bearing bushes; free machining.

DELTA HIGH TENSILE BRASS HT14

Delta Metal (BW) Ltd.
Cu 58.25, Pb 1.5, Sn 0.75, Fe 0.75, Al 0.3, Mn 0.75, bal Zn.
For high pressure gas fittings; free machining.

DELTA HIGH TENSILE BRASS HT16

Delta Metal (BW) Ltd.
Cu 65.8, Sn 0.85, Fe 1.4, Al 3, Mn 0.6, bal Zn.
Light drawn: 85,000 TS; 42,000 YS; 20 El; 155 Brin. Wear and corrosion resistant brass.

DELTA HIGH TENSILE BRASS HT17

Delta Metal (BW) Ltd.
Cu 67.5, Fe 1.4, Al 4.8, Mn 0.5, bal Zn.
Light drawn: 98,000 TS; 50,000 YS; 15 El; 185 Brin. For bearing plates, wear strip.

DELTA HIGH TENSILE BRASS HT18

Delta Metal (BW) Ltd.
Cu 70, Fe 2, Al 6, Mn 3, bal Zn.
Light drawn: 110,000 TS; 60,000 YS; 12 El; 220 Brin. For gears, cams, press slides, etc.

DELTA HIGH TENSILE BRASS HT19

Delta Metal (BW) Ltd.
Cu 57.25, Pb 1, Sn 0.25, Fe 0.5, Al 0.75, Mn 1, Ni 2, bal Zn.
For gas water heating equipment; good hot corrosion resistance.

DELTA HIGH TENSILE BRASS HT2

Delta Metal (BW) Ltd.
Cu 57.5, Pb 1, Sn 0.5, Fe 0.75, Al 1, Mn 1.5, bal Zn.
Light drawn: 80,000 TS; 38,000 YS; 20 El; 120 Brin. For pump rods, valve spindles; not suitable for plating.

DELTA HIGH TENSILE BRASS HT20

Delta Metal (BW) Ltd.
Cu 57.25, Pb 0.5, Al 1.5, Mn 2.5, Si 0.75, bal Zn.
Light drawn: 90,000 TS; 40,000 YS; 15 El; 155 Brin. For bearings and bushings.

DELTA HIGH TENSILE BRASS HT21

Delta Metal (BW) Ltd.
Cu 58, Al 1.75, Mn 3, Si 1, bal Zn.
Light drawn: 80,000 TS; 40,000 YS; 15 El; 130 Brin. For gearbox components.

DELTA HIGH TENSILE BRASS HT22

Delta Metal (BW) Ltd.
Cu 60, Pb 3, Mn 2.5, bal Zn.
Resistance alloy; resistivity 17.

DELTA HIGH TENSILE BRASS HT23

Delta Metal (BW) Ltd.
Cu 56, Pb 3, Mn 6.3, bal Zn.
Resistance alloy; resistivity 20.

DELTA HIGH TENSILE BRASS HT3

Delta Metal (BW) Ltd.
Cu 58.25, Pb 0.65, Sn 0.9, Fe 0.9, Al 0.6, Mn 0.5, bal Zn.
Light drawn: 82,000 TS; 40,000 YS; 20 El; 135 Brin. For pump rods, valve spindles; readily machinable. *Obsolete*

DELTA HIGH TENSILE BRASS HT4

Delta Metal (BW) Ltd.
Cu 58.25, Sn 0.9, Fe 0.9, Al 0.6, Mn 0.5, bal Zn.
Light drawn: 82,000 TS; 40,000 YS; 20 El; 135 Brin. For valve spindles, general marine work.

DELTA HIGH TENSILE BRASS HT5

Delta Metal (BW) Ltd.
Cu 58, Pb 0.25, Sn 1, Fe 0.75, Mn 1.5, bal Zn.
Light drawn: 82,000 TS; 43,000 YS; 20 El; 150 Brin. For general purpose, plating and soldering.

DELTA HIGH TENSILE BRASS HT7

Delta Metal (BW) Ltd.
Cu 57.5, Sn 0.75, Fe 0.75, Mn 1.25, bal Zn.
Light drawn: 74,000 TS; 37,000 YS; 25 El; 135 Brin. For parts subject to shock; 20-25 ft·lb IS (Izod).

DELTA HIGH TENSILE BRASS HT8

Delta Metal (BW) Ltd.
Cu 57, Sn 1, Fe 1, Al 0.75, Mn 1.5, bal Zn.
Light drawn: 85,000 TS; 42,000 YS; 25 El; 150 Brin. For high pressure marine fittings.

DELTA HIGH TENSILE BRASS HT9

Delta Metal (BW) Ltd.
Cu 57, Pb 0.75, Sn 0.3, Fe 1, Al 0.75, Mn 0.75, bal Zn.
Light drawn: 75,000 TS; 40,000 YS; 15 El; 120 Brin. For special hot stampings.

DELTA HT1

Delta (Manganese Bronze) Ltd.
Copper. Cu 57, Pb 1, Sn 0.75, Fe 0.75, Mn 1.5, bal Zn.
Standard high tensile brass suitable for hot stamping and machining. It is aluminum restricted making it suitable for plating, soldering, etc. Forms available: rod, bar, coil, section, hollow rod, rolled. Extruded: 430-500 N/mm^2 TS; 25-30 El. Rolled: 430-500 N/mm^2 TS; 25-30 El. Drawn: 500-590 N/mm^2 TS; 20-25 El.

DELTA HT1 HIGH SPEED MACHINING ROD
Delta Extruded Metals Co., Ltd.
Copper. Cu 57, Pb 1.25, Sn 0.7, Fe 0.75, Mn 1.5, bal Zn.
Aluminum restricted high tensile brass, suitable for use where plating and solderability are a prerequisite. 450-460 N/mm^2 min TS; 12-18 min El.

DELTA HT16
Delta (Manganese Bronze) Ltd.
Copper. Cu 65.8, Sn 0.85, Fe 0.9, Al 3, Mn 0.6, bal Zn.
Higher copper and aluminum contents. Alloy has good wearing properties and excellent corrosion resistance to sea, mine, and acid moorland waters. Can be used for steam up to 200°C. Forms available: rod, rolled, bar, shaft. Extruded: 520-570 N/mm^2 TS; 25-30 El. Rolled: 550-600 N/mm^2 TS; 30-35 El. Drawn: 560-620 N/mm^2 TS; 20-25 El.

DELTA HT17
Delta (Manganese Bronze) Ltd.
Copper. Cu 67.5, Fe 1, Al 4.8, Mn 0.5, bal Zn.
Suitable for applications requiring high strength with ductility. Forms available: rod, bar, section, hollow rod, rolled products. Extruded: 590-650 N/mm^2 TS; 20-25 El. Rolled: 590-650 N/mm^2 TS; 15-20 El. Drawn: 650-710 N/mm^2 TS; 12-18 El.

DELTA HT18
Delta (Manganese Bronze) Ltd.
Copper. Cu 70, Fe 2, Al 6, Mn 3, bal Zn.
High percentage of copper and aluminum make this alloy suitable for highly stressed parts subject to wear. Forms available: rod, bar, hollow rod, rolled products. Extruded: 660-740 N/mm^2 TS; 15-20 El. Rolled: 660-740 N/mm^2 TS; 15-20 El. Drawn: 740-800 N/mm^2 TS; 12-18 El.

DELTA HT2
Delta (Manganese Bronze) Ltd.
Copper. Cu 58, Pb 0.8, Sn 0.6, Fe 0.75, Al 0.75, Mn 0.75, bal Zn.
Suitable for hot stamping and machining, this alloy contains aluminum which gives a bright surface finish to the extruded product and added corrosion resistance. Forms available: rod, bar, coil, section, hollow rod, rolled. Extruded: 430-500 N/mm^2 TS; 25-30 El. Rolled: 430-500 N/mm^2 TS; 25-30 El. Drawn: 540-620 N/mm^2 TS; 18-25 El.

DELTA HT20
Delta (Manganese Bronze) Ltd.
Copper. Cu 58.25, Pb 0.5, Al 1.5, Mn 2.5, Si 0.75, bal Zn.
Specially developed for bearings and similar applications, this alloy contains silicon to improve its wear resistance. Forms available: rod, bar, section, coil, hollow rod. Extruded: 500-650 N/mm^2 TS; 12-20 El. Rolled: 500-650 N/mm^2 TS; 12-20 El. Drawn: 540-700 N/mm^2 TS; 12-18 El.

DELTA HT20 GENERAL PURPOSE ROD AND BAR
Delta Extruded Metals Co., Ltd.
Copper. Cu 58.5, Pb 0.6, Al 1.45, Mn 1.75, Si 0.45, bal Zn.
Silicon bearing high tensile brass specially developed for the production of bearings, synchromesh rings, etc., where high wear resistance is essential. 550 N/mm^2 min TS; 12 min El.

DELTA HT22
Delta (Manganese Bronze) Ltd.
Copper. Cu 60, Pb 3, Mn 2.5, bal Zn.
Relatively high manganese and lead contents. This alloy is supplied as a resistance alloy. Forms available: rod, section. Drawn: 460-520 N/mm^2 TS; 15-20 El.

DELTA HT26
Delta (Manganese Bronze) Ltd.
Copper. Cu 66, Fe 2, Al 5, Mn 3, bal Zn.
High-tensile brass containing significant additions of aluminum, manganese and iron, suitable for highly stressed parts subject to wear. Meets French specification UZ23A4. Forms available: rod, bar, hollow rod, rolled products. Drawn: 720-800 N/mm^2 TS; 15-20 El. Rolled: 690-770 N/mm^2 TS; 15-20 El.

DELTA HT27
Delta (Manganese Bronze) Ltd.
Copper. Cu 66, Fe 1, Al 5, Mn 2, bal Zn.
High percentage of copper and aluminum makes this alloy suitable for highly stressed parts subject to wear. Forms available: rod, bar, hollow rod, rolled products. Drawn: 690-770 N/mm^2 TS; 15-20 El.

DELTA HT28
Delta (Manganese Bronze) Ltd.
Copper. Cu 58, Sn 0.75, Fe 1.25, Mn 0.2, bal Zn.
Lead- and aluminum-free American high-tensile brass suitable for brazing, soldering and plating. Forms available: rod, bar rolled products. Rolled: 400-500 N/mm^2 TS; 30-40 El. Drawn: 520-590 N/mm^2 TS; 25-35 El.

DELTA HT29
Delta (Manganese Bronze) Ltd.
Copper. Cu 57.25, Pb 0.75, Sn 1, Fe 0.7, Mn 0.35, bal Zn.
Leaded version of Delta HT28 to enhance machining properties. Drawn: 430-520 N/mm^2 TS; 25-35 El.

DELTA HT3 FORGING ROD
Delta Extruded Metals Co., Ltd.
Copper. Cu 58, Pb 1.25, Sn 0.6, Fe 0.75, Mn 0.75, Al 1, bal Zn.
Aluminum bearing high tensile brass having an excellent combination of hot forgeability, machinability and corrosion resistance. 460 N/mm^2 min TS; 15 min El.

DELTA HT3 GENERAL PURPOSE ROD AND BAR
Delta Extruded Metals Co., Ltd.
Copper. Cu 58, Pb 1.25, Sn 0.6, Fe 0.75, Al 1, Mn 0.75, bal Zn.
Aluminum bearing high tensile brass having an excellent combination of hot forgeability, machinability and corrosion resistance. 440-520 N/mm^2 min TS; 12-18 min El.

DELTA HT3 HIGH SPEED MACHINING ROD
Delta Extruded Metals Co., Ltd.
Copper. Cu 58.5, Pb 1.25, Sn 0.6, Fe 0.75, Al 1, Mn 0.75, bal Zn.
Aluminum bearing high tensile brass having an excellent combination of hot forgeability, machinability, and corrosion resistance. 440-520 N/mm^2 min TS; 12-18 min El.

DELTA HT5
Delta (Manganese Bronze) Ltd.
Copper. Cu 58, Pb 0.25, Sn 1, Fe 0.75, Mn 1.5, bal Zn.
Aluminum-free, high-tensile brass, with a low lead content. Suitable for hot stamping, soldering and general applications. Forms available: rod, bar, section, hollow rod, coil. Extruded: 500-540 N/mm^2 TS; 25-30 El. Rolled: 500-540 N/mm^2 TS; 25-30 El. Drawn: 530-590 N/mm^2 TS; 20-25 El.

DELTA HT6
Delta (Manganese Bronze) Ltd.
Copper. Cu 58, Sn 1.2, Fe 0.7, Al 0.3, Mn 0.2, bal Zn.
Lead-free, high-tensile brass, that conforms to Lloyd's and Ministry of Transport requirements for shafting and is specially adapted for rolled bars. Forms available: rolled, shaft, bar, sheet. Rolled: 460 N/mm^2 min TS; 25 El min.

DELTA II
Wieland-Werke AG Metallwerke
Cu 56-61, Ni 0-2, Mn 0.2-3, Fe 0.5-1.5, Sn 0-0.5, Al 1.3-2.5, Si 0-0.8, Pb 0-0.8, bal Zn.
Hard: 85,300 TS; 39,800 YS; 10 El; 140 Brin. For ship propellers, hardware, bolts, valve stems, fittings. High strength and corrosion resistance. *Obsolete*

DELTA III
Wieland-Werke AG Metallwerke
Cu 56-61, Ni 0-2, Mn 0.2-3, Fe 0-1.5, Sn 0-0.5, Al 0.4-1.3, Si 0-0.8, Pb 0-1, bal Zn.
Hard: 71,000 TS; 31,300 YS; 12 El; 120 Brin. For ship propellers, marine hardware, bolts, valve stems, fittings. *Obsolete*

DELTA IMMADIUM I
Delta Metal (BW) Ltd.
Zn, Al, Mn, bal Cu.
Cast: 72,400 TS; 19.5 El; 12 RA; 152 Brin. Rolled: 70,000 TS; 35,000 YS; 33 El; 125 Brin. For pump rods, valve spindles, staybolts, propeller shafts; high strength and corrosion resistance.

DELTA IMMADIUM II
Delta Metal (BW) Ltd.
Zn, Al, Mn, bal Cu.
Cast: 68,400 TS; 10 El; 9 RA; 149 Brin. Rolled: 81,000 TS; 40,000 YS; 28 El; 135 Brin. For pump rods, valve spindles, staybolts, shafts, impellers; high strength and corrosion resistance.

DELTA IV
Wieland-Werke AG Metallwerke
Cu 56-61, Ni 0-2, Mn 0.5-3, Fe 0.5-1.5, Sn 0-0.5, bal Zn.
Hard: 64,000 TS; 40,000 YS; 20 El; 110 Brin. For ship propellers, marine hardware, bolts, valve stems. High strength, corrosion resistant. *Obsolete*

DELTA MANGANESE BRASS MB1
Delta Metal (BW) Ltd.
Cu 57, Mn 1.25, Pb 2.25, bal Zn.
Light drawn: 67,000 TS; 27,000 YS; 30 El; 95 Brin. Architectural uses; good machinability.

DELTA MANGANESE BRASS MB2
Delta Metal (BW) Ltd.
Cu 57, Mn 1.25, Pb 1.25, bal Zn.
Light drawn: 67,000 TS; 27,000 YS; 30 El; 95 Brin. Architectural uses; "warmer" appearance.

DELTA MANGANESE BRASS MB3
Delta Metal (BW) Ltd.
Cu 55.5, Mn 1.25, bal Zn.
Architectural uses.

DELTA MANGANESE BRASS MB4
Delta Metal (BW) Ltd.
Cu 57.75, Mn 1.25, bal Zn.
Architectural uses; good cold bending.

DELTA MANGANESE BRONZE MB5
Delta Metal (BW) Ltd.
Cu 58.5, Mn 0.35, Fe 1.25, Sn 1, bal Zn.
Light drawn: 70,000 TS; 35,000 YS; 15 El; 115 Brin. American standard CA675.

DELTA MB1 GENERAL PURPOSE ROD AND BAR
Delta Extruded Metals Co., Ltd.
Copper. Cu 57, Pb 2.25, Mn 1.25, bal Zn.
General purpose manganese brass (bronze) having a good combination of corrosion resistance and machinability. 350 N/mm^2 min TS; 20-25 min El.

DELTA MB2 FORGING ROD
Delta Extruded Metals Co., Ltd.
Copper. Cu 57, Pb 1.25, Mn 1.25, bal Zn.
General purpose manganese brass (bronze) ideally suited for hot forging where some degree of machinability is required. 350 N/mm^2 min TS; 25 min El.

DELTA N1
Delta (Manganese Bronze) Ltd.
Copper. Cu 61.5, Sn 1.2, Pb 0.2, bal Zn.
Standard naval brass for Admiralty requirements suitable for limited machining, forging and hot stamping. Forms available: rod, bar, coil, section, hollow rod, rolled bar. Extruded: 340-390 N/mm^2 TS; 30-40 El. Rolled: 340-390 N/mm^2 TS; 30-40 El. Drawn: 370-430 N/mm^2 TS; 20-30 El.

DELTA N1 GENERAL PURPOSE ROD AND BAR
Delta Extruded Metals Co., Ltd.
Copper. Cu 61.5, Pb 0.2, Sn 1.2, bal Zn.
Standard naval brass suitable for limited machining, hot and cold working. 350-400 N/mm^2 min TS; 15-20 min El.

DELTA N5
Delta (Manganese Bronze) Ltd.
Copper. Cu 60.25, Sn 0.75, bal Zn.
American naval brass with restricted lead and iron content. The alloy has limited machinability but is suitable for hot stamping. Forms available: rod, bar, coil, section, hollow rod, rolled bar. Extruded: 350-400 N/mm^2 TS; 30-40 El. Rolled: 350-400 N/mm^2 TS; 30-40 El. Drawn: 400-460 N/mm^2 TS; 20-30 El.

DELTA N6
Delta (Manganese Bronze) Ltd.
Copper. Cu 60.25, Sn 0.75, Pb 0.6, bal Zn.
With a higher lead content this American naval brass is suitable for simple machining operations as well as hot stamping. The iron content is restricted. Forms available: rod, bar, coil, section, hollow rod. Extruded: 350-400 N/mm^2 TS; 30-40 El. Drawn: 400-460 N/mm^2 TS; 20-30 El.

DELTA N7
Delta (Manganese Bronze) Ltd.
Copper. Cu 60.25, Sn 0.75, Pb 1.5, bal Zn.
Leaded American naval brass with even higher lead content and good machining properties. The iron content is restricted. Available forms: rod, bar, coil, section, hollow rod. Extruded: 350-400 N/mm^2 TS; 25-35 El. Drawn: 400-460 N/mm^2 TS; 15-25 El.

DELTA NICKEL SILVER NS2
Delta Metal (BW) Ltd.
Cu 40, Ni 13.25, Pb 1.9, Mn 2, bal Zn.
Light drawn: 90,000 TS; 58,000 YS; 15 El; 150 Brin. For decorative and ornamental parts; colour of sterling silver.

DELTA NICKEL SILVER NS6
Delta Metal (BW) Ltd.
Cu 49.5, Ni 9.8, Mn 0.35, bal Zn.
Rolled: 83,000 TS; 42,000 YS; 18 El; 145 Brin. Rolled bar and sheet; lead free.

DELTA NS1
Delta (Manganese Bronze) Ltd.
Copper. Cu 45.5, Mn 0.3, Pb 2, Ni 9.5, bal Zn.
Leaded nickel brass suitable for stamping, machining, and engraving. Silver in color with warmer appearance than stainless steel; mainly used for ornamental and decorative purposes. Extruded: 550-620 N/mm^2 TS; 20-30 El. Drawn: 590-680 N/mm^2 TS; 15-20 El. Forms available: rod, section, hollow rod.

DELTA NS6
Delta (Manganese Bronze) Ltd.
Copper. Cu 49.5, Mn 0.35, Ni 9.8, bal Zn.
Lead-free alloy specially designed to have an appearance similar to that of Delta NS1 after rolling. Forms available: rolled bar, plate. Rolled: 560-590 N/mm^2 TS; 18-25 El.

DELTA S1 FORGING ROD
Delta Extruded Metals Co., Ltd.
Copper. Cu 57.5, Pb 2, bal Zn.
Standard general purpose hot forging brass, characterized by excellent hot forging combined with good machinability. 350 N/mm^2 min TS; 25 min El.

DELTA S1 GENERAL PURPOSE ROD AND BAR
Delta Extruded Metals Co., Ltd.
Copper. Cu 57.5, Pb 2, bal Zn.
Standard general purpose hot forging brass with good machinability. 350-425 N/mm^2 min TS; 18-25 min El.

DELTA S1 HIGH SPEED MACHINING ROD
Delta Extruded Metals Co., Ltd.
Copper. Cu 57.5, Pb 2, bal Zn.
Standard general purpose hot forging brass with good machinability. 350-425 N/mm^2 min TS; 18-25 min El.

DELTA S10 FORGING ROD
Delta Extruded Metals Co., Ltd.
Copper. Cu 61, Pb 2.25, As 0.12, Zn 36, bal others.
Arsenic inhibited dezincification resistant leaded brass (DZR) for use in the water and gas service industries. Forgings manufactured of this alloy require a heat treatment to optimize the dezincification properties. 280 N/mm^2 min TS; 30 min El.

DELTA S10 HIGH SPEED MACHINING ROD
Delta Extruded Metals Co., Ltd.
Copper. Cu 61, Pb 2.25, As 0.12, Zn 36, bal others.
Arsenic inhibited dezincification resistant leaded brass (DZR) for use in the water and gas service industries. Machining rod is supplied in the fully heat treated condition in which the dezincification resistant properties have been optimized. 350-380 N/mm^2 min TS; 20-25 El.

DELTA S2 FORGING ROD
Delta Extruded Metals Co., Ltd.
Copper. Cu 59, Pb 2.25, bal Zn.
General hot forging and machining brass with limited cold working properties. 350 N/mm^2 min TS; 25 min El.

DELTA S2 GENERAL PURPOSE ROD AND BAR
Delta Extruded Metals Co., Ltd.
Copper. Cu 59, Pb 2.25, bal Zn.
General machining and hot working brass with limited cold working properties (i.e., bending, swaging, and riveting.) 350-380 N/mm^2 min TS; 22-25 min El.

DELTA S2 HIGH SPEED MACHINING ROD
Delta Extruded Metals Co., Ltd.
Copper. Cu 59, Pb 2.25, bal Zn.
General machining/hot working brass with limited cold working properties (i.e., bending and riveting) combined with good machinability. 350-380 N/mm^2 min TS; 22-25 min El.

DELTA S5 GENERAL PURPOSE ROD AND BAR
Delta Extruded Metals Co., Ltd.
Copper. Cu 59, Pb 1.25, bal Zn.
General purpose alloy having a good combination of machining, cold and hot working properties. 350-380 N/mm^2 min TS; 25-28 min El.

DELTA S5 HIGH SPEED MACHINING ROD
Delta Extruded Metals Co., Ltd.
Copper. Cu 59, Pb 1.25, bal Zn.
Alloy having a combination of good machining, cold working and hot working properties. 350-380 N/mm^2 min TS; 25-28 min El.

DELTA S8 FORGING ROD
Delta Extruded Metals Co., Ltd.
Copper. Cu 58.5, Pb 2.25, bal Zn.
General hot forging and machining brass with limited cold formability.

DELTA S8 HIGH SPEED MACHINING ROD
Delta Extruded Metals Co., Ltd.
Copper. Cu 58.5, Pb 2.25, bal Zn.
General machining and hot forging brass with limited cold formability.

DELTA T II
Wieland-Werke AG Metallwerke
Al 10, Mn 2, Fe 2, bal Cu.
Rolled: 85,000 TS; 42,000 YS; 15 El; 150 Brin. For paper and chemical industries, pump shafts, valve components; aluminum bronze, wear and corrosion resistant. *Obsolete*

DELTA T12 GENERAL PURPOSE ROD AND BAR
Delta Extruded Metals Co., Ltd.
Copper. Cu 57, Pb 3.8, bal Zn.
Extra high speed free machining brass allowing operations at increased speed with reduced tool wear. 350-425 N/mm^2 min TS; 15-25 min El.

DELTA T12 HIGH SPEED MACHINING ROD
Delta Extruded Metals Co., Ltd.
Copper. Cu 57, Pb 3.8, bal Zn.
Extra high speed free machining brass giving superior tool life and speed. 350-425 N/mm^2 min TS; 15-25 min El.

DELTA T2 FORGING ROD
Delta Extruded Metals Co., Ltd.
Copper. Cu 58, Pb 3, bal Zn.
Standard free machining brass with good hot forging properties. 350 N/mm^2 min TS; 25 min El.

DELTA T2 GENERAL PURPOSE ROD AND BAR
Delta Extruded Metals Co., Ltd.
Copper. Cu 58, Pb 3, bal Zn.
Standard high speed free machining brass for general applications on high speed automatic machines. 350-425 N/mm^2 min TS; 15-25 min El.

DELTA T2 HIGH SPEED MACHINING ROD
Delta Extruded Metals Co., Ltd.
Copper. Cu 58, Pb 3, bal Zn.
Standard high speed machining brass supplied for general applications on high speed automatic machines. 350-425 N/mm^2 min TS; 15-25 El.

DELTA T5 GENERAL PURPOSE ROD AND BAR
Delta Extruded Metals Co., Ltd.
Copper. Cu 61.5, Pb 3, bal Zn.
Standard American type high speed machining brass suitable for operations involving thread rolling and light riveting. 280-550 N/mm^2 min TS; 3-22 min El.

DELTA T5 HIGH SPEED MACHINING ROD
Delta Extruded Metals Co., Ltd.
Copper. Cu 61.5, Pb 3, bal Zn.
Standard American type high speed machining brass. The higher copper content gives a superior ductility. This alloy is suitable for such operations as thread rolling and light riveting. 280-550 N/mm^2 min TS; 3-22 min El.

DELTA TR1 GENERAL PURPOSE ROD AND BAR
Delta Extruded Metals Co., Ltd.
Copper. Cu 61.6, Pb 2.25, bal Zn.
Standard turning and riveting brass having an ideal combination of good machinability coupled with good cold working properties. 330-350 N/mm^2 min TS; 22-28 min El.

DELTA TR1 HIGH SPEED MACHINING ROD
Delta Extruded Metals Co., Ltd.
Copper. Cu 61.6, Pb 2.25, bal Zn.
Standard turning and riveting brass combining good machinability with good cold working properties. 330-350 N/mm^2 min TS; 22-28 min El.

DELTA TR4 HIGH SPEED MACHINING ROD
Delta Extruded Metals Co., Ltd.
Copper. Cu 62.5, Pb 1.25, bal Zn.
Very ductile free machining brass, commonly used where riveting properties are paramount, e.g., crimping.

DELTA W1
Delta (Manganese Bronze) Ltd.
Copper. Cu 60.5, Si 0.4, Sn 0.4, bal Zn.
Standard brass filler rod containing small tin addition for improved corrosion resistance in the weld. It is suitable for brazing copper and mild steel and for the fusion welding of duplex brasses. Forms available: rod, coil, or knurled. Melting range: 860-880°C.

DELTA W11
Delta (Manganese Bronze) Ltd.
Copper. Cu 89, Al 9.5, Fe 1, Mn 0.5.
Aluminum bronze welding rod similar to CA8, having the same high impact strength and corrosion resistance. Forms available: rod, coil. Melting range: 1000-1030°C.

DELTA W13

Delta (Manganese Bronze) Ltd.
Copper. Cu 57.5, Si 0.1, Sn 0.9, Fe 0.4, Mn 0.3, bal Zn.
High-tensile filler rod with low fuming properties, suitable for brazing copper and cast and malleable iron for the fusion welding of high-tensile brasses of similar composition. Forms available: rod, coil, plain or knurled. Melting range: 860-880°C.

DELTA W14

Delta (Manganese Bronze) Ltd.
Copper. Cu 59, Ni 0.5, Si 0.1, Sn 0.9, Fe 0.4, Mn 0.4, bal Zn.
High-tensile filler rod, suitable for brazing copper and cast and malleable iron, meets AWS R Cu Zn-B. Forms available: rod, coil. Melting range: 850-880°C.

DELTA W15

Delta (Manganese Bronze) Ltd.
Copper. Cu 59.5, Si 0.2, Sn 0.5, Mn 0.8, bal Zn.
Free-flowing, medium-tensile welding rod containing manganese and tin, giving improved corrosion resistance. Forms available: rod, coil. Melting range: 850-880°C.

DELTA W18

Delta (Manganese Bronze) Ltd.
Copper. Cu 91.4, Al 7, Fe 1.25, Mn 0.35.
One of the single-phase alpha alloys with small addition of iron for increased strength. Forms available: rod, coil, spools. Melting range: 1040-1060°C.

DELTA W19

Delta (Manganese Bronze) Ltd.
Copper. Cu 90, Al 7, Fe 3.
Similar to Delta W18 with approximately three times the iron content for higher strength applications. Forms available: rod, coil, spools. Melting range: 1045-1110°C.

DELTA W2

Delta (Manganese Bronze) Ltd.
Copper. Cu 59.5, Si 0.3, bal Zn.
Brass filler rod is similar to Delta W1, but is tin free for use with gas fluxes. Forms available: rod, coil, plain or knurled. Melting range: 860-880°C.

DELTA W22

Delta (Manganese Bronze) Ltd.
Copper. Cu 83.75, Al 8.75, Fe 2, Ni 4.25, Mn 1.25.
One of the complex aluminum bronzes with addition of nickel, iron and manganese for welding high-strength alloys. Forms available: rod, coil, spools. Melting range: 1055-1075°C.

DELTA W23

Delta (Manganese Bronze) Ltd.
Copper. Cu 81, Al 6.2, Fe 0.6, Si 2.2.
Silicon aluminum bronze used for welding CA10 and CA12. Forms available: rod, coil, spools. Melting range: 980-1010°C.

DELTA W4

Delta (Manganese Bronze) Ltd.
Copper. Cu 59.5, Si 0.25, Sn 0.4, Fe 0.4, Mn 0.2, bal Zn.
High-tensile filler rod, suitable for brazing copper and cast and malleable iron for fusion welding of high-tensile brasses of similar composition. Forms available: rod, coil. Melting range: 860-880°C.

DELTA W5

Delta (Manganese Bronze) Ltd.
Copper. Cu 48, Ni 8.5, Si 0.25, bal Zn.
Nickel brass welding rod, suitable for brazing ferrous metals and for the fusion welding of nickel brass with similar composition. Forms available: rod, coil. Melting range: 890-920°C.

DELTA WELDING ALLOY W 12

Delta Metal (BW) Ltd.
Al 9.75, Fe 2.25, Mn 1.75, bal Cu.
As welded: 63,000 TS. For welding aluminum bronzes.

DELTA WELDING ALLOY W 5

Delta Metal (BW) Ltd.
Cu 48.5, Ni 9, Si 0.25, Mn 0.4, bal Zn.
Welded: 60,000 TS. Nickel silver weld metal; for brazing ferrous metals.

DELTA WELDING ALLOY W 6

Delta Metal (BW) Ltd.
Cu 48, Ni 8.5, Si 0.25, Sn 0.3, Mn 0.4, bal Zn.
Nickel silver weld metal; for brazing ferrous metals.

DELTA, KRUPP

Krupp Stahl AG
copper. Cu 54-56, Zn 41-42, Pb 0.7-1.8, Fe 0.86-1.28, Mn 0.8-1.4.
For high strength structural castings; tough. *Obsolete*

DELTA-MU

Magnetic Metals Co.
Ni 45, Cr 5, bal Fe.
For magnetic cores; high electrical resistivity. *Obsolete*

DELTABRA

Alcan-Booth Industries, Ltd.
Cu 77, Al 2, As 0.03, bal Zn.
Cold drawn: 82,880 TS; 76,160 YS; 8 El. For condenser tubes, tanker heater tubes; resists sea water corrosion.

DELTAL

Deutsch Delta Metallgesellschaft
Mg 0.6-1.4, Si 0.6-1.6, Mn 0.6-1, Cr 0-0.3, bal Al.
Annealed: 21,000 TS; 80,000 YS; 24 El; 36 Brin. Hard: 32,000 TS; 29,000 YS; 6 El. For structural members.

DELTALVIT

Deutsch Delta Metallgesellschaft
Aluminum.
Aluminum alloy AlCuMg. *Obsolete*

DELTAMAX

Allegheny Ludlum Steel
Fe 50, Ni 50.
For magnets, choke cores; grain-oriented. *Obsolete*

DELTOKAL

Deutsch Delta Metallgesellschaft
Aluminum.
Aluminum alloy AlMg. *Obsolete*

DELTOXAL

Deutsch Delta Metallgesellschaft
Mg 2-4, Mn 0-0.4, Cr 0-0.3, bal Al.
Annealed: 28,000 TS; 13,000 YS; 30 El; 47 Brin. Hard: 42,000 TS; 37,000 YS. For aircraft, missile and marine parts; corrosion resistant to seawater.

DELTUMIN

Deutsch Delta Metallgesellschaft
Aluminum. Cu 3.5-5.5, Si 0.2-1.5, Mn 0.1-1.5, Mg 0.2-2, bal Al.
Heat treated: 70,000 TS; 60,000 YS; 13 El; 135 Brin. For fittings, hardware, aircraft engine components. Heat treatable, high tensile and fatigue strength. *Obsolete*

DEM NO. 1

Engelhard Industries
Ag 63.3, bal Cu.
For silver solder; MP 1275°F.

DEM NO. 10

Engelhard Industries
Ag 22.2, bal Cu.
For silver solder; MP 1207°F.

DEM NO. 102

Engelhard Industries
Au 41.7, bal Cu.
For gold solder; MP 1436 to 1508°F.

DEM NO. 104

Engelhard Industries
Au 41.7, bal Cu.
For gold solder; MP 1248 to 1318°F.

DEM NO. 106

Engelhard Industries
Au 41.7, bal Cu.
For gold solder; MP 1378 to 1410°F.

DEM NO. 110

Engelhard Industries
Au 58.3, bal Cu.
For gold solder; MP 1395 to 1310°F.

DEM NO. 112

Engelhard Industries
Au 50, bal Cu.
For gold solder; MP 1420 to 1495°F.

DEM NO. 114

Engelhard Industries
Au 33.3, bal Cu.
For gold solder; MP 1360 to 1392°F.

DEM NO. 115

Engelhard Industries
Au 33.3, bal Cu.
For gold solder; MP 1245 to 1310°F.

DEM NO. 119

Engelhard Industries
Au 25, bal Cu.
For gold solder; MP 1350 to 1141°F.

DEM NO. 120

Engelhard Industries
Au 41.7, bal Cu.
For gold solder; MP 1378 to 1418°F.

DEM NO. 2

Engelhard Industries
Ag 66.7, bal Cu.
For silver solder; MP 1363°F.

DEM NO. 22

Engelhard Industries
Ag 50, bal Cu.
For silver solder; MP 1255°F.

DEM NO. 23

Engelhard Industries
Ag 60, bal Cu.
For silver solder; MP 1248°F.

DEM NO. 293

Engelhard Industries
Au 50, bal Cu.
For gold solder; MP 1317 to 1255°F.

DEM NO. 37

Engelhard Industries
Ag 9, bal Cu.
For silver solder; MP 1510°F.

DEM NO. 41

Engelhard Industries
Ag 71.8, bal Cu.
For silver solder; MP 1434°F.

DEM NO. 426

Engelhard Industries
Au 33.3, bal Cu.
For gold solder; MP 1407 to 1522°F.

DEM NO. 429

Engelhard Industries
Au 50, bal Cu.
For gold solder; MP 1392 to 1373°F.

DEM NO. 43
Engelhard Industries
Ag 54, bal Cu.
For silver solder; MP 1223°F.

DEM NO. 430
Engelhard Industries
Au 58.3, bal Cu.
For gold solder; MP 1323 to 1248°F.

DEM NO. 431
Engelhard Industries
Au 58.3, bal Cu.
For gold solder; MP 1495 to 1440°F.

DEM NO. 48
Engelhard Industries
Ag 17, bal Cu.
For silver solder; MP 1382°F.

DEM NO. 64
Engelhard Industries
Ag 19-20, bal Cu.
For silver solder; MP 1430°F.

DEM NO. 65
Engelhard Industries
Ag 40, bal Cu.
For silver solder; MP 1233°F.

DEM NO. 66
Engelhard Industries
Ag 64-66, bal Cu.
For silver solder; MP 1280°F.

DEM NO. 69
Engelhard Industries
Ag 44-46, bal Cu.
For silver solder; MP 1250°F.

DEM NO. 71
Engelhard Industries
Ag 49-51, bal Cu.
For silver solder; MP 1160°F.

DEM NO. 73
Engelhard Industries
Ag 49-51, bal Cu.
For silver solder; MP 1195°F.

DEM NO. SSEI-70
Engelhard Industries
Ag 46, bal Cu.
For silver solder; MP 1171°F.

DEM NO. SSEI-82
Engelhard Industries
Ag 46, bal Cu.
For silver solder; MP 1165°F.

DEMARK
Pipe Machinery Corp.
Ti, C, Co.
For thread gauging dies; sintered carbide.

DEMMLER "D"
Teledyne Firth Sterling
C 0.55, Mn 0.9, Cr 1.2, V 0.2, bal Fe.
For tools, punches, dies, arbors, chisels, collets; water or oil hardened. *Obsolete*

DEMO BRONZE
Manufacturer not listed
Cu 61, Pb 33, Sn 4-6, Ni 1-2.
18,000 TS; 3 El; 0.35 RA; 52 Brin. For bearings, utensils.

DENAVIS
Denman & Davis Co.
C 0.7-1.2, bal Fe
For tools and cutters. Water hardening. *Obsolete*

DENAVIS HIGH SPEED
Denman & Davis Co.
C 0.7, W 18, Cr 4, bal Fe.
For tools, cutters, and reamers. High speed steel. *Obsolete*

DENINE
Atlas Specialty Steels
C 1.2, W 1.5, bal Fe.
For taps, drawing dies; water hardened. *Obsolete*

DENSALLOY
Welded Carbide Co. Inc.
W 90, Ni 6, Cu 4.
Sintered: 85,000-115,000 TS; 2-10 El; 400 Brin. For balance weights, centrifugal clutches; sintered, heavy metal.

DENSCAST
Driver Harris Co.
Ni 80, Cr 20.
For castings; thin wall. *Obsolete*

DENSE ALLOY 112 TYPE E
GTE Sylvania
W-Co-Ag.
For balancing weights; sintered. *Obsolete*

DENSE ALLOY 112 TYPE P
GTE Sylvania
W-Cu-Ni.
For balancing weights; sintered. *Obsolete*

DENSE ALLOY 112 TYPE Y
GTE Sylvania
W-Co-Ni.
For balancing weights; sintered. *Obsolete*

DENSE STEEL
American manufacture
Cr, Si, Mo, C, bal Fe.
For tools.

DENSIFIED CRU-DIE
Slater Steels Corp.
Alloy steel. C 0.55, Ni 0.85, Mn 0.9, Cr 1.05, Si 0.3, Mo 0.4, bal Fe.
Die blocks and die steel; tough alloy steel of a balanced chromium, nickel, molybdenum composition.

DENSIFIED CRU-DIE
Colt Industries
C 0.55, Mn 0.9, Si 0.3, Ni 0.8, Cr 1.05, Mo 0.4, bal Fe.
Furnished in 4 hardness ranges (269-429 Brin) for forging large sections of tool steels, particularly for dies. *Obsolete*

DENSITE
Jamison Steel Corp.
C 1.4, Cr 11-13, Mo 0.8, 0.20 V min, bal Fe.
For cutting, forming and trimming dies; air hardened.

DENSITE AH
Jamison Steel Corp.
C 1.5, Cr 12, Mo 1, bal Fe.
For forming and drawing dies, punches; Type D2; air hardened, nondeforming.

DENTAL BURR (1% W)
Crucible Materials Corp.
Tool material. C 1.15, W 1.25, bal Fe.

DENTAL BURR (3-1/2% W)
Crucible Materials Corp.
Tool material. C 1.05, Cr 0.75, W 3.5, bal Fe.

DENTALLOY
Cerro Metal Products Co.
Bi 38.14, Pb 26.42, Sn 31.67, Cd 2.64, Sb 1, Cu 0.06.
Cast: 15 Brin. For dental castings and models; see "Cerrodent." *Obsolete*

DENTURE CLASP
J.M. Ney Co.
Au 56, Pd 5.
Gold color clasp and orthodontic wire. Fusing temperature: 1650°F.

DEOXIDIZED COPPER
J.W. Harris Co., Inc.
Cu 98.8, Sn 0.75, Si 0.25, Mn 0.2.
Melting point: 1967°F approx. For inert gas and oxyacetylene welding and build up on copper tanks and assemblies. AWS A5.6-57T Class ECU; 5.7-57T Class RCu; ASTM B225-57T Class ECU; B259-57T Class RCu.

DEOXIDIZED COPPER
Chase Brass & Copper Co.
P 0.02, 99.9%+ Cu.
Annealed: 33,000 TS; 10,000 YS; 45 El. Rolled: 50,000 TS; 45,000 YS; 6 El. For condensers, evaporators; corrosion resistant. *Obsolete*

DEOXIDIZED COPPER 105
Olin Brass, Indianapolis
Copper. P 0.02, 99.9 Cu min.
Hard: 55,000 psi TS; 50,000 psi YS; 8 El; 107 Brin. Soft: 32,000 psi TS; 10,000 psi YS; 45 El; 40 Brin. For deep drawing, water tubes; deoxidized copper.

DEOXIDIZED LEADED COPPER-129
Anaconda Co.
Copper. Pb 0.8-1.2, 99.90 Cu + Pb min.
Hard: 51,000-60,000 TS; 46,000-55,000 YS; 10-14 El; 83-100 Brin. For contact pins and inserts, screw machine products. High thermal and electrical conductivity, free cutting, corrosion resistant.

DEOXOLOY
Belmont Metals Inc.
Cu alloy.
Deoxidizer for Cu alloys.

DEPAL
Duval et Poulain
Cu 2, Mn 2, Ni 2, bal Al.
Cast: 20,000-26,000 TS; 10,000-14,000 YS; 1.5-3.5 El; 55-65 Brin. For pistons, cylinder heads; age-hardenable.

DETACLAD SERIES
Du Pont Co.
Explosion-bonded clad metals; plate, heads and tube sheet; variety of metal combinations.

DETACOUPLE
Du Pont Co.
Welding transition joints for shipbuilding, aerospace, cryogenic and electrometallurgical industries.

DEURANCE METAL
English manufacture
Sn 33.3, Sb 44.5, Cu 22.2.
For locomotive bearing; Babbitt.

DEUTRO 18/8
Deutro GmbH
Stainless steel. C 0.2-0.8, Mn 0-2, Cr 17-19, Ni 8-10, bal Fe.
Annealed: 85,000 TS; 35,000 YS; 60 El; 70 RA; 150 Brin. Cold drawn: 125,000 TS; 95,000 YS; 25 El; 55 RA; 277 Brin. For oil refinery and chemical plant equipment; Type 302; stainless, austenitic.

DEUTRO 18/8 MS
Deutro GmbH
Stainless steel. C 0.1, Cr 16-18, Ni 10-14, Mo 2-3, bal Fe.
Annealed: 85,000-95,000 TS; 35,000-45,000 YS; 50-60 El; 60-75 RA; 150-190 Brin. For chemical plant equipment, mixers, agitators, filters; Type 316; stainless, austenitic.

DEUTRO 18/8S
Deutro GmbH
Stainless steel. C 0-0.08, Cr 18-20, Ni 8-11, Mn 0-2, bal Fe.
Annealed: 90,000 TS; 45,000 YS; 60 El; 135 Brin. Cold drawn:
180,000 TS; 150,000 YS; 10 El; 330 Brin. For chemical plant
equipment, welded structures; Type 304; stainless, austenitic.

DEUTRO 20/15
Deutro GmbH
C 0-0.2, Cr 22-24, Ni 12-15, Mn 0-2, Si 0-1, bal Fe.
Annealed: 85,000-95,000 TS; 40,000-50,000 YS; 45-55 El;
150-185 Brin. For heat treating boxes, oil refinery and
chemical plant equipment; Type 309; austenitic, heat
resistant.

DEUTRO 23/20
Deutro GmbH
C 0-0.25, Cr 24-26, Ni 19-22, bal Fe.
Annealed: 95,000 TS; 45,000 YS; 50 El; 65 RA; 180 Brin. At
1200°F: 57,000 TS; 22,000 YS; 32 El; 45 RA. For furnace
parts and equipment, heat treating boxes; Type 310;
austenitic, heat resistant.

DEUTSCHE NICKEL CUBE 2
Vereinigte Deutsche Nickel-Werke AG
Copper. Be 2, bal Cu.
Wire material for electronics, controls, drawing and stamping
parts. C17200

DEUTSCHE NICKEL CUNI 1.5 SI
Vereinigte Deutsche Nickel-Werke AG
Copper. Si 0.6, 1.3 Ni + Co, bal Cu.
Strip, wire, bar, forged and turned parts for electronics,
fasteners, drawing and stamping parts.

DEUTSCHE NICKEL CUNI 10
Vereinigte Deutsche Nickel-Werke AG
Copper. 10 Ni + Co, bal Cu.
Strip, wire, bar, forged and turned parts for electronics and
controls.

DEUTSCHE NICKEL CUNI 10 FE 1 MN
Vereinigte Deutsche Nickel-Werke AG
Copper. Fe 1.5, C 0.005, Mn 0.7, 10.5 Ni + Co, bal Cu.
Strip, wire, bar, forged and turned parts for apparatus, heat
exchangers and radiators, marine and offshore applications
and welding. C70600

DEUTSCHE NICKEL CUNI 15
Vereinigte Deutsche Nickel-Werke AG
Copper. 14 Ni + Co, bal Cu.
Strip and wire for electronics, controls, drawing and stamping
parts.

DEUTSCHE NICKEL CUNI 2
Vereinigte Deutsche Nickel-Werke AG
Copper. 2 Ni + Co, bal Cu.
Strip and wire for electronics, controls, drawing and stamping
parts. C70200

DEUTSCHE NICKEL CUNI 2 SI
Vereinigte Deutsche Nickel-Werke AG
Copper. Si 0.7, 2.1 Ni + Co, bal Cu.
Strip, wire, bar, forged and turned parts for fasteners, drawing
and stamping parts.

DEUTSCHE NICKEL CUNI 2.5 SI
Vereinigte Deutsche Nickel-Werke AG
Copper. Si 0.7, 2.5 Ni + Co, bal Cu.
Strip, wire, bar, forged and turned parts for electronics,
fasteners, drawing and stamping parts.

DEUTSCHE NICKEL CUNI 20
Vereinigte Deutsche Nickel-Werke AG
Copper. 19.5 Ni + Co, bal Cu.
Strip and wire for electronics, controls, fasteners, drawing and
stamping parts.

DEUTSCHE NICKEL CUNI 23
Vereinigte Deutsche Nickel-Werke AG
Copper. 23 Ni + Co, bal Cu.
Strip and wire for electronics, controls, drawing and stamping
parts. C71100

DEUTSCHE NICKEL CUNI 30 FE 2 MN 2
Vereinigte Deutsche Nickel-Werke AG
Copper. Fe 1.8, C 0.04, Mn 1.8, 30.6 Ni + Co, bal Cu.
Strip, wire and bar for electronics, apparatus, heat
exchangers and radiators, marine and offshore applications,
drawing and stamping parts.

DEUTSCHE NICKEL CUNI 30 MN 1 FE
Vereinigte Deutsche Nickel-Werke AG
Copper. Fe 0.6, C 0.03, Mn 0.7, 30.8 Ni + Co, bal Cu.
Strip, wire, bar, forged and turned parts for apparatus, heat
exchangers and radiators, marine and offshore applications,
drawing and stamping parts. C71500

DEUTSCHE NICKEL CUNI 6
Vereinigte Deutsche Nickel-Werke AG
Copper. 5.5 Ni + Co, bal Cu.
Strip and wire for electronics, controls, drawing and stamping
parts. C70300

DEUTSCHE NICKEL NIMN 1
Vereinigte Deutsche Nickel-Werke AG
Nickel. Cu 0.03, Fe 0.1, C 0.05, Mn 0.6, Fe 0.05, 98.5 Ni +
Co.
Strip, wire, bar, forged and turned parts for apparatus,
drawing and stamping parts.

DEUTSCHE NICKEL NIMN1C
Vereinigte Deutsche Nickel-Werke AG
Nickel. Cu 0.05, Fe 0.1, C 0.18, Mn 1.1, Si 0.07, 98 Ni + Co.
Strip, wire, bar, forged and turned parts for apparatus,
drawing and stamping parts.

DEUTSCHE NICKEL NIMN2
Vereinigte Deutsche Nickel-Werke AG
Nickel. Cu 0.04, Fe 0.1, C 0.04, Mn 1.8, Si 0.12, 98 Ni + Co.
Strip, wire, bar, forged and turned parts for electronics,
optical parts, drawing and stamping parts.

DEVA
Degussa AG
Precious metal.
Precious metal alloy for dentistry and dental engineering.

DEVILS IRON
Allegheny Ludlum Steel
Si, bal Fe.
For core transformers, telephone components; soft magnet,
grain oriented. *Obsolete*

DEW CMS
Thyssen Edelstahlwerke AG
C 0.43, Si 0.2, Mn 1.3, Cr 1.8, Mo 0.2, bal Fe.
Heat treated: 170,000-190,000 TS. For extrusion dies, liners,
containers. Hot work steel. Resists high pressure. *Obsolete*

DEW D12L
Thyssen Edelstahlwerke AG
C 0.55, Cr 1.1, Mo 0.75, Ni 1.7, V 0.1, bal Fe.
For press dies and extrusion press stems, heavy duty forging
dies, die inserts, upsetters. Oil or air hardening. Tough, shock
resistant. *Obsolete*

DEW E38W
Thyssen Edelstahlwerke AG
C 0.38, Si 1.2, Cr 5.6, Mo 1.5, V 0.3, W 1.5, bal Fe.
Annealed: 100,000 TS; 60,000 YS; 28 El; B 92 Rock. Heat
treated: 275,000 TS; 235,000 YS; 8 El; C 52 Rock. For
extrusion presses, die casting dies, forging dies, hot
punches, gripper and header dies. Hot work tool steel. Hot
wear resistance. Tough and shock resistant. *Obsolete*

DEW E38MO
Thyssen Edelstahlwerke AG
C 0.38, Si 1, Cr 5.3, Mo 1.5, V 0.45, bal Fe.
Heat treated: 300,000 TS; 250,000 YS; 6 El; C 55 Rock.
Annealed: 102,000 TS; 66,000 YS; 28 El; B 93 Rock. For
aluminum and magnesium die casting dies, cold chamber
die casting cylinders and pistons. Air or oil hardening, good
hot hardness, resists thermal shock. *Obsolete*

DEW E38V
Thyssen Edelstahlwerke AG
C 0.4, Si 1, Cr 5.3, Mo 1.4, V 1, bal Fe.
Annealed: 100,000 TS; 65,000 YS; 30 El; B 92 Rock. Heat
treated: 280,000 TS; 240,000 YS; 8 El; C 52 Rock. For oil or
water cooled mandrels, extrusion press dies, aluminum die
casting dies. Hot wear resistance, tough and shock
resistance. *Obsolete*

DEW ECR
Thyssen Edelstahlwerke AG
C 0.7, Cr 0.3, Mn 0.3, Si 0.25, bal Fe.
Annealed: 110,000 TS; 60,000 YS; 20 El; 190 Brin. For axes,
hammers, crimpers, springs, punches. Water hardening.
Obsolete

DEW GSE
Thyssen Edelstahlwerke AG
C 0.52, Si 1.3, Mn 0.7, Cr 13.5, Ni 13, V 1.2, W 1.3, bal Fe.
For extrusion dies up to 1100 F. Excellent hot hardness.
Resists thermal shock. Hot work steel, austenitic. *Obsolete*

DEW SPEZIAL-MS
Thyssen Edelstahlwerke AG
C 0.28, Cr 1.2, Mo 1.4, V 0.25, W 2.3, Co 2.3, bal Fe.
For die casting dies. Hot work tool steel, insensible to hot
checking. *Obsolete*

DEW SPEZIAL-W5
Thyssen Edelstahlwerke AG
C 0.3, Cr 2.35, V 0.55, W 4.25, bal Fe.
For hot screw, rivet and nut dies, press mandrels, extrusion
dies, hot punches, piercers, shear blades, die casting dies.
Hot work tool steel. Good hot hardness. *Obsolete*

DEW WM559
Thyssen Edelstahlwerke AG
C 0.42, Si 0.6, Mn 0.4, Cr 1.4, Mo 0.45, V 0.8, W 0.45, bal Fe.
Heat treated: 170,000 TS. For extrusion dies, water cooled
dummy blocks. Hot work steel. *Obsolete*

DEW-AMS EXTRA
Thyssen Edelstahlwerke AG
C 0.55, Si 0.2, Mn 0.7, Cr 1.1, Mo 0.45, Ni 1.7, V 0.1, bal Fe.
Heat treated: 230,000-250,000 TS. For extrusion dies, stems,
punches, rivet sets. Hot work steel, tough. *Obsolete*

DEW88MET
Dewramet Ltd.
Nickel. C 0-0.05, Mn 0-1.25, Si 0-0.5, Cr 11-14, Mo 2-3.5, Fe
0-2, Sn 3-5, Bi 3-5, bal Ni.
45,000 psi TS; 38,000 psi YS; 7 El; 150 Brin. For food,
pharmaceutical, nuclear and fluid control.

DEWARD
Allegheny Ludlum Steel
Cu 0.9, Mn 1.5, Si 0.2, Cr 0.08, Mo 0.3, bal Fe.
Oil hardened: 67,000-80,000 TS; 296-650 Brin. For dies, taps,
tools, reamers, cutters, saws, blanking dies; non-deforming,
non-shrinking. *Obsolete*

DEWARD
AL Tech Specialty Steel Corp.
C 0.9, Mn 1.5, Si 0.2, Cr 0.08, Mo 0.3, bal Fe.
Usual working hardness 58/62 Rock C. Oil hardening, non-
deforming tool steel suitable for punches, dies, taps, gages,
cutters, rolls and bushings. AISI 0-2.

DEWRANCE CM50
Dewramet Ltd.
Now ENDEWRANCE 50.

DEWRANCE CM52
Dewramet Ltd.
Now ENDEWRANCE 52.

DEWRANCE CM53
Dewramet Ltd.
Now ENDEWRANCE 53.

DEWRANCE CM56
Dewramet Ltd.
Ni 71, Cr 15, Si 4.5, B 3.5.
For high temperature brazing; flow point 1030°C, corrosion resistant. *Obsolete*

DEX-TUNG
North American Steel Corp.
C 0.32, Cr 0.93, Mo 0.48, W 0.47, bal Fe.
Heat treated: 177,000-265,000 TS; 170,000-208,000 YS; 14-19 El. For gears, shafting, clutch parts, punches, collets, blacksmith tools; oil hardened.

DEXITE AH
North American Steel Corp.
C 1.15, Mn 0.8, Si 0.2, Cr 5.45, V 0.35, Mo 1.25, bal Fe.
Air hardened, cold-work tool steel, high carbon; good wear resistance.

DEXITE AH MODIFIED
North American Steel Corp.
C 0.55, Mn 0.9, Si 0.25, Cr 3.5, Ni 0.25, V 0.25, Mo 1.4, bal Fe.
Air hardened, tool steel.

DEXITE NO. 14
North American Steel Corp.
C 0.75, Mn 0.8, Cr 0.94, Ni 1.8, Mo 0.4, bal Fe.
Oil hardened, tool steel; for dies, drills, shear blades, swaging dies.

DEXITE NO. 15
North American Steel Corp.
C, alloy, bal Fe.
For boring tools, cams, rollers, reamers, gages; deep hardening.

DEXITE NO. 16
North American Steel Corp.
C, alloy, bal Fe.
For chisels, shovel teeth, wrenches, crushers; abrasion resistant, water hardened, tough.

DFO 110 SUPERVITAC
Creusot-Loire
C 0.3, Ni 2, Cr 2, Mo 0.35, bal Fe.
Bar, treated: 1080 N/mm^2 psi TS; 830 N/mm^2 psi YS min; 11 El. For structural purposes.

DFO 95 SUPERVITAC
Creusot-Loire
C 0.4, Ni 0.9, Cr 0.8, Mo 0.25, bal Fe.
Bars, treated: 930 N/mm^2 psi TS; 735 N/mm^2 psi YS min; 13 El. For structural purposes.

DG20C
Texas Instruments Inc./Materials Control
Pd 59.5-60.5, Ag 39.1-40.5, Cu 0-1, Au 0-2.
Diffused gold surface electrical contact material for low noise, stable contact applications.

DH NO. 241
Harrison Alloys Inc.
Nickel.
Now HAI-NICR 40.

DH NO. 242
Harrison Alloys Inc.
Nickel.
Now HAI-NICR 80 CB.

DH NO. 243
Driver Harris Co.
Nickel. Cr 19.5, Si 1.45, Mn 1.8, C 0.03, bal Ni.
For furnace heating elements, furnace belts and conveyors. High heat and oxidation resistance. *Obsolete*

DH NO. 520
Harrison Alloys Inc.
Superalloy.
Now HAI-NICR 35 CB.

DH NO. 525
Driver Harris Co.
Superalloy. Ni 35, Cr 19.5, Si 1.45, C 0.05, bal Fe.
Rolled: 150,000-70,000 TS; 35 El. For high-temperature resistors. Heat resistant to 800°C. *Obsolete*

DH T2 (TYPE KN)
Harrison Alloys Inc.
Nickel.
Now HAI-KN.

DH-27
Chemetron Corp.
Mn 0.72, Si 0.27, Weld metal: 0.11 Cu, bal Fe.
As welded: 73,500 psi TS; 61,000 psi YS; 27 El. AC-DC electrode for repair of heavy equipment, ship building, pressure vessels. AWS Class E6027.

DHP COPPER
Criterion Metals, Inc.
Copper. Cu 99.99, P 0.015-0.04.
Thin gauge sheet, various tempers: 23-52 ksi TS min; 5-51 ksi YS min. C12200

DI HARD
Certified Alloy Products Inc.
C 3.5, Ni 4.2, B 1.1, bal Fe.
As deposited: 578 Brin. Fusable coating for lined cylinders; wear resistant. *Obsolete*

DI HW
Henckels Zwillingwerke, G.A.
C 1.3, Cr 0-0.2, W 4.75, Mn 0.3, bal Fe.
For bearings, liners, forming and blanking dies; water hardened.

DI R35
Henckels Zwillingwerke, G.A.
C 1.42, Mn 0.3, Si 0.25, W, V, bal Fe.
For cutters, bearings, forming dies; water hardened.

DI-ALLOY
S.I.P.I. Metals Corp.
Al, Cu, bal Zn.
For die castings.

DI-HARD
Now XALOY 100.

DI-IRON
Aluminium Union Ltd.
C 3.2, Si 2.2, Ni 1.5, Cr 0.8, Mo 0.15, bal Fe.
Cast: 50,000-60,000 TS; 230-290 Brin. For dies, gears, housings, shafts; cast iron; transverse strength 4100; transverse deflection 320.

DI-METAL NO. 2
American Smelting & Refining Co.
Al 3.5-4.5, Cu 2.5-3.5, Mg 0.02-0.1, Fe 0-1, bal Zn.
Die cast: 48,000-52,000 psi TS; 5-8 El; 100-105 Brin. For die castings; general purpose.

DI-METAL NO. 3
American Smelting & Refining Co.
Al 3.5-4.3, Cu 0.75-1.25, Mg 0.03-0.08, Fe 0-1, bal Zn.
Die cast: 40,000-41,000 psi TS; 5-10 El; 75-85 Brin. For die castings; general purpose.

DI-METAL NO. 5
American Smelting & Refining Co.
Al 3.5-4.3, Cu 0.75-1.25, Mg 0.03-0.08, Fe 0-1, bal Zn.
Die cast: 45,000-48,000 psi TS; 3-7 El; 95-100 Brin. For die castings; general purpose.

DI-MOL
Disston Inc.
C 0.8, Cr 4, V 1, W 1.5, Mo 9, bal Fe.

DIABOLIQUE SATAN NO. 2
Creusot-Loire
C 1.15, W 1.9, Cr 0.45, bal Fe.
For cutters, tools; fast finishing tool steel; 115WC19. *Obsolete*

DIACRO
Weatherly Casting & Machine Co.
C 1.4-1.8, Si 0-2, Ni 1.75-2.5, Cr 27-29, Mn 0-1.5, Mo 1-2.25, bal Fe.
Stress relieved: 90,000 TS; 380-450 Brin; 4 El. Heat treated: 110,000 TS; 500 Brin min; 2 El. Cast alloy for applications requiring abrasion and corrosion resistance.

DIALLIST
English manufacture
Ni-Al-Co-Fe.
For permanent magnets; high coercive force.

DIALLOY
Gorham Tool Industries Inc.
C, Mn, Cr, Ni, Mo, bal Fe.
For battering tools, punches; high impact and fatigue resistance.

DIALOX
Plansee, Metallwerk Gesellschaft
Al$_2$O$_3$.
For cutting tools and wear parts. Ceramic aluminum oxide.

DIALOY
J.M. Ney Co.
Bi 50, bal Pb, Sn, Cd.
For dental tooth impressions, dental bridge work, models; expands on solidifying. *Obsolete*

DIAMANT
Styria-Stahl Steirische Gusstahlwerke AG
C 1.3, W 4.5, bal Fe.
For tools, dies, cutters; fast finishing steel. *Obsolete*

DIAMANT 12
Plettenberger Gusstahlfabrik
C 2.1, Cr 11.5, bal Fe.
For blanking and forming dies; oil or water hardened.

DIAMANT 28
W. Ossenberg & Cie Edelstahlwerke
C 2.7, Mn 0.2, Si 0.3, Cr 26, bal Fe.
For drawing dies.

DIAMANT 3
Thyssen Edelstahlwerke AG
C 1.35, Cr 0.1, V 0.1, bal Fe.
For fluting tools; water or oil hardened. *Obsolete*

DIAMANT 44
Now VEW K400.

DIAMANT 5
Thyssen Edelstahlwerke AG
C 1.3, Cr 0.2, W 4.75, bal Fe.
For drawing dies; oil hardening. *Obsolete*

DIAMANT 5
Plettenberger Gusstahlfabrik
C 1.3, W 4.75, Cr 0-0.2, bal Fe.
For fast finishing cutters, engraver's tools; water hardened.

DIAMANT EXTRA
Thyssen Edelstahlwerke AG
C 1.5, W, Cr, V, bal Fe.
For fast finishing cutters, engravers tools; water hardened.
Obsolete

DIAMANT EXTRA
Thyssen Edelstahlwerke AG
C 0.7, W 18, Cr 4, V 1, bal Fe.
For gun rifling tools, engraving tools; high speed steel.
Obsolete

DIAMANT EXTRA DURO
Vereinigte Edelstahlwerke
C 1.3, W 4.5, bal Fe.
For tools, dies, cutters; fast finishing steel. *Obsolete*

DIAMANT HK
Haeckerstahl GmbH
C 1.42, W, V, bal Fe.
For fast finishing cutters, engravers' tools; water hardened.

DIAMANT HOCHHART
Vereinigte Edelstahlwerke
C 1.42, W, V, bal Fe.
For fast finishing cutters, engravers tools; water hardened.
Obsolete

DIAMANT R
Thyssen Edelstahlwerke AG
C 1.42, W, V, bal Fe.
For fast finishing cutters, engravers tools; water hardened.
Obsolete

DIAMANT R
Thyssen Edelstahlwerke AG
C 1.4, Cr 0.3, V 0.25, W 3, bal Fe.
For grooving tools; cold work steel. *Obsolete*

DIAMANT R6
Henckels Zwillingwerke, G.A.
C 1.42, W, V, bal Fe.
For fast finishing cutters, engravers' tools; water hardened.

DIAMANT SPECIAL
Styria-Stahl Steirische Gusstahlwerke AG
C 1.3, W 5, Cr 1.5, V 0.5, bal Fe.
For punches, dies, upsetters, crimpers. Nondeforming oil or
water hardening, shock and wear resistant. *Obsolete*

DIAMANT SPECIAL
Vereinigte Edelstahlwerke
C 1.3, W 5, Cr 1.5, V 0.5, bal Fe.
For punches, dies, upsetters, crimpers. Nondeforming oil or
water hardening, shock and wear resistant. *Obsolete*

DIAMANT WDSA
Now VEW K400.

DIAMANTBRONZE
Ostermann GmbH & Co.
Copper. Cu 58, Ni 1, Mn 1, Al 1, Fe 1, bal Zn.
20 El. For valve gauge fittings; corrosion resistant.

DIAMAX 19112
Eutectic Corp.
Alloy powder for metal spraying hard final coat. *Obsolete*

DIAMEND 27
Arcos Alloys
C 0.9-1.1, Mn 3.7-4.3, Si 0.5-1, Cr 11-13, bal Fe.
Welded: 260-300 Brin. For hardfacing electrodes; work
hardenable, for applications having severe impact. *Obsolete*

DIAMEND 305
Arcos Alloys
C 0.1, Mn 1.2, Si 0.6, Cr 2, Mo 0.5, bal Fe.
Welded: 37 Rock C. For hardfacing electrodes; pearlitic,
machinable steel for buildup and repair of worn machine
parts. *Obsolete*

DIAMEND 350
Arcos Alloys
C 0.15, Mn 1, Si 0.6, Cr 2, Mo 0.4, bal Fe.
Welded: 37 Rock C. For hardfacing electrodes; pearlitic,
machinable steel for repair of worn machine parts. *Obsolete*

DIAMEND 500
Arcos Alloys
C 0.4, Mn 1, Si 0.6, Cr 3.5, Mo 0.5, bal Fe.
Welded: 55 Rock C. For hardfacing electrodes; martensitic
steel to withstand severe abrasion, heavy impact. *Obsolete*

DIAMEND 600
Arcos Alloys
C 0.6, Mn 1, Si 0.6, Cr 4, Mo 0.5, bal Fe.
Welded: 60 Rock C. For hardfacing electrodes; martensitic
steel to withstand severe abrasion with mild impact. *Obsolete*

DIAMEND 605
Arcos Alloys
C 0.6, Mn 1, Si 0.9, Cr 4.5, V 0.7, bal Fe.
Welded: 57 Rock C. For hardfacing electrodes; martensitic
steel to resist severe abrasion especially on alloy steels.
Obsolete

DIAMEND 63
Arcos Alloys
C 0.9, Mn 1, Si 0.7, Cr 5, Mo 8.5, Co 2, V 1, W 2.5, bal Fe.
Welded: 60 Rock C. For hardfacing electrodes; high speed
tool steel resistant to abrasion, oxidation, light impact to
1000°F. *Obsolete*

DIAMEND B
Arcos Alloys
C 2, Mn 0.4, Si 0.5, Cr 4.5, bal Fe.
Welded: 55 Rock C. For hardfacing electrodes; martensitic
cast iron general purpose electrode. *Obsolete*

DIAMITE
Rennie Tool Co. Ltd.
WC.
For cutting tools; sintered carbide.

DIAMITE
Weatherly Casting & Machine Co.
Ni 3-5, Cr 1-3, Si 1.1-1.5, Mn 0.6-0.9, C 3-3.5, bal Fe.
Cast: 600-725 Brin. For liners for chutes, pulverizer hammers,
sand pump parts, welding rod for hard facing; Alloy white
iron; resistant to abrasion.

DIAMITE MO
Weatherly Casting & Machine Co.
Mo 0-3, Cr 15-18, C, bal Fe.
650 Brin min. For abrasion resistant castings. Abrasion
resistant.

DIAMOND
Spencer Clark Metal Industries Ltd.
Carbon tool steel plus vanadium.

DIAMOND 00
Saville & Co. Ltd., J.J.
C 1.6, Cr 0.6, V 1, W 5.5, bal Fe.
For finishing tools, form tools, rifling tools; for light cuts.

DIAMOND 00
Jessop-Saville Ltd.
C 1.6, Cr 0.6, V 1, W 5.5, bal Fe.
For finishing tools, form tools, rifling tools; for light cuts.

DIAMOND A
Midvale-Heppenstall Co.
C 2.2, V 0.4, W 11.5, bal Fe.
For dies; air hardening.

DIAMOND A.11
Burys & Co. Ltd.
C 1.7, W 7, bal Fe.
For turning and planing tools, cold drawing dies; fast
finishing steel, wear resistant.

DIAMOND ALLOY
Brown Alloy Works
Co 45, Mo 40, Cr 15.
For cutters, tools; wear and abrasion resistant.

DIAMOND B
Midvale-Heppenstall Co.
C 2, Cr 12, V 0.8, Mo 0.4, bal Fe.
For cold work tools, blanking and forming dies; abrasion
resistant, non-deforming, oil hardened.

DIAMOND BRAND
Midvale-Heppenstall Co.
C 1.7, Mn 0.45, Cr 18.57, V 0.55, bal Fe.
For tools, dies, valves, punches; rust and abrasion resisting.

DIAMOND BRONZE
American manufacture
Al 10, Si 2, bal Cu.
For pump parts, valves, marine parts, hardware; high
strength; corrosion resistant.

DIAMOND D-406
Diamond Metal Alloys
Sintered carbide tool material. Cutting tools for cast iron;
blast nozzles, valve trim.

DIAMOND D-406N
Diamond Metal Alloys
Sintered carbide tool material. Hard wear resistant material
for valve trim; corrosion resistant.

DIAMOND D-409
Diamond Metal Alloys
Sintered carbide tool material. Cutting tools for machining
cast iron.

DIAMOND D-409N
Diamond Metal Alloys
Sintered carbide tool material. For valve trim; corrosion
resistant and erosion resistant.

DIAMOND E
Edgcomb Steel Co.
C 0.6-1.4, bal Fe.
For punches, taps, springs, hobs, reamers; water hardened;
Type W1. *Obsolete*

DIAMOND G BRONZE
E.A. Williams & Sons
Pb, Sn, bal Cu.
For bearings, bushings; tough.

DIAMOND M
Teledyne Firth Sterling
Steel. C 1.35, Cr 0.3, bal Fe.
For form cutters, tools, dies; water hardened. *Obsolete*

DIAMOND NO. 1 B.16.D.
Burys & Co. Ltd.
C 1.25, W 3, V 0.15, bal Fe.
For drawing and burnishing dies; water hardened, wear
resistant.

DIAMOND S
Wallace Murray Corp.
P 0-0.025, S 0-0.02, C as desired, V optional, bal Fe.
Water hardening tool steel; AISI W1. *Obsolete*

DIAMOND S-1
Sivyer Steel Corp.
C 0.3, Cr 18, bal Fe.
For corrosion resistant parts; corrosion resistant. *Obsolete*

DIAMOND T
Sivyer Steel Corp.
C 0.3, Cr 12, Ni 25, bal Fe.
For heat resistant parts; heat resistant. *Obsolete*

DIAMONITE
John A. Crowley Inc.
W 95.65, C 3.91, bal Fe.
For cutting tools, dies; W_2C + WC.

DIAPHRAGM BRASS
American manufacture
Cu 95, Sn 3, Zn 2.
For diaphragms; corrosion resistant.

DIAWELD
Krupp Stahl AG
cast iron. C 4.2-4.4, Si 1.8, Mn 5-6, Cr 30, bal Fe.
For hard surfacing electrodes; corrosion resistant. *Obsolete*

DIBRONZE
Ampco Pittsburgh Corp.
Al, Fe, bal Cu.
For bending and forming dies, deep drawing dies; Al bronze, resists wear and galling. *Obsolete*

DICA (FLAME HARD) CAST TO SHAPE
Jessop Steel Co.
C 0.5, Mn 0.75, Si 0.25, Cr 1, Mo 0.2, bal Fe.
Hardness normally 250-300 Brin or can be annealed. For automobile bumper dies. Oil hardening tool steel.

DICA A
Jessop Steel Co.
C 0.42, Mn 0.75, Cr 1, Mo 0.2, bal Fe.
For die casting dies; oil hardening. *Obsolete*

DICA B
Jessop Steel Co.
C 0.33, Cr 4.8, Mn 0.4, Mo 1.5, Si 1, V 0.2, bal Fe.
For Al die casting dies, hot gripper and header dies; hot work steel, air or oil hardened. AISI H 12.

DICA B MOD
Jessop Steel Co.
C 0.37, Si 1, Cr 5, V 0.5, Mo 1.35, bal Fe.
For upsetters, punches, shears, extrusion tools; hot work steel, oil hardened. AISI H11.

DICA B TOOL ROOM
Jessop Steel Co.
C 0.42, Si 1, Cr 4.85, V 0.35, Mo 1.5, bal Fe.
For upsetters, punches, shears; hot work steel, oil hardened. *Obsolete*

DICA B-VANADIUM
Jessop Steel Co.
C 0.35, Cr 5, V 1, Mo 1.5, bal Fe.
For Al and Mg die casting dies; Type H13, oil hardened. AISI H 13.

DICA C
Jessop Steel Co.
C 0.3, Cr 3.5, V 0.4, Mo 0.25, W 10, bal Fe.
For bolt and rivet dies, hot piercing dies, forging mandrels; hot work steel. *Obsolete*

DICA D
Jessop Steel Co.
C, alloy, bal Fe.
For tools, dies, punches; oil hardened. *Obsolete*

DICAL
Die Castings Ltd.
Si 12, bal Cu.
For light alloy castings; die cast.

DICK'S BEARING BRONZE
English manufacture
Cu 80, Pb 10, Sn 9.2, P 0.8.
For bearings.

DICROME
Firth Brown Ltd.
C 0.6, Si 0.25, Mn 0.6, Cr 0.6, bal Fe.
For punches, heading dies, axes; Brit. EN11.

DIDBY'S CUPRITIC ALLOY
Manufacturer not listed
Cu 18, Cr 18, Fe 72.

DIDYMIUM METAL
Ronson Metals Corp.
Ne-Pr.
For alloying.

DIE BLANKING
Midvale-Heppenstall Co.
C 2.2, Cr 10, V 0.2, bal Fe.
For tools. *Obsolete*

DIE BLOCK "C"
Midvale-Heppenstall Co.
C 0.45, Ni 1.5, Cr 0.7, bal Fe.
For tools, dies; oil hardened. *Obsolete*

DIE BLOCK "E"
Midvale-Heppenstall Co.
C 0.85, W 1, bal Fe.
For tools, cutters; water hardened. *Obsolete*

DIE BLOCK (A)
Midvale-Heppenstall Co.
C 0.62, Mn 0.7, bal Fe.
For dies, tools; water hardened. *Obsolete*

DIE BLOCK (B)
Midvale-Heppenstall Co.
C 0.8, Mn 0.25, bal Fe.
For tools, dies; water hardened. *Obsolete*

DIE CAST
Boyd-Wagner Co.
C 0.45, Si 1, Cr 1.4, W 2.2, V 0.2, Mo 0.3, bal Fe.
For aluminum die casting dies; nonwarping.

DIE CAST NO. 1
Jamison Steel Corp.
C 0.35-0.4, Mo 1.3, V 0.5, Si 1, Cr 5.2, bal Fe.
For blanking and forging dies; hot work steel, tough and wear resistant.

DIE CASTING
CCS Braeburn Alloy Steel
C 0.45, Cr 2.5, Mn 0.6, V 0.25, Si 0.3, bal Fe.
For die casting dies; hot work steel. *Obsolete*

DIE CASTING ALLOY "A"
Manufacturer not listed
Sn 84-91, Sb 2-9, Cu 4.5-8.
For high grade bearings; anti-friction.

DIE CASTING ALLOY "B"
Manufacturer not listed
Sn 84-91, Sb 10-17, Cu 0-1.
For light duty bearings; anti-friction.

DIE CASTING ALLOY "C"
Manufacturer not listed
Sn 80, Pb 10, Sb 10.
For light duty bearings; anti-friction.

DIE CASTING ALLOY "D"
Manufacturer not listed
Sn 5-61, Pb 25, Sb 10.5, Cu 3.
For light duty bearings; anti-friction.

DIE CASTING ALLOY "E"
Manufacturer not listed
Zn 74, Sn 15, Al 6, Cu 5.
For soft work die castings; anti-friction.

DIE CASTING ALLOY "F"
Manufacturer not listed
Zn 85, Sn 8, Cu 4, Al 3.
For standard die castings; anti-friction.

DIE CASTING ALLOY "G"
Manufacturer not listed
Zn 75-85, Al 13.75, Cu 3.
For hard die castings.

DIE CASTING ALLOY "H"
Manufacturer not listed
Zn 46, Sn 31, Cu 20, Sb 3.
For hard die castings.

DIE CASTING ALLOY "I"
Manufacturer not listed
Zn 86.5, Cu 10.75, Al 2.75.
For die castings.

DIE CASTING ALLOY "J"
Manufacturer not listed
Zn 83-90, Cu 5-11, Al 2-5, Sn 1-5.
For die castings.

DIE CASTING ALLOY-1026
Anaconda Co.
Cu 81.5, Zn 14.1, Si 4.25, Mn 0.15.
Cast: 85,000 TS; 50,000 YS; 8 El. For die castings. *Obsolete*

DIE CASTING ALLOY-624
Anaconda Co.
Cu 60, Zn 37.65, Sn 1, Pb 1, Al 0.1.
Cast: 55,000 TS; 35,000 YS; 8 El. For die castings. *Obsolete*

DIE FLEX
Wallace Murray Corp.
C 0.4, Cr 1.5, Mo 0.8, Ni 4.25, bal Fe.
Hot work die steel; tough.

DIE L
Wallace Murray Corp.
C 1.45-1.55, Cr 12.5-14, Co 0.65-0.75, V 1.1, bal Fe.
For dies, broaches, reamers; nondeforming.

DIE STEEL NO. 13200
Wallace Murray Corp.
C 0.7, bal Fe.
For dies, tools, punches; water or oil hardening. *Obsolete*

DIE-MATE
U.S. Metalsource
C 0.2, Mn 0.8, P 0.015, S 0.015, Si 0.2, bal Fe.
Replacement for 1018 cold finished flats and squares for non-critical components of tooling and die sets.

DIECARB H6
Teledyne Firth Sterling
C 4.6-5.6, W 65-90, Co 5.7-6.3, Ta 0-5.
For blanking and forming dies; shock resistant, sintered.

DIECRAT
Marshall Steel Co.
C 0.75, Mn 2, Si 0.3, Cr 1, Mo 1.35, bal Fe.
Air hardening cold work tool steel; AISI A6.

DIEHARD BB
Firth Brown Ltd.
C 1.5, Si 0.3, Mn 0.3, Cr 12, V 1.15, Co 1.3, bal Fe.
For dies; nondeforming. *Obsolete*

DIEHARD HCD
Firth Brown Ltd.
C 2, Si 0.3, Mn 0.3, Cr 12.9, bal Fe.
For dies; nondeforming. *Obsolete*

DIEHARD LC
Firth Brown Ltd.
C 0.85, Si 0.3, Mn 0.3, Cr 12, V 1.15, Co 1.3, bal Fe.
For dies; nondeforming. *Obsolete*

DIEHL GR. A
Diehl Steel Co.
C 1, Mn 0.3, Si 0.25, bal Fe.
For drills, taps, reamers, hobs, broaches, cutters; Type W1; water hardened.

DIEMAC
Macauley Foundry Co.
C 3.1-3.3, Si 1.7-1.9, Mn 0.6-0.8, Mo 0.8-1, Ni 1-1.2, P 0-0.05, bal Fe.
Close grained electric furnace cast iron for use in permanent molds. 180-200 Brin in heavy sections.

DIEMAC
Macauley Foundry Co.
C 3.2, Si 2.6, Ni 1, Mo 0.6, S 0.1, P 0.1, bal Fe.
For casting dies, and permanent molds for Al, Zn, Pb, alloys and plastic casting by gravity or injection molding. *Obsolete*

DIENETT GERMAN SILVER
English manufacture
Cu 51, Zn 32, Pb 9.5, Ni 6.4, Sn 1.6.
For ornamental, hardware novelties; free-cutting.

DIESEL BEARINGS
English manufacture
Sn 80, Sb 15, Cu 5.
For bearings for Diesel engines; Babbitt.

DIESEL OIL ENGINE BABBITT
NL Industries
Sb 9-11, Cu 4-6, bal Sn.
4-34 Brin. For bearings for diesel engines; resists impact loading. *Obsolete*

DIETEMPER-1
Gulf Steel Corp.
C 0.4, Cr 1.5, V 0.25, Mo 1, Mn 0.65, bal Fe.
For forging and die casting dies; hot work steel, oil hardened.

DIETEMPER-2
Gulf Steel Corp.
C 0.3, Cr 2.75, Mo 0.3, W 9.5, Ni 1.6, bal Fe.
For extrusion dies, rams, punches; hot work steel, oil hardened.

DIEWEAR
Gorham Tool Industries Inc.
C 0.9, Mn 1, Cr 0.5, W 0.5, bal Fe.
For dies, punches, blanking and forming dies; Type O1; oil hardened, non-deforming.

DIEWELD
Edgcomb Metals Co.
C 0.7, W 18, Cr 4, V 1, bal Fe.
For high speed tool steel welding electrodes; high speed steel. *Obsolete*

DILATHERM 15
Vereinigte Edelstahlwerke
C 0.1, Ni 36, bal Fe.
For minimum expansion; low coefficient of expansion. *Obsolete*

DILATHERM 85
Vereinigte Edelstahlwerke

For heat conductors. *Obsolete*

DILATHERM 90
Vereinigte Edelstahlwerke
C 0.1, Si 0.1, Cr 25, bal Fe.
For glass to metal seals; low coefficient of expansion. *Obsolete*

DILATON 206
Vereinigte Deutsche Nickel-Werke AG
Alloy steel. C 0.05, Mn 6.5, Si 0.2.
Strip, wire, bar, forged and turned parts for controls, drawing and stamping parts.

DILATON 2821
Vereinigte Deutsche Nickel-Werke AG
Alloy steel. C 0.03, Co 21, Mn 0.5, Si 0.1, 28 Ni + Co, bal Fe.
Wire material for electronics, glass-to-metal sealing, controls, drawing and stamping parts.

DILATON 36
Vereinigte Deutsche Nickel-Werke AG
C 0.05, Mn 0.4, Si 0.2, 36 Ni + Co, bal Fe.
Strip, wire, bars, forged and turned parts for electronics, fasteners, drawing and stamping, glass-to-metal sealing, controls and apparatus.

DILATON 41 TI
Vereinigte Deutsche Nickel-Werke AG
Alloy steel. C 0.02, Mn 0.05, Si 0.05, Ti 0.35, 41 Ni + Co, bal Fe.
Strip material for electronics, glass-to-metal sealing, controls, drawing and stamping parts.

DILATON 42
Vereinigte Deutsche Nickel-Werke AG
C 0.05, Mn 0.9, Si 0.15, 42 Ni + Co, bal Fe.
Strip, wire, bars, forged and turned parts for electronics, glass-to-metal sealing, controls, apparatus, drawing and stamping parts.

DILATON 47
Vereinigte Deutsche Nickel-Werke AG
Alloy steel. C 0.02, Mn 0.35, Si 0.1, 47 Ni + Co, bal Fe.
Strip, wire, bar, forged and turned parts for electronics, glass-to-metal sealing, controls, drawing and stamping parts.

DILATON 50
Vereinigte Deutsche Nickel-Werke AG
C 0.05, Mn 0.6, Si 0.15, 50 Ni + Co, bal Fe.
Strip, wire, bars, forged and turned parts for electronics, glass-to-metal sealing, controls, drawing and stamping parts.

DILATON 51
Vereinigte Deutsche Nickel-Werke AG
Alloy steel. C 0.05, Mn 0.6, Si 0.15, 51 Ni + Co, bal Fe.
Strip, wire, bar, forged and turned parts for electronics, glass-to-metal sealing, controls, drawing and stamping parts.

DILATON 60
Vereinigte Deutsche Nickel-Werke AG
Ni 60, bal Fe.
For glass-to-metal seals; controlled expansion. *Obsolete*

DILPHY
Creusot-Loire
Ni, bal Fe.
For thermostats; controlled coefficient of expansion. *Obsolete*

DILVER
Creusot-Loire
Ni 42-46, Fe 54-58.
For heating elements, leads in electric light bulbs; similar to "Platinite." *Obsolete*

DILVER O
Creusot-Loire
Cr 25, bal Fe.
For glass sealing and for high temperature applications; corrosion and heat resistant. *Obsolete*

DILVER P
Creusot-Loire
Ni 29, Co 17, bal Fe.
For glass sealing and for high temperature applications; corrosion and heat resistant. *Obsolete*

DILVER PL
Creusot-Loire
Nickel alloy.
For electronic equipment; glass sealing alloy. *Obsolete*

DILVER T
Creusot-Loire
Nickel alloy.
For electronic equipment; glass sealing alloy. *Obsolete*

DIMONDITE
Teledyne Firth Sterling
W 95.65, C 3.91, bal Co or 87-95% tungsten carbide, bal Co.
To tip tools and cutters for high speed cutting; sintered alloy, hard as diamond. *Obsolete*

DIMPALLOY
English manufacture
Zn, bal Cu.
For brazing of brasses and bronzes.

DIN ARGO-FLO
Johnson Matthey plc
Ag 40, Cu 19, Zn 21, Cd 20.
Cadmium bearing silver brazing alloy; 595-630°C MP; DIN 8513 L-Ag40Cd.

DINERTIA
Deloro Stellite
Reactive & refractory. Co, Cr, Mo, V.
Cast: 90,000 psi TS; 50,000 psi YS; 2.5 El; 200 Brin. Rolled: 190,000 psi TS; 175,000 psi YS; 9 El; 330 Brin. For dental and surgical tools; corrosion resistant.

DIP METAL
Cu 70, Sn 0.5, Pb 0.5, Ni 0.5, bal Zn.
Rolled: 35,000 TS; 50 El. For hardware; free-cutting.

DIPPING BRASS
American manufacture
Cu 67, Zn 33.
For deep drawn, spun or stamped parts; high ductility.

DIPPIT
United American Metals Corp.
Pb 94.5, Sn 5, Cu 0.5.
Babbitt metal. *Obsolete*

DIRIGOLD
Dirigold Corp.
Al 9-10, Ni 1.1-2, Sn 0-1.7, bal Cu.
For strong corrosion resisting parts, ornamental castings. Corrosion resistant.

DIRILYTE METAL
American Art Metals Inc.
Al 5, bal Cu.
For tableware; color of gold Al bronze.

DIRO 703
Karl Diederichs Stahl-Walz-Hammerwerk
C 0.45, Mn 0.7, Si 0.3, Cr 1.5, V 0.3, Mo 0.7, bal Fe.
Hot work tool steel; oil hardening. For pressure punches. W-Nr. 1.2323; 48 Cr Mo V 6.7.

DIRO B 30
Karl Diederichs Stahl-Walz-Hammerwerk
C 0.4, Mn 0.3, Si 0.4, Cr 13, bal Fe.
Hot or cold work tool steel; air or oil hardening; corrosion resistant. For tools, plastics and light metals. W.Nr. 1.2083; X 42 Cr 13.

DIRO C 10
Karl Diederichs Stahl-Walz-Hammerwerk
C 0.1, Mn 0.4, Si 0-0.25, bal Fe.
Low carbon structural steel. W.Nr. 1.0301. Formerly ST.C. 10.61.

DIRO CV 5
Karl Diederichs Stahl-Walz-Hammerwerk
C 0.5, Mn 1, Si 0.25, Cr 1, V 0.1, bal Fe.
Cold work tool steel; oil hardening. For hand tools, screw drivers, shears, shanks for carbide tools. W.Nr. 1.2241; 51 Cr V 4.

DIRO CV 6

Karl Diederichs Stahl-Walz-Hammerwerk
C 0.58, Mn 1, Si 0.25, Cr 1, V 0.1, bal Fe.
Cold work tool steel; oil hardening. For punches, chisels, shear blades. W.Nr. 1.2242; DIN 59 Cr V 4.

DIRO CW

Karl Diederichs Stahl-Walz-Hammerwerk
C 1.05, Mn 1, Si 0.25, Cr 1, W 1.2, bal Fe.
Cold work tool steel; oil hardening. For shears, broaches, reamers, end mills. W.Nr. 1.2419.

DIRO EC 30

Karl Diederichs Stahl-Walz-Hammerwerk
C 0.13, Mn 0.5, Si 0.25, Cr 0.4, bal Fe.
Low alloy carburizing steel. W.Nr. 1.7012.

DIRO EXTRA 1

Karl Diederichs Stahl-Walz-Hammerwerk
C 1.3, Mn 0.25, Si 0.2, P 0.025, S 0.025, bal Fe.
Carbon tool steel; water hardening. For punches, drills, lathe tools. W.Nr. 1.1560; DIN C 125 W 1.

DIRO EXTRA 2

Karl Diederichs Stahl-Walz-Hammerwerk
C 1.1, Mn 0.25, Si 0.2, P 0.025, S 0.025, bal Fe.
Carbon tool steel; water hardening. For hand and pneumatic chisels, drills. W.Nr. 1.1550; similar to AISI W1.

DIRO EXTRA 3

Karl Diederichs Stahl-Walz-Hammerwerk
C 1, Mn 0.25, Si 0.2, P 0.025, S 0.025, bal Fe.
Carbon tool steel; water hardening. For hand chisels, shears, punches, hand dies. W.Nr. 1.1540; DIN C 100 W1.

DIRO EXTRA 4

Karl Diederichs Stahl-Walz-Hammerwerk
C 0.85, Mn 0.3, Si 0.2, P 0.025, S 0.025, bal Fe.
Carbon tool steel; water hardening. For hand tools, rock drills, hammers. W.Nr. 1.1530; DIN C 85 W1.

DIRO EXTRA 5

Karl Diederichs Stahl-Walz-Hammerwerk
C 0.7, Mn 0.3, Si 0.2, P 0.025, S 0.025, bal Fe.
Carbon tool steel; water hardening. For hand tools, wood working tools. W.Nr. 1.1520; DIN C 70 W1.

DIRO FORTIS

Karl Diederichs Stahl-Walz-Hammerwerk
C 2, Mn 0.3, Si 0.3, Cr 12, bal Fe.
Hot or cold work tool steel; oil hardening. For hot pressure casting molds; heavy duty cold punching dies and shear blades. W.Nr. 1.2080; DIN X 210 Cr 12.

DIRO FORTIS EXTRA

Karl Diederichs Stahl-Walz-Hammerwerk
C 1.65, Mn 0.3, Si 0.3, Cr 12, V 0.1, bal Fe.
Cold work tool steel; air or oil hardening. For coining dies, broaches, threading tools. W.Nr. 1.2201; X 165 Cr V 12.

DIRO FORTIS PRIMA

Karl Diederichs Stahl-Walz-Hammerwerk
C 1.65, Mn 0.3, Si 0.3, Cr 12, W 0.5, V 0.1, Mo 0.6, bal Fe.
Cold work tool steel; air or oil hardening. For heavy duty stamping, drawing and trimming dies; coining dies. W.Nr. 1.2601; X Cr Mo V 12.

DIRO FORTIS SPEZIAL

Karl Diederichs Stahl-Walz-Hammerwerk
C 2.1, Mn 0.3, Si 0.3, Cr 12, W 0.7, bal Fe.
Cold work tool steel; oil or air hardening. For heavy duty punching and trimming dies. W.Nr. 1.2436; X 210 Cr W 12.

DIRO HS 4

Karl Diederichs Stahl-Walz-Hammerwerk
C 0.8, Cr 4, W 9, V 0.85, Mo 1.6, bal Fe.
High speed tool steel. For milling cutters, drills, taps. W.Nr. 1.3316.

DIRO KH 20

Karl Diederichs Stahl-Walz-Hammerwerk
C 0.2, Mn 1.2, Si 0.3, Cr 1, bal Fe.
To be carburized and oil hardened for dies and molds for artificial resin molding press. W.Nr. 1.2162; DIN 21 Cr Mn 5.

DIRO KLS

Karl Diederichs Stahl-Walz-Hammerwerk
C 1, Mn 0.3, Si 0.25, Cr 1.5, bal Fe.
Cold work tool steel; oil hardening. For lathe centers, taps, threading tools. W.Nr. 1.2067; DIN 100 Cr 6.

DIRO KONSTANT

Karl Diederichs Stahl-Walz-Hammerwerk
C 1.05, Mn 0.25, Si 0.3, Cr 1.4, bal Fe.
Cold work tool steel; oil hardening. For gages, small dies and stamps. W.Nr. 1.2060; DIN 100 Cr 5.

DIRO KONSTANT 15

Karl Diederichs Stahl-Walz-Hammerwerk
C 1.45, Mn 0.6, Si 0.2, Cr 1.4, bal Fe.
Cold work tool steel; oil or water hardening. For taps, broaches, milling cutters, counter bores. W.Nr. 1.2063; 145 Cr 6.

DIRO KONSTANT M

Karl Diederichs Stahl-Walz-Hammerwerk
C 1, Mn 0.7, Si 0.25, Cr 1, bal Fe.
Cold work tool steel; oil hardening. For short run cutting and stamping tools and dies. W.Nr. 1.2061; DIN 100 Cr 4.

DIRO KONSTANT S

Karl Diederichs Stahl-Walz-Hammerwerk
C 1.25, Mn 0.7, Si 1.2, Cr 1.2, bal Fe.
Cold work tool steel; oil hardening. For drills, punches, end mills, broaches. W.Nr. 1.2109; DIN 125 Cr Si 5.

DIRO L 18

Karl Diederichs Stahl-Walz-Hammerwerk
C 0.9, Mn 2, Si 0.2, V 0.1, bal Fe.
Cold work tool steel; oil hardening. For small cutting and punching tools, taps. W.Nr. 1.2842; DIN 90 Mn V 8.

DIRO MS 53 G

Karl Diederichs Stahl-Walz-Hammerwerk
C 0.53, Mn 1, Si 1, bal Fe.
Hot or cold work tool steel; oil hardening. For forging or trimming dies; short runs. W.Nr. 1.2826; DIN 60 Mn Si 4.

DIRO MV

Karl Diederichs Stahl-Walz-Hammerwerk
C 0.55, Mn 0.7, Si 0.3, Cr 1, Ni 1.7, V 0.1, Mo 0.5, bal Fe.
Hot or cold work tool steel; oil or air hardening. For forging dies, molding dies. W.Nr. 1.2714; DIN 56 Ni Cr Mo V 7.

DIRO MW

Karl Diederichs Stahl-Walz-Hammerwerk
C 0.55, Mn 0.6, Si 0.3, Cr 0.7, Ni 1.7, V 0.1, Mo 0.2, bal Fe.
Hot or cold work tool steel; oil hardening. Forging and extrusion dies. W.Nr. 1.2713; DIN 55 Ni Cr Mo V 6.

DIRO NC 3

Karl Diederichs Stahl-Walz-Hammerwerk
C 0.15, Mn 0.4, Si 0.3, Cr 0.7, Ni 3.5, bal Fe.
To be carburized and hardened for dies for artificial resins. W.Nr. 1.2735; 15 Ni Cr 14.

DIRO NC 4

Karl Diederichs Stahl-Walz-Hammerwerk
C 0.15, Mn 0.4, Si 0.3, Cr 1, Ni 4.5, bal Fe.
To be carburized and hardened for dies for artificial resins. W.Nr. 1.2745; 14 Ni Cr 18.

DIRO PRIMA 00

Karl Diederichs Stahl-Walz-Hammerwerk
C 0.55, Mn 0-0.4, Si 0.15, P 0.03, S 0.03, bal Fe.
Water hardening carbon tool steel. For anvils, hammers, knives, axes, and shears. W.Nr. 1.1820; DIN C 55 W2.

DIRO PRIMA 1

Karl Diederichs Stahl-Walz-Hammerwerk
C 1.3, Mn 0.3, Si 0.25, P 0.03, S 0.03, bal Fe.
Water hardening; carbon tool steel. For hand punches, shears, drills. W.Nr. 1.1660; DIN C 125 W 2.

DIRO PRIMA 2

Karl Diederichs Stahl-Walz-Hammerwerk
C 1.15, Mn 0.3, Si 0.25, P 0.03, S 0.03, bal Fe.
Water hardening; carbon tool steel. For hand and pneumatic chisels, punches, threading tools. W.Nr. 1.1650; DIN C 110 W2.

DIRO PRIMA 3

Karl Diederichs Stahl-Walz-Hammerwerk
C 1, Mn 0.3, Si 0.25, P 0.03, S 0.03, bal Fe.
Water hardening; carbon tool steel. For hand tools, stamping dies, drills. W.Nr. 1.1640; DIN C 100 W2.

DIRO PRIMA 4

Karl Diederichs Stahl-Walz-Hammerwerk
C 0.85, Mn 0.3, Si 0.25, P 0.03, S 0.03, bal Fe.
Water hardening; carbon tool steel. For hand tools, woodworking tools. W.Nr. 1.1630; DIN C 85 W 2.

DIRO PRIMA 5

Karl Diederichs Stahl-Walz-Hammerwerk
C 0.7, Mn 0.3, Si 0.25, P 0.025, S 0.025, bal Fe.
Water hardening; carbon tool steel. For shears, hand tools, woodworking tools. W.Nr. 1.1620; DIN C 70 W 2.

DIRO PS

Karl Diederichs Stahl-Walz-Hammerwerk
C 0.5, Mn 0.5, Si 0.25, Cr 1, Ni 3.5, bal Fe.
Cold work tool steel; air or oil hardening. For stamping dies, cold forming dies, shear blades. W.Nr. 1.2721; 50 Ni Cr 13.

DIRO SC

Karl Diederichs Stahl-Walz-Hammerwerk
C 0.67, Mn 0.5, Si 1.3, Cr 0.5, bal Fe.
Hot or cold work tool steel; oil hardening. For hand tools, shanks, pressure plates. W.Nr. 1.2101; DIN 62 Si Mn Cr 4.

DIRO SD

Karl Diederichs Stahl-Walz-Hammerwerk
C 0.35, Mn 0.3, Si 1, Cr 1, W 2, V 0.2, bal Fe.
Hot work tool steel; water hardening; for chisels, shears. W.Nr. 1.2541; 35 W Cr V 7.

DIRO SD 45

Karl Diederichs Stahl-Walz-Hammerwerk
C 0.45, Mn 0.3, Si 1, Cr 1, W 2, V 0.2, bal Fe.
Hot work tool steel; oil or water hardening. For hot cutting dies and punches, trimming tools. W.Nr. 1.2542.

DIRO SD 50

Karl Diederichs Stahl-Walz-Hammerwerk
C 0.55, Mn 0.3, Si 1, Cr 1, W 2, V 0.2, bal Fe.
Hot or cold work tool steel; oil hardening. For punches, shear blades, chisels. W.Nr. 1.2550; 60 W Cr V 7.

DIRO SEB

Karl Diederichs Stahl-Walz-Hammerwerk
C 0.95, Cr 4, W 3, V 2.8, Mo 2.5, bal Fe.
High speed tool steel. For twist drills, milling cutters, broaches. W.Nr. 1.3333.

DIRO SS

Karl Diederichs Stahl-Walz-Hammerwerk
C 0.8, Cr 4.5, W 18, V 1.7, Mo 1, Co 10, bal Fe.
Co-W high speed steel. For lathe and planer tools. W.Nr. 1.3265; DIN S 18-1-2-10.

DIRO SS 1

Karl Diederichs Stahl-Walz-Hammerwerk
C 1.4, Cr 4.5, W 12, V 4, Mo 1, Co 5, bal Fe.
W-Co-V high speed steel. For lathe tools. W.Nr. 1.3202; similar to AISI T15.

DIRO SS 183
Karl Diederichs Stahl-Walz-Hammerwerk
C 0.9, Cr 4, W 12, V 0.85, Mo 2.5, bal Fe.
High speed steel. For lathe tools, milling cutters. W.Nr. 1.3318.

DIRO SS 2
Karl Diederichs Stahl-Walz-Hammerwerk
C 0.75, Cr 4, W 18, V 1, bal Fe.
Tungsten high speed steel. For drills, lathe tools, milling cutters, taps. W.Nr. 1.3355; AISI T1.

DIRO SS 3
Karl Diederichs Stahl-Walz-Hammerwerk
C 0.8, Cr 4.5, W 12, V 2, Mo 1, Co 3, bal Fe.
W-Co high speed steel. For lathe tools, milling cutters, taps. W.Nr. 1.3211; DIN S 12-1-2-3.

DIRO SS 4
Karl Diederichs Stahl-Walz-Hammerwerk
C 0.8, Cr 4.5, W 12, V 1, Mo 1.7, Co 5, bal Fe.
Co-W high speed steel. For lathe and planing tools. W.Nr. 1.3251.

DIRO SS 45
Karl Diederichs Stahl-Walz-Hammerwerk
C 1.3, Cr 4.5, W 12, V 4, Mo 1, bal Fe.
High speed steel. For machining aluminum, bakelite, other plastics. W.Nr. 1.3302.

DIRO SS 5
Karl Diederichs Stahl-Walz-Hammerwerk
C 0.8, Cr 4.5, W 18, V 1.7, Mo 1, Co 5, bal Fe.
Co-W high speed steel; for lathe tools. W.Nr. 1.3255; DIN S 18-1-2-5.

DIRO SWP
Karl Diederichs Stahl-Walz-Hammerwerk
C 0.3, Mn 0.3, Si 0.2, Cr 2.5, W 9, V 0.4, bal Fe.
Hot work tool steel; oil or air hardening. For pressing dies, pressure castings, molds. W.Nr. 1.2581; X 30 W Cr V 9.3.

DIRO WM 4
Karl Diederichs Stahl-Walz-Hammerwerk
C 0.3, Mn 0.4, Si 1, Cr 1, W 4, V 0.2, bal Fe.
Hot work tool steel; oil or water hardening. Pressing mandrels for non-ferrous alloys. W.Nr. 1.2564; DIN X 30 W Cr V 4.1.

DIRO WM 5
Karl Diederichs Stahl-Walz-Hammerwerk
C 0.3, Mn 0.3, Si 0.2, Cr 2.5, W 4.5, V 0.6, bal Fe.
Hot work tool steel; oil or air hardening. For pressure casting molds, forming dies. W.Nr. 1.2567; DIN X 30 W Cr V 5.3.

DIRO WS 15
Karl Diederichs Stahl-Walz-Hammerwerk
C 0.15, Mn 0.3, Si 0.25, P 0.035, S 0.035, bal Fe.
To be carburized and water hardened. For punches, lathe jaws, molding presses. W.Nr. 1.1805; DIN C WS.

DIRO WS 35
Karl Diederichs Stahl-Walz-Hammerwerk
C 0.35, Mn 0.5, Si 0.3, P 0.035, S 0.0355, bal Fe.
Water hardening; carbon tool steel. For wrenches, tongs, pliers. W.Nr. 1.1720; DIN C 35 W 3.

DIRO WS 45
Karl Diederichs Stahl-Walz-Hammerwerk
C 0.45, Mn 0.7, Si 0.3, P 0.035, S 0.035, bal Fe.
Water or oil hardening; carbon tool steel. For hammers, axes, screw drivers, wrenches. W.Nr. 1.1730; DIN C 45 W 3.

DIRO WS 60
Karl Diederichs Stahl-Walz-Hammerwerk
C 0.6, Mn 0.7, Si 0.35, P 0.035, S 0.035, bal Fe.
Water or oil hardening; tool steel. For lathe tool shanks, hammers, hand tools. W.Nr. 1.1740; DIN C 60 W 3.

DIRO WS 67
Karl Diederichs Stahl-Walz-Hammerwerk
C 0.67, Mn 0.7, Si 0.35, P 0.035, S 0.035, bal Fe.
Oil hardening; carbon tool steel. For hand saws, woodworking tools. W.Nr. 1.1744.

DIRO-10 CRMO 9.10
Dirostahl Karl Diedricks
Alloy steel. C 0.1, Si 0.35, Mn 0.5, P 0-0.035, S 0-0.035, Cr 2.25, Mo 1, bal Fe.
Equivalent to Fr. 10 CD 9.10; Sw. SS 14 2218; GBr. 1501-622; SA 182 Gr.F22; 212 Brin max.

DIRO-100 CR 6
Dirostahl Karl Diedricks
Alloy steel. C 1, Si 0.25, Mn 0.35, P 0-0.03, S 0-0.025, Cr 1.5, bal Fe.
Equivalent to Fr. 100 C 6; Sw. SS 14 2258; GBr. 970-534 A 99; SAE 52100; 207 Brin max.

DIRO-100 CRMN 6
Dirostahl Karl Diedricks
Alloy steel. C 1, Si 0.25, Mn 1.1, P 0-0.03, S 0-0.025, Cr 1.5, bal Fe.
Equivalent to ASTM A 485 Gr.1; 217 Brin max.

DIRO-100 CRMO 73
Dirostahl Karl Diedricks
Alloy steel. C 1, Si 0.25, Mn 0.7, P 0-0.03, S 0-0.025, Cr 1.8, Mo 0.3, bal Fe.
Equivalent to Fr. 100 CD 7; ASTM A485 Gr.3; 217 Brin max.

DIRO-13 CRMO 4.4
Dirostahl Karl Diedricks
Alloy steel. C 0.13, Si 0.25, Mn 0.6, P 0-0.035, S 0-0.035, Cr 1, Mo 0.3-0.45, bal Fe.
Equivalent to Fr. 15 CD 4.05; Sw. SS 14 2216; GBr. 1501-620; SA 182 Gr.F11; 200 Brin max.

DIRO-14 CRMOV 6.9
Dirostahl Karl Diedricks
Alloy steel. C 0.15, Si 0.15, Mn 0.9, P 0-0.03, S 0-0.035, Cr 1.35, Mo 0.9, V 0.25, bal Fe.
170 Brin max.

DIRO-14 NICR 14
Dirostahl Karl Diedricks
Alloy steel. C 0.14, Si 0.25, Mn 0.4, P 0-0.035, S 0-0.035, Cr 0.75, Ni 3.5, bal Fe.
Equivalent to Fr. 14 NC 11; Sw. SS 14 2514; GBr. 655 A 12; SAE 3310; 220 Brin max.

DIRO-15 CRNI 6
Dirostahl Karl Diedricks
Alloy steel. C 0.15, Si 0.25, Mn 0.5, P 0-0.035, S 0-0.035, Cr 1.55, Ni 1.55, bal Fe.
Equivalent to Fr. Z 16 NC 6; Sw. SS 14 2511; GBr. 637 A 16; SAE 3215; 217 Brin max.

DIRO-15 MO 3
Dirostahl Karl Diedricks
Alloy steel. C 0.15, Si 0.25, Mn 0.65, P 0-0.035, S 0-0.035, Mo 0.3, bal Fe.
Equivalent to Fr. 15 D3; Sw. SS 14 2912; GBr. 1504-240; SA 204 Gr.A; 190 Brin max.

DIRO-16 MNCR 5
Dirostahl Karl Diedricks
Alloy steel. C 0.16, Si 0.25, Mn 1.15, P 0-0.035, S 0-0.035, Cr 0.95, bal Fe.
Equivalent to Fr. 16 MC 5; 207 Brin max.

DIRO-17 CRNIMO 6
Dirostahl Karl Diedricks
Alloy steel. C 0.17, Si 0.25, Mn 0.5, P 0-0.035, S 0-0.035, Cr 1.65, Mo 0.3, Ni 1.55, bal Fe.
Equivalent to Fr. 18 NCD 6; GBr. 820 A 16; 822 A 17; 229 Brin max.

DIRO-18 CRNI 8
Dirostahl Karl Diedricks
Alloy steel. C 0.18, Si 0.25, Mn 0.5, P 0-0.035, S 0-0.035, Cr 1.95, Ni 1.95, bal Fe.
Equivalent to SAE 3316; 235 Brin max.

DIRO-19 MN 5
Dirostahl Karl Diedricks
Carbon steel. C 0.19, Si 0.4, Mn 1.2, P 0-0.05, S 0-0.05, bal Fe.
Equivalent to Fr. A 52 C1; Sw. SS 14 2101; GBr. 1503-221; SA 350 LF2; 190 Brin max.

DIRO-20 MN 5
Dirostahl Karl Diedricks
Carbon steel. C 0.2, Si 0.4, Mn 1.2, P 0-0.035, S 0-0.035, bal Fe.
Equivalent to Sw. SS 14 2132; GBr. 970-120M 19; SAE 1518; 200 Brin max.

DIRO-20 MNCR 5
Dirostahl Karl Diedricks
Alloy steel. C 0.2, Si 0.25, Mn 1.25, P 0-0.035, S 0-0.035, Cr 1.15, bal Fe.
Equivalent to Fr. 20 MC 5; 217 Brin max.

DIRO-24 CRMO 5
Dirostahl Karl Diedricks
Alloy steel. C 0.24, Si 0.25, Mn 0.6, P 0-0.035, S 0-0.035, Cr 1.1, Mo 0.25, bal Fe.
Equivalent to Sw. SS 14 2225; 212 Brin max.

DIRO-25 CRMO 4
Dirostahl Karl Diedricks
Alloy steel. C 0.25, Si 0.25, Mn 0.65, P 0-0.035, S 0-0.035, Cr 1.05, Mo 0.25, bal Fe.
Equivalent to Fr. 25 CD 4; Sw. SS 14 2225; SAE 4130; 212 Brin max.

DIRO-26 CRMO 4
Dirostahl Karl Diedricks
Alloy steel. C 0.26, Si 0.25, Mn 0.65, P 0-0.035, S 0-0.035, Cr 1.05, Mo 0.025, bal Fe.
Equivalent to Sw. SS 14 2225; 212 Brin max.

DIRO-28 NICRMOV 10
Dirostahl Karl Diedricks
Alloy steel. C 0.28, Si 0.4, Mn 0.3, P 0-0.035, S 0-0.035, Cr 0.7, Mo 0.6, Ni 2.5, V 0.3, bal Fe.
240 Brin max.

DIRO-30 CRMOV 9
Dirostahl Karl Diedricks
Alloy steel. C 0.3, Si 0.25, Mn 0.55, P 0-0.035, S 0-0.035, Cr 2.5, Mo 0.2, V 0.15, bal Fe.
248 Brin max.

DIRO-30 CRNIMO 8
Dirostahl Karl Diedricks
Alloy steel. C 0.3, Si 0.25, Mn 0.45, P 0-0.035, S 0-0.035, Cr 2, Mo 0.4, Ni 2, bal Fe.
Equivalent to Fr. 30 CND 8; GBr. 970-823 M 30; 248 Brin max.

DIRO-30 MN 5
Dirostahl Karl Diedricks
Carbon steel. C 0.3, Si 0.25, Mn 1.3, P 0-0.035, S 0-0.035, bal Fe.
Equivalent to GBr. 970-120M 36; SAE 1536; 217 Brin max.

DIRO-31 CRMO 12
Dirostahl Karl Diedricks
Alloy steel. C 0.32, Si 0.25, Mn 0.55, P 0-0.03, S 0-0.035, Cr 3.05, Mo 0.4, bal Fe.
Equivalent to Fr. 30 CD 12; Sw. SS 14 2240; GBr. 970-722 M 24; 248 Brin max.

DIRO-31 CRMOV 9
Dirostahl Karl Diedricks
Alloy steel. C 0.3, Si 0.25, Mn 0.55, P 0-0.03, S 0-0.035, Cr 2.5, Mo 0.2, V 0.15, bal Fe.
Equivalent to GBr. 970-905 M 39; 248 Brin max.

DIRO-32 CRMO 12
Dirostahl Karl Diedricks
Alloy steel. C 0.32, Si 0.25, Mn 0.55, P 0-0.035, S 0-0.035, Cr 3, Mo 0.4, bal Fe.
Equivalent to Fr. 30 CD 12; Sw. SS 14 2240; GBr. 970-722 M 24; 248 Brin max.

DIRO-34 CRALNI 7
Dirostahl Karl Diedricks
Alloy steel. C 0.34, Si 0.25, Mn 0.55, P 0-0.03, S 0-0.035, Al 1, Cr 1.65, Mo 0.2, Ni 1, bal Fe.
Equivalent to GBr. 970-905 M 31; 248 Brin max.

DIRO-34 CRMO 4
Dirostahl Karl Diedricks
Alloy steel. C 0.34, Si 0.25, Mn 0.65, P 0-0.035, S 0-0.035, Cr 1.05, Mo 0.25, bal Fe.
Equivalent to Fr. 35 CD 4; Sw. SS 14 2234; GBr. 970-708 A 37; SAE 4135; 217 Brin max.

DIRO-34 CRNIMO 6
Dirostahl Karl Diedricks
Alloy steel. C 0.34, Si 0.25, Mn 0.55, P 0-0.035, S 0-0.035, Cr 1.55, Mo 0.2, Ni 1.55, bal Fe.
Equivalent to Fr. 35 NCD 6; Sw. SS 14 2541; GBr. 970-816 M 40; SAE 4340; 235 Brin max.

DIRO-37 CR 4
Dirostahl Karl Diedricks
Alloy steel. C 0.37, Si 0.25, Mn 0.75, P 0-0.035, S 0-0.035, Cr 1.05, bal Fe.
Equivalent to Fr. 38 C 4; GBr. 970-530 A 36; SAE 5135; 217 Brin max.

DIRO-37 MNSI 5
Dirostahl Karl Diedricks
Carbon steel. C 0.37, Si 1.3, Mn 1.3, P 0-0.035, S 0-0.035, bal Fe.
217 Brin max.

DIRO-41 CR 4
Dirostahl Karl Diedricks
Alloy steel. C 0.41, Si 0.25, Mn 0.65, P 0-0.035, S 0-0.035, Cr 1.05, bal Fe.
Equivalent to Fr. 45 C 4; GBr. 970-530 A 40; SAE 5140; 217 Brin max.

DIRO-41 CRMO 4
Dirostahl Karl Diedricks
Alloy steel. C 0.41, Si 0.25, Mn 0.65, P 0-0.035, S 0-0.035, Cr 1.05, Mo 0.25, bal Fe.
Equivalent to Fr. 42 CD 4 TS; Sw. SS 14 2244; GBr. 970-708 M 40; SAE 4142; 57 Rock .

DIRO-42 CRMO 4
Dirostahl Karl Diedricks
Alloy steel. C 0.42, Si 0.25, Mn 0.65, P 0-0.035, S 0-0.035, Cr 1.05, Mo 0.25, bal Fe.
Equivalent to Fr. 42 CD 4; Sw. SS 14 2244; GBr. 970-708 M 40, 970-709 M 40, 970-711 M 40; SAE 4140, SAE 4142; 217 Brin max.

DIRO-49 CRMO 4
Dirostahl Karl Diedricks
Alloy steel. C 0.49, Si 0.25, Mn 0.65, P 0-0.035, S 0-0.035, Cr 1.05, Mo 0.25, bal Fe.
Equivalent to GBr. 970-709 M 40; SAE 4150; 59 Rock C.

DIRO-50 CRV 4
Dirostahl Karl Diedricks
Alloy steel. C 0.5, Si 0.25, Mn 0.95, P 0-0.035, S 0-0.035, Cr 1.05, V 0.15, bal Fe.
Equivalent to Fr. 50 CV 4; Sw. SS 14 2230; SAE 6145; 235 Brin max.

DIRO-5042 CRMO 4
Dirostahl Karl Diedricks
Alloy steel. C 0.5, Si 0.25, Mn 0.65, P 0-0.035, S 0-0.035, Cr 1.05, Mo 0.25, bal Fe.
Equivalent to SAE 4150; 235 Brin max.

DIRO-53 MNSI 4
Dirostahl Karl Diedricks
Carbon steel. C 0.53, Si 0.9, Mn 1, P 0-0.035, S 0-0.035, bal Fe.
235 Brin max.

DIRO-55 NICRMOV 6
Dirostahl Karl Diedricks
Alloy steel. C 0.55, Si 0.3, Mn 0.6, P 0-0.035, S 0-0.035, Cr 0.7, Mo 0.3, Ni 1.7, V 0.1, bal Fe.
Equivalent to Fr. 55 NCDV 7; 240 Brin max.

DIRO-56 NICRMOV 7
Dirostahl Karl Diedricks
Alloy steel. C 0.56, Si 0.3, Mn 0.7, P 0-0.035, S 0-0.035, Cr 1.1, Mo 0.5, Ni 1.7, V 0.1, bal Fe.
Equivalent to Fr. 55 NCDV 7; 250 Brin max.

DIRO-58 CRV 4
Dirostahl Karl Diedricks
Alloy steel. C 0.58, Si 0.25, Mn 0.95, P 0-0.035, S 0-0.035, Cr 1.05, V 0.15, bal Fe.
235 Brin max.

DIRO-A ST 35
Dirostahl Karl Diedricks
Carbon steel. C 0.14, Si 0.25, Mn 0.5, P 0-0.045, S 0-0.045, 0.025 Al min, bal Fe.
Equivalent to Fr. A 37 P1; ASTM A442 Gr.55; 146 Brin max.

DIRO-A ST 41
Dirostahl Karl Diedricks
Carbon steel. C 0.18, Si 0.25, Mn 0.8, P 0-0.045, S 0-0.045, 0.025 Al min, bal Fe.
Equivalent to Fr. A 42 P1; Sw. SS 14 1434; GBr. 1501-221Gr28LTD; ASTM A442 Gr.60; 156 Brin max.

DIRO-A ST 52
Dirostahl Karl Diedricks
Carbon steel. C 0.18, Si 0.25, Mn 1.3, P 0-0.045, S 0-0.045, 0.025 Al min, bal Fe.
Equivalent to Fr. A 52 P1; Sw. SS 14 2103; GBr. 1501-224Gr30LTD; ASTM A455; 190 Brin max.

DIRO-C 10
Dirostahl Karl Diedricks
Carbon steel. C 0.1, Si 0.25, Mn 0.45, P 0-0.045, S 0-0.045, bal Fe.
Equivalent to Fr. CC 10; GBr. 970-040 A10; SAE 1010; 131 Brin max.

DIRO-C 15
Dirostahl Karl Diedricks
Carbon steel. C 0.15, Si 0.25, Mn 0.45, P 0-0.045, S 0-0.045, bal Fe.
Equivalent to Fr. CC 20; Sw. SS 14 1350; GBr. 970-040 A15; SAE 1015; 146 Brin max.

DIRO-C 22
Dirostahl Karl Diedricks
Carbon steel. C 0.22, Si 0.25, Mn 0.45, P 0-0.045, S 0-0.045, bal Fe.
Equivalent to Fr. CC 20; Sw. SS 14 1450; GBr. 970-040 A20; SAE 1020; 156 Brin max.

DIRO-C 22.8
Dirostahl Karl Diedricks
Carbon steel. C 0.22, Si 0.25, Mn 0.45, P 0-0.045, S 0-0.045, bal Fe.
Equivalent to GBr. 1501-161 Gr.28A; SA 181 Gr.1; 156 Brin max.

DIRO-C 35
Dirostahl Karl Diedricks
Carbon steel. C 0.35, Si 0.25, Mn 0.65, P 0-0.045, S 0-0.045, bal Fe.
Equivalent to Fr. CC 35; Sw. SS 14 1550; GBr. 970-060 A35; SAE 1038; 183 Brin max.

DIRO-C 45
Dirostahl Karl Diedricks
Carbon steel. C 0.45, Si 0.25, Mn 0.65, P 0-0.045, S 0-0.045, bal Fe.
Equivalent to Fr. CC 45; Sw. SS 14 1650; GBr. 970-080 M46; SAE 1045; 207 Brin max.

DIRO-C 55
Dirostahl Karl Diedricks
Carbon steel. C 0.55, Si 0.25, Mn 0.75, P 0-0.045, S 0-0.045, bal Fe.
Equivalent to Sw. SS 14 1655; GBr. 970-070 M55; SAE 1055; 229 Brin max.

DIRO-C 60
Dirostahl Karl Diedricks
Carbon steel. C 0.6, Si 0.25, Mn 0.75, P 0-0.045, S 0-0.045, bal Fe.
Equivalent to Sw. SS 14 1655; GBr. 970-080 A62; SAE 1060; 241 Brin max.

DIRO-C 67
Dirostahl Karl Diedricks
Carbon steel. C 0.67, Si 0.3, Mn 0.75, P 0-0.045, S 0-0.045, bal Fe.
Equivalent to GBr. 970-080 A67; SAE 1070; 250 Brin max.

DIRO-CF 35
Dirostahl Karl Diedricks
Carbon steel. C 0.35, Si 0.25, Mn 0.65, P 0-0.025, S 0-0.035, bal Fe.
Equivalent to Fr. XC 38 TS; Sw. SS 14 1572; GBr. 970-060 A 35; SAE 1038; 48 Rock .

DIRO-CF 45
Dirostahl Karl Diedricks
Carbon steel. C 0.45, Si 0.25, Mn 0.65, P 0-0.025, S 0-0.035, bal Fe.
Equivalent to Fr. XC 48 TS; Sw. SS 14 1672; GBr. 970-080 M 46; SAE 1045; 53 Rock C.

DIRO-CF 53
Dirostahl Karl Diedricks
Carbon steel. C 0.53, Si 0.3, Mn 0.55, P 0-0.025, S 0-0.035, bal Fe.
Equivalent to Sw. SS 14 1674; GBr. 970-070 M 55; SAE 1055; 55 Rock C.

DIRO-CF 70
Dirostahl Karl Diedricks
Carbon steel. C 0.7, Si 0.25, Mn 0.7, P 0-0.025, S 0-0.035, bal Fe.
Equivalent to GBr. 970-080 A 67; SAE 1070; 59 Rock C.

DIRO-CK 10
Dirostahl Karl Diedricks
Carbon steel. C 0.1, Si 0.25, Mn 0.45, P 0-0.035, S 0-0.035, bal Fe.
Equivalent to Fr. XC 10; SAE 1010; 131 Brin max.

DIRO-CK 15
Dirostahl Karl Diedricks
Carbon steel. C 0.15, Si 0.25, Mn 0.45, P 0-0.035, S 0-0.035, bal Fe.
Equivalent to Fr. XC 18 S; Sw. SS 14 1370; SAE 1015; 146 Brin max.

DIRO-CK 22
Dirostahl Karl Diedricks
Carbon steel. C 0.22, Si 0.25, Mn 0.45, P 0-0.035, S 0-0.035, bal Fe.
Equivalent to Fr. XC 18 S; SAE 1020; 156 Brin max.

DIRO-CK 35
Dirostahl Karl Diedricks
Carbon steel. C 0.35, Si 0.25, Mn 0.65, P 0-0.035, S 0-0.035, bal Fe.
Equivalent to Fr. XC 38; Sw. SS 14 1572; SAE 1038; 183 Brin max.

DIRO-CK 45
Dirostahl Karl Diedricks
Carbon steel. C 0.45, Si 0.25, Mn 0.65, P 0-0.035, S 0-0.035, bal Fe.
Equivalent to Fr. XC 48; Sw. SS 14 1672; SAE 1045; 207 Brin max.

DIRO-CK 55
Dirostahl Karl Diedricks
Carbon steel. C 0.55, Si 0.25, Mn 0.75, P 0-0.035, S 0-0.035, bal Fe.
Equivalent to Fr. XC 55; Sw. SS 14 1674; SAE 1055; 229 Brin max.

DIRO-CK 60
Dirostahl Karl Diedricks
Carbon steel. C 0.6, Si 0.25, Mn 0.75, P 0-0.035, S 0-0.035, bal Fe.
Equivalent to Fr. XC 60; Sw. SS 14 1678; SAE 1060; 241 Brin max.

DIRO-CK 67
Dirostahl Karl Diedricks
Carbon steel. C 0.67, Si 0.3, Mn 0.75, P 0-0.035, S 0-0.035, bal Fe.
Equivalent to Fr. XC 68; Sw. SS 14 1770; SAE 1070; 250 Brin max.

DIRO-H 11
Dirostahl Karl Diedricks
Carbon steel. C 0.18, Si 0.25, Mn 0.8, P 0-0.05, S 0-0.05, bal Fe.
Equivalent to Fr. A 42 C1; Sw. SS 14 1430; GBr. 1503-161 Gr.26; SA 515 Gr.60; 156 Brin max.

DIRO-R ST 34-2
Dirostahl Karl Diedricks
Carbon steel. C 0.1, Si 0.25, Mn 0.45, P 0-0.05, S 0-0.05, bal Fe.
Equivalent to Fr. A 34-2; Sw. SS A 330-2; ASTM A283 Gr.B; 131 Brin max.

DIRO-R ST 37-2
Dirostahl Karl Diedricks
Carbon steel. C 0.15, Si 0.25, Mn 0.45, P 0-0.05, S 0-0.05, bal Fe.
Equivalent to Fr. A37-2; Sw. SS 14 1312; ASTM A283 Gr.C; 146 Brin max.

DIRO-R ST 42-2
Dirostahl Karl Diedricks
Carbon steel. C 0.22, Si 0.25, Mn 0.45, P 0-0.05, S 0-0.05, bal Fe.
Equivalent to Sw. SS A 410-2; GBr. 4360-40 A; ASTM A283 Gr.D; 156 Brin max.

DIRO-ST 33
Dirostahl Karl Diedricks
Carbon steel. C 0.1, Si 0.25, Mn 0.45, P 0-0.06, S 0-0.05, bal Fe.
Equivalent to Fr. A33-2; Sw. SS 14 1300; ASTM A283 Gr.A; 131 Brin max.

DIRO-ST 33-2
Dirostahl Karl Diedricks
Carbon steel. C 0.1, Si 0.25, Mn 0.45, P 0-0.06, S 0-0.05, bal Fe.
Equivalent to Fr. A 33-2; Sw. SS A 330-2; ASTM A283 Gr.A; 131 Brin max.

DIRO-ST 37-2
Dirostahl Karl Diedricks
Carbon steel. C 0.1, Si 0.25, Mn 0.45, P 0-0.05, S 0-0.05, bal Fe.
Equivalent to Fr. A34-2; Sw. SS 14 1232; ASTM A283 Gr.B; 131 Brin max.

DIRO-ST 44-2
Dirostahl Karl Diedricks
Carbon steel. C 0.22, Si 0.25, Mn 0.45, P 0-0.05, S 0-0.05, bal Fe.
Equivalent to Sw. SS 14 1412; GBr. 4360-40A; ASTM A283 Gr.D; 156 Brin max.

DIRO-ST 50-2
Dirostahl Karl Diedricks
Carbon steel. C 0.35, Si 0.25, Mn 0.65, P 0-0.05, S 0-0.05, bal Fe.
Equivalent to Fr. A50-2; ASTM A235 Gr.E; 183 Brin max.

DIRO-ST 52-3
Dirostahl Karl Diedricks
Carbon steel. C 0.18, Si 0.35, Mn 1.3, P 0-0.045, S 0-0.045, 0.015 Al min, bal Fe.
Equivalent to Fr. A52-4; GBr. 4360-50D; Sw. SS 14 2172; ASTM A299/A537; 190 Brin max.

DIRO-ST 60-2
Dirostahl Karl Diedricks
Carbon steel. C 0.45, Si 0.25, Mn 0.65, P 0-0.05, S 0-0.05, bal Fe.
Equivalent to Fr. A60-2; ASTM A235 Gr.G; 207 Brin max.

DIRO-ST 70-2
Dirostahl Karl Diedricks
Carbon steel. C 0.6, Si 0.25, Mn 0.75, P 0-0.05, S 0-0.05, bal Fe.
Equivalent to Fr. A70-2; 241 Brin max.

DIRO-TT ST 35
Dirostahl Karl Diedricks
Carbon steel. C 0.14, Si 0.25, Mn 0.5, P 0-0.045, S 0-0.045, bal Fe.
Equivalent to Sw. SS 14 1313; ASTM A516 Gr.55; 146 Brin max.

DIRO-TT ST 41
Dirostahl Karl Diedricks
Carbon steel. C 0.18, Si 0.25, Mn 0.8, P 0-0.045, S 0-0.045, bal Fe.
Equivalent to Fr. A 42 FP1; Sw. SS 14 1414; GBr. 1503-224Gr.28LT; ASTM A516 GR.60; 156 Brin max.

DIRO-TT ST E 36
Dirostahl Karl Diedricks
Carbon steel. C 0.16, Si 0.35, Mn 1.3, P 0-0.03, S 0-0.03, 0.015 Al min, bal Fe.
Equivalent to Fr. A 52 FP1; Sw. SS 14 2107; GBr. 1503-224Gr.28LT; ASTM A 350 LF2; 190 Brin max.

DIRO-X 10 CR 13
Dirostahl Karl Diedricks
Stainless steel. C 0.1, Si 0.5, Mn 0.5, P 0-0.045, S 0-0.045, Cr 13, bal Fe.
Equivalent to Fr. Z 10 C 13; Sw. SS 14 2302; GBr. 970-410 S 21; SAE 51410; 180 Brin max.

DIRO-X 20 CR 13
Dirostahl Karl Diedricks
Stainless steel. C 0.2, Si 0.5, Mn 0.5, P 0-0.045, S 0-0.045, Cr 13, bal Fe.
Equivalent to Fr. Z 20 C 13; Sw. SS 14 2303; GBr. 970-420 S 37; SAE 51420; 220 Brin max.

DIRO-X 40 CR 13
Dirostahl Karl Diedricks
Stainless steel. C 0.4, Si 0.5, Mn 0.5, P 0-0.045, S 0-0.045, Cr 13, bal Fe.
Equivalent to Fr. Z 40 C 14; Sw. SS 14 2304; GBr. 970-420 S 45; 225 Brin max.

DIRON
British Steel plc
Mild steel with one coat vitreous enamel.

DIROSTAHL
Now DIRO ALLOYS.

DISC ALLOY
Teledyne Allvac
C 0.04, Mn 0.9, Si 0.8, Cr 13.5, Ni 26, Mo 2.75, Ti 1.75, Al 0.1, bal Fe.
For furnace parts, retorts, combustion chambers, salt pots, heat treating equipment. Heat and corrosion resistant. *Obsolete*

DISCALOY
Cannon-Muskegon Corp.
C 0-0.08, Al 0-0.35, Si 0.6-1, Mn 0.6-1, Cr 12-15, Ni 24-28, Ti 1.35-1.85, Mo 2.5-3.5, bal Fe.
At 70°F: 145,000 TS; 106,000 YS; 19 El; 23 RA. At 1200°F: 104,000 TS; 91,000 YS; 19 El; 24 RA. At 1350°F: 82,000 TS; 74,000 YS; 14 El; 26 RA. For turbine discs, bolts, extrusion dies, nonmagnetic gears and pinions; austenitic, heat resistant.

DISCALOY 24
Westinghouse Electric Corp.
C 0.36, Cr 13.5, Ni 26, Mo 4, Ti 1.6, Si 1, Mn 1.4, Al 0.1, bal Fe.
For jet and gas turbine components; corrosion and heat resistant. *Obsolete*

DISCUS
British Steel plc
High strength galvanized corrugated sheet; mild steel.

DISQUE E
Societe Nouvelle des Acieries de Pompey
C 0.45, Mn 0.7, Si 1.75, bal Fe.
For agricultural equipment, plows; oil hardened, shock resistant. *Obsolete*

DISQUE H
Societe Nouvelle des Acieries de Pompey
C 0.55, Mn 0.7, Si 1.75, bal Fe.
For agricultural equipment, plows; oil hardened, shock resistant. *Obsolete*

DISSTEEL 812
Disston Inc.
C 1.8, Cr 12, bal Fe.
For dies; nondeforming. *Obsolete*

DISSTON 6-N-6
Disston Inc.
C 0.8, Cr 4, V 1, W 6, Mo 6, bal Fe.
For cutting tools, taps, reamers, broaches; high speed steel. *Obsolete*

DISSTON 66
Disston Inc.
W 6, Mo 6, Cr 4, V 1, Cu 2, C, bal Fe.
For cutting tools; high speed steel. *Obsolete*

DISSTON 6N6-M2
Disston Inc.
C 0.82, Cr 4.25, W 6.35, Mo 5, V 2, bal Fe.
For tools, cutters; high speed steel. *Obsolete*

DISSTON D-12-CO
Disston Inc.
C, alloy, bal Fe.

DISSTON HRW
CCS Braeburn Alloy Steel
C 0.95, Cr 3.75, bal Fe.
For hot work dies, headers, punches; hot work steel, oil hardened. *Obsolete*

DISSTON NO. 51110
Disston Inc.
C 1.1, Mn 0.3, bal Fe.

DISSTON NO. 5170
Disston Inc.
C 0.7, Mn 0.3, bal Fe.

DISSTON NO. 5190
Disston Inc.
C 0.9, Mn 0.3, bal Fe.

DISSTON NO. 5390
Disston Inc.
C 0.9, Mn 0.7, bal Fe.

DISSTON NO. 819
Disston Inc.
C 1.25, Mn 0.3, Cr 0.4, bal Fe.
For lawn mower blades; tough. *Obsolete*

DISSTON NO. 821
Disston Inc.
C 1.2, V 0.2, bal Fe.
For tools; water hardening. *Obsolete*

DISSTON NO. 825
Disston Inc.
C 0.9, Mn 0.4, Cr 2.25, bal Fe.
For lawn mower blades; tough. *Obsolete*

DISSTON NO. 826
Disston Inc.
C 1.3, V 0.2, bal Fe.
For tools; water hardening. *Obsolete*

DISSTON NO. 827
Disston Inc.
C 0.7, Ni 1.6, Cr 1, Cu 0.3.
bal Fe. *Obsolete*

DISSTON NO. 841
Disston Inc.
C 0.4, Mn 0.3, Cr 0.7, Mo 0.2, bal Fe.
For lawn mower blades; tough. *Obsolete*

DISSTON NO. 842
Disston Inc.
C 1, Mn 0.3, Cr 1, bal Fe.
For lawn mower blades; oil hardening. *Obsolete*

DISSTON NO. 844
Disston Inc.
C 0.85, Cr 0.8, Si 0.25, bal Fe.

DISSTON NO. 871
Disston Inc.
C 0.55, Mn 0.65, Mo 0.2, bal Fe.

DISSTON NO. 872
Disston Inc.
C 0.35, Cr 3.25, W 9.25, V 0.3, bal Fe.

DISSTON NO. 873
Disston Inc.
C 0.38, Cr 5, W 1.25, Mo 1.35, V 0.4, Si 1, bal Fe.

DISSTON STAINLESS A
Disston Inc.
C 0.3, Cr 14, bal Fe.

DISSTON STAINLESS B
Disston Inc.
C 0.6, Cr 17, bal Fe.

DISSTON STAINLESS D
Disston Inc.
C 0.45, Si 1, Cr 8.75, Mo 1.4, Cu 2, bal Fe.

DISSTON STAINLESS IRON
Disston Inc.
C 0.1, Cr 12-14, bal Fe.

DISTALOY 4600A
Hoeganaes Corp.
Ni 1.8, Mo 0.55, Cu 1.6.
Diffusion bonded powder. Responds well to heat treatment.

DISTALOY 4800A
Hoeganaes Corp.
Ni 4, Mo 0.55, Cu 1.6.
Diffusion bonded powder. Suitable for heat treating.

DIVCO 40/60 NO. 180 SOLDER
Division Lead Co.
Sn 40, Pb 60.
For solder; rosin core.

DIXI NO. 3
Hidalgo Steel Co. Inc.
C, Cr, bal Fe.
For tools and dies.

DIXIE BRAND
A. Milne & Co.
C 0.7-1.4, bal Fe.
For tools, dies, drills; water hardening.

DIXOILBRONZ
Seitzinger's Inc.
Sn 10, Zn 2, bal Cu.
For bearings, bushings, gears, liners. Corrosion resistant.

DLP COPPER
Criterion Metals, Inc.
Copper. Cu 99.99, P 0.004-0.012.
Thin gauge sheet, various tempers: 23-52 ksi TS min; 5-51 ksi YS min. C12000

DLP COPPER 120
Anaconda Co.
P 0.01, 99.9%+ Cu.
Hard: 45,000 TS; 40,000 YS; 10 El; 75 Brin. Soft: 33,000 TS; 10,000 YS; 45 El; 42 Brin. For electrical apparatus; high conductivity. *Obsolete*

DM
Southern Malleable Iron Co.
Cu 19.76, Fe 55.12, Si 0.57, Ni 0.87, C 0.29, Cr 15.18.
155 Brin. For corrosion resistant parts; formerly "Durbin."

DM 7
Acciaierie Valbruna s.p.a.
C 0.95-1.05, Cr 3.25-4.25, Mo 8-9, W 1.25-2.25, V 1.8-2.2, bal Fe.
Molybdenum high speed tool steel. W. Nr. 1.3348; AISI M 7.
Obsolete

DM-35
Timken Co.
C 0.33-0.4, Cr 1-1.5, Mo 0.4-0.6, Si 0.55-0.75, Mn 0.45-0.65, bal Fe.
Heat treated: 125,000 TS; 105,000 YS; 16 El; 50 RA; 250 Brin. For high temperature bolting, flanges, fittings; heat resistant. *Obsolete*

DM-45
Timken Co.
C 0.45, Mn 0.55, Cr 1.25, Mo 0.5, Si 0.65, bal Fe.
Normalized: 146,000 TS; 108,000 YS; 17.5 El; 47.4 RA; 285 Brin. For gears, forgings, high temperature bolts; heat resistant to 950 F. *Obsolete*

DM2-45
Timken Co.
C 0.3-0.45, Mn 0.7-1, Cr 0.8-1.1, Mo 0.5-0.7, bal Fe.
Heat treated: 125,000 TS; 105,000 YS; 10 El; 50 RA; 250 Brin. For high temperature bolting; resists heat to 950 F. *Obsolete*

DM; DM2
Now TIMKEN DM, DM-2.

DMAT 13
Creusot-Loire
C 0.55, Ni 2, Cr 0.85, Mo 0.35, V 0.04, bal Fe.
Oil hardening hot work tool steel; for hammer forging and upsetting dies; good resistance to shock. AFNOR 55 NCDV 7.

DMO 5
W. Ossenberg & Cie Edelstahlwerke
C 0.85, Cr 4.5, Mo 5, W 6.5, V 1.8, bal Fe.
High speed steel for lathe tools, milling tools, drills. W.-Nr. 1.3343; similar to AISI M2.

DMO 5 CO
W. Ossenberg & Cie Edelstahlwerke
C 0.9, Cr 4, Mo 5, W 6.5, V 1.5, Co 5, bal Fe.
High speed steel for heavy duty lathe tools, milling cutters, drills. W.-Nr. 1.3243.

DMOC WEICH
Thyssen Edelstahlwerke AG
C, Cr, Mo, bal Fe.
14 El. For machinery parts. *Obsolete*

DN 1
Thyssen Edelstahlwerke AG
C, Mn, Ni, bal Fe.
16-20 El. For machinery parts. *Obsolete*

DN 15
Thyssen Edelstahlwerke AG
C, Ni, Mn, bal Fe.
For non-magnetic parts; non-magnetic. *Obsolete*

DN 25
Thyssen Edelstahlwerke AG
C, Ni, Mn, bal Fe.
For non-magnetic parts; non-magnetic. *Obsolete*

DN 3
Thyssen Edelstahlwerke AG
C, Ni, bal Fe.
17-21 El. For machinery parts. *Obsolete*

DN 33
Thyssen Edelstahlwerke AG
C, Ni, bal Fe.
For wire for electrical apparatus. *Obsolete*

DN 36
Thyssen Edelstahlwerke AG
C, Ni, bal Fe.
For wire for electrical apparatus. *Obsolete*

DN 42
Thyssen Edelstahlwerke AG
C, Ni, bal Fe.
For wire for electrical apparatus. *Obsolete*

DN 5
Thyssen Edelstahlwerke AG
C, Ni, bal Fe.
16-20 El. For machinery parts. *Obsolete*

DN-15
German manufacture
C 0.4-0.6, Ni 11-13, Mn 4-6, Cr 3, Mo 0.5, bal Fe.
Austenitic steel, stainless. For aeronautical structures, gas turbines and internal combustion engines; coefficient of expansion similar to that of aluminum.

DO 40
DoAll Co.
Titanium coated carbide. For general purpose machining of steels and metal that tend to alloy or weld to carbides.

DO 80
DoAll Co.
Ceramic cutting material. For high speed machining of abrasive superalloys, hardened steel and cast iron.

DO ALL AIR HARDENING AISI-A2
DoAll Co.
C 1, Mn 0.6, Cr 5.25, Mo 1.1, V 0.25, S 0.12, bal Fe.
Free machining, precision ground. For dies, punches, taps, shear blades, reamers, gages.

DO ALL AIR HARDENING AISI-A6
DoAll Co.
C 0.7, Mn 2.1, Cr 1, Mo 1.3, Si 0.3, S 0.12, bal Fe.
Free machining, precision ground. For blanking and coining dies, shear blades, jigs, and fixtures, punches.

DO ALL LOW CARBON AISI-1018
DoAll Co.
C 0.18, Mn 0.6, Si 0.2, bal Fe.
Precision ground flat stock. For patterns, jigs, fixtures, machine parts that do not require heat treatment.

DO ALL M41
DoAll Co.
C 1.1, Cr 4.25, V 2, W 6.75, Mo 3.75, Co 5, bal Fe.
High speed steel for cutting tools, tungsten-molybdenum-cobalt type; AISI M-41.

DO ALL O.H.
DoAll Co.
C 1.2, Cr 0.55, Mo 0.33, Mn 0.75, bal Fe.
For dies, punches, gauges, machine parts; oil hardened.

DO ALL OIL HARDENING AISI-O1
DoAll Co.
C 0.9, Mn 1.2, Si 0.3, W 0.5, Cr 0.5, V 0.2, bal Fe.
Precision ground flat stock. For blanking and trimming dies, cutters, punches, special gauges.

DO ALL T1
DoAll Co.
C 0.73, Cr 4, V 1.1, W 18, bal Fe.
High speed steel, general purpose cutting tools, tungsten type; AISI T1.

DO ALL T15
DoAll Co.
C 1.5, W 12, Cr 4, V 5, Co 5, bal Fe.
Tungsten type high speed steel. Recommended for extreme abrasion resistance.

DO ALL T5
DoAll Co.
C 0.8, W 18, Cr 4, V 2, Co 8, bal Fe.
Tungsten-cobalt high speed steel.

DO ALL T8
DoAll Co.
C 0.75, W 14, Cr 4, V 2, Co 5, bal Fe.
Tungsten-cobalt high speed steel for hogging of hard, tough material.

DO ALLOY
DoAll Co.
Co, Cr, W.
Cast alloy for general purpose machining, high red hardness and shock resistance.

DO-1
DoAll Co.
Sintered carbide.
For rough cutting and chip removal on cast iron and non-ferrous metals.

DO-10
DoAll Co.
Sintered carbide.
For wear and shock resistance.

DO-11
DoAll Co.
Sintered carbide.
For light impact duty.

DO-13
DoAll Co.
Sintered carbide.
For medium impact duty.

DO-14
DoAll Co.
Sintered carbide.
For heavy impact duty.

DO-15
DoAll Co.
Sintered carbide.
Cutting tools for heavy roughing and interrupted cuts on steel.

DO-16
DoAll Co.
Sintered carbide.
For general purpose cutting tools.

DO-17
DoAll Co.
Sintered carbide.
Cutting tools for finish cutting steel.

DO-18
DoAll Co.
Sintered carbide.
Cutting tools for finish cutting and precision boring steel.

DO-2
DoAll Co.
Sintered carbide.
General purpose cutting tool for cast iron and non-ferrous metals; good resistance to wear.

DO-3
DoAll Co.
Sintered carbide.
Cutting tool for finish cutting on cast iron and non-ferrous metals.

DO-30
DoAll Co.
Sintered carbide.
Cutting tools for rough machining of cast iron and non-ferrous metal.

DO-34
DoAll Co.
Sintered carbide.
Cutting tools for general purpose machining and finishing of steel.

DO-35
Manufacturer not listed
Sintered carbide.
Cutting tools for heavy, roughing, interrupted cuts on steel.

DO-36
DoAll Co.
Sintered carbide.
Cutting tools for general purpose machining of steel.

DO-4
DoAll Co.
Sintered carbide.
Cutting tool for finishing and precision boring of cast iron and non-ferrous metals.

DO-DI
NL Industries
Cu 60, Sn 0.65, Pb 0.82, Fe 0.35, Al 0.53, bal Cu.
Cast: 56,000 TS; 20,100 YS; 26 El; 32 RA; 86 Brin. For marine parts, hardware; free-cutting. *Obsolete*

DO-IT
Ziv Steel & Wire Co.
C 0.7, Mn 0.8, Si 2, Mo 0.2, bal Fe.
For shear blades, punches, chisels; oil hardened. *Obsolete*

DOBLINSCHE ALLOY
German manufacture
Si 0-50, bal Co.
For chemical construction; corrosion resistant.

DOCO
Saarstahl AG
C 1.75, Co 2.5, Cr 12.5, Mo 0.9, V 0.25, bal Fe.
Cold work tool steel. For heavy duty cutting and stamping dynamo and transformer sheets. Similar to W.-Nr. 2880. *Obsolete*

DOCTOR METAL
English manufacture
Cu 88, Zn 9.5, Sn 2.5.
For corrosion resistant brass parts.

DODGE
Dodge Foundry & Machine Co.
All alloys are now ASTM grades.

DODGE D-1
Dodge Foundry & Machine Co.
C 0.3, Mn 1.2, Cr 0.3, Mo 0.3, bal Fe.
Cast: 80,000 TS; 55,000 YS; 22 El; 40 RA; 163 Brin. For structural parts; water hardening. *Obsolete*

DODGE D-10
Dodge Foundry & Machine Co.
C 0-0.12, Mn 0-0.3, Si 0-0.6, bal Fe.
Cast: 48,000 TS; 29,000 YP; 30 El; 45 RA. For electrical applications, motor and generator housings, pole pieces, magnetic chucks. High permeability. *Obsolete*

DODGE D-11
Dodge Foundry & Machine Co.
C 0-0.1, Mn 0.5-0.8, Ni 3-4, bal Fe.
Cast: 70,000 TS; 40,000 YP; 28 El; 50 RA. For gears, housings, bolts, shafts, fittings, valves, pumps. Tough, oil hardening, shock resistant. Suitable for use to minus 150 F. *Obsolete*

DODGE D-12
Dodge Foundry & Machine Co.
C 0.18-0.2, Mn 0.55-0.65, Ni 2.25-2.75, bal Fe.
Cast: 75,000 TS; 48,000 YS; 24 El; 35 RA; 153 Brin. For low temperature applications; tough. *Obsolete*

DODGE D-14
Dodge Foundry & Machine Co.
C 0.4, Ni 1.8, Cr 0.8, Mo 0.25, bal Fe.
Annealed: 120,000 TS; 80,000 YS; 240 Brin. Heat treated: 185,000 TS; 165,000 YS; 380 Brin. Uses: gears, pressure vessels, die blocks, bushings, shafts, dies, fittings. Type 4340 steel castings; tough, oil hardening. *Obsolete*

DODGE D-2
Dodge Foundry & Machine Co.
C 0.15-0.23, Mn 0.5-0.7, Mo 0.4-0.6, bal Fe.
Cast: 65,000 TS; 37,000 YS; 24 El; 36 RA; 143 Brin. For turbine castings, steam valves and fittings; creep resistant. *Obsolete*

DODGE D-2 A
Dodge Foundry & Machine Co.
C 0.2-0.28, Mn 0.5-0.7, Mo 0.4-0.6, bal Fe.
Cast: 70,000 TS; 45,000 YS; 22 El; 35 RA; 149 Brin. For valves, pump casings, turbines; water hardening. *Obsolete*

DODGE D-2 B

Dodge Foundry & Machine Co.
C 0.2-0.28, Mn 0.5-0.7, Cr 0.5-0.8, Mo 0.4-0.6, bal Fe.
Cast: 70,000 TS; 45,000 YS; 22 El; 35 RA; 143 Brin. For high temperature applications; resists graphitization at high temperature. *Obsolete*

DODGE D-2 C

Dodge Foundry & Machine Co.
C 0-0.15, Mn 0.5-0.7, Si 0.5-1, Cr 1-1.5, Mo 0.45-0.65, bal Fe.
Cast: 70,000 TS; 40,000 YS; 22 El; 35 RA; 153 Brin. For high temperature applications. *Obsolete*

DODGE D-3

Dodge Foundry & Machine Co.
C 0.3, Cr 0.95, Mo 0.2, bal Fe.
Annealed: 88,000 TS; 60,000 YS; 180 Brin. Heat treated: 137,000 TS; 109,000 YS; 13.5 El; 31.2 RA. Normalized: 115,000 TS; 95,000 YS; 16 El; 35 RA. Uses: gears, bolts, shafts, housings, cylinders, fittings. Type 4130 steel castings. *Obsolete*

DODGE D-4

Dodge Foundry & Machine Co.
C 0.38-0.44, Mn 0.8, bal Fe.
Cast: 80,000-95,000 TS; 43,000-60,000 YS; 15-25 El; 20-35 RA; 143-174 Brin. For gears, dies, wear plates, structural castings; water hardened. *Obsolete*

DODGE D-5

Dodge Foundry & Machine Co.
C 0.38-0.43, Mn 0.75-1, Cr 0.8-1.1, Mo 0.15-0.25, bal Fe.
Normalized: 100,000 TS; 65,000 YS; 15 El; 30 RA. For gears, shafts; suitable for nitriding. *Obsolete*

DODGE D-6

Dodge Foundry & Machine Co.
C 0.6-0.75, Mn 0.7-0.9, bal Fe.
Cast: 85,000-110,000 TS; 45,000-66,000 YS; 10-20 El; 15-25 RA; 175-220 Brin. For bending, blanking and forming dies; wear resistant. *Obsolete*

DODGE D-7

Dodge Foundry & Machine Co.
C 0.15-0.2, Mn 0.4-0.6, Cr 4.5-6, Mo 0.4-0.6, bal Fe.
Cast: 90,000 TS; 60,000 YS; 18 El; 30 RA; 197-240 Brin. For valve bodies, fittings, pump cases; resists corrosion. *Obsolete*

DODGE D-8

Dodge Foundry & Machine Co.
C 0-0.2, Mn 0.5-0.7, Cr 1-1.5, Mo 0.4-0.6, bal Fe.
Cast: 70,000 TS; 45,000 YS; 22 El; 35 RA; 153 Brin. For high temperature applications; creep resistant. *Obsolete*

DODGE D-9

Dodge Foundry & Machine Co.
C 0-0.15, Mn 0.7, Si 0.6, Cr 2-3, Mo 0.8-1.1, bal Fe.
Cast: 70,000 TS; 45,000 YS; 22 El; 45 RA; 163 Brin. For high temperature applications; creep resistant. *Obsolete*

DODGE DTS 100

Dodge Foundry & Machine Co.
C 0.95-1.05, Cr 5-5.5, Mo 1-1.25, V 0.35-0.45, bal Fe.
Heat treated: 250,000 TS; 190,000 YS; C 50 Rock. For trimming and deep drawing dies, blanking and forming dies, cams, gauges, master tools, extrusion dies. Type A2 tool steel. Air hardening. Tough castings. *Obsolete*

DODGE DTS 150

Dodge Foundry & Machine Co.
C 1.45-1.55, Cr 11.5-12.5, Mo 0.7-0.9, V 0.18-0.24, bal Fe.
Heat treated: 278,000 TS; 214,000 YS; 1 El; C 57 Rock. For blanking and drawing dies, threading dies, burnishing tools, cams, collars, coining and forming dies. Type D 2 tool steel castings. Air hardening, abrasion resistant. *Obsolete*

DODGE DTS 165

Dodge Foundry & Machine Co.
C 1.55-1.65, Cr 12.5-14, V 0.3-0.45, Mo 0.7-0.9, Co 0.7-0.9, bal Fe.
Heat treated: 280,000 TS; 210,000 YS; C 57 Rock. For drawing and blanking dies, cams, punches, gauges, trimming dies. Deep hardening, non-deforming castings. High wear and compression resistance. Type D2 air hardening tool steel. *Obsolete*

DODGE DTS 30W8

Dodge Foundry & Machine Co.
C 0.3-0.35, Cr 3-3.5, W 8-9, V 0.2-0.3, bal Fe.
At 80 F: 200,000 TS; 173,000 YS; 8 El; 30 RA. At 1200 F: 108,000 TS; 85,000 YS; 4 El; 9 RA. Uses: brass forging dies and inserts, extrusion dies, punches, gripper and swaging dies, permanent molds. Type H21 hot work steel castings; red-hard. *Obsolete*

DODGE DTS 40

Dodge Foundry & Machine Co.
C 0.35-0.45, Si 0.9-1.2, Cr 4.75-5.25, Mo 1.2-1.5, V 0.75-1, bal Fe.
At 80 F: 217,000 TS; 184,000 YS; 8 El; 14 RA. At 1000 F: 155,000 TS; 125,000 YS; 8.8 El; 24 RA. Uses: extrusion dies, gripper and header dies, hot punches, mandrels, trimming dies. Type H13 hot work steel. Will not heat check, air hardening. Tough castings. *Obsolete*

DODGE DTS 60

Dodge Foundry & Machine Co.
C 0.55-0.65, Ni 0.4-0.5, Cr 1.2-1.3, Mo 0.5-0.7, V 0.12-0.18, bal Fe.
Heat treated: C 42-65 Rock. For brake dies, bushings, cams, gears, jigs, chucks, shear blades, pinions, trimmer dies. Tough, oil hardening, wear resistant. *Obsolete*

DODGE DX

Dodge Foundry & Machine Co.
C 0.10-0.24, Mn 0.4-0.8, bal Fe.
Castings: 60,000 TS; 30,000 YS; 26 El; 38 RA; 163 Brin. For castings for locomotives, valves, fittings; weldable. *Obsolete*

DODGE DZ

Dodge Foundry & Machine Co.
C 0.26-0.32, Mn 0.6-0.8, bal Fe.
Cast: 70,000 TS; 30,000 YS; 24 El; 36 RA; 143 Brin. For general castings; water hardening. *Obsolete*

DOE RUN COPPERIZED LEAD

Manufacturer not listed
Cu 0.05, Ag 0.0006, Cd 0.0003, bal Pb.
Cast: 2400-2600 TS; 40-50 El; 5 Brin. 618°F melting point; 0.41 lbs/in^3 density. For cable sheathing, chemical equipment.

DOE RUN LEAD

Manufacturer not listed
Ag 0.0005, Cd 0.0003, Pb 99.99-0.
Cast: 1700-2000 TS; 35-50 El; 4 Brin. 621°F melting point; 0.41 lbs/in^3 density. For battery oxide, solder, chemical pigments anti-knock compounds. Exceeds ASTM B 29-55.

DOFASCO

Dofasco Inc.
C 0.3, Ni 2-3, Cr 0.9-1.1, Mo 0.3-0.4, bal Fe.
For valves, fittings, oil crusher roll shells; high pressure service to 1100°F. *Obsolete*

DOFASCOLOY 1

Dofasco Inc.
C 0-0.22, Mn 0-1.65, P 0-0.04, S 0-0.05, Si 0-0.45, Cu 0-0.6, Ni 0-0.9, 0.005 Cb min, bal Fe.
70,000 TS min; 50,000 YS min; 22 El min. For exposed structural members used in bridges, railway and agricultural equipment, siding and tubing. Good weathering, welding, impact properties. ASTM A242/A242M, A606; SAE J1392, J1442.

DOFASCOLOY 42 W

Dofasco Inc.
C 0-0.26, Mn 0-1.65, P 0-0.04, S 0-0.05, 0.005 Cb min, bal Fe.
60,000 TS min; 42,000 YS min; 20 El min. For automobile components, railroad and agricultural equipment, structural members in bridges and buildings. ASTM A572/A572M, A607, A816/A816M; SAE J1392, J1442.

DOFASCOLOY 45 W

Dofasco Inc.
C 0-0.26, Mn 0-1.65, P 0-0.04, S 0-0.05, 0.005 Cb min, bal Fe.
60,000 TS min; 45,000 YS min; 22 El min. For automobile components, railroad and agricultural equipment, structural members in bridges and buildings. ASTM A572/A572M, A607, A816/A816M; SAE J1392, J1442.

DOFASCOLOY 45F

Dofasco Inc.
C 0-0.15, Mn 0-1.65, P 0-0.025, S 0-0.035, Si 0-0.9, 0.005 Cb min, bal Fe.
55,000 TS min; 45,000 YS min; 24 El min. For automotive, railroad and agricultural equipment; weldable, formable. ASTM A607, A656/A656M, A715, A816/A816M; SAE J1392, J1442.

DOFASCOLOY 50 W

Dofasco Inc.
C 0-0.26, Mn 0-1.65, P 0-0.04, S 0-0.05, 0.005 Cb min, bal Fe.
65,000 TS min; 50,000 YS min; 22 El min. For automobile components, railroad and agricultural equipment, structural members in bridges and buildings. ASTM A572/A572M, A607, A816/A816M; SAE J1392, J1442.

DOFASCOLOY 50F

Dofasco Inc.
C 0-0.15, Mn 0-1.65, P 0-0.025, S 0-0.035, Si 0-0.9, 0.005 Cb min, bal Fe.
60,000 TS min; 50,000 YS min; 22 El min. For automotive, railroad and agricultural equipment; weldable, formable. ASTM A607, A656/A656M, A715, A816/A816M; SAE J1392, J1442.

DOFASCOLOY 55 W

Dofasco Inc.
C 0-0.26, Mn 0-1.65, P 0-0.04, S 0-0.05, 0.005 Cb min, bal Fe.
70,000 TS min; 55,000 YS min; 18 El min. For automobile components, railroad and agricultural equipment, structural members in bridges and buildings. ASTM A572/A572M, A607, A816/A816M; SAE J1392, J1442.

DOFASCOLOY 55F

Dofasco Inc.
C 0-0.15, Mn 0-1.65, P 0-0.025, S 0-0.035, Si 0-0.9, 0.005 Cb min, bal Fe.
65,000 TS min; 55,000 YS min; 22 El min. For automotive, railroad and agricultural equipment; weldable, formable. ASTM A607, A656/A656M, A715, A816/A816M; SAE J1392, J1442.

DOFASCOLOY 60F

Dofasco Inc.
C 0-0.15, Mn 0-1.65, P 0-0.025, S 0-0.035, Si 0-0.9, 0.005 Cb min, bal Fe.
70,000 TS min; 60,000 YS min; 20 El min. For automotive, railroad and agricultural equipment; weldable, formable. ASTM A607, A656/A656M, A715, A816/A816M; SAE J1392, J1442.

DOFASCOLOY 60W

Dofasco Inc.
C 0-0.26, Mn 0-1.65, P 0-0.04, S 0-0.05, 0.005 Cb min, bal Fe.
75,000 TS min; 60,000 YS min; 16 El min. For automobile components, railroad and agricultural equipment structural members in bridges and buildings. ASTM A572/A572M, A607, A816/A816M; SAE J1392, J1442.

DOFASCOLOY 65W

Dofasco Inc.
C 0-0.26, Mn 0-1.65, P 0-0.04, S 0-0.05, 0.005 Cb min, bal Fe.
80,000 TS min; 65,000 YS min; 14 El min. For automobile
components, railroad and agricultural equipment structural
members in bridges and buildings. ASTM A572/A572M,
A607, A816/A816M; SAE J1392, J1442.

DOFASCOLOY 70 W

Dofasco Inc.
C 0-0.26, Mn 0-1.65, P 0-0.04, S 0-0.05, 0.005 Cb min, bal Fe.
85,000 TS min; 70,000 YS min; 12 El. For automobile
components, railroad and agricultural equipment structural
members in bridges and buildings. ASTM A572/A572M,
A607, A816/A816M; SAE J1392, J1442.

DOFASCOLOY 70F

Dofasco Inc.
C 0-0.15, Mn 0-1.65, P 0-0.025, S 0-0.035, Si 0-0.9, 0.005 Cb
min, bal Fe.
80,000 TS min; 70,000 YS min; 18 El min. For automotive,
railroad and agricultural equipment; weldable, formable.
ASTM A607, A656/A656M, A715, A816/A816M; SAE J1392,
J1442.

DOFASCOLOY 80 W

Dofasco Inc.
C 0-0.26, Mn 0-1.65, P 0-0.04, S 0-0.05, 0.005 Cb min, bal Fe.
95,000 TS min; 80,000 YS min; 12 El min. For automobile
components, railroad and agricultural equipment structural
members in bridges and buildings. ASTM A572/A572M,
A607, A816/A816M; SAE J1392, J1442.

DOFASCOLOY 80F

Dofasco Inc.
C 0-0.15, Mn 0-1.65, P 0-0.025, S 0-0.035, Si 0-0.9, 0.005 Cb
min, bal Fe.
90,000 TS min; 80,000 YS min; 16 El min. For automotive,
railroad and agricultural equipment; weldable, formable.
ASTM A607, A656/A656M, A715, A816/A816M; SAE J1392,
J1442.

DOFASCOLOY F

Dofasco Inc.
See separate grades.

DOFASCOLOY M

Dofasco Inc.
C 0-0.28, Mn 0-1.6, 0.30 Cu min, bal Fe.
Plate: 70,000 TS min; 60,000 YS min; 18 El min. For tanks,
trailers, trucks, auto frames, cranes, booms, buckets, stokers.
High strength low alloy steel. *Obsolete*

DOFASCOLOY MV

Dofasco Inc.
C 0-0.22, Mn 0.85-1.25, P 0-0.04, S 0-0.05, Si 0-0.4, 0.20 Cu
min, 0.02 V min, bal Fe.
65,000 TS min; 50,000 YS min; 20 El min. For welded
structural members used in railroad equipment trailers,
trucks, buildings, cranes, derricks. High strength low alloy
steel.

DOFASCOLOY NO. 1

Dominion Foundries & Steel Ltd.(Canada)
C 0-0.18, Mn 0-1, Ni 0-0.9, Cu 0-0.6, bal Fe.
Plate: 63,000 min TS; 42,000 min YS; 24 min El. For tanks,
trailers, trucks, auto frames, mine cars, buckets, stokers,
cranes. High strength, low alloy steel. *Obsolete*

DOFASCOLOY NO. 2

Dofasco Inc.
C 0-0.18, Mn 0-1, Ni 0-0.9, Cu 0-0.6, bal Fe.
Plate: 65,000 TS min; 45,000 YS min; 19 El min. For tanks,
trailers, trucks, auto frames, mine cars, buckets, stokers,
cranes, booms. High strength low alloy steel. *Obsolete*

DOFASCOLOY P

Dofasco Inc.
C 0-0.22, Mn 0-1.25, P 0-0.15, S 0-0.05, Si 0.15-0.35, Cu
0-0.6, Ni 0-0.9, Cr 0-0.6, bal Fe.
70,000 TS min; 50,000 YS min; 22 El min. For exposed
structural members used in buildings, railway, agricultural
equipment and other structures. ASTM A242/A242M, A606;
SAE J1392, J1442.

DOFASCOLOY W

Dofasco Inc.
C 0-0.25, Mn 0.5-1.25.
See separate grades.

DOHLEN CSV45

Dohlen-Stahl Gusstahl-Handels GmbH
C 0.45, Cr, V, Si, bal Fe.
For machine tool parts, springs, gears, bolts; oil hardened,
tough.

DOHLEN DELTA PEB

Dohlen-Stahl Gusstahl-Handels GmbH
C 0.55, Si 0.9, Cr 1.05, V 0.18, W 1.85, bal Fe.
For header dies, upsetters, punches; oil hardened, tough.

DOHLEN DP10 EXTRA

Dohlen-Stahl Gusstahl-Handels GmbH
C 1, Si 0-0.25, Mn 0-0.25, bal Fe.
Annealed: 100,000 TS; 53,000 YS; 21 El; 42 RA; 200 Brin. For
springs, tools, cutters, dies, drills, taps; Type W1; water
hardened.

DOHLEN DP10 PRIMA

Dohlen-Stahl Gusstahl-Handels GmbH
C 1, Si 0-0.25, Mn 0-0.25, bal Fe.
Annealed: 100,000 TS; 53,000 YS; 21 El; 42 RA; 200 Brin. For
springs, tools, taps, drills, reamers; Type W1; water hardened.

DOHLEN DP11 EXTRA

Dohlen-Stahl Gusstahl-Handels GmbH
C 1.1, Si 0-0.25, Mn 0-0.25, bal Fe.
Annealed: 110,000 TS; 58,000 YS; 18 El; 40 RA; 210 Brin. For
springs, tools, drills, taps, reamers; Type W1; water hardened.

DOHLEN DP11 PRIMA

Dohlen-Stahl Gusstahl-Handels GmbH
C 1.15, Si 0-0.25, Mn 0-0.25, bal Fe.
Annealed: 110,000 TS; 58,000 YS; 18 El; 40 RA; 210 Brin. For
springs, tools, reamers, broaches; Type W1; water hardened.

DOHLEN DP12 PRIMA

Dohlen-Stahl Gusstahl-Handels GmbH
C 1.3, Si 0-0.25, Mn 0-0.25, bal Fe.
For cutters, tools, drills, taps, reamers; Type W1; water
hardened.

DOHLEN DP1W

Dohlen-Stahl Gusstahl-Handels GmbH
C 0.15, Si 0.15-0.35, Mn 0.25-0.5, bal Fe.
Annealed: 70,000 TS; 40,000 YS; 25 El; 60 RA; 145 Brin. For
gears, cams, camshafts, machine tool parts; case hardening
steel.

DOHLEN DP3W

Dohlen-Stahl Gusstahl-Handels GmbH
C 0.35, Si 0.25-0.5, Mn 0.3-0.8, bal Fe.
Hot rolled: 85,000 TS; 55,000 YS; 30 El; 53 RA; 185 Brin. For
shears, bolts, machine tool parts, fasteners; water hardened.

DOHLEN DP4W

Dohlen-Stahl Gusstahl-Handels GmbH
C 0.45, Si 0.25-0.5, Mn 0.3-0.8, bal Fe.
Hot rolled: 98,000 TS; 60,000 YS; 24 El; 45 RA; 212 Brin. For
gears, bolts, machine tool parts, fasteners; water hardened.

DOHLEN DP6W

Dohlen-Stahl Gusstahl-Handels GmbH
C 0.6, Si 0.25-0.5, Mn 0.3-0.8, bal Fe.
Heat treated: 160,000 TS; 115,000 YS; 12 El; 40 RA; 325 Brin.
For gears, bolts, springs, rails, axles, shafts; water hardened.

DOHLEN DP7 EXTRA

Dohlen-Stahl Gusstahl-Handels GmbH
C 0.7, S 0-0.25, Mn 0-0.25, bal Fe.
Heat treated: 175,000 TS; 128,000 YS; 12 El; 37 RA; 355 Brin.
For springs, tools, rails, punches, mandrels; water hardened;
Type W1.

DOHLEN DP7 PRIMA

Dohlen-Stahl Gusstahl-Handels GmbH
C 0.7, Si 0-0.25, Mn 0-0.25, bal Fe.
Heat treated: 175,000 TS; 128,000 YS; 12 El; 37 RA; 355 Brin.
For springs, tools, rails, punches, mandrels; water hardened;
Type W1.

DOHLEN DP7W

Dohlen-Stahl Gusstahl-Handels GmbH
C 0.7, Si 0.25-0.5, Mn 0.3-0.8, bal Fe.
Heat treated: 175,000 TS; 128,000 YS; 12 El; 37 RA; 355 Brin.
For springs, tools, punches, mandrels; water hardened; Type
W1.

DOHLEN DP8

Dohlen-Stahl Gusstahl-Handels GmbH
C 0.85, Si 0-0.25, Mn 0-0.25, bal Fe.
Heat treated: 190,000 TS; 145,000 YS; 12 El; 33 RA; 400 Brin.
For tools, springs, drills, taps, cutters; Type W1; water
hardened.

DOHLEN DP8 PRIMA

Dohlen-Stahl Gusstahl-Handels GmbH
C 0.85, Si 0-0.25, Mn 0-0.25, bal Fe.
Heat treated: 190,000 TS; 145,000 YS; 12 El; 33 RA; 400 Brin.
For springs, tools, cutters, taps, drills; water hardened; Type
W1.

DOHLEN DP8W

Dohlen-Stahl Gusstahl-Handels GmbH
C 0.9, Si 0.25-0.5, Mn 0.3-0.8, bal Fe.
Heat treated: 195,000 TS; 150,000 YS; 10 El; 30 RA; 410 Brin.
For springs, tools, cutters, dies, drills, taps; Type W1; water
hardened.

DOHLEN DSW1

Dohlen-Stahl Gusstahl-Handels GmbH
C 0.3, Cr 2.65, V 0.35, W 8.5, bal Fe.
For extrusion press liners and rams; oil hardened, hot work
steel.

DOHLEN DSW2

Dohlen-Stahl Gusstahl-Handels GmbH
C 0.3, Cr 2.35, V 0.6, W 4.25, bal Fe.
For hot work tools, dies, punches; hot work steel, oil
hardened.

DOHLEN GSR

Dohlen-Stahl Gusstahl-Handels GmbH
C 0.67, Si 1.3, Mn 0.5, Cr 0.5, bal Fe.
For punches, dies, shears, upsetters; oil hardened, tough.

DOHLEN KE

Dohlen-Stahl Gusstahl-Handels GmbH
C 0.2, Cr 1.15, Mn 1.25, bal Fe.
For camshafts, cams, bolts; case hardened, tough.

DOHLEN P 53

Dohlen-Stahl Gusstahl-Handels GmbH
C 0.5, Cr 1.05, Ni 3.25, Mn 0.5, bal Fe.
For gears, bolts, crankshafts; oil hardened, shock resistant.

DOHLEN SAR

Dohlen-Stahl Gusstahl-Handels GmbH
C 1.15, Cr 0.65, V 0.1, Mn 0.3, bal Fe.
For blanking and header dies; oil hardened, abrasion
resistant.

DOHLEN STANDARD

Dohlen-Stahl Gusstahl-Handels GmbH
C 0.9, Mn 1.9, V 0.1, bal Fe.
For punches, blanking and forming dies; oil hardened, non-
deforming.

DOHLEN-CSV35
Dohlen-Stahl Gusstahl-Handels GmbH
C 0.38, Cr, V, Si, bal Fe.
For machine tool parts, gears, shafts, fasteners, crankshafts.
Oil hardening, tough, shock resisting.

DOLER ALUMIN NO. 133
NL Industries
Cu 13, Si 3, bal Al.
Die cast: 42,000 TS; 24,000 YS; 0.5 El; 105 Brin. For general
die castings. *Obsolete*

DOLER ALUMIN NO. 308
NL Industries
Si 8, bal Al.
Die cast: 38,000 TS; 20,000 YS; 5 El; 80 Brin. For die
castings. *Obsolete*

DOLER ALUMIN NO. 99
NL Industries
Al 99.
Die cast: 10,000 TS; 5,000 YS; 25 El; 23 Brin. For general die
castings. *Obsolete*

DOLER BRASS 1
NL Industries
Cu 65, Zn 34, Si 1.
Die cast: 65,000 TS; 35,000 YS; 15 El; 120 Brin. For die
castings; silicon brass. *Obsolete*

DOLER BRASS 4
NL Industries
Cu 81.5, Zn 14.5, Si 4.
Die cast: 85,000 TS; 50,000 YS; 10 El; 170 Brin. For die
castings; corrosion resistant. *Obsolete*

DOLER BRASS 5
NL Industries
Zn 10, Si 5, Mn 1, Al 1, bal Cu.
Die cast: 105,000 TS; 60,000 YS; 5 El; 190 Brin. For die
castings. *Obsolete*

DOLER NO. 112
NL Industries
Cu 6-8, Si 2.5-3.5, Zn 1.5, bal Al.
Die cast: 30,000 TS; 1.5 El; 85 Brin. *Obsolete*

DOLER-BRASS
NL Industries
Cu 63-65, Si 0.5-1.5, bal Zn.
Die cast: 65,000-75,000 TS; 30,000-40,000 YS; 25-20 El;
110-120 Brin. For die castings; hardware, plumbing.
Obsolete

DOLER-BRASS 2
NL Industries
Cu 60, Zn 38, Sn 0.6, Pb 0.6.
Cast: 55,000 TS; 30,000 YS; 10 El; 110 Brin. For hardware;
yellow brass. *Obsolete*

DOLER-BRASS 3
NL Industries
Cu 81.5, Zn 15, Si 3.5.
Cast: 80,000 TS; 35,000 YS; 20 El; 140 Brin. For hardware;
silicon-brass. *Obsolete*

DOLER-MAG
NL Industries
Al 9, Cu 0.1-0.3, Zn 0.6, bal Mg.
Die cast: 34,000 TS; 23,000 YS; 4 El; 60 Brin. For general
purpose die castings. *Obsolete*

DOLER-MAG NO. 10
NL Industries
Al 10, Mn 0.2, Si 0.5, bal Mg.
For die castings. *Obsolete*

DOLER-MAG NO. 2
NL Industries
Al 2, Mn 0.2, Si 0.2, bal Mg.
For general die castings. *Obsolete*

DOLER-MAG NO. 6
NL Industries
Al 6, Mn 0.2, Si 0.2, bal Mg.
For general die castings. *Obsolete*

DOLER-ZINC
NL Industries
Cu 2.5-3, Al 3.75-4.3, bal Zn.
Die cast: 45,000-48,000 TS; 5-6 El; 80-90 Brin. For
carburetors, valve parts, fuel pumps. *Obsolete*

DOLER-ZINC 2
NL Industries
Cu 2.7, Al 4.1, bal Zn.
Die cast: 48,000 TS; 5 El; 83 Brin. For die castings. *Obsolete*

DOLER-ZINC 3
NL Industries
Al 4, Mg 0.04, Cu 0.2, bal Zn.
Die cast: 41,000 TS; 10 El; 82 Brin. For die castings,
instrument housings, ornamental grill; C.S. 60,500, S.S.
30,900. *Obsolete*

DOLER-ZINC 5
NL Industries
Cu 1, Al 4, Mg 0.04, bal Zn.
Die cast: 47,000 TS; 7 El; 90 Brin. For die castings; C.S.
87,300, S.S. 38,400. *Obsolete*

DOLER-ZINC 8
NL Industries
Mn 0.5, Al 0.25, bal Zn.
Cast: 20,000 TS; 60 Brin. For die castings, ornaments,
instrument cases; solders readily. *Obsolete*

DOLLAR BLUE CHIP
English manufacture
C 0.62, Cr 4.33, W 17.82, V 1.66, Co 3.93, bal Fe.
For tools, cutters, reamers; high speed steel.

DOM
Vallourec S.A.
C 0.37, Cr 0.95, Mo 0.2, bal Fe.
High elongation; for mechanical and tool joints.

DOMAL AM100A
Dominion Magnesium Ltd.
Al 10, Mn 0.1, bal Mg.
Cast: 22,000 TS; 12,000 YS; 1-3 El; 55 Brin. Heat treated:
38,000 TS; 20,000 YS; 1-4 El; 80 Brin. For high strength
castings; age-hardenable.

DOMAL AM80A
Dominion Magnesium Ltd.
Al 8, Mn 0.2, bal Mg.
Cast: 24,000-27,000 TS; 10,000-12,500 YS; 4-7 El; 48-52 Brin.
Forged: 42,000-48,000 TS; 28,000-32,000 YS; 8-12 El; 54-58
Brin. For aircraft castings, housings, crankcases; age-
hardenable, high strength.

DOMAL AS100
Dominion Magnesium Ltd.
Al 10, Mn 0.5, bal Mg.
For die castings, instrument cases; high strength.

DOMAL AZ21X
Dominion Magnesium Ltd.
Al 1.2-2, Zn 0.4-0.75, 0.20 Mn min, bal Mg.
Extruded: 32,000 TS; 18,000 YS; 5 El. Tube: 32,000 TS;
15,000 YS; 5 El. Forged: 32,000 TS; 17,000 YS; 5 El. For
airframes, cowls, tanks, structures; good weldability and
workability.

DOMAL AZ31B
Dominion Magnesium Ltd.
Al 9, Mn 0.3, bal Mg.
Extruded: 37,000 TS; 28,000 YS; 12 El; 49 Brin. For aircraft
structures; high ductility.

DOMAL AZ31X
Dominion Magnesium Ltd.
Al 2.5-3.5, Zn 0.6-1.4, 0.2 Mn min, bal Mg.
Extruded: 32,000-34,000 TS; 16,000-20,000 YS; 6-8 El. Plate
1/2 Hard: 33,000-39,000 TS; 16,000-29,000 YS; 4-10 El. For
aircraft parts; good formability.

DOMAL AZ61A
Dominion Magnesium Ltd.
Al 6.5, Zn 1, bal Mg.
Extruded: 44,000 TS; 30,000 YS; 14 El; 60 Brin. For aircraft
structures; strong and tough.

DOMAL AZ61X
Dominion Magnesium Ltd.
Al 5.8-7.2, Zn 0.4-1.5, 0.20 Mn min, bal Mg.
Extruded: 39,000 TS; 24,000 YS; 9 El. Forged: 38,000 TS;
22,000 YS; 6 El. Tube: 36,000 TS; 16,000 YS; 7 El. For aircraft
parts; good weldability.

DOMAL AZ63A
Dominion Magnesium Ltd.
Zn 2.7-3.3, 0.20 Mn min, bal Mg.
Cast: 27,000 TS; 14,000 YS; 16 El; 50 Brin. Heat treated:
40,000 TS; 24,000 YS; 10 El; 60 Brin. For cylinder heads,
pistons, high strength castings; age-hardenable.

DOMAL AZ80A
Dominion Magnesium Ltd.
Al 8.5, Zn 0.5, bal Mg.
Extruded: 48,000 TS; 32,000 YS; 12 El; 60 Brin. Heat treated:
52,000 TS; 36,000 YS; 5 El; 82 Brin. For cylinder heads, valve
and pump bodies; age-hardenable.

DOMAL AZ80X
Dominion Magnesium Ltd.
Al 7.5-9.2, Zn 0.2-0.8, 0.20 Mn min, bal Mg.
F-temper: 43,000 TS; 28,000 YS; 6 El. T6-temper: 47,000 TS;
30,000 YS; 2 El. For aircraft parts; age hardenable.

DOMAL AZ91B
Dominion Magnesium Ltd.
Al 9, Zn 0.6, bal Mg.
Die cast: 33,000 TS; 22,000 YS; 3 El; 60 Brin. For die
castings, instrument housings; good strength.

DOMAL AZ91C
Dominion Magnesium Ltd.
Al 9, Zn 0.5, bal Mg.
Die cast: 38,000 TS; 19,000 YS; 9 El; 59 Brin. Heat treated:
40,000 TS; 20,000 YS; 4 El; 66 Brin. For general high strength
castings; age-hardenable.

DOMAL AZ92A
Dominion Magnesium Ltd.
Al 9, Zn 2, bal Mg.
Cast: 24,000 TS; 14,000 YS; 2 El; 65 Brin. Heat treated:
40,000 TS; 23,000 YS; 2 El; 84 Brin. For aircraft engine
components; age-hardenable.

DOMAL E733
Dominion Magnesium Ltd.
Zn 2-3.5, Zr 0.5-1, 2.5-4.0 rare earths, bal Mg.
T5-temper: 20,000 TS; 13,000 YS; 3 El. For engine castings,
diffuser and compressor casings; heat and creep resistant to
500°F.

DOMAL HZ32
Dominion Magnesium Ltd.
Th 3, Zn 2.3, Mn 0.7, bal Mg.
At 20°C: 30,000 TS; 13,000 YS; 7 El. At 250°C: 14,900 TS;
9000 YS; 50 El. At 350°C: 10,000 TS; 40 El. For jet engine
and missile components; high creep resistance to 660°F.

DOMAL M1B
Dominion Magnesium Ltd.
Mn 1.2, bal Mg.
F-temper: 33,000 TS; 27,000 YS; 44 El. H24-temper: 37,000
TS; 28,000 YS; 7 El; 56 Brin. For parts subjected to high
stresses; good weldability and corrosion resistance.

DOMAL TA54A
Dominion Magnesium Ltd.
Sn 5, Al 3, bal Mg.
Forged: 40,000 TS; 28,000 YS; 12 El; 52 Brin. For pistons; high ductility.

DOMAL ZH62
Dominion Magnesium Ltd.
Zn 5.8, Th 1.8, Zr 0.7, bal Mg.
Cast: 40,000 TS; 24,000 YS; 8.5 El. Sheet: 48,000 TS; 42,000 YS; 15 El. Extruded: 49,000 TS; 42,000 YS; 20 El. For aircraft engine and missile components; good formability and weldability.

DOMAL ZK31
Dominion Magnesium Ltd.
Zn 2.5-3.5, Zr 0.5-1, bal Mg.
Extruded: 40,000 TS; 28,000 YS; 5 El. Forged: 38,000 TS; 26,000 YS; 5 El. For airframe structural parts, landing wheels, gear casings; heavy duty service.

DOMAL ZK60
Dominion Magnesium Ltd.
Al 4.8-6.5, Fe 0-0.003, 0.50 Zr min, bal Mg.
F-temper: 42,000 TS; 28,000 YS; 5 El. T5-temper: 44,000 TS; 32,000 YS; 4 El. Cast: 35,000 TS; 18,000 YS; 8 El. T6-temper: 42,000 TS; 26,000 YS; 5 El. For aircraft structural members; age hardenable, shock resistant.

DOMAL-AZ91X
Dominion Magnesium Ltd.
Al 8.3-9.7, Zn 0.4-1, Mn 0.2-0.4, Si 0-0.1, bal Mg.
Die cast: 33,000 TS; 22,000 YS; 3 El; 60 Brin. Sand cast: 24,000 TS; 14,000 YS; 2 El; 52 Brin. Heat treated: 40,000 TS; 19,000 YS; 4 El; 66 Brin. For instrument housings, portable tools, casings. Age-hardenable.

DOMESTIC
Edgar T. Ward's Sons Co.
C 0.7-1.2, bal Fe.
For tools, punches, taps; water hardening.

DOMINIAL 170
Kind & Co. Edelstahlwerk
C 0.45, Cr, Ni, bal Fe.
For gears, bolts, crankshafts, fasteners. Oil hardened, shock resistant. *Obsolete*

DOMINIAL ACM 1
Kind & Co. Edelstahlwerk
C 0.33, Si 0.2, Mn 0.8, Cr 1.1, Mo 0.2, Al 1, bal Fe.
Alloyed tool steel. W. Nr. 2852.

DOMINIAL ACM 2
Kind & Co. Edelstahlwerk
C 0.34, Si 0.2, Mn 0.5, Cr 1.7, Mo 0.2, Ni 1, Al 1, bal Fe.
Alloyed tool steel. W. Nr. 8550.

DOMINIAL AWS
Kind & Co. Edelstahlwerk
C 0.5, Si 1.4, Mn 0.7, Cr 13, Ni 13, V 0.6, W 2.4, bal Fe.
Alloyed tool steel. W. Nr. 2731.

DOMINIAL BA
Kind & Co. Edelstahlwerk
C 1.05, Cr 1, Si 0.25, Mn 1, W 1.1, bal Fe.
For oil refinery equipment. Creep and wear resistant. W. Nr. 2419.

DOMINIAL BB
Kind & Co. Edelstahlwerk
C 1.2, Si 0.45, Mn 12, bal Fe.
Alloyed tool steel. W. Nr. 3401.

DOMINIAL BZM
Kind & Co. Edelstahlwerk
C, alloy, bal Fe.
For gears, shafts, bolts. Oil hardening. *Obsolete*

DOMINIAL C 2 MO
Kind & Co. Edelstahlwerk
C 1, Si 0.2, Mn 0.3, Cr 1.2, Mo 0.3, bal Fe.
Alloyed tool steel. W. Nr. 2303.

DOMINIAL CH
Kind & Co. Edelstahlwerk
C 2, Cr 12, Si 0.3, Mn 0.3, bal Fe.
For blanking and forming dies. Oil or air hardened, nondeforming. W. Nr. 2080.

DOMINIAL CH 10
Kind & Co. Edelstahlwerk
C 0.96, Si 0.3, Mn 0.3, Cr 11.5, Mo 0.9, V 0.9, bal Fe.
Alloyed tool steel. W. Nr. 2376.

DOMINIAL CH 16 V
Kind & Co. Edelstahlwerk
C 1.6, Si 0.4, Mn 0.4, Cr 12.5, Mo 0.7, V 1, bal Fe.
Alloyed tool steel. W. Nr. 2379.

DOMINIAL CH 160 W
Kind & Co. Edelstahlwerk
C 1.65, Cr 12, Mo 0.6, V 0.12, Si 0.3, Mn 0.4, W 0.6, bal Fe.
For blanking and forming dies, punches. Air hardened, nondeforming. W. Nr. 2601.

DOMINIAL CH 165
Kind & Co. Edelstahlwerk
C 1.7, Cr 12, V 0.12, Si 0.3, Mn 0.3, bal Fe.
For blanking and forming dies, punches. Air hardened, nondeforming. W. Nr. 2201.

DOMINIAL CH 17 CO
Kind & Co. Edelstahlwerk
C 1.65, Si 0.3, Mn 0.4, Co 1.3, Cr 12, Mo 0.6, bal Fe.
Alloyed tool steel. W. Nr. 2880.

DOMINIAL CH 2 V
Kind & Co. Edelstahlwerk
C 2.1, Si 0.25, Mn 0.3, Cr 12.5, Mo 0.9, V 2.1, bal Fe.
Alloyed tool steel. W. Nr. 2378.

DOMINIAL CH 5 M
Kind & Co. Edelstahlwerk
C 1, Si 0.3, Mn 0.6, Cr 5.25, Mo 1.2, V 0.3, bal Fe.
Alloyed tool steel. W. Nr. 2363.

DOMINIAL CH 8 CO
Kind & Co. Edelstahlwerk
C 0.8, Si 0.5, Mn 0.4, Co 1.1, Cr 13.5, bal Fe.
Alloyed tool steel. W. Nr. 2883.

DOMINIAL CH CO
Kind & Co. Edelstahlwerk
C 2.1, Si 0.3, Mn 0.4, Co 1, Cr 12.5, W 0.7, Mo 0.5, bal Fe.
Alloyed tool steel. W. Nr. 2884.

DOMINIAL CH V 4
Kind & Co. Edelstahlwerk
C 2, Si 0.3, Mn 0.4, Cr 12, Mo 0.8, V 3.9, bal Fe.
Alloyed tool steel. W. Nr. 2285.

DOMINIAL CH-SPEZIAL
Kind & Co. Edelstahlwerk
C 1.65, Si 0.4, Mn 0.4, Co 1.5, Cr 13.5, Mo 1.2, bal Fe.
Alloyed tool steel.

DOMINIAL CHK
Kind & Co. Edelstahlwerk
C 2, Si 0.3, Mn 0.3, Cr 12, V 0.2, bal Fe.
Alloyed tool steel.

DOMINIAL CHW
Kind & Co. Edelstahlwerk
C 2.1, Cr 12, W 0.8, Si 0.3, Mn 0.3, bal Fe.
For blanking and forming dies, punches. Oil or air hardened, nondeforming. W. Nr. 2436.

DOMINIAL CHW EXTRA
Kind & Co. Edelstahlwerk
C 2.1, Si 0.3, Mn 0.3, Cr 12, W 1.1, bal Fe.
Alloyed tool steel.

DOMINIAL CHWV
Kind & Co. Edelstahlwerk
C 2.1, Si 0.3, Mn 0.5, Cr 12, Mo 0.6, V 1.1, W 0.8, bal Fe.
Alloyed tool steel.

DOMINIAL CM 167
Kind & Co. Edelstahlwerk
C 0.45, Cr 1.5, Mo 0.75, V 0.3, Si 0.3, Mn 0.7, bal Fe.
For gears, bolts, crankshafts, fasteners. Oil hardened, tough. W. Nr. 2323.

DOMINIAL CMR
Kind & Co. Edelstahlwerk
C 0.35, Si 0.4, Mn 0.3, Cr 17, Mo 1.2, bal Fe.
Alloyed tool steel. W. Nr. 2316.

DOMINIAL CR 7 V
Kind & Co. Edelstahlwerk
C 0.47, Si 0.85, Mn 0.35, Cr 7.8, Mo 1.4, V 1.35, bal Fe.
Alloyed tool steel.

DOMINIAL DAG
Kind & Co. Edelstahlwerk
C 1.5, Mn 0.6, Cr 1.5, Si 0.25, bal Fe.
For bearings, sleeves, blanking dies. Water hardened, wear resistant. W. Nr. 2063.

DOMINIAL DKS
Kind & Co. Edelstahlwerk
C 0.2, Si 0.35, Mn 0.4, Cr 13.2, Mo 1.1, bal Fe.
Alloyed tool steel.

DOMINIAL DS
Kind & Co. Edelstahlwerk
C 0.38, Cr 6, Si 0.9, V 0.1, bal Fe.
For oil refinery equipment. Creep and heat resistant.
Obsolete

DOMINIAL DSW
Kind & Co. Edelstahlwerk
C 0.37, Si 1, Cr 1.2, V 0.2, W 2, Mn 0.3, bal Fe.
For cold heading dies, crimpers, punches, upsetters. Oil hardened, tough. W. Nr. 2541.

DOMINIAL DWK
Kind & Co. Edelstahlwerk
C 1.3, Cr 0-0.2, W 5, Si 0.25, Mn 0.3, bal Fe.
For engravers' tools, cutters. Oil hardened. W. Nr. 2453.

DOMINIAL EC 4
Kind & Co. Edelstahlwerk
C 0.15, Ni 3.5, Cr 0.75, Si 0.25, Mn 0.4, bal Fe.
For bolts, fasteners, gears, cams, camshafts. Case hardened steel, shock resistant. W. Nr. 2735.

DOMINIAL EC 5
Kind & Co. Edelstahlwerk
C 0.15, Cr 1.2, Ni 4.5, Si 0.25, Mn 0.4, bal Fe.
For gears, cams, camshafts. Case hardened steel, shock resistant. W. Nr. 2745.

DOMINIAL ECNL
Kind & Co. Edelstahlwerk
C 0.2, Cr 1.2, Mo 0.2, Ni 4, Si 0.25, Mn 0.4, bal Fe.
For gears, bolts, camshafts, fasteners. Case hardened steel, shock resistant. W. Nr. 2764.

DOMINIAL EDN
Kind & Co. Edelstahlwerk
C 0.18, Si 0.3, Mn 0.5, Cr 2, Ni 2, bal Fe.
Alloyed tool steel. W. Nr. 2722.

DOMINIAL EXTRA
Kind & Co. Edelstahlwerk
C 1.05, Cr 1.5, Mn 0-0.2, Si 0.3, bal Fe.
For bearings, liners, sleeves. Water hardened, wear resistant.
W. Nr. 2060.

DOMINIAL EXTRA E
Kind & Co. Edelstahlwerk
C 1, Si 0.3, Mn 0.3, Cr 1.5, bal Fe.
Alloyed tool steel. W. Nr. 2067.

DOMINIAL EXTRA-MO
Kind & Co. Edelstahlwerk
C 0.98, Si 0.3, Mn 0.7, Cr 1.8, Mo 0.3, bal Fe.
Alloyed tool steel.

DOMINIAL FAM
Kind & Co. Edelstahlwerk
C 0.2, Si 0-1, Mn 0-1, Cr 17, Ni 1.7, bal Fe.
Alloyed tool steel. W. Nr. 2787.

DOMINIAL FGS
Kind & Co. Edelstahlwerk
C 1.25, Si 1.15, Mn 0.7, Cr 1.2, bal Fe.
For bearings, liners, sleeves. Oil hardened, wear resistant. W.
Nr. 2109.

DOMINIAL GH
Kind & Co. Edelstahlwerk
C 2.1, Cr 11.5, bal Fe.
For blanking and forming dies, punches. Oil or air hardened,
nondeforming. *Obsolete*

DOMINIAL H 13
Kind & Co. Edelstahlwerk
C 0-0.08, Si 1.2, Mn 0.7, Cr 13, Al 1, bal Fe.
Heat resistant steel. W. Nr. 4724.

DOMINIAL H 13 SI
Kind & Co. Edelstahlwerk
C 0-0.08, Si 2.1, Mn 0.7, Cr 13, bal Fe.
Heat resistant steel. W. Nr. 4722.

DOMINIAL H 18
Kind & Co. Edelstahlwerk
C 0-0.08, Si 1, Mn 0.7, Cr 18, Al 1, bal Fe.
Heat resistant steel. W. Nr. 4742.

DOMINIAL H 264
Kind & Co. Edelstahlwerk
C 0-0.2, Si 1.1, Mn 0.7, Cr 25, Ni 4, bal Fe.
Heat resistant steel. W. Nr. 4821.

DOMINIAL HBS 1
Kind & Co. Edelstahlwerk
C 0-0.12, Si 2, Mn 0.8, Cr 20, Ni 12, bal Fe.
Heat resistant steel. W. Nr. 4828.

DOMINIAL HBS 2
Kind & Co. Edelstahlwerk
C 0-0.12, Si 2, Mn 0.8, Cr 25, Ni 20, bal Fe.
Heat resistant steel. W. Nr. 4841.

DOMINIAL HBS 36
Kind & Co. Edelstahlwerk
C 0-0.1, Si 1.7, Mn 1.25, Cr 16, Ni 36, bal Fe.
Heat resistant steel. W. Nr. 4864.

DOMINIAL HMOD
Kind & Co. Edelstahlwerk
C 0.45, Si 0.4, Mn 0.4, Co 4.5, Cr 4.5, Mo 3, V 2, bal Fe.
Alloyed tool steel. W. Nr. 2889.

DOMINIAL HW
Kind & Co. Edelstahlwerk
C 0.3, Si 0.4, Mn 0.5, Co 4.5, Cr 1.2, Mo 0.7, V 0.5, W 5, bal
Fe.
Alloyed tool steel.

DOMINIAL HWD
Kind & Co. Edelstahlwerk
C 0.4, Si 0.3, Mn 0.4, Co 4.5, Cr 4.5, Mo 0.5, V 2.1, W 4.5, bal
Fe.
Alloyed tool steel. W. Nr. 2678.

DOMINIAL HWF
Kind & Co. Edelstahlwerk
C 0.05, Si 1, Mn 1.1, Cr 15, Mo 1.5, Ni 26, Ti 2.1, bal Fe.
Alloyed tool steel. W. Nr. 2779.

DOMINIAL KC 15
Kind & Co. Edelstahlwerk
C 0.45, Cr 1.5, Si 1.4, V 0.1, Mn 0.6, bal Fe.
For springs, gears, bolts, crankshafts. Oil hardened, shock
resistant. W. Nr. 2249.

DOMINIAL KHP
Kind & Co. Edelstahlwerk
C 0.15, Cr 0.65, Mn 0.5, bal Fe. .
For gears, pinions, cams, camshafts. Case hardened steel.
Obsolete

DOMINIAL KL
Kind & Co. Edelstahlwerk
C 0.6, Si 0.6, Mn 0.3, Cr 1.1, V 0.2, W 2, bal Fe.
For cold heading dies, crimpers, punches, upsetters. Oil
hardened. W. Nr. 2550.

DOMINIAL KS 60
Kind & Co. Edelstahlwerk
C 0.62, Si 0.9, Mn 0.8, Cr 1.2, V 0.1, bal Fe.
Alloyed tool steel. W. Nr. 2243.

DOMINIAL KS 80
Kind & Co. Edelstahlwerk
C 0.9, Si 1.2, Mn 0.7, Cr 1.2, bal Fe.
Alloyed tool steel. W. Nr. 2108.

DOMINIAL KST
Kind & Co. Edelstahlwerk
C 1, Si 0.2, Mn 0.2, V 0.1, bal Fe.
For cutters, drills, taps, reamers, broaches. Type W2; water
hardened. W. Nr. 2833.

DOMINIAL KSTSW
Kind & Co. Edelstahlwerk
C 0.95, Si 0.3, Mn 0.4, V 0.4, bal Fe.
Alloyed tool steel. W. Nr. 2835.

DOMINIAL KSV
Kind & Co. Edelstahlwerk
C 1.45, Si 0.3, Mn 0.4, V 3.25, bal Fe.
Alloyed tool steel. W. Nr. 2838.

DOMINIAL KTS
Kind & Co. Edelstahlwerk
C 0.42, Si 0.3, Mn 1.5, Cr 2, Mo 0.2, S, bal Fe.
Alloyed tool steel. W. Nr. 2312.

DOMINIAL KTW
Kind & Co. Edelstahlwerk
C 0.42, Si 0.3, Mn 1.5, Cr 2, Mo 0.2, bal Fe.
For gears, shafts, bolts, studs, crankshafts. Oil hardened,
tough. W. Nr. 2311.

DOMINIAL KTW-NI
Kind & Co. Edelstahlwerk
C 0.42, Si 0.3, Mn 1.5, Cr 2, Mo 0.2, Ni 1, bal Fe.
Alloyed tool steel. W. Nr. 2738.

DOMINIAL KZR
Kind & Co. Edelstahlwerk
C 1.35, Cr 0.3, V 0.1, Si 0.25, Mn 0.3, bal Fe.
For engravers' tools, cutters, taps, reamers. Water hardened.
W. Nr. 2206.

DOMINIAL MA
Kind & Co. Edelstahlwerk
C 0.3, Cr 2.7, V 0.35, W 9, Si 0.3, Mn 0.3, bal Fe.
For upsetters, riveters, punches, crimpers. Hot work steel, oil
hardened. W. Nr. 2581.

DOMINIAL MA-REKORD
Kind & Co. Edelstahlwerk
C 0.55, Si 1.4, Mn 0.7, Co 1.5, Cr 4, Mo 0.6, Ni 11.5, V 1.1-1,
W 12, bal Fe.
Alloyed tool steel. W. Nr. 2758.

DOMINIAL MA-SUPRA
Kind & Co. Edelstahlwerk
C 0.3, Co 2.1, Cr 2.4, V 0.25, W 8.5, Si 0.3, Mn 0.3, bal Fe.
For extrusion press rams, hot punches, shears. Hot work
steel, oil hardened. W. Nr. 2662.

DOMINIAL MAK
Kind & Co. Edelstahlwerk
C 0.3, Cr 2.35, V 0.6, W 4.5, Si 0.3, Mn 0.3, bal Fe.
For upsetters, riveters, punches. Hot work steel, oil hardened.
W. Nr. 2567.

DOMINIAL MC
Kind & Co. Edelstahlwerk
C 0.3, Cr 2.5, Mo 0.2, V 0.2, Si 0.25, Mn 0.6, bal Fe.
For gears, bolts, crankshafts, fasteners. Oil hardened, tough.
W. Nr. 2307.

DOMINIAL MK
Kind & Co. Edelstahlwerk
C 1.05, Si 0.25, Mn 1.1, Cr 0.9, bal Fe.
For drills, taps, bearings, races. Water hardened, wear
resistant. W. Nr. 2127.

DOMINIAL MKST
Kind & Co. Edelstahlwerk
C 0.9, V 0.12, Mn 2, Si 0.25, Cr 0.35, bal Fe.
For punches, forming dies, shears, crimpers. Oil hardened,
nondeforming. W. Nr. 2842.

DOMINIAL N 400
Kind & Co. Edelstahlwerk
C 0.4, Si 0.25, Mn 0.4, Cr 1.35, Ni 3.9, Mo 0.25, W 0.5, bal Fe.
For gears, bolts, crankshafts, fasteners. Oil hardened, shock
resistant. W. Nr. 2767.

DOMINIAL NC 18
Kind & Co. Edelstahlwerk
C 0.53, Si 0.25, Mn 0.45, Cr 0.6, Mo 2.7, bal Fe.
Alloyed tool steel. W. Nr. 2718.

DOMINIAL P 120
Kind & Co. Edelstahlwerk
C 0.26, Si 0.4, Mn 0.3, Cr 0.75, Mo 0.3, Ni 1.5, V 0.2, bal Fe.
Alloyed tool steel. W. Nr. 2726.

DOMINIAL P 170
Kind & Co. Edelstahlwerk
C 0.45, Si 0.3, Mn 0.7, Cr 1.4, Ni 1.7, bal Fe.
Alloyed tool steel. W. Nr. 2710.

DOMINIAL P 264
Kind & Co. Edelstahlwerk
C 0.2, Si 1.2, Mn 0.5, Cr 25, Ni 4.5, bal Fe.
Alloyed tool steel. W. Nr. 2789.

DOMINIAL P 430
Kind & Co. Edelstahlwerk
C 2, Si 0.2, Mn 1, Cr 2.1, bal Fe.
Alloyed tool steel. W. Nr. 2129.

DOMINIAL P 50
Kind & Co. Edelstahlwerk
C 0.28, Si 0.3, Mn 0.3, Cr 0.4, Mo 1.2, Ni 4.5, V 0.2, bal Fe.
Alloyed tool steel. W. Nr. 2747.

DOMINIAL P 604
Kind & Co. Edelstahlwerk
C 0-0.07, Si 0-0.2, Mn 0-0.2, Cr 3.7, Mo 0.5, bal Fe.
Alloyed tool steel. W. Nr. 2341.

DOMINIAL P 80
Kind & Co. Edelstahlwerk
C 0.28, Si 0.4, Mn 0.3, Cr 0.75, Mo 0.6, Ni 2.5, V 0.3, bal Fe.
Alloyed tool steel. W. Nr. 2740.

DOMINIAL PD
Kind & Co. Edelstahlwerk
C 0.58, Cr 4, Mo 0.9, V 0.8, W 9, Si 0.25, Mn 0.25, bal Fe.
For lathe and planer tools, reamers, broaches, taps. High speed steel. W. Nr. 2622.

DOMINIAL PK
Kind & Co. Edelstahlwerk
C 0.45, Cr 1.1, V 0.2, W 2, Si 1, Mn 0.3, bal Fe.
For cold heading tools, punches, mandrels. Oil hardened, tough. W. Nr. 2542.

DOMINIAL PK 17
Kind & Co. Edelstahlwerk
C 0.45, Si 0.9, Mn 0.3, Cr 1.7, V 0.2, W 2, bal Fe.
For hot shears and punches, upsetters. Hot work tools, oil hardened. W. Nr. 2547.

DOMINIAL PW
Kind & Co. Edelstahlwerk
C 0.45, Si 1, Mn 0.45, Cr 1.1, V 0.2, W 2.4, bal Fe.
Alloyed tool steel.

DOMINIAL PW 15
Kind & Co. Edelstahlwerk
C 0.55, Cr 0.7, V 0.1, Ni 1.7, Mo 0.3, Si 0.3, Mn 0.8, bal Fe.
For gears, bolts, crankshafts. Oil hardened, shock resistant. W. Nr. 2713.

DOMINIAL PW 75
Kind & Co. Edelstahlwerk
C 0.75, Si 0.2, Mn 0.25, Cr 1.5, Mo 0.7, Ni 0.5, W 0.3, bal Fe.
Alloyed tool steel. W. Nr. 2762.

DOMINIAL PWH
Kind & Co. Edelstahlwerk
C 0.2, Si 0.3, Mn 0.7, Cr 0-0.15, Mo 3.2, Ni 3.2, bal Fe.
Alloyed tool steel. W. Nr. 2777.

DOMINIAL PWM
Kind & Co. Edelstahlwerk
C 0.55, Mn 0.8, Si 0.3, Cr 1.1, Mo 0.5, Ni 1.7, V 0.1, bal Fe.
For gears, bolts, crankshafts, axles. Oil hardened, shock resistant. W. Nr. 2714.

DOMINIAL PWU
Kind & Co. Edelstahlwerk
C 0.55, Si 0.3, Mn 0.7, Cr 1.1, Mo 0.8, Ni 1.7, V 0.12, bal Fe.
Alloyed tool steel. W. Nr. 2744.

DOMINIAL R 13
Kind & Co. Edelstahlwerk
Stainless steel. C 0-0.08, Si 0.4, Mn 0.5, Cr 13, bal Fe.
Stainless steel. W. Nr. 4000.

DOMINIAL R 13 F
Kind & Co. Edelstahlwerk
Stainless steel. C 0-0.08, Si 0.4, Cr 14, Mn 0.5, bal Fe.
Annealed: 75,000 TS; 40,000 YS; 35 El; 70 RA; 155 Brin. Cold drawn: 100,000 TS; 85,000 YS; 17 El; 60 RA; 205 Brin. For turbine blades, surgical instruments, valves, cutlery. Type 410; stainless. W. Nr. 4001.

DOMINIAL R 13 FM
Kind & Co. Edelstahlwerk
Stainless steel. C 0.1, Si 0.4, Mn 0.5, Cr 13, bal Fe.
Stainless steel. W. Nr. 4006.

DOMINIAL R 17 F
Kind & Co. Edelstahlwerk
Stainless steel. C 0-0.08, Si 0.4, Mn 0.4, Cr 17, bal Fe.
Stainless steel. W. Nr. 4016.

DOMINIAL RAN
Kind & Co. Edelstahlwerk
Stainless steel.
Now DOMINIAL RANP.

DOMINIAL RAN 304
Kind & Co. Edelstahlwerk
Stainless steel. C 0-0.06, Si 0.4, Mn 1.5, Cr 18, Ni 10, bal Fe.
Stainless steel. W. Nr. 4301.

DOMINIAL RAN 304 L
Kind & Co. Edelstahlwerk
Stainless steel. C 0-0.03, Si 0.4, Mn 1.5, Cr 19, Ni 11, bal Fe.
Stainless steel. W. Nr. 4306.

DOMINIAL RAN 316 L
Kind & Co. Edelstahlwerk
Stainless steel. C 0-0.03, Si 0.4, Mn 1.5, Cr 17, Mo 2.3, Ni 13, bal Fe.
Stainless steel. W. Nr. 4404.

DOMINIAL RANP
Kind & Co. Edelstahlwerk
Stainless steel. C 0-0.04, Cr 18, Ni 13, Si 0.4, Mn 1.5, bal Fe.
Annealed: 80,000 TS; 35,000 YS; 55 El; 75 RA; 150 Brin. Cold drawn: 180,000 TS; 150,000 YS; 10 El; 250 Brin. For chemical plant equipment, tanks, mixers, filters. Type 302; stainless, austenitic. W. Nr. 3941.

DOMINIAL RAS
Kind & Co. Edelstahlwerk
Stainless steel. C 0-0.08, Cr 18, Ni 10, Si 0.4, Mn 1.5, Ti = 5 x C, bal Fe.
Annealed: 85,000 TS; 35,000 YS; 55 El; 65 RA; 150 Brin. Cold drawn: 95,000 TS; 60,000 YS; 40 El; 60 RA; 185 Brin. For welded chemical plant equipment, tanks, mixers. Type 321; stainless, austenitic. W. Nr. 4541.

DOMINIAL RAS 317
Kind & Co. Edelstahlwerk
Stainless steel. C 0-0.06, Si 0.4, Mn 1.5, Cr 17, Mo 4.5, Ni 14, bal Fe.
Stainless steel. W. Nr. 4449.

DOMINIAL RAS 4
Kind & Co. Edelstahlwerk
Stainless steel. C 0-0.08, Cr 17, Mo 2.3, Ni 12, Si 0.4, Mn 1.3, Ti = 5 x C, bal Fe.
Annealed: 85,000 TS; 35,000 YS; 45 El; 60 RA; 160 Brin. For welded acid resistant chemical plant equipment. Type 316Ti; stainless, austenitic. W. Nr. 4571.

DOMINIAL RAS 400
Kind & Co. Edelstahlwerk
Stainless steel. C 0-0.06, Si 0.4, Mn 1.3, Cr 17, Mo 2.3, Ni 12, bal Fe.
Stainless steel. W. Nr. 4401.

DOMINIAL RAS-CU
Kind & Co. Edelstahlwerk
Stainless steel. C 0-0.07, Si 0-1, Mn 0-1, Cr 16.5, Ni 4, Nb 0.3, bal Fe.
Stainless steel. W. Nr. 4542.

DOMINIAL RAS-EXTRA
Kind & Co. Edelstahlwerk
Stainless steel. C 0-0.06, Si 0.4, Mn 1.3, Cr 18, Mo 2.3, Ni 20, Cu 2, Nb = 8 x C, bal Fe.
Stainless steel. W. Nr. 4505.

DOMINIAL RASA
Kind & Co. Edelstahlwerk
Stainless steel. C 0-0.08, Si 0.4, Mn 1.5, Cr 18, Ni 10, Nb = 8 x C, bal Fe.
Stainless steel. W. Nr. 4550.

DOMINIAL RASA 4
Kind & Co. Edelstahlwerk
Stainless steel. C 0-0.08, Si 0.4, Mn 1.5, Cr 17, Mo 2.3, Ni 12, Nb = 8 x C, bal Fe.
Stainless steel. W. Nr. 4580.

DOMINIAL RF
Kind & Co. Edelstahlwerk
Stainless steel. C 0.4, Cr 13.2, Si 0.4, Mn 0.35, bal Fe.
Annealed: 95,000 TS; 50,000 YS; 25 El; 55 RA; 200 Brin. For valves, cutlery, surgical and dental instruments. Type 420; stainless, hardenable. W. Nr. 2083.

DOMINIAL RF-SPEZIAL
Kind & Co. Edelstahlwerk
C 0.2, Si 0.4, Mn 0.35, Cr 13.2, bal Fe.
Alloyed tool steel. W. Nr. 2082.

DOMINIAL RM 10 CO
Kind & Co. Edelstahlwerk
C 0.2, Si 0.2, Mn 0.5, Co 10, Cr 9.5, Mo 2, W 5.5, bal Fe.
Alloyed tool steel. W. Nr. 2888.

DOMINIAL RM 10 MO
Kind & Co. Edelstahlwerk
C 0.15, Si 0.1, Mn 0.2, Co 10, Cr 10, Mo 5, V 0.5, bal Fe.
Alloyed tool steel. W. Nr. 2886.

DOMINIAL RM 13
Kind & Co. Edelstahlwerk
Stainless steel. C 0.2, Cr 13, Si 0.4, Mn 0.4, bal Fe.
Annealed: 95,000 TS; 50,000 YS; 25 El; 55 RA; 200 Brin. For valves, cutlery, turbine blades, surgical instruments. Type 420; stainless, hardenable. W. Nr. 4021.

DOMINIAL RM 13 H
Kind & Co. Edelstahlwerk
Stainless steel. C 0.46, Si 0-1, Cr 13, Mn 0-1, bal Fe.
Annealed: 95,000 TS; 50,000 YS; 25 El; 55 RA; 200 Brin. For valves, cutlery, surgical and dental instruments. Type 420; stainless, hardenable. W. Nr. 4034.

DOMINIAL RM 13 MO
Kind & Co. Edelstahlwerk
Stainless steel. C 0.2, Si 0.4, Cr 13, Mo 1.2, Mn 0.4, bal Fe.
Annealed: 95,000 TS; 50,000 YS; 25 El; 55 RA; 200 Brin. For turbine blades, valves, cutlery. Type 420Mo; stainless. W. Nr. 4120.

DOMINIAL RM 161 A
Kind & Co. Edelstahlwerk
Stainless steel. C 0.14, Si 0.4, Mn 0.7, Cr 17, Mo 0.25, S 0.2, bal Fe.
Stainless steel. W. Nr. 4104.

DOMINIAL RM 17
Kind & Co. Edelstahlwerk
Stainless steel. C 0.2, Cr 17, Ni 1.7, bal Fe.
Annealed: 125,000 TS; 95,000 YS; 20 El; 55 RA; 260 Brin. Cold drawn: 130,000 TS; 110,000 YS; 15 El; 35 RA; 270 Brin. For pumps, valves, marine hardware. Type 431; stainless. W. Nr. 4057.

DOMINIAL RM 174
Kind & Co. Edelstahlwerk
Stainless steel. C 0.38, Si 0.4, Cr 17, Mo 1.2, Mn 0.4, Ni 0-1, bal Fe.
For chemical plant equipment, furnace parts. Corrosion and heat resistant. W. Nr. 4122.

DOMINIAL RM 18
Kind & Co. Edelstahlwerk
Stainless steel. C 0-0.08, Si 0.4, Mn 0.5, Cr 17, Ti = 7 x C, bal Fe.
Stainless steel. W. Nr. 4510.

DOMINIAL RM 189
Kind & Co. Edelstahlwerk
Stainless steel. C 0.9, Cr 18, Mo 1.2, Si 0.4, Mn 0.4, V 0.1, bal Fe.
For cutlery, valves, ball bearings, surgical instruments. Type 440B; corrosion resistant, hardenable. W. Nr. 4112.

DOMINIAL RM 195
Kind & Co. Edelstahlwerk
Stainless steel. C 1.1, Si 0.4, Mn 0.4, Cr 15, Mo 0.5, V 0.12, bal Fe.
Stainless steel. W. Nr. 4111.

DOMINIAL RP
Kind & Co. Edelstahlwerk
C 0.32, Si 0.4, Mn 0.4, Cr 3, Mo 2.8, V 0.6, bal Fe.
Alloyed tool steel. W. Nr. 2365.

DOMINIAL RP CO
Kind & Co. Edelstahlwerk
C 0.32, Si 0.4, Mn 0.4, Co 3, Cr 3, Mo 2.8, V 0.6, bal Fe.
Alloyed tool steel. W. Nr. 2885.

DOMINIAL RP CO 5
Kind & Co. Edelstahlwerk
C 0.32, Si 0.4, Mn 0.4, Co 4.5, Cr 3, Mo 2.8, V 0.6, bal Fe.
Alloyed tool steel. W. Nr. 2887.

DOMINIAL RPMOH
Kind & Co. Edelstahlwerk
C 0.8, Si 0.2, Mn 0.3, Cr 4, Mo 4.25, V 1, bal Fe.
Alloyed tool steel. W. Nr. 2369.

DOMINIAL RPU
Kind & Co. Edelstahlwerk
C 0.38, Si 0.4, Mn 0.4, Cr 5.2, Mo 2.8, V 0.6, bal Fe.
Alloyed tool steel. W. Nr. 2367.

DOMINIAL RS
Kind & Co. Edelstahlwerk
C 0.62, Si 1.1, Mn 1.1, Cr 0.6, bal Fe.
Alloyed tool steel. W. Nr. 2101.

DOMINIAL SA 50 NI
Kind & Co. Edelstahlwerk
C 0-0.12, Si 0-0.5, Mn 0-0.1, Co 11, Cr 19, Mo 9.5, Ti 3, Al 1.6, others, bal Ni.
Alloyed tool steel. W. Nr. 2.4973.

DOMINIAL SA 718
Kind & Co. Edelstahlwerk
C 0.05, Si 0-0.35, Mn 0-0.35, Cr 19, Mo 3, Nb 5, Ti 0.9, Al 0.5, bal Ni.
Alloyed tool steel. W. Nr. 2.4668.

DOMINIAL SACO 21
Kind & Co. Edelstahlwerk
Si 0-1, Mn 0-1, Cr 28, Mo 5.5, Ni 2, bal Co.
Alloyed tool steel. W. Nr. 2.4979.

DOMINIAL SES
Kind & Co. Edelstahlwerk
C 1.2, Cr 0.7, V 0.1, Mn 0.3, Si 0.2, bal Fe.
For bearings, sleeves, cutters. Water hardened, wear resistant. W. Nr. 2210.

DOMINIAL SGW
Kind & Co. Edelstahlwerk
C 0.9, Mn 0.3, Cr 0.8, Si 0.25, bal Fe.
For cutters, bearings, sleeves. Water hardened, wear resistant. W. Nr. 2056.

DOMINIAL SHRS
Kind & Co. Edelstahlwerk
C 1.65, Si 0.3, Mn 0.4, Cr 12, Mo 0.6, V 0.12, W 0.6, bal Fe.
Alloyed tool steel.

DOMINIAL SN
Kind & Co. Edelstahlwerk
C 0.5, Mn 0.5, Cr 1, Ni 3.3, Si 0.3, bal Fe.
For gears, bolts, crankshafts, axles, fasteners. Oil hardened, shock resistant. W. Nr. 2721.

DOMINIAL SP
Kind & Co. Edelstahlwerk
C 0.6, Si 1, Mn 1, Cr 0.3, bal Fe.
For punches, dies, upsetters, springs, crimpers. Oil hardened, tough. W. Nr. 2826.

DOMINIAL SS 50
Kind & Co. Edelstahlwerk
C 0.5, Mn 1, Cr 1.1, V 0.1, Si 0.3, bal Fe.
For gears, springs, bolts, studs, crankshafts. Oil hardened, shock resistant. W. Nr. 2241.

DOMINIAL SZS
Kind & Co. Edelstahlwerk
C 1.5, Si 0.0, Mn 0.0, Cr 0.3, V 1.5, W 0-0.8, bal Fe.
Alloyed tool steel. W. Nr. 2203.

DOMINIAL TCG
Kind & Co. Edelstahlwerk
C 0.37, Si 0.5, Mn 0.4, Cr 1.5, Mo 0.9, Ni 1.3, V 0.55, W 2.1, bal Fe.
Alloyed tool steel.

DOMINIAL UHF 3
Kind & Co. Edelstahlwerk
C 0-0.03, Si 0-0.1, Mn 0-0.1, Co 9.5, Mo 5.2, Ni 18, Ti 0.95, others, bal Fe.
Alloyed tool steel. W. Nr. 2709.

DOMINIAL UNI
Kind & Co. Edelstahlwerk
C 1, Si 0.3, Mn 1.1, Cr 0.6, V 0.1, W 0.6, bal Fe.
Alloyed tool steel. W. Nr. 2510.

DOMINIAL US
Kind & Co. Edelstahlwerk
C 0.36, Si 1, Mn 0.4, Cr 5.2, Mo 1.4, V 0.3, W 1.4, bal Fe.
Alloyed tool steel. W. Nr. 2606.

DOMINIAL US 6
Kind & Co. Edelstahlwerk
C 0.56, Si 1, Mn 0.4, Cr 5.2, Mo 1.3, V 0.3, W 1.4, bal Fe.
Alloyed tool steel.

DOMINIAL USCO V
Kind & Co. Edelstahlwerk
C 0.4, Si 0.3, Mn 0.3, Co 1.6, Cr 3.3, Mo 1.2, V 1.25, W 2.4, bal Fe.
Alloyed tool steel.

DOMINIAL USD
Kind & Co. Edelstahlwerk
C 0.4, Si 1, Mn 0.4, Cr 5.2, Mo 1.3, V 1, bal Fe.
Alloyed tool steel. W. Nr. 2344.

DOMINIAL USDH
Kind & Co. Edelstahlwerk
C 0.5, Si 1, Mn 0.4, Cr 5.2, Mo 1.3, V 1, bal Fe.
Alloyed tool steel. W. Nr. 2345.

DOMINIAL USN
Kind & Co. Edelstahlwerk
C 0.38, Si 1, Mn 0.4, Cr 5.2, Mo 1.3, V 0.4, bal Fe.
Alloyed tool steel. W. Nr. 2343.

DOMINIAL USNS
Kind & Co. Edelstahlwerk
C 0.4, Si 1, Mn 0.4, Cr 5.2, Mo 1.3, V 1, S, bal Fe.
Alloyed tool steel. W. Nr. 2347.

DOMINIAL VCMO 140
Kind & Co. Edelstahlwerk
C 0.42, Si 0.3, Mn 0.7, Cr 1, Mo 0.2, bal Fe.
Alloyed tool steel.

DOMINIAL VSF
Kind & Co. Edelstahlwerk
C 0.9, Mn 1, Si 0.37, bal Fe.
For drills, taps, reamers, hobs, broaches. Water hardened, wear resistant. *Obsolete*

DOMINIAL W 44
Kind & Co. Edelstahlwerk
C 0.45, Cr 1.6, Mo 0.6, V 0.8, W 0.5, Si 0.6, Mn 0.4, bal Fe.
For cold heading dies, upsetters, riveters. Oil hardened, tough. W. Nr. 2603.

DOMINIAL WEH
Kind & Co. Edelstahlwerk
C 0.21, Mn 1.2, Cr 1.2, Si 0.3, bal Fe.
For gears, cams, camshafts, fasteners. Case hardened steel. W. Nr. 2162.

DOMINIAL WEH 2
Kind & Co. Edelstahlwerk
C 0.16, Si 0.3, Mn 1.2, Cr 1, bal Fe.
Alloyed tool steel. W. Nr. 2161.

DOMINIAL WF
Kind & Co. Edelstahlwerk
C 1.2, Cr 0.2, V 0.1, W 1, bal Fe.
For blanking and forming dies, cutters. Water hardened, wear resistant. *Obsolete*

DOMINIAL WF 2
Kind & Co. Edelstahlwerk
C 1.15, Si 0.2, Mn 0.3, Cr 0.2, W 2, bal Fe.
For blanking dies, cutters, punches. Water hardened, wear resistant. W. Nr. 2442.

DOMINIAL WMK
Kind & Co. Edelstahlwerk
C 0.32, Si 0.8, Cr 1.1, V 0.2, W 4, Mn 0.4, bal Fe.
For cold heading dies, punches, upsetters. Oil hardened, tough. W. Nr. 2564.

DOMINIAL WR
Kind & Co. Edelstahlwerk
C 1.4, Si 0.3, V 0.3, W 3.1, Mn 0.3, Cr 0.3, bal Fe.
For bearings, forming dies, cutters. Water hardened. W. Nr. 2562.

DOMINIAL WSE
Kind & Co. Edelstahlwerk
C 0.54, Si 0.3, Mn 0.7, Cr 0.7, Mo 0.3, Ni 1.7, V 0.12, bal Fe.
Alloyed tool steel. W. Nr. 2711.

DOMINIAL ZCS
Kind & Co. Edelstahlwerk
Alloyed tool steel.

DOMINIAL ZF 1
Kind & Co. Edelstahlwerk
C 0.12, Si 2, Mn 0.9, Cr 20, Ni 12, bal Fe.
Alloyed tool steel. W. Nr. 2780.

DOMINIAL ZF 2
Kind & Co. Edelstahlwerk
C 0.12, Si 2, Mn 0.9, Cr 25, Ni 20, bal Fe.
Alloyed tool steel. W. Nr. 2782.

DOMINIAL ZF 36
Kind & Co. Edelstahlwerk
C 0.1, Si 1.7, Mn 1.25, Cr 16, Ni 36, bal Fe.
Alloyed tool steel. W. Nr. 2786.

DOMITE 30
Dominion Wheel & Foundry Co.
C 3.2, Mn 1.5, Si 2, bal Fe.
Cast: 30,000 TS; 180 Brin. For stamping dies, pulleys; cast iron; transverse strength 2200.

DOMITE 35

Dominion Wheel & Foundry Co.
C 3.5, Si 2.5, bal Fe.
Cast: 35,000 TS; 190 Brin. For pumps, die shoes, evaporators; cast iron; transverse strength 2400.

DOMITE 45

Dominion Wheel & Foundry Co.
C 3.3, Si 2.2, bal Fe.
Cast: 45,000 TS; 220 Brin. For gears, flywheels, pump impellers; cast iron; transverse strength 2800.

DOMITE 55

Dominion Wheel & Foundry Co.
C 3.1, Si 1.5, bal Fe.
Cast: 55,000 TS; 250 Brin. For crankshafts, forming dies, gears; cast iron; transverse strength 3400.

DOMITE NI-HARD

Dominion Wheel & Foundry Co.
Ni 4.2-4.7, C 3-3.6, Cr 1.4-2.5, Si 0.5, Mn 0.4-6, bal Fe.
Cast: 50,000-60,000 TS; 600-725 Brin. For wear resistant castings; wear and abrasion resistant.

DOMITE NODULOY

Dominion Wheel & Foundry Co.
TC 3.3, Si 1.9, bal Fe.
Cast: 115,000 TS; 2 El. Annealed: 75,000 TS; 15 El. For high strength castings; high ductility.

DOMITE WEAR RESISTING A

Dominion Wheel & Foundry Co.
high C, alloy, Si, bal Fe.
Cast: 200 Brin. For wheels, brake shoes, brick dies; pearlitic cast iron.

DOMITE WEAR RESISTING B

Dominion Wheel & Foundry Co.
high C, alloy, Si, bal Fe.
Cast: 300 Brin. For iron liners, plow points, wheels; wear resistant; cast iron.

DOMITE WEAR RESISTING C

Dominion Wheel & Foundry Co.
high C, alloy, Si, bal Fe.
Cast: 475 Brin. For railway wheels, brake shoes; chilled cast iron.

DOMITE WEAR RESISTING D

Dominion Wheel & Foundry Co.
high C, alloy, Si, bal Fe.
Cast: 600 Brin. For railway wheels, wear, parts; chilled cast iron.

DONAL

Constrictor Ltd.
Aluminum. Si 0.3-0.5, Mn 1-2, bal Al.
Soft: 22,000 TS; 35 El; 30 Brin. Half hard: 26,000 TS; 21,000 YS; 5-15 El; 40-50 Brin. For general structures, containers, heat exchangers, trim. Nonhardenable. Corrosion resistant.

DONAL

Wieland-Werke AG Metallwerke
Al 98.5, Mn 1.5.
For light alloy parts; die castings. *Obsolete*

DONEGAL D-1

Donegal Steel Foundry Co.
C 0.2, Mn 0.65, bal Fe.
Annealed: 60,000 TS; 30,000 YS; 24 El; 35 RA; 130 Brin. For machine tool parts, gears, shafts; ASTM Gr. WCA.

DONEGAL D-10

Donegal Steel Foundry Co.
C 0.3, Mn 0.65, Ni 0.6, Cr 0.6, Mo 0.5, bal Fe.
Normalized: 100,000 TS; 60,000 YS; 15 El; 30 RA; 202 Brin. Water hardened: 150,000 TS; 125,000 YS; 10 El; 25 RA; 320 Brin. For machine tool parts.

DONEGAL D-11

Donegal Steel Foundry Co.
C 0.2, Mn 0.65, Mo 0.55, bal Fe.
Normalized: 65,000 TS; 35,000 YS; 24 El; 35 RA; 145 Brin. For machine tool parts, gears, shafts; ASTM Gr. WC1.

DONEGAL D-12

Donegal Steel Foundry Co.
C 0.2, Mn 0.5, Cr 9, Mo 1, bal Fe.
Normalized: 90,000 TS; 60,000 YS; 18 El; 35 RA; 190 Brin. For oil refinery equipment; ASTM Gr. C 12; creep resistant.

DONEGAL D-13

Donegal Steel Foundry Co.
C 0.25, Mn 0.65, Ni 2.5, bal Fe.
Normalized: 65,000 TS; 40,000 YS; 24 El; 35 RA; 179 Brin. For gears, machine tool parts; ASTM Gr. CL2; tough.

DONEGAL D-14

Donegal Steel Foundry Co.
C 0.15, Mn 0.65, Ni 3.5, bal Fe.
Normalized: 65,000 TS; 40,000 YS; 25 El; 35 RA; 179 Brin. For gears, bolts, machine tool parts; ASTM Gr. CL3; shock resistant.

DONEGAL D-15

Donegal Steel Foundry Co.
C, alloy, bal Fe.
For armor; tough.

DONEGAL D-16

Donegal Steel Foundry Co.
C 0.2, Mn 0.6, Cr 5, Mo 0.5, bal Fe.
Normalized: 90,000 TS; 60,000 YS; 18 El; 35 RA; 190 Brin. For oil refinery equipment; ASTM Gr. C 5, creep resistant.

DONEGAL D-2

Donegal Steel Foundry Co.
C 0.27, Mn 0.65, bal Fe.
Annealed: 70,000 TS; 35,000 YS; 24 El; 35 RA; 140 Brin. For machine tool parts, gears, housings; ASTM Gr. WCB.

DONEGAL D-20

Donegal Steel Foundry Co.
C 0.07, Cr 18-22, Ni 21-31, Mo 2.5, Cu 4, bal Fe.
Cast: 65,000 TS; 30,000 YS; 30 El; 150 Brin. For chemical plant equipment; ACI Type CN-7M; austenitic, stainless.

DONEGAL D-21

Donegal Steel Foundry Co.
C 0-0.15, Cr 11-14, Ni 0-1, bal Fe.
Cast: 90,000 TS; 65,000 YS; 18 El; 30 RA; 200 Brin. For chemical plant equipment, valves, cutlery; ACI Type CA15; corrosion resistant.

DONEGAL D-23

Donegal Steel Foundry Co.
C 0.2-0.4, Cr 11.5-14, Ni 0-1, bal Fe.
Cast: 110,000 TS; 75,000 YS; 15 El; 25 RA; 220 Brin. For valves, cutlery, chemical plant equipment; ACI Type CA40; corrosion resistant.

DONEGAL D-24

Donegal Steel Foundry Co.
C 0-0.3, Cr 18-22, Ni 0-2, bal Fe.
Cast: 65,000-95,000 TS; 30,000-60,000 YS; 15 El; 170-195 Brin. For furnace parts, heat treating boxes; ACI Type CB30; corrosion resistant.

DONEGAL D-25

Donegal Steel Foundry Co.
C 0-0.5, Cr 26-30, Ni 0-4, bal Fe.
Cast: 55,000 TS; 190 Brin. Heat treated: 97,000 TS; 65,000 YS; 18 El; 210 Brin. For cylinder liners, bushings, valve seats; ACI Type CC50; corrosion resistant.

DONEGAL D-26

Donegal Steel Foundry Co.
C 0-0.3, Cr 26-30, Ni 8-11, bal Fe.
Cast: 80,000 TS; 40,000 YS; 10-20 El; 20 RA; 170 Brin. For furnace parts, salt pots, heat treating boxes; ACI Type CE30; austenitic, stainless.

DONEGAL D-27

Donegal Steel Foundry Co.
C 0-0.2, Cr 18-21, Ni 8-11, bal Fe.
Cast: 70,000 TS; 30,000 YS; 30 El; 160 Brin. For furnace parts, heat treating boxes, retorts; ACI Type CF20; stainless, austenitic.

DONEGAL D-28

Donegal Steel Foundry Co.
C 0-0.08, Cr 18-21, Ni 8-11, bal Fe.
Cast: 70,000 TS; 28,000 YS; 35 El; 150 Brin. For furnace parts, chemical plant equipment; ACI Type CF8; stainless, austenitic.

DONEGAL D-29

Donegal Steel Foundry Co.
C 0-0.16, Cr 18-21, Ni 9-10, Se 0.2-0.35, bal Fe.
Cast: 70,000 TS; 30,000 YS; 25 El; 150 Brin. For chemical plant equipment; ACI Type CF16F; free-cutting, stainless.

DONEGAL D-3

Donegal Steel Foundry Co.
C 0.4, Mn 0.65, bal Fe.
Normalized: 80,000 TS; 40,000 YS; 18 El; 30 RA; 165 Brin. For gears, housings, shafts, machine tool parts; ASTM Gr. 80-40.

DONEGAL D-30

Donegal Steel Foundry Co.
C 0-0.08, Cr 18-21, Ni 9-12, Cb = 8 x C, bal Fe.
Cast: 70,000 TS; 30,000 YS; 30 El; 150 Brin. For welded chemical plant equipment; ACI Type CF8C; stainless, austenitic.

DONEGAL D-31

Donegal Steel Foundry Co.
C 0-0.08, Cr 18-21, Ni 9-12, Mo 2-3, bal Fe.
Cast: 70,000 TS; 30,000 YS; 30 El; 150 Brin. For acid resistant chemical plant equipment; ACI Type CF8M; stainless, austenitic.

DONEGAL D-32

Donegal Steel Foundry Co.
C 0-0.2, Cr 22-26, Ni 12-15, bal Fe.
Cast: 70,000 TS; 30,000 YS; 30 El; 150 Brin. For furnace parts, heat treating boxes, retorts; ACI Type CH20; corrosion and heat resistant.

DONEGAL D-33

Donegal Steel Foundry Co.
C 0-0.2, Cr 23-27, Ni 19-22, bal Fe.
Cast: 65,000 TS; 28,000 YS; 30 El; 170 Brin. For retorts, pots, furnace equipment and parts; ACI Type CK20; corrosion and heat resistant.

DONEGAL D-4

Donegal Steel Foundry Co.
C 0.2, Mn 0.65, Ni 0.9, Cr 0.65, Mo 0.55, bal Fe.
Normalized: 70,000 TS; 40,000 YS; 20 El; 35 RA; 160 Brin. For gears, bolts, crankshafts, housings; ASTM Gr. WC4; shock resistant.

DONEGAL D-40

Donegal Steel Foundry Co.
C 0.2-0.5, Cr 24-28, Ni 11-14, bal Fe.
Cast: 75,000 TS; 35,000 YS; 15 El; 200 Brin. For furnace shafts, beams, rollers, tube supports; ACI Type HH; corrosion and heat resistant.

DONEGAL D-41

Donegal Steel Foundry Co.
C 0.2-0.5, Cr 26-30, Ni 8-11, bal Fe.
Cast: 85,000 TS; 40,000 YS; 9 El; 170 Brin. For furnace parts, retorts, salt pots; ACI Type HE; corrosion and heat resistant.

DONEGAL D-42

Donegal Steel Foundry Co.
C 0.2-0.6, Cr 24-28, Ni 18-22, bal Fe.
Cast: 75,000 TS; 35,000 YS; 15 El; 170 Brin. For furnace parts, retorts, skids, stack dampers; ACI Type HK; corrosion and heat resistant.

DONEGAL D-43

Donegal Steel Foundry Co.
C 0.35-0.75, Cr 13-17, Ni 33-37, bal Fe.
Cast: 65,000 TS; 40,000 YS; 4-10 El; 12 RA; 180 Brin. For salt pots, furnace parts, heat treating boxes; ACI Type HT; corrosion and heat resistant.

DONEGAL D-5

Donegal Steel Foundry Co.
C 0.2, Ni 0.8, Cr 0.7, Mo 1, Mn 0.6, bal Fe.
Normalized: 70,000 TS; 40,000 YS; 20 El; 35 RA; 160 Brin. For gears, housings, machine tool parts; ASTM Gr. WC5; case hardened, shock resistant.

DONEGAL D-6

Donegal Steel Foundry Co.
C 0.2, Mn 0.65, Cr 1.25, Mo 0.55, bal Fe.
Normalized: 70,000 TS; 40,000 YS; 20 El; 35 RA; 160 Brin. For gears, housings, machine tool parts; ASTM Gr. W 6.

DONEGAL D-7

Donegal Steel Foundry Co.
C 0.3, Mn 1.35, bal Fe.
Normalized: 90,000 TS; 60,000 YS; 20 El; 40 RA; 190 Brin. For gears, bolts, crankshafts, machine tool parts; ASTM Gr. 90-60; water or oil hardened.

DONEGAL D-8

Donegal Steel Foundry Co.
C 0.3, Mn 1.3, Mo 0.25, bal Fe.
Normalized: 90,000 TS; 60,000 YS; 20 El; 40 RA; 190 Brin. For gears, bolts, crankshafts, machine tool parts; tough.

DONEGAL D-9

Donegal Steel Foundry Co.
C 0.18, Cr 0.5, Mo 1, Mn 0.6, bal Fe.
Normalized: 70,000 TS; 40,000 YS; 20 El; 35 RA; 160 Brin. For machine tool parts; ASTM Gr. WC9.

DONEGAL DC-40

Donegal Steel Foundry Co.
C 0.25, Mn 0.75, Si 0.75, Ni 1.5, Cr 19, bal Fe.
Annealed: 95,000 TS; 60,000 YS; 15 El; 195 Brin. For chemical and food processing equipment, furnace parts; ACI-CB30; corrosion resistant.

DONEGAL DC-41

Donegal Steel Foundry Co.
C 0.1, Mn 0.65, Si 0.8, Ni 0.5, Cr 12.5, bal Fe.
Annealed: 95,000 TS; 78,000 YS; 22 El; 185 Brin. Heat treated: 163,000-213,000 TS; 75,000-173,000 YS; 6-28 El; 9-59 RA; 185-415 Brin. For chemical and food processing equipment, ship propellers; ACI-CA15 corrosion resistant.

DONEGAL DC-42

Donegal Steel Foundry Co.
C 0.3, Mn 0.75, Si 0.75, Ni 3, Cr 28, bal Fe.
Cast: 95,000 TS; 60,000 YS; 15 El; 193 Brin. Annealed: 97,000 TS; 65,000 YS; 18 El; 210 Brin. For bushings, impellers, cylinder liners, valve seats and bodies; ACI-CC50; corrosion resistant.

DONEGAL DC-43

Donegal Steel Foundry Co.
C 0.08, Mn 0.75, Si 1.25, Ni 10.5, Cr 19.5, Cb, bal Fe.
Cast: 70,000 TS; 30,000 YS; 30 El; 150 Brin. For chemical and food processing equipment; ACI-CF8C; corrosion resistant, austenitic.

DONEGAL DC-44

Donegal Steel Foundry Co.
C 0-0.08, Mn 0.75, Si 1.25, Ni 9, Cr 19.5, bal Fe.
Cast: 70,000 TS; 28,000 YS; 35 El; 150 Brin. For chemical and food processing equipment; ACI-CF8; corrosion resistant, austenitic.

DONEGAL DC-45

Donegal Steel Foundry Co.
C 0-0.16, Mn 0.75, Si 1.25, Ni 9, Cr 19.5, Se, bal Fe.
Cast: 70,000 TS; 28,000 YS; 30 El; 150 Brin. For chemical and food processing equipment; ACI-CF16F; corrosion resistant, austenitic.

DONEGAL DC-46

Donegal Steel Foundry Co.
C 0-0.08, Mn 0.75, Si 1.25, Ni 10, Cr 19.5, Mo, bal Fe.
Cast: 70,000 TS; 30,000 YS; 30 El; 150 Brin. For chemical and food processing equipment; ACI-CF8M; corrosion resistant, austenitic.

DONEGAL DC-47

Donegal Steel Foundry Co.
C 0-0.2, Mn 0.75, Si 1.25, Ni 9, Cr 19.5, bal Fe.
Cast: 70,000 TS; 30,000 YS; 30 El; 150 Brin. For chemical and food processing equipment; ACI-CF20; corrosion resistant, austenitic.

DONEGAL DC-48

Donegal Steel Foundry Co.
C 0-0.2, Mn 0.75, Si 1.25, Ni 13.5, Cr 24.5, bal Fe.
Cast: 70,000 TS; 30,000 YS; 30 El; 170 Brin. For chemical plant equipment, furnace parts; ACI-CH20; corrosion and heat resistant.

DONEGAL DC-49

Donegal Steel Foundry Co.
C 0-0.2, Mn 0.75, Si 1.25, Ni 20.5, Cr 25, bal Fe.
Cast: 65,000 TS; 28,000 YS; 30 El; 170 Brin. For chemical plant equipment, furnace parts; ACI-CK20; corrosion and heat resistant.

DONEGAL DC-50

Donegal Steel Foundry Co.
C 0.05, Cr 16.5, Ni 4, Cu 4, bal Fe.
Heat treated: 180,000-210,000 TS; 165,000-200,000 YS; 6-15 El; 30-60 RA; 375-440 Brin. For chemical plant equipment; age hardenable, corrosion resistant.

DOPPLOY 30

Sowers Mfg. Co.
Si 2, Mn 0.8, Ni 18.5, Cr 2.5, 3.0 total C, bal Fe.
Cast: 35,000 TS. For casting jacketed kettles, agitators, etc. Corrosion resistant.

DOQUAT

Russian manufacture
Fe, W.
For cutting tools; hard sintered alloy.

DORDENT

Johnson Matthey plc
Au 72.5, bal Pt.
For dental purposes.

DORIUM "D"

Manufacturer not listed
70 Cu, 30 Zn. For condenser tubes.

DORRENBERG A 50

Dorrenberg Edelstahl GmbH
C 0.56, Ni, Cr, Mo, V, bal Fe.
For crankshafts, punches, dies, oil hardened; shock resistant.

DORRENBERG CN60

Dorrenberg Edelstahl GmbH
C 0.15, Ni, Cr, bal Fe.
For gears, bolts, machine tool parts; case hardened; shock resistant.

DORRENBERG CP10V

Dorrenberg Edelstahl GmbH
C 2.1, Si 0.35, Mn 0.3, Cr 11.5, bal Fe.
For blanking and forming dies, gages, punches; oil hardened; nondeforming.

DORRENBERG CPP

Dorrenberg Edelstahl GmbH
C 1.65, Cr 11.5, V 0.1, bal Fe.
For blanking and forming dies, gages, punches; air hardened; nondeforming.

DORRENBERG CPW

Dorrenberg Edelstahl GmbH
C 2.1, Si 0.25, Cr 11.5, W 0.7, bal Fe.
For blanking and forming dies, gages, punches; oil hardened; nondeforming.

DORRENBERG DGS

Dorrenberg Edelstahl GmbH
C 0.58, Mn, Si, bal Fe.
For machine tool parts; oil hardened.

DORRENBERG DM1

Dorrenberg Edelstahl GmbH
C 0.45, Cr 1.4, Mo 0.7, V 0.3, bal Fe.
For gears, machine tool parts; oil hardened; tough.

DORRENBERG DML

Dorrenberg Edelstahl GmbH
C 0.5, Cr, Mo, bal Fe.
For gears, bolts, studs, fasteners; oil hardened; shock resistant.

DORRENBERG ECN4M

Dorrenberg Edelstahl GmbH
C 0.19, Cr 1.25, Mo 0.2, Ni 3.75, bal Fe.
For gears, bolts, camshafts, cams; case hardened.

DORRENBERG EPM1

Dorrenberg Edelstahl GmbH
C 0.19, Cr 1.25, Mo 0.2, Ni 3.75, bal Fe.
For gears, cams, camshafts, fasteners; case hardened; shock resistant.

DORRENBERG EPM2

Dorrenberg Edelstahl GmbH
C 0.2, Cr 1.15, Mn 1.25, bal Fe.
For gears, cams, camshafts; case hardened; tough.

DORRENBERG EXTRA 12

Dorrenberg Edelstahl GmbH
C 6.6, Si 0.25-0.5, Mn 0.3-0.8, bal Fe.
For springs, tools, rails, hammers; water hardened.

DORRENBERG EXTRA 16

Dorrenberg Edelstahl GmbH
C 0.55, Si 0.1-0.4, Mn 0.5-0.7, bal Fe.
Heat treated: 150,000 TS; 110,000 YS; 15 El; 45 RA; 310 Brin. For tools, punches, hammers; water hardened.

DORRENBERG EXTRA NO. 10

Dorrenberg Edelstahl GmbH
C 0.6, Si 0.25-0.5, Mn 0.3-0.8, bal Fe.
Heat treated: 160,000 TS; 113,000 YS; 12 El; 40 RA; 325 Brin. For tools, punches, springs, axes, hammers; water hardened.

DORRENBERG GWS2

Dorrenberg Edelstahl GmbH
C 0.9, Cr 1.2, Si 1.15, Mn 0.7, bal Fe.
For punches, cutters, bearings, bushings; oil hardened; wear resistant.

DORRENBERG GWS4

Dorrenberg Edelstahl GmbH
C 1.25, Cr 1.2, Si 1.15, Mn 0.7, bal Fe.
For cutters, dies, bearings, bushings; oil hardened; wear resistant.

DORRENBERG H24CN
Dorrenberg Edelstahl GmbH
C 0.2, Si 1.2, Cr 25, Ni 4, bal Fe.
For valve seats and bodies, bushings; corrosion and heat resistant.

DORRENBERG HC15
Dorrenberg Edelstahl GmbH
C 0-0.12, Si 2.2, Cr 13, bal Fe.
For oil refinery and chemical plant equipment; corrosion and heat resistant.

DORRENBERG HC17M
Dorrenberg Edelstahl GmbH
C 0.3, Cr 16, Mo, bal Fe.
For oil refinery and chemical plant equipment; corrosion and heat resistant.

DORRENBERG HC22
Dorrenberg Edelstahl GmbH
C 0.2, Cr 18, Al 1, bal Fe.
For oil refinery equipment; corrosion and heat resistant.

DORRENBERG HC25
Dorrenberg Edelstahl GmbH
C 0-0.12, Si 1.5, Al 1.5, Cr 24, bal Fe.
For heat treating boxes, furnace parts and equipment; heat resistant.

DORRENBERG HC30
Dorrenberg Edelstahl GmbH
C 0.2, Cr 29, Al 1, bal Fe.
For heat treating boxes, furnace parts and equipment; heat resistant.

DORRENBERG HC50
Dorrenberg Edelstahl GmbH
C 0.4, Si 0.4, Mn 0.3, Cr 13, bal Fe.
For cutlery, surgical instruments; corrosion resistant.

DORRENBERG HCN
Dorrenberg Edelstahl GmbH
C 0.15, Cr 24, Ni 19, bal Fe.
For furnace parts, pots, retorts; Type 314; corrosion resistant.

DORRENBERG HCNN
Dorrenberg Edelstahl GmbH
C 0.15, Cr 19.5, Ni 9.5, bal Fe.
Annealed: 80,000 TS; 35,000 YS; 55 El; 75 RA; 150 Brin. For chemical plant equipment; Type 302; stainless, austenitic.

DORRENBERG MAB
Dorrenberg Edelstahl GmbH
C 1.1, W, Cr, V, bal Fe.
For cutters, dies, tools; oil hardened.

DORRENBERG MCM
Dorrenberg Edelstahl GmbH
C 0.4, Cr, Mn, Mo, bal Fe.
For gears, bolts, fasteners, shafts; oil hardened; tough.

DORRENBERG MS
Dorrenberg Edelstahl GmbH
C 0.53, Si 0.9, Mn 0.9, bal Fe.
For gears, bolts, shafts, fasteners; water hardened.

DORRENBERG NC15
Dorrenberg Edelstahl GmbH
C 0.45, Cr, Ni, Fe.
For gears, shafts, machine tool parts; oil hardened; shock resistant.

DORRENBERG NC15A
Dorrenberg Edelstahl GmbH
C 0.5, Ni, Cr, bal Fe.
For gears, shafts, machine tool parts; oil hardened; shock resistant.

DORRENBERG NCM1
Dorrenberg Edelstahl GmbH
C 0.55, Mn 0.6, Cr 0.7, Mo 0.18, Ni 1.65, V 0.1, bal Fe.
For forging and heading dies, upsetters; oil hardened; tough.

DORRENBERG P 60
Dorrenberg Edelstahl GmbH
C 0.61, Si 0.85, Mn 0.75, Cr 1.18, V 0.1, bal Fe.
For dies, punches, gages, shear blades; oil hardened; tough.

DORRENBERG PC130
Dorrenberg Edelstahl GmbH
C 0.38, Si, Cr, V, bal Fe.
For gears, shafts, machine tool parts; oil hardened; tough.

DORRENBERG PMV
Dorrenberg Edelstahl GmbH
C 0.58, Mn 0.95, Cr 1, V 0.09, bal Fe.
For springs, gears, bolts; oil hardened; tough.

DORRENBERG PMVW
Dorrenberg Edelstahl GmbH
C 0.5, Cr 1, V 0.09, Mn 0.85, bal Fe.
For springs, gears, bolts; oil hardened; shock resistant.

DORRENBERG PNC EXTRA
Dorrenberg Edelstahl GmbH
C 0.5, Cr 1.05, Ni 3.25, Mn 0.5, bal Fe.
For gears, bolts, crankshafts; oil hardened ; shock resistant.

DORRENBERG PV15
Dorrenberg Edelstahl GmbH
C 1, Cr 1.1, Mn 0.07, bal Fe.
For bearings, bushings, liners; water hardened; wear resistant.

DORRENBERG PV35
Dorrenberg Edelstahl GmbH
C 1.5, Cr, Si, bal Fe.
For tools, dies; oil hardened.

DORRENBERG PV3W
Dorrenberg Edelstahl GmbH
C 0.85, Cr 1, bal Fe.
For bearings, cutters, liners, bushings; water hardened; wear resistant.

DORRENBERG PV4
Dorrenberg Edelstahl GmbH
C 0.85, Cr 1.15, bal Fe.
For bearings, cutters, liners, bushings; water hardened; wear resistant.

DORRENBERG PV5
Dorrenberg Edelstahl GmbH
C 1.15, Cr 0.65, V 0.1, bal Fe.
For bearings, cutters, liners, bushings; water hardened; wear resistant.

DORRENBERG R 15
Dorrenberg Edelstahl GmbH
C 0-0.12, Si 0.4, Cr 13, bal Fe.
Annealed: 75,000 TS; 40,000 YS; 35 El; 70 RA; 155 Brin. For turbine blades, cutlery, valves; Type 410; stainless.

DORRENBERG R 17
Dorrenberg Edelstahl GmbH
C 0.9, Si, Cr, Mo, bal Fe.
For bearings, liners, bushings; oil hardened; wear resistant.

DORRENBERG R 18
Dorrenberg Edelstahl GmbH
C 0.22, Cr 17, Ni 1.5, bal Fe.
Annealed: 125,000 TS; 95,000 YS; 20 El; 55 RA; 260 Brin. For pumps, marine hardware, valves; Type 431; corrosion resistant.

DORRENBERG R 25
Dorrenberg Edelstahl GmbH
C 0.2, Cr 13, Si 0.4, bal Fe.
Annealed: 95,000 TS; 50,000 YS; 25 El; 55 RA; 195 Brin. For turbine blades, valves, cutlery; Type 420; stainless.

DORRENBERG R 45
Dorrenberg Edelstahl GmbH
C 0.4, Mn 0.3, Si 0.4, Cr 13, bal Fe.
Annealed: 95,000 TS; 50,000 YS; 25 El; 55 RA; 195 Brin. For valves, cutlery, turbine blades; Type 420; stainless.

DORRENBERG SA
Dorrenberg Edelstahl GmbH
C 0-0.15, Cr 18, Ni 8.5, bal Fe.
Annealed: 80,000 TS; 35,000 YS; 55 El; 75 RA; 150 Brin. For chemical plant equipment, tanks, vessels; Type 302; stainless, austenitic.

DORRENBERG SAT
Dorrenberg Edelstahl GmbH
C 0-0.12, Cr 18, Ni 9.5, Ti = 4 x C, bal Fe.
Annealed: 85,000 TS; 35,000 YS; 55 El; 65 RA; 150 Brin. For welded chemical plant equipment, tanks, mixers; Type 321; stainless, austenitic.

DORRENBERG SAW
Dorrenberg Edelstahl GmbH
C 0-0.07, Cr 18, Ni 9.5, bal Fe.
Annealed: 85,000 TS; 35,000 YS; 60 El; 70 RA; 150 Brin. For chemical plant equipment, tanks; Type 304; stainless, austenitic.

DORRENBERG SB
Dorrenberg Edelstahl GmbH
C 0-0.1, Cr 18, Ni 9.5, Mo 2, bal Fe.
Annealed: 85,000 TS; 35,000 YS; 50 El; 65 RA; 160 Brin. For acid resistant chemical plant equipment, tanks; Type 316; stainless, austenitic.

DORRENBERG SBT
Dorrenberg Edelstahl GmbH
C 0-0.12, Cr 18, Mo 2, Ni 10.5, Ti = 4 x C, bal Fe.
Annealed: 85,000 TS; 35,000 YS; 50 El; 65 RA; 160 Brin. For welded acid resistant chemical plant equipment; Type 316Ti; austenitic, stainless.

DORRENBERG SBW
Dorrenberg Edelstahl GmbH
C 0-0.07, Cr 18, Ni 10.5, Mo 2, bal Fe.
Annealed: 90,000 TS; 40,000 YS; 45 El; 60 RA; 180 Brin. For acid resistant chemical plant equipment, tanks; Type 317; stainless, austenitic.

DORRENBERG SPEZIAL NR. 2
Dorrenberg Edelstahl GmbH
C 0.4, Si 0.25-0.5, Mn 0.3-0.8, bal Fe.
Hot rolled: 91,000 TS; 58,000 YS; 27 El; 50 RA; 200 Brin. For fishplates, gears, bolts, fasteners; water hardened.

DORRENBERG ST2
Dorrenberg Edelstahl GmbH
C 0.7, Si 1.7, Mn 0.7, bal Fe.
For springs, punches, chisels; oil hardened; shock resistant.

DORRENBERG STP
Dorrenberg Edelstahl GmbH
C 0.67, Si 1.3, Mn 0.5, Cr 0.5, bal Fe.
For springs, punches, chisels; oil hardened; shock resistant.

DORRENBERG VNC4
Dorrenberg Edelstahl GmbH
C 0.4, Ni, Cr, Mo, bal Fe.
For gears, bolts, crankshafts; oil hardened; shock resistant.

DORRENBERG W 4
Dorrenberg Edelstahl GmbH
C 0.3, Si 1, Mn 0.4, Cr 1.1, V 0.18, W 3.75, bal Fe.
For extrusion arms, punches, upsetters; oil hardened; tough.

DORRENBERG W 5
Dorrenberg Edelstahl GmbH
C 0.3, Cr 2.35, V 0.6, W 4.25, bal Fe.
For extrusion arms, punches, upsetters; oil hardened; hot work steel.

DORRENBERG W 9
Dorrenberg Edelstahl GmbH
C 0.3, Cr 2.65, V 0.35, W 8.5, bal Fe.
For extrusion rams and liners, upsetters; oil hardened; hot work steel.

DORRENBERG WC11
Dorrenberg Edelstahl GmbH
C 0.3, Cr 2.4, V 0.25, W 8.5, Co 2, bal Fe.
For extrusion rams and liners, upsetters; oil hardened; hot work steel.

DORRENBERG WD1
Dorrenberg Edelstahl GmbH
C 1.2, Cr 0.2, V 0.1, W 1, bal Fe.
For blanking and forming dies; oil hardened; wear resistant.

DORRENBERG Z 1A
Dorrenberg Edelstahl GmbH
C 1.1, Mo, V, bal Fe.
For cutters, blanking and forming dies; oil hardened; wear resistant.

DORRENBERG Z 1B
Dorrenberg Edelstahl GmbH
C 0.9, Mn 1.9, V 0.1, bal Fe.
For punches, blanking and forming dies; oil hardened; nondeforming.

DORRENBERG Z 3C
Dorrenberg Edelstahl GmbH
C, alloy, bal Fe.
For machine tool parts; oil hardened.

DOS SPEZIAL
Vereinigte Edelstahlwerke
C 0.3, Cr 2.5, Mo 0.2, V 0.15, bal Fe.
For gears, pinions, bolts, studs, crankshafts; oil hardened, shock resistant. *Obsolete*

DOUBLE BRONZE
English manufacture
Al 6.5, skin of pure Cu, bal Cu.
For electrical equipment; wire.

DOUBLE CHROME VANADIUM
Allegheny Ludlum Steel
C 0.5, Cr 1.5, V 0.2, bal Fe.
For tools, drills, dies. *Obsolete*

DOUBLE CLOCHE 0.1
Compagnie Ateliers et Forges de la Loire
C 1.3, W 5, bal Fe.
Hardened: 63-65 Rock C. For cold work tools, punches, bending dies, piercing tools, bearings. Cold working steel, shock and wear resisting. Water or oil hardening.

DOUBLE CLOCHE 2.3
Compagnie Ateliers et Forges de la Loire
C 1.1, Cr 0.5, W 2.1, bal Fe.
Hardened: 60-64 Rock C. For cold working tools, punches, hobs, bending dies, drills, cutters. Oil or water hardening. Cold working steel, shock resisting.

DOUBLE CLOCHE E
Compagnie Ateliers et Forges de la Loire
C 1.45, W 3.25, Cr 0.3, V 0.25, bal Fe.
For cold working tools, bending dies, punches, reamers, cutters. Cold working steel and wear resistant. Oil hardening.

DOUBLE CONQUEROR
Joseph Beardshaw & Son Ltd.
C 1.1-1.4, bal Fe.
For roll turning tools, rock drills, punches; water hardened, wear resistant.

DOUBLE EXTRA
Midvale-Heppenstall Co.
C 0.8, Cr 0.65, bal Fe.
For tools, dies, mandrels; oil hardening. *Obsolete*

DOUBLE EXTRA BEST
Leadbeater & Scott Ltd.
High C, W, bal Fe.
For turning tools; oil or water hardened.

DOUBLE EXTRA LION
Burys & Co. Ltd.
C 0.75, W 19, Cr 4, V 1.2, Co 2, Mo 0.5, bal Fe.
For drills, reamers, lathe and planer tools, hobs, broaches; high speed steel.

DOUBLE EXTRA MINT DIE STEEL
Eagle & Globe Steel Ltd.
Tool material. C 1.05, Mn 0.3, V 0.2, Si 0.25, bal Fe.
Carbon tool steel. Water hardened, shock resistant. For dies for minting and cold heading, draw punches, plugs and dies. AS1239 W2A10; AISI W2; Werkstoff 1.2833.

DOUBLE GEANT
Societe des Acieries de Longwy
C 0.7, W 18, Co 9, Cr 4, V 2, Mo 1, bal Fe.
For tools, dies, cutters; high speed steel.

DOUBLE GRIFFIN
Darwins Alloy Castings
C 0.35, Cr 1.75, Ni 3.5, Mn 0.3, bal Fe.
For shears, punches, pneumatic tools; hot work steel, oil hardened, shock resistant. *Obsolete*

DOUBLE GRIFFIN
Eagle & Globe Steel Ltd.
Tool material. C 0.35, Mn 0.3, Cr 1.75, Ni 3.5, Si 0.2, bal Fe.
Ni-Cr shock resisting steel. For punches, shear blades, chisels, snaps, wedges. AS1239 S100A.

DOUBLE GRIFFIN SUPER CHISEL
Darwins Alloy Castings
C, alloy, bal Fe.
For tools, chisels; tough. *Obsolete*

DOUBLE MUSHET
Osborn Steels Ltd.
W 18, Cr 4, V 1, C, bal Fe.
For tools, cutters; high speed steel. *Obsolete*

DOUBLE RAPID
George Cook & Co., Ltd.
C 0.8, Cr 4.5, W 18.5, Co 5.5, Mo 0.3, V 1.2, bal Fe.
For lathe and planer tools, reamers, drills, hobs; high speed steel.

DOUBLE SATAN
Creusot-Loire
C 0.55, W 5, Cr 3.5, bal Fe.
For hot work tools and dies; hot work steel, oil hardened. *Obsolete*

DOUBLE SEVEN
Edgar Allen Balfour Ltd.
C 1.25, Cr 12.5, Co 2.7.
For shear blades, cold forging dies, blanking dies; air hardened, non-deforming.

DOUBLE SEVEN
A. Milne & Co.
C 1.3, Cr 13.3, Co 3-4.5, Mo 0.25, Ni 1, bal Fe.
For tools, broaches, lamination dies; nondeforming.

DOUBLE SIX
Edgar Allen Balfour Ltd.
C 1.9, Cr 12.5, Mo 0.8, V 0.25, bal Fe.
For dies, plug and ring gages, press tools; oil or air hardened, non-deforming.

DOUBLE SIX
Allegheny Ludlum Steel
C, alloy, bal Fe.
For tools, dies, taps. *Obsolete*

DOUBLE SIX
A. Milne & Co.
C 2.25, Cr 14, bal Fe.
For blanking and drawing dies, gages; non-shrinking.

DOUBLE SIX M-2
Latrobe Steel Co.
C 0.85, W 6.3, Mo 5.05, Cr 4.15, V 1.85, bal Fe.
For cutting tools, drills, taps, broaches, reamers; high speed steel.

DOUBLE SIX M-2XL
Latrobe Steel Co.
C 0.8, Cr 4, W 6, Mo 6, V 2, bal Fe.
For lathe and planer tools, reamers, hobs, taps; high speed steel. *Obsolete*

DOUBLE SIX SUPER
Edgar Allen Balfour Ltd.
C 1.9, Cr 12.5, Mo 0.8, V 0.25, bal Fe.
Heat treated: 275,000 TS; 210,000 YS; 1 El; 58 Rock C. For blanking, coining and drawing dies. Plug and ring gages, mandrels, press tools, reamers, broaches. Non-shrinking; air hardened.

DOUBLE SPECIAL
Hawkridge Bros Co.
Cr 0.25, Mn 0.0, W 3.5, Si 0.45, C, bal Fe.
For burnishing tools; water hardening. *Obsolete*

DOUBLE SPECIAL
Sanderson Kayser Ltd.
C 1.4, W 5, Cr 0.8, bal Fe.
For turning and finishing cutters; fast finishing, water hardened. *Obsolete*

DOUBLE SPECIAL D.S.
Crucible Specialty Metals
C 1.3, Mn 0.25, Cr 0.55, W 3.25, bal Fe.
For tools, finishing tools, rifle barrel drills. *Obsolete*

DOUBLE SPECIAL D.S.
Columbia Tool Steel Co.
C 1.3, Mn 0.25, Cr 0.55, W 3.25, bal Fe.
For tools, finishing tools, rifle barrel drills. *Obsolete*

DOUBLE SUPER EXPRESS
Leadbeater & Scott Ltd.
C 0.7, W 22, Cr 5, V 1, bal Fe.
For cutters for asbestos and rubber; high speed steel.

DOUBLE SUPER HYDRA
Osborn Steels Ltd.
C 0.8, W 18, Cr 4.5, V 1.75, Co 5, bal Fe.
For broaches, drills, reamers, lathe tools; high speed steel. *Obsolete*

DOUBLE TWELVE
Edgar Allen Balfour Ltd.
C 0.35, Cr 12, W 12, V 1, bal Fe.
For master hubs for beryllium-copper, mold inserts in the plastic industry, brass die casting molds, extrusion dies; oil hardened.

DOUBLE TWO
Multicore
Lead. Pb 96, Sn 2, Sb 2.
Solder for high temperature joints. MP 581°F. FP 589°F.

DOUBLE YOU CHROME
Wardlows Ltd.
C 2.25, Cr 13, bal Fe.
For tool, dies; non-shrink.

DOUBLE YOU DIE
Wardlows Ltd.
C 0.9, Cr 3.75, bal Fe.
For tools, dies; air hardened.

DOUBLE YOU HOT STUFF
Wardlows Ltd.
W 9.5, Cr 2.5, V 0.1, C, bal Fe.
For tools, hot dies; hot die steel.

DOUBLE ZEBRA
Sanderson Kayser Ltd.
C 0.4, Cr 0.3, Ni 3.5, bal Fe.
Cold work tool steel for chisels; hardenable to about 600 DPH.

DOUX
Vallourec S.A.
C 0.2, Si 0.25, Mn 0.8, bal Fe.
Annealed: 47,000 TS; 28,000 YS; 35 El; 137 Brin. For oil refinery tubes.

DOVER
Time Steel Service Inc.
C 1, Mn 0.25, Si 0.25, bal Fe.
Water hardened tool steel. AISI W1.

DOW METAL O1
Now AZ80A.

DOWMETAL AM60X1
Dow Chemical Co.
Al 6, Mn 0.25, Be 0.0004, bal Mg.
For cast automobile wheels. Tough and ductile. *Obsolete*

DOWMETAL AZ91B
Now AZ91B.

DOWMETAL B
Dow Chemical Co.
Mg 87.9, Al 12, Mn 0.1.
Sand cast: 19,000-23,000 TS; 15,000-17,000 YS; 0-1 El; 57-61 Brin. For heat treated castings requiring high yield strength and hardness. *Obsolete*

DOWMETAL D
Dow Chemical Co.
Al 8.5, Mn 0.15, Cu 2, Cd 1, Zn 0.5, bal Mg.
Sand cast: 22,000 TS; 14,000 YS; 2 El; 58 Brin. Forged: 47,000 TS; 33,000 YS; 7 El; 62 Brin. For light alloy castings; pistons. *Obsolete*

DOWMETAL E
Dow Chemical Co.
Al 6, Mn 0.3, bal Mg.
Sand cast: 26,000-31,000 TS; 7,000-11,000 YS; 7-12 El; 46-52 Brin. Forged: 40,000-44,000 TS; 27,000-32,000 YS; 10-15 El; 50-58 Brin. For castings requiring good strength without heat treatment; high strength and ductile wrought parts. *Obsolete*

DOWMETAL EX
Dow Chemical Co.
Al 6.5, Si 0.2, bal Mg.
Die cast: 27,000 TS; 17,000 YS; 4 El; 50 Brin. For die castings. *Obsolete*

DOWMETAL F
Dow Chemical Co.
Al 4, Mn 0.3, bal Mg.
Cast: 24,000 TS; 9,000 YS; 8 El; 48 Brin. Forged: 38,000-41,000 TS; 26,000-30,000 YS; 12-16 El; 45-54 Brin. For forging, rolling and extrusion, where maximum ductility is required. *Obsolete*

DOWMETAL G
Now AM100A.

DOWMETAL H
Now AZ63A.

DOWMETAL HK31A
Now HK31A.

DOWMETAL J
Now AZ61A.

DOWMETAL J1
Dow Chemical Co.
Al 6.5, Zn 1, bal Mg.
Now AZ61 alloy. O-temper: 44,000 TS; 30,000 YS; 14 El; 60 Brin. For structural members. *Obsolete*

DOWMETAL K
Dow Chemical Co.
Al 10, Mn 0.1, Si 0.5, bal Mg.
Die cast: 29,000-33,000 TS; 20,000-22,000 YS; 1-3 El; 60-70 Brin. For die castings. *Obsolete*

DOWMETAL L
Dow Chemical Co.
Al 2.5, Mn 0.3, Cd 3.5, bal Mg.
Forged: 34,000 TS; 19,000 YS; 6 El; 51 Brin. For light alloy forgings. *Obsolete*

DOWMETAL O1
Dow Chemical Co.
Al 8.5, Zn 0.5, bal Mg.
Now AZ80A alloy. F temper: 48,000 TS; 32,000 YS; 12 El; 60 Brin. T5-temper: 52,000 TS; 36,000 YS; 15 El; 82 Brin. For cylinder heads, valve and pump bodies, superchargers; age-hardenable. *Obsolete*

DOWMETAL P
Dow Chemical Co.
Al 10, Mn 0.1, Zn 1, bal Mg.
Heat treated: 30,000 TS; 19,000 YS; 1 El; 77 Brin. For castings. *Obsolete*

DOWMETAL RC
Dow Chemical Co.
Al 9, Zn 0.6, bal Mg.
Now AZ91B alloy. Die cast: 33,000 TS; 22,000 YS; 3 El; 60 Brin. For instrument cases, portable tools, general die castings. *Obsolete*

DOWMETAL T
Dow Chemical Co.
Al 2, Mn 0.2, Cd 2, Cu 4, bal Mg.
Sand cast: 22,000-25,000 TS; 6,000-7,000 YS; 4-7 El; 43-48 Brin. Forged: 38,000-42,000 TS; 27,000-31,000 YS; 5-9 El; 45-50 Brin. For pistons and other parts requiring maximum thermal properties. *Obsolete*

DOWMETAL X
Dow Chemical Co.
Al 3, Mn 0.2, Zn 3, Si 0-0.5, bal Mg.
Cast: 38,000 TS; 20,000 YS; 9 El; 59 Brin. For light alloy parts; forgings and extrusions. *Obsolete*

DOWMETAL ZK60A
Now ZK60A.

DP3
Sumitomo Metal America Inc.
C 0-0.03, Si 0-0.75, Mn 0-1, Cu 0.2-0.8, Ni 5.5-7.5, Cr 24-26, Mo 2.5-3.5, W 0.1-0.4, N 0.1-0, bal Fe.
118,000 psi TS; 88,200 psi YS; 36 El. Duplex-phase stainless steel resistant to crevice corrosion by seawater at high temperature. Good weldability.

DRACO
Cytemp Specialty Steel Div.
C 0.66-1.15, V 0.2, bal Fe.
For tools, stamping dies, cold heading dies, taps; high strength and toughness. *Obsolete*

DRACO 3460
Chiers-Chatillon
C 0.15, Cr 23, Ni 14, bal Fe.
Annealed: 90,000 TS; 40,000 YS; 50 El; B 83 Rock. For heat treating boxes, furnace parts, salt pots, brazing fixtures, combustion chambers. Type 309 stainless steel, austenitic, heat resistant. *Obsolete*

DRACO DV
Cytemp Specialty Steel Div.
C 1, Mn 0.3, V 0.45, bal Fe.
For cold header dies, edge tools; shock resistant. *Obsolete*

DRAGON
Allegheny Ludlum Steel
C 0.35, Mn 0.55, Cr 0.7, Mo 0.25, bal Fe.
For bucket teeth, keys, studs, bolts; water hardening. *Obsolete*

DRAGONITE
British Steel plc
Now ZINTEC.

DRAGONZIN
British Steel plc
Now GALVATITE.

DRAWALLOY 240
Welding Equipment & Supply Co.
C 0.2, Cr 18, Ni 8, bal Fe.
Cast: 240-260 Brin. For hard surfacing welding rods; austenitic, nonmagnetic. *Obsolete*

DRAWING BRASS
Anaconda Co.
Copper. Cu 67-70, Zn 30-33.
Hard: 77,000 TS; 67,000 YS; 15 El; 50 RA; 145 Brin. For seamless tubes, condensers and evaporators. Deep drawing.

DREADNAUGHT
Hawkridge Bros Co.
C 0.7, Cr 4, V 1.15, W 18, bal Fe.
For tools, cutters, hacksaw blades; high speed steel. *Obsolete*

DREADNOUGHT
Colt Industries
C 0.8, Mn 0.28, Cr 4, W 18, V 1.03, bal Fe.
Oil or air. For tools, dies, punches, general purpose tools; high speed steel; R.H.C. 64 minimum. *Obsolete*

DREWOSTA
Stahlwerk Stahlschmidt GmbH & Co.
C 1.42, W, V, bal Fe.
For engravers' tools, forming and blanking dies; water hardened, wear resistant.

DRIL-TEC 88
Eutectic Corp.
Hard carbides in non-ferrous matrix.
For hard surfacing using torch. Wear and abrasion resistant. For earth moving equipment, etc. *Obsolete*

DRILL
U.S. Steel Corp.
C 0.85, Mn 0.5, Si 0.25, bal Fe.
For drills, chisels, shear blades, axes, hammers; water hardening. *Obsolete*

DRILL ROD
Peninsular Steel Co.
C 1.2, bal Fe.
For drills, tools; water hardened.

DRILL ROD BRASS

American manufacture
Cu 62, Zn 35.5, Pb 2.5.
For automatic screw machine parts; drills and turns easily.

DRILLALLOY

Delsteel Inc.
C 0.65, Mn 0.7, Mo 0.4, bal Fe.
For mine drills, tools; water hardened.

DRILLEX

Rankin Mfg. Co.
Bulk tungsten sintered carbide screened to -60 + 100 mesh
size. This is then alloyed with small amount of Ni, Si + Mn to
develop a strong impact resistant matrix for welding.

DRITTEL SILVER

German manufacture
Al 66.6, Ag 33.3.
For ornamental trade.

DRIVER 180 ALLOY

Wilbur B. Driver Co.
Copper. Cu 78, Ni 22.
Annealed: 50,000 TS. Cold worked: 100,000 TS. For
resistances, precision resistors, rheostats, instruments.
Maximum operating temperature 1000°F in air.

DRIVER 30 ALLOY

Wilbur B. Driver Co.
Copper. Cu 98, Ni 2.
Annealed: 30,000 TS. Cold worked: 60,000 TS. For
resistances, precision resistors, rheostats, instruments.
Maximum operating temperature 600°F in air.

DRIVER 60 ALLOY

Wilbur B. Driver Co.
Copper. Cu 94, Ni 3.5.
Annealed: 35,000 TS. Cold worked: 70,000 TS. For
resistances, resistors, rheostats, instruments. Maximum
operating temperature 600°F in air.

DRIVER 90 ALLOY

Wilbur B. Driver Co.
Copper. Cu 88, Ni 12.
Annealed: 35,000 TS. Cold worked: 75,000 TS. For
resistances, rheostats, precision resistors, instruments.
Maximum operating temperature 800°F in air.

DRW COPPER

Climax Performance Materials Corp.
Electrolytic tough pitch copper. For electrical applications
and general usage in rod and wire applications.

DS NICKEL

Manufacturer not listed
Ni 98, 2 ThO$_2$, (dispersion strengthened nickel).
Powder for sintering; and sheet or bar. Sintered: 8.9 g/cc.
Room temperature: 71,200 psi TS. 2000°F: 17,500 psi TS.
Good high temperature properties.

DS NICR

Manufacturer not listed
Cr 20, 1.7 ThO$_2$, bal Ni.
Powder for sintering.

DS-LEAD

St. Joseph Lead Co.
Pb 99.9, 0.1 PbO.
Sheet: 6000 TS; 5800 YS; 18 El; 6500 CS. At 300 °F: 3250
TS; 20 El. For chemical construction, storage batteries,
counterweights, roofing, gutters. Dispersion strengthened,
fine-grained, corrosion resistant.

DSC HIGH C

Detroit Alloy Steel Co.
C 0.25-0.8, bal Fe.
For springs; water hardened.

DTC

Acciaierie Valbruna s.p.a.
Tool Material. C 2.1, Si 0-0.45, Mn 0-0.5, Cr 13, V 0.2, bal Fe.
Cold work tool steel. AISI D3; W. Nr. 1.2080.

DTC

Acciaierie Valbruna s.p.a.
Tool Material. C 2.1, Cr 13, V 0.2, Si 0-0.45, Mn 0-0.5, bal Fe.
Cold work tool steel. UX 210 Cr 13 KU; AISI D3.

DTC W

Acciaierie Valbruna s.p.a.
Tool Material. C 2.15, Si 0-0.45, Mn 0-0.4, Cr 11.5, W 0.7, bal
Fe.
Cold work tool steel. W. Nr. 1.2436.

DTC-AR

Acciaierie Valbruna s.p.a.
Tool Material. C 1.5, Cr 12, Mo 0.75, Si 0-0.6, Mn 0-0.5, V
0.25, bal Fe.
Cold work tool steel. W. Nr. 1.2379; AISI D2.

DTC-ARK

Acciaierie Valbruna s.p.a.
Tool Material. C 1.5, Si 0-0.6, Mn 0-0.6, Co 3, Cr 12, Mo 0.95,
V 0-1, bal.
Cold work tool steel. AISI D5; W. Nr. 1.2880.

DTC-ARW

Acciaierie Valbruna s.p.a.
Tool Material. C 1.65, Si 0-0.4, Mn 0-0.4, Cr 11.5, Mo 0.6, V
0.3, W 0.5, bal Fe.
Cold work tool steel. W. Nr. 1.2601.

DTC-B

Acciaierie Valbruna s.p.a.
Tool Material. C 1.65, Si 0-0.4, Mn 0-0.4, Cr 11.5, V 0.1, bal
Fe.
Cold work tool steel. W. Nr. 1.2201.

DTD

English manufacture
C 0.4, Cr 3, Mo 0.8, V 0.2, bal Fe.
For gas turbine parts; creep resistant.

DTD 424

British manufacture
Cu, Si, bal Al.
For castings; age-hardenable.

DU PONT D-14

Du Pont Co.
Zr 5, bal Cb.
Rolled: 75,000 TS; 60,000 YS. For hot surfaces on space re-
entry vehicles; high heat resistance.

DU PONT D-36

Du Pont Co.
Ti 10, Zr 5, bal Cb.
Rolled: 80,000 TS; 70,000 YS. For hot surfaces on space re-
entry vehicles; high heat resistance.

DU PONT FA 22 NO. 1

Du Pont Co.
C 0.07, Ni 29, Cu 3, Mo 2, bal Fe.
For chemical plant equipment; resists H$_2$SO$_4$.

DU PONT FA 22 NO. 2

Du Pont Co.
C 0-0.07, Si 2.5, Mn 0.75, Cr 20.5, Ni 28.5, Cu, bal Fe.
Cast: 72,000 TS; 35,000 YS; 45 El; 150 Brin. For heat and
corrosion resistant casting; heat and corrosion resistant.

DUAL SHIELD 110

Chemetron Corp.
Mn 1.29, Si 0.82, weld metal; 0.10 C, bal Fe.
As welded: 86,000 psi TS; 81,000 psi YS; 30 El. Flux-cored
continuous electrode for welding low and medium carbon
steel in the mill scaled or rusty condition. AWS Class E70T-2.

DUAL SHIELD 111A

Chemetron Corp.
Mn 0.85, Si 0.4, weld metal; 0.08 C, bal Fe.
As welded: 74,800 psi TS; 64,200 psi TS; 25 El. Flux-cored
continuous electrode for single and multipass welding of mild
steel in CO$_2$ atmosphere. AWS Class E70T-1.

DUAL SHIELD 111A-C

Chemetron Corp.
Mn 1.45, Si 0.48, weld metal; 0.07 C, bal Fe.
As welded: 88,900 psi TS; 77,400 psi YS; 26 El. Flux-cored
continuous electrode for single and multipass welding of mild
and medium carbon steels in CO$_2$ atmosphere. AWS C
E70T-1; AWS A5.20.

DUAL SHIELD 111HD

Chemetron Corp.
Mn 1.33, Si 0.58, weld metal; 0.09 C, bal Fe.
As welded: 91,000 psi TS; 80,500 psi YS; 26 El. Flux-cored
continuous electrode for single and multipass welding of mild
and medium carbon steels in CO$_2$ atmosphere. AWS Class
E70T-1.

DUAL SHIELD 150

Chemetron Corp.
Mn 1.4, Si 0.38, Mo 0.62, weld metal; 0.06 C, bal Fe.
As welded: 108,500 psi TS; 97,000 psi YS; 16 El. Flux-cored
continuous electrode for multipass welding of low alloy-high
tensile strength steels. AWS Grade E100T.

DUAL SHIELD 7000

Chemetron Corp.
Mn 1.47, Si 0.74, weld metal; 0.08 C, bal Fe.
As welded: 93,000 psi TS; 86,000 psi YS; 25 El. Flux-cored
continuous electrode for welding mild and medium carbon
steels in all positions with CO$_2$ gas. AWS Class E70T-1.

DUAL SHIELD 7000-A1

Chemetron Corp.
Mn 1.13, Si 0.57, Mo 0.55, weld metal; 0.07 C, bal Fe.
As welded: 102,000 psi TS; 97,000 psi YS; 23 El. Flux-cored
continuous electrode designed for welding low alloy-high
tensile steels in out-of-position work with 75% A-25% CO$_2$
gas. AWS Grade E100T.

DUAL SHIELD 78

Chemetron Corp.
Mn 1.2, Si 0.4, weld metal; 0.06 C, bal Fe.
As welded: 84,000 psi TS; 74,500 psi YS; 28 El. Flux-cored
continuous electrode for single and multipass welding of low
and medium carbon steels. AWS Class E70T-1.

DUAL SHIELD 8000-B2

Chemetron Corp.
Mn 0.85, Si 0.64, Cr 1.41, Mo 0.55, weld metal; 0.05 C, bal
Fe.
As welded: 107,500 psi TS; 97,000 psi YS; 20 El. Flux-cored
continuous electrode for all-position welding of 1% Cr - 1/2%
Mo and similar steels.

DUAL SHIELD 85 NM

Chemetron Corp.
Mn 1.09, Si 0.38, Ni 0.96, Mo 0.5, weld metal; 0.07 C, bal Fe.
As welded: 98,500 psi TS; 89,700 psi YS; 23 El. Used largely
for welding nuclear pressure vessels. AWS Grade E90T.

DUAL SHIELD 85-C1

Chemetron Corp.
Mn 1.17, Si 0.3, Ni 2.75, weld metal; 0.05 C, bal Fe.
As welded: 95,000 psi TS; 85,100 psi YS; 23 El. Flux-cored
continuous electrode for strong welds of good toughness
down to minus 100°F. AWS Grade E90T.

DUAL SHIELD 88 CM

Chemetron Corp.
Mn 0.74, Si 0.34, Cr 1.13, Mo 0.5, weld metal; 0.06 C, bal Fe.
As welded: 106,000 psi TS; 94,000 psi YS; 14 El. Flux-cored
continuous electrode for welding HSLA steel and low alloy
steels.

DUAL SHIELD 88-C3

Chemetron Corp.
Mn 1.22, Si 0.35, Ni 1.07, weld metal; 0.07 C, bal Fe.
As welded: 88,600 psi TS; 79,300 psi YS; 24 El. Flux-cored continuous electrode for welding high tensile steels in the 70,000-80,000 psi tensile strength range. AWS Grade E80T.

DUAL SHIELD 9000-B3

Chemetron Corp.
Mn 0.85, Si 0.7, Cr 2.3, Mo 1.1, weld metal; 0.06 C, bal Fe.
As welded: 124,000 psi TS; 108,000 psi YS; 12 El. Flux-cored continuous electrode for all-position welding of 2 1/4% Cr-Mo steels.

DUAL SHIELD 9000-C1

Chemetron Corp.
Mn 1.3, Si 0.6, Ni 2.4, weld metal; 0.05 C, bal Fe.
As welded: 106,250 psi TS; 101,000 psi YS; 20 El. Flux-cored continuous electrode for all-position welding of 2-3% Ni steels for low temperature operation. AWS Grade E100T.

DUAL SHIELD 9000-D1

Chemetron Corp.
Mn 1.37, Si 0.73, Mo 0.45, weld metal; 0.09 C, bal Fe.
As welded: 100,500 psi TS; 92,000 psi YS; 23 El. Flux-cored continuous electrode for all-position welding of low alloy high strength steels requiring 100,000 psi strength. AWS Grade E100T.

DUAL SHIELD 9000-M

Chemetron Corp.
Mn 1.1, Si 0.65, Ni 1.8, Mo 0.25, weld metal; 0.06 C, bal Fe.
As welded: 103,500 psi TS; 94,000 psi YS; 23 El. Flux-cored continuous electrode for all-position welding (fillets) of HY-80 and similar steels. AWS Grade E100T.

DUAL SHIELD 98

Chemetron Corp.
Mn 1.1, Ni 1.75, Mo 0.2, Si 0.3, weld metal; 0.07 C, bal Fe.
As welded: 94,000 psi TS; 85,000 psi YS; 20 El. Flux-cored continuous electrode for welding HY-80, HY-90 and T-1 and similar high tensile quenched and tempered steels. AWS Grade E90T.

DUAL SHIELD 98-CM

Chemetron Corp.
Mn 0.7, Si 0.4, Cr 2.21, Mo 1.04, weld metal; 0.06 C, bal Fe.
As welded: 128,000 psi TS; 109,500 psi YS; 12 El. (Usually stress annealed to 90,000 min TS). For welding 2 Cr-1 Mo steels.

DUAL SHIELD T-62

Chemetron Corp.
Mn 1.35, Si 0.55, weld metal; 0.08 C, bal Fe.
As welded: 82,000 psi TS; 69,000 psi YS; 26 El. Flux-cored continuous electrode for single and multipass welding of mild and medium carbon steels in CO_2 atmosphere. AWS Class E70T-1.

DUAL SHIELD T-63

Chemetron Corp.
Mn 1.6, Si 0.72, weld metal; 0.08 C, bal Fe.
As welded: 92,000 psi TS; 79,000 psi YS; 24 El. Highly deoxidized, flux-cored electrode for welding steel in spite of oil, scale or rust. AWS E70T-2.

DUAL SHIELD SP

Chemetron Corp.
Mn 1.05, Si 0.61, weld metal; 0.11 C, bal Fe.
As welded: probably 72,000 to 100,000 psi TS. Flux-cored continuous electrode for welding scaly or rusty steel. AWS Class E70T-2.

DUAL SHIELD T-100

Chemetron Corp.
Mn 1.4, Si 0.5, Cr 0.3, Ni 2, weld metal; 0.08 C, 0.30 M bal Fe.
As welded: 104,000 psi TS; 91,000 psi YS; 23 El. Flux-cored electrode for welds requiring 100,000 psi strength. AWS Grade E100T.

DUAL SHIELD T-115

Chemetron Corp.
Mn 2, Si 0.47, Ni 2, Mo 0.5, weld metal; 0.05 C, bal Fe.
As welded: 113,000 psi TS; 107,000 psi YS; 24 El. Flux-cored electrode for welding high strength steels. AWS Grade E110T.

DUAL SHIELD T-4130

Chemetron Corp.
Mn 1.1, Si 0.29, Cr 0.35, Ni 1.25, Fe 0.22, weld metal; 0.20 C.
As welded: 110,000 psi TS; 90,000 psi YS; 20 El. Hardenable (Q + T) to above 170,000 psi YS. For welding AISI 8630, 4130 and similar low alloy steels, especially for reheat treat.

DUAL SHIELD T-75

Chemetron Corp.
Mn 1.4, Si 0.45, weld metal; 0.06 C, bal Fe.
As welded: 79,000 psi TS; 67,000 psi YS; 28 El. Flux-cored continuous electrode producing welds of good impact strength. AWS Class E70T-5.

DUAL SHIELD T-8

Chemetron Corp.
Mn 1.45, Si 0.34, Cr 0.28, Ni 1.85, Mo 0.45, weld metal; 0.06 C, bal Fe.
As welded: 117,000 psi TS; 106,500 psi YS; 20 El. Flux-cored continuous electrode for welding high strength quenched and tempered steels. AWS Grade E110T.

DUAL SHIELD T-90C1

Chemetron Corp.
Mn 1, Si 0.4, Ni 2.5, weld metal; 0.07 C, bal Fe.
As welded: 84,000 psi TS; 70,000 psi YS; 25 El. Flux-cored electrode for welding 2 1/2-3 3/4% Ni steels for low temperature operations. AWS Grade E80T.

DUALLOR

Degussa AG
Precious metal.
Precious metal alloy for dentistry and dental engineering.

DUALOY

Duke Steel Co. Inc.
C, W, bal Fe.
For tools, dies; water hardening.

DUCOL W 21

British Steel Corp.
C 0.23, Mn 0-1.7, Ni 0-0.5, Cr 0-0.25, bal Fe.
Cold drawn: 83,000-97,000 TS; 46,000 YS; 14-18 El. For bridges, rail cars, building structures; good weldability. *Obsolete*

DUCOL W 25

British Steel Corp.
C 0-0.2, Mn 0-1.5, Ni 0-0.5, Cr 0-0.3, Mo 0-0.3, bal Fe.
Tempered: 83,000 min TS; 61,000 min YS; 15 min El. For bridges, rail cars, building structures; good weldability. *Obsolete*

DUCOL W 30

British Steel Corp.
C 0-0.18, Mn 0-1.4, Cr 0-0.8, Ni 0-0.5, Mo 0-0.25, V 0-0.1, Cu 0-0.5, bal Fe.
Tempered: 85,000 min TS; 66,000 min YS; 15 min El. For building structures, rail cars, bridges; good weldability. *Obsolete*

DUCOL W30 GRADE A

British Steel plc
C 0.11-0.17, Si 0-0.4, Mn 1-1.5, Cr 0.4-0.7, Cu 0-0.3, V 0.04-0.12, Mo 0.2-0.28, Ni 0-0.7, bal Fe.
High strength; weldable. For pressure vessel applications. *Obsolete*

DUCOL W30 GRADE B

British Steel plc
C 0.09-0.15, Si 0-0.4, Mn 0.9-1.3, Cr 0.4-0.7, Mo 0.2-0.28, V 0.04-0.12, Cu 0-0.3, Nb 0-0.1, N 0-0.015, Ni 0.7-1, bal Fe.
High strength; weldable; improved notch ductility. For pressure vessel applications. *Obsolete*

DUCTALLOY

American Injector Co.
Cast Iron. TC 3-3.5, Si 2-2.5, Mg 0.15, Mn 0.6, bal Fe.
Cast: 115,000 TS; 1 El; 230 Brin. Annealed: 60,000 TS; 20 El; 175 Brin. For machinery castings, gears, shafts; ductile cast iron.

DUCTALLOY 60

Abex Corp. (American Brake Shoe Co.)
C 3.4-4, Si 2-3, Mn 0.2-0.6, Ni 0-2, bal Fe.
Cast: 60,000-80,000 TS; 45,000-60,000 YS; 10-15 El; 140-200 Brin. For valve and pump bodies, compressor heads, gear housings; nodular iron, ferritic. *Obsolete*

DUCTALLOY 80

Abex Corp. (American Brake Shoe Co.)
C 3.3-3.8, Si 2-3, Mn 0.2-0.6, Ni 0-2, bal Fe.
Cast: 80,000-110,000 TS; 60,000-75,000 YS; 3-10 El; 230-280 Brin. For anvils for forging hammers, impellers, cams, roll dies; nodular iron, pearlitic. *Obsolete*

DUCTALLOY A50

Abex Corp. (American Brake Shoe Co.)
C 2.7-3.3, Si 2.2-3.2, Mn 1.5-2.5, Ni 20-26, bal Fe.
Cast: 50,000-68,000 TS; 25,000-38,000 YS; 20-40 El; 120-150 Brin. For pumps and valve impellers, grids; nodular iron, austenitic, corrosion resistant. *Obsolete*

DUCTALUMINUM 356S

Abex Corp.
Mg 0.2-0.4, Si 6.5-7.5, Cu 0-0.2, Fe 0-0.5, Ti 0-0.2, bal Al.
Heat treated: 42,000 TS; 35,000 YS; 3 El. For aircraft castings; age-hardenable.

DUCTALUMINUM 356T

Abex Corp.
Mg 0.2-0.4, Si 6.5-7.5, Cu 0-0.2, Mn 0-0.5, Ti 0-0.2, bal Al.
Heat treated: 38,000 TS; 28,000 YS; 6 El. For aircraft castings; age-hardenable.

DUCTILE CECOLLOY

Chambersburg Engineering Co.
TC 3.3-3.6, Si 2.2-2.6, Mg 0.5, bal Fe.
Cast and annealed: 60,000-80,000 TS; 40,000-60,000 YS; 10-5 El. For castings, forging hammer components. Good machinability. Ductile cast iron.

DUCTILE IRON

Grede Foundries Inc.
C 3.3, Si 2.5, Mn 0.7, Mg 0.05, bal Fe.
Cast: 60,000-90,000 TS; 45,000-65,000 YS; 2-10 El; 163-277 Brin. For gears, crankshafts, hydraulic castings; high strength ductile iron.

DUCTILE IRON FERRITIC

International Nickel Inc.
Cast iron. TC 3-4, Si 1-3, Mn 0.1-0.6, P 0-0.1, Ni 0-2, bal Fe.
Annealed: 60,000 TS; 45,000 YS; 15 El; 143-207 Brin. Ductile cast iron for pressure castings; tough. *Obsolete*

DUCTILE IRON, HARDENED

International Nickel Inc.
Cast iron. TC 3-4, Si 1-3, Mn 0.1-0.6, P 0-0.1, Ni 0-2, bal Fe.
Heat treated: 120,000 TS; 9,000 YS; 2 El; 269-388 Brin. Ductile cast iron for gears, dies, machine tool parts. *Obsolete*

DUCTILE IRON, PEARLITIC

International Nickel Inc.
Cast iron. TC 3-4, Si 1-3, Mn 0.1-0.6, P 0-0.1, Ni 0-2, bal Fe.
Normalized: 100,000 TS; 70,000 YS; 3 El; 225-302 Brin. Ductile cast iron for gears, crankshafts, agricultural equipment; wear resistant. *Obsolete*

DUCTILE IRON, SEMI-PEARLITIC

International Nickel Inc.
Cast iron. TC 3-4, Si 1-3, Mn 0.1-0.6, P 0-0.1, Ni 0-2, bal Fe.
Cast: 85,000 TS; 55,000 YS; 3 El; 200-280 Brin. Ductile cast iron for heavy machinery castings; wear resistant. *Obsolete*

DUCTILEND 100
Arcos Alloys
C 0.05, Mn 1.4, Si 0.5, Ni 1.8, Mo 0.35, bal Fe.
For welding electrodes; class E10018-M, for joints in HY80 and steels of 100,000-110,000 psi TS range. *Obsolete*

DUCTILEND 110
Arcos Alloys
C 0.05, Mn 1.6, Si 0.4, Ni 2.2, Mo 0.4, bal Fe.
For welding electrodes; class E11018-M, for joints in HY 80 and "T" steel, etc., of 110,000-120,000 psi TS range. *Obsolete*

DUCTILEND 120
Arcos Alloys
C 0.05, Mn 1.9, Si 0.55, Cr 0.5, Ni 2.2, Mo 0.45, bal Fe.
For welding electrodes; class E12018-M for joints in HY100 and "T" steel, etc., of 120,000 psi TS. *Obsolete*

DUCTILEND 70
Arcos Alloys
C 0.06, Mn 0.6, Si 0.4, Mo 0.1, bal Fe.
For welding electrodes; class E7018. *Obsolete*

DUCTILEND 70 MO
Arcos Alloys
C 0.05, Mn 0.08, Si 0.6, Mo 0.6, bal Fe.
For welding electrodes; class E7018-Al carbon-molybenum steel. *Obsolete*

DUCTILEND 80
Arcos Alloys
C 0.6, Mn 0.8, Si 0.5, Ni 0.9, Mo 0.15, bal Fe.
For welding electrodes; class E8018-C3, 1% nickel steel. *Obsolete*

DUCTILEND 85
Arcos Alloys
C 0.06, Mn 1, Si 0.5, Ni 1.5, Mo 0.3, bal Fe.
For welding electrodes; class E9018M, for joints in HY80 and steels of 90,000-100,000 psi TS range. *Obsolete*

DUCTILEND 90
Arcos Alloys
C 0.06, Mn 1.5, Si 0.4, Mo 0.45, bal Fe.
For welding electrodes; class E9018-D1, manganese-molybdenum steel of 90,000 psi TS range. *Obsolete*

DUCTILITE
Holt Equipment Co.
Cast Iron. C 3.3-3.5, Si 2.2-2.5, Mn 0.7, Mg 0.05, bal Fe.
Cast: 70,000-100,000 TS; 5-15 El; 210-250 Brin. For earth moving equipment. Ductile cast iron.

DUCTILLITE
Wheeling-Pittsburgh Steel Corp.
Tinplate.
For machinery; improved tin plate. *Obsolete*

DUCTILOY
Now NAX HIGH TENSILE.

DUCTIMET 10
Alloy Metal Products Inc.
Ni 50, Si 20, Mg 15, bal Fe.
Inoculant for production of nodular iron.

DUCTIMET 21
Alloy Metal Products Inc.
Ni 85, Mg 15.
Inoculant for production of nodular iron.

DUCTIMET 80-20
Alloy Metal Products Inc.
Ni 80, Mg 20.
Inoculant for production of nodular iron.

DUCTIMET 90-10
Alloy Metal Products Inc.
Ni 90, Mg 10.
Inoculant for production of nodular iron.

DUCTIMET 95-5
Alloy Metal Products Inc.
Ni 95, Mg 5.
Inoculant for production of nodular iron.

DUCTLIRON TYPE GS
American manufacture
C 3.3, Mn 0.7, Si 2.2, Mg 0.05, bal Fe.
Cast: 75,000 TS; 55,000 YS; 1-4 El; 217-269 Brin. For gears, shafts, machine tool housings; ductile cast iron.

DUCTLIRON TYPE GSF
American manufacture
C 3.3, Mn 0.7, Si 2.2, Mg 0.05, bal Fe.
Cast: 60,000 TS; 45,000 YS; 5-10 El; 207 Brin. Annealed: 20 El; 170 Brin. For gears, shafts, machine tool housings; ductile cast iron.

DUDLEY'S BEARING METALS
English manufacture
Sn 98, Cu 1.6, Pb 0.3.
For bearings.

DUDLEY'S BRONZE "B"
English manufacture
Cu 77, Sn 8, Pb 15.
For bearings, bushings; heavy duty.

DUDLEY'S BRONZE "K"
English manufacture
Cu 77, Sn 10.5, Pb 12.5.
For bearings, bronze castings; heavy duty.

DUDLEYS PHOSPHOR BRONZE
English manufacture
Cu 80, Sn 10, Pb 9.6, P 0.8.
30,000 TS; 6 El; 55 Brin. For bearings; heavy duty.

DUEX
Compagnie Ateliers et Forges de la Loire
C 0.88, W 17.5, Co 8, V 1.5, Mo 1.2, bal Fe.
Hardened: 64-66 Rock C. For lathe machining high manganese steels, heavy duty lathe and planer tools, broaches, hobs, boring tools. High speed steel, oil hardening. High red hardness.

DUFOUR WHITE GOLD
English manufacture
Au 75, Pb 21.5, Pt 4.5.
For ornaments; corrosion resistant.

DUKANE
Denman & Davis Co.
C 0.4, bal Fe.
For gears and shafts. Water hardening. *Obsolete*

DUKANE DRILL ROD
Teledyne Pittsburgh Tool Steel
C 0.95-1.05, bal Fe.
For dowel pins, mandrels, screw machine parts, shafts, small tools. AISI Type 1; water hardening tool steel.

DUKE'S METAL
English manufacture
Ni 40, Cu 30, Fe 30.
For corrosion resistant and heat resistant parts.

DUKE'S METAL
English manufacture
Cr 12, C 1.5, Co 4, Si 0.6, W 0.4, Mn 0.2, bal Fe.
For tools, dies, corrosion and heat resisting parts; non-deforming.

DUKE-KIDD
Duke Steel Co. Inc.
C 0.9, bal Fe.
For tools, drills, taps; water hardened.

DUKEX
Duke Steel Co. Inc.
C, W, bal Fe.
For cutting tools; water hardening.

DULLRAY
Henry Wiggin & Co. Ltd.
Ni 33-35, Cr 3-5, bal Fe.
46,000 TS. For electrical resistances, motor starter resistances; for temperatures below dull red; known as "Ferrozoid." *Obsolete*

DUMET
Creusot-Loire
Fe 54, Ni 46, Plated with Cu.
82,000 TS; 54,600 YS; 38 El; 65 RA. For resistance alloy for sealing in leads in electric light bulbs and radio tubes; platinite wire with an oxidized Cu plated coating. *Obsolete*

DUMET
Carpenter Technology Corp.
Ni 46, Fe 54, sheath of Cu.
Rolled: 70,000 TS; 18 El. For glass to metal seals, lead-in wires; thermal expansion similar to soft glass.

DUMET
Westinghouse Electric Corp.
Ni 46, Fe 54, sheath of Cu.
Rolled: 70,000 TS; 18 El. For glass to metal seals, lead-in wires; thermal expansion similar to soft glass.

DUMORE
Ziv Steel & Wire Co.
C 0.95-1.05, Cr 5-5.5, Mo 0.95-1.25, bal Fe.
For blanking, forming and drawing dies; air hardened.

DUMOST NO. 1
Republic Steel Corp.
C 0.85, Mn 0.3, V 0.3, bal Fe.
For punching, shearing and forming dies; water hardening. *Obsolete*

DUMOST NO. 16
Republic Steel Corp.
C 1.6, V 0.3, bal Fe.
For reamers, special tools; water hardened, keen cutting edge. *Obsolete*

DUMOST NO. 2
Republic Steel Corp.
C 0.95, Mn 0.3, V 0.3, bal Fe.
For punching, shearing, forming dies; water hardening. *Obsolete*

DUMOST NO. 3
Republic Steel Corp.
C 1.1, V 0.3, bal Fe.
For tools, drills, taps; water hardening. *Obsolete*

DUNDIE
Allegheny Ludlum Steel
C, alloy, bal Fe.
For tools, gears, shafts. *Obsolete*

DUNELT 10
Dunford Hadfields Ltd.
C 0.28-0.35, Ni 2.75-3.25, Cr 0-0.3, bal Fe.
Heat treated: 100,000 TS; 74,000 YS; 22 El; 192 Brin. For axles, shafts, tie rods, bolts, connecting rods; oil hardened, shock resistant. *Obsolete*

DUNELT 100
Dunford Hadfields Ltd.
C 0.35-0.45, Mn 0.45-0.65, Ni 1.3-1.8, Cr 0.9-1.4, bal Fe.
Heat treated: 125,000-145,000 TS; 18 El; 248-302 Brin. For gears, pinions, shafts, axles; oil hardened, shock resistant. *Obsolete*

DUNELT 101
Dunford Hadfields Ltd.
C 0.35-0.45, Cr 0.9-1.4, Mo 0.2-0.35, Ni 1.3-1.8, bal Fe.
Heat treated: 112,000-224,000 TS; 8-20 El; 217-444 Brin. For gears, pinions, shafts, axles, bolts; oil hardened, shock resistant. *Obsolete*

DUNELT 128
Dunford Hadfields Ltd.
C 0.25-0.35, Ni 3.9-4.5, Cr 1.1-1.5, bal Fe.
Heat treated: 224,000 TS; 10 El; 444 Brin. For camshafts, gears, bolts; oil hardened, shock resistant. *Obsolete*

DUNELT 129
Dunford Hadfields Ltd.
C 0.25-0.35, Ni 3.75-4.5, Cr 1-1.5, Mo 0.2-0.4, bal Fe.
Heat treated: 224,000 TS; 12 El; 444 Brin. For gears, pinions, shafts, bolts; oil hardened, shock resistant. *Obsolete*

DUNELT 14
Dunford Hadfields Ltd.
C 0.1-0.18, Si 0-0.3, Mn 0.5-0.9, bal Fe.
Heat treated: 72,000 TS; 20 El; 50 RA. For cams, gears, camshafts; case hardened. *Obsolete*

DUNELT 15
Dunford Hadfields Ltd.
C 0.1-0.15, Ni 2.75-3.5, Cr 0-0.3, bal Fe.
Heat treated: 101,000-135,000 TS; 18 El; 45 RA. For gears, cams, camshafts, valve tappets; case hardened, shock resistant. *Obsolete*

DUNELT 16
Dunford Hadfields Ltd.
C 0.1-0.15, Ni 3-3.5, Cr 0.8-1.1, bal Fe.
Heat treated: 123,000 TS; 18 El. For gears, cams, camshafts, worm gears; case hardened, shock resistant. *Obsolete*

DUNELT 17
Dunford Hadfields Ltd.
C 0.08-0.14, Ni 4.6-5.4, Cr 0-0.3, bal Fe.
Heat treated: 103,000 TS; 20 El. For gears, cams, camshafts, valve tappets; case hardened, shock resistant. *Obsolete*

DUNELT 18
Dunford Hadfields Ltd.
C 0.1-0.16, Ni 4.6-5.4, Cr 0-0.3, Mo 0-0.3, bal Fe.
Heat treated: 145,000 TS; 14 El. For gears, cams, camshafts, valve tappets; case hardened, shock resistant. *Obsolete*

DUNELT 20
Dunford Hadfields Ltd.
C 0.12-0.2, Ni 1.5-2, Mo 0.2-0.3, bal Fe.
Heat treated: 101,000-135,000 TS; 18 El. For gears, cams, camshafts, valve tappets; case hardened, shock resistant. *Obsolete*

DUNELT 21
Dunford Hadfields Ltd.
C 0.15-0.25, Mn 0-1, Si 0-0.35, bal Fe.
Hot rolled: 56,000 TS; 25 El. Cold drawn: 63,000 TS; 17 El. For fasteners, screws, bolts, machinery parts; case hardened. *Obsolete*

DUNELT 26
Dunford Hadfields Ltd.
C 0.35-0.45, Ni 0.5-1, Mn 0.4-1, Si 0-0.3, comment, al Fe.
Heat treated: 90,000-112,000 TS; 22 El. For axles, gears, crankshafts, connecting rods; oil hardened, tough. *Obsolete*

DUNELT 3
Dunford Hadfields Ltd.
C 0.08-0.13, Si 0-0.3, Mn 0.4-0.7, bal Fe.
Normalized: 58,000 TS; 45,000 YS; 30 El; 60 RA; 130 Brin. For staybolts; case hardened. *Obsolete*

DUNELT 33
Dunford Hadfields Ltd.
C 0.3-0.45, Cr 0.8-1.2, Mn 0.6-1, Si 0-0.3, bal Fe.
Heat treated: 100,000-145,000 TS; 18-24 El. For shafts, gears, bolts, studs; oil or water hardened, tough. *Obsolete*

DUNELT 34
Dunford Hadfields Ltd.
C 0.3-0.45, Cr 0.9-1.5, Mo 0.2-0.4, Mn 0.5-0.8, bal Fe.
Heat treated: 112,000-168,000 TS; 15-22 El. For crankshafts, connecting rods, bolts; oil hardened, tough. *Obsolete*

DUNELT 36
Dunford Hadfields Ltd.
C 0.35-0.5, Cr 1-1.5, V 0.15-0.25, bal Fe.
Heat treated: 123,000-168,000 TS; 15-20 El; 241-341 Brin. For crankshafts, connecting rods, bolts; oil or water hardened, tough. *Obsolete*

DUNELT 39
Dunford Hadfields Ltd.
C 0.15-0.3, Mn 1.3-1.8, Si 0-0.3, bal Fe.
Heat treated: 90,000-125,000 TS; 20 El; 163-225 Brin. For gears, cams, camshafts, bolts; case hardened, tough. *Obsolete*

DUNELT 4
Dunford Hadfields Ltd.
C 0.25-0.35, Mn 0.5-1.1, S 0.1-0.2, P 0-0.06, bal Fe.
Heat treated: 78,500-101,000 TS; 15 El. For bolts, nuts, studs; free-cutting, water hardened. *Obsolete*

DUNELT 40
Dunford Hadfields Ltd.
C 0.3-0.4, Mn 1.2-1.8, bal Fe.
Heat treated: 90,000-135,000 TS; 20 El; 163-293 Brin. For gears, axles, crankshafts; oil or water hardened. *Obsolete*

DUNELT 41
Dunford Hadfields Ltd.
C 0.3-0.45, Mn 1.1-1.4, S 0-0.15, bal Fe.
Heat treated: 90,000-125,000 TS; 15 El; 163-293 Brin. For gears, shafts, crankshafts, axles; free-cutting, oil or water hardened. *Obsolete*

DUNELT 5
Dunford Hadfields Ltd.
C 0.1-0.2, Mn 1-1.5, S 0.1-0.2, P 0-0.06, bal Fe.
Heat treated: 78,500 TS; 15 El. For nuts, bolts, studs; case hardened, free-cutting. *Obsolete*

DUNELT 50
Dunford Hadfields Ltd.
C 0.25-0.4, Mn 1.2-1.8, Mo 0.2-0.4, bal Fe.
Heat treated: 100,000-145,000 TS; 16-22 El; 192-352 Brin. For crankshafts, axles, studs, bolts, nuts; oil or water hardened, tough. *Obsolete*

DUNELT 51
Dunford Hadfields Ltd.
C 0.3-0.4, Mn 1.3-1.8, Mo 0.35-0.55, bal Fe.
Heat treated: 123,000-168,000 TS; 16-20 El; 248-341 Brin. For crankshafts, axles, studs, connecting rods; oil hardened, tough. *Obsolete*

DUNELT 53
Dunford Hadfields Ltd.
C 0.5-0.6, Mn 0.5-0.9, Ni 0-0.8, bal Fe.
Heat treated: 112,000-135,000 TS; 15-18 El; 223-286 Brin. For gears, pinions, axles, shafts, bolts; oil hardened. *Obsolete*

DUNELT 55
Dunford Hadfields Ltd.
C 0.25-0.33, Mn 0.45-0.7, Ni 2.75-3.75, Cr 0.5-1, bal Fe.
Heat treated: 112,000-145,000 TS; 18-22 El; 223-311 Brin. For crankshafts, connecting rods, propeller shafts; oil hardened, shock resistant. *Obsolete*

DUNELT 6
Dunford Hadfields Ltd.
C 0.35-0.45, Mn 0.4-1, Si 0-0.3, bal Fe.
Heat treated: 92,000-112,000 TS; 18-22 El. For gears, pinions, shafts, axles, machinery parts; water hardened. *Obsolete*

DUNELT 60
Dunford Hadfields Ltd.
C 0.32-0.38, Ni 3-3.75, Cr 0.5-1, bal Fe.
Heat treated: 135,000-168,000 TS; 17 El; 262-352 Brin. For crankshafts, connecting rods, propeller shafts; oil hardened, shock resistant. *Obsolete*

DUNELT 61
Dunford Hadfields Ltd.
C 0-0.15, Mn 0-1, Ni 0-1, 12% min Cr, bal Fe.
Heat treated: 78,000-103,000 TS; 25 El; 140-223 Brin. For valves, pump spindles, golf club heads; corrosion resistant, oil hardened. *Obsolete*

DUNELT 62
Dunford Hadfields Ltd.
C 0.15-0.35, Ni 0-1, 12% min Cr, bal Fe.
Heat treated: 100,000-125,000 TS; 20 El; 192-269 Brin. For chemical and textile plant equipment; corrosion resistant. *Obsolete*

DUNELT 63
Dunford Hadfields Ltd.
C 0-0.25, Cr 15-20, 1% min Ni, bal Fe.
Heat treated: 125,000 TS; 15 El; 241 Brin. For chemical and textile plant equipment; corrosion resistant. *Obsolete*

DUNELT 64
Dunford Hadfields Ltd.
C 0.27-0.44, Ni 2.3-2.8, Cr 0.5-0.9, Mo 0.4-0.7, bal Fe.
Heat treated: 123,000-225,000 TS; 18-20 El; 255-444 Brin. For axles, gears, crankshafts, connecting rods; oil hardened, shock resistant. *Obsolete*

DUNELT 65
Dunford Hadfields Ltd.
C 0.22-0.32, Ni 2.75-3.25, Cr 1-1.4, Mo 0-0.65, V 0-0.25, bal Fe.
Heat treated: 125,000-168,000 TS; 16-18 El; 248-352 Brin. For axles, gears, crankshafts, shafts, bolts; oil hardened, shock resistant. *Obsolete*

DUNELT 67
Dunford Hadfields Ltd.
C 0.15-0.35, Cr 3-3.5, Mo 0.3-0.8, Ni 0-0.4, bal Fe.
Heat treated: 103,000-168,000 TS; 15-20 El; 203-341 Brin. For cylinders, crankpins, gears, bolts; oil hardened, tough. *Obsolete*

DUNELT 69
Dunford Hadfields Ltd.
C 0.35-0.45, Ni 3.25-3.75, Cr 0-0.3, bal Fe.
Heat treated: 123,000-145,000 TS; 18-20 El; 248-302 Brin. For connecting rods, axles, crankshafts; oil hardened, shock resistant. *Obsolete*

DUNELT 7
Dunford Hadfields Ltd.
C 0.1-0.2, Mn 1-1.5, S 0.2-0.4, P 0.05-0.2, bal Fe.
Cold drawn: 78,000-103,000 TS; 15 El. For fasteners, bolts, screws, nuts, studs; case hardened, free-cutting. *Obsolete*

DUNELT 82M
Dunford Hadfields Ltd.
C 0.12-0.18, Ni 3.8-4.5, Cr 1-1.4, Mo 0.15-0.35, bal Fe.
Heat treated: 190,000 TS; 12 El. For gears, cams, camshafts, fasteners; case hardened, shock resistant. *Obsolete*

DUNELT 83
Dunford Hadfields Ltd.
C 0.4-0.5, Si 3-3.75, Ni 0-0.5, Cr 7.5-9.5, bal Fe.
Annealed: 255 Brin. For heat resistant valves; corrosion and heat resistant. *Obsolete*

DUNELT 84
Dunford Hadfields Ltd.
C 0.3-0.5, Cr 12-16, W 2-4, Mo 0-0.55, bal Fe.
Annealed: 302 Brin. For high duty valves; corrosion and heat resistant. *Obsolete*

DUNELT 88
Dunford Hadfields Ltd.
C 0-0.2, Ni 8-10, Cr 16-20, bal Fe.
Heat treated: 78,000 TS; 30 El. For chemical and dye plant equipment; stainless, austenitic. *Obsolete*

DUNELT 90
Dunford Hadfields Ltd.
C 0.9-1.1, Cr 1-1.5, Mn 0.3-0.75, Ni 0-0.25, bal Fe.
Annealed: 229 Brin. For ball and roller bearings, liners; oil or water hardened, wear resistant. *Obsolete*

DUNKIRK EZ
Allegheny Ludlum Steel
C 1.3, Mn 1, Si 1.4, Cr 0.25, Mo 1.25, bal Fe.
For dies, gauges, forming rolls, bushings, liners; free-machining. *Obsolete*

DUNNLEVIC & JONES ANTIFRICTION METAL
English manufacture
Pb 60, Sb 20, Zn 20.
For bearings, bushings; Babbitt.

DUNNLEVIC & JONES RUSSIAN
Russian manufacture
Zn 80, Cu 8, Sb 12.
For bearings; antifriction.

DUOZINC
DuPont de Nemours & Co., E.I.
Hg 0.25, bal Zn.
Cast. For anodes in zinc plating processes, electric resistance purposes; resistant. *Obsolete*

DUPLEX
Faitout Iron & Steel Co.
C 0.9-1.1, Si 0.2, Mn 0.2, bal Fe.
Water hardened: 200,000-216,000 TS; 138,000-152,000 YS; 11-13 El; 32-35 RA; 388-600 Brin. For taps, drills, reamers, punches, stamps, knurls, mandrels. Type W1 water hardening.

DUPLEX GEAR NO. 1
Colt Industries
Ni 3.5, Cr 1.25, C 0.4, bal Fe.
Annealed: 95,000 TS; 65,000 YS; 29 El; 62 RA; 187 Brin. Heat treated: 275,000 TS; 24,000 YS; 11 El; 37 RA; 490 Brin. For gears, shafts, axles, crankshafts; oil hardening. *Obsolete*

DUPLEX GEAR NO. 2
Colt Industries
Ni 1.5, Cr 1, C 0.5, bal Fe.
Annealed: 98,000 TS; 60,000 YS; 25 El; 58 RA; 196 Brin. Heat treated: 265,000 TS; 220,000 YS; 9 El; 28 RA; 512 Brin. For gears, shafts, axles, crankshafts; oil hardening. *Obsolete*

DUPLEX GEAR NO. 3
Colt Industries
Ni 3, Cr 0.75, C 0.35, bal Fe.
Annealed: 100,000 TS; 60,000 YS; 25 El; 59 RA. Toughened: 135,000 TS; 110,000 YS; 18 El; 56 RA; 311 Brin. For gears, shafts, axles, crankshafts; oil hardening. *Obsolete*

DUPONT METAL
Cutler Hammer Inc.
Copper. Cu 96, Si 3, Mn 1, Fe 0.1.
For bolts, screws, turnbuckles for airplanes; same as "Everdur;" non-corrosive.

DUQUESNE SPECIAL
Continental Foundry & Machine Co.
C, Cr, Mo, bal Fe.
For rolls; subjected to severe service.

DUR 500
Thyssen Edelstahlwerke AG
C 1, Mn 1.8, Si 0.25, Cr 1.8, Al 0.1, Ti 0.2, bal Fe.
For welding electrodes. *Obsolete*

DURA-STRAND
International Wire Products
High strength, non-magnetic core wire, with copper plus silver plate for coaxial transmission designs.

DURABIL
Duke Steel Co. Inc.
C, bal Fe.
For tools and dies; water hardening.

DURACAST NO. 7
Now WEST NO. 7, ETC.

DURACID
Machinenbau A.G.
Cast iron. Si 16, C 3, bal Fe.
Cast: for chemical apparatus; corrosion resistant; cast iron. *Obsolete*

DURAFLEX
Anaconda Co.
Copper. Sn 1-10, P 0.01-0.3, bal Cu.
O temper: 55,000 TS. H temper: 72,000 TS. Spring temper: 91,000 TS. For springs, clips, diaphragms. Phosphor-bronze; high endurance limit.

DURAL
English manufacture
Al 91.7-93.9, Cu 4-6, Mn 0.59-0.63, Mg 0.46, Fe 0.8, Si 0.2, Cr 0.11.
For light alloy parts, aircraft structures; age-hardenable.

DURAL 79
British Alcan Aluminium Ltd.
Aluminum.
Aluminum alloy AlZn4Mg3CuCr.

DURAL A
Darwins Alloy Castings
C, alloy, bal Fe.
For turning tools. *Obsolete*

DURAL B
Darwins Alloy Castings
C, alloy, bal Fe.
For coal cutter picks, for welding of hard facing material; abrasion resistant. *Obsolete*

DURAL C
Darwins Alloy Castings
C, alloy, bal Fe.
For valve parts, dry battery manufacture; chemical and abrasive resistant. *Obsolete*

DURALCAN 2014 (1)
Dural Aluminum Composites Corp.
Aluminum. 0.10 Al_2O_3, bal Al.
73,000-75,000 psi TS; 68,000-70,000 psi YS; 1-3 El; 12-15 million ME. Used for extrusions and forgings. Heat treatable.

DURALCAN 2014 (2)
Dural Aluminum Composites Corp.
Aluminum. 0.15 Al_2O_3, bal Al.
73,000-75,000 psi TS; 68,000-70,000 psi YS; 1-3 El; 12-15 million ME. Used for extrusions and forgings. Heat treatable.

DURALCAN 2014 (3)
Dural Aluminum Composites Corp.
Aluminum. 0.20 Al_2O_3, bal Al.
73,000-75,000 psi TS; 68,000-70,000 psi YS; 1-3 El; 12-15 million ME. Used for extrusions and forgings. Heat treatable.

DURALCAN 2219 (1)
Dural Aluminum Composites Corp.
Aluminum. 0.15 Al_2O_3, bal Al.
60,000 psi TS; 52,000 psi YS; 3 El; 13 million ME. Used for extrusions and forgings. Heat treatable and weldable.

DURALCAN 2219 (2)
Dural Aluminum Composites Corp.
Aluminum. 0.20 Al_2O_3, bal Al.
60,000 psi TS; 52,000 psi YS; 3 El; 13 million ME. Used for extrusions and forgings. Heat treatable and weldable.

DURALCAN 6061 (1)
Dural Aluminum Composites Corp.
Aluminum. 0.10 Al_2O_3, bal Al.
49,000-55,000 psi TS; 43,000-52,000 psi YS; 2-8 El; 12-14 million ME. Used for extrusions and forgings. Heat treatable and weldable.

DURALCAN 6061 (2)
Dural Aluminum Composites Corp.
Aluminum. 0.15 Al_2O_3, bal Al.
49,000-55,000 psi TS; 43,000-52,000 psi YS; 2-8 El; 12-14 million ME. Used for extrusions and forgings. Heat treatable and weldable.

DURALCAN 6061 (3)
Dural Aluminum Composites Corp.
Aluminum. 0.20 Al_2O_3, bal Al.
49,000-55,000 psi TS; 43,000-52,000 psi YS; 2-8 El; 12-14 million ME. Used for extrusions and forgings. Heat treatable and weldable.

DURALCAN A356 (1)
Dural Aluminum Composites Corp.
Aluminum. 0.10 SiC, bal Al.
Sand cast, heat treated: 43,000 psi TS; 42,000 psi YS; 1 El; 16 million ME. Used for sand, permanent mold, and investment castings requiring light weight, improved strength and stiffness, superior wear resistance, and thermal stability.

DURALCAN A356 (2)
Dural Aluminum Composites Corp.
Aluminum. 0.15 SiC, bal Al.
Sand cast, heat treated: 43,000 psi TS; 42,000 psi YS; 1 El; 16 million ME. Used for sand, permanent mold, and investment castings requiring light weight, improved strength and stiffness, superior wear resistance, and thermal stability.

DURALCAN A356 (3)
Dural Aluminum Composites Corp.
Aluminum. 0.20 SiC, bal Al.
Sand cast, heat treated: 43,000 psi TS; 42,000 psi YS; 1 El; 16 million ME. Used for sand, permanent mold, and investment castings requiring light weight, improved strength and stiffness, superior wear resistance, and thermal stability.

DURALFA
Alfa Romeo
Al alloy.
For light alloy parts.

DURALIMIN
Russian manufacture
Cu 4, bal Al.
For light alloy parts; age-hardenable.

DURALINE
Atlas Specialty Steels
C 0.35, Si 1.3, Cr 3.5, V 0.85, Mo 4.25, bal Fe.
At 1000°F: 216,000 TS; 8.5 El; 28.3 RA; 633 Brin. For extrusion tools and dies; high hot strength and hardness. *Obsolete*

DURALINOX
Societe du Duralumin
Mg 7, bal Al.
Rolled: 18-22 El. For floors, partitions.

DURALINOX
French manufacture
Mg 4.5, Al 95.5.
For light alloy parts; formerly "Alumag."

DURALINOX 5
Societe du Duralumin
Mg 5, Mn 0.5, bal Al.
For light alloy parts; corrosion resistant.

DURALINOX P
Societe du Duralumin
Mg 12, Mn 0.5, bal Al.
For light alloy parts; corrosion resistant.

DURALIT
French manufacture
Cu 3, Ni 0.5, Mg 0.5, Si 0.7, Ti 1, bal Al.
For forged connecting rods; age-hardenable.

DURALITE
Alluminio SA
Aluminum. Cu 3, Si 0.7, Ni 0.5, Mg 0.5, Fe 1.5, Ti 0.2, bal Al.
Aged: 55,000-67,000 TS; 41,000-51,000 YS; 10-18 El;
115-140 Brin. For auto and aircraft parts, pistons; age
hardenable.

DURALITE 35
Alluminio SA
Aluminum. Cu 3.3-3.8, Fe 1.3-1.6, Si 0.5-0.7, Mg 0.55-0.75,
bal Al.
Annealed: 28,000 TS; 18,000 YS; 15 El; 55 Brin. TA Temper:
64,000 TS; 50,000 YS; 9 El; 145 Brin. For general engineering
components, structures, fittings, fasteners, bolts. Age
hardenable, high strength.

DURALITE 35
Alluminio SA
Cu 3.3-3.8, Fe 1.3-1.6, Si 0.5-0.7, Mg 0.5-0.7, Ti 0.05-0.15, Ni
0.5-0.8, bal Al.
TA temper: 56,000-65,000 TS; 42,000-50,000 YS; 9-15 El;
125-145 Brin. For pistons, fasteners, bolts. Age-hardenable.
High strength. *Obsolete*

DURALLOY
West Steel Casting Co.
Ni 1-3.5, Cr 1-2, C, bal Fe.
For castings; cast steel.

DURALLOY NON GAMMA 2
Degussa AG
Silver.
Silver alloy for mixing amalgams for dentistry and dental
engineering.

DURALOY
Edgar Allen Balfour Ltd.
C 1.3-1.4, Si 0-0.5, Mn 13-15, Cr 1.5-1.7, bal Fe.
For wear plates, tread plates; wear and abrasion resistant
castings.

DURALOY
Ugine Aciers
C 0.25, Cr 23-30, bal Fe.
Annealed: 95,000 TS; 60,000 YS; 25 El; 200 Brin. For heat
treating boxes, furnace parts, fittings, valves, exhaust
manifolds. Type 446 stainless steel, heat and corrosion
resistant.

DURALOY
Now DBK.

DURALOY "18-8"
Duraloy Blaw-Knox/Union Steel Casting
Mn 0.5, C 0-0.2, Ni 7-10, Cr 17-20, bal Fe.
Rolled: 115,000 TS; 80,000 YS; 50 El. Cast: 80,000 TS; 35,000
YS; 35 El. For stainless articles; stainless, resists heat to 1600
F. *Obsolete*

DURALOY "A"
Duraloy Blaw-Knox/Union Steel Casting
Mn 0.5, C 0.2-1, Cr 27-30, bal Fe.
Rolled: 80,000-90,000 TS; 60,000-70,000 YS; 10-27 El; 15-45
RA. Cast: 40,000-50,000 TS; 30,000-40,000 YS; 1 El; 0-2 RA.
For valves and pump parts, furnace parts, rabble arms; resists
oxidation, corrosion and abrasion. *Obsolete*

DURALOY "B"
Duraloy Blaw-Knox/Union Steel Casting
Mn 0.5, C 0-0.2, Cr 16-18, bal Fe.
Rolled: 95,000 TS; 75,000 YS; 35 El; 150 Brin. Cast:
80,000-90,000 TS; 60,000-70,000 YS; 10-19 El; 160 Brin. For
valves and pump parts, furnace parts, rabble arms; resists
heat to 1550 F; corrosion resisting. *Obsolete*

DURALOY "C"
Duraloy Blaw-Knox/Union Steel Casting
C 0.3, Cr 12-14, bal Fe.
Cast: 100,000-110,000 TS; 75,000-85,000 YS. For furnace
parts; heat resisting to 135 F. *Obsolete*

DURALOY "H"
Duraloy Blaw-Knox/Union Steel Casting
C 2.75, Cr 27-30, bal Fe.
Cast: 50,000-60,000 TS; 0 El; 450-500 Brin. For abrasion
resisting castings, hard surface welding rods; heat resisting,
abrasion resisting. *Obsolete*

DURALOY "N"
Duraloy Blaw-Knox/Union Steel Casting
C 0.25-0.5, Cr 20-30, Ni 8-14, bal Fe.
Rolled: 95,000-130,000 TS; 60,000-105,000 YS; 45-55 El;
50-65 RA; 160-260 Brin. Cast: 85,000-110,000 TS; 30-40 El.
Resists heat to 2100 F; can be rolled or forged; for furnace
parts, grids. *Obsolete*

DURALOY "X"
Duraloy Blaw-Knox/Union Steel Casting
Cr-Ni-W.
Cast: 75,000-80,000 TS; 3-5 El. For chains in furnaces.
Obsolete

DURALOY 18-8 SMO
Duraloy Blaw-Knox/Union Steel Casting
Cr 17-19, Ni 7-9, C 0-0.07, Mo 2-4, Mn 1, bal Fe.
Cast: 65,000 TS; 35,000 YS; 35 El; 180 Brin. For stainless
steel castings; heat and corrosion resistant. *Obsolete*

DURALOY 18-8S
Duraloy Blaw-Knox/Union Steel Casting
Cr 17-19, Ni 7-9, C 0-0.07, Mn 1, bal Fe.
Cast: 65,000 TS; 35,000 YS; 35 El; 180 Brin. For stainless
steel castings; heat and corrosion resistant. *Obsolete*

DURALOY 22H
Duraloy
C 0.4-0.6, Mn 0-1.5, Si 0-1.75, Cr 26-30, Ni 45.5-50, W 4, bal
Fe.
Cast alloy. Good strength and oxidation resistance. For use
at 1800-2200°F.

DURALOY 25-20 MO
Duraloy Blaw-Knox/Union Steel Casting
Cr 24-26, Ni 19-21, C 0.2, Mo 2-3, bal Fe.
Cast: 65,000 TS; 45,000 YS; 20 El; 185 Brin. For stainless
steel castings; heat and corrosion resistant. *Obsolete*

DURALOY 50/50
Duraloy
Stainless steel. C 0-0.1, Mn 0-0.3, Si 0-1, Cr 48-52, bal Ni.
Cast alloy; corrosion and temperature resistant. ASTM A560
Gr. 50 Cr-50 Ni.

DURALOY 50/50 CB
Duraloy
Stainless steel. C 0-0.1, Mn 0-0.3, Si 0-0.5, Cr 48-52, Cb 1, bal
Ni.
Cast alloy; corrosion and heat resistant.

DURALOY CA-15
Duraloy
Stainless steel. C 0-0.15, Cr 11-14, bal Fe.
Heat treated: 100,000-203,000 TS; 75,000-173,000 YS; 6-28
El; 9-62 RA; 185-415 Brin. For pumps, valves, bushings,
gears, impellers. Ferritic, corrosion resistant; Type 410.

DURALOY CA-15, ETC.
Now DBK CA-15, ETC.

DURALOY CA-40
Duraloy
Stainless steel. C 0.3-0.4, Cr 11-14, bal Fe.
Annealed: 95,000 TS; 60,000 YS; 25 El; 190 Brin. Heat
treated: 220,000 TS; 165,000 YS; 1 El; 470 Brin. For fan
blades, oil pump pistons, chemical plant equipment. Heat
resistant to 1200°F.

DURALOY CB
Duraloy Blaw-Knox/Union Steel Casting
C 0.2, Cr 16-18, bal Fe.
Cast: 90,000 TS; 70,000 YS; 15 El. For chemical equipment,
pump parts, valves; corrosion resistant; resists heat to 1500 F.
Obsolete

DURALOY CB-30
Duraloy
Stainless steel. C 0-0.3, Cr 18-22, bal Fe.
For valve sleeves, chemical equipment impellers. Corrosion
resistant.

DURALOY CB-50
Duraloy
Stainless steel. C 0-0.5, Cr 26-30, Ni 0-4, bal Fe.
For lead pots, pump and valve bodies, furnace parts. Heat
and corrosion resistant.

DURALOY CC
Duraloy Blaw-Knox/Union Steel Casting
C 0.2, Cr 27-30, bal Fe.
Cast: 50,000 TS; 40,000 YS; 1 El. For pumps; valves, rabble
arms and blades; corrosion and heat resistant up to 2100 F.
Obsolete

DURALOY CE30
Duraloy Blaw-Knox/Union Steel Casting
C 0-0.3, Cr 26-30, Ni 3-11, bal Fe.
Cast: 95,000 TS; 60,000 YS; 15 El; 170 Brin. Annealed:
97,000 TS; 63,000 YS; 18 El; 170 Brin. For pumps, valves,
pipe fittings, chemical plant equipment; ferritic, corrosion
resistant. *Obsolete*

DURALOY CF
Duraloy Blaw-Knox/Union Steel Casting
C 0.1, Ni 7-9, Cr 17-21, bal Fe.
Heat treated: 75,000 TS; 40,000 YS; 50 El. For pump parts,
valves, digesters; corrosion resistant; resists oxidation up to
1500 F. *Obsolete*

DURALOY CF-10
Duraloy Blaw-Knox/Union Steel Casting
C 0-0.1, Cr 18-20, Ni 8-10, bal Fe.
Cast: for valves, autoclaves, impellers; austenitic, stainless.
Obsolete

DURALOY CF-10M
Duraloy Blaw-Knox/Union Steel Casting
C 0-0.1, Cr 18-20, Ni 8-10, Mo 2-3, bal Fe.
For valves, fittings; resists salt water, austenitic, stainless.
Obsolete

DURALOY CF-16
Duraloy Blaw-Knox/Union Steel Casting
C 0-0.16, Cr 18-20, Ni 8-10, bal Fe.
Annealed: 77,000 TS; 40,000 YS; 52 El; 150 Brin. For valves, impellers, autoclaves; stainless, austenitic, non-magnetic. *Obsolete*

DURALOY CF-16M
Duriron Co. Inc.
C 0.16, Cr 18-20, Ni 8-10, Mo 2-3, bal Fe.
For valves, autoclaves, fittings; acid resistant, austenitic. *Obsolete*

DURALOY CF-20
Duraloy
Stainless steel. C 0-0.2, Mn 0-1.5, Si 0-2, Cr 18-21, Ni 8-11, bal Fe.
Cast austenitic stainless steel. Similar to AISI 302.

DURALOY CF-3
Duraloy
C 0-0.03, Mn 0-1.5, Si 0-2, Cr 17-21, Ni 9-12, bal Fe.
Cast austenitic stainless steel. Ductile and corrosion resistant. Similar to AISI 304L.

DURALOY CF-3M
Duraloy
C 0-0.03, Mn 0-1.5, Si 0-2, Cr 18-21, Ni 9-12, Mo 2-3, bal Fe.
Cast austenitic stainless steel. Ductile and corrosion resistant. Similar to AISI 316L.

DURALOY CF-7
Duraloy Blaw-Knox/Union Steel Casting
C 0-0.07, Cr 18-20, Ni 8-10, bal Fe.
For valves, impellers, autoclaves, pipe fittings; corrosion resistant; austenitic. *Obsolete*

DURALOY CF-7C
Duraloy Blaw-Knox/Union Steel Casting
C 0-0.07, Cr 18-20, Ni 8-10, Cb, bal Fe.
For corrosion resistant parts, welded structures; corrosion resistant, austenitic, stabilized. *Obsolete*

DURALOY CF-7M
Duraloy Blaw-Knox/Union Steel Casting
C 0-0.07, Cr 18-20, Ni 8-10, Mo 2-3, bal Fe.
For valves, autoclaves; acid resistant; austenitic. *Obsolete*

DURALOY CF-7SE
Duraloy Blaw-Knox/Union Steel Casting
C 0-0.07, Cr 18-20, Ni 8-10, Se, bal Fe.
For valves, fittings, autoclaves; free-cutting, stainless, austenitic. *Obsolete*

DURALOY CF-8
Duraloy
Stainless steel. C 0-0.08, Mn 0-1.5, Si 0-2, Cr 18-21, Ni 8-11, bal Fe.
Cast austenitic stainless steel. Ductile and corrosion resistant. Similar to AISI 304.

DURALOY CF-8M
Duraloy
C 0-0.08, Mn 0-1.5, Si 0-2, Cr 17-21, Ni 8-11, Mo 2-3, bal Fe.
Cast austenitic stainless steel. Ductile and corrosion resistant. Similar to AISI 316.

DURALOY CG-10
Duriron Co. Inc.
C 0-0.1, Cr 20-22, Ni 10-12, bal Fe.
For valves, autoclaves, pipe fittings; stainless, austenitic. *Obsolete*

DURALOY CG-16
Duriron Co. Inc.
C 0-0.16, Cr 20-22, Ni 10-12, bal Fe.
For valves, impellers, autoclaves; austenitic, stainless. *Obsolete*

DURALOY CG-16SE
Duriron Co. Inc.
C 0-0.16, Cr 17-20, Ni 10-14, Mo 2-3, bal Fe.
For valves, fittings, autoclaves; austenitic, stainless, free cutting. *Obsolete*

DURALOY CG-7
Duriron Co. Inc.
C 0-0.07, Cr 20-22, Ni 10-12, bal Fe.
For valves, impellers, pump casings; austenitic, stainless.

DURALOY CG-7C
Duraloy Blaw-Knox/Union Steel Casting
C 0-0.07, Cr 17-20, Ni 10, Cb, bal Fe.
For valves, fittings, autoclaves; stabilized, austenitic, stainless. *Obsolete*

DURALOY CG-7M
Duraloy Blaw-Knox/Union Steel Casting
C 0.07, Cr 17-20, Ni 10-14, Mo 2.3, bal Fe.
For paper and rayon industry equipment, valves, rolls, piping; stainless; austentic. *Obsolete*

DURALOY CH-10
Duraloy Blaw-Knox/Union Steel Casting
C 0-0.1, Cr 22-26, Ni 12-15, bal Fe.
For pump bushings, liners, roasting equipment; stainless, austenitic. *Obsolete*

DURALOY CH-20
Duraloy Blaw-Knox/Union Steel Casting
C 0-0.2, Cr 22-26, Ni 12-15, bal Fe.
For chemical roasting equipment; heat and corrosion resistant. *Obsolete*

DURALOY CH-20
Duraloy
Stainless steel. C 0-0.2, Mn 0-1.5, Si 0-2, Cr 22-26, Ni 12-15, bal Fe.
Cast austenitic stainless. Resistant to hot dilute H_2SO_4. Similar to AISI 309.

DURALOY CK-20
Duraloy
Stainless steel. C 0-0.2, Mn 0-2, Si 0-2, Cr 23-27, Ni 19-22, bal Fe.
Cast austenitic stainless. Good corrosion and oxidation resistance at elevated temperatures. Similar to AISI 310.

DURALOY CK25
Duraloy Blaw-Knox/Union Steel Casting
C 0.2, Cr 25, Ni 20, bal Fe.
For carburizing and heat treating boxes; heat resistant. *Obsolete*

DURALOY CM-25
Duraloy Blaw-Knox/Union Steel Casting
Ni 19-22, Cr 8-11, C, bal Fe.
For corrosion resistant parts; for special applications. *Obsolete*

DURALOY CT-7
Duraloy Blaw-Knox/Union Steel Casting
C 0-0.07, Cr 15, Ni 35, bal Fe.
For special food and dairy process equipment; heat and corrosion resistant. *Obsolete*

DURALOY DF-20
Duraloy Blaw-Knox/Union Steel Casting
C 0-0.2, Cr 18-20, Ni 8-10, bal Fe.
For valves, autoclaves, impellers; stainless, austenitic. *Obsolete*

DURALOY H
Duraloy Blaw-Knox/Union Steel Casting
Cr 28, C, bal Fe.
For roll guides, crusher plates; heat and abrasion resistant. *Obsolete*

DURALOY H-2
Duraloy Blaw-Knox/Union Steel Casting
C 1.5, Cr 12, bal Fe.
For dies, conveyor bars and plates, molds, abrasion resistant, heat treatable. *Obsolete*

DURALOY HA
Duraloy Blaw-Knox/Union Steel Casting
C 0.15, Cr 11-14, Mo 0.9-1.2, bal Fe.
Annealed: 95,000 TS; 65,000 YS; 23 El; 60 RA; 180 Brin. Normalized: 107,000 TS; 81,000 YS; 21 El; 56 RA; 220 Brin.
For furnace rollers and conveyors, fan blades; heat and corrosion resistant. *Obsolete*

DURALOY HC
Duraloy
Stainless steel. C 0-0.5, Cr 26-30, Ni 0-4, Mo 0-0.5, bal Fe.
Cast: 70,000 TS; 65,000 YS; 2 El; 190 Brin. Aged: 115,000 TS; 80,000 YS; 18 El; 230 Brin. For roasting furnaces, rabble arms, tuyeres, blowers. Heat resistant.

DURALOY HCA
Duraloy Blaw-Knox/Union Steel Casting
Cr 28, C, bal Fe.
For screen conveyors, pumps; corrosion resistant. *Obsolete*

DURALOY HD
Duraloy
Stainless steel. Cr 26-30, Ni 3-7, C, bal Fe.
Cast. For oil and gas burner parts. Heat and corrosion resistant.

DURALOY HE
Duraloy Blaw-Knox/Union Steel Casting
Cr 26-30, Ni 8-11, C, bal Fe.
For chains, conveyors, furnace parts; resist high sulfur atmosphere, corrosion resistant. *Obsolete*

DURALOY HE
Duraloy
Stainless steel. C 0.2-0.5, Mn 0-2, Si 0-2, Cr 26-30, Ni 8-11, Mo 0-0.5, bal Fe.
Cast alloy. Corrosion and heat resistant. ASTM A297 Gr. HE.

DURALOY HF
Duriron Co. Inc.
Cr 17-20, Ni 8-10, C, bal Fe.
For furnace dampers, glass kiln conveyor rolls; heat and corrosion resistant. *Obsolete*

DURALOY HF
Duraloy
Stainless steel. C 0.2-0.4, Mn 0-2, Si 0-2, Cr 19-23, Ni 9-12, Mo 0-0.5, bal Fe.
Cast austenitic stainless. Heat and corrosion resistant at elevated temperatures. ASTM A297 Gr. HF.

DURALOY HH
Duraloy
Stainless steel. C 0.2, Ni 9-13, Cr 23-28, bal Fe.
Cast. 70,000 TS; 45,000 YS; 50 El. For furnace parts, soot blowers, kiln parts. Resists oxidation up to 2100°F.

DURALOY HI
Duraloy
Stainless steel. Cr 26-33, Ni 14-17, C, bal Fe.
For retorts for distillation of Mg. Heat and corrosion resistant. Cast alloy.

DURALOY HK
Duraloy
Stainless steel. C 0.2-0.6, Mn 0-2, Si 0-2, Cr 24-28, Ni 18-22, Mo 0-0.5, bal Fe.
Cast austenitic stainless. Heat and corrosion resistant; similar to AISI 310. ASTM A297 Gr. HK.

DURALOY HL
Duraloy Blaw-Knox/Union Steel Casting
C 0.2, Ni 9-13, Cr 23-28, bal Fe.
For furnace parts; heat resistant. *Obsolete*

DURALOY HL
Duraloy
Stainless steel. C 0.2-0.6, Mn 0-2, Si 0-2, Cr 28-32, Ni 18-22, Mo 0-0.5, bal Fe.
Cast stainless alloy. ASTM A297 Gr. HL.

DURALOY HN
Duraloy Blaw-Knox/Union Steel Casting
Cr 18-22, Ni 23-26, bal Fe.
For heat treating furnaces and boxes; heat and corrosion resistant. *Obsolete*

DURALOY HN
Duraloy
Stainless steel. C 0.2-0.5, Mn 0-2, Si 0-2, Cr 19-23, Ni 23-27, Mo 0-0.5, bal Fe.
Cast stainless alloy. ASTM A297 Gr. HN.

DURALOY HP
Duraloy Blaw-Knox/Union Steel Casting
Cr 28-32, Ni 29-31, C, bal Fe.
For oil cracking equipment; heat and corrosion resistant. *Obsolete*

DURALOY HP
Duraloy
Superalloy. C 0.35-0.75, Mn 0-2, Si 0-2, Cr 24-28, Ni 33-37, Mo 0-0.5, bal Fe.
Cast stainless alloy. For high temperature equipment such as furnace parts. ASTM A297 Gr. HP.

DURALOY HT
Duraloy Blaw-Knox/Union Steel Casting
C 0.2, Ni 34-37, Cr 13-17, bal Fe.
Cast: 60,000 TS; 35,000 YS; 5 El. For furnace parts; retorts; heat resistant. *Obsolete*

DURALOY HT
Duraloy
Superalloy. C 0.35-0.75, Mn 0-2, Si 0-2, Cr 15-19, Ni 33-37, Mo 0-0.5, bal Fe.
Cast stainless alloy. For high temperatures. ASTM A297 Gr. HT.

DURALOY HU
Duraloy
Superalloy. C 0.35-0.75, Cr 17-21, Ni 37-41, bal Fe.
Cast stainless alloy. For salt pots, retorts, annealing boxes. Heat and corrosion resistant. ASTM A297 Gr. HU.

DURALOY HW
Duraloy Blaw-Knox/Union Steel Casting
C 0.02, Cr 14, Ni 60, bal Fe.
Cast: 68,000 TS; 36,000 YS; 4 El; 179 Brin. Aged: 75,000 TS; 38,000 YS; 4 El; 205 Brin. For furnace parts, retorts, heat treating boxes; corrosion and heat resistant. *Obsolete*

DURALOY HW
Duraloy
Nickel. C 0.35-0.75, Mn 0-2, Si 0-2.5, Cr 10-14, Ni 58-62, Mo 0-0.5, bal Fe.
Cast stainless alloy. For high temperature operation. ASTM A297 Gr. HW.

DURALOY HW
Duraloy
C 0.2, Cr 10-14, Ni 58-62, bal Fe.
Cast: 68,000 TS; 38,000 YS; 4 El; 179 Brin. Aged: 75,000 TS; 38,000 YS; 4 El; 205 Brin. For retorts, heat treating fixtures, salt pots. Type HW; heat resistant.

DURALOY HX
Duraloy Blaw-Knox/Union Steel Casting
C 0.2, Cr 13-17, Ni 34-37, bal Fe.
Cast: 75,000 TS; 50,000 YS; 14 El. For furnace parts, transmission chains; heat resistant. *Obsolete*

DURALOY HX
Duraloy
Nickel. C 0.35-0.75, Mn 0-2, Si 0-2.5, Cr 15-19, Ni 64-68, Mo 0-0.5, bal Fe.
Cast high temperature stainless alloy. Usable to 2100°F. ASTM A297 Gr. HX.

DURALOY JPT
Duraloy
C 0.05-0.15, Cr 19-21, Ni 10-12, W 14-16, Co 46-53, Fe 0-3, Mn 1.4.
Cold drawn wire: 178,500-207,000 TS; 45,000-50,000 YS; 7-24 El. For springs, high temperature applications.

DURALOY MH
Duraloy Blaw-Knox/Union Steel Casting
Cr 28, C, bal Fe.
For sand blast segments, sand pumps; heat and abrasion resistant. *Obsolete*

DURALOY MO-RE 1 (1)
Duraloy
Superalloy. C 0.4-0.5, Cr 25-28, Ni 35-38, Mn 0-1.25, Si 0-1.5, W 1.25-2, Mo 0-0.5, bal Fe.
At 1800°F: 18,800 psi TS; 34 El. Cast austenitic alloy for petrochemical and industrial heating applications at 1600-2100°F.

DURALOY MO-RE 1 (2)
Duraloy
C 0.4-0.5, Mn 0-1.25, Si 0-1.5, Cr 25-28, Ni 35-38, Nb 1-1.5, bal Fe.
Cast austenitic stainless steel. For temperatures up to 1950°F.

DURALOY MO-RE 10
Duraloy
C 0.35-0.45, Mn 0-2, Si 0-2, Cr 23-27, Ni 33-37, Nb 0.75-1.5, bal Fe.
Cast austenitic stainless steel. For temperatures up to 1950°F; used in pyrolysis furnaces.

DURALOY MO-RE 15
Duraloy
Superalloy. C 0.35-0.6, Mn 0-2, Si 0-2.5, Cr 15-19, Ni 33-37, Mo 0-0.5, Cb 0.8-1.2, bal Fe.
Cast stainless alloy. Resistant to carburization at elevated temperatures.

DURALOY MO-RE 2 (1)
Duraloy
Nickel. C 0.15-0.25, Cr 32-34, Ni 48-52, Mn 0-0.3, Si 0-0.3, W 15-17, Al 0.75-1.25, bal Fe.
At 2000°F: 24,000 psi TS. Cast heat resistant alloy with good high temperature creep resistance and oxidation resistance. For use at 2100-2400°F.

DURALOY MO-RE 2 (2)
Duraloy
C 0.15-0.25, Mn 0-0.5, Si 0-0.5, Cr 32-36, Ni 48-52, Nb 15-18, Al 0.75-1.25, bal Fe.
Cast austenitic stainless steel. For temperatures up to 2400°F.

DURALOY MO-RE 21
Duraloy
C 0.06-0.14, Mn 0-1.5, Si 0-1.25, Cr 19-23, Ni 31-35, Nb 1-1.5, bal Fe.
Cast austenitic stainless steel. For elevated temperatures.

DURALOY MO-RE 3
Duraloy
Superalloy. C 0.25-0.5, Cr 24-27, Ni 44-47, Mn 0-1.25, Si 0-1.5, W 2.5-4, Mo 2.5-4, Co 2.5-4, bal Fe.
At 1800°F: 16,500 psi TS; 30.5 El. Cast heat resistant austenitic alloy for use up to 2200°F.

DURALOY MO-RE 31
Duraloy
C 0.4-0.5, Mn 0-1.25, Si 0-2, Cr 24-28, Ni 31-35, Nb 1-1.5, W 1-4, bal Fe.
Cast austenitic stainless steel. For elevated temperatures.

DURALOY MO-RE 5
Duraloy
Stainless steel. C 0-0.15, Cr 26-30, Mn 0-1, Si 0-1, Co 0-50, bal Ni.
At 1650°F: 18,600 psi TS; 9 El. Utilized for anti-pickup rolls in steel mill annealing lines.

DURALOY NA-10
Duraloy
C 1.05-1.35, Mn 12-16, Si 0-1.5, Cr 0-0.5, Ni 0-0.5, bal Fe.
Cast austenitic manganese steel with properties similar to wrought material.

DURALOY NA-3
Duraloy
Stainless steel. C 0.2-0.5, Cr 21-25, Ni 23-27, Mn 0-2, Si 0-2, bal Fe.
At 1800°F: 11,950 psi TS; 51 El. Cast austenitic stainless. For use as radiant tubes, blow pipes at 1800-2000°F.

DURALOY NA-6
Duraloy
Stainless steel. C 0.2-0.5, Cr 24-28, Ni 11-14, Mn 0-2, Si 0-2, bal Fe.
At 1800°F: 9,000 psi TS; 45 El. Cast austenitic stainless. Good high temperature properties to 1800°F. For radiant tubes, furnace parts. *Obsolete*

DURALOY NA-7
Duraloy
superalloy. C 0.2-0.6, Cr 24-28, Ni 18-22, Mn 0-2, Si 0-2, bal Fe.
At 1800°F: 12,400 psi TS; 42 El. Cast austenitic alloy. For use in radiant tubes, carburizing fixtures, cement kiln parts up to 2000°F. *Obsolete*

DURALOY NO. 15-35
Duraloy Blaw-Knox/Union Steel Casting
C 0.3, Ni 35, Cr 15, bal Fe.
Cast: 69,000 TS; 2 El. For furnace parts; resists heat to 2100 F; can be forged or rolled only at high temperature. *Obsolete*

DURALOY NSMO
Duraloy Blaw-Knox/Union Steel Casting
Cr 23-25, Ni 11-13, C 0-0.15, Mo 2-3, bal Fe.
Cast: 75,000 TS; 35,000 YS; 25 El; 195 Brin. For stainless steel castings; heat and corrosion resistant. *Obsolete*

DURALOY R
Duraloy Blaw-Knox/Union Steel Casting
Cr 17, C, bal Fe.
For plungers, tumbler molds and valves; corrosion resistant. *Obsolete*

DURALOY SUPER 22-H
Duraloy
Superalloy. C 0.4-0.6, Cr 26-30, Ni 46-50, Mn 0-1.5, Si 0-1.75, W 4-6, bal Fe.
At 1800°F: 18,000 psi TS; 32 El. Cast alloy. Good strength and oxidation resistance; for use at 1800-2250°F.

DURALOY-HOM
Duraloy Blaw-Knox/Union Steel Casting
Nickel alloy.
At 1600 F: 33,000 TS. At 2000 F: 13,500 TS. For high temperature castings; maintains high working strength to 2200 F. *Obsolete*

DURALPLAT
Durener Metallwerke
base metal of Cu containing Duralumin coated, with a Cu-free Duralumin.
For light alloy parts. British Patent 318999; high corrosion resistance.

DURALUM
German manufacture
Al 79, Mg 11, Cu 10, P 0.5.
For light alloy parts; non-hardened.

DURALUMIN
English manufacture
Cu 3.5-4.5, Mg 0.3-0.6, Mn 0-0.8, Fe 0.4-1, Si 0-1, bal Al.
Annealed: 32,000-36,000 TS; 7000-10,000 YS; 34 El; 45-55
Brin. Heat treated: 63,000 TS; 30,000-40,000 YS; 18-25 El.
For airplane and dirigible construction; same as "Alcoa No.
17"; age-hardening alloy.

DURALUMIN 3L-1
English manufacture
Cu 4.5, bal Al.
For housings, cases, structures, screw machine products;
age-hardenable, high strength.

DURALUMIN 3L-3
English manufacture
Cu 4.5, bal Al.
For housings, cases, structures, screw machine products;
age-hardenable, high strength.

DURALUMIN 681A
Durener Metallwerke
Aluminum. Cu 4.5, Mg 0.5, Mn 0.6, bal Al.
For light alloy parts; age hardenable.

DURALUMIN 681B
Societa Metallurgica Italia
Cu 4.5, Mg 0.5, Mn 0.6, bal Al.
Heat treated: 65,000 TS; 45,000 YS. For light alloy parts; heat
treatable.

DURALUMIN 681B
Durener Metallwerke
Cu 4.5, Mg 0.5, Mn 0.6, bal Al.
Heat treated: 65,000 TS; 45,000 YS. For light alloy parts; heat
treatable.

DURALUMIN 681H
Europa Metalli-LMI SpA
Aluminum. Cu 2.5, Mg 0.5, bal Al.
Heat treated: 51,000 TS; 25 El. For light alloy parts; heat
treatable. *Obsolete*

DURALUMIN 681K
Societe du Duralumin
Mg 0.2, bal Al.
Heat treated: 31,000 TS; 20 El. For light alloy parts; heat
treatable.

DURALUMIN 681K
Durener Metallwerke
Mg 0.2, bal Al.
Heat treated: 31,000 TS; 20 El. For light alloy parts; heat
treatable.

DURALUMIN 681K
Societa Metallurgica Italia
Mg 0.2, bal Al.
Heat treated: 31,000 TS; 20 El. For light alloy parts; heat
treatable.

DURALUMIN 681ZB
Europa Metalli-LMI SpA
Aluminum. Si 0.6, Cu 4.5, Mg 0.5, Mn 0.8, bal Al.
Heat treated: 62,000-88,000 TS; 3-14 El; 128-175 Brin. For
light alloy parts; heat treatable. *Obsolete*

DURALUMIN DM31
Societa Metallurgica Italia
Si 0.6, Cu 4, Mg 1, Mn 1.2, bal Al.
Heat treated: 71,000-74,000 TS; 57,000-60,000 YS; 10-12 El;
125-140 Brin. For light alloy parts; heat treatable.

DURALUMIN DM31
Alcan-Booth Industries, Ltd.
Si 0.6, Cu 4, Mg 1, Mn 1.2, bal Al.
Heat treated: 71,000-74,000 TS; 57,000-60,000 YS; 10-12 El;
125-140 Brin. For light alloy parts; heat treatable.

DURALUMIN E
Alcan-Booth Industries, Ltd.
Cu 1.5, Mg 0-1, Fe 0-1, Si 1, Mn 0.6, bal Al.
Heat treated: 50,000-58,000 TS; 40,000-49,000 YS; 8-15 El;
130-160 Brin. For aircraft structural parts; age-hardenable.
Obsolete

DURALUMIN E
English manufacture
Cu 3-4.5, Si 0-1, Mg 0-1, Mn 0-1.2, bal Al.
60,000 TS; 52,000 YS; 8 El. For light alloy parts; heat
treatable.

DURALUMIN F
Now ALCAN GB-050.

DURALUMIN F
Alcan-Booth Industries, Ltd.
Mg 1, Si 0.5, Cr 0.25, Cu 0.25, bal Al.
WP-temper: 40,400 TS; 33,600 YS; 10 El. Annealed: 18,000
TS; 15 El. For extrusions of intricate sections; age
hardenable.

DURALUMIN G
Now ALCAN GB-24S.

DURALUMIN G
Alcan-Booth Industries, Ltd.
Cu 3.5-4.5, Mn 0.7-1.5, Mg 1.4-1.8, bal Al.
Rolled: 65,000 TS; 40,000 YS; 8-15 El. Extruded: 69,000 TS;
45,000 YS; 15 El. For aircraft structures and panels; age-
hardenable. *Obsolete*

DURALUMIN H
Now ALCAN GB-B51S.

DURALUMIN HX
Alcan-Booth Industries, Ltd.
Cu 0-1, Fe 0-0.7, Mg 0.5-1.2, Mn 0-1, Si 0.7-1.2, bal Al.
Heat treated: 52,000 TS; 40,000 YS; 13 El; 100 Brin. For light
alloy parts; spinnings and pressing. *Obsolete*

DURALUMIN IMITATION
English manufacture
Al 79, Mg 11, Cu 10, P 0.05.
For light alloy parts; age-hardened.

DURALUMIN J
Alcan-Booth Industries, Ltd.
Cu 2.25, Mg 1.5, Si 0.9, Fe 1.2, Ni 1, bal Al.
Aged: 56,000 TS; 38,000 YS; 8 El. For aircraft structures and
parts; age-hardenable. *Obsolete*

DURALUMIN JJ
Now ALCAN GB-D19S.

DURALUMIN K
Durener Metallwerke
Aluminum. Mg 0.5-2, Si 0.3-1.5, Mn 0-1.5, bal Al.
For light alloy parts; corrosion resistant, hardenable.

DURALUMIN K
Now ALCAN GB-D75S.

DURALUMIN K
Alcan-Booth Industries, Ltd.
Cu 1, Mg 2.7, Mn 0.4, Zn 6, bal Al.
Aged: 72,000-78,000 TS; 60,000-67,000 YS; 5-7 El. For
aircraft engine components; age-hardenable, high strength.
Obsolete

DURALUMIN K
Now ALCAN GB-B26S.

DURALUMIN K
Multiple manufacturers
Aluminum. Si 0.3-1.5, Mn 1.5, Mg 0.5-2, bal Al.
Heat treated: 50,000 TS; 35,000 YS; 10-20 El; 100 Brin. For
hardware, fasteners. Age hardenable.

DURALUMIN KC
Alcan-Booth Industries, Ltd.
Cu 1, Mg 2.75, Mn 0.4, Zn 6, Cr, bal Al.
Heat treated: 78,000 TS; 68,000 YS; 7 El. For aicraft engine
components. Heat treatable, high strength. *Obsolete*

DURALUMIN L
Alcan-Booth Industries, Ltd.
Cu 0.15, Mg 2-3.5, Si 0.5, Fe 0.5, Mn 0.25-1, Ti 0.2, Zn
4.5-6.5, bal Al.
WP-temper: 72,000-78,000 TS; 60,500-67,000 YS; 5-7 El. For
aircraft engine components; age-hardened, high strength.
Obsolete

DURALUMIN LC
Alcan-Booth Industries, Ltd.
Cu 0.6, Mg 2.6, Mn 0.5, Zn 6, Cr 0.05, bal Al.
Heat treated: 73,000 TS; 62,000 YS; 5 El. For aircraft engine
components. Heat treatable, high strength. *Obsolete*

DURALUMIN M
Alcan-Booth Industries, Ltd.
Cu 1.5-3, Fe 0-0.7, Mg 0.2-0.5, Si 0-0.7, bal Fe.
Heat treated: 40,000 TS; 18,000 YS; 25 El; 90 Brin. For rivets.
Obsolete

DURALUMIN Q
Alcan-Booth Industries, Ltd.
Cu 4.2, Mg 0.9, Mn 0.5, bal Al.
For aircraft structures and components; age-hardenable.
Obsolete

DURALUMIN S
Alcan-Booth Industries, Ltd.
Cu 3.5-4.8, Si 0.8, Mg 0-1, Fe 0-1, bal Al.
Heat treated: 62,000-75,000 TS; 54,000-62,000 YS; 8-12 El;
110-130 Brin. For aircraft engine components; age-
hardenable, high strength. *Obsolete*

DURALUMIN SPECIAL "Y"
English manufacture
Al 92.5, Cu 4, Ni 2, Mg 1.5.
For dirigible, airplane and motor car light alloy parts; age
hardening.

DURALUMIN SUPER
German manufacture
Cu 5.5, Fe 0.3, Mg 1, Mn 0.6, Si 0.6, bal Al.
71,000-78,000 TS; 10-20 El; 130-140 Brin. For light alloy
parts; heat treatable.

DURALUMIN T
Durener Metallwerke
Aluminum. Cu 2.1, Fe 0.5-1.5, Mg 0.3-1.5, Ni 0-2, Mn 0-1, bal
Al.
Heat treated: 62,000 TS; 41,000 YS; 12 El; 130 Brin. For
aircraft engine forgings; age hardenable.

DURALUMIN W
Multiple manufacturers
Aluminum. Cu 3.5-4.5, Mg 1-1.8, Ni 1.8-2.2, bal Al.
Heat treated: 60,000 TS; 35,000 YS; 10 El; 100-200 Brin. For
fasteners, hardware, engine components. Heat treatable,
high strength.

DURALUMIN W
Farbenindustrie Atkiengesellschaft
Cu 4, Fe 0.3, Mg 1.5, Ni 1, Si 0.2, bal Al.
For light alloy parts, aircraft; age-hardenable.

DURALUMIN W
Durener Metallwerke
Cu 4, Fe 0.3, Mg 1.5, Ni 1, Si 0.2, bal Al.
For light alloy parts, aircraft; age-hardenable.

DURALUMIN X

Now ALCAN GB-51S.

DURALUMIN X

Alcan-Booth Industries, Ltd.
Mg 0.75, Si 1, bal Al.
WP-temper: 40,400 TS; 33,600 YS; 10 El. Annealed: 18,000 TS; 15 El. For extrusions of intricate sections; age-hardenable. *Obsolete*

DURALUMIN Z

German manufacture
Cu 4.3-5, Fe 0.3, Mg 0.9, Mn 0.6, Si 0.6, bal Al.
68,000-71,000 TS; 12-14 El; 130-135 Brin. For light alloy parts; heat treatable.

DURALUMIN ZC

Now ALCAN GB-E74S.

DURALUMIN, VICKERS

Alcan-Booth Industries, Ltd.
Cu 3.5-4.5, Mn 0.4-0.7, Mg 0.4-0.7, Fe 0-0.75, Al 0-94.5.
Sheets: 56,000 TS; 33,500 YS; 8-15 El. Bars: 45,000-56,000 TS; 22,500-33,500 YS; 15 El; 20 RA. For airplane and dirigible parts, crankcases, cylinder heads, pistons, airscrews; light alloy, age-hardening. *Obsolete*

DURALUMIN-C

Alcan-Booth Industries, Ltd.
Cu 5.5, Pb 0.4, Bi 0.4, bal Al.
Heat treated: 42,000 TS; 30,000 YS; 12 El. For machined components, fasteners; good machinability, age-hardenable. *Obsolete*

DURALUMIN-L

Alcan-Booth Industries, Ltd.
Zn 5.8, Mg 2.8, Cu 0.5, Mn 0.4, bal Al.
Heat treated: 71,700 TS; 60,500 YS; 5 El. For aircraft structures; age-hardened, high strength. *Obsolete*

DURAMIUM

Duke Steel Co. Inc.
C 0.7, W 18, Cr 4, V 1, bal Fe.
For reamers, hobs, drills, taps; high speed steel.

DURAMOLD A

Bethlehem Steel Corp.
C 0.07, Mn 0.35, Cr 4.5, Mo 0.5, bal Fe.
For hobbing cavity molds; air hardening, wear resistant, high core hardness. AISI P4. *Obsolete*

DURAMOLD B

Bethlehem Steel Corp.
C 0.07, Ni 0.25, Cr 0.95, Mo 0.25, bal Fe.
For cold hobbing cavities, molds; oil hardened, medium core hardness. AISI P2. *Obsolete*

DURAMOLD C

Bethlehem Steel Corp.
C 0.07, Mn 0.15, bal Fe.
For hobbing cavity molds; water hardening, low core hardness *Obsolete*

DURAMOLD N

Bethlehem Steel Corp.
C 0.1, Cr 1.5, Ni 3.25, bal Fe.
Heat treated: 130,000-173,000 TS; 88,000-142,000 YS; 15-20 El; 262-340 Brin. For plastic molds. Type P6; case-hardened tool steel. *Obsolete*

DURAMOLD NI-CR

Bethlehem Steel Corp.
C 0.1, Cr 0.6, Ni 1.25, bal Fe.
Heat treated: 100,000 TS; 70,000 YS; 200 El. For plastic molds, cold hubbed. Good core strength and toughness. Type P3 tool steel. Case hardened. *Obsolete*

DURANA-1

German manufacture
Cu 64-78, Zn 29.5, Sn 2.2, Al 1.7, Fe 1.5.
60,000-65,000 TS; 14 El. For chemical equipment construction; also known as "Duranametal"; corrosion resistant.

DURANA-2

German manufacture
Cu 65, Zn 30, Al 1.5, Fe 1.5, Pb 2.
For chemical equipment; also known as "Duranametal"; corrosion resistant, high strength.

DURANA-3

German manufacture
Cu 59, Zn 40, Sn 1, Fe 0.3, Pb 0.4.
For chemical equipment; high strength.

DURANAL

German manufacture
Mg 5-10, Mn 0.6, Si 0.2, bal Al.
For light alloy parts.

DURANALIUM

Durener Metallwerke
Aluminum. Mg 5-10, bal Al.
For light alloy parts, aircraft castings; corrosion resistant.

DURANALIUM

Durener Metallwerke
Aluminum. Mn 0.3-0.6, Mg 2.5-9, bal Al.
Bar: 26,000-57,000 TS; 8500-34,000 YS; 15-25 El; 45-100 Brin. For hardware, fittings, fasteners. Age hardenable. *Obsolete*

DURANALIUM 2 S

Durener Metallwerke
Aluminum. Mg 2-2.5, Mn 1-2, Sb 0-0.2, bal Al.
Soft: 26,000 TS; 12,000 YS; 25 El; 45 Brin. Hard: 45,000 TS; 36,000 YS; 3 El; 75 Brin. For light alloy parts; corrosion resistant.

DURANALIUM 3

Durener Metallwerke
Aluminum. Mg 2-4, Mn 0-0.4, Cr 0-0.3, bal Al.
Soft: 28,000 TS; 13,000 YS; 30 El; 47 Brin. Hard: 42,000 TS; 37,000 YS; 8 El; 77 Brin. For light alloy parts; resists sea water corrosion.

DURANALIUM 3.5

Durener Metallwerke
Aluminum. Mg 2-6, Mn 0.25-0.75, bal Al.
Cast: 20,000-22,000 TS; 11,000-15,000 YS; 3-5 El; 54-58 Brin. For rivets, bolts, screws; corrosion resistant.

DURANALIUM 5

Durener Metallwerke
Aluminum. Mg 4-6, bal Al.
Soft: 35,000 TS; 18,000 YS; 26 El; 60 Brin. Hard: 55,000 TS; 45,000 YS; 4 El; 105 Brin. For light alloy parts; corrosion resistant.

DURANALIUM 7

Durener Metallwerke
Aluminum. Mg 7, Mn 0.45, bal Al.
Rolled: 52,000 TS; 32,000 YS; 12 El; 115 Brin. For seaplane pontoons, propellers; resists sea water corrosion.

DURANALIUM 9

Durener Metallwerke
Aluminum. Mg 8-10, bal Al.
Soft: 50,000 TS; 25,000 YS; 15 El; 80 Brin. Hard: 65,000 TS; 45,000 YS; 10 El; 110 Brin. For light alloy parts; corrosion resistant.

DURANCE BEARINGS

German manufacture
Sb 44.5, Sn 33.3, Cu 22.2.
For bearings; Babbitt.

DURAND'S ALLOY

English manufacture
Al 67, Zn 33.
For strong and tough light weight parts; non-hardenable.

DURANIC

French manufacture
Al 96, Mn 2, Ni 2.
23,000-26,000 TS; 1.0-2.5 El; 60-70 Brin. For light alloy parts; non-hardenable.

DURANICKEL

Now DURANICKEL ALLOY.

DURANICKEL 301

Criterion Metals, Inc.
Nickel. Ni 94, C 0.15, Mn 0.25, Fe 0.15, S 0.005, Si 0.55, Cu 0.05, Ti 0.5, Al 4.5.
Thin gauge sheet, various tempers. Annealed: 90-120 ksi TS; 35-60 ksi YS. Age hardened: 160-190 ksi TS.

DURANICKEL ALLOY 301

Inco Alloys International Inc.
Ni 96.5, C 0.15, Mn 0.25, Fe 0.3, S 0.005, Si 0.5, Ti 0.63, Cu 0.13, Al 4.4.
Annealed: 90,000-120,000 TS; 30,000-60,000 YS; 35-55 El; 135-185 Brin. Annealed and aged: 150,000-190,000 TS; 110,000-140,000 YS; 20-30 El; 285-360 Brin. For extrusion press parts; molds used in glass industry; clips, diaphragms and springs. Corrosion resistant, high strength and hardness.

DURANIT ATS 354

Thyssen Edelstahlwerke AG
C 0.04, Cr 19, Mo 6, W 1, Al 2, Co 12, Ti 3, bal Ni.
Coated electrodes for overlaying on forging tools. Meets DIN 8555. 721-746 N/mm^2 YS; 826-830 N/mm^2 TS.

DURANIT M 10

Thyssen Edelstahlwerke AG
C 0.85, Cr 3.7, Mo 8.5, W 1.8, V 1.1, bal Fe.
Coated electrodes for overlaying on turning and planing tools. Werkstoff Nr. 1.3346.

DURANIT M 18

Thyssen Edelstahlwerke AG
C 1.2, Cr 6, Mo 18, Co 13, V 1.5, Nb 0.5, bal Fe.
Coated electrodes for overlaying on plough blades. Meets DIN 8555.

DURANIT NIMO CA

Thyssen Edelstahlwerke AG
C 0.1, Cr 17, Mo 15-17.5, W 4-4.5, Fe 0-5, bal Ni.
Coated electrodes for overlaying on trimming dies. Werkstoff Nr. 2.4537.

DURANIT W 4

Thyssen Edelstahlwerke AG
C 0.3, Cr 2.3, W 4.2, V 0.5, bal Fe.
Coated electrodes for overlaying on tools. Werkstoff Nr. 1.2567.

DURANIT W 5

Thyssen Edelstahlwerke AG
C 0.5, Si 0.5, Mn 1.5, Cr 4.5, Mo 3.5, W 4, bal Fe.
Coated electrodes for overlaying of steel casting rods. Meets DIN 8555.

DURANIT W 6

Thyssen Edelstahlwerke AG
C 0.5, Si 0.5, Mn 2.5, Cr 6, Mo 1.5, W 1.5, bal Fe.
Coated electrodes for overlaying of steel casting rods. Meets DIN 8555.

DURANIT W 8

Thyssen Edelstahlwerke AG
C 0.3, Cr 2.3, Co 2, W 8.8, V 0.4, bal Fe.
Coated electrodes for overlaying of cutting tools. Werkstoff Nr. 1.2662.

DURANMIUM

Duke Steel Co. Inc.
C, W, bal Fe.
For high speed cutting tools; high speed steel.

DURAPERM

Hamilton Technology Inc.
Fe 84.5, Si 9.5, Al 6.
Hard, brittle, soft-magnetic, alloy. Saturation induction: 9 K Gausses. Coercive force: 0.05 oersteds (H max = 2.0 oersteds). For pole tips in magnetic recording heads which operate from audio through the video frequency range. *Obsolete*

DURAPLAT

Durener Metallwerke
Aluminum. Cu 3.5-5.5, Mg 0.2-2, Si 1, Mn 1, bal Al.
For light alloy parts; Aluminum ply alloy, bonded; corrosion resistant.

DURASINT GRADE RM 4

Sintered Products Ltd.
Iron, sintered. Moderate duty friction material; for main drive clutches.

DURASINT GRADE S.1

Sintered Products Ltd.
Bronze, sintered. Good general purpose friction material for power shift clutches, disc brake pads.

DURASINT GRADE S.106

Sintered Products Ltd.
Iron ceramic, sintered. Friction material for low friction level, low wear; aircraft brakes.

DURASINT GRADE S.14

Sintered Products Ltd.
Bronze ceramic, sintered. Friction material for heavy duty dry applications as tractor main drive clutches.

DURASINT GRADE S.193

Sintered Products Ltd.
Bronze ceramic. Friction material; high coefficient of friction with good wear; for heavy duty main clutches.

DURASINT GRADE S.2

Sintered Products Ltd.
Bronze, sintered. Friction material for tractor steering and industrial clutches.

DURASINT GRADE S.200

Sintered Products Ltd.
Bronze, sintered. Friction material for semi-automatic gear boxes, aircraft actuators.

DURASINT GRADE S.206

Sintered Products Ltd.
Bronze, sintered. Friction material for torque limiting clutches on farm machinery.

DURASINT GRADE S.208

Sintered Products Ltd.
Bronze, sintered. Friction material for tension and overload clutches.

DURASINT GRADE S.210

Sintered Products Ltd.
Bronze, sintered. Friction material for steering clutches and steering band brakes for track laying vehicles.

DURASINT GRADE S.214

Sintered Products Ltd.
Iron ceramic, sintered. High coefficient of friction and good wear resistance under high temperature conditions. For heavy duty main drive clutches.

DURASINT GRADE S.215

Sintered Products Ltd.
Bronze, sintered. High coefficient of friction. For automotive and farm equipment clutches.

DURASINT GRADE S.217

Manufacturer not listed
Bronze, sintered (friction material). Good energy absorption and wear resistance. For clutches for construction equipment.

DURASINT GRADE S.22

Sintered Products Ltd.
Iron, sintered. High friction level material. For main drive clutches.

DURASINT GRADE S.220

Sintered Products Ltd.
Iron, sintered. Friction material for heavy duty brake material.

DURASINT GRADE S.3

Sintered Products Ltd.
Bronze, sintered. Friction material for automatic and power shift transmissions.

DURASINT GRADE S.71

Sintered Products Ltd.
Bronze graphite, sintered. Friction material for medium duty dry friction applications.

DURASINT GRADE S.72

Sintered Products Ltd.
Bronze ceramic, sintered. Friction material for moderate to high duty, multidisc brake material.

DURASINT GRADE S.73

Sintered Products Ltd.
Bronze graphite, sintered. Friction material for low wear conditions; safety clutches.

DURASINT GRADE S112

Sintered Products Ltd.
Iron ceramic, sintered. Friction material for moderate friction, low wear; aircraft brakes.

DURASINT GRADES RM5, RM6

Sintered Products Ltd.
Iron, sintered. Friction material for brake drums for passenger cars.

DURASINT GRADES S.113, S.114

Sintered Products Ltd.
Iron, sintered. Friction material for aircraft brakes.

DURASINT RF7

Sintered Products Ltd.
Nonmetallic friction material. For oil cooled clutches.

DURASPUN

Duraloy Blaw-Knox/Union Steel Casting
C 0.2, Cr 18, Ni 8, bal Fe.
For sleeves, liners; centrifugally cast tubes, corrosion resistant. *Obsolete*

DURATHERM 200

Vacuumschmelze GmbH
Co, Ni, Cr.
Temperature and corrosion resistant spring nonmagnetic alloy; can be age hardened. *Obsolete*

DURATHERM 2602

Vacuumschmelze GmbH
Co, Ni, Cr.
Temperature and corrosion resistant spring nonmagnetic alloy; can be age hardened. *Obsolete*

DURATHERM 418

Vacuumschmelze GmbH
Spring alloy in hard, age hardened state. 500 Vickers min; 1700 N/mm^2 YS min; 1800 N/mm^2 TS min.

DURATHERM 600

Vacuumschmelze GmbH
Co, Ni, Cr.
Temperature and corrosion resistant spring nonmagnetic alloy; can be age hardened. 520 Vickers; 1650 N/mm^2 YS; 1720 N/mm^2 TS. R30600

DURATHERM 700

Vacuumschmelze GmbH
Spring alloy in hard, age hardened state. 550 Vickers min; 1800 N/mm^2 YS min; 1900 N/mm^2 TS min. R30700

DURAWELD

Plansee, Metallwerk Gesellschaft
WC.
For hard facing electrode. Sintered.

DURAWELD NO. 1

Duraweld Metal Products Corp.
For hard facing welding rods; wear resistant.

DURAWELD NO. 2

Duraweld Metal Products Corp.
For hard facing welding rods; wear resistant.

DURAWELD NO. 3

Duraweld Metal Products Corp.
For hard facing welding rods; wear resistant.

DURAWELD NO. 4

Duraweld Metal Products Corp.
For hard facing welding rods; wear resistant.

DURAX

Swedish American Steel Corp.
C 0.5, Cr 1.5, W 2.5, bal Fe.
For tools, dies; Type S1; oil hardened, shock resistant.

DURAX D

Thyssen Edelstahlwerke AG
C 0.35, Si 1, Cr 1, V 0.2, W 2, bal Fe.
For pneumatic snaps, punches; oil hardened, shock resistant. *Obsolete*

DURAX EXTRA HART

Thyssen Edelstahlwerke AG
C 0.8, W, Cr, V, bal Fe.
For cutters, tools, dies; oil hardened. *Obsolete*

DURAX H

Thyssen Edelstahlwerke AG
C 0.55, Cr 1.2, V 0.4, W 2, bal Fe.
For heading and forming dies, upsetters; oil hardened, tough *Obsolete*

DURAX MO3

Marathon Specialty Steels Inc.
C 0.3, Mn 0.3, Si 0.3, Cr 3, Mo 2.8, V 0.5, bal Fe.
Heat treated: 215,000-230,000 TS. For extrusion dies, mandrels, die casting dies, hot work tools. Hot work steel. Thermal shock resistant.

DURAX N

Thyssen Edelstahlwerke AG
C 0.35, Cr 1, Ni 3.5, V 0.2, W 5.3, bal Fe.
For extrusion and piercing mandrels; hot work steel, tough, oil hardened. *Obsolete*

DURAX SPECIAL

Thyssen Edelstahlwerke AG
C 0.25, Cr 1.1, V 0.1, W 3.7, bal Fe.
For hot dies, knives, shears, punches; oil hardened, tough. *Obsolete*

DURAX W2

Thyssen Edelstahlwerke AG
C 0.45, Cr 1.2, Si 0.9, V 0.2, W 2, bal Fe.
For pneumatic rivet tools, shear cutters, tools. Oil hardening, shock resisting. *Obsolete*

DURAX W3

Thyssen Edelstahlwerke AG
C 0.5, Cr 1.5, W 2.2, V 0.2, bal Fe.
For riveting sets, hot chisels; oil hardened, tough. *Obsolete*

DURAX W5

Thyssen Edelstahlwerke AG
C 0.3, Cr 2.3, V 0.6, W 4.25, bal Fe.
For cold work tools, upsetters, punches; oil hardened, tough.
Obsolete

DURAZIT

Plansee, Metallwerk Gesellschaft
WC.
For hard facing electrode. Sintered.

DURAZIT I

Plansee, Metallwerk Gesellschaft
Co-Cr-W.
Cast: 580 Brin. For chemical equipment; hard facing.

DURAZIT II

Plansee, Metallwerk Gesellschaft
Co-Cr-W.
Cast: 460 Brin. For chemical equipment; hard facing.

DURBAR

British Steel plc
Slip resisting raised pattern steel floor plate.

DURBAR

Buffalo Bronze Die Casting Co.
Cu 70-72, Pb 20-24, Sn 4-10.
For bearings, ornaments. *Obsolete*

DURBAR-HARD

Buffalo Bronze Die Casting Co.
Cu 70, Pb 20, Sn 10.
22,100 TS; 14,450 YS; 12 El; 11 RA; 55 Brin. For bearings for
cranes, excavating machinery, steel mill machinery, heavy
duty. *Obsolete*

DURBAR-STANDARD

Buffalo Bronze Die Casting Co.
Cu 72, Pb 24, Sn 4.
19,100 TS; 13,250 YS; 10 El; 15 RA; 46 Brin. For bearings for
machine tools, electric motors, pumps, cranks, gas engines.
Obsolete

DURCILIUM

Trefileries & Laminoirs du Havre
Al 94.3, Cu 4, Mg 0.7, Mn 0.5, Si 0.5.
Annealed: 34,000 TS; 17,000 YS; 16 El; 45 Brin. Hardened:
61,000 TS; 37,000 YS; 16 El; 110 Brin. For light alloy parts,
aircraft structures; age-hardenable.

DURCILIUM B

Pechiney/E. & E. Kaye Ltd.
Mg 3.5, Cr 0.25, bal Al.
Aluminium wire. AA 5154.

DURCILIUM C

Pechiney/E. & E. Kaye Ltd.
Drawn aluminium tubing.

DURCILIUM E

Pechiney/E. & E. Kaye Ltd.
Si 0.5, Mg 0.6, bal Al.
Aluminium bar for electrical purposes. AA 6101.

DURCILIUM ET

Pechiney/E. & E. Kaye Ltd.
Cu 0-0.5, bal Al.
1/4 or hard drawn bar for electrical purposes. BS 2898 E1E;
BS 215.

DURCILIUM F

Trefileries & Laminoirs du Havre
Cu 3.4-4.8, Mg 0.7, Si 0.8, Ni 0.2, Fe 1, Mn 0.4-1.2, Zn 0.2, Ti
0.2, bal Al.
W-temper: 58,200 TS; 42,600 YS; 8-13 El. For structural
members; age-hardened.

DURCILIUM FD

Pechiney/E. & E. Kaye Ltd.
Si 5.2, bal Al.
Aluminium alloy wire; mostly for welding. AA 4043.

DURCILIUM J

Pechiney/E. & E. Kaye Ltd.
Mg 5, Mn 0.12, Cr 0.12, bal Al.
Aluminium wire. AA 5056.

DURCILIUM K

Trefileries & Laminoirs du Havre
Cu 3.5-5, Mg 0.4-1.2, Si 0.7, Fe 0.7, Mn 0.4-1.2, Ti 0.3, bal Al.
Heat treated: 53,800-56,000 TS; 31,400-33,600 YS; 15-10 El.
For structural members; age-hardened, corrosion resistant.

DURCILIUM K

E. & E. Kaye Ltd.
Cu 3.5-5, Mg 0.4-1.2, Si 0.7, Fe 0.7, Mn 0.4-1.2, Ti 0.3, bal Al.
Heat treated: 53,800-56,000 TS; 31,400-33,600 YS; 15-10 El.
For structural members; age-hardened, corrosion resistant.

DURCILIUM L

Trefileries & Laminoirs du Havre
Cu 3.5-5.8, Mg 0.6, Si 1.5, Fe 1, Mn 1.2, Ti 0.3, bal Al.
W-temper: 53,800-56,000 TS; 31,400-33,600 YS; 10-15 El.
WP-temper: 62,000-67,200 TS; 53,800-58,200 YS; 6-8 El. For
structural members; age-hardened.

DURCILIUM M

Trefileries & Laminoirs du Havre
Cu 1-2, Mg 0.5-1.2, Si 0.7-1.2, Fe 0.7, Mn 1, Ti 0.2, bal Al.
W-temper: 42,100 TS; 32,400 YS; 18 El. WP-temper: 62,000
TS; 49,800 YS; 13 El. For aircraft and other structural
members; age hardenable.

DURCILIUM MX

Pechiney/E. & E. Kaye Ltd.
Cu 2.6, Mg 0.35, bal Al.
Aluminium wire. AA 2117.

DURCILIUM P

Pechiney/E. & E. Kaye Ltd.
Special aluminium wire to meet DTD 5074 A.

DURCILIUM R

Trefileries & Laminoirs du Havre
Cu 0.15, Mg 0.4-1.5, Si 0.7-1.3, Fe 0.6, Mn 1, Ti 0.2, bal Al.
W-temper: 27,000 TS; 15,700 YS; 18 El. WP-temper: 40,400
TS; 33,600 YS; 10 El. For structural members; age-hardened,
corrosion resistant.

DURCILIUM S

Pechiney/E. & E. Kaye Ltd.
Si 0.7-1.3, Fe 0-0.5, Mg 0.6, Mn 0.6, bal Al.
Aluminium wire, bar, tubing. AA 6351.

DURCILIUM T

Pechiney/E. & E. Kaye Ltd.
Si 0-0.25, Fe 0-0.4, 99.5 Al min.
Aluminium wire, bar, tube. 1050 A.

DURCILIUM TA

Pechiney/E. & E. Kaye Ltd.
Si 0-0.15, Fe 0-0.15, 99.8 Al min.
Aluminium wire, bar, tube. 1080 A.

DURCILIUM V

Pechiney/E. & E. Kaye Ltd.
Mg 2, bal Al.
Aluminium wire, bar and tube. 5251.

DURCILIUM W

Trefileries & Laminoirs du Havre
Cu 0.15, Mg 0.4-0.9, Si 0.3-0.7, Fe 0.6, Ti 0.2, bal Al.
M-temper: 15,700 TS; 11,200 YS; 15 El. W-temper: 24,600 TS;
15,700 YS; 23 El. WP-temper: 31,400 TS; 26,900 YS; 14 El.
For structural members; age-hardened.

DURCILIUM W

E. & E. Kaye Ltd.
Cu 0.15, Mg 0.4-0.9, Si 0.3-0.7, Fe 0.6, Ti 0.2, bal Al.
M-temper: 15,700 TS; 11,200 YS; 15 El. W-temper: 24,600 TS;
15,700 YS; 23 El. WP-temper: 31,400 TS; 26,900 YS; 14 El.
For structural members; age-hardened.

DURCILIUM WT

Pechiney/E. & E. Kaye Ltd.
Si 0.4, Mg 0.7, bal Fe.
Aluminium bar. AA 6463.

DURCILIUM XT

Pechiney/E. & E. Kaye Ltd.
Aluminum wire specially processed for use in telephone
cable.

DURCILIUM-F

Pechiney World Trade (USA) Inc.
Cu 3.4-3.8, Mg 0.55-0.85, Si 0.6-0.9, Fe 1, Mn 0.4-1.2, Ni 0.2,
Zn 0.2, Ti 0.3, bal Al.
WP-temper: 74,000 TS; 65,000 YS; 12 El. W-temper: 65,000
TS; 36,000 YS; 16 El. For aircraft structures; age-hardenable.
Obsolete

DURCILIUM-M

Pechiney World Trade (USA) Inc.
Cu 1.2, Mg 0.5-1.25, Si 0.75-1.25, Fe 0.75, Mn 1, Ti 0.2, bal
Al.
WP-temper: 62,700 TS; 49,300 YS; 13 El. W-temper: 42,600
TS; 31,400 YS; 18 El. For aircraft structures; age-hardenable.
Obsolete

DURCILIUM-R

Pechiney World Trade (USA) Inc.
Cu 0.15, Mg 0.4-1.5, Si 0.75-1.3, Fe 0.6, Mn 1, Ti 0.2, bal Al.
WP-temper: 47,100 TS; 40,400 YS; 12 El. P-temper: 31,400
TS; 26,900 YS; 20 El. For aircraft and light alloy parts; age-
hardenable, corrosion resistant. *Obsolete*

DURCILIUM-W

Pechiney World Trade (USA) Inc.
Cu 0.15, Mg 0.4-0.9, Si 0.3-0.7, Fe 0.6, Ti 0.2, bal Al.
WP-temper: 31,400 TS; 26,900 YS; 14 El. P-temper: 24,600
TS; 15,700 YS; 23 El. For aircraft and light alloy parts; age-
hardenable, corrosion resistant. *Obsolete*

DURCILIUM-Y

Pechiney World Trade (USA) Inc.
Al alloy.
Rolled: 15,700 TS; 12 El. For light alloy parts; corrosion
resistant, good weldability. *Obsolete*

DURCILIUM-Z

Pechiney World Trade (USA) Inc.
Cu, Mg, Si, Ti, bal Al.
WP-temper: 76,500-88,500 TS; 67,000-80,000 YS; 1 El. For
aircraft structures; age-hardenable. *Obsolete*

DURCO

Duriron Co. Inc.
C 0.2, Cr 18, Ni 8, bal Fe.
For heat and corrosion resisting parts; heat and corrosion
resistant. *Obsolete*

DURCO 18-8-S

Duriron Co. Inc.
Cr 17-19, Ni 7-9, Mn 0.5, Si 1, <0.07% C, bal Fe.
Cast: 70,000 TS; 30,000 YS; 40-50 El; 40-50 RA; 150 Brin. For
food handling and chemical plant equipment; corrosion
resisting to HNO 3 . *Obsolete*

DURCO 18-8-S-CB

Duriron Co. Inc.
C 0-0.07, Cr 18-22, Ni 9-12, Cb 0-1, Si 0-2, Mn 1.5, bal Fe.
Annealed: 70,000 TS; 30,000 YS; 30 El; 40 RA; 150 Brin. For corrosion resistant castings; corrosion resistant. *Obsolete*

DURCO 18-8-S-MO-CB

Duriron Co. Inc.
C 0-0.07, Cr 18-22, Ni 9-12, Mo 2-3, Cb 0-1, bal Fe.
Annealed: 70,000 TS; 30,000 YS; 30 El; 40 RA; 150 Brin. For corrosion resistant castings; corrosion resistant. *Obsolete*

DURCO 181

Duriron Co. Inc.
C 1.1, Cr 18, bal Fe.
For corrosion and heat resistant parts; corrosion and heat resistant. *Obsolete*

DURCO 25-12

Duriron Co. Inc.
C 0.2, Cr 22-26, Ni 12-15, Si 0-2, Mn 0-1.5, bal Fe.
Annealed: 70,000 TS; 30,000 YS; 30 El; 40 RA; 150 Brin. For heat and corrosion resistant castings; corrosion and heat resistant. *Obsolete*

DURCO 25-12-S

Duriron Co. Inc.
C 0-0.07, Cr 22-26, Ni 12-15, Si 0-1.5, Mn 0-1.5, bal Fe.
Annealed: 65,000 TS; 28,000 YS; 30 El; 40 RA; 150 Brin. For heat and corrosion resistant castings; heat and corrosion resistant; ACI-CH-10. *Obsolete*

DURCO 26-10

Duriron Co. Inc.
C 0.15, Cr 26, Ni 10, Si 1, Mn 0.6, bal Fe.
70,000 TS; 30,000 YS; 40-45 El; 150 Brin. For heat and corrosion resistant parts; chemical plant and food handling equipment; heat and corrosion resistant. *Obsolete*

DURCO 26-12

Duriron Co. Inc.
Cr 26, Ni 12, Si 1, Mn 0.6, 0.15 T.C., bal Fe.
Annealed: 70,000 TS; 30,000 YS; 40-45 El; 40-45 RA; 150 Brin. For chemical plant equipment; food handling equipment; apparatus for nitric acid; stainless, corrosion resistant. *Obsolete*

DURCO 28-12

Duriron Co. Inc.
C 0-0.07, Cr 28, Ni 12, bal Fe.
For heat and corrosion resistant parts; heat and corrosion resistant. *Obsolete*

DURCO 30-15

Duriron Co. Inc.
Cr 30, Ni 15, Si 1, Mn 0.6, 0.15 T.C., bal Fe.
For chemical plant equipment; food handling equipment; apparatus for nitric acid; stainless, corrosion resistant. *Obsolete*

DURCO CF-3

Duriron Co. Inc.
Stainless steel. Cr 18-21, Ni 8-12, C 0-0.03, bal Fe.
Heat treated: 70,000 psi TS; 30,000 psi YS; 35 El; 160 Brin. For stainless parts; chemical process and food handling equipment; chemical resistant cast alloy.

DURCO CF-3M

Duriron Co. Inc.
Stainless steel. Cr 18-21, Ni 9-12, Mo 2-3, C 0-0.03, bal Fe.
Heat treated: 70,000 psi TS; 30,000 psi YS; 30 El; 160 Brin. For stainless parts; chemical process and food handling equipment; chemical resistant cast alloy.

DURCO CF-8

Duriron Co. Inc.
Stainless steel. Cr 18-21, Ni 8-11, C 0-0.08, bal Fe.
Heat treated: 70,000 psi TS; 30,000 psi YS; 35 El; 160 Brin. For stainless parts; chemical process and food handling equipment; chemical resistant cast alloy.

DURCO CF-8M

Duriron Co. Inc.
Stainless steel. Cr 18-21, Ni 9-12, Mo 2-3, C 0-0.07, bal Fe.
Heat treated: 70,000 psi TS; 30,000 psi YS; 30 El; 160 Brin. For stainless parts; chemical process and food handling equipment; corrosion resistant cast alloy.

DURCO CY-40

Duriron Co. Inc.
Nickel. Cr 14-17, Fe 0-11, C 0-0.4, bal Ni.
Heat treated: 70,000 psi TS; 28,000 psi YS; 30 El; 150 Brin. For chemical process equipment handling caustic and some salt solutions; cast alloy.

DURCO CZ-100

Duriron Co. Inc.
Nickel. Ni 95, Fe 0-3, C 0-1.
As cast: 50,000 psi TS; 18,000 psi YS; 10 El; 130 Brin. For chemical process equipment handling caustic alkalis, anhydrous HF, and organic acids.

DURCO D-10

Duriron Co. Inc.
Ni 56, Cr 23, Mo 4, W 2, Mn 1, Cu 8.
Cast: 60,000 TS; 5 El; 5 RA; 180 Brin. For heat and corrosion resistant parts; nonmagnetic. *Obsolete*

DURCO D-12

Duriron Co. Inc.
C 0-0.12, Cr 12, bal Fe.
For valve and pump parts; corrosion resistant. *Obsolete*

DURCO D-18

Duriron Co. Inc.
C 0-0.12, Cr 18, bal Fe.
For pumps, valves; corrosion resistant. *Obsolete*

DURCO D-181

Duriron Co. Inc.
C 0.2, Cr 18, bal Fe.
For valve trim, pump parts; corrosion resistant. *Obsolete*

DURCO D-28

Duriron Co. Inc.
C 0.2, Cr 28, bal Fe.
For pumps, valves; corrosion resistant. *Obsolete*

DURCO D-6

Duriron Co. Inc.
C 0.1, Cr 8, Ni 19, bal Fe.
For petroleum and chemical industry equipment; stainless. *Obsolete*

DURCO D-7

Duriron Co. Inc.
Cr 28, Ni 4, Mo 1, C, bal Fe.
For wear and corrosion resistant parts; high resistance to oxidation. *Obsolete*

DURCO DC-8

Duriron Co. Inc.
Reactive and refractory.
Cobalt base alloy. For use as shaft sleeve material; resists wear and corrosion.

DURCO KA-2-MO

Duriron Co. Inc.
Cr 18-22, Ni 7-10.5, Mo 2-4, Si 0-1, 0.15% min C, bal Fe.
70,000 YS; 30,000 YS; 40-45 El; 40-45 RA; 150 Brin. For stainless parts, chemical plant and paper mill equipment; corrosion resistant. *Obsolete*

DURCO M-35

Duriron Co. Inc.
Nickel. Cu 28-32, Fe 0-3.5, C 0-0.35, bal Ni.
As cast: 65,000 psi TS; 30,000 psi YS; 25 El; 150 Brin. For chemical process equipment handling hydrofluoric acid and alkaline brines.

DURCO WCB

Duriron Co. Inc.
Alloy steel. Cr 0-0.5, Ni 0-0.5, Mo 0-0.25, Cu 0-0.5, Si 0-0.6, Mn 0-1, C 0-0.3, bal Fe.
Heat treated: 70,000 psi TS; 36,000 psi YS; 22 El; 150 Brin. For process equipment handling weak caustics, many types of water, and a wide variety of organics.

DURCOMET 100

Duriron Co. Inc.
Stainless steel. Cr 24.5-26.5, Ni 4.75-6, Mo 1.75-2.25, Cu 2.75-3.25, bal Fe.
Heat treated: 100,000 psi TS; 70,000 psi YS; 16 El; 250 Brin. For equipment handling HNO_3, dilute H_2SO_4, H_3PO_4, fatty acids, and erosion-corrosion services.

DUREDGE

Boyd-Wagner Co.
C 0.55, Mn 0.75, Mo 0.35, Si 2, bal Fe.
For punches, shears, dies; water or oil hardening.

DUREDGE

Swedish American Steel Corp.
C 0.55, Si 1.6, Mo 0.4, V 0.6, bal Fe.
For cutting tools, chisels, rivet sets, shear blades; will maintain keen cutting edge, hard and tough.

DUREHETE

British Steel Corp.
C 0.4, Cr 0.9, Mo 1, bal Fe.
For high temperature use, turbine valves, bolts, pipe flanges; creep resisting steel; resists high pressure superheated steam. *Obsolete*

DUREHETE 1050

British Steel Corp.
C 0.1-0.25, Cr 0.9-1.3, Mo 0.85-1.1, V 0.6-0.8, bal Fe.
Heat treated: 130,000 TS; 115,000 YS; 20 El. For bolting, fasteners, high temperature applications; heat resistant to 1050 F. *Obsolete*

DUREHETE 1055

United Engineering Steels Ltd.
C 0.15-0.25, Cr 0.9-1.3, Mo 0.85-1.1, V 0.6-0.8, Ti 0.05-0.2, B 0.001-0.01, bal Fe.
Heat treated: 122,000 TS min; 100,000 YS min; 43 El min. For bolts required to operate at 500-565°C. For use in steam power plants.

DUREHETE 900

United Engineering Steels Ltd.
C 0.35-0.45, Cr 1-1.5, Mo 0.5-0.8, Si 0-0.5, bal Fe.
Heat treated: 147,000 TS; 136,000 YS; 22 El. For high temperature bolts, studs, and fasteners. High creep resistance. B.S. 970-En20B; ASTM A193-61T Gr 7A.

DUREHETE 950

United Engineering Steels Ltd.
C 0.3-0.45, Cr 1-1.5, Mo 0.5-0.8, V 0.25-0.3, bal Fe.
Heat treated: 144,000 TS; 135,000 YS; 25 El; 64 RA. For high temperature bolts, studs, and fasteners. Heat resistant to 950°F. B.S. 1506-661; ASTM A193-61T Gr B16.

DURENER MN20

Durener Metallwerke
Aluminum. Mn 0.5-1.5, Cr 0-0.3, bal Al.
Soft: 16,000 TS; 6000 YS; 40 El. Hard: 29,000 TS; 27,000 YS; 10 El. For cooking utensils, heat exchangers, tanks, furniture; good forming and welding properties.

DURENER Z8

Durener Metallwerke
Aluminum. Zn, Cu, Mg, bal Al.
For light alloy parts.

DUREX

Darwins Alloy Castings
C 0.7, W 18.5, Cr 4.5, V 1, bal Fe.
For shaping and planing tools, broaches, hobs; high-speed steel. *Obsolete*

DUREX

Moraine Mfg., Inc.
Sn 10, Cu 83.22, 0.6 impurities, 4.4-4.7 total C.
30-40 Brin. For bearings (porous to absorb oil, 25.0 oil by volume); carbon present as graphite.

DUREX BRONZE

Moraine Mfg., Inc.
Cu 95, Sn 5.
Cast: 30,000-40,000 TS; 65-100 Brin. For bearings; sintered.

DUREX BRONZE NO. 46

Moraine Mfg., Inc.
Sn 10, 1.2 graphite, bal Cu.
Sintered: 13,000-25,000 TS; 2.0 El; 25-40 Brin. For bearings; sintered; Crushing strength 79,000.

DUREX IRON 362

Moraine Mfg., Inc.
Cu 20, C 1, bal Fe.
75,000 TS; 1 El; 185 Brin. For gears, cams; copper infiltrated sintered steel.

DUREX IRON NO. 62

Moraine Mfg., Inc.
C 1, bal Fe.
Sintered: 28,000 TS; 0.5 El. For bearings; sintered, Crushing strength 160,000.

DUREX IRON NO. 76

Moraine Mfg., Inc.
Fe.
For gears, pole shoes; sintered iron parts. *Obsolete*

DUREX IRON NO. 93

Moraine Mfg., Inc.
Cu 7.5, bal Fe.
Sintered: 15,000-35,000 TS; 1-7 El; 90 Brin. For bearings; sintered alloy.

DURFACE

Eutectic Corp.
C, Ni, Mo, bal Fe.
For hard facing alloy. *Obsolete*

DURICHLOR 51

Duriron Co. Inc.
Alloy steel. Cr 4-5, Si 14.2-14.7, C 0.95-1.1, bal Fe.
For pumps handling H_2SO_4, HNO_3, HCl; and impressed current anodes.

DURIMET

Duriron Co. Inc.
Ni 22, Cr 19, Mo 3, Si 1, Cu 1, C 0.07.
Heat treated: 65,000-75,000 TS; 30,000-40,000 YS; 35-65 El; 130-150 Brin. For heat and corrosion resistant parts; heat and corrosion resistant. *Obsolete*

DURIMET 20

Duriron Co. Inc.
Superalloy. Cr 19-22, Ni 27.5-30.5, Mo 2-3, Cu 3-4, C 0-0.07, bal Fe.
Cast, heat treated: 62,000 psi TS; 25,000 psi YS; 35 El; 130 Brin. For chemical process equipment handling H_2SO_4, HNO_3, mixed acids, and caustics.

DURIMET A

Duriron Co. Inc.
Ni 25, Si 5, C 0.25, bal Fe.
Rolled: 100,000 TS; 45,750 YS; 62 El; 76 RA; 160 Brin. For chemical apparatus, valves, pumps, kettles; austenitic, abrasion and corrosion resistant. *Obsolete*

DURIMET B

Duriron Co. Inc.
Ni 35, Cr 12, Si 5, C 0.25, bal Fe.
110,000 TS; 20 El; 35 RA. For chemical apparatus, valves, pumps, kettles; austenitic, abrasion and corrosion resistant. *Obsolete*

DURIMET D

Duriron Co. Inc.
Ni 15, Cr 2.5, Si 3, C 0.6, bal Fe.
75,000 TS; 48,000 YS; 62 El; 160 Brin. For chemical apparatus, valves, pumps, kettles; austenitic, abrasion and corrosion resistant. *Obsolete*

DURIMET K

Duriron Co. Inc.
C 0.25, Ni 16, Cr 11, Cu 1.25, bal Fe.
100,000-110,000 TS; 35,000-40,000 YS; 40-45 El; 40-45 RA. For ventilation ducts; resists corrosive fumes. *Obsolete*

DURIMET L

Duriron Co. Inc.
C 0.07, Ni 19.5-23, Cr 18.8-21.5, Si 2.75-3.75, Mn 0.5-0.75, Cu 0.25-0.45, bal Fe.
70,000 TS; 30,000 YS; 40-50 El; 40-50 RA; 150 Brin. For heat and corrosion resistant parts, chemical apparatus, pickling tanks, valves, pumps; heat and corrosion resistant; austenitic. *Obsolete*

DURIMET T

Duriron Co. Inc.
Ni 22, Cr 19, Si 1, Cu 1, Mo 3, C 0.07.
100,000-110,000 TS; 35,000-40,000 YS; 40-45 El. For fume ducts; heat and corrosion resistant. *Obsolete*

DURIMET-20

Duriron Co. Inc.
Ni 29, Cr 20, Mo 2-3, Cu 4, Si 1, C 0.07, bal Fe.
Cast: 65,000-75,000 TS; 28,000-38,000 YS; 35-50 El; 40-50 RA; 120-150 Brin. Wrought: 85,000 TS; 35,000 YS; 50 El; 150 Brin. For chemical plant equipment; corrosion resistant to mixed acids. *Obsolete*

DURINVAL

Creusot-Loire
Ni 42, Al 2, Ti 2, bal Fe.
For chronometer springs; age-hardenable, corrosion resistant. *Obsolete*

DURIRON

Duriron Co. Inc.
Alloy steel. Si 14.2-14.7, C 0.8-1.1, bal Fe.
Cast alloy for chemical process equipment handling H_2SO_4 and HNO_3 acids.

DURITAS

Saville & Co. Ltd., J.J.
WC.
For dies, nibs; cemented carbides.

DURITAS

Jessop-Saville Ltd.
WC.
For dies, nibs; cemented carbides.

DURITE

Columbia Tool Steel Co.
C, W, bal Fe.
For tools, dies; hot work steel. *Obsolete*

DURMES

Durener Metallwerke
Aluminum. Cu 2.5-5, Mg 0.2-1.8, Mn 0.3-1.5, Pb 0.5-2.5, Sn, Cd, Bi, bal Al.
For screw machine products; free cutting.

DURO

McInnes Steel Co.
C 0.8, bal Fe.
For roll turning tools, dies for drawing brass tubing. *Obsolete*

DURO

Jessop-Saville Ltd.
C, alloy, bal Fe.
For tools. *Obsolete*

DURO 6

Thyssen Edelstahlwerke AG
C 0.71, Si 0.22, Mn 0.27, bal Fe.
For springs, tools, axes, rails, crimpers; Type W1; water hardened. *Obsolete*

DURO 70

Thyssen Edelstahlwerke AG
C 0.56, Si 0.3, Mn 0.55, bal Fe.
For shafts, rails, axles, gears, crankshafts; water hardened. *Obsolete*

DURO ANTIFRICTION

Federal-Mogul Corp.
Cu 3-5, Sb 7-8, bal Sn.
For bearings, bushings, liners; Babbitt. *Obsolete*

DURO GLOSS "C-2"

Jessop Steel Co.
C 0.1, Cr 16-18, bal Fe.
Annealed: 80,000 TS; 50,000 YS; 27 El; 70 RA; 150 Brin. Hard rolled: 90,000 TS; 65,000 YS; 25 El; 55 RA; 190 Brin. For turbine blades, pump rods, gate valves, fireplace tools, steam fittings; heat and corrosion resistant. *Obsolete*

DURO GLOSS "C-3"

Jessop Steel Co.
C 0.1-0.15, Cr 20-22, bal Fe.
Annealed: 80,000 TS; 50,000 YS; 30 El; 55 RA; 160 Brin. For turbine blades, pump rods, gate valves, fireplace tools, steam fittings; heat and corrosion resistant. *Obsolete*

DURO GLOSS "C-4"

Jessop Steel Co.
C 0.1-0.15, Cr 25-29, bal Fe.
Annealed: 80,000 TS; 55,000 YS; 35 El; 60 RA; 170 Brin. For turbine blades, pump rods, gate valves, fireplace tools, steam fittings; heat and corrosion resistant. *Obsolete*

DURO GLOSS C-1

Jessop Steel Co.
C 0.12, Cr 15, bal Fe.
Annealed: 75,000 TS; 37,000 YS; 35 El; 65 RA; 150 Brin. Hardened: 170,000 TS; 150,000 YS; 15 El; 50 RA; 364 Brin. For machine parts, turbine blades, pump rods, gate valves, fireplace tools, steam fittings; stainless, corrosion resisting. *Obsolete*

DURO GLOSS C-12

Jessop Steel Co.
C 0.13-0.2, Cr 12, bal Fe.
For corrosion resistant parts; corrosion resistant. *Obsolete*

DURO GLOSS S

Jessop Steel Co.
C 0-0.12, Cr 12-15, S, bal Fe.
For corrosion resistant parts; corrosion resistant. *Obsolete*

DURO HIGH SPEED

McInnes Steel Co.
C 0.8, W 18.5, Mo 4, Cr 4, Co 7, bal Fe.
For heavy duty cutters; high speed steel. *Obsolete*

DURO-CHIP

Allied Steel & Tractor Products Inc.
C 0.6, Mn 0.7, Si 1.85, Mo 0.5, V 0.25, bal Fe.
For hand and pneumatic chisels, caulking tools; shock resistant; oil hardening.

DURO-FORM 50

Gulf States Steel, Inc.
HSLA steel. C 0.1, Mn 0.8, Si 0.5, Cu 0.5, Cr 0.9, Cb 0-0.02, bal Fe.
High strength steel with atmospheric corrosion resistance.
50,000 YS min; 60,000 TS min; 22 El min.

DURO-FORM 60

Gulf States Steel, Inc.
HSLA steel. C 0.1, Mn 0.8, Si 0.5, Cu 0.5, Cr 0.9, Cb 0-0.02, bal Fe.
High strength steel with atmospheric corrosion resistance.
60,000 YS min; 70,000 TS min; 21 El min.

DURO-GLOSS FREE MACHINING

Jessop Steel Co.
C 0-0.12, Mn 0-0.6, Cr 12-15, bal Fe.
Annealed: 80,000 TS; 45,000 YS; 35 El; 65 RA; 160 Brin. For bolts, nuts, screws, spindles, bushings. *Obsolete*

DURO-PLATE 50

Gulf States Steel, Inc.
HSLA steel. C 0.17, Mn 0.5-1.2, Si 0.25-0.5, Cu 0.3-0.5, Mo 0.1, Cr 0.4-0.7, Ni 0.4, Cb 0.005, bal Fe.
High strength steel with atmospheric corrosion resistance.
50,000 YS min; 70,000 TS min.

DURO-PLATE 60

Gulf States Steel, Inc.
HSLA steel. C 0.17, Mn 1.2, Si 0.5, Cu 0.5, Cr 0.7, Cb 0.005, V 0-0.07, bal Fe.
High strength steel with atmospheric corrosion resistance.
60,000 YS min; 75,000 TS min.

DURO-PLATE 65

Gulf States Steel, Inc.
HSLA steel. C 0.17, Mn 1.4, Si 0.5, Cu 0.5, Cr 0.7, Cb 0.005, V 0-0.07, bal Fe.
High strength steel with atmospheric corrosion resistance.
65,000 YS min; 80,000 TS min.

DURO-PLATE 70

Gulf States Steel, Inc.
HSLA steel. C 0.17, Mn 1.4, Si 0.5, Cu 0.5, Cr 0.7, Cb 0.005, V 0-0.07, bal Fe.
High strength steel with atmospheric corrosion resistance.
70,000 YS min; 85,000 TS min.

DUROCHROME

TRW Inc.
C, Si, Cr, Mo, bal Fe.
For valve seat inserts; cast iron, wear resistant. *Obsolete*

DUROCYL

Midland Motor Cylinder Co.
C 3, Si 2, Mn 0.8, Cr 0.3, bal Fe.
Cast: 38,000 TS; 240 Brin. For cylinder liners; wear plates; centrifugally cast iron.

DURODE

Thomas Bolton Ltd.
Cu alloy.
For spot welding electrodes; high conductivity. *Obsolete*

DURODE XH

Thomas Bolton Ltd.
Cu alloy.
For resistance welding electrode; high electrical conductivity. *Obsolete*

DURODI

A. Finkl & Sons Co.
Alloy steel. C 0.5-0.6, Mo 0.7-0.9, Mn 0.5-0.6, Cr 0.85-1.15, Ni 1.4-1.75, bal Fe.
Heat treated: 130,000-225,000 TS; 120,000-200,000 YS; 15-26 El; 43-51 RA; 275-460 Brin. For hot work die steel, punches, upsetting and gripping dies, shear knives; tough steel; air hardening, non-distorting.

DURODIE

Jessop-Saville Ltd.
C 0.4, Cr 3.3, W 2.7, V 0.3, bal Fe.
For bolt dies, hot extrusion dies, swaging dies; hot work steel.

DUROLITH

German manufacture
Zn base alloy.
For die castings.

DUROMAX 250

Ambo-Stahl-Gesellschaft
C 0.1-0.2, Si 0.15-0.35, Mn 0.8-1, Cr 0.4-0.65, Mo 0.4-0.6, Ni 0.7-1, V, Cu, B, bal Fe.
High strength low alloy steel.

DUROMAX 321

Ambo-Stahl-Gesellschaft
C 0.15, Si 0.25, Mn 0.85, Cr 0.5, Mo 0.5, Ni 0.5, V 0.05, B, bal Fe.
High strength, low alloy steel; 321 Brin.

DURON

Thyssen Edelstahlwerke AG
C 0.5, Si, Cr, V, bal Fe.
For shafts, gears, springs, crankshafts; oil hardened, shock resistant. *Obsolete*

DURON H

Thyssen Edelstahlwerke AG
C 0.6, Si 0.85, Mn 0.75, Cr 1.10, V 0.1, bal Fe.

DURON W

Thyssen Edelstahlwerke AG
C 0.4, Si, Mn, Cr, V, bal Fe.

DURONZE 606

Olin Brass, Indianapolis
Copper. Si 3, Mn 0.95, Cu 96.05.
Hard: 90,000 psi TS; 45,000 psi YS; 18 El; 85 Rock B. Soft: 55,000 psi TS; 22,000 psi YS; 55 El; 50 Rock B. For bolts, gears, valve stems, marine hardware, and clutch disks. Corrosion resistant; tough; wear resistant.

DURONZE 609

Olin Brass, Indianapolis
Copper. Si 2, Cu 98, bal Cu.
Hard: 75,000 psi TS; 40,000 psi YS; 10 El; 138 Brin. Soft: 50,000 psi TS; 20,000 psi YS; 40 El; 63 Brin. For bolts, cold headed parts, gears, and shafts; formerly Duronze V; corrosion resistant.

DURONZE 632

Olin Brass, Indianapolis
Copper. Si 2.95, Fe 0.1, bal Cu.
Annealed: 63,000 psi TS; 30,000 psi YS; 55 El; 90 Rock F. Hard: 90,000 psi TS; 45,000 psi YS; 18 El; 85 Rock B. For poleline hardware, nuts, bolts, wire and cable connectors, cap screws, springs, conduits, and rivets. 7% electrical conductivity; corrosion resistant.

DURONZE 707

Olin Brass, Indianapolis
Copper. Cu 91, Si 2, Al 7.
For bolts, gears, valve stems, and marine hardware; tough; corrosion resistant.

DURONZE 708

Olin Brass, Indianapolis
Copper. Al 6.75, Si 1.75, Cu 91.5.
Annealed: 85,000 psi TS; 40,000 psi YS; 30 El; 165 Brin. For valve parts, bolts, and fasteners; corrosion and wear resistant; hot forgeable.

DURONZE I

Olin Brass, Indianapolis
Copper. Cu 97.5, Si 1, Sn 1.5.
Annealed: 45,000 psi TS. Cold drawn: 135,000 psi TS. For general use, bolts, cable, hardware, and cap screws; corrosion resistant.

DURONZE II (BRIDGEPORT 632)

Olin Brass, Indianapolis
Copper. Si 3, Cu 97.
Sand cast: 50,000 psi TS; 40 El; 50 RA; 93 Rock D. Annealed: 70,000 psi TS; 55 El; 65 RA; 85 Rock F. For poleline hardware, marine hardware, water storage tanks, chemical vats, and range boilers; strong, hot forging alloy; corrosion resistant.

DURONZE III

Olin Brass, Indianapolis
Copper. Si 2.5, Al 7, Cu 90.5.
Forged: 90,000 psi TS; 50,000 psi YS; 27 El; 35 RA; 165 Brin.
Annealed: 90,000 psi TS; 30 El; 35 RA; 160 Brin. For boat shafting, pump rods, pinions, and gears; corrosion resistant.

DURONZE IV

Olin Brass, Indianapolis
Copper. Al 5, Cu 95.
Annealed: 65,000 psi TS; 45 El. For condenser tubes; sea water corrosion resistant.

DUROPLAT

German manufacture
Al coated Duralumin.
For bimetal parts.

DUROTEC 19910

Eutectic Corp.
Alloy powder for metal spraying; makes super hard, grindable final coat. For bearing areas subject to severe friction.

DURSIL SI5

Durener Metallwerke
Aluminum.
Aluminum alloy AlSi5. *Obsolete*

DURSILIUM

French manufacture
Cu 4-5, bal Al.
For light alloy parts; age-hardenable.

DUTCH BOY 111

NL Industries
Pb 50, Sn 50.
Cast: 6400 TS; 40 El; 12.7 Brin. For solder; MP 361-414°F; 5800 shearing strength. *Obsolete*

DUTCH BOY 222

NL Industries
Pb 55, Sn 45.
Cast: 6600 TS; 40 El; 12.7 Brin. For solder; MP 361-424°F; 5700 shearing strength. *Obsolete*

DUTCH BOY 333

NL Industries
Pb, Sn.
Cast: 6400 TS; 41 El; 12.6 Brin. For solder; MP 361-437°F; 5500 shearing strength. *Obsolete*

DUTCH BOY 444

NL Industries
Pb, Sn.
Cast: 6600 TS; 40 El; 13.3 Brin. For solder; MP 361-437°F; 5625 shearing strength. *Obsolete*

DUTCH BOY 555

NL Industries
Pb, Sn.
Cast: 6700 TS; 39 El; 13.3 Brin. For solder; MP 361-448°F; 5525 shearing strength. *Obsolete*

DUTCH BOY 666

NL Industries
Pb, Sn.
Cast: 7000 TS; 35 El; 14.0 Brin. For solder; MP 361-460°F; 5450 shearing strength. *Obsolete*

DUTCH BOY 777

NL Industries
Pb, Sn.
Cast: 7000 TS; 33 El; 13.9 Brin. For solder; MP 361-460°F; 5375 shearing strength. *Obsolete*

DUTCH BOY 888

NL Industries
Pb, Sn.
Cast: 70,000 TS; 30 El; 13.9 Brin. For solder; MP 361-488°F; "Plumbers' solder." *Obsolete*

DUTCH BOY BEARING METAL

NL Industries
Sb, Cu, bal Sn.
Cast: 23.5 Brin. For bearings for machinery and general mill work; Babbitt, for slow moving machinery. *Obsolete*

DUTCH BOY GENUINE BABBITT

NL Industries
Sn 88.9, Sb 7.4, Cu 3.7.
Cast: 26.7 Brin. For bearings; Babbitt. *Obsolete*

DUTCH BOY NO. 1 JOURNAL

NL Industries
Sb, Cu, bal Sn.
Cast: 32.4 Brin. For heavy pressure bearings; Babbitt, resists heavy crushing strains. *Obsolete*

DUTCH BOY P-A-F

NL Industries
Pb-Sn-Sb-Cu.
For bearings; Babbitt. *Obsolete*

DUTCH BOY STERLING JOURNAL

NL Industries
Sb, Cu, bal Sn.
Cast: 23.3 Brin. For bearings for stationary gas engines; Babbitt, resists heavy crushing strains. *Obsolete*

DUTCH METAL

Manufacturer not listed
Cu 76-80, Zn 20-24.
Cast: 38,400 TS; 22,750 YS; 33 El; 33 RA; 45 Brin. Hard rolled: 66,900 TS; 66,900 YS; 12 El; 42 RA; 145 Brin. For jewelry, gold leaf; red brass.

DUTCH SILVER

Manufacturer not listed
Sn 81.5, Sb 8.8, Cu 9.6.
For tableware, novelties, art metal pieces; corrosion resistant.

DUTCH WHITE METAL

English manufacture
Sn 81.5, Sb 8.8, Cu 9.6.
For bearings, antifriction purposes; Babbitt.

DUTRAX

Duke Steel Co. Inc.
C, bal Fe.
For machinery parts.

DUTRAX 3

Swedish American Steel Corp.
C 0.45, Cr 0.8, W 1.1, Mo 0.25, bal Fe.
For plastic mold dies; oil hardening.

DUTRAX 7

Swedish American Steel Corp.
C 0.45, Cr 1.25, W 2.15, Si 0.2, bal Fe.
For plastic mold dies; oil hardening.

DUVAL AU

Aubert & Duval
C 0.95, Mn 0.3, Si 0.25, bal Fe.
For drills, taps, punches, springs, tools; Type W1; water hardened. *Obsolete*

DUVAL DU-MUV

Aubert & Duval
C 0.8-1, Mn 0.3, Si 0.2, bal Fe.
For cutters, drills, reamers, taps, broaches, springs; Type W1; water hardened. *Obsolete*

DUVAL EXTRA F

Aubert & Duval
C 0.7, W 18, Co 5, Cr 4, V 1.3, bal Fe.
For lathe and planer tools, reamers, drills, taps, broaches; high speed steel, oil hardened. *Obsolete*

DUVAL MU

Aubert & Duval
C 1.1, Mn 0.3, Si 0.25, bal Fe.
For taps, drills, cutters, reamers; Type W1; water hardened. *Obsolete*

DUVAL SF1

Aubert & Duval
C 0.9-1.1, Cr 0.5-1, W 3.5-4.5, V 0.1, bal Fe.
For cold work tools, headers, forming dies; cold work steel, oil hardened. *Obsolete*

DUVAL SF2

Aubert & Duval
C 0.5, Cr 1.5, W 2.25, V 0.2, bal Fe.
For tools, dies, shears, punches; hot work steel, oil hardened. *Obsolete*

DUVAL U7

Aubert & Duval
C 0.7, Mn 0.3, Si 0.25, bal Fe.
For springs, punches, tools, crimpers; Type W1; water hardened. *Obsolete*

DUVAL U8

Aubert & Duval
C 0.75, Mn 0.3, Si 0.25, bal Fe.
For punches, drills, springs, dies, tools; Type W1; water hardened. *Obsolete*

DUX 4

British Steel Corp.
C 0.5, Cr 1.1, W 2.25, bal Fe.
For pneumatic chisels, punches, shear blades; oil hardened, shock resistant. *Obsolete*

DV2A

Crucible Materials Corp.
Stainless steel. C 0.53, Mn 11.5, Si 0-0.28, Cr 20.5, W 2, Cb 1, V 0.4, N 0.45, bal Fe.
For diesel engine valves.

DY-KROME

Allied Steel & Tractor Products Inc.
C 1.5, Cr 12, V 0.2, Mo 0.8, bal Fe.
For blanking dies, broaches, cold extrusion dies; air hardening.

DY-KROME SPECIAL

U.S. Steel Corp.
C 2.15, V 1, Cr 12.5, bal Fe.
For dies, forming and drawing dies; punches; oil hardened, non-deforming. *Obsolete*

DYCAST

Latrobe Steel Co.
C 0.35, Si 0.95, W 1.25, Cr 5, Mo 1.25, bal Fe.
For die casting dies. *Obsolete*

DYCAST NO. 1

Latrobe Steel Co.
Tool material. C 0.4, Cr 5, Mn 0.3, V 0.5, Mo 1.3, Si 1, bal Fe.
For die casting dies and cores; resists heat checking, hot work steel.

DYCRO

Pennsylvania Steel Corp.
C 1.5, Cr 12, V 0.25, Mo 0.8, bal Fe.
For blanking and forming dies, broaches; air hardened, non-deforming.

DYCRO-1

Gulf Steel Corp.
C 1, Cr, Mn, Mo, bal Fe.
For tools, dies; air hardened.

DYMAL

Allied Steel & Tractor Products Inc.
C 0.9, Cr 0.5, Mn 1.15, W 0.5, bal Fe.
For bakelite molding dies, gages, hobs, forming tools; nondeforming.

DYMAL AH

Allied Steel & Tractor Products Inc.
C 1.65-2.1, Mn 0.25, Cr 11-13, V 0.35, Mo 0.9, bal Fe.
For dies, tools; nondeforming.

DYMAL OH

Allied Steel & Tractor Products Inc.
C 0.9, Mn 1.1, Cr 0.5, W 0.5, V 0.2-0.3, bal Fe.
For dies, hobs, milling cutters, gages; oil hardening.

DYMONHARD NO. 55

Dymonhard Corp. of America
Low C, Cr, Ni, bal Fe.
For hard facing electrodes; work hardening. *Obsolete*

DYMONHARD NO. 65

Dymonhard Corp. of America
Mo, W, Cr.
For hard surfacing electrodes; air hardening. *Obsolete*

DYMONHARD NO. 90

Dymonhard Corp. of America
Cr-W-Mo.
700 Brin. For hard facing electrodes; heat and corrosion resistant. *Obsolete*

DYMONHARD NO. 91

Dymonhard Corp. of America
Cr-W-Mo.
600 Brin. For hard facing electrodes; heat and abrasion resistant. *Obsolete*

DYMONHARD NO. 92

Dymonhard Corp. of America
Cr, Ni, Mo, W.
550 Brin. For hard facing electrodes; tough and shock resistant. *Obsolete*

DYMONHARD NO. 93

Dymonhard Corp. of America
Cr, Ni, W.
For hard facing electrodes; corrosion and acid resistant. *Obsolete*

DYMONHARD NO. 94

Dymonhard Corp. of America
Cr, Ni, Mo.
600 Brin. For hard facing electrodes; abrasion resistant. *Obsolete*

DYMONHARD NO. 95

Dymonhard Corp. of America
C, Mo, Mn, Cr.
500 Brin. For hard facing electrodes; abrasion resistant. *Obsolete*

DYMONHARD NO. 96

Dymonhard Corp. of America
Cr, Mo, W.
385-444 Brin. For hard facing electrodes; wear resistant. *Obsolete*

DYN GZ 10

Dyn-Metal Ltd.
Copper.
Phosphor bronze for impact loads. Applications include bearings, glands for refrigerants, slides for food processing machinery, clutch thrust rings and cones. Sand cast: 120-150 N/mm^2 proof stress (0.2%); 220-280 N/mm^2 TS; 15-20 El; $250^\circ C$ maximum operating temperature. Continuous and centrifugal cast: 140-170 N/mm^2 proof stress (0.2%); 280-320 N/mm^2 TS; 12-20 El.

DYN GZ 14

Dyn-Metal Ltd.
Copper.
Phosphor bronze for use in high load and impact conditions. Applications include crank and eccentric bearings, wormwheels and gears, feed traverse and elevating nuts, thrust bearings, wear pads and slides, and bearings for cranes. Sea water resistant. Sand cast: 130-180 N/mm^2 proof stress (0.2%); 240-300 N/mm^2 TS; 8-20 El; 250°C maximum operating temperature. Continuous and centrifugal cast: 150-200 N/mm^2 proof stress (0.2%); 280-350 N/mm^2 TS; 7-15 El.

DYN GZ 15

Dyn-Metal Ltd.
Copper.
Phosphor bronze for continuous high load operation and for bearings with small clearances. Applications include knuckle and crank bearings for presses, machine tool feed nuts, and wear pads and plates against steel. Sea water resistant. Sand cast: 140-170 N/mm^2 proof stress (0.2%); 200-250 N/mm^2 TS; 3-5 El; 250°C maximum operating temperature. Continuous and centrifugal cast: 160-200 N/mm^2 proof stress (0.2%); 250-320 N/mm^2 TS; 3-4 El.

DYN PAN 7

Dyn-Metal Ltd.
Copper.
Lead bronze alloy for moderately high loads with poor lubrication. Applications include bearings for machine tools. Sand cast: 70-100 N/mm^2 proof stress (0.2%); 150-210 N/mm^2 TS; 6-10 El; 250°C maximum operating temperature. Continuous and centrifugal cast: 90-120 N/mm^2 proof stress (0.2%); 170-240 N/mm^2 TS; 8-12 El.

DYN PAN AL0

Dyn-Metal Ltd.
Copper.
Aluminum bronze alloy for high load with impact and high temperature service. Good lubrication and hard, fine finished mating surface required. Erosion, sea water, and acid resistant. Applications include wormwheels and gears, knuckle and crank bearings for presses, high pressure pistons for compressors and hydraulic systems, and guides for steam valves. Sand cast: 270-330 N/mm^2 proof stress (0.2%); 550-650 N/mm^2 TS; 10-18 El; 450°C maximum operating temperature. Rolled bar: 350-450 N/mm^2 proof stress (0.2%); 750-850 N/mm^2 TS; 4-6 El.

DYN PAN AL7

Dyn-Metal Ltd.
Copper.
Aluminum bronze alloy for high load with impact and high temperature service; fine finished, hard mating surface required for bedding-in ease. Erosion, sea water, and acid resistant. Applications include wormwheels and gears. Sand cast: 180-240 N/mm^2 proof stress (0.2%); 450-550 N/mm^2 TS; 13-25 El; 350°C maximum operating temperature.

DYN PAN B

Dyn-Metal Ltd.
Copper.
Lead bronze alloy for use with poor lubrication and such fluids as fresh or sea water, gasoline, and kerosene as lubricants. Applications include bearings, machine slides, and seal rings. Compatible with cast iron, austenitic steel and stainless steel, useful in dusty environments, and beds in well, low starting torque. Sand cast: 80-110 N/mm^2 proof stress (0.2%); 160-220 N/mm^2 TS; 6-12 El; 200°C maximum operating temperature. Continuous and centrifugal cast: 100-130 N/mm^2 proof stress (0.2%); 180-260 N/mm^2 TS; 8-14 El.

DYN PAN H

Dyn-Metal Ltd.
Copper.
Bronze alloy with high lead content for use in dusty environments with poor lubrication and with nonhardened mating surfaces. Applications include bearings for pumps, mixers, and marine propeller shafts, and seals for compressors and hydraulic motors. Sand cast: 100-180 N/mm^2 proof stress (0.2%); 140-200 N/mm^2 TS; 6-10 El; 200°C maximum operating temperature.

DYN PAN SOMS 140

Dyn-Metal Ltd.
Copper.
Manganese bronze alloy for use under high load, impact and low speed conditions. Applications include bearings, screw down nuts in strip and plate mills, and slippers for rolling mills. Sand cast: 150-200 N/mm^2 proof stress (0.2%); 450-550 N/mm^2 TS; 15-22 El; 150°C maximum operating temperature. Good electrical conductivity; 2 m/s maximum rubbing speed with grease lubrication.

DYN PAN SOMS 210

Dyn-Metal Ltd.
Copper.
Manganese bronze alloy for use under high load with impact at low speed conditions; hard mating surfaces required. Applications include bearings for shears in steel mills, and slippers in slab mills. Sand cast: 450-550 N/mm^2 proof stress (0.2%); 750-800 N/mm^2 TS; 8-12 El; 150°C maximum operating temperature; 1 m/s maximum rubbing speed with grease lubrication.

DYN RM

Dyn-Metal Ltd.
Copper.
Lead bronze alloy for high loads and for pulsating loads. Applications include crosshead bearings in compressors and diesel engines, and wear pads and plates. Sea water resistant. Sand cast: 80-100 N/mm^2 proof stress (0.2%); 150-220 N/mm^2 TS; 2-6 El; 250°C maximum operating temperature. Continuous and centrifugal cast: 100-140 N/mm^2 proof stress (0.2%); 180-250 N/mm^2 TS; 2-5 El.

DYN-EL

Alan Wood Steel Co.
C 0.12, Mn 0.6, Cu 0.4, bal Fe.
Heat treated: 72,000 TS; 58,000 YS; 28 El; 60 Brin. For railroads, trucks, buses. *Obsolete*

DYN-GZ10

Machinery & Machine Supplies Co. Inc.
Cu 89.25, Sn 10, Ni 0.5, P 0.25.
Cast: 29,000-50,000 TS; 16,000-27,000 YS; 8-29 El; 70-85 Brin. For nuts, bearings, valve guides, bushings; phosphor bronze, tough.

DYN-GZ14

Machinery & Machine Supplies Co. Inc.
Cu 86.15, Sn 11.5, Ni 2, P 0.3.
Cast: 27,000-42,500 TS; 16,000-25,000 YS; 6-25 El; 90-110 Brin. For worm wheels, toggle pads, bearings; phosphor bronze, shock resistant.

DYN-RM

Machinery & Machine Supplies Co. Inc.
Cu 83.2, Sn 10, Ni 2.5, Pb 4, P 0.3.
Cast: 25,000-41,000 TS; 16,000-22,500 YS; 6-20 El; 80-100 Brin. For mill bearings, bushings, liners; phosphor bronze, shock resistant.

DYNA-KUT DK1211

Stanadyne
C 0-0.13, Mn 0.6-0.9, S 0.08-0.15, P 0.07-0.12, bal Fe.
Free machining steel bars.

DYNA-KUT DK1212

Stanadyne
C 0-0.13, Mn 0.7-1, P 0.07-0.12, S 0.08-0.15, bal Fe.
Free machining steel bars; 47% greater production capability over DK1211.

DYNA-KUT DK1213

Stanadyne
C 0-0.09, Mn 0.7-1, P 0.07-0.12, S 0.24-0.36, bal Fe.
Free machining steel bars; 87% greater production capability over DK1211.

DYNA-KUT DK1216

Stanadyne
C 0-0.15, Mn 0.85-1.15, P 0.06-0.11, S 0.37-0.46, bal Fe.
Free machining steel bars; 131% greater production capability over DK1211.

DYNA-KUT DK12L13

Stanadyne
C 0-0.09, Mn 0.7-1, P 0.07-0.12, S 0.24-0.33, Pb 0.15-0.35, bal Fe.
Free machining steel bars; 154% greater production capability over DK1211.

DYNA-KUT DK12L16

Stanadyne
C 0-0.15, Mn 0.85-1.15, P 0.06-0.11, S 0.37-0.46, Pb 0.15-0.35, bal Fe.
Free machining steel bars; 200% greater production capability over DK1211.

DYNA-LOY DK1210

Stanadyne
C 0-0.09, Mn 0.9-1.2, P 0.06-0.1, S 0.37-0.45, Bi 0.1, bal Fe.
Cold drawn, free machining carbon steel.

DYNA-LOY DL41L43

Stanadyne
C 0.39-0.45, Mn 0.75-1.05, P 0-0.035, S 0.06-0.1, Si 0.15-0.35, Cr 0.8-1.1, Mo 0.15-0.25, Pb 0.15-0.35, bal Fe.
Free machining alloy steel. Hardenable to 33-55 Rock C; 150,000-280,000 psi TS.

DYNA-LOY DL41L49

Stanadyne
C 0.46-0.52, Mn 0.75-1.05, P 0-0.035, S 0.06-0.1, Si 0.15-0.35, Cr 0.8-1.1, Mo 0.15-0.25, Pb 0.15-0.35, bal Fe.
Free machining alloy steel. Hardenable to 43-56 Rock C; 175,000-300,000 psi TS.

DYNA-LOY HS-32

Stanadyne
Mn 0.7-1.1, Si 0.2-0.35, Cr 0.8-1.2, Mo 0.15-0.2, Pb 0.15-0.35, S 0.06-0.1, 0.40 C min, bal Fe.
Hardened, free machining steel. Heat treated: 150,000 psi TS min; 130,000 psi YS min; 10 El; 37 RA; 32 Rock C min.

DYNABRAZE 145 FC

Eutectic Corp.
Flux coated bronze rod for brazing steel and cast iron. Bonding temperature 1400-1600°F; 65,000 TS. *Obsolete*

DYNACUT

Latrobe Steel Co.
C 1.2, W 2.7, Cr 3.75, V 1.6, Mo 8, Co 8.2, bal Fe.
Hardened: 65-66 Rock C. For cutters in heavy duty machining, broaches, chasers, drills. Type M-43 high-speed steel; wear resistant. High red hardness.

DYNAFLEX

Latrobe Steel Co.
Tool material.
Now DYNAFLEX VAC ARC.

DYNAFLEX VAC ARC

Latrobe Steel Co.
Tool material. C 0.4, Mn 0.3, Si 0.9, Cr 5, Mo 1.3, V 0.45, bal Fe.
Air hardened: 300,000 TS; 250,000 YS; 7 El; 30 RA; 525 Brin. For punches, upsetters, extrusion dies; hot work steel; AISI Type H 11.

DYNALLOY 600

Now DYNALLOY 6741 (600).

DYNALLOY 601
Now DYNALLOY 6680 (601).

DYNALLOY 602
Now DYNALLOY 6730 (602).

DYNALLOY 603
Now DYNALLOY 6731 (603).

DYNALLOY 604
Now DYNALLOY 6732 (604).

DYNALLOY 605
Now DYNALLOY 6733 (605).

DYNALLOY 6680 (601)
Mueller Brass Co.
Cu 61, Si 0.8, Mn 2.5, bal Zn.
HH temper: 70,000-85,000 TS; 40,000-65,000 YS; 15-25 El; 70-87 Rock B. For bearings, worm wheels, bushings. Mate with soft member. Corrosion resistant.

DYNALLOY 6730 (602)
Mueller Brass Co.
Cu 61, Pb 1, Si 1, Mn 2.5, bal Zn.
HH temper: 70,000-85,000 TS; 40,000-65,000 YS; 15-25 El; 70-87 Rock B. For bearings, bushings, gears. Leaded forgeable bearing alloy, corrosion resistant.

DYNALLOY 6731 (603)
Mueller Brass Co.
Mn 2.75, Pb 1, Si 1, Cu 59.7-60, bal Zn.
Forged: 50,000 TS; 30,000 YS; 15 El; 65 Rock B. Heat treated: 75,000 TS; 45,000 YS; 20 El; 82 Rock B. For gears, bushings, bearings, pump barrels, machinery parts. Leaded forging brass.

DYNALLOY 6732 (604)
Mueller Brass Co.
Cu 61, Pb 2.5, Si 1, Mn 2.5, bal Zn.
HH temper: 65,000-80,000 TS; 40,000-60,000 YS; 10-20 El; 75-86 Rock B. For bearings, pump gears, valve stems. High leaded bearing alloy, corrosion resistant. *Obsolete*

DYNALLOY 6733 (605)
Mueller Brass Co.
Cu 62, Pb 0.6, Si 1, Mn 2.5, bal Zn.
HH temper: 70,000-75,000 TS; 45,000-55,000 YS; 18-25 El; 75-82 Rock B. For bearings, pump gears, valve stems. Low leaded bearing alloy, corrosion resistant. *Obsolete*

DYNALLOY 6736 (606)
Mueller Brass Co.
Al 4-4.7, Fe 0-0.5, 99.5 Cu min + Al + Fe.
Aluminum bronze forgings. CDA 606. *Obsolete*

DYNALLOY 6741 (600)
Mueller Brass Co.
Cu 58, Si 0.8, Al 1.5, Mn 2.5, bal Zn.
HH temper: 75,000-100,000 TS; 40,000-65,000 YS; 12-18 El; 82-88 Rock B. Forged: 68,000 TS; 34,000 YS; 18 El; 78 Rock B. For bearings, worms and worm wheels, bushings. Forgeable bearing alloy, corrosion resistant.

DYNALLOY II
Alan Wood Steel Co.
C 0.15, Mn 0.8, Si 0.3, Cu 0.45, Mo 0.1, Ni 0.55, bal Fe.
Rolled: 62,000 TS; 45,000 YS; 25 El. For railroad and agriculture equipment, mine cars; high strength, low alloy construction steel. *Obsolete*

DYNAMAX
Now ELECTRITE DYNAMAX.

DYNAMAX
Latrobe Steel Co.
C 1.08, W 1.5, Cr 3.75, V 1.1, Mo 9.5, Co 8, bal Fe.
Hardened: 65-70 Rock C. For cutting tools requiring long production runs or heavy duty machining; very high wear resistance; good red hardness. AISI Type M-42 high speed tool steel.

DYNAMAX
Manufacturer not listed
Ni 65, Fe 33, Mo 2.
For toroidal cores; high permeability.

DYNAMIC C-1
Continental Foundry & Machine Co.
C 0.2-0.3, Mn 0.5-0.8, bal Fe.
For cast steel parts; pearlitic.

DYNAMIC C-1-3
Continental Foundry & Machine Co.
C 0.25-0.35, Mn 0.6-0.8, bal Fe.
For castings; water hardened.

DYNAMIC C-1-4
Continental Foundry & Machine Co.
C 0.35-0.45, Mn 0.65-0.85, bal Fe.
For gears, pinions, coupling boxes; water hardened.

DYNAMIC C-1-5
Continental Foundry & Machine Co.
C 0.4-0.5, Mn 0.65-0.85, bal Fe.
For gears, pinions, power shovels; water hardened.

DYNAMIC C-1-6
Continental Foundry & Machine Co.
C 0.5-0.6, Mn 0.65-0.85, bal Fe.
For cams, bottom plates; wear hardened.

DYNAMIC C-1-7
Continental Foundry & Machine Co.
C 0.7-0.8, Mn 0.65-0.85, bal Fe.
For castings; abrasion resistant.

DYNAMIC C-10
Continental Foundry & Machine Co.
C, Cr, Mo, bal Fe.
Cast. For die blocks, crane wheels; wear resistant.

DYNAMIC C-2
Continental Foundry & Machine Co.
C 0.25-0.35, Mn 0.9-1.4, Ni 0.25-0.45, Mo 0.05-0.15, bal Fe.
Cast: 85,000 TS; 52,500 YS; 30 El; 57 RA. For tractor frames, locomotive castings; resists shock and impact; IZ 45.

DYNAMIC C-2-A
Continental Foundry & Machine Co.
C 0.35-0.45, Mn 1-1.5, Ni 0.25-0.45, Mo 0.05-0.15, bal Fe.
For crawler shoes, gears, sprockets; wear resistant.

DYNAMIC C-3
Continental Foundry & Machine Co.
C 0.25-0.35, Mn 1-1.5, Mo 0.1-0.2, bal Fe.
Cast: 96,000 TS; 68,000 YS; 24 El; 48 RA. For sprockets, spindles, wheel centers, crossheads, gears; wear, shock and impact resistant.

DYNAMIC C-3A
Continental Foundry & Machine Co.
C 0.4, Mn 1.2, Mo 0.15, bal Fe.
Cast. For gears, racks, sprockets; wear resistant.

DYNAMIC C-4
Continental Foundry & Machine Co.
C 0.25, Mn 0.9, Cr 0.8, Ni 1.9, Mo 0.5, bal Fe.
Heat treated: 120,000 TS; 100,000 YS; 15 El; 30 RA; 321 Brin. For rolling mill pinions; impact resistant.

DYNAMIC C-5
Continental Foundry & Machine Co.
C 0.3, Mn 1, Ni 0.7, Mo 0.4, bal Fe.
Cast. For rams, saw blocks.

DYNAMIC C-6
Continental Foundry & Machine Co.
C 0.4-0.7, Cr 3, Mo 0.4, bal Fe.
Cast. For sand mills, rock crushers; abrasion resistant.

DYNAMIC C-7
Continental Foundry & Machine Co.
C 0.4, Mn 0.9, Cr 0.5, Ni 1.5, Mo 0.5, bal Fe.
Cast. For steel castings, crane wheels; high strength.

DYNAMIC C-8
Continental Foundry & Machine Co.
C 0.25-0.35, Mn 1-1.5, Cr 0.3-0.7, Ni 0.4-0.8, V 0.2-0.4, bal Fe.
For gears, sprockets; tough, high impact resistant.

DYNAMIC C-9
Continental Foundry & Machine Co.
C, Mn, V, bal Fe.
Cast. For steel castings; air hardened.

DYNAMO
Patriarche & Bell
C 0.7-0.9, bal Fe.
For tools, dies, jigs; water hardening.

DYNAMO
Johnson Matthey plc
C 0.12, Si 0.2, Mn 0.1, bal Fe.
For electrical purposes; high magnetic permeability.
Obsolete

DYNAMO SHEET STEEL
American manufacture
Si 3-4, C 0-0.1, Mn 0-0.3, 0.03 (P + S) max, bal Fe.
For laminated sheets for dynamos, electrical equipment; Hadfields silicon steel type.

DYNAMO STEEL
Otis Elevator
C 0.09, bal Fe.
For dynamo parts; high permeability.

DYNATRODE 666
Eutectic Corp.
Electrode for AC-DC welding of mild steel; 70,000 psi TS.
Obsolete

DYNATRODE 777
Eutectic Corp.
Mild steel electrode for AC-DC welding of mild steel. 76,000 psi TS.

DYNAVAN
Latrobe Steel Co.
Tool material. C 1.57, W 12.5, Cr 4, V 5, Co 5, bal Fe.
Hardened: 64-66 Rock C. For cutting tools, blanking dies, broaches, drills, milling cutters. Type T-15 high-speed steel. Heat and abrasion resistant. High red-hardness.

DYNAVAR
Hamilton Technology Inc.
C 0.15, Mn 2, Ni 15, Mo 7, Cr 20, Be 0.04, Co 40, bal Fe.
Aged: 330,000-360,000 TS; 260,000-280,000 YS; 560-600 Brin. Rolled: 260,000-290,000 TS; 200,000-220,000 YS; 480-500 Brin. For springs, flapper valves, valve parts, electronic components; corrosion resistant, age-hardenable, non-magnetic. *Obsolete*

DYNELEC
Empire Sheet & Tin Plate Co.
Si 2.5, bal Fe.
For armatures, electric motors and generators; high permeability.

DYSOID
English manufacture
Cu 63, Pb 18, Sn 10, Zn 10.
For hardware; free-cutting.

DYSPROSIUM
Atomergic Chemetals Corp.
Dy.
Purities: 99.9% special distilled grade, 99.5 + %. Forms: ingot, sponge, wire, sheet, rod, turnings, powder, foil, single crystals.

DZ
Thyssen Edelstahlwerke AG
C, bal Fe.
For scales, rules; water hardened. *Obsolete*

DZISTALOY 6971
Anaconda Co.
Copper. Cu 78, Pb 1, Zn 18, Si 2.85, Mn 0.15.
70,000 TS; 40,000 YS; 25 El; 80 Rock B. For automatic screw machine parts. Resistant to corrosion and dezincification; free machining.

DZISTALOY DZ6943
Anaconda Co.
Copper. Cu 81.5, Pb 0.25, Zn 14.25, Si 4.
85,000 TS; 50,000 YS; 20 El; 90 Rock B. Forgeable, free machining. For valve stems, valve seats and parts needing wear resistance and resistance to dezincification by fresh water.

DZR
American Smelting & Refining Co.
Cu, Zn (yellow brass).
Sand cast: 53,000 psi TS; 22,000 psi YS; 23 El; 80 Brin.

E 1010, ETC. (M-1736)
Now AISI 1010, ETC. (STEELS).

E 10B46
Electrometal SA Metals Especials
C 0.46, Si 0.22, Mn 1.25, B 0.0017, bal Fe.
Boron carbon steel.

E 21 S
Thyssen Edelstahlwerke AG
C, Cr, Mo, Ni, bal Fe.
Oil treated: 150,000-177,000 TS; 128,000-157,000 YS; 9-12
El. For machinery parts; gears, shafts; oil hardening, shock
resistant. *Obsolete*

E 22 S
Thyssen Edelstahlwerke AG
C, Cr, Ni, Mo, bal Fe.
Oil treated: 171,000-199,000 TS; 150,000-177,000 YS; 1-8 El.
For machinery parts; oil hardening. *Obsolete*

E 2365
Electrometal SA Metals Especials
C 0.35, Si 0.4, Mn 0.4, Cr 2.95, Mo 2.85, V 0.5, bal Fe.
Hot work tool steel. W.-Nr. 1.2365.

E 2419
Electrometal SA Metals Especials
C 1.05, Si 0.25, Mn 0.95, Cr 1, W 1.25, bal Fe.
Cold work tool steel. W.-Nr. 1.2419.

E 2542
Electrometal SA Metals Especials
C 0.51, Si 0.95, Mn 0.35, Cr 1.25, Mo 0.2, V 0.2, W 2.35, bal
Fe.
Shock resisting tool steel. W.-Nr. 1.2542; similar to AISI S1.

E 2550
Electrometal SA Metals Especials
C 0.6, Si 0.6, Mn 0.3, Cr 1.1, V 0.2, W 2, bal Fe.
Cold work tool steel. W.-Nr. 1.2550.

E 2601
Electrometal SA Metals Especials
C 1.65, Si 0.3, Mn 0.3, Cr 12, V 0.1, Mo 0.6, W 0.5, bal Fe.
Cold work tool steel. W.-Nr. 1.2601.

E 2603
Electrometal SA Metals Especials
C 0.46, Si 0.75, Mn 0.45, Cr 1.6, Mo 0.7, V 0.8, W 0.7, bal Fe.
Hot work tool steel. W.-Nr. 1.2603.

E 2713
Electrometal SA Metals Especials
C 0.55, Si 0.32, Mn 0.6, Cr 0.7, Mo 0.35, Ni 1.5, bal Fe.
Alloy structural steel. For shafts, gears, and springs. W.-Nr.
1.2713.

E 2714
Electrometal SA Metals Especials
C 0.56, Si 0.25, Mn 0.6, Cr 1, Mo 0.45, Ni 1.7, V 0.15, bal Fe.
Alloy structural steel. For shafts, gears, and springs. W.-Nr.
1.2714.

E 2721
Electrometal SA Metals Especials
C 0.54, Si 0.3, Mn 0.35, Cr 0.95, Mo 0.3, Ni 3.25, bal Fe.
Alloy structural steel; deep hardening. W.-Nr. 1.2721.

E 2764
Electrometal SA Metals Especials
C 0.2, Si 0.3, Mn 0.4, Cr 1.25, Mo 0.2, Ni 3.75, bal Fe.
Cold work tool steel or alloy carburizing steel. W.-Nr. 1.2764.

E 2766
Electrometal SA Metals Especials
C 0.35, Si 0.25, Mn 0.5, Cr 1.3, Mo 0.25, Ni 4, bal Fe.
Alloy structural steel; deep hardening. For heavy sections;
hot work tool steel. W.-Nr. 1.2766.

E 2767
Electrometal SA Metals Especials
C 0.41, Si 0.3, Mn 0.4, Cr 1.2, Mo 0.45, Ni 1, W 0.45, bal Fe.
Hot work tool steel. W.-Nr. 1.2767.

E 2842
Electrometal SA Metals Especials
C 0.91, Si 0.25, Mn 2.05, V 0-0.1, bal Fe.
Cold work tool steel. W.-Nr. 1.2842; similar to AISI O2.
Obsolete

E 300M
Electrometal SA Metals Especials
C 0.43, Si 1.63, Mn 0.78, Cr 0.83, Mo 0.4, Ni 1.83, bal Fe.
Cold work tool steel. For shock resistant tools.

E 302, ETC. (M-1736)
Now AISI 302, ETC. (STAINLESS).

E 7131
Electrometal SA Metals Especials
C 0.16, Si 0.25, Mn 1.15, Cr 0.95, bal Fe.
Alloy carburizing steel. W.-Nr. 1.7131.

E 7147
Electrometal SA Metals Especials
C 0.2, Si 0.25, Mn 1.25, Cr 1.15, bal Fe.
Alloy carburizing steel. W.-Nr. 1.7147.

E 7218
Electrometal SA Metals Especials
C 0.25, Si 0.25, Mn 0.65, Cr 1.05, Mo 0.2, bal Fe.
Chrome-moly structural steel. W.-Nr. 1.7218. *Obsolete*

E 7228
Electrometal SA Metals Especials
C 0.5, Si 0.25, Mn 0.65, Cr 1.05, Mo 0.2, bal Fe.
Alloy structural steel. W.-Nr. 1.7228; AISI 4150. *Obsolete*

E 724Z
Thyssen Edelstahlwerke AG
C, Cr, Ni, bal Fe.
Oil treated: 171,000-193,000 TS; 150,000-171,000 YS; 10-13
El. For gears, shafts; oil hardening, tough. *Obsolete*

E 8550
Electrometal SA Metals Especials
C 0.35, Si 0.3, Mn 0.55, Cr 1.7, Mo 0.2, Ni 1, Al 1.05, bal Fe.
Steel for nitriding. W.-Nr. 1.8550.

E 86B15
Electrometal SA Metals Especials
C 0.15, Si 0.25, Mn 0.8, Cr 0.6, Mo 0.5, Ni 0.85, V 0.06, B
0.004, bal Fe.
Boron, low alloy steel. For carburizing.

E 9335
Electrometal SA Metals Especials
C 0.35, Si 0.3, Mn 0.52, Cr 1.2, Mo 0.11, Ni 3.25, bal Fe.
Alloy structural steel; deep hardening. *Obsolete*

E 9840
Electrometal SA Metals Especials
C 0.4, Si 0.3, Mn 0.8, Cr 0.8, Mo 0.25, Ni 1, bal Fe.
Alloy structural steel. *Obsolete*

E ALLOY
National Physical Laboratory
Aluminum. Zn 20, Cu 2.5-3, Si 0.2-1, Mg 0.5, Mn 0.5, Fe, bal
Al.
Hot rolled: 67,000 TS; 53,000 YS; 15 El. For light weight
structures; non-hardenable.

E C S
United States Steel Corp.
C 0.06, Mn 0.75, Si 0.75, Cr 11.5, Ti = 6 x C min (0.75 max),
bal Fe.
Annealed: 60 ksi TS; 30 ksi YS; 40 El. Stainless sheet steel for
automotive emission controls.

E CO 3
German manufacture
C 0.8, W 12, Cr 4, V 1.9, Mo 0.85, Co 3, bal Fe.
Hardened: 64-67 Rock C. For lathe and planer tools, reamers,
broaches, drills, form cutters, taps, hobs; high red-hardness,
high speed steel, Type T8.

E NICKEL
Now NICKEL 212.

E NO. 1
Latrobe Steel Co.
C 0.75, Cr 4.1, W 18, V 1.1, bal Fe.
For twist drills, reamers, milling cutters, tools; high speed
steel. *Obsolete*

E NO. 25
Jessop Steel Co.
C 1, Mn 0.35, V 0.25, bal Fe.
For tools, concrete breakers; water hardened. *Obsolete*

E PHOSPHOR BRONZE
Criterion Metals, Inc.
Copper. Cu 96.82, Zn 0-0.3, Sn 1.3, P 0.03, Fe 0-0.1, Pb
0-0.05.
Thin gauge sheet, various tempers: 37-75 ksi TS min; 9-73 ksi
YS min. ASTM B-103. C50500

E-132
General Motors Corp./Central Foundry Div
Si 11-13, Fe 0-1, Cu 1.8-2.8, Mn 0-0.5, Mg 0.9-1.3, Pb 0-0.1,
Ni 0.5-1.5, Zn 0-1, Ti 0-0.25, Sn 0-0.1, bal Al.
Cast aluminum alloy. *Obsolete*

E-52100 WIRE
U.S. Steel Corp.
C 0.95-1.1, Mn 0.2-0.5, P 0-0.03, S 0-0.035, Si 0.15-0.3, Cr
1.3-1.6, bal Fe.
For roller and ball bearings; water hardened. *Obsolete*

E-AL (957) 1057
J. & A. Erbsloh Aluminium
Aluminum. Si 0.25, Fe 0.35, Cu 0.02, Mg 0.05, Zn 0.05, 0.50
others, bal Al.
Pure aluminum. Pressed: 65 N/mm^2 TS; 25 N/mm^2 YS;
20-23 Brin. Werkstoff Nr. 3.0257.08.

E-ALLOY
NL Industries
Cu 2, Ni 2, bal Al.
Cast: 30,000 TS; 16,000 YS; 5 El; 60 Brin. For castings; non-
hardenable. *Obsolete*

E-ALMGSI 0.5 (491) 6047
J. & A. Erbsloh Aluminium
Aluminum. Si 0.35-0.5, Fe 0.16-0.22, Cu 0.05, Mn 0.03, Mg
0.4-0.55, Zn 0.05, Ti 0.03, 0.10 others, bal Al.
Hardenable alloy. Warm hardened: 170-215 N/mm^2 TS;
120-160 N/mm^2 YS; 55-70 Brin. Werkstoff Nr. 3.3207,
3.3207.71.

E-BRITE
Allegheny Ludlum Steel
Stainless steel. Cr 26, Mo 1, Ni 0.15, Cu 0.02, Mn 0.05, P
0.01, S 0.01, Si 0.2, C 0.002, N 0.01, Cb 0.1, bal Fe.
Sheet, plate, annealed: 70 ksi TS; 50 ksi YS; 30 El; 83 Rock B.
For solenoid valves, glass sealing, vacuum deposition of foil;
corrosion resistant.

E-BRITE 26-1
AL Tech Specialty Steel Corp.
Cr 26, Mo 1, C 0-0.01, N 0.015, bal Fe.
High purity ferritic alloy with superior corrosion resistance,
especially to stress corrosion cracking. For instrument parts,
screening.

E-BRITE 26-1

Bishop Tube Co.
C 0-0.005, Si 0-0.4, Mn 0-0.4, Cr 25-27.5, Mo 0.75-1.25, N 0-0.015, 0.5 Cu max + Ni, bal Fe.
Annealed: 70,000 TS; 45,000 YS; 42 El; 86 RA. (Supplied as ingots for remelting.) Good resistance to impact, intergranular corrosion, pitting and crevice corrosion, and immune to chloride stress cracking. For parts for marine and desalting equipment, food processing, petroleum industries.

E-STEEL

Jones & Laughlin Steel Corp.
C 0-0.06, Mn 0.3, Si 0.25, bal Fe.
For screw machine products, screws, bolts, nuts; free-cutting. *Obsolete*

E-VAN

Midvale-Heppenstall Co.
C 0.95, Mn 0.25, V 0.25, Si 0.3, bal Fe.
For cold work dies; water hardening. *Obsolete*

E-Z DIE V

Columbia Tool Steel Co.
Tool material. C 2.15-2.3, W 1-1.3, Mo 0.95-1.2, Cr 5-5.5, V 4.6-5, bal Fe.
Hardened: 60-66 Rock C. For lamination and forming dies, burnishing rolls, blanking and drawing dies, powder metal dies. High wear and abrasion resistance. Cold work tool steel; Type A7. Air hardened.

E-Z STAINLESS STEEL

Latrobe Steel Co.
C 0.3-0.4, Cr 13-14, bal Fe.
For cutlery, table-ware; stainless, free machining. *Obsolete*

E-Z-CUT 20

Joseph T. Ryerson & Son Inc.
C 0.15, S, Mn, bal Fe.
Free machining; for screw machine products, and case hardened parts. *Obsolete*

E-Z-DIE SMOOTHCUT

Columbia Tool Steel Co.
Tool material. C 1, Mn 0.8, Cr 5, Mo 1, V 0.25, S, bal Fe.
For forming and blanking dies, extrusion equipment, air hardenable; tough and wear resistant. AISI A2.

E-Z-TEM

St. Lawrence Steel Co.
C 0.35, Mn 0.8, Si 0.45, Cr 1, Mo 0.4, Cu 0.5, bal Fe.
Shock resistant tool steel.

E. H. W. NO. 2

Latrobe Steel Co.
W 10-15, C, bal Fe.
For hot work tools and dies; hot work steel. *Obsolete*

E. H. W. NO. 3

Latrobe Steel Co.
W 6.5-9.5, C, bal Fe.
For hot work tools and dies; hot work steel. *Obsolete*

E. I. S. 14

Heppenstall Co.
C 0.75, V 0.15, bal Fe.
For shears; water hardened. *Obsolete*

E. I. S. 15

Heppenstall Co.
C 0.9, V 0.15, bal Fe.
For shears; water hardened. *Obsolete*

E. I. S. 31 "HOTKUT"

Heppenstall Co.
C, Si, Ni, Cr, V, bal Fe.
For hot cut shear knife. *Obsolete*

E. I. S. 42 "KLEENKUT"

Heppenstall Co.
C 2, Cr 12, Co, Mo, bal Fe.
For dies, cold cutting shear blades, trimmer and blanking dies. *Obsolete*

E.B. ALLOY

Atlas Specialty Steels
C 0.5, Cr 4, V 0.4, Mo 0.5, bal Fe.
Heat treated: 217,000-308,000 TS; 176,000-229,000 YS; 6-13 El; 6-34 RA; 44-56 Rock C. For shears, header dies; hot work steel, air or oil hardening. *Obsolete*

E.B.D. BEARING

English manufacture
Cu 88-90, Sn 10, Zn 2, 5 P + Sn.
For bearings; strong.

E.C. 124

English manufacture
Si 12, Cu 1, Mg 0.5, N 1, bal Al.
For light alloy parts; cast or pressed.

E.C.N. ALLOY

Sutcliff, Speakman & Co.
C, Ni, Cr, bal Fe.
At 20°C: 67,000-78,000 TS. At 1000°C: 33,000 TS. For case-hardening boxes, glass molds, retorts, stills, fire boxes, muffles, crucibles, pumps; highly resistant to heat and corrosion.

E.H.W. NO. 1

Latrobe Steel Co.
Tool material. C 0.25, W 15, Cr 4, bal Fe.
For hot work tools and dies; hot work steel. *Obsolete*

E.H.W. STEEL

Latrobe Steel Co.
C 0.25-0.35, W 8.9-9.5, Cr 3-3.6, bal Fe.
For hot working dies, gripper and header dies for automatic bolt machines; oil hardening. *Obsolete*

E.I.S. 49

Heppenstall Co.
Tool material. C 0.85, Cr 5, Mn 0.25, bal Fe.
Hardened: 560 Brin. For trimmers, dies, shears; for heavy shearing.

E.I.S. B-76

Heppenstall Co.
Tool material. C 1, Cr 0.6, Mn 1.2, V 0.2, W 0.5, bal Fe.
For dies, cold blanking and forming dies; nondeforming; oil hardened.

E.I.S. C-57

Heppenstall Co.
Tool material. C 0.6, Cr 1, Ni 2, V 0.12, Mo 0.8, bal Fe.
For shear blades, knives, punches, upsetters; hot work steel; oil hardened.

E.I.S. H-41

Heppenstall Co.
Tool material. C 1, Cr 4, V 0.15, Mo 0.25, bal Fe.
For hot work dies, trimmers; oil or air hardened.

E.I.S. R-43

Heppenstall Co.
Tool material. C 1.5, Cr 12, bal Fe.
For shear blades, trimmers; air hardened, light shearing.

E.I.S. R-97

Heppenstall Co.
Tool material. C 0.5, Cr 0.4, Mn 1, Mo 0.4, Si 2.1, bal Fe.
For dies, inserts, shear knives; oil hardened; heavy shearing.

E.I.S. T-51

Heppenstall Co.
Tool material. C 0.6, Mn 0.7, Ni 1.4, Cr 0.7, W 2.2, V 0.2, bal Fe.
For trimming and forming dies; oil hardened; hot or cold work steel.

E.I.S. T-717

Heppenstall Co.
Tool material. C 0.35, Cr 5, Mo 1.35, W 1.2, Si 1, bal Fe.
For shears, knives; oil hardened.

E.I.S. T-73

Heppenstall Co.
Tool material. C 0.3, Cr 3.5, V 0.35, W 10, bal Fe.
For inserts, dies, punches; hot work brass forgings.

E.I.S. T-79

Heppenstall Co.
Tool material. C 0.4, Cr 5, V 0.2, W 4, Mo 0.6, bal Fe.
For hot work dies, inserts; heavy shearing.

E.R.1

Eagle & Globe Steel Ltd.
Tool material. C 0.32, Cr 3, Mo 2.75, V 0.4, bal Fe.
Hot work steel similar to C.T.U. For copper and brass die casting and forging. AS1239 H10A; AISI H10.

E.R.A. "A.T.V."

English manufacture
Cr 13.4, Ni 22.6, W 3.3, Mn 1.5, Si 1.8, bal Fe.
For heat and corrosion resisting parts; stainless and corrosion resistant. *Obsolete*

E.S.A.

Latrobe Steel Co.
C 1.4, W 4, Cr 0.5, V 0.3, bal Fe.
For finishing tools, reamers; abrasion resistant. *Obsolete*

E.S.C. "C.H.-2 N."

British Steel Corp.
C 0.1-0.15, Ni 2-2.5, bal Fe.
Normalized: 56,000 TS; 30,200 YS; 30 El; 55 RA; 103 Brin.
For general case-hardened objects; IZ-50. *Obsolete*

E.S.C. ALLOYS

Now B.S.C. ALLOYS.

E.S.D. 31

Stackpole Magnet Division
Fe 20.7, Co 11.6, Pb 67.7.
Permanent magnet. Residual flux density: 5000 Gauss.
Coercive force: 1000 Oersted. Anisotropic. *Obsolete*

E.S.D. 32

Stackpole Magnet Division
Fe 18.3, Co 10.3, Pb 72.4.
Permanent magnet. Residual flux density: 6800 Gauss.
Coercive force: 960 Oersted. Anisotropic. *Obsolete*

E.S.D. 41

Stackpole Magnet Division
Fe 20.7, Co 11.6, Pb 67.7.
Permanent magnet. Residual flux density: 3600 Gauss.
Coercive force, 970 Oersted. *Obsolete*

E.S.D. 42

Stackpole Magnet Division
Fe 18.3, Co 10.3, Pb 72.4.
Residual flux density: 4800 Gauss. Coercive force: 830 Oersted. *Obsolete*

E.T.D.

Now COLFORM E.T.D.

E.T.D. 150

LaSalle Steel Co.
C 0.4, Mn 0.7-1, Si 0.15-0.35, Cr 0.8-1.1, Mo 0.15-0.25, bal Fe.
Bar: 150,000 TS min; 130,000 YS min; 10 El; 32 Rock C min; 75% machinability. For shafts, studs, bolts, gears; high strength without heat treatment.

E.T.D. 180

LaSalle Steel Co.
Mn 0.75-1, Si 0.2-0.35, Cr 0.8-1.1, Mo 0.15-0.25, 0.40 C min, bal Fe.
Bar: 180,000 TS min; 165,000 YS min; 5-10 El; 38 Rock C min; 56% machinability. For shafts, gears, studs, bolts; high strength without heat treatment. *Obsolete*

E.V. ALLOY

Follsain-Wycliffe Foundries, Ltd.
Ni 10-12, Cr 22-24, C, bal Fe.
Annealed: 65,000-75,000 TS; 5 El. For carburizing boxes, furnace parts; resists heat to 1175°C.

E.V. STEEL

Follsain-Wycliffe Foundries, Ltd.
Ni-Cr.
Six alloys ranging from 12 to 60% Ni. For heat resistant parts as furnace parts, heat treat boxes, retorts and trays, burner nozzles.

E.V.300

Follsain-Wycliffe Foundries, Ltd.
30% Cr cast iron (no nickel).
For corrosion resistance at room temperature and at elevated temperature. *Obsolete*

E.V.M.

Teledyne Vasco
C 0.82, Si 0.35, Mn 0.25, W 18.5, Cr 4.3, V 2.1, Mo 0.65, bal Fe.
For drills, chasers, lathe tools, forming tools; high speed steel. Type T2.

E/130C

Eagle & Globe Steel Ltd.
Steel. C 1.2-1.3, S 0-0.02, P 0-0.02, Mn 0.2-0.4, Cr 0.15-0.25, bal Fe.
Cold rolled strip steel. For hacksaw blades

E/CVM

Eagle & Globe Steel Ltd.
Steel. C 0.78-0.85, S 0-0.02, P 0-0.02, Mn 0.2-0.4, Cr 0.2-0.3, Mo 0.13-0.23, V 0.2-0.3, bal Fe.
Hardened and tempered strip steel. For high duty handsaws.

E/NC

Eagle & Globe Steel Ltd.
Steel. C 0.75-0.82, S 0-0.02, P 0-0.02, Mn 0.35-0.55, Ni 0.3-0.45, Cr 0.15-0.25, bal Fe.
Hardened and tempered strip steel. For bandknives, narrow bandsaws, handsaws.

E/NM

Eagle & Globe Steel Ltd.
Steel. C 0.75-0.82, S 0-0.02, P 0-0.02, Mn 0.3-0.5, Ni 0.6-0.75, Mo 0.15-0.25, bal Fe.
Hardened and tempered strip steel. For high duty handsaws.

E1 DYNAMO

Russian manufacture
C 0-0.1, Si 1, Mn 0-0.03, bal Fe.
For motor stators and rotors; high permeability.

E100

Friedr. Lohmann GmbH
C 1, V 0.1, Mn 0.25, bal Fe.
For header dies, punches, bearings. Type W2; water or oil hardened. T72301

E1318

Russian manufacture
C 0-0.6, Cr 16-18, Al 4.5-6.5, Ni 0-0.6, bal Fe
For electrical resistance and heating elements; applications to 1100°C.

E1340

Russian manufacture
C 0 0.15, Cr 23-27, Al 4.7, Ni 0-0.6, bal Fe.
For electrical resistance and heating elements; applications to 1200°C.

E1503

Russian manufacture
C 0.5-0.6, Mn 7.5-9.5, Si 0.7, Cr 3.8-4.5, Ni 8-10, bal Fe.
For instrument parts; nonmagnetic.

E18 CO10

German manufacture
C 0.75, W 18, Cr 4, V 1.5, Mo 0.7, Co 10, bal Fe.
Hardened: 64-66 Rock C. For lathe and planer tools, drills, reamers, form cutters, broaches, form tools, gear cutters; for heavy duty work, high speed steel, Type T5, red-hard.

E18 CO5

German manufacture
C 0.78, Cr 4.2, Mo 0.6, W 18.8, V 1.53, Co 5.72, bal Fe.
Hardened: 64-66 Rock C. For tools and cutters, reamers, taps, broaches, lathe and planer tools; high speed steel, type T4, high red-hardness.

E1919

Russian manufacture
C 0.83, Co 10, Cr 4, V 1.8, W 10, bal Fe.
Heat treated: 64-66 Rock C. For cutters, lathe and planer tools, form cutters, drills; high speed steel, high red-hardness.

E1931

Russian manufacture
C 1.49, Cr 4, V 4.4, W 10.6, Co 4.9, bal Fe.
Hardened: 65-68 Rock C. For cutters, drills, broaches, milling cutters, taps, reamers; high red-hardness, abrasion resistant, high speed steel, Type T15.

E1940

Russian manufacture
C 0.83, Cr 4, V 2.1, W 18, Co 5, bal Fe.
Hardened: 62-65 Rock C. For cutters, boring tools, lathe and planer tools, gun-barrel drills, roll turning tools; high speed steel, type T4, red hard.

E2 DYNAMO

Russian manufacture
C 0-0.1, Mn 0-0.03, Si 2, bal Fe.
For motor stators and rotors; high permeability.

E20M

Thyssen Edelstahlwerke AG
C 0.2, Mn 1.2, Cr 1.2, bal Fe.
For bakelite molds; water or oil hardened. *Obsolete*

E3 TRANSFORMER

Russian manufacture
C 0-0.1, Mn 0-0.3, Si 3, bal Fe.
For transformers; high permeability.

E38 MO

Marathon Specialty Steels Inc.
C 0.38, Si 1, Cr 5.3, Mo 1.5, V 0.45, bal Fe.
Annealed: 102,000 TS; 66,000 YS; 28 El; 197 Brin. Hardened: 300,000 TS; 250,000 YS; 6 El; 55 Rock C. For Al and Mg die casting dies, components for die casting machines. Air or oil hardening. Thermal shock resistant, Type H11.

E38 V

Marathon Specialty Steels Inc.
C 0.4, Si 1, Cr 5.3, Mo 1.4, V 1, bal Fe.
Annealed: 98,000 TS; 74,000 YS; 28 El; 210 Brin. Hardened: 220,000 TS; 182,000 YS; 12 El; 430 Brin. For extrusion press mandrels, aluminum die casting dies. Air or oil hardening. Thermal shock resistant, Type H13.

E38 W

Marathon Specialty Steels Inc.
C 0.38, Si 1.2, Cr 5.6, Mo 1.5, V 0.3, W 1.5, bal Fe.
Heat treated: 216,000 TS; 185,000 YP; 14 El; 53 RA. For metal extrusion press components, die casting dies, forging dies, hot shear blades. Air or oil hardening. Thermal shock resistance. Type H12.

E4 TRANSFORMER

Russian manufacture
C 0-0.1, Mn 0-0.3, Si 4, bal Fe.
For transformers; high permeability.

E75

Friedr. Lohmann GmbH
C 0.8, Si 0-0.25, Mn 0-0.25, bal Fe.
Annealed: 210 Brin. For springs, rails, punches, tools. Type W1; water hardened. T72301

EAGLE & GLOBE 20CB-3

Eagle & Globe Steel Ltd.
Stainless steel. C 0-0.06, Mn 0-2, Ni 33.5, Cr 20, Mo 2.5, Cu 3.5, Nb + Ta = 8 x C min, bal Fe.
331 MPa YS; 627 MPa TS; 45 El; 163 Brin. Precipitation hardening and special grades. Austenitic, resistant to hot sulfuric acid. Immune to stress corrosion cracking; weldable, machinable and cold formable. ASTM B473-20Cb-3.

EAGLE & GLOBE 227 HOT DIE STEEL

Eagle & Globe Steel Ltd.
Tool material. C 0.32, Cr 3, W 9, V 0.3, bal Fe.
Annealed: 248 Brin max. For applications where red hardness, high compressive strength and wear resistance at elevated temperatures are required. For hot forging and blanking dies and punches for making nuts, bolts and other similar small components. Dies, cores, inserts, pins, for die casting of copper base alloys; forming dies, shear blades, hot extrusion dies, mandrels, punches, die holders, ejector discs and extrusion liners. AS1239 H21A; AISI H21; BS4659 BH21; Werkstoff 1.2581.

EAGLE & GLOBE 302

Eagle & Globe Steel Ltd.
Stainless steel. C 0.08, Mn 1.7, Ni 9, Cr 18, bal Fe.
280 MPa YS; 620 MPa TS; 50 El; 170 Brin. General purpose austenitic grade for food, chemical and architectural applications. AS2837 302; AISI 302; BS970 302S31; ASTM A276-302.

EAGLE & GLOBE 303

Eagle & Globe Steel Ltd.
Stainless steel. C 0.08, Mn 1.9, Ni 9, Cr 18, S 0.25, bal Fe.
Annealed: 230 MPa YS; 590 MPa TS; 50 El; 170 Brin. Austenitic. For nuts, bolts, screws, fasteners, machine shafts, machine spindles, screw machine parts, pump parts, components in drink dispensing, food handling and chemical handling equipment. AS2837 303; AISI 303; BS970 303S31; ASTM A582-303.

EAGLE & GLOBE 304

Eagle & Globe Steel Ltd.
Stainless steel. C 0.06, Mn 1.7, Ni 9.5, Cr 18, bal Fe.
Annealed: 230 MPa YS; 590 MPa TS; 50 El. 170 Brin. Austenitic. For chemical handling, hospital and dairy equipment, hardware, kitchenware, architectural equipment. AS2837 304; AISI 304; BS970 304S31; ASTM A276-304.

EAGLE & GLOBE 304L

Eagle & Globe Steel Ltd.
Stainless steel. C 0.03, Mn 1.7, Ni 9.5, Cr 18, bal Fe.
Annealed: 180 MPa YS; 480 MPa TS; 50 El; 165 Brin. Austenitic. For chemical handling, hospital and dairy equipment, hardware, kitchenware, architectural equipment. AS2837 304; AISI 304; BS970 304S31; ASTM A276-304.

EAGLE & GLOBE 310

Eagle & Globe Steel Ltd.
Stainless steel. C 0.08, Mn 1.3, Ni 20, Cr 25, Si 1, bal Fe.
310 MPa YS; 620 MPa TS; 45 El; 170 Brin. Austenitic. For high temperature range 850-1000°C. AS2837 310; AISI 310; BS970 310S31; ASTM A276-310.

EAGLE & GLOBE 316

Eagle & Globe Steel Ltd.
Stainless steel. C 0.06, Mn 1.7, Ni 12, Cr 17, Mo 2.25, bal Fe.
Annealed: 230 MPa YS; 590 MPa TS; 50 El; 180 Brin. Austenitic. For chemical equipment, marine fittings, textile machinery, paper industry and petroleum industry. AS2837 316; AISI 316; BS970 316S31; ASTM A276-316.

EAGLE & GLOBE 316L
Eagle & Globe Steel Ltd.
Stainless steel. C 0.03, Mn 1.7, Ni 12, Cr 17, Mo 2.25, bal Fe.
Annealed: 205 MPa YS; 520 MPa TS; 50 El; 190 Brin.
Austenitic. For chemical equipment, marine fittings, textile machinery, paper industry and petroleum industry, pulp handling equipment, marine and petroleum industries. Weldable, corrosion resistant. AS2837 316; AISI 316; BS970 316S31; ASTM A276-316.

EAGLE & GLOBE 321
Eagle & Globe Steel Ltd.
Stainless steel. C 0.06, Mn 1.7, Ni 10, Cr 18, Ti 0.5, bal Fe.
250 MPa YS; 620 MPa TS; 45 El; 180 Brin. Austenitic. For industrial equipment not requiring polished finish, welding without heat treatment involved. Resists temperatures to 800-850°C. AS2837 321; AISI 321; BS970 321S31; ASTM A276-321.

EAGLE & GLOBE 410
Eagle & Globe Steel Ltd.
Stainless steel. C 0.1, Mn 0.4, Cr 12.5, bal Fe.
Heat treated: 415 MPa YS; 600 MPa TS; 26 El; 179 Brin.
Annealed: 310 MPa YS; 520 MPa TS; 30 El; 156 Brin.
Martensitic. General purpose hardenable grade for use in mildly corrosive environments. AS2837 410; AISI 410; BS970 410S21; ASTM A276-410.

EAGLE & GLOBE 416
Eagle & Globe Steel Ltd.
Stainless steel. C 0.1, Mn 1.1, Cr 12.5, S 0.2, bal Fe.
Heat treated: 490 MPa YS; 680 MPa TS; 24 El; 202 Brin.
Annealed: 310 MPa YS; 540 MPa TS; 25 El; 160 Brin.
Martensitic. For general engineering applications where free machining properties are required. AS2837 416; AISI 416; BS970 416S21; ASTM A582-416.

EAGLE & GLOBE 420
Eagle & Globe Steel Ltd.
Stainless steel. C 0.25, Mn 0.4, Cr 12.5, bal Fe.
Heat treated: 650 MPa YS; 880 MPa TS; 18 El; 248 Brin.
Annealed: 340 MPa YS; 650 MPa TS; 25 El; 195 Brin.
Martensitic. For general engineering applications; pump and valve parts. AS2837 420; AISI 420; BS970 420S37; ASTM A276-420.

EAGLE & GLOBE 420C
Eagle & Globe Steel Ltd.
Stainless steel. C 0.3, Mn 0.3, Cr 12.5, bal Fe.
1540 MPa TS; 450-500 Brin. Spring temper: 400-450 Brin.
770-1390 MPa TS; 250-400 Brin. Martensitic. For mildly corrosive conditions. High hardness after heat treatment. BS970 420S45.

EAGLE & GLOBE 430
Eagle & Globe Steel Ltd.
Stainless steel. C 0.08, Mn 0.5, Cr 17, bal Fe.
310 MPa YS; 540 MPa TS; 30 El; 160 Brin. Ferritic. Non-hardenable, good resistance to scaling up to 800°C. AS2837 430; AISI 430; BS970 430S17; ASTM A276-430.

EAGLE & GLOBE 431
Eagle & Globe Steel Ltd.
Stainless steel. C 0.18, Mn 0.7, Ni 2.2, Cr 16, bal Fe.
Annealed: 630 MPa YS; 880 MPa TS; 18 El; 262 Brin. Heat treated (except for forging billet): 715 MPa YS; 920 MPa TS; 17 El; 269 Brin. Martensitic. For pump shafts, valve stems, outboard motor shafts, marine boat shafts, aircraft fittings, dam gate rails, hydraulic fittings, studs, bolts, fasteners. AS2837 431; AISI 431; BS970 431S29; ASTM A276-431.

EAGLE & GLOBE 440C
Eagle & Globe Steel Ltd.
Stainless steel. C 1.1, Mn 0.6, Cr 17, Mo 0.4, bal Fe.
Martensitic. Wear and abrasion resistant. Corrosion resistant. AS2837 440C; AISI 440C; ASTM A276-440C.

EAGLE & GLOBE 52100
Eagle & Globe Steel Ltd.
Tool material. C 1, Si 0.25, Mn 0.35, Cr 1.5, bal Fe.
Annealed: 190 Brin max. Oil hardening steel, machinable. For roller bearings, rolls for metals, plastics, linoleum, crushing dies and tools for paint and allied trades, blanking dies, forming tools, lathe centers, gauges, cams, bushes, marking dies, master hobs. AS1239 L1A; AISI 52100; BS970 EN31; Werkstoff 1.3505.

EAGLE & GLOBE 630
Eagle & Globe Steel Ltd.
Stainless steel. C 0.05, Mn 0.4, Ni 4, Cr 16, Cu 3.7, Nb 0.25, Si 0.4, bal Fe.
710-1275 MPa YS; 1000-1370 MPa TS; 12-18 El; 311-415 Brin. Precipitation hardening and special grades. Martensitic, age hardenable. Weldable and corrosion resistant. For pump shafts, boat shafts, valve stems, studs, bolts. AS2837 630; AISI 630; ASTM A564-630.

EAGLE A BABBITT
NL Industries
Sn, Sb, bal Pb.
At 70°F: 23.5 Brin. At 212°F: 11.9 Brin. For bearings, bushings; Babbitt. *Obsolete*

EAGLE MUSIC WIRE
Washburn Wire Co.
C 0.9, bal Fe.
For wire, springs; water hardened.

EARLUMIN
Spanish manufacture
Cu 4-5, bal Al.
For light alloy parts; age-hardenable.

EASTERN "Z" METAL
Alloy Foundries
C 2-2.6, Si 1, Mn 1, bal Fe.
Cast: 75,000 TS; 50,000 YS; 12 El; 180 Brin. Malleable cast iron for air drill parts, sprockets, gears, unions.

EASTERN 12R
Alloy Foundries
Aluminum. Cu 6-8, Si 1-4, Zn 2.5, Fe 1-4, Mg 0.6, bal Al.
Cast: 19,000 TS; 14,000 YS; 2 El. For light alloy parts; Aluminum 113, SAE33.

EASTERN 19-9 DL
Now 19-9 DL.

EASTERN 29-9
Alloy Foundries
Cr 29, Ni 9, C, bal Fe.
Cast: 97,000 TS; 49,000 YS; 28 El; 90-95 Brin. For valves, pumps, headers, corrosion resistant.

EASTERN 321 SW
Cyclops Industries/Eastern Stainless
C 0-0.08, Mn 0-2, Si 0-1, Cr 17-19, Ni 9-12, Ti = 6 x C min, bal Fe.
"Slag-washed" grade of AISI 321 stainless steel for improved spinning and forming.

EASTERN 32510
Alloy Foundries

Malleable iron casting with good machinability. Cast: 50,000 TS; 32,500 YS; 10 El; 110-135 Brin. Meets ASTM A47-52 Grade 32510.

EASTERN 35018
Alloy Foundries
Malleable iron casting with good machinability. Cast: 53,000 TS; 35,000 YS; 18 El; 110-135 Brin. Meets ASTM A47-52 Grade 35018.

EASTERN 36
Alloy Foundries
Aluminum. C 0.1, Si 2.3-3.1, Mg 0.03, Zn 0.2, Fe 0.6, Mn 0.2, bal Al.
Cast: 17,000 TS; 5500 YS; 20 El. For light alloy parts.

EASTERN 500T
Alloy Foundries
Aluminum. Cu 0.4-1, Si 0.35, Mg 0.2-0.5, Zn 7-8, Fe 1, Mn 0.6, bal Al.
Aged: 30,000 TS; 22,000 YS; 3 El. For light alloy parts; Alloy ZC81A.

EASTERN ES-18-11
Eastmet Corp.
C 0-0.12, Cr 17-19, Ni 10-13, S 0-0.03, P 0-0.04, bal Fe.
For spun and formed parts; Type 305; stainless, austenitic. *Obsolete*

EASTERN ES-18-12-3MO
Eastmet Corp.
C 0-0.08, Cr 16-18, Ni 10-14, Mo 2-3, bal Fe.
Annealed: 85,000 TS; 35,000 YS; 50 El; 65 RA; 160 Brin. For chemical plant equipment; Type 316; stainless, austenitic. *Obsolete*

EASTERN ES-18-12-3MO-L
Eastmet Corp.
C 0-0.03, Cr 16-18, Ni 10-14, Mo 2-3, bal Fe.
Annealed: 85,000 TS; 35,000 YS; 55 El; 70 RA; 155 Brin. For chemical plant equipment, welded structures; Type 316L; stainless, austenitic. *Obsolete*

EASTERN ES-18-12-4MO
Eastmet Corp.
C 0-0.08, Cr 18-20, Ni 11-14, Mo 3-4, bal Fe.
Annealed: 85,000 TS; 35,000 YS; 50 El; 65 RA; 160 Brin. For chemical plant equipment, mixers, agitators; Type 317; stainless, austenitic. *Obsolete*

EASTERN ES-18-8FM
Eastmet Corp.
C 0-0.15, Cr 17-19, Ni 8-10, Zr 0-0.6, 0.07% min P or S, bal Fe.
Annealed: 80,000 TS; 35,000 YS; 50 El; 65 RA; 150 Brin. For shafts, gears, chemical plant equipment, Type 303; free-cutting, stainless. *Obsolete*

EASTERN G-35
Alloy Foundries
C, bal Fe.
Gray cast iron for pressure fittings, cams, castings. 35,000 TS; 0 El; 215 Brin.

EASTERN G-40
Alloy Foundries
C, alloy, bal Fe.
Heat resistant cast iron for die blocks, stoker parts, castings. 40,000 TS; 0 El; 240 Brin.

EASTERN GR. P1
Alloy Foundries
C 3, Si, Mn, bal Fe.
Cast: 70,000 TS; 48,000 YS; 10 El; 187 Brin. Pearlitic malleable iron for gears, shafts, housings.

EASTERN GR. P2
Alloy Foundries
C 3, Si, Mn, bal Fe.
Cast: 75,000 TS; 50,000 YS; 8 El; 201 Brin. Pearlitic malleable iron for gears, shafts, housings.

EASTERN GR. P3
Alloy Foundries
C 3, Si, Mn, bal Fe.
Cast: 80,000 TS; 60,000 YS; 5 El; 217 Brin. Pearlitic malleable iron for gears, shafts, housings.

EASTERN GR. P4

Alloy Foundries
C 3, Si, Mn, bal Fe.
Cast: 85,000 TS; 55,000 YS; 4 El; 207 Brin. Pearlitic malleable iron for gears, shafts, housings.

EASTERN GR. P5

Alloy Foundries
C 3, Si, Mn, bal Fe.
Cast: 75,000 TS; 50,000 YS; 4 El; 201 Brin. Pearlitic malleable iron for gears, shafts, housings.

EASTERN GR. P6

Alloy Foundries
C 3, Si, Mn, bal Fe.
Cast: 65,000 TS; 45,000 YS; 10 El; 187 Brin. Pearlitic malleable iron for gears, shafts, housings.

EASTERN H-55

Alloy Foundries
C 2.2, Si 1.3, alloy, bal Fe.
Cast: 55,000 TS; 0 El; 270 Brin; 3200 transverse strength. Gray cast iron for lapping plates, rolls, pressure fittings, castings.

EASTERN H-60

Alloy Foundries
C 2.2, Si 1, Mn 0.5, bal Fe.
Cast: 60,000 TS; 0 El; 270 Brin. Wear resistant gray cast iron for manhole covers, cylinders, rolls, brake drums, die blocks.

EASTERN H-65

Alloy Foundries
C 2.2, Si 1, alloy, bal Fe.
Cast: 65,000 TS; 0 El; 300 Brin. Cast iron for die blocks, stoker parts, castings; heat resistant to 1200°F.

EASTERN N

Alloy Foundries
Cr 16, C, bal Fe.
Annealed: 50,000 TS; 0 El; 195 Brin. Cast iron for carburizing and annealing boxes, castings, furnaces; heat resistant to 1500°F.

EASTERN N-1

Alloy Foundries
Cr 28, Ni 12, C, bal Fe.
Cast: 75,000 TS; 60,000 YS; 2 El; 260 Brin. Heat resistant cast iron for rabble arms, pump parts, furnace parts.

EASTERN N-2

Alloy Foundries
Cr 28, C, bal Fe.
Cast: 50,000 TS; 40,000 YS; 1 El; 225 Brin. Heat resistant cast iron for annealing boxes, lead pots, pump and valve bodies; resistant to sulfur atmosphere.

EASTERN N-3

Alloy Foundries
Cr 23, C, bal Fe.
Cast: 65,000 TS; 50,000 YS; 1 El; 260 Brin. For glass industry, stainless castings, glass molds; stainless, resists scaling to 1600°F.

EASTERN N-4

Alloy Foundries
Cr 20, Ni 64, C, bal Fe.
Cast: 60,000 TS; 45,000 YS; 2 El; 250 Brin. For carburizing boxes, glass molds, pyrometer tubes; heat resistant to 1800°F.

EASTERN N-5

Alloy Foundries
Cr 20, Ni 38, C, bal Fe.
Cast: 65,000 TS; 45,000 YS; 4 El; 235 Brin. For furnace parts, grids, rollers, hearth plates; heat resistant to 1650°F.

EASTERN N-6

Alloy Foundries
Cr 8, Ni 22, C, bal Fe.
Cast: 75,000 TS; 40,000 YS; 20 El; 160 Brin. For ship propellers, pump and valve bodies; stainless, heat resistant to 1800°F.

EASTERN N-8

Alloy Foundries
Cr 23, C, bal Fe.
Cast: 65,000 TS; 55,000 YS; 0 El; 555 Brin. For heat resistance; abrasion resistant.

EASTERN N7

Alloy Foundries
C 0.95-1.2, Mn 0-1, Si 0-1, Cr 16-18, Mo 0-0.75, bal Fe. High carbon martensitic stainless. AISI 440C.

EASTERN NO. 1030

Eastmet Corp.
Co 12-15, Mo 3.5-5, Ti 2.75-0.25, Al 1-1.5, Zr 0.02-0.15, B 0.001-0.01, Cr 18-21, bal Ni.
Heat treated: 188,000 TS; 115,000 YS; 28 El; 25 RA. At 1400 F: 117,000 TS; 99,000 YS; 28 El; 41 RA. For jet engine turbine buckets, discs, fasteners, combustion chambers. Similar to Waspaloy, high temperature strength, oxidation resistant. *Obsolete*

EASTERN NO. 155

Eastmet Corp.
C 0.15, Si 0.65, Mn 1-2, Cr 22, Ni 20, Mo 3, Cb 1, W 3, Co 20, bal Fe.
Sheet: 118,000 TS; 56,000 YS; 49 El. At 1000 F: 94,000 TS; 40,000 YS; 54 El. For aircraft cabin heaters, exhaust stacks, afterburners, rocket chambers. Similar to Multimet N-155. High heat and corrosion resistance. *Obsolete*

EASTERN NO. 20

Eastmet Corp.
C 0-0.07, Ni 24-30, Cr 19-21, Mo 2-3, Cu 3-4, Mn 0-2, bal Fe.
Sheet: 90,000 TS; 50,000 YS; 35 El; B 90 Rock. At 1000 F: 76,000 TS; 32,000 YS; 48 El; 60 RA. For acid handling equipment, pickling racks, mixing tanks. Austenitic, stainless. Carpenter 20. Acid resistant. *Obsolete*

EASTERN NO. 200

Eastmet Corp.
Cu 0-0.25, Mn 0-0.35, C 0-0.15, Fe 0-0.4, 99% min Ni.
Hard: 115,000 TS; 105,000 YS; 2 El; B 90 min Rock. Soft: 55,000 TS; 15,000 YS; 55 El; B 70 max Rock. For production and handling of fluorine and bromine, food processing equipment, chemical shipping drums. Similar to Inco 200, high corrosion resistance, good cryogenic ductility. *Obsolete*

EASTERN NO. 286

Eastmet Corp.
C 0.05, Mn 1.5, Cr 15, V 0.3, Ni 26, Mo 1.2, Ti 2.2, Al 0.2, bal Fe.
Aged: 155,000 TS; 105,000 YS; 24 El; 302 Brin. At 1200 F: 110,000 TS; 88,000 YS; 13 El; 14 RA. For jet engine and supercharger components, fasteners, turbine parts. Similar to A-286 alloy, age-hardenable. High heat and corrosion resistance. *Obsolete*

EASTERN NO. 41

Eastmet Corp.
C 0.09, Cr 19, Co 11, Mo 10, Ti 3, Al 1.5, bal Ni.
Heat treated: 206,000 TS; 154,000 YS; 14 El. At 1400 F: 160,000 TS; 136,000 YS; 11 El. For jet engine components, afterburners, turbine casings, combusion liners, turbine buckets. Similar to Rene 41, high strength at 1600-1800 F. *Obsolete*

EASTERN NO. 530

Eastmet Corp.
C 0-0.08, W 3-4.5, Cr 14-16, Mo 15-17, bal Ni.
Sheet: 129,000 TS; 64,000 YS; 37 El; 31 RA. At 1200 F: 95,000 TS; 46,000 YS; 22 El; 22 RA. For chemical and petroleum equipment, oil refinery components, exhaust stacks, combustion chambers. Similar to Hastelloy C, corrosion and heat resistant. *Obsolete*

EASTERN NO. 600

Eastmet Corp.
Cr 14-17, Fe 6-10, bal Ni.
Annealed: 90,000 TS; 40,000 YS; 45 El; B 75 Rock. Cold drawn: 150,000 TS; 125,000 YS; 10 El; C 30 Rock. For regenerators, coolers, heaters, furnace muffles, combustion liners, thermocouple tubes. Similar to Inconel. High oxidation and corrosion resistance. *Obsolete*

EASTERN NO. 605

Eastmet Corp.
C 0.05-0.15, Mn 1-2, Cr 19-21, W 14-16, Ni 9-11, Fe 0-3, bal Co.
Annealed: 155,000 TS; 75,000 YS; 60 El; C 22 Rock. At 1000 F: 82,000 TS; 38,000 YS; 24 El. For gas turbine and jet engine components, fasteners. Similar to Haynes 25 and L-605 alloy. High heat and corrosion resistant. *Obsolete*

EASTERN NO. 702

Eastmet Corp.
Cr 14-17, Al 2.7-3.7, Ti 0.25-1, bal Ni.
Age-hardened: 142,000 TS; 82,000 YS; 38 El. At 1000 F: 120,000 TS; 74,000 YS; 40 El. For combustion liners, jet engine turbines. Age-hardenable. Similar to Inconel 702. High oxidation and corrosion resistant. *Obsolete*

EASTERN NO. 718

Eastmet Corp.
Cr 17-21, Cb 4.5-5.7, Mo 2.8-3.3, Al 0.2-1, Ti 0.3-1.3, Ni 50-55, bal Fe.
Annealed: 130,000 TS; 60,000 YS; 45 El. Age-hardened: 190,000 TS; 150,000 YS; 25 El. For aircraft and jet engine structural components. Similar to Inconel 718, age-hardenable. Oxidation and corrosion resistant. *Obsolete*

EASTERN NO. 750

Eastmet Corp.
Cr 14-17, Fe 5-9, Ti 2.2-2.7, Al 0.4-1, Cb 0.7-1.2, bal Ni.
Heat treated: 170,000 TS; 115,000 YS; 18 El; 25 RA. For gas turbine wheels, and rotor blades, turbo superchargers, jet engine afterburners, pressure vessels. Similar to Inconel X-750, tough, age-hardenable, corrosion and oxidation resistant. *Obsolete*

EASTERN NO. 801

Eastmet Corp.
Ni 32, Ti 1, C 0.04, Mn 0.75, Fe 44.5, S 0.007, Cu 0.15, Cr 20.5.
Sheet: 90,000 TS; 40,000 YS; 40 El. At 1000 F: 75,000 TS; 30,000 YS; 35 El. For combustion liners, furnace parts, heaters, high temperature parts. Excellent oxidation resistance to 1700 F. Similar to Incoloy T. High tensile and rupture strength. *Obsolete*

EASTON PM 216

Easton Metal Powder Inc.
Mn 0.43, Ni 0.41, Mo 0.73, Cu 2.05, 0.80 C (including 0.75 graphite), bal Fe.
140-300 mesh metal powder for sintering. Produces hardenable steel.

EASTON RZ 365

Easton Metal Powder Inc.
Fe 98.7, Mn 0.22, 0.06 C (typical).
140-300 mesh iron powder for sintering.

EASTON RZ 365 MM

Easton Metal Powder Inc.
C 0.11-0.14, Mn 0.55, Fe 98.6.
140-300 mesh powder for sintering.

EASTON RZ 365 S

Easton Metal Powder Inc.
C 0.08, Mn 0.17, S 0.31, Fe 98.5.
140-300 mesh iron powder for sintering. For free machining material.

EASTON RZ 4600

Easton Metal Powder Inc.
C 0.08, Ni 1.84, Mo 0.3, Mn 0.21, bal Fe.
140-300 mesh metal powder for sintering. Mixing with varying amounts of graphite before sintering will permit product similar in composition to AISI 46XX.

EASY

Engelhard Corp.
Ag 65, Cu 20, Zn 15.
For silversmithing solder for sterling; 1280-1325°F MP.
Obsolete

EASY-FLO

Handy & Harmon
Precious metal. Ag 50, Cu 15.5, Zn 16.5, Cd 18.
MP: 1175°F; FP: 1200°F; corrosion resistant. For brazing.
AMS 4770C; AWS BAg-la.

EASY-FLO

Johnson Matthey plc
Ag 50, Zn, Cd, Cu.
Silver solder; melt range 620-630°C.

EASY-FLO 2

Johnson Matthey plc
Ag 42, Cu, Zn, Cd.
Silver solder; 608-617°C MP.

EASY-FLO 25

Handy & Harmon
Ag 25, Cu 35, Zn 26.5, Cd 13.5.
MP: 1125°F; FP: 1375°F. Same as Easy-Flo 30, but used for economical joints.

EASY-FLO 25HC

Handy & Harmon
Ag 25, Cu 30, Zn 27.5, Cd 17.5.
MP: 1110°F; FP: 1315°F. Same as Easy-Flo 35, but used for economical joints.

EASY-FLO 3

Handy & Harmon
Precious metal. Ag 50, Cu 15.5, Zn 15.5, Cd 16, Ni 3.
MP: 1170°F; FP: 1270°F. For brazing stainless steel with corrosion resisting joints; brazing carbide tips. AMS 4771A; AWS BAg-3.

EASY-FLO 3

Johnson Matthey plc
Ag 50, Cu, Zn, Cd, Ni.
Silver solder; 634-656°C MP. For brazing stainless steels and carbide tips.

EASY-FLO 30

Handy & Harmon
Precious metal. Ag 30, Cu 27, Zn 23, Cd 20.
MP: 1125°F; FP: 1310°F. 60,000 TS expected when joining 1020 steel. General purpose brazing alloys, particularly for leak-tight joints. AWS BAg-2a.

EASY-FLO 35

Handy & Harmon
Precious metal. Ag 35, Cu 26, Zn 21, Cd 16.
MP: 1125°F; FP: 1295°F. General purpose low melting brazing alloy for brazing ferrous and non-ferrous alloys. AMS 4768A; AWS BAg-2.

EASY-FLO 45

Handy & Harmon
Precious metal. Ag 45, Cu 15, Zn 16, Cd 24.
MP: 1125°F; FP: 1145°F. For brazing ferrous and non-ferrous alloys; lowest FP of silver braze alloys. AMS 4769A; AWS BAg-1.

EASY-FLO TRI-FOIL "C"

Johnson Matthey plc
Ag 50, Cu 15, Zn 16, Cd 19-19.05.
Silver brazing alloys for tungsten carbide; 620-630°C MP; DIN 8513.

EASYARC

Now AIRCO EASYARC.

EATONITE

Eaton Corp.
C 2-2.75, Ni 37-41, W 14, Cr 27-31, Co 9-11, Si 0-1, Fe 0-8.
At 1300°F: 302 Brin. For valve facings, valve seats and faces; retains hardness at high temperature.

EB SILVER BRAZING ALLOY

Handy & Harmon
Ag 57, Mn, Sn, bal Cu.
For silver brazing of carbides; MP 1120-1345 F. *Obsolete*

EC 100

Saarstahl AG
Now SAARSTAHL 20 MNCR 5.

EC 100 B

Saarstahl AG
C 0.18, Si 0.25, Mn 1.15, Cr 1.15, B, bal Fe.
Carburizing steel; for steering shafts, camshafts, and gears.
Obsolete

EC 60

Saarstahl AG
Now SAARSTAHL 17 CR 3.

EC 80

Saarstahl AG
Now SAARSTAHL 16 MNCR 5.

EC 80 B

Saarstahl AG
Now SAARSTAHL 16 MNCRB 5.

ECCENTRIC RING

English manufacture
Cu 84, Sn 14, Zn 2.
For bearings, bells.

ECLAIR

Creusot-Loire
C 0.62, W 14, bal Fe.
For hot work tools; oil hardened. *Obsolete*

ECLAIR 33

Creusot-Loire
C 0.82, W 6, Mo 4.5, V 1.5, bal Fe.
For lathe and planer tools, reamers, cutters; oil hardened.
Obsolete

ECLAIR 44

Creusot-Loire
C 0.82, W 10, V 1.5, Mo 0.65, bal Fe.
For lathe and planer tools; oil hardened. *Obsolete*

ECLAIR 55

Creusot-Loire
C 0.82, W 10, Co 3, V 1.5, Mo 0.7, bal Fe.
For lathe and planer tools, reamers; oil hardened. *Obsolete*

ECLAIR 88

Creusot-Loire
C 0.82, W 6.5, Mo 5, V 2, bal Fe.
For lathe and planer tools, drills, hobs, taps; oil hardened.
Obsolete

ECLAIR 99

Creusot-Loire
C 0.92, W 10, Co 9, V 2, Mo 0.6, bal Fe.
For dies; oil hardened. *Obsolete*

ECLIPSALLOY

Eclipse-Pioneer Foundries
Aluminum.
Aluminum alloy AlCu(NiMg) (S).

ECLIPSE

J.M. Ney Co.
Au 52, Pd 37.5, Zn 3.8.
White color alloy for porcelain to gold restorations subject to stress; hard. 117,000 psi TS; 71,000 psi YS (0.1% offset); 30 El; 254 HV.

ECLIPSE 75

Manufacturer not listed
C, alloy, bal Fe.
For cold hobbing tools; water hardened.

ECLIPSE BRONZE

Sargent & Co.
Sn, bal Cu.
For hardware; white nickel bronze. *Obsolete*

ECO G 100

Johnson Matthey plc
Au 20, Ag 41, Pd 20, In 15, bal Zn.
Economy yellow dental casting gold; Type 3.

ECO J

Johnson Matthey plc
Au 20, Ag 40, Pd 20, In 18, bal Zn.
Economy yellow dental casting gold; Type 3.

ECOCER

Johnson Matthey plc
Pd 79, Cu 10, Ga 9, bal Au.
Low cost white dental bonding alloy.

ECONO

CCS Braeburn Alloy Steel
C 0.4, Mn 0.2, Cr 3.75, V 0.75, W 1.1, Mo 5.75, bal Fe.
For tools, cutters, hot work dies; hot work steel. *Obsolete*

ECONO NO. 1-V

Champion Steel Co.
C 1.05, V 0.25, bal Fe.
Water hardening tool steel; AISI W2.

ECONO-2

Champion Steel Co.
C 0.75, Ni 1.25, Cr 0.8, Mo 0.25, V 0.15, bal Fe.
Heat treated: 121,000-208,000 TS; 90,000-180,000 PL; 29-45 Rock C. For brake dies, bushings, cams, shear blades, gears, set screws, swaging and forming dies, punches. Type L6, tough, oil hardening.

ECONO-5

Champion Steel Co.
C 1, Cr 5, Mo 1, V 0.4, Mn 0.4, bal Fe.
Annealed: 103,000 TS; 52,000 YS; 26 El; 18 Rock C.
Hardened: 178,000-253,000 TS; 146,000-200,000 YS; 3-12 El; 7-32 RA; 41-53 Rock C. For blanking and trimming dies, shears, cutters, gauges, lamination dies. Type A2, air hardening, nondeforming, cold work tool steel, tough.

ECONO-KROME STAINLESS

Associated Steel Corp.
C 0.08, Cr 18-20, Ni 11-12, Mn 0-2, Mo 0-0.8, Zn 0-0.2, bal Fe.
Austenitic, stainless steel; good resistance to intergranular corrosion.

ECONOLOY

United States Steel Corp.
C alloy, bal Fe.
For welding rod, hard surfacing. *Obsolete*

ECONOMET

Alloy Engineering & Casting Co.
C 0.35-0.75, Cr 8-12, Ni 28-32, bal Fe.
For heat treating boxes and furnace parts. Heat and abrasion resistant.

ECONOMET 4

Alloy Engineering & Casting Co.
C 0.36-0.5, Cr 16-23, Ni 7-11, bal Fe.
For heat and corrosion resistant parts. Stainless; heat and corrosion resistant.

ECONOMO NO. 20

Metalsource Corp.
C 0.18, Mo 0.15, Mn 0.8, bal Fe.
Rolled: 73,000 TS; 50,500 YS; 32 El; 61 RA; 143 Brin. Heat treated: 100,000-124,000 TS; 81,000-86,000 YS; 17-21 El; 51-64 RA; 241-293 Brin. For machinery steel for case-hardening, gears, worms, sprockets, clutches, cams, bolts, gages; free machining carburizing steel. *Obsolete*

ECONOMO NO. 50

Metalsource Corp.
C 0.5, Mo, bal Fe.
Water quenched: 110,000-168,000 TS; 96,000-150,000 YS; 16-24 El; 51-61 RA; 217-293 Brin. For shafts, spindles, axles, worms, gears, pinions, bolts; machinery steel. *Obsolete*

ECONOMY BRONZE

Anaconda Co.
Zn, Sn, bal Cu.
For welding rods; for steel or cast iron. *Obsolete*

ECONOMY HARDFACE

Abex Corp.
C 1, Cr 5, Mo 1.7, bal Fe.
Welded: 450-550 Brin. For hard facing electrode; abrasion resistant, air hardened.

ECONOMY HARDFACE C

Abex Corp.
C 0.5, Cr 3, Mo 1.7, bal Fe.
Welded: 500 Brin. For hard facing electrode; wear and abrasion resistant.

ED 2

Electrometal SA Metals Especiais
C 1.5, Si 0.35, Mn 0.5, Cr 12, Mo 0.8, V 0.2, W 0-0.15, bal Fe.
Cold work tool steel. W.-Nr. 1.2379; similar to AISI D2.

ED 3

Electrometal SA Metals Especiais
C 2, Si 0.35, Mn 0.47, Cr 12.5, V 0.2, bal Fe.
Cold work tool steel. Similar to AISI D3.

ED 6

Electrometal SA Metals Especiais
C 2, Si 0.57, Mn 0.47, Cr 12.5, V 0.2, W 1, bal Fe.
Cold work tool steel. W.-Nr. 1.2436.

EDAL

German manufacture
Mg 5, bal Al.
Cast: 40,000-64,000 TS; 21,000-43,000 YS; 5-31 El; 1 RA. For welding rods, light alloy parts; heat treatable.

EDCO PHOSPHOR BRONZE

Eccles & Davis Machinery Co.
Cu 91.5, Sn 8.25, P 0.25.
Cast: 30,000 TS; 7-10 El; 163 Brin. For welding rod; M.P. 1010°C.

EDELBRONZE

Eisenwerke Neubrandenburg G.m.b.H.
Sn, bal Cu.
For chemical plant and textile industry equipment; corrosion resistant.

EDELMESSING

German manufacture
Zn, alloy, bal Cu.
For ornamental hardware; high tensile brass.

EDELSTAHL-MTS5

Thyssen Edelstahlwerke AG
C 0.2, Cr 11.5, Mo 1, Ni 0.4, V 0.3, bal Fe.
Annealed: 92,000 TS; 38,000 YS; 26 El; B 90 Rock.
Hardened: 235,000 TS; 195,000 YS; 10 El; C 46 Rock. For valves, bearings, cutlery, surgical instruments. Corrosion resistant, hardenable. *Obsolete*

EDELSTAHL-MTS6

Thyssen Edelstahlwerke AG
C 0.17, Cr 11.5, Mo 0.6, Ni 0.6, V 0.25, bal Fe.
Annealed: 80,000 TS; 40,000 YS; 25 El; B 92 Rock.
Hardened: 200,000 TS; 165,000 YS; 12 El; C 45 Rock. For table hardware, surgical instruments, springs, cutlery. Corrosion resistant, hardenable. *Obsolete*

EDELWEISS AKRO 6

Vereinigte Edelstahlwerke
C 0-0.12, Si 0.8, Al 0.8, Cr 6.5, bal Fe.
For oil refining equipment; creep and heat resistant. *Obsolete*

EDELWEISS ALSICRO 8

Vereinigte Edelstahlwerke
C 0-0.12, Si 0.8, Al 0.8, Cr 6.5, bal Fe.
For oil refining equipment; creep and heat resistant. *Obsolete*

EDELWEISS R1

Vereinigte Edelstahlwerke
C 0.15, Si 2, Cr 24, Ni 19, bal Fe.
For furnace parts, heat treating boxes; Type CK-20; corrosion and heat resistant. *Obsolete*

EDELWEISS R16

Vereinigte Edelstahlwerke
C 0.15, Cr 19.5, Ni 9.5, Si 2, bal Fe.
Cast: 72,000 TS; 36,000 YS; 35 El; 40 RA; 145 Brin. For chemical plant equipment, mixers, filters; Type CF-16F; corrosion and heat resistant. *Obsolete*

EDELWEISS R1G

Vereinigte Edelstahlwerke
C 0.4, Si 2, Cr 25, Ni 19, bal Fe.
Cast: 75,000 TS; 50,000 YS; 17 El; 170 Brin. Aged: 85,000 TS; 50,000 YS; 10 El; 190 Brin. For furnace parts, retorts, heat treating boxes; Type HK; corrosion and heat resistant. *Obsolete*

EDELWEISS R25

Vereinigte Edelstahlwerke
C 0-0.12, Si 1.5, Al 1.5, Cr 24, bal Fe.
For furnace parts, oil refinery equipment; heat and creep resistant. *Obsolete*

EDELWEISS R25N

Vereinigte Edelstahlwerke
C 0.2, Si 1.2, Cr 25, Ni 4, bal Fe.
Cast: 90,000 TS; 65,000 YS; 2 El; 212 Brin. For furnace parts, retorts, heat treating boxes; corrosion and heat resistant. *Obsolete*

EDELWEISS R25NG

Vereinigte Edelstahlwerke
C 0.4, Si 1.3, Cr 27, Ni 4, bal Fe.
Cast: 90,000 TS; 65,000 YS; 2 El; 212 Brin. Heat treated: 97,000 TS; 65,000 YS; 18 El; 210 Brin. For cylinder liners, bushings, valve seat and bodies; Type CC-50; heat resistant. *Obsolete*

EDELWEISS R9

Vereinigte Edelstahlwerke
C 0-0.15, Cr 20, Ni 14.5, Mo 0.9, Cb 1.2, Ti 0.8, bal Fe.
For furnace parts, heat treating boxes, retorts; corrosion and heat resistant. *Obsolete*

EDELWEISS R9G

Vereinigte Edelstahlwerke
C 0.4, Cr 22, Si 2, Ni 9.5, bal Fe.
Cast: 85,000 TS; 45,000 YS; 35 El; 165 Brin. For heat treating boxes, chains, conveyor belts; corrosion and heat resistant. *Obsolete*

EDGAR ALLEN 66

Edgar Allen Balfour Ltd.
C 2.25, Cr 13, bal Fe.
For dies; non-deforming.

EDGAR ALLEN A.100

Edgar Allen Balfour Ltd.
C 0.32, Cr 0.65, Ni 2.5, Mo 0.55, bal Fe.
Heat treated: 160,000-275,000 TS; 138,000-230,000 YS; 11-18 El; 32-58 RA; 300-450 Brin. For arbors, boring bars, collets, dies, gears; tough, oil hardened.

EDGAR ALLEN A.13

Edgar Allen Balfour Ltd.
C 0.4, Cr 1.15, Ni 1.5, Mo 0.3, Mn 0.6, bal Fe.
Heat treated: 224,000-251,000 TS; 215,000-224,000 YS; 10-12 El; 477-512 Brin. For arbors, axles, beaters, bolts, connecting rods; shock resistant, oil hardened.

EDGAR ALLEN AM.1

Edgar Allen Balfour Ltd.
C 0.35, W 1.35, Cr 5, Mo 1.5, V 0.45, bal Fe.
For Al and brass extrusion dies, heading tools; hot-work steel, resists heat checking.

EDGAR ALLEN AM.3

Edgar Allen Balfour Ltd.
C 0.4, Cr 5, Mo 1.35, V 1.1, bal Fe.
For Al die casting dies, extrusion and forming tools; hot-work steel, resists heat checking.

EDGAR ALLEN DOUBLE 12

Edgar Allen Balfour Ltd.
C 0.3, Cr 12, W 12, bal Fe.
Heat treated: 38-42 Rock C. For liquid hobbing of beryllium copper cavity molds for plastics, brass and copper extrusion dies. Air or oil hardened. Hot-work steel; Type H23.

EDGAR ALLEN DOUBLE SHEAR TEMPER STEEL

Edgar Allen Balfour Ltd.
C, bal Fe.
For shears, cutting tools; water hardening. *Obsolete*

EDGAR ALLEN DRIFT STEEL

Edgar Allen Balfour Ltd.
C, bal Fe.
For tools; water hardened. *Obsolete*

EDGAR ALLEN DRILL STEEL

Edgar Allen Balfour Ltd.
C, bal Fe.
For drills; resists abrasion. *Obsolete*

EDGAR ALLEN DUNTER-PICK STEEL

Edgar Allen Balfour Ltd.
C, bal Fe.
For dunter picks, drills; water hardened. *Obsolete*

EDGAR ALLEN FEATHER STEEL

Edgar Allen Balfour Ltd.
C, bal Fe.
For feather tools; usually requires no heat treatment. *Obsolete*

EDGAR ALLEN FIVE PLY STEEL

Edgar Allen Balfour Ltd.
C, bal Fe.
For engineering construction; manufactured in plates containing 3 layers of soft steel and 2 layers of hard steel.

EDGAR ALLEN GENUINE DOUBLE SHEAR STEEL
Edgar Allen Balfour Ltd.
C, bal Fe.
For carving knives; water hardened.

EDGAR ALLEN GRANITE-CHISEL STEEL
Edgar Allen Balfour Ltd.
C, bal Fe.
For stone chisels; to cut granite. *Obsolete*

EDGAR ALLEN IMP. C.S.
Edgar Allen Balfour Ltd.
C, W, Cr, bal Fe.
For cutting tools; water hardened. *Obsolete*

EDGAR ALLEN IMPERIAL
Edgar Allen Balfour Ltd.
C 1.4, Cr 0.75, W 4.5, bal Fe.
For cutting tools, drills, reamers, textile needles. Water hardened, wear resistant.

EDGAR ALLEN IRON FIBERED STEEL
Edgar Allen Balfour Ltd.
C, bal Fe.
For screwed studs for cylinder and boiler mountings, piston rods, axles, chains, crank pins; resists shock and impact. *Obsolete*

EDGAR ALLEN K-9
Edgar Allen Balfour Ltd.
C 0.95, Cr 0.75, Mn 0.85, W 0.5, bal Fe.
For dies, stay taps, broaches, milling cutters, plug gages; oil hardened, tough.

EDGAR ALLEN MINERS DRILL STEEL
Edgar Allen Balfour Ltd.
C 0.9, bal Fe.
For drills; water hardened. *Obsolete*

EDGAR ALLEN N.I.F.E. 30
Edgar Allen Balfour Ltd.
Fe 70, Ni 30.
For compensating shunts for electrical equipment; thermo-sensitive, magnetic.

EDGAR ALLEN NO. 4 HOT DIE
Edgar Allen Balfour Ltd.
C 0.9, Cr 3.7, bal Fe.
Air hardened. For pressing and squeezing dies, hot nut dies, piercers, punches, gripper dies; resistant to heat and shock. *Obsolete*

EDGAR ALLEN NO. 5 HOT DIE
Edgar Allen Balfour Ltd.
C 0.32, Cr 3.25, W 9.5, V 0.35, bal Fe.
For hot shear blades, blades, press and form dies, piercers; hot-work steel, oil hardened.

EDGAR ALLEN PLUG STEEL
Edgar Allen Balfour Ltd.
C, bal Fe.
For plugs; resists heavy blows. *Obsolete*

EDGAR ALLEN RED LABEL
Edgar Allen Balfour Ltd.
W, C, bal Fe.
For twist drills, taps, milling cutters, reamers; water hardening. *Obsolete*

EDGAR ALLEN SAFE-PLATE
Edgar Allen Balfour Ltd.
C, bal Fe.
For safes. *Obsolete*

EDGAR ALLEN SHEAR TEMPER
Edgar Allen Balfour Ltd.
C, bal Fe.
For carving knives. *Obsolete*

EDGAR ALLEN SILVER STEEL
Edgar Allen Balfour Ltd.
C 0.8-1, bal Fe.
For borers, punches, reamers, twist drills, taps; water hardened. *Obsolete*

EDGAR ALLEN SPECIAL CHROME
Edgar Allen Balfour Ltd.
C, Cr, bal Fe.
For thread rolling dies; retains hardness at high temperatures.

EDGAR ALLEN STAGPICK
Edgar Allen Balfour Ltd.
C, Cr, V, bal Fe.
For coal cutter picks; tough, hard and shock resistant.

EDGAR ALLEN TOOL CLASS "C"
Edgar Allen Balfour Ltd.
C 0.6-1.35, bal Fe.
For spiral springs, chisels, punches, shear blades, drills, reamers, taps, dies; water hardened.

EDGAR ALLEN TOOL CLASS "E"
Edgar Allen Balfour Ltd.
C 0.6-1.35, bal Fe.
For wood working tools, chisels, punches, shear-blades, drills, reamers, taps; water hardening. *Obsolete*

EDGAR ALLEN TOOL CLASS "F"
Edgar Allen Balfour Ltd.
C 0.6-1.35, bal Fe.
For spiral springs chisels, punches, shear-blades, drills, reamers, taps, dies; water hardening. *Obsolete*

EDGAR ALLEN TOOL CLASS "H"
Edgar Allen Balfour Ltd.
C 0.6-1.35, bal Fe.
For dies, taps, reamers, drills, punches; water hardening. *Obsolete*

EDGAR ALLEN TOOL CLASS "P"
Edgar Allen Balfour Ltd.
C 0.6-1.35, bal Fe.
For chisels, punches, shear-blades, snaps, drills, reamers, taps, milling cutter; water hardening. *Obsolete*

EDGAR ALLEN TOOL CLASS "V"
Edgar Allen Balfour Ltd.
C 0.6-1.35, bal Fe.
For spiral springs, chisels, punches, shear blades, drills, reamers, taps; water hardened.

EDGAR ALLEN TURNERS SAFE-PLATE
Edgar Allen Balfour Ltd.
C, bal Fe.
For safes; layers of very hard steel. *Obsolete*

EDGAR ALLEN V.S. 4
Edgar Allen Balfour Ltd.
C 1.05, Mn 0.3, V 0.25, bal Fe.
For cold heading dies; Type W2; water hardened.

EDGAR ALLEN WELDING STEEL
Edgar Allen Balfour Ltd.
C 0.3-0.5, bal Fe.
For welding. *Obsolete*

EDGAR ALLEN YWA
Edgar Allen Balfour Ltd.
C 0.28, Si 0.45, Cr 1.3, Ni 3.4, W 5.8, V 0.25, bal Fe.
For extension dies, mandrels; oil hardened, shock resistant.

EDM ALLOY "A"
Teledyne Firth Sterling
Cu, bal W.
For electric discharge machining electrodes. Copper infiltrated tungsten. *Obsolete*

EDM ALLOY "S"
Teledyne Firth Sterling
W 85.91, Ag 9.15.
Sintered: 100,000 transverse strength; 100 Rock B. For EDM electrodes for electric discharge machining. Silver infiltrated tungsten. *Obsolete*

EDM ALLOY "SA"
Teledyne Firth Sterling
W 75, Ag 25.
Sintered: 125,000 TR.S.; 100 Rock B. For electric discharge machining electrodes. Silver infiltrated tungsten. *Obsolete*

EDWARDS SPECULUM
German manufacture
Cu 63-70, Sn 25-32, As 1.6-2.4, Zn 0-2.6.
For reflectors, mirrors; corrosion resistant.

EE
Now DARWIN EE.

EEL BRAND ANTIFRICTION METAL
Murex Ltd.
Pb 75, Sn 6, Cu 0.9, Cd 1.5, Ni 1, P 0.1, As 0.5, Sb.
For bearings, antifriction metal. Babbitt.

EGALITE
Egal Metal Products Co.
Aluminum. Al alloy.
For light alloy parts; subject to special process while in molten state.

EH 0000
W. Ossenberg & Cie Edelstahlwerke
C 1.2, Cr 4.5, Mo 1, W 12, V 4, bal Fe.
High speed steel, usually for finishing tools. W.-Nr. 1.3302.

EH 0018
W. Ossenberg & Cie Edelstahlwerke
C 0.75, Cr 4.5, W 18, V 1.1, bal Fe.
High speed steel; for lathe tools, thread cutters, broaches, hobs. W.-Nr. 1.3355.

EH 10
Electrometal SA Metals Especials
C 0.4, Si 1, Mn 0.55, Cr 3.25, Mo 2.5, V 0.32, bal Fe.
Hot work tool steel. AISI H10; W.-Nr. 1.2365.

EH 11
Electrometal SA Metals Especials
C 0.4, Si 0.85, Mn 0.35, Cr 5.25, Mo 1.5, V 0.35, bal Fe.
Hot work tool steel. AISI H11; W.-Nr. 1.2343.

EH 12
Electrometal SA Metals Especials
C 0.3, Si 0.85, Mn 0.35, Cr 5.25, Mo 1.5, V 0.25, W 1.2, bal Fe.
Hot work tool steel. W.-Nr. 1.2606; similar to AISI H12.

EH 13
Electrometal SA Metals Especials
C 0.4, Si 0.95, Mn 0.4, Cr 5.25, Mo 1.4, V 0.95, bal Fe.
Hot work tool steel. AISI H13; W.-Nr. 1.2344.

EH 20
Electrometal SA Metals Especials
C 0.3, Si 0.3, Mn 0.4, Cr 2.75, W 9.1, V 0.3, bal Fe.
Tungsten hot work tool steel. W.-Nr. 1.2581.

EH-4
Polish manufacture
C 1, Si 1, Cr 4, bal Fe.
For magnets.

EHRHARDS B.M
English manufacture
Zn 84.4, Sn 0.2, Cu 10.9, Pb 1.2, Al 2.5.
For bearings; will not resist heat or live steam.

EHRHARDS TYPE
English manufacture
Zn 89, Cu 3, Sn 6, Pb 2.
For type metal.

EHRHARDT'S METAL
German manufacture
Zn 89, Cu 4, Sn 4, Pb 3.
For bearings, bushings; Babbitt.

EI 435
Russian manufacture
C 0.1, Cr 20, Ti 0.2, bal Ni.
Heat treated: 103,000 TS; 60,000 YS; 50 El; 50 RA. For gas turbine and jet engine components, combustion chambers; heat and oxidation resistant.

EI 437
Russian manufacture
C, alloy, bal Fe.
For machine tool parts; oil hardened.

EI-435
Russian manufacture
C 0-0.12, Cr 21, Ti 0.25, Al 0-0.15, bal Ni.
Wrought, high temperature, oxidation resistant nickel-base alloy for tailpipes and parts of combustion equipment.

EI-437
Russian manufacture
C 0.03-0.05, Cr 20-21, Ti 2.5-2.8, Al 0.8-1, bal Ni.
Heat treated: 147,000 TS; 41 El; 35 RA; 35 Rock C. For turbine buckets and discs, rotor blades; high creep strength, high heat resistant, similar to Nimonic 80A.

EI-437A
Russian manufacture
C 0.08, Cr 21, Ti 2.5, Al 0.75, bal Ni.
For aircraft rotating turbine blades.

EI-437B
Russian manufacture
C 0.08, Cr 20, Al 0.7, Ti 2.6, B 0.005, bal Ni.
Heat treated: 132,000 TS; 80,000 YS; 45 El; 36 RA. At 800°C: 62,000 TS; 53,000 YS; 8 El; 10 RA. For turbine buckets and discs; high creep strength, boron modified Nimonic 80A, high heat and corrosion resistant.

EI-437R
Russian manufacture
C 0-0.08, Cr 18.5, Ti 2.5, Al 0.7, W 4.5, Mo 4.5, Ce 0-0.01, Fe 0-4, B 0-0.01, bal Ni.
For turbine blades.

EI-559A
Russian manufacture
C 0-0.1, Cr 16.5, Al 3.2, Ni 55-60, bal Fe.
Wrought, high temperature, oxidation resistant nickel-base alloy for combustion equipment.

EI-598
Russian manufacture
C 0-0.12, Cr 17.5, Ti 2.4, Al 1.4, W 2.8, Mo 5, B 0-0.01, Ce 0-0.02, Cb 0.9, bal Ni.
For aircraft turbine blades.

EI-602
Russian manufacture
C 0-0.08, Cr 20.5, Ti 0.55, Al 0.55, Mo 2, Cb 1.5, Fe 0-3, Cu 0-0.2, bal Ni.
Wrought, high temperature, oxidation resistant nickel-base alloy for combustion equipment.

EI-607
Russian manufacture
C 0-0.08, Cr 16, Ti 2, Al 0.7, Cb 1.3, Fe 0-3, bal Ni.
For turbine blades.

EI-607A
Russian manufacture
C 0-0.08, Cr 16, Ti 1.6, Al 0.7, Cb 1.3, Fe 0-3, bal Ni.
For turbine blades.

EI-612K
Russian manufacture
C 0.09, Cr 15, Ni 36.5, Co 4.1, W 3.2, Ti 1.5, B 0.12.
For gas turbine discs, jet engine components.

EI-617
Russian manufacture
C 0-0.12, Cr 15, Ti 2, Al 2, W 6, Mo 3, B 0.005, V 0.3, bal Ni.
For turbine blades.

EI-652
Russian manufacture
C 0.05, Cr 27, Al 3, Fe 0-2, bal Ni.
Wrought, high temperature, oxidation resistant nickel-base alloy for combustion equipment.

EI-747
Russian manufacture
C 0.15, Cr 11.2, Mo 0.6, Ni 0.6, V 0.3, bal Fe.
Corrosion resistant structural steel.

EI-755
Russian manufacture
C 0.13, Cr 10.85, Mo 0.75, Ni 0.3, V 0.09, W 0.2, bal Fe.
Corrosion resistant structural steel.

EI-756
Russian manufacture
C 0.1-0.15, Cr 10.5-12, Mo 0.7, Ni 0-0.8, V 0.25, W 3.7-4.3, bal Fe.
Corrosion resistant structural steel.

EI-765
Russian manufacture
C 0-0.12, Cr 14.5, Ti 1.2, Al 2, W 5, Mo 4, B 0.008, Fe, bal Ni.
For turbine blades.

EI-802
Russian manufacture
C 0.11-0.18, Cr 11-13, Mo 0.5, Ni 0.5-1, V 0.15-0.3, W 0.7-1, bal Fe.
Annealed: 85,000 TS; 40,000 YS; 22 El; 95 Rock B. For cutlery, valves, bearings, surgical instruments; corrosion resistant, hardenable.

EI-826
Russian manufacture
C 0-0.12, Cr 14.5, Ti 2, Al 2.6, W 6, Mo 3.2, B 0-0.01, Ce 0-0.02, Fe 0-5, V 0.03, bal Ni.
For turbine blades.

EI-827
Russian manufacture
C 0-0.47, Cr 10, Al 4.3, W 5, Mo 7.5, Fe 0-4, bal Ni.
High temperature alloy.

EI-867
Russian manufacture
C 0-0.1, Cr 9, Al 4.5, W 5, Mo 10.3, B 0-0.02, Ce 0-0.02, Fe 0-4, Co 5, bal Ni.
Wrought nickel-base superalloy; for turbine blades.

EI-868
Russian manufacture
C 0-0.1, Cr 25.5, Ti 0.5, Al 0-0.5, W 15, Fe 0-4, bal Ni.
Wrought, high temperature, oxidation resistant nickel-base alloy for combustion equipment.

EI-869
Russian manufacture
C 0-0.08, Cr 15.5, Ti 1.7, Al 1.2, B 0-0.005, Cb 1.3, Fe 0-3, Zr 0-0.005, bal Ni.
Wrought nickel-base superalloy; for turbine blades.

EI-893
Russian manufacture
C 0-0.08, Cr 16, Ti 1.4, Al 1.4, W 9, Mo 4.2, B 0-0.01, Ce 0-0.025, bal Ni.
Wrought nickel-base superalloy; for turbine blades.

EI-894
Russian manufacture
C 0.09, Cr 22.5, Ti 1.1, Al 3.1, W 5.7, Fe 9.7, bal Ni.
Wrought nickel-base superalloy.

EI-929
Russian manufacture
C 0-0.12, Cr 10.5, Ti 1.7, Al 4, W 5.5, Mo 5, B 0-0.02, Fe 0-5, Co 14, V 0.6, Ba 0-0.1, bal Ni.
Wrought nickel-base superalloy; for turbine blades.

EI-961
Russian manufacture
C 0.1-0.16, Cr 10.5-12.5, Mo 0.35-0.5, Ni 1.5-1.8, V 0 18-0.3, W 1.5-2, bal Fe.
Corrosion resistant structural steel.

EI-993
Russian manufacture
C 0.2, Cr 12, Cb 0.4, Mo 0.04, Ni 0.1, V 0.3, W 0.6, bal Fe.
Annealed: 95,000 TS; 42,000 YS; 24 El; 95 Rock B. Hardened: 240,000 TS; 195,000 YS; 10 El; 48 Rock C. For bearings, cutlery, hardware, surgical instruments, valves; corrosion resistant, hardenable.

EIFFEL 32/32T
Societe Nouvelle des Acieries de Pompey
C 0.15, Mn 1.1, Si 0.15, bal Fe.
44-54 kg/mm^2 TS; 32 kg/mm^2 YS min; 25 El. For beams, frames, chains.

EIFFEL 36/36T
Societe Nouvelle des Acieries de Pompey
C 0.18, Mn 1.25, Si 0.2, bal Fe.
52-62 kg/mm^2 TS; 36 kg/mm^2 YS min; 24 El. For beams, frames, chains.

EIFFEL 42/42T
Societe Nouvelle des Acieries de Pompey
C 0.16, Mn 1.5, Si 0.15, V 0.1, bal Fe.
50-65 kg/mm^2 TS; 42 kg/mm^2 YS min; 20 El. For beams, frames, chains.

EIFFEL 48/48T
Societe Nouvelle des Acieries de Pompey
C 0.17, Mn 1.55, Si 0.2, V 0.12, Nb 0.03, bal Fe.
60-80 kg/mm^2 TS; 48 kg/mm^2 YS min; 18 El. For beams, frames, chains.

EIFFEL 55T
Societe Nouvelle des Acieries de Pompey
C 0.17, Mn 1.55, Si 0.2, Mo 0.3, V 0.1, bal Fe.
60-80 kg/mm^2 TS; 55 kg/mm^2 YS min; 16 El. For beams, frames, chains.

EIFFEL 60T
Societe Nouvelle des Acieries de Pompey
C 0.18, Mn 1.6, Si 0.2, Mo 0.4, V 0.1, bal Fe.
Plate: 63-85 kg/mm^2 TS; 58 kg/mm^2 YS min; 15 El. For beams, frames, crane booms, girders.

EIGHTEEN PER CENT
English manufacture
Cu 65, Zn 17, Ni 18.
For base for silver plated flatware; nickel silver.

EINHEITSMETALL
German manufacture
Pb 79, Sb 14, Sn 5.3, Cu 1.5.
For bearings, bushings; Babbitt.

EIS 718
Heppenstall Co.
Tool material. C 0.35, Cr 4.95, W 1.25, Mo 1.5, Si 1, bal Fe.
For hot work dies, shears, punches; oil or air hardened.

EIS 77

Heppenstall Co.
Tool material. C 0.35, Cr 4, W 12, V 0.25, bal Fe.
For hot work dies, shears, punches; oil or air hardened.

EIS 96

Heppenstall Co.
Tool material. C 0.55, Mn 0.8, Cr 1, Mo 0.45, V 0.08, bal Fe.
For plastic molding dies; oil hardened.

EIS-H-721

Heppenstall Co.
Tool material. C 0.35, Mn 0.35, Cr 5, Mo 1.55, W 1.25, Si 1, V 0.2, bal Fe.
Air hardened: 550 Brin. For dies, inserts, punches, liners, rams; hot work, press, upsetters, extrusions.

EIS-H5

Heppenstall Co.
Tool material. C 0.55, Cr 1, Mo 0.45, V 0.08, Mn 0.8, bal Fe.
For plastic mold dies; oil hardened.

EIS-H720

Heppenstall Co.
Tool material. C 0.4, Cr 5.25, Mo 1.15, V 1, bal Fe.
For die casting dies and inserts; Type H11; air hardened.

EIS-R-718

Heppenstall Co.
Tool material. C 0.35, Cr 5, Mo 2, Si 1, bal Fe.
Air hardened: 550 Brin. For dies, inserts, punches; hot work, press, upsetters, extrusions.

EIS-R-720

Heppenstall Co.
Tool material. C 0.4, Mn 0.4, Cr 5.25, Mo 1.15, V 1, bal Fe.
Air hardened: 560 Brin. For dies and inserts; die casting.

EIS-T-71

Heppenstall Co.
Tool material. C 0.5, Mn 0.3, Cr 1.2, Ni 2.4, Si 1, bal Fe.
Oil hardened: 560 Brin. For shear knives, punches; cold heavy duty, cold and semi-hot.

EIS-T721

Heppenstall Co.
Tool material. C 0.3, Cr 5, Mo 1.5, W 1.25, V 0.2, bal Fe.
For die casting dies; oil hardened; hot work steel.

EIS-T77

Heppenstall Co.
Tool material. C 0.33, Co 4, W 12, V 0.25, Mn 0.3, bal Fe.
For hot shearing dies, extrusion liners; oil hardened; hot work steel.

EIS-V2

Heppenstall Co.
Tool material. C 0.8, Mn 0.4, V, Si, bal Fe.
Water hardened: 600 Brin. For shear knives; cold heavy duty.

EIS-V3

Heppenstall Co.
Tool material. C 0.95, Mn 0.4, Si, V, bal Fe.
Water hardened: 620 Brin. For shear knives, trimmers; cold light shearing.

EISENBRONZE-1

German manufacture
Cu 57.5, Zn 39.5, Sn 1, Pb 0.3, Fe 1.3, Al 0.3.
For marine parts, hardware; iron bronze.

EISENBRONZE-2

German manufacture
Cu 82.5, Zn 4.45, Sn 8.55, Fe 3.95.
For marine parts, hardware; iron bronze.

EISENNICKEL 75

VDM Nickel-Technologie AG
Ni 78.5, bal Fe.
For magnetic and electrical equipment, motors; soft magnet, high permeability. *Obsolete*

EISLERS BRONZE

American manufacture
Zn 6, bal Cu.
For electrical contacts, springs; good conductivity.

EK-81

Cytemp Specialty Steel Div.
C 1.35, Cr 12.75, Mo 0.8, Co 3, bal Fe.
For blanking and forming dies, punches; air hardening.
Obsolete

EK15

Russian manufacture
C 0.9-1, Cr 7.5-8.5, Co 14-16, Mo 1.5, bal Fe.
For permanent magnet; heat treated.

EK30A

Various foundries
Zr 0.3, 3.0 Rare Earths, bal Mg.
T6-Temper: 20,000-23,000 TS; 14,000-16,000 YS; 2-3 El. At 600°F: 12,000 TS; 8,000 YS; 70 El. Magnesium sand casting; good pressure tightness, weldability and corrosion resistance. For parts operating at 350-500°F. ASTM B80-57T.

EK41A

Various foundries
Zr 0.6, 4.0 Rare Earths, bal Mg.
T6-Temper: 22,000-25,000 TS; 16,000-18,000 YS; 1-3 El. At 600°F: 13,000 TS; 9,000 YS; 53 El. Magnesium sand casting with good pressure tightness, weldability and corrosion resistance. For parts operating at 350-600°F. ASTM B80-57T, B199-57T; ASM 4440, 4441.

EK5

Russian manufacture
C 0.9-1, Cr 5.5-6.5, Co 5.5-6.5, bal Fe.
For permanent magnet; heat treated.

EKATIT 1013

Ekatit Stahl GmbH
Stainless steel. C 0.08-0.12, Si 0-1, Mn 0-1, Cr 12-14, bal Fe.
Chromium stainless steel. Werkstoff Nr. 1.4006.

EKATIT 1217

Ekatit Stahl GmbH
Stainless steel. C 0.1-0.17, Si 0-1, Mn 0-1.5, Cr 15.5-17.5, Mo 0.2-0.3, bal Fe.
Stainless steel. Werkstoff Nr. 1.4104.

EKATIT 13

Ekatit Stahl GmbH
Stainless steel. C 0-0.08, Si 0-1, Mn 0-1, Cr 12-14, bal Fe.
Ferritic type stainless steel. Werkstoff Nr. 1.4000.

EKATIT 13 A1

Ekatit Stahl GmbH
Stainless steel. C 0-0.08, Si 0-1, Mn 0-1, Cr 12-14, Al 0.1-0.3, bal Fe.
Ferritic type stainless steel; nonhardenable. Werkstoff Nr. 1.4002; similar to AISI 405.

EKATIT 14

Ekatit Stahl GmbH
Stainless steel. C 0-0.08, Si 0-1, Mn 0-1, Cr 13-15, bal Fe.
Ferritic type stainless steel. Werkstoff Nr. 1.4001.

EKATIT 1513 MO

Ekatit Stahl GmbH
Stainless steel. C 0.12-0.17, Si 0-1, Mn 0-1, Cr 12-14, Mo 1-1.3, bal Fe.
Martensitic stainless steel. Werkstoff Nr. 1.4119.

EKATIT 17

Ekatit Stahl GmbH
Stainless steel. C 0-0.1, Si 0-1, Mn 0-1, Cr 15.5-17.5, bal Fe.
Ferritic type stainless steel. Werkstoff Nr. 1.4016; AISI 430.

EKATIT 17 E

Ekatit Stahl GmbH
Stainless steel. C 0-0.1, Si 0-1, Mn 0-1, Cr 16-18, Ti = 7 x C min, bal Fe.
Ferritic stainless steel. Werkstoff Nr. 1.4510; titanium stabilized.

EKATIT 17 EMO

Ekatit Stahl GmbH
Stainless steel. C 0-0.1, Si 0-1, Mn 0-1, Cr 16.5-18.5, Mo 1.5-2, Ti = 7 x C min, bal Fe.
Ferritic type stainless steel. Werkstoff Nr. 1.4523; titanium stabilized.

EKATIT 17 N

Ekatit Stahl GmbH
Stainless steel. C 0-0.1, Si 0-1, Mn 0-1, Cr 16-18, Nb = 12 x C, bal Fe.
Ferritic type stainless steel. Werkstoff Nr. 1.4511; niobium stabilized.

EKATIT 17 VCO

Ekatit Stahl GmbH
Stainless steel. C 0.85-0.95, Si 0-1, Mn 0-1, Cr 15.5-17.5, Mo 0.4-0.6, Co 1.2-1.8, V 0.2-0.3, bal Fe.
Martensitic stainless steel. Werkstoff Nr. 1.4535.

EKATIT 17-13 S

Ekatit Stahl GmbH
Stainless steel. C 0-0.07, Si 0-1, Mn 0-2, Cr 16-18, Ni 12.5-14.5, Mo 4-5, bal Fe.
Austenitic stainless steel. Werkstoff Nr. 1.4449.

EKATIT 17-7

Ekatit Stahl GmbH
Stainless steel. C 0-0.12, Si 0-1, Mn 0-2, Cr 16-18, Ni 7-9, bal Fe.
Austenitic stainless steel. Werkstoff Nr. 1.4310; similar to AISI 301.

EKATIT 18-10 E

Ekatit Stahl GmbH
Stainless steel. C 0-0.1, Si 0-1, Mn 0-2, Cr 16.5-18.5, Ni 10.5-13.5, Mo 2-2.5, Ti = 5 x C min, bal Fe.
Stabilized austenitic stainless steel. Werkstoff Nr. 1.4571.

EKATIT 18-10 NB

Ekatit Stahl GmbH
Stainless steel. C 0-0.1, Si 0-1, Mn 0-2, Cr 16.5-18.5, Ni 10.5-13.5, Mo 2-2.5, Nb = 8 x C min, bal Fe.
Stabilized stainless steel. Werkstoff Nr. 1.4580; similar to AISI 347.

EKATIT 18-10 S

Ekatit Stahl GmbH
Stainless steel. C 0-0.07, Si 0-1, Mn 0-2, Cr 16.5-18.5, Ni 10.5-13.5, Mo 2-2.5, bal Fe.
Austenitic stainless steel. Werkstoff Nr. 1.4401; similar to AISI 316.

EKATIT 18-10 SW

Ekatit Stahl GmbH
Stainless steel. C 0-0.03, Si 0-1, Mn 0-2, Cr 16.5-18.5, Ni 11-14, Mo 2-2.5, bal Fe.
Austenitic stainless steel. Werkstoff Nr. 1.4404; AISI 316L.

EKATIT 18-12

Ekatit Stahl GmbH
Stainless steel. C 0-0.03, Si 0-1, Mn 0-2, Cr 16.5-18.5, Ni 12.5-15, Mo 2.5-3, bal Fe.
Austenitic stainless steel. Werkstoff Nr. 1.4435.

EKATIT 18-12 E

Ekatit Stahl GmbH
Stainless steel. C 0-0.1, Si 0-1, Mn 0-2, Cr 16.5-18.5, Ni 12-14.5, Mo 2.5-3, Ti = 5 x C min, bal Fe.
Stabilized austenitic stainless steel. Werkstoff Nr. 1.4573.

EKATIT 18-12 S

Ekatit Stahl GmbH
Stainless steel. C 0-0.07, Si 0-1, Mn 0-2, Cr 16.5-18.5, Ni 11.5-14, Mo 2.5-3, bal Fe.
Austenitic stainless steel. Werkstoff Nr. 1.4436.

EKATIT 18-16

Ekatit Stahl GmbH
Stainless steel. C 0-0.03, Si 0-1, Mn 0-2, Cr 17-19, Ni 15-17, Mo 3-4, bal Fe.
Austenitic stainless steel. Werkstoff Nr. 1.4438.

EKATIT 18-16

Ekatit Stahl GmbH
Stainless steel. C 0-0.03, Si 0-0.5, Mn 0-0.9, Cr 15.5-16.5, Ni 17.5-18.5, bal Fe.
Stainless steel. Werkstoff Nr. 1.4321.

EKATIT 18-8 A

Ekatit Stahl GmbH
Stainless steel. C 0-0.15, Si 0-1, Mn 0-2, Cr 17-19, Ni 8-10, bal Fe.
Austenitic stainless steel. Werkstoff Nr. 1.4305; AISI 302.

EKATIT 18-8 E

Ekatit Stahl GmbH
Stainless steel. C 0-0.1, Si 0-1, Mn 0-2, Cr 17-18, Ni 9-11.5, Ti = 5 x C min, bal Fe.
Titanium stabilized austenitic stainless steel. Werkstoff Nr. 1.4541; similar to AISI 321.

EKATIT 18-8 N

Ekatit Stahl GmbH
Stainless steel. C 0-0.07, Si 0-1, Mn 0-2, Cr 17-20, Ni 9-11.5, Mo 0-0.2, Nb = 10 x C min, bal Fe.
Niobium stabilized austenitic stainless steel. Werkstoff Nr. 1.4543; similar to AISI 347.

EKATIT 18-8 N

Ekatit Stahl GmbH
Stainless steel. C 0-0.12, Si 0-1, Mn 0-2, Cr 17-19, Ni 8-10, bal Fe.
Austenitic stainless steel. Werkstoff Nr. 1.4300; similar to AISI 302.

EKATIT 18-8 NB

Ekatit Stahl GmbH
Stainless steel. C 0-0.1, Si 0-1, Mn 0-2, Cr 17-19, Ni 9-11.5, Nb = 8 x C min, bal Fe.
Niobium stabilized austenitic stainless steel. Werkstoff Nr. 1.4550; similar to AISI 347.

EKATIT 18-8 S

Ekatit Stahl GmbH
Stainless steel. C 0-0.07, Si 0-1, Mn 0-2, Cr 17-20, Ni 8.5-10, bal Fe.
Austenitic stainless steel. Werkstoff Nr. 1.4301; similar to AISI 304.

EKATIT 18-8 SW

Ekatit Stahl GmbH
Stainless steel. C 0-0.03, Si 0-1, Mn 0-2, Cr 17-20, Ni 10-12.5, bal Fe.
Austenitic stainless steel. Werkstoff Nr. 1.4306; similar to AISI 304L.

EKATIT 20-18

Ekatit Stahl GmbH
Stainless steel. C 0-0.07, Si 0-1, Mn 0-2, Cr 16.5-18.5, Ni 19-21, Mo 2-2.5, Cu 2.2, Nb = 8 x C min, bal Fe.
Austenitic stainless steel. Werkstoff Nr. 1.4505.

EKATIT 20-18 T

Ekatit Stahl GmbH
Stainless steel. C 0-0.07, Si 0-1, Mn 0-2, Cr 16.5-18.5, Ni 19-21, Mo 2-2.5, Cu 2.2, Ti = 7 x C min, bal Fe.
Austenitic stainless steel. Werkstoff Nr. 1.4506.

EKATIT 2013

Ekatit Stahl GmbH
Stainless steel. C 0.17-0.22, Si 0-1, Mn 0-1, Cr 12-14, bal Fe.
Martensitic stainless steel. Werkstoff Nr. 1.4021; similar to AISI 420.

EKATIT 22-18

Ekatit Stahl GmbH
Stainless steel. C 0-0.07, Si 0-1, Mn 0-2, Cr 16.5-18.5, Ni 21.5-23.5, Mo 3-3.5, Cu 2, Nb = 8 x C min, bal Fe.
Stabilized stainless steel. Werkstoff Nr. 1.4586.

EKATIT 2217

Ekatit Stahl GmbH
Stainless steel. C 0.15-0.23, Si 0-1, Mn 0-1, Cr 16-18, Ni 1.5-2.5, bal Fe.
Martensitic stainless steel. Werkstoff Nr. 1.4057; similar to AISI 431.

EKATIT 25-12 NB

Ekatit Stahl GmbH
Stainless steel. C 0-0.1, Si 0-1, Mn 0-2, Cr 16.5-18.5, Ni 12-14.5, Mo 2.5-3, Nb = 8 x C min, bal Fe.
Stabilized stainless steel. Werkstoff Nr. 1.4583.

EKATIT 25-25

Ekatit Stahl GmbH
Stainless steel. C 0-0.07, Si 0-1, Mn 0-2, Cr 24-26, Ni 24-26, Mo 2-2.5, Ti = 10 x C min, bal Fe.
Stabilized stainless steel. Werkstoff Nr. 1.4577.

EKATIT 25-7

Ekatit Stahl GmbH
Stainless steel. C 0-0.06, Si 0-1, Mn 0-1.5, Cr 24-26, Ni 6.5-7.5, Mo 1.3-2, Nb = 10 x C min, bal Fe.
Stabilized stainless steel. Werkstoff Nr. 1.4582.

EKATIT 27-5

Ekatit Stahl GmbH
Stainless steel. C 0-0.1, Si 0-1, Mn 0-2, Cr 26-28, Ni 4-5, Mo 1.3-2, 0.10 N (optional), bal Fe.
Stainless steel. Werkstoff Nr. 1.4460.

EKATIT 330

Ekatit Stahl GmbH
Stainless steel. C 0.17-0.22, Si 0-1, Mn 0-1, Cr 12-14, Mo 1-1.3, Ni 0-1, bal Fe.
Martensitic stainless steel. Werkstoff Nr. 1.4120.

EKATIT 3517

Ekatit Stahl GmbH
Stainless steel. C 0.33-0.43, Si 0-1, Mn 0-1, Cr 15.5-17.5, Mo 1-1.3, Ni 0-1, bal Fe.
Martensitic stainless steel. Werkstoff Nr. 1.4122.

EKATIT 4013

Ekatit Stahl GmbH
Stainless steel. C 0.4-0.5, Si 0-1, Mn 0-1, Cr 12-14, bal Fe.
Martensitic stainless steel. Werkstoff Nr. 1.4034.

EKATIT 4024

Ekatit Stahl GmbH
Stainless steel. C 0.12-0.17, Si 0-1, Mn 0-1, Cr 12-14, bal Fe.
Stainless steel. Werkstoff Nr. 1.4024.

EKATIT 4108

Ekatit Stahl GmbH
Stainless steel. C 1-1.1, Si 0-1, Mn 0-1, Cr 12-14, Mo 0.4-0.6, bal Fe.
Martensitic stainless steel. Werkstoff Nr. 1.4108.

EKATIT 4110

Ekatit Stahl GmbH
Stainless steel. C 0.5-0.6, Si 0-1, Mn 0-1, Cr 13-15, Mo 0.5-0.6, bal Fe.
Martensitic stainless steel. Werkstoff Nr. 1.4110.

EKATIT 4113

Ekatit Stahl GmbH
Stainless steel. C 0-0.07, Si 0-1, Mn 0-1, Cr 16-18, Mo 0.9-1.2, bal Fe.
Ferritic stainless steel. Werkstoff Nr. 1.4113.

EKATIT 9018

Ekatit Stahl GmbH
Stainless steel. C 0.85-0.95, Si 0-1, Mn 0-1, Cr 17-19, Mo 1-1.3, V 0.07-0.12, bal Fe.
Martensitic stainless steel. Werkstoff Nr. 1.4112; similar to AISI 440C.

EKATIT G 1

Ekatit Stahl GmbH
Stainless steel. C 0.18-0.25, Si 0-1, Mn 0-1, Cr 12.5-14.5, bal Fe.
Martensitic stainless steel casting. Werkstoff Nr. 1.4027.

EKATIT G 10

Ekatit Stahl GmbH
Stainless steel. C 0-0.08, Si 0-1.5, Mn 0-2, Cr 19-21, Ni 24-26, Mo 2.5-3.5, Cu 2.5, 1.0 Nb optional, bal Fe.
Stainless steel casting. Werkstoff Nr. 1.4500.

EKATIT G 11

Ekatit Stahl GmbH
Stainless steel. C 0.5-0.9, Si 0-2, Mn 0-1, Cr 27-29, bal Fe.
Stainless steel casting. Werkstoff Nr. 1.4085.

EKATIT G 17-13 S

Ekatit Stahl GmbH
Stainless steel. C 0-0.07, Si 0-1, Mn 0-2, Cr 16-18, Ni 12.5-14.5, Mo 4-5, bal Fe.
Austenitic stainless steel casting. Werkstoff Nr. 1.4448.

EKATIT G 18-10 N

Ekatit Stahl GmbH
Stainless steel. C 0-0.08, Si 0-1.5, Mn 0-1.5, Cr 17-19.5, Ni 10.5-12.5, Mo 2-2.5, Nb = 8 x C min, bal Fe.
Stabilized austenitic stainless steel casting. Werkstoff Nr. 1.4581.

EKATIT G 18-10 S

Ekatit Stahl GmbH
Stainless steel. C 0-0.07, Si 0-2, Mn 0-1.5, Cr 17-19.5, Ni 10-12, Mo 2-2.5, bal Fe.
Austenitic stainless steel casting. Werkstoff Nr. 1.4408.

EKATIT G 18-12 NB

Ekatit Stahl GmbH
Stainless steel. C 0-0.1, Si 0-1.5, Mn 0-1.5, Cr 16.5-18.5, Ni 12-14.5, Mo 2.5-3, Nb = 8 x C min, bal Fe.
Stabilized austenitic stainless steel casting. Werkstoff Nr. 1.4583.

EKATIT G 18-8 N

Ekatit Stahl GmbH
Stainless steel. C 0-0.08, Si 0-1.5, Mn 0-1.5, Cr 17.5-20, Ni 9-11, Nb = 8 x C min, bal Fe.
Stabilized austenitic stainless steel casting. Werkstoff Nr. 1.4552.

EKATIT G 18-8 S

Ekatit Stahl GmbH
Stainless steel. C 0-0.07, Si 0-2, Mn 0-1.5, Cr 17.5-20, Ni 9-11, bal Fe.
Austenitic stainless steel casting. Werkstoff Nr. 1.4308.

EKATIT G 2

Ekatit Stahl GmbH
Stainless steel. C 0.2-0.27, Si 0-1, Mn 0-1, Cr 16-18, Ni 1-2, bal Fe.
Martensitic stainless steel casting. Werkstoff Nr. 1.4059.

EKATIT G 3
Ekatit Stahl GmbH
Stainless steel. C 0.9-1.3, Si 0-2, Mn 0-1, Cr 27-30, bal Fe.
Stainless steel casting. Werkstoff Nr. 1.4086.

EKATIT G 4
Ekatit Stahl GmbH
Stainless steel. C 0.5-0.9, Si 0-2, Mn 0-1, Cr 27-30, Mo 2-2.5, bal Fe.
Stainless steel casting. Werkstoff Nr. 1.4136.

EKATIT G 5
Ekatit Stahl GmbH
C 0.9-1.3, Si 0-2, Mn 0-1, Cr 27-29, Mo 2-2.5, bal Fe.
Stainless steel casting. Werkstoff Nr. 1.4138.

EKATIT G 6
Ekatit Stahl GmbH
Stainless steel. C 0-0.12, Si 0-2, Mn 0-1.5, Cr 17-19.5, Ni 8-10, bal Fe.
Austenitic stainless steel casting. Werkstoff Nr. 1.4312.

EKATIT G 7
Ekatit Stahl GmbH
Stainless steel. C 0.3-0.5, Si 0-2, Mn 0-1.5, Cr 26-28, Ni 3.5-5.5, bal Fe.
Stainless steel casting. Werkstoff Nr. 1.4340.

EKATIT G 8
Ekatit Stahl GmbH
Stainless steel. C 0-0.12, Si 0-2, Mn 0-1.5, Cr 17-19.5, Ni 9-11, Mo 2-2.5, bal Fe.
Austenitic stainless steel casting. Werkstoff Nr. 1.4410.

EKATIT G 9
Ekatit Stahl GmbH
Stainless steel. C 0-0.08, Si 0-1.5, Mn 0-2, Cr 16.5-18.5, Ni 19-21, Mo 2-2.5, Cu 2.4, 1.0 Nb optional, bal Fe.
Stainless steel casting. Werkstoff Nr. 1.4585.

EKH3
Russian manufacture
C 0.9-1.1, Mn 0-0.4, Cr 2.8-3.8, bal Fe.
For permanent magnet; heat treated.

EKL
W. Ossenberg & Cie Edelstahlwerke
C 0.6, Mn 0.3, Si 0.6, Cr 1.1, W 2, V 0.2, bal Fe.
Hot forging dies for light metals. Cold punches for sheet metal. W.-Nr. 1.2550.

EKL SPEZIAL
W. Ossenberg & Cie Edelstahlwerke
C 0.6, Mn 0.3, Si 0.7, Cr 1.1, W 3, V 0.2, bal Fe.
Hot or cold work tool steel. Hot shears, punches, hot stamps. Cold stamping and forming.

EL 3
Electrometal SA Metals Especials
C 0.85, Si 0.25, Mn 0.3, Cr 1.7, Mo 0.12, V 0.1, bal Fe.
Special purpose tool steel. W.-Nr. 1.2237; similar to AISI L3. Obsolete

ELASTIC
Haeckerstahl GmbH
C 0.5, Cr 1, V 0.09, bal Fe.
For gears, springs, bolts, studs, countershafts; oil hardened, shock resistant.

ELASTIC 12
J.M. Ney Co.
Au 15, Pt 9, Pd 37.
Platinum color clasp and orthodontic wire. Fusing temperature: 2010°F.

ELASTIC NO. 3
J.M. Ney Co.
84 Au-Pt, Ag, Zn, bal Cu.
Soft: 109,500 TS; 70,500 YS; 17 El. Heat treated: 146,000 TS; 113,500 YS; 11 El. For wire clasps, cribs, tooth crowns, pinlays; M.P. 1865 F. Obsolete

ELASTIC NO. 4
J.M. Ney Co.
Au 81, Pt, Ag, Zn, bal Cu.
Soft: 117,500 TS; 86,500 YS; 15 El. Heat treated: 173,000 TS; 131,500 YS; 7 El. For dental wire clasps, orthopedic appliances; MP 1925°F. Obsolete

ELASTIQUE SPRING
Republic Steel Corp.
C 0.4-0.6, Mn 0.6-1.3, Si 0.1-2, bal Fe.
For springs. Obsolete

ELASTUF 44
Horace T. Potts Co.
C 0.5, Cr 0.8, Ni 1.5, Mo 0.2, bal Fe.
Heat treated: 185,000-220,000 TS; 160,000-200,000 YS; 10-17 El; 30-50 RA; 388-460 Brin. For gears, shafts, compression dies; preheat treated bars. Obsolete

ELASTUF CHRO-MOLY (MODIFIED)
Horace T. Potts Co.
C 0.5, Cr 0.9, Mo 0.2, bal Fe.
For gears, shafts; oil hardened. Obsolete

ELASTUF JJ
Horace T. Potts Co.
C 0.18-0.23, bal Fe.
Hot rolled: 55,000-70,000 TS; 30,000-45,000 YS; 25-35 El; 45-60 RA; 150-180 Brin. For forgings, welded parts, carburized parts; uniform quality controlled. Obsolete

ELASTUF MEDIA PRECISION FINISH
Brown-Wales Co.
C 0.35-0.45, bal Fe.
As rolled: 70,000-90,000 TS; 35,000-55,000 YS; 15-25 El; 30-45 RA; 170-210 Brin. For general purposes, shafts, spindles; tough. Obsolete

ELASTUF MEDIA PRECISION FINISH
Horace T. Potts Co.
C 0.35-0.45, bal Fe.
As rolled: 70,000-90,000 TS; 35,000-55,000 YS; 15-25 El; 30-45 RA; 170-210 Brin. For general purposes, shafts, spindles; tough. Obsolete

ELASTUF MEDIA STEEL
Brown-Wales Co.
C 0.4-0.5, bal Fe.
Hot rolled: 75,000-95,000 TS; 40,000-60,000 YS; 16-25 El; 30-45 RA; 170-210 Brin. For general purposes, machine parts; tough, controlled quality. Obsolete

ELASTUF MEDIA STEEL
Horace T. Potts Co.
C 0.4-0.5, bal Fe.
Hot rolled: 75,000-95,000 TS; 40,000-60,000 YS; 16-25 El; 30-45 RA; 170-210 Brin. For general purposes, machine parts; tough, controlled quality. Obsolete

ELASTUF PB
Horace T. Potts Co.
C 0.5, Mn 0.92, Cr 0.93, Mo 0.2, Pb 0.15-0.35, bal Fe.
Hard: 150,000 TS; 115,000 YS; 17 El; 52 RA; 302 Brin. For gears, shafting; free-cutting, preheat treated.

ELASTUF PENN MACHINERY STEEL
Horace T. Potts Co.
C 0.4-0.48, Mn 1.25, bal Fe.
Hot rolled: 100,000-125,000 TS; 50,000-75,000 YS; 16-15 El; 32-30 RA; 197-217 Brin. For spindles, lead screws, shafts, chuck and jack screws, bolts, arbors, gears; tough, excellent machinability.

ELASTUF PENN MACHINERY STEEL
Brown-Wales Co.
C 0.4-0.48, Mn 1.25, bal Fe.
Hot rolled: 100,000-125,000 TS; 50,000-75,000 YS; 16-15 El; 32-30 RA; 197-217 Brin. For spindles, lead screws, shafts, chuck and jack screws, bolts, arbors, gears; tough, excellent machinability.

ELASTUF SAE 2335
Brown-Wales Co.
C 0.3-0.4, Ni 3.5, bal Fe.
Hot rolled: 90,000-100,000 TS; 55,000-65,000 YS; 18-28 El; 42-52 RA; 170-212 Brin. For machine parts; tough. Obsolete

ELASTUF SAE 2335
Horace T. Potts Co.
C 0.3-0.4, Ni 3.5, bal Fe.
Hot rolled: 90,000-100,000 TS; 55,000-65,000 YS; 18-28 El; 42-52 RA; 170-212 Brin. For machine parts; tough. Obsolete

ELASTUF SAE 3140
Brown-Wales Co.
C 0.35-0.45, Cr, Ni, bal Fe.
Heat treated: 115,000-135,000 TS; 95,000-115,000 YS; 16-22 El; 45-55 RA; 262-320 Brin. For shafts, axles, bolts, studs, gears, machine parts; tough. Obsolete

ELASTUF SAE 3140
Horace T. Potts Co.
C 0.35-0.45, Cr, Ni, bal Fe.
Heat treated: 115,000-135,000 TS; 95,000-115,000 YS; 16-22 El; 45-55 RA; 262-320 Brin. For shafts, axles, bolts, studs, gears, machine parts; tough. Obsolete

ELASTUF STRAINFREE
Horace T. Potts Co.
C 0.44, Mn 1.45, P 0.03, S 0.28, bal Fe.
Cold drawn: 130,000 TS; 109,000 YS; 13 El; 32 RA; 248 Brin. For gears, shafting; free-cutting, stress tempered. Obsolete

ELASTUF TYPE A
Horace T. Potts Co.
C 0.45-0.55, Cr 0.9, V 0.2, bal Fe.
Heat treated: 125,000-160,000 TS; 105,000-135,000 YS; 15-22 El; 42-64 RA; 255-340 Brin. For gears, shafts, preheat treated. Obsolete

ELASTUF TYPE A STEEL NO. 14
Horace T. Potts Co.
C 0.3-0.35, Cr, V, bal Fe.
Heat treated: 125,000-135,000 TS; 115,000-125,000 YS; 12-18 El; 59-64 RA; 262-325 Brin. For shafts, gears, worms, pinions, axles, machine parts; strong and tough. Obsolete

ELASTUF TYPE A STEEL NO. 9
Brown-Wales Co.
C 0.45-0.55, Cr 0.9, V 1.2, bal Fe.
Heat treated: 140,000-160,000 TS; 120,000-135,000 YS; 15-20 El; 42-60 RA; 302-340 Brin. For high pressure bolts, gears, worms, arbors, hammer pistons, spindles, armature shafts; heat, wear and shock resisting. Obsolete

ELASTUF TYPE A STEEL NO. 9
Horace T. Potts Co.
C 0.45-0.55, Cr 0.9, V 1.2, bal Fe.
Heat treated: 140,000-160,000 TS; 120,000-135,000 YS; 15-20 El; 42-60 RA; 302-340 Brin. For high pressure bolts, gears, worms, arbors, hammer pistons, spindles, armature shafts; heat, wear and shock resisting. Obsolete

ELASTUF TYPE A2
Horace T. Potts Co.
C 0.5, Mn 0.85, Cr 1.03, Mo 0.35, V 0.07, bal Fe.
Pre-hardened: 125,000 YS min. Average hardness: 321 Brin. For gears, shafts, axles, boring bars, crankshafts.

ELASTUF TYPE K
Horace T. Potts Co.
C, Cr, bal Fe.
For tools. Obsolete

ELASTUF TYPE STEEL NO. 14
Brown-Wales Co.
C 0.3-0.35, Cr, V, bal Fe.
Heat treated: 125,000-135,000 TS; 115,000-125,000 YS; 12-18 El; 59-64 RA; 262-325 Brin. For shafts, gears, worms, pinions, axles, machine parts; strong and tough. Obsolete

ELB

W. Ossenberg & Cie Edelstahlwerke
C 0.5, Mn 0.5, Si 0.2, Cr 1, Ni 3.5, bal Fe.
Cold work tool steel, for heading dies. W.-Nr. 1.2721.

ELB SPEZIAL

W. Ossenberg & Cie Edelstahlwerke
C 0.45, Mn 0.5, Si 0.2, Cr 1.3, Ni 4, Mo 0.2, bal Fe.
Cold work tool steel, for dies, heading tools, shear blades.
W.-Nr. 1.2767.

ELBEBRONZE LB5

Le Bronze Industriel
Zn 36, Al 0.5, Ni 2, Fe 0.5, Mn 2, Pb 0.5, bal Cu.
High tensile brass; cast or wrought: 110-200 Brin; free-machining. AFNOR UZ36 N3.

ELBEBRONZE LBA 3

Le Bronze Industriel
Zn 36, Al 2, Mn 3, Si 0.6, Pb 0.5, bal Cu.
High strength brass; wrought: 165 Brin. AFNOR UZ36 M3
A2S.

ELBEBRONZE LBR 1

Le Bronze Industriel
Zn 34, Al 2.5, Ni 0.4, Fe 0.5, Mn 3.5, bal Cu.
High tensile brass; wrought: 160-195 Brin.

ELBEBRONZE LBU

Le Bronze Industriel
Zn 39, Pb 2, bal Cu.
Free machining brass, wrought: 90-150 Brin. AFNOR UZ39
Pb2.

ELBEBRONZE NAB

Le Bronze Industriel
Zn 39, Sn 1, bal Cu.
Naval brass; wrought. SAE 73.

ELBRODUR B/BN

KM-kabelmetal AG
Copper. Be 0.5, Co 2, bal Cu.
Electrode material for welding, mold tube and plate. Product
forms: sheet, plate, rods, profiles, forgings, finished parts.
Heat treated: 100-128 ksi TS; 83-107 ksi YS; 10-20 El.

ELBRODUR G

KM-kabelmetal AG
Copper. Cr 0.65, Zr 0.1, bal Cu.
Casting molds, crucibles, casting wheels. Product forms:
forgings, finished parts. Heat treated: 58 ksi TS; 43 ksi YS; 20
El.

ELBRODUR HF

KM-kabelmetal AG
Copper. Cr 0.65, Zr 0.08, bal Cu.
Electrode material for welding and contact tubes for welding
heads. Product forms: strip, sheet, plate, rods, profiles, tubes,
wires, finished parts. Heat treated: 71-93 ksi TS; 64-83 ksi YS;
10-20 El.

ELBRODUR N/RS

KM-kabelmetal AG
Copper. Cr 0.65, Zr 0.05, bal Cu.
Electrode material for spot and roller seam welding. Product
forms: strip, sheet, plate, rods, profiles, tubes, wires, forgings,
finished parts. Heat treated: 54-64 ksi TS; 37-51 ksi YS; 18-25
El.

ELBRODUR N4

KM-kabelmetal AG
Copper. Zr 0.15, bal Cu.
Electrode material for welding, electronics, casting wheels.
Product forms: strip, sheet, plate, rods, profiles, tubes, wires,
forgings, finished parts. Heat treated: 48-60 ksi TS; 40-52 ksi
YS; 12 El.

ELCOMET "F"

La Bour Pump Co.
Ni 20, Cr 20, Si 5, Cu 4, Mo 3, Mn 0.3, C 0-0.15, bal Fe.
For valves, pumps, for the rayon industry; corrosion and
abrasion resistant. *Obsolete*

ELCOMET "L"

La Bour Pump Co.
Ni 10, Cr 20, C 0-0.15, bal Fe.
For valves, pumps; corrosion and abrasion resistant.
Obsolete

ELCOMET "M"

La Bour Pump Co.
Ni 20, Cr 20, Si 1, Cu 4, Mo 2-3, Mn 0-0.3, C 0-0.15, bal Fe.
For valves, pumps; corrosion and abrasion resistant.
Obsolete

ELCOMET K

La Bour Pump Co.
Ni 22, Cr 23, Si 1-1.25, Cu 3.5-4, Mo 2, Mn 0.3, C 0-0.12, bal
Fe.
For spinner heads, valves, pumps; resists abrasion and
corrosion.

ELECKTROBRONZE

Ostermann GmbH & Co.
Al 10, Mn, Fe, bal Cu.
For valve gages, fittings; Al-bronze, corrosion resistant.

ELECTALLOY NO. 2

McQuay-Norris Mfg. Co.
Cast Iron. Si 2-2.9, Ni 0.8-1.2, Mo 0.9-1.3, Cr 0.4, GC 2.7-3.2,
CC 0.4-0.7, bal Fe.
Cast: 60,000 TS. For piston rings. Cast iron. Wear resistant.

ELECTALLOY NO. 3

McQuay-Norris Mfg. Co.
Cast Iron. Si 2.1-2.7, Mn 0.5-1.2, Mo 0.6-1.2, Cr 0.8-1.2, TC
2.8-3.2, CC 0.7-1.15, bal Fe.
Cast: 70,000 TS. For piston rings. Cast iron. Wear resistant.

ELECTALLOY NO. 4

McQuay-Norris Mfg. Co.
Cast Iron. Si 0-2.8, Mn 0.7, Ni 1, Mo 0.9, Cr 0.3, TC 3.75-3.85,
bal Fe.
Cast: 50,000 TS. For piston rings. Cast iron. Wear resistant.

ELECTRA JAP

Cytemp Specialty Steel Div.
C 1.2, Mn 0.6, Cr 0.55, bal Fe.
For taps, edge tools; water or oil hardening. *Obsolete*

ELECTRA JAP M

Cytemp Specialty Steel Div.
C 1.2, Mn 0.85, Cr 0.5, Mo 0.6, bal Fe.
For taps, edge tools; water or oil hardening. *Obsolete*

ELECTRAL CB 4

Le Bronze Industriel
Co 2, Be 0.5, bal Cu.
Wrought or cast berylium-copper alloy. AFNOR UK2 Be.

ELECTRAL CD 1

Le Bronze Industriel
Cd 1, bal Cu.
Wrought cadmium-copper. Good electrical conductivity.

ELECTRAL CRM 16M

Le Bronze Industriel
Zr 0.1-0.2, Cr 0-0.4, bal Cu.
Copper castings; quenched and tempered. Good electrical
conductivity. AFNOR UC1 Zr.

ELECTRAL CRM 16N

Le Bronze Industriel
Zr 0.1-0.2, Cr 0-0.4, bal Cu.
Wrought copper bar, quenched and tempered. Good
electrical conductivity. AFNOR UC1 Zr.

ELECTREX

Columbia Tool Steel Co.
C 0.9, Mn 0.35, bal Fe.
For brake dies, shafts, arbors; water hardening. *Obsolete*

ELECTRIC RAILWAY BABBITT

Hoyt Metal Co. of London Ltd.
Sn, Sb, bal Pb.
At 70 F: 29 Brin. At 212 F: 14 Brin. For bearings, bushings;
Babbitt. *Obsolete*

ELECTRIC SPINDLE

Colt Industries
C 0.7, Mn 1.15, Cr 0.5, bal Fe.
For machine tool spindles; water or oil hardening. *Obsolete*

ELECTRICAL GOLD ALLOY

English manufacture
Au 70, Ag 25, Ni 5.
For contacts.

ELECTRITE CO-12

Latrobe Steel Co.
C 0.86, W 4, Cr 4.1, V 1.8, Mo 5, Co 12, bal Fe.
Hardened: C 64-68 Rock. For high speed lathe tools; very
good red hardness. AISI Type M6 high speed tool steel.
Obsolete

ELECTRITE CO-6

Latrobe Steel Co.
Tool material.
Now CO-6.

ELECTRITE CO-6XL

Latrobe Steel Co.
C 0.9, W 6, Cr 4, V 2, Mo 5, Co 9, S, bal Fe.
For lathe and planer tools, reamers, broaches, taps; high
speed steel. *Obsolete*

ELECTRITE COBALT

Latrobe Steel Co.
Tool material.
Now COBALT.

ELECTRITE COBALT XL

Latrobe Steel Co.
C 0.7, Co 5, W 18, Cr 4, V 1, S, bal Fe.
For lathe and planer tools, reamers, broaches; high speed
steel. *Obsolete*

ELECTRITE CORSAIR

Latrobe Steel Co.
Tool material.
Now CORSAIR.

ELECTRITE CROMOLON

Latrobe Steel Co.
C 0.9, W 1.5, Mo 9, Cr 8, V 1.5, Co 4, bal Fe.
For taps, cutters; high speed steel. *Obsolete*

ELECTRITE DOUBLE SIX

Latrobe Steel Co.
C 0.85, W 5.5, Cr 4, V 1.5, Mo 4.5, bal Fe.
For countersink and counterbore tools; high speed steel.
Obsolete

ELECTRITE DOUBLE SIX M-2

Latrobe Steel Co.
Now DOUBLE SIX M-2.

ELECTRITE DOUBLE SIX M-2XL

Latrobe Steel Co.
Now DOUBLE SIX M-2XL.

ELECTRITE DYNACUT

Latrobe Steel Co.
Now DYNACUT.

ELECTRITE DYNAMAX
Latrobe Steel Co.
Now DYNAMAX.

ELECTRITE DYNAVAN XL
Latrobe Steel Co.
C 1.5, Cr 4.5, W 13.5, V 4.75, Co 5, Mo 0.5, S, bal Fe.
For inserted blade cutting tools; high speed steel. *Obsolete*

ELECTRITE HS 29 XL
Latrobe Steel Co.
Now HS29XL.

ELECTRITE HS-12
Latrobe Steel Co.
Now HS-12.

ELECTRITE KELVAN
Latrobe Steel Co.
Now KELVAN.

ELECTRITE LACOMO
Latrobe Steel Co.
Now LACOMO.

ELECTRITE MCH
Latrobe Steel Co.
Now MCH.

ELECTRITE MCL
Latrobe Steel Co.
Now MCL.

ELECTRITE MV-1
Latrobe Steel Co.
Now MV-1.

ELECTRITE MV-2
Latrobe Steel Co.
Now MV-2.

ELECTRITE MV-4
Latrobe Steel Co.
C 1.4, Cr 4.1, Mo 4.25, V 4.15. *Obsolete*

ELECTRITE NO. 1
Latrobe Steel Co.
Now E NO. 1.

ELECTRITE NO. 19XL
Latrobe Steel Co.
C 0.84, Cr 4, V 2, W 18, S, bal Fe.
For lathe and planer tools, hobs, drills, taps; high speed steel. *Obsolete*

ELECTRITE NO. 2
Latrobe Steel Co.
W 18, Cr 4, V 1, High C, bal Fe.
For cutting tools, insert saw teeth, woodworking tools, dies, drills; high speed steel. *Obsolete*

ELECTRITE NO. 5
Latrobe Steel Co.
Tool material.
Now LATROBE NO. 5.

ELECTRITE NO. 5XL
Latrobe Steel Co.
C 0.6, W 18, Cr 4, V 1, S, bal Fe.
For bolt trimmers, header inserts, punches, cutters; high speed steel. *Obsolete*

ELECTRITE NO. 7
Latrobe Steel Co.
Now LATROBE NO. 7.

ELECTRITE NO. 7XL
Latrobe Steel Co.
C 0.65, W 6.5, Mo 5, Cr 4, V 2, S, bal Fe.
For extrusion dies, header inserts; high speed steel. *Obsolete*

ELECTRITE STARK
Latrobe Steel Co.
Now STARK.

ELECTRITE STEEL
Latrobe Steel Co.
W 14-18, C, bal Fe.
For heavy lathe tools. *Obsolete*

ELECTRITE SUPER COBALT
Latrobe Steel Co.
Now SUPER COBALT.

ELECTRITE SUPER COBALT XL
Latrobe Steel Co.
C 0.8, V 2, Co 9, W 18, Cr 4, S, bal Fe.
For lathe and planer tools, reamers, broaches, taps; Type T5; high speed steel. *Obsolete*

ELECTRITE T.N.W.
Latrobe Steel Co.
Now TNW.

ELECTRITE TATMO
Latrobe Steel Co.
Now TATMO.

ELECTRITE TATMO COBALT
Latrobe Steel Co.
Now TATMO COBALT.

ELECTRITE TATMO V
Latrobe Steel Co.
Now TATMO V.

ELECTRITE TATMO XL
Latrobe Steel Co.
C 0.8, Cr 3.25-4.25, W 1.2-2, V 0.7-1.5, Mo 7.5-9.5, Co, S, bal Fe.
For reamers, drills, lathe and planer tools, hobs; Type M1; high speed steel. *Obsolete*

ELECTRITE TNW-XL
Latrobe Steel Co.
C 0.85, Cr 4, V 2, Mo 8, S, bal Fe.
For form cutters, reamers, taps, drills; high speed steel. *Obsolete*

ELECTRITE U
Latrobe Steel Co.
C 0.7, Mn 0.3, Cr 4, W 14, V 0.2, bal Fe.
For tools, dies, gages, reamers; high speed steel. *Obsolete*

ELECTRITE UB
Latrobe Steel Co.
C 75, Cr 4, Co 5, V 2, W 14, bal Fe.
For reamers, blades, cutting tools; high speed steel. *Obsolete*

ELECTRITE ULTRA COBALT
Latrobe Steel Co.
Now ULTRA COBALT.

ELECTRITE ULTRAVAN
Latrobe Steel Co.
Now ULTRAVAN.

ELECTRITE ULTRAVAN XL
Latrobe Steel Co.
C 1.5, W 6.3, Cr 4.2, V 4.7, Mo 5, Co 5, S, bal Fe.
For inserted blade cutting tools; high speed steel. *Obsolete*

ELECTRITE VANADIUM
Latrobe Steel Co.
Now VANADIUM.

ELECTRITE VANADIUM XL
Latrobe Steel Co.
C 1.1, W 18, Cr 4, V 3.5, S, bal Fe.
For cutters for hard materials; Type T3; high speed steel. *Obsolete*

ELECTRO
Electro-Steel Co.
C 0.7, W 6, Mo 6, bal Fe.
For high speed cutting tools; high speed steel.

ELECTRO HIGH SPEED
English manufacture
Cu 0.76, Mn 0.45, Cr 2.95, W 13.24, V 1.51, bal Fe.
For tools, cutters, drills; high speed steel.

ELECTRO SHEET COPPER FOIL
Anaconda Co.
99.5%+ Cu.
Commercial quality. Foils in rolls; 1/2 oz-7 oz/ft 2 . 40,00 TS (nominal); 4.5-13 El; 94.3-99.5 electrical conductivity. For electrical coil windings, transformer windings, flexible cable, electrostatic shielding, die stamped circuits. *Obsolete*

ELECTRO-ALLOY H3
English manufacture
Ni, Cr.
For high temperature applications; heat and corrosion resistant.

ELECTROLOY
Molecu-Wire Corp.
Cr 15, Ni 60, bal Fe.
Wire. 95,000-175,000 TS. For resistors; high electrical resistance.

ELECTROLOY
Electroloy Co. Inc.
Copper. Cu alloy.
For welding equipment.

ELECTROLOY 1
Electroloy Co. Inc.
Copper. Cu, W.
Bar: 135,000 TS; 130 Brin. For resistance welding electrode; 35% electrical conductivity, for stainless steel.

ELECTROLOY 10
Electroloy Co. Inc.
Copper. Cu, W.
Bar: 160,000 TS; 205 Brin. For resistance welding electrode; 28% electrical conductivity, for inserts and facings.

ELECTROLOY 100
Electroloy Co. Inc.
Copper. Cu, W.
Bar: 200,000 TS; 76 Rock A. For resistance welding electrode; 30% electrical conductivity, for Red Brass.

ELECTROLOY 20
Electroloy Co. Inc.
Copper. Cu, W.
Bar: 170,000 TS; 228 Brin. For resistance welding electrode; for heavy projection welding.

ELECTROLOY GRADE B
Electroloy Co. Inc.
Copper. Cu alloy.
175,000 TS; 5 El; 365 Brin. For facing and inserts on projection welding electrodes; electrical conductivity 40-45%.

ELECTROLOY GRADE BX
Electroloy Co. Inc.
Copper. Cu alloy.
95,000 TS; 4-8 El; 210-228 Brin. For flash and butt welding dies; electrical conductivity 50%.

ELECTROLOY GRADE C

Electroloy Co. Inc.
Copper. Cu alloy.
60,000 TS; 20 El; 150 Brin. For spot welding tips; electrical conductivity 55-65%.

ELECTROLOY GRADE XX

Electroloy Co. Inc.
Copper. Cu alloy.
70,000 TS; 20 El; 150 Brin. For spot welding tips; electrical conductivity 90%.

ELECTROLOY MATCH PLATE ALUMINUM

FERRO Corp.
Al alloy.
For match plates; non-hardenable. *Obsolete*

ELECTROLOY MOLIN 2

Electroloy Co. Inc.
Copper. Cu alloy.
Cast: 65,000-75,000 TS; 12,000-16,000 YS; 2-10 El; 116-165 Brin. For resistance welding electrode; 10-15% electrical conductivity.

ELECTROLOY TX

Electroloy Co. Inc.
Copper. Cu alloy.
Bar: 100,000 TS; 50,000 YS; 10 El; 185 Brin. Cast: 85,000 TS; 45,000 YS; 10 El; 180 Brin. For resistance welding electrode; for stainless steel.

ELECTROLYTIC IRON

Western Electric Co.
C 0.006, S 0.004, Si 0.005, Cu 0.015, Fe 99.965.
Forged: 55,000 TS; 48,500 YS; 33 El; 83 RA. For thin seamless tubing, magnetic cores, electrical instruments; brittle as deposited.

ELECTROLYTIC TOUGH PITCH COPPER

Chase Brass & Copper Co.
Cu 99.9, 0.04% oxygen.
Annealed: 30,000 TS; 10,000 YS; 45 El; 48 Brin. Cold drawn: 50,000 TS; 45,000 YS; 6 El; 106 Brin. For electric conductors, anodes, bus bars, radiators; high conductivity. *Obsolete*

ELECTROLYTIC TOUGH PITCH COPPER 100

Anaconda Co.
Copper. 99.9 Cu min.
Hard: 55,000 TS; 45,000 YS; 5 El. For electric conductors, pipes, tubes. High conductivity.

ELECTROLYTIC TOUGH PITCH COPPER 110

Anaconda Co.
Copper. 99.9 Cu min.
Hard: 48,000 TS; 40,000 YS; 15 El; 45 Rock B. Soft: 32,000 TS; 10,000 YS; 50 El; 45 Rock F. For roofing, flashing, gutters, architectural shapes, fasteners, plating anodes. Electrical conductivity: 101.

ELECTROMATIC

American manufacture
C 0.75, bal Fe.
For springs; oil tempered wire.

ELECTROMET CALCIUM-ALUMINUM-SILICON

Union Carbide Corp.
Ca 10-14, Al 8-12, Si 50-53.
For deoxidizing and degasifying steel.

ELECTROMET CALCIUM-MANGANESE-SILICON

Union Carbide Corp.
Ca 16-20, Mn 14-18, Si 53-59, bal Fe.
Used as a scavenger for steel; cleanser of oxides and gases.

ELECTROMET CALCIUM-SILICON

Union Carbide Corp.
Ca 30-33, Si 65, Fe 1.5-3.
For metallurgical applications, deoxidizer and degasifier; graded according to Ca content.

ELECTROMET CHROMIUM (HIGH CARBON)

Union Carbide Corp.
C 9-11, Cr 87-90, bal Fe.
For production of nonferrous Cr alloys, deoxidized; graded according to carbon content.

ELECTROMET CHROMIUM CARBIDE

Union Carbide Corp.
Cr 86.1-86.6, 12.8 13.3 total C.
Sintered: 88 HRA. For extrusion dies, gage blocks; pressed or slip-cast and sintered.

ELECTROMET CHROMIUM METAL (LOW C)

Union Carbide Corp.
Cr 97-0, Fe 0-1, 0.2 or 0.5 max C.
For Cr bearing alloy production; Cr-additions.

ELECTROMET CHROMIUM-COPPER

Union Carbide Corp.
Cr 8-11, Cu 88-90, Fe 0-1, Si 0-0.5.
For metallurgical applications.

ELECTROMET COLUMBIUM METAL

Union Carbide Corp.
C 0.27, N 0.01, O 0.04, Ta 0.1, 0.5 ppm B, bal Cb.
Cold drawn: 82,500 TS; 72,000 YS; 17 El; 68 RA. Annealed: 50,000 TS; 36,000 YS; 49 El; 82 RA. For atomic reactors, chemical processing equipment, high temperature alloys; high heat and corrosion resistance.

ELECTROMET DISTILLED CALCIUM

Union Carbide Corp.
Ca 99.48, Fe 0.045, Mn 0.025, Ni 0.03, C 0.03, Cr 0.004.
Rolled: 16,700 TS; 12,300 YS; 7 El; 35 RA. Annealed: 6960 TS; 1990 YS; 51 El; 58 RA. For alloying agent for Al, Pb, Si, copper deoxidizer; getter for electronic tubes.

ELECTROMET FERRO-BORON

Union Carbide Corp.
B 17-25, Al 0-0.5, Si 0-1, C 0-0.5, bal Fe.
For deoxidizer and degasifier additions to steel; boron; boron additions.

ELECTROMET FERRO-CHROME (FOUNDRY)

Union Carbide Corp.
C 5-7, Cr 62-66, Si 7-10, bal Fe.
For metallurgical applications, ladle additions of Cr to cast iron; dissolves readily.

ELECTROMET FERRO-CHROME (HIGH CARBON)

Union Carbide Corp.
C 4.5-7, Si 1-3, Cr 65-70, bal Fe.
For production of Cr steels and cast irons; graded according to carbon content.

ELECTROMET FERRO-CHROME (LOW CARBON)

Union Carbide Corp.
Cr 67-73, C 0.02-2, Si 0.2-1, bal Fe.
For stainless steels; graded according to carbon content.

ELECTROMET FERRO-CHROME (NITROGEN)

Union Carbide Corp.
C 0-0.1, Cr 67-71, Si 0.3-1, approximately 0.75 N, bal Fe.
For metallurgical applications for high Cr steels; graded according to N_2 content.

ELECTROMET FERRO-COLUMBIUM

Union Carbide Corp.
Cb 50-60, Si 0-8, C 0-0.4, bal Fe.
For additions to stainless steel and high temperature alloys of Cb; increases strength at elevated temperatures.

ELECTROMET FERRO-MANGANESE (LOW CARBON)

Union Carbide Corp.
C 0.07-0.5, Mn 85-90, bal Fe.
For high Mn additions to low C steels, metallurgical applications; grade according to carbon content.

ELECTROMET FERRO-MANGANESE (LOW IRON)

Union Carbide Corp.
Mn 85-90, Fe 0-2, Si 0-3, approx 7.0 C, bal Fe.
For metallurgical applications in non-ferrous alloys; where high Mn, low Fe are needed.

ELECTROMET FERRO-MANGANESE (MED. CARBON)

Union Carbide Corp.
C 1.5, Si 0-1.5, Mn 80-85, bal Fe.
For high Mn additions to medium C steels, metallurgical applications; fo C-Mn steels.

ELECTROMET FERRO-MANGANESE (STANDARD)

Union Carbide Corp.
Mn 74-76, Si 0-1, P 0-0.3, approx 7 C, bal Fe.
For metallurgical applications, alloy deoxidizer.

ELECTROMET FERRO-SILICON (15%)

Union Carbide Corp.
Si 14-20, C 0-1, Fe 74-84, P 0-0.05, S 0-0.04.
For addition of Si to steel, metallurgical applications, graded according to Si content.

ELECTROMET FERRO-SILICON (50%)

Union Carbide Corp.
Si 47-51, bal Fe.
For metallurgical applications, graphitizer and deoxidizer; graded according to Si content.

ELECTROMET FERRO-SILICON (75%)

Union Carbide Corp.
Si 73-78, bal Fe.
For metallurgical applications, silicon additions; graded according to Si content.

ELECTROMET FERRO-SILICON (85%)

Union Carbide Corp.
Si 83-88, bal Fe.
For high silicon alloys, metallurgical applications; inoculant for cast iron.

ELECTROMET FERRO-SILICON (90%)

Union Carbide Corp.
Si 92-95, bal Fe.
For high silicon alloys, metallurgical applications; graded according to Si content.

ELECTROMET FERRO-TUNGSTEN

Union Carbide Corp.
W 70-80, C 0-0.6, bal Fe.
For steel alloys, alloying W to steel; graded according to C content.

ELECTROMET FERRO-VANADIUM

Union Carbide Corp.
V 55-50, Si 0-10, C 0-3, bal Fe.
For metallurgical applications, steel alloys, V additions; graded according to C and Si content.

ELECTROMET FERROCHROME-SILICON

Union Carbide Corp.
Cr 33-57, Si 23-48, C 0-0.05, bal Fe.
For production of stainless steel; to reduce metal value in slag.

ELECTROMET FERROMANGANESE SILICON
Union Carbide Corp.
Mn 63-66, Si 28-32, C 0-0.1, P 0-0.05, bal Fe.
For slag reducing agent and Mn additions to steel.

ELECTROMET FERROSILICON (50% LOW AL)
Union Carbide Corp.
Si 47-51, Al 0-0.4, bal Fe.
For additions of Si to steel and cast iron; deoxidizer.

ELECTROMET FERROSILICON (65% LOW AL)
Union Carbide Corp.
Si 64.5-69.5, Al 0-0.5, bal Fe.
For production of electrical sheet steel; deoxidizer; Si-additions.

ELECTROMET FERROSILICON (75% LOW AL)
Union Carbide Corp.
Si 73-78, Al 0-0.5, bal Fe.
For deoxidizer and alloy for steel, silicon additions; inoculant for cast iron.

ELECTROMET FERROSILICON (85% LOW AL)
Union Carbide Corp.
Si 83-88, Al 0-0.5, bal Fe.
For deoxidizer and alloy for steel; inoculant for cast iron.

ELECTROMET FERROSILICON (90% LOW AL)
Union Carbide Corp.
Si 92-95, Al 0-0.5, bal Fe.
For deoxidizer and alloy for steel; inoculant for cast iron.

ELECTROMET FERROSILICON-CHROME
Union Carbide Corp.
Cr 50-54, Si 28-32, C 0-1.25, bal Fe.
For ladle additions of Cr and Si to steel; alloying.

ELECTROMET LOW PHOSPHORUS FERROMANGANESE
Union Carbide Corp.
Mn 90-0, P 0-0.06, C 0-0.07, bal Fe.
For Mn additions to austenitic stainless steels; alloying.

ELECTROMET MANGANESE
Union Carbide Corp.
Fe 0-2.5, Si 0-1.5, C 0-0.2, Mn 96-0.
For metallurgical applications, alloying, deoxidizer; graded according to and Fe content.

ELECTROMET MANGANESE-BORON
Union Carbide Corp.
Mn 75, B 17.5-0, Fe 0-5, Si 0-1.5, C 0-3.
For deoxidizer, degasifier and cleanser; for nonferrous alloys.

ELECTROMET MANGANESE-COPPER
Union Carbide Corp.
Mn 28-32, Cu 65-70, Fe 0-4, Si 0-0.75.
For non-ferrous metallurgical applications; graded according to Fe content.

ELECTROMET MANGANESE-NICKEL TITANIUM
Union Carbide Corp.
C 0-0.1, Mn 6-8, Ni 29-31, Ti 46.5-48.5, Al 12.5-14.5, Fe 0-1, Si 0-0.2.
For Ti additions to high temperature alloys.

ELECTROMET NICKEL BORON
Union Carbide Corp.
B 15-18, Fe 0-3, C 0-0.5, bal Ni.
For addition agent for Ni and Al alloys; for alloying in hard abrasion resistant alloys.

ELECTROMET NICKEL-ZIRCONIUM
Union Carbide Corp.
Ni 43-50, Zr 25-30, Fe 0-2, Si 0-10, approx 15 Al.
For deoxidizer and degasifier for Ni and Fe alloys.

ELECTROMET SILICO-MANGANESE
Union Carbide Corp.
C 0-3, Mn 65-68, Si 12-21, bal Fe.
For metallurgical applications, alloying, and deoxidizing; graded according to carbon content.

ELECTROMET SILICON (REFINED)
Union Carbide Corp.
Si 96-0, Fe 0-2.
For metallurgical applications for Cu and Al silicon alloys; graded according to Fe content.

ELECTROMET SILICON TITANIUM
Union Carbide Corp.
Ti 40-50, Si 45-50, Fe 0-3.
For tool and die steels, also nonferrous alloys; titanium additions.

ELECTROMET SILICON-SPIEGEL
Union Carbide Corp.
Mn 25-30, Si 5-8, C 2-4, bal Fe.
For metallurgical applications; furnished in standard and special grades.

ELECTROMET SPIEGELEISEN
Union Carbide Corp.
Mn 19-28, Si 0-1, bal Fe, C.
For metallurgical applications; graded according to Mn content.

ELECTROMET TANTALUM METAL
Union Carbide Corp.
C 0.015, H 0.001, Fe 0.01, bal Ta.
Cold drawn: 114,200 TS; 104,500 YS; 19 El; 76 RA; 205 BHN.
Recrystallized: 54,900 TS; 37,700 YS; 54 El; 92 RA; 103 BHN.
For heat exchangers, bayonet heaters, pickling tanks; high corrosion resistance.

ELECTROMET TITANIUM CARBIDE
Union Carbide Corp.
Fe 0-0.2, 19.2 min combined C, 0.3 max free C, 95.0 TiC min.
For additions to tungsten carbide; pressed or slip-cast and sintered.

ELECTROMET VANADIUM METAL
Union Carbide Corp.
C 0.1, H 0.1, N 0.1, O 0.1, bal V.
Annealed: 60,000-70,000 TS; 49,100 YS; 30-20 El; 80-70 RA; 140 BHN. For atomic reactor structures.

ELECTROMET ZIRCONIUM (12-15%)
Union Carbide Corp.
Zr 12-15, Si 39-43, C 0-0.3, bal Fe.
For ferrous metallurgical applications, steel deoxidizer; nitride and sulfide former.

ELECTROMET ZIRCONIUM (35-40%)
Union Carbide Corp.
Zr 35-40, Si 47-52, C 0-0.5, bal Fe.
For ferrous metallurgical applications, deoxidizer; nitride and sulfide former.

ELECTROMETALL
Magnesium Elektron Ltd.
Al 6, bal Mg.
For light alloy parts. *Obsolete*

ELECTRON "A.Z.D."
Hellefors Bruks Aktiebolag
Al 5, Zn 3, Mn 0.2-0.5, Cd 3, bal Mg.
Extruded: 42,000 TS; 28,000 YS; 11 El; 20 RA; 53 Brin. For light alloy parts. *Obsolete*

ELECTRON "A.Z.D."
Farbenindustrie Atkiengesellschaft
Al 5, Zn 3, Mn 0.2-0.5, Cd 3, bal Mg.
Extruded: 42,000 TS; 28,000 YS; 11 El; 20 RA; 53 Brin. For light alloy parts. *Obsolete*

ELECTRON "A.Z.F."
Hellefors Bruks Aktiebolag
Al 4, Zn 3, Mn 0.2-0.5, Si 0.3, bal Mg.
Cast: 24,000-30,000 TS; 13,000 YS; 5-9 El; 9 RA; 45 Brin. For airplane wheels, castings to resist high impact stresses. *Obsolete*

ELECTRON "A.Z.F."
Farbenindustrie Atkiengesellschaft
Al 4, Zn 3, Mn 0.2-0.5, Si 0.3, bal Mg.
Cast: 24,000-30,000 TS; 13,000 YS; 5-9 El; 9 RA; 45 Brin. For airplane wheels, castings to resist high impact stresses. *Obsolete*

ELECTRON "A.Z.G"
Hellefors Bruks Aktiebolag
Al 6, Zn 3, Mn 0.35, Si 0.3, bal Mg.
Cast: 23,000-29,000 TS; 14,500-19,000 YS; 6-3 El; 7 RA; 55 Brin. For automobile and airplane engine crankshafts; high fatigue strength.

ELECTRON "A.Z.G"
Farbenindustrie Atkiengesellschaft
Al 6, Zn 3, Mn 0.35, Si 0.3, bal Mg.
Cast: 23,000-29,000 TS; 14,500-19,000 YS; 6-3 El; 7 RA; 55 Brin. For automobile and airplane engine crankshafts; high fatigue strength.

ELECTRON "A.Z.M"
Farbenindustrie Atkiengesellschaft
Al 6, Zn 1, Mn 0.2-0.5, Si 0.2, bal Mg.
Rolled: 49,000 TS; 31,000 YS; 14 El; 27 RA; 70 Brin.
Forgings: 45,000 TS; 29,000 YS; 8 El; 70 Brin. For extruded and forged light alloy parts, airplane propellers, bus body frames; general structural work.

ELECTRON "A.Z.M"
Alcan-Booth Industries, Ltd.
Al 6, Zn 1, Mn 0.2-0.5, Si 0.2, bal Mg.
Rolled: 49,000 TS; 31,000 YS; 14 El; 27 RA; 70 Brin.
Forgings: 45,000 TS; 29,000 YS; 8 El; 70 Brin. For extruded and forged light alloy parts, airplane propellers, bus body frames; general structural work.

ELECTRON "A.Z.M"
Hellefors Bruks Aktiebolag
Al 6, Zn 1, Mn 0.2-0.5, Si 0.2, bal Mg.
Rolled: 49,000 TS; 31,000 YS; 14 El; 27 RA; 70 Brin.
Forgings: 45,000 TS; 29,000 YS; 8 El; 70 Brin. For extruded and forged light alloy parts, airplane propellers, bus body frames; general structural work.

ELECTRON "V-1"
Farbenindustrie Atkiengesellschaft
Al 10, Mn 0.3-0.5, bal Mg.
Chill cast: 20,500 TS; 15,000 YS; 2-3 El; 1.2 RA. Extruded: 49,000 TS; 35,000 YS; 8 El; 11 RA; 70 Brin. For light alloy parts with highest possible hardness. *Obsolete*

ELECTRON "V-1"
Maulbronn Aluminiumwerke
Al 10, Mn 0.3-0.5, bal Mg.
Chill cast: 20,500 TS; 15,000 YS; 2-3 El; 1.2 RA. Extruded: 49,000 TS; 35,000 YS; 8 El; 11 RA; 70 Brin. For light alloy parts with highest possible hardness. *Obsolete*

ELECTRON "Z-3"
Farbenindustrie Atkiengesellschaft
Zn 3, Al 5, bal Mg.
Soft: 28,000-34,000 TS; 18,000-21,000 YS; 10-12 El; 42 Brin.
Hard: 37,000-46,000 TS; 2-5 El; 60 Brin. For stampings, structures. *Obsolete*

ELECTRON A Z
Maulbronn Aluminiumwerke
Al 8.5, Zn 3.5, Mn 0.5, bal Mg.
Sand cast: 25,000-29,000 TS; 4-6 El; 55 Brin. For light alloys.
Obsolete

ELECTRON AZ-102
Farbenindustrie Atkiengesellschaft
Al 11, Zn 2.25, Mn 0.7, 1.5% max impurities, bal Mg.
For extrusions, welding, rods. *Obsolete*

ELECTRON AZ-102
Sterling Metals Ltd.
Al 11, Zn 2.25, Mn 0.7, 1.5% max impurities, bal Mg.
For extrusions, welding, rods. *Obsolete*

ELECTRON AZ-31
Farbenindustrie Atkiengesellschaft
Magnesium. Al 3.3, Mn 0.3-0.5, Zn 1.25, bal Mg.
Cast: 36,000-40,000 TS; 25,000-29,000 YS; 12-8 El; 50 BHN.
For light alloy parts, structures; C.S.-49000-52000.

ELECTRON AZ-855
Farbenindustrie Atkiengesellschaft
Magnesium. Al 8.5-9, Zn 0.2-0.6, Mn 0.1-0.3, bal Mg,
impurities 0.5.
Forged: 40,500-49,000 TS; 22,500-31,000 YS; 15-8 El; 15-10
RA; 65-7 BHN. For forgings, aircraft components; Iz 4-6.

ELECTRON AZ61
English manufacture
Al 5.5-7, Zn 0.1-1, Mn 0.1-0.5, bal Mg.
Used to hold broken bones together.

ELECTRON AZMQ
Farbenindustrie Atkiengesellschaft
Magnesium. Al 6.5, Zn 1, Mn 0.3-0.5, bal Mg.
Cast: 41,000-47,000 TS; 31,000-33,000 YS; 16-10 El; 55 BHN.
Extruded: 40,000 TS, 32,000 YS; 13 El; 27 RA; 55 BHN. For
light alloy parts; now obsolete; C.S.-50000-56000.

ELECTRON CM-SI
Farbenindustrie Atkiengesellschaft
Si 1.5, bal Mg.
Cast: 13,000-14,000 TS; 8,500 YS; 1.5 El; 3 RA; 45 Brin. For
light alloy parts. *Obsolete*

ELECTRON NO. 23
Farbenindustrie Atkiengesellschaft
Al 6, bal Mg.
For light alloy parts. *Obsolete*

ELECTRON Z-1-B
Farbenindustrie Atkiengesellschaft
Al 6.5, Zn 1, Mn 0.3-0.5, bal Mg.
Extruded: 36,500 TS; 24,000 YS; 16 El; 27 RA; 45 Brin. For
light alloy parts. *Obsolete*

ELECTRON ZS-32
Farbenindustrie Atkiengesellschaft
Magnesium. Al 5, Zn 1, bal Mg.
Forged: 7 El; 15-5 RA; 53 BHN. For forgings, pistons;
hardenable.

ELECTROPLATE
English manufacture
Cu 50-70, Ni 10-20, Zn 5-30.
Base for silver plate, tableware.

ELECTROTYPE (STANDARD)
English manufacture
Pb 93, Sb 4, Sn 3.
For electrotype.

ELECTRUM
English manufacture
Au 55-58, Ag 15-45.
For jewelry, ornaments; corrosion resistant.

ELECTRUM
English manufacture
Cu 52, Ni 26, Zn 23.
For corrosion resisting parts; nickel silver.

ELEFANT EXTRA ZH
Dorrenberg Edelstahl GmbH
C 1, Si 0-0.25, Mn 0-0.25, bal Fe.
Annealed: 100,000 TS; 53,000 YS; 21 El; 42 RA; 200 Brin. For
drills, taps, reamers, hobs, broaches; Type W1; water
hardened.

ELEFANT HART
Dorrenberg Edelstahl GmbH
C 1.3, Si 0-0.25, Mn 0-0.25, bal Fe.
For engravers' tools, taps, milling cutters, hobs; Type W1;
water hardened.

ELEFANT MH
Dorrenberg Edelstahl GmbH
C 1.15, Si 0-0.25, Mn 0-0.25, bal Fe.
Annealed: 110,000 TS; 60,000 YS; 18 El; 40 RA; 210 Brin. For
drills, taps, reamers, hobs, broaches; Type W1; water
hardened.

ELEFANT NR 3
Dorrenberg Edelstahl GmbH
C 1, Si 0-0.25, Mn 0-0.25, bal Fe.
Annealed: 100,000 TS; 53,000 YS; 21 El; 42 RA; 200 Brin. For
drills, taps, reamers, hobs, cutters, punches; Type W1; water
hardened.

ELEFANT NR 4
Dorrenberg Edelstahl GmbH
C 0.85, Si 0-0.25, Mn 0-0.25, bal Fe.
Heat treated: 190,000 TS; 145,000 YS; 12 El; 35 RA; 400 Brin.
For drills, taps, springs, hobs; Type W1; water hardened.

ELEFANT NR 5
Dorrenberg Edelstahl GmbH
C 0.7, Si 0-0.25, Mn 0-0.25, bal Fe.
Heat treated: 175,000 TS; 130,000 YS; 12 El; 36 RA; 360 Brin.
For springs, tools, hammers, rails, axes; Type W1; water
hardened.

ELEFANT ZAH
Dorrenberg Edelstahl GmbH
C 0.7, Si 0-0.25, Mn 0-0.25, bal Fe.
Heat treated: 175,000 TS; 130,000 YS; 12 El; 36 RA; 360 Brin.
For springs, tools, rails, axes, hammers; Type W1; water
hardened.

ELEFANT ZH
Dorrenberg Edelstahl GmbH
C 0.85, Si 0-0.25, Mn 0-0.25, bal Fe.
Heat treated: 190,000 TS; 145,000 YS; 12 El; 35 RA; 400 Brin.
For springs, cutters, drills, taps, reamers; Type W1; water
hardened.

ELEKTRO
Friedr. Lohmann GmbH
C 0.85, Si 0.3, Mn 0.6, bal Fe.
Heat treated: 190,000 TS; 145,000 YS; 10 El; 30 RA; 400 Brin.
For springs, tools, drills, taps, reamers. Type W1; water
hardened. *Obsolete*

ELEKTROBRONZE
Ostermann GmbH & Co.
Copper. Cu 81, Al 10, Fe, bal Mn.
20 El; 160-170 Brin. For valve gauge fittings; corrosion
resistant.

ELEKTRON 3Z3
Magnesium Elektron Ltd.
Zn 3, Zr 0.7, bal Mg.
Cast: 33,000-35,500 TS; 13,000-15,500 YS; 30 El. For light
alloy castings; non-hardenable. *Obsolete*

ELEKTRON A 8
Alcan-Booth Industries, Ltd.
Al 8, Zn 0.5, Mn 0.3, bal Mg.
Heat treated: 28,000 TS; 10,000 YS; 5 El; 50 Brin. Annealed:
20,000-39,000 TS; 10,000-12,000 YS; 2-5 El; 50-60 Brin. For
forgings. *Obsolete*

ELEKTRON A10
Farbenindustrie Atkiengesellschaft
Al 9.5, Zn 0.5, Mn 0.2, bal Mg.
Cast. For sand castings. *Obsolete*

ELEKTRON A11V
Birmingham Aluminium Casting Co.
Al 11, bal Mg.
For light alloy parts. *Obsolete*

ELEKTRON A4
F.A. Hughes & Co., Ltd.
Al 9, Zn 1.5, Mn 1, bal Mg.
Rolled: 32,000-36,000 TS; 14,000-18,000 YS; 10-12 El. For
light alloy parts; age-hardenable. *Obsolete*

ELEKTRON A4
Magnesium Elektron Ltd.
Al 9, Zn 1.5, Mn 1, bal Mg.
Rolled: 32,000-36,000 TS; 14,000-18,000 YS; 10-12 El. For
light alloy parts; age-hardenable. *Obsolete*

ELEKTRON A5
Birmingham Aluminium Casting Co.
Magnesium. Al 5, bal Mg.
For light alloy parts.

ELEKTRON A6
Alcan-Booth Industries, Ltd.
Al 6, Zn 1, Mn 0.2, bal Mg.
Rolled: 40,000 TS; 23,000 YS; 8 El. For high strength parts;
age-hardenable. *Obsolete*

ELEKTRON A7
Birmingham Aluminium Casting Co.
Al 7, bal Mg.
For light alloy parts. *Obsolete*

ELEKTRON A8
Magnesium Elektron Ltd.
Magnesium. Al 7.5, Zn 0.7, Mn 0.3, bal Mg.
Cast: 34.0 ksi TS; 11.0 ksi YS; 7 El; 50-60 Brin (T4 temper).
General purpose alloy.

ELEKTRON A8
Birmingham Aluminium Casting Co.
Magnesium. Al 7.9-9, Zn 0.3-1, Mn 0.15-0.4, Cu 0-0.15, bal
Mg.
Sand cast: 22,800 TS; 11,400 YS; 4 El; 55 Brin. Permanent
mold: 27,000 TS; 10,000 YS; 4.5 El; 60 Brin. Heat treated:
36,000 TS; 13,000 YS; 12 El; 60 Brin. For aircraft parts,
general purpose castings; age-hardenable.

ELEKTRON A8
Farbenindustrie Atkiengesellschaft
Magnesium. Al 0-8, Zn 0-0.4, Mn 0.4, bal Mg.
Sand cast: 29,400 TS; 115,000 YS; 4 El; 7 RA; 55 BHN. Aged:
36,000 TS; 21,000 YS; 13 El; 9 RA; 60 BHN. For engine
components, housings, gear cases; sand and permanent
mold castings, age-hardenable.

ELEKTRON A8
F.A. Hughes & Co., Ltd.
Magnesium. Al 8, Zn 0.5, Mn 0.3, Si 0.2, bal Mg.
Cast: 22,000 TS; 11,000 YS; 5 El; 50 Brin. Heat treated:
38,000 TS; 11,000 YS; 9 El; 60 Brin. For gear and blower
casings, aero engines; sand or die castings.

ELEKTRON A8 HP
Sterling Metals Ltd.
Al 7.5-9, Zn 0.3-1, Mn 0.15-0.4.
s sand cast: 18,000-22,400 TS; 10,000-12,400 YS; 0-2 L; 55 Brin; 0.5 IZ. Chill cast, heat treated: 9,000-36,000 TS; 15,500-18,000 YS; 2-4 El; 75 Brin; 0.6 Z. General purpose sand or gravity die magnesium alloy asting. Gen. Eng. BS 2970: MAG 2; MAG 7. OMPOSITION al Mg (low residuals). *Obsolete*

ELEKTRON A8HT
Farbenindustrie Atkiengesellschaft
Magnesium. Al 8, Zn 1, Mn 0.5, bal Mg.
Heat treated: 26,000-32,000 TS; 6-10 El; 50 BHN. For light alloy parts; sand and permanent mold castings.

ELEKTRON A9
Farbenindustrie Atkiengesellschaft
Al 8.5, Zn 0.5, Mn 0.2, bal Mg.
Cast: 21,200 TS; 15,700 YS; 2 El; 54 Brin. For sand castings. *Obsolete*

ELEKTRON A9V
Birmingham Aluminium Casting Co.
Magnesium. Al 8.5, Zn 0.5, Mn 0.3, bal Mg.
For light alloy parts; heat treatable.

ELEKTRON AM-503
Farbenindustrie Atkiengesellschaft
Magnesium. Al 0.2, Zn 0.2, Mn 2.5, Cu 0.2, Si 0.4, bal Mg.
Cast: 14,000-18,000 TS; 3 El; 30 BHN. Rolled: 40,000 TS; 22,000 YS; 5 El; 20 RA. For light alloy parts, aircraft panelling, cowling, fairing; easy to weld, rolled sheets.

ELEKTRON AM503
F.A. Hughes & Co., Ltd.
Magnesium. Mn 1.8, Si 0.2, bal Mg.
Cast: 14,000 TS; 40,000 YS; 3 El; 45 Brin. For light alloy parts; corrosion resistant.

ELEKTRON AM503
Magnesium Elektron Ltd.
Magnesium. Mn 1.5, bal Mg.
Extruded: 230 N/mm^2 TS; 130 N/mm^2 proof stress; 4 El; 45-55 VHN. Low strength, general purpose; weldable; good corrosion resistance.

ELEKTRON AM503
Birmingham Aluminium Casting Co.
Magnesium. Mn 1-2, bal Mg.
Rolled: 29,000-40,000 TS; 13,500-22,500 YS; 5-12 El; 35-45 Brin. For sheets, shapes, aircraft; for welding.

ELEKTRON AM537
F.A. Hughes & Co., Ltd.
Mn 1.8, Ce 0.4, bal Mg.
Rolled: 32,000 TS; 19,000 YS; 17 El. For fuselages, wing ribs and coverings. *Obsolete*

ELEKTRON AM537
Magnesium Elektron Ltd.
Mn 1.8, Ce 0.4, bal Mg.
Rolled: 32,000 TS; 19,000 YS; 17 El. For fuselages, wing ribs and coverings. *Obsolete*

ELEKTRON AM6
Birmingham Aluminium Casting Co.
Mn 2, Ce 6, bal Mg.
Forged: 36,400-40,500 TS; 30,800-31,900 YS; 3-7 El; 61-67 Brin. For supercharger impellers, housings; high temperature applications. *Obsolete*

ELEKTRON AZ 551
Birmingham Aluminium Casting Co.
Al 5.5, Zn 1, bal Mg.
For light alloy parts, engine. *Obsolete*

ELEKTRON AZ31
Magnesium Elektron Ltd.
Magnesium. Al 3, Zn 1, Mn 0.3, bal Mg.
Extruded: 230 N/mm^2 TS; 150 N/mm^2(10 mm) proof stress; 8 El; 50-65 VHN. Medium strength; good formability, weldable. BS: 3373 MAG-E-101M.

ELEKTRON AZ31
Alcan-Booth Industries, Ltd.
Al 3, Zn 1, Mn 0.3, bal Mg.
Rolled: 32,000-42,000 TS; 16,000-29,000 YS; 4-20 El; 45-50 Brin. For aircraft structures, truck bodies, oil tanks; extrusion alloy. *Obsolete*

ELEKTRON AZ31
Birmingham Aluminium Casting Co.
Magnesium. Al 2.5-3.5, Zn 0.5-1.5, Mn 0.2-0.4, bal Mg.
Cast: 21,000 TS; 11,000 YS; 7 El; 45 Brin. Chilled: 30,000 TS; 11,000 YS; 15 El; 45 Brin. For brake shoes; sand or gravity die cast.

ELEKTRON AZ31
F.A. Hughes & Co., Ltd.
Al 3, Zn 1, Mn 0.3, bal Mg.
Rolled: 32,000-42,000 TS; 16,000-29,000 YS; 4-20 El; 45-50 Brin. For aircraft structures, truck bodies, oil tanks; extrusion alloy. *Obsolete*

ELEKTRON AZ855
Magnesium Elektron Ltd.
Al 7.8, Zn 0.4, Mn 0.3, Si 0.1, bal Mg.
Forged: 44,000-49,900 TS; 28,000-32,000 YS; 12-8 El; 65-75 Brin. For propeller blades, forgings; press forgings.

ELEKTRON AZ855
F.A. Hughes & Co., Ltd.
Al 7.8, Zn 0.4, Mn 0.3, Si 0.1, bal Mg.
Forged: 44,000-49,900 TS; 28,000-32,000 YS; 12-8 El; 65-75 Brin. For propeller blades, forgings; press forgings.

ELEKTRON AZ91
Magnesium Elektron Ltd.
Magnesium. Al 9, Zn 0.7, Mn 0.3, bal Mg.
Diecast (F): 34.0 ksi TS; 23.0 ksi YS; 3 El; 63 Brin. General purpose die casting alloy. AMS 4490E.

ELEKTRON AZ91
Magnesium Elektron Ltd.
Al 9.5, Zn 0.5, Mn 0.3, Si 0.2, bal Mg.
Cast: 20,000 TS; 11,000 YS; 3 El; 50 Brin. Heat treated: 34,000 TS; 18,000 YS; 1 El; 80 Brin. For aircraft engine parts, crankcases; sand or die castings.

ELEKTRON AZ91
Birmingham Aluminium Casting Co.
Magnesium. Al 9-10.5, Zn 0.1-1, Mn 0.2-0.4, bal Mg.
Sand cast: 20,000 TS; 16,000 YS; 2 El; 60 Brin. Permanent mold: 28,000 TS; 16,000 YS; 5 El; 70 Brin. For engine casings, aircraft parts; age-hardenable.

ELEKTRON AZ91
F.A. Hughes & Co., Ltd.
Al 9.5, Zn 0.5, Mn 0.3, Si 0.2, bal Mg.
Cast: 20,000 TS; 11,000 YS; 3 El; 50 Brin. Heat treated: 34,000 TS; 18,000 YS; 1 El; 80 Brin. For aircraft engine parts, crankcases; sand or die castings.

ELEKTRON AZ91
Farbenindustrie Atkiengesellschaft
Magnesium. Al 9.5, Zn 0.5, Mn 0.2, bal Mg.
Cast: 24,000 TS; 13,500 YS; 2 El; 65 BHN. For die castings, engine parts; thin sections.

ELEKTRON AZ92A
Magnesium Elektron Ltd.
Magnesium. Al 9, Zn 2, Mn 0.3, bal Mg.
Cast: 34.0 ksi TS; 18.0 ksi YS; 1 El (T6 temper). General purpose alloy.

ELEKTRON AZF
Birmingham Aluminium Casting Co.
Al 4, Zn 3, Mn 0.2, bal Mg.
For light alloy sand castings. *Obsolete*

ELEKTRON AZG
Birmingham Aluminium Casting Co.
Al 6, Zn 2.25-3.5, Mn 0.2-0.5, bal Mg.
Cast: 22,000 TS; 14,000 YS; 5 El; 55 Brin. Chilled: 30,000 TS; 15,000 YS; 10 El; 60 Brin. For light alloy parts. *Obsolete*

ELEKTRON AZG
Magnesium Elektron Ltd.
Al 5.5, Zn 2.9, Mn 0.3, Si 0.2, bal Mg.
Cast: 22,000 TS; 11,000 YS; 3 El; 50 Brin. Heat treated: 34,000 TS; 11,000 YS; 8 El; 55 Brin. For light alloy castings. *Obsolete*

ELEKTRON AZG
F.A. Hughes & Co., Ltd.
Al 5.5, Zn 2.9, Mn 0.3, Si 0.2, bal Mg.
Cast: 22,000 TS; 11,000 YS; 3 El; 50 Brin. Heat treated: 34,000 TS; 11,000 YS; 8 El; 55 Brin. For light alloy castings. *Obsolete*

ELEKTRON AZM
Magnesium Elektron Ltd.
Magnesium. Mn 0.3, Al 6, Zn 1, bal Mg.
Extruded: 270 N/mm^2 TS; 180 N/mm^2 proof stress; 8 El; 7-60 VHN. General purpose alloy; gas and arc weldable. BS: 3373 MAG-E-121M.

ELEKTRON AZM
F.A. Hughes & Co., Ltd.
Magnesium. Al 6.2, Zn 1, Mn 0.3, Si 0.2, bal Mg.
Forged: 40,000 TS; 24,000 YS; 10 El; 70 Brin. Extruded: 49,000 TS; 31,000 YS; 10 El; 70 Brin. For aircraft wings and fuselage; general construction.

ELEKTRON C
F.A. Hughes & Co., Ltd.
Magnesium. Al 8, Zn 0.4, Mn 0.3, bal Mg.
Cast: 24,000 TS; 12,000 YS; 2 El; 60 Brin. Heat treated: 36,000 TS; 17,000 YS; 2 El; 80 Brin. For light parts, engine parts; heat treatable.

ELEKTRON C
Magnesium Elektron Ltd.
Magnesium. Al 7.5-9.5, Zn 0.3-1.5, 0.15 Mn min, bal Mg.
Cast: 24,100 TS; 11,200 YS; 4 El; 60 Brin. T4-temper: 35,000 TS; 11,200 YS; 8 El; 60 Brin. T6-temper: 35,000 TS; 14,000 YS; 2 El; 80 Brin. For aircraft engine components, rear axle casings; age hardenable.

ELEKTRON M2
Alcan-Booth Industries, Ltd.
Mn 1.5, bal Mg.
Rolled: 34,000 TS; 18,000 YS; 4 El. For housings, tanks, cold formed parts; corrosion resistant, ductile. *Obsolete*

ELEKTRON MCZ
Magnesium Elektron Ltd.
Magnesium. Zr 0.4-1, 2.5-4.0 Mischmetal, bal Mg.
Annealed: 20,000-25,000 TS; 11,000-13,500 YS; 3-6 El; 45-55 Brin. For engine components, aircraft structural parts; high creep strength at 500°F.

ELEKTRON MCZ
Sterling Metals Ltd.
Zr 0.4-1, 2.5-4.0% rare earth, bal Mg.
Sand cast: 20,000-25,000 TS; 12,200-13,400 YS; 3-5 El; 50-60 Brin; 1 IZ. Sand and gravity die cast magnesium alloy casting; creep resistant. *Obsolete*

ELEKTRON MG5
Sterling Metals Ltd.
Cu 0.15, Si 0.6, Fe 0.75, Mn 1, Ti 0.2, bal Al.
Extruded: 36,000 TS; 14,000 YS; 22 RA; 70 Brin. 1/4 H-temper: 42,600 TS; 31,400 YS; 8 El; 90 Brin. For seaplane propellers, rivets; resists sea water corrosion. *Obsolete*

ELEKTRON MSR/QE22

Magnesium Elektron Ltd.
Magnesium. Ag 2.5, RE 2, Zr 0.6, bal Mg.
Cast: 35.0 ksi TS; 25.0 ksi YS; 2 El; 70-90 Brin (T6 temper).
Heat treatable casting alloy. AMS 4418C.

ELEKTRON MTZ

Magnesium Elektron Ltd.
Magnesium. Th 3, Zr 0.7, bal Mg.
Cast: 27.0 ksi TS; 13.0 ksi YS; 4 El; 50-60 Brin (T6 temper).
Pressure tight and weldable; creep-resistant up to 650°F for short time applications. AMS 4445C.

ELEKTRON QH21A

Magnesium Elektron Ltd.
Magnesium. Ag 2-3, Th 0.4-1.6, RE 0.6-1.5, Zr 0.5-1, 75 Nd min, 1.6-2.2 RE + Th, bal Mg.
Rare earth casting alloy. Good mechanical properties to 250°C.

ELEKTRON RZ5

Magnesium Elektron Ltd.
Magnesium. Zn 4.2, RE 1.3, Zr 0.7, bal Mg.
Cast: 29.0 ksi TS; 19.5 ksi YS; 2.5 El; 55-70 Brin (T6 temper).
Easily cast; weldable, pressure tight; useful strength at elevated temperatures. AMS 4439.

ELEKTRON RZ5

Birmingham Aluminium Casting Co.
Zn 3.5-5, Zr 0.4-1, 0.75-1.75 rare earths, bal Mg.
Heat treated: 30,000-32,000 TS; 20,000-22,000 YS; 5-3 El; 65-75 Brin. For structural and engine parts; age-hardenable, high temperature use to 400°F.

ELEKTRON RZ5

Stone Manganese - J. Stone & Co. Ltd.
Zn 3.5-5, Zr 0.4-1, 0.75-1.75 rare earths, bal Mg.
Heat treated: 30,000-32,000 TS; 20,000-22,000 YS; 5-3 El; 65-75 Brin. For structural and engine parts; age hardenable, high temperature use to 400°F.

ELEKTRON RZ5

Sterling Metals Ltd.
Zn 3.5-5, Zr 0.4-1, 0.75-1.75% rare earth, bal Mg.
Sand cast: 24,600-27,000 TS; 12,400-14,500 YS; 3-5 El; 55-65 Brin; 1.5 IZ. Sand and gravity die cast magnesium alloy casting; weldable; for casting "spidery" castings. Gen. Eng. BS 2970 MAG 5. *Obsolete*

ELEKTRON TZ6

Magnesium Elektron Ltd.
Magnesium. Zn 5.5, Th 1.8, Zr 0.7, bal Mg.
Cast: 35.0 ksi TS; 22.0 ksi YS; 5 El; 65-75 Brin (T5 temper).
Weldable, pressure tight castings. AMS 4438B.

ELEKTRON TZ6

Sterling Metals Ltd.
Th 1.5-2.3, Zn 5-5.8, Zr 0.4-1, bal Mg.
Sand cast: 38,000-42,500 TS; 21,000-24,600 YS; 5-12 El; 65-75 Brin; 1.5 IZ. High strength thorium alloy, magnesium base casting; for use up to 150 C. AIRCRAFT DID 5015 (England). *Obsolete*

ELEKTRON TZ6

Magnesium Elektron Ltd.
Zn 5.5, Zr 0.7, Th 1.8, bal Mg.
Cast: 38,000 TS; 23,000 YS; 5 El; 65 Brin. Stabilized: 43,500 TS; 26,500 YS; 15 El; 75 Brin. For gear casings, engine castings; used to 400 F. *Obsolete*

ELEKTRON Z5Z

Magnesium Elektron Ltd.
Magnesium. Zn 4.5, Zr 0.7, bal Mg.
Cast: 34.0 ksi TS; 20.0 ksi YS; 5 El; 65-75 Brin (T5 temper).
General purpose casting with useful properties to about 300°F. AMS 4443B.

ELEKTRON Z5Z

Sterling Metals Ltd.
Zn 3.5-5.5, Zr 0.4-1, bal Mg.
Sand cast: 29,000-33,500 TS; 15,700-17,900 YS; 7-10 El; 55-65 Brin; 3 IZ. Sand and gravity die cast magnesium alloy castings; high yield strength; airframe applications. Gen. Eng. BS 2970 MAG 4. *Obsolete*

ELEKTRON Z5Z

Birmingham Aluminium Casting Co.
Magnesium. Zn 3.5-5.5, Zr 0.4-1, Mn 0-0.15, bal Mg.
Cast: 33,600 TS; 17,500 YS; 10 El; 55 Brin. Heat treated: 38,200 TS; 22,000 YS; 7 El; 65 Brin. For high strength castings; age-hardenable.

ELEKTRON ZE63

Magnesium Elektron Ltd.
Magnesium. Zn 5.8, RE 2.5, Zr 0.7, bal Mg.
Cast: 39.0 ksi TS; 26.0 ksi YS; 5 El; 60-85 Brin (T6 temper).
Excellent castability; pressure tight and weldable, good in thin wall castings. AMS 4425.

ELEKTRON ZM21

Magnesium Elektron Ltd.
Magnesium. Zn 2, Mn 1, bal Mg.
Extruded: 230 N/mm^2 TS; 150 N/mm^2 (10 mm) proof stress; 8 El; 50-65 VHN. Weldable by Argon arc process; sheet is easily formed. BS: 3373 MAG-E-131M.

ELEKTRON ZRE0

Magnesium Elektron Ltd.
Magnesium. Zn 0.5, Zr 0.6, 2.7 rare earths, bal Mg.
Annealed: 20,000-24,500 TS; 12,000-15,000 YS; 3-6 El; 50-60 Brin. For jet and turbine parts; creep resistant, used to 500°F.

ELEKTRON ZRE1

F.A. Hughes & Co., Ltd.
Magnesium. Ce 2.75, Zr 0.6, Zn 2.25, bal Mg.
Annealed: 22,000 TS; 12,000 YS; 5 El; 55 Brin. Heat treated: 40,500 TS; 23,500 YS; 5 El; 75 Brin. For light parts, aircraft engine parts; heat resistant, operating to 500°F.

ELEKTRON ZRE1

Sterling Metals Ltd.
Zn 0.8-3, Zr 0.4-1, 2.5-4.0% rare earth, bal Mg.
Sand cast: 20,000-22,400 TS; 12,300-13,400 YS; 3-5 El; 50-60 Brin; 1 IZ. Sand and gravity die magnesium alloy casting; pressure tight, heat treatable, weldable, good creep resistance up to 250 C. Gen. Eng. BS 2970 MAG 6. *Obsolete*

ELEKTRON ZRE1

Birmingham Aluminium Casting Co.
Magnesium. Zn 0.8-3, Zr 0.4-1, 2.5-4.0 rare earths, bal Mg.
At 20°C: 21,000 TS; 14,000 YS; 5 El; 50 Brin. At 200°C: 20,200 TS; 8960 YS; 30 El. For engine and turbine castings; high creep and heat resistance.

ELEKTRON ZRE1

Magnesium Elektron Ltd.
Magnesium. Zn 2.5, RE 3, Zr 0.6, bal Mg.
Cast: 20.0 ksi TS; 14.0 ksi YS; 2 El; 50-60 Brin (T5 temper).
Excellent castability; pressure tight and weldable; creep resistant to 500°F. AMS 4442B.

ELEKTRON ZRE2

Birmingham Aluminium Casting Co.
Zn 4, Ce 2, Zr 0.7, bal Mg.
Heat treated: 27,000-30,000 TS; 19,000-21,000 YS; 1-3 El. For engine castings; operating to 400 F. *Obsolete*

ELEKTRON ZRE2

Stone Manganese - J. Stone & Co. Ltd.
Zn 4, Ce 2, Zr 0.7, bal Mg.
Heat treated: 27,000-30,000 TS; 19,000-21,000 YS; 1-3 El. For engine castings; operating to 400 F. *Obsolete*

ELEKTRON ZRE2

Magnesium Elektron Ltd.
Zn 4, Ce 2, Zr 0.7, bal Mg.
Heat treated: 27,000-30,000 TS; 19,000-21,000 YS; 1-3 El. For engine castings; operating to 400 F. *Obsolete*

ELEKTRON ZRE3

Birmingham Aluminium Casting Co.
Zn 0.5, Ce 2.75, Zr 0.6, bal Mg.
Annealed: 20,000-23,500 TS; 11,000-13,500 YS; 3-6 El; 45-55 Brin. For pressure tight castings; creep resistant to 500 F. *Obsolete*

ELEKTRON ZRE3

Stone Manganese - J. Stone & Co. Ltd.
Zn 0.5, Ce 2.75, Zr 0.6, bal Mg.
Annealed: 20,000-23,500 TS; 11,000-13,500 YS; 3-6 El; 45-55 Brin. For pressure tight castings; creep resistant to 500 F. *Obsolete*

ELEKTRON ZRE3

Magnesium Elektron Ltd.
Zn 0.5, Ce 2.75, Zr 0.6, bal Mg.
Annealed: 20,000-23,500 TS; 11,000-13,500 YS; 3-6 El; 45-55 Brin. For pressure tight castings; creep resistant to 500 F. *Obsolete*

ELEKTRON ZRE0

Sterling Metals Ltd.
Zn 0.3-0.7, Zr 0.4-1, 2.5-4.0% rare earth, bal Mg.
Sand cast: 20,000-22,400 TS; 12,300-13,400 YS; 3-5 El; 50-60 Brin; 1 IZ. Sand and gravity die magnesium alloy casting; improved resistance to sea water; hardenable, weldable. AIRCRAFT DTD 718 (England). *Obsolete*

ELEKTRON ZT1

Magnesium Elektron Ltd.
Magnesium. Th 3, Zn 2.2, Zr 0.7, bal Mg.
Cast: 27.0 ksi TS; 13.0 ksi YS; 4 El; 50-60 Brin (T5 temper).
Pressure tight and weldable; creep-resistant up to 650°F. AMS 4447B.

ELEKTRON ZT1

Sterling Metals Ltd.
Th 2.5-4, Zn 1.7-2.5, Zr 0.4-1, bal Mg.
Sand cast: 27,000-31,400 TS; 12,300-13,400 YS; 5-8 El; 50-60 Brin; 1.5 IZ. Sand and gravity die cast magnesium alloy casting; free from porosity; creep resistant to 350 C. AIRCRAFT DTD 5005 (England). *Obsolete*

ELEKTRON ZT1

Howard Foundry Co.
Th 3, Zn 2.3, Zr 0.7, bal Mg.
Heat treated: 27,000-31,500 TS; 12,500-15,000 YS; 5-10 El; 50-60 Brin. For jet engine and aircraft castings; age hardenable, used up to 600 F. *Obsolete*

ELEKTRON ZTX

Magnesium Elektron Ltd.
Magnesium. Th 2.5, Zn 1, Zr 0.6, bal Mg.
33,600 TS; 20,200 YS; 15 El (at 20°C). 14,600 TS; 9500 YS; 32 El (at 225°C). For light weight parts, housings, castings.
Extrusion and forging alloy.

ELEKTRON ZTY

Magnesium Elektron Ltd.
Magnesium. Th 0.8, Zn 0.6, Zr 0.6, bal Mg.
Extruded: 230 N/mm^2 TS; 130 N/mm^2 proof stress; 6 El; 6-50 VHN. Weldable; creep resistant to 350°C.

ELEKTRON ZW 2

F.A. Hughes & Co., Ltd.
Magnesium. Zn 2, Zr 0.7, bal Mg.
Tube: 38,000 TS; 5 El; 75 Brin. For light parts; nonhardenable.

ELEKTRON ZW 3

F.A. Hughes & Co., Ltd.
Magnesium. Zn 3, Zr 0.7, bal Mg.
Rolled: 37,000 TS; 24,000 YS; 18 El; 60 Brin. Forgings: 49,000 TS; 33,000 YS; 8 El; 80 Brin. For light parts, aircraft structures; heavy duty.

ELEKTRON ZW1
Magnesium Elektron Ltd.
Magnesium. Zn 1.3, Zr 0.6, bal Mg.
Extruded: 250 N/mm^2 TS; 170 N/mm^2 (10 mm) proof stress; 6-8 El; 60-75 VHN. Can be forged and welded; sheet and plate available. BS: 3373 MAG-E-141M.

ELEKTRON ZW1
Alcan-Booth Industries, Ltd.
Zn 1.5, Zr 0.7, bal Mg.
Extruded: 45,000 TS; 31,500 YS; 5 El. Rolled: 40,000 TS; 27,000 YS; 8 El. For aircraft structures; heavy duty. *Obsolete*

ELEKTRON ZW3
Magnesium Elektron Ltd.
Magnesium. Zn 3, Zr 0.6, bal Mg.
Extruded: 280 N/mm^2 TS; 200 N/mm^2 (10 mm) proof stress; 8 El; 65-75 VHN. Can be forged or welded; sheet, plate available. BS: 3373 MAG-E-151M.

ELEKTRON ZW6
Magnesium Elektron Ltd.
Magnesium. Zn 5.5, Zr 0.06, bal Mg.
Extruded and precipitation hardened: 315 N/mm^2 TS; 230 N/mm^2 proof stress; 8 El; 60-75 VHN. High strength alloy; not weldable. BS: 3373-MAG-E-161TE.

ELEKTRON ZZ
Magnesium Elektron Ltd.
Magnesium. Zr 0.5-1, Zn 1-5, Cd 0-4, bal Mg.
For aircraft extrusions; high temperature use.

ELEKTRON-MSR
Magnesium Elektron Ltd.
Ag 2, Zr 0.7, 0.7-2.5% rare earths, bal Mg.
Heat treated: 78,000-79,000 TS; 50,400-55,000 YS; 2-4 El; 70-90 Brin. For transmission cases; heat treatable. *Obsolete*

ELEKTRON-MTZ
Magnesium Elektron Ltd.
Th 3, Zr 0.7, bal Mg.
T6 temper: 26,000-32,000 TS; 11,000-14,000 YS; 5-10 El; 55 Brin. For airframes, rockets and missile components. Good creep resistance and high temperature strength to 700 F. *Obsolete*

ELEPHANT BRAND HARDENING NO. 1
Seymour Products Co.
P 4, Sn 50, bal Cu.
For hardener for Cu alloys.

ELEPHANT BRAND HARDENING NO. 2
Seymour Products Co.
P 7, Sn 10, bal Cu.
For hardener for Cu alloys.

ELEPHANT BRAND NO. 113
Seymour Products Co.
Cu 90, Al 10.
42,500 TS; 19,000 YS; 9 El. For hardware; corrosion resistant.

ELEPHANT BRAND NO. 126R METAL
Seymour Products Co.
Cu 97.2, Sn 2.75, P 0.05.
For springs, diaphragms, friction plates. Phosphor Bronze Grade E.

ELEPHANT BRAND NO. 13 D
Seymour Products Co.
Sn 4-6, Pb 4-6, P 0.1, bal Cu.
55,000 TS; 30,000 YS; 12 El. For hardware; free-cutting.

ELEPHANT BRAND NO. 133 METAL
Seymour Products Co.
Cu 93.75, Sn 5, Pb 1, P 0.25.
Rod: 55,000-70,000 TS; 5-25 El. For screw machine parts. Phosphor Bronze Grade B1.

ELEPHANT BRAND NO. 16 METAL
Seymour Products Co.
Cu 95.7, Sn 4, P 0.3.
For springs, knife switches, diaphragms, electrical contacts. Phosphor Bronze Grade A.

ELEPHANT BRAND NO. 170 METAL
Seymour Products Co.
Cu 91.85, Sn 8, P 0.15.
For springs, diaphragms, gears, welding rod. Phosphor Bronze Grade C.

ELEPHANT BRAND NO. 175 METAL
Seymour Products Co.
Cu 93.8, Sn 6, P 0.2.
For fuse clips, welding rod, wire cloth. Phosphor Bronze Grade F.

ELEPHANT BRAND NO. 192 METAL
Seymour Products Co.
Cu 98.95, Sn 1, P 0.05.
For springs, hot forgings, signal wire. Phosphor Bronze Grade E.

ELEPHANT BRAND NO. 2 METAL
Seymour Products Co.
Cu 97.85, Sn 2, P 0.15.
For flexible hose, welding rod, line wire. Phosphor bronze Grade E.

ELEPHANT BRAND NO. 22 METAL
Seymour Products Co.
Cu 94.7, Sn 5, P 0.3.
For diaphragms, welding rods, springs. Phosphor Bronze Grade A.

ELEPHANT BRAND NO. 25 METAL
Seymour Products Co.
Cu 94.45, Sn 4, Pb 0.25, P 0.3.
Rod: 55,000-70,000 TS; 5-25 El. For bolts, nuts, hardware; Phosphor Bronze Grade B.

ELEPHANT BRAND NO. 28 METAL
Seymour Products Co.
Sn 9, Zn 0.3, P 0.01-0.35, bal Cu.
Rolled: 70,000-100,000 TS; 55,000-80,000 YS; 15-20 El. For diaphragms; P-Bronze Gr. D.

ELEPHANT BRAND NO. 3 METAL
Seymour Products Co.
Cu 96.95, Sn 3, P 0.05.
For electrical contacts, cold heading stock. Phosphor bronze Grade E.

ELEPHANT BRAND PHOSPHOR BRONZE NO. 130
Seymour Products Co.
Cu 90, Sn 5, Pb 5, P.
Cast: 55,000 TS; 45,000 YS; 20 El. For bearings; heavy loads.

ELEPHANT BRAND PHOSPHOR BRONZE NO. 146L
Seymour Products Co.
Cu 88, Sn 10, Pb 2.
For castings, hardware; free-cutting.

ELEPHANT BRAND PHOSPHOR BRONZE NO. 187
Seymour Products Co.
Cu 88, Sn 8, Pb 4, P.
For bearings; heavy duty.

ELEPHANT BRAND PHOSPHOR BRONZE NO. 190
Ste des Acieries de Micheville
Sn 10, Pb 5, bal Cu.
For bearings. Heavy duty.

ELEPHANT BRAND PHOSPHOR BRONZE NO. 64
Seymour Products Co.
Cu 79.8, Sn 10, Pb 10, P 0.2.
For castings; pressure tight.

ELEPHANT BRAND PHOSPHOR BRONZE NO. GK
Seymour Products Co.
Cu 83.5, Sn 8, Pb 8, P 0.5.
For castings, hardware; free-cutting.

ELEPHANT BRAND PHOSPHOR BRONZE NO. SN 1
Seymour Products Co.
Cu 78.9, Sn 10.5, Pb 9, Ni 0.75, P 0.85.
For bearings; heavy duty.

ELEPHANT BRAND PHOSPHOR BRONZE, F-2
Seymour Products Co.
Sn 18.5, P 0.5, bal Cu.
Chill cast: 43,000 TS; 0.2 El; 170 Brin. For slow moving bearings under extreme pressures; hard.

ELEPHANT BRAND PHOSPHOR BRONZE, GRADE A
Seymour Products Co.
Cu 95.5, Sn 3.9, P 0.3.
Rolled rod: 45,000-80,000 TS; 25-50 El. For springs, knife switches, diaphragms; very tough.

ELEPHANT BRAND PHOSPHOR BRONZE, GRADE B
Seymour Products Co.
Cu 96, Sn 3.25, Pb 0.25, P 0.5.
Rolled: 45,000-80,000 TS; 12-40 El. For machine castings, pinions, cogwheels, propeller screws, piston rods, hardware; corrosion resistant, tough and hard.

ELEPHANT BRAND PHOSPHOR BRONZE, GRADE B-
Seymour Products Co.
Cu 94, Sn 5, Pb 1, P 0.5.
Rolled rod: 60,000-70,000 TS; 25 Brin. For rods; rolled or drawn.

ELEPHANT BRAND PHOSPHOR BRONZE, GRADE B-
Delta Metal (BW) Ltd.
Cu 94, Sn 5, Pb 1, P 0.5.
Rolled rod: 60,000-70,000 TS; 25 Brin. For rods; rolled or drawn.

ELEPHANT BRAND PHOSPHOR BRONZE, GRADE C
Seymour Products Co.
Cu 92, Sn 8, P 0.5.
Rolled rod: 75,000-106,000 TS; 20-52 El. For valves, cocks, cylinder liners; wear and corrosion resistant; hard and durable.

ELEPHANT BRAND PHOSPHOR BRONZE, GRADE D
Seymour Products Co.
Cu 90, Sn 10, P 0.5.
Rolled rod: 89,000-135,000 TS; 16-29 El. For gears, valves, pumps, plungers, slides, powder mill tools; hard.

ELEPHANT BRAND PHOSPHOR BRONZE, GRADE E
Seymour Products Co.
Sn 2, P 0.15, bal Cu.
For valves, bearings of heated rolls; very hard.

ELEPHANT BRAND PHOSPHOR BRONZE, GRADE G

Seymour Products Co.
Cu 90, Sn 5, Pb 5, P 0.35.
For rods, bolts; very tough.

ELEPHANT BRAND PHOSPHOR BRONZE, GRADE H

Seymour Products Co.
Cu 97.5, Sn 2.5, P 0.08.
For sheets, wire; rolled or drawn.

ELEPHANT BRAND PHOSPHOR BRONZE, GRADE H

Delta Metal (BW) Ltd.
Cu 97.5, Sn 2.5, P 0.08.
For sheets, wire; rolled or drawn.

ELEPHANT BRAND PHOSPHOR BRONZE, GRADE S

Seymour Products Co.
P 0.7-1, Sn 9-11, Pb 8-11, bal Cu.
Cast: 35,000-40,000 TS; 19,000-31,000 YS; 3-4 El; 57-90 Brin. For bearings for locomotive, marine and stationary engines, roll neck bearings, piston rings; very hard and durable.

ELEPHANT BRAND PHOSPHOR BRONZE, GRADE V

Seymour Products Co.
Cu 87, Sn 7, Pb 6, P 0.5.
For rods; rolled or drawn.

ELEPHANT BRAND PHOSPHOR BRONZE, GRADE V

Delta Metal (BW) Ltd.
Cu 87, Sn 7, Pb 6, P 0.5.
For rods; rolled or drawn.

ELEPHANT BRAND PHOSPHOR-COPPER

Ste des Acieries de Micheville
P, bal Cu.
For hardener for Cu alloys. *Obsolete*

ELEPHANT BRAND PHOSPHORIZED ANTIFRICTION

Ste des Acieries de Micheville
Sn, Sb, P, bal Cu.
For antifriction lining, bearings; for heavy pressure. *Obsolete*

ELFUR IRON NO. 1

Baldwin-Lima-Hamilton Corp.
C 3.44-3.4, Si 1.3, Mn 0.6, Ni 1.5, bal Fe.
Cast: 35,000-56,000 TS; 180-275 Brin. For turbine casings, cylinder heads, liners, pistons; cast iron. *Obsolete*

ELFUR IRON NO. 2

Baldwin-Lima-Hamilton Corp.
C 3.3, Si 1.3, P 0.3, Mn 0.8, Ni 1.5, bal Fe.
Cast: 35,000-40,000 TS; 35,000-40,000 YS; 0 El; 0 RA; 185-215 Brin. For engine liners and cylinders; alloy cast iron. *Obsolete*

ELFUR IRON NO. 3

Baldwin-Lima-Hamilton Corp.
C 3.2, Si 1.5, Ni 3, Cr 0.3, bal Fe.
Cast: 35,000-45,000 TS; 225-275 Brin. For valve bushings, brake drums, camshafts; cast iron. *Obsolete*

ELGILOY

Elgiloy Limited Partnership
Cobalt base superalloy. Co 40, Cr 20, Ni 15, Mo 7, Mn 2, Fe 15, C 0.15, Be 0.05.
Heat treated: 380,000 TS; 280,000 YS; 702 Brin. Available in wire, rod, strip, cable, tubing for springs, instrumentation, seal components. Nonmagnetic, corrosion resistant, high strength.

ELGINITE

Elgin Watch Co.
Cr, Ni, Fe, Mo, Mn, bal Co.
For temperature compensating hair springs; constant flexibility from -35 to 122°F.

ELHANCO

English manufacture
C 0.7-1.2, V 0.2, bal Fe.
For general tools; water hardening.

ELIANITE I

Italian manufacture
Si 15, Mn 0.6, bal Fe.
For pumps, valves, drains, chemical plant equipment; acid resistant, brittle.

ELIANITE II

Italian manufacture
Si 15, Mn 0.5, Ni 2.2, C 0.82, bal Fe.
For pumps, valves, chemical plant equipment; acid resistant, brittle.

ELINVAR

Wallace Murray Corp.
Ni 36, Cr 12, Si 1-2, C 0.8, W, bal Fe.
Rolled: 107,000 TS; 64,500 YS. For watches, chronometers, hair springs, resistances; slightly magnetic.

ELINVAR

Telcon Metals Ltd.
Ni 36, Cr 12, Si 1-2, C 0.8, W, bal Fe.
Rolled: 107,000 TS; 64,500 YS. For watches, chronometers, hair springs, resistances; slightly magnetic.

ELINVAR EXTRA

Hamilton Technology Inc.
Ni 43, Cr 5, Ti 2.75, C 0.04, Al 0.3, Si 0.5, Co 0.35, bal Fe.
Precipitation hardening alloy; aged 160,000 psi TS; cold worked and aged 200,000 psi TS. For springs, orthodontic wires, flexures.

ELINVAR EXTRA

International Nickel Inc.
Nickel. Ni 41-43, Ti 2.4, Cr 5.5, C 0-0.6, Mn 0.5, Si 0.4, bal Fe.
Heat treated: 90,000 TS; 35,000 YS; 40 El; 125 Brin. Aged: 180,000 TS; 115,000 YS; 18 El; 330 Brin. Cold drawn: 200,000 TS; 180,000 YS; 7 El; 420 Brin. For springs, diaphragms, bourdon tubes; constant modulus. *Obsolete*

ELITE

Stahlwerk Stahlschmidt GmbH & Co.
C 1.35, W, Co, bal Fe.
For blanking and forming dies, punches; oil hardening, wear resistant.

ELKALOY A

CMW Inc.
Cu, Cd.
Bar, rod, strip, shapes: 65,000 TS; 60,000 YS; 15 El; 70 Rock B. Electrical conductivity: 90% IACS. Resistance welding electrodes (RWMA Class 1), electrical conducting parts, and contacts.

ELKALOY A

CMW Inc.
Cd, bal Cu.
Bars: 65,000 TS; 60,000 YS; 15 El; 50 RA; 120 Brin. For electrodes for resisting welding of steel; 90% conductivity. *Obsolete*

ELKALOY D

CMW Inc.
Cu, Al.
Cast: 70,000 TS; 30,000 YS; 12 El; 75 Rock B. Electrical conductivity: 18% IACS. For corrosion resistant jigs and fixtures, resistance welding electrodes (RWMA Class 5); flash welding dies for steel.

ELKALOY D-120

CMW Inc.
Al 11-12.2, Fe 3.25-4.5, bal Cu.
Cast: 80,000-85,000 TS; 35,000-70,000 YS; 2-10 El; 200-350 RA. For bushings, bearings, forming and flash welder dies; Al bronze, heat treatable, wear resistant. *Obsolete*

ELKALOY D110

CMW Inc.
Cu, Al, Fe.
Cast: 90,000 TS; 12 El; 88 Rock B. Electrical conductivity: 12% IACS. For corrosion and wear resistant jigs and fixtures; flash welding dies for steel.

ELKALOY D120

CMW Inc.
Cu, Al, Fe.
Cast: 85,000 TS; 4 El; 95 Rock B. Electrical conductivity: 10% IACS. For corrosion and wear resistant jigs and fixtures; heavy duty flash welding dies for steel.

ELKALOY D130

CMW Inc.
Al, Fe, bal Cu.
Cast: 85,000 TS; 0.5 El; 350 Brin. For forming dies and rolls, guide posts; 9% conductivity, Al bronze.

ELKALOY D140

CMW Inc.
Cu, Al, Fe, Ni.
Cast: 80,000 TS; 0.1 El; 42 Rock C. Electrical conductivity: 5% IACS. For corrosion and very wear resistant dies, rolls and guide posts where impact is minimal. *Obsolete*

ELKALOY DX

CMW Inc.
Cu alloy.
Cast: 120,000 TS; 240 Brin. For welding electrodes. *Obsolete*

ELKONITE (10-W-3)

CMW Inc.
W 77, Cu 23.
Sintered: 90,000 TS; 225 Brin. For resistance welding electrode dies, valve seats, switches, bearing metals; sintered alloy; does not respond to heat treatment. *Obsolete*

ELKONITE 10-W-53

CMW Inc.
Cu, bal W.
Heat treated: 160,000 TS. For resistance welding dies; heat treatable. *Obsolete*

ELKONITE 100M

CMW Inc.
Mo.
Wrought: 80,000 TS. For resistance welding electrodes; conductivity 31%. *Obsolete*

ELKONITE 100M MOLYBDENUM

CMW Inc.
Molybdenum.
Electrical conductivity: 31% IACS. 110,000 psi TS; 90 Rock B. RWMA Class 14 welding electrodes.

ELKONITE 100W

CMW Inc.
W.
Wrought: 50,000 TS; 280 Brin. For electrodes for spot welding copper and silver; sintered. *Obsolete*

ELKONITE 100W TUNGSTEN

CMW Inc.
tungsten (W).
100,000 psi TS; 70 Rock A hardness. RWMA Class 13 welding electrodes and electrical contacts.

ELKONITE 1062

CMW Inc.
Ag-WC-Co.
230 Brin. For electrical contacts; electrical conductivity 35%. *Obsolete*

ELKONITE 10W3
CMW Inc.
Cu 25, bal W.
Electrical conductivity: 45% IACS. 98 Rock B hardness. Density; 14.9 g/cm^3; transverse strength: 150,000 psi. For electrical contacts resistant to arcing, power transformer switches: RWMA Class 11, resistance welding electrodes. Semiconductor heat sink; EDM electrodes.

ELKONITE 10W53
CMW Inc.
W, Cu, Ni, Si.
Electrical conductivity: 27% IACS. 109 Rock B hardness. Density: 14.75 g/cm^3. Heat treated: 160,000 psi TS. For resistance welding products.

ELKONITE 1W
CMW Inc.
Cu, W.
For electrodes for welding steel. *Obsolete*

ELKONITE 1W3
CMW Inc.
Cu 45, bal W.
Electrical conductivity: 53% IACS. 75 Rock B hardness. Density: 12.5 g/cm^3; transverse rupture strength: 110,000 psi. For arcing and current carrying electrical contacts. RWMA Class 10 resistance welding electrodes.

ELKONITE 1W3
CMW Inc.
Cu 44, W 56.
Bar: 63,000 TS. For resistance electrodes for spot welding steel, electrical contacts. Electrical conductivity 55%. *Obsolete*

ELKONITE 2050C
CMW Inc.
Cu 50, bal W.
Electrical conductivity: 58% IACS. 70 Rock B hardness. Density: 12.0 g/cm^3. For current carrying and arcing electrical contacts.

ELKONITE 2052
CMW Inc.
Ag 48, W 51.75, C 0.25.
Sintered: Rock B 55. Electrical conductivity 65%. For electrical contacts. *Obsolete*

ELKONITE 20S
CMW Inc.
Ag 26, W 74.
Electrical conductivity: 50% IACS. 90 Rock B hardness. Density: 15.85 g/cm^3. Transverse strength: 130,000 psi. For electrical contacts resistant to arcing and welding for circuit breakers, etc.

ELKONITE 20W3
CMW Inc.
Cu, W.
For welding electrodes; sintered, 36% conductivity. *Obsolete*

ELKONITE 2110
CMW Inc.
Ag 90, bal W.
Wrought, annealed: 25 Rock B. Electrical conductivity: 92% IACS. For electrical contacts.

ELKONITE 2125C
CMW Inc.
Cu 75, W 25.
For electrical contacts. *Obsolete*

ELKONITE 2140C
CMW Inc.
Cu 58, Ni 2, bal W.
Density: 10.5 g/cm^3. For structural rotors.

ELKONITE 2150
CMW Inc.
Ag 50, W 50.
For electrical contacts. 60% electrical conductivity.

ELKONITE 2165
CMW Inc.
Ag 35, W 65.
Sintered: Rock B 65 (hard); B 55 (annealed). Electrical conductivity 50%. For electrical contacts. *Obsolete*

ELKONITE 2173
CMW Inc.
Ag 27, W 73.
Sintered: Rock B 68 (hard); B 58 (annealed). Electrical conductivity 41%. For electrical contacts. *Obsolete*

ELKONITE 2650
CMW Inc.
W 50, Ag 50.
Annealed: 33,000 TS; 21,500 YS; 3.5 El; 46 R-30T. Cold worked: 48,000 TS; 45,000 YS; 1.2 El; 55 R-30T. Electrical conductivity 68-72%; density 13.41. For thin strip, formable, electrical contacts. *Obsolete*

ELKONITE 2665
CMW Inc.
W 65, Ag 35.
Annealed: 56,500 TS; 39,500 YS; 1.8 El; 67 R-30 T. Cold worked: 60,000 TS; 58,000 YS; 0.5 El; 70 R-30 T. Electrical conductivity 57-59%; density 14.7. For thin strip, formable, electrical contacts. *Obsolete*

ELKONITE 2673
CMW Inc.
W 73, Ag 27.
Annealed: 55,000 TS; 48,000 YS; 0.6 El; 70 R-30 T. Cold worked: 64,000 TS; 62,000 YS; 74 R-30 T. Electrical conductivity 53-54%; density 15.49. For thin strip, formable, electrical contacts. *Obsolete*

ELKONITE 3042
CMW Inc.
Ag 58, bal WC.
Electrical conductivity: 52% IACS. 80 Rock B hardness. Density: 11.95 g/cm^3. Transverse strength: 80,000 psi. For arcing and current carrying electrical contacts resistant to welding and sticking.

ELKONITE 30S
CMW Inc.
Ag 30, bal W.
Electrical conductivity: 50% IACS. 85 Rock B hardness. Density: 15.25 g/cm^3. Transverse strength: 125,000 psi. For electrical contacts, circuit breakers. Resistant to arcing and welding.

ELKONITE 30W3
CMW Inc.
Cu 20, bal W.
Electrical conductivity: 40% IACS. 40 Rock B hardness. Density: 15.55 g/cm^3; 98,000 psi TS. For projection welding dies, die facing. RWMA Class 12 resistance welding electrode; low expansion structural members.

ELKONITE 3135
CMW Inc.
Ag 65, WC 35.
Bar: 75 Brin. For electrical contacts. Electrical conductivity 55%. *Obsolete*

ELKONITE 3150
CMW Inc.
Ag 50, bal WC.
Electrical conductivity: 45% IACS. Annealed: 60 Rock B hardness. Density: 11.93 g/cm^3. Transverse strength: 62,000 psi. For arcing and current carrying electrical contacts resistant to welding and sticking.

ELKONITE 3165
CMW Inc.
Ag 35, WC 65.
Hard: Rock B 85. Annealed: Rock B 62. Electrical conductivity 35%. For electrical contacts. *Obsolete*

ELKONITE 3250-C
CMW Inc.
Cu, WC.
190 Brin. For electrical contacts; electrical conductivity 45%.

ELKONITE 35S
CMW Inc.
Ag 34, bal W.
Electrical conductivity: 55% IACS. 78 Rock B hardness. Density: 14.85 g/cm^3. Transverse strength: 120,000 psi. For arcing and current carrying electrical contacts; household and power circuit breaker contacts.

ELKONITE 3W3
CMW Inc.
Cu 32, bal W.
Electrical conductivity: 50% IACS. 90 Rock B hardness. Density: 13.93 g/cm^3; transverse rupture strength: 130,000 psi. For arcing and current carrying electrical contacts, power transformer contacts, resistance welding electrodes.

ELKONITE 3W53
CMW Inc.
W, Cu, Ni, Si.
Electrical conductivity: 30% IACS. 105 Rock B hardness. Density: 14.00 g/cm^3; heat treated: 120,000 psi TS. For resistance welding electrodes.

ELKONITE 4050
CMW Inc.
Ag 50, bal W.
Electrical conductivity: 62% IACS. 65 Rock B hardness. Density: 13.4 g/cm^3. Transverse strength: 110,000 psi. For current carrying and arcing electrical contacts.

ELKONITE 4055
CMW Inc.
Ag 46, WC 19.6, bal W.
Electrical conductivity: 55% IACS. 85 Rock B hardness. Density: 13.3 g/cm^3. Transverse strength: 90,000 psi. For current carrying and arcing electrical contacts.

ELKONITE 40S
CMW Inc.
Ag 40, bal W.
Electrical conductivity: 58% IACS. 70 Rock B hardness. Density: 14.25 g/cm^3. For arcing and current carrying electrical contacts.

ELKONITE 40W3
CMW Inc.
Cu 13.35, bal W.
Electrical conductivity: 40% IACS. 25 Rock C hardness. Density: 16.6 g/cm^3. For vacuum switch contacts, low expansion structural members.

ELKONITE 45S
CMW Inc.
Ag 45, bal W.
Electrical conductivity: 62% IACS. 62 Rock B hardness. Density: 13.8 g/cm^3. For arcing and current carrying electrical contacts.

ELKONITE 5-S
CMW Inc.
W 88, Ag 12.
Sintered: 55,000 TS; 30 Rock C. Density: 17.53; electrical conductivity: 38%. For rocket nozzles; elevated temperature erosion resistant material. *Obsolete*

ELKONITE 50S

CMW Inc.

Ag 49, bal W.

Electrical conductivity: 65% IACS. 55 Rock B hardness. Density: 13.5 g/cm^3. Transverse strength: 80,000 psi. For arcing and current carrying electrical contacts.

ELKONITE 50W3

CMW Inc.

Cu 10.36, bal W.

Electrical conductivity: 35% IACS. 27 Rock C hardness. Density: 17.15 g/cm^3. For vacuum switch contacts.

ELKONITE 5W-53

CMW Inc.

Ag, bal W.

Heat treated: 120,000 TS. For resistance welding dies; heat treatable. *Obsolete*

ELKONITE 5W3

CMW Inc.

Cu 30, bal W.

Electrical conductivity: 48% IACS. 95 Rock B hardness. Density: 14.2 g/cm^3; transverse strength: 140,000 psi. For arcing and current carrying electrical contacts; power transformer contacts.

ELKONITE 7130

CMW Inc.

Ag, Mo.

Sintered: 121 Brin. For electrical contacts. *Obsolete*

ELKONITE 7150

CMW Inc.

Mo-Ag.

For electric contacts; sintered; 50% conductive. *Obsolete*

ELKONITE 7160

CMW Inc.

Ag 40, Mo 60.

Hard: Rock B 68. Annealed: Rock B 50. Electrical conductivity 45%. For electrical contacts. *Obsolete*

ELKONITE G-12

CMW Inc.

Ag 65, bal WC.

Electrical conductivity: 57% IACS. 57 Rock B hardness. Density: 11.55 g/cm^3. Transverse strength: 65,000 psi. For current carrying and arcing electrical contacts resistan to welding and sticking.

ELKONITE G-13

CMW Inc.

Ag 50, bal WC.

Electrical conductivity: 47% IACS. 91 Rock B hardness. Density: 12.35 g/cm^3. Transverse strength: 95,000 psi. For current carrying and arcing electrical contacts resistant to welding and sticking.

ELKONITE G-14

CMW Inc.

Ag 40, bal WC.

Electrical conductivity: 37% IACS. 100 Rock B hardness. Density: 12.9 g/cm^3. Transverse strength: 120,000 psi. For current carrying and arcing electrical contacts resistant to welding and sticking.

ELKONITE G-17

CMW Inc.

Ag 39.5, Cu 0.7, bal Mo.

Electrical conductivity: 47% IACS. 82 Rock B hardness. Density: 10.2 g/cm^3. Transverse strength: 135,000 psi. For arcing electrical contacts especially where low mass is required.

ELKONITE G-18

CMW Inc.

Ag 49.5, Cu 0.7, bal Mo.

Electrical conductivity: 55% IACS. 67 Rock B hardness. Density: 10.23 g/cm^3. Transverse strength: 110,000 psi. For current carrying and arcing electrical contacts, especially where low mass is required.

ELKONITE G12

CMW Inc.

Ag 65, WC 35.

Sintered: 35,000 TS; B 57 Rock. For electrical contacts. Electrical conductivity 57%. *Obsolete*

ELKONITE G14

CMW Inc.

Ag 40, WC 60.

Sintered: 60,000 TS; B 100 Rock. For electrical contacts. Electrical conductivity 37%. *Obsolete*

ELKONITE G18

CMW Inc.

Ag 50, Mo 50.

Sintered: 42,000 TS; B 75 Rock. For electrical contacts. Electrical conductivity 52%. *Obsolete*

ELKONITE TC-10

CMW Inc.

Cu 44, W 3.3, bal WC.

Electrical conductivity: 42% IACS. 100 Rock B hardness. Density: 11.65 g/cm^3. Transverse strength: 180,000 psi. For wiping and arcing electrical contacts and production welding tips.

ELKONITE TC-20

CMW Inc.

WC, Cu.

Electrical conductivity: 30% IACS. 37 Rock C hardness. Density. 12.65 g/cm^3. 85,000 psi TS. For electroforging and electrical upsetting dies.

ELKONITE TC-5

CMW Inc.

Cu 50, W 5, bal WC.

Electrical conductivity: 45% IACS. 94 Rock B hardness. Density: 11.25 g/cm^3. Transverse strength: 160,000 psi. For wiping and arcing electrical contacts and light duty welding tips.

ELKONITE TC-53

CMW Inc.

WC, Cu, Ni, Si.

Electrical conductivity: 18% IACS. 47 Rock C hardness. Density: 12.65 g/cm^3. Heat treated: 150,000 psi TS. For electroforging and electrical upsetting dies.

ELKONITE TC10

CMW Inc.

Cu 44, WC 56.

Sintered: 75,000 TS; B 99 Rock. For resistance welding electrodes and dies. Electrical conductivity 43%. *Obsolete*

ELKONITE TC5

CMW Inc.

Cu 50, WC 50.

Sintered: 75,000 TS; B 94 Rock. For resistance welding electrodes and dies. Electrical conductivity 47%. *Obsolete*

ELKONITE, OW3

CMW Inc.

Cu 50, W 50.

For resisting welding electrode; easy to machine. *Obsolete*

ELKONIUM 1

CMW Inc.

Ag 75, Cu 24.5, Ni 0.5.

Annealed: 45,000 TS; 32 El; 78 Rock 15T. Cold worked: 80,000 TS; 4 El; 88 Rock 15 T. Electrical conductivity: 75% IACS. Density: 5.27. For electrical contacts.

ELKONIUM 117

CMW Inc.

Ag-Cu-Cd-Ni.

Wrought: 86,000 TS; 2 El. For electrical contacts; conductivity 30%. *Obsolete*

ELKONIUM 12

CMW Inc.

Ag-Cu-Ni.

Wrought: 81,000 TS; 3 El. For electrical contacts; electrical conductivity 70%. *Obsolete*

ELKONIUM 15

CMW Inc.

Ag-Cu.

Wrought: 78,000 TS; 2 El. For electrical contacts; electrical conductivity 85%. *Obsolete*

ELKONIUM 17

CMW Inc.

Ag 77, Cd 22.6, Ni 0.4.

Annealed: 35,000 TS; 50 El. Wrought: 76,000 TS; 3 El. For electrical contacts. 31% electrical conductivity.

ELKONIUM 18

CMW Inc.

Ag 86.8, Cd 5.5, Ni 0.2, Cu 7.5.

Wrought: 75,000 TS; 50,000 YS; 3 El. For current carrying reeds and springs, make and break electrical contacts. 43% electrical conductivity.

ELKONIUM 21

CMW Inc.

Ag-Cu.

Wrought: 77,000 TS; 4 El. For electrical contacts; electrical conductivity 83%. *Obsolete*

ELKONIUM 217

CMW Inc.

Ag 85, Cd 15.

Annealed: 28,000 TS; 55 El. Wrought: 58,000 TS; 5 El. For electrical contacts. 35% electrical conductivity.

ELKONIUM 22

CMW Inc.

Ag 72, Cu 28.

Annealed: 53,000 TS; 20 El; 79 Rock 15T. Cold worked: 80,000 TS; 5 El; 85 Rock 15T. Electrical conductivity: 84. For electrical contacts.

ELKONIUM 23

CMW Inc.

Cu 1.5, bal Ag.

Wrought: 53,000 TS; 3 El. For electrical contacts; 97% conductivity

ELKONIUM 24

CMW Inc.

Ag, Be.

Wrought. For electrical contacts. *Obsolete*

ELKONIUM 26

CMW Inc.

Ag-Zn.

Wrought: 52,000 TS; 10 El. For electrical contacts; electrical conductivity 15%. *Obsolete*

ELKONIUM 28

CMW Inc.

Mn 5, bal Ag.

Wrought: 33,000 psi TS; 7 El; 73 Rock 15T. Electrical conductivity; 10% IACS. For electrical contacts.

ELKONIUM 30

CMW Inc.

Pt 99.9.

Annealed: 20,000 TS; 35 El. Wrought: 35,000 TS; 5 El. For electrical contacts. 15% electrical conductivity.

ELKONIUM 301

CMW Inc.
Pt 85, Ir 15.
Annealed: 75,000 TS; 12 El. Wrought: 120,000 TS; 5 El. For electrical contacts. 6% electrical conductivity.

ELKONIUM 302

CMW Inc.
Pt 73.4, Pd 18.4, Ru 8.2.
Wrought: 125,000 TS; 2 El; 240 Brin. For electrical contacts. 4% electrical conductivity.

ELKONIUM 305

CMW Inc.
Ru 5, bal Pt.
Wrought, worked: 115,000 TS; 5 El; 89 Rock 15T. Annealed: 60,000 TS; 18 El; 84 Rock 15T. Electrical conductivity: 5% IACS. For electrical contacts.

ELKONIUM 306

CMW Inc.
Ru 6, bal Pt.
For electrical contacts.

ELKONIUM 31

CMW Inc.
Pt 90, Ir 10.
Annealed: 55,000 TS; 20 El. Wrought: 90,000 TS; 5 El. For electrical contacts. Electrical conductivity 7%. *Obsolete*

ELKONIUM 32

CMW Inc.
Pt 80, Ir 20.
Annealed: 100,000 TS; 12 El. Wrought: 145,000 TS; 5 El. For electrical contacts. Electrical conductivity 5.5%. *Obsolete*

ELKONIUM 323

CMW Inc.
Pt-Pd.
Wrought: 56,000 TS; 5 El. For electrical contacts; electrical conductivity 8%. *Obsolete*

ELKONIUM 323

CMW Inc.
Pt, Pd, Ni.
Wrought: 70,000 TS; 5 El; 123 Brin. For electrical contacts; 10% electrical conductivity. *Obsolete*

ELKONIUM 33

CMW Inc.
Pt 89, Ru 11.
Annealed: 85,000 TS; 12 El. Wrought: 150,000 TS; 5 El. For electrical contacts. 3% electrical conductivity.

ELKONIUM 34

CMW Inc.
Pt 92, Ru 8.
Annealed: 70,000 TS; 15 El. Wrought: 130,000 TS; 5 El. For electrical contacts. 4% electrical conductivity.

ELKONIUM 35

CMW Inc.
Pt 65, Ir 35.
Annealed: 100,000 TS; 8 El. Wrought: 195,000 TS; 3 El. For electrical contacts. Electrical conductivity 5%. *Obsolete*

ELKONIUM 36

CMW Inc.
Pt 3, bal Ag.
Wrought, worked: 47,000 psi TS; 3 El; 77 Rock 15T. Annealed: 25,000 psi TS; 37 El; 45 Rock 15T. Electrical conductivity: 45% IACS. For electrical contacts.

ELKONIUM 37

CMW Inc.
Pt 86, Ru 14.
Annealed: 95,000 TS; 10 El. Wrought: 170,000 TS; 3 El. For electrical contacts. Electrical conductivity 4%. *Obsolete*

ELKONIUM 38

CMW Inc.
Pt 75, Ir 25.
Annealed: 125,000 TS; 10 El. Wrought: 170,000 TS; 5 El. For electrical contacts. Electrical conductivity 5.5%. *Obsolete*

ELKONIUM 39

CMW Inc.
Pt-Ir-Os.
For electrical contacts; electrical conductivity 4.5%. *Obsolete*

ELKONIUM 40

CMW Inc.
Pd 99.9.
Wrought: 47,000 TS; 5 El. Annealed: 28,000 TS; 28 El. For electrical contacts. 16% electrical conductivity.

ELKONIUM 401

CMW Inc.
Pd 35, Pt 9.5, Au 9, Cu 14, Ag 32.5.
Wrought: 150,000 TS; 2 El. For electrical contacts. Electrical conductivity 5.5%. *Obsolete*

ELKONIUM 404

CMW Inc.
Pd-Ru.
For electrical contacts; electrical conductivity 6.5%. *Obsolete*

ELKONIUM 405

CMW Inc.
Pd 89, Ru 11.
Wrought: 100,000 TS; 2 El. For electrical contacts. 6% electrical conductivity.

ELKONIUM 41

CMW Inc.
Ag 26, Ni 2, bal Pd.
Wrought, worked: 100,000 psi TS; 2 El; 90 Rock 15T. Annealed: 68,000 psi TS; 13 El; 82 Rock 15T. Electrical conductivity: 4% IACS. For electrical contacts.

ELKONIUM 42

CMW Inc.
Ag 40, bal Pd.
Wrought, worked: 100,000 TS; 5 El; 91 Rock 15T. Annealed: 54,000 TS; 20 El; 65 Rock 15T. Electrical conductivity: 4% IACS. For electrical contacts.

ELKONIUM 43

CMW Inc.
Pd 23, Ag 60, Cu 12, Ni 5.
Wrought: 110,000 TS; 3 El. For electrical contacts. 11.5% electrical conductivity

ELKONIUM 44

CMW Inc.
Ag 90, Pd 10.
Wrought: 56,000 TS; 3 El. For electrical contacts. Electrical conductivity 27%. *Obsolete*

ELKONIUM 45

CMW Inc.
Pd 3, bal Ag.
Wrought, worked: 48,000 TS; 3 El; 77 Rock 15T. Annealed: 27,000 TS; 37 El; 45 Rock 15T. Electrical conductivity: 58% IACS. For electrical contacts.

ELKONIUM 46

CMW Inc.
Pd 1, bal Ag.
Wrought, worked: 47,000 TS; 3 El; 76 Rock 15T. Annealed: 26,000 TS; 42 El; 44 Rock 15T. Electrical conductivity: 79% IACS. For electrical contacts.

ELKONIUM 460

CMW Inc.
Pd 60, Cu 40.
Wrought: 103,000 TS; 2 El. For electrical contacts. Electrical conductivity 8%. *Obsolete*

ELKONIUM 47

CMW Inc.
Pd 95, Ru 5.
Wrought: 75,000 TS; 2 El. For electrical contacts. Electrical conductivity 8%. *Obsolete*

ELKONIUM 49

CMW Inc.
Ag, Pd, Cu, Ni.
Wrought: 91,000 TS; 2 El; 185 Brin. For electrical contacts; 21% electrical conductivity. *Obsolete*

ELKONIUM 60

CMW Inc.
Ag alloy.
For springs; corrosion resistant. *Obsolete*

ELKONIUM 63

CMW Inc.
Ag 99.55, Mg 0.25, Ni 0.2.
Wrought: 50,000 TS; 6 El; 100 Brin. For electrical contacts. 71% electrical conductivity.

ELKONIUM 70

CMW Inc.
Au 72, Ag 2.62, Ni 1.8.
Wrought: 50,000 TS; 5 El; 81 Rock 15 T. For electrical contacts. 14% electrical conductivity.

ELKONIUM 71

CMW Inc.
Ag 90, Au 10.
Annealed: 29,000 TS; 28 El. Wrought: 46,000 TS; 3 El. For electrical contacts. 40% electrical conductivity.

ELKONIUM 72

CMW Inc.
Au 69, Ag 25, Pt 6.
Wrought: 55,000 TS; 4 El; 66 Brin. For electrical contacts. 11% electrical conductivity.

ELKONIUM 73

CMW Inc.
Cu, Cd, Ag, Au.
Wrought, worked: 65,000 TS; 2 El; 84 Rock 15T. Annealed: 37,000 TS; 40 El; 51 Rock 15T. Electrical conductivity: 90% IACS. For electrical contacts.

ELKONIUM 74

CMW Inc.
Au 50, Ag 50.
Electrical conductivity: 16%. For electrical contacts. Corrosion resistant. *Obsolete*

ELKONIUM 75

CMW Inc.
Au 71, Ag 5, Pt 9, Cu 15.
Annealed: 101,000 TS; Rock 15T-88.5. Cold worked: 170,000 TS; Rock 15N-75. Electrical conductivity 8%. For electrical contacts, brush contacts. *Obsolete*

ELKONIUM 76

CMW Inc.
Au 90, Cu 10.
Annealed: 58,000 TS; 76 Rock 15T. Cold worked: 102,000 TS; 91 Rock 15T. Electrical conductivity: 16%. For electrical contacts.

ELKONIUM NO. 25

CMW Inc.
Ag 93, Ti 3.
Wrought: 54,000 TS; 4 El. For electrical contacts; 60% conductivity. *Obsolete*

ELKONIUM NO. 57

CMW Inc.
Cu alloy.
For electrical contacts. *Obsolete*

ELKONIUM NO.13
CMW Inc.
Pd-alloy.
240 Brin. For electrical contacts. *Obsolete*

ELKRO
Disston Inc.
C 0.5, Cr 0.95, V 0.2, bal Fe.
For dies; oil hardening. *Obsolete*

ELMARID
English manufacture
C 5.9, W 83, Co 4.5, Fe 0.4, W_2C + WC.
For tips for high speed tools, dies.

ELMEDUR
Plansee, Metallwerk Gesellschaft
W alloy.
For resistance welding electrodes. Sintered alloy.

ELMET
Plansee, Metallwerk Gesellschaft
TiC, Co, Cr, Ni.
For electrical contacts. Sintered.

ELMET 2H
Compound Electro Metals Ltd.
Mo-Ag.
130 Brin. For electrical contacts; sintered.

ELMET 3C
Compound Electro Metals Ltd.
W-Cu-Ni.
198 Brin. For electrical contacts; sintered.

ELMET 4
Compound Electro Metals Ltd.
W-Cu-Ni.
275 Brin. For electrical contacts; sintered.

ELMET 5D
Compound Electro Metals Ltd.
W-Ag.
180 Brin. For electrical contacts; sintered.

ELMET 5K
Compound Electro Metals Ltd.
W-Ag.
110 Brin. For electrical contacts; sintered.

ELMET C
Compound Electro Metals Ltd.
Cu alloy.
120 Brin. For welding electrode; 82% conductivity; for spot welding iron.

ELMET CU
Compound Electro Metals Ltd.
Cu alloy.
350 Brin. For welding electrodes; 36% conductivity; spot welding of copper.

ELMET H-1
Compound Electro Metals Ltd.
Cu alloy.
120 Brin. For welding electrodes; butt welding, 59% conductivity.

ELMET H-2
Compound Electro Metals Ltd.
Cu alloy.
180 Brin. For butt welding electrodes; 45% conductivity.

ELMET H-3
Plansee, Metallwerk Gesellschaft
W alloy.
For resistance welding electrodes. Sintered alloy.

ELMET H-3
Compound Electro Metals Ltd.
W, bal Cu.
For butt welding electrodes; 36% conductivity.

ELMET HEAVY METAL
Plansee, Metallwerk Gesellschaft
W alloy.
For flywheel weights and protection against radioactivity.

ELMET HR
Plansee, Metallwerk Gesellschaft
TiC, Ni, Co, Cr.
For dies, tools, and jet engine components. High oxidation resistance; sintered.

ELMET S17
Plansee, Metallwerk Gesellschaft
W, Cu, Ni.
For balancing weights and isotope containers. Heavy metal.

ELMET S18
Plansee, Metallwerk Gesellschaft
W, Ni.
For balancing weights and isotope containers. Heavy metal.

ELMET U
Compound Electro Metals Ltd.
Cu alloy.
180 Brin. For spot welding electrodes; 68% conductivity.

ELMET-ROTUNG
Plansee, Metallwerk Gesellschaft
W, Cu.
For electrical contacts. Sintered.

ELMET-SILVUNG
Plansee, Metallwerk Gesellschaft
W, Ag.
For electrical contacts. Sintered alloy.

ELMET-W
Plansee, Metallwerk Gesellschaft
Cu, Cr.
For resistance welding electrodes. Sintered alloy.

ELNERS GERMAN SILVER
English manufacture
Cu 57.4, Zn 26.6, Ni 13, Fe 3.
For ornaments; nickel silver.

ELOXAL
German manufacture
Al alloy.
For light alloy parts; anodically oxidized.

ELPHAL
British Steel plc
Aluminum coated mild steel.

ELRAN
Vereinigte Metall. Ranshofen-Berndorf
Aluminum.
Aluminum alloy AlMgSi0.5.

ELSOLD
Bleiwerk Goslar GmbH & Co. KG
Solder wire (solid and filled with fluxes), special solders, electronic pastes and fluxes.

ELVERITE "B"
Babcock & Wilcox Co.
C 3.3, Si 2.4, Ni 1.5, Cr 0.7, bal Fe.
For wear resisting cast parts in pulverizers, crushers, stamp mills, conveyors; chilled cast iron. *Obsolete*

ELVERITE A
Babcock & Wilcox Co.
C 3.4-3.7, Si 0.5-1, bal Fe.
Cast: 500 Brin. For tube mill linings, wheels, jaw crusher sprockets; wear resisting; chilled cast iron. *Obsolete*

ELVERITE C
Babcock & Wilcox
C 3.5, Si 0.6, Cr 1.5, Ni 4.5, Mo 0.6, bal Fe.
Cast: 400-650 Brin. For coal and cement pulverizers, crushers, mixers; abrasion resistant.

ELVERITE C
Babcock & Wilcox Co.
Cr 1-1.8, C 3.5, Si 0.25-1, Ni 4.5, Mo 0.6, bal Fe.
Sand cast: 40,000-50,000 TS; 650-700 Brin. For coal and cement pulverizers, crushers, mixers; abrasion resistant. *Obsolete*

ELVERITE D
Babcock & Wilcox Co.
C 3.3, Si 2.2, Mo 0.4, Mn 0.7, bal Fe.
For castings; abrasive and wear resistant, cast iron. *Obsolete*

EM 15
Allegheny Ludlum Steel
Ni, bal Fe.
For glass to metal seals; controlled expansion. *Obsolete*

EM 2
Electrometal SA Metals Especials
C 0.88, Si 0.25, Mn 0.25, Cr 4.15, Mo 5.15, W 6.3, V 1.95, bal Fe.
High speed steel. AISI M2; W.-Nr. 1.3343.

EM 35
Electrometal SA Metals Especials
C 0.9, Si 0.3, Mn 0.0, Cr 4, Mo 5, W 6.3, V 2, Co 5, bal Fe.
High speed steel. W.-Nr. 1.3243.

EM 44
Electrometal SA Metals Especials
C 1.2, Si 0.25, Mn 0.3, Cr 4.15, Mo 5, W 6.4, V 2.75, Ni 0-0.3, Co 10, bal Fe.
Cobalt high speed steel. W.-Nr. 1.3207; similar to AISI M44.

EM 51
Manufacturer not listed
Ce 5-6, Mn 1.2, bal Mg.
For aircraft engine parts; heat treatable.

EM IV
Thyssen Edelstahlwerke AG
C 0.14, Mn 3, Si 0.12, bal Fe.
For welding electrode; general purpose. *Obsolete*

EM-62
German manufacture
Ce 6, Mn 2, bal Mg.
For engine cylinder heads, high temperature use.

EM10
Vallourec S.A.
C 0-0.15, Si 0.25-1, Cr 8-10, Mo 0.8-1.2, bal Fe.
Annealed: 60,000 TS; 30,000 YS; 30 El; 179 Brin. For oil refinery tubes and equipment; ASTM-A-212; AFNOR Z10CD9.

EM11
Vallourec S.A.
C 0-0.15, Cr 1-1.5, Mo 0.44-0.65, bal Fe.
Annealed: 60,000 TS; 30,000 YS; 30 El; 163 Brin. For oil refinery tubes and equipment; ASTM-A-188.

EM110
Vallourec S.A.
C 0.32, Cr 0.5, Mo 0.3, V 0.15, bal Fe.
High tensile strength; for casting.

EM12

Vallourec S.A.
C 0.15, Cr 9.5, Mo 2, Nb 0.4, bal Fe.
Good creep rupture properties up to 650°C. For boilers, superheaters.

EM17

Vallourec S.A.
C 0.34, Cr 1, Mo 0.3, bal Fe.
High tensile strength and high resilience under low temperature. For aeronautics, automotive, gas cylinders.

EM18

Vallourec S.A.
C 0.25, Cr 1, Mo 0.22, bal Fe.
Good strength; for automotive, aeronautics.

EM2

Vallourec S.A.
C 0-0.17, Ni 2.13-2.67, Si 0.13-0.32, bal Fe.
Annealed: 65,000-76,000 TS; 37,000 YS; 25 El. For low temperature tubing; AFNOR 9N10, minimum operating temperature -60°C.

EM24

Vallourec S.A.
C 0.3, Cr 0.75, Ni 2.75, bal Fe.
Good tensile strength; for aeronautics.

EM26

Vallourec S.A.
C 0.1, Cr 1.25, Mo 0.55, bal Fe.
Good elongation; for boilers, refineries.

EM29

Vallourec S.A.
C 0.4, Cr 0.95, Mo 0.2, bal Fe.
High tensile strength; for car industry, aeronautics.

EM35

Vallourec S.A.
C 0-0.19, Ni 3.18-3.82, bal Fe.
Annealed: 65,000 TS; 35,000 YS; 30 El; 190 Brin. For low temperature tubing; AFNOR 10N14, minimum operating temperature -100°C.

EM36A

Vallourec S.A.
C 0.05, Ni 9, bal Fe.
Good properties at low temperature; for liquefied gas lines, low temperature heat exchangers.

EM5

Vallourec S.A.
C 0-0.15, Cr 6-8, Mo 0.45-0.65, bal Fe.
Annealed: 60,000 TS; 24,000 YS; 30 El; 179 Brin. For oil refinery tubes and equipment; ASTM-A-335; AFNOR Z8CD7.

EM6

Vallourec S.A.
C 0-0.15, Si 1-2, Cr 4-6, Mo 0.45-0.65, bal Fe.
Annealed: 60,000 TS; 24,000 YS; 30 El; 179 Brin. For oil refinery tubes and equipment; ASTM-A-213.

EM7

Vallourec S.A.
C 0-0.15, Cr 4-6, Mn 0-0.5, Mo 0.45-0.65, bal Fe.
Annealed: 60,000 TS; 24,000 YS; 30 El; 163 Brin. For oil refinery tubes; ASTM-A-199.

EM75

Vallourec S.A.
C 0.44, Cr 0.15, Mo 0.2, bal Fe.
Good tensile strength; for oil industry, special casing.

EM8

Vallourec S.A.
C 0-0.15, Si 0-1.6, Cr 4-6, Mo 0.45-0.65, Al 0-1, bal Fe.
Annealed: 70,000-88,000 TS; 33,000 YS; 18 El; 179 Brin. For oil refinery tubes and equipment; AFNOR Z7CSAD5.

EM80

Vallourec S.A.
C 0.36, Mo 0.15, V 0.04, bal Fe.
Good tensile strength; for drill pipes, casing and tubing.

EM9

Vallourec S.A.
C 0.15-0.2, Cr 5-7, Mo 0.5-0.6, V 0.35-0.45, bal Fe.
Rolled: 92,000-114,000 TS; 65,000 YS; 18 El. For oil refinery tubes and equipment; AFNOR Z17CDV6.

EMBEEBUSH METAL

Delta Metal (BW) Ltd.
Cu alloy.
For castings; corrosion resistant. *Obsolete*

EME

Now UNITEMP EME.

EME

Midvale-Heppenstall Co.
C 0.15, Cr 19, Ni 12, W 3, Cb 1, bal Fe.
Rolled: 135,000 TS; 100,000 YS; 20 El; 45 RA. Forged: 115,000 TS; 80,000 YS; 25 El; 50 RA. For jet engine and missile components; stainless and heat resistant, creep resistant.

EMK

English manufacture
Fe 71, Mo 19, Co 19, Si 0.8.
For alloy for sealing in glass; coefficient of expansion 92 x 10^{-7}.

EMOCW

Thyssen Edelstahlwerke AG
C, Cr, Mo, bal Fe.
Oil treated: 114,000-143,000 TS; 85,000-114,000 YS; 11-17 El. For gears, shafts; oil hardening, tough. *Obsolete*

EMP 300 M

A.O. Smith-Inland Inc.
C 0-0.01, Mn 0.15, P 0.005, S 0.01, Si 0.02, 99%+ Fe, H_2 loss = 0.12%.
High grade iron powder for sintering. Particle size: 97% through 100 mesh screen, 52% through 250 mesh screen. To be blended, usually with graphite and zinc stearate, compressed and sintered. *Obsolete*

EMP 400 MS

A.O. Smith-Inland Inc.
Fe 99.3, C 0.02, Mn 0.25, P 0.007, S 0.2, Si 0-0.02, 0.10% O_2.
High grade iron powder for sintering. Particle size: 97% through 100 mesh screen, 51% through 250 mesh screen. To be blended, usually with graphite and zinc stearate, compressed and sintered. This grade produces free-machining material. *Obsolete*

EMP 4600

A.O. Smith-Inland Inc.
C 0-0.01, Mn 0.2, P 0.008, S 0.015, Si 0.02, Ni 1.9, Mo 0.5, 0.09% O_2, bal Fe.
High grade iron alloy powder for sintering. Particle size: 97% through 100 mesh screen, 53% through 250 mesh screen. To be blended, usually with graphite and zinc stearate, compressed and sintered. This grade permits oil hardenable parts. *Obsolete*

EMP 8600

A.O. Smith-Inland Inc.
C 0.02, Mn 0.2, P 0.01, S 0.015, Cr 0.47, Si 0.1, Ni 0.62, Mo 0.4, 0.24% O_2, bal Fe.
High grade iron alloy powder for sintering. Particle size: 97% through 100 mesh screen, 52% through 250 mesh screen. To be blended, usually with graphite and zinc stearate, compressed and sintered. This grade produces 8600 alloy steel. *Obsolete*

EMP 9400

A.O. Smith-Inland Inc.
C 0.02, Mn 0.2, P 0.01, S 0.015, Si 0.1, Cr 0.27, Ni 0.28, Mo 0.21, 0.15% O_2, bal Fe.
High grade iron alloy powder for sintering. Particle size: 97% through 100 mesh screen, 52% through 250 mesh screen. To be blended, usually with graphite and zinc stearate, compressed and sintered. This grade produces 9400 alloy steel. *Obsolete*

EMPEDUR

Plansee, Metallwerk Gesellschaft
WC, Fe.
For hard facing electrode. Sintered.

EMPEROR

Sutcliff, Speakman & Co.
C 0.2, Cr 20, Ni 10, bal Fe.
For hardening boxes, carburizing pans, furnaces, crucibles, glass molds; heat and corrosion resistant.

EMPEROR

German manufacture
Sn, Sb, Pb.
For bearings; anti-friction.

EMPEROR BRASS

English manufacture
Cu 60, Al 20, Zn 20.
For corrosion resistant parts.

EMPEROR CHROME NICKEL ALLOY

Sutcliff, Speakman & Co.
Stainless Steel. C 0.2, Ni 9, Cr 19, bal Fe.
For furnace parts; heat resistant, stainless, austenitic.

EMPIRE "D"

Empire Steel Castings Co.
Cr 26, Mo 4, Ni 25, V 1, C, bal Fe.
For engineering applications; heat and corrosion resistant. *Obsolete*

EMPIRE 12

Empire Steel Castings Co.
C 0.25, Ni 2-3, Cr 1-1.5, Mn 0.7, Mo 0.3, bal Fe.
For castings; tough. *Obsolete*

EMPIRE 1214

Empire Steel Castings Co.
Cr 14, C, bal Fe.
For corrosion resistant castings. *Obsolete*

EMPIRE 18

Empire Steel Castings Co.
C 0-0.25, Cr 4-6, Mn 0.5, bal Fe.
For castings; for oil well and refineries. *Obsolete*

EMPIRE 2010

Empire Steel Castings Co.
Cr 20, Ni 10, C, with or without Ti or Se, bal Fe.
For heat and corrosion resistant castings. *Obsolete*

EMPIRE 22 C

Empire Steel Castings Co.
Cr 22, Cu 1, C, bal Fe.
For heat and corrosion resisting castings. *Obsolete*

EMPIRE 2511

Empire Steel Castings Co.
Cr 25, Ni 11, C, with or without Ti or Se, bal Fe.
For heat and corrosion resistant castings. *Obsolete*

EMPIRE 2810

Empire Steel Castings Co.
Cr 28, Ni 10, C, Se, W, bal Fe.
For heat and corrosion resistant castings. *Obsolete*

EMPIRE 2815
Empire Steel Castings Co.
Cr 28, Ni 15, C, Se, W, bal Fe.
For heat and corrosion resistant castings. *Obsolete*

EMPIRE 30 H
Empire Steel Castings Co.
Cr 30, 1.5 min C, special alloys, bal Fe.
For heat and corrosion resisting castings. *Obsolete*

EMPIRE 46
Empire Steel Castings Co.
C 0.13-0.2, Cr 5, Mo or W, bal Fe.
For castings. *Obsolete*

EMPIRE CF-16FA
Empire Steel Castings Co.
C 0-0.16, Cr 18-21, Ni 9-12, Se 0.2-0.3, Mo 0-0.15, bal Fe.
Cast: 75,000 TS; 35,000 YS; 40 El; 160 Brin. For chemical, paper, textile and bleaching equipment; stainless, austenitic, free-cutting. *Obsolete*

EMPIRE CF-3
Empire Steel Castings Co.
C 0-0.03, Cr 18-21, Ni 9-12, bal Fe.
Cast: 75,000 TS; 36,000 YS; 45 El; 140 Brin. For welded fabrications, chemical plant equipment; stainless, austenitic; Type 304L. *Obsolete*

EMPIRE CF-30
Empire Steel Castings Co.
C 0-0.3, Cr 17-20, Ni 8-11, bal Fe.
Cast: 80,000 TS; 40,000 YS; 40 El; 160 Brin. For chemical plant and electrical equipment; stainless, austenitic. *Obsolete*

EMPIRE CF-3M
Empire Steel Castings Co.
C 0-0.03, Cr 18-21, Ni 10-13, Mo 2-3, bal Fe.
Cast: 78,000 TS; 40,000 YS; 44 El; 150 Brin. For welded acid resistant castings; stainless, austenitic; Type 316L. *Obsolete*

EMPIRE MACHINERY STEEL
Colt Industries
C 0.3, bal Fe.
For machinery parts. *Obsolete*

EMPIRE NO. 11
Empire Steel Castings Co.
C 0.25-0.35, Mn 0.6-0.8, Ni 2.25-2.75, Cr 1-1.25, Mo 0.3, bal Fe.
For castings. *Obsolete*

EMPIRE NO. 13
Empire Steel Castings Co.
C 0.25-0.35, Mn 0.6-0.8, Si 2.25-2.5, Ni 2.25-2.5, Cr 7-9, bal Fe.
For castings. *Obsolete*

EMPIRE NO. 16A
Empire Steel Castings Co.
C 0.1, Mn 0.4-0.7, Ni 7-9, Cr 17-21, bal Fe.
For stainless castings; stainless. *Obsolete*

EMPIRE NO. 18
Empire Steel Castings Co.
C 0.2, Cr 18, bal Fe.
80,000-100,000 TS; 65,000-75,000 YS; 15-25 El; 30-40 RA; 170-200 Brin. For corrosion resisting parts; corrosion resistant. *Obsolete*

EMPIRE NO. 18 CU
Empire Steel Castings Co.
C 0.21-0.35, Cr 18, Cu, bal Fe.
For heat and corrosion resistant parts; heat and corrosion resistant. *Obsolete*

EMPIRE NO. 18-8
Empire Steel Castings Co.
C 0.15, Cr 18, Ni 8, bal Fe.
80,000-100,000 TS; 35,000-50,000 YS; 5-15 El; 10-20 RA. For stainless and corrosion resisting parts; stainless, corrosion resistant. *Obsolete*

EMPIRE NO. 18B
Empire Steel Castings Co.
C 0.13-0.2, Cr 18, Si, bal Fe.
For heat and corrosion resistant parts; heat and corrosion resistant. *Obsolete*

EMPIRE NO. 18C
Empire Steel Castings Co.
C 0.21-0.35, Cr 18, bal Fe.
For heat and corrosion resistant parts; heat and corrosion resistant. *Obsolete*

EMPIRE NO. 23
Empire Steel Castings Co.
C 0.15-0.25, Mn 0.4-0.7, Ni 34-37, Cr 13-17, bal Fe.
For corrosion and heat resisting castings; corrosion and heat resistant. *Obsolete*

EMPIRE NO. 24-12
Empire Steel Castings Co.
C 0.25, Cr 24, Ni 12, bal Fe.
70,000-90,000 TS; 30,000-45,000 YS; 5-15 El; 10-20 RA. For heat and corrosion resisting parts; heat and corrosion resisting. *Obsolete*

EMPIRE NO. 25-5
Empire Steel Castings Co.
C 0.2, Cr 25, Ni 5, bal Fe.
For heat and corrosion resisting parts; heat and corrosion resisting. *Obsolete*

EMPIRE NO. 30
Empire Steel Castings Co.
C 0.3, Cr 30, bal Fe.
40,000-50,000 TS; 30,000-40,000 YS; 0-1 El; 0-2 RA. *Obsolete*

EMPIRE NO. 35-15
Empire Steel Castings Co.
C 0.35, Cr 15, Ni 35, bal Fe.
60,000-70,000 TS; 2-8 El; 1-3 RA; 180 Brin. For heat and corrosion resistant parts; heat and corrosion resistant. *Obsolete*

EMPIRE NO. 5
Empire Steel Castings Co.
C 0.13-0.2, Cr 5, bal Fe.
For corrosion resistant parts; corrosion resistant. *Obsolete*

EMPIRE NO. 60-20
Empire Steel Castings Co.
C 0.5, Cr 20, Ni 60, bal Fe.
80,000-90,000 TS; 50,000-60,000 YS; 20-25 El; 40-50 RA. For heat and corrosion resisting parts; heat and corrosion resistant. *Obsolete*

EMPIRE NO. 75-20
Empire Steel Castings Co.
Ni 75, Cr 20, Fe 5.
For heat and corrosion resistant parts; heat and corrosion resistant. *Obsolete*

EMPIRE RESISTA
Empire Steel Castings Co.
C 0-0.3, Cr 10, Ni 2.5, Al 1.5, Si 1.75, bal Fe.
For heat and corrosion resistant castings. *Obsolete*

EMPIRE SHA
Empire Steel Castings Co.
Cr 28, Ni 36, Mo 3, Cu 1.5, C, bal Fe.
For heat and corrosion resistant castings. *Obsolete*

EMPIRE STEEL
Colt Industries
C 0.08-0.18, bal Fe.
For screw machine products; case hardening. *Obsolete*

EMRO
Pyramid Steel Company
C, alloy, bal Fe.
For gears, shafts, hooks, chains, chisels; shock and fatigue resistant.

EMS-1
Eaton Corp.
C 0.45, Mn 0.4, Cr 8.5, Si 3.25, bal Fe.
Exhaust valve alloy. Known commercially as SIL-1; SAE HNV-3.

EMS-10
Eaton Corp.
C 0.52, Mn 9, Cr 27, Ni 4, N 0.44, bal Fe.
Exhaust valve alloy. Known commercially as 21-4N; SAE EV-5.

EMVAC 20
Electrometal SA Metals Especials
C 0-0.07, Si 0-1, Mn 0-2, Cr 19-21, Mo 2-3, Ni 24-30, Cu 3-4, bal Fe.
Corrosion resistant. Similar to alloy 20.

EMVAC 200
Electrometal SA Metals Especials
C 0-0.15, Si 0-0.35, Mn 0-0.35, 99.0 Ni min.
Corrosion resistant. DIN 2.4066. Similar to Nickel 200.

EMVAC 286
Electrometal SA Metals Especials
C 0-0.08, Si 0-1, Mn 0-2, Cr 13.5-16, Mo 1-1.5, Ni 24-27, V 0.1-0.5, Al 0-0.35, Ti 1.9-2.35, bal Fe.
High-temperature corrosion-resistant superalloy. DIN 1.4980. Similar to alloy 286.

EMVAC 400
Electrometal SA Metals Especials
C 0-0.15, Si 0-0.5, Mn 0-1.25, Fe 1-2.5, Cu 28-34, 63.0 Ni min.
Corrosion resistant. DIN 2.4360. Similar to Monel 400.

EMVAC 600
Electrometal SA Metals Especials
C 0-0.06, Si 0-0.5, Mn 0-1, Cr 14-17, Fe 6-10, 73.0 Ni min.
High-temperature corrosion-resistant alloy. Similar to Inconel 600.

EMVAC 601
Electrometal SA Metals Especials
C 0-0.1, Si 0-0.5, Mn 0-1, Cr 21-25, Ni 58-63, Al 1-1.7, Cu 0-1.
High-temperature corrosion-resistant alloy. Similar to Inconel 601.

EMVAC 625
Electrometal SA Metals Especials
C 0-0.05, Si 0-0.2, Mn 0-0.2, Cr 20-23, Mo 8-10, Al 0-0.4, Ti 0-0.4, Fe 0-5, Co 0-0.5, 3.15-4.15 Nb + Ta, bal Ni.
Corrosion resistant. DIN 2.4856. Similar to Inconel 625.

EMVAC 718
Electrometal SA Metals Especials
C 0.02-0.08, Si 0-0.35, Mn 0-0.35, Cr 17-21, Mo 2.8-3.3, Al 0.3-0.7, Ti 0.8-1.15, Fe 17-20, Co 0-1, 5.0-5.5 Nb + Ta, bal Ni.
High-temperature corrosion-resistant superalloy. Similar to Inconel 718.

EMVAC 751
Electrometal SA Metals Especials
C 0.03-0.1, Si 0-0.5, Mn 0-0.5, Cr 14-17, Mo 0-0.5, V 0-0.15, Al 1.1-1.35, Ti 2-2.6, Fe 5-9, Co 0-1, Cu 0-0.5, W 0-0.5, 0.7-1.2 Nb + Ta, bal Ni.
High-temperature corrosion-resistant superalloy. Similar to Inconel 751.

EMVAC 800 H
Electrometal SA Metals Especials
C 0.05-0.1, Si 0-1, Mn 0-1.5, Cr 19-23, Ni 30-35, Al 0.15-0.6, Ti 0.15-0.6, Cu 0-0.75, 39.5 Fe min.
Corrosion resistant. Similar to Incoloy 800 H.

EMVAC 825
Electrometal SA Metals Especials
C 0-0.045, Si 0-0.5, Mn 0-1, Cr 19.5-23.5, Mo 2.5-3.5, Ni 38-46, Al 0-0.2, Ti 0.6-1.2, Cu 1.5-3.
Corrosion resistant. DIN 2.4858. Similar to Incoloy 825.

EMVAC 90
Electrometal SA Metals Especials
C 0-0.13, Si 0-1, Mn 0-1, Cr 19-21, Al 0.8-2, Ti 1.8-3, Fe 0-2, Co 17-21, bal Ni.
High-temperature corrosion-resistant superalloy. DIN 2.4632. Similar to Nimonic 90.

EMVAC HB2
Electrometal SA Metals Especials
C 0-0.02, Si 0-0.01, Mn 0-1, Mo 26-30, Fe 0-2, bal Ni.
High-temperature corrosion-resistant Ni-base alloy. DIN 2.4615. Similar to Hastelloy B2.

EMVAC HC
Electrometal SA Metals Especials
C 0-0.02, Si 0-0.05, Mn 0-1, Cr 14-16, Mo 15-17, V 0-0.35, Fe 4-7, Co 0-2.5, bal Ni.
High-temperature corrosion-resistant Ni-base alloy. Similar to Hastelloy C 276.

EMVAC K500
Electrometal SA Metals Especials
C 0-0.25, Si 0-1, Mn 0-1.5, Ni 63-70, Al 2-4, Ti 0.25-1, Fe 0-2, bal Cu.
Corrosion resistant. DIN 2.4375. Similar to Monel K500.

EMVAC NIFE 37
Electrometal SA Metals Especials
Ni 37, bal Fe.
Soft magnetic alloy. Similar to Permenorm 3601 K5.

EMVAC NIFE 50
Electrometal SA Metals Especials
Ni 48, bal Fe.
Soft magnetic alloy. Similar to Permenorm 5000 H2.

EMVAC NIFE 75
Electrometal SA Metals Especials
Cr 0-1.5, Ni 75, Cu 4-5, bal Fe.
Soft magnetic alloy.

EMVAC WP
Electrometal SA Metals Especials
C 0.03-0.1, Si 0-0.15, Mn 0-0.5, Cr 18-21, Mo 3.5-5, Co 12-15, bal Ni.
High-temperature corrosion-resistant superalloy. Similar to Waspalloy.

EMVACAST 235
Electrometal SA Metals Especials
C 0.1-0.2, Si 0.25-0.6, Mn 0.15-0.4, Cr 14-17, Mo 4.5-6, Al 3.25-4, Ti 1.5-2.5, B 0.025-0.07, Fe 8-12, bal Ni.
Vacuum produced alloy; Ni-base. For gas turbine components. Similar to GMR 235.

EMVACAST 713C
Electrometal SA Metals Especials
C 0.08-0.2, Si 0-0.5, Mn 0-0.25, Cr 12-14, Mo 3.8-5.2, Al 5.5-6.5, Ti 0.5-1, Fe 0-2.5, 1.8-2.8 Nb + Ta, bal Ni.
Vacuum produced alloy; Ni-base. For gas turbines, blades, and rotors. DIN 2.4670. Similar to Inconel 713C.

EN12KHG
Russian manufacture
C 0.5-0.6, Mn 4.5-5.5, Cr 3-4.5, Ni 11-13, bal Fe.
For instrument parts; paramagnetic.

EN25
Russian manufacture
C 0.3-0.6, Cr 2-3, Ni 22-25, bal Fe.
For instrument parts; paramagnetic.

EN25 HIGH TENSILE STEEL
Eagle & Globe Steel Ltd.
Alloy steel. C 0.3, Si 0.25, Mn 0.6, Ni 2.5, Cr 0.5, Mo 0.5, bal Fe.
Heat treated. For high stress shafts in heavy truck axles, tractor axles and transmission shafts. BS970 826M31.

EN26 HIGH TENSILE STEEL
Eagle & Globe Steel Ltd.
Alloy steel. C 0.4, Si 0.25, Mn 0.6, Ni 2.5, Cr 0.5, Mo 0.5, bal Fe.
Heat treated. For severely loaded gears and axles. BS970 826M40.

EN36
Russian manufacture
C 0-0.26, Mn 0-0.7, Si 0-0.35, Ni 35-37, bal Fe.
For electrical equipment, instruments; low coefficient of expansion.

EN36A CASE HARDENING STEEL
Eagle & Globe Steel Ltd.
Alloy steel. C 0.12, Si 0.25, Mn 0.5, Ni 3.2, Cr 0.9, bal Fe.
Annealed, case hardened for gears in trucks, agricultural and mining machinery, pins and bushes, heavily loaded shafts and applications requiring a hard surface with a tough, shock resisting core. AS1444 X3312 (H); BS970 EN36A; BS970 655M13/655H13; SAE 9310.

EN39B CASE HARDENING STEEL
Eagle & Globe Steel Ltd.
Alloy steel. C 0.15, Si 0.25, Mn 0.5, Ni 4.2, Cr 1.2, Mo 0.2, bal Fe.
Annealed, case hardened for large gears. AS1444 X9315; BS970 EN39B; BS970 835M15/835H15; SAE 9315.

EN41B
Eagle & Globe Steel Ltd.
Alloy steel. C 0.4, Mn 0.6, Cr 1.5, Mo 0.2, Al 1, bal Fe.
Up to and including 150 mm diameter; 850-1000 MPa TS.
Nitriding steel for pivot balls, rocking levers, pneumatic hammer parts, lathe centers, gears and pinions. BS970 Part 2 S05M39.

EN42
Russian manufacture
C 0-0.3, Mn 0-0.8, Si 0-0.4, Ni 42-44, bal Fe.
For calibrating instruments; controlled expansion coefficient.

EN60
Russian manufacture
C 0-0.25, Mn 0.7-1.5, Cr 12-15, Al 3.5-5.5, bal Fe.
For electrical resistance and heating elements; applications to 1100°C.

ENCORE
J.M. Ney Co.
Au 48, Pd 40.5.
White color gold alloy for porcelain to gold restorations subject to stress; hard. 115,000 psi TS; 59,000 psi PL; 20 El; 140 Brin.

ENDEWRANCE 202
Dewramet Ltd.
C 0.05-0.12, Ti 0.18, Cb 0.6, Cr 26-30, Co 47-52, bal Fe.
Cast: 135,000 TS; 48,000 YS; 10 El; 10 RA; 250 Brin.
Wrought: 132,000 TS; 61,000 YS; 7 El; 6 RA; 350 Brin. For furnace baffles, burner tips, sintering grates, quench baskets. Corrosion and heat resistant. Resists thermal shock.
Obsolete

ENDEWRANCE 206T
Dewramet Ltd.
Nickel. Cr 19, Si 2.5, B 1.5, C 0.5, P 0-0.02, Co 50, bal Ni.
45 Rock C. For hot trimming; dies, tongue bits, hot punches; steam valve discs.

ENDEWRANCE 50
Dewramet Ltd.
Nickel. Si 3.5, B 1.9, Fe 0-1.5, C 0-0.06, P 0-0.02, bal Ni.
40 Rock C. For high temperature brazing; flow point 1055°C, high ductility.

ENDEWRANCE 50B
Dewramet Ltd.
Nickel. Si 2.4, B 1.4, Fe 0-0.5, C 0-0.06, P 0-0.02, bal Ni.
15 Rock C. For blades, knives and the glass industry.

ENDEWRANCE 50CU
Dewramet Ltd.
Nickel. Si 1.75, B 0.95, Fe 0-0.75, C 0-0.06, P 0-0.02, Cu 50, bal Ni.
35 Rock C. For rebuilding worn surfaces or undersize parts.

ENDEWRANCE 52
Dewramet Ltd.
Nickel. Si 4.5, B 2.9, Cr 7, Fe 0-0.5, C 0-0.06, P 0-0.02, bal Ni.
59 Rock C. For high temperature brazing; flow point 992°C, resists molten Ni, Hg, and Hg vapor.

ENDEWRANCE 53
Dewramet Ltd.
Nickel. Si 4.5, B 2.9, Fe 3, C 0-0.06, P 0-0.06, bal Ni.
59 Rock C. For high temperature brazing; flow point 996°C, corrosion resistant.

ENDEWRANCE 53WC
Dewramet Ltd.
Nickel. Cr 4.2, Si 2.75, B 1.75, Fe 2, P 0-0.03, WC 40, bal Ni.
Matrix: 600 Vickers. For gages, seal rings. Wear resistant.

ENDEWRANCE 54
Dewramet Ltd.
Nickel. Cr 10, Si 2.2, B 2, Fe 2.5, C 0.45, P 0-0.02, bal Ni.
40 Rock C. For dies, valves, shafts and the glass industry.

ENDEWRANCE 55
Dewramet Ltd.
Nickel. Cr 11, Si 4, B 2.5, Fe 4, C 0.6, P 0-0.02, bal Ni.
48 Rock C. For steam valve seats and discs, pump sleeves, extrusion screws.

ENDEWRANCE 56HC
Dewramet Ltd.
Nickel. Cr 16, Si 4.5, B 3.5, Fe 4, C 0.6, P 0-0.02, bal Ni.
59 Rock C. For exhaust valve stems, pump sleeves, pump shafts, thrust collars.

ENDEWRANCE 56HCWC
Dewramet Ltd.
Nickel. Cr 9.6, Si 2.7, B 2.1, Fe 2.5, P 0-0.02, WC 40, bal Ni.
Matrix: 600 Vickers. For exhaust valve stems, pump sleeves, pump shafts, thrust collars. Wear resistant.

ENDEWRANCE 56LC
Dewramet Ltd.
Nickel. Cr 15, Si 5, B 3.5, Fe 4, C 0-0.06, P 0-0.02, bal Ni.
59 Rock C. For exhaust valve stems, pump sleeves, pump shafts, thrust collars.

ENDEWRANCE 60
Dewramet Ltd.
Nickel. Cr 19, Si 10, C 0-0.15, P 0-0.02, bal Ni.
Boron-free alloy with low neutron cross section structure. For nuclear high radiation applications.

ENDEWRANCE-51
Dewramet Ltd.
B 1.8-5, Si 3-5, Cr 0-2, Fe 0-3.5, bal Ni.
Cast: 40-62 Rock C. For hard facing electrodes for pump sleeves, steam valves, gages. Heat, corrosion and wear resistant. *Obsolete*

ENDEWRANCE-CM.18
Dewramet Ltd.
Cr 25, Ni 15, Mo 8, Co 25, bal Fe.
Cast: 40-42 Rock C; 62,000 TS. For welding rods, steam valves, punches, pump sleeves, hot trimming dies. High wear resistance. Heat and corrosion resistant. *Obsolete*

ENDEWRANCE-CM.40
Dewramet Ltd.
Cr 30, Ni 10, Co 10, W 14, bal Fe.
Cast: 64,000 TS; 38-42 Rock C. For welding rods, valve and valve seats, gas turbine parts, superchargers. Heat, corrosion and wear resistant. *Obsolete*

ENDEWRANCE-CM2
Dewramet Ltd.
Cr 23, Mo 28, Co 10, bal Fe.
Cast: 112,000 TS; 50-54 Rock C. For hard facing welding rods, rollers, steel mill guides, pug mill paddles, coal pulverizing hammers. Heat and corrosion resistant. High wear resistance. *Obsolete*

ENDOTEC DO.04
Eutectic Corp.
Continuous electrode alloy for maintenance and repair welding, joining, build-up and removal of metal. Slag-free, Fe-base superalloy. Resists metal/metal friction, corrosion, and oxidation up to 1200°F (650°C). For high-temperature extrusion plungers, forging and trimming dies, valve heads and seats, and continuous casting drive rolls.

ENDOTEC DO.05
Eutectic Corp.
Continuous electrode alloy for maintenance and repair welding, joining, build-up and removal of metal. For mill wobblers, shovel teeth, crusher rolls, buckets, augers, and hydroelectric turbine runners. 20 Rock C (typical).

ENDOTEC DO.10
Eutectic Corp.
Continuous electrode alloy for maintenance and repair welding, joining, build-up and removal of metal. For coating of dragline buckets, shovel buckets, and cutter heads. Resists severe abrasion under high pressure. 52 Rock C (typical, 1 layer); 56 Rock C (typical, 2 layers).

ENDOTEC DO.14
Eutectic Corp.
Continuous electrode alloy for maintenance and repair welding, joining, build-up and removal of metal. Protective coating applications on plain carbon steels and ferritic and martensitic low and medium alloy steels. For the automotive, construction, and agricultural industries. 45 Rock C (typical).

ENDOTEC DO.29
Eutectic Corp.
Continuous electrode alloy for maintenance and repair welding, joining, build-up and removal of metal. Multi-component iron-base alloy containing Ni, Mo, and Cr to prevent corrosion. For pump casings, chemical vats, and pulp digesters. 80,000 psi TS (typical).

ENDOTEC DO.29S
Eutectic Corp.
Ni, Mo, Cr, bal Fe.
Continuous electrode alloy for maintenance and repair welding, joining, build-up and removal of metal. For pump casings, chemical vats, and pulp digesters. Contains special fluxing (slag) system.

ENDOTEC DO.30
Eutectic Corp.
Continuous electrode alloy for maintenance and repair welding, joining, build-up and removal of metal. For all steels, nickel alloys and stainless steel; for mining, crushing, construction, lumber, and cement industries. 62 Rock C (typical).

ENDOTEC DO.66
Eutectic Corp.
Ni, Mo, bal Fe.
Continuous electrode alloy for maintenance and repair welding, joining, build-up and removal of metal. For augers, bulldozers, mixer blades, tanks, and cranes. 85,0000 psi TS (typical); 26 El (typical).

ENDOTEC DO.68
Eutectic Corp.
Continuous electrode alloy for maintenance and repair welding, joining, build-up and removal of metal. For frames, gears, cylinders, dies, springs, and shafts. 105,000 psi TS (typical).

ENDOTEC DO.80
Eutectic Corp.
Continuous electrode alloy for maintenance and repair welding, joining, build-up and removal of metal. For valve plugs and seats, hot work dies, and hot punches. Resists high-temperature oxidation, corrosion, and impact. Work hardens at 45 Rock C.

ENDURANCE AA
Cu 87, Sn 12, P 1.
For aero and motor bronze castings; wear resistant, low coefficient of friction.

ENDURANCE BB
A. Cohn Ltd.
Cu 89, Sn 10.5, P 0.5.
For bushings; resists abrasive wear.

ENDURANCE CC
A. Cohn Ltd.
Cu 65-89, Sn 10, P 0.35.
For pumps, collars, bearings; corrosion resistant.

ENDURANCE DD
A. Cohn Ltd.
Cu 80, Sn 10, Pb 9.75, P 0.25.
For railway bearings, connecting rod bearings; heavy duty.

ENDURANCE EE
A. Cohn Ltd.
Cu 84, Sn 9.5, Pb 6.3, P 0.2.
For railway bearings, connecting rod bearings; heavy duty.

ENDURIA
Bethlehem Steel Corp.
C 0.7, bal Fe.
For springs. *Obsolete*

ENDURO "A"
Republic Steel Corp.
Cr 16.5-18.5, Ni 0.25, C 0.1, Si 0.5-1.25, Mn 0.5, bal Fe.
78,000 TS; 50,000 YS; 32 El; 70 RA. For cooking utensils, chemical plant apparatus, tanks; corrosion resistant. *Obsolete*

ENDURO "A.A."
Now ENDURO 430.

ENDURO "F.C."
Now ENDURO 416.

ENDURO "H C"
Now ENDURO 446.

ENDURO "HCN"
Now ENDURO 309.

ENDURO "K A2"
Republic Steel Corp.
Cr 16.5-19.5, Ni 7-10, 0.15% min C, 0.50% min Mn, 0.75% min Si, bal Fe.
Annealed: 85,000 TS; 35,000 YS; 50 El; 60 RA; 163 Brin. Heat treated: 85,000-95,000 TS; 30,000-40,000 YS; 57-60 El; 70-75 RA; 130-145 Brin. For cutlery, condensers, tanks, pipes, stills; stainless and corrosion resisting. *Obsolete*

ENDURO "K M-1"
Republic Steel Corp.
Cr 13-15, Ni 2, C 0.12, Mn 0.3-0.6, bal Fe.
130,000 TS; 90,000 YS; 14 El; 60 RA. For piston rods, valves; stainless and corrosion resisting. *Obsolete*

ENDURO "K.A.2 S"
Republic Steel Corp.
C 0.07, Cr 16.5-20, Mn 0.6, Ni 7-10.5, Si 0.75, bal Fe.
90,000 TS; 45,000 YS; 60 El; 65 RA; 140 Brin. For use in oil industries for combination of high temperature and high pressures and where severe acid corrosion is encountered; stainless; used for temperatures of 1000-1350 F. *Obsolete*

ENDURO "K.N.C.-3"
Republic Steel Corp.
C 0-0.2, Cr 23-27, Ni 17-21, Mn 0-1.5, Si 0-2, bal Fe.
Annealed: 100,000 TS; 50,000 YS; 45 El; 50 RA; 179 Brin. For corrosion and heat resistant parts stainless; resists heat to 2100 F. *Obsolete*

ENDURO "N.C.T.-3"
Republic Steel Corp.
C 0.2, Cr 23-27, Ni 17-21, bal Fe.
For furnace parts, retorts, grates, heat treating boxes; stainless. *Obsolete*

ENDURO "S"
Republic Steel Corp.
C 0-0.1, Cr 14-16, Mn 0.3-0.5, bal Fe.
Normalized: 85,000 TS; 48,000 YS; 24 El; 18 RA. Heat treated: 109,000-204,300 TS; 91,300-183,400 YS; 11-23 El; 36-69 RA; 207-340 Brin. For pump shafts, steam turbine parts, railroad equipment, cutlery, valve seats, scissors, valve parts; stainless and corrosion resisting. *Obsolete*

ENDURO "S" HIGH CARBON
Now ENDURO 420.

ENDURO "S" TURBINE QUALITY
Now ENDURO 403.

ENDURO "S-F.C"
Republic Steel Corp.
C 0.12, Cr 12.5-15, bal Fe.
Annealed: 75,000 TS; 40,000 YS; 25 El; 55 RA; 159 Brin. For corrosion resisting parts; stainless. *Obsolete*

ENDURO 17-7
Now ENDURO 301.

ENDURO 18-8
Now ENDURO 002.

ENDURO 18-8 CB
Now ENDURO 347.

ENDURO 18-8 B
Now ENDURO 302B.

ENDURO 18-8 FM
Now ENDURO 303.

ENDURO 18-8 FS
Now ENDURO 305.

ENDURO 18-8 HT
Republic Steel Corp.
C 0.08-0.2, Cr 17-20, Ni 7-10, Mn 0-1.5, bal Fe.
Annealed: 85,000 TS; 35,000 YS; 50 El; 80 Rb. Hard: 185,000 TS; 140,000 YS; 5 El; 40 Rc. For bus bodies, airplanes, trucks; stainless, austenitic. *Obsolete*

ENDURO 18-8 S
Now ENDURO 304.

ENDURO 18-8 S-CB
Now ENDURO 348.

ENDURO 18-8 S-MO
Now ENDURO 316.

ENDURO 18-8 S-MO-CB
Now ENDURO 318.

ENDURO 18-8 S-TI
Now ENDURO 321.

ENDURO 19-9
Republic Steel Corp.
C 0.08-0.2, Cr 18-20, Ni 8-10, bal Fe.
For stainless parts; austenitic. *Obsolete*

ENDURO 19-9 S-MO
Now ENDURO 317.

ENDURO 19-9S
Republic Steel Corp.
C 0-0.07, Cr 19, Ni 9, bal Fe.
For stainless parts; stainless. *Obsolete*

ENDURO 19-9S TI
Republic Steel Corp.
C 0-0.07, Cr 19, Ni 9, Ti, bal Fe.
For stainless parts; stainless. *Obsolete*

ENDURO 20-10
Republic Steel Corp.
C 0.08-0.2, Cr 19-22, Ni 9-12, bal Fe.
Annealed: 85,000 TS; 40,000 YS; 45 El. For heat and corrosion resisting purposes; heat and corrosion resistant. *Obsolete*

ENDURO 20-10 S
Now ENDURO 308.

ENDURO 20-23
Republic Steel Corp.
C 0-0.2, Cr 18-23, Si 0-0.5, bal Fe.
For stainless parts, still and condenser tubes, pump parts, fittings, valves; stainless, heat and corrosion resistant. *Obsolete*

ENDURO 20-25
Republic Steel Corp.
C 0-0.25, Cr 19-21, Ni 24-26, bal Fe.
For heat and corrosion resistant parts; heat and corrosion resistant. *Obsolete*

ENDURO 20-25
Republic Steel Corp.
C 0-0.25, Cr 19-21, Ni 24-26, bal Fe.
For heat and corrosion resistant parts; heat and corrosion resistant. *Obsolete*

ENDURO 4-6 CHROME
Republic Steel Corp.
Cr 4-6, C 0.1-0.25, Mn 0.5, Si 0.5, bal Fe.
Annealed: 65,000-77,000 TS; 28,000-41,000 YS; 31-34 El; 73-75 RA; 131-149 Brin. For still tubes, condenser tubes, hot oil and vapor lines, preheaters, pump parts; resists oxidation. *Obsolete*

ENDURO 4-6 CR-"MO"
Republic Steel Corp.
C 0-0.25, Mn 0-0.6, Cr 4-6, Mo 0.4-0.6, bal Fe.
Rolled: 170 Brin. For corrosion resisting parts; stainless and heat resistant. *Obsolete*

ENDURO 4-6 CR-"W"
Republic Steel Corp.
C 0-0.25, Mn 0-0.5, Cr 4-6, W 0.75-1.25, bal Fe.
For corrosion resisting parts; stainless and heat resistant. *Obsolete*

ENDURO 201
Republic Steel Corp.
C 0-0.15, Mn 5.5-7.5, Cr 16-18, Ni 3.5-5.5, N 0.25, bal Fe.
Rolled: 115,000 TS; 55,000 YS; 55 El; 185 Brin. For railroad car bodies; Type 201; stainless, austenitic. *Obsolete*

ENDURO 202
Republic Steel Corp.
C 0-0.15, Mn 7.5-10, Cr 17-19, Ni 4-6, N 0-0.25, bal Fe.
Rolled: 105,000 TS; 55,000 YS; 55 El; 185 Brin. For kitchen utensils, storage and pasteurizing equipment; Type 202; stainless, austenitic. *Obsolete*

ENDURO 301
Republic Steel Corp.
C 0-0.15, Mn 0-2, Si 0-1, Cr 16-18, Ni 6-8, bal Fe.
Annealed: 110,000 TS; 40,000 YS; 60 RA; 165 Brin. 1/2 H-temper: 150,000 TS; 110,000 YS; 16 El; 320 Brin. Hard: 185,000 TS; 140,000 YS; 9 El; 410 Brin. For transportation equipment and aircraft parts; stainless, austenitic; Type 301. *Obsolete*

ENDURO 302
Republic Steel Corp.
C 0-0.15, Mn 2, Cr 17-19, Ni 8-10, bal Fe.
Annealed: 85,000-90,000 TS; 35,000-40,000 YS; 50-60 El; 65-70 RA; 1 Brin. For evaporators, vessels, chemical engineering equipment, sterilizers, marine parts; stainless; resists scaling to 1450°F; Type 302. *Obsolete*

ENDURO 302 B
Republic Steel Corp.
C 0-0.15, Mn 0-2, Si 2-3, Cr 17-19, Ni 8-10, bal Fe.
Annealed: 90,000-95,000 TS; 40,000 YS: 55-50 El; 65 RA; 156-160 Brin. For annealing boxes, furnace parts, pump parts, pre-heaters; stainless, resists oxidation to 1700°F; Type 302B. *Obsolete*

ENDURO 303
Republic Steel Corp.
C 0-0.15, Cr 17-19, Ni 8-10, P, S, or Se 0.07, Mo 0-0.6, bal Fe.
Heat treated: 101,700 TS; 46,900 YS; 55 El; 62 RA; 179 Brin. Annealed: 90,000 TS; 35,000 YS; 50 El; 55 RA: 160 Brin. For tanks, vessels, chemical engineering equipment, pump shafts, valves; stainless and heat resistant; free machining. *Obsolete*

ENDURO 304
Republic Steel Corp.
C 0-0.08, Mn 0-2, Cr 18-20, Ni 8-11, bal Fe.
Annealed: 85,000 TS; 35,000 YS; 55 El; 70 RA; 150 Brin. For welded stainless structures where severe acid corrosion and high temperatures are encountered; stainless, corrosion resistant; Type 304. *Obsolete*

ENDURO 305
Republic Steel Corp.
C 0-0.12, Cr 17-19, Ni 10-13, bal Fe.
Annealed: 85,000 TS; 38,000 YS: 50 El; 880 Brin. For spinners, stainless parts; stainless, austenitic; Type 305. *Obsolete*

ENDURO 308
Republic Steel Corp.
C 0-0.08, Cr 19-21, Ni 10-12, Mn 0-0.2, bal Fe.
Annealed: 85,000-95,000 TS; 40,000-60,000 YS; 45-50 El; 60 RA; 150 Brin. For working electrodes and filler rods; heat and corrosion resistant. *Obsolete*

ENDURO 309
Republic Steel Corp.
C 0-0.20, Mn 0-20, Si 0-1, Cr 22-24, Ni 12-15, bal Fe.
Annealed: 90,000 TS; 45,000 YS; 50 El; 60 RA; 170 Brin. For furnace parts, oil refining and chemical plant equipment, skid rails; stainless, corrosion and heat resistant; resists heat to 1950°F. *Obsolete*

ENDURO 309 S
Republic Steel Corp.
C 0-0.08, Mn 0-2, Si 0-1, Cr 22-24, Ni 12-15, bal Fe.
Annealed: 90,000 TS; 45,000 YS; 50 El; 60 RA; 170 Brin. For furnace parts, skid rails, oil refining and chemical equipment; stainless and heat resistant to 1950°F. *Obsolete*

ENDURO 310
Republic Steel Corp.
C 0-0.25, Mn 0-2, Si 1.5, Cr 24-26, Ni 19-22, bal Fe.
Annealed: 95,000 TS; 45,000 YS; 45 El; 65 RA; 183 Brin. For retorts, valves, furnace parts, pump parts, rails; heat and corrosion resistant to 2000°F. *Obsolete*

ENDURO 316
Republic Steel Corp.
C 0-0.10, Mn 2, Cr 16-18, Ni 10-14, Mo 2-3, Si 0-0.1, bal Fe.
Annealed: 80,000-90,000 TS; 30,000-40,000 YS; 50-60 El; 65-70 RA; 150-156 Brin. For textile and dye equipment, chemical apparatus, paper and pulp mill equipment; stainless, corrosion resistant; Type 316. *Obsolete*

ENDURO 317
Republic Steel Corp.
C 0-0.10, Mn 0-2, Si 0-1, Cr 18-20, Ni 11-14, Mo 3-4, bal Fe.
Annealed: 90,000 TS; 40,000 YS; 45 El; 165 Brin. For corrosion and heat resistant parts; corrosion resistant, high temperature strength, austenitic. *Obsolete*

ENDURO 318
Republic Steel Corp.
C 0-0.08, Mn 0-2, Si 0-1, Cr 17-19, Ni 13-15, Mo 2-3, Cb = 10 x C min, bal Fe.
Annealed: 90,000 TS; 40,000 YS; 40 El; 172 Brin. For corrosion and heat resistant parts; corrosion resistant, high temperature strength, austenitic. *Obsolete*

ENDURO 321
Republic Steel Corp.
C 0-0.08, Mn 2, Si 0-1, Cr 17-19, Ni 7-9, Ti = 5 x C, bal Fe.
Annealed: 85,000-90,000 TS; 35,000 YS; 55-50 El; 65 RA; 150-156 Brin. For welded stainless structures where severe acid corrosion and high temperatures are encountered; stainless, corrosion resistant; Type 321. *Obsolete*

ENDURO 347
Republic Steel Corp.
C 0.15, Cr 17-19, Ni 7-9.5, Cb = 6-10 x C, bal Fe.
Annealed: 80,000-90,000 TS; 35,000-45,000 YS; 60-55 El; 60-55 RA; 135-185 Brin. For stainless parts, welded structures, tanks; non-magnetic, austenitic. *Obsolete*

ENDURO 348
Republic Steel Corp.
C 0-0.08, Cr 17-19, Ni 9-12, Mn 0-2, Cb = 10 x C, bal Fe.
Annealed: 80,000-95,000 TS; 35,000-45,000 YS; 60-55 El; 60-55 RA; 160 Brin. For airplane exhaust stacks, collector rings, heat resistors; stainless, austenitic, corrosion and heat resistant; Type 347. *Obsolete*

ENDURO 403
Republic Steel Corp.
C 0-0.15, Cr 11.5-13.0, bal Fe.
Annealed: 70,000 TS; 40,000 YS; 30 El; 60 RA; 155 Brin. Hardened: 110,000 TS; 85,000 YS; 20 El; 55 RA; 225 Brin. For turbine blades and parts; corrosion and heat resistant; Type 410. *Obsolete*

ENDURO 405
Republic Steel Corp.
C 0-0.08, Cr 11.5-13.5, Al 0.10-0.30, bal Fe.
Annealed: 65,000 TS; 40,000 YS; 25 El; 170 Brin. Cold drawn: 85,000 TS; 70,000 YS; 20 El; 60 RA; 185 Brin. For oil refining equipment; creep and heat resistant. *Obsolete*

ENDURO 410

Republic Steel Corp.
C 0-0.15, Cr 11.5-13.5, bal Fe.
Annealed: 65,000 TS; 35,000 YS; 35 El; 70 RA; 155 Brin. Heat treated: 110,000 TS; 85,000 YS; 23 El; 65 RA; 225 Brin. For cutlery, pump shafts, steam turbine parts, valve seats, rifle and revolver barrels; stainless and heat resistant; Type 410. *Obsolete*

ENDURO 414

Republic Steel Corp.
C 0-0.15, Mn 0-1, Si 0-1, Cr 11.5-13.0, Ni 1.25-2.50, bal Fe.
Heat treated: 210,000 TS; 105,000 YS; 15 El; 228 Brin. For heat treated springs; corrosion resistant; Type 414. *Obsolete*

ENDURO 416

Republic Steel Corp.
C 0-0.15, Mn 0-1.25, Cr 12-14, P 0-0.07, S 1, Se, Mo 0-0.6, Zr, bal Fe.
Annealed: 75,000 TS; 40,000 YS; 30 El; 60 RA; 160 Brin. Hardened: 110,000 TS; 85,000 YS; 18 El; 55 RA; 230 Brin. For turbine blades, valve stems, chemical engineering equipment, pump shafts, bushings; stainless, corrosion and heat resistant; free machining; Type 416. *Obsolete*

ENDURO 420

Republic Steel Corp.
C 0.15 min, Cr 12-14, bal Fe.
Annealed: 95,000 TS; 50,000 YS; 25 El; 55 RA; 195 Brin. For corrosion resistant parts, valves, cutlery; corrosion resistant; Type 420. *Obsolete*

ENDURO 430

Republic Steel Corp.
C 0-0.12, Cr 14-18, Mn 0-1.0, bal Fe.
Annealed: 80,000 TS; 50,000 YS; 25 El; 60 RA; 163 Brin. For valves heat transfer parts, condenser tubes, nitrogen fixation apparatus, pumps; non-hardening, corrosion resistant; Type 430; heat resistant to 1400°F. *Obsolete*

ENDURO 430 F

Republic Steel Corp.
C 0-0.12, Cr 14-18, Mn 0-0.60, P, S or Se 0.07 min, Mo or Zr 0.60, bal Fe.
Annealed: 80,000 TS; 55,000 YS; 25 El; 60 RA; 170 Brin. Rolled: 90,000 TS; 80,000 YS; 15 El; 55 RA; 190 Brin. For pump shafts, trim, oil burner parts, hardware, tanks, valves, screw machine parts; free machining, corrosion resistant. *Obsolete*

ENDURO 431

Republic Steel Corp.
C 0-0.20, Mn 0-1, Si 0-1, Cr 15-17, Ni 1.25-2.50, bal Fe.
Wire: 135,000 TS; 115,000 YS; 10 El; 50 RA; 290 Brin. Bars: 125,000 TS; 95,000 YS; 20 El; 55 RA; 200 Brin. For spring temper applications; corrosion resistant; Type 431. *Obsolete*

ENDURO 440C

Republic Steel Corp.
C 0.9-1.2 Cr 16-18, bal Fe.
Annealed: 125,000 TS; 90,000 YS; 10 El; 30 RA; 265 Brin. Hardened: 240,000 TS; 200,000 YS; 5 El; 10 RA; 550 Brin. For corrosion resistant parts, valves, bearings; corrosion and wear resistant; Type 440 C. *Obsolete*

ENDURO 446

Republic Steel Corp.
C 0-0.35, Cr 23-27, Mn 0-1.0, bal Fe.
Annealed: 80,000 TS; 45,000 YS; 25 El; 45 RA; 179 Brin. For annealing boxes, glass molds, valves, fittings, rabble arms, furnace parts; stainless, corrosion resistant; resists heat to 1900°F; Type 44. *Obsolete*

ENDURO AA HIGH CARBON

Now ENDURO 440 C.

ENDURO AA-FM

Now ENDURO 430F.

ENDURO AA-NI

Now ENDURO 431.

ENDURO FREE CUTTING (ENDURO F.C.)

Republic Steel Corp.
C 0-0.1, Cr 13.5-15, Mo 0.45-0.65, bal Fe.
For corrosion resisting parts; stainless. *Obsolete*

ENDURO HCN LOW CARBON

Now ENDURO 309S.

ENDURO K.A.-2-MO

Republic Steel Corp.
C 0-0.16, Mn 0-1.5, Ni 7-10.5, Mo 2-4, bal Fe.
For heat and corrosion resisting parts; stainless and heat resistant. *Obsolete*

ENDURO K.A.-2-S-MO

Republic Steel Corp.
C 0-0.07, Mn 0-1.5, Si 0-0.75, Cr 18-22, Ni 7-10.5, Mo 2-4, bal Fe.
For heat and corrosion resisting parts; stainless and heat resistant. *Obsolete*

ENDURO K.A.-2-S-TI

Republic Steel Corp.
C 0-0.07, Mn 0-0.6, Si 0-0.75, Cr 16.5-20, Ni 7-10.5, Mo 3.5, Ti, bal Fe.
For stainless parts, welded structures; stainless and heat resistant. *Obsolete*

ENDURO K.A.-2-TI

Republic Steel Corp.
C 0-0.16, Mn 0-0.6, Si 0.75, Cr 16.5-20, Ni 7-10.5, Mo 0.8, Ti, bal Fe.
For stainless parts, welded structures; stainless and heat resistant. *Obsolete*

ENDURO K.A.-28

Republic Steel Corp.
C 0.12-0.18, Mn 0-1.5, Si 2-2.5, Cr 16.5-19.5, Ni 8-10.5, bal Fe.
For heat and corrosion resisting parts; stainless and heat resistant. *Obsolete*

ENDURO NC-3

Now ENDURO 310.

ENDURO S-1

Now ENDURO 410.

ENDURO S-1 ALUMINUM

Now ENDURO 405.

ENDURO S-1-NI

Now ENDURO 414.

ENDURO S-15

Republic Steel Corp.
Cr 12.5-14.5, Ni 0.5, C 0.12-0.18, Si 0.5, bal Fe.
Rolled: 100,000 TS; 80,000 YS; 25 El; 55 RA. For pump shafts; corrosion resisting. *Obsolete*

ENDURON

Sheepbridge Engineering Ltd.
C 1.9, Si 1.9, P 0.06, Cr 16, bal Fe.
Cast: 78,000 TS; 269-321 Brin. For piston rings, cylinder liners; heat and abrasion resistant.

ENDURON

Sheepbridge Alloy Castings Ltd.
C 1.9, Si 1.9, P 0.06, Cr 16, bal Fe.
Cast: 78,000 TS; 269-321 Brin. For piston rings, cylinder liners; heat and abrasion resistant

ENDWELDUR

American Chain & Cable
C 0.2, bal Fe.
Normalized: 55,000 TS. For chains. *Obsolete*

ENDWELDUR 125

American Chain & Cable
C 0.4, Ni 3.5, bal Fe.
Heat treated: 125,000 TS. For chains. *Obsolete*

ENDWELDUR 85

American Chain & Cable
C 0.4, bal Fe.
Normalized: 85,000 TS. For chains. *Obsolete*

ENEBRA BRONZE

Eisenwerke Neubrandenburg G.m.b.H.
Sn, bal Cu.
For high static and dynamic parts; corrosion resistant.

ENEBRA EDELBRONZE NI

Eisenwerke Neubrandenburg G.m.b.H.
Sn, bal Cu.
Sand cast: 50,000-65,000 TS; 21,000-36,000 YS; 20-30 El; 110-130 Brin. For chemical industries; corrosion resistant.

ENEBRA EDELBRONZE TO

Eisenwerke Neubrandenburg G.m.b.H.
Sn, bal Cu.
Sand cast: 71,000-85,000 TS; 26,000-36,000 YS; 20-30 El; 120-150 Brin. For chemical and textile industries; corrosion resistant.

ENEBRA EDELBRONZE TU-I

Eisenwerke Neubrandenburg G.m.b.H.
Sn, bal Cu.
Sand cast: 92,000-107,500 TS; 43,000-56,000 YS; 8-15 El; 150-190 Brin. For chemical and textile industries; corrosion resistant.

ENEBRA EDELBRONZE TU-II

Eisenwerke Neubrandenburg G.m.b.H.
Sn, bal Cu.
Sand cast: 78,500-92,000 TS; 28,000-43,000 YS; 20-30 El; 120-150 Brin. For chemical and textile industries; corrosion resistant.

ENEBRA LAGERBRONZE B-33

Eisenwerke Neubrandenburg G.m.b.H.
Sn, bal Cu.
For machine parts, automobile and diesel engine parts; high strength.

ENEBRA LAGERBRONZE BZ-20

Eisenwerke Neubrandenburg G.m.b.H.
Sn, bal Cu.
For machine parts, automobile and diesel engine parts; high strength.

ENEBRA LAGERBRONZE BZ-25

Eisenwerke Neubrandenburg G.m.b.H.
Sn, bal Cu.
For machine parts, automobile and diesel engine parts; high strength.

ENEBRA LAGERBRONZE PN

Eisenwerke Neubrandenburg G.m.b.H.
Sn, bal Cu.
36,000-51,000 TS; 23,000-28,000 YS; 10-18 El; 70-90 Brin. For pump parts; corrosion resistant.

ENEBRA SONDERMESSINGE S-IIM

Eisenwerke Neubrandenburg G.m.b.H.
Zn, bal Cu.
56,000-72,000 TS; 19,000-25,000 YS; 25-35 El; 80-120 Brin. For general construction.

ENEBRA SONDERMESSINGE S-IM

Eisenwerke Neubrandenburg G.m.b.H.
Zn, bal Cu.
78,400-93,000 TS; 36,000-43,000 YS; 10-15 El; 130-160 Brin. For general construction.

ENEBRA SPEZIALBRONZE EH-I
Eisenwerke Neubrandenburg G.m.b.H.
Sn, bal Cu.
85,000-100,800 TS; 36,000-50,000 YS; 10-15 El; 150-180 Brin. For electrical industries.

ENEBRA SPEZIALBRONZE EH-II
Eisenwerke Neubrandenburg G.m.b.H.
Sn, bal Cu.
72,000-85,000 TS; 28,000-36,000 YS; 15-25 El; 130-150 Brin. For electrical industries.

ENEBRA SPEZIALBRONZE EH-III
Eisenwerke Neubrandenburg G.m.b.H.
Sn, bal Cu.
72,000-78,500 TS; 25,000-31,000 YS; 20-30 El; 110-130 Brin. For electrical industries.

ENGALOY 129
Engelhard Corp.
Pd 60, Ni 40.
For brazing electronic components. 2260°F MP; eutectic alloy. Good wettability.

ENGALOY 135
Engelhard Corp.
Pd 21, Ni 48, Mn 31.
Shear strength: 38,000 psi at 1250°F; 20,000 psi at 1500°F; 50,000 psi at RT. For brazing electronic components. Eutectic alloy; 2050°F MP. Good wettability and ductility.

ENGALOY 142
Engelhard Corp.
Pd 60, Ni 39.8, B 0.05, Li 0.15.
For brazing electronic components. 1630°F MP. Eutectic alloy. Good wettability and ductility.

ENGALOY 14597
Engelhard Corp.
Cu 55, Pd 20, Ni 15, Mn 10.
Shear strength: 27,000 psi at 1000°F; 21,000 psi at 1250°F; 18,000 psi at 1500°F. For brazing electronic components. 1940-2020°F MP. Good wettability and ductility.

ENGALOY 238
Engelhard Corp.
Au 80, Cu 20.
Annealed: 67,200 psi TS; 17 El. For brazing electronic components. 1630°F MP. Eutectic alloy. Good wettability and ductility.

ENGALOY 241
Engelhard Corp.
Au 50, Cu 50.
Annealed wire: 77,800 psi TS; 32 El. For brazing filler metal for electronic applications. 950-975°C MP.

ENGALOY 242
Engelhard Corp.
Au 37.5, Cu 62.5.
Annealed: 53,400 psi TS; 35 El. For brazing electronic components. 1814-1859°F MP. Good wettability and ductility.

ENGALOY 243
Engelhard Corp.
Au 35, Cu 62, Ni 3.
Annealed: 93,700 psi TS; 9 El. For brazing electronic components. 1787-1886°F MP. Good wettability and ductility.

ENGALOY 255
Engelhard Corp.
Au 82, Ni 18.
Annealed: 100,800 psi TS; 5 El. For brazing electronic components. 1742°F MP. Eutectic alloy. Good wettability and ductility.

ENGALOY 259
Engelhard Corp.
Au 94, Cu 6.
Annealed wire: 39,600 psi TS; 31 El. For brazing filler metal for electronic applications. 965-990°C MP.

ENGALOY 260
Engelhard Corp.
Au 35, Cu 65.
Annealed: 53,900 psi TS; 40 El. For brazing electronic components. 1832-1870°F MP. Good wettability.

ENGALOY 261
Engelhard Corp.
Au 75, Ag 5, Cu 20.
Annealed wire: 78,900 psi TS; 36 El. Uses: Brazing alloy for electronic components. 1625-1640°F MP. Good wettability and ductility.

ENGALOY 265
Engelhard Corp.
Au 72, Ni 22, Cr 6.
Brazing alloy for electronic components. 1785-1835°F MP. Good wettability and ductility.

ENGALOY 269
Engelhard Corp.
Au 81.5, Cu 15.5, Ni 3.
Annealed wire: 90,000 psi TS; 25 El. For brazing filler metal for electronic applications. 900-910°C MP.

ENGALOY 428
Engelhard Corp.
Pd 5, Ag 95.
Annealed wire: 31,500 psi TS; 28 El. Uses: Brazing alloy for electronic components. 1780-1850°F MP. Good wettability and ductility.

ENGALOY 431
Engelhard Corp.
Pd 10, Ag 90.
Annealed wire: 31,600 psi TS; 44 El. For brazing alloy for electronic components. 1835-1950°F MP. Good wettability and ductility.

ENGALOY 440
Engelhard Corp.
Pd 20, Ag 75, Mn 5.
Shear strength: 19,000 psi at 1250°F; 14,000 psi at 1500°F. For brazing electronic components. 1830-2050°F MP. Good wettability.

ENGALOY 441
Engelhard Corp.
Pd 33, Ag 64, Mn 3.
Shear strength: 27,000 psi at 1250°F; 18,000 psi at 1500°F. For brazing electronic components. 2100-2190°F MP. Good wettability.

ENGALOY 447
Engelhard Corp.
Pd 20, Ag 80.
Annealed wire: 39,000 psi TS; 34 El. For brazing alloy for electronic components. 1960-2150°F MP.

ENGALOY 478
Engelhard Corp.
Pd 5, Ag 68, Cu 27.
Annealed wire: 65,100 psi TS; 23 El. For brazing alloy for electronic components. 1480-1490°F MP. Good wettability and ductility.

ENGALOY 485
Engelhard Corp.
Pd 30, Ag 70.
Uses: Brazing alloy for electronic components. 2120-2250°F MP. Good wettability and ductility.

ENGALOY 490
Engelhard Corp.
Pd 15, Ag 65, Cu 20.
Annealed wire: 68,300 psi TS; 20 El. Uses: Brazing alloy for electronic components. 1565-1650°F MP. Good wettability and ductility.

ENGALOY 491
Engelhard Corp.
Pd 10, Ag 58, Cu 32.
Annealed wire: 69,500 psi TS; 22 El. Brazing alloy for electronic components. 1520-1565°F MP. Good wettability and ductility.

ENGALOY 492
Engelhard Corp.
Pd 25, Ag 54, Cu 21.
Annealed wire: 73,000 psi TS; 15 El. Brazing alloy for electronic components. 1650-1740°F MP. Good wettability and ductility.

ENGALOY 493
Engelhard Corp.
Pd 20, Ag 52, Cu 28.
Annealed wire: 71,700 psi TS; 25 El. Brazing alloy for electronic components. 1615-1650°F MP. Good wettability and ductility.

ENGALOY 845
Engelhard Corp.
Ag 84.6, Cu 7.5, Pb 2.2, In 5.5, Li 0.2.
For brazing electronic components. 1400-1615°F MP. Good wettability and ductility.

ENGELHARD 15065
Toughite Process Co.
Mg 0.2, Ni 0.2, bal Ag.
Annealed: 32,000-38,000 TS; 12-27 El; 75 Knoop Hard. Hardened: 68,000-74,000 TS; 2-6 El; 169 Knoop Hard. For springs, switches, relays. Oxidation hardened. Good spring properties. 75% elec. cond. as hardened.

ENGELHARD 15065
Engelhard Minerals & Chemicals Corp.
Mg 0.2, Ni 0.2, bal Ag.
Annealed: 32,000-38,000 TS; 12-27 El; 75 Knoop Hard. Hardened: 68,000-74,000 TS; 2-6 El; 169 Knoop Hard. For springs, switches, relays. Oxidation hardened. Good spring properties. 75% elec. cond. as hardened.

ENGELHARD 16527
Engelhard Industries
Mg 0.15, Ni 0.2, bal Ag.
Annealed: 28,000-34,000 TS; 20-38 El; 82 Knoop. Hardened: 58,000-64,000 TS; 9-16 El; 143 Knoop. For springs, switches. Oxidation hardened. 75% elec. cond. as hardened.

ENGELHARD 16527
Engelhard Minerals & Chemicals Corp.
Mg 0.15, Ni 0.2, bal Ag.
Annealed: 28,000-34,000 TS; 20-38 El; 82 Knoop. Hardened: 58,000-64,000 TS; 9-16 El; 143 Knoop. For springs, switches. Oxidation hardened. 75% elec. cond. as hardened.

ENGELHARD 4556
Toughite Process Co.
Mg 1.7, bal Ag.
Annealed: 46,000 TS; 15 El; Rock (30T) 55. Cold drawn: 62,000 TS; 5 El; Rock (30T) 66. For secondary emitter in television camera tubes. Elec. cond. 35 annealed, 32 cold rolled.

ENGELHARD 4556
Engelhard Minerals & Chemicals Corp.
Mg 1.7, bal Ag.
Annealed: 46,000 TS; 15 El; Rock (30T) 55. Cold drawn: 62,000 TS; 5 El; Rock (30T) 66. For secondary emitter in television camera tubes. Elec. cond. 35 annealed, 32 cold rolled.

ENGELHARD ALLOY NO. 10355
Engelhard Corp.
Au 35, Cu 65.
For brazing filler metal; 1778-1850°F MP.

ENGELHARD ALLOY NO. 109
Engelhard Corp.
Pt 90, Ir 10.
For thermocouple wire, magnetos, electrical contacts, thermostats, laboratory ware, surgical tools, voltage regulators, potentiometer wire, and jewelry.

ENGELHARD ALLOY NO. 11098
Engelhard Corp.
Pt 1.5, Pd 44, Ag 38, Au 0.5, Cu 15, Ni 1.
For contacts in low voltage DC circuits, switches, and relays.

ENGELHARD ALLOY NO. 13516
Engelhard Corp.
Pt 10, Pd 35, Au 10, Ag 30, Cu 15.
For contacts in miniature relays and spring arms in miniature relays.

ENGELHARD ALLOY NO. 141
Engelhard Corp.
Pt 6, Au 69, Ag 25.
For electrical contacts.

ENGELHARD ALLOY NO. 14282
Engelhard Corp.
Pt 1, Pd 44, Ag 39, Cu 14, Ni 1, Zn 1.
For wiper contact with nickel-chromium resistance wires.

ENGELHARD ALLOY NO. 1453
Engelhard Corp.
Pd 50, Ag 50.
Dental material.

ENGELHARD ALLOY NO. 1560
Engelhard Corp.
Pt 10, Pd 35, Au 10, Ag 30, Cu 14, Zn 1.
For electrical contacts.

ENGELHARD ALLOY NO. 158
Engelhard Corp.
Pt 85, Ir 15.
For thermocouple wire, magnetos, electrical contacts, thermostats, laboratory ware, surgical tools, voltage regulators, potentiometer wire, and jewelry.

ENGELHARD ALLOY NO. 17632
Engelhard Corp.
Pt 8.5, Au 72.5, Ag 4.75, Cu 14, Zn 0.25.
For electrical contacts.

ENGELHARD ALLOY NO. 18
Engelhard Corp.
Pt 75, Ir 25.
For thermocouple wire, magnetos, electrical contacts, thermostats, laboratory ware, surgical tools, voltage regulators, potentiometer wire, and jewelry.

ENGELHARD ALLOY NO. 190
Engelhard Corp.
Pt 3, Ag 97.
For electrical contacts.

ENGELHARD ALLOY NO. 19182
Engelhard Corp.
Pt 77.7, Co 22.3.
For magnets.

ENGELHARD ALLOY NO. 1990
Engelhard Corp.
Pd 40, Au 60.
For thermocouples and resistance wire.

ENGELHARD ALLOY NO. 2006
Engelhard Corp.
100% Ag.
For solder, electrical contacts, and chemical processing equipment.

ENGELHARD ALLOY NO. 208
Engelhard Corp.
Pt 80, Ir 20.
For thermocouple wire, magnetos, electrical contacts, thermostats, laboratory ware, surgical tools, voltage regulators, potentiometer wire, and jewelry.

ENGELHARD ALLOY NO. 2246
Engelhard Corp.
Pt 90, Pd 5, Rh 5.
For nitric acid catalyst.

ENGELHARD ALLOY NO. 2352
Engelhard Corp.
Ag 60, Cu 20, Zn 20.
Annealed: 78,600 psi TS; 18 El. For coined parts, bearings, Silver solder; 680°C MP.

ENGELHARD ALLOY NO. 24067
Engelhard Corp.
Pt 90, Ir 10.
Medical grade alloy. For pacemaker leads, electrodes, and feed-throughs.

ENGELHARD ALLOY NO. 24737 (Z-ALLOY)
Engelhard Corp.
Pt 1, bal Ag.

ENGELHARD ALLOY NO. 257
Engelhard Corp.
Ag 22, Cu 44, Zn 34.
For silver solder; corrosion resistant.

ENGELHARD ALLOY NO. 2576
Engelhard Corp.
Pt 30, Au 70.
For spinnerettes.

ENGELHARD ALLOY NO. 26
Engelhard Corp.
Pt 70, Ir 30.
For thermocouple wire, magnetos, electrical contacts, thermostats, laboratory ware, surgical tools, voltage regulators, potentiometer wire, and jewelry.

ENGELHARD ALLOY NO. 273
Engelhard Corp.
Ag 40, Cu 36, Zn 24.
For silver solder; corrosion resistant.

ENGELHARD ALLOY NO. 27987 ODS
Engelhard Corp.
Pt 99.9, 0.10 Y_2O_3.
Dispersion strengthened alloy. For crystal growing, spark plug electrodes, thermocouple sheaths, and glass manufacturing vessels.

ENGELHARD ALLOY NO. 28159 ODS
Engelhard Corp.
Pt 89.9, Rh 10, 0.10 Y_2O_3.
Dispersion strengthened alloy. For base plates for glass fiber production, containers for crystal growing, furnace wire windings, and thermocouple sheaths.

ENGELHARD ALLOY NO. 28225 ODS
Engelhard Corp.
Pt 94.9, Au 5, 0.10 Y_2O_3.
Dispersion strengthened alloy. For crucibles, casting molds, and bushings.

ENGELHARD ALLOY NO. 3004
Engelhard Corp.
Au 100.
For brazing filler metal; 1945°F MP.

ENGELHARD ALLOY NO. 3004
Engelhard Corp.
100% Au.
For chemical equipment.

ENGELHARD ALLOY NO. 3202
Engelhard Corp.
Au 82, Ni 18.
For brazing filler metal; 1742°F MP.

ENGELHARD ALLOY NO. 33043 ODS
Engelhard Corp.
Pt 95.9, W 4, 0.10 Y_2O_3.
Dispersion strengthened alloy. For high-performance spark plug electrodes, and sensors for electronic engine management systems.

ENGELHARD ALLOY NO. 349
Engelhard Corp.
Pd 75, Ag 25.
For hydrogen purification.

ENGELHARD ALLOY NO. 3558
Engelhard Corp.
Ag 56, Cu 22, Zn 17, Sn 5.
For silver solder, brazing; 1145-1205°F MP.

ENGELHARD ALLOY NO. 356
Engelhard Corp.
Ag 90, Pd 10.
For high temperature brazing for honeycomb assemblies; 1850-1975°F MP. High oxidation resistance.

ENGELHARD ALLOY NO. 3947
Engelhard Corp.
Au 75, Ag 12, Cu 13.
18 karat gold jewelry.

ENGELHARD ALLOY NO. 4002
Engelhard Corp.
100% Ir.
For crucibles, spark plug electrodes, high temperature apparatus.

ENGELHARD ALLOY NO. 430
Engelhard Corp.
Pt 95, Ru 5.
For contact material, resistance wire, and jewelry.

ENGELHARD ALLOY NO. 4606
Engelhard Corp.
Au 80, Cu 20.
For brazing filler metal; 1620-1630°F MP.

ENGELHARD ALLOY NO. 5009
Engelhard Corp.
100% Os.
Alloying element for pen nibs, record player needles, electrical contacts and instrument pivots.

ENGELHARD ALLOY NO. 5033
Engelhard Corp.
Au 76, Ni 16.5, Cu 1, Zn 6.5.
18 karat white gold jewelry.

ENGELHARD ALLOY NO. 5124
Engelhard Corp.
Pt 4, Pd 26, Au 25, Ag 28, Cu 16, Zn 1.
Dental material.

ENGELHARD ALLOY NO. 5264
Engelhard Corp.
Au 40, Fe 10, Pd 50, bal Pd.
Wire: 175,000-210,000 psi TS. For resistors; high electrical resistance.

ENGELHARD ALLOY NO. 5330
Engelhard Corp.
Pt 13.95, Pd 83, Au 3.05.
For thermocouple wire.

ENGELHARD ALLOY NO. 5348
Engelhard Corp.
Pt 8.5, Au 72.5, Ag 4, Cu 14, Zn 1.
Wire: 182,500 psi TS; 1.3 El. For electrical resistances; resistance wire.

ENGELHARD ALLOY NO. 5355
Engelhard Corp.
Pt 31.4, Pd 55, Au 13.6.
For thermocouple wire.

ENGELHARD ALLOY NO. 5496
Engelhard Corp.
Au 75, Cu 20, Ag 5.
For brazing filler metal; 1640-1650°F MP.

ENGELHARD ALLOY NO. 5538
Engelhard Corp.
Pt 84, Pd 10, Ru 6.
For spark plug electrodes.

ENGELHARD ALLOY NO. 5637 (S-ALLOY)
Engelhard Corp.
Pt 95.8, W 4, 0.2 ThO$_2$.
Dispersion strengthened alloy. For spark plug electrodes.

ENGELHARD ALLOY NO. 5769
Engelhard Corp.
Au 37.5, Cu 62.5.
For brazing filler metal; 1755-1815°F MP.

ENGELHARD ALLOY NO. 6007
Engelhard Corp.
100% Pd.
For telephone relay contacts.

ENGELHARD ALLOY NO. 6031
Engelhard Corp.
Pt 92, W 8.
300,000 psi TS; 2 El. For electrical resistances; resistance wire.

ENGELHARD ALLOY NO. 6346
Engelhard Corp.
Au 56.25, Cu 24.96, Ni 12.64, Zn 6.15.
14 karat jewelry alloy.

ENGELHARD ALLOY NO. 6361
Engelhard Corp.
Au 56.25, Ag 8.75, Cu 30.63, Zn 4.37.
14 karat jewelry alloy.

ENGELHARD ALLOY NO. 6395
Engelhard Corp.
Pt 97, Ru 3.
For contact material, resistance wire, and jewelry.

ENGELHARD ALLOY NO. 6429
Engelhard Corp.
Au 72.9, Ag 16.1, Cu 11.
For electrical contacts.

ENGELHARD ALLOY NO. 6551
Engelhard Corp.
Pt 85, Ni 15.
For cathode emitter wire.

ENGELHARD ALLOY NO. 661
Engelhard Corp.
Pd 60, Cu 40.
For sliding contacts and slip rings.

ENGELHARD ALLOY NO. 6833
Engelhard Corp.
Ag 65, Cu 25, Zn 10.
For silver solder; corrosion resistant.

ENGELHARD ALLOY NO. 7005
Engelhard Corp.
100% Rh.
For plated reflectors, sliding electrical contact plate, and jewelry plate.

ENGELHARD ALLOY NO. 7070
Engelhard Corp.
Pt 95, Au 5.
For electrical ribbon, spinnerettes, brazing wire, laboratory ware.

ENGELHARD ALLOY NO. 7146
Engelhard Corp.
Pt 70, Rh 30.
For thermocouples, resistance heaters, spinnerettes, and glass manufacturing; nitric acid catalyst.

ENGELHARD ALLOY NO. 7294
Engelhard Corp.
Pd 40.5, Fe 10, Au 49.5.
Wire: 200,000-220,000 psi TS. For resistors; high electrical resistance.

ENGELHARD ALLOY NO. 7450
Engelhard Corp.
Pt 95, Rh 5.
For thermocouples, resistance heaters, spinnerettes, and glass manufacturing; nitric acid catalyst.

ENGELHARD ALLOY NO. 7500, 7526, 7542
Engelhard Corp.
Pt 90, Rh 10.
For thermocouples, resistance heaters, spinnerettes, and glass manufacturing; nitric acid catalyst.

ENGELHARD ALLOY NO. 7609; 11,775
Engelhard Corp.
Pt 80, Rh 20.
For thermocouples, resistance heaters, spinnerettes, and glass manufacturing; nitric acid catalyst.

ENGELHARD ALLOY NO. 7658
Engelhard Corp.
Pt 91, W 9.
For spark plug electrodes, electrical contacts, and strain gages.

ENGELHARD ALLOY NO. 7666
Engelhard Corp.
Au 50, Cu 50.
For brazing filler metal; 1697-1733°F MP.

ENGELHARD ALLOY NO. 7674
Engelhard Corp.
Pd 34.95, Au 65.05.
For thermocouple wire.

ENGELHARD ALLOY NO. 7872
Engelhard Corp.
Pt 90, Ru 10.
For contact material, resistance wire, and jewelry.

ENGELHARD ALLOY NO. 7989
Engelhard Corp.
Pt 79, Rh 15, Ru 6.
Wire: 278,000 psi TS; 1.8 El. For electrical contact equipment. Corrosion resistant.

ENGELHARD ALLOY NO. 8003
Engelhard Corp.
100% Ru.
Hardener for Pt and Pd; alloyed with 90% Pt for aircraft magneto contacts.

ENGELHARD ALLOY NO. 8169
Engelhard Corp.
Ag 75, Pd 20, Mn 5.
For high temperature brazing; 1960-2050°F MP; high strength.

ENGELHARD ALLOY NO. 8177
Engelhard Corp.
Mn 3, Ag 64, Pd 33.
For high temperature brazing; 2100-2250°F MP; high strength.

ENGELHARD ALLOY NO. 8227
Engelhard Corp.
Au 75, Ag 22, Ni 3.
For thermostats.

ENGELHARD ALLOY NO. 8292
Engelhard Corp.
Au 35, Cu 62, Ni 3.
For brazing filler metal; 1815-1875°F MP.

ENGELHARD ALLOY NO. 8318
Engelhard Corp.
Pd 92, W 8.
Potentiometer resistor wire.

ENGELHARD ALLOY NO. 8383
Engelhard Corp.
Pt 60, Rh 40.
For thermocouples, resistance heaters, spinnerettes, and glass manufacturing; nitric acid catalyst.

ENGELHARD ALLOY NO. 8391
Engelhard Corp.
Pd 95.5, Ru 4.5.
For electrical contacts and jewelry.

ENGELHARD ALLOY NO. 8409
Engelhard Corp.
Pd 60, Ni 40.
For high temperature brazing; 2260°F MP; good oxidation resistance.

ENGELHARD ALLOY NO. 8466
Engelhard Corp.
Pt 50, Pd 26, Au 20, Ag 4.
Annealed: 107,000 psi TS; 85,000 psi YS; 25 El. Hardened: 171,000 psi TS; 163,000 psi YS; 3 El. For electrical contacts; heat resistant.

ENGELHARD ALLOY NO. 8904
Engelhard Corp.
Pt 96, Ru 4.
For contact material, resistance wire, and jewelry.

ENGELHARD ALLOY NO. 8938
Engelhard Corp.
Pt 94, Ru 6.
For contact material, resistance wire, and jewelry.

ENGELHARD ALLOY NO. 8961
Engelhard Corp.
Pt 87, Rh 13.
For thermocouples, resistance heaters, spinnerettes, and glass manufacturing; nitric acid catalyst.

ENGELHARD ALLOY NO. 8987
Engelhard Corp.
Pd 93.5, Ru 6.5.
For resistor wire; white jewelry alloy.

ENGELHARD ALLOY NO. 9001
Engelhard Corp.
100% Pt.
For electrical contacts, thermocouples, resistance thermometers, metal to glass seals, laboratory ware, and cathodic protection.

ENGELHARD ALLOY NO. 9142
Engelhard Corp.
Pt 96.5, Rh 3.5.
For heavy duty laboratory ware.

ENGINEERING
Dunford Hadfields Ltd.
C 0.18-0.25, Cr 12-14, bal Fe.
Heat treated: 100,000-200,000 TS; 8-20 El; 201-460 Brin. For hydraulic and steam equipment, valves, valve seats; corrosion resistant; Type 420. *Obsolete*

ENGLISH ALLOY
English manufacture
Sn 53, Pb 33, Sb 11, Cu 2.4, Zn 1.
For bearings, bushings; Babbitt.

ENGLISH B.M
English manufacture
Zn 80, Sn 14.5, Cu 5.5.
For bearings; will not resist heat or live steam.

ENGLISH BRASS
English manufacture
Cu 70, Zn 29, Pb 0.3, Sn 0.2.
For hardware, novelties; free-cutting.

ENGLISH GERMAN SILVER
English manufacture
Cu 59.4, Zn 25, Ni 13, Fe 3.
For ornamental parts; corrosion resistant.

ENGLISH LINOTYPE
English manufacture
Pb 83, Sn 5, Sb 12.
For type metal for linotype machines.

ENGLISH PENNY BRONZE
English manufacture
Sn 3, Zn 1.5, bal Cu.
For springs; corrosion resistant.

ENGLISH PEWTER
English manufacture
Sn 81.2, Sb 5.7, Cu 1.6, Pb 11.5.
For bearings, ornaments, household utensils; corrosion resistant.

ENGLISH PHOSPHOR BRONZE
English manufacture
Sn 10, Pb 9.6, P 1, bal Cu.
For bearings, gear wheels; heavy duty.

ENGLISH SPECULUM
English manufacture
Cu 67, Zn 33.
For deep drawing, spinning, stamping; high ductility.

ENGLISH STEREOTYPE (STANDARD)
English manufacture
Pb 82.5, Sn 4.5, Sb 13.
For type metal.

ENGLISH TYPE METAL "D"
English manufacture
Pb 58, Sn 15, Sb 26, Cu 1.
For type metal.

ENGLISH TYPE METAL "E"
English manufacture
Pb 77.5, Sn 6.5, Sb 16.
For type metal.

ENGLISH TYPE METAL (A)
English manufacture
Pb 63.2, Sn 12, Sb 24, Cu 0.8.
For type metal.

ENGLISH TYPE METAL (B)
English manufacture
Pb 60.5, Sn 14.5, Sb 24.2, Cu 0.8.
For type metal.

ENGLISH TYPE METAL OLD "C"
English manufacture
Pb 69.2, Sn 9.1, Sb 19.5, Cu 1.7.
For type metal.

ENGLISH-1
English manufacture
Cu 61.3, Zn 19.1, Ni 19.1.
For knives, forks, etc. for silver plating; nickel silver.

ENGLISH-2
English manufacture
Cu 70.3, Zn 29.3, Sn 0.17, Pb 0.26.
For hardware; free-cutting.

ENGRAVER PLATE
Colt Industries
C 0.35, bal Fe.
For engraver's plates. *Obsolete*

ENGRAVER'S BRASS
Century Brass Products Inc.
Cu 62.5-64.5, Pb 1.5-2.5, bal Zn.
Annealed: 45,000 TS; 50 El. Hard: 73,000 TS; 7 El. For gears, watch and clock backs; free cutting. *Obsolete*

ENGRAVERS BRASS
American manufacture
Cu 66, Zn 33, Pb 1.
For engravings; free-cutting.

ENGRAVERS BRASS 63
Olin Brass, Indianapolis
Copper. Cu 65.5, Zn 33.4, Pb 1.1.
Hard: 74,000 psi TS; 60,000 psi YS; 7 El; 150 Brin. Soft: 49,000 psi TS; 17,000 psi YS; 54 El; 61 Brin. For engraved items and dials; brass.

ENORM S
Remystahl
C 0.82, Cr 4.1, Mo 0.85, V 1.6, W 8.7, bal Fe.
For lathe and planer tools, drills, reamers; high speed steel. *Obsolete*

ENTECROD 192
Eutectic Corp.
Cd, Ag alloy.
Cast: 11,000 TS. For solder; hard, 650-700 F bonding temperature. *Obsolete*

EO 1
Electrometal SA Metals Especials
C 0.95, Si 0.3, Mn 1.25, Cr 0.52, V 0.1, W 0.55, bal Fe.
Oil hardening tool steel. AISI O1; W.-Nr. 1.2510.

EP 20
Electrometal SA Metals Especials
C 0.36, Si 0.3, Mn 0.6, Cr 1.8, Mo 0.2, V 1, bal Fe.
Special purpose tool steel. For molds. W.-Nr. 1.2330; similar to AISI P20.

EP 65
Russian manufacture
C 0.23, Cr 13, Mo 1, Ni 1, V 1, W 1, bal Fe.
Hardenable to 50 Rock C min. For cutlery, surgical instruments.

EPCO-OIL HARD
Williamson Bros. Inc.
C 1, Cr 0.5, Mn 1, W 0.5, bal Fe.
For dies, gages, master tools; non-deforming.

EPCO-WATER HARD
Williamson Bros. Inc.
C 1.05, Mn 0.25, bal Fe.
For blanking, forming, and trimming dies; water hardened.

EPD
W. Ossenberg & Cie Edelstahlwerke
C 0.6, Si 0.3, Mn 0.3, Cr 3.8, Mo 0.9, V 0.7, W 9, bal Fe.
Hot work tool steel. For tube pressing mandrels. W.-Nr. 1.2622.

EPD MO
W. Ossenberg & Cie Edelstahlwerke
C 0.6, Si 0.3, Mn 0.3, Cr 3.6, Mo 8.5, V 1.75, W 1.5, bal Fe.
Hot work tool steel. For hot pressing dies and blocks.

EPM
W. Ossenberg & Cie Edelstahlwerke
C 0.45, Si 1, Mn 0.3, Cr 1, V 0.2, W 2, bal Fe.
Hot or cold work forming, punching and trimming dies. W.-Nr. 1.2542.

EPS
W. Ossenberg & Cie Edelstahlwerke
C 0.35, Si 0.2, Mn 0.3, Cr 2.5, V 0.4, W 8.5, bal Fe.
Hot work tool steel. Pressing and forging dies. W.-Nr. 1.2581; similar to AISI H21.

EPS 25
W. Ossenberg & Cie Edelstahlwerke
C 0.45, Si 1, Mn 0.3, Cr 1.7, V 0.2, W 2, bal Fe.
Hot work tool steel. Pressing dies for lead, zinc and light metals. W.-Nr. 1.2547.

EPS 33
W. Ossenberg & Cie Edelstahlwerke
Mn 0.4, Cr 3, Mo 2.8, V 0.5, C 0.3, Si 0.4, bal Fe.
Hot work tool steel. Piercing mandrels; pressing dies. W.-Nr. 1.2365.

EPS 35
W. Ossenberg & Cie Edelstahlwerke
Mn 0.4, Cr 1, V 0.2, W 4, C 0.3, Si 1, bal Fe.
Hot work tool steel. For water cooled forming tools for bolts, screws. W.-Nr. 1.2564.

EPS 45
W. Ossenberg & Cie Edelstahlwerke
Mn 0.3, Cr 2.2, V 0.6, W 4.5, C 0.3, Si 0.2, bal Fe.
Hot work tool steel. Heading tools for screws, bolts, non-ferrous alloys. W.-Nr. 1.2567.

EPS 51
W. Ossenberg & Cie Edelstahlwerke
C 0.38, Si 1, Mn 0.4, Cr 5.5, Mo 1.3, V 0.3, bal Fe.
Hot work tool steel. Forging dies; casting molds for light metals. W.-Nr. 1.2343.

EPS 51 V
W. Ossenberg & Cie Edelstahlwerke
C 0.38, Mn 0.4, Si 1.1, Cr 5.5, V 1, Mo 1.5, bal Fe.
Hot work tool steel. Forging dies; casting molds; extrusion and piercing mandrels. W.-Nr. 1.2344.

EPSK 2
W. Ossenberg & Cie Edelstahlwerke
C 0.3, Si 0.2, Mn 0.3, Co 2, Cr 2.5, V 0.3, W 9, bal Fe.
Hot work tool steel; forging tools. W.-Nr. 1.2662.

EPSK SPEZIAL
W. Ossenberg & Cie Edelstahlwerke
C 0.65, Si 0.3, Mn 0.3, Cr 4, V 0.5, W 15, Co 5, bal Fe.
Hot work tool steel. Forging and pressing dies.

EPSNI
W. Ossenberg & Cie Edelstahlwerke
C 0.45, Mn 0.7, Si 1.4, Cr 13.5, W 2.5, V 0.5, Ni 13, bal Fe.
Hot work tool steel; pressing dies. W. Nr. 1.2731.

EPW
W. Ossenberg & Cie Edelstahlwerke
C 0.35, Si 1, Mn 0.3, Cr 1, V 0.2, W 2, bal Fe.
Hot or cold work tool steel. For water-cooled hot work tools as hot shears; also cold punches. W.-Nr. 1.2541.

ER 347, ETC.
Now ARCALOY ER 347, ETC.

ERA 131
Dunford Hadfields Ltd.
C, Mo, Cr, Cu, bal Fe.
For tubes for steam turbines; creep resistant. *Obsolete*

ERA 1414
Dunford Hadfields Ltd.
Cr 14, Ni 14, C, W, bal Fe.
For exhaust valves; heat resistant. *Obsolete*

ERA 156
Dunford Hadfields Ltd.
C 0.3-0.4, Mn 1.3-1.8, Mo 0.35-0.55, bal Fe.
Bar: 100,000-145,000 TS; 72,000-112,000 YS; 16-22% El;
201-341 Brin. For gears, shafts, axles, bolts, crankshafts,
countershafts. Oil hardening, shock resisting. *Obsolete*

ERA 165
Dunford Hadfields Ltd.
C 0.15-0.25, Mn 1.4-1.8, Ni 0.4-0.7, Mo 0.15-0.35, bal Fe.
Bar: 90,000 TS; 60,000 YS; 22% El; 200 Brin. For gears, bolts,
fasteners, shafts. Oil or water harden. *Obsolete*

ERA 171
Dunford Hadfields Ltd.
C 0.3-0.5, Cr 2.5-3.5, Ni 0-0.4, Mo 0.7-1.2, V 0.1-0.3, bal Fe.
Oil hardened: 190,000 TS; 152,000 YS; 10% El. For gears,
bolts, cams, camshafts, rolls. Nitriding steel, wear and
abrasion resistant. *Obsolete*

ERA 59
Dunford Hadfields Ltd.
C 0.25-0.4, Mn 1.3-1.8, Mo 0.2-0.35, bal Fe.
Bar: 100,000-145,000 TS; 72,000-112,000 YS; 16-22% El;
201-341 Brin. For gears, crankshafts, fasteners, bolts. Water
harden, shock resistant. *Obsolete*

ERA 60 M
Dunford Hadfields Ltd.
C 0.18, Si 0.05-0.35, Mn 1.1-1.5, S 0.1-0.18, P 0-0.05, bal Fe.
Bar: 90,000 TS; 20 El; 52,000 YS; 190 Brin. Annealed: 62,000
TS; 40,000 YS: 32 El. 121 Brin. Heat treated: 102,000 TS;
74,000 YS; 26 El; 269 Brin. For gears, bolts, cams, camshafts,
bushings, clips, jaws, fasteners. SAE 1118 case hardening
steel. Free cutting, shock resistant. *Obsolete*

ERA 7393
Dunford Hadfields Ltd.
Cr 18, Ni 8, C, Mo, W, Ti, bal Fe.
For gas turbine rotors, blades; heat resistant. *Obsolete*

ERA A.T.V.
Dunford Hadfields Ltd.
C 0.3, Cr 14, Ni 28, W 4, Mn 1.5, Si 1.8, bal Fe.
108,000 TS; 60,000 YS; 25 El; 45 RA. For high duty I.C.
engine valves, supercharger rotors; heat resisting. *Obsolete*

ERA BORON 1
Dunford Hadfields Ltd.
C 0.1, Si 0.5, Mn 0.3, B 1, bal Fe.
For nuclear reactors; for neutron absorption. *Obsolete*

ERA BORON 2
Dunford Hadfields Ltd.
C 0.1, Si 0.5, Mn 0.3, B 2, bal Fe.
For nuclear reactors; for neutron absorption. *Obsolete*

ERA BORON 3
Dunford Hadfields Ltd.
C 0.1, Si 0.5, Mn 0.3, B 3, bal Fe.
For nuclear reactors; for neutron absorption. *Obsolete*

ERA BORON 4
Dunford Hadfields Ltd.
C 0.1, Si 0.5, Mn 0.3, B 4, bal Fe.
For nuclear reactors; for neutron absorption. *Obsolete*

ERA C.R. 2
Dunford Hadfields Ltd.
Cr 20, Ni 8, C, Cu, W, bal Fe.
For impellers, castings; stainless. *Obsolete*

ERA C.R. 3
Dunford Hadfields Ltd.
Cr 10, Ni 20, C, Cu, W, bal Fe.
For stainless parts; corrosion resistant. *Obsolete*

ERA C.R. BORON
Dunford Hadfields Ltd.
C 0.08, Si 0.5, Mn 0.3, B 2-4, Cr 18, Ni 10, bal Fe.
For nuclear reactors; for neutron absorption. *Obsolete*

ERA C.R.I.
Dunford Hadfields Ltd.
C 0.1, Cr 18, Ni 8, Mn 0.5-1.5, bal Fe.
88,000-132,000 TS; 66,000-88,000 YS; 37-75 El; 150-230
Brin. For domestic utensils, fittings, chemical plant
equipment; corrosion resistant; austenitic. *Obsolete*

ERA C.R.I.L.
Dunford Hadfields Ltd.
C 0.08, Cr 18, Ni 8, bal Fe.
For chemical equipment; austenitic, stainless. *Obsolete*

ERA C.R.I.S.
Dunford Hadfields Ltd.
C 0.1, Cr 18, Ni 8, Ti 0.1, bal Fe.
Annealed: 95,000 TS; 45,000 YS; 50 El; 60 RA; 175 Brin. For
chemical equipment, welded structures; Type 321; austenitic,
stainless. *Obsolete*

ERA C.R.I.S. (CB)
Dunford Hadfields Ltd.
Cr 18, Ni 8, C, Cb, bal Fe.
For stainless equipment to be welded; stabilized. *Obsolete*

ERA CR4
Dunford Hadfields Ltd.
C 0-0.1, Cr 16-18, Ni 10-14, Mo 2-3, bal Fe.
Annealed: 95,000 TS; 45,000 YS; 50 El; 60 RA; 190 Brin. For
acid resistant parts, chemical plant equipment; Type 316;
stainless, austenitic. *Obsolete*

ERA CR4S
Dunford Hadfields Ltd.
C 0.1, Cr 18, Ni 9, Mo 2, Ti, bal Fe.
Annealed: 95,000 TS; 45,000 YS; 50 El; 60 RA; 190 Brin. For
chemical plant equipment; Type 316Ti; stainless, austenitic.
Obsolete

ERA H.R. 4
Dunford Hadfields Ltd.
Cr 28, C, bal Fe.
For heat resistant castings; resists scaling to 1150 C.
Obsolete

ERA H.R. 5
Dunford Hadfields Ltd.
Cr 12, Ni 60, C, bal Fe.
For high temperature equipment; heat resistant. *Obsolete*

ERA H.R. 6
Dunford Hadfields Ltd.
Cr 25, Ni 12, C, bal Fe.
For furnace parts; heat resistant to 1050 C in sulfur
atmosphere. *Obsolete*

ERA H.R.-1
Dunford Hadfields Ltd.
Cr 22, Ni 12, Si 0.2, W 0.4, C 0.2, bal Fe.
At 20 C: 130,000 TS; 85,000 YS; 37 El; 37 RA; 200 Brin. At
700 C: 69,000 TS; 35 El; 57 RA. At 1000 C: 45,000 TS; 30 El;
64 RA. For engine valves, crucibles, furnace conveyors,
dredger parts, carburizing boxes; corrosion resisting,
stainless; range of alloys. *Obsolete*

ERA H.R.I.S.
Dunford Hadfields Ltd.
Cr 21, Ni 7, Si 3.5, bal Fe.
For heat resistant castings; heat resistant. *Obsolete*

ERA HR2
Dunford Hadfields Ltd.
C 0.25, Cr 21, Ni 7, Si 1.5, bal Fe.
Annealed: 85,000 TS; 42,000 YS; 45 El; 60 RA; 180 Brin. For
furnace parts, heat treat boxes; heat and corrosion resistant.
Obsolete

ERA HR3
Dunford Hadfields Ltd.
C 0-0.2, Cr 25, Ni 20, bal Fe.
Annealed: 95,000 TS; 45,000 YS; 50 El; 65 RA; 180 Brin. For
furnace parts, heat treat boxes, engine components; Type
310; austenitic, heat resistant. *Obsolete*

ERA HR7
Dunford Hadfields Ltd.
C 0.3, Cr 15, Ni 35, bal Fe.
Cast: 70,000 TS; 40,000 YP; 10 El; 170 Brin. At 1600 F:
22,000 TS; 28 El; 36 RA. For carburizing boxes, trays, furnace
parts, grills. Oxidation and heat resistant. *Obsolete*

ERA MANGANESE STEEL
Dunford Hadfields Ltd.
Alloy steel. C 1.25, Mn 11-14, bal Fe.
85,000-100,000 TS; 60-80 El. For crushers, dredging
machines, soldiers' helmets; wear resistant, non-magnetic.

ERAYDO
Illinois Zinc Co.
Zinc. Cu 1-3, Ag 0.03-0.1, bal Zn.
Rolled: 30,000-50,000 TS; 12,000-22,000 YS; 15-40 El. For
panels, indoor trim, electrical appliances, drawn cases;
rustless, nonmagnetic.

ERBIUM
Atomergic Chemetals Corp.
Er.
Purities: 99.9% special distilled grade, 99.5+%. Forms: ingot,
lump, wire, sheet, foil, turnings, sponge, powder, single
crystals.

ERCO
Erie Steel Co.
C 0.8, bal Fe.
For dies to resist alternate heating and cooling; water
hardened.

ERFTAL
German manufacture
Fe 0.04, Al 99.9.
For reflectors; a high gloss mill aluminum.

ERGAL 55
Alluminio SA
Aluminum. Zn 5.8, Mg 2.5, Cu 1.6, Mn 0.2, Cr 0.15, Ti 0.08,
bal Al.
TA-Temper: 90,000 TS; 80,000 YS; 6 El; 160 Brin. For cast
aircraft components, hardware.

ERGAL 60
Soc. Alluminio Veneto per Azioni
Aluminum. Zn 6.8, Mg 2.5, Cu 1.6, Ti 0.1, Mn 0.2, Cr 0.15, bal
Al.
Cast. For aircraft castings; self-aging.

ERGAL 60
Feltrina Societa Metallurgica
Zn 6.8, Mg 2.5, Cu 1.6, Ti 0.1, Mn 0.2, Cr 0.15, bal Al.
Cast. For aircraft castings; self-aging.

ERGAL 65

Alluminio SA

Aluminum. Zn 7.8, Mg 2.5, Cu 1.6, Mn 0.2, Cr 0.15, Ti 0.1, bal Al.

TA-Temper: 82,000-96,000 TS; 72,000-85,000 YS; 4-9 El; 150-180 Brin. For cast aircraft components. Self aging, high strength cast alloy.

ERGALPLAT 65

Alluminio SA

Aluminum. Cu 1.5-1.7, Mg 2.5, Mn 0.2, Zn 8, bal Al.

R-Temper: 27,000-37,000 TS; 23,000 YS; 8-20 El. TA-Temper: 80,000-94,000 TS; 71,000-85,000 YS; 4-14 El. For aircraft constructional parts; self aging. Heat and corrosion resistant. High strength and fatigue resistant.

ERGES 4

Schmole GmbH

Mg 0.6-1.4, Si 0-1.6, Mn 0.6-1, Cr 0-0.3, bal Al.

For chemical equipment; corrosion resistant. *Obsolete*

ERGSTE 120

Ergst, Stahlwerke

C 0-0.1, Cr 12.5, Ni 13, bal Fe.

For chemical plant equipment; corrosion resistant. *Obsolete*

ERGSTE 13, ETC.

Ergst, Stahlwerke

C 0-0.12, Si 0.4, Cr 13, bal Fe.

Annealed: 75,000 TS; 40,000 YS; 35 El; 70 RA; 155 Brin. Cold drawn: 100,000 TS; 85,000 YS; 17 El; 60 RA; 205 Brin. For turbine blades, valve trim, chemical plant equipment; Type 410; corrosion resistant.

ERGSTE 135P

Ergst, Stahlwerke

C 0.07, Cr 17, Ni 13, Mo 2, bal Fe.

Annealed: 85,000 TS; 35,000 YS; 50 El; 65 RA; 160 Brin. Cold drawn: 150,000 TS; 135,000 YS; 6 El; 300 Brin. For acid resistant equipment, mixers, agitators; Type 316; stainless, austenitic. *Obsolete*

ERGSTE 17

Ergst, Stahlwerke

C 0.8, Cr 17, bal Fe.

Annealed: 107,000 TS; 62,000 YS; 18 El; 35 RA; 220 Brin. Heat treated: 280,000 TS; 270,000 YS; 3 El; 15 RA; 555 Brin. For bearings, valves, surgical instruments; Type 440B; corrosion resistant, hardenable. *Obsolete*

ERGSTE 17L

Ergst, Stahlwerke

C 0.1, Cr 17, Mo 2, bal Fe.

Annealed: 125,000 TS; 95,000 YS; 20 El; 55 RA; 260 Brin. For chemical plant equipment; corrosion resistant. *Obsolete*

ERGSTE 17LT

Ergst, Stahlwerke

C 0-0.1, Cr 17, Mo 1.8, Ti = 7 x C, bal Fe.

Annealed: 130,000 TS; 98,000 YS; 20 El; 53 RA; 260 Brin. For chemical plant welded structures; corrosion resistant, stabilized. *Obsolete*

ERGSTE 17T

Ergst, Stahlwerke

C 0-0.1, Ti = 7 x C, bal Fe.

Annealed: 130,000 TS; 98,000 YS; 20 El; 53 RA; 260 Brin. For chemical plant welded equipment; corrosion resistant, stabilized. *Obsolete*

ERGSTE 17U

Ergst, Stahlwerke

C 0-0.12, Cr 16.5, Mo 0.25, S 0.2, bal Fe.

Annealed: 80,000 TS; 50,000 YS; 25 El; 50 RA; 160 Brin. For screw machine products, chemical plant equipment; Type 430F; corrosion resistant. *Obsolete*

ERGSTE 182RT

Ergst, Stahlwerke

C 0-0.07, Cr 17.5, Ni 17.5, Mo 2, Cu 2, Ti = 7 x C, bal Fe.

Annealed: 85,000 TS; 35,000 YS; 50 El; 65 Brin. For welded acid resistant, equipment, tanks, mixers; stainless, austenitic, stabilized. *Obsolete*

ERGSTE 25L

Ergst, Stahlwerke

C 0.15, Cr 16, Mo 0.2, S 0.2, bal Fe.

Annealed: 80,000 TS; 50,000 YS; 25 El; 50 RA; 160 Brin. For screw machine products, chemical plant equipment; Type 430F; corrosion resistant. *Obsolete*

ERGSTE 80K

Ergst, Stahlwerke

C 0-0.15, Cr 18, Ni 8.5, bal Fe.

Annealed: 80,000 TS; 35,000 YS; 55 El; 75 RA; 150 Brin. For chemical plant equipment, tanks; Type 302; stainless, austenitic. *Obsolete*

ERGSTE 80P

Ergst, Stahlwerke

C 0-0.07, Cr 18, Ni 9.5, bal Fe.

Annealed: 85,000 TS; 35,000 YS; 60 El; 70 RA; 150 Brin. Cold drawn: 180,000 TS; 125,000 YS; 10 El; 330 Brin. For chemical plant equipment, welded tanks, mixers; Type 304; stainless, austenitic. *Obsolete*

ERGSTE 80S

Ergst, Stahlwerke

C 0-0.1, Cr 18, Ni 8.5, bal Fe.

Annealed: 85,000 TS; 35,000 YS; 60 El; 70 RA; 150 Brin. Cold drawn: 180,000 TS; 125,000 YS; 10 El; 330 Brin. For chemical plant equipment, welded tanks, mixers; Type 304; stainless, austenitic. *Obsolete*

ERGSTE 80T

Ergst, Stahlwerke

C 0-0.12, Cr 18, Ni 9.5, Ti = 4 x C, bal Fe.

Annealed: 85,000 TS; 35,000 YS; 55 El; 65 RA; 150 Brin. Cold drawn: 95,000 TS; 60,000 YS; 40 El; 60 RA; 185 Brin. For welded structures, chemical plant equipment; Type 321; stainless, austenitic. *Obsolete*

ERGSTE 80U

Ergst, Stahlwerke

C 0.12, Ni 8.5, Cr 18, bal Fe.

Annealed: 80,000 TS; 35,000 YS; 55 El; 75 RA; 150 Brin. For chemical plant equipment, fasteners; Type 302; stainless, austenitic. *Obsolete*

ERGSTE 80X

Ergst, Stahlwerke

C 0-0.12, Cr 18, Ni 9.5, Cb = 8 x C, bal Fe.

Annealed: 90,000 TS; 35,000 YS; 56 El; 65 RA; 160 Brin. Cold drawn: 100,000 TS; 65,000 YS; 40 El; 60 RA; 202 Brin. For welded structures, chemical plant equipment; Type 347; stainless, austenitic. *Obsolete*

ERGSTE 82P

Ergst, Stahlwerke

C 0-0.07, Cr 18, Ni 10.5, Mo 2, bal Fe.

Annealed: 80,000 TS; 35,000 YS; 55 El; 70 RA; 150 Brin. Cold drawn: 140,000 TS; 130,000 YS; 7 El; 280 Brin. For acid resistant equipment, mixers, agitators; Type 316L; stainless, austenitic. *Obsolete*

ERGSTE 82RT

Ergst, Stahlwerke

C 0.05, Cr 18, Ni 10, Cu, bal Fe.

Annealed: 80,000 TS; 35,000 YS; 55 El; 70 RA; 150 Brin. For chemical plant equipment; stainless, austenitic. *Obsolete*

ERGSTE 82S

Ergst, Stahlwerke

C 0-0.1, Cr 18, Ni 9.5, Mo 2, bal Fe.

Annealed: 85,000 TS; 35,000 YS; 50 El; 65 RA; 160 Brin. Cold drawn: 150,000 TS; 135,000 YS; 6 El; 300 Brin. For acid resistant equipment, mixers, agitators; Type 316; stainless, austenitic. *Obsolete*

ERGSTE 82T

Ergst, Stahlwerke

C 0-0.12, Cr 18, Ni 10.5, Mo 2, Ti = 4 x C, bal Fe.

Annealed: 85,000 TS; 35,000 YS; 50 El; 65 RA; 160 Brin. Cold drawn: 150,000 TS; 135,000 YS; 6 El; 300 Brin. For welded acid resistant equipment; Type 316Ti; stainless, austenitic. *Obsolete*

ERGSTE 82X

Ergst, Stahlwerke

C 0-0.12, Cr 18, Ni 10.5, Mo 2, Cb = 8 x C, bal Fe.

Annealed: 85,000 TS; 35,000 YS; 50 El; 65 RA; 160 Brin. Cold drawn: 150,000 TS; 135,000 YS; 6 El; 300 Brin. For welded acid resistant equipment; Type 316Cb; stainless, austenitic. *Obsolete*

ERGSTE CN20

Ergst, Stahlwerke

C 0.13, Cr, Mo, bal Fe.

For gears, cams, camshafts; case hardening, tough. *Obsolete*

ERGSTE CN30

Ergst, Stahlwerke

C 0.15, Ni, Cr, bal Fe.

For gears, bolts, machine tool parts; case hardening, tough. *Obsolete*

ERGSTE CN60

Ergst, Stahlwerke

C 0.15, Ni, Cr, bal Fe.

For gears, bolts, machine tool parts; case hardened, shock resistant. *Obsolete*

ERGSTE CN80

Ergst, Stahlwerke

C 0.15, Ni, Cr, bal Fe.

For gears, bolts, machine tool parts; case hardened, shock resistant. *Obsolete*

ERGSTE E110S

Ergst, Stahlwerke

C 1.15, Si 0-0.25, Mn 0-0.25, bal Fe.

Annealed: 110,000 TS; 58,000 YS; 18 El; 40 RA; 210 Brin. For springs, tools, drills, taps, reamers; Type W1; water hardened. *Obsolete*

ERGSTE ESCV110

Ergst, Stahlwerke

C 1.15, Cr 0.65, V 0.1, Mn 0.3, bal Fe.

For bearings, bushings, cutters; oil hardened, tough. *Obsolete*

ERGSTE ESWV

Ergst, Stahlwerke

C 1.2, Cr 0.2, V 0.1, W 1, bal Fe.

For forming dies, shears; oil hardened. *Obsolete*

ERGSTE K13

Ergst, Stahlwerke

C 0.13, Si 0.3-0.5, Mn 0.25-0.5, bal Fe.

Annealed: 70,000 TS; 40,000 YS; 25 El; 60 RA; 145 Brin. For gears, bolts, machine tool parts; case hardening steel. *Obsolete*

ERGSTE K20

Ergst, Stahlwerke

C 0.2, Si 0.4, Cr 13, bal Fe.

Annealed: 95,000 TS; 50,000 YS; 25 El; 55 RA; 195 Brin. Cold drawn: 105,000 TS; 85,000 YS; 17 El; 50 RA; 215 Brin. For cutlery, valves, turbine blades, surgical instruments; Type 420; stainless. *Obsolete*

ERGSTE K20L

Ergst, Stahlwerke

C 0.2, Si 0.4, Cr 13, Mo 1.15, bal Fe.

Annealed: 95,000 TS; 50,000 YS; 25 El; 55 RA; 195 Brin. For turbine blades, cutlery, valves; Type 420 Mo; stainless. *Obsolete*

ERGSTE K20N

Ergst, Stahlwerke
C 0.22, Si 0.4, Cr 17, Ni 1.5, bal Fe.
Annealed: 125,000 TS; 95,000 YS; 20 El; 55 RA; 260 Brin. For pumps, marine hardware, valves; Type 431; corrosion and heat resistant. *Obsolete*

ERGSTE K40

Ergst, Stahlwerke
C 0.4, Si 0.4, Cr 13, bal Fe.
Annealed: 95,000 TS; 50,000 YS; 25 El; 55 RA; 195 Brin. For cutlery, valves, surgical instruments; Type 420; stainless. *Obsolete*

ERGSTE K90

Ergst, Stahlwerke
C 0.85, Cr, V, bal Fe.
For tools, cutters, drills, taps, broaches; water or oil hardened. *Obsolete*

ERGSTE K90L

Ergst, Stahlwerke
C 0.9, Cr, Mo, bal Fe.
For bearings, punches, cutters; water or oil hardened. *Obsolete*

ERHARD BRONZE

German manufacture
Zn, Cu, Al.
For ornamental fittings.

ERIE

Time Steel Service Inc.
C 1, Mn 0.35, Si 0.2, bal Fe.
Water hardened tool steel. AISI W1.

ERIE AA

Erie Steel Co.
C 0.7, W 18, Cr 4, V 1, bal Fe.
For lathe and planer tools, drills, reamers, taps; high speed steel.

ERIE EXTRA

Time Steel Service Inc.
C 1, Mn 0.35, Si 0.2, bal Fe.
Water hardened tool steel. AISI W1.

ERIE NO. 5

Erie Steel Co.
C 0.7, alloy, bal Fe.
For shear blades, punches, blacksmith tools; water hardened.

ERIE RA

Erie Steel Co.
C 0.7, W 18, Cr 4, V 1, Co 5, bal Fe.
For tools, drills, lathe and planer cutters; high speed steel.

ERIE SPECIAL

Time Steel Service Inc.
C 1, Mn 0.35, Si 0.2, bal Fe.
Water hardened tool steel. AISI W1.

ERK ADG

Erkenzweig & Schwemann
C 2.1, Cr 11.5, Mn 0.3, bal Fe.
For blanking and forming dies, gauges; oil hardened, nondeforming.

ERKENZWEIG BRA

Erkenzweig & Schwemann
C 0.2, Si 0.4, Mn 0.3, Cr 13, bal Fe.
Annealed: 95,000 TS; 50,000 YS; 25 El; 55 RA; 196 Brin. For turbine blades, valves, cutlery, dental instruments; Type 420; stainless.

ERKENZWEIG EX

Erkenzweig & Schwemann
C 1.4, Cr 0.3, V 0.1, Mn 0.3, bal Fe.
For forming dies, engraving tools, cutters; water hardened, wear resistant.

ERKENZWEIG EXTRA BEST MH

Erkenzweig & Schwemann
C 1.1, Si 0-0.25, Mn 0-0.25, bal Fe.
Annealed: 110,000 TS; 58,000 YS; 20 El; 40 RA; 210 Brin. For springs, tools, cutters, drills, broaches; Type W1; water hardened.

ERKENZWEIG EXTRA BEST Z

Erkenzweig & Schwemann
C 0.85, Si 0-0.25, Mn 0-0.25, bal Fe.
Heat treated: 190,000 TS; 145,000 YS; 10 El; 30 RA; 400 Brin. For punches, drills, taps, reamers, cutters; Type W1; water hardened.

ERKENZWEIG EXTRA BEST ZH

Erkenzweig & Schwemann
C 1, Mn 0-0.25, Si 0-0.25, bal Fe.
Annealed: 100,000 TS; 53,000 YS; 21 El; 42 RA; 200 Brin. For springs, taps, drills, cutters; Type W1; water hardened.

ERKENZWEIG HB

Erkenzweig & Schwemann
C 1.4, Cr 0.3, V 0.1, bal Fe.
For header dies, cutters, bearings; oil hardened, tough.

ERKENZWEIG KM45

Erkenzweig & Schwemann
C 0.5, Cr 1.05, Ni 3.25, bal Fe.
For gears, pinions, crankshafts, forging dies; oil hardened, shock resistant.

ERKENZWEIG KWB

Erkenzweig & Schwemann
C 1.05, Cr 1, W 1.15, Mn 0.9, bal Fe.
For cutters, shears, header dies; oil hardened, tough.

ERKENZWEIG NH

Erkenzweig & Schwemann
C 0.95, W, Mo, Cr, bal Fe.
For lathe and planer tools, drills; high speed steel.

ERKENZWEIG NHA

Erkenzweig & Schwemann
C 1.35, W, Co, Cr, V, bal Fe.
For engravers' tools, forming and blanking dies; high speed steel.

ERKENZWEIG NHB

Erkenzweig & Schwemann
C 0.79, Co 4.75, Cr 4.3, Mo 0.75, V 1.5, W 18, bal Fe.
For lathe and planer tools, broaches, reamers, drills; high speed steel.

ERKENZWEIG NHC

Erkenzweig & Schwemann
C 0.76, Co 10, Cr 4.2, Mo 0.8, V 1.8, W 18, bal Fe.
For lathe and planer tools, drills, taps, hobs; high speed steel.

ERKENZWEIG NHE

Erkenzweig & Schwemann
C 1.3, Cr 4.3, Mo 0.85, V 3.8, W 12, bal Fe.
For engravers' tools, blanking and forming dies; high speed steel.

ERKENZWEIG NHH

Erkenzweig & Schwemann
C 0.82, Cr 4.1, Mo 0.85, V 1.6, W 8.7, bal Fe.
For lathe and planer tools, reamers, broaches; high speed steel.

ERKENZWEIG NHPC

Erkenzweig & Schwemann
C 0.86, Cr 4.3, Mo 0.85, V 2.1, W 12, bal Fe.
For lathe and planer tools, reamers, broaches; high speed steel.

ERKENZWEIG NHUC

Erkenzweig & Schwemann
C 0.86, Mo 0.85, V 2.5, W 12, Cr 4.1, bal Fe.
For lathe and planer tools, drills, taps, hobs; high speed steel.

ERKENZWEIG PKU

Erkenzweig & Schwemann
C 1, V 0.1, Mn 0.25, bal Fe.
Heat treated: 200,000 TS; 130,000 YS; 8 El; 25 RA; 400 Brin. For reamers, drills, drawing and forming dies; oil or water hardened, wear resistant.

ERKENZWEIG PLD III

Erkenzweig & Schwemann
C 0.35, Si 0.9, Cr 1.05, V 0.18, W 1.85, bal Fe.
For heading and forging dies, upsetters, punches; oil hardened, tough.

ERKENZWEIG PLDI

Erkenzweig & Schwemann
C 0.55, V 0.18, W 1.85, Si 0.9, Cr 1.05, bal Fe.
For heading and forging dies, upsetters, punches; oil hardened, tough.

ERKENZWEIG PSA

Erkenzweig & Schwemann
C 1.45, Cr 1.4, Mn 0.6, bal Fe.
For blanking and forming dies, cutters, punches; oil hardened, wear resistant.

ERKENZWEIG PSX

Erkenzweig & Schwemann
C 1.4, Cr 0.3, V 0.1, Mn 0.3, bal Fe.
For blanking and forming dies, bearings; oil or water hardened.

ERKENZWEIG SPEZIAL K

Erkenzweig & Schwemann
C 2.1, Cr 11.5, Mn 0.3, bal Fe.
For blanking and forming dies, punches; oil hardened, nondeforming.

ERKENZWEIG SPWC

Erkenzweig & Schwemann
C 0.3, Co 2, Cr 2.4, V 0.25, W 8.5, bal Fe.
For extrusion rams, dies, liners; oil hardened, hot work steel.

ERKENZWEIG SWK III

Erkenzweig & Schwemann
C 0.67, Si 1.3, Mn 0.5, Cr 0.5, bal Fe.
For springs, shear blades, punches; oil hardened, shock resistant.

ERKENZWEIG SWK V

Erkenzweig & Schwemann
C 0.45, Cr, Ni, bal Fe.
For gears, bolts, crankshafts, fasteners; oil hardened, shock resistant.

ERKENZWEIG SWK VI

Erkenzweig & Schwemann
C 2.1, Cr 11.5, W 0.7, bal Fe.
For blanking and forming dies, punches; oil hardened, nondeforming.

ERKENZWEIG SWK VII

Erkenzweig & Schwemann
C 0.45, W, Cr, V, bal Fe.
For shear blades, forging and heading dies; oil hardened, tough.

ERKENZWEIG SWK VIII

Erkenzweig & Schwemann
C 0.3, Cr 2.65, V 0.35, W 8.5, bal Fe.
For extrusion rams, dies and liners; hot work steel, oil hardened.

ERKENZWEIG WCR
Erkenzweig & Schwemann
C 1.42, W, V, bal Fe.
For blanking and heading dies; oil hardened, wear resistant.

ERKENZWEIG WNC 40
Erkenzweig & Schwemann
C 0.5, Cr 1.05, Ni 3.25, bal Fe.
For gears, bolts, crankshafts; oil hardened, shock resistant.

ERM 3A
Enfield Rolling Mills Ltd.
Ni 0.9-1.1, P 0.18-0.25, S 0.1-0.15, bal Cu.
Bar: 62,000-78,000 TS; 18-25 El; 150-175 Brin. For electrical switchgear and resistance welding equipment, electrical contacts. 55-60% electrical conductivity. *Obsolete*

ERM ALW
Delta Enfield Metals Ltd.
Now DELTA ENFIELD ALW. *Obsolete*

ERM ALW
Enfield Rolling Mills Ltd.
Te 0.5-0.7, Ni 0.08-0.12, bal Cu.
Bar: 38,000-45,000 TS; 15-18 El; 95-110 Brin. For spot and seam welding of aluminum alloys. High conductivity resistance welding electrode. *Obsolete*

ERM CCS
Enfield Rolling Mills Ltd.
Cr 0.5-0.8, bal Cu.
Bar: 66,000-78,000 TS; 20-30 El; 135-160 Brin. For switchgear and resistance welding equipment, rotor bars. High electrical conductivity both at room and elevated temperatures. *Obsolete*

ERM CCS (M3)
Delta Enfield Metals Ltd.
Cr 0.5-1.2, Si 0.01-0.1, S 0.02-0.05, bal Cu.
Bar: 54,000-77,000 TS; 15-30 El; 110-165 Brin. Forged: 52,000-67,000 TS; 17-32 El; 105-135 Brin; 78-85% electrical conductivity. For switchgear and resistance welding equipment, rotor bars. ISO 5182 and RWMA alloy A/2/1. C18200

ERM CUS
Delta Enfield Metals Ltd.
S 0.3-0.6, bal Cu.
Bar: 36,000-47,000 TS; 16-21 El; 75-90 Brin. Forged: 25,000-38,000 TS; 20-30 El; 45-55 Brin. 95-98% electrical conductivity. Free machining copper. C14700

ERM HSM
Enfield Rolling Mills Ltd.
Te 0.5-0.8, 0.02-0.05% oxygen, bal Cu.
Bar: 37,000-41,000 TS; 14-25 El; 90 Brin. For commutator bars, gas welding nozzles. 95-98% electrical conductivity. *Obsolete*

ERM N.S. (M100)
Delta Enfield Metals Ltd.
Co 2-2.8, Be 0.4-0.7, Si 0.08-0.15, bal Cu.
Bar: 90,000-130,000 TS; 9-20 El; 170-230 Brin. Forged: 90,000-130,000 TS; 9-20 El; 170-230 Brin. For machine parts and resistance welding equipment. High strength and hardness. Age hardenable. BS 4577 and ISO 5182 alloy A/3/1.

ERM NS
Enfield Rolling Mills Ltd.
Be 0.3-0.5, Co 2-2.5, bal Cu.
Bar: 112,000-123,000 TS; 10 El; 240-310 Brin. For machine parts and resistance welding equipment. High strength and hardness. Age-hardenable. *Obsolete*

ERM SC65
Enfield Rolling Mills Ltd.
Ag 5.5-6.5, 0.02-0.04% oxygen, bal Cu.
Forged: 63,000-67,000 TS; 12.5 El; 125-150 Brin. For high duty seam welding electrode. 82-86% conductivity. *Obsolete*

ERMAL
Erie EMI Company
Cast iron. C 2.2-2.4, Si 0.9-1, Mn 0.7-0.8, bal Fe.
Obsolete

ERMALITE
Erie Malleable Iron Co.
2.4-2.6 total C, 1.6-2.0 graphitic C, 0.6-0.8 combined, C, 1.9-2.0 Si, 0.4-0.45 Si, 0.35-0.45 Ni, or Mo, bal Fe.
Cast: 60,000-65,000 TS; 60,000-65,000 YS; 1-2 El; 240-250 Brin. For brake drums; wear and shock resisting. *Obsolete*

ERODUR 15 3
Bergische Stahl Industrie
C 3.5, Cr 15, Mo 2.7, bal Fe.
Abrasion resistant cast steel. G-X 350 CrMo 15 3.

ERODUR 16 SR
Bergische Stahl Industrie
C 1.6, Cr 12, Mo 0.75, V 0.25, bal Fe.
Hardenable to 500-600 Brin. Castings for wear resistant parts or tools. DIN G-X160CrMoV12.

ERODUR 16 SRV
Bergische Stahl Industrie
C 1.6, Cr 12, Mo 0.75, V 1, bal Fe.
Abrasion resistant cast steel. SEW 1.2379; G-X 155 CrVMo 12 1.

ERODUR 28
Bergische Stahl Industrie
C 2.7, Cr 28, bal Fe.
Hardenable to 500-600 Brin. Castings for wear resistant parts or tools. DIN G-X270Cr29.

ERODUR CM 1H
Bergische Stahl Industrie
C 0.42, Cr 1, Mo 0.2, bal Fe.
Cast, heat treated: 118,000-184,000 TS; 100,000-145,000 YS; 4-9 El. Castings for wear resistant parts or tools. DIN GS-42CrMo4; W.-Nr. 1.7225.

ERODUR E 2
Bergische Stahl Industrie
C 2, Cr 2.1, bal Fe.
Hardenable to 500-600 Brin. Castings for wear resistant parts or tools. DIN GS-200CrMn8; Werkstoff Nr. 1.2129. *Obsolete*

ERODUR E-C1 H
Bergische Stahl Industrie
C 0.8, Cr 1.8, bal Fe.
Hardenable to 400 Brin. Castings for wear resisting parts or tools. DIN GS-80Cr7. *Obsolete*

ERODUR E-VMS
Bergische Stahl Industrie
C 0.3, Mn 1.25, bal Fe.
Heat treated: 106,000-128,000 TS; 78,000 minimum YS; 12 minimum El. Castings for small wear resisting parts or tools. DIN GS-30Mn5; Werkstoff Nr. 1.5066. *Obsolete*

ERODUR MNA 1
Bergische Stahl Industrie
C 1.2, Cr 13, bal Fe.
Solution annealed and aged: 84,000 TS min; 42,500 YS min; 25 El min. Austenitic manganese steel casting; work hardens to hard, wear resistant martensite. DIN G-X120Mn12; W.-Nr. 1.3401.

ERODUR MNA 1 U
Bergische Stahl Industrie
C 1.05, Mn 12, bal Fe.
Solution annealed & aged: 84,000 minimum TS; 42,500 minimum YS; 25 minimum El. Austenitic manganese steel casting; work-hardens to hard, wear resistant martensite. DIN G-X100Mn12. *Obsolete*

ERODUR MNA 2
Bergische Stahl Industrie
C 1.4, Mn 17, Cr 2.25, bal Fe.
Solution annealed & aged: 100,000 minimum TS; 57,000 minimum YS; 15 minimum El. Austenitic wear resistant casting. DIN G-X140MnCr162. *Obsolete*

ERODUR V 2 Z
Bergische Stahl Industrie
C 0.32-0.39, Si 0.3-0.5, Mn 0.5-0.8, Cr 2.2-2.7, Mo 0.3-0.5, V 0.05-0.15, bal Fe.
Abrasion resistant cast steel. GS-35 CrMoV 10 4; SEW 1.1755.

ERW 3
W. Ossenberg & Cie Edelstahlwerke
C 1.4, Mn 0.3, Si 0.2, Cr 0.3, V 0.25, W 3, bal Fe.
Hardenable to 67 Rock C. For scrapers, rifflers, and engraving tools. W.-Nr. 1.2562.

ERZBERG WEICH
Vereinigte Edelstahlwerke
C 0.35, Si 0.25-0.5, Mn 0.2-0.8, bal Fe.
Hot rolled: 85,000 TS; 54,000 YS; 30 El; 53 RA; 185 Brin. For gears, bolts, shafts, machine tool parts; water hardened. *Obsolete*

ERZBERG ZAH
Vereinigte Edelstahlwerke
C 0.45, Si 0.25-0.5, Mn 0.2-0.8, bal Fe.
Hot rolled: 98,000 TS; 60,000 YS; 24 El; 45 RA; 215 Brin. For gears, bolts, shafts, machine tool parts; water hardened. *Obsolete*

ERZBERG ZAHHART
Vereinigte Edelstahlwerke
C 0.6, Si 0.25-0.5, Mn 0.2-0.8, bal Fe.
Heat treated: 160,000 TS; 113,000 YS; 12 El; 40 RA; 325 Brin. For gears, axles, rails, shafts, punches; water hardened. *Obsolete*

ES 1
Electrometal SA Metals Especials
C 0.45, Si 0.95, Mn 0.3, Cr 1, Mo 0.2, W 1.95, bal Fe.
Shock resisting tool steel. W.-Nr. 1.2542; similar to AISI S1.

ES 17-7
Eastmet Corp.
C 0.09-0.15, Ni 6-8, Mn 0-2, Cr 16-18, bal Fe.
Annealed: 100,000 TS; 40,000 YS; 55 El; 80 Rb. Worked: 185,000 TS; 140,000 YS; 16 El; 43 Rc. For refrigerator and kitchen equipment trim; Type 301; corrosion resistant, non-magnetic. *Obsolete*

ES 18-8SI
Eastmet Corp.
C 0.08-0.15, Si 2-3, Cr 17-19, Ni 8-10, bal Fe.
Annealed: 95,000 TS; 50,000 YS; 45 El; B90 Brin. For heat resistant parts, furnace equipment; stainless, non-magnetic. *Obsolete*

ES 19-12 MOL
Eastmet Corp.
C 0-0.03, Mn 0-2, Si 0-1, Cr 18-20, Ni 11-15, Mo 3-4, bal Fe.
Stainless sheet, plate, strip, coil. Annealed: 80,000-90,000 TS; 30,000-50,000 YS; 45-52 El; 77-83 Rb. Excellent corrosion resistance; resistant to scaling up to 1600 F. For food processing equipment, paper pulp, chemical plant equipment, fountain pens. AISI 317L. *Obsolete*

ES EXTRA
Thyssen Edelstahlwerke AG
C 0.9-1.2, bal Fe.
For tools; water hardening. *Obsolete*

ES PRIMA
Vereinigte Edelstahlwerke
C 0.12, Ni 2.5, Cr 0.8, bal Fe.
Hardened: 130,000-160,000 TS; 100,000 YS; 13 El; 40 RA; 270-330 Brin. For gears, shafts, pinions; case-hardening. *Obsolete*

ES-12
Eastmet Corp.
C 0-0.15, Cr 11-14, Mn 0.25-0.45, bal Fe.
Annealed: 75,000 TS; 48,000 YS; 20 El; 180 Brin. For corrosion resistant parts, turbine blades; Type 410; magnetic, corrosion resistant. *Obsolete*

ES-15-35
Cyclops Industries/Eastern Stainless
C 0-0.15, Ni 17-20, Cr 34-37, bal Fe.
Annealed: 80,000 TS; 40,000 YS; 40 El; 150 Brin. For salt pots, furnace equipment, heat treating boxes; heat resistant, austenitic; Type 330.

ES-18-10 CB
Eastmet Corp.
C 0-0.1, Mn 0-2, Cr 17-19, Ni 9-13, Cb = 8 x C, bal Fe.
Annealed: 90,000 TS; 45,000 YS; 45 El; 180 Brin. For welded tanks, manifolds, stainless parts; Type 347; heat and corrosion resistant. *Obsolete*

ES-18-10 TI
Eastmet Corp.
C 0-0.1, Mn 0-2, Cr 17-19, Ni 9-12, Ti = 5 x C, bal Fe.
Annealed: 85,000 TS; 40,000 YS; 50 El; 180 Brin. For exhaust manifolds, welded tanks, stainless parts; Type 321; heat and corrosion resistant. *Obsolete*

ES-18-10CB-LO-TA
Eastmet Corp.
C 0-0.08, Cr 17-19, Ni 9-13, Ta 0-0.1, Cb = 8 x C, bal Fe.
Annealed: 88,000 TS; 45,000 YS; 45 El; 160 Brin. For chemical plant equipment, welded structures; stainless, austenitic; Type 348. *Obsolete*

ES-18-8
Eastmet Corp.
C 0.09-0.2, Mn 0-2, Cr 17-19, Ni 8-10, bal Fe.
Annealed: 90,000 TS; 40,000 YS; 55 El; 180 Brin. For stainless parts, chemical plant equipment; Type 302; austenitic, non-magnetic. *Obsolete*

ES-18-8 LC
Eastmet Corp.
C 0-0.08, Mn 0-2, Cr 18-20, Ni 8-10, bal Fe.
Annealed: 85,000 TS; 35,000 YS; 60 El; 180 Brin. For stainless parts, chemical plant equipment; austenitic, non-magnetic; Type 304. *Obsolete*

ES-19-9L
Eastmet Corp.
C 0-0.03, Cr 18-20, Ni 8-12, bal Fe.
Annealed: 78,000 TS; 35,000 YS; 52 El; 144 Brin. For chemical plant equipment; stainless, austenitic; Type 304L. *Obsolete*

ES-25-12
Eastmet Corp.
C 0-0.2, Mn 0-2, Si 0-1.25, Cr 22-24, Ni 12-15, bal Fe.
Annealed: 90,000 TS; 45,000 YS; 45 El; 185 Brin. For furnace parts, heat resistant parts, heat treating boxes; heat and corrosion resistant; Type 309. *Obsolete*

ES-25-20
Eastmet Corp.
C 0-0.08, Mn 0-2, Si 1.5, Cr 24-26, Ni 19-22, bal Fe.
Annealed: 85,000 TS; 45,000 YS; 45 El; 185 Brin. For furnace parts, carburizing and annealing boxes; Type 310; heat and corrosion resistant. *Obsolete*

ES-FENI 5060
Vereinigte Deutsche Nickel-Werke AG
Alloy steel. Cu 0.1, C 0.05, Mn 0.7, Si 0.15, 50-60 Ni + Co, bal Fe.
Strip and wire for the welding industry.

ES3
Thyssen Edelstahlwerke AG
C 1.15, Si 0-0.25, Mn 0-0.25, bal Fe.
Annealed: 110,000 TS; 58,000 YS; 18 El; 40 RA; 210 Brin. For drills, taps, reamers, broaches; Type W1; water hardened. *Obsolete*

ES4
Thyssen Edelstahlwerke AG
C 1, Si 0-0.25, Mn 0-0.25, bal Fe.
Annealed: 100,000 TS; 53,000 YS; 21 El; 42 RA; 200 Brin. For drills, taps, hobs, reamers, spring; Type W1; water hardened. *Obsolete*

ES5
Thyssen Edelstahlwerke AG
C 0.85, Si 0-0.25, Mn 0-0.25, bal Fe.
Heat treated: 190,000 TS; 145,000 YS; 10 El; 30 RA; 400 Brin. For springs, drills, taps, reamers; Type W1; water hardened. *Obsolete*

ES6
Thyssen Edelstahlwerke AG
C 0.8-1.2, Si 0.25, Mn 0.25, bal Fe.
Water hardened: 166,000-216,000 TS; 100,000-150,000 YS; 11-15 El; 32-37 RA; 330-600 Brin. For taps, drills, reamers, punches, stamps, knurls, mandrels. Type W1; water hardening. *Obsolete*

ESC
W. Ossenberg & Cie Edelstahlwerke
C 2.2, Mn 0.4, Si 0.3, Cr 12, V 0.12, bal Fe.
Hot or cold work tool steel. For zinc die casting dies, hammer cores. For cold heavy duty stamping dies. W.-Nr. 1.2080; similar to AISI D3.

ESCO 12M
ESCO Corp.
C 0.25-0.35, Mo 0.2-0.3, Mn 1-1.3, Si 0.5, bal Fe.
Cast: 102,000 TS; 81,000 YS; 21 El; 225 Brin. Heat treated: 240,000 TS; 190,000 YS; 5 El; 495 Brin. For earth moving equipment, dredging and cement equipment; wear, shock and abrasion resistant.

ESCO 14
ESCO Corp.
C 1.15-1.25, Mn 12-14, Si 0.4-1, bal Fe.
Cast: 120,000 TS; 55,000 YS; 20 El; 200 Brin. For rock and ore crushers, earth moving equipment; impact, shock and abrasion resistant.

ESCO 16W
ESCO Corp.
C 0-0.2, Cr 1-1.5, Mo 0.45-0.65, Mn 0.5-0.8, bal Fe.
Cast: 80,000-90,000 TS; 50,000-70,000 YS; 22 El; 210-220 Brin. For structural and steam turbine components; high temperature use.

ESCO 16Z
ESCO Corp.
C 0-0.18, Cr 2-2.7, Mo 0.9-1.2, bal Fe.
Cast: 70,000-100,000 TS; 40,000-75,000 YS; 20-21 El; 200-220 Brin. For structural and steam turbine components; high temperature use.

ESCO 18CW
ESCO Corp.
C 0.65-0.75, Cr 1.9-2.4, Ni 0.4-0.6, Mo 0-0.25, bal Fe.
Cast: 135,000-180,000 TS; 118,000-130,000 YS; 4-12 El; 290-390 Brin. For liners for abrasion slurries; wear resistant.

ESCO 1B
ESCO Corp.
C 0.3, Mn 1, Si 0.4-0.6, bal Fe.
Cast: 79,000 TS; 46,000 YS; 30 El; 47 RA; 163 Brin. For housings, gears, shafts, machine tool parts; water hardened.

ESCO 20
ESCO Corp.
C 0-0.07, Cr 19-22, Ni 27.5-31, Mo 1.8-2.5, Cu 3-3.5, bal Fe.
Cast: 68,000 TS; 30,000 YS; 47 El; 60 RA; 125 Brin. For acid resistant equipment; Type CN7M; austenitic, corrosion resistant.

ESCO 20E
ESCO Corp.
C 0-0.07, Cr 19-21, Ni 30-38, Mo 2-3, Cu 3-4, bal Fe.
Cast: 68,000 TS; 28,000 YS; 47 El; 60 RA; 160 Brin. Austenitic; for acid resistant equipment; specific for H_2SO_4.

ESCO 22
ESCO Corp.
C 0-0.3, Ni 62-68, Fe 0-3, bal Cu.
Cast: 75,000 TS; 32,000 YS; 50 El; 50 RA; 120 Brin. For corrosion resistant parts; cast Monel, corrosion resistant.

ESCO 22H
ESCO Corp.
C 0-0.35, Mn 0.5-1.5, Fe 0-2.5, Si 1, Ni 62-68, bal Cu.
Cast: 85,000-120,000 TS; 50,000-80,000 YS; 10-20 El; 175-250 Brin. For chemical plant equipment; cast H-Monel, corrosion resistant.

ESCO 22S
ESCO Corp.
C 0-0.35, Mn 0.5-1.5, Fe 0-2.5, Si 1-2.2, Ni 62-68, bal Cu.
Cast: 120,000-145,000 TS; 80,000-130,000 YS; 1-4 El; 275-350 Brin. For chemical plant equipment; age-hardenable, cast S-Monel; corrosion resistant.

ESCO 23
ESCO Corp.
C 0-0.2, Cr 12-15, Ni 76-79, Fe 0-6.
Cast: 80,000 TS; 37,000 YS; 16 El; 5 RA; 150 Brin. At 1200°F: 63,000 TS; 25,000 YS; 39 El. At 1800°F: 72,000 TS; 3800 YS; 118 El. For nitriding containers; cast Inconel, corrosion and heat resistant.

ESCO 26
ESCO Corp.
C 0-0.12, Mo 26-30, Fe 4-7, bal Ni.
Cast: 78,000 TS; 56,000 YS; 12 El; 12 RA; 196 Brin. For corrosion resistant equipment; Hastelloy B; corrosion resistant.

ESCO 27
ESCO Corp.
C 0-0.15, Cr 15.5-17.5, Mo 16-18, W 3.7-4.7, Fe 4-7, bal Ni.
Cast: 70,000 TS; 50,000 YS; 15 El; 22 RA; 200 Brin. For corrosion resistant equipment; Hastelloy C.

ESCO 28
ESCO Corp.
Si 0-11, Cu 0-4, Mn 1, Al 1, bal Ni.
Cast: 36,000-40,000 TS; 35,000-40,000 YS; 0 El; 0 RA; 484-545 Brin. For pump valves; Hastelloy D; resists HCl and H_2SO_4.

ESCO 32
ESCO Corp.
C 0.2-0.4, Cr 12, Ni 0-1, bal Fe.
Heat treated: 200,000 TS; 150,000 YS; 5 El; 7 RA; 450 Brin. For cutlery, valves, furnace parts; Type CA40; corrosion resistant.

ESCO 32B
ESCO Corp.
C 0.06-0.15, Cr 11.5-13.5, Ni 0-0.25, bal Fe.
Heat treated: 160,000 TS; 145,000 YS; 12 El; 25 RA; 360 Brin. For turbine blades, valves, cutlery; Type 403; corrosion and heat resistant.

ESCO 32C

ESCO Corp.

C 0.06-0.15, Cr 11.5-14, Ni 0-1, Mo 0-0.5, bal Fe.
Heat treated: 135,000-200,000 TS; 115,000-150,000 YS; 7-17 El; 6-43 RA; 390-260 Brin. Cast: 100,000-115,000 TS; 75,000-100,000 YS; 20-30 El; 52 RA; 185-225 Brin. For cutlery, valves, pump parts; Type CA15; corrosion resistant.

ESCO 33 G

ESCO Corp.

C 0-0.06, Cr 11.5-14, Ni 3.5-4.5, Mo 0.5-1, bal Fe.
Heat treated: 110,000 TS min; 80,000 YS min; 15 El; 35 RA; 363 Brin. Type CA6NM; corrosion resistant, for large propellers, water wheels, impellers, valves, pump parts.

ESCO 33-A

ESCO Corp.

C 0.6-0.75, Mn 1, Cr 16-18, Mo 0.75, bal Fe.
Heat treated: 400 Brin. For bearings, races, pivots, valves, cutlery; Type 440A; stainless, hardenable.

ESCO 33-B

ESCO Corp.

C 0.75-0.95, Cr 16-18, Mo 0-0.75, Mn 1, Si 1, bal Fe.
Annealed: 107,000 TS; 62,000 YS; 18 El; 35 RA; 220 Brin. For bearings, cutlery, valves; Type 440B; stainless, hardenable.

ESCO 33C

ESCO Corp.

C 0.95-1.2, Cr 16-18, Mo 0-0.75, bal Fe.
For bearings, valves; Type 440C; corrosion resistant.

ESCO 35-T

ESCO Corp.

C 0.5, Cr 26-30, Ni 4-7, Mo 0-0.5, bal Fe.
Cast: 97,000 TS; 65,000 YS; 18 El; 212 Brin. For cylinder liners, valves, furnace parts; Type CC50; corrosion and heat resistant.

ESCO 35AW

ESCO Corp.

C 2.3-2.8, Mn 1.25, Cr 24-28, bal Fe.
Cast: 630 Brin. Annealed: 370 Brin. For roasters, impellers; high abrasion resistance.

ESCO 35H

ESCO Corp.

C 0-0.5, Cr 26-30, Ni 0-4, bal Fe.
At 1400°F: 10,500 TS; 8700 YS; 65 El. At 1800°F: 250 TS; 2100 YS; 110 El. At 70°F: 75,000 TS; 66,000 YS. For furnace rabble arms, sintering bars, grates, dampers; heat resistant to 2000°F in Sulfur atmosphere.

ESCO 36

ESCO Corp.

C 0-0.2, Cr 26-30, Ni 3-4, Mo 1.5, bal Fe.
For corrosion and heat resisting parts; acid resistant.

ESCO 36F

ESCO Corp.

C 0-0.07, Cr 14-15.5, Ni 3.5-5.5, Cu 2.5-4.5, bal Fe.
Heat treated (age hardened at 925°F, aged): 175,000 TS min; 150,000 YS min; 5 El; 375 Brin. High strength cast, corrosion resistant steel. CB 7Cu2155PH.

ESCO 36PH

ESCO Corp.

C 0.07, Cr 16, Ni 4, Cu 2.5, bal Fe.
Heat treated: 160,000 TS; 145,000 YS; 12 El; 26 RA; 360 Brin. For corrosion resistant castings; age-hardenable, high strength.

ESCO 37 PH

ESCO Corp.

C 0.04, Mn 1, Si 1, Cr 25-27, Ni 4.75-6, Mo 1.75-2.25, Cu 2.75-3.25, bal Fe.
Heat treated: 130,000-144,000 TS; 95,000-105,000 YS; 10-22 El; 15-41 RA; 285-321 Brin. For chemical plant equipment, digesters, autoclaves; high corrosion resistance.

ESCO 40B

ESCO Corp.

C 0-0.08, Cr 18-21, Ni 8-11, B 1.75-2.25, bal Fe.
Cast: 80,000 TS; 50,000 YS; 2 El; 302 Brin. Neutron absorbing stainless steel; for control rods in nuclear reactors.

ESCO 40F

ESCO Corp.

C 0-0.16, Cr 18-21, Ni 9-12, Mo 0-1.5, Se 0.2-0.35, bal Fe.
Cast: 72,000 TS; 35,000 YS; 58 El; 60 RA; 140 Brin. For chemical plant equipment; Type CF16F; corrosion resistant, free-cutting.

ESCO 40H

ESCO Corp.

C 0.2-0.4, Cr 18-23, Ni 9-12, Mo 0-0.5, bal Fe.
At 1200°F: 57,000 TS; 16 El. At 1600°F: 22,000 TS; 2 El. At 70°F: 90,000 TS; 50,000 YS; 25 El. For furnace dampers, oil still supports, annealing furnaces; strength and scale resistant to 1600°F; Type HF.

ESCO 40L

ESCO Corp.

C 0-0.03, Cr 18-21, Ni 8-11, bal Fe.
Cast: 65,000 TS; 28,000 YS; 55 El; 60 RA; 140 Brin. For chemical plant equipment, tanks, mixers; Type 304L; austenitic, stainless.

ESCO 40S

ESCO Corp.

C 0-0.08, Cr 18-21, Ni 8-11, bal Fe.
Cast: 72,000 TS; 32,000 YS; 60 El; 65 RA; 140 Brin. For chemical plant equipment; Type CF8; austenitic, corrosion resistant.

ESCO 40T

ESCO Corp.

C 0-0.08, Cr 17-20, Ni 9-12, bal Fe.
Cast: 65,000 TS; 28,000 YS; 45 El; 50 RA; 125 Brin. For chemical plant equipment; Type 305; austenitic, stainless.

ESCO 41

ESCO Corp.

C 0-0.08, Cr 18-21, Ni 9-12, Cb = 8 x C, bal Fe.
Cast: 72,000 TS; 36,000 YS; 35 El; 35 RA; 145 Brin. For welded chemical plant equipment; Type CF8C; austenitic, stainless.

ESCO 43C

ESCO Corp.

C 0-0.2, Cr 22-26, Ni 12-15, bal Fe.
Cast: 70,000 TS; 30,000 YS; 35 El; 60 RA; 150 Brin. For furnace parts, heat treat boxes; Type CH20; austenitic, heat resistant.

ESCO 43H

ESCO Corp.

C 0.2-0.5, Cr 24-28, Ni 11-14, Mo 0-1.5, N 0.2, bal Fe.
At 1400°F: 35,000 TS; 18,000 YS; 12 El. At 1800°F: 11,000 TS; 7000 YS; 30 El. At 70°F: 85,000 TS; 45,000 YS; 15 El. For furnace shafts, beams and rollers, tube supports; Type HH; scale resistant at 2000°F.

ESCO 44

ESCO Corp.

C 0-0.03, Cr 17.5-18.5, Ni 12-13, Mo 2-3, Cb 0.2-0.3, Mn 0.5-1, bal Fe.
Cast: 71,500-76,500 TS; 42,750-48,250 YS; 39-47 El; 45-50 RA; 170 Brin. For heat exchanger tubes; corrosion and heat resistant.

ESCO 45L

ESCO Corp.

C 0-0.03, Cr 18-21, Ni 10-13, Mo 2-3, bal Fe.
Cast: 70,000 TS; 35,000 YS; 40 El; 55 RA; 170 Brin. For acid resistant chemical plant equipment; Type 316L; austenitic, stainless.

ESCO 45M

ESCO Corp.

C 0-0.08, Cr 18-21, Ni 9-13, Mo 3-4, bal Fe.
Cast: 75,000 TS min; 35,000 YS min; 25 El; 170 Brin. For chemical plant equipment, sulfite pulp mills, acid resistant service. Type CF8M; austenitic, stainless.

ESCO 45S

ESCO Corp.

C 0-0.08, Cr 18-21, Ni 9-12, Mo 2-3, bal Fe.
Cast: 80,000 TS; 42,000 YS; 45 El; 65 RA; 170 Brin. For acid resistant equipment; Type CF8M; stainless.

ESCO 45T

ESCO Corp.

C 0-0.08, Cr 16-19, Ni 13-15, Mo 2-3, bal Fe.
Cast: 70,000 TS; 30,000 YS; 50 El; 62 RA; 125 Brin. For acid resistant equipment; Type 316; stainless, austenitic.

ESCO 46

ESCO Corp.

C 0.2-0.5, Cr 26-30, Ni 14-18, bal Fe.
For heat resistant parts, furnace equipment; stainless and heat resistant, austenitic.

ESCO 48

ESCO Corp.

C 0.2-0.6, Cr 28-32, Ni 18-22, bal Fe.
Cast: 85,000 TS; 50,000 YS; 10 El; 190 Brin. For furnace parts and equipment, salt pots; Type HL; heat resistant.

ESCO 49

ESCO Corp.

C 0-0.05, Mn 8-11, Cr 14.5-17.5, Ni 18-27, Mo 1.75-2.75, bal Fe.
Cast, RT: 65,000 TS; 30,000 YS; 50 El; At 423°F: 125,000 TS; 80,000 YS; 40 El. Non-magnetic; very low temperature service; liquid hydrogen, liquid nitrogen; excellent weldability. Kromarc 55.

ESCO 5

ESCO Corp.

C 0.05, Cr 16.05, Ni 4.02, Cu 2.82, Mn 0.56, Si 0.61, bal Fe.
Annealed: 138,000 TS; 113,000 YS; 14 El; 315 Brin. Heat treated: 180,000 TS; 156,000 YS; 3 El; 6 RA; 418 Brin. Uses: Pump impellers, valves. Precipitation hardening stainless steel. *Obsolete*

ESCO 50

ESCO Corp.

C 0-0.12, Si 2.5-3, Cr 20-23, Ni 25-28, Mo 2.5, Cu 1.5, bal Fe.
Cast: 70,000 TS; 35,000 YS; 50 El; 60 RA; 140 Brin. For salt pots, heat and corrosion resistant parts; Type CNMCu; austenitic, corrosion resistant.

ESCO 51

ESCO Corp.

C 0-0.3, Cr 26-30, Ni 8-11, Mo 0-0.5, bal Fe.
Cast: 90,000 TS; 57,000 YS; 25 El; 34 RA; 200 Brin. For furnace parts, heat treat boxes; Type CE30; austenitic, stainless.

ESCO 51 H

ESCO Corp.

C 0.2-0.5, Cr 26-30, Ni 8-11, bal Fe.
Cast: 85,000 TS min; 40,000 YS min; 9 El. Heat resistant service, good resistance to high sulfur content gases, ore roasting equipment. ACI Type HE.

ESCO 52C

ESCO Corp.

C 0.1, Cr 14-17, Ni 33-37, bal Fe.
Cast: 70,000 TS; 40,000 YS; 10 El; 12 RA; 170 Brin. For salt pots, furnace pots, heat treat boxes; Type HT; heat resistant.

ESCO 52H
ESCO Corp.
C 0.35-0.75, Cr 13-17, Ni 33-37, Mo 0-0.5, bal Fe.
At 1200°F: 42,000 TS; 28,000 YS; 5 El. At 1800°F: 11,000 TS; 8000 YS; 28 El. At 70°F: 67,000 TS; 40,000 YS; 10 El. For retorts, radiant tubes, salt pots, hearth plates, carburizing boxes; resists oxidation and thermal cycling.

ESCO 53C
ESCO Corp.
C 0-0.2, Cr 23-27, Ni 19-22, bal Fe.
Cast: 75,000 TS; 35,000 YS; 35 El; 42 RA; 140 Brin. For furnace parts, salt pots, heat treat boxes; Type CK20; austenitic, heat resistant.

ESCO 53H
ESCO Corp.
C 0.2-0.6, Cr 24-28, Ni 18-22, Mo 0-0.5, bal Fe.
At 1600°F: 23,000 TS; 21 El. At 70°F: 75,000 TS; 40,000 YS; 18 El. For gas dissociation equipment, fixtures, baskets, calcining tubes; good strength and oxidation resistance to 2100°F; Type HK.

ESCO 54
ESCO Corp.
C 0.2-0.5, Cr 19-23, Ni 23-27, bal Fe.
For heat treating boxes, furnace parts; Type HN; heat resistant.

ESCO 55
ESCO Corp.
C 0.3-0.5, Cr 17-21, Ni 37-41, bal Fe.
For salt pots, heat treating boxes, furnace parts; Type HU; heat resistant.

ESCO 56
ESCO Corp.
C 0.35-0.75, Cr 10-14, Ni 58-62, bal Fe.
At 1400°F: 32,000 TS; 23,000 YS. At 1800°F: 10,000 TS; 8000 YS; 40 El. At 70°F: 70,000 TS; 40,000 YS; 6 El. For carburizing and hardening fixtures, retorts; Type HW; resists thermal cycling.

ESCO 57
ESCO Corp.
C 0.3-0.7, Cr 15-19, Ni 64-68, bal Fe.
For heat treat boxes, furnace equipment; heat and corrosion resistant.

ESCO 58
ESCO Corp.
C 0.4-0.6, Cr 24-28, Ni 33-37, W 4.5-5.5, bal Fe.
For heat resistant service at 2000-2200°F.

ESCO 5T
ESCO Corp.
C 0-0.2, Cr 8-10, Mo 0.9-1.2, bal Fe.
Cast: 90,000-105,000 TS; 65,000-85,000 YS; 18-22 El; 40-55 RA; 200-240 Brin. For oil refinery equipment; Type HA; corrosion resistant.

ESCO 5W
ESCO Corp.
C 0-0.2, Cr 4-6.5, Mo 0.45-0.65, bal Fe.
Cast: 90,000-100,000 TS; 75,000-80,000 YS; 18-24 El; 45-60 RA; 190-248 Brin. For oil refinery equipment; Type 501; corrosion resistant.

ESCO 5X
ESCO Corp.
C 0-0.2, Cr 4-6.5, Mo 0.45-0.65, bal Fe.
For oil refinery equipment; Type 502; corrosion resistant.

ESCO 6-T
ESCO Corp.
C 0-0.15, Ni 3-4, Mn 0.6, Si 0.5, bal Fe.

ESCO 61A
ESCO Corp.
C 0-3, Cr 1.7-2.5, Ni 13.5-17.5, Cu 5.5-7.5, bal Fe.
Cast: 25,000 TS; 135 Brin. For corrosion and wear resistant parts; Ni-Resist Type 1; corrosion resistant cast iron.

ESCO 625
ESCO Corp.
C 0-0.1, Cr 20-23, Mo 8-10, Cb 3.15-4.15, Fe 0-4.5, bal Ni.
Electroslag remelted billet, forging stock and shaped forms. Inconel 625.

ESCO 62A
ESCO Corp.
C 0-3, Cr 1.7-2.5, Ni 18-22, bal Fe.
Cast: 25,000 TS; 135 Brin. For corrosion and wear resistant parts; Ni-Resist Type 2; corrosion resistant cast iron.

ESCO 63
ESCO Corp.
C 0-3, Cr 2, Ni 20, bal Fe.
For corrosion and wear resistant parts; Ni-Resist Type 2B; corrosion resistant cast iron.

ESCO 70
ESCO Corp.
C 0.08-0.16, Mn 1-2, Cr 20-22.5, Ni 19-21, Mo 2.5-3.5, Co 18.5-21, W 2-3, Cb 0.75-1.25, N 0.15, bal Fe.
For high temperature applications; Multimet N-155.

ESCO 718C
ESCO Corp.
C 0.08, Cr 17-21, Mo 2-4, Cb 4.5-5.75, Ti 0.4-1.3, Fe 19-21, bal Ni.
Electroslag remelted billet, forging stock. Inconel 718C.

ESCO 72
ESCO Corp.
C 0.1, Ti 0.18, Cb 0.6, Cr 26-30, Co 47-52, bal Fe.
Cast: 135,000 TS; 48,000 YS; 10 El; 10 RA; 250 Brin. For furnace baffles, burner tips, sintering grates, quench baskets. Corrosion, heat and thermal shock resistant.

ESCO 800
ESCO Corp.
C 0.05, Cr 21, Ni 32.5, Ti 0.4, Al 0.4, bal Fe.
Electroslag remelted billet, forging stock and shaped forms. Incoloy 800.

ESCO 802
ESCO Corp.
C 0.35, Cr 21, Ni 32.5, Al 0.55, Ti 0.75, bal Fe.
Electroslag remelted billet, forging stock and shaped forms. Incoloy 802.

ESCO A286
ESCO Corp.
C 0-0.08, Cr 13.5-16, Ni 24-27, Mo 1-1.5, Ti 1.9-2.35, bal Fe.
Electroslag remelted billet, forging stock, and shaped forms. AISI 660.

ESCO N155 (ESCO 70 CASTINGS)
ESCO Corp.
C 0.8-0.16, Cr 20-22.5, Ni 19-21, Co 18.5-21, Mo 2.5-3.5, W 0.75-1.25, bal Fe.
Electroslag remelted billet, forging stock and shaped forms. AISI 661, N155.

ESCO NIROSTA KNC-3
ESCO Corp.
C 0.2, Cr 25, Ni 19, bal Fe.
For furnace parts, retorts; stainless. *Obsolete*

ESCO NO. 16-N
ESCO Corp.
C 0.25, Cr 2.5-3, Mo 0.4-0.65, Mn 1, bal Fe.
For oil refinery equipment; creep resistant. *Obsolete*

ESCO NO. 20
Crane Co.
C 0.07, Ni 29, Cr 20, Mo 2-3, Cu 4, bal Fe.
Cast: 65,000-75,000 TS; 28,000-38,000 YS; 35-50 El; 40-50 RA; 120-150 Brin. For chemical plant equipment, valves; resists mixed acids, austenitic. *Obsolete*

ESCO NO. 21
ESCO Corp.
C 0-0.7, Si 0-2, Mn 0-0.75, 96.5 Ni min, bal Fe.
Cast: 60,000 TS; 25,000 YS; 25 El; 30 RA; 120 Brin. For corrosion and heat resistant castings; cast nickel, corrosion and heat resistant.

ESCO NO. 35C
ESCO Corp.
C 0.5, Cr 28, Ni 0-3, bal Fe.
For furnace parts, heat resistant parts; Type CC50.

ESCO NO. 40
ESCO Corp.
C 0.16, Cr 20, Ni 9, bal Fe.
Cast: 80,000 TS; 35,000 YS; 15 El; 150 Brin. For chemical and plastic plant equipment; Type CF-8; stainless, austenitic. *Obsolete*

ESCO NO. 41W
ESCO Corp.
C 0-0.08, Cr 18-21, Ni 9-12, Cb + Ti = 10 x C, bal Fe.
Cast: 72,000 TS; 36,000 YS; 35 El; 35 RA; 145 Brin. For welded chemical plant equipment; Type CF-8C; austenitic, stainless. *Obsolete*

ESCO NO. 43
ESCO Corp.
C 0-0.2, Cr 22-26, Ni 10-13, bal Fe.
For corrosion and heat resisting parts; heat and corrosion resistant. *Obsolete*

ESCO NO. 45
ESCO Corp.
C 0-0.2, Cr 18-22, Ni 7-9, bal Fe.
For turbine castings, chemical plant equipment; Type CF-20; stainless, austenitic. *Obsolete*

ESCO STAINLESS NO. 33
ESCO Corp.
C 0-0.25, Cr 12-15, bal Fe.
For corrosion resistant parts; corrosion resistant. *Obsolete*

ESCO STAINLESS NO. 34
ESCO Corp.
C 0-0.25, Cr 16-20, bal Fe.
For heat and corrosion resistant parts; heat and corrosion resistant. *Obsolete*

ESCO STAINLESS NO. 35
ESCO Corp.
C 0-0.25, Cr 24-30, bal Fe.
For heat and corrosion resistant parts; heat and corrosion resistant. *Obsolete*

ESCO STAINLESS NO.49
ESCO Corp.
C 0.13-0.2, Cr 16-23, Ni 24-30, Si, bal Fe.
For heat and corrosion resistant parts. *Obsolete*

ESMO
W. Ossenberg & Cie Edelstahlwerke
C 1.65, Mn 0.3, Si 0.3, Cr 12, Mo 0.6, W 0.5, V 0.1, bal Fe.
For broaches, coining, punching dies for thin sheet. W.-Nr. 1.2601; similar to AISI D2.

ESMO 2
W. Ossenberg & Cie Edelstahlwerke
C 1.65, Cr 12, Mo 0.8, V 0.9, bal Fe.
Air or oil hardenable to 58-63 Rock C. For punching and coining dies for sheet metal. Similar to AISI D2.

ESS

W. Ossenberg & Cie Edelstahlwerke
C 2.1, Mn 0.3, Si 0.3, Cr 12, W 0.8, V 0.1, bal Fe.
Cold work tool steel for heavy duty shears, punching and trimming dies. W.-Nr. 1.2436.

ESS 1

W. Ossenberg & Cie Edelstahlwerke
C 2.1, Mn 0.35, Si 0.3, Cr 12.5, Co 1.2, Mo 0.5, W 0.7, bal Fe.
Cold work tool steel for stamping transformer laminations. W.-Nr. 1.2884.

ESS SPEZIAL

W. Ossenberg & Cie Edelstahlwerke
C 1.65, Mn 0.3, Si 0.3, Cr 12, Co 1.3, Mo 0.5, bal Fe.
Cold work tool steel for stamping transformer laminations. W.-Nr. 1.2880.

ESSHETE 1250

United Engineering Steels Ltd.
C 0-0.15, Si 0.2-1, Mn 5.5-7, Ni 9-11, Cr 14-16, Mo 0.8-1.2, V 0.15-0.4, Nb 0.75-1.25, B 0.003-0.009, bal Fe.
Solution treated: 72,000 TS min; 25,800 YS min; 30 El. For superheater tubes, pressure vessels, steam piping, and parts operating up to 650°C.

ESSHETE 316

United Engineering Steels Ltd.
C 0.04-0.09, Si 0-0.8, Mn 1-2, Ni 11-14, Cr 16-18, Mo 2-2.75, B, bal Fe.
Solution treated: 74,000 TS min; 30,000 YS min; 30 El. For superheater tubes and power plant components. B.S. 3605-855 (1963); ASTM 213-64T TP316H.

ESSHETE 321

United Engineering Steels Ltd.
C 0.04-0.09, Si 0-1, Mn 0-2, Ni 9-11, Cr 17-19, Ti 0-0.7, bal Fe.
Solution treated: 74,000 TS min; 30,000 YS min; 30 El. For superheater tubes and power plant components. B.S. 970-En58B; ASTM 213-64T TP321H; AISI 321.

ESSHETE 347

United Engineering Steels Ltd.
C 0.04-0.09, Si 0-0.8, Mn 0.5-2, Ni 9-13, Cr 17-19, Nb = 10 x C min, bal Fe.
Solution treated: 74,000 TS min; 32,600 YS min; 30 El. For superheater tubes and power plant components. B.S. 3605-822 Nb (1963): AISI 347; ASTM 213-64T TP347H.

ESSHETE 600

United Engineering Steels Ltd.
C 0-0.15, Si 0-0.5, Mn 0-1, Cr 14-17, Fe 6-10, Cu 0-0.5, bal Ni + Co.
Good high temperature strength and corrosion properties above 600°C.

ESSHETE 800

United Engineering Steels Ltd.
C 0-0.1, Si 0-1, Mn 0-1.5, Cr 19-23, Al 0.15-0.6, Ti 0.15-0.6, Ni 30-35, Cu 0-0.75, bal Fe.
Good high temperature strength and corrosion properties.

ESSHETE 800L

United Engineering Steels Ltd.
C 0-0.03, Si 0-1, Mn 0-1.5, Cr 20-23, Al 0.15-0.6, Ti 0.15-0.6, N 0-0.03, Ni 32-35, Cu 0-0.75, bal Fe.
Good high temperature strength and improved corrosion properties.

ESSHETE CML

United Engineering Steels Ltd.
C 0-0.12, Mo 0.4-0.7, Cr 0.75-1.25, bal Fe.
Normalized: 62,000-78,000 TS; 22,000-25,000 YS; 45-41 El; 70-63 RA. For superheater tubes and steam piping. Resists heat to 1000°F; creep resistant. BS 1507, 1508; ASTM A199-64T-Gr T11.

ESSHETE CRM 12

United Engineering Steels Ltd.
C 0-0.15, Cr 12, Mo 1, V 0.25, bal Fe.
For long time rupture strength up to about 600°C. For power plant components, especially superheater tubes.

ESSHETE CRM1

United Engineering Steels Ltd.
C 0.1-0.15, Cr 1, Mo 0.5, bal Fe.
For applications up to 540°C.

ESSHETE CRM2

United Engineering Steels Ltd.
C 0-0.15, Cr 2-2.5, Mo 0.9-1.1, bal Fe.
Normalized: 93,000 TS; 80,000 YS; 30 El. Annealed: 72,500 TS; 34,500 YS; 40 El. For steam piping and superheater tubes. Creep resistance to 1100°F. B.S. 1503, 1508; ASTM A200-62T Gr T22.

ESSHETE CRM5

United Engineering Steels Ltd.
C 0-0.12, Cr 4-6, Mo 0.45-0.65, bal Fe.
Annealed: 67,000 TS; 30,000 YS; 40 El. For heat exchangers, valves, superheaters, and steam and oil tubes. Resists scaling to 1100°F.

ESSHETE CRM9

United Engineering Steels Ltd.
C 0-0.15, Si 0.25-1, Mn 0.3-0.6, Cr 8-10, Mo 0.9-1.1, bal Fe.
32,500-39,000 psi. 0.2% proof stress at 540°C. B.S. 1607-GrP17; ASTM 199 Gr T9; DIN X9CrMo91; Werkstoff Spec 213.

ESSHETE D4C

United Engineering Steels Ltd.
C 0.15, Mo 0.5, bal Fe.
Normalized: 56,000-68,000 TS. For superheater tubes and steam piping. Resists heat to 950°F; creep resistant.

ESSHETE MV

United Engineering Steels Ltd.
C 0.08-0.15, Mn 0.4-0.7, Cr 0.25-0.5, Mo 0.5-0.7, V 0.22-0.3, bal Fe.
Normalized and tempered: 67,000 TS min; 42,200 YS min; 17 El. For steam piping. B.S. 3604 CD.660 (1963).

ESSW

W. Ossenberg & Cie Edelstahlwerke
C 1.65, Mn 0.4, Si 0.3, Cr 12, W 0.8, V 0.1, bal Fe.
Oil or air hardenable to 60-63 Rock C. For cold punching and forming dies.

ESTHONIA 5 MARK

English manufacture
Cu 70, Zn 20, Ni 10.
For coinage; corrosion resistant.

ESV

W. Ossenberg & Cie Edelstahlwerke
C 2.2, Si 0.4, Mn 0.4, Cr 12.5, Mo 1.1, V 4, bal Fe.
Hardenable to 65-66 Rock C. For cold stamping and coining; for precision rollers.

ESW

W. Ossenberg & Cie Edelstahlwerke
C 1.65, Mn 0.3, Si 0.3, Cr 12, V 0.1, bal Fe.
Cold work tool steel for punching and shearing dies, trimming and coining dies. W.-Nr. 1.2201; similar to AISI D2.

ET-21

Forjas Alavesas S.A.
C 0.2, Mn 0.4, Si 0-0.08, bal Fe.
Steel for cold heading and extrusion.

ET-8

Forjas Alavesas S.A.
C 0-0.1, Mn 0.35, Si 0-0.06, bal Fe.
Low carbon steel for cold heading.

ETERNOS 10

Now ETERNOS N110.

ETERNOS 10

Remystahl
C 0.15, Cr 19.5, Ni 9.5, bal Fe.
Annealed: 80,000 TS; 35,000 YS; 55 El; 75 RA; 150 Brin. For chemical plant equipment, tanks, mixers, filters; Type 302; stainless, austenitic. *Obsolete*

ETERNOS 11

Remystahl
C 0.15, Cr 24, Ni 19, bal Fe.
Annealed: 100,000 TS; 45,000 YS; 50 El; 65 RA; 185 Brin. For furnace parts, valves, turbine and jet engine parts; Type 310; stainless, austenitic. *Obsolete*

ETERNOS 23

Now ETERNOS A1 24.

ETERNOS AL 13

Remystahl
C 0-0.12, Si 0.7-1.4, Mn 0-1, Cr 12-14, Al 0.7-1.2, bal Fe.
For furnace equipment; elevated temperature to 950°C. W.-Nr. 1.4724.

ETERNOS AL 18

Remystahl
C 0-0.12, Si 0.7-1.4, Mn 0-1, Cr 17-19, Al 0.7-1.2, bal Fe.
For furnace parts; elevated temperature to 1050°C. W.-Nr. 1.4742.

ETERNOS AL 24

Remystahl
C 0-0.12, Si 0.7-1.4, Mn 0-1, Cr 23-25, Al 1.2-1.7, bal Fe.
For high temperature equipment to 1200°C. W.-Nr. 1.4762.

ETERNOS AL 7

Remystahl
C 0-0.12, Si 0.5-1, Mn 0-1, Cr 6-8, Al 0.5-1, bal Fe.
For annealing boxes; elevated temperature to 800°C. W.-Nr. 1.4713.

ETERNOS CS

Remystahl
C 0-0.1, Si 1.5-1.8, Mn 0-1, Cr 1.5-2, bal Fe.
Non-highly-stressed parts for elevated temperature to 600°C. W.-Nr. 1.4700.

ETERNOS NAT 32

Remystahl
C 0-0.12, Si 0-1, Mn 0-2, Cr 19-23, Ni 30-34, Al 0.15-0.6, Ti 0.15-0.6, bal Fe.
Parts for furnace and steam boiler equipment to 1100°C. W.-Nr. 1.4876.

ETERNOS NI 10

Remystahl
C 0-0.12, Cr 17-19, Ni 9-11.5, Ti = 4 x C min (0.80 max), bal Fe.
For furnace and boiler parts; high temperature. W.-Nr. 1.4878; similar to AISI 321.

ETERNOS NI 12

Remystahl
C 0-0.2, Si 1.5-2.5, Mn 0-2, Cr 19-21, Ni 11-13, bal Fe.
For annealing boxes and bells. Heat resistant to 1000°C. W.-Nr. 1.4828; similar to AISI 309.

ETERNOS NI 20

Remystahl
C 0-0.2, Si 1.5-2.5, Mn 0-2, Cr 24-26, Ni 19-22, bal Fe.
For annealing and carburizing equipment. Heat resistant to 1150°C. W.-Nr. 1.4841; AISI 310.

ETERNOS NI 35

Remystahl
C 0-0.15, Si 1-2, Mn 0-2, Cr 15-17, Ni 33-37, bal Fe.
For furnace parts for high temperature to 1100°C. W.-Nr. 1.4864.

ETERNOS NI 4
Remystahl
C 0.1-0.2, Si 0.8-1.5, Mn 0-2, Cr 24-27, Ni 3.5-5.5, bal Fe.
For high temperature equipment to 1100°C. W.-Nr. 1.4821.

ETERNOS SI 13
Remystahl
C 0-0.12, Si 1.9-2.4, Mn 0-1, Cr 12-14, bal Fe.
Furnace rails, grates; thermocouple tubes. Heat resistant to 950°C. W.-Nr. 1.4722.

ETERNOS SI 18
Remystahl
C 0-0.12, Si 1.9-2.4, Mn 0-1, Cr 17-19, bal Fe.
For furnace parts; elevated temperature. Heat resistant to 1050°C. W.-Nr. 1.4741.

ETERNOS SI 6
Remystahl
C 0-0.12, Si 2-2.5, Mn 0-1, Cr 5.5-6.5, bal Fe.
For carburizing boxes. Heat resistant to 850°C. W.-Nr. 1.4712.

ETIRAL
Creusot-Loire
Cr 10, C, bal Fe.
For wire drawing dies; tough and hard. *Obsolete*

ETP COPPER
Criterion Metals, Inc.
Copper. Cu 99.99.
Thin gauge sheet, various tempers: 23-52 ksi TS min; 5-51 ksi YS min. C11000

EUGENE VADERS
English manufacture
Cu 57.5, Fe 0.3, Mn 2.5, Al 1.5, Si 0.5, bal Zn.
For propellers, gears, high strength castings; corrosion resistant.

EUREKA
Manufacturer not listed
Pb, Sb, bal Sn.
For Babbitt bearings; Babbitt metal.

EUREKA
Columbia Tool Steel Co.
C, bal Fe.
For machinery parts; water hardened. *Obsolete*

EUREKA
Manufacturer not listed
Cu 60, Ni 40.
Hot rolled: 60,500-65,000 TS; 20,000-22,400 YS; 45-51 El; 65-67 RA. For resistance materials and thermocouples; similar to "Constantan" and "Ferry Metals." *Obsolete*

EUREKA "BU" BARE ROD-TIG
Eureka Welding Alloys, Inc.
For joining or heavy buildup of low alloy steels and general maintenance.

EUREKA "FORGEWELD" LOW ALLOY ELECTRODE
Eureka Welding Alloys, Inc.
Now EUREKA PRESSALLOY LOW ALLOY ELECTRODE.

EUREKA "HAMMERWELD" ELECTRODE
Eureka Welding Alloys, Inc.
Now EUREKA HAMMERALLOY ELECTRODE.

EUREKA 1 ELECTRODE
Eureka Welding Alloys, Inc.
C 2.5, Cr, W, high Co.
For surfacing parts subject to severe wear, heat and abrasion.

EUREKA 100 ELECTRODE
Eureka Welding Alloys, Inc.
Ni 99, bal Fe.

EUREKA 100-A PURE NICKEL BARE ROD
Eureka Welding Alloys, Inc.
Ni 99, bal others.

EUREKA 1000-A H.W. BARE ROD
Welding Equipment & Supply Co.
C 0.4, Cr 5, Mo, V, Si, bal Fe.
Tungsten free rod for welding die cast dies made from H-11 or H-13. *Obsolete*

EUREKA 12 ELECTRODE
Eureka Welding Alloys, Inc.
C 1.4-1.45, Cr, Ni, W, Mo, high Co.
For surfacing parts subject to severe wear, heat and abrasion.

EUREKA 1215 A. H. ELECTRODE
Eureka Welding Alloys, Inc.
C 0.75-1, Cr 4.75, W 1.05, Mo 1.85, V, Mn, bal Fe.
For welding air hardening tool and die steels.

EUREKA 1215-A A. H. BARE ROD
Eureka Welding Alloys, Inc.
C 1, Cr 5, Mo, W, Mn, bal Fe.
For welding air hard tool steel by GTAW.

EUREKA 1216 H. S. ELECTRODE
Eureka Welding Alloys, Inc.
C 0.6-0.75, Cr 4.6, Mo 8.75, W, V, Mn, bal Fe.
For welding high speed steel and other hot work units.

EUREKA 1216-A HIGH SPEED BARE ROD
Eureka Welding Alloys, Inc.
C 0.93, Mo 8.5, Cr 3.5, V, W, bal Fe.
High Mo high speed for welding tools of the M1-M2 type.

EUREKA 1220-A A. H. BARE ROD
Eureka Welding Alloys, Inc.
C 2.15, Cr 11.5, Mo, V, Mn, bal Fe.
For welding high carbon, high chromium tool steel by GTAW. *Obsolete*

EUREKA 130 ALLOY ELECTRODE
Eureka Welding Alloys, Inc.
C 0.25-0.35, Cr 1.15, Mo, Mn, bal Fe.
For welding SAE 4130 and flame hardening units.

EUREKA 130-A BARE ROD
Eureka Welding Alloys, Inc.
C 0.32, Mo, Cr, Mn, Si, bal Fe.
Low alloy for welding flame hardened dies and SAE 4130 steel.

EUREKA 145 ALLOY ELECTRODE
Eureka Welding Alloys, Inc.
C 0.4-0.5, Cr 1, Mn, Mo, V, bal Fe.
For welding SAE 6145 and flame hardening dies.

EUREKA 145-A BARE ROD
Eureka Welding Alloys, Inc.
C 0.45-0.5, Cr, Mo, Mn, Si, bal Fe.
Medium alloy for welding flame hardened dies and 6145 steel.

EUREKA 2 HIGH SPEED BARE ROD
Eureka Welding Alloys, Inc.
C 0.75, Cr 4, Mo 4, W, plus high Co.
For welding high speed cutting tools by GTAW.

EUREKA 200 HARD SURFACING ELECTRODE
Eureka Welding Alloys, Inc.
C 0.06-0.12, Cr, Ni, Mo, Mn, Cb, bal Fe.
Work hardening; 200 Brin min; for surfacing parts subject to wear, heat and impact; austenitic. *Obsolete*

EUREKA 240 DRAWALLOY ELECTRODE
Eureka Welding Alloys, Inc.
C 0.1-0.2, Cr, Ni, Mo, Mn, bal Fe.
Work hardening, 240 Brin min; for surfacing parts subject to wear, heat and impact; austenitic.

EUREKA 31-A
Eureka Welding Alloys, Inc.
Same composition as AISI H 13 steel, W free.
For welding die cast dies, hot work steel in general.

EUREKA 340 DRAWALLOY ELECTRODE
Eureka Welding Alloys, Inc.
C 0.6-0.75, Mn, Cr, W, Mo, Cb, bal Fe.
Work hardening; 340 Brin min; for surfacing parts subject to wear, heat and impact; austenitic. *Obsolete*

EUREKA 35 H. W. ELECTRODE
Eureka Welding Alloys, Inc.
C 0.07, Cr 4.35, Ni 0.85, Mo, Mn, bal Fe.
For welding hot work die units, forging and upsetter dies. *Obsolete*

EUREKA 350-A ALLOY BARE ROD
Eureka Welding Alloys, Inc.
C 0.06, Cr 12.5, Ni 1.85, bal Fe.
For joining or repairing cracked die sections.

EUREKA 4 HIGH SPEED BARE ROD
Eureka Welding Alloys, Inc.
C 1.55, W 12, V 4.85, Cr, Co, bal Fe.
For welding high speed tool where 63-65 Rock C hardness is required. *Obsolete*

EUREKA 40-A A.H. BARE ROD
Welding Equipment & Supply Co.
C 0.7, Cr 1, Mn, Mo, bal Fe.
For welding the low air hardening and graphitic tool steels by G.T.A. method. *Obsolete*

EUREKA 400 BUILD UP ELECTRODE
Eureka Welding Alloys, Inc.
C 0.55-0.65, Mn, Cr, Si, bal Fe.
For build up on water hardening tool and die steels. *Obsolete*

EUREKA 440 DRAWALLOY ELECTRODE
Eureka Welding Alloys, Inc.
C 0.75-1, Mn, Cr, Ni, W, Mo, Cb, bal Fe.
Work hardening; 440 Brin min; for surfacing parts subject to wear, heat and impact; austenitic.

EUREKA 45 H. W. ELECTRODE
Eureka Welding Alloys, Inc.
C 0.1, Cr 4.25, W 2, Mn, Mo, Ni, V, bal Fe.
For welding hot work die units, forging and upsetter dies.

EUREKA 45-A H. W. BARE ROD
Eureka Welding Alloys, Inc.
C 0.08, Cr, Ni, Mo, Si, bal Fe.
Very low carbon hot work for welding hot work units that must be machinable, forging and die cast dies.

EUREKA 45-N H.W. ELECTRODE
Eureka Welding Alloys, Inc.
C 0.1, Cr 4.25, W 2, Ni 25-30, Mn, Mo, V, bal Fe.
Tougher weld metal for welding hot work die units, forging and upsetter dies.

EUREKA 500 HIGH ALLOY ELECTRODE
Now EUREKA 590 HIGH ALLOY ELECTRODE.

EUREKA 500-AND 580 HIGH ALLOY ELECTRODES
Welding Equipment & Supply Co.
Low C, Cr, Ni, Mn, Mo, Si, bal Fe.
For joining and repairing high alloys and tool steel. *Obsolete*

EUREKA 52 ELECTRODE
Eureka Welding Alloys, Inc.
Pure deoxidized copper.
Obsolete

EUREKA 5545 ELECTRODE
Eureka Welding Alloys, Inc.
Ni 45, Cu 55.
Electrode. *Obsolete*

EUREKA 590 HIGH ALLOY ELECTRODE
Eureka Welding Alloys, Inc.
Low C, Cr, Ni, Mn, Mo, Si, bal Fe.
For welding high carbon steel and joining hard to weld steels.

EUREKA 6 ELECTRODE
Eureka Welding Alloys, Inc.
C 1.45, Cr 28.5, W 3.9, Ni, Si, high Co.
For surfacing punches, dies, shear blades subject to severe heat and impact.

EUREKA 60 NICKEL ELECTRODE
Eureka Welding Alloys, Inc.
Ni 60, bal Fe.
For strong welds on all cast iron.

EUREKA 70-A O. H. BARE ROD
Eureka Welding Alloys, Inc.
C 0.9, Cr, V, W, Mo, Mn, bal Fe.
For welding oil hard tool steel by GTAW.

EUREKA 70-W O. H. ELECTRODE
Eureka Welding Alloys, Inc.
C 0.65-0.75, Cr 1.4, W 2.15, Mn, V, Si, bal Fe.
For welding "W" type oil hardening tool steels. *Obsolete*

EUREKA 71-M O. H. ELECTRODE
Eureka Welding Alloys, Inc.
C 0.65-0.75, Cr 1.4, Mo 1.15, Mn, V, bal Fe.
For welding oil hardening tool and die steels.

EUREKA 72 H. W. ELECTRODE
Eureka Welding Alloys, Inc.
C 0.25-0.3, Cr 5, W 2.75, Mo 2.25, Mn, V, bal Fe.
For welding hot work die steels.

EUREKA 72-A H. W. BARE ROD
Eureka Welding Alloys, Inc.
C 0.36, Cr 5, W, Mo, Si, bal Fe.
For welding hot work tool steel punches, shear blades, etc.

EUREKA 73 H. W. ELECTRODE
Eureka Welding Alloys, Inc.
C 0.4-0.5, Cr 4.25, W 9.5, Mo 1.4, bal Fe.
For welding extreme hot work die steels.

EUREKA 73-A H. W. BARE ROD
Eureka Welding Alloys, Inc.
C 0.5, Cr 7.5, V 1.48, Mo, Si, bal Fe.
For welding hot work shear blades and punches; heat resistant.

EUREKA 75-X H. W. ELECTRODE
Eureka Welding Alloys, Inc.
C 0.75-1, Cr 0.8, V 0.4, Mo, Mn, bal Fe.
For welding water hardening tools and dies.

EUREKA 75-XA W. H. BARE ROD
Eureka Welding Alloys, Inc.
C 1.15, Si, Mn, V, bal Fe.
For welding water hard tool steel by GTAW.

EUREKA 78-A H. W. BARE ROD
Eureka Welding Alloys, Inc.
C 0.5, Cr 7.5, V 1.48, Mo, Si, bal Fe.
For welding high C, high Cr and similar alloys; hard. *Obsolete*

EUREKA 8510 W. H. ELECTRODE
Eureka Welding Alloys, Inc.
C 0.5-0.75, Cr 1.15, Mo 0.45, V, Mn, bal Fe.
For welding water hardening tools and dies. *Obsolete*

EUREKA 88 H. W. ELECTRODE
Eureka Welding Alloys, Inc.
C 0.4-0.5, Cr 5, W 5, Co 3.5, Mn, Mo, V, bal Fe.
For welding hot work die steels.

EUREKA 88-A H. W. BARE ROD
Eureka Welding Alloys, Inc.
C 0.39, Cr 4.25, W 4.1, V, Co, bal Fe.
For welding hot work dies; heat and impact resistant.

EUREKA 99 NICKEL ELECTRODE
Eureka Welding Alloys, Inc.
Ni 97, bal Fe.

EUREKA EXP-10 SURFACING ELECTRODE
Eureka Welding Alloys, Inc.
For overlaying sheet metal, drawing and forming dies.

EUREKA EXP-10, EXP-20
Welding Equipment & Supply Co.
Low C, Cr, Ni, Mo, Si, Mn, bal Fe.
For welding cast iron and alloy drawing and forming dies. Surfacing electrode. *Obsolete*

EUREKA EXP-20 SURFACING ELECTRODE
Eureka Welding Alloys, Inc.
For overlaying sheet metal, drawing and forming dies. Harder than EXP-10. *Obsolete*

EUREKA HAMMERALLOY COATED ELECTRODE
Eureka Welding Alloys, Inc.
C, Cr, Mo, Ni alloy.
For repair of forging dies, rams, sow blocks and die holders; hard.

EUREKA HAMMERALLOY ELECTRODE
Eureka Welding Alloys, Inc.
Low C, Cr, Ni, Mo, Si, Mn, bal Fe.
For welding and repairing forging dies.

EUREKA MARWELD 250A BARE ROD
Eureka Welding Alloys, Inc.
Maraging steel for repairing dies of maraging steel or overlaying hot work dies.

EUREKA MARWELD BARE ROD
Eureka Welding Alloys, Inc.
Now EUREKA MARWELD 250A BARE ROD.

EUREKA MARWELD NO. 250-A BARE ROD-TIG
Eureka Welding Alloys, Inc.
Ni, Co, Mo alloy.
For repair of die casting dies, plastic molds, extrusion dies, mandrels, dummy blocks and cold heading dies.

EUREKA NO. 1 HARD FACING BARE RODS
Eureka Welding Alloys, Inc.
C 2.5, Cr, W, high Co.
For surfacing parts subject to severe wear, heat and abrasion.

EUREKA NO. 12 HARD FACING BARE ROD
Eureka Welding Alloys, Inc.
C 1.4, Ni, W, Mo, high Co.
For surfacing parts subject to severe wear, heat and abrasion.

EUREKA NO. 140 COATED ELECTRODE
Eureka Welding Alloys, Inc.
C, Cr, Mo alloy.
For repair or joining of 4100 or 4300 steels such as forgings, castings, plastics molds and composite dies.

EUREKA NO. 2 ELECTRODE
Eureka Welding Alloys, Inc.
C, Cr, Mo alloy.
For repair of forging components such as rams, sow blocks, die shoes, die shanks, etc.

EUREKA NO. 20-A BARE ROD-TIG
Eureka Welding Alloys, Inc.
Cr, Mo, Ni alloy.
For joining underlayment or build up of AISI 8600 series steels.

EUREKA NO. 240-A BARE ROD-TIG
Eureka Welding Alloys, Inc.
For repair or joining of furnace parts, bearing surfaces, billet hooks and forging tongs.

EUREKA NO. 26 COATED ELECTRODE
Eureka Welding Alloys, Inc.
A general purpose joining and fabrication electrode for low alloy steels.

EUREKA NO. 31 COATED ELECTRODE
Eureka Welding Alloys, Inc.
C, Cr, Mo, V alloy.
For repair of forging dies, die casting, extrusion and plastics molds constructed of H-13 tool steels.

EUREKA NO. 350 COATED ELECTRODE
Eureka Welding Alloys, Inc.
Cr, Ni alloy.
Heat treatable alloy for repair or joining of die casting dies, arbors, shafts, and fractured tools and dies in general.

EUREKA NO. 450 COATED ELECTRODE
Eureka Welding Alloys, Inc.
C, Cr, Mo, Ni alloy.
For repair or reclamation of forging hammer dies possessing medium depth impressions.

EUREKA NO. 500 COATED ELECTRODE
Eureka Welding Alloys, Inc.
Cr, Ni alloy.
For repair or joining of all types of tool steels and general maintenance.

EUREKA NO. 500-A BARE ROD-TIG
Eureka Welding Alloys, Inc.
Cr, Ni alloy.
For repair or joining of all types of tool steels and general maintenance.

EUREKA NO. 5545-A CUPRO-NICKEL BARE ROD
Eureka Welding Alloys, Inc.
Ni 45, Cu 55.
For repairing cast iron patterns.

EUREKA NO. 6 HARD FACING BARE ROD
Eureka Welding Alloys, Inc.
C 1.45, Cr 28.5, W 3.9, Ni, Si, high Co.
For surfacing punches, dies, shear blades subject to severe heat and impact.

EUREKA NO. 635 COATED ELECTRODE
Eureka Welding Alloys, Inc.
C, Cr, Mo, Ni alloy.
For repair or reclamation of forging dies, rams, sow blocks, steel gears, arbors and couplings.

EUREKA NO. 650 COATED ELECTRODE
Eureka Welding Alloys, Inc.
C, Cr, Mo, V, Ni alloy.
For repair of shallow and medium depth press forging dies.

EUREKA NO. 74 COATED ELECTRODE
Eureka Welding Alloys, Inc.
C, Cr, Mo, V alloy.
For repair or fabrication of shock resisting tool steels. Excellent for trimming, shearing, slitting, and punching operations.

EUREKA NO. 74-A BARE ROD-TIG
Eureka Welding Alloys, Inc.
C, Cr, Mo, V alloy.
For repair or fabrication of shock resisting tool steels. Excellent for trimming, shearing, slitting and punching operations.

EUREKA PRESSALLOY COATED ELECTRODE
Eureka Welding Alloys, Inc.
C, Cr, Mo, Ta, Ni alloy.
For repair of medium to shallow impressions forging dies.

EUREKA PRESSALLOY LOW ALLOY ELECTRODE
Eureka Welding Alloys, Inc.
Low C, Mo, Si, Mn, bal Fe.
For repairing press forging type dies.

EUREKA T.G.A. LOW AIR ELECTRODE
Eureka Welding Alloys, Inc.
C 0.7-0.75, Cr 0.75, Ni 1.7, Mo 1.65, Mn 1.15, bal Fe.
For welding low air tool steels. *Obsolete*

EUREKALLOY "C" ELECTRODE
Eureka Welding Alloys, Inc.
C 0.05-0.1, Ni 55, Cr 16.25, Mo 17.5, W, Mn, V, bal Fe.
For surfacing parts subject to heat, impact, oxidation and certain chemicals.

EUREKALLOY "X" ELECTRODE
Eureka Welding Alloys, Inc.
C 0.3-0.35, Ni, Mo, high Co.
For surfacing units subject to pressure, impact, heat and galling, hot extrusion and forge dies.

EUREKAMATIC "BUILD-UP" SOLID WIRE
Eureka Welding Alloys, Inc.
Low C, bal Fe.
Alloy for preliminary build up on die sections before facing with harder material.

EUREKAMATIC 1216 SOLID WIRE
Eureka Welding Alloys, Inc.
C, Cr, high Mo, bal Fe.
Type M1 and M2.

EUREKAMATIC 130 CORED WIRE
Eureka Welding Alloys, Inc.
Similar to AISI 4130.
For welding and build up of AISI 4130 and 4140 steel. *Obsolete*

EUREKAMATIC 130 SOLID WIRE
Eureka Welding Alloys, Inc.
C 0.32, Cr, Mo, Mn, Si, bal Fe.
For overlay with mild alloy; meets SAE 4130.

EUREKAMATIC 145 SOLID WIRE
Eureka Welding Alloys, Inc.
C 0.45-0.5, Cr, V, bal Fe.
Medium carbon; meets SAE 6145.

EUREKAMATIC 2 CORED WIRE
Eureka Welding Alloys, Inc.
Low C, bal Fe.
Medium alloy for repairing rams and sow blocks.

EUREKAMATIC 3 CORED WIRE
Eureka Welding Alloys, Inc.
Now EUREKAMATIC 635 CORED WIRE.

EUREKAMATIC 350 ALLOY SOLID WIRE
Eureka Welding Alloys, Inc.
C 0.06, Mn, Mo, high Cr, bal Fe.
For joining tool steel, repairing crane wheels and forge dies.

EUREKAMATIC 350 CORED WIRE
Eureka Welding Alloys, Inc.
Very low C, bal Fe.
Alloy for joining tool steels and high alloys that must be heat treated; 38-40 Rock C.

EUREKAMATIC 45 CORED WIRE
Eureka Welding Alloys, Inc.
Low C, Cr, Ni, Mn, Mo, W, bal Fe.
For filling forge die impressions.

EUREKAMATIC 45 SOLID WIRE
Eureka Welding Alloys, Inc.
C 0.08, Medium Cr, Mo, bal Fe.
Semi-hot work grade, for repairing upsetter and forge dies.

EUREKAMATIC 450
Eureka Welding Alloys, Inc.
Now EUREKAMATIC 450 ALLOY CORED WIRE.

EUREKAMATIC 450 ALLOY CORED WIRE
Eureka Welding Alloys, Inc.
Low carbon, bal Fe.
Alloy for repairing forge die blocks; 44-46 Rock C.

EUREKAMATIC 5 CORED WIRE
Eureka Welding Alloys, Inc.
Low C, bal Fe.
Medium alloy for forge die impression and shanks. *Obsolete*

EUREKAMATIC 635 CORED WIRE
Eureka Welding Alloys, Inc.
Analysis similar to typical forge block. For repair and flooding forge die blocks; 38-40 Rock C.

EUREKAMATIC 650
Eureka Welding Alloys, Inc.
Low carbon, bal Fe.
Chromium alloy for repairing forge die blocks; 50 Rock C approx.

EUREKAMATIC 72 CORED WIRE
Eureka Welding Alloys, Inc.
Similar to AISI H12.
For repair welding dies, shear blades; 51-54 Rock C.

EUREKAMATIC 72 SOLID WIRE
Eureka Welding Alloys, Inc.
Low C, Cr, Mo, W, bal Fe.
For repairing shear blades, forge dies, and building composite units, hot work.

EUREKAMATIC 726 ALLOY CORED WIRE-MIG
Eureka Welding Alloys, Inc.
C, Cr, Mo, W, V alloy.
For repair of shallow impression press forging dies.

EUREKAMATIC 73 SOLID WIRE
Eureka Welding Alloys, Inc.
Low C, medium Cr, high W, bal Fe.
For repairing shear blades and other hot work steel subject to severe heat. *Obsolete*

EUREKAMATIC 78 SOLID WIRE
Eureka Welding Alloys, Inc.
Medium C, high Cr, W free, bal Fe.
Hot work grade for repairing hot or cold die units. *Obsolete*

EUREKAMATIC EXP-10 ALLOY CORED WIRE-MIG
Eureka Welding Alloys, Inc.
Cr, Mo, Ni alloy.
For repair of all grades of nodular and gray cast irons.

EUREKAMATIC FORGEWELD CORED WIRE
Eureka Welding Alloys, Inc.
Now EUREKAMATIC PRESSALLOY CORED WIRE.

EUREKAMATIC FORGEWELD CORED WIRE
Eureka Welding Alloys, Inc.
Low alloy wire for repairing forge shop components; can be applied without covering gas. *Obsolete*

EUREKAMATIC HAMMERALLOY ALLOY CORED WIRE
Eureka Welding Alloys, Inc.
C, Cr, Mo, Ni alloy.
For repair of forging dies, rams, sow blocks, and die holders; hard.

EUREKAMATIC MARWELD NO. 250 SOLID WIRE
Eureka Welding Alloys, Inc.
Ni, Co, Mo alloy.
For repair of die casting dies, plastic molds, extrusion dies, mandrels, dummy blocks and cold heading dies.

EUREKAMATIC NO. 1215 SOLID WIRE-MIG
Eureka Welding Alloys, Inc.
C, Cr, Mo, V alloy.
For repair of air hardening grades of tool steels.

EUREKAMATIC NO. 1216 ALLOY CORED WIRE
Eureka Welding Alloys, Inc.
C, Mo, W, Cr alloy.
For repair of AISI types M1/M2 and D2.

EUREKAMATIC NO. 2 HSS - SOLID WIRE-MIG
Eureka Welding Alloys, Inc.
C, Mo, Co, Cr, W, V alloy.
For repair of broaches, drills, milling cutters, end mills and cutting tools.

EUREKAMATIC NO. 650 ALLOY CORED WIRE-MIG
Eureka Welding Alloys, Inc.
C, Cr, Mo, V, Ni alloy.
For repair of shallow and medium depth press forging dies.

EUREKAMATIC NO. 70 SOLID WIRE-MIG
Eureka Welding Alloys, Inc.
C, Mg, Cr, W alloy.
For repair of oil hardening type tool steels.

EUREKAMATIC NO. 74 ALLOY CORED WIRE-MIG
Eureka Welding Alloys, Inc.
C, Cr, Mo, V alloy.
For repair or fabrication of shock resisting tool steels. Excellent for trimming, shearing, slitting, and punching operations.

EUREKAMATIC NO. 74 SOLID WIRE-MIG
Eureka Welding Alloys, Inc.
C, Cr, Mo, V alloy.
For repair or fabrication of shock resisting tool steels. Excellent for trimming, shearing, slitting and punching operations.

EUREKAMATIC NO. 75X SOLID WIRE-MIG
Eureka Welding Alloys, Inc.
C, V alloy.
For repair of water hardening tool steel types.

EUREKAMATIC PRESSALLOY CORED WIRE
Eureka Welding Alloys, Inc.
High alloy wire for repairing forge dies.

EUREKAMOLD BARE ROD
Welding Equipment & Supply Co.
Low C, Cr, Ni, Si, bal Fe.
For mold repair-P6 type. *Obsolete*

EUREKAMOLD P-20 BARE ROD

Eureka Welding Alloys, Inc.
Same as AISI P20.
For welding molds in general, those that must be grained or etched.

EUREKAMOLD P-6 BARE ROD

Eureka Welding Alloys, Inc.
Similar to AISI P6.
For repairing plastic molds; high luster. *Obsolete*

EUROPEAN "REAMUR"

English manufacture
C 2.8-3.5, Si 0.6-0.8, graphite, bal Fe.
For castings; malleable cast iron.

EUROPIUM

Atomergic Chemetals Corp.
Eu.
Purities: 99.9% special distilled grade, 99.5+%. Forms: ingot, lump, wire, sheet, foil.

EUT-O-MAT 3010A

Eutectic Corp.
Now OA 3010. *Obsolete*

EUT-O-MAT 3205

Eutectic Corp.
Now OA 3205. *Obsolete*

EUT-O-MAT 3220A

Eutectic Corp.
Continuous electrode for joining and coating manganese steel; base for harder coating. AC-DC; 20 Rock C; work hardens to 50 Rock C. *Obsolete*

EUT-O-MAT 4601A

Eutectic Corp.
Now OA 4601. *Obsolete*

EUT-O-MAT 4625A

Eutectic Corp.
Now OA 4525. *Obsolete*

EUT-O-MAT AN690

Eutectic Corp.
Now OA 690. *Obsolete*

EUTALLITE UNIVERSAL 10092

Eutectic Corp.
45-50 Rock C. Powder for spray welding of steel. Resists scaling and softening at elevated temperatures.

EUTALLITE-6

Eutectic Corp.
Cr 29, C 1.1, W 4, bal Co.
As cast: Rock. C45. For hard facing overlay on engine valves, tractor treads, fuel pump shafts, forging dies. Alloy powders deposited by oxy-acetylene. Wear resistant. *Obsolete*

EUTEC ROD 14

Eutectic Corp.
C 3, Si 2, bal Fe.
For gas welding rod; for cast iron. *Obsolete*

EUTEC ROD 1700

Eutectic Corp.
Ag, bal Cu.
For silver brazing. *Obsolete*

EUTEC ROD 1850FC

Eutectic Corp.
Cu alloy.
Cast: 90,000 TS; 180 Brin. For torch welding rod for copper; discontinued. *Obsolete*

EUTEC ROD 189

Eutectic Corp.
Cu alloy.
Welded: 51,000 TS. For welding rod for steel, bronze and copper. *Obsolete*

EUTEC TRODE 2100

Eutectic Corp.
Al alloy.
For welding rod (gas or electric); for Al sheet and castings. *Obsolete*

EUTEC TRODE 24

Eutectic Corp.
Ni alloy.
For arc welding rod; machinable welds on cast iron. *Obsolete*

EUTEC TRODE 24/49

Eutectic Corp.
C, Si, bal Fe.
For welding rod for cast iron; machinable. *Obsolete*

EUTEC-SILWELD 1618

Eutectic Corp.
For ferrous and non-ferrous joining. 1125°F MP; 85,000 psi TS.

EUTEC-TINWELD I

Eutectic Corp.
For soldering iron, nickel, copper base alloys; M.P. 450 F; 7,000 TS. *Obsolete*

EUTEC-TINWELD II

Eutectic Corp.
For lead free soldering; M.P. 450 F; 15,000 TS. *Obsolete*

EUTEC-TINWELD III

Eutectic Corp.
For soldering electronic and radio components; M.P. 360 F; 8,000 TS. *Obsolete*

EUTEC-TINWELD IV

Eutectic Corp.
For soldering copper base alloys M.P. 452 F; 10,000 TS. *Obsolete*

EUTECBOR 9

Eutectic Corp.
For hard facing overlays using torch. Operates at 1800-1900 F; wear and corrosion resistant. *Obsolete*

EUTECROD 14 FC

Eutectic Corp.
Bonding temperature 1400-1600°F. Hard: 200 Brin. Electrode for torch brazing or building up cast iron, as machine bases, motor and gear housings. *Obsolete*

EUTECROD 141

Eutectic Corp.
Rod for torch brazing cast iron; easily machinable. Bonding temperature 1400-1600°F; 40,000 psi TS. *Obsolete*

EUTECROD 146

Eutectic Corp.
Electrode for torch brazing of cast iron, steels and copper base alloys at low heat; Bonding temperature 1400-1600°F; 65,000 psi TS. *Obsolete*

EUTECROD 146 FC

Eutectic Corp.
Flux coated modification of EUTECROD 146. *Obsolete*

EUTECROD 15

Eutectic Corp.
Bonding temperature 450-600°F; 1500 psi YS. Electrode for low temperature torch build up and sealing of cast iron, malleable, steel, copper and nickel alloys. For crankcases, water jackets. *Obsolete*

EUTECROD 155

Eutectic Corp.
Rod for joining most ferrous and non-ferrous metals at low temperature. Bonding temperature 725 F; 16,000 TS. *Obsolete*

EUTECROD 157

Eutectic Corp.
Sn 95.5, Ag 3.46.
Solder type alloy free from lead, zinc, antimony and cadmium. 425°F MP; 15,000 psi TS.

EUTECROD 157-B

Eutectic Corp.
Flux cored solder type alloy free from lead, zinc, antimony and cadmium; 425°F MP; 15,000 psi TS. *Obsolete*

EUTECROD 158-B

Eutectic Corp.
Flux-cored solder; Melts 001 F, 7,000 TS. *Obsolete*

EUTECROD 16

Eutectic Corp.
Copper base with high nickel content.
Rod for torch brazing ferrous and non-ferrous metals. Bonding temperature 1400-1600°F; 100,000 TS. *Obsolete*

EUTECROD 16 FC

Eutectic Corp.
Flux coated modification of EUTECROD 16. *Obsolete*

EUTECROD 1600 SUPER

Eutectic Corp.
Cadmium free silver solder type for high strength braze joints on steels, copper, brass, and dissimilar metals. Bonding temperature 1375°F; 60,000 psi TS. *Obsolete*

EUTECROD 1601

Eutectic Corp.
Rod for torch, furnace or induction brazing; ferrous and non-ferrous. Brazing temperature 1225°F; 60,000 psi TS.

EUTECROD 1702

Eutectic Corp.
Electrode for torch brazing steel, stainless steel, and carbide tips. 1200°F MP; 85,000 psi TS. *Obsolete*

EUTECROD 18

Eutectic Corp.
Zn, bal Cu.
For gas welding rod; for bronze and brass. *Obsolete*

EUTECROD 18 FC

Eutectic Corp.
Flux coated grade of EUTECROD 18. *Obsolete*

EUTECROD 180

Eutectic Corp.
Electrode for torch brazing copper base alloys; 1290°F MP; 42,000 psi TS.

EUTECROD 1800

Eutectic Corp.
Ag, bal Cu.
For gas welding rod; for steel and copper alloys.

EUTECROD 1801

Eutectic Corp.
Ag 51, Cu 22, Zn 21, Cd 5, Sn 1.
For brazing ferrous and non-ferrous. 1120°F MP; 90,000 psi TS.

EUTECROD 1803

Eutectic Corp.
For torch brazing copper and copper alloys; Melts 1185 F; 55,000 TS. *Obsolete*

EUTECROD 1804
Eutectic Corp.
For torch brazing copper and copper alloys; 1185°F MP; 50,000 psi TS.

EUTECROD 1810
Eutectic Corp.
Silver solder type for joining copper, brass, steels and dissimilar metals. Bonding temperature 1195°F; 78,000 psi TS.

EUTECROD 185
Eutectic Corp.
Bronze electrode for torch build-up; machinable and wear resistant. Bonding temperature 1400-1600°F; 130-200 Brin. *Obsolete*

EUTECROD 185 FC
Eutectic Corp.
Machinable overlays on cast iron, steel, copper and nickel alloys. Flux coated; for torch. Bonding temperature 1400-1600°F; 200 Brin.

EUTECROD 186 FC
Eutectic Corp.
For machinable overlays on ferrous and non-ferrous parts; bonding temperature 1400-1600 F, using torch; hardness 200-300 Brin. *Obsolete*

EUTECROD 19
Eutectic Corp.
Zn base with Al.
For surface build up on aluminum. *Obsolete*

EUTECROD 190
Eutectic Corp.
For torch welding aluminum. 1070°F MP; 34,000 psi TS.

EUTECROD 1900
Eutectic Corp.
For torch welding magnesium. 1100°F MP; 30,000 psi TS. *Obsolete*

EUTECROD 1909
Eutectic Corp.
For soldering aluminum and magnesium; M.P. 315 F; 10,000 TS. *Obsolete*

EUTECROD 196
Eutectic Corp.
For torch build-up and joining zinc die castings. 700°F MP; 28,000 TS. *Obsolete*

EUTECROD 21
Eutectic Corp.
For torch welding fillet and bead joints on sheet, tubular, extruded and cast aluminum. Bonding temperature 1090°F; 33,000 psi TS. *Obsolete*

EUTECROD 21-FC-E
Eutectic Corp.
For torch joining of aluminum; 1090°F MP; 33,000 psi TS.

EUTECROD 80
Eutectic Corp.
Copper base with nickel.
Rod for torch brazing steel as bicylces, ladders, steel furniture. Bonding temperature 1400-1600 F; 80,000 TS. *Obsolete*

EUTECROD 80 FC
Eutectic Corp.
Flux coated modification of Eutecrod 80. *Obsolete*

EUTECROD 800
Eutectic Corp.
For torch brazing copper base alloys; Melts 1305 F; 40,000 TS; *Obsolete*

EUTECROD 91
Eutectic Corp.
Cobalt-Chromium-Tungsten.
Hard facing alloy for torch or TIG for build up on check valves, exhaust valves, hot punches. *Obsolete*

EUTECSIL 1020 FC
Eutectic Corp.
Now EUTECSIL 1020XFC. *Obsolete*

EUTECSIL 1020XFC
Eutectic Corp.
Flux coated alloy for torch brazing of ferrous and non-ferrous alloys. 1050°F MP; 85,000 psi TS.

EUTECSIL 1030 FC
Eutectic Corp.
Flux coated rod, silver braze type, for joining ferrous and non-ferrous metals. Bonding temperature 1150°F; 60,000 psi TS.

EUTECSIL 1801 FC
Eutectic Corp.
Silver solder type alloy for joining ferrous and non-ferrous metals and dissimilar metals. Flux coated. Bonding temperature 1120°F; 90,000 psi TS.

EUTECSILVERWELD ECON 1
Eutectic Corp.
General purpose silver alloy and flux in unitized container. Bonding temperature: 1295°F; 58,000 psi TS. *Obsolete*

EUTECSILVERWELD ECON 2
Eutectic Corp.
Cadmium-free silver alloy and flux in unitized container. Bonding temperature 1130°F; 85,000 psi TS.

EUTECTAL
Alais Forges et Camargue
Cu 1.5, Mn 0.8, Mg 1.58, Ti 0.35, Si 0.25, bal Al.
Heat treated: 38,000 TS; 4 El; 98 Brin. For light alloy parts; age hardenable.

EUTECTIC ALLOY-1
American manufacture
Bi 50, Sn 13, Cd 10, Pb 27.
For fusible alloys, fuses.

EUTECTIC ALLOY-2
American manufacture
Bi 52, Pb 40, Cd 8.
For fusible alloys, fuses.

EUTECTIC ALLOY-3
American manufacture
Bi 53, Pb 32, Sn 15.
For fusible alloys, fuses.

EUTECTIC ALLOY-4
American manufacture
Bi 54, Sn 26, Cd 20.
For fusible alloys, fuses.

EUTECTIC ALLOY-5
Manufacturer not listed
Sn 50, Cd 18, Pb 32.

EUTECTRODE 1851 (DC)
Eutectic Corp.
For arc welding copper base alloys.

EUTECTRODE 2101
Eutectic Corp.
For DC reverse welding of aluminum alloys. 34,000 psi TS.

EUTECTRODE 232
Eutectic Corp.
For AC-DC metallic arc machinable welds on alloy cast irons including ductile and high phosphorus types. 55,000 psi TS. *Obsolete*

EUTECTRODE 240
Eutectic Corp.
Now EUTECTRODE 244. *Obsolete*

EUTECTRODE 244
Eutectic Corp.
For AC-DC metallic arc build up and fill gray cast iron with machinable deposits. 53,000 psi TS.

EUTECTRODE 27
Eutectic Corp.
For repair welding cast iron; AC-DC. For machine bases, frames, and supports. 60,000 psi TS.

EUTECTRODE 280
Eutectic Corp.
Electrode for reverse DC welding copper alloys to ferrous metals; 60,000 psi TS. *Obsolete*

EUTECTRODE 300
Eutectic Corp.
For DC reverse welding copper; 35,000 TS. *Obsolete*

EUTECTRODE 4
Eutectic Corp.
Electrode for AC-DC metallic arc tough coating on manganese and alloy steels; good shock resistance; 90 Rock B; work hardens to 45 Rock C.

EUTECTRODE 40
Eutectic Corp.
Electrode for AC-DC metallic arc build-up and join manganese steels; deposits can be flame cut; 80-90 Rock B; work hardens to 45-50 Rock C.

EUTECTRODE 501
Eutectic Corp.
Electrode for AC-DC metallic arc welding of mild steel sheet, plate, angle iron and pipe. 68,000 psi TS. *Obsolete*

EUTECTRODE 518
Eutectic Corp.
Electrode for AC-DC metallic arc joining of low alloy and medium carbon steels in all positions. 80,000 psi TS. *Obsolete*

EUTECTRODE 526
Eutectic Corp.
Electrode for AC-DC metallic arc welding of most steels including dissimilar combinations. 120,000 psi TS. *Obsolete*

EUTECTRODE 53
Eutectic Corp.
For AC-DC metallic arc welding 316 stainless steel; moly bearing; 85,000 psi TS. *Obsolete*

EUTECTRODE 53-L
Eutectic Corp.
For AC-DC metallic arc welding of 316 L; moly bearing with low carbon; 85,000 psi TS.

EUTECTRODE 54
Eutectic Corp.
For AC-DC metallic arc joining of 301, 302, 304, 305 and 308 stainless steels; 80,000 psi TS. *Obsolete*

EUTECTRODE 54-L
Eutectic Corp.
308L-16 stainless; low carbon content; for joining types 304, 304L, 308, 347; 75,000 psi TS.

EUTECTRODE 554-L
Eutectic Corp.
All purpose extra low carbon stainless for AC-DC welding of dissimilar combinations. 75,000 psi TS. *Obsolete*

EUTECTRODE 57
Eutectic Corp.
Now CEC 57L. *Obsolete*

EUTECTRODE 65

Eutectic Corp.
Cr, Ni, Mo steel, coated.
Electrode for DC-Rev. welding and overlaying to give harder and stronger weld metal. 140,000 TS; 320 Brin. *Obsolete*

EUTECTRODE 66

Eutectic Corp.
Electrode for AC-DC welding of mild steel, beams, channel iron, and pipes. 80,000 TS; machinable.

EUTECTRODE 670

Eutectic Corp.
Cr, Ni steel.
Electrode for AC-DC welding of low alloy steel and some stainless steel. 95,000 psi TS.

EUTECTRODE 680

Eutectic Corp.
Electrode for AC-DC welding of alloy steels, leaf and coil springs, some stainless grades. 120,000 psi TS.

EUTECTRODE 6800

Eutectic Corp.
Electrode for AC-DC metallic arc welding and build-up of ferrous and nickel alloys. Heat and corrosion resistant; 85,000 psi TS.

EUTECTRODE 691

Eutectic Corp.
Electrode for AC-DC welding of miscellaneous steels; 105,000 TS. *Obsolete*

EUTECTRODE 71

Eutectic Corp.
Electrode for AC-DC metallic arc welding thin sections of 4130, 4140 and 8620 steels; similar response to heat treatment; 100,000 psi TS. *Obsolete*

EUTECTRODE N 2

Eutectic Corp.
Now XHD N102. *Obsolete*

EUTECTRODE N 4

Eutectic Corp.
Hard overlays on manganese steels; AC-DC; for railroad, construction equipment, rails and frogs. Hardness as deposited 98 Rb, work hardens to C 45 Rock. *Obsolete*

EUTECTRODE N 40

Eutectic Corp.
For hard facing or joining austenitic manganese steels; AC-DC; for bucket teeth and lips, crushers, hammer mills and similar parts subject to severe impact; work hardenable. *Obsolete*

EUTECTRODE N 61

Eutectic Corp.
For heavy build-ups of hard and abrasion resistant overlays on steel and cast iron; AC-DC; Hardness C 50-55 Rock. *Obsolete*

EUTECTRODE N 6800

Eutectic Corp.
Electrode for AC-DC overlays on steel as forging dies, rams, piercing tools, tongs. 85,000 TS; 210 Brin. *Obsolete*

EUTECTRODE N 90

Eutectic Corp.
Hard overlay for ferrous metals; AC-DC; deposits retain high strength and hardness at elevated temperatures. For ash remover impellers, exhaust valves, and hot punches.

EUTECTRODE N5005

Eutectic Corp.
Electrode for AC-DC metallic arc coating on carbon, alloy and manganese steel for combined hardness and toughness; 57-60 Rock C.

EUTECTRODE N9080

Eutectic Corp.
Electrode for AC-DC metallic arc build-up to give high hardness, corrosion resistance and resistance to high temperature and impact. 30 Rock C; work hardens to 45-50 Rock C.

EUTECTRODE SUPER 110

Eutectic Corp.
Electrode for AC-DC welding steel, angle iron, I-beams, and channel iron; 110,000 psi TS.

EUTHERM A11

Otto Wolff Handelsgesellschaft
C 0.15, Cr 19.5, Ni 9.5, bal Fe.
Annealed: 80,000 TS; 35,000 YS; 55 El; 75 RA; 150 Brin. For chemical plant equipment, tanks, mixers, filters; Type 302; stainless austenitic.

EUTHERM A22

Otto Wolff Handelsgesellschaft
C 0.15, Cr 24, Ni 19, bal Fe.
Annealed: 100,000 TS; 45,000 YS; 50 El; 65 RA; 185 Brin. For furnace parts, pumps, valves, turbine parts; Type 310; stainless, austenitic.

EUTHERM F10

Otto Wolff Handelsgesellschaft
C 0-0.12, Si 2, Cr 18, bal Fe.
Annealed: 80,000 TS; 50,000 YS; 25 El; 50 RA; 150 Brin. For shafts, gears, chemical plant equipment, sinks; Type 430; stainless.

EUTHERM F30

Otto Wolff Handelsgesellschaft
C 0-0.12, Al 1.5, Cr 24, bal Fe.
Annealed: 85,000 TS; 50,000 YS; 30 El; 55 RA; 180 Brin. For oil refinery and chemical plant equipment; heat resistant.

EUTHERM F8

Otto Wolff Handelsgesellschaft
C 0-0.12, Si 2.3, Cr 6, bal Fe.
Annealed: 70,000 TS; 30,000 YS; 28 El; 65 RA; 160 Brin. For oil refinery equipment; heat and creep resistant.

EUTHERM FA25

Otto Wolff Handelsgesellschaft
C 0.2, Cr 25, Ni 4, bal Fe.
Cast: 80,000 TS; 55,000 YS; 200 Brin. For furnace parts, heat treating boxes, valves, pumps. Type CC; corrosion and heat resistant.

EUZONIT

Mannesmannrohren-Werke AG
Ni 55, Mo 20, Fe 20.
For corrosion and heat resistant equipment; corrosion and heat resistant. *Obsolete*

EUZONIT

Now MARKER EUZONIT.

EUZONIT 60

Mannesmannrohren-Werke AG
C 0.1, Cr 17, Ni 60, Mo 20, bal Fe.
For corrosion and heat resistant parts; corrosion and heat resistant. Resists sulfuric acid. *Obsolete*

EUZONIT 70

Mannesmannrohren-Werke AG
Ni 67, Mo 30, bal Fe.
For high temperature applications; corrosion and heat resistant. Resists sulfuric acid. *Obsolete*

EUZONIT 85

Mannesmannrohren-Werke AG
Ni 85, Si 9, Cu 4.
For high temperature applications; corrosion and heat resistant. *Obsolete*

EV. 12

Follsain-Wycliffe Foundries, Ltd.
C 0.2-0.5, Si 0-1.75, Cr 23-26, Ni 11-14, W 1.5-2, bal Fe.
Stainless alloy casting; good hot strength, resistance to oxidizing and scaling. BS 1648/1957, Gr. E; ASTM A297, Gr. HH.

EV. 20

Follsain-Wycliffe Foundries, Ltd.
C 0.2-0.5, Si 2.25-3, Cr 24-26, Ni 17-20, bal Fe.
Austenitic stainless casting; good creep resistance at elevated temperatures. BS 1648/1957, Gr. F; ASTM A297, Gr. HK.

EV. 25

Follsain-Wycliffe Foundries, Ltd.
C 0-0.5, Si 0-3, Cr 17-23, Ni 23-28, bal Fe.
Alloy casting; for elevated temperature work. BS. 1648/1957, Gr. G; ASTM A297, Gr. HN.

EV. 300

Follsain-Wycliffe Foundries, Ltd.
C 0.5-1, Cr 27-30, Si 0-2, bal Fe.
Ferritic type alloy casting; good resistance to scaling and sulfurous atmospheres to 1150°C. BS. 1648/1957, Gr. B; ASTM A297, Gr. HC.

EV. 37

Follsain-Wycliffe Foundries, Ltd.
C 0-0.5, Si 0-2.5, Cr 17-21, Ni 37-41, bal Fe.
Austenitic stainless casting; for high temperature equipment as furnace parts. BS 1648/1957, Gr. H; ASTM A297, Gr. HU.

EV. 500

Follsain-Wycliffe Foundries, Ltd.
C 0.05-0.12, Si 0.5-1, Cr 26-30, Co 47-52, bal Fe.
Alloy casting; resistant to oxidization and sulfurous atmospheres to 1200°C. Good thermal shock and abrasion resisting properties. Equivalent to UMCO 50.

EV. 60

Follsain-Wycliffe Foundries, Ltd.
C 0.35-0.75, Si 0-2.5, Cr 15-20, Ni 58-62, bal Fe.
Austenitic stainless casting; good hot strength and scale resistance to 1100°C. BS. 1648/1957, Gr. K; ASTM A297, Gr. HW.

EV4CO

German manufacture
C 1.3, Co 5, Mo 1, V 4, W 12, Cr 4, bal Fe.
Hardened: 64-66 Rock C. For cast cutting tools, reamers, drills, broaches, engravers tools; high speed steel, Type T15, high hot-hardness and wear resistance.

EV5

Russian manufacture
C 0.7-0.8, Mn 0-0.4, Cr 0.3-0.5, W 5-6.5, bal Fe.
For permanent magnets; heat treated.

EVANOHM

Gilby-Fodor S.A.
Cr 20, Al 3, Cu 2, bal Ni.
Annealed: 90,000 TS; 65,000 YS; 40 El. For precision resistors. Heat resistant to 300°C.

EVANOHM

Wilbur B. Driver Co.
Nickel. Cr 20, Ni 75, Al 2.75, Cu 2.75.
Annealed: 150,000 TS; 117,000 YS; 20 El. For precision resistors; high electrical resistance, low resistance change with temperature.

EVANOHM ALLOY R

Carpenter Technology Corp.
Ni 75, Cr 20, Al 2.5, Cu 2.5.
Annealed: 689 MPa TS. Cold worked: 1379 MPa TS. Resistance wire with high corrosion resistance and tensile strength. Nonmagnetic. Used in fine sizes for precision wound resistors.

EVANOHM ALLOY S
Carpenter Technology Corp.
Ni 72, Cr 20, Al 3, Mn 4, Si 1.
145-185 ksi TS; 75-155 ksi YS; 10-30 El. Resistance alloy with high tensile strength and corrosion resistance. Nonmagnetic. Used in fine wire sizes for precision wound resistors from -55 to +220 °C.

EVANS PEERLESS
Evans Steel Co.
C 0.9-1.2, bal Fe.
For tools, dies, jigs; water hardened.

EVANSTEEL, GRADE I
Chicago Steel Foundry
Ni 1.5-2, Cr 0.75-1, C 0.3, bal Fe.
Annealed: 95,000-100,000 TS; 50,000-60,000 YS; 22-27 El; 30-40 RA; 200 Brin. For sprockets, gears, high pressure valves; resists abrasion and wear; resists temperature to 1000°F.

EVANSTEEL, GRADE II
Chicago Steel Foundry
C 0.3, Ni 1.5-2, Cr 0.75-1, bal Fe.
Annealed: 115,000-125,000 TS; 65,000-75,000 YS; 15-20 El; 25-30 RA; 275 Brin. For tractor shoes, bucket lips, dipper teeth, sheaves; abrasion resistant; heat resistant to 1000°F.

EVANSTEEL, GRADE III
Chicago Steel Foundry
C 0.3, Ni 1.5-2, Cr 0.75-1, bal Fe.
Annealed: 145,000-160,000 TS; 120,000 YS; 12 El; 10 RA; 400-500 Brin. For dipper teeth, liner plates, guard rails, pulverizer hammers; abrasion resistant; heat resistant to 1000°F.

EVERARD
Everard Tap & Die Co.
C 1-1.2, bal Fe.
For dies, tools; water hardening. *Obsolete*

EVERBRITE
American Manganese Bronze Co.
Copper. Cu 60, Ni 30, Fe 3, Si 3, Cr 3.
Cast: 75,000 TS; 45,000 YS; 14 El; 170 Brin. For valves, chemical plants; corrosion resistant.

EVERBRITE
Allis-Chalmers Mfg. Co.
Ni 29.5-31.5, Cu 59-65, Fe 5-8, Mn 0-0.6, C 0-0.25, Si 0-0.6.
Cast: 75,000 TS; 50,000 YS; 14 El; 20 RA; 200 Brin. For valve disks, valve seat rings, steam turbine nozzle blocks and valves; machinable; resists action of steam.

EVERBRITE NO. 82
Curtis Bay Copper & Iron Works
Ni 30.5, Cu 62, Fe 7, Mn 0.5.
For steam applications; high corrosion resistance. *Obsolete*

EVERBRITE NO. 90
Curtis Bay Copper & Iron Works
Ni 35, Cu 60.
For valves, chemical plants. *Obsolete*

EVERBRITE NO. 92
Curtis Bay Copper & Iron Works
Ni 33.5, Cu 56.75, Co 0.35, Mn 2.5, Fe 1, Cr 3.75.
Cast: 95,000 TS; 60,000 YS; 20 El; 137 Brin. For high pressure steam valves. *Obsolete*

EVERCLAD
British Steel plc
Galvanized plastic-coated roofing and cladding steel; mild steel base.

EVERDUR
Now IMI 705.

EVERDUR 1000
Now EVERDUR 6552.

EVERDUR 1010
Now EVERDUR 655.

EVERDUR 1012
Now EVERDUR 661.

EVERDUR 1014
Now EVERDUR 637.

EVERDUR 1015
Now EVERDUR 651.

EVERDUR 637
Anaconda Co.
Copper. Si 2, Al 7.25, bal Cu.
Annealed: 88,000 TS; 44,000 YS; 25 El; 85 Rock B. For screw machine products and forgings, studs, bolts, switchgear, electrical hardware, valve stems. Corrosion resistant, nonmagnetic.

EVERDUR 651
Anaconda Co.
Copper. Si 1.5, Mn 0.25, bal Cu.
Soft: 40,000 TS; 15,000 YS; 50 El; 55 Rock F. Hard: 70,000 TS; 55,000 YS; 15 El; 80 Rock B. For cold headed and roll threaded bolts, hardware marine fittings, pole line hardware, rigid conduit, welding rod. Corrosion resistant; good for severe cold working.

EVERDUR 655
Anaconda Co.
Copper. Cu 95.8, Si 3.1, Mn 1.1.
Annealed rod: 55,000 TS; 20,000 YS; 65 El; 35 Rock B. Hard: 90,000 TS; 55,000 YS; 18 El; 90 Rock B. For bearing plates, bolts, screws, forgings, hardware, springs, tanks, marine fittings, oil storage tanks. High corrosion resistance and strength.

EVERDUR 6552
Anaconda Co.
Copper. Si 4, Mn 1.1, bal Cu.
Cast: 50,000 TS; 20,000 YS; 25 El. For castings, pipe fittings, valves. 1000°C MP.

EVERDUR 656
Anaconda Co.
Copper. Cu 95.8, Si 3.1, Mn 1.1.
For welding rods to weld copper and copper alloys, plain and galvanized steel. Good for carbon-arc, acetylene welding and inert gas welding.

EVERDUR 661
Anaconda Co.
Copper. Si 3, Mn 1, Pb 0.4, bal Cu.
Soft: 58,000 TS; 22,000 YS; 70 El. Hard: 90,000 TS; 60,000 YS; 18 El. For screw machine parts. Free cutting, wear and corrosion resistant.

EVERDUR A
Now EVERDUR 655.

EVERDUR NO. 1026 (WEBERT ALLOY)
Anaconda Co.
Cu 81.5, Si 4.25, Zn 14.25, Trace Mn.
90,000 TS; 10 El. For pressure die castings. *Obsolete*

EVERIT 35
Thyssen Edelstahlwerke AG
C 0.15, Si 1, Mn 1.5, Cr 2.5, Mo 0.5, bal Fe.
Coated electrodes for overlaying of brake drums. Meets DIN 8555.

EVERIT 48
Thyssen Edelstahlwerke AG
C 3, Si 1.5, Mn 2, Cr 15.5, bal Fe.
Coated electrodes for overlaying of crusher jaws. Meets DIN 8555.

EVERIT 50
Thyssen Edelstahlwerke AG
C 2.5, Cr 25, Mo 3.2, V 0.55, bal Fe.
Coated electrodes for overlaying of valves. Meets DIN 8555.

EVERIT 55
Thyssen Edelstahlwerke AG
C 3.6-5, Si 1-2, Mn 0.5-3, Cr 27-31, bal Fe.
Coated electrodes for overlaying of coal plows. Meets DIN 8555.

EVERIT 60
Thyssen Edelstahlwerke AG
C 0.5, Si 1-2.8, Mn 0.4-1.5, Cr 6-9, Mo 1.5, W 1.5, bal Fe.
Coated electrodes for overlaying of mangle rolls. Werkstoff Nr. 1.4718.

EVERIT 60 TI
Thyssen Edelstahlwerke AG
C 2, Si 0.5, Mn 1.5, Cr 6, Mo 1, Ti 5.5, bal Fe.
Coated electrodes for overlaying of pick hammers. Meets DIN 8555.

EVERIT 65
Thyssen Edelstahlwerke AG
C 4.5-5.5, Si 1, Mo 7-9, Cr 22-23, Nb 7-8.5, W 2-2.5, V 1-1.4, bal Fe.
Coated electrodes for overlaying of spike breakers. Meets DIN 8555.

EVERIT 68
Thyssen Edelstahlwerke AG
C 4.8-5.2, Cr 32-38, B 2-2.4, Nb 0-5.2, bal Fe.
Coated electrodes for overlaying of coal plows. Meets DIN 8555.

EVERIT A 43
Thyssen Edelstahlwerke AG
C 5, Si 0.5, Mn 1, Cr 22, Nb 7, V 0.5, bal Fe.
Coated electrodes for overlaying of spike breakers. Meets DIN 8555.

EVERIT A 45
Thyssen Edelstahlwerke AG
C 4.5-5.5, Si 1, Mo 7-9, Cr 22-23, Nb 7-8.5, W 2-2.5, V 1-1.4, bal Fe.
Coated electrodes for overlaying of spike breakers. Meets DIN 8555.

EVERIT FECR 50
Thyssen Edelstahlwerke AG
C 2.2, Cr 27, bal Fe.
Welding consumables for gas welding on valve stem ends. Meets DIN 8555.

EVERSHYNE
Osborn Steels Ltd.
C 0.2, Cr 18, Ni 8, bal Fe.
For stainless parts; stainless. *Obsolete*

EW
Thyssen Edelstahlwerke AG
C, bal Fe.
Water treated: 72,000-85,000 TS; 42,000-56,000 YS; 22-29 El. *Obsolete*

EW 1
Electrometal SA Metals Especials
C 0.7, Si 0.25, Mn 0.27, bal Fe.
Carbon tool steel; water hardening. W.-Nr. 1.1620; AISI W1.

EW 108
Electrometal SA Metals Especials
C 0.85, Si 0.25, Mn 0.27, bal Fe.
Carbon tool steel; water hardening. W.-Nr. 1.1620; AISI W1. *Obsolete*

EW 110

Electrometal SA Metals Especials
C 1, Si 0.25, Mn 0.27, Cr 0-0.2, bal Fe.
Carbon tool steel; water hardening. W.-Nr. 1.2005; AISI W1.
Obsolete

EW 2

Electrometal SA Metals Especials
C 1, Si 0.25, Mn 0.27, V 0.2, bal Fe.
Carbon tool steel; water hardening. W.-Nr. 1.1640; AISI W2.

EW 711

Emil Weingartner & Co.
C 0-0.13, Mn 0.9, P 0-0.1, S 0.18-0.25, bal Fe.
Free machining steel for automatic screw machines.
Werkstoff Nr. 1.0711.

EW AERO

Emil Weingartner & Co.
C 1.05, Mn 1, Cr 1, W 1.2, bal Fe.
For punching and trimming dies. Werkstoff Nr. 1.2419.

EW APM

Emil Weingartner & Co.
C 0.9, Mn 2, Cr 0.35, V 0.1, bal Fe.
For shear blades, reamers. Werkstoff Nr. 1.2842.

EW METEORIT I

Emil Weingartner & Co.
Cr 1.05, W 2, V 0.2, C 0.6, Si 0.6, bal Fe.
Hot or cold work punches, shears. Werkstoff Nr. 1.2550.

EW METEORIT II

Emil Weingartner & Co.
C 0.45, Si 1.5, Cr 1.5, V 0.1, bal Fe.
Hot or cold punches, shears. Werkstoff Nr. 1.2249.

EW METEORIT III

Emil Weingartner & Co.
C 0.45, Si 1, Cr 1.1, W 2, V 0.2, bal Fe.
For punches, trimming dies, pneumatic tools. Werkstoff Nr. 1.2542.

EW ORION III

Emil Weingartner & Co.
C 1.45, Si 0.2, Mn 0.6, Cr 1.5, bal Fe.
Oil hardening cold work tool steel. Werkstoff Nr. 1.2063.

EW ORION SPEZIAL

Emil Weingartner & Co.
C 0.9, Si 1.15, Mn 0.7, Cr 1.2, bal Fe.
Cold work tool steel; for shear blades, milling cutters.
Werkstoff Nr. 1.2108.

EW SATURN

Emil Weingartner & Co.
C 0.3, Cr 3, Mo 2.8, V 0.5, bal Fe.
Hot work tool steel; for pressure casting molds. Werkstoff Nr. 1.2365.

EW SATURN W

Emil Weingartner & Co.
C 0.3, Cr 2.7, W 8.5, V 0.35, bal Fe.
Hot work tool steel; for extrusion dies, pressure casting molds, forging dies. Werkstoff Nr. 1.2581.

EW URUS

Emil Weingartner & Co.
C 0-0.12, P 0.1, S 0.27, Mn 0.7, Pb 0.2, bal Fe.
Leaded free machining steel. Werkstoff Nr. 1.0716; DIN 9SPb23.

EW URUS MN

Emil Weingartner & Co.
C 0-0.14, Mn 1.1, P 0-0.1, S 0.24-0.32, Pb 0.15-0.3, bal Fe.
Free machining steel for automatic screw machines.

EW ZIRKON 1

Emil Weingartner & Co.
C 2, Cr 11.5, bal Fe.
Pressure casting molds; cold work sheet punching dies.
Werkstoff Nr. 1.2080.

EW ZIRKON W

Emil Weingartner & Co.
C 2.1, Cr 11.5, W 0.7, bal Fe.
Heavy duty sheet punching dies. Werkstoff Nr. 1.2436.

EWP

Styria-Stahl Steirische Gusstahlwerke AG
W 9.5, Cr 2.5, V 0.1, C, bal Fe.
For hot dies; hot work steel. *Obsolete*

EX-2

Manufacturer not listed
C 0.65-0.75, Mn 0.25-0.45, Si 0.2-0.35, Ni 0.7-1, Cr 0.15-0.3, Mo 0.08-0.15, bal Fe.
Hardened: 60 Rock C min. Annealed: 200 Brin max. For spindles, rolls, bearings, plungers, valves, meters, pins, cylinders. Wear resistant, deep hardening.

EX-CELL-O 10A

Ex-Cell-O
Co 11, WC 73.5, 7.5 TiC, 8.0 TaC.
90.8 Rock A. For heavy cuts at slow speeds, interrupted cuts, and severe milling of steels and alloy cast irons. *Obsolete*

EX-CELL-O 509

Ex-Cell-O
Co 8, WC 76, 16.0 TiC.
92.5 Rock A. For precision boring and turning of steel and alloy cast iron at high speed and low feeds. *Obsolete*

EX-CELL-O 606

Ex-Cell-O
Co 8, WC 78, 10.0 TiC, 4.0 TaC.
91.8 Rock A. For medium roughing cuts on steels and alloy cast irons. *Obsolete*

EX-CELL-O 6A

Ex-Cell-O
Co 7, WC 76, 7.0 TiC, 10.0 TaC.
91.9 Rock A. General purpose medium roughing grade for steels and alloy cast irons. *Obsolete*

EX-CELL-O 6AX

Ex-Cell-O
Co 5, WC 71.5, 12.5 TiC, 11.0 TaC.
92.6 Rock A. For light roughing and finishing of steels and alloy cast irons. *Obsolete*

EX-CELL-O 8A

Ex-Cell-O
Co 8.5, WC 72, 7.5 TiC, 12.0 TaC.
91.2 Rock A. General purpose heavy duty roughing grade for steels and alloy cast irons. *Obsolete*

EX-CELL-O E16

Ex-Cell-O
Co 16, WC 84.
87.5 Rock A. For moderate impact use where good wear resistance is required. *Obsolete*

EX-CELL-O E20

Ex-Cell-O
Co 20, WC 80.
86 Rock A. For high impact applications. *Obsolete*

EX-CELL-O E25

Ex-Cell-O
Co 25, WC 75.
85 Rock A. For applications requiring maximum shock resistance. *Obsolete*

EX-CELL-O E3

Ex-Cell-O
Co 3, WC 96.5, 0.5 TaC.
93.3 Rock A. For precision turning and boring of cast iron and nonferrous materials. *Obsolete*

EX-CELL-O E5

Ex-Cell-O
Co 5, WC 94.5, 0.5 TaC.
92.8 Rock A. For general purpose semi-finishing of cast iron and nonferrous materials. *Obsolete*

EX-CELL-O E6

Ex-Cell-O
Co 6.5, WC 93, 0.5 TaC.
92.2 Rock A. For wear applications with little shock and general purpose roughing of cast iron and nonferrous materials. *Obsolete*

EX-CELL-O E8

Ex-Cell-O
Co 6, WC 94.
91 Rock A. For heavy duty roughing cuts of cast iron and nonferrous materials. *Obsolete*

EX-CELL-O E9

Ex-Cell-O
Co 9, WC 91.
90.7 Rock A. General purpose wear grade with medium shock and roughing cuts on high temperature alloys. *Obsolete*

EX-CELL-O W12C

Ex-Cell-O
Co 12, WC 88.
89 Rock A. For wear applications where heavy shock is present. *Obsolete*

EX-CELL-O XL028

Ex-Cell-O
Co 6, WC 91, 3.0 TaC.
92.8 Rock A. For milling, broaching, reamer applications on cast iron and nonferrous materials. *Obsolete*

EX-CELL-O XL061

Ex-Cell-O
Co 8, WC 80, 6.0 TiC, 6.0 TaC.
92 Rock A. General purpose milling grade for steels and alloy cast irons. *Obsolete*

EX-CELL-O XL620

Ex-Cell-O
Co 6, WC 74, 20.0 TaC.
91.4 Rock A. Special purpose grade where good hot hardness and high temperature lubricity are required. *Obsolete*

EX-CELL-O XL85

Ex-Cell-O
Ni 21, Mo 9, 70 TiC.
91 Rock A. For roughing use at moderate speeds on steels. *Obsolete*

EX-CELL-O XL86

Ex-Cell-O
Ni 18, Mo 9, 73 TiC.
91.5 Rock A. For general purpose machining of steels. *Obsolete*

EX-CELL-O XL88

Ex-Cell-O
Ni 12.5, Mo 11, 76.5 TiC.
93 Rock A. For finish and semi-finish machining of steels and alloy cast irons at high speeds. *Obsolete*

EX-TEN 42

United States Steel Corp.
C 0-0.21, Cb 0.01, Mn 0-1.35, Si 0-0.3, 0.02 V min, bal Fe.
Plate: 42,000 min YS; 63,000 min TS; 24 El; 42,000 Yield point in compression in pounds per square in. For structural applications, automotive and truck parts, storage tanks, pipelines, cargo containers and construction machinery. Good formability and weldability. Tough.

EX-TEN 45

United States Steel Corp.
C 0-0.22, Mn 0-1.35, Si 0.1, Cb 0.02, 0.02 V min, bal Fe.
Rolled: 60,000 TS; 45,000 YS; 25 El. For railroad car and bus bodies, cargo containers. High strength low alloy construction steel.

EX-TEN 50

United States Steel Corp.
C 0-0.26, Mn 0-1.35, Si 0.1, 0.01 Cb min, 0.02 V min, bal Fe.
Rolled: 65,000 TS; 50,000 YS; 22 El. For railroad car and bus bodies, cargo containers. High strength low alloy construction steel.

EX-TEN 55

United States Steel Corp.
C 0-0.25, Mn 0-1.35, Si 0.1, Cb 0.02, 0.02 V min, bal Fe.
Rolled: 70,000 TS; 55,000 YS; 20 El. For automotive and truck bodies, cargo containers, tote boxes. High strength, low alloy steel.

EX-TEN 60

United States Steel Corp.
C 0-0.26, Mn 0-1.35, N 0-0.012, 0.02 V min, 0.02 Cb min, bal Fe.
Bar: 75,000 min TS; 60,000 min YP; 18 El. For automotive and truck parts, cargo containers, tote boxes, gas cylinders, construction machinery. Low-alloy, high strength steel.

EX-TEN 65

United States Steel Corp.
C 0-0.26, Mn 0-1.35, N 0-0.012, 0.01 Cb min, 0.02 V min, bal Fe.
Bar: 80,000 min TS; 65,000 min YP; 16 El. For automotive and truck parts, cargo containers, tote boxes, gas cylinders, construction machinery. Low-alloy, high strength steel.

EX-TEN 70

United States Steel Corp.
C 0-0.26, Mn 0-1.35, Si 0-0.4, C, V, bal Fe.
590 MPa (85 ksi) TS; 485 MPa (70 ksi) YS; 14 El. For railroad freight cars, bridges, towers.

EX. "B" METAL

English manufacture
Cu 77, Pb 15, Sn 8, P.
For bearings, bushings; heavy duty.

EX. "K" METAL

English manufacture
Cu 77, Pb 12.5, Sn 10.5, P.
For bearings, bushings; heavy duty.

EXCELITE

Excelite Co.
Cr 32, W 18, Co 45, C 2, B 0.2, bal Fe.
For cutting tools; cast.

EXCELLO

H. Boker & Co.
Ni 85, Cr 14, Fe 0.5, Mn 0.5.
Rolled: 95,000 TS. For resistance wires, heating elements; oxidation resistant. *Obsolete*

EXCELO

Carpenter Technology Corp.
C 0.5, Mn 0.3, Si 0.3, Cr 1.5, W 2.6, V 0.25, bal Fe.
Shock resisting tool steel used for heading dies, punches, chisels and impact tooling.

EXCELSIOR

H.K. Porter Co., Inc.
Cu 55, Ni 45, Fe 0.3.
For resistance alloy; heat resistant. *Obsolete*

EXCELSIOR

Riverside Metals Corp.
Ni 45, bal Cu.
Annealed: 70,000 TS; 25,000 YS; 35 El. For resistors, rheostats, heating elements; low electrical resistance, for service up to 800 F. *Obsolete*

EXD1

Eagle & Globe Steel Ltd.
Tool material. C 0.33, Mo 5.5, W 1.5, Ni 3.75, bal Fe.
Hot work steel for copper alloy extrusion tooling, press forging stainless steel. AS1239 H101A.

EXD5

Eagle & Globe Steel Ltd.
Tool material. C 0.33, Si 0.3, Mo 12, W 12, V 1.1, bal Fe.
Hot work steel resists softening and wear at high temperatures. Hot extrusion dies for brass and copper alloys. AS1239 H23A; AISI H23.

EXELLOY

Now CRANE EXELLOY.

EXL-DIE

Columbia Tool Steel Co.
Tool material. C 0.95, Mn 1.3, Cr 0.5, W 0.5, V 0.1, bal Fe.
Oil hardened, cold work tool steel. For taps, reamers, broaches, extrusion and engraving dies, form tools; nondeforming. AISI O1.

EXL-DIE SMOOTHCUT

Columbia Tool Steel Co.
C 0.9, Mn 1, W 0.5, Cr 0.5, bal Fe.
For tools, dies, chisels, shear blades, drills, vise jaws; non-deforming, oil hardened. *Obsolete*

EXLO

Foote Mineral Co.
C 0-0.025, Cr, Si, bal Fe.
Used as addition for stainless steel melting. *Obsolete*

EXOCUT

Allegheny Ludlum Steel
C 1.09, Cr 3.75, Co 8, Mo 9.5, W 1.6, V 1.15, Mn 0.2, Si 0.3, bal Fe.
Hardened: Rock C 68-70. For cutters to machine heat treated superalloys. Super high speed steel. Not shock resistant. *Obsolete*

EXOCUT

AL Tech Specialty Steel Corp.
C 1.09, Cr 3.75, Co 8, Mo 9.5, W 1.6, V 1.15, Mo 0.2, Si 0.3, bal Fe.
Usual heat-treated hardness: 67/70 Rock C. Heavy duty high speed steel designed to machine high temperature alloys, titanium, and prehardened alloy steels. AISI M-42.

EXOHARD

Allegheny Ludlum Steel
C 1.1, Mn 0.2, Si 0.3, Cr 3.75, Co 5, Mo 9.5, W 1.6, P 0.015, S 0.02, V 1.25, bal Fe.
Hardened: Rock C 64-68. For cutting tools for hard metals, lathe and planer tools, broaches. Super high speed steel. Not shock resistant. *Obsolete*

EXOTEC 29904

Eutectic Corp.
Alloy powder for metal spraying bond coat on all base metals except copper. *Obsolete*

EXP. 99

Sterling Metals Ltd.
Cu 4.5-5.5, Mn 0.2-0.3, Ni 0.8-1.2, Ti 0.15-0.25, Sb 0.1-0.4, Co 0.1-0.4, bal Al.
Sand cast: 31,400-35,800 TS; 29,000 YS; 2 El; 90-100 Brin; 0.6 Iz. Modified "Y" alloy for service up to 250 C. *Obsolete*

EXPANDAL

Now LESCALLOY EXPANDAL.

EXPANDED METAL

American manufacture
C 0.15-0.25, bal Fe.
For reinforced concrete or plaster walls; low carbon steel mesh.

EXPANDING ALLOY

English manufacture
Pb 67, Sn 25, Bi 8.3.
For type metal, mounting purposes.

EXPANSIVE METAL

Cerro Metal Products Co.
Pb 75, Sb 16.7, Bi 8.3.
For plugging castings. Expands on freezing. *Obsolete*

EXPORT

Paul Bergsoe & Son
Sn 92, Sb 3.9, Cu 3.9, Ni 0.2.
Cast: 7000 TS; 19 Brin. MP: 440-600°F. For engine bearings and bushings. Shock resistant.

EXPRESS

Leadbeater & Scott Ltd.
C 0.7, W 14, Cr 4, V 1, bal Fe.
For lathe and planer tools, reamers, drills, broaches; high speed steel.

EXPRESS COBALT FIVE

Leadbeater & Scott Ltd.
C 0.7, Cr 4, W 18, V 1, Co 5, bal Fe.
For lathe and planer tools, reamers, broaches; high speed steel.

EXPRESS COBALT TEN

Leadbeater & Scott Ltd.
C 0.7, Cr 4, Co 10-12, W, Mo, V, bal Fe.
For cutters for heat treated steels; high speed steel.

EXPRESS E.Z.

Compagnie Ateliers et Forges de la Loire
C 0.72, W 10.5, V 1.3, Mo 0.5, Co 0.25, bal Fe.
Hardened: 64-66 Rock C. For lathe and planer tools, hobs, broaches, end mills, form cutters. High-speed steel, oil hardening, high red hardness.

EXTENDO-DIE HOT WORK DIE STEEL

Carpenter Technology Corp.
Tool material. C 0.44, Mn 0.45, S 0-0.005, Si 1, Cr 6, Mo 1.9, V 0.8, bal Fe.
As treated: 58-59 Rock C. Air hardening hot work die steel with extreme toughness and good red-hardness. For dies, mandrels and inserts for hot working aluminum.

EXTRA

Agawam Tool Co.
C 1.15, Mn 0.3, V 0.5, Si 0.35, bal Fe.
For drills, taps, water hardened.

EXTRA

Carpenter Technology Corp.
C 0.6-1.2, bal Fe.
For taps, reamers, drills, dies, cold heading dies; water hardening. *Obsolete*

EXTRA

CCS Braeburn Alloy Steel
C 0.8-1.2, Mn 0.25, Si 0.25, bal Fe.
Water hardening tool steel. AISI W1. *Obsolete*

EXTRA "L"
Teledyne Vasco
C 0.6-1.4, bal Fe.
For tools, drills, razor blades. *Obsolete*

EXTRA (DISSTON)
Disston Inc.
C 1, bal Fe.
For tools, drills, cutters; water hardened. *Obsolete*

EXTRA BEST
Manufacturer not listed
C 1, bal Fe.
For tools, taps, reamers; water hardening.

EXTRA BEST CAST
Edgar T. Ward's Sons Co.
C, Cr, V, bal Fe.
For tools. *Obsolete*

EXTRA BEST MH
Otto Wolff Handelsgesellschaft
C 1.1, Si 0-0.25, Mn 0-0.25, bal Fe.
Annealed: 110,000 TS; 57,000 YS; 18 El; 40 RA; 210 Brin. For drills, taps, springs, hobs, reamers. Type W1; water hardened.

EXTRA BEST VZH
Otto Wolff Handelsgesellschaft
C 1, Mn 0.25, V 0.1, bal Fe.
For springs, taps, reamers, broaches; Type W2; water hardened.

EXTRA BEST Z
Otto Wolff Handelsgesellschaft
C 0.85, Si 0-0.25, Mn 0-0.25, bal Fe.
Heat treated: 190,000 TS; 145,000 YS; 10 El; 30 RA; 400 Brin. For springs, taps, drills, reamers, punches. Type W1; water hardened.

EXTRA BEST ZH
Otto Wolff Handelsgesellschaft
C 1, Si 0-0.25, Mn 0-0.25, bal Fe.
Annealed: 100,000 TS; 53,000 YS; 21 El; 42 RA; 200 Brin. For springs, taps, drills, hobs, reamers. Type W1; water hardened.

EXTRA BEST ZW
Otto Wolff Handelsgesellschaft
C 0.7, Si 0-0.25, Mn 0-0.25, bal Fe.
Heat treated: 175,000 TS; 128,000 YS; 12 El; 37 RA; 355 Brin. For springs, rails, punches, axes, crimpers. Type W1; water hardened.

EXTRA CARBON
Cytemp Specialty Steel Div.
C 0.7-1.4, bal Fe.
For tools, dies, punches; water hardened. *Obsolete*

EXTRA CARBON VANADIUM
Allegheny Ludlum Steel
C 0.8, V 0.2, bal Fe.
For tools, drills, taps. *Obsolete*

EXTRA OO3T
Sanderson Kayser Ltd.
C 1.05-1.15, bal Fe.
For tools, drills; water hardened. *Obsolete*

EXTRA D.R.
Latrobe Steel Co.
C 1.2, bal Fe.
For drills, taps, cutters; drill rod, water hardening. *Obsolete*

EXTRA DRACO
Cytemp Specialty Steel Div.
C 0.7-1.2, V 0.2, bal Fe.
For tools, cold header dies; water hardened. *Obsolete*

EXTRA DRACO DV
Cytemp Specialty Steel Div.
C 1, V 0.5, bal Fe.
For tools, dies, taps, drills, cutters; Type W2; water hardened. *Obsolete*

EXTRA DURO MX4
Vereinigte Edelstahlwerke
C, alloy, bal Fe.
For tools; oil hardened. *Obsolete*

EXTRA EXTRA
Thyssen Edelstahlwerke AG
C 1.2-1.3, bal Fe.
For tools, water hardened. *Obsolete*

EXTRA G
D.G. Gautier & Co.
C 0.7-1.1, bal Fe.
For tools, drills, taps; water hardened.

EXTRA H
Raven Steel & Tool Co.
C 1, V 0.5, bal Fe.
For taps, cutters, drills, tools; Type W2; water hardened.

EXTRA H
Latrobe Steel Co.
C 1, Mn 0.25, Si 0.25, bal Fe.
For general tools, cutters, drills, hobs, taps; Type W1; water hardened. *Obsolete*

EXTRA HEADER DIE
Columbia Tool Steel Co.
C 0.95, Mn 0.35, Si 0.25, bal Fe.
For header dies; water hardened. *Obsolete*

EXTRA HIGH LEADED BRASS 356
Anaconda Co.
Copper. Cu 62.5, Zn 35, Pb 2.5.
Hard sheet: 73,000 TS; 60,000 YS; 7 El; 80 Rock B. Soft sheet: 45,000 TS; 17,000 YS; 50 El; 15 Rock B. For clocks, instruments, screw machine parts, fasteners. Electrical conductivity: 26. Hard and strong; good machinability.

EXTRA L
Teledyne Vasco
C 0.6-1.25, bal Fe.
For tools, taps, punches; varying tempers. *Obsolete*

EXTRA LION
Burys & Co. Ltd.
C 0.7, W 18, Cr 4, V 1, bal Fe.
For drills, cutters, hobs, broaches, reamers; high speed steel.

EXTRA M.G.
Houghton & Richards Inc.
C 1.05, Mn 1.1, Cr 0.5, Si 0.25, bal Fe.
For tools and dies, long taps, reamers, screw dies, calipers; nondeforming.

EXTRA MH
Otto Wolff Handelsgesellschaft
C 1.1, Si 0-0.25, Mn 0-0.25, bal Fe.
Annealed: 110,000 TS; 58,000 YS; 18 El; 40 RA; 210 Brin. For springs, taps, reamers, cutters. Type W1; water hardened.

EXTRA NO. 5
Peter A. Frasse & Co.
C 0.7-0.9, bal Fe.
For tools, drills, taps, water hardened. *Obsolete*

EXTRA PUNCH
Latrobe Steel Co.
C 0.7, W 18, Cr 4, V 1, bal Fe.
For punches, dies; oil hardening. *Obsolete*

EXTRA QUALITY
Houghton & Richards Inc.
C 0.7-1.2, bal Fe.
For punches, dies, general tools; water hardening.

EXTRA RAPID 300
Vereinigte Edelstahlwerke
C 1.3, Cr 4, W 12, Mo 1, V 4, bal Fe.
For cutting tools, engravers' tools; high speed steel. *Obsolete*

EXTRA RAPID 300A
Vereinigte Edelstahlwerke
C 1.2, W 10.5, Cr 4.3, V 3.8, bal Fe.
For tools, dies, cutters; high speed steel. *Obsolete*

EXTRA S
Raven Steel & Tool Co.
C 0.6-1.4, V 0.25, bal Fe.
For tools, taps, cutters, reamers; Type W2, water hardened.

EXTRA SO
Bohler Gesellschaft M.B.H.
C 1.05, bal Fe.
For tools, drills, taps; water hardened.

EXTRA SPECIAL
Columbia Tool Steel Co.
C, Cr, bal Fe.
For tools. *Obsolete*

EXTRA SPECIAL
Manufacturer not listed
C 1.2, bal Fe.
For tools, finishing tools, drills; water hardening.

EXTRA SPECIAL
Lebanon Steel Foundry
C 1.3-1.5, Mn 0.2-0.4, Cr 0.15-0.4, W 3.5-4.5, V 0.35, bal Fe.
For tools, form cutters, reamers; fast finishing steel. *Obsolete*

EXTRA SPECIAL (POTTS)
Horace T. Potts Co.
C, alloy, bal Fe.
For tools. *Obsolete*

EXTRA SPECIAL ALLOY
Latrobe Steel Co.
C, W, bal Fe.
For tools, drills, taps; water hardened. *Obsolete*

EXTRA SPECIAL HIGH SPEED
Bethlehem Steel Corp.
C 0.7, Cr 4, W 14, V 2.2, bal Fe.
For tools, extrusion dies, cutters; high speed steel. *Obsolete*

EXTRA SPECIAL POWER TRANSFORMER 52-58
Follansbee Steel Co.
C 0.03, Si 4.5, bal Fe.
Annealed: 75,000 TS; 68,000 YS; 2 El. For power and radio transformers; high permeability. *Obsolete*

EXTRA SUPER INCOMPARABLE
Flockton, Tompkin & Co., Ltd.
C 0.8, W 22, Cr 4.5, V 1.8, Co 10, Mo 0.7, bal Fe.
For lathe and planer tools, cutters; high speed steel.

EXTRA SUPERDURALUMIN
Japanese manufacture
Al-Cu-Zn-Mn-Mg.
For light alloy parts.

EXTRA TENAZ DURO
Vereinigte Edelstahlwerke
C 0.9-1, bal Fe.
For tools, drills, taps, water hardened. *Obsolete*

EXTRA TOOL
John A. Crowley Inc.
C 1.1, bal Fe.
For tools, fixtures; water hardened.

EXTRA TOOL
Midvale-Heppenstall Co.
C 0.7-1.2, bal Fe.
For cutters, dies, general tools; water hardened.

EXTRA TOUGH AND HARD
Vereinigte Edelstahlwerke
C 1.1, bal Fe.
For general purpose tools and dies; water hardening. *Obsolete*

EXTRA TOUGH NO. 4
Jessop Steel Co.
C 0.7, Ni 1.5, Cr 0.5, Mo 0.2, bal Fe.
For punches, shear blades; tough. AISI L 6.

EXTRA TOUGH NO. 6
Jessop Steel Co.
C 0.5, Mn 0.8, Cr 0.95, V 0.2, bal Fe.
For gears, bolts, springs, crankshafts; oil hardened, tough. AISI L 2.

EXTRA TOUGH NO. 7
Jessop Steel Co.
C 0.7, Mn 0.7, Mo 0.3, bal Fe.
For rails, hammers, axles, springs; oil or water hardened. *Obsolete*

EXTRA TOUGH NO. 7 H.C.
Jessop Steel Co.
C 0.83, Mn 0.83, Mo 0.25, bal Fe.
For rails, hammers, axles; oil or water hardened. *Obsolete*

EXTRA TRIPLE CONQUEROR
Joseph Beardshaw & Son Ltd.
C 1.6, Cr 0.6, W 6, Si 0.15, Mn 0.4, bal Fe.
For turning tools, drawing dies; finishing steel, wear resistant.

EXTRA TRIPLE GRIFFIN
Darwins Alloy Castings
C 1.5, Cr 0.35, W 6, Mn 0.35, bal Fe.
For drawing dies, cutters; water hardened, hard case and tough core. *Obsolete*

EXTRA V
Teledyne Firth Sterling
Steel. C 0.9-1, V 0.2, bal Fe.
For tools, drills, taps, cutters, reamers; Type W2; water hardened. *Obsolete*

EXTRA VALTOOL
Disston Inc.
C 0.6-1.4, V 0.25, bal Fe.
For tools, drills, taps, cutters; Type W2; water hardened. *Obsolete*

EXTRA WARRANTED
Crucible Specialty Metals
C 1-1.2, bal Fe.
For taps, dies, reamers, drills, knives; Type W1; water hardened. *Obsolete*

EXTRA ZH
Otto Wolff Handelgesellschaft
C 1, Si 0-0.25, Mn 0-0.25, bal Fe.
Annealed: 100,000 TS; 53,000 YS; 21 El; 42 RA; 200 Brin. For springs, taps, drills, hobs. Type W1; water hardened.

EXTRA ZW
Otto Wolff Handelgesellschaft
C 1, Si 0-0.25, Mn 0-0.25, bal Fe.
Annealed: 100,000 TS; 53,000 YS; 21 El; 42 RA; 200 Brin. For springs, taps, drills, hobs, reamers. Type W1; water hardened.

EXTRAD
Midvale-Heppenstall Co.
C 0.35, Si 1, Cr 5, W 1.5, Mo 1.65, V 0.2, bal Fe.
For hot forming dies; air hardening. *Obsolete*

EXTRARD
Midvale-Heppenstall Co.
C 0.35, Cr 5, Mo 1.4, W 1.25, V 0.2, Si 1, bal Fe.
For heat resistant parts, hot work tools; Type H12; air or oil hardened.

EXTRUDAL
Aluminium Industrie Aktiengesellschaft
Si 0.4-0.8, Mg 0.4-0.8, bal Al.
Forged: 36,000-43,000 TS; 29,000-36,000 YS; 10-18 El; 75-90 Brin. For light alloy parts; good formability.

EXTRUDAL 12
Alluminio SA
Aluminum. Cu 0.8-1.3, Si 11-13, Mg 0.8-1.2, Fe 0-0.8, Ni 0.6-1.2, bal Al.
TA Temper: 55,000-60,000 TS; 42,000-48,000 YS; 5-12 El; 110-135 Brin. For forged pistons. High corrosion resistance.

EYA-1 CAST IRON
Russian manufacture
C 1.3-3.2, Cr 17.8-19.2, Ni 8.5-9.2, Si 1.7-2.6, bal Fe.
For furnace parts and equipment; corrosion resistant, cast iron.

EYELET BRASS
American manufacture
Cu 65-68, Zn 32-35.
Hard: 85,000 TS; 60,000 YS; 4 El; 156 Brin. Soft: 46,000 TS; 20,000 YS; 58 El; 53 Brin. For eyelets, drawn shells; high ductility.

EZ 14W
Boyd-Wagner Co.
C 0.7, W 14, Cr 3, V 0.2, bal Fe.
240 Brin. For hot dies, tools, extrusion and hot forging dies; high speed steel.

EZ 14W
Boyd-Wagner Co.
C 0.7, W 14, Cr 3, V 0.2, bal Fe.
240 Brin. For hot dies, tools, extrusion and hot forging dies; high speed steel.

EZ 33A
Dow Chemical Co.
Magnesium. Re 2.5-4, Zn 2-3.1, Zr 0.45-1, bal Mg.
Melting point: 1189°F. For welding rod and electrodes. *Obsolete*

EZ 33A
Various foundries
Zn 2.7, Zr 0-7, 3.0 Rare Earths, bal Mg.
T5-temper: 20,000-23,000 TS; 14,000-15,000 YS; 2-3 El. At 600°F: 12,000 TS; 8000 YS; 50 El. Magnesium sand and permanent mold casting with good pressure tightness and weldability. For parts operating at 350-500°F. ASTM B80-69, B199-68; AMS 4442; QQ-M-55, QQ-M-56; SAE 506.

EZ 9W
Boyd-Wagner Co.
C 0.7, W 9.5, Cr 3.5, V 0.2, bal Fe.
For gripper and hot forming dies, punches, piercers; hot work steel.

EZ 9W
Boyd-Wagner Co.
C 0.7, W 9.5, Cr 3.5, V 0.2, bal Fe.
For gripper and hot forming dies, punches, piercers; hot work steel.

EZ CARB
Wallace Murray Corp.
C 0.35, Cr 0.85, Mn 0.75, Mo 0.5, W 0.5, bal Fe.
Shock resistant; cold work tool steel.

EZ CUT 20
Joseph T. Ryerson & Son Inc.
C 0.18, Mn 1.2, P 0.08, S 0.29, Si 0.23, bal Fe.
67 ksi TS; 38 ksi YS; 31 El; 33 RA. Free-cutting steel for die bases, jigs, molds; case hardenable.

EZ CUT 45
Joseph T. Ryerson & Son Inc.
C 0.46, Mn 1.2, P 0.03, S 0.26, Si 0.18, bal Fe.
92 ksi TS; 45 ksi YS; 20 El; 42 RA. Direct hardenable, free-cutting steel for dies, jigs, molds.

EZ HEAD
U.S. Steel Corp.
C 0-0.08, Cr 16-18, Ni 17.5-19, bal Fe.
For cold headed parts; stainless. *Obsolete*

EZ-DIE
Columbia Tool Steel Co.
Tool material. C 1, Cr 5.2, Mo 1.15, V 0.25, bal Fe.
For blanking, coining and forming dies; air hardened; Type A2; nondeforming.

EZ33A
Fansteel/Wellman Dynamics
Magnesium. Zn 2-3, RE 2.5-4, Zr 0.5-1, bal Mg.
Magnesium casting alloy. Cast: 20,000 psi TS; 14,000 psi YS; 2 El; 50 Brin.

F & G 1
Aluminium-Zentrale e.V.
Aluminum. Cu 3-5.5, Si 0.3-0.9, Mn 0.3-1, Mg 0.5-1.5, bal Al.
Soft: 23,000-31,000 TS; 15-20 El; 50-60 Brin. Heat treated: 68,000 TS; 48,000 YS; 10 El; 140 Brin. For rivets, hydraulic fittings, hardware, bridges, heavy duty structures. Age hardenable, high strength. *Obsolete*

F & G 3
Aluminium-Zentrale e.V.
Aluminum. Cu 5-6, Si 1, Mn 0.5, bal Al.
Soft: 28,000 TS; 15-25 El; 50-60 Brin. Heat treated: 71,000 TS; 57,000 YS; 2 El; 140 Brin. For hydraulic fittings, hardware, aircraft and engine components. Age hardenable, high strength and endurance limit. *Obsolete*

F & G 4
Aluminium-Zentrale e.V.
Aluminum. Si 0.5-1, Mn 0.8-1.5, Mg 0.5-1, bal Al.
Heat treated: 50,000 TS; 36,000 YS; 12 El; 95 Brin. For hardware, structures, tanks, aircraft components. Age hardenable, corrosion resistant. *Obsolete*

F & G 5
Aluminium-Zentrale e.V.
Aluminum. Mn 1.5, Mg 5-9, bal Al.
Soft: 32,000 TS; 12,800 YS; 25 El; 55 Brin. Hard: 68,000 TS; 48,000 YS; 10 El; 120 Brin. For heat exchangers, fixtures, tanks. Nonhardenable, corrosion resistant. *Obsolete*

F & G 8
Aluminium-Zentrale e.V.
Aluminum. Mn 1-2, bal Al.
Half hard: 26,000 TS; 21,000 YS; 5 El; 45 Brin. Hard: 35,000 TS; 28,000 YS; 2 El; 60 Brin. For commercial roofing, heat exchangers, ducts, fixtures, tanks. Nonheat treatable, good weldability. *Obsolete*

F 1 K
Saarstahl AG
Now SAARSTAHL 55 CR 3.

F 1 KB
Saarstahl AG
C 0.58, Si 0.3, Mn 1, Cr 0.9, B, bal Fe.
For flat or coil springs. W.-Nr. 7163. *Obsolete*

F 2 K
Saarstahl AG
Now SAARSTAHL 50 CRV 4.

F 2 KH
Saarstahl AG
Now SAARSTAHL 58 CRV 4.

F 4 K
Saarstahl AG
Now SAARSTAHL 51 CRMOV 4.

F ALLOY
Otis Elevator Co.
Cu 2.5, Zn 20, Mg 0.5, Mn 0.5, Si 0.75, bal Al.
Heat treated: 8100 TS; 57,000 YS; 19 El. For aircraft parts, structures; die casting.

F EXTRA
Thyssen Edelstahlwerke AG
C, bal Fe.
For files; water hardening. *Obsolete*

F QUALITY
Delta Metal (BW) Ltd.
Cu 58, Pb 1.5-2, bal Zn.
Extruded: 56,000 TS; 30-35 El; 25-30 RA; 95-110 Brin. Drawn: 64,000 TS; 25-30 El; 20-25 RA; 110-125 Brin. For hot stampings; free-cutting. *Obsolete*

F SPEZIAL
Thyssen Edelstahlwerke AG
C, bal Fe.
For files; water hardening. *Obsolete*

F-16
Koppers Co. Inc.
TC 3.4-3.8, Ni 1-1.5, Cr 0-0.5, Mo 0-0.5, bal Fe.
Heat treated: 40,000 TS; 400-450 Brin. For hardened liners and sleeves. *Obsolete*

F-17 TRIMETAL
Now CLEVITE F-17.

F-17 TRIMETAL
Gould Inc.
Pb 8-12, Sn 8-12, Sb 7-8, Overlay 3-5% Cu, bal Sn.
For heavy duty bearings; with steel backing. *Obsolete*

F-23
Now CLEVITE F-23.

F-23
Gould Inc.
Sn 0.9-1.25, Sb 14.75-15.5, As 0.8-1.1, Cu 0-0.6, bal Pb.
For bearings; SAE-15. *Obsolete*

F-3034
Olin Brass, Indianapolis
S 0.1, P 0.25, Cu 99.65.
For hardware; phosphor bronze; corrosion resistant. *Obsolete*

F-37
Olin Brass, Indianapolis
Copper. Zn 30, Cu 70.
For bearing retainer cages; cartridge brass.

F-37
Koppers Co. Inc.
TC 2.9-3.9, Si 2.2-3.1, Mn 0.4-0.8, Mo 0.9-1.3, Cr 0.2-0.4, Ni 0.8-1.25, bal Fe.
Cast: 65,000 TS; 290-350 Brin. For aero and auto piston rings; impact resisting. *Obsolete*

F-8 IRON
Koppers Co. Inc.
TC 3.4-3.8, Ni 1-1.5, Cr 0-0.5, Si, bal Fe.
Cast: 35,000 TS; 300 Brin. For piston rings, wear resistant parts for high speed engines; wear and corrosion resistant. *Obsolete*

F-80-S
Forjas Alavesas S.A.
C 0-0.1, Mn 0-0.7, Si 0-0.06, P 0-0.03, S 0-0.12, bal Fe.
For cold stamping, cold heading and free machining.

F-95
Kidd Drawn Steel Co.
TC 3.2, Si 1.6, Mn 0.7, Ni 1, Cr 0.25, bal Fe.
Cast: 108,800 TS; 330 Brin. For piston rings; wear and fatigue resistant.

F.A.S. STEEL
Firth-Vickers Stainless Steels Ltd.
C 0.4, Cr 11.5, bal Fe.
Oil treated: 112,000-123,000 TS; 78,500-90,000 YS; 18-22 El; 35-43 RA; 210-260 Brin. For aircraft and automobile valves; heat and corrosion resistant; Firth Brown J-180.

F.B.D
German manufacture
C 0.15, Mn 1, Si 1.2, Cr 17.5, Ni 16, Mo 1.75, Cb 2, V 0.12, bal Fe.
For supercharger parts, gas engines; high heat resistant.

F.C.C. AIR HARDENING
Allegheny Ludlum Steel
C 1.45-1.6, Mn 0.4-0.5, Co 0.7, Cr 11.5-12.5, Mo 0.8, V 0.5, bal Fe.
For blanking, forming, drawing, embossing dies, punches, hot press work, rotary shears; air hardening; non-deforming; cast to shape; resists abrasion. *Obsolete*

F.C.C. NO. 3 1/2
Allegheny Ludlum Steel
C, 12 Cr + W + Mo, bal Fe.
For heavy duty hot work dies; resists wear at high temperature. *Obsolete*

F.C.C.-P.R.K.-33 (CAST-TO-SHAPE)
Allegheny Ludlum Steel
C 1.45-1.6, Cr 11.5-12.5, Mo 0.9-1, Co 3-3.5, Ni 0.4-0.5, bal Fe.
Air hardened. For cutting edge forming dies; cast to shape. *Obsolete*

F.C.C.25
Allegheny Ludlum Steel
C 0.3-0.35, Cr 3.75-4.25, Mo 1.75-2.25, W 13-15, Ni 2.5-3, bal Fe.
For brass die casting dies; oil hardened. *Obsolete*

F.C.C.26
Allegheny Ludlum Steel
C 0.35-0.42, Cr 1-1.5, Si 0.6-1.1, Mo 0.45-0.55, W 2.5-3, V 0.15-0.3, bal Fe.
For hot work dies; oil hardened. *Obsolete*

F.C.C.27
Allegheny Ludlum Steel
C 0.35-0.45, Cr 2.75-3.25, Si 1.75-2, Mo 2.75-3, W 5-5.5, V 0.5-0.6, Co 1-1.5, bal Fe.
For tools, chisels; oil hardened. *Obsolete*

F.C.C.34 M
Allegheny Ludlum Steel
C 0.9-1, Mn 0.25-0.35, Si 0.15-0.2, bal Fe.
For general purpose tools; water hardened. *Obsolete*

F.C.C.39
Allegheny Ludlum Steel
C 0.85-0.95, Cr 0.4-0.6, V 0.15-0.2, bal Fe.
For draw rings; water hardened. *Obsolete*

F.C.C.41 M
Allegheny Ludlum Steel
C 0.65-0.75, Cr 3.5-4.5, W 17-18.5, V 1-1.35, bal Fe.
For cutters, hobs; high speed steel. *Obsolete*

F.C.C.44
Allegheny Ludlum Steel
C 0.44-0.48, Cr 4.75-5.5, Si 1.5-1.75, Mo 5-6, V 0.15-0.25, Ni 1.4-1.6, bal Fe.
For brass forging dies; oil hardened. *Obsolete*

F.C.C.46 M
Allegheny Ludlum Steel
C 0.9-1, Cr 0.4-0.6, Mn 1-1.25, W 0.4-0.6, V 0.2-0.3, bal Fe.
For punches, dies; oil hardened. *Obsolete*

F.C.C.47 M
Allegheny Ludlum Steel
C 1.15-1.25, Cr 0.4-0.5, W 1.5-1.75, V 0.15-0.25, bal Fe.
For fast finishing tools; oil hardened. *Obsolete*

F.C.C.49
Allegheny Ludlum Steel
C 0.3-0.45, Cr 0.75-1, W 5.25-6.25, Ni 3.5-4.25, bal Fe.
For hot piercing dies; oil hardened. *Obsolete*

F.C.C.53
Allegheny Ludlum Steel
C 0.5-0.53, Cr 3.5-4.5, W 17-19, V 0.75-1.25, bal Fe.
For hot drawing dies; oil hardened. *Obsolete*

F.C.I. FREE-CUTTING STAINLESS STEEL
Firth-Vickers Stainless Steels Ltd.
C 0-0.12, Si 0-0.8, Mn 0-1, Cr 11.5-13.5, Mo 0.3-0.6, S 0.15-0.3, bal Fe.
Free machining martensitic stainless steel. BS 970 (PT4) 416S21; similar to AISI 416.

F.C.S. STAINLESS STEEL, FREE CUTTING
Firth-Vickers Stainless Steels Ltd.
C 0.2-0.28, Si 0.6, Mn 1.2, S 0.23, Mo 0.25, Cr 13.5, bal Fe.
Heat treated: 92,000 TS; 65,000 YS; 25 El; 54 RA; 220 Brin.
For hardware, screw machine products, bolts, shafts; free
cutting, corrosion resistant.

F.I. 20 STEEL
Firth-Vickers Stainless Steels Ltd.
C 0.06, Si 0.3, Mn 0.7, Cr 21, bal Fe.
Heat treated: 72,000 TS; 55,000 YS; 34 El; 60 RA; 175 Brin.
For corrosion resistant parts; corrosion resistant.

F.J.A.B. EXTRA
A. Milne & Co.
C 0.9-1.4, bal Fe.
For general tools and dies; water hardening.

F.J.A.B. HOLLOW DRILL
A. Milne & Co.
C 1-1.2, bal Fe.
For mining, quarrying and construction tools; water
hardening.

F.J.A.B. REGULAR
A. Milne & Co.
C 0.9-1.2, bal Fe.
For general tools and dies; water hardening.

F.J.A.B. ROOLED AUGER
A. Milne & Co.
C 0.7-0.9, bal Fe.
For tools for mining soft ores, coal, etc.; water hardening.

F.J.A.B. SOLID DRILL
A. Milne & Co.
C 1.1-1.2, bal Fe.
For mining, quarrying and construction work tools; water
hardening.

F.J.A.B. SPECIAL
A. Milne & Co.
C 0.9-1.2, bal Fe.
For special die work tools.

F.U.G. 1
Felten & Guillaume Calswerke AG
Cu 2.5-5, Mg 0.2-1.8, Mn 0.3-1.5, bal Al.
Annealed: 27,000 TS; 11,000 YS; 22 El; 47 Brin. Heat treated:
72,000 TS; 57,000 YS; 130 Brin. For aircraft structures and
fittings, fasteners; age-hardenable.

F.U.G. 4
Felten & Guillaume Calswerke AG
Mg 0.6-1.4, Si 0.6-1.6, Mn 0.6-1, Cr 0.3, bal Al.
Annealed: 21,000 TS; 8000 YS; 24 El. For window frames, fan
blades, gutters, boats; good forming and welding properties.

F.U.G. 6
Felten & Guillaume Calswerke AG
Mg 2-4, Mn 0-0.4, Cr 0-0.3, bal Al.
Soft: 28,000 TS; 13,000 YS; 30 El; 47 Brin. Hard: 40,000 TS;
35,000 YS; 10 El; 73 Brin. For aircraft tanks and fittings, fuel
lines, marine parts; resists seawater corrosion.

F.U.G. 8
Felten & Guillaume Calswerke AG
Mn 1-1.5, Cr 0-0.3, bal Al.
Soft: 16,000 TS; 6000 YS; 40 El. Hard: 29,000 TS; 27,000 YS;
10 El. For cooking utensils, heat exchangers, tanks, furniture;
good forming and welding properties.

F.V.S. STEEL
Firth-Vickers Stainless Steels Ltd.
C 0.42, Si 1.5, Mn 0.7, Cr 14, Ni 14, W 2.5, bal Fe.
Heat treated: 130,000 TS; 74,000 YS; 23 El; 24 RA. For
stainless parts; austenitic, stainless.

F11
Keystone Carbon Co.
Iron copper 99 10. Sintered: 20,000 psi TS; 18% porosity;
5.7-6.1 density; 70 Rock H. Low cost bearing for moderate
loading.

F20, BEARING GRADE
Keystone Carbon Co.
Iron copper 90 10. Sintered: 30,000 psi TS; 18% porosity;
5.8-6.2 density; 80 Rock H. Low cost iron bearing for
moderate loading. ASTM B439-83, Grade 3.

F20, STRUCTURAL GRADE
Keystone Carbon Co.
Cu 10, bal Fe.
Sintered: 5.8-6.2 density; 31,000 psi TS; 85 Rock H. Standard
grade for medium stressed structural parts. ASTM B-222-83A.

F58
Sintered Products Ltd.
Cu (sintered), bal Fe.
5.7-5.9 g/cm^3 density; 70 Vickers; 1-3 El; 25-28 porosity.
Hardenable to 250 Vickers. Meets BSS A200; SAE 850.

F58/1
Sintered Products Ltd.
Cu + C (sintered), bal Fe.
5.7-5.9 g/cm^3 density; 120 Vickers; 1 El; 25-28 porosity.
Hardenable to 250 Vickers. Meets SAE 855; MPIE F-0010-N.

F68
Sintered Products Ltd.
bal Fe.
Sintered. 6.6-6.9 g/cm^3 density; 12-16 porosity. Can be
carbonitrided to 650 Vickers and 25 Rock C. Meets BSS
A-202; MPIE F-0000-R.

F80/0
Sintered Products Ltd.
Sintered iron. 6.2 g/cm^3 density; 12.5 kg/mm^2 TS; 70
Vickers. For motor industry bearings; timing chain adjusters.

F80/1
Sintered Products Ltd.
Sintered iron with 1% carbon. 6.2 g/cm^3 density; 23.5
kg/mm^2 TS; 140 Vickers. For brackets, levers, cams,
business machine parts.

FA 20
Cooper Alloy Corp.
Stainless steel. Mo 3.5, Cr 19-21, Ni 28-30, Cu 4-5, C 0-0.07,
Si 0-1.5, Mn 0-2.5, bal Fe.
Water quenched: 65,000-85,000 psi TS; 40,000-50,000 psi
YS; 35-45 El; 40-50 RA; 140-150 Brin. For fittings, valves, and
pumps. Corrosion resistant; austenitic.

FA-20CB3
Cooper Alloy Corp.
Stainless steel. C 0.07, Si 1.5, Mn 1.5, Cr 19-22, Ni 32.5-35,
Mo 2-3, Cu 3-4, Cb = 8 x C min (1.0 max), bal Fe.
Corrosion resistant casting.

FA-3
Howment Corp. (Carbide Div.)
Sintered carbide tool material. For heavy roughing cuts and
interrupted cuts on cast iron, non-ferrous and non-metallic
materials. *Obsolete*

FA-4
Howment Corp. (Carbide Div.)
Sintered carbide tool material. For roughing cuts and light
interrupted cuts on cast iron, non-ferrous and non-metallic
materials. *Obsolete*

FA-5
Howment Corp. (Carbide Div.)
Sintered carbide tool material. For general purpose
machining of cast iron, non-ferrous and non-metallic
materials, and for wear resisting applications. *Obsolete*

FA-6
Howment Corp. (Carbide Div.)
Sintered carbide tool material. For general purpose
machining of cast iron, non-ferrous metals and non-metallics.
Obsolete

FA-62
Howment Corp. (Carbide Div.)
Sintered carbide tool material. For light roughing and
finishing high temperature alloys and non-ferrous metals.
Obsolete

FA-7
Howment Corp. (Carbide Div.)
Sintered carbide tool materials. For finish cuts on cast iron,
non-ferrous metals and non-metallics. *Obsolete*

FA-8
Howment Corp. (Carbide Div.)
Sintered carbide tool material. For fine finishing cuts on cast
iron, non-ferrous metals and non-metallics. *Obsolete*

FA-9
Howment Corp. (Carbide Div.)
Sintered carbide tool material. For finish cutting cast iron,
non-ferrous metals and non-metallics. *Obsolete*

FA20CB
Cooper Alloy Corp.
Stainless steel. C 0-0.07, Cr 19-22, Mo 2-3, Cu 3-4, Cb 0-1, Ni
27.5-30.5, bal Fe.
Annealed: 75,000 psi TS; 45,000 psi YS; 40 El; 145 Brin. For
fittings, valves, and pumps. ACI-CN7MCb; corrosion resistant;
austenitic.

FABCO 115
Hobart Welding Products
C 0.05, Mn 1.7, Si 0.5, Cr 0.35, Ni 2.4, Mo 0.45, bal Fe.
Weld metal. As welded: 110,500 psi TS; 98,500 psi YS; 22 El.
Tubular flux cored electrode for CO_2 welding.

FABCO 1CM
Hobart Welding Products
C 0.1, Mn 0.7, Si 0.25, Cr 1.25, Mo 0.5, bal Fe.
Weld metal. As welded: 88,500 psi TS; 78,500 psi YS; 21 El.
Tubular flux cored electrode for CO_2 welding. *Obsolete*

FABCO 1MN
Hobart Welding Products
C 0.07, Mn 1.33, Si 0.26, Mo 0.47, bal Fe.
Weld metal. As welded: 113,000 psi TS; 103,000 psi YS; 16
El. Tubular flux cored electrode for CO_2 welding. *Obsolete*

FABCO 80
Hobart Welding Products
C 0.1, Mn 1.9, Si 0.94, bal Fe.
Weld metal. As welded: 101,000 TS; 86,000 YS; 24 El.
Tubular flux cored electrode for CO_2 welding. AWS-E70T-2.
Obsolete

FABCO 801
Hobart Welding Products
C 0.06, Mn 0.7, Si 0.22, Ni 2.47, bal Fe.
Weld metal. As welded: 73,500 psi TS; 62,000 psi YS; 26 El.
Tubular flux cored electrode for CO_2 welding. AWS E-70T-G;
ABS H-3. *Obsolete*

FABCO 802
Hobart Welding Products
C 0.03, Mn 1.25, Si 0.6, P 0.011, S 0.013, bal Fe.
Weld metal. As welded: 87,100 psi TS; 75,400 psi YS; 27.5 El;
67.1 RA; Charpy V-notch 50 ft·lb (average) at 0°F. Tubular
welding wire for CO_2 welding. AWS A5.20, Class E71T-1.

FABCO 81
Hobart Welding Products
C 0.077, Mn 1.45, Si 0.72, 0.10 other metals, bal Fe.
Weld metal. As welded: 83,000 psi TS; 70,000 psi YS; 28 El.
Tubular flux cored electrode for CO_2 welding. AWS-E70T-1.

FABCO 82

Hobart Welding Products
C 0.08, Mn 1.4, Si 0.5, 0.12 other metals, bal Fe.
Weld metal. As welded: 88,000 TS; 77,000 YS; 25 El. Tubular flux cored electrode for CO_2 welding. AWS-E70T-1. *Obsolete*

FABCO 85

Hobart Welding Products
C 0.07, Mn 1.4, Si 0.5, 0.10 other metals, bal Fe.
Weld metal. As welded: 77,000 psi TS; 63,500 psi YS; 28 El. Tubular flux cored electrode for CO_2 welding. AWS-E70T-5.

FABIS

Sanderson Kayser Ltd.
C, bal Fe.
Water hardening tool steel. *Obsolete*

FABLOY 1CM

Hobart Welding Products
C 0.07, Mn 0.8, Si 0.65, Cr 1.2, Mo 0.5, bal Fe.
Weld metal. Welded, stress relieved 1275°F: 81,000 TS; 67,000 YS; 22 El. Tubular flux cored electrode for CO_2 welding. AWS E-8018-B2; ASME A-3.

FABLOY 2CM

Hobart Welding Products
C 0.06, Mn 0.82, Si 0.67, Cr 2.25, Mo 1, bal Fe.
Weld metal. Welded, stress relieved 1275°F: 91,000 TS; 78,000 YS; 20 El. Tubular flux cored electrode for CO_2 welding. AWS E-9018-B3; ASME A-4.

FABLOY 2N

Hobart Welding Products
C 0.05, Mn 1.03, Si 0.68, Ni 2.39, bal Fe.
Weld metal. Welded, stress relieved 1150°F: 82,000 TS; 70,000 YS; 28 El. Tubular flux cored electrode for CO_2 welding on 2-1/2 Ni steels requiring impact strength. AWS E-8018-C1.

FABLOY 3N

Hobart Welding Products
C 0.04, Mn 0.97, Si 0.56, Ni 3.62, bal Fe.
Weld metal. Welded, stress relieved 1150°F: 81,000 TS; 69,000 YS; 25 El. Tubular flux cored electrode for CO_2 welding on 3-1/2 Ni steels used at very low temperatures. AWS E-8018-C2.

FABLOY 410

Hobart Welding Products
C 0.057, Mn 0.86, Si 0.64, Cr 12.71, Ni 0.42, bal Fe.
Weld metal. As welded; 87,000 TS; 65,000 YS; 25.5 El. Tubular flux cored electrode for CO_2 welding of corrosion resistant steels. AWS E410.

FABLOY 4130

Hobart Welding Products
C 0.3, Mn 0.8, Si 0.51, Cr 0.98, Mo 0.34, bal Fe.
Weld metal. Welded, quenched and tempered 750°F: 181,000 psi TS. Tubular flux cored electrode for CO_2 welding of SAE 4130 and similar steels.

FABLOY 5CM

Hobart Welding Products
C 0.05, Mn 0.92, Si 0.7, Cr 5.3, Mo 0.57, bal Fe.
Weld metal. Welded, annealed: 71,000 TS; 40,000 YS; 33 El. Tubular flux cored electrode for CO_2 welding on 4-6 Cr steels used at high temperature. AWS E502; ASME A-4.

FABLOY 9CM

Hobart Welding Products
C 0.05, Mn 0.91, Si 0.79, Cr 10.15, Mo 0.93, bal Fe.
Weld metal. As welded: 72,000 TS; 37,500 YS; 29 El. Tubular flux cored electrode for CO_2 welding on 9-11 Cr steels. AWS E505.

FABLOY T

Hobart Welding Products
C 0.05, Mn 2, Si 0.4, Ni 2.5, Mo 0.4, bal Fe.
Weld metal. As welded: 123,000 TS; 108,000 YS; 14 El. Tubular flux cored electrode for CO_2 welding. AWS E-12018G.

FABRIALLOY

Industrial Overlay Metals Corp.
C 0.5-0.6, bal Fe.
For welding electrodes. Shielded.

FABSHIELD 31

Hobart Welding Products
C 0.16, Mn 0.63, Si 0.1, Al 0.06, bal Fe.
Weld metal. As welded: 82,000 TS; 64,000 YS; 22.5 El. Self shielded, flux cored welding electrode wire. AWS E60T-7.

FABSHIELD 4

Hobart Welding Products
C 0.15, Mn 1.25, Si 0.28, Al 1.15, 0.10 other metals, bal Fe.
Weld metal. As welded: 84,000 TS; 64,000 YS; 24 El. Self shielded, flux cored welding electrode wire. AWS E70T-4.

FABSHIELD 55

Hobart Welding Products
C 0.18, Mn 1.35, Si 0.5, bal Fe.
Weld metal. As welded: 101,000 psi TS (transverse). For high speed single pass welding. AWS E70T-G.

FABSHIELD 8

Hobart Welding Products
C 0.08, Mn 1.38, Si 0.21, Al 0.73, bal Fe.
Weld metal. As welded: 88,000 TS; 75,000 YS; 23 El. Self shielded, flux cored welding electrode wire. AWS E70T-G. *Obsolete*

FABSHIELD 8 NI

Hobart Welding Products
C 0.07, Mn 1.31, Si 0.21, Ni 2, Al 0.59, bal Fe.
Weld metal. As welded: 90,000 TS; 76,000 YS; 23 El. Self shielded, flux cored welding electrode wire. AWS E70T-G. *Obsolete*

FABTUF 250

Hobart Welding Products
C 0.26, Mn 1.73, Cr 1.85, Si 0.32, bal Fe.
Weld metal. 24 Rock C. For build-up, magnetic, low abrasion resistance, high impact property.

FABTUF 960

Hobart Welding Products
C 0.7, Mn 2, Cr 8, Si 1, bal Fe.
Weld metal. 55-60 Rock C. For build-up, magnetic, high abrasion resistance.

FACE-COR 2200

Airco Vacuum Metals
Mn 15, Cr 4, Ni 4, bal Fe.
Wire for joining manganese steel and dissimilar steels and as build-up under base for hard coatings. Weld deposit: 0.80 C. For crusher jaws, railroad rails, frogs, and shovel buckets. Tough; non-magnetic.

FACE-COR 2500 RB

Airco Vacuum Metals
Mn 17.5, V 0.5, bal Fe.
DC reverse polarity wire for build-up on crusher hammers, dredge pump parts, etc. Tough, non-magnetic. Weld deposit: 0.90 C.

FACE-COR 2800

Airco Vacuum Metals
Mn 17.5, V 0.5, bal Fe.
DC reverse polarity wire for build-up on crusher rolls and jaws, dipper lips and teeth. Tough; non-magnetic. Weld deposit: 0.90 C.

FACE-COR 5200

Airco Vacuum Metals
Mn 1.5, Si 1.5, Cr 16, Mo 1, bal Fe.
DC reverse polarity wire for hard, non-machinable alloy deposits. Weld deposit: 3.0 C; 50-54 Rock C (two passes); magnetic. For crusher rolls, and impactor breaker bars.

FACE-COR 5500 R.F

Airco Vacuum Metals
Mn 1.5, Si 1.5, Cr 16, Mo 1, bal Fe.
DC reverse polarity wire for hard, non-machinable alloy deposits. Weld deposit: 3.0 C; 50-54 Rock C (two passes); magnetic.

FACE-COR 6500

Airco Vacuum Metals
Mn 1.5, Si 1.5, Cr 30, bal Fe.
Electrode wire for hard build-up. Weld deposit: 5.0 C; 58-61 Rock C (2 passes on mild steel). Non-machinable and cannot be flame cut. For breaker bars, sand pump parts, mixing plows and scrapers.

FACEWELD 1

Lincoln Electric Co.
C 4.2, Mn 5.4, Si 0.71, Mo 0.25, Cr 22.5, V 0.12, bal Fe.
Hard surfacing, arc welding electrodes; resistance to severe abrasion.

FACEWELD 12

Lincoln Electric Co.
C 4.5, Mn 1, Si 1, Mo 6, Cr 18.5, V 0.7, bal Fe.
Hard surfacing, arc welding electrodes; resistance to severe abrasion.

FAGERSTA A540

Fagersta Bruks Aktiebolag
C 0.8, Si 0.2, Mn 0.3, V 0.1, bal Fe.
For chisels, axles, shafts, drills; Type W1; water hardened. *Obsolete*

FAGERSTA AD 95

Swedish Steel Mills, A.A.
C 0.7, Cr 0.25, V 0.15, W 1, bal Fe.
For coining dies; oil hardened.

FAGERSTA ALLOY "30" DRILL

Achorn Steel Co.
C 1, Cr 1.1, Mn 0.3, Mo 0.35, bal Fe.
For tools, drills; water hardened.

FAGERSTA ALLOY CHISEL

Brukskoncernen AB
Cr, W, Mo, C, bal Fe.
For punches, chisels, rivet sets and busters; shock resistant.

FAGERSTA ALLOY CHISEL

Fagersta Stainless AB
Cr, W, Mo, C, bal Fe.
For punches, chisels, rivet sets and busters; shock resistant. *Obsolete*

FAGERSTA ALLOY SHOE DIE

Achorn Steel Co.
C 0.5, Cr 0.6, Mn 0.6, Mo 0.4, bal Fe.
For tools, dies; oil hardened.

FAGERSTA ALLOY SHOE DIE

Fagersta Stainless AB
C 0.55, Cr 0.65, Mo 0.35, bal Fe.
For shoe dies, cutting dies for leather and paper and rubber; oil hardened. *Obsolete*

FAGERSTA ALLOY SHOE DIE

Brukskoncernen AB
C 0.55, Cr 0.65, Mo 0.35, bal Fe.
For shoe dies, cutting dies for leather and paper and rubber; oil hardened.

FAGERSTA B 103

Fagersta Bruks Aktiebolag
C 0.45, Si 0.25, Mn 0.8, bal Fe.
Normalized. 85,500-107,000 TS; 46,000 minimum YS; 18 El; 180-230 Brin. For surface hardened gears, levers, shifter for forks, shafts. Water hardening. *Obsolete*

FAGERSTA B 805

Fagersta Bruks Aktiebolag
C 0.4, Si 0.25, Mn 1.3, bal Fe.
Hardened: 100,000-118,000 TS; 65,000 minimum YS; 17 minimum El; 50 minimum RA; 210-250 Brin. For surface hardened gears and shafts. Oil or water hardening. *Obsolete*

FAGERSTA B85

Fagersta Bruks Aktiebolag
C 0.4, Mn 1.25, bal Fe.
For gears, machinery parts; tough. *Obsolete*

FAGERSTA BRILLIANT H.H.

Achorn Steel Co.
C 0.7, Cr 3.5, W 14, bal Fe.
For tools, cutters, drills; high speed steel.

FAGERSTA BRILLIANT WKE EXTRA

Achorn Steel Co.
C 0.7, Cr 4.5, Co 9, V 1.5, W 19, bal Fe.
For tools, cutters, drills; high speed steel.

FAGERSTA BRILLIANT WW

Achorn Steel Co.
C 0.7, Cr 4.5, V 1.5, W 18, bal Fe.
For tools, cutters, reamers; high speed steel.

FAGERSTA BROACH

Brukskoncernen AB
C 0.7, W 18, Cr 4, V 1, bal Fe.
For broaches, cutters, drills, taps; high speed steel.

FAGERSTA BROACH

Fagersta Stainless AB
C 0.7, W 18, Cr 4, V 1, bal Fe.
For broaches, cutters, drills, taps; high speed steel. *Obsolete*

FAGERSTA BROACH

Achorn Steel Co.
C 1.15, Cr 0.4, Mn 1.25, bal Fe.
For tools, broaches, reamers; tough.

FAGERSTA C 345

Fagersta Bruks Aktiebolag
C 0.42, Si 0.25, Mn 0.8, Cr 1.1, Mo 0.2, bal Fe.
Hardened: 112,000-117,000 TS; 72,000 minimum YS; 10 minimum El; 40 minimum RA; 210-370 Brin. For gears, shafts, power transmission components. Oil hardening. *Obsolete*

FAGERSTA C 642

Fagersta Bruks Aktiebolag
C 0.25, Mn 0.65, Cr 1.05, Mo 0.2, Ni 0-0.3, bal Fe.
Hardened: 156,000-185,000 TS; 128,000 minimum YS; 8 minimum El; 330-390 Brin. For shafts, gears, machinery parts; oil or water hardenable. *Obsolete*

FAGERSTA C 643

Fagersta Bruks Aktiebolag
C 0.34, Si 0.25, Mn 0.7, Cr 1, Mo 0.2, bal Fe.
Hardened: 128,000-150,000 TS; 98,000 minimum YS; 12 minimum El; 50 minimum RA; 270-310 Brin. For gears, shafts, power transmission components. Oil hardening. *Obsolete*

FAGERSTA C143

Fagersta Bruks Aktiebolag
C 0.5, Mn 0.85, Cr 1.05, Mo 0.3, V 0.15, bal Fe.
For gears, machinery parts; oil hardened, tough. *Obsolete*

FAGERSTA C182

Fagersta Bruks Aktiebolag
C 0.9, Cr 0.5, bal Fe.
For wood working tools; water hardened. *Obsolete*

FAGERSTA C265

Fagersta Bruks Aktiebolag
C 1.5, Cr 12, Mo 0.8, V 0.2, bal Fe.
For drawing and blanking dies, punches; air hardened, non-deforming. *Obsolete*

FAGERSTA C424

Fagersta Bruks Aktiebolag
C 0.5, Cr 1.05, V 0.15, Mn 0.85, bal Fe.
For gears, shafts, springs; shock resistant. *Obsolete*

FAGERSTA C46

Fagersta Bruks Aktiebolag
C 1.3, Cr 0.5, bal Fe.
For files, razors; water hardened. *Obsolete*

FAGERSTA C50

Fagersta Bruks Aktiebolag
C 1.8, Cr 12, Mo 0.6, Co 3.25, bal Fe.
For cold work tools, cutters; oil or air hardened, non-deforming. *Obsolete*

FAGERSTA C550

Fagersta Bruks Aktiebolag
C 0.9, Cr 5.2, Mo 1.1, V 0.2, bal Fe.
For finishing tools, blanking dies; non-deforming, air hardened. *Obsolete*

FAGERSTA C71

Fagersta Bruks Aktiebolag
C 0.95, Si 1.5, Mn 0.75, Cr 1, bal Fe.
For drawing and forming dies; oil hardened. *Obsolete*

FAGERSTA CARBON

Brukskoncernen AB
C, bal Fe.
For general tools; water hardening.

FAGERSTA CARBON

Fagersta Stainless AB
C, bal Fe.
For general tools; water hardened. *Obsolete*

FAGERSTA CIRCULAR PLATES

Brukskoncernen AB
C, Cr, Ni, bal Fe.
For saws; oil hardening. *Obsolete*

FAGERSTA CIRCULAR PLATES

Fagersta Bruks Aktiebolag
C, Cr, Ni, bal Fe.
For saws; oil hardening. *Obsolete*

FAGERSTA COLD HEADER

Brukskoncernen AB
C, V, bal Fe.
For cold heading and swedging dies; tough. *Obsolete*

FAGERSTA COLD HEADER

Fagersta Bruks Aktiebolag
C, V, bal Fe.
For cold heading and swedging dies; tough. *Obsolete*

FAGERSTA CUTLERY STEEL

Brukskoncernen AB
C, bal Fe.
For knives, saws; water hardening. *Obsolete*

FAGERSTA CUTLERY STEEL

Fagersta Bruks Aktiebolag
C, bal Fe.
For knives, saws; water hardening. *Obsolete*

FAGERSTA D 161

Fagersta Stainless AB
Tool material. C 0.9, Mn 1.2, Cr 0.5, W 0.5, V 0.2, bal Fe.
For drawing and forming dies, punches; oil hardened, non-deforming. *Obsolete*

FAGERSTA D-921

Fagersta Stainless AB
Tool material. C 0.72, Cr 4, W 17.8, V 1.1, bal Fe.
High speed steel for cutting tools. AISI T1. *Obsolete*

FAGERSTA D-927

Fagersta Stainless AB
Tool material. C 1.25, Cr 4.1, Mo 3.2, W 9, V 3, Co 9, bal Fe.
Cobalt high speed steel; heat treated: 67-68 Rock C. Good red hardness for tough cutting tool operations (WKE 4). *Obsolete*

FAGERSTA D-930

Fagersta Stainless AB
Tool material. C 1.4, Cr 4.2, Mo 3.5, W 8.5, V 3.4, Co 11, bal Fe.
Cobalt high speed steel; heat treated: 67-68 Rock C. Good red hardness and wear resistance for cutting tools on special alloys (WKE 45). *Obsolete*

FAGERSTA D-933

Fagersta Stainless AB
Tool material. C 0.88, Cr 3.7, Mo 9.5, W 1.7, V 1.2, Co 8.3, bal Fe.
Co-Mo high speed steel for cutting tools; good red hardness for tough machining jobs. AISI M33. *Obsolete*

FAGERSTA D-941

Fagersta Stainless AB
Tool material. C 0.86, Cr 4, Mo 5, W 6.5, V 1.9, bal Fe.
Mo-W high speed steel for cutting tools; good for lathe rough machining. AISI M2. *Obsolete*

FAGERSTA D-942

Fagersta Stainless AB
Tool material. C 0.89, Cr 4.2, Mo 3, W 6.4, V 1.9, bal Fe.
High speed steel. *Obsolete*

FAGERSTA D-943

Fagersta Stainless AB
Tool material. C 0.83, Cr 3.9, Mo 8.7, W 1.8, V 1.2, bal Fe.
Mo high speed steel for drills, cutting tools. AISI M1. *Obsolete*

FAGERSTA D-946

Fagersta Stainless AB
Tool material. C 0.89, Cr 4.1, Mo 5.1, W 6.2, V 1.9, Co 5, bal Fe.
Co-Mo-W high speed steel for cutting tools; good red hardness (WKE-46). *Obsolete*

FAGERSTA D-948

Fagersta Stainless AB
Tool material. C 1.8, Cr 3.7, Mo 9.4, W 1.5, V 1.1, Co 8, bal Fe.
Special high speed steel; good abrasion resistance and red hardness. For special applications. Similar to AISI M42. *Obsolete*

FAGERSTA D-950

Fagersta Stainless AB
Tool material. C 0.95, Cr 4, Mo 5, W 1.8, V 1.2, bal Fe.
High speed steel. For twist drills, end mills. *Obsolete*

FAGERSTA D-952

Fagersta Stainless AB
Tool material. C 0.89, Cr 4.1, Mo 4.4, W 1.2, V 1.8, bal Fe.
High speed steel. *Obsolete*

FAGERSTA D-954

Fagersta Stainless AB
Tool material. C 0.99, Cr 3.9, Mo 8.8, W 1.7, V 2, bal Fe.
Mo-high carbon high speed steel; good abrasion resistance. AISI M7. *Obsolete*

FAGERSTA D-960

Fagersta Stainless AB
Tool material. C 0.97, Cr 3.9, Mo 7.9, W 0.6, V 1.9, bal Fe.
Mo-high carbon, high speed steel; good abrasion resistance. Similar to AISI M10. *Obsolete*

FAGERSTA D110

Fagersta Bruks Aktiebolag
C 0.35, Si 1.25, Cr 5, W 5, Co 0.65, bal Fe.
For Al die casting dies, hot work tools; hot work steel, oil hardened. *Obsolete*

FAGERSTA D249

Fagersta Bruks Aktiebolag
C 0.45, Cr 1.2, Mo 0.25, W 2.2, V 0.15, Si 0.9, bal Fe.
For pneumatic tools, chisels, shears; hot work steel, oil hardened. *Obsolete*

FAGERSTA D366

Fagersta Bruks Aktiebolag
C 0.3, Mn 0.3, Cr 3, Ni 1.7, W 9, V 0.3, bal Fe.
For die casting dies, hot work tools; hot work steel, oil hardened. *Obsolete*

FAGERSTA D61

Now FAGERSTA D-161.

FAGERSTA D65

Fagersta Bruks Aktiebolag
C 2, Cr 13, W 1.2, Mn 0.75, bal Fe.
For forming and drawing dies, cold work tools; oil or air hardened, non-deforming. *Obsolete*

FAGERSTA D66

Fagersta Bruks Aktiebolag
C 0.3, Cr 3.25, W 9, V 0.25, bal Fe.
For die casting dies, hot work tools; hot work steel, oil hardened. *Obsolete*

FAGERSTA DIE CASTING

Achorn Steel Co.
C 0.5, Cr 3, Mn 0.35, V 0.25, bal Fe.
For die casting dies, tools; oil hardened.

FAGERSTA DRAWING DIE STEEL

Brukskoncernen AB
C, Mn, Cr, W, bal Fe.
For wire drawing dies; oil hardening.

FAGERSTA DRAWING DIE STEEL

Fagersta Stainless AB
C, Mn, Cr, W, bal Fe.
For wire drawing dies; oil hardened. *Obsolete*

FAGERSTA ENGRAVER PLATE

Achorn Steel Co.
C 0.35, Mn 0.3, Si 0.1, bal Fe.
For engraver plates; water hardened.

FAGERSTA ENGRAVERS PLATE

Brukskoncernen AB
C 0.35, bal Fe.
For engravers plates.

FAGERSTA ENGRAVERS PLATE

Fagersta Stainless AB
C 0.35, bal Fe.
For engravers' plate. *Obsolete*

FAGERSTA FAST FINISHING

Brukskoncernen AB
C 1.2, W 2, bal Fe.
For finishing tools; water hardening.

FAGERSTA FAST FINISHING

Fagersta Stainless AB
C 1.2, W 2, bal Fe.
For finishing tools; water hardened. *Obsolete*

FAGERSTA FB-01

Now FAGERSTA D-161.

FAGERSTA FB-06 GRAPHITIC

Fagersta Bruks Aktiebolag
C 1.45, Mn 0.8, Si 1.15, Mo 0.25, bal Fe.
AISI Type O6 oil hardening tool steel. *Obsolete*

FAGERSTA FB-52

Fagersta Stainless AB
Tool material. C 0.89, Cr 4.1, Mo 4.4, W 1.2, V 1.8, bal Fe.
Low cost high speed steel for cutting tools, drills, wood working tools. *Obsolete*

FAGERSTA FB-A-2

Fagersta Bruks Aktiebolag
C 1, Mn 0.7, Cr 5.2, V 0.2, Mo 1.1, bal Fe.
Heat treated: 250,000 TS; 195,000 YS; 5 El; C 50 Rock. For blanking and trimming dies, gauges, shear knives, punches, forming dies, thread rolling dies. Air hardening tool steel. AISI Type A2; nondeforming. *Obsolete*

FAGERSTA FB-A-4

Fagersta Bruks Aktiebolag
C 0.95, Mn 2, Cr 2.2, Mo 1.1, bal Fe.
Heat treated: C 62-65 Rock. For punches, forming and blanking tools, shear blades, broaches, drills, reamers. Air hardening tool steel. AISI Type A4 nondeforming. *Obsolete*

FAGERSTA FB-A-6

Fagersta Bruks Aktiebolag
C 0.7, Mn 2.2, Cr 1, Mo 1.4, bal Fe.
Heat treated: 290,000 TS; 264,000 YS; 1 El; C 55 Rock. For blanking and forming dies, shear blades, mandrels, master hobs. Air hardening tool steel. Type AISI-A6 non-deforming. *Obsolete*

FAGERSTA FB-BRILLIANT DIE

Fagersta Bruks Aktiebolag
C 0.5, Mn 1, Si 0.3, Cr 1.1, Mo 0.25, bal Fe.
Prehardened low alloy steel for molds. *Obsolete*

FAGERSTA FB-D2

Fagersta Bruks Aktiebolag
C 1.5, Cr 11.5, V 0.2, Mo 0.8, bal Fe.
Heat treated: 278,000 TS; 214,000 YS; 1 El; C 57 Rock. For blanking and drawing dies, gauges, burnishing tools, thread rolling dies. Air hardening tool steel. AISI Type D2 nondeforming. *Obsolete*

FAGERSTA FB-D2 HIGH PRODUCTION

Fagersta Bruks Aktiebolag
C 1.55-1.7, Mn 0.25-0.35, Si 0.25-0.35, Cr 11.5-12.5, V 0.7-1, Mo 0.7-0.9, bal Fe.
Air or oil hardening cold work tool steel, chromium type; AISI D2. *Obsolete*

FAGERSTA FB-D3 HIGH PRODUCTION "H"

Fagersta Bruks Aktiebolag
C 2.05, Mn 0.4-4, Cr 11.5, V 0.6, bal Fe.
Air or oil hardening cold work tool steel, chromium type; AISI D3. *Obsolete*

FAGERSTA FB-D6

Fagersta Bruks Aktiebolag
C 2.1, Mn 0.7, Si 0.3, Cr 13, W 1.3, bal Fe.
Hardened: Rock C 62-64. For punching and drawing dies, shear blades, knives, rotary slitters, gauges, wear plates and cams, impact extrusion dies. Abrasion and wear resistant, non-deforming. AISI Type D6; Air or oil hardening. *Obsolete*

FAGERSTA FB-H11

Fagersta Bruks Aktiebolag
C 0.4, Mn 0.4, Si 1.05, Cr 5, V 0.5, Mo 1.35, bal Fe.
Air or oil hardening hot work tool steel, chromium type; for forging dies and hot forming dies; AISI H11. *Obsolete*

FAGERSTA FB-H12

Fagersta Bruks Aktiebolag
C 0.35, Si 1, W 1.5, Mo 1.6, Cr 5, bal Fe.
Heat treated: 216,000 TS; 185,000 YS; 14 El; 53 RA. For trimmer dies, die casting dies, forging and bolt heading dies. Hot work tool steel. AISI Type H12. Tough, shock resistant. *Obsolete*

FAGERSTA FB-H13

Fagersta Bruks Aktiebolag
C 0.35, Cr 5, V 1, Mo 1.5, bal Fe.
Heat treated: 290,000 TS; 228,000 YS; 3 El; C 55 Rock. For forging and heading dies, bulldozer dies, piercing and forming punches. Hot work, air hardening tool steel. AISI Type H13, shock resisting. *Obsolete*

FAGERSTA FB-M1

Now FAGERSTA D-943.

FAGERSTA FB-M1

Fagersta Bruks Aktiebolag
C 0.78, Cr 3.8, V 1.15, W 1.5, Mo 8.75, bal Fe.
High speed steel for lathe tools, milling cutters, drills, broaches; AISI M1. Molybdenum type. *Obsolete*

FAGERSTA FB-M10

Now FAGERSTA D-960.

FAGERSTA FB-M10

Fagersta Bruks Aktiebolag
C 0.9, Cr 4, V 2, Mo 8, bal Fe.
High speed steel for cutting tools; AISI M10. *Obsolete*

FAGERSTA FB-M2

Now FAGERSTA D-941.

FAGERSTA FB-M2

Fagersta Bruks Aktiebolag
C 0.83, Cr 4.15, V 1.9, W 6.5, Mo 5, bal Fe.
High speed steel, molybdenum-tungsten type, for general purpose cutting tools, as drills, lathe tools, milling cutters; AISI M2. *Obsolete*

FAGERSTA FB-M3 (CLASS 2)

Fagersta Bruks Aktiebolag
C 1.2, Cr 4, V 3, W 6, Mo 6, bal Fe.
High speed steel for cutting tools, molybdenum-tungsten type; for lathe tools, gun drills, reamers, thread chasers; high hardness. AISI M3 Class 2. *Obsolete*

FAGERSTA FB-M7

Now FAGERSTA D-954.

FAGERSTA FB-M7

Fagersta Bruks Aktiebolag
C 1, Cr 3.75, V 2.05, W 1.75, Mo 8.75, bal Fe.
High speed steel for cutting tools; AISI M7. *Obsolete*

FAGERSTA FB-O1

Fagersta Bruks Aktiebolag
C 0.95, Cr 0.5, W 0.5, bal Fe.
Heat treated: 250,000 TS; 225,000 YS; 8 El; C 50 Rock. For blanking and forming dies, punches, reamers, plug and ring gauges, taps. Oil hardening tool steel. AISI Type O1, shock resisting. *Obsolete*

FAGERSTA FB-S1

Fagersta Bruks Aktiebolag
C 0.55, W 2, Cr 1.7, V 0.25, bal Fe.
Heat treated: 295,000 TS; 258,000 YS; 4 El; C 54 Rock. For chisels, punches, shear blades, bolt heading dies, die casting dies. Shock resisting tool steel. AISI Type S1, tough. *Obsolete*

FAGERSTA FB-S1 MIRYCAL

Fagersta Bruks Aktiebolag
C 0.45-0.5, Mn 0.2-0.5, Si 0.75-1, Cr 1.15-1.25, V 0.15-0.25, W 2.25-2.6, bal Fe.
AISI Type S1 shock resisting tool steel. *Obsolete*

FAGERSTA FB-S4

Fagersta Bruks Aktiebolag
C 0.55, Mn 0.8, Si 2, bal Fe.
AISI Type S4 shock resisting tool steel. *Obsolete*

FAGERSTA FB-S5 TUFCUT

Fagersta Bruks Aktiebolag
C 0.5-0.55, Mn 0.6-0.8, Si 1.3-1.6, Mo 0.3-0.4, bal Fe.
AISI Type S5 shock resisting tool steel. *Obsolete*

FAGERSTA FB-S7 TUF DIE

Fagersta Bruks Aktiebolag
C 0.5, Mn 0.7, Si 0.25, Cr 3.25, Mo 1.4, bal Fe.
Air or oil hardening, shock resisting tool steel; AISI S7. *Obsolete*

FAGERSTA FB-T1

Now FAGERSTA D-921.

FAGERSTA FB-T1
Fagersta Bruks Aktiebolag
C 0.7-0.75, Cr 3.75-4.25, V 1.05, W 18, bal Fe.
High speed steel for general purpose machining; AISI T1.
Obsolete

FAGERSTA FB-T4
Fagersta Bruks Aktiebolag
C 0.68-0.75, Cr 4, V 1.05, W 17.25, Mo 0.5, Co 4.5, bal Fe.
High speed steel for cutting tools for machining tough
metals; AISI T4. *Obsolete*

FAGERSTA FB-W1
Fagersta Stainless AB
Tool material. C 1.05, Mn 0.25, Si 0.25, bal Fe.
Heat treated: 208,000 TS; 146,000 YS; 12 El; 460 Brin. For
cold working tools, dies, blanking and coining dies, knurling
and threading dies, punches, stamps, taps, reamers. AISI
Type W1, water hardened. *Obsolete*

FAGERSTA FB-W1 COLD HEADER DIE
STEEL
Fagersta Bruks Aktiebolag
Composition by arrangement. AISI Type W1 water hardening
tool steel. *Obsolete*

FAGERSTA FB-W1 EXTRA CARBON
Fagersta Bruks Aktiebolag
C 1-1.1, Mn 0.25-0.35, Si 0.25-0.35, bal Fe.
AISI Type W1 water hardening tool steel. *Obsolete*

FAGERSTA FB-W1 SPECIAL
Fagersta Stainless AB
Tool material. C 1-1.2, Mn 0.25-0.35, Si 0.2-0.3, bal Fe.
AISI W1; water hardened tool steel. *Obsolete*

FAGERSTA FB-W1 SPECIAL CARBON
Fagersta Stainless AB
Tool material. C 1-1.2, Mn 0.25-0.35, Si 0.2-0.3, bal Fe.
AISI W1; water hardened tool steel. *Obsolete*

FAGERSTA FB-W2
Fagersta Bruks Aktiebolag
C 1.05, Mn 0.25, Si 0.25, V 0.2, bal Fe.
Heat treated: 210,000 TS; 148,000 YS; 10 El; 470 Brin. For
cold working tools, dies, blanking and coining dies, knurling
and threading dies, punches, stamps, taps, reamers. AISI
Type W2, water hardening. *Obsolete*

FAGERSTA HACK SAW
Brukskoncernen AB
C, W, bal Fe.
For hacksaws; water hardening.

FAGERSTA HACK SAW
Fagersta Stainless AB
C, W, bal Fe.
For hack saws, water hardened. *Obsolete*

FAGERSTA HIGH PRODUCTION
Brukskoncernen AB
C, Cr, Mo, V, bal Fe.
For production dies and tools; non-deforming. *Obsolete*

FAGERSTA HIGH PRODUCTION
Fagersta Bruks Aktiebolag
C, Cr, Mo, V, bal Fe.
For production dies and tools; non-deforming. *Obsolete*

FAGERSTA HOLLOW DRILL
Achorn Steel Co.
C 0.8, Mn 0.3, bal Fe.
For hollow drills; water hardened.

FAGERSTA HOLLOW DRILL
Brukskoncernen AB
C 0.8, Mn 0.3, bal Fe.
For drills for mining and quarrying; water hardening.
Obsolete

FAGERSTA HOLLOW DRILL
Fagersta Bruks Aktiebolag
C 0.8, Mn 0.3, bal Fe.
For drills for mining and quarrying; water hardening.
Obsolete

FAGERSTA HOT DIE
Brukskoncernen AB
C, Cr, W, bal Fe.
For hot work dies; hot work steel.

FAGERSTA HOT DIE
Fagersta Stainless AB
C, Cr, W, bal Fe.
For hot work dies; hot work steel. *Obsolete*

FAGERSTA HX
Fagersta Stainless AB
Tool material.
Sintered carbide for turning difficult metals. *Obsolete*

FAGERSTA K291
Fagersta Bruks Aktiebolag
C 0.12, Cr 0.75, Ni 3, bal Fe.
For gears, camshafts, crankshafts; case hardened, shock
resistant. *Obsolete*

FAGERSTA K333
Fagersta Bruks Aktiebolag
C 0.15, Cr 0.65, Ni 1.25, bal Fe.
For gears, camshafts, crankshafts; case hardened, shock
resistant. *Obsolete*

FAGERSTA K336
Fagersta Bruks Aktiebolag
C 0.18, Cr 0.75, Ni 3, bal Fe.
For gears, camshafts, crankshafts; case hardened, heavy
duty. *Obsolete*

FAGERSTA K669
Fagersta Bruks Aktiebolag
C 0.3, Cr 1.25, Ni 4.25, bal Fe.
For gears, shafts; oil hardened, tough. *Obsolete*

FAGERSTA K825
Fagersta Bruks Aktiebolag
C 0.4, Cr 1.4, Ni 1.25, Mo 0.2, bal Fe.
Heat treated: 165,000 TS; 145,000 YS; 14 El; 350 Brin. For
gears, shafts, connecting rods, bolts, fasteners, hardware.
Shock resistant, oil hardening. *Obsolete*

FAGERSTA K845
Fagersta Bruks Aktiebolag
C 0.35, Cr 1.15, Ni 2.6, bal Fe.
For gears, shafts, crankshafts; oil hardened, tough. *Obsolete*

FAGERSTA L441
Fagersta Bruks Aktiebolag
C 0.2, Mn 0.55, Ni 1.8, Mo 0.25, bal Fe.
For gears, shafts; case hardened. *Obsolete*

FAGERSTA L536
Fagersta Bruks Aktiebolag
C 0.3, Cr 1.05, Ni 3.25, Mo 0.25, bal Fe.
Hardened: 213,000 min TS; 170,000 min YS; 8 min El; 25 min
RA; 440-510 Brin. Oil hardening steel for gears, shafts,
machinery parts. *Obsolete*

FAGERSTA L97
Fagersta Bruks Aktiebolag
C 0.55, Ni 3, Mo 0.3, Mn 0.4, Cr 1, bal Fe.
For drop forgings dies, cold work tools; oil hardened, shock
resistant. *Obsolete*

FAGERSTA NON-DEFORMING
Achorn Steel Co.
C 0.95, Cr 0.45, Mn 1.05, W 0.45, bal Fe.
For tools, dies; non-deforming.

FAGERSTA OIL HARDENING
Brukskoncernen AB
C, Cr, W, bal Fe.
For blanking, stamping and forming dies and tools; non-
deforming.

FAGERSTA OIL HARDENING
Fagersta Stainless AB
C, Cr, W, bal Fe.
For blanking, stamping and forming dies and tools; non-
deforming. *Obsolete*

FAGERSTA OVERCOAT AXE
Achorn Steel Co.
C 1, Mn 0.25, bal Fe.
For axes, tools, water hardened. *Obsolete*

FAGERSTA OVERCOAT AXE
Brukskoncernen AB
C 1, Mn 0.25, bal Fe.
For axes; water hardening.

FAGERSTA OVERCOAT AXE
Fagersta Inc.
C 1, Mn 0.25, bal Fe.
For axes; water hardening.

FAGERSTA P10
Fagersta Bruks Aktiebolag
C 0.7, Cr 4.5, W 18.5, V 1.2, Co 0-0.6, bal Fe.
For twist drills, hobs, cutters, reamers; high speed steel.
Obsolete

FAGERSTA P15
Fagersta Bruks Aktiebolag
C 0.8, Cr 4.5, W 18.5, Co 2.5, V 1.6, bal Fe.
For planing and shaping tools, cutters, reamers; high speed
steel. *Obsolete*

FAGERSTA PAVEMENT BREAKER
Achorn Steel Co.
C 0.65, Mn 0.3, bal Fe.
For pavement breaking tools; water hardened.

FAGERSTA PAVEMENT BREAKER
Brukskoncernen AB
C 0.65, Mn 0.3, bal Fe.
For spades, moil points for concrete breaking; water
hardening. *Obsolete*

FAGERSTA PAVEMENT BREAKER
Fagersta Bruks Aktiebolag
C 0.65, Mn 0.3, bal Fe.
For spades, moil points for concrete breaking; water
hardening. *Obsolete*

FAGERSTA POLHEIM WIRE
Brukskoncernen AB
C 1.8, Cr 2, Mn 2, W 13, bal Fe.
For wire drawing dies; oil hardened. *Obsolete*

FAGERSTA POLHEM WIRE DRAWING
Achorn Steel Co.
C 1.8, Cr 2, Mn 2, W 13, bal Fe.
For wire drawing dies; oil hardened.

FAGERSTA Q10
Fagersta Bruks Aktiebolag
C 0.8, Mo 1, Cr 4.5, W 18.5, Co 10, V 1.6, bal Fe.
For lathe and planer tools, reamers, broaches, hobs; high
speed steel. *Obsolete*

FAGERSTA Q5
Fagersta Bruks Aktiebolag
C 0.8, Mo 1.2, Cr 4.5, W 18.5, Co 5.5, V 1.6, bal Fe.
For milling cutters, lathe and planer tools; high speed steel.
Obsolete

FAGERSTA QRO-45 HOT WORK
Fagersta Bruks Aktiebolag
C 0.3, Cr 2.8, V 0.5, Mo 2.8, Co 2.8, bal Fe.
Air or oil hardening tool and die steel for hot working
operations. *Obsolete*

FAGERSTA R-250
Fagersta Stainless AB
Stainless steel. C 0-0.06, Cr 17.5, Ni 0-0.5, bal Fe.
Corrosion resistant steel, ferritic. Non-hardenable; magnetic.
AISI 430. *Obsolete*

FAGERSTA R-300
Fagersta Stainless AB
Stainless steel. C 0-0.11, Cr 18, Ni 8.2, bal Fe.
Austenitic; stainless steel; non-magnetic; non-hardenable.
AISI 302. *Obsolete*

FAGERSTA R-326
Fagersta Stainless AB
Stainless steel. C 0-0.06, Cr 20, Ni 10, bal Fe.
Austenitic; stainless steel; non-magnetic; non-hardenable.
AISI 308. *Obsolete*

FAGERSTA R-358
Fagersta Stainless AB
Stainless steel. C 0-0.08, Cr 18, Ni 9.5, Cb, bal Fe.
Stabilized, austenitic; stainless steel for welded assemblies.
Similar to AISI 347. *Obsolete*

FAGERSTA R-360
Fagersta Stainless AB
Stainless steel. C 0-0.03, Cr 18.5, Ni 9.5, bal Fe.
For welded, austenitic; stainless equipment as sinks,
chemical tanks, food and beverage processing equipment.
AISI 304L. *Obsolete*

FAGERSTA R-366
Fagersta Stainless AB
Stainless steel. C 0-0.02, Cr 19.5, Ni 10, bal Fe.
Low carbon grade of AISI 308 for welded equipment, or for
welding 18-8 type stainless. Good corrosion resistance and
usable in food processing equipment. *Obsolete*

FAGERSTA R-380
Fagersta Stainless AB
Stainless steel. C 0-0.05, Cr 18, Ni 9, Mo 0.3, S or Se, bal Fe.
Free machining grade of austenitic, stainless; Type 304. For
screw machine parts. Similar to AISI 303. *Obsolete*

FAGERSTA R-390
Fagersta Stainless AB
Stainless steel. C 0-0.05, Cr 18, Ni 11.5, bal Fe.
Austenitic, stainless steel; low work hardening rate; for cold
heading, severe drawing, spinning and forming. AISI 305.
Obsolete

FAGERSTA R-440
Fagersta Stainless AB
Stainless steel. C 0-0.05, Cr 17, Ni 13, Mo 2.8, bal Fe.
Better corrosion resistance than 304 alloy; for chemical
equipment, food processing. AISI 316. *Obsolete*

FAGERSTA R-460
Fagersta Stainless AB
Stainless steel. C 0-0.25, Cr 17, Ni 13, Mo 2.8, bal Fe.
Low carbon type of AISI 316; particularly for welded
assemblies. AISI 316L. *Obsolete*

FAGERSTA R-470
Fagersta Stainless AB
Stainless steel. C 0-0.02, Cr 18.5, Ni 14, Mo 3.7, bal Fe.
Low carbon grade of AISI 317; particularly for welded
assemblies. AISI 317L. *Obsolete*

FAGERSTA R-575
Fagersta Stainless AB
Stainless steel. C 0-0.08, Cr 18, Ni 9, Cu 3.5, bal Fe.
Austenitic, stainless steel. *Obsolete*

FAGERSTA R-806
Fagersta Stainless AB
Stainless steel. C 0-0.08, Cr 22.5, Ni 13.5, bal Fe.
For high temperature parts and equipment. AISI 309.
Obsolete

FAGERSTA R-820
Fagersta Stainless AB
Stainless steel. C 0.12, Cr 26, Ni 21, bal Fe.
For high temperature parts and equipment. AISI 310.
Obsolete

FAGERSTA R-823
Fagersta Stainless AB
Stainless steel. C 0.1, Cr 23.5, Ni 20, bal Fe.
For elevated temperature operation; annealing and
carburizing boxes; heat treat fixtures. AISI 314. *Obsolete*

FAGERSTA R-860
Fagersta Stainless AB
Stainless steel. C 0-0.08, Cr 19, Ni 35, bal Fe.
For furnace parts, heat treat fixtures, carburizing boxes. AISI
330. *Obsolete*

FAGERSTA R140
Fagersta Bruks Aktiebolag
C 0.1, Cr 13.5, Mo 1.1, bal Fe.
Annealed: 70,000 TS; 35,000 YS; 30 El; B 80 Rock. Cold
drawn: 95,000 TS; 80,000 YS; 15 El; B 92 Rock. For oil
refinery and chemical plant equipment, valve trim, shafts.
Corrosion resistant. *Obsolete*

FAGERSTA R320
Fagersta Stainless AB
Stainless steel. C 0-0.08, Cr 17.2-18.7, Ni 8.5-10, bal Fe.
Annealed: 78,000 TS min; 28,000 YS min; 40 El min; 55 RA
min; 90 Rock B. For equipment in the food and chemical
industries, washing machines, fittings, kitchen sinks. Type
304; stainless, austenitic. *Obsolete*

FAGERSTA R350
Fagersta Bruks Aktiebolag
C 0-0.06, Cr 18-19.5, Ni 8.5-10, Mn 0.8-1.5, Si 0.25-0.75, bal
Fe.
Annealed: 80,000 TS; 35,000 YS; 60 El; B 80 Rock. Cold
drawn: 240,000 TS; 212,000 YS; 5 El. For food, chemical,
textile and plastic industries, storage tanks, vats, containers.
Austenitic, stainless, oxidation resistant. Type 304. *Obsolete*

FAGERSTA R740
Fagersta Bruks Aktiebolag
C 0.3, Cr 14, Mo 0.6, Ni 0.4, bal Fe.
Annealed: 95,000 TS; 50,000 YS; 25 El; B 92 Rock. Heat
treated: 250,000 TS; 215,000 YS; 8 El; C 52 Rock. For cutlery,
surgical instruments, gauges, shafts, hardware, needle
valves. Corrosion resistant, hardenable. *Obsolete*

FAGERSTA REGULAR
Achorn Steel Co.
C 1, Mn 0.35, bal Fe.
For tools, drills, taps; water hardened.

FAGERSTA RO 8155 HOT WORK
Fagersta Bruks Aktiebolag
C 0.4, Cr 3, Mo 0.5, bal Fe.
Air or oil hardening tool and die steel for hot working
operations. *Obsolete*

FAGERSTA ROLLED AUGER
Achorn Steel Co.
C 0.85, Mn 0.3, bal Fe.
For augers, drills, punches; water hardened.

FAGERSTA RRJ10
Fagersta Bruks Aktiebolag
C 0-0.08, Cr 11.5-13, Al 0.1-0.3, bal Fe.
Annealed: 71,000 TS; 42,600 YS; 22 El; 70 RA; 150 Brin. Heat
treated: 175,000 TS; 145,000 YS; 21 El; 64 RA; 352 Brin. For
oil refinery and chemical plant equipment; Type 405;
corrosion resistant. *Obsolete*

FAGERSTA RRJ10
Fagersta Bruks Aktiebolag
C 0-0.08, Cr 13, bal Fe.
Annealed: 75,000 TS; 40,000 YS; 35 El; 70 RA; 155 Brin. For
turbine blades, valves, cutlery; Type 403; corrosion resistant.
Obsolete

FAGERSTA RRJ11
Fagersta Bruks Aktiebolag
C 0.09-0.12, Cr 13, Si 0-0.5, bal Fe.
Annealed: 75,000 TS; 40,000 YS; 35 El; 70 RA; 155 Brin. For
turbine blades, valves, cutlery, surgical instruments; Type 403
and 410; corrosion resistant. *Obsolete*

FAGERSTA RRJ11
Kay-Brunner Steel Products Inc.
C 0-0.15, Cr 11.5-13.5, Mn 0-1, Si 0-1, bal Fe.
Annealed: 80,000 TS; 40,000 YS; 35 El; 70 RA; 155 Brin. Cold
drawn: 100,000 TS; 85,000 YS; 17 El; 60 RA; 205 Brin. For
valve parts, turbine blades, cutlery, knives; Type 410,
corrosion resistant. *Obsolete*

FAGERSTA RRJ14
Fagersta Bruks Aktiebolag
C 0.08, Cr 13, Ni 2, Mo 1, bal Fe.
For oil refinery equipment; corrosion resistant. *Obsolete*

FAGERSTA RRM20
Now FAGERSTA R-250.

FAGERSTA RRM20
Fagersta Bruks Aktiebolag
C 0-0.12, Cr 14-18, bal Fe.
Annealed: 70,000 TS; 40,000 YS; 30 El; 55 RA; 150 Brin. Cold
drawn: 130,000 TS; 120,000 YS; 2 El; 185 Brin. For oil refinery
equipment; bolts, kitchen sinks; Type 430; stainless, ferritic.
Obsolete

FAGERSTA RRM22
Fagersta Bruks Aktiebolag
C 0.2, Cr 15-17, Ni 1.25-2.5, bal Fe.
Heat treated: 100,000-135,000 TS; 78,000-112,000 YS; 15-25
El; 45-60 RA; 240-280 Brin. For pump spindles, propeller
shafts, aircraft structures; Type 431; stainless, hardenable.
Obsolete

FAGERSTA RRM23
Fagersta Bruks Aktiebolag
C 0.1, Cr 17, Ni 8, Mo, bal Fe.
Annealed: 80,000 TS; 35,000 YS; 60 El; B 80 Rock. For
chemical plant equipment, evaporators, digesters, acid tanks.
Stainless steel, acid resistant, austenitic. *Obsolete*

FAGERSTA RRM24
Fagersta Bruks Aktiebolag
C 0-0.1, Cr 17, Mo 1.5, bal Fe.
Annealed: 80,000 TS; 50,000 YS; 25 El; 50 RA; 150 Brin. For
oil refinery equipment, oil burners and heaters; Type 430Mo;
corrosion resistant. *Obsolete*

FAGERSTA RRM28
Fagersta Bruks Aktiebolag
C 0-0.12, Cr 17, S, Se, bal Fe,.
Annealed: 80,000 TS; 50,000 YS; 25 El; 50 RA; 150 Brin. For
screw machine products, shafts; Type 430F; corrosion
resistant, free-cutting. *Obsolete*

FAGERSTA RRNJ30
Now FAGERSTA R-320.

FAGERSTA RRNJ30
Fagersta Bruks Aktiebolag
C 0.06, Cr 18, Ni 8, bal Fe.
Annealed: 80,000 TS; 35,000 YS; 55 El; 75 RA; 150 Brin. For
chemical plant equipment, tanks, mixers; Type 304;
austenitic, stainless. *Obsolete*

FAGERSTA RRNJ30.32
Fagersta Bruks Aktiebolag
C 0.08-0.2, Mn 0-2, Cr 17-19, Ni 8-10, bal Fe.
Annealed: 85,000 TS; 35,000 YS; 60 El; 70 RA; 150 Brin. Cold drawn: 125,000 TS; 95,000 YS; 22 El; 55 RA; 277 Brin. For oil refinery chemical plant equipment; Type 302; stainless, austenitic. *Obsolete*

FAGERSTA RRNJ31
Now FAGERSTA R-320.

FAGERSTA RRNJ31
Fagersta Bruks Aktiebolag
C 0.05, Cr 18, Ni 8, bal Fe.
Annealed: 80,000 TS; 35,000 YS; 55 El; 75 RA; 150 Brin. For chemical plant equipment, tanks, filters, mixers; Type 304; stainless, austenitic. *Obsolete*

FAGERSTA RRNJ32
Now FAGERSTA R-320.

FAGERSTA RRNJ32
Fagersta Bruks Aktiebolag
C 0.06, Cr 18, Ni 8, bal Fe.
Annealed: 80,000 TS; 35,000 YS; 55 El; 75 RA; 150 Brin. For chemical plant equipment, tanks; Type 304; stainless. *Obsolete*

FAGERSTA RRNJ33
Now FAGERSTA R-300.

FAGERSTA RRNJ33
Fagersta Bruks Aktiebolag
C 0.1, Cr 18, Ni 8, bal Fe.
Annealed: 80,000 TS; 35,000 YS; 55 El; 75 RA; 150 Brin. For chemical plant equipment, tanks, mixers; Type 302; stainless, austenitic. *Obsolete*

FAGERSTA RRNJ35
Fagersta Bruks Aktiebolag
C 0.05, Cr 18, Ni 8, bal Fe.
Annealed: 80,000 TS; 35,000 YS; 55 El; 75 RA; 150 Brin. For chemical plant equipment, tanks, mixers; Type 304; stainless, austenitic. *Obsolete*

FAGERSTA RRNJ36
Now FAGERSTA R-360.

FAGERSTA RRNJ36
Fagersta Bruks Aktiebolag
C 0-0.03, Cr 19, Ni 10, bal Fe.
Annealed: 80,000 TS; 30,000 YS; 60 El; B 76 Rock. For welded structures, chemical plant equipment, tanks, evaporators, digesters, agitators. Type 304L stainless steel, austenitic. *Obsolete*

FAGERSTA RRNJ38
Fagersta Bruks Aktiebolag
C 0.08-0.15, Cr 18, Ni 9, P, S, Se, Zr, bal Fe.
Annealed: 80,000 TS; 35,000 YS; 55 El; 75 RA; 150 Brin. For screw machine products, shafts; Type 303; stainless, austenitic. *Obsolete*

FAGERSTA RRNJ39
Now FAGERSTA R-300.

FAGERSTA RRNJ40
Fagersta Bruks Aktiebolag
C 0-0.08, Cr 18, Ni 10, Mo 1.5, bal Fe.
Annealed: 85,000 TS; 35,000 YS; 50 El; 65 RA; 160 Brin. For acid resistant chemical plant equipment, tanks; Type 316; stainless, austenitic. *Obsolete*

FAGERSTA RRNJ41
Fagersta Bruks Aktiebolag
C 0-0.06, Cr 18, Ni 9, Mo 1.5, bal Fe.
Annealed: 85,000 TS; 35,000 YS; 50 El; 65 RA; 160 Brin. For acid resistant chemical plant equipment, tanks; Type 316; stainless, austenitic. *Obsolete*

FAGERSTA RRNJ42
Fagersta Bruks Aktiebolag
C 0-0.08, Cr 17, Ni 12, bal Fe.
Annealed: 80,000 TS; 32,000 YS; 55 El; B 80 Rock. For chemical plant equipment, evaporators, digesters, acid tanks. Stainless steel; austenitic. *Obsolete*

FAGERSTA RRNJ44
Fagersta Bruks Aktiebolag
C 0-0.1, Cr 16-18, Ni 10-14, Mo 2-3, bal Fe.
Annealed: 85,000-95,000 TS; 35,000-45,000 YS; 50-60 El; 60-75 RA; 150-190 Brin. For acid-resistant chemical plant equipment; Type 316; stainless, austenitic. *Obsolete*

FAGERSTA RRNJ45
Fagersta Bruks Aktiebolag
C 0-0.06, Cr 18, Ni 9, Mo 1.5, bal Fe.
Annealed: 85,000 TS; 35,000 YS; 50 El; 65 RA; 160 Brin. For acid resistant chemical plant equipment, tanks; Type 316; stainless, austenitic. *Obsolete*

FAGERSTA RRNJ46
Fagersta Bruks Aktiebolag
C 0.03, Cr 17, Ni 12, bal Fe.
Annealed: 78,000 TS; 30,000 YS; 60 El; B 78 Rock. For chemical plant equipment, tanks, evaporators, digesters, mixers. Stainless steel, austenitic. *Obsolete*

FAGERSTA RRNJ47
Fagersta Bruks Aktiebolag
C 0.1, Cr 18, Ni 15, bal Fe.
Annealed: 85,000 TS; 36,000 YS; 55 El; B 84 Rock. For chemical and textile plant equipment, evaporators. Stainless steel, austenitic. Heat and corrosion resistant. *Obsolete*

FAGERSTA RRNJ50
Fagersta Bruks Aktiebolag
C 0.06, Cr 18, Ni 12, Cb 0.6, bal Fe.
Annealed: 88,000 TS; 38,000 YS; 55 El; B 82 Rock. For welded structures, chemical plant equipment, evaporators, mixers, digesters. Type 347 stainless steel, austenitic, stabilized, welding grade. *Obsolete*

FAGERSTA RRNJ51
Fagersta Bruks Aktiebolag
C 0-0.08, Cr 17-19, Ni 8-11, Ti = 5 x C, bal Fe.
Normalized: 93,000 TS; 36,000 YS; 45 El; 60 RA; 165 Brin.
Annealed: 87,000 TS; 33,000 YS; 57 El; 73 RA; 155 Brin. For chemical plant equipment; Type 321; stainless, austenitic. *Obsolete*

FAGERSTA RRNJ52
Fagersta Bruks Aktiebolag
C 0.12, Cr 18, Ni 5, Mn 9, bal Fe.
Annealed: 100,000 TS; 50,000 YS; 60 El; B 90 Rock. For pumps, valves, valve bodies, food processing equipment, digesters, hardware, spray nozzles. Type 202 and 204 stainless steel non-hardenable. *Obsolete*

FAGERSTA RRNJ59
Fagersta Bruks Aktiebolag
C 0.08-0.16, Cr 12, Ni 12, bal Fe.
For valves; corrosion resistant. *Obsolete*

FAGERSTA RRO4
Fagersta Bruks Aktiebolag
C 0-0.1, Cr 5, bal Fe.
Annealed: 70,000 TS; 30,000 YS; 28 El; 65 RA; 160 Brin. For oil refinery equipment; Type 501; creep and heat resistant. *Obsolete*

FAGERSTA RRO6
Fagersta Bruks Aktiebolag
C 0-0.1, Cr 5, bal Fe.
Annealed: 70,000 TS; 30,000 YS; 28 El; 65 RA; 160 Brin. For oil refinery equipment; Type 501; creep and heat resistant. *Obsolete*

FAGERSTA RRS70
Fagersta Bruks Aktiebolag
C 0-0.15, Cr 11.5-13.5, bal Fe.
Heat treated: 120,000-135,000 TS; 110,000-117,000 YS; 15-16 El; 58-63 RA; 220-240 Brin. For cutlery, valves, turbine blades; Type 410; corrosion resistant. *Obsolete*

FAGERSTA RRS70
Fagersta Bruks Aktiebolag
C 0.13-0.18, Cr 13, bal Fe.
Annealed: 80,000 TS; 42,000 YS; 33 El; 65 RA; 165 Brin. For turbine blades, valves, cutlery, surgical instruments; Type 410; corrosion resistant. *Obsolete*

FAGERSTA RRS70(72)
Fagersta Bruks Aktiebolag
Cr 12-14, 0.15% C min, bal Fe.
Annealed: 88,000 TS; 40,000 YS; 32 El; 68 RA; 170 Brin. For cutlery, valve trim, turbine blades; Type 420; stainless, hardenable. *Obsolete*

FAGERSTA RRS71(7.3)
Fagersta Bruks Aktiebolag
Cr 12-14, 0.15% C min, bal Fe.
Annealed: 88,000 TS; 40,000 YS; 32 El; 68 RA; 170 Brin. For cutlery, valve trim, turbine blades; Type 420; stainless, hardenable. *Obsolete*

FAGERSTA RRS72
Fagersta Bruks Aktiebolag
C 0.26-0.37, Cr 13, bal Fe.
Annealed: 95,000 TS; 50,000 YS; 25 El; 55 RA; 195 Brin. For valves, cutlery, surgical and dental instruments; Type 420; corrosion resistant. *Obsolete*

FAGERSTA RRS73
Fagersta Bruks Aktiebolag
C 0.38-0.45, Cr 13, bal Fe.
Annealed: 100,000 TS; 55,000 YS; 20 El; 50 RA; 210 Brin. For valves, cutlery, surgical and dental instruments; Type 420; corrosion resistant. *Obsolete*

FAGERSTA RRS74
Fagersta Bruks Aktiebolag
C 0.2, Cr 13, Mo 1, bal Fe.
For oil refinery and chemical plant equipment; corrosion resistant. *Obsolete*

FAGERSTA RRT80
Fagersta Bruks Aktiebolag
C 0-0.12, Cr 20, Ni 20, bal Fe.
For valves; corrosion and heat resistant. *Obsolete*

FAGERSTA RRT83
Now FAGERSTA R-820.

FAGERSTA RRT83
Fagersta Bruks Aktiebolag
C 0-0.25, Cr 24-26, Ni 19-22, bal Fe.
Annealed: 95,000 TS; 45,000 YS; 50 El; 65 RA; 185 Brin. At 1200 F: 57,000 TS; 22,000 YS; 32 El; 45 RA. For furnace parts, valves, pumps, heat treating boxes; Type 310; austenitic, heat resistant. *Obsolete*

FAGERSTA RRT85
Fagersta Bruks Aktiebolag
C 0.15, Cr 23, Ni 14, S, bal Fe.
Annealed: 90,000 TS; 40,000 YS; 50 El; B 83 Rock. For heat treating boxes, furnaces, gas and oil heater baffles, salt pots. Free-machining. Type 309S stainless steel, austenitic, heat and corrosion resistant. *Obsolete*

FAGERSTA RRV60
Fagersta Bruks Aktiebolag
C 0.35, Cr 23-27, bal Fe.
Annealed: 90,000 TS; 60,000 YS; 20 El; 45 RA; 180 Brin. For furnace parts, heat treating boxes; Type 446; heat resistant. *Obsolete*

FAGERSTA RRV61

Fagersta Bruks Aktiebolag
C 0.35, Cr 23-27, bal Fe.
Annealed: 90,000 TS; 60,000 YS; 20 El; 45 RA; 180 Brin. For furnace parts, heat treating boxes; Type 446; heat resistant. *Obsolete*

FAGERSTA RRV61

Fagersta Bruks Aktiebolag
C 0-0.15, Cr 25, Mo 2.5, Ti 1.8, bal Fe.
For welded furnace parts and equipment; corrosion and heat resistant. *Obsolete*

FAGERSTA RRV62

Fagersta Bruks Aktiebolag
C 0.25, Cr 23-30, Ni 3-5, bal Fe.
Normalized: 93,000 TS; 71,000 YS; 20 El; 48 RA; 210 Brin. For valves, pumps, furnace parts; Type 327; heat resistant. *Obsolete*

FAGERSTA RRV62

Fagersta Bruks Aktiebolag
C 0-0.12, Cr 26, Ni 4, bal Fe.
Cast: 90,000 TS; 65,000 YS; 2 El; 212 Brin. For furnace parts and equipment, heat treating boxes; Type CC-50; corrosion and heat resistant. *Obsolete*

FAGERSTA RRV64

Fagersta Bruks Aktiebolag
C 0.2, Cr 23-28, Ni 2.5-5, Mo 1-2, bal Fe.
Normalized: 103,000 TS; 78,000 YS; 18 El; 45 RA; 235 Brin. Annealed: 95,000 TS; 41,000 YS; 29 El; 60 RA; 225 Brin. For valves, pumps, furnace parts; Type 327; heat resistant. *Obsolete*

FAGERSTA RRV64

Fagersta Bruks Aktiebolag
C 0-0.1, Cr 26, Ni 5, Mo 1.5, bal Fe.
For furnace parts and equipment, salt pots; Type 329; heat resistant. *Obsolete*

FAGERSTA RRV66

Fagersta Bruks Aktiebolag
C 0-0.35, Cr 20, bal Fe.
For cutlery, valves, surgical instruments; Type 442; corrosion resistant. *Obsolete*

FAGERSTA S25M

Fagersta Stainless AB
Tool material.
Sintered carbide cutting tools for milling cast iron and non-ferrous materials. *Obsolete*

FAGERSTA S310

Fagersta Bruks Aktiebolag
C 0.55, Si 1.75, Mn 0.75, bal Fe.
For springs; oil hardened. *Obsolete*

FAGERSTA SHOE DIE

Brukskoncernen AB
C 0.55, Cr 0.65, Mo 0.35, bal Fe.
For cutting dies for leather, paper and rubber; water hardened.

FAGERSTA SHOE DIE

Fagersta Stainless AB
C 0.55, Cr 0.65, Mo 0.35, bal Fe.
For cutting dies for leather, paper and rubber; water hardened. *Obsolete*

FAGERSTA SOLID DRILL

Brukskoncernen AB
C 1, Si 0.3, Mn 0.3, bal Fe.
For drills, chisels, taps; water hardened.

FAGERSTA SOLID DRILL

Fagersta Stainless AB
C 1, Si 0.3, Mn 0.3, bal Fe.
For drills, chisels, taps; water hardened. *Obsolete*

FAGERSTA SPECIAL

Achorn Steel Co.
C 1, Mn 0.35, bal Fe.
For tools, drills, taps; water hardened.

FAGERSTA SSL

Fagersta Stainless AB
Tool material. C 0.8, W 18.5, Mo 1.2, Co 2.5, Cr 4.5, V 1.6, bal Fe.
For cutting and planing tools, milling cutters; high speed steel. *Obsolete*

FAGERSTA TWISTED AUGER

Brukskoncernen AB
C 0.7, Si 0.3, Mn 0.35, bal Fe.
For augers for coal and ore mining; water hardened.

FAGERSTA TWISTED AUGER

Fagersta Stainless AB
C 0.7, Si 0.3, Mn 0.35, bal Fe.
For augers for coal and ore mining; water hardened. *Obsolete*

FAGERSTA W221

Fagersta Bruks Aktiebolag
C 1.2, W 0.55, V 0.1, Mn 0.3, bal Fe.
For twist drills, hobs, taps, reamers; water hardened. *Obsolete*

FAGERSTA W401

Fagersta Bruks Aktiebolag
C 1.1, Cr 0.3, W 1, V 0.1, bal Fe.
For twist drills, reamers, taps; water hardened. *Obsolete*

FAGERSTA W406

Fagersta Bruks Aktiebolag
C 1.4, Cr 0.4, W 5.5, V 0.5, bal Fe.
For finishing cutter; water hardened. *Obsolete*

FAGERSTA WKE-4

Fagersta Stainless AB
Tool material. C 1.25, Cr 4.1, V 3.1, W 9, Mo 3.1, Co 9, bal Fe.
Heat treated: 67 Rock C. For cutters, thread tools, lathe and planer tools. Super high speed steel. *Obsolete*

FAGERSTA WKE-45

Fagersta Stainless AB
Tool material. C 1.4, Cr 4.2, V 3.5, W 9, Mo 3.5, Co 11, bal Fe.
High speed steel for finish machining steel; high hardness, good wear resistance, good red hardness. 68 Rock C. *Obsolete*

FAGERSTA-16

Fagersta Bruks Aktiebolag
C 0.8, Si 0.2, Mn 0.3, bal Fe.
Heat treated: 188,000 TS; 143,000 YS; 12 El; 35 RA; 388 Brin. For clutch discs, girders, rails, tie rods, drills; Type W1; water hardened. *Obsolete*

FAGERSTA-20

Fagersta Bruks Aktiebolag
C 1, Si 0.2, Mn 0.3, bal Fe.
Annealed: 100,000 TS; 53,000 YS; 21 El; 42 RA; 200 Brin. For drills, reamers, hobs, taps; Type W1; water hardened. *Obsolete*

FAGERSTA-24

Fagersta Bruks Aktiebolag
C 1.2, Si 0.2, Mn 0.3, bal Fe.
Heat treated: 200,000 TS; 125,000 YS; 8 El; 28 RA; 400 Brin. For drills, cutters, reamers, broaches; Type W1; water hardened. *Obsolete*

FAHLUN BRILLIANTS (FALUNER DIAMENTEN)

German manufacture
Sn 60, Pb 40.
For ornaments.

FAHR

Manufacturer not listed.
Fe-Ni-Cr.
For heat resistant parts; heat resistant.

FAHRALLOY C

Fahralloy Co.
Ni 56, Cr 14, Mo 17, W 5, bal Fe.
For pumps, valves, plastic and chemical plant equipment. Heat and corrosion resistant.

FAHRALLOY CD

Fahralloy Co.
C 0.25, Si 1, Cr 27-30, Ni 3-6, bal Fe.
Cast: 85,000 TS; 48,000 YS; 18 El; 190 Brin. At 1600°F: 23,000 TS; 18 El. For salt pots, furnaces, sintering bars, cracking equipment, and recuperators. High oxidation resistance in sulfur atmosphere. Corrosion resistant casting alloy. Type HD.

FAHRALLOY CF-4

Fahralloy Co.
C 0.04, Si 1.5, Cr 17-20, Ni 9-12, bal Fe.
Annealed: 79,000 TS; 34,000 YS; 71 El. For computer parts, engine mountings, filter press plates, hardware, oil burner parts, and spray nozzles. Stainless casting alloy, austenitic, non-hardenable.

FAHRALLOY CG-12

Fahralloy Co.
C 0.12, Si 1.5, Cr 20-23, Ni 10-13, bal Fe.
Annealed: 77,000 TS; 36,000 YS; 50 El; 163 Brin. For oil refinery and chemical processing equipment, power plants, pumps, and valve bodies. Heat and corrosion resistant casting alloy; austenitic; non-hardenable.

FAHRALLOY CR-W

Fahralloy Co.
C 2.25, Cr 11, Si 0-1, W 1, bal Fe.
For heat resistant castings. Corrosion and heat resistant.

FAHRALLOY D-1

Fahralloy Co.
C 2.5-3, Cr 10-12, Mo 4-5, Si 0-1, Mn 0-1, bal Fe.
For heat resistant castings. Corrosion and heat resistant.

FAHRALLOY F-0726

Fahralloy Co.
C 0.5, Cr 6-8, Ni 25-27, Si 0-1, Mn 0-1.25, bal Fe.
For heat resistant castings. Corrosion and heat resistant.

FAHRALLOY F-0821

Fahralloy Co.
C 0.5, Cr 7-10, Ni 21-23, Si 0-1.5, Mn 0-1, bal Fe.
For heat resistant castings. Corrosion and heat resistant.

FAHRALLOY F-1

Fahralloy Co.
C 0.35-0.75, Ni 37-41, Cr 17-21, bal Fe.
For heat resistant castings. Heat resistant; Type HU.

FAHRALLOY F-10

Fahralloy Co.
C 0.2-0.5, Cr 23-27, Ni 11-14, bal Fe.
For heat resistant castings. Heat resistant, Type HH.

FAHRALLOY F-10LC

Fahralloy Co.
C 0.2, Cr 22-26, Ni 12-15, Si 2, Mn 1.5, bal Fe.
Annealed: 88,000 TS; 50,000 YS; 38 El; 190 Brin. For chemical plant equipment, valves, pumps, and digester fittings. Type CH 20; stainless; austenitic.

FAHRALLOY F-11

Fahralloy Co.
Cr 14-16, Ni 49-51, Mo 19-21, W, bal Fe.
For heat resistant castings. Corrosion and heat resistant.

FAHRALLOY F-12
Fahralloy Co.
C 1, Cr 28, Ni 10, bal Fe.
For furnace parts. Heat resistant.

FAHRALLOY F-1260
Fahralloy Co.
C 0.35-0.75, Cr 10-14, Ni 58-62, Si 0-2.5, Mn 0-2.
For heat resistant castings. Corrosion and heat resistant;
Type HW.

FAHRALLOY F-1400
Fahralloy Co.
C 0-0.15, Cr 12-14, Ni 0-1.5, bal Fe.
Heat treated: 100,000-216,000 TS; 75,000-173,000 YS; 6-28
El; 9-62 RA; 185-415 Brin. For chemical plant equipment,
heat treating boxes. Corrosion resistant; Type CA-15.

FAHRALLOY F-1535
Fahralloy Co.
C 0.35-0.75, Cr 13-17, Ni 33-37, Si 0-2.5, bal Fe.
Cast: 60,000 TS; 38,000 YS; 20 El; 21 RA; 156 Brin.
Annealed: 84,000 TS; 50,000 YS; 10 El; 18 RA; 200 Brin. For
furnace parts, muffles, heat treat boxes, and pots. Type HT;
heat and corrosion resistant.

FAHRALLOY F-1800
Fahralloy Co.
C 0-0.3, Cr 18-22, Ni 2, Si 1.5, Mn 1, bal Fe.
For castings. Stainless; Type CB-30.

FAHRALLOY F-1808
Fahralloy Co.
C 0-0.2, Cr 18-21, Ni 8-11, Si 2, Mn 1.5, bal Fe.
For castings. Stainless; austenitic; Type CF-20.

FAHRALLOY F-1808CB
Fahralloy Co.
C 0-0.08, Cr 18-21, Ni 9-12, Cb = 8 x C, bal Fe.
Annealed: 85,000 TS; 45,000 YS; 45 El; 165 Brin. For welded
construction and chemical plant equipment. Type 347;
stainless; stabilized.

FAHRALLOY F-1808LC
Fahralloy Co.
C 0-0.08, Cr 18-21, Ni 8-11, Si 2, Mn 1.5, bal Fe.
Annealed: 78,000-85,000 TS; 42,000-45,000 YS; 45-50 El;
155-165 Brin. For valves, pumps, mixers, spray nozzles, and
evaporators. Type CF-8; stainless; austenitic.

FAHRALLOY F-1808MO
Fahralloy Co.
C 0-0.08, Cr 18-21, Ni 9-12, Si 2, Mn 1.5, Mo 2-3, bal Fe.
Annealed: 80,000-88,000 TS; 45,000-48,000 YS; 42-48 El;
160-170 Brin. For valves, pumps, mixers, spray nozzles, and
evaporators. Type CF-8M; stainless; austenitic.

FAHRALLOY F-1824
Fahralloy Co.
C 0.08, Cr 16-19, Ni 21-25, Si 2, Mn 1.5, Mo 2-3, bal Fe.
For heat resistant castings. Corrosion and heat resistant;
Type CN-7.

FAHRALLOY F-2
Fahralloy Co.
C 0.5, Cr 21-24, Ni 3-6, Si 0-1.75, Mn 0-2, bal Fe.
For heat resistant castings. Corrosion and heat resistant.

FAHRALLOY F-2-B (1)
Fahralloy Co.
Cr 22, Ni 9, C, bal Fe.
For castings. Heat resistant.

FAHRALLOY F-2-B (2)
Fahralloy Co.
Cr 16-23, Ni 7-11, 0.8 C min, Si, Al, bal Fe.
For heat and corrosion resistant parts. Heat and corrosion
resistant.

FAHRALLOY F-2210
Fahralloy Co.
Cr 20-23, Ni 9-11, C, Al, bal Fe.
For heat resistant castings. Corrosion and heat resistant.

FAHRALLOY F-2520
Fahralloy Co.
C 0.25, Cr 23-27, Ni 19-22, Si 2, Mn 2, bal Fe.
For heat resistant castings. Corrosion and heat resistant;
Type CK-25.

FAHRALLOY F-2520-HK
Fahralloy Co.
C 0.2-0.6, Cr 24-28, Ni 18-22, Si 0-3, Mn 0-2, bal Fe.
For heat resistant castings. Corrosion and heat resistant;
Type HK.

FAHRALLOY F-2802
Fahralloy Co.
C 0.5, Cr 26-30, Ni 0-4, Si 0-2, bal Fe.
Cast: 70,000-110,000 TS; 65,000-75,000 YS; 2-19 El; 190-223
Brin. For ore roasting furnaces, rabble arms, and dampers.
Type HC; heat resistant.

FAHRALLOY F-2808
Fahralloy Co.
C 0-0.3, Cr 26-30, Ni 8-11, Si 2, Mn 1.5, bal Fe.
Cast: 95,000 TS; 55,000 YS; 16 El; 18 RA; 212 Brin.
Annealed: 90,000 TS; 50,000 YS; 18 El; 20 RA; 197 Brin. For
valves, pumps, and fittings. Type CE-30; corrosion resistant.

FAHRALLOY F-2810
Fahralloy Co.
C 0.2-0.5, Cr 26-30, Ni 8-11, Si 0-2, Mn 0-2, bal Fe.
For heat resistant castings. Corrosion and heat resistant;
Type HE.

FAHRALLOY F-2817
Fahralloy Co.
C 0.25, Cr 27-30, Ni 16-18, Si 1.5, Mn 1.5, bal Fe.
For heat resistant castings. Corrosion and heat resistant.

FAHRALLOY F-2817-HI
Fahralloy Co.
C 0.2-0.5, Cr 26-30, Ni 14-18, Si 0-2, bal Fe.
Cast: 80,000 TS; 45,000 YS; 12 El; 180 Brin. Aged: 90,000 TS;
65,000 YS; 6 El; 200 Brin. For valves, fittings, pumps, and
furnace parts. Type HI; corrosion and heat resistant.

FAHRALLOY F-3
Fahralloy Co.
C 0.5, Cr 26-30, Ni 4, Si 1.5, Mn 1, bal Fe.
For castings. Heat resistant.

FAHRALLOY F-3020
Fahralloy Co.
C 0.2-0.6, Cr 28-32, Ni 18-22, Si 0-0.3, Mn 0-2, bal Fe.
For heat resistant castings. Corrosion and heat resistant;
Type HL.

FAHRALLOY F-35
Fahralloy Co.
C 1, Cr 33-36, bal Fe.
For heat resistant castings. Heat resistant.

FAHRALLOY F-35 N
Fahralloy Co.
C 2, Cr 33-36, Ni 9-11, Si 0-2, Mn 0-2, bal Fe.
For heat resistant castings. Corrosion and heat resistant.

FAHRALLOY F-5
Fahralloy Co.
C 0.35-0.75, Cr 15-19, Ni 64-68, Si 0-2.5, Mn 0-2.
For heat resistant castings. Corrosion and heat resistant;
Type HX.

FAHRALLOY F-5-B
Fahralloy Co.
C 0.51-0.8, Ni 49-57, Cr 16-23, Mn, Si, bal Fe.
For heat and corrosion resistant parts. Heat and corrosion
resistant.

FAHRALLOY F3X
Fahralloy Co.
C 3.1-3.5, Cr 26.5-28.5, Ni 2.5-4, Si 0-1.25, Mn 0-0.5, bal Fe.
For heat resistant castings. Corrosion and heat resistant; cast
iron.

FAHRALLOY-HB
Fahralloy Co.
C 0.4, Si 1.5, Cr 18-22, Ni 0-2, bal Fe.
Annealed: 95,000 TS; 60,000 YS; 15 El; 195 Brin. For furnace
brackets and hangers, rabble arms, valve parts and bodies.
Heat and corrosion resistant casting alloy. Type CB-30.

FAHRALLOY-HT75
Fahralloy Co.
C 0.75, Si 1.75, Cr 13-17, Ni 33-37, bal Fe.
Cast: 70,000 TS; 40,000 YS; 10 El; 180 Brin. At 1400°F:
35,000 TS; 26,000 YS; 10 El. For air ducts, brazing trays,
cyanide pots, fan blades, and glass molds. Heat and
abrasion resistant casting alloy. Type HT. High fatigue
strength.

FAHRALLOY-HTCB
Fahralloy Co.
C 0.5, Si 1.75, Cr 13-17, Ni 33-37, Cb 1, bal Fe.
Cast: 68,000 TS; 39,000 YS; 12 El; 175 Brin. For glass molds,
gear spacers, heat treating fixtures, and muffles. Heat and
corrosion resistant casting alloy. Type HT. High fatigue
strength.

FAHRALLOY-HU CB
Fahralloy Co.
C 0.5, Si 1.75, Cr 17-21, Ni 37-41, Cb 1, bal Fe.
Cast: 70,000 TS; 40,000 YS; 9 El; 170 Brin. For carburizing
retorts, lead pots, resistor guides, muffles, pouring spouts,
and conveyor screws. Heat and corrosion resistant casting
alloy. Type HU.

FAHRENWALD RESISTING GOLD ALLOY
English manufacture
Au 60-90, Pd 10-40.
For chemical equipment; white gold, high acid resistance.

FAHRIG ANTIFRICTION METAL
English manufacture
Sn 90, Cu 10.
For bearings; antifriction.

FAHRITE ALLOYS
Now OHIOLOY ALLOYS.

FAHRITE C-7
Teledyne Ohiocast
C 0-0.2, Cr 24-27, Ni 10-13, Mn 0-1.25, bal Fe.
Cast: 84,000 TS; 40,000 YS; 25 El; 160 Brin. For heat and
corrosion resisting parts; heat and corrosion resistant.
Obsolete

FAHRITE C-7M
Teledyne Ohiocast
C 0-0.2, Cr 24-27, Ni 10-13, Mo 2.5-3.5, bal Fe.
Cast: 85,000 TS; 42,000 YS; 27 El; 179 Brin. For heat and
corrosion resistant parts; resists most acids and alkalis.
Obsolete

FAHRITE C-8A
Teledyne Ohiocast
C 0.07, Cr 18-21, Ni 8-10, Mn 1, bal Fe.
Heat treated: 80,000-90,000 TS; 36,000-45,000 YS; 50 El;
160-170 Brin. For stainless parts, castings; stainless.
Obsolete

FAHRITE C-8AM
Teledyne Ohiocast
C 0.07, Cr 18-21, Ni 8-10, Mo 2.5-3.5, Mn 1, bal Fe.
Cast. For stainless castings; austenitic. *Obsolete*

FAHRITE N-10

Teledyne Ohiocast
Ni 19-21, Cr 23-27, C, bal Fe.
For heat and corrosion resistant parts; heat and corrosion resistant. *Obsolete*

FAHRITE N-11

Teledyne Ohiocast
Ni 29-32, Cr 8-12, C, bal Fe.
For heat and corrosion resistant parts; heat and corrosion resistant. *Obsolete*

FAHRITE N-17

Teledyne Ohiocast
C 0.2, Cr 16-20, Ni 2, bal Fe.
Cast: 80,000 TS; 30 El; 175 Brin. For stainless castings, valves, pumps; corrosion resistant. *Obsolete*

FAHRITE N-2

Teledyne Ohiocast
C 0.15-0.25, Cr 17-23, Ni 7-10, Mn 0.5, bal Fe.
Cast: 80,000-80,000 TS; 22,000-38,000 YS; 20-30 El; 20-30 RA; 140-180 Brin. For oil cracking units, furnace parts; heat resistant to 1700 F. *Obsolete*

FAHRITE N-31

Teledyne Ohiocast
C 0-0.3, Ni 23-26, Cr 19-23, bal Fe.
For heat resisting castings; heat resisting. *Obsolete*

FAHRITE N-9

Teledyne Ohiocast
Ni 19-21, Cr 28-32, C, bal Fe.
For heat and corrosion resistant parts; heat and corrosion resistant. *Obsolete*

FAHRITE N-9 A

Teledyne Ohiocast
Ni 29-31, Cr 28-32, C, bal Fe.
For heat and corrosion resistant parts; heat and corrosion resistant. *Obsolete*

FAIRLEYS

Jas. Fairley & Sons Ltd.
C 0.7, W 18, Cr 4, V 1, bal Fe.
For high speed tools and cutters. High speed steel.

FAKIR FAS2

Fakirstahl Hoffmanns GmbH & Co. KG
C 0.53, Si 0.9, Mn 0.9, bal Fe.
For gears, bolts, punches, crimpers; water hardened.

FAKIR FC1

Fakirstahl Hoffmanns GmbH & Co. KG
C 1.05, Cr 1, Mn 0.3, bal Fe.
For bearings, bushings, liners; water hardened, wear resistant.

FAKIR FGS

Fakirstahl Hoffmanns GmbH & Co. KG
C 1.05, Cr 1, W 1.15, Mn 0.9, bal Fe.
For bearings, cutters, punches; water hardened, wear resistant.

FAKIR FJS

Fakirstahl Hoffmanns GmbH & Co. KG
C 0.9, Mn 1.9, V 0.1, bal Fe.
For punches, upsetters, blanking dies; oil hardened, nondeforming.

FAKIR FKL

Fakirstahl Hoffmanns GmbH & Co. KG
C 0.55, Cr 1.05, V 0.18, W 1.85, bal Fe.
For cold header and upsetter dies, crimpers; oil hardened, tough.

FAKIR FKO-10

Fakirstahl Hoffmanns GmbH & Co. KG
C 0.78, Co 10, Cr 4.2, Mo 0.9, V 1.0, bal Fe.
For lathe and planer tools, form cutters, hobs; high speed steel, heavy duty.

FAKIR FKP

Fakirstahl Hoffmanns GmbH & Co. KG
C 0.5, Cr 1.05, Ni 3.25, bal Fe.
For gears, bolts, crankshafts, fasteners; oil hardened, shock resistant.

FAKIR FP

Fakirstahl Hoffmanns GmbH & Co. KG
C 0.85, Cr 4.1, Mo 0.85, V 1.6, W 8.7, bal Fe.
For lathe and planer tools, reamers, broaches, taps; high speed steel.

FAKIR FRS27MO

Fakirstahl Hoffmanns GmbH & Co. KG
C 0.95, Cr 4, W, Mo, V, bal Fe.
For form and milling cutters, reamers, hobs; high speed steel.

FAKIR FSC

Fakirstahl Hoffmanns GmbH & Co. KG
C 2.1, Cr 11.5, W 0.7, bal Fe.
For blanking and forming dies, punches; oil hardened, nondeforming.

FAKIR FSD

Fakirstahl Hoffmanns GmbH & Co. KG
C 0.35, V 0.18, Cr 1.05, W 1.85, bal Fe.
For header dies, upsetters, punches, crimpers; oil hardened, tough.

FAKIR FSM

Fakirstahl Hoffmanns GmbH & Co. KG
C 0.35, V 0.18, Cr 1.05, W 1.85, bal Fe.
For header dies, shears, punches, upsetters; oil hardened, tough.

FAKIR FSP

Fakirstahl Hoffmanns GmbH & Co. KG
C 0.3, Co 2, Cr 2.4, V 0.25, W 8.5, bal Fe.
For upsetters, punches, rivet sets; hot work steel, oil hardened.

FAKIR FSS

Fakirstahl Hoffmanns GmbH & Co. KG
C 1.05, Cr 1, W 1.15, Mn 0.9, bal Fe.
For bearings, cutters, bushings; water hardened, wear resistant.

FAKIR FSZ

Fakirstahl Hoffmanns GmbH & Co. KG
C 1.4, Cr 0.3, V 0.1, Mn 0.3, bal Fe.
For engravers' tools, milling cutters, reamers; water hardened.

FAKIR FW2

Fakirstahl Hoffmanns GmbH & Co. KG
C 1.3, Si 0-0.25, Mn 0-0.25, bal Fe.
For cutters, drills; Type W1; water hardened.

FAKIR FW3

Fakirstahl Hoffmanns GmbH & Co. KG
C 1.15, Si 0-0.25, Mn 0-0.25, bal Fe.
For cutters, drills, hobs; Type W1; water hardened.

FAKIR FW4

Fakirstahl Hoffmanns GmbH & Co. KG
C, bal Fe.
For tools, cutters; water hardened.

FAKIR FW5

Fakirstahl Hoffmanns GmbH & Co. KG
C, bal Fe.
For tools, cutters; water hardened.

FAKIR FWG

Fakirstahl Hoffmanns GmbH & Co. KG
C, bal Fe.
For tools, cutters; water hardened.

FAKIR FWM

Fakirstahl Hoffmanns GmbH & Co. KG
C, bal Fe.
For tools, cutters; water hardened.

FAKIR FWN

Fakirstahl Hoffmanns GmbH & Co. KG
C 0.55, Ni, Cr, Mo, V, bal Fe.
For punches, crimpers, upsetters; oil hardened, shock resistant.

FALC223

Krupp Stahl AG
Stainless Steel. C 0-0.03, Cr 21-23, Mo 2.5-3.5, Ni 4.5-6.5, N 0.08-0.2, bal Fe.
Ferritic-austenitic. DIN X2CrNiMoN2253; W. Nr. 1.4462. 450 N/mm^2 YS, 650-900 N/mm^2 TS, 30 El.

FALCON 4

Atlas Specialty Steels
C 0.45, W 2.25, Cr 1.5, V 0.25, bal Fe.
For chisels, pneumatic tools; shock resistant.

FALCON 6

Atlas Specialty Steels
C 0.55, W 2.25, Cr 1.25, V 0.25, bal Fe.
For punches, shear blades, dies; shock resistant.

FALCON EXTRA

Farrelloy Co.
high C, bal Fe.
For chisels, punches; see Seminole.

FALCON-14

Forjas Alavesas S.A.
C 0.14, Mn 0.7, Si 0.25, Ni 1, Cr 1, Mo 0.15, bal Fe.
Ni-Cr-Mo carburizing steel; deep hardening. IHA F-159.

FALCON-15

Forjas Alavesas S.A.
C 0.15, Mn 0.45, Si 0.25, Ni 4, Cr 1, Mo 0.25, bal Fe.
Ni-Cr-Mo carburizing steel; very deep hardening. IHA F-156; BS EN39B.

FALCON-16

Forjas Alavesas S.A.
C 0.15, Mn 0.5, Si 0.25, Ni 3.3, Cr 1, Mo 0.25, bal Fe.
Ni-Cr-Mo carburizing steel; very deep hardening. AFNOR 16NCD13; BS EN36B.

FALCON-18

Forjas Alavesas S.A.
C 0.18, Mn 0.9, Si 0.25, Ni 1, Cr 1, Mo 0.15, bal Fe.
Ni-Cr-Mo carburizing steel. IHA F-158; AFNOR 18NCD6; UNI 19NCD4.

FALCON-19

Forjas Alavesas S.A.
C 0.2, Mn 0.8, Si 0.25, Ni 0.55, Cr 0.5, Mo 0.2, bal Fe.
Ni-Cr-Mo carburizing steel. AFNOR 20NCD2; AISI 8620.

FALCON-20

Forjas Alavesas S.A.
C 0.36, Mn 0.7, Si 0.25, Ni 1.1, Cr 1, Mo 0.25, bal Fe.
Ni-Cr-Mo structural steel; hardenable to about 48-56 Rock C. IHA F-128; DIN 36 CrNiMo 4; BS EN100.

FALCON-25

Forjas Alavesas S.A.
C 0.32, Mn 0.6, Si 0.25, Ni 2.5, Cr 0.7, Mo 0.4, bal Fe.
Ni-Cr-Mo structural steel; deep hardening. IHA F-127; BS En25; UNI 30NCD12.

FALCON-27

Forjas Alavesas S.A.
C 0.4, Mn 0.7, Si 0.25, Ni 0.9, Cr 0.75, Mo 0.25, bal Fe.
Ni-Cr-Mo structural steel; hardenable to about 49-58 Rock C; 170-209 kg/mm^2 TS. CFNIM F-1282/40 NiCrMo4; UNI 38NCD4.

FALCON-29
Forjas Alavesas S.A.
C 0.4, Mn 0.7, Si 0.25, Ni 1.8, Cr 0.8, Mo 0.25, bal Fe.
Ni-Cr-Mo structural steel; hardenable to about 50-59 Rock C;
171-209 kg/mm² TS. CENIM F-1272/40NiCrMo7; AISI 4340.

FALCON-30
Forjas Alavesas S.A.
C 0.33, Mn 0.5, Si 0.25, Ni 4, Cr 1.25, Mo 0.35, bal Fe.
Ni-Cr-Mo structural steel; deep hardening for highly stressed
heavy equipment. IHA F-126; AFNOR 35 NCD16; BS EN30B.

FALCON-40T
Forjas Alavesas S.A.
C 0.4, Mn 0.85, Si 0.2, Ni 0.55, Cr 0.5, Mo 0.2, bal Fe.
Ni-Cr-Mo structural steel. AFNOR 40 NCD2; AISI 8640.

FALK NO. 1 GEARALLOY
Falk Corp.
C 0.27-0.37, Mn 0.7-1, Si 0-0.6, Cr 0.6-0.9, Ni 0-0.9, Mo
0.3-0.4, bal Fe.
Alloy cast steel; hardenable to 335-375 Brin up to 6 in.
sections. For large heavily loaded gearing, flywheels,
construction machinery.

FALK NO. 3 GEARALLOY
Falk Corp.
C 0.3-0.37, Mn 0.7-1, Si 0-0.6, Cr 0.6-0.9, Ni 0-0.9, Mo 0.4-0.5,
bal Fe.
Alloy cast steel; hardenable to 350-390 Brin up to 4 in.
sections. For large, heavily loaded construction machinery as
flywheels, coupling hubs, gearing.

FAMA 100
Swedish manufacture
Ni 24, Al 13, Cu 4, bal Fe.
For permanent magnets, electrical and magnetic equipment;
high permeability.

FAMA 600
Swedish manufacture
Ni 24, Al 13, Cu 4, bal Fe.
For permanent magnets, electrical and magnetic equipment;
high permeability.

FAMA 700
Swedish manufacture
Ni 21, Co 12, Al 10, Cu 6, bal Fe.
For permanent magnets, electrical and magnetic equipment;
high permeability.

FAN BLADES
English manufacture
Cu 61, Zn 37, Pb 1.5.
For fan blades; free-cutting.

FANITE
Fansteel Metals
Zn 55, bal Cu.
For brazing metal. *Obsolete*

FANSTEEL 103 METAL
Fansteel Metals
Miscellaneous nonferrous. Hf 10, Ti 1, bal Cb.
Recrystallized: 62,000 psi TS; 47,000 psi YS; 35 El; 8.8
density. Good fabricability and weldability; moderate strength
and excellent stability at elevated temperatures; can be
coated to provide resistance to oxidizing atmospheres. For
rocket motor skirts. *Obsolete*

FANSTEEL 17
Fansteel Metals
Ni 6, Cu 4, bal W.
Sintered: 90,000 TS; 1 El; 230-255 Brin. For rotors, flywheels,
governors, radiation shields. Machinable. Powder metallurgy.
Obsolete

FANSTEEL 21
Fansteel Metals
W 85-90, Ag 10-15.
Sintered: 75,000 TS. For high temperature rocket hardware.
Silver infiltrated tungsten. High density. 33% electrical
conductivity. *Obsolete*

FANSTEEL 222
Fansteel Metals
W 8.5-9.5, Hf 2-2.8, C 0-0.0135, Cb 0-0.06, bal Ta.
Stress relieved (90% cold worked): 220,000 TS; 190,000 YS.
Recrystallized: 180,000 TS; 170,000 YS; 28 El. For aerospace
equipment, supersonic air and space craft, missile hardware,
liquid metal reactors. Low transition temperature, good creep
properties. Resists extreme environmental conditions.
Obsolete

FANSTEEL 291 METAL
Fansteel Metals
Miscellaneous nonferrous. W 10, Ta 10, bal Cb.
Recrystallized: 66,000 psi TS; 53,000 psi YS; 32 El; 9.6
density. Good fabricability: high strength up to 3000°F; for
rocket nozzle and other aerospace applications. *Obsolete*

FANSTEEL 42
Fansteel Metals
Ti 0.5, Zr 0.08, bal Mo.
Sheet: 120,000-145,000 TS; 105,000-135,000 YS; 7-16 El. At
2400 F: 50,000 TS. For high temperature applications, aircraft
skins and structural parts. High recrystallization temperature,
good hot strength. *Obsolete*

FANSTEEL 60
Fansteel Metals
W 10, Ta 90.
Sheet: 165,000 TS; 165,000 YS; 3 El; Rock B 97. At 1500 F:
104,000 TS; 96,000 YS; 3 El. For chemical equipment to
resist corrosion and high temperatures, missile nozzles,
electronic components. High corrosion resistance. Work
hardens. *Obsolete*

FANSTEEL 60 METAL
Fansteel Metals
Reactive/Refractory. W 10, bal Ta.
Recrystallized: 94,000 psi TS; 81,000 psi YS; 35 El; 78
Rock-30T; 16.9 density. High strength at elevated
temperatures. Aerospace and chemical process applications;
electron beam melted. *Obsolete*

FANSTEEL 601
Fansteel Metals
Cu alloy.
Heat treated: 65,000 TS; 55,000 YS; 20 El; 125 Brin. For
soldering iron tips; electrical conductivity 85%. *Obsolete*

FANSTEEL 602
Fansteel Metals
Cu alloy.
Heat treated: 75,000 TS; 70,000 YS; 15 El; 150 Brin. For
collector rings, circuit breaker parts, torch tips, commutator
segments; heat treatable, electrical conductivity 80%.
Obsolete

FANSTEEL 603
Fansteel Metals
Cu alloy.
Heat treated: 120,000 TS; 100,000 YS; 8 El; 190 Brin. For
springs, slip rings, circuit breakers, switch blades; heat
treatable, electrical conductivity 50%. *Obsolete*

FANSTEEL 606
Fansteel Metals
Cu alloy.
Heat treated: 95,000-115,000 TS; 95,000 YS; 10-15 El; 240
Brin. For springs, vibrators, circuit breakers, switch blades;
heat treatable, corrosion resistant, electrical conductivity 53%.
Obsolete

FANSTEEL 607
Fansteel Metals
Cu alloy.
Heat treated: 175,000-195,000 TS; 170,000 YS; 2-5 El; 420
Brin. For springs, vibrator arms, contact brushes, fuse clips;
heat treatable, electrical conductivity 24%. *Obsolete*

FANSTEEL 61 METAL
Fansteel Metals
Reactive/refractory. W 7.5, bal Ta.
Cold-worked: 165,000 psi TS; 160,000 psi YS; 5 El; 35 Rock
C; 16.8 density. High strength for springs or other elastic
parts for operation under chlorinating or severe acid corrosive
conditions. P/M material; also called TANTALOY and
FANSTEEL TA W ALLOY. *Obsolete*

FANSTEEL 63 METAL
Fansteel Metals
Reactive/refractory. W 2.5, Cb 0.15, bal Ta.
Recrystallized: 56,000 psi TS; 34,000 psi YS; 38 El; 54
Rock-30T; 16.7 density. Excellent fabricability and weldability;
outstanding corrosion resistance; electron beam melted.
Obsolete

FANSTEEL 77
W base alloy.
High density alloy; 90,000 TS; 1 El; 230-255 Brin. For rotors,
flywheels, governor weights, radiation shields; machinable.
Obsolete

FANSTEEL 77 METAL
Fansteel Metals
Reactive/Refractory. W 89, Ni 7, Cu 4.
Sintered: 100,000 psi TS; 2.5 El; 300-370 DPH; 17.1 density.
For rotors, gyroscopes, governors, and radiation shields.
Machinable; excellent resistance to electrical arc erosion in
non-oxidizing atmospheres. *Obsolete*

FANSTEEL 80
Fansteel Metals
Zr, bal Cb.
At 2000 F: 30,000 TS. At 2400 F: 12,000 TS. Rolled: 60,000
TS; 3 El; 300 Brin. For missile and rocket components; high
heat and corrosion resistance. *Obsolete*

FANSTEEL 80 METAL
Fansteel Metals
Miscellaneous nonferrous. Zr 1, bal Cb.
Recrystallized: 42,000 psi TS; 27,000 psi YS; 34 El; 8.6
density. Low neutron absorption cross section; resistance to
liquid metals; fabricated readily at room temperature.
Obsolete

FANSTEEL 82
Fansteel Metals
Ta 33, Zr 0.7, bal Cb.
At 2000 F: 30,000 TS. At 2400 F: 12,000 TS. For missile and
other high temperature applications; heat and oxidation
resistant. *Obsolete*

FANSTEEL 85
Fansteel Metals
Zr 0.6-1.1, Ta 26-29, W 10-12, bal Cb.
At 70 F: 109,000 TS; 92,000 YS; 14 El. At 2000 F: 55,000 TS;
43,000 YS; 20 El. For space vehicles, nuclear reactors. Good
combination of density, strength, and oxidation resistance at
high temperatures. *Obsolete*

FANSTEEL 85 METAL
Fansteel Metals
Miscellaneous nonferrous. W 11, Ta 28, Zr 1, bal Cb.
Recrystallized: 86,000 psi TS; 66,000 psi YS; 30 El; 10.3
density. Exceptional resistance to creep up to 2500°F;
fabricable, and can be coated to provide resistance to
oxidizing environments. *Obsolete*

FANSTEEL 99
Fansteel Metals
W alloy.
For rotors, shields, counterweights; high density. *Obsolete*

FANSTEEL TAW ALLOY
Now FANSTEEL 61 METAL.

FANSTEEL BL2
Fansteel Metals
Ni 1, bal W.
For counterweights, gyros; high density. *Obsolete*

FANSTEEL COLUMBIUM
Fansteel Metals
Miscellaneous nonferrous. Cb, commercially pure, unalloyed. Recrystallized: 38,000 psi TS; 24,000 psi YS; 37 El; 38 Rock-30T; 8.6 density. Good fabricability and weldability; corrosion resistance similar to Ta in many environments; low neutron absorption cross section; electron beam melted. *Obsolete*

FANSTEEL CS FOIL
Fansteel Metals
Reactive/Refractory.
Ta material, special grade. For foil type Ta electric capacitors; provides relatively constant capacitance over a wide range of operating temperatures. *Obsolete*

FANSTEEL NO. 1
Fansteel Metals
Cu alloy.
For soldering tips; for low P/M solders. *Obsolete*

FANSTEEL NO. 2
Fansteel Metals
Cu alloy.
For soldering tips; for heavy duty service. *Obsolete*

FANSTEEL SCB-291
Now FANSTEEL 291 METAL.

FANSTEEL TANTALUM
Fansteel Metals
Reactive/Refractory
Ta, commercially pure, unalloyed. Recrystallized: 40,000 psi TS; 26,000 psi YS; 38 El; 38 Rock-30T; 16.6 density. Excellent fabricability and weldability; corrosion resistance similar to glass; stable, high dielectric oxide useful in capacitors; electron beam melted. *Obsolete*

FANSTEEL TPX
Fansteel Metals
Reactive/Refractory.
Ta material, special grade. For embedded lead wire in solid Ta electrolytic capacitors and for vacuum furnace hardware; provides resistance to embrittlement during sintering. *Obsolete*

FANTASTEC
Eutectic Corp.
Electrode for AC-DC metallic arc super-fast welding of all steels including low, medium and high alloy types; 100,000 psi TS. *Obsolete*

FANWELD
Fansteel Metals
W, Cr, Co, TaC, CbC.
Cast: 530 Brin. For hard facing welding rod, hot work dies; wear resistant, high red-hardness. *Obsolete*

FARCO
Farrelloy Co. (American Solder&Flux)
For solder for aluminum, stainless steel, Monel; corrosion resistant. *Obsolete*

FARM-ALLOY
Resisto-Loy Company, Inc.
C 2, Cr 12, Mo 4, V 1, Ni 3, Mn 2, B 0.5, bal Fe.
Welded: 70,000 TS; 580 Brin. For hard facing electrodes; wear resistant. *Obsolete*

FARMFACE
Abex Corp.
C 4.5, Mn 6, Si 2, Cr 30, bal Fe.
Welded: 600 Brin. For hard facing rod; austenitic, wear resistant.

FARRELL F-85-B
Farrell-Cheek Steel Co.
C 0.25-0.35, Mn 1.2-1.4, Si 0.35-0.6, Mo 0.2-0.3, bal Fe.
Hardened: 120,000 TS; 100,000 YS; 14 El; 30 RA; 240-310 Brin. For gears, pinions, shafts, crankshafts; wear and abrasion resistant. *Obsolete*

FARRELL F-85-BKN
Farrell-Cheek Steel Co.
C 0.33-0.43, Mn 0.6-0.8, Si 0.35-0.6, Mo 0.2-0.3, Ni 1.5-2, Cr 0.6-0.9, bal Fe.
Hardened: 175,000 TS; 150,000 YS; 6 El; 12 RA; 350-400 Brin. For gears, pinions, shafts, crankshafts; air or oil hardening, tough, fatigue resistant. *Obsolete*

FARRELL F-85-HCM
Farrell-Cheek Steel Co.
C 0.33-0.43, Mn 0.65-1, Si 0.35-0.6, Mo 0.15-0.3, Cr 0.8-1, bal Fe.
Normalized: 120,000 TS; 100,000 YS; 16 El; 30 RA; 250-300 Brin. For gears, pinions, shafts, crankshafts; high hardenability, fatigue and abrasion resistant. *Obsolete*

FARRELL F-85-L
Farrell-Cheek Steel Co.
C 0.25-0.35, Mn 1.2-1.4, Si 0.35-0.6, Mo 0.2-0.3, bal Fe.
Normalized: 100,000 TS; 65,000 YS; 18 El; 30 RA; 160-240 Brin. For gears, pinions, shafts, housings; wear and abrasion resistant. *Obsolete*

FARRELL'S "85" ALLOY (A)
Farrell-Cheek Steel Co.
C 0.4, Mo 0.2, Cr 0.8, Mn 0.7, Ni 1.5, bal Fe.
100,000-170,000 TS; 60,000-140,000 YS; 7-21 El; 15-36 RA; 200-400 Brin. For chains, shovels, crusher rolls, gears, car wheels, sprockets, coal cutters; resists wear, shock, and abrasion. *Obsolete*

FARRELL-CHEEK F-85
Farrell-Cheek Steel Co.
C 0.25-0.35, Mn 1.2-1.4, Si 0.35-0.6, Mo 0.2-0.3, bal Fe.
Annealed: 85,000 TS; 53,000 YS; 22 El; 35 RA; 195 Brin. Heat treated: 105,000-150,000 TS; 85,000-125,000 YS; 10-17 El; 25-35 RA; 240-325 Brin. For gears, shafts, housings; wear and abrasion resistant. *Obsolete*

FARRELL-CHEEK F-85-ACM
Farrell-Cheek Steel Co.
C 0.33-0.43, Mn 0.65-1, Mo 0.15-0.3, Cr 0.8-1.1, bal Fe.
Normalized: 100,000 TS; 75,000 YS; 18 El; 35 RA; 225 Brin. For gears, shafts, housings; fatigue and abrasion resistant. *Obsolete*

FARRELL-CHEEK F-85-AKN
Farrell-Cheek Steel Co.
C 0.33-0.43, Ni 1.5-2, Cr 0.6-0.9, Mo 0.2-0.3, bal Fe.
Normalized: 100,000 TS; 75,000 YS; 18 El; 35 RA; 225 Brin. For gears, shafts, housings; fatigue and abrasion resistant. *Obsolete*

FARRELL-CHEEK F85-4135C
Farrell-Cheek Steel Co.
C 0.37, Mn 0.8, Si 0.45, Cr 1, Mo 0.25, bal Fe.
Cast, normalized and tempered: 105,000 psi TS; 85,000 psi YS; 17 El; 210-260 Brin. ASTM A-148 105-85; SAE 0105.

FARRELL-CHEEK F85-4135D
Farrell-Cheek Steel Co.
C 0.37, Mn 0.8, Si 0.45, Cr 1, Mo 0.25, bal Fe.
Cast, quenched and tempered: 105,000 psi TS; 85,000 psi YS; 17 El; 240-290 Brin.

FARRELL-CHEEK F85-4135E
Farrell-Cheek Steel Co.
C 0.37, Mn 0.8, Si 0.45, Cr 1, Mo 0.25, bal Fe.
Cast, quenched and tempered: 105,000 psi TS; 85,000 psi YS; 17 El; 300-350 Brin.

FARRELL-CHEEK F85-4135F
Farrell-Cheek Steel Co.
C 0.37, Mn 0.8, Si 0.45, Cr 1, Mo 0.25, bal Fe.
Cast, quenched and tempered: 105,000 psi TS; 85,000 psi YS; 17 El; 350-400 Brin.

FARRELL-CHEEK F85-4330F
Farrell-Cheek Steel Co.
C 0.37, Mn 0.7, Si 0.45, Ni 1.75, Cr 0.75, Mo 0.25, bal Fe.
Cast, quenched and tempered: 105,000 psi TS; 85,000 psi YS; 17 El; 350-400 Brin.

FARRELL-CHEEK F85-4335C
Farrell-Cheek Steel Co.
C 0.07, Mn 0.7, Si 0.45, Ni 1.75, Cr 0.75, Mo 0.25, bal Fe.
Cast, normalized and tempered: 105,000 psi TS; 85,000 psi YS; 17 El; 210-260 Brin. ASTM A-148 105-85; SAE 0105.

FARRELL-CHEEK F85-4335D
Farrell-Cheek Steel Co.
C 0.37, Mn 0.7, Si 0.45, Ni 1.75, Cr 0.75, Mo 0.25, bal Fe.
Cast, quenched and tempered: 105,000 psi TS; 85,000 psi YS; 17 El; 240-290 Brin.

FARRELL-CHEEK F85-4335E
Farrell-Cheek Steel Co.
C 0.37, Mn 0.7, Si 0.45, Ni 1.75, Cr 0.75, Mo 0.25, bal Fe.
Cast, quenched and tempered: 105,000 psi TS; 85,000 psi YS; 17 El; 300-350 Brin.

FARRELL-CHEEK F85-8620A
Farrell-Cheek Steel Co.
C 0.2, Mn 0.7, Si 0.45, Ni 0.55, Cr 0.5, Mo 0.2, bal Fe.
Cast, normalized and tempered: 85,000 psi TS; 50,000 psi YS; 22 El; 170-220 Brin. AISI 8620.

FARRELL-CHEEK F85-8635A
Farrell-Cheek Steel Co.
C 0.35, Mn 0.7, Si 0.45, Ni 0.55, Cr 0.5, Mo 0.2, bal Fe.
Cast, normalized and tempered: 85,000 psi TS; 50,000 psi YS; 22 El; 170-220 Brin. AISI 8635.

FARRELL-CHEEK F85-8635B
Farrell-Cheek Steel Co.
C 0.35, Mn 0.7, Si 0.45, Ni 0.55, Cr 0.5, Mo 0.2, bal Fe.
Cast, normalized and tempered: 85,000 psi TS; 50,000 psi YS; 22 El; 180-230 Brin.

FARRELL-CHEEK F85-8635C
Farrell-Cheek Steel Co.
C 0.35, Mn 0.7, Si 0.45, Ni 0.55, Cr 0.5, Mo 0.2, bal Fe.
Cast, heat treated: 85,000 psi TS; 50,000 psi YS; 22 El; 210-260 Brin.

FARRELL-CHEEK F85-8635D
Farrell-Cheek Steel Co.
C 0.35, Mn 0.7, Si 0.45, Ni 0.55, Cr 0.5, Mo 0.2, bal Fe.
Cast, heat treated: 85,000 psi TS; 50,000 psi YS; 22 El; 240-290 Brin.

FARRELL-CHEEK F85-8635E
Farrell-Cheek Steel Co.
C 0.35, Mn 0.7, Si 0.45, Ni 0.55, Cr 0.5, Mo 0.2, bal Fe.
Cast, heat treated: 85,000 psi TS; 50,000 psi YS; 22 El; 300-350 Brin.

FARRELL-CHEEK F85LCA
Farrell-Cheek Steel Co.
C 0.24, Mn 1.35, Si 0.45, Mo 0.25, bal Fe.
Cast, normalized and tempered: 85,000 psi TS; 50,000 psi YS; 22 El; 170-220 Brin. ASTM A-148 90-60; SAE 090.

FARRELL-CHEEK F85MMB
Farrell-Cheek Steel Co.
C 0.3, Mn 1.35, Si 0.45, Mo 0.25, bal Fe.
Cast, normalized and tempered: 90,000 psi TS; 60,000 psi YS; 20 El; 180-230 Brin.

FARRELL-CHEEK F85MMC
Farrell-Cheek Steel Co.
C 0.3, Mn 1.35, Si 0.45, Mo 0.25, bal Fe.
Cast, quenched and tempered: 105,000 psi TS; 85,000 psi YS; 17 El; 210-260 Brin.

FARRELL-CHEEK F85MMD
Farrell-Cheek Steel Co.
C 0.3, Mn 1.35, Si 0.45, Mo 0.25, bal Fe.
Cast, quenched and tempered: 120,000 psi TS; 100,000 psi YS; 14 El; 240-290 Brin.

FARRELL-CHEEK F85MME
Farrell-Cheek Steel Co.
C 0.3, Mn 1.35, Si 0.45, Mo 0.25, bal Fe.
Cast, quenched and tempered: 150,000 psi TS; 125,000 psi YS; 10 El; 300-350 Brin.

FARRELL-CHEEK FC 1045 Q
Farrell-Cheek Steel Co.
C 0.45, Mn 0.7, Si 0.45, bal Fe.
Cast, quenched and tempered: 90,000 psi TS; 60,000 psi YS; 20 El; 180-230 Brin. ASTM A-148 90-60; SAE 090.

FARRELL-CHEEK FC-1020
Farrell-Cheek Steel Co.
C 0.15-0.25, Mn 0.65-0.85, Si 0.35-0.6, bal Fe.
Annealed: 60,000 TS; 30,000 YS; 26 El; 38 RA; 120-170 Brin. For machine tool castings, gears, shafts, housings; SAE1020, water hardened. *Obsolete*

FARRELL-CHEEK FC-1022
Farrell-Cheek Steel Co.
C 0.22, Mn 0.7, Si 0.45, bal Fe.
Cast steel, normalized or normalized and tempered; 135-170 Brin. AISI 1022; ASTM A-216 WCA; SAE 0025.

FARRELL-CHEEK FC-1025
Farrell-Cheek Steel Co.
C 0.2-0.3, Mn 0.65-0.85, Si 0.35-0.6, bal Fe.
Annealed: 78,000 TS; 38,000 YS; 24 El; 36 RA; 150 Brin. Heat treated: 80,000 TS; 55,000 YS; 22 El; 35 RA; 200 Brin. For machine tool castings, gears, housings, shafts; SAE 1025, water hardened.

FARRELL-CHEEK FC-1030
Farrell-Cheek Steel Co.
C 0.25-0.35, Mn 0.65-0.85, Si 0.35-0.6, bal Fe.
Annealed: 75,000 TS; 40,000 YS; 24 El; 35 RA; 170 Brin. Heat treated: 90,000 TS; 60,000 YS; 20 El; 40 RA; 210 Brin. For machine tools castings, gears, shafts, housings; SAE 1030, water hardened.

FARRELL-CHEEK FC-1045
Farrell-Cheek Steel Co.
C 0.4-0.5, Mn 0.65-0.85, Si 0.35-0.6, bal Fe.
Annealed: 80,000 TS; 40,000 YS; 18 El; 30 RA; 175 Brin. Normalized: 85,000 TS; 50,000 YS; 18 El; 26 RA; 195 Brin. For machine tool castings, gears, shafts; SAE 1045, water hardened.

FARRELL-CHEEK FC1030Q
Farrell-Cheek Steel Co.
C 0.3, Mn 0.7, Si 0.45, bal Fe.
Cast, quenched and tempered: 80,000 psi TS; 50,000 psi YS; 22 El; 160-210 Brin. ASTM A-148 80-50; SAE 80-50.

FARRELLS HARD EDGE
Farrell-Cheek Steel Co.
C 0.6, Ni 1.5, Cr 0.8, bal Fe.
Heat treated: 650-750 Brin. For crane wheels, rollers, sheaves, sprockets, gears; abrasion resistant. *Obsolete*

FASAL-40
Forjas Alavesas S.A.
C 0.4, Mn 0.6, Si 0.25, Cr 1.5, Mo 0.25, Al 1.1, bal Fe.
Nitriding steel. IHA F-174; AFNOR 40CAD6-12; BS EN41.

FASALOY
Fansteel Metals
Ag, Cu, Co.
For electrical contacts. *Obsolete*

FASALOY 115
Fansteel Metals
Precious metal. Ag 77, Cd 22.6, Ni 0.4.
Annealed: 68 Rock-15T; electrical conductivity 30% IACS. For electrical contacts. *Obsolete*

FASALOY 13
Fansteel Metals
Precious metal. Pt 90, Ru 10.
Annealed: 90 Rock-15T; electrical conductivity 4% IACS. For electrical contacts. *Obsolete*

FASALOY 130
Fansteel Metals
Precious metal. Ag 75, Cu 19.5, Cd 5, Ni 0.5.
Annealed: 78 Rock-15T; electrical conductivity 50% IACS. For electrical contacts. *Obsolete*

FASALOY 136
Fansteel Metals
Precious metal. Ag 90, Ni 10.
Annealed: 56 Rock-15T; electrical conductivity 90% IACS. For electrical contacts. *Obsolete*

FASALOY 137
Fansteel Metals
Precious metal. Ag 85, Ni 15.
Annealed: 60 Rock-15T; electrical conductivity 85% IACS. For electrical contacts. *Obsolete*

FASALOY 138
Fansteel Metals
Precious metal. Ag 80, Ni 20.
Annealed: 61 Rock-15T; electrical conductivity 81% IACS. For electrical contacts. *Obsolete*

FASALOY 139
Fansteel Metals
Precious metal. Ag 50, Ni 40.
Annealed: 75 Rock-15T; electrical conductivity 65% IACS. For electrical contacts. *Obsolete*

FASALOY 14
Fansteel Metals
Precious metal. Pt 90, Ir 10.
Annealed: 67 Rock-15T; electrical conductivity 7% IACS. For electrical contacts. *Obsolete*

FASALOY 142
Fansteel Metals
Precious metal. Ag 90, Fe 10.
Annealed: 55 Rock-15T; electrical conductivity 92% IACS. For electrical contacts. *Obsolete*

FASALOY 16
Fansteel Metals
Precious metal. Pt 85, Ir 15.
Annealed: 88 Rock-15T; electrical conductivity 6% IACS. For electrical contacts. *Obsolete*

FASALOY 19
Fansteel Metals
Precious metal. Pt 73.4, Pd 18.4, Ru 8.2.
Annealed: 90 Rock-15T; electrical conductivity 4% IACS. For electrical contacts. *Obsolete*

FASALOY 24
Fansteel Metals
Precious metal. Ag 10, Au 10.
Annealed: 48 Rock-15T; electrical conductivity 47% IACS. For electrical contacts. *Obsolete*

FASALOY 31
Fansteel Metals
Precious metal. Pt 92, Ru 8.
Annealed: 91 Rock-15T; electrical conductivity 8% IACS. For electrical contacts. *Obsolete*

FASALOY 35
Fansteel Metals
Precious metal. Pd 72, Ag 26.2, Cu 1.4, Ni 0.4.
Annealed: 85 Rock-15T; electrical conductivity 4% IACS. For electrical contacts. *Obsolete*

FASALOY 37
Fansteel Metals
Precious metal. Pd 72, Ag 26, Ni 2.
Annealed: 82 Rock-15T; electrical conductivity 4% IACS. For electrical contacts. *Obsolete*

FASALOY 38
Fansteel Metals
Precious metal. Pd 40, Ag 30, Cu 30.
Annealed: 81 Rock-15T; electrical conductivity 8% IACS. For electrical contacts. *Obsolete*

FASALOY 41
Fansteel Metals
Precious metal. Ag 97, Pd 3.
Annealed: 55 Rock-15T; electrical conductivity 59% IACS. For electrical contacts. *Obsolete*

FASALOY 42
Fansteel Metals
Precious metal. Ag 90, Pd 10.
Annealed: 64 Rock-15T; electrical conductivity 29% IACS. For electrical contacts. *Obsolete*

FASALOY 43
Fansteel Metals
Precious metal. Ag 80, Pd 20.
Annealed: 70 Rock-15T; electrical conductivity 14% IACS. For electrical contacts. *Obsolete*

FASALOY 5
Fansteel Metals
Precious metal. Ag 75, Cu 24.5, Ni 0.5.
Annealed: 78 Rock-15T; electrical conductivity 79% IACS. For electrical contacts. *Obsolete*

FASALOY 51
Fansteel Metals
Precious metal. Ag 97, Pt 3.
Annealed: 56 Rock-15T; electrical conductivity 47% IACS. For electrical contacts. *Obsolete*

FASALOY 7
Fansteel Metals
Precious metal. Ag 85, Cd 15.
Annealed: 56 Rock-15T; electrical conductivity 34% IACS. For electrical contacts. *Obsolete*

FASALOY 72
Fansteel Metals
Precious metal. Au 72, Ag 26.2, Ni 1.8.
Annealed: 61 Rock-15T; electrical conductivity 14% IACS. For electrical contacts. *Obsolete*

FASALOY 72
Fansteel Metals
Au 72, Ag 26.2, Ni 1.8.
For brazing tantalum to steel. Melting range: 1904-1940 F. *Obsolete*

FASALOY 73
Fansteel Metals
Precious metal. Au 68.8, Ag 25.9, Pt 5.3.
Annealed: 74 Rock-15T; electrical conductivity 14% IACS. For electrical contacts. *Obsolete*

FASALOY 99
Fansteel Metals
Precious metal. Ag 99.87, Ge 0.13.
Annealed: 55 Rock-15T; electrical conductivity 80% IACS. For electrical contacts. *Obsolete*

FASALOY 99
Fansteel Metals
Ag alloy.
For contacts. *Obsolete*

FASALOY GAH
Fansteel Metals
Precious metal. Ag 99.58, Ni 0.2, 0.2200 MgO.
Sintered: 60 Rock-30T; electrical conductivity 75% IACS. For electrical contents. *Obsolete*

FASALOY NO. 106
Fansteel Metals
Ag, Cu.
B 82 Rockwell. For electrical contacts, springs. *Obsolete*

FASALOY NO. 12
Fansteel Metals
Pt, Ru.
For electrical contacts. *Obsolete*

FASALOY NO. 136
Fansteel Metals
Ag, Ni.
For electrical contacts. *Obsolete*

FASALOY NO. 138
Fansteel Metals
Ag, Ni.
For electrical contacts. *Obsolete*

FASALOY NO. 139
Fansteel Metals
Ag, Ni.
For electrical contacts. *Obsolete*

FASALOY NO. 14
Fansteel Metals
Pt, Ir.
For electrical contacts. *Obsolete*

FASALOY NO. 24
Fansteel Metals
Ag, Au.
For electrical contacts. *Obsolete*

FASALOY NO. 3
Fansteel Metals
Ag, Cu.
For electrical contacts. *Obsolete*

FASALOY NO. 4
Fansteel Metals
Ag, Cu.
For electrical contacts. *Obsolete*

FASALOY NO. 5
Fansteel Metals
Ag, Cu, Ni.
For electrical contacts. *Obsolete*

FASALOY NO. 6
Fansteel Metals
Ag, Cu.
For contact parts, electrical contacts; high electrical conductivity. *Obsolete*

FASALOY NO. 7
Fansteel Metals
Ag, Cd.
For electrical contacts. *Obsolete*

FASALOY RJA
Fansteel Metals
Precious metal. Ag 90, 10.0 CdO.
Annealed: 50 Rock F; electrical conductivity 79% IACS. *Obsolete*

FASALOY RJC
Fansteel Metals
Precious metal. Ag 90, 10.0 CdO.
Annealed: 55 Rock F; electrical conductivity 79% IACS. For electrical contacts. *Obsolete*

FASALOY ROA
Fansteel Metals
Precious metal. Ag 85, 15.0 CdO.
Annealed: 50 Rock; electrical conductivity 72% IACS. For electrical contacts. *Obsolete*

FASALOY RRA
Fansteel Metals
Precious metal. Ag 82, 18.0 CdO.
Annealed: 50 Rock F; electrical conductivity 67% IACS. For electrical contacts. *Obsolete*

FAST FINISHING
A. Milne & Co.
C 1.25-1.35, W 3.5-5, bal Fe.
For tools, reamers, arbors; oil hardened. *Obsolete*

FAST FINISHING
Carpenter Technology Corp.
C 1.3, W 3.5, bal Fe.
For tools, fast finishing cutters. *Obsolete*

FASTBOR HOLLOW DRILL
Hoyland Steel Co.
C 0.8, bal Fe.
For hollow drill; water hardening. *Obsolete*

FASTBOR HOLLOW DRILL
Great Western Steel Co.
C 0.8, bal Fe.
For hollow drills; tough.

FASTBOR HOLLOW DRILL
Hoyland Steel Co.
C 0.8, bal Fe.
For hollow drills; tough.

FASTELL 01103 (UB)
Fansteel Metals
Precious metal. Ag 90, W 10.
Coined: 65 Rock F; electrical conductivity 92% IACS. For electrical contacts; special compositions also available. *Obsolete*

FASTELL 01153 (UC)
Fansteel Metals
Precious metal. Ag 85, W 15.
Coined: 70 Rock F; electrical conductivity 88% IACS. For electrical contacts. *Obsolete*

FASTELL 01253 (UE)
Fansteel Metals
Precious metal. Ag 75, W 25.
Sintered: electrical conductivity 77% IACS. For electrical contacts. *Obsolete*

FASTELL 01501 (UJ)
Fansteel Metals
Precious metal. Ag 50, W 50.
Sintered: 55 Rock B; electrical conductivity 65% IACS. For electrical contacts. *Obsolete*

FASTELL 01651 (UM)
Fansteel Metals
Reactive/Refractory. W 65, Ag 35.
Sintered: 88 Rock B; electrical conductivity 54% IACS. For electrical contacts. *Obsolete*

FASTELL 01655 (UM-5)
Fansteel Metals
Reactive/Refractory. W 65, Ag 35.
Sintered: 88 Rock F; electrical conductivity 58% IACS. For electrical contacts. *Obsolete*

FASTELL 01751 (UO)
Fansteel Metals
Reactive/Refractory. W 75, Ag 25.
Sintered: 93 Rock B; electrical conductivity 46% IACS. For electrical contacts. *Obsolete*

FASTELL 01755 (UO-5)
Fansteel Metals
Reactive/Refractory. W 75, Ag 25.
Sintered: 96 Rock B; electrical conductivity 48% IACS. For electrical contacts. *Obsolete*

FASTELL 01801 (UP)
Fansteel Metals
Reactive/Refractory. W 80, Ag 20.
Sintered: 100 Rock B; Electrical. conductivity: 42% IACS. For electrical contacts. *Obsolete*

FASTELL 01901 (UR)
Fansteel Metals
Reactive/Refractory. W 90, Ag 10.
Sintered: 100 Rock B; electrical conductivity 32% IACS. For electrical contacts; special compositions also available. *Obsolete*

FASTELL 02253 (NE)
Fansteel Metals
Copper. W 25, Cu 75.
Annealed: 60 Rock F; electrical conductivity 46% IACS. For electrical contacts. *Obsolete*

FASTELL 02501 (NJ)
Fansteel Metals
Copper. W 50, Cu 50.
Sintered: 70 Rock B; electrical conductivity 38% IACS. For electrical contacts. *Obsolete*

FASTELL 02505 (NJ-5)
Fansteel Metals
Copper. W 50, Cu 50.
Sintered: 80 Rock B; electrical conductivity 55% IACS. For electrical contacts. *Obsolete*

FASTELL 02601 (NL)
Fansteel Metals
Reactive/Refractory. W 60, Cu 40.
Sintered: 81 Rock B; electrical conductivity 35% IACS. For electrical contacts. *Obsolete*

FASTELL 02651 (NM)
Fansteel Metals
Reactive/Refractory. W 65, Cu 35.
Sintered: 87 Rock B; electrical conductivity 32% IACS. For electrical contacts. *Obsolete*

FASTELL 02701 (NN)
Manufacturer not listed.
W 70, Cu 30.
Sintered: 91 Rock B; electrical conductivity: 30% IACS. For electrical contacts.

FASTELL 02751 (N)
Fansteel Metals
Reactive/Refractory. W 75, Cu 25.
Sintered: 98 Rock; electrical conductivity 28% IACS. For electrical contacts. *Obsolete*

FASTELL 02785 (NP-5)
Fansteel Metals
Reactive/Refractory. W 78, Cu 22.
Sintered: 100 Rock B; electrical conductivity 44% IACS. For electrical contacts. *Obsolete*

FASTELL 06501 (BJ)
Fansteel Metals
Precious metal. Mo 50, Ag 50.
Sintered: 68 Rock B; electrical conductivity 55% IACS. For electrical contacts. *Obsolete*

FASTELL 06601 (BL)
Fansteel Metals
Reactive/Refractory. Mo 60, Ag 40.
Sintered: 85 Rock B; electrical conductivity 45% IACS. For electrical contacts. *Obsolete*

FASTELL 06651 (E)
Fansteel Metals
Reactive/Refractory. Mo 65, Ag 35.
Sintered: 97 Rock B; electrical conductivity 42% IACS. For electrical contacts. *Obsolete*

FASTELL 07103 (RJ)
Fansteel Metals
Precious metal. Ag 90, 10.0 CdO.
Annealed: 45 Rock F; electrical conductivity 71% IACS. For electrical contacts. *Obsolete*

FASTELL 07153 (RO)
Fansteel Metals
Precious metal. Ag 85, 15.0 CdO.
Annealed: 50 Rock F; electrical conductivity 62% IACS. For electrical contacts. *Obsolete*

FASTELL 07173 (RR)
Fansteel Metals
Precious metal. Ag 82, 18.0 CdO.
Annealed: 55 Rock F: electrical conductivity 59% IACS. *Obsolete*

FASTELL 2900
Fansteel Metals
Cu.
Sintered: 28,000 TS; 8,000 YS; 34 El; 15 Rock F. Coined: 37,000 TS; 30,000 YS; 15 El; 75 Rock F. For electrical parts, contacts, terminals, connectors, conductors, welding electrodes. High conductivity, optimum density. Sintered electrical copper material. *Obsolete*

FASTELL BH
Fansteel Metals
Mo-Ag.
For electrical contacts; sintered. *Obsolete*

FASTELL E
Now FASTELL 06651 (E).

FASTELL E
Fansteel Metals
Ag, Mo.
For electrical contacts; sintered. *Obsolete*

FASTELL N
Now FASTELL 02751 (N).

FASTELL NL
Now FASTELL 02601 (NL).

FASTELL NO. 130
Fansteel Metals
Ag, Cd.
For electrical contacts; formerly No. 3653. *Obsolete*

FASTELL NO. 142
Fansteel Metals
Ag, Ni.
For electrical contacts; formerly No. 2647. *Obsolete*

FASTELL NO. 31
Fansteel Metals
Pd, Ru.
For electrical contacts; formerly No. 928. *Obsolete*

FASTELL NO. 35
Fansteel Metals
Pd, Ag, Ni.
B 85 Rockwell. For electrical contacts; formerly No. 2. *Obsolete*

FASTELL NO. 41
Fansteel Metals
Ag, Pd.
B 80 Rockwell. For electrical contacts; formerly No. 1. *Obsolete*

FASTELL NO. 42
Fansteel Metals
Ag, Pd.
For electrical contacts; formerly No. 1090. *Obsolete*

FASTELL NO. 51
Fansteel Metals
Ag, Pt.
For electrical contacts; formerly No. 10. *Obsolete*

FASTELL NO. 72
Fansteel Metals
Au 72, Ag 26.2, Ni 1.8.
For electrical contacts; formerly No. 11. *Obsolete*

FASTELL UJ-5
Fansteel Metals
Precious metal. Ag 50, W 50.
Sintered: 55 Rock B; electrical conductivity 70% IACS. For electrical contacts. *Obsolete*

FASTELL UM
Now FASTELL 01651 (UM).

FASTELL UP
Now FASTELL 01801 (UP).

FASTELL UR
Now FASTELL 01901 (UR).

FATIGUE-PROOF
LaSalle Steel Co.
C 0.4-0.48, Mn 1.35-1.65, P 0-0.04, S 0.24-0.33, Si 0.15-0.3, bal Fe.
Steel bar. 140,000 psi TS (min); 125,000 psi YS min (2% offset); 10 El (mean); 26 RA (mean); 30 Rock C min; 280 Brin min. 80% of 1212 machining characteristics.

FATIGUE-PROOF
LaSalle Steel Co.
C 0.4-0.48, Mn 1.35-1.65, P 0-0.04, S 0.24-0.33, Si 0.15-0.3, bal Fe.
Cold finished: 140,000 min TS; 125,000 min YS; 8 El; 30 min Rock C; 80% machinability. For studs, shafts, bolts, axles, machined parts.

FATIGUE-PROOF
Steel Co. of Canada Ltd.
C 0.4-0.48, Mn 1.35-1.65, P 0-0.04, S 0.24-0.33, Si 0.15-0.3, bal Fe.
Cold finished: 140,000 min TS; 125,000 min YS; 8 El; 30 min Rock C; 80% machinability. For studs, shafts, bolts, axles, machined parts.

FATIGUE-PROOF
LaSalle Steel Co.
C 0.45, Mn 1.5, S 0.28, Si 0.21, bal Fe.
Heat treated: 140,000-150,000 TS; 135,000 YS; 7 El; 25 RA; 300 Brin. For gears, shafts, axles, machine tool parts; preheat treated, free-cutting. *Obsolete*

FAULTLESS "A" BABBITT
Hoyt Metal Co. of London Ltd.
Sn, Sb, bal Pb.
At 70 F: 21.8 Brin. At 212 F: 8.2 Brin. For bearings, bushings; Babbitt. *Obsolete*

FAULTLESS-A BABBITT
NL Industries
Sn, Sb, Pb, Cu.
Cast: 21.8 Brin. For bearings for saw mills and wood working equipment; Babbitt, resists sudden strain. *Obsolete*

FAVORIT
Bohler Gesellschaft M.B.H.
C 0.9, Mn 1.9, V 0.1, bal Fe.

FAVORITO SC1
Vereinigte Edelstahlwerke
C 0.95, Mn 1.2, Cr 0.5, W 0.5, bal Fe.
For tools, dies, cutters; non-deforming. *Obsolete*

FB-3
Howment Corp. (Carbide Div.)
Sintered carbide tool material. For blanking dies and punches for heavy shock in punching operations. *Obsolete*

FB-4
Howment Corp. (Carbide Div.)
Sintered carbide tool material. For general purposes, blanking dies and punches. *Obsolete*

FB-5
Howment Corp. (Carbide Div.)
Sintered carbide tool material. For blanking dies and punches, slitter knives for thicker material. *Obsolete*

FB-6
Howment Corp. (Carbide Div.)
Sintered carbide tool material. For blanking dies and punches, for punching thin material. *Obsolete*

FB-M1 (AND OTHER "FB" ALLOYS)
Now FAGERSTA FB-M1.

FC 85
Now FARRELL-CHEEK FC85.

FC AIR HARDENING CAST TO SHAPE
Allegheny Ludlum Steel
C 1.5, Mn 0.4, Si 0.4, Cr 12, Ni 0.25, V 0.5, Mo 0.9, Co 0.75, bal Fe.
AISI Type D2 cold work tool steel. *Obsolete*

FC AIR HARDENING FORGING
Allegheny Ludlum Steel
C 1.5, Mn 0.4, Si 0.4, Cr 12, Ni 0.25, V 0.5, Mo 0.9, Co 0.75.
AISI Type D2 cold work tool steel. *Obsolete*

FC CAST TOOL STEEL
Allegheny Ludlum Steel
C 0.9, Mn 0.35, Si 0.3, bal Fe.
AISI Type W1 water hardening tool steel. *Obsolete*

FC NO 55
Allegheny Ludlum Steel
C 1.5, Cr 11.5-13, Mo 0.95, V 0.5, Co 2, Ni 0.4, bal Fe.
For cold work rolls, tools, dies; air hardened, nondeforming. *Obsolete*

FC NO. 1
Allegheny Ludlum Steel
C 0.5, Cr 0.7, Ni 1.6, Mn 0.5, Si 0.5, bal Fe.
For stripper dies, backing blocks, die blocks; shock resistant, oil hardened. *Obsolete*

FC NO. 10
Allegheny Ludlum Steel
C 1-1.2, Cr 1.1-1.3, Mo 0.3-0.5, bal Fe.
For cold rolling dies; oil hardened. *Obsolete*

FC NO. 14
Allegheny Ludlum Steel
C 0.35, Cr 4-5.5, V 0.25, W 3.25-4.25, Si 1.25-1.75, bal Fe.
For hot die inserts, dummy blocks, press forging dies; oil or air hardened, non-deforming. *Obsolete*

FC NO. 19

Allegheny Ludlum Steel
C 0.35, Cr 5, Si 1.5, V 0.3, W 4, bal Fe.
For upset form punches, press forging dies, container liners; hot work steel, oil hardened. *Obsolete*

FC NO. 2

Allegheny Ludlum Steel
C 0.18-0.23, Cr 1.5, Mo 0.5, bal Fe.
For upset insert dies; oil hardened, shock resistant. *Obsolete*

FC NO. 20

Allegheny Ludlum Steel
C 1.4, Si 0.6, W 3.5, V 0.2, bal Fe.
For cold drawing dies; water hardened. *Obsolete*

FC NO. 23

Allegheny Ludlum Steel
C 0.35-0.42, Cr 1-1.5, Si 0.8, W 3.75-4.25, V 0.25, bal Fe.
For extrusion liners and rams; oil hardened, hot work steel. *Obsolete*

FC NO. 38M

Allegheny Ludlum Steel
C 0.9, Mn 0.25, V 0.15-0.25, bal Fe.
For special purpose tools; water hardened. *Obsolete*

FC NO. 5

Allegheny Ludlum Steel
C 1.9-2.2, Cr 11.5-12.5, Mo 0.6-0.9, bal Fe.
For draw dies, forming and blanking dies; non-deforming, oil hardened. *Obsolete*

FC NO. 5XI

Allegheny Ludlum Steel
C 0.3, Cr 5, Si 1.2, Mo 1.4, V 0.3, W 1.3, bal Fe.
For tools, dies, mandrels, extrusion rams; hot work steel, oil hardened. *Obsolete*

FC NO. 66

Allegheny Ludlum Steel
C 1.4-1.6, Cr 11.5-12.5, Mo 1, V 0.6, Co 3.5, Ni 0.4, bal Fe.
For forming and blanking dies, punches; air hardened, non-deforming. *Obsolete*

FC SPEZIAL

Thyssen Edelstahlwerke AG
C, Cr, bal Fe.
For files; water hardening. *Obsolete*

FC-1025

Farrell-Cheek Steel Co.
C 0.2-0.3, Mn 0.7, Si 0.4, S 0.5, P 0.5, bal Fe.
Annealed: 70,000 TS; 38,000 YS; 24 El; 36 RA; 143 Brin. For gears, shafts, machine tool housings; SAE 1020 to 1027 steel castings. *Obsolete*

FC-1025V

Farrell-Cheek Steel Co.
C 0.2-0.3, Mn 0.7, Si 0.4, V 0.1-0.2, bal Fe.
Normalized: 75,000 TS; 48,000 YS; 25 El; 40 RA; 153 Brin.
For gears, cases, machine tool castings; shock resistant castings. *Obsolete*

FC-1040

Farrell-Cheek Steel Co.
C 0.4-0.5, Mn 0.7, Si 0.5, S 0.05, P 0.05, bal Fe.
Annealed: 80,000 TS; 45,000 YS; 17 El; 25 RA; 187 Brin. For gears, shafts, machine tool housings; SAE 1040 to 1046 steel castings. *Obsolete*

FC-1040 N

Farrell-Cheek Steel Co.
C 0.4-0.5, Mn 0.7, Si 0.5, S 0.5, P 0.05, bal Fe.
Normalized: 85,000 TS; 50,000 YS; 18 El; 26 RA; 170 Brin. For gears, shafts, machine tool housings; SAE 1040 to 1046 steel castings. *Obsolete*

FC-1040 Q

Farrell-Cheek Steel Co.
C 0.4-0.5, Mn 0.7, Si 0.5, S 0.05, P 0.05, bal Fe.
Heat treated: 95,000 TS; 65,000 YS; 19 El; 35 RA; 200 Brin.For gears, shafts, machine tool housings; SAE 1040 and 1046 steel castings. *Obsolete*

FC-1084

Allegheny Ludlum Steel
C 0.25-0.35, Mn 0.4, Si 1-1.5, Cr 8-10, Mo 1-1.25, bal Fe.
For oil refinery equipment, stills; resists oil corrosion. *Obsolete*

FC-440C

Allegheny Ludlum Steel
C 0.95-1.2, Cr 16-18, Mo 0.4-0.65, bal Fe.
For surgical instruments, ball bearings, liners; Type 440C; corrosion resistant. *Obsolete*

FC-5XIV

Allegheny Ludlum Steel
C 0.37-0.42, Cr 5, Mo 1.2, V 1, Si 1.2, bal Fe.
For die casting dies, hot work dies; oil or air hardened. *Obsolete*

FC-5X1 SPEC

Allegheny Ludlum Steel
C 0.3-0.35, Cr 2.5-3, W 9-10.5, V 0.25, Mo 0.25, Ni 1.7, bal Fe.
For cast hot work tools; hot work steel, oil hardened. *Obsolete*

FC-85 T

Farrell-Cheek Steel Co.
C 0.2-0.3, Mn 0.7, Si 0.4, S 0.5, P 0.5, bal Fe.
Heat treated: 82,000 TS; 46,000 YS; 21 El; 32 RA; 180 Brin.
For gears, shafts, machine tool housings; heat treated castings. *Obsolete*

FC 85 AKN

Farrell-Cheek Steel Co.
C 0.33-0.4, Mn 0.8, Si 0.4, Mo 0.2-0.3, Ni 1.5-2, Cr 0.45-0.9, bal Fe.
Normalized: 100,000 TS; 75,000 YS; 18 El; 35 RA; 220 Brin.
For gears, shafts, machine tool castings; air hardened, immune to temper brittleness. *Obsolete*

FC-85-AN

Farrell-Cheek Steel Co.
C 0.16-0.23, Mn 0.8, Si 0.4, Ni 1.75-2.5, bal Fe.
Normalized: 80,000 TS; 50,000 YS; 26 El; 55 RA; 160 Brin.
For gears, cases, machine tool castings; shock and fatigue resistant. *Obsolete*

FC-85-AV

Farrell-Cheek Steel Co.
C 0.25-0.35, Mn 1.3, Si 0.4, Mo 0.25, V 0.15, bal Fe.
Normalized: 90,000 TS; 60,000 YS; 22 El; 42 RA; 193 Brin.
For gears, cases, machine tool castings; shock resistant castings. *Obsolete*

FC-85-BN

Farrell-Cheek Steel Co.
C 0.25-0.3, Mn 0.8, Si 0.4, Mo 0.2-0.3, Ni 1.75-2.5, bal Fe.
Heat treated: 125,000 TS; 110,000 YS; 14 El; 32 RA; 275 Brin.
For gears, cases, machine tool castings; low temperature applications, tough. *Obsolete*

FC-85-HN

Farrell-Cheek Steel Co.
C 0.25-0.3, Mn 0.7, Si 0.4, Ni 1.75-2.5, bal Fe.
Normalized: 90,000 TS; 60,000 YS; 25 El; 50 RA; 190 Brin.
For gears, casings, machine tool castings; immune to temper brittleness. *Obsolete*

FC-85A

Farrell-Cheek Steel Co.
C 0.25-0.35, Mn 1.2-1.4, Si 0.4, Mo 0.3, bal Fe.
Normalized: 90,000 TS; 60,000 YS; 20 El; 40 RA; 193 Brin.
For gears, shafts, machine tool housings; tough and wear resistant. *Obsolete*

FC-85AC

Farrell-Cheek Steel Co.
C 0.25-0.35, Mn 1.2-1.4, Si 0.4, Mo 0.3, Cu 1.2, bal Fe.
Normalized: 90,000 TS; 60,000 YS; 22 El; 42 RA; 193 Brin.
For gears, cases, machine tool castings; tough and wear resistant. *Obsolete*

FC-8740

Allegheny Ludlum Steel
C 0.38-0.43, Cr 0.5, Mo 0.3, Ni 0.5, bal Fe.
For cast gears, shafts, machine tools; oil hardened. *Obsolete*

FC-AH

Allegheny Ludlum Steel
C 1.45-1.6, Cr 11.5-12.5, Mo 0.8, V 0.5, Ni 0.2, bal Fe.
For forming and blanking dies; air hardened, non-deforming. *Obsolete*

FC-ALX

Allegheny Ludlum Steel
C 1.75-2, Cr 32-34, W 16-17, Mo 0.7, Mn 0.5, bal Co.
For tools, dies; corrosion and abrasion resistant. *Obsolete*

FC-ALX NO. 6

Allegheny Ludlum Steel
C 1, Cr 33.5, W 6, bal Co.
For high temperature applications; heat resistant. *Obsolete*

FC-CMS

Allegheny Ludlum Steel
C 0.55-0.65, Cr 0.8-1.2, Mn 0.7, Mo 0.4, Ni 0.5, V 0.15, bal Fe.
For forming dies, oil hardened; tough.

FC-CNS

Allegheny Ludlum Steel
C 1.6, Cr 11-13, V 0.5, Mo 0.9, Co 0-0.6, bal Fe.
For blanking and forming dies, castings; air hardened, non-deforming. *Obsolete*

FC-CTS

Allegheny Ludlum Steel
C 0.8-1.1, bal Fe.
For cold work dies, inserts, machinery parts; water hardened. *Obsolete*

FC-CV

Allegheny Ludlum Steel
C 0.45, Cr 1, Mn 0.7, V 0.2, bal Fe.
For machinery parts, dies; oil or water hardened. *Obsolete*

FC-EZ

Allegheny Ludlum Steel
C 1.4, Cr 0.25, Mo 0.25, Si 1.3, Mn 1, bal Fe.
For cutters; oil or water hardened. *Obsolete*

FC-FH

Allegheny Ludlum Steel
C 0.5, Mn 1.1, Cr 1.2, Mo 0.4, V 0.1, bal Fe.
For machinery parts, cold work dies; oil hardened. *Obsolete*

FC-FLAMHARD

Allegheny Ludlum Steel
C 0.5, Mn 1.15, Si 0.5, Cr 1.2, V 0.12, Mo 0.4, bal Fe.
Prehardened: 300-325 Brin. For die casting dies, die holder blocks, frames. Hot work die steel. *Obsolete*

FC-HE

Farrell-Cheek Steel Co.
C 0.25-0.35, Mn 1.2, Si 0.4, Mo 0.2-0.3, bal Fe.
Normalized: 110,000 TS; 90,000 YS; 20 El; 45 RA. For hard edge tools, gears; wear and abrasion resistant.

FC-NCI

Allegheny Ludlum Steel
C 2.75, Cr 1.2, Mo 0.25, Al 1, Si 2.6, bal Fe.
For liners, sleeves, machinery parts; nitriding cast iron. *Obsolete*

FC-ROLOY

Allegheny Ludlum Steel
C 1.5, Cr 12, Mo 1.1, V 0.5, Co 0.7, Ni 0.25, bal Fe.
For cold working rolls, dies, blanking and forming dies; air hardened, non-deforming. *Obsolete*

FC-ROLOY NO. 2

Allegheny Ludlum Steel
C 1, Cr 5.25, Mo 1.15, V 0.5, bal Fe.
For blanking and forming dies, cold work dies; air hardened. *Obsolete*

FC2/0

Sintered Products Ltd.
Iron-copper, sintered. 6.2 g/cm^3 density; 22 kg/mm^2 TS; 70 Vickers. Road transport shock absorber parts.

FC2/63/0

Sintered Products Ltd.
6.2-6.4 g/cm^3 density; 23.5 kg/mm^2 TS. Hardenable to 25 Rock C. Meets BSS A301; MPIE F-0200-P.

FC2/63/1

Sintered Products Ltd.
Cu + C (sintered), bal Fe.
6.2-6.4 g/cm^3 density; 30 kg/mm^2 TS; 1 El; 17-21 porosity. Hardenable to 25 Rock C. Meets BSS A350; SAE 864B.

FC2/63/1/2

Sintered Products Ltd.
Cu + C (sintered), bal Fe.
6.2-6.4 g/cm^3 density; 27.5 kg/mm^2 TS; 1 El; 17-21 porosity. Meets MPIE FC-0210P.

FC3/0

Sintered Products Ltd.
Iron-copper, sintered. 6.2 g/cm^3 density; 25 kg/mm^2 TS; 80 Vickers. Shock absorber parts, valves, pistons, levers, spacers.

FC3/1

Sintered Products Ltd.
Iron-copper, 1% carbon, sintered. 6.2 g/cm^3 density; 37 kg/mm^2 TS; 145 Vickers. Medium duty structural parts.

FC3/58/1

Sintered Products Ltd.
Cu + C (sintered), bal Fe.
5.7-6.0 g/cm^3 density; 27.5 kg/mm^2 TS; 1 El; 24-28 porosity. Hardenable to 22 Rock C. Meets BSS A350; MPIE FC-0310N.

FC3/61/1

Sintered Products Ltd.
Cu + C (sintered), bal Fe.
6.0-6.2 g/cm^3 density; 31.5 kg/mm^2 TS; 1 El; 21-24 porosity. Hardenable to 25 Rock C. Meets BSS A350; MPIE FC-0310N.

FC3/63/0

Sintered Products Ltd.
Cu + C (sintered), bal Fe.
6.2-6.4 g/cm^3 density; 25 kg/mm^2 TS; 1 El; 17-21 porosity. Hardenable to 25 Rock C. Meets BSS A301; MPIE FC-0300P.

FC3/63/1

Sintered Products Ltd.
Cu + C (sintered), bal Fe.
6.2-6.4 g/cm^3 density; 35 kg/mm^2 TS; 1 El; 17-21 porosity. Hardenable to 25 Rock C. Meets BSS A350; MPIE FC-0310P.

FC3/63/1/2

Sintered Products Ltd.
Cu + C (sintered), bal Fe.
6.2-6.4 g/cm^3 density; 31.5 kg/mm^2 TS; 1 El; 17-21 porosity. Meets BMSA 31B; MPIE FC-0305P.

FC3/66/1

Sintered Products Ltd.
Cu + C (sintered), bal Fe.
6.5-6.8 g/cm^3 density; 35 kg/mm^2 TS; 1 El; 13-16 porosity. Hardenable to 30 Rock C. Meets MPIE FC-0310R.

FC7/63/0

Sintered Products Ltd.
Cu + C (sintered), bal Fe.
6.2-6.4 g/cm^3 density; 31 kg/mm^2 TS; 1 El; 17-21 porosity. Hardenable to 25 Rock C. Meets BSS A303; MPIE FC-0700P; SAE 862.

FC7/63/1

Sintered Products Ltd.
Cu + C (sintered), bal Fe.
6.2-6.4 g/cm^3 density; 32.5 kg/mm^2 TS; 1 El; 17-21 porosity. Hardenable to 25 Rock C. Meets BSS A352; MPIE FC-0710P.

FC7/63/1/2

Sintered Products Ltd.
Cu + C (sintered), bal Fe.
6.2-6.4 g/cm^3 density; 31.5 kg/mm^2 TS; 1 El; 17-21 porosity. Meets BMSA 33; MPIE FC-0705P.

FCB(T)

English manufacture
C 0.12, Cr 17.5, Ni 12, Cb 1, bal Fe.
For gas turbine parts; stainless, austenitic, stabilized.

FCC 11

Allegheny Ludlum Steel
C 0.4-0.5, Cr 1.45, Ni 3.5, bal Fe.
For gears, pinions, shafts; tough. *Obsolete*

FCC 168

Allegheny Ludlum Steel
C 0.65, Cr 0.9, Mo 0.8, V 0.1, Ni 1.8, bal Fe.
For hot forming dies; oil and air hardened, hot work steel. *Obsolete*

FCC 16B CAST-TO-SHAPE

Allegheny Ludlum Steel
C 0.65, Mn 0.55, Cr 0.9, Mo 0.8, V 0.1, Si 0.3, Ni 1.8, bal Fe.
For hot forming dies; hot work steel. *Obsolete*

FCC 17

Allegheny Ludlum Steel
C 0.2-0.5, Mn 7-9, Si 0.8-1.4, Ni 19-20, Cr 0.6, bal Fe.
For resistance grids. *Obsolete*

FCC 21

Allegheny Ludlum Steel
C 0.5, Cr 1.4, Si 0.9, W 1.5-2.25, V 0.25, bal Fe.
For extruding dies, dummy blocks, blanking dies; oil hardened. *Obsolete*

FCC 4

Allegheny Ludlum Steel
C 1, W 0.8, bal Fe.
For finishing tools. *Obsolete*

FCC 6

Allegheny Ludlum Steel
C 0.4, Cr 1, Mn 0.7, V 0.2, bal Fe.
For zinc die casting dies; water hardened. *Obsolete*

FCC 7

Allegheny Ludlum Steel
C 0.37-0.42, Cr 4.5-5.5, Si 0.8, Mo 0.5, bal Fe.
For die casting dies for aluminum; oil hardened. *Obsolete*

FCC AIR HARDENING NO. 48M

Allegheny Ludlum Steel
C 1.6, Cr 11.5-12.5, Mo 0.8, V 0.2, Co 0.8, bal Fe.
For forming and blanking dies; air hardened. *Obsolete*

FCC OIL HARDENING

Allegheny Ludlum Steel
C 0.85-0.95, Mn 1-1.2, Cr 0.5, W 0.6, V 0.3, bal Fe.
For forming dies, punches; oil hardened. *Obsolete*

FCC OIL HARDENING

Allegheny Ludlum Steel
C 0.85, Mn 1.2, Cr 0.5, W 0.5, V 0.25, bal Fe.
For cams, dies, gauges, punches, machinery parts; oil hardened. *Obsolete*

FCC-6

Allegheny Ludlum Steel
C 0.37, Mn 0.8, Cr 0.95, V 0.17, Si 0.3, bal Fe.
For die casting dies; oil hardened. *Obsolete*

FCC-R3

Allegheny Ludlum Steel
C 0.8, Cr 1, Mo 0.4, V 0.1, Ni 0.45, bal Fe.
For forming and swaging rolls; oil hardened. *Obsolete*

FCC-TMS

Allegheny Ludlum Steel
C 0.45-0.55, Cr 1.6, Ni 2.7, bal Fe.
For forming and hot work dies; oil hardened. *Obsolete*

FCC-TS

Allegheny Ludlum Steel
C 0.8-1, Mn 0.5, Si 0.3, bal Fe.
For inserts, dies, pressure pads. *Obsolete*

FCF SPECIAL

Thyssen Edelstahlwerke AG
C, Cr, W, bal Fe.
For files; water hardening. *Obsolete*

FCHD

Kapfenberg
C 0-0.14, Mn 17-19, Cr 10-13, Si 0.5-2, 0.3-0.8 Ti or 0.1-0.3 N, bal Fe.
For turbine blades; high heat resistant.

FCI STAINLESS IRON

Firth-Vickers Stainless Steels Ltd.
C 0.12, Si 0.6, Mn 1.2, Cr 13.5, Mo 0.28, S 0.23, bal Fe.
Heat treated: 85,000 TS; 58,000 YS; 28 El; 62 RA; 180 Brin.
For hardware, screw machine products; free-cutting, corrosion resistant.

FCW SPECIAL

Thyssen Edelstahlwerke AG
C, Cr, W, bal Fe.
For files; water hardening. *Obsolete*

FD-3

Howment Corp. (Carbide Div.)
Sintered carbide tool material. For draw dies, for drawing wire, tube and bar stock. *Obsolete*

FD-4

Howment Corp. (Carbide Div.)
Sintered carbide tool material. For draw dies, for drawing wire, tube and bar stock. *Obsolete*

FD-5

Howment Corp. (Carbide Div.)
Sintered carbide tool material. For draw dies, for drawing wire, tube and bar stock. *Obsolete*

FD65 STEEL

English manufacture
C 0.25, Mn 0.5, Ni 3, Cr 1.2, Mo 0.45, V 0.2, bal Fe.
Hardened: 150,000 TS; 20 El; 65 RA; 320 Brin. For aircraft engine parts; case hardened.

FE5

Teledyne Firth Sterling
Tool material.
Cubic boron nitride for machining ferrous materials of varying hardness; 45-65 Rock C.

FE74
Sintered Products Ltd.
Fe, sintered.
7.2-7.9 g/cm³ density; 4-8 porosity. High purity iron; magnetic properties. Hardenable and case hardenable. Meets BSS A203; MPIE F-0000-T.

FE88
Sintered Products Ltd.
Sintered iron with 1% carbon. 6.6 g/cm³ density; 20 kg/mm² TS; 6-12% El; 70 Vickers.

FEAL
French manufacture
Al 25, bal Fe.
For atomic reactors; outstanding resistance to oxidation and excellent strength at high temperatures, particularly transparent to neutrons.

FECHRAL
Russian manufacture
C 0.06-0.15, Cr 17-25, Al 4-7, bal Fe.
For resistance wire; heat resistant.

FECRALOY
Wilbur B. Driver Co.
Stainless steel. C 0.2, Cr 13, Al 4, bal Fe.
Rolled: 128,000 TS; 104,000 YS; 5 El. For rheostats, heating elements; high heat resistance.

FEDERAL HIGH SPEED
Swedish Iron & Steel Corp.
C 0.7, W 18, Cr 4, V 1, bal Fe.
For high speed cutting tools; high speed steel. *Obsolete*

FEDERAL-MOGUL B-10025
Federal-Mogul Corp.
Cu 3-5, Pb 0-0.25, Sb 7.3-8, bal Sn.
Cast: 12,075 TS; 15.5 El; 26 Brin. For bearings, connecting rods cast on mandrels; die castings. *Obsolete*

FEDERAL-MOGUL B-101
Federal-Mogul Corp.
Cu 3.5-5.5, Sb 4-5, Sn 88.5-91.5.
At 70 F: 9,500 TS; 9 El; 20 Brin. At 212 F: 6,000 TS; 17 El; 11 Brin. At 300 F: 4,100 TS; 22 El; 6 Brin. For lining for steel and bronze back bearings; Babbitt. *Obsolete*

FEDERAL-MOGUL B-10105
Federal-Mogul Corp.
Cu 3.5-5, Pb 0-0.25, Sb 4-5, bal Sn.
Cast: 11,000 TS; 11 El; 23 Brin. For bearings; Babbitt. *Obsolete*

FEDERAL-MOGUL B-103
Federal-Mogul Corp.
Cu 5-6.5, Sb 6.5-7.5, Sn 85-87.5.
At 70 F: 10,500 TS; 7 El; 25 Brin. At 212 F: 6,500 TS; 10 El; 14 Brin. At 300 F: 4,000 TS; 15 El; 8 Brin. For lining for steel and bronze back bearings; Babbitt. *Obsolete*

FEDERAL-MOGUL B-10325
Federal-Mogul Corp.
Cu 5-6.5, Pb 0-0.25, Sb 6-7.5, bal Sn.
Cast: 12,600 TS; 4.8 El; 30 Brin. For lined bearings; Babbitt. *Obsolete*

FEDERAL-MOGUL B-104
Federal-Mogul Corp.
Cu 6.5-8.5, Sb 7-8.5, Sn 82-85.5.
At 70 F: 10,750 TS; 4.8 El; 29 Brin. At 212 F: 6,800 TS; 10 El; 16 Brin. At 300 F: 4,400 TS; 16 El; 9 Brin. For lining for steel and bronze back bearings; Babbitt. *Obsolete*

FEDERAL-MOGUL B-10425
Federal-Mogul Corp.
Cu 6.5-8, Pb 0-0.25, Sb 7.25-8, bal Sn.
Cast: 10,760 TS; 2 El; 32 Brin. For main and connecting rods, engine bearings, heavy duty. *Obsolete*

FEDERAL-MOGUL B-109
Federal-Mogul Corp.
Cu 1.5-2.5, Sb 8.5-9.5, Sn 88-90.
At 70 F: 11,650 TS; 15 El; 26 Brin. At 212 F: 6,600 TS; 20 El; 13 Brin. At 300 F: 3,650 TS; 24 El; 8 Brin. For lining for steel and bronze back bearings; Babbitt. *Obsolete*

FEDERAL-MOGUL B-10925
Federal-Mogul Corp.
Cu 1.75-2.75, Sb 8.5-9.5, 0.25% min Pb, bal Sn.
Cast: 11,650 TS; 17 El; 26 Brin. For bearings; Babbitt. *Obsolete*

FEDERAL-MOGUL B-11125
Federal-Mogul Corp.
Cu 1.75-2.75, Pb 0-0.25, Sb 7.5-8.5, bal Sn.
Cast: 11,600 TS; 17 El; 26 Brin. For bearings; Babbitt. *Obsolete*

FEDERAL-MOGUL CS-50
Federal-Mogul Corp
Ag 0.75, Cu 0.5, bal Cd.
At 70 F: 18,000 TS; 50 El; 40 Brin. At 212 F: 11,800 TS; 44 El; 19 Brin. At 400 F: 4,350 TS; 68 El. For bearings; Babbitt. *Obsolete*

FEDERAL-MOGUL DURO
Federal-Mogul Corp.
Cu 3.5, Sb 7-8, bal Sn.
Cast. For slow moving pulley and axle bearings; antifriction. *Obsolete*

FEDERAL-MOGUL F-1
Federal-Mogul Corp.
Cu 88, Sn 10, Zn 2.
Cast: 48,400 TS; 48 El; 41 RA; 65 Brin. For bearings, gears, piston pin bushings; S.A.E. 62. *Obsolete*

FEDERAL-MOGUL F-11
Federal-Mogul Corp.
Cu 87-89, Sn 9-11, Pb 1.5-2.5, 1% others.
Cast: 46,000 TS; 40 El; 31 RA; 69 Brin. For bearings, bushings; S.A.E. 63. *Obsolete*

FEDERAL-MOGUL F-13
Federal-Mogul Corp.
Cu 86-89, Sn 9-10, Pb 1-3, Zn 0.75-1.5, 1% others.
Cast: 44,000 TS; 37 El; 31 RA; 64 Brin. For bearings; heavy duty. *Obsolete*

FEDERAL-MOGUL F-15
Federal-Mogul Corp.
Cu 70, Sn 10, Pb 20.
Cast 31,300 TS; 16 El; 14.6 RA; 59 Brin. For bearings. *Obsolete*

FEDERAL-MOGUL F-16
Federal-Mogul Corp.
Cu 70, Sn 5, Pb 25.
Cast: 21,000 TS; 15.7 El; 15.7 RA; 42 Brin. For bearings for bushings under water shafts; "Journal Bearing Metal". *Obsolete*

FEDERAL-MOGUL F-18
Federal-Mogul Corp.
Cu 72-74, Sn 6-8, Pb 18-20, 1.5% others.
Cast: 25,000 TS; 16 El; 14 RA; 47 Brin. For bearings; heavy duty. *Obsolete*

FEDERAL-MOGUL F-19
Federal-Mogul Corp.
Cu 85-89, Sn 7.5-8.5, Pb 0.75-1.5, Zn 2.25-4.2.
Cast: 45,000 TS; 42 El; 31 RA; 58 Brin. For bearings; "Ford Metal". *Obsolete*

FEDERAL-MOGUL F-2
Federal-Mogul Corp.
Cu 80, Sn 10, Zn 10.
Cast 40,000 TS; 29 El; 27 RA; 64 Brin. For bushings, bearings; S.A.E. 64. *Obsolete*

FEDERAL-MOGUL F-20
Federal-Mogul Corp.
Cu 88.5-89.5, Sn 10.5-11.5, P 0-0.25.
Sand cast: 52,000 TS; 47 El; 70 Brin. Chill cast: 51,000 TS; 17 El; 100 Brin. For bearings, gears, bushings; S.A.E. 65 "Gear Bronze". *Obsolete*

FEDERAL-MOGUL F-22
Federal-Mogul Corp.
Al 5-12, Fe, Ni, Mn, Zn, Sn, bal Cu.
Cast 70,000 TS; 25 El; 130 Brin. For marine propellers; resists sea water corrosion. *Obsolete*

FEDERAL-MOGUL F-23
Federal-Mogul Corp.
Cu 88-90, Al 9.5-10.5, Fe 0.75-1.25.
Cast: 65,000 TS; 15 El; 125 Brin. For gears, bushings, high strength castings; good corrosion resistance. *Obsolete*

FEDERAL-MOGUL F-26
Federal-Mogul Corp.
Sn 2-4, Pb 30-37, bal Cu.
Cast: 19,500 TS; 10.5 El; 7.5 RA; 36 Brin. For bearings, bushings. *Obsolete*

FEDERAL-MOGUL F-27
Federal-Mogul Corp.
Sn 11.5-13.5, Pb 13.5-16.5, bal Cu.
Cast: 30,000 TS; 5 El; 69 Brin. For heavily loaded bearings. *Obsolete*

FEDERAL-MOGUL F-28
Federal-Mogul Corp.
Sn 4.5-5.5, Pb 17-19, 1.75% others, bal Cu.
Cast: 25,000 TS; 12 El; 50 Brin. For compressor seals for refrigerators; resists Freon and sulfur dioxide. *Obsolete*

FEDERAL-MOGUL F-3
Federal-Mogul Corp.
Cu 83-86, Sn 4-6, Pb 4-6, Zn 4-6.
Cast: 30,400 TS; 20 El; 22 RA; 56 Brin. For bearings, Babbitt lined bearings; "Ounce Metal", "Steam Metal". *Obsolete*

FEDERAL-MOGUL F-32
Federal-Mogul Corp.
Al 10.8-11.1, Fe 4-4.5, 1.0% others, bal Cu.
185 Brin. For rollers, safety tools, forming and drawing dies; wear resistant. *Obsolete*

FEDERAL-MOGUL F-33
Federal-Mogul Corp.
Al 11.5-12, Fe 4-4.5, 1.0% others, bal Cu.
225 Brin. For rollers, safety tools, dies; wear resistant. *Obsolete*

FEDERAL-MOGUL F-34
Federal-Mogul Corp.
Al 13, Fe 4-4.5, 1.0% max others, bal Cu.
290 Brin. For rollers, safety tools, drawing dies; wear resistant. *Obsolete*

FEDERAL-MOGUL F-5
Federal-Mogul Corp.
Cu 85, Sn 5, Pb 9, Zn 1.
Cast: 31,000 TS; 16.5 El; 20 RA; 52 Brin. For bearings, bushings, Babbitt lined bearings; S.A.E. 66. *Obsolete*

FEDERAL-MOGUL F-6
Federal-Mogul Corp.
Cu 76-78, Sn 7-9, Pb 14-16, 1.5% others.
Cast: 27,000 TS; 17 El; 11 RA; 50 Brin. For bearings for locomotive rods; S.A.E. 67. *Obsolete*

FEDERAL-MOGUL F-8
Federal-Mogul Corp.
Cu 83, Sn 7, Pb 7, Zn 3.
Cast 34,000 TS; 23 El; 24 RA; 53 Brin. For bushings, bearings. *Obsolete*

FEDERAL-MOGUL L-100
Federal-Mogul Corp.
Sn 5-7, Sb 9-11, Cu 0-0.5, bal Pb.
Cast: 11,000 TS; 4.5 El; 20 Brin. For bearings; "Bermax".
Obsolete

FEDERAL-MOGUL NO. 407
Federal-Mogul Corp.
Ni 1, Cu 3, Pb 2, Sb 6, bal Sn.
For heavy duty bearings; Babbitt. *Obsolete*

FEDERAL-MOGUL NO. 408
Federal-Mogul Corp.
Cu 4-5, Sb 8-10, bal Sn.
For bearings for motor pumps, rock crushers; Babbitt.
Obsolete

FEDERAL-MOGUL SPECIAL "B"
Federal-Mogul Corp.
Cu 3, Sb 7-9, bal Pb.
For slow speed bearings; antifriction. *Obsolete*

FEDERALOY A-200
Federal-Mogul Corp.
Cu 1, Ni 1, Cd 3, Al 95.
Aluminum base bearing liner strip for extra heavy duty
gasoline and diesel engines; 40 Brin.

FEDERALOY A-300
Federal-Mogul Corp.
Cu 1, Sn 6.25, Ni 0.5, Si 1.5, Al 91.25.
Aluminum base bearing lining strip for connecting rod and
main bearings, thrust washers for heavy duty gasoline and
diesel engines; 50 Brin.

FEDERALOY AF-6
Federal-Mogul Corp.
Sn, bal Al.
Cast. For journal and crankpin bearings; heavy duty.
Obsolete

FEDERALOY AT 4
Federal-Mogul Corp.
Al 91, Sn 4, Si 4, Cu 1.
Aluminum base bearing liner strip for medium duty on
connecting rod and flange main bearings; 70 Rock 15T.

FEDERALOY AT-20
Federal-Mogul Corp.
Cu 1, Sn 20, Al 79.
Aluminum base bearing liner for connecting rod and main
bearings for passenger cars, electric motors. 38 Brin.

FEDERALOY AT-7
Federal-Mogul Corp.
Sn 5.5-7, Cu 0.7-1.3, Ni 0.7-1.3, bal Al.
Cast: 21,000 TS; 10,500 YS; 11 El; 44 Brin. For journal and
crankpin bearings; heavy duty.

FEDERALOY B-100
Federal-Mogul Corp.
Cu 3-5, Sb 7-8, bal Sn.
At 70°F: 105,000 TS; 11 El; 24 Brin. At 300°F: 42,000 TS; 22
El; 8 Brin. For reciprocating engine bearings, pump bearings;
Babbitt.

FEDERALOY B-30
Federal-Mogul Corp.
Pb 25-32, Ag 0-1.5, bal Cu.
Cast: 8,000-9,000 TS; 6-8 El; 22-32 Brin. For journal and
crankpin bearings; SAE 48; heavy duty. *Obsolete*

FEDERALOY B-35
Federal-Mogul Corp.
Pb 32-37, Ag 0-1.5, bal Cu.
Cast: 7,750-8,750; 6-8 El; 20-30 Brin. For journal and
crankpin bearings; SAE 480; heavy duty. *Obsolete*

FEDERALOY B-40
Federal-Mogul Corp.
Pb 37-42, Ag 0-2, bal Cu.
Cast: 75,000-85,000 TS; 6-8 El; 20-30 Brin. For bearings,
bushings; superior to Babbitt. *Obsolete*

FEDERALOY F-1
Federal-Mogul Corp.
Cu 87, Sn 10, Zn 2, Ni 0-1.
Cast bronze bearing metal.

FEDERALOY F-12
Federal-Mogul Corp.
Cu 72, Sn 7, Pb 19, Zn 0-1.25, Ni 0-1.
Cast bronze bearing metal.

FEDERALOY F-12
Federal-Mogul Corp.
Sn 6-8, Pb 18-20, bal Cu.
Cast: 25,000 TS; 16 El; 14 RA; 47 Brin. For bearings,
bushings; resists adverse lubrication. *Obsolete*

FEDERALOY F-15
Federal-Mogul Corp.
Cu 69, Sn 9, Pb 20, Zn 0-1, Ni 0-1.
Cast bronze bearing metal.

FEDERALOY F-2
Federal-Mogul Corp.
Cu 80, Sn 10, Pb 10.
Cast bronze bearing metal; SAE 64.

FEDERALOY F-20
Federal-Mogul Corp.
Cu 89, Sn 11.
Cast bronze bearing metal.

FEDERALOY F-3
Federal-Mogul Corp.
Cu 84, Sn 5, Pb 5, Zn 5, Ni 0-1.
Cast bronze bearing metal. SAE 40.

FEDERALOY F-5
Federal-Mogul Corp.
Cu 84, Sn 5, Pb 9, Zn 0-2.
Cast bronze bearing metal. SAE 66.

FEDERALOY F-8
Federal-Mogul Corp.
Cu 83, Sn 7, Pb 7, Zn 3.
Cast bronze bearing metal. SAE 660.

FEDERALOY H-116
Federal-Mogul Corp.
Cu 73, Sn 3.25, Pb 23.75.
Copper-lead bearing liner strip for connecting rod and main
bearings for extra heavy duty gasoline and diesel engines; 75
Rock 15T.

FEDERALOY H-24
Federal-Mogul Corp.
Pb 21-27, 1.0 Sn min, bal Cu.
Cast. For journal and crankpin bearings; SAE 49; heavy duty.

FEDERALOY H-35
Federal-Mogul Corp.
Pb 32-37, bal Cu.
Cast. For journal and crankpin bearings; SAE 480; heavy
duty.

FEDERALOY H-35-LT
Federal-Mogul Corp.
Sn 7, Pb 28, Cu 65.
Bronze-lead bearing metal liner.

FEDERALOY HF-16
Federal-Mogul Corp.
Sn 3-4, Zn 0-3, Pb 21-25, bal Cu.
Cast. For heavy duty bushings, bearings; SAE 799.

FEDERALOY HF-2
Federal-Mogul Corp.
Sn 9-11, Pb 9-11, Zn 0-0.75, Ni 0-0.5, bal Cu.
Cast. For heavy duty bushings, bearings; SAE-797.

FEDERALOY HF-24
Federal-Mogul Corp.
Cu 76, Pb 24.
Cast. For bearings, liners, bushings; plastic bronze. *Obsolete*

FEDERALOY HF-3
Federal-Mogul Corp.
Sn 4, Zn 4, Pb 8, Sb 0-0.5, P 0-0.5, bal Cu.
Cast. For thrust washers, bearings, bushings; SAE-798.

FEDERALOY L-200
Federal-Mogul Corp.
Sb 14-16, Sn 0.75-1.25, Cu 0-0.6, As 1, bal Pb.
At 70°F: 8800 TS; 2 El; 21 Brin. For bearings for engines,
compressors, pumps and turbines; Babbitt.

FEDERALOY L-300
Federal-Mogul Corp.
Sn 9.25-10.75, Sb 14-16, Cu 0-0.5, bal Pb.
For bearings, bushings; Babbitt.

FEDERALOY L-301
Federal-Mogul Corp.
Sn 2-4, Hg 0.3-1, Ca 0.05-1, Al 0-0.15, bal Pb.
At 70°F: 10,300 TS; 13 El; 23 Brin. For bearings; Babbitt.

FEDERALOY RS-11
Federal-Mogul Corp.
Sn 0.25-0.75, Fe 0-0.1, Cu 88-92, bal Zn.
For architectural trim; commercial bronze; SAE 795.

FEDERALOY RS-13
Federal-Mogul Corp.
Cu 66, Zn 34.
For ornaments, fixtures, hardware; SAE 70C; yellow brass.

FEDERALOY RS-21
Federal-Mogul Corp.
Sn 3.5-4.5, Zn 3.5-4.5, Pb 3.5-4.5, bal Cu.
Cast. For piston pins, bushings, thrust washers; SAE 791;
leaded bronze.

FEDERALOY RS-25
Federal-Mogul Corp.
Sn 0.5, Zn 8, Pb 1.5, bal Cu.
Cast. For hardware; free-cutting. *Obsolete*

FEDERALOY RS-27
Federal-Mogul Corp.
Zn 36, Pb 1.5, bal Cu.
Cast. For hardware; free-cutting; SAE 72. *Obsolete*

FEDERALOY RS-29
Federal-Mogul Corp.
99.0% Al min.
Cast. For electrical machinery conductors. *Obsolete*

FEDERATED F 430
American Smelting & Refining Co.
Cu 3.5-4.5, Si 3, Mg 0.05, Mn 0.5, Fe 1, Ni 0.3, Zn 0.3, bal Al.
Cast: 25,000 psi TS; 12,000 psi YS; 3 El; 57 Brin. Aged:
28,000 psi TS; 21,000 psi YS; 2 El; 74 Brin. For light alloy
castings; general purpose alloy.

FEDERATED F-250
American Smelting & Refining Co.
Cu 2, Si 5, bal Al.
For light alloy castings; age-hardenable.

FEDERATED F-480
American Smelting & Refining Co.
Cu 4, Si 8, bal Al.
For die castings; high strength.

FEDERATED F-720
American Smelting & Refining Co.
Cu 7, Si 2, bal Al.
For castings; general use.

FEDERATED F004
American Smelting & Refining Co.
Mg 4, bal Al.
Sand cast: 22,000-26,000 psi TS; 12,000-14,000 psi YS; 6-10 El; 50-60 Brin. For corrosion resistant castings. Tough; Aluminum 214 alloy. Not heat treatable.

FEDERATED F004ZN
American Smelting & Refining Co.
Mg 4, Zn 2, bal Al.
Cast: 24,000-29,000 psi TS; 16,000-19,000 psi YS; 3-7 El; 70 Brin. For cookware, food handling equipment, fittings, marine hardware. Pressure tight, similar to Aluminum A214.

FEDERATED F008
American Smelting & Refining Co.
Mg 8, bal Al.
Cast: 45,000 psi TS; 28,000 psi YS; 5 El; 75 Brin. For general purpose die castings, brake shoes, propellers, cylinder blocks, motor brackets. Tough, strong, high corrosion resistance. Similar to Aluminum 218.

FEDERATED F0120
American Smelting & Refining Co.
Si 12, bal Al.
Sand cast: 27,000 psi TS; 12,000 psi YS; 8 El; 55 Brin. Permanent mold: 26,000 psi TS; 12,000 psi YS; 3 El. For meter cases, switch boxes, thin walled castings. Similar to Aluminum 13. Not heat treatable, pressure tight, high fluidity.

FEDERATED F024
American Smelting & Refining Co.
Si 2, Mg 4, bal Al.
Cast: 20,000-24,000 psi TS; 13,000-16,000 psi YS; 1-2 El; 60 Brin.For pressure tight castings, cooking utensils, marine fittings. Aluminum B214; excellent corrosion resistance.

FEDERATED F050
American Smelting & Refining Co.
Si 5, bal Al.
Cast: 22,000 psi TS; 10,000 psi YS; 4 El; 50 Brin. For pressure tight castings, marine hardware, architectural parts. Aluminum 43 alloy; tough and corrosion resistant.

FEDERATED F070.3
American Smelting & Refining Co.
Si 7, Mg 0.3, bal Al.
Heat treated: 33,000 psi TS; 24,000 psi YS; 3 El; 70 Brin. Sand cast: 23,000 psi TS; 12,000 psi YS; 3 El; 55 Brin. For complex pressure tight castings. Aluminum 356. Age hardenable; corrosion resistant.

FEDERATED F090.5
American Smelting & Refining Co.
Si 9.5, Mg 0.5, bal Al.
Cast: 46,000 psi TS; 24,000 psi YS; 4 El; 80 Brin. For general purpose castings, cover plates, instrument housings, castings, corrosion resistant die castings. Similar to Aluminum 360.

FEDERATED F100.4
American Smelting & Refining Co.
Cu 0.8, Zn 8, Mn 0.4, bal Al.
Cast: 29,000 psi TS; 16,000 psi YS; 6-7 El; 60 Brin. For permanent mold castings.

FEDERATED F1000.3
American Smelting & Refining Co.
Cu 10, Mg 0.3, bal Al.
Heat treated: 44,000 psi TS; 40,000 psi YS; 0.5 El; 120 Brin. For cylinder heads, pistons. Aluminum 122 alloy. Heat treatable, good high temperature strength.

FEDERATED F1040.3
American Smelting & Refining Co.
Cu 10, Si 4, Mg 0.4, bal Al.
Heat treated: 46,000 psi TS; 44,000 psi YS; 125 Brin. For cylinder heads, pistons. Aluminum 138 alloy, age-hardenable. Good high temperature strength.

FEDERATED F1121 NI
American Smelting & Refining Co.
Cu 1, Si 12, Mg 1, Ni 2.5, bal Al.
Stress-relieved: 32,000-36,000 psi TS; 28,000-30,000 psi YS; 0.5 El; 105 Brin. For general purpose castings. Aluminum A132 alloy. Not heat treatable.

FEDERATED F150.5
American Smelting & Refining Co.
Cu 1, Si 5, Mg 0.5, bal Al.
Heat treated: 44,000 psi TS; 42,000 psi YS; 1 El; 95 Brin. Cast: 26,000 TS; 17,000 psi YS; 2 El; 75 Brin. For complex pressure tight castings, cylinders, housings, hardware. Age-hardenable, good castability, pressure tight.

FEDERATED F360
American Smelting & Refining Co.
Cu 3, Si 6, bal Al.
Heat treated: 35,000 psi TS; 20,000 psi YS; 3 El; 80 Brin. Sand cast: 28,000 psi TS; 15,000 psi YS; 4 El; 60 Brin. For housings, oil pans, casings. Heat treatable, high strength.

FEDERATED F391 NI
American Smelting & Refining Co.
Cu 3.5, Si 9, Mg 1, Ni 1, bal Al.
Cast: 32,000 psi TS; 26,000 psi YS; 2 El; 85 Brin. For pistons. Aluminum D 132 alloy.

FEDERATED F401.5 NI
American Smelting & Refining Co.
Cu 4, Mg 1.5, Ni 2, bal Al.
Heat treated: 45,000 psi TS; 42,000 psi YS; 1 El; 100 Brin. Cast: 34,000 TS; 32,000 psi YS; 1 El; 86 Brin. For cylinder heads, pistons. Aluminum 142 alloy, wear and heat resistant. Age hardenable.

FEDERATED F410
American Smelting & Refining Co.
Cu 4, Si 1, bal Al.
Heat treated: 47,000 psi TS; 42,000 psi YS; 1 El; 110 Brin. Cast: 25,000 TS; 11,000 psi YS; 4 El; 65 Brin. For high strength castings, housings, hardware, gear cases, oil pans. Age-hardenable. Aluminum 195 alloy.

FEDERATED F4110
American Smelting & Refining Co.
Si, bal Al.
For die castings; high fluidity.

FEDERATED F450
American Smelting & Refining Co.
Cu 4, Si 5, bal Al.
Heat treated: 45,000 psi TS; 40,000 psi YS; 1 El; 105 Brin. Cast: 25,000 TS; 15,000 psi YS; 2 El; 70 Brin. For general purpose castings. Aluminum A108, heat treatable. Good castability.

FEDERATED F460
American Smelting & Refining Co.
Cu 3-4, Si 5.5-6.5, bal Al.
Heat treated: 45,000 psi TS; 40,000 psi YS; 1 El; 105 Brin. Cast: 25,000 TS; 15,000 psi YS; 2 El; 70 Brin. For general purpose castings, grills, reflectors, housings, castings. Heat treatable, Aluminum 319 alloy.

FEDERATED F750.3
American Smelting & Refining Co.
Cu 6.5, Si 5.5, Mg 0.3, bal Al.
Heat treated: 50,000 psi TS; 48,000 psi YS; 0.5 El; 130 Brin. For pistons. Aluminum 152 alloy, age-hardenable, high strength.

FEDERATED NO. 14
American Smelting & Refining Co.
Pb 83.75, Sn 0.75, Sb 12.5, As 3.
Cast: 9800 psi TS; 1.5 El; 22 Brin. For bearings; antifriction, Babbitt.

FEDERATED NO. 15
American Smelting & Refining Co.
Pb 83, Sn 1, As 1, Sb 15.
Cast: 10,350 psi TS; 2 El; 20 Brin. For bearings; antifriction, Babbitt.

FEDERATED NO. 2
American Smelting & Refining Co.
Sb 7.5, Cu 3.5, Sn 89.
Cast: 10,900 psi TS; 8 El; 24 Brin; 466°F MP.

FEDERATED NO. 3
American Smelting & Refining Co.
Sn 83.3, Sb 8.3, Cu 8.3.
Cast: 12,300 psi TS; 2 El; 26 Brin. For bearings; antifriction, Babbitt.

FEDERATED NO. 7
American Smelting & Refining Co.
Pb 75, Sn 10, Sb 15.
Cast: 10,500 psi TS; 5 El; 20 Brin. For bearings; antifriction, Babbitt.

FEDERATED NO. 8
American Smelting & Refining Co.
Pb 80, Sn 5, Sb 15.
Cast: 10,000 psi TS; 5 El; 20 Brin. For bearings; antifriction, Babbitt.

FEDERATED STEEL MILL ALUMINUM GR. 4
American Smelting & Refining Co.
Al 85-90, bal Cu, Fe, Mn.
For deoxidizer in steel mills.

FEDERATED XXXX NICKEL
American Smelting & Refining Co.
Sn alloy.
For bearings; Babbitt.

FEDERSTAHLDRAHT FD
Krupp Stahl AG
P 0-0.03, S 0-0.03, Cu 0-0.12, bal Fe.
DIN 17223 BL.2/84; W. Nr. 1.1230.

FEDERSTAHLDRAHT VD
Krupp Stahl AG
C 0.6-0.7, Si 0-0.25, Mn 0.5-0.9, P 0-0.03, S 0-0.02, Cu 0-0.06, bal Fe.
DIN 17223 BL.2/64; W. Nr. 1.1250.

FEDOL
Swedish Iron & Steel Corp.
C 1.2, V 0.2, bal Fe.
For tools, dies; water hardening. *Obsolete*

FELAX-45
Forjas Alavesas S.A.
C 0.47, Mn 0.7, Si 1.75, bal Fe.
Si-Mn spring steel; hardenable to 56-61 Rock C. IHA F-145; AFNOR 4658; DIN 46Si7.

FELAX-55
Forjas Alavesas S.A.
C 0.56, Mn 0.75, Si 1.75, bal Fe.
Si Mn spring steel, hardenable to 57-62 Rock C. IHA F-144; DIN 55Si7; AISI 9255; BSEN45.

FELAX-60
Forjas Alavesas S.A.
C 0.6, Mn 0.8, Si 1.75, bal Fe.
Si-Mn spring steel; hardenable to 58-63 Rock C. DIN 65Si7; AISI 9280; BS EN45A.

FELTMETAL 347-10-20-AC3-A
Brunswick Technetics, ECS
Stainless steel.
Fiber metal acoustic product made by diffusion bonding randomly oriented metal fibers. Fiber type 8 micrometers; 10 rayls (dyne/cm^2 per cm/s); 9000 psi TS; Type 347 stainless steel.

FELTMETAL 347-10-30-AC3-A
Brunswick Technetics, ECS
Stainless steel.
Fiber metal acoustic product made by diffusion bonding randomly oriented metal fibers. Fiber type 8 micrometers; 10 rayls (dyne/cm^2 per cm/s); 6500 psi TS; Type 347 stainless steel.

FELTMETAL 347-50-20-AC3-A
Brunswick Technetics, ECS
Stainless steel.
Fiber metal acoustic product made by diffusion bonding randomly oriented metal fibers. Fiber type 8 micrometers; 50 rayls (dyne/cm^2 per cm/s); 9000 psi TS; Type 347 stainless steel.

FELTMETAL 347-70-15-AC3-B1
Brunswick Technetics, ECS
Stainless steel.
Fiber metal acoustic product made by diffusion bonding randomly oriented metal fibers. Fiber type 8 micrometers; 70 rayls (dyne/cm^2 per cm/s); 4500 psi TS; Type 347 stainless steel.

FELTMETAL FM122
Brunswick Technetics, ECS
Stainless steel.
Fiber metal acoustic product made by diffusion bonding randomly oriented metal fibers. Fiber type A04; 50 rayls (dyne/cm^2 per cm/s); 5500 psi TS; Type 347 stainless steel.

FELTMETAL FM123
Brunswick Technetics, ECS
Stainless steel.
Fiber metal acoustic product made by diffusion bonding randomly oriented metal fibers. Fiber type C40; 50 rayls (dyne/cm^2 per cm/s); 17,000 psi TS; Type 347 stainless steel.

FELTMETAL FM125
Brunswick Technetics, ECS
Stainless steel.
Fiber metal acoustic product made by diffusion bonding randomly oriented metal fibers. Fiber type C40; 10 rayls (dyne/cm^2 per cm/s); 12,000 psi TS; Type 347 stainless steel.

FELTMETAL FM126
Brunswick Technetics, ECS
Stainless steel.
Fiber metal acoustic product made by diffusion bonding randomly oriented metal fibers. Fiber type C30; 50 rayls (dyne/cm^2 per cm/s); 20,000 psi TS; Type 347 stainless steel.

FELTMETAL FM127
Brunswick Technetics, ECS
Stainless steel.
Fiber metal acoustic product made by diffusion bonding randomly oriented metal fibers. Fiber type C40; 10 rayls (dyne/cm^2 per cm/s); 13,000 psi TS; Type 347 stainless steel.

FELTMETAL FM128
Brunswick Technetics, ECS
Stainless steel.
Fiber metal acoustic product made by diffusion bonding randomly oriented metal fibers. Fiber type C20; 50 rayls (dyne/cm^2 per cm/s); 22,000 psi TS; Type 347 stainless steel.

FELTMETAL FM134
Brunswick Technetics, ECS
Stainless steel.
Fiber metal acoustic product made by diffusion bonding randomly oriented metal fibers. Fiber type C40; 35 rayls (dyne/cm^2 per cm/s); 20,000 psi TS; Type 347 stainless steel.

FELTMETAL FM1802
Brunswick Technetics, ECS
Stainless steel.
Fiber metal acoustic product made by diffusion bonding randomly oriented metal fibers. Fiber type 8 micrometers; 20 rayls (dyne/cm^2 per cm/s); 2500 psi TS; Type 347 stainless steel; web overlay material.

FELTMETAL FM1808
Brunswick Technetics, ECS
Stainless steel.
Fiber metal acoustic product made by diffusion bonding randomly oriented metal fibers. Fiber type 8 micrometers; 70 rayls (dyne/cm^2 per cm/s); 8000 psi TS; Type 347 stainless steel.

FELTMETAL FM1810
Brunswick Technetics, ECS
Stainless steel.
Fiber metal acoustic product made by diffusion bonding randomly oriented metal fibers. Fiber type 8 micrometers; 70 rayls (dyne/cm^2 per cm/s); 11,000 psi TS; Type 347 stainless steel.

FELTMETAL FM1812
Brunswick Technetics, ECS
Stainless steel.
Fiber metal acoustic product made by diffusion bonding randomly oriented metal fibers. Fiber type 8 micrometers; 30 rayls (dyne/cm^2 per cm/s); 13,000 psi TS; Type 347 stainless steel.

FELTMETAL FM1813
Brunswick Technetics, ECS
Stainless steel.
Fiber metal acoustic product made by diffusion bonding randomly oriented metal fibers. Fiber type 12 micrometers; 35 rayls (dyne/cm^2 per cm/s); 20,000 psi TS; Type 347 stainless steel.

FELTMETAL FM1814
Brunswick Technetics, ECS
Stainless steel.
Fiber metal acoustic product made by diffusion bonding randomly oriented metal fibers. Fiber type 8 micrometers; 55 rayls (dyne/cm^2 per cm/s); 13,000 psi TS; Type 347 stainless steel.

FELTMETAL FM1815
Brunswick Technetics, ECS
Stainless steel.
Fiber metal acoustic product made by diffusion bonding randomly oriented metal fibers. Fiber type: wire cloth; 75 rayls (dyne/cm^2 per cm/s); 20,000 psi TS; Type 347 stainless steel.

FELTMETAL FM1816
Brunswick Technetics, ECS
Stainless steel.
Fiber metal acoustic product made by diffusion bonding randomly oriented metal fibers. Fiber type 12 micrometers; 75 rayls (dyne/cm^2 per cm/s); 20,000 psi TS; Type 347 stainless steel.

FELTMETAL FM185
Brunswick Technetics, ECS
Stainless steel.
Fiber metal acoustic product made by diffusion bonding randomly oriented metal fibers. Fiber type C20; 10 rayls (dyne/cm^2 per cm/s); 12,000 psi TS; Type 347 stainless steel.

FELTMETAL FM189
Brunswick Technetics, ECS
Stainless steel.
Fiber metal acoustic product made by diffusion bonding randomly oriented metal fibers. Fiber type C20; 30 rayls (dyne/cm^2 per cm/s); 16,000 psi TS; Type 347 stainless steel.

FELTMETAL FM190
Brunswick Technetics, ECS
Stainless steel.
Fiber metal acoustic product made by diffusion bonding randomly oriented metal fibers. Fiber type C20; 50 rayls (dyne/cm^2 per cm/s); 18,000 psi TS; Type 347 stainless steel.

FELTMETAL FM802
Brunswick Technetics, ECS
Superalloy.
Fiber metal acoustic product made by diffusion bonding randomly oriented metal fibers. Fiber type A08 Hastelloy X; 20 rayls (dyne/cm^2 per cm/s); 7300 psi TS.

FELTMETAL FM806
Brunswick Technetics, ECS
Stainless steel.
Fiber metal acoustic product made by diffusion bonding randomly oriented metal fibers. Fiber type C40; 50 rayls (dyne/cm^2 per cm/s); 18,000 psi TS; Type 347 stainless steel.

FELTMETAL FM817
Brunswick Technetics, ECS
Stainless steel.
Fiber metal acoustic product made by diffusion bonding randomly oriented metal fibers. Fiber type C40; 100 rayls (dyne/cm^2 per cm/s); 27,500 psi TS; Type 347 stainless steel.

FELTMETAL FM818
Brunswick Technetics, ECS
Stainless steel.
Fiber metal acoustic product made by diffusion bonding randomly oriented metal fibers. Fiber type B62 430 SS; 35 rayls (dyne/cm^2 per cm/s); 18,000 psi TS; Type 347 stainless steel.

FELTMETAL FM827
Brunswick Technetics, ECS
Superalloy. Fe-Cr-Al-Si alloy.
Fiber metal acoustic product made by diffusion bonding randomly oriented metal fibers. Fiber type C40 Hoskins 875; 35 rayls (dyne/cm^2 per cm/s); 12,000 psi TS.

FENICOLOY
Molecu-Wire Corp.
Ni 29, Co 17, Mn 0.2, bal Fe.
Resistance wire: 294 ohms per cir. mil. ft. Annealed: 75,000 psi; hard drawn; 150,000 psi. For resistors; max. operation range; 450 C. *Obsolete*

FENTON'S ALLOY
English manufacture
Zn 80, Cu 8.5, Sb 11.5.
For bearings; Babbitt.

FER
A. Milne & Co.
C 0.7-1.2, bal Fe.
For tools, dies, punches; water hardening.

FERAL
German manufacture
C 0.06, Mn 0.6, bal Fe, clad with Al alloy.
For aircraft fire walls; aluminum-clad steel.

FERALITE
Chapman Valve Mfg. Co.
Cast Iron. C 3, Si, Mn, bal Fe.
For housings, fittings, gears, shafts; malleable iron.

FERALSI

Unexcelled Mfg. Co.
Fe-Al-Si.
For magnetic applications; high magnetic permeability.

FERALUN

American Abrasive Metals Co.
Special iron base with embedded abrasive grains.
For stairs, platforms; castings; wear resistant.

FERAN

Hoesch Stahl AG
Fe-Al.
For bi-metals in the canning industry. *Obsolete*

FERAN

English manufacture
sheet iron coated with Al and rolled together.
For engineering construction; corrosion resistance of Al and strength of Fe.

FERAN

Massillon Steel Casting Co.
Sheet iron coated with Al and rolled together. For engineering construction; corrosion resistance of Al and strength of Fe. *Obsolete*

FERCHROMIT

Skoda Works National Corp.
Cr 22-30, Si 0.5-2.5, C 0.1-1.5, bal Fe.
Resists oxidation at high temperatures.

FEREX

American manufacture
C 0.2, bal Fe.
For welding rod for armor; coated.

FERHOLZER

Creusot-Loire
C 0.005, Fe (very pure).
Annealed, slow cooled: 275 N/mm^2 psi TS. Very soft magnetic material; for solenoids, relays, electromagnets.

FERIMPHY

Creusot-Loire
Iron alloy.
For armatures, transformers; soft magnet for strong or moderate field. *Obsolete*

FERINOX

Sandvik Steel Co.
C 0.09, Si 1.2, Mn 1.3, Cr 17, Ni 8, Mo 0.7, bal Fe.
Rolled: 299,000 TS; 250,000 YS; 8 El. For watch mainsprings. Stainless, austenitic, high fatigue strength.

FERMET

English manufacture
Cr 4, Ni 18, Mn 2.2, W 1, Cu 0.3, C 0.35, bal Fe.
For corrosion resisting parts; stainless.

FERMO SPECIAL

British Steel Corp.
C 0.2, bal Fe.
Heat treated: 72,000-80,000 TS; 44,000-54,000 YS; 27-32 El; 60-65 RA; 156-183 Brin. For case hardened parts. *Obsolete*

FERNITE 17A

Allegheny Ludlum Steel
C 0.3-0.5, Cr 8-9, Ni 18.7-19.7, bal Fe.
For high temperature applications; heat resistant. *Obsolete*

FERNITE 17B

Allegheny Ludlum Steel
C 0.2-0.3, Cr 6.7-7.7, Ni 18.7-19.7, bal Fe.
For high temperature applications; heat resistant. *Obsolete*

FERNITE 24

Allegheny Ludlum Steel
C 2.3, Cr 24, Mn 0.3, Si 0.2, W 5.3, V 0.3, Ni 0.3, B 0.9, bal Fe.
For forming and sizing dies; cast iron, corrosion resistant. *Obsolete*

FERNITE 5C

Allegheny Ludlum Steel
C 0.45-0.55, Mo 1.2, Co 4, Cr 19-21, Ni 11-13, Se 0.2, bal Fe.
For heat resistant parts; heat and corrosion resistant. *Obsolete*

FERNITE NO. 2

Allegheny Ludlum Steel
C 0.2, Cr 15, Ni 35, bal Fe.
For heat resistant parts; heat and corrosion resistant. *Obsolete*

FERNITE NO. 3

Allegheny Ludlum Steel
C 0.3, Cr 12, Ni 24, bal Fe.
For heat resistant parts; heat and corrosion resistant. *Obsolete*

FERNITE NO. 4

Allegheny Ludlum Steel
C 0.3, Cr 28, Ni 22, bal Fe.
For heat resistant parts; heat and corrosion resistant. *Obsolete*

FERNITE NO. 6

Allegheny Ludlum Steel
C 0.2, Cr 18, Ni 8, bal Fe.
For stainless parts; stainless, austenitic. *Obsolete*

FERNITE NO. 7

Allegheny Ludlum Steel
C 0.2, Cr 28, bal Fe.
For heat resistant castings. *Obsolete*

FERNITE NO. 8

Allegheny Ludlum Steel
C 0.2, Cr 14, bal Fe.
For corrosion resistant castings. *Obsolete*

FERNITE NO. 9

Allegheny Ludlum Steel
C 0.5-0.7, Cr 28, bal Fe.
For heat and corrosion resistant castings. *Obsolete*

FERNITE NO. L

Allegheny Ludlum Steel
C 0.1, Cr 15, Ni 65, bal Fe.
For heat resistant parts; heat and corrosion resistant. *Obsolete*

FERNO

Lehigh Steel Corp.
C 0.7, Cr 3.75, V 0.55, Mo 0.7, bal Fe.
For hot work dies, punches, shear blades; good red hardness.

FERNO EXTRA

Lehigh Steel Corp.
C 0.45, Cr 3.75, Mn 0.4, V 0.7, W 12, bal Fe.
For hot work dies, punches, shear blades; tough and abrasion resistant.

FEROBA-III

Edgar Allen Balfour Ltd.
BaO, 6 Fe$_2$O$_3$.
3500 Br, 2500 Hc, 2,800,000 BH max. For permanent magnets in electrical and magnetic equipment. High permeability.

FEROVAN

SMC (Shieldalloy Metallurgical Corp.)
V 42-48, C 0-1, P 0-0.4, S 0-0.4, Si 0-7, Al 0-0.1, Mn 0-5.5, Ni 0-8, Cr 0-7.5, bal Fe.
Master alloy for ladle additions.

FEROVAN

Cyprus Foote Mineral Co.
V 43, C 0.6, Cr 6.6, Mn 4, Si 6.3, Al 0-0.1, bal Fe.
Ferrovanadium alloy for additive in steel melting. *Obsolete*

FERRAL

Russian manufacture
C 0-0.25, Cr 12-15, Al 3.5-5.5, bal Fe.
For electrical and heat resistance elements; applications to 1100°C.

FERRALIUM

Langley Alloys Ltd.
C 0-0.08, Cr 24-27, Ni 4.5-6.5, Cu 1.3-4, Mo 2-4, 0.1 N min, bal Fe.
Cast or wrought ferritic, austenitic stainless. 820 N/mm^2 TS; 540 N/mm^2 YS; 18-20 El. Improved strength and corrosion resistance.

FERRALIUM ALLOY NO. 255

Haynes International, Inc.
Stainless steel. Ni 5.5, Cr 26, Mo 3.1, Si 0-1, Mn 0-1.5, C 0-0.04, Cu 1.7, N 0.17.
Ferritic-austenitic with strength and resistance to localized corrosion.

FERRIMAG 5

Colt Industries
BaO, 6 Fe 2 O 3.
Bar: 3,950 Br, 2,400 Hc, 2,500,000 BH max. For electrical and magnetic equipment. Permanent magnet, high permeability. *Obsolete*

FERRIMAG I

Colt Industries
Barium oxide ferrate.
For permanent magnets; sintered ferrite, high coercive force *Obsolete*

FERRO BRONZE

American manufacture
Fe 8, Cr 0.7, bal Cu.
For hardware; corrosion resistant.

FERRO MOLYBDENUM (LOW CARBON)

Molycorp, Inc.
C 0-0.5, Mo 60-70.
For metallurgical applications in steel making; Mo-additions.

FERRO MOLYBDENUM (STANDARD)

Molycorp, Inc.
C 1.6-2.25, Mo 58-65, bal Fe.
For metallurgical applications in steel making; Mo-additions.

FERRO TUNGSTEN (ASTM)

Molycorp, Inc.
C 0.6-0.7, W 70-80, Si 0-1, bal Fe.
For metallurgical applications in steel making; H$_2$S metals less than 0.25.

FERRO TUNGSTEN (LOW MELTING)

Molycorp, Inc.
C 2.5, W 65-75, Si 0-1.5, bal Fe.
For metallurgical applications in steel making; H$_2$S metals less than 0.25.

FERRO TUNGSTEN (STANDARD)

Molycorp, Inc.
C 0.6-0.7, W 76-80, Si 0-0.75, bal Fe.
For metallurgical applications in steel making; H$_2$S metals less than 0.25.

FERRO-MANGANESE

Now ELECTROMET FERR-MANGANESE.

FERRO-TIC GR SK

Alloy Technology International, Inc.
Mo 4, Ni 0.5, C 0.5, 40% TiC (by volume), matrix = 5 Cr, bal Fe.
Annealed: 37 Rock C. Hardened: 64 Rock C. For applications involving mechanical and thermal shock.

FERRO-TIC GR. A

Alloy Technology International, Inc.
Ti 26, C 7, Cr 2, Mo 2, bal Fe.
Annealed: 400 Brin. Heat treated: 700 Brin. For drawing and forming dies, rolls, cutters; sintered carbide, heat treatable. *Obsolete*

FERRO-TIC GR. C

Alloy Technology International, Inc.
Ti 26, C 7, Cr 2, Mo 2, bal Fe (45% carbide by volume).
Annealed: 43 Rock C. Heat treated: 300,000 transverse strength; 70 Rock C. For tools, dies, rolling and blanking dies, valves, wear parts. Sintered carbides with alloy steel matrix, wear resistant; maximum operating temperature 400°F.

FERRO-TIC GR. CM

Alloy Technology International, Inc.
Fe 56.5, Ti 27.6, Cr 6.6, Mo 2, 7.4 C (45% carbide by volume).
Annealed: 45 Rock C. Heat treated: 310,000 transverse strength; 66-70 Rock C. For tools, dies, valves, forming dies. Steel-bonded carbide; maximum operating temperature 1000°F.

FERRO-TIC GR. CN-5

Alloy Technology International, Inc.
45% WC (by volume).
For tools and dies, and for wear resistant parts exposed to sea water; corrosion resistant, excellent resistance to sea water. Age hardenable copper-nickel alloy. *Obsolete*

FERRO-TIC GR. CS-40

Alloy Technology International, Inc.
Stainless steel. 45% TiC (by volume).
Annealed: 50 Rock C. Hardened: 225,000 transverse strength; 68-70 Rock C. Good wear and corrosion resistance; for can closing tools, chemical plant valve seats and wear parts. Martensitic, stainless steel matrix.

FERRO-TIC GR. DN-1

Alloy Technology International, Inc.
45% TiC (by volume).
Annealed: 40-41 Rock C. Age-hardened: 52-53 Rock C. Good corrosion resistance toward strong chlorides. Age hardenable nickel-aluminum alloy matrix. *Obsolete*

FERRO-TIC GR. HT-2

Alloy Technology International, Inc.
45% TiC (by volume).
Annealed: 44 Rock C. Aged: 54 Rock C; 250,000 transverse strength. Ni-Fe-Cr age hardenable alloy.

FERRO-TIC GR. HT-6A

Alloy Technology International, Inc.
45% TiC (by volume).
Solutionized: 46 Rock C. Aged (1st step): 51 Rock C. Aged (fully): 54 Rock C; 310,000 transverse strength. Good oxidation and corrosion resistance and high-temperature strength. Ni-Cr age hardenable steel matrix.

FERRO-TIC GR. J

Alloy Technology International, Inc.
40% W TiC$_2$ (by volume).
Annealed: 48 Rock C. Hardened and tempered: 70 Rock C; 250,000 transverse strength. For tools for hot working applications as piercing, forming, machining of non-ferrous materials. High-speed steel matrix. *Obsolete*

FERRO-TIC GR. M-6

Alloy Technology International, Inc.
45% TiC (by volume).
Solutionized: 51 Rock C. Aged: 65 Rock C; 280,000 transverse strength. Can be nitrided for 75 Rock C surface. Maraging steel matrix. *Obsolete*

FERRO-TIC GR. M-6A

Alloy Technology International, Inc.
50% TiC (by volume).
Solutionized: 54 Rock C. Aged: 67 Rock C; 375,000 transverse strength. Good abrasion resistance; can be nitrided to surface hardness of 74 Rock C. Maraging steel matrix. *Obsolete*

FERRO-TIC GR. M-6B

Alloy Technology International, Inc.
55% TiC (by volume).
Annealed: 58 Rock C. Hardened: 68 Rock C; 400,000 transverse strength. Good wear resistance. Maraging steel matrix. *Obsolete*

FERRO-TIC GR. MS-5A

Alloy Technology International, Inc.
Stainless steel. 45% TiC (by volume).
Solutionized: 51 Rock C. Aged: 63-64 Rock C; 280,000 transverse strength. Good corrosion resistance; can be nitrided for high surface hardness; for abrasion resistance parts in the food, chemical and aerospace industries. Age hardenable, martensitic, stainless steel matrix.

FERRO-TIC GR. S-45

Alloy Technology International, Inc.
Cr 11, Ni 7.3, Fe 42.7, 39.0 TiC.
450 Brin. For valve parts, seal rings, knives, molds; oxidation and heat resistant. *Obsolete*

FERRO-TIC GR. S-55

Alloy Technology International, Inc.
Cr 8.6, Ni 5.7, Fe 33.7, 52.0 TiC.
550 Brin. For valve parts, seal rings, knives, molds; oxidation and heat resistant. *Obsolete*

FERRO-TIC PK

Alloy Technology International, Inc.
45% TiC (by volume), bal maraging steel matrix.
Solutionized: 50-51 Rock C. Aged: 64-65 Rock C. 280,000 transverse strength.

FERRO-TITANIUM 18-2.5

Foote Mineral Co.
C 2.5, Ti 18, bal Fe.
For metallurgical applications in steel making. *Obsolete*

FERRO-TITANIUM HIGH CARBON

Foote Mineral Co.
C 6-8, Ti 15-18, bal Fe.
For metallurgical applications in steel. *Obsolete*

FERRO-TITANIUM LOW CARBON GRADE

Foote Mineral Co.
C 0-0.1, Ti 20-25, Al 0-7.5, bal Fe.
For metallurgical applications in stainless steel. *Obsolete*

FERROCAL

Aluminiumwerke Maulbronn
Al alloy.
Chill cast: 23,000-28,000 TS; 2-4 El. For permanent mold castings; general machinery parts, automobile and motorcycle parts.

FERROCHROME

Now ELECTROMET FERROCHROME.

FERROCHRONIN 15-60

Vereinigte Deutsche Nickel-Werke AG
Cr 15, Ni 60, bal Fe.
For electrical equipment, heating elements, resistors; high heat and electrical resistance. *Obsolete*

FERROCHRONIN 20-30

Vereinigte Deutsche Nickel-Werke AG
Cr 20, Ni 30, bal Fe.
For electrical equipment, heating elements, resistors; high heat and electrical resistance. *Obsolete*

FERROCHRONIN 75-15

Vereinigte Deutsche Nickel-Werke AG
Cr 14-16, Ni 72, Fe 6-9.
For heaters, combustion liners, manifolds, regenerators; high heat and corrosion resistance. *Obsolete*

FERROCITE

Charles Hardy & Co.
C 0.5, bal Fe.
For valves, guides; powder metals.

FERROCOLUMBIUM-VACUUM

Reading Alloys, Inc.
Reactive/refractory. C 0.1, Cb 60-67, Mn 0.05, Ni 0.1, Si 0.2.
Master alloy.

FERRODUR

Janney Cylinder Co.
C 3, Mo 0.5, Ni 1.2, Mn 0.9, Si 2, Cr 0.2, bal Fe.
Heat treated: 50,000 TS; 0 El; 500 Brin. For pump liners, shaft sleeves; centrifugal casting.

FERROLUM

Knapp Mills Inc.
Lead clad steel.
For chemical plant equipment; acid resistant.

FERROMET

Plansee, Metallwerk Gesellschaft
Fe alloy.
For machine parts, bearings, and magnets. Sintered alloy.

FERRON

Fansteel Metals
Fe 50, Ni 35, Cr 15.
For heating elements, heat and corrosion resisting parts. *Obsolete*

FERROPERM

NKK Corp.
C 0-0.004, Si 0-0.03, Mn 0-0.1, P 0-0.005, S 0-0.003, Al 0.7-1.3, bal Fe.
Soft magnetic plate with good DC magnetic shield performance.

FERROPYR-I

English manufacture
Fe 86, Cr 7, Al 7, 1 Mn + Si.
For electrical resistances; used in high temperature range of 2100-2460°F.

FERROSAD

Metalltechnik Schmidt GmbH & Co.
C 0.1, Si 0.15, Mn 1-1.3, S 0-0.03, P 0-0.03, bal Fe.
Blasting shot, bainitic structure.

FERROSILICON

Now ELECTROMET FERROSILICON.

FERROSILID

Russian manufacture
C 0.5-0.7, Si 12-18, bal Fe.
For cast pumps and conduits; acid resistant, cast.

FERROSTEEL

Now CRANE FERROSTEEL.

FERROSTEEL

Crane Co.
C 3, Si 1.8, bal Fe.
Cast: 35,000 TS. For valves, fittings; transverse strength 3600; transverse deflection 0.14 in.

FERROTHERM 4700

Krupp Stahl AG
C 0-0.1, Si 1.5-1.8, Mn 0-1, P 0-0.045, S 0-0.03, Cr 1.5-2, bal Fe.
SEL 90; W. Nr. 1.4700.

FERROTHERM 4712
Krupp Stahl AG
C 0-0.12, Si 2-2.5, Mn 0-1, P 0-0.045, S 0-0.03, Cr 5.5-6.5, bal Fe.
SEL 90; W. Nr. 1.4712.

FERROTHERM 4713
Krupp Stahl AG
C 0-0.12, Si 0.5-1, Al 0.5-1, Cr 6-8, bal Fe.
Ferritic. DIN X10CrAl7; W. Nr. 1.4713. 220 N/mm^2 YS, 420-620 N/mm^2 TS, 20 El.

FERROTHERM 4720
Krupp Stahl AG
C 0-0.08, Si 0-1, Mn 0-1, P 0-0.04, S 0-0.03, Cr 10.5-12.5, Ti 0-1, bal Fe.
SEW 470/76; W. Nr. 1.4720.

FERROTHERM 4724
Krupp Stahl AG
C 0-0.12, Si 0.7-1.4, Al 0.7-1.2, Cr 12-14, bal Fe.
Ferritic. DIN X10CrAl13; W. Nr. 1.4724. 250 N/mm^2 YS, 450-650 N/mm^2 TS, 15 El.

FERROTHERM 4742
Krupp Stahl AG
C 0-0.12, Si 0.7-1.4, Al 0.7-1.2, Cr 17-19, bal Fe.
Ferritic. DIN X10CrAl18; W. Nr. 1.4742. 270 N/mm^2 YS, 500-700 N/mm^2 TS, 12 El.

FERROTHERM 4762
Krupp Stahl AG
C 0-0.12, Si 0.7-1.4, Al 1.2-1.7, Cr 23-26, bal Fe.
Ferritic. DIN X10CrAl24; W. Nr. 1.4762. 280 N/mm^2 YS, 520-720 N/mm^2 TS, 10 El.

FERROTHERM 4821
Krupp Stahl AG
C 0.1-0.2, Si 0.8-1.5, Cr 24-27, Ni 3.5-5.5, bal Fe.
Ferritic-austenitic. DIN X20CrNiSi254; W. Nr. 1.4821. 400 N/mm^2 YS, 600-850 N/mm^2 TS, 16 El.

FERROTHERM 4828
Krupp Stahl AG
C 0-0.2, Si 1.5-2.5, Cr 19-21, Ni 11-13, bal Fe.
Austenitic. DIN X15CrNiSi2012; W. Nr. 1.4828. 230 N/mm^2 YS, 500-750 N/mm^2 TS, 30 El.

FERROTHERM 4833
Krupp Stahl AG
C 0-0.08, Si 0-1, Mn 0-2, P 0-0.045, S 0-0.03, Cr 21-23, Ni 12-15, bal Fe.
SEW 470/76; W. Nr. 1.4833.

FERROTHERM 4841
Krupp Stahl AG
C 0-0.2, Si 1.5-2.5, Cr 24-26, Ni 19-22, bal Fe.
Austenitic. DIN X15CrNiSi2520; W. Nr. 1.4841. 230 N/mm^2 YS, 550-800 N/mm^2 TS, 30 El.

FERROTHERM 4845
Krupp Stahl AG
C 0-0.15, Si 0-0.75, Cr 24-26, Ni 19-22, bal Fe.
Austenitic. DIN X12CrNi2521; W. Nr. 1.4845. 210 N/mm^2 YS, 500-750 N/mm^2 TS, 35 El.

FERROTHERM 4847
Krupp Stahl AG
C 0-0.08, Si 0-1, Mn 0-1, P 0-0.03, S 0-0.015, Cr 18-22, Ni 18-22, Al 0-0.6, Ti 0-0.6, bal Fe.
SEL 90; W. Nr. 1.4847.

FERROTHERM 4860
Krupp Stahl AG
C 0-0.2, Si 2-3, Mn 0-1.5, P 0-0.045, S 0-0.03, Cr 20-22, Ni 28-31, bal Fe.
DIN 17740/83; W. Nr. 1.4860.

FERROTHERM 4861
Krupp Stahl AG
C 0-0.12, Si 0-1, Mn 0-1.5, P 0-0.045, S 0-0.03, Cr 19-22, Ni 30-34, bal Fe.
SEL 90; W. Nr. 1.4861.

FERROTHERM 4864
Krupp Stahl AG
C 0-0.15, Si 1-2, Cr 15-17, Ni 33-37, bal Fe.
Austenitic. DIN X12NiCrSi3616; W. Nr. 1.4864. 230 N/mm^2 YS, 550-750 N/mm^2 TS, 30 El.

FERROTHERM 4876
Krupp Stahl AG
C 0.04-0.12, Si 0-1, Al 0.15-0.6, Cr 19-23, Ni 30-34, Ti 0.15-0.6, bal Fe.
Austenitic. DIN X10NiCrAlTi3220; W. Nr. 1.4876. 170 N/mm^2 YS, 450-700 N/mm^2 TS, 30 El.

FERROTHERM 4877
Krupp Stahl AG
C 0.04-0.08, Si 0-0.3, Mn 0-1, P 0-0.02, S 0-0.01, Cr 26-28, Ni 31-33, Al 0-0.025, Ce 0.05-0.1, Nb 0.6-1, bal Fe.
SEL 90; W. Nr. 1.4877.

FERROTHERM 4878
Krupp Stahl AG
C 0-0.12, Si 0-1, Cr 17-19, Ni 9-11.5, Ti 0-0.8, 4 x C less than or equal to Ti, bal Fe.
Austenitic. DIN X12CrNiTi189; W. Nr. 1.4878. 210 N/mm^2 YS, 500-750 N/mm^2 TS, 40 El.

FERROTHERM 5310
Krupp Stahl AG
C 0-0.1, Si 0.7-1.1, Mn 0.7-1, P 0-0.035, S 0-0.035, bal Fe.
SEL 90; W. Nr. 1.5310.

FERROTHERM FF 112
Krupp Stahl AG
stainless steel. C 1, Cr 12, bal Fe.
For furnace parts, crucibles, autoclaves, recuperators; heat and corrosion resistant to 850°C. *Obsolete*

FERROTHERM FF 118
Krupp Stahl AG
stainless steel. C 1, Cr 18, bal Fe.
For furnace parts, crucibles, autoclaves, recuperators; heat and corrosion resistant to 1000°C. *Obsolete*

FERROTHERM FF 128
Krupp Stahl AG
stainless steel. C 1, Cr 28, bal Fe.
For furnace parts, crucibles, autoclaves, recuperators; heat and corrosion resistant to 1100°C. *Obsolete*

FERROTHERM FF 18
Krupp Stahl AG
stainless steel. C 0.15, Cr 18, bal Fe.
Annealed: 78,000 TS; 50,000 YS; 12 El; 12 RA. For furnace parts, crucibles, autoclaves, recuperators; heat and corrosion resistant to 1000°C. *Obsolete*

FERROTHERM FF 228
Krupp Stahl AG
stainless steel. C 2, Cr 28, bal Fe.
For furnace parts, crucibles, autoclaves, recuperators; heat and corrosion resistant to 1100°C. *Obsolete*

FERROTHERM FF 30
Krupp Stahl AG
stainless steel. C 0.15, Cr 30, bal Fe.
Annealed: 78,000 TS; 58,000 YS; 12 El; 8 RA. For furnace parts, crucibles, autoclaves, recuperators; heat and corrosion resistant to 1200°C. *Obsolete*

FERROTHERM FF 6
Krupp Stahl AG
alloy steel. C 0.15, Cr 6, bal Fe.
Heat treated: 78,000 TS; 50,000 YS; 18 El; 50 RA. For furnace parts, crucibles, autoclaves, recuperators; heat and corrosion resistant to 800°C. *Obsolete*

FERROTITE
Pacific Foundry Co.
C 0.2, bal Fe.
For welding rod.

FERROTROD 2-B
Eutectic Corp.
Electrode for AC-DC metallic arc for machinable build-up on steel; resists severe impact; 28-31 Rock C.

FERROTRODE N2B
Eutectic Corp.
For machinable overlays, and mild abrasion on ferrous metals; AC-DC; for hammers, sprockets, rollers, excavator equipment. Hardness C 28-31 Rock. *Obsolete*

FERROVAC 1020
Crucible Materials Corp.
C 0.2, Ni 0.01, Mn 0.003, P 0.005, Si 0.01, bal Fe.
For pole pieces for magnetrons, klystron components; vacuum melted. *Obsolete*

FERROVAC 4340
Crucible Specialty Steel
C 0.4, Cr 0.8, Ni 1.8, Mo 0.25, bal Fe.
Heat treated: 185,000-285,000 TS; 166,000-215,000 YS; 16-12 El; 60-41 RA; 410-540 Brin. For bolts, gears, shafts, crankshafts; vacuum melted, shock resistant.

FERROVAC 4340
Colt Industries
C 0.4, Cr 0.8, Ni 1.8, Mo 0.25, bal Fe.
Heat treated: 185,000-285,000 TS; 166,000-215,000 YS; 16-12 El; 60-41 RA; 410-540 Brin. For bolts, gears, shafts, crankshafts; vacuum melted, shock resistant.

FERROVAC 52100
Crucible Specialty Steel
C 0.95-1.1, Cr 1.3-1.6, bal Fe.
Annealed: 95,000 TS; 65,000 YS; 27 El; 62 RA; 180 Brin. Heat treated: 200,000 TS; 185,000 YS; 3 El; 30 RA; 400 Brin. For plug and ring gages, jet engine bearings; vacuum melted, wear resistant.

FERROVAC 52100
Colt Industries
C 0.95-1.1, Cr 1.3-1.6, bal Fe.
Annealed: 95,000 TS; 65,000 YS; 27 El; 62 RA; 180 Brin. Heat treated: 200,000 TS; 185,000 YS; 3 El; 30 RA; 400 Brin. For plug and ring gages, jet engine bearings; vacuum melted, wear resistant.

FERROVAC E
Crucible Materials Corp.
Fe 99.9.
Gas free, high purity iron. High maximum permeability. For relays, solenoid plungers, armatures, pole pieces, magnetic core devices.

FERROVAC HALMO
Crucible Materials Corp.
Tool material. C 0.58, Cr 4.75, Si 1.15, Mo 5.25, V 0.55, bal Fe.
For high temperature bearings, shears, punches; vacuum melted hot work steel.

FERROWELD
Pacific Foundry Co.
C 0.1, bal Fe.
For welding rods for cast iron; coated. AWS ESt.

FERROWELD
Lincoln Electric Co.
C 0.15, Si 0.03, Mn 0.3, P 0.04, S 0.04, bal Fe.
Low carbon steel, arc welding electrode. AWS Class ESt.

FERROXDURE
Dutch manufacture
BaO + 6 Fe$_2$O$_3$.
For permanent magnets.

FERROXDURE I

North American Philips Co. Inc.
$BaO + 6Fe_2O_3$.
For magnets for motors and oil filters; magnetic ceramic.

FERROXDURE II

North American Philips Co. Inc.
$BaFe_{12}O_{19}$.
For magnets for motors and oil filters; magnetic ceramic.

FERROZOID

Henry Wiggin & Co. Ltd.
Ni 33-35, Cr 3.5, bal Fe.
Now known as "Dullray". For electrical resistances; resistance alloy. *Obsolete*

FERRUL

English manufacture
Cu 54.6, Zn 40, Pb 5, Al 0.4.
For bearings, intricate castings; free-cutting.

FERRY

English manufacture
Ba 2, Ca 1, Hg 0.25, bal Pb.
For bearings, solder.

FERRY

Mond Nickel Co. Ltd.
Cu 55-56, Ni 44-45.
Annealed: 73,000 TS; 34,000 YS; 45 El; 77 RA; 160 Brin. For electrical resistances, thermocouples. Max. operating temperature 440°C. Corrosion resistant.

FERRY

Henry Wiggin & Co. Ltd.
Cu 55-56, Ni 44-45.
Annealed: 73,000 TS; 34,000 YS; 45 El; 77 RA; 160 Brin. For electrical resistances, thermocouples. Max. operating temperature 440°C. Corrosion resistant.

FERRY ALLOY

Inco Alloys International Inc.
Copper. Cu 55, Fe 0-1, C 0-0.1, Mn 0-1, Si 0-0.5, bal Ni.
Annealed: 60,000 TS; 21,000 YS; 32 El. For electrical properties; medium-range electrical resitivity, low temperature coefficient of resistance. For wire-wound precision resistors having optimum temperatures up to 750°F.

FERRYDUR

Ferry-Capitain, Usines De Bussy
C 1.8-2, Cr 12-13, C, others, bal Fe.
Up to 48-58 Rock C hardness. Excellent wear resistance. For castings to resist wear. AFNOR Z 180 CD 12 (approximately).

FERRYNOX 10

Ferry-Capitain, Usines De Bussy
C 0.05-0.1, Cr 16-20, Ni 8-10, Ti, bal Fe.
Corrosion resistant; temperature resistant; austenitic. For food equipment, sugar refineries. AFNOR Z 6 CN 18.10. *Obsolete*

FERRYNOX 12

Ferry-Capitain, Usines De Bussy
C 0.18-0.25, Cr 22-27, Ni 11-13, bal Fe.
Good high temperature properties. Austenitic, corrosion resistant. For furnace and heat treat equipment. AFNOR Z 20 CN 25.13. *Obsolete*

FERRYNOX 15

Ferry-Capitain, Usines De Bussy
C 0.18-0.25, Cr 23-27, Ni 13-16, bal Fe.
Good high temperature properties. Furnace equipment, jet engine parts. AFNOR Z 20 CN 25.15. *Obsolete*

FERRYNOX 20

Ferry-Capitain, Usines De Bussy
C 0.18-0.25, Cr 24-26, Ni 19-22, bal Fe.
Good high temperature properties. For jet engine, gas turbines, heat exchangers. AFNOR Z 20 CN 25.20. AISI 310 Stainless. *Obsolete*

FERRYNOX 35

Ferry-Capitain, Usines De Bussy
C 0.18-0.25, Cr 14-16, Ni 34-36, bal Fe.
Good high temperature properties in carburizing atmospheres. For heat treat equipment, petroleum industry. AFNOR Z 20 NC 35.15. *Obsolete*

FERRYNOX 65

Ferry-Capitain, Usines De Bussy
C 0.18-0.25, Cr 14-16, Ni 62-66, bal Fe.
Good high temperature properties, good corrosion resistance. For chemical plant equipment. AFNOR NC 15 Fe. *Obsolete*

FERRYNOX 8

Ferry-Capitain, Usines De Bussy
C 0.18-0.25, Cr 17-19, Ni 7-9, bal Fe.
Corrosion resistant, good resistance to scaling; austenitic. For ventilator fans, heat exchangers. AFNOR Z 20 CN 18.08. *Obsolete*

FERRYNOX 8M

Ferry-Capitain, Usines De Bussy
C 0.18-0.25, Cr 16-20, Ni 8-10, Mn 3, bal Fe.
Good corrosion and heat resistance; austenitic. For pumps for chemical plants. AFNOR Z 20 CNM 18.10. *Obsolete*

FERRYNOX 8S

Ferry-Capitain, Usines De Bussy
C 0.18-0.25, Cr 16-20, Ni 8-10, Mo 2-3, bal Fe.
Excellent corrosion resistance, good resistance to scaling at elevated temperatures; austenitic. For pump rotors, compressors. AFNOR Z 20 CND 18.10. *Obsolete*

FERRYNOX CO 50

Ferry-Capitain, Usines De Bussy
Co 50, Cr 28, bal Fe.
RT: 60-70 kg/mm^2 TS; 2-4 El. 1000°C: 11-13 kg/mm^2 TS; 15-17 El. Good strength at high temperatures. Resistant to oxidation and many chemicals. For furnace equipment. *Obsolete*

FESTEL METAL

Cabot Corporation
Fe 55, Co 23, Cr 21, C 0.7.
For weld metal. *Obsolete*

FESTEL STELLITE

Cabot Corporation
Co-Cr-W.
Cast: 101,000 TS; 1 El. Hammered: 154,000 TS; 5.5 El. For welding rod, fabrication of cutlery. *Obsolete*

FH "62"

Joseph T. Ryerson & Son Inc.
C 0.58, Mn 0.8, Si 0.25, bal Fe.
Induction or flame hardenable to 60-66 Rock C. For machine ways, gibs, rails, slides, wear strips.

FH-2

Howment Corp. (Carbide Div.)
Sintered carbide tool material. For heading dies and punches. *Obsolete*

FH-3

Howment Corp. (Carbide Div.)
Sintered carbide material. For heading dies and punches. *Obsolete*

FH-4

Howment Corp. (Carbide Div.)
Sintered carbide material. For heading dies and punches. *Obsolete*

FH-5

Howment Corp. (Carbide Div.)
Sintered carbide tool material. For heading dies and punches *Obsolete*

FH-6

Howment Corp. (Carbide Div.)
Sintered carbide tool material. For heading dies and punches *Obsolete*

FIBRO

British Steel Corp.
C 0.15-0.25, bal Fe.
Heat treated: 66,000-76,000 TS; 44,000-54,000 YS; 27-32 El; 60-65 RA; 153-166 Brin. For case hardened parts, gears, shafts. *Obsolete*

FIBRTOUGH

Atlas Steels Ltd.
C 0.15, Mn 0.7, Si 0.25, bal Fe.
For boiler staybolts; high ductility. *Obsolete*

FIELD

English manufacture
Si 10, Cu 3, Al 1.5-2, bal Ni.
For corrosion resisting parts; stainless.

FIELD ROLLED

English manufacture
Ni 60, Fe 29, Mo 20.
Stainless and corrosion resistant.

FIFTEEN PER CENT

English manufacture
Cu 56.66, Zn 28.33, Ni 15.
For ornamental parts; corrosion resistant.

FIFTY N

United States Steel Corp.
C 0.18, Mn 1.3, Si 0.35, Cb 0.05, bal Fe.
Normalized: 70 ksi TS; 50 ksi YS; 23 El. Readily formed and welded. For stressed structures at low temperature; arctic and marine structures. ASTM A633 Grade C.

FIFTY-FIVE

Lake Erie Engineering Corp.
C 2.9-3.1, Ni 1.2-1.5, Mo 0.4-0.6, Si 1.4-1.6, bal Fe.
Cast: 55,000-62,000 TS. For hydraulic press and diesel engine parts; cast iron.

FIL-SODER

Swiss Laboratory Inc.
Sn-Pb.
For solder for Al; no flux required.

FILBRITE

English manufacture
Cr 17, Ni 35, C, bal Fe.
For heat and corrosion resisting parts.

FILE BRONZE

American manufacture
Sn 18-31, Pb 7-8.5, Zn 0-10, bal Cu.
For plumbing, hardware; free-cutting.

FILE METAL (GENFER)

Manufacturer not listed.
Cu 62-64.6, Sn 18-20, Zn 10, Pb 7.6-8.
For nail file; free-cutting.

FILE METAL (VOGEL)

Manufacturer not listed.
Cu 51-73, Sn 19-28.5, Zn 0-7, Pb 7-8.5.
For nail file; free-cutting.

FILIPOFF LEAD-CALCIUM

Manufacturer not listed.
Cu 1.37, Ca 1.9, Sr 1, Ba 1.1, Na 0.1, bal Pb.
For bearings; anti-friction.

FILNIC F

Driver Harris Co.
Ni-Co-Ti-Fe.
For resistance metal, filaments. *Obsolete*

FILNIC H
Driver Harris Co.
Ni 99.9.
For filaments. *Obsolete*

FILNIC I
Driver Harris Co.
Si, bal Ni.
For filaments. *Obsolete*

FILNIC J
Driver Harris Co.
Si 3, al N i.
For filaments. *Obsolete*

FILNIC K
Driver Harris Co.
Al-Ni.
For filaments; heat resistant. *Obsolete*

FILNIC M
Driver Harris Co.
Co 45, Ni 55.
For filaments; heat resistant. *Obsolete*

FILNIC T
Driver Harris Co.
Ni alloy.
For filaments. *Obsolete*

FINE SILVER
Fansteel Metals
Precious metal. Ag 99.9.
Annealed: 30 Rock-15T; Electrical conductivity: 106% IACS.
For electrical contacts. *Obsolete*

FINE SILVER
Handy & Harmon
Precious metal. Ag 99.9.
Hard: 41,000 TS; 30,000 YS; 3 El. Annealed: 23,000 TS;
10,000 YS; 35 El; 38 RA. MP: 1761°F. For jewelry, alloys.

FINGAL
Uddeholm Corp.
C 0.5, Cr 0.6, W 0.5, Mo 0.2, bal Fe.
For pneumatic tools; oil or water hardened. *Obsolete*

FINIS
Apex Steel Co.
C 0.7, W 18, Cr 4, V 1, bal Fe.
For high speed tools and cutters; high speed steel.

FINISHING SPECIAL
Midvale-Heppenstall Co.
C, Cr, W, bal Fe.
For tools.

FINKL COLD-HOT
A. Finkl & Sons Co.
C 0.9, Cr 0.7, Mo 0.2, Ni 0.5, bal Fe.
For cold and hot trimming dies; heat resistant; tough.
Obsolete

FINKL CUPRODIE
A. Finkl & Sons Co.
Tool material. C 0.5-0.6, Mn 0.65-0.9, Ni 1.4-1.75, Cr 0.8-1.1,
Mo 0.25-0.35, Cu 0.6-0.9, bal Fe.
Prehardened (as required) from 37-50 Rock C. Mechanical
properties depend on hardness. For hammer dies, forging
press dies, drop hammer dies. Resists heat checking.

FINKL DC
A. Finkl & Sons Co.
Tool material. C 0.37-0.42, Mn 0.2-0.42, Si 0.85-1.2, Cr 5-5.5,
Mo 1.2-1.75, V 0.88-0.92, bal Fe.
Vacuum electric furnace degassed. As annealed: 229 Brin.
For die casting dies for Al, Mg, Zn, lead and tin base alloys;
for plastic mold dies. AISI H 13; hot work steel.

FINKL F
A. Finkl & Sons Co.
Tool material. C 0.55, Cr-Ni-Mo steel.
Prehardened to ordered hardness. Economy hot work steel
for short runs as hammer dies, saw blocks and other hot work
tools. *Obsolete*

FINKL FI-CARBO
A. Finkl & Sons Co.
High C, bal Fe.
For dies for hot trimming drop forgings of simple sections;
water hardened. *Obsolete*

FINKL FS
A. Finkl & Sons Co.
Tool material. C 0.55-0.58, Mn 0.75-0.95, Si 0.15-0.35, Ni
0.85-1.05, Cr 0.85-1.15, Mo 0.33-0.43, bal Fe.
Vacuum degassed. FS is the annealed condition, see FINKL
FX for prehardened condition. For forging dies and other hot
working tools.

FINKL FX
Now FX-XTRA.

FINKL GRADE F
A. Finkl & Sons Co.
C 0.7, Mn 1.2, bal Fe.
229-352 Brin. For dies for short runs of simple forgings, sow
blocks; treated dies. *Obsolete*

FINKL HB
A. Finkl & Sons Co.
Tool material. C 0.47-0.55, Mn 0.95-1.3, Si 0.27, Cr 0.5-0.9, S
0.06-0.1, Mo 0.1-0.25, bal Fe.
Prehardened: 28-34 Rock C (or as required). For holder
blocks, forging and extrusion dies, die casting and plastic
molding dies. Prehardened; good machinability.

FINKL MD
A. Finkl & Sons Co.
Tool material. C 0.3-0.35, Mn 0.7-0.9, Si 0.35-0.55, Cr
1.55-1.85, Mo 0.4-0.5, bal Fe.
Prehardened: 28-34 Rock C. Vacuum degassed. For zinc die
casting dies, plastic molds. Prehardened; Type P20; tool
steel.

FINKL MO-CRO-NI
A. Finkl & Sons Co.
C, Cr, Mo, Ni, bal Fe.
For dies for hot trimming medium and heavy section drop
forgings; heat resistant. *Obsolete*

FINKL TYPE R
A. Finkl & Sons Co.
C, Ni, Cr, Mo, bal Fe.
Forged: 110,000 TS; 86,000 YS; 23 El; 66 Brin. For
locomotive side rods; forgings. *Obsolete*

FINKL W4X
A. Finkl & Sons Co.
Tool material. C 0.37, Mn 0.45, Si 1, Cr 5, Mo 1.45, V 0.35, W
1.25.
Annealed or hardened as required: 37-46 Rock C. For forging
press dies, inserts, extrusion press rams, liners, forging
machine headers. AISI H-12; hot work steel. *Obsolete*

FINKL WF
A. Finkl & Sons Co.
Tool material. C 0.33-0.38, Mn 0.45-0.65, Si 0.4-0.6, Ni
0.5-0.8, Cr 2.2-2.6, Mo 0.9-1.1, bal Fe.
Prehardened (as required): 28-50 Rock C. Mechanical
properties depend on hardness. For hammer dies, press
dies, punches, headers, gripper dies, insert dies, saw blocks.

FIREARMOR "A"
Michiana Products Corp.
Nickel. Ni 65, Cr 15, Mn 1.75, C 0.5, Si 1, bal Fe.
Cast: 60,000 TS; 46,000 YS; 2.9 El; 1.2 RA; 190 Brin. For
carburizing boxes; heat and abrasion resisting.

FIREARMOR B
Michiana Products Corp.
Nickel. Cr 15, Ni 60, C 0.5, Mn 1.7, Si 1.2, bal Fe.
Cast: 62,000 TS; 33,000 YS; 1.5 El; 3.0 RA; 200 Brin. For
furnace parts, grates; heat resistant.

FIREDIE
Columbia Tool Steel Co.
Tool material. C 0.38, Cr 5, Mn 0.45, V 0.5, Mo 1.4, Si 1, bal
Fe.
For hot work dies, die casting dies; hot work die steel. AISI
H11.

FIREDIE 13
Columbia Tool Steel Co.
Tool material. C 0.38, Mo 1.4, Cr 5.2, V 1.05, Si 1, bal Fe.
Heat treated: 285,000 TS; 225,000 YS; 4 El; 54 Rock C. For
hot extrusion tools and die casting dies for Al, Mg, and Zn
alloys. Hot work tool steel. Type H13; tough; shock resistant.

FIREDIE 13 SMOOTHCUT
Columbia Tool Steel Co.
C 0.38, Si 1, Mo 1.4, Cr 5.2, V 1.05, Sulphides, bal Fe.
Heat treated: 286,000 TS; 226,000 YS; 3 El; C 54 Rock. For
hot sizing dies for forming titanium alloys, hot extrusion and
die casting dies, scroll shear dies. Free machining, oil
hardening. Hot work steel Type H13. *Obsolete*

FIREDIE 9
Columbia Tool Steel Co.
Tool material. C 0.38, Mn 0.3, Si 0.9, Cr 3.6, Mo 3, V 0.65, Co
2, bal Fe.
Hot work tool steel; for forging and extrusion dies.

FIREX
Darwin & Milner Inc.
Tool material. C 0.4, Ni 4.5, Cr 1.1, Si 0.3, Mn 0.6, bal Fe.
For cutting tools, hot shears, shock resisting tools, chisels; air
or oil hardened.

FIREX SPECIAL
Darwin & Milner Inc.
Tool material. C 0.4-0.55, Cr 0.4-1, Ni 2.75-4, Mn 0.4-0.9, V
0.15-0.2, Mo 0.4-0.7, bal Fe.
For tools, dies, punches; shock resistant, tough.

FIRINIT
Metallo-Chemische Fabrik
For welding and soldering rods.

FIRMILAY
J.F. Jelenko & Co.
Precious metal. Au 74.5, Pd 3.5, Ag 11.
Quenched: 63,000 psi TS; 30,000 psi YS; 39 El; 110 Brin.
Hardened: 77,000 psi TS; 40,000 psi YS; 19 El; 165 Brin.
Type III hard dental alloy. For inlays, crowns, and fixed
bridgework.

FIRMINY A.C.M.
Creusot-Loire
C 0.55, Ni 1.4, Cr 0.6, Mo 0.4, V 0.1, bal Fe.
For forging dies; oil hardened. *Obsolete*

FIRMINY A.D.F.
Creusot-Loire
C 0.28, Ni 3.5, Cr 1.1, bal Fe.
For gears, bolts, crankshafts, machine tool parts; oil
hardened, tough. *Obsolete*

FIRMINY A.D.F.M.
Creusot-Loire
C 0.3, Cr 1.3, Ni 3.5, Mo 0.4, bal Fe.
For forging dies, crankshafts; oil hardened, tough. *Obsolete*

FIRMINY A.L.S.
Creusot-Loire
C 1.45, W 5.5, Cr 0.55, V 0.5, bal Fe.
For header dies, engraving tools; oil hardened. *Obsolete*

FIRMINY A.M.3
Creusot-Loire
C 0.35, Mn 1.1, bal Fe.
For gears, bolts, crankshafts; oil hardened, tough. *Obsolete*

FIRMINY B.L.N.
Creusot-Loire
C 0.28, W 10, Cr 3, Mo 0.3, V 0.6, bal Fe.
For extrusion rams, dies and liners, punches; hot work steel, oil hardened. *Obsolete*

FIRMINY B.O.L.
Creusot-Loire
C 0.38, W 4.2, Cr 3.4, Mo 0.6, V 1, bal Fe.
For shear blades, punches, pneumatic dies; oil hardened. *Obsolete*

FIRMINY B.T.R.
Creusot-Loire
C 0.35, Ni 4.7, Cr 0.35, Mo 1.15, bal Fe.
For forging and plastic mold dies; oil hardened. *Obsolete*

FIRMINY C.L.D.
Creusot-Loire
C 0.9, Cr 1.5, bal Fe.
For bearings, dies; oil hardened, wear resistant. *Obsolete*

FIRMINY C.M.Y.1
Creusot-Loire
C 0.5, Cr 1, Mo 0.25, bal Fe.
For bolts, gears, crankshafts; oil hardened, tough. *Obsolete*

FIRMINY C.M.Y.16
Creusot-Loire
C 0.13, Cr 5.5, Mo 0.6, bal Fe.
For oil refinery equipment; creep resistant. *Obsolete*

FIRMINY C.M.Y.2
Creusot-Loire
C 0.35, Cr 1, Mo 0.25, bal Fe.
For bolts, gears, crankshafts; oil hardened, tough. *Obsolete*

FIRMINY C.M.Y.4
Creusot-Loire
C 0.25, Cr 1, Mo 0.25, bal Fe.
For bolts, gears, crankshafts, machine tool parts; oil hardened, tough. *Obsolete*

FIRMINY C.M.Y.5
Creusot-Loire
C 0.2, Cr 1, Mo 0.25, bal Fe.
For crankshafts, machine tool parts; case hardened, tough. *Obsolete*

FIRMINY C.M.Y.6
Creusot-Loire
C 0.12, Cr 0.5, Mo 0.5, bal Fe.
For machine tool parts, cams; case hardened, tough. *Obsolete*

FIRMINY C.M.Y.7
Creusot-Loire
C 0.08, Cr 1, Mo 0.25, bal Fe.
For machine tool parts, cams, camshafts; case hardened, tough. *Obsolete*

FIRMINY C.N.W.
Creusot-Loire
C 0.38, W 3, Cr 2.5, Mo 0.7, V 0.25, bal Fe.
For pneumatic tools, chisels, upsetting dies; oil hardened. *Obsolete*

FIRMINY C.R.V.
Creusot-Loire
C 1.5, Cr 1.8, V 0.25, bal Fe.
For bearings, dies; oil hardened, wear resistant. *Obsolete*

FIRMINY C.T.
Creusot-Loire
C, alloy, bal Fe.
For machine tool parts; oil hardened. *Obsolete*

FIRMINY C.T.N.2
Creusot-Loire
C 0.1, Ni 2, bal Fe.
For camshafts, cams, fasteners; case hardened, tough. *Obsolete*

FIRMINY C.T.N.6
Creusot-Loire
C 0.1, Ni 6, bal Fe.
For hobbing dies. *Obsolete*

FIRMINY C.T.N.C.4
Creusot-Loire
C 0.16, Cr 1.8, Ni 1, Mo 0.1, bal Fe.
For cams, camshafts; case hardened, tough. *Obsolete*

FIRMINY C.T.N.C.5
Creusot-Loire
C 0.1, Ni 1, Cr 1.4, Mo 0.1, bal Fe.
For machine tool parts; case hardened, tough. *Obsolete*

FIRMINY C.T.N.M.
Creusot-Loire
C 0.1, Ni 1.2, Mo 0.12, Cr 0.1, bal Fe.
For gears, bolts, fasteners, cams; case hardened, tough. *Obsolete*

FIRMINY C.T.N.V.
Creusot-Loire
C 0.1, Ni 3, Cr 0.3, bal Fe.
For camshafts, cams, gears, fasteners; case hardened, shock resistant. *Obsolete*

FIRMINY C.V.A.1
Creusot-Loire
C 0.5, Cr 1, V 0.15, bal Fe.
For gears, bolts, crankshafts; oil hardened, shock resistant. *Obsolete*

FIRMINY C.V.A.2
Creusot-Loire
C 0.4, Cr 1, V 0.15, bal Fe.
For gears, bolts, crankshafts; oil hardened, shock resistant. *Obsolete*

FIRMINY C.V.A.3
Creusot-Loire
C 0.3, Cr 2.5, V 0.2, bal Fe.
For oil refinery equipment; creep resistant. *Obsolete*

FIRMINY C.V.A.4
Creusot-Loire
C 0.3, Cr 1, V 0.15, bal Fe.
For gears, bolts, machine tool parts; oil hardened, tough. *Obsolete*

FIRMINY C.V.A.6
Creusot-Loire
C 0.18, Cr 1, V 0.1, bal Fe.
For gears, bolts, camshafts; case hardened, shock resistant. *Obsolete*

FIRMINY C.V.A.7
Creusot-Loire
C 0.12, Cr 1, V 0.1, bal Fe.
For gears, bolts, camshafts; case hardened, shock resistant. *Obsolete*

FIRMINY C.W.2
Creusot-Loire
C 0.55, W 1.8, Cr 1, Si 0.9, bal Fe.
For header dies; oil or water hardened. *Obsolete*

FIRMINY C.W.3
Creusot-Loire
C 0.45, W 1.8, Cr 1, Si 0.8, bal Fe.
For header dies; water or oil hardened. *Obsolete*

FIRMINY F.C.K.
Creusot-Loire
C 1.9, Cr 13, bal Fe.
For blanking and forming dies; oil or air hardened. *Obsolete*

FIRMINY F.G.O.
Creusot-Loire
C 2, W 3.75.
bal Fe. *Obsolete*

FIRMINY G.R.N.
Creusot-Loire
C 0.42, Si 2, bal Fe.
For machine tool parts; heat resistant. *Obsolete*

FIRMINY ICN001
Creusot-Loire
C 0.11-0.16, Cr 18, Ni 8, bal Fe.
Annealed: 80,000 TS; 35,000 YS; 35 El; 75 RA; 150 Brin. For chemical plant equipment, tanks, mixers, filters; Type 302; stainless, austenitic. *Obsolete*

FIRMINY ICN164
Creusot-Loire
C 0.12, Cr 18-19, Ni 10-12, Mo 2.5-3.5, bal Fe.
Annealed: 85,000 TS; 35,000 YS; 50 El; 65 RA; 160 Brin. For acid resistant chemical plant equipment, mixers; Type 316 and 317; stainless, austenitic. *Obsolete*

FIRMINY ICN164T
Creusot-Loire
C 0-0.12, Cr 18, Ni 10, Mo 2.5, Ti, bal Fe.
Annealed: 85,000 TS; 35,000 YS; 50 El; 65 RA; 160 Brin. For acid resistant chemical plant equipment, mixers; Type 316; stainless, austenitic. *Obsolete*

FIRMINY ICN472
Creusot-Loire
C 0.08, Cr 18, Ni 8, bal Fe.
Annealed: 85,000 TS; 35,000 YS; 60 El; 70 RA; 150 Brin. For chemical plant equipment, tanks, mixers; Type 304; stainless, austenitic. *Obsolete*

FIRMINY ICN472T
Creusot-Loire
C 0-0.08, Cr 18, Ni 8, Ti, bal Fe.
Annealed: 85,000 TS; 35,000 YS; 55 El; 65 RA; 150 Brin. For welded chemical plant equipment, tanks, mixers; Type 321; stainless, austenitic. *Obsolete*

FIRMINY ICN583
Creusot-Loire
C 0-0.05, Cr 18, Ni 8, bal Fe.
Annealed: 85,000 TS; 35,000 YS; 60 El; 70 RA; 150 Brin. For chemical plant equipment, tanks, mixers; Type 304; stainless, austenitic. *Obsolete*

FIRMINY L.F.
Creusot-Loire
C 1.15, Cr 0.7, bal Fe.
For bearings, cutters, dies; oil or water hardened, wear resistant. *Obsolete*

FIRMINY M.12A.F.Y.
Creusot-Loire
C 1, Mn 12, bal Fe.
For wear resistant parts; wear and abrasion resistant. *Obsolete*

FIRMINY M.14A.F.Y.
Creusot-Loire
C 1.2, Mn 14, Cr 1.5, bal Fe.
For wear resistant parts; wear and abrasion resistant. *Obsolete*

FIRMINY M.A.S.
Creusot-Loire
C 0.38, Mn 1.2, Si 1.2, bal Fe.
For machine tool parts; tough. *Obsolete*

FIRMINY M.C.R.1
Creusot-Loire
C 0.45, Cr 0.9, Mn 0.7, bal Fe.
For machine tool parts, gears; oil or water hardened.
Obsolete

FIRMINY M.C.R.2
Creusot-Loire
C 0.35, Cr 0.9, Mn 0.7, bal Fe.
For machine tool parts, gears; oil or water hardened.
Obsolete

FIRMINY M.C.R.3
Creusot-Loire
C 0.15, Cr 0.9, Mn 1.2, bal Fe.
For camshafts, cams, bolts; case hardened, tough. *Obsolete*

FIRMINY M.C.R.4
Creusot-Loire
C 0.17, Cr 0.8, Mn 0.7, bal Fe.
For camshafts, cams, gears; case hardened, tough. *Obsolete*

FIRMINY M.C.R.5
Creusot-Loire
C 0.1, Cr 0.8, Mn 0.7, bal Fe.
For camshafts, cams, machine tool parts; case hardened,
tough. *Obsolete*

FIRMINY M.L.
Creusot-Loire
C 0.92, Mn 1.3, Cr 0.5, bal Fe.
For punches, blanking and forming dies; oil hardened, non-
deforming. *Obsolete*

FIRMINY N.2
Creusot-Loire
C 0.28, Ni 2, bal Fe.
For gears, bolts, fasteners; oil hardened, tough. *Obsolete*

FIRMINY N.3
Creusot-Loire
C 0.28, Ni 3, bal Fe.
For gears, bolts, fasteners; oil hardened, tough. *Obsolete*

FIRMINY N.5
Creusot-Loire
C 0.28, Ni 5, bal Fe.
For gears, bolts, fasteners, machine tool parts; oil hardened,
tough. *Obsolete*

FIRMINY N.C.1
Creusot-Loire
C 0.4, Ni 2.8, Cr 0.7, bal Fe.
For gears, bolts, crankshafts, fasteners; oil hardened, shock
resistant. *Obsolete*

FIRMINY N.C.2
Creusot-Loire
C 0.3, Ni 2.8, Cr 0.5, bal Fe.
For machine tool parts, bolts, gears; oil hardened, shock
resistant. *Obsolete*

FIRMINY N.C.3
Creusot-Loire
C 0.25, Ni 2.8, Cr 0.5, bal Fe.
For machine tool parts, bolts, gears; oil hardened, shock
resistant. *Obsolete*

FIRMINY N.C.4
Creusot-Loire
C 0.18, Ni 2.8, Cr 0.5, bal Fe.
For camshafts, cams, gears; case hardened, tough. *Obsolete*

FIRMINY N.C.M.2
Creusot-Loire
C 0.32, Ni 2.6, Cr 0.7, Mo 0.6, bal Fe.
For gears, bolts, crankshafts, fasteners; oil hardened, shock
resistant. *Obsolete*

FIRMINY N.C.M.4
Creusot-Loire
C 0.16, Ni 2.9, Cr 0.45, Mo 0.3, bal Fe.
For machine tool parts, cams; case hardened, shock
resistant. *Obsolete*

FIRMINY P.F.
Creusot-Loire
C 0.14, Ni 4, Cr 0.9, bal Fe.
For camshafts, cams, gears, fasteners; case hardened, shock
resistant. *Obsolete*

FIRMINY P.F.3
Creusot-Loire
C 0.1, Ni 3.5, Cr 0.6, bal Fe.
For camshafts, machine tool parts, gears; case hardened,
shock resistant. *Obsolete*

FIRMINY S.A.M.
Creusot-Loire
C 1.15, W 1.9, Cr 0.45, bal Fe.
For heading dies; oil hardened. *Obsolete*

FIRMINY T.R.D.
Creusot-Loire
C 1.3, W 0.8, Cr 0.5, bal Fe.
For cutters, dies; oil hardened. *Obsolete*

FIRMINY T.S.W.
Creusot-Loire
C 0.42, Si 2, W 0.4, bal Fe.
For machine tool parts; heat resistant. *Obsolete*

FIRMINY V.D.L.
Creusot-Loire
C 0.14, Ni 5.5, Cr 1.5, bal Fe.
For camshafts, cams, gears, bolts; case hardened, shock
resistant. *Obsolete*

FIRMINY V.D.L.D.
Creusot-Loire
C 0.35, Ni 3.8, Cr 1.5, bal Fe.
For gears, bolts, crankshafts; oil hardened, shock resistant.
Obsolete

FIRST QUALITY
Houghton & Richards Inc.
C 0.8-1.2, bal Fe.
For tools, punches, dies; water hardening.

FIRTH
Firth Brown Ltd.
C 1.05, Mn 0.22, bal Fe.
For tools, dies, taps, drills; water hardened. *Obsolete*

FIRTH "AW"
Firth Brown Ltd.
C 1.2, Mn 0.25, bal Fe.

FIRTH "DIEHARD"
Firth Brown Ltd.
C 1.5, Cr 13, bal Fe.
Hardened: 550-660 Brin. For forging, blanking and drawing
dies, cold trimming and punching dies; abrasion resisting
"Firth Brown M-254." Air-hardening. *Obsolete*

FIRTH "F-65"
Firth Brown Ltd.
C 0.2-0.3, Ni 2.75-3.5, Cr 1-1.4, V 0-0.25, Mo 0.65, bal Fe.
Oil treated: 146,000-168,000 TS; 127,000-135,000 YS; 17 El;
40 RA. For crank shafts, propeller shafts, rear axles, gear box
shafts, connecting rods; "Firth Brown E-81"; for high duty
work. *Obsolete*

FIRTH "NONVAR" CAST STEEL
Firth Brown Ltd.
C, bal Fe, high Mn.
Hardened: 444-600 Brin. For taps, boiler stay bolts, punches,
dies, jigs, gages; "Firth Brown M-240." *Obsolete*

FIRTH "RBD" STEEL
Firth Brown Ltd.
C, high W, Cr, bal Fe.
Hardened: 286-418 Brin. For hot rivet dies, die blocks, shear
blades, hot iron punches; air hardening "Firth Brown M-261."
Obsolete

FIRTH 16-25-6
Teledyne Firth Sterling
Steel. C 0.08, Cr 17, Ni 25, Mo 6, N 0.15, bal Fe.
Rolled: 115,000 TS; 60,000 YS; 35 El; 50 RA; 201 Brin. Cold
drawn: 150,000 TS; 120,000 YS; 19 El; 40 RA; 300 Brin. For
jet engine components, fasteners, bolts; ductile, high
temperature applications to 1300°F. *Obsolete*

FIRTH 19-9DL
Teledyne Firth Sterling
Steel. C 0.28-0.35, Cr 18-21, Ni 8-11, Mo 1.5, W 1.4, Cb 0.5,
Ti 0.4, Mn 1.2, bal Fe.
Hot rolled: 118,000 TS; 69,000 YS; 58 El; 55 RA; 216 Brin. At
1000°F: 89,000 TS; 42,000 YS; 43 El; 52 RA. For turbine
wheels, bolts, turbo superchargers; high temperature uses to
1200°F. *Obsolete*

FIRTH 19-9DX
Teledyne Firth Sterling
Steel. C 0.3, Cr 19, Ni 9, Mo 1.5, W 1.2, Ti 0.55, bal Fe.
Hot rolled: 118,000 TS; 69,000 YS; 58 El; 55 RA; 216 Brin. At
1000°F: 89,000 TS; 42,000 YS; 43 El; 52 RA. For jet engine
fasteners; high temperature uses to 1200°F. *Obsolete*

FIRTH A-286
Teledyne Firth Sterling
Steel. C 0.05, Mn 1.35, Cr 15, Ni 26, Mo 1.2, Ti 2, V 0.3, bal
Fe.
Rolled: 135,000-160,000 TS; 85,000-115,000 YS; 20-30 El;
30-55 RA. For gas turbine discs, jet engine components;
austenitic, heat treatable. *Obsolete*

FIRTH BEST TOOL
Teledyne Firth Sterling
Cr 0.1, Mn 0.28, C, bal Fe.
For drills, tools, dies; water hardening. *Obsolete*

FIRTH BROWN 50CT
Firth Brown Ltd.
C 0.48, Si 0.25, Mn 0.7, bal Fe.
For streetcar and railway wheel tires; water hardened.
Obsolete

FIRTH BROWN 60CT
Firth Brown Ltd.
C 0.6, Si 0.25, Mn 0.7, bal Fe.
For streetcar and railway wheel tires; water hardened.
Obsolete

FIRTH BROWN 65CF
Firth Brown Ltd.
C 0.65, Si 0.35, Mn 0.7, bal Fe.
For machine tool parts, springs; water hardened. *Obsolete*

FIRTH BROWN 65CT
Firth Brown Ltd.
C 0.65, Si 0.25, Mn 0.7, bal Fe.
For streetcar and railway wheel tires; water hardened.
Obsolete

FIRTH BROWN CAST STEEL WARRANTED
Firth Brown Ltd.
C 0.7-1.4, Si 0.15, Mn 0.3, bal Fe.
For drills, reamers, broaches, hobs; Type W1; water hardened
Obsolete

FIRTH BROWN CHROMVA-W
Firth Brown Ltd.
C 0.17, Si 0.3, Mn 0.3, Cr 2.75, Mo 0.6, V 0.7, W 0.5, bal Fe.
For elevated temperature applications. *Obsolete*

FIRTH BROWN CRM5
Firth Brown Ltd.
C 0.13, Si 0.25, Mn 0.4, Cr 5.25, Mo 0.55, bal Fe.
For oil refinery equipment; creep and corrosion resistant, high temperature use. *Obsolete*

FIRTH BROWN CRM6
Firth Brown Ltd.
C 0.12, Cr 2.25, Mo 1, Mn 0.45, Si 0.2, bal Fe.
For oil refinery equipment; creep and corrosion resistant. *Obsolete*

FIRTH BROWN CRMO
Firth Brown Ltd.
C 0.1, Si 0.4, Mn 0.5, Cr 0.85, Mo 0.5, bal Fe.
For oil refinery equipment; creep resistant, high temperature use. *Obsolete*

FIRTH BROWN F60C
Firth Brown Ltd.
C 0.6, Si 0.25, Mn 0.65, bal Fe.
Heat treated: 160,000 TS; 113,000 YS; 12 El; 40 RA; 320 Brin. For die blocks, girders, clutch discs; water hardened. *Obsolete*

FIRTH BROWN F65CH
Firth Brown Ltd.
C 0.18, Si 0.25, Mn 0.7, Ni 1.3, Cr 1, Mo 0.1, bal Fe.
For gears, pinions, camshafts, cams, machinery parts, case hardened; Brit.EN353; shock resistant. *Obsolete*

FIRTH BROWN GK3
Firth Brown Ltd.
C 0.35, Si 0.3, Mn 0.5, Cr 2, Mo 0.25, V 0.15, bal Fe.
For crankshafts, oil refinery equipment; oil hardened, creep resistant. *Obsolete*

FIRTH BROWN GK5
Firth Brown Ltd.
C 0.25, Si 0.3, Mn 0.5, Cr 2, Mo 0.25, V 0.15, bal Fe.
For crankshafts, oil refinery equipment; oil hardened, creep resistant. *Obsolete*

FIRTH BROWN GK7
Firth Brown Ltd.
C 0.18, Si 0.3, Mn 0.5, Cr 2, Mo 0.25, V 0.15, bal Fe.
For oil refinery equipment; creep resistant, high temperature use. *Obsolete*

FIRTH BROWN HACKSAW
Firth Brown Ltd.
C 1.2, W 1.25, Mn 0.25, Si 0.25, bal Fe.
For hacksaws, cutters; water hardened. *Obsolete*

FIRTH BROWN J-183 (VESUVIUS FORGINGS)
Firth Brown Ltd.
C 0.1, Cr 30, bal Fe.
For furnace parts, annealing and carburizing parts, trays; corrosion and heat resistant. *Obsolete*

FIRTH BROWN J-185
Firth Brown Ltd.
C, high Ni, Cr, W, bal Fe.
For furnace parts, mechanical stokers, exhaust valves, heat treating boxes; heat and corrosion resistant. *Obsolete*

FIRTH BROWN LCMO
Firth Brown Ltd.
C 0.17, Si 0.25, Mn 0.45, Mo 0.5, bal Fe.
For gears, bolts, camshafts, cams; case hardened, tough. *Obsolete*

FIRTH BROWN MOVA(1)
Firth Brown Ltd.
C 0.1, Si 0.25, Mn 0.5, Mo 0.7, V 0.25, bal Fe.
For oil refinery equipment; elevated temperature use. *Obsolete*

FIRTH BROWN MOVA(2)
Firth Brown Ltd.
C 0.2, Si 0.25, Mn 0.5, Mo 0.7, V 0.25, bal Fe.
For high temperature applications. *Obsolete*

FIRTH BROWN STANDARD
Firth Brown Ltd.
C 1.5, Cr 12, Mo 0.9, V 0.9, bal Fe.
For blanking and forming dies, punches; oil hardened, non-deforming. *Obsolete*

FIRTH BROWN TREBLE EXTRA
Firth Brown Ltd.
C 0.7-1.4, Si 0.15, Mn 0.3, bal Fe.
For drills, reamers, taps, broaches; Type W1; water hardened. *Obsolete*

FIRTH DISCALOY
Teledyne Firth Sterling
Steel. C 0.04, Cr 13, Ni 25, Mo 3, Ti 1.8, Al 0.2, bal Fe.
At 70°F: 145,000 TS; 116,000 YS; 19 El; 23 RA. At 1000°F: 125,000 TS; 98,000 YS; 16 El; 34 RA. For gas turbine blades, rotors, jet engine parts; age hardenable, high temperature applications. *Obsolete*

FIRTH F.S.M.-10
Teledyne Firth Sterling
Steel. C 0.7, Mo 8, V 2, bal Fe.
For cutting tools, drills, hobs, reamers; high speed steel. *Obsolete*

FIRTH F.S.M.-10 MOD
Teledyne Firth Sterling
Steel. C 0.65, Cr 4, Mo 8.2, V 2, bal Fe.
For hot work tools and dies, punches, shears; hot work steel, oil hardened. *Obsolete*

FIRTH FCN-5 STEEL
Firth Brown Ltd.
C 0.08-0.15, Ni 4.75-5.25, bal Fe.
Heat treated: 101,000 TS; 65,700 YS; 20 El; 50 RA. For motor car gears; case-hardening "Firth Brown F-102", "Atlas NCH 5." *Obsolete*

FIRTH FCNC STEEL
Firth Brown Ltd.
C 0.12-0.18, Ni 4-4.5, Cr 1-1.4, bal Fe.
Heat treated: 180,000 TS; 144,000 YS; 12 El; 35 RA. For gears for aero-engines, racing car gears; case-hardening "Firth Brown F-104", "Browns Atecor." *Obsolete*

FIRTH FN-3 STEEL
Firth Brown Ltd.
C 0.25-0.4, Ni 2.75-3.25, bal Fe.
Oil hardened: 101,000-123,000 TS; 65,700-80,000 YS; 22 El; 50 RA. For valves, propeller shafts, rear axles, gear box shafts, connecting rods; "Firth Brown D-61", "Atlas ATN." *Obsolete*

FIRTH FNC STEEL
Firth Brown Ltd.
C 0.28-0.34, Ni 3-3.75, Cr 0.5-1, bal Fe.
Oil hardened: 123,000-146,000 TS; 92,000-110,000 YS; 20 El; 50 RA. For crank shafts, propeller shafts, rear axle connecting rods, motor cars; "Firth Brown D-64", "Atlas NC-45." *Obsolete*

FIRTH FNC-3
Firth Brown Ltd.
C 0.08-0.15, Ni 2.75-3.5, bal Fe.
Heat treated: 94,100 TS; 56,500 YS; 18 El; 45 RA; 140 Brin. For gears, camshafts, steering gear parts; for case hardening "Firth Brown F-101", "Atlas NCH-3." *Obsolete*

FIRTH FS-2
Teledyne Firth Sterling
Steel. Ni 29.6, Cr 7.4, 63.0 TiC.
Sintered: 87 Rock A. For jet engine components; super refractory, sintered carbides. *Obsolete*

FIRTH FS-26
Teledyne Firth Sterling
Steel. Ni 40, Cr 5.7, 54.3 TiC.
At 80°F: 33,000 TS; 0 El. At 1200°F: 48,000 TS; 0 El. At 1800°F: 41,000 TS; 0.8 El. For jet engine components, nozzles, dies; oxidation, corrosion and heat resistant. *Obsolete*

FIRTH FS-27
Teledyne Firth Sterling
Ni 50, Cr 7.1, 42.9 TiC.
At 80°F: 75,000 TS; 0 El. At 800°F: 76,500 TS; 0 El. At 1800°F: 35,000 TS; 1.7 El. High temperature steel for jet engine components, cutting tools; oxidation, corrosion, and heat resistant. *Obsolete*

FIRTH FS-5
Teledyne Firth Sterling
Co 25.9, Cr 11.1, 63.0 TiC.
Sintered: 90 Rock A. High temperature steel for jet engine parts; sintered carbides. *Obsolete*

FIRTH FS-8
Teledyne Firth Sterling
Ni 22.2, Co 7.4, Cr 7.4, 63.0 TiC.
At 75°F: 26,200 TS; 0 El. At 800°F: 43,000 TS; 1.3 El. At 1800°F: 37,800 TS; 2.4 El. High temperature steel for jet engine components, cutting tools; oxidation, corrosion and heat resistant. *Obsolete*

FIRTH FS-9
Teledyne Firth Sterling
Ni 30, Co 10, Cr 10, 50.0 TiC.
At 75°F: 30,500 TS; 0 El. High temperature steel for jet engine components, nozzles, dies; oxidation, corrosion and heat resistant. *Obsolete*

FIRTH FT-30 STEEL
Firth Brown Ltd.
C 0.28-0.4, Si 0.3, Mn 0.45-0.8, bal Fe.
Normalized: 67,000-90,000 TS; 35,500-45,000 YS; 25 El; 45 RA. For general engineering use; "Firth Brown C-41", "Atlas ATC-30." *Obsolete*

FIRTH FT-40 STEEL
Firth Brown Ltd.
C 0.45-0.55, Si 0.3, Mn 0.5-0.7, bal Fe.
Normalized: 90,000-112,000 TS; 45,000-56,000 YS; 20 El; 40 RA. For general engineering use; "Firth Brown C-43", "Atlas ATC-50." *Obsolete*

FIRTH GREEK ASCOLOY
Teledyne Firth Sterling
C 0.15, Si 0.3, Mn 0.3, Cr 13, Ni 2, W 3, Al 0.12, bal Fe.
Heat treated: 137,000-195,000 TS; 108,000-160,000 YS; 14-23 El; 51-63 RA; 290-400 Brin. High temperature steel for compressor wheels and blades, bolts, fasteners; heat resistant to 1000°F. *Obsolete*

FIRTH H.W.D. MOD
Teledyne Firth Sterling
Tool steel. C 0.55, Si 1, W 1.4, Cr 5.25, bal Fe.
For dies; air hardened, nondeforming. *Obsolete*

FIRTH HEAVY METAL
Teledyne Firth Sterling
W 95, bal Ni + Cu.
Sintered: 120,000 TS; 2 El; 300 Brin; 17.5-18.2 density. For counterweights, X-ray tube screens. *Obsolete*

FIRTH PNUSNAP
Firth Brown Ltd.
C 0.35, Si 0.5, Mn 0.25, Cr 1, W 1.8, bal Fe.
Heat treated: 512-555 Brin. For pneumatic snaps, for hot riveting chisels, punches; tough, shock resistant. *Obsolete*

FIRTH RT

Firth Brown Ltd.
C 1.2, Mn 0.28, W 3, bal Fe.
For draw dies, punches, burnishing tools; fast finishing tool steel. *Obsolete*

FIRTH STERLING METEOR

Teledyne Firth Sterling
C 1.15, W 1.4, bal Fe.
For taps, punches, stamps, drills. *Obsolete*

FIRTH STERLING NIROSTA "K2A"

Teledyne Firth Sterling
C 0.23, Si 0.37, Cr 19, bal Fe.
Rolled: 255 Brin. For stainless steel parts, sheets, tubes, cooking utensils; stainless, corrosion resistant. *Obsolete*

FIRTH VC

Teledyne Firth Sterling
Tool steel. C 1.1, Cr 4, Mo 2.6, W 2.5, V 4, bal Fe.
For twist drills, reamers, broaches; abrasion resistant.
Obsolete

FIRTH VICKERS F.I.

Firth-Vickers Stainless Steels Ltd.
C 0-0.08, Cr 11.5-13, Al 0.1-0.3, bal Fe.
Normalized: 71,000 TS; 42,600 YS; 22 El; 70 RA; 150 Brin.
Heat treated: 175,000 TS; 145,000 YS; 21 El; 64 RA; 352 Brin.
For oil refinery and chemical plant equipment; Type 405; corrosion resistant. *Obsolete*

FIRTH W.C.R.

Teledyne Firth Sterling
Tool steel. C 0.5, W 7.5, Cr 7.5, bal Fe.
For hot work tools and dies; hot work steel, oil hardened.
Obsolete

FIRTH XDL

Teledyne Firth Sterling
Tool steel. C 0.4, W 15, Cr 3.5, V 0.5, bal Fe.
For hot work dies, extrusion rams and liners; hot work steel, oil hardened. *Obsolete*

FIRTH-BRONW FVS

Firth Brown Ltd.
C 0.37-0.47, Si 1-2, Mn 0.5-1, Cr 13-15, Mo 0.4-0.7, Ni 13-15, W 2.2-3, bal Fe.
Valve steel.

FIRTH-BROWN 20CF

Firth Brown Ltd.
C 0.2, Si 0.35, Mn 0.9, bal Fe.
For gears, shafts; case hardening. *Obsolete*

FIRTH-BROWN 21/4/N

Firth Brown Ltd.
C 0.48-0.58, Si 0-0.25, Mn 8-10, Cr 20-22, Ni 3.5-4.5, N 0.38-0.5.
Valve steel. C + N = 0.90 min, bal Fe.

FIRTH-BROWN 30CF

Firth Brown Ltd.
C 0.3, Si 0.35, Mn 0.9, bal Fe.
For gears, shafts; water hardening. *Obsolete*

FIRTH-BROWN 40CF

Firth Brown Ltd.
C 0.42, Si 0.4, Mn 0.8, bal Fe.
For gears, shafts; water hardening. *Obsolete*

FIRTH-BROWN 50CF

Firth Brown Ltd.
C 0.52, Si 0.4, Mn 0.8, bal Fe.
For gears, shafts; water hardening. *Obsolete*

FIRTH-BROWN 50CT

Firth Brown Ltd.
C 0.48, Si 0.25, Mn 0.7, bal Fe.
For railway tires, wheels. *Obsolete*

FIRTH-BROWN 60CT

Firth Brown Ltd.
C 0.6, Si 0.25, Mn 0.7, bal Fe.
For railway tires, wheels. *Obsolete*

FIRTH-BROWN 65CT

Firth Brown Ltd.
C 0.65, Si 0.25, Mn 0.7, bal Fe.
For railway tires, wheels. *Obsolete*

FIRTH-BROWN ANC1

Firth Brown Ltd.
C 0.3, Mn 0.45, Ni 4.25, Cr 1.25, bal Fe.
For shafts, gears; formerly "Brown ATG."

FIRTH-BROWN ANC2

Firth Brown Ltd.
C 0.4, Mn 0.5, Ni 3.5, Cr 1.6, bal Fe.
For shafts, gears; formerly "Brown ATG." *Obsolete*

FIRTH-BROWN ANCM

Firth Brown Ltd.
C 0.3, Mn 0.45, Ni 4.25, Cr 1.25, Mo 0.25, bal Fe.
For shafts, gears; formerly "Atlas AHNC."

FIRTH-BROWN B45CH

Firth Brown Ltd.
C 0.15, Ni 0.6, Cr 0.65, Mo 0.12, bal Fe.
For gears, cams, camshafts, fasteners; case hardened, EN361.

FIRTH-BROWN B55CH

Firth Brown Ltd.
C 0.2, Cr 0.65, Ni 0.6, Mo 0.12, bal Fe.
For gears, cams, camshafts, machinery parts; case hardened, EN362.

FIRTH BROWN B65CH

Firth Brown Ltd.
C 0.24, Ni 0.6, Cr 0.65, Mo 0.12, bal Fe.
For gears, cams, camshafts; case hardened, EN363.

FIRTH-BROWN BEST C.S.

Firth Brown Ltd.
C 0.7-1.4, bal Fe.
Hardened: 500-700 Brin. For hammers, tools, drills, chisels; cheap tool steel "Firth Brown L-222." *Obsolete*

FIRTH-BROWN BRB

Firth Brown Ltd.
C 1.4, Si 0.35, Mn 0.4, Cr 0.75, W 5, bal Fe.
For tools, dies, drawing dies; oil hardening. *Obsolete*

FIRTH-BROWN CAST STEEL

Firth Brown Ltd.
C 0.7-1.4, bal Fe.
Hardened: 500-700 Brin. For turning, planing and slotting dies, shaping tools, wood saws, axes, scythes, drills; tool steel intermediate quality "Firth Brown L-221." *Obsolete*

FIRTH-BROWN CCR 1

Firth Brown Ltd.
C 0.85, Mn 0.3, Cr 1.8, bal Fe.
For tools, dies; oil hardening. *Obsolete*

FIRTH-BROWN CCRF

Firth Brown Ltd.
C 0.42, Si 0.3, Mn 0.7, Cr 0.9, bal Fe.
For gears, shafts, bolts; water hardening. *Obsolete*

FIRTH-BROWN CCW

Firth Brown Ltd.
C 0.52, Si 0.7, Mn 0.55, Cr 1.2, W 6, bal Fe.
For tools, dies; hot work steel. *Obsolete*

FIRTH-BROWN CDD

Firth Brown Ltd.
C 1.9, Si 0.15, Mn 0.25, Cr 0.6, W 2.3, bal Fe.
For tools, dies, drawing dies; oil hardening. *Obsolete*

FIRTH-BROWN CMCR

Firth Brown Ltd.
C 0.85, Mn 0.8, Cr 0.8, bal Fe.
For tools, dies; oil hardened. *Obsolete*

FIRTH-BROWN CMN 2

Firth Brown Ltd.
C 0.35, Mn 1.5, bal Fe.
For gears, shafts; water hardening.

FIRTH-BROWN CMN1

Firth Brown Ltd.
C 0.25, Mn 1.5, bal Fe.
For gears, shafts; water hardening.

FIRTH-BROWN CR1T

Firth Brown Ltd.
C 0.52, Si 0.25, Mn 0.7, Cr 0.45, bal Fe.
For railway ties, wheels. *Obsolete*

FIRTH-BROWN CR2T

Firth Brown Ltd.
C 0.6, Si 0.25, Mn 0.7, Cr 0.45, bal Fe.
For railway tires, wheels. *Obsolete*

FIRTH-BROWN CR3T

Firth Brown Ltd.
C 0.65, Si 0.25, Mn 0.7, Cr 0.45, bal Fe.
For railway tires, wheels. *Obsolete*

FIRTH-BROWN CR4T

Firth Brown Ltd.
C 0.72, Si 0.25, Mn 0.7, Cr 0.45, bal Fe.
For railway tires, wheels. *Obsolete*

FIRTH-BROWN CRM1

Firth Brown Ltd.
C 0.4, Mn 0.65, Cr 1.2, Mo 0.25, bal Fe.
For gears, shafts, bolts; oil hardening.

FIRTH-BROWN CRM2

Firth Brown Ltd.
C 0.4, Mn 0.6, Cr 1.3, Mo 0.85, bal Fe.
For gears, shafts, bolts; tough, oil hardening.

FIRTH-BROWN CRM3

Firth Brown Ltd.
C 0.27, Mn 0.6, Cr 3.25, Mo 0.6, bal Fe.
For gears, shafts, machinery parts; oil hardened, tough.

FIRTH-BROWN CRM4

Firth Brown Ltd.
C 0.3, Mn 0.6, Ni 2.75, Mo 0.6, bal Fe.
For gears, shafts; tough. *Obsolete*

FIRTH-BROWN CRV1

Firth Brown Ltd.
C 0.47, Si 0.25, Mn 0.65, Cr 0.85, V 0.2, bal Fe.
For laminated springs; oil hardening.

FIRTH-BROWN CRV2

Firth Brown Ltd.
C 0.45, Si 0.25, Mn 0.65, Cr 1.25, V 0.2, bal Fe.
For laminated springs; oil hardening.

FIRTH-BROWN CRV3

Firth Brown Ltd.
C 0.35, Cr 1.25, V 0.2, bal Fe.
For forgings; formerly Atlas ATCV. *Obsolete*

FIRTH-BROWN CRWC

Firth Brown Ltd.
C 0.34, Si 0.9, Mn 0.25, Cr 5.5, W 4.7, Co 0.5, bal Fe.
For tools, dies; hot work steel. *Obsolete*

FIRTH-BROWN CVW3

Firth Brown Ltd.
C 1.35, Si 0.15, Mn 0.3, V 0.5, W 3.75, bal Fe.
For tools, dies, drawing dies; oil hardening. *Obsolete*

FIRTH-BROWN DDQ
Firth Brown Ltd.
C 0.16, Mn 2, Ni 11-14, Cr 11-14, 0.2 Si min, bal Fe.
Acid, rust and heat resisting steel.

FIRTH-BROWN DICROME
Firth Brown Ltd.
C 0.6, Cr 0.6, Mn 0.6, Si 0.25, bal Fe.
For machine tool parts; oil hardened.

FIRTH-BROWN EMS
Firth Brown Ltd.
C 0-0.12, Si 0.2-1, Mn 0.5-2, Cr 17-19, Ni 8-11, Ti = 5 x C min (0.90 max), bal Fe.
Stabilized austenitic stainless.

FIRTH-BROWN EXTRA CS
Firth Brown Ltd.
C 0.7-1.4, bal Fe.
Hardened: 500-700 Brin. For finishing tools, drills, milling cutters, reamers, razors, punches, surgical instruments, taps, needles; tool steel "Firth Brown L-220." *Obsolete*

FIRTH-BROWN F10C
Firth Brown Ltd.
C 0.1, Si 0.25, Mn 0.65, bal Fe.
Annealed: 64,000 psi TS; 40,000 psi YS; 28 El; 65 RA; 130 Brin. For nails, rivets, fan blades, case hardened parts; case hardened.

FIRTH-BROWN F117
Firth Brown Ltd.
C 0.12, Si 1, Mn 1, Ni 0.5, Cr 16-18, bal Fe.
Ferritic rust resisting steel.

FIRTH-BROWN F120
Firth Brown Ltd.
C 0.12, Si 1, Mn 1, Ni 0.5, Cr 20-22, bal Fe.
Ferritic rust resisting steel.

FIRTH-BROWN F15C
Firth Brown Ltd.
C 0.15, Si 0.25, Mn 0.65, bal Fe.
Annealed: 70,000 psi TS; 45,000 psi YS; 25 El; 60 RA; 145 Brin. For screws, bolts, case hardened parts; case hardened.

FIRTH-BROWN F20C
Firth Brown Ltd.
C 0.2, Si 0.25, Mn 0.65, bal Fe.
Annealed: 73,000 psi TS; 45,000 psi YS; 22 El; 58 RA; 150 Brin. For screws, bolts, gears, case hardened parts; case hardened.

FIRTH-BROWN F25C
Firth Brown Ltd.
C 0.25, Si 0.25, Mn 0.65, bal Fe.
Hot rolled: 70,000 psi TS; 45,000 psi YS; 31 El; 58 RA; 145 Brin. For crankshafts, gears, bolts, fasteners; water hardened.

FIRTH-BROWN F27C
Firth Brown Ltd.
C 0.25-0.35, Si 0.05-0.35, Mn 0.7-0.9, bal Fe.
Carbon steel.

FIRTH-BROWN F30C
Firth Brown Ltd.
C 0.32, Si 0.25, Mn 0.65, bal Fe.
For axles.

FIRTH-BROWN F32C
Firth Brown Ltd.
C 0.3-0.35, Si 0.05-0.35, Mn 0.7-0.9, bal Fe.
Carbon steel.

FIRTH-BROWN F35C
Firth Brown Ltd.
C 0.35, Si 0.25, Mn 0.65, bal Fe.
Hot rolled: 85,000 psi TS; 54,000 psi YS; 30 El; 53 RA; 183 Brin. For gears, shafts, axles, bolts; water hardened.

FIRTH-BROWN F40C
Firth Brown Ltd.
C 0.4, Si 0.25, Mn 0.65, bal Fe.
Hot rolled: 91,000 psi TS; 58,000 psi YS; 27 El; 50 RA; 200 Brin. For gears, shafts, axles, bolts, crankshafts; water hardened.

FIRTH-BROWN F45C
Firth Brown Ltd.
C 0.45, Si 0.25, Mn 0.65, bal Fe.
Hot rolled: 98,000 psi TS; 59,000 psi YS; 24 El; 45 RA; 212 Brin. For gears, shafts, bolts, axles, crankshafts; water hardened.

FIRTH-BROWN F45CH
Firth Brown Ltd.
C 0.15, Si 0.25, Mn 0.7, Ni 0.8, Cr 0.6, bal Fe.
For gears, pinions, shafts, cams, camshafts; case hardened; Brit. EN351; shock resistant.

FIRTH-BROWN F47C
Firth Brown Ltd.
C 0.45-0.5, Si 0.05-0.35, Mn 0.7-1, bal Fe.
Carbon steel.

FIRTH-BROWN F50C
Firth Brown Ltd.
C 0.5, Si 0.25, Mn 0.65, bal Fe.
Annealed: 96,000 psi TS; 52,000 psi YS; 16 El; 23 RA; 170 Brin. For gears, shafts, bolts, axles, crankshafts; water hardened.

FIRTH-BROWN F52C
Firth Brown Ltd.
C 0.45-0.6, Si 0.1-0.4, Mn 0.6-0.8, bal Fe.
Carbon steel.

FIRTH-BROWN F543
Firth Brown Ltd.
C 0.09, Si 0.6, Mn 0.6, Ni 4.8, Cr 3.9, Mo 3, bal Fe.
For tools, dies. *Obsolete*

FIRTH-BROWN F55C
Firth Brown Ltd.
C 0.55, Mo 0.7, bal Fe.
For springs; water hardening.

FIRTH-BROWN F55CH
Firth Brown Ltd.
C 0.18, Si 0.25, Mn 0.7, Ni 1, Cr 0.8, bal Fe.
For gears, pinions, cams, camshafts, machine tool parts; case hardened; Brit. EN352; shock resistant.

FIRTH-BROWN F62C
Manufacturer not listed.
C 0.6-0.65, Si 0.05-0.35, Mn 0.4-0.6, bal Fe.
Carbon steel; for springs and hand tools.

FIRTH-BROWN F65C
Firth Brown Ltd.
C 0.65, Si 0.25, Mn 0.65, bal Fe.
Heat treated: 170,000 psi TS; 116,000 psi YS; 10 El; 38 RA; 350 Brin. For die blocks, girders, clutch discs; water hardened.

FIRTH-BROWN F67C
Firth Brown Ltd.
C 0.65-0.7, Si 0.05-0.35, Mn 0.7-0.9, bal Fe.
Carbon steel; for springs and hand tools.

FIRTH-BROWN F70C
Firth Brown Ltd.
C 0.7, Si 0.25, Mn 0.65, bal Fe.
Heat treated: 174,000 psi TS; 128,000 psi YS; 12 El; 37 RA; 352 Brin. For springs, rails, die blocks, girders; water hardened.

FIRTH-BROWN F75C
Firth Brown Ltd.
C 0.75, Mn 0.65, bal Fe.
For springs; water hardening.

FIRTH-BROWN F75CH
Firth Brown Ltd.
C 0.18, Si 0.25, Mn 0.7, Ni 1.75, Cr 1, Mo 0.15, bal Fe.
For gears, cams, camshafts, machine tool parts; case hardened; Brit. EN354; shock resistant.

FIRTH-BROWN F80C
Firth Brown Ltd.
C 0.7-0.85, Si 0.1-0.4, Mn 0.55-0.75, bal Fe.
Carbon tool and spring steel.

FIRTH-BROWN F85C
Firth Brown Ltd.
C 0.85, Si 0.25, Mn 0.65, bal Fe.
Heat treated: 190,000 psi TS; 145,000 psi YS; 12 El; 35 RA; 390 Brin. For drills, punches, springs, reamers, hobs; Type W1; water hardened.

FIRTH-BROWN F85CH
Firth Brown Ltd.
C 0.18, Si 0.25, Mn 0.7, Cr 1.6, Mo 0.2, Ni 2, bal Fe.
For gears, cams, machine tool parts; case hardened; Brit. EN355; shock resistant.

FIRTH-BROWN F95C
Firth Brown Ltd.
C 0.95, Mn 0.6, bal Fe.
For springs; water hardening. *Obsolete*

FIRTH-BROWN FCB
Firth Brown Ltd.
C 0-0.08, Si 0.2-1, Mn 0.5-2, Cr 17-19, Ni 9-12, Nb = 10 x C min (1.0 C max), bal Fe.
Stabilized austenitic stainless steel. AISI 347.

FIRTH-BROWN FCCH
Firth Brown Ltd.
C 0.14, Mn 0.6, bal Fe.
For shafts, gears; formerly "Atlas CCH;" case hardening.

FIRTH-BROWN FCCR
Firth Brown Ltd.
C 0.45, Si 0.25, Mn 0.9, Cr 0.95, bal Fe.
For gears, bolts, crankshafts; water hardened.

FIRTH-BROWN FCI
Firth Brown Ltd.
C 0.08-0.15, Si 0-1, Mn 0-1.5, Ni 0-1, Cr 11.5-13.5, S 0.15-0.3, P 0-0.04, bal Fe.
Free-machining martensitic stainless steel. AISI 416.

FIRTH-BROWN FCMO
Firth Brown Ltd.
C 0.3, Mn 0.75, Mo 0.6, bal Fe.
For gears, shafts; water hardening. *Obsolete*

FIRTH-BROWN FCS
Firth Brown Ltd.
C 0.2-0.28, Si 0-1, Mn 0-1.5, Ni 0-1, Cr 12-14, Mo 0-0.6, bal Fe.
Martensitic stainless steel. Similar to AISI 420.

FIRTH-BROWN FCVA
Firth Brown Ltd.
C 1.08, Si 0.1, Mn 0.2, V 0.15, bal Fe.
For tools, dies, drills, taps; water hardening. *Obsolete*

FIRTH-BROWN FCW5
Firth Brown Ltd.
C 1.35, Si 0.15, Mn 0.35, W 5, bal Fe.
For tools, dies, drawing dies; oil hardening. *Obsolete*

FIRTH-BROWN FDP
Firth Brown Ltd.
C 0-0.12, Si 0.2-1, Mn 0.5-2, Cr 17-19, Ni 8-11, Ti = 5 x C min (0.90 C max), bal Fe.
Stabilized austenitic stainless steel. AISI 321.

FIRTH-BROWN FDP(L)
Firth Brown Ltd.
C 0-0.08, Si 0.2-1, Mn 0.5-2, Cr 17-19, Ni 9-12, Ti = 5 x C min (0.70 C max), bal Fe.
Stabilized austenitic stainless steel. AISI 321.

FIRTH-BROWN FG
Firth Brown Ltd.
C 0.2-0.28, Si 0-0.8, Mn 0-1, Ni 0-1, Cr 12-14, bal Fe.
Martensitic stainless steel. AISI 420.

FIRTH-BROWN FG(L)
Firth Brown Ltd.
C 0.14-0.2, Si 0-0.8, Mn 0-1, Ni 0-1, Cr 11.5-13.5, bal Fe.
Martensitic stainless steel.

FIRTH-BROWN FH
Firth Brown Ltd.
C 0.28-0.36, Si 0-0.8, Mn 0-1, Ni 0-1, Cr 12-14, bal Fe.
Martensitic stainless steel. Similar to AISI 420.

FIRTH-BROWN FI
Firth Brown Ltd.
C 0-0.08, Si 0-0.8, Mn 0-1, Ni 0-0.5, Cr 12-14, bal Fe.
Ferritic type stainless steel. AISI 403.

FIRTH-BROWN FI
Firth Brown Ltd.
C 0.09-0.15, Si 0-0.8, Mn 0-0.8, Ni 0-0.5, Cr 11.5-13.5, bal Fe.
Martensitic stainless steel.

FIRTH-BROWN FI17
Firth Brown Ltd.
C 0-0.1, Si 0-0.8, Mn 0-1, Ni 0-0.5, Cr 16-18, bal Fe.
Ferritic type stainless steel. AISI 430.

FIRTH-BROWN FMB
Firth Brown Ltd.
C 0-0.07, Si 0.2-1, Mn 0.5-2, Cr 16.5-18.5, Ni 10-13, Mo 2.25-3, bal Fe.
Austenitic stainless steel. AISI 316.

FIRTH-BROWN FMB (LC)
Firth Brown Ltd.
C 0-0.03, Si 0.2-1, Mn 0.5-2, Cr 16.5-18.5, Ni 11-14, Mo 2.25-3, bal Fe.
Austenitic stainless steel. AISI 316L.

FIRTH-BROWN FMBTI
Firth Brown Ltd.
C 0-0.08, Si 0.2-1, Mn 0.5-2, Cr 16.5-18.5, Ni 11-14, Mo 2.25-3, Ti = 4 x C min (0.60 max), bal Fe.
Stabilized austenitic stainless steel.

FIRTH-BROWN FML
Firth Brown Ltd.
C 0-0.07, Si 0.2-1, Mn 0.5-2, Cr 16.5-18.5, Ni 9-11, Mo 1.25-1.75, bal Fe.

FIRTH-BROWN FN10
Firth Brown Ltd.
C 0.34-0.46, Si 0.1-0.35, Mn 0.65-1.05, Ni 0.7-1, bal Fe.
Structural steel.

FIRTH-BROWN FN15
Firth Brown Ltd.
C 0.6, Mn 0.65, Ni 1.25, bal Fe.
For shafts; formerly "Atlas ND." *Obsolete*

FIRTH-BROWN FN30
Firth Brown Ltd.
C 0.3, Mn 0.65, Ni 3, bal Fe.
For gears, shafts; tough.

FIRTH-BROWN FN35
Firth Brown Ltd.
C 0.3, Mn 0.6, Ni 3.6, bal Fe.
For gears, shafts; oil hardening, tough.

FIRTH-BROWN FN50
Firth Brown Ltd.
C 0.2, Mn 0.5, Ni 5, bal Fe.
For gears, bolts, shafts; case hardening. *Obsolete*

FIRTH-BROWN FNCF
Firth Brown Ltd.
C 0.35, Si 0.35, Mn 0.8, Ni 1.75, Cr 0.75, bal Fe.
For gears, punches, shafts; oil hardening. *Obsolete*

FIRTH-BROWN FNCR
Firth Brown Ltd.
C 0.35, Ni 1.25, Cr 0.6, bal Fe.
For gears, pinions, bolts, machinery parts; Brit. EN111; oil hardened, tough.

FIRTH-BROWN FNMC
Firth Brown Ltd.
C 0.65, Si 0.3, Mn 5.2, Ni 12, Cr 3.7, bal Fe.
For heat and corrosion resistant parts; non-magnetic, austenitic.

FIRTH-BROWN FPRF
Firth Brown Ltd.
C 1.1, Si 0.4, Mn 1, Ni 0.9, Cr 3.25, Mo 0.35, bal Fe.
For tools, dies; oil hardening. *Obsolete*

FIRTH-BROWN FSL
Firth Brown Ltd.
C 0.08, Si 0.2-1, Mn 2, Cr 17.5-20, Ni 8-11, bal Fe.
Austenitic stainless steel for wire.

FIRTH-BROWN FSL (LC)
Firth Brown Ltd.
C 0-0.03, Si 0.2-1, Mn 0.5-2, Cr 17-19, Ni 9-12, bal Fe.
Low carbon austenitic stainless steel. AISI 304L.

FIRTH-BROWN FST
Firth Brown Ltd.
C 0-0.12, Cr 17-19, Ni 8-11, bal Fe.
Austenitic stainless steel. AISI 302.

FIRTH-BROWN FST (FC)
Firth Brown Ltd.
C 0-0.12, Mn 1-2, Cr 17-19, Ni 8-11, S 0.15-0.3, P 0-0.045, bal Fe.
Free machining austenitic stainless steel. AISI 303.

FIRTH-BROWN FST (L)
Firth Brown Ltd.
C 0-0.06, Si 0.2-1, Mn 0.5-2, Cr 17.5-19, Ni 8-11, bal Fe.
Low carbon austenitic stainless steel. AISI 304.

FIRTH-BROWN FV 520B
Firth Brown Ltd.
C 0-0.07, Si 0-0.6, Mn 0-1, Cr 13.2-14.7, Ni 5-5.8, Mo 1.2-2, Cu 1.2-2, Nb 0.2-0.5, bal Fe.
Precipitation hardening stainless.

FIRTH-BROWN FV 535 VR
Firth Brown Ltd.
C 0.06-0.11, Si 0.1-0.7, Mn 0.6-1.1, Ni 0.2-0.8, Cr 9.8-11.2, Mo 0.5-1, B 0.004-0.012, Co 5-7, Nb 0.2-0.45, V 0.1-0.35, N 0.01-0.035, bal Fe.
Heat resisting steel.

FIRTH-BROWN FV317
Firth Brown Ltd.
C 0-0.06, Si 0.2-1, Mn 0.5-2, Cr 17.5-19.5, Ni 12-15, Mo 3-4, bal Fe.
Austenitic stainless steel. AISI 317.

FIRTH-BROWN FV317 (LC)
Firth Brown Ltd.
C 0-0.03, Si 0.2-1, Mn 0.5-2, Cr 17.5-19.5, Ni 14-17, Mo 3-4, bal Fe.
Austenitic stainless steel.

FIRTH-BROWN FV448
Firth Brown Ltd.
C 0.08-0.16, Si 0.15-0.6, Mn 0.3-1.2, Ni 0.6-1.2, Cr 9.8-11.2, Mo 0.4-0.8, Nb 0.15-0.45, V 0.1-0.25, N 0.03-0.075, bal Fe.
Heat resisting steel.

FIRTH-BROWN FVS
Firth Brown Ltd.
C 0.35-0.5, Si 1-2, Mn 0.5-1, Cr 12-15, Ni 12-15, W 2-3, bal Fe.
Valve steel.

FIRTH-BROWN G110
Firth Brown Ltd.
C 0-0.015, Si 0-0.1, Mn 0-0.1, Ni 17-19, Cr 0-0.25, Mo 4.6-5.2, Al 0.05-0.15, Co 7-8.5, Ti 0.3-0.6, bal Fe.
Maraging steel.

FIRTH-BROWN HCM3
Firth Brown Ltd.
C 0.4, Cr 3, Mo 1, V 0.25, bal Fe.
For oil refinery equipment; Brit. EN40C; creep resistant.

FIRTH-BROWN HCM5
Firth Brown Ltd.
C 0.3, Cr 3, Mo 0.4, bal Fe.
For oil refinery equipment; Brit. EN40B; creep resistant.

FIRTH-BROWN HCM7
Firth Brown Ltd.
C 0.2, Cr 3, Mo 0.4, bal Fe.
For oil refinery equipment; Brit. EN40A; creep resistant.

FIRTH-BROWN HDBS
Firth Brown Ltd.
C 0.36, Si 0.35, Mn 0.3, Cr 3.6, V 0.25, W 3, bal Fe.
For tools, dies; hot work steel. *Obsolete*

FIRTH-BROWN HRC MAX
Firth Brown Ltd.
Mn 1, Ni 6-12, W 2-4, C 0.18-0.45, Si 1-2, 17.0 Cr min, bal Fe.
High Cr-Ni-W valve steel.

FIRTH-BROWN LK3
Firth Brown Ltd.
C 0.35-0.43, Si 0.1-0.45, Mn 0.4-0.65, Cr 1.4-1.8, Mo 0.15-0.25, Al 0.9-1.3, bal Fe.
Steel to be nitrided.

FIRTH-BROWN LK5
Firth Brown Ltd.
C 0.27-0.35, Si 0.1-0.45, Mn 0.4-0.65, Cr 1.4-1.6, Mo 0.15-0.25, Al 0.9-1.3, bal Fe.
Steel to be nitrided.

FIRTH-BROWN M13F
Firth Brown Ltd.
C 1.1, Si 0.7-0, Mn 13, Cr 1, bal Fe.
For wear resistant parts; wear resistant, austenitic. *Obsolete*

FIRTH-BROWN MCMT
Firth Brown Ltd.
C 0.36, Si 0.25, Mn 1.25, Cr 0.7, Mo 0.45, bal Fe.
For railway tires, wheels. *Obsolete*

FIRTH-BROWN MCT
Firth Brown Ltd.
C 0.95, Si 0.25, Mn 1.1, W 0.5, Cr 0.5, bal Fe.
For bearings, punches, dies, crimpers, cutters; oil hardened, non-deforming. *Obsolete*

FIRTH-BROWN MMO1
Firth Brown Ltd.
C 0.32, Mn 1.6, Mo 0.27, bal Fe.
For gears, shafts; oil hardening.

FIRTH-BROWN MMO2
Firth Brown Ltd.
C 0.33, Mn 1.6, Mo 0.4, bal Fe.
For gears, shafts; oil hardening.

FIRTH-BROWN MMOF
Firth Brown Ltd.
C 0.26, Si 0.35, Mn 1.5, Mo 0.35, bal Fe.
For gears, shafts; water hardening. *Obsolete*

FIRTH-BROWN MN1F
Firth Brown Ltd.
C 0.26, Si 0.35, Mn 1.5, bal Fe.
For gears, shafts. *Obsolete*

FIRTH-BROWN MN3F
Firth Brown Ltd.
C 0.42, Si 0.35, Mn 1.4, bal Fe.
For gears, shafts, punches; oil hardening. *Obsolete*

FIRTH-BROWN MO1F
Firth Brown Ltd.
C 0.2, Si 0.35, Mn 0.8, Mo 0.65, bal Fe.
For gears, shafts; case hardening. *Obsolete*

FIRTH-BROWN N2MCH
Firth Brown Ltd.
C 0.17, Mn 0.5, Ni 1.8, Mo 0.25, bal Fe.
For gears, shafts; case-hardening.

FIRTH-BROWN N3CCH
Firth Brown Ltd.
C 0.12, Mn 0.4, Ni 3.3, Cr 0.8, bal Fe.
For gears, shafts; case-hardening, tough.

FIRTH-BROWN N3CH
Firth Brown Ltd.
C 0.14, Si 0.25, Mn 0.4, Ni 3.1, bal Fe.
For gears, shafts; case-hardening.

FIRTH-BROWN N3CMCH
Firth Brown Ltd.
C 0.12, Ni 3.3, Cr 0.8, Mo 0.2, bal Fe.
For gears, shafts, machinery parts, cams; case-hardened, shock resistant.

FIRTH-BROWN N4CCH
Firth Brown Ltd.
C 0.14, Mn 0.4, Ni 4.25, Cr 1.25, bal Fe.
For gears, shafts; case-hardening, tough.

FIRTH-BROWN N4CMCH
Firth Brown Ltd.
C 0.14, Ni 4.25, Cr 1.25, Mo 0.25, bal Fe.
For gears, shafts, camshafts, cams, bolts; case-hardened, shock resistant.

FIRTH-BROWN N5CH
Firth Brown Ltd.
C 0.1, Mn 0.3, Ni 5, bal Fe.
For gears, shafts; case-hardening, tough.

FIRTH-BROWN N5MCH
Firth Brown Ltd.
C 0.12, Mn 0.4, Ni 5.1, Mo 0.2, bal Fe.
For gears, shafts; case-hardening, shock resistant.

FIRTH-BROWN NCM1
Firth Brown Ltd.
C 0.4, Mn 0.6, Ni 1.6, Cr 1, Mo 0.5, bal Fe.
For gears, shafts, bolts; shock resistant, heavy duty.

FIRTH-BROWN NCM2
Firth Brown Ltd.
C 0.6, Mn 0.65, Ni 1.5, Cr 0.7, Mo 0.25, bal Fe.
For gears, shafts; heavy duty. *Obsolete*

FIRTH-BROWN NCM3
Firth Brown Ltd.
C 0.3, Mn 0.6, Ni 2.4, Cr 0.6, Mo 0.45, bal Fe.
For gears, shafts; shock resistant.

FIRTH-BROWN NCM4
Firth Brown Ltd.
C 0.4, Mn 0.6, Cr 0.65, Mo 0.6, Ni 0.3, bal Fe.
For gears, shafts, bolts, machinery parts; oil hardened, shock resistant.

FIRTH-BROWN NCM5
Firth Brown Ltd.
C 0.3, Mn 0.6, Ni 3.25, Cr 0.6, Mo 0.6, bal Fe.
For gears, shafts, bolts; shock resistant.

FIRTH-BROWN NCM6
Firth Brown Ltd.
C 0.3, Mn 0.6, Ni 3.25, Cr 0.8, Mo 0.3, bal Fe.
For bolts, gears, shafts; tough, shock resistant.

FIRTH-BROWN NCM7
Firth Brown Ltd.
C 0.3, Mn 0.6, Ni 3.5, Cr 1.25, Mo 0.6, bal Fe.
For bolts, gears, shafts; tough, shock resistant. *Obsolete*

FIRTH-BROWN NCMCH
Firth Brown Ltd.
C 0.14, Mn 0.4, Ni 4.25, Cr 1.25, Mo 0.25, bal Fe.
For gears, shafts; case hardening, tough. *Obsolete*

FIRTH-BROWN NCMF
Firth Brown Ltd.
C 0.35, Si 0.35, Mn 0.8, Ni 1.25, Cr 0.75, Mo 0.35, bal Fe.
For gears, shafts, dies; oil hardening. *Obsolete*

FIRTH-BROWN NCMO
Firth Brown Ltd.
C 0.38, Si 0.25, Mn 0.65, Ni 1.4, Cr 1.2, Mo 0.15, bal Fe.
For gears, bolts, machine tool parts; Brit. EN110; oil hardened, tough.

FIRTH-BROWN NCMV
Firth Brown Ltd.
C 0.25, Mn 0.55, Ni 3, Cr 1.2, Mo 0.45, V 0.2, bal Fe.
For bolts, gears, shafts; tough, shock resistant. *Obsolete*

FIRTH-BROWN NCR1
Firth Brown Ltd.
C 0.55, Mn 0.65, Ni 1.5, Cr 0.65, bal Fe.
For gears, shafts; formerly "Atlas NCD." *Obsolete*

FIRTH-BROWN NCR2
Firth Brown Ltd.
C 0.3, Mn 0.6, Ni 3.25, Cr 0.75, bal Fe.
For gears, shafts; tough.

FIRTH-BROWN NCR3
Firth Brown Ltd.
C 0.65, Mn 0.15, Ni 1.95, Cr 1.9, bal Fe.
For gears, shafts; tough. *Obsolete*

FIRTH-BROWN NIMCH
Firth Brown Ltd.
C 0.1, Ni 0.5, Mo 0.2, Mn 0.3, bal Fe.
For gears, shafts; formerly "Fibratlas"; case hardening. *Obsolete*

FIRTH-BROWN NMCM
Firth Brown Ltd.
C 0.4, Ni 0.75, Cr 0.45, Mo 0.2, Mn 1.35, Si 0.25, bal Fe.
For gears, bolts, machine tool parts, fasteners; Brit. EN100; oil hardened, tough.

FIRTH-BROWN NMCW
Firth Brown Ltd.
C 0.7, Si 0.3, Mn 9.2, Ni 7.7, Cr 3.3, W 0.6, bal Fe.
For heat and corrosion resistant parts; nonmagnetic, austenitic. *Obsolete*

FIRTH-BROWN NMM1
Firth Brown Ltd.
C 0.18, Mn 1.4, Ni 0.5, Mo 0.3, bal Fe.
For axles; tough.

FIRTH-BROWN NMM2
Firth Brown Ltd.
C 0.28, Mn 1.2, Ni 0.5, Mo 0.25, bal Fe.
For axles; tough. *Obsolete*

FIRTH-BROWN O5CF
Firth Brown Ltd.
C 0.07, Si 0.35, Mn 0.07, bal Fe.
For hubbing dies; case hardening. *Obsolete*

FIRTH-BROWN RBD
Firth Brown Ltd.
C 0.4, Si 0.15, Mn 0.25, Cr 3, V 0.5, W 10, bal Fe.
For tools, dies; hot work steel. *Obsolete*

FIRTH-BROWN RBD-E
Firth Brown Ltd.
C 0.22, Si 0.25, Mn 0.25, Ni 2.25, Cr 2.25, Mo 0.45, V 0.2, W 10, bal Fe.
For tools, dies; hot work steel. *Obsolete*

FIRTH-BROWN S80
Firth Brown Ltd.
C 0.12-0.2, Si 0-0.8, Mn 0-1, Ni 2-3, Cr 15-18, bal Fe.
Martensitic stainless steel. AISI 431.

FIRTH-BROWN SHC1
Firth Brown Ltd.
C 0.44, Si 1, Mn 0.65, Ni 3, Cr 0.65, bal Fe.
For tools, dies, gears, shafts; tough. *Obsolete*

FIRTH-BROWN SLV
Now S.L.V. STEEL.

FIRTH-BROWN SLV
Firth Brown Ltd.
C 0.4-0.5, Si 3-3.75, Mn 0.3-0.75, Cr 7.5-9.5, Ni 0-0.5, bal Fe.
Valve steel.

FIRTH-BROWN SP
Firth Brown Ltd.
C 1.1, Si 0.15, Mn 0.25, Cr 0.6, W 0.35, bal Fe.
For tools, dies; water hardening. *Obsolete*

FIRTH-BROWN T-374 (FOUNDRY MN 12)
Firth Brown Ltd.
Mn 12, C, bal Fe.
200-240 Brin.

FIRTH-BROWN TDC
Firth Brown Ltd.
C 1.2, Si 0.25, Mn 0.25, W 1.25, bal Fe.
For tools, dies, hacksaws; water hardening. *Obsolete*

FIRTH-BROWN TMS
Firth Brown Ltd.
C 0.92, Si 0.3, Mn 1.75, bal Fe.
For tools, dies; oil hardening, non-deforming. *Obsolete*

FIRTH-BROWN XB
Firth Brown Ltd.
C 0.74-0.84, Si 1.75-2.25, Ni 1.15-1.65, Cr 19-20.5, bal Fe.
Hardened and tempered to 47 Rock C. For valves.

FIRTH-CHQ
Teledyne Firth Sterling
C, alloy, bal Fe.
For cold heading dies; cold work steel. *Obsolete*

FIRTH-VICKERS "VIKRO"
Firth-Vickers Stainless Steels Ltd.
Ni 60-65, Cr 15-25, C 1-0.1, Si 0.5-1, Mn 0-1, bal Fe.
Normalized: 102,000-134,000 TS; 65,000-78,000 YS; 25-15 El; 35-25 RA; 180-240 Brin. At 1000°C: 17,600 TS. For case hardening boxes, gas or oil burners, furnace muffles, stay bolts; heat resisting; stainless.

FIRTH-VICKERS "VIKRO"

British Steel Corp.
Ni 60-65, Cr 15-25, C 1-0.1, Si 0.5-1, Mn 0-1, bal Fe.
Normalized: 102,000-134,000 TS; 65,000-78,000 YS; 25-15 El;
35-25 RA; 180-240 Brin. At 1000°C: 17,600 TS. For case
hardening boxes, gas or oil burners, furnace muffles, stay
bolts; heat resisting; stainless.

FIRTH-VICKERS 21/4N STEEL

Firth-Vickers Stainless Steels Ltd.
C 0.48-0.58, Si 0-0.25, Mn 8-10, Cr 20-22, Ni 3.25-4.5, N
0.38-0.5, bal Fe.
Non-magnetic, precipitation hardenable stainless steel. Heat
treat tested at 800°C: 53,000 TS; 35,800 YS. Good oxidation
resistance up to 850°C. For valves in internal combustion
engines.

FIRTH-VICKERS 309

Firth-Vickers Stainless Steels Ltd.
C 0-0.15, Si 0-1, Mn 0-2, Cr 22-24, Ni 13-15, bal Fe.
RT: 90,000 TS; 35,000 YS; 50 El. At 800°C: 20,000 TS; 10,800
YS (0.2% proof). Good resistance to corrosion and to scaling
at elevated temperatures; for superheater supports, furnace
parts, heat treat boxes and annealing racks. Weldable. AISI
309.

FIRTH-VICKERS 448 STEEL

Firth-Vickers Stainless Steels Ltd.
C 0.1-0.13, Si 0.5, Mn 1, Cr 11, Ni 0.75, Mo 0.65, V 0.15-0.3,
Nb 0.4, bal Fe.
Air or oil hardened, tempered 650-700°C. At RT: 150,000 TS;
127,000 YS; 20 El. At 700°C: 51,500 TS; 44,000 YS; 28 El.
For aircraft gas turbine discs and similar stressed parts at
elevated temperatures.

FIRTH-VICKERS 467

Firth-Vickers Stainless Steels Ltd.
C 0.2, Ni 9.5, Cr 14, Mo 2, Cu 2.5, Ti 0.7, bal Fe.
Annealed: 98,000 TS; 43,500 YS; 52 El; 56 RA. At 500°F:
90,000 TS; 33,000 YS; 31 El; 36 RA. At 750°F: 61,000 TS;
33,000 YS; 24 El; 27 RA. For jet engine components; heat
and creep resistant, austenitic.

FIRTH VICKERS 507

Firth-Vickers Stainless Steels Ltd.
C 0.12, Cb 0.4, Cr 10.5, Mo 0.75, V 0.15, bal Fe.
Annealed: 75,000 TS; 35,000 YS; 30 El; B 82 Rock. Cold
drawn: 95,000 TS; 80,000 YS; 15 El; B 92 Rock. For springs,
table flatware, knives, oil refinery equipment. Corrosion
resistant. *Obsolete*

FIRTH-VICKERS 535 STEEL

Firth-Vickers Stainless Steels Ltd.
C 0.07, Si 0.4, Mn 0.85, Cr 10.5, Ni 0.3, Co 6, Mo 0.75, V 0.2,
Nb 0.45, bal Fe.
Air or oil hardened, tempered 600-650°C. At RT: 160,000 TS;
142,000 TS; 22 El. At 550°C: 118,000 TS; 103,000 YS; 22 El.
For aircraft turbine and compressor disc forgings and
wrought rings.

FIRTH-VICKERS 548 STEEL

Firth-Vickers Stainless Steels Ltd.
C 0.08, Si 0.4, Mn 1, Cr 16.5, Ni 11.5, Mo 1.5, Nb 1, bal Fe.
Austenitic, stainless steel; weldable. RT: 80,000 TS; 33,000
YS; 55 El. At 700°C: 53,000 TS; 18,000 YS; 41 El. For
structural parts and welded assemblies operating at elevated
temperatures.

FIRTH-VICKERS 555 STEEL

Firth-Vickers Stainless Steels Ltd.
C 0.05, Si 0.4, Mn 1.5, Cr 16.5, Ni 10.5, Mo 2.4, bal Fe.
Austenitic, stainless; weldable. At 700°C: 45,600 TS; 13,800
YS. For steam pipe, chemical plant equipment.

FIRTH-VICKERS 566

Firth-Vickers Stainless Steels Ltd.
C 0.12, Si 0.4, Mn 0.8, Cr 11.5, Ni 2.3, Mo 1.4, V 0.15, Nb 0.3,
bal Fe.
Air hardened, tempered at 630-650°C. RT: 147,000 TS;
124,000 YS; 22 El. At 600°C: 80,000 TS; 66,000 YS; 28 El.
For turbine blading for use up to 450°C.

FIRTH-VICKERS 607 STEEL

Firth-Vickers Stainless Steels Ltd.
C 0.15, Si 0.3, Mn 0.8, Cr 11, Ni 0.6, Mo 0.8, V 0.25, bal Fe.
Air or oil hardenable, corrosion resistant steel, primarily for
sheet production. For stamped or formed parts stressed at
temperatures up to 600°C.

FIRTH-VICKERS F 17

Firth-Vickers Stainless Steels Ltd.
C 0.1, Cr 14-18, bal Fe.
Annealed: 80,000 TS; 45,000 YS; 28 El; 145 Brin. For
automotive trim, kitchen sinks, oil burners, fasteners, furnace
parts. Type 430 stainless steel, nonhardenable, ferritic, heat
resistant. *Obsolete*

FIRTH-VICKERS F.A.L.

Firth-Vickers Stainless Steels Ltd.
C 0-0.12, Si 0-0.8, Mn 0-10, Cr 12-14, Al 3.8-4.8, bal Fe.
Heat treated: 90,000 TS; 67,000 YS; 25 El. Corrosion resistant
steel with high electrical resistivity; for resistance and
rheostats.

FIRTH-VICKERS F.A.S.

Firth-Vickers Stainless Steels Ltd.
C 0.43, Si 0.3, Mn 0.5, Cr 11.5, bal Fe.
Annealed: 110,000 TS; 79,000 YP; 20 El; 240 Brin. For
stainless ball bearings. Corrosion resistant.

FIRTH-VICKERS F.C.I.

Firth-Vickers Stainless Steels Ltd.
C 0-0.12, Si 0-0.8, Mn 1-2, S 0.15-0.3, Cr 11.5-13.5, Mn
0.3-0.6, bal Fe.
Martensitic, free-machining, hardenable, corrosion resistant
steel. Air or oil hardenable to about 28-38 Rock C. For shafts,
hardware, fasteners, cap screws. BS EN.56AM; AISI 416.

FIRTH-VICKERS F.C.S.

Firth-Vickers Stainless Steels Ltd.
C 0.2-0.28, Si 0-0.8, Mn 1-2, S 0.15-0.3, Cr 12-14, Mo 0.3-0.6,
bal Fe.
Martensitic, free-machining, hardenable, corrosion resistant
steel. Air or oil hardenable to about 44-52 Rock C. For spline
shafts, gears, studs, hardware. BS EN.56 CM, 2S.124; AISI
420F.

FIRTH-VICKERS F.G.

Firth-Vickers Stainless Steels Ltd.
C 0.2-0.28, Si 0-0.8, Mn 0-1, Cr 12-14, bal Fe.
Martensitic, corrosion resistant steel. Heat treated and
tempered at 400°C: 218,000 TS; 196,000 YS; 18 El; 480
Vickers. For cutlery, surgical instruments, dental tools, turbine
equipment, valves. BS 3S.62, EN56C; AISI 420.

FIRTH-VICKERS F.G. (L)

Firth-Vickers Stainless Steels Ltd.
C 0.12-0.2, Si 0-0.8, Mn 0-1, Cr 11.5-13.5, bal Fe.
Martensitic, corrosion resistant steel; hardenable to about
42-48 Rock C. For intricate cutlery and surgical instruments
where hot forging may be difficult. (AISI 420-low carbon
range).

FIRTH-VICKERS F.H.

Firth-Vickers Stainless Steels Ltd.
C 0.28-0.36, Si 0-0.8, Mn 0-1, Cr 12-14, bal Fe.
Martensitic, corrosion resistant steel. Air or oil hardenable to
240,000 psi TS and 598 Vickers. For knives, cutlery, surgical
and dental instruments.

FIRTH-VICKERS F.H.M.

Firth-Vickers Stainless Steels Ltd.
C 0.7-0.9, Si 0-0.8, Mn 0-1, Cr 15.5-17.5, Mo 0.3-0.7, bal Fe.
Martensitic, corrosion resistant. Air or oil hardenable to
about 50-56 Rock C. For cutlery, dental and surgical
instruments, valves, stainless ball and roller bearings. AISI
440B.

FIRTH-VICKERS F.I. (AL)

Firth-Vickers Stainless Steels Ltd.
C 0-0.08, Cr 11.5-14.5, Al 0.1-0.3, bal Fe.
Annealed: 65,000 TS; 38,000 YS; 36 El; 145 Brin. Cold drawn:
85,000 TS; 70,000 YS; 20 El; 185 Brin. For heat treating
boxes, quenching racks, oil refining and chemical plant
equipment. Heat and corrosion resistant. AISI 405.

FIRTH-VICKERS F.I. (P) STEEL

Firth-Vickers Stainless Steels Ltd.
C 0-0.08, Cr 12-14, bal Fe.
Ferritic-martensitic type stainless steel.

FIRTH-VICKERS F.I. (TI) STEEL

Firth-Vickers Stainless Steels Ltd.
C 0-0.06, Si 0-0.8, Mn 0-1.5, Cr 10.5-12, Ni 1, Ti 0.2-0.8, bal
Fe.
Soft: 63,000 TS; 34,000 YS; 32 El. Hard: 72,000 TS; 52,000
YS; 25 El. For lightly loaded parts to resist scaling up to
700-750°C. Similar to AISI 405.

FIRTH-VICKERS F.I. 17

Firth-Vickers Stainless Steels Ltd.
C 0.07, Cr 17, Mn 0.8, bal Fe.
Annealed: 78,400 TS; 49,300 YS; 32 El; 50 RA. 160 Brin. For
automotive trim, window frames, grills; corrosion resistant;
Type 430.

FIRTH-VICKERS F.I. 17 MO STEEL

Firth-Vickers Stainless Steels Ltd.
C 0.07, Si 0.4, Mn 0.6, Cr 16.3, Ni 0.4, Mo 1, bal Fe.
Ferritic, stainless steel; for motor car hub caps and trim,
domestic flatware and holloware.

FIRTH-VICKERS F.I. MO

Now FIRTH-VICKERS MOLYBDENUM STAINLESS STEEL.

FIRTH-VICKERS F.L.

Firth-Vickers Stainless Steels Ltd.
C 0-0.08, Cr 11.5-14.5, Al 0.1-0.3, bal Fe.
Annealed: 70,000 TS; 40,000 YS; 30 El; 60 RA; 160 Brin. For
heat treating boxes, quenching racks, high temperature
units. Type 405; stainless steel.

FIRTH-VICKERS F.M.B.

Firth-Vickers Stainless Steels Ltd.
C 0-0.08, Cr 16-18, Ni 10-14, Mo 2-3, bal Fe.
Annealed: 85,000-95,000 TS; 35,000-45,000 YS; 50-60 El;
60-75 RA; 150-190 Brin. For chemical plant equipment,
mixers, agitators, filters; Type 316; stainless, austenitic.

FIRTH-VICKERS F.S.M.1

Firth-Vickers Stainless Steels Ltd.
C 0.05, Cr 17.2, Mn 7.15, Ni 4.5, N 0.12, bal Fe.
Annealed: 104,000 TS; 56,000 YS; 51 El; B 90 Rock. For
aeronautical applications, kitchen equipment, fasteners. Type
204, corrosion and heat resistant. Stainless, austenitic, high
strength. *Obsolete*

FIRTH-VICKERS F.V. 507

Firth-Vickers Stainless Steels Ltd.
C 0.12, Cr 11, Mo 0.8, V 0.15, Nb 0.4, bal Fe.
Martensitic corrosion resistant steel. Air or oil
hardenable to about 35-44 Rock C. In harden conditions has
good strength and corrosion resistance at room temperature
and is used to about 700 C; for housings, brackets. *Obsolete*

FIRTH-VICKERS F.V. 520 (B) STEEL

Firth-Vickers Stainless Steels Ltd.
C 0-0.07, Si 0-0.7, Mn 0-1, Cr 13.2-14.7, Ni 5-6, Cu 1.2-2, Mo
1.2-2, Nb 0.2-0.7, bal Fe.
Corrosion resistant maraging type steel. Aged:
122,000-212,000 TS; 78,000-150,000 YS; 10-20 El; 270-450
Vickers (varying treatment). Weldable; for blading, bolts,
fasteners, valves, pumps, water impellers, assemblies.

FIRTH-VICKERS F.V. 520 (S) STEEL
Firth-Vickers Stainless Steels Ltd.
C 0.04-0.07, Si 0-0.6, Mn 0.8-1.8, Cr 15.3-16, Ni 5-5.8, Cu 1.4-2.1, Mo 1.2-2, Ti 0.5-0.15, bal Fe.
Corrosion resistant maraging type steel. Aged: 142,000-196,000 TS; 116,000-142,000 YS; 6 El min; 320-400 Vickers (varying treatment). Weldable; for bolts, fasteners, valves, pump parts, water impellers, assemblies.

FIRTH-VICKERS F.V. 702
Firth-Vickers Stainless Steels Ltd.
C 0.03, Si 0.5, Mn 0.6, Cr 15.7, Ni 2.5, Mo 1, Nb 0.5, bal Fe.
Good resistance to corrosion and stress corrosion; equipment for chemical plants. *Obsolete*

FIRTH-VICKERS FHM NO. 2
Firth-Vickers Stainless Steels Ltd.
C 0.95-1.2, Cr 16-18, Mo 0.75, bal Fe.
Martensitic, air-hardening, stainless steel; AISI 440C.

FIRTH-VICKERS FVS
Firth-Vickers Stainless Steels Ltd.
C 0.4, Cr 13.6, Ni 14, W 2.6, bal Fe.
Valve steel.

FIRTH-VICKERS H.R.CROWN 1 STEEL
Firth-Vickers Stainless Steels Ltd.
C 0.2, Si 0.8, Mn 0.9, Cr 22.5, Ni 11, W 0.5, bal Fe.
Austenitic stainless steel casting. For room temperature and elevated temperature parts subject to corrosive conditions; good high temperature strength. AISI 309; BS; 1648-E. *Obsolete*

FIRTH-VICKERS H.R.CROWN MAX STEEL
Firth-Vickers Stainless Steels Ltd.
C 0.2, Si 1, Mn 0.8, Cr 23, Ni 11.5, W 3, bal Fe.
Austenitic stainless steel casting. For room temperature and elevated temperature parts subject to corrosive conditions; good high temperature strength. AISI 309; BS; 1648-E. *Obsolete*

FIRTH-VICKERS MO-V STAINLESS
Firth-Vickers Stainless Steels Ltd.
C 0.12, Si 0.3, Mn 0.6, Cr 12.25, Ni 0.8, Mo 0.6, V 0.18, bal Fe.
Air or oil hardenable, tempered at 675-700°C. At RT: 115,000 TS; 100,000 YS; 29 El. At 650°C: 48,000 TS; 40,000 YS (0.2% proof). For stressed parts operating up to 650°C.

FIRTH-VICKERS MOLYBDENUM STAINLESS STEEL
Firth-Vickers Stainless Steels Ltd.
C 0.1, Si 0.3, Mn 0.3, Cr 12.5, Mo 0.75, bal Fe.
Air or oil hardenable; good resistance to scaling and creep at elevated temperatures; for turbine blades and similar applications.

FIRTH-VICKERS S. 80
Firth-Vickers Stainless Steels Ltd.
C 0.16, Cr 16.5, Ni 2.5, bal Fe.
Oil treated: 100,000-135,000 TS; 78,000-112,000 YS; 15-25 El; 45-60 RA; 240-280 Brin. For pump spindles, propeller shafts, seaplane and aircraft construction; magnetic, stainless and rust resistant. AISI 431.

FIRTH-VICKERS S.L.V.
Firth-Vickers Stainless Steels Ltd.
C 0.4-0.5, Si 3-3.75, Mn 0.3-0.6, Cr 7.5-9.5, bal Fe.
Heat treated and tempered, 650°C: 134,000-154,000 TS; 100,000-112,000 YS; 10-20 El. Good mechanical properties and scale resistance at elevated temperatures. For valves in internal combustion engines. BS EN.52.

FIRTH-VICKERS STAINLESS STEEL F.I.
Firth-Vickers Stainless Steels Ltd.
C 0.08, Si 0.25, Mn 0.25, Cr 13.3, bal Fe.
Stainless steel, for elevated temperature operation; as steam turbine blades.

FIRTH-VICKERS XB
Firth-Vickers Stainless Steels Ltd.
C 0.78, Si 2, Mn 0.45, Cr 19.5, Ni 1.35, bal Fe.
Hardened: 140,000 TS; 127,000 YS; 16 El; 21 RA. At 400°C: 110,000 TS; 78,000 YS; 15 El; 28 RA. For valves. Stainless, heat and corrosion resistant.

FIRTHAG
Firth Brown Ltd.
C 0.45, Mn 0.9, Cr 0.95, Si 0.25, bal Fe.
For punches, gears, bolts; Brit. EN18.

FIRTHALOY
Ferranti Ltd.
Tool material. W.C.
For wire drawing and extrusion dies; cemented tungsten carbide.

FIRTHITE
Teledyne Firth Sterling
Co 5-13, TC 87-97.
For tipped cutting tools; cemented tungsten carbide. *Obsolete*

FIRTHITE GR. TXL
Teledyne Firth Sterling
Co 6.3-6.8, 12.5 TiC, 81.0 WC min.
Sintered: 790 Brin. For cutting tools; carbide cement.

FIRTHITE H
Teledyne Firth Sterling
Now FIRTHITE H6.

FIRTHITE H-23
Teledyne Firth Sterling
WC, Co.
Sintered: 240,000 transverse strength; 91.8 Rock A; 14.8 density. Cutting tools for roughing cuts; for machining superalloys. *Obsolete*

FIRTHITE H6
Teledyne Firth Sterling
Co 5.7-6.3, 94.0 WC min.
78 Rock C to 91 Rock A. For cutters on cast iron, nonferrous metals and nonmetallics; cemented WC.

FIRTHITE HA
Teledyne Firth Sterling
Co 5.2-5.8, 1.0 TaC, 91.0 WC min.
80 Rock C to 92.0 Rock A. For tool bits, cutters; for semi-finishing hard irons.

FIRTHITE HB
Teledyne Firth Sterling
Co 8, WC 92.
77 Rock C. For tool bits, cutters; for roughing cast iron. *Obsolete*

FIRTHITE HE
Teledyne Firth Sterling
WC 95.5, Co 4.5.
Sintered: 92 Rock A. For cutting tools; sintered carbide. *Obsolete*

FIRTHITE HF
Teledyne Firth Sterling
Co 2.7-3.3, bal WC.
93.5 Rock A. For tools, cutters; finishing cuts.

FIRTHITE NHA
Teledyne Firth Sterling
WC.
For cutters for machining cast iron; cemented carbide. *Obsolete*

FIRTHITE NTA
Teledyne Firth Sterling
WC, Co.
Sintered: 250,000 transverse strength; 91.0 Rock A; 12.9 density. Cutting tools for rough machining steel.

FIRTHITE T 0 4
Teledyne Firth Sterling
WC 85, Co 10, 1.0 TaC, 4.0 TiC.
For tipped tools for rough cuts; cemented.

FIRTHITE T-22
Teledyne Firth Sterling
WC, Co.
Sintered: 240,000 transverse strength; 92.0 Rock A; 12.8 density. Cutting tool for general purpose machining steel.

FIRTHITE T-25
Teledyne Firth Sterling
WC, Co.
Sintered: 190,000 transverse strength; 92.4 Rock A; 11.72 density. For semifinish cuts on ferrous and nonferrous metals; crater resistant grade.

FIRTHITE T-31
Teledyne Firth Sterling
WC 68.5, Co 6.5, 25.0 TiC.
For bits for cutting tools; hard, wear resistant, sintered. *Obsolete*

FIRTHITE T-41
Teledyne Firth Sterling
Co 8, bal WC.
79 Rock C. For tools, dies, cutters; cemented WC. *Obsolete*

FIRTHITE T-66
Teledyne Firth Sterling
WC 60, Co 12, Ta 28.
For heavy duty cutters and tools; sintered.

FIRTHITE T-89
Teledyne Firth Sterling
Co 12, WC 59, 22.0 TaC, 7.0 TiC.
77 Rock C. For rough and semifinish cuts, tipped tools; cemented. *Obsolete*

FIRTHITE TXH
Teledyne Firth Sterling
WC 83, Co 8, 9.0 TiC.
Sintered: 91.7 Rock A. For cutting tools; sintered carbide.

FIRTHITE TXL
Teledyne Firth Sterling
Co 6.2-6.8, 12.5 TiC, 81.0 WC min.
Sintered: 92 Rock A. For cutting tools; sintered carbide.

FIRTHITE WF
Teledyne Firth Sterling
Ni 12, 70.0 TiC, 18.0 MoC.
Sintered: 92 Rock A; 125,000 transverse strength. Sintered carbide tool inserts for cutting steel; high speed fine finishing. *Obsolete*

FIRTHITE WL
Teledyne Firth Sterling
WC 73, Co 5.7-6.3, 21.0 TaC.
Sintered: 91.3 Rock A. For cutting tools; sintered carbide.

FIRTHOB
Firth Brown Ltd.
C 0.09, Mn 0.5, Ni 1.25, Cr 0.5, bal Fe.
For gears, shafts, cams, machinery parts; case hardened, shock resistant. *Obsolete*

FISCHER VIS-11
Georg Fisher AG
C 0.25, Cr 10.8, Mo 1.1, Ni 0.6, V 0.4, W 0.4, bal Fe.
Annealed: 95,000 TS; 45,000 YS; 25 El; 92 Rock B.
Hardened: 245,000 TS; 210,000 YS; 8 El; 50 Rock C. For valves, gears, cutlery, shafts, bearings, surgical instruments. Corrosion resistant, hardenable. Cast only.

FISCO 66
Faitout Iron & Steel Co.
C 0.8, Cr 4, W 5.5, V 1.5, Mo 5, bal Fe.
For tools, cutters; high speed steel.

FISCO AIRDIE
Faitout Iron & Steel Co.
C 0.6, Cr 5, bal Fe.
For tools, cutters; oil hardening.

FISCO AIRQUENCH
Faitout Iron & Steel Co.
C 0.95, Mn 2.04, Cr 1.9, Mo 1, bal Fe.
For dies, rolls, gauges; air hardening.

FISCO CARBON
Faitout Iron & Steel Co.
C 1, bal Fe.
For tools, dies, cutters; water hardening.

FISCO CHROMDIE
Faitout Iron & Steel Co.
C 1.6, Cr 12, Mo 0.75, V 0.25, bal Fe.
For dies, gauges, mandrels, shear blades; air hardening; abrasion resistant.

FISCO COBALT
Faitout Iron & Steel Co.
C 0.74, Cr 4.5, V 1.4, W 18.8, Mo 0.6, Co 5, bal Fe.
For tools, cutters, dies; high speed steel.

FISCO DUPLEX
Faitout Iron & Steel Co.
C 0.8, bal Fe.
For tools; water hardening.

FISCO EXCELL
Faitout Iron & Steel Co.
C 0.8, W 18, Cr 4, V 2, Mo 0.7, bal Fe.
For tools, cutters; high speed steel.

FISCO HIGH SPEED
Faitout Iron & Steel Co.
C 0.7, W 18, Cr 4, V 1, bal Fe.
For tools, cutters; high speed steel.

FISCO HOT WORK
Faitout Iron & Steel Co.
C 0.6, Mo 8.5, V 1.7, Cr 3.6, bal Fe.
For hot work tools and dies; hot work steel.

FISCO MOLY
Faitout Iron & Steel Co.
C 0.7, W 1.5, Mo 9.5, Cr 4, V 1, bal Fe.
For tools, cutters; high speed steel.

FISCO NO. 1
Faitout Iron & Steel Co.
C 0.15, Mn 0.5, Ni 1.8, Mo 0.25, bal Fe.
For arbors, gears, pinions; carburizing steel.

FISCO NO. 2
Faitout Iron & Steel Co.
C 0.3-0.4, Mn 0.9-1.2, bal Fe.
Rolled: 90,000 TS; 65,000 YS; 18 El; 45 RA; 212 Brin. For gears, shafts, bolts, gears, lead screws, spindles; excellent machining qualities.

FISCO NO. 3
Faitout Iron & Steel Co.
C, Mo, Cr, bal Fe.
For axles, bolts, gears, keys, pinions; oil hardening.

FISCO NO. 4
Faitout Iron & Steel Co.
C 0.45-0.55, Mn 0.6-0.9, Cr 0.8-1.1, Mo 0.15-0.25, bal Fe.
Oil quenched: 130,000-150,000 TS; 100,000-130,000 YS; 17-18 El; 50-55 RA; 230-300 Brin. For shafts, gears, axles; high impact and creep strength, pre-heat-treated.

FISCO OILDIE
Faitout Iron & Steel Co.
C, bal Fe.
For tools, dies; oil hardening.

FISCO OILHARD
Faitout Iron & Steel Co.
C 0.9, Mn 1.1, Cr 0.5, V 0.25, W 0.4, bal Fe.
For tools, cutters, dies, gauges; non-deforming.

FISCO OMEGA
Faitout Iron & Steel Co.
C 0.55, Mn 0.8, Mo 0.5, V 0.25, Si 2.3, bal Fe.
For impact tools; shock resistant; water hardened.

FISCO PNEUMATIC
Faitout Iron & Steel Co.
C, bal Fe.
For tools; oil hardening.

FISCO PRECISION
Faitout Iron & Steel Co.
C, bal Fe.
For tools; water hardening.

FISCO SEVEN
Faitout Iron & Steel Co.
C, bal Fe.

FISCO SPECIAL
Faitout Iron & Steel Co.
C 1, bal Fe.
For tools, dies, cutters; water hardening.

FISCO STAR TUNG
Faitout Iron & Steel Co.
C, W, bal Fe.
For tools, cutters; oil hardening.

FISCO SUPERIOR
Faitout Iron & Steel Co.
C 0.8, Co 12, W 20, Cr 4, V 1, Mo 0.6, bal Fe.
For tools, cutters; high speed steel.

FISCO TIGER
Faitout Iron & Steel Co.
C 1, W 18, Cr 4, V 2, bal Fe.
For tools, cutters; high speed steel.

FISCO VANADIUM
Faitout Iron & Steel Co.
C 0.9, V 0.16, bal Fe.
For tools, dies, cutters; water hardening.

FISCO-MO NO. 1
Faitout Iron & Steel Co.
C 0.1-0.2, Mn 0.4-0.7, Ni 1.65-2, Mo 0.2-0.3, bal Fe.
Oil quenched: 120,000 TS; 90,000 YS; 22 El; 60 RA; 248 Brin.
For carburized parts, arbors, gears, pinions, cams, clutches; minimum distortion.

FISCO-MO NO. 3
Faitout Iron & Steel Co.
C 0.45-0.55, Mn 0.6-0.9, Cr 0.9-1.1, Mo 0.15-0.25, bal Fe.
Annealed: 100,000 TS; 68,000 YS; 25 El; 55 RA; 183 Brin.
Heat treated: 240,000 TS; 225,000 YS; 12 El; 30 RA; 460 Brin.
For heavy duty shafts, gears, clutches; good machining qualities.

FIVE POINT DEEP HARD
Foote Bros. Gear & Machine Corp.
C 0.4, Ni 1.5, Cr 0.8, Mo 0.2, bal Fe.
For gears, axles, wrist pins, wheels, valves; oil hardening.

FIVE STAR
Midvale-Heppenstall Co.
C 0.85, Cr 4, Co 10.5, W 18, Mo 0.55, bal Fe.
For tools, cutters; high speed steel. *Obsolete*

FIXAMPER
Creusot-Loire
Ni-Fe.
For heating elements, electrical resistances; good to 600 C. *Obsolete*

FIXINVAR
Creusot-Loire
Ni 36, bal Fe.
For clocks. *Obsolete*

FK(D)M-10
German manufacture
C 0.25, Cr 3, W 0.4, Mo 0.4, V 0.2, bal Fe.
For turbine blades for jet engines; oil hardened.

FK10F
Teledyne Firth Sterling
Tool material. WC 94, Co 6.
Abrasion resistant grade to cut high temperature alloys and abrasive material. 92.7 Rock A.

FK20M
Teledyne Firth Sterling
Tool material. WC 93.5, Co 6, 0.5 TiC + TaC + NbC.
Abrasion resistant grade for use in cutting metal to yield a short or powder-like chip. 92.1 Rock A.

FK30F
Teledyne Firth Sterling
Tool material. WC 90, Co 10.
Abrasion resistant grade to cut high temperature alloys and abrasive material. 91.4 Rock A.

FK40B
Teledyne Firth Sterling
Tool material. WC 88.5, Co 11, 0.50 TiC + TaC + NbC.
For use as base material for coated cutting grades. 89.7 Rock A.

FLAMALOY
Detroit Alloy Steel Co.
C, alloy, bal Fe.
For castings; castings, wear resistant.

FLAME HARD
Now ISC FLAME HARD.

FLAME RESISTING ALLOY
English manufacture
Cr 14, Ni 9.7, Mn 0.8, Si 0.2, C 0.2, bal Fe.
For heat resisting alloy parts; corrosion resistant.

FLANGE METAL, FRENCH
French manufacture
Cu 94, Sn 5.6, Pb 0.05.
For bushings, fittings, flanges; tough.

FLANGE METAL, GERMAN
German manufacture
Cu 92, Zn 5, Sn 2.5.
For pipes, fittings, flanges; corrosion resistant.

FLASH ALLOY
S.G. Taylor Chain Co.
C 0.17-0.22, Ni 1.65-2, Mo 0.2-0.3, bal Fe.
For sling chains.

FLECTO IRON
Ohio Brass Co.
C 2.4-2.6, Mn 0.45, Si 1.1, bal Fe.
Cast: 52,000 TS; 35,000 YS; 14 El; 135 Brin. For castings to be hot galvanized; specially annealed malleable cast iron. *Obsolete*

FLEETWELD
Lincoln Electric Co.
C, bal Fe.
For welding electrodes; coated electrode. *Obsolete*

FLEETWELD 180
Lincoln Electric Co.
C, bal Fe.
Steel arc welding electrode. AWS Class E6011.

FLEETWELD 35
Lincoln Electric Co.
C, bal Fe.
Steel arc welding electrodes. AWS Class E6011.

FLEETWELD 35LS
Lincoln Electric Co.
C, bal Fe.
Steel arc welding electrode. AWS Class E6011.

FLEETWELD 37
Lincoln Electric Co.
C, bal Fe.
Steel arc welding electrode. AWS Class E6013.

FLEETWELD 47
Lincoln Electric Co.
C, bal Fe.
Steel arc welding electrode. AWS Class E7014.

FLEETWELD 5
Lincoln Electric Co.
C, bal Fe.
Steel arc welding electrode. AWS Class E6010.

FLEETWELD 57
Lincoln Electric Co.
C, bal Fe.
Steel arc welding electrode. AWS Class E6013.

FLEETWELD 5P
Lincoln Electric Co.
C, bal Fe.
Steel arc welding electrode. AWS Class E6010.

FLEETWELD 7
Lincoln Electric Co.
C, bal Fe.
Steel arc welding electrode. AWS Class E6012.

FLEETWELD NO. 10
Lincoln Electric Co.
C, bal Fe.
For welding rods; flux coated. *Obsolete*

FLEETWELD NO. 11
Lincoln Electric Co.
C, bal Fe.
For welding rods; coated, shielded arc. *Obsolete*

FLEETWELD NO. 8
Lincoln Electric Co.
C, bal Fe.
For welding rods; coated, shielded arc. *Obsolete*

FLEETWELD NO. 9
Lincoln Electric Co.
C, bal Fe.
For arc welding electrodes; coated, shielded arc. *Obsolete*

FLEETWELD NO. 9 HT
Lincoln Electric Co.
C, bal Fe.
For welding rods; coated, shielded arc. *Obsolete*

FLETCHER & EMPERER BEARING
English manufacture
Al 92, Cu 7.5, Sn 0.3.
For bearings; non-hardenable.

FLETCHER'S ALLOY
English manufacture
Al 96, Cu 3, Sn 1, Sb 0.5, P.
For aircraft and light weight parts; non-hardenable.

FLETCHERS BEARING METAL
Manufacturer not listed.
Al 90, Cu 7, Zn 1.
For bearings.

FLEX
Now VEW F180.

FLEX-O-LOY
Bridgeport Rolling Mills Co.
Cu 68.5-71.5, Pb 0.03, Fe 0.05, Zn 30.
Hard: 76,000 TS; 63,000 YS; 8 El; Rock B 82. Spring: 95,000 TS; 85,000 YS; 3 El; Rock B 91. For electrical, electronic, and wiring device components. Good formability. *Obsolete*

FLEXAL
Aluminium Norf GmbH
Aluminum.
Aluminum alloy AlFe.

FLEXALOY
Bergstrom Alloys Corp.
C, Cr, Mn, Ni, bal Fe.
550 Brin. For hard surfacing electrodes; wear resistant against earth.

FLEXARC ACP-MO
Westinghouse Electric Corp.
C, bal Fe.
For welding electrodes for steel. *Obsolete*

FLEXARC AP
Westinghouse Electric Corp.
C 0.1, bal Fe.
Welded: 62,000 psi TS; 52,000 psi YS; 22 El. For welding electrodes for low carbon steel. Meets E6012. *Obsolete*

FLEXARC AP-MO
Westinghouse Electric Corp.
C, alloy, bal Fe.
For welding electrodes. Meets E-7010. *Obsolete*

FLEXARC DH
Westinghouse Electric Corp.
C 0.2, bal Fe.
Welded: 62,000 psi TS; 55,000 psi YS; 25 El. For welding electrodes; coated, for medium carbon steel. Meets E6020. *Obsolete*

FLEXARC DH-MO
Westinghouse Electric Corp.
C, alloy, bal Fe.
Welded: 70,000 psi TS; 57,000 psi YS; 25 El. For welding electrodes; high tensile. Meets E-7020. *Obsolete*

FLEXARC FP
Westinghouse Electric Corp.
C 0.3, bal Fe.
Welded: 68,000 psi TS; 55,000 psi YS; 17 El. For welding electrodes; coated, for medium carbon steel. Meets E6012. *Obsolete*

FLEXARC FP-2
Westinghouse Electric Corp.
C, bal Fe.
Welded: 68,000 psi TS; 55,000 psi YS; 17 El. For arc welding electrodes. Meets AWS-E-6012. *Obsolete*

FLEXARC GRADE 18
Westinghouse Electric Corp.
C, bal Fe.
For welding rods; Sulcoat. Meets E-4510. *Obsolete*

FLEXARC LOH-2
Westinghouse Electric Corp.
C, alloy, bal Fe.
Welded: 70,000 psi TS; 57,000 psi YS; 22 El. H_2 coated electrode for alloy steel. Meets E-7016. *Obsolete*

FLEXARC SW-2
Westinghouse Electric Corp.
Fe alloy.
Welded: 68,000 psi TS; 55,000 psi YS; 17 El. For welding electrodes. Meets E-6013. *Obsolete*

FLEXIBLE HACK BAND
Disston Inc.
C 1.3, Cr 0.15, bal Fe.
For tools, dies; water hardening. *Obsolete*

FLEXITE
St. Lawrence Steel Co.
C 0.45, Mn 1.1, Cr 1.1, Ni 1.75, V 0.2, Mo 0.35, bal Fe.
Oil hardened, low alloy tool steel.

FLEXO STEEL
Carpenter Technology Corp.
C 0.6, Mn 0.75, Si 2, bal Fe.
Heat treated: 135,000-325,000 TS; 100,000-300,000 YS; 4-20 El; 10-50 RA; 250-600 Brin. For leaf springs, recoil springs, shuttle springs; great toughness for high service stresses. *Obsolete*

FLEXOGRAIN PHOSPHOR BRONZE
Riverside Metals Corp.
Sn, P, bal Cu.
For springs, contacts, clips; high fatigue life. *Obsolete*

FLEXOR
Pennsylvania Steel Corp.
C 0.34, Cr 0.85, Mo 0.5, W 0.45, bal Fe.
Heat treated: 160,000-250,000 TS; 125,000-200,000 YS; 9-14 El; 37-44 RA; 300-495 Brin. For chuck jaws, arbors, motor shafts, gears, racks, arbors, pre-treated machinery steel.

FLINSO
Grant & West Ltd.
Pb, bal Sn.
For solder for Al.

FLINT ALLOY
English manufacture
Fe 83, Cr 12.5, Si 0.5, C 0.3.
For cutlery, stainless parts, corrosion resisting parts.

FLINTCAST
Pacific Foundry Co.
TC 3.3, Si 2.4, Cr 1, bal Fe.
For abrasion resisting iron castings; abrasion resisting.

FLINTMETAL
Morris Machine Works
C 3-3.6, Ni 4-4.75, Cr 1.4-3.5, Si 0.4-0.7, bal Fe.
Sand cast: 45,000 TS; 600 Brin. Permanent mold: 50,000 TS; 675 Brin. For cams, dies, rollers, bearing races; white cast iron, hard, corrosion resistant. *Obsolete*

FLINTUFF
Teledyne Ohiocast
C 3, Ni, bal Fe.
For rolls; hard alloy. *Obsolete*

FLINTYPE
Grand Northern Products Ltd.
C 0.9, Si 1, Mn 0.3, Cr 9.5, Mo 0.6, Ni 3.2, bal Fe.
Hardfacing electrode. As arc welded: 55,000 psi TS; 560 Brin. For overlay or buildup on plow points, heavy earth moving machinery; abrasion resistant, not designed for edge impact. *Obsolete*

FLO-KOTE
Stulz Sickles Steel Co.
C 0.2, Mn 11-13, Ni, bal Fe.
For welding electrode; tough, wear resistant.

FLOCKTON 4.0 TUNGSTEN
Flockton, Tompkin & Co., Ltd.
C 1.3, W 4-5, Cr 1-1.2, V 0.25, bal Fe.
For reamers; oil hardened.

FLOCKTON C. T.
Flockton, Tompkin & Co., Ltd.
C 0.3, W 9, Cr 2.5, V 0.4, bal Fe.
For heading and piercing dies, punches; hot work steel, oil hardened.

FLOCKTON H.C.C.
Flockton, Tompkin & Co., Ltd.
C, Co, bal Fe.
For shear blades, nibblers; air hardened.

FLOCKTON H.D.M.
Flockton, Tompkin & Co., Ltd.
C 0.35, Cr 5, Mo 1.5, Si 1, Mn 0.4, V 0.4, bal Fe.
For extrusion dies, die casting dies; hot work steel, oil hardened.

FLOCKTON I.E.-P
Flockton, Tompkin & Co., Ltd.
C 1, bal Fe.
Annealed: 100,000 TS; 53,000 YS; 21 El; 42 RA; 200 Brin. For tools, dies, drills, taps; water hardened; Type W1.

FLOCKTON N.T.C.
Flockton, Tompkin & Co., Ltd.
C, alloy, bal Fe.
For chisels; nontempering steel, air hardened.

FLOCKTON R.H.D.
Flockton, Tompkin & Co., Ltd.
C 0.38, Cr 5, Mo 1.5, W 1.5, Si 1, Mn 0.3, bal Fe.
For extrusion and heading dies, rams, mandrels; oil or air hardened, hot work steel.

FLOCKTON T.C.
Flockton, Tompkin & Co., Ltd.
C 1.5, W 3.2, Cr 0.15, bal Fe.

FLOCTON P-NI
Flockton, Tompkin & Co., Ltd.
C 0.8, Ni 1.8, Cr 1, Mn 0.35, bal Fe.
For machinery parts, cold heading dies; oil hardened, shock resistant.

FLOCTON P.B.7
Flockton, Tompkin & Co., Ltd.
C, alloy, bal Fe.
For taps, dies, drills, reamers; resists severe compression and abrasion.

FLOTECTIC SILVER 2
Eutectic Corp.
Silver solder type alloy for torch, furnace or induction brazing, ferrous and non-ferrous alloys. 1295°F MP; 58,000 psi TS. *Obsolete*

FLOTRET PROCESS ALLOY
Globe Metallurgical, Inc.
Miscellaneous nonferrous. Si 43-47, Mg 3-3.75, Ca 0.7-1.1, Al 0-1.2, 1.25-1.75 total rare earth elements.
Magnesium ferrosilicon alloy with controlled chemistry for use in the FLOTRET Process.

FLUGINOX 51
Ugine Aciers
C 0.05, Al 0.2, Cr 12.5, Mo 0.5, bal Fe.
Annealed: 70,000 TS; 35,000 YS; 40 El; 80 RA; 145 Brin. Cold drawn: 90,000 TS; 80,000 YS; 20 El; 65 RA; 200 Brin. For oil refinery and chemical plant equipment, table flatware, hardware. Corrosion resistant. *Obsolete*

FLUGINOX 51
Ugine Aciers
C 0.06, Ni 0.4, Cr 12.8, Mo 0.5, Al 0.25, bal Fe.
Stainless steel, resistant to corrosion and oxidation at elevated temperature; resistant to creep.

FLUGINOX 60
Ugine Aciers
C 0-0.05, Cr 13, Mo 0.5, bal Fe.
Annealed: 70,000 TS; 35,000 YS; 40 El; 80 RA; 145 Brin. Cold drawn: 90,000 TS; 80,000 YS; 20 El; 65 RA; 200 Brin. For oil refinery and chemical plant equipment, table flatware, hardware, Corrosion resistant. *Obsolete*

FLUGINOX 61
Ugine Aciers
C 0.2, Ni 0.5, Cr 12, Mo 1, V 0.3, bal Fe.
Martensitic stainless steel; resistant to creep up to 600°C, for turbines and hydrocarbon industries. Modified AISI 420.

FLUGINOX 62
Ugine Aciers
C 0.2, Cr 12, Mo 1, Ni 0.8, V 0.25, W 1, bal Fe.
Annealed: 75,000 TS; 40,000 YS; 30 El; 70 RA; B 82 Rock.
Heat treated: 230,000 TS; 195,000 YS; 8 El; 30 RA; C 50 Rock. For cutlery, surgical instruments, valves, bearings, shafts. Corrosion resistant, hardenable. *Obsolete*

FLUGINOX 65
Ugine Aciers
C 0.22, Cr 11, Mo 1, V, Cb, bal Fe.
Annealed: 75,000 TS; 35,000 YS; 30 El; 70 RA; B 82 Rock.
Cold drawn: 90,000 TS; 80,000 YS; 15 El; 60 RA; B 92 Rock.
For knives, cutlery, valves, bearings, shafts, gears, surgical instruments. Corrosion resistant, hardenable. *Obsolete*

FLUGINOX 65
Ugine Aciers
C 0.22, Ni 0.7, Cr 11, Mo 0.9, Nb 0.3, V 0.3, S 0-0.015, bal Fe.
Martensitic stainless steel, resistant to oxidation and corrosion and creep up to 500-650°C, steam and hot oil.

FLUGINOX 71
Ugine Aciers
C 0.1, Ni 2.3, Cr 12, Mo 1.7, V 0.3, bal Fe.
Martensitic stainless steel; resistant to creep up to 500°C.

FLUKS
Paul Bergsoe & Son
Sn 55, Sb 10, Cu 2.5, Pb 32.5.
Cast: 11,400 TS; 22 Brin. MP: 355-620°F. For refrigerator and electric motor bearings. Good castability.

FLUSH PLATE
English manufacture
Cu 65.75, Zn 32.75, Pb 1.5.
For hardware; free-cutting.

FLUX-COR NO. 5
Airco Vacuum Metals
Mn 1.27, Si 0.39, Ni 0.06, Cu 0.06, bal Fe.
Welded: 81,400 psi TS; 64,000 psi YS; 30 El. Good impact strength. Weld deposit: 0.072 C. MIL-E-24403/1-A CLASS MIL-70T-5.

FLUXCOR 1
Airco Vacuum Metals
C 0.11, Mn 1.6, Si 0.84, bal Fe.
Welded: 101,700 psi TS; 89,200 psi YS; 24 El. Wire for welding or build-up on mild steel. AWS E 70T-2.

FLUXCOR 1/2 MO
Airco Vacuum Metals
C 0.08, Mn 0.8, Si 0.5, Mo 0.5, bal Fe.
Welded, stress annealed: 92,000 psi TS; 81,000 psi YS; 26 El. Wire for welding and build-up of low alloy steel using CO_2 shielded gas.

FLUXCOR 2
Airco Vacuum Metals
For welding steel having residual oil from machining and forming. AWS E 70T-1.

FLUXCOR CR-MO 1
Airco Vacuum Metals
C 0.07, Mn 0.84, Si 0.46, Cr 1.23, Mo 0.51, bal Fe.
Welded, stress annealed: 103,000 psi TS; 93,000 psi YS; 25 El. Wire to join or repair Cr-Mo steel castings and similar Cr-Mo steels.

FLUXCOR CR-MO 2
Airco Vacuum Metals
C 0.04, Mn 0.83, Si 0.48, Cr 2.23, Mo 1.08, bal Fe.
Welded, stress annealed: 110,000 psi TS; 100,000 psi YS; 20 El. Wire for welding or build-up of Cr-Mo castings and wrought steel of similar composition. Used with CO_2 shielded gas.

FLUXCOR NO. 1
Airco Vacuum Metals
Mn 1.46, Si 0.74, Ni 0.06, Cr 0.07, Mo 0.07, bal Fe.
Welded: 87,000 psi TS; 74,700 psi YS; 26.5 El. Low alloy cored wire for welding low alloy steel. Weld deposit: 0.60 C. AWS A5.20 Class E 70T-2.

FLUXRITE
NL Industries
Sn, bal Pb.
For solder; soft. *Obsolete*

FLYLIGHT NO. 5
Howard Foundry Co.
Al 7.5-8.5, Zn 0.2-0.6, Mn 0.3, Si 0.2, bal Mg.
Cast: 20,000 TS; 11,000 YS; 3 El; 50 Brin. Heat treated: 32,000 TS; 11,000 YS; 9 El; 60 Brin. For gear and blower casing for aero engines; heat treatable. *Obsolete*

FLYLITE NO. 4
Howard Foundry Co.
Al 8.3-9.7, Zn 1.7-2.3, Mn 0-0.1, bal Mg.
Cast: 20,000 TS; 10,000 YS; 1 El; 65 Brin. T-temper: 34,000 TS; 18,000 YS; 1 El; 84 Brin. For sand castings; heat treatable. *Obsolete*

FLYLITE NO. 8
Howard Foundry Co.
Al 5.3-6.7, Zn 2.5-3.5, Mn 0-0.15, bal Mg.
Cast: 24,000 TS; 10,000 YS; 4 El; 50 Brin. T-temper: 34,000 TS; 18,000 YS; 3 El; 73 Brin. For sand castings; heat treatable. *Obsolete*

FLYLITE NO. 9
Howard Foundry Co.
Al 9.2-9.8, Zn 0.2-0.6, Mn 0.2-0.4, bal Mg.
Cast: 18,000 TS; 11,000 YS; 1 El; 55 Brin. T-temper: 34,000 TS; 17,000 YS; 1 El; 75 Brin. For sand castings; heat treatable. *Obsolete*

FM-100
Muskegon Piston Ring Co.
Cu 3, Si 0.2, Mn 0.5, Mo 0.9, TC 1.25, CC 0.92, bal Fe.
Sintered: 100,000 TS; 95 Rock B. For piston rings; wear resistant.

FM-2
Howment Corp. (Carbide Div.)
Sintered carbide tool material. For percussion type mining tools. *Obsolete*

FM-20
Forjas Alavesas S.A.
C 0.19, Mo 1.25, Si 0.2, bal Fe.
Manganese structural or carburizing steel. AFNOR 20M5; SAE 1522; BS EN7A.

FM-3
Howment Corp. (Carbide Div.)
Sintered carbide tool material. For percussion and rotary drilling type mining tools. *Obsolete*

FM-30T
Forjas Alavesas S.A.
C 0.34, Mn 1.25, Si 0.2, bal Fe.
Manganese structural steel. IHA F-411; DIN 30 Mn5; AISI 1536.

FM-5
Howment Corp. (Carbide Div.)
Sintered carbide tool material. For coal mining tools, percussion, rotary drilling, auger bits. *Obsolete*

FM-6
Howment Corp. (Carbide Div.)
Sintered carbide tool material. For percussion and rotary drilling type mining tools. *Obsolete*

FM10B
Teledyne Firth Sterling
Tool material. WC 87.05, Co 5.5, 7.45 TiC + TaC + NbC. For use as base material for coated cutting grades. 91.5 Rock A.

FM202
Contact Technologies, Inc.
99.0 Mo min.
Density: 9.7 g/cm^3; specific resistivity: 0.067 micro-ohm-m; hardness: 30N31; strength: 110,000 psi. Heat sinks for semiconductor applications. *Obsolete*

FM26
Contact Technologies, Inc.
Ag 52.
Silver molybdenum. Density: 10.2 g/cm^3; specific resistivity: 0.030 micro-ohm-m; hardness: 30T60; strength: 90,000 psi. Electrical contacts for circuit breaker applications.

FMP 035
F.M. Parkin Ltd.
C 1.05, Si 0.25, Mn 0.35, V 0.35, bal Fe.
Water hardening tool steel. AISI W2.

FMP 1850
F.M. Parkin Ltd.
C 0.55, Si 0.3, Mn 0.3, W 18, Cr 4.1, V 0.7, Mo 0-1, bal Fe.
Tungsten type hot work tool steel. For hot forming dies. AISI H26.

FMP 200
F.M. Parkin Ltd.
C 0.95, Si 0.3, Mn 1.25, W 0.5, Cr 0.5, V 0.17, bal Fe.
Oil hardening cold work tool steel. For blanking and forming tools, punches. AISI O1.

FMP 328
F.M. Parkin Ltd.
C 0.4, Si 1, Mn 0.3, Cr 5.25, V 0.4, Mo 1.35, bal Fe.
Chromium type hot work tool steel. For hot forming and forging dies. AISI H11.

FMP 329
F.M. Parkin Ltd.
C 0.4, Si 1, Mn 0.35, Cr 5.25, V 1, Mo 1.35, bal Fe.
Chromium type hot work tool steel. For hot forming dies, light metal casting dies. AISI H13.

FMP 336
F.M. Parkin Ltd.
C 1.5, Si 0.4, Mn 0.4, Cr 12, V 0.9, Mo 0.8, bal Fe.
High carbon-high chrome type cold work tool steel. For forming and blanking dies, thread rolling dies, plastic molds. AISI D2.

FMP 338
F.M. Parkin Ltd.
C 2.05, Si 0.4, Mn 0.4, Cr 13, bal Fe.
High carbon-high chrome type cold work tool steel; air or oil hardening. For blanking and forming dies, gauges, brick mold liners. AISI D3.

FMP 348
F.M. Parkin Ltd.
C 0.43, Si 0.3, Mn 0.4, Cr 1.55, Mo 0.3, Ni 4, bal Fe.
Shock resisting tool steel. For heavy duty shear blades, trimming dies.

FMP 379
F.M. Parkin Ltd.
C 1, Si 0.3, Mn 0.5, Cr 5, V 0.3, Mo 1.1, bal Fe.
Air hardening cold work tool steel. For punching, drawing and coining dies. AISI A2.

FMP 399
F.M. Parkin Ltd.
C 0.5, Si 0.9, Mn 0.35, W 2.25, Cr 1.45, V 0.2, bal Fe.
Oil hardening shock resisting tool steel. For chisels, shear blades, pneumatic chisels. AISI S1.

FMP 455
F.M. Parkin Ltd.
C 0.8, Si 0.3, Mn 0.3, W 18.5, Cr 4.25, V 1.3, Co 5, Mo 0-0.75, bal Fe.
Tungsten-cobalt type high speed tool steel. AISI T4.

FMP 470
F.M. Parkin Ltd.
C 0.8, Si 0.3, Mn 0.3, W 22, Cr 4.5, V 1.5, Mo 0-1, bal Fe.
Tungsten type high speed steel.

FMP 501
F.M. Parkin Ltd.
C 0.8, Si 0.3, Mn 0.3, W 2, Cr 3.9, V 1.25, Mo 9, bal Fe.
Molybdenum type high speed tool steel. For drills, rough cutting lathe tools. AISI M1.

FMP 504
F.M. Parkin Ltd.
C 1.3, Si 0.3, Mn 0.3, W 5.75, Cr 4.4, V 4, Mo 4.6, bal Fe.
Molybdenum type high speed steel. AISI M4.

FMP 505
F.M. Parkin Ltd.
C 0.32, Si 0.3, Mn 0.3, W 9.25, Cr 3.25, V 0.5, Mo 0.5, bal Fe.
Tungsten type hot work tool steel. For hot forming and swaging dies. AISI H21.

FMP 507
F.M. Parkin Ltd.
C 0.3, Si 0.3, Mn 0.3, W 9, Cr 3, V 0.3, Mo 0.5, Ni 2.5, bal Fe.
Tungsten type hot work tool steel. For hot punches and hot heading dies.

FMP 513
F.M. Parkin Ltd.
C 0.35, Si 1, Mn 0.3, W 1.25, Cr 5.25, V 0.35, Mo 1.5, bal Fe.
Chromium type hot work tool steel. For hot forming, blanking and extrusion dies. AISI H12.

FMP 526
F.M. Parkin Ltd.
C 1.1, Si 0.3, Mn 0.3, W 6.75, Cr 4.25, V 2, Co 5, Mo 3.75, bal Fe.
Molybdenum-cobalt type high speed steel. AISI M41.

FMP 530
F.M. Parkin Ltd.
C 0.8, Si 0.3, Mn 0.3, W 2, Cr 4, V 1.2, Co 5, Mo 8.25, bal Fe.
Molybdenum-cobalt type high speed steel. AISI M30.

FMP 536
F.M. Parkin Ltd.
C 1.5, Si 0.3, Mn 0.3, W 6.5, Cr 4, V 5, Co 5, Mo 3.5, bal Fe.
Molybdenum type high speed steel. AISI M15.

FMP 542
F.M. Parkin Ltd.
C 1.05, Si 0.3, Mn 0.3, W 1.5, Cr 3.75, V 1.15, Co 8, Mo 9.5, bal Fe.
Molybdenum-cobalt type high speed steel. AISI M42.

FMP 555
F.M. Parkin Ltd.
C 1.5, Si 0.3, Mn 0.3, W 12.75, Cr 4.75, V 5, Co 5, Mo 0-1, bal Fe.
Tungsten-vanadium-cobalt high speed steel. AISI T15.

FMP 562
F.M. Parkin Ltd.
C 0.82, Si 0.3, Mn 0.3, W 6.25, Cr 4.1, V 2, Mo 5, bal Fe.
Molybdenum type high speed steel. AISI M2.

FMP 563
F.M. Parkin Ltd.
C 1.05, Si 0.3, Mn 0.3, W 6, Cr 4, V 2.35, Mo 5, bal Fe.
Molybdenum type high speed steel. AISI M3 Class 1.

FMP 599
F.M. Parkin Ltd.
C 0.7, Si 0.3, Mn 0.3, W 14, Cr 4.25, V 0.8, bal Fe.
Tungsten type high speed tool steel. For hacksaws, slitting saws, cold punches.

FMP 622
F.M. Parkin Ltd.
C 0.75, Si 0.3, Mn 0.3, W 18.2, Cr 4.1, V 1.1, Mo 0-1, bal Fe.
Tungsten type high speed tool steel. For cutting tools. AISI T1.

FMP 682
F.M. Parkin Ltd.
C 0.6, Si 0.3, Mn 0.3, Cr 4, V 2, Mo 8.25, bal Fe.
Molybdenum type hot work tool steel. For hot working tools and dies. AISI H43.

FMP 808
F.M. Parkin Ltd.
C 0.8, Si 0.3, Mn 0.3, W 18.5, Cr 4.25, V 1.6, Co 10, Mo 1, bal Fe.
Tungsten-cobalt type high speed tool steel. Similar to AISI T5 or T6.

FMP 828
F.M. Parkin Ltd.
C 0.8, Si 0.3, Mn 0.3, W 18.5, Cr 4.5, V 2, Co 8, Mo 0-1, bal Fe.
Tungsten-cobalt type high speed tool steel. AISI T5.

FMP 842
F.M. Parkin Ltd.
C 0.84, Si 0.3, Mn 0.3, W 18.5, Cr 4.25, V 2.25, Mo 0-1, bal Fe.
Tungsten type high speed tool steel. For cutting tools. AISI T2.

FMP 922
F.M. Parkin Ltd.
C 1, Si 0.3, Mn 0.3, W 1.75, Cr 4, V 2, Mo 8.75, bal Fe.
Molybdenum type high speed steel. AISI M7.

FMP 928
F.M. Parkin Ltd.
C 0.9, Si 0.3, Mn 0.3, W 2, Cr 4, V 2, Co 8, Mo 8.5, bal Fe.
Molybdenum-cobalt type high speed steel. AISI M34.

FMP 929
F.M. Parkin Ltd.
C 1.25, Si 0.3, Mn 0.3, W 1.8, Cr 3.75, V 2, Co 8.25, Mo 8.75, bal Fe.
Molybdenum-cobalt type high speed steel. AISI M43.

FMP 933
F.M. Parkin Ltd.
C 1.3, Si 0.3, Mn 0.3, W 9.25, Cr 4.25, V 3.5, Mo 3.75, Co 10, bal Fe.
Tungsten type high speed tool steel with cobalt.

FMP 948
F.M. Parkin Ltd.
C 0.88, Si 0.3, Mn 0.3, Cr 4, V 2, Mo 8.25, bal Fe.
Molybdenum type high speed steel. AISI M10.

FMS-35
Forjas Alavesas S.A.
C 0.36, Mn 1.3, Si 1.3, bal Fe.
Mn-Si structural steel; for gears, axles. Hardenable, water quench to 48-56 Rock C. DIN 37 Mn Si 5; AFNOR 38MS5.

FO 18-9
Acieries du Forez
C 0.2-0.28, Cr 17-18, Ni 8.5-9.5, Mn 0-1.5, Si 0-0.5, bal Fe.
Austenitic stainless steel; good corrosion resistance. AFNOR Z 25 CN 18.9.

FO 25-12
Acieries du Forez
Stainless steel; similar to AISI 309.

FO 25-20
Acieries du Forez
Stainless steel; similar to AISI 310.

FO 25-20 U.
Acieries du Forez
C 0-0.03, Cr 19-21, Ni 24-28, Mo 4.5, Mn 0-2, Si 0-1, Cu 1.5, bal Fe.
Stainless steel for petroleum refineries, paper making, ammonium sulfate. AFNOR Z 2 NCDU 25.20.

FO 36-18
Acieries du Forez
Stainless steel; similar to AISI 330.

FO 90 SUPERVITAC
Creusot-Loire
C 0.42, Cr 1, Mo 0.2, bal Fe.
Bar, treated: 880 N/mm^2 psi TS; 685 N/mm^2 psi YS min; 15 El; 262 Brin. For structural purposes. Similar to AISI 4142.

FOB METAL
English manufacture
Cu 87.5, Zn 12, Sn 0.5.
For fobs, ornaments; red brass.

FOBALLOY
Plessey Inc.
Sn 95, Ag 3.5, Sb 1.5.
For hermetic sealing of integrated ceramic packages.

FOBES METAL
English manufacture
Zn 54, Cu 46.
For ornamental castings; low strength.

FOC 3 POINTS
Compagnie Ateliers et Forges de la Loire
C 0.9, bal Fe.
Annealed: 100,000 TS; 53,000 YP; 2 El; 197 Brin. Hardened: 216,000 TS; 152,000 YP; 11 El; 600 Brin. For chisels, bolts, cutting tools, springs, punches, drills. Water hardening, wear resistant.

FOC 4 POINTS
Compagnie Ateliers et Forges de la Loire
C 0.8, bal Fe.
Annealed: 120,000 TS; 65,000 YS; 15 El; 223 Brin. Hardened: 190,000 TS; 142,000 YS; 12 El; 390 Brin. For chisels, bolts, springs, bearings, liners. Water hardening, wear resistant.

FOC O POINTS
Compagnie Ateliers et Forges de la Loire
C 1.25, bal Fe.
Hardened: 65-66 Rock C. For files, cutting tools, drills. Water hardening, wear resistant.

FOCARBO 10
Forjas Alavesas S.A.
C 0.1, Mn 0.5, Si 0.25, P 0-0.03, S 0-0.03, bal Fe.
Low carbon steel. DIN CK 10; AFNOR XC 10; AISI 1010.

FOCARBO 15
Forjas Alavesas S.A.
C 0.15, Mn 0.5, Si 0.25, bal Fe.
Low carbon steel, sometimes for case hardening. DIN CK-15; AFNOR XC 18. AISI 1015.

FOCARBO 20
Forjas Alavesas S.A.
C 0.2, Mn 0.55, Si 0.25, bal Fe.
Low carbon structural steel. DIN CK-22; AFNOR XC 18. AISI 1020.

FOCARBO 25
Forjas Alavesas S.A.
C 0.25, Mn 0.6, Si 0.25, bal Fe.
Structural steel. IHA F-112; AISI 1025; BS EN 4.

FOCARBO 32
Forjas Alavesas S.A.
C 0.32, Mn 0.65, Si 0.25, bal Fe.
Structural steel. AFNOR XC-32; AISI 1030; BS EN 5C.

FOCARBO 35
Forjas Alavesas S.A.
C 0.35, Mn 0.65, Si 0.25, bal Fe.
Structural steel. DIN-CK 35; IHA F-113; AISI 1035; UNI C-35.

FOCARBO 37
Forjas Alavesas S.A.
C 0.37, Mn 0.65, Si 0.25, bal Fe.
Medium carbon structural steel. AFNOR XC 38; AISI 1038; BS EN 6A.

FOCARBO 42
Forjas Alavesas S.A.
C 0.42, Mn 0.65, Si 0.2, bal Fe.
Medium carbon structural steel. AFNOR XC 42; AISI 1042; BS EN 8D, UNI C-40.

FOCARBO 45
Forjas Alavesas S.A.
C 0.45, C 0.65, Si 0.25, bal Fe.
Medium carbon structural steel. IHA F-114; DIN CK 45; AISI 1045; UNI C-45.

FOCARBO 47
Forjas Alavesas S.A.
C 0.47, Mn 0.65, Si 0.25, bal Fe.
Medium carbon structural steel. CENIM F-1142/C47K; AFNOR XC48; AISI 1049.

FOCARBO 55
Forjas Alavesas S.A.
C 0.55, Mn 0.65, Si 0.25, bal Fe.
Medium-high carbon structural steel. IHA F-115; DIN CK 55; AFNOR XC 55; AISI 1055.

FOCARBO-35T
Forjas Alavesas S.A.
C 0.36, Mn 0.8, Si 0.25, bal Fe.
For cold forming, threading and subsequent heat treatment of small screws. IHA F-113; DIN Cq-35; AISI 1037.

FOCEM-12
Forjas Alavesas S.A.
C 0.1, Mn 0.5, Si 0.25, bal Fe.
Low carbon steel for carburizing. AFNOR XC-10; DIN CK-10; AISI 1010.

FOCEM-17
Forjas Alavesas S.A.
C 0.17, Mn 0.5, Si 0.25, bal Fe.
Carburizing steel. IHA F-111; AFNOR XC-18; DIN CK-15; AISI 1017.

FOLDER DIE STEEL
McInnes Steel Co.
C, bal Fe.
For forming dies, brakes. *Obsolete*

FOLLANSBEE ARMATURE ELECTRIC NO. 1
Follansbee Steel Co.
C 0.4, Si 0.5, bal Fe.
Annealed: 45,000 TS; 25,000 YS; 25 El. For laminated pole pieces, motors, transformers; high permeability. *Obsolete*

FOLLANSBEE EXT.SPEC. TRANSFORMER NO. 7
Follansbee Steel Co.
Si 3, bal Fe.
For power and distribution transformers. *Obsolete*

FOLLANSBEE IMPROVED ELECTRIC NO. 2
Follansbee Steel Co.
For motor armatures. *Obsolete*

FOLLANSBEE RADIO A-1 NO. 13
Follansbee Steel Co.
Si 2, bal Fe.
For radio transformers, chokes and loudspeaker transformers. *Obsolete*

FOLLANSBEE RADIO A-2 NO. 12
Follansbee Steel Co.
Si 2, bal Fe.
For radio, choke, loud-speaker transformers. *Obsolete*

FOLLANSBEE RADIO A-3 NO. 11
Follansbee Steel Co.
Si 2, bal Fe.
For radio, choke and loud speaker transformers. *Obsolete*

FOLLANSBEE RADIO B NO. 10
Follansbee Steel Co.
Si 2, bal Fe.
For choke coils, transformers. *Obsolete*

FOLLANSBEE RADIO C NO. 9
Follansbee Steel Co.
Si 2, bal Fe.
For choke and filters, audio transformers. *Obsolete*

FOLLANSBEE RADIO D NO. 8
Follansbee Steel Co.
Si 2, bal Fe.
For choke coils. *Obsolete*

FOLLANSBEE REGULAR TRANSFORMER NO. 5
Follansbee Steel Co.
Si 3, bal Fe.
For transformers, generators, motors. *Obsolete*

FOLLANSBEE SPECIAL DYNAMO NO. 4
Follansbee Steel Co.
Si 3.5, bal Fe.
For induction motors, generators, motors. *Obsolete*

FOLLANSBEE SPECIAL MOTOR NO. 3
Follansbee Steel Co.
Si 1, bal Fe.
For motors. *Obsolete*

FOLLANSBEE SPECIAL TRANSFORMER NO. 6
Follansbee Steel Co.
Si 3.5, bal Fe.
For transformers; low watt loss. *Obsolete*

FOMO-12
Forjas Alavesas S.A.
C 0.13, Mn 0.7, Si 0.25, Cr 1, Mo 0.2, bal Fe.
Cr-Mo carburizing steel. IHA F-155; DIN 12CrMo4; AFNOR 12Cd4.

FOMO-18
Forjas Alavesas S.A.
C 0.18, Mn 0.75, Si 0.25, Cr 1, Mo 0.2, bal Fe.
Cr-Mo carburizing steel. AFNOR 18CD4; DIN 16CrMo4.

FOMO-20
Forjas Alavesas S.A.
C 0.2, Mn 0.75, Si 0.25, Cr 0.4, Mo 0.45, bal Fe.
Mo-Cr carburizing steel. CENIM F-1523/20 CrMo2; DIN 20 MoCr4.

FOMO-25
Forjas Alavesas S.A.
C 0.27, Mn 0.6, Si 0.25, Cr 1, Mo 0.2, bal Fe.
Cr-Mo structural steel. IHA F-222; AFNOR 25 CD4; UNI 25CD4.

FOMO-30
Forjas Alavesas S.A.
C 0.3, Mn 0.7, Si 0.25, Cr 1, Mo 0.2, bal Fe.
Cr-Mo structural steel. UNI 30CD4; AISI 4130.

FOMO-35
Forjas Alavesas S.A.
C 0.35, Mn 0.7, Si 0.25, Cr 1, Mo 0.2, bal Fe.
Cr-Mo structural steel. IHA F-125; AFNOR 35CD4; AISI 4135.

FOMO-35T
Forjas Alavesas S.A.
C 0.37, Mn 0.7, Si 0.25, Cr 1, Mo 0.2, bal Fe.
Cr-Mo structural steel. IHA F-125; AFNOR 38CD4; AISI 4137.

FOMO-40
Forjas Alavesas S.A.
C 0.4, Mn 0.75, Si 0.25, Cr 1, Mo 0.2, bal Fe.
Cr-Mo structural steel, for shafts, axles. DIN 42 CrMo4; AISI 4140; BS En19.

FONI-30
Forjas Alavesas S.A.
C 0.32, Mn 0.55, Si 0.25, Ni 3, Cr 0.7, bal Fe.
Ni-Cr structural steel; deep hardening. IHA F-123; AFNOR 30NC11; BS EN23.

FONIX-10
Forjas Alavesas S.A.
C 0-0.12, Mn 0-0.7, Si 0-0.7, Cr 13, bal Fe.
Ferritic or martensitic stainless steel. IHA F-311; DIN X10 Cr13; similar to AISI 410.

FONIX-16
Forjas Alavesas S.A.
C 0-0.1, Mn 0-1, Si 0-1, Cr 17, bal Fe.
Ferritic type stainless steel; not hardenable by heat teatment. DIN X8Cr17; AISI 430; BS EN60.

FONIX-17-2
Forjas Alavesas S.A.
C 0.17, M 0-0.7, Si 0-0.7, Cr 17, Ni 2, Mn 0-0.7, bal Fe.
Martensitic stainless steel; hardenable to 100-135 kg/mm^2 TS. For aircraft fittings, marine equipment.

FONIX-18-12-2-TI
Forjas Alavesas S.A.
C 0-0.08, Mn 0-2, Si 0-1, Cr 18, Ni 12, Mo 2.25, Ti = 5 x C min, bal Fe.
Weldable grade of AISI 316 Stainless. AFNOR Z 8CNDT 18-12; BS En 58 J.

FONIX-18-8
Forjas Alavesas S.A.
C 0-0.08, Mn 0-2, Si 0-1, Cr 18, Ni 8, bal Fe.
Austenitic stainless steel; for food, dairy, textile and photographic equipment. IHA F-314; AISI 304; BS EN58E.

FONIX-18-8-2
Forjas Alavesas S.A.
C 0-0.08, Mn 0-2, Si 0-1, Cr 18, Ni 12, bal Fe.
Austenitic stainless steel; improved corrosion resistance for food, photographic, and beverage equipment. DIN X5CrNiMo 18-10; AISI 316; BS EN58H.

FONIX-18-8-S
Forjas Alavesas S.A.
C 0-0.12, Mn 0-2, Si 0-1, S 0.2, Cr 18, Ni 9, bal Fe.
Free-machining, austenitic stainless steel. For threaded parts. AISI 303; AFNOR Z10CNF 1809.

FONIX-18-8-TI
Forjas Alavesas S.A.
C 0-0.08, Mn 0-2, Si 0-1, Cr 18, Ni 11, Ti = 5 x C min, bal Fe.
Weldable austenitic stainless steel. IHA F-332; AISI 321; BS EN58C.

FONIX-20
Forjas Alavesas S.A.
C 0.2, Mn 0-0.7, Si 0-0.7, Cr 13, bal Fe.
Martensitic stainless steel; hardenable to 44-50 Rock C. For valves, arbors, plastic molds. DIN X20Cr13; AFNOR Z20C13; AISI 420.

FONIX-30
Forjas Alavesas S.A.
C 0.3, Mn 0-0.7, Si 0-0.7, Cr 13.5, bal Fe.
Martensitic stainless steel; hardenable to 47-54 Rock C. For valves, parts for chemical equipment. AFNOR Z30C13; BS EN56D; AISI 420.

FONIX-35
Forjas Alavesas S.A.
C 0.38, Mn 0-0.7, Si 0-0.7, Cr 13.5, bal Fe.
Martensitic stainless steel; hardenable to 48-54 Rock. For pump shafts, cutlery, shears, valves. IHA F-312; DIN X40Cr13; AISI 420.

FONTAINMOREAU BRONZE
English manufacture
Cu 0-8, Fe 0-1, Pb 0-1, bal Zn.
For ornaments, fittings, die castings.

FONTANOR 15.3
Acieries et Forges d'Anor
C 2.4-2.6, Cr 14-16, Mo 2.5-3, bal Fe.
High Cr-Mo white iron.

FONTANOR 2
Acieries et Forges d'Anor
C 3.3-3.5, Cr 1.7-1.9, Ni 3.5-3.7, bal Fe.
Low alloy white cast iron type Ni-hard.

FONTANOR 30
Acieries et Forges d'Anor
C 3.2, Cr 30, bal Fe.
Alloy iron casting. *Obsolete*

FONTANOR 300
Acieries et Forges d'Anor
C 3.3-3.5, Cr 15-17, W 0.5-0.7, Mo 0.5-0.7, bal Fe.
High alloy white iron.

FONTANOR F30B
Acieries et Forges d'Anor
C 2.8-2.9, Cr 24-25, bal Fe.
High chromium white iron.

FONTANOR F4
Acieries et Forges d'Anor
C 3.1-3.3, Cr 8-9, Ni 5.5-6.5, bal Fe.
High alloy white cast iron; type Ni-hard 4.

FONTANOR NI-RESIST
Acieries et Forges d'Anor
C 2.7-2.9, Cr 0-1.5, Ni 14.5-15.5, Cu 5-5.5, bal Fe.
Ni-Resist type gray cast iron.

FONTE R. A.
Ste de Produits Metallurgiques
C 0.2, bal Fe.
For structural parts; carburizing steel.

FOOL-PROOF
P.F. McDonald & Co.
C 0.7, W 4, Cr 2, bal Fe.
For hand and pneumatic chisels; oil hardening.

FOOTE ELECTROMANGANESE NO. 1
Cyprus Foote Mineral Co.
Mn 99.5, S 0.028, C 0.008, Si 0.001, H 0.0005, N 0.013, O 0.48.
For alloying in stainless and special steels. *Obsolete*

FOOTE ELECTROMANGANESE, HYDROGEN REMOVED
Cyprus Foote Mineral Co.
Mn 99.5, S 0.028, C 0.008, Si 0.001, H 0.0005, N 0.013, O 0.48.
For alloying in stainless and special steels. *Obsolete*

FOOTE ELECTROMANGANESE, LOW OXYGEN
Manufacturer not listed.
Mn 99.8, S 0.028, C 0.008, Si 0.001, H 0.024, N 0.002, O 0.1.
For alloying in stainless and special steels.

FOOTE NITREMANG, GRADE "A"
Cyprus Foote Mineral Co.
Mn 93.7, N 6, C 0.03, S 0.032, Fe 0.005, P 0.001, O 0.35, S 0.001, H 0.0005.
Alloying agent for introducing manganese and nitrogen into special steels. *Obsolete*

FOOTE NITREMANG, GRADE "B"
Cyprus Foote Mineral Co.
Mn 94.4, N 5, C 0.03, S 0.032, Fe 0.005, P 0.001, Si 0.001, H 0.005, O 0.35.
Alloying agent for introducing manganese and nitrogen into special steels. *Obsolete*

FOOTE NODULOY 3
Now NODULOY 3.

FOR TEN 50
Ford Motor Co.
HSLA steel, semi-killed or killed. Meets SAE 950.

FOR TEN 60
Ford Motor Co.
HSLA steel; meets SAE 960.

FORAL
Forjas Alavesas S.A.
C 0.1, Mn 1.1, Si 0-0.06, P 0-0.05, S 0.3, Pb 0.2, Se 0-0.1, bal Fe.
Free machining, low carbon steel. IHA F-212; AISI 12L14.

FORD 406
Cannon-Muskegon Corp.
Cr 6, Co 10, Mo 1, W 8.5, Cb 2, Ti 2, Al 4.5, Ta 6, bal Ni.
For integrally cast turbine wheels.

FOREZ 2 AS BC
Acieries du Forez
C 0.6-0.7, Cr 4-4.3, W 6-6.75, Mo 5.25-5.75, V 1.85-2.15, bal Fe.
For punches and cold chisels, extrusion mandrels. AFNOR Z 65 WDV 06.05.02; AISI H42.

FOREZ 2 AS CO
Acieries du Forez
C 0.8-0.85, Cr 4-4.3, Mo 5.25-5.75, W 6-6.75, V 1.85-2.15, Co 4.75-5.25, bal Fe.
Cobalt-molybdenum-tungsten high speed steel. Cutting tool for higher speed operation. AFNOR Z 85 WDKV 06.06.05.02; AISI M35.

FOREZ 2AS
Acieries du Forez
C 0.85, W 6, V 2, Mo 6, Cr 4.5, bal Fe.
For lathe and planer tools, drills, reamers, hobs; high speed steel (Z85WD06-06).

FOREZ 333
Acieries du Forez
C 1.5-1.6, Cr 4.5-5, W 12-13, V 4.75-5.25, Co 4.75-5.25, bal Fe.
Tungsten-vanadium-cobalt high speed steel. Good wear resistant cutting tool. AFNOR Z 150 WKV 12.05.05; AISI T 15.

FOREZ 3AS
Acieries du Forez
C 0.8, W 18, Cr 5, Mo 1, V 1.5, bal Fe.
For lathe and planer tools, drills, taps, hobs; high speed steel (Z80W18).

FOREZ 3ASR
Acieries du Forez
C 0.7, W 12, Mo 2.5, V 2, Cr 5, bal Fe.
For lathe and planer tools, reamers, milling cutters; high speed steel (Z70WD12). *Obsolete*

FOREZ 444
Acieries du Forez
C 1.25, W 20, Cr 4, V, Co, bal Fe.
For lathe and planer tools, reams, taps, drills; high speed steel, oil hardened (Z125WK20-15).

FOREZ 4AS
Acieries du Forez
C 0.85, W 18, Cr 5, V 1.5, Mn 1, Co 5, bal Fe.
For lathe and planer tools, milling cutters, hobs; high speed steel (Z85WK-18-05).

FOREZ CH1
Acieries du Forez
C 1, Mn 0.3, Si 0.3, bal Fe.
For cutters; water hardened (100C2). *Obsolete*

FOREZ CH3
Acieries du Forez
C 0.9, Mn 0.3, Si 0.3, bal Fe.
For milling cutters, reamers, drills, punches, files; water hardened (90C2). *Obsolete*

FOREZ CH4
Acieries du Forez
C 1.2, Mn 0.3, Si 0.3, bal Fe.
For nail dies, hot snaps, punches, shears; hard and tough (120C2).

FOREZ CH5
Acieries du Forez
C 0.8, Mn 1.2-1.5, Si 0.4, bal Fe.
For reamers, taps, gauges, special dies; oil hardened, nondeforming (80M5).

FOREZ CX100
Acieries du Forez
C 1, bal Fe.
Annealed: 100,000 TS; 53,000 YS; 21 El; 42 RA; 200 Brin. For reamers, taps, drills, punches; water hardened. *Obsolete*

FOREZ CX125
Acieries du Forez
C 1.25, bal Fe.
For drills, reamers, milling cutters; water hardened. *Obsolete*

FOREZ CX50
Acieries du Forez
C 0.5, bal Fe.
Annealed: 96,000 TS; 52,000 YS; 18 El; 23 RA; 170 Brin. For impact and shear tools, pneumatic chisels; water hardened. *Obsolete*

FOREZ CX75
Acieries du Forez
C 0.75, bal Fe.
Heat treated: 180,000 TS; 132,000 YS; 11 El; 36 RA; 360 Brin. For wire drawing dies, reamers, punches; water hardened. *Obsolete*

FOREZ DT
Acieries du Forez
C 2-2.2, Cr 12-14, bal Fe.
High carbon, high chrome cold work tool steel for dies for punching and embossing. AFNOR Z 200 C 12; AISI D3.

FOREZ DT
Acieries du Forez
C 2, Cr 12, bal Fe.
For shear blades, wire drawing and piercing dies; nondeforming, oil hardened (Z200 C12).

FOREZ DT VA
Acieries du Forez
C 1.5-1.6, Cr 11.5-12.5, Mo 0.75-0.85, V 0.7-0.9, bal Fe.
High carbon, high chrome cold work tool steel for punching and embossing dies. AFNOR Z 150 CDV 12; AISI D2.

FOREZ E3
Acieries du Forez
C 1.25, W 11, V 3, Mo 0.6, Cr 4.5, bal Fe.
For lathe and planer tools, reams, taps, drills; high speed steel, oil hardened (Z125WV15-03). *Obsolete*

FOREZ MO6
Acieries du Forez
C 0.35, Alloy, bal Fe.
For injection nozzles, molds, die casting dies; hot work steel, tough (35CD06-04). *Obsolete*

FOREZ MT18
Acieries du Forez
C 0.5, Cr 4, W 18, Mo 1, V 1, bal Fe.
For cold work tools, punches, crimpers; cold impact steel (Z50WV18-01). *Obsolete*

FOREZ NI5
Acieries du Forez
C 0.35, Ni 3, Cr 0.8, bal Fe.
Annealed: 92,000 TS; 20 El; 50 RA. Heat treated: 165,000 TS; 10 El; 40 RA. For arbors, gears, machine tool parts, shafts; oil hardened (35NC11). *Obsolete*

FOREZ NID
Acieries du Forez
C 0.4, Ni 3, Cr 0.8, bal Fe.
Annealed: 103,000 TS; 15 El; 50 RA. Heat treated: 235,000 TS; 3 El; 20 RA. For gears, machine tool parts, shafts, axles; oil hardened (40NC11). *Obsolete*

FOREZ NVM
Acieries du Forez
C 0.4, Ni, Cr, bal Fe.
Annealed: 128,000 TS; 18 El; 40 RA. Heat treated: 275,000 TS; 6 El; 20 RA. For hot work punches and dies; tough and hard (40NW5). *Obsolete*

FOREZ SM
Acieries du Forez
C, Si, Mn, bal Fe.
For crankshafts, propeller shafts, axles, gears; oil hardened. *Obsolete*

FOREZ SMX
Acieries du Forez
C 0.45, Si 1.9, Mn 0.6, bal Fe.
For springs, washers, looms, machine tool parts; oil or water hardened (45S8). *Obsolete*

FOREZ SUPRA EXTRA 4AS
Acieries du Forez
C 0.85, W 18, Cr 5, V 1.5, Mo 1, Co 10, bal Fe.
For lathe and planer tools, broaches, reamers, hobs; high speed steel (Z85WK-18-10).

FOREZ TE
Acieries du Forez
C 0.3, Cr 3, W 9, bal Fe.
For punches, dies, bolt and nail dies; hot work steel (Z30WC09).

FOREZ TES
Acieries du Forez
C 0.3, W, Cr, V, bal Fe.
For punches, dies, wire drawing dies; cold work steel (Z30WCV10). *Obsolete*

FOREZ TFR
Acieries du Forez
C 0.85, W 6, V 2, Mo 1, Cr 4.5, V 2, bal Fe.
For lathe and planer tools, reamers, hobs, taps; high speed steel (Z85WD06-02). *Obsolete*

FOREZ TM
Acieries du Forez
C 0.4, alloy, bal Fe.
For hot and cold stamping tools, pneumatic chisels; shock resistant (40WNCD).

FORGE DIE C-30
Teledyne Vasco
C 0.4, W 15, bal Fe.
For hot work tools, extrusion rams and liners; hot work steel. *Obsolete*

FORGE-DIE
Teledyne Vasco
C 0.25, W 13.5-14.5, Cr 3.5, V 0.5, bal Fe.
For upsetter heads and dies, piercing punches, extrusion dies; resists heat checking. Hot work steel.

FORGE-WELL
St. Lawrence Steel Co.
C 0.27, Mn 1, Cr 1.2, V 0.25, Mo 0.3, bal Fe.
Oil hardened, low alloy tool steel, for molds.

FORGEMASTER DIE BLOCK STEEL
Eagle & Globe Steel Ltd.
Tool material. C 0.55, Si 0.3, Mn 0.65, Cr 0.65, Mo 0.3, Ni 1.5, bal Fe.
Heat treated. For drop forging die blocks, shear blades, die casting dies for zinc, extrusion tooling for lead and its alloys, and supporting tools used in copper and aluminum extrusion to back-up or carry the higher alloy steel tools in direct contract with the extrusion product. AS1239 L104A; AISI L104A; BSS224 No. 5; Werkstoff 1.2714.

FORGING BRASS
English manufacture
Cu 57-60, Zn 40-43.
For corrosion resisting forgings; high strength.

FORGING BRASS NO. 377
Chase Brass & Copper Co., Inc.
Copper. Cu 60, Zn 38, Pb 2.
Extruded: 52,000 psi TS; 20,000 psi YS; 45 El; 32,000 psi shear strength; 78 Rock F. For forgings, tire valve stems. ASTM B 124, Alloy 2, B 283.

FORGING BRASS-377
Anaconda Co.
Copper. Cu 60, Zn 38, Pb 2.
Soft rod: 54,000 TS; 20,000 YS; 45 El; 45 Rock B. For hardware, gears, fasteners, bolts. Free machining leaded brass.

FORGING RUSSIAN
Russian manufacture
Cu 53.5, Zn 42, Mn 4.5.
For corrosion resisting forgings.

FORMA
Uddeholm Corp.
C 0-0.05, bal Fe.
For plastic molding dies; case hardened. *Obsolete*

FORMA 2
Uddeholm Corp.
C 0.08, Si 0.1, Mn 0.15, bal Fe.
For deep drawn parts; deep drawing steel. *Obsolete*

FORMA-1

Uddeholm Corp.
C 0-0.05, Si 0-0.1, Mn 0-0.15, bal Fe.
Annealed: 58,000 TS; 45,000 YS; 30 El; 66 RA; 125 Brin. For plastic mold dies; case hardened. *Obsolete*

FORMA-1

Uddeholm Corp.
C 0.05, Si 0.05, Mn 0.1, bal Fe.
For bakelite molding dies; deep hobbing, case hardened. *Obsolete*

FORMABLE-40 K (PHOS)

Armco
HSLA steel.
High strength steel. Cold-rolled: 55 ksi (379 MPa) TS; 40 ksi (276 MPa) YS; 25 El (in 2 in.); 67 Rock B.

FORMABLE-45 C

Armco
HSLA steel.
High strength steel. Hot-rolled: 60 ksi (414 MPa) TS; 45 ksi (310 MPa) YS; 25 El (in 2 in.); 70 Rock B.

FORMABLE-50 B

Armco
HSLA steel.
High strength steel. Hot-rolled: 70 ksi (483 MPa) TS; 50 ksi (345 MPa) YS; 22 El (in 2 in.); 75 Rock B.

FORMABLE-50 C

Armco
HSLA steel.
High strength steel. Hot-rolled: 65 ksi (448 MPa) TS; 50 ksi (345 MPa) YS; 22 El (in 2 in.); 75 Rock B. Cold-rolled: 65 ksi (448 MPa) TS; 50 ksi (345 MPa) YS; 22 El (in 2 in.); 72 Rock B.

FORMABLE-50 F

Armco
HSLA steel.
High strength steel. Hot-rolled: 60 ksi (414 MPa) TS; 50 ksi (345 MPa) YS; 24 El (in 2 in.); 77 Rock B. Cold-rolled: 60 ksi (414 MPa) TS; 50 ksi (345 MPa) YS; 24 El (in 2 in.); 72 Rock B.

FORMABLE-50 K

Armco
HSLA steel.
High strength steel. Cold-rolled: 65 ksi (448 MPa) TS; 50 ksi (345 MPa) YS; 22 El (in 2 in.); 72 Rock B.

FORMABLE-50 V

Armco
HSLA steel.
High strength steel. Hot-rolled: 60 ksi (414 MPa) TS; 50 ksi (345 MPa) YS; 24 El (in 2 in.); 77 Rock B.

FORMABLE-55 C

Armco
HSLA steel.
High strength steel. Hot-rolled: 70 ksi (483 MPa) TS; 55 ksi (379 MPa) YS; 20 El (in 2 in.); 77 Rock B.

FORMABLE-55 F

Armco
HSLA steel.
High strength steel. Hot-rolled: 65 ksi (448 MPa) TS; 55 ksi (379 MPa) YS; 24 El (in 2 in.); 77 Rock B.

FORMABLE-55 V

Armco
HSLA steel.
High strength steel. Hot-rolled: 65 ksi (448 MPa) TS; 55 ksi (379 MPa) YS; 24 El (in 2 in.); 77 Rock B.

FORMABLE-70 F

Armco
HSLA steel.
High strength steel. Hot-rolled: 80 ksi (552 MPa) TS; 70 ksi (483 MPa) YS; 20 El (in 2 in.); 88 Rock B.

FORMABLE-70 V

Armco
HSLA steel.
High strength steel. Hot-rolled: 80 ksi (552 MPa) TS; 70 ksi (483 MPa) YS; 20 El (in 2 in.); 88 Rock B.

FORMABLE-80 F

Armco
HSLA steel.
High strength steel. Hot-rolled: 90 ksi (621 MPa) TS; 80 ksi (552 MPa) YS; 18 El (in 2 in.); 90 Rock B.

FORMABLE-80 V

Armco
HSLA steel.
High strength steel. Hot-rolled: 90 ksi (621 MPa) TS; 80 ksi (552 MPa) YS; 18 El (in 2 in.); 90 Rock B.

FORMALOY

Boyd-Wagner Co.
C, Mn, Ni, Cr, Mo, bal Fe.
For dies, shearing tools; oil hardening.

FORMALOY

American Smelting & Refining Co.
Cu 3, Mg 0.4, bal Zn.
Aged: 40,300 psi TS; 33,100 psi YS; 2.8 El; 103 Brin. Cast: TS: 4.7 El; 90 Brin. Chilled: 36,000 psi TS; 2.7 El; 93 Brin For dies for forming aluminum sheet, drop hammer punches and dies. Die casting alloy; high strength.

FORMBRITE

Anaconda Co.
Copper, Zn, bal Cu.
For deep drawn parts. Good formability.

FORMDIE

Columbia Tool Steel Co.
Tool material. C 0.51, Mn 0.45, Si 1, Cr 5.2, Ni 1.5, V 1.05, Mo 1.4, bal Fe.
For shear blades, extrusion dies, inserts, heavy punches. Type A9; air hardened tool steel.

FORMDIE SMOOTHCUT

Columbia Tool Steel Co.
C 0.48-0.54, Si 0.9-1.1, Mo 1.25-1.5, Cr 5-5.4, V 0.9-1.1, Ni 1.35-1.65, S, bal Fe.
For shear blades, hot upsetting and gripper dies, drop forging and hot heading dies, punches, air hardening. Cold work tool steel, Type A9. *Obsolete*

FORMITE MO. 2

Columbia Tool Steel Co.
C 0.3, Cr 3.25, W 9.25, V 0.5, bal Fe.
For forming, extrusion and die casting dies; hot work steel, oil hardened. *Obsolete*

FORMITE NO. 1

Columbia Tool Steel Co.
C 0.35, W 13.5, Cr 3.7, Si 0.4, V 0.4, Mn 0.3, bal Fe.
For hot punches, dies, shear blades; heat resistant, oil hardened; "Durite V-40". *Obsolete*

FORMITE NO. 2

Columbia Tool Steel Co.
C 0.35, Cr 3.5, W 9, bal Fe.
For punches, shears, crimpers, upsetters; hot work steel, oil hardened. *Obsolete*

FORMITE NO. 21

Columbia Tool Steel Co.
Tool material. C 0.33, W 9.25, Cr 3.3, V 0.5, bal Fe.
For hot temperature springs, forging dies, forming dies, gripping dies. Type H21; hot work tool steel.

FORMITE NO. 24

Columbia Tool Steel Co.
Tool material. C 0.51, W 15, Cr 3, V 0.5, bal Fe.
For die casting dies, punches, piercing tools, blanking dies. Type H24; hot work tool steel.

FORMITE NO. 3

Columbia Tool Steel Co.
C 0.51, W 15, Cr 3, V 0.5, bal Fe.
Hardened: C 52-56 Rock. For hot shear blades and punches, hot forming and extrusion rolls and dies. Hot work tool steel Type H24; High temperature hardness. *Obsolete*

FORMOLD

Crucible Materials Corp.
Tool material. C 0-0.07, Ni 0.55, Cr 1.35, Mo 0.2, bal Fe.
For plastic mold dies, cold hubbed; oil hardened, high core strength.

FORNANC SPECIAL

British Steel Corp.
C 0.15-0.25, Mn 1.4-1.8, Ni 0.4-0.7, Mo 0.15-0.35, bal Fe.
Hardened: 90,000-107,000 TS; 67,000-80,000 YS; 22-30 El; 50-70 Brin. For gears, shafts, axles; oil hardened. *Obsolete*

FORTAL

Cie Francais des Metaux
Al 94.3, Cu 4, Mg 0.5, Mn 0.5, Si 0.7.
For light alloy parts; age hardenable.

FORTE-100

Forjas Alavesas S.A.
C 0.97, Mn 0.25, Si 0-0.25, bal Fe.
Water hardening tool steel, for hand tools. IHA F-515; AFNOR XC100; AISI W1.

FORTE-115

Forjas Alavesas S.A.
C 1.12, Mn 0.25, Si 0-0.25, bal Fe.
Water hardening tool steel. IHA F-516; DIN C110W1; AISI W1.

FORTE-130

Forjas Alavesas S.A.
C 1.3, Mn 0.25, Si 0-0.25, bal Fe.
Water hardening tool steel. IHA F-517; AFNOR XC-120.

FORTE-50 M

Forjas Alavesas S.A.
C 0.5, Mn 0.7, Si 0-0.4, bal Fe.
For hand tools as hammers, hatchets, etc. DIN C45W3.

FORTE-55M

Forjas Alavesas S.A.
C 0.54, Mn 0.7, Si 0-0.4, bal Fe.
For hand tools, axes, hammers.

FORTE-65

Forjas Alavesas S.A.
C 0.65, Mn 0.25, Si 0-0.25, bal Fe.
Water hardening tool steel. IHA F-512; AFNOR XC-60; AISI W1.

FORTE-65 M

Forjas Alavesas S.A.
C 0.65, Mn 0.65, Si 0-0.4, bal Fe.
For hand tools, saws, knives, wood-working tools. DIN C67W3.

FORTE-75

Forjas Alavesas S.A.
C 0.75, Mn 0.25, Si 0-0.25, bal Fe.
Water hardening tool steel, as carpenter tools, hand tools. IHA-513; AFNOR XC-75.

FORTE-75M

Forjas Alavesas S.A.
C 0.75, Mn 0.8, Si 0-0.4, bal Fe.
For hand tools, farm tools, chisels, DIN C75W3.

FORTE-90

Forjas Alavesas S.A.
C 0.88, Mn 0.25, Si 0-0.25, bal Fe.
Water hardening tool steel. IHA F-514; AFNOR XC-85; UNI UC-85.

FORTEX-4W
Forjas Alavesas S.A.
C 1.35, Mn 0.25, Si 0.25, Cr 0.7, W 3.75, V 0.2, bal Fe.
For broaches, finish turning tools. IHA F-531; DIN 130W19.

FORTEX-C
Forjas Alavesas S.A.
C 0.9, Mn 0.3, Si 0-0.4, Cr 0.75, V 0.12, bal Fe.
Cold work tool steel; for stamping and coining dies.

FORTEX-W
Forjas Alavesas S.A.
C 1.15, Mn 0.25, Si 0.2, W 1, bal Fe.
For files, saws, threading tools. IHA F-532; AFNOR 100 WC 15-04.

FORTICAST
J.F. Jelenko & Co.
Precious metal. Au 42, Pd 9, Ag 26.
Quenched: 85,000 psi TS; 65,000 psi YS; 18 El; 175 Brin.
Hardened: 128,500 psi TS; 122,500 psi YS; 3 El; 265 Brin.
Type IV extra-hard dental alloy. For hard inlays, thin crowns, fixed bridgework, and partial dentures.

FORTINOX
British Steel plc
Ni-Cr-Mn.
Semi-austenitic stainless steel. For watch springs. *Obsolete*

FORTISSIMUS
Rudolf Schmidt Stahlwerke
C 0.7, W 18, Cr 4, V 1, bal Fe.
For cutters, tools; high speed steel.

FORTIWELD
British Steel plc
C 0.1-0.16, Mo 0.4-0.6, Mn 0-0.6, Si 0-0.4, B 0-0.005, bal Fe.
Normalized: 97,500-88,000 TS; 75,000-69,000 YS; 26-17 El; 64 RA. For bridge cranes, railroad cars, and material handling equipment. Good formability and weldability.

FORTIWELD PRESSURE VESSEL STEEL
British Steel plc
C 0.1-0.17, Si 0.1-0.4, Mn 0.4-0.8, Mo 0.4-0.6, Cr 0-0.25, Ni 0-0.3, Cu 0-0.3, B 0.001-0.005, bal Fe.
Weldable. For high yield strength, elevated temperature pressure vessels. *Obsolete*

FORTIWELD STRUCTURAL STEEL
British Steel plc
C 0-0.17, Si 0-0.4, Mn 0-0.8, Mo 0-0.6, B 0-0.005, bal Fe.
High yield strength structural steel; weldable. *Obsolete*

FORTUNA 12M
Stahlwerke Sudwestfalen
C 1.2, Mn 12, bal Fe.
For crusher jaws, impact tools, liners, grab-teeth; wear resistant, work hardened. *Obsolete*

FORTUNA A12
Stahlwerke Sudwestfalen
C 0-0.1, Cr 12.5, Ni 12, bal Fe.
Annealed: 72,000-92,000 TS; 30,000 YS; 60 El; 115-150 Brin.
For clock cases, chemical plant equipment; EN58D Special; stainless, good cold workability. *Obsolete*

FORTUNA A18
Stahlwerke Sudwestfalen
C 0-0.15, Cr 18, Ni 9, bal Fe.
Annealed: 78,000-108,000 TS; 36,000 YS; 55 El; 130-180 Brin. For chemical plant equipment, vessels, tanks, agitators; Type 302; stainless, austenitic. *Obsolete*

FORTUNA A182ZN
Stahlwerke Sudwestfalen
C 0-0.1, Cr 18, Ni 11, Ti 2.3, Cb + Ta, bal Fe.
Annealed: 78,000-108,000 TS; 38,000 YS; 45 El; 140-190 Brin. For acid resistant chemical plant equipment; Type 316 + Cb, stainless, austenitic. *Obsolete*

FORTUNA A182Z
Stahlwerke Sudwestfalen
C 0-0.1, Cr 18, Ni 11, Mo 2.3, Ti, bal Fe.
Annealed: 78,000-108,000 TS; 38,000 YS; 50 El; 140-190 Brin. For acid resistant chemical plant equipment; mixers; Type 316 + Ti; stainless, austenitic. *Obsolete*

FORTUNA A183Z
Stahlwerke Sudwestfalen
C 0-0.1, Cr 18, Ni 12, Mo 2.8, Ti, bal Fe.
Annealed: 78,000-108,000 TS; 38,000 YS; 50 El; 140-190 Brin. For chemical and pharmaceutical equipment; EN58J Special; stainless, austenitic. *Obsolete*

FORTUNA A183ZN
Stahlwerke Sudwestfalen
C 0-0.1, Cr 18, Ni 12, Mo 2.8, Ta + Cb, bal Fe.
Annealed: 78,000-108,000 TS; 38,000 YS; 45 El; 140-190 Brin. For pharmaceutical and chemical equipment, resists sulfide lye; Type 318; stainless, austenitic. *Obsolete*

FORTUNA A18Z
Stahlwerke Sudwestfalen
C 0-0.1, Cr 18, Ni 10, Ti, bal Fe.
Annealed: 78,000-108,000 TS; 38,000 YS; 50 El; 140-190 Brin. For welded chemical plant equipment, tanks, vessels; Type 321; stainless, austenitic. *Obsolete*

FORTUNA A18ZN
Stahlwerke Sudwestfalen
C 0-0.1, Cr 18, Ni 10, Cb + Ta, bal Fe.
Annealed: 78,000-108,000 TS; 38,000 YS; 45-50 El; 140-190 Brin. For acid resistant welded chemical plant equipment; Type 347; stainless, austenitic. *Obsolete*

FORTUNA A2182ZN
Stahlwerke Sudwestfalen
C 0-0.06, Cr 18, Ni 18, Mo 2.3, Cu, Ta, Cb, bal Fe.
Annealed: 78,000-108,000 TS; 34,000 YS; 45 El; 140-190 Brin. For equipment for pickling and sulfuric acid plants; acid resistant, austenitic. *Obsolete*

FORTUNA AS175
Stahlwerke Sudwestfalen
C 0-0.06, Cr 17, Ni 14, Mo 4.5, bal Fe.
Annealed: 78,000-108,000 TS; 34,000 YS; 50 El; 140-190 Brin. For equipment for salt works and soda plants, cooling coils; Type 317; stainless, austenitic. *Obsolete*

FORTUNA AS18
Stahlwerke Sudwestfalen
C 0-0.06, Cr 18, Ni 10, bal Fe.
Annealed: 72,000-101,000 TS; 32,000 YS; 55 El; 130-180 Brin. For chemical plant equipment, tanks, breweries; Type 304; stainless, austenitic. *Obsolete*

FORTUNA AS182
Stahlwerke Sudwestfalen
C 0-0.06, Cr 18, Ni 11, Mo 2.3, bal Fe.
Annealed: 78,000-108,000 TS; 32,000 YS; 45 El; 140-180 Brin. For chemical plant equipment, agitators, mixers; tanks; Type 316; stainless, austenitic. *Obsolete*

FORTUNA AS183
Stahlwerke Sudwestfalen
C 0-0.06, Cr 18, Ni 12, Mo 2.8, bal Fe.
Annealed: 78,000-108,000 TS; 32,000 YS; 50 El; 140-180 Brin. For acid resistant chemical and textile plant equipment; Type 316; stainless, austenitic. *Obsolete*

FORTUNA BS 45
Stahlwerke Sudwestfalen
C 0.4, Cr 1.4, Ni 4, W 0.5, bal Fe.
Annealed: 225 Brin. For plastic mold dies; oil or air hardened. *Obsolete*

FORTUNA BS50
Stahlwerke Sudwestfalen
C 0.5, Cr 1.1, Ni 3.5, bal Fe.
For cutlery, blanking dies, hobbing dies, punches; air or oil hardened, shock resistant. *Obsolete*

FORTUNA C 1215
Stahlwerke Sudwestfalen
C 1.65, Cr 12, V 0.1, bal Fe.
For blanking and drawing dies, piercing punches; air hardened, non-deforming. *Obsolete*

FORTUNA C 1215 SUPRA
Stahlwerke Sudwestfalen
C 1.65, Cr 12.5, Co 1.4, Mo 1.2, bal Fe.
For blanking and forming dies, die casting dies; air hardened, non-deforming. *Obsolete*

FORTUNA C 1220
Stahlwerke Sudwestfalen
C 2.1, Cr 12, bal Fe.
For drawing blanking dies, thread rolling dies; oil hardened, non-deforming. *Obsolete*

FORTUNA CA 1215
Stahlwerke Sudwestfalen
C 1.65, Cr 12, Mo 0.8, W 0.5, bal Fe.
For blanking and forming dies, die casting dies; air hardened, non-deforming. *Obsolete*

FORTUNA CA 1220
Stahlwerke Sudwestfalen
C 2.1, Cr 12, W 0.1, bal Fe.
For blanking and forming dies, punches, shears; oil hardened, non-deforming. *Obsolete*

FORTUNA CA 1220 SUPRA
Stahlwerke Sudwestfalen
C 2.1, Cr 12, W 0.1, Co, Mo, bal Fe.
For blanking and forming dies, shearing knives; oil hardened, non-deforming. *Obsolete*

FORTUNA CO1000
Stahlwerke Sudwestfalen
C 0.75, Co 10, W 18, Cr 4.2, Mo 0.7, V 1.5, bal Fe.
For cutting tools, broaches, reamers; high speed steel. *Obsolete*

FORTUNA CO300
Stahlwerke Sudwestfalen
C 0.8, Co 3, W 12, Cr 4.2, Mo 0.5, V 1.8, bal Fe.
For lathe and planer tools, reamers, broaches; high speed steel. *Obsolete*

FORTUNA CO500
Stahlwerke Sudwestfalen
C 0.8, Co 5, W 18, Cr 4.2, Mo 0.7, V 1.5, bal Fe.
For lathe and planer tools, drills, taps, reamers; high speed steel. *Obsolete*

FORTUNA CSV 4
Stahlwerke Sudwestfalen
C 0.3, Si 1.5, Cr 1.5, V 0.1, bal Fe.
For shear blades, dies, stamps, piercers; hot work steel, water hardened. *Obsolete*

FORTUNA CSV5
Stahlwerke Sudwestfalen
C 0.45, Si 1.5, Cr 1.5, V 0.1, bal Fe.
Anneald: 225 Brin. For chisels, shear blades, punches, pneumatic tools; oil hardened, shock resistant. *Obsolete*

FORTUNA CSV6
Stahlwerke Sudwestfalen
C 0.6, Si 1, Cr 1.2, V 0.1, bal Fe.
Annealed: 235 Brin. For punches, blanking dies, cold shear blades; oil hardened, shock resistant. *Obsolete*

FORTUNA DMO10
Stahlwerke Sudwestfalen
C 0.24, Si 0.25, Mn 0.6, Cr 1.2, Mo 0.25, bal Fe.
Heat treated: 85,000-115,000 TS; 65,000-72,000 YS; 19-21 El.
For bolts and nuts, oil refinery and chemical plant equipment; creep resistant to 500°C. *Obsolete*

FORTUNA DMO11
Stahlwerke Sudwestfalen
C 0.15, Si 0.25, Mn 0.6, Mo 0.3, bal Fe.
Heat treated: 62,000-78,000 TS; 38,000 YS; 24 El. For flanges, welded collars, oil refinery equipment; creep resistant to 500°C. *Obsolete*

FORTUNA DMO14
Stahlwerke Sudwestfalen
C 0.13, Si 0.25, Mn 0.6, Cr 1, Mo 0.45, bal Fe.
Heat treated: 62,000-78,000 TS; 41,000 YS; 24 El. For flanges, welded collars, oil refinery equipment; creep resistant to 500°C. *Obsolete*

FORTUNA DMO20
Stahlwerke Sudwestfalen
C 0.24, Si 0.25, Mn 0.45, Cr 1.4, Mo 0.55, V 0.2, bal Fe.
Heat treated: 100,000-135,000 TS; 78,000-85,000 YS; 15-19 El. For bolts, nut, oil refinery and chemical plant equipment; creep resistant to 500°C. *Obsolete*

FORTUNA DMO22
Stahlwerke Sudwestfalen
C 0.21, Si 0.47, Mn 0.4, Cr 1.35, Mo 1.1, V 0.3, bal Fe.
Heat treated: 101,000-121,000 TS; 78,500 YS; 19 El. For bolts and nuts, oil refinery and chemical plant equipment; creep resistant to 500°C. *Obsolete*

FORTUNA EC3
Stahlwerke Sudwestfalen
C 0.15, Si 0.25, Mn 0.5, Cr 0.6, bal Fe.
Heat treated: 85,000-120,000 TS; 58,000 YS; 16 El; 45 RA. For gears, pinions, camshafts; case hardened. *Obsolete*

FORTUNA ECMO2H
Stahlwerke Sudwestfalen
C 0.2, Si 0-0.35, Mn 0.7, Cr 0.6, Mo 0.35, bal Fe.
Heat treated: 128,000-170,000 TS; 92,000 YS; 12 El; 45 RA. For gears, pinions, camshafts, cams; case hardened. *Obsolete*

FORTUNA ECMO4
Stahlwerke Sudwestfalen
C 0.15, Si 0-0.35, Mn 0.95, Cr 1.15, Mo 0.25, bal Fe.
Heat treated: 121,000-157,000 TS; 92,000 YS; 12 El; 40 RA. For gears, pinions, camshafts; case hardened. *Obsolete*

FORTUNA ECMO5
Stahlwerke Sudwestfalen
C 0.2, Si 0-0.35, Mn 1, Cr 1.2, Mo 0.25, bal Fe.
Heat treated: 156,000-192,000 TS; 107,000 YS; 9 El; 30 RA. For gears, cams, camshafts, bolts, fasteners; case hardened, tough. *Obsolete*

FORTUNA EMC5
Stahlwerke Sudwestfalen
C 0.16, Si 0.25, Mn 1.2, Cr 1, bal Fe.
Heat treated: 114,000-157,000 TS; 85,000 YS; 12 El; 40 RA. For gears, cams, camshafts, crankshafts; case hardened, tough. *Obsolete*

FORTUNA EMC5H
Stahlwerke Sudwestfalen
C 0.2, Si 0.25, Mn 1.3, Cr 1.2, bal Fe.
Heat treated: 144,000-186,000 TS; 101,000 YS; 10 El; 35 RA. For gears, bolts, crankshafts; case hardened, tough. *Obsolete*

FORTUNA ENC10
Stahlwerke Sudwestfalen
C 0.13, Si 0-0.35, Mn 0-0.5, Cr 0.75, Ni 2.5, bal Fe.
Heat treated: 114,000-144,000 TS; 78,000 YS; 17 El; 50 RA. For gears, pinions, camshafts, cams, fasteners; case hardened, shock resistant. *Obsolete*

FORTUNA ENC14
Stahlwerke Sudwestfalen
C 0.13, Si 0-0.35, Mn 0-0.5, Cr 0.75, Ni 3.5, bal Fe.
Heat treated: 128,000-170,000 TS; 92,000 YS; 11 El; 45 RA. For gears, pinions, camshafts, cams; case hardened, tough. *Obsolete*

FORTUNA ENC18
Stahlwerke Sudwestfalen
C 0.13, Si 0-0.35, Mn 0-0.5, Cr 1.1, Ni 4.5, bal Fe.
Heat treated: 177,000-200,000 TS; 128,000 YS; 9 El; 40 RA. For gears, shafts, camshafts, cams; case hardened, tough. *Obsolete*

FORTUNA ENC6
Stahlwerke Sudwestfalen
C 0.13, Si 0-0.35, Mn 0-0.5, Cr 0-0.2, Ni 1.5, bal Fe.
Heat treated: 85,000-115,000 TS; 58,000 YS; 12 El; 50 RA. For gears, pinions, camshafts, cams, bolts; case hardened, shock resistant. *Obsolete*

FORTUNA EW 15 SPECIAL
Stahlwerke Sudwestfalen
C 0.19, Cr 1.3, Mo 0.2, Ni 4, bal Fe.
Annealed: 245 Brin. For plastic mold dies; case hardened. *Obsolete*

FORTUNA EW 52H
Stahlwerke Sudwestfalen
C 0.2, Mn 1, Cr 1.2, Mo 0.25, bal Fe.
Annealed: 217 Brin. For plastic mold dies; case hardened. *Obsolete*

FORTUNA EW 5H
Stahlwerke Sudwestfalen
C 0.2, Mn 1.3, Cr 1.2, bal Fe.
Annealed: 217 Brin. For plastic mold dies; case hardened. *Obsolete*

FORTUNA EWX 50
Stahlwerke Sudwestfalen
C 0-0.07, Cr 5, Mo 1, V 0.2, bal Fe.
Annealed: 140 Brin. For plastic mold dies; case hardened. *Obsolete*

FORTUNA EX15
Stahlwerke Sudwestfalen
C 0.15, Si 0.25, Mn 0.5, Cr 1.5, Ni 1.5, bal Fe.
Heat treated: 128,000-170,000 TS; 92,000 YS; 11 El; 40 RA. For gears, fasteners, camshafts, cams, bolts; case hardened. *Obsolete*

FORTUNA EX17
Stahlwerke Sudwestfalen
C 0.15, Si 0.25, Mn 0.5, Cr 1.7, Ni 1.5, Mo 0.3, bal Fe.
Heat treated: 156,000-192,000 TS; 107,000 YS; 10 El; 40 RA. For gears, axles, shafts, cams, camshafts; case hardened, tough. *Obsolete*

FORTUNA EX20
Stahlwerke Sudwestfalen
C 0.18, Si 0.25, Mn 0.5, Cr 2, Ni 2, bal Fe.
Heat treated: 170,000-206,000 TS; 114,000 YS; 9 El; 35 RA. For machine tool parts, gears, crankshafts, camshafts; case hardened, tough. *Obsolete*

FORTUNA EX8H
Stahlwerke Sudwestfalen
C 0.2, Si 0.25, Mn 0.8, Cr 0.5, Ni 0.6, Mo 0.2, bal Fe.
Heat treated: 135,000-177,000 TS; 92,000 YS; 11 El; 40 RA. For gears, fasteners, camshafts, cams; case hardened. *Obsolete*

FORTUNA F13
Stahlwerke Sudwestfalen
C 0-0.1, Cr 13, bal Fe.
Heat treated: 85,000-107,000 TS; 63,000 YS; 26 RA; 170-210 Brin. For cutlery, fittings, appliances, pump and valve parts; corrosion resistant; Type 410. *Obsolete*

FORTUNA F17
Stahlwerke Sudwestfalen
C 0.08, Cr 17, bal Fe.
Annealed: 62,000-85,000 TS; 43,000 YS; 24 El; 140-180 Brin. For cutlery, fittings, kitchen appliances and sinks; Type 430; corrosion resistant. *Obsolete*

FORTUNA F17A
Stahlwerke Sudwestfalen
C 0-0.15, Cr 17, Mo 0.25, S, bal Fe.
Heat treated: 100,000-123,000 TS; 63,000 YS; 16 El; 190-240 Brin. For screw machining products, screws, bolts, gears; free-cutting, corrosion resistant. *Obsolete*

FORTUNA F17Z
Stahlwerke Sudwestfalen
C 0.08, Cr 17, Ti, bal Fe.
Annealed: 68,000-85,000 TS; 43,000 YS; 24 El; 130-170 Brin. For welded acid resistance and chemical plant equipment; stabilized, welding grade, corrosion resistant. *Obsolete*

FORTUNA FU
Stahlwerke Sudwestfalen
C 1.3, Mn 0.25, bal Fe.
For files, rasps; water hardened. *Obsolete*

FORTUNA FU33
Stahlwerke Sudwestfalen
C 0.37, Mn 0.5, bal Fe.
Hot rolled: 85,000 TS; 54,000 YS; 30 El; 53 RA; 185 Brin. For files, rasps; case hardened tools. *Obsolete*

FORTUNA FU43
Stahlwerke Sudwestfalen
C 0.47, Mn 0.67, bal Fe.
Hot rolled: 98,000 TS; 60,000 YS; 24 El; 54 RA; 212 Brin. For files, rasps; water hardened. *Obsolete*

FORTUNA GB0
Stahlwerke Sudwestfalen
C 0.1, Si 0.25, Mn 0.4, 0.035 max S and P, bal Fe.
Heat treated: 60,000-74,000 TS; 36,000 YS; 22 El; 55 RA. For machine and construction parts, fasteners; case hardened. *Obsolete*

FORTUNA GB1
Stahlwerke Sudwestfalen
C 0.15, Si 0.25, Mn 0.4, 0.035 max S and P, bal Fe.
Heat treated: 72,000-94,000 TS; 42,000 YS; 19 El; 50 RA. For bolts, levers, journals, gears; case hardened. *Obsolete*

FORTUNA GB10
Stahlwerke Sudwestfalen
C 0.1, Si 0.25, Mn 0.4, 0.045 max S and P, bal Fe.
Heat treated: 60,000-74,000 TS; 36,000 YS; 22 El; 55 RA. For bushings, bolts, levers, gears; case hardened. *Obsolete*

FORTUNA GB11
Stahlwerke Sudwestfalen
C 0.15, Si 0.25, Mn 0.4, 0.045 max S and P, bal Fe.
Heat treated: 72,000-94,000 TS; 42,000 YS; 19 El; 50 RA. For machine and construction parts, gears, bolts; case hardened. *Obsolete*

FORTUNA GB12
Stahlwerke Sudwestfalen
C 0.22, Si 0.25, Mn 0.45, 0.045 max S and P, bal Fe.
Heat treated: 60,000-92,000 TS; 33,000-56,000 YS; 24-30 El; 40-45 RA. For machine tools parts, gears, bolts, shafts; water hardened. *Obsolete*

FORTUNA GB13
Stahlwerke Sudwestfalen
C 0.35, Si 0.25, Mn 0.55, 0.045 max S and P, bal Fe.
Heat treated: 72,000-114,000 TS; 40,000-62,000 YS; 19-25 El; 35-45 RA. For shafts, gears, axles, machine tool parts; oil or water hardened. *Obsolete*

FORTUNA GB14
Stahlwerke Sudwestfalen
C 0.45, Si 0.25, Mn 0.65, 0.045 max S and P, bal Fe.
Heat treated: 85,000-128,000 TS; 49,000-70,000 YS; 17-21 El; 30-40 RA. For shafts, axles, gears, bolts; oil or water hardened. *Obsolete*

FORTUNA GB16

Stahlwerke Sudwestfalen

C 0.6, Si 0.25, Mn 0.65, 0.045 max S and P, bal Fe.
Heat treated: 100,000-150,000 TS; 56,000-80,000 YS; 15-18 El; 25-35 RA. For axles, shafts, spindles, bolts, gears; oil or water hardened. *Obsolete*

FORTUNA GB2

Stahlwerke Sudwestfalen

C 0.22, Si 0.25, Mn 0.45, 0.035 max S and P, bal Fe.
Heat treated: 72,000-92,000 TS; 42,000-52,000 YS; 24-25 El; 45-50 RA. For machine tool parts, gears, shafts, bolts; water hardened. *Obsolete*

FORTUNA GB3

Stahlwerke Sudwestfalen

C 0.35, Si 0.25, Mn 0.55, 0.035 max S and P, bal Fe.
Heat treated: 78,000-115,000 TS; 47,000-61,000 YS; 19-24 El; 40-50 RA. For machine tool parts, gears, fasteners; water hardened. *Obsolete*

FORTUNA GB4

Stahlwerke Sudwestfalen

C 0.45, Si 0.25, Mn 0.65, 0.035 max S and P, bal Fe.
Heat treated: 85,000-128,000 TS; 52,000-67,000 YS; 17-21 El; 35-45 RA. For pinions, bolts, transmission gears, shafts; for surface hardening. *Obsolete*

FORTUNA GB5H

Stahlwerke Sudwestfalen

C 0.56, Si 0.3, Mn 0.55, bal Fe.
Heat treated: 92,000-135,000 TS; 58,000-76,000 YS; 16-19 El; 35-40 RA. For pistons, machine tool parts; for surface hardening. *Obsolete*

FORTUNA GB6

Stahlwerke Sudwestfalen

C 0.6, Si 0.25, Mn 0.65, 0.035 max S and P, bal Fe.
Heat treated: 100,000-150,000 TS; 60,000-80,000 YS; 15-18 El; 30-40 RA. For clutch levers, gears, axles, shafts; water hardened. *Obsolete*

FORTUNA GB6H

Stahlwerke Sudwestfalen

C 0.68, Si 0.32, Mn 0.7, 0.040 max S and P, bal Fe.
Annealed: 98,000 TS. For shafts, piston rods, valve stems; wear resistant, oil or water hardened. *Obsolete*

FORTUNA GB9

Stahlwerke Sudwestfalen

C 0.9, Si 0.32, Mn 1.1, 0.040 max S and P, bal Fe.
Annealed: 145,000 TS; 85,000 YS; 12 El. Heat treated: 242,000 TS; 199,000 YS; 7 El. For molds for briquette presses; wear resistant, oil hardened. *Obsolete*

FORTUNA GFD2

Stahlwerke Sudwestfalen

C 0.38, Si 1.5, Mn 0.65, bal Fe.
Heat treated: 170,000-200,000 TS; 150,000 YS; 7-10 El; 30 RA; 350-400 Brin. For railway and auto springs; water hardened. *Obsolete*

FORTUNA GFD3

Stahlwerke Sudwestfalen

C 0.48, Si 0.25, Mn 1.75, 0.050 max S and P, bal Fe.
Heat treated: 186,000-213,000 TS; 157,000 YS; 7-10 El; 30 RA; 380-435 Brin. For railway and auto springs; oil hardened. *Obsolete*

FORTUNA GFD4

Stahlwerke Sudwestfalen

C 0.46, Si 1.65, Mn 0.65, 0.050 max S and P, bal Fe.
Heat treated: 186,000-213,000 TS; 157,000 YS; 7-10 El; 30 RA; 380-435 Brin. For railway and auto springs; water hardened. *Obsolete*

FORTUNA GFD5

Stahlwerke Sudwestfalen

C 0.66, Si 1.65, Mn 0.7, 0.050 max S and P, bal Fe.
Heat treated: 186,000-213,000 TS; 157,000 YS; 7-10 El; 30 RA; 380-435 Brin. For railway and auto springs; oil hardened. *Obsolete*

FORTUNA GFD530

Stahlwerke Sudwestfalen

C 0.3, Si 0.22, Mn 0.3, Cr 2.35, V 0.6, W 4.35, bal Fe.
Heat treated: 202,000-248,000 TS; 177,000 YS; 6-9 El; 30 RA. For high temperature springs; oil hardened, heat resistant. *Obsolete*

FORTUNA GFD5W

Stahlwerke Sudwestfalen

C 0.56, Si 1.65, Mn 0.7, 0.050 max S and P, bal Fe.
Heat treated: 186,000-213,000 TS; 157,000 YS; 7-10 El; 30 RA; 380-435 Brin. For railway and auto springs; oil hardened. *Obsolete*

FORTUNA GFD7

Stahlwerke Sudwestfalen

C 0.65, Si 1.1, Mn 1, 0.050 max S and P, bal Fe.
Heat treated: 190,000-225,000 TS; 157,000 YS; 7-10 El; 30 RA; 390-450 Brin. For railway and auto springs; oil hardened. *Obsolete*

FORTUNA GFD8

Stahlwerke Sudwestfalen

C 0.67, Si 1.3, Mn 0.5, Cr 0.5, bal Fe.
Annealed: 112,000 TS; 65,000 YS; 16 El. Heat treated: 232,000 TS; 190,000 YS; 8 El. For highly stressed springs; wear resistant, heavy duty. *Obsolete*

FORTUNA GFD8R

Stahlwerke Sudwestfalen

C 0.6, Si 1.1, Mn 1, Cr 0.5, bal Fe.
Heat treated: 200,000-238,000 TS; 177,000 YS; 7-10 El; 30 RA; 410-435 Brin. For highly stressed springs; oil hardened. *Obsolete*

FORTUNA GK

Stahlwerke Sudwestfalen

C 1.05, Si 0.25, Mn 0.32, Cr 0.5, bal Fe.
Annealed: 215 Brin. For ball bearings, roller and needle bearings; oil or water hardened. *Obsolete*

FORTUNA GKL

Stahlwerke Sudwestfalen

C 1, Si 0.25, Mn 0.32, Cr 1.5, bal Fe.
Annealed: 215 Brin. For ball bearings, ball races and discs; oil hardened. *Obsolete*

FORTUNA GKL0

Stahlwerke Sudwestfalen

C 1, Si 0.25, Mn 0.32, Cr 1.5, bal Fe.
Annealed: 207 Brin. Heat treated: 600 Brin. For ball and roller bearings; wear resistant, heavy duty. *Obsolete*

FORTUNA GMKL

Stahlwerke Sudwestfalen

C 1, Si 0.6, Mn 1.1, Cr 1.5, bal Fe.
Annealed: 226 Brin. For ball bearing races; oil or water hardened. *Obsolete*

FORTUNA GRL

Stahlwerke Sudwestfalen

C 1.05, Si 0.25, Mn 0.32, Cr 1, bal Fe.
Annealed: 215 Brin. For ball bearings, balls and rollers, races; oil or water hardened. *Obsolete*

FORTUNA GSB10

Stahlwerke Sudwestfalen

C 1, Mn 0.25, bal Fe.
Annealed: 100,000 TS; 53,000 YS; 21 El; 42 RA; 200 Brin. For stone working tools; water hardened. *Obsolete*

FORTUNA GSB11

Stahlwerke Sudwestfalen

C 1.1, Mn 0.25, bal Fe.
Annealed: 110,000 TS; 56,000 YS; 19 El; 40 RA; 210 Brin. For stone working tools; water hardened. *Obsolete*

FORTUNA GSB63

Stahlwerke Sudwestfalen

C 0.62, Mn 0.7, bal Fe.
Heat treated: 160,000 TS; 113,000 YS; 12 El; 40 RA; 325 Brin. For stone working tools; water hardened. *Obsolete*

FORTUNA GSB7

Stahlwerke Sudwestfalen

C 0.7, Mn 0.3, bal Fe.
Heat treated: 174,000 TS; 128,000 YS; 12 El; 37 RA; 335 Brin. For stone working tools; water hardened. *Obsolete*

FORTUNA GSB8

Stahlwerke Sudwestfalen

C 0.85, Mn 0.3, bal Fe.
Heat treated: 185,000 TS; 142,000 YS; 12 El; 32 RA; 390 Brin. For stone working tools; water hardened. *Obsolete*

FORTUNA GSP5

Stahlwerke Sudwestfalen

C 0.53, Si 1, Mn 1, bal Fe.
Rolled: 235-290 Brin. For hammer and press saddles, trimming tools; hot work steel, oil hardened. *Obsolete*

FORTUNA GSP6

Stahlwerke Sudwestfalen

C 0.65, Si 1, Mn 1, bal Fe.
For spring collets, screw drivers, clamping jaws; oil hardened. *Obsolete*

FORTUNA GSP8

Stahlwerke Sudwestfalen

C 0.67, Si 1.31, Cr 0.5, bal Fe.
For spring collets, screw drivers, clamping jaws; oil hardened. *Obsolete*

FORTUNA HSB1

Stahlwerke Sudwestfalen

C 0.85, bal Fe.
For mechanical and hard wood saws; water hardened; Type W1. *Obsolete*

FORTUNA HSB2

Stahlwerke Sudwestfalen

C 0.75, Mn 0.7, bal Fe.
Heat treated: 180,000 TS; 140,000 YS; 14 El; 38 RA; 375 Brin. For agricultural equipment, saws, cement scrapers; oil or water hardened. *Obsolete*

FORTUNA HSB4

Stahlwerke Sudwestfalen

C 0.8, Cr 0.5, V 0.4, bal Fe.
For wood working tools, circular saws; oil or water hardened. *Obsolete*

FORTUNA M-2171

Stahlwerke Sudwestfalen

C 0.9, Cr 17, Mo 0.5, V, Co, bal Fe.
Heat treated: 580-600 Brin. For bearing rolls and races, cutters; hardenable, corrosion resistant. *Obsolete*

FORTUNA M13

Stahlwerke Sudwestfalen

C 0.4, Cr 13, bal Fe.
Heat treated: 540-560 Brin. For cutlery, knives, springs, dies, instruments; corrosion resistant, hardenable. *Obsolete*

FORTUNA M131

Stahlwerke Sudwestfalen

C 0.5, Cr 14, Mo 0.5, bal Fe.
Heat treated: 550-570 Brin. For cutlery, knives, surgical instruments; corrosion resistant, hardenable. *Obsolete*

FORTUNA M171
Stahlwerke Sudwestfalen
C 0.9, Cr 18, Mo 1.2, V, bal Fe.
Heat treated: 560-580 Brin. For gears, bearings, cutters; Type 440B; hardenable, corrosion resistant. *Obsolete*

FORTUNA M171H
Stahlwerke Sudwestfalen
C 1.1, Cr 17, Mo 1, V, bal Fe.
Heat treated: 590-620 Brin. For bearing rolls and rings, cutters; Type 440C; hardenable, corrosion resistant. *Obsolete*

FORTUNA MO 500
Stahlwerke Sudwestfalen
C 0.85, Mo 5, V 1.8, W 6.5, Cr 4, bal Fe.
For planing and turning tools, milling cutters, saws; high speed steel. *Obsolete*

FORTUNA MO 503
Stahlwerke Sudwestfalen
C 1.2, Mo 5, V 3.3, W 6.5, Cr 4.5, bal Fe.
For drills, reamers, broaches, saw teeth; high speed steel. *Obsolete*

FORTUNA MO 550
Stahlwerke Sudwestfalen
C 0.8, Co 5, Mo 5, V 1.8, W 6.5, Cr 4.25, bal Fe.
For heavy milling and planing cutters; high speed steel. *Obsolete*

FORTUNA MOG 110
Stahlwerke Sudwestfalen
C 0.45, Cr 1.5, Mo 0.7, V 0.3, bal Fe.
For die casting dies, forging tools; hot work steel, oil hardened. *Obsolete*

FORTUNA MOG 111
Stahlwerke Sudwestfalen
C 0.45, Cr 1.5, Mo 0.5, V 0.8, W 0.5, bal Fe.
For shear blades, forging tools, gripper dies; hot work steel, oil hardened. *Obsolete*

FORTUNA MOG 330
Stahlwerke Sudwestfalen
C 0.3, Cr 3, Mo 3, V 0.6, bal Fe.
For piercing mandrels, die inserts, heading dies; hot work steel, oil hardened. *Obsolete*

FORTUNA MOG 510
Stahlwerke Sudwestfalen
C 0.4, Si 1, Cr 5, Mo 1.5, V 0.6, bal Fe.
For forging dies, die inserts, gripping dies; hot work steel, oil or air hardened. *Obsolete*

FORTUNA MOG 511
Stahlwerke Sudwestfalen
C 0.35, Si 1, Cr 5, Mo 1.5, V 0.3, W 1.5, bal Fe.
For forging dies, die inserts, shear blades; hot work steel, air or oil hardened. *Obsolete*

FORTUNA NC6
Stahlwerke Sudwestfalen
C 0.27, Cr 1.5, Al 1, bal Fe.
Heat treated: 92,000-114,000 TS; 65,000 YS; 19 El; 55 RA. For gears, shafts, camshafts, cams; nitriding steel. *Obsolete*

FORTUNA NC6H
Stahlwerke Sudwestfalen
C 0.34, Cr 1.5, Al 1, bal Fe.
Heat treated: 114,000-144,000 TS; 85,000 YS; 15 El; 50 RA. For gears, shafts, camshafts, cams; nitriding steel. *Obsolete*

FORTUNA NC7 EXTRA
Stahlwerke Sudwestfalen
C 0.33, Cr 1.7, Ni 1, Al 1, bal Fe.
Heat treated: 114,000-144,000 TS; 85,000 YS; 17 El; 50 RA. For gears, shafts, camshafts, cams, piston rods; nitriding steel. *Obsolete*

FORTUNA NCM04
Stahlwerke Sudwestfalen
C 0.32, Cr 1, Mo 0.2, Al 1, bal Fe.
Heat treated: 114,000-135,000 TS; 85,000 YS; 15 El; 50 RA. For gears, shafts, camshafts, cams, nitriding steel. *Obsolete*

FORTUNA NCS5
Bochum Stahlwerke AG
C 0.3, Cr 1.2, Al 0.9, S 0.1, bal Fe.
Heat treated: 85,000-115,000 TS; 65,000 YS; 15 El; 50 RA. For gears, shafts, camshafts; nitriding steel. *Obsolete*

FORTUNA NCS5
Sudwestfalen Stahlwerke AG.
C 0.3, Cr 1.2, Al 0.9, S 0.1, bal Fe.
Heat treated: 85,000-115,000 TS; 65,000 YS; 15 El; 50 RA. For gears, shafts, camshafts; nitriding steel. *Obsolete*

FORTUNA NCS5
Stahlwerke Sudwestfalen
C 0.3, Cr 1.2, Al 0.9, S 0.1, bal Fe.
Heat treated: 85,000-115,000 TS; 65,000 YS; 15 El; 50 RA. For gears, shafts, camshafts; nitriding steel. *Obsolete*

FORTUNA NCV9 SPEC
Stahlwerke Sudwestfalen
C 0.3, Cr 2.5, Mo 0.12, V 0.15, bal Fe.
Heat treated: 128,000-164,000 TS; 100,000-114,000 YS; 13-16 El; 45-50 RA. For gears, shafts, cams, valve spindles, camshafts; nitriding steel. *Obsolete*

FORTUNA NG
Stahlwerke Sudwestfalen
C 0.55, Cr 0.7, Mo 0.2, Ni 1.7, V 0.1, bal Fe.
For forging dies, crankshafts, camshafts; hot work steel, oil hardened. *Obsolete*

FORTUNA NG2 SUPRA
Stahlwerke Sudwestfalen
C 0.55, Cr 1, Mo 0.5, Ni 1.7, V 0.1, bal Fe.
For forging and gripping dies, die inserts; hot work steel, oil or air hardened. *Obsolete*

FORTUNA NG3 SUPRA
Stahlwerke Sudwestfalen
C 0.55, Cr 1, Mo 0.7, Ni 2.2, bal Fe.
For drop forging dies, hot mandrels; hot work steel, oil or air hardened. *Obsolete*

FORTUNA OB7
Stahlwerke Sudwestfalen
C 0.71, Si 0.2, Mn 0.22, bal Fe.
Heat treated: 107,000-144,000 TS; 58,000-80,000 YS; 11-15 El; 30-35 RA. For machine tool parts, springs, hammers, axes; for surface hardening. *Obsolete*

FORTUNA PH10
Stahlwerke Sudwestfalen
C 0.21, Cr 2.35, Mo 0.4, V 0.8, W 0.37, bal Fe.
Heat treated: 114,000-135,000 TS; 78,000 YS; 17 El. For parts resistant to 520°C; creep and heat resistant. *Obsolete*

FORTUNA PH20
Stahlwerke Sudwestfalen
C 0-0.15, Cr 5, Mo 0.55, bal Fe.
Heat treated: 65,000-107,000 TS; 36,000-58,000 YS; 20-23 El. For pipes for oil refinery plants; creep and heat resistant. *Obsolete*

FORTUNA PH5B
Stahlwerke Sudwestfalen
C 0.1, Cr 2.85, bal Fe.
Heat treated: 65,000-76,000 TS; 32,000 YS; 25 El. For pipes for oil refineries and high pressure washers; creep and heat resistant. *Obsolete*

FORTUNA PH5C
Stahlwerke Sudwestfalen
C 0.15, Cr 2.35, Mn 0.4, Si 0.25, bal Fe.
Heat treated: 72,000-85,000 TS; 43,000 YS; 22 El. For parts for core pipes for high pressure windings; creep and heat resistant. *Obsolete*

FORTUNA PH8N
Stahlwerke Sudwestfalen
C 0.17, Cr 2.65, Mo 0.25, V 0.15, bal Fe.
Heat treated: 92,000-115,000 TS; 65,000 YS; 19 El. For parts resistant to 400°C; creep and heat resistant. *Obsolete*

FORTUNA PH9
Stahlwerke Sudwestfalen
C 0.2, Cr 3.15, Mo 0.55, V 0.5, bal Fe.
Heat treated: 114,000-135,000 TS; 78,000 YS; 17 El. For parts resistant to 480°C; creep and heat resistant. *Obsolete*

FORTUNA R3
Stahlwerke Sudwestfalen
C 1.42, Cr 0.3, V 0.25, W 3, bal Fe.
For thread cutters, scrapers, chisels; water hardened. *Obsolete*

FORTUNA R5
Stahlwerke Sudwestfalen
C 1.3, Cr 0.2, W 5, bal Fe.
For thread cutters, scrapers, chisels, engraving needles; water hardened. *Obsolete*

FORTUNA SC 150
Stahlwerke Sudwestfalen
C 1.45, Cr 1.4, bal Fe.
For taps, milling cutters, reamers, broaches; oil hardened, wear resistant *Obsolete*

FORTUNA SICV
Stahlwerke Sudwestfalen
C 1.15, Cr 0.7, V 0.1, bal Fe.
For drills, taps, cutters, files, punches; oil or water hardened. *Obsolete*

FORTUNA SIW
Stahlwerke Sudwestfalen
C 1.2, V 0.1, W 1, bal Fe.
For drills, taps, cutters, files, punches; oil or water hardened. *Obsolete*

FORTUNA SMV 200
Stahlwerke Sudwestfalen
C 0.9, Mn 2, V 0.1, bal Fe.
For blanking tools, punches, tap drills, chasers; oil hardened, cold work steel. *Obsolete*

FORTUNA SW 100
Stahlwerke Sudwestfalen
C 1.2, W 1, bal Fe.
For drills, taps, cutters, files, punches; oil or water hardened. *Obsolete*

FORTUNA SW 111
Stahlwerke Sudwestfalen
C 1.05, Mn 1, Cr 1, W 1.2, bal Fe.
For milling cutters, swages, blanking tools, reamers; oil hardened, cold work steel. *Obsolete*

FORTUNA SW 55
Stahlwerke Sudwestfalen
C 0.95, Mn 1.3, Cr 0.5, W 0.6, bal Fe.
For milling cutters, blanking dies; oil hardened, cold work steel. *Obsolete*

FORTUNA T13
Stahlwerke Sudwestfalen
C 0.2, Cr 13, bal Fe.
Heat treated: 107,000-130,000 TS; 78,000 YS; 18 El; 210-250 Brin. For pumps, valves, piston rods, cutlery; Type 420; corrosion resistant, hardenable. *Obsolete*

FORTUNA T131
Stahlwerke Sudwestfalen
C 0.2, Cr 13, Mo 1.2, bal Fe.
Heat treated: 107,000-130,000 TS; 78,000 YS; 18 El; 220-260 Brin. For steam turbine blades, molds; corrosion resistant, hardenable. *Obsolete*

FORTUNA T17
Stahlwerke Sudwestfalen
C 0.2, Cr 17, Ni 2, bal Fe.
Heat treated: 114,000-137,000 TS; 85,000 YS; 18 El; 225-280 Brin. For shafts, axles, valves, pump parts; Type 431; corrosion resistant to sea water. *Obsolete*

FORTUNA T171
Stahlwerke Sudwestfalen
C 0.35, Cr 17, Mo 1.2, bal Fe.
Heat treated: 114,000-137,000 TS; 85,000 YS; 18 El; 225-265 Brin. For axles, spindles, valve seats and cones; corrosion resistant. *Obsolete*

FORTUNA TC4
Stahlwerke Sudwestfalen
C 0.34, Si 0.25, Mn 0.7, Cr 1, bal Fe.
Heat treated: 114,000-170,000 TS; 78,000-114,000 YS; 13-17 El; 40-50 RA. For aircraft and auto engine parts; oil hardened, tough. *Obsolete*

FORTUNA TC4 SPEC
Stahlwerke Sudwestfalen
C 0.41, Si 0.25, Mn 0.7, Cr 1, bal Fe.
Heat treated: 219,000-250,000 TS; 185,000 YS; 9 El; 30 RA. For gears, shafts, crankshafts; oil hardened, shock resistant. *Obsolete*

FORTUNA TC4B
Stahlwerke Sudwestfalen
C 0.37, Si 0.25, Mn 0.65, Cr 1, bal Fe.
Heat treated: 114,000-170,000 TS; 78,000-114,000 YS; 13-17 El; 40-50 RA. For gears, shafts, machinery parts; for surface hardening. *Obsolete*

FORTUNA TC6
Stahlwerke Sudwestfalen
C 0.36, Si 0.25, Mn 0.5, Cr 1.5, bal Fe.
Heat treated: 100,000-135,000 TS; 65,000-78,000 YS; 17-18 El; 55-60 RA. For levers, axles, steering parts, gears, shafts; oil hardened, tough. *Obsolete*

FORTUNA TCM04
Stahlwerke Sudwestfalen
C 0.34, Si 0.25, Mn 0.65, Cr 1, Mo 0.2, bal Fe.
Heat treated: 100,000-176,000 TS; 65,000-114,000 YS; 13-18 El; 45-60 RA. For crankshafts, axles, gears, connecting rods; for surface hardening. *Obsolete*

FORTUNA TCM04H
Stahlwerke Sudwestfalen
C 0.42, Si 0.25, Mn 0.65, Cr 1, Mo 0.2, bal Fe.
Heat treated: 107,000-186,000 TS; 78,000 YS; 17 El; 55 RA. For axle swivels, connecting rods, gears, shafts; oil hardened, shock resistant. *Obsolete*

FORTUNA TCM04W
Stahlwerke Sudwestfalen
C 0.25, Si 0.25, Mn 0.65, Cr 1, Mo 0.2, bal Fe.
Heat treated: 92,000-150,000 TS; 60,000-92,000 YS; 15-19 El; 50-65 RA. For axles, shafts, steering parts; oil hardened, tough. *Obsolete*

FORTUNA TCM05
Stahlwerke Sudwestfalen
C 0.5, Cr 1, Mo 0.2, bal Fe.
Heat treated: 114,000-185,000 TS; 85,000-128,000 YS; 12-18 El; 40-50 RA. For gears, shafts, crankshafts; for surface hardening. *Obsolete*

FORTUNA TCV4
Stahlwerke Sudwestfalen
C 0.5, Si 0.25, Mn 0.9, Cr 1, V 0.1, bal Fe.
Heat treated: 212,000-250,000 TS; 185,000 YS; 10 El; 35 RA. For axles, levers, gears, shafts; oil hardened, shock resistant. *Obsolete*

FORTUNA TCV5
Stahlwerke Sudwestfalen
C 0.59, Si 0.25, Mn 1, Cr 1.1, V 0.1, bal Fe.
Heat treated: 212,000-242,000 TS; 190,000 YS; 7-10 El; 30 RA; 435-495 Brin. For highly stressed springs; oil hardened. *Obsolete*

FORTUNA TCV6
Stahlwerke Sudwestfalen
C 0.42, Si 0.25, Mn 0.6, Cr 1.5, V 0.1, bal Fe.
Heat treated: 107,000-128,000 TS; 78,000 YS; 17 El; 55 RA. For gears, shafts, crankshafts, axles; oil hardened, shock resistant. *Obsolete*

FORTUNA TCV9 SPEC
Stahlwerke Sudwestfalen
C 0.3, Si 0.25, Mn 0.55, Cr 2.5, Mo 0.2, V 0.15, bal Fe.
Heat treated: 175,000-208,000 TS; 145,000-157,000 YS; 11 El; 35-40 RA. For crankshafts, bolts, gears, screws; oil hardened, shock resistant. *Obsolete*

FORTUNA TM4
Stahlwerke Sudwestfalen
C 0.4, Si 0.35, Mn 1, bal Fe.
Heat treated: 100,000-150,000 TS; 65,000-92,000 YS; 15-18 El; 40-50 RA. For bolts, spindles, shafts; oil or water hardened. *Obsolete*

FORTUNA TM5
Stahlwerke Sudwestfalen
C 0.3, Si 0.25, Mn 1.4, bal Fe.
Heat treated: 92,000-105,000 TS; 60,000-75,000 YS; 17-19 El; 45-55 RA. For forgings, axles, shafts, gears; water hardened. *Obsolete*

FORTUNA TMCV4
Stahlwerke Sudwestfalen
C 0.27, Si 0.25, Mn 1.1, Cr 0.7, V 0.1, bal Fe.
Heat treated: 92,000-150,000 TS; 60,000-92,000 YS; 15-19 El; 50-65 RA. For crankshafts, axles, levers, gears; oil hardened, tough. *Obsolete*

FORTUNA TMS4
Stahlwerke Sudwestfalen
C 0.53, Si 0.9, Mn 0.9, bal Fe.
Heat treated: 100,000-170,000 TS; 65,000-114,000 YS; 13-18 El; 35-50 RA. For gears, pinions, camshafts, connecting rods; for surface hardening. *Obsolete*

FORTUNA TMS5
Stahlwerke Sudwestfalen
C 0.37, Si 1.25, Mn 1.25, bal Fe.
Heat treated: 100,000-170,000 TS; 65,000-114,000 YS; 13-18 El; 35-50 RA. For shafts, gears; tough and abrasion resistant. *Obsolete*

FORTUNA TMV7
Stahlwerke Sudwestfalen
C 0.42, Si 0.25, Mn 1.8, V 0.1, bal Fe.
Heat treated: 127,000-186,000 TS; 100,000-128,000 YS; 12-15 El; 30-40 RA. For axles, shafts, gears, bolts, connecting rods; oil hardened, tough. *Obsolete*

FORTUNA TNC10H
Stahlwerke Sudwestfalen
C 0.35, Si 0.3, Mn 0.5, Cr 0.75, Ni 2.5, bal Fe.
Heat treated: 114,000-135,000 TS; 78,000 YS; 12 El; 50 RA. For wheels, hubs, axles, swivels, gears; oil hardened, tough. *Obsolete*

FORTUNA TNC10W
Stahlwerke Sudwestfalen
C 0.3, Si 0.3, Mn 0.5, Cr 0.75, Ni 2.5, bal Fe.
Heat treated: 101,000-121,000 TS; 72,000 YS; 17 El; 50 RA. For wheel hubs, axle swivels, gears, shafts; oil hardened, shock resistant. *Obsolete*

FORTUNA TNC14H
Stahlwerke Sudwestfalen
C 0.3, Si 0.3, Mn 0.5, Cr 0.75, Ni 3.5, bal Fe.
Heat treated: 128,000-150,000 TS; 101,000 YS; 12 El; 50 RA. For axles, shafts, levers, gears, connecting rods; oil hardened, shock resistant. *Obsolete*

FORTUNA TNC14W
Stahlwerke Sudwestfalen
C 0.25, Si 0.3, Mn 0.5, Cr 0.75, Ni 3.5, bal Fe.
Heat treated: 108,000-128,000 TS; 78,000 YS; 17 El; 50 RA. For axles, shafts, levers, gears, connecting rods; oil hardened, shock resistant. *Obsolete*

FORTUNA TNC18
Stahlwerke Sudwestfalen
C 0.35, Si 0.3, Mn 0.5, Cr 1.3, Ni 4.5, bal Fe.
Heat treated: 144,000-166,000 TS; 114,000 YS; 11 El; 50 RA. For shafts, axles, axle tubes; oil hardened, shock resistant. *Obsolete*

FORTUNA TNC6H
Stahlwerke Sudwestfalen
C 0.35, Si 0.3, Mn 0.5, Cr 0.5, Ni 1.5, bal Fe.
Heat treated: 108,000-121,000 TS; 74,000 YS; 18 El; 50 RA. For axle swivels, axles, shafts, gears; oil hardened, shock resistant. *Obsolete*

FORTUNA TNC6W
Stahlwerke Sudwestfalen
C 0.3, Si 0.3, Mn 0.5, Cr 0.5, Ni 1.5, bal Fe.
Heat treated: 92,000-108,000 TS; 61,000 YS; 21 El; 50 RA. For construction and machine parts, shafts, gears; oil hardened, shock resistant. *Obsolete*

FORTUNA TT13
Stahlwerke Sudwestfalen
C 0.15, Cr 13, bal Fe.
Heat treated: 100,000-123,000 TS; 72,000 YS; 20 El; 190-420 Brin. For turbine blades, cutlery; Type 420; corrosion resistant, hardenable. *Obsolete*

FORTUNA TX10
Stahlwerke Sudwestfalen
C 0.36, Si 0.25, Mn 0.65, Cr 1, Ni 1, Mo 0.2, bal Fe.
For airplane and auto parts, axles, shafts, gears, countershafts; oil hardened, tough. *Obsolete*

FORTUNA TX15
Stahlwerke Sudwestfalen
C 0.34, Si 0.25, Mn 0.55, Cr 1.5, Ni 1.5, Mo 0.2, bal Fe.
Heat treated: 152,000-186,000 TS; 123,000-135,000 YS; 12 El; 40-45 RA. For gears, shafts, aircraft and auto parts, machine tool components; oil hardened, tough. *Obsolete*

FORTUNA TX20
Stahlwerke Sudwestfalen
C 0.3, Si 0.25, Mn 0.45, Cr 2, Mo 0.3, Ni 2, bal Fe.
Heat treated: 175,000-208,000 TS; 145,000-157,000 YS; 11 El; 35-40 RA. For aircraft and auto parts, machine tool parts, gears, shafts; subjected to high stresses, oil hardened. *Obsolete*

FORTUNA V 300
Stahlwerke Sudwestfalen
C 0.85, W 12, V 2.5, Cr 4, Mo 0.8, bal Fe.
For milling cutters, lathe and planer tools, saws; high speed steel. *Obsolete*

FORTUNA V400
Stahlwerke Sudwestfalen
C 1.25, W 12, V 4, Cr 4.2, Mo 0.8, bal Fe.
For milling cutters, lathe and planer tools, drills; high speed steel. *Obsolete*

FORTUNA V450
Stahlwerke Sudwestfalen
C 1.35, Co 5, W 12, V 4, Cr 4.2, Mo 0.8, bal Fe.
For cutters for abrasive material; high speed. *Obsolete*

FORTUNA VC12
Stahlwerke Sudwestfalen
C 2.1, Si 0.3, Mn 0.3, Cr 11.5, bal Fe.
Heat treated: 114,000-135,000 TS; 72,000 YS; 9 El; 10 RA.
For valve seats; heat and corrosion resistant. *Obsolete*

FORTUNA VCN 18-8
Stahlwerke Sudwestfalen
C 0.42, Si 2.25, Cr 18, Ni 9, W 1, bal Fe.
Heat treated: 114,000-144,000 TS; 58,000 YS; 27 El; 35 RA.
For exhaust valves; high oxidation and heat resistance.
Obsolete

FORTUNA VCN 235
Stahlwerke Sudwestfalen
C 0.45, Si 1.15, Mn 1.05, Cr 23, Mo 2.7, Ni 5, bal Fe.
Heat treated: 107,000-135,000 TS; 85,000 YS; 16 El; 25 RA.
For exhaust valves; for service up to 700°C. *Obsolete*

FORTUNA VCS 20
Stahlwerke Sudwestfalen
C 0.8, Si 2.15, Mn 0.4, Cr 20, Ni 1.37, bal Fe.
Heat treated: 127,000-150,000 TS; 101,000 YS; 7 El. For
exhaust valves; good strength, oxidation resistant. *Obsolete*

FORTUNA VCS2
Stahlwerke Sudwestfalen
C 0.45, Si 4, Mn 0.45, Cr 2.85, bal Fe.
Heat treated: 128,000-151,000 TS; 101,000 YS; 16 El; 40 RA.
For exhaust and inlet valves; oil hardened. *Obsolete*

FORTUNA VCS9
Stahlwerke Sudwestfalen
C 0.45, Si 3.05, Mn 0.45, Cr 9, bal Fe.
Heat treated: 128,000-151,000 TS; 101,000 YS; 16 El; 40 RA.
For exhaust valves; high stressed, oil hardened. *Obsolete*

FORTUNA W10 EXTRA
Stahlwerke Sudwestfalen
C 1, bal Fe.
Annealed: 95,000 TS; 50,000 YS; 23 El; 45 RA; 200 Brin. For
jaws, shear blades, snaps, cold headers; water hardened.
Obsolete

FORTUNA W10 PRIMA
Stahlwerke Sudwestfalen
C 1, bal Fe.
Annealed: 100,000 TS; 53,000 YS; 21 El; 42 RA; 200 Brin. For
cold forging dies, knives, milling cutters, reamers; water
hardened. *Obsolete*

FORTUNA W10V
Stahlwerke Sudwestfalen
C 1, V 0.1, bal Fe.
For cold forging dies, shear blades, punches; water
hardened; Type W2. *Obsolete*

FORTUNA W11 EXTRA
Stahlwerke Sudwestfalen
C 1.1, bal Fe.
Annealed: 100,000 TS; 53,000 YS; 21 El; 42 RA; 210 Brin. For
jaws, shear blades, cold heading dies; water hardened, wear
resistant. *Obsolete*

FORTUNA W11 PRIMA
Stahlwerke Sudwestfalen
C 1.15, bal Fe.
Annealed: 110,000 TS; 56,000 YS; 19 El; 39 RA; 210 Brin. For
wood and leather tools, files, saws, drills; water hardened.
Obsolete

FORTUNA W13 PRIMA
Stahlwerke Sudwestfalen
C 1.3, bal Fe.
Annealed: 210 Brin. For lathe and planer tools, taps, draw
punches; water hardened. *Obsolete*

FORTUNA W18
Stahlwerke Sudwestfalen
C 0.75, W 18, Cr 4, V 1, bal Fe.
For lathe and planer tools, reamers, hobs, drills; high speed
steel. *Obsolete*

FORTUNA W23
Stahlwerke Sudwestfalen
C 0.15, Mn 0.4, bal Fe.
Annealed: 70,000 TS; 40,000 YS; 25 El; 60 RA; 130 Brin. For
plastic mold dies, rollers; water hardened. *Obsolete*

FORTUNA W3
Stahlwerke Sudwestfalen
C 0.45, Mn 0.7, bal Fe.
Hot rolled: 98,000 TS; 59,000 YS; 24 El; 45 RA; 212 Brin. For
hammers, axes, pliers, shears, anvils, stamps; water
hardened. *Obsolete*

FORTUNA W33
Stahlwerke Sudwestfalen
C 0.35, Mn 0.5, bal Fe.
Hot rolled: 85,000 TS; 54,000 YS; 30 El; 53 RA; 185 Brin. For
pliers, screw drivers, augers, forks; water hardened. *Obsolete*

FORTUNA W63
Stahlwerke Sudwestfalen
C 0.6, Mn 0.7, bal Fe.
Heat treated: 160,000 TS; 113,000 YS; 12 El; 40 RA; 325 Brin.
For hammers, hatches, knives, vise-jaws, tool holders; water
hardened. *Obsolete*

FORTUNA W63K
Stahlwerke Sudwestfalen
C 0.65, Mn 1, bal Fe.
For shackles, bolts, liners, hammers; oil hardened, wear
resistant. *Obsolete*

FORTUNA W7 EXTRA
Stahlwerke Sudwestfalen
C 0.7, bal Fe.
Heat treated: 174,000 TS; 128,000 YS; 12 El; 37 RA; 352 Brin.
For cold shears, rivet snaps, blanking tools; water hardened.
Obsolete

FORTUNA W7 PRIMA
Stahlwerke Sudwestfalen
C 0.7, bal Fe.
Heat treated: 174,000 TS; 128,000 YS; 12 El; 37 RA; 350 Brin.
For hot and cold work tools, knives, hammers; water
hardened. *Obsolete*

FORTUNA W73
Stahlwerke Sudwestfalen
C 0.7, Mn 0.7, bal Fe.
Heat treated: 180,000 TS; 140,000 YS; 14 El; 38 RA; 375 Brin.
For axes, wood working tools, knives; water hardened.
Obsolete

FORTUNA W8 EXTRA
Stahlwerke Sudwestfalen
C 0.85, bal Fe.
Heat treated: 190,000 TS; 145,000 YS; 10 El; 30 RA; 400 Brin.
For cold punches, leather stampers, gauge tools; water
hardened. *Obsolete*

FORTUNA W8 PRIMA
Stahlwerke Sudwestfalen
C 0.85, bal Fe.
Annealed: 190 Brin. For shear blades, knives, hammer tools;
Type W1; water hardened. *Obsolete*

FORTUNA W83K
Stahlwerke Sudwestfalen
C 0.8, Mn 1, bal Fe.
For shackles, bolts, liners, hammers; oil hardened, wear
resistant. *Obsolete*

FORTUNA W8N
Stahlwerke Sudwestfalen
C 0.85, Ni 0.6, V 0.1, bal Fe.
For coining, gripping and heading dies; water hardened.
Obsolete

FORTUNA W93
Stahlwerke Sudwestfalen
C 0.9, Mn 0.5, bal Fe.
Heat treated: 190,000 TS; 145,000 YS; 10 El; 30 RA; 400 Brin.
For cement crushers, knives, pressure plates; oil or water
hardened. *Obsolete*

FORTUNA W93K
Stahlwerke Sudwestfalen
C 0.9, Mn 1, bal Fe.
For shackles, bolts, liners, drag rails, hammers; oil hardened,
wear resistant. *Obsolete*

FORTUNA WA 235
Stahlwerke Sudwestfalen
C 0.35, Si 1, Cr 1, V 0.2, W 2, bal Fe.
For hand chisels, cold shears, rivet snaps; hot work steel,
water hardened. *Obsolete*

FORTUNA WA 245
Stahlwerke Sudwestfalen
C 0.45, Si 1, Mn 0.3, Cr 1, V 0.2, W 2, bal Fe.
For hot piercing and trimming tools, cutters; hot work steel,
oil hardened. *Obsolete*

FORTUNA WA 2930
Stahlwerke Sudwestfalen
C 0.3, Co 2, Cr 2.5, V 0.3, W 9, bal Fe.
For extrusion dies, liners, rams; hot work steel, oil hardened.
Obsolete

FORTUNA WA 430
Stahlwerke Sudwestfalen
C 0.3, Si 1, Cr 1, V 0.2, W 4, bal Fe.
For shear blades, gripping dies, mandrels; hot work steel, oil
hardened. *Obsolete*

FORTUNA WA 530
Stahlwerke Sudwestfalen
C 0.3, Cr 2.5, V 0.6, W 4.5, bal Fe.
For extrusion and tube press tools, liners; hot work steel, oil
hardened. *Obsolete*

FORTUNA WA 930
Stahlwerke Sudwestfalen
C 0.3, Cr 2.5, V 0.4, W 9, bal Fe.
For extrusion dies, liners and rams, die casting dies; hot work
steel, oil or air hardened. *Obsolete*

FORTUNA WA255
Stahlwerke Sudwestfalen
C 0.55, Si 1, Cr 1, V 0.2, W 2, bal Fe.
For blanking dies, shear blades, punches, knives; oil
hardened, tough. *Obsolete*

FORTUNA WC 6H
Stahlwerke Sudwestfalen
C 0.34, Al 1, Cr 1.5, bal Fe.
Annealed: 110 Brin. For plastic mold dies; nitrided. *Obsolete*

FORTUNA WF8
Stahlwerke Sudwestfalen
C 0.67, Si 1.3, Cr 0.5, bal Fe.
For crusher parts, hammer mills, brick molds; wear resistant,
tough. *Obsolete*

FORTUNA WGKL
Stahlwerke Sudwestfalen
C 1, Cr 1.5, bal Fe.
For blanking tools, punches, thread cutting tools; oil
hardened, wear resistant. *Obsolete*

FORTUNA WM 13
Stahlwerke Sudwestfalen
C 0.4, Cr 13, bal Fe.
Annealed: 225 Brin. For plastic mold dies; corrosion resistant.
Obsolete

FORTUNA WO 3
Stahlwerke Sudwestfalen
C 0.1, Mn 0.4, bal Fe.
Annealed: 140 Brin. For plastic mold dies; case hardened.
Obsolete

FORTUNA WRL
Stahlwerke Sudwestfalen
C 0.9, Cr 0.8, bal Fe.
For stamping and coining dies, cold rolls; water hardened.
Obsolete

FORTUNA WSB EXTRA
Stahlwerke Sudwestfalen
C 1.15, W 2, bal Fe.
For saws; oil hardened. *Obsolete*

FORTUNA WT 131
Stahlwerke Sudwestfalen
C 0.3, Cr 13, Mo 0.4, bal Fe.
For dies, casting dies, piercing mandrels; hot work steel, corrosion resistant. *Obsolete*

FORTUNA ZW
Stahlwerke Sudwestfalen
C 1.4, Cr 0.3, V 0.1, bal Fe.
For drawing dies; water hardened, abrasion and wear resistant. *Obsolete*

FORTY-TWO N
United States Steel Corp.
C 0.16, Mn 1.2, Si 0.35, Cb 0.05, bal Fe.
Normalized: 63 ksi TS; 42 ksi YS; 23 El. Readily formed and welded. For stressed structures at low temperatures; arctic and marine structures. ASTM A633 Grade A.

FORVA-50
Forjas Alavesas S.A.
C 0.5, Mn 0.85, Si 0.25, Cr 1, V 0.15, bal Fe.
Cr-V spring steel; hardenable to 58-63 Rock C. IHA F-143; DIN 50CrV4; AISI 6150; BS EN47.

FOS-FLO 5
Handy & Harmon
Cu 95, P 5.
MP: 1310°F; FP: 1695°F. Filler metal with wide melting range and low fluidity. Used for preplacing in resistance welding applications or to supplement spot welding operations. Recommended joint clearance: 0.003-0.005 in. Slow flow.

FOS-FLO 7
Handy & Harmon
Copper. Cu 92.75, P 7.25.
MP: 1310°F; FP: 1350°F. For joining copper and copper alloys where joint does not involve critical impact or vibration. AWS BCuP-2.

FOUR STAR
Now CARPENTER FOUR STAR.

FOUR STAR
Midvale-Heppenstall Co.
C 0.7-1.2, bal Fe.
For tools, drills, taps; water hardened. *Obsolete*

FOURDINIER WIRE
English manufacture
Cu 80-85, Zn 15-20, Sn 0-0.4, Pb 0.01.
For wire for Fourdinier screens used in paper manufacture.

FOURDRINIER BRASS
Chase Brass & Copper Co.
Cu 83, Zn 17.
Annealed: 43,000 TS; 14,000 YS; 49 El. Cold drawn: 95,000 TS; 9 El. For Fourdrinier cloth; red brass, wire. *Obsolete*

FOURTEEN PER CENT
English manufacture
Cu 58-60, Zn 26-28, Ni 14.
For ornamental flatware; corrosion resistant.

FOX NO. 002 "MANCASE"
British Steel Corp.
Ni 3, C 0.1-0.25, bal Fe.
Heat treated: 85,000-100,000 TS; 25-30 El; 45-55 RA. Normalized: 67,000 TS; 41,000 YS; 36 El; 65 RA. For case hardened parts; free cutting case-hardening steel; IZ-60-90. *Obsolete*

FOX NO. 031 (531)
British Steel Corp.
C 0.2, Cr 0.9, Ni 1.5, bal Fe.
Hardened: 130,000-170,000 TS; 110,000-150,000 YS; 14-20 El; 50-60 RA. For gears, pinions, axles; case hardening steel. *Obsolete*

FOX NO. 040 (540)
British Steel Corp.
C 0.12-0.18, Ni 3.8-4.5, Cr 1-1.4, bal Fe.
Hardened: 170,000-200,000 TS; 130,000-160,000 YS; 13-18 El; 45-60 RA. For gears, pinions, shafts; case hardening steel. *Obsolete*

FOX NO. 1038
British Steel Corp.
C 0.5, Si 3.2, Cr 8, bal Fe.
Hardened: 134,000 TS; 98,000 YS; 27 El; 59 RA; 262 Brin. For exhaust valves; heat resistant. *Obsolete*

FOX NO. 120
British Steel Corp.
C 0.3-0.4, Mn 1.3-1.7, bal Fe.
Normalized: 80,000-90,000 TS; 48,000-60,000 YS; 27-30 El; 50-60 RA. For automotive parts, nuts, bolts; oil hardening. *Obsolete*

FOX NO. 130
British Steel Corp.
C 0.25-0.4, Mn 1.3-1.8, Mo 0.2-0.35, bal Fe.
Normalized: 90,000-110,000 TS; 50,000-90,000 YS; 20-25 El; 50-60 RA. For crankshafts, axles, bolts, connecting rods; tough. *Obsolete*

FOX NO. 131
British Steel Corp.
C 0.3-0.4, Mn 1.3-1.8, Mo 0.35-0.55, bal Fe.
Hardened: 90,000-160,000 TS; 80,000-140,000 YS; 12-22 El; 40-60 RA. For crankshafts, axles, bolts; tough. *Obsolete*

FOX NO. 135
British Steel Corp.
C 0.15-0.25, Mn 1.4-1.8, Ni 0.4-0.7, Mo 0.15-0.35, bal Fe.
Hardened: 80,000-100,000 TS; 60,000-80,000 YS; 22-30 El; 50-70 RA. For gears, shafts, axles; oil hardening. *Obsolete*

FOX NO. 525
British Steel Corp.
C 0.2, Ni 5, bal Fe.
Hardened: 100,000 TS; 70,000 YS; 20 El; 45 RA. For gears, shafts, axles; case hardening steel. *Obsolete*

FOX NO. 540
British Steel Corp.
C 0.14-0.18, Ni 4-4.5, Cr 1-1.5, Mo 0-0.35, bal Fe.
Hardened: 170,000-200,000 TS; 130,000 YS; 13 El; 45 RA. For gears, shafts, axles; case hardening steel. *Obsolete*

FOX NO. 671
British Steel plc
C 0.35-0.45, Ni 1.25-1.75, Cr 0.8-1.4, bal Fe.
Obsolete

FOX S.F. 10
British Steel Corp.
C 0.09-0.12, Cr 13, Si 0-0.5, bal Fe.
Annealed: 75,000 TS; 40,000 YS; 35 El; 70 RA; 155 Brin. For turbine blades, valves, cutlery; Type 410; corrosion resistant. *Obsolete*

FOX S.F. 11
British Steel Corp.
C 0.13-0.18, Cr 13, bal Fe.
Annealed: 80,000 TS; 45,000 YS; 32 El; 68 RA; 170 Brin. For turbine blades, valves, cutlery, surgical instruments; Type 420; corrosion resistant. *Obsolete*

FOX S.F. 13
British Steel Corp.
C 0.26-0.37, Cr 13, bal Fe.
Annealed: 95,000 TS; 50,000 YS; 25 El; 55 RA; 195 Brin. For turbine blades, valves, cutlery, surgical instruments; Type 420; corrosion resistant. *Obsolete*

FOX S.F. 17
British Steel Corp.
C 0.07-0.12, Cr 17, bal Fe.
Annealed: 80,000 TS; 50,000 YS; 25 El; 50 RA; 150 Brin. For oil refinery equipment, sinks, oil burners; Type 430; corrosion resistant. *Obsolete*

FOX S.F. 20
British Steel Corp.
C 0.11-0.16, Cr 18, Ni 8, bal Fe.
Annealed: 80,000 TS; 35,000 YS; 55 El; 75 RA; 150 Brin. For chemical plant equipment, tanks, mixers, filters; Type 302; stainless, austenitic. *Obsolete*

FOX S.F. 22
British Steel Corp.
C 0-0.08, Cr 18, Ni 8, Ti, bal Fe.
Annealed: 85,000 TS; 35,000 YS; 55 El; 65 RA; 150 Brin. For welded chemical plant equipment, tanks; Type 321; stainless, austenitic. *Obsolete*

FOX S.F. 25
British Steel Corp.
C 0-0.12, Cr 18, Ni 9, Mo 1.5, bal Fe.
Annealed: 85,000 TS; 35,000 YS; 50 El; 65 RA; 160 Brin. For acid resistant chemical plant equipment, tanks; Type 316; stainless, austenitic. *Obsolete*

FOX S.F. 35
British Steel Corp.
C 0-0.1, Cr 19, Ni 12, Mo 3.5, bal Fe.
Annealed: 85,000 TS; 35,000 YS; 50 El; 65 RA; 160 Brin. For acid resistant chemical plant equipment; Type 317; stainless, austenitic. *Obsolete*

FOX S.F. 620
British Steel Corp.
C 0-0.05, Cr 18, Ni 8, bal Fe.
Annealed: 85,000 TS; 35,000 YS; 60 El; 70 RA; 150 Brin. For chemical plant equipment, tanks, mixers, filters; Type 304; stainless, austenitic. *Obsolete*

FOX S.F.12
British Steel Corp.
C 0.19-0.25, Cr 13, bal Fe.
Annealed: 95,000 TS; 50,000 YS; 25 El; 55 RA; 195 Brin. For turbine blades, valves, cutlery, surgical instruments; Type 420; corrosion resistant. *Obsolete*

FOX SF23
British Steel Corp.
C 0.05, Cr 17-19, Ni 9-12, Ti 0.3, bal Fe.
Annealed: 85,000 TS; 30,000 YS; 55 El; B 80 Rock. For welded structures, vessels, digesters, evaporators, tanks. Type 321 stainless steel, stabilized, austenitic, welding grade. *Obsolete*

FOX SF23

British Steel Corp.
C 0.05, Cr 17-19, Ni 9-12, Ti 0.3, bal Fe.
Annealed: 85,000 TS; 30,000 YS; 55 El; B 80 Rock. For welded structures, vessels, digesters, evaporators, tanks. Type 321 stainless steel, stabilized, austenitic, welding grade. *Obsolete*

FOX SF301

British Steel Corp.
C 0.12, Cr 16-18, Ni 6-8, bal Fe.
Annealed: 110,000 TS; 40,000 YS; 60 El; B 85 Rock. For aircraft structural members, diaphragms, household utensils, trailer bodies, springs. Type 301 stainless steel, austenitic. *Obsolete*

FOX SF301

British Steel Corp.
C 0.12, Cr 16-18, Ni 6-8, bal Fe.
Annealed: 110,000 TS; 40,000 YS; 60 El; B 85 Rock. For aircraft structural members, diaphragms, household utensils, trailer bodies, springs. Type 301 stainless steel, austenitic. *Obsolete*

FOX SF304 ELC

British Steel Corp.
C 0.06, Cr 18-20, Ni 8-12, bal Fe.
Annealed: 85,000 TS; 35,000 YS; 60 El; 150 Brin. For processing chemical plant equipment, welded structures, evaporators. Type 304 stainless steel, austenitic, non-hardenable. *Obsolete*

FOX SF304 ELC

British Steel Corp.
C 0.06, Cr 18-20, Ni 8-12, bal Fe.
Annealed: 85,000 TS; 35,000 YS; 60 El; 150 Brin. For processing chemical plant equipment, welded structures, evaporators. Type 304 stainless steel, austenitic, non-hardenable. *Obsolete*

FOX SF316 TI

British Steel Corp.
C 0.06, Cr 18, Ni 11, Ti 0.4, bal Fe.
Annealed: 85,000 TS; 30,000 YS; 55 El; B 82 Rock. For welded structures, tanks, evaporators, digesters. Type 321 stainless steel, austenitic, stabilized. *Obsolete*

FOX SF316 TI

British Steel Corp.
C 0.06, Cr 18, Ni 11, Ti 0.4, bal Fe.
Annealed: 85,000 TS; 30,000 YS; 55 El; B 82 Rock. For welded structures, tanks, evaporators, digesters. Type 321 stainless steel, austenitic, stabilized. *Obsolete*

FOX SF318

British Steel Corp.
C 0.1, Cr 17, Ni 13, Mo 3, Cb 0.6, bal Fe.
Annealed: 85,000 TS; 40,000 YS; 50 El; B 83 Rock. For chemical plant equipment, welded structures and tanks, mixers, agitators. Type 318 stainless steel, austenitic, welding grade. *Obsolete*

FOX SF318

British Steel Corp.
C 0.1, Cr 17, Ni 13, Mo 3, Cb 0.6, bal Fe.
Annealed: 85,000 TS; 40,000 YS; 50 El; B 83 Rock. For chemical plant equipment, welded structures and tanks, mixers, agitators. Type 318 stainless steel, austenitic, welding grade. *Obsolete*

FOX SF320

British Steel Corp.
C 0.03, Cr 18-20, Ni 8-12, bal Fe.
Annealed: 77,000 TS; 30,000 YS; 60 El; 140 Brin. For welded structures, evaporators, digesters, mixers, chemical plant equipment. Type 304L stainless steel, austenitic, non-hardenable. *Obsolete*

FOX SF321

British Steel Corp.
C 0.06, Cr 17-19, Ni 9-12, Ti 0.3, bal Fe.
Annealed: 85,000 TS; 30,000 YS; 55 El; B 80 Rock. For welded structures, mixers, evaporators, digesters, chemical plant equipment. Type 321 stainless steel, austenitic, stabilized, welding grade. *Obsolete*

FOX SF6

British Steel Corp.
C 0.06, Cr 13, Al 0.2, bal Fe.
Annealed: 65,000 TS; 38,000 YS; 30 El; 160 Brin. For annealing boxes, furnace parts, oil refinery equipment. Type 405 stainless steel, heat and corrosion resistant. *Obsolete*

FOX SF80T

British Steel Corp.
C 0.06, Ni 4, Cr 16, Mo 2, Co 2, bal Fe.
Annealed: 130,000 TS; 100,000 YS; 16 El; C 26 Rock. For marine hardware, pump shafts, valve parts, aircraft structures. High strength, stainless. *Obsolete*

FOX SF835

British Steel Corp.
C 0.08, Cr 18.5, Ni 11, Ti 0.4, bal Fe.
Annealed: 85,000 TS; 35,000 YS; 55 El; B 82 Rock. For welded structures, chemical plant equipment, evaporators, tanks, digesters. Type 321 stainless steel, stabilized, austenitic, welding grade. *Obsolete*

FOX SF920

British Steel Corp.
C 0.1, Cr 17-19, Ni 9-10, bal Fe.
Annealed: 90,000 TS; 40,000 YS; 40 El; B 85 Rock. For chemical plant and food processing equipment, tanks, mixers. Type 302 stainless steel, austenitic. *Obsolete*

FP20B

Teledyne Firth Sterling
Tool material. WC 71.5, Co 8.5, 20.0 TiC + TaC + NbC. For use as base material for coated cutting grades. 91.4 Rock A.

FP20M

Teledyne Firth Sterling
Tool material. WC 80.5, Co 7, 12.5 TiC + TaC + NbC. Crater resistant grade to cut materials that yield a continuous chip and tend to produce a crater on top rake surfaces. 92.0 Rock A.

FP25B

Teledyne Firth Sterling
Tool material. WC 85.5, Co 5.5, 9.0 TiC + TaC + NbC. For use as base material for coated cutting grades. 90.5 Rock A.

FP25M

Teledyne Firth Sterling
Tool material. WC 77.8, Co 8.5, 13.7 TiC + TaC + NbC. Crater resistant grade to cut materials that yield a continuous chip and tend to produce a crater on top rake surfaces. 91.4 Rock A.

FP30B

Teledyne Firth Sterling
Tool material. WC 81.7, Co 8.5, 9.8 TiC + TaC + NbC. For use as base material for coated cutting grades. 90.5 Rock A.

FP30M

Teledyne Firth Sterling
Tool material. WC 68.5, Co 10.5, 21.0 TiC + TaC + NbC. Crater resistant grade to cut materials that yield a continuous chip and tend to produce a crater on top rake surfaces. 91.1 Rock A.

FPC NONTEMPERING

A. Milne & Co.
C 0.3-0.4, Cr 0.7-0.9, Mn 0.7, Mo 0.3-0.6, bal Fe.
For chisels, punches, caulking tools; water hardening.

FRAMDIE

Columbia Tool Steel Co.
C 1, Cr 1.4, Mo 0.4, Mn 0.35, bal Fe.
For drive shafts, spindles, arbors; oil hardened. *Obsolete*

FRANCO

E. Frank Atkinson & Sons Ltd.
C 0.7, W 18, Cr 4, V 1, bal Fe.
For high speed tools and cutters. High speed steel.

FRANKITE E-212

Frank Foundries Corp.
C, bal Fe.
For hydraulic bodies, compressor cylinders. *Obsolete*

FRANKITE E-450

Frank Foundries Corp.
Ni 12-15, Cu 5-7, Cr, 1.1 C min, bal Fe.
25,000-35,000 TS; 145-170 Brin. For heat and corrosion resistant parts; heat and corrosion resistant. *Obsolete*

FRANKITE E-604

Frank Foundries Corp.
Ni 4.5, Cr 1.5, C, bal Fe.
For mixer blades, ash chutes, scrapers; "Ni-Hard;" abrasion resistant. *Obsolete*

FRANKITE E-821

Frank Foundries Corp.
C 0.7-1.1, Cr 22-25, bal Fe.
42,000-50,000 TS; 200-230 Brin. For heat and corrosion resistant parts; heat and corrosion resistant. *Obsolete*

FRANKITE E-822

Frank Foundries Corp.
C 1.75-2, Cr 22-25, bal Fe.
80,000-90,000 TS; 320-350 Brin. For heat and corrosion resistant parts; heat and corrosion resistant. *Obsolete*

FRANKITE E-830N

Frank Foundries Corp.
Cr 30, Ni 3, C, bal Fe.
For furnace supports, kilns; corrosion and heat resistant. *Obsolete*

FRANKITE E-831

Frank Foundries Corp.
C 0.8-1.1, Cr 30-34, bal Fe.
45,000-55,000 TS; 280-300 Brin. For heat and corrosion resistant parts; heat and corrosion resistant. *Obsolete*

FRANKITE E-832

Frank Foundries Corp.
Cr 31-39, 1.1 C min, bal Fe.
For heat and corrosion resistant parts; heat and corrosion resistant. *Obsolete*

FRANXA 9

Compagnie des Mines
C 0.58-0.63, Si 1.5-1.9, Mn 0.5-0.9, bal Fe.
Annealed: 120,000 TS; 78,000 YS; 12 El. Hardened: 225,000 TS; 141,000 YS; 6 El. For tools, dies, springs; shock resistant.

FRANZ MAYER NO. 2

Vereinigte Edelstahlwerke
C 0.55, Si 0.1-0.4, Mn 0.5-0.7, bal Fe.
Heat treated: 160,000 TS; 115,000 YS; 13 El; 42 RA; 320 Brin. For gears, bolts, machine tool parts; water hardened. *Obsolete*

FRAPIMPHY 1

Creusot-Loire
C 0-0.03, Cr 17.5, Ni 9.5, Cu 3.25, bal Fe.
Wrought austenitic stainless steel designed for cold heading hexagonal socket heat screws (HSH screw). AFNOR Z 6 CNU 18.10 DF.

FRAPIMPHY 304 BC

Creusot-Loire
C 0-0.03, Cr 18.5, Ni 11.5, bal Fe.
Wrought austenitic stainless steel designed for cold heading bolts and nut for chemical and shipbuilding industries. AFNOR Z2CN 18.10 DF; AISI 304L.

FRAPIMPHY 316 BC

Creusot-Loire
C 0-0.03, Cr 17, Ni 13, Mo 2.3, bal Fe.
Wrought austenitic stainless steel designed for cold heading bolts and nuts for chemical and shipbuilding industries. AFNOR Z 2 CND 17.13 DF; AISI 316L.

FRAPIMPHY A3

Creusot-Loire
C 0-0.2, Cr 12.5, bal Fe.
Wrought, annealed, martensitic stainless steel for cold-heading balls, pins, hinges. AFNOR Z 12C13DF; AISI 410-420.

FRAPIMPHY A5

Creusot-Loire
C 0-0.32, Cr 13.5, bal Fe.
Wrought, annealed, martensitic stainless steel for cold heading balls, pins, hinges. AFNOR Z30 C13DF; AISI 420.

FRAPIMPHY B4

Creusot-Loire
C 0-0.08, Cr 17, bal Fe.
Wrought, annealed, ferritic stainless steel for cold heading wood screws and automobile bolts and nuts. AFNOR Z8C17 DF; AISI 430.

FRAPPANT

Otto Wolff Handelgesellschaft
C 0.9, Mn 1.9, V 0.1, bal Fe.
For blanking and forming dies, punches. Nondeforming, oil hardened.

FRARY METAL

NL Industries
Hg 0.24, Cu 0.04, Ba 0.83, Ca 0.42, Sn 0.12, bal Pb.
Hard: 1.15 El; 0.92 RA; 27 Brin. For bearings. *Obsolete*

FRARY METAL

NL Industries
Cu 0.04, Ba 2.02, Ca 0.71, Sn 0.26, bal Pb.
Medium: 2.3 El; 1.8 RA; 23 Brin. For bearings. *Obsolete*

FRARY METAL

NL Industries
Hg 0.3, Ba 0.2-2, Ca 0.1-1, Sn 0.1-0.3, bal Pb.
Cast: 5 El; 1 RA. For bearings; non-corrosive die casting. *Obsolete*

FRASSE GRADE A

Peter A. Frasse & Co.
C 1.1-1.2, Mn 0.15-0.25, bal Fe.
For augers, broaches, cutters, tools; water hardened. *Obsolete*

FRASSE GRADE B

Peter A. Frasse & Co.
C 1-1.1, Mn 0.15-0.25, bal Fe.
For cold chisels, drills, knives, reamers; water hardened. *Obsolete*

FRASSE GRADE C

Peter A. Frasse & Co.
C 0.9-1.1, Mn 0.15-0.25, bal Fe.
For chisels, drills, hammers, picks, punches; water hardened. *Obsolete*

FRASSE GRADE H

Peter A. Frasse & Co.
C 1-1.1, W 1.25-1.79, bal Fe.
For cutters, tools; oil hardened. *Obsolete*

FRASSE TEMPER "A"

Peter A. Frasse & Co.
C 0.7-0.9, bal Fe.
For general tools. *Obsolete*

FRASSE TEMPER "B"

Peter A. Frasse & Co.
C 0.8-1.1, bal Fe.
For general tools. *Obsolete*

FRASSE TEMPER "C"

Peter A. Frasse & Co.
C 0.9-1.2, bal Fe.
For general tools. *Obsolete*

FREE CUTTING BRASS, COPPER ALLOY NO. 360

Chase Brass & Copper Co., Inc.
Copper. Cu 62, Zn 34.75, Pb 3.25.
Half hard (1.0 inch): 58,000 psi TS; 45,000 psi YS; 25 El; 34,000 psi shear strength; 78 Rock B. Excellent machinability; for automatic machine parts as studs, bolts, nuts shafts. ASTM B16; CDA 360.

FREE CUTTING BRASS, SCOVILL 276

Century Brass Products Inc.
Cu 60-63, Pb 2.5-3.7, bal Zn.
Annealed: 49,000 TS; 53 El; 58 RA. 1/2 Hard: 58,000 TS; 25 El; 50 RA; 144 Brin. For screw machine products; free-cutting. *Obsolete*

FREE CUTTING BRASS-360

Anaconda Co.
Copper. Cu 61.5, Zn 35.25, Pb 3.25.
Hard: 58,000 TS; 42,000 YS; 20 El; 70 Rock B. Soft: 47,000 TS; 18,000 YS; 55 El; 20 Rock B. For hardware, gears, pinions, screw machine products, fasteners. Free machining leaded brass.

FREE CUTTING BRONZE

American manufacture
Cu 89, Zn 10, Pb 1.5.
For screw stock, bolts, automobile radiators; free-cutting.

FREE CUTTING MUNTZ METAL-293

Anaconda Co.
Copper. Cu 60, Zn 39, Pb 1.
Soft: 54,000 TS; 20,000 YS; 40 El. Hard: 80,000 TS; 60,000 YS; 6 El. For screw machine parts. Free cutting.

FREE CUTTING MUNTZ METAL-3711

Anaconda Co.
Copper. Cu 60, Zn 36.75, Pb 3.25.
Hard: 60,000 TS; 45,000 YS; 18 El; 72 Rock B. Soft: 48,000 TS; 20,000 YS; 54 El; 20 Rock B. For butt hinges, lock bodies, mechanical devices, screw machine products, forging rods. Free cutting, extrudable.

FREE CUTTING PHOSPHOR BRONZE 610

Anaconda Co.
Copper. Sn 4, Zn 4, Pb 4, bal Cu.
Hard: 60,000 TS; 50,000 YS; 20 El. For bearings, bushings, valve and pump parts, gears, pinions. Free cutting, wear resistant.

FREE CUTTING STAYBITE F.S.T. STEEL

Now STAYBRITE F.S.T. (FC).

FREE CUTTING TUBE BRASS-332

Anaconda Co.
Copper. Cu 66.5, Zn 31.9, Pb 1.6.
Hard tube: 73,000 TS; 60,000 YS; 10 El; 80 Rock B. Soft tube: 45,000 TS; 17,000 YS; 50 El; 15 Rock B. For screw machine products, ball point pens, plumbing fixtures, musical instruments. Free machining tubes.

FREE CUTTING YELLOW BRASS 271

Anaconda Co.
Copper. Pb 3.25, Zn 35.25, bal Cu.
Soft: 47,000 TS; 32,000 YS; 60 El; 60 Brin. Hard: 62,000 TS; 50,000 YS; 20 El; 140 Brin. For machined parts, hardware. Free cutting.

FREE-CUTTING TUBE BRASS-282

Anaconda Co.
Copper. Cu 66.5, Zn 31.9, Pb 1.6.
Soft: 45,000 TS; 17,000 YS. Hard: 73,000 TS; 60,000 YS. For screw machine products. Free cutting.

FREECUT 15

Peninsular Steel Co.
C 0.2, Mn 1.25, Si 0.05, S 0.25, P 0.02, bal Fe.
Free machining steel plate stock.

FREECUT 45

Peninsular Steel Co.
C 0.45, Mn 1.25, Si 0.05, S 0.25, P 0.02, bal Fe.
Free machining, flame hardenable plate stock.

FREEMACHINEWELD

Westinghouse Electric Corp.
bal Ni.
For welding electrodes for cast iron; machinable welds. *Obsolete*

FREMAX 15

United States Steel Corp.
C 0.13, Mn 1.1, S 0.25, bal Fe.
Free machining steel. 143 Brin. For molds, dies, seals, gears.

FREMAX 45

United States Steel Corp.
C 0.45, Mn 1.1, S 0.25, bal Fe.
Free machining steel. 187 Brin. Hardenable; for molds, dies, seals, gears.

FRENCH

French manufacture
Cu 50, Zn 31, Ni 18.
For ornamental parts, fittings; corrosion resistant.

FRENCH ALLOY

French manufacture
Cu 58-50, Zn 25-30, Ni 17-20.
For electrical resistors, ornamental parts; nickel silver.

FRENCH AUTO

English manufacture
Pb 75, Sn 10, Sb 15.
For automobile bearings; Babbitt.

FRENCH NAVY ANTIFRICTION METAL

French manufacture
Cu 7, Sn 7.5, Zn 78.5.
For bearings, bushings; Babbitt.

FRENCH SILVER SOLDER

French manufacture
Ag 66, Cu 24, Zn 10.
For silver solder.

FRENCH TYPE METAL "A"

French manufacture
Pb 55, Sn 22, Sb 23.
For type metal.

FRENCH TYPE METAL "B"

French manufacture
Pb 55, Sn 15, Sb 30.
For type metal.

FREUND STEEL

German manufacture
Si 0.7-1.3, Mn 0.3-0.6, C 0.1-0.15, Cr, bal Fe.
For highly stressed structural members; high elastic steel.

FRICKE'S HARDER
Manufacturer not listed.
Cu 69, Zn 30, Ni 10.
For strong corrosion resistant parts; corrosion resistant.

FRICKE'S SILVERY
Manufacturer not listed.
Cu 50, Zn 18.8, Ni 31.2.
For ornamental and decorative parts; corrosion resistant.

FRICKS ALLOY
English manufacture
Cu 50-55, Zn 30-31, Ni 17-19.
For resistances, decorative parts; corrosion resistant.

FRICKS ALLOY
English manufacture
Cu 50-69, Zn 18-38, Ni 5.5-31.
For ornamental, base for plated ware; nickel silver.

FRICKS BLUISH YELLOW, HARD
Manufacturer not listed.
Cu 55.5, Zn 39, Ni 5.5.
For hardware, decorative parts; corrosion resistant.

FRICKS PALE YELLOW, DUCTILE
Manufacturer not listed.
Cu 62.5, Zn 31.2, Ni 6.3.
For hardware, decorative parts; corrosion resistant.

FRICTION ALLOY (STANDARD)
English manufacture
Pb 50, Sn 40, Sb 10.
For antifriction metal; Babbitt.

FRICTIONLESS
American Smelting & Refining Co.
Pb alloy.
For bearings, linings; Babbitt.

FRIDUCTIL 5622, ETC.
Thyssen Edelstahlwerke AG
See Werkstoff Nr. 1.5622, etc.
Steels for toughness at sub-zero temp; 9 grades.

FRIGIDAL
Vereinigte Deutsche Nickel-Werke AG
Ni 35, low Cr, bal Fe.
86,000 TS; 32 El. For electrical resistances. Coefficient of expansion (0-40°C): 0.0000014. *Obsolete*

FRILOLIT RF
Friedr. Lohmann GmbH
C 0.4, Si 0.4, Cr 13, bal Fe.
Annealed: 95,000 TS; 50,000 YS; 25 El; 55 RA; 195 Brin. Cold drawn: 105,000 TS; 85,000 YS; 17 El; 50 RA; 215 Brin. For valves, cutlery, surgical instruments. Type 420; corrosion resistant. *Obsolete*

FRILOLIT RF SPEZIAL
Friedr. Lohmann GmbH
C 0.85, Cr, V, bal Fe.
For cutters, tools. Oil or water hardened. *Obsolete*

FRILOLIT RFOO
Friedr. Lohmann GmbH
C 0-0.12, Cr 13, Si 0.4, bal Fe.
Annealed: 75,000 TS; 40,000 YS; 35 El; 70 RA; 155 Brin. Cold drawn: 100,000 TS; 85,000 YS; 17 El; 60 RA; 205 Brin. For turbine blades, surgical instruments. Type 410; corrosion resistant. *Obsolete*

FRILOLIT RFW
Friedr. Lohmann GmbH
C 0.2, Si 0.4, Cr 13, bal Fe.
Annealed: 95,000 TS; 50,000 YS; 25 El; 55 RA; 195 Brin. Cold drawn: 105,000 TS; 85,000 YS; 17 El; 50 RA; 215 Brin. For turbine blades, surgical instruments. Type 420; corrosion resistant. *Obsolete*

FRISMUTH ALUMINUM SOLDER
English manufacture
Sn 67, Pb 27, Al 3.
For aluminum solder.

FRIXTEC 19850
Eutectic Corp.
Aluminum bronze powder for metal spraying. Machinable; smooth, bright finish.

FRKIT
Frank Foundries Corp.
C 3.25-3.5, Si 2-2.25, alloy, bal Fe.
For cylinder liners in diesel and gasoline engines and compressors.

FROGALLOY
Now MCKAY FROGALLOY.

FROGALLOY
Teledyne McKay
C 0.4, Mn 4.1, Cr 19.5, Ni 10, Mo 1.4, bal Fe.
Welded: 116,000 TS; 91,000 YS; 14 El; 550 Brin. For hard surfacing electrodes; work hardenable; wear resistant.

FROGALLOY M
Teledyne McKay
C 0.7, Mn 3.9, Cr 18.6, Ni 10.4, bal Fe.
Welded: 116,000 TS; 91,000 YS; 14 El; 550 Brin. For welding electrodes, hard surfacing; stainless, work hardenable; shielded arc. *Obsolete*

FRONTIER 40E
Frontier Bronze Corp.
Copper. Zn 5.5, Mg 0.55, Cr 0.5, Ti 0.2, bal Al.
Aged: 32,000-38,000 TS; 22,000-26,000 YS; 3-10 El; 75-90 Brin. For aircraft parts, pressure tight castings; age hardenable at room temperature.

FRONTIER ALLOY NO. 24
Frontier Bronze Corp.
Cu 70, Sn 1, Zn 27, Pb 2.
29,500 TS; 10,500 YS; 26 El; 29 RA; 53 Brin. For bearings; formerly Bronze No. 24. *Obsolete*

FRONTIER BRONZE
Sterling International Technology Ltd.
Mg 0.5-0.75, Ni 0.1, Zn 4.8-5.7, Pb 0.1, Sn 0.05, Ti 0.15-0.25, Cr 0.4-0.6, bal Al.
Sand cast: 31,200-35,800 TS; 22,400-24,700 YS; 4-6 El; 60-70 Brin; 3 Izod. Medium strength, reasonably shock resistant sand or gravity die cast aluminum alloy. Ages 21-30 days at room temperature. Somewhat hot short; for simple castings only.

FRONTIER NO. 1
Frontier Bronze Corp.
Cu 90, Al 10.
65,000-80,000 TS; 22,000-26,000 YS; 15-25 El; 120-130 Brin. For worm gears; IZ-20-30. *Obsolete*

FRONTIER NO. 10
Frontier Bronze Corp.
Copper. Sn 8, Zn 4, bal Cu.
Cast: 30,000-45,000 TS; 18,000-23,000 YS; 15-30 El; 65-74 Brin. For bearings, screw down nuts, worm gears, hydraulic castings; gun metal bronze.

FRONTIER NO. 11
Frontier Bronze Corp.
Copper. Cu 88, Ni 5, Sn 5, Zn 2.
Cast: 70,000 TS; 50,000 YS; 15 El; 160 Brin. For bearings, gears, screws, bolts; age hardened; tough.

FRONTIER NO. 12
Frontier Bronze Corp.
Sn 10, Pb 2, P 0.25, bal Cu.
Cast: 32,000-40,000 TS; 19,000-23,000 YS; 15-5 El; 60-75 Brin. For work gears, bearings; free-cutting. *Obsolete*

FRONTIER NO. 14
Frontier Bronze Corp.
Copper. Sn 10, Pb 5, P 0.25, bal Cu.
Cast: 30,000-40,000 TS; 19,000-22,000 YS; 5-15 El; 50-61 Brin. For bearings against soft steel; for moderate speeds.

FRONTIER NO. 15Y
Frontier Bronze Corp.
Copper. Cu 79.75, Sn 10, Pb 10, P 0.25.
28,000-35,000 TS; 19,000-22,000 YS; 5-10 El; 52-70 Brin. For small machine tool bearings; Izod 3-8.

FRONTIER NO. 18
Frontier Bronze Corp.
Copper. Cu 85, Sn 5, Pb 5, Zn 5.
27,000-33,000 TS; 15,000-20,000 YS; 15-20 El; 50-70 Brin. For general utility bronze; not recommended for bearings. Izod 13-18.

FRONTIER NO. 20
Frontier Bronze Corp.
Copper. Cu 77.75, Sn 7, Pb 15, P 0.25.
27,000-31,000 TS; 17,000-18,000 YS; 12-18 El; 43-63 Brin. For bearings, acid resisting alloy in mining machinery; for high speeds; Izod 6-8.

FRONTIER NO. 28
Frontier Bronze Corp.
Copper. Cu 99.6-99.9.
17,000-20,000 TS; 6000-9000 YS; 40-50 El; 40-50 Brin. For high electrical conductivity bearings.

FRONTIER NO. 29
Frontier Bronze Corp.
Copper. Sn 0.5, Zn 41, Fe 1, Al 1, Mn 0.5, bal Cu.
Cast: 65,000-85,000 TS; 25,000-35,000 YS; 35-20 El; 104-119 Brin. For propeller blades and hubs, valves; manganese bronze.

FRONTIER NO. 3
Frontier Bronze Corp.
Copper. Cu 88.5, Sn 11, Pb 0.25, P 0.25.
35,000-40,000 TS; 20,000-25,000 YS; 6-10 El; 67-85 Brin. For worm gears, bearings.

FRONTIER NO. 30
Frontier Bronze Corp.
Copper. Zn 20, Fe 2.5, Al 6.5, Mn 2.5, bal Cu.
Cast: 90,000-110,000 TS; 50,000-65,000 YS; 10-15 El; 196-220 Brin. For propeller blades, gears; manganese bronze, high strength.

FRONTIER NO. 32
Frontier Bronze Corp.
Al 82, Cu 8.
16,000-22,000 TS; 11,000-13,000 YS; 1-2 El; 50-55 Brin. For general castings. *Obsolete*

FRONTIER NO. 33
Frontier Bronze Corp.
Al 82, Zn 15, Cu 3.
18,000-25,000 TS; 12,000-15,000 YS; 1-3 El; 60-65 Brin. For free machining Al castings; IZ-1-3. *Obsolete*

FRONTIER NO. 38
Frontier Bronze Corp.
Copper. Cu 67, Sn 8, Pb 24, Zn 1.
Cast: 22,500-27,500 TS; 14,500-19,000 YS; 10-15 El; 45-55 Brin. For castings, bearings; heavy duty.

FRONTIER NO. 39
Frontier Bronze Corp.
Sn, bal Cu.
Cast: 45,000 TS; 28,000 YS; 6-10 El; 100-120 Brin. For castings. *Obsolete*

FRONTIER NO. 40
Frontier Bronze Corp.
Ti 0.15, Cr 0.5, Zn 7-8, Mg 0.6, bal Al.
Cast: 30,000-38,000 TS; 28,000-32,000 YS; 1-5 El; 69-73 Brin. For light alloy castings. *Obsolete*

FRONTIER NO. 40X

Frontier Bronze Corp.
Al alloy.
For casting. *Obsolete*

FRONTIER NO. 5

Frontier Bronze Corp.
Copper. Al 10, Cu 89, Fe 1, trace Ti.
Untreated: 60,000-80,000 TS; 22,000-28,000 YS; 30-15 El;
120-130 Brin. Heat treated: 80,000-95,000 TS; 50,000-65,000
YS; 4-10 El; 180-200 Brin. For housing nuts, worm gears,
spur gears, bearings, spacer boxes; wear and shock resistant;
heat resistant.

FRONTIER NO. 6

Frontier Bronze Corp.
Copper. Cu 86, Sn 9.5, Pb 2.5, Zn 2.
30,000-45,000 TS; 18,000-20,000 YS; 15-20 El; 55-70 Brin.
For general utility bearing bronze; Izod 8-10.

FRONTIER NO. 7

Frontier Bronze Corp.
Copper. Cu 79.5, Al 11, Fe 6, Mn 3.5.
70,000-80,000 TS; 35,000-45,000 YS; 3-5 El; 150-170 Brin.
For paper mill machinery and valve seats; Izod 3-5.

FRONTIER NO. 8

Frontier Bronze Corp.
Copper. Cu 83.5, Sn 10, Pb 2.5, Ni 3.5, P 0.5.
35,000-45,000 TS; 25,000-28,000 YS; 10-15 El; 80-93 Brin.
For heavy duty bearings, worm gears, lead screws; high
strength bearing; Izod 3-5.

FRONTIER NO. 88

Frontier Bronze Corp.
Al alloy.
For light alloy castings. *Obsolete*

FRONTIER NO. 9

Frontier Bronze Corp.
Copper. Cu 90, Sn 10.
30,000-35,000 TS; 15,000-20,000 YS; 10-15 El; 63-70 Brin.
For acid resisting work; Izod 3-10.

FRONTIER NO.37

Frontier Bronze Corp.
Cu 10, Zn 86, Al 4.
30,000-35,000 TS; 105-119 Brin. For electric motor and
machine tool bearings. *Obsolete*

FS

Now FINKL FS.

FS 85

Now FANSTEEL FS 85.

FS M-10

Teledyne Firth Sterling
Tool steel. C 0.85, Cr 4.25, Mo 8.25, V 2, bal Fe.
For drills, taps, dies, chasers, broaches; high speed steel.
Obsolete

FSX-414

Cannon-Muskegon Corp.
C 0.25, Mn 0-1, Si 0-1, Cr 29.5, Ni 10.5, W 7, B 0.012, Fe 0-2,
bal Co.
Cast alloy for gas turbine vanes.

FSX-418

Cannon-Muskegon Corp.
C 0.25, Mn 0-1, Si 0-1, Cr 29.5, Ni 10.5, W 7, B 0.012, Fe 0-2,
Y 0.15, bal Co.
Cast alloy for gas turbine vanes; improved oxidation
resistance.

FSX-430

Cannon-Muskegon Corp.
C 0.4, Cr 29.5, Ni 10, W 7.5, B 0.027, Zr 0.9, Y 0.5, bal Co.
Cast alloy for gas turbine vanes.

FT-2

Howment Corp. (Carbide Div.)
Sintered carbide tool material. For heavy roughing and
interrupted cuts on carbon and alloy steels. *Obsolete*

FT-21

Howment Corp. (Carbide Div.)
Sintered carbide tool material. For heavy roughing and
interrupted cuts on carbon and alloy steels; for gauge blocks.
Obsolete

FT-3

Howment Corp. (Carbide Div.)
Sintered carbide tool materials. For heavy roughing or
interrupted cuts on carbon and alloy steels. *Obsolete*

FT-35

Howment Corp. (Carbide Div.)
Sintered carbide tool material. For heavy duty milling and
interrupted cuts on carbon and alloy steels. *Obsolete*

FT-4

Howment Corp. (Carbide Div.)
Sintered carbide tool materials. For general machine shop
cutting of carbon and alloy steels. *Obsolete*

FT-42

Howment Corp. (Carbide Div.)
Sintered carbide tool material. For general purpose milling of
carbon and alloy steels. *Obsolete*

FT-5

Howment Corp. (Carbide Div.)
Sintered carbide tool material. For general purpose
machining of carbon and alloy steels including heavy duty,
interrupted cuts and elevated temperatures. *Obsolete*

FT-6

Howment Corp. (Carbide Div.)
Sintered carbide tool material. For finish cutting carbon and
alloy steels. *Obsolete*

FT-62

Howment Corp. (Carbide Div.)
Sintered carbide tool material. For general purpose cutting of
carbon and alloy steels; wide range. *Obsolete*

FT-7

Howment Corp. (Carbide Div.)
Sintered carbide tool material. For light finishing and
precision boring of carbon and alloy steels. *Obsolete*

FT-8

Howment Corp. (Carbide Div.)
Sintered carbide tool material. For cutting hot flash trim,
special purpose wire drawing. *Obsolete*

FT-9

Howment Corp. (Carbide Div.)
Sintered carbide tool material. For heavy duty machining at
elevated temperatures. *Obsolete*

FUCHS 2000

Otto Fuchs Metallwerke
Copper. 99.90 Cu (contains O_2).
Good electrical conductivity and formability. For electrical
parts as sockets. Type E-Cu.

FUCHS 2020

Otto Fuchs Metallwerke
Copper. Ag 0.025-0.25, bal Cu.
For armature windings and commutator segments.

FUCHS 2050

Otto Fuchs Metallwerke
Copper. Te 0.4-1.1, bal Cu.
High electrical conductivity and good machinability.

FUCHS 2061

Otto Fuchs Metallwerke
Copper. Pb 0.8-1.2, bal Cu.
Corrosion resistant with good machinability.

FUCHS 2155

Otto Fuchs Metallwerke
Cu 53.5-56, Pb 1-2.5, bal Zn.
For extrusion, hot forming, thin walled sections. Type
CuZn44Pb2; similar to CDA 380.

FUCHS 2156

Otto Fuchs Metallwerke
Cu 53.5-56, Pb 1-2.5, Al 0.5, bal Zn.
For hot formability. Type CuZn44Pb2; similar to CDA 380.

FUCHS 2157

Otto Fuchs Metallwerke
Cu 57.5-59, bal Zn.
For extrusions, hot forming, or cold bending. Type CuZn42.

FUCHS 2158

Otto Fuchs Metallwerke
Cu 57.5-59, Pb 2.5-3.3, bal Zn.
For hot pressing and forming, good machinability. Type
CuZn39Pb3; similar to CDA 385, Architectural Bronze.

FUCHS 2159

Otto Fuchs Metallwerke
Cu 57.5-59, Pb 2.5-3, bal Zn.
Processed to give improved toughness. Type CuZn39Pb3;
similar to CDA 385.

FUCHS 2160

Otto Fuchs Metallwerke
Cu 59-61.5, bal Zn.
Muntz metal; for hot or cold forming, riveting. Type CuZn40;
similar to CDA 280.

FUCHS 2161

Otto Fuchs Metallwerke
Cu 59.5-61.5, Pb 0.5-2, bal Zn.
For hot or cold forming; lock brass. Type CuZn38Pb1; similar
to CDA 377.

FUCHS 2163

Otto Fuchs Metallwerke
Cu 62-64, bal Zn.
For cold forming, deep drawing; costume jewelry, holloware,
zip fasteners. Type CuZn37; CDA 274. 63% yellow brass.

FUCHS 2167

Otto Fuchs Metallwerke
Cu 66-68.5, bal Zn.
Very good cold formability; for radiator strips, tube rivets, wire
netting. Type CuZn33; CDA 268. 66% yellow brass.

FUCHS 2170

Otto Fuchs Metallwerke
Cu 69-71, bal Zn.
Excellent cold formability, deep drawing and solderability; for
instruments, tubes, bushings, cartridge cases. Type CuZn30;
CDA 260; cartridge brass.

FUCHS 2171

Otto Fuchs Metallwerke
Cu 70-72.5, Sn 0.9-1.3, bal Zn.
Admiralty bronze, corrosion and dezincification resistant, for
tubes, condenser and heat exchanger plates. Type
CuZn28Sn; CDA 442.

FUCHS 2180

Otto Fuchs Metallwerke
Cu 79-81, bal Zn.
Cold formability for electrical parts and installations, flexible
hose. Type CuZn20; 80.0 low brass.

514 / WOLDMAN'S ENGINEERING ALLOYS

FUCHS 2181

Otto Fuchs Metallwerke
Cu 79-83, Sn 0.6-1.2, bal Zn.
Corrosion and dezincification resistant trumpet brass for tubes, and musical instruments such as horns. Type CuZn20Sn; CDA 435.

FUCHS 2185

Otto Fuchs Metallwerke
Cu 84-86, bal Zn.
85% red brass for cold formability such as spinning and embossing for jewelry, and bushings for spring assembly. Type CuZn15; CDA 230.

FUCHS 2190

Otto Fuchs Metallwerke
Cu 89-91, bal Zn.
10% commercial bronze with cold formability for electrical components and tubes. Type CuZn10; CDA 230.

FUCHS 2195

Henning Bros. & Smith Inc.
Copper. Cu 95-96, bal Zn.
For handicraft articles, tubes. Type CuZn 5.

FUCHS 2201

Otto Fuchs Metallwerke
Cu 57-59, Pb 1-2, Mn 0.4-1.8, bal Zn.
Good hot formability and machinability. For hot pressing, valves. Type CuZn40MnPb.

FUCHS 2202

Otto Fuchs Metallwerke
Cu 58-61, Mn 1.5-2.5, Al 0.3-1.5, Ni 2-3, bal Zn.
Good strength for structural applications such as appliances, marine engineering. Type CuZn35Ni.

FUCHS 2203

Otto Fuchs Metallwerke
Cu 56.5-59.5, Mn 0.4-1.8, Al 0.4-1.6, bal Zn.
Weather resistant for architecture and appliances. Type CuZn40Al1.

FUCHS 2204

Otto Fuchs Metallwerke
Cu 57-59, Mn 1-2.5, bal Zn.
Weather resistant, can be soldered and brazed for appliances, architectural sections, and handrails. Type CuZn40Mn.

FUCHS 2205

Otto Fuchs Metallwerke
Cu 55.5-59, Pb 0-0.8, Fe 0-1, Sn 0.05, Mn 1-2.4, Al 1.3-2, Si 0-0.8, Ni 0-2, bal Zn.
High strength bronze; weather resistant for structural components. Type CuZn40Al2.

FUCHS 2206

Otto Fuchs Metallwerke
Cu 55.5-59, Pb 0-0.8, Fe 0-1, Sn 0-0.5, Mn 1-2.4, Al 1.3-2.3, Si 0-0.8, Ni 0-2, bal Zn.
Good strength and bearing properties for worm gears, sprockets. Type CuZn40Al2.

FUCHS 2210 THROUGH 2214

Now FUCHS 2206.

FUCHS 2264

Otto Fuchs Metallwerke
Cu 61-66, Fe 0.5-3.5, Mn 2-5, Al 2.5-7.5, Si 0-0.5, Ni 0-0.5, bal Zn.
Strong, wear resistant alloy. Similar to CDA 670, manganese bronze B.

FUCHS 2356

Otto Fuchs Metallwerke
Cu 56-58, Pb 1.5-2.5, bal Zn.
For extruded sections. Type CuZn41Pb2; similar to CDA 380.

FUCHS 2357

Otto Fuchs Metallwerke
Cu 57.5-59, Pb 1.5-2.5, bal Zn.
Good hot formability and punchability. Type CuZn40Pb2; similar to CDA 380.

FUCHS 2358

Otto Fuchs Metallwerke
Cu 57.5-59, Pb 2.5-3.3, bal Zn.
For free cutting applications. Type CuZn39Pb3; similar to CDA 385. Architectural bronze.

FUCHS 2360

Otto Fuchs Metallwerke
Pb 1.5-2.5, Cu 59.5-61.5, bal Zn.
Forging brass for hot pressed, punched, and machined parts. Type CuZn38Pb 2; similar to CDA 377.

FUCHS 2361

Otto Fuchs Metallwerke
Cu 59.5-61.5, Pb 1.5-2.5, bal Zn.
For hot or cold forming, threaded parts, screws. Type CuZn37Pb2; similar to CDA 377.

FUCHS 2362

Otto Fuchs Metallwerke
Cu 60-62, Pb 2.5-3.5, bal Zn.
Free cutting brass for screws, and machined parts; hot or cold formed. Type CuZn36Pb3; similar to CDA 360.

FUCHS 2363

Otto Fuchs Metallwerke
Cu 62-63.5, Pb 2-3, bal Zn.
High leaded brass, cold forming and machinable; for machined parts, tubes. Type CuZn35Pb2; CDA 256.

FUCHS 2458

Otto Fuchs Metallwerke
Cu 57.5-59, Pb 2.5-3.3, bal Zn.
For pressure tight fittings, valve sections. Type CuZn39Pb3; similar to CDA 385.

FUCHS 2560

Otto Fuchs Metallwerke
Cu 58-62, Sn 0-0.5, Si 0.1-0.5, bal Zn.
Good formability for drawn wire, welding wire.

FUCHS A1

Otto Fuchs Metallwerke
Aluminum.
Unalloyed aluminum. 60-130 N/mm^2 TS; 20-110 N/mm^2 YS; 5-27 El. Type Al99.8; similar to AA 1080.

FUCHS A2

Otto Fuchs Metallwerke
Aluminum.
Unalloyed aluminum. 60-130 N/mm^2 TS; 20-110 N/mm^2 YS; 5-27 El. Type Al99.7; similar to AA 1070.

FUCHS ACID RESISTING ALLOY

English manufacture
Ag 73-80, Ni 13-15, Au 13.5-15.
For chemical equipment, corrosion resisting parts; acid resistant.

FUCHS ACID RESISTING GOLD ALLOY

English manufacture
Au 75, Ni 10-15, W 10-15.
For chemical equipment; acid resistant.

FUCHS AM, AG, AK, AS, AZ

Now AM, AG, AK, AS, AZ.

FUCHS E

Otto Fuchs Metallwerke
Aluminum. 99.9 Al.
Unalloyed aluminium. Type Al99.9; similar to AA 1090.

FUCHS E 05

Otto Fuchs Metallwerke
Aluminum. Mg 0.5, bal Al.
Good formability and corrosion resistance. Type Al99.9Mg0.5.

FUCHS E 2

Otto Fuchs Metallwerke
Aluminum. Mg 1.8, bal Al.
Good formability and corrosion resistance. Type Al99.9Mg2.

FUCHS E1

Otto Fuchs Metallwerke
Aluminum. Mg 1, bal Al.
Good formability and corrosion resistance. Type Al99.9Mg1; similar to AA 5657.

FUCHS ES 70

Otto Fuchs Metallwerke
Aluminum. Cu 0.2, Mg 0.65, Si 0.65, bal Al.
130-235 N/mm^2 TS; 80-155 N/mm^2 YS; 14-17 El. Good formability, surface finish. Type Al99.7MgSiCu.

FUCHS ES 90

Otto Fuchs Metallwerke
Aluminum. Mg 0.6, Si 0.5, bal Al.
130-235 N/mm^2 TS; 80-155 N/mm^2 YS; 14-17 El. Good formability, surface finish. Type Al99.9MgSi.

FUCHS M1

Otto Fuchs Metallwerke
Magnesium.
Unalloyed magnesium. Good hot formability and weldability. Type H Mg99.8; similar to ASTM B92-45.

FUCHS MA3

Gottingen Aluminiumwerke GmbH
Al 3, Zn 1, Mn 0.15, bal Mg.
25 kp/mm^2 TS; 16 kp/mm^2 YS; 10 El. Medium strength, good hot formability, ductility, weldability and machinability. Type Mg Al 3 Zn; Bs 3371 MAG-T-111 M. (AZ 31 B).

FUCHS MA39

Otto Fuchs Metallwerke
Magnesium. Al 3, Zn 1, Mn 0.35, bal Mg.
25 kp/mm^2 TS; 16 kp/mm^2 YS; 10 El. Similar to MA 3 but with lower iron content. Type MgAl3Zn. Meets ASTM B107. (AZ 31).

FUCHS MA6

Otto Fuchs Metallwerke
Magnesium. Al 6.3, Zn 1, Mn 0.15, bal Mg.
26-28 kp/mm^2 TS; 18-20 kp/mm^2 YS; 8-10 El. High strength alloy, medium weldability and machinability. Type MgAl6Zn; ASTM B107. (AZ 61A).

FUCHS MA74

Otto Fuchs Metallwerke
Magnesium. Al 7.2, Zn 1.2, Mn 0.3, bal Mg.
28-32 kp/mm^2 TS; 20-23 kp/mm^2 YS; 8-10 El. Heat treatable, high strength alloy, good machinability; limited weldability. Type MgAl7Zn.

FUCHS MA8

Otto Fuchs Metallwerke
Magnesium. Al 8, Zn 0.5, Mn 0.2, bal Mg.
28-32 kp/mm^2 TS; 20-23 kp/mm^2 YS; 6-10 El. Heat treatable, high strength, limited weldability. Type MgAl8Zn. Meets AZ 80A.

FUCHS MG2

Otto Fuchs Metallwerke
Magnesium. Mn 1.2-2, bal Mg.
20-23 kp/mm^2 TS; 15-17 kp/mm^2 YS; 2 El. Hot formability and corrosion resistance. Type MgMn2; meets ASTM B107.

FUCHS MZ64
Otto Fuchs Metallwerke
Magnesium. Zn 5.5, Zr 0.6, bal Mg.
28-32 kp/mm^2 TS; 18-25 kp/mm^2 YS; 4-7 El. Heat treatable, high strength, limited weldability. Type MgZn6Zr. Meets ZK 60A.

FUCHS R
Otto Fuchs Metallwerke
Aluminum. 99.99 Al.
High purity aluminum. Type Al 99.99R; AFNOR A9.

FUCHS R 05
Otto Fuchs Metallwerke
Aluminum. Mg 0.5, bal Al.
Good formability, corrosion resistance. Type AlRMg0.5.

FUCHS R 1
Otto Fuchs Metallwerke
Aluminum. Mg 1, bal Al.
Good formability and corrosion resistance. Type AlHMg1; AFNOR A 9-G 1.

FUCHS T 2
Otto Fuchs Metallwerke
Titanium. Fe 0-0.2, C 0.08, Ti 99.5, 0.10 O$_2$, 0.0125 H$_2$ max, 0.05 N$_2$.
30-42 kp/mm^2 TS; 20 kp/mm^2 YS min; 25 El min. Corrosion resistant with good formability and weldability. Grade Ti 99.5.

FUCHS T 3
Otto Fuchs Metallwerke
Titanium. C 0.08, Ti 99.4, Fe 0-0.25, 0.20 O$_2$, 0.0125 H$_2$ max, 0.06 N$_2$.
40-55 kp/mm^2 TS; 28 kp/mm^2 YS min; 20 El min. Similar to FUCHS T2. Grade Ti 99.4; meets AMS 4902; DTD 5003 B.

FUCHS T 6
Otto Fuchs Metallwerke
Titanium. Fe 0-0.35, C 0.1, Ti 99.2, 0.30 O$_2$, 0.0125 H$_2$ max, 0.07 N$_2$.
55-75 kp/mm^2 TS; 45 kp/mm^2 YS min; 15 El min. Moderate weldability; high strength. Grade Ti 99.2; meets AMS 4921 A.

FUCHS TA 44
Otto Fuchs Metallwerke
Titanium. Al 3-5, Mo 3-5, Sn 1.5-2.5, Si 0.3-0.7, Fe 0-0.2, 0.0150 H$_2$ max, bal Ti.
Heat treated: 107-130 kp/mm^2 TS; 92 kp/mm^2 YS min; 9 El min. Used in British aircraft projects. Grade TiAl4Mo4Sn2; DTD 5153.

FUCHS TA 52
Otto Fuchs Metallwerke
Al 4-6, Sn 2-3, Fe 0-0.25, C 0.08, 0.20 O$_2$, 0.020 H$_2$ max, 0.07 N$_2$, bal Ti.
80 kp/mm^2 TS min; 77 kp/mm^2 YS min; 10 El min. Weldable, good elevated temperature strength. Grade TiAl5Sn2.5; meets AMS 4966 B; DTD 5083.

FUCHS TA 64
Otto Fuchs Metallwerke
Al 5.75-6.75, V 3.5-4.5, Fe 0-0.25, C 0.08, 0.20 O$_2$, 0.125 H$_2$ max, 0.07 N$_2$, bal Ti.
90 kp/mm^2 TS min; 84 kp/mm^2 YS min; 10 El min. Forgeable, heat treatable titanium alloy with limited weldability. Hardenable to about 110 kp/mm^2 TS. Grade TiAl6V4; meets AMS 4928 E; DTD 5173.

FUCHS TA 66
Otto Fuchs Metallwerke
Titanium. Al 5-6, V 5-6, Sn 1.5-2.5, Cu 0.35-1, Fe 0.35-1, C 0.05, 0.20 O$_2$, 0.015 H$_2$ max, 0.04 N$_2$, bal Ti.
Heat treated: 112-126 kp/mm^2 TS; 98-119 kp/mm^2 YS; 5 El min. Titanium alloy with good forgeability. Grade TiAl6V6Sn2; meets AMS 4971.

FUCHS TA 74
Otto Fuchs Metallwerke
Titanium. Al 6.5-7.3, Mo 3.5-4.5, Fe 0-0.25, C 0.08, 0.20 O$_2$, 0.0125 H$_2$ max, 0.07 N$_2$, bal Ti.
Heat treated: 105-119 kp/mm^2 TS; 98-117 kp/mm^2 YS; 6 El min. For heavy sections requiring heat treatment. Grade TiAl7Mo4; meets AMS 4970 A.

FUCHS TC 2
Otto Fuchs Metallwerke
Titanium. Cu 2-3, Fe 0-0.2, C 0.1, 0.20 O$_2$, 0.010 H$_2$ max, 0.05 N$_2$, bal Ti.
55 kp/mm^2 TS min; 39 kp/mm^2 YS min; 16 El min. Weldable, heat treatable to about 80 kp/mm^2 TS. Grade TiCu2; DTD 5123.

FUCHS TP 02
Otto Fuchs Metallwerke
Titanium. Pd 0.2, Fe 0-0.25, Ti 99.4, 0.20 O$_2$, 0.0125 H$_2$ max, 0.08-0.06 N$_2$.
45 kp/mm^2 TS min; 35 kp/mm^2 YS min; 16 El min. Improved resistance to hydrochloric acid and other reducing acids. Grade TiPd0.15.

FUCHSAL
Otto Fuchs Metallwerke
Mg 0.6-1.4, Si 0.6-1.6, Mn 0.6-1, Cr 0-0.3, bal Al.
Annealed: 21,000 TS; 8,000 YS; 24 El. For window frames, gutters, fan blades, boats; good forming and welding properties. *Obsolete*

FUCHSDUR
Otto Fuchs Metallwerke
Cu 2.5-5, Mg 0.2-1.8, Mn 0.3-1.5, bal Al.
Annealed: 27,000 TS; 11,000 YS; 22 El; 47 Brin. Heat treated: 72,000 TS; 57,000 YS; 130 Brin. For aircraft structures and fittings, fasteners; age-hardenable. *Obsolete*

FUCHSMAN
Otto Fuchs Metallwerke
Mn 1-1.5, Cr 0-0.3, bal Al.
Soft: 16,000 TS; 6,000 YS; 40 El. Hard: 29,000 TS; 27,000 YS; 10 El. For cooking utensils, heat exchangers, tanks, furniture; good forming and welding characteristics. *Obsolete*

FUEGO
Hidalgo Steel Co. Inc.
C 0.5, Si 0.9, Mn 0.5, Cr, V, bal Fe.
For punches, shears, chisels; water hardened.

FUGI-HIZ
Fugi Iron & Steel Co., Ltd.
C 0.1-0.18, Mn 0.6-1, Cu 0.15-0.5, Ni 0.7-1, V 0.03-0.1, B 0.002-0.006, Cr 0.4-0.8, Mo 0.4-0.6, bal Fe.
Heat treated: 114,000-135,000 TS; 100,000 YP min; 20.0 El min. For car and railroad bodies, agricultural equipment. High strength-low alloy constructional steel.

FUJI-FTW 42
Fugi Iron & Steel Co., Ltd.
C 0.18, Mn 0-1.5, Si 0-0.55, bal Fe.
Rolled: 74,000-88,000 TS; 50,000 YP min; 20 El min. For building structures, derricks, booms, bridges. Structural low carbon steel.

FUJI-FTW 58
Fugi Iron & Steel Co., Ltd.
C 0.18, Mn 0-1.5, Si 0-0.55, bal Fe.
Heat treated: 82,000-97,000 TS; 65,000 YP min; 16 El min. For mine cars, bus bodies, booms, derricks, bridges. Structural low carbon steel.

FULLER ALLOY
Fuller & Dasche Co.
Zn 6, Mg 1.2, Fe 2, bal Al.
Heat treated: 43,000-48,000 TS; 0-2 El; 110-115 Brin. For light alloy castings; heat treatable.

FULTALLOY
Fulton Iron Works Co.
C 0.5, C, bal Fe.
For castings. *Obsolete*

FULTON EGR
Fulton Gold Refineries Corp.
Ag 56, Cu 22, Zn 18, Sn 4.
For silver solder, brazing; MP 1145-1205°F.

FULTON NO. 110-A
Fulton Gold Refineries Corp.
Ag 5, Cu 88.5, P 6.5.
For silver solder, self-fluxing; MP 1185-1300°F.

FULTON NO. 111
Fulton Gold Refineries Corp.
Ag 9, Cu 51, Zn 40.
For silver solder; MP 1510°F.

FULTON NO. 111A
Fulton Gold Refineries Corp.
Ag 15, Cu 80, P 5.
For silver solder; MP 1185°F.

FULTON NO. 112
Fulton Gold Refineries Corp.
Ag 20, Cu 45, Zn 30, Cd 5.
For silver solder, brazing; MP 1140-1410°F.

FULTON NO. 112-A
Fulton Gold Refineries Corp.
Ag 25, Cu 52.5, Zn 22.5.
For silver solder; MP 1500-1575°F.

FULTON NO. 113
Fulton Gold Refineries Corp.
Ag 30, Cu 38, Zn 32.
For silver solder; MP 1370°F.

FULTON NO. 113-A
Fulton Gold Refineries Corp.
Ag 35, Cu 26, Zn 21, Cd 18.
For silver solder, brazing; MP 1125-1295°F.

FULTON NO. 114
Fulton Gold Refineries Corp.
Ag 40, Cu 30, Zn 28, Ni 2.
For silver solder, brazing; MP 1240°F.

FULTON NO. 114-A
Fulton Gold Refineries Corp.
Ag 45, Cu 30, Zn 25.
For silver solder; MP 1250°F.

FULTON NO. 114-AN
Fulton Gold Refineries Corp.
Ag 45, Cu 15, Zn 16, Cd 24.
For silver solder; MP 1120°F.

FULTON NO. 114-B
Fulton Gold Refineries Corp.
Ag 40, Cu 35, Zn 25, Ni 5.
For silver solder; MP 1240-1560°F.

FULTON NO. 114-N
Fulton Gold Refineries Corp.
Ag 40, Cu 18, Zn 15, Cd 27.
For silver solder; MP 1076°F.

FULTON NO. 115
Fulton Gold Refineries Corp.
Ag 50, Cu 15.5, Zn 16.5, Cd 18.
For silver solder; MP 1160°F.

FULTON NO. 115-4
Fulton Gold Refineries Corp.
Ag 54, Cu 40, Zn 5, Ni 1.
For silver solder; MP 1325-1575°F.

FULTON NO. 117-2
Fulton Gold Refineries Corp.
Ag 72, Cu 28.
For silver solder; MP 1435-1435°F.

FULTON NO. 118-A
Fulton Gold Refineries Corp.
Ag 85, Mn 15.
For silver solder; MP 1760-1778°F.

FULTON NO. 216
Fulton Gold Refineries Corp.
Ag 60, Cu 25, Zn 15.
For silver solder; MP 1260°F.

FULTON NO. 216-A
Fulton Gold Refineries Corp.
Ag 65, Cu 20, Zn 15.
For silver solder; MP 1280-1325°F.

FULTON NO. 217
Fulton Gold Refineries Corp.
Ag 70, Cu 20, Zn 10.
For silver solder; MP 1335°F.

FULTON NO. 218
Fulton Gold Refineries Corp.
Ag 80, Cu 16, Zn 4.
For silver solder; MP 1360°F.

FULTON NO. A-114
Fulton Gold Refineries Corp.
Ag 40, Cu 36, Zn 24.
For silver solder; MP 1330°F.

FULTON NO. A-115
Fulton Gold Refineries Corp.
Ag 50, Cu 34, Zn 16.
For silver solder; MP 1275°F.

FULTON NO. G-112
Fulton Gold Refineries Corp.
Ag 20, Cu 45, Zn 35.
For silver solder; MP 1430-1500°F.

FULTON NO. G4-115
Fulton Gold Refineries Corp.
Ag 50, Cu 15.5, Zn 16.5, Cd 18.
For silver solder; MP 1160-1175°F.

FULTON NO. G5-115
Fulton Gold Refineries Corp.
Ag 50, Cu 15.5, Zn 15.5, Cd 16, Ni 3.
For silver solder; MP 1195-1270°F.

FULTON NO. LA-115
Fulton Gold Refineries Corp.
Ag 50, Cu 28, Zn 22.
For silver solder; MP 1250-1340°F.

FURBALOI
English manufacture
Cu 0.45, Mn 0.05, C 0.13, Cr 13.35, Ni 0.08, bal Fe.
77,000 TS; 38 El; 75 RA. For corrosion resisting parts.

FURIOUS
Osborn Steels Ltd.
C 1.25, W 4.5, Cr 1.2, V 0.25, bal Fe.
For blanking and heading dies, punches; oil hardened,
abrasion resistant. *Obsolete*

FURODIT "S"
Rochling Burbach GmbH
C 0.2, Cr 27, Mo 1, bal Fe.
52,700-64,000 TS; 18-30 El; 50-60 RA. For heat and corrosion
resisting parts; heat and corrosion resistant. *Obsolete*

FURODIT "SS"
Rochling Burbach GmbH
C 0.2, Cr 29, Mo 2, Ta 1-2, bal Fe.
57,000-71,000 TS; 16-30 El; 50-60 RA. For heat and corrosion
resisting parts; heat and corrosion resistant. *Obsolete*

FURODIT "Z"
Rochling Burbach GmbH
C 0.2, Cr 25, bal Fe.
52,700-64,000 TS; 18-30 El; 50-60 RA. For heat resisting
parts; heat resistant to 2010 F. *Obsolete*

FURODIT 10
Rochling Burbach GmbH
C 0.12, Si 2, Cr 18, bal Fe.
Annealed: 80,000 TS; 50,000 YS; 25 El; 50 RA. 150 Brin. For
oil refinery and food processing equipment, sinks, bolts,
burners; corrosion and heat resistant; Type 430. *Obsolete*

FURODIT 10G
Rochling Burbach GmbH
C 0.6, Si 1.5, Cr 22, bal Fe.
For oil refinery equipment, furnace parts, burners; heat
resistant. *Obsolete*

FURODIT 12G
Rochling Burbach GmbH
C 0.6, Si 1.5, Cr 29, bal Fe.
For oil refinery and chemical plant equipment, furnace parts;
Type CC-60; heat resistant. *Obsolete*

FURODIT 25
Rochling Burbach GmbH
C 0.2, Ni 6, Cr 25, bal Fe.
Cast: 85,000 TS; 48,000 YS; 16 El; 190 Brin. For salt pots,
furnace parts, grate bars, dampers, rabble arms and blades;
Type HD; heat resistant. *Obsolete*

FURODIT 30
Rochling Burbach GmbH
C 0-0.12, Si 1.5, Al 1.5, Cr 24, bal Fe.
Annealed: 85,000 TS; 50,000 YS; 30 El; 55 RA. 180 Brin. For
oil refinery equipment, heat exchangers; heat and creep
resistant. *Obsolete*

FURODIT 7G
Rochling Burbach GmbH
C 0.3, Si 2.2, Cr 6, bal Fe.
For oil refinery equipment; creep and heat resistant.
Obsolete

FURODIT 8G
Rochling Burbach GmbH
C 0.3, Si 2.3, Cr 13, bal Fe.
Annealed: 95,000 TS; 50,000 YS; 25 El; 55 RA. 195 Brin. For
oil refinery equipment, valves; creep and heat resistant.
Obsolete

FURODIT 9G
Rochling Burbach GmbH
C 0.5, Si 1.5, Cr 17, bal Fe.
For furnace parts, conveyors, heat treating boxes, retorts;
corrosion and heat resistant. *Obsolete*

FURODIT N6
Rochling Burbach GmbH
C 0.15, Si 1.5, Cr 19, Mo 2, Ni 5, bal Fe.
Annealed: 90,000 TS; 40,000 YS; 45 El; 60 RA. 170 Brin. For
acid resistant, chemical plant equipment; stainless. *Obsolete*

FURODIT NH11
Rochling Burbach GmbH
C 0.15, Si 2, Cr 19.5, Ni 9.5, bal Fe.
Annealed: 80,000 TS; 35,000 YS; 55 El; 75 RA. 150 Brin. For
chemical plant equipment, tanks, mixers, filters; Type 302;
stainless, austenitic. *Obsolete*

FURODIT NH12
Rochling Burbach GmbH
C 0.15, Si 2, Cr 24, Ni 19, bal Fe.
Annealed: 100,000 TS; 45,000 YS; 50 El; 65 RA. 185 Brin. For
furnaces, valves, pumps, turbine and jet components; Type
310; stainless, austenitic. *Obsolete*

FURODIT NO. 8
Rochling Burbach GmbH
C 0.2, Cr 6, Al 2, bal Fe.
49,800-57,000 TS; 27-35 El; 70 RA. For heat resisting parts;
heat resistant to 1470 F. *Obsolete*

FURODIT Z SPECIAL
Rochling Burbach GmbH
C 0.2, Cr 25, Ta 2, bal Fe.
Annealed: 57,000-68,500 TS; 15-25 El; 50-60 RA. For furnace
parts, heat treating boxes, pots; heat resistant to 2100 F.
Obsolete

FUSE-WELL NO. 28
Chicago Hardware Foundry Co.
C 3.2, Si 2, bal Fe.
For welding rod for cast iron; flux coated, gas welding.

FUSIBLE
English manufacture
Bi 45, Sn 17, Pb 30, 10.0 HgS.
For solders, binding plugs; melting point 85°C.

FUSIBLE ALLOY-1
American manufacture
Bi 50, Sn 34, Cd 17.
For fusible alloy; melting point 65°C.

FUSIBLE ALLOY-2
American manufacture
Bi 15, Sn 42, Pb 43.
For fusible alloy; low melting point.

FUSIBLE ALLOY-3
American manufacture
Bi 33.3, Sn 33.3, Pb 33.3.
For fusible alloy; low melting point.

FUSIBLE METAL "D"
Manufacturer not listed.
Bi 33.3, Sn 66.7.
For fusible alloy; M.P. 166°F.

FUSIBLE METAL "E"
Manufacturer not listed.
Bi 13, Sn 40, bal Pb.
For fuses, fire extinguishers; M.P. 172°F.

FUSIBLE METAL "F"
Manufacturer not listed.
Bi 12.5, Sn 39.5, bal Pb.
For fuses, fire extinguishers; M.P. 178°F.

FUSIBLE METAL "G"
Manufacturer not listed.
Bi 20, Sn 80.
For fuses, fire extinguishers; M.P. 200°F.

FUSION NO. 1000
Fusion Inc.
Ag 45, Cu 15, Zn 16, Cd 24, plus flux.
Solidus 1125°F; liquidus 1145°F. Paste type braze alloy.
AWS A5.8-69 BAg1.

FUSION NO. 1050
Fusion Inc.
Ag 50, Cu 15.5, Zn 16.5, Cd 18, plus flux.
Solidus 1160°F; liquidus 1175°F. Paste type braze alloy.
AWS A5.8-69 BAg1a; AMS-4770.

FUSION NO. 1100

Fusion Inc.
Ag 35, Cu 26, Zn 21, Cd 18, plus flux.
Solidus 1125°F; liquidus 1295°F. Paste type braze alloy.
AWS A5.8-69 BAg2.

FUSION NO. 1115

Fusion Inc.
Ag 60, Cu 30, Sn 10, plus flux.
Solidus 1095°F; liquidus 1325°F. Paste type braze alloy.
AWS A5.8-69 BAg 18.

FUSION NO. 1120

Fusion Inc.
Ag 30, Cu 27, Zn 23, Cd 20, plus flux.
Solidus 1125°F; liquidus 1310°F. Paste type braze alloy.
AWS A5.8-69 BAg Za.

FUSION NO. 1190

Fusion Inc.
Cu 75, P 7.25, Ag 17.75, plus flux.
Solidus 1190°F; liquidus 1190°F. Paste type braze alloy.

FUSION NO. 1200

Fusion Inc.
Ag 50, Cu 15.5, Zn 15.5, Cd 16, Ni 3, plus flux.
Solidus 1195°F; liquidus 1270°F. Paste type braze alloy.
AWS A5.8-69 BAg3.

FUSION NO. 1205

Fusion Inc.
Ag 56, Cu 22, Zn 17, Sn 5, plus flux.
Solidus 1152°F; liquidus 1203°F. Paste type braze alloy.
AWS A5.8-69 BAg 7.

FUSION NO. 1230

Fusion Inc.
Ag 60, Cu 25, Zn 15, plus flux.
Solidus 1260°F; liquidus 1325°F. Paste type braze alloy.

FUSION NO. 1235

Fusion Inc.
Ag 65, Cu 20, Zn 15, plus flux.
Solidus 1280°F; liquidus 1325°F. Paste type braze alloy.

FUSION NO. 1240

Fusion Inc.
Ag 40, Cu 30, Zn 28, Ni 2, plus flux.
Solidus 1220°F; liquidus 1435°F. Paste type braze alloy.
AWS A5.8-69 BAg 4.

FUSION NO. 1245

Fusion Inc.
Ag 40, Cu 30, Zn 25, Ni 5, plus flux.
Solidus 1260°F; liquidus 1550°F. Paste type braze alloy.

FUSION NO. 1250

Fusion Inc.
Ag 45, Cu 30, Zn 25, plus flux.
Solidus 1250°F; liquidus 1370°F. Paste type braze alloy.
AWS A5.8-69 BAg 5.

FUSION NO. 1260

Fusion Inc.
Ag 50, Cu 20, Zn 28, Ni 2, plus flux.
Solidus 1220°F; liquidus 1305°F. Paste type braze alloy.

FUSION NO. 1300

Fusion Inc.
Cu 92.75, P 7.25, plus flux.
Solidus 1310°F; liquidus 1456°F. Paste type braze alloy.
AWS A5.8-69 BCuP2.

FUSION NO. 1306

Fusion Inc.
Cu 86.75, P 7.25, Ag 6, plus flux.
Solidus 1190°F; liquidus 1320°F. Paste type braze alloy.
AWS A5.8-69 BCuP4.

FUSION NO. 1400

Fusion Inc.
Ag 72, Cu 28, plus flux.
Solidus 1435°F; liquidus 1435°F. Paste type braze alloy;
eutectic alloy. AWS A5.8-69 BAg8.

FUSION NO. 1440

Fusion Inc.
Cu 27.25, Zn 64.75, Sn 7.5, Pb 0.5, plus flux.
Solidus 1385°F; liquidus 1440°F. Paste type braze alloy.

FUSION NO. 1450

Fusion Inc.
Ag 50, Cu 34, Zn 16, plus flux.
Solidus 1260°F; liquidus 1410°F. Paste type braze alloy.
AWS A5.8-69 BAg6.

FUSION NO. 1565

Fusion Inc.
Cu 53, Ag 9, Zn 38, plus flux.
Solidus 1450°F; liquidus 1565°F. Paste type braze alloy.

FUSION NO. 1600

Fusion Inc.
Cu 51.5, Ag 4.5, Zn 44, plus flux.
Solidus 1410°F; liquidus 1635°F. Paste type braze alloy.

FUSION NO. 1610

Fusion Inc.
Ni 89, P 11, plus flux.
Solidus 1610°F; liquidus 1610°F. Paste type braze alloy.
AWS A5.8-69 BNi6.

FUSION NO. 1630

Fusion Inc.
Ni 77, Cr 12, P 10, plus flux.
Solidus 1630°F; liquidus 1630°F. Paste type braze alloy.
AWS A5.8-69 BNi7.

FUSION NO. 1650

Fusion Inc.
Cu 55, Zn 44.75, Mn 0.25, plus flux.
Solidus 1610°F; liquidus 1635°F. Paste type braze alloy.

FUSION NO. 1740

Fusion Inc.
Ag 54, Cu 21, Pd 25, plus flux.
Paste type braze alloy.

FUSION NO. 1742

Fusion Inc.
Au 82, Ni 18, plus flux.
Solidus 1740°F; liquidus 1740°F. Paste type braze alloy.
AWS A5.8-69 BAu4.

FUSION NO. 1761

Fusion Inc.
Ag 100, plus flux.
Solidus 1761°F; liquidus 1761°F. Paste type braze alloy;
pure silver.

FUSION NO. 1800

Fusion Inc.
Cu 80, Sn 20, plus flux.
Solidus 1470°F; liquidus 1635°F. Paste type braze alloy.

FUSION NO. 1830

Fusion Inc.
Cu 90, Sn 10, plus flux.
Solidus 1750°F; liquidus 1830°F. Paste type braze alloy.

FUSION NO. 1850

Fusion Inc.
100 Cu_2O, plus flux.
Solidus 2040°F; liquidus 2100°F. Paste type braze alloy.

FUSION NO. 1900

Fusion Inc.
Cu 100, plus flux.
Solidus 1980°F; liquidus 1980°F. Paste type braze alloy;
pure copper. AWS A5.8-69 BCu 1a.

FUSION NO. 1900-C

Fusion Inc.
Cu 90, 10 Cu_2O, plus flux.
Liquidus 1980°F. Paste type braze alloy.

FUSION NO. 1900-F

Fusion Inc.
Cu 95, 5 Fe_2O_3, plus flux.
Paste type braze alloy.

FUSION NO. 1900-FC

Fusion Inc.
Cu 90, 7 Cu_2O, 3 Fe_2O_3, plus flux.
Liquidus 1980°F. Paste type braze alloy; improved filleting.

FUSION NO. 2412

Fusion Inc.
Au 48, Ag 29, Cu 17, Zn 6, plus flux.
Solidus 1350°F; liquidus 1450°F. Paste type braze alloy; for
jewelry.

FUSION NO. 2460

Fusion Inc.
Au 42, Ag 32, Cu 16, Zn 10, plus flux.
Solidus 1335°F; liquidus 1380°F. Paste type braze alloy; for
jewelry.

FUSION NO. 2466

Fusion Inc.
Au 23, Ag 32, Cu 25, Zn 1, Cd 19, plus flux.
Solidus 1200°F; liquidus 1285°F. Paste type braze alloy; for
jewelry.

FUSION NO. 2468

Fusion Inc.
Au 29, Ag 31, Cu 20, Zn 1, Cd 19, plus flux.
Solidus 1280°F; liquidus 1400°F. Paste type braze alloy; for
jewelry.

FUSION NO. 24695

Fusion Inc.
Au 38, Ag 26, Cu 19, Zn 1, Cd 16, plus flux.
Solidus 1175°F; liquidus 1300°F. Paste type braze alloy; for
jewelry.

FUSION NO. 300

Fusion Inc.
Sn 43, Pb 43, Bi 14, plus flux.
Solidus 290°F; liquidus 310°F. Paste type solder for very low
temperature soldering.

FUSION NO. 360

Fusion Inc.
Sn 60, Pb 40, plus flux.
Solidus 361°F; liquidus 374°F. Paste type solder. ASTM
B32-70 At60 B.

FUSION NO. 361

Fusion Inc.
Sn 62.5, Pb 36, Ag 1.5, plus flux.
Solidus 350°F; liquidus 372°F. Paste type solder.

FUSION NO. 365

Fusion Inc.
Sn 63, Pb 37, plus flux.
Solidus 361°F; liquidus 361°F. Paste type solder, eutectic
alloy. ASTM B32-70AT 63B.

FUSION NO. 430

Fusion Inc.
Sn 96.5, Ag 3.5, plus flux.
Solidus 430°F; liquidus 430°F. Paste type solder; eutectic
alloy. ASTM B32-70AT 96.5 TS.

FUSION NO. 440

Fusion Inc.
Sn 45, Pb 55, plus flux.
Solidus 361°F; liquidus 441°F. Paste type solder. ASTM B 32-70AT 45B.

FUSION NO. 450

Fusion Inc.
Sn 50, Pb 50, plus flux.
Solidus 361°F; liquidus 421°F. Paste type solder. ASTM B32-70AT 50B.

FUSION NO. 455

Fusion Inc.
Sn 40, Pb 60, plus flux.
Solidus 360°F; liquidus 460°F. Paste type solder. ASTM B32-70AT 40B.

FUSION NO. 460

Fusion Inc.
Sn 95, Sb 5, plus flux.
Solidus 452°F; liquidus 464°F. Paste type solder. ASTM B32-70AT 95 TA.

FUSION NO. 470

Fusion Inc.
Sn 30, Pb 70, plus flux.
Solidus 361°F; liquidus 491°F. Paste type solder. ASTM B32-70AT 30B.

FUSION NO. 4765

Fusion Inc.
Ag 56, Cu 42, Ni 2, plus flux.
Solidus 1420°F; liquidus 1640°F. Paste type braze alloy.

FUSION NO. 4772

Fusion Inc.
Ag 54, Cu 40, Zn 5, Ni 1, plus flux.
Solidus 1325°F; liquidus 1575°F. Paste type braze alloy. AWS A5.8-69 BAg 13; AMS 4772.

FUSION NO. 4774

Fusion Inc.
Ag 63, Cu 28.5, Ni 2.5, Sn 6, plus flux.
Solidus 1275°F; liquidus 1475°F. Paste type braze alloy. AMS 4774.

FUSION NO. 4775

Fusion Inc.
Ni 74, Cr 14, Fe 4.5, Si 4.5, B 3, plus flux.
Solidus 1780°F; liquidus 1900°F. Paste type braze alloy. AWS A5.8-69 BNi1.

FUSION NO. 4776

Fusion Inc.
Ni 74.5, Cr 15, Fe 3, Si 4.5, B 3, plus flux.
Solidus 1780°F; liquidus 1970°F. Paste type braze alloy. AMS 4776. For stainless steels.

FUSION NO. 4777

Fusion Inc.
Ni 82.6, Cr 7, Fe 3, Si 4.5, B 2.9, plus flux.
Solidus 1780°F; liquidus 1830°F. Paste type braze alloy. AWS A5.8-69 BNi2.

FUSION NO. 4778

Fusion Inc.
Ni 92.5, Si 4.5, B 3, plus flux.
Solidus 1800°F; liquidus 1900°F. Paste type braze alloy. AWS A5.8-69 BNi 3.

FUSION NO. 4779

Fusion Inc.
Ni 94.5, Si 3.5, B 2, plus flux.
Solidus 1810°F; liquidus 1935°F. Paste type braze alloy. AWS A5.8-69 BNi4.

FUSION NO. 490

Fusion Inc.
Sn 25, Pb 75, plus flux.
Solidus 361°F; liquidus 511°F. Paste type solder. ASTM B32-70AT 25B.

FUSION NO. 500

Fusion Inc.
Sn 100, plus flux.
Solidus 450°F; liquidus 450°F. Paste type solder; pure tin.

FUSION NO. 505

Fusion Inc.
Sn 95, Ag 5, plus flux.
Solidus 430°F; liquidus 473°F. Paste type solder.

FUSION NO. 560

Fusion Inc.
Sn 5, Pb 93, Ag 2, plus flux.
Solidus 530°F; liquidus 568°F. Paste type solder.

FUSION NO. 570

Fusion Inc.
Sn 10, Pb 88, Ag 2, plus flux.
Solidus 514°F; liquidus 554°F. Paste type solder.

FUSION NO. 575

Fusion Inc.
Sn 10, Pb 90, plus flux.
Solidus 527°F; liquidus 572°F. Paste type solder. ASTM B32-70AT 10A.

FUSION NO. 8100

Fusion Inc.
Ni 70.8, Cr 19, Si 10.2, plus flux.
Solidus 1975°F; liquidus 2075°F. Paste type braze alloy. AWS A5.8-69 BNi5.

FUSION NO. 8300

Fusion Inc.
Ni 61, Cr 19, Si 10, Mn 10, plus flux.
Solidus 1975°F; liquidus 2075°F. Paste type braze alloy; for stainless steel.

FUSION NO. NBA-1040

Fusion Inc.
Al 76, Si 10, Zn 10, Cu 4.
Solidus 960°F; liquidus 1040°F. Fusion paste alloy for joining aluminum. Aluminum Association 4245.

FUSION NO. NBA-1070

Fusion Inc.
Al 88, Si 12.
Solidus 1070°F; liquidus 1080°F. Fusion paste alloy for joining aluminum. AWS BAlSi-4.

FW 45

Bergische Stahl Industrie
C 0.42-0.5, Si 0.15-0.4, Mn 0.6-0.9, bal Fe.
Carbon tool steel. W.-Nr. 1.1730.

FW 63

Bergische Stahl Industrie
C 0.58-0.64, Si 0.15-0.4, Mn 0.6-0.9, bal Fe.
Carbon tool steel. W.-Nr. 1.1740.

FW 75 C

Bergische Stahl Industrie
C 0.7-0.8, Si 0.25-0.5, Mn 0.5-0.7, Cr 0.25-0.4, bal Fe.
Cold work tool steel. W.-Nr. 1.2003.

FW 90

Bergische Stahl Industrie
C 0.85-0.95, Si 0.15-0.4, Mn 0.4-0.6, bal Fe.
Carbon tool steel. W.-Nr. 1.1760.

FX XTRA

A. Finkl & Sons Co.
Tool material. C 0.5-0.58, Mn 0.75-0.95, Ni 0.85-1.05, Cr 0.85-1.15, Mo 0.33-0.43, bal Fe.
Vacuum arc degassed; hot work die steel. Good high temperature properties and toughness.

G 6 AG CU ZN
Johnson Matthey plc
Ag 67, Cu, Zn.
Brazing alloy for silver; 705-723°C MP. *Obsolete*

G-192 ALLOY
Manufacturer not listed.
C 0.6, Cr 22, Mn 8.5, N 0.35, bal Fe.
For high temperature valves, gas turbine parts, nozzles, afterburners; super strength, high temperature alloy.

G-97
Aluminium Industrie Aktiengesellschaft
Si 0.5, Fe 0.85, Mn 1.4, Mg 0.25, Cu 12.75, bal Al.
Cast: 24,000-32,000 TS; 21,000-26,000 YS; 0.5-1.0 El. For pistons; cast alloy. *Obsolete*

G-ALCU4AG1MG K01
Titan-Aluminium-Feinguss GmbH
Aluminum. Mg 0.3, Ti 0.25, Cu 4.5, Mn 0-0.25, Ag 0-0.7, bal Al.
Annealed: 345 N/mm^2 YS; 410 N/mm^2 TS; 3-5 El; 115 Brin. Can be made to AMS 4228, AMS 4229.

G-ALCU4TI
Titan-Aluminium-Feinguss GmbH
Aluminum. Ti 0.26, Cu 4.9, Mn 0-0.05, Zn 0-0.07, bal Al.
Annealed: 185 N/mm^2 YS; 265 N/mm^2 TS; 2 El; 95 Brin.

G-ALCU4TIMG
Titan-Aluminium-Feinguss GmbH
Aluminum. Mg 0.28, Ti 0.26, Cu 4.6, Mn 0-0.5, Ni 0-0.3, Zn 0-0.1, bal Al.
Annealed: 195 N/mm^2 YS; 295 N/mm^2 TS; 3-6 El; 85 Brin. Can be made to A601B, A-U5GT.

G-ALCU5NI1.5ZRCOSB RR350
Titan-Aluminium-Feinguss GmbH
Aluminum. Ti 0.2, Cu 5, Ni 0-1.5, bal Al.
Annealed: 160-165 N/mm^2 YS; 180-220 N/mm^2 TS; 1 El; 85 Brin. Can be made to AMS 4225.

G-ALLOY
National Physical Laboratory
Aluminum. Zn 18, Cu 2.5, Mg 0.35, Mn 0.35, Fe 0.2, Si 0.75, bal Al.
Hot rolled: 72,000-78,000 TS; 58,000-69,500 YS; 17-19 El. For cast parts for trucks, airplanes and boats; light weight.

G-ALLOY
American Smelting & Refining Co.
Pb alloy.
10,000 TS; 22 Brin. For bearings.

G-ALMG3SI
Titan-Aluminium-Feinguss GmbH
Aluminum. Si 1, Mg 3, bal Al.
Annealed: 120 N/mm^2 YS; 180 N/mm^2 TS; 2 El; 55 Brin. Can be made to A-G3T.

G-ALSI7MGO.3 A356
Titan-Aluminium-Feinguss GmbH
Aluminum. Si 7, Mg 0.3, Mn 0-0.1, Zn 0-0.1, bal Al.
Annealed: 80-195 N/mm^2 YS; 135-265 N/mm^2 TS; 2-4 El; 45-80 Brin. Can be made to AMS 4218B, SAE 336, and MIL-C-21180.

G-ALSI7MGO.6 A357
Titan-Aluminium-Feinguss GmbH
Aluminum. Si 7, Mg 0.6, Ti 0.15, Be 0-0.07, Mn 0-0.1, Zn 0-0.1, bal Al.
Annealed: 120-270 N/mm^2 YS; 165-320 N/mm^2 TS; 2-5 El; 55-80 Brin. Can be made to MIL-C-21180.

G-ALZN2MG
Titan-Aluminium-Feinguss GmbH
Aluminum. Si 0.6, Mg 1.1, Zn 0-2, bal Al.
Annealed: 220 N/mm^2 YS; 300 N/mm^2 TS; 3 El; 100 Brin.

G-B NO. 1300
NL Industries
Ag 30, Cu 29, Zn 25, Cd 16.
For silver solder, brazing; MP 1125-1320°F. *Obsolete*

G-B NO. 1445
NL Industries
Ag 40, Cu 18, Zn 15, Cd 27.
For silver solder, brazing; MP 1120-1205°F. *Obsolete*

G-B NO. 35
NL Industries
Ag 40, Cu 18, Zn 15, Cd 27.
For silver solder, brazing; MP 1120-1205°F. *Obsolete*

G-B NO. 40
NL Industries
Ag 40, Cu 20, Zn 16, Cd 24.
For silver solder, brazing; MP 1125-1235°F. *Obsolete*

G-B NO. 45
NL Industries
Ag 45, Cu 15, Zn 16, Cd 24.
For silver solder, brazing; MP 1125-1145°F. *Obsolete*

G-B NO. 50
NL Industries
Ag 50, Cu 17.5, Zn 11.5, Cd 21.
For silver solder, brazing; MP 1170-1220°F. *Obsolete*

G-B SH 7
Texas Instruments Inc./Materials Control
Ag, bal Cu.
For silver solder. *Obsolete*

G-IRON
Tonawanda Electric Steel Casting
C 4.2, Si 2.4, Mn 0.6, bal Fe.
For making gray iron castings; graphitized pig iron.

G-TI99.4
Titan-Aluminium-Feinguss GmbH
Fe 0-0.3, C 0-0.1, bal Ti.
Investment casting alloy. 280 N/mm^2 YS; 350 N/mm^2 TS; 20 El. ASTM B367-38, 3.7034 Grade C-2.

G-TI99.4PD
Titan-Aluminium-Feinguss GmbH
Fe 0-0.3, C 0-0.1, bal Ti.
Investment casting alloy. 280 N/mm^2 YS; 350 N/mm^2 TS; 20 El. ASTM B367-38, Grade C-7B.

G-TIAL6V4
Titan-Aluminium-Feinguss GmbH
Fe 0-0.4, C 0-0.1, Al 5.5-6.75, V 3.5-4.5, bal Ti.
Investment casting alloy. 825 N/mm^2 YS; 895 N/mm^2 TS; 5 El. ASTM B367-38, 3.7264 Grade C-5.

G. BABBITT
American Smelting & Refining Co.
Sb 12.5, As 3, Sn 0.75, bal Pb.
At 70°F: 9800 psi TS; 1.5 El; 22 Brin. At 392°F: 190 70 El. For Babbitts, bearings; 486-549°F MP; heavy duty.

G.A. PERCIT EXTRA
Krupp Stahl AG
stainless steel. C 1.2-1.4, Si 2-3, Cr 26-28, Fe 57-63.
For hard surfacing electrodes; wear and corrosion resistant. *Obsolete*

G.C.C. CERIUM METAL
General Cerium Co.
Fe 0.7, C 0.03, Si 0.02, 0.70% other rare earth metals, bal Ce.
For alloying, flints; pyrophoric. *Obsolete*

G.C.C. DIDYMIUM
General Cerium Co.
Ce 2.3, La 44.7, Nd 37.4, Pr 10.5, Sm 3.4, Fe 0.7.
Didymium metal. *Obsolete*

G.C.C. MISCHMETAL
General Cerium Co.
Ce 50-55, La 24-26, Nd 15-16, Pr 5-6, Sm 2-3.
For desulfurizer and deoxidizer in alloying; pyrophoric. *Obsolete*

G.C.C. NO. 112
General Cerium Co.
Ce alloy.
Addition agent to stainless steels. *Obsolete*

G.C.C. PURE CERIUM METAL
General Cerium Co.
Fe 0.7, C 0.03, Si 0.02, 0.70% other rare earths, bal Ce.
For desulfurizer and deoxidizer in alloying; pyrophoric. *Obsolete*

G.D.H. BLUE LABEL
Grammer, Dempsey & Hudson Co.
C 1.55-1.7, Cr 11.5-12.5, V 0.15-0.25, Mo 0.7-0.9, bal Fe.
For tools, gauges, punches, dies; oil hardened.

G.D.H. BRONZE LABEL
Grammer, Dempsey & Hudson Co.
C 0.9-0.95, Mn 1-1.1, Cr 0.4-0.5, W 0.4-0.5, bal Fe.
Heat treated: 640 Brin. For tools, gauges, broaches, taps; oil hardened.

G.D.H. RED LABEL
Grammer, Dempsey & Hudson Co.
C 0.65-0.75, Cr 3-4, V 1-1.5, W 18-20, bal Fe.
For tools, cutters; high speed steel.

G.D.H. SILVER LABEL
Grammer, Dempsey & Hudson Co.
C 1, bal Fe.
Heat treated: 660 Brin. For tools, punches, shears, dies; water hardened.

G.E. 33
Manufacturer not listed.
Cu 84.5, Zn 15, Sn 0.5.
For slip rings. *Obsolete*

G.E. CO. NO. 1 SILVER SOLDER
Manufacturer not listed.
Ag 65, Cu 20, Zn 15.
64,800 TS; 34 El. For solder, brazing of stainless steels; malleable, ductile, corrosion resistant. *Obsolete*

G.E. CO. NO. 2 SILVER SOLDER
Manufacturer not listed.
Ag 20, Cu 45, Zn 35.
For solder, brazing of stainless steels; malleable, ductile, corrosion resistant. *Obsolete*

G.E. CO. NO. 3 SILVER SOLDER
Manufacturer not listed.
Ag 55.5, Cu 4.5, Cd 40.
For solder, brazing of stainless steels; malleable, ductile, corrosion resistant. MP 725 C. *Obsolete*

G.E. ELECTRODE TYPE W-85
Manufacturer not listed.
C, alloy, bal Fe.
For electrode for manganese welding to repair worn manganese steel castings. *Obsolete*

G.E.C. HEAVY ALLOY
British General Electric Co. Ltd.
W 90, Ni 7.5, Cu 2.5.
Sintered: 90,000 TS; 83,000 YS; 4 El; 290 Brin. For screens in X-ray tubes.

G.F.A. NICKEL
English manufacture
99.0 min Ni + Co.
For anode support wires, valve components; corrosion and heat resistant.

G.S.N.
Latrobe Steel Co.
Now GSN.

G.S.N. + MO
Latrobe Steel Co.
C 1.5, Cr 12, Mo 0.8, V 1, bal Fe.
For stamping and forming dies; non-deforming. *Obsolete*

G.W. "C.V.M."
Great Western Steel Co.
C 1, Cr 5, Mo 1, bal Fe.
For tools, dies, punches; Type A2; air hardened, nondeforming.

G.W. 422 MIRYCAL
Great Western Steel Co.
C 0.5, Cr 0.95, W 1, Mo 0.2, bal Fe.
For dies, punches, shear blades; Type S1; oil hardened.

G.W. 6-6-2 HIGH SPEED
Great Western Steel Co.
C 0.8, Cr 4, W 6, Mo 4.5, V 1.5, bal Fe.
For tools, cutters, chasers; high speed steel.

G.W. C.W. "OIL"
Great Western Steel Co.
C 0.9, Mn 1, Cr 0.5, W 0.5, bal Fe.
For dies, punches, rollers, mandrels; Type O1; oil hardened, nondeforming.

G.W. COLD HEADER
Great Western Steel Co.
C 0.95, Mn 0.3, Si 0.3, bal Fe.
For cold heading tools; water hardening.

G.W. EXTRA
Great Western Steel Co.
C 1, bal Fe.
Annealed: 100,000 TS; 53,000 YS; 21 El; 42 RA; 200 Brin. For drills, taps, reamers, lathe and planer tools; Type W1; water hardened.

G.W. NO. 265 (HIGH PRODUCTION)
Great Western Steel Co.
C 1.6, Mn 0.3, Cr 12, Mo 0.8, V 0.2, bal Fe.
For tools, dies, blanking, forming and drawing dies; abrasion resistant, air or oil hardening.

G.W. NO. 310
Great Western Steel Co.
C 0.3, Mn 0.25, Cr 3.25, W 10, V 0.35, bal Fe.
For hot work dies, brass forming dies; hot work steel.

G.W. NO. 313
Great Western Steel Co.
C 0.35, Mn 0.3, Cr 3, W 13.5, bal Fe.
Hot work tools and dies; hot work steel, high abrasion resistant.

G.W. NO. 350
Great Western Steel Co.
C 1.3, Mn 0.3, W 3.5, bal Fe.
For cutting tools; fast finishing tool steel.

G.W. NO. 515
Great Western Steel Co.
C 0.35, Si 1, Cr 5, W 5, Mo 0.2, bal Fe.
For hot work tools and dies; hot work steel, air hardening, tough.

G.W. NO. 99 HOT WORK
Great Western Steel Co.
C 0.35, Mn 0.35, Si 0.9, Cr 4.7, W 1.1, Mo 1.5, V 0.25, bal Fe.
For hot work tools and dies, aluminum die casting dies; hot work steel.

G.W. OIL HARDENING (G.W.O.H.)
Great Western Steel Co.
C 0.9, Mn 1.1, Cr 0.5, W 0.5, bal Fe.
For tools, dies; oil hardening.

G.W. PAVEMENT BREAKER
Great Western Steel Co.
C 0.65, Mn 0.3, bal Fe.
For tools, hammers, crushers; water hardening.

G.W. REGULAR
Great Western Steel Co.
C 0.7-1.2, Mn 0.3, bal Fe.
For tools, punches; water hardening.

G.W. SOLID DRILL
Great Western Steel Co.
C 0.85, Mn 0.3, bal Fe.
For tools, pivots; water hardening.

G.W. SPECIAL
Great Western Steel Co.
C 1-1.1, Mn 0.3, bal Fe.
For tools, drills, taps; water hardening.

G.W.32
William Guertler GmbH
Zn 7.6, Mg 1.7, bal Al.
Heat treated: 43,000-50,000 TS; 406 El; 130-140 Brin. For light alloy castings, housings; age-hardenable.

G7
Sheffield Smelting Co. Ltd.
Ag 35, W.
Annealed: 165 Vickers. Electrical resistivity: 3.5 micro-ohm/cm. Electrical conductivity: 49% IACS. For electrical contacts; good erosion resistance.

G97
Aluminium Industrie Aktiengesellschaft
Cu 12, Mn 1.4, Mg 0.25, Fe 0.8, bal Al.
Cast: 24,000-32,000 TS; 21,000-26,000 YS; 0.5-1.0 El. For pistons. Cast alloy.

GADOLINIUM
Atomergic Chemetals Corp.
Gd.
Purities: 99.9% special distilled grade, 99.5 +%. Forms: ingot, lump, sponge, wire, sheet, foil, turnings, powder, single crystals.

GAINEX-38 SA
Armco
HSLA steel.
High strength steel. Hot-rolled: 53 ksi (365 MPa) TS; 38 ksi (262 MPa) YS; 29 El (in 2 in.); 58 Rock B.

GAINEX-40 C
Armco
HSLA steel.
High strength steel. Hot-rolled: 53 ksi (365 MPa) TS; 40 ksi (276 MPa) YS; 26 El (in 2 in.); 60 Rock B. Cold-rolled: 40 ksi (276 MPa) YS.

GAINEX-40 SA
Armco
HSLA steel.
High strength steel. Cold-rolled: 52 ksi (359 MPa) TS; 40 ksi (276 MPa) YS; 25 El (in 2 in.); 62 Rock B.

GAINEX-45 C
Armco
HSLA steel.
High strength steel. Hot-rolled: 63 ksi (434 MPa) TS; 45 ksi (310 MPa) YS; 26 El (in 2 in.); 70 Rock B.

GAINEX-45 SA
Armco
HSLA steel.
High strength steel. Cold-rolled: 58 ksi (400 MPa) TS; 45 ksi (310 MPa) YS; 24 El (in 2 in.); 64 Rock B.

GAINEX-50 C
Armco
HSLA steel.
High strength steel. Hot-rolled: 63 ksi (434 MPa) TS; 50 ksi (345 MPa) YS; 24 El (in 2 in.); 75 Rock B.

GAINEX-60 R
Armco
HSLA steel.
High strength steel. Cold-rolled: 65 ksi (448 MPa) TS; 60 ksi (414 MPa) YS; 12 El (in 2 in.); 82 Rock B.

GAINEX-70 R
Armco
HSLA steel.
High strength steel Cold-rolled: 75 ksi (517 MPa) TS; 70 ksi (483 MPa) YS; 10 El (in 2 in.); 85 Rock B.

GALAHAD A
Dunford Hadfields Ltd.
Cr 13, C 0.1, bal Fe.
For corrosion resisting parts; cutlery; rustless and wear resistant. *Obsolete*

GALAHAD A.C.
Dunford Hadfields Ltd.
C 0-0.15, Cr 11.5-13.5, bal Fe.
Heat treated: 120,000-135,000 TS; 110,000-117,000 YS; 15-16 El; 58-63 RA; 220-240 Brin. For cutlery, valves, turbine blades; Type 410; corrosion resistant. *Obsolete*

GALAHAD A.F.C.
Dunford Hadfields Ltd.
C 0.12, Cr 12-14, S 0-0.75, Ni 0-1, Mo 0-0.6, bal Fe.
Bar: 78,000-112,000 TS; 12-20 El; 280 Brin. For cutlery, turbine blades, gears, oil refinery equipment. Type 416 F stainless, free-cutting. *Obsolete*

GALAHAD B
Dunford Hadfields Ltd.
C 0.15, Cr 13, bal Fe.
For corrosion resistant parts; corrosion resistant, hardenable. *Obsolete*

GALAHAD B.F.C.
Dunford Hadfields Ltd.
C 0.12-0.18, S 0-0.75, Cr 12-14, Ni 0-1, Mo 0-0.6, bal Fe.
Bar: 78,000-112,000 TS; 12-20 El; 150-277 Brin. For cutlery, tableware, turbine blades, structural parts. Free machining. Type 416F stainless steel. Hardenable. *Obsolete*

GALAHAD C
Dunford Hadfields Ltd.
C 0.2, Cr 13, bal Fe.
For corrosion resistant parts; stainless, hardenable. *Obsolete*

GALAHAD C.F.C.
Dunford Hadfields Ltd.
C 0.18-0.25, S 0-0.75, Cr 12-14, Ni 0-1, Mo 0-0.6, bal Fe.
Bar: 78,000-120,000 TS; 12-20 El; 150-277 Brin. For cutlery, tableware, turbine blades, structural parts. Type 420F stainless steel. Hardenable. Free machining. *Obsolete*

GALAHAD D
Dunford Hadfields Ltd.
C 0.25, Cr 13, bal Fe.
Heat treated: 250,000 TS; 185,000 YS; 9 El; 15 RA; 500 Brin. For cutlery, knives, surgical instruments; Type 420; corrosion resistant. *Obsolete*

GALAHAD D.F.C.
Dunford Hadfields Ltd.
C 0.25-0.35, S 0-0.75, Cr 12-14, Ni 0-1, Mo 0-0.6, bal Fe.
Bar: 78,000-112,000 TS; 12-20 El; 150-277 Brin. For cutlery, tableware, valves, knives, turbine blades. Free machining. Type 420 F stainless steel. Hardenable. *Obsolete*

GALAHAD E

Dunford Hadfields Ltd.
C 0.3, Cr 17, Ni 2, bal Fe.
Heat treated: 135,000 TS; 112,000 YS; 15 El; 45 RA; 280 Brin. For propeller shafts, marine hardware; Type 431; corrosion resistant. *Obsolete*

GALAHAD F

Dunford Hadfields Ltd.
C 0-0.12, Cr 16-18, Ni 0-0.5, bal Fe.
Annealed: 75,000 TS; 40,000 YS; B 80 Rock. For nitric acid storage tanks, heat treatment equipment, furnace parts. Heat resisting Type 442. *Obsolete*

GALAVAN

United States Steel Corp.
Low C, Mn, bal Fe.
82 ksi TS; 80 ksi YS. Pre-painted galvanized steel sheet for truck and trailer bodies.

GALICAR

Charles Carr Ltd.
Pb, Sb, bal Sn.
For antifriction alloy; Babbitt.

GALLIMORE METAL

Wm. Gallimore & Sons, Ltd.
Ni 45, Cu 28, Zn 25, Fe 2, Si, Mn.
Hard rolled: 156,000 TS; 2.5 El. Soft rolled: 98,000 TS; 42 El. For airplane parts, noncorrosive stampings. Noncorrosive in sea water.

GALLIUM

Atomergic Chemetals Corp.
Ga.
Purities: 99.99999%, 99.9999%, 99.999%, 99.99%, 99.9%. Packed in sealed poly packets, poly or glass bottles, quartz boats or sealed ampules.

GALV-WELD

Galv-Weld Inc.
Zinc. Pb, Zn, Sn, Bi.
For solder for regalvanizing. Zn-base alloy.

GALVA-ONE

United States Steel Corp.
Low C, Mn, bal Fe.
Electro-galvanized steel sheet; zinc coated on one side only.

GALVALLOY

Metalloy Products Co.
Zn, Pb-Sn.
For Al solder; no flux required.

GALVAMATT

British Steel Corp.
C 0.2, bal Fe.
For stoves, metal signs; galvanized sheets. *Obsolete*

GALVANNEALED

United States Steel Corp.
Low C, Mn, bal Fe.
Steel sheet with remelted zinc coating.

GALVAPRIME

British Steel plc
Mild steel. Galvanized and prepainted roofing and cladding steel.

GALVATITE

British Steel plc
Hot dipped galvanized steel; mild steel.

GALVOBRITE

American Smelting & Refining Co.
Zn alloy.
For galvanizing industry; improves fluidity of Zn.

GALVOMAG

Dow Chemical Co.
Magnesium. Al 0-0.01, Mn 0.5-1.3, Cu 0-0.02, Ni 0-0.001, Fe 0-0.03, bal Mg.
Extruded anode.

GALVOROD

Dow Chemical Co.
Magnesium. Al 2.5-3.5, Zn 0.7-1.3, Si 0-0.05, Cu 0-0.01, Ni 0-0.001, Fe 0-0.002, 0.2 Mn min, bal Mg.
Extruded anode.

GAMA

English manufacture
Cu 12, Ni 2, Fe 0.5, Si 0.5, bal Al.
For pistons; low friction.

GAMAN H

Crucible Materials Corp.
Stainless steel. C 0.51, Mn 12, Si 2.7, Cr 21.25, N 0.45, bal Fe.
For diesel engine valves.

GAMAN H

Colt Industries
C 0.45, Mn 12, Cr 23, N 0.45, bal Fe.
Obsolete

GAMAN L

Colt Industries
C 0.4, Mn 13, Cr 12, N 0.2, bal Fe.
Obsolete

GAMAN R

Colt Industries
C 0.18, Mn 12, Cr 18, N 0.4, bal Fe.
Obsolete

GAMMA COLUMBIUM

Allegheny Ludlum Steel
C 0.4, Mn 1, Si 1, Cr 15, Ni 25, Mo 2-4, Cb 4, bal Fe.
Annealed: 108,000 TS; 32,000 YS 37 El. For superchargers, wheels, aircraft engine components; high heat resistance. *Obsolete*

GAMMA COLUMBIUM

Universal Cyclops
C 0.4, Mn 1, Si 1, Cr 15, Ni 25, Mo 2-4, Cb 4, bal Fe.
Annealed: 108,000 TS; 32,000 YS; 37 El. For superchargers, wheels, aircraft engine components; high heat resistance. *Obsolete*

GAMMA NICKEL STEEL

Midvale-Heppenstall Co.
Ni 36, bal Fe.
Annealed: 70,000 TS; 24,000 YS; 36 El; 68 RA; 143 Brin. Cold drawn: 90,000 TS; 70,000 YS; 20 El; 60 RA; 185 Brin. For instruments, chronometers; low thermal expansion.

GANNALOY

Midvale-Heppenstall Co.
C 0.03, Mn 1.4, Si 0.4, Cr 5.5, Ni 24.5, Ti 2.25, Al 0.68, B 0.003, bal Fe.
For corrosion and heat resisting parts.

GANNALOY

Titanium Metals Corp.
C 0.03, Mn 1.4, Si 0.4, Cr 5.5, Ni 24.5, Ti 2.25, Al 0.68, B 0.003, bal Fe.
For corrosion and heat resisting parts.

GAPASIL 9

GTE Products Corp./Wesgo Div.
Ag 82, Pd 9, Ga 9.
Brazing alloy available in foil, flexibraze, wire, powder, extrudable paste and preform. Liquidus 1616°F. Solidus 1553°F.

GAPASIL 9

Western Gold & Platinum Co.
Pd 9, Ga 9, Ag 82.
Brazing range: 1616-1688°F (880-920°C). Uses include brazing titanium in partial vacuum (500-1000 microns of He or argon).

GARANT DOMO

Haeckerstahl GmbH
C 0.85, Cr 4, W, Mo, V, bal Fe.
For lathe and planer tools, drills, reamers taps; high speed steel.

GARANT DOMO CO

Haeckerstahl GmbH
C 0.85, Cr 4, Co, Mo, W, V, bal Fe.
For lathe and planer tools, hobs, reamers, taps, drills; high speed steel.

GARANT DOMO V

Haeckerstahl GmbH
C 0.85, Cr 4, W, Mo, V, bal Fe.
For lathe and planer tools, reamers, broaches, taps; high speed steel.

GARANT EXTRA

Haeckerstahl GmbH
C 0.74, Cr 4, V 1.1, W 18.5, bal Fe.
For lathe and planer tools, reamers, hobs, drills, taps; high speed steel.

GARANT EXTRA 333

Haeckerstahl GmbH
C 0.95, Cr 4, W, Mo, bal Fe.
For lathe and planer tools, reamers, hobs, drills; high speed steel.

GARANT PRIMA

Haeckerstahl GmbH
C 0.82, Cr 4, Mo 0.85, V 1.6, W 8.7, bal Fe.
For lathe and planer tools, reamers, hobs, taps; high speed steel.

GARANT REKORD 10

Haeckerstahl GmbH
C 0.76, Co 10, Cr 4, Mo 0.8, V 1.8, W 18, bal Fe.
For lathe and planer tools, hobs, broaches; high speed steel.

GARANT REKORD 3

Haeckerstahl GmbH
C 0.86, Co 2.8, Cr 4.3, Mo 0.85, V 2.1, W 12, bal Fe.
For lathe and planer tools, hobs, taps, broaches; high speed steel.

GARANT REKORD 5

Haeckerstahl GmbH
C 0.79, Co 4.7, Cr 4.2, Mo 0.8, V 1.6, W 18, bal Fe.
For lathe and planer tools, drills, reamers, taps; high speed steel.

GARANT SONDERKLASSE 500

Haeckerstahl GmbH
C 1.3, Cr 4.3, Mo 0.85, V 3.8, W 12, bal Fe.
For engravers' tools, form cutters, reamers; high speed steel.

GARANT SONDERKLASSE CO

Haeckerstahl GmbH
C 1.35, Cr 4.2, Co, Mo, W, V, bal Fe.
For engravers' tools, form cutters, reamers; high speed steel.

GARANT SPEZIAL 275

Haeckerstahl GmbH
C 0.86, Cr 4, Mo 0.85, V 2.5, W 12, bal Fe.
For lathe and planer tools, broaches, reamers, taps; high speed steel.

GARBA

Ekstrand & Tholand Co.
C 1, bal Fe.
For drills, taps, drill rod. Water hardening.

GARFIELD
Time Steel Service Inc.
C 1, Mn 0.3, Si 0.25, Cr 0.5, V 0.2, bal Fe.
Water hardened tool steel, AISI W7.

GAS ENGINE BABBITT
Hoyt Metal Co. of London Ltd.
Sn, Sb, Bal Pb.
At 70 F: 24.1 Brin. At 212 F: 11.7 Brin. For bearings for gas engines; Babbitt. *Obsolete*

GASID NO. 12
Manufacturer not listed.
Copper. Sn 9, Zn 3, bal Cu.
30,000 TS; 22 El; 20 RA; 125 Brin. For high pressure acid pumps in oil refineries; corrosion resistant to hot hydrochloric acid. *Obsolete*

GASID NO. 15
Manufacturer not listed.
Copper. Al 7, Fe 2.5, bal Cu.
42,000 TS; 20-30 El; 32 RA; 193 Brin. For chemical apparatus and equipment; corrosion and acid resistant. *Obsolete*

GASITE
Georgia Iron Works Co.
C 3-3.6, Ni 4-4.75, Cr 1.4-3.5, Si 0.4-0.7, bal Fe.
Sand cast: 45,000 TS; 600 Brin. Permanent mold: 55,000 TS; 675 Brin. For heavy cams, dies, roller bearing races; white cast iron, corrosion resistant.

GAUSSIT 180
German manufacture
Co 50, V 14, bal Fe.
8000 Br, 1.4 BH max. For electromechanical devices, hysteresis motors, digital computers; precipitation hardening permanent magnet.

GBN
W. Ossenberg & Cie Edelstahlwerke
C 0.38, Mn 0.4, Si 1.5, V 0.1, bal Fe.
Cold work tool steel for chisels, punches, shearing tools. W.-Nr. 1.2248.

GBV
W. Ossenberg & Cie Edelstahlwerke
C 0.45, Mn 0.1, Si 1.5, Cr 1.5, V 0.1, bal Fe.
Hot work tool steel. Pressure casting molds for lead, tin, zinc alloys. Also cold work punches. W.-Nr. 1.2249.

GBV 6
W. Ossenberg & Cie Edelstahlwerke
C 0.6, Mn 0.8, Si 0.9, Cr 1.2, V 0.1, bal Fe.
Cold work tool steel for punches and stamping tools. W.-Nr. 1.2243.

GBZ 10 BRONZE
German manufacture
Cu 90, Sn 10.
Cast: 45,500 TS; 20,000 YS; 45 El; 62 Brin. For castings.

GBZ 14 BRONZE
German manufacture
Cu 86, Sn 14.
Cast: 32,000 TS; 27,000 YS; 7 El; 79 Brin. For castings.

GBZ 20
German manufacture
Cu 80, Sn 20.
Cast: 27,000 TS; 0 El; 120 Brin. For castings.

GCR15
China Metallurgical Import&Export Corp.
Alloy steel. C 0.95-1.05, Mn 0.2-0.4, Si 0.15-0.35, Cr 1.3-1.65, S 0-0.02, P 0-0.027, bal Fe.
170-207 Brin, 52100 (SAE). Used for heavy-duty ball or roller bearings, races, cams and rolls.

GCR15SIMN
China Metallurgical Import&Export Corp.
Alloy steel. C 0.9-1.05, Mn 0.9-1.2, Si 0.4-0.65, Cr 1.3-1.65, S 0-0.02, P 0-0.027, bal Fe.
179-217 Brin. Used for heavy-duty ball or roller bearings with thick sections.

GCR6
China Metallurgical Import&Export Corp.
Alloy steel. C 1.05-1.15, Mn 0.2-0.4, Si 0.15-0.35, Cr 0.4-0.7, S 0-0.02, P 0-0.027, bal Fe.
170-207 Brin, 50100 (SAE). Used for ball or roller bearings.

GCR9
China Metallurgical Import&Export Corp.
Alloy steel. C 1-1.1, Mn 0.2-0.4, Si 0.15-0.35, Cr 0.9-1.2, S 0-0.02, P 0-0.027, bal Fe.
170-207 Brin, 51100 (SAE). Used for ball or roller bearings.

GCR9SIMN
China Metallurgical Import&Export Corp.
Alloy steel. C 1-1.1, Mn 0.9-1.2, Si 0.4-0.7, Cr 0.9-1.2, S 0-0.02, P 0-0.027, bal Fe.
179-217 Brin. Used for ball bearings with thick sections.

GCX
National Intergroup Inc.
C 0.08, Mn 0.45, Si 0.05, Cb 0.04, bal Fe.
For parts to be carbonitrided, as cams, ratchets, transmission levers.

GD15
Sheffield Smelting Co. Ltd.
Ag 85, CdO.
Annealed: 60 Vickers. Electrical resistivity: 2.1 micro-ohm/cm. Electrical conductivity: 82% IACS. For electrical contacts; good anti-weld properties.

GDH BRAKE DIE
Grammer, Dempsey & Hudson Co.
C 0.5, Mo 0.2, Mn 0.9, Cr 1, bal Fe.
Oil hardening tool steel for shafts, arbors, lathe centers, forming tools.

GDH GRAPH-AIR
Grammer, Dempsey & Hudson Co.
C 1.35, Mn 1.8, Si 1.2, Ni 1.85, Mo 1.5, bal Fe.
Air or oil hardening, medium alloy, cold work tool steel; AISI A10.

GDH GRAPH-TUNG
Grammer, Dempsey & Hudson Co.
C 1.5, Mn 1, Si 0.9, Ni 0.5, W 2.75, bal Fe.
Water or oil hardening tool steel; good wear resistance and high hardness.

GDH LUSTRE DIE
Grammer, Dempsey & Hudson Co.
C 0.5, Mn 0.9, Cr 1, bal Fe.
Oil hardening tool steel for shafts, arbors.

GDH NO. 212
Grammer, Dempsey & Hudson Co.
C 0.45-0.5, Cr 0.85-1.05, W 0.9-1.2, Mo 0.15-0.25, bal Fe.
For chisels, shear blades, die blocks; tough.

GDH NO. 280
Grammer, Dempsey & Hudson Co.
C 0.5, Si 1.3-1.6, Mo 0.3-0.4, bal Fe.
For tools, punches, shear blades; tough.

GDH NO. 33
Grammer, Dempsey & Hudson Co.
C 0.35, Si 0.8-1, Cr 4.5-5, W 1-1.2, Mo 1.3-1.5, bal Fe.
For forging dies, hot piercing and punching dies; hot work steel.

GDH NO. 350
Grammer, Dempsey & Hudson Co.
C 1.25-1.35, W 3.5-3.7, bal Fe.
For drills, dies, gear cutters, taps; keen cutting edge.

GDH NON TEMPERING
Grammer, Dempsey & Hudson Co.
C 0.35, Mn 0.7, Si 0.45, Cr 0.8, Mo 0.3, Cu 0.3, bal Fe.
Oil hardening tool, shock resisting type.

GDH-07
Grammer, Dempsey & Hudson Co.
C 1.25, Mn 0.3, Si 0.35, Cr 0.4, V 0.2, W 1.4, bal Fe.
Oil hardening tool steel, high hardness and water resistance; AISI O7.

GDH-10
Grammer, Dempsey & Hudson Co.
C 0.4, Mn 0.55, Si 1, Cr 3.25, V 0.35, Mo 2.5, bal Fe.
Air or oil hardening hot work tool and die steel for forging dies, hot forming tools; AISI H10.

GDH-11
Grammer, Dempsey & Hudson Co.
C 0.4, Mn 0.4, Si 1, Cr 5, V 0.5, Mo 1.35, bal Fe.
Air or oil hardening hot work tool and die steel for forging dies, hot forming tools; AISI H11.

GDH-12
Grammer, Dempsey & Hudson Co.
C 0.35, Mn 0.35, Si 1.05, Cr 5, V 0.35, W 1.25, Mo 1.35, bal Fe.
Air or oil hardening hot work tool and die steel for forging dies, hot forming tools; AISI H12.

GDH-13
Grammer, Dempsey & Hudson Co.
C 0.4, Mn 0.4, Si 1.1, Cr 5, V 1.1, Mo 1.35, bal Fe.
Air or oil hardening hot work tool steel, chromium type; AISI H13.

GDH-3
Grammer, Dempsey & Hudson Co.
C 2.15, Mn 0.25, Si 0.25, Cr 12, V 0.8, bal Fe.
Air or oil hardening tool steel for cold working dies, punches; AISI D3.

GDH-3SP
Grammer, Dempsey & Hudson Co.
C 2.1, Cr 12.5, Ni 0.5, bal Fe.
Air or oil hardening cold work tool steel for dies, punches, gages; AISI D3.

GDH-A2
Grammer, Dempsey & Hudson Co.
C 1, Mn 0.7, Si 0.3, Cr 5.25, V 0.3, Mo 1.15, bal Fe.
Air hardening, medium alloy, cold work tool steel; AISI A2.

GDH-A4
Grammer, Dempsey & Hudson Co.
C 0.95, Mn 2, Cr 2.2, Mo 1.1, Pb added, bal Fe.
Air hardening, medium alloy, cold work tool steel; AISI A4.

GDH-CHW
Grammer, Dempsey & Hudson Co.
C 0.23, Mn 0.6, Si 1.25, Cr 10, Ni 0.75, V 1, W 0.45, Mo 1, N 0.1, bal Fe.
Air or oil hardening hot work tool steel, chromium type.

GDH-D2
Grammer, Dempsey & Hudson Co.
C 1.55, Mn 0.35, Si 0.45, Cr 11.5, V 0.9, Mo 0.8, bal Fe.
Air or oil hardening cold work tool steel, chromium type; for lamination dies, punches, thread forming tools; AISI D2.

GDH-D2 FM
Grammer, Dempsey & Hudson Co.
C 1.55, Mn 0.35, Si 0.45, Cr 11.5, V 0.9, Mo 0.8, S, bal Fe.
Air or oil hardening cold work tool steel, chromium type; free machining grade; AISI D2.

GDH-D4
Grammer, Dempsey & Hudson Co.
C 2.25, Mn 0.35, Si 0.5, Cr 11.5, V 0.2, Mo 0.8, bal Fe.
Oil or air hardening cold work tool steel, chromium type; AISI D4.

GDH-D5

Grammer, Dempsey & Hudson Co.
C 1.5, Mn 0.4, Si 0.4, Cr 12, Ni 0.35, V 0.5, Mo 0.9, Co 3.2, bal Fe.
Air or oil hardening cold work tool steel, high carbon, high chromium type; AISI D5.

GDH-H24

Grammer, Dempsey & Hudson Co.
C 0.45, Mn 0.3, Si 0.3, Cr 3.5, V 0.7, W 14, bal Fe.
Air or oil hardening hot work tool steel, tungsten type; AISI H24.

GDH-L2

Grammer, Dempsey & Hudson Co.
C 0.5, Mn 0.8, Si 0.3, Cr 1, V 0.2, bal Fe.
Oil hardening tool steel for shafts, arbors, lathe centers; AISI L2; similar to AISI 4150.

GDH-L6

Grammer, Dempsey & Hudson Co.
C 0.75, Mn 0.75, Cr 0.9, Ni 1.75, Mo 0.35, bal Fe.
Oil hardening tool steel for shafts, arbors, lathe centers, drill bushings; AISI L6.

GDH-M2

Grammer, Dempsey & Hudson Co.
C 0.85, Cr 4.15, V 1.95, W 6.4, Mo 5, bal Fe.
High speed tool steel, molybdenum-tungsten type. For lathe tools, milling cutters, drills, broaches; AISI M2.

GDH-M41

Grammer, Dempsey & Hudson Co.
C 1.1, Mn 0.45, Si 0.3, Cr 4.25, V 2, W 6.75, Mo 3.75, Co 5, bal Fe.
High speed steel, molybdenum-tungsten-cobalt type; AISI M41.

GDH-O1

Grammer, Dempsey & Hudson Co.
C 0.9, Mn 1.2, Cr 0.5, V 0.2, W 0.5, bal Fe.
Oil hardening tool steel; AISI O1.

GDH-O2

Grammer, Dempsey & Hudson Co.
C 0.9, Mn 1.5, Si 0.25, Mo 0.3, bal Fe.
Oil hardening tool steel; AISI O2.

GDH-S1

Grammer, Dempsey & Hudson Co.
C 0.45-0.5, Mn 0.2-0.5, Si 0.75-1, Cr 1.15-1.25, V 0.15-0.25, W 2.25-2.6, bal Fe.
Oil hardening tool, shock resisting type; AISI S1.

GDH-S5

Grammer, Dempsey & Hudson Co.
C 0.6, Mn 0.8, Si 2, Cr 0.25, V 0.2, Mo 0.25, bal Fe.
Water or oil hardening tool steel, shock resisting type; AISI S5.

GDH-S7

Grammer, Dempsey & Hudson Co.
C 0.5, Mn 0.7, Cr 3.25, Mo 1.4, bal Fe.
Oil hardening tool steel, shock resisting type; AISI S7.

GDH-T1

Grammer, Dempsey & Hudson Co.
C 0.75, Mn 0.3, Si 0.3, Cr 4, V 1.15, W 18.25, bal Fe.
High speed steel, tungsten type, general purpose cutting tool; AISI T1.

GDH-W1

Grammer, Dempsey & Hudson Co.
C 1.1, Mn 0.3, Si 0.25, bal Fe.
Water hardening tool steel; AISI W1.

GDH-W2

Grammer, Dempsey & Hudson Co.
C 1.05, Mn 0.25, Si 0.25, V 0.2, bal Fe.
Water hardening tool steel; AISI W2.

GE 1570

General Electric Co.
C 0.2, Cr 20, Ni 29, Co 37.5, Mo 7, Ti 4.2, Fe 1.5.
High temperature alloy.

GE 218

General Electric Co.
Tungsten base wire. For operation to 3000°F.

GE 473

General Electric Co.
W 7, Re 3, bal Ta.

GE-17 PS

General Electric Co.
1.8-2.2 ThO$_2$, bal W.
For lamp filaments, heaterwire, rocket and missile high temperature parts.

GEANT 22

Forges et Acieries de Voelkingen
C 0.8, W 10, Cr 4, V 1.7, Mo 0.9, bal Fe.
For lathe and planer tools, hobs, reamers; high speed steel.

GEANT 3

Societe des Acieries de Longwy
C 0.7, W 18, Co 5, Cr 4, V 1.3, bal Fe.
For tools, dies, cutters, hobs, drills; high speed steel.

GEANT 50

Forges et Acieries de Voelkingen
C 0.75, W 18, Cr 4, V 1, Mo 0.7, bal Fe.
For lathe and planer tools, reamers, broaches, taps; high speed steel.

GEANT 55

Forges et Acieries de Voelkingen
C 0.85, W 12, Cr 4, V 2, Mo 0.9, Co 2.8, bal Fe.
For lathe and planer tools, drills, reamers; high speed steel.

GEANT 60

Forges et Acieries de Voelkingen
C 0.8, W 18, Cr 4, V 1.5, Mo 0.75, bal Fe.
For lathe and planer tools, drills, reamers, hobs; high speed steel.

GEANT 66

Forges et Acieries de Voelkingen
C 1.3, Cr 4, W 12, V 4.5, Mo 0.9, Co 4.5, bal Fe.
For blanking and forming dies, engravers' tools; high speed steel.

GEANT 77

Forges et Acieries de Voelkingen
C 0.85, W 18, Cr 4, V 1.5, Mo 0.7, Co 5, bal Fe.
For lathe and planer tools, milling cutters, drills; high speed steel.

GEANT 88

Forges et Acieries de Voelkingen
C 0.85, W 18, Cr 4, V 1.5, Mo 0.7, Co 10, bal Fe.
For lathe and planer tools, cutters, broaches; high speed steel.

GEANT M5

Forges et Acieries de Voelkingen
C 0.85, W 6, Cr 4, V 2.4, Mo 5, bal Fe.
For lathe and planer tools, reamers; high speed steel.

GEAR BRONZE-1

American manufacture
Cu 78-91, Sn 10-12, Zn 0-3, Pb 0.2, P 0.1-0.3.
Cast: 38,000 TS; 20,000 YS; 10 El; 80 Brin. For gears, worm wheels; tough.

GEAR BRONZE-2

American manufacture
Cu 88, Sn 10, Pb 2.
For gears; free-cutting.

GEAR BRONZE-3

American manufacture
Cu 85, Sn 13, Zn 2.
For gears; tough.

GEAR STEEL

English manufacture
Ni 3.5, Cr 1.5, C 0.45-0.5, bal Fe.
For gears and pinions; oil-hardening.

GEAR STEEL HIGH DUTY

English manufacture
C, alloy, bal Fe.
For gears; oil-hardening.

GEAR-WHEEL BRONZE

Stone Manganese - J. Stone & Co. Ltd.
Sn 11-12, Zn 0-1.5, P 0.1-0.3, bal Cu.
Cast: 30,000-37,000 TS; 20,000-23,000 YS; 1-4 El; 90-125 Brin. For high duty worm wheels; working with hardened steel worms.

GEARING BRONZE

English manufacture
Cu 91.3, Sn 8.7.
For gears; tough.

GEARS BRONZE

English manufacture
Cu 85, Sn 10, Zn 3, Pb 2.
For gears; free-cutting.

GECOR

Manufacturer not listed.
Cobalt rare earth compounds. Anisotropic per magnet material. Highest energy product and coercive force. *Obsolete*

GEDGE'S METAL

English manufacture
Cu 60, Zn 38.2, Fe 1.8.
For ship sheathing, cylinders for hydraulic presses; malleable at red heat.

GEMCO

GTE Products Corp./Wesgo Div.
Cu 87.75, Ge 12, Ni 0.25.
Brazing alloy available in foil, flexibraze, wire, powder, extrudable paste and preform. Liquidus 1769°F. Solidus 1508°F.

GEMCO

Western Gold & Platinum Co.
Cu 87.75, Ge 12, Ni 0.25.
Melting point: 1508-1769°F. For high temperature brazing.

GEMINOL N

Driver Harris Co.
Nickel. Si 3, bal Ni.
Negative nuclear grade thermocouple. *Obsolete*

GEMINOL P

Driver Harris Co.
Nickel. Ni 80, Cr 20, Si 1.
Positive nuclear-grade thermocouple. *Obsolete*

GEMMA

Creusot-Loire
Ni 35, bal Fe.
For magnets; high permeability. *Obsolete*

GEMPCO

General Metals Powder Co.
Sn-Pb-graphite, bal Cu.
For friction material, clutch discs, motors, brushes; sintered, self lubricating.

GENALLOY

Alloy Engineering & Casting Co.
Ni 37-40, Cr 17-21, C, Si, bal Fe.
Cast: 67,000 TS; 38,000 YS; 6 El; 207 Brin. For furnace parts.
High resistance to heat and corrosion. *Obsolete*

GENALLOY B

Alloy Engineering & Casting Co.
C 0.36-0.5, Ni 31-39, Cr 12-15, bal Fe.
For heat and corrosion resistant parts. Heat and corrosion
resistant. *Obsolete*

GENCALLOY

American manufacture
Pb alloy.
For cable sheathing; corrosion resistant.

GENCALOY

American manufacture
Pb alloy.
For cable sheath.

GENELITE

British Thomson Houston Co. Ltd.
Cu 70-73, Sn 12-14, Pb 9-10, 4.5- 5.5 graphite.
For bearings for aero engines; synthetic bronze, self-
lubricating.

GENERAL ALLOYS CN-1-H

Alloy Engineering & Casting Co.
C 0.2-0.5, Cr 26-30, Ni 8-11, bal Fe.
Cast: 95,000 TS; 45,000 YS; 20 El; 200 Brin. For oil burner
parts, rabble arms, and tube supports. Heat resistant,
stainless. *Obsolete*

GENERAL ALLOYS Q-10

Alloy Engineering & Casting Co.
C 0-0.2, Cr 18-21, Ni 8-11, bal Fe.
Annealed: 77,000 TS; 36,000 YS; 50 El; 163 Brin. For pumps,
rolls, and valve bodies in chemical plant equipment.
Stainless; austenitic; ACI CF-20. *Obsolete*

GENERAL ALLOYS Q-11

Alloy Engineering & Casting Co.
C 0-0.08, Cr 18-21, Ni 8-11, bal Fe.
Annealed: 77,000 TS; 37,000 YS; 55 El; 140 Brin. For spray
nozzles, sanitary fittings and chemical plant equipment.
Stainless; austenitic; ACI CF-8. *Obsolete*

GENERAL ALLOYS Q-12

Alloy Engineering & Casting Co.
C 0-0.08, Cr 18-21, Ni 9-12, Mo 2-3, bal Fe.
Annealed: 80,000 TS; 42,000 YS; 50 El; 168 Brin. For mixing
propellers and fittings in chemical plant equipment.
Stainless; austenitic; ACI CF-8C.

GENERAL ALLOYS Q-13

Alloy Engineering & Casting Co.
C 0-0.08, Cr 19, Ni 11, Mo 3-4, bal Fe.
Annealed: 80,000 TS; 42,000 YS; 50 El; 168 Brin. For acid
resistant casting. Type 317; corrosion resistant. For fittings
and mixing propellers.

GENERAL ALLOYS Q-14

Alloy Engineering & Casting Co.
C 0-0.08, Cr 18-21, Ni 9-12, Cb, bal Fe.
Annealed: 77,000 TS; 38,000 YS; 39 El; 149 Brin. For pump
parts and valve bodies in chemical plant equipment.
Stainless; austenitic; ACI CF-8C.

GENERAL ALLOYS Q-15

Alloy Engineering & Casting Co.
C 0-0.16, Cr 18-21, Ni 9-12, Se 0.2-0.35, bal Fe.
Annealed: 77,000 TS; 40,000 YS; 52 El; 150 Brin. For pumps,
castings, and valves in chemical plant equipment. Free-
machining austenitic stainless; ACI CF-16F.

GENERAL ALLOYS Q-16

Alloy Engineering & Casting Co.
C 0-0.2, Cr 19, Ni 9, Mo 3, bal Fe.
Annealed: 70,000 TS; 30,000 YS; 35 El; 40 RA; 160 Brin. For
chemical plant equipment. ACI-CH20. *Obsolete*

GENERAL ALLOYS Q-21

Alloy Engineering & Casting Co.
C 0-0.2, Cr 22-26, Ni 12-15, bal Fe.
Annealed: 83,000 TS; 50,000 YS; 38 El; 190 Brin. For pumps,
roasting equipment, furnace parts, and heat treating boxes.
Heat resistant. ACI CH-20. *Obsolete*

GENERAL ALLOYS Q-22

Alloy Engineering & Casting Co.
C 0-0.1, Cr 22-26, Ni 12-15, bal Fe.
Annealed: 86,000 TS; 40,000 YS; 40 El; 170 Brin. For furnace
parts, heat treating boxes, and chemical plant equipment.
Heat resistant; ACI CH-20.

GENERAL ALLOYS Q-23

Alloy Engineering & Casting Co.
C 0-0.1, Cr 24, Ni 12, Mo 3, bal Fe.
Annealed: 70,000 TS; 30,000 YS; 45 El; 55 RA; 150 Brin. For
chemical plant equipment. ACI-CH10M; heat and corrosion
resistant. *Obsolete*

GENERAL ALLOYS Q-30

Alloy Engineering & Casting Co.
C 0-0.2, Cr 23-27, Ni 19-22, bal Fe.
Annealed: 75,000 TS; 45,000 YS; 20 El; 180 Brin. For valves,
digesters, jet engine parts, furnace parts, and heat treating
boxes. Heat resistant; ACI CK-20. *Obsolete*

GENERAL ALLOYS Q-35

Alloy Engineering & Casting Co.
C 0.2, Cr 23-27, Ni 19.22, bal Fe.
Annealed: 65,000 TS; 30,000 YS; 30 El; 165 Brin. For furnace
parts, heat treating boxes, and chemical plant parts. ACI
CK-20. *Obsolete*

GENERAL ALLOYS Q-41

Alloy Engineering & Casting Co.
C 0-0.07, Cr 16, Ni 35, Mo, Cu, bal Fe.
Annealed: 70,000 TS; 35,000 YS; El; 45 RA; 150 Brin. For
chemical plant equipment and furnace fixtures. Corrosion
and heat resistant. *Obsolete*

GENERAL ALLOYS Q-50

Alloy Engineering & Casting Co.
C 0.08-0.15, Cr 12, Ni 0-1, bal Fe.
Normalized: 100,000-200,000 TS; 75,000-150,000 YS; 7-30 El;
40 RA; 185-390 Brin. For chemical plant equipment. ACI-
CA15; corrosion resistant.

GENERAL ALLOYS Q-51

Alloy Engineering & Casting Co.
C 0.2-0.4, Cr 12, Ni 0-1, bal Fe.
Heat treated: 110,000-220,000 TS; 67,000-165,000 YS; 1-16
El; 7 RA; 212-470 Brin. For chemical plant equipment. ACI-
CA40; corrosion resistant.

GENERAL ALLOYS Q-56

Alloy Engineering & Casting Co.
C 0.6-0.75, Cr 17, Ni 0-1, bal Fe.
Heat treated: 400 Brin. AISI 440A; corrosion resistant.

GENERAL ALLOYS Q-57

Alloy Engineering & Casting Co.
C 0.95-1.2, Cr 17, Ni 0-1, bal Fe.
Heat treated: 550 Brin. AISI 440C; corrosion resistant.

GENERAL ALLOYS Q-58

Alloy Engineering & Casting Co.
C 0-0.12, Cr 16, Ni 0-1, bal Fe.
Annealed: 80,000 TS; 45,000 YS; 10 El; 10 RA; 200 Brin. For
chemical plant equipment. AISI 430; corrosion resistant.

GENERAL ALLOYS Q-60

Alloy Engineering & Casting Co.
C 0-0.3, Cr 20, Ni 0-2, bal Fe.
Annealed: 95,000 TS; 60,000 YS; 15 El; 10 RA; 200 Brin. For
chemical plant equipment. ACI-CB20; corrosion resistant.

GENERAL ALLOYS Q-65

Alloy Engineering & Casting Co.
C 0.3-0.5, Cr 28, Ni 0-4, bal Fe.
Annealed: 97,000 TS; 65,000 YS; 18 El; 212 Brin. For furnace
parts and heat treating boxes. ACI-CC50; corrosion and heat
resistant.

GENERAL ALLOYS Q-66

Alloy Engineering & Casting Co.
C 0-0.2, Cr 28, Ni 0-3, bal Fe.
Annealed: 55,000 TS; 45,000 YS. For furnace parts, heat
treating boxes, and chemical plant equipment. ACI-CC20;
heat and corrosion resistant.

GENERAL ALLOYS Q-67

Alloy Engineering & Casting Co.
C 0-0.35, Cr 25, Ni 0-1, bal Fe.
Annealed: 55,000 TS; 40,000 YS. For chemical plant
equipment. AISI 446; corrosion resistant.

GENERAL ALLOYS Q-80

Alloy Engineering & Casting Co.
C 0-0.1, Cr 26-30, Ni 3-6, Mo, bal Fe.
Annealed: 60,000 TS; 40,000 YS. For chemical plant
equipment. Resists high temperature and high temperature
corrosion; AISI 329.

GENERAL ALLOYS Q-81

Alloy Engineering & Casting Co.
C 0-0.07, Cr 20, Ni 25, Mo 3, Cu 1, bal Fe.
Cast. For chemical plant equipment; corrosion resistant.
Obsolete

GENERAL ALLOYS Q-82

Alloy Engineering & Casting Co.
C 0-0.3, Cr 29, Ni 9, bal Fe.
Cast: 80,000 TS; 40,000 YS; 20 El; 20 RA. For furnace parts
and chemical plant equipment. ACI-CE30; corrosion and
heat resistant.

GENERAL ALLOYS Q-85

Alloy Engineering & Casting Co.
C 0.07, Cr 19-22, Ni 27-31, Mo 1.75-2.5, 3.0 Cu min, bal Fe.
Annealed: 65,000-75,000 TS; 28,000-38,000 YS; 35-50 El;
120-150 Brin. For towers and pickling hooks in chemical
plant equipment. Resists sulfuric acid.

GENERAL ALLOYS Q-86

Alloy Engineering & Casting Co.
C 0-0.12, Cr 0-1, Ni 64, Mo 28, bal Fe.
Annealed: 80,000 TS; 57,000 YS; 8 El; 179-235 Brin. For
casting. Corrosion and heat resistant; Hastelloy B. For
chemical plant parts.

GENERAL ALLOYS Q-87

Alloy Engineering & Casting Co.
C 0-0.12, Cr 0-1, Ni 55, Mo 21, bal Fe.
Annealed: 71,000 TS; 47,000 YS; 12 El; 149-187 Brin. For
castings. Corrosion and heat resistant; Hastelloy A.

GENERAL ALLOYS Q-88

Alloy Engineering & Casting Co.
C 0-0.15, Cr 16, Ni 56, W 4, Mo 17, bal Fe.
Annealed: 78,000 TS; 57,000 YS; 10 El; 187-248 Brin. For
castings; corrosion and heat resistant to chlorine. Hastelloy
C.

GENERAL ALLOYS Q-89

Alloy Engineering & Casting Co.
C 0-0.12, Cr 0-1, Ni 84, Si 9, Cu 4, bal Fe.
Annealed: 118,000 TS; 118,000 YS; 0-2 El. Acid resistant
casting; maximum resistance to sulfuric acid. Reference:
Hastelloy D.

GENERAL ALLOYS Q-90
Alloy Engineering & Casting Co.
C 0-0.2, Cr 22, Ni 45, Mo, Co, W, bal Fe.
Annealed: 70,000 TS; 48,000 YS; 11 El; 170 Brin. For casting. Corrosion and heat resistant; Hastelloy X. Oxidation resistant to 2200°F.

GENERAL ALLOYS X-3
Alloy Engineering & Casting Co.
C 0.2-0.6, Cr 28-32, Ni 18-22, bal Fe.
Cast: 82,000 TS; 52,000 YS; 19 El; 192 Brin. For carrier fingers, enameling fixtures, and stack dampers. Heat resistant; stainless. *Obsolete*

GENERAL ALLOYS X-4
Alloy Engineering & Casting Co.
C 0.2-0.6, Cr 24-28, Ni 18-22, bal Fe.
Cast: 75,000 TS; 50,000 YS; 19 El; 192 Brin. For cement kiln segments, pier caps, and skid rails. Heat resistant; stainless. *Obsolete*

GENERAL ALLOYS X-5
Alloy Engineering & Casting Co.
C 0-0.5, Cr 26-30, Ni 4-7, bal Fe.
Cast: 85,000 TS; 48,000 YS; 16 El; 190 Brin. For brazing furnace parts, cracking equipment, and rabble shoes. Heat resistant. *Obsolete*

GENERAL ALLOYS X-6
Alloy Engineering & Casting Co.
C 0.2-0.5, Cr 19-23, Ni 23-27, bal Fe.
Cast: 68,000 TS; 38,000 YS; 17 El; 160 Brin. For chains, furnace beams, and tube supports. Heat resistant; stainless. *Obsolete*

GENERAL ALLOYS X-7
Alloy Engineering & Casting Co.
C 0.2-0.5, Cr 24-28, Ni 11-14, bal Fe.
Cast: 80,000 TS; 50,000 YS; 25 El; 185 Brin. For exhaust manifolds, rabble arms, and tube hangers. Heat resistant; stainless. *Obsolete*

GENERAL ALLOYS X-8
Alloy Engineering & Casting Co.
Cr 19-23, Ni 9-12, C 0.2-0.4, bal Fe.
Cast: 85,000 TS; 45,000 YS; 35 El; 165 Brin. For burner tips, fan housings, and wear plates. Heat resistant; stainless.

GENERAL ALLOYS X-9
Alloy Engineering & Casting Co.
C 0.2-0.3, Cr 26-30, Ni 14-18, bal Fe.
Cast: 80,000 TS; 45,000 YS; 12 El; 180 Brin. For furnace rails, lead pots, and tube spacers. Heat resistant; stainless. *Obsolete*

GENERAL ALLOYS X-B
Alloy Engineering & Casting Co.
Ni 33.37, Cr 13-17, C 0.35-0.75, bal Fe.
Cast: 70,000 TS; 40,000 YS; 10 El; 180 Brin. For furnace parts, brazing trays, heat treating trays and baskets. Heat and corrosion resistant. *Obsolete*

GENERAL ALLOYS X-B CB
Alloy Engineering & Casting Co.
C 0.35-0.75, Ni 35, Cr 15, Cb, bal Fe.
Cast: 70,000 TS; 40,000 YS; 10 El; 180 Brin. For furnace parts, heat treating boxes, and salt pots. Heat and corrosion resistant. *Obsolete*

GENERAL ALLOYS-Q91
Alloy Engineering & Casting Co.
C 0-0.04, Cr 26, Ni 5, Mo 2, Cu 3, bal Fe.
Aged: 140,000 TS; 115,000 YS; 18 El; 310 Brin. Cast: 120,000 TS; 83,000 YS; 27 El; 260 Brin. For acid resistant and chemical plant equipment, pumps, valves, and fittings. ACI-CD4MCu; corrosion resistant.

GENERAL PLATE 154
Texas Instruments Inc./Materials Control
Ag 45, Cu 17, Zn 18.5, Cd 20.5, Ni 0.5.
For silver brazing; MP 1135-1150°F. *Obsolete*

GENERAL PLATE 715
Texas Instruments Inc./Materials Control
Mn 15, Ni 15, bal Cu.
Hardened: 100,000-133,000 TS; 40,000-77,000 YS; 12-20 El; 100-320 Brin. For springs, diaphragms, watch cases; age hardened, nonmagnetic, corrosion resistant. *Obsolete*

GENERAL PLATE 720
Texas Instruments Inc./Materials Control
Mn 20, Ni 20, bal Cu.
Wrought: 88,000-130,000 TS; 41,000-113,000 YS; 0.5-37 El; 150-300 Brin. For springs, diaphragms, watch cases; age hardened, nonmagnetic, corrosion resistant. *Obsolete*

GENERAL PLATE BH-1
Texas Instruments Inc./Materials Control
Ag 10, Cu 50, Zn 40.
For silver brazing; MP 1495-1590°F. *Obsolete*

GENERAL PLATE BH-2
Texas Instruments Inc./Materials Control
Ag 10, Cu 52, Zn 38.
For silver brazing. *Obsolete*

GENERAL PLATE CH-1
Texas Instruments Inc./Materials Control
Ag 40, Cu 36, Zn 24.
For silver brazing; MP 1340-1405°F. *Obsolete*

GENERAL PLATE CK-4
Texas Instruments Inc./Materials Control
Ag 45, Cu 30, Zn 25.
For silver brazing. *Obsolete*

GENERAL PLATE KA-1
Super Tool Co.
Ag 52.5, Cu 20, Zn 14, Cd 11.5, Ni 2.
For brazing carbide tips; melting point 1115-1105°F.

GENERAL PLATE KC-4
Texas Instruments Inc./Materials Control
Ag 54, Cu 40, Zn 5, Ni 1.
For silver brazing; MP 1430-1470°F. *Obsolete*

GENERAL PLATE KH-105
Texas Instruments Inc./Materials Control
Ag 50, Cu 15.5, Zn 25, Cd 10.
For silver brazing. *Obsolete*

GENERAL PLATE KH-2
Texas Instruments Inc./Materials Control
Ag 50, Cu 34, Zn 16.
For silver brazing. *Obsolete*

GENERAL PLATE KH-4
Texas Instruments Inc./Materials Control
Ag 50, Cu 15.5, Zn 15.5, Cd 16, Ni 3.
For brazing carbide tips; MP 1195-1270°F. *Obsolete*

GENERAL PLATE KH-7
Texas Instruments Inc./Materials Control
Ag 50, Cu 15.5, Zn 16.5, Cd 18.
For brazing; MP 1160-1175°F. *Obsolete*

GENERAL PLATE KK-5
Texas Instruments Inc./Materials Control
Ag 55, Cu 31.5, Zn 11.7, Ni 1.8.
For furnace brazing of steel parts; MP 1300-1355°F. *Obsolete*

GENERAL PLATE LH-1
Texas Instruments Inc./Materials Control
Ag 20, Cu 45, Zn 30, Cd 5.
For silver brazing; MP 1430-1500°F. *Obsolete*

GENERAL PLATE LH-3
Texas Instruments Inc./Materials Control
Ag 19.45, Cu 47.75, Zn 32.8.
For silver brazing; MP 1440-1500°F. *Obsolete*

GENERAL PLATE LH-4
Texas Instruments Inc./Materials Control
Ag 20, Cu 45, Zn 35.
For silver brazing. *Obsolete*

GENERAL PLATE LM-1
Texas Instruments Inc./Materials Control
Ag 27, Cu 40.15, Zn 38.25.
For silver brazing; MP 1350-1430°F. *Obsolete*

GENERAL PLATE MA-1
Texas Instruments Inc./Materials Control
Ag 72.15, Cu 22.8, Zn 5.05.
For silver brazing; MP 1345-1400°F. *Obsolete*

GENERAL PLATE MH-4
Texas Instruments Inc./Materials Control
Ag 70, Cu 20, Zn 10.
For silver brazing. *Obsolete*

GENERAL PLATE ML
Texas Instruments Inc./Materials Control
Ag 72, Cu 28.
For silver brazing; MP 1435°F; eutectic alloy. *Obsolete*

GENERAL PLATE SB-2
Texas Instruments Inc./Materials Control
Ag 60.5, Cu 22.5, Zn 7, Cd 10.
For silver brazing; MP 1285-1335°F. *Obsolete*

GENERAL PLATE SH-7
Texas Instruments Inc./Materials Control
Ag 60, Cu 20, Zn 7, Sn 3.
For silver brazing; MP 1270-1300°F. *Obsolete*

GENERAL PLATE SI-1
Texas Instruments Inc./Materials Control
Ag 68, Cu 27, Sn 5.
For silver brazing; MP 1370-1400°F. *Obsolete*

GENERAL PLATE SK-4
Texas Instruments Inc./Materials Control
Ag 65, Cu 20, Zn 15.
For silver brazing; MP 1285-1325°F. *Obsolete*

GENERAL PLATE SM-1
Texas Instruments Inc./Materials Control
Ag 66.7, Cu 28.25, Zn 5.05.
For silver brazing; MP 1360-1395°F. *Obsolete*

GENESEE 100
Symington-Gould Corp.
C 0.4, Cr, Mo, bal Fe.
Heat treated: 105,000-135,000 TS; 85,000-115,000 YS; 11-15 El; 25-30 RA. For gears, shafts; oil hardened.

GENESEE 180
Symington-Gould Corp.
C 0.3, Mn, bal Fe.
Normalized: 80,000 TS; 45,000 YS; 24 El; 45 RA. For structures; water hardened.

GENESEE 185
Symington-Gould Corp.
C 0.27, Mn, Mo, bal Fe.
Normalized: 85,000 TS; 53,000 YS; 22 El; 35 RA. For structures; water hardened.

GENESEE 191
Symington-Gould Corp.
C 0.3, Mn, Cr, Mo, bal Fe.
Rolled: 90,000 TS; 60,000 YS; 22 El; 45 RA. For abrasion resisting parts; shock resisting.

GENESEE 194
Symington-Gould Corp.
C 0.2, Mo, bal Fe.
Normalized: 65,000 TS; 35,000 YS; 24 El; 35 RA. For structures; case hardened.

GENESEE 194B
Symington-Gould Corp.
C 0.25, Mo, bal Fe.
Normalized: 70,000 TS; 45,000 YS; 22 El; 35 RA. For structures; water hardened.

GENESEE 195
Symington-Gould Corp.
C 0.27, Mn, Mo, Cr, bal Fe.
Normalized: 100,000 TS; 65,000 YS; 17 El; 30 RA. For structures; water hardened.

GENESEE 212
Symington-Gould Corp.
C 0-0.2, Cr 12-16, bal Fe.
For chemical engineering equipment; corrosion resistant.

GENESEE 255
Symington-Gould Corp.
Cr 4-6, Mo 0.5, C, bal Fe.
For chemical engineering equipment; corrosion resistant.

GENESEE 280
Symington-Gould Corp.
Cr 28, Ni 10, C, bal Fe.
For corrosion and heat resistant parts; heat and corrosion resistant.

GENESEE 303
Symington-Gould Corp.
C 0.2, Cr 25, Ni 20, bal Fe.
75,000 TS; 150 Brin. For corrosion and heat resistant parts; heat and corrosion resistant to sulfur fuels.

GENESEE 304
Symington-Gould Corp.
C 35, Cr 17, Ni 35, Mn 0.65, bal Fe.
For hearth supports, heat treating baskets, valves, salt pots; heat and corrosion resistant.

GENESEE 305
Symington-Gould Corp.
C 0.15, Cr 17, Ni 65, Mn 0.6, bal Fe.
For hearth supports, heat treating baskets, valves, salt pots; heat and corrosion resistant.

GENESEE 315
Symington-Gould Corp.
C 2.8, Mo, Si, bal Fe.
Cast: 40,000 TS; 250 Brin. For chemical engineering equipment, stoves, grates; alloy gray iron.

GENESEE 405
Symington-Gould Corp.
C 0.45, Cr 1, Ni 0.75, Mn 1.5, Mo 0.3, bal Fe.
For chemical engineering equipment; oil hardening.

GENESEE 412
Symington-Gould Corp.
Mn 11-14, C, Cr, bal Fe.
200 Brin. For chemical engineering equipment, crusher plates; wear and abrasion resistant.

GENESEE 460
Symington-Gould Corp.
C 0.6, Cr 1, Ni 3, Mo 0.3, bal Fe.
For chemical engineering equipment; oil hardening.

GENESEE 500
Symington-Gould Corp.
C 3.1, Cr 2, Ni 4.5, bal Fe.
550 Brin. For castings, pulverizers; corrosion and abrasion resisting.

GENESEE KA-2
Symington-Gould Corp.
C 0.2, Cr 18, Ni 8, bal Fe.
For chemical engineering equipment, pump parts; stainless.

GENESEE KA4
Symington-Gould Corp.
C 0-0.2, Cr, Ni, Mo, bal Fe.
Annealed: 75,000 TS; 160 Brin. For paper mill equipment; resists sulfites; austenitic, corrosion resistant.

GENESEE NI-HARD
Symington-Gould Corp.
C 3.2, Cr 2, Ni 4.5, bal Fe.
550 Brin. For chemical engineering equipment; wear resistant.

GENESEE NI-RESIST
Symington-Gould Corp.
Cr 2, Ni 13, Cu 6, C, bal Fe.
For chemical engineering equipment; heat resistant.

GENESEE NICKEL N-RESIST
Symington-Gould Corp.
Cr 2, Ni 16, C, bal Fe.
For castings; corrosion resisting.

GENESIS II
J.F. Jelenko & Co.
Au 53, Pt 27, W 10, Ru 3.
110,000 psi TS; 61,000 psi YS; 9 El; 265 Brin. Dental alloy for fusing porcelain to metal.

GENESSEE 110
Symington-Gould Corp.
C, alloy, bal Fe.
For magnetic applications. *Obsolete*

GENESSEE 130
Symington-Gould Corp.
C 0.4, Ni 1.5, Cr 0.8, bal Fe.
For gears, shafts. *Obsolete*

GENESSEE 190
Symington-Gould Corp.
C, alloy, bal Fe.
For abrasion resisting parts; shock resisting. *Obsolete*

GENSTEEL
General Malleable Corp.
C, alloy, bal Fe.
For castings.

GENUINE "A" BABBITT METAL
Hoyt Metal Co. of London Ltd.
Sn, Sb, Pb.
At 70 F: 28.6 Brin. At 212 F; 12.8 Brin. For bearings, bushings; M.P. 437 F; Babbitt. *Obsolete*

GENUINE BABBITT
Belmont Metals Inc.
Sb 7.5, Cu 3.5, bal Sn.
10,900 psi TS. For high speed bearings; melting point 669°F.

GENUINE SILVERINE
Glacier Metal Co.
Pb, Sb, bal Sn.
For bearings; Babbitt.

GENUINE SOVEREIGN BABBITT
Glacier Metal Co.
Sn 86-90, bal Cu, Sb.
Cast: 28.4 Brin. For internal combustion engine bearings; maximum freezing point 306°C.

GENUINE WROUGHT IRON
A.M. Byers Co.
C 0-0.05, Mn 0-0.05, P 0.1-0.15, S 0-0.03, Si 0.1-0.15, bal Fe.
45,000-50,000 TS; 25,000-30,000 YS; 35 El. For staybolts, rivets, roofing sheets, water and steam pipes, boiler tubes; ductile, easily welded. *Obsolete*

GEOMANT
Thyssen Edelstahlwerke AG
bal Fe.
Coated electrodes for overlaying of coal plows. Meets DIN 8555.

GEORO
GTE Products Corp./Wesgo Div.
Au 88, Ge 12.
Brazing alloy available in foil, flexibraze, wire, powder, extrudable paste and preform. Liquidus 673°F. Solidus 673°F.

GEORO
Western Gold & Platinum Co.
Au 88, Ge 12.
Melting point: 673°F. For brazing. Eutectic alloy.

GERMAN ALUMINUM ALLOY
German manufacture
Al 93, Mg 7.
For pistons; heat treatable.

GERMAN NAVY ANTIFRICTION METAL
German manufacture
Cu 7.5, Sn 85, Sb 7.5.
For bearings, bushings; Babbitt.

GERMAN SILVER BERLIN
German manufacture
Cu 54, Zn 28, Ni 18.
For ornamental purposes, electrical resistances; corrosion resistant.

GERMAN SILVER EXTRA WHITE
Century Brass Products Inc.
Ni 30, Cu 50, Zn 20.
For corrosion resistant parts; German Silver. *Obsolete*

GERMAN SILVER FOUNDRY ALLOY
Century Brass Products Inc.
Ni 35, Cu 45, Zn 20.
For white meal castings; German Silver. *Obsolete*

GERMAN SILVER RUSSIAN
Russian manufacture
Cu 45-56, Zn 18-23.5, Ni 20-36.
For ornamental purposes, electrical resistances; corrosion resistant.

GERMAN SILVER SOLDER-1
German manufacture
Ag 20, Cu 30, Zn 30, Cd 20.
For silver solder.

GERMAN SILVER SOLDER-2
German manufacture
Ag 10, Cu 40, Zn 30, Cd 20.
For silver solder.

GERMAN SILVER SOLDER-3
German manufacture
Ag 10, Cu 3, Zn 2, Sn 75.
For silver solder; corrosion resistant.

GERMAN SILVER WHITE
Century Brass Products Inc.
Ni 24, Cu 54, Zn 22.
For corrosion resistant parts; German Silver. *Obsolete*

GERMAN SILVER, AUSTRIAN
Austrian manufacture
Zn 50, Cu 4, Sn 3.3, Pb 1.2.
For solder, brazing; nearly white.

GERMAN SILVER, AUSTRIAN
Austrian manufacture
Cu 50-60, Zn 20-25, Ni 20-25.
For ornamental purposes, electrical resistances; corrosion resistant.

GERMAN SILVER, BEST
German manufacture
Cu 46-50, Ni 20-31, Zn 29-34.
For springs and contact points in electrical work; German Silver.

GERMAN SILVER, BIRMINGHAM
English manufacture
Cu 50-62, Zn 20-32, Ni 12-30.
For ornamental purposes, electrical resistances; corrosion resistant.

GERMAN SILVER, COMMON FORMULA
German manufacture
Cu 55, Zn 25, Ni 20.
For ornamental purposes, electrical resistances; corrosion resistant.

GERMAN SILVER, FIFTHS
Century Brass Products Inc.
Ni 7, Cu 57, Zn 36.
Soft: 55,000 TS; 20,000 YS; 50 El. For ornamental parts; German Silver. *Obsolete*

GERMAN SILVER, FIRSTS "A" SPECIAL
Century Brass Products Inc.
Ni 17, Cu 56, Zn 27.
For electrical resistances, flat work; German Silver. *Obsolete*

GERMAN SILVER, FIRSTS "A-1" BEST
Century Brass Products Inc.
Ni 16, Cu 56, Zn 28.
Soft: 60,000 TS; 50 El. For cheap flat work; German Silver. *Obsolete*

GERMAN SILVER, FOURTHS
Century Brass Products Inc.
Ni 10, Cu 55, Zn 35.
For ornamental parts; German Silver. *Obsolete*

GERMAN SILVER, FRENCH (ARCET)
French manufacture
Cu 50, Zn 30-31.3, Ni 18.7-20.
For ornamental purposes, electrical resistances; corrosion resistant.

GERMAN SILVER, FRENCH (CHAVAL)
French manufacture
Cu 58.3, Zn 25, Ni 16.7.
For ornamental purposes, electrical resistances; corrosion resistant.

GERMAN SILVER, SECONDS
Century Brass Products Inc.
Ni 14, Cu 62, Zn 24.
Hard: 120,000 TS; 2 El. For ornamental and decorative parts; German Silver. *Obsolete*

GERMAN SILVER, SHEFFIELD
English manufacture
Cu 55.2, Zn 24.1, Ni 20.7.
For ornamental purposes, electrical resistances; Silver White.

GERMAN SILVER, SHEFFIELD
English manufacture
Cu 45.7, Zn 20, Ni 31.3.
For ornamental purposes, electrical resistances; Hard Alloy.

GERMAN SILVER, SHEFFIELD
English manufacture
Cu 59.3, Zn 25.9, Ni 14.8.
For ornamental purposes, electrical resistances; Common Yellow.

GERMAN SILVER, SHEFFIELD
English manufacture
Cu 51.6, Zn 22.6, Ni 25.8.
For ornamental purposes, electrical resistances; Electrum, Bluish.

GERMAN SILVER, SPECIAL THIRDS
Century Brass Products Inc.
Ni 11, Cu 56.5, Zn 32.5.
For white metal trim; German Silver. *Obsolete*

GERMAN SILVER, THIRDS
Century Brass Products Inc.
Ni 12, Cu 56, Zn 32.
For white metal trim; German Silver. *Obsolete*

GERMAN TYPE METAL "A"
German manufacture
Pb 75, Sn 2, Sb 23.
For type metal.

GERMAN TYPE METAL "B"
German manufacture
Pb 60, Sn 35, Sb 5.
For type metal.

GERMAN TYPE METAL "C"
German manufacture
Pb 60, Sn 34.6, Sb 5.4.
For type metal.

GERMAN TYPE METAL "D"
German manufacture
Pb 60, Sn 15, Sb 25.
For type metal.

GERMANIA B. BRONZE
English manufacture
Zn 80.4, Sn 9.6, Cu 4.4, Pb 4.7, Fe 0.8.
For bearings; will not resist heat or live stream.

GERMANIUM
Atomergic Chemetals Corp.
Ge.
Grades: intrinsic 30, 40, 50 ohm/cm, First reduction 5 ohm/cm, epitaxial. Forms: semicircular, trapezoidal, circular rod, lump, powder, single crystals, slices.

GEROWAL 11
G. Robert Wilms A.G.
Cu 4-7, Si 2-4, Zn 0-2.5, Fe 0-1.1, bal Al.
For light alloy parts; age hardenable.

GEROWAL I
G. Robert Wilms A.G.
Si 5.5-7, Cu 2-4, Mn 0.4-0.6, Zn 0-2, Fe 0-1, bal Al.
Heat treated: 36,000 TS; 24,000 YS; 2 El; 80 Brin. For light alloy parts; age hardenable.

GEROWI SILBERIT
G. Robert Wilms A.G.
Mg 1.5-3.5, Si 0-1.3, Mn 0-0.6, Cr 0.2, Ti 0.2, bal Al.
Soft: 20,000 TS; 13,000 YS; 30 El; 47 Brin. For aircraft tanks and fittings, fuel lines, marine parts; resists seawater corrosion.

GEROWI SILBERIT-K
G. Robert Wilms A.G.
Mg 2-4, Mn 0-0.4, Cr 0-0.3, bal Al.
Soft: 28,000 TS; 13,000 YS; 30 El; 47 Brin. Hard: 40,000 TS; 35,000 YS; 10 El; 73 Brin. For aircraft tanks and fittings, fuel lines, marine parts; resists seawater corrosion.

GEWA 235
Stahlwerke Sudwestfalen
C 0.35, Si 0.9, Mn 0.3, Cr 1, V 0.2, W 1.8, bal Fe.
For cold work tools, upsetters, crimpers, tough, shock resistant. *Obsolete*

GEWA 250
Stahlwerke Sudwestfalen
C 0.45, Si 0.9, Mn 0.3, W 2, Cr 1, V 0.4, bal Fe.
For punches, chisels, riveters, crimpers, upsetters; tough, shock resistant. *Obsolete*

GEWA 255
Stahlwerke Sudwestfalen
C 0.55, Si 0.9, Mn 0.3, Cr 1, V 0.2, W 1.8, bal Fe.
For punches, chisels, riveters, crimpers, upsetters; tough, shock resistant. *Obsolete*

GEWA 2930
Stahlwerke Sudwestfalen
C 0.3, Co 2, Cr 2.4, V 0.25, W 8.5, bal Fe.
For extrusion rams and liners, punches, shears; hot work steel, oil hardened. *Obsolete*

GEWA 430
Stahlwerke Sudwestfalen
C 0.3, Si 1, Mn 0.4, Cr 1.1, V 0.18, W 3.75, bal Fe.
For punches, chisels, riveters, crimpers, upsetters; tough, shock resistant. *Obsolete*

GEWA 530
Stahlwerke Sudwestfalen
C 0.3, Mn 0.3, Cr 2.3, V 0.6, W 4.2, bal Fe.
For punches, upsetters, riveters, shears, crimpers; hot work steel, oil hardened. *Obsolete*

GEWA 930
Stahlwerke Sudwestfalen
C 0.3, Si 0.2, Mn 0.3, Cr 2.6, V 0.35, W 8.5, bal Fe.
For extrusion rams and liners, punches, upsetters; hot work steel, oil hardened. *Obsolete*

GEWA 960
Stahlwerke Sudwestfalen
C 0.65, Cr 3.7, Mo 0.85, V 0.7, W 8.5, bal Fe.
For drills, reamers, taps, hobs, broaches; high speed steel, oil hardened. *Obsolete*

GIANT SPECIAL
Champion Steel Co.
C 0.35, Cr 0.9, Mo 0.3, bal Fe.
Oil or water hardening steel designed for plastic molds.

GIBRALTAR
H. Boker & Co.
C 0.95-1.05, Si 0.3, Mn 0.25, bal Fe.
For cutters, reamers, drills, punches, broaches, taps; Type W1; water hardened. *Obsolete*

GIBSILOY A 3
Gibson Electric Co.
Ag 50.95, Ni 5-50.
For electric contacts. Sintered powders, ductile. *Obsolete*

GIBSILOY A-1
Gibson Electric Co.
Ag 95, Ni 5.
For contact rivets, discs, screws. Low contact resistance, high ductility. *Obsolete*

GIBSILOY A-4
Gibson Electric Co.
Ag, Ni.
For electrical contacts, circuit breakers. *Obsolete*

GIBSILOY A-6
Gibson Electric Co.
Ag, Ni.
For electrical contacts, circuit breakers. *Obsolete*

GIBSILOY A8
Gibson Electric Co.
Ag, Ni.
Sintered: 36,000-58,000 TS. For electrical contacts, disconnecting switch contacts. *Obsolete*

GIBSILOY C 4
Gibson Electric Co.
Ag 93-99, 1.0-7.0 graphite.
For electric contacts, sliding. Mechanically processed.
Obsolete

GIBSILOY C-1
Gibson Electric Co.
Ag 99, 1.0 graphite.
For contact discs. Non-sticking. *Obsolete*

GIBSILOY C-2
Gibson Electric Co.
Ag, C.
For electrical contacts, disconnecting switches. *Obsolete*

GIBSILOY C-7
Gibson Electric Co.
Ag 93, graphite.
For contact discs. Non-sticking. *Obsolete*

GIBSILOY C5
Gibson Electric Co.
Ag, C.
Sintered. For electrical contacts, circuit breakers. Non-welding. *Obsolete*

GIBSILOY CW-42
Gibson Electric Co.
Ag, W, C.
For electrical contacts, circuit breakers. *Obsolete*

GIBSILOY CW-52
Gibson Electric Co.
Ag, W, C.
For electrical contacts, air circuit breakers. *Obsolete*

GIBSILOY M 10
Gibson Electric Co.
Ag 30-90, Mo 10-70.
For electrical contacts. Arcing. *Obsolete*

GIBSILOY M-12
Gibson Electric Co.
Ag, Mo.
Sintered: 55,000 TS; 165 Brin. For electrical contacts for air circuit breakers. Arcing. *Obsolete*

GIBSILOY NC 22
Gibson Electric Co.
Ag, Ni, graphite.
For electrical contacts, rheostat contacts. *Obsolete*

GIBSILOY NC-43
Gibson Electric Co.
Ag, Ni, C.
For electrical contacts, rheostats, contactors. *Obsolete*

GIBSILOY ND
Gibson Electric Co.
Ni 5-25, Cd 1-10, bal Ag.
For electrical contacts. *Obsolete*

GIBSILOY NM
Gibson Electric Co.
Ni 1-35, Mo 1-30, bal Ag.
For electrical contacts; for high current densities. *Obsolete*

GIBSILOY NW
Gibson Electric Co.
Ni 1-35, W 1-30, bal Ag.
For electrical contacts; for high current densities. *Obsolete*

GIBSILOY NW-54
Gibson Electric Co.
Ag, Ni, W.
For electrical contacts; DC service. *Obsolete*

GIBSILOY NW-55
Gibson Electric Co.
Ag, Ni, W.
For electrical contacts, circuit breakers. *Obsolete*

GIBSILOY UC-5
Gibson Electric Co.
Ag, Cu, C.
For electrical contacts, rheostats, contactors. *Obsolete*

GIBSILOY UT-10
Gibson Electric Co.
Cu, WC.
Sintered: 95 Rock B; density 11.1. For electrical contacts, switches. *Obsolete*

GIBSILOY UT-6
Gibson Electric Co.
Cu, WC.
Sintered: 100 Rock B; density 12.3. For electrical contacts, switches. *Obsolete*

GIBSILOY UT8
Gibson Electric Co.
Cu, WC.
Sintered: 230 Brin. For electrical contacts and facings for resistance welding electrodes. *Obsolete*

GIBSILOY UW-5
Gibson Electric Co.
Cu, W.
Sintered: 100 Rock B; density 14.3. For electrical contacts, switches. *Obsolete*

GIBSILOY UW-8
Gibson Electric Co.
Cu, W.
Sintered: 85 Rock B; density 12.6. For electrical contacts, switches. *Obsolete*

GIBSILOY UW4
Gibson Electric Co.
Cu, W.
For electrical contacts and facings for resistance welding electrodes; sintered. *Obsolete*

GIBSILOY UW6
Gibson Electric Co.
Cu, W.
Sintered: 90 Rock B; 50% electrical conductivity. For electric contacts, switches. *Obsolete*

GIBSILOY W 10
Gibson Electric Co.
Ag, W.
For electrical contacts. Sintered powders. *Obsolete*

GIBSILOY W 2
Gibson Electric Co.
Ag 30-90, W 10-70.
For electrical contacts. *Obsolete*

GIBSILOY W-13
Gibson Electric Co.
Ag, W.
Sintered: 43,000 TS; 150 Brin. For electrical contacts for air circuit breakers. Arcing. *Obsolete*

GIBSILOY W-15
Gibson Electric Co.
Ag, W.
Sintered: 63,000 TS; 200 Brin. For electrical contacts for air circuit breakers. Arcing. *Obsolete*

GIBSILOY W-4
Gibson Electric Co.
Ag, W.
For electrical contacts, contactor contacts. *Obsolete*

GIBSILOY W12
Gibson Electric Co.
Ag, W.
For electrical contacts, air circuit breakers. *Obsolete*

GIBSILOY WC-10
Gibson Electric Co.
Ag, WC.
For electrical contacts for air circuit breakers. *Obsolete*

GIBSILOY WID
Gibson Electric Co.
Ag, W.
For electrical contacts for air circuit breakers. *Obsolete*

GIBSON S-10
Gibson Electric Co.
Ag, Cd.
For thermostats, contactors, switches, relays, electrical contacts. *Obsolete*

GIBSON S-15
Gibson Electric Co.
Ag, Pd.
For thermostats, contactors, switches, relays, electrical contacts. *Obsolete*

GIBSON S-16
Gibson Electric Co.
Ag, Pd.
For electrical contacts, thermostats, switches, relays. *Obsolete*

GIBSON S-20
Gibson Electric Co.
Ag, Pt.
For thermostats, contactors, switches, relays, electrical contacts. *Obsolete*

GIBSON S-25
Gibson Electric Co.
Ag, Cu.
For electrical contacts. *Obsolete*

GIBSON S-26
Gibson Electric Co.
Ag, Cu, Ni.
For electrical contacts, relays, switches. *Obsolete*

GIBSON S-5
Gibson Electric Co.
Ag, Zn.
For thermostats, contactors, switches, relays, electrical contacts. *Obsolete*

GIESCHE Z13
Zinkberatungsstelle GmbH
Al 4, Cu 0.5-1, Mg 0.03, bal Zn.
Rolled: 44,000-71,000 TS; 75-140 Brin. For die castings, instrument cases, and housings. Rolled or cast.

GIESCHE Z14
Zinkberatungsstelle GmbH
Al 0.8-1, Cu 0.3-0.4, bal Zn.
Rolled: 31,000-42,500 TS; 40-70 Brin. For die castings, housings, gears. Rolled or cast.

GIESCHE ZL 2
German manufacture
Al 4, Cu 1-1.5, bal Zn.
For die or sand castings.

GIESCHE ZL 3
German manufacture
Al 4, Cu 0.5, bal Zn.
For pressed or drawn parts.

GIESCHE ZL7
Zinkberatungsstelle GmbH
Cu 4, Al 0.2, bal Zn.
Rolled: 28,000-59,600 TS; 45-100 Brin. For die castings, housings, gears. Rolled or cast.

GIESCHE ZO4
Zinkberatungsstelle GmbH
Cu 4, bal Zn.
For die castings. Rolled or cast.

GIGANT
Rochling Burbach GmbH
C 1.05-1.15, Cr 4.5, W 11-12, Mo 0.6, V 2.5, bal Fe.
For tools, dies, cutters; high speed steel. *Obsolete*

GIGANT
Edelstahlwerk Rochling AG
C 0.7, W 18, Cr 4, V 1, bal Fe.
For cutters, tools; high speed steel.

GIGANT 10
Rochling Burbach GmbH
W 18, Co 9, Mo 1, Cr 4, V 2, C, bal Fe.
For tools, dies, cutters; high speed steel. *Obsolete*

GIGANT 100
Saarstahl AG
C 1.25, Cr 4.25, Mo 3.75, V 3.25, W 10, Co 12.5, bal Fe.
High speed steel, particularly for automatic lathe tools. W.-Nr. 3207; DIN S 10-4-3-10. *Obsolete*

GIGANT 11
Rochling Burbach GmbH
C 0.92-0.98, Cr 4, W 1.2-1.5, Mo 2.2-2.5, V 2.2-2.5, bal Fe.
For tools, dies, cutters; high speed steel. *Obsolete*

GIGANT 201
Saarstahl AG
Now SAARSTAHL 1.3246.

GIGANT 22
Rochling Burbach GmbH
C 0.78-0.85, Cr 3.5-4, W 7.3-8, V 1.8-2, bal Fe.
For tools, dies, cutters; high speed steel. *Obsolete*

GIGANT 301
Saarstahl AG
C 1.45, Cr 4.25, Mo 5, V 2.4, W 6.35, Co 8, bal Fe.
High speed steel; for finishing tools and for machining abrasive materials. W.-Nr. 3222; DIN S 6-5-2-8. *Obsolete*

GIGANT 33
Rochling Burbach GmbH
C 0.85-0.95, Cr 4.5, W 13, Mo 0.7, V 2, Co 2.4, bal Fe.
For tools, dies, cutters; high speed steel. *Obsolete*

GIGANT 44
Saarstahl AG
C 1.3, Cr 4.25, Mo 0.85, V 3.75, W 12, bal Fe.
High speed steel for broaches, lathe tools for finishing cuts. W.-Nr. 3302; DIN S 12-1-4. *Obsolete*

GIGANT 44
Rochling Burbach GmbH
C 1.3-1.4, Cr 4-4.5, W 9-10, V 3.5-4, bal Fe.
For tools, dies, cutters; high speed steel. *Obsolete*

GIGANT 5
Rochling Burbach GmbH
C 0.9, Cr 4, W 18, Co 5, V 1.3, bal Fe.
For tools, dies, cutters; high speed steel. *Obsolete*

GIGANT 50
Saarstahl AG
C 0.75, Cr 4.25, V 1.1, W 18, bal Fe.
High speed steel for lathe tools, milling cutters, taps. W.-Nr. 3355; AISI T1. *Obsolete*

GIGANT 50
Rochling Burbach GmbH
C 0.75, Cr 4, V 1, W 18, bal Fe.
For lathe and planer tools, drills, taps, hobs; high speed steel. *Obsolete*

GIGANT 55
Rochling Burbach GmbH
C 0.86, Co 2.8, Cr 4.3, Mo 8.5, V 2.1, W 12, bal Fe.
For lathe and planer tools, reamers, broaches, hobs; high speed steel. *Obsolete*

GIGANT 60
Rochling Burbach GmbH
C 0.7, W 18, Mo 1, V 1.5, bal Fe.
For lathe and planer tools, drills, taps, reamers; high speed steel. *Obsolete*

GIGANT 66
Saarstahl AG
C 1.4, Cr 4.25, Mo 0.85, V 3.75, W 12, Co 4.8, bal Fe.
High speed steel for cutting tools and special operations. Resistant to wear; W.-Nr. 3202. Similar to AISI T15. *Obsolete*

GIGANT 66
Rochling Burbach GmbH
C 1.4-1.5, Cr 4-4.5, W 12-13, Mo 0.4-0.7, V 4.2-5, Co 4.2-5, bal Fe.
For tools, dies, cutters; high speed steel. *Obsolete*

GIGANT 70
Rochling Burbach GmbH
C 0.8, Cr 4, V 1, W, Co, Mo, bal Fe.
For lathe and planer tools, drills, taps, reamers; high speed steel. *Obsolete*

GIGANT 77
Saarstahl AG
C 0.8, Cr 4.25, Mo 0.65, V 1.55, W 18, Co 4.8, bal Fe.
High speed steel for lathe tools and other cutting tools for special operations. Resistant to elevated temperatures in difficult cutting. W.-Nr. 3255; similar to AISI T4. *Obsolete*

GIGANT 77
Rochling Burbach GmbH
C 0.8, Co 4.7, Cr 4, Mo 0.7, V 1.5, W 18, bal Fe.
For lathe and planer tools, reamers, taps, hobs; high speed steel. *Obsolete*

GIGANT 88
Saarstahl AG
C 0.75, Cr 4.25, Mo 0.65, V 1.55, W 18, Co 9.5, bal Fe.
High speed steel; best red-hardness. For lathe tools. W.-Nr. 3265; similar to AISI T5. *Obsolete*

GIGANT 88
Rochling Burbach GmbH
C 0.76, Co 10, Cr 4, Mo 0.8, V 1.8, W 18, bal Fe.
For lathe and planer tools, reamers, broaches; high speed steel. *Obsolete*

GIGANT BST
Rochling Burbach GmbH
C 0.85, Cr 4, W, Mo.
For lathe and planer tools, reamers, hobs, taps; high speed steel. *Obsolete*

GIGANT DUPLO
Rochling Burbach GmbH
C 1.35, W, Co, Cr, bal Fe.
For fast finishing tools, engraving cutters; high speed steel; keen edge. *Obsolete*

GIGANT M 5 CO
Saarstahl AG
Now SAARSTAHL 1.3243.

GIGANT M 5 H
Saarstahl AG
Now SAARSTAHL 1.3247.

GIGANT M 5 S
Saarstahl AG
C 0.88, S 0.12, Cr 4.25, Mo 5, V 1.85, W 6.35, bal Fe.
Free machining Mo-W high speed steel. For special tools or intricate shapes. *Obsolete*

GIGANT M 5 V
Saarstahl AG
Now SAARSTAHL 1.3344.

GIGANT M 9
Saarstahl AG
Now SAARSTAHL 1.3346.

GIGANT M 9 V
Saarstahl AG
Now SAARSTAHL 1.3348.

GIGANT M5
Saarstahl AG
Now SAARSTAHL 1.3343.

GIGANT M5
Rochling Burbach GmbH
C 0.85, Cr 4, W, Mo, bal Fe.
For lathe and planer tools, drills, taps, hobs; high speed steel. *Obsolete*

GIGANT N
Rochling Burbach GmbH
C 0.72-0.8, Cr 4.5, W 13, Mo 0.7, V 1.7, Co 2.4, bal Fe.
For tools, dies, cutters; high speed steel. *Obsolete*

GIGANT UNO
Rochling Burbach GmbH
C 0.72-0.8, W 13, Mo 0.7, V 1.7, Co 5, Cr 4, bal Fe.
For tools, dies, cutters; high speed steel. *Obsolete*

GILDING 210
Anaconda Co.
Copper. Cu 95, Zn 5.
Hard sheet: 56,000 TS; 46,000 YS; 5 El; 64 Rock B. Soft sheet: 36,000 TS; 11,000 YS; 42 El; 50 Rock F. For coins, costume jewelry, medallions, ornamental trim. 56% electrical conductivity; malleable.

GILDING 4
Anaconda Co.
Copper. Cu 95, Zn 5.
Soft: 35,000 TS; 11,000 YS; 38 El. Hard: 55,000 TS; 44,000 YS; 5 El. For jewelry and bullet jackets, fuse caps, primers. Gilding metal.

GILDING 95%
Chase Brass & Copper Co., Inc.
Copper. Cu 95, Zn 5.
Annealed: 35,000 psi TS; 11,000 psi YS; 45 El. Hard rolled: 56,000 TS; 50,000 psi YS; 5 El. For angles, channels; red brass.

GILDING FOIL
American manufacture
Cu 98, Cu 2.2, Fe 0.1.
For cheap jewelry; corrosion resistant.

GILDING METAL
Chase Brass & Copper Co.
Cu 95-97, Zn 3-5.
35,000-60,000 TS; 10,000-30,000 YS; 5-40 El. For cheap jewelry; corrosion resistant. *Obsolete*

GILDING METAL
English manufacture
Cu 64-72, Zn 23-34, Sn 0.3-2.5, P 0.3.
For cheap jewelry, bullet jackets; corrosion resistant.

GILDING METAL 26

Olin Brass, Indianapolis
Copper. Cu 95, Zn 5.
Hard: 56,000 psi TS; 50,000 psi YS; 5 El; 114 Brin. Soft:
35,000 psi TS; 11,000 psi YS; 45 El; 50 Brin. For jewelry;
gilding metal.

GILDING METAL CLAD STEEL 747

Texas Instruments Inc./Materials Control
5 brass, 80 steel, 15 brass (nominal).
Small arms ammunition, coinage, decorative brass clad steel.

GILDING METAL, 95%

Chase Brass & Copper Co.
Cu 95, Zn 5.
Annealed: 35,000 TS; 11,000 YS; 45 El. Rolled hard: 56,000
TS; 50,000 YS; 5 El. For angles, coinage, medallions. Good
corrosion resistance.

GILDING METAL, SCOVILL ALLOY 110

Century Brass Products Inc.
Cu 94-96, bal Zn.
Annealed: 35,000 TS 45 El. Hard: 55,000 TS; 5 El; 114 Brin.
For drawn and spun parts; corrosion resistant. *Obsolete*

GILGRID 10

Gilby-Fodor S.A.
Ni 45, Fe 45, Mo 10.
Annealed: 90,000 TS; 72,000 YS; 33 El. For grid wires, and
electronic tubes. Corrosion and heat resistant.

GILGRID 30

Gilby-Fodor S.A.
Ni 67, Mo 30, Cr 1, Si 1, Mn 1.
Annealed: 90,000 TS; 63,000 YS; 15 El; 18 RA. For grid wires
and electronic tubes. Corrosion and heat resistant.

GILMORE NICKEL ALLOY

F.F. Gilmore & Co.
Cu, bal Ni.
For diamond setting tools, holders.

GILMORE TOOL

Bissett Steel Co.
C 0.85, Cr 3.5, W 1, bal Fe.
For tools and dies; hot or cold work steel. *Obsolete*

GILSON BM12C20

Gilson Ltd.
C 0-0.2, Cr 22-24, Ni 12-15, Mn 0-2, Si 0-3.5, bal Fe.
Annealed: 85,000-95,000 TS; 40,000-50,000 YS; 45-55 El;
150-185 Brin. For heat treat boxes, furnace parts and
equipment; Type 309; austenitic, heat resistant.

GILSON UC-BC14

Gilson Ltd.
C 0-0.15, Cr 11.5-13.5, bal Fe.
Heat treated: 120,000-135,000 TS; 110,000-117,000 YS;
15-16 El; 58-63 RA; 220-240 Brin. For cutlery, valves, turbine
blades; Type 410; corrosion resistant.

GILSON UG

Gilson Ltd.
C 0-0.2, Cr 22-24, Ni 12-15, Mn 0-2, Si 0-3.5, bal Fe.
Annealed: 85,000-95,000 TS; 40,000-50,000 YS; 45-55 El;
150-185 Brin. For heat treat boxes, furnace parts, refinery
equipment; Type 309; austenitic, heat resistant.

GILSON UG-AM

Gilson Ltd.
C 0-0.1, Cr 16-18, Ni 10-14, Mo 1.7-2.7, Cb = 10 x C, bal Fe.
Annealed: 85,000-95,000 TS; 35,000-45,000 YS; 50-60 El;
60-75 RA; 150-190 Brin. For acid resistant chemical plant
equipment; Type 316 Cb; stainless, austenitic.

GILSON UG-AM 8C, 18MO

Gilson Ltd.
C 0-0.08, Cr 17-19, Ni 8-11, Ti = 5 x C, bal Fe.
Normalized: 93,000 TS; 36,000 YS; 45 El; 60 RA; 165 Brin.
Annealed: 87,000 TS; 33,000 YS; 57 El; 73 RA; 155 Brin. For
chemical plant equipment, muffles, cowls; Type 321;
stainless, austenitic.

GILSON UG-CC14

Gilson Ltd.
Cr 12-14, 0.15 C min, bal Fe.
Annealed: 88,000 TS; 40,000 YS; 32 El; 68 RA; 170 Brin. Heat
treated: 256,000 TS; 190,000 YS; 6 El; 10 RA; 540 Brin. For
cutlery, valve trim, springs, turbine blades; Type 420;
stainless, hardenable.

GILSON-UGAC14

Gilson Ltd.
C 0.08, Cr 11.5-13, Al 0.1-0.3, bal Fe.
Normalized: 71,000 TS; 42,600 YS; 22 El; 150 Brin. Heat
treated: 175,000 TS; 145,000 YS; 21 El; 352 Brin. For
injectors, table flatware, valves, pumps, oil refinery
equipment. Type 405, stainless.

GIMAP

Pechiney/SOFREM
Powdered manganese alloys for coating of welding rods and
wire for welding.

GIMEL

Pechiney/SOFREM
Mn 96-98, Fe 1-3.
For alloying in steel, copper alloys and light alloys;
electrothermic manganese metal.

GITTERMETALL

German manufacture
Sn 10, Sb 15, Cu 1.75, 0.20 graphite, bal Pb.
For bearings; graphitic.

GKM Z 400

G.K.M. Greuter & Kerscher GmbH & Co. KG
Zinc. Al 3.9-4.3, Mg 0.03, Cu 0-0.03, Fe 0-0.02, Sn 0-0.001,
0.006 Pb + Cd max, bal Zn.
Zinc alloy addition for pressure casting. Pouring temperature
400-420°C; MP 390-385°C; 25-30 kg/mm^2 TS; 3-6 El;
specific gravity 6.6; bending strength 8-12 cmkg/cm^2. DIN
1743.

GKM Z 410

G.K.M. Greuter & Kerscher GmbH & Co. KG
Zinc. Al 3.9-4.3, Cu 0.75-1.25, Mg 0.03-0.06, Fe 0-0.02, Sn
0-0.001, 0.006 Pb + Cd max, bal Zn.
Zinc alloy addition for pressure casting. Pouring temperature
410-420°C; MP 395-380°C; 28-35 kg/mm^2 TS; 2-5 El;
specific gravity 6.7; bending strength 7-11 cmkg/mm^2. DIN
1743.

GKM Z 430

G.K.M. Greuter & Kerscher GmbH & Co. KG
Zinc. Al 3.9-4.3, Cu 2.5-3.2, Mg 0.03-0.06, Fe 0-0.05, Sn
0-0.001, 0.006 Pb + Cd max, bal Zn.
Zinc alloy addition for sand and chill casting. Sand casting:
21-26 kg/mm^2 TS; 0.5-2.5 El; specific gravity 6.75; bending
strength 6-8 cmkg/mm^2. Chill casting: 24-30 kg/mm^2 TS; 1-3
El; specific gravity 6.75; bending strength 6-8 cmkg/mm^2.

GKM Z 610

G.K.M. Greuter & Kerscher GmbH & Co. KG
Zinc. Al 5.6-6, Cu 1.2-1.6, Fe 0-0.05, Mg 0-0.005, Sn 0-0.001,
0.006 Pb + Cd max, bal Zn.
Zinc alloy addition for sand and chill casting. Sand casting:
18-23 kg/mm^2 TS; 1.5-3.5 El; specific gravity 6.5; bending
strength 4-6 cmkg/mm^2. Chill casting: 23-27 kg/mm^2 TS;
1.5-3.5 El; specific gravity 6.5; bending strength 4-6
cmkg/mm^2.

GKN 58-L

Presmet Corp.
Fe 97, 3.0 graphite max.
Sintered, oil impregnated (26%); 10,400 psi compressive
strength. Bearings for light duty motors. *Obsolete*

GKN 59-FM

Presmet Corp.
Fe 96.25, S 0.4, 3.0 other elements max.
Sintered, oil impregnated (25%); 17,000 TS; 11,500 psi
compressive strength. Excellent machining, ductile enough
for staking or spinning. For low and medium loaded
bearings. ASTM B439-67 Grade 1; ASTM B310-67 Type I
Class A2, SAE 850; PMPMA F-0000-N. *Obsolete*

GKN 59-I

Presmet Corp.
Fe 97, 3.0 other elements max.
Sintered, oil impregnated (25%); 12,500 psi compressive
strength. For low and medium load bearings. ASTM B439-67
Grade 1; ASTM B310-67 Type I Class A; SAE 850; PMPMA
F-0000-N. *Obsolete*

GKN 59-PC

Presmet Corp.
C 0.7, Cu 4, Fe 95.3.
Sintered and oil impregnated (20%); 31,000 psi compressive
strength. Good strength, for products subject to frequent
impact or shock loading. ASTM B426-65 Grade 2 Type 1;
SAE 865 A. *Obsolete*

GKN 61-A

Presmet Corp.
Cu 87, Sn 9.5, 3.5 graphite.
Sintered, oil impregnated (20%); 10,000 psi compressive
strength. Bearings for light, non-shock loads. *Obsolete*

GKN 61-P

Presmet Corp.
Cu 10, 87.0 Fe min, 0.0-3.0 other elements.
Sintered, oil impregnated (25%); 30,000 psi compressive
strength. Good load bearing qualities for wheel bearings,
automotive and farm equipment. ASTM B439-67 Grade 3;
ASTM B222-61; SAE 862; PMPMA; F-1000-N.

GKN 62-E

Presmet Corp.
Cu 38, N 4, 55.75 Fe min, 1.0 graphite.
Sintered, oil impregnated (23%); 12,500 psi compressive
strength. For self-aligning bearings in motors, fan motors,
agricultural equipment.

GKN 63-H

Presmet Corp.
Cu 89, Sn 9.75, 1.25 graphite.
Sintered, oil impregnated (27%); 10,500 psi compressive
strength. For self-aligning bearings in fractional horse power
motors.

GKN 63-PC

Presmet Corp.
C 0.7, Cu 4, Fe 95.3.
Sintered, oil impregnated (8%); 48,000 TS. High strength
grade for units subject to frequent shock or impact loading.
Obsolete

GKN 63-PZ

Presmet Corp.
C 0.7, Cu 20, 76.8 Fe min, 0.0-2.5 other elements.
As sintered: 47,000 TS; 33,000 psi compressive strength;
density 6.3. For bearings having intermittent or oscillating
loading; needs supplementary lubrication. ASTM B426 Grade
4 Type II; SAE 867B. *Obsolete*

GKN 66-H

Presmet Corp.
Cu 89, Sn 9.75, 1.25 graphite.
Sintered, oil impregnated (23%); 12,000 psi compressive
strength. Good load bearing qualities for consumer products
and industrial equipment. ASTM B438-70 Grade 1 Type II;
SAE 841; PMPMA BT-0010-R.

GKN 66-Q

Presmet Corp.
Cu 90, Sn 10.
Sintered, oil impregnated (25%); 12,000 psi compressive strength. Good bearing qualities; can be machined and staked. ASTM B438-70 Grade 1 Type II; SAE 841; PMPMA BT-0010-R. *Obsolete*

GKN 66-R

Presmet Corp.
Cu 87, Sn 9.5, 3.5 graphite.
Sintered, oil impregnated (18%); 12,000 psi compressive strength. For oscillating and reciprocating loads. *Obsolete*

GKN 70-H

Presmet Corp.
Cu 89, Sn 9.75, 1.25 graphite.
Sintered, oil impregnated (15%); 15,000 psi compressive strength; density 7.0. Bearings for heavy duty service such as construction equipment, farming equipment. ASTM B255-70 Type II; SAE 842; PMPMA BT-0010-S. *Obsolete*

GKN 70-Q

Presmet Corp.
Cu 90, Sn 10.
Sintered, oil impregnated (20%); 15,000 psi compressive strength. Bearings for shock and high loading. *Obsolete*

GKN 70-R

Presmet Corp.
Cu 87, Sn 9.5, 3.5 graphite.
Sintered, oil impregnated (5%); 18,000 psi compressive strength. For oscillating and reciprocating loads. *Obsolete*

GKN C-000

Presmet Corp.
99.0 Cu min.
As sintered: 20,000-34,000 TS; 10-17 El; 60-90 Rock H; density 7.7-8.5. Excellent electrical and thermal conductivity. *Obsolete*

GKN CT-100

Presmet Corp.
Cu 87.5-90.5, Sn 9.5-10.5.
As sintered: 13,500-20,000 TS; 1.0-3.0 El; 50-90 Rock H; density 6.4-7.2. For bearings and structural parts; may be coined for dimensional control. ASTM B255 Type I or II; CT-0010-R; CT-0010-S.

GKN CZ-103

Presmet Corp.
Cu 87-90, Pb 1-2, 7.35 Zn min.
As sintered: 16,500-19,400 TS; 10-12 El; 45-70 Rock H; density: 7.2-8.0. Commercial bronze for ornamental parts and lock hardware with good machinability. CZP-0210-T; CZP-0210-U. *Obsolete*

GKN CZ-203

Presmet Corp.
Cu 77-80, Pb 1-2, 17.35 Zn min.
As sintered: 20,000-30,000 TS; 10-19 El; 50-85 Rock H; density 7.2-8.0. For lock hardware, plates, latches, electrical equipment. Good machinability. ASTM B282 Type I or II; CZP-0200-T; CZP-0220-U.

GKN D-1030

Presmet Corp.
Cu 91-94, Sn 3.5-4.5, Zn 0.4-1.4.
As sintered: 20,000-35,000 TS; 12-22 El; 50-90 Rock H; density 7.4-8.2. High ductility; good wear and corrosion properties. May be coined to 8.2 density for increased strength and wear.

GKN I-000

Presmet Corp.
C 0-0.25, 97.75 Fe min.
Iron type sintered metal. 21,000-27,000 TS; 1.5-3.0 El; 40 Rock F; density 6.5-6.9. For ductile, light duty mechanical components. ASTM B310 Class A Type III; F-0000-R.

GKN NS-180

Presmet Corp.
Cu 62.5-65.5, Ni 16.5-19.5, 14.0 Zn min.
As sintered: 27,000-33,000 TS; 13-16 El; 75-85 Rock H; density 7.3-7.7. For ornamental parts, lock hardware, structural components. Nickel silver type. ASTM B458 Grade I Type I; CZN-1818-U.

GKN S-005

Presmet Corp.
C 0.25-0.6, 97.4 Fe min.
As sintered: 17,000-22,000 psi TS; 1.0-1.5 El; 40-65 Rock F; density 5.7-6.1. Carbon steel grade for bearings and low cost structural parts. ASTM B310 Class B Type 1; F-0005-N.

GKN SA-4200

Presmet Corp.
C 0.5-0.9, Mn 0.35, Ni 0.45, Mo 0.55, 97.6 Fe min.
As sintered: 55,000-78,000 TS; 1.0-2.0 El; 70-90 Rock B. Hardened: 90,000-145,000 TS; 0.5-1.0 El; 25 Rock C min. Good hardenability; for structural parts.

GKN SA-4600

Presmet Corp.
C 0.5-0.9, Mn 0.2, Ni 1.9, Mo 0.5, 96.25 Fe min.
As sintered: 50,000-90,000 TS; 2-10 El; 75-95 Rock B. Hardened: 85,000-145,000 TS; 0.5-1.0 El; 25 Rock C min. Excellent hardenability and wear resistance. For heavy duty cams, ratchets, gears, pawls.

GKN SC-210

Presmet Corp.
C 0.6-1, Cu 1.5-2.5, 94.5 Fe min.
As sintered: 60,000-100,000 TS; 0.5-1.5 El; 65-85 Rock B; density 6.4-7.2. Copper steel grade; hardenable to 72 Rock 15-N min. For heavy duty structural parts, gears, ratchets. ASTM B426 Grade 1 Type III or IV; FC-0208-R; FC-0208-S.

GKN SC-310

Presmet Corp.
C 0.6-1, Cu 2.5-3.5, 92.5 Fe min.
As sintered: 37,000-42,000 TS; 0.5 El; 40-60 Rock B; density 5.9-6.2. Copper steel grade, hardenable to 90 Rock B min. For structural components such as gears, cams, pawls. ASTM B426 Grade 2 Type I or II.

GKN SC-705

Presmet Corp.
C 0.25-0.6, Cu 6-8, 88.4 Fe min.
As sintered: 29,000-34,000 TS; 0.5-1.0 El; 50-80 Rock F; density 5.9-6.2. Copper steel grade, hardenable. For structural components. ASTM B222.

GKN SC-710

Presmet Corp.
C 0.6-1, Cu 6-8, 88.0 Fe min.
As sintered: 40,000-46,000 TS; 0.5 El; 40-60 Rock B; density 5.9-6.2. Copper steel grade; hardenable to 90 Rock B min. For structural components. ASTM B426 Grade 3 Type I or II.

GKN SN-200

Presmet Corp.
Cu 0-1.25, Ni 1.75-2.75, 93.75 Fe min.
As sintered: 30,000-47,000 TS; 4.0-10.0 El; 60-85 Rock F; density 6.4-7.2. Nickel steel with good ductility and impact resistance. For firearm parts. Higher density parts can be case carburized for wear resistance. ASTM B484 Grade 1 Class A Type I or II; FN-0200-R; FN-0200-S.

GKN SN-208

Presmet Corp.
C 0.6-0.9, Ni 1.75-2.75, 94.35 Fe min.
As sintered: 46,000-75,000 TS; 1.5-3.5 El; 60-85 Rock B; density 6.4-7.2. Nickel steel; hardenable to 72 Rock 15-N min and 94,000-155,000 psi TS. High strength; usually heat treated. ASTM B484 Grade 1 Class C Type I or II; FN-0208-R; FN-0208-S.

GKN SN-405

Presmet Corp.
C 0.4-0.7, Cu 0.75-1.25, Ni 3.5-4.5, 91.55 Fe min.
As sintered: 45,000-74,000 TS; 2.5-5.5 El; 60-75 Rock B; density 6.4-7.2. Nickel steel; hardenable to 75 Rock 15-N min and 100,000-170,000 psi TS. Good combination of strength and impact resistance; superior heat treated properties. ASTM B484 Grade 2 Class B Type I or II; FN-0405-R; FN-0405-S.

GKN SS-304

Presmet Corp.
Cr 18-20, Ni 8-12, 62.8 Fe min.
As sintered: 36,000-50,000 TS; 2.0-4.0 El; 40-65 Rock B. AISI 304 composition. Nonmagnetic; good corrosion resistance. ASTM B525 Grade 1 Type I.

GKN SS-316

Presmet Corp.
Cr 16-18, Ni 10-14, Mo 2-3, 59.0 Fe min.
As sintered: 40,000-63,500; 3.0-7.0 El; 50-75 Rock B. AISI 316 composition; nonmagnetic, ductile, very good corrosion resistance. ASTM B525 Grade 2 Type 1; SS-316-R.

GKN SS-410

Presmet Corp.
Cr 11.5-13.5, 83.8 Fe min.
Sintered and heat treated: 65,000-85,000 TS; 55-70 Rock A. Martensitic, heat treatable stainless; corrosion resistant. SS-410-P.

GKN SX-000

Presmet Corp.
C 0-0.25, Cu 15-25, Zn 0-3, Fe 69.75-85.
As sintered: 60,000 psi TS; 5.0 El; 40 Rock B min; density 7.1 min. Infiltrated steel; hardenable to 70 Rock 15-N min. Good impact resistance, excellent machinability. ASTM B303 Class A; FX-2000-T.

GKN SX-010

Presmet Corp.
C 0.6-1, Cu 15-25, Zn 0-3, Fe 69-85.
As sintered: 100,000 TS; 0.5 El; 75 Rock B min; density 7.1 min. Infiltrated steel: hardenable to 75 Rock 15-N min. High strength steel; uniform density in complex designs. ASTM B303 Class C; FX-2008-T.

GKN-S-010

Presmet Corp.
97.0 Fe min, C.
As sintered: 22,000-28,000 TS; 0.5 El; 20-40 Rock B; density 5.6-6.1. Carbon steel type, hardenable to 85 Rock B min. For low cost, wear resistant machine parts. ASTM B310 Class C Type 1; F-0008-N.

GLACIER PRECISION FINISH LEAD BRONZE

Glacier Metal Co., Ltd.
Sn, Pb, bal Cu.
For machinery parts; free cutting.

GLASS MOLD ALLOY

English manufacture
Cu 55-65, Ni 12-18, Zn 11-17, Fe 8-12, Si 0.5-1.
For glass molds; U. S. Pat. 1360773.

GLASS SEALING

Now CARPENTER GLASS SEALING.

GLASS SEALING 42

Carpenter Technology Corp.
C 0-0.15, Mn 1, Ni 41.5, bal Fe.
For sealing into glass. Low expansion.

GLASS SEALING 49

Carpenter Technology Corp.
C 0.1, Ni 49, bal Fe.
For glass sealing. Same coefficient of expansion as glass.

GLASSEAL

Now TECHALLOY GLASSEAL.

GLIDCOP AL-10
SCM Metal Products Inc.
Cu 99.8, 0.2 Al_2O_3.
Pre-alloyed metal powder for sintering. Good conductivity; recommended for wire and strip. Meets RWMA specifications for class I material.

GLIDCOP AL-20
SCM Metal Products Inc.
Cu 99.6, 0.4 Al_2O_3.
Pre-alloyed metal powder for sintering. General purpose grade; good electrical and thermal conductivity.

GLIDCOP AL-35
SCM Metal Products Inc.
Cu 99.3, 0.7 Al_2O_3.
Pre-alloyed metal powder for sintering. For good conductivity; good strength and wear for long term elevated operation. Meets RWMA specifications for class II materials.

GLIDCOP AL-60
SCM Metal Products Inc.
Cu 98.9, 1.1 Al_2O_3.
Pre-alloyed metal powder for sintering. Good strength, hardness and wear resistance. Retains strength at elevated temperatures. Conductivity: 78% IACS at 68°F.

GLIDDEN 303-L
SCM Metal Products Inc.
Cr 17.5, Ni 12.5, S 0.2, Si 0.7, Mn 0.2, C 0.02, bal Fe.
Pre-alloyed metal powders; free-machining austenitic stainless. Typical sintered properties: 50,000-60,000 psi TS; 3-12 El; 50-64 Rock B.

GLIDDEN 304L
SCM Metal Products Inc.
C 0.02, Cr 18, Ni 10, bal Fe.
Pre-alloyed metal powders; stainless. Typical sintered properties: 50,000-62,000 TS; 3.5-12.4 El; 52-62 Rock B.

GLIDDEN 316L
SCM Metal Products Inc.
C 0.03, Cr 17, Ni 13, Mo 2.5, bal Fe.
Pre-alloyed metal powders; stainless. Typical sintered properties: 49,000-69,000 TS; 3-14 El; 50-60 Rock B.

GLIDDEN 410L (ANNEALED)
SCM Metal Products Inc.
C 0.02, Cr 12, bal Fe.
Pre-alloyed metal powders; stainless. Typical sintered properties: 58,000-95,000 TS; 2-3 El; 90-96 Rock B.

GLIDDEN 434-L
SCM Metal Products Inc.
Cr 17, Mo 1, Si 0.8, Mo 0.2, C 0.02, bal Fe.
Pre-alloyed metal powders; ferritic type stainless. Typical sintered properties: 450-550 N/mm^2 TS; 350-400 N/mm^2 YS; 3-5 El; 70-80 Rock B.

GLIDDEN 4600
SCM Metal Products Inc.
C 0.04, Mn 0.6, Ni 1.9, Mo 0.25, bal Fe.
Pre-alloyed metal powders; low alloy steel. Sintered properties depend on carbon pick-up from graphite and zinc stearate and on heat treatment.

GLIDDEN 830
SCM Metal Products Inc.
Cr 20.2, Ni 30, Mo 2.5, Cu 3.5, Si 1, Mn 0.2, C 0.02, bal Fe.
Pre-alloyed metal powders; corrosion resistant austenitic grade. Meets chemistry of ACI CN-7M.

GLIDDEN A-210; B-214
SCM Metal Products Inc.
Iron powder for sintering into pole pieces and other soft P/M parts.

GLIDDEN A-220
SCM Metal Products Inc.
Iron base powder for sintering into permanent magnets.

GLIDDEN CN-1
SCM Metal Products Inc.
Ni 70, Cu 30.
Pre-alloyed metal powders; Ni-Cu alloy. Typical sintered properties: 39,000-44,000 TS; 12.5-17.1 El; 34-38 Rock B. For parts requiring corrosion resistance.

GLIDDEN CN-1; CN-4
SCM Metal Products Inc.
Cu-Ni alloy powders to be sintered for corrosion resistant filters.

GLIDDEN H-120
SCM Metal Products Inc.
Fe 50, 50 Al alloy powder for sintering into permanent magnets.

GLIDDEN HN-1; HN-4
SCM Metal Products Inc.
Ni-base alloy powders to be sintered for corrosion resistant and high temperature filters; and for hard facing.

GLIDDEN J-8100
SCM Pigments (Glidden Metals)
Cr 19, Si 10, bal Ni (boron-free).
Vacuum melted brazing material; brazing temperature 2125-2175 F. Oxidation resistance to 1900 F; for general purpose and nuclear brazing applications. *Obsolete*

GLIDDEN J-8102
SCM Pigments (Glidden Metals)
Cr 15.2, Si 8, bal Ni.
Vacuum melted brazing material; brazing temperature 2150-2200 F. Boron free for thin gauge (0.005 in.) honeycomb assemblies; good filleting and low erosion tendencies. *Obsolete*

GLIDDEN J-8103
SCM Pigments (Glidden Metals)
Cr 17.1, Si 9.2, B 0.1, bal Ni.
Vacuum melted brazing material; brazing temperature 2100-2150 F. Low boron grade, good flow, less fillet and more erosion than J-8102. *Obsolete*

GLIDDEN J-8104
SCM Pigments (Glidden Metals)
Cr 11.4, Si 6.8, B 0.3, bal Ni.
Vacuum melted brazing material; brazing temperature 2125-2150 F. For bridging gaps up to 0.060 inch, and for two or more brazing operations. *Obsolete*

GLIDDEN J-8202
SCM Pigments (Glidden Metals)
Cr 2.2, Si 2.6, B 1.5, bal Ni.
Vacuum melted brazing material; brazing temperature 2000-2110 F. For bridging wide gaps; service temperature up to 1700 F. *Obsolete*

GLIDDEN J-8207
SCM Pigments (Glidden Metals)
Cr 5.6, Si 3.2, B 2.4, bal Ni.
Vacuum melted brazing material; brazing temperature 2050-2120 F. For general wide gap brazing. *Obsolete*

GLIDDEN J-8300
SCM Pigments (Glidden Metals)
Cr 19, Si 10, Mn 10, bal Ni.
Vacuum melted brazing material; brazing temperature 2100-2150 F in hydrogen (not vacuum). For brazing stainless steel honeycombs. *Obsolete*

GLIDDEN J-8400
SCM Pigments (Glidden Metals)
Cr 19, Ni 16.5, Si 8, W 4, B 0.8, bal Co.
Vacuum melted brazing material; brazing temperature 2150-2200 F. For brazing L-605, J-1570 and other cobalt base materials. *Obsolete*

GLIDDEN J-8600
SCM Pigments (Glidden Metals)
Cr 33, Si 4, Pd 24, bal Ni.
Vacuum melted brazing material; brazing temperature 2140-2170 F. For joining Rene 41 and similar superalloys. *Obsolete*

GLIDDEN J-8950
SCM Pigments (Glidden Metals)
Cr 32, Pd 18, Si 4.8, Ti 2.8, Al 1.8, bal Ni.
Vacuum melted brazing material; brazing temperature 2125-2175 F. Braze joint has exceptional high temperature mechanical properties. *Obsolete*

GLIDDEN LA-100
SCM Metal Products Inc.
Pb-Sn-Sb eutectic alloy powder for use in solder paste for low temperature type metal soldering.

GLIDDEN NF-1
SCM Metal Products Inc.
Ni 50, Fe 50.
Pre-alloyed metal powders. Typical sintered properties: 32,000-38,000 TS; 8.4-12.1 El; 26-29 Rock B. For soft magnetic and electrical applications.

GLIDDEN SF-9
SCM Metal Products Inc.
Si 9, bal Fe.
Pre-alloyed metal powders. Master alloy for soft magnetic parts.

GLIDDEN SF-9; SF-17
SCM Metal Products Inc.
Pre-alloyed Si-Fe powder for blending into silicon-iron alloys for soft magnetic applications.

GLIDDEN TL-15
SCM Metal Products Inc.
One of several Pb-Sn alloys in powder form; for use as solders.

GLIEVOR BEARING
English manufacture
Sb 14, Sn 8, Pb 77, Fe 1.5.
For bearings; anti-friction.

GLIEVOR BEARING
English manufacture
Zn 74, Sb 9, Sn 6.7, Pb 5, Cu 4.4, Cd 1.4.
For bearings; anti-friction.

GLIXEY NO. 1
Die Casting Appliance Corp., Ltd.
Pb, Sb, bal Sn.
For transmission bearings, pump bearings; anti-friction metal.

GLIXEY NO. 2
Die Casting Appliance Corp., Ltd.
Pb, Sb, bal Sn.
For bearings; anti-friction metal.

GLIXEY NO. 2A
Die Casting Appliance Corp., Ltd.
Pb, Sb, bal Sn.
For bearings; anti-friction metal.

GLIXEY NO. 3
Die Casting Appliance Corp., Ltd.
Pb, Sb, bal Sn.
For bearings; anti-friction metal.

GLM
Bergische Stahl Industrie
C 1.05-1.15, Si 0.15-0.3, Mn 0.2-0.4, Cr 1.1-1.3, W 1.2-1.4, V 0.15-0.25, bal Fe.
Cold work tool steel. W.-Nr. 1.2519.

GLO C-10
Globe Metallurgical, Inc.
Miscellaneous nonferrous. Si 36-40, Ce 9-11, 11.0-15.0 total rare earth elements.
Inoculant with RE content, for cast iron.

GLO C-30
Globe Metallurgical, Inc.
Miscellaneous nonferrous. Si 30-35, Ce 28-32, 40.0 max total rare earth elements.
Inoculant with RE content, for cast iron.

GLO SB2
Globe Metallurgical, Inc.
Miscellaneous nonferrous. Si 72-77, Al 0.8-1.2, Ba 1.75-2.25.
Inoculant for cast iron.

GLO SLC-5
Globe Metallurgical, Inc.
Miscellaneous nonferrous. Si 46-51, Sr 0.6-1, Al 0-0.5, Ca 0-0.1.
Graphitizing inoculant for use in gray iron.

GLO SLC-7
Globe Metallurgical, Inc.
Miscellaneous nonferrous. Si 72-77, Sr 0.6-1, Al 0-0.5, Ca 0-0.1.
Silicon based inoculant for cast iron.

GLO SMB
Globe Metallurgical, Inc.
Miscellaneous nonferrous. Si 60-65, Mn 9-11, Al 1-1.5, Ca 1-3, Ba 4-6.
Inoculant for gray and ductile iron.

GLO STC-1
Globe Metallurgical, Inc.
Miscellaneous nonferrous. Si 50-55, Ca 0.5-1, Ti 9-11.
Inoculant for cast iron.

GLO ZAC
Globe Metallurgical, Inc.
Miscellaneous nonferrous. Si 74-80, Al 1-2.5, Ca 2-3, Zr 1-2.
Zr and Ca containing inoculant based on 75 Fe Si; for cast iron.

GLOBE .35 MAX FE (1)
Globe Metallurgical, Inc.
Miscellaneous nonferrous. Fe 0-0.35, Ca 0-0.07, 98.50 Si min.
Alloying element.

GLOBE .35 MAX FE (2)
Globe Metallurgical, Inc.
Miscellaneous nonferrous. Fe 0-0.35, Ca 0-0.03, 98.50 Si min.
Alloying element.

GLOBE .50 MAX FE (1)
Globe Metallurgical, Inc.
Miscellaneous nonferrous. Fe 0-0.5, Ca 0.15-0.4, 98.40 Si min.
Alloying element.

GLOBE .50 MAX FE (2)
Globe Metallurgical, Inc.
Miscellaneous nonferrous. Fe 0-0.5, Ca 0-0.07, 98.40 Si min.
Alloying element.

GLOBE .50 MAX FE (3)
Globe Metallurgical, Inc.
Miscellaneous nonferrous. Fe 0-0.5, Ca 0-0.03, 98.40 Si min.
Alloying element.

GLOBE 1.00 MAX FE (1)
Globe Metallurgical, Inc.
Miscellaneous nonferrous. Fe 0-1, Ca 0.15-0.4, 98.25 Si min.
Alloying element.

GLOBE 1.00 MAX FE (2)
Globe Metallurgical, Inc.
Miscellaneous nonferrous. Fe 0-1, Ca 0-0.07, 98.25 Si min.
Alloying element.

GLOBE 1.00 MAX FE (3)
Globe Metallurgical, Inc.
Miscellaneous nonferrous. Fe 0-1, Ca 0-0.03, 98.25 Si min.
Alloying element.

GLOBE SPECIAL (1.00% MAX FE)
Globe Metallurgical, Inc.
Miscellaneous nonferrous. Al 0.2-0.5, 98.00 Si min.
Used in the production of silicones.

GLOBE SPECIAL LOW 0.10% IRON GRADE (1)
Globe Metallurgical, Inc.
Miscellaneous nonferrous. Fe 0-0.1, Ca 0-0.07, 99.00 Si min.
Alloying element.

GLOBE SPECIAL LOW 0.10% IRON GRADE (2)
Globe Metallurgical, Inc.
Miscellaneous nonferrous. Fe 0-0.1, Ca 0-0.03, 99.00 Si min.
Alloying element.

GLOBE SPECIAL LOW 0.15% IRON GRADE (1)
Globe Metallurgical, Inc.
Miscellaneous nonferrous. Fe 0-0.15, Ca 0-0.07, 99.00 Si min.
Alloying element.

GLOBE SPECIAL LOW 0.15% IRON GRADE (2)
Globe Metallurgical, Inc.
Miscellaneous nonferrous. Fe 0-0.15, Ca 0-0.03, 99.00 Si min.
Alloying element.

GLOBE SPECIAL LOW 0.25% IRON GRADE (1)
Globe Metallurgical, Inc.
Miscellaneous nonferrous. Fe 0-0.25, Ca 0-0.07, 98.50 Si min.
Alloying element.

GLOBE SPECIAL LOW 0.25% IRON GRADE (2)
Globe Metallurgical, Inc.
Miscellaneous nonferrous. Fe 0-0.25, Ca 0-0.03, 98.50 Si min.
Alloying element.

GLOBLE DRILL STEEL
Teledyne Firth Sterling
Steel. C 1.25, Cr 0.1, Mn 0.3, bal Fe.
For drills, reamers, cutters; water hardened. *Obsolete*

GLOMAG C3
Globe Metallurgical, Inc.
Miscellaneous nonferrous. Si 43-47, Mg 3-3.75, Ca 0.7-1.1, Al 0-1.2.
Used in the production of ductile iron; can be either cerium or total rare earth based.

GLOMAG C5
Globe Metallurgical, Inc.
Miscellaneous nonferrous. Si 43-47, Mg 5-6, Ca 0.7-1.1, Al 0-1.2.
Used in the production of ductile iron; can be either cerium or total rare earth based.

GLOMAG C5-10
Globe Metallurgical, Inc.
Miscellaneous nonferrous. Si 43-47, Mg 5-6, Ce 0.9-1.2, Ca 0.7-1.1, Al 0-1.2.
Used in the production of ductile iron; can be either cerium or total rare earth based.

GLOMAG C5-3
Globe Metallurgical, Inc.
Miscellaneous nonferrous. Si 43-47, Mg 5-6, Ce 0.25-0.35, Ca 0.7-1.1, Al 0-1.2.
Used in the production of ductile iron; can be either cerium or total rare earth based.

GLOMAG C5-6
Globe Metallurgical, Inc.
Miscellaneous nonferrous. Si 43-47, Mg 5-6, Ce 0.5-0.7, Ca 0.7-1.1, Al 0-1.2.
Used in the production of ductile iron; can be either cerium or total rare earth based.

GLOMAG C9
Globe Metallurgical, Inc.
Miscellaneous nonferrous. Si 43-47, Mg 8.5-10, Ca 0.7-1.1, Al 0-1.2.
Used in the production of ductile iron; can be either cerium or total rare earth based.

GLOMAG C9-3
Globe Metallurgical, Inc.
Miscellaneous nonferrous. Si 43-47, Mg 8.5-10, Ce 0.3-0.5, Ca 0.7-1.1, Al 0-1.2.
Used in the production of ductile iron; can be either cerium or total rare earth based.

GLOMAG C9-5
Globe Metallurgical, Inc.
Miscellaneous nonferrous. Si 43-47, Mg 8.5-10, Ce 0.5-0.7, Ca 0.7-1.1, Al 0-1.2.
Used in the production of ductile iron; can be either cerium or total rare earth based.

GLOMAG R5-10
Globe Metallurgical, Inc.
Miscellaneous nonferrous. Si 43-47, Mg 5-6, Ce 0.45-0.6, Ca 0.7-1.1, Al 0-1.2, 0.90-1.20 total rare earth elements.
Used in the production of ductile iron; can be either cerium or total rare earth based.

GLOMAG R5-8
Globe Metallurgical, Inc.
Miscellaneous nonferrous. Si 43-47, Mg 5-6, Ce 0.3-0.45, Ca 0.7-1.1, Al 0-1.2, 0.60-0.90 total rare earth elements.
Used in the production of ductile iron; can be either cerium or total rare earth based.

GLOMAG R9-10
Globe Metallurgical, Inc.
Miscellaneous nonferrous. Si 43-47, Mg 8.5-10, Ca 0.7-1.1, Al 0-1.2, 0.70-1.00 total rare earth elements.
Used in the production of ductile iron; can be either cerium or total rare earth based.

GLOWRAY
Henry Wiggin & Co. Ltd.
Ni 65, Cr 15, Fe 20.
Annealed: 99,000 TS; 41,000 YS; 44 El; 66 RA; 190 Brin. For heating elements, furnace parts; heat resistant. *Obsolete*

GLUHFEST ALLOYS
Now JUNKER.

GLX-42W
National Intergroup Inc.
C 0.21, Mn 1.35, Cb, bal Fe.
Aluminum killed. Hot rolled plate: 60,000 psi TS; 42,000 psi YS min; 24 El. For structural parts.

GLX-45W

National Intergroup Inc.
C 0.22, Mn 1.2, Si 0-0.1, Cb 0.01-0.04, bal Fe.
Rolled (min): 60,000 TS; 45,000 YS; 22 El. For structural parts; weldable.

GLX-50W

National Intergroup Inc.
C 0.22, Mn 1.2, Si 0-0.1, Cb 0.01-0.04, bal Fe.
Rolled (min): 65,000 TS; 50,000 YS; 22 El. For structural parts; weldable.

GLX-55W

National Intergroup Inc.
C 0.22, Mn 1.2, Si 0-0.1, Cb 0.01-0.04, bal Fe.
Rolled (min): 70,000 TS; 55,000 YS; 20 El. For structural parts; weldable.

GLX-60W

National Intergroup Inc.
C 0.22, Mn 1.2, Si 0-0.1, Cb 0.01-0.04, bal Fe.
Rolled (min): 75,000 TS; 60,000 YS; 18 El. For structural parts; weldable.

GLX-65W

Manufacturer not listed.
C 0.22, Mn 1.2, Si 0-0.1, Cb 0.01-0.04, bal Fe.
Rolled (min): 80,000 TS; 65,000 YS; 16 El. For structural parts; weldable.

GLYCO

Joseph T. Ryerson & Son Inc.
Sb 15, Sn 4.5, As 0.5, bal Pb.
For bearings, bushings; anti-friction; Babbitt.

GLYCO 24

Glyco-Metallwerke Daelen & Hofmann KG
Lead. Sb 14, Sn 1, As 1, Cu 2, bal Pb.
200 Brin min. Steel backed lead base babbitt. Anti-friction properties and conformability. For camshaft bearings, transmission bushes, compressor bearings; corrosion resistant. SAE 15; DIN 1703.

GLYCO 38

Glyco-Metallwerke Daelen & Hofmann KG
Tin. Sb 7.5, Cu 3.5, Pb 0-0.5, bal Sn.
200 Brin min. Steel backed tin base babbitt. Used in refrigeration engineering and with high-alloyed lubricants. Corrosion resistant due to high tin content. SAE 12; DIN 1703.

GLYCO 40

Glyco-Metallwerke Daelen & Hofmann KG
Pb 22, Sn 1.5, bal Cu.
250 Brin min. Trimetal bearing consisting of steel backed cast copper-lead, nickel dam and overlay. For high loads and running speeds. Standard material for main and big end bearings in automobile and diesel engine applications. Corrosion resistant. Available as GLYCO 41 without overlay. SAE 49; SAE 19. Overlay composition: 10.0 Sn, 2.0 Cu, bal Pb.

GLYCO 46

Glyco-Metallwerke Daelen & Hofmann KG
Copper. Pb 10, Sn 10, bal Cu.
45 Rock C min. Steel backed lead bronze. Used for piston pin bushes, valve rocker bushes, transmission part applications, journal bushes, thrust washers. Shock and load carrying capacities; wear resistant; corrosion resistant. SAE 792; DIN 1716.

GLYCO 48

Glyco-Metallwerke Daelen & Hofmann KG
Copper. Pb 22, Sn 3, bal Cu.
40 Rock C min. Steel backed lead bronze. Used for pump, transmission and camshaft bushes, thrust washers. High load carrying capacity. SAE 794; DIN 1716.

GLYCO 60

Glyco-Metallwerke Daelen & Hofmann KG
Lead. Cu 49, Sn 1, Pb 50.
250 Brin min. Steel backed sintered copper-lead. Used for thrust washers, transmission, pump and camshaft bushes. SAE 485.

GLYCO 66

Glyco-Metallwerke Daelen & Hofmann KG
Copper. Pb 10, Sn 10, bal Cu.
45 Rock C min. Steel backed lead bronze. Used for valve rocker, transmission and journal bushes, thrust washers. High load carrying capacity. Wear resistant; corrosion resistant. SAE 797.

GLYCO 68

Glyco-Metallwerke Daelen & Hofmann KG
Copper. Pb 23, Sn 3, bal Cu.
40 Rock C min. Steel backed lead bronze. For transmission bush applications, thrust washers. SAE 799.

GLYCO 73

Glyco-Metallwerke Daelen & Hofmann KG
Sn 6, Cu 1, bal Al.
250 Brin min. Trimetal bearing consisting of steel backed aluminum alloy and overlay. Used for main and big end bearings. SAE 780; SAE 19. Overlay composition: 10.0 Sn, 2.0 Cu, bal Pb.

GLYCO 74

Glyco-Metallwerke Daelen & Hofmann KG
Aluminum. Sn 20, Cu 1, bal Al.
250 Brin min. Steel backed aluminum alloy. Used for main and big end bearings, transmission and pump applications, thrust washers. Wear resistant; corrosion resistant.

GLYCO METAL

Manufacturer not listed.
Zn 85.5, Sn 5, Pb 4.7, Cu 0.4, Al 2.
For bearings, bushings; Babbitt.

GLYCO TURBO

English manufacture
Pb 70, Sb 22, Sn 8.
For turbine bearings; Babbitt.

GLYCODUR A

Glyco-Metallwerke Daelen & Hofmann KG
Copper. Cu 90, Sn 10.
150 Brin min. Porous tin bronze sintered onto steel, pores filled with acetal resin and surface layer made up of same material. Bearing surface provided with lubrication pockets. Used for building machinery, hydraulic excavators and cylinders, mechanical handling plants. DIN 1494.

GLYCODUR F

Glyco-Metallwerke Daelen & Hofmann KG
Copper. Cu 90, Sn 10.
150 Brin min. Porous tin bronze sintered onto steel, pores filled with PTFE and MoS_2 additives, running-in layer of the same composition. Used for shock absorbers, pumps, agricultural equipment, mechanical handling units, transmissions. DIN 1494.

GM 14M

General Motors Corp./Central Foundry Div
C 3.2-3.5, Si 2.15-2.45, Mn 0.5-0.8, Cr 0.1-0.4, P 0-0.15, S 0-0.15, bal Fe.
Ferritic-pearlitic type gray cast iron for easily machinable automotive castings. 20,000 psi TS; 187 Brin max. Similar to ASTM/SAE Gr. G-1800.

GM 6129 GR 5203

General Motors Corp./Central Foundry Div
C 3.6-3.9, Si 1.8-2.7, Mn 0.3-1, Mg 0.03-0.07, Cu 0-0.7, bal Fe.
Nodular cast iron, as cast, pearlitic: 52,000 psi YS; 3 El; 187-269 Brin; 50.0 ferrite max, 50.0 pearlite min, 10.0 carbide max.

GM 6129 M, GR 3815

General Motors Corp./Central Foundry Div
C 3.6-3.9, Si 2.2-2.7, Mn 0.25-0.9, Mg 0.03-0.05, bal Fe.
Nodular cast iron, annealed (ferritic). 38,000 psi YS; 15 El; 126-179 Brin; 90.0 ferrite min, 10.0 pearlite max, 3.0 carbide max.

GM 6129M GR 4010

General Motors Corp./Central Foundry Div
C 3.6-3.9, Si 2.2-2.7, Mn 0.25-0.9, Mg 0.03-0.07, bal Fe.
Nodular cast iron, as cast, ferritic: 40,000 psi YS; 10 El; 143-217 Brin; 40.0 ferrite min, 60.0 pearlite max, 3.0 carbide max.

GM-4199M

General Motors Corp./Central Foundry Div
Si 7.5-9.5, Fe 0-1.3, Cu 3-4, Mn 0-0.5, Mg 0-0.1, Ni 0-0.5, Zn 0-3, Sn 0-0.35, bal Al.
Cast aluminum alloy. UNS 13800 (380).

GM-4227 M

General Motors Corp./Central Foundry Div
Si 5.5-7, Fe 0-1, Cu 3-4, Mn 0-0.5, Mg 0-0.6, Ni 0-0.35, Zn 0-1, Ti 0-0.25, bal Al.
Cast aluminum alloy. UNS A03190 (319).

GM-4334M

General Motors Corp./Central Foundry Div
Si 8.5-10.5, Fe 0-1.2, Cu 2-4, Mn 0-0.5, Mg 0.5-1.5, Ni 0-0.5, Zn 0-1, Ti 0-0.25, Sn 0-0.2, bal Al.
Cast aluminum alloy. UNS A63320 (F-132).

GM-6209 M

General Motors Corp./Central Foundry Div
Cr 0.1-0.4, S 0-0.15, P 0-0.15, TC 3.2-3.45, Si 2.15-2.55, Mn 0.5-0.8, bal Fe.
Pearlite gray cast iron for small or medium size automotive castings. 30,000 psi TS; 170-241 Brin. Similar to ASTM/SAE Gr.G-3000.

GM-6213 M

General Motors Corp./Central Foundry Div
Si 2.1-2.4, Mn 0.5-0.7, Cr 0.2-0.4, S 0-0.15, P 0-0.15, C 3.1-3.4, bal Fe.
Pearlite gray cast iron for cylinder heads, cylinder blocks, and engine bearing caps. 35,000 psi TS; 179-255 Brin. Similar to ASTM/SAE Gr.G-3500.

GM-O2 DIE

Columbia Tool Steel Co.
C 0.9, Mn 1.6, bal Fe.
For dies, punches, crimpers, upsetters; Type O2; oil hardened. *Obsolete*

GM6214 M

General Motors Corp./Central Foundry Div
Si 2.15-2.55, Mn 0.5-0.8, Cr 0.1-0.4, P 0-0.15, S 0-0.15, C 3.2-3.5, bal Fe.
Ferritic-pearlite gray cast iron for lightly loaded automotive castings. 18,000 psi TS; 187 Brin max. Similar to ASTM/SAE Gr.G-1800.

GMOODIE

General Motors Corp./Central Foundry
Al 4, Cu 3.25, Ni 0.8, Ti 0.2, Mg 0.15, bal Zn.
Dies for sheet forming.

GMR 235 D

General Motors Corp./Central Foundry
C 0.15, Cr 15.5, Mo 5, Ti 2.5, Al 3.5, B 0.05, bal Ni.
For jet engine parts.

GMR 235 D

Cannon-Muskegon Corp.
C 0.15, Cr 15.5, Mo 5, Ti 2.5, Al 3.5, B 0.05, bal Ni.
For jet engine parts.

GMR 236

General Motors Corp./Central Foundry Div
C 0.15, Cr 15.5, Mo 5, Ti 2.2, Al 3.25, Fe 25, B 0.06, bal Ni.
Corrosion and heat resistant. *Obsolete*

GMR-235

Crucible Specialty Steel
C 0.15, Fe 8-12, Cr 14-17, Mo 4.5-6, B 0.1, Ti 2, Al 2.5-3.5, bal Ni.
At 70°F: 147,000 TS; 43 El. At 1500°F: 90,000 TS; 5 El. At 1800°F: 21,800 TS; 38 El. For jet engine components, gas turbine parts; high heat resistant to 1500°F.

GMR-235

Cannon-Muskegon Corp.
C 0.15, Fe 8-12, Cr 14-17, Mo 4.5-6, B 0.1, Ti 2, Al 2.5-3.5, bal Ni.
At 70°F: 147,000 TS; 43 El. At 1500°F: 90,000 TS; 5 El. At 1800°F: 21,800 TS; 38 El. For jet engine components, gas turbine parts; high heat resistant to 1500°F.

GMR-235

General Motors Corp./Central Foundry
C 0.15, Fe 8-12, Cr 14-17, Mo 4.5-6, B 0.1, Ti 2, Al 2.5-3.5, bal Ni.
At 70°F: 147,000 TS; 43 El. At 1500°F: 90,000 TS; 5 El. At 1800°F: 21,800 TS; 38 El. For jet engine components, gas turbine parts; high heat resistant to 1500°F.

GMS 58

German manufacture
Zn, bal Cu.
For castings; brass.

GMS 63

German manufacture
Zn, bal Cu.
For castings; brass.

GMS 67

German manufacture
Zn, bal Cu.
For castings; brass.

GN3

Sheffield Smelting Co. Ltd.
Ag 80, Ni.
Annealed: 70 Vickers. Electrical resistivity: 2.2 micro-ohm/cm. Electrical conductivity: 78% IACS. For electrical contacts; low contact resistance.

GN5

Sheffield Smelting Co. Ltd.
Ag 90, Ni.
Annealed: 60 Vickers. Electrical resistivity: 2.0 micro-ohm/cm. Electrical conductivity: 86% IACS. For electrical contacts; low contact resistance.

GOHI IRON

Newport Steel Co.
C 0.02, Cu 0.25, Mn 0.025, Si 0.003, bal Fe.
45,000-50,000 TS; 40-50 El. For sheet-metal construction work, roofing, culverts, pipe, ventilators, skylights; rust resistant.

GOLD

Atomergic Chemetals Corp.
Au.
Purities, zone refined: 99.9999%, 99.999%, 99.99%. Forms: sponge, powder, wire, foil, sheet, ingot, single crystals.

GOLD ANCHOR

Teledyne Allvac
C 0.68-0.73, Cr 3.75-4.25, V 0.95-1.15, W 17.5-18.5, bal Fe.
For chasers, dies, punches, special tools; high speed steel.

GOLD ANCHOR DRILL ROD

Teledyne Allvac
C 0.68-0.73, W 17.5-18.5, Cr 3.75-4.25, V 0.95-1.15, bal Fe.
For twist drills, reamers, punches, small tools; high speed steel.

GOLD BRONZE

American manufacture
Sn 6.5, Zn 3, bal Cu.
For jewelry, ornaments; corrosion resistant.

GOLD BUTTONS

English manufacture
Zn 33, Sn 5, Pb 3, bal Cu.
For buttons, jewelry; free-cutting.

GOLD COLOR ELASTIC

J.M. Ney Co.
Gold color clasp and orthodontic wire. Fusing temperature: 1675°F. *Obsolete*

GOLD IMITATION

English manufacture
Cu 97.8, Al 2, Au 0.2.
For gold solder; corrosion resistant.

GOLD LABEL

Edgar T. Ward's Sons Co.
C, W, bal Fe.
For tools. *Obsolete*

GOLD LABEL

Heller Bros. Co.
C 0.6-0.7, Cr 3-4, W 18-20, Co 2-3, V 1.25-1.5, bal Fe.
For cutting tools; high speed steel, "Hellers Peerless."

GOLD LABEL

Hidalgo Steel Co. Inc.
C 0.6-0.7, Cr 3-4, W 18-20, Co 2-3, V 1.25-1.5, bal Fe.
For cutting tools; high speed steel, "Hellers Peerless."

GOLD LABEL (H. & R.)

Houghton & Richards Inc.
C 1.2, W 2.5, bal Fe.
For finishing tools, cutters; water or oil hardening.

GOLD LEAF

English manufacture
Cu 66-80, Zn 34-20.
For gold leaf substitute, signs.

GOLD LEAF METAL

English manufacture
Cu 84, Zn 16.
For gold leaf substitute, signs.

GOLD LEAF, AIX

English manufacture
Cu 64.8, Zn 32.8, Sn 2, Pb 0.4.
For gold leaf substitute, signs.

GOLD LEAF, JEMMAPES

English manufacture
Cu 64.6, Zn 33.7, Sn 1.4, Pb 0.2.
For gold leaf substitute, signs.

GOLD METAL LEAF

English manufacture
Cu 66-84, Zn 16-34, Pb 0-4.
For gold solder; corrosion resistant.

GOLD SOLDER

English manufacture
Ag, Au, Cu, Zn.
For gold solder; corrosion resistant.

GOLD SOLDER 10 KT NO. 10

Texas Instruments Inc./Materials Control
Au alloy.
For gold soldering; hard flowing, yellow. *Obsolete*

GOLD SOLDER 10 KT. CADMIUM NO. 1

Texas Instruments Inc./Materials Control
Au alloy.
For gold soldering; easy flowing, yellow. *Obsolete*

GOLD SOLDER 10 KT. WHITE NO. 5

Texas Instruments Inc./Materials Control
Au alloy.
For gold soldering; medium flowing, white. *Obsolete*

GOLD SOLDER 8 KT. CADMIUM NO. 1

Texas Instruments Inc./Materials Control
Au alloy.
For gold soldering; easy running, yellow. *Obsolete*

GOLD SOLDER 8 KT. NO. 10

Texas Instruments Inc./Materials Control
Au alloy.
For gold soldering; hard running, yellow. *Obsolete*

GOLD SOLDER 8 KT. WHITE NO. 5

Texas Instruments Inc./Materials Control
Au alloy.
For gold soldering; medium running, white. *Obsolete*

GOLD SOLDER 9-1/2 KT. CADMIUM NO. 1

Texas Instruments Inc./Materials Control
Au alloy.
For gold soldering; easy running, yellow. *Obsolete*

GOLD SOLDER 9-1/2 KT. NO. 10

Texas Instruments Inc./Materials Control
Au alloy.
For gold soldering; hard running, yellow. *Obsolete*

GOLD SOLDER 9-1/2 KT. WHITE NO. 5

Texas Instruments Inc./Materials Control
Au alloy.
For gold soldering; medium running, white. *Obsolete*

GOLD SOLDER, 10 CARAT

Manufacturer not listed.
Au 41, Ag 37, Cu 21, Br 0.6.
For gold solder.

GOLD SOLDER, 12 CARAT

Manufacturer not listed.
Au 50, Cu 35, Ag 15.
For gold solder; corrosion resistant.

GOLD SOLDER, 14 CARAT

Manufacturer not listed.
Au 50, Ag 33, Cu 17.
For gold solder.

GOLD SOLDER, 16 CARAT

Manufacturer not listed.
Au 75, Ag 17, Cu 8.3.
For gold solder.

GOLD SOLDER, 18 CARAT

Manufacturer not listed.
Au 63-75, Ag 13-31, Cu 6.3-12.
For gold solder.

GOLD SOLDER, 8 CARAT

Manufacturer not listed.
Au 40, Ag 37, Cu 23.
For gold solder.

GOLD SOLDER, BEST

Manufacturer not listed.
Au 63, Ag 23, Cu 15.
For gold solder.

GOLD SOLDER, EASY MELT

Manufacturer not listed.
Ag 55, Ag 32, Cu 14.
For gold solder.

GOLD SOLDER, VERY EASY MELT

Manufacturer not listed.
Ag 55, Cu 29, Au 12, Zn 5.5.
For gold solder.

GOLD SPEZIAL

Plettenberger Gusstahlfabrik
C 1.35, Co 2, Cr 4.3, V, W, bal Fe.
For engravers' tools, milling and form cutters, taps; high speed steel.

GOLD STAR

Carpenter Technology Corp.
C 0.77, W 13.7, Co 5, Cr 3.75, V 2, bal Fe.
For high speed cutting tools. High speed steel for heavy duty service. *Obsolete*

GOLD SUPER

Plettenberger Gusstahlfabrik
C 0.79, Cr 4.3, Mo 0.75, V 1.5, W 18, bal Fe.
For lathe and planer tools, reamers, broaches, taps; high speed steel.

GOLD SUPER KOBALT 10

Plettenberger Gusstahlfabrik
C 0.76, Co 10, Cr 4.2, Mo 0.8, V 1.8, W 18, bal Fe.
For lathe and planer tools, drills, taps, hobs; high speed steel.

GOLD TIP

LaSalle Steel Co.
C 0.08-0.13, Mn 0.6-0.9, S 0.24-0.33, bal Fe.
Cold drawn: 80,000 TS; 75,000 YS; 15 El; 48 RA; 180 Brin.
For screw machine parts; SAE-X-1112. *Obsolete*

GOLD, 10 CARAT

Manufacturer not listed.
Cu 38-46, Au 42, Ag 12-20.
For jewelry; corrosion resistant.

GOLD, 14 CARAT

Manufacturer not listed.
Cu 14-28, Au 58, Ag 4-28.
For jewelry; corrosion resistant.

GOLD, 14 CARAT

Manufacturer not listed.
Cu 12, Au 58, Ag 30.
For dental fillings and crowns; corrosion resistant.

GOLD, 15 CARAT

Manufacturer not listed.
Cu 13, Au 62, Ag 11.
For jewelry; corrosion resistant.

GOLD, 16 CARAT

Manufacturer not listed.
Cu 8-27, Au 67, Ag 6.6-26.
For jewelry.

GOLD, 18 CARAT

Manufacturer not listed.
Cu 5-15, Au 75, Ag 10-20.
For jewelry.

GOLD, 20 CARAT

Manufacturer not listed.
Cu 6-8.3, Au 84, Ag 8.3-11.
For jewelry.

GOLD, 22 CARAT

Manufacturer not listed.
Cu 3.4, Au 92, Ag 4.9.
For dental crown and fillings; corrosion resistant.

GOLD, 22 CARAT

Manufacturer not listed.
Cu 4.2, Au 92, Ag 4.2.
For jewelry.

GOLD, 8 CARAT

Manufacturer not listed.
Cu 47, Au 33, Ag 20.
For jewelry; corrosion resistant.

GOLD-1

Plettenberger Gusstahlfabrik
C 0.86, Co 2.8, Cr 4.3, Mo 0.85, V 2.1, W 12, bal Fe.
For lathe and planer tools, reamers, drills, taps; high speed steel.

GOLD-2

Plettenberger Gusstahlfabrik
C 1.3, Cr 4.3, Mo 0.85, V 3.8, W 12, bal Fe.
For engravers' tools, milling cutters, taps, reamers; high speed steel.

GOLDAL

English manufacture
Au, alloy.
For gold solder, dental alloy; corrosion resistant.

GOLDMASTER GM15

Herbert Cutanit Ltd.
Sintered carbide tool material. Coated throwaway tips for light cutting on steel and all cutting on cast iron and non-ferrous metal. *Obsolete*

GOLDMASTER GM35

Herbert Cutanit Ltd.
Sintered carbide tool material. Coated throwaway tips for heavier cutting on steel. *Obsolete*

GOLDPUNKT

SWB Stahlformguss Gesellschaft mbH
C 0.86, Cr 4.1, Mo 0.85, V 2.5, W 12, bal Fe.
For lathe and planer tools, drills, taps, hobs; high speed steel. *Obsolete*

GOLDSMITH GB-02

NL Industries
Ag 2, Cu 91, P 7.
For silver solder; MP 1180-1270°F; self-fluxing. *Obsolete*

GOLDSMITH GB-05

NL Industries
Ag 5, Cu 58, Zn 37.
For silver solder; MP 1575-1600°F. *Obsolete*

GOLDSMITH GB-06

NL Industries
Ag 6, Cu 86.5, P 7.5.
For silver solder; MP 1175-1350°F; self-fluxing. *Obsolete*

GOLDSMITH GB-07

NL Industries
Ag 7, Cu 85, Sn 8.
For silver solder; MP 1450-1565°F. *Obsolete*

GOLDSMITH GB09

NL Industries
Ag 9, Cu 53, Zn 38.
Melting range: 1450-1565°F. Brazing alloy. *Obsolete*

GOLDSMITH GB110

NL Industries
Ag 10, Cu 52, Zn 38.
For silver solder; MP 1450-1565°F. *Obsolete*

GOLDSMITH GB120

NL Industries
Ag 20, Cu 45, Zn 35.
For silver solder; MP 1430-1500°F. *Obsolete*

GOLDSMITH GB1270

NL Industries
Ag 91, P 7.
Melting range: 1185-1450°F. For brazing copper and copper alloys. *Obsolete*

GOLDSMITH GB131

NL Industries
Ag 31.5, Cu 34, Zn 15.5, Cd 19.
For silver solder; MP 1165-1390°F. *Obsolete*

GOLDSMITH GB140

NL Industries
Ag 40, Cu 18, Zn 15, Cd 27.
For silver solder; MP 1135-1205°F. *Obsolete*

GOLDSMITH GB145

NL Industries
Ag 45, Cu 30, Zn 25.
For silver solder; MP 1250-1370°F. *Obsolete*

GOLDSMITH GB15

Marrel Freres, S.A.
Ag 15, Cu 80, P 5.
For brazing; melting point: 1185-1300°F. Silver solder.

GOLDSMITH GB150

NL Industries
Ag 50, Cu 15, Zn 25, Cd 10.
For silver solder; MP 1166-1190°F. *Obsolete*

GOLDSMITH GB160

NL Industries
Cu 30, Sn 10, Ag 60.
Melting range: 1095-1325°F. Brazing alloy; AMS 4773. *Obsolete*

GOLDSMITH GB165

NL Industries
Ag 65, Cu 28, Mn 5, Ni 2.
Melting range: 1385-1445°F. Brazing alloy. *Obsolete*

GOLDSMITH GB175

NL Industries
Ag 75, Cu 20, Zn 5.
For silver solder; MP 1350-1425°F. *Obsolete*

GOLDSMITH GB20

NL Industries
Ag 20, Cu 45, Zn 30, Cd 5.
For brazing; MP 1140-1500°F; silver solder. *Obsolete*

GOLDSMITH GB240

NL Industries
Ag 40, Cu 30, Zn 28, Ni 2.
For silver solder; MP 1240-1435°F. *Obsolete*

GOLDSMITH GB245

NL Industries
Ag 45, Cu 18, Zn 21, Cd 16.
For silver solder; MP 1140-1185°F. *Obsolete*

GOLDSMITH GB25

NL Industries
Ag 25, Cu 32.5, Zn 22.5.
Melting range: 1300-1575°F. Brazing alloy. *Obsolete*

GOLDSMITH GB250

NL Industries
Ag 50, Cu 34, Zn 16.
For brazing; MP 1260-1410°F; silver solder. *Obsolete*

GOLDSMITH GB260

NL Industries
Ag 60, Cu 27, Zn 13.
Melting range: 1125-1310°F. Brazing alloy. *Obsolete*

GOLDSMITH GB275

NL Industries
Ag 75, Zn 25.
For silver solder; MP 1300-1325°F. *Obsolete*

GOLDSMITH GB30

NL Industries
Ag 30, Cu 38, Zn 32.
For silver solder; MP 1370-1410°F. *Obsolete*

GOLDSMITH GB31

NL Industries
Cu 34, Cd 19, Ag 31.5, Zn 15.5.
Melting range: 1165-1390°F. Brazing alloy. *Obsolete*

GOLDSMITH GB340
NL Industries
Ag 40, Cu 36, Zn 24.
For silver solder; MP 1330-1445°F. *Obsolete*

GOLDSMITH GB35
NL Industries
Ag 35, Cu 26, Zn 21, Cd 18.
For brazing; MP 1125-1295°F; silver solder. *Obsolete*

GOLDSMITH GB350
NL Industries
Ag 50, Cu 15.5, Zn 15.5, Cd 16, Ni 3.
For brazing; MP 1195-1275°F, silver solder. *Obsolete*

GOLDSMITH GB41
NL Industries
Ag 41, Cu 17, Zn 18, Cd 24.
For brazing; MP 1125-1160°F; silver solder. *Obsolete*

GOLDSMITH GB440
NL Industries
Ag 40, Cu 30.5, Zn 29.5.
Melting range: 1150-1350°F. Brazing alloy. *Obsolete*

GOLDSMITH GB45
NL Industries
Ag 45, Cu 15, Zn 16, Cd 24.
For brazing; MP 1125-1145°F; silver solder. *Obsolete*

GOLDSMITH GB50
NL Industries
Ag 50, Cu 15.5, Zn 16.5, Cd 18.
For brazing; MP 1160-1175°F; silver solder. *Obsolete*

GOLDSMITH GB54
NL Industries
Ag 54, Cu 40, Zn 5, Ni 1.
Melting range: 1075-1575°F. Brazing alloy, AMS 4772.
Obsolete

GOLDSMITH GB540
NL Industries
Ag 40, Cu 30, Zn 25, Ni 5.
For brazing; MP 1260-1550°F; silver solder. *Obsolete*

GOLDSMITH GB56
NL Industries
Ag 56, Cu 22, Zn 17, Sn 5.
For silver solder; MP 1152-1203°F. *Obsolete*

GOLDSMITH GB57
NL Industries
Ag 57, Cu 33, Mn 3, Sn 7.
Melting range: 1120-1345°F. Brazing alloy. *Obsolete*

GOLDSMITH GB60
NL Industries
Ag 60, Cu 25, Zn 15.
For brazing; MP 1260-1325°F; silver solder. *Obsolete*

GOLDSMITH GB65
NL Industries
Ag 65, Cu 20, Zn 15.
For silver solder; MP 1280-1325°F. *Obsolete*

GOLDSMITH GB650
NL Industries
Ag 50, Cu 28, Zn 22.
For silver solder; MP 1250-1340°F. *Obsolete*

GOLDSMITH GB70
NL Industries
Ag 70, Cu 20, Zn 10.
For silver solder; MP 1335-1390°F. *Obsolete*

GOLDSMITH GB72
NL Industries
Ag 72, Cu 28.
For silver solder; MP 1335-1435°F. *Obsolete*

GOLDSMITH GB75
NL Industries
Ag 75, Cu 0.22, Zn 3.
For silver solder; MP 1365-1490°F. *Obsolete*

GOLDSMITH GB80
NL Industries
Ag 80, Cu 16, Zn 4.
For brazing; MP 1360-1490°F; silver solder. *Obsolete*

GOLDSMITH GB85
NL Industries
Ag 85, Mn 15.
For silver solder; MP 1760-1778°F. *Obsolete*

GOLDSMITH GB95-5
NL Industries
Ag 5, Cd 95.
Melting range: 620-750°F. Solder alloy. *Obsolete*

GOLDSTAR
J.F. Jelenko & Co.
Precious metal. Au 2, Pd 60, Ag 26.
95,000 psi TS; 67,000 psi YS; 20 El; 172 Brin. Dental alloy for fusing porcelain to metal.

GOLFALLOY
Certified Alloy Products Inc.
C 0.08-0.2, Cr 15-16.5, Ni 1.5-2.2, Mn 0-1, Si 0-1, Cu 0-0.5, Mo 0-0.5, bal Fe.
Cast, heat treated: 118,000 TS; 87,000 YS; 21 El; 18-25 Rock C. For golf club heads, investment cast; corrosion resistant

GOLFALLOY
Centrifugal Products Inc.
C 0.08-0.2, Cr 15-16.5, Ni 1.5-2.2, Mn 0-1, Si 0-1, Cu 0-0.5, Mo 0-0.5, bal Fe.
Cast, heat treated: 118,000 TS; 87,000 YS; 21 El; 18-25 Rock C. For golf club heads, investment cast; corrosion resistant

GOLIATH
Thyssen Edelstahlwerke AG
C, alloy, bal Fe.
For dies; oil hardened. *Obsolete*

GOLIATH G
Thyssen Edelstahlwerke AG
C, alloy, bal Fe.
For dies; oil hardened. *Obsolete*

GOLIATH GNV
Thyssen Edelstahlwerke AG
C 0.55, Cr 0.7, Mo 0.2, Ni 1.5, V 0.1, bal Fe.
For die blocks, gears, mandrels; oil hardened, tough. *Obsolete*

GOLIATH SPECIAL M
Thyssen Edelstahlwerke AG
C, alloy, bal Fe.
For dies; oil hardened. *Obsolete*

GOLIATH SPEZIAL
Thyssen Edelstahlwerke AG
C 2.2, Cr 13, bal Fe.
For drawing dies; oil or air hardened. *Obsolete*

GOLIATH SPEZIAL M
Thyssen Edelstahlwerke AG
C, Cr, Ni, Mo, bal Fe.
For swaging dies; oil hardened. *Obsolete*

GOLIATH V
Thyssen Edelstahlwerke AG
C, alloy, bal Fe.
For dies; oil hardened. *Obsolete*

GOLLET STEEL
Bethlehem Steel Corp.
C 0.95, Mn 0.7, bal Fe.
For collets, tools; water hardened. *Obsolete*

GOMAK 2
French manufacture
Al 3.5-5, Cu 2.5-3.5, Mg 0.02-0.1, bal Zn.
Die cast: 46,000-53,000 TS; 5-14 El; 80-100 Brin. For motor frames, gear housings, hardware; similar to Zamak 2.

GOMAK 3
French manufacture
Al 3.5-4.3, Mg 0.03-0.08, bal Zn.
Die cast: 35,000-43,000 TS; 4-11 El; 60-85 Brin. For gear housings, motor frames, fuel pumps; similar to Zamak 3.

GOMAK 5
French manufacture
Al 3.5-4.3, Cu 0.75-1.25, Mg 0.03-0.06, bal Zn.
Die cast: 42,000-50,000 TS; 13-3 El; 70-95 Brin. For motor frames, gear housings, hardware; similar to Zamak 5.

GONG METAL
English manufacture
Cu 78, Zn 22.
For gongs.

GORDON
Latrobe Steel Co.
C 0.6, Cr 0.6, Mn 0.3, Mo 0.4, V 0.2, Si 0.7, bal Fe.
For arbors, tools, cold cutters, cold bending dies; tough. *Obsolete*

GORDON DIE STEEL
Latrobe Steel Co.
C 0.57, Si 0.7, Cr 0.6, Mo 0.35, bal Fe.
For pneumatic tools, swedging dies, cams, chisels; tough. *Obsolete*

GORHAM IMPERIAL
Gorham Tool Industries Inc.
C, Co, W, bal Fe.
For tool bits, cutters; for heavy cuts. *Obsolete*

GORHAM IMPERIAL 9
Gorham Tool Industries Inc.
C 0.7, Mo 8-9, Co 8-9, bal Fe.
For tools, cutters; high speed steel. *Obsolete*

GORHAM M-40-C
Gorham Tool Industries Inc.
C 0.7, Mo 9, Co 5, V 1, W 1, Cr 4, B, bal Fe.
For special cutters for nonferrous alloys; cast high speed steel. *Obsolete*

GORHAM M-40-T
Gorham Tool Industries Inc.
C 0.8, Mo 9, Co 5, V 3, W 1, Cr 4, B, bal Fe.
For cutters, tool bits; cast high speed steel. *Obsolete*

GORHAM M-40-U
Gorham Tool Industries Inc.
C 0.7, Mo 9, Co 3, W 1, Cr 4, V 1, bal Fe.
Cast. For blades, centers, gauges, wear parts; cast, abrasion and wear resistant. *Obsolete*

GORHAM M-40-U12
Gorham Tool Industries Inc.
C 0.7, Cr 4, Mo, B, Co, bal Fe.
For tools, dies, cutters; high speed steel, abrasion resistant.

GORHAM MOLYBDENUM
Gorham Tool Industries Inc.
C 0.7, W, Mo, Cr, V, bal Fe.
For tools, cutters; high speed steel. *Obsolete*

GORMET
Gorham Tool Industries Inc.
C, Cr, Co, Mo, W, bal Fe.
For cutters for iron, brass and bakelite, form tools; cast. *Obsolete*

GORMET FMC
Gorham Tool Industries Inc.
W, Cr, Co.
For cutters for cast iron, brass and bakelite; cast, abrasion resistant. *Obsolete*

GOST
Manufacturer not listed.
Prefix for many USSR alloys.

GOST-0C18N9
Russian manufacture
C 0.06, Cr 18, Ni 9, bal Fe.
Annealed: 80,000 TS; 35,000 YS; 80 Rock B. For tanks, evaporators, agitators, mixers, chemical plant equipment; Type 304 stainless steel, austenitic.

GOST-0X18H9
Russian manufacture
C 0.06, Cr 19, Ni 10, bal Fe.
Annealed: 85,000 TS; 35,000 YS; 150 Brin. For chemical plant equipment, evaporators, tanks, mixers; Type 304 stainless steel austenitic.

GOST-1C18N
Russian manufacture
C 0.05, Cr 18, Ni 12, Mo 2, bal Fe.
Annealed: 85,000 TS; 35,000 YS; 80 Rock B. For chemical plant equipment, acid tanks, evaporators, digesters; Type 316 stainlesss steel, austenitic, acid resistant.

GOST-1C18N9
Russian manufacture
C 0.1, Cr 18, Ni 9, bal Fe.
Annealed: 80,000 TS; 35,000 YS; 80 Rock B. For chemical and pharmaceutical plant equipment, evaporators, trim, molding, digesters; Type 302 stainless steel, austenitic.

GOST-1C18N9T
Russian manufacture
Cr 18, Ni 10, Ti 0.3, C, bal Fe.
Annealed: 85,000 TS; 35,000 YS; 185 Brin. For welded structures, and tanks, mixers, evaporators, agitators; Type 321 stainless steel, austenitic, stabilized, welding grade.

GOST-1X18H
Russian manufacture
C 0.06, Cr 17, Ni 12, Mo 2, bal Fe.
Annealed: 80,000 TS; 30,000 YS; 80 Rock B. For chemical plant equipment, valve trim, digesters, evaporators; Type 316 stainless steel, austenitic, corrosion resistant.

GOST-1X18H9
Russian manufacture
C 0.12, Cr 18, Ni 9, bal Fe.
Annealed: 90,000 TS; 40,000 YS; 85 Rock B. For tanks, chemical and pharmaceutical plant equipment, evaporators; Type 302 stainless steel, austentic.

GOST-1X18H9T
Russian manufacture
C 0.06; Cr 18, Ni 11, Cb 0.3, bal Fe.
Annealed: 85,000 TS; 35,000 YS; 55 El; 150 Brin. For welded structures and tanks, mixers, agitators, digesters; Type 321 stainless steel, stabilized, austenitic, welding grade.

GOST-2C13
Russian manufacture
C 0.2, Cr 13, bal Fe.
Annealed: 95,000 TS; 50,000 YS; 92 Rock B. For cutlery, surgical instruments, gears, shafts, ball bearings; Type 420 stainless steel, heat treatable.

GOST-2X13
Russian manufacture
C 0.2, Cr 12-14, bal Fe.
Annealed: 95,000 TS; 50,000 YS; 92 Rock B. For gears, shafts, cutlery, surgical instruments; Type 420 stainless steel, hardenable.

GOST-C17
Russian manufacture
C 0.1, Cr 17, bal Fe.
Annealed: 70,000 TS; 40,000 YS; 140 Brin. For dairy and chemical plant equipment, oil refinery accessories, automobile trim, oil burners, fasteners; Type 430 stainless steel, non-hardenable.

GOST-C18N11B
Russian manufacture
C 0.06, Cr 18, Ni 12, Cb 0.6, bal Fe.
Annealed: 90,000 TS; 35,000 YS; 85 Rock B. For welded structures and tanks, chemical plant equipment, agitators, mixers, evaporators; Type 348 stainless steel, austenitic, heat and corrosion resistant.

GOST-C18N12M2T
Russian manufacture
C 0.08, Cr 18, Ni 10, Mo 2, bal Fe.
For chemical plant equipment, evaporators, agitators, tanks; Type 18-10 Mo stainless steel, acid resistant, austenitic.

GOST-C20N14S2
Russian manufacture
C 0.16, Cr 24, Ni 14, bal Fe.
Annealed: 90,000 TS; 40,000 YS; 84 Rock B. For salt pots, furnace parts, heat treating fixtures; Type 309B stainless steel, austenitic, heat and corrosion resistant.

GOST-C23N13
Russian manufacture
C 0.15, Cr 23, Ni 14, bal Fe.
Annealed: 90,000 TS; 40,000 YS; 83 Rock B. For heat treating boxes, combustion chambers, salt pots, kilns; Type 309S stainless steel, heat and corrosion resistant.

GOST-C23N18
Russian manufacture
C 0.2, Cr 25, Ni 21, bal Fe.
Annealed: 95,000 TS; 45,000 YS; 90 Rock B. For salt pots, furnace parts, heat treating equipment; Type 310S stainless steel, heat and corrosion resistant.

GOST-C25
Russian manufacture
C 0.2, Cr 25, bal Fe.
Annealed: 85,000 TS; 50,000 YS; 180 Brin. For salt pots, furnace parts, heat treating equipment and fixtures; Type 446 stainless steel, heat and corrosion resistant.

GOST-C25N20S2
Russian manufacture
C 0.2, Cr 25, Ni 20, bal Fe.
Annealed: 100,000 TS; 50,000 YS; 90 Rock B. For furnace parts, salt pots, heat treating fixtures; Type 314 stainless steel, austenitic, heat and corrosion resistant.

GOST-X17
Russian manufacture
C 0.1, Cr 14-18, bal Fe.
Cold rolled: 100,000 TS; 70,000 YS; 190 Brin. For automotive trim, hardware, oil burners, fasteners, soot blowers; Type 430 stainless steel, corrosion resistant.

GOST-X18H115
Russian manufacture
C 0.06, Cr 18, Ni 12, Cb 0.8, bal Fe.
Annealed: 90,000 TS; 40,000 YS; 82 Rock B. For exhaust manifolds, storage containers, welded structures, chemical plant equipment; Type 347 stainless steel, heat and corrosion resistant, austenitic.

GOST-X18H12M2T
Russian manufacture
C 0.06, Cr 17, Ni 12, Mo 2, bal Fe.
Annealed: 85,000 TS; 35,000 YS; 82 Rock B. For chemical plant equipment, fixtures, mixers, agitators; Type 18-10 Mo stainless steel, acid resistant, austenitic.

GOST-X20H14C2
Russian manufacture
C 0.15, Cr 18, Ni 10, Si 3, bal Fe.
Annealed: 90,000 TS; 40,000 YS; 85 Rock B. Heat treating fixtures and equipment, furnace parts, tube supports; Type 302B stainless steel, heat and corrosion resistant.

GOST-X23H13
Russian manufacture
C 0.18, Cr 23, Ni 14, bal Fe.
Annealed: 90,000 TS; 40,000 YS; 84 Rock B. For heat treating fixtures, furnace parts, salt pots, oil burners; Type 309S stainless steel, heat and corrosion resistant.

GOST-X23H18
Russian manufacture
C 0.2, Cr 25, Ni 21, bal Fe.
Annealed: 95,000 TS; 45,000 YS; 89 Rock B. For furnace parts and equipment, heat treating boxes and fixtures; Type 310S stainless steel, heat and corrosion resistant.

GOST-X25
Russian manufacture
C 0.25, Cr 24-30, bal Fe.
Annealed: 80,000 TS; 48,000 YS; 85 Rock B. For heat treating boxes, furnace parts, fittings, hardware, valves; Type 446 stainless steel, heat and corrosion resistant.

GOST-X25H20C2
Russian manufacture
C 0.2, Cr 24, Ni 20, bal Fe.
Annealed: 100,000 TS; 50,000 YS; 90 Rock B. For furnace parts, radiant tubes, heat treating boxes, fixtures; Type 314 Stainless steel, heat and corrosion resistant.

GOTTINGEN AGG2
Gottingen Aluminiumwerke GmbH
Al 99.5.
Soft: 13,000 TS; 5000 YS; 45 El; 23 Brin. Hard: 24,000 TS; 22,000 YS; 15 El; 44 Brin. For roofing, culverts, architectural trim; corrosion resistant.

GOTTINGEN AGG3
Gottingen Aluminiumwerke GmbH
Mn 1-1.5, Cr 0-0.3, bal Al.
Soft: 16,000 TS; 6000 YS; 40 El. Hard: 29,000 TS; 27,000 YS; 10 El. For cooking utensils, heat exchangers, tanks, furniture; good forming and welding properties.

GOTTINGEN AGG4
Gottingen Aluminiumwerke GmbH
Mg 1.5-3, Mn 0.5-1.5, Cr 0-0.3, bal Al.
Soft: 26,000 TS; 10,000 YS; 20 El; 45 Brin. Hard: 41,000 TS; 36,000 YS; 5 El; 77 Brin. For roofing, hydraulic tubing, architectural trim; good forming and welding properties.

GOTTINGEN AGG51
Gottingen Aluminiumwerke GmbH
Mg 0.6-1.4, Si 0.6-1.6, Cr 0-0.3, Mn 0.6-1, bal Al.
Annealed: 21,000 TS; 8000 YS; 24 El. For window frames, fan blades, gutters, boats; good formability and weldability.

GOTTINGEN AGG57
Gottingen Aluminiumwerke GmbH
Mg 2-4, Mn 0-0.4, Cr 0-0.3, bal Al.
Soft: 28,000 TS; 13,000 YS; 30 El; 47 Brin. Hard: 40,000 TS; 35,000 YS; 10 El; 73 Brin. For aircraft tanks and fittings, fuel lines, marine parts; resists seawater corrosion.

GOVERNMENT BRONZE
Century Brass Products Inc.
Cu 86-89, Sn 7.5-11, Zn 1.5-4.5, Ni 0.75, Pb 0.3.
35,000 TS; 17,000 YS; 14-16 El. For valves, gears, fittings, steam valves; composition "G." *Obsolete*

GOVERNMENT BRONZE SUBSTITUTE
Manufacturer not listed.
Cu 90, Sn 6.5, Zn 3, Pb 0.5.
For structural and marine fittings; corrosion resistant.

GOVERNMENT BRONZE-H
American manufacture
Cu 82-84, Sn 12.5-14.5, Zn 2.5-4.5, Pb 0.8-1, Fe 0.1.
For journals, bushings, bearings; high strength.

GOVERNMENT GENUINE BABBITT
United American Metals Corp.
Pb, Sb, Sn.
Cast: 7100 TS; 0.75 El; 2.8 RA. For bearings; good shock
resistance; Babbitt metal.

GRAC
Graphitized Alloy Corp.
Sn 10, Sb, As, Cd, Ni, graphite, bal Pb.
27 Brin. For bearings; M.P. 453-576°F.

GRACITE
Grayborn Steel Co.
C 0.9, Mn 1.25, Cr 0.5, W 0.5, bal Fe.
For punches, forming and beading dies, upsetters; oil
hardened, nondeforming

GRADE "A" (GRADE 1)
Diehl Steel Co.
C 0.95, Mn 0.3, Si 0.2, bal Fe.
Water hardening tool steel; AISI W1.

GRADE "A" PHOSPHOR BRONZE 351 ALLOY
Anaconda Co.
Cu 95, 5% Sn + P.
55,000-70,000 TS. For sleeves, bushings. *Obsolete*

GRADE "H"
Peter A. Frasse & Co.
C, W, bal Fe.
For general tools. *Obsolete*

GRADE S
Oil Well Supply Co.
C, Ni, bal Fe.
For sucker rods in oil wells; resists H_2S in water.

GRADE X
Midvale-Heppenstall Co.
C 1.2, bal Fe.
For drills; taps, cutters; drill rod, water hardening. *Obsolete*

GRAFIDIN
Pechiney Electrometallurgie
Miscellaneous nonferrous. Al 1.5, Ba 4.5, Ca 2.5, Mn 9, Si 63,
bal Fe.
Cast iron inoculant.

GRAINAL 100
SMC (Shieldalloy Metallurgical Corp.)
B 1, Al 13, Ti 20, Si 5, Zr 4, Mn 8, bal Fe.
Master alloy for ladle additions.

GRAINAL 100
Cyprus Foote Mineral Co.
Ti 20, Al 13, Zr 4, Mn 8, Si 0-5, B 1, bal Fe.
For addition in steel melting; increases hardenability.
Obsolete

GRAINAL 79
SMC (Shieldalloy Metallurgical Corp.)
B 0.5, Al 13, Ti 20, Si 5, Zr 3.5, Mn 8, bal Fe.
Master alloy for ladle additions.

GRAINAL NO. 1
Now VANADIUM GRAINAL NO. 1.

GRAINAL NO. 79
Cyprus Foote Mineral Co.
Ti 20, Al 13, Zr 4, Mn 8, Si 0-5, B 0.5, bal Fe.
Used in steel manufacturing; increases hardenability.
Obsolete

GRAINLOY
Birdsboro Corp.
C 3, Si 1.3, Mn, bal Fe.
For rolls; cast iron. *Obsolete*

GRALUR
Duke Steel Co. Inc.
C, bal Fe.
For drills.

GRAMIX
U.S. Graphite Inc.
Cu-Sn-graphite.
For light machine bearings, bushings; finely powdered
mixture; self-lubricating. *Obsolete*

GRAMIX GRADE NO. 138
U.S. Graphite Inc.
Copper. Cu.
Sintered: 50,000 psi (345 MPa) crushing strength; 23,000 psi
(158 MPa) TS; 11,000 psi (75 MPa) YS COMP; 12 El; 7.6-8.0
density (dry). For commutator rings. *Obsolete*

GRAMIX GRADE NO. 183
U.S. Graphite Inc.
Copper. Sn, Pb, Ni, bal Cu.
Bronze type. Sintered: 30,000 psi (207 MPa) crushing
strength; 15,000 psi (103 MPa) TS, 13,000 psi (189 MPa) YS
COMP; 2 El; 6.4-6.8 density (dry). Bearings for shaded pole
motors. *Obsolete*

GRAMIX GRADE NO. 266
U.S. Graphite Inc.
Cu, Pb, C, bal Fe.
Iron alloy. Sintered: 46,000 psi (317 MPa) crushing strength;
24,000 psi (165 MPa) TS; 20,000 psi (138 MPa) YS COMP; 0.5
El; 5.7-6.1 density (dry); 88-90 Rock F. For bearings,
structural parts, rotors. *Obsolete*

GRAMIX GRADE NO. 272
U.S. Graphite Inc.
Copper. Zn, bal Cu.
Brass type. Sintered: 42,000 psi (289 MPa) crushing strength;
20,000 psi (138 MPa) TS; 10,000 psi YS COMP; 11 El; 7.0-7.4
density (dry). For brass parts as packing glands. *Obsolete*

GRAMIX GRADE NO. 273
U.S. Graphite Inc.
Copper. Ni, Zn, bal Cu.
Nickel-Silver. Sintered: 49,000 psi (338 MPa) crushing
strength; 27,000 psi (186 MPa) TS; 22,000 psi (151 MPa) YS
COMP; 10 El; 6.7-7.1 density (dry). Light duty, corrosion
resistant seal face applications. *Obsolete*

GRAMIX GRADE NO. 278
U.S. Graphite Inc.
Copper. Sn, C, bal Cu.
Bronze type. Sintered: 25,000 psi (172 MPa) crushing
strength; 13,000 psi (89 MPa) TS; 10,000 psi (69 MPa) YS
COMP; 2 El; 6.1-6.5 density (dry). Bearings for small
appliances. *Obsolete*

GRAMIX GRADE NO. 353
U.S. Graphite Inc.
Copper. Ni, Sn, bal Cu.
Nickel-Bronze type. Sintered: 44,000 psi (303 MPa) crushing
strength; 24,000 psi (165 MPa) TS; 20,000 psi (138 MPa) YS
COMP; 1.5 El; 6.6-7.0 density (dry). High load bearing and
high strength applications. *Obsolete*

GRAMIX GRADE NO. 361
U.S. Graphite Inc.
Fe.
Machinable iron. Sintered: 37,000 psi (255 MPa) crushing
strength; 18,000 psi (124 MPa) TS; 12,500 psi (86 MPa) YS
COMP; 1.5 El; 5.8-6.2 density (dry); 43-49 Rock F. For
machinable structural parts and pole shoes. *Obsolete*

GRAMIX GRADE NO. 363
U.S. Graphite Inc.
Copper. Sn, C, bal Cu.
Bronze type. Sintered: 23,000 psi (158 MPa) crushing
strength; 11,000 psi (75 MPa) TS; 9,000 psi (62 MPa) YS
COMP; 0.5 El; 6.3-6.7 density (dry). Used on bearing
applications where quiet operation is major consideration.
Obsolete

GRAMIX GRADE NO. 400
U.S. Graphite Inc.
Copper. Sn, C, bal Cu.
Bronze type. Sintered: 28,000 psi (193 MPa) crushing
strength; 14,000 psi (96 MPa) TS; 11,000 psi (75 MPa) YS
COMP; 2 El; 6.4-6.8 density (treated). ASTM B-438-73 Grade
1, Type 2. *Obsolete*

GRAMIX GRADE NO. 401
U.S. Graphite Inc.
Copper. Sn, Pb, C, bal Cu.
Bronze type. Sintered: 28,000 psi (193 MPa) crushing
strength; 14,000 psi (96 MPa) TS; 11,000 psi (75 MPa) YS
COMP; 2 El; 6.5-6.9 density (treated). ASTM B-438-73 Grade
2, Type 1. *Obsolete*

GRAMIX GRADE NO. 402
U.S. Graphite Inc.
C 0-0.25, bal Fe.
Iron type. Sintered: 29,000 psi (200 MPa) crushing strength;
14,000 psi (96 MPa) TS; 10,000 psi (69 MPa) YS COMP; 1.5
El; 5.7-6.1 density (treated); 44-52 Rock F. ASTM B-439-70
Grade 1. *Obsolete*

GRAMIX GRADE NO. 403
U.S. Graphite Inc.
C 0.25-0.6, bal Fe.
Iron-carbon type. Sintered: 35,000 psi (241 MPa) crushing
strength; 17,000 psi (117 MPa) TS; 15,000 psi (103 MPa) YS
COMP; 1 El; 5.7-6.1 density (treated); 5.8-6. Rock F. ASTM
B-439-70 Grade 2. *Obsolete*

GRAMIX GRADE NO. 404
U.S. Graphite Inc.
C 0.6-1, bal Fe.
Iron-carbon type. Sintered: 45,000 psi (310 MPa) crushing
strength; 22,000 psi (151 MPa) TS; 20,000 psi (138 MPa) YS
COMP; 0.5 El; 5.7-6.1 density (treated); 67-80 Rock F. ASTM
B-310-70 Class C, Type 1. *Obsolete*

GRAMIX GRADE NO. 405
U.S. Graphite Inc.
Cu, C, bal Fe.
Iron alloy. Sintered: 53,000 psi (365 MPa) crushing strength;
29,000 psi (200 MPa) TS; 25,000 psi (172 MPa) YS COMP; 0.5
El; 5.8-6.2 density (treated); 80-90 Rock F. ASTM B-439-70
Grade 3. *Obsolete*

GRAMIX GRADE NO. 406
U.S. Graphite Inc.
Copper. Zn, Pb, bal Cu.
Brass type. Sintered: 45,000 psi (310 MPa) crushing strength;
21,000 psi (144 MPa) TS; 11,000 psi (75 MPa) YS COMP; 9
El; 7.2-7.7 density (dry). Lock parts. ASTM B-282-70 Type 1.
Obsolete

GRAMIX GRADE NO. 407
U.S. Graphite Inc.
Copper. Zn, Pb, bal Cu.
Brass type. Sintered: 52,000 psi (358 MPa) crushing strength;
24,000 psi (165 MPa) TS; 13,000 psi (89 MPa) YS COMP; 11
El; 7.7 min density (dry). Lock parts. ASTM B-282-70 Type 2.
Obsolete

GRAMIX GRADE NO. 408
U.S. Graphite Inc.
Cu, C, bal Fe.
Iron alloy. Sintered: 68,000 psi (469 MPa) crushing strength;
34,000 psi (234 MPa) TS; 29,000 psi (200 MPa) YS COMP; 1
El; 5.8-6.2 density (dry); 77-87 Rock F. ASTM B-222-70.
Obsolete

GRAMIX GRADE NO. 410

U.S. Graphite Inc.
Copper. Sn, C, bal Cu.
Bronze type. Sintered: 36,000 psi (248 MPa) crushing strength; 18,000 psi (124 MPa) TS; 17,000 psi (117 MPa) YS COMP; 3 El; 6.8-7.2 density (dry). ASTM B-255-70 Type 2. *Obsolete*

GRAMIX GRADE NO. 411

U.S. Graphite Inc.
C 0-0.25, bal Fe.
Iron. Sintered: 31,000 psi (213 MPa) crushing strength; 15,000 psi (106 MPa) TS; 12,500 psi (86 MPa) YS COMP; 2 El; 5.7-6.1 density (dry); 44-52 Rock F. ASTM B-310-70 Type 1, Class A. *Obsolete*

GRAMIX GRADE NO. 412

U.S. Graphite Inc.
C 0.25-0.6, bal Fe.
Iron-carbon. Sintered: 39,000 psi (269 MPa) crushing strength; 19,000 psi (131 MPa) TS; 17,000 psi (117 MPa) YS COMP; 1 El; 5.7-6.1 density (dry); 58-68 Rock F. ASTM B-310-70 Type 1, Class B. *Obsolete*

GRAMIX GRADE NO. 413

U.S. Graphite Inc.
C 0.6-1, bal Fe.
Iron-carbon. Sintered: 48,000 psi (331 MPa) crushing strength; 24,000 psi (166 MPa) TS; 22,500 psi (155 MPa) YS COMP; 0.5 El; 6.1-6.5 density (dry). For valve plates. *Obsolete*

GRAMIX GRADE NO. 414

U.S. Graphite Inc.
C 0-0.25, bal Fe.
Iron type. Sintered: 41,000 psi (282 MPa) crushing strength; 20,500 psi (141 MPa) TS; 17,500 psi (120 MPa) YS COMP; 3.5 El; 6.1-6.5 density (dry); 60-70 Rock F. ASTM B-310-70 Type 2, Class A. *Obsolete*

GRAMIX GRADE NO. 416

U.S. Graphite Inc.
C 0.6-1, bal Fe.
Iron-carbon type. Sintered: 61,000 psi (420 MPa) crushing strength; 31,000 psi (213 MPa) TS; 26,000 psi (179 MPa) YS COMP; 0.5 El; 6.1-6.5 density (dry); 80-90 Rock F. ASTM B-310-70 Type 2, Class C. *Obsolete*

GRAMIX GRADE NO. 417

U.S. Graphite Inc.
C 0-0.25, Cu, bal Fe.
Infiltrated iron. Sintered: 122,000 psi (841 MPa) crushing strength; 65,000 psi (448 MPa) TS; 70,000 psi (483 MPa) YS COMP; 1 El; 7.1-7.6 density (dry); 89-93 Rock F. ASTM B-303-70 Class A. *Obsolete*

GRAMIX GRADE NO. 419

U.S. Graphite Inc.
C 0.6-1, Cu, bal Fe.
Infiltrated iron. Sintered: 165,000 psi (1138 MPa) crushing strength; 85,000 psi (586 MPa) TS; 90,000 psi (621 MPa) YS COMP; 0.5 El; 7.1-7.6 density (dry); 98-101 Rock F. ASTM B-303-70 Class C. Note: On all Gramix grades YS COMP is yield in compression. *Obsolete*

GRAMIX GRADE NO. 510

U.S. Graphite Inc.
Copper. Pb, Sn, bal Cu.
High lead bronze. Sintered: 16,000 psi (110 MPa) crushing strength; 8000 psi (55 MPa) TS; 5000 psi (34 MPa) YS COMP; 1 El; 6.6-7.0 density (dry). For bearings requiring a leaded bronze material. *Obsolete*

GRAMIX GRADE NO. 514

U.S. Graphite Inc.
C 0.6-1, Cu, bal Fe.
Iron alloy. Sintered: 80,000 psi (552 MPa) crushing strength; 45,000 psi (310 MPa) TS; 42,000 psi (289 MPa) YS COMP; 0.5 El; 5.8-6.2 density (dry); 83-84 Rock F. For high strength structural parts. *Obsolete*

GRAMIX GRADE NO. 548

U.S. Graphite Inc.
Fe, C.
Iron-graphite. Sintered: 9000 psi (62 MPa) crushing strength; 4500 psi (31 MPa) TS; 3000 psi (20 MPa) YS COMP; 1 El; 5.4-5.8 density (dry). Specially developed iron bearing material for high speed, light duty applications. *Obsolete*

GRAMIX GRADE NO. 560

U.S. Graphite Inc.
Cu, Fe, Mn, Si, bal Ni.
Monel type. Sintered: 55,000 psi (379 MPa) crushing strength; 30,000 psi (207 MPa) TS; 22,000 psi (151 MPa) YS COMP; 5 El; 7.0-7.4 density (dry). For corrosion resistant parts. *Obsolete*

GRAMIX GRADE NO. 562

U.S. Graphite Inc.
Ni, C, bal Fe.
Nickel steel. Sintered: 120,000 psi (828 MPa) crushing strength; 60,000 psi (414 MPa) TS; 50,000 psi (345 MPa) YS COMP; 2 El; 6.6-7.0 density (dry); 81-83 Rock F. For high strength structural parts. *Obsolete*

GRAMIX GRADE NO. 562-P3

U.S. Graphite Inc.
Ni, C, bal Fe.
Hardened nickel steel. Sintered, HT; 175,000 psi (1207 MPa) crushing strength; 100,000 psi (690 MPa) TS; 100,000 (690 MPa) YS COMP; 0.5 El; 6.6-7.0 density (dry); 25-30 Rock C. For high strength structural parts. Note: On all Gramix grades YS COMP is yield in compression. *Obsolete*

GRAMIX GRADE NO. 563

U.S. Graphite Inc.
C 0-0.25, bal Fe.
Machinable iron. Sintered: 36,000 psi (248 MPa) crushing strength; 18,000 psi (124 MPa) TS; 11,000 psi (75 MPa) YS COMP; 3 El; 5.6-6.0 density (dry); 34-40 Rock F. For pole pieces, complex parts. *Obsolete*

GRAMIX GRADE NO. 564

U.S. Graphite Inc.
C 0.25-0.6, bal Fe.
Machinable iron. Sintered: 48,000 psi (331 MPa) crushing strength; 24,000 psi (165 MPa) TS; 18,000 psi (124 MPa) YS COMP; 1 El; 5.7-6.1 density (dry); 57-58 Rock F. For shock absorber pistons. *Obsolete*

GRAMIX GRADE NO. 565

U.S. Graphite Inc.
C 0.6-1, bal Fe.
Machinable iron-carbon. Sintered: 75,000 psi (517 MPa) crushing strength; 38,000 psi (262 MPa) TS; 24,000 psi (165 MPa) YS COMP; 0.5 El; 5.8-6.2 density (dry); 79-83 Rock F. For structural parts. *Obsolete*

GRAMIX GRADE NO. 567

U.S. Graphite Inc.
Copper. Ni, Sn, bal Cu.
Nickel bronze. Sintered: 75,000 psi (517 MPa) crushing strength; 40,000 psi (276 MPa) TS; 32,000 psi (220 MPa) YS COMP; 3 El; 6.8-7.2 density (dry). For structural parts. *Obsolete*

GRAMIX GRADE NO. 568

U.S. Graphite Inc.
C 0-0.25, bal Fe.
Iron. Sintered: 53,000 psi (365 MPa) crushing strength; 20,000 psi (138 MPa) TS; 21,000 psi (144 MPa) YS COMP; 4 El; 6.5-6.9 density (dry); 66-72 Rock F. Magnetic applications. *Obsolete*

GRAMIX GRADE NO. 569

U.S. Graphite Inc.
C 0.6-1, bal Fe.
Iron-carbon alloy. Sintered: 100,000 psi (690 MPa) crushing strength; 52,000 psi (358 MPa) TS; 41,000 psi (282 MPa) YS COMP; 1 El; 6.4-6.8 density (dry); 88-95 Rock F. For gears and structural parts. *Obsolete*

GRAMIX GRADE NO. 571

U.S. Graphite Inc.
Copper. Zn, bal Cu.
Brass type. Sintered: 26,700 psi (184 MPa) crushing strength; 12,700 psi (87 MPa) TS; 6000 psi (41 MPa) YS COMP; 7 El; 6.8-7.2 density (dry). *Obsolete*

GRAMIX GRADE NO. 572

U.S. Graphite Inc.
C 0-0.25, bal Fe.
Machinable iron. Sintered: 48,000 psi (331 MPa) crushing strength; 19,000 psi (131 MPa) TS; 17,000 psi (117 MPa) YS COMP; 3 El; 6.6-7.0 density (dry); 64-72 Rock F. For machinable and magnetic parts. *Obsolete*

GRAMIX GRADE NO. 574

U.S. Graphite Inc.
C 0.6-1, Cu, bal Fe.
Infiltrated iron-carbon. Sintered: 175,000 psi (1207 MPa) crushing strength; 90,000 psi (621 MPa) TS; 75,000 psi (517 MPa) YS COMP; 1 El; 7.3-7.7 density (dry) 78-84 Rock F. For high strength structural parts. *Obsolete*

GRAMIX GRADE NO. 575

U.S. Graphite Inc.
Cu, bal Fe.
Iron-copper. Sintered: 50,000 psi (345 MPa) crushing strength; 16,500 psi (113 MPa) TS; 21,000 psi (144 MPa) YS COMP; 0.5 El; 6.0-6.4 density (dry); 48-60 Rock F. For structural parts. Note: On all Gramix grades YS COMP is yield in compression. *Obsolete*

GRAMIX GRADE NO. 576

U.S. Graphite Inc.
C 0-0.25, bal Fe.
Iron. Sintered: 96,000 psi (662 MPa) crushing strength; 27,600 psi (190 MPa) TS; 24,000 psi (167 MPa) YS COMP; 4 El; 7.0-7.4 density (dry); 89-93 Rock F. High density, magnetic applications. *Obsolete*

GRAMIX GRADE NO. 577

U.S. Graphite Inc.
C 0.25-0.6, Cu, bal Fe.
Iron alloy. Sintered: 73,200 psi (505 MPa) crushing strength; 38,100 psi (262 MPa) TS; 33,000 psi (227 MPa) YS COMP; 1 El; 5.8-6.2 density (dry); 75-80 Rock F. *Obsolete*

GRAMIX GRADE NO. 578

U.S. Graphite Inc.
C 0-0.25, Ni, bal Fe.
Iron-nickel. Sintered: 69,000 psi (476 MPa) crushing strength; 30,000 psi (207 MPa) TS; 20,000 psi (138 MPa) YS; 2 El; 6.6-7.0 density (dry); 71-75 Rock F. *Obsolete*

GRAMIX GRADE NO. 580

U.S. Graphite Inc.
C 0.2-0.4, bal Fe.
Machinable iron-carbon. Sintered: 54,000 psi (372 MPa) crushing strength; 16,000 psi (110 MPa) TS; 14,000 psi (96 MPa) YS COMP; 1 El; 6.2-6.6 density (dry); 79-85 Rock F. For structural parts; good machinability. *Obsolete*

GRAMIX GRADE NO. 581

U.S. Graphite Inc.
C 0.25-0.6, bal Fe.
Machinable iron-carbon. Sintered: 58,000 psi (400 MPa) crushing strength; 23,300 psi (160 MPa) TS; 19,000 psi (131 MPa) YS COMP; 2 El; 5.7-6.1 density (dry); 64-73 Rock F. Structural parts; good machinability. *Obsolete*

GRAMIX GRADE NO. 582

U.S. Graphite Inc.
C 0.6-1, bal Fe.
Iron-carbon. Sintered: 70,000 psi (483 MPa) crushing strength; 36,000 psi (248 MPa) TS; 21,000 psi (144 MPa) YS COMP; 2 El; 6.4-6.8 density (dry); 33-50 Rock F. Structural parts. *Obsolete*

GRAMIX GRADE NO. 61
U.S. Graphite Inc.
Copper. Sn, bal Cu.
Bronze type. Sintered: 29,000 psi (200 MPa) crushing strength; 15,000 psi (103 MPa) TS; 12,000 psi (82 MPa) YS COMP; 2 El; 6.4-6.8 density (treated). Standard grade for bronze parts. *Obsolete*

GRANADA TOOL
Colt Industries
C 1, bal Fe.
For mining sledges, tools, dies; water hardened. *Obsolete*

GRANADA VANADIUM
Crucible Materials Corp.
Tool material. C 1, Mn 0.3, V 0.2, bal Fe.
For blacksmith tools, shears, punches, forming dies; water hardening.

GRANALEC
National Intergroup Inc.
Si, bal Fe.
For laminations; high magnetic permeability.

GRANATOR
National Intergroup Inc.
Si, bal Fe.
For laminations; high magnetic permeability.

GRANATURE
National Intergroup Inc.
Si, bal Fe.
For laminations; high magnetic permeability.

GRANDIOS
Now VEW S305.

GRANDIOS 3VN
Vereinigte Edelstahlwerke
C 0.86, Co 2.8, Cr 4.3, Mo 0.8, V 2, W 12, bal Fe.
For lathe and planer tools, drills, hobs, reamers; high speed steel. *Obsolete*

GRANDIOS 5V
Now VEW S308.

GRANDIOS 5VN
Vereinigte Edelstahlwerke
C 0.8, Cr 4.3, Co, Mo, V, W, bal Fe.
For lathe and planer tools, broaches; high speed steel. *Obsolete*

GRANDIOS EXTRA
Bohler Gesellschaft M.B.H.
C 0.76, Co 10, Cr 4.2, Mo 0.8, V 1.8, W 18, bal Fe.
For lathe and planer tools, milling cutters; high speed steel.

GRANE 1
Uddeholm Corp.
C 0.55, Cr 1, Ni 3, Mo 0.3, bal Fe.
For drop forging dies; tough.

GRANE 2
Uddeholm Corp.
C 0.55, Cr 1.5, Ni 3, bal Fe.
For drop forging dies; tough.

GRANFIN CAST IRON
English manufacture
C 2.2-2.6, Mn 0.7-1, Si 2.3-2.5, 1.7-2.1 graphite, bal Fe.
For frames, housings, general castings; cast iron.

GRANFORMER
National Intergroup Inc.
Si, bal Fe.
For laminations; high magnetic permeability.

GRANFORMER 52
National Intergroup Inc.
Si 4.7, bal Fe.
For transformers; high permeability.

GRANFORMER 58
National Intergroup Inc.
Si 4-4.4, bal Fe.
For transformers; high permeability.

GRANFORMER 65
National Intergroup Inc.
Si 4-4.5, bal Fe.
For transformers; low core loss.

GRANFORMER 72
National Intergroup Inc.
Si 3-3.4, bal Fe.
For transformers for radios; high permeability.

GRANIMO
National Intergroup Inc.
Si, bal Fe.
For laminations; high magnetic permeability.

GRANISIL
National Intergroup Inc.
Si, bal Fe.
For laminations; high magnetic permeability.

GRANIT
Haeckerstahl GmbH
C 2.1, Cr 11.5, bal Fe.
For blanking and forming dies; oil hardening, nondeforming.

GRANIT 0
Bergische Stahl Industrie
C 0.55, Cr 0.7, Mo 0.18, Ni 1.65, V 0.1, bal Fe.
For gears, pinions, shafts, crankshafts; oil hardened, shock resistant. *Obsolete*

GRANIT 1
Bergische Stahl Industrie
C 0.1, Mn 0.37, bal Fe.
For plastic mold dies; hubbable, case hardening. *Obsolete*

GRANIT 2
Bergische Stahl Industrie
C 0.18-0.24, Si 0.15-0.35, Mn 1.1-1.4, Cr 1-1.3, bal Fe.
Cold work tool steel. W.-Nr. 1.2162.

GRANIT 2
Bergische Stahl Industrie
C 1.65, Si 0.35, Cr 11.5, V 0.1, bal Fe.
For blanking and forming dies; air hardening, nondeforming. *Obsolete*

GRANIT 3
Bergische Stahl Industrie
C 0.56, Mn, Cr, Mo, bal Fe.
For gears, crankshafts; oil hardened, shock resistant. *Obsolete*

GRANIT 30
Bergische Stahl Industrie
C 1.05, Mn 0.9, Cr 1, W 1.15, bal Fe.
For cold work tools, dies, bearings; water hardened. *Obsolete*

GRANIT CO
Haeckerstahl GmbH
C 0.65, Cr, Co, bal Fe.
For tools, dies; oil hardening.

GRANIT IV
Bergische Stahl Industrie
C 0.4, Cr 13, bal Fe.
Annealed: 90,000 TS; 45,000 YS; 40 El; 50 RA; 190 Brin. For cutlery, surgical instruments; corrosion resistant; Type 420. *Obsolete*

GRANIT V
Bergische Stahl Industrie
C 0.2, Mn 1.25, Cr 1.15, bal Fe.
For bearings, liners; case hardening steel, water or oil hardening. *Obsolete*

GRANIT W
Haeckerstahl GmbH
C 2.1, Cr 11.5, W 0.7, bal Fe.
For thread rolling dies, blanking and forming dies; oil hardened, nondeforming.

GRANITE
Midvale-Heppenstall Co.
C, alloy, bal Fe.
For tools, oil hardening. *Obsolete*

GRANITE CITY HIGH-YIELD NO. 1
National Intergroup Inc.
Mn 0.7-0.9, P 0-0.01, S 0.035, Si 0.15-0.2, 0.2 Cu min, 0.15 Cr min, C, bal Fe.
Annealed: 70,000 TS; 55,000 YS; 25 El. For transportation equipment, cars, buses, trucks; corrosion resistant.

GRANITE CITY HIGH-YIELD NO. 2
National Intergroup Inc.
Mn 1.2-1.6, P 0-0.01, S 0.035, Si 0.15-0.2, 0.2 Cu min, 0.15 Cr min, C, bal Fe.
Annealed: 85,000 TS; 70,000 YS; 18 El. For bridges, transportation equipment, cars, buses, trucks; corrosion resistant.

GRANT
Time Steel Service Inc.
C 0.45, Mn 0.55, Si 0.2, Cr 0.95, V 0.2, bal Fe.
Tool steel for miscellaneous applications as hammers, hatchets. AISI L2.

GRAPH M.N.S.
Alexander Benecke, Inc.
C 1.5, Mn 1.25, Si 1.2, Ni 1.75, Mo 0.5, Cr 0.5, bal Fe.
Annealed: 135,000 TS; 78,000 YS; 17 El; 34 RA; 262 Brin. For blanking and forming dies, punches, rolls; graphitic steel. *Obsolete*

GRAPH M.N.S.
Timken Co.
C 1.5, Mn 1.25, Si 1.2, Ni 1.75, Mo 0.5, Cr 0.5, bal Fe.
Annealed: 135,000 TS; 78,000 YS; 17 El; 34 RA; 262 Brin. For blanking and forming dies, punches, rolls; graphitic steel. *Obsolete*

GRAPH-AIR
Latrobe Steel Co.
C 1.35, Si 1.2, Mn 1.8, Ni 1.85, Mo 1.5.
Air hardening tool steel. For gauges, arbors, forming rolls, shear blades, and metal working dies and punches. Cold work die steel, AISI A10.

GRAPH-AIR
Peninsular Steel Co.
C 1.35, Mn 1.85, Si 1.2, Ni 1.85, Mo 1.5, bal Fe.
Air hardening tool steel, minimal distortion.

GRAPH-AIR
Timken Co.
C 1.35, Mn 1.85, Si 1.2, Ni 1.85, Mo 1.5, bal Fe.
Air hardening tool steel, minimal distortion.

GRAPH-MO
Latrobe Steel Co.
C 1.45, Si 0.9, Mn 1, Mo 0.25.
Oil hardening, cold work die steel. For dies and punches in drawing, forming and shaping operations. AISI O6.

GRAPH-MO

Peninsular Steel Co.
C 1.5, Mo 0.25, Si 0.8, Mn 0.4, bal Fe.
Annealed: 85,000 TS; 50,000 YS; 25 El; 40 RA; 197 Brin. Heat treated: 218,000 TS; 170,000 YS; 9 El; 14 RA; 388 Brin. For plug and ring gauge dies, taps, spindles, rolls; wear and abrasion resistant graphitic steel, non-seizing.

GRAPH-MO

Timken Co.
C 1.5, Mo 0.25, Si 0.8, Mn 0.4, bal Fe.
Annealed: 85,000 TS; 50,000 YS; 25 El; 40 RA; 197 Brin. Heat treated: 218,000 TS; 170,000 YS; 9 El; 14 RA; 388 Brin. For plug and ring gauge dies, taps, spindles, rolls; wear and abrasion resistant graphitic steel, non-seizing.

GRAPH-SIL

Timken Co.
C 1.5, Si 0.75-0.95, CC 0.35-0.4, bal Fe.
Annealed: 100,000 TS; 57,000 YS; 25 El; 47 RA; 171 Brin. Heat treated: 220,000 TS; 160,000 YS; 9 El; 18 RA; 400 Brin. For cold header dies, bushings, engine sleeves; graphitic steel, abrasion resistant. *Obsolete*

GRAPH-TUNG

Timken Co.
C 1.5, Mn 0.5, Si 0.65, Mo 0.5, W 2.8, bal Fe.
For blanking and drawing dies, coining dies, shear blades; abrasion resistant graphitic steel, non-seizing. *Obsolete*

GRAPHALLOY BABBITT

Graphite Metallizing Corp.
graphite impregnated with Pb base Babbitt.
Sintered: 5000 TS. For self-lubricating bearings, seal rings, contact shoes. Crushing strength: 19,000 psi.

GRAPHALLOY BRONZE

Graphite Metallizing Corp.
graphite impregnated with bronze.
Sintered: 5500 TS. For self-lubricating bearings, seal rings, contact shoes. Crushing strength: 24,000 psi.

GRAPHALLOY CADMIUM

Graphite Metallizing Corp.
graphite impregnated with Cd.
Sintered: 5000 TS. For bearings, contacts for controllers, switches, seal rings; maximum temperature 250°F; self lubricating.

GRAPHALLOY COPPER

Graphite Metallizing Corp.
graphite impregnated with Cu.
Sintered: 6000 TS. For slip ring brushes, contacts on controllers, relays, bearings; self lubricating, maximum temperature 700°F. Crushing strength: 25,000 psi.

GRAPHALLOY GRADE A

Graphite Metallizing Corp.
55 graphite, bal Pb Babbitt.
For bearings of light service; Babbitt soft grade.

GRAPHALLOY GRADE N

Graphite Metallizing Corp.
55 graphite, bal Pb Babbitt.
For bearings of medium service; Babbitt medium grade.

GRAPHALLOY GRADE O

Graphite Metallizing Corp.
55 graphite, bal Pb Babbitt.
For bearings; Babbitt hard grade.

GRAPHALLOY GRADE S

Graphite Metallizing Corp.
55 graphite, bal Pb Babbitt.
For bearings; medium hardness.

GRAPHALLOY IRON

Graphite Metallizing Corp.
graphite impregnated with Fe.
Sintered: 6000 TS. For self-lubricating bearings, submerged bearings. Crushing strength: 25,000 psi.

GRAPHALLOY SILVER 8

Graphite Metallizing Corp.
graphite impregnated with Ag.
Sintered: 5500 TS. For contacts on relays, circuit breakers, bearings, brushes; maximum temperature 700°F; self lubricating. Crushing strength: 24,000 psi.

GRAPHEX

Wakefield Corp.
Cu 87-92, Sn 9-11, 4.0 graphite.
13,000 TS; 11,000 YS; 1 El; 30 Brin. For bearings, bushings; self lubricating. *Obsolete*

GRAPHIDOX

Cyprus Foote Mineral Co.
Ca 5-7, Si 50-55, Ti 9-11, Al 1-1.3, C 0.15-0.25, bal Fe.
Ferroalloy for inoculant for cast iron; also for supplementary deoxidation of steel. *Obsolete*

GRAPHIDOX NO. 4

Foote Mineral Co.
Ti 10, Si 48, Ca 6, bal Fe.
Used in iron manufacturing; for graphitization, reduces chill. *Obsolete*

GRAPHITE METAL-1

English manufacture
Pb 80, Sb 20.
For crucibles, lubricants, lead pencils.

GRAPHITE METAL-2

English manufacture
Pb 68, Sb 17, Sn 15.
For foundry facings, electric brush carbons.

GRAPHITE NITRALLOY

Bethlehem Steel Corp.
C 1.25-1.5, Si 1.25-1.5, Mn 0.4-0.6, Al 1-1.5, Cr 0.2-0.4, Mo 0.3, bal Fe.
Heat treated: 85,000 TS; 70,000 YS; 19 El; 24 RA. For wear resistant parts; wear resistant after nitriding. *Obsolete*

GRAPHITE NITRALLOY

Nitralloy Corp.
C 1.25-1.5, Si 1.25-1.5, Mn 0.4-0.6, Al 1-1.5, Cr 0.2-0.4, Mo 0.3, bal Fe.
Heat treated: 85,000 TS; 70,000 YS; 19 El; 24 RA. For wear resistant parts; wear resistant after nitriding. *Obsolete*

GRAPHO BABBITT METAL NO. 1

Lehigh Babbitt Co.
Graphite and Babbitt.
19 Brin. For Babbitt, graphite bearings, transmission and railroad axle bearings, engine bearings; low bearing temperature, reduced friction losses.

GRAPHO BABBITT METAL NO. 2

Lehigh Babbitt Co.
0.3 graphite, S, Pb, Sb.
26 Brin. For bearings for electrical motors, blowers, high speed transmissions; high velocity and high pressures.

GRAPHO BABBITT METAL NO. 3

Lehigh Babbitt Co.
0.3 graphite, Sn, Pb, Sb.
29 Brin. For bearings for crankshafts in internal combustion engines, mining machinery, machine tools, reciprocating pumps; high pressure and high velocity with alternating load.

GRAPHO BABBITT METAL NO. 4

Lehigh Babbitt Co.
0.4 graphite, Sn, Pb, Sb.
34 Brin. For bearings for diesel engines, heavy duty compressors; heavy duty; all kinds of shocks and velocities.

GRAY CUT COBALT

Teledyne Vasco
C 0.8, V 1.6, W 20.5, Cr 4.25, Co 12.25, Mo 0.6, bal Fe.
For cutting tools, shapers, planers; high speed steel. *Obsolete*

GRAY DEVIL

Champion Rivet Co.
C 0.09, bal Fe.
Welded: 77,000 TS; 65,000 YS; 19 El; 30 RA. For welding rods for steel; E-6012.

GRAY GOLD

English manufacture
Au 86, Fe 5.7-17, Ag 0-8.6.
For jewelry; corrosion resistant.

GRAY LABEL

Peninsular Steel Co.
C 0.95, bal Fe.
For cold beading dies; water hardening.

GRAY LABEL STAYPUT

Heller Bros. Co.
C 0.8-0.9, Mn 1.2, Cr 0.5, W 0.5, bal Fe.
For molding and threading dies, chasers, taps, reamers; oil hardened, nondeforming. *Obsolete*

GRAY LABEL STYRIAN

Skoda Works National Corp.
C 0.88, Mn 0.19, bal Fe.
For tools, drills, taps; water hardening.

GRAYDAC

Champion Rivet Co.
C 0.08, Mn 0.31, Si 0.29, bal Fe.
Welded: 64,000-72,000 TS; 55,000-62,000 YS; 19-26 El; 30-45 RA. For welding electrodes; low spatter; E-6013.

GREAT WESTERN GW

Great Western Steel Co.
C 0.85, Mn 1, Si 0.35, Cr 0.5, W 0.5, bal Fe.
For milling cutters, drills, dies, punches, mandrels; oil hardened, nondeforming.

GREEK ASCOLOY

Teledyne Firth Sterling
C 0-0.12, Cr 12-14, Ni 1.8-2.2, W 2.5-3.5, Mo 0-0.5, Cu 0-0.5, bal Fe.
Heat treated: 137,000-195,000 TS; 108,000-160,000 YS; 14-23 El; 51-63 RA; 290-400 Brin. High temperature steel for compressor blades and vanes, fasteners, jet engine components; heat resistant to 1000°F. *Obsolete*

GREEN GOLD

English manufacture
Au 75, Ag 11-25, Cd 0-13.
For jewelry; corrosion resistant.

GREEN LABEL

Ackerlind Steel Co., Inc.
C 0.9, Mn 1.15, Cr 0.5, W 0.5, V 0.2, bal Fe.
For tools, dies, and punches. Nondeforming; oil hardening.

GREEN LABEL

Wallace Murray Corp.
C 0.7-1.4, bal Fe.
For tools, drills, cutters; water hardening. *Obsolete*

GREEN LABEL

A. Milne & Co.
C 0.7-0.8, bal Fe.
For tools.

GREEN LABEL

Wallace Murray Corp.
C 0.7-1.4, bal Fe.
For tools, drills, cutters; water hardening.

GREEN LABEL

Wallace Murray Corp.
C 0.8-1.1, bal Fe.
For tools depending on temper; "Heller's Extra Tool."

GREEN LABEL
Wallace Murray Corp.
C 0.8-0.9, Mn 1.2-1.4, W 0.4-0.6, Cr 0.4-0.6, bal Fe.
For precision tools and files, taps, reamers, chasters, gages; nondeforming; "Stayput" Oil Hardening.

GREEN LABEL
Edgar T. Ward's Sons Co.
C 1-1.2, bal Fe.
For tools. *Obsolete*

GREEN LABEL
Peninsular Steel Co.
C 0.5, Cr 1, Mn 0.75, V 0.18, bal Fe.
For gears, crankshafts, springs; oil hardening; shock resisting.

GREEN LABEL EXTRA
Wallace Murray Corp.
C 0.8-1.1, bal Fe.
For tools, dies; water hardening

GREENLEAF 94AL$_2$O$_3$
Greenleaf Corp.
Cold pressed and sintered Al$_2$O$_3$.
280 MPa transverse rupture strength. Density 3.67 g/cm^3.

GREENLEAF 97AL$_2$O$_3$
Greenleaf Corp.
Cold pressed and sintered Al$_2$O$_3$.
280 MPa transverse rupture strength. Density 3.80 g/cm^3.

GREENLEAF 99AL$_2$O$_3$
Greenleaf Corp.
Cold pressed and sintered Al$_2$O$_3$.
340 MPa transverse rupture strength. Density 3.90 g/cm^3.

GREENLEAF AB40
Greenleaf Corp.
Pressed and sintered Al$_2$O$_3$-40% B$_4$C composite.
620 MPa transverse rupture strength. Density 3.39 g/cm^3.

GREENLEAF AB50
Greenleaf Corp.
Pressed and sintered Al$_2$O$_3$-50% B$_4$C composite.
620 MPa transverse rupture strength. Density 3.28 g/cm^3.

GREENLEAF GEM 1
Greenleaf Corp.
Pressed and sintered Al$_2$O$_3$.
620 MPa transverse rupture strength. Density 3.99 g/cm^3.

GREENLEAF GEM 2
Greenleaf Corp.
Pressed and sintered Al$_2$O$_3$-TiC composite.
760 MPa transverse rupture strength. Density 4.26 g/cm^3.

GREENLEAF GEM 4
Greenleaf Corp.
Pressed and sintered Si$_3$N$_4$-hardened Al$_2$O$_3$.
350 MPa transverse rupture strength. Density 3.78 g/cm^3.

GREENLEAF S-6
Greenleaf Corp.
WC + Co.
For tipped cutting tools. Cemented carbides.

GRIDALOY M
Molecu-Wire Corp.
Ni 94, Mn 5.
Wire: 50,000-140,000 TS. For resistors; high heat resistance. *Obsolete*

GRIDALOY P
Molecu-Wire Corp.
Ni 98, Plus additions.
Wire: 90,000-190,000 TS. For resistors; high heat resistance. *Obsolete*

GRIDNIC A
Driver Harris Co.
Ni-Fe.
For radio sets and tube parts. *Obsolete*

GRIDNIC B
Driver Harris Co.
Ni-Cr-Fe.
For radio parts and tube parts. *Obsolete*

GRIDNIC C
Driver Harris Co.
Ni 85, Cr 15.
For radio sets and tube parts; heat resistant. *Obsolete*

GRIDNIC D
Driver Harris Co.
Ni-Cr.
For radio sets and tube parts. *Obsolete*

GRIDNIC E
Driver Harris Co.
Mn 4.5, bal Ni.
For radio sets and tube parts; heat resistant. *Obsolete*

GRIDNIC F
Driver Harris Co.
Co 17, Ti 9, Si 0.5, Al 0.25, Mn 16, bal Fe.
For radio sets and tube parts; heat resistant. *Obsolete*

GRIDNIC T
Driver Harris Co.
C 0-0.4, Mn 0.2-0.6, Mg 0.2-0.5, bal Ni.
For grid wire; heat resistant. *Obsolete*

GRIDUR C25
Griesogen Griesheimer Autogen
C 2.6, Cr 27, bal Fe.
Weld: 50-52 Rock C. For worm conveyors, guide bars, slides, drawing dies. Hard facing electrode. Wear and corrosion resistant.

GRIDUR C35
Griesogen Griesheimer Autogen
C 3.5, Si 1, Cr 35, bal Fe.
Weld: 60-62 Rock C. For machine parts and tools subject to abrasion by sand, gravel, coal, cement. Corrosion and wear resistant. Hard facing electrode.

GRIDUR G10
Griesogen Griesheimer Autogen
Granular WC in steel tubes.
For tools and machine parts, rotary borers, rock drills, dredger teeth. Hard facing electrode. Wear resistant.

GRIDUR G15
Griesogen Griesheimer Autogen
Granular WC in steel tubes.
For tool and machine parts, rotary borers, rock drills, dredger teeth, crushing jaws. Hard facing electrode. Wear resistant.

GRIDUR G5
Griesogen Griesheimer Autogen
Granular WC in steel tubes.
For tools and machine parts, rotary borers, rock drills, dredger teeth. Hard facing electrode. Wear resistant.

GRIDUR K40
Griesogen Griesheimer Autogen
C 1.2, Si 1.5, Co 65, Cr 27, W 5.5, bal Fe.
Weld: 40-45 Rock C. For steam valves, exhaust valves, bearings, trimming dies. Hard facing electrode. Corrosion, heat and wear resistant.

GRIDUR K50
Griesogen Griesheimer Autogen
C 2.2, Co 55, Cr 27, W 14, bal Fe.
Weld: 50-54 Rock C. For steam valves, exhaust valves, bearings, trimming dies. Corrosion, heat and wear resistant. Hard facing electrode.

GRIDUR K60
Griesogen Griesheimer Autogen
C 2.7, Co 53, Cr 25, W 19, bal Fe.
Weld: 56-60 Rock C. For needle valves, impeller parts. Hard facing electrode. Wear and corrosion resistant.

GRIDUR SA1
Griesogen Griesheimer Autogen
C 1, Cr 4.2, Mo 2.8, V 2.5, W 3, bal Fe.
Weld: 48 Rock C. For hot working tools, air cooling systems. Hard facing electrode. High speed steel.

GRIDUR SA2
Griesogen Griesheimer Autogen
C 0.85, Cr 4.2, V 1.2, W 18, bal Fe.
Weld: 58-62 Rock C. For tough and impact resistant cutting tools. Hard facing electrode. High speed steel.

GRIDUR SA3
Griesogen Griesheimer Autogen
C 1, Cr 4.2, Mo 9, V 1.2, W 1.8, bal Fe.
Weld: 58-62 Rock C. For turning tools, milling cutters, reamers, drills, hot shear blades. Hard facing electrode. High speed steel.

GRIDUR SA4
Griesogen Griesheimer Autogen
C 1.5, Co 8, Cr 4.5, Mo 6, V 3.2, W 8, bal Fe.
Weld: 62-65 Rock C. For cutting tools with highest wear resistance and hardness at elevated temperatures. Hard facing electrode. High speed steel.

GRIDUR SCS-E
Griesogen Griesheimer Autogen
C 2, Cr 12, bal Fe.
Weld: 63-65 Rock C. For hard facing cutting and blanking dies, punches, shears, broaches. Hard facing electrode.

GRIDUR WA1
Griesogen Griesheimer Autogen
C 0.5, Cr 2.3, Mo 0.8, Ni 1.7, V 0.1, bal Fe.
Weld: 300-400 Brin. For welding on tools requiring high toughness and impact resistance, die blocks, upsetters, piercers to 450°C. Hard facing electrode. Hot work steel.

GRIDUR WA2
Griesogen Griesheimer Autogen
C 0.3, Cr 2.3, Mo 0.5, V 0.6, W 4.5, bal Fe.
Weld: 45 Rock C. For welding on hot working tools, forging dies, mandrels, punches, bushings. Hard facing electrode. Hot work steel.

GRIDUR WA3
Griesogen Griesheimer Autogen
C 0.3, Co 2.3, Cr 2.7, V 0.4, W 9, bal Fe.
Weld: 48 Rock C. For hot working tools, air cooling systems. Hard facing electrode. Hot work steel.

GRIDUR-SCS
Griesogen Griesheimer Autogen
C 2, Si 0.3, Mn 0.3, Cr 12, W 0.8, bal Fe.
Weld: 450 Brin. For welding cutting and blanking dies, punches, shears, broaches, drawing tools. Hard facing electrode.

GRIESHEIM "RAUCHLOS"
Griesogen Griesheimer Autogen
C 0.2, bal Fe.
For welding rods.

GRIESHEIM "SINI"
Griesogen Griesheimer Autogen
Alloy, bal Cu.
For welding rods for copper welding.

GRIESHEIM EA-B
Griesogen Griesheimer Autogen
C 0.22, Si 1.2, Mn 0.8, bal Fe.
Weld: 250 Brin. For welding for rails, switch tongues, wheel tires, worms, axes, cams. Welding electrode, hard surfacing.

GRIESHEIM N

Griesogen Griesheimer Autogen
C 0.7, Si 0.6, Mn 13, Ni 3.5, bal Fe.
Weld: 200 Brin. For welding crusher jaws, mill beaters, dredger teeth, rolling mill parts. Hard facing electrode. Austenitic. Work hardens to 450 Brin.

GRIESHEIM S

Griesogen Griesheimer Autogen
C 1, Si 0.4, Mn 13, bal Fe.
Weld: 200 Brin. For welding mill beaters, crushing jaws, dredger teeth, rolling mill parts. Austenitic hard facing electrode. Work hardens to 450 Brin.

GRIESHEIM-EA-V

Griesogen Griesheimer Autogen
C 0.3, Si 1.2, Mn 1.2, Cr 1, bal Fe.
Weld: 350 Brin. For welding rails, switch tongues, wheel tires, worms, axes, shafts, cams. Welding electrode, hard facing.

GRIESHEIM-EA2

Griesogen Griesheimer Autogen
C 0.5, Si 0.4, Mn 5, bal Fe.
Weld: 500 Brin. For welding skid rails, crusher jaws, baffle plates, dredger parts. Hard facing electrode.

GRIESHEIM-EA3

Griesogen Griesheimer Autogen
C 0.45, Si 0.45, Mn 0.4, Cr 5.8, bal Fe.
Weld: 600 Brin. For welding crushing jaws, dies, cams, worms, breakers, chain wheels. Hard facing electrode.

GRIESHEIM-EA600

Griesogen Griesheimer Autogen
C 0.5, Si 2.4, Mn 0.4, Cr 9, bal Fe.
Welded: 560-600 Brin. For welding crushing jaws, dies, cams, worms, breakers, chain wheels. Hard facing electrode.

GRIESHEIM-EA600W

Griesogen Griesheimer Autogen
C 0.5, Mn 0.5, Si 2.5, Cr 8, bal Fe.
Weld: 560 Brin. For welding crusher jaws, dies, breakers, cams, worms, chain wheels. Hard facing electrode.

GRIFFIN

Adams & Osgood Steel Co.
C, Cr, Mn, bal Fe.
For dies, punches, thread rolling tools; non-deforming.

GRILLO 2105

Zinkberatungsstelle GmbH
Al 5, Cu 1, Ga 0.02, bal Zn.
For die castings.

GRILLO 31010

Zinkberatungsstelle GmbH
Al 10, Cu 1, Ga 0.02, bal Zn.
For castings.

GRIMM 4S4-ZH

Gustav Grimm Edelstahl-Werk GmbH
C 0.85, Si 0-0.25, Mn 0-0.25, bal Fe.
Heat treated: 190,000 TS; 150,000 YS; 10 El; 30 RA; 390 Brin. For springs, taps, drills, hobs, reamers; Type W1; water hardened.

GRIMM BRF

Gustav Grimm Edelstahl-Werk GmbH
C 0.2, Cr 13, Mn 0.3, bal Fe.
Annealed: 95,000 TS; 50,000 YS; 25 El; 55 RA; 196 Brin. For turbine blades, valves, cutlery, surgical instruments; Type 420; stainless.

GRIMM CWL

Gustav Grimm Edelstahl-Werk GmbH
C 2.1, Cr 11.5, W 0.7, Mn 0.3, bal Fe.
For blanking and forming dies, punches; oil hardened, non-deforming.

GRIMM DSA

Gustav Grimm Edelstahl-Werk GmbH
C 0.38, Cr, V, Si, bal Fe.
For gears, bolts, machine tool parts; oil hardened, tough.

GRIMM DSW

Gustav Grimm Edelstahl-Werk GmbH
C 0.45, W, Cr, V, bal Fe.
For header dies, upsetters, crimpers; oil hardened.

GRIMM G4

Gustav Grimm Edelstahl-Werk GmbH
C 0.75, Si 0.25-0.5, Mn 0.3-0.8, bal Fe.
Heat treated: 185,000 TS; 140,000 YS; 15 El; 40 RA; 375 Brin.
For rails, springs, tools, axes, hammers; water hardened.

GRIMM GC120

Gustav Grimm Edelstahl-Werk GmbH
C 2.1, Cr 11.5, Mn 0.3, bal Fe.
For blanking and forming dies; oil hardened, non-deforming.

GRIMM GC15

Gustav Grimm Edelstahl-Werk GmbH
C 1.05, Cr 1, bal Fe.
For bearings, liners, bushings; water hardened, wear and abrasion resistant.

GRIMM GCK

Gustav Grimm Edelstahl-Werk GmbH
C 1.45, Cr 1.4, Mn 0.6, bal Fe.
For bearings, bushings, liners; water hardened, wear resistant.

GRIMM GCM

Gustav Grimm Edelstahl-Werk GmbH
C 0.9, Mn 1.9, V 0.1, bal Fe.
For punches, dies, shears, crimpers; oil hardened, non-deforming.

GRIMM GCS

Gustav Grimm Edelstahl-Werk GmbH
C 0.9, Cr, Si, bal Fe.
For bearings, bushings, liners; water hardened, wear resistant.

GRIMM GCZ

Gustav Grimm Edelstahl-Werk GmbH
C 0.9, Cr 0.8, Mn 0.3, Si 0.25, bal Fe.
For bearings, cutters, liners, bushings; water hardened, wear resistant.

GRIMM GEE

Gustav Grimm Edelstahl-Werk GmbH
C 0.15, Si 0.25-0.5, Mn 0.3-0.8, bal Fe.
Annealed: 70,000 TS; 40,000 YS; 25 El; 60 RA; 145 Brin. For gears, bolts, machine tool parts; case hardening steel.

GRIMM GH4

Gustav Grimm Edelstahl-Werk GmbH
C 0.75, Si 0.25-0.5, Mn 0.3-0.8, bal Fe.
Heat treated: 180,000 TS; 132,000 YS; 10 El; 34 RA; 375 Brin.
For springs, rails, hammers, tools, cutters; Type W1; water hardened.

GRIMM GH5

Gustav Grimm Edelstahl-Werk GmbH
C 0.6, Si 0.25-0.5, Mn 0.3-0.8, bal Fe.
Heat treated: 160,000 TS; 113,000 YS; 12 El; 40 RA; 325 Brin.
For gears, springs, fasteners; water hardened.

GRIMM GHM

Gustav Grimm Edelstahl-Werk GmbH
C 0.56, Ni, Cr, Mo, V, bal Fe.
For gears, bolts, crankshafts; oil hardened, shock resistant.

GRIMM GK10

Gustav Grimm Edelstahl-Werk GmbH
C 0.76, Co 10, Cr 4.2, Mo 0.8, V 1.8, W 18, bal Fe.
For lathe and planer tools, reamers, milling cutters; high speed steel.

GRIMM GK3

Gustav Grimm Edelstahl-Werk GmbH
C 0.86, Co 2.8, Cr 4.3, Mo 0.85, V 2.1, W 12, bal Fe.
For lathe and planer tools, reamers, broaches; high speed steel.

GRIMM GK5

Gustav Grimm Edelstahl-Werk GmbH
C 0.79, Co 4.75, Cr 4.3, Mo 0.75, V 1.55, W 18, bal Fe.
For lathe and planer tools, reamers; high speed steel.

GRIMM GKS

Gustav Grimm Edelstahl-Werk GmbH
C 1, V 0.1, Si 0.25, Mn 0.25, bal Fe.
For drills, taps, reamers, broaches; Type W2; water hardened.

GRIMM GLS

Gustav Grimm Edelstahl-Werk GmbH
C 0.55, W, Cr, V, bal Fe.
For upsetters, shears, punches; oil hardened, tough.

GRIMM GM

Gustav Grimm Edelstahl-Werk GmbH
C 0.55, Cr 0.7, Mo 0.18, Ni 1.65, V 0.1, bal Fe.
For gears, bolts, crank shafts; oil hardened, shock resistant.

GRIMM GME

Gustav Grimm Edelstahl-Werk GmbH
C 0.4, Cr, Mo, Mn, bal Fe.
For gears, bolts, machine tool parts; oil hardened, tough.

GRIMM GR

Gustav Grimm Edelstahl-Werk GmbH
C 0.5, Cr 1.05, Mn 0.95, V 0.1, bal Fe.
For springs, bolts, gears, studs; oil hardened, shock resistant.

GRIMM GSM

Gustav Grimm Edelstahl-Werk GmbH
C 0.53, Si 0.9, Mn 0.9, bal Fe.
For springs, bolts, crankshafts; water or oil hardened.

GRIMM GWP

Gustav Grimm Edelstahl-Werk GmbH
C 0.5, Cr 1.05, Ni 3.25, bal Fe.
For gears, pinions, countershafts, axles; oil hardened, shock resistant.

GRIMM GWV112B

Gustav Grimm Edelstahl-Werk GmbH
C 0.74, Cr 4.1, V 1.1, W 18.5, bal Fe.
For lathe and planer tools, reamers, drills, taps; high speed steel.

GRIMM GWV122 SPEZIAL

Gustav Grimm Edelstahl-Werk GmbH
C 0.82, Cr 4, Mo 0.8, V 1.6, W 8.7, bal Fe.
For lathe and planer tools, reamers, broaches; high speed steel.

GRIMM GWV132 RAPID

Gustav Grimm Edelstahl-Werk GmbH
C 0.86, Cr 0.4, Mo 0.8, V 2.5, W 12, bal Fe.
For lathe and planer tools, reamers, broaches; high speed steel.

GRIMM GWV145 RECORD

Gustav Grimm Edelstahl-Werk GmbH
C 1.3, Cr 4.3, Mo 0.8, V 3.8, W 12, bal Fe.
For form cutters, broaches, reamers; high speed steel.

GRIMM HPM

Gustav Grimm Edelstahl-Werk GmbH
C 0.3, Co 2, Cr 2.4, V 0.25, W 8.5, bal Fe.
For extrusion press dies, rams and liners; oil hardened, tough.

GRIMM HWP
Gustav Grimm Edelstahl-Werk GmbH
C 0.3, Cr 2.65, V 0.35, W 8.5, bal Fe.
For extrusion press liners and rams, punches, shears; oil hardened, tough.

GRIMM KSA
Gustav Grimm Edelstahl-Werk GmbH
C 0.45, Si, Cr, V, bal Fe.
For dies, bolts, crankshafts, gears, shears; oil hardened, tough.

GRIMM NS12
Gustav Grimm Edelstahl-Werk GmbH
C 0.34, Al 1.1, Cr 1.4, bal Fe.
For oil refinery equipment; creep resistant.

GRIMM NS13
Gustav Grimm Edelstahl-Werk GmbH
C 0.32, Al 1.1, Cr 1.1, Mo 0.18, bal Fe.
For oil refinery equipment; creep resistant.

GRIMM NS15
Gustav Grimm Edelstahl-Werk GmbH
C 0.27, Al 1.1, Cr 1.4, bal Fe.
For oil refinery equipment; creep resistant.

GRIMM NS23
Gustav Grimm Edelstahl-Werk GmbH
C 0.33, Al 1.1, Cr 1.7, Ni 1, bal Fe.
For oil refinery equipment; creep resistant.

GRIMM NS48
Gustav Grimm Edelstahl-Werk GmbH
C 0.3, Cr 1.1, V 0.2, bal Fe.
For gears, pinions, bolts, fasteners; oil hardened, shock resistant.

GRIMM NS54
Gustav Grimm Edelstahl-Werk GmbH
C 0.31, Cr 2.5, Mo 0.18, V 0.13, bal Fe.
For gears, pinions, bolts, fasteners, shafts; oil hardened, shock resistant.

GRIMM RSA
Gustav Grimm Edelstahl-Werk GmbH
C 1.42, W, V, bal Fe.
For forming and heading dies; oil or water hardened.

GRIMM SB
Gustav Grimm Edelstahl-Werk GmbH
C 1.05, Cr 1, W 1.15, Mn 0.9, bal Fe.
For forming and blanking dies, punches; oil hardened, tough.

GRIMM SE
Gustav Grimm Edelstahl-Werk GmbH
C 0.2, Mn 1.25, Cr 1.15, bal Fe.
For gears, bolts, camshafts, cams; case hardened.

GRIMM SH4-ZH
Gustav Grimm Edelstahl-Werk GmbH
C 0.75, Si 0.4, Mn 0.6, bal Fe.
Heat treated: 180,000 TS; 130,000 YS; 10 El; 35 RA; 360 Brin.
For springs, rails, clutch discs; Type W1; water hardened.

GRIMM SH5-ZAH
Gustav Grimm Edelstahl-Werk GmbH
C 0.6, Si 0.4, Mn 0.6, bal Fe.
Heat treated: 160,000 TS; 114,000 YS; 12 El; 40 RA; 325 Brin.
For wheels, die blocks, rails, axles, bolts; water hardened.

GRIMM SH6-ZW
Gustav Grimm Edelstahl-Werk GmbH
C 0.45, Si 0.4, Mn 0.6, bal Fe.
Hot rolled: 98,000 TS; 58,000 YS; 24 El; 45 RA; 212 Brin. For axles, gears, bolts, bushings; water hardened.

GRIMM SSA
Gustav Grimm Edelstahl-Werk GmbH
C 0.95, W, Mo, bal Fe.
For tools, dies, cutters; oil hardened.

GRIMM SSCV
Gustav Grimm Edelstahl-Werk GmbH
C 1.35, W, Co, bal Fe.
For engravers' tools, blanking dies; oil hardened, wear resistant.

GRIMM WA5
Gustav Grimm Edelstahl-Werk GmbH
C 0.55, Si 0.1-0.4, Mn 0.5-0.7, bal Fe.
Heat treated: 155,000 TS; 110,000 YS; 14 El; 42 RA; 315 Brin.
For axles, tie rods, bushings, bolts, shafts; water hardened.

GRIMM WE2 HARD
Gustav Grimm Edelstahl-Werk GmbH
C 1.3, Si 0-0.25, Mn 0-0.25, bal Fe.
For engravers' tools, reamers, blanking dies; Type W1; water hardened.

GRIMM WE3-MH
Gustav Grimm Edelstahl-Werk GmbH
C 1, Si 0-0.25, Mn 0-0.25, bal Fe.
Heat treated: 200,000 TS; 125,000 YS; 8 El; 27 RA; 400 Brin.
For springs, taps, reamers, drills, cutters; Type W1; water hardened.

GRIMM WE4-ZH
Gustav Grimm Edelstahl-Werk GmbH
C 0.85, Si 0-0.25, Mn 0-0.25, bal Fe.
Heat treated: 190,000 TS; 140,000 YS; 10 El; 30 RA; 390 Brin.
For springs, taps, drills, hobs, reamers; Type W1; water hardened.

GRIMM WE5-ZAH
Gustav Grimm Edelstahl-Werk GmbH
C 0.7, Si 0-0.25, Mn 0-0.25, bal Fe.
Heat treated: 174,000 TS; 128,000 YS; 12 El; 37 RA; 350 Brin.
For springs, clutch disks, rails; Type W1; water hardened.

GRIMM WPN
Gustav Grimm Edelstahl-Werk GmbH
C 0.3, Si 1, Cr 1.1, V 0.18, W 3.75, bal Fe.
For extrusion rams and liners, punches, upsetters; hot work steel, oil hardened.

GRIMM WPN EXTRA
Gustav Grimm Edelstahl-Werk GmbH
C 0.3, Cr 2.35, V 0.6, W 4.25, bal Fe.
For shears, punches, upsetters, dies; hot work steel, oil hardened.

GRIMM WS2
Gustav Grimm Edelstahl-Werk GmbH
C 1.1, Si 0-0.25, Mn 0-0.25, bal Fe.
Annealed: 110,000 TS; 58,000 YS; 19 El; 40 RA; 210 Brin. For springs, taps, drills, hobs, reamers; Type W1; water hardened.

GRIMM WS3-MH
Gustav Grimm Edelstahl-Werk GmbH
C 1, Si 0-0.25, Mn 0-0.25, bal Fe.
Heat treated: 200,000 TS; 125,000 YS; 10 El; 30 RA; 400 Brin.
For springs, taps, drills, hobs, reamers; Type W1; water hardened.

GRIMM WS5-ZAH
Gustav Grimm Edelstahl-Werk GmbH
C 0.7, Si 0-0.25, Mn 0-0.25, bal Fe.
Heat treated: 175,000 TS; 128,000 YS; 12 El; 37 RA; 350 Brin.
For wheels, springs, die blocks, girders, rails; Type W1; water hardened.

GRIMM ZR SPEZIAL
Gustav Grimm Edelstahl-Werk GmbH
C 1.4, Cr 0.3, V 0.1, bal Fe.
For bearings, cutters, liners; oil hardened, wear resistant.

GRIMM'S ALUMINUM SOLDER-1
English manufacture
Sn 69.1, Pb 28.8, Zn 1.44, Ag 0.72.
For aluminum solder.

GRIMM'S ALUMINUM SOLDER-2
English manufacture
Sn 50, Pb 25, Zn 25.
For aluminum solder.

GRIPMORE
Bissett Steel Co.
C 0.35, Cr 4, V 0.25, Mo 0.5, bal Fe.
For hot blanking and forming dies, gripper dies; non-deforming, oil hardened. *Obsolete*

GRIPMORE NO. 1
Bethlehem Steel Corp.
C 0.35, Cr 5, Mo 1, V 0.4, bal Fe.
For dies, punches, shears, extrusion rams; hot work steel, oil hardened. *Obsolete*

GRIPMORE NO. 1
Bissett Steel Co.
C 0.35, Mo 1, V 0.4, Si 1, bal Fe.
For hot blanking and forming dies, gripper dies; oil hardened, hot work steel. *Obsolete*

GRIPMORE NO. 1-V
Bissett Steel Co.
C 0.35, Cr 5, Mo 1, V 1.1, Si 1, bal Fe.
For hot forming and blanking dies, punches; air hardened, hot work steel. *Obsolete*

GRIPMORE NO. 2
Bissett Steel Co.
C 0.35, W 1.35, Mo 1.75, Si 1, Cr 5, bal Fe.
For punches, gripping dies, hot work tools; air hardened, hot work steel. *Obsolete*

GRIPMORE TOOL
Bissett Steel Co.
C 0.95, Cr 3.5, W 1, bal Fe.
For hot work tools; hot work steel. *Obsolete*

GRITALLOY 10011
Eutectic Corp.
Tungsten carbide and nickel base alloy powder for hard build up.

GRM 235D
Universal Cyclops
C 0.15, Cr 15.5, Mo 5.2, Al 3.0, Ti 2.5, Fe 4.2, B 0.075, bal Ni.
For high temperature cast components in jet engines; high heat and stress-rupture properties.

GRM 235D
Cannon-Muskegon Corp.
C 0.15, Cr 15.5, Mo 5.2, Al 3.6, Ti 2.5, Fe 4.2, B 0.075, bal Ni.
For high temperature cast components in jet engines; high heat and stress-rupture properties.

GROMMET BRASS
American manufacture
Cu 70, Zn 30.
For grommets; high ductility.

GROSSMAN
Manufacturer not listed.
Al alloy. For light alloy parts.

GS
Thyssen Edelstahlwerke AG
C 1.6, Cr 12, bal Fe.
For valves for internal combustion engines; heat resistant. *Obsolete*

GS 62
Bergische Stahl Industrie
C 0.4, bal Fe.
Carbon steel casting. GS-62.3.

GS-38
Bergische Stahl Industrie
C 0.15, bal Fe.
Annealed: 54,000 TS min; 28,000 YS min; 25 El min. Low carbon, unalloyed steel casting. W.-Nr. 1.0416.

GS-45
Bergische Stahl Industrie
C 0.22, bal Fe.
Annealed: 64,000 TS min; 35,000 YS min; 22 El min. Low carbon steel casting. W.-Nr. 1.0443.

GS-52
Bergische Stahl Industrie
C 0.3, bal Fe.
Annealed: 74,000 TS min; 40,000 YS min; 18 El min. Carbon steel casting. W.-Nr. 1.0551.

GS-60
Bergische Stahl Industrie
C 0.37, bal Fe.
Annealed: 75,000 TS min; 42,000 YS min; 15 El min. Carbon steel casting.

GS-70
Bergische Stahl Industrie
C 0.45, bal Fe.
Annealed: 88,000 TS min; 50,000 YS min; 12 El min. Carbon steel casting. W.-Nr. 1.0553.

GSC
Thyssen Edelstahlwerke AG
C, Cr, Mo, Co, bal Fe.
For valves for internal combustion engines; heat resistant. *Obsolete*

GSE
Thyssen Edelstahlwerke AG
C 0.45, Si 1.5, Ni 13, Cr 15, W 1.2, bal Fe.
For valves; austenitic. *Obsolete*

GSM
Thyssen Edelstahlwerke AG
C 0.4, Si 2.25, Mn 15.5, Cr 12.5, bal Fe.
For valves; corrosion and heat resistant. *Obsolete*

GSN
Latrobe Steel Co.
C 2.15, Cr 12.25, Si 0.4, Mn 0.4, bal Fe.
For stamping and forming dies; non-deforming.

GSNI
Friedr. Lohmann GmbH
C 0.56, Ni, Cr, Mo, V, bal Fe.
For forging dies, mold frames, machine tool parts. Type L6; oil hardened, shock resistant. T61206

GT-45
Armco
C 0.08, Cr 16.7, Ni 14, Mo 2.7, Cu 3, Ti 0.25, Cb 0.35, Si 0.5, Mn 1.25, bal Fe.
For high temperature applications, jet engines and turbo superchargers; heat and corrosion resistant. *Obsolete*

GT2A METAL
French manufacture
C, Cr, Ni, bal Fe.
For chemical equipment; corrosion resistant.

GT4A METAL
French manufacture
C, Cr, Ni, bal Fe.
For chemical equipment; corrosion resistant.

GUETTIERES BUTTON
English manufacture
Cu 61.5, Zn 29-32, Sn 6.5-9.7.
For buttons.

GUETTIERES BUTTON
English manufacture
Cu 56, Zn 44.
For buttons.

GUILLAUME'S METAL
English manufacture
Cu 64.3, Ni 35.7.
For tape, chemical equipment; name also applied to "Invar" tape.

GUINEA GOLD
English manufacture
Cu 88, Zn 12.
For ornamental parts, cheap jewelry; red brass.

GUISHIBUICHI
Japanese manufacture
Cu 51-67, Ag 32-49, Au, Fe.
For jewelry, ornaments; v. Shibu-ichi.

GULF AIR
Gulf Steel Corp.
C 1, Cr 5, Mo 1, bal Fe.
For tools and dies; Type A2; air hardened, nondeforming.

GULF GH-VAN
Gulf Steel Corp.
C, alloy, bal Fe.
For machine tool parts; oil hardened.

GULF GH14
Gulf Steel Corp.
C, alloy, bal Fe.
For machine tool parts; oil hardened.

GULF GH9
Gulf Steel Corp.
C, alloy, bal Fe.
For machine tool parts; oil hardened.

GULF GHW
Gulf Steel Corp.
C, alloy, bal Fe.
For machine tool parts; oil hardened.

GULF H.S.
Gulf Steel Corp.
C 0.7, Cr 4, V 1, W 18, bal Fe.
For lathe and planer tools, drills, hobs, reamers; Type T-1; high speed steel.

GULF H.S.5
Gulf Steel Corp.
C 0.8, Cr 4, V 2, W 18, Co 8, bal Fe.
For lathe and planer tools, reamers, broaches; Type T-5; high speed steel.

GULF M-2
Gulf Steel Corp.
C 0.8, Cr 4, V 2, W 6, Mo 5, bal Fe.
For thread rolling dies, broaches, reamers, cutters; Type M2; high speed steel.

GULF O.H.
Gulf Steel Corp.
C 0.9, Mn 1, Cr 0.5, W 0.5, bal Fe.
For dies, punches, rollers, mandrels; Type O1; oil hardening, nondeforming.

GUN IRON
Power Products Inc.
Si 1.5, Mn 0.85, Ni 1.9, Cr 0.4, Mo 0.65, S 0.075, P 0.18, TC 3, bal Fe.
Cast: 35,000-48,000 TS; 200-600 Brin. For cylinder liners, castings; Tr.S. 2700; Tr.D 0.265. *Obsolete*

GUN METAL
McCallum-Hatch Bronze Co. Inc.
Copper. Cu 88, Sn 8, Zn 4.
Cast: 40,000-48,000 TS; 18,000 YS; 20-50 El; 70 Brin. For steam and structural bronze, gears, bolts; resists salt water corrosion.

GUN METAL COCHIN CHINA
Chinese manufacture
Cu 93.19, Sn 5.43, Fe 1.38.
For early cannons and guns.

GUN METAL COCHIN CHINA
Chinese manufacture
Cu 77.18, Sn 3.42, Zn 5.02, Pb 13.22, Fe 1.16.
For early cannons and guns.

GUN METAL MODIFIED
McCallum-Hatch Bronze Co. Inc.
Copper. Cu 86, Sn 9.5, Pb 2.5, Zn 2.
Cast: 32,000-40,000 TS; 15-25 El; 12-23 RA; 63-72 Brin. For gears, bearings; "Barr Alloy No. 2."

GUN METAL NO. 1
Knowsley Cast Metal Co., Ltd.
Cu 88, Sn 10, Zn 2.
Cast: 36,000 TS; 12 El; 50-65 Brin. For bearings, gears, marine parts; Gun Metal.

GUN METAL NO. 2
Knowsley Cast Metal Co., Ltd.
Cu 85, Sn 5, Zn 5, Pb 5.
Cast: 27,000 TS; 12 El; 45-55 Brin. For steam valves, bearings, gears; pressure tight, creep resistant.

GUN METAL NO. 3
Knowsley Cast Metal Co., Ltd.
Cu 86, Sn 7, Zn 5, Pb 2.
Cast: 31,500 TS; 12 El; 45-60 Brin. For backing for white metal lined bearings; pressure tight castings, creep resistant.

GUN METAL NO. 4
Knowsley Cast Metal Co., Ltd.
Sn, Zn, bal Cu.
Cast: 38,100 TS; 25 El; 45-60 Brin. For bearings, pressure tight castings; pressure tight, not good above 450°F.

GUN METAL PRUSSIAN
Prussian manufacture
Cu 90.9, Sn 9.1.
For cannons and guns; modern.

GUN METAL, CHINESE
Chinese manufacture
Cu 93.2, Sn 5.05, Fe 1.72.
For early cannons and guns.

GUN METAL, CHINESE
Chinese manufacture
Cu 71.16, Zn 27.36, Fe 1.4.
For early cannons and guns.

GUN METAL, ENGLISH
English manufacture
Cu 89-92, Sn 8-11.
For early cannons and guns.

GUN METAL, FRENCH
French manufacture
Cu 89.44, Sn 8.91, Zn 1.31, Pb 0.16.
For early cannons and guns; old.

GUN METAL, FRENCH

French manufacture
Cu 90.1, Sn 9.9.
For early cannnons and guns; modern.

GUN METAL, RUSSIAN

Russian manufacture
Cu 88-91, Sn 9-12, Fe 0.07.
For early cannons and guns.

GUN METAL, SWISS

Swiss manufacture
Cu 88.93, Sn 10.37, Zn 0.42.
For early cannons and guns; "lucern."

GUN METAL, TURKISH

Turkish manufacture
Cu 91-95, Sn 4.7-8.8, Fe 0.02.
For early cannons and guns.

GUN METAL-A

English manufacture
Cu 90, Sn 10.
For electric contacts, valve disks; corrosion resistant.

GUN METAL-B

English manufacture
Cu 90, Sn 6.5, Zn 2, Pb 1.5.
For steam fittings, valves, cocks; free-cutting.

GUN MOUNT BRONZE

American manufacture
Zn 17, Sn 3, bal Cu.
For gun mounts; high strength.

GUNIT 50/GUNIT 100

Thyssen Edelstahlwerke AG
bal Fe.
Coated electrodes for welding on cast iron.

GUNITE A

Kelsey Hayes Co.
Si 2, Mo 0.25, 0.8% CC, 2.2% GC, bal Fe.
Cast: 40,000 TS; 223 Brin. For cylinder sleeves, camshafts; Tr.S. 3200; Tr.D. 32; to be hardened. *Obsolete*

GUNITE B-1

Kelsey Hayes Co.
Si 2, Mo 0.75, Cr 0.65, Ni 0.9, 0.8% C.C., 2.2% G.C., bal Fe.
Cast: 55,000 TS; 269 Brin. For heavy forming dies; Tr.S. 3400; Tr.D. 30; wear resistant. *Obsolete*

GUNITE C

Kelsey Hayes Co.
Si 2, CC 0.8, GC 2.2, Bal Fe.
Cast: 35,000 TS; 212 Brin. For automotive brake drums, cams, rolls, gears, worms; wear resistant under friction Tr.S.-3,000. *Obsolete*

GUNITE D

Kelsey Hayes Co.
Si 2, Cr 0.55, Ni 1.2, 0.8% CC, 2.2% GC, bal Fe.
Cast: 45,000 TS; 235 Brin. For pump bodies; Tr.S. 3300; Tr.D. 02; abrasion resistant. *Obsolete*

GUNITE E

Kelsey Hayes Co.
Mo 0.4, Cu 0.4, 0.8% CC, 2.2% GC, bal Fe.
Cast: 45,000 TS; 229 Brin. For brake drums, crankshafts, gears; Tr.S. 3300; Tr.D. 34. *Obsolete*

GUNITE E-1

Kelsey Hayes Co.
Si 2, Mo 0.1, Cu 0.2, 0.8% CC, 2.2% GC, bal Fe.
Cast: 40,000 TS; 223 Brin. For brake drums, cylinder sleeves; Tr.S. 3200; Tr.D. 32; heavy duty service. *Obsolete*

GUNITE F

Kelsey Hayes Co.
Si 2, Mo 0.8, Cu 0.6, Cr 0.35, 0.8% CC, 2.2% GC, bal Fe.
Cast: 50,000 TS; 248 Brin. For crankshafts; Tr.D. 34; Tr.S. 3500; wear resistant. *Obsolete*

GUNITE R

Kelsey Hayes Co.
Si 2, Mo 1.25, Ni 1.25, 0.8% CC, 2.2% GC, bal Fe.
Cast: 60,000 TS; 269 Brin. For heavy crankshafts; Tr.S. 3800; Tr.D. 34; tough. *Obsolete*

GUNITE-K

Kelsey Hayes Co.
C 2.3, Si 1, S 0.08, P 0.16, bal Fe.
Cast: 105,000 TS; 85,000 YS; 4 El; 210 Brin. For gears, shafts, housings; cast iron, high strength. *Obsolete*

GURLEY'S METAL

English manufacture
Cu 86.5, Zn 5.4, Sn 5.4, Pb 2.7.
For steam fittings, valves, cocks; free-cutting.

GURNEYS BRONZE

American manufacture
Cu 76, Sn 9, Pb 15.
For machine parts; also called "U.S. Government Bronze."

GURONIT

Guronitwerke Vervoort GmbH
C, alloy, bal Fe.
For machine tool products; oil hardened.

GURONIT 1.1 H

Guronitwerke Vervoort GmbH
C 19, Si 1.3, Cr 29, bal Fe.
For wear plates, crushers, heat and abrasion resistant.

GURONIT 14

Guronitwerke Vervoort GmbH
C 0.25, Cr 14.5, Ni 0-1, bal Fe.
Annealed: 85,000 TS; 40,000 YS; 50 El; 65 RA; 185 Brin. For surgical instruments, gages, valves; Type 420; stainless, austenitic.

GURONIT 18

Guronitwerke Vervoort GmbH
C 0.25, Cr 17, Ni 0-1.8, bal Fe.
For chemical plant equipment; corrosion resistant.

GURONIT 18-8

Guronitwerke Vervoort GmbH
C 0.15, Cr 18, Ni 8.5, bal Fe.
Annealed: 75,000 TS; 35,000 YS; 55 El; 75 RA; 140 Brin. For chemical plant equipment, tanks, vessels; Type 302; stainless, austenitic.

GURONIT 18-8E

Guronitwerke Vervoort GmbH
C 0-0.12, Cr 18, Ni 9.5, Ti = 4 x C, bal Fe.
Annealed: 85,000 TS; 35,000 YS; 55 El; 65 RA; 170 Brin. Cold drawn: 95,000 TS; 60,000 YS; 40 El; 60 RA; 185 Brin. For welded structures, chemical plant equipment; Type 321; stainless, austenitic.

GURONIT 18-8MO

Guronitwerke Vervoort GmbH
C 0.15, Cr 18, Ni 9.5, Mo 2, bal Fe.
Annealed: 85,000 TS; 35,000 YS; 50 El; 75 RA; 150 Brin. Cold drawn: 150,000 TS; 135,000 YS; 6 El; 300 Brin. For acid resistant equipment; Type 316; stainless, austenitic.

GURONIT 18-8MOE

Guronitwerke Vervoort GmbH
C 0-0.12, Mo 2, Cr 18, Ni 10.5, Cb = 8 x C, bal Fe.
Annealed: 85,000 TS; 35,000 YS; 50 El; 75 RA; 150 Brin. Cold drawn: 150,000 TS; 135,000 YS; 6 El; 300 Brin. For welded acid resistant equipment; Type 316Cb; stainless, austenitic.

GURONIT 2.2 H

Guronitwerke Vervoort GmbH
C 1.2, Si 1.3, Cr 29, Mo 2, bal Fe.
For wear plates, crushers; corrosion, heat and wear resistant.

GURONIT 20-25MO

Guronitwerke Vervoort GmbH
C 0.1, Ni 24, Cr 20, Mo, Cu, bal Fe.
Annealed: 100,000 TS; 45,000 YS; 50 El; 65 RA; 185 Brin. For acid resistant equipment; corrosion resistant, austenitic.

GURONIT 28.4

Guronitwerke Vervoort GmbH
C 0.4, Si 1.3, Cr 27, Ni 4, bal Fe.
Cast: 115,000 TS; 80,000 YS; 15 El; 190 Brin. For furnace equipment, heat treat boxes; heat resistant.

GURONIT GS22-10

Guronitwerke Vervoort GmbH
C 0.4, Cr 22, Ni 9.5, bal Fe.
Cast: 85,000 TS; 45,000 YS; 35 El; 165 Brin. For heat treating boxes, furnace parts, conveyor belts; Type HF; corrosion and heat resistant.

GURONIT GS25-15

Guronitwerke Vervoort GmbH
C 0.4, Cr 26, Ni 14, bal Fe.
Cast: 75,000 TS; 47,000 YS; 17 El; 25 RA; 200 Brin. For heat treating boxes, salt pots, retorts, chains; Type HH; corrosion and heat resistant.

GURONIT GS25-35

Guronitwerke Vervoort GmbH
C 0.5, Cr 25, Ni 30, bal Fe.
For furnace parts, retorts, pots, heat treating boxes; corrosion and heat resistant.

GURONIT GS28-10

Guronitwerke Vervoort GmbH
C 0.4, Ni 10, Cr, bal Fe.
For furnace parts, heat treating boxes; Type HE; heat resistant.

GURONIT GS28-20

Guronitwerke Vervoort GmbH
C 0.4, Cr 25, Ni 19, bal Fe.
Cast: 75,000 TS; 50,000 YS; 17 El; 170 Brin. For furnace parts, retorts, stack dampers; Type HK; heat resistant.

GURONIT GS28-4

Guronitwerke Vervoort GmbH
C 0.4, Si 1.3, Cr 27, Ni 4, bal Fe.
Cast: 90,000 TS; 65,000 YS; 2 El; 212 Brin. For cylinder liners, bushings, valve seats and bodies; Type CC-50; heat resistant.

GURONIT H10

Guronitwerke Vervoort GmbH
C 0.6, Si 1.5, Cr 22, bal Fe.
For oil refinery equipment, rollers, crushers; heat and wear resistant.

GURONIT H11

Guronitwerke Vervoort GmbH
C 1.3, Si 1.5, Cr 29, bal Fe.
For crushers, rollers, grate bars; heat and abrasion resistant.

GURONIT H7

Guronitwerke Vervoort GmbH
C 0.3, Cr 6, Si 2.2, bal Fe.
For oil refinery equipment; creep and heat resistant.

GURONIT H8

Guronitwerke Vervoort GmbH
C 0.5, Si 1.5, Cr 17, bal Fe.
For furnace parts, heat treating boxes; heat resistant.

GURONIT H9

Guronitwerke Vervoort GmbH
C 0.3, Si 2.2, Cr 17, bal Fe.
For furnace parts, heat treating boxes, oil refinery equipment; heat and creep resistant.

GURONITE

Guronitwerke Vervoort GmbH
C 0.7-2.3, Cr 30-35, Si 0.7, Mo, bal Fe.
For heat treating furnaces, pump fittings, valves, grates; heat and corrosion resistant.

GUSS PANTAL

German manufacture
Mg 2-4, Mn 1.2-1.5, 0 0.2 Sb or Ti, bal Al.
For light alloy castings; resists sea water corrosion.

GUSS-KORROFESTAL

Aluminium-Werke Wutoeschingen GmbH
Aluminum. Si 9-13, Mg 0.25-0.4, Mn 0.3-0.5, bal Al.
For light alloy parts; high corrosion resistance. *Obsolete*

GUSSBRONZE 10

Oederlin & Co. Ltd.
Copper. Sn, bal Cu.
Cast: 32,000-36,000 TS; 15 El; 70 Brin. For hardware; bronze.

GUSSBRONZE 12

Oederlin & Co. Ltd.
Copper. Sn, bal Cu.
Cast: 32,000-36,000 TS; 12 El; 85 Brin. For hardware; bronze.

GUSSBRONZE 14

Oederlin & Co. Ltd.
Copper. Sn, bal Fe.
Cast: 29,000-32,000 TS; 3 El; 110 Brin. For hardware; bronze.

GUSSBRONZE 20

Oederlin & Co. Ltd.
Copper. Sn, bal Cu.
Cast: 15,000-21,000 TS; 0 El; 180 Brin. For hardware; bronze.

GUSSBRONZE 6

Oederlin & Co. Ltd.
Copper. Sn, bal Cu.
Cast: 21,000-29,000 TS; 18 El; 60 Brin. For hardware; bronze.

GUSSMESSING 63

Oederlin & Co. Ltd.
Copper. Zn, bal Cu.
Cast: 22,000-25,000 TS; 6-10 El; 45 Brin. For air conditioners, housings, machinery parts; brass.

GUSSMESSING 67

Oederlin & Co. Ltd.
Copper. Zn, bal Cu.
Cast: 25,000-29,000 TS; 15-12 El; 40 Brin. For air conditioners, housings, machinery parts; brass.

GUSSMESSING 90

Oederlin & Co. Ltd.
Copper. Zn, bal Cu.
Cast: 23,000-25,000 TS; 18-20 El; 60 Brin. For current carrying equipment, conductors; brass.

GUSSTAHL 4H

Now VEW K970.

GUSSTAHL 4W

Now VEW K960.

GUSSTAHL 5H

Now VEW K945.

GUSSTAHL 5W

Now VEW K935.

GUSSTAHL D8518

Gusstahl-Handels GmbH
C 0.22, Mn 0.6, Cr 11.7, Mo 1.05, Ni 0.6, V 0.3, bal Fe.
Annealed: 95,000 TS; 40,000 YS; 25 El; 94 Rock B.
Hardened: 245,000 TS; 205,000 YS; 9 El; 50 Rock C. For cutlery, surgical instruments, hardware, gears, and shafts. Corrosion resistant; hardenable.

GUSSTAHL-D8514

Gusstahl-Handels GmbH
C 0.17, Cb 0.2, Cr 11, Mo 0.6, Ni 0.55, V 0.6, bal Fe.
Annealed: 75,000 TS; 40,000 YS; 30 El; 70 RA; 82 Rock B.
Cold drawn: 95,000 TS; 80,000 YS; 15 El; 60 RA; 92 Rock B.
For springs, table flatware, knives, oil refinery and chemical plant equipment. Corrosion resistant; hardenable.

GUSSTAHL-D8518W

Gusstahl-Handels GmbH
C 0.21, Cr 12, Mo 1, Ni 0.6, V 0.3, W 0.5, bal Fe.
Annealed: 90,000 TS; 38,000 YS; 25 El; 92 Rock B.
Hardened: 240,000 TS; 205,000 YS; 9 El; 50 Rock C. For surgical instruments, valves, cutlery, gears, and shafts. Corrosion resistant; hardenable.

GUTHRIES ALLOY

English manufacture
Bi 47, Sn 20, Pb 19, Cd 13.
For fire extinguisher plugs; fusible alloy.

GUYS ALLOY

English manufacture
Ni 82.5, Al 17.5.
At 70°F: 83,800 TS; 0.6 El. At 1500°F: 52,900 TS; 2. El. For turbine blades, jet engine components; high stress-rupture strength.

GW 280 TUF KUT

Great Western Steel Co.
C 0.6, Mn 0.85, Si 2, Cr 0.25, Mo 0.25, V 0.2, bal Fe.
Annealed: 107,000 TS; 64,000 YS; 27 El; 212 Brin. Oil hardened: 145,000-340,000 TS; 127,000-283,000 YS; 5-24 El; 20-44 RA; 293-611 Brin. For shear blades, pneumatic tools, punches, chisels, caulking tools. Type S5 shock resisting tool steel. Oil or water hardening.

GYRO

CCS Braeburn Alloy Steel
C 0.7, Cr 4, V 2, W 14, bal Fe.
For tools, cutters; water hardened. *Obsolete*

GYROCAST

Youngstown Alloy Castings Co.
C 0.4, Cr 1, bal Fe.
For straightening rolls, bar mill guides; centrifugally cast castings.

GYROMET

CMW Inc.
Ni 3, Mo 3, Cu 3, Fe 1.5, bal W.
140,000 psi TS; 120,000 psi YS; 2 El; 68 Rock A. Density: 16.95 g/cm^3. For rotating inertia members, weights requiring high strength.

GYROMET

CMW Inc.
W-alloy.
For gyros, counterweights; high density. *Obsolete*

GYROMET 1100

CMW Inc.
Ni 4, Mo 4, Fe 2, bal W.
165,000 psi TS; 160,000 psi YS; 1.0 El; 38 Rock C. Density: 17.25 g/cm^3. For rotating inertia members, weights requiring high strength.

GYROMET 1100

CMW Inc.
W 90, Ni 4, Fe 2, Mo 4.
Sintered: 150,000 TS; 140,000 YS; 2 El; density: 17.25; hardness 36 Rock C. For gyroscope rotors, governor counter weights, radiation shielding. *Obsolete*

H & R O6 GRAPHITIC

Houghton & Richards Inc.
C 1.45, Mn 1, Si 1.25, Mo 0.25, bal Fe.
AISI Type O6; oil hardening; tool steel.

H & R A4

Houghton & Richards Inc.
C 0.95, Mn 2, Si 0.35, Cr 2.2, Mo 1.1, Lead added, bal Fe.
Free machining grade of air hardening cold work tool steel.
AISI A4.

H & R BRAKE DIE

Houghton & Richards Inc.
C 0.5, Mn 1, Cr 0.95, Mo 0.2, bal Fe.
For brake dies; pre-heat treated.

H & R CARBON

Houghton & Richards Inc.
C 0.9-1.05, bal Fe.
For punches, dies, tools; water hardening.

H & R COBALT HIGH SPEED STEEL

Houghton & Richards Inc.
C 0.75-0.8, Cr 4, W 18, Co 5, V 1, bal Fe.
For cutting tools; high speed steel.

H & R COBALT MOLY

Houghton & Richards Inc.
C 0.88, Cr 4.1, W 6, Mo 6, V 1.9, Co 9, bal Fe.
For cutters, tools; high speed steel.

H & R GOLD LABEL

Houghton & Richards Inc.
C 1.33, Mo 0.35, W 4.25, bal Fe.
For drawing dies, finishing tools; keen cutting edge.

H & R GRAY LABEL

Houghton & Richards Inc.
C 0.9-1.05, bal Fe.
For punches, dies, stamps, headers.

H & R HEADING DIE STEEL

Houghton & Richards Inc.
C 0.9-1, bal Fe.
For embossing dies, cold heading dies; water hardening.

H & R HOT WORK

Houghton & Richards Inc.
C 0.3-0.35, Cr 3-3.5, V 0.3-0.5, W 10-11.5, bal Fe.
For extrusion and swedging dies, shear blades; hot work steel.

H & R HOT WORK 7

Houghton & Richards Inc.
C 0.55, W 1.2, Mo 1.2, Si 0.95, Mn 0.3, bal Fe.
For dies, hot work tools, punches; air or oil hardening, shock resistant.

H & R HOT WORK NO. 10

Houghton & Richards Inc.
C 0.23, Mn 0.6, Si 1.25, Cr 10, Ni 0.75, V 1, W 0.45, Mo 1, N 0.1, bal Fe.
Air or oil hardening; hot work tool and die steel; chromium type.

H & R HOT WORK NO. 12

Houghton & Richards Inc.
C 0.3, Mn 0.35, Si 0.5, Cr 12, V 0.9, W 12, bal Fe.
Air or oil hardening; hot work tool and die steel; tungsten type; AISI H23.

H & R HOT WORK NO. 15A

Houghton & Richards Inc.
C 0.25, W 14-16, Cr 3.75-4.25, V 0.4-0.6, bal Fe.
For hot heading dies, gripper dies, high temperature springs; hot work steel, oil hardening.

H & R HOT WORK NO. 4

Houghton & Richards Inc.
C 0.97, Cr 3.9, bal Fe.
For upsetting dies, gripper dies; oil or air hardening.

H & R HOT WORK NO. 5

Houghton & Richards Inc.
C 0.35, Si 1, Cr 5, V 0.4, Mo 1, bal Fe.
For die casting dies, tools, hot punches; hot work steel.

H & R HOT WORK NO. 6

Houghton & Richards Inc.
C 0.35, Si 1, Cr 5, W 1.35, Mo 1.75, bal Fe.
For extrusion dies, piercing mandrels; hot work steel.

H & R K2

Houghton & Richards Inc.
C 1.5, Cr 11.5, Mo 0.75, V 0.25, bal Fe.
For blanking and forming dies, lamination dies; nondeforming.

H & R K2L

Houghton & Richards Inc.
C 0.85, Ni 1, Mo 0.45, Cr 11.5, V 0.3, bal Fe.
For shear blades, punches, dies; shock and abrasion resistant, nondeforming.

H & R K3

Houghton & Richards Inc.
C 2.4, Cr 12.75, V 4, Mo 1.1, bal Fe.
For mold liners, lamination dies; oil hardening, nondeforming.

H & R K4

Houghton & Richards Inc.
C 1.4, Cr 4.1, V 4.1, Mo 4.25, bal Fe.
For cutting tools; machining scaly material.

H & R M4

Houghton & Richards Inc.
C 1.28, Cr 4.5, V 4, W 5.5, Mo 4.5, bal Fe.
Molybdenum-tungsten-vanadium grade of high speed steel for cutting tools; AISI M4.

H & R M42

Houghton & Richards Inc.
C 1.07, Cr 3.75, V 1.15, W 1.5, Mo 9.5, Co 8, bal Fe.
High speed tool steel; molybdenum-cobalt type, AISI M42.

H & R M43

Houghton & Richards Inc.
C 1.2, Cr 3.75, V 1.6, W 2.7, Mo 8, Co 8.2, bal Fe.
High speed tool steel; molybdenum-cobalt type; AISI M-43.

H & R M7

Houghton & Richards Inc.
C 1, Cr 3.75, V 2.05, W 1.75, Mo 8.75, bal Fe.
High speed tool steel; molybdenum type; AISI M7.

H & R MOLY VAN

Houghton & Richards Inc.
C 0.8, Cr 4, Mo 9, V 2.2, bal Fe.
For cutters, tools; high speed steel.

H & R MOLYHI

Houghton & Richards Inc.
C 0.8, Cr 4, W 1.5, Mo 8.5, V 1.15, bal Fe.
For cutters, tools; high speed steel.

H & R MULTIMOLD

Houghton & Richards Inc.
C 0.35, Cr 0.8, Mo 0.3, Mn 0.7, bal Fe.
For plastic mold dies; for high mold pressures.

H & R MY

Vereinigte Edelstahlwerke
C 0.42, Cr 1.4, Si 1.5, V 0.25, bal Fe.
For pneumatic tools, chisels, snaps, hot dies, hot shears; shock resistant.

H & R MY

Houghton & Richards Inc.
C 0.42, Cr 1.4, Si 1.5, V 0.25, bal Fe.
For pneumatic tools, chisels, snaps, hot dies, hot shears; shock resistant.

H & R MY-A

Houghton & Richards Inc.
C 0.4, Cr 1.5, Si 1.5, V 0.25, bal Fe.
For rivet sets, hand chisels, punches, shears; tough; shock resistant.

H & R N175

Houghton & Richards Inc.
C 0.75, Mn 0.75, Si 0.3, Cr 0.9, Ni 1.75, Mo 0.42, bal Fe.
Oil hardening; low alloy tool steel for shafts, arbors, lathe centers; AISI L6.

H & R NO. 1

Houghton & Richards Inc.
C 0.55-0.75, Cr 3.8-4.3, V 0.9-1.25, W 17.75-18.85, bal Fe.
For lathe and planer tools, cutters; high speed steel.

H & R NO. 14

Houghton & Richards Inc.
C 0.75, Cr 4, V 2, W 14, Co 5, bal Fe.
For lathe and planer tools, reamers, broaches; Type T8; high speed steel.

H & R NO. 15

Houghton & Richards Inc.
C 0.37, Mo 3, W 15.5, Cr 4, Ni 2, bal Fe.
For extrusion dies, hot piercing punches; hot work steel.

H & R NO. 150

Houghton & Richards Inc.
C 0.7, Cr 0.85, Mo 0.42, Ni 1.4, bal Fe.
For carbide tool shanks; oil hardening.

H & R NO. 19

Houghton & Richards Inc.
C 0.9, Mn 1.5, Mo 0.3, bal Fe.
For precision tools, dies; oil hardening.

H & R NO. 2

Houghton & Richards Inc.
C 0.8-0.85, Cr 4-4.5, V 2-2.25, W 18-19, Mo 0.6-0.8, bal Fe.
For lathe tools, drills, taps, reamers; high speed steel.

H & R NO. 225

Houghton & Richards Inc.
C 0.5, Cr 1.5, W 2.5, V 0.25, bal Fe.
For dies, shear blades, cutters; oil hardening.

H & R NO. 3

Houghton & Richards Inc.
C 1, Mn 0.25, Cr 4, W 18, Mo 0.8, V 3.4, bal Fe.
For cutters, tools; high speed steel.

H & R NO. 4

Houghton & Richards Inc.
C 0.8, W 18.5, Cr 4.5, V 1.75, Co 7.5, Mo 0.8, bal Fe.
For cutters, reamers; high speed steel.

H & R NO. 434

Houghton & Richards Inc.
C 1.18, Mo 4.25, V 3.1, Cr 4.1, bal Fe.
For cutting tools; for cutting abrasive materials.

H & R NO. 44

Houghton & Richards Inc.
C 0.8, Cr 4, Mo 4.2, V 1.1, W 6, bal Fe.
For cutting tools, drills, hobs, reamers; high speed steel.

H & R NO. 444

Houghton & Richards Inc.
C 1.4, Cr 4.1, Mo 4.25, V 4.15, bal Fe.
For cutting tools; abrasion resistant.

H & R NO. 445
Houghton & Richards Inc.
C 1.5, W 13.5, Cr 4.5, V 4.75, Co 5, Mo 0.5, bal Fe.
For cutting tools; high speed steel.

H & R NO. 45
Houghton & Richards Inc.
C 0.65, Cr 4, V 2, W 6, Mo 5, bal Fe.
For drills, reamers, lathe and planer tools; Type M2; high speed steel.

H & R NO. 50
Houghton & Richards Inc.
C 0.57, W 18, Cr 4, V 1.15, bal Fe.
For drills, reamers, taps, milling cutters; high speed steel.

H & R NO. 55
Houghton & Richards Inc.
C 0.35, Cr 5.2, W 5.2, Mo 0.2, V 0.2, Co 0.5, Si 0.9, bal Fe.
For die casting dies; oil hardening.

H & R NO. 57
Houghton & Richards Inc.
C 0.8, Cr 3.75, W 5.75, Mo 4.5, V 1.5, bal Fe.
For taps, chasers, broaches, reamers; high speed steel.

H & R NO. 59
Houghton & Richards Inc.
C 0.88, Cr 4.1, V 1.8, Mo 4.25, W 6, bal Fe.
For thread chasers, drills, pipe taps, tools; high speed steel.

H & R NO. 60
Houghton & Richards Inc.
C 1.2, Cr 0.7, Mo 0.25, V 1.6-0.2, bal Fe.
For taps, broaches, punches, gages; non-distorting.

H & R NO. 61
Houghton & Richards Inc.
C 1.3, Cr 12, Mo 0.6, Co 3, Si 0.5, bal Fe.
For blanking and forming dies, cold heading dies; air hardening.

H & R NO. 7
Houghton & Richards Inc.
C 1.3, W 5.5, Cr 4.5, Mo 4.5, V 4, bal Fe.
For broaches, reamers, chasers; high speed steel.

H & R NO. 8
Houghton & Richards Inc.
C 0.55, Mn 0.85, Cr 0.25, V 0.25, Si 2.1, bal Fe.
For carbide tool shanks; water or oil hardening.

H & R NO. 80
Houghton & Richards Inc.
C 1, Cr 0.5, Mn 0.7, V 0.25, Mo 1.25, bal Fe.
For lamination dies, gages, reamers; oil hardening.

H & R NO. 85
Houghton & Richards Inc.
C 0.5, Cr 0.95, V 0.2, bal Fe.
For die casting dies, shear blades; shock resistant.

H & R NO. 8M
Houghton & Richards Inc.
C 0.55, Mn 0.75, Mo 0.2, Si 2, bal Fe.
For carbide tool shanks; water or oil hardening.

H & R NON-TEMPERING
Houghton & Richards Inc.
C 0.35, Mn 0.7, Si 0.45, Cr 0.8, Mo 0.3, Cu 0.3, bal Fe.
Oil hardening tool steel; shock resistant.

H & R OIL HARDENING
Houghton & Richards Inc.
C 0.9-1, Mn 0.9-1, Cr 0.5-0.6, V 0.2, bal Fe.
For cold forming tools and dies, punches, broaches, gages.

H & R PISTON
Houghton & Richards Inc.
C 1.14, Cr 0.58, V 0.19, Si 0.2, Mn 0.3, bal Fe.
For pistons; water hardening.

H & R PLASTIC MOLD L
Houghton & Richards Inc.
C 0.5, Mn 1, Si 0.3, Cr 1.1, Mo 0.25, bal Fe.
Oil hardening; tool and die steel designed for plastic molds.

H & R PLASTIC MOLD-A
Houghton & Richards Inc.
C 0.07, Cr 4.5, Mo 0.45, Mn 0.4, Si 0.25, bal Fe.
For plastic mold dies; hobbing die steel, air hardening.

H & R PLASTIC MOLD-B
Houghton & Richards Inc.
C 0.06, Cr 1, Mo 0.25, B, bal Fe.
For plastic mold dies; for cold hobbing, oil hardening.

H & R PLASTIC MOLD-C
Houghton & Richards Inc.
C 0.1, bal Fe.
For plastic mold dies; case hardening.

H & R SILICO 1
Houghton & Richards Inc.
C 0.5, Mo 0.5, V 0.2, Si 1.1, bal Fe.
For impact tools, punches, chisels; shock resistant: oil or water hardening.

H & R SILICO 2
Houghton & Richards Inc.
C 0.55, Mn 0.7, Mo 0.45, V 0.2, Si 2.15, bal Fe.
For impact tools, punches, chisels; shock resistant, oil or water hardening.

H & R SPECIAL CARBON
Houghton & Richards Inc.
C 1-1.2, Mn 0.3, Si 0.2, bal Fe.
For drills, taps, hobs, reamers; Type W1; water hardening.

H & R SPECIAL HARDENING DIE
Houghton & Richards Inc.
C 0.9-1, Mn 0.2, Si 0.3, bal Fe.
For hardening dies; water hardening.

H & R SPECIAL HEADING DIE
Houghton & Richards Inc.
C 0.9-1, Mn 0.2, Si 0.3, bal Fe.
AISI Type W1; water hardening tool steel.

H & R SUPER COBALT
Houghton & Richards Inc.
C 0.8, W 20-21, Cr 4-4.5, V 1.3, Co 12, Mo 0.6, bal Fe.
For cutters, hobs, millers; high speed steel.

H & R SUPER MOLYHI
Houghton & Richards Inc.
C 0.8, Cr 4, W 1.5, Mo 8.5, V 1.2, Co 5, bal Fe.
For cutters, tools; high speed steel.

H & R TUNGSTEN OIL HARDENING
Houghton & Richards Inc.
C 0.9, Mn 1.1, W 0.5, V 0.2, Cr 0.5, bal Fe.
For hobs, reamers, broaches, gages; fast finishing tool steel.

H & R VANADIUM
Houghton & Richards Inc.
C 1.05-1.15, V 0.18, bal Fe.
For striking and heading dies, swedging tools; water hardening.

H & R VH
Houghton & Richards Inc.
C 1.4, Mn 0.4, Si 0.35, Cr 0-0.15, V 3.5, Mo 0-0.1, bal Fe.
Water hardening tool steel.

H & R WORK 5V
Houghton & Richards Inc.
C 0.35, Cr 5, V 1, Mo 1.5, bal Fe.
For hot work steel, extrusion rams and liners; Type H13; hot work steel.

H & R-K
Houghton & Richards Inc.
C 2.25-2.4, Mn 0.25-0.4, Cr 12.75-13.25, V 0.15-0.25, bal Fe.
For plug gages, punches, forming dies; nondeforming.

H & R-MY EXTRA
Houghton & Richards Inc.
C 0.4, Si 0.9, Cr 1, W 2, bal Fe.
For pneumatic tools, chisels, snaps, hot dies, hot shears; water hardening.

H 12
Titanium Metals Corp.
C 0.35, Mn 0.35, Si 1.05, Cr 5, V 0.35, W 1.25, Mo 1.35, bal Fe.
Hot work tool and die steel, chromium type; AISI H12.
Obsolete

H BRAND
Darwin & Milner Inc.
Tool material. C 0.95, Cr 0.35, Si 0.2, Mn 1, bal Fe.
For dies, punches; nondeforming.

H P D
Ziv Steel & Wire Co.
C 0.35, Si 1.05, Cr 5.15, V 0.3, W 1.25, Mo 1.55, bal Fe.
Air or oil hardening hot work tool and die steel; AISI H12.

H R C MAX
Firth-Vickers Stainless Steels Ltd.
C 0-0.25, Cr 18-20, Ni 24-26, bal Fe.
Annealed: 85,000-95,000 TS; 35,000-45,000 YS; 50-60 El; 60-75 RA; 150-190 Brin. For furnace parts, heat treating boxes, valves, pumps; Type 311; austenitic, heat resistant.

H RAPID 199
Handler GmbH
C 0.86, Co 2.8, Cr 4.3, Mo 0.85, V 2.1, W 12, bal Fe.
For lathe and planer tools, reamers, broaches; high-speed steel.

H RAPID 200
Handler GmbH
C 1.35, Cr 4, W, Co, bal Fe.
For blanking and forming dies, engravers' tools; high-speed steel.

H RAPID 201
Handler GmbH
C 0.79, Cr 4.3, Co 4.75, Mo 0.75, V 1.5, W 18, bal Fe.
For lathe and planer tools, cutters, reamers, taps; high-speed steel.

H REKORD EMINENT
Handler GmbH
C 0.76, Co 10, Cr 4.3, Mo 0.8, V 1.8, W 18, bal Fe.
For lathe and planer tools, reamers, broaches, taps; high-speed steel.

H REKORD SUPERIOR
Handler GmbH
C 0.74, V 1.1, W 18.5, Cr 4.1, bal Fe.
For lathe and planer tools, reamers; high-speed steel.

H S C 18-4-1
Hoyland Steel Co.
C 0.7, Cr 4, W 18, V 1, bal Fe.
For tools, taps, cutters, chasers, punches; high speed steel.

H S C 350
Hoyland Steel Co.
C 1.3, W 3.5, Mn 0.3, Si 0.45, bal Fe.
For cutters, tools, cold extrusion dies; fast finishing steel.

H S C 6-6-2
Manufacturer not listed.
C 0.8, Cr 4, W 5.75, Mo 5, V 1.5, bal Fe.
For tools, drills, cutters, chasers, broaches; high speed steel.

H S C ALLOY SHOE DIE
Hoyland Steel Co.
C 0.5, Cr 0.6, Mn 0.6, Mo 0.4, bal Fe.
For shoe dies; wear resistant; oil hardening.

H S C COBALT 5
Hoyland Steel Co.
C 0.7, Cr 4, W 18, Mo 0.5, Co 5, V 1, bal Fe.
For tools, cutters, broaches, chasers, reamers; high speed steel.

H S C COLD HEADER
Hoyland Steel Co.
C 0.95, Mn 0.3, bal Fe.
For cold headed dies, tools.

H S C CUTLERY
Hoyland Steel Co.
C 1.1, Mn 0.2, bal Fe.
For tools, cutlery, pocket knives; water hardening.

H S C NO. 310
Hoyland Steel Co.
C 0.3, Cr 3.25, W 10, V 0.35, bal Fe.
For hot work dies, brass forging dies; oil and air hardening.

H S C NO. 313
Hoyland Steel Co.
C 0.35, Cr 3, W 13.5, bal Fe.
For hot work dies, hot punching and extruding dies; hot work steel.

H S C NO. 515
Hoyland Steel Co.
C 0.35, Si 1, Cr 5, W 5, Mo 0.2, bal Fe.
For hot work dies; hot work steel, air hardening.

H S C OVERCOAT AXE
Hoyland Steel Co.
C 1, Mn 0.25, Si 0.15, bal Fe.
For axes, tools; water hardening.

H S C PAVEMENT BREAKER
Hoyland Steel Co.
C 0.65, Mn 0.3, Si 0.1, bal Fe.
For tools, sledges, pavement breakers; tough.

H S C REGULAR
Hoyland Steel Co.
C 0.7-1.1, Mn 0.3, bal Fe.
For tools, cold work tools; water hardened.

H S C SHOE DIE
Hoyland Steel Co.
C 0.9, Mn 0.35, Si 0.15, bal Fe.
For shoe dies; water hardening.

H S C SOLID DRILL
Hoyland Steel Co.
C 0.85, bal Fe.
For drills; tough.

H S C SPECIAL
Hoyland Steel Co.
c 1.1, Mn 0.3, Si 0.3, bal Fe.
For tools, cold work tools, taps, blanking dies; water hardened.

H SUPER REKORD
Handler GmbH
C 1.3, Cr 4.3, Mo 0.85, V 3.8, W 12, bal Fe.
For engravers' tools, blanking and forming tools; high-speed steel.

H SUPER REKORD CO
Handler GmbH
C 1.35, Cr, Co, V, bal Fe.
For engravers' tools, blanking and forming dies; high-speed steel.

H T S 33
Central Pattern & Foundry Co.
Al alloy.
For castings; not hardenable.

H V BLUE CHIP
Teledyne Firth Sterling
Tool steel. C 0.82, Cr 4, V 2, W 18, bal Fe.
For finishing tools and cutters; high speed steel. *Obsolete*

H W A
Darwin & Milner Inc.
Tool material. C 0.6, Cr 5, Mo 1, V 1, bal Fe.
Heat treated: 440-550 Brin. For die casting dies, hot working tools and dies; hot work steel; air or oil hardened.

H W D 2
Teledyne Firth Sterling
Steel. C 0.38, Cr 5.25, V 0.5, Mo 1.35, Mn 0.4, bal Fe.
For hot punches and dies; hot work steel, oil hardened. *Obsolete*

H W D 3
Teledyne Firth Sterling
Steel. C 0.4, V 1.05, Cr 5.25, Mo 1.25, bal Fe.
At 950°F: 153,000 TS; 138,000 YS; 17 El; 63 RA; 260 Brin. At 1150°F: 86,000 TS; 78,000 YS; 27 El; 78 RA; 155 Brin. For hot punches, shear blades, die casting dies; Type H13; hot work steel, oil hardened. *Obsolete*

H W D NO. 1
Teledyne Firth Sterling
Steel. C 0.4, Cr 0.5, V 0.15, W 1.1, Mo 1.5, Si 0.8, bal Fe.
At 950°F: 141,000 TS; 130,000 YS; 19 El; 61 RA; 275 Brin. At 1150°F: 80,000 TS; 68,000 YS; 29 El; 80 RA; 165 Brin. For hot punches, mandrels, shear blades; Type H12; hot work steel. *Obsolete*

H Y C C
Crucible Materials Corp.
Tool material. C 2.2, Cr 12, V 0.25, Mo 0.8, bal Fe.
For forming and crimping rolls, perforating dies, plug gages; wear resistant; air or oil hardened.

H-11
Titanium Metals Corp.
C 0.4, Mn 0.4, Si 1.05, Cr 5, V 0.5, Mo 1.35, bal Fe.
Air or oil hardening hot work tool steel, chromium type, AISI H11. *Obsolete*

H-17
Teledyne Firth Sterling
Tool material. WC 90, Co 10.
Submicron grain structure yields abrasion resistance and strength for machining abrasive materials and high temperature alloys. 91.4 Rock A.

H-3
Atrax Cemented Carbide
Sintered carbide. 400,000 transverse strength; 12.9-13.1 g/cm^3 density; 83.5-84.5 RA. Industry code: C-13.

H-36
Teledyne Firth Sterling
Tool material. WC 94, Co 6.
Submicron grain structure yields abrasion resistance and strength for machining abrasive materials and high temperature alloys. 92.7 Rock A.

H-4
Atrax Cemented Carbide
Sintered carbide. 400,000 transverse strength; 13.4-13.6 g/cm^3 density; 85.5-86.5 RA. Industry code: C-12.

H-4-X
Allied Steel & Tractor Products Inc.
C 0.35, Cr 0.8, Mo 0.5, Mn 0.7, Si 0.2, Ni 0-0.15, bal Fe.
Annealed: 100,000 TS; 80,000 YS; 2 El; 300 Brin. For gears, shafts, pinions; water hardening.

H-46
Now CARPENTER H-46.

H-9 DOUBLE HEADER
Carpenter Technology Corp.
C 0.9, Mn 0.35, Si 0.35, bal Fe.
For header dies, punches, coining and striking dies. Water hardened, wear resistant.

H-E IRON
Hansell-Elcock Co.
C, Si, Mn, bal Fe.
Cast: 30,000-60,000 TS; 180-260 Brin. For machine tool castings, gears, housings; gray cast iron.

H-M-BLUE CHIP
Teledyne Firth Sterling
C 0.7, W 18, Cr 4, V 1, bal Fe.
For tools, cutters; high speed steel. *Obsolete*

H.A.L.H. STEEL
English manufacture
C, alloy, bal Fe.
For tools, cutters; same as "Stellite."

H.C. PICK STEEL
Colt Industries
C, alloy, bal Fe.
For mining picks; water hardened. *Obsolete*

H.C. POKER BAR STEEL
Colt Industries
C, bal Fe.
For mining poker bars; water hardened. *Obsolete*

H.C.A.
CCS Braeburn Alloy Steel
C 0.3, Cr 12, W 1.2, V 0.9, bal Fe.
For master hobs, brass extrusion dies, die casting dies. Hot work steel, oil hardened. *Obsolete*

H.D.A. 8151
Cabot Corporation
Co 66, Cr 20, W 12.5, B 0.05, C 0.5, Mn 0.6, Si 0.6.
For jet engine components. High temperature alloys, heat and corrosion resistant. *Obsolete*

H.P.A. NICKEL
Henry Wiggin & Co. Ltd.
C 0-0.1, Cu 0-0.04, Fe 0-0.05, Mn 0-0.02, 99.5% Ni + Co min.
For electronic valves, cathodes; high temperature applications. *Obsolete*

H.P.B. NICKEL
Henry Wiggin & Co. Ltd.
C 0-0.1, Cu 0-0.04, Fe 0.05-1, Si 0.15-0.25, bal Fe.
For electronic valves, cathodes; high temperature applications. *Obsolete*

H.R.S.
British Steel Corp.
C, Si, Mn, bal Fe.
Heat treated: 170,000-190,000 TS; 150,000-170,000 YS; 10-12 El; 20-25 RA; 387-444 Brin. For automobile springs; good resiliency. *Obsolete*

H.R.W. DIE STEEL
Disston Inc.
C 0.9, Cr 3.7, V 0.15, bal Fe.
For dies, tools, forging mandrels; hot work steel. *Obsolete*

H.T. SILVER FUSE
Fusion Inc.
For solder, MP 600-700 F. *Obsolete*

H.T.C. NICKEL
Henry Wiggin & Co. Ltd.
Ni.
Cold drawn: 67,000 TS; 40-50 El; 60-75 RA; 110 Brin. For resistance thermometers; high temperature coefficient. *Obsolete*

H.T.M. STEEL
Joseph T. Ryerson & Son Inc.
C 0.35-0.45, Ni 2, Cr 0.9, Mo 0.4, bal Fe.
Heat treated: 158,000-237,000 TS; 142,000-207,000 YS; 11-18 El; 44-58 RA; 321-461 Brin. For gears, spindles, shafts, rolls; tough, shock resistant. *Obsolete*

H.T.S.
British Steel Corp.
C, alloy, bal Fe.
Normalized: 74,000 TS; 60,000 YS; 27 El; 45 RA; 217 Brin.
Heat treated: 110,000 TS; 74,000 YS; 25 El; 50 RA; 255 Brin.
For gears, shafts. *Obsolete*

H1
Sheffield Smelting Co. Ltd.
Ag 61, Cu, Zn.
Melting range: 690-735°C. Maximum stress: 39.4 kgf/mm^2; 23 El. For silver brazing, high quality.

H14
Sheffield Smelting Co. Ltd.
Ag 74, Cu, Zn.
Melting range: 745-778°C. For silver brazing, sterling silver.

H15
Sheffield Smelting Co. Ltd.
Ag 75, Cu, Zn.
Melting range: 740-788°C. For silver brazing, sterling silver.

H2
Atrax Cemented Carbide
Sintered carbide. 400,000 transverse strength; 12.8-13.0 g/cm^3 density; 82.0-83.5 RA. Industry code: C-14.

H20
Sheffield Smelting Co. Ltd.
Ag 80, Cu, Zn.
Melting range: 740-780°C. For silver brazing, sterling silver.

H44
Now HEPPENSTALL H44.

H5DM
Manufacturer not listed.
C 0.15, Mn 1.5, Cr 0.5, Mo 0.4, Ni 4, V 0.1, bal Fe.
Plate: 107,000 min TS; 92,000 min YS; 13 min El. For pressure vessels, bridges, mine cars, power shovels, cranes, trucks, trailers. Shock and wear resistant.

H6
Sheffield Smelting Co. Ltd.
Ag 66.5, Cu, Zn.
Melting range: 700-720°C. For silver brazing, sterling silver.

HA-0
Belgian manufacture
C 0.9, Cr 3, Mn 0.3, bal Fe.
H max 300; B max 13,500; Br 9900; Coercive force 66; (BdHd) max 320,000. For permanent magnets in electrical and magnetic equipment. Hardened, high permeability.

HA-1
Belgian manufacture
C 1, Cr 9, Co 3, Mo 1.5, bal Fe.
For permanent magnets in electrical and magnetic equipment. Hardened, high permeability.

HA-2
Belgian manufacture
C 1, Cr 9, Co 6, Mo 1.5, bal Fe.
For permanent magnets in electrical and magnetic equipment. Hardened. High permeability.

HA-25 ALLOY
Now HAYNES ALLOY NO. 25.

HA-3
Belgian manufacture
C 1, Co 15, Cr 9, Mo 1.5, bal Fe.
For permanent magnets in electrical and magnetic equipment. Hardened. High permeability.

HACKER CNBVO
Haeckerstahl GmbH
C 0.36, Cr 1, Mn 0.2, Ni 1, bal Fe.
For gears, bolts, machine tool parts; oil hardened, shock resistant.

HACKER CNBVO/2H
Haeckerstahl GmbH
C 0.3, Ni 2, Cr 2, Mo 0.3, bal Fe.
For gears, bolts, machine tool parts; oil hardened, shock resistant.

HACKER CNBVO/H
Haeckerstahl GmbH
C 0.34, Mn 0.55, Cr 1.55, Ni 1.55, Mo 0.2, bal Fe.
For gears, bolts, machine tool parts; oil hardened, shock resistant.

HACKER HWG
Haeckerstahl GmbH
C 0.55, Cr 0.7, Mo 0.18, V 0.1, Ni 1.65, bal Fe.
For heading and forging dies, upsetters, punches; oil hardened, tough.

HACKER HWG EXTRA
Haeckerstahl GmbH
C 0.56, Cr 0.85, Mo 0.2, Ni 1.8, V 0.1, bal Fe.
For forging dies, upsetting dies, punches; oil hardened, tough.

HACKER LD EXTRA EXTRA ZH
Haeckerstahl GmbH
C 1, Si 0-0.25, Mn 0-0.25, bal Fe.
Heat treated: 130,000-200,000 TS; 90,000-150,000 YS; 10-20 El; 30-50 RA; 270-420 Brin. For drills, reamers, taps, hobs; Type W1; water hardened.

HACKER LD EXTRA H
Haeckerstahl GmbH
C 1.3, Mn 0-0.25, Si 0-0.25, bal Fe.
For engravers' tools, blanking and forming dies; Type W1; water hardened.

HACKER LD EXTRA MH
Haeckerstahl GmbH
C 1.15, Si 0-0.25, Mn 0-0.25, bal Fe.
For springs, taps, drills, reamers, broaches; Type W1; water hardened.

HACKER LD EXTRA SEHR ZAH
Haeckerstahl GmbH
C 0.7, Si 0-0.25, Mn 0-0.25, bal Fe.
Heat treated: 122,000-175,000 TS; 82,000-130,000 YS; 12-22 El; 37-52 RA; 240-360 Brin. For springs, rails, punches, clutch discs; Type W1; water hardened.

HACKER LD EXTRA ZAH
Haeckerstahl GmbH
C 0.85, Si 0-0.25, Mn 0-0.25, bal Fe.
Heat treated: 130,000-200,000 TS; 90,000-150,000 YS; 10-20 El; 30-50 RA; 260-400 Brin. For drills, reamers, taps, hobs, cutters; Type W1; water hardened.

HACKER LD PRIMA EXTRA ZH
Haeckerstahl GmbH
C 0.67, Si 0.4, Mn 0.7, bal Fe.
Heat treated: 120,000-170,000 TS; 80,000-125,000 YS; 15-25 El; 40-55 RA; 240-350 Brin. For springs, clutch discs, girders, rails, die blocks; water hardened.

HACKER LD PRIMA H
Haeckerstahl GmbH
C 0.9, Si 0.4, Mn 0.6, bal Fe.
Heat treated: 130,000-200,000 TS; 90,000-150,000 YS; 10-20 El; 30-50 RA; 260-400 Brin. For drills, taps, reamers, cutters; Type W1; water hardened.

HACKER LD PRIMA MH
Haeckerstahl GmbH
C 0.75, Si 0.4, Mn 0.6, bal Fe.
Heat treated: 125,000-175,000 TS; 85,000-130,000 YS; 10-20 El; 35-50 RA; 250-360 Brin. For wheels, die blocks, rails, springs, clutches; Type W1; water hardened.

HACKER LD PRIMA SEHR ZAH
Haeckerstahl GmbH
C 0.35, Si 0.6, Mn 0.6, bal Fe.
Hot rolled: 85,000 TS; 54,000 YS; 30 El; 53 RA; 185 Brin. For gears, shafts, axles, bolts, fishplates; water hardened.

HACKER LD PRIMA ZAH
Haeckerstahl GmbH
C 0.6, Si 0.35, Mn 0.6, bal Fe.
Heat treated: 115,000-160,000 TS; 77,000-113,000 YS; 12-23 El; 40-54 RA; 230-320 Brin. For wheels, die blocks, girders, rails; water hardened.

HACKER LD PRIMA ZW
Haeckerstahl GmbH
C 0.15, Si 0.4, Mn 0.6, bal Fe.
Annealed: 70,000 TS; 55,000 YS; 25 El; 60 RA; 145 Brin. For screws, bolts, nuts, camshafts, rivets; case hardening steel.

HACKER LD SPECIAL ZAH
Haeckerstahl GmbH
Mn 0.2, C 0.85, Si 0.2, bal Fe.
Heat treated: 128,000-180,000 TS; 85,000-135,000 YS; 12-22 El; 35-50 RA; 250-400 Brin. For springs, tools, cutters, taps, drills, reamers; Type W1; water hardened.

HACKER LD SPEZIAL
Haeckerstahl GmbH
C 0.75, Si 0.25, Mn 0.25, bal Fe.
Heat treated: 125,000-180,000 TS; 85,000-135,000 YS; 12-22 El; 36-50 RA; 250-370 Brin. For springs, rails, punches, girders, crimpers; Type W1; water hardened.

HACKER LD SPEZIAL EXTRA ZH
Haeckerstahl GmbH
C 1, Si 0.25, Mn 0.25, bal Fe.
Heat treated: 190,000 TS; 120,000 YS; 10 El; 30 RA; 400 Brin. For springs, tools, reamers, drills; Type W1; water hardened.

HACKER LD SPEZIAL MH
Haeckerstahl GmbH
C 1.1, Si 0.25, Mn 0.25, bal Fe.
Annealed: 110,000 TS; 58,000 YS; 20 El; 40 RA; 210 Brin. For springs, drills, taps, reamers, broaches; Type W1; water hardened.

HACKER LDS
Haeckerstahl GmbH
C 0.9, Mn 1.9, V 0.1, bal Fe.
For punches, crimpers, forming dies; oil hardened, nondeforming.

HACKER LDX
Haeckerstahl GmbH
C 1.65, Cr 11.5, V 0.1, bal Fe.
For blanking and forming dies, punches; air hardened, nondeforming.

HACKER MN180
Haeckerstahl GmbH
C 0.14, Cr 1.1, Ni 4.5, bal Fe.
For gears, bolts, crankshafts, cams; case hardened, shock resistant.

HACKER ON130
Haeckerstahl GmbH
C 0.15, Si 0.25, Mn 0.5, Cr 0.65, bal Fe.
For gears, bolts, camshafts, cams; case hardened, tough.

HACKER ON160
Haeckerstahl GmbH
C 0.16, Si 0.25, Mn 1.15, Cr 0.95, bal Fe.
For gears, bolts, camshafts, cams; case hardened, tough.

HACKER ON200
Haeckerstahl GmbH
C 0.2, Cr 1.15, Mn 1.25, bal Fe.
For gears, bolts, camshafts, cams; case hardened, tough.

HACKER PS
Haeckerstahl GmbH
C 0.45, V 0.2, W 1.85, Cr 1.05, bal Fe.
For punches, rivet sets, upsetters, forging dies; oil hardened, tough.

HACKER PS SPEZIAL
Haeckerstahl GmbH
C 0.45, W, Cr, V, bal Fe.
For forging and header dies, upsetters; oil hardened, tough.

HACKER PS505
Haeckerstahl GmbH
C 0.55, Si 0.9, Mn 0.3, Cr 1.05, V 0.18, W 1.85, bal Fe.
For forging and header dies, upsetters; oil hardened, tough.

HACKER PSN
Haeckerstahl GmbH
C 0.45, Si, Cr, V, bal Fe.
For springs, gears, bolts, crankshafts; oil hardened, shock resistant.

HACKER TE
Haeckerstahl GmbH
C 1.05, Mn 0.9, Cr 1, W 1.15, bal Fe.
For bearings, cutters, bushings; oil or water hardened, tough.

HACKER WGNH
Haeckerstahl GmbH
C 0.6, Si 0.4, Mn 0.6, bal Fe.
Heat treated: 160,000 TS; 113,000 YS; 12 El; 40 RA; 320 Brin.
For wheels, die blocks, clutch discs, rails; water hardened.

HACKER WGNH EXTRA
Haeckerstahl GmbH
C 0.9, Si 0.4, Mn 0.6, bal Fe.
Heat treated: 190,000 TS; 145,000 YS; 10 El; 30 RA; 400 Brin.
For springs, taps, drills, hobs, reamers; Type W1; water hardened.

HACKER WGNH SPEZIAL
Haeckerstahl GmbH
C 0.5, Si 0.9, Mn 0.9, bal Fe.
For pneumatic tools, punches, dies, shear blades; oil hardened, tough.

HACKER WGW
Haeckerstahl GmbH
C 0.85, Si 0-0.25, Mn 0-0.25, bal Fe.
Heat treated: 188,000 TS; 145,000 YS; 12 El; 35 RA; 400 Brin.
For springs, drills, reamers, taps, tools, hammers; Type W1; water hardened.

HACKER WHH
Haeckerstahl GmbH
C 1.3, W, bal Fe.
For cutters, engravers' tools, cold headers; oil or water hardened.

HACKER WHH1
Haeckerstahl GmbH
C 1.1, Cr 0.4, Mn 0.3, Si 0.2, bal Fe.
For bearings, cutters, liners, drills; Type W5; water hardened.

HACKER WHHS
Haeckerstahl GmbH
C 1.2, Cr 0.2, V 0.1, W 1, bal Fe.
For heading and forming dies, punches; oil hardened, wear resistant.

HACKETT K-COPPER
Hackett Brass Foundry
Alloy Cu.
Cast: 70,000 TS; 20 El; 50 RA; 150 Brin. For welding tips and holders; 75-80% conductivity of Cu.

HACKSAW
Adams & Osgood Steel Co.
C 0.7, W 2, bal Fe.
For tools, hacksaws; oil hardened.

HACKSAW "A"
Jessop-Saville Ltd.
bal Fe.
For hack saws; water hardening.

HACKSAW "B"
Jessop-Saville Ltd.
C, W, bal Fe.
For hack saws; water hardening.

HACKSAW REG C
Adams & Osgood Steel Co.
C 0.7-1, bal Fe.
For tools, hacksaws; water hardened.

HACKSAW STEEL
Colt Industries
C, alloy, bal Fe.
For tools, hack saws. *Obsolete*

HADFIELD MANGANESE STEEL
Now MANGANESE GR.B.

HADFIELD MANGANESE STEEL
U.S. Steel Corp.
C 1-1.35, Mn 10.5-15, P 0-0.1, Si 0.1-0.3, bal Fe.
For perforated screens, dipper teeth, journal box liners, wear plates, screens, rails; abrasion resistant. *Obsolete*

HADFIELD SILICON STEEL
Dunford Hadfields Ltd.
Si 4, C, bal Fe.
For transformer cores, armatures for dynamos and electric motors, loading coils; low magnetic hysteresis with high permeability. *Obsolete*

HADFIELDS CR-MN STEEL
Bonney-Floyd Co.
C 1-1.4, Mn 10-14, Cr 2.8-3.1, bal Fe.
Heat treated: 80,000-140,000 TS; 35,000-55,000 YS; 20-30 El; 15-30 RA; 201-255 Brin. For wear resisting castings; non-magnetic. *Obsolete*

HADFIELDS MANGANESE STEEL
English manufacture
C 1-1.4, Mn 10-14, bal Fe.
Heat treated: 80,000-120,000 TS; 30,000-50,000 YS; 25-35 El; 20-35 RA; 170-201 Brin. For rails, wear plates, castings; non-magnetic, wear resisting.

HADURA
Dunford Hadfields Ltd.
C, Cr, bal Fe.
For rolls for cold rolling; water hardening. *Obsolete*

HAECKER 135H
Haeckerstahl GmbH
C 0.4, Cr, bal Fe.
For gears, machine tools parts; water hardened.

HAECKER 135M
Haeckerstahl GmbH
C 0.2, Cr, bal Fe.
For gears, cams, machine tool parts; case hardening.

HAECKER 135W
Haeckerstahl GmbH
C 0.1, Cr, bal Fe.
For gears, cams, shafts; case hardening.

HAECKER ADS
Haeckerstahl GmbH
C 0.38, Cr, Si, V, bal Fe.
For gears, bolts, crankshafts; oil hardened, shock resistant.

HAECKER AE13C
Haeckerstahl GmbH
C 2.1, Cr 11.5, bal Fe.
For blanking and forming dies; oil or air hardened, nondeforming.

HAECKER BSL-SPEZIAL
Haeckerstahl GmbH
C 0.4, Ni, Cr, Mo, bal Fe.
For gears, bolts, shafts, studs; oil hardened, shock resistant.

HAECKER F480E
Haeckerstahl GmbH
C 0-0.12, Cr 18, Ni 5, Mo 2, Ti = 4 x C, bal Fe.
Annealed: 90,000 TS; 45,000 YS; 50 El; 60 RA; 180 Brin. For chemical plant equipment, mixers, tanks, vessels; Type 316 Ti; stainless, austenitic.

HAECKER KF17
Haeckerstahl GmbH
C 0.08, Cr, bal Fe.
For gears, cams, camshafts, case hardened.

HAECKER KF17NI
Haeckerstahl GmbH
C 0.22, Cr, Ni, bal Fe.
For gears, cams, camshafts; case hardened.

HAECKER KF17S
Haeckerstahl GmbH
C 0.12, Cr 16.5, Mo 0.25, S 0.2, bal Fe.
Annealed: 80,000 TS; 50,000 YS; 25 El; 50 RA; 150 Brin. For chemical plant equipment; corrosion and heat resistant.

HAECKER KF300
Haeckerstahl GmbH
C 0-0.15, Cr 18, Ni 8.5, bal Fe.
Annealed: 80,000 TS; 35,000 YS; 50 El; 65 RA; 160 Brin. For chemical plant equipment; Type 302; stainless, austenitic.

HAECKER KF300E
Haeckerstahl GmbH
C 0-0.12, Cr 18, Ni 9.5, Ti = 4 x C, bal Fe.
Annealed: 90,000 TS; 45,000 YS; 45 El; 60 RA; 180 Brin. For welded structures, chemical plant equipment; Type 321; stainless, austenitic.

HAECKER KF300S
Haeckerstahl GmbH
C 0-0.07, Cr 18, Ni 9.5, bal Fe.
Annealed: 80,000 TS; 35,000 YS; 55 El; 70 RA; 150 Brin. For welded structures, chemical plant equipment; Type 304; stainless, austenitic.

HAECKER KF480S
Haeckerstahl GmbH
C 0-0.07, Cr 18, Mo 2, Ni 10.5, bal Fe.
Annealed: 90,000 TS; 45,000 YS; 50 El; 60 RA; 180 Brin. For acid resistant equipment, tanks, mixers, fitters; Type 316; stainless, austenitic.

HAECKER LR
Haeckerstahl GmbH
C 0.3, Cr 2.65, V 0.35, W 8.5, bal Fe.
For extrusion press dies, mandrels, liners, punches; oil hardened, hot work steel.

HAECKER MN130
Haeckerstahl GmbH
C 0.14, Cr 0.7, Ni 3.5, bal Fe.
For gears, bolts, camshafts, cams; case hardened, shock resistant.

HAECKER PSBG
Haeckerstahl GmbH
C 1, V 0.1, Mn 0.3, Si 0.25, bal Fe.
Heat treated: 185,000 TS; 120,000 YS; 10 El; 30 RA; 400 Brin. For knives, drills, drawing and stamping dies; oil or water hardened, wear resistant.

HAECKER PSX
Haeckerstahl GmbH
C 1.4, Cr 0.3, V 0.1, bal Fe.
For bearings, liners, blanking and forming dies; oil hardened, wear resistant.

HAFNIA
Paul Bergsoe & Son
Sn 24.5, Sb 13, Cu 0.5, Pb 62.
Cast: 12,500 TS; 28 Brin. MP: 355-535°F. For refrigerator and electric motor bearings. Good castability.

HAFNIUM
Atomergic Chemetals Corp.
Hf.
Purities: spectrographic grade 99.9%, reactor grade, commercial grade. Forms: sponge, crystal bar, ingot, powder, wire, sheet, foil, rod, arc melted buttons, zone refined rod, single crystals.

HAGESTA 0060 M
J.C. Soding & Halbach
C 0.1, Mo 28, Ni 65, bal Fe.
Cast, quenched: 64,000-92,000 TS; 50,000 YS; 10 El; 190 Brin approx. Highly corrosion-resistant cast alloy. DIN G-X10NiMo6528.

HAGESTA 1855 MW
J.C. Soding & Halbach
C 0.1, Cr 16, Mo 17, Ni 60, bal Fe.
Cast, quenched: 64,000-92,000 TS; 50,000 YS; 10 El; 190 Brin approx. Highly corrosion-resistant cast alloy. DIN G-X10NiMoCr601716.

HAI-11 ALLOY
Harrison Alloys Inc.
Copper. Cu.
Lead wire for precious metal thermocouple.

HAI-180 ALLOY
Harrison Alloys Inc.
Copper. Ni 23, bal Cu.
For electrical resistance; 180 ohms/circular mil ft.

HAI-30 ALLOY
Harrison Alloys Inc.
Copper. Ni 2, bal Cu.
Rolled: 30,000-60,000 TS. 30 ohms/circular mil ft. For corrosion resistant parts; electrical and corrosion resistant.

HAI-36 ALLOY
Harrison Alloys Inc.
Nickel. Ni 36, bal Fe.
Wrought: 150,000-70,000 TS. For bimetals, measuring tapes, length standards, thermostats; low-expansion alloy; maximum operating temperature 200°C.

HAI-373 ALLOY
Harrison Alloys Inc.
Superalloy. Ni 28.5-29.5, Co 16.5-17.5, bal Fe.
Sheet: 89,700 TS; 50,500 YP; 200-250 Brin. For electrical resistors, glass to metal seals, grid glow tubes. Heat resistant. Can be brazed or welded.

HAI-380 ALLOY
Harrison Alloys Inc.
Nickel. Ni 70, bal Fe.
Wrought: 150,000-70,000 TS. For immersion heaters, heater pads, electrical resistances; maximum operating temperature 500°C.

HAI-400 ALLOY
Harrison Alloys Inc.
Ni 63-70, bal Cu.
Good corrosion resistance in a wide range of environments.

HAI-42 ALLOY
Harrison Alloys Inc.
Nickel. Ni 41, bal Fe.
Rolled: 70,000-150,000 TS. For metal to glass seals, thermostats; matches 8160 glass expansion.

HAI-46 ALLOY
Harrison Alloys Inc.
Nickel. Ni 46, bal Fe.
Rolled: 70,000-150,000 TS. For seals with ceramic coated materials; special expansion properties.

HAI-52 ALLOY
Harrison Alloys Inc.
Nickel. Ni 50-51, bal Fe.
Wrought: 150,000-70,000 TS. For glass to metal seals, grid wire; maximum operating temperature 500°C.

HAI-60 ALLOY
Harrison Alloys Inc.
Copper. Ni 5, bal Cu.
For electrical resistances; 60 ohms/circular mil ft.

HAI-600 ALLOY
Harrison Alloys Inc.
Cr 14-17, Fe 6-10, 72.0 Ni min.
Good corrosion resistance and high temperature oxidation resistance. Nonmagnetic.

HAI-601 ALLOY
Harrison Alloys Inc.
Ni 58-63, Cr 21-25, Al 1-1.7, bal Fe.
Good corrosion and oxidation resistance. Good strength and toughness from cryogenic temperatures to 2000°F.

HAI-63 ALLOY
Harrison Alloys Inc.
Nickel. Mn 4, Si 1, bal Ni.
Wrought: 175,000-70,000 TS. For spark plug electrodes, resistances; heat resistant to 750°C.

HAI-66 ALLOY
Harrison Alloys Inc.
Nickel. Cr 11-15, Fe 0-10, 70.0 Ni min.
For heat resisting parts; stainless, heat resistant.

HAI-90 ALLOY
Harrison Alloys Inc.
Copper. Ni 11, bal Cu.
For electrical resistance; load banks, power resistors; 90 ohms/circular mil ft.

HAI-CUNI 102
Harrison Alloys Inc.
Ni 45, bal Cu.
Used for wire-wound resistors. Constantan.

HAI-FECRAL 15
Harrison Alloys Inc.
Cr 15, Al 4.5, bal Fe.
For electrical resistors; 750 ohms/circular mil ft.

HAI-FECRAL 20
Harrison Alloys Inc.
Cr 22, Al 5, bal Fe.
Rolled: 110,000 TS; 60,000 YS; 20 El. For electric heating elements and resistors; magnetic; 815 ohms/circular mil ft.

HAI-FECRAL-25
Harrison Alloys Inc.
Stainless steel. Cr 22, Al 5.5, bal Fe.
Rolled: 110,000 TS; 60,000 YS; 20 El. 875 ohms/circular mil ft. For electric heating elements and resistances; magnetic.

HAI-JN
Harrison Alloys Inc.
Ni 45, bal Cu.
Negative leg for ISA Type S, T and E thermocouple. Constantan, nonmagnetic.

HAI-JP
Harrison Alloys Inc.
Fe 99, bal Fe.
Positive leg for ISA Type S thermocouple. Magnetic.

HAI-KN
Harrison Alloys Inc.
Nickel. Si 1, Al 1, Mn 1, bal Ni.
Negative Type K extension and thermocouple wire and ribbon.

HAI-KP
Harrison Alloys Inc.
Nickel. Ni 90, Cr 10.
Positive Type K extension and thermocouple wire and ribbon.

HAI-MANGANIN 12
Harrison Alloys Inc.
Mn 12.5, Ni 4-5, bal Cu.
Annealed: 70,000 TS. For wire-wound resistors; low thermal EMF; low TCR.

HAI-MANGANIN 13
Harrison Alloys Inc.
Copper. Mn 9.5-10.5, Ni 4-5, bal Cu.
Annealed: 70,000 TS. For shunts; low thermal EMF, low TCR.

HAI-NI 200
Harrison Alloys Inc.
Nickel. 99.0 Ni min.
For magnetostriction, jewelry, mechanical parts.

HAI-NI 211
Harrison Alloys Inc.
Nickel. Mn 4-5, bal Ni.
For heating element lead wire, lamp fuse leads.

HAI-NI 634
Harrison Alloys Inc.
Nickel. W 4, bal Ni.
Ribbons and strip for high-strength electron tube cathodes.

HAI-NI 99
Harrison Alloys Inc.
Nickel. Ni 99.5.
For electrical equipment, resistance thermometers; heat and corrosion resistant.

HAI-NICR 35
Harrison Alloys Inc.
Superalloy. Ni 35, Cr 20, Fe 45, Si 1.
Wire, rod or ribbon for heating elements up to 1700°F.

HAI-NICR 35 CB
Harrison Alloys Inc.
Superalloy. Ni 35, Cr 21.5, Si 2, Cb 1, C 0.05, bal Fe.
For furnace heating elements, furnace belts and conveyors. High heat and oxidation resistance.

HAI-NICR 40
Harrison Alloys Inc.
Nickel. Ni 37, Cr 21, Si 0.75-1.7, C 0-0.15, Mn 0-1, bal Fe.
For open heating elements. High heat and oxidation resistant.

HAI-NICR 60
Harrison Alloys Inc.
Nickel. Ni 60, Cr 15, Fe 25, Si 1.
Wire and ribbon for heating elements up to 1900°F.

HAI-NICR 70
Harrison Alloys Inc.
Nickel. Ni 70, Cr 30.
Wire, rod, ribbon for heating elements up to 2200°F.

HAI-NICR 80
Harrison Alloys Inc.
Nickel. Ni 80, Cr 20.
95,000-175,000 TS; 50,000 YS; 35 El. For heating elements in electric furnaces, electric ranges; maximum operating temperature 1150°C; high heat and oxidation resistant.

HAI-NICR 80 CB
Harrison Alloys Inc.
Nickel. C 0.03, Si 1.2, Cr 19.5, Cb 1.1, bal Ni.
For thermocouples, conveyor belts; high heat resistance; stable in reducing atmosphere.

HAIRSPRINGS
American manufacture
Sn 4.5-6.5, P 0.2, bal Cu.
For hair springs, phosphor bronze.

HALBERLAND ALLOY
German manufacture
Cu 87, Zn 13.
For hardware fittings; Red Brass.

HALCO
Crucible Specialty Metals
C 0.9, Cr 3.6, bal Fe.
For hot work dies, hot punches and shear blades; tough and heat resistant. *Obsolete*

HALCO
Colt Industries
C 0.9, Cr 3.6, bal Fe.
For hot work dies, hot punches and shear blades; tough and heat resistant. *Obsolete*

HALCOMB
Evans Steel Co.
C 0.9, Cr 1, bal Fe.
For tools, dies; hardened.

HALCOMB "S.R.B."
Colt Industries
C 1.08, Mn 0.37, Cr 1.37, bal Fe.
For tools, dies; oil-hardened. *Obsolete*

HALCOMB 1370
Colt Industries
C 0.43, Cr 2.2, Mo 0.7, W 1, V 0.65, bal Fe.
For hot work dies; hot work steel. *Obsolete*

HALCOMB 218
Crucible Materials Corp.
Tool material. C 0.4, Si 1, Cr 5, V 0.3, Mo 1.4, bal Fe.
Air hardened: 400 Brin. For brass extrusion mandrels, hot heading and forging dies, All die casting dies, hot punches; tough, resists heat checking, hot work steel. AISI H11.

HALCOMB 236
Crucible Materials Corp.
Tool material. C 0.3, Cr 13, W 13, V 0.9, bal Fe.
For permanent molds, dies, extension dies; corrosion and heat resistant, hot work steel.

HALCOMB 425
Crucible Materials Corp.
Tool material. C 0.4, Mn 0.3, Cr 4.25, V 2.1, W 4.25, Mo 0.4, Co 4.25, bal Fe.
At 1200°F: 110,000 TS; 16 El; 36 RA; 44 Rock C. At 800°F: 180,000 TS; 9 El; 36 RA; 45 Rock C. For hot work tools, brass extrusion dies and dummy blocks, forging dies and inserts. Shock and abrasion resistant. Hot work tool steel, Type H10.

HALCOMB 440
Hawkridge Bros. Co.
C 0.8, Cr 4.5, V 1.5, W 20, Co 12, bal Fe.
For lathe and planer tools, reamers, broaches; Type T6; high speed steel. *Obsolete*

HALCOMB 777
Colt Industries
C 0.7, Mn 0.7, Mo 0.7, bal Fe.
For tools; oil or water hardening. *Obsolete*

HALCOMB 999
Hawkridge Bros. Co.
C 0.75, Cr 4, V 2, W 14, Co 5, bal Fe.
For lathe and planer tools, reamers, broaches; Type T8; high speed steel.

HALCOMB 999
Hawkridge Bros. Co.
C 0.75, Cr 4, V 2, W 14, Co 5, bal Fe.
For lathe and planer tools, reamers, broaches; Type T8; high speed steel. *Obsolete*

HALCOMB BRAKE DIE
Colt Industries
C 0.35, Cr 0.9, Mo 0.2, bal Fe.
Heat treated: 260-300 Brin. For brake dies; furnished heat treated. *Obsolete*

HALCOMB C.C.S.
Crucible Specialty Metals
C 0.4, Si 1.1, Cr 5, W 4, bal Fe.
Air-oil: 400 Brin. For tools, die casting dies, press forging dies; tough, wear resistant. *Obsolete*

HALCOMB C.C.S.
Colt Industries
C 0.4, Si 1.1, Cr 5, W 4, bal Fe.
Air-oil: 400 Brin. For tools, die casting dies, press forging dies; tough, wear resistant. *Obsolete*

HALCOMB CHROME-MOLY HOG KNIFE
Colt Industries
C 0.6, Cr 0.9, bal Fe.
For knives; water hardening. *Obsolete*

HALCOMB CSM NO. 2
Colt Industries
C 0.3, Cr 0.8, Mo 0.3, bal Fe.
200 Brin. For dies for zinc base die castings. *Obsolete*

HALCOMB DOUBLE SPECIAL W
Crucible Specialty Metals
C 1.19, Mn 0.22, Cr 0.12, W 2.84, bal Fe.
Brine harden: For tools, finishing tools, drawing dies; finishing steel. *Obsolete*

HALCOMB DOUBLE SPECIAL W
Colt Industries
C 1.19, Mn 0.22, Cr 0.12, W 2.84, bal Fe.
Brine harden: For tools, finishing tools, drawing dies; finishing steel. *Obsolete*

HALCOMB FIRST QUALITY
Colt Industries
C 1.25, bal Fe.
For tools; drill rod. *Obsolete*

HALCOMB FM-2 FREE MACHINING
Crucible Specialty Metals
C 0-0.12, Cr 15, S, Mo, bal Fe.
Oil hardened. For valves, bolts, nuts, golf heads. *Obsolete*

HALCOMB FM-2 FREE MACHINING
Colt Industries
C 0-0.12, Cr 15, S, Mo, bal Fe.
Oil hardened. For valves, bolts, nuts, golf heads. *Obsolete*

HALCOMB GRADE A STAINLESS
Crucible Specialty Metals
C 0.35, Cr 13, bal Fe.
At 200 C: 225,000 TS; 186,000 YS; 9 El; 25 RA; 548 Brin. At 700 C: 125,000 TS; 91,000 YS; 22 El; 59 RA; 294 Brin. For ball bearings, paper knives, cutlery, surgical instruments; corrosion resistant. *Obsolete*

HALCOMB GRADE A STAINLESS
Colt Industries
C 0.35, Cr 13, bal Fe.
At 200 C: 225,000 TS; 186,000 YS; 9 El; 25 RA; 548 Brin. At 700 C: 125,000 TS; 91,000 YS; 22 El; 59 RA; 294 Brin. For ball bearings, paper knives, cutlery, surgical instruments; corrosion resistant. *Obsolete*

HALCOMB GRADE B HIGH CARBON STAINLESS
Crucible Specialty Metals
C 1, Cr 17, bal Fe.
For corrosion resistant parts; corrosion resistant. *Obsolete*

HALCOMB GRADE B HIGH CARBON STAINLESS
Colt Industries
C 1, Cr 17, bal Fe.
For corrosion resistant parts; corrosion resistant. *Obsolete*

HALCOMB GRADE B STAINLESS
Crucible Specialty Metals
C 0.75, Cr 17, bal Fe.
For cutlery and dental instruments; corrosion resistant. *Obsolete*

HALCOMB GRADE B STAINLESS
Colt Industries
C 0.75, Cr 17, bal Fe.
For cutlery and dental instruments; corrosion resistant. *Obsolete*

HALCOMB HIGH SPEED
Colt Industries
C 0.75, Mn 0.27, Cr 1.7, W 17.8, V 1.07, bal Fe.
For tools, cutters, drills; high speed steel. *Obsolete*

HALCOMB HOT WORK
Crucible Specialty Metals
C 0.4, W 8.7, Cr 3.5, V 0.25, bal Fe.
Oil hardened: 400 Brin. For hot forging dies, punches, grippers; resists heat checking. *Obsolete*

HALCOMB HOT WORK
Colt Industries
C 0.4, W 8.7, Cr 3.5, V 0.25, bal Fe.
Oil hardened: 400 Brin. For hot forging dies, punches, grippers; resists heat checking. *Obsolete*

HALCOMB L.C.T.
Crucible Specialty Metals
C 0.4, Cr 3, V 0.3, W 15, bal Fe.
For hot work dies and tools, shear blades, nut piercers; tough and heat resistant. *Obsolete*

HALCOMB L.C.T.
Colt Industries
C 0.4, Cr 3, V 0.3, W 15, bal Fe.
For hot work dies and tools, shear blades, nut piercers; tough and heat resistant. *Obsolete*

HALCOMB L.C.T. NO. 2
Crucible Specialty Metals
C 0.45, Cr 2, V 0.3, W 11, bal Fe.
Oil hardened: 400 Brin. For brass forging dies, extrusion dies, dummies for brass; resists heat checking. *Obsolete*

HALCOMB L.C.T. NO. 2
Colt Industries
C 0.45, Cr 2, V 0.3, W 11, bal Fe.
Oil hardened: 400 Brin. For brass forging dies, extrusion dies, dummies for brass; resists heat checking. *Obsolete*

HALCOMB LO CHRO W STUD
Crucible Specialty Metals
C 0.25, Cr 6, W 1, bal Fe.
For corrosion resistant parts; corrosion resistant. *Obsolete*

HALCOMB LO CHRO W STUD
Colt Industries
C 0.25, Cr 6, W 1, bal Fe.
For corrosion resistant parts; corrosion resistant. *Obsolete*

HALCOMB NU-DIE
Colt Industries
C 0.38, Si 1, Cr 5, V 0.4, Mo 1.4, bal Fe.
Heat treated: 200,000 TS; 180,000 YS; 13.5 El; 47 RA; 420 Brin. For die casting dies, sleeves, slides, plungers; resists heat checking. *Obsolete*

HALCOMB PYRO
Crucible Specialty Metals
C 0.7, Cr 0.9, V 0.2, bal Fe.
For hot and cold work impact tools, bull-dozing tools; resists shock and wear. *Obsolete*

HALCOMB PYRO
Colt Industries
C 0.7, Cr 0.9, V 0.2, bal Fe.
For hot and cold work impact tools, bull-dozing tools; resists shock and wear. *Obsolete*

HALCOMB SPECIAL
Crucible Specialty Metals
C 1.35, Mn 0.24, Cr 0.25, bal Fe.
For tools, dies, taps, reamers; water hardened. *Obsolete*

HALCOMB SPECIAL
Crucible Specialty Metals
C 0.6, Cr 1, V 0.2, bal Fe.
For tools, taps, dies, reamers; deep hardened. *Obsolete*

HALCOMB SS TOOL STEEL
Colt Industries
C 1.05, Cr 1.2, Mo 0.3, bal Fe.
Hardened: C 58-65 Rock. Oil hardening tool steel. For broaches, dies, drills, gauges, thread rolling dies. *Obsolete*

HALCOMB STAINLESS IRON NO. 12
Crucible Specialty Metals
C 0-0.12, Cr 13, bal Fe.
Annealed: 67,500 TS; 34,500 YS; 39 El; 72 RA; 137 Brin. Water quenched: 174,000 TS; 135,000 YS; 14 El; 42 RA; 332 Brin. For cutlery, dental instruments; corrosion resistant. *Obsolete*

HALCOMB STAINLESS IRON NO. 12
Colt Industries
C 0-0.12, Cr 13, bal Fe.
Annealed: 67,500 TS; 34,500 YS; 39 El; 72 RA; 137 Brin. Water quenched: 174,000 TS; 135,000 YS; 14 El; 42 RA; 332 Brin. For cutlery, dental instruments; corrosion resistant. *Obsolete*

HALCOMB STAINLESS NO. 16
Crucible Specialty Metals
C 0-0.1, Cr 17, bal Fe.
For corrosion resistant parts; corrosion resistant. *Obsolete*

HALCOMB STAINLESS NO. 16
Colt Industries
C 0-0.1, Cr 17, bal Fe.
For corrosion resistant parts; corrosion resistant. *Obsolete*

HALCOMB STAINLESS NO. 18
Crucible Specialty Metals
C 0-0.1, Cr 19, bal Fe.
Annealed: 72,000 TS; 35 El; 70 RA; 170 Brin. For cooking utensils, table tops, hardware, automobile head lamps; corrosion resistant. *Obsolete*

HALCOMB STAINLESS NO. 18
Colt Industries
C 0-0.1, Cr 19, bal Fe.
Annealed: 72,000 TS; 35 El; 70 RA; 170 Brin. For cooking utensils, table tops, hardware, automobile head lamps; corrosion resistant. *Obsolete*

HALCOMB STAINLESS NO. 24
Crucible Specialty Metals
C 0.24, Cr 26, bal Fe.
For heat and corrosion resistant parts, chemical plant equipment; corrosion and heat resistant. *Obsolete*

HALCOMB STAINLESS NO. 24
Colt Industries
C 0.24, Cr 26, bal Fe.
For heat and corrosion resistant parts, chemical plant equipment; corrosion and heat resistant. *Obsolete*

HALCUT
Crucible Specialty Metals
C 0.5, Cr 1.3, W 2.7, bal Fe.
Oil hardened: 400 Brin. For hot work tools, chisels, shear blades, forging dies; tough and abrasion resistant. *Obsolete*

HALCUT
Colt Industries
C 0.5, Cr 1.3, W 2.7, bal Fe.
Oil hardened: 400 Brin. For hot work tools, chisels, shear blades, forging dies; tough and abrasion resistant. *Obsolete*

HALDI
Crucible Specialty Metals
C 2.25, Cr 11.5, bal Fe.
For dies, swages, punches, shears, drawing dies; air hardening; resists wear and abrasion. *Obsolete*

HALDI
Colt Industries
C 2.25, Cr 11.5, bal Fe.
For dies, swages, punches, shears, drawing dies; air hardening; resists wear and abrasion. *Obsolete*

HALDI NO. 2
Colt Industries
C 1.5, Cr 11.5, bal Fe.
For dies, swages, punches, shears, drawing dies, high abrasion resistant. *Obsolete*

HALGRAPH
Crucible Materials Corp.
Tool material. C 1.5, Mo 0.25, Si 1, Mn 0.75, bal Fe.
For machine tool parts; oil hardened; Type O6.

HALLAMITE
Hallamshire Steel Co.
C 0.7, W 14, Cr 4, V 1, bal Fe.
For turning and planing tools, shear blades, saws; high speed steel, oil hardened.

HALLAMITIER
Hallamshire Steel Co.
C 0.7, W 18, Cr 4, V 1, bal Fe.
For saws, saw blades, taps, broaches, reamers; high speed steel, oil hardened.

HALLAMITIEST
Hallamshire Steel Co.
C 0.7, W 22, Cr 4, V 1, bal Fe.
For hacksaw blades, saws, lathe and planer tools; high speed steel, oil hardened.

HALLAMSTEEL NO. 1
Hallamshire Steel Co.
C 1-1.2, bal Fe.
For circular saws, turning and shaping tools; Type W1; water hardened.

HALLAMSTEEL NO. 2
Hallamshire Steel Co.
C 0.9-1, bal Fe.
For saws, drills, pneumatic chisels, shear blades; Type W1; water hardened.

HALLAMSTEEL NO. 3
Hallamshire Steel Co.
C 0.8-0.9, bal Fe.
For band saws, caulking tools, shear blades; Type W2; water hardened.

HALLAMSTEEL NO. 4
Hallamshire Steel Co.
C 0.7-0.8, bal Fe.
For cutlery, drills, wood saws, stamping dies; Type W1; water hardened.

HALLAMSTEEL NO. 5
Hallamshire Steel Co.
C 0.6-0.7, bal Fe.
For awl blades, blacksmith and boilermaker tools; water hardened.

HALLAMSTEEL NO. 6
Hallamshire Steel Co.
C 0.5-0.6, bal Fe.
For scissors, hammers, chisels; water hardened.

HALLIMAX I
Hallamshire Steel Co.
Ni 65, Cr 15, bal Fe.
Annealed: 100,000 TS; 25-35 El; 55 RA; 147-157 Brin. For resistances, heating elements; high heat resistant.

HALLIMAX II
Hallamshire Steel Co.
Ni 80, Cr 20.
Annealed: 95,000 TS; 25-35 El; 55 RA; 142-157 Brin. For resistances, heating elements; high heat resistance.

HALMO
Colt Industries
C 0.35, Cr 5, Mo 5, bal Fe.
For tools and dies; oil or air hardening. *Obsolete*

HALVAN
Crucible Materials Corp.
Tool material. C 0.5, Cr 1, Mn 0.7, V 0.2, bal Fe.
Hardened: 540 Brin. For chisels, die holders, fixtures, jigs, punches; shock resistant, tough, AISI L2.

HAMILOY
Hamilton Die Cast
Si, bal Al.
Die cast: 37,000 TS; 24,000 YS; 5-7 El. For die castings; corrosion resistant.

HAMILTON ALLOYED IRON
Hamilton Foundry
Alloy cast iron. As cast: 25,000-60,000 TS; 20,000-48,000 YS; 145-350 Brin. Higher strength castings.

HAMILTON DUCTILE IRON
Hamilton Foundry
Cast and heat treated: 60,000-175,000 TS; 40,000-150,000 YS; 143-338 Brin. High strength and ductility. Several grades.

HAMILTON DUCTILE NI-RESIST
Hamilton Foundry
18-25 Ni castings.
Cast: 55,000-60,000 TS; 28,000-32,000 YS; 130-240 Brin. Austenitic, ductile, corrosion resistant, with good strength; 8 grades. *Obsolete*

HAMILTON GRAY IRON
Hamilton Foundry
Gray iron castings. As cast: 20,000-25,000 TS; 16,000-20,000 YS; 130-220 Brin. For general purpose cast iron.

HAMILTON MEEHANITE
Hamilton Foundry
Cast: 30,000-70,000 TS; 24,000-56,000 YS; 174-600 Brin. For high quality gray iron castings; 24 types. *Obsolete*

HAMILTON METAL-A
English manufacture
Zn 93, Cu 3.5, Pb 3.1, Sb 1.5, 0.5 P + Sn.
For ornamental castings; similar to "Chrysorin."

HAMILTON METAL-B
English manufacture
Zn 33.3, Cu 66.7.
For drawn, spun and stamped articles; high ductility.

HAMILTON NI-HARD
Hamilton Foundry
Ni-Cr cast iron.
Cast: 40,000-75,000 TS; 525-600 Brin. White cast iron, good wear properties, not machinable; for pump parts, metal working rolls, coffee grinding burrs. *Obsolete*

HAMILTON NI-RESIST
Hamilton Foundry
Cast: 20,000-30,000 TS; 100-250 Brin. Ni or Ni-Cu cast iron; good resistance to corrosion and erosion; 8 grades. *Obsolete*

HAMMOND ALLOY NO. 1
Hammond Brass Works
Cu 80-82, Sn 2.75-3.5, Pb 6.5-7.5, Zn 8-10.
Cast: 29,000-39,000 TS; 13,700-17,000 YS; 16-30 El; 12-27 RA; 50-60 Brin. For valve and plumbing castings.

HAMMOND ALLOY NO. 10
Hammond Brass Works
Cu 86, Sn 7, Pb 7.
Cast: 31,000 TS; 19,000 YS; 15 El; 55 Brin. For bearings. Acid resisting.

HAMMOND ALLOY NO. 11
Hammond Brass Works
Cu 84.5-86.5, Sn 8.5-9.5, Pb 3.5-4.5, Zn 1-2.
Cast: 36,000-38,000 TS; 20,000-26,000 YS; 24-30 El; 65-75 Brin. For bearings. Shock resisting.

HAMMOND ALLOY NO. 12
Hammond Brass Works
Cu 82, Sn 17, P 0-1.
Cast. For journals and trunions.

HAMMOND ALLOY NO. 13
Hammond Brass Works
Cu 80-82, Sn 10.5-11.5, Pb 5.5-6.5, Ni 1.5-2.5, P 0.4-0.5.
Cast: 38,000-42,000 TS; 20,000-24,000 YS; 15-22 El; 75-85 Brin. For bearings. Shock resisting.

HAMMOND ALLOY NO. 14
Hammond Brass Works
Cu 80, Sn 8, Pb 10, P 0.5, Ni 1.5.
Cast: 32,000 TS; 18,000 YS; 15 El; 70 Brin. For bearings. High speed.

HAMMOND ALLOY NO. 15
Hammond Brass Works
Si, bal Cu.
Cast: 45,000 TS; 22,000 YS; 18 El; 90 Brin. For worms, gears, and pinions. Acid resisting.

HAMMOND ALLOY NO. 2
Hammond Brass Works
Cu 83, Sn 4, Pb 6, Zn 7.
Cast: 30,000 TS; 14,000 YS; 16 El; 55 Brin. For hydraulic pumps. Free-cutting.

HAMMOND ALLOY NO. 3
Hammond Brass Works
Cu 84-86, Sn 4-6, Pb 4-6, Zn 4-6.
Cast: 33,000-48,000 TS; 17,000-24,000 YS; 18-35 El; 12-32 RA; 55-65 Brin. For steam valves and bearings. Free cutting.

HAMMOND ALLOY NO. 4
Hammond Brass Works
Cu 78.5-81.5, Sn 9-11, Pb 9-11, Zn 0-0.75.
Cast: 32,000-38,000 TS; 17,000-22,000 YS; 15-20 El; 6-17 RA; 60-70 Brin. For motor bearings and castings. Tough.

HAMMOND ALLOY NO. 5
Hammond Brass Works
Cu 86-89, Sn 9-11, Pb 1, Zn 1-2.5.
Cast: 36,000-46,000 TS; 20,000-26,000 YS; 17-25 El; 12-26 RA; 70-80 Brin. For bearings, bushings, and gears. Tough.

HAMMOND ALLOY NO. 6
Hammond Brass Works
Cu 86-89, Sn 9-11, Pb 0-0.2, Zn 1-3.
Cast: 38,000-45,000 TS; 21,000-26,000 YS; 22-28 El; 20-28 RA; 70-80 Brin. For gears, worms, and hydraulic castings. Tough.

HAMMOND ALLOY NO. 7
Hammond Brass Works
Cu 87-88.5, N 7.75-8.5, Pb 0-0.3, Zn 3-5, Ni 0.4-0.6.
Cast: 40,000-50,000 TS; 18,000-23,000 YS; 25-40 El; 32-37 RA; 65-75 Brin. For gears, bolts, pipe fittings, and valves. Steam metal.

HAMMOND ALLOY NO. 8
Hammond Brass Works
Cu 88, Sn 10, Pb 2.
Cast: 36,000 TS; 19,000 YS; 22 El; 65 Brin. For gears and bearings. Free-cutting.

HAMMOND ALLOY NO. 9
Hammond Brass Works
Cu 75, Sn 8, Pb 15, Zn 2.
Cast: 26,000 TS; 15,000 YS; 10 El; 50 Brin. For bearings and bushings. Tough.

HAMMOND ALLOY NO. A-21
Hammond Brass Works
Cu 74.5, Fe 8, Al 13, Mn 0.5, Ni 4.
Cast: 60,000 TS; 60,000 YS; 0 El; 295 Brin. For guides, dies, and pistons. Al-bronze.

HAMMOND ALLOY NO. A-22
Hammond Brass Works
Cu 85, Fe 3, Al 12, Mn 0.25.
Cast: 68,000 TS; 45,000 YS; 1 El; 225 Brin. For castings. Al-bronze.

HAMMOND ALLOY NO. A-23
Hammond Brass Works
Cu 87, Fe 4, Al 9, Mn 0.1, Ni 0.1.
Cast: 78,000 TS; 27,000 YS; 32 El; 100 Brin. For castings. Al-bronze.

HAMMOND ALLOY NO. A-24
Hammond Brass Works
Cu 89, Fe 1, Al 10, Mn 0.1, Ni 0.25.
Cast: 80,000 TS; 28,000 YS; 25 El; 115 Brin. For castings. Al-bronze.

HAMMOND ALLOY NO. A-25
Hammond Brass Works
Cu 86, Fe 4, Al 10, Mn 0.1, Ni 0.25.
Cast: 87,000 TS; 28,000 YS; 18 El; 125 Brin. For gears and bushings. Al-bronze.

HAMMOND ALLOY NO. A-26
Hammond Brass Works
Cu 83.5, Fe 4.5, Al 10.75, Mn 1.25.
Cast: 90,000 TS; 35,000 YS; 10 El; 165 Brin. For gears. Al-bronze.

HAMMOND ALLOY NO. A-27
Hammond Brass Works
Cu 79, Fe 5, Al 11, Mn 0.1, Ni 5.
Cast: 105,000 TS; 50,000 YS; 5 El; 175 Brin. For castings. Al-bronze.

HAMMOND ALLOY NO. M-16
Hammond Brass Works
Cu 55-60, Zn 38-42, Sn 0-1.5, Mn 0-3.5.
Cast: 60,000 TS; 28,000 YS; 30 El; 90 Brin. For valve stems, and hubs. Free machining.

HAMMOND ALLOY NO. M-17
Hammond Brass Works
Zn, Fe, Mn, bal Cu.
Cast: 70,000 TS; 30,000 YS; 2 El; 125 Brin. For castings. Corrosion resistant.

HAMMOND ALLOY NO. M-18
Hammond Brass Works
Zn, Fe, Mn, bal Cu.
Cast: 80,000 TS; 40,000 YS; 20 El; 150 Brin. For castings. Corrosion resistant.

HAMMOND ALLOY NO. M-19
Hammond Brass Works
Zn, Fe, Mn, bal Cu.
Cast: 90,000 TS; 45,000 YS; 15 El; 170 Brin. For castings. Corrosion resistant.

HAMMOND ALLOY NO. M-20
Hammond Brass Works
Zn, Fe, Mn, bal Cu.
Cast: 100,000 TS; 60,000 YS; 10 El; 235 Brin. For castings. High strength.

HAMMONIA METAL
German manufacture
Sn 65, Zn 32, Cu 3.3.
For ornamental parts.

HAMPDEN
Now CARPENTER HAMPDEN.

HANDLER BHS
Handler GmbH
C 0.6, Si 0.25-0.5, Mn 0.3-0.8, bal Fe.
Heat treated: 160,000 TS; 113,000 YS; 12 El; 40 RA; 325 Brin. For rails, punches, gears, bolts; water hardened.

HANDLER BSD
Handler GmbH
C 0.9, Cr 1.9, V 0.1, bal Fe.
For bearings, cutters, liners, sleeves; water hardened, wear resistant.

HANDLER BSD SPEZIAL
Handler GmbH
C 1.45, Cr 1.4, Mn 0.0, bal Fe.
For bearings, cutters, liners, sleeves; water hardened, wear resistant.

HANDLER H18
J.C. Soding & Halbach
C 0.95, Cr 4, W, V, Mo, bal Fe.
For lathe and planer tools, reamers, broaches; high-speed steel.

HANDLER H20
Handler GmbH
C 0.82, Cr 4.1, Mo 0.85, V 1.6, W 8.7, bal Fe.
For lathe and planer tools, reamers, broaches; high-speed steel.

HANDLER HC EXTRA AH
Handler GmbH
C 0.85, Si 0-0.25, Mn 0-0.25, bal Fe.
Heat treated: 190,000 TS, 145,000 YS; 10 El; 30 RA; 400 Brin. For springs, cutters, taps, hobs, reamers; Type W1; water hardened.

HANDLER HC EXTRA HART
Handler GmbH
C 1.3, Mn 0-0.25, Si 0-0.25, bal Fe.
For engraving tools, cutters, broaches; Type W1; water hardened.

HANDLER HC EXTRA MH
Handler GmbH
C 1.1, Si 0-0.25, Mn 0-0.25, bal Fe.
Annealed: 110,000 TS; 55,000 YS; 20 El; 40 RA; 210 Brin. For springs, cutters, taps, broaches; Type W1; water hardened.

HANDLER HC EXTRA ZH
Handler GmbH
C 1, Si 0-0.25, Mn 0-0.25, bal Fe.
Annealed: 100,000 TS; 53,000 YS; 21 El; 42 RA; 200 Brin. For springs, taps, hobs, drills, reamers; Type W1; water hardened.

HANDLER HC PRIMA H
Handler GmbH
C 1.3, Mn 0-0.25, Si 0-0.25, bal Fe.
For engraving tools, cutters, broaches; Type W1; water hardened.

HANDLER HC PRIMA MH
Handler GmbH
C 1.15, Mn 0-0.25, Si 0-0.25, bal Fe.
Annealed: 110,000 TS; 55,000 YS; 18 El; 40 RA; 210 Brin. For springs, taps, reamers, broaches; Type W1; water hardened.

HANDLER HC PRIMA WEICH
Handler GmbH
C 0.7, Si 0-0.25, Mn 0-0.25, bal Fe.
Heat treated: 175,000 TS; 130,000 YS; 12 El; 37 RA; 355 Brin. For rails, axes, punches, crimpers; Type W1; water hardened.

HANDLER HC PRIMA ZAH
Handler GmbH
C 0.85, Si 0-0.25, Mn 0-0.25, bal Fe.
Heat treated: 190,000 TS; 145,000 YS; 10 El; 30 RA; 400 Brin. For springs, tools, cutters, taps, drills, hobs; Type W1; water hardened.

HANDLER HC PRIMA ZAHHART
Handler GmbH
C 1, Mn 0-0.25, Si 0-0.25, bal Fe.
Annealed: 100,00 TS; 53,000 YS; 21 El; 42 RA; 200 Brin. For springs, tools, cutters, taps, drills, hobs; Type W1; water hardened.

HANDLER IIA GUSSTAHL
Handler GmbH
C 0.35, Si 0.4, Mn 0.6, bal Fe.
Hot rolled: 85,000 TS; 54,000 YS; 30 El; 53 RA; 185 Brin. For gears, bolts, machine tool parts; water hardened; SAE 1035.

HANDLER MV2
Handler GmbH
C 0.42, V 0.1, Mn 1.75, Si 0.25, bal Fe.
For punches, gears, upsetters, header dies; oil hardened, tough.

HANDLER NI 10
Handler GmbH
C 0.33, Al 1.1, Cr 1.7, Ni 1, Mn 0.5, bal Fe.
For oil refinery equipment; heat and creep resistant.

HANDLER NI 19
Handler GmbH
C 0.31, Cr 2.35, Mo 0.18, V 0.13, Mn 0.6, bal Fe.
For oil refinery equipment; heat and creep resistant.

HANDLER NI 3
Handler GmbH
C 0.27, Al 1.1, Cr 1.4, Mn 0.6, bal Fe.
For oil refinery equipment; heat and creep resistant.

HANDLER NI 4
Handler GmbH
C 0.34, Al 1.1, Cr 1.4, Mn 0.6, bal Fe.
For oil refinery equipment; heat and creep resistant.

HANDLER NI 7
Handler GmbH
C 0.32, Al 1.1, Cr 1.1, Mo 0.18, bal Fe.
For oil refinery equipment; heat and creep resistant.

HANDLER SF 29
Handler GmbH
C 0.85, W, Mo, bal Fe.
For upsetters, punches, dies; hot-work steel.

HANDLER SF 566
Handler GmbH
Stainless steel. C 0-0.12, Cr 18, Mo 2, Ni 10.5, Cb = 8 x C, bal Fe.
For welded acid resistant chemical plant equipment, Type 316Cb; stainless, austenitic.

HANDLER SF10
Handler GmbH
C 1, Cr 1.1, Mn 0.07, Si 0.25, bal Fe.
For bearings, cutters, liners, sleeves; water hardened, wear resistant.

HANDLER SF11
Handler GmbH
C 0.85, Si 0.2, Mn 0.3, Cr 0.3, bal Fe.
Heat treated: 190,000 TS; 145,000 YS; 10 El; 30 RA; 400 Brin. For cutters, bearings, liners, sleeves; water hardened, wear resistant.

HANDLER SF12
Handler GmbH
C 1, V 0.1, Mn 0.25, Si 0.2, bal Fe.
For bearings, dies, cutters, taps; water hardened, wear resistant.

HANDLER SF15
Handler GmbH
C 0.9, Cr, Si, bal Fe.
For bearings, dies, cutters, drills, taps; water hardened, wear resistant.

HANDLER SF16
Handler GmbH
C 1.05, Cr 1, W 1.15, Mn 0.9, bal Fe.
For drawing and forming dies, cutters; oil hardened, tough.

HANDLER SF18
Handler GmbH
C 0.35, Cr 1.05, V 0.18, W 1.85, Si 0.9, bal Fe.
For heading and forming dies, punches; oil hardened, tough.

HANDLER SF1951
Handler GmbH
C 0.85, V 2.5, Mo 0.85, W 12, bal Fe.
For cutters, forming and drawing dies; high-speed steel.

HANDLER SF2
Handler GmbH
C 0.7, Si 0-0.25, Mn 0-0.25, bal Fe.
Heat treated: 174,000 TS; 128,000 YS; 12 El; 37 RA; 350 Brin. For springs, rails, punches; Type W1; water-hardened tool steel.

HANDLER SF21
Handler GmbH
C 1.42, W, V, bal Fe.
For bearings, cutters, liners, sleeves; oil or water hardened.

HANDLER SF22
Handler GmbH
C 0.3, Cr 2.65, V 0.35, W 8.5, bal Fe.
For extrusion rams, dies, liners, punches; hot-work steel, oil hardened.

HANDLER SF25
Handler GmbH
C 0.74, Cr 4.1, V 1.1, W 18.5, bal Fe.
For lathe and planer tools, drills, taps, reamers; high-speed steel.

HANDLER SF27
Handler GmbH
C 0.79, Co 4.75, Cr 4.3, Mo 0.75, V 1.5, W 18, bal Fe.
For lathe and planer tools, drills, taps, reamers; high-speed steel.

HANDLER SF28
Handler GmbH
C 0.76, Co 10, Cr 4.2, Mo 0.8, V 1.8, W 18, bal Fe.
For lathe and planer tools, drills; high-speed steel.

HANDLER SF3
Handler GmbH
C 0.85, Si 0-0.25, Mn 0-0.25, bal Fe.
Heat treated: 190,000 TS; 145,000 YS; 10 El; 30 RA; 400 Brin. For springs, taps, reamers, drills, cutters; Type W1; water-hardened tool steel.

HANDLER SF4
Handler GmbH
C 1, Si 0-0.25, Mn 0-0.25, bal Fe.
Annealed: 100,000 TS; 53,000 YS; 21 El; 42 RA; 200 Brin. For springs, taps, reamers; Type W1; water-hardened tool steel.

HANDLER SF41
Handler GmbH
C 0.6, Si 0.25-0.5, Mn 0.3-0.8, bal Fe.
Heat treated: 160,000 TS; 113,000 YS; 12 El; 40 RA; 320 Brin. For wheels die blocks, rails, girders, springs; water hardened.

HANDLER SF43
Handler GmbH
C 0.7, Si 1.7, Mn 0.7, bal Fe.
For springs, punches, pneumatic tools; oil hardened, tough.

HANDLER SF46
Handler GmbH
C 0.5, Mn 0.95, Cr 1.05, V 0.1, bal Fe.
For gears, bolts, crankshafts, springs; oil hardened, shock resistant.

HANDLER SF5
Handler GmbH
C 1.1, Si 0-0.25, Mn 0-0.25, bal Fe.
Annealed: 100,000 TS; 53,000 YS; 21 El; 42 RA; 200 Brin. For springs, taps, reamers, cutters; Type W1; water hardened.

HANDLER SF502
Handler GmbH
Stainless steel. C 0-0.12, Si 0.4, Cr 13, bal Fe.
Annealed: 75,000 TS; 40,000 YS; 35 El; 70 RA; 155 Brin. For turbine blades, gages, valves, cutlery; Type 410; stainless.

HANDLER SF511
Handler GmbH
Stainless steel. C 0.2, Si 0.4, Cr 13, bal Fe.
Annealed: 95,000 TS; 50,000 YS; 25 El; 55 RA; 196 Brin. For cutlery, valves, surgical instruments; Type 420; stainless.

HANDLER SF512
Handler GmbH
Stainless steel. C 0.4, Si 0.4, Cr 13, bal Fe.
Annealed: 100,000 TS; 55,000 YS; 22 El; 52 RA; 200 Brin. For cutlery, valves, surgical and dental instruments; Type 420; stainless.

HANDLER SF522
Handler GmbH
C 0.2, Si 1.2, Cr 25, Ni 4, bal Fe.
Aged: 115,000 TS; 80,000 YS; 15 El. Cast: 70,000 TS; 65,000 YS; 12 El; 190 Brin. For furnace parts, grate bars, salt pots; heat and corrosion resistant.

HANDLER SF527
Handler GmbH
C 0.12, Cr 16.5, Mo 0.25, S 0.2, bal Fe.
Annealed: 80,000 TS; 50,000 YS; 25 El; 50 RA; 150 Brin. For screw machine products, soot blowers; Type 430F; corrosion resistant.

HANDLER SF553
Handler GmbH
Stainless steel. C 0-0.07, Cr 18, Ni 9.5, bal Fe.
Annealed: 85,000 TS; 35,000 YS; 60 El; 70 RA; 150 Brin. For chemical plant equipment, tanks, agitators; Type 304; stainless, austenitic.

HANDLER SF555
Handler GmbH
Stainless steel. C 0-0.15, Cr 18, Ni 8.5, bal Fe.
Annealed: 80,000 TS; 35,000 YS; 55 El; 75 RA; 150 Brin. For chemical plant equipment, tanks, mixers, agitators; Type 302; stainless, austenitic.

HANDLER SF559
Handler GmbH
C 0-0.1, Cr 12.5, Ni 12, bal Fe.
For valves, furnace parts; corrosion resistant.

HANDLER SF564
Handler GmbH
Stainless steel. C 0-0.07, Cr 18, Mo 2, Ni 10.5, bal Fe.
Annealed: 85,000 TS; 35,000 YS; 50 El; 65 RA; 160 Brin. For acid resistant chemical plant equipment, tanks; Type 316; stainless, austenitic.

HANDLER SF62
Handler GmbH
C 2.1, Cr 11.5, W 0.7, bal Fe.
For blanking and forming dies, punches; oil hardened, non-deforming.

HANDLER SF85
Handler GmbH
C 0.45, Ni, Cr, W, bal Fe.
For forging and heading dies; oil hardened, tough.

HANDLER SS212
Handler GmbH
C 2.1, Cr 11.5, bal Fe.
For blanking and forming dies, punches; oil hardened, non-deforming.

HANDLER SS308
Handler GmbH
C 1.25, Si 1.15, Mn 0.7, Cr 1.2, bal Fe.
For blanking and forming dies, headers; oil hardened, tough.

HANDLER SS511
Handler GmbH
C 0.56, Ni, Cr, Mo, bal Fe.
For forging and heading dies, shears, punches; oil hardened, shock resistant.

HANDLER SS513
Handler GmbH
C 0.45, Si, Cr, V, bal Fe.
For springs, gears, bolts, fasteners, crankshafts; oil hardened, shock resistant.

HANDLER SS514
Handler GmbH
C 0.3, Cr 2.35, V 0.6, W 4.25, bal Fe.
For shear blades, pneumatic tools, chisels; oil hardened, shock resistant.

HANDLER ST MN 2
Handler GmbH
C 0.9, Mn 1.9, V 0.1, bal Fe.
For punches, shears, blanking and forming dies; oil hardened, non-deforming.

HANDY 85 AG-15 MN
Now PERMABRAZE 130.

HANDY 94 CU-6 AG ALLOY
Handy & Harmon
Cu 94, Ag 6.
Heat treated: 140,000-165,000 TS; For spring parts, watch and instrument parts; 70% conductivity, high strength. *Obsolete*

HANDY AT SPECIAL
Now BRAZE 202.

HANDY ATT
Now BRAZE 200.

HANDY BT
Now BRAZE 720.

HANDY COIN SILVER
Handy & Harmon
Precious metal. Ag 90, Cu 10.
For coins; corrosion resistant.

HANDY DE
Now BRAZE 450.

HANDY DT
Now BRAZE 400.

HANDY E.T.C.
Handy & Harmon
Ag 50, Cu 16.5, Zn 10.5, Cd 18.
For silver solder, brazing; superseded by "Easy-Flo". *Obsolete*

HANDY EXT
Now BRAZE 501.

HANDY HI-TEMP 095
Handy & Harmon
Cu 52.5, Ni 9.5, B 0-0.1, Fe 0-0.1, bal Mn.
MP 1615 F; flow P 1700 F. For brazing copper, steel, stainless and nickel base alloys; for combined braze and heat treat. Ductile, fair corrosion resistance. AMS 4764. *Obsolete*

HANDY HI-TEMP 710
Handy & Harmon
Ni 70.5, Cr 19, Si 10.1, C 0-0.1, Co 0-0.1.
MP 1975 F; flow P 2075 F. For brazing stainless steel and high temperature alloys. Good oxidation resistance at high temperature-including against NaK in nuclear reactors. AMS 4782; AWS BNi-5. *Obsolete*

HANDY HI-TEMP 720
Handy & Harmon
Cr 14, Si 4, B 3.35, Fe 4.5, C 0.75, 72% Ni (plus Co).
MP 1790 F; flow P 1900 F. High temperature brazing alloy for brazing high nickel, cobalt and other high temperature alloys and stainless steels. Good corrosion resistance and high temperature strength. AMS 4775A and AWS B Ni-1. *Obsolete*

HANDY HI-TEMP 721 (LOW CARBON)
Handy & Harmon
Cr 14, Si 4, B 3.35, Fe 4.5, C 0-0.6, 72% Ni (plus Co).
MP 1780 F; flow P 1950 F. High temperature brazing alloy for brazing high nickel, cobalt and other high temperature alloys and stainless steels. Good corrosion resistance and high temperature strength. AMS 4776. *Obsolete*

HANDY HI-TEMP 820
Handy & Harmon
Cr 7, Si 4.5, B 3.1, Fe 3, C 0 0.15, 82% Ni (plus Co).
MP 1780 F; flow P 1830 F. For brazing stainless steel and high temperature alloys; flows into narrow joints. Good corrosion resistance and high temperature strength. AMS 4777; AWS B Ni2. *Obsolete*

HANDY HI-TEMP 90
Handy & Harmon
Cu 67.5, Ni 9, Mn 23.5, B 0-0.1, Fe 0-0.1.
MP 1670 F; flow P 1710 F. For brazing copper, steel, stainless and nickel base alloys; for combined braze and heat treat. Ductile, fair corrosion resistance. *Obsolete*

HANDY HI-TEMP 910
Handy & Harmon
Si 4.5, B 3.1, Fe 1.5, C 0-0.06, 91% Ni (plus Co).
MP 1800 F; flow P 1900 F. For brazing stainless steels and high temperature alloys; flows into narrow joints. Good corrosion resistance and high temperature strength. AMS 4778A; AWS BNi-3. *Obsolete*

HANDY HI-TEMP 930
Handy & Harmon
Si 3.5, B 1.6, Fe 1.5, C 0-0.06, 93% Ni (plus Co).
MP 1800 F; flow P 1950 F. For brazing stainless steels and high temperature alloys; will bridge wide and irregular joints. Good oxidation and corrosion resistance. AMS 4779; AWS BNi-4. *Obsolete*

HANDY IT
Now BRAZE 800.

HANDY NE
Now BRAZE 250.

HANDY NT
Now BRAZE 655.

HANDY RE-MN
Now BRAZE 655.

HANDY RE-MN6
Now BRAZE 655.

HANDY RSNI
Handy & Harmon
Ag 60-63, Cu 28-28.5, Sn 6-10.
For brazing alloy for Type 430 stainless steel; immune to crevice corrosion; MP 1245-1450 F. *Obsolete*

HANDY RSNI
Handy & Harmon
Ag 63, Cu 28.5, Sn 6, Ni 2.5.
For silver brazing of 400 series stainless steels; MP 1325-1475 F; overcomes crevice corrosion. *Obsolete*

HANDY RT
Now BRAZE 600.

HANDY RT-SN
Now BRAZE 603.

HANDY SILVER SOLDER
Now BRAZE 650.

HANDY SILVER SOLDER HARD
Now BRAZE 750.

HANDY SILVER SOLDER MEDIUM
Now BRAZE 700.

HANDY SN NO. 7
Now BRAZE 071.

HANDY SS
Now BRAZE 403.

HANDY SS-5
Now BRAZE 404.

HANDY TE SPECIAL
Now BRAZE 051.

HANDY TL
Now BRAZE 090.

HANFORD 20
Hanford Foundry Co.
C 0-0.07, Ni 29, Cr 20, Mo 2-3, Cu 4, bal Fe.
Cast: 65,000-75,000 TS; 28,000-38,000 YS; 35-50 El; 40-50 RA; 120-150 Brin. For chemical plant equipment, mixers; resists mixed acids, austenitic.

HANOVER WHITE METAL
American manufacture
Sn 86.8, Sb 7.6, Cu 5.6.
For bearings; anti-friction.

HANSA ESPECIAL K10
Vereinigte Edelstahlwerke
C 0.7, W 18, Co 9, Mo 1, Cr 4, V 2, bal Fe.
For tools, dies, cutters, reamers, hobs; high speed steel.
Obsolete

HANSA ESPECIAL K5
Vereinigte Edelstahlwerke
C 0.7, W 18, Co 4, Cr 4, V 1.3, bal Fe.
For tools, dies, cutters, drills, taps; high speed steel.
Obsolete

HANSA ESPECIAL T17 EXTRA
Vereinigte Edelstahlwerke
C 0.7, W 19, Cr 4, V 2, bal Fe.
For lathe and planer tools, reamers, drills, broaches, hobs; high speed steel, oil hardened. *Obsolete*

HANSA SPEZIAL
Vereinigte Edelstahlwerke
C 0.74, Cr 4.1, V 1.1, W 18.5, bal Fe.
For lathe and planer tools, reamers, broaches, drills; high speed steel. *Obsolete*

HANSA SPEZIAL 325D
Vereinigte Edelstahlwerke
C 0.95, Cr 4.1, W, Mo, bal Fe.
For lathe and planer tools, form cutters, drills, taps, hobs; high speed steel. *Obsolete*

HANSA SPEZIAL K3
Vereinigte Edelstahlwerke
C 0.86, Co 2.8, Cr 4.3, Mo 0.85, V 2.1, W 12, bal Fe.
For lathe and planer tools, reamers, broaches; high speed steel. *Obsolete*

HANSA SPEZIAL K6
Vereinigte Edelstahlwerke
C 0.79, Co 4.75, Cr 4.3, Mo 0.75, V 1.55, W 18, bal Fe.
For drills, taps, hobs, reamers, lathe and planer tools; high speed steel. *Obsolete*

HANSA SPEZIAL T17
Vereinigte Edelstahlwerke
C 0.82, Cr 4.1, Mo 0.85, V 1.6, W 8.7, bal Fe.
For lathe and planer tools, reamers, broaches, drills; high speed steel. *Obsolete*

HANSA SPEZIAL T50
Vereinigte Edelstahlwerke
C 1.3, Cr 4.3, Mo 0.85, V 3.8, W 12, bal Fe.
For engravers' tools, forming cutters, blanking dies; high speed steel. *Obsolete*

HARBRONZE
Arthur Harris & Co.
Sn, bal Cu.
For bearings.

HARD
Electro-Steel Co.
C 1, CO 8.27544e+008-0, Fo 0.750007-0, CO 8.90405e+011-3.37774e+009, ba 0.

HARD
Engelhard Corp.
Ag 75, Cu 22, Zn 3.
Silversmithing solder for sterling; 1365-1450°F MP. *Obsolete*

HARD BABBITT
English manufacture
Sn 83, Cu 8.4, Sb 8.3.
For bearings; anti-friction.

HARD BEARING
English manufacture
Pb 98.7, Mg 1.3.
For bearings.

HARD BRONZE-1
General Motors Corp./Central Foundry
Cu 88, Pb 2, Sn 7, Zn 3.
30,000 psi TS; 12 El. For gears, bushings; tough.

HARD BRONZE-2
General Motors Corp./Central Foundry
Cu 88, Zn 2, Sn 10.
For gears, bushings; GM No. 4048 M.

HARD DEVIL
Champion Rivet Co.
C 0.85-1, bal Fe.
250-400 Brin. For hard facing welding electrodes.

HARD FACING R 459
Abex Corp.
C, Cr, Mn, Mo, bal Fe.
500-600 Brin. For hard facing welding rods; wear and heat resistant.

HARD HEAD
English manufacture
Sn 90, Sb 8, Cu 2.
For bearings; anti-friction.

HARD KOTE "H"
Bonney-Floyd Co.
W 24-30, Cr 16-23, 1.1% C min, Ni, bal Fe.
For heat and corrosion resistant parts, tools, dies; heat and corrosion resistant. *Obsolete*

HARD KOTE "T"
Bonney-Floyd Co.
Ni 24-30, Cr 16-23, W 16-23, 1.1% C min, bal Fe.
For heat and corrosion resistant parts, tools, dies; heat and corrosion resistant. *Obsolete*

HARD PHOSPHOR BRONZE
American manufacture
Sn 7, P 0.2, bal Cu.
For springs, bearings, electrical parts, gears; strong, corrosion resistant.

HARD SILVER SOLDER
English manufacture
Ag 80, Cu 13, Zn 6.8.
For solder.

HARD SOLDER
English manufacture
Cu 50-57, Zn 43-50.
For solder.

HARD SURFACE WELDING A-6120
U.S. Steel Corp.
C 0.17-0.22, Mn 0.3-0.6, P 0-0.04, S 0-0.05, Si 0.15-0.3, Cr 0.8-1.1, 0.15% V min, bal Fe.
Deposited metal: 200-350 Brin. For welding wire (arc or acetylene); hard surfacing for resistance to abrasive wear. *Obsolete*

HARD YELLOW SOLDER
English manufacture
Cu 58, Zn 43, Sn 1.3, Pb 0.3.
For solder.

HARD ZINC (HARTZINK)
German manufacture
Zn 92, Fe 5.3, Pb 2.4, Cu 0.1.
For wash boards.

HARD-ROD NO. 100
Marquette Corp.
C 0.9, Mn 0.6, Mo 0.3, Cr 1, bal Fe.
For hard facing rod; for carbon steels.

HARD-ROD NO. 450
Marquette Corp.
C 0.5-0.6, Cr 6-8, Si 0.6, bal Fe.
Welded: 500 Brin. For hard facing electrode; for high impact and abrasion.

HARD-ROD NO. 550
Marquette Corp.
C 0.6. Cr 8, Mn 0.4, Si 0.6, bal Fe.
Welded: 550 Brin. For hard facing electrode; for high abrasion and impact.

HARD-TRAK
Joseph T. Ryerson & Son Inc.
C 0.31, Mn 1.4, Si 0.24, Cr 0.5, Mo 0.11, B 0.0004, bal Fe.
Heat treated: 360-400 Brin. For wear bars and rails.

HARDALLOY 118
Teledyne McKay
C 0.8, Mn 16.5, Si 0.5, Cr 5, Ni 0.3, bal Fe.
Covered surfacing electrodes. 127,000 psi TS; 78,000 psi YS; 50 El; 18-22 Rock C (as deposited). Work hardening austenitic manganese steel alloy. For crusher rolls, impact hammers, impactor bars, gyratory cones, bucket teeth, and railroad frogs and crossings.

HARDALLOY 119
Teledyne McKay
C 1, Mn 19.5, Si 0.5, Cr 5, bal Fe.
Covered surfacing electrode. 135,000 psi TS; 97,000 psi YS; 31 El; 21 Rock C (as deposited). Work hardening austenitic manganese steel alloy. For shovel and dredge parts, shovel teeth, wear plates, frogs, crossovers, and rail ends.

HARDALLOY 120
Teledyne McKay
C 0.07, Mn 1.3, Si 0.45, Cr 23.5, Ni 9.7, bal Fe.
Covered surfacing electrode. 97,000 psi TS; 71,000 psi YS; 37 El; 17-21 Rock C (as deposited). Stainless steel alloy; 20 FN ferrite content.

HARDALLOY 140
Teledyne McKay
C 3, Mn 0.4, Si 2, Cr 30, Mo 0.7, bal Fe.
Covered surfacing electrodes. 50-58 Rock C. For final overlay on crusher rolls and hammermill hammers, bucket teeth, coke pusher shoes, augers, and muller tires.

HARDALLOY 32
Teledyne McKay
C 0.18, Mn 1.3, Si 0.6, Cr 0.7, Mo 0.3, bal Fe.
Covered surfacing electrodes. 17-30 Rock C. Heat treatable; for build-up of carbon and low alloy steels. For steel mill wobblers, coupling boxes, tractor rollers, mine car wheels, and gear teeth.

HARDALLOY 40 TIC
Teledyne McKay
C 3, Mn 1.1, Si 0.8, Cr 8.2, Ti 1.5, bal Fe.
Covered surfacing electrode. 39-50 Rock C (as deposited). High alloy cast iron. For final overlay on crusher rolls and hammermill hammers, bucket teeth, coke pusher shoes, augers, and muller tires.

HARDALLOY 42
Teledyne McKay
C 0.17, Mn 1.8, Si 0.5, Cr 2, Mo 0.7, V 0.3, bal Fe.
Covered surfacing electrode. 29-45 Rock C (as deposited). For tractor rollers, trailer idlers, crane wheels, carbon steel rolls, and steel shafts.

HARDALLOY 48
Now MCKAY HARDALLOY 48.

HARDALLOY 48
Teledyne McKay
C 1.8, Mn 1.2, Si 1.5, Cr 30, Ni 3, Mo 1.5, bal Fe.
Covered surfacing electrode. 35-40 Rock C (as deposited). For rolling mill guides, pulleys, ingot tongs, and furnace skid pipes.

HARDALLOY 55
Teledyne McKay
C 4.6, Mn 1, Si 1.1, Cr 27, Mo 3.5, bal Fe.
Covered surfacing electrodes. 43-56 Rock C. Extra-high carbon-chromium alloy steel. For final overlay on crusher rolls and hammermill hammers, bucket teeth, coke pusher shoes, augers, and muller tires.

HARDALLOY 55 TIC
Teledyne McKay
C 6, Mn 2.8, Si 1, Cr 13, Ti 5.5, bal Fe.
Covered surfacing electrode. 41-56 Rock C (as deposited). High chromium, high titanium cast iron for overlay of surfaces subjected to severe abrasion; bucket teeth, coke pusher shoes, augers, and muller tires.

HARDALLOY 58
Teledyne McKay
C 0.6, Mn 1.2, Si 0.7, Cr 5.5, Mo 0.5, bal Fe.
Covered surfacing electrodes. 45-60 Rock C. Martensitic alloy for hard, tough overlays on carbon and low alloy steel parts. For machine components, tools, and sliding metal parts.

HARDALLOY 61
Teledyne McKay
C 0.8, Mn 0.5, Si 0.7, Cr 4, W 1.1, Mo 8, V 1.1, bal Fe.
Covered surfacing electrode. 53-63 Rock C. Martensitic; similar to high speed tool steel deposit. For shear blades, trimming dies, punching dies, and sliding metal parts.

HARDALLOY CHROME-MANG
Teledyne McKay
Stainless steel. C 0.4, Mn 14.5, Si 0.6, Cr 14, Ni 1, Mo 1.5, V 0.55, bal Fe.
Covered surfacing electrodes. 130,000 psi TS; 94,000 psi YS; 40 El; 18-22 Rock C (as deposited). Work hardening chromium-manganese austenitic stainless steel alloy. For crusher rolls, impact hammers, impactor bars, and railroad frogs, and crossings.

HARDALLOY M-932
Teledyne McKay
C 0.13, Mn 0.8, Si 0.4, Cr 2.2, Mo 1, bal Fe.
Covered surfacing electrode. 145,000 psi TS; 38-39 Rock C (as deposited). For railroad track repair applications.

HARDENITE
Hardenite Steel Co. Ltd.
C 0.9, Mn 1.2, bal Fe.
For tools, dies. Oil hardened.

HARDENTOUGH 1
Westinghouse Electric Corp.
C 0.85, bal Fe.
300-425 Brin. For welding electrodes; hard facing wear resistant. *Obsolete*

HARDENTOUGH 2
Westinghouse Electric Corp.
C, Mn, Ni, bal Fe.
347 Brin. For welding electrodes; hard facing wear resistant. *Obsolete*

HARDENTOUGH 250
Westinghouse Electric Corp.
C, alloy, bal Fe.
Weld: 230-300 Brin. For arc welding electrode; hard facing. *Obsolete*

HARDENTOUGH 350
Westinghouse Electric Corp.
C, alloy, bal Fe.
Weld: 500 Brin. For arc welding electrode; hard facing. *Obsolete*

HARDENTOUGH 4
Westinghouse Electric Corp.
C, alloy, bal Fe.
500-600 Brin. For hard facing welding electrodes; wear resistant. *Obsolete*

HARDENTOUGH 450
Westinghouse Electric Corp.
C, alloy, bal Fe.
Welded: 600 Brin. Welding electrodes; hard surfacing. *Obsolete*

HARDENTOUGH 5
Westinghouse Electric Corp.
C, alloy, bal Fe.
600-675 Brin. For hard facing welding electrodes; wear resistant. *Obsolete*

HARDENTOUGH 550
Westinghouse Electric Corp.
C, alloy, bal Fe.
Welded: 400-600 Brin. For arc welding electrodes; hard surfacing. *Obsolete*

HARDENTOUGH 6
Westinghouse Electric Corp.
C, alloy, bal Fe.
700-800 Brin. For hard facing welding electrodes; wear resistant. *Obsolete*

HARDENTOUGH FORMWEAR
Westinghouse Electric Corp.
C, alloy, bal Fe.
Welded: 500-650 Brin. For arc welding electrodes; hard surfacing. *Obsolete*

HARDENTOUGH HA
Westinghouse Electric Corp.
C, alloy, bal Fe.
Welded: 575-650 Brin. For arc welding eletrodes; hard surfacing. *Obsolete*

HARDENTOUGH MOLYMANG
Westinghouse Electric Corp.
Mn 13, C, bal Fe.
Weld hardened: 475 Brin. For arc welding electrodes; abrasion resistant, work hardens. *Obsolete*

HARDENTOUGH WH
Westinghouse Electric Corp.
C, alloy, bal Fe.
Cast: 270-380 Brin. For welding electrodes; hard surfacing. *Obsolete*

HARDFACE 200
Champion Rivet Co.
C, alloy, bal Fe.
Welded: 250 Brin. For hard facing electrodes. *Obsolete*

HARDFACE 400
Champion Rivet Co.
C, alloy, bal Fe.
Welded: 400 Brin. For hard facing electrodes. *Obsolete*

HARDFACE 600
Champion Rivet Co.
C, alloy, bal Fe.
Welded: 600 Brin. For welding electrodes; hard facing. *Obsolete*

HARDFACER NO. 11
Hobart Bros. Co.
C, Cr, Ni, bal Fe.
Welded: 520 Brin. For welding electrodes, hard facing; arc, resists abrasion. *Obsolete*

HARDFACER NO. 112
Hobart Bros. Co.
C, Cr, Ni, Mo, Ti, bal Fe.
Welded: 570 Brin. For welding electrodes, hard facing; arc, resists abrasion. *Obsolete*

HARDFLEX-11
Sandvik Steel Co.
C 0.5-0.6, Mn 0.6-0.9, bal Fe.
Heat treated: 128,000-190,000 TS; 107,000-142,000 YS; 10-14 El; 350 diamond pyramid hardness; 36 Rock C. For office and business machine components, electric razors, sewing machines, appliances. AISI-C1055. Prehardened strip steel, austempered.

HARDFLEX-13M
Sandvik Steel Co.
Mn 0.7-0.8, bal Fe.
Hardened: 190,000-228,000 TS; 149,000-200,000 YS; 6-9 El. For office and business machine components, electric razors, sewing machines, appliances. Prehardened strip steel, austempered. AISI-C1074.

HARDITE A
Hardite Metals Inc.
Ni 55, Cr 13, Si 1.5-2, Mn 2, bal Fe.
At 70°F: 103,000 TS. At 1800°F: 43,000 TS. For carburizing boxes, heating elements, lead pots, retorts; heat resistant.

HARDITE B
Hardite Metals Inc.
Ni 33-35, Cr 12-14, Si 1.5-2, Mn 2, bal Fe.
For carburizing boxes, heating elements, lead pots, retorts; heat resistant.

HARDITE C
Hardite Metals Inc.
Ni 93, Si 5, Mn 2.
For heating elements, retorts; high heat resistance.

HARDITE S
Hardite Metals Inc.
Ni 88, Si 4.5, bal Fe.
For carburizing boxes, heating elements, lead pots, retorts; heat resistant.

HARDITE X
Hardite Metals Inc.
Ni 82-86, Cr 10.13, Si 2-4, Mn 2.
For heating elements, retorts; heat resistant.

HARDITE Z
Hardite Metals Inc.
Stainless steel. Ni 6-12, Cr 15-25, Si 0.4, Mn 1-3, C, bal Fe.
For heat treating boxes, furnace parts; stainless.

HARDNAIR
Atlantic Steel Corp.
C 1, Cr 5, Mo 1, Mn 0.4, V 0.4, bal Fe.
Annealed: 103,000 TS; 52,000 YS; 26 El; 18 Rock C.
Hardened: 178,000-253,000 TS; 146,000-200,000 YS; 3-12 El; 7-32 RA; 41-53 Rock C. For blanking and trimming dies, cams, cutters, gages, lamination dies, shears. Type A2; air hardened tool steel, nondeforming.

HARDRITE
Universal Cyclops
C 1.1, W 1.7, Cr 0.6, V 0.25, bal Fe.
For light and heavy press dies, punches, taps, reamers, thread rolling dies, knives, cams, forming dies; non-shrinkable; AISI O7. Was Vulcan Hardrite.

HARDRITE
H.K. Porter Co.
C 1.1, W 1.7, Cr 0.6, V 0.25, bal Fe.
For light and heavy press dies, punches, taps, reamers, thread rolling dies, knives, cams, forming dies; non-shrinkable; AISI O7. Was Vulcan Hardrite.

HARDROCK 33
Industries Trading Co.
Co 7, W 18, Cr 4, V 1, C, bal Fe.
For tool bits; high speed steel.

HARDSTEEL

Black Drill Co. Inc.
Co 40, Cr 32, W 18, Ni 2, Fe 2, Trace B, trace Mn.
For drills, tool bits and wear parts.

HARDTEM

Creusot-Loire
C 0.55, Mn 0.75, Ni 0.55, Cr 1, Mo 0.45, V 0.05, bal Fe.
Oil hardening hot work tool steel; for hot forming dies.
AFNOR 55CNDV 4.

HARDTEM

Heppenstall Co.
Tool material. C 0.6, Mn 0.7, Cr 0.61, V 0.17, Mo 0.28, Ni, bal Fe.
Oil treated: 123,000 TS; 101,500 YS; 53 El; 23 RA; 262 Brin.
For die blocks; drop forged and pressed.

HARDWARE BRONZE

Chase Brass & Copper Co.
Cu 89, Zn 9, Pb 2.
Annealed: 37,000 TS; 12,000 YS; 45 El. Drawn (20%): 50,000 TS; 45,000 YS; 25 El. For hardware, bolts, screws; free-cutting. *Obsolete*

HARDWARE BRONZE 267

Anaconda Co.
Copper. Cu 85, Zn 13.25, Pb 1.75.
Soft: 40,000 TS; 15,000 YS; 45 El; 72 RA; 62 Brin. Hard: 50,000 TS; 43,000 YS; 20 El; 49 RA; 218 Brin. For bolts, screws, nuts, tie rods.

HARDWARE BRONZE-320

Anaconda Co.
Copper. Cu 85, Zn 13.25, Pb 1.75.
Hard: 52,000 TS; 43,000 YS; 20 El; 55 Rock B. Soft: 40,000 TS; 15,000 YS; 45 El; 5 Rock B. For screw machine products, hardware, architectural applications. Electrical conductivity: 36. Corrosion resistant.

HARDWELD

Lincoln Electric Co.
C 1, bal Fe.
For hard facing welding rod for manganese steel; hard, wear resistant. *Obsolete*

HARDWELD 100

Lincoln Electric Co.
C 1, Alloy, bal Fe.
For hard facing welding rod; shock resistant. *Obsolete*

HARDWELD 50

Lincoln Electric Co.
C 1, Alloy, bal Fe.
For hard facing welding rod; shielded arc, shock resistant. *Obsolete*

HARDY NICKEL IRON

Atlas Specialty Steels
C 0.08, Mn 0.2, Ni 2, bal Fe.
Annealed: 50,000 TS; 35,000 YS; 35 El; 75 RA; 112 Brin. For tinning rolls; sub-zero applications. *Obsolete*

HARDYNE

Charles Hardy & Co.
Oxides bound with thermoplastic resin.
For permanent magnets; pressed compound oxides.

HARGUS

Agawam Tool Co.
C 0.9, Mn 1.2, bal Fe.
For punching and blanking dies; non-deforming.

HARGUS

Ziv Steel & Wire Co.
C 0.9, Mn 1.2, Cr 0.5, W 0.5, bal Fe.
For taps, dies, punches, gauges, reamers; non-deforming, oil hardened.

HARLINGTON BRONZE

English manufacture
Cu 56, Zn 43, Sn 0.9, Fe 0.6.
For sheating, condenser tubes, bolts, nuts; high strength.

HARMONIA BRONZE

English manufacture
Cu 55.7-57, Zn 40-41.2, Pb 0.4-0.46, Fe 1.29-1.8, Sn 0-0.5, Al 0-0.86.
For fittings, hardware.

HARRIS 308 L STAINLESS WELD WIRE

J.W. Harris Co., Inc.
C 0-0.03, Cr 19.5-22, Si 0.3-0.65, Ni 9.5-11, Mn 1-2.5, bal Fe.
For MIG, TIG, manual and submerged arc welding AISI 304L, 308L, 321 and 347 stainless. AWS A5.9 ER308L.

HARRIS 308 STAINLESS WELD WIRE

J.W. Harris Co., Inc.
C 0-0.08, Cr 19.5-22, Ni 9-11, Mn 1-2.5, Si 0.3-0.65.
AWS A5.9 ER308.

HARRIS 309 STAINLESS WELD WIRE

J.W. Harris Co., Inc.
C 0-0.12, Cr 23-25, Si 0.3-0.65, Ni 13-14, Mn 1-2.5, bal others.
For MIG, TIG, manual and submerged arc welding AISI 309 stainless and stainless to mild steel. AWS A5.9 ER309.

HARRIS 310 STAINLESS WELD WIRE

J.W. Harris Co., Inc.
C 0.08-0.15, Cr 25-28, Ni 20-22.5, Mn 1-2.5, Si 0.3-0.65, bal others.
For MIG, TIG, manual and submerged arc welding AISI 310, 304, and some dissimilar stainless and carbon or alloy steels. AWS A5.9 ER 310.

HARRIS 312 STAINLESS WELD WIRE

J.W. Harris Co., Inc.
C 0.08-0.15, Cr 28-32, Ni 8-10.5, Mn 1-2.5, Si 0.3-0.65, bal others.
For MIG, TIG, manual and submerged arc welding stainless to mild steels and high strength steels. AWS A5.9 ER 312.

HARRIS 316 L STAINLESS WELD WIRE

J.W. Harris Co., Inc.
C 0-0.03, Cr 18-20, Ni 11-14, Mn 1-2.5, Si 0.3-0.65, Mo 2-3, bal others.
For MIG, TIG, manual and submerged arc welding AISI 316 L stainless. AWS A5.9 ER316L.

HARRIS 316 STAINLESS WELD WIRE

J.W. Harris Co., Inc.
C 0-0.08, Cr 18-20, Ni 11-14, Mn 1-2.5, Si 0.3-0.65, Mo 2-3, bal others.
For MIG, TIG, manual and submerged arc welding AISI 316 stainless steel. AWS A5.9 ER 316.

HARRIS 347 STAINLESS WELD WIRE

J.W. Harris Co., Inc.
C 0-0.08, Cr 19.5-21.5, Ni 9-11, Mn 1-2.5, Si 0.3-0.65, Cb + Ta, bal others.
For MIG, TIG, manual and submerged arc welding AISI 321, 347, 304 and 304 L stainless. AWS A5.9 ER 347.

HARRIS 404 ALUM ELECTRODE

J.W. Harris Co., Inc.
Aluminum electrode extruded coating, DC reverse polarity. For miscellaneous maintenance repair of aluminum equipment. AWS A5.2 E 4043.

HARRIS 410 STAINLESS WELD WIRE

J.W. Harris Co., Inc.
C 0.07-0.12, Cr 11.5-13.5, Ni 0-0.6, Mn 0-0.6, Si 0-0.5, Mo 0-0.75, bal others.
For MIG, TIG, manual and submerged arc welding of AISI 410, 403, 405, 414 and 416 stainless steels. AWS S.9 ER 410.

HARRIS 430 STAINLESS WELD WIRE

J.W. Harris Co., Inc.
C 0-0.01, Cr 15.5-17, Ni 0-0.6, Mn 0-0.6, Si 0-0.5, bal Fe.
For MIG, TIG, manual and submerged arc welding of AISI 430 stainless steel. AWS A5.9-62T; ASTM A371-62T. *Obsolete*

HARRIS ALUMINUM WELDING ROD

J.W. Harris Co., Inc.
Available in all aluminum alloy weldable grades. For gas welding and inert tungsten gas (TIG) welding of aluminum alloys. AWS A5-10.

HARRIS ALUMINUM WELDING WIRE

J.W. Harris Co., Inc.
Available in all aluminum alloy weldable grades. For inert gas metal arc (MIG) welding of aluminum alloys. *Obsolete*

HARRIS AMERICAN LOW FUMING BRONZE

J.W. Harris Co., Inc.
Cu 57.8, Zn 40.3, Sn 0.95, Fe 0.85, Si 0.1, Mn 0.03.
Bronze brazing rod; flux coated and bare. For joining and build-up on bronze, steel, cast iron, and miscellaneous alloys. AWS A5.27 RCuZn-C.

HARRIS CUT ROD

J.W. Harris Co., Inc.
Fast working electrode without use of oxygen. For cutting, beveling, gouging, piercing on most commercial metals. AC-DC; reverse polarity on DC.

HARRIS MAGNESIUM WELDING WIRE

J.W. Harris Co., Inc.
Meets AWS. ASTM and government specifications. Available as AZ92A, AZ61A, or E233A.

HARRIS NAVAL BRONZE

J.W. Harris Co., Inc.
Cu 57-61, Sn 0.25-1, Al 0-0.01, Pb 0-0.05, 0.05 others max, bal Zn.
Melting point: 1625°F approx. For oxyacetylene braze welding and build-up on cast iron and steel. AWS A5.27 RBCuZn-A.

HARRIS NICKEL BRONZE

J.W. Harris Co., Inc.
Cu 56-60, Sn 0.8-1.1, Ni 0.2-0.8, Fe 0.25-1.2, Mn 0.01-0.5, Si 0.04-0.15, Al 0-0.01, Pb 0-0.05, 0.50 others max, bal Zn.
Melting point: 1680°F approx. For oxyacetylene braze welding on malleable iron and steel. AWS A5.27 RBCuZn-D.

HARRIS NICKEL SILVER

J.W. Harris Co., Inc.
Cu 46-50, Ni 9-11, Si 0.04-0.25, P 0-0.25, Al 0-0.01, Pb 0-0.05, 0.50 others max, bal Zn.
Melting point: 1680°F approx. For oxyacetylene braze welding iron and steel for color match. AWS A5.27 RBCuZn-D.

HARRIS PHOSPHOR BRONZE A

J.W. Harris Co., Inc.
Sn 4-6, P 0.1-0.35, 93.5 Cu min, bal others.
For inert gas and carbon arc welding of phosphor bronze and copper, and repair of castings. AWS A5.7 ERCuSn-A.

HARRIS PHOSPHOR BRONZE C

J.W. Harris Co., Inc.
Sn 7-9, P 0.1, Al 0-0.01, Pb 0-0.02, 0.50 others max, bal Cu.
For joining and repair of ferrous and non-ferrous parts, and for build-up. AWS A5.6 ECuSn-C.

HARRIS SILICON BRONZE

J.W. Harris Co., Inc.
Si 2.8-4, Mn 0-1.5, Zn 0-1.5, Sn 0-1.5, Fe 0-0.5, Al 0-0.01, Pb 0-0.02, 0.05 others max.
For inert gas, carbon-arc and oxyacetylene welding of copper base alloys and ferrous metals. AWS A5.7 ERCuSi-A.

HARRISON FERRITIC
Harrison Steel Castings Co.
Iron. C 3.55, Mn 0.35, Si 2.5, B 0.035, bal Fe.
As cast: 48,000 TS; 67,000 YS; 20 El; 160 Brin. Meets ASTM A536 Gr. 65-45-12.

HARRISON PEARLITIC
Harrison Steel Castings Co.
Iron. C 3.55, Mn 0.47, Si 2.5, Cu 0.3, B 0.035, bal Fe.
Tempered at 1150°F: 61,000 TS; 94,000 YS; 12 El; 207 Brin. Meets ASTM A536 Gr. 80-55-06.

HARTALUMIN
Manufacturer not listed.
Al alloy. For light alloy parts.

HARTMETALLE
German manufacture
For tools, dies; sintered WC.

HARVEY 66S
Martin-Marietta Corp.
Cu 0.7-1.2, Si 0.9-1.8, Mg 0.8-1.4, Mn 0.6-1.1, bal Al.
O temper: 22,000 TS; 12,000 YS; 18 El; 43 Brin. T6 temper: 62,000 TS; 55,000 YS; 8 El; 120 Brin. T4 temper: 52,000 TS; 30,000 YS; 18 El; 90 Brin. For structures, bus and truck bodies, boom scaffolds; age hardenable, good weldability.

HARVEY HA-1900
Martin-Marietta Corp.
H 0-0.0125, N 0-0.07, 0 0-0.25, C 0-0.1, Fe 0-0.3, bal Ti.
Bar: 80,000 TS min; 70,000 YS min; 15 El; 30 RA; 10-11 Charpy impact. For non-structural and moderately stressed aircraft parts, corrosion applications. Alpha titanium, corrosion resistant.

HARVEY HA-1930
Martin-Marietta Corp.
Ti 99.5.
Annealed: 48,000 TS; 25,000 YS; 35 El. At 600°F: 20,000 TS; 10,000 YS; 50 El. For chemical plant equipment, high temperature applications. Highest formability. Corrosion resistant.

HARVEY HA-1940
Martin-Marietta Corp.
Ti 99.2.
Annealed: 65,000 TS; 43,000 YS; 30 El. At 600°F: 28,000 TS; 13,000 YS; 45 El. For aerospace equipment, chemical and marine applications. Pipe and tubing. Corrosion resistant.

HARVEY HA-1940 PD
Martin-Marietta Corp.
Pd 0.15-0.2, bal Ti.
Annealed: 62,000 TS; 46,000 YS; 27 El. At 600°F: 28,000 TS; 13,000 YS; 30 El. At 1000°F: 15,400 TS; 90,000 YS; 31 El. For chemical industry equipment. Corrosion resistant grade.

HARVEY HA-1950
Martin-Marietta Corp.
Ti 99.
Annealed: 75,000 TS; 63,000 YS; 25 El. At 600°F: 33,000 TS; 19,000 YS; 33 El. For marine applications, high temperature applications. Corrosion resistant.

HARVEY HA-1970
Martin-Marietta Corp.
Ti 99.
Annealed: 95,000 TS; 80,000 YS; 22 El. At 600°F: 43,000 TS; 27,000 YS; 28 El. For marine applications, high temperature components. Commercially pure titanium. Corrosion resistant.

HARVEY HA-4145
Martin-Marietta Corp.
Al 4, Mn 4, bal Ti.
Annealed: 148,000 TS; 135,000 YS; 15 El. At 600°F: 110,000 TS; 90,000 YS; 17 El. Aged: 162,000 TS; 143,000 YS; 10 El. For heavy section aircraft components, fasteners, jet engine and guided missile parts. Alpha-beta titanium alloy.

HARVEY HA-5137
Martin-Marietta Corp.
Al 4-6, Sn 2-3, bal Ti.
Bar: 125,000 TS; 120,000 YS; 18 El; 25 RA; 35 Rock C. At 600°F: 82,000 TS; 65,000 YS; 19 El. For structural aircraft parts, missiles. Weldable, high strength; Alpha titanium, corrosion resistant.

HARVEY HA-5137 ELI
Martin-Marietta Corp.
Al 5, Sn 2.5, low O, bal Ti.
Annealed: 110,000 TS; 95,000 YS; 20 El. At 600°F: 78,000 TS; 60,000 YS; 20 El. For cryogenic applications, missile and aircraft components. Alpha titanium, corrosion resistant.

HARVEY HA-5158
Martin-Marietta Corp.
Al 6, V 6, Sn 2, Fe 1, Cu 1, bal Ti.
Annealed: 160,000 TS; 155,000 YS; 15 El. Aged: 190,000 TS; 180,000 YS; 10 El. For ordnance applications and aircraft components, pressure vessels. Alpha-beta titanium alloy. Hardenable.

HARVEY HA-6148
Martin-Marietta Corp.
Al 5.5-6.5, V 3.5-4.5, bal Ti.
Bar: 130,000 TS min; 120,000 YS min; 10 El; 25 RA; 38 Rock C. For airframes and jet engine compressor parts, ordnance equipment. Alpha-beta titanium; corrosion resistant.

HARVEY HA-6510
Martin-Marietta Corp.
Al 6, V 4, bal Ti.
Annealed: 140,000 TS; 130,000 YS; 18 El. At 600°F: 105,000 TS; 95,000 YS; 11 El. Aged: 165,000 TS; 155,000 YS; 13 El. For airframes and jet engine components, ordnance equipment. Heat treatable to high strength. Alpha-beta type.

HARVEY HA-6510 ELI
Martin-Marietta Corp.
Al 6, V 4, low O, bal Ti.
Annealed: 135,000 TS; 125,000 YS; 20 El. At 600°F: 105,000 TS; 95,000 YS; 12 El. For cryogenic applications. Aircraft and missile components. Heat treatable. Alpha-beta alloy.

HARVEY HA-7146
Martin-Marietta Corp.
Al 7, Mo 4, bal Ti.
Annealed: 160,000 TS; 150,000 YS; 16 El. At 600°F: 127,000 TS; 108,000 YS; 18 El. Aged: 185,000 TS; 175,000 YS; 15 El. For engine and airframe applications, jet engine and missile components. Heat treatable; high creep strength.

HARVEY HA-8116
Martin-Marietta Corp.
Al 8, Mo 1, V 1, bal Ti.
Annealed: 140,000 TS; 130,000 YS; 15 El. At 1000°F: 95,000 TS; 73,000 YS; 14 El; 30 RA. For aircraft and jet components, discs, spacers and blades. High strength, long time creep resistant, high temperature stability.

HARVILL NO. 1
Harville Machine Inc.
Al 0-1.5, Pb 0-0.4, Sn 0-1.5, Zn 38-42, Mn 3.5-4.5, Cu 55-60.
Cast: 65,000 TS; 30,000 YS; 10 El; 100-130 Brin. For casting; corrosion resistant. *Obsolete*

HARVILL NO. 2
Harville Machine Inc.
Al 7-9, Ni 2.5-4.5, Cu 85-87.
Cast: 75,000 TS; 25 El; 110 Brin. For die castings; Al Bronze. *Obsolete*

HARVILL NO. 3
Harville Machine Inc.
Ni 18, Cu 55-64, bal Zn.
Cast: 55,000 TS; 35,000 YS; 10 El; 115 Brin. For die castings; white Ni brass. *Obsolete*

HARVILL NO. 4
Harville Machine Inc.
Cu 57-59, Sn 0.5-1.5, Zn 40-42.
Cast: 55,000 TS; 40,000 YS; 15 El; 130 Brin. For die castings; yellow brass. *Obsolete*

HARZ REFINED LEAD
Manufacturer not listed.
Pb 99.9.
For lead tubes and sheet for chemical industry; corrosion resistant. *Obsolete*

HAS-ALL
Wm. Hassall & Sons
C 0.7, W 18, Cr 4, V 1, bal Fe.
For high speed tools. High speed steel.

HASCROME ALLOY
Haynes International, Inc.
C 1.25, Cr 12, Mn 3, bal Fe.
Covered electrode for AC-DC welding and build-up. As welded: 26 Rock C; work hardens to 43 Rock C. *Obsolete*

HASS
English manufacture
Cu 4.5, Mn 0.75, Si 1, bal Al.
For aircraft structures; age-hardening alloy.

HAST B (1)
Ancast, Inc.
Nickel. C 0-0.12, Mn 0-1, Si 0-1, Cr 0-1, Mo 26-30, Fe 4-6, P 0-0.04, S 0-0.03, V 0.2-0.6, bal Ni.
ASTM A-494 GR N-12 MV.

HAST B (2)
Ancast, Inc.
Nickel. C 0-0.07, Mn 0-1, Si 0-1, Cr 0-1, Mo 30-33, Fe 0-3, P 0-0.04, S 0-0.03, bal Ni.
ASTM A-494 GR N-7M.

HAST B (3)
Ancast, Inc.
Nickel. C 0-0.12, Mn 0-1, Si 0-1, Cr 0-1, Mo 26-33, Co 0-2.5, Fe 0-6, P 0-0.04, S 0-0.03, V 0-0.6, bal Ni.
ASTM A-743 GR N12M.

HAST C (1)
Ancast, Inc.
Nickel. C 0-0.12, Mn 0-1, Si 0-1.5, Cr 15.5-20, Mo 16-20, W 0-5.25, Co 0-2.5, Fe 0-7.5, P 0-0.04, S 0-0.03, V 0-0.4, bal Ni.
ASTM A-743 GR CW-12M.

HAST C (2)
Ancast, Inc.
Nickel. C 0-0.12, Mn 0-1, Si 0-1, Cr 15.5-17.5, Mo 16-18, W 3.75-5.25, Fe 4.5-7.5, P 0-0.04, S 0-0.03, V 0.2-0.4, bal Ni.
ASTM A-494 GR CW-12MW.

HAST C (3)
Ancast, Inc.
Nickel. C 0-0.07, Mn 0-1, Si 0-1, Cr 17-20, Mo 17-20, Fe 0-3, P 0-0.04, S 0-0.03, bal Ni.
ASTM A-494 GR CW-7M.

HAST G
Ancast, Inc.
Nickel. C 0-0.12, Mn 1-2, Si 0-1, Cr 21-23.5, Mo 5.5-7.5, W 0-1, Co 0-2.5, Fe 18-21, P 0-0.04, S 0-0.03, Cu 1.5-2.5, CO 8.27544e+008-525621, CO 8.90405e+011-3.37774e+009, 1.1.0082e-008-0.

HAST X (1)
Ancast, Inc.
Nickel. C 0-0.2, Mn 0-1, Si 0-1, Cr 20.5-23, Mo 8-10, W 0.2-1, Co 0.5-2.5, Fe 17-20, P 0-0.04, S 0-0.03, bal Ni.
AMS 5390B.

HAST X (2)

Ancast, Inc.
Nickel. C 0-0.2, Mn 0-1, Si 0-1, Cr 20.5-23, Mo 8-10, W 0.2-1, Co 0.5-2.5, Fe 17-20, P 0-0.04, S 0-0.04, bal Ni.
ASTM A-567 GR5.

HASTELLOY ALLOY 500

Cabot Corporation
Co 16-20, Cr 16-20, Mo 3-5, Fe 0-2, C 0-0.1, Al 2.5-3.5, Ti 2.5-3.2, B 0.01, bal Ni.
At 70 F: 195,000 TS; 125,000 YS; 18 El; 23 RA; 400 Brin. At 1000 F: 185,000 TS; 118,000 YS; 10 El; 15 RA. At 1400 F: 150,000 TS; 110,000 YS; 20 El; 27 RA. For high temperature applications, turbine buckets, jet engine components; vacuum melted, age-hardenable, heat resistant to 1700 F. *Obsolete*

HASTELLOY ALLOY 700

Cabot Corporation
C 0.1, B 0.007, Cr 2.9, Ti 2.9, Co 18, Cr 18, Mo 4, Fe 0-2, bal Ni.
Aged: 171,000 TS; 104,000 YS; 25 El; 27 RA. At 1600 F: 84,000 TS; 56,000 YS; 7 El; 8 RA. For turbine buckets, jet engine components; heat resistant to 1700 F, age-hardenable. *Obsolete*

HASTELLOY ALLOY B

Stellite Division
Mo 26-30, Fe 4-6, C 0-0.12, Ni 62.
Cast: 82,000 TS; 57,000 YS; 9 El; 13 RA; 230 Brin. Rolled: 140,000 TS; 65,000 YS; 45 El; 45 RA; 235 Brin. For agitators, heating and cooling coils, pump parts, valves; resists HCl and boiling H_3PO_4.

HASTELLOY ALLOY B

Langley Alloys Ltd.
Mo 26-30, Fe 4-6, C 0-0.12, Ni 62.
Cast: 82,000 TS; 57,000 YS; 9 El; 13 RA; 230 Brin. Rolled: 140,000 TS; 65,000 YS; 45 El; 45 RA; 235 Brin. For agitators, heating and cooling coils, pump parts, valves; resists HCl and boiling H_3PO_4.

HASTELLOY ALLOY B

Teledyne Rodney Metals
Mo 26-30, Fe 4-6, C 0-0.12, Ni 62.
Cast: 82,000 TS; 57,000 YS; 9 El; 13 RA; 230 Brin. Rolled: 140,000 TS; 65,000 YS; 45 El; 45 RA; 235 Brin. For agitators, heating and cooling coils, pump parts, valves; resists HCl and boiling H_3PO_4.

HASTELLOY ALLOY B-2

Haynes International, Inc.
Nickel. C 0-0.01, Mo 26-30, Fe 0-2, Cr 0-1, Mn 0-1, Co 0-1, Si 0-0.1, bal Ni.
Plate, solution treated: 129,000 psi (894 MPa) TS; 59,800 psi (412 MPa) YS; 61 El. Good strength to 800°F; corrosion resistant (except for ferric or cupric salts).

HASTELLOY ALLOY C

Haynes International, Inc.
Nickel. Cr 17, C 0.1, Mo 17, Fe 6, W 5, bal Ni.
Covered electrode; tube wire sub-arc; heat and impact resistant; for use in hot-forge dies.

HASTELLOY ALLOY C

Stellite Division
Mo 16-18, Cr 13-17.5, W 3.7-5.3, Fe 4.5-7, bal Ni.
Cast: 72,000-80,000 TS; 45,000-48,000 YS; 15-10 El; 16-11 RA; 175-215 Brin. Rolled: 130,000 TS; 65,000 YS; 25 El; 210 Brin. For pumps and valves for H_2SO_4 at high temperature; resists SO_3, P_2O_5 and Cl_2.

HASTELLOY ALLOY C

Teledyne Rodney Metals
Mo 16-18, Cr 13-17.5, W 3.7-5.3, Fe 4.5-7, bal Ni.
Cast: 72,000-80,000 TS; 45,000-48,000 YS; 15-10 El; 16-11 RA; 175-215 Brin. Rolled: 130,000 TS; 65,000 YS; 25 El; 210 Brin. For pumps and valves for H_2SO_4 at high temperature; resists SO_3, P_2O_5 and Cl_2.

HASTELLOY ALLOY C (POWDER)

Haynes International, Inc.
Nickel. C 0-0.12, Cr 16.5, Mo 17, Fe 5.5, Co 0-2.5, W 4.5, Si 0-1, Mn 0-1, bal Ni.
For plasma spray build-up.

HASTELLOY ALLOY C-22

Haynes International, Inc.
Nickel. Co 0-2.5, Cr 22, Mo 13, W 2-3, Fe 3, Si 0-0.08, Mn 0-0.5, C 0-0.01, V 0-0.35, bal Ni.
Resistance to pitting, crevice corrosion and stress-corrosion cracking. Plate, sheet, strip, billet, bar, wire, electrodes, pipe and tubing.

HASTELLOY ALLOY C-276

Haynes International, Inc.
Nickel. Co 0-2.5, Cr 15.5, Mo 16, W 4, Fe 5.5, Si 0-0.08, Mn 0-1, C 0-0.01, V 0-0.35, bal Ni.
Resistant to oxidizing and reducing corrosive acids and chlorine contaminated hydrocarbons.

HASTELLOY ALLOY C-276

Stellite Division
Cr 14.5-16.5, Mo 15-17, Co 0-2.5, W 3-4.5, Fe 4-7, C 0-0.02, Mn 0-1, bal Ni.
Hardenable by cold work plus aging from 90 Rock B to max of 49 Rock C. Excellent corrosion resistance even to such chemicals as wet chlorine gas and hypochlorites.

HASTELLOY ALLOY C-276

Teledyne Rodney Metals
Cr 14.5-16.5, Mo 15-17, Co 0-2.5, W 3-4.5, Fe 4-7, C 0-0.02, Mn 0-1, bal Ni.
Hardenable by cold work plus aging from 90 Rock B to max of 49 Rock C. Excellent corrosion resistance even to such chemicals as wet chlorine gas and hypochlorites.

HASTELLOY ALLOY C-4

Haynes International, Inc.
C 0-0.015, Co 0-2, Cr 14-18, Mo 14-17, Ti 0-0.7, Fe 0-3, Mn 0-1, Si 0-0.08, bal Ni.
1/2 in. plate, welded: 112,700 psi TS; 68,300 psi YS; 40 El. For corrosion resistance and high-temperature stability.

HASTELLOY ALLOY D

Cabot Corporation
Si 0-11, Cu 0-4, bal Ni.
Cast: 36,000-40,000 TS; 35,000-40,000 YS; 0 El; 0 RA; 484-547 Brin. Pump valves; resists HCl and H 2 SO 4 , but HNO 3 or Cl 2 . *Obsolete*

HASTELLOY ALLOY F

Haynes International, Inc.
C 0-0.05, Mn 1-2, Si 0-1, Cr 21-23, Mo 5.5-7.5, W 0-1, Ni 44-47, Co 0-2.5, 1.75-2.5 Cb + Ta, bal Fe.
Obsolete

HASTELLOY ALLOY G

Haynes International, Inc.
Cr 21-23.5, Mo 5.5-7.5, W 0-1, Mn 1-2, Co 0-2.5, Si 0-1, Fe 18-21, C 0-0.05, Cu 1.5-2.5, 1.75-2.50 Cb + Ta, bal Ni.
Alloy with excellent resistance to hot sulfuric and phosphoric acids. For chemical plant equipment.

HASTELLOY ALLOY G-3

Haynes International, Inc.
Nickel. Co 0-5, Cr 21-23.5, Mo 6-8, W 0-1.5, Fe 18-21, Si 0-1, Mn 0-1, C 0-0.015, Cu 1.5-2.5, P 0-0.04, S 0-0.03, bal Ni.
Improved wrought version of Hastelloy alloy G. Corrosion resistance; resistance to heat affected zone; used in flue gas desulfurization systems.

HASTELLOY ALLOY G-30

Haynes International, Inc.
Nickel. Co 0-5, Cr 28-31.5, Mo 4-6, W 1.5-4, Fe 13-17, Si 0-0.08, Mn 0-1.5, C 0-0.03, Cu 1-2.4, 0.3-1.5 Cb + Ta, bal Ni.
Corrosion resistance in oxidizing acids; chemical process applications.

HASTELLOY ALLOY H

Haynes International, Inc.
Nickel. Cr 22, Fe 18, Mo 9, W 2, bal Ni.
Equivalent or better localized corrosion resistance compared to alloy 625.

HASTELLOY ALLOY N

Haynes International, Inc.
Cr 6-8, Mo 15-18, Fe 5, C 0.04-0.08, Al 0.5, B 0.01, bal Ni.
Annealed: 86,400 TS; 37,300 YS; 17 El. For containers for molten fluorides; oxidation resistant to 1800°F.

HASTELLOY ALLOY S

Haynes International, Inc.
Nickel. C 0-0.02, Cr 14.5-17, Mo 14-16.5, Fe 0-3, Al 0.1-0.5, Si 0.2-0.75, Mn 0.3-1, B 0-0.015, Co 0-2, bal Ni.
High temperature wrought alloy for applications involving severe cyclic heating conditions. Weldable, can be hot or cold worked.

HASTELLOY ALLOY W

Haynes International, Inc.
Nickel. C 0-0.12, Cr 5, Mo 24.5, Co 2.5, Fe 6, Mn 0-1, Si 1, bal Ni.
For high temperature applications; corrosion and heat resistant; for welding of dissimilar alloys.

HASTELLOY ALLOY X

Haynes International, Inc.
Nickel. Co 0.5-2.5, Cr 20.5-23, Mo 8-10, W 0.2-1, Fe 17-20, C 0.05-0.15, Si 0-1, Mn 0-1, bal Ni.
Oxidation resistance, fabricability and high temperature strength; for aircraft, furnace and chemical process components.

HASTELLOY ALLOY X

Stellite Division
C 0.05-0.2, Cr 20.5-23, Fe 17-20, Mo 8-10, Co 0.5-2.5, W 0.2-1, bal Ni.
At 70°F: 113,000 TS; 55,800 YS; 44 El. At 1200°F: 83,000 TS; 40,700 YS; 37 El. At 1800°F: 21,000 TS; 17,000 YS; 43 El. For aircraft and jet engine components; good creep and stress rupture properties.

HASTELLOY ALLOY X

Cannon-Muskegon Corp.
C 0.05-0.2, Cr 20.5-23, Fe 17-20, Mo 8-10, Co 0.5-2.5, W 0.2-1, bal Ni.
At 70°F: 113,000 TS; 55,800 YS; 44 El. At 1200°F: 83,000 TS; 40,700 YS; 37 El. At 1800°F: 21,000 TS; 17,000 YS; 43 El. For aircraft and jet engine components; good creep and stress rupture properties.

HASTELLOY B

Otto Junker GmbH
C 0-0.1, Mo 26-30, Fe 4-6, V 0.2-0.6, bal Ni.
Cast: 71,000 TS; 38,000 YS; 20 El; 180-240 Brin. Corrosion resistant casting.

HASTELLOY B

Enpar Sonderwerkstoffe GmbH
Nickel.
Alternate manufacturer.

HASTELLOY B, C

Now JUNKER HASTELLOY B & C.

HASTELLOY B-2

Enpar Sonderwerkstoffe GmbH
Nickel.
Alternate manufacturer.

HASTELLOY B2

Otto Junker GmbH
C 0-0.02, Mo 26-30, bal Ni.
71,000 psi TS min; 35,000 psi YS min; 30 El min. Corrosion resistant casting.

HASTELLOY C
Driver Harris Co.
Nickel. Cr 15, Fe 5, W 4, Mo 15, bal Ni.
Wire, rod, ribbon for welding wire and fastener stock.
Obsolete

HASTELLOY C
Enpar Sonderwerkstoffe GmbH
Nickel.
Alternate manufacturer.

HASTELLOY C
Otto Junker GmbH
C 0-0.1, Cr 15.5-17.5, Mo 16-18.5, V 0.2-0.4, Fe 4.5-7, W 3.75-5.25, bal Ni.
Cast: 71,000 TS; 40,000 YS; 10 El; 180-240 Brin. Corrosion resistant casting.

HASTELLOY C WIRE
National Standard Co.. C 0-0.08, Cr 14.5-16.5, Mo 15-17, W 3-4.5, Co 0-2.5, V 0-0.35, Fe 4-7, bal Ni.
For springs requiring resistance to chlorine and sulphur attack.

HASTELLOY C-276
Enpar Sonderwerkstoffe GmbH
Nickel.
Alternate manufacturer.

HASTELLOY C-4
Enpar Sonderwerkstoffe GmbH
Nickel.
Alternate manufacturer.

HASTELLOY C22
Otto Junker GmbH
C 0-0.02, Cr 20-22.5, Mo 12.5-14.5, W 2.5-3.5, Fe 2-6, bal Ni.
64,000 psi TS min; 33,000 psi YS min; 20 El min. Corrosion resistant casting.

HASTELLOY C4
Otto Junker GmbH
C 0-0.02, Cr 15-17.5, Mo 15-17.5, bal Ni.
68,000 psi TS min; 35,000 psi YS min; 40 El min. Corrosion resistant casting.

HASTELLOY CHF
Cabot Corporation
Ni-Cr.
Cast: 80,000 TS; 48,000 YS; 10 El; 11 RA; 215 Brin. For hard facing welding rod; deposits 16-18 Mo, 13-17.5 Cr, 3.7-5.3 W, 4.5-7 Fe, bal Ni; heat resistant. *Obsolete*

HASTELLOY DEVELOPMENT ALLOY C-455
Now HASTELLOY ALLOY C-4.

HASTELLOY G-3
Enpar Sonderwerkstoffe GmbH
Nickel.
Alternate manufacturer.

HASTELLOY G3
Otto Junker GmbH
C 0-0.02, Cr 21-23.5, Mo 6-8, Fe 18-21, Cu 1.5-2.5, bal Ni.
71,000 psi TS min; 30,000 psi YS min; 20 El min. Corrosion resistant casting.

HASTELLOY R-235
Cannon-Muskegon Corp.
C 0.15, Cr 15.5, Co 0-2.5, Mo 5.5, Ti 2.5, Al 2, Fe 10, bal Ni.
High temperature alloy; for gas turbine and engine parts, sheet.

HASTELLOY T
Haynes International, Inc.
C 0.02, Mn 0.2, Cr 12, Mo 9, W 14, Al 0.2, La 0.02, bal Ni.
Corrosion resistant, low expansion alloy. *Obsolete*

HASTELLOY W
Driver Harris Co.
Nickel. Cr 5, Fe 5, Mo 24, bal Ni.
Wire, rod, ribbon for welding wire and fastener stock.
Obsolete

HASTELLOY X
Driver Harris Co.
Nickel. Cr 21, Co 1, Fe 18, W 1, Mo 9, bal Ni.
Wire, rod, ribbon for welding wire and fastener stock.
Obsolete

HASTELLOY X
Enpar Sonderwerkstoffe GmbH
Nickel.
Alternate manufacturer.

HATHAL-A
Aluminium-Zentrale e.V.
Aluminum. Cu 3-5, Si 0.4, Mn 0.4-1, Mg 0.3-1, bal Al.
Heat treated: 23,000-65,000 TS; 10-25 El; 40-120 Brin. For rivets, hydraulic fittings, hardware. Age hardenable. Aluminum 2014. Good machinability and weldability. *Obsolete*

HATHAL-B
Aluminium-Zentrale e.V.
Aluminum. Si 0.2-1, Mn 0.2-0.6, Mg 2-8, bal Al.
Heat treated: 31,000-54,000 TS; 5-20 El; 50-85 Brin. For structural members, marine hardware, fasteners. Age hardenable. Corrosion resistant. *Obsolete*

HATHAL-C
Aluminium-Zentrale e.V.
Aluminum. Si 0.5-1, Mn 0.3-1.5, Mg 0.8-1.5, bal Al.
Heat treated: 14,000-43,000 TS; 6-22 El; 30-80 Brin. For structures, marine parts, screw machine products. Age hardenable. Corrosion resistant. *Obsolete*

HAVAR
Hamilton Technology Inc.
Co 42.5, Ni 13, Cr 20, Mo 2, C 0.2, Be 0.04, Mn 1.6, W 2.8, bal Fe.
Rolled: 260,000-290,000 TS; 200,000-220,000 YS; 48-50 El.
Aged: 330,000-360,000 TS; 260,000-280,000 YS; 56-60 El.
For watch and power springs, valves, flapper valves, electronic parts; age hardenable, high strength and fatigue resistance.

HAVOC
Wallace Murray Corp.
C 0.5, Si 1, V 0.2, Mo 0.5, bal Fe.
For plug gauges, reamers, arbors, punches; water hardening, shock resistant. *Obsolete*

HAWK "H"
Hawkridge Bros. Co.
C 0.85-1.1, Cr 1-1.5, V 2, Mn 0.3-0.6, Si 0.15-0.3, bal Fe.
For general tools, dies, rolls; deep hardening, fine grained. *Obsolete*

HAWK 3110
Hawkridge Bros. Co.
C 0.1, Mn 0.7, Ni 1.25, Cr 0.65, bal Fe.
For plastic mold and die casting dies; case hardened. *Obsolete*

HAWK 3312
Hawkridge Bros. Co.
C 0.08-0.13, Mn 0.45-0.6, Cr 1.4-1.75, Ni 3.25-3.75, bal Fe.
Deep hardening carburizing grade alloy steel.

HAWK 777
Hawkridge Bros. Co.
C 0.7, Cr 0.09, Mn 0.8, Mo 0.7, bal Fe.
For drills, cutters, punches, mandrels; oil hardened. *Obsolete*

HAWK 977
Hawkridge Bros. Co.
C 0.9, Cr 0.09, Mo 0.72, bal Fe.
For rivet punches; oil hardened; tough. *Obsolete*

HAWK A-2
Hawkridge Bros. Co.
C 1, Mn 0.7, Si 0.3, Cr 5.25, V 0.3, Mo 1.15, bal Fe.
AISI Type A2 air hardening tool steel.

HAWK A-2-S
Hawkridge Bros. Co.
C 1, Mn 0.7, Si 0.3, Cr 5.25, V 0.3, Mo 1.15, S 0.15, bal Fe.
Free machine grade. AISI Type A2 air hardening tool steel.

HAWK BRAND
Hawkridge Bros. Co.
C 0.6-1.4, Mn 0.15-0.3, Si 0.15-0.3, bal Fe.
For general tools, dies; shallow hardening, fine grained.

HAWK COLD HEADER DIE
Hawkridge Bros. Co.
C 0.95, bal Fe.
For heading dies; tough, water hardened. *Obsolete*

HAWK D2
Hawkridge Bros. Co.
C 1.55, Mn 0.35, Si 0.45, Cr 11.5, V 0.9, Mo 0.8, bal Fe.
Air or oil hardening cold work tool and die steel, chromium type; AISI D2.

HAWK D2S
Hawkridge Bros. Co.
C 1.55, Mn 0.35, Si 0.45, Cr 11.5, V 0.9, Mo 0.8, S 0.15, bal Fe.
Air or oil hardening cold work tool and die steel, chromium type, free machining grade; AISI D2.

HAWK IMPACTO
Hawkridge Bros. Co.
C 0.9-1.1, bal Fe.
For silverware blanking and striking dies; water hardened. *Obsolete*

HAWK O1
Hawkridge Bros. Co.
C 0.9, Mn 1.25, Cr 0.5, W 0.5, bal Fe.
AISI Type O1 oil hardening tool steel.

HAWK PREFAK
Hawkridge Bros. Co.
C 0.9, Cr 0.5, Mn 1.25, W 0.5, bal Fe.
For tools, cutters, dies; nondeforming. *Obsolete*

HAWK SPECIAL
Hawkridge Bros. Co.
C 0.9-1.4, Cr 0.2-0.3, Mn 0.15-0.3, Si 0.4-0.6, bal Fe.
For general tools, dies, cutters, taps, drills; holds its cutting edge. *Obsolete*

HAWK SS EXTRA
Hawkridge Bros. Co.
C 1.05, Cr 1.2, Mo 0.3, bal Fe.
For tools, dies; Type L7.

HAWK STANDARD
Hawkridge Bros. Co.
C 0.7-1.2, bal Fe.
For tools, dies; water hardening.

HAWK VANADIUM
Hawkridge Bros. Co.
C 0.7-1.4, V 0.15-0.25, Mn 0.15-0.3, Si 0.15-0.3, bal Fe.
For general tools, dies, cutters, punches; tough.

HAWK W1
Hawkridge Bros. Co.
C 0.95-1.05, Mn, Si, bal Fe.
AISI Type W1 water hardening tool steel.

HAWKS BRAND WHITE METAL
Billington & Newton Ltd.
Sn 53.2, Sb 7.87, Cu 0.48, Pb 38.48.
For bearings, castings.

HAYNES 11
Haynes International, Inc.
C 3.9, Cr 15, Si 1.4, Mn 1, Mo 0.5, bal Fe.
Covered electrode for AC-DC welding. For general purpose and maintenance; 55 Rock C. *Obsolete*

HAYNES 11-0
Haynes International, Inc.
C 3.5, Cr 16, Si 1, Mn 1.3, Mo 0.5, bal Fe.
Tube wire, open-arc welding. For general purpose and maintenance; 48 Rock C. *Obsolete*

HAYNES 25
Enpar Sonderwerkstoffe GmbH
Cobalt.
Alternate manufacturer.

HAYNES 42
Cabot Corporation
Ni alloy.
For hard facing electrode; corrosion resistant. *Obsolete*

HAYNES 420-S
Haynes International, Inc.
C 0.3, Cr 12, Si 1, Mn 2, bal Fe.
Tube wire, sub-arc welding. For general purpose and maintenance; 54 Rock C. *Obsolete*

HAYNES 4560
Cabot Corporation
C 0.6, Cr 5, Mn 0.8, Mo 0.8, bal Fe.
Welded: 450 Brin. For hard facing rod; resists cold abrasion. *Obsolete*

HAYNES 4561 ALLOY
Cabot Corporation
Cr 2.5, C 0.55, Mn 1.1, Mo 0.4, Si 0.5, bal Fe.
Welded: 550 Brin. For hard facing electrode; heavy impact and moderate abrasion resistance. *Obsolete*

HAYNES 5
Haynes International, Inc.
C 0.2, Cr 2.3, Si 0.8, Mn 2, Mo 0.3, bal Fe.
Covered electrode for AC-DC welding or build-up. Good impact resistance; 45 Rock C. *Obsolete*

HAYNES 5-0
Haynes International, Inc.
C 0.1, Cr 3.5, Si 0.3, Mn 2, Mo 0.2, bal Fe.
Tube wire, open-arc welding or build-up. Good impact resistance; 44 Rock C. *Obsolete*

HAYNES 5-S
Haynes International, Inc.
C 0.15, Cr 3, Si 0.6, Mn 2, Mo 0.4, bal Fe.
Tube wire, sub-arc welding. Good impact resistance; 40 Rock C. *Obsolete*

HAYNES 52
Haynes International, Inc.
C 0.8, Cr 6.8, Si 0.3, Mn 1.4, bal Fe.
Covered electrode for AC-DC welding. For general purpose and maintenance; 54 Rock C. *Obsolete*

HAYNES 52-S
Haynes International, Inc.
C 0.4, Cr 5.4, Si 0.9, W 1.4, Mn 3, Mo 0.8, bal Fe.
Tube wire, sub-arc welding. For general purpose and maintenance; 52 Rock C. *Obsolete*

HAYNES 5260
Cabot Corporation
Cr 9, C 1, Mn 0.6, Mo 1.15, bal Fe.
Welded: 520 Brin. For hard facing rod; low crack sensitivity. *Obsolete*

HAYNES 5261 ALLOY
Cabot Corporation
Cr 5.5, C 0.7, Mn 1.15, Mo 0.5, Si 0.8, bal Fe.
Welded: 570 Brin. For hard facing electrode; wear resistant. *Obsolete*

HAYNES 5461 ALLOY
Cabot Corporation
C 3, Cr 12, Mn 2.5, Si 1.5, Mo 1.5, bal Fe.
Welded: 550 Brin. For hard facing electrode; corrosion and erosion resistant. *Obsolete*

HAYNES 6-S
Haynes International, Inc.
C 3.1, Si 0.9, Mn 2, Mo 4.5, Ni 5, bal Fe.
Tube wire, sub-arc welding. Good impact resistance; 40 Rock C. *Obsolete*

HAYNES 7-S
Haynes International, Inc.
C 0.1, Cr 1.7, Si 0.5, Mn 1.8, Mo 0.5, bal Fe.
Tube wire, sub-arc welding. Good impact resistance; 38 Rock C. *Obsolete*

HAYNES 90
Haynes International, Inc.
C 2.5, Cr 26, Si 0.25, Mn 0.8, bal Fe.
Covered electrode for AC-DC welding. For general purpose and maintenance; 52 Rock C. *Obsolete*

HAYNES 92
Haynes International, Inc.
C 3.75, Mo 10, bal Fe.
Bare cast rod for AC-CD welding. For severe wear, light impact; 64 Rock C. *Obsolete*

HAYNES 93
Haynes International, Inc.
C 3, Cr 17, Mo 19, Co 6.3, bal Fe.
Bare cast rod for DC welding. For severe wear, but light impact; 62 Rock C. Note: HAYNES 93 covered electrode, same but 57 Rock C. *Obsolete*

HAYNES 94
Haynes International, Inc.
C 3.5, Cr 31, Si 0.5, Mn 1.2, Mo 1.5, bal Fe.
Covered electrode for AC-DC welding. For severe wear, light impact; 57 Rock C. *Obsolete*

HAYNES 94-0
Haynes International, Inc.
C 3.25, Cr 27, Si 0.95, Mn 0.75, Co 3, bal Fe.
Tube wire, open-arc welding. For severe wear, light impact; 56 Rock C. *Obsolete*

HAYNES 94-G
Haynes International, Inc.
C 2.3, Cr 29, Si 1.4, Mn 1, bal Fe.
Bare tube rod for welding. For severe wear, but light impact; 60 Rock C. *Obsolete*

HAYNES ALLOY 150
Haynes International, Inc.
C 0.08, Mn 0.65, Si 0.75, Cr 28, Fe 2, bal Co.
For parts subject to intermittent heating and SO_2 gas.

HAYNES ALLOY 31
Haynes International, Inc.
C 0.5, Cr 25, Fe 0-2, Ni 10.5, W 7.5, bal Co.
Powder for plasma spray.

HAYNES ALLOY 556
Haynes International, Inc.
C 0.1, Cr 22, Ni 20, Co 18, Mo 3, W 2.5, La 0.02, N 0.2, Ta 0.6, Al 0.2, Si 0.4, Mn 1, Zr 0.02, bal Fe.
High temperature alloy; for stressed components operating to 2000°F (1095°C). Useful in corrosive environments.

HAYNES ALLOY 6B
Haynes International, Inc.
C 1.1, Cr 30, W 4.5, Fe 0-3, Ni 0-3, bal Co.
Plate: 148,000 TS; 88,000 YS; 7 El; 9 RA; 380 Brin. Sheet: 165,000 TS; 110,000 YS; 5 El; 430 Brin. For high temperature applications, valve parts, liners; heat and corrosion resistant.

HAYNES ALLOY NO. 13
Cabot Corporation
Cr, W, bal Co.
Cast: 101,000 TS; 1 El. For hard facing rod. *Obsolete*

HAYNES ALLOY NO. 151
Cabot Corporation
C 0.48, B 0.06, Cr 20, W 12.8, 3% max Fe + Ni, bal Co.
Cast: 103,000 TS; 74,000 YS; 8 El; 330 Brin. For turbine vanes, high temperature applications; good creep-rupture properties to 1800 F. *Obsolete*

HAYNES ALLOY NO. 188
Haynes International, Inc.
Refractory. Cr 21-23, Ni 20-24, W 13-15, Fe 0-3, C 0.05-0.15, Si 0.2-0.5, Mn 0-1.25, La 0.03-0.12, B 0-0.015, bal Co.
High temperature strength, oxidation resistance to 2100°F, good post-aging ductility; gas turbine applications.

HAYNES ALLOY NO. 188
Stellite Division
Cr 20-24, Ni 20-24, W 13-16, Fe 0-3, C 0.05-0.15, Si 0.2-0.5, Mn 0-1.25, La 0.03-0.15, bal Co.
Sheet, HtTr, RT.: 139,400 TS; 69,500 YS; 56 El. Sheet, HtTr, 1800°F: 36,800 TS; 23,500 YS; 72 El. For high temperature applications in aircraft, space, gas turbine, nuclear industries. Weldable, oxidation and corrosion resistant.

HAYNES ALLOY NO. 188
Teledyne Rodney Metals
Cr 20-24, Ni 20-24, W 13-16, Fe 0-3, C 0.05-0.15, Si 0.2-0.5, Mn 0-1.25, La 0.03-0.15, bal Co.
Sheet, H.T., RT: 139,400 TS; 69,500 YS; 56 El. Sheet, HtTr, 1800°F: 36,800 TS; 23,500 YS; 72 El. For high temperature applications in aircraft, space, gas turbine, nuclear industries. Weldable, oxidation and corrosion resistant.

HAYNES ALLOY NO. 188
Cannon-Muskegon Corp.
Cr 20-24, Ni 20-24, W 13-16, Fe 0-3, C 0.05-0.15, Si 0.2-0.5, Mn 0-1.25, La 0.03-0.15, bal Co.
Sheet, HtTr, RT: 139,400 TS; 69,500 YS; 56 El. Sheet, HtTr, 1800°F: 36,800 TS; 23,500 YS; 72 El. For high temperature applications in aircraft, space, gas turbine, nuclear industries. Weldable, oxidation and corrosion resistant.

HAYNES ALLOY NO. 190
Haynes International, Inc.
C 3, Cr 26, W 14, B 1, Fe 0-5, bal Co.
Cobalt base superalloy; for hard facing and powder metallurgy. *Obsolete*

HAYNES ALLOY NO. 20
Haynes International, Inc.
Cr 21-23, Mo 4-6, Ni 25-27, bal Fe.
Very good corrosion resistance. *Obsolete*

HAYNES ALLOY NO. 20 MOD
Haynes International, Inc.
C 0-0.05, Ni 25-27, Cr 21-23, Mo 4-6, Si 0-1, Mn 0-2.5, Ti = 4 x C min, bal Fe.
Plate, solution treated: 94,900 TS; 42,000 YS; 52 El.
Weldable; resistant to corrosion and to chloride pitting; good low temperature toughness; good high temperature strength. For boilers and pressure vessels. *Obsolete*

HAYNES ALLOY NO. 200
Haynes International, Inc.
Ni 99.2, Cu 0-0.25, Fe 0-0.4, Mn 0-0.35, C 0-0.1, Si 0-0.15.
Wrought nickel for chemical equipment. UNS NO2200; ASTM B160 and B162. *Obsolete*

HAYNES ALLOY NO. 201

Haynes International, Inc.
Ni 99, Cu 0-0.25, Fe 0-0.4, Mn 0-0.35, Mg 0-0.15, C 0-0.02, Si 0-0.15.
Wrought nickel for chemical equipment. *Obsolete*

HAYNES ALLOY NO. 208 PM

Haynes International, Inc.
C 2.1-2.7, Si 0-1, Co 9-11, Cr 25-27, Mo 9-11, W 9-11, Fe 11.5-13.5, Mn 0-0.75, B 0-1, bal Ni.
Sintered: 8.4 density; 117,000 psi transverse strength; 100,000 psi TS; hardness 44 Rock C. For high temperature operation. *Obsolete*

HAYNES ALLOY NO. 214

Haynes International, Inc.
Nickel. Cr 16, Fe 3, Al 4.5, Y present, bal Ni.
High temperature alloy with resistance to oxidation, carburization, and chlorination. Plate, sheet, strip, bar, wire, tubing.

HAYNES ALLOY NO. 230

Haynes International, Inc.
Nickel. Cr 22, W 14, Mo 2, Fe 0-3, Co 0-5, Mn 0.5, Si 0.4, Al 0.3, C 0.1, La 0.02, B 0.005, bal Ni.
High temperature strength and oxidation resistance; aerospace applications. Plate, sheet, strip, foil, billet, bar, wire, welding, pipe and tubing.

HAYNES ALLOY NO. 25

Haynes International, Inc.
Refractory. Cr 19-21, W 14-16, Ni 9-11, C 0.05-0.15, Fe 0-2, Mn 1-2, bal Co.
Annealed: 155,000 TS; 70,000 YS; 65 El; 225 Brin. For turbine blades and discs, combustion chambers, jet stack; corrosion and heat resistant.

HAYNES ALLOY NO. 263

Haynes International, Inc.
Nickel. C 0.04-0.08, Cr 19-21, Mo 5.6-6.1, Co 19-21, Al 0.3-0.6, Ti 1.9-2.4, Fe 0-0.7, Cu 0-0.2, Si 0-0.4, Mn 0-0.6, bal Ni.
Sheet: 144,000 psi (993 MPa) TS; 87,000 psi (600 MPa) YS; 37 El. Good strength at elevated temperatures.

HAYNES ALLOY NO. 31

Haynes International, Inc.
C 0.45-0.55, Si 0-1, Ni 9.5-11.5, Fe 0-2, Mn 0-1, Cr 24.5-26, W 7-8, bal Co.
Sintered: 8.45 density; 120,000 psi TS; 72,000 psi YS; 4 El. For high temperature operation.

HAYNES ALLOY NO. 40 (POWDER)

Haynes International, Inc.
C 0.75, Cr 14, Si 4, Fe 4, B 3.4, 2.0 others, bal Ni.
Powder for flame spray/plasma arc surfacing; 52-56 Rock C. Good erosion and corrosion resistance. *Obsolete*

HAYNES ALLOY NO. 40 (ROD)

Haynes International, Inc.
C 0.75, Cr 15, Si 4, Fe 4, 3.5 others, bal Ni.
Bare cast rod for hard-facing; 57 Rock C. Good erosion and corrosion resistance. *Obsolete*

HAYNES ALLOY NO. 400

Haynes International, Inc.
Ni 63-70, Cu 28-34, Fe 1-2.5, Mn 0-1.25, C 0-0.15, Al 0-0.5.
Cu-Ni alloy for marine equipment and chemical equipment.
UNS N04400; ASTM B127 and B164. *Obsolete*

HAYNES ALLOY NO. 41 (POWDER)

Haynes International, Inc.
C 0.45, Cr 12, Si 3.5, Fe 3, B 2.5, bal Ni.
Powder for flame spray/plasma arc surfacing; 46-52 Rock C. Good erosion and corrosion resistance. *Obsolete*

HAYNES ALLOY NO. 41 (ROD)

Haynes International, Inc.
C 0.05, Cr 12, Si 3.5, Fe 3, 2.50 others, bal Ni.
Bare cast rod for hard-facing; 51 Rock C. Good erosion and corrosion resistance. *Obsolete*

HAYNES ALLOY NO. 43

Haynes International, Inc.
C 0.85, Cr 17, Si 3.9, Fe 2, B 3.3, bal Ni.
Powder for flame spray or plasma arc hard surfacing; 54-60 Rock C. For abrasion resistance. *Obsolete*

HAYNES ALLOY NO. 44

Haynes International, Inc.
C 0.45, Cr 9, Si 3, Fe 3.8, B 2, bal Ni.
Powder for flame spray/plasma arc surfacing; 40-44 Rock C. *Obsolete*

HAYNES ALLOY NO. 45

Haynes International, Inc.
C 0-0.1, Cr 0-0.5, Si 2.5, Fe 0-1, B 1.5, bal Ni.
Powder for manual torch build-up; 20 Rock C. *Obsolete*

HAYNES ALLOY NO. 46

Haynes International, Inc.
C 0-0.1, Cr 0-0.5, Si 3, Fe 0-1, B 1.8, bal Ni.
Powder for manual torch build-up; 30 Rock C. *Obsolete*

HAYNES ALLOY NO. 48

Haynes International, Inc.
C 0.4, Cr 16, Si 4, Mo 2.5, Fe 3, B 4, Cu 2.5, bal Ni.
Powder for flame spray surfacing; 55 Rock C. For abrasion resistance. *Obsolete*

HAYNES ALLOY NO. 520

Haynes International, Inc.
C 1, Cr 5, Mo 6.3, V 2.7, W 7.5, bal Fe.
Powder for plasma arc surfacing; 64 Rock C. Good abrasion resistance. *Obsolete*

HAYNES ALLOY NO. 525

Haynes International, Inc.
C 7, Cr 41, Si 1, Mn 1, B 0.08, bal Fe.
Powder for plasma arc surfacing; 62 Rock C. Abrasion resistant. *Obsolete*

HAYNES ALLOY NO. 589

Haynes International, Inc.
C 2.9-3.4, Si 0.5-1.5, Mn 0-0.5, Cr 15.5-18.5, Mo 14.5-17.5, V 1.65-2.1, 3.0 others max, bal Fe.
Sintered: 7.55 density; 170,000 psi transverse strength; 125,000 psi TS; hardness 58 Rock C. *Obsolete*

HAYNES ALLOY NO. 600

Haynes International, Inc.
Ni 72, Fe 6-10, Cr 14-17, C 0-0.08, Ti 0-0.5.
Ni-Cr-Fe wrought alloy; good oxidation resistance to 2150°F. For heat treated, furnace, and tank equipment. UNS N06600; ASTM B166 and B168. *Obsolete*

HAYNES ALLOY NO. 713C

Cabot Corporation
C 0.08-0.2, Cr 11-14, Mo 3.5-5.5, Al 6, Ti 0.6, Cb 2, Fe 0-2.5, Zr 0.04, B 0.01, bal Ni.
At 70 F: 123,000 TS; 108,000 YS; 7.6 El; 9.9 RA. At 1800 F: 72,000 TS; 51,300 YS; 14 El; 28 RA. For jet engine blades; vacuum melted, vacuum cast, age-hardenable. *Obsolete*

HAYNES ALLOY NO. 718

Haynes International, Inc.
Nickel. C 0-0.08, Co 0-1, Cr 17-21, Mo 2.8-3.3, Al 0.2-0.8, Ti 0.65-1.15, Ni 50-55, B 0-0.006, Si 0-0.35, Cu 0-0.1, Mn 0-0.35, 4.75-5.50 Cb + Ta, bal Fe.
Solution treated and aged: 180,000 psi TS; 150,000 psi YS; 15 min El. Excellent high temperature properties; weldable; used also as filler metal.

HAYNES ALLOY NO. 88

Cabot Corporation
C 0.07, Mn 1.4, Cr 12, Ni 15, Mo 2, W 0.5, Ti 0.5, B 0.1, bal Fe.
Cold drawn: 104,000 TS; 85,000 YS; 32 El; 63 RA; 233 Brin. For high temperature bolts, turbine discs and blades; heat resistant, good relaxation properties. *Obsolete*

HAYNES ALLOY NO. 90

Haynes International, Inc.
C 2.75, Cr 27, bal Fe.
Powder for plasma arc surfacing; 52 Rock C. Good cold abrasion resistance. *Obsolete*

HAYNES ALLOY NO. 92

Haynes International, Inc.
C 3.75, Cr 0-1.5, Mo 10, bal Fe.
Powder for plasma arc surfacing; 60 Rock C. Good erosion and cold abrasion resistance. *Obsolete*

HAYNES ALLOY NO. 93

Haynes International, Inc.
C 3, Cr 15-19, Mo 13-17, Co 4-7, V 0.5-3, bal Fe.
Heat treated: 90,000 TS; 90,000 YS; 0 El; 0 RA; 650-745 Brin. For hard surfacing electrode; abrasion resistant, high cold hardness. *Obsolete*

HAYNES ALLOY NO. 94

Haynes International, Inc.
Cr 32, C 3.5, Mn 1.6, Mo 1, Si 0-0.3, bal Fe.
Welded: 580 Brin. For hard facing electrode; resists corrosion and oxidation. *Obsolete*

HAYNES ALLOY R-41

Haynes International, Inc.
C 0-0.12, Cr 18-20, Fe 0-5, Co 10-12, Ti 3-3.3, Mo 9-10.5, Al 1.4-1.8, B, bal Ni.
Heat treated: 140,000-206,000 TS; 97,000-154,000 YS; 9-18 El; 400 Brin. For high temperature applications, jet engine components; age-hardenable, high oxidation resistance.

HAYNES ALLOY X750

Haynes International, Inc.
C 0.08, Cr 15.5, Fe 7, Cu 0.5, Al 0.7, Ti 2.5, bal Ni.
Nickel base high temperature alloy; high creep strength to 1500°F. ASTM A461.

HAYNES CRUSHER

Haynes International, Inc.
C 4.1, Cr 2.5, Si 0.75, Mn 0.25, Mo 3, bal Fe.
Covered electrode for AC-DC welding. For severe wear, light impact; 57 Rock C. *Obsolete*

HAYNES CRUSHER-O

Haynes International, Inc.
C 3.2, Cr 23, Si 0.2, Mn 0.1, Mo 2, bal Fe.
Tube wire, open-arc welding. For severe wear, but light impact; 52 Rock C. *Obsolete*

HAYNES DEVELOPMENT ALLOY 556

Haynes International, Inc.
Ni 20, Co 20, Cr 22, Mo 3, W 2.5, La 0.02, C 0.1, N 0.2, Al 0.3, Si 0.4, Mn 1.5, Zr 0.02, 1.0 Cb + Ta, bal Fe.
Good oxidation resistance to 2000°F. Weldable, hot or cold formed.

HAYNES DEVELOPMENT ALLOY NO. 8117

Haynes International, Inc.
C 2.5-3.8, Si 0-1, Ni 0 2.5, Fe 2-5, Cr 27.5-31.5, W 15-18.5, Mn 1-3, B 0-1, 3.0-6.5 Cb + Ta, bal Co.
Sintered: 8.8 density; 160,000 psi transverse strength; 95,000 psi TS; hardness 61 Rock C. For high temperature operations. *Obsolete*

HAYNES L-605

Now HAYNES ALLOY NO. 25.

HAYNES MULTIPASS

Haynes International, Inc.
C 0.1, Cr 2, Si 0.4, Mn 0.9, bal Fe.
Covered electrode for AC-DC welding or build-up. Good impact resistance; 32 Rock C. *Obsolete*

HAYNES MULTIPASS-O

Haynes International, Inc.
C 0.2, Cr 0.9, Si 1.4, Mn 2, Mo 0.4, bal Fe.
Tube wire, open arc welding or build-up. Good impact resistance; 34 Rock C. *Obsolete*

HAYNES MULTIPASS-S

Haynes International, Inc.
C 0.1, Si 0.6, Mn 1.9, Mo 0.4, bal Fe.
Tube wire, sub-arc welding. Good impact resistance; 32 Rock C. *Obsolete*

HAYNES NI-MANG-O

Haynes International, Inc.
C 0.8, Cr 3.8, Si 0.4, Mn 14.5, Ni 3.4, bal Fe.
Tube wire, open-arc welding and build-up. As welded: 86 Rock B; work hardens to 44 Rock C. Severe impact resistance; moderate to slight abrasion. *Obsolete*

HAYNES NICKEL-MANGANESE

Cabot Corporation
Cr 0.5, C 0.7, Mn 13.5, Ni 3.5, Si 0.8, bal Fe.
Welded: 127,000 TS; 185-450 Brin. For hard facing electrode; work hardens, abrasion and impact resistant. *Obsolete*

HAYNES PATENT

American manufacture
Cr 8-60, C 0-1, bal Fe.
For tools, dies, corrosion and heat resisting parts; stainless and corrosion resistant.

HAYNES STA-MANG

Haynes International, Inc.
C 0.6, Cr 15, Si 0.3, Mn 15, Mo 0.3, Ni 1.5, bal Fe.
Covered electrode for AC-DC welding and build-up. As welded: 16 Rock C; work hardens to 47 Rock C. Severe impact resistance; moderate to slight abrasion. *Obsolete*

HAYNES STA-MANG-O

Haynes International, Inc.
C 0.1, Cr 12, Si 0.2, Mn 14, Mo 0.1, Ni 1.4, bal Fe.
Tube wire, open-arc welding and build-up. As welded: 19 Rock C; work hardens to 48 Rock C. Severe impact resistance; moderate to slight abrasion. *Obsolete*

HAYNES STELLITE 98M2 ALLOY

Deloro Stellite, Inc.
Now STELLITE 98M2 ALLOY.

HAYNES STELLITE ALLOY NO. 1 (POWDER)

Deloro Stellite, Inc.
Now STELLITE ALLOY NO. 1 (POWDER).

HAYNES STELLITE ALLOY NO. 1 (SOLID)

Deloro Stellite, Inc.
Now STELLITE ALLOY NO. 1 (SOLID).

HAYNES STELLITE ALLOY NO. 1016 (POWDER)

Deloro Stellite, Inc.
Now STELLITE ALLOY NO. 1016 (POWDER).

HAYNES STELLITE ALLOY NO. 1016 (SOLID)

Deloro Stellite, Inc.
Now STELLITE ALLOY NO. 1016 (SOLID).

HAYNES STELLITE ALLOY NO. 12 (POWDER)

Deloro Stellite, Inc.
Now STELLITE ALLOY NO. 12 (POWDER).

HAYNES STELLITE ALLOY NO. 12 (SOLID)

Deloro Stellite, Inc.
Now STELLITE ALLOY NO. 12 (SOLID).

HAYNES STELLITE ALLOY NO. 156

Deloro Stellite, Inc.
Now STELLITE ALLOY NO. 156.

HAYNES STELLITE ALLOY NO. 157

Deloro Stellite, Inc.
Now STELLITE ALLOY NO. 157.

HAYNES STELLITE ALLOY NO. 158

Deloro Stellite, Inc.
Now STELLITE ALLOY NO. 158.

HAYNES STELLITE ALLOY NO. 19

Deloro Stellite, Inc.
Now STELLITE ALLOY NO. 19.

HAYNES STELLITE ALLOY NO. 190 PM

Deloro Stellite, Inc.
Now STELLITE ALLOY NO. 190 PM.

HAYNES STELLITE ALLOY NO. 21 (SOLID)

Deloro Stellite, Inc.
Now STELLITE ALLOY NO. 21 (SOLID).

HAYNES STELLITE ALLOY NO. 3

Now HAYNES STELLITE ALLOY NO. 3 PM.

HAYNES STELLITE ALLOY NO. 3 PM

Deloro Stellite, Inc.
Now STELLITE ALLOY NO. 3 PM.

HAYNES STELLITE ALLOY NO. 4

Cabot Corporation
C 0.6, Co 51, Cr 30, W 14, Ni 0-3.
Cast: 39,100 TS; 39,100 YS; 1 El; 1 RA; 444 Brin. For castings for wear resistant cutting tools and dies; heat, wear and corrosion resistant. *Obsolete*

HAYNES STELLITE ALLOY NO. 6 (POWDER)

Deloro Stellite, Inc.
Now STELLITE ALLOY NO. 6 (POWDER).

HAYNES STELLITE ALLOY NO. 6 (SOLID)

Deloro Stellite, Inc.
Now STELLITE ALLOY NO. 6 (SOLID).

HAYNES STELLITE ALLOY NO. 6 PM

Deloro Stellite, Inc.
Now STELLITE ALLOY NO. 6 PM.

HAYNES STELLITE ALLOY NO. 6K

Haynes International, Inc.
C 1.6, Cr 31, W 4.5, Fe 0-3, Ni 0-3, bal Co.
Sheet: 176,500 TS; 3.5 El; 460 Brin. At 1500°F: 70,200 TS; 17.0 El. For high temperature applications, knives, scrapers; heat and corrosion resistant. *Obsolete*

HAYNES STELLITE NO. 21 (POWDER)

Deloro Stellite, Inc.
Now STELLITE NO. 21 (POWDER).

HAYNES STELLITE NO. 23

Cabot Corporation
C 0.35-0.5, Cr 23-29, W 4-7, Ni 0-1.5, Fe 0-2, bal Co.
Cast: 110,000 TS; 86,000 YS; 9 El; 12 RA; 300 Brin. For turbine blades, jet engine components; high oxidation resistance. *Obsolete*

HAYNES STELLITE NO. 27

Cabot Corporation
C 0.35-0.5, Cr 23-29, Mo 5-7, Ni 15, 30 min Co.
Cast: 90,000 TS; 65,000 YS; 12 El; 13 RA; 200 Brin. For turbine blades, jet engine components; high oxidation resistance. *Obsolete*

HAYNES STELLITE NO. 30

Cabot Corporation
C 0.35-0.5, Cr 23-29, Mo 5-7, Ni 13-17, 2% min Fe, bal Co.
Cast: 97,000 TS; 81,000 YS; 10 El; 12 RA; C-27 Brin. For turbine blades, jet engines; high heat resistant. *Obsolete*

HAYNES STELLITE STAR J-METAL PM

Deloro Stellite, Inc.
Now STELLITE STAR J-METAL PM.

HAYNES-NI-MANG

Haynes International, Inc.
C 0.6, Cr 2.5, Si 0.2, Mn 13.5, Ni 3.5, bal Fe.
Covered electrode for AC-DC welding and build-up. As welded: 88 Rock B; work hardens to 46 Rock C. Severe impact resistance; moderate to slight abrasion. *Obsolete*

HAYNES-STELLITE NO. 2

Cabot Corporation
Co-Cr-W-C.
Cast: 67,050 TS. For tools for metal cutting; heat and corrosion resistant. *Obsolete*

HAYNES-STELLITE NO. 8

Cabot Corporation
C-Cr-Co-W.
Cast: 387 Brin. For welding rod. *Obsolete*

HAYSTELLITE 954

Haynes International, Inc.
Co 10, 90.0 tungsten carbide.
Powder; may be blended with other metal powders. *Obsolete*

HAYSTELLITE 956

Haynes International, Inc.
Tungsten carbide.
Powder; for blending with other metal powders. *Obsolete*

HAYSTELLITE 967

Haynes International, Inc.
C 3.9, Fe 0-2, Co 12, bal W.
For plasma spray; 91 Rock A. *Obsolete*

HAYSTELLITE COMPOSITE NO. 1

Haynes International, Inc.
50.0 tungsten carbide, 50.0 nickel-alloy matrix.
Powder for flame spray hard surfacing; 91 Rock A. To resist hot and cold abrasion. *Obsolete*

HAYSTELLITE COMPOSITE NO. 3

Haynes International, Inc.
15.0 tungsten carbide, 85.0 nickel-alloy matrix.
Powder for flame spray hard surfacing; 91 Rock A. Resists hot and cold abrasion. *Obsolete*

HAYSTELLITE COMPOSITE NO. 4

Haynes International, Inc.
60.0 tungsten carbide, 40.0 nickel-alloy matrix.
Powder for manual torch hard surfacing; 91 Rock A. To resist hot and cold abrasion. *Obsolete*

HAYSTELLITE COMPOSITE NO. 6

Deloro Stellite, Inc.
35.0 tungsten carbide, 65.0 nickel-alloy matrix.
Powder for flame spray hard surfacing; 91 Rock A. Resists hot and cold abrasion. *Obsolete*

HAYWOOD'S HIGH TENSILE BRONZE GRADE 8

Haywoods NCA Metal Ltd.
Zn 35, Al 7, bal Cu.
Forged: 86,000 TS; 61,000 YS; 36 El; 115 Brin. For corrosion resisting, high strength cast and forged bronze parts. Noncorrosive.

HAZEL BRONZE

Thomas Bolton Ltd.
Cu alloy.
For nuts, bolts, studs. *Obsolete*

HB

Now FINKL HB.

HB 5

Thyssen Edelstahlwerke AG
Al 10.5, Fe 4, Mn 4, Zn 1, Cr 0.5, bal Cu.
For gears; corrosion resistant Al bronze. *Obsolete*

HB 5
Thyssen Edelstahlwerke AG
Al 10.5, Fe 4, Mn 4, Zn 1, Cr 0.5, bal Cu.
For gears; corrosion resistant aluminum bronze.

HB-18
Hobart Welding Products
C 0.1, Mn 1.3, Si 0.3, Mo 0.41, bal Fe.
Weld metal. As welded: 94,000 psi TS; 78,000 psi YS; 20 El.
Welding wire electrode, for welding pipe. AWS E 70S-1B.

HB-25
Hobart Welding Products
C 0.09, Mn 0.64, Si 0.28, bal Fe.
Weld metal. As welded: 75,000 psi TS; 60,000 psi YS; 24 El.
Welding wire for welding killed or semi-killed steels. AWS
E70S-3.

HB-28
Hobart Welding Products
C 0.08, Mn 1.18, Si 0.62, bal Fe.
Weld metal. As welded: 89,000 psi TS; 70,000 psi YS; 22 El.
Good weldability with CO_2 gas. AWS E70S-6.

HC 5
Thyssen Edelstahlwerke AG
C 0.1, Cr 2.5-2.7, Si 0.35, Mn 0.4, bal Fe. *Obsolete*
For case hardened parts, dies; carburizing steel. *Obsolete*

HC 8
Thyssen Edelstahlwerke AG
C 0.25, Cr 2.5-2.7, Si 0.4, Mn 0.35, bal Fe.
For case hardened parts; carburizing steel. *Obsolete*

HC 9
Thyssen Edelstahlwerke AG
C 0.2, Cr 3-3.2, Mo 0.25, V 5, bal Fe.
For case hardened parts; carburizing steel. *Obsolete*

HC QUALITY
Delta Metal (BW) Ltd.
60% min Cu, 2% min Pb, bal Zn.
Extruded: 55,000 TS; 30-35 El; 25 RA; 100 Brin. Drawn:
60,000 TS; 25 El; 22 RA; 110 Brin. For spun parts, hardware;
free-cutting. *Obsolete*

HC-250
Abex Corp.
Cast iron. C 2.5, Ni 0-0.3, Cr 26, Mn 0.75, Si 1, bal Fe.
Cast heat resistant alloy; good high temperature strength and
creep strength; abrasion resistant.

HC-HC
Delsteel Inc.
C 1.85, Cr 12.25, bal Fe.
For punches and dies; non-deforming.

HC250
Abex Corp.
C, Cr, bal Fe.
Annealed: 85,000 TS; 70,000 YS; 3 El; 350 Brin. Heat treated:
100,000 TS; 80,000 YS; 750 Brin. For cylinder liners, fan
blades, molds, nozzles; abrasion resistant, hot gas resistant.

HCM9M
Sumitomo Metal America Inc.
C 0-0.08, Si 0-0.5, Mn 0.3-0.7, Cr 8-10, Mo 1.8-2.2, bal Fe.
High strength boiler tube alloy; intermediate high
temperature strength between austenitic stainless steels and
commercial low alloy steels.

HCR ALLOY
Telcon Metals Ltd.
Ni 50, Fe 50.
Soft magnetic alloy, rectangular hysteresis loop, for magnetic
amplifers.

HCS16
Kawasaki Steel Corp.
Stainless steel. C 0.32, Mn 0.3, P 0-0.035, S 0-0.009, Si 0.8,
Cr 15.6, bal Fe.
97,000 psi TS; 27 El. Ferritic stainless steel for automotive
and motorcycle disc brakes.

HCS27
Kawasaki Steel Corp.
C 0.05, Mn 0.5, P 0-0.03, S 0-0.01, Si 0.4, Cr 26.5, bal Fe.
76,000 psi TS; 28 El. Glass-to-metal sealing alloy.

HD10
Spencer Clark Metal Industries Ltd.
C 0.32, Cr 3, W 9.5, V 0.3, Mo 0.3, bal Fe.
Tungsten hot work steel for gripper and heading dies.

HDA-188
Union Carbide Corp.
C 0.08, Cr 22.5, Ni 22, W 15, La 0.08, bal Co.
Cobalt base alloy for high temperature operation. Same as
HAYNES ALLOY No 188.

HDNC-400
Remington Arms Co. Inc.

Ni-Cu alloy, sintered. Medium strength, high ductility with
outstanding corrosion resistance. Nonmagnetic; can be
used for filter applications. Similar to Monel 400.

HDNC-500
Remington Arms Co. Inc.
Ni-Cu-Si alloy.
Sintered. High strength: 110,000 psi TS; 20 Rock C. Anti-
galling properties up to 1100°F; good corrosion resistance;
nonmagnetic. Similar to Monel 505.

HDW 2
Teledyne Firth Sterling
Steel. C 0.4, Cr 5, V 0.25, Mo 1.5, Si 0.8, bal Fe.
For dies, mandrels, shear blades; hot work steel, tough.
Obsolete

HDW 3
Teledyne Firth Sterling
Steel. C 0.4, Cr 5, Mo 1.5, W 1.4, Si 0.8, bal Fe.
For dies, mandrels, shear blades; hot work steel, tough.
Obsolete

HE 2048
American manufacture
C 0.4, Co 15, Ni 30, Fe 21, Cr 26, Mo 4, W 2, B 0.15.
Cast. For high temperature applications; high strength, heat
resistant.

HE-30
Hansell-Elcock Co.
C 3.2-3.3, Si 2.2, Mn 0.8, bal Fe.
Cast: 30,000-40,000 TS; 180-200 Brin. For motor blocks,
machine tool beds, gibs; cast iron.

HE-40
Hansell-Elcock Co.
C 3.1-3.25, Si 1.6-1.8, Mn 0.8, bal Fe.
Cast: 40,000-50,000 TS; 200-240 Brin. For radical drill, arms
and bases; cast iron.

HE-50
Hansell-Elcock Co.
C 2.9-3.1, Si 1.3-1.5, Mn 0.8, bal Fe.
Cast: 50,000-60,000 TS; 230-265 Brin. For machine tool
runways, saddles, gears; cast iron.

HE-60
Hansell-Elcock Co.
C 2.9-3.1, Si 1.3-1.5, Mn 0.8, Ni 1.25-1.5, Mo 0.5, bal Fe.
Cast: 60,000-70,000 TS; 240-280 Brin. For crankshafts,
hydraulic rams; cast iron, tough.

HEADER
Uddeholm Corp.
C 0.9, V 0.2, bal Fe.
For tools, cold heading dies; water hardened. *Obsolete*

HEADER DIE DX
Carpenter Technology Corp.
C 1.35, Mn 0.3, Si 1, Cr 6.25, Mo 1, V 6, bal Fe.
Header dies, brick mold liners.

HEADMORE
Bissett Steel Co.
C 0.6, Cr 1, V 0.2, bal Fe.
For header dies; oil hardening. *Obsolete*

HEAT RESISTANT STEEL
English manufacture
Fe 70, Cr 15, Co 14, Mn 0.5, C 0.5, Si 0.5.
For corrosion and heat resisting parts; U. S. Pat. 1357549.

HEAT RESISTING ACID METAL
Belmont Metals Inc.
Cu 55, Pb 6, Zn 9, Ni 30.
For chemical apparatus; acid resistant.

HEAT-RESISTING STEELS, NO. 4
Jessop Steel Co.
C 0-0.3, Cr 8-9, Ni 21-22, bal Fe.
Annealed: 90,000-110,000 TS; 30,000-40,000 YS; 25-35 El;
45-60 RA; 160 Brin. For furnace parts; corrosion, heat and
abrasion resistant. *Obsolete*

HEAVY ALLOY
General Electric Ltd.
W 76-90, Ni 1-16, Cu 3-20.
Sintered: 80,000 TS; 74,000 YS; 1 El; 290 Brin. For balancing
crankshafts of aero motors; powder metals. *Obsolete*

HEAVY AXLE BEARING
English manufacture
Zn 47, Sn 38, Sb 6, Pb 4, Cu 1.
For axle bearings; anti-friction.

HEAVY BEARING
English manufacture
Sn 85, Cu 7.5, Sb 7.5.
For bearings, bushings; anti-friction.

HEAVY DUTY BABBITT
Belmont Metals Inc.
Sb 8.33, Cu 8.33, bal Sn.
12,300 psi TS. For high speed bearings; melting point 792°F.

HEAVY DUTY SW-16
Solar Basic Industries
C 0.2, bal Fe.
For welding electrodes; for light gage steel.

HEAVY METAL TM-17
Surahammars Bruk
Cu, Ni, bal W.
Sintered: 100,000 TS; 90,000 YS; 4 El; 250-300 Brin. For
radioactive shields, pendulums, counterbalance weights;
density 16.7. *Obsolete*

HEAVY METAL TM-170
Surahammars Bruk
Cu, Ni, bal W.
Sintered: 110,000 TS; 95,000 YS; 4 El; 250-300 Brin. For
radioactive shields, pendulums, counterbalance weights;
density 17.0. *Obsolete*

HEAVY METAL TM-175
Surahammars Bruk
Cu, Ni, bal W.
Sintered: 117,000 TS; 102,000 YS; 3.5 El; 250-300 Brin. For
radioactive shields, pendulums, counterbalance weights;
density 17.5. *Obsolete*

HEAVY METAL TM-18

Surahammars Bruk
Cu, Ni, bal W.
Sintered: 117,000 TS; 108,000 YS; 2.5 El; 250-300 Brin. For radioactive shields, pendulums, counterbalance weights; density 18.0. *Obsolete*

HEAVY PRESSURE

NL Industries
Sb, Cu, bal Sn.
Cast: 30 Brin. For bearings for crushers; heavy duty Babbitt. *Obsolete*

HEAVY PRESSURE MILL GLYCO

Joseph T. Ryerson & Son Inc.
Sn, Sb, bal Pb.
For bearings; Babbitt.

HEAVY TUNGSTEN ALLOY

Now KENNERTIUM W-2.

HEAVY TUNGSTEN ALLOY CLASS 2

Kennametal Inc.
W 95, 5% Ni + Cu.
Sintered: 50,000 TS; 270-290 Brin. For radioactive shielding, gyroscope components; sintered, density 18-18.5. *Obsolete*

HEAVY TUNGSTEN ALLOY CLASS 3

Kennametal Inc.
W 90, 10% Ni + Cu.
Sintered: 80,000 TS; 3 El; 305 Brin. For radioactive shielding, gyroscope components; density 17-17.5. *Obsolete*

HEAVY TUNGSTEN-W2

Kennametal Inc.
W 97.5, Ni 2.5.
For shielding, isotope containers, gyro parts, counterweights, electrical contacts. Heavy metal. *Obsolete*

HEAVY TUNGSTEN-W5

Kennametal Inc.
W 95, Ni 3.3, Cu 1.7.
For rocket nozzles, jet and guide vanes. Heavy metal, heat resistant. *Obsolete*

HEBONITE

English manufacture
C, alloy, bal Fe.
For heat resistant parts.

HEC

Pose-Marre Edelstahlwerk G.m.b.H.
C 2, Cr 2, bal Fe.
As tempered: 60 Rock C. For wear resistance against abrasion.

HECKFORD "HP" NICKEL

Arthur E. Heckford Ltd.
Nickel. Ni 99.98.
Electrical resistance alloy. High purity grade made by powder metallurgy.

HECKFORD ALLOY 400

Arthur E. Heckford Ltd.
Copper. Cu 32, Ni 66, Fe 1, Mn 1.
Electrical resistance alloy. Used for terminations up to a maximum operating temperature of 450°C.

HECKFORD N/C 37/18

Arthur E. Heckford Ltd.
Nickel. Ni 37, Mn 1, Cr 20, bal Fe.
Electrical resistance alloy. Suitable for continuous operation up to a maximum temperature of 1050°C. For furnace use with atmospheres that might otherwise cause dry corrosion of higher nickel content materials.

HECKFORD N/C 60/16

Arthur E. Heckford Ltd.
Nickel. Ni 60, Fe 22, Cr 16.
Electrical resistance alloy. Suitable for use up to 1100°C.

HECKFORD N/C 76/16

Arthur E. Heckford Ltd.
Nickel. Ni 76, Fe 8, Cr 15.5.
Electrical resistance alloy. Suitable for use up to 1150°C with good oxidation resistance. Used in furnace construction, chemical, food processing and nuclear engineering.

HECKFORD N/C 80/20C

Arthur E. Heckford Ltd.
Nickel. Fe 1, Mn 0.25, Cr 19.5, bal Ni.
Electrical resistance alloy. For applications subject to frequent switching and wide temperature fluctuations.

HECKFORD N/C 80/20W

Arthur E. Heckford Ltd.
Nickel. Fe 1, Mn 1.2, Cr 20, bal Ni.
Electrical resistance alloy. Suitable for continuous operation up to a maximum temperature of 1150°C.

HECLA

Cerro Metal Products Co.
Cu 60, Pb 1, Sn 0.7, bal Zn.
Drawn: 74,000 TS; 43,000 YS; 21 El; 25 RA; 137 Brin. Normalized: 63,000 TS; 51,000 YS; 43 El; 46 RA; 100 Brin. For screw machine parts, and hot pressings. Corrosion resistant; free-cutting; CA 482.

HECLA 10

Dunford Hadfields Ltd.
C 0.5-0.6, Mn 0.5-0.8, Si 0.05-0.35, Ni 0.5-0.8, bal Fe. Annealed: 100,000 TS; 55,000 YS; 22 El; 190 Brin. Heat treated: 160,000 TS; 110,000 YS; 12 El; 330 Brin. For structural parts, shafts, gears, axles, countershafts, punches. Water or oil harden, tough, shock resistant. *Obsolete*

HECLA 100

Dunford Hadfields Ltd.
C 0.35-0.45, Mn 1.2-1.5, Cr 0.3-0.6, Ni 0.5-1, Mo 0.15-0.25, bal Fe.
Bar: 101,000 TS; 72,000 YS; 22 El; 201-255 Brin. For structural parts, gears, shafts, bolts, fasteners. Oil hardening, shock resisting. *Obsolete*

HECLA 104

Dunford Hadfields Ltd.
C 0.5-0.7, Cr 0.5-0.8, Mn 0.5-0.8, bal Fe.
Bar: 123,000-145,000 TS; 12-15 El; 248-341 Brin. For structural parts, bolts, crankshafts, punches. SAE 5160. *Obsolete*

HECLA 105

Dunford Hadfields Ltd.
C 0.25-0.45, Mn 0.6-0.95, Cr 0.85-1.15, bal Fe.
Bar: 100,000-123,000 TS; 72,000-92,000 YS; 18-22 El; 201-302 Brin. For shafts, bolts, gears, fasteners, countershafts. SAE 5130-5140 steel, oil hardening. *Obsolete*

HECLA 110

Dunford Hadfields Ltd.
C 0.35-0.45, Cr 0.9-1.4, Ni 1.2-1.6, Mo 0.1-0.2, bal Fe.
Bar: 112,000 TS; 81,000 YS; 20 El; 223-277 Brin. For gears, bolts, countershafts, fasteners, crankshafts. Oil hardening, tough. *Obsolete*

HECLA 115

Dunford Hadfields Ltd.
C 0.3-0.45, Si 0.1-0.35, Mn 0.7-0.9, Ni 0.6-1, bal Fe. Normalized: 78,000 TS; 20 El; 152-207 Brin. For structural parts, gears, axles, shafts, bolts. Water hardening. *Obsolete*

HECLA 116

Dunford Hadfields Ltd.
C 0.25-0.35, Cr 0.5-1, Ni 2.75-3.5, Mo 0-0.65, bal Fe.
Bar: 112,000-146,000 TS; 80,000-112,000 YS; 16-20 El; 223-341 Brin. For gears, shafts, axles, bolts, countershafts. Oil hardening, shock resisting. *Obsolete*

HECLA 120

Dunford Hadfields Ltd.
C 0.6, Cr 9, Si, bal Fe.
For valves; heat resistant. *Obsolete*

HECLA 135

Dunford Hadfields Ltd.
Alloy steel. C 0.6, Si 0.3, Mn 0.3, Cr 2, Ni 2, Mo 0.45, bal Fe. For hot-work extrusion components.

HECLA 138

Dunford Hadfields Ltd.
C 0.27-0.35, Cr 0.5-0.8, Ni 2.3-2.8, Mo 0.4-0.7, bal Fe. Bar: 123,000-224,000 TS; 92,000-180,000 YS; 10-18 El; 248-444 Brin. For gears, shafts, axles, bolts, countershafts. Shock resisting, oil hardening. *Obsolete*

HECLA 138H

Dunford Hadfields Ltd.
Alloy steel. C 0.4, Si 0.3, Mn 0.6, Cr 0.6, Ni 2.6, Mo 0.4, bal Fe.
For hot-work extrusion components.

HECLA 142

Dunford Hadfields Ltd.
C 0.25-0.4, Cr 0.75-1.5, Ni 3-4.5, Mo 0.2-0.65, bal Fe. Bar: 134,000-180,000 TS; 103,000-144,000 YS; 14-17 El; 269-415 Brin. For structural parts, gears, pinions, shafts, bolts, countershafts. Oil hardening, shock resisting. *Obsolete*

HECLA 143

Dunford Hadfields Ltd.
C 0.12-0.18, Si 0.1-0.35, Mn 0.3-0.6, Cr 0.6-1.1, Ni 3-3.75, bal Fe.
Bar: 123,000-145,000 TS; 13-15 El. For gears, bolts, cams, camshafts, fasteners. Case hardening steel. Tough, shock resistant. *Obsolete*

HECLA 143B

Dunford Hadfields Ltd.
C 0.12-0.18, Cr 0.6-1.1, Ni 3-3.75, Mo 0.1-0.25, bal Fe. Bar: 145,000 TS; 13 El; 88,000 YS. Heat treated: 180,000 TS; 145,000 YP; 15 El; 370 Brin. For gears, bolts, cams, camshafts. Case hardening steel. Tough, shock resistant. *Obsolete*

HECLA 146

Dunford Hadfields Ltd.
C 0.12-0.18, Cr 1-1.4, Ni 3.8-4.5, bal Fe.
Bar: 190,000 TS; 12 El; 145,000 YP; 50 RA; 380 Brin. For gears, bolts, cams, camshafts, clutch parts, pneumatic tools. Tough, shock resisting, case hardening; abrasion resistant. *Obsolete*

HECLA 146B

Dunford Hadfields Ltd.
C 0.12-0.18, Cr 1-1.4, Ni 3.8-4.5, Mo 0.15-0.35, bal Fe. Bar: 190,000 TS; 12 El; 145,000 YP; 50 RA; 375 Brin. For gears, bolts, cams, camshafts, clutch parts, pneumatic tools. Tough, shock resisting, case hardening; abrasion resistant. *Obsolete*

HECLA 147

Dunford Hadfields Ltd.
C 0.3-0, Cr 5, Mo 0.5, bal Fe.
For oil refinery castings and tubes; creep and corrosion resistant. *Obsolete*

HECLA 149C

Dunford Hadfields Ltd.
Alloy steel. C 0.3, Si 0.3, Mn 0.35, Cr 3.5, Mo 0.45, W 9, V 0.3, bal Fe.
For hot-work extrusion components.

HECLA 150

Dunford Hadfields Ltd.
C 0.35-0.45, Cr 0.9-1.5, Mo 0.2-0.4, bal Fe.
Bar: 100,000-180,000 TS; 72,000-112,000 YS; 16-22 El; 201-341 Brin. For gears, shafts, bolts, fasteners, countershafts. SAE 4137-4140 steel, oil hardening. *Obsolete*

HECLA 151

Dunford Hadfields Ltd.

C 0.14-0.2, Si 0.1-0.35, Mn 0.3-0.6, Ni 1.5-2, Mo 0.2-0.3, bal Fe.

Bar: 101,000 TS; 18 El; 88,000 YP; 245 Brin. Heat treated: 150,000 TS; 106,000 YP; 16 El; 300 Brin. For gears, bolts, cams, camshafts. SAE 4617 case hardening steel. Tough, wear resistant. *Obsolete*

HECLA 151B

Dunford Hadfields Ltd.

C 0.2-0.28, Si 0.1-0.35, Mn 0.3-0.6, Ni 1.5-2, Mo 0.2-0.3, bal Fe.

Bar: 123,000 TS; 15 El; 93,000 YP; 260 Brin. Heat treated: 186,000 TS; 156,000 YP; 9 El; 360 Brin. Gears, bolts, cams, camshafts. SAE 4617 and 4620 case hardening steel. Tough, abrasion resistant. *Obsolete*

HECLA 152

Dunford Hadfields Ltd.

C 0.35-0.45, Cr 0.9-1.4, Ni 1.3-1.8, Mo 0.2-0.35, bal Fe.

Bar: 112,000-224,000 TS; 80,000-180,000 YS; 8-20 El; 223-444 Brin. For gears, shafts, axles, bolts, countershafts, fasteners. Oil hardening, shock resisting. *Obsolete*

HECLA 153

Dunford Hadfields Ltd.

C 0.2-0.45, Cr 0.5-1.5, Mo 0.5-0.9, bal Fe.

Bar: 123,000-146,000 TS; 92,000-112,000 YS; 16-18 El; 248-341 Brin. For bolts, fasteners, gears, shafts, countershafts. Oil or water hardening. *Obsolete*

HECLA 157

Dunford Hadfields Ltd.

C 0.35-0.45, Cr 0.9-1.5, Mo 0.2-0.4, bal Fe.

Bar: 100,000-180,000 TS; 72,000-112,000 YS; 16-22 El; 201-341 Brin. For gears, shafts, bolts, fasteners, countershafts. SAE 4137-4140 steel, oil hardening. *Obsolete*

HECLA 159

Dunford Hadfields Ltd.

C 1.5, Cr 12, Mo, bal Fe.

For tools, dies, cutters; air hardening. *Obsolete*

HECLA 160C

Dunford Hadfields Ltd.

Alloy steel. C 0.35, Si 1, Mn 0.35, Cr 1.5, Mo 0.45, W 3.5, V 0.2, bal Fe.

For hot-work extrusion components.

HECLA 163

Dunford Hadfields Ltd.

C 0-0.16, Cr 0-0.3, Ni 4.5-5.5, Mo 0.15-0.3, bal Fe.

Bar: 146,000 TS; 13 El; 115,000 YP; 275 Brin. Oil hardened: 193,000 TS; 155,000 YP; 14 El; 362 Brin. For gears, bolts, cams, camshafts, pawls, machinery parts. Tough, shock resisting, case hardening, oil harden. *Obsolete*

HECLA 167G

Dunford Hadfields Ltd.

Alloy steel. C 0.4, Si 0.3, Mn 0.6, Cr 1.5, Ni 0.7, Mo 0.4, bal Fe.

For hot-work extrusion components.

HECLA 172

Dunford Hadfields Ltd.

Alloy steel. C 0.25, Si 0.4, Mn 0.3, Cr 3, Mo 0.6, W 0.6, Mo 0.9.

For hot-work extrusion components.

HECLA 174

Dunford Hadfields Ltd.

Alloy steel. C 0.35, Si 1, Mn 0.5, Cr 5, Mo 1.3, V 0.85, bal Fe.

For hot-work extrusion components.

HECLA 174

Dunford Hadfields Ltd.

C 0.4, Cr 5, Mo 1.3, V 0.8, bal Fe.

Heat treated: 270,000 TS; 230,000 YS; 6 El. For hot forging dies for brass and aluminum, extrusion dies; hot work steel, air hardened. *Obsolete*

HECLA 177

Dunford Hadfields Ltd.

Alloy steel. C 0.35, Si 1, Mn 0.35, Cr 5, Mo 1.3, W 1.2, V 0.2, bal Fe.

For hot-work extrusion components.

HECLA 179

Dunford Hadfields Ltd.

Alloy steel. C 0.3, Si 0.3, Mn 0.35, Cr 3, Mo 3, V 0.4, bal Fe.

For hot-work extrusion components.

HECLA 18

Dunford Hadfields Ltd.

C 0.65-0.9, Si 0-0.35, Mn 0.3-0.6, bal Fe.

Annealed: 120,000 TS; 66,000 YS; 15 El; 223 Brin. Oil hardened: 190,000 TS; 142,000 YP; 12 El; 388 Brin. For hammers, tools, punches, springs, structural parts, thrust washers, lock washers. SAE 1064-1078 steel, water hardening. *Obsolete*

HECLA 181

Dunford Hadfields Ltd.

C 0-0.2, Cr 0.4-0.8, Ni 0.6-1, Mo 0-0.1, bal Fe.

Bar: 101,000 TS; 18 El; 55,000 YP; 200 Brin. Oil hardened: 145,000 TS; 115,000 YP; 17 El; 302 Brin. For gears, bolts, cams, camshafts. Tough, shock resisting, case hardening. *Obsolete*

HECLA 182

Dunford Hadfields Ltd.

C 0-0.2, Cr 0.6-1, Ni 0.85-1.25, Mo 0-0.1, bal Fe.

Bar: 123,000 TS; 15 El. Heat treated: 215,000 TS; 170,000 YP; 15 El; 425 Brin. For gears, bolts, cams, camshafts, chuck jaws. Tough, shock resisting, case hardening, wear and abrasion resistant. *Obsolete*

HECLA 183

Dunford Hadfields Ltd.

C 0-0.2, Cr 0.75-1.25, Ni 1-1.5, Mo 0.08-0.15, bal Fe.

Bar: 146,000 TS; 12 El. Heat treated: 218,000 TS; 178,000 YP; 13 El; 430 Brin. For gears, bolts, cams, camshafts, chuck jaws. Tough, shock resisting, case hardening, wear and abrasion resistant. *Obsolete*

HECLA 184

Dunford Hadfields Ltd.

C 0-0.2, Cr 0.75-1.25, Ni 1.5-2, Mo 0.1-0.2, bal Fe.

Bar: 168,000 TS; 12 El. Heat treated: 200,000 TS; 150,000 YP; 15 El; C 42 Rock. For gears, bolts, cams, camshafts. Tough, shock resisting, case hardening, wear resistant. *Obsolete*

HECLA 185

Dunford Hadfields Ltd.

C 0.2, Cr 1.4-1.7, Ni 1.8-2.2, Mo 0.15-0.25, bal Fe.

Bar: 120,000 TS; 18 El; 230 Brin. Heat treated: 220,000 TS; 160,000 YP; 13 El; C 40 Rock. For gears, bolts, cams, camshafts, arbors, bearings, chuck jaws. Tough, shock resisting, case hardening, wear resistant. *Obsolete*

HECLA 190

Dunford Hadfields Ltd.

Alloy steel. C 0.35, Si 1, Mn 0.35, Cr 1.5, Mo 1.5, W 1.5, V 0.2, bal Fe.

For hot-work extrusion components.

HECLA 191

Dunford Hadfields Ltd.

C 0.13-0.17, Cr 0.55-0.8, Ni 0.4-0.7, Mo 0.08-0.15, bal Fe.

Bar: 101,000 TS; 18 El; 80,000 YS; 220 Brin. Heat treated: 200,000 TS; 155,000 YS; 15 El; 375 Brin. For gears, bolts, cams, camshafts. Tough, shock resisting, case hardening, oil harden, wear resistant. *Obsolete*

HECLA 192

Dunford Hadfields Ltd.

C 0.18-0.23, Cr 0.55-0.8, Ni 0.4-0.7, Mo 0.08-0.15, bal Fe.

Bar: 123,000 TS; 15 El; 77,000 YS; 240 Brin. Heat treated: 220,000 TS; 180,000 YS; 12 El; 430 Brin. For gears, bolts, cams, camshafts. Tough, shock resisting, case hardening, oil harden, wear resistant. *Obsolete*

HECLA 193

Dunford Hadfields Ltd.

C 0.22-0.26, Cr 0.55-0.8, Ni 0.4-0.7, Mo 0.08-0.15, bal Fe.

Bar: 146,000 TS; 85,000 YS; 260 Brin. Heat treated: 220,000 TS; 180,000 YP; 12 El; 440 Brin. For gears, bolts, cams, camshafts, pawls, chuck jaws, die bodies. Tough, shock resisting, case hardening, oil hardening. *Obsolete*

HECLA 196

Dunford Hadfields Ltd.

C 0.25-0.45, Cr 1.4-1.8, Ni 0-0.4, Mo 0.1-0.25, Al 0.9-1.3, bal Fe.

Oil hardened: 100,000-123,000 TS; 72,000-92,000 YS; 17-20 El. For gears, bolts, cams, camshafts, rolls, crankshafts, anvils. *Obsolete*

HECLA 197

Dunford Hadfields Ltd.

C 0-0.22, Cr 0.4-0.6, Ni 1.5-2, Mo 0.2-0.3, bal Fe.

Bar: 123,000 TS; 15 El; 78,000 YS; 248 Brin. Heat treated: 218,000 TS; 178,000 YP; 13 El; 430 Brin. For gears, cams, camshafts, bolts. Tough, shock resisting, case hardening. SAE 4320, oil hardening. *Obsolete*

HECLA 206

Dunford Hadfields Ltd.

C 0.12-0.17, Cr 0.3-0.5, bal Fe.

Annealed: 75,000 TS; 48,000 YP; 38 El; 145 Brin. Hardened: 145,000 TS; 120,000 YP; 16 El; 310 Brin. For gears, bolts, cams, camshafts, pawls, machinery parts. Case hardening, water harden. SAE 5115, wear resistant. *Obsolete*

HECLA 207

Dunford Hadfields Ltd.

C 0.16-0.21, Cr 0.6-0.8, bal Fe.

Annealed: 75,000 TS; 50,000 YP; 35 El; 150 Brin. Water hardened: 190,000 TS; 170,000 YP; 13 El; 370 Brin. For gears, bolts, cams, camshafts, fasteners, pawls, machinery parts. Case hardening. SAE 5120, water or oil hardening. *Obsolete*

HECLA 317

Dunford Hadfields Ltd.

C 0.1-0.3, Cr 2.9-3.5, Ni 0-0.4, Mo 0.4-0.7, bal Fe.

Bar: 101,000-135,000 TS; 72,000-103,000 YS; 17-22 El. For gears, bolts, cams, camshafts, rolls, fasteners, crankshafts. Tough, shock resisting, case hardening and nitriding steel. *Obsolete*

HECLA 34

Dunford Hadfields Ltd.

C 0.9-1.25, Si 0-0.35, Mn 0.3-0.7, bal Fe.

Annealed: 100,000 TS; 55,000 YP; 20 El; 200 Brin. Heat treated: 216,00 TS; 152,000 YP; 11 El; 600 Brin. For springs, cutting tools, drills, punches, jigs, and fixtures. SAE 1095 steel. Water harden, wear resistant. *Obsolete*

HECLA 35

Dunford Hadfields Ltd.

C 0.15-0.3, Mn 0.4-0.6, bal Fe.

Forged: 56,000-78,000 TS; 25 El. Normalized: 62,000-85,000 TS; 25 El; 126-180 Brin. For structural parts, fasteners, gears, shafts. SAE 1017-1020 steel, water harden. *Obsolete*

HECLA 36

Dunford Hadfields Ltd.

C 0.7-0.9, Si 0-0.35, Mn 0.55-0.8, bal Fe.

Annealed: 120,000 TS; 66,000 YS; 15 El; 223 Brin. Water harden: 183,000 TS; 135,000 YS; 13 El; 390 Brin. For tools, cutters, punches, drills, hammers, springs, bushings, bearings. SAE-1080 steel, water harden, wear resistant. *Obsolete*

HECLA 37

Dunford Hadfields Ltd.

C 0.35-0.45, Mn 0.6-1, bal Fe.

Normalized: 78,000-100,000 TS; 20-22 El; 152-255 Brin. For gears, bolts, fasteners, axles, shafts, structural parts. SAE 1035-1040 steel, water harden. *Obsolete*

HECLA 40
Dunford Hadfields Ltd.
C 0.25-0.35, Si 0.05-0.35, Mn 0.7-0.9, bal Fe.
Normalized: 72,000-100,000 TS; 20-25 El; 143-255 Brin. For structural parts, bolts, gears, shafts, axles. SAE 1026-1033 steel, water harden. *Obsolete*

HECLA 41
Dunford Hadfields Ltd.
C 0.45-0.6, Si 0.05-0.4, Mn 0.3-1, bal Fe.
Normalized: 90,000-112,000 TS; 63,000-69,000 YS; 18 El. For structural parts, gears, shafts, crankshafts, bolts. SAE 1046-1050 steels, water harden. *Obsolete*

HECLA 42
Dunford Hadfields Ltd.
C 0.6-0.9, Si 0-0.35, Mn 0.55-0.8, bal Fe.
Annealed: 120,000 TS; 65,000 YS; 15 El; 223 Brin. Water harden: 183,000 TS; 135,000 YS; 13 El; 390 Brin. For structural parts, gears, shafts, springs, cutlery, punches. SAE 1064 steel-SAE 1084 steel, water harden. *Obsolete*

HECLA 46
Dunford Hadfields Ltd.
C 0.15, Si 0.05-0.35, Mn 0.4-0.7, bal Fe.
Bar: 72,000 TS; 20 El; 64,000 YS; 155 Brin. Water hardened: 105,000 TS; 80,000 YS; 6 El; 205 Brin. For case hardened gears, shafts, cams. SAE 1015 case hardening steel. Water harden. *Obsolete*

HECLA 66
Dunford Hadfields Ltd.
C 0.1-0.15, Si 0.1-0.35, Mn 0.3-0.6, Cr 0.3, Ni 2.75-3.5, bal Fe.
Bar: 101,000 TS; 18 El; 78,000 YP; 200 Brin. Heat treated: 180,000 TS; 155,000 YS; 14 El; 360 Brin. For gears, bolts, cams, camshafts, clutch parts, piston pins. Tough, shock resistant case hardening. Abrasion resistant. *Obsolete*

HECLA 67
Dunford Hadfields Ltd.
C 0.26-0.34, Cr 1.1-1.4, Ni 3.9-4.3, bal Fe.
Heat treated: 224,000 TS; 180,000 YS; 10 El; 444 Brin. For structural parts, gears, pinions, shafts, bolts, countershafts. Shock resisting, oil hardening. *Obsolete*

HECLA 67B
Dunford Hadfields Ltd.
C 0.26-0.34, Cr 1.1-1.4, Ni 3.9-4.3, Mo 0.2-0.4, bal Fe.
Heat treated: 224,000 TS; 180,000 YS; 10 El; 444 Brin. For structural parts, gears, pinions, countershafts, axles, bolts. Oil hardening, shock resisting. *Obsolete*

HECLA 70
Dunford Hadfields Ltd.
C 0.35-0.45, Cr 0-0.3, Ni 3.25-3.75, bal Fe.
Bar: 112,000 TS; 80,000 YS; 20 El; 223-277 Brin. For gears, shafts, bolts, fasteners, countershafts, pinions. Oil hardening, shock resisting. *Obsolete*

HECLA 73
Dunford Hadfields Ltd.
C 0.2-0.35, Cr 0-0.3, Ni 2.5-3.5, bal Fe.
Bar: 101,000-112,000 TS; 72,000-81,000 YS; 20-22 El; 201-277 Brin. For gears, shafts, bolts, fasteners, countershafts. Oil hardening, tough. *Obsolete*

HECLA 78
Dunford Hadfields Ltd.
C 0.16, Mn 0.45, Cr 0.3, Ni 4.5-5.2, bal Fe.
Bar: 90,000 TS; 20 El; 72,000 YP; 180 Brin. Heat treated: 182,000 TS; 160,000 YP; 15 El; 375 Brin. For gears, bolts, cams, camshafts, pawls, arbors, pneumatic tools. Tough, shock resisting, case hardening. *Obsolete*

HECLA 98
Dunford Hadfields Ltd.
C 0.3-0.4, Cr 0.45-0.75, Ni 1-1.5, bal Fe.
Bar: 100,000-135,000 TS; 72,000-103,000 YS; 17-22 El; 201-321 Brin. For bolts, gears, countershafts, structural parts, fasteners. Oil hardening, tough. *Obsolete*

HECLA A.T.G.
Dunford Hadfields Ltd.
C 0.2, Cr 12, Ni 60, W 4, bal Fe.
For glass industry equipment; heat resistant. *Obsolete*

HECLA A.T.V.
Dunford Hadfields Ltd.
C 0.2, Cr 11, Ni 35, bal Fe.
Treated: 110,000 TS; 66,000 YS; 20 El; 35 RA. For turbine blades and nozzles; heat resistant. *Obsolete*

HECLA BRONZE
Thomas Paulson & Sons Inc.
Sn, Pb, bal Cu.
For bearings; heavy duty. *Obsolete*

HECLA CH31
Dunford Hadfields Ltd.
C 0.1-0.18, Si 0.05-0.35, Mn 0.6-1, bal Fe.
Bar: 72,000 TS; 20 El; 64,000 YS; 155 Brin. Water hardened: 105,000 TS; 80,000 YS; 6 El; 205 Brin. For gears, cams, shafts, bolts, fasteners, pawls, machinery parts. SAE 1016 case hardening steel, water harden. *Obsolete*

HECLA D17
Dunford Hadfields Ltd.
C 0.6-0.7, Si 0-0.35, Mn 0.55-0.8, bal Fe.
Oil hardened: 180,000 TS; 130,000 YS; 12 El; 360 Brin. Annealed: 112,000 TS; 60,000 YS; 17 El; 207 Brin. For structural parts, axles, shafts, springs, cutlery, punches, heavy machine parts, hatchets. SAE 1064 steel. Water harden, wear resistant. *Obsolete*

HECLA EM20
Dunford Hadfields Ltd.
C 0.18, Mn 1.5, Ni 15, Cr 17, Co 12, Mo 2.9, W 2.5, Cb 1, N 0.09, bal Fe.
For jet engine components, turbine rotor blades; heat resistant. *Obsolete*

HECLA EM35
Dunford Hadfields Ltd.
C 0.35, Mn 1.5, Ni 15, Cr 17, Co 12, Mo 2.9, W 2.5, N 0.9, Cb 1, bal Fe.
For jet engine components, turbine rotor blades; heat resistant. *Obsolete*

HECLA HCT-4
Dunford Hadfields Ltd.
C 0.17, Si 0.5, Mn 1, Cr 11.5, Mo 0.6, Cb 0.2, V 0.2, N 0.07, B 0.3, bal Fe.
For jet engine components, turbine rotor blades; heat resistant. *Obsolete*

HECLA MM20
Dunford Hadfields Ltd.
C 0.18, Mn 1.5, Ni 20, Cr 21, Co 20, Mo 2.9, W 2.5, Cb 1, N 0.12, bal Fe.
For jet engine components, turbine rotor blades; heat resistant. *Obsolete*

HECLA MM35
Dunford Hadfields Ltd.
C 0.35, Mn 1.5, Ni 20, Cr 21, Co 20, Mo 2.9, W 2.5, Cb 1, N 0.12, bal Fe.
For jet engine components, turbine rotor blades; heat resistant. *Obsolete*

HECLA NO. 13
Cerro Metal Products Co.
Cu 60, Pb 1.8, Sn 0.7, bal Zn.
Soft: 58,000 TS; 28,000 YS; 35 El; 83 Brin. 1/2 H-temper: 65,000 TS; 38,000 YS; 25 El; 125 Brin. H-temper: 75,000 TS; 52,000 YS; 20 El; 160 Brin. For screw machine products and marine hardware. Free cutting; C 485.

HECLA NO. 69
Cerro Metal Products Co.
Cu 60, Pb 0.7, Sn 0.75, bal Zn.
Soft: 58,000 TS; 28,000 YS; 40 El; 83 Brin. 1/2 H-temper: 65,000 TS; 38,000 YS; 30 El; 125 Brin. H-temper: 75,000 TS; 52,000 YS; 22 El; 160 Brin. For screw machine products and marine hardware. Free cutting; CA 482.

HECLA NO. 7
Cerro Metal Products Co.
Cu 60, Pb 1, Sn 0.7, bal Zn.
For screw machine products. Free cutting, corrosion resistant; C 482.

HECLA S55
Dunford Hadfields Ltd.
C 0.55-0.65, Si 1.5-2, Mn 0.7-1, bal Fe.
Annealed: 118,000 TS; 68,000 YS; 22 El; 240 Brin. Oil harden: 325,000 TS; 280,000 YS; 9 El; 620 Brin. For springs, pneumatic tools, punches. SAE 9255-9260 spring steel, oil harden, tough. *Obsolete*

HECLA SNS
Dunford Hadfields Ltd.
C 0.74-0.84, Si 1.75-2.25, Mn 0.2-0.6, Cr 19-20.5, Ni 1.15-1.65, bal Fe.
Annealed: 105,000 TS; 65,000 YS; Rock B 95. For cutlery, valve parts, ball bearings, surgical instruments. Corrosion resistant. Hardenable. *Obsolete*

HECLA-H GT4
Dunford Hadfields Ltd.
C 0.15-0.19, Cr 10.5-12, Cb 0.15-0.25, Mo 0.5-0.7, V 0.15-0.25, B 0.02, N 0.08, Ti 0.2, bal Fe.
Annealed: 90,000 TS; 38,000 YS; 26 El; B 90 Rock. Hardened: 230,000 TS; 195,000 YS; 10 El; C 47 Rock. For bearings, valves, cutlery, surgical instruments. Corrosion resistant, hardenable. *Obsolete*

HECNUM
Arthur E. Heckford Ltd.
Copper. Cu 55, Ni 45.
Annealed: 70,000 TS; 30,000 YS; 50 El; 110 Brin. For electrical resistance thermocouples. Negligible temperature coefficient.

HECORROS 76
VDM Nickel-Technologie AG
Cu 76-78, Zn 18-22, Al 1.8-2.3, As 0.02-0.035.
Minimum values at 20°C: 340 N/mm^2 TS; 120 N/mm^2 YS; 55 El. For heat exchanger tubes and piping systems on ships and in power plants. Material No. 2.0460. C68700

HEDDAL
VDM Nickel-Technologie AG
C 0.2, Mn 0.3, Si 0.25, bal Fe.
For machinery parts; water hardened. *Obsolete*

HEDDAL
VDM Aluminium GmbH
Aluminum. Mn 1.5, bal Al.
Soft: 14,000 TS; 5690 YS; 30-40 El; 28-32 Brin. Hard: 35,600 TS; 28,500 YS; 3 El; 60 Brin. For heat exchangers, pressure and storage tanks, chemical equipment. Similar to Aluminum 3003, not heat treatable.

HEDDENAL 2
VDM Nickel-Technologie AG
Mg 1.5-3, Mn 0.5-1.5, Cr 0-0.3, bal Al.
Soft: 26,000 TS; 10,000 YS; 20 El; 45 Brin. Hard: 41,000 TS; 36,000 YS; 5 El; 77 Brin. For roofing, hydraulic tubing, architectural trim; good forming and welding properties. *Obsolete*

HEDDENAL 3
VDM Nickel-Technologie AG
Mg 2-4, Mn 0-0.4, Cr 0-0.3, bal Al.
Soft: 28,000 TS; 13,000 YS; 30 El; 47 Brin. Hard: 40,000 TS; 35,000 YS; 10 El; 73 Brin. For aircraft tanks and fittings, fuel lines, marine parts; resists sea water corrosion. *Obsolete*

HEDDENAL 5

VDM Nickel-Technologie AG
Mg 4-5.5, Mn 0-0.8, Cr 0-0.3, bal Al.
Soft: 42,000 TS; 22,000 YS; 35 El; 65 Brin. Hard: 60,000 TS;
50,000 YS; 10 El; 105 Brin. For aircraft and marine parts;
good corrosion resistance. *Obsolete*

HEDDENAL 7

VDM Nickel-Technologie AG
Mg 5.5-7.5, Mn 0-0.8, Cr 0-0.3, bal Al.
For aircraft and marine parts; good corrosion resistance.
Obsolete

HEDDRONAL 3.5

VDM Aluminium GmbH
Aluminum. Mn 0.5, Mg 3.5, bal Al.
Rolled: 35,000 TS; 20-25 El; 70-100 Brin. For structural
towers, unfired pressure vessels, rocket motor parts, missile
containers. Similar to Aluminum 5086, nonheat treatable.
Corrosion resistant.

HEDDRONAL 5

VDM Aluminium GmbH
Aluminum. Mn 0.5, Mg 5, bal Al.
Rolled: 32,000-38,000 TS; 15-22 El; 70-100 Brin. For deck
housing, overhead cranes, ship unloaders, heavy duty
structures. Similar to Aluminum 5456. Not heat treatable;
corrosion resistant.

HEDDRONAL 7

VDM Aluminium GmbH
Aluminum. Mn 0.5, Mg 7, bal Al.
Rolled: 45,000-52,000 TS; 20-25 El; 80-100 Brin. For heavy
duty structures, overhead cranes, ship unloaders. Corrosion
resistant.

HEDDROXAL

VDM Nickel-Technologie AG
Mg 2-4, Mn 0.4, Cr 0-0.3, bal Al.
Soft: 28,000 TS; 13,000 YS; 30 El; 47 Brin. Hard: 40,000 TS;
35,000 YS; 10 El; 73 Brin. For aircraft tanks and fittings, fuel
lines, marine parts; resists sea water corrosion. *Obsolete*

HEDDUR

Daido Steel Co. Ltd.
Cu 3.5-5, Si 0.3-0.8, Mn 0.2-1.2, Mg 0.4-1, bal Al.
Heat treated: 68,000-78,000 TS; 50,000-57,000 YS; 12-16 El;
115-130 Brin. For structural applications, screw machine
products, fasteners. Heat treatable. Similar to Aluminum
2017.

HEDDUR

VDM Nickel-Technologie AG
Cu 2.5-5, Mg 0.2-1.8, Mn 0.3-1.5, bal Al.
Annealed: 27,000 TS; 11,000 YS; 22 El; 47 Brin. Heat treated:
72,000 TS; 57,000 YS; 130 Brin. For aircraft structures and
fittings, fasteners; age hardenable. *Obsolete*

HEDERVAN

Latrobe Steel Co.
C 0.9, V, bal Fe.
For cold heading and striking dies, punches, draw dies; water
hardened; Type W2. *Obsolete*

HEERDT T10F

Dusseldorf-Heerdt GmbH & Co., KG
C 1, Si 0-0.25, Mn 0-0.25, bal Fe.
Heat treated: 121,000-213,000 TS; 84,000-150,000 YS; 11-20
El; 33-47 RA; 235-535 Brin. For drills, hobs, cutting tools;
Type W1; water hardened.

HEERDT T10P

Dusseldorf-Heerdt GmbH & Co., KG
C 1, Si 0-0.25, Mn 0-0.25, bal Fe.
Heat treated: 121,000-213,000 TS; 84,000-150,000 YS; 11-20
El; 33-47 RA; 235-535 Brin. For drills, hobs, cutting tools,
reamers; Type W1; water hardened.

HEERDT T11E

Dusseldorf-Heerdt GmbH & Co., KG
C 1.1, Mn 0-0.25, Si 0-0.25, bal Fe.
Heat treated: 122,000-215,000 TS; 85,000-155,000 YS; 10-18
El; 30-45 RA; 240-555 Brin. For reamers, drills, taps, hobs,
lathe tools; Type W1; water hardened.

HEERDT T11P

Dusseldorf-Heerdt GmbH & Co., KG
C 1.15, Mn 0-0.25, Si 0-0.25, bal Fe.
Heat treated: 122,000-215,000 TS; 85,000-155,000 YS; 10-18
El; 30-45 RA; 240-555 Brin. For reamers, drills, taps, hobs,
lathe tools; Type W1; water hardened.

HEERDT T12E

Dusseldorf-Heerdt GmbH & Co., KG
C 1.3, Mn 0-0.25, Si 0-0.25, bal Fe.
For reamers, drills, taps, hobs, lathe tools; Type W1; water
hardened.

HEERDT T12P

Dusseldorf-Heerdt GmbH & Co., KG
C 1.3, Si 0-0.25, Mn 0-0.25, bal Fe.
For reamers, drills, hobs, lathe tools; Type W1; water
hardened.

HEERDT T7P

Dusseldorf-Heerdt GmbH & Co., KG
C 0.7, Si 0-0.25, P 0-0.25, bal Fe.
Heat treated: 120,000-174,000 TS; 82,000-128,000 YS; 12-22
El; 37-52 RA; 241-352 Brin. For crimpers, punches, springs,
rails, dies; Type W1; water hardened.

HEERDT T8E

Dusseldorf-Heerdt GmbH & Co., KG
C 0.7, Si 0-0.2, P 0-0.2, bal Fe.
Heat treated: 120,000-174,000 TS; 82,000-128,000 YS; 12-22
El; 37-52 RA; 241-352 Brin. For crimpers, punches, dies,
springs, rails; Type W1; water hardened.

HEERDT T8V

Dusseldorf-Heerdt GmbH & Co., KG
C 0.8, Si 0.2, Mn 0.2, bal Fe.
Heat treated: 129,000-188,000 TS; 87,000-143,000 YS; 12-21
El; 35-50 RA; 255-388 Brin. For tools, dies, springs, cutters;
Type W1; water hardened.

HEERDT T9E

Dusseldorf-Heerdt GmbH & Co., KG
C 0.85, Si 0-0.25, Mn 0-0.25, bal Fe.
Heat treated: 130,000-190,000 TS; 88,000-145,000 YS; 11-20
El; 32-48 RA; 260-400 Brin. For tools, dies, springs, cutters,
drills; Type W1; water hardened.

HEERDT T9P

Dusseldorf-Heerdt GmbH & Co., KG
C 0.85, Si 0-0.25, Mn 0-0.25, bal Fe.
Heat treated: 130,000-190,000 TS; 88,000-145,000 YS; 11-20
El; 32-48 RA; 260-400 Brin. For tools, dies, springs, cutters,
drills; Type W1; water hardened.

HELBIMPHY

Creusot-Loire
C, bal Fe.
For tools; superior quality. *Obsolete*

HELBIMPHY 10

Henry A. Kries & Sons Co.
C 1.05-1.15, bal Fe.
For tools, taps, drills; water hardened.

HELBIMPHY 12

Henry A. Kries & Sons Co.
C 1.2, bal Fe.
For tools, drills, taps; water hardened.

HELBIMPHY 14

Henry A. Kries & Sons Co.
C 1.4, bal Fe.
For cutting tools, drills; water hardened.

HELBIMPHY 6

Henry A. Kries & Sons Co.
C 0.7-0.8, bal Fe.
For tools, springs, punches; water hardened.

HELBIMPHY 7

Burndy Corp.
C 0.7-0.8, bal Fe.
For tools, springs, drills; water hardening.

HELBIMPHY 8

Burndy Corp.
C 0.9-1, bal Fe.
For tools, drills, punches, shears; water hardening.

HELBIMPHY 9

Henry A. Kries & Sons Co.
C 1.05-1.15, bal Fe.
For tools, reamers, punches; water hardened.

HELBIMPHY SPECIAL

Creusot-Loire
C, bal Fe.
For tools; extra-superior quality. *Obsolete*

HELCO H-40

Hansell-Elcock Co.
C 3.2, Si 2.2, Ni 1.5, bal Fe.
228 Brin. For hydraulic press cylinders; cast iron, pressure
resistant. *Obsolete*

HELCO HD-50

Hansell-Elcock Co.
C 3.2, Si 2.4, Ni 2, bal Fe.
Cast: 50,000 TS. For castings; cast iron. *Obsolete*

HELCO HD-60

Hansell-Elcock Co.
C 3, Si 2, Ni 2, bal Fe.
Cast: 60,000 TS. For castings; cast iron. *Obsolete*

HELCO N-140

Hansell-Elcock Co.
C 3.1, Si 2.3, Ni 1, bal Fe.
For heavy castings; cast iron. *Obsolete*

HELIOS D

CCS Braeburn Alloy Steel
C, alloy, bal Fe.
For tools, dies. *Obsolete*

HELIOS G

CCS Braeburn Alloy Steel
C, alloy, bal Fe.
For tools, dies. *Obsolete*

HELIOS H

CCS Braeburn Alloy Steel
C, alloy, bal Fe.
For tools, dies. *Obsolete*

HELIOS K

CCS Braeburn Alloy Steel
C, alloy, bal Fe.
For tools, dies. *Obsolete*

HELIOTITAL

Cegedur Pechiney
Aluminum.
Aluminum alloy Al99.85MgTi.

HELLEFORS 123H

Ovako Steel Hellefors AB
C 0.4, Si 0.3, Mn 1.25, bal Fe.
For gears, machinery parts; tough. *Obsolete*

HELLEFORS 134 HF

Ovako Steel Hellefors AB
C 0.55, Si 1.7, Mn 0.7, bal Fe.
For leaf and coil springs, shock resistant. *Obsolete*

HELLEFORS 206A
Ovako Steel Hellefors AB
C 0.9, Cr 0.5, bal Fe.
For punches, forming dies, headers; water hardened.
Obsolete

HELLEFORS 209A
Ovako Steel Hellefors AB
C 1.3, Cr 0.5, bal Fe.
For blanking and forming dies, engravers' tools; water or oil hardened. *Obsolete*

HELLEFORS 209H
Ovako Steel Hellefors AB
C 1.5, Cr 2, bal Fe.
For bearings, liners, sleeves; oil or water hardened. *Obsolete*

HELLEFORS 216 HM
Ovako Steel Hellefors AB
C 0.95, Si 1.5, Cr 1, bal Fe.
For metal and wood working, cutting tools; non-deforming, oil hardening. *Obsolete*

HELLEFORS 249A
Ovako Steel Hellefors AB
C 1.35, Si 0.25, Mn 0.25, Cr 0.3, bal Fe.
For blanking and forming dies, engravers' tools; water or oil hardened. *Obsolete*

HELLEFORS 264C
Ovako Steel Hellefors AB
C 0.18-0.23, Cr 0.2, Mn 0.45, Ni 1, B 0.45, bal Fe.
For chains; water hardened. *Obsolete*

HELLEFORS 267B
Ovako Steel Hellefors AB
C 1, Mn 0.3, Cr 0.5, V 0.1, bal Fe.
For bearings, punches, liners, forming dies; water hardened. *Obsolete*

HELLEFORS 288 J
Ovako Steel Hellefors AB
C 1.1, Cr 0.3, W 1.1, V 0.1, bal Fe.
For structural parts, cutting tools, drill rod. *Obsolete*

HELLEFORS 288A
Ovako Steel Hellefors AB
C 1.2, Si 0.2, Mn 0.3, W 0.55, V 0.1, bal Fe.
For cold-work tools, punches, heading dies; oil or water hardened. *Obsolete*

HELLEFORS 352H
Ovako Steel Hellefors AB
C 0.25, Mn 0.65, Cr 1.05, Mo 0.2, bal Fe.
For gears, shafts; tough. *Obsolete*

HELLEFORS 371 L
Ovako Steel Hellefors AB
C 0.15, Cr 0.6, Ni 3, bal Fe.
Hardened: 128,000 TS; 107,000 YS; 16 El; 55 RA; 270 Brin.
For gears, crankshafts, chain balls; case hardened. *Obsolete*

HELLEFORS 371H
Ovako Steel Hellefors AB
C 0.15, Mn 0.55, Cr 0.65, Ni 1.25, bal Fe.
For gears, camshafts, crankshafts; case hardened. *Obsolete*

HELLEFORS 371L2
Ovako Steel Hellefors AB
C 0.18, Mn 0.55, Cr 0.75, Ni 3, bal Fe.
For gears, camshafts, crankshafts; case hardened. *Obsolete*

HELLEFORS 371M
Ovako Steel Hellefors AB
C 0.2, Mn 0.55, Cr 1.8, Mo 0.25, bal Fe.
For gears, shafts; case hardened. *Obsolete*

HELLEFORS 372H
Ovako Steel Hellefors AB
C 0.4, Mn 0.8, Cr 0.8, Ni 1.25, bal Fe.
For gears, shafts; shock resistant. *Obsolete*

HELLEFORS 372J
Ovako Steel Hellefors AB
C 0.35, Cr 1.2, Ni 2.6, bal Fe.
Oil hardened: 143,000 TS; 130,000 YS; 19-22 El; 55-60 RA; 300 Brin. For shafts, axles, bolts, gears; tough. *Obsolete*

HELLEFORS 372S
Ovako Steel Hellefors AB
C 0.3, Mn 0.55, Cr 1.25, Ni 4.25, bal Fe.
For gears, shafts, machinery parts; tough, shock resistant. *Obsolete*

HELLEFORS 372Y
Ovako Steel Hellefors AB
C 0.3, Mn 0.55, Cr 1.05, Ni 3.25, Mo 0.25, bal Fe.
For gears, shafts, machinery parts; tough, shock resistant. *Obsolete*

HELLEFORS 374 H
Ovako Steel Hellefors AB
C 0.5, Cr 0.6, Ni 1.7, Mo 0.3, bal Fe.
For hot and cold working tools; oil hardening. *Obsolete*

HELLEFORS 374R
Ovako Steel Hellefors AB
C 0.55, Cr 1.5, Ni 3, Mn 0.4, bal Fe.
For cold work dies and tools; oil hardened, shock resistant. *Obsolete*

HELLEFORS 374RK
Ovako Steel Hellefors AB
C 0.55, Cr 1, Ni 3, Mo 0.3, Mn 0.4, bal Fe.
For drop forge dies, cold working tools; oil hardened, shock resistant. *Obsolete*

HELLEFORS 394T
Ovako Steel Hellefors AB
C 0.55, Cr 0.7, Ni 1.75, Mo 0.7, Mn 0.7, bal Fe.
For forging dies, headers, upsetters; oil hardened, shock resistant. *Obsolete*

HELLEFORS 408C
Ovako Steel Hellefors AB
C 1.2, Si 0.2, Mn 0.3, W 1, bal Fe.
For cold work tools, punches, heading dies; oil or water hardened. *Obsolete*

HELLEFORS 605 F
Ovako Steel Hellefors AB
C 0.8, V 0.1, bal Fe.
For chisels, punches, dies; hard and tough. *Obsolete*

HELLEFORS 606F
Ovako Steel Hellefors AB
C 0.9, Si 0.2, Mn 0.3, V 0.1, bal Fe.
Heat treated: 185,000 TS; 118,000 YS; 10 El; 30 RA; 375 Brin. For springs, taps, reamers, hobs, broaches; Type W2; water hardened. *Obsolete*

HELLEFORS 607C
Ovako Steel Hellefors AB
C 1, Si 0.2, Mn 0.3, V 0.1, bal Fe.
Annealed: 100,000 TS; 55,000 YS; 20 El; 42 RA; 200 Brin. For cold work dies and tools, drills, taps; Type W2; water hardened. *Obsolete*

HELLEFORS 666
Ovako Steel Hellefors AB
C 1.2, bal Fe.
For structural parts, tools; water hardening. *Obsolete*

HELLEFORS 677R
Ovako Steel Hellefors AB
C 1, Si 0.2, Mn 0.6, Cr 5.2, Mo 1.1, V 0.2, bal Fe.
For dies, punches, crimpers; air hardened, wear resistant. *Obsolete*

HELLEFORS 782R
Ovako Steel Hellefors AB
C 0.35, Cr 1, W 2.5, bal Fe.
For pneumatic chisels, shears, hot work tools; oil hardened, shock resistant. *Obsolete*

HELLEFORS 783 R
Ovako Steel Hellefors AB
C 0.45, Si 0.9, Cr 1.1, W 2.1, bal Fe.
For chisels, punches; oil hardening, tough. *Obsolete*

HELLEFORS 786 AM
Ovako Steel Hellefors AB
C 0.9, Mn 1.2, Cr 0.5, W 0.5, bal Fe.
For punches, dies, shears; non-deforming. *Obsolete*

HELLEFORS 920-AL
Ovako Steel Hellefors AB
Stainless steel. C 0.1, Cr 14, Ni 10, bal Fe.
For fasteners, chemical plant equipment; stainless steel. *Obsolete*

HELLEFORS 920-H
Ovako Steel Hellefors AB
Stainless steel. C 0.08, Cr 17, Ni 10, bal Fe.
Annealed: 80,000 TS; 38,000 YS; 80 Rock B. For chemical and pharmaceutical plant equipment, mixers, agitators, tanks, digesters; Type 17-10; stainless steel, austenitic. *Obsolete*

HELLEFORS 920-M
Ovako Steel Hellefors AB
Stainless steel. C 0-0.15, Cr 11.5-13.5, bal Fe.
Annealed: 75,000 TS; 40,000 YS; 35 El; 155 Brin. Cold drawn: 100,000 TS; 85,000 YS; 17 El; 205 Brin. For flat springs, knives, table flatware; Type 410; stainless, hardenable. *Obsolete*

HELLEFORS 920A
Ovako Steel Hellefors AB
C 0-0.08, Cr 11.5-13, Al 0.1-0.3, bal Fe.
Annealed: 71,000 TS; 42,600 YS; 22 El; 70 RA; 150 Brin. Heat treated: 175,000 TS; 145,000 YS; 21 El; 64 RA; 352 Brin. For oil refinery and chemical plant equipment; Type 405; corrosion resistant. *Obsolete*

HELLEFORS 920B
Ovako Steel Hellefors AB
Stainless steel. C 0-0.12, Cr 14-18, bal Fe.
Annealed: 70,000 TS; 40,000 YS; 30 El; 55 RA; 150 Brin. Cold drawn: 130,000 TS; 120,000 YS; 2 El; 185 Brin. For oil refinery equipment, bolts, kitchen sinks; Type 430; stainless, ferritic. *Obsolete*

HELLEFORS 921A
Ovako Steel Hellefors AB
C 0-0.15, Cr 11.5-13.5, Mn 0-1, bal Fe.
Annealed: 80,000 TS; 40,000 YS; 35 El; 70 RA; 155 Brin. Cold drawn: 100,000 TS; 85,000 YS; 17 El; 60 RA; 205 Brin. For valve parts, turbine blades, cutlery, knives; Type 410; corrosion resistant. *Obsolete*

HELLEFORS 921C
Ovako Steel Hellefors AB
Stainless steel. Cr 12-14, 0.15 C min, bal Fe.
Annealed: 88,000 TS; 40,000 YS; 32 El; 68 RA; 170 Brin. For cutlery, valve trim, turbine blades; Type 420; stainless, hardenable. *Obsolete*

HELLEFORS 922-A
Ovako Steel Hellefors AB
Stainless steel. Cr 12-14, 0.15 C min, bal Fe.
Annealed: 95,000 TS; 50,000 YS; 25 El; 92 Rock B. Oil hardened: 250,000 TS; 192,000 YS; 10 El; 49 Rock C. For cutlery, surgical instruments, scissors, ball bearings, valves, gears; Type 420; stainless steel, hardenable. *Obsolete*

HELLEFORS 923A

Ovako Steel Hellefors AB
C 0.38-0.45, Cr 13, bal Fe.
Annealed: 95,000 TS; 50,000 YS; 25 El; 55 RA; 196 Brin. For valves, cutlery, surgical and dental instruments; Type 420; corrosion resistant. *Obsolete*

HELLEFORS 930-B

Ovako Steel Hellefors AB
Stainless steel. C 0-0.15, Cr 17-19, Ni 8-10, bal Fe.
Cold rolled: 140,000 TS; 100,000 YS; 30 El; 100 Rock B.
Annealed: 90,000 TS; 40,000 YS; 50 El; 85 Rock B. For chemical and pharmaceutical plant equipment, valve trim, fasteners; Type 302; stainless steel, austenitic. Hardenable only by cold work. *Obsolete*

HELLEFORS 930A

Ovako Steel Hellefors AB
Stainless steel. C 0.08-0.2, Mn 0-2, Cr 17-19, Ni 8-10, bal Fe.
Annealed: 85,000 TS; 35,000 YS; 60 El; 70 RA; 150 Brin. Cold drawn: 125,000 TS; 95,000 YS; 22 El; 55 RA; 277 Brin. For oil refinery and chemical plant equipment; Type 302; stainless, austenitic. *Obsolete*

HELLEFORS S-7

Ovako Steel Hellefors AB
C 0.15, bal Fe.
Hardened: 100,000 TS; 78,000 YS; 19 El; 40 RA; 200 Brin. For gears, bolts, pins; case hardened. *Obsolete*

HELLEFORS V100

Ovako Steel Hellefors AB
C 1, Si 0.2, Mn 0.3, bal Fe.
Annealed: 100,000 TS; 53,000 YS; 21 El; 42 RA; 200 Brin. For drills, reamers, taps, hobs, broaches; water hardened; Type W1. *Obsolete*

HELLEFORS V110

Ovako Steel Hellefors AB
C 1.1, Si 0.2, Mn 0.3, bal Fe.
Annealed: 105,000 TS; 55,000 YS; 20 El; 40 RA; 210 Brin. For drills, springs, reamers, taps, broaches; water hardened; Type W1. *Obsolete*

HELLEFORS V120

Ovako Steel Hellefors AB
C 1.2, Si 0.2, Mn 0.3, bal Fe.
Annealed: 110,000 TS; 55,000 YS; 18 El; 38 RA; 220 Brin. For cutters, taps, reamers, broaches; Type W1; water hardened. *Obsolete*

HELLEFORS V140

Ovako Steel Hellefors AB
C 1.4, Si 0.3, Mn 0.3, bal Fe.
For engravers' tools, blanking dies, cutters; Type W1; water hardened. *Obsolete*

HELLEFORS V60

Ovako Steel Hellefors AB
C 0.6, Si 0.2, Mn 0.3, bal Fe.
Heat treated: 115,000-160,000 TS; 77,000-113,000 YS; 12-23 El; 40-54 RA; 229-320 Brin. For wheels, die blocks, rails, girders; water hardened. *Obsolete*

HELLEFORS V70

Ovako Steel Hellefors AB
C 0.7, Si 0.2, Mn 0.3, bal Fe.
Heat treated: 122,000-174,000 TS; 82,000-128,000 YS; 12-22 El; 37-52 RA; 240-350 Brin. For springs, rails, girders, clutch discs; water hardened. *Obsolete*

HELLEFORS V80

Ovako Steel Hellefors AB
C 0.8, Si 0.2, Mn 0.3, bal Fe.
Heat treated: 130,000-188,000 TS; 87,000-143,000 YS; 12-21 El; 35-50 RA; 285-355 Brin. For drills, reamers, hobs, taps, cutters; water hardened; Type W1. *Obsolete*

HELLEFORS W406

Ovako Steel Hellefors AB
C 1.4, Cr 0.4, W 5.5, V 0.5, bal Fe.
For finishing tools, cutters; oil hardened. *Obsolete*

HELLER 70-20 V

Wallace Murray Corp.
C 0.6-0.7, V 0.15-0.25, bal Fe.
For pneumatic tools; tough.

HELLER 70-20-M

Wallace Murray Corp.
C 0.6-0.7, Mo 0.15-0.25, bal Fe.
For pneumatic tools; tough.

HELLER AIR HARDENING DIE STEEL

Wallace Murray Corp.
C 0.95-1.05, Mn 0.5-0.7, Si 0.3-0.5, Cr 5-5.5, Mo 0.9-1.1, V 0.2-0.3, bal Fe.
Flat ground die steel. For punches, dies, gauges, wear resisting tools and parts. AISI type A2 tool steel.

HELLER OIL HARDENING DIE STEEL

Wallace Murray Corp.
C 0.85-0.95, Mn 1-1.25, Si 0.2-0.4, Cr 0.4-0.6, W 0.4-0.6, V 0.1-0.2, bal Fe.
Flat ground die steel. For dies, punches, jigs, gauges, templates, stamps, shims, machine parts. AISI Type 01.

HELLER'S CHROME DIE

Wallace Murray Corp.
C 0.85-1, Cr 3-4, bal Fe.
For tools, dies, shears; now BROWN LABEL.

HELLERS'S HOLLOW DRILL

Heller Bros. Co.
High C, bal Fe.
For tools, dies. *Obsolete*

HELMENT BRONZE

English manufacture
Cu 70, Zn 30.
For water pipes, cartridges, shell cases, deep drawn parts; max ductility.

HELMET METAL

English manufacture
Cu 70-72, Zn 28-30.
For cartridges, shell cases, helmets; deep drawn.

HELUMIN

Transleteur & Co.
Cu 1.8, Fe 1.5, bal Al.
Heat treated: 50,000 TS; 15 El; 50 Brin. For light alloy parts; non-hardenable.

HELVE

Hardenite Steel Co. Ltd.
C 0.7-1.26, bal Fe.
For tools, dies. Water hardening.

HENCKELS AFH

Henckels Zwillingwerke, G.A.
C 0.6, Si 0.4, Mn 0.6, bal Fe.
Heat treated: 160,000 TS; 113,000 YS; 12 El; 40 RA; 325 Brin. For crimpers, axes, rails; water hardened.

HENCKELS G5

Henckels Zwillingwerke, G.A.
C 0.6, Si 0-0.35, Mn 0.3-0.8, bal Fe.
Heat treated: 165,000 TS; 118,000 YS; 12 El; 40 RA; 330 Brin. For gears, springs, shafts, axes, hammers; water hardened.

HENCKELS NI 187

Henckels Zwillingwerke, G.A.
C 0.56, Ni, Cr, Mo, V, bal Fe.
For forging and heading dies, punches; oil hardened, shock resistant.

HENCKELS NI 35

Henckels Zwillingwerke, G.A.
C 0.5, Cr 1.05, Ni 3.25, Mn 0.5, bal Fe.
For gears, bolts, crankshafts, forging dies; oil hardened, tough.

HENCKELS NI 44

Henckels Zwillingwerke, G.A.
C 0.4, Ni, Cr, Mo, bal Fe.
For gears, bolts, crankshafts; oil hardened, shock resistant.

HENCKELS SPBA

Henckels Zwillingwerke, G.A.
C, alloy, bal Fe.
For machine tool parts; oil hardened, tough.

HENCKELS SPDM-H

Henckels Zwillingwerke, G.A.
C 0.55, Si 0.9, Cr 1.05, V 0.18, V 1.85, W 1.85, bal Fe.
For pneumatic tools, chisels, punches; oil hardened, shock resistant.

HENCKELS SPDN

Henckels Zwillingwerke, G.A.
C 0.45, W, Cr, V, bal Fe.
For forging and heading dies, pneumatic tools; oil hardened, tough.

HENCKELS SPDN-A

Henckels Zwillingwerke, G.A.
C 0.45, Si, Cr, V, bal Fe.
For gears, springs, crankshafts, bolts; oil hardened, shock resistant.

HENCKELS SPDN-W

Henckels Zwillingwerke, G.A.
C 0.35, Si 0.9, Cr 1.05, V 0.18, W 1.85, bal Fe.
For pneumatic tools, chisels, rivet sets; oil hardened, shock resistant.

HENCKELS SPEZIAL ZH

Henckels Zwillingwerke, G.A.
C 1.45, Cr 1.4, Mn 0.6, bal Fe.
For blanking and forming dies; oil or water hardened.

HENCKELS SPG/A

Henckels Zwillingwerke, G.A.
C 1.05, Cr, bal Fe.
For cutters, bearings, liners; oil or water hardened, wear resistant.

HENCKELS SPHF

Henckels Zwillingwerke, G.A.
C 1.3, W 4.75, Cr 0-0.2, bal Fe.
For blanking and forming dies, cutters; water hardened.

HENCKELS SPHF-A

Henckels Zwillingwerke, G.A.
C 1, Cr, bal Fe.
For bearings, cutters, liners, sleeves; water hardened, wear resistant.

HENCKELS SPIII VA

Henckels Zwillingwerke, G.A.
C, alloy, bal Fe.
For machine tool parts; oil hardened, tough.

HENCKELS SPJ

Henckels Zwillingwerke, G.A.
C 1.4, Cr 0.3, V 0.1, Mn 0.3, bal Fe.
For blanking and forming dies, engravers' tools; oil or water hardened, wear resistant.

HENCKELS SPM

Henckels Zwillingwerke, G.A.
C 0.9, V 0.1, Mn 1.9, bal Fe.
For blanking and forming dies, punches, cutters; oil hardened, non-deforming.

HENCKELS SPS
Henckels Zwillingwerke, G.A.
C 2.1, Cr 11.5, Mn 0.3, bal Fe.
For blanking and forming dies, punches; oil hardened, non-deforming.

HENCKELS SPSB
Henckels Zwillingwerke, G.A.
C 1.65, Cr, Mo, V, bal Fe.
For blanking and forming dies, punches; oil hardened, non-deforming.

HENCKELS SPSE
Henckels Zwillingwerke, G.A.
C 1.05, W 1.15, Mn 0.9, Cr 1, bal Fe.
For blanking and heading dies; oil hardened, tough.

HENCKELS SPSE/A
Henckels Zwillingwerke, G.A.
C 1.25, Si 1.15, Mn 0.7, Cr 1.2, bal Fe.
For cold work tools; oil hardened, tough.

HENCKELS SPSWO
Henckels Zwillingwerke, G.A.
C 2.1, Cr 11.5, W 0.7, bal Fe.
For blanking and forming dies, shears, punches; oil hardened, non-deforming.

HENCKELS SPWO4
Henckels Zwillingwerke, G.A.
C 0.3, Cr 1.1, V 0.18, W 3.75, bal Fe.
For pneumatic tools, shears, punches, chisels; oil hardened, tough.

HENCKELS SPWO5VA
Henckels Zwillingwerke, G.A.
C 0.3, Cr 2.35, V 0.6, W 4.25, bal Fe.
For pneumatic tools, punches, chisels, shears; oil hardened, tough.

HENCKELS SPWO9
Henckels Zwillingwerke, G.A.
C 0.3, W 8.5, V 0.35, Cr 2.65, bal Fe.
For extrusion dies, rams, liners, punches; oil hardened, hot work steel.

HENCKELS SPWON
Henckels Zwillingwerke, G.A.
C 0.3, Co 2, W 8.5, Cr 2.4, V 0.25, bal Fe.
For extrusion dies, rams, liners, punches; oil hardened, hot work steel.

HENCKELS SPZII
Henckels Zwillingwerke, G.A.
C 1, V 0.1, Mn 0.25, Si 0.25, bal Fe.
For blanking and heading dies, cutters; water hardened; Type W2.

HENNEQUIN METAL
Manufacturer not listed.
C, bal Fe.

HENRICOT BMES
Usines Emile Henricot, S.A.
C 0.05, Mn, Si, bal Fe.
For plastic mold dies; case hardened. *Obsolete*

HENRICOT BO
Usines Emile Henricot, S.A.
C 0.1, Mn, Si, bal Fe.
For rivets, nails, carburized parts; case hardened. *Obsolete*

HENRICOT FAMO
Usines Emile Henricot, S.A.
C, Mn, Si, Ni, Cr, Mo, bal Fe.
Cast: 55,000-60,000 TS; 270-310 Brin. For wear plates; acicular cast iron. *Obsolete*

HENRICOT FC
Usines Emile Henricot, S.A.
C 3, Cr 0.7, Mn, Si, bal Fe.
Cast: 220-270 Brin. For wear resistant castings; refractory cast iron. *Obsolete*

HENRICOT FC2
Usines Emile Henricot, S.A.
C 3, Cr 2, Si, Mn, bal Fe.
Cast: 200-300 Brin. For abrasion resistant castings; cast iron. *Obsolete*

HENRICOT FP1
Usines Emile Henricot, S.A.
C 3, Mn, Si, bal Fe.
Cast: 25,000-30,000 TS; 180-250 Brin. For gears, shafts, housings, brackets; pearlitic cast iron. *Obsolete*

HENRICOT FP2
Usines Emile Henricot, S.A.
C 3, Mn, Si, bal Fe.
Cast: 40,000-50,000 TS; 230-250 Brin. For gears, shafts, housings, brackets; pearlitic cast iron. *Obsolete*

HENRICOT FPNC
Usines Emile Henricot, S.A.
C 3, Mn, Si, Ni, Cr, bal Fe.
Cast: 35,000-45,000 TS; 200-250 Brin. For gears, shafts, housings, brackets; pearlitic cast iron. *Obsolete*

HENRICOT H1F
Usines Emile Henricot, S.A.
C, Ni, Cr, Mo, bal Fe.
Heat treated: 130,000-140,000 TS; 110,000-130,000 YS; 8-10 El. For gears, shafts, housings; air hardened, tough. *Obsolete*

HENRICOT HA0
Usines Emile Henricot, S.A.
C 1, Cr 6, bal Fe.
For permanent magnets; hardened. *Obsolete*

HENRICOT HA1
Usines Emile Henricot, S.A.
C 1, Cr 6, Co 3, bal Fe.
For permanent magnets; heat treated. *Obsolete*

HENRICOT HA2
Usines Emile Henricot, S.A.
C 1, Cr 8, Co 12, bal Fe.
For permanent magnets; hardened. *Obsolete*

HENRICOT HA3
Usines Emile Henricot, S.A.
C 1, Cr 10, Co 15, bal Fe.
For permanent magnets; hardened. *Obsolete*

HENRICOT HA4
Usines Emile Henricot, S.A.
C 0.7, Cr 6, W 4, Co 35, bal Fe.
For permanent magnets; high coercive force. *Obsolete*

HENRICOT HA5
Usines Emile Henricot, S.A.
C, alloy, bal Fe.
For permanent magnets; heat treated *Obsolete*

HENRICOT HBA
Usines Emile Henricot, S.A.
C 0.4, Ni 4.5, Cr 1.5, W 0.75, bal Fe.
Hardened: 500 Brin. For bakelite molding dies; air hardening. *Obsolete*

HENRICOT HBE
Usines Emile Henricot, S.A.
C 0.1, Cr 0.4, bal Fe.
For dies and molds for plastics; hubbing die steel, water hardened. *Obsolete*

HENRICOT HBNO
Usines Emile Henricot, S.A.
C 0.5, Cr 12, bal Fe.
Hardened: 500 Brin. For bakelite molding dies; oil or air hardening. *Obsolete*

HENRICOT HD2
Usines Emile Henricot, S.A.
C 1, Cr 5, Mo 1, bal Fe.
For dies; oil or air hardening. *Obsolete*

HENRICOT HDK
Usines Emile Henricot, S.A.
C 2, Cr 12, bal Fe.
Hardened: 600 Brin. For dies; oil hardening, non-deforming. *Obsolete*

HENRICOT HDKW
Usines Emile Henricot, S.A.
C 1.5, Cr 12, Mo 0.75, W 0.75, bal Fe.
Hardened: 600 Brin. For dies; air hardening, non-deforming. *Obsolete*

HENRICOT HDZ
Usines Emile Henricot, S.A.
C 1, Mn 1, Cr 0.5, V 0.15, bal Fe.
Hardened: 650 Brin. For dies, punches, cold forming dies; non-deforming, oil hardening. *Obsolete*

HENRICOT HEB
Usines Emile Henricot, S.A.
C 0.52, Cr, Si, bal Fe.
For shear blades, cutters, pneumatic chisels; oil hardened. *Obsolete*

HENRICOT HEM3
Usines Emile Henricot, S.A.
C 0.5, Cr 1, V 0.15, bal Fe.
Hardened: 142,000 TS; 128,000 YS; 10 El; 50 RA. For chisels, springs; tough, shock resistant. *Obsolete*

HENRICOT HF
Usines Emile Henricot, S.A.
C 0.5-1.25, bal Fe.
For hand chisels, shear blades, hammer dies; water hardening. *Obsolete*

HENRICOT HF1
Usines Emile Henricot, S.A.
Cr 12-14, 0.15% C min, bal Fe.
Annealed: 88,000 TS; 40,000 YS; 32 El; 68 RA; 170 Brin. Heat treated: 256,000 TS; 190,000 YS; 6 El; 10 RA; 540 Brin. For cutlery, valve trim, springs, surgical instruments; Type 420; stainless, hardenable. *Obsolete*

HENRICOT HF2
Usines Emile Henricot, S.A.
C 0.7, bal Fe.
Heat treated: 174,000 TS; 128,000 YS; 12 El; 37 RA; 352 Brin. For short shear blades, clutch discs; water hardened; Type W1. *Obsolete*

HENRICOT HF3
Usines Emile Henricot, S.A.
C 0.85, bal Fe.
Heat treated: 188,000 TS; 143,000 YS; 12 El; 35 RA; 388 Brin. For short shear blades, tools; water hardened; Type W1. *Obsolete*

HENRICOT HF4
Usines Emile Henricot, S.A.
C 1, bal Fe.
Annealed: 100,000 TS; 53,000 YS; 21 El; 42 RA; 200 Brin. For cutting tools, drills, hobs, cold punches; Type W1; water hardened. *Obsolete*

HENRICOT HIB
Usines Emile Henricott, SA
C 0-0.06, Mn 0-1, Si 0-1, Cr 18-21, Mo 14-16, Fe 0-2, bal Ni.
Cast: 45-60 kg/mm^2 TS; 32 kg/mm^2 YS min; 8 El min. Good corrosion resistance.

HENRICOT HIF
Usines Emile Henricot, S.A.
C 0.3-0.35, Cr 1.4, Ni 3.2, Mo, bal Fe.
For gears, pinions, shafts, crimpers; oil hardened, shock resistant. *Obsolete*

HENRICOT HIF2
Usines Emile Henricott, SA
C 0.25-0.35, Mn 0-0.75, Ni 3-4, Cr 1-1.5, Mo 0.16-0.35, Si 0-0.6, bal Fe.
Cast alloy; hardenable to 420-500 Brin. For highly stressed parts.

HENRICOT HIF4
Usines Emile Henricott, SA
C 0.3-0.4, Mn 0-0.75, Si 0-0.6, Ni 3-4, Cr 1.2-1.7, Mo 0.15-0.35, bal Fe.
Cast alloy; hardenable to 500-550 Brin. For highly stressed parts.

HENRICOT HIG
Usines Emile Henricott, SA
C 0-0.06, Mn 0-2, Si 0-1, Ni 46-50, Cr 23-27, Mo 5-7, Cu 2-4, Nb, bal Fe.
Cast: 45-60 kg/mm^2 TS; 20 kg/mm^2 YS min; 25 El min. Good corrosion resistance. Similar to Hastelloy G.

HENRICOT HIL
Usines Emile Henricott, SA
C 0-0.06, Mn 0-1, Si 0-1, Cr 23-27, Mo 8-10, Fe 0-2, Co 7-9, bal Ni.
Corrosion resistant casting; for chemical equipment.

HENRICOT HIM2
Usines Emile Henricott, SA
C 0-0.06, Mn 0-1, Si 0-1, Mo 26-30, Fe 0-7, bal Ni.
Corrosion resistant casting; for chemical equipment. Similar to Hastelloy B.

HENRICOT HIT
Usines Emile Henricott, SA
C 0-0.06, Cr 21-24, Mo 6-8, Fe 12-14, Co 2-3.5, bal Ni.
Corrosion resistant casting; for chemical equipment. Similar to Hastelloy F.

HENRICOT HIW
Usines Emile Henricott, SA
C 0-0.06, Mn 0-1, Si 0-1, Cr 15.5-17.5, Mo 16-18, Fe 0-7, W 3.75-5.25, bal Ni.
Cast: 45-60 kg/mm^2 TS; 32 kg/mm^2 YS min; 5 El min. Corrosion resistant alloy; for chemical equipment. Similar to Hastelloy C.

HENRICOT HKZ
Usines Emile Henricott, S.A.
C 0.7, Cr 1, V 0.15, bal Fe.
Hardened: 170,000 TS; 156,000 YS; 7 El; 40 RA. For tools, shear blades; oil hardening. *Obsolete*

HENRICOT HL
Usines Emile Henricott, S.A.
C, Cr, Ni, Mo, bal Fe.
For gears, shafts; heat treatable. *Obsolete*

HENRICOT HL1
Usines Emile Henricott, S.A.
Ni 1, C, Cr, bal Fe.
For construction and automotive parts; oil hardened, shock resistant. *Obsolete*

HENRICOT HL10
Usines Emile Henricott, SA
C 0.3-0.4, Mn 0-0.75, Si 0-0.6, Ni 1.25-1.75, Cr 1.25-1.75, Mo 0.15-0.25, bal Fe.
Medium carbon alloy steel casting.

HENRICOT HL11
Usines Emile Henricott, SA
C 0.2-0.3, Mn 0-0.75, Si 0-0.6, Ni 1.25-1.75, Cr 1.25-1.75, Mo 0.15-0.25, bal Fe.
Alloy steel casting.

HENRICOT HL13
Usines Emile Henricott, SA
C 0.14-0.2, Mn 0.75-0.95, Si 0.4-0.6, Ni 1.5-2, Cr 0.55-0.7, Mo 0.35-0.45, V 0.03-0.06, bal Fe.
As cast: 286-321 Brin. Alloy carburizing steel casting. Similar to AISI 4320.

HENRICOT HL1A
Usines Emile Henricot, S.A.
Ni 1.5, Cr 0.75, Mo 0.15, C, bal Fe.
Normalized: 100,000-115,000 TS; 70,000-85,000 YS; 8-10 El. For cast housings, gears, shafts; oil hardened, tough. *Obsolete*

HENRICOT HL1B
Usines Emile Henricott, SA
C 0.25-0.36, Mn 0.6-1, Si 0.2-0.5, Ni 0.5-1, Cr 0.5-1, Mo 0.15-0.25, bal Fe.
Cast: 286-321 Brin. Similar to AISI 8630.

HENRICOT HL1C
Usines Emile Henricott, SA
C 0.13-0.18, Mn 0.7-1.1, Si 0.15-0.4, Ni 0.8-1.2, Cr 0.6-1, Mo 0-0.1, bal Fe.
As cast: 220 Brin max. Low alloy carburizing steel casting.

HENRICOT HL1E
Usines Emile Henricott, SA
C 0.14-0.2, Mn 0.5-1, Si 0-0.6, Ni 0.5-1, Cr 0.4-0.8, Mo 0.4-0.6, V 0-0.1, bal Fe.
Low alloy carburizing steel casting.

HENRICOT HL1X
Usines Emile Henricot, S.A.
Ni 1.5, Cr 0.5, Mo 0.15, C, bal Fe.
Heat treated: 90,000-110,000 TS; 70,000-85,000 YS; 12-14 El. For cast housings, flanges, gears, shafts; water or oil hardened, tough. *Obsolete*

HENRICOT HL2
Usines Emile Henricott, S.A.
Ni 2, C, Cr, bal Fe.
For construction and automotive parts; oil hardened, shock resistant. *Obsolete*

HENRICOT HL27
Usines Emile Henricot, S.A.
C 0.3, Ni 22, Cr 5, bal Fe.
Annealed: 85,000-107,000 TS; 50,000-72,000 YS; 20 El; 50 RA. For electrical industry; nonmagnetic. *Obsolete*

HENRICOT HL2X
Usines Emile Henricot, S.A.
Ni 2.5, Cr 1, Mo 0.2, C, bal Fe.
Heat treated: 115,000-130,000 TS; 90,000-110,000 YS; 12 El. For gears, shafts, housings; oil hardened, shock resistant. *Obsolete*

HENRICOT HL3
Usines Emile Henricott, S.A.
C 0.4, Cr 1.5, Ni 3, bal Fe.
For construction and automotive parts; gears; tough, oil hardened. *Obsolete*

HENRICOT HL3C
Usines Emile Henricott, S.A.
C 0.15, Ni 3, Cr 1, Mo 0.2, bal Fe.
For gears, pinions, shafts; case hardening steel. *Obsolete*

HENRICOT HL3X
Usines Emile Henricot, S.A.
Ni 3, Cr 1, Mo 0.2, C, bal Fe.
Heat treated: 115,000-130,000 TS; 100,000-115,000 YS; 10-13 El. For gears, shafts, housings; oil hardened, shock resistant. *Obsolete*

HENRICOT HL5C
Usines Emile Henricott, S.A.
C 0.15, Cr 1.5, Ni 5, Mo 0.2, bal Fe.
Heat treated: 130,000-200,000 TS. For camshafts, gears, pinions, bolts; case hardened, shock resistant. *Obsolete*

HENRICOT HL6
Usines Emile Henricott, S.A.
C 0.4, Ni 4, Cr, bal Fe.
For gears, pinions, shafts, auto parts; oil hardened, shock resistant. *Obsolete*

HENRICOT HM
Usines Emile Henricott, S.A.
C, Cr, Mo, bal Fe.
For gears, shafts; heat treatable. *Obsolete*

HENRICOT HM1
Usines Emile Henricott, SA
C 0-0.35, Mn 1-2, Si 0-0.6, Cr 0.5-1, Mo 0.4-0.6, bal Fe.
Alloy steel casting; hardenable to 450-500 Brin.

HENRICOT HM3
Usines Emile Henricott, SA
C 0.4-0.5, Mn 0-1, Si 0-0.6, Cr 2.5-3.5, Mo 0.35-0.45, bal Fe.
Alloy steel casting; deep hardening.

HENRICOT HM4
Usines Emile Henricott, SA
C 0.6-0.7, Mn 0.75-1.25, Si 0-0.6, Ni 0.5-1, Cr 1.25-2, Mo 0.35-0.45, bal Fe.
Alloy steel casting; good hardenability.

HENRICOT HM51
Usines Emile Henricott, SA
C 0.9-1, Mn 0.75-1.25, Si 0-0.6, Cr 2-3, Mo 0.25-0.4, bal Fe.
HT: 450-500 Brin in heavy sections.

HENRICOT HM7
Usines Emile Henricott, SA
C 0.9-1.1, Mn 0.4-0.6, Si 0-0.6, Cr 4.5-5.5, Mo 0.9-1.1, V 0.5, bal Fe.
Alloy steel casting. Similar to AISI A2 tool steel.

HENRICOT HM9
Usines Emile Henricott, SA
C 0.9-1.1, Mn 0.4-0.6, Si 0-0.6, Cr 4.5-5.5, Mo 0.9-1.1, V, bal Fe.
Alloy steel casting. Similar to AISI A2 or A3.

HENRICOT HMA
Usines Emile Henricott, SA
C 0-0.25, Mn 0.5-0.8, Si 0-0.6, Ni 0-0.5, Cr 0-0.35, Mo 0.45-0.65, Cu 0-0.5, W 0-0.1, bal Fe.
As cast: 48 kg/mm^2 TS; 25 kg/mm^2 YS; 22-24 El. Creep resistant to 450°C; for oil refineries. BS. 1398; GS 22 Mo 4.

HENRICOT HMA1
Usines Emile Henricott, SA
C 0.13-0.18, Mn 0.5-0.8, Si 0-0.5, Ni 0-0.2, Cr 0-0.2, Mo 0.3-0.5, bal Fe.
Low carbon, low alloy steel casting.

HENRICOT HMA2
Usines Emile Henricott, SA
C 0.2-0.26, Mn 1.1-1.5, Si 0-0.6, Ni 0-0.5, Cr 0-0.3, Mo 0-0.25, Cu 0-0.5, bal Fe.
Low alloy casting.

HENRICOT HMA3
Usines Emile Henricott, SA
C 0.16-0.2, Mn 1.3-1.7, Si 0.3-0.6, Mo 0.25-0.35, bal Fe.
Manganese alloy steel casting.

HENRICOT HMA4
Usines Emile Henricott, SA
C 0.32-0.37, Mn 0.5-0.8, Si 0-0.6, Mo 0.35, bal Fe.
Low alloy steel casting.

HENRICOT HMA5
Usines Emile Henricott, SA
C 0.1, Mn 1.5, Si 0-0.6, Mo 0.2, V 0.05, bal Fe.
Steel casting.

HENRICOT HMA6
Usines Emile Henricott, SA
C 0.21-0.28, Mn 1.15-1.35, Mo 0.15-0.25, B 0.003-0.006, Al 0.05-0.15, Ti 0.05-0.15, bal Fe.
Manganese-boron steel casting.

HENRICOT HMA7
Usines Emile Henricott, SA
C 0.18-0.25, Mn 0.5-0.8, Si 0.3-0.6, Cr 0-0.3, Mo 0.35-0.4, bal Fe.
Casting for elevated structural parts. Similar to W.-Nr. 1.5419.

HENRICOT HME
Usines Emile Henricott, SA
C 0.2-0.3, Mn 0.4-0.8, Si 0-0.6, Cr 0.9-1.2, Mo 0.15-0.25, bal Fe.
Alloy steel casting. Similar to AISI 4130.

HENRICOT HMES
Usines Emile Henricott, S.A.
C 0.2, Cr 0.5, Mo 0.1, bal Fe.
Heat treated: 90,000-100,000 TS; 60,000-70,000 YS; 8-10 El. For cast housings, flanges, supports; water hardened. *Obsolete*

HENRICOT HMEV
Usines Emile Henricott, SA
C 0.25-0.35, Mn 0-1.3, Si 0-0.6, Cr 0-0.6, Mo 0.25-0.35, V 0.03-0.07, bal Fe.
Alloy casting; hardenable to 425-515 Brin.

HENRICOT HMH
Usines Emile Henricott, S.A.
C, Mo, bal Fe.
For surface hardened gears, shafts; wear resistant. *Obsolete*

HENRICOT HMH
Usines Emile Henricott, S.A.
C, Cr, Mo, bal Fe.
For construction and automotive parts; oil hardened, shock resistant. *Obsolete*

HENRICOT HMH1
Usines Emile Henricott, SA
C 0.3-0.4, Mn 0.5-0.8, Si 0.6, Cr 0.9-1.2, Mo 0.2-0.3, bal Fe.
Alloy steel casting. Similar to AISI 4135.

HENRICOT HMH2
Usines Emile Henricott, SA
C 0.35-0.45, Mn 0.5-0.8, Si 0-0.6, Cr 0.9-1.2, Mo 0.2-0.3, bal Fe.
Alloy steel casting. Similar to AISI 4140.

HENRICOT HMH3
Usines Emile Henricott, SA
C 0.48-0.53, Mn 0.8-0.95, Si 0-0.6, Cr 0.8-1.1, Mo 0.15-0.2, bal Fe.
Alloy steel casting. Similar to AISI 4150.

HENRICOT HMO
Usines Emile Henricott, SA
C 0-0.35, Mn 1-2, Si 0-0.6, Cr 0.5-1, Mo 0.4-0.6, bal Fe.
Alloy steel casting; hardenable to 425 Brin min.

HENRICOT HMV1
Usines Emile Henricott, SA
C 0-0.2, Mn 0.5-0.8, Si 0-0.5, Cr 1-1.5, Mo 0.45-0.65, V 0.15-0.25, bal Fe.
Alloy steel casting; for oil refinery equipment.

HENRICOT HMV3
Usines Emile Henricott, SA
C 0.15-0.2, Mn 0.5-0.8, Si 0-0.6, Cr 1.2-1.5, Mo 0.9-1.1, V 0.2-0.3, bal Fe.
Alloy steel casting; for oil refinery equipment.

HENRICOT HMVX
Usines Emile Henricot, S.A.
C 0.15, Cr 1, Mo 0.5, V, bal Fe.
Annealed: 85,000-100,000 TS; 60,000-70,000 YS; 10-12 El; 150-170 Brin. For oil refinery equipment; creep resistant to 575 C. *Obsolete*

HENRICOT HMX
Usines Emile Henricot, S.A.
C 0.25, Cr 1, Mo 0.75, bal Fe.
Heat treated: 142,000 TS; 128,000 YS; 10 El; 45 RA; 250 Brin. For oil refinery equipment; creep resistant. *Obsolete*

HENRICOT HMX1
Usines Emile Henricott, SA
C 0-0.2, Mn 0.5-0.8, Si 0-0.5, Cr 1.01-1.2, Mo 0.45-0.7, V 0-0.5, 0.50 max residual Ni, bal Fe.
Alloy steel casting; for oil refinery equipment.

HENRICOT HMX3
Usines Emile Henricott, SA
C 0-0.18, Mn 0.4-0.7, Si 0-0.5, Cr 2-2.5, Mo 0.9-1.2, bal Fe.
Alloy steel casting; for oil refinery equipment.

HENRICOT HMX4
Usines Emile Henricott, SA
C 0.16-0.2, Mn 0.6-0.9, Si 0.3-0.5, Ni 0-0.3, Cr 0.6-0.9, Mo 0.3-0.4, bal Fe.
Alloy steel casting.

HENRICOT HN
Usines Emile Henricot, S.A.
C 0-0.2, Ni 30-50, bal Fe.
Annealed. For controlled expansion properties; water quenched from 1100 C. *Obsolete*

HENRICOT HN11
Usines Emile Henricott, SA
C 0.13-0.15, Mn 0.45-0.65, Si 0-0.6, Ni 1.65-2, Mo 0.2-0.3, bal Fe.
Alloy steel casting. Similar to AISI 4615.

HENRICOT HN12
Usines Emile Henricott, SA
C 0-0.25, Mn 0.5-0.8, Si 0-0.6, Ni 2-3, bal Fe.
As cast: 49 kg/mm^2 TS; 35 kg/mm^2 YS; 20 El.

HENRICOT HN13
Usines Emile Henricott, SA
C 0-0.15, Mn 0.5-0.8, Si 0-0.6, Ni 3-4, bal Fe.
Alloy steel casting. Similar to SAE 2317 steel.

HENRICOT HN14
Usines Emile Henricott, SA
C 0-0.12, Mn 0.5-0.8, Si 0-0.6, Ni 4-5, bal Fe.
As cast: 52 kg/mm^2 TS; 35 kg/mm^2 YS; 20 El. Similar to SAE 2512.

HENRICOT HN50
Usines Emile Henricot, S.A.
Ni 50-51, bal Fe.
Annealed: 60,000-75,000 TS; 30,000-40,000 YS; 30-35 El. For electrical equipment; controlled expansion. *Obsolete*

HENRICOT HNF
Usines Emile Henricot, S.A.
Ni 45, bal Fe.
Annealed: 90,000-130,000 TS; 40,000-70,000 YS; 20-30 El. For electrical equipment, motors; high permeability, soft magnet. *Obsolete*

HENRICOT HO
Usines Emile Henricot, S.A.
C 0.05, Mn, Si, bal Fe.
For plastic mold dies; case hardened. *Obsolete*

HENRICOT HO10
Usines Emile Henricott, SA
C 0-0.12, Mn 0-0.6, Si 0-0.6, bal Fe.
Cast, normalized: 40-45 kg/mm^2 TS; 20 kg/mm^2 YS; 24 El; GS 38/38-3.

HENRICOT HO20
Usines Emile Henricot, SA
C 0-0.15, Mn 0-0.9, Si 0-0.6, bal Fe.
Cast, normalized: 40-45 kg/mm^2 TS; 25 kg/mm^2 YS; 24 El.

HENRICOT HO30
Usines Emile Henricott, SA
C 0-0.18, Mn 0-0.85, Si 0-0.6, Ni 0.5, Cr 0.4, Mo 0.25, Cu 0.5, bal Fe.
Cast, normalized: 45-50 kg/mm^2 TS; 28 kg/mm^2 YS; 22 El. AAR Gr. A.

HENRICOT HO34
Usines Emile Henricott, SA
C 0-0.18, Mn 0-0.85, Si 0-0.6, Cu 0.35-0.6, bal Fe.
Cast, normalized: 45-50 kg/mm^2 TS; 28 kg/mm^2 YS; 22 El.

HENRICOT HO40
Usines Emile Henricott, SA
C 0-0.25, Mn 0-1, Si 0-0.6, Ni 0.5, Cr 0.4, Mo 0.25, Cu 0.5, bal Fe.
Cast, normalized: 45-55 kg/mm^2 TS; 30 kg/mm^2 YS; 20 El. ASTM A27-654 Class 70-40.

HENRICOT HO41
Usines Emile Henricott, SA
C 0-0.28, Mn 0-0.85, Si 0-0.6, Ni 0.5, Cr 0.4, Cu 0.25-0.5, bal Fe.
Cast, normalized: 49 kg/mm^2 TS; 27 kg/mm^2 YS; 20 El. ASTM A27-65 Class 70-35.

HENRICOT HO42
Usines Emile Henricott, SA
C 0.25, Mn 1.1-1.3, Si 0-0.6, Ni 0.5, Cr 0.4, Mo 0.25, Cu 0.5, bal Fe.
Cast, normalized: 49 kg/mm^2 TS; 27 kg/mm^2 YS; 20 El. ASTM A27-65 Class 70-40.

HENRICOT HO43
Usines Emile Henricott, SA
C 0.2-0.28, Mn 0.5-1.2, Si 0.6, Ni 0.5, Cr 0.4, Mo 0.25, Cu 0.5, bal Fe.
Cast, normalized: 50-60 kg/mm^2 TS; 30 kg/mm^2 YS; 15 El.

HENRICOT HO44
Usines Emile Henricott, SA
C 0-0.25, Mn 0-1, Si 0-0.6, Cu 0.35-0.6, bal Fe.
Cast, normalized: 48-55 kg/mm^2 TS; 30 kg/mm^2 YS; 20 El.

HENRICOT HO45
Usines Emile Henricott, SA
C 0.13, Mn 0.2-0.55, Si 0.25-0.75, P 0.07-0.09, Ni 0.6, Cr 0.3-1.25, Cu 0.2-0.55, bal Fe.
Cast, normalized: 45-55 kg/mm^2 TS; 30 kg/mm^2 YS; 18 El.

HENRICOT HO46
Usines Emile Henricott, SA
C 0.21-0.29, Mn 0.5-0.8, Si 0.3-0.6, bal Fe.
Cast, normalized: 45.5 kg/mm^2 TS; 24.5 kg/mm^2 YS; 20 El.

HENRICOT HO50
Usines Emile Henricott, SA
C 0-0.3, Mn 0-1.5, Si 0.3-0.6, Ni 0.5, Cr 0.4, Mo 0.25, Cu 0.5, bal Fe.
Cast, normalized: 55-60 kg/mm^2 TS; 35 kg/mm^2 YS; 15 El. Similar to ASTM A148-65 Class 80-50.

HENRICOT HO51
Usines Emile Henricott, SA
C 0-0.3, Mn 0-1, Si 0.3-0.6, Ni 0.5, Cr 0.4, Mo 0.25, Cu 0.5, bal Fe.
Cast, normalized: 55-60 kg/mm^2 TS; 28 kg/mm^2 YS. Similar to Bs 592 Gr. C.

HENRICOT HO52
Usines Emile Henricott, SA
C 0-0.28, Mn 0-1.2, Si 0.3-0.6, Ni 0.5, Cr 0.4, Mo 0.25, Cu 0.5, bal Fe.
Cast, normalized: 52-60 kg/mm^2 TS; 30 kg/mm^2 YS; 18 El.

HENRICOT HO53
Usines Emile Henricott, SA
C 0.25-0.35, Mn 0.5-0.78, Si 0.25-0.5, bal Fe.
Carbon steel casting.

HENRICOT HO54
Usines Emile Henricott, SA
C 0-0.3, Mn 0-1.5, Si 0-0.6, Cu 0.35-0.6, bal Fe.
Cast, normalized: 55-60 kg/mm^2 TS; 35 kg/mm^2 YS; 15 El.

HENRICOT HO60
Usines Emile Henricott, SA
C 0-0.25, Mn 0-2, Si 0-0.6, bal Fe.
Cast: 60-70 kg/mm^2 TS; 35-45 kg/mm^2 YS; 18 El.

HENRICOT HO61
Usines Emile Henricott, SA
C 0-0.45, Mn 0-1.2, Si 0-0.6, bal Fe.
Cast: 60-70 kg/mm^2 TS; 35-45 kg/mm^2 YS; 16 El.

HENRICOT HO62
Usines Emile Henricott, SA
C 0.2-0.28, Mn 0-1.5, Si 0-0.6, Cr 0.2-0.5, bal Fe.
Cast: 60-65 kg/mm^2 TS; 40 kg/mm^2 YS; 18 El. Similar to
ASTM A148 Class 60-90.

HENRICOT HO63
Usines Emile Henricott, SA
C 0-0.3, Mn 0-1.6, Si 0-0.6, Cr 0.2-0.5, bal Fe.
Cast: 68-78 kg/mm^2 TS; 50 kg/mm^2 YS; 18 El.

HENRICOT HO64
Usines Emile Henricott, SA
C 0-0.3, Mn 0-1.5, Si 0-0.6, Ni 0.2-0.5, Cr 0.4-0.8, bal Fe..
Cast, tempered: 270-320 Brin.

HENRICOT HO65
Usines Emile Henricott, SA
C 0-0.35, Mn 0-1.5, Si 0-0.6, bal Fe.
Cast: 65-75 kg/mm^2 TS; 35-45 kg/mm^2 YS; 15 El.

HENRICOT HO66
Usines Emile Henricott, SA
C 0-0.3, Mn 0-1.6, Si 0-0.6, bal Fe.
Cast: 63 kg/mm^2 TS; 42 kg/mm^2 YS; 22 El.

HENRICOT HO70
Usines Emile Henricott, SA
C 0.32-0.4, Mn 1.25-1.65, Si 0.3-0.6, bal Fe.
Cast: 70-75 kg/mm^2 TS; 40-45 kg/mm^2 YS; 16 El.

HENRICOT HO71
Usines Emile Henricott, SA
C 0.4-0.5, Mn 0.6-0.9, Si 0.3-0.6, bal Fe.
Cast carbon steel.

HENRICOT HO72
Usines Emile Henricott, SA
C 0.35-0.45, Mn 1-1.25, Si 0-0.6, bal Fe.
Cast carbon steel.

HENRICOT HO80
Usines Emile Henricott, SA
C 0.35-0.45, Mn 0-1.75, Si 0-0.6, bal Fe.
Normalized and tempered: 80-85 kg/mm^2 TS; 48 kg/mm^2
YS. Carbon-manganese steel casting.

HENRICOT HO81
Usines Emile Henricott, SA
C 0-0.5, Mn 0-1, S 0-0.6, bal Fe.
Normalized and tempered: 75-85 kg/mm^2 TS; 45 kg/mm^2
YS. Carbon-manganese steel casting.

HENRICOT HO82
Usines Emile Henricott, SA
C 0-0.5, Mn 0-1.5, Si 0-0.6, bal Fe.
Normalized and tempered: 75-85 kg/mm^2 TS; 45 kg/mm^2
YS. Carbon-manganese steel casting.

HENRICOT HO83
Usines Emile Henricott, SA
C 0.35-0.45, Mn 0-1.4, Si 0-0.6, Mo 0.1-0.2, bal Fe.
Normalized and tempered: 75-85 kg/mm^2 TS; 45 kg/mm^2
YS. Carbon-manganese steel casting.

HENRICOT HO85
Usines Emile Henricott, SA
C 0-0.6, Mn 0-0.85, Si 0-0.6, bal Fe.
Normalized and tempered: 75-85 kg/mm^2 TS; 45 kg/mm^2
YS; 10 El. Carbon steel casting.

HENRICOT HO90
Usines Emile Henricott, SA
C 0-0.7, Mn 0-0.85, Si 0-0.6, bal Fe.
Normalized and tempered: 80-90 kg/mm^2 TS; 48 kg/mm^2
YS; 8 El. Carbon steel casting.

HENRICOT HOT
Usines Emile Henricott, SA
Mn 0-0.7, Si 0-0.6, 0.30 C min, bal Fe.
Normalized and tempered: 50-60 kg/mm^2 TS; 30 kg/mm^2
YS; 16 El. Carbon steel casting. SIS 1505.

HENRICOT HOT1
Usines Emile Henricott, SA
Si 0-0.6, Mn 0-0.65, 0.35 C min, bal Fe.
Normalized and tempered: 60-70 kg/mm^2 TS; 35 kg/mm^2
YS; 15 El. Carbon steel casting.

HENRICOT HOT2
Usines Emile Henricott, SA
Mn 0-0.6, Si 0-0.6, 0.45 C min, bal Fe.
Normalized and tempered: 65-75 kg/mm^2 TS; 38 kg/mm^2
YS; 14 El. Carbon steel casting.

HENRICOT HOT3
Usines Emile Henricott, SA
Mn 0-0.55, Si 0-0.6, 0.50 C min, bal Fe.
Normalized and tempered: 75-85 kg/mm^2 TS; 45 kg/mm^2
YS. Carbon steel casting.

HENRICOT HP2
Usines Emile Henricot, S.A.
C 0.35, Si 1, Cr 5, Mo 1.25, W 1.25, bal Fe.
Hardened: 450 Brin. For dies; air hardening. *Obsolete*

HENRICOT HP93
Usines Emile Henricott, S.A.
C 0.55, Cr, Mo, bal Fe.
For pneumatic hammers, chisels; oil hardened, tough.
Obsolete

HENRICOT HPB
Usines Emile Henricott, S.A.
W 14, C, V, Cr, bal Fe.
For drawing and forming dies, spindles; oil hardened, tough.
Obsolete

HENRICOT HPC
Usines Emile Henricott, S.A.
Cr 5, C, Mo, V, bal Fe.
For drawing dies and spindles, Al die casting dies; oil
hardened, non-deforming. *Obsolete*

HENRICOT HPL
Usines Emile Henricott, SA
C 0.5-0.6, Mn 0-1, Si 0-0.6, Ni 2-2.5, Cr 1.4-2, Mo 0.15-0.25,
bal Fe.
Alloy cast steel; wear resistant.

HENRICOT HPNR
Usines Emile Henricot, S.A.
C 0.35, Cr 2.5, W 9, bal Fe.
Hardened: 400 Brin. For extrusion and drawing dies; hot work
steel. *Obsolete*

HENRICOT HPO
Usines Emile Henricot, S.A.
W 5, C, Ni, Cr, bal Fe.
For bolt bushing, drawing dies for Cu alloy; oil hardened.
Obsolete

HENRICOT HPT
Usines Emile Henricott, S.A.
C 0.3, Cr 2.5, W 4.5, V 0.5, bal Fe.
Hardened: 450 Brin. For dies for brass tubing; oil hardening.
Obsolete

HENRICOT HR.33
Usines Emile Henricott, SA
C 0-0.15, Mn 0-1, Si 0-1, Ni 2-3, Cr 17-19, bal Fe.
Similar to ASTM A296 CB-30.

HENRICOT HR.34
Usines Emile Henricott, SA
C 0-0.08, Mn 0-1, Si 0-1, Ni 3-5, Cr 15.5-17.5, Cu 3-5, Nb
0.1-0.5, bal Fe.
Precipitation hardening stainless steel casting; 17-4 PH Type.

HENRICOT HR.3M3
Usines Emile Henricott, SA
C 0.95-1.2, Mn 0-1, Si 0-1, Cr 16-18, Mo 0-0.75, bal Fe.
Corrosion resistant steel casting. AISI 440C.

HENRICOT HR.3M5
Usines Emile Henricott, SA
C 3-3.5, Cr 15-18, Mo 3-3.2, bal Fe.
Corrosion resistant alloy casting.

HENRICOT HR.43
Usines Emile Henricott, SA
C 0-0.5, Mn 0-1, Si 0-3, Ni 2-4, Cr 26-30, 0.50 Mo residual
max, bal Fe.
Heat resistant alloy casting. Similar to ASTM A297 HC.

HENRICOT HR.44
Usines Emile Henricott, SA
C 0.3-0.5, Mn 0-1, Si 0-1.5, Ni 3.5-4.5, Cr 26-28, bal Fe.
Heat resistant alloy casting. Similar to ASTM A297 HC; W.-Nr.
1.4340.

HENRICOT HR.46
Usines Emile Henricott, SA
C 0.5, Mn 0-1, Si 0-3, Ni 5-7, Cr 26-30, 0.50 Mo residual max,
bal Fe.
Heat resistant alloy casting. Similar to ASTM A297 HD.

HENRICOT HR.4M
Usines Emile Henricott, SA
C 0-0.1, Mn 0-2, Si 0-1, Ni 4.5-6, Cr 25-27, Mo 1.3-2, bal Fe.
Corrosion resistant alloy steel casting. Similar to AISI 329; SIS
2324.

HENRICOT HR.51
Usines Emile Henricott, SA
C 0-2.5, Mn 0-0.75, Si 0-0.75, Cr 27-30, bal Fe.
Alloy steel casting for high temperature application, as
furnace parts.

HENRICOT HR.56
Usines Emile Henricott, SA
C 1-1.5, Mn 0-1, Si 1-2.5, Cr 27-30, bal Fe.
Alloy steel casting for high temperature operation. W.-Nr.
1.4777.

HENRICOT HR.5A
Usines Emile Henricott, SA
C 2.5-3.5, Mn 0-1.5, Si 0-1, Cr 25-29, bal Fe.
Alloy steel casting for high temperature operation.

HENRICOT HR.5M
Usines Emile Henricott, SA
Mn 0-1, Si 0-1, Cr 27-31, Mo 4-5, bal Fe.
Alloy steel casting for chemical equipment and high
temperature operation.

HENRICOT HR1
Usines Emile Henricot, S.A.
C 0-0.08, Cr 11.5-13, Al 0.1-0.3, bal Fe.
Normalized: 71,000 TS; 42,600 YS; 22 El; 70 RA; 150 Brin.
Heat treated: 175,000 TS; 145,000 YS; 21 El; 64 RA; 352 Brin.
For injectors, table flatware, valves, pumps; Type 405;
corrosion resistant. *Obsolete*

HENRICOT HR2
Usines Emile Henricot, S.A.
C 0.2, Cr 12, Ni 0.5, bal Fe.
Heat treated: 256,000 TS; 190,000 YS; 6 El; 10 RA; 540 Brin.
For turbine blading, coal screens; corrosion resistant,
hardenable. *Obsolete*

HENRICOT HR24
Usines Emile Henricott, SA
C 0-0.15, Mn 0-1, Si 0-1.5, Ni 0-1, Cr 11.5-14.5, Mo 0-0.5, bal
Fe.
Quenched and tempered: 70-85 kg/mm^2 TS; 50 kg/mm^2 YS
min; 15 El min (at 720°C). ASTM A296 Type CA-15.

HENRICOT HR25
Usines Emile Henricott, SA
C 0.08-0.12, Mn 0-1, Si 0-1.5, Ni 0.8-1.2, Cr 12.2-13.2, bal Fe.
Quenched and tempered: 65-75 kg/mm^2 TS; 50 kg/mm^2 YS
min; 15 El min (at 720°C). Corrosion resistant steel casting.

HENRICOT HR26
Usines Emile Henricott, SA
C 0-0.18, Mn 0-1, Si 0-0.6, Cr 12-14, bal Fe.
Quenched and tempered: 70-80 kg/mm^2 TS; 45 kg/mm^2 YS
min; 15 El min; 50 RA min (at 720°C). Corrosion resistant
steel casting.

HENRICOT HR27
Usines Emile Henricott, SA
C 0.12, Ni 1, Cr 13, Mo 1, bal Fe.
Corrosion resistant steel casting.

HENRICOT HR28
Usines Emile Henricott, SA
C 0-0.07, Mn 0-0.75, Si 0-0.6, Ni 3.25-4.25, Cr 11.5-14, Mo
0-1, bal Fe.
Quenched and tempered: 80-95 kg/mm^2 TS; 60 kg/mm^2 YS
min; 14 El min (at 630°C). Corrosion resistant steel casting.

HENRICOT HR2F
Usines Emile Henricot, S.A.
C 3, Cr 14, bal Fe.
For abrasion and heat resistant castings; cast iron, wear and
heat resistant. *Obsolete*

HENRICOT HR2F4
Usines Emile Henricott, SA
C 1.35-1.4, Cr 12-13, Mo 1.2-1.4, bal Fe.
Corrosion resistant steel casting.

HENRICOT HR2L
Usines Emile Henricot, S.A.
Cr 14, C, bal Fe.
Annealed: 90,000-100,000 TS; 40,000-60,000 YS; 12-15 El;
150-170 Brin. For kitchen equipment, tableware; corrosion
resistant. *Obsolete*

HENRICOT HR2R
Usines Emile Henricott, SA
C 0.25-0.35, Mn 0-0.85, Si 0-1, Cr 12-14, Mo 0-0.5, 1.0 Ni
residual, bal Fe.
Quenched and tempered: 85-95 kg/mm^2 TS; 68 kg/mm^2 YS;
10 El (at 650°C). ASTM A296 Type CA-40.

HENRICOT HR2R2
Usines Emile Henricott, SA
C 0.4-0.5, Mn 0-1, Si 0-1, Cr 12-14, 1.0 Ni residual, bal Fe.
Corrosion resistant steel casting.

HENRICOT HR2X
Usines Emile Henricot, S.A.
C 0.08, Cr 12, Al 0.2, Mn 0-1, Si 0-1, bal Fe.
Annealed: 85,000 TS; 60,000 YS; 20 El; 150 Brin. Heat
treated: 170,000 TS; 145,000 YS; 21 El; 64 RA; 350 Brin. For
injectors, flatware, valve seats; heat resistant, hardenable.
Obsolete

HENRICOT HR3
Usines Emile Henricot, S.A.
C 0-0.1, Cr 18, bal Fe.
Normalized: 86,000 TS; 57,000 YS; 20 El; 50 RA. For
nonsealing purposes up to 900 C; corrosion resistant.
Obsolete

HENRICOT HR30
Usines Emile Henricott, SA
C 0-0.15, Mn 0-1, Si 0-1, Cr 17-19, 1.0 Ni residual max, bal
Fe.
Corrosion resistant steel casting.

HENRICOT HR3L
Usines Emile Henricot, S.A.
C 0.1, Cr 18, Ni 3, bal Fe.
Heat treated: 121,000 TS; 100,000 YS; 15 El; 50 RA; 200-220
Brin. For valves, valve seats, turbine blades; stainless,
hardenable. *Obsolete*

HENRICOT HR4
Usines Emile Henricot, S.A.
C 0-0.2, Cr 24-26, Si 0-1, bal Fe.
Annealed: 75,000-85,000 TS; 50,000-60,000 YS; 12 El; 180
Brin. For furnace parts, heat treating boxes; Type 442;
corrosion and heat resistant. *Obsolete*

HENRICOT HR4LA
Usines Emile Henricot, S.A.
Cr 24, Ni 12, C, N, bal Fe.
Annealed: 100,000-115,000 TS; 40,000-60,000 YS; 4-6 El. For
furnace parts and accessories, heat treating boxes; heat
resistant to 1050 C. *Obsolete*

HENRICOT HR4LB
Usines Emile Henricot, S.A.
C 0.2, Cr 23-24, Ni 12-14, Si 1, bal Fe.
Annealed: 85,000-100,000 TS; 40,000-60,000 YS; 12-15 El.
For furnace parts, heat treat equipment, burners; heat
resistant to 1050 C. *Obsolete*

HENRICOT HR4LX
Usines Emile Henricot, S.A.
C 0.15, Si 0.75, Ni 12, Cr 23, bal Fe.
Annealed: 93,000 TS; 50,000 YS; 40 El; 50 RA; 180 Brin. For
heat treating equipment; austenitic, stainless. *Obsolete*

HENRICOT HR5
Usines Emile Henricot, S.A.
C 2, Cr 30, bal Fe.
Cast: 350 Brin. For furnace parts; operating temperature to
1050 C. *Obsolete*

HENRICOT HR5II
Usines Emile Henricot, S.A.
C 2, Cr 30, Mo, bal Fe.
Cast: 350 Brin. For heat and wear resistant castings; cast
iron, heat resistant to 1050 C. *Obsolete*

HENRICOT HR5III
Usines Emile Henricott, SA
C 2, Cr 30, Mo, bal Fe.
Cast: 350 Brin. For heat and wear resistant castings; cast
iron, heat resistant to 1050 C. *Obsolete*

HENRICOT HR5M
Usines Emile Henricott, SA
C 2, Cr 30, Mo, bal Fe.
For chemical plant equipment; cast iron; corrosion and wear
resistant.

HENRICOT HR5VI
Usines Emile Henricot, S.A.
C 2, Cr 30, bal Fe.
For chemical plant equipment; cast iron, corrosion and wear
resistant. *Obsolete*

HENRICOT HR6
Usines Emile Henricot, S.A.
C 0.35, Cr 23-27, N 0-0.25, bal Fe.
Annealed: 90,000 TS; 60,000 YS; 20 El; 45 RA; 180 Brin. Cold
drawn: 175,000 TS; 155,000 YS; 2 El; 25 RA; 250 Brin. For
furnace parts, preheaters, annealing boxes; Type 446;
corrosion resistant. *Obsolete*

HENRICOT HR60
Usines Emile Henricot, S.A.
C 0.2, Mn 1, Si 1, Cr 27-30, bal Fe.
Corrosion resistant alloy steel casting.

HENRICOT HR6P
Usines Emile Henricot, S.A.
C 0-0.1, Cr 27-30, bal Fe.
Annealed: 75,000-85,000 TS; 50,000-60,000 YS; 12 El. For
glass and petroleum industry equipment; corrosion and heat
resistant to 1200 C. *Obsolete*

HENRICOT HR80
Usines Emile Henricott, SA
C 0-0.2, Mn 0.4-0.7, Si 0.75, Ni 0.5, Cr 4-6.5, Mo 0.45-0.65,
Cu 0.5, W 1, bal Fe.
Quenched and tempered: 65-80 kg/mm^2 TS; 45 kg/mm^2 YS;
15 El (at 720°C). Alloy steel casting for pressure vessels for
high temperature service. ASTM A217 C5.

HENRICOT HR8S
Usines Emile Henricot, S.A.
C 0.2, Cr 5, Mo 0.5, bal Fe.
Annealed: 100,000 TS; 85,000 YS; 12 El; 160 Brin. For oil
refinery equipment; creep resistant to 600 C. *Obsolete*

HENRICOT HRC
Usines Emile Henricot, S.A.
C 0.3, Cr 12, bal Fe.
Hardened: 142,000 TS; 121,000 YS; 8 El; 40 RA. For cutlery;
corrosion resistant, hardenable. *Obsolete*

HENRICOT HRM
Usines Emile Henricot, S.A.
C 0.3, Cr 14, Mo 1.2, bal Fe.
Heat treated: 129,000 TS; 107,000 YS; 13 El; 45 RA. For
paper industry beaters and blades; corrosion resistant,
hardenable. *Obsolete*

HENRICOT HS1
Usines Emile Henricott, SA
C 0.95-1.15, Mn 11.5-14, Si 0-0.65, Cr 1-1.25, bal Fe.
Annealed: 160 Brin. Cold worked: 375 Brin. Austenitic
manganese steel castings. ASTM A128.

HENRICOT HS10
Usines Emile Henricott, SA
C 1-1.15, Mn 11.5-14, Si 0.65, P 0.07, bal Fe.
Annealed: 160 Brin. Cold worked: 375 Brin. Austenitic
manganese steel castings. ASTM A128.

HENRICOT HS2
Usines Emile Henricott, SA
C 1.1-1.3, Mn 11.5-14, Si 1, Cr 1.1-1.3, bal Fe.
Annealed: 190 Brin. Cold worked: 425 Brin. Austenitic
manganese steel castings. ASTM A128.

HENRICOT HS20
Usines Emile Henricott, SA
C 1.1-1.3, Mn 11.5-14, Si 0-1, P 0-0.07, bal Fe.
Annealed: 180 Brin. Cold worked: 400 Brin. Austenitic
manganese steel castings. ASTM A128.

HENRICOT HS25
Usines Emile Henricott, SA
C 1.3-1.5, Mn 18-20, Si 0-1, Cr 1.7-2.3, bal Fe.
Stainless steel casting.

HENRICOT HS26

Usines Emile Henricott, SA
C 1.3-1.5, Mn 18-20, Si 0-1, Cr 2.3-2.9, bal Fe.
Stainless steel casting.

HENRICOT HS3

Usines Emile Henricott, SA
C 1.2-1.4, Mn 12-14, Si 0-1, Cr 1.5-2, bal Fe.
Annealed: 200 Brin. Cold worked: 450 Brin. Austenitic
manganese steel castings. ASTM A128.

HENRICOT HS30

Usines Emile Henricott, SA
C 1.25-1.4, Mn 12-14, Si 1, P 0-0.07, bal Fe.
Annealed: 200 Brin. Cold worked: 450 Brin. Austenitic
manganese steel castings. ASTM A128.

HENRICOT HSO

Usines Emile Henricott, SA
C 0.95-1.15, Mn 11.5-14, Si 0-0.65, bal Fe.
Austenitic manganese steel castings; ASTM A128.

HENRICOT HT1

Usines Emile Henricott, S.A.
C 0.15, Si 2, Ni 15, Cr 18, bal Fe.
Annealed: 86,000 TS; 43,000 YS; 35 El; 50 RA. For heat
resistant applications; heat resistant to 1000 C. *Obsolete*

HENRICOT HT10

Usines Emile Henricott, SA
C 0.2-0.4, Mn 0-2, Si 0-2, Ni 15-17, Cr 18-20, bal Fe.
Cast: 48-65 kg/mm^2 TS; 25 kg/mm^2 YS; 15 El. Stainless
steel casting for high temperature operation. Similar to W.-Nr.
1.4832.

HENRICOT HT12

Usines Emile Henricott, SA
C 0.2-0.4, Mn 0-2, Si 0-2, Ni 8-12, Cr 18-23, Mo 0-0.5, bal Fe.
Cast: 49 kg/mm^2 TS min; 24 kg/mm^2 YS; 20 El. Austenitic
stainless steel casting for high temperature operation. Similar
to W.-Nr.1.4826; ASTM A297-HF.

HENRICOT HT14

Usines Emile Henricott, SA
C 0.2-0.4, Mn 0-1, Si 0-1, Ni 7.5-9.5, Cr 17-19, bal Fe.
Cast: 50-65 kg/mm^2 TS; 25 kg/mm^2 YS; 20 El. Stainless
steel casting; B.S. 1648.D.

HENRICOT HT2

Usines Emile Henricott, S.A.
C 0.15, Si 2, Ni 25, Cr 25, bal Fe.
Annealed: 93,000 TS; 50,000 YS; 30 El; 50 RA. For heat
resistant applications; heat resistant to 1100 C. *Obsolete*

HENRICOT HT20

Usines Emile Henricott, SA
C 0.2-0.4, Mn 0-2, Si 0-2, Ni 18-22, Cr 24-28, 0.50 Mo residual
max, bal Fe.
Cast: 48 kg/mm^2 TS min; 26 kg/mm^2 YS; 10 El. Heat
resistant cast alloy. W.-Nr. 1.4848; ASTM A297 HK.

HENRICOT HT21

Usines Emile Henricott, SA
C 0-0.3, Mn 0-2, Si 0-2.5, Ni 24-28, Cr 24-28, bal Fe.
Cast: 43 kg/mm^2 TS min; 28 kg/mm^2 YS min; 12 El. Heat
resistant alloy casting.

HENRICOT HT23

Usines Emile Henricott, SA
C 0.35-0.45, Mn 0-2, Si 0-1.5, Ni 19-22, Cr 24-27, 0.50 Mo
residual max, bal Fe.
Cast: 45.5 kg/mm^2 TS min; 25 kg/mm^2 YS min; 10 El min.
Heat resistant cast alloy. Similar to W.-Nr. 1.4848; B.S. 1648-
F; ASTM A297-HK.

HENRICOT HT24

Usines Emile Henricott, SA
C 0.35-0.45, Mn 0-2, Si 0-1.5, Ni 19-22, Cr 24-27, 0.50 Mo
residual max, bal Fe.
Cast: 45 kg/mm^2 TS min; 25 kg/mm^2 min YS; 8 El min. Heat
resistant cast alloy. Similar to W.-Nr. 1.4848; B.S. 1648-F;
ASTM A297-HK.

HENRICOT HT25

Usines Emile Henricott, SA
C 0.25-0.35, Mn 0-2, Si 0-1.5, Ni 23-25, Cr 23-25, Cu 1-2, bal
Fe.
Cast: 45-65 kg/mm^2 TS; 25 kg/mm^2 YS; 10 El. Heat resistant
cast alloy.

HENRICOT HT3

Usines Emile Henricot, S.A.
C 0.15, Si 2, Ni 35, Cr 20, bal Fe.
Annealed: 100,000 TS; 57,000 YS; 25 El; 45 RA. For heat
resistant applications; heat resistant to 1200 C. *Obsolete*

HENRICOT HT30

Usines Emile Henricott, SA
C 0.35-0.45, Mn 0-2, Si 0-2.5, Ni 33-37, Cr 13-17, 0.50 Mo
residual max, bal Fe.
Cast: 45-65 kg/mm^2 TS; 25 kg/mm^2 YS; 4 El. Heat resistant
alloy; for furnace parts. Similar to W.-Nr. 1.4865; ASTM A297-
HT.

HENRICOT HT31

Usines Emile Henricott, SA
C 0-0.2, Mn 0-2, Si 0-0.8, Ni 33-37, Cr 19-23, 0.50 Mo residual
max, bal Fe.
Cast, annealed: 50 kg/mm^2 TS; 21 kg/mm^2 YS; 25 El. Heat
and oxidation resistant alloy. Similar to W.-Nr. 1.4861; Incoloy
800.

HENRICOT HT32

Usines Emile Henricott, SA
C 0.35-0.45, Mn 0-2, Si 0-2.5, Ni 37-41, Cr 17-21, 0.50 Mo
residual max, bal Fe.
Cast, annealed: 45-65 kg/mm^2 TS; 25 kg/mm^2 YS; 4 El.
Heat and oxidation resistant alloy. Similar to W.-Nr. 1.4865;
ASTM A297-HU.

HENRICOT HT33

Usines Emile Henricott, SA
C 0.45-0.55, Mn 0-1, Si 0-2.5, Ni 45-55, Cr 26-30, Mo 0-0.5, W
4-6, bal Fe.
42-60 kg/mm^2 TS. Heat and oxidation resistant casting.
Similar to W.-Nr. 2.4879.

HENRICOT HT35

Usines Emile Henricott, SA
C 0-0.15, Mn 0-2, Si 0-1.5, Ni 31-33, Cr 19-22, Co 0.5-1.5, bal
Fe.
Heat and oxidation resistant casting.

HENRICOT HT36

Usines Emile Henricott, SA
C 0-0.12, Mn 0-1, Si 0-1, Cr 28-30, Co 48-52, bal Fe.
50-65 kg/mm^2 TS; 28 kg/mm^2 YS; 10 El. Heat and oxidation
resistant casting.

HENRICOT HT37

Usines Emile Henricott, SA
C 0.25-0.4, Mn 0.5-1, Si 0.5-1, Cr 27-29, Co 48-52, Nb 1.8-2.5,
bal Fe.
Heat and oxidation resistant casting.

HENRICOT HT3S

Usines Emile Henricot, S.A.
C 0.3, Ni 30, Cr 12, bal Fe.
Water quenched: 85,000-93,000 TS; 35,000-50,000 YS; 20 El;
40 RA. For the steam turbine industry; corrosion and heat
resistant. *Obsolete*

HENRICOT HT42

Usines Emile Henricott, SA
C 0.2-0.5, Mn 0-1.5, Si 0-2, Ni 11-14, Cr 24-28, Mo 0-0.5, bal
Fe.
Cast: 52 kg/mm^2 TS; 24 kg/mm^2 YS; 25 El. Heat and
oxidation resistant castings. W.-Nr. 1.4837; ASTM A297-HH.

HENRICOT HT43

Usines Emile Henricott, SA
C 0.2-0.45, Mn 0-2, Si 0-1.75, Ni 11-14, Cr 23-28, Mo 0-0.5, N
0.2, bal Fe.
Cast: 56 kg/mm^2 TS; 4 El. Heat and oxidation resistant alloy
casting. Similar to ASTM A447 Type 2.

HENRICOT HT46

Usines Emile Henricott, SA
C 0-0.3, Mn 0-1.5, Si 0-2, Ni 8-11, Cr 26-30, Mo 0-0.5, bal Fe.
Cast, tempered: 50 kg/mm^2 TS; 28 kg/mm^2 YS; 10 El. Heat
and oxidation resistant alloy casting. Similar to ASTM A296-
CE30.

HENRICOT HT50

Usines Emile Henricott, SA
C 0-0.05, Mn 0-0.6, Si 0-0.45, Cr 48-52, Fe 0-2, bal Ni.
Cast: 55-70 kg/mm^2 TS; 35 kg/mm^2 YS; 15 El. Heat and
oxidation resistant alloy casting.

HENRICOT HT51

Usines Emile Henricott, SA
C 0-0.05, Mn 0-0.6, Si 0-0.45, Cr 48-52, Fe 0-2, Nb 0-2, bal Ni.
Cast: 50-75 kg/mm^2 TS; 40 kg/mm^2 YS; 25 El. Heat and
oxidation resistant alloy casting.

HENRICOT HT6

Usines Emile Henricott, S.A.
C 0.15, Si 2, Ni 65, Cr 18, bal Fe.
Annealed: 100,000 TS; 57,000 YS; 20 El; 45 RA. For heat
resistant applications; heat resistant to 1250 C. *Obsolete*

HENRICOT HT60

Usines Emile Henricott, SA
C 0.25-0.35, Mn 0-2, Si 0-2.5, Ni 58-62, Cr 16-20, Mo 0-0.5,
bal Fe.
Cast: 45 kg/mm^2 TS. Heat and oxidation resistant alloy
casting. Similar to ASTM A297-HW.

HENRICOT HT70

Usines Emile Henricott, SA
C 0-0.1, Mn 0-1, Si 0-1.5, Cr 14-17, Fe 6-10, Nb 1-2, Co 0-0.1,
Cu 0-0.5, bal Ni.
Cast: 48 kg/mm^2 TS; 18 kg/mm^2 YS; 35 El. Heat and
oxidation resistant alloy casting.

HENRICOT HT75

Usines Emile Henricott, SA
C 0-0.2, Mn 0-2, Si 0-2, Ni 75-79, Cr 12.5-14.5, bal Fe.
Heat and oxidation resistant casting. Similar to W.-Nr. 2.4640.

HENRICOT HTL

Usines Emile Henricott, S.A.
C 0.2, Cr 19, Ni 9, Mo 1.5, W 1.5, bal Fe.
For blades for gas turbines; austenitic, stainless. *Obsolete*

HENRICOT HTY

Usines Emile Henricott, S.A.
C 0.15, Cr 1, Mo 0.4, V 0.4, W 0.5, bal Fe.
Annealed: 80,000-90,000 TS; 60,000-70,000 YS; 14-16 El;
150-170 Brin. For oil refinery equipment; creep resistant to
575 C. *Obsolete*

HENRICOT HV12

Usines Emile Henricott, SA
C 0-0.08, Mn 0-1.5, Si 0-0.75, Ni 8-11, Cr 18-20, bal Fe.
Cast: 48 kg/mm^2 TS; 20 kg/mm^2 YS; 35 El. Corrosion
resistant steel casting. W.-Nr. 1.4308; ASTM A296-CF8.

HENRICOT HV13
Usines Emile Henricott, SA
C 0-0.03, Mn 0-1.5, Si 0-1, Ni 8-12, Cr 18-20, bal Fe.
Cast: 49 kg/mm² TS; 21 kg/mm² YS; 35 El. Corrosion resistant steel casting. Similar to W.-Nr. 1.4306; ASTM A296-CF3.

HENRICOT HV181
Usines Emile Henricott, SA
C 0-0.06, Mn 0-2, Si 0-1.5, Ni 8-11, Cr 18-21, Co 0.2, Cu 1, Nb 0.15, N 0.02, bal Fe.
Cast: 50 kg/mm² TS; 21 kg/mm² YS; 30 El. Corrosion resistant steel casting. Similar to ASTM A296-CF8. For nuclear energy.

HENRICOT HV182
Usines Emile Henricott, SA
Mn 0-1.5, Si 0-2, Ni 8-11, Cr 18-21, Co 0.2, bal Fe.
Cast: 54 kg/mm² TS; 24.5 kg/mm² YS; 35 El. Corrosion resistant steel casting. Similar to ASTM A296-CF8. For nuclear energy.

HENRICOT HV1S
Usines Emile Henricot, S.A.
C 0-0.08, Ni 8, Cr 18, bal Fe.
Annealed: 79,000 TS; 36,000 YS; 40 El; 50 RA; 150 Brin. For chemical plant equipment; Type 304; stainless, austenitic. *Obsolete*

HENRICOT HV2S
Usines Emile Henricot, S.A.
C 0-0.15, Cr 18, Ni 9, bal Fe.
Annealed: 75,000-85,000 TS; 30,000-40,000 YS; 40-45 El; 150-160 Brin. For chemical and textile plant equipment, tanks, mixers; stainless, austenitic; Type 302. *Obsolete*

HENRICOT HV3
Usines Emile Henricot, S.A.
C 0-0.08, Ni 11, Cr 18, Mo 2.5, bal Fe.
Annealed: 85,000-95,000 TS; 36,000-50,000 YS; 50-60 El; 60-75 RA; 150-190 Brin. For chemical and plastic plant equipment; Type 316; stainless, austenitic. *Obsolete*

HENRICOT HV30
Usines Emile Henricott, SA
C 0-0.08, Mn 0-1.5, Si 0-1.5, Ni 10.5-13.5, C 16.5-18.5, Mo 2-3, bal Fe.
Cast, annealed: 49 kg/mm² TS; 21 kg/mm² YS; 30 El (at 1100°C). Corrosion resistant cast steel. Similar to W.-Nr. 1.4580.

HENRICOT HV30B
Usines Emile Henricott, SA
C 0-0.06, Mn 0-2, Si 0-1, Ni 11-14, Cr 16-19, Mo 1.5-3, bal Fe.
Corrosion resistant steel casting.

HENRICOT HV31
Usines Emile Henricott, SA
C 0-0.08, Mn 0-1.5, Si 0-2, Ni 9-12, Cr 18-21, Mo 2-3, bal Fe.
Cast, annealed: 49 kg/mm² TS; 21 kg/mm² YS; 30 El (at 1100°C). Corrosion resistant steel casting. Similar to W.-Nr. 1.4408; ASTM A296-CF8M; AISI 316.

HENRICOT HV32
Usines Emile Henricott, SA
C 0-0.06, Mn 0-2, Si 0-1, Ni 10-13.5, Cr 16.5-19, Mo 2.5-3, bal Fe.
Corrosion resistant steel casting. Similar to W.-Nr. 1.4437.

HENRICOT HV36
Usines Emile Henricott, SA
C 0-0.06, Mn 0-2, Si 0-1, Ni 13-14, Cr 18-20, Mo 3-4, bal Fe.
Cast, annealed: 47 kg/mm² TS; 20 kg/mm² YS; 30 El (at 1100°C). Corrosion resistant steel casting. Similar to W.-Nr. 1.4448; B.S. 1632-A; AISI 317.

HENRICOT HV37
Usines Emile Henricott, SA
C 0-0.03, Mn 0-2, Si 0-1, Ni 10-14, Cr 16-18, Mo 2-3, bal Fe.
Corrosion resistant steel casting; Similar to ASTM A296-CF3M; AISI 316L.

HENRICOT HV382
Usines Emile Henricott, SA
C 0-0.06, Mn 0-1.5, Si 0-1.5, Ni 9-12, Cr 18-21, Mo 2.3-2.8, Co 0.2, C 1, 0.15 Nb + Ta, bal Fe.
Corrosion resistant steel casting.

HENRICOT HV3P
Usines Emile Henricot, S.A.
C 0.2, Cr 18, Ni 8, Mo 1.5, bal Fe.
Annealed: 70,000-80,000 TS; 30,000-40,000 YS; 20-25 El; 150-160 Brin. For marine ornamental trim, hardware; Type 316; stainless, austenitic. *Obsolete*

HENRICOT HV4
Usines Emile Henricot, S.A.
C 0.08, Cr 18-20, Ni 8-11, Mn 0-2, bal·Fe.
Annealed: 90,000 TS; 45,000 YS; 60 El; 135 Brin. Cold drawn: 18,000 TS; 15,000 YS; 10 El; 330 Brin. For chemical plant equipment, welded structures; Type 304; stainless, austenitic. *Obsolete*

HENRICOT HV5
Usines Emile Henricot, S.A.
C 0-0.25, Cr 24-26, Ni 19-22, bal Fe.
Annealed: 95,000 TS; 45,000 YS; 50 El; 65 RA; 180 Brin. At 1200 F: 57,000 TS; 22,000 YS; 32 El; 45 RA. For engine components, valves, furnace equipment; Type 310; austenitic, heat resistant. *Obsolete*

HENRICOT HV6
Usines Emile Henricot, S.A.
C 0-0.08, Ni 9, Cr 18, Ti 0.6, bal Fe.
Annealed: 86,000 TS; 43,000 YS; 40 El; 50 RA. For welded stainless construction; austenitic, stainless, stabilized. *Obsolete*

HENRICOT HV60
Usines Emile Henricott, SA
C 0-0.08, Mn 0-1.5, Si 0-0.75, Ni 9-12, Cr 18-21, Nb = 8 x C, bal Fe.
Cast, annealed: 49 kg/mm² TS; 21 kg/mm² YS; 35 El. Corrosion resistant steel casting. Similar to W.-Nr. 1.4552; ASTM A296-CF8C.

HENRICOT HV6S
Usines Emile Henricot, S.A.
C 0-0.08, Cr 17-19, Ni 9-12, Cb = 10 x C, bal Fe.
Annealed: 85,000-95,000 TS; 35,000-45,000 YS; 50-55 El; 175 Brin. For welded structures, chemical plant equipment; Type 347, corrosion and heat resistant. *Obsolete*

HENRICOT HV7
Usines Emile Henricot, S.A.
C 0-0.08, Ni 20, Cr 15, Mo 4.5, bal Fe.
Annealed: 85,000-95,000 TS; 35,000-50,000 YS; 50-60 El; 60-75 RA; 150-190 Brin. For chemical plant equipment, furnace parts; stainless, austenitic. *Obsolete*

HENRICOT HV70
Usines Emile Henricott, SA
C 0-0.07, Mn 0-2, Si 0-1, Ni 19-21, Cr 15.5-17.5, Mo 4, Nb, bal Fe.
Cast, annealed: 48 kg/mm² TS; 23 kg/mm² YS; 30 El. Corrosion resistant steel casting.

HENRICOT HV8
Usines Emile Henricot, S.A.
C 0-0.07, Cr 17, Ni 13, Mo 4.5, bal Fe.
Annealed: 90,000 TS; 40,000 YS; 45 El; 60 RA; 180 Brin. For acid resistant chemical plant equipment; Type 317; stainless, austenitic. *Obsolete*

HENRICOT HV9
Usines Emile Henricott, SA
C 0-0.04, Mn 0-2, Si 0-1, Ni 24-26, C 20-23, Mo 4, Nb, bal Fe.
Heat and corrosion resistant wrought steel.

HENRICOT HV90
Usines Emile Henricott, SA
C 0-0.06, Mn 0-2, Si 0-1.5, Ni 27-29, Cr 20-23, Mo 4-5, Cu 2-4, Nb, bal Fe.
Corrosion resistant steel casting.

HENRICOT HV91
Usines Emile Henricott, SA
C 0-0.07, Mn 0-1.5, Si 0-1.3, Ni 27-32, Cr 19-22, Mo 2-3, Cu 3-4, bal Fe.
Cast, annealed: 45 kg/mm² TS; 21 kg/mm² YS; 35 El. Corrosion resistant steel casting. Similar to ASTM A296-CN7M.

HENRICOT HV98
Usines Emile Henricott, SA
C 0-0.04, Mn 0-2, Si 0-1, Ni 24-26, Cr 20-23, Mo 4, Nb 0.3, bal Fe.
Corrosion resistant steel casting.

HENRICOT HV9A
Usines Emile Henricott, SA
C 0-0.04, Mn 0-2, Si 0-1, Ni 24-26, C 20-23, Mo 4-5, Cu, Nb, bal Fe.
Heat and corrosion resistant wrought steel.

HENRICOT HVD1
Usines Emile Henricott, SA
C 0-0.08, Mn 0-2, Si 0-1, Ni 7-9, Cr 24-26, Mo 2-3, Cu 1-2, bal Fe.
Heat and corrosion resistant wrought steel.

HENRICOT HVD10
Usines Emile Henricott, SA
C 0-0.08, Mn 0-2, Si 0-1, Ni 7-9, Cr 24-26, Mo 2-3, Cu 1-2, bal Fe.
Cast: 70-85 kg/mm² TS; 50 kg/mm² YS; 20 El. Heat and corrosion resistant steel casting.

HENRICOT HVD13
Usines Emile Henricott, SA
C 0-0.1, Mn 0-2, Si 0-1, Ni 7-9, Cr 19.5-22, Mo 2-3, Cu 1-2, bal Fe.
Cast: 65 kg/mm² TS; 35 kg/mm² YS; 30 El. Corrosion resistant steel casting.

HENRICOT HVD20
Usines Emile Henricott, SA
C 0-0.08, Mn 0-2, Si 0-1.5, Ni 7-9, Cr 25-27, Mo 2-3, Cu 1-2, bal Fe.
Heat and corrosion resistant steel casting.

HENRICOT HVI
Usines Emile Henricot, S.A.
C 0.08-0.2, Mn 0-2, Cr 17-19, Ni 8-10, bal Fe.
Annealed: 85,000 TS; 35,000 YS; 60 El; 70 RA; 150 Brin. Cold drawn: 25,000 TS; 95,000 YS; 25 El; 55 RA; 277 Brin. For oil refinery and chemical plant equipment; Type 302; stainless, austenitic. *Obsolete*

HENRICOT HVIR
Usines Emile Henricot, S.A.
C 0.3-0.4, Cr 18, Ni 8, bal Fe.
For beating trough blades in paper refining machines; austenitic, stainless. *Obsolete*

HENRICOT HVIX
Usines Emile Henricot, S.A.
C 0.05-0.08, Cr 18, Ni 9.5, bal Fe.
Annealed: 75,000-85,000 TS; 25,000-35,000 YS; 40-45 El; 150-160 Brin. For chemical plant equipment, trim, tanks; stainless, austenitic; Type 304. *Obsolete*

HENRICOT HW2
Usines Emile Henricot, S.A.
C 0.5, Cr 1, W 2, bal Fe.
For punches, chisels; tough, shock resistant. *Obsolete*

HENRICOT HW2E
Usines Emile Henricot, S.A.
C 0.38, Cr 16-18, bal Fe.
For oil refinery equipment, bolts; corrosion resistant.
Obsolete

HENRICOT HW2E
Usines Emile Henricot, S.A.
C 0.37, Cr 1, W 1.75, bal Fe.
For punches, chisels; water hardening. *Obsolete*

HENRICOT HW3
Usines Emile Henricot, S.A.
C 1.25, W 3, bal Fe.
Hardened: 650 Brin. For punches, chucks, wood working
tools; water hardening. *Obsolete*

HENRICOT HWC
Usines Emile Henricot, S.A.
C 0.8, Cr 4, W 19, V 1.6, Co 13, Mo 1, bal Fe.
For cutters, lathes and planer tools; high speed steel.
Obsolete

HENRICOT HWD
Usines Emile Henricot, S.A.
C 0.7, W 21, Cr, V, bal Fe.
For lathe and planer tools, cutters; high speed steel.
Obsolete

HENRICOT HWF
Usines Emile Henricot, S.A.
C 0.7, Cr 4, W 18, V 1.25, bal Fe.
For cutters, hobs, drills; high speed steel. *Obsolete*

HENRICOT HWF1
Usines Emile Henricot, S.A.
C 0.7, W 18, Cr 4, V 1, bal Fe.
For cutters, drills, taps, hobs, broaches; high speed steel, oil
hardened. *Obsolete*

HENRICOT HWF2
Usines Emile Henricot, S.A.
C 0.7, Cr 4, V 1, W 18, bal Fe.
For cutters, drills, reamers, taps, broaches; high speed steel.
Obsolete

HENRICOT HWIE
Usines Emile Henricot, S.A.
C 1.2, W 1, bal Fe.
For punches; water hardening. *Obsolete*

HENRICOT HWMO4
Usines Emile Henricot, S.A.
C 0.85, Cr 4, V 2, W 6, Mo 5, bal Fe.
Hardened: 640 Brin. For twist drills, reamers, milling cutters;
high speed steel. *Obsolete*

HENRICOT HWMO7
Usines Emile Henricot, S.A.
C 0.85, Cr 4, V 1.5, W 1.5, Mo 8, bal Fe.
Hardened: 650 Brin. For gear cutters, broaches, boring tools;
high speed steel. *Obsolete*

HENRICOT HWS
Usines Emile Henricot, S.A.
C 0.7, Cr 4, W 14, V 1.2, bal Fe.
Hardened: 610 Brin. For tools, cutters, hobs, drills, reamers;
high speed steel. *Obsolete*

HENRICOT HWT
Usines Emile Henricot, S.A.
C 0.8, Cr 4, W 19, V 1.6, Co 6, Mo 1, bal Fe.
For cutters, broaches, lathe tools, drills; high speed steel
Obsolete

HENRICOT HWT
Usines Emile Henricot, S.A.
C 0.7, W 18, Co 5-6, bal Fe.
For milling cutters, lathe and planer tools; high speed steel.
Obsolete

HENRICOT MONEL
Usines Emile Henricott, SA
C 0.1-0.3, Mn 0.5-1.5, Si 1.25-2, Cu 28-34, Fe 3, bal Ni.
Cast: 150 Brin.

HENRICOT NI-AL-0
Usines Emile Henricot, S.A.
Ni 22, Al 10, bal Fe.
For permanent magnets; high permeability. *Obsolete*

HENRICOT NI-AL-1
Usines Emile Henricot, S.A.
Ni 30, Al 12, bal Fe.
For permanent magnets; high permeability. *Obsolete*

HENRICOT NI-AL-CO-2
Usines Emile Henricot, S.A.
Ni 28, Al 10, Co 6, bal Fe.
For permanent magnets; high permeability. *Obsolete*

HENRICOT NI-AL-CO-2B
Usines Emile Henricot, S.A.
Ni 26, Al 12, Co 8, bal Fe.
For permanent magnets; high permeability. *Obsolete*

HENRICOT NI-AL-CO-6
Usines Emile Henricot, S.A.
Ni 20, Al 10, Co 12, Cu 6, bal Fe.
For permanent magnets; high permeability. *Obsolete*

HENRICOT-HRM2M1
Usines Emile Henricot, S.A.
C 0.15, Cr 13, Mo 1.15, bal Fe.
Annealed: 70,000 TS; 35,000 YS; 40 El; 80 RA; 145 Brin. Cold
drawn: 90,000 TS; 80,000 YS; 20 El; 65 RA; 205 Brin. For
chemical plant and oil refinery equipment, table flatware,
knives, cutlery. Corrosion resistant. *Obsolete*

HENRICOT-HRM2M2
Usines Emile Henricot, S.A.
C 0.2, Cr 13, Mo 1.15, bal Fe.
Annealed: 95,000 TS; 40,000 YS; 25 El; 55 RA; B 92 Rock.
Heat treated: 240,000 TS; 210,000 YS;; 8 El; 25 RA; C 50
Rock. For cutlery, shafts, surgical instruments, gears, knives.
Corrosion resistant, hardenable. *Obsolete*

HEPOSIL
German manufacture
Si 21-22, Cu 1.5, Ni 1.5, Mn 0.7, Co 1.2, Mg 0.5, bal Al.
For pistons in engines, liners; low expansivity, high corrosion
resistance.

HEPPENSTALL 2 C 30
Heppenstall Co.
Alloy steel. C 0.35, Mn 0.7, Cr 0.8, Mo 0.45, Ni 1.75, bal Fe.
Hardened: 135,000 TS; 115,000 YS; 20 El; 50 RA; 293 Brin.
For piston rods, crankshafts, rams; upsetters.

HEPPENSTALL 2V72
Heppenstall Co.
Tool material. C 0.75, Mn 0.4, V 0.1, Si 0.6, bal Fe.
For coining and crimping dies, striking dies, shear knives,
trimmers; water hardened tool steel; AISI W2.

HEPPENSTALL 3C40
Heppenstall Co.
Tool material. C 0.43, Ni 2.6, Cr 1.35, Mo 0.55, V 0.18, bal Fe.
Annealed: 250 Brin. Heat treated: 470 Brin. For die blocks; oil
hardened; shock resistant.

HEPPENSTALL 3N24
Heppenstall Co.
C, Ni, bal Fe.
For forgings. *Obsolete*

HEPPENSTALL 3N25
Heppenstall Co.
C 0.2-0.3, Mn 0.7-1, Ni 2.5-3, bal Fe.
Normalized: 88,000 TS; 64,000 YS; 25 El; 52 RA; 160 Brin.
Heat treated: 102,000 TS; 77,000 YS; 28 El; 58 RA; 255 Brin.
For general forgings. *Obsolete*

HEPPENSTALL 5H50
Heppenstall Co.
Tool material. C 0.55, Cr 0.9, Mo 0.43, V 0.06, bal Fe.
Annealed: 210 Brin. Heat treated: 450 Brin. For die blocks,
inserts; oil hardened; tough.

HEPPENSTALL 6H55
Heppenstall Co.
Tool material. C 0.55, Cr 1.05, Mo 0.47, V 0.13, bal Fe.
Annealed: 210 Brin. Heat treated: 500 Brin. For die blocks,
inserts; oil hardened; tough.

HEPPENSTALL 9C60
Heppenstall Co.
Tool material. C 0.6, Ni 1.25, Cr 0.65, Mo 0.45, bal Fe.
Annealed: 250 Brin. Heat treated: 540 Brin. For
miscellaneous forgings, shear knives; oil hardened; shock
resistant.

HEPPENSTALL B76
Heppenstall Co.
Tool material. C 0.97, Mn 1.15, Cr 0.53, V 0.2, W 0.5, bal Fe.
Annealed: 250 Brin. Heat treated: 620 Brin. For stamping,
forming and blanking dies; shear knives; oil hardened; tough.
AISI W1.

HEPPENSTALL C-55
Heppenstall Co.
C 0.5-0.6, Ni 1-1.5, Cr 0.5-0.8, Mo 0.25-0.35, bal Fe.
Normalized: 108,000 TS; 73,000 YS; 19 El; 48 RA; 275 Brin.
Heat treated: 130,000 TS; 110,000 YS; 21 El; 54 RA; 305 Brin.
For forgings. *Obsolete*

HEPPENSTALL C45
Heppenstall Co.
C 0.4-0.55, Ni 1-1.5, Cr 0.5-0.8, Mo 0.25-0.35, bal Fe.
Normalized: 106,000 TS; 71,000 YS; 20 El; 48 RA; 230 Brin.
Heat treated: 125,000 TS; 105,000 YS; 22 El; 50 RA; 285 Brin.
For forgings. *Obsolete*

HEPPENSTALL C50
Heppenstall Co.
Alloy steel. C 0.5, Mn 0.55, Cr 1, Ni 1.5, bal Fe.
Oil hardened: 115,000-125,000 TS; 85,000-100,000 YS; 18-20
El; 45-50 RA; 248-293 Brin. For large mill pinions, couplings;
high tensile forgings.

HEPPENSTALL C55
Heppenstall Co.
Tool material. C 0.55, Cr 0.9, Mo 0.3, Ni 1.5, bal Fe.
For hot work dies, punches; oil hardened; tough.

HEPPENSTALL C93
Heppenstall Co.
Tool material. C 0.57, Ni 1.2, Cr 2.65, Mo 0.45, bal Fe.
Annealed: 250 Brin. Heat treated: 580 Brin. For cold shear
knives; oil hardened; shock resistant.

HEPPENSTALL CR T77
Heppenstall Co.
Tool material. C 0.3-0.35, V 0.2-0.25, Mn 0.2-0.4, Cr
3.75-4.15, W 11-12, bal Fe.
Heat treated: 40-53 Rock C; 220,000-240,000 TS;
180,000-215,000 YS; 10-12 El; 37-42 RA. For die casting dies,
forging and press dies, extrusion and gripper dies, piercers,
punches, extrusion mandrels. Hot work steel. Type H21; high
resistance to heat. Wear and abrasion resistant.

HEPPENSTALL GR 14A60
Heppenstall Co.
Tool material. C 0.55-0.65, Mn 0.7-0.9, Si 0.15-0.35, Cr
0.2-0.3, bal Fe.
Heat treated: 285-461 Brin. For trimmers, die and saw blocks,
axles, crankpins, rollers, shafting. Shallow hardening; wear
resistant. Type W4; water hardened tool steel.

HEPPENSTALL GR 2V90
Heppenstall Co.
Tool material. C 0.9-0.95, Mn 0.3-0.5, Si 0.5-0.7, V 0.08-0.1, bal Fe.
Water hardened: 216,000 TS; 152,000 YS; 11 El; 600 Brin. For coining and crimping dies, embossing dies, and rolls, heading and forming dies, swaging and trimming dies. Type W2; water hardened tool steel, wear and abrasion resistant.

HEPPENSTALL GR 5M21
Heppenstall Co.
Tool material. C 0.18-0.23, Ni 3-3.25, Mo 3.25-3.5, Mn 0.7, Si 0.3, bal Fe.
For press and upsetter dies, piercers, punches, die casting dies; supplied in two grades, hardened and annealed.

HEPPENSTALL GR 6E14GV
Heppenstall Co.
Alloy steel. C 0.15, Ni 3.5, Cr 1.6, V 0.05, bal Fe.
Heat treated: 320 Brin. Annealed: 230 Brin. For piston rods, cams, camshafts. Case hardened and shock resistant.

HEPPENSTALL GR 9C68
Heppenstall Co.
Tool material. C 0.65-0.7, Mo 0.4-0.5, Ni 1-1.5, Cr 0.5-0.8, Mn 0.5-0.8, Si 0.2-0.35, bal Fe.
Oil hardened: 208,000 TS; 180,000 PL; 45 Rock C. For spindles, punches, brake dies, collets, plastic mold dies, idler rolls, rocker plates, stamping dies, shear blades. Type L6 tool steel. Deep oil hardening; tough and wear resistant.

HEPPENSTALL GR 9R40
Heppenstall Co.
Alloy steel. C 0.4, Cr 0.95, Mo 0.45, bal Fe.
Annealed: 220 Brin. Heat treated: 370 Brin. For piston rods, shafts, counter shafts, gears. Oil hardened; shock resistant.

HEPPENSTALL GR A110
Heppenstall Co.
Tool material. C 1.05-1.15, Mn 0.25-0.35, Si 0.25-0.35, bal Fe.
Annealed: 100,000 TS; 54,000 YS; 21 El; 197 Brin. Water hardened: 216,000 TS; 152,000 YS; 11 El; 600 Brin. For cold header and striking dies, arbors, coining and broaching tools, mandrels, pipe cutters, tube drawing dies. Type W1 tool steel; wear and abrasion resistant.

HEPPENSTALL GR C57
Heppenstall Co.
Tool material. C 0.52-0.57, Ni 2-2.2, Mn 0.53-0.68, Cr 0.8-0.95, Si 0.5-0.7, Mo 0.7-0.75, bal Fe.
For hammer dies, press and upsetter dies, gripper and header dies, piercing mandrels, shear blades. Hot work steel. High compressive strength.

HEPPENSTALL GR C58
Heppenstall Co.
Tool material. C 0.42-0.47, Ni 4.1-4.5, Cr 1.25-1.65, Mo 0.7-0.8, V 0.12-0.16, bal Fe.
Heat treated: 218,000 TS; 190,000 YS; 12 El; 33 RA; 45 Rock C. At 1000°F: 144,000 TS; 133,000 YS; 9 El; 39 RA. For press and hammer dies, insert and upsetter dies, deep drawing dies. Resists softening at elevated temperatures. High compressive strength.

HEPPENSTALL GR H230
Heppenstall Co.
Tool material. C 0.32, Cr 3.3, Mo 2.4, V 0.37, bal Fe.
Air hardened: 180,000-296,000 TS; 160,000-247,000 YS; 10-14 El; 16-4 RA; 40-59 Rock C. For hot extrusion dies, punches, shear blades, gages. Hot work tool and die steel. Type H10.

HEPPENSTALL GR H44
Heppenstall Co.
Tool material. C 0.95-1.05, Si 0.2-0.4, Mn 0.5-0.7, Mo 0.95-1.05, Cr 4.9-5.3, V 0.2-0.3, bal Fe.
Annealed: 104,000 TS; 51,000 YS; 26 El; 18 Rock C. Air hardened: 253,000 TS; 200,000 YS; 3 El; 53 Rock C. For cold work trimming and blanking dies, forming and drawing dies, gages, rolls, punches, thread rolling dies. Type A2 cold work tool steel. Deep air hardening; tough; wear resistant.

HEPPENSTALL GR H720
Heppenstall Co.
Tool material. C 0.37-0.42, Cr 5-5.5, Mn 0.23-0.38, Mo 1-1.3, Si 0.85-1.1, V 0.9-1, bal Fe.
At 80°F: 217,000 TS; 184,000 YS; 13 El; 40 RA. At 1100°F: 110,000 TS; 88,000 YS; 21 El; 69 RA. For extrusion rams and liners, mandrels, punches, piercers, upsetting and forging dies, die casting dies. Type H13; hot work steel.

HEPPENSTALL GR H722
Heppenstall Co.
Tool material. C 0.37-0.42, Cr 5-5.5, Mn 0.23-0.38, Mo 1-1.2, Si 0.85-1.1, V 0.45-0.55, bal Fe.
At 80°F: 217,000 TS; 184,000 YS; 13 El; 40 RA. At 1100°F: 110,000 TS; 88,000 YS; 21 El; 69 RA. For extrusion rams and liners, mandrels, punches, piercers, upsetting dies, die casting dies. Air hardening and nondeforming. Type H11. High resistance to wash and erosion and heat checking.

HEPPENSTALL GR T716
Heppenstall Co.
Tool material. C 0.44-0.5, Cr 0.7-0.9, Mn 0.4-0.6, Si 0.2-0.35, bal Fe.
Heat treated: 415-601 Brin. For punches, chisels, pneumatic tools, rivet busters, hubbing dies, hot heading and swaging dies, hot forming dies. Type S3 tool steel for cold and hot working.

HEPPENSTALL GR T719
Heppenstall Co.
Tool material. C 0.62-0.67, Si 0.9-1.1, Mo 1.4-1.6, Mn 0.2-0.4, Cr 4.5-5, W 1-1.2, bal Fe.
For cold work plastic dies and forming dies. For hot work gripper and compression dies, trimming dies, aluminum extrusion dies. Air or oil hardened. Tough and wear resistant.

HEPPENSTALL GR T72
Heppenstall Co.
Tool material. C 0.45-0.5, Cr 1.2-1.5, W 1.8-2.05, V 0.2-0.25, bal Fe.
For chisels, punches, compression and trimming dies, hot swaging and heading dies, insert dies. Hot and cold work tool steel. Oil hardened. Fatigue resistant.

HEPPENSTALL GR T721
Heppenstall Co.
Tool material. C 0.34-0.4, Mn 0.25-0.45, Si 0.85-1.15, Cr 4.75-5.25, Mo 1.25-1.65, W 1.05-1.45, V 0.25-0.45, bal Fe.
At 80°F: 217,000 TS; 184,000 YS; 13 El; 40 RA. At 1100°F: 110,000 TS; 88,000 YS; 21 El; 69 RA. For extrusion rams and liners, mandrels, punches, piercers, dummy blocks, shear blades, aluminum die casting dies. Type H12 hot work steel. Resists heat checking.

HEPPENSTALL GR T74
Heppenstall Co.
Tool material. C 0.47-0.53, Cr 1.3-1.5, W 2.25-2.5, V 0.18-0.23, Mo 0-0.4, bal Fe.
For chisels, punches, compression and trimming dies, hot swaging and heading dies, insert dies. Hot and cold work tool steel. Oil hardened. Fatigue resistant.

HEPPENSTALL GRADE C
Heppenstall Co.
Tool material. C 0.55, Ni, Cr, Mo, C, bal Fe.
Oil treated: 206,000-305,000 TS; 195,000-288,000 YS; 6-18 El; 17-52 RA; 555 Brin. For die blocks, inserts; drop forged.

HEPPENSTALL H340
Heppenstall Co.
Tool material. C 0.41, Cr 3.3, Mo 2.25, V 0.38, bal Fe.
Annealed: 250 Brin. For extrusion parts; oil hardened; tough.

HEPPENSTALL H41
Heppenstall Co.
Tool material. C 0.95, Cr 3.75, Mo 0.23, V 0.16, bal Fe.
Annealed: 220 Brin. Heat treated: 510 Brin. For hot and cold trimmers; oil hardened.

HEPPENSTALL R43
Heppenstall Co.
Tool material. C 1.55, Cr 11.5, Mo 0.55, W 0.55, bal Fe.
Annealed: 250 Brin. Heat treated: 610 Brin. For shear knives, cold work dies; air hardened, nondeforming.

HEPPENSTALL R45
Heppenstall Co.
Tool material. C 0.83, Cr 11.5, Mo 0.43, bal Fe.
Annealed: 250 Brin. Heat treated: 560 Brin. For shear knives, cold work dies; air hardened, nondeforming.

HEPPENSTALL R718
Heppenstall Co.
Tool material. C 0.33, Cr 4.75, Mo 1.87, bal Fe.
Annealed: 250 Brin. Heat treated: 510 Brin. For extrusion parts, shear knives; oil hardened.

HEPPENSTALL R97
Heppenstall Co.
Tool material. C 0.55, Mn 1.05, Si 2.1, Cr 0.3, Mo 0.38, bal Fe.
Annealed: 250 Brin. Heat treated: 570 Brin. For shear knives, cold work dies; oil hardened; shock resistant. AISI S5.

HEPPENSTALL SPECIAL
Heppenstall Co.
Tool material. C 0.45, Ni 4.3, Cr 1.45, Mo 0.75, V 0.14, bal Fe.
For hot work dies, trimmers, punches. Tough; oil hardened.

HEPPENSTALL T51
Heppenstall Co.
Tool material. C 0.63, Ni 1.35, Cr 0.7, W 2.12, bal Fe.
Annealed: 250 Brin. Heat treated: 580 Brin. For hot and cold trimmers; oil hardened; tough.

HEPPENSTALL T71
Heppenstall Co.
Tool material. C 0.5, Cr 1.12, W 2.35, bal Fe.
Annealed: 250 Brin. Heat treated: 560 Brin. For cold shear knives, hot heading dies; oil hardened; tough.

HEPPENSTALL T717
Heppenstall Co.
Tool material. C 0.33, Cr 4.75, Mo 1.3, W 1.1, bal Fe.
Annealed: 250 Brin. Heat treated: 530 Brin. For shear knives, dies, punches, mandrels; oil hardened.

HEPPENSTALL T73
Heppenstall Co.
Tool material. C 0.28, Cr 3.4, V 0.23, W 8.75, bal Fe.
Annealed: 250 Brin. Heat treated: 500 Brin. For shear knives, inserts, hot work dies; oil hardened; tough.

HEPPENSTALL T745
Heppenstall Co.
Tool material. C 0.45, Cr 8.5, Mo 1.1, W 1.1, bal Fe.
Annealed: 250 Brin. Heat treated: 580 Brin. For chipper knives; oil hardened.

HEPPENSTALL T746
Heppenstall Co.
Tool material. C 0.68, Cr 9.5, Mo 1.1, W 1.1, bal Fe.
Annealed: 250 Brin. Heat treated: 620 Brin. For flaking knives; oil hardened.

HEPPENSTALL T75
Heppenstall Co.
Tool material. C 0.68, Cr 4, V 1, W 18, bal Fe.
Heat treated: 650 Brin. For shear knives; oil hardened; high speed steel.

HEPPENSTALL T78
Heppenstall Co.
Tool material. C 0.85, Cr 4.25, Mo 5, V 1.85, W 6.25, bal Fe.
Annealed: 250 Brin. Heat treated: 630 Brin. For shear knives; high speed steel; oil hardened.

HEPPENSTALL T79
Heppenstall Co.
Tool material. C 0.43, Cr 5.25, V 0.22, W 4, bal Fe.
Annealed: 250 Brin. Heat treated: 600 Brin. For rams, die casting dies, shear knives; oil hardened.

HEPPENSTALL X-60
Heppenstall Co.
Ni 1.5, Cr 0.7, Mo 0.2, V 0.17, C 0.6, 15-18 IZ, bal Fe.
Heat treated: 150,000 TS; 119,000 YS; 18 El; 49 RA; 302 Brin. For mandrel sleeves. *Obsolete*

HERBOHN BELL METAL
German manufacture
Cu 60-71.43, Sn 26.4-35, Zn 2.7-5.
For bells; corrosion resistant.

HERC-ALLOY
Columbus McKinnon Corp.
Ni, Cr and/or Mo alloy.
Heat treated alloy for chains, fittings, joining links, sling chains; wear and impact resistant.

HERCULES
Cytemp Specialty Steel Div.
C 1, Cr 0.5, V 0.2, bal Fe.
For blanking, forming and drawing dies; fatigue resistant. AISI W7. *Obsolete*

HERCULES BRONZE-A
American manufacture
Cu 86, Sn 10, Al 2.5, Zn 2.
For hardware, worm gears; high strength.

HERCULES BRONZE-B
American manufacture
Cu 54, Zn 36, Fe 7.5, Al 2.5.
For hardware.

HERCULITE
Midvale-Heppenstall Co.
C 0.54, Ni 2.6, Mo 0.45, bal Fe.
For tools. *Obsolete*

HERCULOY
Revere Copper Products, Inc.
Cu 96.25, Si 3.25, Sn 0.5.
Annealed: 60,000 TS; 80 El. Hot rolled: 68,000 TS; 60 El. Cold drawn: 120,000 TS; 2 El. For bolts, nuts, screws, turn buckles, storage tanks, boilers, blowers, fans, hardware; non-magnetic; corrosion resisting. *Obsolete*

HERCULOY - HIGH SILICON BRONZE A
Revere Copper Products, Inc.
Si 2.8-3.3, Mn 0.75-1.3, bal Cu.
Half hard: 78,000 TS; 45,000 YS; 17 El. Marine applications, pressure vessels, bearing plates, tanks, sculpture. Corrosion resistant. C65500

HERCULOY 418
Revere Copper Products, Inc.
Si 3-3.25, Sn 0.5, bal Cu.
Hot rolled: 67,000 TS; 60 El. Cold rolled: 120,000 TS; 5 El. For bolts, hardware; tough. *Obsolete*

HERCULOY 419
Revere Copper Products, Inc.
Si 2-2.5, Sn 0.25, bal Cu.
Hot rolled: 50,000 TS; 80 El. Cold rolled: 100,000 TS; 10 El. For bolts, nuts, studs; tough. *Obsolete*

HERCULOY 420
Now HERCULOY 655.

HERCULOY 421
Now HERCULOY 651.

HERCULOY 651
Revere Copper Products, Inc.
Si 1.5-2, Mn 0.25, bal Cu.
Hard: 70,000 TS; 55,000 YS; 15 El; 880 Brin. For bolts, hardware; corrosion resistant. *Obsolete*

HERCULOY 655
Revere Copper Products, Inc.
Now HERCULOY - HIGH SILICON BRONZE A. C65500

HERGERMUHL BRASS
English manufacture
Cu 62-72, Sn 0.2-1, Pb 0-0.8, bal Zn.
For hardware, fittings, sheathing; good workability.

HERKULES BS
Westa-Westdeutsche
C 2.1, Si 0.3, Mn 0.35, Cr 11.5, bal Fe.
For forming and blanking dies, punches; oil or air hardened, nondeforming.

HERKULES FZ
Westa-Westdeutsche
C 2.1, Cr 11.5, Mn 0.3, Si 0.3, bal Fe.
For forming and blanking dies, punches; oil or air hardened, nondeforming.

HERKULES ME
Westa-Westdeutsche
C 2.1, Si 0.3, Mn 0.35, Cr 11.5, bal Fe.
For forming and blanking dies, punches; oil or air hardened, nondeforming.

HERMES 35
Vereinigte Edelstahlwerke
C 0.35, Si 0.25-0.5, Mn 0.3-0.8, bal Fe.
For gears, shafts, machine tool parts; water hardened. *Obsolete*

HERMES 45
Vereinigte Edelstahlwerke
C 0.45, Si 0.25-0.5, Mn 0.3-0.8, bal Fe.
For gears, bolts, shafts, machine tool parts; water hardened. *Obsolete*

HERMES 60
Vereinigte Edelstahlwerke
C 0.6, Si 0.25-0.5, Mn 0.3-0.8, bal Fe.
For axes, hammers, crimpers; water hardened. *Obsolete*

HERMES 90
Bohler Gesellschaft M.B.H.
C 0.9, Si 0.25-0.5, Mn 0.3-0.8, bal Fe.
For drills, taps, hobs, reamers, springs, cutters; water hardened. Type W1.

HETZEL L SN 25
Hetzel & Co. Metallhuttenwerk GmbH
Sn 25, bal Pb.
For solder; soft.

HETZEL L SN 40
Hetzel & Co. Metallhuttenwerk GmbH
Sn 40, bal Pb.
For solder; soft.

HETZEL L ZN 98
Hetzel & Co. Metallhuttenwerk GmbH
Zn 98, Cu 2.
For solder; 790°F MP. *Obsolete*

HEUSLER ALLOY-1
German manufacture
Al 11.1, Cu 66.5, Mn 22.4.
For electrical machinery; ferromagnetic.

HEUSLER ALLOY-2
German manufacture
Al 10, Pb 4, Cu 68, Mn 18.
For electrical machinery; ferromagnetic.

HEUSLER ALLOY-3
German manufacture
Al 4-15, Cu 54-76, Mn 16-30.
For electrical machinery; ferromagnetic.

HEUSLER ALLOY-4
German manufacture
Cu 61, Al 13, Mn 26.
For electrical machinery; ferromagnetic.

HEVA-12A
ACENOR, S.A.
C 1.6, Mn 0.35, Si 0.3, Cr 11.5, Mo 0.7, V 0.45, W 0.5, bal Fe.
Air hardenable to 58-63 Rock C. Cold work tool steel. Similar to AISI D2. *Obsolete*

HEVA-17CN
ACENOR, S.A.
C 0.1-0.2, Ni 3-15, bal Fe.
Hardenable stainless steel for combination of high strength for paper machinery and sea water resistance. Similar to AISI 431. *Obsolete*

HEVA-5B4
ACENOR, S.A.
C 0.4, Mn 0.4, Si 1, Cr 5.3, Mo 1.5, V 1, bal Fe.
Heat treated: 48-57 Rock C. For tools, dies, hot working mandrels for extrusion. *Obsolete*

HEVA-ACROM 35
ACENOR, S.A.
See ACROM-35.

HEVA-ALS
ACENOR, S.A.
See ACENOR ALS.

HEVA-ALSP
ACENOR, S.A.
See ACENOR ALSP.

HEVA-BT1
ACENOR, S.A.
C 0.42, Mn 0.3, Si 1, Cr 1.2, W 2, bal Fe.
Heat treated: 52-56 Rock C. For punches, pneumatic tools; shock resistant. *Obsolete*

HEVA-CNE
ACENOR, S.A.
See ACENOR CNE.

HEVA-CTM
ACENOR, S.A.
C 0.92, Mn 1.3, Si 0.3, Cr 0.5, V 0.12, W 0.5, bal Fe.
Oil hardenable to 58-62 Rock C. Cold work tool steel. Similar to AISI O1. *Obsolete*

HEVA-DF-100
ACENOR, S.A.
C 0.35-0.41, Mn 0.5-0.8, Si 0.15-0.4, Cr 0.85-1.15, Mo 0.15-0.25, bal Fe.
Annealed: 68 kg/mm^2 TS max. Heat treated: 100 kg/mm^2 TS min. For cold headed, oil hardened bolts and parts. *Obsolete*

HEVA-DF-120
ACENOR, S.A.
C 0.37-0.43, Mn 0.65-0.95, Si 0.15-0.4, Cr 0.35-0.65, Ni 0.4-0.7, Mo 0.15-0.25, bal Fe.
Annealed: 68 kg/mm^2 TS max. Heat treated: 120 kg/mm^2 TS min. For cold headed, oil hardened bolts and parts. *Obsolete*

HEVA-DF-35-B
ACENOR, S.A.
C 0.32-0.38, Mn 0.45-0.75, Si 0.15-0.4, 0.003 B min, bal Fe.
Annealed: 60 kg/mm^2 TS max. Heat treated: 80 kg/mm^2 min TS. For cold headed bolts and parts. *Obsolete*

HEVA-DF-35-M
ACENOR, S.A.
C 0.32-0.38, Mn 1.1-1.35, Si 0.15-0.4, bal Fe.
Annealed: 62 kg/mm^2 TS max. Heat treated: 80 kg/mm^2 TS min. For cold headed bolts and parts. *Obsolete*

HEVA-DTA
ACENOR, S.A.
See ACENOR DTA.

HEVA-EM-3
ACENOR, S.A.
C 0.32, Mn 0.3, Si 0.3, Cr 3, Mo 2.8, V 0.5, bal Fe.
Heat treated: 46-50 Rock C. Containers and mandrels for extrusion of nonferrous metals. Similar to AISI H10. *Obsolete*

HEVA-ENIMO
ACENOR, S.A.
C 0.2, Mn 0.7, Si 0.3, Ni 3.15, Mo 3.4, bal Fe.
Precipitation hardening steel for tools and dies; heat treated: 36-42 Rock C. *Obsolete*

HEVA-EST-EXTRA
ACENOR, S.A.
C 0.55, Mn 0.7, Si 0.22, Cr 1.1, Ni 1.7, Mo 0.5, V 0.1, bal Fe.
Heat treated: 35-45 Rock C. For tools, containers for rod and tube extrusion, hot work steel for drop forging dies. *Obsolete*

HEVA-F-1282
ACENOR, S.A.
See ACENOR F-1282.

HEVA-FCA
ACENOR, S.A.
C 2.2, Mn 0.35, Si 0.5, Cr 11.5, Mo 0.8, V 0.2, bal Fe.
Air hardenable to 60-65 Rock C. Cold work tool steel. Similar to AISI D4. *Obsolete*

HEVA-HI
ACENOR, S.A.
C 0-0.15, Cr 11.5-14, bal Fe.
Quenched and treated: 70-90 kg/mm^2 TS min; 55 kg/mm^2 YS min. General purpose heat treatable type for machine parts, pump shafts. Similar to AISI 403 and 410. *Obsolete*

HEVA-IDF
ACENOR, S.A.
C 0-0.1, Cr 17-19, Ni 11-13, bal Fe.
50-70 kg/mm^2 TS; 20 kg/mm^2 YS min; 40 El min. For corrosion parts. Austenitic stainless steel for cold heading. Similar to AISI 305. *Obsolete*

HEVA-IF
ACENOR, S.A.
C 0-0.1, Cr 16-18, bal Fe.
45-65 kg/mm^2 TS; 26 kg/mm^2 YS min; 18 El min. General purpose nonhardenable stainless steel, chromium type. AISI 430. *Obsolete*

HEVA-IM-3
ACENOR, S.A.
C 0-0.03, Cr 16-18, Ni 10-14, Mo 2-3, bal Fe.
45-65 kg/mm^2 TS; 20 kg/mm^2 YS min; 40 El min. For corrosion parts. Low carbon for restriction of carbide precipitation during welding. AISI 316L. *Obsolete*

HEVA-IM-8
ACENOR, S.A.
C 0-0.08, Cr 16-18, Ni 10-14, Mo 2-3, bal Fe.
50-70 kg/mm^2 TS; 20 kg/mm^2 YS min; 40 El min. For corrosion parts. Austenitic stainless steel. AISI 316. *Obsolete*

HEVA-INOX 20
ACENOR, S.A.
C 0.16-0.25, Cr 12-16, bal Fe.
70-90 kg/mm^2 TS; 55 kg/mm^2 YS min; 13 El min. Higher carbon modification of HEVA-HI, often used for cutlery; hardenable to 40-50 Rock C. Similar to AISI 420. *Obsolete*

HEVA-INOX 42
ACENOR, S.A.
C 0.36-0.45, Cr 12.5-14.5, bal Fe.
Hardenable to 45-55 Rock C. Martensitic stainless, high carbon modification of HEVA-HI and Inox 20, often used for cutlery and surgical instruments. *Obsolete*

HEVA-KLT
ACENOR, S.A.
C 0.3, Mn 0.3, Si 0.3, Cr 2.6, W 8.8, V 0.35, bal Fe.
Heat treated: 48-52 Rock C. For tools, dies, hot work tools. *Obsolete*

HEVA-LCH-3
ACENOR, S.A.
C 0-0.03, Cr 18-20, Ni 8-12, bal Fe.
45-65 kg/mm^2 TS; 18 kg/mm^2 YS min; 40 El min. For corrosion parts. Low carbon for restriction of carbide precipitation during welding. AISI 304L. *Obsolete*

HEVA-LCH-8
ACENOR, S.A.
C 0-0.08, Cr 18-20, Ni 8-10.5, bal Fe.
50-70 kg/mm^2 TS; 20 kg/mm^2 YS min; 40 El min. For corrosion parts. Austenitic stainless. AISI 304. *Obsolete*

HEVA-MCV
ACENOR, S.A.
See ACENOR MCV.

HEVA-MOLICORT
Johnson Matthey
C 0.85, Mn 0.3, Si 0.25, Cr 4, Mo 5, W 6.25, V 1.85, bal Fe.
Heat treated: 63-66 Rock C. High speed steel, for tools, drills, cutters, cold forming dies. AISI M2. *Obsolete*

HEVA-MP
ACENOR, S.A.
See ACENOR MP2311.

HEVA-PERFOR
ACENOR, S.A.
See ACENOR PERFOR.

HEVA-RAPID
ACENOR, S.A.
C 0.9, Mn 0.3, Si 0.25, Cr 4, Mo 5, W 6.25, V 1.85, Co 4.75, bal Fe.
Heat treated: 63-66 Rock C. High speed steel, for tools, cutters. *Obsolete*

HEVA-SM-75
ACENOR, S.A.
C 0.32-0.38, Mn 1.1-1.4, Si 1.1-1.4, bal Fe.
Quenched and treated: 70-100 kg/mm^2 TS; 50 kg/mm^2 YS min; 12 El. For gears, shafts; water hardening. *Obsolete*

HEVA-SPT
ACENOR, S.A.
See ACENOR SPT.

HEVA-SUPER BONO
ACENOR, S.A.
C 0.78, Mn 0.3, Si 0.25, Cr 4.25, W 18, V 1, bal Fe.
Heat treated: 63-65 Rock C. High speed steel, for tools, dies, cutters. AISI T1. *Obsolete*

HEVA-TC1
ACENOR, S.A.
C 0.47, Mn 0.5, Si 0.22, Cr 1.3, Ni 4, Mo 0.2, bal Fe.
Heat treated: 40-50 Rock C. For cold tools, shock resistant; air hardening. *Obsolete*

HEVA-TSD
ACENOR, S.A.
C 0.29-0.35, Mn 0.45-0.75, Si 0.15-0.4, Cr 0.5-0.8, Ni 2.8-3.25, bal Fe.
Quenched and treated: 95-125 kg/mm^2 TS; 60 kg/mm^2 YS min; 10 El. For gears, shafts; oil hardening. *Obsolete*

HEVA-V-214
ACENOR, S.A.
C 0.49-0.57, Mn 8-10, Cr 20-22, Ni 3.3-4.3, N 0.36-0.5, bal Fe.
100-120 kg/mm^2 TS; 60 kg/mm^2 YS; 8 El min. For auto exhaust valves; resists leaded fuels. Precipitation hardening steel. *Obsolete*

HEVA-VSC
ACENOR, S.A.
C 0.38-0.48, Si 2.5-3.2, Cr 8.6-9.5, bal Fe.
90-105 kg/mm^2 TS; 70 kg/mm^2 YS min; 15 El min. For auto inlet and exhaust valves; heat resistant. *Obsolete*

HEVA-XKW
ACENOR, S.A.
C 0.76, Mn 0.3, Si 0.25, Cr 4.25, W 18.5, Mo 1, V 1.55, Co 9.5, bal Fe.
Heat treated: 63-66 Rock C. High speed steel for tools, dies, cutters. Similar to AISI T5. *Obsolete*

HEVI-DUTY SW-14
Solar Basic Industries
C 0.3, bal Fe.
For welding rod; high tensile.

HEVIMET
Now CARBOLOY HEVIMET.

HEWITT COPPER-HARD
Hewitt Metals Corp.
Cu 4.5, Sb 4.5, bal Sn.
Cast: 11,050 TS; 24 Brin. For heavy duty bearings, and bushings. Babbitt; 365-590°F MP.

HEWITT GENUINE
Hewitt Metals Corp.
Sb 7.5, Cu 3.5, bal Sn.
Cast: 15,150 TS; 29 Brin. For heavy duty bearings and bushings. Babbitt; 460-682°F MP.

HEWITT MILL
Hewitt Metals Corp.
Pb, Sn, Sb.
For steel mill bearings; Babbitt. *Obsolete*

HEWMET
Hewitt Metals Corp.
Sn 5, Sb 15, bal Pb.
Cast: 7580 TS; 23.7 Brin. For bearings and bushings. Babbitt; heavy loads.

HFI
German manufacture
Co 2.5, 24.0 TiC, 73.5 WC.
Sintered: 92 Rock A. For cutters to machine steel, wear dies; sintered carbides; hard and abrasion resistant.

HG-NICKEL
Henry Wiggin & Co. Ltd.
99.0% Ni min, 0.06% C min.
For forming gassy seals, terminal pins, electronic valves. Heat and corrosion resistant. *Obsolete*

HGT3
Dunford Hadfields Ltd.
C 0.2, Cr 3, Mo 0.5, W 0.5, V 0.75, bal Fe.
For gas turbine discs and rotors; creep and heat resistant. *Obsolete*

HH1
German manufacture
Co 2.5, 97.5 WC.
Sintered: 92 Rock A. For cutters to machine cast iron, form tools; sintered carbide, hard and abrasion resistant.

HI 1
W. Ossenberg & Cie Edelstahlwerke
C 0.1, Si 0.8, Al 0.8, Cr 6.5, bal Fe.
For elevated temperature operations, as pyrometer sheath tubes; heat resisting steel. W.-Nr. 1.4713.

HI 10

W. Ossenberg & Cie Edelstahlwerke
C 0.18, Si 1.2, Mn 0.7, Cr 25, Ni 4, bal Fe.
For high temperature equipment; furnace parts. W.-Nr.
1.4821.

HI 2

W. Ossenberg & Cie Edelstahlwerke
C 0.1, Si 1, Al 1, Cr 13, bal Fe.
For elevated temperature operation, as furnace rails and
supports. W.-Nr. 1.4724.

HI 20

W. Ossenberg & Cie Edelstahlwerke
C 0.15, Si 0.4, Mn 1.6, Cr 18, Ni 10.5, Ti 0.4, bal Fe.
For high temperature operation, as annealing boxes. W.-Nr.
1.4878.

HI 21

W. Ossenberg & Cie Edelstahlwerke
C 0.15, Si 2, Mn 0.7, Cr 20, Ni 12, bal Fe.
For high temperature operation, as annealing boxes. W.-Nr.
1.4828; similar to AISI 309.

HI 22

W. Ossenberg & Cie Edelstahlwerke
C 0.15, Si 2, Mn 0.7, Cr 25, Ni 20, bal Fe.
For high temperature equipment; annealing pots,
thermocouple housings. W.-Nr. 1.4841; similar to AISI 310.

HI 23

W. Ossenberg & Cie Edelstahlwerke
C 0.1, Si 1.8, Mn 1.4, Cr 16, Ni 35.5, bal Fe.
For high temperature equipment, as furnace parts. W.-Nr.
1.4864.

HI 24

W. Ossenberg & Cie Edelstahlwerke
Cr 0.1, Si 2, Mn 1.8, Cr 21, Mo 1.8, Ni 16.5, Nb 1.2, bal Fe.
For high temperature equipment, as support bars for bright
annealing. W.-Nr. 1.4885.

HI 3

W. Ossenberg & Cie Edelstahlwerke
C 0.12, Si 1, Mn 0.7, Al 1, Cr 18, bal Fe.
For elevated temperature equipment as furnace fittings. W.-
Nr. 1.4742.

HI 4

W. Ossenberg & Cie Edelstahlwerke
C 0.12, Si 1.4, Mn 0.7, Al 1.45, Cr 24, bal Fe.
For high temperature operation as steam boiler equipment.
W.-Nr. 1.4762.

HI 50

W. Ossenberg & Cie Edelstahlwerke
C 0.1, Si 1, Mn 1, Cr 28, Co 48, bal Fe.
For high temperature equipment. W.-Nr. 2.4778.

HI MAG PERM

Erie Steel Co.
C 0.03-0.05, P 0.005-0.009, Si 0.01-0.02, Cr 0.03-0.07, Al
0.006-0.01, Mo 0.04-0.07, bal Fe.
Annealed: 40,000 TS; 20,000 YS; 40 El; 78 RA; 69 Brin. For
electrical applications, magnetic control devices, magnetic
clutches and chucks, pole pieces, armatures. High magnetic
permeability. Vacuum degassed.

HI PROOF 304L

British Steel Corp.
C 0-0.03, Mn 0-2, Cr 18-20, Ni 8-11, N 0.2, bal Fe.
Room temperature: 85,000-114,000 TS; 42,500 YS; 35 El. At
400 C: 20,000 minimum YS. Weldable; for stainless parts and
assemblies. *Obsolete*

HI SHOCK 60

Now CARPENTER HI SHOCK 60.

HI TM 900

Standard Pressed Steel Co.
C 0.37-0.43, Cr 5, Mo 1.2, V 0.5, Si 0.9, Mn 0.3, bal Fe.
Heat treated: 220,000 TS; 185,000 YS; 10 El; 35 RA; 460 Brin.
For fasteners, bolts; high strength at elevated temperature.
Obsolete

HI-10

Riverside Metals Corp.
Sn 3.75, P 0.25, bal Cu.
For springs; wire, phosphor bronze. *Obsolete*

HI-120

Westinghouse Electric Corp.
Cb, Ti.
For superconducting wires. High magnetic fields at low
reduced power use. *Obsolete*

HI-ALLOY

Dow Chemical Co.
Al 5.8-0.7, Mn 0.18, Zn 2.5-3.5, Si 0.3, Cu 0.05, bal Mg.
For anodes; for cathodic protection from corrosion. *Obsolete*

HI-C-SUPER HY-TUF

Colt Industries
C 0.47, Mn 1.38, Si 2.4, Cr 1.1, Mo 0.4, V 0.25, bal Fe.
Heat treated: 325,000 TS. For aircraft structures, landing
gears; tough, shock resistant. *Obsolete*

HI-CHROME

Xaloy Inc.
C 2.6, Cr 27, Mn 1, Si 1, Mo 0.6, V 0.3, bal Fe.
Hardened: 60 Rock C. For pump and cylinder liners; abrasion
resistant.

HI-COBALT

Indiana General
Co 40, bal Fe.
For permanent magnets; Br 9800, Hc 240. *Obsolete*

HI-DI 5

Boyd-Wagner Co.
C 1, Cr 5, Mo 1, Mn 0.4, V 0.4, bal Fe.
Annealed: 103,000 TS; 51,000 YS; 26 El; 18 Rock C.
Hardened: 178,000-255,000 TS; 145,00-200,000 YS; 3-12 El;
7-32 RA; 41-53 Rock C. For blanking and trimming dies,
shear blades, punches gauges, master tools, rolling dies,
broaches. Type A2 air hardening, nondeforming cold work
tool steel, tough.

HI-FORM 40

Inland Steel Co.
HSLA steel. C 0-0.1, P 0-0.15, S 0-0.035, Cb 0-0.05, Mn 0-0.6,
bal Fe.
Sheet: 50 ksi TS; 40 ksi YS; 28 El. For transportation and
mobile equipment parts requiring good formability and
weldability.

HI-FORM 50

Inland Steel Co.
HSLA steel. C 0-0.15, Mn 0-1.4, Cb 0.005-0.15, bal Fe.
Sheet: 60 ksi TS; 50 ksi YS; 25 El. For transportation and
mobile equipment parts requiring good formability and
weldability.

HI-FORM 60

Inland Steel Co.
HSLA steel. C 0-0.15, Mn 0-1.4, Cb 0.005-0.15, bal Fe.
Sheet: 70 ksi TS; 60 ksi YS; 21 El. For transportation and
mobile equipment parts requiring good formability and
weldability.

HI-FORM 70

Inland Steel Co.
HSLA steel. C 0-0.15, Mn 0-1.4, Cb 0.005-0.15, bal Fe.
Sheet: 80 ksi TS; 70 ksi YS; 18 El. For transportation and
mobile equipment parts requiring good formability and
weldability.

HI-FORM 80

Inland Steel Co.
HSLA steel. C 0-0.15, Mn 0-1.4, Cb 0.005-0.15, bal Fe.
Sheet: 90 ksi TS; 80 ksi YS; 16 El. For transportation and
mobile equipment parts requiring good formability and
weldability.

HI-GLOSS "C"

Jessop Steel Co.
C 0.1, Cr 18, Ni 8, Mn 0.35, bal Fe.
Rolled: 105,000-120,000 TS; 60,000-90,000 YS; 35-45 El;
45-65 RA; 196-228 Brin. Annealed: 88,000 TS; 40,000 YS; 60
El; 70 RA; 135 Brin. For cooking utensils, sterilizing
equipment, bumpers, piston rods, pumps, valves, radiator
shells; stainless, austenitic, corrosion resisting. *Obsolete*

HI-GLOSS "D.D."

Jessop Steel Co.
C 0.1, Cr 12, Ni 12, Mn 0.35, bal Fe.
Annealed: 90,000-100,000 TS; 45,000-55,000 YS; 55-65 El;
60-70 RA; 150-160 Brin. For stainless and corrosion resistant
parts; corrosion resistant. *Obsolete*

HI-GLOSS FREE MACHINING

Jessop Steel Co.
C 0-0.12, Mn 0-1.2, Se 0.25, Cr 18, Ni 8, bal Fe.
Annealed: 95,000 TS; 45,000 YS; 53 El; 65 RA; 160 Brin. For
stainless bolts, nuts, screws, spindles, bushings; stainless.
Obsolete

HI-MAN

Inland Steel Co.
Alloy steel. C 0.25, Mn 1.35, Si 0.3, Cu 0.2, bal Fe.
Rolled: 75,000 TS; 50,000 YS; 20 El. For railroad and
agriculture equipment, mine cars, automobile bodies; high
strength, low alloy construction steel.

HI-MAN 440

Inland Steel Co.
Alloy steel. C 0.28, Mn 1.35, Si 0.3, Cu 0.2, bal Fe.
Rolled: 70,000 TS; 50,000 YS. For railroad and agriculture
equipment, mine cars; high strength, low alloy construction
steel.

HI-MO

Teledyne Firth Sterling
Steel. C 0.83, W 1.5, Mo 8.75, Cr 4, V 1.25, bal Fe.
For cutting tools, reamers; high speed steel. *Obsolete*

HI-NICKEL ALLOY

Wallace Murray Corp.
C, Ni, bal Fe.
For tools. *Obsolete*

HI-NICKEL ALLOY

Wallace Murray Corp.
C, Ni, bal Fe.
For corrosion and heat resistant parts; corrosion and heat
resistant. *Obsolete*

HI-PHY KIRKSITE

NL Industries
Cu, Al, Mg, bal Zn.
For forming dies; refined grain.

HI-PROOF 304

British Steel Corp.
C 0-0.06, Mn 0-2, Cr 18-20, Ni 8-11, N 0.2, bal Fe.
Room temperature: 85,000-114,000 TS; 42,500 YS; 35 El. At
400 C: 20,000 minimum YS. For stainless parts in food
industry. *Obsolete*

HI-PROOF 316

British Steel Corp.
C 0-0.07, Mn 0-2, Cr 16-18, Ni 10-14, Mo 2-3, N 0.2, bal Fe.
Room temperature: 92,000-114,000 TS; 46,000 YS; 35 El. At
400 C: 22,400 minimum YS. For stainless steel parts.
Obsolete

HI-PROOF 316L
British Steel Corp.
C 0-0.03, Mn 0-2, Cr 16-18, Ni 10-14, Mo 2-3, N 0.2, bal Fe.
Room temperature: 92,000-114,000 TS; 46,000 YS; 35 El. At 400 C: 22,400 minimum YS. Weldable; for stainless steel parts and assemblies. *Obsolete*

HI-PROOF 347
British Steel Corp.
C 0-0.08, Mn 0-2, Cr 17-19, Ni 9-11, N 0.2, Nb = 10 x C min, bal Fe.
Room temperature: 92,000-114,000 TS; 49,000 YS; 35 El. At 400 C: 31,000 minimum YS. Weldable; for stainless steel parts and assemblies. *Obsolete*

HI-QUA-LED
Alco Products
Leaded alloy steels.
For screw machine products; free-cutting.

HI-QUA-LED 10L45
Alco Products
C 0.43-0.5, Mn 0.6-0.9, Pb, bal Fe.
Heat treated: 110,000 TS; 68,500 YS; 22 El; 44 RA; 230 Brin. For gears, bolts, machine tool parts; free-cutting, leaded steel.

HI-QUA-LED 10L50
Alco Products
C 0.48-0.5, Mn 0.6-0.9, Pb, bal Fe.
Heat treated: 117,000 TS; 72,500 YS; 22 El; 43 RA; 235 Brin. For crankshafts, gears, tie rods, bushings; free-cutting, leaded steel.

HI-QUA-LED 10L60
Alco Products
C 0.55-0.65, Mn 0.7-1, Pb, bal Fe.
Heat treated: 142,000 TS; 81,000 YS; 16 El; 32 RA; 270 Brin. For crankshafts, die blocks, girders; free-cutting, leaded steel.

HI-QUA-LED 10L70
Alco Products
C 0.65-0.75, Mn 0.9, Pb, bal Fe.
Heat treated: 145,000-157,000 TS; 87,000-104,000 YS; 14-15.5 El; 26.5-37.0 RA; 285-321 Brin. For punches, hammers, tools; free-cutting leaded steel.

HI-QUA-LED 41L30
Alco Products
C 0.28-0.35, Mn 0.4-0.6, Cr 0.8-1.1, Mo 0.15-0.25, Pb, bal Fe.
Heat treated: 126,000 TS; 105,000 YS; 18 El; 52 RA; 262 Brin. For gears, bolts, crankshafts; free-cutting, leaded steel.

HI-QUA-LED 41L37
Alco Products
C 0.35-0.4, Cr 0.8-1.1, Mo 0.15-0.25, Pb, bal Fe.
Heat treated: 122,000 TS; 100,000 YS; 23 El; 62 RA; 248 Brin. For gears, bolts, crankshafts; free-cutting, leaded steel.

HI-QUA-LED 41L40
Alco Products
C 0.38-0.43, Mn 0.75-1, Cr 0.8-1.1, Mo 0.2, Pb, bal Fe.
Heat treated: 118,000-132,000 TS; 92,000-106,000 YS; 17-20 El; 46-57 RA; 235-277 Brin. For gears, bolts, machine tool parts; free-cutting, leaded steel.

HI-QUA-LED 43L40
Alco Products
C 0.38-0.43, Ni 1.6-2, Cr 0.7-0.9, Mo 0.3, Pb, bal Fe.
Heat treated: 145,000 TS; 125,000 YS; 16 El; 48 RA; 310 Brin. For machine tool parts; free-cutting leaded steel.

HI-RUN
Manufacturer not listed.
C 1.55, Cr 11.5, Mo 0.8, V 0.9, bal Fe.
Air hardening. For tools, dies. AISI D2 Hi Chrome- Hi Carbon tool steel.

HI-TEM-IRON NO. 7
Bethlehem Foundry & Machine Co.
C 3, Si 1.5, Ni, Cr, bal Fe.
For retorts, dye and paint equipment; corrosion and heat resisting, cast iron.

HI-TEMP 080
Handy & Harmon
Cu 54.85, Ni 8, Zn 25, Mn 12, Si 0.15.
MP: 1575°F; FP:1675°F. Economical high strength filler metal for joining carbides to alloy steel.

HI-TEMP 095
Handy & Harmon
Cu 52.5, Ni 9.5, Mn 38.
MP: 1615°F; FP: 1700°F. High strength filler metal for joining carbides, steel and heat resistant alloys.

HI-TEMP 548
Handy & Harmon
Cu 55, Ni 6, Zn 35, Mn 4.
MP: 1615°F; FP: 1685°F. Tough, moderate strength, low melting improved nickel silver filler metal for carbides, tool steel, stainless steel and nickel alloys.

HI-TEMP 575
Handy & Harmon
Cu 57.5, Zn 38.5, Mn 2, Co 2.
MP: 1635°F; FP: 1705°F. Low melting, free flowing bronze filler metal for brazing carbides and tool steel.

HI-TEMP 870
Handy & Harmon
Cu 87, Mn 10, Co 3.
MP: 1760°F; FP: 1885°F. Free flowing, high melting filler metal with good high temperature strength, for brazing carbides, tool steel, stainless steel and nickel alloys.

HI-TEMP AF-183
Allegheny Ludlum Steel
C 0.25-0.4, Cr 11.5-13.5, Mo 2.75-3.25, V 0.6-0.95, Mn 17-19, N 0.2, bal Fe.
Heat treated: 136,800 TS; 81,600 YS; 33.5 El; 23.4 RA; 258 Brin. For gas turbine parts, afterburners, nozzles; heat treatable, stainless, high heat resistance. *Obsolete*

HI-TENSILE
R. Lavin & Sons, Inc.
Zn 7-8, Mg 0.2-0.5, Cu 0.4-1, Ti 0.2, bal Al.
T5-temper: 35,000 TS; 26,000 YS; 6 El; 75 Brin. For aircraft and machine tool casting. Room temperature aging properties.

HI-TENSO 1622
Anaconda Co.
Copper. Cd 1, bal Cu.
Hard wire: 90,000 TS; 1.5 El. For low load transmission lines, trolley wire, spring contracts. Corrosion and wear resistant. Copper Alloy No. 1622.

HI-TENSO 961
Now HI-TENSO 162.

HI-TENSO 965
Now HI-TENSO 165.

HI-THORIA
Manufacturer not listed.
W, ThO 2.
For welding electrodes; stable arc. *Obsolete*

HI-TOP
British Steel plc
Electrolytic chromium and chromium oxide steel.

HI-VAN
Peninsular Steel Co.
Si 1.1, Cr 5.25, Mo 1.2, V 0.9, bal Fe.
For punches, upsetters, forming dies; hot work steel, oil hardened.

HI-VAN NO. 28
Midvale-Heppenstall Co.
Cr 4, W 19, V 2, Mo 0.75, C, bal Fe.
For tools, cutters, reamers, taps, gauges; high speed steel.

HI-WEAR 64
Now CARPENTER HI-WEAR 64.

HI-YAW-TEN
Yawata Iron & Steel Co., Ltd.
C 0-0.12, Si 0.25-0.75, Mn 0.2-0.5, P 0.06-0.12, Cu 0.25-0.5, Ni 0-0.65, Cr 0.4-1, Ti 0-0.15, bal Fe.
Annealed: 67,000 TS min; 50,000 YS min. Rolled: 71,000 TS min; 57,000 YS min. For buildings, rolling stock, buses, mine cars, bridges. Resists atmospheric corrosion.

HI-YIELD 42
National Intergroup Inc.
C 0.21, Mn 0.9, 0.01 Cb min, bal Fe.
Plate: 63,000 TS; 42,000 YS; 24 El. For trucks, mine cars, derricks, bridges, plow frames, pressure vessels. Good fabricability and weldability.

HI-YIELD 45
National Intergroup Inc.
C 0.22, Mn 1.25, 0.01 Cb min, bal Fe.
Plate: 60,000 TS; 45,000 YS; 22 El. For trucks, mine cars, derricks, bridges, plow frames, pressure vessels. Good fabricability and weldability.

HI-YIELD 50
National Intergroup Inc.
C 0.22, Mn 1.25, 0.01 Cb min, bal Fe.
Plate: 65,000 TS; 50,000 YS; 20 El. For trucks, mine cars, derricks, bridges, plow frames, pressure vessels. Good fabricability and weldability.

HI-YIELD 55
National Intergroup Inc.
C 0.25, Mn 1.35, 0.01 Cb min, bal Fe.
Plate: 70,000 TS; 55,000 YS; 18 El. For trucks, mine cars, derricks, bridges, plow frames, pressure vessels. Good fabricability and weldability.

HI-Z
Fugi Iron & Steel Co., Ltd.
C 0.1-0.18, Mn 0.6-1, Si 0.15-0.35, B 0.002-0.006, Cr 0.4-0.8, Cu 0.15-0.5, Mo 0.4-6, Ni 0.7-1, V 0.03-0.1, bal Fe.
Plate: 114,000-135,000 TS; 100,000 YS min; 20 El min. For pressure vessels, bridges, mine cars, power shovels, cranes, trucks, trailers. Shock and wear resistant. Heat treated by mill.

HIA
Allegheny Ludlum Steel
C 0.4, Cr 0.45, Mo 4, Ni 1.6, bal Fe.
For tools, gears; oil hardening. *Obsolete*

HIBBO NO. 100
Central Brass & Aluminum Foundry Co.
Al, bal Cu.
50,000 TS; 100 Brin. For light bearings and bushings; heavy duty.

HIBBO NO. 125
Central Brass & Aluminum Foundry Co.
Al, bal Cu.
Cast: 81,000 TS; 29,000 YS; 24 El; 22 RA; 120 Brin. For gears, worm wheels, bearings; for general service.

HIBBO NO. 150
Central Brass & Aluminum Foundry Co.
Al, bal Cu.
Cast: 83,000 TS; 35,000 YS; 13 El; 19 RA; 140 Brin. For bearings, gears, bushings, pinions; heavy duty, shock resistant.

HIBBO NO. 175
Central Brass & Aluminum Foundry Co.
Al, bal Cu.
Cast: 82,000 TS; 36,000 YS; 9 El; 10 RA; 180 Brin. For gears, pinions, shafts; heavy compressive loading.

HIBBO NO. 200
Central Brass & Aluminum Foundry Co.
Al, bal Cu.
Cast: 96,4000 TS; 48,000 YS; 6 El; 65 RA; 190-220 Brin. For bushings, bearings, gears, welder dies and jaws; wear and corrosion resistant, Al-bronze.

HIBBO NO. 225
Central Brass & Aluminum Foundry Co.
Al, bal Cu.
85,000 TS; 190-240 Brin. For forming dies, bearing slides, cam rollers, non-sparking tools; resists shock, fatigue and corrosion.

HIBBO NO. 250
Central Brass & Aluminum Foundry Co.
Al, bal Cu.
Cast: 81,000 TS; 76,000 YS; 0.5 El; 0 RA; 230-260 Brin. For bearing slides, cam rollers, wear strips; high wear resistance.

HIBBO NO. 275
Central Brass & Aluminum Foundry Co.
Al, bal Cu.
90,000 TS; 250 Brin. For dies, rolls, pins, cams, bushings; resists shock, fatigue and corrosion.

HIBBO NO. 300
Central Brass & Aluminum Foundry Co.
Al, bal Cu.
270-300 Brin. For forming and drawing dies; very hard, tough to machine.

HICKORY NO. 7
Jamison Steel Corp.
C 0.5, Cr 1.5, W 2.5, bal Fe.
For punches, rivet sets; Type S1; shock resistant.

HICORE 75
British Steel Corp.
C 0.13, Cr 0.9, Ni 3.1, bal Fe.
Heat treated: 155,000-180,000 TS; 123,000-156,000 YS; 15-20 El; 50-60 RA. For gears, pinions, differential axles; case-hardening steel; IZ 30-50. *Obsolete*

HICORE 90
British Steel Corp.
C 0.16, Ni 4.2, Cr 1.2, Mo, bal Fe.
Heat treated: 190,000-213,000 TS; 145,000-180,000 YS; 13-18 El; 45-60 RA. For gears, pinions; case-hardening steel; IZ 25-35. *Obsolete*

HICRO
Diehl Steel Co.
C 1.5, Cr 12, Mo 0.8, V 0.35, bal Fe.
For cold forming dies; oil hardened.

HICRO 200
Diehl Steel Co.
C 2, Mn 0.7, Si 0.3, Cr 13, W 1.2, bal Fe.
Air or oil hardening cold work tool steel, chromium type; AISI D3.

HICRO T
Diehl Steel Co.
C 2.1, Mn 0.7, Si 0.3, Cr 13, W 1.3, bal Fe.
Air or oil hardenable cold work tool and die steel, chromium type; AISI D6.

HICRO-150
Diehl Steel Co.
C 1.5, Cr 12, V 0.4, Mo 1, bal Fe.
Hardened: 278,000 TS; 214,000 YS; 1 El; 567 Brin. For blanking and drawing dies, wire drawing and stamping dies, punches, gauges, hobs. Type D2 air hardening, nondeforming, cold work tool steel; tough.

HIDALGO
Dorrenberg Edelstahl GmbH
C 0.7, W 14, Cr 4, V 2, bal Fe.
For cutters, tools; high speed steel.

HIDALGO I
Dorrenberg Edelstahl GmbH
C 0.55, Si 0.9, Cr 1, V 0.18, W 1.85, bal Fe.
For cold work tools, punches, headers; oil hardened; tough.

HIDALGO II
Dorrenberg Edelstahl GmbH
C 0.4, Si 0.9, Cr 1, V 0.18, W 1.85, bal Fe.
For cold work tools, punches; oil hardened; tough.

HIDALGO III
Dorrenberg Edelstahl GmbH
C 0.35, Si 0.9, Cr 1.05, V 0.18, W 1.85, bal Fe.
For cold work tools, headers, punches, upsetters; oil hardened; tough.

HIDALGO NO. 48 ALLOY
Hidalgo Steel Co. Inc.
C 0.55, Cr 0.2, Mn 0.75, V 0.2, Si 2, bal Fe.
For chisels, shear blades, punches; shock resistant.

HIDALGO RL
Hidalgo Steel Co. Inc.
C 1.25, Cr 0.4-0.5, W 5.5, bal Fe.
For cutting tools; for finishing cuts on hard material.

HIDUMATIC
High Duty Alloys Ltd.
Aluminum.
Aluminum alloy AlCuBiPb.

HIDUMINIUM 00
High Duty Alloys Ltd.
Cu 0.75-2.5, Si 9-11.5, Ni 0-3, bal Al.
Sand cast: 16,000 TS. Die cast: 29,000 TS; 13,500 YS; 2 El; 85 Brin. For general purpose castings of thin wall sections; high fluidity. *Obsolete*

HIDUMINIUM 01
High Duty Alloys Ltd.
Cu 4, Mg 0.6, Mn 0.5, bal Al.
Forged: 54,000 TS; 30,500 YS; 15 El. Extruded: 54,000 TS; 31,500 YS; 12 El. Tube: 62,000 TS; 47,000 YS; 12 El. For general engineering components; age-hardenable.

HIDUMINIUM 01
Birmingham Aluminium Casting Co.
Cu 4, Mg 0.6, Mn 0.5, bal Al.
Forged: 54,000 TS; 30,500 YS; 15 El. Extruded: 54,000 TS; 31,500 YS; 12 El. Tube: 62,000 TS; 47,000 YS; 12 El. For general engineering components; age-hardenable.

HIDUMINIUM 02
High Duty Alloys Ltd.
Cu 4, Mg 1.5, Ni 2, bal Al.
Heat treated: 49,000-58,000 TS; 31,500-32,000 YS; 8-20 El; 115 Brin. For pistons, cylinder heads; age hardenable, high temperature uses. *Obsolete*

HIDUMINIUM 03
Rolls-Royce Mfg. Co.
Cu 1, Mg 1, Si 1, bal Al.
T4-temper: 38,000 TS; 22,400 YS; 15 El. T6-temper: 56,000 TS; 45,000 YS; 10 El. For aircraft forgings and extrusions.

HIDUMINIUM 03
High Duty Alloys Ltd.
Cu 1, Mg 1, Si 1, bal Al.
T4-temper: 38,000 TS; 22,400 YS; 15 El. T6-temper: 56,000 TS; 45,000 YS; 10 El. For aircraft forgings and extrusions.

HIDUMINIUM 05
Rolls-Royce Mfg. Co.
Mg 5, bal Al.
Annealed: 38,000 TS; 18,000 YS; 18 El. 1/2 Hard: 40,500 TS; 31,500 YS; 5 El. For marine structures; corrosion resistant.

HIDUMINIUM 05
High Duty Alloys Ltd.
Mg 5, bal Al.
Annealed: 38,000 TS; 18,000 YS; 18 El. 1/2 Hard: 40,500 TS; 31,500 YS; 5 El. For marine structures; corrosion resistant.

HIDUMINIUM 07
Rolls-Royce Mfg. Co.
Mg 7, Mn 0.6, Fe 0-0.6, bal Al.
Annealed: 45,000 TS; 18,000 YS; 18 El; 20 RA. 1/2 Hard: 56,000 TS; 36,000 YS; 5 El. For hardware, sporting goods, machine tool parts; high corrosion resistance.

HIDUMINIUM 07
High Duty Alloys Ltd.
Mg 7, Mn 0.6, Fe 0-0.6, bal Al.
Annealed: 45,000 TS; 18,000 YS; 18 El; 20 RA. 1/2 Hard: 56,000 TS; 36,000 YS; 5 El. For hardware, sporting goods, machine tool parts; high corrosion resistance.

HIDUMINIUM 08
High Duty Alloys Ltd.
Cu 1, Mg 1, Si 11, Ni 1, bal Al.
Heat treated: 47,000 TS; 33,500 YS; 5 El; 115 Brin. For pistons, cylinder heads; age hardenable. *Obsolete*

HIDUMINIUM 10
High Duty Alloys Ltd.
Si 10-13, Ti 0-0.2, Cu 0-0.1, Fe 0-0.6, bal Al.
Sand cast: 26,000 TS; 9,000 YS; 8 El; 55 Brin. Die cast: 30,500 TS; 10,000 YS; 10 El; 65 Brin. For general purpose castings of intricate shape; good fluidity. *Obsolete*

HIDUMINIUM 100
Rolls-Royce Mfg. Co.
Al_2O_3 + Al.
Sintered. For components operating above 250°C; high temperature resistance.

HIDUMINIUM 100
Mond Nickel Co. Ltd.
Al_2O_3 + Al.
Sintered. For components operating above 250°C; high temperature resistance.

HIDUMINIUM 100
High Duty Alloys Ltd.
Al_2O_3 + Al.
Sintered. For components operating above 250°C; high temperature resistance.

HIDUMINIUM 11
High Duty Alloys Ltd.
Cu 0-0.15, Si 0-0.6, Fe 0-0.75, Mn 1-1.5, bal Al.
Soft temper: 15,000 TS; 12,000 YS; 25 El; 27 Brin. 3/4 H temper: 22,000 TS; 19,000 YS; 5 El; 55 Brin. For structures; not heat treatable, corrosion resistant. *Obsolete*

HIDUMINIUM 1A
High Duty Alloys Ltd.
Al 98.8.
O-temper: 11,200 TS; 35 El. 1/2 H-temper: 13,500 TS; 8 El. H-temper: 18,000 TS; 5 El. For light alloy parts. *Obsolete*

HIDUMINIUM 1B
High Duty Alloys Ltd.
Al 99.5.
O-temper: 13,500 TS; 30 El. 1/2 H-temper: 14,500 TS; 8 El. H-temper: 19,000 TS; 5 El. For light alloy parts. *Obsolete*

HIDUMINIUM 1C
High Duty Alloys Ltd.
Al 99.
O-temper: 14,500 TS; 5600 YS; 30 El. 1/2 H-temper: 15,700 TS; 14,500 YS; 7 El. H-temper: 20,000 TS; 19,000 YS; 3 El. For light alloy parts. *Obsolete*

HIDUMINIUM 20

High Duty Alloys Ltd.
Cu 2-4, Si 3-6, Mn 0.3-0.7, Zn 0-0.5, Fe 0-0.8, bal Al.
Sand cast: 18,000 TS; 8000 YS; 2 El; 55 Brin. Die cast: 20,000 TS; 9000 YS; 2 El; 80 Brin. For marine castings; age-hardenable. *Obsolete*

HIDUMINIUM 22

High Duty Alloys Ltd.
Cu 0-0.015, Mg 1.5-2.5, Si 0-0.6, Fe 0-0.75, bal Al.
Soft temper: 24,000 TS; 12,000 YS; 18 El; 45 Brin. 1/2 H temper: 35,000 TS; 27,000 YS; 5 El; 85 Brin. For fuel pipes; high corrosion resistance. *Obsolete*

HIDUMINIUM 24

High Duty Alloys Ltd.
Mg 2, bal Al.
O-temper: 24,500-27,000 TS; 10,500 YS; 18 El. For deep drawn and formed parts; high ductility. *Obsolete*

HIDUMINIUM 29

High Duty Alloys Ltd.
Sn 6.5, Cu 1, Ni 0.8, bal Al.
Sand cast: 18,000 TS; 7800 YS; 8 El. Permanent mold: 20,000 TS; 18,000 YS; 15 El. For bearings; shock resistant. *Obsolete*

HIDUMINIUM 33

High Duty Alloys Ltd.
Mg 3-4, Mn 0-1, Cu 0-0.15, Fe 0-0.7, Cr 0-0.5, bal Al.
Soft temper: 28,000 TS; 14,000 YS; 18 El; 55 Brin. 1/2 temper: 36,000 TS; 30,000 YS; 5 El; 100 Brin. For marine parts; highest corrosion resistance. *Obsolete*

HIDUMINIUM 35

High Duty Alloys Ltd.
Mg 4.25, bal Al.
O-temper: 32,000 TS; 15,700 YS; 18 El. M-temper: 38,000 TS; 18,000 YS; 12 El. For light alloy parts; high corrosion resistance and ductility. *Obsolete*

HIDUMINIUM 40

High Duty Alloys Ltd.
Mg 0.2-0.8, Si 4.5-6, Ti 0-0.2, bal Al.
Cast: 17,000 TS; 10,000 YS; 2.5 El; 50 Brin. W-temper: 22,000 TS; 14,000 YS; 2.5 El; 50 Brin. WP-temper: 30,000 TS; 26,000 YS; 100 Brin. For marine castings; high fluidity, pressure tight. *Obsolete*

HIDUMINIUM 42

High Duty Alloys Ltd.
Mg 0.75, Si 1, bal Al.
W-temper: 26,700 TS; 15,700 YS; 18 El. WP-temper: 43,600 TS; 40,300 YS; 10 El. For light alloy parts; good corrosion resistance and formability; age-hardenable. *Obsolete*

HIDUMINIUM 44

High Duty Alloys Ltd.
Mg 0.5-1.2, Si 0.75-1.3, Mn 0-1, Fe 0-0.6, bal Al.
Annealed and tempered: 18,000 TS; 10,000 YS; 27 El; 30 Brin. WP-temper: 46,000 TS; 36,000 YS; 8 El; 110 Brin. For marine applications; heat treatable, resists seawater corrosion. *Obsolete*

HIDUMINIUM 46

High Duty Alloys Ltd.
Mg 0.7, Si 0.5, Ti 0.05, bal Al.
W-temper: 27,100 TS; 16,200 YS; 25 El. WP-temper: 35,800 TS; 27,000 YS; 18 El. For structural extrusions; heat treatable; corrosion resistant. *Obsolete*

HIDUMINIUM 51

High Duty Alloys Ltd.
Cu 0.8-2, Mg 0.05-0.2, Si 0.75-2.8, Fe 0.25-1.4, Ni 0.8-1.75, bal Al.
Sand cast: 23,000 TS; 11,000 YS; 2 El; 55 Brin. Die cast: 25,000 TS; 13,000 YS; 2 El; 80 Brin. For general purpose castings; fair corrosion resistance. *Obsolete*

HIDUMINIUM 55

High Duty Alloys Ltd.
Cu 1.75-2.7, Mg 0.5-1.3, Si 0.4-1.5, Fe 0-1.4, Ni 0-1.4, bal Al.
Forged: 50,000 TS; 36,000 YS; 8 El; 148 Brin. WP-temper: 44,000 TS; 36,000 YS; 8 El. For high strength parts, aircraft structures; heat treatable. *Obsolete*

HIDUMINIUM 66

High Duty Alloys Ltd.
Cu 3.5-4.8, Mg 0.3-0.6, Si 0.1-1.5, Mn 0-1.2, bal Al.
WP-temper: 64,000 TS; 52,000 YS; 8 El; 160 Brin. For high strength parts, aircraft structures; heat treatable. *Obsolete*

HIDUMINIUM 72

High Duty Alloys Ltd.
Cu 4.6, Mg 1.3, Fe 0.25, Mn 0.7, bal Al.
Heat treated: 67,000 TS; 51,000 YS; 8 El. For aircraft and general engineering parts; age-hardenable. *Obsolete*

HIDUMINIUM 80

High Duty Alloys Ltd.
Cu 4-5, Ti 0.1-0.25, Si 0-0.25, bal Al.
W-temper: 33,000 TS; 18,000 YS; 10 El; 60 Brin. WP-temper: 48,000 TS; 27,000 YS; 10 El; 95 Brin. For brackets, levers, housings, aircraft castings; age-hardenable, hot-short. *Obsolete*

HIDUMINIUM 90

High Duty Alloys Ltd.
Mg 9.5-11, Cu 0-0.15, Si 0-0.25, bal Al.
Sand cast: 45,000 TS; 22,000 YS; 15 El; 90 Brin. Die cast: 47,000 TS; 25,000 YS; 15 El; 90 Brin. For gasoline flow-meters; age-hardenable, corrosion resistant. *Obsolete*

HIDUMINIUM DU BRAND

High Duty Alloys Ltd.
Cu 3.5-4.8, Mg 0.6, Mn 0.5, Ti 0.3, bal Al.
Rolled: 54,000 TS; 35,000 YS; 18 El; 120 Brin. For light alloy parts, seamless tubes, airscrews; age-hardened. *Obsolete*

HIDUMINIUM R R 53

Birmingham Aluminium Casting Co.
Aluminum. Cu 1.5-2.5, Ni 0.5-2, Mg 1.4-1.8, Fe 1.2-1.5, Ti 0.2-0.12, Si 0.4-2, bal Al.
Heat treated: 50,000 TS; 44,000 YS; 2 El; 140 Brin. For pistons, cylinder heads; die casting.

HIDUMINIUM R.R. 50

Mond Nickel Co. Ltd.
Cu 0.8-2, Mg 0.05-2, Ti 0.2, Si 1.5-2.8, Fe 0.8-1.4, Ni 0.8-1.7, bal Al.
Heat treated: 25,000-30,000 TS; 3 El; 5 RA; 72 Brin. For cylinder blocks, cylinder heads, pistons; crankcases; heat treatable.

HIDUMINIUM R.R. 50

Rolls-Royce Mfg. Co.
Cu 0.8-2, Mg 0.05-2, Ti 0.2, Si 1.5-2.8, Fe 0.8-1.4, Ni 0.8-1.7, bal Al.
Heat treated: 25,000-30,000 TS; 3 El; 5 RA; 72 Brin. For cylinder blocks, cylinder heads, pistons; crankcases; heat treatable.

HIDUMINIUM R.R. 50

Birmingham Aluminium Casting Co.
Cu 0.8-2, Mg 0.05-2, Ti 0.2, Si 1.5-2.8, Fe 0.8-1.4, Ni 0.8-1.7, bal Al.
Heat treated: 25,000-30,000 TS; 3 El; 5 RA; 72 Brin. For cylinder blocks, cylinder heads, pistons; crankcases; heat treatable.

HIDUMINIUM R.R. 50

High Duty Alloys Ltd.
Cu 0.8-2, Mg 0.05-2, Ti 0.2, Si 1.5-2.8, Fe 0.8-1.4, Ni 0.8-1.7, bal Al.
Heat treated: 25,000-30,000 TS; 3 El; 5 RA; 72 Brin. For cylinder blocks, cylinder heads, pistons; crankcases; heat treatable.

HIDUMINIUM R.R. 56

Rolls-Royce Mfg. Co.
Cu 2, Ni 1.25, Fe 1.2, Mg 0.8, Ti 0.08, Si 0.6, bal Al.
Heat treated: 60,000-72,000 TS; 54,000-58,000 YS; 10-20 El; 14-25 R 120-160 Brin. For automobile forgings, connecting rods, supercharger rotors; forgings; Rolls-Royce automobile.

HIDUMINIUM R.R. 56

High Duty Alloys Ltd.
Cu 2, Ni 1.25, Fe 1.2, Mg 0.8, Ti 0.08, Si 0.6, bal Al.
Heat treated: 60,000-72,000 TS; 54,000-58,000 YS; 10-20 El; 14-25 R 120-160 Brin. For automobile forgings, connecting rods, supercharger rotors; forgings; Rolls-Royce automobile.

HIDUMINIUM R.R. 57

High Duty Alloys Ltd.
Cu 5.75-6.25, Si 0-0.2, Mn 0.2-0.3, Ti 0.1-0.15, bal Al.
At 70°F: 54,000 TS; 33,000 YS; 8 El; 121 Brin. At 650°F: 14,000 TS; 9000 YS; 27 El. For high temperature applications, aircraft structures; heat treatable. *Obsolete*

HIDUMINIUM R.R. 58

High Duty Alloys Ltd.
Cu 1.5-3, Mg 1.2-1.8, Fe 1-1.5, Ni 0.5-1.5, Ti 0-0.2, bal Al.
At 70°F: 60,000 TS; 44,000 YS; 10 El; 148 Brin. At 650°F: 12,500 TS; 10,000 YS; 27.5 El. For high temperature applications, aircraft structures; heat treatable. *Obsolete*

HIDUMINIUM R.R. 59

Rolls-Royce Mfg. Co.
Cu 2.2, Ni 1.35, Mg 1.25, Fe 1.35, Si 0.08, bal Al.
Forged: 52,000-65,000 TS; 50,000-56,000 YS; 6-10 El; 10-20 RA; 120-150 Brin. Heat treated: 55,000 TS; 47,000 YS; 8 El; 17 RA 127 Brin. For pistons, cylinder heads; forgings; Rolls Royce automobile; aero-engines.

HIDUMINIUM R.R. 59

High Duty Alloys Ltd.
Cu 2.2, Ni 1.35, Mg 1.25, Fe 1.35, Si 0.08, bal Al.
Forged: 52,000-65,000 TS; 50,000-56,000 YS; 6-10 El; 10-20 RA; 120-150 Brin. Heat treated: 55,000 TS; 47,000 YS; 8 El; 17 RA 127 Brin. For pistons, cylinder heads; forgings; Rolls Royce automobile; aero-engines.

HIDUMINIUM R.R. 60

High Duty Alloys Ltd.
Cu 1.5-3, Si 0.3-1.5, Mg 0.5-1.5, Fe 0.5-1.5, Ni 0-2, Ce 0.3, bal Al.
For light alloy parts; antifriction properties. *Obsolete*

HIDUMINIUM R.R.53

Rolls-Royce Mfg. Co.
Cu 2.2, Ni 1.3, Mg 1.6, Fe 1.4, Ti 0.08, Si 1.2, bal Al.
Die cast: 31,300 TS; 28,000 YS; 3 El; 4 RA; 80 Brin. Heat treated: 56,000 TS; 50,400 YS; 1 El; 1.5 RA; 130 Brin. For pistons for automobile and aircraft engines, cylinder heads; die cast; Rolls-Royce automobile.

HIDUMINIUM R.R.53

High Duty Alloys Ltd.
Cu 2.2, Ni 1.3, Mg 1.6, Fe 1.4, Ti 0.08, Si 1.2, bal Al.
Die cast: 31,300 TS; 28,000 YS; 3 El; 4 RA; 80 Brin. Heat treated: 56,000 TS; 50,400 YS; 1 El; 1.5 RA; 130 Brin. For pistons for automobile and aircraft engines, cylinder heads; die cast; Rolls-Royce automobile.

HIDUMINIUM RR 53 B

Birmingham Aluminium Casting Co.
Aluminum. Cu 1.5, Ni 1-2, Mg 0.6-1, Fe 0.8-1.5, Si 0.7-0.9, bal Al.
Cast: 21,000 TS; 14,000 YS; 3 El; 78 Brin. Heat treated: 42,000 TS; 37,000 YS; 2 El; 138 Brin. For levers, brackets, textile and food machinery, pistons; high temperature strength.

HIDUMINIUM RR 53 C

High Duty Alloys Ltd.
Cu 0.8-2, Ni 0.5-1.5, Mg 0.3-0.8, Fe 0.8-1.4, Si 2-3, Ti 0.3, bal Al.
Cast: 30,000 TS; 20,000 YS; 2 El; 75 Brin. Heat treated: 44,000 TS; 40,000 YS; 2 El; 115 Brin. For stand and die castings; light alloy. *Obsolete*

HIDUMINIUM RR 75

High Duty Alloys Ltd.
Al 97.6, Cu 2, Mg 0.4.
For wire, rivets. *Obsolete*

HIDUMINIUM RR 77

High Duty Alloys Ltd.
Cu 1.5, Zn 4-6, Mg 2-4, Fe 0.6, Si 0.6, Ni 0-1, Ti 0.3, bal Al.
Annealed: 32,000 TS; 19,000 YS; 20 El; 65 Brin. Heat treated: 85,000 TS; 74,000 YS; 16 El; 180 Brin. For extruded and drawn tubing, rolled sheet; light alloy, high strength.
Obsolete

HIDUMINIUM RR 82

High Duty Alloys Ltd.
Cu, Si, bal Al.
For fuel pipes. *Obsolete*

HIDUMINIUM RR-66

High Duty Alloys Ltd.
Cu 0.5, Ni 0.33, Mg 4.8, Fe 0.33, bal Al.
Annealed: 45,000 TS; 22,000 YS; 24 El; 80 Brin. Cold worked: 60,500 TS; 49,000 YS; 10 El; 121 Brin. For automobile parts, airplanes, boats; corrosion resistant. *Obsolete*

HIDUMINIUM RR250

Rolls-Royce Mfg. Co.
Cu 5, Ti 0.2, Mn 0.25, Ce 0.25, Ni 1, Sb 0.25, bal Al.
Heat treated: 36,000 TS; 22,500 YS; 2 El. For engine components; age hardenable; sand castings.

HIDUMINIUM RR250

High Duty Alloys Ltd.
Cu 5, Ti 0.2, Mn 0.25, Ce 0.25, Ni 1, Sb 0.25, bal Al.
Heat treated: 36,000 TS; 22,500 YS; 2 El. For engine components; age hardenable; sand castings.

HIDUMINIUM RR257

High Duty Alloys Ltd.
Cu 6, Co 0.25, Ni 1, Sb 0.25, Mn 0.25, Ti 0.2, bal Al.
WP-temper: 54,000 TS; 33,500 YS; 8 El. For engine castings; age hardenable. *Obsolete*

HIDUMINIUM RR72

High Duty Alloys Ltd.
Al 94, Cu 4, Mg 1.2, Mn 0.8.
Extruded: 60,000 TS; 15 El; 148 Brin. For extruded sections, solid drawn tubes; heat treatable. *Obsolete*

HIDUMINIUM RR88

High Duty Alloys Ltd.
Cu 0-3, Mg 0-4, Zn 4-8.5, Cr 0-1, bal Al.
Aged: 89,000 TS; 78,000 YS; 7 El; 180 Brin. For aircraft construction, high strength applications; age hardenable.
Obsolete

HIDUMINIUM RRAC9A

Rolls-Royce Mfg. Co.
Al alloy.
For light alloy parts.

HIDUMINIUM RRAC9A

High Duty Alloys Ltd.
Al alloy.
For light alloy parts.

HIDUMINIUM S 12

High Duty Alloys Ltd.
Si 12, Cu 1, Ni 1, Mg 1.2, bal Al.
Forged: 38,000 TS; 7 El; 105 Brin. For aircraft pistons; forging alloy. *Obsolete*

HIDUMINIUM SR

High Duty Alloys Ltd.
Cu 2.5, Mg 0.1, Fe 1, Ni 0.75, Ti 0.1, bal Al.
60-85 Brin. For gasket sealing rings; corrosion resistant.
Obsolete

HIDUMINIUM Y

High Duty Alloys Ltd.
Cu 4, Mg 1.5, Ni 2, Fe 0.6, Mn 0.5, bal Al.
Forged: 55,000 TS; 35,000 YS; 2 El; 120 Brin. Cast: 40,000 TS; 34,000 YS; 1 El; 105 Brin. *Obsolete*

HIDUMINIUM-12

High Duty Alloys Ltd.
Mg 1.3, bal Al.
Hard drawn: 36,000 TS; 35,000 YS; 4 El; 105 Rock H. For deep drawn parts. For trim molding, washing machine tubs, vacuum cleaner hoods, fan blades. Corrosion resistant, low strength. Similar to Aluminum 5050. *Obsolete*

HIDUMINIUM-14

High Duty Alloys Ltd.
Mg 1, Mn 0.5, bal Al.
Hard drawn: 40,000 TS; 36,000 YS; 6 El; 77 Brin. Annealed: 26,000 TS; 10,000 YS; 25 El; 45 Brin. For general purpose applications, commercial roofing, vessels, tanks. Corrosion resistant. Similar to Aluminum 3004. *Obsolete*

HIDUMINIUM-21

High Duty Alloys Ltd.
Cu 3, Si 5, Mn 0.5, Mg 0.25, bal Al.
Heat treated: 36,000 TS; 24,000 YS; 2 El; 80 Brin. Cast: 27,000 TS; 18,000 YS; 2 El; 70 Brin. For sand and permanent mold castings, crankcases, housings, cylinder heads, oil pans. Heat treatable, pressure-tight. *Obsolete*

HIDURAX 1 CAST

Langley Alloys Ltd.
Al 8.5-10.5, Fe 4-5.5, Ni 4-5.5, bal Cu.
Cast: 90,00-100,000 TS; 40,000-45,000 YS; 12-20 El; 170-180 Brin. For shafts, gears, spindles, valve seats; resists corrosion and cavitation erosion.

HIDURAX 1 WROUGHT

Langley Alloys Ltd.
Al 8.5-10.5, Fe 4-6, Ni 4-6, Mn 0.5, bal Cu.
Forged: 100,000-116,000 TS; 56,000-76,000 YS; 15-25 El; 180-240 Brin. For shafts, gears, spindles, valve seats; resists corrosion and cavitation.

HIDURAX 1/12A

Langley Alloys Ltd.
Al 10.5-11, Fe 4-6, Ni 4-6, Mn 0.5, bal Cu.
Heat treated: 125,000-150,000 TS; 70,000-100,000 YS; 10-20 El. For axial flow compressor blades; high temperature strength and fatigue resistance. *Obsolete*

HIDURAX 1A

Langley Alloys Ltd.
Al 10, Ni 5, Fe 5, Mn 1.5, bal Cu.
Rolled: 85,000-94,000 TS; 36,000-54,000 YS; 25-35 El; 150-200 Brin. For valves, pumps, chemical plant equipment; BS-2032; corrosion resistant, aluminum bronze. *Obsolete*

HIDURAX 2-CAST

Langley Alloys Ltd.
Al 8.5-10.5, Fe 1.5-3.5, Mn 0.5, bal Cu.
Cast: 72,000-85,000 TS; 25,000-30,000 YS; 20-30 El; 110-140 Brin. For pumps, valves, gears, bushings; resists wear, corrosion.

HIDURAX 2-WROUGHT

Langley Alloys Ltd.
Al 8.5-10.5, Fe 0.5-2.5, Ni 1-3, bal Cu.
Rolled: 85,000-108,000 TS; 45,000-56,000 YS; 18-25 El; 149-212 Brin. For shafts, spindles, gears, liners, bushings; resists wear, corrosion and erosion.

HIDURAX 21A

Langley Alloys Ltd.
Ni 4-7, Al 8-11, Mn 0.5-2, Fe 1.5-3.5, bal Cu.
Rolled: 90,000 TS; 23,000 YS; 10 El. For valves, pumps; ASTM-B171 Alloy E; aluminum bronze, corrosion resistant.
Obsolete

HIDURAX 26A

Langley Alloys Ltd.
Fe 1.5-3.5, Al 6-8, bal Cu.
Rolled: 72,000 TS; 36,000 YS; 35 El. For valves, pumps; ASTM-B171 Alloy D; aluminum bronze, corrosion resistant.
Obsolete

HIDURAX 3

Langley Alloys Ltd.
Al 8.5-9.5, Fe 4-6, Mn 4-6, bal Cu.
Sand cast: 70,000-85,000 TS; 25,000-30,000 YS; 20-30 El; 110-140 Brin. For pumps, valves, liners, sheaves, gears; resists corrosion in H_2SO_4. *Obsolete*

HIDURAX 4

Langley Alloys Ltd.
Al 11, Fe 4-5, Ni 4, bal Cu.
Heat treated: 110,000-135,000 TS; 67,000-90,000 YS; 5-12 El; 218-270 Brin. For valve inserts, plastic molding dies, nozzles; tough, hardenable, corrosion resistant.

HIDURAX 4/16A

Langley Alloys Ltd.
Al, bal Cu.
Rolled: 107,000-121,000 TS; 67,000-79,000 YS; 4-10 El; 250-280 Brin. For valves, pumps; aluminum bronze, corrosion resistant. *Obsolete*

HIDURAX 4/17A

Langley Alloys Ltd.
Al, bal Cu.
Rolled: 112,000-135,000 TS; 67,000-90,000 YS; 5-12 El; 218-270 Brin. For valves, pumps; aluminum bronze, corrosion resistant. *Obsolete*

HIDURAX 5

Langley Alloys Ltd.
Al 9-9.8, bal Cu.
Forged: 72,000-85,000 TS; 34,000-45,000 YS; 15-30 El; 126-179 Brin. For valve seat inserts, rotors, shafts; aluminum bronze, corrosion resistant. *Obsolete*

HIDURAX 5/27A

Langley Alloys Ltd.
Al, bal Cu.
Rolled: 76,000 TS; 38,000 YS; 35 El; 130-180 Brin. For valves, pumps; aluminum bronze, corrosion resistant. *Obsolete*

HIDURAX 5/2A

Langley Alloys Ltd.
Al, bal Cu.
Rolled: 72,000-85,000 TS; 33,000-45,000 YS; 15-30 El; 126-179 Brin. For valves, pumps; DTD 160; aluminum bronze, corrosion resistant. *Obsolete*

HIDURAX SPECIAL

Langley Alloys Ltd.
Al 2-4, Fe 1-3, Ni 12-16, bal Cu.
Forged: 123,000-134,000 TS; 90,000-112,000 YS; 12-20 El; 240-270 Brin. For pump shafts, valve cones, valve seats; high strength, corrosion resistant.

HIDUREL 5-CAST

Langley Alloys Ltd.
Ni 1.8-2.6, Si 0.3-5, bal Cu.
Heat treated: 63,000-72,000 TS; 40,000-54,000 YS; 15-18 El; 140-170 Brin. For switchgears, bushings, electrical conductors; high strength and conductivity.

HIDUREL 6
Langley Alloys Ltd.
Cr 0.4-1.2, 0.2 others, bal Cu.
Heat treated: 40,000-80,000 TS; 27,000-78,000 YS; 15-30 El;
140-170 Brin. For switchgears, resistance welding electrodes;
high strength at high temperature.

HIDUREL 640
Rasmussen Mfg. Co.
Cr, Zr, bal Cu.
Annealed: 21-24 tons/in.^{2} TS; 100-120 Brin. Cold worked
28-34 tons/in.^{2} TS; 125-170 Brin. High conductivity bronze
bar. For bus bars.

HIDUREL 7
Langley Alloys Ltd.
Zn, bal Cu.
Cast: 78,000-90,000 TS; 36,000-49,000 YS; 10-20 El; 130-160
Brin. For general castings; high conductivity bronze.
Obsolete

HIDUREL-5
Langley Alloys Ltd.
Ni 0-3.5, Si 0.4-0.8, bal Cu.
Bar: 85,000-117,000 TS; 61,000-105,000 YS; 15-25 El;
160-210 Brin. Strip: 85,000-108,000 TS; 61,000-94,000 YS;
15-25 El; 160-210 Brin. For switch gears, contacts, gears,
current carrying parts; high conductivity and high strength.

HIDUREX 7
Langley Alloys Ltd.
Al 6-6.4, Si 2-2.4, Fe 0.5-0.7, bal Cu.
Wrought: 34 tsi TS; 130-180 Brin. Cast: 30 tsi TS; 110-150
Brin. For pump shafts, bolts, studs, impellers.

HIDURIT 10
Langley Alloys Ltd.
Zn, bal Cu.
Cast: 100,000-112,000 TS; 56,000-67,000 YS; 12-18 El;
180-220 Brin. For general castings; high strength brass.
Obsolete

HIDURIT 11
Langley Alloys Ltd.
Zn, bal Cu.
Cast: 85,000-95,000 TS; 38,000-54,000 YS; 20-30 El; 130-160
Brin. For general castings; BS-1400; high strength brass.
Obsolete

HIDURIT 12
Langley Alloys Ltd.
Zn, bal Cu.
Cast: 67,000-75,000 TS; 27,000-41,000 YS; 20-35 El; 110-130
Brin. For general castings; BS-1400; high strength brass.
Obsolete

HIDURIT 15
Langley Alloys Ltd.
Zn, bal Cu.
Rolled: 67,000-85,000 TS; 34,000-45,000 YS; 15-30 El;
120-180 Brin. For general castings; BS-250; high strength
brass.

HIDURON 102
Langley Alloys Ltd.
Ni 10, Cu 90.
Wrought: 18 tsi; 30 El min. BS. 2875-CN 102.

HIDURON 107
Langley Alloys Ltd.
Ni 30, Cu 70.
Wrought: 20 tsi; 30 El min. BS. 2875-CN 107.

HIDURON 191
Langley Alloys Ltd.
Wrought cupro-nickel + Al, Fe, Mn.
46 tsi TS; 27 tsi proof strength. For improved strength over
normal cupro-nickel.

HIDURON 501
Langley Alloys Ltd.
Ni 12, Al, Fe, Mn, Cr, bal Cu.
Cast cupro-nickel with additions. 30 tsi TS; 17 tsi proof stress.
For improved strength; marine use.

HIFLEX
Gulf Steel Corp.
C, alloy, bal Fe.
For tools, dies, machinery parts; fatigue resistant.

HIGH BRASS
Anaconda Co.
Copper. Cu 65, Zn 35.
Rolled: 47,000-75,000 TS; 20,000-60,000 YS; 50-60 El; 50-75
RA; 45-180 Brin. For spun parts, tanks, vessels. High
strength, good ductility.

HIGH CARBON ALLOY
Edgar T. Ward's Sons Co.
C 0.9, V 0.2, bal Fe.
For tools, drills, taps; water hardening.

HIGH CHROME-NICKEL
Johnson Matthey plc
C 0.33-0.45, Cr 0.4-0.75, Mn 1-1.35, Ni 0.9-1.2, Si 0.4-0.75,
bal Fe.
Annealed: 85,000-100,000 TS; 60,000 YS; 18 El; 23 RA. For
gears, shafts; tough. *Obsolete*

HIGH DOUBLE EXTRA
Midvale-Heppenstall Co.
C, W, Co, V, bal Fe.
For high speed cutters. *Obsolete*

HIGH DUTY
Flockton, Tompkin & Co., Ltd.
C 2.2, W 0.6, Cr 12.8, Mn 0.4, bal Fe.
For blanking and thread rolling dies, molds; oil or air
hardened, abrasion resistant.

HIGH DUTY GREY IRON
Sterling International Technology Ltd.
Si 2.1-2.3, Mn 0.6-0.9, Cr 0.2-0.6, P 0-0.15, S 0-0.12, 3.2-3.4
total C, 0.55-0.75 combined C, bal Fe.
230-280 N/mm^2 TS; 385-485 N/mm^2 transverse strength (1.2
in. diam test bar). Mainly for cylinder blocks.

HIGH DYNAMIC
Timken Co.
C 0.4, Ni 1.8, Cr 0.8, Mo 0.3, bal Fe.
Heat treated: 127,000 TS; 110,000 YS; 18 El; 55 RA. For
piston rods; tough. *Obsolete*

HIGH EXPANSION
Carpenter Technology Corp.
C 0.1, Mn 0.5, Si 0.25, Cr 3.1, Ni 2.2, bal Fe.
Annealed: 70,000 TS; 40,000 YS; 35 El; 74 Rock B. For
applications requiring high thermal expansion, bimetals.
High thermal expansion, austenitic, nonmagnetic. *Obsolete*

HIGH GRAPHITIC IRON
Janney Cylinder Co.
TC 3.5, CC 0.4, Si 3, Mo 0.5, bal Fe.
Cast: 38,000 TS; 220 Brin. For steam cylinder liners, packing
rings; centrifugal castings. *Obsolete*

HIGH LEADED BRASS 353
Anaconda Co.
Copper. Cu 61.5, Zn 36.7, Pb 1.8.
Annealed: 49,000 TS; 17,000 YS; 52 El; 68 Rock F. Half hard:
61,000 TS; 50,000 YS; 20 El; 70 Rock B. For plaques, hinges,
gears, pinions, wheels, valve stems, rivets. Corrosion
resistant. Electrical conductivity: 26. Free cutting.

HIGH LEADED BRASS 3531
Anaconda Co.
Copper. Zn 37, Pb 2, bal Cu.
Hard: 58,000 TS; 42,000 YS; 22 El; 125 Brin. Soft: 47,000 TS;
18,000 YS; 58 El; 62 Brin. For hardware, clock parts, screw
machine parts. Free cutting.

HIGH LEADED BRASS 3532
Anaconda Co.
Copper. Zn 38, Pb 1.5, bal Cu.
Hard: 73,000 TS; 60,000 YS; 7 El; 150 Brin. Soft: 45,000 TS;
17,000 YS; 50 El; 60 Brin. For hardware, clock parts. Free
cutting.

HIGH LEADED BRASS 62
Olin Brass, Indianapolis
Copper. Cu 62.25, Zn 35.75, Pb 2.
Hard: 74,000 psi TS; 60,000 psi YS; 7 El; 150 Brin. Soft:
49,000 psi TS; 17,000 psi YS; 52 El; 61 Brin. For clock gears
and key stock; high leaded brass.

HIGH LEADED BRASS 62%
Chase Brass & Copper Co., Inc.
Copper. Cu 62, Zn 36, Pb 2.
Annealed: 49,000 psi TS; 18,00 psi YS; 50 El. Hard rolled:
74,000 psi TS; 63,000 psi YS; 9 El. For hardware, clocks,
gears, locks, meters; good machinability.

HIGH LEADED BRASS, COPPER ALLOY NO. 342
Chase Brass & Copper Co.
Cu 63.5, Zn 34.6, Pb 1.9.
Half hard strip: 61,000 TS; 50,000 YS; 20 El; 40,000 shear; 70
Rock B. For clock and watch gears, plates and cases, keys,
nuts. *Obsolete*

HIGH LEADED BRASS-342
Anaconda Co.
Copper. Cu 64, Zn 34, Pb 2.
Annealed: 49,000 TS; 17,000 YS; 52 El; 68 Rock F. Half hard:
61,000 TS; 50,000 YS; 20 El; 70 Rock B. For plaques, hinges,
gears, pinions, wheels, valve stems, rivets.

HIGH LEADED TUBE BRASS
Chase Brass & Copper Co., Inc.
Copper. Cu 66, Zn 32.25, Pb 1.75.
Annealed: 52,000 psi TS; 20,000 psi YS; 50 El. Drawn: 75,000
psi TS; 60,000 psi YS; 7 El. For screw machine products; free
cutting.

HIGH MANGANESE NICKEL
English manufacture
Ni 94-98, Mn 2-6.
For spark plug wire; heat resistant.

HIGH PERMEABILITY 49
Carpenter Technology Corp.
C 0-0.1, Ni 49, bal Fe.
Rolled: 25 El; 62 RA; 200 Brin. For transformer cores. High
permeability.

HIGH PHOSPHORUS COPPER
Now COPPER DEOXIDIZED D H P.

HIGH SILICON BRONZE A
Chase Brass & Copper Co., Inc.
Copper. Cu 97, Si 3.
Annealed: 58,000 psi TS; 22,000 psi YS; 60 El. Drawn (36%):
92,000 psi TS; 55,000 psi YS; 22 El. For marine and pole lin
hardware, bolts, shafts.

HIGH SPEED BRASS
Buckeye Brass & Mfg. Co.
Cu 88, Sn 7, Zn 5.
For high speed bearings; heavy duty. *Obsolete*

HIGH SPEED EXTRA DRILL
Midvale-Heppenstall Co.
C 0.7, Cr 3.5, W 18, V 1, bal Fe.
For drills, tools; high speed steel. *Obsolete*

HIGH SPEED MO EXTRA
Midvale-Heppenstall Co.
C, W, Mo, V, bal Fe.
For high speed cutters, tools; high speed steel. *Obsolete*

HIGH SPEED N M HOT ROD
Grand Northern Products Ltd.
Mild steel welding rod with alloy coating. ASW 6010.

HIGH SPEED NO. 2
Edgar Allen Balfour Ltd.
C, W, bal Fe.
For tools; high speed steel. *Obsolete*

HIGH SPEED TOOL-A
English manufacture
W 5-12, Ti 6-12, Si 3-6, Al 3-5, B 0-1, bal Ni.
For cutting tools, dies; cast nonferrous.

HIGH STRENGTH
Manufacturer not listed.
Zn 18.5-21.5, Fe 2-3, Al 6-7, Ni 1.5-2.25, bal Cu.
For high strength corrosion resistant parts; corrosion resistant.

HIGH STRENGTH ALUMINUM BRONZE
Criterion Metals, Inc.
Copper. Cu 73.3, Zn 22.7, Ni 0.6, Al 0-3.4.
Thin gauge sheet, various tempers: 77-130 ksi TS min; 44-117 ksi YS min. ASTM B-592. C69000

HIGH STRENGTH COMMERCIAL BRONZE
Anaconda Co.
Copper. Zn 7.65, Pb 1.75, P 0.1, Ni 1, bal Cu.
Hard: 68,000 TS; 58,000 YS; 10 El; 75 Rock B. For cable clamps, pole line hardware, screw machine products. High strength, corrosion resistance. Electrical conductivity: 32.

HIGH TEMPERATURE BRONZE
American manufacture
Zn 6.3, Sn 2.7, bal Cu.
For fittings, hardware; high strength.

HIGH TENSILE ALLOY
National Steel Corp.
C 0.4-0.5, bal Fe.
For gears, shafts; water hardening. *Obsolete*

HIGH TENSILE BRASS NO. 1
Knowsley Cast Metal Co., Ltd.
Zn, Sn, bal Cu.
Cast: 67,200 TS; 20 El; 120-165 Brin. For high strength castings; corrosion resistant.

HIGH TENSILE BRASS NO. 2
Knowsley Cast Metal Co., Ltd.
Zn, Sn, bal Cu.
Cast: 85,200 TS; 15 El; 170-220 Brin. For heavy castings; high strength.

HIGH TENSILE BRASS NO. 3
Knowsley Cast Metal Co., Ltd.
Zn, Sn, bal Cu.
Cast: 107,500 TS; 12 El; 180-229 Brin. For machine tool parts; high strength.

HIGH TENSILE BRONZE
American manufacture
Cu 68.5, Al 6.5, Mn 2.2, Fe 2.5.
Forged: 101,000-110,000 TS; 50,000-57,000 YS; 10-5 El; 200-220 Brin. For propellers, nuts, bolts; corrosion resistant.

HIGH TEST
International Nickel Inc.
Cast iron. TC 2.75-3.15, Ni 1-1.25, Mn 0.6-1, Si 0.9-1.1, bal Fe.
For brake drums, valve bodies; cast iron. *Obsolete*

HIGH TIN COMMERCIAL BRONZE
Now TIN BRASS 2% TIN.

HIGH YIELD
Now RODNEY 270 & 290.

HIGH YIELD
National Intergroup Inc.
C, bal Fe.
For transportation equipment, buses, bridges.

HIGH YIELD 70/30
Olin Brass, Indianapolis
Copper. Ni 31, Fe 0.5, Mn 0.75, Cu 67.75.
Cold drawn: 72,000 psi TS; 50,000 psi YS; 15 El; 125 Brin. For condenser and heat exchanger tubes and feed-water heaters; high strength; corrosion resistant.

HIGH-STRENGTH GRAY IRON
Frank Foundries Corp.
C 2.8-3.4, Si 1.5-2.4, Mn 0.6-1.25, Ni 0.5-1.5, Cr 0.2-0.4, bal Fe.

HIGHTENSILE BRONZE NO. 1
Janney Cylinder Co.
Cu 60-70, Zn 21-25, Al 3-7, Mn 3-7, Fe 2-5.
Cast: 120,000 TS; 55,000 YS; 15 El; 220 Brin. For pump liners, sleeves, bushings, gears, pinions; pressure tight castings.

HIGHTENSILE BRONZE NO. 2
Janney Cylinder Co.
Cu 60-70, Zn 21-25, Al 3-7, Mn 3-7, Fe 2-5.
Cast: 100,000 TS; 50,000 YS; 20 El; 200 Brin. For pump liners, sleeves, bushings; pressure tight castings.

HIGHTENSILE BRONZE NO. 3
Janney Cylinder Co.
Cu 60-70, Zn 21-25, Mn 3-7, Fe 2-5.
Cast: 90,000 TS; 45,000 YS; 25 El; 180 Brin. For pump liners, sleeves, bushings; centrifugal castings.

HIGHWAY
Apollo Steel Co.
C, Cu, bal Fe.
For building construction.

HILANIC
Gilby-Fodor S.A.
Ni 79, Co 18, Si 1, Fe 2.
Annealed: 85,000 TS; 60,000 YS; 35 El. For cathodes and filaments in electronic tubes. Heat resistant.

HILLS-MCCANNA NO. 1
Hills-McCanna Co.
Copper. Cu 89, Sn 11.
35,000-40,000 TS; 22,000-25,000 YS; 6-10 El; 7-9 RA; 75-85 Brin. For gears, worm wheels; known as "Stones English Worm Gear Bronze."

HILLS-MCCANNA NO. 10
Hills-McCanna Co.
Copper. Cu 80, Sn 10, Pb 10.
28,000-33,000 TS; 19,000-22,000 YS; 5-10 El; 6-11 RA; 52-60 Brin. For bearings for heavy duty; resists shock and vibration.

HILLS-MCCANNA NO. 102
Hills-McCanna Co.
Copper. Cu 88, Sn 10, Zn 2.
Cast: 48,000 TS; 24,000 YS; 42 El; 35 RA; 85 Brin. For castings; SAE-62.

HILLS-MCCANNA NO. 102 P
Hills-McCanna Co.
Copper. Si 2, Fe 1, Special hardener, bal Cu.
Cast: 52,000 TS; 26,000 YS; 17 El; 19 RA; 119 Brin. For castings; Si bronze.

HILLS-MCCANNA NO. 105
Hills-McCanna Co.
Copper. Cu 85.75, Sn 10, Pb 2.5, Ni 1.75.
Cast: 44,000 TS; 33,500 YS; 12 El; 15 RA; 96 Brin. For gears.

HILLS-MCCANNA NO. 11
Hills-McCanna Co.
Copper. Cu 88, Sn 10, Pb 2.
30,000-40,000 TS; 19,000-23,000 YS; 15-25 El; 12-20 RA; 65-70 Brin. For centrifugal pump parts, mine pump bodies, rods, gears; C.E.L. 15000.

HILLS-MCCANNA NO. 110
Hills-McCanna Co.
Copper. Cu 80, Sn 10, Pb 10.
Cast: 45,000 TS; 25,000 YS; 32 El; 24 RA; 64 Brin. For bearings; SAE-64.

HILLS-MCCANNA NO. 111
Hills-McCanna Co.
Copper. Cu 88, Sn 10, Pb 2.
Cast: 49,000 TS; 24,000 YS; 49 El; 36 RA; 81 Brin. For bearings; SAE-63.

HILLS-MCCANNA NO. 2
Hills-McCanna Co.
Copper. Cu 88, Sn 10, Zn 2.
32,000-45,000 TS; 19,000-23,000 YS; 15-30 El; 12-25 RA; 65-75 Brin. For gears, superheated steam and hydraulic castings, bearings; known as "Gun Metal" or "G-Bronze."

HILLS-MCCANNA NO. 20
Hills-McCanna Co.
Copper. Cu 85, Sn 5, Pb 5, Zn 5.
27,000-33,000 TS; 15,000-19,000 YS; 16-20 El; 15-20 RA; 50-60 Brin. For pump bodies, valves, bearings; known as "Ounce Metal" or "Red Composition."

HILLS-MCCANNA NO. 20 P
Hills-McCanna Co.
Copper. Si 2, Fe 0.5, bal Cu.
Cast: 40,000 TS; 19,000 YS; 35 El; 30 RA; 70 Brin. For bronze castings; corrosion resistant.

HILLS-MCCANNA NO. 21
Hills-McCanna Co.
Copper. Cu 82, Sn 4, Pb 6, Zn 8.
28,000-33,000 TS; 14,000-16,000 YS; 15-20 El; 20-26 RA; 55-60 Brin. For general use, bearings; C.E.L. 10000.

HILLS-MCCANNA NO. 22
Hills-McCanna Co.
Copper. Cu 65, Pb 2, Zn 30, 1 special element.
Cast: 30,000-35,000 TS; 25-35 El; 20-30 RA; 40-50 Brin. For plumbing fixtures; SAE-41.

HILLS-MCCANNA NO. 25
Hills-McCanna Co.
Copper. Cu 99.6-99.9.
17,000-20,000 TS; 6000-9000 YS; 50-40 El; 70-60 RA; 30-40 Brin. For electrical installations, parts of electric welding machines; C.E.L. 4000.

HILLS-MCCANNA NO. 30
Hills-McCanna Co.
Copper. Cu 56, Zn 41, Fe 1, Al 1, Mn 0.5, Sn 0.5.
65,000-85,000 TS; 26,000-33,000 YS; 20-35 El; 18-30 RA; 104-119 Brin. For propeller blades, hubs, valve stems, engine frames, bearings, gears; C.E.L. 28000.

HILLS-MCCANNA NO. 30 A
Hills-McCanna Co.
Copper. Zn 39, Mn 0.5, Al 1.5, bal Cu.
Cast: 57,000-62,000 TS; 32,000-35,000 YS; 24-35 El; 27-30 RA; 96-99 Brin. For bronze castings; Mn bronze.

HILLS-MCCANNA NO. 30 B
Hills-McCanna Co.
Copper. Zn 40, Al 1, Fe 1.5, bal Cu.
Cast: 71,000-80,000 TS; 30,000-35,000 YS; 25-30 El; 25-30 RA; 130-135 Brin. For bronze castings; SAE-43.

HILLS-MCCANNA NO. 30 C
Hills-McCanna Co.
Copper. Al 1, Mn 0.5, Zn 3, bal Cu.
Cast: 115,000-119,000 TS; 99,000-102,000 YS; 9-9.5 El;
8.4-8.6 RA; 220-235 Brin. For Mn bronze castings; corrosion
resistant.

HILLS-MCCANNA NO. 35
Hills-McCanna Co.
Aluminum. Al 92, Cu 8.
16,000-22,000 TS; 11,000-13,000 YS; 1-2 El; 1-2 RA; 50-55
Brin. For crankcases, housings, automobile castings; C.E.L.
10000.

HILLS-MCCANNA NO. 39
Hills-McCanna Co.
Copper. Si 3, Fe 1, Special hardener, bal Cu.
Cast: 57,000 TS; 26,000 YS; 27 El; 27 RA; 100 Brin. For
castings; same as "Everdur D."

HILLS-MCCANNA NO. 40
Hills-McCanna Co.
Copper. Cu 90, Al 10.
65,000-80,000 TS; 23,000-28,000 YS; 20-30 El; 21-29 RA. For
structural parts; non-magnetic; wear resistant.

HILLS-MCCANNA NO. 41
Hills-McCanna Co.
Copper. Cu 77, Al 11, Fe 5, Mn 7.
80,000-97,800 TS; 26,000-28,200 YS; 10-30 El; 12-29 RA; 190
Brin. For structural parts; wear resistant; C.E.L. 28300.

HILLS-MCCANNA NO. 42
Hills-McCanna Co.
Copper. Cu 88, Al 8, Fe 4.
65,000-80,000 TS; 23,000-28,000 YS; 20-30 El; 21-29 RA;
92-100 Brin. For gears, bushings, bolts; wear resistant; C.E.L.
19000.

HILLS-MCCANNA NO. 43
Hills-McCanna Co.
Copper. Al 10, Fe 1, Special hardener, bal Cu.
Cast: 84,000 TS; 33,000 YS; 22 El; 25 RA; 134 Brin. For
bearings, bushings; Al bronze.

HILLS-MCCANNA NO. 45
Hills-McCanna Co.
Copper. Cu 89, Al 10, Fe 1.
65,000-80,000 TS; 23,000-28,000 YS; 20-30 El; 21-29 RA;
92-100 Brin. For equipment in oil refineries; wear and
corrosion resistant; C.E.L. 19000.

HILLS-MCCANNA NO. 46
Hills-McCanna Co.
Copper. Cu 81, Al 11, Fe 5, Mn 3.
80,000-94,300 TS; 26,400-28,400 YS; 10-30 El; 11-29 RA; 176
Brin. For packing tools in the explosive industry; wear
resistant; C.E.L. 26100.

HILLS-MCCANNA NO. 47
Hills-McCanna Co.
Copper. Cu 80, Al 12, Fe 8.
65,000-80,000 TS; 23,000-28,000 YS; 20-30 El; 21-29 RA. For
bearings, gears to resist heavy loads and shocks; wear
resistant; C.E.L. 22100.

HILLS-MCCANNA NO. 48
Hills-McCanna Co.
Copper. Al 9, Special hardener, bal Cu.
Cast: 72,000 TS; 55,000 YS; 3 El; 3 RA; 229 Brin. For
construction parts; Al bronze.

HILLS-MCCANNA NO. 49
Hills-McCanna Co.
Copper. Al 10, Fe 4, Special hardener, bal Cu.
Cast: 55,000 TS; 55,000 YS; 0 El; 0 RA; 300 Brin. For dies for
forming tools; Al bronze.

HILLS-MCCANNA NO. 50
Hills-McCanna Co.
Copper. Cu 52, Ni 31, Fe 12, Cr 5.
75,000-92,300 TS; 26,400-42,000 YS; 4.5-8.5 El; 7.8-8.5 RA;
190-201 Brin. For hydraulic valve seats; corrosion resistant;
C.E.L. 33040.

HILLS-MCCANNA NO. 50A
Hills-McCanna Co.
Copper. Cu 60.5, Ni 23.5, Sn 2, Pb 3, Zn 11.
38,000-45,000 TS; 19,000-22,000 YS; 13-21 El; 15-20 RA;
85-90 Brin. For vessels to resist H_2SO_4; corrosion resistant.

HILLS-MCCANNA NO. 52
Hills-McCanna Co.
Copper. Cu 74.5, Ni 16.5, Al 9.
85,000 TS; 36,000 YS; 4 El; 3.5 RA; 213 Brin. For vessels to
resist hot oils; corrosion resistant; C.E.L. 29610.

HILLS-MCCANNA NO. 53
Hills-McCanna Co.
Copper. Cu 80, Al 10, Ni 5, Fe 5.
Untreated: 95,000 TS; 44,000 YS; 9 El; 18 RA; 170 Brin. Heat
treated: 110,000 TS; 77,000 YS; 3 El; 9 RA; 235 Brin. For
spark proof tools; corrosion resistant.

HILLS-MCCANNA NO. 54
Hills-McCanna Co.
Nickel. Ni 60-65, Cu 23-28, Fe 3.5, Mn 2, Si 0.75, C 0.25.
65,000-70,000 TS; 35,000-40,000 YS; 25-30 El; 100 Brin. For
chemical apparatus; corrosion resistant; C.E.L. 29610.

HILLS-MCCANNA NO. 55
Hills-McCanna Co.
Nickel. Ni 65-67, Cu 30-32, Si 2.5-3.5, Fe 2.5-3.5.
75,000-80,000 TS; 44,000-48,000 YS; 10-10.5 El; 13-16 RA;
170 Brin. For still plugs in oil industries; corrosion resistant.

HILO
Gilby-Fodor S.A.
Ni 75, Co 18, Ti 2, Fe 5.
Annealed: 105,000 TS; 75,000 YS; 30 El. For cathodes and
filaments in electronic tubes. Heat resistant.

HILO
Wilbur B. Driver Co.
Nickel. Ni 75, Co 18, Ti 2, bal Fe.
Annealed: 90,000-106,000 TS; 20-30 El. For radio tube
filaments; magnetic.

HILO, MODIFIED
Wilbur B. Driver Co.
Ni 79, Co 18, Fe 2.
Annealed: 82,000 TS. For vacuum tube filaments; heat and
corrosion resistant. *Obsolete*

HINGE
English manufacture
Cu 62-63.5, Pb 1-2, bal Zn.
Soft: 50,000 TS; 50 El; 65 Brin. Hard: 85,000 TS; 0 El; 180
Brin. Engraving brass.

HIOLOY "CU"
Teledyne Ohiocast
C, alloy, bal Fe.
For corrosion resistant parts; corrosion resisting. *Obsolete*

HIOLOY 0-3
Teledyne Ohiocast
C, Ni, Cr, bal Fe.
For castings for oil refinery equipment. *Obsolete*

HIOLOY 0-6
Teledyne Ohiocast
C 0.55-0.75, Si 0.3-0.55, Cr 0.8-1.2, Mn 0.5-0.8, bal Fe.
For liner plates, screen plates, cement mills; wear resistant;
formerly "Hioloy W." *Obsolete*

HIOLOY-0-2
Teledyne Ohiocast
C 0.2-0.35, Si 0.3-0.55, Mn 0.6-0.9, Mo 0.4-0.6, bal Fe.
85,000 TS; 55,000 YS; 20 El; 30 RA. For oil refinery
equipment; water hardened. *Obsolete*

HIOLOY-0-4
Teledyne Ohiocast
C 0.15-0.35, Si 0.3-0.55, Mn 0.4-0.7, Cr 4-6, Mo 0.4-0.65, bal
Fe.
110,000 TS; 75,000 YS; 16 El; 30 RA. For oil refinery
equipment; corrosion resistant. *Obsolete*

HIOLOY-0-7
Teledyne Ohiocast
C 0.2-0.35, Si 0.5-0.8, Ni 1.75-2.25, Cr 0.6-0.9, Mo 0.2-0.3, bal
Fe.
100,000 TS; 65,000 YS; 18 El; 30 RA. For oil refinery
equipment; oil hardened. *Obsolete*

HIPERCO
Westinghouse Electric Corp.
Co 35, Cr 1, bal Fe.
Annealed: 25,000 TS; 20,000 YS; 1-2 El. Unannealed: 80,000
TS; 60,000 YS; 20 El. For core material, magnets for electric
motors and generators; magnetic alloy of high permeability.
Obsolete

HIPERCO 27
Westinghouse Electric Corp.
Co 27, Cr 0-0.5, Ni 0-0.3, Mn 0-0.3, C 0-0.03, bal Fe.
Annealed: 80,000 TS; 45,000 YS; 20 El; 48.8 RA; 150 Brin.
For aircraft generators, current transformers, rectifiers; high
magnetic saturation. *Obsolete*

HIPERCO 35
Westinghouse Electric Corp.
Co 35, Mn 0-0.3, Cr 0-0.5, Ni 0-0.3, C 0-0.03, bal Fe.
Annealed: 36,000 TS; 36,000 YS; 190 Brin. For servo
mechanism, torque motors, magnetic clutches; high
magnetic saturation. *Obsolete*

HIPERCO 50-FM
Now CARPENTER HIPERCO 50-FM.

HIPERCO-50
Westinghouse Electric Corp.
V 2, Co 49, Fe 49.
Cold rolled: 195,000 TS; 185,000 YS; 1 El; C 35 Rock.
Annealed: 80,000 TS; 48,000 YS; 0.5 El; B 97 Rock. For
applications requiring high magnetic inductions at low fields
of 10-50 oersteds, diaphragms, motors, generators, relays.
High permeability at high induction. *Obsolete*

HIPERNIK
Westinghouse Electric Corp.
Ni 50, Fe 50, traces Mn.
Annealed: 55,000 TS; 20,000 YS; 50 El; 50 RA. For core
material for radio and current transformers, motors, relays;
soft and ductile magnetic alloy. *Obsolete*

HIPERNIK V
Westinghouse Electric Corp.
Ni 50, Fe 50, Trace Mn.
Annealed: 55,000 TS; 20,000 YS; 33 El. For magnetic core
material for transductor applications; magnetic amplifiers,
saturated reactors. *Obsolete*

HIPERNOM V
Westinghouse Electric Corp.
Ni 79, Mo 4, bal Fe.
40,000 to 50,000 permeability in the 40 gauss range. 250,000
to 500,000 permeability at 3,500 gauss. For sensitive
saturable core devices where extreme sensitivity is required.
Audio transformers, relays, amplifiers. Processed to exhibit
square loop properties. High permeability. *Obsolete*

HIPERSIL
Westinghouse Electric Corp.
Si 3-4, bal Fe.
Cast: 50,000 psi TS; 45,000 psi YS; 6 El. For transformer cores; high permeability. *Obsolete*

HIPROOF 304
British Steel plc
C 0-0.06, Cr 17.5-19, Ni 8-11, N 0-0.25, bal Fe.
High proof stress version of 304. For cryogenic, storage, and pressure vessels. Similar to AISI 304.

HIPROOF 316
British Steel plc
C 0-0.07, Cr 16.5-18.5, Ni 10-13, Mo 2.25-3, N 0-0.25, bal Fe.
High proof stress version of 316. For cryogenic storage and pressure vessels.

HIPROOF 347
British Steel plc
C 0-0.08, Cr 17-19, Ni 9-12, N 0.15-0.25, Nb = 10 x C min, bal Fe.
High proof strength version of 347.

HIROX
Westinghouse Electric Corp.
Al 6-10, Cr 3-9, Mn 0-4, B, Zr, bal Fe.
At 70°F: 118,850 psi TS; 111,650 psi YS; 15 El; 42 RA. At 1300°F: 15,000 psi TS; 14,950 psi YS; 94 El; 98 RA. For resistances, heating elements; high electrical resistivity and oxidation resistance to 2300°F. *Obsolete*

HITEM IRON NO. 5
Bethlehem Foundry & Machine Co.
C 3.2, Si 2, Mn 1, bal Fe.
For chemical plant apparatus; corrosion resistant. *Obsolete*

HITEM IRON NO. 7
Bethlehem Foundry & Machine Co.
C 3.2, Si 2, Mn 0.6, bal Fe.
For dye and paint equipment, retorts; corrosion and heat resistant; (Hi-Tem). *Obsolete*

HITEN-SPEED 45
British Rolling Mills Ltd.
C 0.33-0.41, Mn 1.3-1.6, bal Fe.
For gears, bolts, crankshafts, axles; oil hardened, tough.

HITEN-SPEED 55
British Rolling Mills Ltd.
C 0.3-0.41, Mn 1.3-1.9, Mo 0.25-0.34, bal Fe.
For gears, bolts, crankshafts, axles; oil hardened, tough.

HITENSILOY BRONZE
Cerro Wire & Cable Company
Cu 56.75, Ni 1.5, Pb 1, Al 0.5, bal Zn.
Annealed: 80,000 TS; 40,000 YS; 30 El; 25 RA; 125 Brin. Cold drawn: 95,000 TS; 50,000 YS; 20 El; 15 RA; 130 Brin. For welding rod, pump rods, valve stems, bolts, nuts; corrosion resisting. *Obsolete*

HITENSILOY NO. 11
Cerro Metal Products Co.
Cu 57, Pb 1, Ni 1.75, bal Zn.
Rolled: 80,000 TS; 45,000 YS; 165 Brin. For pump rods, valve stems, bolts, nuts, and die castings. Nickel brass. *Obsolete*

HITENSO 162
Anaconda Co.
Copper. Cu 99, Cd 1.
Soft rod: 37,000 TS; 12,000 YS; 50 El; 47 Rock F. Hard rod: 58,000 TS; 50,000 YS; 15 El; 65 Rock B. For electrical equipment, trolley wire, contact shoes. High electrical conductivity, wear resistant. Copper Alloy No. 162. Formerly Hitenso 961.

HITENSO 165
Anaconda Co.
Copper. Cd 0.8, Si 0.02, Sn 0.6, bal Cu.
Hard: 65,000 TS; 55,000 YS; 15 El; 75 Rock B. Soft: 40,000 TS; 14,000 YS; 55 El; 6 Rock B. Wire: 95,000 TS; 1.5 El. For electrical equipment, trolley wire, contact shoes, electrical welding tips and wheels. 58% electrical conductivity. High strength. Copper Alloy No. 165.

HITEST
Medart Engineering & Equipment Co.
C 3, Si, Mn, bal Fe.
For gears, shafts, housings, machinery parts; cast iron. *Obsolete*

HIZUTIT 700
Ekatit Stahl GmbH
C 0-0.1, Si 1.5-1.8, Mn 0-1, Cr 1.5-2, bal Fe.
Heat resisting steel. Werkstoff Nr. 1.4700.

HIZUTIT 701
Ekatit Stahl GmbH
C 0-0.12, Si 2-2.5, Mn 0-1, Cr 5.5-6.5, bal Fe.
Heat resisting steel. Werkstoff Nr. 1.4712.

HIZUTIT 702
Ekatit Stahl GmbH
C 0-0.12, Si 0.5-1, Mn 0-1, Cr 6-7, Al 1, bal Fe.
Heat resisting steel. Werkstoff Nr. 1.4713.

HIZUTIT 703
Ekatit Stahl GmbH
C 0-0.12, Si 1.9-2.4, Mn 0-1, Cr 12-14, bal Fe.
Heat resisting steel. Werkstoff Nr. 1.4722.

HIZUTIT 704
Ekatit Stahl GmbH
C 0-0.12, Si 0.9-1.4, Mn 0-1, Cr 12-14, Al 1.2, bal Fe.
Heat resisting steel. Werkstoff Nr. 1.4724.

HIZUTIT 705
Ekatit Stahl GmbH
C 0-0.12, Si 1.9-2.4, Mn 0-1, Cr 17-19, bal Fe.
Heat resisting steel. Werkstoff Nr. 1.4741.

HIZUTIT 706
Ekatit Stahl GmbH
C 0-0.12, Si 0.7-1.2, Mn 0-1, Cr 17-19, Al 1.2, bal Fe.
Heat resisting steel. Werkstoff Nr. 1.4742.

HIZUTIT 707
Ekatit Stahl GmbH
C 0-0.12, Si 1.2-1.5, Mn 0-1, Cr 23-25, Al 1.7, bal Fe.
Heat resisting steel. Werkstoff Nr. 1.4762.

HIZUTIT 709
Ekatit Stahl GmbH
C 0.15-0.25, Si 0.8-1.3, Mn 0-2, Cr 24-26, Ni 3.3-5.5, bal Fe.
Heat resisting steel. Werkstoff Nr. 1.4821.

HIZUTIT 710
Ekatit Stahl GmbH
Si 1.8-2.3, Mn 0-2, Cr 19-21, Ni 11-13, C 0-0.2, bal Fe
Heat resisting steel. Werkstoff Nr. 1.4828.

HIZUTIT 711
Ekatit Stahl GmbH
C 0-0.2, Si 1.8-2.3, Mn 0-2, Cr 24-26, Ni 19-21, bal Fe.
Heat resisting steel. Werkstoff Nr. 1.4841.

HIZUTIT 712
Ekatit Stahl GmbH
C 0-0.15, Si 0-0.75, Mn 0-2, Cr 24-26, Ni 19-22, bal Fe.
Heat resisting steel. Werkstoff Nr. 1.4845.

HIZUTIT 713
Ekatit Stahl GmbH
C 0-0.15, Si 1.5-2, Mn 0-2, Cr 15-17, Ni 34-37, bal Fe.
Heat resisting steel. Werkstoff Nr. 1.4864.

HIZUTIT 714
Ekatit Stahl GmbH
C 0.15, Si 0-1, Mn 0-2, Cr 17-19, Ni 9-11, Ti 0.7, bal Fe.
Heat resisting steel. Werkstoff Nr. 1.4878.

HIZUTIT G 1
Ekatit Stahl GmbH
C 0.2-0.4, Si 1-2.5, Mn 0-1, Cr 6-8, bal Fe.
Heatproof steel castings. Werkstoff Nr. 1.4710.

HIZUTIT G 10
Ekatit Stahl GmbH
C 0.2-0.5, Si 1-2.5, Mn 0-1.5, Cr 25-28, Ni 13-16, bal Fe.
Heatproof steel castings. Werkstoff Nr. 1.4846.

HIZUTIT G 11
Ekatit Stahl GmbH
C 0.2-0.5, Si 1-2.5, Mn 0-1.5, Cr 24-27, Ni 19-21, bal Fe.
Heatproof steel castings. Werkstoff Nr. 1.4848.

HIZUTIT G 12
Ekatit Stahl GmbH
C 0.2-0.5, Si 1-2.5, Mn 0-1.5, Cr 16-19, Ni 38-39, bal Fe.
Heatproof steel castings. Werkstoff Nr. 1.4865.

HIZUTIT G 13
Ekatit Stahl GmbH
C 1.4-1.8, Si 1-2.5, Mn 0-1, Cr 17-19, bal Fe.
Heatproof steel castings. Werkstoff Nr. 1.4743.

HIZUTIT G 14
Ekatit Stahl GmbH
C 0.2-0.5, Si 1-2.5, Mn 0-1.5, Cr 24-26, Ni 11-14, bal Fe.
Heatproof steel castings. Werkstoff Nr. 1.4837.

HIZUTIT G 15
Ekatit Stahl GmbH
C 0.15-0.35, Si 1-2.5, Mn 0-1.5, Cr 17-19, Ni 8-10, bal Fe.
Heatproof steel castings. Werkstoff Nr. 1.4825.

HIZUTIT G 16
Ekatit Stahl GmbH
C 0.1-0.2, Si 1-2, Mn 0-1.5, Cr 24-27, Ni 19-21, bal Fe.
Heatproof steel castings. Werkstoff Nr. 1.4849.

HIZUTIT G 2
Ekatit Stahl GmbH
C 0.3-0.6, Si 1-2.5, Mn 0-1, Cr 12-14, bal Fe.
Heatproof steel castings. Werkstoff Nr. 1.4729.

HIZUTIT G 4
Ekatit Stahl GmbH
C 0.3-0.6, Si 1-2.5, Mn 0-1, Cr 22-24, bal Fe.
Heatproof steel castings. Werkstoff Nr. 1.4745.

HIZUTIT G 5
Ekatit Stahl GmbH
C 0.3-0.6, Si 1-2.5, Mn 0-1, Cr 27-30, bal Fe.
Heatproof steel castings. Werkstoff Nr. 1.4776.

HIZUTIT G 6
Ekatit Stahl GmbH
C 1.2-1.4, Si 1-2.5, Mn 0-1, Cr 27-30, bal Fe.
Heatproof steel castings. Werkstoff Nr. 1.4777.

HIZUTIT G 7
Ekatit Stahl GmbH
C 0.3-0.5, Si 1-2, Mn 0-1, Cr 26-28, Ni 3.5-5.5, bal Fe.
Heatproof steel castings. Werkstoff Nr. 1.4823.

HIZUTIT G 8
Ekatit Stahl GmbH
C 0.3-0.5, Si 1-2.5, Mn 0-1.5, Cr 21-23, Ni 9-11, bal Fe.
Heatproof steel castings. Werkstoff Nr. 1.4826.

HIZUTIT G 9
Ekatit Stahl GmbH
C 0.15-0.35, Si 1-2.5, Mn 0-1.5, Cr 18-21, Ni 13-15, bal Fe.
Heatproof steel castings. Werkstoff Nr. 1.4832.

HK 4 M

Sumitomo Metal America Inc.
C 0.2-0.3, Si 0-0.75, Mn 0-1.5, Cr 24-26, Ni 24-26, Ti 0.2-0.6, Al 0.2-0.6, bal Fe.
75,400 TS min; 34,100 YS min; 25 El min. For high temperature service on petrochemical plants, and as reformer tubes and cracking tubes.

HK-40CB

Certified Alloy Products Inc.
C 0.4, Cr 23, Ni 21, Cb 1.5, bal Fe.
As cast: 86,000 psi TS; 44,000 psi YS; 16 El; 19 RA. Cast high temperature furnace fittings; creep strength and thermal fatigue resistant. *Obsolete*

HK-HITEN 80

American Carbide Corp.
C 0.18, Mn 1, Si 0.15-0.35, B 0.006, Cr 0.8, Cu 0.15-0.5, Mo 0.6, Ni 1, V 0.1, bal Fe.
Plate: 114,000 TS min; 100,000 YS min; 18 El min. For mine cars, bus bodies, trailers, pressure vessels, cranes, bridges. Wear and shock resistant.

HK30

Cooper Alloy Corp.
Stainless steel. C 0.25-0.35, Si 1.75, Mn 1.5, Cr 23-27, Ni 19-22, bal Fe.
Cast: 448 MPa TS; 241 MPa YS; 10 El. Heat resistant alloy; ASTM A-351 HK30.

HK31A

Fansteel/Wellman Dynamics
Magnesium. Zn 0-0.3, Zr 0.5-1, Th 2.5-4, bal Mg.
Magnesium casting alloy. Cast: 27,000 psi TS; 13,000 psi YS; 4 El; 55 Brin.

HK31A

Various foundries
Zr 0.7, Th 3, bal Mg.
T6-temper: 27,000-32,000 TS; 13,000-15,000 YS; 4-8 El. At 700°F: 13,000 TS; 8000 YS; 26 El. Magnesium sand castings with good pressure tightness, weldability and corrosion resistance. For parts operating at 400-700°F. ASTM B80-69; AMS 4445; SAE 507; QQ-M-56.

HK31A

Dow Chemical Co.
Magnesium. Th 3, Zr 0.7, bal Mg.
O-temper: 30,000-33,000 psi TS; 18,000-20,000 psi YS; 12-23 El. H 24-temper: 34,000-39,000 psi TS; 24,000-31,000 psi YS; 4-14 El. Sheet and plate have good formability and weldability; used at 400°F to 600°F in aircraft and missiles. ASTM B90-69; AMS 4384, 4385. *Obsolete*

HK32A

Fansteel/Wellman Dynamics
Magnesium. Zn 1.7-2.5, RE 0-0.1, Zr 0.5-1, Th 2.5-4, bal Mg.
Magnesium casting alloy. Cast: 27,000 psi TS; 13,000 psi YS; 4 El; 57 Brin.

HK40

Cooper Alloy Corp.
Stainless steel. C 0.35-0.45, Si 1.75, Mn 1.5, Cr 23-27, Ni 19-22, bal Fe.
Cast: 431 MPa TS; 241 MPa YS; 10 El. Heat resistant alloy; ASTM A-351 HK40.

HM

Thyssen Edelstahlwerke AG
C, Mn, bal Fe.
For hand chisels; water hardened. *Obsolete*

HM21A

Dow Chemical Co.
Magnesium. Mn 0.6, Th 2, bal Mg.
T8-temper: 33,000-35,000 psi TS; 18,000-23,000 psi YS; 6-11 El; 56 Brin. Sheet and plate have good properties at 400-750°F; good formability and weldability. *Obsolete*

HM31A (EXTRUSIONS)

Dow Chemical Co.
Magnesium. Mn 1.2, Th 3, bal Mg.
T5-temper: 37,000-44,000 psi TS; 26,000-38,000 psi YS; 4-13 El; 63 Brin. Extrusions: good elevated temperature properties, 400-800°F; used in aircraft and missiles. AMS 4388, 4389. *Obsolete*

HMS STEELS

Thyssen Edelstahlwerke AG
C, bal Fe.
For general machine parts; water hardened. *Obsolete*

HMS-CHROME

Cyprus Foote Mineral Co.
Cr 22-24, C 1.4-1.55, Si 29-31, bal Fe.
Ferroalloy for source of chromium and silicon, particularly in stainless steels. *Obsolete*

HNS 40

Creusot-Loire
C 0.18, Mn 1.1, Si 0-0.3, bal Fe.
Normalized (23 mm max): 510-610 N/mm^2 psi TS; 355 N/mm^2 psi YS min. For chains. AFNOR 20 Mn 5.

HNS 50

Creusot-Loire
C 0.18, Mn 1.1, Si 0-0.3, bal Fe.
Quenched and tempered (23 mm max): 785-880 N/mm^2 TS; 685 N/mm^2 min YS; 15 El. For chains. AFNOR 20 Mn 5.

HNS 60

Creusot-Loire
C 0.2, Mn 1.1, Si 0-0.3, Cr 1, bal Fe.
Quenched and tempered (23 mm max): 930-1030 N/mm^2 psi TS; 835 N/mm^2 YS; 16 El. For chains. AFNOR 20 MC 4.

HNS 80

Creusot-Loire
C 0.22, Mn 1.1, Si 0-0.3, Cr 1, Mo 0.2, bal Fe.
Quenched and tempered (23 mm max): 1280-1380 N/mm^2 psi TS; 1175 N/mm^2 YS min. For chains. AFNOR 22 MCD 4.

HO 124

Honsel-Werke Aktiengesellschaft
AlSi12CuNiMg T6.
Chill cast: 180 N/mm^2 TS; 130 Brin. Sand cast: 170 N/mm^2 TS; 85 Brin. Aluminum castings with good elevated temperature strength and wear resistance, as engine pistons.

HOB-A-DIE

Ziv Steel & Wire Co.
C 0.06, Mn 0.3, Si 0.2, Mo 0.25, Cr 1, bal Fe.
For Zn casting dies; resists heat checking.

HOB-A-FORM

Manufacturer not listed.
C 0-0.06, Mn 0.15, Si 0.1, bal Fe.
For plastic mold dies; water hardening, hobbed cavity.

HOBALITE

Agawam Tool Co.
C 0.05, Mn 0.16, bal Fe.
For mold dies; water hardened.

HOBALITE

Darwin & Milner Inc.
Tool material. C, alloy, bal Fe.
For mold dies, hobs; air hardened.

HOBALITE

Ziv Steel & Wire Co.
C 0.05, Mn 0.15, bal Fe.
For plastic mold dies; case hardening steel.

HOBALITE 500A

Ovako Steel Hellefors AB
C 0.05, Si 0.05, Mn 0.1, bal Fe.
For deep hobbing steel for bakelite dies; case hardened. *Obsolete*

HOBALLOY

Colt Industries
C 0-0.1, Ni 1.3, Cr 0.6, bal Fe.
For Bakelite molds; shock resistant. *Obsolete*

HOBART 10

Hobart Welding Products
C 0.07, Mn 0.5, Si 0.25, P 0.016, S 0.02, bal Fe.
Weld metal. As welded: 62,000-72,000 psi TS; 51,000-67,000 psi YS. AWS E-6010 all position coated electrode; DCEP. *Obsolete*

HOBART 101P

Hobart Welding Products
C 0.11, Mn 0.5, Si 0.2, P 0.1, S 0.022, bal Fe.
Weld metal. As welded: 64,000-74,500 psi TS; 53,000-64,000 psi YS; 22 El; 35-60 RA. AWS E-6010 all position electrode; DCEP. *Obsolete*

HOBART 1100

Hobart Welding Products
Cu 0.05-0.2, Mn 0.05, Si 1, Zn 0.1, B 0.0008, Fe, 99.0 Al min.
Solid aluminum welding electrode wire; 25,000 psi TS min. AWS A5.10-69 ER-1100. *Obsolete*

HOBART 1139

Hobart Welding Products
C 0.06, Mn 0.33, Si 0.31, P 0.02, S 0.022, bal Fe.
Weld metal. As welded: 60,000 psi TS min. Designed for spot welding of roof decking. AWS E-6022; DCEN or AC.

HOBART 12

Hobart Welding Products
C 0.08, Mn 0.38, Si 0.2, P 0.023, S 0.026, bal Fe.
Weld metal. As welded: 67,000 psi TS; 55,000 psi YS; 17 El; 25-50 RA. AWS E-6012 general purpose electrode; AC or DCEN.

HOBART 12A

Hobart Welding Products
C 0.07, Mn 0.5, Si 0.13, P 0.016, S 0.025, bal Fe.
Weld metal. As welded: 67,000 psi TS; 55,000 psi YS; 17 El. AWS E-6012 electrode; DCEN.

HOBART 13A

Hobart Welding Products
C 0.09, Mn 0.5, Si 0.41, P 0.015, S 0.024, bal Fe.
Weld metal. As welded: 67,000 psi TS; 55,000 psi YS; 17 El. AWS E-6013; AC, DCEN or DCEP. *Obsolete*

HOBART 14A

Hobart Welding Products
C 0.09, Mn 0.46, Si 0.32, P 0.018, S 0.022, Ni 0.039, Cr 0.054, Mo 0.02, V 0.023, bal Fe.
Weld metal. As welded: 72,000 psi TS; 60,000 psi YS; 17 El. AWS E-7014; DCEN or AC. *Obsolete*

HOBART 16

Hobart Welding Products
C 0.08, Mn 0.8, Si 0.35, P 0.015, S 0.02, bal Fe.
Weld metal. As welded: 72,000 psi TS; 60,000 psi YS; 22 El. AWS E-7016; DCEP or AC. *Obsolete*

HOBART 212A

Hobart Welding Products
C 0.06, Mn 0.25, Si 0.15, P 0.018, S 0.019, bal Fe.
Weld metal. As welded: 67,000 psi TS; 55,000 psi YS; 17 El. AWS E-6012; DCEN or AC. *Obsolete*

HOBART 24

Hobart Welding Products
C 0.04, Mn 0.6, Si 0.5, P 0.035, S 0.035, Ni 0.052, Cr 0.055, Mo 0.022, V 0.018, bal Fe.
Weld metal. As welded: 72,000 psi TS; 60,000 psi YS; 17 El. AWS E-7024; AC or DCEN.

HOBART 24H

Hobart Welding Products
C 0.084, Mn 0.76, Si 0.61, P 0.018, S 0.01, bal Fe.
Weld metal. As welded: 83,500 psi TS; 71,000 psi YS; 25 El.
AWS E-7024. Electrode for shielded metal arc welding.
Obsolete

HOBART 25 P

Hobart Welding Products
C 0.11, Mn 1.12, Si 0.5, bal Fe.
Solid wire consumable electrode for electroslag welding of
mild steel. *Obsolete*

HOBART 27

Hobart Welding Products
C 0.08, Mn 0.62, Si 0.17, P 0.01, S 0.015, bal Fe.
Weld metal. As welded: 62,000 psi TS; 50,000 psi YS; 25 El.
AWS E-6027; AC or DCEN. *Obsolete*

HOBART 27H

Hobart Welding Products
C 0.08, Mn 1, Si 0.42, P 0.017, S 0.016, bal Fe.
Weld metal. As welded: 71,000 psi TS; 61,000 psi YS; 28 El.
AWS E-6027. Electrode for shielded metal arc welding.
Obsolete

HOBART 308-15 & 308-16

Hobart Welding Products
C 0-0.07, Cr 19, Ni 9.5, Mn 1.6, Si 0.5, bal Fe.
Weld metal. As welded: 85,000-95,000 psi TS; 40-50 El. For
welding AISI 301, 302, 304, and 308 stainless. *Obsolete*

HOBART 308L

Hobart Welding Products
C 0-0.25, Mn 1.8, Si 0.4, Cr 20.5, Ni 10, bal Fe.
80,000 psi TS; 40 El. Solid stainless wire. For welding types
304, 308, 321, and 347 steels. AWS A5.9-69 ER 308L.

HOBART 308L HISIL

Hobart Welding Products
C 0-0.25, Mn 1.8, Si 0.85, Cr 20.5, Ni 10, bal Fe.
80,000 psi TS; 40 El. Solid stainless steel welding electrode
wire. ER-308L Si.

HOBART 308L-15 & 308L-16

Hobart Welding Products
C 0-0.04, Cr 19, Ni 9.5, Mn 1, Si 0.3, bal Fe.
Weld metal. As welded: 80,000-90,000 psi TS; 40-50 El. For
welding AISI 304 ELC stainless steel. *Obsolete*

HOBART 309

Hobart Welding Products
C 0-0.25, Mn 1.8, Si 0.4, Cr 24, Ni 13.5, bal Fe.
80,000 psi TS; 38 El. Solid stainless steel welding electrode
wire. For welding 309 steels; ER-309.

HOBART 309-15 & 309-16

Hobart Welding Products
C 0-0.1, Cr 23, Ni 13, Mn 1.6, Si 0.5, bal Fe.
Weld metal. As welded: 85,000-95,000 psi TS; 34-45 El. For
welding AISI 309 stainless steel. *Obsolete*

HOBART 309CB-15 & 309CB-16

Hobart Welding Products
C 0-0.1, Cr 23, Ni 13, Mn 1.6, Cb 0.8, Si 0.6, bal Fe.
Weld metal. As welded: 85,000-95,000 psi TS; 30-40 El. For
welding AISI 347 and 321 stainless clad steels. *Obsolete*

HOBART 309MO-15

Hobart Welding Products
C 0-0.1, Cr 23, Ni 13, Mn 1.7, Mo 2.2, Si 0.5, bal Fe.
Weld metal. As welded: 85,000-95,000 psi TS; 35-45 El. For
welding AISI 316 stainless clad steel. *Obsolete*

HOBART 310

Hobart Welding Products
C 0.12, Mn 1.8, Si 0.45, Cr 26, Ni 21, bal Fe.
80,000 psi TS; 30 El. Solid stainless steel welding electrode
wire. For welding AISI 310 stainless; ER-310.

HOBART 310-15 & 310-16

Hobart Welding Products
C 0-0.2, Cr 26, Ni 21, Mn 1.8, Si 0.4, bal Fe.
Weld metal. As welded: 85,000-95,000 psi TS; 35-45 El. For
welding AISI 410, 430, and 502 stainless steels. *Obsolete*

HOBART 310CB-15 & 310CB-16

Hobart Welding Products
C 0-0.12, Cr 26, Ni 21, Mn 1.8, Cb 0.8, Si 0.4, bal Fe.
Weld metal. As welded: 85,000-95,000 psi TS; 30-40 El. For
welding AISI 310, 321, and 347 stainless steels. *Obsolete*

HOBART 310MO-15 & 310MO-16

Hobart Welding Products
C 0-0.12, Cr 26, Ni 21, Mn 1.8, Mo 2, Si 0.4, bal Fe.
Weld metal. As welded: 85,000-95,000 psi TS; 35-45 El. For
welding AISI 310 and 316 stainless steels. *Obsolete*

HOBART 312-15 & 312-16

Hobart Welding Products
C 0-0.15, Cr 29, Ni 9.5, Mn 1.9, Si 0.5, bal Fe.
Weld metal. As welded: 110,000-120,000 psi TS;
80,000-90,000 psi YS; 22-75 El. Highest "as-welded" strength
of any austenitic stainless weld. *Obsolete*

HOBART 316-15 & 316-16

Hobart Welding Products
C 0-0.07, Cr 18, Ni 13, Mo 2.25, Mn 1.7, Si 0.4, bal Fe.
Weld metal. As welded: 85,000-95,000 psi TS; 35-45 El. For
welding AISI 316 stainless steel. *Obsolete*

HOBART 316L

Hobart Welding Products
C 0-0.025, Mn 1.8, Si 0.35, Cr 19.5, Ni 13, Mo 2.2, bal Fe.
78,000 psi TS; 30 El. Solid stainless steel welding electrode
wire. For welding AISI 316 and 316L stainless; ER-316L.

HOBART 316L HISIL

Hobart Welding Products
C 0-0.025, Mn 1.8, Si 0.85, Cr 1.5, Ni 13, Mo 2.7, bal Fe.
78,000 psi TS; 30 El. Solid stainless steel welding electrode
wire. For improved welding 316 and 316L steels; ER-316L-Si.

HOBART 316L-15 & 316L-16

Hobart Welding Products
C 0-0.04, Cr 18, Ni 13, Mo 2.25, Mn 1, Si 0.3, bal Fe.
Weld metal. As welded: 80,000-90,000 psi TS; 35-45 El. For
welding AISI 316L stainless steel. *Obsolete*

HOBART 317-15 & 317-16

Hobart Welding Products
C 0-0.07, Cr 19, Ni 13, Mo 3.5, Mn 1.7, Si 0.5, bal Fe.
Weld metal. As welded: 85,000-95,000 psi TS; 35-45 El. For
welding AISI 317 stainless steel. *Obsolete*

HOBART 318-15 & 318-16

Hobart Welding Products
C 0-0.07, Cr 18, Ni 12, Mo 2.25, Mn 1.6, Cb 0.8, Si 0.6, bal
Fe.
Weld metal. As welded: 85,000-95,000 psi TS; 30-40 El. For
welding Type 318 stainless steel. *Obsolete*

HOBART 320CB-15

Hobart Welding Products
C 0-0.07, Cr 20, Ni 29, Cu 3, Mo 2, Cb 0.5, Mn 1.5, Si 0.4, bal
Fe.
Weld metal. As welded: 75,000-85,000 psi TS; 35-40 El. For
welding Carpenter No. 20 or Durimet 20 and similar stainless
steels. *Obsolete*

HOBART 330-15 & 330-16

Hobart Welding Products
C 0-0.25, Cr 15, Ni 35, Mn 1.6, Si 0.3, bal Fe.
Weld metal. As welded: 75,000-85,000 psi TS; 25-35 El. For
welding Type 330 stainless steel. *Obsolete*

HOBART 330HC-15

Hobart Welding Products
C 0.3, Cr 15, Ni 35, Mo 2, Si 0.3, bal Fe.
Weld metal. As welded: 75,000-85,000 psi TS; 25-35 El. For
welding high carbon 330HC stainless. *Obsolete*

HOBART 335A

Hobart Welding Products
C 0.1, Mn 0.45, Si 0.2, P 0.015, S 0.035, bal Fe.
Weld metal. As welded: 62,000 psi TS; 50,000 YS; 22 El;
22-63 RA. AWS E-6011 all position electrode; AC, DCEP or
DCEN.

HOBART 347

Hobart Welding Products
C 0.06, Mn 1.3, Si 0.4, Cr 19.5, Ni 9.5, Cb 0.8, bal Fe.
82,000 psi TS; 30 El. Solid stainless steel welding electrode
wire. For welding AISI 347 and 321 stainless steels. ER-347.

HOBART 347-15 & 347-16

Hobart Welding Products
C 0-0.07, Cr 19, Ni 9.5, Cb 0.8, Mn 1.6, Si 0.6, bal Fe.
Weld metal. As welded: 85,000-95,000 psi TS; 35-45 El. For
welding AISI 347 stainless steels. *Obsolete*

HOBART 349-15 & 349-16

Hobart Welding Products
C 0-0.13, Cr 19, Ni 9, W 1.4, Mo 0.5, Cb 1, Mn 1.5, Si 0.7, bal
Fe.
Weld metal. As welded: 105,000-110,000 psi TS;
80,000-85,000 psi YS; 27-37 El. For welding material of
similar analysis on turbojet engines. *Obsolete*

HOBART 4043

Hobart Welding Products
Cu 0.3, Mg 0.05, Mn 0.05, Si 4.5-6, Fe 0.8, Zn 0.1, Ti 0.2, B
0.0008, bal Al.
Solid aluminum welding electrode wire; 27,000 psi TS min.
ER-4043.

HOBART 4047

Hobart Welding Products
Cu 0.3, Mg 0.1, Mn 0.15, Si 11-13, Fe 0.08, Zn 0.2, B 0.0008,
bal Al.
Solid aluminum welding electrode wire; 30,000 psi TS min.
ER-4047. *Obsolete*

HOBART 410-15 & 410-16

Hobart Welding Products
C 0-0.1, Cr 12.5, Mn 0.6, Si 0.4, bal Fe.
Weld metal. Welded and annealed: 80,000-90,000 psi TS;
55,000-60,000 psi YS; 30-35 El For welding AISI 410 stainless
steel. *Obsolete*

HOBART 413

Hobart Welding Products
C 0.08, Mn 0.5, Si 0.56, P 0.019, S 0.02, bal Fe.
Weld metal. As welded: 67,000 psi TS; 55,000 psi YS; 17 El.
AWS E-6013; AC, DCEN or DCEP. *Obsolete*

HOBART 430-15 & 430-16

Hobart Welding Products
C 0-0.1, Cr 16, Mn 0.6, Si 0.6, bal Fe.
Weld metal. Welded and annealed: 75,000-80,000 psi TS;
40,000-45,000 psi YS; 30-35 El. For welding AISI 430 stainless
steels. *Obsolete*

HOBART 447A

Hobart Welding Products
C 0.12, Mn 0.41, Si 0.3, P 0.019, S 0.022, bal Fe.
Weld metal. As welded: 67,000 psi TS; 55,000 YS; 17 El. AWS
E-6013; AC, DCEN or DCEP.

HOBART 5183

Hobart Welding Products
Cu 0.1, Mg 4.3-5.2, Mn 0.5-1, Cr 0.05-0.25, Si 0.4, Fe 0.4, Zn
0.25, Ti 0.15, B 0.0008, bal Al.
Solid aluminum welding electrode wire; 42,000 psi TS min.
ER 5100. *Obsolete*

HOBART 5356

Hobart Welding Products
Cu 0.1, Mg 4.5-5.5, Mn 0.05-0.2, Cr 0.05-0.2, Si 0.5, Zn 0.1, Ti
0.06-0.2, B 0.0008, Fe, bal Al.
Solid aluminum welding electrode wire; 42,000 psi TS min.
ER-5356.

HOBART 5554
Hobart Welding Products
Cu 0.1, Mg 2.4-3, Mn 0.5-1, Cr 0.05-0.2, Si 0.4, Zn 0.25, Ti 0.05-0.2, B 0.0008, Fe, bal Al.
Solid aluminum welding electrode wire; 30,000 psi TS min. ER-5554. *Obsolete*

HOBART 5654
Hobart Welding Products
Cu 0.05, Mg 3.1-3.9, Mn 0.01, Cr 0.15-0.35, Si 0.45, Zn 0.2, Ti 0.05-0.15, B 0.0008, Fe, bal Al.
Solid aluminum welding electrode wire; 30,000 psi TS min. ER-5654. *Obsolete*

HOBART 70AP
Hobart Welding Products
C 0.1, Mn 0.3, Si 0.15, P 0.015, S 0.01, Ni 1.5, Mo 0.17, bal Fe.
Weld metal. As welded: 70,000 psi TS; 57,000 psi YS; 22 El. AWS E-7010-G; DCEP. For welding pipe steels, storage tanks, drill platforms, ships.

HOBART 718
Hobart Welding Products
C 0.08, Mn 0.65, Si 0.37, P 0.011, S 0.013, Ni 0.01, Cr 0.088, Mo 0.018, V 0.014, Cu 0.011, bal Fe.
Weld metal. As welded: 72,000 psi TS; 60,000 psi YS; 22 El. AWS E-7018; DCEP or AC.

HOBART 718 SR
Hobart Welding Products
C 0.04, Mn 0.78, Si 0.32, P 0.21, S 0.013, Ni 0.015, C 0.038, Mo 0.16, V 0.011, Cu 0.013, bal Fe.
Weld metal. As welded: 72,000 psi TS; 60,000 psi YS; 22 El. AWS E-7018; DCEP. *Obsolete*

HOBART HI-CARBON NO. 40-HC
Hobart Bros. Co.
High C, bal Fe.
For welding electrodes for high C steels; coated. *Obsolete*

HOBART HOBRONZE
Hobart Bros. Co.
Cu alloy.
For welding electrodes for Cu alloys; carbon or metallic arc. *Obsolete*

HOBART LH-1018
Hobart Welding Products
C 0.06, Mn 1.77, Si 0.68, Mo 0.44, bal Fe (low hydrogen).
Weld metal. As welded: 106,000 psi TS; 101,00 psi YS; 22 El. E10018-D2; for welding high strength steels. *Obsolete*

HOBART LH-1018M
Hobart Welding Products
C 0.05, Mn 1.2, Si 0.54, Ni 1.73, Mo 0.33, bal Fe (low hydrogen).
Weld metal. As welded: 103,000 psi TS; 96,000 psi YS; 24 El. E10018-M; for strong welds in highly stressed military equipment. *Obsolete*

HOBART LH-1118
Hobart Welding Products
C 0.06, Mn 1.53, Si 0.27, Cr 0.31, Mo 0.42, Ni 1.88, bal Fe (low hydrogen).
Weld metal. As welded: 115,000 psi TS; 103,000 psi YS; 22 El. E11018-M; for welding high strength steel. *Obsolete*

HOBART LH-1218
Hobart Welding Products
C 0.05, Mn 1.9, Si 0.25, Cr 0.85, Mo 0.5, Ni 2, bal Fe (low hydrogen).
Weld metal. As welded: 132,000 psi TS; 120,000 psi YS; 20 El. E12018-M; for welding high strength steels. DIN high stressed assemblies. *Obsolete*

HOBART LH-4130
Hobart Welding Products
C 0.25, Mn 0.6, Si 0.22, Cr 0.8, Mo 0.25, bal Fe (low hydrogen).
Weld metal. As welded: 121,000 psi TS; 93,500 psi YS; 16 El. For welding SAE 4130 and 8630 and similar steels that are heat treated after welding. *Obsolete*

HOBART LH-4340
Hobart Welding Products
C 0.37, Mn 0.83, Si 0.33, Cr 0.67, Mo 0.24, Ni 1.68, bal Fe (low hydrogen).
Weld metal. As welded: 137,000 psi TS; 95,000 psi YS; 6 El. For welding SAE 4140, 4330 and 4340 and similar steels that are to be heat treated after welding; weld metal will harden to same value as base metal. *Obsolete*

HOBART LH-718-MO
Hobart Welding Products
C 0.05, Mn 0.75, Si 0.56, Mo 0.53, bal Fe (low hydrogen).
Weld metal. As welded: 79,000 psi TS; 68,000 YS; 31 El. E7018-A1; for welding low alloy, high tensile steels of 50,000 psi YS min. *Obsolete*

HOBART LH-818-CM
Hobart Welding Products
C 0.05, Mn 0.68, Si 0.6, Cr 1.24, Mo 0.49, bal Fe (low hydrogen).
Weld metal. Welded, stress relieved 1275°F: 92,000 psi TS; 83,000 psi YS; 27 El; 62 RA. E8018-B2; for welding low alloy Mo and Cr-Mo steels. *Obsolete*

HOBART LH-818-N1
Hobart Welding Products
C 0.04, Mn 1.04, Si 0.31, Ni 2.37, bal Fe (low hydrogen).
Weld metal. As welded: 88,500 psi TS; 73,800 psi YS; 28 El. E8018-C1; for welding nickel steels that operate at low temperatures. *Obsolete*

HOBART LH-818-N2
Hobart Welding Products
C 0.05, Mn 0.84, Si 0.37, Ni 3.3, bal Fe (low hydrogen).
Weld metal. As welded: 94,000 psi TS; 83,000 psi YS; 25 El. E8018-C2; for welding nickel steels that operate at low temperatures. *Obsolete*

HOBART LH-818-N3
Hobart Welding Products
C 0.05, Mn 1.06, Si 0.38, Ni 1.04, bal Fe (low hydrogen).
Weld metal. As welded: 84,000 psi TS; 73,500 psi YS; 30 El. E8018-C3; for welding 70,000-80,000 psi steels for operation down to -60°F. *Obsolete*

HOBART LH-918-CM
Hobart Welding Products
C 0.05, Mn 0.78, Si 0.6, Cr 2.2, Mo 1.05, bal Fe (low hydrogen).
Weld metal. Welded, stress relieved 1275°F: 96,000 psi TS; 83,000 psi YS; 25 El; 67 RA. E9018-B3; for welding steels up to 2% Cr and 1% Mo. *Obsolete*

HOBART LH-918M
Hobart Welding Products
C 0.05, Mn 1.11, Si 0.32, i 1.7, Mo 0.28, bal Fe (low hydrogen).
Weld metal. As welded: 94,400 psi TS; 84,600 psi YS; 27 El. E9018-M; for welding Hy-80 and Hy-90 and similar high tensile and heat treated steels. *Obsolete*

HOBART MANGANICK
Hobart Bros. Co.
Mn 13-15, Ni 3.5-4.5, C, bal Fe.
Welded: 475 Brin. For welding electrodes for Mn steel; arc, coated. *Obsolete*

HOBART MANGANOL
Hobart Bros. Co.
Mn 13-16, C, bal Fe.
For welding electrode for Mn steels; arc, coated or bare. *Obsolete*

HOBART NO. 111
Hobart Bros. Co.
C 0.1, Mn 0.4, Si 0.1, bal Fe.
Welded: 61,000-69,000 TS; 52,000-60,000 YS; 25-31 El. For welding electrodes for steel; arc. *Obsolete*

HOBART NO. 111 H.T.
Hobart Bros. Co.
bal Fe.
Welded: 75,000 TS; 67,000 YS; 25 El. For welding electrodes for low alloy steels; arc, coated. *Obsolete*

HOBART NO. 13
Hobart Bros. Co.
C 0.1, bal Fe.
For welding electrodes; for light gauge steel. *Obsolete*

HOBART NO. 55
Hobart Bros. Co.
C 0.1, Mn 0.5, Si 0.2, al Fe.
Welded: 65,000-75,000 TS; 50,000-60,000 YS; 22-33 El. For welding electrodes for mild steel; arc, coated. *Obsolete*

HOBART NO. 77
Hobart Bros. Co.
C 0.08, Mn 0.5, Si 0.12, bal Fe.
Welded: 68,000-80,000 TS; 58,000-65,000 YS; 17-21 El. For welding electrodes for mild steel; arc. *Obsolete*

HOBART NO. 885
Hobart Bros. Co.
C 0.1, Mn 0.5, Si 0.2, Mo 0.5, bal Fe.
Welded: 80,000 TS; 64,000 YS; 22 El. For welding electrodes for high tensile steel; arc, coated. *Obsolete*

HOBART NO. 90-PL
Hobart Bros. Co.
C, bal Fe.
For welding electrodes; for high tensile, low alloy steels. *Obsolete*

HOBART NO. 912
Hobart Bros. Co.
Mn, Zn, bal Cu.
For welding rod for cast iron and steel; low fuming for gas welding, MP 1600 F. *Obsolete*

HOBART NO. 914
Hobart Bros. Co.
Mn, Zn, bal Cu.
For welding rod for cast iron and steel; gas welding. *Obsolete*

HOBART NO. 932
Hobart Bros. Co.
Si, bal Cu.
For welding electrodes; MP 1010 C. *Obsolete*

HOBART PS-588
Hobart Welding Products
C 0.07, Mn 1.1, Si 0.5, Cr 0.5, Ni 0.8, Cu 0.45, bal Fe.
Weld metal. Tubular welding wire for welding weathering steels. *Obsolete*

HOBART QUIKFIL 525
Hobart Welding Products
C 0.02, Mn 0.2, Si 0.01, Ni 1.8, Mo 0.25, bal Fe.
Granular material to enrich weld deposit in one pass submerged arc welding. *Obsolete*

HOBART SOFTCAST A
Hobart Bros. Co.
C, bal Fe.
Welded: 200 Brin. For welding electrodes of cast iron; coated, arc, machinable. *Obsolete*

HOBART SOFTCAST C
Hobart Bros. Co.
C, bal Fe.
For welding electrodes for cast iron; coated cast electrode, machinable. *Obsolete*

HOBART STRONGCAST
Hobart Bros. Co.
C, bal Fe.
For welding electrodes for cast iron; coated, arc. *Obsolete*

HOBART SULKOTE
Hobart Bros. Co.
C 0.13-0.18, bal Fe.
Welded: 48,000 TS; 8 El. For welding electrodes for mild steel; arc, iron oxide coating. *Obsolete*

HOBART TOOLFACER
Hobart Bros. Co.
C, Cr, Mo, W, V, bal Fe.
Welded: 600 Brin. For welding electrodes for tools, dies; high speed steel, arc. *Obsolete*

HOBB DIE STEEL
Faitout Iron & Steel Co.
C 0-0.07, Mn 0.15, bal Fe.
For plastic molds, hobs; carburizing steel.

HOBBING DIE STEEL
Bethlehem Steel Corp.
C 0.06, Mn 0.15, Si 0.1, bal Fe.
For plastic mold dies; water or oil hardened. *Obsolete*

HOBRITE
Swedish American Steel Corp.
C 0-0.05, Mn 0.11, bal Fe.
For hobbing molds; plastic mold dies; water hardening.

HOBSON NON-SHRINK
Hobson, Houghton & Co.
C 0.9, Mn 1.2, V 0.2, bal Fe.
For dies; and tools; non-deforming.

HOBSON'S CHOICE
Hobson, Houghton & Co.
C 1.48, Mn 0.34, bal Fe.
For tools, cutters; water hardened.

HOBSON'S CHOICE DRILL RODS
Hobson, Houghton & Co.
C 1.1, bal Fe.
For tools, drills, taps; water hardened.

HOBSON'S CHOICE XX
Hobson, Houghton & Co.
C 0.7-1.2, Mn 0.3, Si 0.3, bal Fe.
For drills, punches, taps, reamers, hobs, broaches; water hardened; Type W1.

HOBSON'S FAST FINISHING
Hobson, Houghton & Co.
C 1.2, V 0.2, bal Fe.
For fast finishing tools, cutters; water or oil hardened.

HOBSON'S OIL HARDENING
Hobson, Houghton & Co.
C 0.9, Mn 1.2, bal Fe.
For tools, dies, punches; non-deforming.

HOBSON'S SPECIAL TAP RODS
Hobson, Houghton & Co.
C 1, bal Fe.
For tools, taps; water hardened.

HOBSON'S WARRANTED BEST
Hobson, Houghton & Co.
C 0.7-1.2, bal Fe.
For general tools; water hardening.

HOCKLEISTEUNGSTAHL
German manufacture
Cr 10, W 2, V 0.5, C 1.5, Si 1.2, Mn 0.4, bal Fe.
For shear blades, dies.

HOCO EXTRA EXTRA MH
Hoffman & Co. KG
C 1.1, Si 0-0.25, Mn 0-0.25, bal Fe.
Annealed: 110,000 TS; 58,000 YS; 18 El; 40 RA; 210 Brin. For springs, taps, cutters, drills, hobs; Type W1; water hardened.

HOCO EXTRA EXTRA SEHR ZAH
Hoffman & Co. KG
C 0.7, Mn 0-0.25, bal Fe.
Heat treated: 175,000 TS; 128,000 YS; 12 El; 37 RA; 355 Brin. For springs, axes, punches, crimpers; type W1; water hardened.

HOCO EXTRA EXTRA ZAH
Welded Carbide Co. Inc.
C 0.85, Si 0-0.25, Mn 0-0.25, bal Fe.
Heat treated: 190,000 TS; 145,000 YS; 10 El; 30 RA; 400 Brin. For springs, taps, drills, hobs, reamers; type W1; water hardened.

HOCO EXTRA EXTRA ZH
Hoffman & Co. KG
C 1, Si 0-0.25, Mn 0-0.25, bal Fe.
Annealed: 100,000 TS; 53,000 YS; 21 El; 42 RA; 200 Brin. For springs, drills, reamers, broaches; type W1; water hardened.

HOCO EXTRA HART
Hoffman & Co. KG
C 1.3, Si 0-0.25, Mn 0-0.25, bal Fe.
For engravers tools, forming dies, reamers; type W1; water hardened.

HOCO EXTRA MH
Hoffman & Co. KG
C 1.15, Si 0-0.25, Mn 0-0.25, bal Fe.
Annealed: 110,000 TS; 58,000 YS; 20 El; 40 RA; 210 Brin. For springs, tools, reamers, hobs, drills, taps; type W1; water hardened.

HOCO EXTRA SEHR HART
Hoffman & Co. KG
C 0.7, Si 0-0.25, Mn 0-0.25, bal Fe.
Heat treated: 175,000 TS; 128,000 YS; 12 El; 37 RA; 355 Brin. For springs, rails, punches, axes, hammers; type W1; water hardened.

HOCO EXTRA WEICH SCHWEISSBAR
Hoffman & Co. KG
C 0.55, Si 0.1-0.4, Mn 0.5-0.7, bal Fe.
Heat treated: 160,000 TS; 115,000 YS; 12 El; 40 RA; 325 Brin. For gears, bolts, shafts, machine tool parts; water hardened.

HOCO EXTRA ZAH
Hoffman & Co. KG
C 0.85, Si 0-0.25, Mn 0-0.25, bal Fe.
Heat treated: 190,000 TS; 145,000 YS; 10 El; 30 RA; 400 Brin. For springs, taps, reamers, drills, broaches; type W1; water hardened.

HOCO EXTRA ZH
Hoffman & Co. KG
C 1, Si 0-0.25, Mn 0-0.25, bal Fe.
Annealed: 100,000 TS; 53,000 YS; 21 El; 42 RA; 200 Brin. For springs, tools, cutters, drills; type W1; water hardened.

HOCO HSSP
Hoffman & Co. KG
C 0.7, Si 1.7, Mn 0.7, bal Fe.
For springs; oil hardened, tough.

HOCO HWA
Hoffman & Co. KG
C 0.3, Si 0.25, Mn 0.25, Cr 2.5, Mo 0.15, V 0.1, bal Fe.
For gears, bolts, crankshafts; oil hardened, tough.

HOCO HWA100M
Hoffman & Co. KG
C 0.65, Cr 3.75, Mo 0.85, V 0.7, W 8.5, bal Fe.
For lathe planer tools, taps, reamers; high speed steel.

HOCO HWA50
Hoffman & Co. KG
C 0.3, Cr 2.35, V 0.6, W 4.25, bal Fe.
For extrusion dies, rams and liners; oil hardened, tough.

HOCO HWA90
Hoffman & Co. KG
C 0.3, Cr 2.65, V 0.35, W 8.5, bal Fe.
For extrusion dies, rams and liners; hot work steel, oil hardened.

HOCO HWA95K
Hoffman & Co. KG
C 0.3, Cr 2.4, Co 2, V 0.25, W 8.5, bal Fe.
For extrusion dies and liners; hot work steel, oil hardened.

HOCO HWAA40
Hoffman & Co. KG
C 0.3, Si 1, Cr 1.1, W 3.75, V 0.18, bal Fe.
For punches, upsetters, riveters; oil hardened, tough.

HOCO HWAMV
Hoffman & Co. KG
C 0.45, Cr 1.4, Mo 0.7, V 0.3, bal Fe.
For header and forging dies; oil hardened, tough.

HOCO HWR4
Hoffman & Co. KG
C 1.42, W, V, bal Fe.
For bearings, bushing, cutters; water or oil hardened, wear resistant.

HOCO HZR
Hoffman & Co. KG
C 1.4, Cr 0.3, V 0.1, bal Fe.
For bearings, bushings, cutters; water or oil hardened, wear resistant.

HOCO KG
Hoffman & Co. KG
C 1, Cr 1.55, Mn 0.35, bal Fe.
For bearings, cutters, liners; water or oil hardened, shock resistant.

HOCO PRIMA SEHR WEICH
Hoffman & Co. KG
C 0.35, Si 0.25-0.5, Mn 0.3-0.8, bal Fe.
Hot rolled: 84,000 TS; 54,000 YS; 30 El; 53 RA; 185 Brin. For gears, shafts, axles, bolts, screws; water hardened.

HOCO PRIMA ZAH
Hoffman & Co. KG
C 0.6, Si 0.25-0.5, Mn 0.3-0.8, bal Fe.
Heat treated: 160,000-115,000 TS; 113,000-77,000 YS; 12-23 El; 40-54 RA; 320-230 Brin. For wheels, die blocks, girders, rails; water hardened.

HOCO PRIMA-MH
Hoffman & Co. KG
C 0.9, Si 0.25-0.5, Mn 0.3-0.8, bal Fe.
Heat treated: 185,000 TS; 118,000 YS; 10 El; 30 RA; 375 Brin. For springs, taps, drills, reamers; Type W1; water hardened.

HOCXO PRIMA WEICH
Hoffman & Co. KG
C 0.45, Si 0.25-0.5, Mn 0.3-0.8, bal Fe.
Hot rolled: 98,000 TS; 60,000 YS; El; 54 RA; 212 Brin. For axles, gears, bolts, shafts; water hardened.

HODI
Atlas Specialty Steels
C 0.28, W 9, Cr 4, V 0.4, bal Fe.
For hot gripping and swaging dies. *Obsolete*

HODUR
Honsel-Werke Aktiengesellschaft
Cu 2.5-5, Mg 0.2-1.8, Mn 0.3-1.5, bal Al.
Annealed: 27,000 TS; 11,000 YS; 22 El; 47 Brin. Heat treated: 72,000 TS; 57,000 YS; 130 Brin. For aircraft structures, fittings, fasteners; age-hardenable. *Obsolete*

HODUR
Honsel-Werke Atkiengesellschaft
Aluminum.
Aluminum alloy AlCuMg.

HOESCH ARK 35
Hoesch Stahl AG
C 0-0.12, Si 0-0.35, Mn 0-0.7, Ni 0-3.8, P 0-0.025, S 0-0.025, Al, bal Fe.
Normalized: 440-610 N/mm^2 TS; 345 N/mm^2 YS min; 24 El min. Welded construction for low temperatures.

HOESCH ARK 50
Hoesch Stahl AG
C 0-0.1, Si 0-0.35, Mn 0-0.7, Ni 0-5.25, P 0-0.025, S 0-0.025, Al, bal Fe.
Normalized: 490 N/mm^2 TS; 390 N/mm^2 YS min; 25 El min. Welded construction for low temperatures.

HOESCH ARK 6-29
Hoesch Stahl AG
C 0-0.13, Si 0-0.35, Mn 0-1.5, Ni 0-0.8, P 0-0.025, S 0-0.025, Al, bal Fe.
Normalized: 410-530 N/mm^2 TS; 285 N/mm^2 YS min; 24 El min. Containers for liquefied gas; for low temperatures.

HOESCH ARK 6-32
Hoesch Stahl AG
C 0-0.14, Si 0-0.35, Mn 0-1.6, Ni 0-0.8, P 0-0.025, S 0-0.025, Al, bal Fe.
Normalized: 440-560 N/mm^2 TS; 315 N/mm^2 YS min; 23 El min. Tanks and ships; good at low temperatures.

HOESCH ARK 6-36
Hoesch Stahl AG
C 0-0.15, Si 0-0.35, Mn 0-1.65, Ni 0-0.8, P 0-0.025, S 0-0.025, Al, bal Fe.
Normalized: 490-610 N/mm^2 TS; 355 N/mm^2 YS min; 22 El min. Containers for use at low temperatures.

HOESCH ARK 90
Hoesch Stahl AG
C 0-0.1, Si 0-0.3, Mn 0-0.7, Ni 0-10, P 0-0.025, S 0-0.025, Al, bal Fe.
Normalized: 640-830 N/mm^2 TS; 490 N/mm^2 YS min; 18 El min. Welded construction for low temperatures.

HOESCH NOVAR K 260 PR
Hoesch Stahl AG
C 0-0.1, Si 0-0.5, Mn 0-0.8, Nb, V, Ti, Al, bal Fe.
Cold rolled and process annealed: 350-480 N/mm^2 TS; 260 N/mm^2 YS; 26 El min. For cold forming.

HOESCH NOVAR K 300 PR
Hoesch Stahl AG
C 0-0.1, Si 0-0.5, Mn 0-1, Nb, V, Ti, Al, bal Fe.
Cold rolled and process annealed: 380-510 N/mm^2 TS; 300 N/mm^2 YS min; 24 El min. For cold forming.

HOESCH NOVAR K 340 PR
Hoesch Stahl AG
C 0-0.1, Si 0-0.5, Mn 0-1.2, Nb, V, Ti, Al, bal Fe.
Cold rolled and process annealed: 410-540 N/mm^2 TS; 340 M/mm^2 YS min; 22 El min. For cold forming.

HOESCH NOVAR K 380 PR
Hoesch Stahl AG
C 0-0.1, Si 0-0.5, Mn 0-1.3, Nb, V, Ti, Al, bal Fe.
Cold rolled and process annealed: 440-580 N/mm^2 TS; 380 N/mm^2 YS min; 20 El min. For cold forming.

HOESCH NOVAR K 420 PR
Hoesch Stahl AG
C 0-0.1, Si 0-0.5, Mn 0-1.4, Nb, V, Ti, Al, bal Fe.
Cold rolled and process annealed: 470-620 N/mm^2 TS; 420 N/mm^2 YS min; 18 El min. For cold forming.

HOESCH NOVAR W 340 PR
Hoesch Stahl AG
C 0-0.1, Si 0-0.5, Mn 0-0.8, V 0-0.08, 0.02 Nb min, Al, bal Fe.
Treated: 420-540 N/mm^2 TS; 340 N/mm^2 YS min; 19 El min. For cold forming.

HOESCH NOVAR W 38D PR
Hoesch Stahl AG
C 0-0.1, Si 0-0.5, Mn 0-0.9, V 0-0.08, 0.02 Nb min, Al, bal Fe.
Treated: 450-590 N/mm^2 TS; 380 N/mm^2 YS min; 18 El min. For cold forming.

HOESCH NOVAR W 420 PR
Hoesch Stahl AG
C 0-0.1, Si 0-0.5, Mn 0-1, V 0-0.08, 0.02 Nb min, Al, bal Fe.
Treated: 480-620 N/mm^2 TS; 420 N/mm^2 YS min; 16 El min. For cold forming.

HOESCH NOVAR W 460 PR
Hoesch Stahl AG
C 0.1, Si 0.5, Mn 1.2, V 0.08, 0.02 Nb min, Al, bal Fe.
Treated: 520-670 N/mm^2 TS; 460 N/mm^2 YS min; 14 El min. For cold forming.

HOESCH NOVAR W 500 PR
Hoesch Stahl AG
C 0-0.1, Si 0-0.5, Mn 0-1.4, V 0-0.08, 0.02 Nb min, Al, bal Fe.
Treated: 550-700 N/mm^2 TS; 500 N/mm^2 YS min; 12 El min. For cold forming.

HOESCH NOVAR X 52 PR
Hoesch Stahl AG
C 0-0.12, Si 0-0.25, Mn 0-1.4, V 0-0.08, Nb 0-0.05, Al, bal Fe.
Treated: 510-630 N/mm^2 TS; 365 YS min; 20 El min. For welded pipes.

HOESCH NOVAR X 56 PR
Hoesch Stahl AG
C 0-0.12, Si 0-0.25, Mn 0-1.4, V 0-0.08, Nb 0-0.05, Al, bal Fe.
Treated: 530-680 N/mm^2 TS; 385 N/mm^2 YS min; 19 El min. For welded pipes.

HOESCH NOVAR X 60 PR
Hoesch Stahl AG
C 0-0.12, Si 0-0.3, Mn 0-1.45, V 0-0.08, Nb 0-0.05, Al, bal Fe.
Treated: 550-710 N/mm^2 TS; 410 N/mm^2 YS min; 18 El min. For welded pipes.

HOESCH NOVAR X 65 PR
Hoesch Stahl AG
C 0-0.12, Si 0-0.35, Mn 0-1.5, V 0-0.08, Nb 0-0.05, Al, bal Fe.
Treated: 560-710 N/mm^2 TS; 445 N/mm^2 YS min; 18 El min. For welded pipe.

HOESCH NOVAR X 70 PR
Hoesch Stahl AG
C 0-0.12, Si 0-0.35, Mn 0-1.5, V 0-0.08, Nb 0-0.05, Al, bal Fe.
Treated: 600-750 N/mm^2 TS; 480 N/mm^2 YS min; 18 El min. For welded pipe.

HOESCH RESISTASTAHL 37
Hoesch Stahl AG
C 0-0.13, Si 0-0.4, Mn 0-0.5, Cu 0-0.5, Cr 0-0.8, Ni 0-0.4, Al, bal Fe.
360-440 N/mm^2 TS; 235 N/mm^2 YS; 25 El min. For bridges, contraction, vehicles; resistant to atmospheric corrosion.

HOESCH RESISTASTAHL 37 EXTRA
Hoesch Stahl AG
C 0-0.13, Si 0-0.4, Mn 0-0.5, P 0-0.045, Cu 0-0.5, Cr 0-0.8, N 0-0.4, Al, bal Fe.
360-440 N/mm^2 TS; 235 N/mm^2 YS min; 25 El min. For steel construction, penstock pipes; resistant to atmospheric corrosion.

HOESCH RESISTASTAHL 52
Hoesch Stahl AG
C 0-0.12, Si 0-0.75, Mn 0-0.5, Cu 0-0.55, Cr 0-1.25, Ni 0-0.65, Al, bal Fe.
510-610 N/mm^2 TS; 355 N/mm^2 YS min; 22 El min. For bridge construction, industrial plants; resistant to atmospheric corrosion.

HOESCH RESISTASTAHL 52 EXTRA
Hoesch Stahl AG
C 0.15, Si 0.4, Mn 1.3, P 0.45, Cu 0.5, Cr 0.8, Ni 0-0.4, V 0-0.1, Al, bal Fe.
510-610 N/mm^2 TS; 355 N/mm^2 YS min; 22 El min. For construction machinery, vehicles; resistant to atmospheric corrosion.

HOESCH UNION 26 AK
Hoesch Stahl AG
C 0-0.16, Si 0-0.4, Mn 0-1.3, Al, bal Fe.
Normalized: 360-480 N/mm^2 TS; 255 N/mm^2 YS min; 25 El min. Special non-aging quality with low temperature toughness.

HOESCH UNION 26 G
Hoesch Stahl AG
C 0-0.18, Si 0-0.4, Mn 0-1.3, Al, bal Fe.
Normalized: 360-480 N/mm^2 TS; 255 N/mm^2 YS min; 25 El min. Standard quality weldable structural steel.

HOESCH UNION 26 W
Hoesch Stahl AG
C 0-0.18, Si 0-0.4, Mn 0-1.3, Al, bal Fe.
Normalized: 360-480 N/mm^2 TS; 255 N/mm^2 YS min; 25 El min. Weldable structural steel; for resisting high temperatures.

HOESCH UNION 29 AK
Hoesch Stahl AG
C 0-0.16, Si 0-0.4, Mn 0-1.4, Al, bal Fe.
Normalized: 390-510 N/mm^2 TS; 285 N/mm^2 YS min; 24 El min. Special non-aging quality with low temperature toughness.

HOESCH UNION 29 G
Hoesch Stahl AG
C 0-0.18, Si 0-0.4, Mn 0-1.4, Al, bal Fe.
Normalized: 390-510 N/mm^2 TS; 285 N/mm^2 YS min; 24 El min. Standard quality, weldable, fine-grained structural steel.

HOESCH UNION 29 W
Hoesch Stahl AG
C 0-0.18, Si 0-0.4, Mn 0-1.4, Al, bal Fe.
Normalized: 390-510 N/mm^2 TS; 285 N/mm^2 YS min; 24 El min. Weldable structural steel; for resisting high temperatures.

HOESCH UNION 32 AK
Hoesch Stahl AG
C 0-0.16, Si 0-0.45, Mn 0-1.5, Al, bal Fe.
Normalized: 440-560 N/mm^2 TS; 315 N/mm^2 YS min; 23 El min. Non-aging quality with low temperature toughness for liquefied gas containers, etc.

HOESCH UNION 32 G
Hoesch Stahl AG
C 0-0.18, Si 0-0.45, Mn 0-1.5, Al, bal Fe.
Normalized: 440-560 N/mm^2 TS; 315 N/mm^2 YS min; 23 El min. For bridges, steel structures, vehicles.

HOESCH UNION 32 W
Hoesch Stahl AG
C 0-0.18, Si 0-0.45, Mn 0-1.5, Al, bal Fe.
Normalized: 440-560 N/mm^2 TS; 315 N/mm^2 YS min; 23 El min. For boiler construction, engine parts, etc.; for resisting high temperatures.

HOESCH UNION 36 AK
Hoesch Stahl AG
C 0-0.18, Si 0-0.5, Mn 0-1.6, Al, bal Fe.
Normalized: 490-630 N/mm^2 TS; 355 N/mm^2 YS min; 22 El min. Non-aging quality for low temperature operations.

HOESCH UNION 36 G
Hoesch Stahl AG
C 0-0.2, Si 0-0.5, Mn 0-1.6, Al, bal Fe.
Normalized: 490-630 N/mm^2 TS; 355 N/mm^2 YS min; 22 El min. For vehicles, steel structures, spheres.

HOESCH UNION 36 NB
Hoesch Stahl AG
C 0-0.18, Si 0-0.5, Mn 0-1.6, 0.02 Nb min, Al, bal Fe.
Normalized: 490-630 N/mm^2 TS; 355 N/mm^2 YS min; 22 El min. For tanks and containers.

HOESCH UNION 36 W
Hoesch Stahl AG
C 0-0.2, Si 0-0.5, Mn 0-1.6, Al, bal Fe.
Normalized: 490-630 N/mm^2 TS; 355 N/mm^2 YS min; 22 El min. For equipment to operate at elevated temperatures.

HOESCH UNION 39 AK
Hoesch Stahl AG
C 0-0.18, Si 0-0.6, Mn 0-1.6, Ni 0-0.8, 0.20 V or Ti, bal Fe.
Normalized: 500-650 N/mm^2 TS; 380 N/mm^2 YS min; 20 El min. Storage and transport of liquefied gas.

HOESCH UNION 39 G
Hoesch Stahl AG
C 0-0.18, Si 0-0.6, Mn 0-1.6, Ni 0-0.8, 0.20 V or Ti, bal Fe.
Normalized: 500-650 N/mm^2 TS; 380 N/mm^2 YS min; 20 El min. Steel structures, vehicles, tanks.

HOESCH UNION 39 W
Hoesch Stahl AG
C 0-0.18, Si 0-0.6, Mn 0-1.6, Ni 0-0.8, 0.20 V or Ti, bal Fe.
Normalized: 500-650 N/mm^2 TS; 380 N/mm^2 YS min; 20 El min. Equipment for elevated temperatures.

HOESCH UNION 43 AK
Hoesch Stahl AG
C 0-0.18, Si 0-0.6, Mn 0-1.7, Ni 0-0.8, 0.20 V or Ti, bal Fe.
Normalized: 530-680 N/mm^2 TS; 420 N/mm^2 YS min; 19 El min. Non-aging quality with low temperature toughness.

HOESCH UNION 43 G
Hoesch Stahl AG
C 0-0.18, Si 0-0.6, Mn 0-1.7, Ni 0-0.8, 0.20 V or Ti, bal Fe.
Normalized: 530-680 N/mm^2 TS; 420 N/mm^2 YS min; 19 El min. For structures, transport equipment.

HOESCH UNION 43 W
Hoesch Stahl AG
C 0-0.18, Si 0-0.6, Mn 0-1.7, Ni 0-0.8, 0.20 V or Ti, bal Fe.
Normalized: 530-680 N/mm^2 TS; 420 N/mm^2 YS min; 19 El min. Equipment to resist high temperatures.

HOESCH UNION 47 AK
Hoesch Stahl AG
C 0-0.2, Si 0-0.6, Mn 0-1.7, Ni 0-0.8, 0.20 V or Ti, bal Fe.
Normalized: 560-730 N/mm^2 TS; 460 N/mm^2 YS min; 17 El min. Non-aging quality with low temperature toughness.

HOESCH UNION 47 G
Hoesch Stahl AG
C 0-0.2, Si 0-0.6, Mn 0-1.7, Ni 0-0.8, 0.20 V or Ti, bal Fe.
Normalized: 560-730 N/mm^2 TS; 460 N/mm^2 YS min; 17 El min. For penstock construction, bridges, vehicles.

HOESCH UNION 47 W
Hoesch Stahl AG
C 0-0.2, Si 0-0.6, Mn 0-1.7, Ni 0-0.8, 0.20 V or Ti, bal Fe.
Normalized: 560-730 N/mm^2 TS; 460 N/mm^2 YS min; 17 El min. Equipment for resisting high temperatures.

HOESCH UNION H 60 L
Hoesch Stahl AG
C 0.24, Si 0.6, Mn 1.6, Cr 1.2, Al, bal Fe.
Normalized: 590 N/mm^2 min TS; 175 Brin min. Wear resistant, hardenable alloy for construction, agricultural equipment and mixers.

HOESCH UNION H 90 L
Hoesch Stahl AG
C 0.22, Si 0.4, Mn 1.3, Cr 0.7, Ti 0.2, Al, bal Fe.
Normalized: 880 N/mm^2 min TS; 265 Brin min. For parts not subject to abrasion; either normalized or hardened.

HOESCH UNION H 90 U
Hoesch Stahl AG
C 0.5, Si 0.6, Mn 2, Al, bal Fe.
Normalized: 880 N/mm^2 min TS; 265 Brin min. Wear resistant and hardenable alloy for agricultural equipment, chutes.

HOESCH UNION Q 260
Hoesch Stahl AG
C 0-0.16, Si 0-0.5, Mn 0-1.2, Ti 0-0.2, Al, bal Fe.
Normalized: 370-490 N/mm^2 TS; 260 N/mm^2 YS min; 30 El min. For cold forming and bending; aging resistant.

HOESCH UNION Q 300
Hoesch Stahl AG
C 0-0.16, Si 0-0.5, Mn 0-1.4, Ti 0-0.2, Al, bal Fe.
Normalized: 400-520 N/mm^2 TS; 300 N/mm^2 YS min; 27 El min. For cold forming and bending; aging resistant.

HOESCH UNION Q 340
Hoesch Stahl AG
C 0-0.16, Si 0-0.5, Mn 0-1.5, Ti 0-0.2, Al, bal Fe.
Normalized: 460-580 N/mm^2 TS; 340 N/mm^2 YS min; 25 El min. For cold forming and bending; aging resistant.

HOESCH UNION Q 340 TM
Hoesch Stahl AG
C 0-0.12, Si 0-0.5, Mn 0-1.3, Al, V, Nb, Ti, bal Fe.
Treated: 420-540 N/mm^2 TS; 340 N/mm^2 YS min; 25 El min. For cold forming and bending; aging resistant.

HOESCH UNION Q 360
Hoesch Stahl AG
C 0-0.18, Si 0-0.5, Mn 0-1.6, Ti 0-0.2, Al, bal Fe.
Normalized: 480-610 N/mm^2 TS; 360 N/mm^2 YS min; 25 El min. For cold forming and bending; aging resistant.

HOESCH UNION Q 380
Hoesch Stahl AG
C 0-0.18, Si 0-0.5, Mn 0-1.6, Ti 0-0.2, Al, bal Fe.
Normalized: 500-640 N/mm^2 TS; 380 N/mm^2 YS min; 25 El min. For cold forming and bending; aging resistant.

HOESCH UNION Q 380 TM
Hoesch Stahl AG
C 0-0.12, Si 0-0.5, Mn 0-1.4, Al, V, Nb, Ti, bal Fe.
Treated: 450-580 N/mm^2 TS; 380 N/mm^2 YS min; 23 El min. For cold forming and bending; aging resistant.

HOESCH UNION Q 420
Hoesch Stahl AG
C 0-0.2, Si 0-0.5, Mn 0-1.6, Ti 0-0.2, Al, bal Fe.
Normalized: 530-670 N/mm^2 TS; 420 N/mm^2 YS min; 23 El min. For cold forming and bending; aging resistant.

HOESCH UNION Q 420 TM
Hoesch Stahl AG
C 0-0.12, Si 0-0.5, Mn 0-1.5, Al, V, Nb, Ti, bal Fe.
Treated: 480-620 N/mm^2 TS; 420 N/mm^2 YS min; 21 El min. For cold forming and bending; aging resistant.

HOESCH UNION Q 460
Hoesch Stahl AG
C 0-0.21, Si 0-0.5, Mn 0-1.7, Ti 0-0.2, Al, bal Fe.
Normalized: 550-700 N/mm^2 TS; 460 N/mm^2 YS min; 21 El min. For cold forming and bending; aging resistant.

HOESCH UNION Q 460 TM
Hoesch Stahl AG
C 0-0.12, Si 0-0.5, Mn 0-1.6, Al, V, Nb, Ti, bal Fe.
Treated: 520-670 N/mm^2 TS; 460 N/mm^2 YS min; 19 El min. For cold forming and bending; aging resistant.

HOESCH UNION Q 500
Hoesch Stahl AG
C 0-0.22, Si 0-0.5, Mn 0-1.7, Ti 0-0.2, Al, bal Fe.
Normalized: 580-730 N/mm^2 TS; 500 N/mm^2 YS min; 19 El min. For cold forming and bending; aging resistant.

HOESCH UNION Q 500 TM
Hoesch Stahl AG
C 0-0.12, Si 0-0.5, Mn 0-1.7, Al, V, Nb, Ti, bal Fe.
Treated: 550-700 N/mm^2 TS; 500 N/mm^2 YS min; 17 El min. For cold forming and bending, aging resistant.

HOESCH UNION Q 550
Hoesch Stahl AG
C 0-0.24, Si 0-0.5, Mn 0-1.7, Ti 0-0.2, Al, bal Fe.
Normalized: 590-750 N/mm^2 TS; 550 N/mm^2 YS min; 18 El min. For cold forming and bending; aging resistant.

HOFORS KN-13
Hofors Steel Works
C 0.18, Si 0.3, Mn 0.55, Cr 0.75, Ni 3, bal Fe.
For gears, camshafts, crankshafts; case hardened.

HOFORS KN-14
Hofors Steel Works
C 0.4, Si 0.3, Mn 0.8, Cr 0.8, Ni 1.25, bal Fe.
For gears, shafts; shock resistant.

HOFORS KN-16
Hofors Steel Works
C 0.15, Si 0.3, Mn 0.55, Cr 0.65, Ni 1.25, bal Fe.
For gears, camshafts, crankshafts; case hardened.

HOFORS KN-17
Hofors Steel Works
C 0.12, Cr 0.75, Ni 3, bal Fe.
For gears, camshafts, crankshafts; case hardened.

HOFORS KN-18
Hofors Steel Works
C 0.35, Si 0.3, Mn 0.55, Cr 1.15, Ni 2.6, bal Fe.
For gears, machinery parts; shock resistant.

HOFORS KN-19
Hofors Steel Works
C 0.3, Si 0.3, Mn 0.55, Cr 1.25, Ni 4.25, bal Fe.
For gears, shafts; tough, oil hardened.

HOFORS KNY-1
Hofors Steel Works
C 0.3, Mn 0.55, Cr 1.05, Ni 3.25, Mo 0.25, bal Fe.
For gears, machinery parts, shafts; tough.

HOFORS KY-2
Hofors Steel Works
C 0.25, Mn 0.65, Cr 1.05, Mo 0.2, bal Fe.
For gears, machinery parts; tough.

HOFORS M-11
Hofors Steel Works
C 0.4, Si 0.3, Mn 1.25, bal Fe.
For machinery parts; tough.

HOFORS NY-1
Hofors Steel Works
C 0.2, Mn 0.55, Ni 1.8, Mo 0.25, bal Fe.
For gears, shafts; case hardened.

HOFORS SM-5
Hofors Steel Works
C 0.55, Si 1.75, Mn 0.75, bal Fe.
For springs; oil hardened.

HOFORS-1
Hofors Steel Works
C 1, Mn 1.1, Si 0.6, Cr 1, bal Fe.
For large ball and roller bearing rings; oil hardened; non-deforming.

HOFORS-100

Hofors Steel Works
C 1, bal Fe.
For twist drills, punches, shear blades; water hardened.

HOFORS-13

Hofors Steel Works
C 1.05, Cr 1, bal Fe.
For ball bearing rings, rollers, punches, taps.

HOFORS-22

Hofors Steel Works
C 1, Cr 1.1, Mo 0.35, bal Fe.
For cutting tools, chisels, drill, rollers, taps; wear resistant.

HOFORS-3

Hofors Steel Works
C 1, Cr 1.5, bal Fe.
For ball bearings, master gauges, taps.

HOFORS-46

Hofors Steel Works
C 0.9, Mn 1.15, Cr 0.5, W 0.5, V 0.1, bal Fe.
For blanking dies, broaches, form tools, punches, taps; oil hardened; tough.

HOFORS-7

Hofors Steel Works
C 1.1, Cr 0.6, bal Fe.
For taps, twist drills, dies, punches, reamers.

HOFORS-711

Hofors Steel Works
C 0.5, Mn 0.7, Cr 1.15, W 2.5, V 0.15, bal Fe.
For broaches, milling cutters, blanking and heading dies; oil hardened.

HOFORS-9

Hofors Steel Works
C 1.1, Cr 0.5, bal Fe.
For bearings, rollers, reamers, taps.

HOHENZOLLERN BRASS

German manufacture
Zn 30-40, bal Cu.
For gears, worm wheels; a high tensile brass.

HOLDAX

Uddeholm Corp.
C 0.4, Si 0.32, Mn 1.6, Cr 1, Mo 0.2, S 0.08, bal Fe.
At 68°F: 146,000 psi TS; 116,000 psi YS; 18 El; 50 RA.
Vacuum-degassed chromium-molybdenum alloy steel supplied in the hardened and tempered condition. For molds, dies and plates. AISI 4140, modified.

HOLDER BLOCK

Ackerlind Steel Co., Inc.
C 0.3, Cr, Ni, Mo, bal Fe.
Prehardened steel; 250-310 Brin. For die casting dies and plastic molds.

HOLDER BLOCK STEEL

Slater Steels Corp.
Alloy steel. C 0.5, Mn 1.15, S 0.08, Cr 0.65, Mo 0.18, bal Fe.
Medium carbon, chromium-molybedenum alloy steel that exhibits superior machinability due to its sulfur content. Typical applications include backers for forging dies, brake dies, die holders, frames for plastic molds, holders for forging dies, molds requiring noncritical finish.

HOLDERTEM

Heppenstall Co.
Alloy steel. C 0.4, Mn 0.85, Cr 0.95, Mo 0.2, bal Fe.
Prehardened low alloy steel for shafts, arbors, structural parts, and special purpose tools.

HOLFOS 10% LEAD BRONZE

Holcroft Castings & Forgings Ltd.
Cu 80, Sn 10, Pb 10.
Cast: 29,000-36,000 TS; 22,500-27,000 YS; 2-6 El; 70-85 Brin.
For bushings, bearings; free cutting, SAE 64. *Obsolete*

HOLFOS 20% LEAD BRONZE

Holcroft Castings & Forgings Ltd.
Cu 75, Sn 5, Pb 20.
Cast: 22,500-29,500 TS; 13,500-18,000 YS; 7-10 El; 50-60 Brin. For bushings, bearings; operates best with soft shafts
Obsolete

HOLFOS AB1 AS CAST

Holcroft Castings & Forgings Ltd.
Copper. Cu 87.8, Al 9.5, Fe 2.5.
Cast: 72,000-78,000 TS; 25,000-31,000 YS; 20 El; 120-140 Brin. For high strength components in food handling, acid resistant fittings, flanges.

HOLFOS AB1 HEAT TREATED

Holcroft Castings & Forgings Ltd.
Cu 88, Al 9.5, Fe 2.5.
Cast and heat treated: 85,000-94,000 TS; 43,000-50,000 YS; 12 El; 130-180 Brin. Gears, shafts, splined couplings for hydraulic systems, food processing. *Obsolete*

HOLFOS AB2 AS CAST

Holcroft Castings & Forgings Ltd.
Copper. Cu 80, Al 10, Fe 4.5, Ni 5.5.
Cast: 90,000-95,000 TS; 36,000 YS; 15 El; 165 Brin. For gears, shafts, components in marine equipment, food and oil industry. BS 1400 AB2-C; BS 1073; ASTM B148-65T-9D.

HOLFOS AB2 HEAT TREATED

Holcroft Castings & Forgings Ltd.
Cu 80, Al 10, Fe 4.5, Ni 5.5.
Cast and heat treated: 87,000-100,000 TS; 38,000-54,000 YS; 8-12 El; 155-190 Brin. For highly stressed equipment in marine, food, petroleum industries. Near ASTM B148-65T-9D-HT *Obsolete*

HOLFOS B O

Holcroft Castings & Forgings Ltd.
Copper. Cu 87.8, Sn 11.5, P 0.7.
Cast: 38,000-54,000 TS; 27,000-47,000 YS; 5-7 El; 90-130 Brin. Properties vary with method of casting. Good wear and abrasion characteristics. For gears, shafts, wear parts. BS 1400 PB1-C and PB2-C; BS 1059.

HOLFOS B O H T

Holcroft Castings & Forgings Ltd.
Copper. Cu 87.8, Sn 11.5, P 0.7.
Centrifugally-spun and continuously cast. As cast: 45,000-54,000 TS; 22,400-23,500 YS; 20-40 El; 85-90 Brin. Good wear and abrasion resistance, resistant to sea water. For elevating nuts, bearings, small worm gears. DTD 900/4454/A.

HOLFOS CT1

Holcroft Castings & Forgings Ltd.
Copper. Cu 89.85, Sn 10, P 0-0.15.
Cast tin bronze; 70-130 Brin depending on casting method. For bearings, pump bodies, piston shafts, marine equipment. BS 1400 CT1-C.

HOLFOS G 3-TF (FULLY HEAT TREATED)

Holcroft Castings & Forgings Ltd.
Copper. Cu 85.2, Sn 7, Zn 2.3, Ni 5.5.
Heat treated cast tin-nickel bronze; improved strength and wear properties. 160-200 Brin. BS 1400 G3-TF.

HOLFOS G1

Holcroft Castings & Forgings Ltd.
Copper. Cu 88, Sn 10, Zn 2.
Cast: 38,000-50,000 TS; 18,000-27,000 YS; 3-15 El; 85-100 Brin. Properties vary with method of casting. Known as Admiralty Gunmetal. For pumps, shaft liners, hydraulic parts. BS 1400 G1-C; BS 383: SAE 62; AMS 4845D.

HOLFOS G2

Holcroft Castings & Forgings Ltd.
Cu 88, Sn 8, Zn 4.
Cast: 31,000-45,000 TS; 17,000-22,400 YS; 3-15 El; 85-100 Brin. Properties vary with method of casting: continuous cast is strongest, shell or sand is most ductile. For pump fittings, valve nuts, bolts. BS 1400 G2-C; BS 1022; SAE 620. *Obsolete*

HOLFOS G3 (AS CAST)

Holcroft Castings & Forgings Ltd.
Copper. Cu 85.2, Sn 7, Zn 2.3, Ni 5.5.
Cast tin-nickel bronze (gunmetal); 70-130 Brin depending on casting methods and conditions. Good wear resistance. For gear wheels, actuating nuts, spindles. BS 1400 G3-C.

HOLFOS G3-WP

Holcroft Castings & Forgings Ltd.
Cu 84.75, Sn 7, Zn 2.25, P 0-0.5, Ni 5.5.
Cast, then age hardened: 63,000 TS; 40,000 YS; 3 El; 160-200 Brin. High strength bronze for critical applications. BS 1400 G3-W. *Obsolete*

HOLFOS HTB 1

Holcroft Castings & Forgings Ltd.
Copper. Cu 60, Zn 36, Al 1, Mn 2, Fe 1.
Cast: 67,000-70,000 TS; 25,000-31,000 YS; 20 El; 110-120 Brin. For screw down nuts in rolling mills and screw presses, neck bushings and stuffing boxes. BS 1400 HTB1-C; SAE 43; AMS 4860A.

HOLFOS HTB 2

Holcroft Castings & Forgings Ltd.
Cu 60, Zn 33, Al 3.5, Mn 2, Fe 1.5.
Cast: 85,000-87,000 TS; 40,000-42,000 YS; 15 El; 130-140 Brin. For valve and control parts, seatings, cones, marine hardware. BS 1400 HTB2-C. *Obsolete*

HOLFOS HTB 3

Holcroft Castings & Forgings Ltd.
Copper. Cu 60, Zn 30, Al 5, Mn 3, Fe 2.
Cast: 107,000 TS; 60,000 YS; 12-15 El; 180-185 Brin. For highly stressed hardware and components. BS 1400 HTB3-C.

HOLFOS J H 17

Holcroft Castings & Forgings Ltd.
Copper. Cu 84.7, Sn 14, P 1.3.
Centrifugally spun and continuously cast. As cast: 49,000-63,000 TS; 43,000-58,000 YS; 1-1.5 El; 130-160 Brin. Good wear resistance and resistant to sea water. For aircraft auxiliary driving gears, timing gears. (Ger) DIN 1705 G-SnBz14.

HOLFOS JHR42

Holcroft Castings & Forgings Ltd.
Cu 92.5, Zn 2, Mn 0.5, Fe 1.5, Si 3.5.
Cast: 49,000 TS; 29,000 YS; 18 El; 110 Brin. For bushings and bearings, esp. for top end bushings for diesel engines. BS 1400 SB1-C (1948); BS 1030. *Obsolete*

HOLFOS LB1

Holcroft Castings & Forgings Ltd.
Copper. Cu 76, Sn 9, Pb 15.
Cast: 25,000-34,000 TS; 11,200-22,400 YS; 3-8 El; 65-90 Brin. Properties vary with method of casting. For bearings and bushings, acid resistant fittings. BS 1400 LB1-C; ASTM B144-52-3D.

HOLFOS LB2

Holcroft Castings & Forgings Ltd.
Copper. Cu 80, Sn 10, Pb 10.
Cast: 27,000-40,000 TS; 11,200-25,000 YS; 3-10 El; 70-95 Brin. Properties vary with method of casting. Good sliding properties, corrosion resistant. For bearings for heavy duty. BS 1400 LB2-C; BS 963: SAE 64.

HOLFOS LB3

Holcroft Castings & Forgings Ltd.
Cu 85, Sn 10, Pb 5.
Cast: 27,000-45,000 TS; 11,200-34,000 YS; 3-15 El; 80-100 Brin. Properties vary with method of casting. For bearings, acid resistant fittings. BS 1400 LB3-C; BS 961. *Obsolete*

HOLFOS LB4

Holcroft Castings & Forgings Ltd.
Copper. Cu 85, Sn 5, Pb 10.
Cast: 22,400-38,000 TS; 9,000-22,400 YS; 5-25 El; 65-75 Brin. Properties vary with method of casting. For shafts, bushes and bearings. BS 1400 LB4-C; SAE 66.

HOLFOS LB5

Holcroft Castings & Forgings Ltd.
Copper. Cu 75, Sn 5, Pb 20.
Cast: 22,400-27,000 TS; 9,000-22,400 YS; 6-8 El; 60-70 Brin.
Properties vary with method of casting. For bearings for agricultural and railroad equipment, flour mill equipment. BS 1400 LB5-C; ASTM B144-52-3E.

HOLFOS LG1

Holcroft Castings & Forgings Ltd.
Copper. Cu 83, Sn 3, Zn 9, Pb 5.
Cast: 25,000-38,000 TS; 13,500-18,000 YS; 12-40 El; 55-70 Brin. Properties vary with method of casting. For lightly loaded marine hardware. BS 1400 LGI-C; BS 1159; ASTM B145-63-5A.

HOLFOS LG2

Holcroft Castings & Forgings Ltd.
Copper. Cu 85, Sn 5, Zn 5, Pb 5.
Cast: 25,000-38,000 TS; 13,500-22,400 YS; 7-25 El; 65-80 Brin. Properties may vary with method of casting. For pump castings, water pump impellers, hose couplings. BS 1400 LG2-C; SAE 40; AMS 4855 B.

HOLFOS LG3

Cu 86, Sn 7, Zn 5, Pb 2.
Cast: 31,000-45,000 TS; 13,500-25,000 YS; 5-15 El; 70-85 Brin. Properties vary with method of casting. For valves, hydraulic systems, super heated steam components. BS 1400 LG3-C; BS 1024. *Obsolete*

HOLFOS LG4

Holcroft Castings & Forgings Ltd.
Copper. Cu 87, Sn 7, Zn 3, Pb 3.
Cast: 36,000-43,000 TS; 18,000-22,400 YS; 5-20 El; 65-75 Brin. Properties vary with method of casting. For hydraulic pressure control equipment, petrol pump meters, valve bodies, compressor parts. BS 1400 LS4-C.

HOLFOS LG773

Holcroft Castings & Forgings Ltd.
Copper. Cu 83, Sn 7, Zn 3, Pb 7.
Cast: 31,000-54,000 TS; 14,5000-22,400 YS; 12-18 El; 75-90 Brin. Properties vary with methods of casting. For shafts, bearings, bushings, sliding surfaces, marine equipment. SAE 660; ASTM B144-52-3B.

HOLFOS LPB1

Holcroft Castings & Forgings Ltd.
Copper. Cu 88.2, Sn 7.5, Pb 4, 0.03 P min.
Cast: 27,000-46,000 TS; 11,200-27,000 YS; 2-30 El; 85-90 Brin. Properties vary with method of casting. Free machining bronze for bushings, valves, steam pressure fittings. BS 1400 LPB1-C; BS 1061.

HOLFOS PB1

Holcroft Castings & Forgings Ltd.
Copper. Cu 89.5, Sn 10, 0.5 P min.
Cast: 34,000-66,000 TS; 18,000-45,000 YS; 2-20 El; 70-120 Brin. Properties vary with method of casting. Good bearing properties. For shafts, guide wheels, blade wheels for pumps and water turbines. BS 1400 PB1-C, BS 2B8.

HOLFOS PB2

Holcroft Castings & Forgings Ltd.
Copper. Cu 88.35, Sn 11.5, P 0.15-0.
Cast: 36,000-54,000 TS; 20,000-47,000 YS; 6-10 El; 85-130 Brin. Properties vary with method of casting. Good wear and abrasion resistance. For bearings, feed nuts operating under load, coupling blocks, worm wheels. BS 1400 PB2-C; BS 421; SAE 65.

HOLFOS PB3

Holcroft Castings & Forgings Ltd.
Cu 90.2, Sn 9.5, P 0.3.
Cast: 34,000-50,000 TS; 20,000-36,000 YS; 5-20 El; 90-110 Brin. Properties vary with method of casting. For pressure vessels, pump bodies, casings, impellers, piston shafts. BS 1400 PB3-C; ASTM D135-66, Alloy 524 (Chem). *Obsolete*

HOLFOS PB4

Holcroft Castings & Forgings Ltd.
Copper. Cu 89.5, Sn 10, P 0.5.
Cast tin bronze; 70-130 Brin depending on casting method. For bearings with hard shafts, medium duty gears, marine equipment. BS 1400 PB4-C.

HOLFOS W W

Holcroft Castings & Forgings Ltd.
Copper. Cu 88.3, Sn 11.5, P 0.2.
Sand cast: 36,000 TS; 20,000 YS; 10 El; 895 Brin. Good wear properties. For gears, shafts, splined couplings. BS 1400 PB2-C; BS 421 SAE 65.

HOLLO

Hidalgo Steel Co. Inc.
C, bal Fe.
For rock drills; water hardening.

HOLLOBAR

Peninsular Steel Co.
C 1, Cr 1.2, Mo 0.3, bal Fe.
For tools; water hardened.

HOLLOBAR

Diehl Steel Co.
C 1.05, Mn 0.38, Cr 1.46, Si 0.28, bal Fe.
For cold work dies; oil hardened.

HOLLOMEK 510

British Steel plc
C 0-0.2, Si 0-0.4, Mn 0-1.5, P 0-0.04, S 0-0.04, 0.025 Al min, bal Fe.
490-630 N/mm^2 TS; 355 N/mm^2 YS. For hollows for machining of angular components, fluid power cylinders, and rollers. Weldable steel tube.

HOLLOMEK 6V

British Steel plc
C 0.17-0.22, Si 0.2-0.35, Mn 1.4-1.6, P 0-0.025, S 0-0.025, Cr 0-0.25, Mo 0-0.1, Ni 0-0.35, Cu 0-0.35, V 0.08-0.12, 0.025 Al min, bal Fe.
600 N/mm^2 TS min; 500 N/mm^2 YS min. Hollow bar.

HOLLOW BAR

P.F. McDonald & Co.
C, Cr, bal Fe.
For dies, punches; oil hardening.

HOLLOW BLUE BAND

Hollup Corp.
C 0-0.1, Cr 25, Ni 12, bal Fe.
For welding electrodes for stainless steels; stainless; coated.

HOLLOW DRILL

Agawam Tool Co.
Cr 1-1.1, Mn 0.35, Si 0.3, C, bal Fe.
For rock drills; water hardened.

HOLLOW DRILL

Horace T. Potts Co.
C 1-1.3, bal Fe.
For hollow drills. *Obsolete*

HOLLOW DRILL

Ziv Steel & Wire Co.
C 1.2, bal Fe.
For hollow drills; water hardened.

HOLLUP GR. 30-S

Hollup Corp.
C 0.1-0.15, bal Fe.
Welded: 50,000 TS; 8 El. For welding electrodes; coated.

HOLLUP GR. 30XL

Societe Nouvelle des Acieries de Pompey
C, bal Fe.
Welded: 60,000 psi TS; 12 El. For welding rods; coated.

HOLLUP GR. NO. 120S

Hollup Corp.
C 0.1, Cr 18, Ni 8, bal Fe.
For welding rod; stainless.

HOLLUP GR. NO. 150

Hollup Corp.
C 0.08-0.1, bal Fe.
Welded: 45,000 TS; 13 El. For welding electrodes; coated.

HOLLUP GR. NO. 30-X

Hollup Corp.
C 0.1-0.15, bal Fe.
For welding electrodes; coated.

HOLLUP GR. NO. 30SB

Hollup Corp.
C 0.1-0.15, bal Fe.
Welded: 55,000 TS; 10 RA. For welding electrodes; coated.

HOLLUP GR. NO. 30XB

Hollup Corp.
C 0.1-0.15, bal Fe.
For welding electrodes; coated.

HOLLUP GR. NO. 33

Societe Nouvelle des Acieries de Pompey
C 0.08-0.1, bal Fe.
For welding electrodes; copper coated.

HOLLUP GR. NO. 50

Hollup Corp.
C 0.18-0.25, bal Fe.
Welded: 70,000 TS. For welding electrodes; coated.

HOLLUP GR. NO. 50A

Hollup Corp.
C 0.18-0.25, bal Fe.
For welding electrodes; coated.

HOLLUP GR. NO. 55

Hollup Corp.
C 0.2, bal Fe.
For welding rod; for pipe and pressure weld.

HOLLUP GR. NO. 70A

Hollup Corp.
C 0.85-1.1, bal Fe.
For welding electrodes; coated.

HOLLUP GR. YOLOY

Hollup Corp.
C 0.15, Cu 0.5, bal Fe.
For welding electrodes; coated.

HOLLUP GRADE 1

Hollup Corp.
C 0-0.06, bal Fe.
Cast: 45,000 TS; 13 El. For welding electrodes for low carbon steel; coated.

HOLLUP GRADE 10

Hollup Corp.
C 0-0.06, bal Fe.
Cast: 45,000 TS; 13 El. For welding electrodes; coated.

HOLLUP GRADE 10X

Hollup Corp.
C 0-0.06, bal Fe.
For welding electrodes; coated, ductile.

HOLLUP GRADE 11

Hollup Corp.
C 0-0.06, bal Fe.
For welding rod for wrought iron and steel; Cu coated.

HOLLUP GRADE 15

Hollup Corp.
C 0.08-0.1, bal Fe.
Cast: 45,000 TS; 10 El. For welding electrodes; coated.

HOLLUP GRADE 20
Hollup Corp.
Cu alloy.
Cast: 40,000 TS. For coated bronze welding rod for copper alloys.

HOLLUP GRADE 20B
Hollup Corp.
Cu alloy.
Cast: 40,000 TS. For coated bronze welding rod, for light gauge sheet.

HOLLUP GRADE 3
Hollup Corp.
C 0.1-0.15, bal Fe.
Cast: 60,000 TS; 12 El. For welding electrodes for boiler flues and fire boxes; coated.

HOLLUP GRADE 3R
Hollup Corp.
C 0.1-0.15, bal Fe.
For welding electrodes for boiler tubes; special coating.

HOLLUP GRADE 5
Hollup Corp.
C 0.18-0.25, bal Fe.
Cast: 70,000 TS. For welding electrodes for building up bearing surfaces; coated, wear resistant.

HOLLUP GRADE 6
Hollup Corp.
C, Ni, bal Fe.
For welding electrodes; coated; shock resistant.

HOLLUP GRADE 7
Hollup Corp.
C 0.85-1.1, bal Fe.
For welding electrodes for high carbon steel parts; coated.

HOLLUP GRADE 9
Hollup Corp.
C 0.75-0.95, Mn 12-14, Ni 3-4, bal Fe.
190-500 Brin. For welding electrodes for high manganese steels; coated.

HOLLUP GRADE CI-10
Hollup Corp.
C, bal Fe.
435 Brin. For welding electrodes for hard surfacing cast iron; coated.

HOLLUP GRADE CI-3
Hollup Corp.
C, bal Fe.
For welding electrodes for cast iron; coated.

HOLLUP GRADE CI-8
Hollup Corp.
C, bal Fe.
For welding electrodes for cast iron; coated.

HOLLUP GRADE CI-9
Hollup Corp.
C, Ni, Cu, bal Fe.
For welding electrodes for cast iron; coated.

HOLLUP GRADE NO. 22
Hollup Corp.
C, bal Fe.
For welding electrodes.

HOLLUP GRADE NO. 30
Hollup Corp.
C 0.1-0.15, bal Fe.
Cast: 60,000 TS; 12 El. For welding electrodes for flues, boilers and fire boxes; coated.

HOLLUP H.S.V.
Hollup Corp.
C 0.2, Ni 1, Cr 0.6, V 0.15, bal Fe.
For coated welding electrodes.

HOLLUP I.H.S.
Hollup Corp.
C, alloy, bal Fe.
For coated welding electrodes.

HOLLUP MCM
Hollup Corp.
C 0.2, Cr 1.2, Mo 0.2, bal Fe.
Cast: 85,000 TS; 27 El. For coated welding electrodes; for aircraft tubing.

HOLLUP N.C.
Hollup Corp.
C, bal Fe.
For coated welding electrodes.

HOLLUP NO. 1217
Hollup Corp.
Al alloy.
For coated Al welding rod.

HOLLUP NO. 2320
Hollup Corp.
C 0.1, Ni 3.5, bal Fe.
For coated welding electrodes for Ni steels; shock resistant.

HOLLUP NO. 2512
Hollup Corp.
C 0.2, Ni 5, bal Fe.
Cast: 100,000 TS; 25 El. For coated welding electrodes for Ni steels; tough.

HOLLUP NO. 6723
Hollup Corp.
Fe 0-3.5, 23.0 Cu min, bal Ni.
50,000 TS. For Monel metal welding rod; corrosion resistant.

HOLLUP PHOS-COPPER
Hollup Corp.
P 7-9, bal Cu.
For P-Cu brazing rod for copper alloys; self-fluxing.

HOLLUP R.D.S.
Hollup Corp.
C 0.2, bal Fe.
For coated welding electrodes.

HOLLUP RED BAND
Hollup Corp.
C 0.1, Cr 18, Ni 8, bal Fe.
Cast: 90,000 TS; 30 El. For stainless steel welding electrodes; stainless.

HOLLUP YELLOW BAND
Hollup Corp.
C 0.1, Cr 4-6, Mo 0.5, bal Fe.
Cast: 75,000 TS; 35 El. For welding electrodes for 5 Cr steel.

HOLMIUM
Atomergic Chemetals Corp.
Ho.
Purities: 99.9% special distilled grade, 99.5+%. Forms: ingot, lump, sheet, foil, wire, fillings, sponge, powder, single crystals.

HOLTITE
Belmont Metals Inc.
Cu 50, Zn 50, granular.
For brazing all metals except aluminum; melting point 1610°F.

HOLTO 1948
Compagnie Ateliers et Forges de la Loire
C 0.28, Cr 1.5, Mo 0.6, V 0.15, Co 5, bal Fe.
Heat treated: 128,000-270,000 TS. For die casting dies, brass forging dies. Hot work tool steel, deep hardening.

HOLTO A.C.M.
Compagnie Ateliers et Forges de la Loire
C 0.6, Ni 1.6, Cr 0.8, Mo 0.3, V 0.1, bal Fe.
Heat treated: 150,000-220,000 TS; 130,000-200,000 YS; 12-20 El; 40-58 RA; 150-400 Brin. For hot work tools, punches, hot shears, upsetters, caulking tools. Hot work steel, oil hardening, shock resistant.

HOLTO A.F.L.M.
Compagnie Ateliers et Forges de la Loire
C 0.6, Ni 2.5, Cr 0.9, Mo 0.35, V 0.15, bal Fe.
For hot work tolls, punches, hot shears, upsetters. Hot work steel, oil hardening, tough.

HOLTO C.R.V.
Compagnie Ateliers et Forges de la Loire
C 1.5, Cr 1.8, V 0.25, bal Fe.
For gages, taps, threading dies, cold cutting and stamping tools, plastic molds. Cold working tool steel, oil hardening.

HOLTO D.R.B.
Compagnie Ateliers et Forges de la Loire
C 1, Cr 1.25, bal Fe.
Oil hardened: 288,000 TS; 278,000 YS; 540 Brin. For rollers, bearings, taps punches, cutters, pivot pins, stamping dies, plug gages. Cold working steel, oil hardening, deep hardening, medium toughness.

HOLTO N.B.2.
Compagnie Ateliers et Forges de la Loire
C 0.9, Mn 1.3, Cr 0.5, bal Fe.
Annealed: 85,000 TS; 60,000 YS; 26 El; 185 Brin. Oil hardened: 280,000 TS; 272,000 El; 535 Brin. For dies, punches, gauges, cutters, cold headers, stamping dies, broaches. Non-deforming tool steel, oil hardening, shock resistant, tough.

HOLTO R.B.L
Compagnie Ateliers et Forges de la Loire
C 1.4, Cr 0.7, bal Fe.
For cutters, bearings, cold heading dies, hand taps, bearings, reamers, boring tools. Cold working steel. Oil hardening.

HOLTO R.B.O.
Compagnie Ateliers et Forges de la Loire
C 0.95, Cr 0.75, bal Fe.
Annealed: 85,000 TS; 46,000 YS; 175 Brin. Hardened: 260,000 TS; 240,000 YS; 525 Brin. For bearings, rollers, bushings, gauges, mandrels, punches, cold heading tools. Cold work tool steel, oil or water harden, wear resistant.

HOLTO S.P.2
Compagnie Ateliers et Forges de la Loire
C 0.28, Mo 5, V 0.25, Cr 6, W 5, Co 0.6, bal Fe.
Hot rolled: 114,000 TS. Heat treated: 47-56 Rock C. For die casting dies, extrusion rams and liners, brass forging dies. Hot work steel. High hot hardness and resistance to thermal shock.

HOLTO SPECIAL B.M.
Compagnie Ateliers et Forges de la Loire
C 0.7, Cr 1.4, Mo 0.7, V 0.15, bal Fe.
For punches, shears, die casting dies. Hot and cold work tool steel, oil hardening, shock resisting.

HOLTO SPECIAL D
Compagnie Ateliers et Forges de la Loire
C 0.28, W 9.5, Cr 3.1, V 0.3, bal Fe.
Heat treated: 142,000-350,000 TS; 120,000-180,000 YS; 6-13 El. For drawing dies, extrusion rams and dies, valve heads, stamping dies. Hot work tool steel. High hot hardness above 650°C.

HOLTO SPECIAL D.A.
Compagnie Ateliers et Forges de la Loire
C 0.4, Cr 3.25, W 2.5, V 0.4, bal Fe.
Hardened: 250,000 TS; 200,000 YS; 8 El; 12 RA. For press dies, upsetters, rivet sets, forging and punching dies, dummy blocks. Hot work steel up to 600°C applications.

HOLTO SPECIAL D.C.
Compagnie Ateliers et Forges de la Loire
C 0.21, W 8, Cr 2.6, Co 2, Mo 1, V 0.4, bal Fe.
Heat treated: 135,000-340,000 TS; 110,000-214,000 YS; 10-14 El. For wire drawing dies, hot upsetters, stamping dies, forging dies. Hot work tool steel. High hot hardness.

HOLTO SPECIAL E.X.
Compagnie Ateliers et Forges de la Loire
C 0.45, Ni 4.6, W 1.75, Cr 0.4, Mo 0.3, bal Fe.
For forging and riveting dies, shears, punches. Hot and cold work steel. Shock resistant. Oil hardening.

HOLTO SPECIAL F.A.41
Compagnie Ateliers et Forges de la Loire
C 0.45, Cr 4.75, W 3.75, V 0.45, bal Fe.
Air hardened: 270,000 TS; 208,000 YS; 5 El; 9 RA. For forging dies and inserts, extrusion dies and rams, dummy blocks. Hot work steel, oil or air hardening, non-deforming, deep hardening.

HOLTO SPECIAL F.A.48
Compagnie Ateliers et Forges de la Loire
C 0.42, W 4.75, Co 2, Cr 4.75, V 0.45, bal Fe.
Heat treated: 170,000-300,000 TS; 156,000-256,000 YS; 4-8 El. For thread rolling dies, spinning tools, dies. Hot work steel, oil hardening. Resists cyclic heating.

HOLTO SPECIAL F.A.82
Compagnie Ateliers et Forges de la Loire
C 0.38, Cr 5, Mo 1.3, Si 1, V 0.5, bal Fe.
Annealed: 90,000 TS; 46,000 YS; 18 El. Air hardened: 217,000 TS; 184,000 YS; 13 El; 40 RA; 53-55 Rock C. For die casting dies, forging dies and inserts, stamping dies, extrusion dies, hot punches. Hot work steel, oil hardening, non-deforming, shock resistant.

HOLTO SPECIAL M
Compagnie Ateliers et Forges de la Loire
C 0.42, Si 1.6, Cr 0.8, Mo 0.8, bal Fe.
Water hardened: 190,000-300,000 TS; 6-16 El; 15-50 RA; 40-55 Rock C. For cold headers, punches, chisels, rivet sets, upsetters, pneumatic tools. Cold working steel. Shock resistant. Oil hardening.

HOLTO SPECIAL M.D.
Compagnie Ateliers et Forges de la Loire
C 0.55, Cr 1.1, Se 0.9, bal Fe.
For chisels, rivet sets, upsetters, pneumatic tools. Cold work steel, shock resistant.

HOLTO SPECIAL M.O.V.
Compagnie Ateliers et Forges de la Loire
C 0.45, Cr 1.9, Mo 0.6, V 0.35, bal Fe.
Annealed: 100,000 TS; 48,000 YS; 22 El; 197 Brin. Hardened: 290,000 TS; 245,000 YS; 10 El; 580 Brin. For hot work tools, hot upsetters, rivet sets, punches, forging dies, inserts, molds for plastics. Hot work steel, oil hardening, shock resistant.

HOLTO SPECIAL M.O.V.2
Compagnie Ateliers et Forges de la Loire
C 0.37, W 3, Cr 2.5, Mo 0.7, V 0.25, bal Fe.
For wire cutters, forging die inserts, die casting dies, hot heading bolts. Hot work steel, oil hardening, wear resistant.

HOLTO SPECIAL P
Compagnie Ateliers et Forges de la Loire
C 0.42, Ni 4.6, Mo 1.15, Cr 0.3, bal Fe.
For hot work tools, punches, upsetters. Hot work steel, oil hardening.

HOLTO SPECIAL R
Compagnie Ateliers et Forges de la Loire
C 1, Cr 1.65, V 0.15, bal Fe.
Hardened: 280,000 TS; 255,000 YS; 555 Brin. Annealed: 90,000 TS; 50,000 YS; 28 El; 185 Brin. For rolling mill cylinders, cold headers, gages, bearings, bushings, drawing dies, punching, drills. Cold work tools, oil hardening, deep hardening.

HOLTO SPECIAL R.K.
Compagnie Ateliers et Forges de la Loire
C 1.4, Cr 1.45, Mo 0.3, V 0.15, bal Fe.
For gages, headers, drawing dies, punches. Cold work tool steel, oil hardening.

HOLTO SPECIAL R.U.
Compagnie Ateliers et Forges de la Loire
C 0.95, Cr 0.9, Mo 0.45, bal Fe.
For bearings, bushings, gauges, rollers, cold header dies. Cold work steel, oil hardening.

HOLTO SPECIAL S.C.1
Compagnie Ateliers et Forges de la Loire
C 1.2, W 1.25, bal Fe.
Heat treated: 280,000 TS; 270,000 YS; 540 Brin. Annealed: 85,000 TS; 22 El; 45 RA; 185 Brin. For fast finishing cutters, drills, taps, cold cutting tools, augers, paper knives. Cold work steel, oil or water hardening, wear resistant.

HOLTO SPECIAL-U
Compagnie Ateliers et Forges de la Loire
C 0.9, Mn 2, bal Fe.
Hardened: 64-65 Rock C. For stamping dies, gages, punches, taps, cold headers, broaches, trimming dies. Non-deforming, oil hardening, shock resistant, Type O2 cold work tool steel.

HOLTO V.D.L.D.M.
Compagnie Ateliers et Forges de la Loire
C 0.35, Ni 4, Cr 1.7, Mo 0.5, bal Fe.
For hot work tools, punches, hot shears, upsetters. Hot work steel, oil hardening, tough, shock resistant.

HOMBERGS ALLOY
Manufacturer not listed.
Pb 33.3, Sn 33.3, Bi 33.3.
For sprinkler plugs, fusible metal; M.P. 251°F.

HONALIUM 31
Honsel-Werke Aktiengesellschaft
AlMg3Si T6.
Chill cast: 220 N/mm² TS; 90 Brin. Sand cast: 180 N/mm² TS; 60 Brin. For corrosion resistant aluminum castings; resists seawater. Good decorative anodized finish.

HONALIUM 411
Honsel-Werke Aktiengesellschaft
AlMg4Si1Mn.
Chill cast, F: 160 N/mm² TS; 50 Brin. Chill cast, T6: 320 N/mm² TS; 100 Brin. Aluminum casting for elevated temperature, as cylinder heads.

HONALIUM 5
Honsel-Werke Aktiengesellschaft
AlMg5.
Chill cast: 150 N/mm² TS; 55 Brin. Sand cast: 140 N/mm² TS; 50 Brin. For corrosion resistant aluminum castings; resists seawater. Good decorative anodized finish.

HONALIUM 51
Honsel-Werke Aktiengesellschaft
AlMg5Si.
Chill cast: 150 N/mm² TS; 60 Brin. Sand cast: 140 N/mm² TS; 55 Brin. Aluminum castings; resistant to seawater; better castability than HONALIUM 5. Good decorative anodized finish.

HONALIUM S
Honsel-Werke Aktiengesellschaft
AlMg10Si3 T2.
Chill cast: 190 N/mm² TS; 105 Brin. Sand cast: 170 N/mm² TS; 80 Brin. For aluminum castings with good corrosion resistance.

HONALIUM-ELEKTRAL
Honsel-Werke Aktiengesellschaft
AlSi5Mg T6.
Casting: 220 N/mm² TS; 60-85 Brin. Aluminum casting with high electrical conductivity.

HONDA NEW
Japanese manufacture
Ti 6.7, Al 3.7, Ni 18, Co 27, Fe 45.
For permanent magnets.

HONSEL HO 3
Honsel-Werke Aktiengesellschaft
AlMn.
Soft: 100-120 N/mm² TS; 30-35 Brin. Hard: 160-230 N/mm² TS; 40-55 Brin. Sheet and plate for cooking utensils, chemical equipment, storage tanks, builders hardware.

HONSEL HO E 10
Honsel-Werke Aktiengesellschaft
AlMg1.
Soft: 100-130 N/mm² TS; 30-35 Brin. Hard: 160-240 N/mm² TS; 50-65 Brin. Sheet and plate for architectural trim, builders hardware, appliances.

HONSEL HO E 20
Honsel-Werke Aktiengesellschaft
AlMg2.
Soft: 150-180 N/mm² TS; 40-55 Brin. Hard: 210-280 N/mm² TS; 60-75 Brin. Aluminum alloy sheet and plate for architectural and furniture trim, traffic signs.

HONSEL HO E 25
Honsel-Werke Aktiengesellschaft
AlMg2.5.
Soft: 170-215 N/mm² TS; 50 Brin. Hard: 250-290 N/mm² TS; 80 Brin. Aluminum sheet and plate for metal work, auto and appliance trim.

HONSEL HO S
Honsel-Werke Aktiengesellschaft
AlMgMn.
Soft: 180-220 N/mm² TS; 45-60 Brin. Hard: 260-350 N/mm² TS; 75-90 Brin. Aluminum sheet and plate for moderate strength structures, storage tanks, chemical equipment.

HONTAL
Honsel-Werke Aktiengesellschaft
AlCu4Ti T6.
Chill cast: 280 N/mm² TS; 90 Brin. Sand cast: 250 N/mm² TS; 85 Brin. Aluminum casting with good strength and high pressure tightness; good castability.

HONTAL S
Honsel-Werke Aktiengesellschaft
AlCu4TiMg T6.
Chill cast: 300 N/mm² TS; 95 Brin. Sand cast: 280 N/mm² TS; 90 Brin. Aluminum casting with good strength and high pressure tightness.

HONTRON A6
Honsel-Werke Aktiengesellschaft
MgAl6.
Sand cast: 150 N/mm² TS; 50-65 Brin. Magnesium casting alloy with high tear resistance, as for motor car wheels.

HOOKER BRASS
English manufacture
Cu 61, Pb 2, Zn 37.
For hot forgings, brass parts; free-cutting.

HOPKINSON ALLOY
Manufacturer not listed.
Fe 75, Ni 24.5, trace Si.
160,000 TS; 88,000 YS; 45 El; 68 RA. For corrosion and heat resistant parts; Sc.-21.

HORBACH HH100
Horbach & Schmitz GmbH
C 0-0.12, Al 1, Cr 18, bal Fe.
For oil refinery equipment; heat and creep resistant.

HORBACH HH110A
Horbach & Schmitz GmbH
Stainless steel. C 0.15, Cr 19.5, Ni 9.5, bal Fe.
Annealed: 80,000 TS; 35,000 YS; 55 El; 75 RA; 160 Brin. For chemical plant equipment, tanks, mixers; Type 302; austenitic.

HORBACH HH120
Horbach & Schmitz GmbH
C 0-0.12, Al 1.5, Cr 24, bal Fe.
For oil refinery equipment; heat and creep resistant.

HORBACH HH120A
Horbach & Schmitz GmbH
Stainless steel. C 0.15, Cr 24, Ni 19, bal Fe.
Annealed: 100,000 TS; 45,000 YS; 50 El; 65 RA; 185 Brin. For valves, pumps, furnace parts, turbine and jet parts; Type 310; austenitic.

HORBACH HH80
Horbach & Schmitz GmbH
C 0.12, Cr 13, Si 2.2, bal Fe.
Annealed: 75,000 TS; 40,000 YS; 35 El; 70 RA; 155 Brin. For valves, cutlery, pump bodies, bolts; Type 410; corrosion resistant.

HORBACH HN1
Horbach & Schmitz GmbH
Stainless steel. C 0-0.15, Cr 18, Ni 8, bal Fe.
Annealed: 80,000 TS; 35,000 YS; 55 El; 75 RA; 150 Brin. For chemical plant equipment tanks, mixers, agitators; Type 302; austenitic.

HORBACH HN1 EXTRA
Horbach & Schmitz GmbH
Stainless steel. C 0-0.12, Cr 18, Ni 9.5, Ti = 4 x C, bal Fe.
Annealed: 85,000 TS; 35,000 YS; 55 El; 65 RA; 150 Brin. For welded chemical plant equipment, tanks, mixers; Type 321; austenitic.

HORBACH HN1 SUPRA
Horbach & Schmitz GmbH
Stainless steel. C 0-0.07, Cr 18, Ni 9.5, bal Fe.
Annealed: 85,000 TS; 35,000 YS; 60 El; 70 RA; 150 Brin. For chemical plant equipment, tanks, vessels; Type 304; austenitic.

HORBACH HN2 EXTRA
Horbach & Schmitz GmbH
Stainless steel. C 0-0.12, Cr 18, Mo 2, Ni 10.5, Ti = 4 x C, bal Fe.
Annealed: 85,000 TS; 35,000 YS; 50 El; 65 RA; 160 Brin. For chemical plant equipment, welded structures; Type 316 Ti; austenitic.

HORBACH HN2 SUPRA
Horbach & Schmitz GmbH
Stainless steel. C 0.07, Cr 18, Ni 10.5, Mo 2, bal Fe.
Annealed: 85,000 TS; 35,000 YS; 50 El; 65 RA; 160 Brin. For acid resistant chemical plant equipment; Type 316; austenitic.

HORBACH HNH
Horbach & Schmitz GmbH
Stainless steel. C 0.4, Cr 13, Si 0.4, bal Fe.
Annealed: 95,000 TS; 50,000 YS; 25 El; 55 RA; 196 Brin. For valves, cutlery, surgical and dental instruments; Type 420.

HORBACH HNM
Horbach & Schmitz GmbH
Stainless steel. C 0.2, Si 0.4, Cr 13, bal Fe.
Annealed: 95,000 TS; 50,000 YS; 25 El; 55 RA; 196 Brin. For valves, cutlery, surgical instruments; Type 420.

HORBACH HNW
Horbach & Schmitz GmbH
Stainless steel. C 0-0.12, Si 0.4, Cr 13, bal Fe.
Annealed: 75,000 TS:; 40,000 YS; 35 El; 70 RA; 155 Brin. For turbine blades, cutlery, valves; Type 410.

HORSEHEAD ZAMAK NO. 3
Now ZAMAK NO. 3.

HOSKINS 717
Now CHROMEL 1A.

HOSKINS 827
Hoskins Mfg. Co.
Cr 20, Fe 8, Si 2, bal Ni.
For thermoelements. Heat and oxidation resistant. *Obsolete*

HOSKINS ALLOY 400
Hoskins Mfg. Co.
Nickel. Ni 67, Cu 31, Fe 1, Mn 1.
Drawn wire: 60,000 psi TS. For corrosion resistance.

HOSKINS ALLOY 600
Hoskins Mfg. Co.
Nickel. Ni 76, Cr 15.5, Fe 8, Mn 0.5, Si 0.2.
For corrosion resistance; heat and oxidation resistant.

HOSKINS ALLOY 750
Hoskins Mfg. Co.
Stainless steel. Cr 15, Al 4, Si 0.5, C 0.1, bal Fe.
Elec. resistance: 750 ohms/circular mil ft. Resistance wire for furnaces, kilns, other heating devices; requires adequate support at high temperatures. Useful to 2050°F.

HOSKINS ALLOY 800
Hoskins Mfg. Co.
Superalloy. Ni 35, Cr 21, Si 1.6, bal Fe.
Drawn wire: 100,000 psi TS. For furnace fixtures; heat resistance applications; heat and oxidation resistant.

HOSKINS ALLOY 815
Hoskins Mfg. Co.
Stainless steel. C 0.1, Si 0.5, Al 4.6, Cr 22.5, bal Fe.
Hot rolled: 143,000 TS; 12 El; 234 BHN. Annealed: 108,000 TS; 27 El; 190 BHN. For resistances, heating elements, rheostats; heat resistant to 2150°F.

HOSKINS ALLOY 831
Hoskins Mfg. Co.
Nickel. Cr 15, Fe 7.5, bal Ni.
Drawn: 100,000 TS. For spark plugs. Electrode alloy, oxidation resistant.

HOSKINS ALLOY 875
Hoskins Mfg. Co.
Stainless steel. Cr 22.5, Al 5.5, Si 0.5, C 0.1, bal Fe.
Hot rolled: 143,000 TS; 12 El; 185 BHN. Annealed: 108,000 TS; 27 El; 190 BHN. For electrical resistances, heating elements; high heat resistance to 2350°F.

HOT
Sheffield Smelting Co. Ltd.
Ag 60, Cu, Zn.
Melting range: 600-720°C. Maximum stress: 34.6 kgf/mm^2; 5 El. For silver brazing in protective atmospheres.

HOT DIE
Atlas Specialty Steels
C 0.3, W 11, Cr 3, V 0.5, bal Fe.
For hot die tools, punches; hot die steel. *Obsolete*

HOT DIE
Adams & Osgood Steel Co.
C, W, bal Fe.
For tools; oil hardened.

HOT DIE "C"
CCS Braeburn Alloy Steel
C 0.3, Cr 2.5, W 12, Co 3.6, Cr 4, bal Fe.
For hot punches, forming and extrusion dies; hot work steel, oil hardened. *Obsolete*

HOT DIE 593
Allegheny Ludlum Steel
C 0.38, Mo 0.25, W 5, Cr 5, V 0.2, Mn 0.25, Co 0.45, Si 0.9, bal Fe.
Heat treated: 270,000 TS; 208,000 YS; 5 El; 9 RA. At 900°F: 217,000 TS; 172,000 YS; 10 El; 31 RA. Uses: extrusion dies, hot upset punches, insert forging dies, dummy blocks. Type H14; hot work steel.

HOT DIE NO. 2
CCS Braeburn Alloy Steel
C 1, Cr 4, bal Fe.
For hot work dies; hot work steel. *Obsolete*

HOT DIE STEEL
Teledyne Vasco
C 0.5, Cr 1.5, V 0.3, W 2.4, bal Fe.
For punches, hot upsetters; hot work steel. *Obsolete*

HOT FORM DRILL ROD
Teledyne Allvac
C 0.35, Si 1, W 1.25, Cr 5, V 0.5, Mo 1.45, bal Fe.
For die casting die tools, high temperature fasteners; hot work steel, oil hardened.

HOT HEADER NO. 3
Teledyne Vasco
C 0.9, Cr 5, bal Fe.
For hot work tools and dies. *Obsolete*

HOT STAMPING ALLOY DIE
Edgar T. Ward's Sons Co.
C 0.5, Cr 3, bal Fe.
For hot stamping dies; hot work steel.

HOT STAMPING ALLOY DIE
Jessop-Saville Ltd.
C, alloy, bal Fe.
For hot stamping dies, tools; oil hardened.

HOT WORK 8
Bethlehem Steel Corp..
C 0.6, Mo 8.5, V 1.5, Cr 3.2, bal Fe.
For hot-work tools and dies, punches, blanking dies; hot-work steel, wear resistant. *Obsolete*

HOT WORK NO. 2
Abex Corp.
C 0.3, Cr 27, Mo 5.5, Ni 2.8, Co 62.
Weld hardness: 32 Rock C; work hardened. Hard facing rod and electrode; resistant to heat, corrosion and impact. For buildup and repair valves and dies.

HOT WORK NO. 22
Manufacturer not listed.
C 0.4, Cr 3, W 9, V 0.2, bal Fe.
For tools, dies; hot work steel.

HOT WORK NO. 23
Manufacturer not listed.
C 0.45, Cr 3, W 14, V 0.3, bal Fe.
For tools, cutters, taps, gauges, reamers; high speed steel.

HOT-DIE (CHROME)
American Saw & Mfg. Co.
Cr 3.5-4, C 0.8-0.9, bal Fe.
For dies for hot metal, tools; shock and wear resistant. *Obsolete*

HOT-DIE (CHROME)

Latrobe Steel Co.
Cr 3.5-4, C 0.8-0.9, bal Fe.
For dies for hot metal, tools; shock and wear resistant.
Obsolete

HOT-DIE (TUNGSTEN)

American Saw & Mfg. Co.
Cr 2-3, C 0.3-0.45, V 1, W 10-14, bal Fe.
For dies for hot metal, tools; shock and wear resistant.
Obsolete

HOT-DIE (TUNGSTEN)

Latrobe Steel Co.
Cr 2-3, C 0.3-0.45, V 1, W 10-14, bal Fe.
For dies for hot metal, tools; shock and wear resistant.
Obsolete

HOTFORM NO. 2

Teledyne Vasco
C 0.35, Si 0.9, Cr 4.7, Mo 1.3, V 0.5, bal Fe.
For punches, upsetters, hot work tools and dies, extrusion press; Type H11; hot work steel.

HOTFORM NO. 3

Teledyne Vasco
C 0.55, Si 0.9, Mo 1.4, Cr 4.75, bal Fe.
For punches, upsetters, hot dies and tools; Type H12; air hardened.

HOTFORM ULTRA

Teledyne Vasco
C 0.33, Mo 1.3, Cr 5, V 0.45, Co 3, Si 0.9, Mn 0.3.
54-56 Rock C. Hot work die steel; heat treated. For die casting dies, core pins and inserts, piercing points, hot forging dies, and hot extrusion dies.

HOTFORM V

Teledyne Vasco
C 0.37-0.43, Si 1, Cr 5-5.5, Mo 1.2, V 1.1, bal Fe.
For Al die casting dies, shear blades, forging dies; hot work steel; oil hardened. Type H13.

HOTPRESS

Teledyne Vasco
C 0.35, Cr 2, W 9.5, V 0.5, bal Fe.
For hot work dies, upsetter dies; tough and heat resistant; hot work steel. Type 20.

HOTSPUR

Dunford Hadfields Ltd.
C 0.2, Cr 12-14, Ni 7-25, bal Fe.
Cast: 89,600 TS. For furnace parts and equipment; heat and oxidation resistant. *Obsolete*

HOUGHTON'S SHAVING PACKING

English manufacture
Sn 5.97, Pb 94.02, trace Cu.
For piston rod packing for steam engines.

HOVER 123

Hover, Gebruder, Edelstahlwerk
C 1.45, Cr 1.4, bal Fe.
For bearings, liners, races, sleeves; water hardened.

HOVER 151

Hover, Gebruder, Edelstahlwerk
C 0.61, Cr 1.18, V 0.1, bal Fe.
For springs, shafts, crankshafts, punches, crimpers; oil or water hardened.

HOVER 210

Hover, Gebruder, Edelstahlwerk
C 1.25, Si 1.15, Mn 0.7, Cr 1.2, bal Fe.
For dies, punches, shock resistant cutters; oil hardened, tough.

HOVER 350

Hover, Gebruder, Edelstahlwerk
C 1, V 0.1, Mn 0.25, bal Fe.
For cutters, tools, dies, taps, drills; Type W2; water hardened.

HOVER 401

Hover, Gebruder, Edelstahlwerk
C 1.05, Mn 0.9, Cr 1, W 1.15, bal Fe.
For fast finishing cutters, reamers; water hardened.

HOVER 550

Hover, Gebruder, Edelstahlwerk
C 0.7, Si 1.7, Mn 0.7, bal Fe.
For punches, upsetters, riveters, springs; oil hardened, tough.

HOVER 702

Hover, Gebruder, Edelstahlwerk
C 0.28, Ni, Cr, Mo, V, bal Fe.
For gears, bolts, machine tool parts; oil hardened, shock resistant.

HOVER 91C

Hover, Gebruder, Edelstahlwerk
C 0.55, Cr 1, V 0.18, W 1.85, bal Fe.
For cold work tools, hammers, upsetters; oil or water hardened.

HOVER 91KA

Hover, Gebruder, Edelstahlwerk
C 0.45, Cr 1, V 0.2, bal Fe.
For gears, pinions, shafts, bolts, crankshafts; oil hardened, shock resistant.

HOVER A18A

Hover, Gebruder, Edelstahlwerk
C 0.12, Cr 18, Ni 8, bal Fe.
Annealed: 75,000 TS; 35,000 YS; 50 El; 65 RA; 150 Brin. For chemical plant equipment; Type 302; stainless, austenitic.

HOVER A18N

Hover, Gebruder, Edelstahlwerk
C 0-0.12, Cr 18, Ni 9.5, Cb = 8 x C, bal Fe.
Annealed: 90,000 TS; 35,000 YS; 50 El; 65 RA; 160 Brin. Cold drawn: 100,000 TS; 65,000 YS; 40 El; 60 RA; 212 Brin. For welded structures, chemical plant equipment; Type 347; stainless, austenitic.

HOVER A18T

Hover, Gebruder, Edelstahlwerk
C 0-0.12, Cr 18, Ni 9.5, Ti = 4 x C, bal Fe.
Annealed: 85,000 TS; 35,000 YS; 55 El; 65 RA; 150 Brin. Cold drawn: 95,000 TS; 60,000 YS; 40 El; 60 RA; 185 Brin. For welded structures, chemical plant equipment; Type 321; stainless, austenitic.

HOVER A18Z

Hover, Gebruder, Edelstahlwerk
C 0-0.07, Cr 18, Cr 9.5, Ni, bal Fe.
Annealed: 85,000 TS; 35,000 YS; 60 El; 70 RA; 150 Brin. Cold drawn: 180,000 TS; 125,000 YS; 10 El; 330 Brin. For welded structures, chemical plant equipment; Type 304; stainless, austenitic.

HOVER A18ZS

Hover, Gebruder, Edelstahlwerk
C 0-0.1, Cr 18, Ni 8.5, bal Fe.
Annealed: 75,000 TS; 35,000 YS; 55 El; 75 RA; 155 Brin. For chemical plant equipment, welded tanks and vessels; Type 304; stainless, austenitic.

HOVER A20T

Hover, Gebruder, Edelstahlwerk
C 0-0.12, Cr 18, Ni 10.5, Mo 2, Ti = 4 x C, bal Fe.
Annealed: 85,000 TS; 35,000 YS; 50 El; 75 RA; 150 Brin. Cold drawn: 150,000 TS; 130,000 YS; 6 El; 300 Brin. For acid resistant equipment, welded structures; Type 316 Ti; stainless, austenitic.

HOVER A20Z

Hover, Gebruder, Edelstahlwerk
C 0.07, Cr 18, Ni 10.5, Mo 2, bal Fe.
Annealed: 85,000 TS; 35,000 YS; 50 El; 75 RA; 150 Brin. Cold drawn: 150,000 TS; 130,000 YS; 6 El; 300 Brin. For acid resistant equipment; Type 316; stainless; austenitic.

HOVER A20ZS

Hover, Gebruder, Edelstahlwerk
C 0-0.1, Cr 18, Ni 9.5, Mo 2, Si 2.2, bal Fe.
Annealed: 85,000 TS; 35,000 YS; 50 El; 75 RA; 150 Brin. Cold drawn: 150,000 TS; 135,000 YS; 6 El; 300 Brin. For acid resistant equipment, chemical plant apparatus; Type 316; stainless, austenitic.

HOVER A21Z13

Hover, Gebruder, Edelstahlwerk
C 0-0.07, Cr 17, Mo 4.75, Ni 13, bal Fe.
Annealed: 90,000 TS; 40,000 YS; 45 El; 70 RA; 160 Brin. For acid resistant equipment, chemical plant apparatus; Type 317; stainless, austenitic.

HOVER GHK243

Hover, Gebruder, Edelstahlwerk
C 0.4, Cr, Mn, Mo, bal Fe.
For gears, bolts, machine tool parts; oil hardened, tough.

HOVER H10

Hover, Gebruder, Edelstahlwerk
C 1.1, Si 0-0.25, Mn 0-0.25, bal Fe.
Annealed: 110,000 TS; 56,000 YS; 20 El; 40 RA; 210 Brin. For springs, cutters, drills, taps, reamers; Type W1; water hardened.

HOVER H1000

Hover, Gebruder, Edelstahlwerk
C 0.15, Si 2, Cr 19.5, Ni 9.5, bal Fe.
Annealed: 85,000 TS; 40,000 YS; 50 El; 65 RA; 160 Brin. For chemical and oil refinery equipment; Type 302; stainless, austenitic.

HOVER H1050-ON

Hover, Gebruder, Edelstahlwerk
C 0-0.12, Cr 18, V 0.95, bal Fe.
Annealed: 80,000 TS; 35,000 YS; 50 El; 65 RA; 160 Brin. For chemical plant and oil refinery equipment; corrosion and heat resistant.

HOVER H1200

Hover, Gebruder, Edelstahlwerk
C 0.15, Cr 24, Ni 19, bal Fe.
Annealed: 90,000 TS; 40,000 YS; 55 El; 70 RA; 165 Brin. For heat treating boxes, furnace parts and equipment; heat resistant, austenitic.

HOVER H127

Hover, Gebruder, Edelstahlwerk
C 1, Mn 0-0.25, Si 0-0.25, bal Fe.
Annealed: 100,000 TS; 53,000 YS; 21 El; 42 RA; 200 Brin. For drills, taps, reamers, hobs; Type W1; water hardened.

HOVER H15

Hover, Gebruder, Edelstahlwerk
C 0.15, Cr 1.55, Ni 1.55, bal Fe.
For gears, cams, camshafts; case hardening.

HOVER H20

Hover, Gebruder, Edelstahlwerk
C 1, Si 0-0.25, Mn 0-0.25, bal Fe.
For springs, tools, drills, taps, reamers; Type W1; water hardened.

HOVER H20

Hover, Gebruder, Edelstahlwerk
C 0.18, Ni 2, Cr 2, Mn 0.5, bal Fe.
For gears, bolts, camshafts, cams; case hardened, tough.

HOVER H391

Hover, Gebruder, Edelstahlwerk
C 0.9, Cr 0.8, Mn 0.3, bal Fe.
For drills, punches, springs, taps, reamers; water hardened.

HOVER H680

Hover, Gebruder, Edelstahlwerk
C 0.5, Mn 0.85, Cr 1, V 3.1, bal Fe.
For gears, shafts, crankshafts; oil hardened, shock resistant.

HOVER HED16
Hover, Gebruder, Edelstahlwerk
C 0.15, Si 0.25, Mn 0.37, bal Fe.
Annealed: 70,000 TS; 40,000 YS; 25 El; 60 RA; 145 Brin. For gears, bolts, machine tool parts; case hardened.

HOVER HK904
Hover, Gebruder, Edelstahlwerk
C 0.3, Cr 2.35, V 0.6, W 4.25, bal Fe.
For extrusion press dies, rams, liners; hot work steel, oil hardened.

HOVER HNRO 11
Hover, Gebruder, Edelstahlwerk
C 0.2, Cr, Mo, bal Fe.
For chemical plant equipment; stainless.

HOVER HNRO 12
Hover, Gebruder, Edelstahlwerk
C 0-0.1, Ni 9.5, Mo 2, Cr 18, bal Fe.
Annealed: 85,000 TS; 50 El; 65 RA; 160 Brin. For acid resistant chemical plant equipment, tanks; Type 316; stainless, austenitic.

HOVER HNRO 14
Hover, Gebruder, Edelstahlwerk
C 0-0.12, Cr 18, Ni 9.5, Ti = 4 x C, bal Fe.
Annealed: 85,000 TS; 35,000 YS; 55 El; 65 RA; 150 Brin. For welded chemical plant equipment, tanks; Type 321; stainless, austenitic.

HOVER HNRO 16
Hover, Gebruder, Edelstahlwerk
C 0.22, Cr 17, Ni 1.5, bal Fe.
Annealed: 125,000 TS; 90,000 YS; 20 El; 55 RA; 260 Brin. For pumps, marine hardware, valves; Type 321; stainless.

HOVER HNRO 17
Hover, Gebruder, Edelstahlwerk
C 0-0.12, Cr 18, Mo 2, Ni 10.5, Ti = 4 x C, bal Fe.
Annealed: 85,000 TS; 35,000 YS; 50 El; 65 RA; 160 Brin. For welded acid resistant chemical plant equipment; Type 316 Ti; stainless, austenitic.

HOVER HNRO 19
Hover, Gebruder, Edelstahlwerk
C 0.35, Mo 1.15, Cr 16.5, bal Fe.
For oil refinery equipment; heat resistant.

HOVER HNRO 20
Hover, Gebruder, Edelstahlwerk
C 0.8, Cr 17, bal Fe.
Heat treated: 280,000 TS; 270,000 YS; 3 El; 15 RA; 555 Brin. For bearings, surgical instruments, cutlery; corrosion resistant, wear resistant; Type 440B.

HOVER HNRO 24
Hover, Gebruder, Edelstahlwerk
C 0-0.1, Cr 17, Mo 1.8, Ti = 7 x C, bal Fe.
Annealed: 125,000 TS; 95,000 YS; 20 El; 55 RA; 260 Brin. For welded structures, pump parts; corrosion resistant; Type 431 Ti.

HOVER HNRO 27
Hover, Gebruder, Edelstahlwerk
C 0-0.12, Cr 18, Mo 2, Ni 10.5, Cb = 8 x C, bal Fe.
Annealed: 90,000 TS; 45,000 YS; 55 El; 65 RA; 160 Brin. For welded chemical plant equipment, tanks, mixers; Type 347; stainless, austenitic.

HOVER HNRO 8
Hover, Gebruder, Edelstahlwerk
C 0-0.07, Cr 18, Ni 9.5, bal Fe.
Annealed: 85,000 TS; 35,000 YS; 60 El; 70 RA; 150 Brin. For chemical plant equipment, tanks, vessels; Type 304; stainless, austenitic.

HOVER HNRO 9
Hover, Gebruder, Edelstahlwerk
C 0-0.07, Cr 18, Mo 2, Ni 10.5, bal Fe.
Annealed: 85,000 TS; 35,000 YS; 50 El; 65 RA; 160 Brin. For acid resistant chemical plant equipment, tanks; Type 316; stainless, austenitic.

HOVER HNRO7
Hover, Gebruder, Edelstahlwerk
C 0-0.15, Ni 8.5, Cr 18, bal Fe.
Annealed: 80,000 TS; 35,000 YS; 55 El; 75 RA; 150 Brin. For chemical plant equipment, tanks; Type 302; stainless, austenitic.

HOVER HSE
Hover, Gebruder, Edelstahlwerk
C 0.2, Mn 1.25, Cr 1.15, bal Fe.
For camshafts, cams, gears, bolts; case hardened.

HOVER HVU 2 1/2
Hover, Gebruder, Edelstahlwerk
C 0.22, Si 0.25, Mn 0.45, bal Fe.
Annealed: 75,000 TS; 42,000 YS; 20 El; 55 RA; 150 Brin. For gears, bolts, machine tool parts; case hardened.

HOVER HVU 3 1/2
Hover, Gebruder, Edelstahlwerk
C 0.35, Si 0.25, Mn 0.55, bal Fe.
Hot rolled: 85,000 TS; 54,000 YS; 30 El; 53 RA; 185 Brin. For gears, bolts, machine tool parts; water hardened.

HOVER HVU 4 1/2
Hover, Gebruder, Edelstahlwerk
C 0.45, Cr, Mo, bal Fe.
For chemical plant equipment; stainless.

HOVER HVU 6
Hover, Gebruder, Edelstahlwerk
C 0.45, Si 0.25, Mn 0.65, bal Fe.
Hot rolled: 98,000 TS; 60,000 YS; 24 El; 45 RA; 212 Brin. For gears, bolts, machine tool parts; water hardened.

HOVER HW
Hover, Gebruder, Edelstahlwerk
C 0.53, Si 0.9, Mn 0.9, bal Fe.
For machine tool parts, punches; oil hardened.

HOVER I-EBN
Hover, Gebruder, Edelstahlwerk
C 0.15, Cr 0.65, Mn 0.5, bal Fe.
For gears, cams, camshafts, fasteners; case hardening steel.

HOVER I-EC
Hover, Gebruder, Edelstahlwerk
C 0.15, Cr 1.15, Mo 0.25, bal Fe.
For gears, cams, camshafts, fasteners; case hardening steel.

HOVER I-ECN
Hover, Gebruder, Edelstahlwerk
C 0.16, Mn 1.15, Cr 0.95, bal Fe.
For gears, cams, camshafts, fasteners; case hardening steel.

HOVER I-ED
Hover, Gebruder, Edelstahlwerk
C 0.2, Cr 1.15, Mo 0.25, bal Fe.
For gears, cams, camshafts, fasteners; case hardening steel.

HOVER I-EDN
Hover, Gebruder, Edelstahlwerk
C 0.2, Mn 1.25, Cr 1.15, bal Fe.
For gears, cams, camshafts; case hardening steel.

HOVER II-VA
Hover, Gebruder, Edelstahlwerk
C 0.25, Mn 0.65, Cr 1, bal Fe.
For gears, pinions, bolts, fasteners; water hardened.

HOVER II-VB
Hover, Gebruder, Edelstahlwerk
C 0.33, Mn 0.65, Cr 1, bal Fe.
For gears, shafts, crankshafts; water hardened.

HOVER II-VC
Hover, Gebruder, Edelstahlwerk
C 0.33, Mn 0.65, Cr 1, bal Fe.
For gears, shafts, crankshafts; water hardened.

HOVER II-VD
Hover, Gebruder, Edelstahlwerk
C 0.42, Cr 1, Mo 0.2, bal Fe.
For gears, shafts, crankshafts; oil hardened, shock resistant.

HOVER II-VE
Hover, Gebruder, Edelstahlwerk
C 0.5, Cr, bal Fe.
For gears, shafts, crankshafts, bolts; oil or water hardened.

HOVER ILO-105H
Hover, Gebruder, Edelstahlwerk
C 0.76, Co 11, Cr 4.2, Mo 0.8, V 1.8, W 18, bal Fe.
For lathe and planer tools, drills, taps, hobs, reamers; high speed steel.

HOVER ILO-35H
Hover, Gebruder, Edelstahlwerk
C 0.86, Co 2.8, Cr 4.3, Mo 0.85, V 2.1, W 12, bal Fe.
For lathe and planer tools, reamers, drills, taps; high speed steel.

HOVER ILO-45H
Hover, Gebruder, Edelstahlwerk
C 0.8, Cr 4, Co, V, W, Mo, bal Fe.
For lathe and planer tools, reamers, broaches, taps; high speed steel.

HOVER ILO-55H
Hover, Gebruder, Edelstahlwerk
C 1.35, W, Cr, Co, bal Fe.
For blanking and forming dies, engravers' tools; high speed steel.

HOVER ILO-ABC II
Hover, Gebruder, Edelstahlwerk
C 0.82, Cr 4.1, Mo 0.85, V 1.6, W 8.7, bal Fe.
For lathe and planer tools, reamers, hobs, taps, drills; high speed steel.

HOVER ILO-BCO
Hover, Gebruder, Edelstahlwerk
C 0.79, Co 4.7, Cr 4.3, Mo 0.75, V 1.5, W 18, bal Fe.
For drills, taps, hobs, reamers, lathe tools; high speed steel.

HOVER ILO-EXTRA
Hover, Gebruder, Edelstahlwerk
C 0.86, Cr 4.1, Mo 0.85, V 2.5, W 18, bal Fe.
For lathe and planer tools, drills, taps, reamers; high speed steel.

HOVER ILO-PRIMA
Hover, Gebruder, Edelstahlwerk
C 0.95, Cr 4.1, Mo 0.85, V, W, bal Fe.
For lathe and planer tools, drills, reamers; high speed steel.

HOVER ILO-SUPER B
Hover, Gebruder, Edelstahlwerk
C 1.3, Cr 4.3, Mo 0.85, V 3.8, W 12, bal Fe.
For blanking and forming dies, form cutters; high speed steel.

HOVER ILO-ULTRA W
Hover, Gebruder, Edelstahlwerk
C 0.74, Cr 4.1, V 1.1, W 18.5, bal Fe.
For lathe and planer tools, drills, taps, hobs; high speed steel.

HOVER IV-SF EXTRA QUALITY
Hover, Gebruder, Edelstahlwerk
C 0.55, Si 0.3, Mn 0.6, bal Fe.
Heat treated: 110,000-160,000 TS; 75,000-112,000 YS; 13-25 El; 42-55 RA; 230-325 Brin. For gears, axles, crankshafts; water hardened.

HOVER M48
Hover, Gebruder, Edelstahlwerk
C 0.38, Si, Cr, V, bal Fe.
For gears, bolts, machine tool parts; oil hardened, tough.

HOVER P12
Hover, Gebruder, Edelstahlwerk
C 0.56, Ni, Cr, Mo, V, bal Fe.
For forging and heading dies; oil hardened, tough.

HOVER P14
Hover, Gebruder, Edelstahlwerk
C 0.55, Cr 0.7, Mo 0.18, Ni 1.65, V 0.1, bal Fe.
For forging and heading dies; oil hardened, tough.

HOVER P15
Hover, Gebruder, Edelstahlwerk
C 0.45, Cr, Ni, bal Fe.
For gears, bolts, crankshafts; oil hardened, tough.

HOVER P16
Hover, Gebruder, Edelstahlwerk
C 1.1, Si 0.2, Cr 0.4, Mn 0.3, bal Fe.
For bearings, cutters, liners, sleeves; water hardened, wear resistant.

HOVER P35
Hover, Gebruder, Edelstahlwerk
C 0.5, Cr 1.05, Ni 3.25, bal Fe.
For gears, bolts, crankshafts; oil hardened, shock resistant.

HOVER PRIMA SEHR ZAH
Hover, Gebruder, Edelstahlwerk
C 0.6, Si 0.25-0.5, Mn 0.3-0.8, bal Fe.
Heat treated: 115-000-160,000 TS; 77,000-113,000 YS; 12-23 El; 40-54 RA; 230-320 Brin. For wheels, die blocks, rails, girders; water hardened.

HOVER PRIMA WEICH
Hover, Gebruder, Edelstahlwerk
C 0.75, Si 0.25-0.5, Mn 0.3-0.8, bal Fe.
Heat treated: 125,000-180,000 TS; 85,000-140,000 YS; 12-20 El; 38-52 RA; 240-360 Brin. For springs, clutch discs, rails, hammers; Type W1; water hardened.

HOVER PRIMA ZH
Hover, Gebruder, Edelstahlwerk
C 0.9, Si 0.25-0.5, Mn 0.3-0.8, bal Fe.
Heat treated: 180,000 TS; 118,000 YS; 10 El; 30 RA; 375 Brin. For springs, cutters, hobs, drills, taps; Type W1; water hardened.

HOVER R10
Hover, Gebruder, Edelstahlwerk
C 0.45, Cr 1.4, Mo 0.7, V 0.3, Mn 0.7, bal Fe.
For die casting dies, forging and heading dies; oil hardened, tough.

HOVER R11
Hover, Gebruder, Edelstahlwerk
C 1.4, V 0.1, Cr 0.3, Mn 0.3, bal Fe.
For engravers' tools, textile needles, cutters; oil or water hardened.

HOVER R5
Hover, Gebruder, Edelstahlwerk
C 0.3, Si 0.25, Mn 0.55, Cr 2.5, Mo 0.2, V 0.15, bal Fe.
For die casting dies, upsetters, rivet sets; oil hardened, tough.

HOVER SBL1
Hover, Gebruder, Edelstahlwerk
C 0-0.12, Si 0.4, Cr 13, bal Fe.
Annealed: 75,000 TS; 40,000 YS; 35 El; 70 RA; 155 Brin. For turbine blades, valves, cutlery; Type 410; stainless.

HOVER SBL2
Hover, Gebruder, Edelstahlwerk
C 0.2, Si 0.4, Cr 13, bal Fe.
Annealed: 95,000 TS; 50,000 YS; 25 El; 55 RA; 195 Brin. For turbine blades, valves, cutlery, knives; Type 420; stainless.

HOVER SBL4
Hover, Gebruder, Edelstahlwerk
C 0.4, Si 0.4, Cr 13, bal Fe.
Annealed: 100,000 TS; 55,000 YS; 20 El; 50 RA; 210 Brin. For valves, cutlery, surgical and dental instruments; Type 420; stainless.

HOVER SBL9
Hover, Gebruder, Edelstahlwerk
C 0.9, Si 0.4, Cr 18, Mo 1.15, V 1, bal Fe.
Annealed: 110,000 TS; 65,000 YS; 16 El; 30 RA; 230 Brin. For cutlery, ball bearings; Type 440BMo; stainless.

HOVER SP50
Hover, Gebruder, Edelstahlwerk
C 1.15, Cr 0.65, V 0.1, Mn 0.3, bal Fe.
For heading and forming dies, cutters; oil hardened, wear resistant.

HOVER SPEZIAL MH
Hover, Gebruder, Edelstahlwerk
C 1.1, Si 0-0.25, Mn 0-0.25, bal Fe.
Annealed: 100,000 TS; 53,000 YS; 21 El; 42 RA; 200 Brin. For springs, cutters, reamers, drills, broaches; Type W1; water hardened.

HOVER SPEZIAL SEHR HART
Hover, Gebruder, Edelstahlwerk
C 0.7, Si 0-0.25, Mn 0-0.25, bal Fe.
Heat treated: 174,000 TS; 128,000 YS; 12 El; 37 RA; 352 Brin. For springs, rails, girders, dies; Type W1; water hardened.

HOVER SPEZIAL ZAH
Hover, Gebruder, Edelstahlwerk
C 0.85, Si 0-0.25, Mn 0-0.25, bal Fe.
Heat treated: 188,000 TS; 145,000 YS; 10 El; 30 RA; 400 Brin. For drills, taps, reamers, broaches, hobs; Type W1; water hardened.

HOVER SPEZIAL ZH
Hover, Gebruder, Edelstahlwerk
C 1, Si 0-0.25, Mn 0-0.25, bal Fe.
Annealed: 100,000 TS; 53,000 YS; 21 El; 42 RA; 200 Brin. For drills, springs, taps, reamers, broaches; Type W1; water hardened.

HOVER WA18
Hover, Gebruder, Edelstahlwerk
C 0.55, Si 0.9, Mn 0.3, Cr 1.05, V 0.18, W 1.85, bal Fe.
For heading and forming dies, punches; cold work steel, tough.

HOWAL
Honsel-Werke Aktiengesellschaft
Mg 0.6-1.4, Si 0.6-1.6, Mn 0.6-1, Cr 0.3, bal Al.
Annealed: 21,000 TS; 8,000 YS; 24 El. For window frames, gutters, fan blades, boats; good formability and weldability. *Obsolete*

HOWARD ALLOY NO. 1
Howard Foundry Co.
Th 2.5-3.25, Zn 2-2.5, 0.5% Zr min, bal Mg.
T5 temper: 32,000 TS; 14,000 YS; 10 El; 60 Brin. For engine parts, structural components. Good creep properties to 650 F. Requires no solution treatment. *Obsolete*

HOWARD ALLOY NO. 3
Howard Foundry Co.
Zn 4.5, Zr 0.7, bal Mg.
T5 temper: 40,000 TS; 24,000 YS; 8 El; 60 Brin. For structural parts or engine sections. High strength. ASTM B80; alloy ZK51. Requires no solution treatment. *Obsolete*

HOWARD ALLOY NO. 4
Howard Foundry Co.
Al 9, Zn 2, Mn 0.1.
s cast: 25,000 TS; 14,000 YS; 2 El; 65 Brin. T6 temper: 9,000 TS; 21,000 YS; 2 El; 85 Brin. F temper: 24,000 TS; 14,000 YS; 2 El; 65 Brin. For aircraft engine parts, rankcases, housings, pressure-tight castings. Age ardenable, ASTM B80; Alloy AZ92. OMPOSITION al Mg. *Obsolete*

HOWARD ALLOY NO. 5
Howard Foundry Co.
Al 8.6, Zn 0.7, Mn 0.15, bal Mg.
As cast: 24,000 TS; 14,000 YS; 3 El; 55 Brin. T6-temper: 38,000 TS; 18,000 YS; 5 El; 80 Brin. For casings, housings, portable tools, leakproof die and sand castings, aircraft and automotive parts. Pressure tight. Heat treatable. ASTM B80; Alloy AZ91. *Obsolete*

HOWARD ALLOY NO. 6
Howard Foundry Co.
Zn 5.5, Th 1.7, Zr 0.7, bal Mg.
T5 temper: 40,000 TS; 26,000 YS; 9 El; 60 Brin. For airframe castings, engine casings, gear casings. Good castability and weldability. Requires no solution treatment. *Obsolete*

HOWARD ALLOY NO. 7
Howard Foundry Co.
Zn 2.5, Zr 0.6, 3.0% rare earths, bal Mg.
T5 temper: 23,000 TS; 16,000 YS; 3 El; 58 Brin. For engine castings operating between 300 and 500 F. Good creep properties to 500 F. Requires no solution treatment. *Obsolete*

HOWARD ALLOY NO. 8
Howard Foundry Co.
Al 6, Zn 3, Mn 0.2, bal Mg.
As cast: 26,000 TS; 12,000 YS; 5 El; 55 Brin. T6-temper: 38,000 TS; 18,000 YS; 4 El; 80 Brin. For airplane wheels and brakes, gear housings, crankcases, oil pumps. Age hardenable. ASTM B80; Alloy AZ63. Good properties to 300 F. *Obsolete*

HOWE BROWN EXTRA
Colt Industries
C, alloy, bal Fe.
For tools, dies; water hardened. *Obsolete*

HOWE BROWN SPECIAL
Colt Industries
C, alloy, bal P.
For tools, dies; water hardened. *Obsolete*

HOWEGE EXTRA
Dohlen-Stahl Gusstahl-Handels GmbH
C 0.55, Cr 0.7, Mo 0.18, W 1.85, V 0.1, bal Fe.
For cold work tools, headers, upsetters; oil hardened, tough.

HOWEGE EXTRA I
Dohlen-Stahl Gusstahl-Handels GmbH
C 0.56, Cr 0.7, Mo 0.18, W 1.85, V 0.1, bal Fe.
For cold work tools, headers, upsetters; oil hardened, tough.

HOWEGE SPEZIAL
Dohlen-Stahl Gusstahl-Handels GmbH
C 0.4, Cr, Mo, bal Fe.
For gears, shafts, bolts, studs, crankshafts; oil hardened, tough.

HOWES SHEAR BLADE
Crucible Materials Corp.
Tool material. C 0.9, Mo 0.3, bal Fe.
For shear blades; water hardening.

HOWMET FA
Now FA.

HOWMET NO. 3
Waimet Alloys Co.
C 2.45, Cr 31, W 12.5, Fe 0-3, Ni 0-3, bal Co.
Cast: 64,000 TS; 53 Rock C. For cutting tools, ball bearing rolls, sleeves, bushings, wear strips, valve seat inserts, scraper blades, valves. High heat and abrasion resistance, non-galling; corrosion and oxidation resistant.

HOWMET STANDARD
Waimet Alloys Co.
C 2.45, Cr 31, W 12.5, Fe 0-3, Ni 0-3, bal Co.
Cast: 64,000 TS; 530 Brin. For cutting tools, high temperature parts. Oxidation and wear resistant.

HOWMET STANDARD NO. 6
Waimet Alloys Co.
C 0.9-1.4, Fe 0-3, Ni 0-3, Mo 0-1.5, Cr 27-31, W 3.5-5.5, bal Co.
Cast: 115,000 TS; 96,000 YS; 3 El; 410 Brin. At 1500°F: 70,000 TS; 5 El; 8 RA. For turbine blades, valve parts, hot work punches. Heat and thermal shock resistant.

HOWMET SUPER-3
Waimet Alloys Co.
C 2.45, Cr 31, W 12.5, Fe 0-6, Ni 0-6, bal Co.
Cast: 62,000 TS; 560 Brin. For cutting tools, high temperature parts. Oxidation and wear resistant.

HOWMET SUPER-6
Waimet Alloys Co.
C 0.9-1.4, Fe 0-6, Ni 0-6, Mo 0-1.5, Cr 27-31, W 3.5-5.5, bal Co.
Cast: 114,000 TS; 101,000 YS; 1.5 El; 430 Brin. At 1500°F: 72,000 TS; 7 El; 7 RA; 230 Brin. For turbine blades, valve parts, hot work punches. Heat and thermal shock resistant.

HOWMET-25
Waimet Alloys Co.
C 0.12, Mn 1.5, Si 1, Cr 20, Ni 10, Co 51, W 15, Fe 1.
Cast: 106,800 TS; 24.1 El; 18-25 Rock C. Heat treated: 145,000-165,000 TS; 65,000-80,000 YS; 55-70 El; 95 Rock B; 2 Rock C. For gas turbine rotors and buckets, nozzles, afterburners, valves. Similar to L605 Alloy. Investment cast.

HOWMET-50
Waimet Alloys Co.
C 0.05-0.12, Ti 0.18, Cb 0.6, Cr 26-30, Co 47-52, bal Fe.
Cast: 135,000 TS; 48,000 YS; 10 El; 10 RA; 250 Brin. For furnace baffles, burner tips, sintering grates, quench baskets. Corrosion, heat and thermal shock resistant.

HOWORD A
Wallace Murray Corp.
C 0.35, Cr 5, V 0.4, Mo 1.5, bal Fe.
Air or oil hardening; hot work tool and die steel for forging dies, hot forming tools. AISI H11.

HOWORD B
Wallace Murray Corp.
C 0.35, Cr 5, V 0.4, W 1.5, W 1.5, Mo 1.5, bal Fe.
Air or oil hardening; hot work tool and die steel for forging dies, hot work forming tools. AISI H12.

HOWORD C
Wallace Murray Corp.
C 0.35, Cr 5, V 1, Mo 1.5, bal Fe.
Air or oil hardening; hot work tool and die steel; AISI H13.

HOYLES METAL
American manufacture
Sn 46, Sb 12, Pb 42.
For bearings; Babbitt.

HOYT
NL Industries
Sn, Sb, bal Fe.
For bearings; Babbitt. *Obsolete*

HOYT 11
NL Industries
Pb, Cu, Sb, bal Sn.
For bearings; Babbitt. *Obsolete*

HOYT ARROW
Hoyt Darchem Ltd.
Sb, Cu, Pb, bal Sn.
For bearings for heavy loads and high speeds; antifriction metal; tough.

HOYT I C E
Hoyt Darchem Ltd.
Cu, Pb, Sb, bal Sn.
For bearings for internal combustion engines; antifriction metal.

HOYT METAL
NL Industries
Cu 10, Sn 78, Pb 12, Br 26.
For bearings for internal combustion engines and pumps; bronze Babbitt; MP 360°C. *Obsolete*

HOYT METAL
Hoyt Metal Co. of London Ltd.
Cu 10, Sn 78, Pb 12.
26 Brin. For bearings for internal combustion engines and pumps; bronze Babbitt; M.P. 360 C. *Obsolete*

HOYT METAL
Hoyt Metal Co. of London Ltd.
Pb 90-94, Sb 6-10.
For bearings; hard. *Obsolete*

HOYT NO. 1
Hoyt Darchem Ltd.
Cu, Sn, Sb, bal Pb.
For bearings for heavy loads and medium speeds; antifriction metal, tough.

HOYT NO. 1 PHOSPHOR BRONZE
Hoyt Metal Co. of London Ltd.
P, Sn, bal Cu.
For chill castings; tough, wear resistant. *Obsolete*

HOYT NO. 11 D
Hoyt Metal Co. of London Ltd.
Ni 0.55, Pb 0.18, Sb 3.5, Cu 4.3, bal Sn.
Cast: 11,000 TS; 7,500 YS; 13.5 El; 27 Brin. For connecting rod bearings; for severe duty. *Obsolete*

HOYT NO. 133C
Hoyt Darchem Ltd.
Cu, Sn, Sb, bal Pb.
For bearings; Babbitt.

HOYT NO. 14
Hoyt Darchem Ltd.
Cu, Sn, Sb, bal Pb.
For stern tubes; zinc-free Babbitt.

HOYT NO. 156B
Hoyt Metal Co. of London Ltd.
Sn, Sb, bal Pb.
For bearings; Babbitt. *Obsolete*

HOYT NO. 175
Hoyt Darchem Ltd.
Cu, Pb, Sb, bal Sn.
Cast: 12,000 TS; 10,500 YS; 1.6 El; 35 Brin. For bearings for locomotives, crushers and tube mills; Babbitt.

HOYT NO. 3M
Hoyt Darchem Ltd.
Cu, Sn, Sb, bal Pb.
For bearings; antifriction metal.

HOYT NO. 40
Hoyt Darchem Ltd.
Cu, Sn, Sb, Pb, Zn.
For stern tubes and outer bearings; Babbitt.

HOYT NO. 400
Hoyt Darchem Ltd.
Sb, Cu, Cd, bal Sn.
For bearings operating in sea water; Babbitt.

HOYT NO. 71
Hoyt Darchem Ltd.
Cu, Sn, Sb, bal Pb.
For engine bearings; Babbitt.

HOYT NO. ELEVEN D
Hoyt Darchem Ltd.

Tin base, fine grain White metal. Compressive proof strength 56.37 N/mm². For bearings for marine steam engines and heavy duty compressors.

HOYT NO. ELEVEN R
Hoyt Darchem Ltd.
Sb, Cu, Ni, Pb, bal Sn.
Cast: 12,200 TS; 9000 YS; 11.7 El; 32 Brin. For bearings for diesel engines and turbines; Babbitt. Grain refined, fatigue resistant.

HOYT NO. ELEVEN Z3
Hoyt Darchem Ltd.
Sb, Cu, Ni, Pb, bal Sn.
Cast: 13,500 TS; 9300 YS; 13.9 El; 30 Brin. For bearings for aero engines and gasoline engines; Babbitt.

HOYT STAR BRAND
Hoyt Darchem Ltd.
Sn, Sb, bal Pb.
For bearings for fans, stone crushers, mining machinery; antifriction metal.

HOYT STAR M
Hoyt Darchem Ltd.
Sb, Cu, Pb, bal Sn.
Cast: 11,500 TS; 11,200 YS; 1.5 El; 30 Brin. For machine tool bearings; Babbitt.

HP 9-4-20, ETC.
Now REPUBLIC HP 9-4-20.

HP NICKEL
Now JELLIFF HP NICKEL.

HP-3
J.F. Jelenko & Co.
Precious metal. Au 86, Pt 10, Pd 2.
85,000 psi TS; 75,000 psi YS; 5 El; 170 Brin. Dental alloy for fusing porcelain to metal.

HPM-NICKEL
Henry Wiggin & Co. Ltd.
Ni 99.9.
For components of special valves, hydrogen thyratrons, electric resistance thermometers and controls. Oxidation and corrosion resistant. *Obsolete*

HPM1
Eagle & Globe Steel Ltd.
Tool material. C 0.15, Mn 1.2, Cr 0.6, Cu 2, bal Fe.
Machining properties and strength for mass production dies and molds, die plates and holders, molds for automotive parts and electrical home appliances. AISI P21.

HPM2
Eagle & Globe Steel Ltd.
Tool material. C 0.33, Cr 2.2, Mo 0.5, bal Fe.
Machining qualities and mechanical characteristics. Hardened for large automotive molds. For general purpose molds and forming dies. AISI P20.

HPW-NICKEL
Henry Wiggin & Co. Ltd.
Ni 96.5, W 3.5.
For oxide-coated cathodes of special valves, submarine repeaters. Heat resistant. *Obsolete*

HR MONEL
International Nickel Inc.
C 0-0.35, Mn 0.5-1.5, Fe 0-2.5, Si 2, Ni 62-68, bal Cu.
Cast: 100,000 TS; 70,000 YS; 15 El; 200 Brin. For corrosion resistant parts; free-cutting. *Obsolete*

HRA 376
English manufacture
C 0.03-0.15, Cr 4-8, W 3.5-10.5, Al 5-7, Ta 3-9.5, B 0.005-0.04, Co 10-14, Mo 0-4, Cb 0-3.5, Zr 0-0.15, bal Ni.
For cast turbine blades; vacuum cast, high creep fatigue resistance.

HS 100
Latrobe Steel Co.
C 1.08, Si 0.3, W 1.5, Cr 3.75, V 1.1, Mo 9.5, Co 8, bal Fe.
Super high-speed steel, hardenable to 68 Rock C. For long production runs or heavy duty machining, particularly for lathe tools. AISI M 42.

HS 188
Now HAYNES ALLOY 188.

HS 31
Now HAYNES STELLITE ALLOY NO. 31.

HS PRIMA
Thyssen Edelstahlwerke AG
C 1-1.2, bal Fe.
For tools; water hardening. *Obsolete*

HS QUALITY
Delta Metal (BW) Ltd.
Cu 58, Zn 40, Pb 2-2.5.
Extruded: 60,000 TS; 28-32 El; 25-30 RA; 120 Brin. Drawn: 68,000 TS; 25-30 El; 20-25 RA; 130 Brin. For screw machine products, hardware; free-cutting. *Obsolete*

HS-1
Japan Steel & Tube Co.Ltd.
C 0-0.18, Mn 0-0.14, Si 0-0.55, bal Fe.
Rolled: 71,000-85,000 TS; 47,000 min YP; 23 min El. For bridges, booms, buildings, bus and truck bodies. Constructional steel.

HS-12
Latrobe Steel Co.
C 0.99, W 0.72, Cr 3.9, V 1.96, Mo 8-11, bal Fe.
Hardened: 63-66 Rock C. For lathe tools and other cutting tools and dies requiring extra hardness and wear resistance. AISI Type M-7 (modified) high speed tool steel.

HS-151
Now HAYNES ALLOY 151.

HS-151
Cannon-Muskegon Corp.
C 0.5, Mn 0-1, Si 0-1, Cr 20, W 12.7, B 0.05, bal Co.
Cast alloy for gas turbine blades, vanes.

HS-21
Cannon-Muskegon Corp.
C 0.25, Mn 0.6, Si 0.6, Cr 27, Ni 3, Mo 5, Fe 1, bal Co.
Cast alloy for gas turbine parts.

HS-220
Timken Co.
C 0.3, Ni 2, Cr 1.2, Mo 0.45, bal Fe.
Heat treated: 237,000-250,000 TS; 201,000-212,000 YS; 11-13 El; 36-51 RA; 500-540 Brin. For aircraft structures, landing gears; tough, shock resistant, oil hardened. *Obsolete*

HS-220-07
Timken Co.
C 0.27-0.33, Ni 1.85-2.25, Cr 1-1.35, Mo 0.35-0.55, bal Fe.

HS-220-18
Timken Co.
C 0.23-0.28, Ni 1.65-2, Cr 0.2-0.4, Mo 0.35-0.45, bal Fe.

HS-220-27
Timken Co.
C 0.28-0.33, Ni 1.65-2, Cr 0.75-1, Mo 0.35-0.5, V 0.05-0.1, bal Fe.

HS-220-27 CA2
Timken Co.
C 0.28-0.33, Ni 1.65-2, Cr 0.75-1, Mo 0.35-0.5, V 0.05-0.1, S 0-0.015, bal Fe.

HS-220-28
Timken Co.
C 0.32-0.38, Ni 1.65-2, Cr 0.65-0.9, Mo 0.3-0.4, V 0.17-0.23, bal Fe.

HS-25
Cannon-Muskegon Corp.
C 0.1, Mn 1.5, Si 0.5, Cr 20, Ni 10, W 15, bal Co.
For jet engine parts, sheet.

HS-260
Timken Co.
C 0.38-0.43, Cr 1-1.35, Ni 1.85-2.25, Mo 0.35-0.5, bal Fe.
Heat treated: 271,000-280,000 TS; 221,000-235,000 YS; 7-10.5 El; 24.5-38.5 RA; 510-530 Brin. For aircraft structures, landing gears; tough, shock resistant, oil hardened.

HS-31
Cannon-Muskegon Corp.
C 0.5, Mn 0.5, Si 0.5, Cr 25, Ni 10, W 7.5, Fe 1.5, bal Co.
Cast alloy for gas turbine parts, nozzle vans.

HS-35
Inland Steel Co.
Carbon steel. C 0-0.25, Mn 0-0.9, bal Fe.
Sheet, plate: 50 ksi TS; 35 ksi YS; 22 El. Structural equipment where little or no forming is required.

HS-40
Inland Steel Co.
Carbon steel. C 0-0.25, Mn 0-0.9, bal Fe.
Sheet, plate: 55 ksi TS; 40 ksi YS; 21 El. Structural equipment where little or no forming is required.

HS-45
Inland Steel Co.
HSLA steel. C 0-0.25, Mn 0-0.9, N 0.01-0.013, bal Fe.
Sheet, plate: 60 ksi TS; 45 ksi YS; 20 El. Structural equipment where little or no forming is required.

HS-50
Inland Steel Co.
HSLA steel. C 0-0.25, Mn 0-0.9, N 0.1-0.13, bal Fe.
Sheet, plate: 65 ksi TS; 50 ksi YS; 18 El. Structural equipment where little or no forming is required.

HS100
Harrison Steel Castings Co.
Alloy steel. C 0.26-0.34, Mn 1.2-1.5, Si 0.3-0.6, Mo 0.22-0.32, bal Fe.
Normalized or quenched and tempered: 69,000-115,000 TS; 97,000-127,000 YS; 17-23 El; 201-277 Brin. Meets ASTM A148.

HS142
Harrison Steel Castings Co.
Alloy steel. C 0.16-0.22, Mn 0.7-1, Si 0.35-0.65, Cr 0.45-0.75, Mo 0.28-0.43, bal Fe.
Normalized, quenched, and tempered: 94,000-141,000 TS; 111,000-165,000 YS; 11-21 El; 248-352 Brin. Meets ASTM A148.

HS144
Harrison Steel Castings Co.
Alloy steel. C 0.22-0.28, Mn 0.7-1, Si 0.3-0.6, Cr 0.4-0.6, Mo 0.15-0.25, bal Fe.
Normalized, quenched, and tempered: 92,000-141,000 TS; 110,000-172,000 YS; 10-22 El; 248-363 Brin. Meets ASTM A148.

HS145
Harrison Steel Castings Co.
Alloy steel. C 0.19-0.26, Mn 0.75-1.1, Si 0.4-0.7, Cr 0.4-0.6, Mo 0.2-0.3, bal Fe.
Normalized, quenched, and tempered: 100,000-154,000 TS; 114,000-175,000 YS; 10-20 El; 255-363 Brin. Meets ASTM A148.

HS146
Harrison Steel Castings Co.
Alloy steel. C 0.26-0.33, Mn 0.75-1.1, Si 0.4-0.7, Cr 0.4-0.6, Mo 0.2-0.3, B 0.003-0.005, bal Fe.
Normalized, quenched, and tempered: 100,000-207,000 TS; 117,000-232,000 YS; 6-19 El; 255-495 Brin. Meets ASTM A148.

HS150
Harrison Steel Castings Co.
Alloy steel. C 0.16-0.22, Mn 0.7-1, Si 0.3-0.6, Ni 0.4-0.6, Cr 0.4-0.6, Mo 0.2-0.3, bal Fe.
Normalized and tempered or normalized, quenched, and tempered: 58,000-104,000 TS; 85,000-124,000 YS; 17-25 El; 183-277 Brin. Meets ASTM A148.

HS154
Harrison Steel Castings Co.
Alloy steel. C 0.22-0.28, Mn 0.7-1, Si 0.3-0.6, Ni 0.4-0.6, Cr 0.4-0.6, Mo 0.15-0.25, bal Fe.
Normalized, quenched, and tempered: 83,000-109,000 TS; 102,000-124,000 YS; 18-23 El; 223-269 Brin. Meets ASTM A148.

HS180
Harrison Steel Castings Co.
Alloy steel. C 0.26-0.32, Mn 0.7-1, Si 0.3-0.6, Ni 0.4-0.6, Cr 0.4-0.6, Mo 0.18-0.28, bal Fe.
Normalized, quenched, and tempered: 99,000-198,000 TS; 116,000-226,000 YS; 6.5-20 El; 255-477 Brin. Meets ASTM A148.

HS180B
Harrison Steel Castings Co.
Alloy steel. C 0.26-0.33, Mn 0.65-1.05, Si 0.4-0.7, Ni 0.4-0.6, Cr 0.4-0.6, Mo 0.2-0.3, B 0.003-0.005, bal Fe.
Normalized, quenched, and tempered: 104,000-202,000 TS; 120,000-230,000 YS; 6.5-20 El; 255-477 Brin. Meets ASTM A148.

HS180BA
Harrison Steel Castings Co.
Alloy steel. C 0.19-0.26, Mn 0.65-1.05, Si 0.4-0.7, Ni 0.4-0.6, Cr 0.4-0.6, Mo 0.2-0.3, B 0.003-0.005, bal Fe.
Normalized, quenched, and tempered: 151,000 TS; 172,000 YS; 10 El; 363 Brin. Meets ASTM A148 Gr. 165-145.

HS193
Harrison Steel Castings Co.
Alloy steel. C 0.25-0.33, Mn 0.5-0.9, Si 1.4-1.8, Cr 1.6-2, Mo 0.3-0.4, bal Fe.
Normalized, quenched, and tempered: 213,000 TS; 243,000 YS; 5.5 El; 495 Brin. Abrasion resistant.

HS29XL
Latrobe Steel Co.
C 0.98, Cr 4.15, W 0.3, V 1.85, Mo 5, bal Fe, plus alloy sulfides.
Hardened: 64-67 Rock C; free machining. For lathe tools and other cutting tools requiring extra high hardness and wear resistance. AISI Type M2 (modified) high speed tool steel.

HS65
Harrison Steel Castings Co.
Carbon steel. C 0.19-0.29, Mn 0.4-0.8, Si 0.3-0.6, bal Fe.
Normalized: 45,000 TS; 77,000 YS; 30 El; 153 Brin. Meets ASTM A27 Gr. 65-35

HS70
Harrison Steel Castings Co.
Carbon steel. C 0.18-0.26, Mn 1.2-1.5, Si 0.3-0.6, bal Fe.
Normalized: 54,000 TS; 86,000 YS; 28 El; 172 Brin. Meets ASTM A148 Gr. 80-50.

HS70V
Harrison Steel Castings Co.
Alloy steel. C 0.18-0.26, Mn 1.2-1.5, Si 0.3-0.6, V 0.08-0.13, bal Fe.
Normalized: 61,000 TS; 91,000 YS; 27 El; 183 Brin. Meets ASTM A148 Gr. 80-50.

HS80
Harrison Steel Castings Co.
Carbon steel. C 0.4-0.5, Mn 0.5-0.9, Si 0.3-0.6, bal Fe.
Normalized: 57,000 TS; 100,000 YS; 21 El; 197 Brin. Meets ASTM A148 Gr. 80-40.

HS80B
Harrison Steel Castings Co.
Alloy steel. C 0.3-0.38, Mn 0.9-1.15, Si 0.3-0.6, B 0.003-0.005, bal Fe.
Normalized: 52,000 TS; 94,000 YS; 24 El; 183 Brin. Meets ASTM A148 Gr. 80-40; Gr. 80-50.

HS90A
Harrison Steel Castings Co.
Carbon steel. C 0.32-0.4, Mn 1.2-1.5, Si 0.3-0.6, bal Fe.
Normalized or quenched and tempered: 63,000-115,000 TS; 102,000-130,000 YS; 16-23 El; 207-286 Brin. Meets ASTM A148.

HSC NO. 265
Hoyland Steel Co.
C 1.55-1.7, Cr 11.5-12.5, Mo 0.7-0.9, V 0.15-0.25, bal Fe.
For dies, tools, blanking and forming dies; non-deforming.

HSC NO. 280
Hoyland Steel Co.
C 0.55, Si 1.5, Mo 0.35, bal Fe.
For tools, punches, chisels; shock resistant.

HSC NO. 33
Hoyland Steel Co.
C 0.3-0.35, Si 0.9, Cr 5, W 1.2, Mo 1.5, bal Fe.
For hot work tools, Al-die casting dies; hot work steel.

HSC NO. 422
Hoyland Steel Co.
C 0.45-0.5, Cr 0.85-1.05, W 0.9-1.2, Mo 0.15-0.25, bal Fe.
For tools, punches, shear blades; shock resistant.

HSC-CVM
Hoyland Steel Co.
C 1, Mn 0.7, Cr 5.25, Mo 1.1, V 0.25, bal Fe.
For cold work dies, shears, punches; air hardening.

HSC-SS EXTRA
Hoyland Steel Co.
C 1-1.1, Mn 0.3, Si 0.3, bal Fe.
For drills, taps, hobs, reamers, broaches; Type 1; water hardened.

HSM 260
Hoesch Hohenlimburg AG
HSLA steel. C 0-0.12, Si 0-0.5, Mn 0-1.1, P 0-0.03, S 0-0.03, Nb 0-0.09, Ti 0-0.22, 0.015 Al min, bal Fe.
Hot rolled fine grain structural steel for cold forming. 260 N/mm^2 YP min; 380-500 N/mm^2 TS min; 23 El.

HSM 260 N
Hoesch Hohenlimburg AG
HSLA steel. C 0-0.18, Si 0-0.5, Mn 0-1.2, P 0-0.03, S 0-0.03, Nb 0-0.09, Ti 0-0.22, 0.015 Al min, bal Fe.
Hot rolled fine grain structural steel for cold forming. 260 N/mm^2 YP min; 370-490 N/mm^2 TS min; 24 El.

HSM 300
Hoesch Hohenlimburg AG
HSLA steel. C 0-0.12, Si 0-0.5, Mn 0-1.2, P 0-0.03, S 0-0.03, Nb 0-0.09, Ti 0-0.22, 0.015 Al min, bal Fe.
Hot rolled fine grain structural steel for cold forming. 300 N/mm^2 YP min; 400-520 N/mm^2 TS min; 21 El.

HSM 340
Hoesch Hohenlimburg AG
HSLA steel. C 0-0.12, Si 0-0.5, Mn 0-1.3, P 0-0.03, S 0-0.03, Nb 0-0.09, Ti 0-0.22, 0.015 Al min, bal Fe.
Hot rolled fine grain structural steel for cold forming. 340 N/mm^2 YP min; 420-540 N/mm^2 TS min; 19 El.

HSM 340 N
Hoesch Hohenlimburg AG
HSLA steel. C 0-0.18, Si 0-0.5, Mn 0-1.5, P 0-0.03, S 0-0.03, Nb 0-0.09, Ti 0-0.22, 0.015 Al min, bal Fe.
Hot rolled fine grain structural steel for cold forming. 340 N/mm^2 YP min; 460-580 N/mm^2 TS min; 21 El.

HSM 380
Hoesch Hohenlimburg AG
HSLA steel. C 0-0.12, Si 0-0.5, Mn 0-1.4, P 0-0.03, S 0-0.03, Nb 0-0.09, Ti 0-0.22, 0.015 Al min, bal Fe.
Hot rolled fine grain structural steel for cold forming. 380 N/mm^2 YP min; 450-590 N/mm^2 TS min; 18 El.

HSM 380 N
Hoesch Hohenlimburg AG
HSLA steel. C 0-0.18, Si 0-0.5, Mn 0-1.6, P 0-0.03, S 0-0.03, Nb 0-0.09, Ti 0-0.22, 0.015 Al min, bal Fe.
Hot rolled fine grain structural steel for cold forming. 380 N/mm^2 YP min; 500-640 N/mm^2 TS min; 19 El.

HSM 420
Hoesch Hohenlimburg AG
HSLA steel. C 0-0.12, Si 0-0.5, Mn 0-1.5, P 0-0.03, S 0-0.03, Nb 0-0.09, Ti 0-0.22, 0.015 Al min, bal Fe.
Hot rolled fine grain structural steel for cold forming. 420 N/mm^2 YP min; 480-620 N/mm^2 TS min; 16 El.

HSM 420 N
Hoesch Hohenlimburg AG
HSLA steel. C 0-0.2, Si 0-0.5, Mn 0-1.6, P 0-0.03, S 0-0.03, Nb 0-0.09, Ti 0-0.22, 0.015 Al min, bal Fe.
Hot rolled fine grain structural steel for cold forming. 420 N/mm^2 YP min; 530-670 N/mm^2 TS min; 18 El.

HSM 460
Hoesch Hohenlimburg AG
HSLA steel. C 0-0.12, Si 0-0.5, Mn 0-1.6, P 0-0.03, S 0-0.03, Nb 0-0.09, Ti 0-0.22, 0.015 Al min, bal Fe.
Hot rolled fine grain structural steel for cold forming. 460 N/mm^2 YP min; 520-670 N/mm^2 TS min; 14 El.

HSM 500
Hoesch Hohenlimburg AG
HSLA steel. C 0-0.12, Si 0-0.5, Mn 0-1.7, P 0-0.03, S 0-0.03, Nb 0-0.09, Ti 0-0.22, 0.015 Al min, bal Fe.
Hot rolled fine grain structural steel for cold forming. 500 N/mm^2 YP min; 550-700 N/mm^2 TS min; 12 El.

HSM 550
Hoesch Hohenlimburg AG
HSLA steel. C 0-0.18, Si 0-0.5, Mn 0-1.8, P 0-0.03, S 0-0.03, Nb 0-0.09, Ti 0-0.22, 0.015 Al min, bal Fe.
Hot rolled fine grain structural steel for cold forming. 550 N/mm^2 YP min; 600-760 N/mm^2 TS min; 10 El.

HSM 600
Hoesch Hohenlimburg AG
HSLA steel. C 0-0.18, Si 0-0.5, Mn 0-1.9, P 0-0.03, S 0-0.03, Nb 0-0.09, Ti 0-0.22, 0.015 Al min, bal Fe.
Hot rolled fine grain structural steel for cold forming. 600 N/mm^2 YP min; 650-800 N/mm^2 TS min; 10 El.

HSM 650
Hoesch Hohenlimburg AG
HSLA steel. C 0-0.18, Si 0-0.5, Mn 0-2, P 0-0.03, S 0-0.03, Nb 0-0.09, Ti 0-0.22, 0.015 Al min, bal Fe.
Hot rolled fine grain structural steel for cold forming. 650 N/mm^2 YP min; 700-850 N/mm^2 TS min; 10 El.

HSM 700
Hoesch Hohenlimburg AG
HSLA steel. C 0-0.18, Si 0-0.5, Mn 0-2.1, P 0-0.03, S 0-0.03, Nb 0-0.09, Ti 0-0.22, 0.015 Al min, bal Fe.
Hot rolled fine grain structural steel for cold forming. 700 N/mm^2 YP min; 750-900 N/mm^2 TS min; 10 El.

HSM/S
Delta Enfield Metals Ltd.
Now ERM CUS.

HST.100
British Steel Corp.
C 0.4, 3.0% Cr + Mo + V, bal Fe.
For high temperature service. *Obsolete*

HST.120
British Steel Corp.
C 0.3, 3% Cr + Mo + V, bal Fe.
For high temperature service. *Obsolete*

HST.140
British Steel Corp.
C 0.4, 5% Cr + Mo + V, bal Fe.
For high temperature service. *Obsolete*

HT IRON-40
Ingersoll-Rand Co.
TC 3.1, Si 0.2, Mn 0.7, Ni 1.5, bal Fe.
Cast: 40,000 TS; 220 Brin. For roller bearing cages; cast iron; transverse strength 3800.

HT MOLYBDENUM
Plansee, Metallwerk Gesellschaft
Mo 100.
For electronic, nuclear, chemical, aerospace, metallurgical and electrical industries; vacuum equipment. Ductile after exposure to 2700-3600°F.

HT30
Cooper Alloy Corp.
Stainless steel. C 0.25-0.35, Si 2.5, Mn 2, Cr 13-17, Ni 33-37, Mo 0.5, bal Fe.
Cast: 448 MPa TS; 193 MPa YS; 15 El. Heat resistant alloy; ASTM A-351 HT30.

HT850
Eagle & Globe Steel Ltd.
Alloy steel. C 0.4, Mn 1.6, bal Fe.
850-1000 MPa TS; high tensile fasteners, shafts, applications requiring high tensile strength and good impact toughness. AS1442 XK1340; SAE 1541.

HTB-3
Allegheny Ludlum Steel
C 0.57, Mn 0.3, Si 1.15, Cr 4.75, Mo 5.25, V 0.55, bal Fe.
For jet engine bearings; for high temperature operation. *Obsolete*

HTM
Capitol Castings Inc.
C 3, Mn, Si, bal Fe.
For tractor parts; high strength malleable iron. *Obsolete*

HTP 50
Kawasaki Steel Corp.
C 0.12, Mn 1.1, P 0.011, S 0.007, Ti 0.025, bal Fe.
74,700 psi YS; 33 El. High-strength low-alloy steel for structural purposes.

HTP 60
Kawasaki Steel Corp.
C 0.16, Mn 1.4, P 0.01, S 0.006, Ti 0.045, bal Fe.
92,000 psi TS; 74,700 psi YS; 28 El. High-strength low- alloy steel for structural purposes.

HTP 60 W
Kawasaki Steel Corp.
C 0.16, Mn 1.48, P 0.02, S 0.009, Si 0.47, Cr 0.21, V 0.043, Cb 0.025, bal Fe.
96,000 psi TS; 76,700 psi YS; 40 El. High-strength low- alloy normalized steel for weldable structures and construction equipment.

HTP 80 E
Kawasaki Steel Corp.
C 0.12, Mn 1.98, P 0.01, S 0.006, Si 0.22, Ti 0.138, Cb 0.044, bal Fe.
119,200 psi TS; 108,200 psi YS; 23 El. High-strength low-alloy steel for structural purposes.

HTP100E
Kawasaki Steel Corp.
C 0.09, Mn 2.23, P 0.006, S 0.005, Si 0.2, Ti 0.22, Cb 0.039, Mo 0.65, bal Fe.
147,000 psi TS; 120,000 psi YS; 18 El. High-strength low-alloy steel for structural purposes.

HTW
Bethlehem Steel Corp.
C 0.9, Cr 1.7, V 0.2, bal Fe.
For bearing races; wear resistant. *Obsolete*

HUB BUSHING-HB-1
Lunkenheimer Co.
Cu 83, Pb 0.9, Ni 11, Fe 4.5, Si 0.8.
Cast: 63,000 psi TS; 90-145 Brin. For valve components where iron is cast around the bronze bushing in core. Good corrosion resistance; wears well.

HUB METAL-H-1
Lunkenheimer Co.
Cu 80.5, Sn 5, Zn 4.5, Pb 7.
Cast: 30,000 TS; 45-50 Brin. For pressure containing parts at temperatures to 400°F; wears well with silicon brass stems.

HUBARD SPECIAL
Hubbard Steel Co.
C, Ni, Cr, bal Fe.
For rolls, guides, castings; high wear resistance.

HUBARD SPECIAL
Continental Foundry & Machine Co.
C, Ni, Cr, bal Fe.
For rolls, guides, castings; high wear resistance.

HUBBING DIE
Bethlehem Steel Corp.
C 0.06, Mn 0.15, bal Fe.
For hubbing dies; case hardened. *Obsolete*

HUCKINGER KI
Mannesmann-Huttenwerk AG
C 0-0.17, Si 0-0.35, Mn 0.35, bal Fe.
Annealed: 70,000 TS; 55,000 YS; 25 El; 60 RA; 143 Brin. Cold drawn: 72,000 TS; 60,000 YS; 22 El; 58 RA; 146 Brin. For screws, bolts, gears, shafts, nails, bushings; case hardened.

HUCKINGER KII
Mannesmann-Huttenwerk AG
C 0-0.23, Si 0-0.25, Mn 0-0.35, bal Fe.
Hot rolled: 70,000 TS; 45,000 YS; 31 El; 58 RA; 143 Brin. Cold drawn: 80,000 TS; 69,000 YS; 18 El; 48 RA; 162 Brin. For crankshafts, gears, bolts, armature shafts; water hardened.

HUCKINGER KIIIW
Mannesmann-Huttenwerk AG
C 0-0.22, Si 0-0.35, Mn 0-0.55, bal Fe.
Cold drawn: 78,000 TS; 68,000 YS; 20 El; 55 RA; 159 Brin. Annealed: 73,000 TS; 61,000 YS; 22 El; 58 RA; 149 Brin. For gears, bolts, crankshafts, fan blades; water hardened.

HUCKINGER KIIW
Mannesmann-Huttenwerk AG
C 0-0.2, Si 0-0.35, Mn 0-0.5, bal Fe.
Cold drawn: 78,000 TS; 68,000 YS; 20 El; 55 RA; 149 Brin. For fan blades, bushings, gears, bolts; water hardened.

HUCKINGER KIVW
Mannesmann-Huttenwerk AG
C 0-0.26, Si 0-0.35, Mn 0-0.6, bal Fe.
Hot rolled: 70,000 TS; 45,000 YS; 31 El; 58 RA; 143 Brin. Cold drawn: 80,000 TS; 69,000 YS; 18 El; 48 RA; 162 Brin. For crankshafts, gears, bolts, armature shafts; water hardened.

HUCKINGER KIW
Mannesmann-Huttenwerk AG
C 0-0.16, Si 0-0.35, Mn 0-0.4, bal Fe.
Annealed: 70,000 TS; 55,000 YS; 25 El; 60 RA; 143 Brin. Cold drawn: 72,000 TS; 60,000 YS; 22 El; 58 RA; 146 Brin. For screws, bolts, gears, shafts, nails, bushings; case hardened.

HUCKINGERHUTTE HH428U
Mannesmann-Huttenwerk AG
C 0.15, Mn 0.6, Mg 0.3, Mo 0.3, bal Fe.
For gears, bolts, shafts, cams, fasteners; case hardened.

HUCKINGERHUTTE HH432
Mannesmann-Huttenwerk AG
C 0.19, Si 0.5, Mn 1.15, bal Fe.
For gears, pinions, camshafts, cams, bolts; case hardened, tough.

HUGHES ALLOY
Egal Metal Products Co.
Lead. Sn 17, Pb 70, Sb 13, Br 30.
For metallic packing.

HUNDRED METAL
English manufacture
Ni, Cr.
For high temperature applications; heat and corrosion resistant.

HUNT-SPILLER HIGH CARBON GUN IRON
Power Products Inc.
Si 1.5, 3.6 % TC, bal Fe.
Cast: 35,000 TS. For brake drums, clutch pressure plates; resists heat or thermal checking. *Obsolete*

HUNT-SPILLER NO. 101
Power Products Inc.
C 0.3, bal Fe.
Cast: 70,000-75,000 TS; 40,000-45,000 YS; 25-30 El; 45-50 RA. For castings, gears, shafts; water hardening. *Obsolete*

HUNT-SPILLER NO. 101A
Power Products Inc.
C 0.4, bal Fe.
Cast: 70,000-75,000 TS; 40,000-45,000 YS; 25-30 El; 45-50 RA. For castings, gears, shafts; water hardening. *Obsolete*

HUNT-SPILLER NO. 104
Power Products Inc.
C 0-0.25, Mo 0.4-0.6, Cu 0-0.3, Ni 0-1, Cr 0-0.2, bal Fe.
Cast: 65,000 TS; 35,000 YS; 20 El; 30 RA. For valves, fittings; water hardening. *Obsolete*

HUNTER-DOUGLAS ZG73
Bridgeport Brass Corp.
Al alloy.
For extrusion; heat treatable, resists fuming nitric acid.

HURBENIUM B
Hurbenium Co. of America
Pb 91, Sn 8, Bi 1.
For hot dip plating; corrosion resistant.

HURON
AL Tech Specialty Steel Corp.
C 2.1, Cr 12.5, V 1, Mn 0.32, Si 0.45, bal Fe.
Heat treated: 744-418 Brin. For high performance blanking and cold forming dies, punches, rolls and gages. High resistance to abrasion. AISI D-3.

HURON ALUMINUM ALLOY
English manufacture
Cu 3-7, Mn 0-0.6, Mg 0.5, Ni 0-1.25, Cr, Co, Cd, bal Al.
For light alloy parts; heat treatable.

HURON ALUMINUM ALLOY A ROLLED
Manufacturer not listed.
Cu 3.5, Mn 0.6, Mg 0.5, Cr 0.1, bal Al.
For light alloy parts; heat treatable.

HURON ALUMINUM ALLOY B ROLLED
Manufacturer not listed.
Cu 4, Mn 0.6, Mg 0.5, Cr 0.1, bal Al.
For light alloy parts; heat treatable.

HURON ALUMINUM ALLOY D ROLLED
Manufacturer not listed.
Cu 4, Mg 0.5, Cr 0.6, bal Al.
For light alloy parts; heat treatable.

HURON ALUMINUM CASTINGS A-5 ALLOY
Manufacturer not listed.
Cu 6.6, Mg 0.5, Ni 1.25, Co 0.25, Mn 0.5, Sn, Cd, bal Al.
For light alloy parts; heat treatable.

HURON METAL CAST
Manufacturer not listed.
Cu 6.6, Ni 1.3, Mg 0.5, Co 0.3, bal Al.
For light alloy parts; heat treatable.

HURON METAL ROLLED
Manufacturer not listed.
Cu 3.5-4, Mg 0.5, Mn 0-0.6, Cr 0.1-0.6, bal Al.
For light alloy parts; heat treatable.

HURON V
Allegheny Ludlum Steel
C 2.45, Cr 12, V 4, Mo 1.8, bal Fe.
Annealed: 269 Brin. Heat treated: 640 Brin. For brick mold liners, blanking and lamination dies; oil hardened, nondeforming. *Obsolete*

HURON-EZ
Allegheny Ludlum Steel
C 2.1, Cr 12.5, V 1, Mn 0.3, Si 0.4, bal Fe.
For blanking and forming dies; rolls, gages; Type D3; oil hardened, abrasion resistant.

HUSMANN METAL
English manufacture
Sn 74, Sb 11, Pb 11, Cu 4, Fe 0.2, Zn 0.2.
For bearings, bushings; anti-friction.

HUSQUARNA HVA
Husquarna Vapenfabrik Aktiebolag
C 0-0.05, Si 0-0.02, Mn 0-0.06, S 0-0.01, bal Fe.
Sintered. For sintered parts; electrolytic powdered iron.

HUZUTIT G 3
Ekalit Stahl GmbH
C 0.4-0.6, Si 1-2.5, Mn 0-1, Cr 16-18, bal Fe.
Heatproof steel castings. Werkstoff Nr. 1.4740.

HV 5
Spencer Clark Metal Industries Ltd.
C 1.5, W 12.5, Cr 4.75, Mo 0.5, V 5, Co 5, bal Fe.
Cobalt-vanadium high speed steel; extra high abrasion resistance.

HW-8
Bethlehem Steel Corp.
C 0.55, Cr 4, Mo 8, V 2, bal Fe.
Hot-work tool steel (same as HOT WORK 8). *Obsolete*

HW10C
Sheffield Smelting Co. Ltd.
Ag 49.75, WC, Co.
Annealed: 130 Vickers. Electrical resistivity: 2.8 micro-ohm/cm. Electrical conductivity: 62% IACS. For electrical contacts; good erosion resistance.

HWD MOD
Teledyne Firth Sterling
Steel. C 0.55, W 1.4, Cr 5, Mo 1.5, V 0.3, bal Fe.
For shear blades, forging dies, punches; air hardened, nondeforming. *Obsolete*

HWF 22
Chr. Hover & Sohn Edelstahlwerk
C 0.2, Cr 12, Mo 0-1.2, Ni 0-0.8, V 0-0.35, bal Fe.
For high temperature parts as steam power plant equipment.

HWF 81
Chr. Hover & Sohn Edelstahlwerk
C 0-0.1, Mn 0-1.5, Si 0-0.6, Cr 16.5, Mo 0-2, Ni 16, Nb = 10 x C, bal Fe.
High temperature stainless; for steam and gas turbines.

HWS
Now DARWIN HWS.

HY 43
Vereinigte Leichtmetallwerke G.m.b.H.
Aluminum. Zn 4.5, Mg 3.5, Mn 0.3, Cr 0-0.1, Cu 0.4, V 0.2-0.6, bal Al.
Aged: 71,000-77,000 TS; 64,000-68,000 YS; 8 El. For light alloy parts; high strength, age-hardenable.

HY MU 400
Carpenter Technology Corp.
Fe-Ni.
For laminations for motor cores; high permeability. *Obsolete*

HY-100
Lukens Steel
C 0.2, Mn 0.1-0.4, Ni 2.25-3.5, Cr 1-1.8, Mo 0.2-0.6, Ti 0.02, V 0.02, Cu 0.25, Cr, bal Fe.
Plate: 100,000-120,000 YS; 18 min El; 50 min RA; 105,700 CYS. For warship hulls, structures, deck railings, heat exchangers, submarine hulls, pressure vessels. Supplied in quenched and tempered condition.

HY-100
Earle M. Jorgensen
C 0.2, Mn 0.1-0.4, Ni 2.25-3.5, Cr 1-1.8, Mo 0.2-0.6, Ti 0.02, V 0.02, Cu 0.25, Cr, bal Fe.
Plate: 100,000-120,000 YS; 18 min El; 50 min RA; 105,700 CYS. For warship hulls, structures, deck railings, heat exchangers, submarine hulls, pressure vessels. Supplied in quenched and tempered condition.

HY-100
United States Steel Corp.
C 0.2, Mn 0.1-0.4, Ni 2.25-3.5, Cr 1-1.8, Mo 0.2-0.6, Ti 0.02, V 0.02, Cu 0.25, Cr, bal Fe.
Plate: 100,000-120,000 YS; 18 min El; 50 min RA; 105,700 CYS. For warship hulls, structures, deck railings, heat exchangers, submarine hulls, pressure vessels. Supplied in quenched and tempered condition.

HY-130
United States Steel Corp.
C 0.12, Mn 0.6-0.9, Ni 4.75-5.25, Cr 0.4-0.7, Mo 0.3-0.65, Ti 0.02, V 0.05-0.1, Cu 0.15, bal Fe.
Plate: 130,000 TYS; 141,000 CYS; 15 El; 50 RA. For submarine hulls, pressure vessels. Tough, high impact resistance.

HY-130
Lukens Steel
C 0.12, Mn 0.6-0.9, Ni 4.75-5.25, Cr 0.4-0.7, Mo 0.3-0.65, Ti 0.02, V 0.05-0.1, Cu 0.15, bal Fe.
Plate: 130,000 TYS; 141,000 CYS; 15 El; 50 RA. For submarine hulls, pressure vessels. Tough, high impact resistance.

HY-130/140 MOD
Earle M. Jorgensen
C 0.1-0.17, Mn 0.6-0.9, Si 0.15-0.35, Cr 0.45-0.65, Mo 0.45-0.65, Ni 4.75-5.25, V 0.05-0.1, bal Fe.
Plate, quenched and tempered: 145,000 psi YS; 15 El. For large, highly stressed structural equipment.

HY-140
U.S. Steel Corp.
C 0-0.12, Mn 0.6-0.9, Si 0.2-0.35, Ni 4.75-5.25, Cr 0.4-0.7, Mo 0.3-0.65, V 0.05-0.1, Ti 0.02, Cu 0-0.15, bal Fe.
Plate: 140,000-155,000 YS; 50 min Charpy. For pressure vessels, submarine hulls. Tough, shock resistant. *Obsolete*

HY-150 STEEL
U.S. Steel Corp.
C 0.16-0.2, Ni 3.5-4, Cr 1.2-1.7, Mo 0.3-0.5, V 0.07-0.12, Mn 0.4-0.6, 0.01% P and S, bal Fe.
Heat treated: 125,000-180,000 YS. For ship hulls, structures, deck railings, heat exchangers. Supplied in quenched and tempered condition. *Obsolete*

HY-180
United States Steel Corp.
C 0.1, Ni 10, Co 8, Cr 2, Mo 1, bal Fe.
Plate: 180,000 YS; 95 Charpy V-notch.

HY-511
German manufacture
Mg 5, Si 1, Ti 0.2, Mn, Fe, bal Al.
Cast. For cylinder heads; heat resistant.

HY-65
American manufacture
C 0.12, Mn 0.48, Ni 2.16, Mo 0.39, Cu 0.7, bal Fe.
For pressure vessels; tough, corrosion resistant.

HY-80
Now BIRDSBORO HY80.

HY-80
Titanium Metals Corp.
C 0.18, Mn 0.1-0.4, Ni 2-3.25, Cr 1-1.8, Mo 0.2-0.6, Ti 0.002, V 0.02, Cu 0.25, bal Fe.
Plate: 80,000 YS; 84,800 YS; 20 El; 55 RA; 50 IS (Charpy, at-120°F). For pressure vessels, submarine hulls. Tough, good weldability.

HY-80
Earle M. Jorgensen
C 0.18, Mn 0.1-0.4, Ni 2-3.25, Cr 1-1.8, Mo 0.2-0.6, Ti 0.002, V 0.02, Cu 0.25, bal Fe.
Plate: 80,000 YS; 84,800 YS; 20 El; 55 RA; 50 IS (Charpy, at-120°F). For pressure vessels, submarine hulls. Tough, good weldability.

HY-80
United States Steel Corp.
C 0.18, Mn 0.1-0.4, Ni 2-3.25, Cr 1-1.8, Mo 0.2-0.6, Ti 0.002, V 0.02, Cu 0.25, bal Fe.
Plate: 80,000 YS; 84,800 YS; 20 El; 55 RA; 50 IS (Charpy, at-120°F). For pressure vessels, submarine hulls. Tough, good weldability.

HY-80
Lukens Steel
C 0.18, Mn 0.1-0.4, Ni 2-3.25, Cr 1-1.8, Mo 0.2-0.6, Ti 0.002, V 0.02, Cu 0.25, bal Fe.
Plate: 80,000 YS; 84,800 YS; 20 El; 55 RA; 50 IS (Charpy, at-120°F). For pressure vessels, submarine hulls. Tough, good weldability.

HY-80
English manufacture
C 0-0.22, Ni 1.9-3.3, Cr 0.8-1.9, Mo 0.13-0.63, bal Fe.
Heat treated: 105,000 TS; 80,000 YS; 20 El; 72 RA. For structural and pressure vessels; good weldability.

HY-BB
Jessop Steel Co.
C 0.25-0.33, Mn 0.6-0.9, Si 0.15-0.3, Cr 0.9-1.1, Mo 0.4-0.6, V 0.1-0.15, 2.5 Ni min, bal Fe.
For bleach rings and breech blocks. MIL-S-10185A. High yield and impact strength.

HY-DI 5
Manufacturer not listed.
C 1, Si 0.3, Mn 0.7, Cr 5.2, Mo 1, V 0.2, bal Fe.
For dies, cutters, cold working tools; air hardened, nondeforming.

HY-DI 5
Boyd-Wagner Co.
C 1, Si 0.3, Mn 0.7, Cr 5.2, Mo 1, V 0.2, bal Fe.
For dies, cutters, cold working tools; air hardened, nondeforming.

HY-LO
Boyd-Wagner Co.
C 0.85, Mn 2.24, V 0.1, bal Fe.
For dies; nonshrinking.

HY-LO
Champion Rivet Co.
C 0.12, Mn 0.77, Si 0.72, Mo 0.8, bal Fe.
Welded: 79,000-85,000 TS; 67,000-72,000 YS; 20-26 El; 35-50 RA. For welding; AWS-E7015, low H_2 content.

HY-PRESS 20
Sheepbridge Alloy Castings Ltd.
C 0.06, Mn 0.58, Cu 0.16, Nb 0.02, bal Fe.
As rolled: 63,000-72,000 TS; 47,000-54,000 YS; 30-34 El. Sheet and strip for stamped and formed parts; weldable.

HY-PRESS 20
British Steel Corp.
C 0.06, Mn 0.58, Cu 0.16, Nb 0.02, bal Fe.
As rolled: 63,000-72,000 TS; 47,000-54,000 YS; 30-34 El. Sheet and strip for stamped and formed parts; weldable.

HY-SPEED
Illinois Zinc Co.
Copper. Cu 88, Sn 10, Pb 2.
For bushings, bearings; heavy duty.

HY-TEN "A" NO. 1
Metalsource Corp.
C 0.2, Mn 1, Cr 1.5, bal Fe.
Hot rolled: 80,000 TS; 50,000 YS; 30 El; 58 RA; 163 Brin. Heat treated: 134,000 TS; 116,000 YS; 11 El; 36 RA; 302 Brin. For parts to be carburized, pins, cams, gears, gages, bolts, shafts; case-hardening steel. *Obsolete*

HY-TEN "A" NO. 15
Metalsource Corp.
C 0.15, Ni, Mo, bal Fe.
Heat treated: 118,000 TS; 85,000 YS; 25 El; 65 RA; 235 Brin. For carburized arbors, camshafts, piston pins, ratchets, worms, valve tappets, pinions; similar to SAE 4615 *Obsolete*

HY-TEN "A" NO. 15
U.S. Metalsource
C 0.15, Ni, Mo, bal Fe.
Heat treated: 118,000 TS; 85,000 YS; 25 El; 65 RA; 235 Brin. For carburized arbors, camshafts, piston pins, ratchets, worms, valve tappets, pinions; similar to SAE 4615. *Obsolete*

HY-TEN "A" TEMPER NO. 20
U.S. Metalsource
C 0.2, Ni 0.6, Cr 0.55, Mo 0.2, bal Fe.
Heat treated: 167,000 TS; 138,000 YS; 11 El; 52 RA; 352 Brin. For gears, worms, clutches, arbors, spindles, rolls, toggle pins. Carburizing alloy steel, shock resistant. *Obsolete*

HY-TEN "B" NO. 2

U.S. Metalsource
C 0.4, Mn, Cr, bal Fe.
Hot rolled: 100,000 TS; 62,000 YS; 25 El; 47 RA; 201 Brin. For shafts, spindles, tap shanks, sleeves, worms, arbors, gears; free machining; tough. *Obsolete*

HY-TEN "B" NO. 4

Aluminium Co. of Canada, Ltd.
C 0.5, Mn, Cr, bal Fe.
Annealed: 95,000 TS; 60,000 YS; 28 El; 60 RA; 179 Brin. Heat treated: 240,000 TS; 222,000 YS; 10 El; 43 RA; 477 Brin. For gears, clutches, shafts, pins, springs; oil hardened, tough. *Obsolete*

HY-TEN "B" NO. 5

Metalsource Corp.
C 0.95, low Cr, bal Fe.
Heat treated. For cams, arbors, springs, hammers, reamers, piston rings, tools; water hardening steel. *Obsolete*

HY-TEN "B" TEMPER NO. 3-X-40

Metalsource Corp.
C 0.4, Mn 1.1, Cr 0.75, Mo 0.2, S 0.08, bal Fe.
Heat treated: 230,000 TS; 215,000 YS; 11 El; 43 RA; 460 Brin. Annealed: 95,000 TS; 64,000 YS; 25 El; 57 RA; 183 Brin. For gears, clutches, arbors, cams, mandrels; oil hardened, impact resistant. *Obsolete*

HY-TEN "M" TEMPER

U.S. Metalsource
C 0.7, Mn 0.6, Ni 1.35, Cr 0.65, Mo 0.25, bal Fe.
Oil quenched from 1475°F, tempered at 325°F: 305 ksi TS; 280 ksi YS; 18 RA; 653 Brin; 62 Rock C. Oil hardenable alloy steel for rolls, bushings, cams, collets, stamps, blades, gages, and dies.

HY-TEN "MC" TEMPER

U.S. Metalsource
C 0.7, Cr 0.8, Ni 0.5, Mo 0.25, V 0.08, bal Fe.
Annealed: 96,000 TS; 71,000 YS; 24 El; 53 RA; 200 Brin. Heat treated: 302,000 TS; 279,000 YS; 653 Brin. For gears, arbors, punches, clutches; oil hardened, shock resistant. *Obsolete*

HY-TEN A NO. 1 X

Metalsource Corp.
C 0.2, Mo, Mn, bal Fe.
Rolled: 85,000 TS; 55,000 YS; 35 El; 65 RA; 163 Brin. Heat treated: 140,000 TS; 120,000 YS; 10 El; 35 RA; 311 Brin. For carburized pins, cams, gears, gauges, bolts, shafts; case hardening steel. *Obsolete*

HY-TEN A-2

Metalsource Corp.
C 1, Cr 5.25, Mo 1, V 0.25, bal Fe.
For punches, dies, blanking and forming tools; air hardened, non-deforming. *Obsolete*

HY-TEN AH-70-FM

U.S. Metalsource
C 0.7, Mn 1.15, Ni 3, Cr 1.5, Mo 0.25, Nb 0.05, bal Fe.
Air-hardened tool steel; non-deforming, tough, good machinability. *Obsolete*

HY-TEN ALLOY CHISEL STEEL

Metalsource Corp.
C 0.75, Cr 0.5, Mn 0.35, bal Fe.
For battering tools, chipping chisels, cold chisels; water hardened. *Obsolete*

HY-TEN B340

Metalsource Corp.
C 0.43, Mn 0.95, Cr 0.75, Mo 0.25, bal Fe.
Heat treated: 167,000-265,000 TS; 153,000-235,000 YS; 10-17 El; 32-54 RA; 341-555 Brin. For zinc die casting dies, gears, shafts, cutters; impact resistant, deep hardening. *Obsolete*

HY-TEN B350

Metalsource Corp.
C 0.45, Mn 1.05, Cr 0.85, Mo 0.25, bal Fe.
Heat treated: 185,000-290,000 TS; 170,000-255,000 YS; 10-16 El 32-52 RA; 321-555 Brin. For plastic molds, stripper and bolster plates; impact resistant, deep hardening. *Obsolete*

HY-TEN B3X

U.S. Metalsource
C 0.5, Mn 1.05, S 0.08, Cr 0.7, Mo 0.2, bal Fe.
Rolled: 135,000 TS; 70,000 YS; 18 El; 45 RA; 293 Brin. Heat treated: 240,000 TS; 220,000 YS; 10 El; 40 RA; 477 Brin. For clutches, gears, drive shafts, pinions, connecting rods, mandrels; readily machined at high hardness.

HY-TEN B3X BRAKE DIE

U.S. Metalsource
C 0.5, Mn 1.05, Cr 0.7, Mo 0.2, 0.04 S min, bal Fe.
145 ksi TS; 105 ksi YP; 15 El; 40 RA; 302 Brin. For press brake dies and press brake tooling applications, for guides, roller tracks, machine stops, tool holders, ways, tool bodies, and trim dies.

HY-TEN D-2

U.S. Metalsource
C 1.5, Cr 11.5, V 0.9, Mn 0.4, Mo 0.8, bal Fe.
For forming and drawing dies, punches; air hardened, non-deforming. *Obsolete*

HY-TEN MOLD

U.S. Metalsource
C 0.35, Mn, Cr, Mo, bal Fe.
Heat treated: 160,000-240,000 TS; 132,000-205,000 YS; 10-19 El; 37-58 RA; 311-475 Brin. For plastic molds, Zn die cast molds, tie rods; shock and impact resistant. *Obsolete*

HY-TEN PNEUMATIC CHISEL

U.S. Metalsource
C 0.6, Mn 0.85, Si 2, Cr 0.3, bal Fe.
For chisels, punches, cold cutters, pick points; oil hardened, shock resistant. *Obsolete*

HY-TEN SAE 2315

Metalsource Corp.
C 0.15, Ni 3.5, bal Fe.
Heat treated: 100,000-110,000 TS; 70,000-80,000 YS; 32-33 El; 60-65 RA; 200-220 Brin. For carburized automotive parts, cams, camshafts, arbors, gears, spindles, worms, clutches; case hardening steel; IZ-70-72. *Obsolete*

HY-TEN SAE 3140

Metalsource Corp.
C 0.4, Ni 1.5, Cr 0.75, bal Fe.
Heat treated: 125,000-190,000 TS; 100,000-150,000 YS; 11-21 El; 44-60 RA; 250-380 Brin. For heavy duty shafts, axles, tough gears, spindles, bolts; IZ-57-39. *Obsolete*

HY-TEN SAE 4150

Metalsource Corp.
C 0.5, Cr, Mo, bal Fe.
Heat treated: 132,000-152,000 TS; 118,000-136,000 YS; 10-18 El; 33-59 RA; 318-580 Brin. For shafts, axles, gears, spindles, clutches; resists shock and torsion stresses. *Obsolete*

HY-TEN SAE 4615

Metalsource Corp.
C 0.1-0.2, Ni 1.5-2, Mo 0.2-0.3, bal Fe.
For carburized gears, worms, spindles, clutches, automotive parts; case hardening steel. *Obsolete*

HY-TEN SAE 6145

Metalsource Corp.
C 0.45, Cr, V, bal Fe.
Heat treated: 210,000-260,000 TS; 180,000-225,000 YS; 8-17 El; 25-43 RA; 420-525 Brin. For axles, shafts, clash gears, clutches; shock-resistant. *Obsolete*

HY-TEN-SL BRONZE GR. 2

American Manganese Bronze Co.
Copper. Cu 60-68, Zn 20-24, Al 3-7, Mn 2.5-5, Fe 2-4.
Cast: 100,000 TS; 55,000 YS; 16 El; 16 RA; 200 Brin. For pump and fan impellers, gears, valves; wear and corrosion resistant.

HY-TEN-SL BRONZE GR. 3

American Manganese Bronze Co.
Copper. Cu 60-68, Zn 20-24, Al 3-7, Mn 2.5-5, Fe 2-4.
Cast: 90,000 TS; 45,000 YS; 20 El; 20 RA; 175 Brin. For impellers, gears, valves; wear and corrosion resistant.

HY-TEN-SL BRONZE GR. 4

American Manganese Bronze Co.
Copper. Cu 60-68, Zn 20-24, Al 3-7, Mn 2.5-5, Fe 2-4.
Rolled: 85,000 TS; 40,000 YS; 25 El; 25 RA; 150 Brin. For impellers, gears, valves; wear and corrosion resistant.

HY-TEN-SL BRONZE GRADE 1

American Manganese Bronze Co.
Copper. Cu 60-68, Zn 20-24, Al 3-7, Mn 2.5-5, Fe 2-4.
Cast: 108,000 TS; 65,000 YS; 14 El; 14 RA; 220 Brin. For pump and fan impellers, gears, valves; wear and corrosion resistant.

HY-TEN-SL GR. 1 A

American Manganese Bronze Co.
Copper. Cu 60-68, Zn 20-24, Al 3-7, Mn 2-5, Fe 2-4.
Sand cast: 115,000 TS; 70,000 YS; 12 El; 12 RA; 210 Brin. Forged: 120,000 TS; 73,000 YS; 8 El; 8 RA; 240 Brin. For gears, worm wheels, bearings; wear and corrosion resistant.

HY-TENSO M-50

Alloy Foundries
TC 2.2, Si 1, Mn 1, bal Fe.
Cast: 75,000 TS; 50,000 YS; 3-7 El; 200 Brin. For caster plates, chain links, clutches, cams; case hardenable. *Obsolete*

HY-TENSO M-51

Alloy Foundries
C 2.3, Si 1.2, bal Fe.
Cast: 65,000 TS; 40,000 YS; 4-8 El; 185 Brin. Refined pearlitic malleable iron for chain links, connecting rods, gears.

HY-TENSO M20

Alloy Foundries
C 2.4, Si 1.2, alloy, bal Fe.
Cast: 75,000 TS; 50,000 YS; 5 El; 200 Brin. Pearlitic cast iron for stoker parts, stove and grate parts, dampers.

HY-TEST-ROD NO. 110

Marquette Corp.
C 0.12, Mn 0.6, Mo 0.3, Cr 0.6-1, bal Fe.
Welded: 110,000 TS; 100,000 YS; 15 El; 20 RA; 200 Brin. For hard facing electrode; for moderate shock resistance.

HY-TEST-ROD NO. 85

Marquette Corp.
C 0.1-0.15, Mn 0.95-1.05, Si 0.25-0.35, Mo 0.4-0.6, bal Fe.
Welded: 85,000-95,000 TS; 65,000-75,000 YS; 18-22 El; 25-35 RA. For welding rod; AWS-E8011.

HY-TUF

Crucible Materials Corp.
Alloy steel. C 0.25, Mn 1.3, Si 1.5, Ni 1.8, Mo 0.4, bal Fe.
Heat treated: 234,000 TS; 193,000 YS; 13 El; 50 RA; 470 Brin. For engineering parts, gears, shafts; tough.

HY-TUF

Latrobe Steel Co.
C 0.25, Si 1.5, Ni 1.8, Mn 1.3, Mo 0.4, bal Fe.
Quenched, 1600°F: 112,000 YS; 186,000 TS; 16.8 El; 48.6 RA; 41 Rock C. Low alloy, tough, through-hardening, high tensile strength steel.

HY-TUF
Eagle & Globe Steel Ltd.
Alloy steel. C 0.25, Mn 1.3, Ni 1.8, Cr 0.35, Mo 0.4, Si 1.5, bal Fe.
Normalized and tempered to 248 Brin. Hardened and tempered to 1500-1650 MPa TS. For aircraft landing gear, earth moving equipment, pneumatic tool parts, rock drill bodies. AMS 6418.

HYBNICKEL A
Hybnickel Alloys Co.
Ni 12-25, Cr 20-24, bal Fe.
Cast: 50,000 TS; 40,000 YS; 0-10 El. Forged: 100,000 TS; 70,000 YS; 20 El. For carburizing boxes, hearth plated, furnace conveyor parts; heat resisting; maximum operating temperature 2100°F.

HYBNICKEL A-1
Hybnickel Alloys Co.
Ni 31-39, Cr 16-23, C 0.51-0.65, bal Fe.
For furnace parts; heat resistant.

HYBNICKEL B
Hybnickel Alloys Co.
Ni 5-7, Cr 18-22, bal Fe.
Cast: 50,000 TS; 40,000 YS; 0-15 El. Forged: 100,000 TS; 70,000 YS; 25 El. For carburizing boxes, hearth plates, furnace conveyor parts; heat resisting; maximum operating temperature 1800°F.

HYBNICKEL C
Hybnickel Alloys Co.
Ni 15-35, Cr 25-35, bal Fe.
Cast: 50,000 TS; 40,000 YS; 0.15 El. Forged: 100,000 TS; 70,000 YS; 25 El. For carburizing boxes, hearth plates, furnace conveyor parts; acid resisting; maximum operating temperature 2200°F.

HYBNICKEL D
Hybnickel Alloys Co.
C 0.25-0.5, Cr 20-30, Ni 5-10, bal Fe.
For heat and corrosion resistant parts; heat and corrosion resistant.

HYBNICKEL R
Hybnickel Alloys Co.
C 0.2, Ni 31-39, Cr 7-11, bal Fe.
For heat and corrosion resistant parts; heat resistant.

HYBNICKEL S
Hybnickel Alloys Co.
Ni 25, Cr 20, bal Fe.
Cast: 50,000 TS; 40,000 YS; 0-15 El. Forged: 100,000 TS; 20,000 YS; 25 El. For furnace parts; acid and alkali resistant.

HYBOR
British Steel plc
Stainless steel.
Special nuclear grades containing high boron levels for increased neutron absorption.

HYCHROME 5616
Latrobe Steel Co.
C 0.18, Mn 0.35, Si 0.35, Cr 13, Ni 2, W 3, bal Fe.
Heat treated: 125,000-218,000 TS; 98,000-180,000 YS; 16-22 El; 53-62 RA. For steam turbine buckets and blades, compress parts; hardenable, martensitic, heat resistant to 1200 F. *Obsolete*

HYCLAD 304
British Steel plc
Stainless steel. Cr 18, Ni 9, bal Fe.
For roof and wall cladding in flat or profiled form. Corrosion resistant.

HYCLAD 316
British Steel plc
Stainless steel. Cr 17, Ni 11, Mo 2.25, bal Fe.
Similar to HyClad 304. Corrosion resistant.

HYCLAD 430
British Steel plc
Stainless steel. Cr 17, bal Fe.
Similar to HyClad 304.

HYCO
Lehigh Steel Corp.
C 2, Cr 12.5, V 1, Mo 0.75, bal Fe.
For dies, punches, rivet sets, rolls; wear and abrasion resistant.

HYCO
A. Milne & Co.
C 0.7, W 18, Cr 4, V 1-2, Co 3-5.5, bal Fe.
For tools, cutters; high speed steel.

HYCO 1
Lehigh Steel Corp.
C 1.5, Mn 0.35, Si 0.3, Cr 12, V 0.9, Mo 0.8, bal Fe.
Air or oil hardening cold work tool steel, chromium type; AISI D2.

HYCO 2
Lehigh Steel Corp.
C 2.25, Si 0.3, Cr 11.5, V 0.2, Mo 0.8, bal Fe.
Air or oil hardening cold work tool and die steel; chromium type; AISI D3.

HYCO-SPAN
U.S. Steel Corp.
C 0-0.12, Cr 17-19, Ni 10-13, bal Fe.
For aircraft control cables; stainless. *Obsolete*

HYCOMAX-I
Gibson Electric Co.
Ni 21, Co 20, Al 9.5, Cu 2, bal Fe.
Annealed: 850 coercive force; 9000 residual induction. For electrical and magnetic equipment. Permanent magnet. High permeability. *Obsolete*

HYCOMAX-II
Edgar Allen Balfour Ltd.
Ni 21, Co 20, Al 9.5, Cu 2, bal Fe.
For permanent magnets in electrical and magnetic equipment. High permeability. Permanent magnet.

HYCOMAX-II
Gibson Electric Co.
Ni 14, Co 29, Cu 3, Al 7, Ti 4, Cb 2, bal Fe.
8500 residual induction; 1200 coercive force; 4,000,000 BH max. For electrical and magnetic equipment. Permanent magnet. *Obsolete*

HYCOMAX-III
Gibson Electric Co.
Ni 14, Co 34, Cu 4, Al 7, Ti 5, bal Fe.
8800 residual induction; 1500 coercive force; 5,300,000 BH max. For electrical and magnetic equipment. Permanent magnet. Similar to Alnico VIII. High permeability. *Obsolete*

HYCOMAX-IV
Gibson Electric Co.
Ni 14, Co 40, Cu 3, Al 7.5, Ti 8, bal Fe.
7400 residual induction; 2100 coercive force; 6,000,000 BH max. For electrical and magnetic equipment. Permanent magnet. *Obsolete*

HYCRO V80
English manufacture
C 0.2, Cr 3, Mo 0.6, W 0.5, V 0.8, bal Fe.
For gas turbine components; creep resistant.

HYCV
Crucible Materials Corp.
Tool material. C 2.45, V 3.75, Mo 1, Cr 12.25, bal Fe.
Heat treated: 660 Brin. For brick mold liners, porcelain molds, punches; air or oil hardened, nondeforming.

HYDRA
Osborn Steels Ltd.
C 0.65, W 15, Cr 4, V 0.5, bal Fe.
For lathe and planer tools, drills, hobs, taps; high speed steel. *Obsolete*

HYDRA HUSKY
Osborn Steels Ltd.
C 0.8, W 22, Cr 4.5, V 1.5, bal Fe.
For broaches, hobs, reamers, drills, high speed steel. *Obsolete*

HYDRA MULTICO
Osborn Steels Ltd.
C 0.85, W 22, Cr 4.5, V 1.5, Co 11, Mo 0.75, bal Fe.
For broaches, drills, reamers, hobs, lathe tools; high speed steel. *Obsolete*

HYDRA VANTAGE
Osborn Steels Ltd.
C 1.25, W 13.5, Cr 4.5, V 4, bal Fe.
For drills, taps, hobs, milling cutters; high speed steel. *Obsolete*

HYDRA-E
Osborn Steels Ltd.
C 0.3, W 8.5, Cr 3.5, V 0.35, bal Fe.
For hot extrusion dies for brasses; hot work steel, oil hardened. *Obsolete*

HYDRA-H.D.
Osborn Steels Ltd.
C 0.4, W 14, Cr 4, V 0.7, bal Fe.
For die casting dies; hot work steel, oil hardened. *Obsolete*

HYDRA-M
Osborn Steels Ltd.
C 0.35, W 4, Cr 1.5, Mo 0.5, Ni 4, bal Fe.
For mandrels, pilger bars, hot punches; hot work steel, oil hardened. *Obsolete*

HYDRA-V.K.
Osborn Steels Ltd.
C 0.4, Cr 5, V 0.6, Mo 1.4, Si 1, bal Fe.
For Al die casting dies; hot work steel, oil hardened. *Obsolete*

HYDRA-Z
Osborn Steels Ltd.
C 0.26, W 8.5, Cr 3, V 0.25, Mo 0.5, Ni 2.5, bal Fe.
For hot forging dies, punches; hot work steel, oil hardened. *Obsolete*

HYDRAULIC 56
Carpenter Technology Corp.
C 0.85, Mn 0.3, Si 0.3, Cr 1.75, V 0.2, bal Fe.
Oil hardening tool steel for shafts, arbors, drill bushings, lathe centers; AISI L3.

HYDRAULIC BRONZE
American manufacture
Cu 83, Sn 11, Zn 6, Pb 0.1.
For pumps, cocks, valves; pressure tight.

HYDRAULIC BRONZE
American manufacture
Cu 72.5, Zn 19.5, Sn 1.75, Pb 6.5.
For pumps, cocks, valves; pressure tight.

HYDREX 1212
Osborn Steels Ltd.
C 0-0.12, Cr 12-13, Ni 11.5-12.5, bal Fe.
For auto trim, kitchenware, architectural uses; austenitic, stainless, malleable. *Obsolete*

HYDREX 188
Osborn Steels Ltd.
C 0-0.12, Cr 17.5-18.5, Ni 7.5-9.5, bal Fe.
Annealed: 80,000 TS; 35,000 YS; 55 El; 75 RA; 150 Brin. For fittings, chemical plant equipment; Type 302; stainless, austenitic. *Obsolete*

HYDREX 367
Osborn Steels Ltd.
C 0.07, Cr 17.2-19, Ni 7.2-9, bal Fe.
Annealed: 85,000 TS; 35,000 YS; 60 El; 70 RA; 150 Brin. For fittings, chemical plant equipment; Type 304; stainless, austenitic. *Obsolete*

HYDREX 368
Osborn Steels Ltd.
C 0-0.07, Cr 17-19, Ni 7-9, Ti 0.25-0.4, bal Fe.
Annealed: 85,000 TS; 35,000 YS; 55 El; 65 RA; 150 Brin. For acid resistant chemical plant equipment; Type 321; stainless, austenitic. *Obsolete*

HYDREX 369
Osborn Steels Ltd.
C 0-0.12, Cr 18-19, Ni 12-13, bal Fe.
Annealed: 90,000 TS; 40,000 YS; 55 El; 65 RA; 160 Brin. For chemical plant equipment, tanks, mixers; stainless, austenitic. *Obsolete*

HYDREX 372
Osborn Steels Ltd.
C 0-0.1, Cr 18-20, Ni 8-9.5, Cb 0.8, bal Fe.
Annealed: 90,000 TS; 45,000 YS; 56 El; 65 RA; 160 Brin. For welded chemical plant equipment; Type 347; stainless, austenitic. *Obsolete*

HYDREX 375
Osborn Steels Ltd.
C 0-0.12, Cr 19-21, Ni 10.5-12, bal Fe.
Annealed: 85,000 TS; 35,000 YS; 60 El; 70 RA; 150 Brin. For core wire for welding electrodes; stainless, austenitic. *Obsolete*

HYDREX 399
Osborn Steels Ltd.
C 0-0.1, Cr 18.5-20, Ni 8.5-10, Mo 1-1.5, bal Fe.
Annealed: 85,000 TS; 35,000 YS; 60 El; 70 RA; 150 Brin. For core wire for welding electrodes; stainless, austenitic. *Obsolete*

HYDREX CTY
Osborn Steels Ltd.
C 0.25-0.35, Cr 12-14, Ni 0-1, bal Fe.
Annealed: 100,000 TS; 55,000 YS; 20 El; 50 RA; 210 Brin. For cutlery, scissors, surgical instruments; Type 420; stainless, hardenable. *Obsolete*

HYDREX ENGG
Osborn Steels Ltd.
C 0.18-0.25, Cr 12-14, Ni 0-1, bal Fe.
Annealed: 95,000 TS; 50,000 YS; 25 El; 55 RA; 195 Brin. For pump and piston rods, surgical instruments, cutlery; Type 420; stainless, hardenable. *Obsolete*

HYDREX ENGG-FM
Osborn Steels Ltd.
C 0.2-0.27, Cr 12-14, Ni 0-1, Mo 0.4, S 0.2, bal Fe.
Annealed: 100,000 TS; 55,000 YS; 20 El; 50 RA; 200 Brin. For fittings, piston and pump rods, valves and valve seats; Type 420F; stainless, free-cutting. *Obsolete*

HYDREX HCi
Osborn Steels Ltd.
C 0-0.12, Cr 19.5-22.5, Ni 0-0.25, bal Fe.
For valve seats, stokers, fire grids; heat and scale resistant to 1000 C. *Obsolete*

HYDREX HR 6015
Osborn Steels Ltd.
C 0-0.2, Ni 60, 15% min Cr, bal Fe.
For furnace parts, recuperators, heat treating boxes; resists heat and scaling to 1100 C. *Obsolete*

HYDREX HR2520
Osborn Steels Ltd.
C 0-0.12, Cr 24-26, Ni 19-21, bal Fe.
For heat treating boxes, furnace muffles; oxidation and heat resistant to 1150 C. *Obsolete*

HYDREX HR356
Osborn Steels Ltd.
C 0-0.15, Cr 20-22, Ni 7-8, bal Fe.
For welding electrodes; resists heat and scaling to 900 C. *Obsolete*

HYDREX HR393
Osborn Steels Ltd.
C 0.25, Cr 22-24, Ni 10-12, W 3, bal Fe.
For exhaust inlet valves, furnace impellers; stainless, austenitic, resists scaling to 1050 C. *Obsolete*

HYDREX HRV
Osborn Steels Ltd.
C 0.39-0.45, Cr 13-14, Ni 13.2-14.2, W 2.5-3, bal Fe.
For exhaust and inlet valves, furnace impellers; high heat and corrosion resistance. *Obsolete*

HYDREX MC1
Osborn Steels Ltd.
C 0-0.12, Cr 12-14, Ni 0-0.25, bal Fe.
Annealed: 75,000 TS; 40,000 YS; 35 El; 70 RA; 155 Brin. For turbine blades, cutlery, valves, fittings; Type 410; stainless. *Obsolete*

HYDREX MC1-FM
Osborn Steels Ltd.
C 0-0.12, Cr 12-14, Ni 0-1, Mo 0.4, S 0.2, bal Fe.
Annealed: 75,000 TS; 40,000 YS; 35 El; 70 RA; 155 Brin. For screw machine products, turbine blades; Type 410 F; free-cutting, stainless. *Obsolete*

HYDREX MODIFIED WIM
Osborn Steels Ltd.
C 0-0.12, Cr 19-20, Ni 9-10, Mo 1.5-2, bal Fe.
Annealed: 85,000 TS; 35,000 YS; 50 El; 65 RA; 160 Brin. For paper and pulp equipment, dye and bleaching equipment; Type 316; stainless, austenitic. *Obsolete*

HYDREX S80
Osborn Steels Ltd.
C 0.13-0.18, Cr 16.5-17.5, Ni 1.6-2, bal Fe.
Annealed: 80,000 TS; 50,000 YS; 25 El; 50 RA; 150 Brin. For valve spindles, impeller shafts, Type 430; stainless, martensitic. *Obsolete*

HYDREX SPECIAL HR2520
Osborn Steels Ltd.
C 0-0.12, Cr 25-28, Ni 19-21, bal Fe.
For welding electrodes; high creep and oxidation resistance. *Obsolete*

HYDREX W
Osborn Steels Ltd.
C 0-0.12, Cr 18-20, Ni 8-10, Ti 0.5, bal Fe.
Annealed: 85,000 TS; 35,000 YS; 55 El; 65 RA; 150 Brin. For welded chemical plant equipment; Type 321; stainless, austenitic. *Obsolete*

HYDREX W-FM
Osborn Steels Ltd.
C 0.15, Cr 17-20, Ni 7-10, Mo 0.2-0.4, Ti 0.5, bal Fe.
Annealed: 85,000 TS; 35,000 YS; 50 El; 65 RA; 160 Brin. For acid resistant chemical plant equipment; Type 316Ti; stainless, austenitic. *Obsolete*

HYDREX WIM
Osborn Steels Ltd.
C 0-0.12, Cr 19-20, Ni 9-10, Mo 2.7-3.2, bal Fe.
Annealed: 90,000 TS; 40,000 YS; 45 El; 60 RA; 170 Brin. For paper and pulp industry equipment; Type 317; stainless, austenitic. *Obsolete*

HYDREX XB (IIR329)
Osborn Steels Ltd.
C 0.74-0.89, Cr 19-20.5, Ni 1.15-1.65, bal Fe.
For exhaust and inlet valves; resists exhaust gases containing lead. *Obsolete*

HYDRO-T-METAL
Hydrometals Inc.
Cu 0.4-0.7, Ti 0.08-0.16, Mn 0.002-0.01, Cr 0.003-0.02, bal Zn.
Rolled: 26,500 TS; 20,000 YS; 36 El. For auto trim, furniture, rain gutters, hardware, electrical fixtures; stable, corrosion resistant.

HYDRO-T-METAL 200
Illinois Zinc Co.
Cu 0.3-1.25, Ti 0.1-0.4, bal Zn.
Sheet: 24,000 TS; 10 El; 55 Brin. For roofing, gutters, trim, fuses, housings, curtain walls. High creep resistance.

HYDRO-T-METAL 200
Hydrometals Inc.
Cu 0.3-1.25, Ti 0.1-0.4, bal Zn.
Sheet: 24,000 TS; 10 El; 55 Brin. For roofing, gutters, trim, fuses, housings, curtain walls. High creep resistance.

HYDRONALIUM
Oederlin & Co. Ltd.
Aluminum. Mg 1.5-3.5, bal Al.
Sand cast: 19,900-27,000 TS; 11,400-14,000 YS; 3-8 El; 50-60 Brin. For marine castings; corrosion resistant.

HYDRONALIUM
Westfalische Leichtmetallwerke GmbH
Mg 6-10, Mn 0.2-0.7, Fe 0-1.5, bal Al.
Heat treated: 48,000 TS; 26,000 YS; 16 El; 75 Brin. For light alloy parts; age hardenable.

HYDRONALIUM
Vesevorder Metallwerke GmbH
Mg 6-10, Mn 0.2-0.7, Fe 0-1.5, bal Al.
Heat treated: 48,000 TS; 26,000 YS; 16 El; 75 Brin. For light alloy parts; age hardenable.

HYDRONALIUM 10
Farbenindustrie Atkiengesellschaft
Aluminum. Mg 1, Mn 0.5, Si 0.2-1.5, bal Al.
Forged: 54,000 TS; 32,000 YS; 15 El; 93 BHN. For seaplane pontoons and propellers; resists sea water corrosion.

HYDRONALIUM 2
Westfalische Leichtmetallwerke GmbH
Mg 1.3, Mn 0-0.4, Cr 0-0.3, bal Al.
Soft: 28,000 TS; 13,000 YS; 30 El; 47 Brin. Hard: 42,000 TS; 37,000 YS; 8 El; 77 Brin. For marine products, aircraft structures; resists seawater corrosion.

HYDRONALIUM 21
Westfalische Leichtmetallwerke GmbH
Mg 0.6-1.4, Si 0.6-1.6, Mn 0.6-1, Cr 0-0.3, bal Al.
Soft: 28,000 TS; 13,000 YS; 30 El; 47 Brin. For marine products, aircraft structures; corrosion resistant.

HYDRONALIUM 3
Westfalische Leichtmetallwerke GmbH
Mg 2-4, Mn 0-0.4, Cr 0-0.3, bal Al.
Soft: 30,000 TS; 15,000 YS; 28 El; 50 Brin. For marine and aircraft parts; resists seawater corrosion.

HYDRONALIUM 43
Farbenindustrie Atkiengesellschaft
Aluminum. Zn 4.5, Mg 3.5, bal Al.
Wire: 70,000 TS; 45,000 YS; 16 El. Sheet: 63,000 TS; 40,500 YS; 18.5 El. For light alloy parts; heat treatable.

HYDRONALIUM 43
Westfalische Leichtmetallwerke GmbH
Mg, Zn, bal Al.
For aircraft and light alloy parts.

HYDRONALIUM 5
Farbenindustrie Atkiengesellschaft
Aluminum. Al 95, Mg 5.
Extruded: 33,000-36,000 TS; 13,000-14,000 YS; 16-22 El; 60-70 BHN. For seaplane pistons, propellers; resists sea water.

HYDRONALIUM 5
Westfälische Leichtmetallwerke GmbH
Mg 4-5.5, Mn 0-0.8, Cr 0-0.3, bal Al.
Soft: 42,000 TS; 22,000 YS; 35 El; 65 Brin. Hard: 60,000 TS; 50,000 YS; 15 El; 100 Brin. For marine hardware, aircraft parts; corrosion resistant.

HYDRONALIUM 51
Farbenindustrie Atkiengesellschaft
Aluminum. Si 0.2-1.5, Mg 5-12, Mn 0.2-0.5, bal Al.
Cast: 23,000-27,000 TS; 13,000-18,000 YS; 2-5 El; 60-70 BHN. For light alloy parts, aircraft components; corrosion resistant.

HYDRONALIUM 51
Westfälische Leichtmetallwerke GmbH
Mg 5-12, Si 0.2-1.5, Mn 0.2-0.5, Cr, Ti, bal Al.
Sand cast: 23,000-27,000 TS; 13,000-18,000 YS; 2-5 El; 60-70 Brin. For gasoline flow meters, engine components; corrosion resistant.

HYDRONALIUM 511
Farbenindustrie Atkiengesellschaft
Aluminum. Fe 0.3, Mn 0.2, Mg 5, Si 1, Ti 0.1, bal Al.
For castings; sand.

HYDRONALIUM 5112
Farbenindustrie Atkiengesellschaft
Aluminum. Cu 1, Fe 0.3, Mn 0.1, Mg 5, Si 1, Ti 0.1, Be 0.18, bal Al.
For sand castings; heat treatable.

HYDRONALIUM 7
Farbenindustrie Atkiengesellschaft
Aluminum. Mg 7, Mn 0.45, bal Al.
Rolled: 52,000 TS; 32,000 YS; 12 El; 115 BHN. For light alloy parts, seaplane pontoons and propellers; resists sea water corrosion.

HYDRONALIUM 7
Westfälische Leichtmetallwerke GmbH
Mg 5.5-7.5, Mn 0-0.8, Cr 0-0.3, bal Al.
For light alloy parts; corrosion resistant.

HYDRONALIUM 71
Farbenindustrie Atkiengesellschaft
Aluminum. Mg 7, bal Al.
Cast: 32,000-36,000 TS; 17,000-20,000 YS; 5-8 El; 70-80 BHN. For gasoline flow meters and aircraft parts; corrosion resistant.

HYDRONALIUM 71
Westfälische Leichtmetallwerke GmbH
Mg 7, Si, Mn, Cr, Ti, bal Al.
Permanent mold: 32,000-36,000 TS; 17,000-20,000 YS; 5-8 El; 70-80 Brin. For marine and aircraft parts; corrosion resistant.

HYDRONALIUM 9
Farbenindustrie Atkiengesellschaft
Aluminum. Mg 9, bal Al.
Rolled: 60,000 TS; 40,000 YS; 10 El. Extruded: 54,000 TS; 25,000 YS; 20 El; 45 RA; 82 BHN. For seaplane pontoons and propellers; resists sea water corrosion.

HYDRONALIUM 91
Westfälische Leichtmetallwerke GmbH
Mg, Si, Mn, Cr, Ti, bal Al.
For light alloy parts.

HYDRONALIUM D B A
Farbenindustrie Atkiengesellschaft
Aluminum. Al alloy.
For light alloy parts, aircraft parts, corrosion resistant.

HYDRONALIUM HY25
Farbenindustrie Atkiengesellschaft
Aluminum. Si 0.2-1, Mn 0.2-0.5, Mg 3-12, bal Al.
Soft: 27,000 TS; 12,800 YS; 24 El; 50 Brin. Half hard: 36,000 TS; 26,000 YS; 10 El; 60 Brin. For hardware, aircraft parts. Not heat treatable. *Obsolete*

HYDRONALIUM HY51
Oederlin & Co. Ltd.
Aluminum. Mg 5-12, Si 0.2-1.5, Mn 0.2-0.5, bal Al.
Sand cast: 23,000-28,000 TS; 11,400-14,400 YS; 2.5-5.0 El; 55-70 Brin. For aircraft and marine equipment; corrosion resistant to seawater.

HYDRONALIUM HY71
Oederlin & Co. Ltd.
Aluminum. Mg 7, Si, Mn, bal Al.
Permanent mold: 32,000-36,000 TS; 17,000-20,000 YS; 8-5 El; 70-80 Brin. For aircraft and marine equipment; corrosion resistant to seawater.

HYDRONE
Cutler Hammer Inc.
Lead. Na 31, bal Pb.
Used as a deoxidizer; for non-ferrous metals.

HYDRONE
English manufacture
Pb 67, Sb 33.
For bearings; light shocks only.

HYFAB 3/12
British Steel plc
Stainless steel. Cr 11.5, bal Fe.
Weldable chromium steel. Corrosion resistant.

HYFLOR DURBAR 304L
British Steel plc
Stainless steel. Cr 18, Ni 10, bal Fe.
Stainless raised pattern floor plate.

HYFLOR DURBAR 316L
British Steel plc
Stainless steel. Cr 17, Ni 12, Mo 2.25, bal Fe.
Stainless floor plate. Corrosion resistant.

HYFLOW 420R
British Steel plc
Stainless steel. Cr 12, bal Fe.
Hard stainless steel plate. Wear resistant.

HYFLOW COMPOSITE
British Steel plc
Stainless steel.
Composite plate.

HYFLUX ALNICO K-7
Electronic Memories & Magnetics Corp.
Al 8, Ni 14, Co 24, bal Fe.
Energy product 7.5×10^6 max; residual induction 13,400 gauss; coercive force 3000 oersted. Permanent magnet.

HYFLUX ALNICO V
Indiana General
Al 8, Ni 14, Co 24, bal Fe.
For permanent magnets; high energy product. *Obsolete*

HYFLUX ALNICO-9
Electronic Memories & Magnetics Corp.
Al 7, Ni 15, Co 35, Fe 34, Cu 4, Ti 5.
Energy product 10,000,000 gauss-oersted. Coercive force 1600 oersted. For straight field focusing devices, repulsion and torque transmitters, motors, space age applications. Permanent magnet. Resists demagnetization.

HYFORM 409
British Steel plc
Stainless steel. Cr 11, bal Fe.
Corrosion and oxidation resistant. For car exhaust systems and canisters containing exhaust catalysts.

HYKRO
British Steel plc
Stainless steel. Cr 3, Mo 0.5, bal Fe.
Armor plate steel for applications requiring plate greater than 80 mm thick.

HYKRO
British Steel Corp.
C 0.25, Mn 0.55, Cr 3.2, Mo 0.5, bal Fe.
Oil hardened: 104,000-145,000 TS; 83,000-121,000 YS; 20-28 El; 65-72 RA; 75-110 IZ. For crankshafts, cylinder liners, gears, machine parts. Nitriding steel. Wear resistant. *Obsolete*

HYKRO V-80
English manufacture
C 0.2, Cr 3, Mo 0.6, W 0.5, V 0.8, bal Fe.
For oil refinery equipment; corrosion and heat resistant.

HYKRO-V
British Steel Corp.
C 0.4, Mn 0.6, Cr 3, Mo 0.9, V 0.2, bal Fe.
Oil hardened: 184,000-260,000 TS; 157,000-197,000 YS; 15-20 El; 54 RA; 23-40 IZ. For gears, cams, camshafts, rollers. Oil or nitride hardening. *Obsolete*

HYLASTIC
Manufacturer not listed.
C 0.3, Mn 1.6, V 0.1, Si 0.4, bal Fe.
Normalized: 100,000-90,000 TS; 76,000-60,000 YS; 28-25 El; 58-50 RA; 180 Brin, 40 IS (Izod). For railroad and structural work where high physicals are required.

HYLITE 10
Jessop-Saville Ltd.
Ti.
Bar: 60,500 TS; 36,000 YS; 30 El; 150 Brin. Sheet: 61,600 TS; 44,800 YS; 20 El; 160 Brin. For aircraft parts; good form- and weldability.

HYLITE 15
Jessop-Saville Ltd.
Ti.
Bar: 71,700 TS; 44,800 YS; 25 El. Sheet: 76,200 TS; 56,000 YS; 18 El. For aircraft parts; alpha type.

HYLITE 20
Jessop-Saville Ltd.
Al 5, Sn 2.5, bal Ti.
Bar: 125,500 TS; 103,000 YS; 18 El. For aircraft and missile components; alpha type.

HYLITE 30
Jessop-Saville Ltd.
Mn 2, Al 2, bal Ti.
Bar: 98,600 TS; 80,600 YS; 20 El. For compressor discs and blades, fasteners, fuel systems; heat treatable.

HYLITE 40
Jessop-Saville Ltd.
Mn 4, Al 4, bal Ti.
Bar: 138,900 TS; 130,000 YS; 18 El. For compressor discs and blades, fasteners, fuel systems; heat treatable.

HYLITE 45
Jessop-Saville Ltd.
Al 6, V 4, bal Ti.
Bar: 141,200 TS; 130,000 YS; 22 El. For compressor discs and blades, fasteners, fuel systems; heat treatable.

HYLITE-1
Jessop-Saville Ltd.
Ti 100.
Rolled: 78,000 TS; 38,000 YS; 30 El. For the chemical industries. Commercially pure Ti.

HYLITE-15H
Jessop-Saville Ltd.
Ti 100.
Rolled: 98,000 TS; 78,000 YS; 27 El. For the chemical industry. Commercially pure Ti with addition of oxygen to raise strength.

HYLITE-25
Jessop-Saville Ltd.
Cu 2.5, bal Ti.
Rolled: 85,000 TS; 64,000 YS; 27 El. For the chemical industry. Alpha-eutectoid alloy. Good weldability, and ductility.

HYLITE-50
Jessop-Saville Ltd.
Al 4, Mo 4, Sn 2, Si 0.5, bal Ti.
Bar: 161,300 TS; 143,400 YS; 15 El. For compressor discs and blades, fuel systems, fasteners. Heat treatable.

HYLITE-51
Jessop-Saville Ltd.
Al 4, Mo 4, Sn 4, Si 0.5, bal Ti.
Heat treated: 200,000 TS; 179,000 YS; 15 El. For heavy duty service aircraft structural members. Alpha-beta alloy. Good creep properties.

HYLITE-55
Jessop-Saville Ltd.
Sn 6, Zr 5, Al 3, Si 0.5, bal Ti.
Heat teated: 130,000 TS; 116,000 YS; 16 El. For tubine compressor blades and discs. Alpha titanium alloy. Good properties, to over 500°C. High temperature strength.

HYLITE-60
Jessop-Saville Ltd.
Al 3, Sn 6, Zr 5, Mo 2, Si 0.4, bal Ti.
At 70°F: 157,000 TS; 135,000 YS; 14 El. At 1000°F: 90,000 TS; 81,000 YS. For turbine compressor blades and discs up to 500°C. Good creep and tensile properties. Alpha-beta titanium alloy.

HYLITE-65
Jessop-Saville Ltd.
Sn 6, Zr 5, Al 3, Mo 0.5, Si 0.5, bal Ti.
Heat treatd: 155,000 TS; 136,000 YS; 13 El; 27 RA. At 600°C: 91,400 TS; 69,500 YS; 16.5 El; 49 RA. For jet engine compressor components, discs, spacers, blades. High creep resistance, weldable. Near alpha alloy, heat treatable.

HYMAN
English manufacture
Ag 58.3, Cu 16.7, Zn 16.7, Ni 8.3.
For jewelry, ornaments; Nickel silver.

HYMAN ALLOY
Manufacturer not listed.
Si 0.8, Cu 3, Mg 0.5, Ni 0.5, bal Al.
Heat treated: 36,000-43,000 TS; 6-9 El. For light alloy parts; age-hardenable.

HYMAX 15% COBALT MAGNET
Edgar Allen Balfour Ltd.
Co 15, C 1.15, Cr 10, Mo 1.5, bal Fe.
Cast. For magnet steel; 7850-7600 Br; 200-220 Hc; 648,000 BH max.

HYMAX 3% COBALT MAGNET
Edgar Allen Balfour Ltd.
Co 3, C 1.15, Cr 10, Mo 1.5, bal Fe.
Cast. For magnet steel; 7000-6750 Br; 140-130 Hc; 410,000 BH max.

HYMAX 35% COBALT MAGNET
Edgar Allen Balfour Ltd.
Co 35, C 0.85, W 4.5, Cr 6, bal Fe.
Cast. For magnet steel; 8800-8350 Br; 265-240 Hc.

HYMAX 6% COBALT MAGNET
Edgar Allen Balfour Ltd.
C 1.15, Co 6, Cr 10, Mo 1.5, bal Fe.
Cast. For magnet steel; 7250-7000 Br; 160-145 Hc; 460,000 BH max.

HYMAX 9% COBALT MAGNET
Edgar Allen Balfour Ltd.
Co 9, C 1.15, Cr 10, Mo 1.5, bal Fe.
Cast. For magnet steel; 7550-7250 Br; 175-165 Hc; 527,000 BH max.

HYMU 80
Carpenter Technology Corp.
C 0-0.1, Ni 80, Mo 4, bal Fe.
Cold drawn: 97,000 TS; 69,000 YS; 37 El; 71 RA; 220 Brin.
For transformer cores, magnetic shields. Hydrogen annealed.

HYMU 800
Carpenter Technology Corp.
Mo 4, Ni 79, bal Fe.
For laminations for motor cores, toroids. High magnetic permeability, soft magnet.

HYNICAL
English manufacture
Ni 31, Al 12.5, Ti 0.4, bal Fe.
For permanent magnets, electrical and magnetic equipment; high permeability.

HYNICO II
Johnson Matthey
Ni, Co.
For magnet; permanent. *Obsolete*

HYNICO-I
Johnson Matthey
Ni 19, Co 12, Al 10, Cu 6, bal Fe.
For permanent magnets, electrical and magnetic equipment. High permeability. *Obsolete*

HYPAR
Universal Cyclops
C, alloy, bal Fe.
For tools, dies. *Obsolete*

HYPERM 36
Krupp Stahl AG
alloy steel. C, alloy, bal Fe.
For magnets, electrical apparatus; magnetic alloy, high permeability. *Obsolete*

HYPERM 4
Krupp Stahl AG
alloy steel. Si 2.5-4.5, C, bal Fe.
For precision transformers; magnetic alloy. *Obsolete*

HYPERM 50 Y
Krupp Stahl AG
alloy steel. C, high Ni, bal Fe.
For magnets, electrical apparatus; magnetic alloy, high permeability. *Obsolete*

HYPERM 6
Krupp Stahl AG
alloy steel. C, Si, bal Fe.
For magnets, electrical apparatus; magnetic alloy, high permeability. *Obsolete*

HYPERM-702
J.C. Soding & Halbach
C 0.1, Si 0.1, Mn 1.4, Mo 3.25, Cu 9, Ni 74, bal Fe.
For electrical and magnetic equipment magnets. High permeability. *Obsolete*

HYPERM-766
J.C. Soding & Halbach
C 0.1, Si 0.1, Mn 0.65, Cr 2, Mo 3.25, C 5, Ni 76, bal Fe.
For electrical and magnetic equipment magnets. High permeability. *Obsolete*

HYPERM-CO 27
Krupp Stahl AG
alloy steel. C 0.006, Co 27.5, bal Fe, 0.002 oxy.
Annealed: 80,000 TS; 45,000 YS; 15 El; 85 Rock B. Cold Rolled: 100,000 TS; 90,000 YS; 1 El; 30 Rock C. For torque motors, generators, pole pieces, relays, alternators, transformer rectifiers. Soft magnetic alloy, high magnetic saturation. Operates at high flux levels. *Obsolete*

HYPERM-CO 50
Krupp Stahl AG
react and refract. Co 49, V 2, bal Fe.
Annealed: 80,000 TS; 48,000 YS; 0.5 El; B 97 Rock. Cold rolled: 195,000 TS; 185,000 YS; 1 El; C 35 Rock. For electric motors, computers, transformers, magnetic amplifiers, magnetostriction reducers. Magnetically soft for high flux levels. High permeability at high inductions. *Obsolete*

HYPERM-O
Krupp Stahl AG
alloy steel. Ni, bal Fe.
For electrical equipment. *Obsolete*

HYPERNOM
Now CARPENTER HYPERNOM.

HYPERNOM
Westinghouse Electric Corp.
Ni 79, Mo 4, bal Fe.
Hot rolled: 125,000 psi TS; 112,000 psi YS; 17.5 El; 165 Brin; 85 Rock B. For telephone loading coils, instrument transformers, electronic shields, magnetometers; high permeability, low fields. *Obsolete*

HYPERSILICIE
English manufacture
Cu 3, Fe 0.5, Si 18, bal Al.
For pistons; low density, cast.

HYPERSILID
Bradley Laboratories
Si 14-16, C, bal Fe.
For castings; cast iron, acid resistant.

HYPERTHERMOS 150
Vallourec S.A.
C 0.25-0.35, Mn 0-1, Cr 27-29, Mo 5-6, Ni 1.5-3, Si 0-1, bal Co.
For high temperature equipment.

HYPERTHERMOS 190
Vallourec S.A.
C 0.07-0.12, Si 0-1, Mn 0-1, Cr 16-18, Ni 12-14, W 2.5-4, Ti = 4 x C min, bal Fe.
Austenitic stainless steel; for elevated temperature service.

HYPERTHERMOS 304
Vallourec S.A.
C 0-0.07, Si 0-1, Mn 0-2, Cr 17-19, Ni 8-10, bal Fe.
Austenitic stainless steel. AFNOR Z 6 CN 18 09; AISI 304.

HYPERTHERMOS 316
Vallourec S.A.
C 0-0.07, Si 0-1, Mn 0-2, Cr 16-18, Ni 10-12, Mo 2-2.5, bal Fe.
Austenitic stainless steel; for chemical equipment. AFNOR Z 6 CND 17-11; AISI 316.

HYPERTHERMOS 316 ST
Vallourec S.A.
C 0-0.1, Si 0-1, Mn 0-2, Cr 16-18, Ni 11-13, Mo 2-2.5, Ti = 5 x C (up to 0.80), bal Fe.
Stabilized austenitic stainless steel; for welded chemical equipment. AFNOR Z 8 CNDT 17-12.

HYPERTHERMOS 321
Vallourec S.A.
C 0-0.08, Si 0-1, Mn 0-2, Cr 17-19, Ni 10-12, Ti = 5 x C (up to 0.80), bal Fe.
Stabilized austenitic stainless steel. AFNOR Z 6 CNT 18-11; AISI 321.

HYPERTHERMOS 347
Vallourec S.A.
C 0-0.08, Si 0-1, Mn 0-2, Cr 17-19, Ni 10-12, Nb + Tb = 10 x C (up to 1.0), bal Fe.
Stabilized austenitic stainless steel. AFNOR Z 8 CNNb 18-11; AISI 347.

HYPLUS 23
British Steel plc
C 0-0.2, Si 0-0.5, Mn 0-1.5, Nb 0-0.1, bal Fe.
High strength structural steel. For bridges, ships, cranes, and pipelines. *Obsolete*

HYPLUS 29
British Steel plc
C 0-0.22, Si 0.15-0.5, Mn 0-1.6, V 0-0.2, P 0-0.05, S 0-0.03, bal Fe.
High strength structural steel.

HYPOFLEX
Superior Tube Company
Needle drawn stainless tubing. Meets AISI 304 composition.

HYPRESS 20
British Steel plc
C 0-0.12, Mn 0-1.2, bal Fe.
300 N/mm^2 YS min. Cold formable and weldable; hot and cold rolled strip.

HYPRESS 23
British Steel plc
C 0.08-0.12, Mn 0-1.2, bal Fe.
As rolled: 67,000-76,000 TS; 51,000-58,000 YS; 28-34 El. Sheet and strip for stamped and formed parts. Weldable.

HYPRESS 26
British Steel plc
C 0.1-0.12, Mn 0-1.2, bal Fe.
As rolled: 74,000-85,000 TS; 56,000-65,000 YS; 25-31 El. Sheet and strip for stamped and formed parts. Weldable.

HYPRESS 29
British Steel plc
C 0-0.12, Mn 0-1.2, bal Fe.
As rolled: 78,000-87,000 TS; 65,000-72,000 YS; 24-29 El. Sheet and strip for stamped and formed parts. Weldable.

HYPRESS 35
British Steel plc
C 0-0.12, Mn 0-1.2, bal Fe.
550 N/mm^2 YS min.

HYPRESS 40
British Steel plc
C 0-0.12, Mn 0-1.2, bal Fe.
Hot rolled strip: 620 N/mm^2 YS min.

HYPRO
Bissett Steel Co.
C 1.5, Cr 14, bal Fe.
For punches, dies, forming rolls, shear blades; non-deforming. *Obsolete*

HYPRO 61 AIR HARDENING
Boyd-Wagner Co.
C 1.5, Cr 11.5, Mo 0.8, V 2, bal Fe.
For gages, drawing dies, trimmers; air hardening.

HYPRO 63
Boyd-Wagner Co.
C 1.7, Cr 18, bal Fe.
For dies, punches, rotary shears, gauges; nonshrink, deep hardening.

HYPRO A
Bissett Steel Co.
C 0.35, Cr 7.5, W 7.5, Si 1.5, bal Fe.
For dies, oil or air hardened. *Obsolete*

HYPRO A.
Bethlehem Steel Corp.
C 0.35, Mn 0.6, Cr 7.5, W 7.5, Si 1.5, bal Fe.
For hot work dies; hot work steel, oil or air hardened. *Obsolete*

HYPRO B
Bethlehem Steel Corp.
C 0.35, Cr 5.25, W 5.25, Mo 0.2, V 0.2, Co 0.5, Si 0.9, bal Fe.
For hot work dies; hot work steel, air hardened. *Obsolete*

HYPRO B
Bissett Steel Co.
C 0.35, Cr 5.2, W 5.2, V 0.2, Mo 0.2, Co 0.3, Si 0.9, bal Fe.
For dies, punches; air hardened. *Obsolete*

HYPRO NO. 62
Boyd-Wagner Co.
C 2, Cr 12, Mn 0.6, bal Fe.
122,000 TS; 54,000 YS; 7.6 El; 257 Brin. For general tools, high production tools, stamping and blanking dies; nondeforming.

HYPRO-62-O.H
Boyd-Wagner Co.
C 2.2, Cr 12, V 1, bal Fe.
Hardened: 64-67 Rock C. For blanking and forming dies, burnishing tools and rolls, plug gauges, slitting cutters, drawing dies. Type D3 oil hardening cold work steel, deep hardening, abrasion resistant.

HYPROCRODE
Sheepbridge Engineering Ltd.
Ni 12.5-14.5, Cu 6.5-7.5, Cr 4.75-5.75, C, bal Fe.
25,000-30,000 TS; 150 Brin. For hydraulic parts, pump bodies, impellers, plungers, liners, furnace parts, burners, fire bars. Austenitic cast iron; corrosion, heat and erosion resistant.

HYPROOF 304L
British Steel plc
Stainless steel. Cr 19, Ni 9, N, bal Fe.
High proof stress version of 304L.

HYPROOF 316L
British Steel plc
Stainless steel. Cr 17, Ni 11, Mo 2.5, N, bal Fe.
High proof stress version of 316L.

HYPROOF 317L
British Steel plc
Stainless steel. Cr 18, Ni 12, Mo 3.5, N, bal Fe.
High proof stress version of 317L.

HYPROOF 317LM
British Steel plc
Stainless steel. Cr 17.5, Ni 13.5, Mo 4.5, N, bal Fe.
For marine service. High strength; pitting resistant.

HYREM RADIOMETAL
Telcon Metals Ltd.
Soft magnetic alloy, square hysteresis loop, with high remanance for transformers.

HYRESIST 22/5
British Steel plc
Stainless steel. Cr 22, Ni 5.5, Mo 3, N, bal Fe.
Austenitic-ferritic duplex stainless steel. Resists pitting and stress-corrosion cracking.

HYRESIST 317LM
British Steel plc
Stainless steel. Cr 18, Ni 15, Mo 4.5, bal Fe.
Resists pitting and crevice corrosion.

HYRESIST 94L
British Steel plc
Stainless steel. Cr 20, Ni 25, Mo 4.5, Cu 1.5, bal Fe.
Austenitic stainless steel. High resistance to pitting, crevice, and stress corrosion.

HYRHO RADIOMETAL
Telcon Metals Ltd.
Soft magnetic alloy, high resistivity for AC uses,, low eddy current losses for transformers.

HYSPEED NICKEL BRONZE
American Manganese Bronze Co.
Ni, bal Cu.
45,000 TS; 30,000 YS; 15 El. For housing units, slippers for universals, worm rim wheels, bearings for gear drives, bushings; excellent wear resistance. *Obsolete*

HYSTAL 77
British Steel plc
C 0.08-0.17, Si 0.1-0.35, Mn 1.1-1.4, Mo 0.25-0.45, B 0.002-0.006, Ti 0-0.1, P 0-0.04, S 0-0.04, bal Fe.
Quenched and tempered: 790-930 N/mm^2 TS; 700 N/mm^2 YS min. For structural hollow sections.

HYTEMP 306
British Steel plc
Stainless steel. Cr 19, Ni 11, Si 2, bal Fe.
Good scaling resistance up to 1000°C in air and 950°C in flue gases.

HYTEMP 309
British Steel plc
Stainless steel. Cr 23, Ni 14, bal Fe.
Good scaling resistance up to 1050°C in air.

HYTEMP 310
British Steel plc
Stainless steel. Cr 25, Ni 20, bal Fe.
Austenitic. Good scaling resistance up to 1150°C.

HYTEMPCO
Harrison Alloys Inc.
Nickel.
Now HAI-380 ALLOY.

HYTEMPITE (COMMERCIAL NO. 1)
F.J. Ballard & Co.
Ni 25, Cr 15, bal Fe.
Cast: 50,000 TS; 1.5 El; 1.7 RA; 170 Brin. For carburizing and heat treating boxes, annealing pots, pyrometer tubes. *Obsolete*

HYTEN "B" NO. 3
Metalsource Corp.
C 0.5, Mn, Cr, bal Fe.
Heat treated: 152,000-220,000 TS; 130,000-192,000 YS; 11-15 El; 40-58 RA; 311-477 Brin. Rolled: 100,000 TS; 60,000 YS; 18 El; 45 RA; 212 Brin. For shafts, gears, arbors, pinions, spring clips, connecting rods; shock resisting. *Obsolete*

HYTEN B-43
U.S. Metalsource
C 0.4-0.45, Cr 0.65-0.85, Ni 1.5-2, Mo 0.3, bal Fe.
Heat treated: 132,000-326,000 TS; 117,000-294,000 YS; 10-61 El; 32-69 RA; 285-627 Brin. For heavy-duty gears, shafts, pinions, bolts; oil hardened. *Obsolete*

HYTENAL BRONZE
English manufacture
Cu 60, Al 10, Fe 6, Zn 1, Mn 5.
105,000 TS; 65,000 YS; 20 El. For staybolts, pump rods, valves, valve spindles; corrosion resistant.

HYTENSIL ALUMINUM NO. 375
Hytensil Aluminum Co.
Al 98.
Heat treated: 48,000 TS; 3 El; 100 Brin. For light alloy parts; corrosion resistant.

HYTENSILITE
Hytensil Aluminum Co.
Al 98, bal other elements.
Cast: 32,000 TS; 28,000 YS; 15 El; 40 Brin. Heat treated: 45,000 TS; 41,000 YS; 5 El; 100 Brin. For light alloy parts; corrosion resistant.

HYZED
British Steel plc
Plate with Hyzed properties.

HZ 11
Chr. Hover & Sohn Edelstahlwerk
C 0.1-0.2, Si 1, Cr 25, Ni 4.5, bal Fe.
Heat resisting steel; for high temperature equipment.

HZ 16
Chr. Hover & Sohn Edelstahlwerk
C 0-0.2, Si 2, Cr 25, Ni 20, bal Fe.
Heat resisting stainless steel; for high temperature equipment. Similar to AISI 310.

HZ10
Chr. Hover & Sohn Edelstahlwerk
C 0-0.12, Si 1-1.5, Cr 18, Al 1, bal Fe.
Heat resisting steel; for high temperature equipment.

HZ32A
Various foundries
Zn 2.1, Zr 0.7, Th 3, bal Mg.
T5-temper: 27,000-30,000 TS; 13,000-14,000 YS; 4-7 El; 57 Brln; 68 Rock E. At 700°F: 10,000 TS; 7000 YS; 29 El. Magnesium sand castings with good pressure tightness and weldability; some permanent mold. For parts operating at 350-700°F. ASTM B80-69; AMS 4447; QQ-M-56.

I.A.S. INVINCIBLE

Joseph Beardshaw & Son Ltd.
C 1.2, W 1.6, Cr 0.75, Mo 0.25, V 0.25, bal Fe.
For finishing tools, paper and wood working knives; tough, cold work steel.

I.B.D.

T. Inman & Co. Ltd.
C 0.32, Mn 0.4-0.6, Cr 1-1.5, Ni 4-4.5, Mo 0.3-0.4, Si 0.2, bal Fe.
Nickel chrome mold steel; hardenable to 50 Rock C. Can also be cyanide hardened or nitrided.

I.C.N.

T. Inman & Co. Ltd.
C 0.18-0.25, Si 1, Mn 1, Cr 14, Ni 1, bal Fe.
Plastic extrusion die steel; air or oil hardenable, corrosion resistant.

I.C.S.

T. Inman & Co. Ltd.
Cast steel for tools (0.60-0.70 C, bal Fe). For mason tools, Smith's tool, hand punches (0.80-0.90 C, bal Fe). For drills, taps, punches (0.90-1.0 C, bal Fe). For lathe centers, mandrels (1.1-1.2 C, bal Fe). Keen edge for brass and woodworking tools.

I.C.W.

T. Inman & Co. Ltd.
C 2, Mn 0.5, Cr 13, Si 0.6, bal Fe.
Heavy duty die steel; air hardenable. For lamination dies, blanking dies, master hobs.

I.D.I.

T. Inman & Co. Ltd.
C 1.3, Cr 1-1.2, Mn 1.25-1.5, W 0.5, bal Fe.
Oil hardening, non-distorting press tool steel.

I.R. METAL

Oil Well Supply Co.
C 2.5-3.25, B 0.7-1.1, Si 0.5-1.5, Mn 0.5-1.25, Ni 3.5-4.5, bal Fe.
For oil field accessories; abrasion resistant cast iron.

I.W. 1

Henry A. Kries & Sons Co.
C 0.5, Cr 1.5, W 2.2, V 0.2, bal Fe.
For tools, dies; hot work steel.

I.W. 2

Henry A. Kries & Sons Co.
C 0.6, W 9.5, Cr 2.5, V 0.1, bal Fe.
For hot work tools and dies; hot work steel.

I45X

Acme Steel Co.
Now A45YO/A45YK/A45YF.

I50F

Acme Steel Co.
Now A50XF.

I50X

Acme Steel Co.
Now A50YO/A50YK.

I55X

Acme Steel Co.
Now A55YO/A55YK.

I60F

Acme Steel Co.
Now A60XF.

I60X

Acme Steel Co.
Now A60YO/A60YK.

I70F

Acme Steel Co.
Now A70XF.

I80F

Acme Steel Co.
Now A80XF.

IA-IA

H. Boker & Co.
Cu 58-60, Ni 40-42, 1.0% Fe + Mn.
70,000 TS; 46 El. For resistances, thermocouples; negligible temperature coefficient. *Obsolete*

IA-IA

Vereinigte Deutsch Nickel-Werke AG
Cu 58-60, Ni 40-42, 1.0% Fe + Mn.
70,000 TS; 46 El. For resistances, thermocouples; negligibl temperature coefficient. *Obsolete*

IAT QUALITY

Delta Metal (BW) Ltd.
Pb 0-0.3, Cu, Mn, bal Zn.
Extruded: 75,000 TS; 37,000 YS; 28 El; 20 RA; 130 Brin.
Drawn: 85,000 TS; 42,000 YS; 25 El; 18 RA; 135 Brin. For hardware; manganese bronze. *Obsolete*

IBIS

Phosphor Bronze Co. Ltd.
45.0 Sn white metal.
Heavy pressure, moderate severity.

IBIS WHITE NAVY

Phosphor Bronze Co. Ltd.
75.0 Sn white metal.
Medium-high speed-stone crushing plant; for cold rolling mil bearings.

ICI 13518

IMI Knoch Ltd.
C 0.25, Cr 27.5, Mo 5.5, Ni 2.75, Fe 0-3, bal Co.
Cast: 52,000 TS; 10 El; 340 Brin. For high temperature applications; heat resistant.

ICI 13530

IMI Knoch Ltd.
C 0.5, Cr 25.5, Fe 0-2, Ni 10.5, W 7.5, bal Co.
Cast: 50,000 TS; 10 El; 340 Brin. For high temperature applications; heat resistant.

ICI TITANIUM 115

IMI Knoch Ltd.
Fe 0-0.2, 0.010 H_2 max, bal Ti.
Bar: 58,300 TS max; 30 El min. For high temperature applications, corrosion-resistant parts. Commercially pure titanium.

ICI TITANIUM 120

IMI Knoch Ltd.
Fe 0-0.2, 0.010 H_2 max, bal Ti.
Sheet: 67,000 TS max; 25 El min. For high temperature components, corrosion-resistant parts. Commercially pure titanium.

ICI TITANIUM 125

IMI Knoch Ltd.
Fe 0-0.2, 0.010 H_2 max, bal Ti.
Sheet: 56,000-78,500 TS; 22 El min. For high temperature components, corrosion-resistant parts. Commercially pure titanium.

ICI TITANIUM 130

IMI Knoch Ltd.
Fe 0-0.2, 0.010 H_2 max, bal Ti.
Sheet: 56,000-79,000 TS; 22 El min. For high temperature components, corrosion-resistant parts. Commercially pure titanium.

ICI TITANIUM 150

IMI Knoch Ltd.
Fe 0-0.2, 0.010 H_2 max, bal Ti.
Bar: 79,000-100,000 TS; 18 El. For high temperature components, corrosion-resistant parts. Commercially pure titanium.

ICI TITANIUM 160

IMI Knoch Ltd.
Fe 0-0.2, 0.010 H_2 max, bal Ti.
Bar: 89,000-112,000 TS; 18 El; 40 RA. For high temperature applications, corrosion-resistant parts, non-structural aircraft parts. Commercially pure titanium. DTD 5063.

ICI TITANIUM 230

IMI Knoch Ltd.
Cu 1.8-2.8, bal Ti.
Bar: 71,000-100,000 TS; 20 El min. For high temperature applications, corrosion-resistant parts. Good strength retained up to 300°C.

ICI TITANIUM 314A

IMI Knoch Ltd.
Al 3-5, Mn 3-5, bal Ti.
Bar: 139,000 TS min; 15 El min. For jet engine and guided missile components. High strength and good creep resistance up to 300°C.

ICI TITANIUM 314C

IMI Knoch Ltd.
Al 2, Mn 2, bal Ti.
Bar: 94,000-117,000 TS; 20 El min. For high temperature applications, missile and jet engine components. Heat and corrosion resistant.

ICI TITANIUM 317

IMI Knoch Ltd.
Al 4.5-5.5, Sn 2-3, bal Ti.
Bar: 116,000 TS min; 10 El min. For aircraft and guided missile components. Good retention of strength and creep resistance up to 400°C. Good weldability.

ICI TITANIUM 318A

IMI Knoch Ltd.
Al 5.5-6.5, V 3.5-4.5, bal Ti.
Bar: 139,000 TS min; 10 El min. For jet engine components, airframe forgings and fasteners. Good retention of strength and creep resistance up to 300°C.

ICI TITANIUM 679

IMI Knoch Ltd.
Al 2-2.5, Mo 0.8-1.2, Si 0.1-0.5, Sn 10.5-11.5, Zr 4-6, bal Ti.
Bar: 150,000 TS min; 12 El min. For compressor blades. High creep resistance.

ICL 001

Creusot-Loire
C 0-0.12, Cr 18, Ni 9, bal Fe.
Wrought, annealed: 490-690 N/mm^2 psi TS. Austenitic stainless steel for kitchen equipment, furniture springs, cycle spokes. AFNOR Z10CN 18.09; AISI 302.

ICL 004 BC

Creusot-Loire
C 0.045, Cr 18, Ni 7, Mn 7.5, N 0.2, bal Fe.
Wrought, annealed: 650 N/mm^2 psi TS min. Austenitic stainless steel for containers of gases and liquefied gases. AFNOR Z3CMN 18.8.7 (non-standard).

ICL 164

Creusot-Loire
C 0-0.08, Cr 17.5, Ni 12, Mo 2.2, bal Fe.
Wrought, annealed: 520 N/mm^2 psi TS min. Austenitic stainless steel, for equipment for food industry, chemical industry, petroleum. AFNOR Z8 CND 17.11; AISI 316.

ICL 164 BC

Creusot-Loire

C 0-0.03, Cr 17.5, Ni 12, Mo 2.2, bal Fe.

Wrought, annealed: 490 N/mm^2 psi TS min. Austenitic stainless steel, weldable, for best corrosion resistance in food, petroleum, chemical industries. AFNOR Z2CND 17.12; AISI 316L.

ICL 164 FLUAGE

Creusot-Loire

C 0.06, Cr 17, Ni 13, Mo 2.2, B, bal Fe.

Water quenched: 245 N/mm^2 psi YS min; 590 N/mm^2 min TS; 45 El. AFNOR Z6CND 18-12B.

ICL 164 HE

Creusot-Loire

C 0-0.03, Cr 17.5, Ni 12.5, Mo 2.8, N 0.17, bal Fe.

Wrought, annealed: 600 N/mm^2 psi TS min. Austenitic stainless steel, improved strength, for automotive, rail and ship equipment that requires welding and good corrosion resistance. AFNOR Z2CND 17.13.3 + N; AISI 316N.

ICL 164 NB

Creusot-Loire

C 0-0.08, Cr 17.5, Ni 12, Mo 2.2, Nb = 8 x C, bal Fe.

Wrought, annealed: 520 N/mm^2 psi TS min. Stabilized austenitic stainless steel for welded equipment for food, petroleum and chemical industries. AFNOR Z6CNDNb 17.12.

ICL 164 T

Creusot-Loire

C 0-0.08, Cr 17.5, Ni 12, Mo 2.2, Ti = 5 x C, bal Fe.

Wrought, annealed: 520 N/mm^2 psi TS min. Stabilized austenitic stainless steel for welded equipment for food, petroleum and chemical industries. AFNOR Z6CNDT 17.12; AISI 316T.

ICL 166 BC

Creusot-Loire

C 0-0.03, Cr 17, Ni 13, Mo 2.7, bal Fe.

Wrought, annealed: 490 N/mm^2 psi TS min. Austenitic stainless steel; good resistance to dilute sulfuric acid, bisulfites, alkaline earth chlorides, organic acids; weldable. AFNOR Z2CND 17.13.

ICL 167 CN

Creusot-Loire

C 0-0.04, Cr 17.5, Ni 12, Mo 2.5, Cu 0-0.5, Co 0-0.2, Ta 0-0.15, N 0-0.08, bal Fe.

Wrought, annealed: 540 N/mm^2 psi TS min. Austenitic stainless for parts in light-water nuclear power plants. AFNOR Z3CND 17-12.

ICL 168 BC

Creusot-Loire

C 0-0.03, Cr 18.5, Ni 15, Mo 3.3, bal Fe.

Wrought, annealed: 490 N/mm^2 psi TS min. Austenitic stainless steel; good resistance to dilute sulfuric acid, bisulfites, alkaline earth chlorides, organic acids. AFNOR Z3CND 19.15; AISI 317L.

ICL 472

Creusot-Loire

C 0-0.07, Cr 18.5, Ni 10, bal Fe.

Wrought, annealed: 520 N/mm^2 psi TS min. Austenitic stainless steel for petroleum industry, food and beverage equipment. AFNOR Z6CN 18.09; AISI 304.

ICL 472 BC

Creusot-Loire

C 0-0.03, Cr 18.5, Ni 10.5, bal Fe.

Wrought, annealed: 480 N/mm^2 psi TS min. Low carbon austenitic stainless steel, weldable; for petroleum, food industries. AFNOR Z2CN 18.10; AISI 304L.

ICL 472 HE

Creusot-Loire

C 0-0.03, Cr 18.5, Ni 10, N 0.15, bal Fe.

Wrought, annealed: 550 N/mm^2 psi TS min. Low carbon austenitic stainless steel for household appliances, transportation equipment. AFNOR Z2CN 18.10 plus N; AISI 304 N.

ICL 472 NB

Creusot-Loire

C 0-0.08, Cr 18.5, Ni 10.5, Nb = 8 x C, bal Fe.

Wrought, annealed: 520 N/mm^2 psi TS min. Stabilized austenitic stainless steel; weldable; for chemical industry, food and beverage, petroleum industry. AFNOR Z6CN Nb 18.11; AISI 347.

ICL 472 T

Creusot-Loire

C 0-0.08, Cr 18.5, Ni 10.5, Ti = 5 x C, bal Fe.

Wrought, annealed: 520 N/mm^2 psi TS min. Stabilized austenitic stainless steel; weldable; for chemical industry, food and beverage, petroleum industry. AFNOR Z6CNT 18.11; AISI 321.

ICL 472 U

Creusot-Loire

C 0-0.12, Cr 18, Ni 9, Mn 0-0.6, 0.15 S min, bal Fe.

Wrought, annealed: 510 N/mm^2 psi TS min. Free machining austenitic stainless steel. AFNOR Z10CNF 18.09; AISI 303.

ICL 473 BC

Creusot-Loire

C 0-0.04, Cr 19.9, Ni 9.5, Cu 0-0.5, Co 0-0.2, Ta 0-0.15, N 0-0.08, bal Fe.

Wrought, annealed: 540 N/mm^2 psi TS min. Austenitic stainless for parts in light-water nuclear power plants. AFNOR Z3CN19-10.

ICN 001

Societe des AFY

C 0.08-0.2, Cr 17-19, Ni 8-10, Mn 0-2, bal Fe.

Annealed: 85,000 TS; 35,000 YS; 60 El; 70 RA; 150 Brin. Cold drawn: 125,000 TS; 95,000 YS; 25 El; 55 RA; 277 Brin. For oil refinery and chemical plant equipment; Type 302; stainless, austenitic.

ICN 164

Societe des AFY

C 0-0.1, Cr 16-18, Ni 10-14, Mo 2-3, bal Fe.

Annealed: 85,000-95,000 TS; 35,000-45,000 YS; 50-60 El; 60-75 RA; 150-190 Brin. For acid-resistant chemical plant equipment; Type 316; stainless, austenitic.

ICN 164 T (TI)

Societe des AFY

C 0.1, Cr 16-18, Ni 10-14, Mo 1.7-2.7, Ti, bal Fe.

Annealed: 85,000-95,000 TS; 35,000-45,000 YS; 50-60 El; 60-75 RA; 150-190 Brin. For acid resistant chemical plant equipment; Type 316 Ti; stainless, austenitic.

ICN 472

Societe des AFY

C 0-0.08, Cr 18-20, Ni 8-11, Mn 0-2, bal Fe.

Annealed: 90,000 TS; 45,000 YS; 60 El; 135 Brin. Cold drawn: 180,000 TS; 150,000 YS; 10 El; 330 Brin. For chemical plant equipment, welded structures; Type 304; stainless, austenitic.

ICN 472T

Societe des AFY

C 0-0.08, Cr 17-19, Ni 8-11, Ti = 5 x C, bal Fe.

Normalized: 93,000 TS; 36,000 YS; 45 El; 60 RA; 165 Brin. Annealed: 87,000 TS; 33,000 YS; 57 El; 73 RA; 155 Brin. For chemical plant equipment; Type 321; stainless, austenitic.

IDEAL (ELALCO)

Driver Harris Co.

Cu 53-58, Ni 40-45, Fe 0.6-1, Mn 0.5-1.

For electrical resistors, heating elements; heat resistant.

Obsolete

IDEAL BC12V

Idealstahl Breidenbach KG

C 2.1, Cr 11.5, bal Fe.

For blanking and forming dies, punches; oil or air hardened, nondeforming.

IDEAL BC13V

Idealstahl Breidenbach KG

C 1.65, Cr 11.5, V 0.1, Si 0.35, Mn 0.3, bal Fe.

For blanking and forming dies, punches; nondeforming, air hardened.

IDEAL BCK1

Idealstahl Breidenbach KG

C 1.45, Cr 1.4, bal Fe.

For fast finishing cutters, bearings, liners; water hardened, wear resistant.

IDEAL BCK2

Idealstahl Breidenbach KG

C 1, Cr 1.55, bal Fe.

For bearings, bushings, liners; water hardened, wear resistant.

IDEAL BCK3

Idealstahl Breidenbach KG

C 1.05, Cr 1, bal Fe.

IDEAL BCK4

Idealstahl Breidenbach KG

C 1, Cr 1.1, bal Fe.

For bearings, liners, bushings, sleeves, cutters; water hardened.

IDEAL BCR

Idealstahl Breidenbach KG

C 1.05, Cr 1, Mn 0.3, bal Fe.

For bearings, liners, pivots, sleeves; water hardened, wear resistant.

IDEAL BCS

Idealstahl Breidenbach KG

C 0.9, Cr, Si, bal Fe.

For bearings, liners, sleeves; water hardened, wear resistant.

IDEAL BCZ1

Idealstahl Breidenbach KG

C 1.4, Cr 0.3, V 0.1, bal Fe.

For bearings, liners, bushings, sleeves, cutters; water hardened.

IDEAL BCZ2

Idealstahl Breidenbach KG

C 0.9, Cr 0.8, bal Fe.

For bearings, liners, bushings, sleeves, cutters; water hardened.

IDEAL BDS

Idealstahl Breidenbach KG

C 0.5, Mn 0.9, Cr 1.05, V 0.1, bal Fe.

For gears, shafts, springs, crankshafts; oil hardened, shock resistant.

IDEAL BEE

Idealstahl Breidenbach KG

C 0.15, Si 0.2, Mn 0.4, bal Fe.

Annealed: 70,000 TS; 40,000 YS; 25 El; 60 RA; 145 Brin. For gears, pinions, cams, camshafts; case hardening steel.

IDEAL BFM

Idealstahl Breidenbach KG

C 0.7, Si 1.7, Mn 0.7, bal Fe.

For springs, punches; oil hardened, tough.

IDEAL BG1

Idealstahl Breidenbach KG

C 1.05, Cr 1, W 1.15, bal Fe.

For bearings, cold work tools, drawing dies; water hardened.

IDEAL BG2
Idealstahl Breidenbach KG
C 0.45, Ni, Cr, W, bal Fe.
For cold work tools, upsetters; oil hardened, tough.

IDEAL BGW
Idealstahl Breidenbach KG
C 0.5, Cr 1.05, Ni 3.25, bal Fe.
For gears, shafts, crankshafts, bolts, fasteners; oil hardened, shock resistant.

IDEAL BHM
Idealstahl Breidenbach KG
C 0.56, Ni, Cr, Mo, V, bal Fe.
For gears, shafts, crankshafts, bolts, studs; oil hardened, shock resistant.

IDEAL BKS
Idealstahl Breidenbach KG
C 0.45, Si, Cr, V, bal Fe.
For gears, springs, crankshafts, axles; oil hardened, shock resistant.

IDEAL BKV
Idealstahl Breidenbach KG
C 1, V 0.1, Mn 0.25, bal Fe.
For drills, taps, reamers, broaches; Type W2; water hardened.

IDEAL BME
Idealstahl Breidenbach KG
C 0.4, Cr, Mn, Mo, bal Fe.
For gears, pinions, crankshafts; oil hardened, shock resistant.

IDEAL BMK
Idealstahl Breidenbach KG
C 0.85, Cr, bal Fe.
For tools, dies, cutters, taps, bearings; water hardened.

IDEAL DM3
Idealstahl Breidenbach KG
C 0.53, Si 0.8, Mn 1.05, bal Fe.
For gears, axles, crankshafts; water hardened.

IDEAL BMV
Idealstahl Breidenbach KG
C 0.9, Mn 1.9, V 0.1, bal Fe.
For dies, punches, upsetters, cutters; oil hardened, nondeforming.

IDEAL BPK2
Idealstahl Breidenbach KG
C 0.2, Cr 1.15, Mn 1.25, bal Fe.
For cams, fasteners, camshafts; case hardened.

IDEAL BRF1
Idealstahl Breidenbach KG
C 0.2, Cr 13, bal Fe.
Annealed: 95,000 TS; 50,000 YS; 25 El; 55 RA; 195 Brin. Cold drawn: 105,000 TS; 85,000 YS; 17 El; 50 RA; 215 Brin. For turbine blades, valves, cutlery, surgical instruments; Type 420; stainless.

IDEAL BRF2
Idealstahl Breidenbach KG
C 0.4, Cr 13, Si 0.4, Mn 0.3, bal Fe.
Annealed: 95,000 TS; 50,000 YS; 25 El; 55 RA; 195 Brin. For valves, cutlery, surgical and dental instruments; Type 420; stainless.

IDEAL BSA
Idealstahl Breidenbach KG
C 0.38, Si, Cr, V, bal Fe.
For gears, punches, bolts, machine tool parts; oil hardened, tough.

IDEAL BSE
Idealstahl Breidenbach KG
C 0.15, Cr 0.65, Si 0.25, Mn 0.5, bal Fe.
For gears, cams, camshafts, fasteners; case hardening steel, tough.

IDEAL BW6
Idealstahl Breidenbach KG
C 1.2, Cr 0.2, V 0.1, W 1, bal Fe.
For bearings, cutters, sleeves; water hardened, wear resistant.

IDEAL BWP EXTRA
Idealstahl Breidenbach KG
C 0.3, Cr 2.35, V 0.6, W 4.25, bal Fe.
For extrusion dies, upsetters, header dies; oil hardened, tough.

IDEAL BWP SPEZIAL
Idealstahl Breidenbach KG
C 0.3, Cr 2.65, V 0.35, W 8.5, bal Fe.
For header dies, extrusion dies and rams, shears; hot work steel, oil hardened.

IDEAL BWP-PRIMA
Idealstahl Breidenbach KG
C 0.3, Si 1, Cr 1.1, V 0.18, W 3.75, bal Fe.
For header dies, crimpers, punches; oil hardened, tough.

IDEAL BWPKO
Idealstahl Breidenbach KG
C 0.3, Cr 2.4, Co 2, V 0.25, W 8.5, bal Fe.
For shears, punches, upsetters, header dies; hot work steel, oil hardened.

IDEAL BWPU
Idealstahl Breidenbach KG
C 0.45, Cr 1.4, Mo 0.7, V 0.3, Mn 0.7, bal Fe.
For gears, bolts, crankshafts, fasteners; oil hardened, tough.

IDEAL EXTRA HART 1
Idealstahl Breidenbach KG
C 1.3, Si 0-0.25, Mn 0-0.25, bal Fe.
For engravers tools, forming and blanking dies; Type W1; water hardened.

IDEAL EXTRA HART 2
Idealstahl Breidenbach KG
C 1.15, Si 0-0.25, Mn 0-0.25, bal Fe.
Annealed: 110,000 TS; 58,000 YS; 18 El; 38 RA; 220 Brin. For springs, drills, reamers, taps, tools; Type W1; water hardened.

IDEAL EXTRA HART 3
Idealstahl Breidenbach KG
C 1, Si 0-0.25, Mn 0-0.25, bal Fe.
Annealed: 100,000 TS; 53,000 YS; 21 El; 42 RA; 200 Brin. For springs, tools, drills, taps, cutters; Type W1; water hardened.

IDEAL EXTRA HART 4
Idealstahl Breidenbach KG
C 0.85, Si 0-0.25, Mn 0-0.25, bal Fe.
Heat treated: 190,000 TS; 145,000 YS; 10 El; 30 RA; 400 Brin. For springs, drills, cutters, punches; Type W1; water hardened.

IDEAL EXTRA HART 5
Idealstahl Breidenbach KG
C 0.7, Si 0-0.25, Mn 0-0.25, bal Fe.
Heat treated: 175,000 TS; 128,000 YS; 12 El; 37 RA; 355 Brin. For springs, rails, punches, tools, axes; Type W1; water hardened.

IDEAL NIT 1
Idealstahl Breidenbach KG
C 0.26, Cr 1.4, Al 1.1, Mn 0.6, bal Fe.
For oil refinery equipment; creep and heat resistant.

IDEAL NIT 11
Idealstahl Breidenbach KG
C 0.31, Mn 0.6, Cr 2.35, Mo 0.18, bal Fe.
For gears, dies, crankshafts; oil hardened, tough.

IDEAL NIT 3
Idealstahl Breidenbach KG
C 0.34, Cr 1.4, Al 1.1, Mn 0.6, bal Fe.
For oil refinery equipment; creep and heat resistant.

IDEAL NIT 7
Idealstahl Breidenbach KG
C 0.32, Cr 1.1, Al 1.1, V 0.18, Mn 0.6, bal Fe.
For oil refinery equipment; creep and heat resistant.

IDEAL NIT 9
Idealstahl Breidenbach KG
C 0.33, Cr 1.7, Ni 1, Mn 0.5, bal Fe.
For dies, upsetters, punches, shafts; oil hardened, shock resistant.

IDEALOY
Wellman Dynamics Corp.
Sn-Zn-Pb, bal Cu.
For pressure castings; corrosion resistant. *Obsolete*

IDEOR
Darwin & Milner Inc.
C 0.4, W 2, Cr 1, Si 1, mn 0.3, bal Fe.
For chisels, punches, rivet sets, shears, picks; Type S1; tough.

IDEOR
Balfour Darwins Ltd.
C 0.4, W 2, Cr 1, Si 1, mn 0.3, bal Fe.
For chisels, punches, rivet sets, shears, picks; Type S1; tough.

IDEOR SPECIAL
Darwins Alloy Castings
C 0.5, Cr 1, W 2, V 0.15, bal Fe.
For blanking and piercing dies; shock resistant, oil hardened. *Obsolete*

IDIART
Manufacturer not listed.
C 3, Si 2, bal Fe.
For compressor pistons; cast iron.

IDRONAL
Italian manufacture
Mg 5, Si 1, bal Al.
For marine parts; resists sea water corrosion.

IDRONAL
Italian manufacture
Mg 7, bal Al.
For marine parts; resists sea water corrosion.

IGEDUR
Farbenindustrie Atkiengesellschaft
Aluminum. Mg 0.2-2, Cu 3.5-5.5, Si 0.2-1, Mn 0.1-1.2, bal Al.
For light alloy parts; Duralumin type.

IGETALLOY
Sumitoms Trading Co.
WC.
For cutting tools; sintered carbide.

IGNIDUR PC10
Hoffman Elektrogusstahlwerk, Alb.
C 0.6, Si 1.5, Cr 22, bal Fe.
For cylinder liners, rollers, bushings, furnace parts; heat resistant.

IGNIDUR PC11
Hoffman Elektrogusstahlwerk, Alb.
C 1.3, Si 1.5, Cr 29, bal Fe.
For bushings, valve parts, rollers; heat and water resistant.

IGNIDUR PC12
Hoffman Elektrogusstahlwerk, Alb.
C 0.6, Si 1.5, Cr 29, bal Fe.
Cast: 90,000 TS; 65,000 YS; 2 El; 212 Brin. For cylinder liners, bushings, valve seats, furnace parts; heat resistant.

IGNIDUR PC8
Hoffman Elektrogusstahlwerk, Alb.
C 0.3, Si 2.2, Cr 6, bal Fe.
For oil refinery equipment; creep and heat resistant.

IGNIDUR PC9

Hoffman Elektrogusstahlwerk, Alb.
C 0.5, Si 1.5, Cr 17, bal Fe.
Annealed: 100,000 TS; 60,000 YS; 20 El; 38 RA; 210 Brin. For valve parts, cylinder liners, furnace parts; Type 440; stainless, hardenable.

IGNIDUR PK14

Hoffman Elektrogusstahlwerk, Alb.
C 0.4, Si 2, Cr 26, Ni 14, bal Fe.
Cast: 75,000 TS; 47,000 YS; 17 El; 25 RA; 200 Brin. For furnace parts, heat treating boxes, salt pots, belts; Type HH; corrosion and heat resistant.

IGNIDUR PK19

Hoffman Elektrogusstahlwerk, Alb.
C 0.4, Si 2, Cr 25, Ni 19, bal Fe.
Cast: 75,000 TS; 50,000 YS; 17 El; 170 Brin. Aged: 85,000 TS; 50,000 YS; 10 El; 190 Brin. For furnace parts, retorts, skids, rabble arms; Type HK; corrosion and heat resistant.

IGNIDUR PK30

Hoffman Elektrogusstahlwerk, Alb.
C 0.5, Si 1.8, Cr 25, Ni 30, bal Fe.
For furnace parts, heat treating boxes; heat and corrosion resistant.

IGNIDUR PK4

Hoffman Elektrogusstahlwerk, Alb.
C 0.4, Si 1.3, Cr 27, Ni 4, bal Fe.
Cast: 70,000 TS; 65,000 YS; 2 El; 190 Brin. Heat treated: 115,000 TS; 80,000 YS; 15 El. For furnace parts, grate bars, salt pots, baffles; heat resistant.

IGNIDUR PK9

Hoffman Elektrogusstahlwerk, Alb.
C 0.4, Si 2, Cr 22, Ni 9.5, bal Fe.
Cast: 85,000 TS; 45,000 YS; 35 El; 165 Brin. For heat treating boxes, baskets, chains, conveyors, belts; corrosion and heat resistant; Type HF.

IGNITION PIN ALLOY-1

English manufacture
Ce 61, Fe 39.
For ignition pins; pyrophoric.

IGNITION PIN ALLOY-2

English manufacture
Ce 70-73, Fe 1.6-6, Zn 17-24, Al 0-2.4.
For ignition pins; pyrophoric.

IGS-CHRONIN 110

Vereinigte Deutsche Nickel-Werke AG
Nickel. Cr 20, Cu 0.1, Fe 0.5, C 0.5, Ti 0.4, bal Ni.
For the welding industry.

IGS-CHRONIN A

Vereinigte Deutsche Nickel-Werke AG
Nickel. Fe 9, Cr 15.5, Mn 3, Mo 1, Nb 2, bal Ni + Co.
For the welding industry.

IGS-CUNI 10 FE

Vereinigte Deutsche Nickel-Werke AG
Copper. Fe 0.8, C 0.01, Mn 0.8, Ti 0.4, 10 Ni + Co, bal Cu.
Strip and wire for the welding industry.

IGS-CUNI 30 FE

Vereinigte Deutsche Nickel-Werke AG
Copper. Fe 0.5, C 0.03, Mn 0.8, Ti 0.4, 31 Ni + Co, bal Cu.
Strip and wire for the welding industry. C71581

IH-50

International Harvester Co.
C 0-0.22, Mn 0-1.5, Si 0-0.7, 0.2 Cu min, bal Fe.
Plate: 75,000 TS min; 50,000 YP min; 20 El. For earth moving and agricultural equipment, tractors, trucks, trailers, bridges, bus bodies. Good fabricability and weldability. Good fatigue strength and ductility.

IH-60

International Harvester Co.
C 0.22, Mn 1.65, Si 0.7, 0.005 Cb min or 0.01 V min, bal Fe.
Wrought: 80,000 psi TS; 60,000 psi YS; 18 El. HSLA steel.

IH-65

International Harvester Co.
C 0-0.22, Mn 0-1.5, Si 0-0.7, 0.2 Cu min, 0.01 Cb min, bal Fe.
Plate: 90,000 TS min; 65,000 YP min; 18 El. For earth moving and agricultural equipment, tractors, trucks, trailers. Good fabricability and weldability. Good fatigue strength and ductility.

IH-WX STEEL

International Harvester Co.
C 0-0.09, Mn 0.75-1.05, P 0.05-0.1, S 0.24-0.33, bal Fe.
For screw machine products, fasteners, hardware, cams. Free machining, case-hardened.

IHCP

LaSalle Steel Co.
Cold finished, induction-hardened, chrome-plated carbon steel bar; as per order. For piston rods.

IHX-42

International Harvester Co.
C 0.21, Mn 1.35, 0.01 Cb or V min, bal Fe.
60,000 TS min; 42,000 YS min; 24 El min. For structural parts.

IHX-45

International Harvester Co.
C 0-0.2, Mn 0-1, Si 0-0.1, 0.01 Cb or V min, bal Fe.
Plate: 60,000 TS min; 45,000 YS min; 22 El. For earth moving and agricultural equipment, trucks, trailers, tractors, mine cars. High strength, low alloy steel.

IHX-50

International Harvester Co.
C 0-0.22, Mn 0-1.1, Si 0-0.1, 0.01 Cb or V min, bal Fe.
Plate: 65,000 TS min; 50,000 YP min; 20 El. For earth moving and agricultural equipment, mine cars, trucks, trailers, derricks, booms. Shock resistant.

IHX-55

International Harvester Co.
C 0-0.24, Mn 0-1.4, Si 0-0.3, 0.01 Cb or V min, bal Fe.
Plate: 70,000 TS min; 55,000 YP min; 18 El. For earth moving and agricultural equipment, trucks, trailers, tractors, booms, mine cars. Shock resistant.

IHX-60

International Harvester Co.
C 0-0.26, Mn 0-1.55, Si 0-0.3, 0.01 Cb or V min, bal Fe.
Plate: 75,000 TS min; 60,000 YP; 18 El. For earth moving and agricultural equipment, trucks, trailers, tractors, booms, mine cars. Shock resistant.

IHX-65

International Harvester Co.
C 0-0.26, Mn 0-1.6, Si 0-0.3, 0.01 Cb or V min, bal Fe.
Plate: 80,000 TS min; 65,000 YP min; 17 El. For earth moving and agricultural equipment, tractors, trailers, booms, mine cars. Shock resistant.

IHX-70

International Harvester Co.
C 0-0.26, Mn 0-1.65, Si 0-0.3, 0.01 Cb or V min, bal Fe.
Plate: 85,000 TS min; 70,000 YP min; 16 El. For earth moving and agricultural equipment, tractors, trailers, mine cars, buildings. Shock resistant.

IK

French manufacture
Cu 4.5, bal Al.
64,000 TS; 5 El; 110 Brin. For light alloy parts.

IK STEELS

German manufacture
C, Cr, bal Fe.
For heat and corrosion resistant parts; chromized steel, surface layer.

IK1

Thyssen Edelstahlwerke AG
C 0.1, Ti, bal Fe.
For chromizing surface treatment. *Obsolete*

ILLINOIS CHROME-NICKEL-MOLY

U.S. Steel Corp.
C 0.4, Ni 1.5, Cr 0.8, Mo 0.2, bal Fe.
For axles, shafts, jackshafts; resists high torsion, stresses. *Obsolete*

ILLINOIS NICKEL STEEL

Joseph T. Ryerson & Son Inc.
C 0.3-0.4, Ni 3.5, bal Fe.
Hot rolled: 85,000-100,000 TS; 50,000-60,000 YS; 18 El; 25 RA. For dredging cranes, motor truck frames, elevators, power shovels; oil hardening, tough. *Obsolete*

ILLIUM 98

Stainless Foundry & Engineering Inc.
Ni 55, Cr 28, Mo 8, Cu 5, C 0.05.
Cast: 54,000 TS; 41,000 YS; 18 El; 22 RA; 143 Brin. For acid pumps, hardware, impellers, valves, nozzles; resists sulfuric acid.

ILLIUM 98 HF

Stainless Foundry & Engineering Inc.
Ni 50.25-62.5, Cr 26-30, Mo 7.5-9, Cu 4-6.5, Mn 0-1.5, Si 0-0.5, Fe 0-1.5, C 0-0.07.
Cast: 54,000 TS; 41,000 YS; 18 El; 22 RA; 160 Brin. Solution heat treated: 78,000 TS; 43,000 YS; 43 El; 39 RA; 150 Brin. For fluorine service. Corrosion resistant casting alloy.

ILLIUM 98 LSI

Now ILLIUM 98HF.

ILLIUM CD-4MCU

Stainless Foundry & Engineering Inc.
C 0.03, Cr 26, Ni 5.5, Mo 2, Cu 3, bal Fe.
Cast: 108,000 TS; 80,000 YS; 25 El; 250 Brin. Corrosion resistance greater than CF-8 or CF-8M, with double the yield strength of these alloys.

ILLIUM R

Stainless Foundry & Engineering Inc.
Ni 64, Cu 2.5, Fe 6, Cr 22, Mo 5, C 0.05.
Annealed: 105,000 TS; 55,000 YS; 45 El; 240 Brin. Cold drawn: 150,000 TS; 100,000 YS; 12 El; 365 Brin. For chemical equipment, kettles, pump parts, spray nozzles; resists strong acids and thermal shock. *Obsolete*

ILLIUM S

Stainless Foundry & Engineering Inc.
C 0.1, Mn 0.9, Si 9, Cr 85, Al 2, Cu 3.
Cast corrosion resistant alloy.

ILLIUM-B

Stainless Foundry & Engineering Inc.
Cr 28, C 0.05, Mo 8, Cu 5, Fe 5, bal Ni.
Cast: 60,000 TS min; 41,000 YS min; 0.5 El; 221-420 Brin. Hardenable; maximum wear and galling resistance; resists sulfuric acid. For pumps, valves, chemical plant equipment.

ILLIUM-G

Stainless Foundry & Engineering Inc.
C 0.2, Cr 22, Fe 5, Mo 6, Cu 5, bal Ni.
Cast: 60,000 min TS; 35,000 min YS; 5 El; 159 Brin. Excellent resistance to corrosion from hot H_2SO_4, H_3PO_4, acid-salt mixtures, sea water. For pumps, valves, chemical and marine equipment.

ILLIUM-H

Stainless Foundry & Engineering Inc.
C 0.1, Ti 0.18, Cb 0.6, Cr 26-30, Co 47-52, bal Fe.
Cast: 135,000 TS; 48,000 YS; 10 El; 10 RA; 257 Brin. Wrought: 132,000 TS; 61,000 YS; 7 El; 6 RA; 350 Brin. For furnace baffles, burner tips, sintering grates, quench baskets. Corrosion, heat and thermal shock resistant.

ILLIUM-M

Stainless Foundry & Engineering Inc.
C 0.08, Mo 28, bal Ni.
Cast: 72,000 TS min; 46,000 YS min; 61 El; 190 Brin. Resists halogen chemicals and HCl. For chemical plant equipment.

ILLIUM-P

Stainless Foundry & Engineering Inc.
Cr 28, Ni 8, Cu 3, Mo 2, bal Fe.
Cast: 99,000-110,000 TS; 65,000-81,000 YS; 10-18 El; 217-255 Brin. Resists erosion and corrosion by mixed acids; resists phosphoric acid. For chemical plant equipment.

ILLIUM-PD

Stainless Foundry & Engineering Inc.
C 0.08, Cr 26, Ni 5, Mo 2, bal Fe.
Cast: 95,000 TS min; 65,000 YS min; 30 El; 212 Brin. High erosion-corrosion resistance; for equipment in pulping liquors, dairy and chemical plants.

ILLIUM-W

Stainless Foundry & Engineering Inc.
C 0.15, Fe 6, W 4, Mo 17, Cr 16, bal Ni.
Cast: 72,000 TS; 46,000 YS; 4 El. For chemical equipment; resists Cl_2, H_2SO_4, and HNO_3.

ILLIUM-X

Stainless Foundry & Engineering Inc.
C 0.75, Cr 29, W 14, Co 55.
Cast: 100,000 TS min; 90,000 YS min; 2 El; 293 Brin. Corrosion and abrasion resistant, particularly for equipment for dry battery manufacture.

ILSSA AT-15

Ilssa-Viola SpA
C 0-0.25, Cr 25, Ni 20, Si 0-1.5, bal Fe.
Annealed: 100,000 TS; 45,000 YS; 50 El; 65 RA; 185 Brin. For furnace parts, valves, pumps, jet engine parts; Type 310; corrosion and heat resistant.

ILSSA IC

Ilssa-Viola SpA
C 0.09-0.12, Cr 13, Si 0-0.5, bal Fe.
Annealed: 75,000 TS; 40,000 YS; 35 El; 70 RA; 155 Brin. For turbine blades, valves, cutlery, surgical instruments; Type 403 and 410; corrosion resistant.

ILSSA ICC

Ilssa-Viola SpA
C 0.13-0.25, Cr 17, bal Fe.
Annealed: 90,000 TS; 55,000 YS; 20 El; 45 RA; 170 Brin. For oil refinery equipment, oil burners and heaters; Type 430; corrosion resistant.

ILSSA ICMS

Ilssa-Viola SpA
C 0-0.1, Cr 17, Mo 1.5, bal Fe.
Annealed: 90,000 TS; 55,000 YS; 20 El; 45 RA; 170 Brin. For oil refinery equipment, oil burners and heaters; corrosion resistant.

ILSSA ICS

Ilssa-Viola SpA
C 0.07-0.12, Cr 17, bal Fe.
Annealed: 80,000 TS; 50,000 YS; 25 El; 50 RA; 150 Brin. For oil refinery equipment, oil burners and heaters; Type 430; corrosion resistant.

ILSSA IN

Ilssa-Viola SpA
C 0.11-0.16, Cr 18, Ni 8, bal Fe.
Annealed: 80,000 TS; 35,000 YS; 55 El; 75 RA; 150 Brin. For chemical plant equipment, tanks, mixers, agitators; Type 302; stainless, austenitic.

ILSSA INF 20

Ilssa-Viola SpA
C 0-0.35, Cr 25, bal Fe.
Annealed: 85,000 TS; 50,000 YS; 30 El; 55 RA; 180 Brin. For furnace parts and equipment, heat treating boxes; Type 446; heat resistant.

ILSSA INF 25/5

Ilssa-Viola SpA
C 0-0.12, Cr 26, Ni 4, bal Fe.
Cast: 90,000 TS; 65,000 YS; 2 El; 212 Brin. For cylinder liners, bushings, valve seats and bodies; Type CC50; corrosion and heat resistant.

ILSSA INI

Ilssa-Viola SpA
C 0-0.08, Cr 18, Ni 8, bal Fe.
Annealed: 85,000 TS; 35,000 YS; 60 El; 70 RA; 150 Brin. For chemical plant equipment, tanks, mixers, filters; Type 304; stainless, austenitic.

ILSSA INI/BC

Ilssa-Viola SpA
C 0-0.05, Cr 18, Ni 8, bal Fe.
Annealed: 85,000 TS; 35,000 YS; 60 El; 70 RA; 150 Brin. For chemical plant equipment, tanks, mixers; Type 304; stainless, austenitic.

ILSSA INL

Ilssa-Viola SpA
C 0.08-0.2, Cr 18, Ni 9, Si 2.5, bal Fe.
Annealed: 85,000 TS; 40,000 YS; 50 El; 70 RA; 160 Brin. For chemical plant and oil refinery equipment; Type 302B; stainless, austenitic.

ILSSA INMI

Ilssa-Viola SpA
C 0-0.12, Cr 18, Ni 10, Mo 2.5, bal Fe.
Annealed: 85,000 TS; 35,000 YS; 50 El; 65 RA; 160 Brin. For acid resistant chemical plant equipment; Type 316; stainless.

ILSSA INS

Ilssa-Viola SpA
C 0.08-0.15, Cr 18, Ni 9, P, S, Se, Zr, bal Fe.
Annealed: 80,000 TS; 35,000 YS; 40 El; 60 RA; 160 Brin. For screw machine products, shafts; Type 303; stainless, free-cutting.

ILSSA MIC

Ilssa-Viola SpA
C 0.26-0.37, Cr 13, bal Fe.
Annealed: 95,000 TS; 50,000 YS; 25 El; 55 RA; 195 Brin. For valves, cutlery, surgical and dental instruments; Type 420; corrosion resistant.

ILZRO-14

International Lead-Zinc Research Assoc.
Cu 1-1.5, Ti 0.25-0.3, Al 0.01-0.03, bal Zn.
Die cast: 33,000 TS; 20,000 YS; 5-6 El; 79-86 Vickers. For die castings, carburetors, fuel pumps, valve bodies. Creep resistant.

ILZRO-16

International Lead-Zinc Research Assoc.
Cu 1-1.5, Ti 0.15-0.25, Cr 0.1-0.2, Al 0.01-0.04, bal Zn.
Die cast: 33,000 TS; 5 El; 84 Vickers. For automobile die castings. Good creep resistance at high temperatures.

IM 001

Creusot-Loire
C 0-0.12, Mn 0-1.5, Si 0-2, Ni 8-10, Cr 17-19.5, bal Fe.
Cast, annealed: 490 N/mm^2 psi TS min. For faucets, pump parts; corrosion resistant. AFNOR Z10CN 18.9 M; similar to ACI CF 20.

IM 004 BC

Creusot-Loire
C 0-0.04, Mn 7-9, Si 0-1.5, Ni 6-8, Cr 17-19, N 0.15-0.25, bal Fe.
Cast, annealed: 540 N/mm^2 psi TS min. Austenitic stainless for food industry, cryogenic operations. AFNOR Z3CMN 18.8.7 AZM.

IM 164

Creusot-Loire
C 0-0.08, Mn 0-1.5, Si 0-1.5, Ni 9-12, Cr 18-21, Mo 2-3, bal Fe.
Cast, annealed: 490 N/mm^2 psi TS min. Austenitic stainless for pumps, valves, and faucets; improved resistance to corrosion. AFNOR Z5CND 20.10 M; similar to ACI CF-8M.

IM 164 BC

Creusot-Loire
C 0-0.03, Mn 0-1.5, Si 0-1.5, Ni 9-13, Cr 17-21, Mo 2-3, bal Fe.
Cast, annealed: 490 N/mm^2 psi TS min. Austenitic stainless; weldable; for pumps, valves, faucets; improved resistance to corrosion. AFNOR Z3CND 20.10 M; ACI CF-3M.

IM 164 NB

Creusot-Loire
C 0-0.08, Mn 0-1.5, Si 0-1.5, Ni 10.5-12.5, Cr 17-19.5, Mo 2-2.5, Nb = 8 x C, bal Fe.
Cast, annealed: 490 N/mm^2 psi TS min. Austenitic stainless; weldable; for pumps, valves, faucets; improved resistance to corrosion. AFNOR Z4CND Nb 18.12 M; W. Nr. 1.4581.

IM 167 CN

Creusot-Loire
C 0-0.045, Mn 0-1.5, Si 0-1.5, Ni 10.5-11.5, Cr 17-21, Mo 2.3-2.8, Co 0-0.2, N 0-0.08, Cu 0-0.5, bal Fe.
Cast, annealed: 520 N/mm^2 psi TS min. Austenitic stainless for special applications in nuclear processes. AFNOR Z3CND 19.10 M.

IM 168

Creusot-Loire
C 0-0.05, Mn 0-1.5, Si 0-1.5, Ni 12-15, Cr 17.5-20.5, Mo 3-3.5, bal Fe.
Cast, annealed: 490 N/mm^2 psi TS min. Austenitic stainless for pumps, valves, faucets; improved resistance to corrosion. AFNOR Z4CND 19.13 M; similar to AISI 317.

IM 472

Creusot-Loire
C 0-0.08, Mn 0-1.5, Si 0-2, Ni 8-11, Cr 18-21, bal Fe.
Cast, annealed: 490 N/mm^2 psi TS min. Austenitic stainless for pumps, valves, parts requiring corrosion resistance. AFNOR Z6CN 19.9 M; ACI CF-8.

IM 472 BC

Creusot-Loire
C 0-0.03, Mn 0-1.5, Si 0-2, Ni 8-12, Cr 18-21, bal Fe.
Cast, annealed: 490 N/mm^2 psi TS min. Austenitic stainless steel; weldable; for faucets, valves, pumps. AFNOR Z3CN 19.9 M; ACI CF-3.

IM 472 NB

Creusot-Loire
C 0-0.08, Mn 0-1.5, Si 0-2, Ni 9-12, Cr 18-21, Nb = 8 x C, bal Fe.
Cast, annealed: 490 N/mm^2 psi TS min. Stabilized austenitic stainless, preferred for welded pumps, valves, faucets. AFNOR Z4CN Nb 19.10 M; ACI CF-8C.

IM 473 BC

Creusot-Loire
C 0-0.04, Mn 0-1.5, Si 0-1.5, Ni 9-10, Cr 18.5-21, Co 0-0.2, Cu 0-0.5, Ta 0-0.15, N 0-0.08, bal Fe.
Cast, annealed: 510 N/mm^2 psi TS min. Austenitic stainless steel, weldable, for special nuclear applications. AFNOR Z3CN 19.10 M; similar to ACI CF-3.

IMAGE

J.M. Ney Co.
Gold color alloy for porcelain to gold restorations subject to stress; hard. 81,000 psi TS; 15 El; 159 Brin.

IMI 100
IMI Rolled Metals Ltd.
99.95 Cu (including Ag), oxygen free.
For thermionic valves, transistors, welding and brazing wire, glass-metal seals; high conductivity and ductility; not subject to hydrogen embrittlement during annealing and brazing. ASTM B170; BS 1861 2873.C103 (wire). Bar, sheet, strip, wire.

IMI 100
IMI Rod & Wire
99.95 Cu (including Ag), oxygen free.
For thermionic valves, transistors, welding and brazing wire, glass-metal seals; high conductivity and ductility; not subject to hydrogen embrittlement during annealing and brazing. ASTM B170; BS 1861 2873.C103 (wire). Bar, sheet, strip, wire.

IMI 103
IMI Rolled Metals Ltd.
99.9 min Cu (including Ag).
Electrolytic tough pitch copper wire, sheet, strip. Standard copper for electrical purposes and stampings, spinning, etc. Ductile, high conductivity. BS 2870 C101; 2873.C101 (Wire).

IMI 103
IMI Rod & Wire
99.9 min Cu (including Ag).
Electrolytic tough pitch copper wire, sheet, strip. Standard copper for electrical purposes and stampings, spinning, etc. Ductile, high conductivity. BS 2870 C101; 2873.C101 (Wire).

IMI 111
IMI Rolled Metals Ltd.
99.9 min Cu (including Ag).
Fire refined tough pitch copper. For electrical applications and cold heading wire, and rods for bus bars. BS1037/2870.C102 etc; ASTM B5, B124 etc.

IMI 111
IMI Rod & Wire
99.9 min Cu (including Ag).
Fire refined tough pitch copper. For electrical applications and cold heading wire, and rods for bus bars. BS1037/2870.C102 etc; ASTM B5, B124 etc.

IMI 125
IMI Rolled Metals Ltd.
Sulphur addition to copper. Sulphur copper bar, and rod. Improved machinability; good conductivity.

IMI 125
IMI Rod & Wire
Sulphur addition to copper. Sulphur copper bar, and rod. Improved machinability; good conductivity.

IMI 131
IMI Rolled Metals Ltd.
Cu, phosphorus deoxidized, non arsenical.
Ductile and corrosion resistant; for severe heading and forming. Drawn shapes, rod, bar, sheet and strip. ASTM B152 (DHP); BS 2870.C106.

IMI 131
IMI Rod & Wire
Cu, phosphorus deoxidized, non arsenical.
Ductile and corrosion resistant; for severe heading and forming. Drawn shapes, rod, bar, sheet and strip. ASTM B152 (DHP); BS 2870.C106.

IMI 145
IMI Rolled Metals Ltd.
Cu 99.8, Cd 0.2.
High conductivity copper for radiator fin manufacture. *Obsolete*

IMI 146
IMI Rolled Metals Ltd.
As 0.3-0.5, P 0.013-0.05, 99.85 Cu min.
Used in some chemical plants where resistance to scaling is desirable; rod, strip and sheet. BS 2870-C.107. *Obsolete*

IMI 149
IMI Rolled Metals Ltd.
Sn 0.1, Cu 99.9.
O temper: 55 VHN; H temper: 130 VHN. For fin material in car radiators. *Obsolete*

IMI 151
IMI Rolled Metals Ltd.
Cu 99.85, Ag 0.05.
Higher softening temperature to resist softening during tinning and soldering; for electrical commutators, for printers' engraving plates. ASTM B152-STP. *Obsolete*

IMI 153 (KUFIL)
IMI Rolled Metals Ltd.
Cu 99, Ag 1.
Rod, spooled wire or slittings; for oxyacetylene welding of copper. BS 1453.C1. *Obsolete*

IMI 156
IMI Rolled Metals Ltd.
Cu 99.4, Al 0.3, Ti 0.3.
For nitrogen arc welding of copper. BS 2901.C8.

IMI 156
IMI Rod & Wire
Cu 99.4, Al 0.3, Ti 0.3.
For nitrogen arc welding of copper. BS 2901.C8.

IMI 161 (BOROFIL)
IMI Rolled Metals Ltd.
Cu 99.92, B 0.08.
For argon arc welding of copper, retaining high conductivity. BS 2901.C21.

IMI 161 (BOROFIL)
IMI Rod & Wire
Cu 99.92, B 0.08.
For argon arc welding of copper, retaining high conductivity. BS 2901.C21.

IMI 166
IMI Rolled Metals Ltd.
Cd addition to copper.
Increased strength and hardness while still retaining high conductivity. For line wire, trolley wires, overhead conductors, welding tips. BS 23,175,672,2755,2870.C108,2873.C108.

IMI 166
IMI Rod & Wire
Cd addition to copper.
Increased strength and hardness while still retaining high conductivity. For line wire, trolley wires, overhead conductors, welding tips. BS 23,175,672,2755,2870.C108,2873.C108.

IMI 171 (KUMIUM)
IMI Rolled Metals Ltd.
Chromium addition to copper. Heat treatable alloy retaining strength up to 350°C with high electric conductivity. For electrode tips for resistance welding; for certain electronic applications. ASTM F268 (Chromium copper). *Obsolete*

IMI 176 (ARGOFIL)
IMI Rolled Metals Ltd.
Cu 99.5, Si 0.25, Mn 0.25.
For argon arc welding of copper. BS 2901.C7; ASTM B225,B259.

IMI 176 (ARGOFIL)
IMI Rod & Wire
Cu 99.5, Si 0.25, Mn 0.25.
For argon arc welding of copper. BS 2901.C7; ASTM B225,B259.

IMI 180
IMI Rod & Wire
Cu 99.5, Te 0.5.
1/2 hard: 310 N/mm^2 TS; 12 El. Good machinability; for automatic screw machines. BS 2874 C109; UNS C14500.

IMI 195
IMI Rolled Metals Ltd.
Copper containing small amounts of Pb. Improved machinability while still retaining good electrical and thermal conductivity. Called leaded copper. *Obsolete*

IMI 215
IMI Rod & Wire
Cu 85, Zn 15.
Gilding metal; six grades of hardness; 65-150 Vickers. For jewelry. BS 2873 CZ102; UNS C23000.

IMI 220
IMI Rod & Wire
Cu 80, Zn 20.
Low brass; six grades of hardness; 70-160 Vickers. BS 2873 CZ103; UNS C24000.

IMI 230
IMI Titanium Ltd.
Titanium. Cu 2.5, bal Ti.
Age hardenable, good weldability and mechanical properties up to 350°C. Annealed sheet: 460 MPa YS min; 540-700 MPa TS; 18 El min.

IMI 233
IMI Rod & Wire
Cu 67, Zn 33.
Yellow brass; six grades of hardness; 70-175 Vickers. For stamping and forming. BS 2873 CZ107; UNS C27000.

IMI 235
IMI Rolled Metals Ltd.
Cu 65, Zn 35.
As sheet and strip for general commercial presswork and general coppersmith work. BS 2870 CZ.107; ASTM B36-268.

IMI 235
IMI Rod & Wire
Cu 65, Zn 35.
As sheet and strip for general commercial presswork and general coppersmith work. BS 2870 CZ.107; ASTM B36-268.

IMI 237
IMI Rolled Metals Ltd.
Cu 63, Zn 37.
Wire and drawn shapes for cold forming of screws, rivets and cold headed products. BS 265;2870.CZ108;2873.CZ108;ASTM B19.

IMI 237
IMI Rod & Wire
Cu 63, Zn 37.
Wire and drawn shapes for cold forming of screws, rivets and cold headed products. BS 265;2870.CZ108;2873.CZ108;ASTM B19.

IMI 239
IMI Rod & Wire
Cu 61, Zn 39.
Often called "pin wire" because of use in making brass pins.

IMI 240
IMI Rod & Wire
Cu 60, Zn 40.
Wire and rod for brazing ferrous and non-ferrous materials; yellow brass. BS 1845.C23.

IMI 260
IMI Titanium Ltd.
Titanium. Pd 0.15, bal Ti.
Corrosion resistant titanium alloy. 170 MPa YS min; 330-420 MPa TS; 25 El min.

IMI 260
IMI Rod & Wire
Cu 70, Zn 30.
Known as Cartridge Brass; very ductile when soft. For drawn parts. BS 2874 CZ106; UNS C26000.

IMI 262

IMI Titanium Ltd.
Titanium. Pd 0.15, bal Ti.
Corrosion resistant titanium alloy, slightly stronger than IMI 260. Sheet: 290 MPa YS min; 390-540 MPa TS; 22 El min.

IMI 276

IMI Rod & Wire
Cu 70, Zn 30, As 0.03.
Inhibited deep drawing brass; six grades of hardness; 70-170 Vickers. BS 2871 CZ105.

IMI 307 (ALUMBRO)

IMI Rolled Metals Ltd.
Cu 76, Zn 22, Al 2, As 0.4.
Arsenic inhibited bronze strip. BS 2870 CZ 110; ASTM B111-687; CDA and UNS C68700. *Obsolete*

IMI 318

IMI Titanium Ltd.
Titanium. Al 6, V 4, bal Ti.
For titanium castings. Sheet: 900 MPa YS min; 960-1270 MPa TS; 8 El min.

IMI 329

IMI Rod & Wire
Cu 57.5, Fe 0.5, Pb 2.25, Mn 2, Sn 1, Al 0.5, bal Zn.
Rod and bar: 520-540 N/mm^2 TS; 20-25 El; 170 Vickers; 85% machinability. High tensile brass for valve spindles.

IMI 333

IMI Rod & Wire
Cu 57.5, Fe 0.5, Pb 1.25, Mn 1.5, Sn 0.75, bal Zn.
Rod and bar: 530-580 N/mm^2 TS; 20-30 El; 175-190 Vickers; 65% machinability. High tensile brass; may be soldered. BS 2874 CZ114.

IMI 334

IMI Rod & Wire
Cu 57, Fe 1, Pb 1.25, Mn 1.5, Sn 0.4, Al 1.25, bal Zn.
Rod and bar: 600-660 N/mm^2 TS; 20-25 El; 195-210 Vickers; 60% machinability. High strength bar. BS 2874 CZ114.

IMI 341 (COIN BRONZE)

IMI Rolled Metals Ltd.
Cu 97, Zn 2.5, Sn 0.5.
Strip used for stamping coins in Great Britain, Australia, New Zealand and Finland.

IMI 345

IMI Rolled Metals Ltd.
Zn, Sn, bal Cu.
Wire for safety pins, decorative pins, spectacle frames. *Obsolete*

IMI 360 (NAVAL BRASS)

IMI Rolled Metals Ltd.
Cu 62, Sn 1.25, bal Zn.
Sheet, strip, bar for underwater fittings for ships and other marine hardware. BS 250, 409 2872.CZ112, 2874.CZ112. ASTM B171.

IMI 360 (NAVAL BRASS)

IMI Rod & Wire
Cu 62, Sn 1.25, bal Zn.
Sheet, strip, bar for underwater fittings for ships and other marine hardware. BS 250, 409 2872.CZ112, 2874.CZ112. ASTM B171.

IMI 365

IMI Rolled Metals Ltd.
Cu 60, Sn 0.75, bal Zn.
Sheet and strip for underwater fittings or ships and general marine hardware. BS 2870.CZ.113; ASTM B61A. *Obsolete*

IMI 365

IMI Rod & Wire
Cu 60.5, Sn 0.8, bal Zn.
Rod and bar: 410-450 N/mm^2 TS; 20-30 El; 115-135 Vickers; 50% machinability. American Naval brass; UNS C46400.

IMI 388

IMI Rolled Metals Ltd.
Cu 59.4, Si 0.3, Zn 40.3.
Wire and rod for brazing ferrous and non-ferrous materials; called "silicon brass." BS 1845.C26.

IMI 388

IMI Rod & Wire
Cu 59.4, Si 0.3, Zn 40.3.
Wire and rod for brazing ferrous and non-ferrous materials; called "silicon brass." BS 1845.C26.

IMI 410

IMI Rolled Metals Ltd.
Cu 64, Pb 1, bal Zn.
Leaded brass strip, particularly for stamped and machined watch, clock and instrument parts; called "clock and matrix brass." BS 2870.CZ.116. *Obsolete*

IMI 432

IMI Rod & Wire

Called "nipple wire." For cold heading, riveting and cold forming. BS 2873.CZ119; ASTM B146. Leaded brass wire.

IMI 433

IMI Rod & Wire
Cu 62, Pb 2, bal Zn.
Rod and bar: 360-380 N/mm^2 TS; 30-40 El. 75% machinability; for turning, riveting, spinning and cold swaging. BS 2873; UNS 35300.

IMI 441

IMI Rod & Wire
Cu 61.5, Pb 3.25, bal Zn.
Rod and bar: 370-430 N/mm^2 TS; 22-35 El. 100% machinability; for free turning brass. BS 2874; ASTM B16; UNS C36000.

IMI 442

IMI Rod & Wire
Cu 61.5, Pb 2.5, bal Zn.
Rod and bar: 370-400 N/mm^2 TS; 25-45 El. 90% machinability; for high speed machining and riveting.

IMI 443

IMI Rolled Metals Ltd.
Cu 61, Pb 0.25, bal Zn.
Low leaded wire called "shoe rivet wire," main use is upsetting to shoe rivets, and heading operations where some machining is required. *Obsolete*

IMI 452 (MUNTZ METAL)

IMI Rod & Wire
Cu 60, Pb 0-0.5, bal Zn.
Rod and extruded or drawn sections for cold work, bending, riveting, machining. BS 1949 2872.CZ123; 2874.CZ123.

IMI 453

IMI Rolled Metals Ltd.
Cu 60, Pb 0-0.7, bal Zn.
Rod and extruded or drawn sections, called "special bending brass," for cold forming and machining. BS 1949 2872.CZ123; 2874.CZ123.

IMI 453

IMI Rod & Wire
Cu 60, Pb 0-0.7, bal Zn.
Rod and extruded or drawn sections, called "special bending brass," for cold forming and machining. BS 1949 2872.CZ123; 2874.CZ123.

IMI 455

IMI Rod & Wire
Cu 59.25, Pb 2.25, bal Zn.
Rod and bar: 420-460 N/mm^2 TS; 20-30 El. 85% machinability; free machining brass; may be hot forged. ASTM B124; UNS C37700.

IMI 456

IMI Rolled Metals Ltd.
Cu 59, Pb 0.75, bal Zn.
Rod and extruded or drawn sections called "riveting quality brass." For upsetting, stamping and limited turning. BS CZ9B/12872.CZ119; 2874.CZ119. *Obsolete*

IMI 458

IMI Rolled Metals Ltd.
Cu 59, Pb 2, bal Zn.
Strip, called "clock brass." For stamped parts requiring machining to make clock and instrument parts. BS 2870.CZ120. *Obsolete*

IMI 463

IMI Rod & Wire
Cu 58, Pb 2.25, bal Zn.
Extruded: 330-400 N/mm^2 TS; 30-38 El. Hot stamping brass; 90% machinability. UNS 38000.

IMI 466

IMI Rod & Wire
Cu 57.5, Pb 4.25, bal Zn.
Rod and bar: 450-515 N/mm^2 TS; 19-25 El. 100% plus machinability; for free turning brass.

IMI 467

IMI Rod & Wire
Cu 57.75, Pb 2.25, bal Zn.
Rod and extruded or drawn sections for hot stamping and forging; free machining. BS 218 2872.CZ122; 2874.CZ122.

IMI 469

IMI Rod & Wire
Cu 57.5, Pb 3.25, bal Zn.
Rod and extruded or drawn sections for easy machining to bolts, nuts, screws and fittings. BS 249 2872.CZ121; 2874.CZ121.

IMI 470

IMI Rod & Wire
Cu 58.25, Pb 2.75, bal Zn.
Rod and bar: 435-495 N/mm^2 TS; 21-28 El. 100% machinability; for free turning brass. BS 2874.

IMI 471

IMI Rod & Wire
Cu 57, Pb 1.5, bal Zn.
Rod and shapes: 140 Vickers; 85% machinability. General purpose section brass.

IMI 473

IMI Rod & Wire
Cu 58, Pb 3, bal Zn.
Rod and bar: 435-495 N/mm^2 TS; 20-28 El. 100% machinability; for free turning brass. BS 2874; DIN 17660 CuZn 39 Pb3.

IMI 475

IMI Rod & Wire
Cu 57, Pb 3, Al 0.3, bal Zn.
Sections: 160 Vickers; 90% machinability. General purpose section; not for brazing.

IMI 476

IMI Rod & Wire
Cu 56, Pb 1.5, Al 0.5, bal Zn.
Sections: 150 Vickers; 75% machinability. For thin sections.

IMI 500 (NICKEL BRASS COIN)

IMI Rolled Metals Ltd.
Cu 79, Zn 20, Ni 1.
Sometimes used for coins

IMI 506 (NICKEL SILVER)

IMI Rolled Metals Ltd.
Cu 65, Ni 18, Zn 17.
Wire and drawn shapes for decorative purposes, springs, fishing tackle; ductile, malleable, machinable, corrosion resistant. ASTM B122, B206.

IMI 506 (NICKEL SILVER)
IMI Rod & Wire
Cu 65, Ni 18, Zn 17.
Wire and drawn shapes for decorative purposes, springs, fishing tackle; ductile, malleable, machinable, corrosion resistant. ASTM B122, B206.

IMI 511 (NICKEL SILVER)
IMI Rolled Metals Ltd.
Cu 62, Ni 10, Zn 28.
Wire, drawn shapes, sheet for decorative purposes, jewelry, tableware, fishing tackle; ductile, malleable, machinable, corrosion resistant. ASTM B122, B206.

IMI 511 (NICKEL SILVER)
IMI Rod & Wire
Cu 62, Ni 10, Zn 28.
Wire, drawn shapes, sheet for decorative purposes, jewelry, tableware, fishing tackle; ductile, malleable, machinable, corrosion resistant. ASTM B122, B206.

IMI 512 (NICKEL SILVER)
IMI Rolled Metals Ltd.
Cu 62, Ni 12, Zn 26.
Wire, drawn shapes, sheet for decorative purposes, jewelry, slide fastener nameplates; ductile, malleable, machinable, corrosion resistant. ASTM B122, B206.

IMI 512 (NICKEL SILVER)
IMI Rod & Wire
Cu 62, Ni 12, Zn 26.
Wire, drawn shapes, sheet for decorative purposes, jewelry, slide fastener nameplates; ductile, malleable, machinable, corrosion resistant. ASTM B122, B206.

IMI 513 (NICKEL SILVER)
IMI Rolled Metals Ltd.
Cu 62, Ni 15, Zn 23.
Wire and drawn shapes for decorative purposes, tableware, jewelry; ductile, malleable, machinable, corrosion resistant. ASTM B122, B206.

IMI 513 (NICKEL SILVER)
IMI Rod & Wire
Cu 62, Ni 15, Zn 23.
Wire and drawn shapes for decorative purposes, tableware, jewelry; ductile, malleable, machinable, corrosion resistant. ASTM B122, B206.

IMI 514 (NICKEL SILVER)
IMI Rolled Metals Ltd.
Cu 62, Ni 18, Zn 20.
Wire, drawn shapes and sheet for decorative purposes, tableware, name plates, engraving plates, springs; ductile, malleable, machinable, corrosion resistant. ASTM B122, B206.

IMI 514 (NICKEL SILVER)
IMI Rod & Wire
Cu 62, Ni 18, Zn 20.
Wire, drawn shapes and sheet for decorative purposes, tableware, name plates, engraving plates, springs; ductile, malleable, machinable, corrosion resistant. ASTM B122, B206.

IMI 515 (NICKEL SILVER)
IMI Rolled Metals Ltd.
Cu 62, Ni 20, Zn 18.
Wire and drawn shapes for decorative purposes, tableware, springs for electronics and telecommunications; ductile, malleable, machinable, corrosion resistant. BS 790, 1824, 2870, 2873; ASTM B122, B206. *Obsolete*

IMI 525
IMI Rolled Metals Ltd.
Brass plus 10.0% Ni. Nickel brass rod, stamping grade; good white color for sanitary fittings. ASTM B124-55 alloy 14. *Obsolete*

IMI 550
IMI Titanium Ltd.
Titanium. Al 4, Mo 4, Sn 2, Si 0.5, bal Ti.
Readily forgeable high strength alloy. Sheet: 960 MPa YS min; 1100-1280 MPa TS; 9 El min.

IMI 551
IMI Rolled Metals Ltd.
Free machining brass plus 10.0% Ni. Nickel brass rod, free turning grade; good white color for sanitary fittings. BS (near) 2872.NS101; 2874.NS101. *Obsolete*

IMI 551
IMI Titanium Ltd.
Titanium. Al 4, Mo 4, Sn 4, Si 0.5, bal Ti.
High strength titanium alloy. Sheet: 1095 MPa YS min; 1250-1420 MPa TS; 8 El min.

IMI 575 ("KUTHERM" 3)
IMI Rolled Metals Ltd.
For electric blankets; under-road, under-ramp heating. Resistance alloy wire, maximum continuous operating temperature 150°C.

IMI 575 ("KUTHERM" 3)
IMI Rod & Wire
For electric blankets; under-road, under-ramp heating. Resistance alloy wire, maximum continuous operating temperature 150°C.

IMI 576 ("KUTHERM" 10)
IMI Rolled Metals Ltd.
For Cold "tails" electric blankets; under-road, under-ramp heating. Resistance alloy wire, maximum continuous operating temperature 250°C.

IMI 576 ("KUTHERM" 10)
IMI Rod & Wire
For Cold "tails" electric blankets; under-road, under-ramp heating. Resistance alloy wire, maximum continuous operating temperature 250°C.

IMI 577 (KUTHERM X)
IMI Rod & Wire
Resistivity: 10 micro-ohm·cm. Cu-Ni alloy used in electric blankets, controlled resistors for black heat applications.

IMI 579 ("KUTHERM" 26)
IMI Rolled Metals Ltd.
For electric blanket heating cables, fixed resistance, space heating systems. Resistance alloy wire, maximum continuous operating temperature 250°C.

IMI 579 ("KUTHERM" 26)
IMI Rod & Wire
For electric blanket heating cables, fixed resistance, space heating systems. Resistance alloy wire, maximum continuous operating temperature 250°C.

IMI 581 ("KUTHERM" 14)
IMI Rolled Metals Ltd.
For heating cables, thermocouples, space heating systems. Resistance alloy, intermediate electrical resistivity, outstanding stability up to 350°C.

IMI 581 ("KUTHERM" 14)
IMI Rod & Wire
For heating cables, thermocouples, space heating systems. Resistance alloy, intermediate electrical resistivity, outstanding stability up to 350°C.

IMI 583 (KUTHERM 49)
IMI Rod & Wire
Cupro-nickel alloy with low temperature coefficient. Used for instrument shunts, field regulators and all types of black heat resistances. Resistivity: 49 micro-ohm·cm.

IMI 591 (KUTHERM 15)
IMI Rod & Wire
Cupro-nickel alloy with low temperature coefficient. Resistivity: 15 micro-ohm·cm.

IMI 651
IMI Rolled Metals Ltd.
Cu 99, Sn 1.
Conductivity bronze wire; for high strength conductors, rivets and chains. *Obsolete*

IMI 657
IMI Rolled Metals Ltd.
Cu 95, Sn 5, P.
Phosphor bronze wire, various tempers. For springs, electrical contacts, switch parts, chains, welding wire. ASTM B103, B159, B225.

IMI 657
IMI Rod & Wire
Cu 95, Sn 5, P.
Phosphor bronze wire, various tempers. For springs, electrical contacts, switch parts, chains, welding wire. ASTM B103, B159, B225.

IMI 658
IMI Rolled Metals Ltd.
Cu 93, Sn 7, P.
Phosphor bronze, various tempers; for springs and stampings. BS 2870 PB 103. *Obsolete*

IMI 659
IMI Rolled Metals Ltd.
Cu 93, Sn 7, P.
Phosphor bronze wire, various tempers; for flat and coiled springs, weldin wire, Fourdrinier cloth warp. ASTM B103, B159, B225.

IMI 659
IMI Rod & Wire
Cu 93, Sn 7, P.
Phosphor bronze wire, various tempers; for flat and coiled springs, weldin wire, Fourdrinier cloth warp. ASTM B103, B159, B225.

IMI 660 (CAROBRONZE)
IMI Rolled Metals Ltd.
Cu 92, Sn 8.
Tin bronze wire and drawn shapes for bushes and piston rings. *Obsolete*

IMI 661
IMI Rolled Metals Ltd.
Cu 62, Sn 8, P.
Phosphor bronze wire; for Fourdrinier cloth warp. *Obsolete*

IMI 670
IMI Rod & Wire
Cu 99, Sn 0.75, Cd 0.25.
Sometimes called "Post office bronze."

IMI 675
IMI Rod & Wire
Cu 93, Sn 7.
Wire; for manufacture of hair springs. BS 2873 PB103; UNS C52100.

IMI 685
IMI Titanium Ltd.
Titanium. Al 6, Zr 5, Mo 0.5, Si 0.2, bal Ti.
Titanium alloy for high temperature use; weldable with creep resistance up to 520°C. RT: 850 MPa YS min; 990-1140 MPa TS; 6 El min.

IMI 705 ("EVERDUR" A)
IMI Rolled Metals Ltd.
Cu 96, Si 3, Mn 1.
Good strength bars and rods, weldable, and corrosion resistant to weak acids. For screws, bolts, nuts, particularly marine welding wire. ASTM B98.

IMI 705 ("EVERDUR" A)
IMI Rod & Wire
Cu 96, Si 3, Mn 1.
Good strength bars and rods, weldable, and corrosion resistant to weak acids. For screws, bolts, nuts, particularly marine welding wire. ASTM B98.

IMI 757
IMI Rolled Metals Ltd.
Cu 93, Al 7.
Aluminum bronze welding wire; for welding aluminum bronzes.

IMI 757
IMI Rod & Wire
Cu 93, Al 7.
Aluminum bronze welding wire; for welding aluminum bronzes.

IMI 758 ("SILVABRONZE")
IMI Rolled Metals Ltd.
Cu 92.6, Al 7, Sn 0.3, Ag 0.1.
Bronze strip, good resistance to flue gas condensates and to stress corrosion, particularly in chloride-contaminated waters. *Obsolete*

IMI 829
IMI Titanium Ltd.
Titanium. Al 5.5, Sn 3.5, Zr 3, Nb 1, Mo 0.3, Si 0.3, bal Ti. Titanium alloy for high temperature use; oxidation resistant and weldable, with creep resistance up to 520°C. RT: 820 MPa YS min; 950 MPa TS min; 10 El min.

IMI 830 (CUPRONIC)
IMI Rod & Wire
Cupro-nickel alloy with controlled thermo electromotive force used as compensating leads for thermocouples.

IMI 834
IMI Rolled Metals Ltd.
Al 5.8, Sn 4, Zr 3.5, Nb 0.7, Mo 0.5, Si 0.35, bal Ti. Bar, rod, 20°C: 1030 MPa TS; 910 MPa YS; 6 El. 600°C: 585 MPa TS; 480 MPa YS; 9 El.

IMI 837 (CUPRO-NICKEL)
IMI Rolled Metals Ltd.
Cu 94, Ni 6.
Wire and drawn shapes; good strength, ductility and resistance to corrosion. *Obsolete*

IMI 838 (CUPRO-NICKEL)
IMI Rolled Metals Ltd.
Cu 90, Ni 10. *Obsolete*

IMI 842 (CUPRO-NICKEL)
IMI Rolled Metals Ltd.
Cu 80, Ni 20.
Wire and drawn shapes; good strength, ductility and resistance to corrosion. BS 374; ASTM B122. *Obsolete*

IMI 849 (CUPRO-NICKEL)
IMI Rolled Metals Ltd.
Cu 70, Ni 30.
Wire and drawn shapes; good strength, ductility and resistance to corrosion. BS 374; ASTM B122. *Obsolete*

IMI 850
IMI Rod & Wire
Cupro-nickel with controlled electromotive force used in thermocouples.

IMI 885 ("KUNIFER" 10)
Yorkshire Imperial Metals Ltd.
Cu 87, Ni 10, Fe 2, Mn 1.
Sheet and strip for marine parts; good strength and corrosion resistance, especially against sea water; also for heat exchangers. BS 2870 CN.102.

IMI 891 ("KUNIFER 5T")
IMI Rolled Metals Ltd.
Ni 5, Fe 1.3, Ti 0.2-0.5, bal Cu.
Welding wire, for Argon arc welding of "Kunifer" 5 and 95/5 cupro-nickel. BS 2901.C19. *Obsolete*

IMI 892 ("KUNIFER 10T")
IMI Rolled Metals Ltd.
Ni 10, Fe 1.5, Mn 0.8, Ti 0.2-0.5, bal Cu.
Welding wire, for Argon arc welding of "Kunifer" 10 and 90/10 cupro-nickel.

IMI 892 ("KUNIFER 10T")
IMI Rod & Wire
Ni 10, Fe 1.5, Mn 0.8, Ti 0.2-0.5, bal Cu.
Welding wire, for Argon arc welding of "Kunifer" 10 and 90/10 cupro-nickel.

IMI 893 ("KUNIFER 30 T")
Yorkshire Imperial Metals Ltd.
Ni 30, Fe 0.5, Mn 0.8, Ti 0.2-0.5, bal Cu.
Welding wire for Argon arc welding of "Kunifer" 30 and 70/30 cupro-nickel. BS 2901.C18.

IMI 895 ("KUNIFER" 30)
IMI Rolled Metals Ltd.
Cu 68.25, Ni 30, Fe 0.75, Mn 1.
Sheet and strip for marine parts; good strength and corrosion resistance, especially against sea water. Also for heat exchangers. BS 2870 CN.107. *Obsolete*

IMI 981
IMI Rolled Metals Ltd.
Tin coated tough pitch or oxygen free HC copper. For electrical windings and conductors that have to be soldered. BS 128; ASTM B33. *Obsolete*

IMI TITANIUM 115
IMI Rolled Metals Ltd.
Commercially pure titanium. Sheet, rod, bar, wire, extrusions. 42,500-60,000 TS; 29,000 YS; 25 El. Ductile, formable, corrosion resistant. BS TA 1; DTD 501 3B.

IMI TITANIUM 125
IMI Rolled Metals Ltd.
Commercially pure titanium. Sheet, bar, tubes. 54,000-78,000 TS; 42,500 YS; 22 El. Ductile, formable, corrosion resistant. BS TA 2, 3, 4, 5; AMS 4902 etc.

IMI TITANIUM 130
IMI Rolled Metals Ltd.
Commercially pure titanium. Sheet, rod, bar, wire, extrusions, tubes (welded). 67,000-90,000 TS; 49,000 YS; 20 El. Corrosion resistant, formable. DTD 5003B, 5023B; AMS 4900B.

IMI TITANIUM 155
IMI Rolled Metals Ltd.
Commercially pure titanium. Sheet: 83,000-105,000 TS; 67,000 YS; 15 El. Corrosion resistant; slight forming only. BS TA6; AMS 4901, 4921.

IMI TITANIUM 160
IMI Rolled Metals Ltd.
Commercially pure titanium. Bar, rod, wire, extrusions. 78,000-108,000 TS; 63,000 YS; 15 El. Corrosion resistant. BS TA 7, 8, 9; AMS 4921.

IMI TITANIUM 230
IMI Rolled Metals Ltd.
Cu 2.5, bal Ti.
Sheet, rod, bar, wire, extrusions. Annealed: 78,000-112,000 TS; 67,000 YS; 15 El. Solution treated and aged: 110,000-134 TS; 80,000 YS; 10 El. Fatigue limit: 60-65% of TS. Formable, corrosion resistant. DTD 5233, 5243, 5253, 5263; BS TA21, 22 23, 24.

IMI TITANIUM 260
IMI Rolled Metals Ltd.
Pd (slight), bal Ti.
Sheet, bar, tube, wire and rod: 42,500-60,000 TS; 29,000 YS; 25 El. Improved resistance to non-oxidizing acids.

IMI TITANIUM 262
IMI Rolled Metals Ltd.
Commercially pure titanium with 0.15% Pd min. Sheet, bar, tube, wire and rod: 390-540 MPa TS; 290 MPa YS min; 22 El min.

IMI TITANIUM 315
IMI Rolled Metals Ltd.
Al 2, Mn 2, bal Ti.
Bar, rod: 94,000-116,000 TS; 67,000 YS; 20 El. Fatigue limit: 60-65% of TS. DTD 5043B.

IMI TITANIUM 317
IMI Rolled Metals Ltd.
Al 5, Sn 2.5, bal Ti.
Sheet, rod, bar, extrusions. 118,000-157,000 TS; 112,000 YS; 10 El. Weldable, high strength. BS TA14, 15, 16, 17; AMS 4910, 4926, etc.

IMI TITANIUM 318
IMI Rolled Metals Ltd.
Al 6, V 4, bal Ti.
Sheet, bar, wire, extrusions, rod. Rod: 130,000-168,000 TS; 120,000 YS; 8 El (variable properties with varying heat treatment). Fatigue limit: 50-60% of TS. Most popular titanium alloy; weldable, machinable, corrosion resistant. BS TA10, 11, 12, 13, 28; AMS 4911, 4928, etc.

IMI TITANIUM 550
IMI Rolled Metals Ltd.
Al 4, Mo 4, Sn 2, Si 0.5, bal Ti.
Bar, rod: 164,000 TS min; 145,000 YS; 0 El. High strength alloy, creep resistant up to 400°C. DTD 5103, 5153 (Hylite 50).

IMI TITANIUM 551
IMI Rolled Metals Ltd.
Al 4, Mo 4, Sn 4, Si 0.5, bal Ti.
Bar, rod: 179,000 TS min; 159,000 YS; 8 El. Very high strength titanium alloy. DTD 5203, 5223 (Hylite 51).

IMI TITANIUM 679
IMI Rolled Metals Ltd.
Sn 11, Zr 5, Al 2.25, Mo 1, Si 0.2, bal Ti.
Bar, rod, quenched and aged: 160,000-195,000 TS; 141,000 YS; 8 El. High strength alloy; creep resistant up to 450°C. BS TA18, 19, 20, 25, 26, 27; AMS 4974.

IMI TITANIUM 680
IMI Rolled Metals Ltd.
Sn 11, Mo 4, Al 2.25, Si 0.2, bal Ti.
Bar, rod, quenched and aged: 179,000 TS min; 156,000 YS; 10 El. High strength alloy. DTD 5213, M160.

IMI TITANIUM 684
IMI Rolled Metals Ltd.
Al 6, Zr 5, W 1, Si 0.3, bal Ti.
Bar: 143,000 TS; 127,000 YS; 6 El. Good strength alloy, weldable; creep resistant up to 520°C. DTD M200. *Obsolete*

IMI TITANIUM 685
IMI Rolled Metals Ltd.
Al 6, Zr 5, Mo 0.5, Si 0.3, bal Ti.
Forged, bar, rod, wire; 20°C: 990 MPa TS; 850 MPa YS; 6 El. 520°C: 620 MPa TS; 480 MPa YS; 9 El. Compressor blades for turbofan engine.

IMI TITANIUM 700
IMI Rolled Metals Ltd.
Al 6, Zr 5, Mo 4, Cu 1, Sn 0.2, bal Ti.
Air-cooled and aged, bar: 183,000 TS min; 166,000 YS; 8 El. Quenched and aged, bar: 195,000 TS min; 179,000 YS; 6 El. Fatigue limit: 45-55% of TS. Ultrahigh strength alloy, creep resistant up to 400°C. DTD M201.

IMI TITANIUM 829

IMI Rolled Metals Ltd.
Al 5.5, Sn 3.5, Zr 3, Nb 1, Mo 0.3, Si 0.3, bal Ti.
Room temperature, bar, rod: 960 MPa TS min; 820 MPa YS min; 10 El. 540°C: 600 MPa TS min; 460 MPa YS min; 12 El. Good high temperature creep strength.

IMI-RM 236

IMI Rolled Metals Ltd.
Cu 64, Zn 36.
Ductile alloy; as sheet or strip is used in automobile radiator tanks. BS 2870 CZ108; ASTM B36-272.

IMITA GOLD

English manufacture
Cu 89, Al 10.3, Fe 0.33, Ni 0.25, Sn 0.23, Mn 0.2.
For cheap jewelry, ornaments; corrosion resistant.

IMMACULATE 2W

Firth-Vickers Stainless Steels Ltd.
C 0.3, Si 1.3, Mn 0.8, Cr 20, Ni 7.5, W 2, bal Fe.
Annealed: 119,000 TS; 54,500 YS; 49 El; 56 RA; 220 Brin. For valves, furnace equipment; austenitic, stainless.

IMMACULATE 5

Firth Brown Ltd.
C 0-0.15, Si 0.2-1, Mn 0.5-2, Cr 23-26, Ni 19-22, bal Fe.
Austenitic stainless; AISI 310.

IMMACULATE 5

Firth-Vickers Stainless Steels Ltd.
C 0.1, Cr 23, Ni 21, bal Fe.
Stainless steel for elevated temperature operation.

IMMACULATE 5T

Firth Brown Ltd.
C 0.15, Si 0.2-1, Mn 0.5-2, Cr 22-23, Ni 13-16, Ti = 4 x C min, bal Fe.
Stabilized heat resisting steel.

IMMACULATE 5T

Firth-Vickers Stainless Steels Ltd.
C 0-0.15, Si 0-1, Mn 0-2, Cr 21.5-23.5, Ni 16-17.5, Ti 1, bal Fe.
Annealed, RT: 90,000 TS; 40,000 YS; 45 El. At 800°C: 42,000 TS; 52 El. Good strength and resistance to scaling at elevated temperatures. For radiant tube supports, furnace brackets.

IMMUNIT 14S

SWB Stahlformguss Gesellschaft mbH
C 0-0.07, Cr 17, Mo 4.75, Ni 13, bal Fe.
Annealed: 90,000 TS; 40,000 YS; 40 El; 55 RA; 180 Brin. For acid resistant chemical plant equipment, tanks; Type 317; stainless, austenitic. *Obsolete*

IMMUNIT 16S

SWB Stahlformguss Gesellschaft mbH
C 0-0.07, Cr 17.5, Ni 17.5, Cu 2, Mo 2, Cb = 8 x C, bal Fe.
Annealed: 100,000 TS; 50,000 YS; 40 El; 50 RA; 190 Brin. For acid resistant chemical plant equipment, valves; stainless, austenitic. *Obsolete*

IMMUNIT 2E

SWB Stahlformguss Gesellschaft mbH
C 0-0.12, Cr 18, Ni 9.5, Ti = 4 x C, bal Fe.
Annealed: 85,000 TS; 35,000 YS; 55 El; 65 RA; 150 Brin. For welded chemical plant equipment; Type 321; stainless, austenitic. *Obsolete*

IMMUNIT 2G

SWB Stahlformguss Gesellschaft mbH
C 0.15, Cr 18, Ni 8.5, bal Fe.
Annealed: 80,000 TS; 35,000 YS; 55 El; 75 RA; 150 Brin. Cold drawn: 180,000 TS; 150,000 YS; 10 El; 250 Brin. For chemical plant equipment, mixers, filters, tanks; Type 302; stainless, austenitic. *Obsolete*

IMMUNIT 2N

SWB Stahlformguss Gesellschaft mbH
C 0-0.15, Cr 18, Ni 8.5, bal Fe.
Annealed: 80,000 TS; 35,000 YS; 55 El; 75 RA; 150 Brin. Cold drawn: 180,000 TS; 150,000 YS; 10 El; 250 Brin. For chemical plant equipment, mixers, filters, tanks; Type 302; stainless, austenitic. *Obsolete*

IMMUNIT 2NA

SWB Stahlformguss Gesellschaft mbH
C 0-0.15, Cr 18, Ni 8.5, S 0.2, bal Fe.
Annealed: 80,000 TS; 35,000 YS; 45 El; 65 RA; 150 Brin. For screw machine products, bolts, fasteners, shafts; Type 303; stainless, free-cutting. *Obsolete*

IMMUNIT 2S

SWB Stahlformguss Gesellschaft mbH
C 0-0.07, Cr 18, Ni 9.5, bal Fe.
Annealed: 85,000 TS; 35,000 YS; 60 El; 70 RA; 150 Brin. Cold drawn: 180,000 TS; 135,000 YS; 10 El; 330 Brin. For welded chemical plant equipment, tanks, dryers; Type 304; stainless, austenitic. *Obsolete*

IMMUNIT 2X(G)

SWB Stahlformguss Gesellschaft mbH
C 0-0.12, Cr 18, Ni 9.5, Cb = 8 x C, bal Fe.
Annealed: 90,000 TS; 45,000 YS; 56 El; 65 RA; 160 Brin. Cold drawn: 100,000 TS; 65,000 YS; 40 El; 60 RA; 205 Brin. For welded structures, chemical plant equipment, tanks; Type 347; stainless, austenitic. *Obsolete*

IMMUNIT 4E

SWB Stahlformguss Gesellschaft mbH
C 0-0.12, Cr 18, Mo 2, Ni 10.5, Ti = 4 x C, bal Fe.
Annealed: 85,000 TS; 35,000 YS; 50 El; 65 RA; 160 Brin. For welded chemical plant equipment; Type 316 Ti; stainless, austenitic. *Obsolete*

IMMUNIT 4G

SWB Stahlformguss Gesellschaft mbH
C 0.15, Cr 18, Mo 2, Ni 9.5, bal Fe.
Annealed: 85,000 TS; 35,000 YS; 50 El; 65 RA; 160 Brin. For chemical plant equipment, mixers, agitators, filters; Type 316; stainless, austenitic. *Obsolete*

IMMUNIT 4N

SWB Stahlformguss Gesellschaft mbH
C 0.1, Cr 18, Ni 8, Si, bal Fe.
Annealed: 85,000 TS; 35,000 YS; 50 El; 70 RA; 150 Brin. For chemical plant equipment; corrosion resistant, austenitic. *Obsolete*

IMMUNIT 4S

SWB Stahlformguss Gesellschaft mbH
C 0-0.07, Cr 18, Ni 10.5, Mo 2, bal Fe.
Annealed: 85,000 TS; 35,000 YS; 50 El; 65 RA; 160 Brin. Cold drawn: 150,000 TS; 135,000 YS; 6 El; 300 Brin. For acid resistant chemical plant equipment; Type 316; stainless, austenitic. *Obsolete*

IMMUNIT 4SI

SWB Stahlformguss Gesellschaft mbH
C 0.1, Cr 18, Ni 9.5, Mo 2, bal Fe.
Annealed: 85,000 TS; 35,000 YS; 50 El; 65 RA; 160 Brin. Cold drawn: 150,000 TS; 135,000 YS; 6 El; 300 Brin. For acid resistant chemical plant equipment, tanks; Type 316; stainless, austenitic. *Obsolete*

IMMUNIT 4X(G)

SWB Stahlformguss Gesellschaft mbH
C 0-0.12, Cr 18, Ni 10.5, Mo 2, Cb = 8 x C, bal Fe.
Annealed: 85,000 TS; 35,000 YS; 45 El; 60 RA; 170 Brin. For welded chemical plant equipment, mixers, tanks; Type 316 Cb; stainless, austenitic. *Obsolete*

IMMUNIT 68G

SWB Stahlformguss Gesellschaft mbH
C 0.25, Cr 1.18, V 0.1, Mn 0.7, Si 0.8, bal Fe.
For gears, shafts, cams, bolts, camshafts; case hardening steel, tough. *Obsolete*

IMMUNIT 8E

SWB Stahlformguss Gesellschaft mbH
C 0-0.12, Cr 18, Mo 2, Ni 10.5, Ti = 4 x C, bal Fe.
Annealed: 85,000 TS; 35,000 YS; 45 El; 60 RA; 170 Brin. For welded chemical plant equipment, tanks, mixers; Type 316 Ti; stainless, austenitic. *Obsolete*

IMMUNIT C58G

SWB Stahlformguss Gesellschaft mbH
C 0.25, Cr 14.5, Ni 0-1, bal Fe.
Annealed: 95,000 TS; 50,000 YS; 25 El; 55 RA; 200 Brin. For turbine blades, cutlery, surgical and dental instruments; Type 420; corrosion resistant. *Obsolete*

IMMUNIT-R1013 MONI

SWB Stahlformguss Gesellschaft mbH
C 0.1, Cr 12.5, Mo 0.4, Ni 0.45, bal Fe.
Annealed: 70,000 TS; 35,000 YS; 40 El; 80 RA; 145 Brin. Cold drawn: 90,000 TS; 80,000 YS; 20 El; 65 RA; 205 Brin. For oil refinery and chemical plant equipment, table hardware, flatware, knives. Corrosion resistant. *Obsolete*

IMMUNIT-R1213 MOA

SWB Stahlformguss Gesellschaft mbH
C 0-0.15, Cr 13, Mo 0.3, bal Fe.
Annealed: 80,000 TS; 40,000 YS; 25 El; 93 Rock B. For springs, shafts, table flatware, oil refinery equipment. Corrosion resistant, hardenable. *Obsolete*

IMMUNIT-R1513 MO

SWB Stahlformguss Gesellschaft mbH
C 0.15, Cr 13, Mo 1, bal Fe.
Annealed: 80,000 TS; 40,000 YS; 25 El; 93 Rock B. Cold drawn: 100,000 TS; 80,000 YS; 15 El; 96 Rock B. For springs, table flatware, oil and chemical plant equipment. Corrosion resistant, hardenable. *Obsolete*

IMMUNIT-R2013 MO

SWB Stahlformguss Gesellschaft mbH
C 0.2, Cr 13, Mo 1, bal Fe.
Annealed: 95,000 TS; 40,000 YS; 25 El; 92 Rock B. Heat treated: 240,000 TS; 200,000 YS; 10 El; 50 Rock C. For cutlery, bearings, knives, shafts, surgical instruments, scissors. Corrosion resistant, hardenable. *Obsolete*

IMMUNIT-R2213

SWB Stahlformguss Gesellschaft mbH
C 0.2, Cr 12, Mo 1, V 0.3, W 0.5, bal Fe.
Annealed: 90,000 TS; 40,000 YS; 25 El; 92 Rock B. Hardened: 240,000 TS; 210,000 YS; 9 El; 50 Rock C. For valves, bearings, cutlery, surgical instruments. Corrosion resistant. *Obsolete*

IMPACTO

Atlas Specialty Steels
C 0.15, Mn 0.5, Mo 0.25, Ni 1.75, bal Fe.
Normalized: 80,000 TS; 45,000 YS; 30 El; 65 RA; 163 Brin. For case hardened parts, gears, pinions, cams; carburizing steel.

IMPACTO

Rio Algom Corp.
C 0.2, Mn 0.8, Ni 0.55, Cr 0.5, Mo 0.2, bal Fe.
Heat treated: 121,000 TS; 84,500 YS; 19 El; 248 Brin. For carburized splined shafts, piston pins, cam shafts, guide pins, ratchets. Tough, case hardening steel.

IMPACTO

Colt Industries
C 1, Si 0.4, bal Fe.
For tools, dies; water hardening. *Obsolete*

IMPALCO 740

Alcoa of Great Britain
Cu 0.3-0.7, Mg 2.2-3.2, Si 0-0.5, Fe 0-0.5, Mn 0.3-0.7, Zn 5.2-6.2, bal Al.
Heat treated: 72,000 TS min; 7 El min. For machining applications of high strength parts, structural members. Poor weldability. Heat treatable.

IMPALCO 750
Alcoa of Great Britain
Cu 0.3-0.7, Mg 2.2-3.2, Zn 5.2-6.2, Cr 0.08-0.25, bal Al.
Heat treated: 72,000 TS min; 7 El min. For machining applications of high strength parts, structural members. Poor weldability. Heat treatable.

IMPALCO 760
Alcoa of Great Britain
Cu 0.8-1.4, Mg 2.2-3.2, Mn 0.3-0.7, Zn 5-6, bal Al.
Heat treated: 78,000 TS min; 5 El min. For machining applications of high strength parts, structural members. Heat treatable. Poor weldability.

IMPALCO C66
Alcoa of Great Britain
Cu 3.5-4.8, Mg 0-0.85, Si 0-0.9, Fe 0-1, Mn 0-1.2, bal Al.
Heat treated: 69,000 TS min; 8 El min. At 200°C: 55,000 TS; 7.5 El. For highly stressed structural members. Heat treatable. Not corrosion resistant. Fair ductility and weldability.

IMPALCO C66A
Alcoa of Great Britain
Cu 3.5-4.8, Mg 0-0.85, Si 0-0.9, Fe 0-1, Mn 0-1.2, bal Al.
T-4 temper: 54,000 TS min; 15 El min. T-6 temper: 58,000 TS min; 8 El min. For stressed skin structures where corrosion resistance is important. Clad with pure aluminum. High strength. Heat treatable.

IMPALCO C69
Alcoa of Great Britain
Cu 1-2, Mg 0.5-1.2, Si 0.8-1.3, Fe 0-0.7, Mn 0-1, bal Al.
Heat treated: 58,000 TS min; 8 El min. For medium and high strength applications. Heat treatable. Poor resistance to atmospheric attack.

IMPALCO C80
Alcoa of Great Britain
Cu 5-6, Si 0-0.4, Fe 0-0.7, Pb 0.2-0.6, Bi 0.2-0.6, Zn 0-0.3, bal Al.
Heat treated: 42,000 TS min; 12 El min. For screw machine products, fasteners, shafts, screws. Free-machining. Poor resistance to atmospheric attack. Age-hardenable.

IMPALCO M31
Alcoa of Great Britain
Cu 0.15, Mg 0.4-0.8, Si 0-0.15, Fe 0-0.2, Mn 0.25, bal Al.
O-temper: 18,000 TS max. H-temper: 24,600 TS min. For bright trim, automobile panels, architectural applications. Good formability and weldability. Non-heat treatable.

IMPALCO M32
Alcoa of Great Britain
Cu 0-0.15, Mg 0.8-1.2, Si 0-0.15, Fe 0-0.2, Mn 0.25, bal Al.
O-temper: 22,400 TS max. H-temper: 29,200 TS min. For bright trim, architectural applications, automobile panels. Good ductility, formability and weldability. Non-heat treatable.

IMPALCO M32X
Alcoa of Great Britain
Cu 0-0.15, Mg 0.8-1.2, Si 0-0.15, Fe 0-0.2, Mn 0.25, bal Al.
O-temper: 22,400 TS max. H-temper: 29,200 TS min. For automobile trim, architectural applications, paneling. Good ductility, formability and weldability. Non-heat treatable.

IMPALCO M34
Alcoa of Great Britain
Cu 0-0.1, Mg 0.8-1.2, Si 0-0.35, Fe 0-0.5, Mn 0.15-0.3, bal Al.
O-temper: 18,000 TS min; 20 El min. H-temper: 33,600 TS min; 29,200 min YS; 4 El min. For tubular furniture, television antennae. Non-heat treatable. Good formability and weldability.

IMPALCO M35/1
Alcoa of Great Britain
Cu 0-0.1, Mg 1.8-2.7, Si 0-0.6, Fe 0-0.7, Mn 0-0.5, Cr 0-0.5, bal Al.
At 25°C: 25,700 TS; 10,000 YS; 29 El. At 200°C: 20,700 TS; 10,700 YS; 50 El. For architectural and marine applications, paneling, containers, welded structures. Non-heat treatable. Good formability and weldability.

IMPALCO M35/2
Alcoa of Great Britain
Cu 0-0.1, Mg 3-4, Si 0-0.6, Fe 0-0.7, Mn 0-1, Cr 0-0.5, bal Al.
At 25°C: 32,000 TS; 10,000 YS; 29 El. At 200°C: 23,500 TS; 11,200 YS; 53 El. For welded structures, casings. Good formability and weldability. Work hardens rapidly. Non-heat treatable.

IMPALCO M35/3
Alcoa of Great Britain
Cu 0-0.1, Mg 3.5-5.5, Si 0-0.7, Fe 0-0.7, Mn 0-1, Cr 0-0.5, bal Al.
M-temper: 38,100 TS min; 12 El min. O-temper: 38,100 TS min; 18 El min. For riveted and welded vessels or containers subjected to relatively high stresses. Non-heat treatable. Good formability and weldability. Work hardens rapidly.

IMPALCO M36
Alcoa of Great Britain
Cu 0-0.1, Mg 4.5-5.5, Si 0-0.6, Fe 0-0.7, Mn 0-1, Cr 0-0.5, bal Al.
M-temper: 38,100 TS min; 18 El min. 1/4 H-temper: 42,000 TS min; 8 El min. For marine, chemical and food industries, aircraft, welded structures. Non-heat treatable. Good formability and weldability. Work hardens rapidly.

IMPALCO M38
Alcoa of Great Britain
Cu 0-0.04, Mg 0.4-0.9, Si 0.3-0.7, Fe 0-0.5, bal Al.
At 25°C: 33,000 TS; 17 El. At 150°C: 21,000 TS; 20 El. For bus bars. Heat treatable. Good ductility and weldability. High electrical conductivity of 55% min.

IMPALCO M39/1
Alcoa of Great Britain
Cu 0-0.1, Mg 0.4-0.9, Si 0.3-0.7, Fe 0-0.6, Mn 0-0.3, bal Al.
O-temper: 15,700 TS min; 15 El. Heat treated: 27,000 TS min; 12 El min. For glazing bars, window frames, curtain walls. Heat treatable. Good formability and weldability.

IMPALCO M39/2
Alcoa of Great Britain
Cu 0-0.1, Mg 0.4-1.5, Si 0.6-1.3, Fe 0-0.6, Mn 0.4-1, Cr 0-0.3, bal Al.
Heat treated: 42,600 TS min; 10 El min. O-temper: 15,700 TS min; 15 El min. For automobile and bus bodies, coke skips, crossbearers. Heat treatable. Good weldability and ductility. Resists stress and shock.

IMPALCO M40
Alcoa of Great Britain
Cu 0.15-0.4, Mg 0.8-1.2, Si 0.4-0.8, Fe 0-0.7, Mn 0.2-0.8, Cr 0.15-0.35, bal Al.
Heat treated: 40,400 TS min; 10 El min. O-temper: 15,700 TS min; 15 El min. For tubular furniture. Good formability and weldability. Heat treatable.

IMPALCO M42
Alcoa of Great Britain
Cu 0-0.1, Mg 0.4-1.5, Si 0.6-1.3, Fe 0-0.6, Mn 0-0.2, Cr 0-0.1, bal Al.
Heat treated: 40,400 TS min; 10 El min. O-temper: 15,700 TS min; 15 El min. For general purpose applications requiring good surface finish and medium strength. Heat treatable. Good ductility and weldability. Not shock resistant.

IMPALCO P10
Alcoa of Great Britain
Cu 0-0.1, Si 0-0.5, Fe 0-0.7, Mn 0-0.1, 99.0 Al min.
At 25°C: 12,800 TS; 5200 YS; 45 El. At 200°C: 6000 TS; 2500 YS; 70 El. H-temper: 20,200 TS min; 3 El. For processing equipment for chemical, pharmaceutical, petroleum and food industries. High corrosion resistance, good ductility, low strength.

IMPALCO P3
Alcoa of Great Britain
Cu 0-0.02, Si 0-0.15, Fe 0-0.15, 99.8 Al min.
O-temper: 11,200 TS max; 35 El min. 1/2 H-temper: 13,500-16,800 TS; 8 El min. For chemically brightened and anodized components, costume jewelry. High ductility, poor machinability, low strength, high corrosion resistance.

IMPALCO P5
Alcoa of Great Britain
Cu 0-0.05, Si 0-0.3, Fe 0-0.4, Mn 0-0.05, 99.5 Al min.
M-temper: 9000 TS min; 25 El min. O-temper: 13,500 TS max; 30 El min. For food, pharmaceutical and chemical processing equipment, containers. High corrosion resistance, high ductility, and workability.

IMPALCO P5E
Alcoa of Great Britain
Cu 0-0.05, 99.5 Al min, 0.5 Cu + Si + Fe max.
O-temper: 10,000-14,500 TS; 25 El min. For electrical applications, bus bars. Low strength, high conductivity.

IMPALCO PA15
Alcoa of Great Britain
Si 10-13, Cu 0-0.1, Fe 0-0.6, Mn 0-0.5, bal Al.
For brazing applications. Filler alloy. Non-heat treatable. Good fluidity. Good corrosion resistance.

IMPALCO PA16
Alcoa of Great Britain
Cu 0-0.1, Si 4.5-6, Fe 0-0.6, Mn 0-0.5, bal Al.
For brazing applications. Filler alloy. Non-heat treatable. Good fluidity and corrosion resistance.

IMPALCO PA17
Alcoa of Great Britain
Cu 0-0.1, Si 7-8, Fe 0-0.6, Mn 0-0.5, bal Al.
For brazing applications. Filler alloy. Non-heat treatable. Good fluidity and corrosion resistance.

IMPALCO PA19
Alcoa of Great Britain
Cu 0-0.15, Si 0-0.6, Fe 0-0.7, Mn 1-1.5, bal Al.
At 25°C: 15,200 TS; 5000 YS; 40 El. At 200°C: 11,200 TS; 4480 YS; 43 El. For paneling of buildings, or vehicles, containers, holloware, roofing. Not hardenable by heat treatment. Good ductility, fair machinability.

IMPAX
Uddeholm Corp.
C 0.36, Cr 1.4, Ni 1.4, Mo 0.2, bal Fe.
Heat treated: 300-330 Brin. For molds, dies, tools, die casting tools for zinc, plastic molds. Prehardened; Type P20; tool steel. *Obsolete*

IMPAX SUPREME
Uddeholm Corp.
C 0.33, Si 0.3, Mn 1.4, Cr 1.8, Ni 0.8, Mo 0.2, S 0.008, bal Fe.
At 68°F: 146,000 psi TS; 116,000 psi YS; 20 El; 60 RA. Vacuum-degassed Cr-Ni-Mo alloy steel supplied in the hardened and tempered condition. For molds, dies, forming tools and structural components. AISI P20, improved.

IMPERATOR 15
Westfälische Stahlgesellschaft A.G.
C 0.3, Si 1, Mn 0.4, Cr 1.1, V 0.18, W 3.7, bal Fe.
For upsetters, cold headers, crimpers; oil hardened, tough.

IMPERATOR 200
Westfälische Stahlgesellschaft A.G.
C 0.3, Cr 2.35, V 0.6, W 4.25, bal Fe.
For upsetters, header dies, crimpers, oil hardened, tough.

IMPERIAL

Edgar Allen Balfour Ltd.
C 1.4, Cr 0.75, W 4.5, bal Fe.
For cutting tools, drills, reamers, textile needles. Water hardened. Wear resistant.

IMPERIAL

Edgar Allen Balfour Ltd.
C 1.2-1.3, W 4-5, Cr 1-1.25, V 0.2-0.25, bal Fe.
For reamers, cutters; fast finishing tool steel. *Obsolete*

IMPERIAL

Empire Steel Corp.
Copper. Cu 80, Ni 20.
For condenser tubes; corrosion resistant.

IMPERIAL

Bethlehem Steel Corp.
C 0.5, Mn 0.3, Si 1, Mo 0.45, bal Fe.
For cold work applications as beading tools, pipe cutter wheels, caulking tools, concrete breakers, staking tools. AISI Type S2 shock resisting steel. *Obsolete*

IMPERIAL ALL-METAL SOLDER

Imperial Brass Mfg. Co.
Pb, Sn.
For Al solder; high strength.

IMPERIAL C.R. 17

Edgar Allen Balfour Ltd.
C 0.12, Cr 17, Mn 0.3, bal Fe.
Annealed: 78,000 TS; 49,000 YS; 32 El; 50 RA; 160 Brin. For chemical plant equipment, oil burner parts; corrosion resistant, ferritic.

IMPERIAL C.T.

Edgar Allen Balfour Ltd.
C 0.32, Cr 13, Mn 0.3, bal Fe.
Heat treated: 202,000 TS; 190,000 YS; 15 El; 35 RA; 450 Brin. For surgical instruments, valve bodies, pump parts; impellers; corrosion and heat resistant, hardenable.

IMPERIAL E.Q.

Edgar Allen Balfour Ltd.
Stainless steel. C 0.2, Cr 12-14, bal Fe.
Heat treated: 100,000 TS; 80,000 YS; 30 El; 65 RA: 200 Brin. For steam valves, pump bodies; stainless, hardenable.

IMPERIAL MAJOR

Edgar Allen Balfour Ltd.
C 0.7, W 22, Cr 5, Co 13, bal Fe.
For lathe and planing tools; high speed steel. *Obsolete*

IMPERIAL MAJOR

A. Milne & Co.
W 21-22, Cr 4-5, Co 13, V 1.5, Mo 0.5, bal Fe.
For tools, cutters; high speed steel.

IMPERIAL MALLEABLE STAINLESS STEEL

Edgar Allen Balfour Ltd.
Cr, Si, Cu, Ni, bal Fe.
At 700 C: 94,100 TS. For golf clubs, shears, discharge valves, pump rods, resistance grids, tanks, drying cylinders. *Obsolete*

IMPERIAL MANGANESE

Edgar Allen Balfour Ltd.
Mn 12, C 1.2, bal Fe.
123,000 TS; 200 Brin. For tracks, dredger buckets, dipper teeth, gears, sheave wheels, sprockets, tube mill liners; wear resistant; difficult to machine.

IMPERIAL PERMANENT MAGNET

Edgar Allen Balfour Ltd.
6 W or 6 Cr, bal Fe.
For magnet steel. *Obsolete*

IMPERIAL R. 1

Edgar Allen Balfour Ltd.
C 0.12, Cr 13, Mn 0.3, bal Fe.
Heat treated: 85,000 TS; 56,000 YS; 35 El; 75 RA; 175 Brin.
Annealed: 67,000 TS; 38,000 YS; 38 El; 77 RA; 140 Brin. For shafts, housings, valve bodies, pump casings; corrosion resistant.

IMPERIAL R1O

Edgar Allen Balfour Ltd.
Stainless steel. C, Ni, Cr, bal Fe.
For plastic mold dies; stainless, hobbed dies.

IMPERIAL S.61

Edgar Allen Balfour Ltd.
Stainless steel. C 0.3, Cr 13, bal Fe.
For turbine blades, valves, cutlery; stainless, hardenable.

IMPERIAL S.80

Edgar Allen Balfour Ltd.
Stainless steel. C 0.2, Cr 18, Ni 1.25, bal Fe.
Heat treated: 130,000 TS; 107,000 YS; 20 El; 45 RA; 270 Brin.
Annealed: 112,000 TS; 80,000 YS; 14 El; 245 Brin. For steam valves, pump bodies; stainless, hardenable.

IMPERIAL STAINLESS IRON

Edgar Allen Balfour Ltd.
C 0.1, Cr 13, bal Fe.
Oil treated: 161,000-116,500 TS; 12-20 El; 37-67 RA; 220-360 Brin. For dental mirrors, nitric acid equipment, dye vats, heat exchangers, stove fittings; acid and heat resistant, malleable.

IMPERIAL STAINLESS STEEL

Edgar Allen Balfour Ltd.
0.25% min C, Cr, Si, Mn, bal Fe.
Oil treated: 179,200-235,000 TS; 8-16 El; 22-48 RA; 460-660 Brin. *Obsolete*

IMPERIAL STEEL

English manufacture
W 6.4, Mn 2.1, C 1.6, bal Fe.
For tools, cutters; oil hardening.

IMPERIAL TOOL

Edgar Allen Balfour Ltd.
C 1.4, Cr 0.75, W 4.5, bal Fe.
For fast finishing cutters; water hardened.

IMPERIAL TURNING FINISHING

Sterlite Foundry & Mfg. Co.
C 1.3, W 4.5, bal Fe.
For dies; oil hardened. *Obsolete*

IMPHRAM

Creusot-Loire
W 2-3, Mn 0.5-1, C, bal Fe.
For turning tools, planers, borers, gages, fine cutlery, hard hammers, chisels, razors; non-deforming. *Obsolete*

IMPHY "D"

Creusot-Loire
C 0.25, Si, Mn, bal Fe.
Annealed: 70,000-78,000 TS; 43,000-52,000 YS; 25-28 El. For machine parts, locomotive axles; water hardening. *Obsolete*

IMPHY "D.D."

Creusot-Loire
C 0.2, Si, Mn, bal Fe.
Annealed: 64,000-71,000 TS; 38,000-46,000 YS; 28-31 El. For wagon axles, marine forgings; water hardening. *Obsolete*

IMPHY "D.D.D."

Creusot-Loire
C 0.15, Si, Mn, bal Fe.
Annealed: 58,000-64,000 TS; 39,000-45,000 YS; 28-33 El. For rivets, stamped pieces; easily worked. *Obsolete*

IMPHY "D.T."

Creusot-Loire
C 0.4, Si, Mn, bal Fe.
Annealed: 85,000-92,000 TS; 50,000-56,000 YS; 20-24 El. For fire arms, cutlery, edge tools; water hardening. *Obsolete*

IMPHY "F.F."

Creusot-Loire
C 0.1, Si, Mn, bal Fe.
Annealed: 52,000-58,000 TS; 32,500-38,000 YS; 39-42 El. For welding; replaces Swedish Iron. *Obsolete*

IMPHY "M.D."

Creusot-Loire
C 0.3, Si, Mn, bal Fe.
Annealed: 78,000-85,000 TS; 46,000-52,000 YS; 23-26 El. For fire arms, edge tools; water hardening. *Obsolete*

IMPHY "M.T."

Creusot-Loire
C 0.45, Si, Mn, bal Fe.
Annealed: 91,000-100,000 TS; 52,000-60,000 YS; 17-22 El. For paper cutters, sledges, wedges, mining bars; water hardening. *Obsolete*

IMPHY "T"

Creusot-Loire
C 0.5, Si, Mn, bal Fe.
Annealed: 100,000-107,000 TS; 55,000-64,000 YS; 14-19 El. For hammers, punches, dies, cutlery; oil hardening. *Obsolete*

IMPHY "T.T."

Creusot-Loire
C 0.65, Si, Mn, bal Fe.
Annealed: 107,000-120,000 TS; 56,000-71,000 YS; 10-15 El. For files, saws, dies, ball mills; water hardened. *Obsolete*

IMPHY "T.T.T."

Creusot-Loire
C 0.7, Si, Mn, bal Fe.
Annealed: 120,000-143,000 TS; 64,000-78,000 YS; 7-12 El. For files, saws, tools; water hardening. *Obsolete*

IMPHY 1640

Creusot-Loire
C, Cr, bal Fe.
Annealed: 101,000 TS; 58,000 YS; 18 El. Heat treated: 135,000 TS; 114,000 YS; 14 El. For crankshafts; tough. *Obsolete*

IMPHY 1691

Creusot-Loire
C, Ni, Cr, Mo, bal Fe.
Annealed: 121,000 TS; 92,000 YS; 18 El. Oil treated: 144,000 TS; 128,000 YS; 14 El. For gears, crankshafts, still parts, bolts; resists high temperatures. *Obsolete*

IMPHY 301

Creusot-Loire
C 0-0.2, Si 0-1, Mn 0-2, Ni 7, Cr 17, bal Fe.
Annealed: 110,000 TS; 40,000 YS; 60 El; B 85 Rock. Half hard: 150,000 TS; 110,000 YS; 15 El; C 32 Rock. For aircraft structural members, trailer bodies, household utensils, architectural trim; work hardens rapidly. AISI Type 301. Stainless. *Obsolete*

IMPHY 302

Creusot-Loire
C 0-0.12, Si 0-1, Mn 0-2, Ni 9, Cr 18, bal Fe.
Annealed: 90,000 TS; 40,000 YS; 50 El; B 85 Rock. Cold drawn: 140,000 TS; 100,000 YS; 30 El; B 100 Rock. For chemical plant and food processing equipment, valve trim, fasteners. AISI Type 302. Stainless. *Obsolete*

IMPHY 304

Now ICL 472.

IMPHY 304L

Now ICL 472 BC.

IMPHY 305
Creusot-Loire
C 0-0.12, Si 0-1, Mn 0-2, Ni 11.5, Cr 18, bal Fe.
Annealed: 85,000 TS; 38,000 YS; 55 El; B 80 Rock. For electrical instruments, cold headed fasteners, chemical and textile processing equipment. AISI Type 305. *Obsolete*

IMPHY 308
Creusot-Loire
C 0-0.08, Si 0-1, Mn 0-2, Ni 11, Cr 20, bal Fe.
Annealed: 85,000 TS; 30,000 YS; 55 El; B 80 Rock. For welded structures, valve trim, oil refinery and chemical plant equipment, industrial furnaces, welding rod. Heat and corrosion resistant. AISI Type 308. *Obsolete*

IMPHY 309
Creusot-Loire
C 0-0.15, Ni 14, Cr 23, bal Fe.
Annealed: 90,000 TS; 45,000 YS; 45 El; B 85 Rock. For oil burners, furnace parts, oil refinery and chemical plant equipment. High creep and oxidation resistance. AISI Type 309. Stainless. *Obsolete*

IMPHY 30CD12SP
Creusot-Loire
C 0.3, Cr 3, Mo 0.4, bal Fe.
Annealed: 114,000 TS. Heat treated: 157,000 TS. For casings, liners, cylinder liners; Afnor 30CD12; nitriding steel. *Obsolete*

IMPHY 31 BIS SP
Creusot-Loire
C 0.16, Ni, Cr, Mo, bal Fe.
Annealed: 92,000 TS; 65,000 YS; 20 El. Heat treated: 186,000 TS; 143,000 YS; 10 El. For gears, cams, camshafts; case hardened; Afnor 16NCD13. *Obsolete*

IMPHY 310
Creusot-Loire
C 0-0.15, Ni 20, Cr 25, bal Fe.
Annealed: 90,000 TS; 40,000 YS; 55 El; B 80 Rock. At 1200 F: 57,000 TS; 22,000 YS; 32 El; 45 RA. For carburizing boxes, oil burner parts, baffle plates, oil refinery equipment, recuperators. Heat and corrosion resistant. AISI Type 310, austenitic. *Obsolete*

IMPHY 316
Now ICL 164.

IMPHY 316L
Now ICL 164 BC.

IMPHY 317
Creusot-Loire
C 0-0.08, Ni 14, Cr 19, Mo 3.5, bal Fe.
Annealed: 90,000 TS; 55,000 YS; 50 El; B 90 Rock. Cold rolled: 150,000 TS; 125,000 YS; 15 El; C 30 Rock. For textile and dye plant equipment, paper and pulp mill equipment, chemical plant equipment. Stainless, acid resistant. AISI Type 317, austenitic. *Obsolete*

IMPHY 317L
Now ICL 188 BC.

IMPHY 321
Now ICL 472 T.

IMPHY 32SP
Creusot-Loire
C 0.3, Ni 1.2, Cr 0.6, Mo 0.2, bal Fe.
Normalized: 120,000 TS; 85,000 YS; 16 El. Heat treated: 135,000 TS; 114,000 YS; 14 El. For crankshafts, gears, bits; Afnor 30NCD11; oil hardened. *Obsolete*

IMPHY 347
Creusot-Loire
C 0-0.08, Si 0-1, Mn 0-2, Ni 11, Cr 18, Cb = 10 x C, bal Fe.
Annealed: 90,000 TS; 40,000 YS; 50 El; B 83 Rock. Cold drawn: 100,000 TS; 65,000 YS; 40 El; B 95 Rock. For welded structures, exhaust manifolds, radiant superheaters, low temperature processing. Welding grade, stainless. AISI Type 347, austenitic. *Obsolete*

IMPHY 34R
Creusot-Loire
C 0.32, Ni, Cr, Mo, bal Fe.
Normalized: 121,000 TS; 85,000 YS; 16 El. Heat treated: 242,000 TS; 200,000 YS; 6 El. For crankshafts, gears, bolts; Afnor 32NCD11; oil hardened. *Obsolete*

IMPHY 34SP
Creusot-Loire
C 0.32, Ni 2.2, Cr 2.2, Mo 0.7, bal Fe.
Normalized: 121,000 TS; 92,000 YS; 16 El. Heat treated: 250,000 TS; 220,000 YS; 8 El. For crankshafts, gears, bolts; Afnor 32NCD8; oil hardened. *Obsolete*

IMPHY 3693
Creusot-Loire
Cr 1.5, C, Mo, V, bal Fe.
For gears, shafts; oil hardening. *Obsolete*

IMPHY 38MS5SP
Creusot-Loire
C 0.9, Alloy, bal Fe.
Annealed: 114,000 TS; 58,000 YS; 15 El. Heat treated: 200,000 TS; 177,000 YS; 8 El. For springs; Afnor 38MS5; water hardened. *Obsolete*

IMPHY 4
Creusot-Loire
C 0-0.12, Mn 0-1.25, Cr 14-18, bal Fe.
Annealed: 80,000 TS; 45,000 YP; 25 El; 165 Brln. Cold rolled: 130,000 TS; 125,000 YP; 2 El; 270 Brln. For chemical and textile plant equipment, oil burner and heater parts. Type 430 stainless steel; ferritic. *Obsolete*

IMPHY 410
Now SOLEIL A2.

IMPHY 41SP
Creusot-Loire
C 0.42-0.5, Si 1.25-1.75, Cr 0.7-0.9, Mo 0.2-0.3, bal Fe.
Normalized: 108,000 TS; 58,000 YS; 16 El. Heat treated: 242,000 TS; 221,000 YS; 6 El. For springs; Afnor 45SCD6; oil hardened. *Obsolete*

IMPHY 430
Now SOLEIL B4.

IMPHY 431
Creusot-Loire
C 0-0.2, Si 0-1, Mn 0-1, Ni 2, Cr 16, bal Fe.
Annealed: 125,000 TS; 95,000 YS; 20 El; C 24 Rock. Heat treated: 180,000 TS; 125,000 YS; 17 El; C 38 Rock. For pump shafts, marine hardware, surgical trusses, valve trim, aircraft structural members. Corrosion and heat resistant. AISI Type 431. Hardenable. *Obsolete*

IMPHY 505
Creusot-Loire
Cr 5, Mo 0.5, C, bal Fe.
For petroleum industry equipment; corrosion resistant. *Obsolete*

IMPHY 6
Creusot-Loire
C 0.2, Cr 28, Ni 4, bal Fe.
Cast: 100,000 TS; 61,000 YS; 13 El; 207 Brln. At 1000 F: 64,000 TS; 38,000 YS; 25 El; 27 Brln. For furnace blowers, gas burners, recuperators, ore roasting furnaces. Type 327 stainless steel. Resists high sulphur atmospheres. *Obsolete*

IMPHY 766
Creusot-Loire
C 0.4, Cr 1.5, Si, Mo, V, bal Fe.
For gears, shafts; oil hardening. *Obsolete*

IMPHY A.F.T.
Creusot-Loire
Ni alloy. *Obsolete*

IMPHY A.R.C. 1047
Creusot-Loire
Cr 18, Ni 8, C, W, bal Fe.
For chemical equipment; corrosion resistant. *Obsolete*

IMPHY A.R.C. 2702S
Creusot-Loire
C 0.2, Cr 18, Ni 8, bal Fe.
For corrosion resistant parts; stainless, corrosion resistant. *Obsolete*

IMPHY A.T.E.
Creusot-Loire
Ni 95, Mn 2, Al 2, Si 1.
For thermocouple element to be used with Imphy B.T.E.; maximum operating temperature 1100 C. *Obsolete*

IMPHY A.T.G.
Creusot-Loire
Ni 60, Fe 26, Cr 10, W 4.
Forged: 121,000 TS; 91,000 YS; 20 El. Annealed: 100,000 TS; 52,000 YS; 30 El. For gas turbines, internal combustion engines; heat resisting, corrosion resisting. *Obsolete*

IMPHY A.T.V.-1
Creusot-Loire
C 0.3, Ni 35, Cr 11, bal Fe.
Forged: 107,000 TS; 78,000 YS; 20 El. Annealed: 88,000 TS; 46,000 YS; 25 El. For turbines, exhaust valves for internal combustion engines; corrosion and heat resistant. *Obsolete*

IMPHY A8
Creusot-Loire
C 0.06, Cr 12-14, Al 0.2, bal Fe.
Annealed: 70,000 psi TS; 40,000 psi YS; 30 El; 160 Brln. Cold drawn: 85,000 psi TS; 70,000 psi YS; 20 El; 185 Brln. For heat treating boxes, oil refinery equipment, quenching racks. Type 405 stainless steel, ferritic.

IMPHY ADR
Creusot-Loire
Ni 38-40, Mn 1, bal Fe.
For use in construction of rotary converters, thermostats, geodesic measures; low coefficient of expansion. *Obsolete*

IMPHY AFK
Creusot-Loire
Si, bal Fe.
For magnetic equipment; soft magnet for strong or moderate fields.

IMPHY AMF
Creusot-Loire
Ni 55-60, Mn 0.3, C 0.2-0.4, bal Fe.
Annealed: 100,000 TS; 62,000 YS; 40 El; 50 RA. Liquid air T: 102,000 TS; 50,000 YS; 40 El; 55 RA. For valves of expansion engines used in liquid air plants, refrigeration machines; great ductility at low temperature. *Obsolete*

IMPHY ARC
Creusot-Loire
C 0.2, Cr 18, Ni 8, bal Fe.
For chemical apparatus, nitrogen fixation and synthetic ammonia apparatus, valves, pumps; resists chemical attack; stainless. *Obsolete*

IMPHY ARC 098
Creusot-Loire
C 0.1, Ni 14, Cr 25, Mo 1.2, bal Fe.
For chemical plant equipment; resists nitric acids. *Obsolete*

IMPHY ARC 164
Creusot-Loire
C 0.12, Ni 60, Mo 18, bal Fe.
For chemical plant equipment; resists HCl. *Obsolete*

IMPHY ARC 2266S

Creusot-Loire
C 0-0.08, Cr 20, Ni 10, Mo 3, bal Fe.
Annealed: 85,000 TS; 35,000 YS; 50 El; 65 RA; 160 Brin. For chemical plant equipment, tanks, agitators; acid resistant, austenitic; Type 317. *Obsolete*

IMPHY ARC 2266SP

Creusot-Loire
C 0-0.06, Cr 18, Ni 11, Mo 3, Ti 0.4, bal Fe.
Annealed: 85,000 TS; 35,000 YS; 40 El; 55 RA; 160 Brin. For oil refinery and chemical plant equipment; acid resistant, austenitic. *Obsolete*

IMPHY ARC 2266T

Creusot-Loire
C 0-0.06, Cr 17, Ni 12, Mo 3, bal Fe.
Annealed: 85,000 TS; 35,000 YS; 50 El; 65 RA; 160 Brin. For oil refinery and chemical plant equipment; acid resistant, austenitic; Type 316L. *Obsolete*

IMPHY ARC-2266

Now ARCM 2266, ETC.

IMPHY ARC-2702A

Creusot-Loire
C 0.04, Cr 18, Ni 8, bal Fe.
Annealed: 85,000 TS; 35,000 YS; 60 El; 70 RA; 150 Brin.
Rolled: 125,000 TS; 95,000 YS; 25 El; 55 RA; 277 Brin. For chemical plant equipment; Type 304; stainless, austenitic. *Obsolete*

IMPHY ARC-2702B

Creusot-Loire
C 0 09-0.15, Cr 18, Ni 8, Ti, bal Fe.
Annealed: 85,000 TS; 35,000 YS; 60 El; 70 RA; 150 Brin. For welded parts, chemical plant equipment; Type 321; stainless, austenitic. *Obsolete*

IMPHY ATG-B

Creusot-Loire
C, alloy, bal Fe.
For heat resistant parts, oil refinery equipment; heat resistant. *Obsolete*

IMPHY ATG-M

Creusot-Loire
C, alloy, bal Fe.
For heat resistant parts, oil refinery equipment; heat resistant.

IMPHY ATG-R

Creusot-Loire
C, alloy, bal Fe.
For heat resistant parts, oil refinery equipment; heat resistant. *Obsolete*

IMPHY ATG-S

Creusot-Loire
C, alloy, bal Fe.
For oil refinery equipment; heat resistant.

IMPHY ATG-S4

Creusot-Loire
C, alloy, bal Fe.
For oil refinery equipment; heat resistant.

IMPHY ATG-S7

Creusot-Loire
C, alloy, bal Fe.
For oil refinery equipment; heat resistant.

IMPHY ATG-T

Creusot-Loire
C, alloy, bal Fe.
For oil refinery equipment; heat resistant.

IMPHY ATG-Z

Creusot-Loire
C, alloy, bal Fe.
For oil refinery equipment; heat resistant.

IMPHY ATV-1

Creusot-Loire
C, alloy, bal Fe.
For oil refinery equipment; oxidation resistant.

IMPHY ATV-3

Creusot-Loire
C, alloy, bal Fe.
For oil refinery equipment; oxidation resistant.

IMPHY ATV-R

Creusot-Loire
C, alloy, bal Fe.
For oil refinery equipment, oxidation resistant.

IMPHY ATV-S

Creusot-Loire
C, alloy, bal Fe.
For oil refinery equipment; oxidation resistant.

IMPHY B.T.E.

Creusot-Loire
Ni 91, Cr 9.
For thermocouple element to be used with Imphy A.T.E., C.T.E. or nickel; maximum operating temperature 1100 C. *Obsolete*

IMPHY B.Y. 2

Creusot-Loire
C 0.35, Ni 4.25, Cr 1.2, bal Fe.
121,000 TS; 100,000 YS; 12 El. For machinery parts, gears; shock resistant. *Obsolete*

IMPHY B.Y.-1

Creusot-Loire
C 0.25, Ni 4, Cr 1, bal Fe.
Heat treated: 28,000-107,000 TS; 85,000-213,000 YS; 8-15 El. For shafts; air hardening. *Obsolete*

IMPHY BCM

Creusot-Loire
C 0.1, bal Fe.
Heat treated: 50,000-72,000 TS; 36,000-50,000 YS; 25-30 El. For valves, punches, drivers, gears; case hardened; Afnor XC10. *Obsolete*

IMPHY BTE-ATE

Creusot-Loire
Iron alloy.
For pyrometers; good to 1000 C. *Obsolete*

IMPHY BTE-CTE

Creusot-Loire
Iron alloy.
For pyrometers; good to 700 C. *Obsolete*

IMPHY BTE-NTE

Creusot-Loire
Iron alloy.
For pyrometers; good to 900 C. *Obsolete*

IMPHY C.C.R.-CO

Creusot-Loire
C 1.5, Cr 13, Mo 0.7, Co 3, bal Fe.
For bearings, knives, cutlery; heat resistant, hardenable. *Obsolete*

IMPHY C.T.E.

Creusot-Loire
Ni 45, Cu 55.
For thermocouple element to be used with Imphy B.T.E.; useful range (-220 to +700 C). *Obsolete*

IMPHY CCA 1007

Now SOLEIL 33.

IMPHY CCA 1007

Creusot-Loire
C 0.08, Cr 13, Al 0.2, Mn 0-1, bal Fe.
Annealed: 65,000 TS; 38,000 YS; 30 El; 160 Brin. Cold drawn: 85,000 TS; 70,000 YS; 20 El; 185 Brin. For annealing boxes, quenching racks, heat treating equipment, oil refinery and chemical plant equipment. Type 405 stainless steel, heat and corrosion resistant. *Obsolete*

IMPHY CCR

Creusot-Loire
C 2, Cr 13, bal Fe.
Annealed: 100,000-121,000 TS; 58,000-78,000 YS; 15-20 El. Heat treated: 192,000 TS; 163,000 YS; 5 El. For gages, mirrors; abrasion resistant. *Obsolete*

IMPHY CMO

Creusot-Loire
C, alloy, bal Fe.
For oil refinery equipment; oxidation resistant. *Obsolete*

IMPHY CRB

Creusot-Loire
C 1, Cr 1.5, Si 0.3, Mn 0.4, bal Fe.
For rolls, ball bearings; Afnor 100C6; water hardened. *Obsolete*

IMPHY CVC

Creusot-Loire
Cr 1.5, V 0.25, C, bal Fe.
Annealed: 64,000-73,000 TS; 50,000-58,000 YS; 22-28 El. Water quenched: 163,000-186,000 TS; 156,000-177,000 YS; 10-17 El. For piston pins; case-hardening steel. *Obsolete*

IMPHY ERA HR1

Creusot-Loire
C 0.3, Cr 18, Ni 8, W 3, Si 1.5, bal Fe.
Rolled: 114,000 TS; 71,000 YS; 22 El. For chemical plant equipment; stainless, austenitic. *Obsolete*

IMPHY ERA HR2

Creusot-Loire
C 0.13, Cr 18, Ni 8, bal Fe.
Rolled: 100,000 TS; 55,000 YS; 25 El. For chemical plant equipment; Type 302; stainless, austenitic. *Obsolete*

IMPHY HPM

Creusot-Loire
Si, bal Fe.
For armatures, transformers, relays; soft magnet for strong or moderate field. *Obsolete*

IMPHY I.R.R.-S

Creusot-Loire
C 0.7, W 20, Mo 1, V 1.5, Co 10, bal Fe.
For tools, cutters; high speed steel. *Obsolete*

IMPHY I.R.U.

Creusot-Loire
C 0.7, W 18, Mo 1, V 1, bal Fe.
For tools, cutters; high speed steel. *Obsolete*

IMPHY I.T.R.

Creusot-Loire
C 0.7, W 18, Mo 1, V 1, Co 5, bal Fe.
For tools, cutters; high speed steel. *Obsolete*

IMPHY I.W.3

Creusot-Loire
W 9, Cr 2.5, V 0.5, C, bal Fe.
For hot work dies; hot work steel. *Obsolete*

IMPHY ICR

Creusot-Loire
C, bal Fe.
For welded structures. *Obsolete*

IMPHY IR
Creusot-Loire
C 0.7, W 14, Cr 4, V 1, bal Fe.
Hardened: C 64-66 Rock. For cutting tools, drills, reamers, broaches, lathe and planer tools. High speed steel, high red-hardness. *Obsolete*

IMPHY IRU
Creusot-Loire
C 0.7, W 18, Cr 4, V 1, bal Fe.
Hardened: C 64-66 Rock. For cutting tools, taps, drills, lathe and planer tools, reamers, hobs, milling cutters, thread chasers. High speed steel, Type T1. High red-hardness. *Obsolete*

IMPHY ISO
Creusot-Loire
C, bal Fe.
For welded structures. *Obsolete*

IMPHY ITR
Creusot-Loire
C 0.7, W 18, Cr 4, V 1, Co 5, bal Fe.
Hardened: C 64-66 Rock. For cutting tools, reamers, drills, lathe and planer tools, broaches. High speed steel, high red-hardness. *Obsolete*

IMPHY IVO
Creusot-Loire
C, bal Fe.
For welded structures. *Obsolete*

IMPHY IW1
Creusot-Loire
C, bal Fe.
For welded structures. *Obsolete*

IMPHY M 550
Creusot-Loire
Cr 2.5, C, Mo, V, bal Fe.
For gears, shafts; oil hardening. *Obsolete*

IMPHY M-1628
Creusot-Loire
Ni 65, Mo 25, C, bal Fe.
For chemical plant equipment; resists HCl. *Obsolete*

IMPHY M.C.T. 1
Creusot-Loire
C 0.1-0.2, Cr, Mo, bal Fe.
72,000 TS; 43,000 YS; 26 El. For gun tubes, general parts; water hardening. *Obsolete*

IMPHY M.C.T. 2
Creusot-Loire
C 0.2-0.3, Cr, Mo, bal Fe.
93,000 TS; 58,000 YS; 22 El. For shafts, axles, aircraft fuselage; high workability. *Obsolete*

IMPHY M.C.T. 3
Creusot-Loire
C 0.3-0.4, Cr, Mo, bal Fe.
101,000 TS; 65,000 YS; 20 El. For crankshafts, gears, levers, axles; tough. *Obsolete*

IMPHY MA-42
Creusot-Loire
Ni 20, Cr 25, C, Si, bal Fe.
For stainless and wear resistant parts; corrosion resistant; austenitic. *Obsolete*

IMPHY MCT-19
Creusot-Loire
C, alloy bal Fe.
For oil refinery equipment; oxidation resistant. *Obsolete*

IMPHY MCT2S
Creusot-Loire
C 0.25, Cr 1, Mo 0.25, bal Fe.
Annealed: 94,000 TS; 58,000 YS; 20 El. Heat treated: 157,000 TS; 128,000 YS; 12 El. For axles, gears, spindles; Afnor 25CD4; case hardened. *Obsolete*

IMPHY MCT4
Creusot-Loire
C 0.42, Cr 1, Mo 0.25, bal Fe.
Annealed: 114,000 TS; 72,000 YS; 16 El. Heat treated: 280,000 TS; 230,000 YS; 4 El. For gears, shafts, crankshafts; Afnor 42CD4; oil hardened. *Obsolete*

IMPHY MCTO
Creusot-Loire
C 0.12, Cr 1, Mo 0.2, bal Fe.
Annealed: 72,000 TS; 43,000 YS; 24 El. Heat treated: 150,000 TS; 123,000 YS; 10 El. For case hardened gears, shafts, cams; Afnor 12CD4; case hardened. *Obsolete*

IMPHY MIP1
Creusot-Loire
C, bal Fe.
For welded structures. *Obsolete*

IMPHY MOS
Creusot-Loire
Si 1.5, Mn, C, bal Fe.
Annealed: 105,000 TS; 65,000 YS; 20 El. Heat treated: 186,000 TS; 163,000 YS; 7 El. Heat treated: 186,000 TS; 163,000 YS; 7 El. For gears, springs; silico-manganese steel. *Obsolete*

IMPHY MWS
Creusot-Loire
bal Fe.
Annealed: 107,000 TS; 69,000 YS; 22 El. Heat treated: 100,000 TS, 171,000 YS; 7 El. For gears, crankshafts, wearing parts;, tough. *Obsolete*

IMPHY N.7C.M.
Creusot-Loire
Ni 7, C, bal Fe.
Heat treated: 140,000-164,000 TS; 100,000-114,000 YS; 7-10 El. Case-hardened parts; special Ni steel carburized. *Obsolete*

IMPHY N.A.
Creusot-Loire
Ni 2, bal Fe.
Annealed: 78,000 TS; 50,000 YS; 26 El. Heat treated: 143,000 TS; 116,000 YS; 11 El. For crankshafts, transmission shafts, forged parts; tough. *Obsolete*

IMPHY N.C. 3 H.-1
Creusot-Loire
Ni 3, Cr 1, C, bal Fe.
Annealed: 81,000 TS; 55,000 YS; 25 El. Heat treated: 164,000 TS; 150,000 YS; 10 El. For crankshafts, machine parts, dies; shock resistant. *Obsolete*

IMPHY N.C.-4
Creusot-Loire
Cr 25, bal Fe.
Forged: 121,000 TS; 85,000 YS; 25 El. Annealed: 92,000 TS; 39,000 YS; 40 El. For rotor housings, naval compass construction; non-magnetic. *Obsolete*

IMPHY N.C.M.
Creusot-Loire
C 0.2, Ni 2, bal Fe.
Heat treated: 78,000-100,000 TS; 50,000-71,000 YS; 18-25 El. For case-hardened parts; carburized. *Obsolete*

IMPHY N.D.
Creusot-Loire
Ni 5, C, bal Fe.
Annealed: 81,000 TS; 55,000 YS; 25 El. Heat treated: 177,000 TS; 143,000 YS; 10 El. For turbine buckets; resists fatigue stresses. *Obsolete*

IMPHY N.F.
Creusot-Loire
C 0.2, Ni 3, bal Fe.
Heat treated: 100,000-114,000 TS; 78,000-100,000 YS 8-16 El. For case-hardened parts; carburized. *Obsolete*

IMPHY N.F.C. NO. 1
Creusot-Loire
Ni 3.5, Cr 1, bal Fe.
Annealed: 81,000-91,000 TS; 61,000-71,000 YS; 18-25 El. Oil quenched: 171,000-200,000 TS; 156,000-192,000 YS; 7-10 El. For case-hardened parts, axles, shafts, steering parts; carburized. *Obsolete*

IMPHY N.F.C.-2
Creusot-Loire
Ni 3, Cr 1, C, bal Fe.
83,000-200,000 TS; 61,000-192,000 YS; 7-25 El. For axles, shafts, steering parts; oil hardening, tough. *Obsolete*

IMPHY N.F.C.-3
Creusot-Loire
Ni 4, Cr 1, C, bal Fe.
86,000-214,000 TS; 72,000-199,000 YS; 6-20 El. For axles, shafts, steering parts; shock resistant. *Obsolete*

IMPHY N.M.F.
Creusot-Loire
Ni 8, Mn 8, Cr 4, C, bal Fe.
For corrosion and heat resistant parts; corrosion and heat resistant. *Obsolete*

IMPHY N42
Creusot-Loire
Ni 42, bal Fe.
For electrical equipment; controlled expansion. *Obsolete*

IMPHY N48
Creusot-Loire
Ni 48, bal Fe.
For glass sealing electronic rectifiers. Stepped expansion. *Obsolete*

IMPHY N54
Creusot-Loire
Ni 54, bal Fe.
For glass to metal seals. Stepped expansion. *Obsolete*

IMPHY N58
Creusot-Loire
Ni 58, bal Fe.
For electrical equipment; controlled expansion. *Obsolete*

IMPHY N5CM
Creusot-Loire
Ni 5, C, Si, Mn, bal Fe.
69,000-83,000 TS; 50,000-60,000 YS; 24-30 El. For case hardened parts, gun tubes, locomotive parts; shock resistant. *Obsolete*

IMPHY NB
Creusot-Loire
Ni 3, C, Si, Mn, bal Fe.
81,000-96,000 TS; 54,000-67,000 YS; 20-25 El. For crankshafts, transmission shafts, forgings; shock resistant. *Obsolete*

IMPHY NC-3H2
Creusot-Loire
Ni 3, Cr 1, C, bal Fe.
96,000-108,000 TS; 63,000-77,000 YS; 18-24 El. For pinions, crankshafts, forging dies, auto construction; shock resistant. *Obsolete*

IMPHY NC3H1
Creusot-Loire
C 0.2, Ni 3, Cr 0.7, bal Fe.
Annealed: 92,000 TS; 58,000 YS; 20 El. Heat treated: 157,000 TS; 143,000 YS; 10 El. For arbors, axles, shafts, gears, spindles; Afnor 20NC11; oil hardened, case hardening. *Obsolete*

IMPHY NC3H10R

Creusot-Loire
C 0.1, Ni 1.4, Cr 1, bal Fe.
Annealed: 83,000 TS; 50,000 YS; 22 El. Heat treated: 164,000 TS; 127,000 YS; 9 El. For case hardened gears, shafts; case hardened; Afnor 10NC6. *Obsolete*

IMPHY NC3H16R

Creusot-Loire
C 0.16, Ni 1.4, Cr 1, bal Fe.
Annealed: 85,000 TS; 58,000 YS; 20 El. Heat treated: 193,000 TS; 157,000 YS; 9 El. For case hardened gears, shafts; case hardened; Afnor 16NC6. *Obsolete*

IMPHY NC3H1R

Creusot-Loire
C 0.25, Ni 1.4, Cr 1, bal Fe.
Annealed: 92,000 TS; 58,000 YS; 18 El. Heat treated: 120,000 TS; 107,000 YS; 12 El. For arbors, shafts, axles, spindles; Afnor 25NC6; oil or water hardened. *Obsolete*

IMPHY NC3H2OR

Creusot-Loire
C 0.18, Ni 3.5, Cr 1.5, Mo 0.4, bal Fe.
Annealed: 107,000 TS; 72,000 YS; 15 El. Heat treated: 200,000 TS; 150,000 YS; 8 El. For gears, shafts, cams; case hardened; Afnor 18NCD6. *Obsolete*

IMPHY NC3H2R

Creusot-Loire
C 0.35, Ni 1.4, Cr 1, bal Fe.
Annealed: 100,000 TS; 65,000 YS; 17 El. Heat treated: 242,000 TS; 213,000 YS; 5 El. For arbors, axles, shafts, spindles; Afnor 35NC6; oil hardened. *Obsolete*

IMPHY NC3H8R

Creusot-Loire
C 0.08, Ni 1.4, Cr 1, bal Fe.
Heat treated: 78,000-128,000 TS; 43,000-85,000 YS; 12-25 El. For case hardened parts, gears, shafts; case hardened; Afnor 8NC6. *Obsolete*

IMPHY NC3HO

Creusot-Loire
C 0.14, Ni 2.7, Cr 0.8, bal Fe.
Annealed: 82,000 TS; 52,000 YS; 22 El. Heat treated: 170,000 TS; 143,000 YS; 10 El. For case hardened gears, shafts; case hardened; Afnor 14NC11. *Obsolete*

IMPHY NCM

Creusot-Loire
C 0.08, Ni 1.3-2.3, bal Fe.
Heat treated: 60,000-86,000 TS; 46,000-58,000 YS; 20-32 El. For case hardened parts, gears, shafts; case hardened; Afnor 8N8. *Obsolete*

IMPHY NFCO

Creusot-Loire
C 0.08, Ni, Cr, bal Fe.
Annealed: 78,000 TS; 50,000 YS; 25 El. Heat treated: 143,000 TS; 100,000 YS; 11 El. For case hardened gears, shafts; case hardened; Afnor 8NC13. *Obsolete*

IMPHY NMHG

Creusot-Loire
Ni, bal Fe.
For shunts, tachometers; variable Curie temperature from -5 C to +200 C. *Obsolete*

IMPHY NMHG

Creusot-Loire
Fe 70, Ni 30.
For compensating shunts, electrical devices; controlled expansion. *Obsolete*

IMPHY NY

Creusot-Loire
Ni 6, C, Si, Mn, bal Fe.
78,000-107,000 TS; 58,000-85,000 YS; 18-25 El. For axles, gears, shafts; tough. *Obsolete*

IMPHY R23

Creusot-Loire
C 0.7, W 6.5, Mo 5, Cr 4, V 1, bal Fe.
Hardened: C 64-66 Rock. For cutting tools, lathe and planer tools, drills, milling tools, broaches. High speed steel. High red-hardness. *Obsolete*

IMPHY R24

Creusot-Loire
C 0.7, W 10, Mo 1, Cr 4, V 1, bal Fe.
Hardened: C 64-66 Rock. For cutting tools, lathe and planer tools, drills, reamers. High speed steel, high red-hardness. *Obsolete*

IMPHY RAPIDE ULTRA

Henry A. Kries & Sons Co.
C 0.7, W 19, Cr 4, V 2.2, bal Fe.
For tools, dies, cutters; high speed steel.. *Obsolete*

IMPHY RCA 33

Creusot-Loire
C, Cr, Al, bal Fe.
For resistances, radiators; good to 1200 C application. *Obsolete*

IMPHY RCA 33

Creusot-Loire
Cr 20, Al 5, bal Fe.
For electrical resistances; heat resistant. *Obsolete*

IMPHY RCA 44

Creusot-Loire
C, Cr, Al, bal Fe.
For heating elements, radiators; good to 1200 C. *Obsolete*

IMPHY RCA 44

Creusot-Loire
Cr 32, Al 5, bal Fe.
For electrical resistances; heat resistant. *Obsolete*

IMPHY RES

Creusot-Loire
Si 1.5, C, Mn, bal Fe.
Annealed: 107,000 TS; 65,000 YS; 20 El. Heat treated: 186,000 TS; 164,000 YS; 7 El. For springs; tough. *Obsolete*

IMPHY RNC 2

Creusot-Loire
Ni 60, Cr 11, low C, bal Fe.
For rheostats, electrically heated apparatus; heat resistant. *Obsolete*

IMPHY RNC 3

Creusot-Loire
Ni 80, Cr 20, Low C.
For rheostats, electrically heated apparatus; heat resistant to 1100 C. *Obsolete*

IMPHY RNC 30

Creusot-Loire
Ni 40, Cr 22, bal Fe.
For electrical resistances for industrial equipment; heat and corrosion resistant. *Obsolete*

IMPHY RNC 44

Creusot-Loire
Al 5, Cr 30, bal Fe.
For electrical heating resistances; heat resistant. *Obsolete*

IMPHY RNCI

Creusot-Loire
Ni 36, Cr 10, C, bal Fe.
For rheostats, electrical heating units, metallic heat screens; heat resistant to 600 C. *Obsolete*

IMPHY RNCO

Creusot-Loire
Ni 12, Cr 12, bal Fe.
For electrical resistances, rheostats; heat and corrosion resistant. *Obsolete*

IMPHY RWS

Creusot-Loire
C, Si, Mn, bal Fe.
Annealed: 107,000 TS; 69,000 YS; 22 El. Heat treated: 193,000 TS; 171,000 YS; 7 El. For motor car springs; tough. *Obsolete*

IMPHY TRIPLE RAPIDE

Henry A. Kries & Sons Co.
C 0.7, W 18, Co 5, Cr 4, V 1.3, bal Fe.
For tools, dies, cutters; high speed steel. *Obsolete*

IMPHY TUNGSTENE

Henry A. Kries & Sons Co.
C 1.3, W 4.5, bal Fe.
For tools, dies; oil hardened. *Obsolete*

IMPHY UM CO50

Creusot-Loire
C 0.05-0.12, Ti 0.18, Cb 0.6, Cr 26-30, Co 47-52, bal Fe.
Cast: 135,000 TS; 48,000 YS; 10 El; 10 RA; 250 Brin. Wrought: 132,000 TS; 61,000 YS; 7 El; 6 RA; 350 Brin. For furnace baffles, burner tips, sintering grates, quench baskets. Corrosion, heat and thermal shock resistant. *Obsolete*

IMPHY VY2-SP

Creusot-Loire
C 0.4, Ni 4.6, Cr 1.6, Mo 0.5, bal Fe.
Heat treated: 300,000 TS; 230,000 YS; 5 El. For crankshafts, gears; Afnor 40NCD18; oil hardened. *Obsolete*

IMPHY VYP

Creusot-Loire
C 0.3, Ni 4, Cr 1.4, Mo 0.5, bal Fe.
Annealed: 128,000 TS. Heat treated: 250,000 TS; 212,000 YS; 6 El. For crankshafts, gears, bolts; Afnor 30NCD16; oil hardened. *Obsolete*

IMPHY ZCR 716C

Creusot-Loire
Ni 10-12, Cr 8-11, Mo 2-2.5, Al 2-3.5, C 0.05, Si 0.13, Mn 0.47, bal Fe.
Maraged: 230,000 TS; 219,000 YS; 11.3 El; 56 RA. For aircraft landing gears, special springs, used in marine atmospheres. Maraging steel. High yield strength and corrosion resistant. *Obsolete*

IMPHYCHROME

Creusot-Loire
Cr 2.5, C, bal Fe.
For turning tools, planers, drills, punches, cold dies, files; extremely hard. *Obsolete*

IMPHYCORROYE

Creusot-Loire
C, alloy, bal Fe.
For axles, hooks, cutlery, carpenter tools. *Obsolete*

IMPHYNATUREL

Creusot-Loire
C, bal Fe.
For cutlery, picks, axes, wood tools, pistons. *Obsolete*

IMPHYNUSABLE

Creusot-Loire
C, Cr, bal Fe.
For tools, gages, saws, turning tools, planers; resistance t wear and shock. *Obsolete*

IMPHYRAPIDE

Creusot-Loire
Cr 5.5, W 10-12, C, Mn, Si, bal Fe.
For turning tools, high speed planers. *Obsolete*

IMPHYRAPIDE ULTRA

Creusot-Loire
Cr 6, V 16-20, C, Mn, bal Fe.
For high speed tools. *Obsolete*

IMPHYSIL
Creusot-Loire
Si, bal Fe.
For motors, transformers; soft magnet for strong or moderate field. *Obsolete*

IMPHYSTA
Creusot-Loire
C, bal Fe.
For airplane construction; class of Imphy alloys. *Obsolete*

IMPHYTUNGSTENE
Creusot-Loire
Cr 0.5, W 6-8, C, bal Fe.
For tools, planers, circular saws, borers. *Obsolete*

IMPRESS
Boyd-Wagner Co.
C 0.45, Si 1, Cr 1.4, W 2.2, V 0.2, Mo 0.3, bal Fe.
For impression dies; nondeforming.

IMPROVED ELECTRIC
Follansbee Steel Co.
C 0.04, Si 0.1, bal Fe.
Annealed: 50,000 TS; 30,000 YS; 20 El. For small motors; good magnetic properties. *Obsolete*

IMPROVED KS
Japanese manufacture
Ni 20, Ti 15, Co 20, bal Fe.
For magnetic and electrical equipment; permanent magnet.

IMPROVITE
Illingworth Steel Co.
Ni 66-68, Si 3-4, Cu 28-31.
For introducing Ni and Cu into cast iron. *Obsolete*

IN 738
Huntington Alloys Inc.
C 0.17, Cr 16, Co 8.5, Mo 1.75, W 2.6, Cb 0.9, Ti 3.4, Al 3.4, B 0.01, Zr 0.1, Ta 1.75, bal Ni.
Cast good hot corrosion resistant alloy.

IN 738
Cannon-Muskegon Corp.
C 0.17, Cr 16, Co 8.5, Mo 1.75, W 2.6, Cb 0.9, Ti 3.4, Al 3.4, B 0.01, Zr 0.1, Ta 1.75, bal Ni.
Cast good hot corrosion resistant alloy.

IN 738
International Nickel Inc.
C 0.17, Cr 16, Co 8.5, Mo 1.75, W 2.6, Cb 0.9, Ti 3.4, Al 3.4, B 0.01, Zr 0.1, Ta 1.75, bal Ni.
Cast good hot corrosion resistant alloy.

IN-100
Ishikawajima-Harima Heavy Industries Co.
C 0.16, Si 0.25, Mn 0.85, Ni 1.2, Cr 0.6, Mo 0.55, V 0.07, Cu 0.25, N 0.06, bal Fe.
Heat treated: 144,200 TS; 138,000 YS; 20 El. For mine cars, bus and truck bodies, booms, derricks. High-strength low-alloy steel, tough and shock resistant.

IN-100
Cannon-Muskegon Corp.
Cr 8-11, Co 13-17, Mo 2-4, V 0.7-1.2, Al 5-6, Ti 4.5-5.5, C 0.18, B 0.01, Zr 0.06, bal Ni.
Cast: 142,000 TS; 122,000 YS; 8 El. At 1800°F: 75,000 TS; 60,000 YS; 5 El. For jet engine parts, turbine blades; for operating temperatures up to 1900°F.

IN-100
Yawata Iron & Steel Co., Ltd.
C 0.16, Si 0.25, Mn 0.85, Ni 1.2, Cr 0.6, Mo 0.55, V 0.07, Cu 0.25, N 0.06, bal Fe.
Heat treated: 144,200 TS; 138,000 YS; 20 El. For mine cars, bus and truck bodies, booms, derricks. High strength, low alloy steel, tough and shock resistant.

IN-102
Cannon-Muskegon Corp.
Fe 7, Cr 15, Cb 3, Mo 3, W 3, Al 0.5, Ti 0.5, C 0.06, B, Zr, bal Ni.
Annealed: 130,000 TS; 60,000 YS; 45 El. At 1200°F: 103,000 TS; 50,000 YS; 52 El; 42 RA. For high-temperature, high-pressure steam turbines; for long service up to 1200°F, high strength and ductility.

IN-162
Cannon-Muskegon Corp.
C 0.12, Cr 10, Mo 4, W 2, Ti 1, Cb 1, Al 6.5, Ta 2, B 0.02, Zr 0.1, bal Ni.
Cast superalloy; for gas turbine blades and vanes.

IN-40
Ishikawajima-Harima Heavy Industries Co.
C 0.1, Si 0.25, Mn 1, N 0.04-0.075, bal Fe.
Rolled: 69,100 TS; 48,600 YS; 32 El. Heat treated: 60,700 TS; 45,800 YS; 44 El; 79 RA. For bridges, structures, booms, bus bodies. Tough and readily welded.

IN-40
Yawata Iron & Steel Co., Ltd.
C 0.1, Si 0.25, Mn 1, N 0.04-0.075, bal Fe.
Rolled: 69,100 TS; 48,600 YS; 32 El. Heat treated: 60,700 TS; 45,800 YS; 44 El; 79 RA. For bridges, structures, booms, bus bodies. Tough and readily welded.

IN-50
Ishikawajima-Harima Heavy Industries Co.
C 0.14, Si 0.3, Mn 1.4, N 0.04-0.075, bal Fe.
Rolled: 77,500 TS; 54,400 YS; 29 El. Heat treated: 72,100 TS; 56,000 YS; 38 El; 75.6 RA. For bridges, structures, booms, mine and railroad cars. Good fabricability and weldability.

IN-50
Yawata Iron & Steel Co., Ltd.
C 0.14, Si 0.3, Mn 1.4, N 0.04-0.075, bal Fe.
Rolled: 77,500 TS; 54,400 YS; 29 El. Heat treated: 72,100 TS; 56,000 YS; 38 El; 75.6 RA. For bridges, structures, booms, mine and railroad cars. Good fabricability and weldability.

IN-60
Ishikawajima-Harima Heavy Industries Co.
C 0.17, Si 0.35, Mn 1.3, Cr 0.3, N 0.04-0.075, bal Fe.
Rolled: 87,300 TS; 59,700 YS; 35 El; 69 RA. Heat treated: 90,700 TS; 72,800 YS; 32 El; 68.5 RA. For bridges, structures, bus and truck bodies, booms, mine cars. High-strength low-alloy steel. Good weldability.

IN-60
Yawata Iron & Steel Co., Ltd.
C 0.17, Si 0.35, Mn 1.3, Cr 0.3, N 0.04-0.075, bal Fe.
Rolled: 87,300 TS; 59,700 YS; 35 El; 69 RA. Heat treated: 90,700 TS; 72,800 YS; 32 El; 68.5 RA. For bridges, structures, bus and truck bodies, booms, mine cars. High strength, low alloy steel. Good weldability.

IN-643
International Nickel Inc.
Nickel. C 0.5, Si 0.3, Fe 3, Cr 25, Co 12, W 9, Mo 0.5, Nb 2, Ti, Zr, Mg, bal Ni.
Cast: 560-600 MPa TS; 280-300 MPa YS; 8-15 El. Heat resisting alloy for high temperature operation including carburization equipment. *Obsolete*

IN-731
Huntington Alloys Inc.
C 0.18, Mn 0-0.2, Si 0-0.2, Cr 9.5, Co 10, Mo 2.5, Ti 4.65, Al 5.5, B 0.015, Zr 0.06, Fe 0-0.5, V 0.95, bal Ni.
Modified IN-100. Jet engine blades and wheels.

IN-731
Cannon-Muskegon Corp.
C 0.18, Mn 0-0.2, Si 0-0.2, Cr 9.5, Co 10, Mo 2.5, Ti 4.65, Al 5.5, B 0.015, Zr 0.06, Fe 0-0.5, V 0.95, bal Ni.
Modified IN-100. Jet engine blades and wheels.

IN-731
International Nickel Inc.
C 0.18, Mn 0-0.2, Si 0-0.2, Cr 9.5, Co 10, Mo 2.5, Ti 4.65, Al 5.5, B 0.015, Zr 0.06, Fe 0-0.5, V 0.95, bal Ni.
Modified IN-100. Jet engine blades and wheels.

IN-744
International Nickel Inc.
Nickel. C 0-0.06, Cr 26, Ni 6.5, Mn 0.4, Si 0.4, Ti = 5 x C, bal Fe.
At 1000°F: 53,000 psi TS; 45,000 YS psi; 27 El. Corrosion resistant, high temperature properties, weldable. *Obsolete*

IN-787
Titanium Metals Corp.
C 0.05, Mn 0.55, Si 0.26, Ni 0.91, Cr 0.75, Mo 0.2, Cu 1.23, Cb 0.06, bal Fe.
Age-hardened: 78,000-88,000 psi TS; 65,000-75,000 psi YS; 28 El; 78 RA; 180-200 Charpy (at -50°F). Weld neck flanges for low temperature service.

IN-787
International Nickel Inc.
Alloy steel. C 0.04, Mn 0.5, Si 0.3, Ni 0.9, Cr 0.6, Mo 0.2, Cu 1.2, Cb 0.05, bal Fe.
Precipitation hardened and aged: 90,000 TS; 85,000 YS; 25 El. Plate or welded pipe for gas or liquid transmission; low temperature properties.

IN-792
International Nickel Inc.
Cr 12.7, Co 9, Mo 2, W 3.9, Ta 3.9, Al 3.2, Ti 4.2, B 0.02, Zr 0.1, C 0.21, bal Ni.
1600°F: 115,000 TS; 85,000 YS; 15 El. Good high stress rupture properties. (See also PM IN-792.)

IN-792
Cannon-Muskegon Corp.
Cr 12.7, Co 9, Mo 2, W 3.9, Ta 3.9, Al 3.2, Ti 4.2, B 0.02, Zr 0.1, C 0.21, bal Ni.
1600°F: 115,000 TS; 85,000 YS; 15 El. Good high stress rupture properties. (See also PM IN-792).

IN-80
Ishikawajima-Harima Heavy Industries Co.
C 0.16, Si 0.25, Mn 0.85, Cr 1, Mo 0.4, Cu 0.25, N 0.04-0.075, bal Fe.
Heat treated: 107,000 TS; 102,000 YS; 23 El. For structures, buildings, bridges, mine cars, truck and bus bodies. High-strength low-alloy steel, good weldability.

IN-80
Yawata Iron & Steel Co., Ltd.
C 0.16, Si 0.25, Mn 0.85, Cr 1, Mo 0.4, Cu 0.25, N 0.04-0.075, bal Fe.
Heat treated: 107,000 TS; 102,000 YS; 23 El. For structures, buildings, bridges, mine cars, truck and bus bodies. High strength, low alloy steel, good weldability.

IN-80N
Ishikawajima-Harima Heavy Industries Co.
C 0.15, Si 0.25, Mn 0.4, Ni 2.25, Cr 1.2, Mo 0.35, N 0.04-0.075, bal Fe.
Heat treated: 120,600 TS; 105,200 YS; 27 El; 73 RA. For mine cars, crushers, bridges, truck and bus bodies, booms. High-strength low-alloy steel, good weldability.

IN-80N
Yawata Iron & Steel Co., Ltd.
C 0.15, Si 0.25, Mn 0.4, Ni 2.25, Cr 1.2, Mo 0.35, N 0.04-0.075, bal Fe.
Heat treated: 120,600 TS; 105,200 YS; 27 El; 73 RA. For mine cars, crushers, bridges, truck and bus bodies, booms. High strength, low alloy steel, good weldability.

IN-80V
Ishikawajima-Harima Heavy Industries Co.
C 0.14, Si 0.25, Mn 0.85, Ni 0.85, Cr 0.5, Mo 0.45, V 0.07, Cu 0.3, N 0.06, bal Fe.
Heat treated: 113,000 TS; 103,000 YS; 24 El; 67 RA. For bridges, booms, structures, truck and bus bodies, mine cars. High-strength low-alloy steel, tough and shock resistant.

IN-80V
Yawata Iron & Steel Co., Ltd.
C 0.14, Si 0.25, Mn 0.85, Ni 0.85, Cr 0.5, Mo 0.45, V 0.07, Cu 0.3, N 0.06, bal Fe.
Heat treated: 113,000 TS; 103,000 YS; 24 El; 67 RA. For bridges, booms, structures, truck and bus bodies, mine cars. High strength, low alloy steel, tough and shock resistant.

IN-833
International Nickel Inc.
Stainless steel. C 0-0.03, Si 0-1, Mn 0-0.5, Ni 7, Cr 11.5, Al 0.1, Ti 0.1, bal Fe.
Cast and aged: 120,000-150,000 psi TS; 100,000-137,000 psi YS; 14-19 El. Age hardening, weldable, corrosion resistant; good elevated temperature properties. *Obsolete*

IN-853 (MA-753)
Cannon-Muskegon Corp.
C 0.06, Cr 19.7, Ti 2.3, Al 0.88, B 0.007, Zr 0.07, 1.18 Y_2O_3, 1.12 Al_2O_3, bal Ni.
Mechanically alloyed powder, for turbine blades and vanes.

IN-FLUX 2209-O
Teledyne McKay
Nickel. C 0.02, Mn 1.6, Si 0.4, Cr 22, Ni 8.5, Mo 3.3, N 0.14, bal Fe.
Stainless steel and nickel base flux cored open arc welding wire. 116,000 psi TS; 86,000 psi YS; 30 El. Austenitic-ferritic duplex. For joining duplex stainless Alloy 2205, carbon steel or low alloy steel.

IN-FLUX 2253-O
Teledyne McKay
Nickel.
Now IN-FLUX 2209-O.

IN-FLUX 2553CU-O
Teledyne McKay
Nickel.
Now IN-FLUX 259-O.

IN-FLUX 259-O
Teledyne McKay
Nickel. C 0.02, Mn 1, Si 0.3, Cr 25, Ni 10, Mo 3.2, Cu 2, N 0.14, bal Fe.
Stainless steel and nickel base flux cored open arc welding wire. 122,000 psi TS; 91,000 psi YS; 28 El. Austenitic-ferritic duplex; controlled ferrite. For joining duplex stainless to carbon or low alloy steel and for cladding.

IN-FLUX 308-G/S
Teledyne McKay
Stainless steel. C 0.06, Mn 1.4, Si 0.4, Cr 20.5, Ni 10, bal Fe.
Stainless steel metal cored gas shielded submerged arc welding wire. 86,000 psi TS; 60,000 psi YS; 35 El. AWS ER308.

IN-FLUX 308-O
Teledyne McKay
Stainless steel. C 0.05, Mn 1.2, Si 0.5, Cr 20.5, Ni 9.9, bal Fe.
Stainless steel and nickel base flux cored open arc welding wire. 94,000 psi TS; 70,000 psi YS; 40 El. Austenitic; controlled ferrite. For joining common austenitic steels. AWS E308T-3.

IN-FLUX 308L-G/S
Teledyne McKay
Stainless steel. C 0.02, Mn 1.7, Si 0.4, Cr 20, Ni 10.3, bal Fe.
Stainless steel metal cored gas shielded submerged arc welding wire. 84,000 psi TS; 58,000 psi YS; 37 El. AWS ER308L.

IN-FLUX 308L-O
Stainless steel. C 0.02, Mn 1.2, Si 0.5, Cr 20.3, Ni 10.2, bal Fe.
Stainless steel and nickel base flux cored open arc welding wire. 94,000 psi TS; 70,000 psi YS; 40 El. Austenitic; controlled ferrite. For joining common austenitic steels; intermediate layer for hard surfacing. AWS E308LT-3.

IN-FLUX 308L-T1
Teledyne McKay
Stainless steel. C 0.03, Mn 1.2, Si 0.6, Cr 19.5, Ni 9.5, bal Fe.
Stainless steel and nickel base flux cored gas shielded welding wire. 85,000 psi TS; 58,000 psi YS; 35 El. Austenitic; controlled ferrite. For joining common austenitic stainless steels. AWS E308LT-1.

IN-FLUX 309 L-T1
Teledyne McKay
Stainless steel. C 0.03, Mn 1.2, Si 0.6, Cr 24.2, Ni 12.5, bal Fe.
Stainless steel and nickel base flux cored gas shielded welding wire. 90,000 psi TS; 60,000 psi YS; 34 El. Austenitic; controlled ferrite. For joining common austenitic stainless steels. AWS E309LT-1.

IN-FLUX 309CBL-O
Teledyne McKay
Stainless steel. C 0.02, Mn 1.2, Si 0.4, Cr 24, Ni 12.9, 0.85 Cb + Ta, bal Fe.
Stainless steel and nickel base flux cored open arc welding wire. 92,000 psi TS; 70,000 psi YS; 40 El. Austenitic; controlled ferrite. For first layer substrate for cladding carbon or low-alloy Type 347 steel. AWS E309CbLT-3.

IN-FLUX 309L-G/S
Teledyne McKay
Stainless steel. C 0.02, Mn 1.8, Si 0.5, Cr 24, Ni 13, bal Fe.
Stainless steel metal cored gas shielded submerged arc welding wire. 81,000 psi TS; 61,000 psi YS; 35 El. AWS ER309L.

IN-FLUX 309L-O
Teledyne McKay
Stainless steel. C 0.02, Mn 1.4, Si 0.5, Cr 23.8, Ni 12.7, bal Fe.
Stainless steel and nickel base flux cored open arc welding wire. 91,000 psi TS; 70,000 psi YS; 40 El. Austenitic; controlled ferrite. For joining common austenitic stainless steels. AWS E309LT-3.

IN-FLUX 309LCB-G/S
Teledyne McKay
Stainless steel. C 0.02, Mn 1.8, Si 0.5, Cr 23.1, Ni 12.9, Cb 0.8, bal Fe.
Stainless steel metal cored gas shielded submerged arc welding wire. 82,000 psi TS; 60,000 psi YS; 35 El.

IN-FLUX 309LMO-G/S
Teledyne McKay
Stainless steel. C 0.02, Mn 1.6, Si 0.5, Cr 23, Ni 12.8, Mo 2.5, bal Fe.
Stainless steel metal cored gas shielded submerged arc welding wire. 85,000 psi TS; 63,000 psi YS; 35 El.

IN-FLUX 309LMO-T1
Teledyne McKay
Stainless steel. C 0.03, Mn 1.2, Si 0.6, Cr 23, Ni 12.5, Mo 2.3, bal Fe.
Stainless steel and nickel base flux cored gas shielded welding wire. 90,000 psi TS; 60,000 psi YS; 34 El. Austenitic; controlled ferrite. For joining stainless steels to carbon or low alloy steels.

IN-FLUX 310-G/S
Teledyne McKay
Stainless steel. C 0.12, Mn 1.6, Si 0.5, Cr 26.7, Ni 21.4, bal Fe.
Stainless steel metal cored gas shielded submerged arc welding wire. 85,000 psi TS; 62,000 psi YS; 40 El. AWS ER310.

IN-FLUX 310-O
Teledyne McKay
Stainless steel. C 0.1, Mn 2.3, Si 0.4, Cr 26.1, Ni 20.7, bal Fe.
Stainless steel and nickel base flux cored open arc welding wire. 86,000 psi TS; 63,000 psi YS; 25 El. Austenitic; for joining Type 310 base meal. AWS E310T-3.

IN-FLUX 310HC-O
Teledyne McKay
Stainless steel. C 0.4, Mn 2.3, Si 0.4, Cr 26.1, Ni 20.7, bal Fe.
Stainless steel and nickel base flux cored open arc welding wire. Austenitic; for welding ACI Type HK castings.

IN-FLUX 312-G/S
Teledyne McKay
Stainless steel. C 0.12, Mn 1.8, Si 0.5, Cr 29, Ni 9, bal Fe.
Stainless steel metal cored gas shielded submerged arc welding wire. 110,000 psi TS; 93,000 psi YS; 25 El. AWS ER312.

IN-FLUX 312-O
Teledyne McKay
Stainless steel. C 0.08, Mn 1.5, Si 0.3, Cr 29.5, Ni 9, bal Fe.
Stainless steel and nickel base flux cored open arc welding wire. 115,000 psi TS; 95,000 psi YS; 25 El. Mixed ferritic-austenitic. AWS E312T-3.

IN-FLUX 312-T1
Teledyne McKay
Stainless steel. C 0.11, Mn 1.2, Si 0.6, Cr 29, Ni 9, bal Fe.
Stainless steel and nickel base flux cored gas shielded welding wire. 100,000 psi TS; 77,000 psi YS; 24 El. Austenitic; controlled ferrite. For joining dissimilar steels. AWS E312T-1.

IN-FLUX 316L-G/S
Teledyne McKay
Stainless steel. C 0.02, Mn 1.6, Si 0.4, Cr 19, Ni 12.8, Mo 2.3, bal Fe.
Stainless steel metal cored gas shielded submerged arc welding wire. 79,000 psi TS; 52,000 psi YS; 40 El. AWS ER316L.

IN-FLUX 316L-O
Teledyne McKay
Stainless steel. C 0.02, Mn 1.9, Si 0.5, Cr 18.9, Ni 12, Mo 2.4, bal Fe.
Stainless steel and nickel base flux cored open arc welding wire. 89,000 psi TS; 68,000 psi YS; 40 El. Austenitic; for joining Types 316, 316L, CF-8M, and CF-3M stainless steels. AWS E316LT-3.

IN-FLUX 316L-T1
Teledyne McKay
Stainless steel. C 0.03, Mn 1.2, Si 0.6, Cr 19, Ni 12, Mo 2.3, bal Fe.
Stainless steel and nickel base flux cored gas shielded welding wire. 85,000 psi TS; 58,000 psi YS; 35 El. Austenitic; controlled ferrite. For joining Types 316, 316L, CF-8M, and CF-3M stainless steels. AWS E316LT-1.

IN-FLUX 317L-G/S
Teledyne McKay
Stainless steel. C 0.02, Mn 1.4, Si 0.4, Cr 19.5, Ni 14, Mo 3.5, bal Fe.
Stainless steel metal cored gas shielded submerged arc welding wire. 84,000 psi TS; 55,000 psi YS; 39 El. AWS ER317L.

IN-FLUX 317L-O
Teledyne McKay
Stainless steel. C 0.02, Mn 1.5, Si 0.5, Cr 20, Ni 13.5, Mo 3.3, bal Fe.
Stainless steel and nickel base flux cored open arc welding wire. 90,000 psi TS; 69,000 psi YS; 36 El. Austenitic; controlled ferrite. For joining highly corrosion resistant stainless steels. AWS E317LT-3.

IN-FLUX 347-T1

Teledyne McKay
Stainless steel. C 0.05, Mn 1.2, Si 0.6, Cr 19.5, Ni 9.5, Cb 0.6, bal Fe.
Stainless steel and nickel base flux cored gas shielded welding wire. 85,000 psi TS; 58,000 psi YS; 35 El. Austenitic; controlled ferrite. For joining stabilized stainless steels. AWS E347T-1.

IN-FLUX 347L-G/S

Teledyne McKay
Stainless steel. C 0.04, Mn 1.8, Si 0.4, Cr 21, Ni 10, Cb 0.7, bal Fe.
Stainless steel metal cored gas shielded submerged arc welding wire. 84,000 psi TS; 57,500 psi YS; 42 El. AWS ER347.

IN-FLUX 347L-O

Teledyne McKay
Stainless steel. C 0.02, Mn 1.1, Si 0.6, Cr 19.9, Ni 9.8, 0.6 Cb + Ta, bal Fe.
Stainless steel and nickel base flux cored open arc welding wire. 94,000 psi TS; 70,000 psi YS; 40 El. Austenitic; controlled ferrite. For joining stabilized stainless steels, Types 321 and 347. AWS E347T-3.

IN-FLUX 409-G/S

Teledyne McKay
Stainless steel. C 0.05, Mn 0.5, Si 0.3, Cr 12, Ti 0.7, bal Fe.
Stainless steel metal cored gas shielded submerged arc welding wire. 55,000 psi TS; 42,000 psi YS; 30 El.

IN-FLUX 409-O

Teledyne McKay
Stainless steel. C 0.04, Mn 0.6, Si 0.6, Cr 12, Ti 0.7, bal Fe.
Stainless steel and nickel base flux cored open arc welding wire. Ferritic; for joining Type 409 stainless steel. AWS E409T-3.

IN-FLUX 410-G/S

Teledyne McKay
Stainless steel. C 0.1, Mn 0.4, Si 0.2, Cr 12.3, bal Fe.
Stainless steel metal cored gas shielded submerged arc welding wire. Stress relieved 1 h at 1575°F: 75,000 psi TS; 55,000 psi YS; 22 El. AWS ER410.

IN-FLUX 410-O

Teledyne McKay
Stainless steel. C 0.09, Mn 0.5, Si 0.3, Cr 12, bal Fe.
Stainless steel and nickel base flux cored open arc welding wire. 145,000 psi TS; 110,000 psi YS; 8 El. Martensitic; for joining Types 410 and CA-15 stainless steels. AWS E410T-3.

IN-FLUX 410NIMO-G/S

Teledyne McKay
Stainless steel. C 0.02, Mn 0.5, Si 0.3, Cr 12, Ni 4.5, Mo 0.5, bal Fe.
Stainless steel metal cored gas shielded submerged arc welding wire. Stress relieved 1 h at 1150°F: 120,000 psi TS; 100,000 psi YS; 15 El. AWS E410NiMo.

IN-FLUX 410NIMO-O

Teledyne McKay
Nickel. C 0.03, Mn 0.5, Si 0.3, Cr 11.6, Ni 4.4, Mo 0.5, bal Fe.
Stainless steel and nickel base flux cored open arc welding wire. 131,000 psi TS; 111,000 psi YS; 21 El. Martensitic; for joining Type CA-6NB. AWS E410NiMoT-3.

IN-FLUX 410NIMO-T1

Teledyne McKay
Stainless steel. C 0.03, Mn 0.5, Si 0.4, Cr 11.8, Ni 4.5, Mo 0.6, bal Fe.
Stainless steel and nickel base flux cored gas shielded welding wire. 131,000 psi TS; 111,000 psi YS; 21 El. Low carbon; martensitic. For joining Type CA-6NM stainless steel castings and joining Types 409, 410, 410S, and 405 stainless steels. AWS E410NiMoT-1.

IN-FLUX 430-G/S

Teledyne McKay
Stainless steel. C 0.06, Mn 0.4, Si 0.3, Cr 16.3, bal Fe.
Stainless steel metal cored gas shielded submerged arc welding wire. Stress relieved 2 h at 1425°F: 78,000 psi TS; 58,000 psi YS; 22 El. AWS ER430.

IN-FLUX 4K-O

Teledyne McKay
Stainless steel. C 0.02, Mn 2.25, Si 0.3, Cr 17.8, Ni 13.5, Mo 2.2, bal Fe.
Stainless steel and nickel base flux cored open arc welding wire. 75,000 psi TS; 51,000 psi YS; 39 El. Austenitic; low controlled ferrite. For joining Types 304LN and 316LN for low temperature service. AWS E316KT-3.

IN-FLUX 625-O

Teledyne McKay
Nickel. C 0.02, Mn 0.2, Si 0.2, Cr 20.5, Mo 8.4, Cb 3.5, Fe 2, bal Ni.
Stainless steel and nickel base flux cored open arc welding wire. 103,000 psi TS; 62,000 psi YS; 35 El. For cladding of mild and low alloy steels. AWS ENiCrMo3T-3.

IN-FLUX 82-O

Teledyne McKay
Nickel. C 0.02, Mn 3.2, Si 0.2, Cr 20, Cb 2.4, Fe 1.3, bal Ni.
Stainless steel and nickel base flux cored open arc welding wire. 84,500 psi TS; 53,000 psi YS; 26 El. For cladding of mild and low alloy steels. AWS ENiCrMo3T-3.

IN-FLUX A9-G/S

Teledyne McKay
Stainless steel. C 0.1, Mn 1.6, Si 0.5, Cr 19.5, Ni 9.5, Mo 2, bal Fe.
Stainless steel metal cored gas shielded submerged arc welding wire. 90,000 psi TS; 68,000 psi YS; 38 El.

IN-FLUX A9-O

Teledyne McKay
Stainless steel. C 0.1, Mn 1.7, Si 0.6, Cr 20.5, Ni 9.9, Mo 2.1, bal Fe.
Stainless steel and nickel base flux cored open arc welding wire. 95,000 psi TS; 73,000 psi YS; 38 El. Austenitic; controlled ferrite. For joining armor steel to armor steel, to mild or low alloy steel, or to stainless steel.

IN-FLUX A9-T1

Teledyne McKay
Stainless steel. C 0.09, Mn 1.4, Si 0.6, Cr 21, Ni 9.5, Mo 2.2, bal Fe.
Stainless steel and nickel base flux cored gas shielded welding wire. 95,000 psi TS; 65,000 psi YS; 30 El. Austenitic; controlled ferrite. For joining armor steels and difficult-to-weld steels.

IN-FLUX NICR3-O

Teledyne McKay
Nickel.
Now IN-FLUX 82-O.

IN-FLUX NICRMO3-O

Teledyne McKay
Nickel.
Now IN-FLUX 625-O.

IN-FLUX VERTICLAD 9

Teledyne McKay
Nickel. C 0.02, Mn 1.55, Si 0.3, Cr 27.7, Ni 12.5, bal Fe.
Stainless steel and nickel base flux cored open arc welding wire. 97,800 psi TS; 71,800 psi YS; 37 El. Austenitic; for cladding a vertical wall in a horizontal stringer technique.

IN-STEEL

Ishikawajima-Harima Heavy Industries Co.
C 0-0.18, Ni 0-1.5, Cr 0.4-0.8, Mo 0-0.6, V 0-0.1, Si 0.15-0.35, Mn 0.6-1.2, bal Fe.
Plate: 138,000-163,500 TS; 128,000 YS; 15 El. For heavy-duty welded structures and equipment, mine cars, bus bodies. Ultra-high strength, excellent weldability.

INAFOND C41

Montecatini Settore Alluminio
Cu 4.5, Fe 0-0.8, Si 0-1, Ti 0-0.2, bal Al.
Heat treated: 36,000-43,000 TS; 28,000-33,000 YS; 2-4 El; 75-95 Brin. For light alloy parts; age-hardenable.

INAFOND C4S

Montecatini Settore Alluminio
Cu 4.5, Si 2-3, Fe 0-0.45, Ti 0-0.1, bal Al.
Heat treated: 36,000-40,000 TS; 18,000-23,000 YS; 3-6 El; 75-85 Brin. For light alloy parts; age-hardenable.

INAFOND C5

Montecatini Settore Alluminio
Cu 4.2-5, Mg 0.3, Ti 0.2, bal Al.
Heat treated: 47,000-54,000 TS; 28,000-37,000 YS; 7-14 El; 100-115 Brin. For light alloy parts; age-hardenable.

INAFOND C8

Montecatini Settore Alluminio
Cu 7-8.5, Fe 0-0.6, Si 0-0.5, Ti 0.2, bal Al.
Cast: 20,000-26,000 TS; 11,000-13,500 YS; 2-5 El; 55-70 Brin. For light alloy parts.

INAFOND S12

Montecatini Settore Alluminio
Si 11.5-13, Mg 0.3, Mn 0.5, bal Al.
Cast: 34,000-46,000 TS; 28,000-40,000 YS; 1-4 El; 80-110 Brin. For light alloy parts; corrosion resistant.

INAFOND S13

Montecatini Settore Alluminio
Si 12.75-13.25, bal Al.
Cast: 31,000-34,000 TS; 14,000-17,000 YS; 1.5-3 El; 60-80 Brin. For light alloy parts; corrosion resistant.

INAFOND S131

Montecatini Settore Alluminio
Cu 0.8, Si 12-13.3, Mn 0.3, bal Al.
Cast: 31,000-37,000 TS; 17,000-20,000 YS; 1.5-3 El; 60-80 Brin. For light alloy parts; corrosion resistant.

INAFOND S132

Montecatini Settore Alluminio
Cu 1.75-2.25, Si 12-13.3, Mn 0.3, bal Al.
Cast: 23,000-33,000 TS; 18,000-23,000 YS; 1-2.5 El; 65-85 Brin. For light alloy parts; corrosion resistant.

INAFOND S52

Montecatini Settore Alluminio
Cu 1.3, Si 5, Mg 0.5, bal Al.
Heat treated: 50,000-57,000 TS; 40,000-46,000 YS; 2-5 El; 110-140 Brin. For light alloy parts; age-hardenable.

INAFOND S7

Montecatini Settore Alluminio
Si 7, Mg 0.3, Mn 0.5, Ti 0.15, bal Al.
Cast: 37,000-43,000 TS; 25,000-30,000 YS; 6-10 El; 90-110 Brin. For light alloy parts; corrosion resistant.

INAFOND S71

Montecatini Settore Alluminio
Si 7, Mg 0.3, bal Al.
Cast: 37,000-43,000 TS; 25,000-30,000 YS; 6-10 El; 90-110 Brin. For light alloy parts; corrosion resistant.

INAFOND S9

Montecatini Settore Alluminio
Si 9, Mg 0.35, Mn 0.5, bal Al.
Cast: 35,000-43,000 TS; 28,000-37,000 YS; 3.5-5.5 El; 80-95 Brin. For light alloy parts; corrosion resistant.

INAFOND Z5F

Montecatini Settore Alluminio
Fe 1, Mg 0.6, Zn 5, Ti 0.2, bal Al.
Heat treated: 42,000-50,000 TS; 24,000-30,000 YS; 9-14 El; 90-105 Brin. For light alloy parts; heat treatable.

INALIUM
Compagnie des Alliages
Si 0.5, Mg 1.2, Cd 1.7, bal Al.
26,000-29,000 TS; 10,000-18,500 YS; 18-22 El; 50-60 Brin.
For light alloy parts.

INALIUM
Societe des Brevets Bethelmy
Mg 0.8, Fe 0.25, Si 0.45, Zn 0.18, W 0.08, bal Al.
For light alloy parts.

INAMEL
Inland Steel Co.
Carbon steel. C 0.008, Mn 0-0.6, S 0-0.04, bal Fe.
Decarburized steel intended for one coat enameling application.

INCO "B" MONEL
Now MONEL ALLOY 400.

INCO 101
International Nickel Inc.
C 1.2, Cr 2, Ni 13, Cu 6, bal Fe.
For welding rods for Ni-Resist; corrosion resistant. *Obsolete*

INCO 13
International Nickel Inc.
Nickel. Al 5, bal Ni.
For jet engine components; corrosion and heat resistant.
Obsolete

INCO 220 NICKEL
Now NICKEL 220.

INCO 425
International Nickel Inc.
Superalloy. C 0.04, Mn 1.3, Si 0.75, Cr 5.5, Ni 25.5, Ti 2.4, Al 0.65, Fe 63.8.
Superalloy. *Obsolete*

INCO 546
Now INCONEL ALLOY 721.

INCO 550
Now INCONEL ALLOY 751.

INCO 700
International Nickel Inc.
Nickel. C 0.14, Cr 15, Ni 44, Co 30, Mo 3, Ti 2.35, Al 3.1, Fe 1.5, B 0.005, Zr 0.05.
Nickel alloy. *Obsolete*

INCO 739
International Nickel Inc.
C 0.07, Cr 15, Ni 77, Ti 1.7, Al 2.7, Fe 1.
For jet engine components; high heat and corrosion resistant. *Obsolete*

INCO A-15
International Nickel Inc.
Cast iron. C 3.3, Ni 1.5, Si 2, bal Fe.
Cast: 40,000 TS. Cast iron for machine tool castings.
Obsolete

INCO A-2
International Nickel Inc.
Cast iron. TC 3.4, Si 2.3, Mn 0.5, Ni 1, Cr 1, bal Fe.
For bottom plates in chain drag conveyors; heat resistant.
Obsolete

INCO ALLOY 020
Inco Alloys International Inc.
Nickel. Ni 32-38, Cr 19-21, Cu 3-4, Mo 2-3, Nb 1, C 0-0.07, Mn 0-2, P 0-0.0045, S 0-0.035, Si 0-1, bal Fe.
Annealed: 90,000 TS; 45,000 YS; 40 El. Nickel-iron-chromium alloy. For tanks, piping, heat exchangers, pumps, valves, and other process equipment. 8020

INCO ALLOY 330
Inco Alloys International Inc.
Nickel. Ni 34-37, Cr 17-20, Si 0.75-1.5, C 0-0.08, Mn 0-2, P 0-0.03, S 0-0.03, bal Fe.
Annealed: 7000 rupture strength (1000 h) at 1400°F. Nickel-chromium alloy with silicon added for enhanced oxidation resistance. For industrial heating and conveyor systems and for heat-treating baskets and fixtures. 8330

INCO ALLOY 739
International Nickel Inc.
C 0.07, Cr 15.5, Ni 77.5, Al 2.6, Ti 1.7, Mo 2.
For high temperature applications; age hardenable.
Obsolete

INCO ALLOY C-276
Inco Alloys International Inc.
Nickel. Mo 15-17, Cr 14.5-16.5, Fe 4-7, W 3-4.5, Co 0-2.5, Mn 0-1, C 0-0.01, V 0-0.35, P 0-0.04, S 0-0.03, Si 0-0.08, bal Ni.
Annealed: 115,000 TS; 60,000 YS; 50 El. Nickel-molybdenum-chromium alloy; resistant to pitting and crevice corrosion. For pollution control, chemical processing, pulp and paper production, and waste treatment. 10276

INCO ALLOY G-3
Inco Alloys International Inc.
Nickel. Cr 21-23.5, Fe 18-21, Mo 6-8, Cu 1.5-2.5, Nb 0-0.5, C 0-0.015, W 0-1.5, Si 0-1, Mn 0-1, P 0-0.04, S 0-0.03, Co 0-5, bal Ni.
Annealed: 100,000 TS; 47,000 YS; 50 El. Nickel-chromium-iron alloy. For flue-gas scrubbers and handling phosphoric and sulfuric acids. 6985

INCO ALLOY HX
Inco Alloys International Inc.
Nickel. Cr 20.5-23, Fe 17-20, Mo 8-10, Co 0.5-2.5, W 0.2-1, C 0.05-0.15, Si 0-1, Mn 0-1, P 0-0.04, S 0-0.03, bal Ni.
Solution annealed: 16,000 rupture strength (1000 h) at 1400°F. Nickel-iron-molybdenum alloy. For gas turbines, industrial furnaces, heat-treating equipment, and nuclear engineering. 6002

INCO ALLOY MS 250
Inco Alloys International Inc.
Ni 18-20, Mo 2.75-3.25, Ti 1.3-1.45, Al 0.05-0.15, C 0-0.03, Mn 0-0.1, Si 0-0.1, P 0-0.01, S 0-0.01, bal Fe.
Precipitation hardened: 266,000 TS; 254,000 YS; 12 El. Maraging steel. For missile cases, aircraft forgings, power-transmission shafts and couplings, springs, bolts, punches, and dies.

INCO F-NICKEL
International Nickel Inc.
Nickel. Si 5-6, Fe 1-2, Cu 0.2, S 0.025, Ni 90-92.
For alloying cast iron; nickel additions.

INCO FILLER METAL C-276
Inco Alloys International Inc.
For INCO alloy C-276; other pit-resistant alloys; surfacing of steels. ERNiCrMo-4.

INCO FILLER METAL HX
Inco Alloys International Inc.
For INCO alloy HX. ERNiCrMo-2.

INCO IN-100
International Nickel Inc.
Nickel. Cr 8-11, Co 13-17, Mo 2-4, V 0.7-1.2, Al 5-6, Ti 4.5-5.5, C 0.18, B 0.01, Zr 0.06, bal Ni.
Cast: 142,000 TS; 122,000 YS; 8 El. At 1800°F: 75,000 TS; 60,000 YS; 5 El. For jet engine parts, turbine blades, for operating temperatures up to 1900°F. *Obsolete*

INCO IN-102
International Nickel Inc.
Fe 7, Cr 15, Cb 3, Mo 3, W 3, Al 0.5, Ti 0.5, C 0.06, B, Zr, bal Ni.
Annealed: 130,000 TS; 60,000 YS; 45 El. At 1200°F: 103,000 TS; 50,000 YS; 52 El. For high-temperature, high-pressure steam turbines; for long service up to 1200°F; high strength and ductility.

INCO IN-102
Cannon-Muskegon Corp.
Fe 7, Cr 15, Cb 3, Mo 3, W 3, Al 0.5, Ti 0.5, C 0.06, B, Zr, bal Ni.
Annealed: 130,000 TS; 60,000 YS; 45 El. At 1200°F: 103,000 TS; 50,000 YS; 52 El. For high-temperature, high-pressure steam turbines; for long service up to 1200°F; high strength and ductility.

INCO IN-732
International Nickel Inc.
Copper. Ni 30, Cr 2.8, bal Cu.
Air cooled: 55,000 YS; 150 IS (Charpy); 90,000 TS; 30 El; 60 RA. For marine hardware. High strength, tough, corrosion resistant, good weldability. *Obsolete*

INCO IN-738
International Nickel Inc.
Nickel. C 0.17, Co 8.5, Cr 16, Mo 1.75, W 2.6, Ta 1.75, Cb 0.9, Al 3.4, Ti 3.4, B 0.01, Zr 0.1, bal Ni.
Heat treated: 160,000 TS; 140,000 YS; 4 El; 4 RA. At 1200°F: 155,000 TS; 130,000 YS; 3 El; 5 RA. For jet engine and gas turbine components, blades, vanes, integral-wheel configuration. Precipitation hardenable. Sulfidation resistant. High rupture strength at elevated temperatures. *Obsolete*

INCO WELDING ELECTRODE C-276
Inco Alloys International Inc.
For INCO alloy C-276; other pit-resistant alloys; surfacing of steels. ENiCrMo-4.

INCO-WELD A ELECTRODE
Inco Alloys International Inc.
For INCOLOY alloys 800 and 800HT; dissimilar combinations of steels and nickel alloys; 9.0 nickel steel; surfacing of steels. ENiCrFe-2.

INCO-WELD A ELECTRODE
Henry Wiggin & Co. Ltd.
Ni 70, C 0.03, Mn 2, Fe 9, S 0.008, Si 0.3, Cu 0.06, Cr 15, Cb 2, Mo 1.5, Co 0.05, Ta 0.03.
Electrode for shielded metal-arc welding of dissimilar alloys such as austenitic and ferritic steels to each other and to high-nickel alloys; welding Incoloy alloy 800 to itself. AWS A5.11 Class ENiCrFe-2; ASME SFB 5.11.

INCO-WELD A ELECTRODE
Huntington Alloys Inc.
Ni 70, C 0.03, Mn 2, Fe 9, S 0.008, Si 0.3, Cu 0.06, Cr 15, Cb 2, Mo 1.5, Co 0.05, Ta 0.03.
Electrode for shielded metal-arc welding of dissimilar alloys such as austenitic and ferritic steels to each other and to high-nickel alloys; welding Incoloy alloy 800 to itself. AWS A5.11 Class ENiCrFe-2; ASME SFB 5.11.

INCO-WELD B ELECTRODE
Inco Alloys International Inc.
Ni 70, C 0.1, Mn 2, Fe 9, Si 0.3, Cr 15, Cb 2.5, Mo 2.
For shielded metal-arc welding of 9% nickel steel, using AC.

INCO-WELD C ELECTRODE
Inco Alloys International Inc.
For stainless steels; carbon steels; spring steels; general maintenance welding.

INCOCAL ALLOY 10
International Nickel Inc.
Nickel. C 0.3, Ca 5.5, Si 0.6, bal Ni.
For calcium addition to molten iron and steel.

INCOLOY
Now INCOLOY ALLOY 800.

INCOLOY 800

Criterion Metals, Inc.
Nickel. Ni 32, C 0.04, Mn 0.75, Fe 46, S 0.007, Si 0.35, Cu 0.3, Cr 20.5.
Thin gauge sheet, various tempers. Annealed: 75-100 ksi TS; 30-55 ksi YS. Deep drawing: 75-105 ksi TS; 30-55 ksi YS. ASTM B-409.

INCOLOY 800

Enpar Sonderwerkstoffe GmbH
Nickel.
Alternate manufacturer.

INCOLOY 801

Criterion Metals, Inc.
Nickel. Ni 32, C 0.04, Mn 0.75, Fe 44.5, S 0.007, Si 0.35, Cu 0.15, Cr 20.5, Ti 1.
Thin gauge sheet, various tempers. Annealed: 75-100 ksi TS; 30-55 ksi YS. Deep drawing: 75-105 ksi TS; 30-55 ksi YS.

INCOLOY 825

Criterion Metals, Inc.
Nickel. Ni 41.8, C 0.03, Mn 0.65, Fe 30, S 0.007, Si 0.35, Cu 1.8, Cr 21.5, Ti 0.9, Al 0.15, Mo 3.
Thin gauge sheet, various tempers. Annealed: 75-100 ksi TS; 30-55 ksi YS. Deep drawing: 75-105 ksi TS; 30-55 ksi YS.

INCOLOY 825

Enpar Sonderwerkstoffe GmbH
Nickel.
Alternate manufacturer.

INCOLOY ALLOY 800

Inco Alloys International Inc.
Ni 32.5, C 0-0.1, Mn 0.8, Fe 46, Si 0.008, Cu 0.4, Cr 21, Al 0.4, Ti 0.4.
Annealed: 83,000 psi TS; 45,000 psi YS; 60 El. For corrosive environments and high temperatures such as heat treating equipment, heat exchangers, and steam generators.

INCOLOY ALLOY 800

Huntington Alloys Inc.
Ni 32.5, C 0.05, Mn 0.75, Fe 46, S 0.008, Cu 0.38, Cr 21, Al 0.38, Ti 0.38, Si 0.5.
Cold drawn: 100,000-150,000 TS; 75,000-125,000 YS; 30-10 El; 180-300 For heat exchangers and process piping; carburizing fixtures and retorts, electric range element sheathing; resistant to elevated temperature oxidation and carburization.

INCOLOY ALLOY 800

Henry Wiggin & Co. Ltd.
Ni 32.5, C 0.05, Mn 0.75, Fe 46, S 0.008, Cu 0.38, Cr 21, Al 0.38, Ti 0.38, Si 0.5.
Cold drawn: 100,000-150,000 TS; 75,000-125,000 YS; 30-10 El; 180-300 For heat exchangers and process piping; carburizing fixtures and retorts, electric range element sheathing; resistant to elevated temperature oxidation and carburization.

INCOLOY ALLOY 800

Cannon-Muskegon Corp.
Ni 32.5, C 0.05, Mn 0.75, Fe 46, S 0.008, Cu 0.38, Cr 21, Al 0.38, Ti 0.38, Si 0.5.
Cold drawn: 100,000-150,000 TS; 75,000-125,000 YS; 30-10 El; 180-300 For heat exchangers and process piping; carburizing fixtures and retorts, electric range element sheathing; resistant to elevated temperature oxidation and carburization.

INCOLOY ALLOY 800H

Inco Alloys International Inc.
Now INCOLOY ALLOY 800HT.

INCOLOY ALLOY 800HT

Inco Alloys International Inc.
Ni 32.5, C 0.08, Mn 0.8, Fe 46, S 0.008, Si 0.5, Cu 0.4, Cr 21, Al 0.4, Ti 0.4.
Annealed: 80,000 TS; 30,000 YS; 50 El. For petrochemical process piping with steam-generator components. Variation of INCOLOY 800 with controlled carbon content and grain size.

INCOLOY ALLOY 801

Henry Wiggin & Co. Ltd.
Ni 32, C 0.05, Mn 0.75, Fe 44.5, Cu 0.25, Cr 20.5, Ti 1.13, S 0.008, Si 0.5.
For jet engine combustion liners and transition pieces. High temperature tensile and rupture strength. Formerly INCOLOY T.

INCOLOY ALLOY 801

Huntington Alloys Inc.
Ni 32, C 0.05, Mn 0.75, Fe 44.5, Cu 0.25, Cr 20.5, Ti 1.13, S 0.000, Si 0.5.
For jet engine combustion liners and transition pieces. High temperature tensile and rupture strength. Formerly INCOLOY T.

INCOLOY ALLOY 801

Cannon-Muskegon Corp.
Ni 32, C 0.05, Mn 0.75, Fe 44.5, Cu 0.25, Cr 20.5, Ti 1.13, S 0.008, Si 0.5.
For jet engine combustion liners and transition pieces. High temperature tensile and rupture strength. Formerly INCOLOY T.

INCOLOY ALLOY 802

Huntington Alloys Inc.
Ni 32.5, C 0.35, Mn 0.75, Fe 46, S 0.008, Si 0.38, Cr 21, Al 0.58, Ti 0.75.
Hot finished and annealed: 80,000-105,000 TS; 35,000-50,000 YS; 47-18 El; 140-188 Brin. For hot sizing dies for aerospace industry, connecting pins for cast link heat-treating furnace belts; good high temperature strength and corrosion resistance.

INCOLOY ALLOY 802

Henry Wiggin & Co. Ltd.
Ni 32.5, C 0.35, Mn 0.75, Fe 46, S 0.008, Si 0.38, Cr 21, Al 0.58, Ti 0.75.
Hot finished and annealed: 80,000-105,000 TS; 35,000-50,000 YS; 47-18 El; 140-188 Brin. For hot sizing dies for aerospace industry, connecting pins for cast link heat-treating furnace belts; good high temperature strength and corrosion resistance.

INCOLOY ALLOY 802

Cannon-Muskegon Corp.
Ni 32.5, C 0.35, Mn 0.75, Fe 46, S 0.008, Si 0.38, Cr 21, Al 0.58, Ti 0.75.
Hot finished and annealed: 80,000-105,000 TS; 35,000-50,000 YS; 47-18 El; 140-188 Brin. For hot sizing dies for aerospace industry, connecting pins for cast link heat-treating furnace belts; good high temperature strength and corrosion resistance.

INCOLOY ALLOY 804

Inco Alloys International Inc.
Ni 41, C 0.25, Mn 0.75, Fe 25.4, S 0.008, Cr 29.5, Al 0.3, Ti 0.6, Si 0.38, Cu 0.25.
Annealed: 95,000 TS; 45,000 YS; 40 El. High temperature strength; resistant to carburization and sulfidation. *Obsolete*

INCOLOY ALLOY 804

Cannon-Muskegon Corp.
Ni 41, C 0.25, Mn 0.75, Fe 25.4, S 0.008, Cr 29.5, Al 0.3, Ti 0.6, Si 0.38, Cu 0.25.
Annealed: 95,000 TS; 45,000 YS; 40 El. High temperature strength; resistant to carburization and sulfidation.

INCOLOY ALLOY 805

Inco Alloys International Inc.
Ni 36, C 0.12, Mn 0.8, Fe 54.5, S 0.008, Si 0.5, Cu 0.1, Cr 7.5, Mo 0.5. *Obsolete*

INCOLOY ALLOY 825

Manufacturer not listed.
MIL-E-21562.

INCOLOY ALLOY 825

Cannon-Muskegon Corp.
Ni 42, C 0.03, Mn 0.5, Fe 30, S 0.015, Cu 2.25, Cr 21.5, Al 0.1, Ti 0.9, Mo 3.
Annealed: 85,000-105,000 TS; 35,000-65,000 YS; 50-30 El; 120-180 Brin. For phosphoric acid evaporators, pickling tank heaters, pickling hooks and equipment, propeller shafts, tank trucks; corrosion resistant. Formerly Ni-O-NEL alloy 825.

INCOLOY ALLOY 825

Huntington Alloys Inc.
Ni 42, C 0.03, Mn 0.5, Fe 30, S 0.015, Cu 2.25, Cr 21.5, Al 0.1, Ti 0.9, Mo 3.
Annealed: 85,000-105,000 TS; 35,000-65,000 YS; 50-30 El; 120-180 Brin. For phosphoric acid evaporators, pickling tank heaters, pickling hooks and equipment, propeller shafts, tank trucks; corrosion resistant. Formerly Ni-O-NEL alloy 825.

INCOLOY ALLOY 825

Henry Wiggin & Co. Ltd.
Ni 42, C 0.03, Mn 0.5, Fe 30, S 0.015, Cu 2.25, Cr 21.5, Al 0.1, Ti 0.9, Mo 3.
Annealed: 85,000-105,000 TS; 35,000-65,000 YS; 50-30 El; 120-180 Brin. For phosphoric acid evaporators, pickling tank heaters, pickling hooks and equipment, propeller shafts, tank trucks; corrosion resistant. Formerly Ni-O-NEL alloy 825.

INCOLOY ALLOY 825

Inco Alloys International Inc.
Nickel. Ni 38-46, Cr 19.5-23.5, Mo 2.5-3.5, Cu 1.5-3, Ti 0.6-1.2, C 0-0.05, Mn 0-1, S 0-0.03, Si 0-0.5, Al 0-0.2, 22.0 Fe min.
Annealed: 100,000 TS; 45,000 YS; 45 El. Nickel-iron-chromium alloy; resistant to sulfuric and phosphoric acids. For chemical processing, pollution-control equipment, oil and gas well piping, nuclear fuel reprocessing, acid production, and pickling equipment.

INCOLOY ALLOY 825

Inco Alloys International Inc.
C 0.04, Ni 42, Cu 2.2, Cr 21.5, Fe 30, Mo 3, Nb 0.9.
Good corrosion resistance; equipment for phosphoric and sulfuric acids, sea water, nuclear fuel element recovery plant.

INCOLOY ALLOY 825 EP

Inco Alloys International Inc.
Now INCOLOY ALLOY 825.

INCOLOY ALLOY 840

Inco Alloys International Inc.
Ni 21, Fe 58, Cr 19, Si 0.8, C 0.03.
Annealed: 90,000 TS; 50,000 YS; 35 El. For electric heater element sheathing. *Obsolete*

INCOLOY ALLOY 901

Inco Alloys International Inc.
Now NIMONIC ALLOY 901.

INCOLOY ALLOY 901

Cannon-Muskegon Corp.
C 0.05, Cr 12.8, Ni 43, Mo 5.7, Ti 2.4, Fe 35.
Heat treated: 168,000 TS; 110,000 YS; 23 El. At 1200°F: 125,000 TS; 95,000 YS; 5 El. For aircraft gas turbine blades, turbine discs; age hardenable, up to 1400°F service.

INCOLOY ALLOY 903

Inco Alloys International Inc.
Ni 38, Co 15, Al 0.7, Ti 1.4, Cb 3, Fe 41.
Heat treated: 190,000 TS; 160,000 YS; 14 El. At 1200°F: 145,000 TS; 130,000 YS; 18 El. For rocket-engine thrust chambers, steam-turbine bolts, and ordnance hardware.

INCOLOY ALLOY 903
Techalloy Co. Inc.
C 0.02, Ni 38, Fe 41, Co 15, Ti 1.4, Al 0.7, Nb 3.
Low coefficient of expansion, high strength, constant modulus of elasticity and resistance to thermal fatigue and shock from -240 to + 650°C. For instrumentation, rocket engines.

INCOLOY ALLOY 903
Henry Wiggin & Co. Ltd.
C 0.02, Ni 38, Fe 41, Co 15, Ti 1.4, Al 0.7, Nb 3.
Low coefficient of expansion, high strength, constant modulus of elasticity and resistance to thermal fatigue and shock from -240 to + 650°C. For instrumentation, rocket engines.

INCOLOY ALLOY 904
Inco Alloys International Inc.
C 0.02, Ni 33, Fe 50, Co 14, Ti 1.7.
Low coefficient of expansion; good strength. For compensating members in gas turbine engines. *Obsolete*

INCOLOY ALLOY 907
Inco Alloys International Inc.
Nickel. Ni 38, Fe 42, Co 13, Nb 4.7, Ti 1.5, Al 0.03, Si 0.15.
Precipitation hardened: 100,000 rupture strength (1000 h) at 1100°F. Nickel-iron-cobalt alloy. For components of gas turbines including seals, shafts, and casings. 19907

INCOLOY ALLOY 909
Inco Alloys International Inc.
Nickel. Ni 38, Fe 42, Co 13, Nb 4.7-4.6, Ti 1.5, Si 0.4, Al 0.03, C 0.01.
Precipitation hardened: 130,000 rupture strength (1000 h) at 1000°F. Nickel-iron-cobalt alloy with silicon added. For gas-turbine casings, shrouds, vanes, and shafts. 11109

INCOLOY ALLOY 925
Inco Alloys International Inc.
Nickel. Ni 44, Fe 28, Cr 21, Mo 3, Cu 1.8, Ti 2.1, Al 0.3, C 0.01.
Precipitation hardened: 176,000 TS; 118,000 YS; 24 El. Nickel-iron-chromium alloy. For surface and down-hole hardware in sour gas wells and for oil-production equipment. 9925

INCOLOY ALLOY DS
Inco Alloys International Inc.
C 0-0.15, Si 2-2.5, Ni 36-39, Cr 17-19, bal Fe.
At 68°F: 105,300 TS; 38 El; 50 RA. At 1112°F: 66,100 TS; 45 El; 38 RA. At 1832°F: 12,300 TS; 124 El; 83 RA. For furnace parts, high temperature equipment; useful to 950°C in oxidizing atmosphere.

INCOLOY ALLOY MA 956
Inco Alloys International Inc.
Fe 75, Cr 20, Al 4.5, Ti 0.35, 0.5 Y_2O_3.
For combustion chambers in gas turbines and components in fossil-fuel burners. High-temperature oxidation and sulfidation resistance. Melting point approximately 2700°F.

INCOLOY CARPENTER 20CB3
Enpar Sonderwerkstoffe GmbH
Nickel.
Alternate manufacturer.

INCOLOY DS
Enpar Sonderwerkstoffe GmbH
Nickel.
Alternate manufacturer.

INCOLOY DS
Inco Alloys International Inc.
Now INCOLOY ALLOY DS.

INCOLOY FILLER METAL 65
Inco Alloys International Inc.
Ni 42, C 0.03, Mn 0.7, Fe 30, S 0.007, Si 0.3, Cu 1.7, Cr 21, Ti 1, Mo 3.
For gas-shielded arc welding.

INCOLOY T
Now INCOLOY ALLOY 801.

INCOLOY WELDING ELECTRODE 135
Inco Alloys International Inc.
Ni 36, C 0.05, Mn 2, Fe 26, S 0.008, Si 0.4, Cu 1.8, Cr 29, Mo 3.75.
Electrode for shielded metal-arc welding of Incoloy alloy 825 to itself and to steels.

INCOMAG ALLOY 1
International Nickel Inc.
Nickel. C 2, Mg 14-16, bal Ni.
Formerly NICKEL MAGNESIUM ALLOY NO.1. For production of ductile iron.

INCOMAG ALLOY 3LC
International Nickel Inc.
Nickel. C 0.07, Mg 4.2-4.8, bal Ni.
For production of ductile iron.

INCOMAG ALLOY 4
International Nickel Inc.
Nickel. C 1.7, Mg 4-4.5, Fe 34, bal Ni.
For direct melt addition in the production of ductile iron.

INCOMAP ALLOY AL-9052
Inco Alloys International Inc.
Aluminum. Mg 4, O 0.5, C 1.1, bal Al.
65,000 TS; 55,000 YS; 13 El. High strength, corrosion-resistant; made by the mechanical alloying process.

INCOMAP ALLOY AL-905XL
Inco Alloys International Inc.
Aluminum. Mg 4, Li 1.3, O 0.5, C 1.1, bal Al.
Wrought, longitudinal: 75,000 TS; 65,000 YS; 9 El. Low-density alloy; for forgings where weight is critical.

INCOMET NICKEL
International Nickel Inc.
Nickel. Ni 95, Co 1.3, Cu 0.4, Fe 0.4, 1.1 O_2.
Intermediate quality granular nickel for most commercial alloying uses. *Obsolete*

INCOMPARABLE
Flockton, Tompkin & Co., Ltd.
C 0.8, W 22, Cr 4.5, V 1.5, bal Fe.
For form tools, cutters, lathe and planer tools; high speed steel.

INCONEL
Now INCONEL ALLOY 600.

INCONEL 112
Inco Alloys International Inc.
Now INCONEL ALLOY 112. *Obsolete*

INCONEL 132
Gilby-Fodor S.A.
C 0-0.2, Fe 6-10, Cr 14-17, bal Fe.
For heat resistant parts. High heat resistance.

INCONEL 600
Criterion Metals, Inc.
Nickel. Ni 76, C 0.04, Mn 0.2, Fe 7.2, S 0.007, Si 0.2, Cu 0.1, Cr 15.8.
Thin gauge sheet, various tempers. Annealed: 80,000 psi TS; 30,000 psi YS. Hard: 125,000 psi TS; 90,000 psi YS. ASTM B-168.

INCONEL 600
Enpar Sonderwerkstoffe GmbH
Nickel.
Alternate manufacturer.

INCONEL 601
Enpar Sonderwerkstoffe GmbH
Nickel.
Alternate manufacturer.

INCONEL 604
Techalloy Co. Inc.
C 0.04, Ni 74, Mn 0.2, Fe 7.2, Si 0.2, Cu 0.1, 15.8 Cr, Cb.
For woven belts for furnaces, steam turbine nozzle parts.

INCONEL 610
Ancast, Inc.
Nickel. C 0-0.4, Mn 0-1.5, Si 0-3, Cr 14-17, Fe 0-11, P 0-0.03, S 0-0.03, bal Ni.
ASTM A-494 GR CY40.

INCONEL 611
Ancast, Inc.
Nickel. C 0-0.4, Mn 0-1.5, Si 0-2, Cr 14-17, Fe 0-11, Cu 0-0.5, 1.00-3.00 Cb + Ta, bal Ni.

INCONEL 617
Enpar Sonderwerkstoffe GmbH
Nickel.
Alternate manufacturer.

INCONEL 62
Now INCONEL FILLER METAL 62.

INCONEL 625
Criterion Metals, Inc.
Nickel. Ni 58.4, C 0.1, Mn 0.5, Fe 5, S 0.015, Si 0.5, Cr 21, Ti 0.4, Al 0.4, Cb 3.65, Mo 9.
Thin gauge sheet, various tempers. Annealed: 80,000 psi TS; 30,000 psi YS. Hard: 125,000 psi TS; 90,000 psi YS. ASTM B-446.

INCONEL 625
Enpar Sonderwerkstoffe GmbH
Nickel.
Alternate manufacturer.

INCONEL 69 (FILLER METAL)
Inco Alloys International Inc.
C 0.04, Ni 73, Mn 0.55, Fe 6.5, Cr 15.2, S 0.007, Si 0.3, Cu 0.05, Al 0.7, Ti 2.5, Nb 0.85.
Filler metal for gas tungsten-arc welding INCONEL alloy X-750. AWS.14, Class ERNCrFe-7; AMS 5778. *Obsolete*

INCONEL 702
Criterion Metals, Inc.
Nickel. Ni 79.5, C 0.04, Mn 0.05, Fe 0.35, S 0.007, Si 0.2, Cu 0.1, Cr 15.6, Ti 0.7, Al 3.4.
Thin gauge sheet, various tempers.

INCONEL 718
Criterion Metals, Inc.
Nickel. Ni 52.5, C 0.04, Mn 0.2, Fe 18.5, S 0.007, Si 0.3, Cu 0.07, Cr 18.6, Ti 0.9, Cb 0.5, Mo 3.1.
Thin gauge sheet, various tempers.

INCONEL 718
Enpar Sonderwerkstoffe GmbH
Nickel.
Alternate manufacturer.

INCONEL ALLOY 112
Inco Alloys International Inc.
C 0.05, Mn 0.3, Fe 4, S 0.01, Si 0.4, Cr 21.5, Mo 9, Ni 61, 3.65 Nb + Ta.
Welding electrode for shielded-arc welding INCONEL alloy 625 and dissimilar nickel and iron-base alloys. MIL-E-22200/3. *Obsolete*

INCONEL ALLOY 600
Inco Alloys International Inc.
Nickel. Cr 14-17, Fe 6-10, C 0-0.15, Mn 0-1, S 0-0.015, Si 0-0.5, Cu 0-0.5, 72.0 Ni min.
Annealed: 95,000 TS; 45,000 YS; 40 El. Used for furnace components, chemical and food processing, nuclear engineering, and sparking electrodes. 6600

INCONEL ALLOY 600
Henry Wiggin & Co. Ltd.
Ni 76, C 0.08, Mn 0.5, Fe 8, S 0.008, Cr 15.5, Cu 0.25.
Annealed: 80,000-100,000 TS; 30,000-50,000 YS; 50-35 El;
120-170 Brin. Cold drawn: 105,000-150,000 TS;
80,000-125,000 YS; 30-10 El; 180-300 Brin. For furnace
muffles, electronic components, heat exchanger tubing,
nuclear reactors, springs, jet engine parts; good high
temperature properties.

INCONEL ALLOY 600
Huntington Alloys Inc.
Ni 76, C 0.08, Mn 0.5, Fe 8, S 0.008, Cr 15.5, Cu 0.25.
Annealed: 80,000-100,000 TS; 30,000-50,00 YS; 50-35 El;
120-170 Brin. Cold drawn: 105,000-150,000 TS;
80,000-125,000 YS; 30-10 El; 180-300 Brin. For furnace
muffles, electronic components, heat exchanger tubing,
nuclear reactors, springs, jet engine parts; good high
temperature properties.

INCONEL ALLOY 600
Cannon-Muskegon Corp.
Ni 76, C 0.08, Mn 0.5, Fe 8, S 0.008, Cr 15.5, Cu 0.25.
Annealed: 80,000-100,000 TS; 30,000-50,00 YS; 50-35 El;
120-170 Brin. Cold drawn: 105,000-150,000 TS;
80,000-125,000 YS; 30-10 El; 180-300 Brin. For furnace
muffles, electronic components, heat exchanger tubing,
nuclear reactors, springs, jet engine parts; good high
temperature properties.

INCONEL ALLOY 601
Inco Alloys International Inc.
Nickel. Ni 58-63, Cr 21-25, Al 1-1.7, C 0-0.1, Mn 0-1, Si 0-0.5,
S 0-0.015, Cu 0-1, bal Fe.
Solution annealed: 28,000 rupture strength (1000 h) at
1200°F. For industrial furnaces, heat-treating, petrochemical,
and other process equipment, and gas turbine components.
6601

INCONEL ALLOY 601
Huntington Alloys Inc.
Ni 60.5, C 0.05, Mn 0.5, Fe 14.1, S 0.007, Si 0.25, Cu 0.5, Cr
23, Al 1.35.
Hot Rolled, annealed: 86,000-107,000 TS; 29,000-48,000 YS;
46-60 El. For heat treating baskets and fixtures, radiant
furnace tubes, thermocouple protection tubes, furnace muffle
and retorts; good high temperature strength and corrosion
resistance.

INCONEL ALLOY 601
Cannon-Muskegon Corp.
Ni 60.5, C 0.05, Mn 0.5, Fe 14.1, S 0.007, Si 0.25, Cu 0.5, Cr
23, Al 1.35.
Hot Rolled, annealed: 86,000-107,000 TS; 29,000-48,000 YS;
46-60 El. For heat treating baskets and fixtures, radiant
furnace tubes, thermocouple protection tubes, furnace muffle
and retorts; good high temperature strength and corrosion
resistance.

INCONEL ALLOY 601
Henry Wiggin & Co. Ltd.
Ni 60.5, C 0.05, Mn 0.5, Fe 14.1, S 0.007, Si 0.25, Cu 0.5, Cr
23, Al 1.35.
Hot Rolled, annealed; 86,000-107,000 TS; 29,000-48,000 YS;
46-60 El. For heat treating baskets and fixtures, radiant
furnace tubes, thermocouple protection tubes, furnace muffle
and retorts; good high temperature strength and corrosion
resistance.

INCONEL ALLOY 617
Inco Alloys International Inc.
Ni 54, Cr 22, Co 12.5, Mo 9, Al 1, C 0.07.
Annealed: 110,000 TS; 45,000 YS; 60 El. For gas turbine
engines and nuclear reactor components.

INCONEL ALLOY 617
Henry Wiggin & Co. Ltd.
C 0.07, Ni 54, Cr 22, Co 12.5, Mo 9, Al 1.
At 1000°C: 150 N/mm^2 TS; 120 N/mm^2 YS; 120 El. Good
high temperature strength and oxidation resistance; for
equipment such as gas turbines.

INCONEL ALLOY 617
Techalloy Co. Inc.
C 0.07, Ni 54, Cr 22, Co 12.5, Mo 9, Al 1.
At 1000°C: 150 N/mm^2 TS; 120 N/mm^2 YS; 120 El. Good
high temperature strength and oxidation resistance; for
equipment such as gas turbines.

INCONEL ALLOY 622
Inco Alloys International Inc.
Cr 20.5, Mo 14.2, Fe 2.3, W 3.2, Co 0-2.5, V 0-0.35, C
0-0.015, Mg 0-0.5, S 0-0.02, Si 0-0.08, P 0-0.02, bal Ni.
Plate (0.5 in.): 106,300 TS; 47,900 YS; 69 El. Nickel-
chromium-molybdenum alloy for chemical processing, flue-
gas desulfurization, hazardous waste incineration, bleaching
systems in paper manufacture, and processing of radioactive
waste.

INCONEL ALLOY 625
Inco Alloys International Inc.
Nickel. Cr 20-23, Mo 8-10, Nb 3.15-4.15, Fe 0-5, C 0-0.1, Mn
0-0.5, Si 0-0.5, S 0-0.015, Al 0-0.4, Ti 0-0.4, P 0-0.015, Co 0-1,
58.0 Ni min.
Solution annealed: 52,000 rupture strength (1000 h) at
1200°F. Nickel-chromium-molybdenum alloy; resistant to
pitting and crevice corrosion. For chemical processing,
aerospace and marine engineering, pollution-control
equipment, and nuclear reactors.

INCONEL ALLOY 625
Huntington Alloys Inc.
Ni 61, C 0.05, Mn 0.25, Fe 2.5, S 0.008, Si 0.25, Cr 21.5, Al
0.2, Ti 0.2, Mo 9, 3.65 Cb + Ta.
Solution annealed: 105,000-130,000 TS; 42,000-60,000 YS;
65-40 El; 116-194 Brin. As rolled: 120,000-160,000 TS;
60,000-110,000 YS; 60-30 El; 175-240 Brin. For ducting
systems, combustion systems, fuel nozzles, after burners;
high strength and corrosion resistance.

INCONEL ALLOY 625
Henry Wiggin & Co. Ltd.
Ni 61, C 0.05, Mn 0.25, Fe 2.5, S 0.008, Si 0.25, Cr 21.5, Al
0.2, Ti 0.2, Mo 9, 3.65 Cb + Ta.
Solution annealed: 105,000-130,000 TS; 42,000-60,000 YS;
65-40 El; 116-194 Brin. As rolled: 120,000-160,000 TS;
60,000-110,000 YS; 60-30 El; 175-240 Brin. For ducting
systems, combustion systems, fuel nozzles, after burners;
high strength and corrosion resistance.

INCONEL ALLOY 671
Inco Alloys International Inc.
Ni 52.5, Cr 46.5, Ti 0.45, C 0.07.
Annealed: 125,000 TS; 70,000 YS; 25 El. For boiler
superheater tube shields, soot-blower tubes, and hangers.
High-temperature corrosion resistance, particularly fuel-ash
corrosion in atmospheres containing sulfur and vanadium.
Obsolete

INCONEL ALLOY 690
Inco Alloys International Inc.
Ni 60, Cr 30, Fe 9.5, C 0.03.
Annealed: 105,000 TS; 50,000 YS; 40 El. For equipment for
steel pickling and reprocessing of spent nuclear fuels.
Resistance to oxidizing chemicals, high-temperature sulfur-
containing gases, and nitric or nitric/hydrofluoric acid
solutions.

INCONEL ALLOY 702
Inco Alloys International Inc.
Ni 79.5, C 0.05, Mn 0.5, Fe 1, S 0.005, Si 0.35, Cr 15.5, Al
3.25, Ti 0.63, Cu.
For afterburner liners, furnace components and fixtures;
oxidation resistant to 2400°F. *Obsolete*

INCONEL ALLOY 702
Cannon-Muskegon Corp.
Ni 79.5, C 0.05, Mn 0.5, Fe 1, S 0.005, Si 0.35, Cr 15.5, Al
3.25, Ti 0.63, Cu 0.25.
For after-burner liners, furnace components and fixtures;
oxidation resistant to 2400°F.

INCONEL ALLOY 706
Henry Wiggin & Co. Ltd.
Ni 41.5, C 0.3, Mn 0.18, Fe 40, S 0.008, Si 0.18, Cu 0.15, Cr
16, Al 0.2, Ti 1.75, 2.9 Cb + Ta.
Solution treated and aged: 185,000-193,000 TS;
143,000-161,000 YS; 20-19 El. For gas turbine components;
high strength, machinable and weldable.

INCONEL ALLOY 706
Cannon-Muskegon Corp.
Ni 41.5, C 0.3, Mn 0.18, Fe 40, S 0.008, Si 0.18, Cu 0.15, Cr
16, Al 0.2, Ti 1.75, 2.9 Cb + Ta.
Solution treated and aged: 185,000-193,000 TS;
143,000-161,000 YS; 20-19 El. For gas turbine components;
high strength, machinable and weldable.

INCONEL ALLOY 706
Huntington Alloys Inc.
Ni 41.5, C 0.3, Mn 0.18, Fe 40, S 0.008, Si 0.18, Cu 0.15, Cr
16, Al 0.2, Ti 1.75, 2.9 Cb + Ta.
Solution treated and aged: 185,000-193,000 TS;
143,000-161,000 YS; 20-19 El. For gas turbine components;
high strength, machinable and weldable.

INCONEL ALLOY 717C
Huntington Alloys Inc.
Cr 10-14, Mo 3.5-5, Cb 1-3, Co 7-9, Ti 0.75-1.25, Al 7.2-8, bal
Ni.
For turbine blades; 1800 F, rupture strength. *Obsolete*

INCONEL ALLOY 718
Inco Alloys International Inc.
Nickel. Ni 50-55, Cr 17-21, Nb 4.75-5.5, Mo 2.8-3.3, Ti
0.65-1.15, Al 0.2-0.8, Co 0-1, Mn 0-0.35, P 0-0.015,
S 0-0.015, B 0-0.006, Cu 0-0.3, C 0-0.08, bal Fe.
Precipitation hardened: 110,000 rupture strength (1000 h) at
1100°F. Nickel-chromium alloy. For gas turbines, rocket
motors, spacecraft, nuclear reactors, pumps and tooling.
7718

INCONEL ALLOY 718
Henry Wiggin & Co. Ltd.
Ni 52.5, C 0.04, Mn 0.18, Fe 18.5, S 0.008, Si 0.18, Cu 0.15,
Cr 19, Al 0.5, Ti 0.9, Mo 3.05, 5.13 Cb + Ta.
Hot Rolled and aged: 203,500 TS; 181,000 YS; 16 El; 383
Brin. For jet engines, pump bodies and parts, rocket motors
and thrust reversers, space craft. High strength, weldable in
aged condition.

INCONEL ALLOY 718
Huntington Alloys Inc.
Ni 52.5, C 0.04, Mn 0.18, Fe 18.5, S 0.008, Si 0.18, Cu 0.15,
Cr 19, Al 0.5, Ti 0.9, Mo 3.05, 5.13 Cb + Ta.
Hot Rolled and aged: 203,500 TS; 181,000 YS; 16 El; 383
Brin. For jet engines, pump bodies and parts, rocket motors
and thrust reversers, space craft. High strength, weldable in
aged condition.

INCONEL ALLOY 718
Cannon-Muskegon Corp.
Ni 52.5, C 0.04, Mn 0.18, Fe 18.5, S 0.008, Si 0.18, Cu 0.15,
Cr 19, Al 0.5, Ti 0.9, Mo 3.05, 5.13 Cb + Ta.
Hot Rolled and aged: 203,500 TS; 181,000 YS; 16 El; 383
Brin. For jet engines, pump bodies and parts, rocket motors
and thrust reversers, space craft. High strength, weldable in
aged condition.

INCONEL ALLOY 718
Titanium Metals Corp.
Ni 52.5, C 0.04, Mn 0.18, Fe 18.5, S 0.008, Si 0.18, Cu 0.15,
Cr 19, Al 0.5, Ti 0.9, Mo 3.05, 5.13 Cb + Ta.
Hot Rolled and aged: 203,500 TS; 181,000 YS; 16 El; 383
Brin. For jet engines, pump bodies and parts, rocket motors
and thrust reversers, space craft. High strength, weldable in
aged condition.

INCONEL ALLOY 721

Inco Alloys International Inc.
Ni 71, C 0.04, Mn 2.25, Fe 4, S 0.005, Si 0.08, Cu 0.1, Cr 16, Ti 3.05.
For internal combustion engine valves; age-hardenable; high creep strength. Formerly INCONEL M. *Obsolete*

INCONEL ALLOY 722

Inco Alloys International Inc.
Ni 75, C 0.04, Mn 0.5, Fe 7, S 0.005, Si 0.35, Al 0.7, Ti 2.38, Cu 0.25, Cr 15.5.
For jet engine components; age hardenable; high temperature strength and oxidation and corrosion resistant. Formerly INCONEL W. *Obsolete*

INCONEL ALLOY 751

Inco Alloys International Inc.
Ni 72.5, Cr 0.05, Mn 0.5, Fe 7, S 0.005, Si 0.25, Cu 0.25, Cr 15.5, Al 1.2, Ti 2.3, 0.95 Cb + Ta.
For diesel exhaust valves; high rupture strength at 1600°F.

INCONEL ALLOY 806

International Nickel Inc.
C 0.06, Cr 6-8, Mo 15-18, Fe 0-5, B 0-0.01, bal Ni.
Heat treated: 87,000 TS; 38,000 YS; 17 El. At 1000 F: 63,000 TS; 22,000 YS; 29 El. For equipment handling hot fluoride salts, high temperature applications. Heat and corrosion resistant. *Obsolete*

INCONEL ALLOY MA 754

Inco Alloys International Inc.
Ni 78, Cr 20, Ti 0.5, Al 0.3, C 0.05, 0.6 Y_2O_3.
For vanes in aircraft gas-turbine engines. High stress rupture strength to 2100°F.

INCONEL ALLOY X-750

Inco Alloys International Inc.
Nickel. Cr 14-17, Fe 5-9, Ti 2.25-2.75, Al 0.4-1, Nb 0.7-1.2, C 0-0.08, Mn 0-1, Si 0-0.5, S 0-0.01, Cu 0-0.5, Co 0-1, 70.0 Ni min.
Precipitation hardened: 92,000 rupture strength (1000 h) at 1100 °F. Nickel-chromium alloy; precipitation hardenable by additions of aluminum and titanium. For gas turbines, rocket engines, nuclear reactors, pressure vessels, tooling, and aircraft structures. 7750

INCONEL ALLOY X-750

Huntington Alloys Inc.
Ni 73, C 0.04, Mn 0.5, Fe 7, S 0.005, Si 0.25, Cu 0.25, Cr 15.5, Al 0.7, Ti 2.5, 0.95 Cb + Ta.
Hot finished and aged: 170,000-206,000 TS; 100,000-163,000 YS; 30-15 El; 302-400 Brin. For aviation and industrial gas turbine parts, springs for steam service and nuclear reactors, bolts, heat treating fixtures; corrosion and oxidation resistant; age hardenable; high creep-rupture strength.

INCONEL ALLOY X-750

Cannon-Muskegon Corp.
Ni 73, C 0.04, Mn 0.5, Fe 7, S 0.005, Si 0.25, Cu 0.25, Cr 15.5, Al 0.7, Ti 2.5, 0.95 Cb + Ta.
Hot finished and aged: 170,000-206,000 TS; 100,000-163,000 YS; 30-15 El; 302-400 Brin. For aviation and industrial gas turbine parts, springs for steam service and nuclear reactors, bolts, heat treating fixtures; corrosion and oxidation resistant; age hardenable; high creep-rupture strength.

INCONEL ALLOY X-750

Henry Wiggin & Co. Ltd.
Ni 73, C 0.04, Mn 0.5, Fe 7, S 0.005, Si 0.25, Cu 0.25, Cr 15.5, Al 0.7, Ti 2.5, 0.95 Cb + Ta.
Hot finished and aged: 170,000-206,000 TS; 100,000-163,000 YS; 30-15 El; 302-400 Brin. For aviation and industrial gas turbine parts, springs for steam service and nuclear reactors, bolts, heat treating fixtures; corrosion and oxidation resistant; age hardenable; high creep-rupture strength.

INCONEL CY 40

Alloy Foundries
Nickel. C 0-0.4, Mn 0-1.5, Si 0-3, Cr 14-17, Fe 0-11, bal Ni.
Alloy casting; corrosion and temperature resistant. Meets ACI CY-40; ASTM A-296 CY-40.

INCONEL FILLER METAL 601

Inco Alloys International Inc.
Ni 60.5, C 0.05, Mn 0.5, Fe 14.1, S 0.007, Si 0.25, Al 1.35, C 0.5, Cr 23.
For gas tungsten-arc welding of Inconel alloy 601.

INCONEL FILLER METAL 617

Inco Alloys International Inc.
Ni 54, Cr 22, Co 12.5, Mo 9, Al 1, C 0.07.
For gas-tungsten-arc welding of INCONEL Alloy 617.

INCONEL FILLER METAL 62

Inco Alloys International Inc.
Ni 74, C 0.02, Mn 0.1, Fe 7.5, S 0.005, Si 0.1, Cb 2.25, Cu 0.03, Cr 16.
For gas-shielded arc welding of Inconel alloy 600 and Incoloy alloy 800; overlaying on steel. AWS A 5.14 Class ERNiCrFe-5; AMS 5679.

INCONEL FILLER METAL 625

Inco Alloys International Inc.
Ni 61, C 0.05, Mn 0.25, Fe 2.5, S 0.008, Si 0.25, Ti 0.2, Mo 9, Cr 21.5, Al 0.2, 3.65 Cb + Ta.
For gas-shielded arc welding of Inconel alloy 625. AMS 5837.

INCONEL FILLER METAL 69

Inco Alloys International Inc.
Ni 73, C 0.04, Mn 0.55, Fe 6.5, S 0.007, Si 0.3, Al 0.7, Ti 2.5, Cb 0.85, Cu 0.05, Cr 15.2.
For gas tungsten-arc welding of Inconel alloys 722 and X750. AWS A 5.14 Class ERNiCrFe-7; AMS-5778. *Obsolete*

INCONEL FILLER METAL 718

Inco Alloys International Inc.
Ni 52.5, C 0.04, Mn 0.2, Fe 18.5, S 0.007, Si 0.3, Al 0.4, Ti 0.9, Cb 5, Mo 3.1, C 0.07, Cr 18.6.
For gas tungsten-arc welding of Inconel alloys 718 and X-750. AMS 5832.

INCONEL FILLER METAL 82

Inco Alloys International Inc.
Ni 72, C 0.02, Mn 3, Fe 1, S 0.007, Si 0.2, Cu 0.04, Cr 20, Ti 0.55, Cb 2.5.
For gas-shielded arc welding of Inconel alloy 600, Incoloy alloy 800, Inconel alloy 600 to steel, other dissimilar combinations of nickel-base and iron-base alloys; overlaying on steel. AWS A 5.14 Class ERNiCr-3; ASME SFB 5.14.

INCONEL FILLER METAL 92

Inco Alloys International Inc.
Ni 71, C 0.03, Mn 2.3, Fe 6.6, S 0.007, Si 0.1, Cu 0.04, Cr 16.4, Ti 3.2.
For gas shielded arc welding of dissimilar alloys such as austenitic and ferritic steels, to each other and to high-nickel alloys; overlaying on steel. AWS A 5.14 Class ERNiCrFe-6; ASME SFB 5.14.

INCONEL M

Now INCONEL ALLOY 721.

INCONEL W

Now INCONEL ALLOY 722.

INCONEL WELDING ELECTRODE 112

Inco Alloys International Inc.
Ni 61, C 0.05, Mn 0.3, Fe 4, S 0.01, Si 0.4, Cr 21.5, Mo 9, 3.65 Cb + Ta.
Electrode for shielded metal-arc welding of Inconel alloy 625 and dissimilar nickel-base and iron-base alloys. MIL-E-22200/3.

INCONEL WELDING ELECTRODE 113

Inco Alloys International Inc.
Ni 61, C 0.05, Mn 1, Fe 4, Si 0.4, Cr 21.5, Mo 9, 2.75 Cb + Ta.
For shielded metal-arc welding of nickel steels for cryogenic applications. *Obsolete*

INCONEL WELDING ELECTRODE 117

Inco Alloys International Inc.
For INCONEL alloy 617; INCOLOY alloy 800HT; dissimilar combinations of high temperature alloys. ENiCrCoMo-1.

INCONEL WELDING ELECTRODE 132

Inco Alloys International Inc.
Ni 73, C 0.04, Mn 0.75, Fe 8.5, S 0.006, Si 0.2, Cb 2.1, Co 0.05, Ta 0.05, Cu 0.04, Cr 15.
Electrode for shielded metal-arc welding of Inconel alloy 600. AWS A 5.11 Class ENiCrFe-1; AMS 5684.

INCONEL WELDING ELECTRODE 182

Inco Alloys International Inc.
Ni 67, C 0.05, Mn 7.75, Fe 7.5, S 0.008, Si 0.5, Ti 0.4, Cb 1.75, Co 0.08, Ta 0.03, C 0.1, Cr 14.
Electrode for shielded metal-arc welding of Inconel alloy 600; welding dissimilar combinations of nickel-base and iron-base alloys; overlaying on steel. AWS A 5.11 Class ENiCrFe-3; ASME SFB 5.11.

INCONEL X

Now INCONEL ALLOY X750.

INCONEL X-750

Enpar Sonderwerkstoffe GmbH
Nickel.
Alternate manufacturer.

INCONEL X-750

Criterion Metals, Inc.
Nickel. Ni 73, C 0.04, Mn 0.7, Fe 6.75, S 0.007, Si 0.3, Cu 0.05, Cr 15, Ti 2.5, Al 0.8, Cb 0.85.
Thin gauge sheet, various tempers. Annealed: 140,000 psi TS (<0.010 gauge). Skin hard: 135,000 psi TS (0.010-0.025 gauge).

INCONEL X550

Now INCONEL 751.

INCOR 16

Electronic Memories & Magnetics Corp.
Re 33, Co 67.
Energy product 16.0 x 10^6 max; residual induction 8100 gauss; coercive force 7900 oersted; intrinsic coercive force 18,000 oersted. Permanent magnet.

INCRA 4040-6

Brush Wellman
Ni 13.5-16.5, Al 9.5-10.5, Fe 0.4-1, Co 1-2, bal Cu.
At RT: 91,000 TS; 60,000 YS; 5 El; 195 Brin. For molds and accessory components in the glass industry. Thermal shock and fatigue resistant; good oxidation and growth resistance. *Obsolete*

INCRAMET-800

Copper Development Assoc., Inc.
Ni 13.5-16.5, Al 10.75-11.5, Fe 0.4-1, Co 1-2, bal Cu.
Cast: 91,000 TS; 60,000 YS; 5 El; 195 Brin. At 1100°F: 25,000 TS; 30 Rock A. For glass molds, sheet glass rolls. Heat and corrosion resistant.

INCRAMUTE I AND II

N.C. Ashton Ltd.
Cu, Mn.
Alloy with high damping characteristics.

INCROLOY

Electronic Memories & Magnetics Corp.
Co 15, Cr 28, Si 1, bal Fe.
Energy product 4.0 x 10^6 max; residual induction 13,000 gauss; coercive force 550 oersted. Permanent magnet.

INCURO 20

GTE Products Corp./Wesgo Div.
Au 20, Cu 78, In 2.
Brazing alloy available in foil, flexibraze, wire, powder, extrudable paste and preform. Liquidus 1877°F. Solidus 1787°F.

INCURO 20

Western Gold & Platinum Co.
In, Cu, Au.
For brazing alloy; melting point: 970-1015°C.

INCURO 60
GTE Products Corp./Wesgo Div.
Au 60, Cu 37, In 3.
Brazing alloy available in foil, flexibraze, wire, powder, extrudable paste and preform. Liquidus 1652°F. Solidus 1580°F.

INCURO-60
Western Gold & Platinum Co.
Au 60, Cu 32, In 3.
For brazing Kovar. Melting point: 860-900°C. Low vapor pressure.

INCUS
Uddeholm Corp.
C 0.55, Mn 0.7, Cr 0.7, Ni 1.75, Mo 0.7, bal Fe.
For drop forging dies; tough. *Obsolete*

INCUSIL 10
GTE Products Corp./Wesgo Div.
Ag 63, Cu 27, In 10.
Brazing alloy available in foil, flexibraze, wire, powder, extrudable paste and preform. Liquidus 1346°F. Solidus 1265°F.

INCUSIL 10
Western Gold & Platinum Co.
Ag 63, Cu 27, In 10.
For brazing glass headers. Melting point: 685-730°C.

INCUSIL 15
GTE Products Corp./Wesgo Div.
Ag 61.5, Cu 24, In 14.5.
Brazing alloy available in foil, flexibraze, wire, powder, extrudable paste and preform. Liquidus 1301°F. Solidus 1166°F.

INCUSIL 15
Western Gold & Platinum Co.
Ag 61.5, Cu 24, In 14.
For brazing glass headers. Melting point: 630-705°C.

INCUT 4140
Inland Steel Co.
C 0.38-0.43, Mn 0.75-1, Si 0.15-0.3, Pb 0.15-0.35, Cr 0.8-1.1, Mo 0.15-0.25, 0.035% min Te, bal Fe.
Rolled: 137,600 TS; 96,800 YS; 19 El; 47 RA. Heat treated: 139,000-203,000 TS; 116,000-186,000 YS; 9-19 El; 27-55 RA. For gears, shafts, axles, fittings, collets, forge hammer rams, studs, wrenches. Free-machining, oil hardening. *Obsolete*

INCUT 4142
Inland Steel Co.
C 0.4-0.45, Mn 0.75-1, Si 0.15-0.3, Pb 0.15-0.35, Cr 0.8-1.1, Mo 0.15-0.25, 0.035% min Te, bal Fe.
Rolled: 137,600 TS; 96,800 YS; 19 El; 47 RA. Heat treated: 139,000-203,000 TS; 116,000-186,000 YS; 9-19 El; 27-55 RA. For gears, shafts, axles, fittings, collets, forge hammer rams, studs, wrenches. Free-machining, oil hardening. *Obsolete*

INDALLOY
Indiana General
Co 12, Mo 17, bal Fe.
For permanent magnets; sintered; Br 9000, Hc 240. *Obsolete*

INDALLOY 12
Indium Corp. of America
Filler metal. In 5, Ag 5, Pb 90.
5730 psi TS; liquidus 310°C; solidus 290°C.

INDALLOY 13
Indium Corp. of America
Filler metal. In 70, Sn 15, Pb 9.6, Cd 5.4.
1476 psi TS; MP 125°C.

INDALLOY 146
Indium Corp. of America
Filler metal. In 20, Pb 79, Sb 1.
Liquidus 270°C; solidus 184°C.

INDALLOY 147
Indium Corp. of America
Filler metal. In 4, Bi 48, Pb 25.63, Sn 12.77, Cd 9.6.
Liquidus 65°C; solidus 61°C.

INDALLOY 16
Indium Corp. of America
Filler metal. In 16.1, Bi 44.7, Pb 22.6, Sn 11.3, Cd 5.3.
Liquidus 52°C; solidus 47°C.

INDALLOY 162
Indium Corp. of America
Filler metal. In 66.3, Bi 33.7.
MP 72°C; eutectic.

INDALLOY 17
Indium Corp. of America
Filler metal. In 20.89, Bi 49.14, Pb 17.92, Sn 11.55, Cd 0.5.
Liquidus 56°C; solidus 54°C.

INDALLOY 174
Indium Corp. of America
Filler metal. In 26, Bi 57, Sn 17.
MP 79°C; eutectic.

INDALLOY 178
Indium Corp. of America
Filler metal. In 18, Au 82.
Liquidus 485°C; solidus 451°C.

INDALLOY 179
Indium Corp. of America
Brazing alloy. In 15, Ag 61, Cu 24.
Liquidus 705°C; solidus 630°C.

INDALLOY 18
Indium Corp. of America
Filler metal. In 61.72, Bi 30.78, Cd 7.5.
MP 61.5°C; eutectic.

INDALLOY 1E
Indium Corp. of America
Filler metal. In 52, Sn 48.
1720 psi TS; MP 118°C; eutectic.

INDALLOY 203
Indium Corp. of America
Filler metal. In 95, Bi 5.
Liquidus 150°C; solidus 125°C.

INDALLOY 209
Indium Corp. of America
Filler metal. Ag 25, Sb 10, Sn 65.
Cast: 17,000 psi TS; MP 233°C.

INDALLOY 25
Indium Corp. of America
Filler metal. In 41.5, Bi 48.5, Cd 10.
MP 77.5°C; eutectic.

INDALLOY 253
Indium Corp. of America
Filler metal. In 74, Cd 26.
MP 123°C; eutectic.

INDALLOY 27
Indium Corp. of America
Filler metal. In 29.68, Bi 54.02, Cd 16.3.
MP 81°C; eutectic.

INDALLOY 290
Indium Corp. of America
Filler metal. In 97, Ag 3.
MP 141°C; eutectic.

INDALLOY 5
Indium Corp. of America
Filler metal. In 25, Sn 37.5, Pb 37.5.
5260 psi TS; liquidus 181°C; solidus 134°C.

INDALLOY 51
Indium Corp. of America
Filler metal. In 21.5, Ga 62.5, Sn 16.
MP 10.7°C; eutectic.

INDALLOY 53
Indium Corp. of America
Filler metal. In 33, Bi 67.
MP 109°C; eutectic.

INDALLOY 60
Indium Corp. of America
Filler metal. In 24.5, Ga 75.5.
MP 15.7°C; eutectic.

INDALLOY 71
Indium Corp. of America
Filler metal. In 48, Sn 52.
Liquidus 131°C; solidus 118°C.

INDALLOY 8
Indium Corp. of America
Filler metal. In 44, Sn 42, Cd 14.
2632 psi TS; MP 93°C; eutectic.

INDALLOY 87
Indium Corp. of America
Filler metal. In 42, Sn 58.
Liquidus 145°C; solidus 118°C.

INDALLOY 88
Indium Corp. of America
Filler metal. In 99.3, Ga 0.7.
MP 150°C. *Obsolete*

INDALLOY 9
Indium Corp. of America
Filler metal. In 12, Sn 70, Pb 18.
Cast: 5320 psi TS; MP 162°C.

INDALLOY 90
Indium Corp. of America
Filler metal. In 99.4, Ga 0.6.
MP 152°C. *Obsolete*

INDALLOY 91
Indium Corp. of America
Filler metal. In 99.6, Ga 0.4.
MP 153°C. *Obsolete*

INDALLOY 92
Indium Corp. of America
Filler metal. In 99.5, Ga 0.5.
MP 154°C.

INDALLOY-1
Indium Corp. of America
Filler metal. In 50, Sn 50.
Cast: 1720 TS; liquidus 125°C; solidus 118°C.

INDALLOY-3
Indium Corp. of America
Filler metal. In 90, Ag 10.
Cast: 1650 TS; liquidus 237°C; solidus 141°C.

INDALLOY-4
Indium Corp. of America
Filler metal. In 100.
Cast: 575 TS; 41 El; liquidus 157°C; solidus 157°C. For solder.

INDAR 5003
CMW Inc.
Cu 64, Zn 16.5, Ni 18, Pb 1.5.
Sintered: 27,600 TS; 8 El; H 76 Rock. Repressed: 35,000 TS; 15 El; H 80 Rock. For corrosion resistant and decorative parts, marine components. Nickel silver, stainless. *Obsolete*

INDAR 6003
CMW Inc.
Fe 99.75, C 0.25.
Sintered: 18,500-34,000 TS; 2 El; F 47-81 Rock. For structural parts and bearings. Carbon steel, water hardening. *Obsolete*

INDAR 6005
CMW Inc.
Fe 100.
Sintered: 19,500 TS; 1.5 El; F 54 Rock. For low strength structural parts and bearings. SAE 850. Powder metallurgy. *Obsolete*

INDAR 6012
CMW Inc.
C 1, Fe 99.
Sintered: 34,000-52,000 TS; 1 El; B 48-64 Rock. For structural parts and bearings. Water hardening, wear resistant. *Obsolete*

INDAR 6015
CMW Inc.
Fe 99.5, C 0.5.
Sintered: 27,000-42,000 TS; 1-2 El; B 28-55 Rock. For structural parts and bearings. Water hardening. *Obsolete*

INDAR 6016
CMW Inc.
Fe 100.
Sintered: 30,000 TS; 10 El; F 78 Rock. Repressed: 41,000 TS; 25 El. For magnetic pole pieces, relay armatures. Soft magnetic properties, high permeability. *Obsolete*

INDAR 6022
CMW Inc.
Cu 2, Fe 98.
Sintered: 21,000-39,000 TS; 0-1 El; B 60-68 Rock. For structural parts and bearings. May be oil impregnated. Good corrosion resistance. *Obsolete*

INDAR 6027
CMW Inc.
Fe 100.
Sintered: 26,000-36,000 TS; 1-3 El; F 64-72 Rock. For soft magnetic applications, yokes, pole pieces. High permeability. *Obsolete*

INDAR 6028
CMW Inc.
C 0.5, Mo 0.75, Ni 0.35, Mn 0.4, Si 0.25, bal Fe.
Sintered: 32,000-51,000 TS; 0 El; B 46-68 Rock. Heat treated: 37,000-65,000 TS; 0 El; C 28-48 Rock. For structural parts, gears, fasteners. Oil hardening. *Obsolete*

INDAR 6030
CMW Inc.
Cu 7, Fe 93.
Sintered: 29,000-50,000 TS; 0.5-1.0 El; F 72-86 Rock. For structural parts and bearings. May be oil impregnated. *Obsolete*

INDAR 6031
CMW Inc.
Cu 5, Fe 95.
Sintered: 27,000-48,000 TS; 0.5-1.0 El; F 72-86 Rock. For structural parts and bearings. May be oil impregnated. *Obsolete*

INDAR 6051
CMW Inc.
Cu 25, Fe 75.
Sintered: 20,000-40,000 TS; 1-2 El; F 37-75 Rock. For structural parts and bearings. May be oil impregnated. *Obsolete*

INDAR 6053
CMW Inc.
Ni 4, Cu 1, C 0.7, bal Fe.
Heat treated: 90,000-135,000 TS; 0.5 El; Rock C 25-45. For high strength structural parts. Supplied heat treated; wear resistant. *Obsolete*

INDAR 7004
CMW Inc.
Cu 78.5, Zn 20, Pb 1.5.
Sintered: 28,400 TS; 15 El; H 74 Rock. Repressed: 40,900 TS; 3 El; H 97 Rock. For bolts, fasteners, screw machine products. Leaded brass, corrosion resistant. *Obsolete*

INDAR 8001
CMW Inc.
Cu 100.
Sintered: 35,000 TS; 18 El; F 70 Rock. For power switchgear, heat sinks, current carrying structural parts. High conductivity of 90% min. Powder metallurgy. *Obsolete*

INDEFATIGABLE
French manufacture
C 0.26, Ni 4, Cr 0.6, Mo 0.9, bal Fe.
Heat treated: 230,000 TS; 190,000 YS; 6 El. Annealed: 120,000 TS; 100,000 YS; 16 El. For rail and switch parts, street car tracks; oil hardening.

INDEPENDENCE
Sanderson Kayser Ltd.
C 0.8, bal Fe.
For tools, springs; water hardened. *Obsolete*

INDEX
Westa-Westdeutsche
C 1, Si 0.2, Mn 0.25, V 0.1, bal Fe.
For springs, tools, cutters, reamers; Type W2; water hardened.

INDIE ZP
Heppenstall Co.
Tool material. C 0.35, Mn 0.8, Si 0.45, Cr 1.65, Mo 0.43, bal Fe.
Oil hardened tool and die steel designed for molds; AISI P20.

INDIUM
Atomergic Chemetals Corp.
In.
Purities, zone refined: 99.9999%, 99.999%, 99.99%, 99.9%. Forms: rod, ingot, shot, powder, wire, sheet, foil, single crystals.

INDOX 1
Electronic Memories & Magnetics Corp.
Sr 0.6 Fe_2O_3 (ceramic permanent magnet).
Normal peak energy product 1.0×10^6 max; residual induction 2200 gauss; coercive force 1825 oersted; intrinsic coercive force 3750 oersted.

INDOX 3
Electronic Memories & Magnetics Corp.
Sr 0.6 Fe_2O_3 (ceramic permanent magnet).
Peak normal energy product 2.6×10^6 max; residual induction 3350 gauss; coercive force 2350 oersted; intrinsic coercive force 2550 oersted.

INDOX 5
Electronic Memories & Magnetics Corp.
Sr 0.6 Fe_2O_3 (ceramic permanent magnet).
Peak normal energy product 3.4×10^6 max; residual induction 3800 gauss; coercive force 2400 oersted; intrinsic coercive force 2450 oersted.

INDOX 6
Electronic Memories & Magnetics Corp.
Sr 0.6 Fe_2O_3 (ceramic permanent magnet).
Peak normal energy product 2.45×10^6 max; residual induction 3300 gauss; coercive force 2800 oersted; intrinsic coercive force 3300 oersted.

INDOX 7
Electronic Memories & Magnetics Corp.
Sr 0.6 Fe_2O_3 (ceramic permanent magnet).
Normal peak energy product 2.8×10^6 max; residual induction 3450 gauss; coercive force 3250 oersted; intrinsic coercive force 4000 oersted.

INDOX 8
Electronic Memories & Magnetics Corp.
Sr 0.6 Fe_2O_3 (ceramic permanent magnet).
Normal peak energy product 3.5×10^6 max; residual induction 3850 gauss; coercive force 3050 oersted; intrinsic coercive force 3150 oersted.

INDOX V
Indiana General
BaO + FeO.
For permanent magnet for D.C. motors; ferrite, high peak energy product, cermet. *Obsolete*

INDOX VI-A
Indiana General
Barium ferrite.
Coercive force 3000 oersteds. Peak energy product 2.6×10^6 gauss-oersteds. Residual induction 3300 gausses. For applications requiring good resistance to demagnetization. Highly oriented. Permanent magnet. *Obsolete*

INDURA DRAWING DIE
Horace T. Potts Co.
C, alloy, bal Fe.
For drawing dies. *Obsolete*

INDUSTRIAL CRO-TUNG
Industrial Steels Inc.
C 0-0.3, Cr 4.5-6.5, W 0.75-1, bal Fe.
100,000-193,000 TS; 85,000-167,000 YS; 5-24 El; 6-62 RA; 210-490 Brin. For corrosion resistant parts; corrosion resistant.

INDUSTRIAL NO. 100
Industrial Steels Inc.
C 0.95-1.2, Cr 16-18, Mo 0-0.75, bal Fe.
Annealed: 95,000 TS; 60,000 YS; 30 El; 50 RA; 190 Brin. Heat treated: 300,000 TS; 250,000 YS; 1 El; 3 RA; 620 Brin. For bearings, valves, cutlery; corrosion and wear resistant; Type 440C.

INDUSTRIAL NO. 100FM
Industrial Steels Inc.
C 0.95-1.2, Cr 16-18, Ni 0-0.75, Mo 0-0.6, bal Fe.
Annealed: 95,000 TS; 60,000 YS; 30 El; 50 RA; 190 Brin. Heat treated: 300,000 TS; 250,000 YS; 1 El; 3 RA; 620 Brin. For bearings, cutlery, valves; Type 440F; wear resistant.

INDUSTRIAL NO. 12
Industrial Steels Inc.
C 0-0.15, Cr 11.5-13.5, bal Fe.
Annealed: 85,000 TS; 60,000 YS; 35 El; 75 RA; 135 Brin. For valves, cutlery, turbine blades; Type 410; corrosion resistant.

INDUSTRIAL NO. 12T
Industrial Steels Inc.
C 0-0.15, Cr 11.5-13, Mn 0-1, bal Fe.
Annealed: 85,000-100,000 TS; 60,000-75,000 YS; 25-35 El; 60-75 RA; 135-165 Brin. Heat treated: 100,000-200,000 TS; 60,000-175,000 YS; 10-30 El; 50-75 RA; 200-390 Brin. For chemical plant equipment; Type 403; corrosion resistant.

INDUSTRIAL NO. 17
Industrial Steels Inc.
C 0-0.12, Cr 14-18, Mn 0-1, Si 0-1, bal Fe.
Cold drawn: 130,000 TS; 120,000 YS; 3 El; 10 RA; 270 Brin. Annealed: 75,000 TS; 35,000 YS; 40 El; 60 RA; 145 Brin. For auto trim, kitchen sinks, oil refinery equipment; Type 430; corrosion resistant.

INDUSTRIAL NO. 17-7
Industrial Steels Inc.
C 0.09-0.2, Cr 16-18, Ni 7-9, Mn 0-1.2, bal Fe.
Annealed: 100,000 TS; 40,000 YS; 60 El; 160 Brin. 1/2 H temper: 150,000 TS; 110,000 YS; 15 El; 320 Brin. For chemical plant equipment, structural members; austenitic, stainless.

INDUSTRIAL NO. 18
Industrial Steels Inc.
C 0-0.12, Cr 18, bal Fe.
For corrosion resistant parts; corrosion resistant.

INDUSTRIAL NO. 18-10 TI
Industrial Steels Inc.
C 0-0.08, Cr 17-19, Ni 9-12, Ti = 5 x C, bal Fe.
Annealed: 80,000 TS; 35,000 YS; 60 El; 70 RA; 135 Brin. Cold rolled: 180,000 TS; 120,000 YS; 10 El; 300 Brin. For welded chemical plant equipment; Type 321; austenitic, stabilized.

INDUSTRIAL NO. 18-10CB
Industrial Steels Inc.
C 0-0.08, Cr 17-19, Ni 9-13, Cb = 10 x C, bal Fe.
Annealed: 80,000-90,000 TS; 35,000-45,000 YS; 55-60 El; 60-70 RA; 135-185 Brin. Cold drawn: 100,000-180,000 TS; 50,000-120,000 YS; 10-50 El; 180-300 Brin. For chemical plant equipment, welded structures; Type 347; austenitic, stainless.

INDUSTRIAL NO. 18-12MO
Industrial Steels Inc.
C 0-0.08, Cr 16-18, Ni 10-14, Mo 2-3, bal Fe.
Cold drawn: 100,000-150,000 TS; 50,000-125,000 YS; 15-55 El; 180-200 Brin. Annealed: 80,000-90,000 TS; 35,000-50,000 YS; 60-70 El; 60-70 RA; 135-185 Brin. For chemical plant equipment, mixers, agitators; Type 316; austenitic, acid resistant.

INDUSTRIAL NO. 18-8S
Industrial Steels Inc.
C 0-0.08, Cr 18-20, Ni 8-12, Mn 0-2, bal Fe.
Annealed: 80,000 TS; 35,000 YS; 60 El; 70 RA; 135 Brin. Cold drawn: 180,000 TS; 150,000 YS; 10 El; 330 Brin. For kitchen equipment, welded structures, process equipment; stainless, austenitic; Type 304.

INDUSTRIAL NO. 188
Industrial Steels Inc.
C 0-0.15, Cr 17-19, Ni 8-10, bal Fe.
Annealed: 80,000 TS; 35,000 YS; 60 El; 70 RA; 135 Brin. Cold rolled: 180,000 TS; 150,000 YS; 10 El; 330 Brin. For chemical plant equipment, tanks, mixers; Type 302; austenitic, stainless.

INDUSTRIAL NO. 188-M
Industrial Steels Inc.
C 0.2, Cr 18, Ni 8, Mo 2-4, bal Fe.
For stainless and corrosion resistant parts and equipment.

INDUSTRIAL NO. 188-U
Industrial Steels Inc.
C 0-0.07, Cr 17.5-18.5, Ni 8.5-9.5, bal Fe.
90,000-113,000 TS; 35,000-84,000 YS; 40-61 El; 60-68 RA; 135-223 Brin. For stainless parts, chemical plant equipment; stainless, austenitic.

INDUSTRIAL NO. 19-12-3MO
Industrial Steels Inc.
C 0-0.1, Cr 18-20, Ni 11-15, Mo 3-4, bal Fe.
Cold drawn: 100,000-150,000 TS; 50,000-125,000 YS; 15-50 El; 180-300 Brin. Annealed: 80,000-90,000 TS; 35,000-50,000 YS; 60-70 El; 60-70 RA; 135-185 Brin. For chemical plant equipment, mixers, agitators; Type 317; austenitic, acid resistant.

INDUSTRIAL NO. 20-10S
Industrial Steels Inc.
C 0-0.08, Cr 19-22, Ni 10-12, Mn 0-2, bal Fe.
Annealed: 90,000 TS; 40,000 YS; 50 El; 65 RA; 150 Brin. For chemical plant equipment; stainless, austenitic; Type 308.

INDUSTRIAL NO. 20-25
Industrial Steels Inc.
C 0-0.25, Cr 19-21, Ni 24-26, bal Fe.
Annealed: 100,000 TS; 50,000 YS; 45 El; 60 RA; 180 Brin. For furnace parts, heat treat equipment; Type 311; austenitic, heat resistant.

INDUSTRIAL NO. 21
Industrial Steels Inc.
C 0-0.35, Cr 18-23, bal Fe.
Cold drawn: 70,000-80,000 TS; 45,000-55,000 YS; 40-55 El; 55-60 RA; 150-180 Brin. For furnace parts, chemical plant equipment; Type 442; corrosion resistant.

INDUSTRIAL NO. 25-12
Industrial Steels Inc.
C 0-0.2, Cr 22-24, Ni 12-15, bal Fe.
Cold drawn: 100,000-250,000 TS; 50,000-200,000 YS; 5-45 El; 50-60 RA; 160-400 Brin. Annealed: 90,000-110,000 TS; 35,000-45,000 YS; 40-55 El; 50-65 RA; 170-200 Brin. For chemical plant equipment, furnace parts; Type 309; stainless, austenitic.

INDUSTRIAL NO. 25-12S
Industrial Steels Inc.
C 0-0.08, Cr 22-26, Ni 12-14, bal Fe.
Cold drawn: 100,000-250,000 TS; 50,000-200,000 YS; 5-45 El; 50-60 RA; 160-400 Brin. For heat treat boxes, furnace parts, kiln linings; Type 309S; austenitic heat resistant.

INDUSTRIAL NO. 25-20
Industrial Steels Inc.
C 0-0.25, Cr 24-26, Ni 19-22, Mn 0-2, bal Fe.
Annealed: 95,000 TS; 45,000 YS; 50 El; 65 RA; 170 Brin. For furnace parts, heat treat equipment; Type 310; austenitic, heat resistant.

INDUSTRIAL NO. 27
Industrial Steels Inc.
C 0-0.35, Cr 23-30, bal Fe.
Cold drawn: 75,000-85,000 TS; 50,000-60,000 YS; 35-50 El; 50-55 RA; 150-200 Brin. For furnace parts, chemical plant equipment; Type 443; corrosion resistant.

INDUSTRIAL NO. 35
Industrial Steels Inc.
C 0.3-0.45, Cr 12-14, Ni 0-0.5, Mo 0-0.5, bal Fe.
Annealed: 95,000 TS; 60,000 YS; 30 El; 50 RA; 190 Brin. Heat treated: 300,000 TS; 250,000 YS; 1 El; 3 RA; 620 Brin. For surgical instruments, valves, cutlery; Type 420; corrosion resistant.

INDUSTRIAL NO. 512
Industrial Steels Inc.
C 0-0.15, Cr 12-14, P 0-0.06, Mn 1.2, 0.15 S min, bal Fe.
Heat treated: 100,000-200,000 TS; 60,000-125,000 YS; 10-30 El; 50-75 RA; 200-390 Brin. Annealed: 85,000-100,000 TS; 60,000-75,000 YS; 25-35 El; 60-75 RA; 135-165 Brin. For screw machine products, valve trim, pump shafts; Type 416; corrosion resistant.

INDUSTRIAL NO. 512 FREE MACHINING
Industrial Steels Inc.
C 0-0.12, Cr 11.5-13, S 0.3-0.4, bal Fe.
40,000-175,000 TS; 25,000-160,000 YS; 15-35 El; 60-75 RA; 150-375 Brin. For corrosion resistant parts, new machine products; corrosion resistant; free cutting.

INDUSTRIAL NO. 517
Industrial Steels Inc.
C 0-0.12, Cr 14-10, Mo 0-0.6, 0.07 min S or Se, bal Fe.
Cold drawn: 75,000-100,000 TS; 45,000-90,000 YS; 8-34 El; 58-62 RA; 140-250 Brin. For screw machine products, shafts, pumps; Type 430F; corrosion resistant.

INDUSTRIAL NO. 5188
Industrial Steels Inc.
C 0-0.15, Cr 17-19, Ni 8-10, 0.15 S min, bal Fe.
Cold drawn: 150,000 TS; 120,000 YS; 10 El; 330 Brin. Annealed: 80,000 TS; 35,000 YS; 60 El; 75 RA; 130 Brin. For screw machine products; free cutting, stainless; Type 303.

INDUSTRIAL NO. 530
Industrial Steels Inc.
C 0.21-0.35, Cr 12-15, S, bal Fe
For stainless and corrosion resistant parts; corrosion resistant.

INDUSTRIAL NO. 65
Industrial Steels Inc.
C 0.6-0.7, Cr 16-17, bal Fe.
Heat treated: 143,000-250,000 TS; 115,000-223,000 YS; 4-13 El; 8-42 RA; 302-512 Brin. For corrosion resistant parts, cutlery; corrosion resistant.

INFILOY
PYRON Corp.
Cu 90-93, Fe 3.5-6, Mn 1-1.5, Ni 0.4-0.6, 0.5-2.0% others. Metal powder. Used in copper infiltration of powder metallurgy parts. *Obsolete*

INGACLAD
Ingersoll Steel Co.
20% layer 18-8 stainless welded to carbon steel.
62,000-68,000 TS; 48,000-50,000 YS; 48 El; 50 RA. For tanks, vessels, kettles, bins, table tops, vats; corrosion resistant. *Obsolete*

INGACLAD 309
Ingersoll Steel Co.
C 0.2, Cr 22-26, Ni 12-14, Mn 0-2, bal Fe.
Annealed: 60,000-70,000 TS; 40,000-50,000 YS; 30-50 El; 150-200 Brin. For heat and corrosion resistant parts; heat and corrosion resistant clad steel. *Obsolete*

INGACLAD 316
Ingersoll Steel Co.
C 0.1, Cr 16-18, Ni 0-14, Mo 2-3, bal Fe.
For heat and corrosion resistant parts; stainless clad steel. *Obsolete*

INGACLAD 317
Ingersoll Steel Co.
C 0.1, Cr 18-20, Ni 0-14, Mo 3-4, bal Fe.
For heat and corrosion resistant parts; stainless clad steel. *Obsolete*

INGACLAD 347
Ingersoll Steel Co.
C 0.1, Cr 17-20, Ni 8-12, Cb = 10 x C, bal Fe.
For corrosion resistant parts; stainless clad steel. *Obsolete*

INGACLAD TYPE 304
Ingersoll Steel Co.
C 0.8, Cr 18-20, Ni 8-10, bal Fe.
For stainless parts; stainless clad steel. *Obsolete*

INGERSALL
Ingersoll Steel Co.
C 0.1, Cr 18, Ni 8, bal Fe.
Annealed: 85,000 TS; 165 El. For stainless steel parts; stainless. *Obsolete*

INGERSOLL
Ingersoll Steel Co.
C 0.7-0.9, V 0.2, bal Fe.
For tools, drills, punches; water hardening. *Obsolete*

INGERSOLL DBL
Ingersoll Steel Co.
C 0.75-0.8, Cr 4, W 5, V 1.5, Mo 4, bal Fe.
For tools, cutters; high speed steel. *Obsolete*

INGERSOLL I-50
Ingersoll Steel Co.
C 0.5, Mn 0.75, bal Fe.
For knives, saws; water hardened. *Obsolete*

INGERSOLL I-60
Ingersoll Steel Co.
C 0.6, Mn 0.75, bal Fe.
For knives, saws; water hardened. *Obsolete*

INGERSOLL I-75
Ingersoll Steel Co.
C 0.75, Mn 0.75, bal Fe.
For knives, saws; water hardened. *Obsolete*

INGERSOLL I-85
Ingersoll Steel Co.
C 0.85, Mn 0.85, bal Fe.
For knives, saws; water hardened. *Obsolete*

INGERSOLL IC-125
Ingersoll Steel Co.
C 1.25, Mn 0.2, Cr 0.2, bal Fe.
For saws, knives, circular and hack saws; water hardened.
Obsolete

INGERSOLL IC-3C
Ingersoll Steel Co.
C 0.8, Mn 0.65, Cr 0.35, bal Fe.
For knives, saws; water hardened. *Obsolete*

INGERSOLL IC-4C
Ingersoll Steel Co.
C 0.85, Mn 0.6, Cr 0.25, bal Fe.
For knives, saws; water hardened. *Obsolete*

INGERSOLL IC-66
Ingersoll Steel Co.
C 0.7, Mn 0.65, Cr 0.45, bal Fe.
For knives, saws; water hardened. *Obsolete*

INGERSOLL IC-77
Ingersoll Steel Co.
C 0.75, Mn 0.75, Cr 0.45, bal Fe.
For knives, saws; water hardened. *Obsolete*

INGERSOLL IC-88
Ingersoll Steel Co.
C 0.85, Mn 0.45, Cr 0.45, bal Fe.
For knives, saws; water hardened. *Obsolete*

INGERSOLL IC-M
Ingersoll Steel Co.
C 0.9, Mn 0.9, Cr 0.8, bal Fe.
For knives and saws; water hardened. *Obsolete*

INGERSOLL ICN-100
Ingersoll Steel Co.
C 1, Mn 0.4, Cr 0.75, Ni 1.35, bal Fe.
For circular and hack saws, hog and meat knives; oil
hardened. *Obsolete*

INGERSOLL ICN-70
Ingersoll Steel Co.
C 0.7, Mn 0.25, Cr 0.35, Ni 1.35, bal Fe.
For knives, saws; oil hardened. *Obsolete*

INGERSOLL ICN-75
Ingersoll Steel Co.
C 0.75, Mn 0.35, Cr 0.35, Ni 1.25, bal Fe.
For knives, saws, valve discs; oil hardened. *Obsolete*

INGERSOLL ICN-80
Ingersoll Steel Co.
C 0.8, Mn 0.4, Cr 0.25, Ni 0.6, bal Fe.
For circular saws, hog knives, cordwood saws; water or oil
hardened. *Obsolete*

INGERSOLL ICN-K
Ingersoll Steel Co.
C 0.8, Mn 0.25, Cr 0.25, Ni 2.5, bal Fe.
For saws, knives, valve discs; oil hardened. *Obsolete*

INGERSOLL ICNM-55
Ingersoll Steel Co.
C 0.7, Mn 0.5, Cr 0.6, Ni 0.8, Mo 0.2, bal Fe.
For circular and hack saws, knives, valve discs; oil hardened.
Obsolete

INGERSOLL ICNM-9M
Ingersoll Steel Co.
C 0.75, Mn 0.35, Cr 0.2, Ni 2.65, Mo 0.2, bal Fe.
For circular and hack saws, knives; oil hardened. *Obsolete*

INGERSOLL ICNM-CS
Ingersoll Steel Co.
C 0.75, Mn 0.3, Cr 0.2, Ni 0.7, Mo 0.15, bal Fe.
For circular and hack saws, knives; oil hardened. *Obsolete*

INGERSOLL ICNV-LA
Ingersoll Steel Co.
C 1.25, Mn 0.3, Cr 1, Ni 0.55, V 0.25, bal Fe.
For hack and circular saws, knives; oil hardened. *Obsolete*

INGERSOLL ICV-100
Ingersoll Steel Co.
C 1, Mn 0.4, Cr 1.4, V 0.2, bal Fe.
For circular and hack saws, knives; oil or water hardened.
Obsolete

INGERSOLL ICV-35
Ingersoll Steel Co.
C 0.35, Mn 0.45, Cr 1, V 0.2, bal Fe.
For knives, saws, valve discs; water hardened. *Obsolete*

INGERSOLL ICV-50
Ingersoll Steel Co.
C 0.6, Mn 0.45, Cr 1, V 0.2, bal Fe.
For circular and hack saws, knives; water or oil hardened.
Obsolete

INGERSOLL ICV-98
Ingersoll Steel Co.
C 0.9, Mn 0.35, Cr 0.4, V 0.2, bal Fe.
For circular and hack saws, knives; oil or water hardened.
Obsolete

INGERSOLL ICW-90
Ingersoll Steel Co.
C 0.9, Mn 1.1, Cr 0.5, W 0.5, bal Fe.
For hack and circular saws, hog knives; oil hardened, tough.
Obsolete

INGERSOLL IN-40
Ingersoll Steel Co.
C 0.4, Mn 0.8, Ni 3.5, bal Fe.
For knives, saws; oil hardened. *Obsolete*

INGERSOLL IN-8
Ingersoll Steel Co.
C 0.75, Mn 0.4, Ni 2.05, bal Fe.
For knives, saws; oil hardened. *Obsolete*

INGERSOLL IN-8B
Ingersoll Steel Co.
C 0.75, Mn 0.85, Ni 2.05, bal Fe.
For knives, saws; oil hardened. *Obsolete*

INGERSOLL IN-9
Ingersoll Steel Co.
C 0.75, Mn 0.4, Ni 2.75, bal Fe.
For knives, saws; oil hardened. *Obsolete*

INGERSOLL IN-9B
Ingersoll Steel Co.
C 0.75, Mn 0.85, Ni 2.75, bal Fe.
For knives, saws; oil hardened. *Obsolete*

INGERSOLL INM-45
Ingersoll Steel Co.
C 0.45, Mn 0.5, Ni 1.4, Mo 0.35, bal Fe.
For saws, knives, friction saws; oil hardened. *Obsolete*

INGERSOLL INV-75
Ingersoll Steel Co.
C 0.75, Mn 0.35, Ni 2, V 0.1, bal Fe.
For circular saws, meat knives, drag saws; oil hardened.
Obsolete

INGERSOLL IV-85
Ingersoll Steel Co.
C 0.85, Mn 0.25, V 0.25, bal Fe.
For friction and hack saws, knives; water hardened. *Obsolete*

INGERSOLL IW-120
Ingersoll Steel Co.
C 1.2, Mn 0.3, W 1.15, bal Fe.
For saws, knives, friction saws; water hardened. *Obsolete*

INGERSOLL IWM-65
Ingersoll Steel Co.
C 0.65, Mn 0.7, W 0.45, Mo 0.45, bal Fe.
For valve discs, saws, knives; water hardened. *Obsolete*

INGERSOLL M-2
Ingersoll Steel Co.
C 0.8, W 6, Mo 5, V 2, Cr 4, bal Fe.
For hack saw blades; high-speed steel. *Obsolete*

INHIBITED ADMIRALTY
Now ADMIRALTY BRASS, INHIBITED.

INHIBITED ALUMINUM BRASS
Now ALUMINUM BRASS.

INJECTOR BRONZE
American manufacture
Sn 8.5, Pb 2.5, bal Cu.
For injectors, fittings, hardware; free-cutting.

INKOMO
Sanderson Kayser Ltd.
C 0.32, Ni 2.75-3.15, Cr 0.8, Mo 0.5, bal Fe.
Heat treated: 145,000-225,000 TS; 123,000-235,000 YS;
12-22 El; 33-60 RA; 269-504 Brin. For tappet rods, gears,
shafts, crankshafts; tough, shock resistant. *Obsolete*

INKROMSTAHL IK1
Thyssen Edelstahlwerke AG
C 0.1, Si 0.3, Mn 0.5, Ti 0.45, bal Fe.
For molten salt bath fixtures; suitable for chromizing.
Obsolete

INKROMSTAHL IK25
Thyssen Edelstahlwerke AG
C 0.09, Mn 0.35, Si 1, Cr 2.5, V 0.17, Cu 0.2, bal Fe.
For molten salt bath fixtures; suitable for chromizing.
Obsolete

INKROMSTAHL IK3
Colt Industries
C 0.1, 3% min Mn, bal Fe.
For molten salt bath fixtures; suitable for chromizing.
Obsolete

INKROMSTAHL IK85
Thyssen Edelstahlwerke AG
C 0.15, Si 0.8, Cr 7.5, V 0.5, Mo 1.2, bal Fe.
For molten salt bath fixtures; suitable for chromizing.
Obsolete

INKUS
Uddeholm Corp.
C 0.56, Cr 0.85, Mo 0.2, Ni 1.8, V 0.1, bal Fe.
For forging dies, punches; oil hardened, tough. *Obsolete*

INLAND AA
Inland Steel Co.
Carbon steel. C 0.1, Mn 0.65, P 0.01, S 0.2, Si 0.1, Al 0.005,
0.010 N_2, bal Fe.
Cold rolled: 45,000 YS; 64,000 TS; 26.1 El. Aged: 70,000 YS;
15 El. For formed sheet products, sills, containers, vessels,
trunk hinges. Age-hardened, readily fabricated.

INLAND COPPER BEARING STEEL
Inland Steel Co.
Carbon steel. C 0.05-0.15, Cu 0.2-0.3, bal Fe.
For corrosion resistant applications, roofing; corrosion
resistant.

INLAND HI-STEEL
Inland Steel Co.
C 0.1, Mn 0.6, Si 0.2, Cu 1.1, Ni 0.55, bal Fe.
70,000 TS; 55,000 YS. For structural applications, mining and transportation; weldable and workable. *Obsolete*

INLAND LEDLOY
Inland Steel Co.
Alloy steel. Pb 0.15-0.35, C, bal Fe.
For machinery parts; free cutting.

INLAND ZINC-ALLOY STEEL
Inland Steel Co.
Zn coated SAE 1010 steel. For sheets for fabricated parts; Zn alloyed to steel surface. *Obsolete*

INMANITE
T. Inman & Co. Ltd.
C 0.8, Cr 4.5, W 20, Mo 1, V 1.5, Co 5, bal Fe.
Cobalt-tungsten high speed tool steel. For cutting tools for very high speed and for hard materials.

INMET
Polish manufacture
Cu, Mn, Ni, Fe, bal Al.
For light alloy parts; age hardened.

INMOLD PROCESS ALLOY
Globe Metallurgical, Inc.
Miscellaneous nonferrous. Si 43-47, Mg 5-6.5, Ca 0.35-0.7, Al 0-0.75, 0.35-0.70 total rare earth elements.
A master alloy with grain size distribution for the INMOLD Process.

INNER SHIELD NR-203-NICKEL
Lincoln Electric Co.
Ni 2, C, bal Fe.
Self shielded, flux cored, arc welding electrode. For welding low temperature alloys. AWS E70T-G.

INNERBERG HG
Vereinigte Edelstahlwerke
C 0.55, Si 0.1-0.4, Mn 0.5-0.7, bal Fe.
Normalized: 105,000 TS; 65,000 YS; 17 El; 24 RA; 220 Brin. For gears, bolts, machine tool parts, shafts; water hardened. *Obsolete*

INNERSHIELD NR 431
Lincoln Electric Co.
C, bal Fe.
Self shielded, flux cored, arc welding electrode. As welded: 78,000-82,000 TS approx.

INNERSHIELD NR-1
Lincoln Electric Co.
C, bal Fe.
Self shielded, flux cored, arc welding electrode.

INNERSHIELD NR-131
Lincoln Electric Co.
C, bal Fe
Self shielded, flux cored, arc welding electrode. As welded (single pass): 72,000-75,000 TS. AWS 70T-G.

INNERSHIELD NR-202
Lincoln Electric Co.
C, bal Fe.
Self shielded, flux cored, arc welding electrode. AWS Class E60T-7.

INNERSHIELD NR-203
Lincoln Electric Co.
C, bal Fe.
Self shielded, flux cored, arc welding electrode. AWS Class E60T-7.

INNERSHIELD NR-203M
Lincoln Electric Co.
C, bal Fe.
Self shielded, flux cored, arc welding electrode. As welded, aged: 72,000-78,000 psi TS; 22-23 El; 70-120 Charpy V-Notch. AWS E70T-G.

INNERSHIELD NR-211
Lincoln Electric Co.
C, bal Fe.
Self shielded, flux cored, arc welding electrode. As welded, aged: 72,000-78,000 TS; 60,000-67,000 YS; 22-26 El. AWS E70T-G.

INNERSHIELD NR-301
Lincoln Electric Co.
C, bal Fe.
Self shielded, flux cored, arc welding electrode. As welded, aged: 72,000-81,000 TS; 60,000-69,000 YS; 22-29 El; 20-40 Charpy V-notch. AWS E70T-G.

INNERSHIELD NR-311
Lincoln Electric Co.
C, bal Fe.
Self shielded, flux cored, arc welding electrode. AWS Class E70T-G.

INNERSHIELD NR-5
Lincoln Electric Co.
C, bal Fe.
Self shielded, flux cored, arc welding electrode.

INNERSHIELD NS-3M
Lincoln Electric Co.
C, bal Fe.
Self shielded, flux cored, arc welding electrode. AWS Class E70T-4.

INNOSEIN
French manufacture
Cu 5-6, bal Al.
For light alloy parts; age-hardenable.

INNOTAL
Pechiney Electrometallurgie
Miscellaneous nonferrous. Mg 8-10, Ca 1, Ba 4, Si 45-50, Al 0-1, bal Fe.
Cast iron nodularizer.

INOBAR
Pechiney Electrometallurgie
Miscellaneous nonferrous. Al 1.5, Ba 9, Ca 1, Si 63, bal Fe.
Cast iron inoculant.

INOCARB
Pechiney Electrometallurgie
Miscellaneous nonferrous. Al 0.7, Ba 4.5, C 47, Ca 0.5, Si 32, bal Fe.
Cast iron inoculant.

INOCULOY 63
Cyprus Foote Mineral Co.
Si 60-65, Mn 9-12, Ca 1.5-3, Ba 4-6, Al 1-1.5, bal Fe.
For reducing chill in gray cast iron. Cast iron inoculant; ductile iron inoculant. *Obsolete*

INOFO 1
Acieries du Forez
Stainless steel; similar to AISI 410.

INOFO 10
Acieries du Forez
Stainless steel; similar to AISI 302.

INOFO 10B
Acieries du Forez
Stainless steel; similar to AISI 304.

INOFO 10TI
Acieries du Forez
Stainless steel; similar to AISI 321.

INOFO 11
Acieries du Forez
Stainless steel; similar to AISI 304L.

INOFO 12
Acieries du Forez
Stainless steel; similar to AISI 316.

INOFO 12B
Acieries du Forez
Stainless steel; similar to AISI 316 L.

INOFO 12TI
Acieries du Forez
Stainless steel; similar to AISI 316 Ti.

INOFO 13S
Acieries du Forez
Stainless steel; similar to AISI 303.

INOFO 1S
Acieries du Forez
Stainless steel; similar to AISI 416.

INOFO 2
Acieries du Forez
Stainless steel; similar to AISI 420.

INOFO 4
Acieries du Forez
Stainless steel; similar to AISI 430.

INOFO 4S
Acieries du Forez
Stainless steel; similar to AISI 430F.

INOFO 5
Acieries du Forez
Stainless steel; similar to AISI 431.

INOR-8
Westinghouse Electric Corp.
C 0.06, Cr 7, Mo 17, Fe 0-5, Mn 0-0.8, 0.5 Al + Ti max, bal Ni.
Heat treated: 115,100 psi TS; 45,500 psi YS; 50.7 El. At 1600°F: 35,000 psi TS; 25,000 psi YS; 30 El. For high temperature applications; resists hot fluoride salts, oxidation, aging, and embrittlement. *Obsolete*

INOR-8
Teledyne Allvac
C 0.06, Cr 7, Co 0.2, Mo 16, Ti 0.5, B 0.005, Fe 4, Si 1, bal Ni.
At room temperature: 100,000 UTS; 45,000 YS; 50 El. At 1200 F: 70,000 TS; 30,000 YS; 45 El. For high temperature applications; jet engine and gas turbine components. High heat and corrosion resistant. Resists hot fluoride salts. *Obsolete*

INOSSIDALFA
Alfa Romeo
Al alloy.
For light alloy parts.

INOSTRONG
Pechiney Electrometallurgie
Miscellaneous nonferrous. Al 0-0.5, Sr 0.85, Ca 0-0.1, Si 66, bal Fe.
Cast iron inoculant.

INOX 1
Societe Nouvelle des Acieries de Pompey
C 0-0.15, Cr 13, Ni 0-0.5, bal Fe.
Annealed: 75,000 psi TS; 40,000 psi YS; 35 El; 70 RA; 155 Brin. For turbine blades, cutlery, valves; Type 410 stainless.

INOX 1 F
Societe Nouvelle des Acieries de Pompey
C 0-0.15, Cr 13, S 0.2, Mo, bal Fe.
Free machining martensitic stainless. AISI 416.

INOX 16
Societe Nouvelle des Acieries de Pompey
C 0.08, Cr 16, Ni 0-0.5, bal Fe.
Annealed: 80,000 psi TS; 50,000 psi YS; 25 El; 50 RA; 150 Brin. For bolts, shafts, oil refinery equipment, heaters; Type 430 stainless, ferritic.

INOX 2
Societe Nouvelle des Acieries de Pompey
C 0.2, Cr 13.5, Ni 0-0.5, bal Fe.
Annealed: 95,000 psi TS; 50,000 psi YS; 25 El; 55 RA; 195 Brin. For turbine blades, cutlery, valves, surgical instruments; Type 420 stainless, hardenable.

INOX 3
Societe Nouvelle des Acieries de Pompey
C 0.3, Cr 13.5, Ni 0-0.5, bal Fe.
Annealed: 100,000 psi TS; 55,000 psi YS; 22 El; 52 RA; 210 Brin. For turbine blades, valves, cutlery, surgical instruments; Type 420 stainless, martensitic.

INOX 430 F
Societe Nouvelle des Acieries de Pompey
C 0.1, Cr 16.5, S 0.2, Mo, bal Fe.
Free machining ferritic type stainless steel. AISI 430F.

INOX 431
Societe Nouvelle des Acieries de Pompey
C 0.15, Cr 16, Ni 2, bal Fe.
Annealed: 125,000 TS; 95,000 YS; 20 El; 55 RA; 260 Brin. For pumps, marine hardware, valves; Type 431; stainless.
Obsolete

INOXALIUM
T.L.M. Co.
C 0.2, bal Fe.
For machinery parts; water hardenable.

INOXARGENT
Ugine Aciers
C 0-0.12, Cr 12-14, Ni 12-14, Cu, bal Fe.
Annealed: 70,000-78,000 TS; 28,000-44,000 YS; 45-50 El. For clocks, decorative trim, furniture; corrosion resistant.

INOXESCO 1
Vallourec S.A.
C 0-0.08, Mn 0-2, Cr 18-20, Ni 8-12, bal Fe.
Annealed: 75,000 TS; 30,000 YS; 35 El; 190 Brin. For oil refinery tubes and equipment; stainless, austenitic. *Obsolete*

INOXESCO 13
Vallourec S.A.
C 0.17, Mn 0.6, Cr 13, Mo 0.6, Ni 0.5, V 0.2, bal Fe.
Annealed: 85,000 TS; 40,000 YS; 24 El; B 95 Rock.
Hardened: 200,000 TS; 165,000 YS; 12 El; C 48 Rock. For cutlery, valves, bearings, surgical instruments, gears. Corrosion resistant, hardenable. *Obsolete*

INOXESCO 16
Vallourec S.A.
C 0-0.12, Mn 0-1, Cr 14-18, Ni 0-0.5, bal Fe.
Annealed: 60,000 TS; 35,000 YS; 20 El; 190 Brin. For oil refinery tubes and equipment; stainless, ferritic.

INOXESCO 2
Vallourec S.A.
C 0-0.08, Cr 17-20, Ni 9-13, Ti = 5 x C, bal Fe.
Annealed: 75,000 TS; 30,000 YS; 35 El; 190 Brin. For welded oil refinery equipment and tubes; stainless, austenitic.
Obsolete

INOXIUM 130 AB
Chiers-Chatillon
C 0.06, Cr 13, bal Fe.
Ferritic type stainless steel.

INOXIUM 130 TI
Chiers-Chatillon
C 0.06, Cr 13, Ti, bal Fe.
Ferritic type stainless steel for mufflers.

INOXIUM 170
Chiers-Chatillon
C 0.06, C 16.5, bal Fe.
Ferritic stainless steel; for decoration and trim. AISI 430.

INOXIUM 170 MO
Chiers-Chatillon
C 0.06, Cr 16.5, Mo, bal Fe.
Ferritic stainless steel; for automotive bumpers, wheel covers. AISI 434.

INOXIUM 18.10 N
Chiers-Chatillon
C 0.06, Cr 18.5, Ni 9, bal Fe.
Austenitic stainless steel; AISI 302.

INOXIUM 18.10 R
Chiers-Chatillon
C 0.06, Cr 18, Ni 9.5, bal Fe.
Austenitic type stainless; AISI 304.

INOXIUM 18.10 S
Chiers-Chatillon
C 0.02, Cr 18.5, Ni 10.5, bal Fe.
Austenitic type stainless; AISI 304 LC.

INOXIUM 18.10 T
Chiers-Chatillon
C 0.05, Cr 17.5, Ni 9, Ti, bal Fe.
Stabilized austenitic type stainless; AISI 321.

INOXIUM 18.13 D 25 R
Chiers-Chatillon
C 0.05, Cr 17.5, Ni 11, Mo 2.5, bal Fe.
Austenitic type stainless steel; improved corrosion resistance. AISI 316.

INOXIUM 18.13 D 25 T
Chiers-Chatillon
C 0.05, Cr 17.5, Ni 11, Mo 2.5, Ti, bal Fe.
Austenitic type stainless steel; improved corrosion resistance. AISI 316 Ti.

INOXIUM 18.8 C
Chiers-Chatillon
C 0.1, Cr 18, Ni 7.5, bal Fe.
Austenitic type stainless; cold rolled for increased strength.

INOXIUM 18.8 E
Chiers-Chatillon
C 0.1, Cr 17.5, Ni 8, bal Fe.
Austenitic type stainless; cold rolled for increased strength.

INOXYDA BRONZE
Manufacturer not listed.
Al 1-9, Cu 82-89, Zn 0-1.
For steam fittings; corrosion resistant.

INSTRUMENT BRONZE
American manufacture
Sn 13, Zn 5, bal Cu.
For instruments, utensils; corrosion resistant.

INSULUMINUM
Manufacturer not listed.
Steel with a surface impregnation of Al. For chemical engineering equipment. *Obsolete*

INTAL N 54
International Alloys Ltd.
Mg 2, Mn 0.8, bal Al.
For cast door fittings.

INTAL N 89
International Alloys Ltd.
Zn 50, Al 48, Cu 2.
Cast: 49,000 psi TS; 38,000 psi YS. For bearings, sand and permanent mold castings. Was MAIN METAL.

INTENSIV
Remystahl
C 1.9-2.2, Cr 11-12, bal Fe.
For blanking and forming dies, punches; air hardened, non-deforming. W.-Nr. 1.2080. AISI D3.

INTENSIV L SUPRA
Remystahl
C 0.92-1, Cr 11-12, Mo 0.8-1, V 0.8-1, bal Fe.
Cold work tool steel; similar to AISI D2. W.-Nr. 1.2376.

INTENSIV SPEZIAL
Remystahl
C 1.65, Cr 11.5, Mo 0.6, W 0.5, V 0.3, bal Fe.
For blanking and forming dies, punches; air hardening tool steel. W.-Nr. 1.2601.

INTENSIV W CO
Remystahl
C 2-2.3, Cr 11.5-12.5, Mo 0.3-0.5, W 0.6-0.8, Co 0.8-1.1, bal Fe.
Cold work tool steel. W.-Nr. 1.2884.

INTENSIV WO
Remystahl
C 2-2.25, Cr 11-12, W 0.6-0.8, bal Fe.
For blanking and forming dies, punches; oil hardened. W.-Nr. 1.2436.

INTENSIV Z
Remystahl
C 1.65, Cr 11.5, V 0.1, bal Fe.
For blanking and forming dies, punches; air hardened, non-deforming. W.-Nr. 1.2201.

INTENSIV ZAHDUR
Remystahl
C 0.9-1.05, Cr 4.8-5.5, Mo 0.9-1.2, V 0.1-0.3, bal Fe.
Warm work tool steel. W.-Nr. 1.2363.

INTENSIVE 25
Remystahl
C 2.5, Cr 12, bal Fe.
Cold work tool steel; AISI D3. *Obsolete*

INTENSIVE PLATT
Remystahl
C 1.55-1.75, Cr 11-12, Mo 0.5-0.6, Co 1.2-1.4, bal Fe.
Cold work tool steel. W.-Nr. 1.2880.

INTENSIVE SUPRA
Remystahl
C 1.5-1.6, Cr 11-12, Mo 0.6-0.8, V 0.9-1.1, bal Fe.
Cold work tool steel; similar to AISI D2. W.-Nr. 1.2379.

INTENSIVE SUPRA V
Remystahl
C 2.2, Cr 12.5, Mo 0.9, V 2.2, bal Fe.
Cold work tool steel. W.-Nr. 1.2378.

INTERMEDIATE MANGANESE ABRASION STEEL
U.S. Steel Corp.
C 0.35-0.5, Mn 1.5-2, P 0-0.05, S 0-0.055, Si 0.15-0.3, bal Fe.
Hot rolled: 100,000-125,000 TS; 200-250 Brin. For dredge pipe, screens; abrasion resistant; billets, sheets, plates.
Obsolete

INTERMEDIATE MANGANESE MEDIUM CARBON
U.S. Steel Corp.
C 0.28-0.38, Mn 1.1-1.4, P 0-0.04, Si 0.15-0.3, bal Fe.
Hot rolled: 90,000 TS; 55,000 YS; 18 El; 50 RA. For sucker rods; bars and rods; Iz 45 min. *Obsolete*

INTERMEDIATE MANGANESE RAIL STEEL
U.S. Steel Corp.
C 0.5-0.65, Mn 1.2-1.5, P 0-0.04, Si 0.1-0.23, bal Fe.
For railways, rails; wear resistant. *Obsolete*

INTRA
H. Boker & Co.
C 1.2, W 1.3, bal Fe.
For dies, reamers, blanking dies; water hardening. *Obsolete*

INVAR
Enpar Sonderwerkstoffe GmbH
Nickel.
Alternate manufacturer.

INVAR
Driver Harris Co.
C 0.18, Ni 35.5, Mn 0.42, bal Fe.
Annealed: 72,500 TS; 24,000 YS; 39 El; 67 RA. For measuring tapes and instruments, automobile and aircraft engine parts, thermostats; low coefficient of expansion. *Obsolete*

INVAR
Henry Wiggin & Co. Ltd.
C 0.18, Ni 35.5, Mn 0.42, bal Fe.
Annealed: 72,500 TS; 24,000 YS; 39 El; 67 RA. For measuring tapes and instruments, automobile and aircraft engine parts, thermostats; low coefficient of expansion. *Obsolete*

INVAR
Latrobe Steel Co.
C 0.15, Ni 38, Mn 0.8, bal Fe.
Annealed: 70,000 TS; 24,000 YS; 36 El; 68 RA; 143 Brin. Cold drawn: 90,000 TS; 70,000 YS; 20 El; 60 RA; 185 Brin. For bimetal thermostats, geodetic instruments; low coefficient of expansion. *Obsolete*

INVAR
Creusot-Loire
C 0.18, Ni 35.5, Mn 0.42, bal Fe.
Annealed: 72,500 TS; 24,000 YS; 39 El; 67 RA. For measuring tapes and instruments, automobile and aircraft engine parts, thermostats; low coefficient of expansion. *Obsolete*

INVAR
Now CARPENTER INVAR.

INVAR
VILLARES
C 0-0.08, Ni 36, bal Fe.
Low expansion steel. DIN 1.3912.

INVAR 36
Telcon Metals Ltd.
Ni 36, bal Fe.
Low-expansion alloy.

INVAR 36 FREE-CUT
Universal Cyclops
C 0.12, Mn 0.9, Si 0.35, Ni 36, 0.20 Se or S, bal Fe.
Annealed: 65,000 TS; 40,000 YS; 35 El; 70 Rock B. For geodetic instruments, thermostats, temperature control and indicating devices, component parts requiring minimum expansivity. Free-cutting, low coefficient expansion.

INVAR 36 FREE-CUT
Carpenter Technology Corp.
C 0.12, Mn 0.9, Si 0.35, Ni 36, 0.20 Se or S, bal Fe.
Annealed: 65,000 TS; 40,000 YS; 35 El; 70 Rock B. For geodetic instruments, thermostats, temperature control and indicating devices, component parts requiring minimum expansivity. Free-cutting, low coefficient expansion.

INVAR 42
Telcon Metals Ltd.
Ni 42, bal Fe.
Low-expansion alloy.

INVAR CLAD STEEL CLAD INVAR
Texas Instruments Inc./Materials Control
Invar clad 1008 AK steel clad Invar. Volume ratio 6/88/6.
Surge protectors for telephones, computers, and other electrical items.

INVAR FREE-MACHINING
Latrobe Steel Co.
C 0.12, Ni 38, Mn 0.8, Se 0.25, bal Fe.
Annealed: 70,000 TS; 24,000 YS; 36 El; 68 RA; 143 Brin. Cold drawn: 90,000 TS; 70,000 YS; 20 El; 60 RA; 185 Brin. For bimetal thermostats, geodetic instruments; low coefficient of expansion, free-cutting. *Obsolete*

INVAR M63
Creusot-Loire
Ni 36, bal Fe.
Low coefficient of thermal expansion. Good at low temperatures.

INVARIANT
English manufacture
Ni 47, Fe 53.
For surveyors' tape, piston struts; corrosion resistant.

INVARO NO. 1
Teledyne Firth Sterling
Steel. C 0.88, Mn 1-1.25, Cr 0.5, V 0.2, W 0.5, bal Fe.
For tools, hobs, milling cutters, taps; Type O1; oil hardened. *Obsolete*

INVARO-B
Teledyne Firth Sterling
Steel. C 0.94, Mn 1.2, W 0.53, V 0.22, Cr 0.5, bal Fe.
For punches, forming dies, shears; tough, nondeforming, oil hardened. *Obsolete*

INVAROD
Carpenter Technology Corp.
Ni 36, Mn 2.5, Ti 0.75, bal Fe.
Hot rolled: 77,500 TS; 42,000 YS; 40 El; 71 RA. For welding Invar parts. Weld wire permits crack-free welding. Aids cryogenic joining. *Obsolete*

INVARSTAHL
Vereinigte Deutsch Nickel-Werke AG
Ni 36, bal Fe.
86,000 TS; 32 El. For clocks, balances, measuring devices; same as "Invar." *Obsolete*

INVENTOR
Dunford Hadfields Ltd.
C 0.13-0.2, Cr 12-14, bal Fe.
Heat treated: 90,000-123,000 TS; 25 El; 179-228 Brin. For turbine blades, surgical instruments; Type 420; corrosion resistant. *Obsolete*

INVINCIBLE
Joseph Beardshaw & Son Ltd.
C 0.7, W 18, Cr 4, V 1, bal Fe.
For turning tools, reamers, drills, chasers, taps; high speed steel.

INVINCIBLE 22%
Joseph Beardshaw & Son Ltd.
C 0.75, W 22, Cr 4, V 1.2, bal Fe.
For lathe tools, reamers, drills, hobs; high speed steel.

INVINCIBLE DRILL ROD
Kidd Drawn Steel Co.
C 1-1.2, bal Fe.
For drills, general tools; water hardening.

INX-42
Inland Steel Co.
HSLA steel. C 0.2, Mn 0.9-1.35, Cb 0.01, bal Fe.
Rolled (min): 63,000 TS; 42,000 YS; 20 El. For trucks, plow frames, pressure vessels, building components, automobiles, trailers, railroad accessories. Good weldability, tough.

INX-45
Inland Steel Co.
HSLA steel. C 0.2, Mn 1.2, Cb 0.02, bal Fe.
Rolled: 45,000 YS min; 65,000 min TS; 19 min El. For trucks, plow frames, pressure vessels, automobiles, trailers, buildings. Good weldability, tough.

INX-50
Inland Steel Co.
HSLA steel. C 0.2, Mn 1.3, Cb 0.01, bal Fe.
Rolled (min): 70,000 TS; 50,000 YP; 18 El. For trucks, plow frames, pressure vessels, building components, water heaters, railroad accessories. Good weldability, tough.

INX-55
Inland Steel Co.
HSLA steel. C 0.22, Mn 1.35, Cb 0.01, bal Fe.
Rolled (min): 75,000 TS; 55,000 YP; 17 El. For trucks, plow frames, pressure vessels, construction equipment, water heaters, railroad accessories. Good weldability, tough.

INX-60
Inland Steel Co.
HSLA steel. C 0.25, Mn 1.35, Cb 0.01, bal Fe.
Rolled (min): 80,000 TS; 60,000 YS; 16 El. For trucks, plow frames, pressure vessels, building components, pipe, railroad accessories, automobiles. Good weldability, tough.

INX-65
Inland Steel Co.
HSLA steel. C 0.25, Mn 1.3, Si 2, Cb 0.03, bal Fe.
Rolled: 85,000 TS min; 65,000 YS min; 15 El min. For trucks, plow frames, pressure vessels, automobiles, railroad accessories. Good weldability.

INX-70
Inland Steel Co.
HSLA steel. C 0.28, Mn 1.3, Si 0.2, Cb 0.03, V 0.01, bal Fe.
Plate: 90,000 TS min; 70,000 YP min; 14 El min. For structures, plow shares, pressure vessels, railroad accessories, building components, water heaters. Good weldability and forming properties.

IOCHROME
Alloy Technology International, Inc.
Cr 99.997.
Rolled: 44 El; 78 RA. Alloying ingredient for high Cr alloys; for high temperature service. *Obsolete*

IPM 338-IH
IPM Corp.
High strength, low alloy steel, sintered. Density 7.4 g/cm^3, Sintered, heat treated: 148,000 TS; 2 El; 43 Rock C. Unnotched Charpy 13 ft·lb. For high strength and impact resistance.

IPM FG-982
IPM Corp.
Iron graphite, sintered. As sintered: 5000 TS; K Factor 11,500; density 5.8 g/cm^3. For P/M bearings to also include lubricating properties.

IPM FP-1
IPM Corp.
High phosphorus, soft magnetic material, sintered. Density 6.8 g/cm^3. At 10,000 gauss: coercive force 0.9; residual induction 9200; electrical resistivity 27 micro-ohm·cm. Maximum permeability 5500.

IPM SS-100
IPM Corp.
Austenitic stainless steel, sintered. As sintered: 47,000 TS; 16 El; radiofrequency 67; density 6.4 g/cm^3. For corrosion resistance comparable to AISI 316.

IPM SS-101
IPM Corp.
Austenitic stainless steel, sintered. As sintered: 38,000 TS; 9 El; radiofrequency 70; density 6.4 g/cm^3. For corrosion resistance comparable to AISI 304; excellent machinability.

IRALITE
Gulf & Western Mfg. Co.
C 3.1, Cr 1.2, Mo 0.3, bal Fe.
Cast: 36,000-52,000 TS; 150-180 Brin. For sheaves, sprockets, plungers, glass rolls, brake drums, conveyor parts. Pearlitic cast iron, wear resistant.

IRCAMET
Ingersoll-Rand Co.
C 0.07, Cr 20, Ni 29, Mo 2.75, Si 1, Cu 4, Mn 0.8, bal Fe.
Cast: 65,000 psi TS; 30,000 YS; 30-45 El; 40-50 RA; 140
Rolled: 85,000 psi TS; 35,000 YS; 35-50 El; 50-70 RA; 17
For furnace parts, chemical plant equipment; stainless, austenitic, nonmagnetic.

IRIDAL
Istituto Sperimentali Metalli Leggeri
B 0.3, Si 5, Mg 0.7, bal Al.
For architectural uses; heat treatable.

IRIDIUM
Atomergic Chemetals Corp.
Ir.
Purities: 99.999%, 99.99%, 99.9%. Forms: sponge, powder, rod, sheet, foil, single crystals, crucibles.

IRIDIUM
English manufacture
Zn 77-83, Cu 1.21-1.25, Sn 15.7-21.6, trace Sb.
For bearings; Babbitt.

IRIDIUM B.M
English manufacture
Zn 83, Sn 15.75, Cu 1.25, trace Sb.
For bearings; will not resist heat or live steam.

IRIDIUM EXTRA SPECIAL
Becker Stahlwerk A.G.
C 0.7, W 18, Cr 4, V 2, Co 5, bal Fe.
For cutters, tools; high speed steel.

IRIDIUM II
English manufacture
Zn 77.25, Sn 21.63, Cu 1.12.
For bearings; will not resist heat or live steam.

IRIDOSMINE
English manufacture
Os 57, Ru 8, 34 Rh + Ir.
For pen points; "Osmiridium."

IRON
Atomergic Chemetals Corp.
Fe.
Purities: zone and chemical refined 99.999%, 99.99%, 99.9+%. Forms: sponge, rod, chips, powder, wire, foil, single crystals.

IRON "8-J"
Power Products Inc.
C 2.9-3.3, Si 1.6-2, Mn 0.6, Ni 1-1.25, Mo 0.3, Cr 0.3, bal Fe.
Cast: 42,000 TS; 235 Brin. For cams, gears; alloy cast iron. *Obsolete*

IRON BRONZE
International Development Corp.
Sn 8.6, Zn 4.4, Fe 4, bal Cu.
Cast: 20-35 El; 70-85 Brin. Pressed: 20-35 El; 80-90 Brin. For shafts, piston rods, screws, marine parts; resists sea water corrosion.

IRON BRONZE
Allgemeines Deutsches Metallwerk GmbH
Sn 8.6, Zn 4.4, Fe 4, bal Cu.
Cast: 20-35 El; 70-85 Brin. Pressed: 20-35 El; 80-90 Brin. For shafts, piston rods, screws, marine parts; resists sea water corrosion.

IRON OILITE
Chrysler Corp.
Fe 97.25-0, C 0-0.25.
Sintered: 10,000-15,000 TS; 5.7-6.1 density. Oilite bearing material. ASTM B439-67 Gr 1; SAE 850; PMPMA F-0000-N.

IRON OILITE-1
Chrysler Corp.
Fe 97-0, C 0.6-1.
Sintered: 24,000 TS; 5.7-6.1 density. ASTM B310-67 Type I Cl.C; SAE 852.

IRON OILITE-1 IM
Chrysler Corp.
Cu 15-25, Fe 69-84.4, C 0.6-1.
Sintered: 85,000 TS; 7.1 min density. 40,000 shear; 17,000 fatigue. ASTM B303-67 Cl.C; SAE 872. PMPMA FX-2010-T.

IRON OILITE-M
Chrysler Corp.
C 0.5, S 0.5, Fe 99.
Sintered: 30,000-37,000 TS; 25,000-32,000 YS; 1.0-1.5 El; 30 Rb; density 6.1-6.8. Machinable grade oilite bearing material.

IRONAC
English manufacture
Si 13.2, Mn 0.77, C 1.1, P 0.8, bal Fe.
For pumps, valves, ejectors, nozzles, fans; acid resistant cast iron.

IRONIERS BRONZE
English manufacture
Hg 1, Sn, bal Cu.
For special applications; corrosion resistant.

IRONITE
Kinite Corp.
Cr, Ni, V, bal Fe.
35,000-40,000 TS; 230-390 Brin. For cams, gears; wear resisting. *Obsolete*

IRONWELD
Edgcomb Metals Co.
C 3.2, Si 2.2, bal Fe.
For welding rod for cast iron; flux-coated. *Obsolete*

IROQUOIS
Atlas Specialty Steels
C 1.25, W 1.4, bal Fe.
For taps, cartridge dies, punches; water hardening. *Obsolete*

IROQUOIS SPECIAL
Allegheny Ludlum Steel
C, alloy, bal Fe.
For tools, drills, taps. *Obsolete*

IRRUBIGO 16W
Schmidt & Clemens Edelstahlwerke
C 0.08, Cr 17, bal Fe.
Annealed: 80,000 TS; 50,000 YS; 25 El; 50 RA; 150 Brin. For oil refining equipment, sinks, soot blowers; Type 430; corrosion resistant. *Obsolete*

IRRUBIGO 17A
Schmidt & Clemens Edelstahlwerke
C 0.12, Cr 16.5, Mo 0.25, S 0.2, bal Fe.
Annealed: 80,000 TS; 50,000 YS; 25 El; 50 RA; 150 Brin. For screw machine products, fasteners; Type 430F; corrosion resistant, free-cutting. *Obsolete*

IRRUBIGO 17W
Schmidt & Clemens Edelstahlwerke
C 0-0.1, Cr 17.5, Ti = 4 x C, bal Fe.
Annealed: 80,000 TS; 50,000 YS; 30 El; 55 RA; 160 Brin. For welded oil refinery equipment; corrosion resistant. *Obsolete*

IRRUBIGO EXTRA W
Schmidt & Clemens Edelstahlwerke
C 0-0.1, Cr 17, Mo 1.8, Ti = 7 x C, bal Fe.
Annealed: 85,000 TS; 55,000 YS; 20 El; 45 RA; 160 Brin. For welded oil refinery equipment; heat and corrosion resistant. *Obsolete*

IRRUBIGO GS18
Schmidt & Clemens Edelstahlwerke
C 0.25, Cr 17, Ni 0-1.8, bal Fe.
For welded oil refinery equipment; corrosion and heat resistant. *Obsolete*

IRRUBIGO GS30
Schmidt & Clemens Edelstahlwerke
C 1.2, Si 1.3, Cr 29, bal Fe.
For rollers, crushers, grates, rabble arms; corrosion, heat and abrasion resistant. *Obsolete*

IRRUBIGO MC18
Schmidt & Clemens Edelstahlwerke
C 0.9, Cr 18, Mo 2, bal Fe.
Annealed: 105,000 TS; 62,000 YS; 18 El; 35 RA; 220 Brin. Heat treated: 280,000 TS; 270,000 YS; 3 El; 15 RA; 555 Brin. For cutlery, valves, bearings, surgical instruments; Type 440B-Mo; stainless, hardenable. *Obsolete*

IRRUBIGO S14
Schmidt & Clemens Edelstahlwerke
C 0.25, Cr 14.5, Ni 0-1, bal Fe.
Annealed: 80,000 TS; 45,000 YS; 30 El; 65 RA; 170 Brin. For oil refinery equipment; corrosion resistant. *Obsolete*

IRRUBIGO S18
Schmidt & Clemens Edelstahlwerke
C 0.25, Cr 17, Ni 0-1.8, bal Fe.
Annealed: 125,000 TS; 95,000 YS; 20 El; 55 RA; 260 Brin. For pumps, valves, marine hardware; Type 431; corrosion and heat resistant. *Obsolete*

IRRUBIGO S30
Schmidt & Clemens Edelstahlwerke
C 1.2, Si 1.3, Cr 29, bal Fe.
For crushers, rollers, grates, baffles; corrosion, heat and abrasion resistant. *Obsolete*

IRRUBIGO S30M
Schmidt & Clemens Edelstahlwerke
C 1.2, Si 1.3, Cr 29, Mo 2, bal Fe.
For crushers, rollers, grates, baffles; corrosion, heat and abrasion resistant. *Obsolete*

IRRUBIGO SN12
Schmidt & Clemens Edelstahlwerke
C 0.05, Cr 17, Ni 13, Mo 2, Ta, Cb, bal Fe.
Annealed: 90,000 TS; 40,000 YS; 45 El; 60 RA; 170 Brin. For welded acid resistant chemical plant equipment; Type 316 Cb; stainless, austenitic. *Obsolete*

IRRUBIGO SN18
Schmidt & Clemens Edelstahlwerke
C 0.05, Cr 17, Ni 17, Mo 2, Cu 0.2, Ta, Cb, bal Fe.
For valves, pumps, chemical plant equipment; corrosion and heat resistant, austenitic. *Obsolete*

IRRUBIGO SN5
Schmidt & Clemens Edelstahlwerke
C 0.4, Cr 27, Si 1.3, Ni 4, bal Fe.
Cast: 70,000 TS; 65,000 YS; 2 El; 190 Brin. Aged: 115,000 TS; 80,000 YS; 15 El. For furnace parts, grate bars, salt pots; Type HC; heat resistant. *Obsolete*

IRRUBIGO SN8
Schmidt & Clemens Edelstahlwerke
C 0.15, Si 1.5, Cr 18, Ni 8, bal Fe.
Annealed: 80,000 TS; 35,000 YS; 55 El; 75 RA; 150 Brin. Cold drawn: 180,000 TS; 150,000 YS; 10 El; 250 Brin. For chemical plant equipment, tanks, mixers, filters; Type 302; stainless, austenitic. *Obsolete*

IRRUBIGO SN8S

Schmidt & Clemens Edelstahlwerke
C 0-0.12, Cr 18, Ni 9.5, Ti = 4 x C, bal Fe.
Annealed: 85,000 TS; 35,000 YS; 55 El; 65 RA; 150 Brin. For welded chemical plant equipment, tanks, mixers; Type 321; stainless, austenitic. *Obsolete*

IRRUBIGO SN9

Schmidt & Clemens Edelstahlwerke
C 0.15, Si 2, Cr 18, Mo 2, Ni 9.5, bal Fe.
Annealed: 85,000 TS; 35,000 YS; 50 El; 65 RA; 160 Brin. For acid resistant chemical plant equipment, tanks; Type 316; stainless, austenitic. *Obsolete*

IRRUBIGO SN9S

Schmidt & Clemens Edelstahlwerke
C 0-0.12, Cr 18, Mo 2, Ni 10.5, Ti = 4 x C, bal Fe.
Annealed: 85,000 TS; 40,000 YS; 50 El; 65 RA; 170 Brin. For welded acid resistant chemical plant equipment; Type 316 Ti; stainless, austenitic. *Obsolete*

IRRUBIGO, ETC.

Now MARKER IRR, ETC.

ISA 13

Isabellenhutte
Copper. Cu 97, Mn 3.
290 N/mm^2 TS. For electrical equipment and instruments. Resistance alloy. Maximum working temperature 300°C.

ISA 50

Isabellenhutte
Copper. Cu 81.8, Ni 5, Mn 12, Al 1.2.
400 N/mm^2 TS. For electrical equipment and instruments. Resistance alloy. Maximum working temperature 500°C.

ISA-CHROM 60

Isabellenhutte
Nickel, Ni 65, Cr 15, Fe 20.
Annealed: 600 N/mm^2 TS. For electrical equipment and instruments. Resistance alloy. Maximum working temperature to 1150°C.

ISA-CHROM 80

Isabellenhutte
Nickel. Ni 80, Cr 20.
Annealed: 650 N/mm^2 TS. For electrical equipment and instruments. Resistance alloy. Maximum working temperature to 1200°C.

ISA-NICKEL

Isabellenhutte
Nickel. Cu 31, Ni 67, Mn 1, Fe 1.
45 N/mm^2 TS. For electrical equipment and instruments. Resistance alloy. Maximum operating temperature to 600°C.

ISA-SPRAY

Isabellenhutte
Cu 88, Mn 10, Al 2.
Hard: 128,000-144,000 TS. Soft: 64,000-72,000 TS. For electrical equipment and instruments. Resistance alloy. Maximum working temperature to 400°C. *Obsolete*

ISABELLIN

Isabellenhutte
Cu, Mn, Al.
For electrical resistant parts. *Obsolete*

ISABELLIN-A

Isabellenhutte
Cu 84, Mn 13, Al 3.
Hard: 128,000-144,000 TS. Soft: 72,000-78,000 TS. For electrical equipment and instruments. Resistance alloy. Maximum operating temperature to 400°C. *Obsolete*

ISAOHM

Isabellenhutte
Nickel. Ni 74.5, Cr 20, Mn 0.5, Fe 0.5, Al 3.5, Si 1.
Annealed: 1000 N/mm^2 TS. For electrical equipment and instruments. Resistance alloy. Maximum working temperature 250°C.

ISAZIN

Isabellenhutte
Copper. Cu 75.5, Ni 23, Mn 1.5.
350 N/mm^2 TS. For electrical equipment and instruments. Resistance alloy. Maximum operating temperature to 400°C.

ISC 1020

Induction Steel Castings Co. Inc.
C 0.2, Mn 0.6, Si 0.5, bal Fe.
Cast: 60,000 psi TS; 35,000 psi YS; 24 El; 35 RA. General purpose cast steel.

ISC 1045

Induction Steel Castings Co. Inc.
C 0.45, Mn 0.6, Si 0.5, bal Fe.
Cast: 85,000 psi TS; 40,000 psi YS; 16 El; 24 RA. Steel casting; can be hardened by heat treating.

ISC 15-3

Crucible Steel Castings Co.
C 3, Mn 0.7, Si 0.5, Cr 15, Mo 3, bal Fe.
550-650 Brin; extreme wear resistance, poor impact. *Obsolete*

ISC 4335

Induction Steel Castings Co. Inc.
C 0.35, Mn 0.7, Si 0.5, Ni 1.75, Cr 0.7, Mo 0.35, bal Fe.
Alloy steel casting, hardenable to 200-400 Brin in comparatively heavy sections. For components requiring high strength.

ISC 6145

Induction Steel Castings Co. Inc.
C 0.45, Mn 0.6, Si 0.5, Cr 0.95, V 0.15, bal Fe.
For gears, cams, shafts; flame or induction hardened.

ISC 8630

Induction Steel Castings Co. Inc.
C 0.3, Mn 0.7, Si 0.5, Ni 0.5, Cr 0.5, Mo 0.2, bal Fe.
Steel casting, hardenable to 180-350 Brin.

ISC AIR HARD

Induction Steel Castings Co. Inc.
C 1.5, Mo 0.8, V 0.5, Mn 0.7, Cr 12, bal Fe.
For dies, cams, punches, blanking and forming tools; air hardened, non-deforming.

ISC FLAME HARD

Induction Steel Castings Co. Inc.
C 0.55, Mn 1.2, Cr 1.25, Mo 0.4, V 0.1, bal Fe.
For trim dies; oil or flame hardened.

ISC H-13

Induction Steel Castings Co. Inc.
C 0.35, Mn 0.5, Si 0.5, Cr 5, Mo 1.5, V 0.9, bal Fe.
Cast hot work tool steel; air hardening.

ISC OIL HARD

Induction Steel Castings Co. Inc.
C 0.9, Mn 1.25, Cr 0.5, Mo 0.3, V 0.25, bal Fe.
For forming and blanking dies, punches, headers; oil hardened, non-deforming.

ISC WEAR HARD

Induction Steel Castings Co. Inc.
C 0.3, Mn 1.4, Si 0.5, Cr 1, Mo 0.5, B, bal Fe.
Cast (hardened): 450-500 Brin; high wear resistance; good impact properties.

ISCAR IC 2

Iscar Ltd.
Sintered carbide. Transverse strength: 330,000 psi; 91.9 Rock A. For machining soft cast iron, non-ferrous alloys, wood, plastic and other non-metallics. Code C2.

ISCAR IC 20

Iscar Ltd.
Sintered carbide. Transverse strength: 300,000 psi; 92.5 Rock A. For machining cast irons, silicon aluminum alloys and non-metallics. Code C3.

ISCAR IC 28

Iscar Ltd.
Sintered carbide. Transverse strength: 450,000 psi; 90.5 Rock A. For roughing cuts on gray cast iron, austenitic and high temperature alloys. Code C1.

ISCAR IC 4

Iscar Ltd.
Sintered carbide. Transverse strength: 250,000 psi; 93.0 Rock A. For semi-finishing and finishing abrasive alloys and non-metallics. Code C4.

ISCAR IC 424 (ISCOTIC)

Iscar Ltd.
Sintered TiC coated carbide. For machining cast irons, manganese steels, and high temperature resistant alloys. Code C4, C2.

ISCAR IC 50

Iscar Ltd.
Sintered carbide. Transverse strength: 315,000 psi; 91.5 Rock A. For rough and semi-finish machining of steel, steel castings and malleable iron. Code C5.

ISCAR IC 50M

Iscar Ltd.
Sintered carbide. Transverse strength: 340,000 psi; 91.7 Rock A. For rough and semi-finish milling of steel, steel castings and malleable cast iron. Code C6.

ISCAR IC 54

Iscar Ltd.
Sintered carbide. Transverse strength: 380,000 psi; 90.7 Rock A. For planing, turning, heavy roughing, and difficult cutting of steel and iron and steel castings.

ISCAR IC 636 (ISOTIC)

Iscar Ltd.
Sintered TiC/N coated carbide. Tough grade for roughing of tool steels, iron and steel castings. Code C5, C6.

ISCAR IC 70

Stackpole Magnet Division
Sintered carbide. 275,000 psi transverse strength; 92.6 Rock A. For roughing and semi-finishing steel, steel castings and malleable cast iron. Meets code C7, C6. *Obsolete*

ISCAR IC 757 (ISCOTIC)

Iscar Ltd.
Sintered TiC coated carbide. For rough and finish turning of steel castings, malleable and nodular iron. Code C7, C5.

ISCAR IC 78

Iscar Ltd.
Sintered carbide. Transverse strength: 240,000 psi; 92.4 Rock A. For general machining carbon and alloy steels. Code C7.

ISCAR IC 80 T

Iscar Ltd.
Sintered titanium carbide. Transverse strength: 210,000 psi; 92.9 Rock A. For semi-finish machining carbon and alloy steels. Code C7, C8.

ISERLOHN

Manufacturer not listed.
Cu 64, Zn 34, Sn 2.4, Pb 0.3.
For fittings, hardware; corrosion resistant.

ISERLOHN

English manufacture
Cu 70.1, Zn 29.9.
For cartridges, shell cases, condenser tubes; maximum ductility.

ISERLOHN BRASS

English manufacture
Cu 70, Zn 30.
1/4 H-temper: 54,000 TS; 40,000 YS; 43 El; 100 Brin. 1/2 H-temper: 62,000 TS; 52,000 YS; 23 El; 118 Brin. For cartridge cases, condenser tubes; high ductility and workability.

ISO-450

Empire Steel Castings Co.
Stainless steel. C 0-0.05, Cr 14.5-16.5, Ni 5.5-7, Mo 0.5-1, Cu 1.25-1.75, Cb = 8 x C min, bal Fe.
Solution annealed: 128 ksi TS; 118 ksi YS; 15 El; 53 RA. Martensitic, age-hardenable stainless.

ISO-CAST "MONEL-E"

Empire Steel Castings Co.
C 0.15, Ni 63, Cu 30, Si 1.5, Cb 1.
As cast: 65,000 TS; 32,000 YS; 30 El. Used in contact with reducing compounds as boric acid, Ca, K, Na hydroxides.

ISO-CAST 1

Empire Steel Castings Co.
C 0.2-0.3, Mn 0.6-0.8, Si 0.35-0.5, bal Fe.
Annealed: 70,000 TS; 40,000 YS; 25 El; 40 RA; 130 Brin. For castings; general purpose.

ISO-CAST 1A

Empire Steel Castings Co.
C 0.3-0.4, Mn 0.5-0.7, Si 0.35-0.5, bal Fe.
Annealed: 75,000 TS; 45,000 YS; 25 El; 40 RA; 165 Brin. For castings; general purpose.

ISO-CAST 1B

Empire Steel Castings Co.
C 0.4-0.5, Mn 0.6-0.8, Si 0.35-0.5, bal Fe.
Annealed: 85,000 TS; 55,000 YS; 20 El; 30 RA; 174 Brin. For castings; general purpose.

ISO-CAST 2

Empire Steel Castings Co.
C 0.5-0.6, Mn 0.6-0.8, Si 0.35-0.5, bal Fe.
Annealed: 95,000 TS; 60,000 YS; 15 El; 20 RA; 202 Brin. For castings; high strength.

ISO-CAST 2A

Empire Steel Castings Co.
C 0.6-0.7, Mn 0.6-0.8, bal Fe.
Annealed: 105,000 TS; 65,000 YS; 12 El; 15 RA; 222 Brin. For castings; high strength.

ISO-CAST 2B

Empire Steel Castings Co.
C 0.7-0.8, Mn 0.6-0.8, Si 0.35-0.5, bal Fe.
Annealed: 112,000 TS; 75,000 YS; 10 El; 12 RA; 235 Brin. For castings; abrasion resistant.

ISO-CAST 3 S

Empire Steel Castings Co.
C 0.33-0.47, Mn 1.25-1.5, bal Fe.
Cast steel.

ISO-CAST 6A

Empire Steel Castings Co.
C 0.15-0.25, Mn 0.6-0.8, Ni 2.75-3.5, bal Fe.
Tempered: 85,000 TS; 55,000 YS; 20 El; 40 RA; 159 Brin. For castings, machinery parts.

ISO-CAST 7A

Empire Steel Castings Co.
C 0.53-0.67, Mn 0.5, Si 0.5, Cr 2.5, V 0.2, bal Fe.
Cast, low-alloy steel; for mixing blades, crushing equipment, ball mills. Hard and abrasion resistant.

ISO-CAST 85

Empire Steel Castings Co.
C 0-0.2, Cr 4-6.5, Mo 0.45-0.65, bal Fe.
Normalized: 90,000-105,000 TS; 60,000-70,000 YS; 18-24 El; 35-50 RA; 192-200 Brin. For pump valves; for strength at elevated temperatures.

ISO-CAST 8A

Empire Steel Castings Co.
C 0.35-0.45, Cr 1, Mo 0.3, Mn 0.7, Si 0.5, bal Fe.
Normalized: 90,000-105,000 TS; 60,000-80,000 YS; 18-24 El; 35-50 RA; 207-220 Brin. For gears, shafts, machine tool housings; SAE 4142; tough and wear resistant.

ISO-CAST 8B

Empire Steel Castings Co.
C 0-0.2, Cr 1-1.5, Mo 0.45-0.65, bal Fe.
Normalized: 75,000-90,000 TS; 45,000-65,000 YS; 25-30 El; 40-50 RA; 159-170 Brin. For steam service castings; for strength at elevated temperatures.

ISO-CAST 8C

Empire Steel Castings Co.
C 0-0.18, Mn 0.4-0.7, Si 0-0.6, Cr 2-2.75, Mo 0.9-1.2, bal Fe.
Cast: 75,000-95,000 TS; 45,000-65,000 YS; 25-30 El; 40-50 RA. For strength at elevated temperatures, steam service.

ISO-CAST B

Empire Steel Castings Co.
C 0-0.12, Cr 0-1, Ni 63, Mo 28, Fe 6.
Cast: 80,000 TS; 57,000 YS; 8 El; 11 RA; 200 Brin. For equipment handling HCl in all concentrations. Corrosion resistant to H_2SO_4 and H_3PO_4.

ISO-CAST C

Empire Steel Castings Co.
C 0-0.12, Cr 17, Ni 55, Mo 17, Fe 6, W 4.25, Co 0.8, V 0.3.
Cast: 78,000 TS; 57,000 YS; 7 El; 10 RA; 200 Brin. For equipment handling nitric acid, solutions containing Cl_2. Corrosion resistant.

ISO-CAST CA-15

Empire Steel Castings Co.
C 0.15, Ni 1, Cr 11-14, bal Fe.
Quenched and treated: 95,000 TS; 65,000 YS; 20 El; 45 RA; 200 Brin. For valves, pumps, castings; corrosion resistant, hardenable.

ISO-CAST CA-40

Empire Steel Castings Co.
C 0.2-0.4, Ni 1, Cr 11.5-14, bal Fe.
Quenched and treated: 110,000 TS; 75,000 YS; 18 El; 35 RA; 220 Brin. For valves, pumps, castings; corrosion resistant, hardenable.

ISO-CAST CA6NM

Empire Steel Castings Co.
C 0.06, Cr 11.5, Ni 3.5, Mo 0.4, bal Fe.
Corrosion-resistant alloy cast iron. ASTM A 296 CA-6NM.

ISO-CAST CB-30

Empire Steel Castings Co.
C 0.3, Ni 2, Cr 18-22, bal Fe.
Cast: 90,000 TS; 40,000 YS; 200 Brin. For valves and heat resisting castings; corrosion and heat resistant.

ISO-CAST CB7CU

Empire Steel Castings Co.
C 0-0.07, Mn 0-1, Si 0-1, Cr 15.5-17, Ni 3.6-4.6, Cu 2.3-3.3, bal Fe.
Cast grade of 17-4 PH; age hardenable to about 200,000 psi TS.

ISO-CAST CE-30

Empire Steel Castings Co.
C 0.3, Ni 8-11, Cr 26-30, bal Fe.
Cast: 90,000 TS; 55,000 YS; 25 El; 25 RA; 200 Brin. For corrosion and heat-resistant castings; corrosion and heat resistant.

ISO-CAST CF-8

Empire Steel Castings Co.
Stainless steel. C 0.08, Ni 8-11, Cr 18-21, bal Fe.
Annealed: 75,000 TS; 40,000 YS; 45 RA; 140 Brin. For stainless castings, pumps, valves, processing equipment; stainless, austenitic.

ISO-CAST CF-8M

Empire Steel Castings Co.
Stainless steel. C 0.08, Ni 9-12, Cr 18-21, Mo 2-3, bal Fe.
Annealed: 80,000 TS; 40,000 YS; 50 El; 156 Brin. For stainless castings, valves, pumps; stainless, austenitic.

ISO-CAST CG-8M

Empire Steel Castings Co.
Stainless steel. C 0-0.08, Ni 10-14, Cr 18-21, Mo 3-4, bal Fe.
Cast: 80,000 TS; 44,000 YS; 33 El; 165 Brin. For paper and textile equipment, bleaching equipment; Type 317; stainless, austenitic.

ISO-CAST HA

Empire Steel Castings Co.
C 0-0.2, Mn 0.5, Si 0-1, Cr 8-10, Mo 0.9-1.2, bal Fe.
Cast, corrosion-resistant steel; wear and abrasion resistant. ACI HA.

ISO-CAST HD

Empire Steel Castings Co.
C 0-0.5, Mn 1.5, Si 0-2, Cr 26-30, Ni 4-7, Mo 0-0.5, bal Fe.
Cast, corrosion and heat-resistant alloy. ACI HD.

ISO-CAST HH

Empire Steel Castings Co.
C 0.25-0.5, Cr 24-28, Ni 11-14, bal Fe.
Cast: 85,000 TS; 50,000 YS; 35 El; 170 Brin. For corrosion and heat-resistant castings, furnace parts; corrosion and heat resistant.

ISO-CAST HI

Empire Steel Castings Co.
C 0.2-0.5, Mn 2, Si 2, Cr 26-30, Ni 14-18, Mo 0-0.5, bal Fe.
Cast, corrosion and heat-resistant alloy. ACI HI.

ISO-CAST HK

Empire Steel Castings Co.
C 0.2-0.6, Ni 18-22, Cr 24-28, bal Fe.
Cast: 80,000 TS; 45,000 YS; 25 El; 190 Brin. For corrosion and heat-resistant castings, furnace parts; corrosion and heat resistant, austenitic.

ISO-CAST HL

Empire Steel Castings Co.
C 0.2-0.6, Mn 0-2, Si 0-2, Cr 28-32, Ni 18-22, Mo 0-0.5, bal Fe.
Cast, corrosion and heat-resistant alloy. ACI HL.

ISO-CAST HN

Empire Steel Castings Co.
C 0.2-0.5, Mn 0-2, Si 0-2, Cr 19-23, Ni 23-27, Mo 0-0.5, bal Fe.
Cast, corrosion and heat-resistant alloy. ACI HN.

ISO-CAST HT

Empire Steel Castings Co.
C 0.35-0.75, Ni 33-37, Cr 13-17, bal Fe.
Cast: 75,000 TS; 7 El; 180 Brin. For corrosion and heat-resistant castings, furnace parts; corrosion and heat resistant.

ISO-CAST HU

Empire Steel Castings Co.
C 0.35-0.75, Mn 0-2, Si 0-2.5, Cr 17-21, Ni 37-41, Mo 0-0.5, bal Fe.
Cast, corrosion and heat-resistant alloy. ACI HU.

ISO-CAST HW

Empire Steel Castings Co.
C 0.35-0.75, Ni 58-62, Cr 10-14, bal Fe.
Cast: 65,000 TS; 180 Brin. At 1800°F: 10,000 TS; 8000 YS; 40 El. For furnace equipment, retorts, pots; heat and corrosion resistant.

ISO-CAST HX

Empire Steel Castings Co.
C 0.35-0.75, Mn 0-2, Si 0-2.5, Cr 15-19, Ni 64-68, Mo 0-0.5, bal Fe.
Cast, corrosion and heat-resistant alloy. ACI HX.

ISO-CAST M

Empire Steel Castings Co.
C 0.1-0.2, Mn 0.6-0.8, bal Fe.
Annealed: 55,000 TS; 30,000 YS; 35 El; 55 RA; 120 Brin. For castings, instrument parts. *Obsolete*

ISO-CAST NICKEL

Empire Steel Castings Co.
C 0-1, Ni 97, Si 0-2, Mn 0-1.5, Fe 0-3.
As cast: 53 ksi TS; 25 ksi YS; 25 El. Used for resistance to strong caustic solutions.

ISO-CAST-89

Empire Steel Castings Co.
C 0-0.2, Mn 0.35-0.65, Cr 8-10, Mo 0.9-1.2, Si 0-1, bal Fe.
Cast: 90,000-105,000 TS; 60,000-70,000 YS; 18-24 El; 35-50 RA. For strength at elevated temperatures, steam service.

ISOCAST 1EL

Empire Steel Castings Co.
C 0-0.1, Mn 0.5-0.7, Si 0-0.6, bal Fe.
For electric motors and generators; good magnetic permeability. *Obsolete*

ISOCAST 1L

Empire Steel Castings Co.
C 0-0.2, Mn 0.5-0.7, Si 0-0.6, bal Fe.
For electric motors and generators. *Obsolete*

ISOCAST 20

Empire Steel Castings Co.
Stainless steel. C 0-0.07, Cr 19-22, Ni 27.5-30, 2.0 Mo min, 3.0 Cu min, bal Fe.
Cast: 65,000 TS; 30,000 YS; 35 El; 140 Brin. For pumps, valves, chemical plant equipment; austenitic, stainless, resists H_2SO_4.

ISOCAST 24-12B

Empire Steel Castings Co.
C 0.21-0.35, Cr 24, Ni 12, bal Fe.
For heat and corrosion resistant parts; heat and corrosion resistant. *Obsolete*

ISOCAST 3

Empire Steel Castings Co
C 0.25-0.35, Mn 1.25 1.5, bal Fe.
Heat treated: 90,000 TS; 55,000 YS; 25 El; 40 RA; 200 Brin. For machine and railroad parts; castings.

ISOCAST 30-20

Empire Steel Castings Co.
C 0.13-0.2, Cr 30, Ni 20, bal Fe.
For heat and corrosion resistant parts; heat and corrosion resistant. *Obsolete*

ISOCAST 30A

Empire Steel Castings Co.
C 0.13-0.2, Cr 30, bal Fe.
For heat and corrosion resistant parts; heat and corrosion resistant. *Obsolete*

ISOCAST 31

Empire Steel Castings Co.
C 0-0.12, Cr 0-1, Ni 63, Mo 28, Fe 6.
Cast: 80,000 TS; 57,000 YS; 8 El; 11 RA; 200 Brin. For agitators, valves, pump parts; resists hydrochloric acid and phosphoric acid. *Obsolete*

ISOCAST 32

Empire Steel Castings Co,
C 0-0.15, Cr 17, Mo 17, Fe 6, W 4.25.
Cast: 78,000 TS; 57,000 YS; 7 El; 10 RA; 210 Brin. For pump and valves, for strongly oxidizing acids; highly acid resistant. *Obsolete*

ISOCAST 35

Empire Steel Castings Co.
C 0-0.15, Cr 24, Ni 56, Cu 8, Mo 4, bal Fe.
Cast: 60,000 TS; 50,000 YS; 5 El; 5 RA; 180 Brin For chemical plant equipment, valves, pumps; resists sulfuric acid and thermal shock. *Obsolete*

ISOCAST 4

Empire Steel Castings Co.
C 0.15-0.25, Mn 0.5-0.7, Mo 0.4-0.6, bal Fe.
Annealed: 75,000 TS; 45,000 YS; 25 El; 40 RA. For turbines, locomotives, valves, pumps; castings.

ISOCAST 40

Empire Steel Castings Co.
C 0-0.04, Cr 25-57, Ni 4.7-6, Mo 1.7-2.2, Cu 2.7-3.2, bal Fe.
Annealed: 100,000 TS; 80,000 YS; 30 El; 250 Brin. Heat treated: 145,000 TS; 120,000 YS; 10 El; 350 Brin. For high temperature applications; age hardenable, corrosion resistant.

ISOCAST 5

Empire Steel Castings Co.
C 0.25-0.35, Mn 1-1.2, Mo 0.35-0.45, bal Fe.
Heat treated: 80,000 TS; 55,000 YS; 25 El; 40 RA. For cars, excavating machinery sheaves; castings.

ISOCAST 6

Empire Steel Castings Co.
C 0.25-0.35, Mn 0.5-0.7, Ni 1.75-2.25, bal Fe.
Heat treated: 95,000 TS; 60,000 YS; 22 El; 40 RA; 179 Brin. For mining, excavating and ship castings; castings. *Obsolete*

ISOCAST 60-20A

Empire Steel Castings Co.
C 0.13-0.2, Ni 60, Cr 20, bal Fe.
For heat and corrosion resistant parts; heat and corrosion resistant. *Obsolete*

ISOCAST 7

Empire Steel Castings Co.
C 0.85, Mn 0.5, Si 0.5, Cr 2.5, V 0.2, bal Fe.
Heat treated: 95,000-165,000 TS; 60,000-135,000 YS; 15 El; 30 RA; 179-302 Brin. For machine tools, sheaves, brick making equipment; wear-resistant castings.

ISOCAST 7A

Empire Steel Castings Co.
C 0.8-0.9, Mn 0.5-0.7, Cr 1-1.5, bal Fe.
Annealed: 150,000 TS; 110,000 YS; 3 El; 10 RA; 349 Brin. For cement mills, paddles, crushers; castings, abrasion resistant. *Obsolete*

ISOCAST 8

Empire Steel Castings Co.
C 0.25-0.35, Mn 0.6-0.9, Cr 1-1.25, Mo 0.2-0.3, bal Fe.
Heat treated: 90,000-105,000 TS; 55,000-65,000 YS; 15-25 El; 30-45 RA; 210 Brin. For railroads, connecting rods, valves; castings.

ISOCAST 9

Empire Steel Castings Co.
C 0.2 0.3, Ni 0.5-0.7, Cr 0.5-0.7, Mo 0.1-0.2, bal Fe.
Heat treated: 110,000-120,000 TS; 70,000-100,000 YS; 12-18 El; 30-35 RA; 230-270 Brin. For valve bodies, turbines, gears, shafts, housings; SAE 8630; castings.

ISOCAST 9A

Empire Steel Castings Co.
C 0.25-0.35, Ni 1-1.2, Cr 1-1.5, V 0.1-0.2, bal Fe.
Heat treated: 115,000 TS; 65,000 YS; 20 El; 40 RA; 235 Brin. For impellers, pump casings, structural castings; ASTM 4C4; abrasion resistant.

ISOCAST B-75

Empire Steel Castings Co.
C 0.0 0.75, Cr 16-19, Ni 0-2, bal Fe.
Cast: 105,000 TS; 75,000 YS; 240 Brin. For valves, pumps; hardenable, heat resistant. *Obsolete*

ISOCAST CC-50

Empire Steel Castings Co.
Stainless steel. C 0-0.5, Cr 26-30, Ni 0-4, bal Fe.
Cast: 60,000 TS; 190 Brin. For ore roasting equipment; stainless.

ISOCAST CF-12M

Empire Steel Castings Co.
C 0-0.12, Cr 18-21, Ni 9-12, Mo 2-3, bal Fe.
Cast: 80,000 TS; 50,000 YS; 45 El; 170 Brin. For chemical processing equipment; austenitic, corrosion resistant. *Obsolete*

ISOCAST CF-16F

Empire Steel Castings Co.
Stainless steel. C 0-0.08, Cr 18-21, Ni 9-12, Cb = 8 x C, bal Fe.
Cast: 80,000 TS; 44,000 YS; 40 El; 156 Brin. For stainless and heat resistant parts; stainless; heat resistant; welding grade.

ISOCAST CF-20

Empire Steel Castings Co.
C 0-0.2, Cr 20, Ni 8-11, bal Fe.
Cast: 80,000 TS; 40,000 YS; 45 El; 160 Brin. For valves, pumps; austenitic, corrosion resistant.

ISOCAST CF-30

Empire Steel Castings Co.
Stainless steel. C 0-0.3, Cr 17-20, Ni 8-11, bal Fe.
Annealed: 80,000 TS; 40,000 YS; 40 El; 160 Brin. For chemical plant equipment; stainless, austenitic.

ISOCAST CF-8C

Empire Steel Castings Co.
Stainless steel. C 0-0.08, Cr 18-21, Ni 9-12, Cb 0-1, bal Fe.
Cast: 80,000 TS; 44,000 YS; 40 El; 160 Brin. For dairy and chemical plant equipment, valves, tanks; Type 347; stainless, austenitic.

ISOCAST CF16FA

Empire Steel Castings Co.
Stainless steel. C 0-0.16, Cr 18-21, Ni 9-12, 0.20-0.35 Se or 0.2-0.4 S, bal Fe.
Annealed: 75,000 TS; 35,000 YS; 40 El; 160 Brin. For chemical plant equipment; free-cutting, stainless, austenitic.

ISOCAST CF3

Empire Steel Castings Co.
Stainless steel. C 0-0.03, Cr 18-21, Ni 9-12, bal Fe.
Annealed: 75,000 TS; 36,000 YS; 45 El; 140 Brin For chemical plant equipment, food processing equipment; Type 304L; stainless, austenitic.

ISOCAST CF3M

Empire Steel Castings Co.
Stainless steel. C 0-0.03, Cr 18-21, Ni 10-13, Mo 2-3, bal Fe.
Annealed: 78,000 TS; 40,000 YS; 44 El; 150 Brin. For acid resistant chemical plant castings; Type 316L; stainless, austenitic.

ISOCAST CG-12

Empire Steel Castings Co.
Stainless steel. C 0-0.12, Cr 20-23, Ni 10-13, bal Fe.
Cast: 80,000 TS; 40,000 YS; 40 El; 156 Brin. For paper and textile processing equipment; stainless.

ISOCAST CH-20

Empire Steel Castings Co.
Stainless steel. C 0-0.2, Cr 24-26, Ni 12-15, bal Fe.
Cast: 85,000 TS; 50,000 YS; 30 El; 170 Brin. For paper and pulp and chemical plant equipment; Type 309; stainless, austenitic.

ISOCAST CK-20

Empire Steel Castings Co.
Stainless steel. C 0-0.2, Cr 23-27, Ni 19-22, bal Fe.
Cast: 75,000 TS; 45,000 YS; 30 El; 190 Brin. For paper and pulp and chemical plant equipment; Type 310; stainless, austenitic.

ISOCAST HC

Empire Steel Castings Co.
C 0-0.5, Cr 26-30, Ni 0-4, bal Fe.
Cast: 60,000 TS; 200 Brin. For ore roasting equipment; austenitic, heat resistant.

ISOCAST HE

Empire Steel Castings Co.
C 0.2-0.5, Cr 26-30, Ni 8-11, bal Fe.
Cast: 90,000 TS; 55,000 YS; 20 El; 200 Brin. For oil still parts; austenitic, corrosion and heat resistant.

ISOCAST HF
Empire Steel Castings Co.
C 0.2-0.4, Cr 18-23, Ni 8-12, bal Fe.
Cast: 85,000 TS; 40,000 YS; 45 El; 160 Brin. At 1400°F: 37,200 TS; 19,750 YS; 20 El. For valves, pumps, chemical plant equipment; corrosion resistant, austenitic.

ISOCAST N
Empire Steel Castings Co.
Ni 97, bal Fe, Mn, Si.
Cast: 50,000 TS; 25,000 YS; 20 El; 100 Brin. For caustic soda equipment; corrosion resistant. *Obsolete*

ISOCAST WC1
Empire Steel Castings Co.
C 0.2-0.3, Mn 0.5-0.75, bal Fe.
Heat treated: 95,000 TS; 60,000 YS; 22 El; 35 RA; 210 Brin. For valve bodies; castings. *Obsolete*

ISOCAST WC2
Empire Steel Castings Co.
C 0.2-0.3, Mn 0.5-0.7, bal Fe.
Heat treated: 85,000 TS; 55,000 YS; 22 El; 35 RA; 200 Brin. For valves; castings. *Obsolete*

ISOCAST WC3
Empire Steel Castings Co.
C 0-0.3, Mn 0.5-0.7, Mo 0.4-0.6, Cr 0.5-0.8, bal Fe.
Heat treated: 90,000 TS; 60,000 YS; 18 El; 30 RA; 220 Brin. For valves; castings. *Obsolete*

ISOCAST WC5
Empire Steel Castings Co.
C 0.55-0.85, Mo 0.85-1.05, Cr 0.55-0.85, Ni 0.65-0.95, bal Fe.
Heat treated: 85,000 TS; 60,000 YS; 22 El; 35 RA. For valves; castings. *Obsolete*

ISOCAST WC6
Empire Steel Castings Co.
C 0-0.2, Cr 1-1.5, Mo 0.4-0.6, bal Fe.
Heat treated: 75,000 TS; 45,000 YS; 22 El; 35 RA; 163 Brin. For valves; castings. *Obsolete*

ISOCAST WC9
Empire Steel Castings Co.
C, Ni, Cr, Mo, bal Fe.
For general engineering castings; oil hardened, tough. *Obsolete*

ISODISC-1
Vereinigte Edelstahlwerke
C 0.35, Cr 5, V 1, Mo 1, bal Fe.
Heat treated: 276,000 TS; 224,000 YS; 8 El; C 54 Rock. For extrusion and forging dies. Modified Type H13 tool steel. Hot work steel, shock resistant. *Obsolete*

ISODISC-1
U.N. Alloy Steel Corp.
C 0.35, Cr 5, V 1, Mo 1, bal Fe.
Heat treated: 276,000 TS; 224,000 YS; 8 El; C 54 Rock. For extrusion and forging dies. Modified Type H13 tool steel. Hot work steel, shock resistant. *Obsolete*

ISODISC-2
Vereinigte Edelstahlwerke
C 0.35, Si 1, Mn 0.4, Cr 5, W 1.3, Mo 1.4, V 0.35, bal Fe.
Forged: 224,000 TS; 207,800 YS; 8.4 El; 30 RA. For aluminum and brass extrusion dies and press components, forging dies, mandrels, punches, rivet sets. Type H12 hot work steel. *Obsolete*

ISODISC-2
U.N. Alloy Steel Corp.
C 0.35, Si 1, Mn 0.4, Cr 5, W 1.3, Mo 1.4, V 0.35, bal Fe.
Forged: 224,000 TS; 207,800 YS; 8.4 El; 30 RA. For aluminum and brass extrusion dies and press components, forging dies, mandrels, punches, rivet sets. Type H12 hot work steel. *Obsolete*

ISODISC-4
Vereinigte Edelstahlwerke
C 0.33, V 0.5, Cr 3, Mo 3, bal Fe.
For extrusion and forging dies, rams, upsetters. Similar to Bohler WMD alloy. Hot work tool steel, shock resistant. *Obsolete*

ISODISC-4
U.N. Alloy Steel Corp.
C 0.33, V 0.5, Cr 3, Mo 3, bal Fe.
For extrusion and forging dies, rams, upsetters. Similar to Bohler WMD alloy. Hot work tool steel, shock resistant. *Obsolete*

ISOLOY 302
Riverside Metals Corp.
C 0-0.15, Cr 18, Ni 8, bal Fe.
Cold drawn: 250,000-350,000 TS. For springs, wire rope; stainless, austenitic, non-magnetic. *Obsolete*

ISOLOY 304
Riverside Metals Corp.
C 0-0.08, Cr 18, Ni 10, bal Fe.
Cold drawn: 250,000-350,000 TS. For springs, wire rope; stainless, austenitic, non-magnetic. *Obsolete*

ISOLOY 316
Riverside Metals Corp.
C 0-0.08, Cr 17, Ni 12, Mo 2, bal Fe.
Cold drawn: 200,000-250,000 TS. For springs, wire rope; stainless, austenitic, non-magnetic. *Obsolete*

ISOPERM 2
Allgemeine Elektrizitats-Gesellschaft
Ni 35-50, Cu 9-15, bal Fe.
For magnets; high permeability.

ISOPERM 3
Allgemeine Elektrizitats-Gesellschaft
Ni 40-60, Al 3-4, bal Fe.
For magnets; high permeability.

ISOPERM-1
Allgemeine Elektrizitats-Gesellschaft
Ni 40-55, bal Fe.
For magnets; high permeability.

ISOROD
Grand Northern Products Ltd.
C 0.2, Si 0.5, Mn 0.2, Cr 15, Mo 8, Ti 1, bal Fe.
Coated, hardfacing electrode. As arc welded: 185,000 psi TS; 520-580 Brin. For overlay or buildup on tractor sprockets, shovel teeth, dredge pump impellers, heavy hammers; strong and tough.

ISOTAN
Isabellenhutte
Copper. Cu 55, Ni 44, Mn 1.
420 N/mm^2 TS. For electrical equipment and instruments. Resistance alloy. Maximum working temperature 600°C.

ISTEG STEEL
Isteg Steel Products Co.
C, bal Fe.
For reinforcing steel. Mild steel, pre-stressed.

ITALSIL
Italian manufacture
Si 5, bal Al.
For light alloy parts; die and sand castings.

IWI
Now DARWIN IWI.

IX QUALITY
Delta Metal (BW) Ltd.
Cu, Mn, Pb, bal Zn.
Extruded: 72,000 TS; 35,000 YS; 26 El; 20 RA; 130 Brin.
Drawn: 80,000 TS; 38,000 YS; 25 El; 18 RA; 135 Brin. For hardware; manganese bronze. *Obsolete*

IZETT STEEL NO. 1
Krupp Stahl AG
alloy steel. C 0.125, Mn 0.56, P 0.018, S 0.15, Si 0.1, bal Fe.
At 20°C: 62,600 TS; 39,000 YS; 43 El; 68 RA. At 400°C: 50,300 TS; 22,000 YS; 39 El; 70 RA. For boilers; tough, Al deoxidized; non-aging. *Obsolete*

IZETT STEEL NO. 2
Krupp Stahl AG
alloy steel. C 0.15, Mn 0.54, P 0.025, S 0.039, Si 0.06, Al 10.06, bal Fe.
At 20°C: 91,800 TS; 44,500 YS. At 400°C: 68,000 TS; 23,000 YS. For boilers; tough, Al deoxidized; non-aging. *Obsolete*

IZETT STEEL NO. 3
Krupp Stahl AG
alloy steel. C 0.355, Mn 0.49, P 0.024, S 0.024, Si 0.1, bal Fe.
At 20°C: 98,500 TS; 51,000 YS. At 400°C: 77,000 TS; 29,000 YS. For boilers; tough, Al deoxidized; non-aging. *Obsolete*

IZETT STEEL NO. 4
Krupp Stahl AG
alloy steel. C 0.34, Mn 0.48, P 0.025, S 0.025, Si 0.11, bal Fe.
At 20°C: 105,000 TS; 54,000 YS. At 400°C: 81,000 TS; 30,000 YS. For boilers; tough, Al deoxidizing; non-aging. *Obsolete*

J & L 1113 BESSEMER-SELENIUM LEADED

Jones & Laughlin Steel Corp.
C 0-0.09, Mn 0.7-1, P 0.07-0.12, S 0.25-0.35, Pb 0.15-0.35, bal Fe.
For screw machine products, fasteners, cams, camshafts. Free machining, case hardening. *Obsolete*

J & L 1215

LTV Steel
C 0-0.09, Mn 0.75-1.05, P 0.04-0.09, S 0.25-0.35, bal Fe.
For screw machining products, fasteners, cams, gears, and camshafts. Free-machining, case hardening.

J & L COR-TEN

LTV Steel
C 0-0.12, Mn 0.2-0.5, P 0.7-0.15, S 0-0.05, Cu 0.25-0.55, Cr 0.3-1.25, Ni 0-0.65, Si 0.25-0.75, bal Fe.
50,000 psi YS min. Good weldability, atmosphere corrosion resistance. For guard rails and building materials. ASTM A-242.

J & L TYPE 303SM

Jones & Laughlin Steel Corp.
C 0.08, Mn 1.5, Si 0.8, Cr 18, Ni 9, 0.15% min S, bal Fe.
Annealed: 90,000 TS; 35,000 YS; 50 El; 55 RA; 160 Brin. Cold drawn: 125,000 TS; 90,000 YS; 20 El; 50 RA; 277 Brin. For stainless screw machine products, fasteners, instrument gears, trim. Free-machining, stainless, austenitic, nonmagnetic. *Obsolete*

J & L TYPE 304SM

Jones & Laughlin Steel Corp.
C 0-0.08, Mn 0-2, P 0-0.045, S 0-0.03, Si 0-1, Cr 18-20, Ni 8-12, bal Fe.
Annealed: 82,000 TS; 35,000 YS; 60 El; 70 RA; B 80 Rock.
For beer barrels, chemical equipment, evaporators, milking machines, sanitary fittings, vacuum pump parts. Stainless, austenitic. *Obsolete*

J & L TYPE 316SM

Jones & Laughlin Steel Corp.
C 0-0.08, Mn 0-2, P 0-0.045, S 0-0.03, Si 0-1, Cr 16-18, Ni 10-14, Mo 2-3, bal Fe.
Annealed: 80,000 TS; 30,000 YS; 60 El; 70 RA; B 78 Rock.
For chemical and food industries, screens, storage and transportation tanks, digesters. Stainless, austenitic. *Obsolete*

J & L TYPE 416SM

Jones & Laughlin Steel Corp.
C 0-0.15, Mn 0-1.25, P 0-0.06, Si 0-1, Cr 12-14, Mo 0-0.6, 0.15% min S, bal Fe.
Annealed: 80,000 TS; 45,000 YS; 30 El; B 85 Rock. For stainless screw machine products; bolts, fasteners, pump parts, shafts. Free machining, corrosion resistant, hardenable. *Obsolete*

J & L TYPE A

Jones & Laughlin Steel Corp.
C 0-0.15, Mn 0.8-1.2, P 0.04-0.09, S 0.25-0.35, Pb 0.15-0.35, bal Fe.
Cold drawn: 83,000 TS; 75,000 YS; 16 El; 37 RA; 174 Brin.
For automatic screw machine products, shafts, bolts, fasteners, cams. Free-machining. AISI C12L14. *Obsolete*

J & L TYPE A-SELENIUM

Jones & Laughlin Steel Corp.
C 0-0.15, Mn 0.8-1.2, P 0.04-0.09, S 0.25-0.35, Pb 0.15-0.35, Se, bal Fe.
Cold drawn: 83,000 TS; 75,000 YS; 16 El; 37 RA; 175 Brin.
For automatic screw machine products, bolts, fasteners, cams, shafts. Free-machining. *Obsolete*

J & L TYPE B

Jones & Laughlin Steel Corp.
C 0-0.15, Mn 0.85-1.35, P 0.04-0.09, Pb 0.15-0.35, 0.40% min S, bal Fe.
Cold drawn: 87,000 TS; 79,000 YS; 12.5 El; 36.5 RA; 174 Brin.
For automatic screw machine products, screws, bolts, fasteners, shafts. Free-machining. *Obsolete*

J & L-1211

LTV Steel
C 0.08-0.13, P 0.07-0.12, S 0.08-0.15, Mn 0.6-0.9, bal Fe.
Cold drawn: 82,000 psi TS; 80,000 psi YS; 14 El; 47 RA; 174 Brin. For screw machine products, bolts, and shafts; AISI-C1211, free-cutting.

J & L-1212

LTV Steel
C 0-0.13, S 0.16-0.23, P 0.07-0.12, Mn 0.7-1, bal Fe.
Cold drawn: 82,000 psi TS; 80,000 psi YS; 14 El; 47 RA; 174 Brin. For screw machine products, bolts, shafts, and screws; AISI-C1212, free cutting.

J & L-1213

LTV Steel
C 0-0.09, S 0.24-0.33, P 0.07-0.12, Mn 0.7-1, bal Fe.
Cold drawn: 82,000 psi TS; 80,000 psi YS; 14 El; 47 RA; 174 Brin. For screw machine products, bolts, screws, and shafts; AISI-C1213, free cutting.

J & L-12L13

LTV Steel
C 0-0.09, Mn 0.7-1, P 0.07-0.12, S 0.25-0.35, Pb 0.15-0.35, bal Fe.
For screw machine products, fasteners, hardware cams, and camshafts. Free machining, case hardening.

J & L-B11L13

Jones & Laughlin Steel Corp.
C 0-0.09, Mn 0.7-1, P 0.07-0.12, S 0.25-0.35, Pb 0.15-0.35, bal Fe.
Cold drawn: 82,000 TS; 75,000 YS; 16 El; 37 RA; 174 Brin.
For automatic screw machine products, bolts, shafts, fasteners. Free-machining. *Obsolete*

J 1300

General Electric Co.
C 0.08, Cr 14, Ni 33, Mo 4, W 6.5, Ti 2, Al 0.25, Zr 0.25, bal Fe.
For gas turbine blades and parts.

J 1300

Cannon-Muskegon Corp.
C 0.08, Cr 14, Ni 33, Mo 4, W 6.5, Ti 2, Al 0.25, Zr 0.25, bal Fe.
For gas turbine blades and parts.

J HOT WORKING DIE

Jessop Steel Co.
C 0.4, Cr, bal Fe.
For hot working tools and dies; hot work steel; oil hardened. *Obsolete*

J L

Alais Forges et Camargue
Cu 4.5, bal Al.
64,000 TS; 5 El; 120 Brin. For light alloy parts; age-hardenable.

J-100

Now JORGENSEN 100.

J-1500

Darwin & Milner Inc.
C 0.15, Cr 20, Co 10, Mo 10, Ti 3, Al 1, bal Ni.
For gas turbine blades and parts.

J-1500

Cannon-Muskegon Corp.
C 0.15, Cr 20, Co 10, Mo 10, Ti 3, Al 1, bal Ni.
For gas turbine blades and parts.

J-1500

General Electric Co.
C 0.15, Cr 20, Co 10, Mo 10, Ti 3, Al 1, bal Ni.
For gas turbine blades and parts.

J-1530

General Electric Co.
C 0.08, Cr 19.5, Ni 57, Co 13.5, Mo 4.3, Ti 3.1, Al 1.3.
For gas turbine parts.

J-1530

Cannon-Muskegon Corp.
C 0.08, Cr 19.5, Ni 57, Co 13.5, Mo 4.3, Ti 3.1, Al 1.3.
For gas turbine parts.

J-1570

General Electric Co.
C 0.2, Cr 20, Ni 28, W 6, Ti 4, bal Co.
For jet engine parts.

J-1570

Cannon-Muskegon Corp.
C 0.2, Cr 20, Ni 28, W 6, Ti 4, bal Co.
For jet engine parts.

J-1600

General Electric Co.
C 0.1, Cr 19, Co 19, Mo 4, Ti 3, Al 3, bal Ni.
For high temperature equipment.

J-1600

Cannon-Muskegon Corp.
C 0.1, Cr 19, Co 19, Mo 4, Ti 3, Al 3, bal Ni.
For high temperature equipment.

J-1650

General Electric Co.
C 0.2, Cr 19, Ni 27, W 12, Ti 3.8, B 0.02, Ta 2, bal Co.
For jet engine parts.

J-1650

Cannon-Muskegon Corp.
C 0.2, Cr 19, Ni 27, W 12, Ti 3.0, B 0.02, Ta 2, bal Co.
For jet engine parts.

J-4 CHISEL

Ackerlind Steel Co., Inc.
C 0.44, Mn 0.25, Si 0.7, Cr 1.3, W 2.3, V 0.15, bal Fe.
For dies; hot work steel. *Obsolete*

J-4 CHISEL

Jessop-Saville Ltd.
C 0.44, Cr 1.3, W 2.3, V 0.15, bal Fe.
For chisels, shear blades, punches; shock resistant.

J-4 CHISEL

Saville & Co. Ltd., J.J.
C 0.44, Cr 1.3, W 2.3, V 0.15, bal Fe.
For chisels, shear blades, punches; shock resistant.

J-ALLOY

Manufacturer not listed.
C 0.76, Cr 23, Ni 6, Mo 6, Ta 2, bal Co.
For jet engine and gas turbine parts; high heat resistant.

J-S STEEL

Teledyne Firth Sterling
C 0.5, W 2.25, Mn 0.3, Cr 1.4, V 0.2, bal Fe.
Annealed: 80,000 TS; 50,000 YS; 32 El; 58 RA; 156 Brin. Oil hardened: 295,000 TS; 250,000 YS; 6.5 El; 12 RA; 578 Brin.
For chisels, punches, shears, tools, rivet sets, beading tools, blanking dies; tough, hot and cold work tools. *Obsolete*

J. & L. CORRECT BALANCE

Jones & Laughlin Steel Corp.
bal Fe.
For forgings. *Obsolete*

J. & L E-23

Jones & Laughlin Steel Corp.
C 0-0.13, Mn 0.7-0.9, S 0.16-0.23, P 0.07-0.12, bal Fe.
For screw machine products; free-cutting. *Obsolete*

J. & L. E-33
Jones & Laughlin Steel Corp.
C 0-0.13, Mn 0.7-0.9, P 0.07-0.12, S 0.24-0.33, bal Fe.
For screw machine products; free-cutting. *Obsolete*

J. C. DAIRY METAL
Janney Cylinder Co.
Ni 30-35, Cu 55-60, bal special alloy.
For dairy equipment. *Obsolete*

J.& L. E-15
Jones & Laughlin Steel Corp.
C 0-0.13, Mn 0.6-0.9, S 0.08-0.15, P 0.07-0.12, bal Fe.
For screw machine products; free-cutting. *Obsolete*

J.A.C.-ACM
Allard, Soc. Anon. Usines et Acieries
Cr 28, C, Mo, bal Fe.
Normalized: 100,000 TS. For glass, chemical and dyeing
industries; acid resistant.

J.A.C.-ACN 1
Allard, Soc. Anon. Usines et Acieries
Cr 18, Ni 8, C, bal Fe.
Normalized: 100,000-120,000 TS; 50,000-60,000 YS; 25 El.
For food, chemical, petroleum industries; stainless.

J.A.C.-ACN 2
Allard, Soc. Anon. Usines et Acieries
Cr 26, Ni 10, C, bal Fe.
Heat treated: 110,000-120,000 TS; 50,000-60,000 YS; 18 El.
For furnace parts, grates; heat resistant.

J.A.C.-ACN 3
Allard, Soc. Anon. Usines et Acieries
C, Cr, Ni, Mo, bal Fe.
Heat treated: 110,000-130,000 TS; 50,000-70,000 YS; 18 El.
For furnace parts, grates; heat resistant, scale resistant.

J.A.C.-ACR
Allard, Soc. Anon. Usines et Acieries
Cr 30, C, bal Fe.
Heat treated: 84,000-100,000 TS. For furnace parts; scale and
heat resistant.

J.A.C.-AMH
Allard, Soc. Anon. Usines et Acieries
Mn 12-14, C, bal Fe.
Heat treated: 180,000 TS; 80,000 YS; 35 El; 40 RA. For
crusher plates, hammers, harrowers, mixers; wear resistant.

J.A.C.-AMH 2
Allard, Soc. Anon. Usines et Acieries
Mn 12-14, C, Cr, bal Fe.
Heat treated: 180,000-200,000 TS. For crusher plates,
hammers, harrowers, mixers; wear and abrasion resistant.

J.A.C.-ARF
Allard, Soc. Anon. Usines et Acieries
C, Ni, Cr, Mo, bal Fe.
Heat treated:; 160,000-180,000 TS; 120,000-140,000 YS; 12
El; 25 RA. For crusher plates, hammers, harrowers, mixers;
easily machined and welded.

J.A.C.-B 1
Allard, Soc. Anon. Usines et Acieries
C, bal Fe.
Annealed: 80,000 TS; 50,000 YS; 18 El; 35 RA. For general
usage; Bessemer steel.

J.A.C.-B 1 S
Allard, Soc. Anon. Usines et Acieries
C, bal Fe.
Annealed: 80,000 TS; 50,000 YS; 22 El; 40 RA. For general
usage; Bessemer steel.

J.A.C.-B 2
Allard, Soc. Anon. Usines et Acieries
C, bal Fe.
Annealed: 90,000 TS; 60,000 YS; 16 El; 30 RA. For general
usage; Bessemer steel.

J.A.C.-B 2 S
Allard, Soc. Anon. Usines et Acieries
C, bal Fe.
Annealed: 90,000 TS; 60,000 YS; 20 El; 35 RA. For general
usage; Bessemer steel.

J.A.C.-B 3
Allard, Soc. Anon. Usines et Acieries
C, bal Fe.
Annealed: 110,000 TS; 66,000 YS; 12 El; 25 RA. For general
usage; Bessemer steel.

J.A.C.-E 1
Allard, Soc. Anon. Usines et Acieries
P 0.03, S 0.3, C, bal Fe.
Normalized: 71,000 TS; 48,000 YS; 25 El; 50 RA. For motor
frames, rotors, dynamo parts; high magnetic permeability.

J.A.C.-E 2
Allard, Soc. Anon. Usines et Acieries
C, bal Fe.
Normalized: 80,000 TS; 52,000 YS; 22 El; 45 RA. For wheels,
gears, axles; high yield strength.

J.A.C.-E 3
Allard, Soc. Anon. Usines et Acieries
C, bal Fe.
Normalized: 90,000 TS; 60,000 YS; 20 El; 35 RA. For bridge
bearings, rudder frames, gears, naval construction.

J.A.C.-E 4
Allard, Soc. Anon. Usines et Acieries
C, bal Fe.
Normalized: 100,000 TS; 64,000 YS; 20 El; 35 RA. For
machine tool parts, bridge gear boxes.

J.A.C.-E 5
Allard, Soc. Anon. Usines et Acieries
C, bal Fe.
Normalized: 110,000 TS; 72,000 YS; 17 El; 32 RA. For gears,
axles, couplings; water hardening.

J.A.C.-E 6
Allard, Soc. Anon. Usines et Acieries
C, bal Fe.
Normalized: 120,000 TS; 80,000 YS; 15 El; 25 RA. For
pinions, block dies, gears; wear resistant.

J.A.C.-EA 1
Allard, Soc. Anon. Usines et Acieries
C, Cr, Mn, bal Fe.
Normalized: 120,000-160,000 TS; 80,000-110,000 YS; 7-15 El.
For gears, excavator parts, tractors, cranes, dies; tough, wear
resistant.

J.A.C.-EA 2
Allard, Soc. Anon. Usines et Acieries
C, Ni, bal Fe.
Normalized: 120,000 TS; 96,000 YS; 14 El. For drills, pinions,
roller mill cylinders, brake drums; wear and shock resistant.

J.A.C.-EA 3
Allard, Soc. Anon. Usines et Acieries
C, Cr, Ni, bal Fe.
Normalized: 150,000 TS; 120,000 YS; 12 El. For drills,
pinions, roller mill cylinders; wear and shock resistant.

J.A.C.-EA 4
Allard, Soc. Anon. Usines et Acieries
C, Cr, Ni, Mo, bal Fe.
Normalized: 160,000-180,000 TS; 110,000-150,000 YS; 10 El.
For rollers, gears, valves; heat resistant to 500°F.

J.A.C.-EA 5
Allard, Soc. Anon. Usines et Acieries
C, Cr, Mo, bal Fe.
Normalized: 130,000-150,000 TS; 90,000-100,000 YS; 10 El.
For rollers, gears, valves; shock and wear resistant.

J.A.C.-EA 6
Allard, Soc. Anon. Usines et Acieries
C, Cr, Mo, bal Fe.
Normalized: 100,000-120,000 TS; 80,000-90,000 YS; 15 El.
For valves, collectors, piping; maximum operating
temperature 550°F.

J.C./C
Ekstrand & Tholand Co.
For general tools; water hardening.

J.C./C
Degefors Iron & Steel Works
For general tools; water hardening.

J.P.S. STEEL
Jessop-Saville Ltd.
C 0.9-1, W 1.1-1.3, Cr 0.4-0.6, bal Fe.
For shock tools, threading, taps, reamers, broaches; oil
hardening.

J.S. PUNCH
Teledyne Firth Sterling
Steel. C 0.55, Cr 1.25, W 2.75, V 0.2, bal Fe.
Heat treated: 231,000-315,000 TS; 209,000-275,000 YS; 9-11
El; 27-42 RA; 48-56 Rock C. For header and swaging dies,
chipping chisels, rivet busters, track tools. Type S1 shock and
fatigue resistant tool steel. *Obsolete*

J.S.B.
Johnson Bronze Co.
Bronze on steel.
For bushings, bearings, washers; bronze on steel in finished
bearing. *Obsolete*

JACKMANIZED STEEL
Joseph Jackman & Co. Ltd.
C 0.5, bal Fe.
For dredger bucket pins, bushes, tin rolls.

JACKSBERG
Babcock & Wilcox
C 0.7, W 18, Cr 4, V 1, bal Fe.
For high speed tools and cutters; high speed steel. *Obsolete*

JACKSONS BUTTON ALLOY
English manufacture
Zn 30-35, Sb 2-5, bal Cu.
For ornamental and architectural parts; high strength.

JACOBY METAL
English manufacture
Sn 85, Sb 10, Cu 5.
For bearings; liners; Babbitt.

JACOL 1886
Compagnie Ateliers et Forges de la Loire
C 0.08, Cr 1.4, Mo 1.1, V 0.6, bal Fe.
Heat treated: 107,000 TS; 93,000 YS; 15 El. For gas turbine
and steam turbine rotors and discs. Holds properties up to
600°C. Resists oxidation up to 550°C.

JACOL 9
Creusot-Loire
C 0-0.02, Mn 7, Si 0-0.6, Cr 16.5, Ni 12.5, W 3.8, bal Fe.
Austenitic stainless steel; weldable. AFNOR Z 2 CNMW
17-13-7.

JACOL-CVM6
Compagnie Ateliers et Forges de la Loire
C 0.17, Cr 5.5, Mo 0.55, V 0.25, bal Fe.
Annealed: 64,000 TS; 40,000 YS; 32 El. Heat treated: 107,000
TS; 88,000 YS; 17 El. For piping equipment in oil refineries.
Resistant to oxidation to 750°C.

JACONA METAL
English manufacture
Pb 70, Sb 20, Sn 10.
For bearings, liners; Babbitt.

JADE

Ugine Aciers
C 1, bal Fe.
For reamers, taps, punches, cold work tools; Type W1, water hardened.

JADOT J1

Usines de Jadot
C 0-0.08, Cr 13, bal Fe.
Annealed: 75,000 TS; 40,000 YS; 35 El; 70 RA; 155 Brin. For valves, cutlery, surgical instruments; Type 403; corrosion resistant.

JADOT J1

Usines de Jadot
C 0-0.15, Cr 11.5-13.5, bal Fe.
Heat treated: 120,000-135,000 TS; 110,000-117,000 YS; 15-16 El; 58-63 RA; 220-240 Brin. For cutlery, valves, turbine blades, fasteners; Type 410; corrosion resistant.

JADOT J2

Usines de Jadot
C 0.26-0.45, Cr 13, bal Fe.
Annealed: 95,000 TS; 50,000 YS; 25 El; 55 RA; 196 Brin. For valves, cutlery, surgical and dental instruments; Type 420; corrosion resistant.

JADOT JA

Usines de Jadot
C 0.08-0.2, Cr 17-19, Ni 8-10, Mn 0-2, bal Fe.
Annealed: 85,000 TS; 35,000 YS; 60 El; 70 RA; 150 Brin. Cold drawn: 125,000 TS; 95,000 YS; 25 El; 55 RA; 277 Brin. For oil refinery and chemical plant equipment; Type 302; stainless, austenitic.

JADOT JA1

Usines de Jadot
C 0-0.1, Cr 16-18, Ni 10-14, Mo 2-3, bal Fe.
Annealed: 85,000-95,000 TS; 35,000-45,000 YS; 50-60 El; 60-75 RA; 160-100 Brin. For chemical plant equipment, mixers, agitators, filters; Type 316; austenitic, stainless.

JADOT JA2

Usines de Jadot
C 0.1, Cr 18, Ni 10, Mo 2.5, bal Fe.
Annealed: 85,000 TS; 35,000 YS; 50 El; 65 RA; 160 Brin. For acid resistant chemical plant equipment, mixers; Type 316; stainless, austenitic.

JADOT JA2E

Usines de Jadot
C 0.1, Cr 19, Ni 12, Mo 3.5, bal Fe.
Annealed: 90,000 TS; 40,000 YS; 45 El; 60 RA; 180 Brin. For acid resistant chemical plant equipment; mixers; Type 317; stainless, austenitic.

JADOT JAA

Usines de Jadot
C 0.07, Cr 17, Ni 13, Mo 4.5, bal Fe.
Annealed: 90,000 TS; 40,000 YS; 45 El; 60 RA; 180 Brin. For acid resistant chemical plant equipment, mixers; Type 317; stainless, austenitic.

JADOT JAS

Usines de Jadot
C 0.12, Cr 18, Ni 8, bal Fe.
Annealed: 85,000 TS; 35,000 YS; 60 El; 70 RA; 150 Brin. For chemical plant equipment, tanks, mixers, filters; Type 304; stainless, austenitic.

JADOT JD

Usines de Jadot
C 0-0.25, Cr 15, Ni 35, bal Fe.
For furnace parts and equipment, salt pots; Type 330; heat resistant.

JADOT JH

Usines de Jadot
C 0-0.12, Cr 14-18, bal Fe.
Annealed: 70,000 TS; 40,000 YS; 30 El; 55 RA; 150 Brin. Cold drawn: 130,000 TS; 120,000 YS; 2 El; 185 Brin. For oil refinery equipment, bolts, kitchen sinks; Type 430; stainless, ferritic.

JADOT JH1

Usines de Jadot
Cr 24-26, C 0-0.2, Si 0-1, bal Fe.
Annealed: 75,000-85,000 TS; 50,000-60,000 YS; 12 El; 180 Brin. For furnace parts, heat treating boxes; Type 442; corrosion and heat resistant.

JADOT JH1A

Usines de Jadot
C 0.35, Cr 23-27, N 0-0.25, bal Fe.
Annealed: 90,000 TS; 60,000 YS; 20 El; 45 RA; 180 Brin. Cold drawn: 175,000 TS; 155,000 YS; 2 El; 25 RA; 250 Brin. For furnace parts, preheaters, annealing boxes; Type 446; corrosion resistant.

JADOT JH2

Usines de Jadot
C 0-0.2, Cr 22-24, Ni 12-15, Mn 0-2, Si 0-3.5, bal Fe.
Annealed: 85,000-95,000 TS; 40,000-50,000 YS; 45-55 El; 150-185 Brin. For heat treat boxes, furnace parts and equipment; Type 309; austenitic, heat resistant.

JADOT JH3

Usines de Jadot
C 0-0.25, Cr 24-26, Ni 19-22, bal Fe.
Annealed: 95,000 TS; 45,000 YS; 50 El; 65 RA; 180 Brin. At 1200°F: 57,000 TS; 22,000 YS; 32 El; 45 RA. For furnace parts, valves, pumps, engine parts; Type 310; heat resistant, austenitic.

JADOT JH3AS

Usines de Jadot
C 0.12, Cr 20, Ni 20, bal Fe.
For furnace parts, salt pots; corrosion and heat resistant.

JADOT JH4

Usines de Jadot
C 0-0.25, Cr 24-26, Ni 19-22, bal Fe.
Annealed: 95,000 TS; 45,000 YS; 50 El; 65 RA; 180 Brin. At 1200°F: 57,000 TS; 22,000 YS; 32 El; 45 RA. For furnace parts, valves, pumps, engine components; Type 310; austenitic heat resistant.

JADOT JH6

Usines de Jadot
C 0-0.1, Cr 18-20, Ni 11-14, Mo 3-4, bal Fe.
Annealed: 85,000-95,000 TS; 35,000-45,000 YS; 50-60 El; 60-75 RA; 150-190 Brin. For acid resistant chemical plant equipment; Type 317; stainless, austenitic.

JAE METAL

Inco Alloys International Inc..
Ni 70, Cu 30.
Annealed: 63,000 TS; 22,000 YS; 52 El; 81 RA; 110 Brin. For compensating shunts on magnetic instruments; high regularity temperature coefficient of magnetic permeability. *Obsolete*

JALCASE 1

Jones & Laughlin Steel Corp.
C 0.1-0.16, Mn 1-1.3, bal Fe.
Water hardened: 138,000 TS; 100,000 YS; 10 El; 30 RA; 315 Brin. Cold rolled: 65,000 TS; 55,000 YS; 17 El; 45 RA; 137 Brin. For case-hardened parts, screw stock; free cutting C-1113. *Obsolete*

JALCASE 10

Jones & Laughlin Steel Corp.
C 0.4-0.48, Mn 1.35-1.65, S 0.24-0.33, bal Fe.
Cold rolled: 115,000 TS; 100,000 YS; 7 El; 20 RA; 235-285 Brin. For gears, shafts; C-1144; free-cutting. *Obsolete*

JALCASE 100

LTV Steel
C 0.4-0.48, Mn 1.35-1.65, S 0.24-0.33, bal Fe.
Cold drawn: 115,000 psi; 100,000 psi YS; 10-20 El; 30-50 RA; 241 Brin. For screw machine products, gears, and bolts; free-cutting stress stabilized.

JALCASE 2

Jones & Laughlin Steel Corp.
C 0.1-0.16, Mn 1-1.3, bal Fe.
Water hardened: 150,000 TS; 135,000 YS; 12 El; 50 RA; 325 Brin. Cold rolled: 65,000 TS; 55,000 YS; 17 El; 45 RA; 137 Brin. For general use, shafts, gears; free cutting C-1114. *Obsolete*

JALCASE 3

Jones & Laughlin Steel Corp.
C 0.14-0.2, Mn 1.1-1.4, bal Fe.
Water hardened: 190,000 TS; 167,000 YS; 10 El; 38 RA; 375 Brin. Cold rolled: 70,000 TS; 60,000 YS; 15 El; 35 RA; 143 Brin. For general use, shafts, gears, free cutting C-1116. *Obsolete*

JALCASE 4

Jones & Laughlin Steel Corp.
C 0.14-0.2, Mn 1-1.3, bal Fe.
Oil hardened: 150,000 TS; 128,000 YS; 12 El; 38 RA; 310 Brin. Cold rolled: 70,000 TS; 60,000 YS; 15 El; 35 RA; 143 Brin. For general use, shafts; free cutting C-1117. *Obsolete*

JALCASE 5

Jones & Laughlin Steel Corp.
C 0.14-0.2, Mn 1.3-1.6, bal Fe.
Oil hardened: 166,000 TS; 138,000 YS; 10 El; 38 RA; 380 Brin. Cold rolled: 70,000 TS; 60,000 YS; 15 El; 35 RA; 149 Brin. For general use, shafts, gears; free cutting C-1118. *Obsolete*

JALCASE 6

Jones & Laughlin Steel Corp.
C 0.14-0.2, Mn 1-1.3, S 0.2-0.33, bal Fe.
Cold rolled: 70,000 TS; 60,000 YS; 15 El; 35 RA; 143-192 Brin. For gears, shafts, cams, camshafts; C-1119; free cutting. *Obsolete*

JALCASE 7

Jones & Laughlin Steel Corp.
C 0.32-0.39, Mn 1.35-1.65, S 0.08-0.13, bal Fe.
Cold rolled: 90,000 TS; 75,000 YS; 10 El; 35 RA; 192-241 Brin. For gears, shafts; C-1137; free-cutting. *Obsolete*

JALCASE 8

Jones & Laughlin Steel Corp.
C 0.37-0.45, Mn 1.35-1.65, S 0.08-0.13, bal Fe.
Cold rolled: 95,000 TS; 80,000 YS; 10 El; 28 RA; 197-248 Brin. For gears, shafts; C-1141; free-cutting. *Obsolete*

JALCASE 9

Jones & Laughlin Steel Corp.
C 0.4-0.48, Mn 1.35-1.65, S 0.24-0.33, bal Fe.
Cold rolled: 100,000 TS; 85,000 YS; 8 El; 25 RA; 207-255 Brin. For gears, shafts; C-1144; free-cutting. *Obsolete*

JALCOLD

Jones & Laughlin Steel Corp.
C 0.2, bal Fe.
For welding rods. *Obsolete*

JALLOY 280

LTV Steel
C 0.25-0.31, Mn 1.35-1.65, Mo 0.1-0.2, Al, B, Ti, 0.2 Cu min, bal Fe.
Heat treated: 115,000 psi TS; 100,000 psi YS; 15 El; 260-300 Brin. For liners, power shovels, mining equipment, and fan blades; resists impact and abrasion.

JALLOY 280

Jones & Laughlin Steel Corp.
C 0.25-0.31, Mn 1.35-1.65, Mo 0.1-0.2, 0.2% min Cu, bal Fe.
Heat treated: 115,000 TS; 100,000 YS; 15 El; 260-300 Brin.
For spouts in chemical and food processing equipment; heat treated plates. *Obsolete*

JALLOY 3

Jones & Laughlin Steel Corp.
C 0.25-0.31, Mn 1.35-1.65, Mo 0.1-0.2, bal Fe.
Rolled: 102,000 TS; 80,000 YS; 21 El; 51 RA; 241 Brin.
Hardened: 150,000 TS; 140,000 YS; 14 El; 55 RA; 364 Brin.
For mining equipment, buckets; tough. *Obsolete*

JALLOY 320

LTV Steel
C 0.25-0.31, Mn 1.35-1.65, Mo 0.1-0.2, Al, B, Ti, 0.2 Cu min, bal Fe.
Heat treated: 300-340 Brin. For liners, power shovels, mining equipment, fan blades, and stone crushers; resists impact and abrasion.

JALLOY 7

Jones & Laughlin Steel Corp.
C 0.5-0.6, Mn 1.35-1.65, Mo 0.1-0.2, bal Fe.
Rolled: 134,000 TS; 100,000 YS; 15 El; 31 RA; 269 Brin. For mining equipment, buckets; tough. *Obsolete*

JALLOY AR

LTV Steel
C 0.25-0.31, Mn 0-1.65, Si 0.15-0.3, Mo 0-0.35, Cr 0-1.2, 0.0005 B min, 0.2 Cu min, bal Fe.
90,000 psi YS min. Good abrasion resistance, atmosphere corrosion resistance. For snowmobile skis and cleats.

JALLOY AR-280

LTV Steel
C 0.25-0.31, Mn 0-1.65, Si 0.15-0.3, Mo 0-0.35, Cr 0-1.2, 0.0005 B min, 0.2 Cu min, bal Fe.
130,000 psi YS (typical). Good abrasion resistance, atmosphere corrosion resistance. For floor plates, coal chutes, and mining equipment.

JALLOY AR-320

LTV Steel
C 0.25-0.31, Mn 0-1.65, Si 0.15-0.3, Mo 0-0.35, Cr 0-1.2, 0.0005 B min, 0.2 Cu min, bal Fe.
140,000 psi YS (typical). Good abrasion resistance, atmosphere corrosion resistance. For mining equipment, off-highway equipment.

JALLOY AR-360

LTV Steel
C 0.25-0.31, Mn 0-1.65, Si 0.15-0.3, Mo 0-0.35, Cr 0-1.2, 0.0005 B min, 0.2 Cu min, bal Fe.
157,000 psi YS (typical). Excellent abrasion resistance. For coal chutes, floor plates, and mining equipment.

JALLOY AR-400

LTV Steel
C 0.25-0.31, Mn 0-1.65, Si 0.15-0.3, Mo 0-0.35, Cr 0-1.2, 0.0005 B min, 0.21 Cu min, bal Fe.
184,000 psi YS (typical). Excellent abrasion resistance. For snowmobile skis and cleats, coal haulers.

JALLOY AR-Q

LTV Steel
C 0.25-0.31, Mn 0-1.65, Si 0.15-0.3, Mo 0-0.35, Cr 0-1.2, 0.0005 B min, 0.2 Cu min, bal Fe.
217,000 psi YS (typical). Good abrasion resistance, atmosphere corrosion resistance. For highly stressed highway equipment.

JALLOY S-100

LTV Steel
C 0.1-0.2, Mn 0-1.5, Si 0-0.5, Mo 0-0.3, Cr 0-1.5, 0.0005 B min, bal Fe.
100,000 psi YS min. Good cold forming, weldability, impact resistance. For dump trucks, and agricultural equipment. ASTM A-514, A-517.

JALLOY S-110

LTV Steel
C 0.1-0.2, Mn 0-1.5, Si 0-0.5, Mo 0-0.3, Cr 0-1.5, 0.0005 B min, bal Fe.
110,000 psi YS min. Good cold forming, weldability, impact resistance. For mobile cranes, booms, tractor components. ASTM A-514, A-517.

JALLOY S-340

LTV Steel
C 0.1-0.2, Mn 0-1.5, Si 0-0.5, Mo 0-0.3, Cr 0-1.5, bal Fe.
150,000 psi YS (typical). Good weldability, abrasion resistance, impact resistance. For dump trucks, coal chutes, and tractor parts.

JALLOY S-90

LTV Steel
C 0.1-0.2, Mn 0-1.5, Si 0-0.5, Mo 0-0.3, Cr 0-1.5, 0.0005 B min, bal Fe.
90,000 psi YS min. Good forming, weldability, impact resistance. For tractor components, railroad car equipment. ASTM A-514.

JALLOY-1

Jones & Laughlin Steel Corp.
C 0.13-0.18, Mn 1-1.3, Mo 0.1-0.2, bal Fe.
Hot rolled: 85,000 TS; 57,000 YS; 30 El; 63 RA; 179 Brin. Cold drawn: 98,500 TS; 98,000 YS; 17.5 El; 65 RA. For tank armor, mine equipment; shock resistant. *Obsolete*

JALTEN NO. 1

LTV Steel
C 0-0.15, Mn 0-1.3, V 0.035-0.065, 0.2 Cu min, bal Fe.
Rolled: 65,000 psi TS; 50,000 psi YS; 22 El. For structural parts, truck bodies, and mine cars; good formability and weldability. ASTM A441.

JALTEN NO. 2

LTV Steel
C 0-0.15, Mn 0-1.4, P 0-0.14, 0.3 Cu min, bal Fe.
Hot rolled: 70,000 psi TS; 50,000 psi YS; 22 El. For structural parts, truck bodies, and railroad cars; good formability and weldability.

JALTEN NO. 3

LTV Steel
C 0-0.25, Mn 0-1.6, 0.2 Cu min, bal Fe.
Hot rolled: 70,000 psi TS; 50,000 psi YS; 22 El. For construction equipment and truck bodies; moderate formability. ASTM A440.

JALWELD

LTV Steel
C 0.15, bal Fe.
For welding rods.

JAMAG

Blackstone Corp.
C 2.2, Si 1.6, Mn 0.3, Mg 0.1, bal Fe.
Heat treated: 50,000-110,000 YS. For electrical parts, where high rotation speeds are involved. High magnetic permeability, malleable iron.

JAMALEX

Blackstone Corp.
C 2.2, Si 1.6, Mn 0.3, Mg 0.05, bal Fe.
Heat treated: 65,000-140,000 TS; 50,000-110,000 YS; 2-9 El. For gears, sprockets, hubs, cams, bearings, wear resistant parts. Malleable iron, high magnetic permeability.

JAMAPPES BRASS

English manufacture
Sn 0.2, Pb 1.4, Zn 33.4, bal Cu.
For hardware, fittings, forgings; good workability, free-cutting.

JAMISON HOBBING IRON

Jamison Steel Corp.
C 0.05, bal Fe.
For plastic molding hobs.

JAMISON K-46 O.H.

Jamison Steel Corp.
C 0.9, Mn 1, Cr 0.5, W 0.5, bal Fe.
For reamers, broaches, punches, cutters; Type O1; oil hardened, nondeforming.

JAMISON K46

Jamison Steel Corp.
C 0.85, Mn 1.2, Cr 1.5, W 0.5, V 0.1, bal Fe.
For blanking and forming dies; reamers, punches; oil hardened, nondeforming.

JAMISON SPECIAL

Jamison Steel Corp.
C 0.7-1.4, Mn 0.3, Si 0.3, bal Fe.
For drills, reamers, hobs, planer tools, springs; Type W1; water hardened.

JANNEY NO. 2 BRONZE

Janney Cylinder Co.
Sn 10, Pb 2, bal Cu.
Cast 90 Brin. For high pressure bearings, pump liners; corrosion resistant. *Obsolete*

JANNEY NO. 20

Janney Cylinder Co.
Ni 2, Sn 5, Pb 14, bal Cu.
Cast: 28,000 TS; 80 Brin. For bushings, bearings, sleeves, pump liners. *Obsolete*

JANNEY NO. 8 BRONZE

Janney Cylinder Co.
Sn 7.5, Pb 7.5, Zn 7.5, bal Cu.
For crusher bearings, bushings; heavy duty. *Obsolete*

JANO

Hidalgo Steel Co. Inc.
Cr 4, V 2, W 1, Mo 0.6, C, bal Fe.
For tools, cutters; high speed steel.

JANUS

Hidalgo Steel Co. Inc.
C, bal Fe.
For cold dies; very tough.

JANUS EXTRA ZH

Dorrenberg Edelstahl GmbH
C 1-1.1, Mn 0-0.25, Si 0-0.25, bal Fe.
For drills, taps, reamers, broaches, hobs; Type W1; water hardened.

JANUS GHC17M

Dorrenberg Edelstahl GmbH
C 0.35, Cr 17, Mo 2, bal Fe.
For acid resistant chemical plant equipment; corrosion resistant.

JANUS GR15

Dorrenberg Edelstahl GmbH
C 0-0.15, Si 0.75, Cr 13, Ni 0-1.8, bal Fe.
Annealed: 75,000 TS; 40,000 YS; 35 El; 70 RA; 150 Brin. Cold drawn: 100,000 TS; 85,000 YS; 17 El; 60 RA; 205 Brin. Type 410; corrosion resistant.

JANUS GR18

Dorrenberg Edelstahl GmbH
C 0-0.25, Si 0.7, Cr 17, Ni 0-1.8, bal Fe.
For furnace parts, chemical plant equipment; corrosion resistant.

JANUS GR25

Dorrenberg Edelstahl GmbH
C 0.25, Si 0.7, Cr 14.5, Ni 0-1.8, bal Fe.
Annealed: 95,000 TS; 50,000 YS; 25 El; 55 RA; 195 Brin. Cold drawn: 105,000 TS; 85,000 YS; 17 El; 50 RA; 215 Brin. For cutlery, turbine blades, surgical instruments; Type 420; corrosion resistant.

JANUS GSA

Dorrenberg Edelstahl GmbH
C 0.15, Si 1.5, Cr 18, Ni 8.5, bal Fe.
Annealed: 85,000 TS; 40,000 YS; 55 El; 65 RA; 160 Brin. For chemical plant equipment, tanks, mixers, vessels; Type 302; stainless, austenitic.

JANUS GSAT

Dorrenberg Edelstahl GmbH
C 0-0.12, Cr 18, Ni 9.5, Cb = 8 x C, bal Fe.
Annealed: 90,000 TS; 45,000 YS; El; 60 RA; 170 Brin. For chemical plant equipment, welded structures; Type 346; stainless, austenitic.

JANUS GSAW

Dorrenberg Edelstahl GmbH
C 0-0.07, Ni 9.5, Cr 18, bal Fe.
Annealed: 85,000 TS; 40,000 YS; 55 El; 65 RA; 160 Brin. For chemical plant equipment, welded structures; Type 304; stainless, austenitic.

JANUS GSB

Dorrenberg Edelstahl GmbH
C 0.15, Si 2, Cr 18, Ni 9.5, Mo 2, bal Fe.
Annealed: 90,000 TS; 45,000 YS; 50 El; 60 RA; 180 Brin. For acid resistant chemical plant equipment; Type 316; stainless, austenitic.

JANUS GSBT

Dorrenberg Edelstahl GmbH
C 0-0.12, Cr 18, Ni 10.5, Mo 2, Cb = 10 x C, bal Fe.
Annealed: 85,000 TS; 40,000 YS; 50 El; 60 RA; 180 Brin. For acid resistant welded structures, mixers, filters; Type 316 Cb; stainless, austenitic.

JANUS GSBW

Dorrenberg Edelstahl GmbH
C 0-0.07, Cr 18, Ni 10.5, Mo 2, bal Fe.
Annealed: 80,000 TS; 35,000 YS; 55 El; 65 RA; 160 Brin. For chemical plant equipment, mixers, agitators, filters; Type 316L; stainless, austenitic.

JANUS HC17M

Dorrenberg Edelstahl GmbH
C 0.35, Cr 17, Mo 2, bal Fe.
For acid resistant chemical plant equipment; corrosion resistant.

JANUS MH

Dorrenberg Edelstahl GmbH
C 1, Si 0-0.25, Mn 0-0.25, bal Fe.
For drills hobs, reamers, taps, punches; Type W1; water hardened.

JANUS R15

Dorrenberg Edelstahl GmbH
C 0-0.15, Si 0.4, Cr 13, bal Fe.
Annealed: 75,000 TS; 40,000 YS; 35 El; 70 RA; 155 Brin. Cold drawn: 100,000 TS; 85,000 YS; 17 El; 60 RA; 205 Brin. For valves, turbine blades, tableware, hardware; Type 410; corrosion resistant.

JANUS R15W

Dorrenberg Edelstahl GmbH
C 0-0.12, Si 0.4, Cr 13, bal Fe.
Annealed: 75,000 TS; 40,000 YS; 70 RA; 155 Brin. Cold drawn: 100,000 TS; 85,000 YS; 17 El; 60 RA; 205 Brin. For valves, turbine blades, tableware, hardware; Type 410; corrosion resistant.

JANUS R17

Dorrenberg Edelstahl GmbH
C 1, Cr 18, W, Mo, V, bal Fe.
Annealed: 110,000 TS; 65,000 YS; 17 El; 32 RA; 22 Brin. Heat treated: 280,000 TS; 270,000 YS; 3 El; 15 RA; 550 Brin. For bearings, sleeves, liners; corrosion resistant; wear resistant.

JANUS R18

Dorrenberg Edelstahl GmbH
C 0.22, Si 0.4, Cr 17, Ni 1.5, bal Fe.
Annealed: 120,000 TS; 90,000 YS; 15 El; 20 RA; 260 Brin. Heat treated: 180,000 TS; 130,000 YS; 10 El; 15 RA; 420 Brin. For pump shafts, surgical trusses, valve trim; Type 431; corrosion resistant.

JANUS R25

Dorrenberg Edelstahl GmbH
C 0.2, Si 0.4, Cr 13, bal Fe.
Annealed: 95,000 TS; 50,000 YS; 25 El; 55 RA; 195 Brin. Cold drawn: 105,000 TS; 85,000 YS; 17 El; 50 RA; 215 Brin. For cutlery, valves, surgical instruments; Type 420; stainless, hardenable.

JANUS R3V

Dorrenberg Edelstahl GmbH
C 0.3, Cr 14, Ni, V, bal Fe.
Annealed: 100,000 TS; 60,000 YS; 40 El; 50 RA; 200 Brin. For oil refinery equipment; corrosion and creep resistant.

JANUS R45

Dorrenberg Edelstahl GmbH
C 0.4, Cr 14, Si 0.4, bal Fe.
Annealed: 90,000 TS; 45,000 YS; 50 El; 65 RA; 180 Brin. For oil refinery equipment, cutlery, valves; Type 420; corrosion resistant.

JANUS R50

Dorrenberg Edelstahl GmbH
C 0.5, Cr 16, Si 0.4, bal Fe.
For oil refinery equipment; corrosion resistant.

JANUS SA

Dorrenberg Edelstahl GmbH
C 0-0.15, Si 0.4, Cr 18, Ni 8.5, bal Fe.
Annealed: 85,000 TS; 40,000 YS; 50 El; 65 RA; 160 Brin. For chemical plant and oil refinery equipment; Type 302; austenitic, stainless.

JANUS SAT

Dorrenberg Edelstahl GmbH
C 0-0.12, Cr 18, Ni 9.5, Cb = 8 x C, bal Fe.
Annealed: 90,000 TS; 45,000 YS; 45 El; 60 RA; 170 Brin. For chemical plant equipment, welded structures; Type 346; stainless, austenitic.

JANUS SAW

Dorrenberg Edelstahl GmbH
C 0-0.07, Cr 18, Ni 9.5, bal Fe.
Annealed: 85,000 TS; 40,000 YS; 55 El; 65 RA; 160 Brin. For chemical plant equipment, welded structures; Type 304; stainless, austenitic.

JANUS SB

Dorrenberg Edelstahl GmbH
C 0-0.15, Cr 18, Ni 8.5, bal Fe.
Annealed: 80,000 TS; 35,000 YS; 55 El; 65 RA; 160 Brin. For chemical plant equipment, tanks; Type 302; stainless, austenitic.

JANUS SBT

Dorrenberg Edelstahl GmbH
C 0-0.12, Cr 18, Ni 10.5, Mo 2, Cb = 8 x C, bal Fe.
Annealed: 85,000 TS; 40,000 YS; 50 El; 60 RA; 180 Brin. For acid resistant welded structures; Type 316Cb; stainless, austenitic.

JANUS SBW

Dorrenberg Edelstahl GmbH
C 0-0.07, Cr 18, Ni 10.5, Mo 2, bal Fe.
Annealed: 80,000 TS; 35,000 YS; 55 El; 65 RA; 160 Brin. For chemical plant equipment, mixers, agitators, tanks; Type 316L; stainless, austenitic.

JANUS SPEZIAL

Dorrenberg Edelstahl GmbH
C 1.4, W 13, Mn 0.3, V, bal Fe.
For cutters, tools; fast finishing.

JANUS SPEZIAL EXTRA

Dorrenberg Edelstahl GmbH
C 1.3, Cr 0-0.2, W 4.7, bal Fe.
For cutters, tools; fast finishing.

JANUS V

Dorrenberg Edelstahl GmbH
C 0.9, Mn 1.9, V 0.1, bal Fe.
For dies; punches, cutters, crimpers; oil hardened; nondeforming.

JANUS ZAH

Dorrenberg Edelstahl GmbH
C 0.7, Mn 0-0.25, Si 0-0.25, bal Fe.
For drills, punches, crimpers, springs; Type W1; water hardened.

JANUS ZH

Dorrenberg Edelstahl GmbH
C 0.85, Si 0-0.25, Mn 0-0.25, bal Fe.
For drills, taps, reamers, hobs; Type W1; water hardened.

JANUSIT-A

Dorrenberg Edelstahl GmbH
C 0.4, Si 1.3, Cr 27, Ni 4, bal Fe.
For furnace parts and equipment, heat treating boxes; heat resistant.

JANUSIT-B

Dorrenberg Edelstahl GmbH
C 0.4, Si 2, Cr 22, Ni 9.5, bal Fe.
For furnace parts, heat treating boxes; corrosion and heat resistant.

JANUSIT-C

Dorrenberg Edelstahl GmbH
C 0.4, Si 2, Cr 26, Ni 14, bal Fe.
For furnace parts and equipment, heat treating boxes; corrosion and heat resistant; austenitic.

JANUSIT-D

Dorrenberg Edelstahl GmbH
C 0.4, Si 2, Cr 25, Ni 19, bal Fe.
For furnace parts and equipment, heat treating boxes; corrosion and heat resistant; austenitic.

JANUSIT-E

Dorrenberg Edelstahl GmbH
C 0.5, Si 1.8, Cr 25, Ni 30, bal Fe.
For furnace parts and equipment, heat treating boxes; corrosion and heat resistant; austenitic.

JAPAN 2H-SUPER

Japan Steel Works Ltd.
C 0.08-0.16, Mn 0.6-1.2, Ni 0-1, Cu 0-0.5, Mo 0-0.4, bal Fe.
Heat treated: 110,000-114,000 TS; 90,000 YP min; 22 El min. For mine cars, agricultural equipment, railroad car and automobile bodies. High-strength low-alloy constructional steel.

JAPAN 2H-ULTRA

Japan Steel Works Ltd.
C 0.08-0.16, Mn 0.6-1.2, Cu 0.15-0.5, Mo 0-0.7, V 0-0.1, B 0-0.006, Ni 1-1.5, Cr 0-0.8, bal Fe.
Heat treated: 114,000-135,000 TS; 100,000 YP min; 20 El min. For mine and railroad cars, booms, agricultural equipment, structures. High-strength low-alloy constructional steel.

JAPAN BRASS

Japanese manufacture
Cu 66.6, Zn 33.4
For hardware, fixtures, radiators; high strength.

JBR ALLOY

Rennie Tool Co. Ltd.
Co 6, C, W, Cr, V, bal Fe.
For tools, cutters; tipped tools.

JD

J.F. Jelenko & Co.
Au 63, Pt 28.5, Mo 6.
110,000 psi TS; 80,000 psi YS; 5 El; 400 Brin. Dental alloy for partial dentures.

JEFFALOY 55M

Dresser Industries
Mn 0.5-0.7, Mo 0.7-0.9, TC 2.85-3.05, Si, bal Fe.
Cast: 50,000-60,000 TS; 248-269 Brin. For castings; wear resistant.

JEFFALOY A

Dresser Industries
TC 2.85-3.5, Mn 0.3-0.7, Si, bal Fe.
Cast: 35,000-50,000 TS; 192-248 Brin. For castings, pump parts, sprockets, housings; wear resistant cast iron, processed.

JEL-3

J.F. Jelenko & Co.
Precious metal. Au 66, Pd 4, Ag 20.
Quenched: 64,000 psi TS; 41,000 psi YS; 38 El; 120 Brin. Hardened: 96,000 psi TS; 76,000 psi YS; 17 El; 165 Brin. Type III hard dental alloy. For inlays, crowns, and fixed bridgework.

JEL-4

J.F. Jelenko & Co.
Precious metal. Au 66.5, Pd 3.5, Ag 14.5.
Quenched: 66,000 psi TS; 43,000 psi YS; 35 El; 150 Brin. Hardened: 111,000 psi TS; 87,500 psi YS; 4 El; 215 Brin. Type IV extra-hard dental alloy. For hard inlays, thin crowns, fixed bridgework, and partial dentures.

JEL-5

J.F. Jelenko & Co.
Precious metal. Pd 54, Ag 38.5.
105,000 psi TS; 67,000 psi YS; 25 El; 170 Brin. Dental alloy for fusing porcelain to metal.

JEL-62

J.F. Jelenko & Co.
Precious metal. Au 62, Pd 3, Ag 25.
Quenched: 71,000 psi TS; 42,000 psi YS; 35 El; 130 Brin. Hardened: 96,000 psi TS; 70,500 psi YS; 15 El; 190 Brin. Type III hard dental alloy. For inlays, crowns, and fixed bridgework.

JEL-O 75

J.F. Jelenko & Co.
Precious metal. Au 75, Pd 18, Ag 1.
90,000 psi TS; 75,000 psi YS; 5 El; 210 Brin. Dental alloy for fusing porcelain to metal.

JELENKO "O"

J.F. Jelenko & Co.
Precious metal. Au 87.5, Pt 4.5, Pd 6, Ag 1.
72,500 psi TS; 65,300 psi YS; 5 El; 165 Brin. Dental alloy for fusing porcelain to metal.

JELENKO NO. 7

J.F. Jelenko & Co.
Precious metal. Au 69, Pt 3, Pd 3.5, Ag 12.5.
Quenched: 71,000 psi TS; 50,000 psi YS; 35 El; 135 Brin. Hardened: 112,500 psi TS; 72,000 psi YS; 7 El; 240 Brin. Type IV extra-hard dental alloy. For hard inlays, thin crowns, fixed bridgework, and partial dentures.

JELLIFF 800 LN (LOW NOISE)

Jelliff Corp.
Ni, Cr, Mn, Mo.
Resistance alloy for potentiometers and power resistors. 800 ohms/circular mill foot resistivity at 25°C. Similar to Jelliff 800.

JELLIFF ALLOY 1000

Jelliff Corp.
Ni, Cr, Mn, Mo.
Rolled: 165,000 TS. For electrical resistances, potentiometers; max operating temperature 260°C, nonmagn *Obsolete*

JELLIFF ALLOY 180

Jelliff Corp.
Copper. Ni 22, Cu 78.
Hard: 100,000 TS; 3 El. Annealed: 50,000 TS; 30 El. For resistance wire, rheostats, potentiometers; maximum operating temperature 450°C; 180 ohms/circular mill foot resistivity at 25°C.

JELLIFF ALLOY 30

Jelliff Corp.
Copper. Ni 2, Cu 98.
Annealed: 30,000 TS; 30 El. 30 ohms/circular mill foot resistivity at 25°C.

JELLIFF ALLOY 45

Jelliff Corp.
Copper. Ni 45, Cu 55.
Annealed: 60,000 TS; 30 El. Hard: 135,000 TS; 3 El. For thermocouples, resistance coils, low temperature heaters; resists oxidation and corrosion at lower temperature; maximum operating temperature 500°C. 294 ohms/circular mill foot resistivity at 25°C.

JELLIFF ALLOY 60

Jelliff Corp.
Copper. Ni 6, Cu 94.
Annealed: 35,000 TS; 30 El. Hard: 70,000 TS; 3 El. 60 ohms/circular mill foot resistivity at 25°C. For relays, low resistance resistors, rheostats; nonmagnetic, easily soldered.

JELLIFF ALLOY 800

Jelliff Corp.
Nickel. Ni 61, Cr 20, Mn 17.5, Mo 1.5.
Annealed: 150,000 TS; 30 El. Hard: 250,000 TS; 3 El. For precision resistors, potentiometers. 800 ohms/circular mill foot resistivity at 25°C.

JELLIFF ALLOY 90

Jelliff Corp.
Copper. Ni 12, Cu 88.
Hard: 75,000 TS; 3 El. Annealed: 35,000 TS; 30 El. 90 ohms/circular mill foot resistivity at 25°C. For resistance wire, rheostats, potentiometers; maximum operating temperature 400°C.

JELLIFF ALLOY A

Jelliff Corp.
Nickel. Ni 80, Cr 20.
Annealed: 95,000 TS; 30 El. Hard: 105,000 TS; 3 El. For industrial furnace elements, voltmeter windings, electrical instruments; maximum operating temperature 1150°C; high resistance to oxidation and corrosion.

JELLIFF ALLOY C

Jelliff Corp.
Nickel. Ni 60, Cr 16, bal Fe.
Annealed: 95,000 TS; 30 El. Hard 200,000 TS; 3 El. For heating elements, rheostats, resistors, potentiometers; maximum operating temperature 1350°C; heat and corrosion resistant.

JELLIFF ALLOY C/O/J

Jelliff Corp.
Ni, Cr, Mn, Mo.
Fine wire with very high electrical resistivity. 1040 ohms/circular mill foot resistivity at 25°C.

JELLIFF ALLOY D

Jelliff Corp.
Ni 35, Cr 20, bal Fe.
Annealed: 85,000 TS; 25 El. Hard: 160,000 TS; 2 El. For heavy rheostats, car heaters, heating elements; maximum operating temperature 1600°F. *Obsolete*

JELLIFF ALLOY K

Jelliff Corp.
Cr 15.5, Al 5.5, bal Fe.
Annealed: 100,000 TS; 20 El. Hard: 200,000 TS; 3 El. For high temperature heating elements, resistors; maximum operating temperature 1000°C, magnetic. *Obsolete*

JELLIFF ALLOY K20

Jelliff Corp.
Cr 15.5, Al 5.5, bal Fe.
Annealed: 100,000 TS; 20 El. Hard: 200,000 TS; 3 El. For precision resistors, power resistors. *Obsolete*

JELLIFF HP (HIGH PURITY) NICKEL

Jelliff Corp.
Nickel. 99.9 Ni min.
For ballast applications and resistance thermometers. 44 ohms/circular mill foot resistivity at 25°C.

JELLIFF MANGANIN

Jelliff Corp.
Copper. Cu 86, Mn 12, Ni 2.
Wire for resistance boxes, potentiometers. 290 ohms/circular mill foot resistivity at 25°C.

JELSTAR

J.F. Jelenko & Co.
Precious metal. Pd 60, Ag 28.
95,000 psi TS; 67,000 psi YS; 20 El; 172 Brin. Dental alloy for fusing porcelain to metal.

JESSAIR

Jessop Steel Co.
C 0.7, Mn 2, Si 0.3, Cr 1, Mo 1.35, bal Fe.
Heat treated: 293,000 TS; 265,000 YS; 1 El; 2 RA; 55 Rock C. For blanking and forming dies; trimming and notching dies; master hubs, shear blades, bending tools, mandrels, heavy duty punches. Type A6 air hardening tool steel.

JESSCO

Jessop Steel Co.
C 0.7, bal Fe.
For hot and cold dies, heading and forging dies, shear blades, punches; high speed steel. *Obsolete*

JESSCO A

Jessop Steel Co.
C 0.7, W, Cr, V, bal Fe.
For tools, cutters; high speed steel. *Obsolete*

JESSCO B

Jessop Steel Co.
C 0.7, W, Cr, V, Co, bal Fe.
For tools, cutters; high speed steel. *Obsolete*

JESSOP "J" HOT WORK

Jessop Steel Co.
C 0.62, Cr 3.8, Mo 0.55, V 0.55, bal Fe.
For hot work dies, bloom shears, shears, trimmer dies; hot die steel. *Obsolete*

JESSOP 181 NICKEL STEEL

Jessop Steel Co.
Ni 36, C, bal Fe.
Hot rolled: 65,000-85,000 TS; 40,000-60,000 YS; 30-45 El. For thermostatic elements; Rockwell "B" 82-86. *Obsolete*

JESSOP 2 B (MC)

Jessop Steel Co.
Cr 3, Mn 0.25, V 0.3, W 13.5, bal Fe.
For tools, bolts and rivet dies; mandrels. Tough, hot work steel.

JESSOP 2B (HC)

Jessop Steel Co.
C 0.48, Cr 2.75, V 0.3, W 11.25, bal Fe.
For punches, mandrels, grippers, heading dies; hot work steel.

JESSOP 2B (LC)
Jessop Steel Co.
C 0.3, Cr 3.15, V 0.3, W 10, bal Fe.
For forming dies; hot work steel.

JESSOP 3 C
Jessop Steel Co.
C 2.25, Cr 12, V 0.2, Mo 0.8, bal Fe.
AISI D2 cast to shape. For coining, crimping and cutting dies, hobs, punches; wear resistant, oil hardened.

JESSOP 3 C SPECIAL
Jessop Steel Co.
C 1.45, Cr 12.5, Mo 0.8, Co 3.2, Si 0.4, bal Fe.
AISI D5 cast to shape. For die casting dies; air hardened.

JESSOP 302
Jessop Steel Co.
C 0.08-0.2, Cr 17-19, Ni 8-10, bal Fe.
Annealed: 85,000 TS; 40,000 YS; 50 El; 65 RA; 150 Brin. For springs, screens, chemical plant equipment; stainless, austenitic. Type 302.

JESSOP 302-B
Jessop Steel Co.
C 0.15, Mn 2, Cr 17-19, Ni 8-10, Si 2-3, bal Fe.
Improved scale resistance over AISI 302; for furnace parts.

JESSOP 302-S
Jessop Steel Co.
C 0-0.08, Cr 17-18, Ni 8-10, bal Fe.
Modified AISI 304.

JESSOP 303
Jessop Steel Co.
C 0-0.15, Cr 17-19, Ni 8-10, 0.07 P min, S or Se, bal Fe.
Annealed: 85,000 TS; 40,000 YS; 40 El; 55 RA; 150 Brin. For screw machine products, bolts, fasteners; Type 303; stainless, free cutting.

JESSOP 304
Jessop Steel Co.
C 0-0.08, Cr 18-20, Ni 8-11, bal Fe.
Annealed: 85,000 TS; 35,000 YS; 60 El; 70 RA; 150 Brin. Cold drawn: 180,000 TS; 125,000 YS; 10 El; 330 Brin. For architectural molding and trim, kitchen equipment; Type 304; stainless, austenitic.

JESSOP 304 N
Jessop Steel Co.
C 0-0.08, Mn 2, Cr 18-20, Ni 8-10.5, Si 1, N 0.1, bal Fe.
Increased strength over AISI 304.

JESSOP 304L
Jessop Steel Co.
C 0-0.03, Mn 0-2, Si 0-1, Cr 18-20, Ni 8-12, bal Fe.
For boilers and pressure vessels; weldable.

JESSOP 309
Jessop Steel Co.
C 0-0.2, Cr 22-24, Ni 12-15, bal Fe.
Annealed: 90,000 TS; 40,000 YS; 50 El; 65 RA; 180 Brin. For furnace parts, tube supports, heat treat boxes; Type 309; stainless, austenitic.

JESSOP 309-CB
Jessop Steel Co.
C 0.2, Cr 22-24, Ni 12-15, Mn 2, Si 1, Cb, bal Fe.
Columbium stabilized AISI 309 to prevent carbide precipitation at elevated temperatures.

JESSOP 309-MOD
Jessop Steel Co.
Similar to AISI 309 but modified to give very hot strength in high temperature applications.

JESSOP 310
Jessop Steel Co.
C 0-0.25, Cr 24-26, Ni 19-22, bal Fe.
Annealed: 100,000 TS; 45,000 YS; 50 El; 65 RA; 180 Brin. For furnace parts and equipment, valves, pumps, baffles; Type 310; austenitic, heat resistant.

JESSOP 316
Jessop Steel Co.
C 0-0.1, Cr 16-18, Ni 10-14, Mo 2-3, bal Fe.
Annealed: 80,000 TS; 30,000 YS; 60 El; 80 RA; 135 Brin. Cold drawn: 150,000 TS; 135,000 YS; 6 El; 300 Brin. For chemical plant equipment, agitators, digesters, valve trim; Type 316; austenitic, stainless.

JESSOP 316-L
Jessop Steel Co.
C 0-0.03, Mn 0-2, Si 0-1, Cr 16-18, Ni 10-14, Mo 2-3, bal Fe.
Increased corrosion and temperature resistance over AISI 316.

JESSOP 317
Jessop Steel Co.
C 0-0.1, Cr 18-20, Ni 11-14, Mo 3-4, bal Fe.
Annealed: 85,000 TS; 40,000 YS; 50 El; 70 RA; 150 Brin. For chemical plant equipment, agitators, digesters, valve trim; Type 317; stainless, austenitic.

JESSOP 317-L
American Saw & Mfg. Co.
Stainless steel. C 0-0.03, Mn 2, Si 1, Cr 18-20, Ni 11-15, Mo 3-4, bal Fe.
Low carbon grade of AISI 317 for welded assemblies.

JESSOP 317CB
Jessop Steel Co.
C 0-0.1, Cr 18-20, Ni 11-14, Mo 3-4, Cb = 8 x C, bal Fe.
Annealed: 85,000 TS; 40,000 YS; 50 El; 70 RA; 150 Brin. For welded equipment, chemical plant equipment; Type 317 Cb; stabilized, stainless. *Obsolete*

JESSOP 318
Jessop Steel Co.
C 0-0.1, Mn 0-2, Cr 16-18, Ni 10-14, Mo 2-3, Cb = 10 x C, bal Fe.
For welded structures and equipment; stainless, austenitic. *Obsolete*

JESSOP 319
Jessop Steel Co.
C 0-0.07, Cr 18, Ni 12, Mo 2.5, bal Fe.
Modified AISI 316 with improved corrosion resistance.

JESSOP 319-L
Jessop Steel Co.
C 0-0.03, Cr 18, Ni 12, Mo 2.5, bal Fe.
Low carbon modification of JESSOP 319.

JESSOP 321
Jessop Steel Co.
C 0-0.08, Cr 17-19, Ni 8-18, Ti = 5 x C, bal Fe.
Annealed: 85,000 TS; 33,000 YS; 58 El; 75 RA; 165 Brin. For jet aircraft, chemical plant equipment, refinery tubes; Type 321; stabilized, stainless.

JESSOP 347
Jessop Steel Co.
C 0-0.08, Cr 17-19, Ni 9-12, Cb = 10 x C, bal Fe.
Annealed: 91,000 TS; 39,500 YS; 50 El; 71 RA; 200 Brin. For welded structures, chemical plant equipment, vessels, tanks; Type 347; stabilized, stainless.

JESSOP 348
Jessop Steel Co.
C 0-0.08, Mn 0-2, Si 0-1, Cr 17-19, Ni 9-13, Cb + Ta = 10 x C, bal Fe.
Stabilized austenitic stainless steel.

JESSOP 409
Jessop Steel Co.
C 0.08, Mn 1, Cr 10.5-11.75, Si 1, Ti = 6 x C, bal Fe.
Chromium stainless steel.

JESSOP 410
Jessop Steel Co.
C 0-0.15, Cr 11.5-13.5, bal Fe.
Annealed: 75,000 TS; 40,000 YS; 35 El; 70 RA; 155 Brin. Cold drawn: 100,000 TS; 85,000 YS; 17 El; 60 RA; 205 Brin. For tableware, hardware, flat springs; Type 410; corrosion resistant.

JESSOP 416
Jessop Steel Co.
C 0-0.15, Cr 12-14, S or Se, 0.07 P min, bal Fe.
Annealed: 75,000 TS; 40,000 YS; 30 El; 60 RA; 155 Brin. Heat treated: 110,000 TS; 85,000 YS; 18 El; 55 RA; 230 Brin. For screw machine products, shafts, fasteners, gears. Type 416; stainless, free cutting.

JESSOP 420
Jessop Steel Co.
Cr 12-14, 0.15% min C, bal Fe.
Annealed: 95,000 TS; 50,000 YS; 25 El; 55 RA; 195 Brin. Heat treated: 250,000 TS; 215,000 YS; 8 El; 25 RA; 512 Brin. For cutlery, surgical instruments, tableware, gauges; Type 420; stainless, hardenable. *Obsolete*

JESSOP 430
Jessop Steel Co.
C 0-0.12, Cr 14-18, bal Fe.
Annealed: 70,000 TS; 40,000 YS; 30 El; 55 RA; 140 Brin. Cold drawn: 130,000 TS; 120,000 YS; 2 El; 185 Brin. For kitchen sinks, fasteners, soot blowers, bolts; Type 430; corrosion resistant. *Obsolete*

JESSOP 440A
Jessop Steel Co.
Cr 16-18, Mo 0-0.75, C 0.6-0.75, bal Fe.
Annealed: 95,000 TS; 55,000 YS; 20 El; 220 Brin. Heat treated: 225,000 TS; 240,000 YS; 2 El; 550 Brin. For needle valves, ball bearings, pivots, shafts, surgical instruments; Type 440A; stainless, hardenable.

JESSOP 440B
Jessop Steel Co.
C 0.75-0.95, Cr 16-18, Mo 0-0.75, bal Fe.
Annealed: 107,000 TS; 62,000 YS; 18 El; 35 RA; 220 Brin. Heat treated: 280,000 TS; 270,000 YS; 3 El; 15 RA; 555 Brin. For dental and surgical instruments, pivots, ball bearings, valves; Type 440B; stainless, hardenable.

JESSOP 440C
Jessop Steel Co.
C 0.95-1.2, Cr 16-18, Mo 0-0.75, bal Fe.
Annealed: 110,000 TS; 70,000 YS; 15 El; 30 RA; 225 Brin. For ball bearings, valve parts, pivots, Type 440C; stainless, hardenable.

JESSOP 446
Jessop Steel Co.
C 0-0.35, Cr 23-27, N 0-0.25, bal Fe.
Annealed: 75,000 TS; 45,000 YS; 35 El; 65 RA; 160 Brin. Cold drawn: 175,000 TS; 155,000 YS; 2 El; 25 RA; 250 Brin. For annealing boxes, oil burner and furnace parts; Type 446; stainless and heat resistant.

JESSOP 494E
Saville & Co. Ltd., J.J.
C 1.1, Mn 0.3, W 0.5, bal Fe.
For taps, drills, reamers; tough core.

JESSOP 494E
Jessop-Saville Ltd.
C 1.1, Mn 0.3, W 0.5, bal Fe.
For taps, drills, reamers; tough core.

JESSOP 6-STAR VANADIUM
Saville & Co. Ltd., J.J.
C 1, Mn 0.3, V 0.2, bal Fe.
For coining and cutlery dies, cold-heading dies; tough.

JESSOP 6-STAR VANADIUM
Jessop-Saville Ltd.
C 1, Mn 0.3, V 0.2, bal Fe.
For coining and cutlery dies, cold-heading dies; tough.

JESSOP O-6
Jessop Steel Co.
C 1.45, Si 1, Mo 0.25, bal Fe.
Oil hardening tool steel for coining dies, trim dies, cams; AISI O-6.

JESSOP A-00
Saville & Co. Ltd., J.J.
C 0.08, bal Fe.
For plastic molding dies; hobbing steel.

JESSOP A-00
Jessop-Saville Ltd.
C 0.08, bal Fe.
For plastic molding dies; hobbing steel.

JESSOP A-1
Saville & Co. Ltd., J.J.
C 0.15, Mn 0.6, S 0.25, bal Fe.
60,000 TS; 15 El. For bolts, screws, shafts, bushes; free-cutting, case-hardening.

JESSOP A-1
Jessop-Saville Ltd.
C 0.15, Mn 0.6, S 0.25, bal Fe.
60,000 TS; 15 El. For bolts, screws, shafts, bushes; free-cutting, case-hardening.

JESSOP A-2A
Saville & Co. Ltd., J.J.
C 0.15, Mn 0.6, bal Fe.
Rolled: 65,000 TS; 20 El. For camshafts, tappets, rollers, spindles; case-hardening.

JESSOP A-2A
Jessop-Saville Ltd.
C 0.15, Mn 0.6, bal Fe.
Rolled: 65,000 TS; 20 El. For camshafts, tappets, rollers, spindles; case-hardening.

JESSOP A-3
Saville & Co. Ltd., J.J.
C 0.22, Mn 0.7, bal Fe.
For turbine casings, engine frames; good moldability.

JESSOP A-3
Jessop-Saville Ltd.
C 0.22, Mn 0.7, bal Fe.
For turbine casings, engine frames; good moldability.

JESSOP A-3C
Jessop Steel Co.
C 0.9, Mn 0.7, Cr 0.27, bal Fe.
For saws; water hardened. *Obsolete*

JESSOP A-4
J.J. Saville & Co. Ltd
C 0.3, Mn 0.5, bal Fe.
Rolled: 76,200-81,000 TS; 45,000-51,000 YS; 30-35 El; 55-60 RA; 150-160 Brin. For bolts, studs, brake drums; water hardening.

JESSOP A-5
J.J. Saville & Co. Ltd.
C 0.4, Mn 0.8, bal Fe.
Rolled: 85,200-89,600 TS; 49,000-65,000 YS; 26 El; 55 RA; 170 Brin. For axles, axle tubes, bolts; low stress applications.

JESSOP ALLOY 104
Jessop Steel Co.
C 0.1-0.2, Cr 5, W 1, bal Fe.
Annealed: 60,000 TS; 25,000 YS; 30 El; 170 Brin. Heat treated: 143,000 TS; 126,000 YS; 20 El; 67 RA; 270 Brin. For bolts, bushes, pipes; corrosion and heat resistant. *Obsolete*

JESSOP ALLOY C
Jessop-Saville Ltd.
C 2.3, Mn 0.35, Cr 13, bal Fe.
For shear blades, liners, thread rolling dies; cold work steel, nondeforming.

JESSOP AM-2
Saville & Co. Ltd., J.J.
C 0.15, Mn 1.5, bal Fe.
Rolled: 80,000 TS; 20 El. For rifle parts, general engineering parts; case-hardening.

JESSOP AM-2
Jessop-Saville Ltd.
C 0.15, Mn 1.5, bal Fe.
Rolled: 80,000 TS; 20 El. For rifle parts, general engineering parts; case-hardening.

JESSOP AM-2S
Saville & Co. Ltd., J.J.
C 0.15, Mn 1.5, S 0.1, bal Fe.
Rolled: 80,000 TS; 18 El. For general engineering parts; case-hardening, free-cutting.

JESSOP AM-2S
Jessop-Saville Ltd.
C 0.15, Mn 1.5, S 0.1, bal Fe.
Rolled: 80,000 TS; 18 El. For general engineering parts; case-hardening, free-cutting.

JESSOP AM-3
Saville & Co. Ltd., J.J.
C 0.25, Mn 1.6, bal Fe.
For welded structures; tough.

JESSOP AM-3
Jessop-Saville Ltd.
C 0.25, Mn 1.6, bal Fe.
For welded structures; tough.

JESSOP AM-4
Saville & Co. Ltd., J.J.
C 0.35, Mn 1.4, bal Fe.
For axles, shafts, gas cylinders; water hardened.

JESSOP AM-4
Jessop-Saville Ltd.
C 0.35, Mn 1.4, bal Fe.
For axles, shafts, gas cylinders; water hardened.

JESSOP AM-4 NICKEL
Saville & Co. Ltd., J.J.
C 0.35, Mn 1.3, Ni 0.5, bal Fe.
For axles, shafts, gas cylinders; tough.

JESSOP AM-4 NICKEL
Jessop-Saville Ltd.
C 0.35, Mn 1.3, Ni 0.5, bal Fe.
For axles, shafts, gas cylinders; tough.

JESSOP ARMAT
Jessop Steel Co.
C 0.85, Cr 3, bal Fe.
300 Brin. For magnets, electrical apparatus; magnet steel. *Obsolete*

JESSOP B-1
Saville & Co. Ltd., J.J.
C 0.55, bal Fe.
For die blocks; oil or air hardening.

JESSOP B-1
Jessop-Saville Ltd.
C 0.55, bal Fe.
For die blocks; oil or air hardening.

JESSOP B-1 NICKEL
Saville & Co. Ltd., J.J.
C 0.55, Mn 0.6, Ni 0.5, bal Fe.
For rifle barrels, gear wheels, housings; water hardening.

JESSOP B-1 NICKEL
Jessop-Saville Ltd.
C 0.55, Mn 0.6, Ni 0.5, bal Fe.
For rifle barrels, gear wheels, housings; water hardening.

JESSOP BLACK LABEL
Jessop-Saville Ltd.
C 0.7-1.4, Mn 0.3, Si 0.3, bal Fe.
For tools, punches, shear blades, taps, drills; water hardened; Type W1.

JESSOP BLUE LABEL
Jessop-Saville Ltd.
C 0.7-1.2, Mn 0.3, bal Fe.
For taps, mandrels, punches, drills; water hardening.

JESSOP BX-3 (AISI A7)
Jessop Steel Co.
C 2.45, Cr 5, V 4.5, Mo 1.1, bal Fe.
For blanking and forming dies; headers; cold work steel. Wear resistant; AISI A7.

JESSOP C
J.J. Saville & Co. Ltd.
C 1.65, Mn 0.45, Cr 13, Mo 0.7, V 0.3, bal Fe.
For burnishing rolls, shear blades, gages; tough, nondeforming.

JESSOP CHROME MAGNET STEEL
Jessop Steel Co.
C 0.8-0.95, Cr 3-3.5, Mn 0.4-0.6, bal Fe.
300-350 Brin. For magnetos, instruments where high permeability and permanency are required; magnet steel. *Obsolete*

JESSOP CNS NO. 1 (AISI D2)
Jessop Steel Co.
C 1.55, Cr 12.5, V 0.8, Mo 0.8, bal Fe.
For blanking and forming dies, punches; air hardened, non-deforming. AISI D2.

JESSOP CNS NO. 2
Jessop Steel Co.
C 2.15, Cr 12, V 0.25, bal Fe.
For blanking and forming dies, punches; oil hardened, non-deforming. *Obsolete*

JESSOP CNS-3 (AISI D4)
Jessop Steel Co.
C 2.25, Cr 12, V 0.2, Mo 0.8, bal Fe.
Hardened: 65-67 Rock C. For coining and crimping dies, forming and extruding dies, punches, hobs, taps, gages, lamination dies. Air hardening, nondeforming, high wear resistance. Type D4 tool steel.

JESSOP COAL AND MINE BIT STEEL
Jessop Steel Co.
C 0.75, bal Fe.
350 Brin. For coal and mine bits; water hardening. *Obsolete*

JESSOP COBALT MAGNET STEEL
Jessop Steel Co.
C 0.87-0.95, Co 35-37, Cr 4-4.5, W 2-2.5, bal Fe.
350 Brin. For magnets, electrical apparatus; magnet steel. *Obsolete*

JESSOP COMPOSITE "R"
Jessop Steel Co.
Tool steel welded to machinery steel. For tools, cutters, dies; composite steel. *Obsolete*

JESSOP CORROSION RESISTANT R-1
Jessop-Saville Ltd.
C 0-0.15, Cr 12, Ni 0-1, bal Fe.
For corrosion resistant parts; corrosion resistant.

JESSOP CORROSION RESISTANT R-2
Jessop-Saville Ltd.
C 0.3, Cr 14, Ni 0-1, bal Fe.
For corrosion resistant parts; corrosion resistant.

JESSOP CORROSION RESISTANT R-3
Jessop-Saville Ltd.
C 0.12, Cr 18, Ni 0-11, bal Fe.
For stainless parts; stainless.

JESSOP CORROSION RESISTANT R-4
Jessop-Saville Ltd.
C 0.25, Cr 18, Ni 1, bal Fe.
For corrosion resistant parts; corrosion resistant.

JESSOP DB NO. 1
Jessop Steel Co.
C 0.5, Mn 0.55, Ni 1.6, Cr 0.7, bal Fe.
For castings. *Obsolete*

JESSOP DRIL-IT DRILL
Jessop Steel Co.
C 0.8, V 0.25, bal Fe.
For drills for rock mining; alloy drill steel. *Obsolete*

JESSOP E-1A
Saville & Co. Ltd., J.J.
C 0.3, Ni 3.25, bal Fe.
At 70°F: 102,000 TS; 25 El. At 1100°F: 32,000 TS; 50 El. At 1300°F: 12,000 TS; 70 El. For inlet valves, exhaust valves; low to medium temperature applications.

JESSOP E-1A
Jessop-Saville Ltd
C 0.3, Ni 3.25, bal Fe.
At 70°F: 102,000 TS; 25 El. At 1100°F: 32,000 TS; 50 El. At 1300°F: 12,000 TS; 70 El. For inlet valves, exhaust valves; low to medium temperature applications.

JESSOP E-25CV
Jessop Steel Co.
C, alloy, bal Fe.
For pneumatic chisel, trimming dies, stamps, shears; oil hardening. *Obsolete*

JESSOP E-4
Saville & Co. Ltd., J.J.
C 0.12, Mn 0.5, Ni 5, bal Fe.
Rolled: 80,000 TS; 20 El. For gears, pinions, shafts, collets; shock resistant, case-hardening.

JESSOP E-4
Jessop-Saville Ltd.
C 0.12, Mn 0.5, Ni 5, bal Fe.
Rolled: 80,000 TS; 20 El. For gears, pinions, shafts, collets; shock resistant, case-hardening

JESSOP E-4 MOLYBDENUM
Saville & Co. Ltd., J.J.
C 0.12, Mn 0.35, Ni 5, Mo 0.25, bal Fe.
Rolled: 130,000 TS; 13 El. For rocker arms, crankshafts, gears, pinions; shock resistant, case-hardening.

JESSOP E-4 MOLYBDENUM
Jessop-Saville Ltd.
C 0.12, Mn 0.35, Ni 5, Mo 0.25, bal Fe.
Rolled: 130,000 TS; 13 El. For rocker arms, crankshafts, gears, pinions; shock resistant, case-hardening.

JESSOP E-5
Saville & Co. Ltd., J.J.
C 0.6, Ni 1.25, bal Fe.
For dies; high core toughness.

JESSOP E-5
Jessop-Saville Ltd.
C 0.6, Ni 1.25, bal Fe.
For dies; high core toughness.

JESSOP E-6
Saville & Co. Ltd., J.J.
C 0.15, Mn 0.5, Ni 2, Mo 0.25, bal Fe.
Rolled: 90,000 TS: 18 El. For pinions, shafts, rifle parts; shock resistant, case-hardening.

JESSOP E-6
Jessop-Saville Ltd.
C 0.15, Mn 0.5, Ni 2, Mo 0.25, bal Fe.
Rolled: 90,000 TS: 18 El. For pinions, shafts, rifle parts; shock resistant, case-hardening.

JESSOP E-8
Saville & Co. Ltd., J.J.
C 0.55, Ni 0.9, Cr 0.25, Mn 0.95, bal Fe.
For mandrel bars, resists heat checking.

JESSOP E-8
Jessop-Saville Ltd.
C 0.55, Ni 0.9, Cr 0.25, Mn 0.95, bal Fe.
For mandrel bars, resists heat checking.

JESSOP E-83
Jessop Steel Co.
C, bal Fe.
For tools; water hardening. *Obsolete*

JESSOP E-9
Jessop Steel Co.
C, bal Fe.
For engineering parts; super high tensile. *Obsolete*

JESSOP E-9
Jessop-Saville Ltd.
C 0.25, Mn 0.5, Ni 2, Mo 0.25, bal Fe.
Rolled: 110,000 TS; 15 El. For rocker arms, camshafts, gears, pinions; shock resistant.

JESSOP E-9
Saville & Co. Ltd., J.J.
C 0.25, Mn 0.5, Ni 2, Mo 0.25, bal Fe.
Rolled: 110,000 TS; 15 El. For rocker arms, camshafts, gears, pinions; shock resistant.

JESSOP E.T. NO. 7
Jessop Steel Co.
C, alloy, bal Fe.
For tools, dies; oil hardened. *Obsolete*

JESSOP ET NO. 4
Jessop Steel Co.
C 0.7, Mn 0.65, Ni 1.3, Cr 0.55, Mo 0.25, bal Fe.
For castings. *Obsolete*

JESSOP ET NO. 6
Jessop Steel Co.
C 0.5, Mn 0.8, Si 0.3, Cr 0.95, V 0.2, bal Fe.
Temper to desired hardness from 200 to 1000 F. For keys, gears, shafts, bushings, springs, etc. AISI Type L2 tool steel. *Obsolete*

JESSOP F 2
Jessop-Saville Ltd.
C 0.57, Cr 0.7, V 0.2, bal Fe.
For die casting dies; hot die steel.

JESSOP F-5
Saville & Co. Ltd., J.J.
C 0.44, Cr 1.3, W 2.3, V 0.15, bal Fe.
For mandrel bars; resists heat checking.

JESSOP F-5
Jessop-Saville Ltd.
C 0.44, Cr 1.3, W 2.3, V 0.15, bal Fe.
For mandrel bars; resists heat checking.

JESSOP F-7
Saville & Co. Ltd., J.J.
C 0.48, Ni 0.7, Cr 1.5, Mo 0.2, V 0.12, W 1.2, bal Fe.
For punches, chisels, shear blades, dies; shock resistant, durable cutting edge.

JESSOP F-7
Jessop-Saville Ltd.
C 0.48, Ni 0.7, Cr 1.5, Mo 0.2, V 0.12, W 1.2, bal Fe.
For punches, chisels, shear blades, dies; shock resistant, durable cutting edge.

JESSOP G-0
Saville & Co. Ltd., J.J.
C 0.33, Mn 0.6, Ni 3.5, Cr 0.75, bal Fe.
For collets, cold punches, dies; oil hardening.

JESSOP G-0
Jessop-Saville Ltd.
C 0.33, Mn 0.6, Ni 3.5, Cr 0.75, bal Fe.
For collets, cold punches, dies; oil hardening.

JESSOP G-11
Saville & Co. Ltd., J.J.
C 0.3, Mn 0.6, Ni 2.5, Cr 0.8, Mo 0.6, bal Fe.
Hardened: 110,000-200,000 TS; 88,000-170,000 YS; 18-10 El; 248-444 Brin. For crankshafts, connecting rods, engine parts; shock resistant.

JESSOP G-11
Jessop-Saville Ltd.
C 0.3, Mn 0.6, Ni 2.5, Cr 0.8, Mo 0.6, bal Fe.
Hardened: 110,000-200,000 TS; 88,000-170,000 YS; 18-10 El; 248-444 Brin. For crankshafts, connecting rods, engine parts; shock resistant.

JESSOP G-12
Saville & Co. Ltd., J.J.
C 0.4, Mn 0.6, Ni 2.5, Cr 0.6, Mo 0.6, bal Fe.
For connecting rods, gears, turbine parts; shock resistant.

JESSOP G-12
Jessop-Saville Ltd.
C 0.4, Mn 0.6, Ni 2.5, Cr 0.6, Mo 0.6, bal Fe.
For connecting rods, gears, turbine parts; shock resistant.

JESSOP G-14
Saville & Co. Ltd., J.J.
C 0.39, Mn 0.5, Ni 4.1, Cr 1.3, Mo 0.3, bal Fe.
For cold punches, shear blades, chisels; air hardening, shock resistant.

JESSOP G-14
Jessop-Saville Ltd.
C 0.39, Mn 0.5, Ni 4.1, Cr 1.3, Mo 0.3, bal Fe.
For cold punches, shear blades, chisels; air hardening, shock resistant.

JESSOP G-15
Saville & Co. Ltd., J.J.
C 0.12, Mn 0.45, Ni 3, Cr 1, bal Fe.
Heat treated: 110,000-160,000 TS; 15-13 El. For gears, cams, pinions, camshafts; case-hardening, shock resistant.

JESSOP G-15
Jessop-Saville Ltd.
C 0.12, Mn 0.45, Ni 3, Cr 1, bal Fe.
Heat treated: 110,000-160,000 TS; 15-13 El. For gears, cams, pinions, camshafts; case-hardening, shock resistant.

JESSOP G-16
Saville & Co. Ltd., J.J.
C 0.15, Mn 0.5, Ni 4.25, Cr 1.3, Mo 0.3, bal Fe.
Heat treated: 170,000 TS; 12 El. For gears, pinions, tappets, valve rockers; shock resistant, case-hardening.

JESSOP G-16
Jessop-Saville Ltd.
C 0.15, Mn 0.5, Ni 4.25, Cr 1.3, Mo 0.3, bal Fe.
Heat treated: 170,000 TS; 12 El. For gears, pinions, tappets, valve rockers; shock resistant, case-hardening.

JESSOP G-18B

J.J. Saville & Co. Ltd.
C 0.4, Mn 0.8, Si 1, Cr 13, Co 10, W 2.5, Mo 2, Cb 3, Ni 13, bal Fe.
At 70°F: 92,000 TS; 40 El. At 1500°F: 44,000 TS; 9 El. For turbine discs, rotor blades, engine exhaust valves; high heat resistant.

JESSOP G-2

Jessop-Saville Ltd.
C 0.4, Ni 13, Cr 13, W 2.5, bal Fe.
At 70°F: 106,000 TS; 31 El. At 1100°F: 66,000 TS; 35 El. For valves, dies for extruding steel tubes, exhaust valves; high heat resistant, shock resistant.

JESSOP G-20

Saville & Co. Ltd., J.J.
C 0.4, Mn 1.35, Ni 0.65, Cr 0.45, Mo 0.2, bal Fe.
For gears, shafts, drop forgings; shock resistant.

JESSOP G-20

Jessop-Saville Ltd.
C 0.4, Mn 1.35, Ni 0.65, Cr 0.45, Mo 0.2, bal Fe.
For gears, shafts, drop forgings; shock resistant.

JESSOP G-21

Saville & Co. Ltd., J.J.
C 0.42, Si 1.3, Cr 13, Ni 13, W 2.5, Nb 1, bal Fe.
106,000 TS; 50,000 YS; 31 El. For furnace parts, bolts, stays, mandrels; high hot strength.

JESSOP G-21

Jessop-Saville Ltd.
C 0.42, Si 1.3, Cr 13, Ni 13, W 2.5, Nb 1, bal Fe.
106,000 TS; 50,000 YS; 31 El. For furnace parts, bolts, stays, mandrels; high hot strength.

JESSOP G-24

Saville & Co. Ltd., J.J.
C 3, Si 1.5, Ni 13, Cr 5, Cu 5, Mn 1.2, bal Fe.
Annealed. For impellers, pump parts; austenitic, castings, stainless.

JESSOP G-24

Jessop-Saville Ltd.
C 3, Si 1.5, Ni 13, Cr 5, Cu 5, Mn 1.2, bal Fe.
Annealed. For impellers, pump parts; austenitic, castings, stainless.

JESSOP G-25

Saville & Co. Ltd., J.J.
C 0.17, Mn 0.5, Ni 2, Cr 2, Mo 0.2, bal Fe.
Heat treated: 170,000 TS; 12 El. For camshafts, gears, bearings; shock resistant, case-hardening.

JESSOP G-25

Jessop-Saville Ltd.
C 0.17, Mn 0.5, Ni 2, Cr 2, Mo 0.2, bal Fe.
Heat treated: 170,000 TS; 12 El. For camshafts, gears, bearings; shock resistant, case-hardening.

JESSOP G-26

Saville & Co. Ltd., J.J.
C 0.4, Mn 0.55, Ni 1.55, Cr 1.15, Mo 0.15, bal Fe.
Heat treated: 92,000-150,000 TS; 70,000-120,000 YS; 22-15 El; 240-360 Brin. For forging bars, stamping; shock resistant.

JESSOP G-26

Jessop-Saville Ltd.
C 0.4, Mn 0.55, Ni 1.55, Cr 1.15, Mo 0.15, bal Fe.
Heat treated: 92,000-150,000 TS; 70,000-120,000 YS; 22-15 El; 240-360 Brin. For forging bars, stamping; shock resistant.

JESSOP G-27

Saville & Co. Ltd., J.J.
C 0.3, Mn 0.55, Ni 3.5, Cr 0.95, Mo 0.4, bal Fe.
Heat treated: 120,000-160,000 TS; 100,000-140,000 YS; 17-14 El; 300-400 Brin. For crankshafts, connecting rods, gears; shock resistant.

JESSOP G-27

C 0.3, Mn 0.55, Ni 3.5, Cr 0.95, Mo 0.4, bal Fe.
Heat treated: 120,000-160,000 TS; 100,000-140,000 YS; 17-14 El; 300-400 Brin. For crankshafts, connecting rods, gears; shock resistant.

JESSOP G-3

Saville & Co. Ltd., J.J.
C 0.55, Ni 1.5, Cr 0.65, bal Fe.
For dies; tough.

JESSOP G-3

Jessop-Saville Ltd.
C 0.55, Ni 1.5, Cr 0.65, bal Fe.
For dies; tough.

JESSOP G-32

J.J. Saville & Co. Ltd.
C 0.27, Cr 19, Ni 10.5, Mo 2.2, Cb 1.4, Mn 0.8, C 3, Fe 16, bal Co.
Rolled: 126,000 TS; 78,000 YS; 10 El. For gas turbine disc shafts, rotor blades; heat and oxidation resistant.

JESSOP G-32

Jessop-Saville Ltd.
C 0.4, Si 1, Cr 13, Ni 13, Co 15, Cb, Mo, W, bal Fe.
For jet engine parts; high heat resistant.

JESSOP G-33

Saville & Co. Ltd., J.J.
C 0.4, Ni 1.4, Cr 0.6, Mo 0.25, bal Fe.
For mandrel bars; resists heat checking.

JESSOP G-33

Jessop-Saville Ltd.
C 0.4, Ni 1.4, Cr 0.6, Mo 0.25, bal Fe.
For mandrel bars; resists heat checking.

JESSOP G-5 SPECIAL

Saville & Co. Ltd., J.J.
C 0.32, Ni 1.55, Cr 1.1, Mo 0.2, bal Fe.
For mandrel bars, gears, connecting rods; resists heat checking.

JESSOP G-5 SPECIAL

Jessop-Saville Ltd.
C 0.32, Ni 1.55, Cr 1.1, Mo 0.2, bal Fe.
For mandrel bars, gears, connecting rods; resists heat checking.

JESSOP G-7

Saville & Co. Ltd., J.J.
C 0.55, Ni 1.75, Cr 0.6, Mo 0.12, bal Fe.
For stamping dies; deep hardening, tough.

JESSOP G-7

Jessop-Saville Ltd.
C 0.55, Ni 1.75, Cr 0.6, Mo 0.12, bal Fe.
For stamping dies; deep hardening, tough.

JESSOP G-8

Saville & Co. Ltd., J.J.
C 0.25, Ni 3, Cr 1.2, Mo 0.4, bal Fe.
For mandrel bars, crankshafts; resists heat checking.

JESSOP G-8

Jessop-Saville Ltd.
C 0.25, Ni 3, Cr 1.2, Mo 0.4, bal Fe.
For mandrel bars, crankshafts; resists heat checking.

JESSOP G.42

Jessop-Saville Ltd.
Cr 19, Ni 15, Co 25, C, bal Fe.
For high temperature applications; corrosion and heat resistant.

JESSOP G.I. SPECIAL

Jessop-Saville Ltd.
Cr 1.3, Mo 0.2, C 4.1 Ni, bal Fe.
Hardened: 215,000 TS; 190,000 YS; 12 El; 477 Brin. For plastic molding dies, mandrels; shock resist, air hardening.

JESSOP G19

Jessop-Saville Ltd.
C 0.4, Mn 0.8, Si 1, ni 13, Cr 1, Mo 2, W 2.5, Cb 3, Co 10, bal Fe.
Heat treated: 102,500 TS; 30,000 YS; 42 El; 50 RA For nozzle guide vanes in gas turbines; high heat and oxidation resistant.

JESSOP G29

Jessop-Saville Ltd.
C 0.35-0.45, Ni 11-13.5, Cr 19-20.5, Co 7-10, bal Fe.
Normalized: 110,000 TS; 43,000 YS; 40 El; 210 Brin. For gas turbine components, furnace parts; high creep resistance, good to 1740°F.

JESSOP G30

Jessop-Saville Ltd.
C, alloy, bal Fe.
For machine tool parts; oil hardened.

JESSOP G32

Jessop-Saville Ltd.
C 0.3, Ni 12, Cr 19, Fe 16, V 2.8, Mo 2, Cb 1.2, bal Co.
144,000 TS; 85,000 YS; 10 El. For gas turbine rotors; high creep and fatigue strength.

JESSOP G34

Jessop-Saville Ltd.
C 0-1, Mn 0.8, Si 0.3, Ni 13, Cr 19, Mo 2, Co 46, Cb 1.2, V 2.8, bal Fe.
Cast: 89,600 TS; 58,200 YS; 2.5 El; 2.5 RA; 300 Brin. For turbine blades, nozzle guide vanes; high creep and heat resistance, shock resistant.

JESSOP G39

Jessop-Saville Ltd.
C 0.5, Cr 20, W 3, Mo 3, Cb 1.5, Ta 1.5, bal Ni.
Cast: 73,200 TS; 5 El; 5 RA ; 190 Brin. For nozzle guide vanes in gas turbines; high heat and oxidation resistant.

JESSOP G42B

Jessop-Saville Ltd.
C 0.26, Co 25, Cr 20, Ni 15, plus carbide formers.
Heat treated: 120,000 TS; 91,000 YS; 10 El; 12 RA; 260 Brin. For turbine blades; high heat and creep resistance, used to 1650°F.

JESSOP G56

Jessop-Saville Ltd.
C 0.2, Cr 13, Ni 23, bal Fe.
Heat treated: 151,000 TS; 94,700 YS; 26 El; 27 RA. For turbine discs; good creep and heat resistance, stainless.

JESSOP GB-18

Jessop-Saville Ltd.
C 0.4, Si 1, Cr 13, Ni 13, Co 10, W 2.5, 3 Cb 2 Mo, bal Fe.
Age-hardened: 107,500 TS; 62,700 YS; 12 El; 15 RA. For high temperature parts, gas turbine discs; heat resistant.

JESSOP H 4

Jessop-Saville Ltd.
C 0.85, Mn 1.9, bal Fe.
For press tools, gauges taps, dies; non-distorting.

JESSOP H 4

Cleveland Brass Corp.
C 0.85, Mn 1.9, bal Fe.
For press tools, gauges taps, dies; non-distorting.

JESSOP H 42

Jessop-Saville Ltd.
Mn 0.45, Cr 13, Mo 0.7, V 0.3, M 486; 1.65 C, bal Fe.
For drawing and extrusion dies, gauges, shear blades; shock resistant.

JESSOP H-18
Saville & Co. Ltd., J.J.
Si 3.3, Cr 8.5, Mo 1.2, bal Fe.
At 70°F: 114,000 TS; 24 El. At 1100°F: 36,000 TS; 60 El. For auto inlet and exhaust valves; resists scaling.

JESSOP H-18
Jessop-Saville Ltd.
Si 3.3, Cr 8.5, Mo 1.2, bal Fe.
At 70°F: 114,000 TS; 24 El. At 1100°F: 36,000 TS; 60 El. For auto inlet and exhaust valves; resists scaling.

JESSOP H-22
Saville & Co. Ltd., J.J.
C 0.2, Cr 1.55, Mo 0.5, bal Fe.
For small stamping dies; shock resistant.

JESSOP H-22
Cerro Metal Products Co.
C 0.2, Cr 1.55, Mo 0.5, bal Fe.
For small stamping dies; shock resistant.

JESSOP H-23
Saville & Co. Ltd., J.J.
C 0.47, Mn 0.6, Si 1, Cr 1.4, bal Fe.
For shear blades, punches, chipping chisels; oil hardening.

JESSOP H-23
Jessop-Saville Ltd.
C 0.47, Mn 0.6, Si 1, Cr 1.4, bal Fe.
For shear blades, punches, chipping chisels; oil hardening.

JESSOP H-24
Saville & Co. Ltd., J.J.
Heat treated: 90,000-140,000 TS; 70,000-120,000 YS; 22-15 El; 220-360 Brin. For gun barrels, superheater tubes; creep and shock resistant.

JESSOP H-24
Jessop-Saville Ltd.
C 0.4, Mn 0.65, Cr 1.2, Mo 0.3, bal Fe.
Heat treated: 90,000-140,000 TS; 70,000-120,000 YS; 22-15 El; 220-360 Brin. For gun barrels, superheater tubes; creep and shock resistant.

JESSOP H-28
Saville & Co. Ltd., J.J.
C 0.4, Mn 0.8, Cr 1, bal Fe.
Heat treated: 90,000-110,000 TS; 64,000-88000 YS; 22-18 El 240-302 Brin. For general engineering components; water hardening.

JESSOP H-28
Jessop-Saville Ltd.
C 0.4, Mn 0.8, Cr 1, bal Fe.
Heat treated: 90,000-110,000 TS; 64,000-88000 YS; 22-18 El 240-302 Brin. For general engineering components; water hardening.

JESSOP H-29
Saville & Co. Ltd., J.J.
C 0.8, Si 2, Ni 1.5, Cr 19.5, bal Fe.
At 70°F: 120,000 TS; 15 El. At 1100°F: 40,000 TS; 40 El. At 1500°F: 10,000 TS; 90 El. For inlet and exhaust valves; resists leaded fuels and scaling.

JESSOP H-29
Jessop-Saville Ltd.
C 0.8, Si 2, Ni 1.5, Cr 19.5, bal Fe.
At 70°F: 120,000 TS; 15 El. At 1100°F: 40,000 TS; 40 El. At 1500°F: 10,000 TS; 90 El. For inlet and exhaust valves; resists leaded fuels and scaling.

JESSOP H-3
J.J. Saville & Co. Ltd.
C 0.6, Si 1.55, Cr 5.8, bal Fe.
At 70°F: 110,000 TS; 24 El. At 1100°F: 40,000 TS; 50 El. At 1500°F: 8000 TS; 140 El. For inlet and exhaust valves; heat resistant; Silchrom type steel.

JESSOP H-30
Saville & Co. Ltd.
C 0.6, Mn 0.65, Cr 0.6, bal Fe.
For chisels, rivet snaps, expander mandrels; wear and shock resistant.

JESSOP H-30
Jessop-Saville Ltd.
C 0.6, Mn 0.65, Cr 0.6, bal Fe.
For chisels, rivet snaps, expander mandrels; wear and shock resistant.

JESSOP H-32
Jessop-Saville Ltd.
Mn 0.6, Mo 0.55, Cr 3.22, M 486; 0.3 C, bal Fe.
Heat treated: 90,000-200,000 TS; 34,000-170,000 YS; 22-10 El; 248-444 Brin. For crankshafts; cylinder liners; shock resistant, nitriding steel.

JESSOP H-33
Saville & Co. Ltd., J.J.
C 1, Mn 0.35, Cr 6.25, Mo 1, bal Fe.
For shear blades, press tools, taps, dies; abrasion resistant.

JESSOP H-33
Jessop-Saville Ltd.
C 1, Mn 0.35, Cr 6.25, Mo 1, bal Fe.
For shear blades, press tools, taps, dies; abrasion resistant.

JESSOP H-34
Saville & Co. Ltd., J.J.
C 0.35, Mn 0.6, Cr 1, Mo 0.7, bal Fe.
Heat treated: 90,000-14,000 TS; 70,000-120,000 YS; 22-15 El; 230-360 Brin. For superheater tubes, gears; creep and shock resistant.

JESSOP H-34
Jessop-Saville Ltd.
C 0.35, Mn 0.6, Cr 1, Mo 0.7, bal Fe.
Heat treated: 90,000-14,000 TS; 70,000-120,000 YS; 22-15 El; 230-360 Brin. For superheater tubes, gears; creep and shock resistant.

JESSOP H-35
Saville & Co. Ltd., J.J.
C 0.3, Mn 0.5, Cr 1, Mo 1.2, bal Fe.
Heat treated: 110,000 TS; 90,000 YS; 18 El; 293 Brin. For crankshafts; shock resistant, nitriding steel.

JESSOP H-35
Jessop-Saville Ltd.
C 0.3, Mn 0.5, Cr 1, Mo 1.2, bal Fe.
Heat treated: 110,000 TS; 90,000 YS; 18 El; 293 Brin. For crankshafts; shock resistant, nitriding steel.

JESSOP H-36
Saville & Co. Ltd., J.J.
C 0.35, Mn 0.5, Cr 1.5, Mo 0.2, Al 1, bal Fe.
Heat treated: 70,000-90,000 TS; 48,000-84,000 YS; 24-17 El; 207-302 Brin. For crankshafts, cylinder liners; shock resistant, nitriding steel.

JESSOP H-36
Jessop-Saville Ltd.
C 0.35, Mn 0.5, Cr 1.5, Mo 0.2, Al 1, bal Fe.
Heat treated: 70,000-90,000 TS; 48,000-84,000 YS; 24-17 El; 207-302 Brin. For crankshafts, cylinder liners; shock resistant, nitriding steel.

JESSOP H-38
Saville & Co. Ltd., J.J.
C 0.5, Si 3, Cr 7.5, Mo 0.5, bal Fe.
At 70°F: 105,000 TS; 25 El. At 1100°F: 44,000 TS; 55 El. At 1500°F: 7000 TS; 110 El. For auto inlet and exhaust valves; resists scaling.

JESSOP H-38
Jessop-Saville Ltd.
C 0.5, Si 3, Cr 7.5, Mo 0.5, bal Fe.
At 70°F: 135,000 TS; 25 El. At 1100°F: 44,000 TS; 55 El. At 1500°F: 7000 TS; 110 El. For auto inlet and exhaust valves; resists scaling.

JESSOP H-4
Jessop Steel Co.
C 0.8-0.9, Si 0.1-0.2, Mn 1.75-2, Cr, bal Fe.
For tools, taps, gauges; oil harden. *Obsolete*

JESSOP H-44
Saville & Co. Ltd., J.J.
C 1, Mn 0.7, Cr 1.45, bal Fe.
For shear blades, punches, press tools; deep hardening, abrasion resistant.

JESSOP H-44
Jessop-Saville Ltd.
C 1, Mn 0.7, Cr 1.45, bal Fe.
For shear blades, punches, press tools; deep hardening, abrasion resistant.

JESSOP H-46
Saville & Co. Ltd., J.J.
C 0.2, Mn 0.6, Cr 12, Mo 0.6, Cb 0.15, V 0.75, bal Fe.
Annealed: 120,000 TS; 90,000 YS; 20 El; 241 Brin. For aircraft gas turbine discs and blades, furnace parts; corrosion and heat resistant, hardenable.

JESSOP H-46
Jessop-Saville Ltd.
C 0.2, Mn 0.6, Cr 12, Mo 0.6, Cb 0.15, V 0.75, bal Fe.
Annealed: 120,000 TS; 90,000 YS; 20 El; 241 Brin. For aircraft gas turbine discs and blades, furnace parts; corrosion and heat resistant, hardenable.

JESSOP H-9
Jessop-Saville Ltd.
C 2-2.25, Si 0.1-0.15, Mn 0.3-0.5, Cr 13-13.5, bal Fe.
For tools, dies. *Obsolete*

JESSOP H-9
Jessop Steel Co.
C 2-2.25, Si 0.1-0.15, Mn 0.3-0.5, Cr 13-13.5, bal Fe.
For tools, dies. *Obsolete*

JESSOP H27
Jessop-Saville Ltd.
C 0.4, Mn 0.6, Si 0.3, Cr 3, Mo 0.8, bal Fe.
Oil hardened: 135,000 TS; 112,000 YS; 18 El; 40 RA. For turbine disks; heat resistant.

JESSOP H31
Jessop-Saville Ltd.
C 0.4, Mn 0.4, Si 0.3, Cr 1.1, Mo 0.7, bal Fe.
Oil hardened: 130,000 TS; 96,000 YS; 20 El; 40 RA. For turbine disks; heat resistant.

JESSOP H39
Jessop-Saville Ltd.
C 1, Cr 1.4, Mo 0.4, bal Fe.
For bearings, heading dies; Type L7.

JESSOP H3A
Jessop-Saville Ltd.
C 0.6, Si 1.2, Cr 6, Mo 0.5, bal Fe.
Heat treated: 135,000 TS; 100,000 YS; 20 El; 40 RA. For turbine disks; heat resistant.

JESSOP H40
Jessop-Saville Ltd.
C 0.25, Mn 0.4, Si 0.4, Cr 3, Mo 0.5, W 0.5, 0.75 bal Fe.
Oil hardened: 135,000 TS; 112,000 YS; 18 El; 45 RA. For turbine disks; heat resistant.

JESSOP H49
Jessop-Saville Ltd.
C 0 0.07, Cr 3, Mo, V, bal Fe.
For hobbed molds; soft, deep hardening, hobbing steel.

JESSOP H50

Jessop-Saville Ltd.
C 0.37, Si 1, Cr 5, V 1.1, Mo 1.35, bal Fe.
For extrusion and die casting dies; hot work steel; Type H13. Hot punches, shear blades.

JESSOP HEAT-RESISTING NO. 5

Jessop Steel Co.
C 0.2, Cr 24-28, Ni 12-14, bal Fe.
Annealed: 100,000-110,000 TS; 45,000-60,000 YS; 50-60 El; 60-70 RA; 160 Brin. For furnace parts; heat and corrosion resistant to 2100 F. *Obsolete*

JESSOP HIGH SPEED STEEL

Jessop Steel Co.
W 13-19, C 0.6-0.77, Cr 3-4, V 0.25-2.25, bal Fe.
For high speed cutting tools, drills, reamers; high speed steel. *Obsolete*

JESSOP HOT WORKING DIE

Jessop Steel Co.
C, alloy, bal Fe.
For hot work dies; hot work steel. *Obsolete*

JESSOP HP

Saville & Co. Ltd., J.J.
C 0.9, Mn 0.9, C 1.15, W 1.5, bal Fe.
For press tools, cold punches, gauges, taps; abrasion resistant.

JESSOP HP

Jessop-Saville Ltd.
C 0.9, Mn 0.9, C 1.15, W 1.5, bal Fe.
For press tools, cold punches, gauges, taps; abrasion resistant.

JESSOP J 23

Jessop-Saville Ltd.
C 0.36, Si 0.4, Cr 2.6, V 1, W 1.8, Mo 4.3, bal Fe.
For die casting dies, extrusion dies; hot die steel.

JESSOP J-13

Saville & Co. Ltd., J.J.
C 1.2, Cr 4.35, Mo 0.25, V 4.5, W 14, bal Fe.
For broaches, reamers, drills, taps; high speed steel.

JESSOP J-13

Jessop-Saville Ltd.
C 1.2, Cr 4.35, Mo 0.25, V 4.5, W 14, bal Fe.
For broaches, reamers, drills, taps; high speed steel.

JESSOP J-18

Saville & Co. Ltd., J.J.
C 0.84, Cr 4.5, Mo 5.5, V 1.4, W 6, bal Fe.
For milling cutters, drills, taps, reamers, broaches; high speed steel.

JESSOP J-18

Jessop-Saville Ltd.
C 0.84, Cr 4.5, Mo 5.5, V 1.4, W 6, bal Fe.
For milling cutters, drills, taps, reamers, broaches; high speed steel.

JESSOP J-27

Saville & Co. Ltd., J.J.
C 0.42, Cr 3.25, V 0.25, W 14, bal Fe.
For die casting dies; resists heat checking; hot die steel.

JESSOP J-27

Jessop-Saville Ltd.
C 0.42, Cr 3.25, V 0.25, W 14, bal Fe.
For die casting dies; resists heat checking; hot die steel.

JESSOP J-32

Saville & Co. Ltd., J.J.
C 0.35, Ni 4, Cr 1.5, W 5.8, V 0.35, bal Fe.
For mandrels, punches, hot stamping dies; hot work steel.

JESSOP J-32

Jessop-Saville Ltd.
C 0.35, Ni 4, Cr 1.5, W 5.8, V 0.35, bal Fe.
For mandrels, punches, hot stamping dies; hot work steel.

JESSOP J.J.

Jessop Steel Co.
C 0.92, Cr 3.8, Mo 0.5, V 0.5, bal Fe.
For hot work tools, piercing punches, hot pressed nut dies; hot work steel, wear resistant. *Obsolete*

JESSOP J34

Jessop-Saville Ltd.
C 0.8, Cr 4, V 2, W 6, Mo 5, bal Fe.
For drills, taps, reamers, broaches, cutters; high speed steel; Type M4.

JESSOP J35

Jessop-Saville Ltd.
C 80, Cr 4, V 1, W 1.5, Mo 8, 0, bal Fe.
For drills, taps, reamers, hobs, lathe cutters; high speed steel; Type M1

JESSOP JS-777

Jessop Steel Co.
C 0-0.03, Cr 21, Ni 25, Mo 4.5, Cu 0.3, bal Fe.
Improved corrosion resistance over JS-700.

JESSOP JS700

Jessop Steel Co.
C 0.03, Ni 25, Cr 21, Mo 4.5, Mn 1.7, Si 0.5, Cb 0.3, bal Fe.
Bar: 85,000 TS; 39,000 YS; 45 El; 50 RA; 170 Brin. For agitators, centrifuges, mixers, filters, separators, seals, valves, tanks; austenitic, stainless, resists phosphoric acid in presence of halogens.

JESSOP K 18

Jessop-Saville Ltd.
C 0.32, Si 0.95, Cr 5, W 1.4, Mo 1.7, bal Fe.
For die casting dies, swaging dies; hot die steel.

JESSOP K-15

Saville & Co. Ltd., J.J.
C 1.25, Cr 1.1, V 0.2, W 4.5, bal Fe.
For broaches, reamers, taps, punches; abrasion resistant.

JESSOP K-15

Jessop-Saville Ltd.
C 1.25, Cr 1.1, V 0.2, W 4.5, bal Fe.
For broaches, reamers, taps, punches; abrasion resistant.

JESSOP K-2

Jessop-Saville Ltd.
C 1.3-1.5, Si 0.1-0.15, Mn 0.3-0.5, Cr 0.75-0.85, W 4-5, bal Fe.
For tools, form cutters. *Obsolete*

JESSOP K-2

Jessop Steel Co.
C 1.3-1.5, Si 0.1-0.15, Mn 0.3-0.5, Cr 0.75-0.85, W 4-5, bal Fe.
For tools, form cutters. *Obsolete*

JESSOP K-6

Jessop-Saville Ltd.
C 1.15-1.25, Si 0.1-0.2, Mn 0.3-0.5, Cr 0-0.5, W 1.25-1.5, bal Fe.
For tools, hack-saws; water hardening. *Obsolete*

JESSOP K-6

Jessop Steel Co.
C 1.15-1.25, Si 0.1-0.2, Mn 0.3-0.5, Cr 0-0.5, W 1.25-1.5, bal Fe.
For tools, hack-saws; water hardening. *Obsolete*

JESSOP K4 SPECIAL

Jessop-Saville Ltd.
C 0.9, Mn 1, Cr 0.5, W 0.5, bal Fe.
For dies, cutters, punches, hobs, broaches; Type 01; oil hardened, non-deforming.

JESSOP M-1

Saville & Co. Ltd., J.J.
C 0.37, Mn 1.5, Mo 0.25, bal Fe.
Heat treated: 90,000-130,000 TS; 70,000-105,000 YS; 22-16 El; 241-341 Brin. For general engineering parts; shock resistant.

JESSOP M-1

Jessop-Saville Ltd.
C 0.37, Mn 1.5, Mo 0.25, bal Fe.
Heat treated: 90,000-130,000 TS; 70,000-105,000 YS; 22-16 El; 241-341 Brin. For general engineering parts; shock resistant.

JESSOP M-10

Jessop Steel Co.
C 0.85-1.1, Mo 8, Cr 4, V 2, bal Fe.
Molybdenum high speed steel for cutting tools. AISI M10.

JESSOP M-2

Saville & Co. Ltd., J.J.
C 0.35, Mn 1.5, Mo 0.5, bal Fe.
Heat treated: 90,000-130,000 TS; 70,000-105,000 YS; 22-16 El; 241-341 Brin. For general engineering components; shock resistant.

JESSOP M-2

Jessop-Saville Ltd.
C 0.35, Mn 1.5, Mo 0.5, bal Fe.
Heat treated: 90,000-130,000 TS; 70,000-105,000 YS; 22-16 El; 241-341 Brin. For general engineering components; shock resistant.

JESSOP M-3 TYPE 1

Jessop Steel Co.
C 1.15, Cr 4.15, V 3.25, W 6, Mo 6, bal Fe.
Hardened: 64-66 Rock C. For broaches, chasers, drills, hobs, lathe and milling tools, reamers, tap and dies. Good red hardness. High speed steel. High abrasion resistance.

JESSOP M-3 TYPE 2

Jessop Steel Co.
C 1.15, Cr 4.15, V 3.25, W 6, Mo 6, bal Fe.
Hardened: 64-66 Rock C. For broaches, chasers, drills, hobs, lathe and planer tools, reamers, taps and dies. Good red hardness. High speed steel. High abrasion resistance.

JESSOP M-4

Jessop Steel Co.
C 1.3, W 5.5, Mo 4.5, Cr 4, V 4, bal Fe.
High speed steel for cutting tools, abrasion and wear resistant. AISI M4.

JESSOP M-7

Jessop Steel Co.
C 1, W 1.75, Mo 8.75, Cr 4, V 2, bal Fe.
Molybdenum high speed steel for cutting tools; good wear resistance. AISI M7.

JESSOP M.D. DRILL

Jessop Steel Co.
C 0.8, bal Fe.
For drills, mining and rock drills; drill steel. *Obsolete*

JESSOP NEW PROCESS

Jessop Steel Co.
C 0.91-0.98, Mn 0.2-0.3, Si 0.2, bal Fe.
For cold header dies and punches; water hardening. *Obsolete*

JESSOP NO. 10

Jessop Steel Co.
C 1-1.1, Si 0.1-0.2, Mn 0.25-0.6, bal Fe.
For large drills, punches, broaches, tools, cutting tools, knives, jigs, gages, mandrels; water hardening. *Obsolete*

JESSOP NO. 11
Jessop Steel Co.
C 1.1-1.2, Si 0.1-0.2, Mn 0.25-0.6, bal Fe.
For small drills, punches, press tools, tools, heading dies, mandrels, pipe cutters, punches, scrapers; water hardening. *Obsolete*

JESSOP NO. 12
Jessop Steel Co.
C 1.2-1.3, Si 0.1-0.2, Mn 0.25-0.6, bal Fe.
For very small tools, engraving tools, tools, gages, planing tools; water harden. *Obsolete*

JESSOP NO. 139B
Jessop Steel Co.
C 0.82, Mn 0.35, Ni 2.6, Cr 0.25, bal Fe.
Heated treated: 320-620 Brin. For woodworking knives, saws; oil hardened, tough.

JESSOP NO. 14
Jessop Steel Co.
C 0.55-0.65, Si 0.1-0.25, Mn 0.3-0.5, Cr 3.5-4, W 13-14, 0.5% min V, bal Fe.
For tools, high speed cutters; high speed. *Obsolete*

JESSOP NO. 157
Jessop Steel Co.
C, Cr, Ni, bal Fe.
For dies for hot stamping; hot stamping die steel. *Obsolete*

JESSOP NO. 18
Jessop Steel Co.
C 0.65-0.75, Si 0.1-0.25, Mn 0.3-0.5, Cr 4-5, W 17-18.5, bal Fe.
For tools, high speed cutters; high speed. *Obsolete*

JESSOP NO. 200
Jessop Steel Co.
C 0.35, Mn 11.5, Ni 7.5, Cr 0-0.5, bal Fe.
Annealed: 80,000-95,000 TS; 30,000-60,000 YS; 30-50 El; 30-60 RA. For circuit breakers, transformers, motors, motor shafts; nonmagnetic, austenitic.

JESSOP NO. 202
Jessop Steel Co.
C 0.25, Mn 17.5, Ni 1.5, bal Fe.
Rolled: 85,000-95,000 TS; 30,000-60,000 YS; 25-50 El; 30-60 RA. For induction furnace parts, switch covers; austenitic, nonmagnetic. *Obsolete*

JESSOP NO. 259 (AISI S5)
Jessop Steel Co.
C 0.54, Mn 0.75, Si 2, Cr 0.15, V 0.2, Mo 0.5, bal Fe.
Shock resisting tool steel; AISI S5.

JESSOP NO. 27
Jessop Steel Co.
C 1.15, Mn 0.55, Cr 0.55, W 0.95, V 0.2, bal Fe.
For cold metal saws; oil or water hardened. *Obsolete*

JESSOP NO. 271
Jessop Steel Co.
C 0.9, Mn 0.7, Cr 0.25, bal Fe.
Annealed: 100,000 TS; 53,100 YS; 21 El; 42 RA; 200 Brin. For cross-cut saws, circular and hand saws; oil or water hardened. *Obsolete*

JESSOP NO. 276
Jessop Steel Co.
C 0.9, Mn 0.35, Si 0.25, Cr 0.5, bal Fe.
Heat treated: 185,000 TS; 120,000 YS; 10 El; 30 RA; 375-400 Brin. For cross-cut saws; oil or water hardened. *Obsolete*

JESSOP NO. 2B
Jessop Steel Co.
C 0.3-0.55, W 9-15, V 0.5, trace Cr, bal Fe.
For header dies, rivets and bolt dies, forging, machines, bull dozing tools, punches; hot die steel; semi-high speed. *Obsolete*

JESSOP NO. 3C
Jessop Steel Co.
C 2.2, Cr 12.5, Mo 0.8, V 0.2, bal Fe.
Annealed: 228 Brin. For dies, blanking dies, punches, taps, reamers; resists corrosion and oxidation. *Obsolete*

JESSOP NO. 4824
Jessop-Saville Ltd.
C 0.4, Cr 1.5, Ni 0.65, Mo 0.5, V 0.27, Mn 1, Si 1.6, bal Fe.
Heat treated: 275,000 TS; 220,000 YS; 12 El; 37 RA. For aircraft parts, gears, shafts; oil hardened, high tensile.

JESSOP NO. 4828
Jessop-Saville Ltd.
C 0.26, Mn 0.6, Si 1.8, Ni 2.9, Mo 0.47, Cr 1.2, bal Fe.
Heat treated: 246,000 TS; 189,000 YS; 12.5 El; 40 RA. For aircraft parts, gears, shafts; oil hardened, high tensile.

JESSOP NO. 5
Jessop Steel Co.
C 0.55-0.65, Si 0.1-0.2, Mn 0.25-0.6, bal Fe.
For tools. *Obsolete*

JESSOP NO. 51
Jessop Steel Co.
C 1.15, Mn 0.3, Si 0.2, W 1.1, bal Fe.
For hacksaws; water or oil hardened. *Obsolete*

JESSOP NO. 5B
Jessop Steel Co.
C 0-0.25, Mn 0-1.5, Si 0-2, Cr 24-26, Ni 19-21, bal Fe.
Annealed: 90,000-110,000 TS; 40,000-60,000 YS; 45-55 El; 50-60 RA; 145-210 Brin. For furnace parts, carburizing boxes; heat resistant, austenitic. *Obsolete*

JESSOP NO. 67
Jessop Steel Co.
C 0.8, Mn 0.25, Ni 1.35, Cr 0.35, bal Fe.
For cross-cut saws, circular wood saws; oil hardened. *Obsolete*

JESSOP NO. 7
Jessop Steel Co.
C 0.66-0.75, Si 0.1-0.2, Mn 0.25-0.6, bal Fe.
For coal cutters, rock drills, tools, hammers, swages, chisels, press tools, picks, cams; water hardening. *Obsolete*

JESSOP NO. 773
Jessop Steel Co.
C 0.68, Mn 0.68, Mo 0.3, bal Fe.
For cutting tools, dies, punches, stamps, shears; tough, oil hardening. *Obsolete*

JESSOP NO. 774
Jessop Steel Co.
C 0.8, Mn 0.8, Si 0.25, Mo 0.15, bal Fe.
For hacksaw blades; oil hardened. *Obsolete*

JESSOP NO. 8
Jessop Steel Co.
C 0.76-0.85, Si 0.1-0.2, Mn 0.25-0.6, bal Fe.
For tools, drills, taps; water hardening. *Obsolete*

JESSOP NO. 9
Jessop Steel Co.
C 0.86-0.95, Si 0.1-0.2, Mn 0.25-0.6, bal Fe.
For gages, cold chisels, reamers, tools, screw dies, drills, milling cutters, valve tappets; water hardening. *Obsolete*

JESSOP NO. 9 NONMAGNETIC
Jessop Steel Co.
C 0.35, Mn 12.4, Ni 3.25, Cr 4.15, Mo 0.5, bal Fe.
Rolled: 115,000-160,000 TS; 55,000-100,000 YS; 35-45 El; 40-50 RA. For armor plate, coal chutes, mine cars; austenitic, tough and wear resistant.

JESSOP NO. 91
Jessop Steel Co.
C 0.7, Mn 0.5, Ni 0.7, Cr 0.5, Mo 0.15, bal Fe.
For circular wood saws, segment plates; oil hardened.

JESSOP NO. 96
Jessop Steel Co.
C 0.75, Mn 0.25, Ni 1.4, Cr 0.2, bal Fe.
For woodworking tools; oil hardened, tough. *Obsolete*

JESSOP NO. 96-KC
Jessop Steel Co.
Ni 1.5, Cr 0.3, C, bal Fe.
Heat treated: 190,000-200,000 TS; 160,000-170,000 YS; 12 El. For wood saws, circular and band saws; Rockwell C 42. *Obsolete*

JESSOP NO. 96C
Jessop Steel Co.
C 0.9, Mn 0.35, Ni 1.4, Cr 0.3, bal Fe.
Heat treated: 380-620 Brin. For crosscut saws, ripsaws; oil hardened, tough. *Obsolete*

JESSOP NO. 96K
Jessop Steel Co.
C 1, Mn 0.4, Ni 1.4, Cr 0.65, bal Fe.
Heat treated: 380-630 Brin. For saw bits; oil hardened, tough. *Obsolete*

JESSOP NON-MAGNETIC STEEL
Jessop Steel Co.
C 0.25-0.4, Mn 10.5-12, Ni 7-8, bal Fe.
80,000-110,000 TS; 35,000-60,000 YS; 25-50 El; 30-60 RA. For transformer covers, switch covers, electric equipment; austenitic, non-magnetic. *Obsolete*

JESSOP OIL BIT DRILL
Jessop Steel Co.
C 0.7, bal Fe.
For oil bit drills; water hardening. *Obsolete*

JESSOP OK
Jessop-Saville Ltd.
C 0.44, Mn 0.35, Cr 1.3, V 0.15, W 2.3, bal Fe.
For punches, chipping chisels, pneumatic tools; tough, shock resistant.

JESSOP R-1
Saville & Co. Ltd., J.J.
C 0.1, Cr 13, bal Fe.
70,000-160,000 TS; 25-15 El; 153-360 Brin. For fittings, golf club heads, hardware; hardenable, corrosion resistant.

JESSOP R-1
Jessop-Saville Ltd.
C 0.1, Cr 13, bal Fe.
70,000-160,000 TS; 25-15 El; 153-360 Brin. For fittings, golf club heads, hardware; hardenable, corrosion resistant.

JESSOP R-10
Saville & Co. Ltd., J.J.
C 0.12, Ni 12, Cr 12, bal Fe.
Annealed: 66,000 TS; 35 El. For cooking utensils, holloware; austenitic, stainless.

JESSOP R-10
Jessop-Saville Ltd.
C 0.12, Ni 12, Cr 12, bal Fe.
Annealed: 66,000 TS; 35 El. For cooking utensils, holloware; austenitic, stainless.

JESSOP R-11
Saville & Co. Ltd., J.J.
C 0.15, Si 1.2, Ni 14, Cr 23, Mo 0.8, bal Fe.
For furnace parts, annealing boxes, dampers; heat resistant, austenitic.

JESSOP R-11
Jessop-Saville Ltd.
C 0.15, Si 1.2, Ni 14, Cr 23, Mo 0.8, bal Fe.
For furnace parts, annealing boxes, dampers; heat resistant, austenitic.

JESSOP R-12
Saville & Co. Ltd., J.J.
C 0.12, Si 1.25, Ni 1.25, Cr 27, Mo 0.7, Ti 0.4, bal Fe.
For furnace linings, flame tubes, tubes, cowling; high scale resistant.

JESSOP R-12
Jessop-Saville Ltd.
C 0.12, Si 1.25, Ni 1.25, Cr 27, Mo 0.7, Ti 0.4, bal Fe.
For furnace linings, flame tubes, tubes, cowling; high scale resistant.

JESSOP R-13
Saville & Co. Ltd., J.J.
C 0.3, Mn 0.5, Ni 6.5, Cr 18, W 2, Si 1.6, bal Fe.
For salt pots, lead baths, recuperators; heat and corrosion resistant.

JESSOP R-13
Jessop-Saville Ltd.
C 0.3, Mn 0.5, Ni 6.5, Cr 18, W 2, Si 1.6, bal Fe.
For salt pots, lead baths, recuperators; heat and corrosion resistant.

JESSOP R-15
Saville & Co. Ltd., J.J.
C 0.17, Cr 14, S and Zr, bal Fe.
90,000 TS; 20 El; 192-269 Brin. For food processing equipment, nuts, bolts, rivets; corrosion resistant.

JESSOP R-15
Jessop-Saville Ltd.
C 0.17, Cr 14, S and Zr, bal Fe.
90,000 TS; 20 El; 192-269 Brin. For food processing equipment, nuts, bolts, rivets; corrosion resistant.

JESSOP R-16
Saville & Co. Ltd., J.J.
C 0.1, Si 1.2, Ni 9.5, Cr 18.5, Ti 0.7, bal Fe.
Annealed: 70,000 TS; 30 El. For exhaust manifolds and cowling; austenitic, stainless.

JESSOP R-16
Jessop-Saville Ltd.
C 0.1, Si 1.2, Ni 9.5, Cr 18.5, Ti 0.7, bal Fe.
Annealed: 70,000 TS; 30 El. For exhaust manifolds and cowling; austenitic, stainless.

JESSOP R-17
Saville & Co. Ltd., J.J.
C 0.35, Mn 0.6, Si 1.4, Ni 25, Cr 18, W 2, bal Fe.
For salt pots, carburizing boxes; austenitic, heat resistant.

JESSOP R-17
Jessop-Saville Ltd.
C 0.35, Mn 0.6, Si 1.4, Ni 25, Cr 18, W 2, bal Fe.
For salt pots, carburizing boxes; austenitic, heat resistant.

JESSOP R-18
Saville & Co. Ltd., J.J.
C 0.35, Si 1.8, Cr 18, Ni 8, W 3.5, bal Fe.
For furnace parts, impellers; scale resistant, austenitic.

JESSOP R-18
Jessop-Saville Ltd.
C 0.35, Si 1.8, Cr 18, Ni 8, W 3.5, bal Fe.
For furnace parts, impellers; scale resistant, austenitic.

JESSOP R-19
Saville & Co. Ltd., J.J.
C 0.3, Cr 14, S and Zr, bal Fe.
100,000 TS; 20 El; 227-277 Brin. For shafts, steam valves, chemical industries; corrosion resistant.

JESSOP R-19
Jessop-Saville Ltd.
C 0.3, Cr 14, S and Zr, bal Fe.
100,000 TS; 20 El; 227-277 Brin. For shafts, steam valves, chemical industries; corrosion resistant.

JESSOP R-1A
Saville & Co. Ltd., J.J.
C 0.08, Cr 13, bal Fe.
Hardened: 125,000 TS; 110,000 YS; 16 El; 277 Brin. For plastic molding dies; corrosion resistant.

JESSOP R-1A
Jessop-Saville Ltd.
C 0.08, Cr 13, bal Fe.
Hardened: 125,000 TS; 110,000 YS; 16 El; 277 Brin. For plastic molding dies; corrosion resistant.

JESSOP R-2
Saville & Co. Ltd., J.J.
C 0.2, Cr 13, bal Fe.
90,000-200,000 TS; 20-10 El; 207-444 Brin. For steam valves, knives, piston rods, bolts; hardenable, corrosion resistant.

JESSOP R-2
Jessop-Saville Ltd.
C 0.2, Cr 13, bal Fe.
90,000-200,000 TS; 20-10 El; 207-444 Brin. For steam valves, knives, piston rods, bolts; hardenable, corrosion resistant.

JESSOP R-20
Saville & Co. Ltd., J.J.
C 0.1, Ni 14, Cr 19, Nb 1.7, bal Fe.
For exhaust manifolds, furnace parts, cowlings; austenitic, heat resistant.

JESSOP R-20
Jessop-Saville Ltd.
C 0.1, Ni 14, Cr 19, Nb 1.7, bal Fe.
For exhaust manifolds, furnace parts, cowlings; austenitic, heat resistant.

JESSOP R-22
Jessop-Saville Ltd.
C 0.3, Cr 25, Ni 14.5, W 3.2, Si 1.3, bal Fe.
For inlet nozzle blades; high heat resistant. *Obsolete*

JESSOP R-22
J.J. Saville & Co. Ltd.
C 0.3, Mn 0.6, Cr 25, Ni 14, Si 1.35, W 3, bal Fe.
For furnace parts, valves, impellers; austenitic, heat resistant.

JESSOP R-23
Saville & Co. Ltd., J.J.
C 0.12, Ni 22, Cr 25, Si 1.7, bal Fe.
For gas burners, pyrometer tubes; resists heat, austenitic.

JESSOP R-23
Jessop-Saville Ltd.
C 0.12, Ni 22, Cr 25, Si 1.7, bal Fe.
For gas burners, pyrometer tubes; resists heat, austenitic.

JESSOP R-24
Saville & Co. Ltd., J.J.
C 0.1, Ni 10.5, Cr 17.5, Zr, S or Se, bal Fe.
Annealed: 70,000 TS; 30 El. For valve and pump parts, chemical plant; austenitic, stainless, free-cutting.

JESSOP R-24
Jessop-Saville Ltd.
C 0.1, Ni 10.5, Cr 17.5, Zr, S or Se, bal Fe.
Annealed: 70,000 TS; 30 El. For valve and pump parts, chemical plant; austenitic, stainless, free-cutting.

JESSOP R-25
Saville & Co. Ltd., J.J.
C 0.12, Ni 9.5, Cr 18.5, Mo 2.75, bal Fe.
Annealed: 70,000 TS; 30 El. For chemical plant and textile equipment; austenitic, stainless.

JESSOP R-25
Jessop-Saville Ltd.
C 0.12, Ni 9.5, Cr 18.5, Mo 2.75, bal Fe.
Annealed: 70,000 TS; 30 El. For chemical plant and textile equipment; austenitic, stainless.

JESSOP R-28
Saville & Co. Ltd., J.J.
C 0.16, Ni 1.9, Cr 17, S and Zr, bal Fe.
110,000 TS; 13 El; 241-302 Brin. For steam valves, seaplane parts, bolts; corrosion resistant.

JESSOP R-28
Jessop-Saville Ltd.
C 0.16, Ni 1.9, Cr 17, S and Zr, bal Fe.
110,000 TS; 13 El; 241-302 Brin. For steam valves, seaplane parts, bolts; corrosion resistant.

JESSOP R-29
Saville & Co. Ltd., J.J.
C 0.12, Si 0.9, Cr 21, Mo 0.45, Ti 0.35, bal Fe.
For nitriding boxes, cowlings for gas turbines; scale resistant.

JESSOP R-29
Jessop-Saville Ltd.
C 0.12, Si 0.9, Cr 21, Mo 0.45, Ti 0.35, bal Fe.
For nitriding boxes, cowlings for gas turbines; scale resistant.

JESSOP R-3
Saville & Co. Ltd., J.J.
C 0.12, Ni 10.5, Cr 17, bal Fe.
Annealed: 70,000 TS; 30 El. For fittings, chemical plant equipment; austenitic, stainless.

JESSOP R-3
Jessop-Saville Ltd.
C 0.12, Ni 10.5, Cr 17, bal Fe.
Annealed: 70,000 TS; 30 El. For fittings, chemical plant equipment; austenitic, stainless.

JESSOP R-4
Saville & Co. Ltd., J.J.
C 0.16, Ni 1.5, Cr 16.5, bal Fe.
110,000-180,000 TS; 15-10 El; 241-400 Brin. For seaplane parts, propeller shafts, steam valves; corrosion resistant, hardenable.

JESSOP R-4
Jessop-Saville Ltd.
C 0.16, Ni 1.5, Cr 16.5, bal Fe.
110,000-180,000 TS; 15-10 El; 241-400 Brin. For seaplane parts, propeller shafts, steam valves; corrosion resistant, hardenable.

JESSOP R-9
Saville & Co. Ltd., J.J.
C 0.12, Si 1.2, Ni 10, Cr 18, Mo 2, Ti 0.5, bal Fe.
Annealed: 70,000 TS; 30 El. For chemical plant equipment, textile equipment; austenitic, stainless.

JESSOP R-9
Jessop-Saville Ltd.
C 0.12, Si 1.2, Ni 10, Cr 18, Mo 2, Ti 0.5, bal Fe.
Annealed: 70,000 TS; 30 El. For chemical plant equipment, textile equipment; austenitic, stainless.

JESSOP RAPID FINISHING STEEL
Jessop Steel Co.
C 1.35, W 3.75, Cr 0.75, bal Fe.
For boring tools, drawing dies, lathe tools, shaving dies; wear resistant. *Obsolete*

JESSOP RT
Jessop Steel Co.
C, bal Fe.
For tools; water hardening. *Obsolete*

JESSOP RTS
Jessop Steel Co.
C 0.55, Si 0.8, Mn 0.5, Mo 0.45, bal Fe.
Heat treated: 290,000 TS; 270,000 YS; 5 El; 15 RA; 60 Rock C. For chisels, punches, dies, sledges; water hardening. Tough; shock resistant.

JESSOP S-7
Jessop Steel Co.
C 0.5, Mo 1.4, Cr 3.25, bal Fe.
Shock resisting tool steel for punches, chisels, rivet sets, shear blades. AISI S-7.

JESSOP SHEET SPRING
Jessop-Saville Ltd.
C 0.95, Mn 0.3, bal Fe.
For tools, springs; water hardened.

JESSOP SPECIAL ALLOY B
Saville & Co. Ltd., J.J.
C 1.6, Cr 0.6, W 5.5, bal Fe.
For drawing dies, working tools, stone cutting tools; wear resistant.

JESSOP SPECIAL ALLOY B
Jessop-Saville Ltd.
C 1.6, Cr 0.6, W 5.5, bal Fe.
For drawing dies, working tools, stone cutting tools; wear resistant.

JESSOP SPECIAL H 4
Jessop-Saville Ltd.
C 0.9, Mn 1.85, Cr 0.45, W 0.45, bal Fe.
For gauges, taps, dies, chasers, cams, pawls; non-deforming, oil hardening.

JESSOP SPECIAL K-4
Saville & Co. Ltd., J.J.
C 0.92, Mn 1.25, Cr 0.45, W 0.45, bal Fe.
For gauges, taps, dies, chasers, cams; non-deforming, oil hardening.

JESSOP SPECIAL K-4
Jessop-Saville Ltd.
C 0.92, Mn 1.25, Cr 0.45, W 0.45, bal Fe.
For gauges, taps, dies, chasers, cams; non-deforming, oil hardening.

JESSOP STEEL A-2
Jessop-Saville Ltd.
C 0.1-0.18, Mn 0.6-0.09, bal Fe.
Case hardened core: 85,000 TS; 56,000 YS; 25 El; 60 RA; 180 Brin. For case hardened parts; case-hardening steel.

JESSOP STEEL A-4 (S M-30)
Jessop-Saville Ltd.
C 0.25-0.35, Mn 0.6-0.8, bal Fe.
76,200-80,100 TS; 45,000-51,500 YS; 30-35 El; 55-60 RA; 150-160 Brin. For shafts, axles, machinery parts; water hardened.

JESSOP STEEL A-5 (S M-40)
Jessop-Saville Ltd.
C 0.35-0.45, Mn 0.6-0.8, bal Fe.
85,200-89,600 TS; 49,000-65,000 YS; 26 El; 55 RA; 170 Brin. For gears, pinions, shafts, machinery parts; water hardening.

JESSOP STEEL B (S M-50)
Jessop-Saville Ltd.
C 0.5-0.6, Mn 0.5-0.7, bal Fe.
89,600-112,000 TS; 51,500-56,000 YS; 20-25 El; 45-50 RA; 187-217 Brin. For axles, shafts, wheels, shock resistant tools; water hardening.

JESSOP STEEL C-1 (S M-60)
Jessop-Saville Ltd.
C 0.6-0.7, Mn 0.4-0.6, bal Fe.
98,600 TS; 51,500 YS; 21 El; 40 RA; 200 Brin. For blanking, drawing and trimming dies, punches, hobs, gages; abrasive resistance, non-deforming.

JESSOP STEEL C-2 (S M-80)
Jessop-Saville Ltd.
C 0.75-0.85, Mn 0.4-0.6, bal Fe.
Heat treated: 134,000-156,800 TS; 100,800-123,200 YS; 12-18 El; 40 RA; 320 Brin. For blanking, drawing and trimming dies, punches, hobs, gages; resists scaling, corrosion and abrasion.

JESSOP STEEL E
Jessop-Saville Ltd.
C 0.1-0.15, Mn 0.2-0.6, Ni 3-3.5, bal Fe.
Case hardened core: 100,800-136,700 TS; 71,700-85,200 YS; 15-20 El; 4 RA; 280-300 Brin. For case-hardened parts, gears; case-hardening steel.

JESSOP STEEL E-1
Jessop-Saville Ltd.
C 0.4-0.45, Mn 0.4-0.85, Ni 3-3.5, bal Fe.
Heat treated: 112,000-125,000 TS; 71,700-85,200 YS; 22-27 El; 47 RA 225 Brin. For gears, pinions, crankshafts, structural parts; high impact strength.

JESSOP STEEL F
Jessop-Saville Ltd.
C 0.5-0.6, Mn 0.7-0.8, Cr 0.6-0.8, V 0.25, bal Fe.
Heat treated: 132,200 TS; 91,800 YS; 25-30 El; 62 RA; 240 Brin. For tools, wheels, jaws, clutches, springs; oil hardening.

JESSOP STEEL G
Jessop-Saville Ltd.
C 0.28-0.35, Mn 0.45-0.6, Ni 2.5-3.5, Cr 0.75-1, bal Fe.
Heat treated: 112,000 TS; 78,400 YS; 20-25 El; 50 RA; 240 Brin. For gears, pinions, shafts; oil hardening.

JESSOP STEEL G-1
Jessop-Saville Ltd.
C 0.28-0.35, Ni 4-5, Cr 1-1.5, Mo 0.2, bal Fe.
Heat treated: 123,000 TS; 100,800 YS; 20 El; 45 RA; 260-270 Brin. For gears, shafts, pinions, clutches, jaws; air hardening.

JESSOP STEEL H-2
Jessop-Saville Ltd.
C 0.95-1.05, Si 0-0.3, Mn 0.25-0.4, Cr 1.3-1.5, bal Fe.
For balls, shear blades; abrasion resistant.

JESSOP STEEL H-3
Jessop-Saville Ltd.
C 0.55-0.65, Si 1.4-1.6, Mn 0.3-0.6, Cr 6-7, bal Fe.
Heat treated: 118,700-130,000 TS; 69,000-80,500 YS; 24-28 El; 50-60 RA; 230 Brin. For valves, inlet and exhaust; air hardening.

JESSOP STEEL NO. 5
Jessop Steel Co.
C 0.2, Cr 20, Ni 12, bal Fe.
Hot rolled: 130,500 TS; 107,300 YS; 31 El; 60 RA; 260 Brin. At 1000 C: 18,000 TS; 26 El; 42 RA. For furnace parts, recuperators, pumps, boiler baffles, annealing boxes, oil stills; heat and corrosion resistant; resists scaling to *Obsolete*

JESSOP SUPERIOR OIL HARDENING
Jessop Steel Co.
C 0.9, Mn, W, bal Fe.
For cutting tools, dies; non-deforming. *Obsolete*

JESSOP T AND V
Jessop Steel Co.
C 2.4, Cr 32, W 19, Co 45.
620 Brin. For cutting tools; cast to shape. *Obsolete*

JESSOP T-15
Jessop Steel Co.
C 1.5, W 12, Cr 4, V 5, Co 5, bal Fe.
High carbon, tungsten-vanadium-cobalt high speed for cutting tools. Good wear resistance. AISI T15.

JESSOP T-8
Jessop Steel Co.
C 0.75, W 14, Cr 4, V 2, Co 5, bal Fe.
Tungsten-cobalt type high speed steel for cutting tools. AISI T8.

JESSOP TMC
Jessop Steel Co.
C, W, Mo, Cr, bal Fe.
For cutting tools; high speed steel. *Obsolete*

JESSOP TOOL
Jessop-Saville Ltd.
C 0.7-1.2, bal Fe.
For tools, drills, taps; water hardened.

JESSOP TTQ
Jessop-Saville Ltd.
C 0.8, Cr 5, W 20, V 1.6, Co 11, bal Fe.
For broaches, reamers, planer and lathe cutters; high speed steel, oil hardened.

JESSOP TUNGSTEN MAGNET STEEL
Jessop Steel Co.
C 0.8-0.95, W 5-5.5, Cr 0.4-0.5, Mn 0.4-0.6, bal Fe.
300-350 Brin. For magnets, electrical apparatus; magnet steel. *Obsolete*

JESSOP TYPE R
Jessop Steel Co.
High speed steel with low C steel backing. For clipper knives, blades, planer knives; composite tool steel. *Obsolete*

JESSOP WEX 491
Jessop-Saville Ltd.
C 1, Cr 2, Mn 2, Mo 1, bal Fe.
For tool shanks, punches; air hardened, nondeforming.

JESSOP-H4 SPECIAL
Jessop-Saville Ltd.
C 0.9, Mn 1.6, bal Fe.
For dies, punches, upsetters, crimpers; Type 02; oil hardened, non-deforming.

JESSOP-SAVILLE 6 STAR VANADIUM
Jessop-Saville Ltd.
C 1, Mn 0.3, Si 0.25, V 0.12, bal Fe.
For coining and medal dies, striking dies, parts for cold heading operations. Water hardening, wear resistant.

JESSOP-SAVILLE G 110
Jessop-Saville Ltd.
C 0-0.02, Cr 0-0.25, Ni 17-19, Co 7-8.5, Mo 4.6-5.2, Ti 0.3-0.6, Al 0.05-0.15, Zr 0.02, B 0.003, Ca 0.05, bal Fe.
Heat treated: 258,000 TS; 247,000 YS; 10 El; 40 RA; 520 Vickers. For rocket motor cases and accessories, bolts, fasteners, pressure vessels. Maraging steel. Tough and ductile at 250,000 YS.

JESSOP-SAVILLE G 125 D V
Jessop-Saville Ltd.
C 0.01, Ni 18.5, Co 9, Mo 4.85, Ti 0.75, Al 0.1, Si 0.05, Mn 0.05, bal Fe.
Aged: 318,000 TS; 294,000 YS; 8 El; 50 RA. For rotor motor cases and accessories, fasteners, aircraft structures, pressure vessels. Maraging stel, weldable, hardened by aging. Good notch toughness.

JESSOP-SAVILLE G 30
Jessop-Saville Ltd.
C 0.05-0.12, Cr 27-29, Co 48-52, Si 0.5-1, Mn 0.3-1, bal Fe.
Forged: 134,000 TS; 88,000 YS; 10 El; 350 DPH. At 1290°F: 47,000 TS; 33,000 YS; 21 El; 18 RA. For furnace parts, burner tips, quenching baskets, sintering machines. Good oxidation resistance at 2200°F.

JESSOP-SAVILLE G 31
Jessop-Saville Ltd.
C 0.25-0.3, Cb 2-2.2, Cr 27-29, Co 48-52, Si 0.5-1, Mn 0.5-1, bal Fe.
Cast: 89,500 TS; 64,000 YS; 4 El; 280 DPH. For furnace baffles, grates, clinker coolers, rolls. High stress rupture properties. High resistance to thermal shock and oxidation.

JESSOP-SAVILLE G 35
Jessop-Saville Ltd.
C 0.15, Si 0.5, Ni 12, Cr 19.5, bal Fe.
Annealed: 90,000 TS; 40,000 YS; 60 El; B 82 Rock. At 1200°F: 45,000 TS; 30,000 YS; 32 El. For gas turbine rotor blades and nozzle guide vanes. Austenitic, heat and corrosion resisting, non hardening.

JESSOP-SAVILLE G 38
Jessop-Saville Ltd.
C 0.15, Si 0.5, Ni 12, Cr 16, bal Fe.
Annealed: 85,000 TS; 35,000 YS; 60 El; B 80 Rock. Cold Drawn: 100,000 TS; 65,000 YS; 40 El; B 95 Rock. For gas turbine and steam turbine rotors and blades, superchargers. Austenitic, high creep and heat resistance.

JESSOP-SAVILLE G 40
Jessop-Saville Ltd.
C 0.2, Si 0.3, Ni 25, Cr 20, bal Fe.
Annealed: 95,000 TS; 45,000 YS; 50 El; B 89 Rock. Normalized: 108,000 max TS; 30 El. For gas turbine rotor blades, jet engine components, heat exchangers. Austenitic. Heat and corrosion resisting.

JESSOP-SAVILLE G 41
Jessop-Saville Ltd.
C 0.17, Ni 12, Cr 16, bal Fe.
Normalized: 98,500 max TS; 36 El. For gas turbine blades, high temperature bolts, jet engine components. Austenitic, heat and corrosion resistant.

JESSOP-SAVILLE G 46
Jessop-Saville Ltd.
C 0.16, Si 0.3, Cr 11.5, Mo, V, Cb, bal Fe.
Heat treated: 146,000 max TS; 20 El. For aircraft and gas turbine discs. High creep strength. Good scale resistance up to 1380°F.

JESSOP-SAVILLE G 47
Jessop-Saville Ltd.
C 0.17, Mn 0.8, Ni 1, Cr 0.8, bal Fe.
Heat treated: 124,000 max TS; 15 El. For gudgeon pins, pinions, shafts, rifle components. Case hardening steel. Shock resisting.

JESSOP-SAVILLE G 48
Jessop-Saville Ltd.
C 0.18, Mn 0.8, Ni 1.25, Cr 0.9, Mo 0.11, bal Fe.
Heat treated: 145,000 max TS; 12 El. For rocker arms and shafts, camshafts, gears, clutch plates, levers, bushings. Case hardening steel, shock resisting.

JESSOP-SAVILLE G 49
Jessop-Saville Ltd.
C 0.18, Mn 0.8, Ni 1.75, Cr 0.9, Mo 0.15, bal Fe.
Heat treated: 168,000 max TS; 12 El. For gears, pinions, cams, camshafts, fasteners. Case hardening steel, shock resisting.

JESSOP-SAVILLE G 50
Jessop-Saville Ltd.
C 0.18, Mn 0.6, Ni 2, Cr 1.6, Mo 0.2, bal Fe.
Heat treated: 190,000 max TS; 12 El. For gears, pinions, shafts, camshafts, levers, clutch plates, magneto drive wheels. Case hardening steel. Shock resisting.

JESSOP-SAVILLE G 51
Jessop-Saville Ltd.
C 0.15, Mn 0.8, Ni 0.6, Cr 0.65, Mo 0.11, bal Fe.
Heat treated: 101,000 max TS; 18 El. For rocker arms and shafts, gears, camshafts, rifle components, levers, bushings, pinions. Case hardening steel. Shock resisting.

JESSOP-SAVILLE G 52
Jessop-Saville Ltd.
C 0.2, Mn 0.8, Ni 0.6, Cr 0.65, Mo 0.11, bal Fe.
Heat treated: 124,000 max TS; 15 El. For high duty gears and pinions, bearings, camshafts, clutch plates, tappets. Case hardening steel. Shock resisting.

JESSOP-SAVILLE G 53
Jessop-Saville Ltd.
C 0.25, Mn 0.8, Ni 0.6, Cr 0.65, Mo 0.11, bal Fe.
Heat treated: 145,000 max TS. For rocker arms and shafts, gears, camshafts, bushings, levers. Case hardening. Shock resisting.

JESSOP-SAVILLE G 64
Jessop-Saville Ltd.
C 0.1, Cr 11, Mo 3, W 3.5, Cb 2, Al 6, B, bal Fe.
For high temperature castings, turbine blades, gas turbine rotors, turbo-blower impellers. Heat and corrosion resistant.

JESSOP-SAVILLE G 68
Jessop-Saville Ltd.
C 0.05, Mn 1.5, Si 0.5, Ni 25.4, Cr 14, Mo 1.4, Ti 2.35, V 0.3, Al 0.35, B 0.005, bal Fe.
Heat treated: 16,000 TS; 115,000 YS; 20 El; 30 RA; 250-325 Brinell. For jet engines and superchargers, turbine wheels and blades, afterburners. Precipitation hardening. Austenitic. Similar to A286.

JESSOP-SAVILLE G 84
Jessop-Saville Ltd.
C 0.14, Cr 10, Co 15, Mo 2.5, W 1.25, Al 5, Ti 5.2, B, bal Ni.
For high temperature castings, turbine blades, gas turbine rotors, turbo-blower impellers. Heat and corrosion resistant.

JESSOP-SAVILLE G 94
Jessop-Saville Ltd.
C 0.06, Al 6, Cr 9, Cb 4, Co 10, W 4, Mo 4, B, Zr, bal Ni.
For high temperature castings, turbine blades, gas turbine rotors, turbo-blower impellers. Heat and corrosion resistant.

JESSOP-SAVILLE H 48
Jessop-Saville Ltd.
C 0.45, Si 0.6, Ni 4.7, Cr 24, Mo 2.9, bal Fe.
At 70°F: 150,000 max TS; 11 El. At 1470°F: 34,000 ma TS; 50 El. For exhaust valves for aircraft and automobile engines. Corrosion and heat resisting. Air hardening.

JESSOP-SAVILLE H 51
Jessop-Saville Ltd.
C 0.19, Si 0.25, Mo 0.55, V 0.25, bal Fe.
Heat treated: 130,000 max TS; 24 El. For gas turbine rotors and discs, high temperature stresses components. High creep strength up to 1020°F.; shock resistant.

JESSOP-SAVILLE H 52
Jessop-Saville Ltd.
C 0.12, Si 0.35, Cr 16, bal Fe.
Bar: 90,000 max TS; 15 El. For steam turbine and gas turbine components. Non-hardenable. Good creep resistance. Suitable for welding.

JESSOP-SAVILLE H 53
Jessop-Saville Ltd.
C 0.06, Cr 11, Mo 0.8, Cb 0.45, V 0.5, Co 6.5, W 0.8, B, bal Fe.
At 75°F: 148,000 TS; 130,000 YS; 18 El; 62 RA. At 1200°F: 56,000 TS; 50,200 YS; 28 El; 76 RA. For gas turbine discs, compressor discs. High heat resistance. Martensitic stainless steel.

JESSOP-SAVILLE H 55
Jessop-Saville Ltd.
C 0.55, Mn 0.8, Cr 1, Mo 0.45, Ni 0.5, V 0.2, bal Fe.
Ht. Tr.: 265,000 TS; 250,000 YS; 10 El; 514 Brin. For die blocks, forging dies, punches, turbine rotors, rollers. High strength, tough, shock resistant.

JESSOP-SAVILLE H 59
Jessop-Saville Ltd.
C 0.07, Cb 0.15, Cr 11, Mo 1.5, Ni 3, V 0.3, bal Fe.
Annealed: 80,000 TS; 38,000 YS; 24 El; B 95 Rock. For table flatware, oil refinery and chemical plant equipment. Corrosion resistant, non-hardenable.

JESSOP-SAVILLE J 12
Jessop-Saville Ltd.
C 0.26, Ni 3, Cr 2.75, W 10, Mo 0.5, V 0.3, bal Fe.
For forging dies for brass, die inserts. Tough, oil hardening, hot work steel.

JESSOP-SAVILLE J O SPECIAL
Jessop-Saville Ltd.
C 0.28, Cr 3.2, W 9.5, V 0.3, bal Fe.
Heat treated: 224,000 TS; 191,000 YS; 12 El; C 51 Rock. At 600°C: 134,000 TS; 103,000 YS; 10 El; 28 RA. For extrusion dies and liners, forging and trimming dies, hot punches, rotary shear blades. Hot work steel, Type H21.

JESSOP-SAVILLE J-36
Jessop-Saville Ltd.
C 1.5, Cr 4.75, V 5, W 12.5, Co 5, bal Fe.
Hardened: C 64-68 Rock. For form and planing tools, roll turning tools, broaches, milling cutters, end mills, gear cutters. High speed steel. Highest wear resistant. High red-hardness.

JESSOP-SAVILLE K.S.A
Jessop-Saville Ltd.
C 0.9, Mn 1.85, Si 0.25, Cr 0.45, W 0.45, bal Fe.
Hardened: C 60-65 Rock. For gages, taps, dies, chasers, cams, blanking and forming dies. Oil hardening, tough, shock resisting, non-deforming.

JESSOP-SAVILLE R 27
Jessop-Saville Ltd.
C 0.42, Cr 3.25, W 14, V 0.25, bal Fe.
For forging dies, extrusion dies, punches, shear blades. Hot work steel. Oil hardening, good red-hardness.

JESSOP-SAVILLE R 34
Jessop-Saville Ltd.
C 0.3, Cr 13, bal Fe.
Heat treated: 246,000 max TS; 10 El; C 52 Rock. Annealed: 95,000 TS; 50,000 YS; 25 El; 196 Brin. For knife blades, cutlery, surgical instruments, needle valves, gauges. Hardenable, corrosion resisting steel.

JESSOP-SAVILLE R 40
Jessop-Saville Ltd.
C 0.21, Si 1, Ni 11.5, Cr 22.5, W 1, bal Fe.
For unstresses furnace parts, valves, impellers, dampers, roller hearths, burner nozzles. Heat and corrosion resistant. Not hardenable, austenitic.

JESSOP-SAVILLE R 41
Jessop-Saville Ltd.
C 0-0.1, Ni 12, Cr 17, Mo 2.35, bal Fe.
Annealed: 78,000 TS; 30 El; B 78 Rock. Cold Rolled: 125,000 TS; 100,000 YS; 25 El; 250 Brin. Chemical plant equipment, paper making and textile plant equipment. Stainless, acid resisting, austenitic, non-hardenable.

JESSOP-SAVILLE R 42
Jessop-Saville Ltd.
C 0-0.08, Ni 8.5, Cr 18, bal Fe.
Annealed: 78,000 max TS; 30 El; B 80 Rock. Cold Rolled; 150,000 TS; 100,000 YS 25 El; B-100 Rock. For chemical plant equipment, aiscrews, fish knives, fittings, cooking utensils. Austenitic, stainless steel, Type 304, non-heat treatable.

JESSOP-SAVILLE S.V.L
Jessop-Saville Ltd.
C 0.32, Cr 1.3, Mo 0.3, Mn 0.5, Si 0.25, bal Fe.
For cold punches, trimming dies, shear blades, chisels. Oil or air hardening. Shock resistant.

JESSOP-SAVILLE SPECIAL-BB
Jessop-Saville Ltd.
C 0.28, Cr 3.2, W 9.5, V 0.3, bal Fe.
Heat treated: 224,000 TS; 191,000 YS; 12 El; C 51 Rock. At 600°C: 134,000 TS; 103,000 YS; 10 El; 28 RA. For extrusion dies and liners, forging dies, hot trimming dies, and punches, rotary shear blades. Hot work steel, Type H21.

JESSOP-SAVILLE W.P.S
Jessop-Saville Ltd.
C 2.3, Mn 0.35, Si 0.3, Cr 13, bal Fe.
Hardened: C 65-67 Rock. For blanking and forming dies, gauges, shear blades, thread rollers, mandrels. Oil or air hardening. Nondeforming, wear and abrasion resistant.

JESSOP-SAVILLE ZIRCONIUM

Jessop-Saville Ltd.
Sn 1.5, Fe 0.12, Cr 0.1, Ni 0.05, bal Zr.
At 70°F: 74,000 TS; 42,700 YS; 28 El; 38 RA. At 930°F: 24,200 TS; 12,000 YS; 40 El; 60 RA. For canning and structural material for use in pressurized water reactors. Corrosion resistant and low neutron absorption.

JESSOP-SAVILLE-H.D.S

Jessop-Saville Ltd.
C 0.28, Cr 3.2, W 9.5, V 0.3, bal Fe.
Hardened: 243,000 TS; 215,000 YP; 12 El; 37 RA; Rock C 50. For extrusion dies and liners, forging dies, trimming dies, punches, rotar shear blades. Hot work steel, AISI Type H21.

JESSOPS "J-4" STEEL

Jessop-Saville Ltd.
C 0.5, Cr 1.5, W 2.2, V 0.25, bal Fe.
For hand and pneumatic chisels, boiler cups, punches, shear blades, tools oil hardened.

JESSOPS E-0

Saville & Co. Ltd., J.J.
C 0.12, Mn 0.4, Ni 3, bal Fe.
Rolled: 90,000 TS; 18 El. For gears, pinions, shafts, collets; tough, case-hardening.

JESSOPS E-0

Jessop-Saville Ltd.
C 0.12, Mn 0.4, Ni 3, bal Fe.
Rolled: 90,000 TS; 18 El. For gears, pinions, shafts, collets; tough, case-hardening.

JESSOPS SUPERIOR OIL HARDENING

Jessop-Saville Ltd.
C 0.95, Mn 1.2, Cr 0.5, W 0.5, bal Fe.
For tools and dies, punches; non-deforming steel.

JET FORGE

Teledyne Vasco
C 0.45-0.5, Mo 1.3, V 1.4, Si 0.9, Mn 0.3, Cr 0.25-0.8, bal Fe.
For forging dies; air hardened, hot work steel.

JET LH 72

Lincoln Electric Co.
C, bal Fe.
Steel arc welding electrode. AWS E-7018.

JET-LH 8018-C1

Lincoln Electric Co.
C 0.12, Mn 1.2, P 0.03, S 0.04, Si 0.6, Ni 2.5, bal Fe.
Steel arc welding electrode. AWS Class E8018-C1.

JET-LH 8018-C3

Lincoln Electric Co.
C 0.12, Mn 0.4, P 0.03, S 0.03, Si 0.8, Ni 1, Cr 0.15, Mo 0.35, V 0.05, bal Fe.
Steel arc welding electrode. AWS Class E8018-C3.

JET-LH BU-90

Lincoln Electric Co.
C 0.14, Mn 1.15, Si 0.6, Cr 1.4, bal Fe.
Hard surfacing, arc welding electrode; resistance to metal to metal wear.

JETALLOY 209

Canadian Quebec Metallurgical Corp.
C 0.02, Cr 20, Ni 10, W 15, Ti 2, Fe 1, bal Co.
For high temperature applications; high creep resistance, oxidation and wear resistant.

JETALLOY 249

Canadian Quebec Metallurgical Corp.
Cobalt alloy.
For corrosion resistant applications; resists corrosion and abrasion.

JETHETE M 153

British Steel Corp.
C 0.1, Mn 1.5, Ni 1.5, Cr 12, Mo 1.25, bal Fe.
Heat treated: 134,000-190,000 TS; 103,000-146,000 YS; 25-12 El; 64-25 RA. For high temperature applications; heat resistant.

JETHETE M 154

British Steel Corp.
C 0.08-0.15, Mn 0.5-1, Ni 2-3, Cr 11-13, N 0.05, Mo 1.5-2, V 0.25-0.4, bal Fe.
Hardened: 147,000 TS; 129,000 YS; 10 El. For turbine blading, bolting, gas turbine components. Good oxidation resistance and high strength at elevated temperatures.

JETHETE M.160

British Steel Corp.
C 0-0.2, Cr 11-13, Mo 0-1, V 0-1, Cb 0-0.7, Ni 0-1.25, bal Fe.
At 70 F: 143,000 TS; 121,000 YS; 22 El; 56 RA. At 570 F: 124,000 TS; 106,000 YS; 18 El; 51 RA. At 750 F: 116,000 TS; 97,000 YS; ?? El; 65 RA. For gas turbine compressor blades, corrosion and heat resistant. *Obsolete*

JETHETE M151

United Engineering Steels Ltd.
C 0.1, Mn 1.5, Ni 1.25, Cr 12, Mg 0.6, V 0.3, bal Fe.
Heat treated: 134,400 TS; 112,000 YS; 15 El. For gas turbine discs and blades. Creep resistant.

JETHETE M152

United Engineering Steels Ltd.
C 0.12, Mn 0.7, Ni 2.5, Cr 12, Mo 1.7, V 0.3, bal Fe.
Heat treated: 134,400 TS; 112,000 YS; 20 El. For gas turbine discs and blades. Creep resistant.

JETHETE M160

United Engineering Steels Ltd.
C 0-0.2, Cr 11-13, Mo 0-1, V 0-1, Cb 0-0.7, Nb 0-1.25, bal Fe.
At 70°F: 143,000 TS; 121,000 YS; 22 El; 56 RA. At 570°F: 124,000 TS; 106,000 YS; 18 El; 51 RA. At 750°F: 116,000 TS; 97,000 YS; 22 El; 65 RA. For gas turbine compressor blades. Corrosion and heat resistant.

JETHETE M160

Luria Steel & Trading Co.
C 0-0.2, Cr 11-13, Mo 0-1, V 0-1, Cb 0-0.7, Ni 0-1.25, bal Fe.
At 70 F: 143,000 TS; 121,000 YS; 22 El; 56 RA. At 570 F: 124,000 TS; 106,000 YS; 18 El; 51 RA. At 750 F: 116,000 TS; 97,000 YS; 22 El; 65 RA. For gas turbine compressor blades; corrosion and heat resistant. *Obsolete*

JETHETE M160

British Steel Corp.
C 0-0.2, Cr 11-13, Mo 0-1, V 0-1, Cb 0-0.7, Ni 0-1.25, bal Fe.
At 70 F: 143,000 TS; 121,000 YS; 22 El; 56 RA. At 570 F: 124,000 TS; 106,000 YS; 18 El; 51 RA. At 750 F: 116,000 TS; 97,000 YS; 22 El; 65 RA. For gas turbine compressor blades; corrosion and heat resistant. *Obsolete*

JETHETE-M.140

British Steel Corp.
C 0.08, Cr 13, Mo 1, bal Fe.
Annealed: 75,000 TS; 35,000 YS; 30 El; B 92 Rock. Cold drawn: 85,000 TS; 70,000 YS; 20 El; 185 Brin. For springs, table flatware, oil refinery equipment, annealing boxes. Corrosion resistant, ferritic, non-hardenable. *Obsolete*

JETWELD 1

Lincoln Electric Co.
C, bal Fe.
Steel arc welding electrode. AWS Class E7024.

JETWELD 2

Lincoln Electric Co.
C, bal Fe.
Steel arc welding electrode. AWS Class E6027.

JETWELD 2HT

Lincoln Electric Co.
C, bal Fe.
Steel arc welding electrode. AWS Class E7027-Al.

JETWELD 3

Lincoln Electric Co.
C, bal Fe.
Steel arc welding electrode. AWS Class E7024.

JETWELD LH-110M

Lincoln Electric Co.
C 0.1, Mn 1.3, P 0.03, S 0.03, Si 0.6, Ni 2, Cr 0.4, Mo 0.35, V 0.05, bal Fe.
Steel arc welding electrode. AWS Class E11018-M.

JETWELD LH-90

Lincoln Electric Co.
C 0.12, Mn 0.9, P 0.03, S 0.04, Si 0.06, Cr 1.25, Mo 0.5, bal Fe.
Steel arc welding electrode. AWS Class E8018-B2.

JETWELD LH3800

Lincoln Electric Co.
C, bal Fe.
Steel arc welding electrode. AWS Class E7028.

JETWELD LH70

Lincoln Electric Co.
C, bal Fe.
Steel arc welding electrode. AWS Class E7018.

JEWEL CAST IRON

English manufacture
Mn 0.14, Si 0.81, CC 0.88, 1.24 graphite, bal Fe.
Cast. For castings, gears, housings; cast iron.

JEWELER'S METAL

English manufacture
Cu 83-91.5, Zn 6.5-17, Sn 0-2.
For jewelry; corrosion resistant.

JEWELERS MANGANESE BRONZE

Belmont Metals Inc.
Al 0.1-3, Mn 0.1-2, Zn 35-40, bal Cu.
65,000 psi TS. For jewelry and decorative castings; melting point 1580-1620°F.

JEWELL ALLOY

Kencroft Malleable Co.
Si 1.25, C 1.6, Mn 0.4, Ni 0.2-1.5, Cr 0.1-0.5, bal Fe.
65,000-95,000 TS; 47,000-70,000 YS; 10-25 El; 130-250 Brin. For high strength and heat resisting castings, grate bars, furnaces, stoker parts, glass molds and dies; heat resisting; resists scaling up to 1450 F. *Obsolete*

JEWELL ALLOY 42

Apollo Steel Co.
C 3, Si 1, bal Fe.
Cast: 70,000 TS; 50,000 YS; 15 El; 180 Brin. For general castings; pearlitic malleable iron. *Obsolete*

JEWELL ALLOY S

Apollo Steel Co.
C 3, Si 1.5, bal Fe.
Cast: 62,000 TS; 42,000 YS; 24 El; 130 Brin. For general castings; pearlitic malleable iron. *Obsolete*

JEWELL ALLOY V

Apollo Steel Co.
C 2.5, Si 1, bal Fe.
Cast: 85,000 TS; 70,000 YS; 5 El; 250 Brin. For general castings; pearlitic malleable iron. *Obsolete*

JEWELL STEEL

Kencroft Malleable Co.
C 3, Si 1, bal Fe.
Cast: 60,000 TS; 42,000 YS; 20 El; 130 Brin. For general castings, staybolts, fittings; ferritic, malleable iron. *Obsolete*

JEWELL SUPER STRENGTH MALLEABLE

Kencroft Malleable Co.
C 3, Si 1.3, bal Fe.
Cast: 57,000 TS; 38,000 YS; 18 El; 120 Brin. For staybolts, fittings, general castings; ferritic, malleable iron. *Obsolete*

JEWELRY BRONZE 87.5
Chase Brass & Copper Co.
Cu 87.5, Zn 12.5.
Annealed: 40,000 TS; 13,000 YS; 46 El. Hard rolled: 66,000 TS; 56,000 YS; 5 El. For costume jewelry, slide fasteners; containers; red brass.

JISCO SILVERY
Jackson Iron & Steel Co.
Si 5-17, Mn 1-2, C 0.8-3, bal Fe.
For foundry steel and iron castings; pig iron.

JISCON
Jackson Iron & Steel Co.
C 2.3-2.5, Si 6-6.25, Cu 1.25, Cr 0.5, bal Fe.
For high temperature castings; resists growth and scaling to 1700°F.

JJ HOT WORK
Now JESSOP JJ.

JLX-36
LTV Steel
C 0-0.15, Mn 0-0.9, bal Fe.
Sheet: 50,000 psi TS; 36,000 psi YS; 28 El min; 63-75 Rock B. For construction applications, bridges, and tanks. Good formability, weldability and notch toughness.

JLX-42
LTV Steel
C 0-0.2, Mn 0-1, Si 0-0.3, 0.01 Cb or V min, bal Fe.
42,000 psi YS min. Good weldability and impact toughness. For bumpers, agricultural equipment.

JLX-45
LTV Steel
C 0-0.2, Mn 0-1.1, Si 0-0.3, 0.01 Cb or V min, bal Fe.
45,000 psi YS min. Good weldability and impact toughness. For bumpers and railroad car components.

JLX-50
LTV Steel
C 0-0.22, Mn 0-1.2, Si 0-0.3, 0.1 Cb or V min, bal Fe.
50,000 psi YS min. Good weldability. For agricultural equipment. ASTM A-572.

JLX-50 CC
LTV Steel
C 0-0.12, Mn 0-0.9, P 0-0.01, S 0-0.25, Si 0-0.02, 0.01 Cb min, bal Fe.
50,000 psi YS min. Good weldability and impact toughness. For automobile bumpers, couplings, and agricultural equipment. ASTM A-572.

JLX-55
LTV Steel
C 0-0.24, Mn 0-1.2, Si 0-0.3, 0.01 Cb or V min, bal Fe.
55,00 psi YS min. Good weldability. For agricultural equipment.

JLX-60
LTV Steel
C 0-0.25, Mn 0-1.35, Si 0-0.3, 0.01 Cb or V min, bal Fe.
60,000 psi YS min. Good weldability. For railroad car components.

JLX-65
LTV Steel
C 0-0.26, Mn 0-1.5, Si 0-0.3, 0.01 Cb or V min, bal Fe.
65,000 psi YS min. Good weldability. For automobile bumpers and agriculture equipment.

JLX-70
LTV Steel
C 0-0.26, Mn 0-1.65, Si 0-0.3, 0.01 Cb or V min, bal Fe.
70,000 psi YS min. Good weldability. For truck frames, dump trucks, and farm equipment.

JLX-W
LTV Steel
C 0-0.2, Mn 0.5-1, Si 0-0.1, 0.01 Cb min, bal Fe.
Gr. 45: 60,000 psi TS; 45,000 psi YS; 24 El; 79 Rock B. Gr. 50: 65,000 psi TS; 50,000 psi YS; 22 El; 82 Rock B. Gr. 55: 70,000 psi TS; 55,000 psi YS; 20 El; 83 Rock B. Gr. 60: 75,000 psi TS; 60,000 psi YS; 18 El; 85 Rock B. For water wheel generators, freight cars, trucks, and storage tanks. Produced to minimum yield points indicated.

JMC 1715 MG NI
Johnson Matthey plc
Ag, Mg, and others.
For electrical spring contact; 55% IACS conductivity. *Obsolete*

JMC 625
Johnson Matthey plc
Cu, Ag, Au.
For wiping contacts. *Obsolete*

JMC 77
Johnson Matthey plc
Cu 40, bal Pd.
For wiping contacts. *Obsolete*

JMM 77
Johnson Matthey plc
Pt, Pd, Au, Ag.
Precious metal electrical contacts. *Obsolete*

JOBBINS 3-6 SUPREME
W.F. Jobbins Inc.
Cu 2.75-3.75, Si 5.5-6.5, bal Al.
Cast: 29,000 TS; 17,000 YS; 2.5 El; 75 Brin. T6 Temper: 40,000 TS; 28,000 YS; 2 El; 90 Brin. For sand and permanent mold castings; heat treatable. *Obsolete*

JOBBINS 3-6-6 SUPREME
W.F. Jobbins Inc.
Cu 2.5-3.5, Si 5-6, Zn 5-6, bal Al.
Cast: 26,000 TS; 14,500 YS; 2 El; 65 Brin. T6 Temper: 45,000 TS; 38,000 YS; 1 El; 100 Brin. For sand and permanent mold castings; hardenable. *Obsolete*

JOBBINS 4-8 SUPREME
W.F. Jobbins Inc.
Cu 3.5-4.5, Si 7.5-9.5, bal Al.
Cast: 32,000 TS; 17,000 YS; 2.5 El; 85 Brin. T6 Temper: 45,000 TS; 28,000 YS; 3.0 El; 100 Brin. For permanent mold castings; hardenable. *Obsolete*

JOBBINS ALMAG 35
Now USCO-JOBBINS ALMAG 35 (535.2).

JOBBINS ALMAG 56
W.F. Jobbins Inc.
Mg 9.5-10.5, bal Al.
Cast T4 Temper: 58,000 TS; 30,000 YS; 18 El; 95 Brin. For sand castings; heat treatable. *Obsolete*

JOHNSON
Manufacturer not listed.
Ni 9-36, Cr 10-30, Si 1-10, bal Fe.
For heat and corrosion resisting parts; heat and corrosion resistant.

JOHNSON BRONZE
Johnson Bronze Co.
Sn 5, bal Cu.
For machine bearings and bushings. *Obsolete*

JOHNSON BRONZE BABBITT NO. 11
Johnson Bronze Co.
Sn 87, Cu 6, Sb 7.
For bearings; bronze Babbitt. *Obsolete*

JOHNSON BRONZE BABBITT NO. 97
Johnson Bronze Co.
Sn 90, Cu 2.5, Sb 7.5.
For bearings; bronze Babbitt. *Obsolete*

JOHNSON BRONZE GRAPHITED BEARINGS
Johnson Bronze Co.
Graphite, Sn, bal Cu.
For bearings; self lubricating. *Obsolete*

JOHNSON BRONZE NO. 12
Johnson Bronze Co.
Sn 90, Sb 7.5, Cu 2.5.
For bearings; high speed and high temperature service. *Obsolete*

JOHNSON BRONZE NO. 29
Now SAE NO. 67.

JOHNSON BRONZE NO. 40
Johnson Bronze Co.
Cu 90, Sn 0.5, Zn 9.5.
For bearings; sheet metal. *Obsolete*

JOHNSON BRONZE NO. 44
Johnson Bronze Co.
Cu 86-90, Sn 3-4.5, Pb 3.5-4, Zn 3.5-4.5.
For bearings; sheet metal. *Obsolete*

JOHNSON BRONZE NO. 53
Now SAE NO. 63.

JOHNSON BRONZE NO. LX
Johnson Bronze Co.
Pb 74.75, Sb 15, Sn 10, Cu 0.25.
For camshaft bearings; Babbitt. *Obsolete*

JOHNSON LOCOMOTIVE BRONZE
English manufacture
Sn 7.8, Zn 5, bal Cu.
For railroad hardware, fittings, injectors; high strength.

JOHNSON NO. 9 BABBITT ALLOY
Johnson Bronze Co.
Pb, Sb, bal Sn.
For automobile bearings; Babbitt metal. *Obsolete*

JOHNSONS (A) STAINLESS STEEL SOLDER
Johnson Mfg. Co.
For solder for stainless steel; M.P. 450°F.

JOHNSTOWN A-1
Johnstown Corp.
C 0.3-0.4, Mn 0-0.85, Cr 0.75-1, Ni 2.5-3, 0.15 Si min, bal Fe.
Coupling boxes, spindles, wheels, pinions, gears, ram heads, peels, clamps, drums, rings, hoppers, discs, etc. 100,000 TS; 65,000 YS; 16 El; 30 RA.

JOHNSTOWN A-3
Johnstown Corp.
C 0.45-0.55, Mn 1.25-1.5, Cr 0.75-1, Ni 0.6-0.8, Mo 0.3-0.4, 0.15 Si min, bal Fe.
Screen plates, liner plates, lifter bars, gears, wheels, rollers, crusher rolls, hammers, face plates, skid bars, shear blades. 350 Brin min.

JOHNSTOWN A-4
Johnstown Corp.
C 0.3-0.4, Mn 0-0.85, Cr 0.75-1, Mo 0.3-0.4, 0.15 Si min, bal Fe.
Guides, shoes, sheaves, wheels, pinions, piercer discs. 100,000 TS; 50,000 YS; 20 El; 40 RA.

JOHNSTOWN A-4D
Johnstown Corp.
C 0.8-0.9, Mn 0-0.85, Cr 1.8-2.4, Mo 0.25-0.4, 0.15 Si min, bal Fe.
Liner plates, skid plates.

JOHNSTOWN A-5
Johnstown Corp.
C 0.3-0.4, Mn 1.6-2, 0.15 Si min, bal Fe.
Bridge shoes, cross heads, gears, sheaves, valves, fittings. 90,000 TS; 50,000 YS; 20 El; 30 RA.

JOHNSTOWN A-6
Johnstown Corp.
C 0.25-0.35, Mn 1.15-1.4, Mo 0.25-0.35, 0.15 Si min, bal Fe.
Sheaves, pinions, bevel gears, chain links, mandrel segments, draw bars, levers, cutter heads, plates, rings, drums. 90,000 TS; 60,000 YS; 20 El; 40 RA.

JOHNSTOWN A-7
Johnstown Corp.
C 0.25-0.35, Mn 0-0.85, Ni 3-3.5, 0.15 Si min, 0.20 Mo min, bal Fe.
Coupling boxes, large mill pinions.

JOHNSTOWN A-9
Johnstown Corp.
C 0.4-0.5, Mn 0-0.85, Cr 0.8-1, V 0.15-0.2, 0.15 Si min, bal Fe.
Machine tool parts, lathe rests, tool posts, gears, pulleys. 100,000 TS; 60,000 YS; 15 El; 30 RA.

JOHNSTOWN A319 F
Johnstown Corp.
Mn 0-1, Si 0.9-2.7, Cr 0.41-0.65, 3.20 C min, bal Fe.
30,000 TS.

JOHNSTOWN A319 N
Johnstown Corp.
C 3.5-3.9, Mn 0.45-0.6, Si 1.35-1.9.
30,000 TS.

JOHNSTOWN A319 S
Johnstown Corp.
C 3.15-3.5, Mn 0.5-0.8, Si 1.25-1.85.
30,000 TS.

JOHNSTOWN A319 T
Johnstown Corp.
C 3-3.5, Mn 0.5-0.8, Si 1.25-1.85, Cr 0.2-0.4.
Coke oven door jambs, heavy castings. 30,000 TS.

JOHNSTOWN GJ-2
Johnstown Corp.
C 0-0.8, Mn 0-1, Si 0-2, Cr 26-30, Ni 8-10, Mo 1.75-2.25.
Sintering plant grate bars with a service temperature to 2000°F. ASTM A297 GR HE.

JOHNSTOWN JC GR I
Johnstown Corp.
C 3.25-3.45, Mn 0-1, Si 1.5-2.5, bal Fe.
Pig machine molds, grate bars, tuyere boxes, blow pipes, bars, pulleys, wheels, gears, housings, racks, boxes, hot top castings. 30,000 TS; 160-200 Brin.

JOHNSTOWN JC GR IV
Johnstown Corp.
C 3-3.5, Mn 0-1, Si 1.5-2.25, 0.50 Cr min, 1.50 Ni min, bal Fe.
Molds, manifolds, door frames, dampers, stoker parts. 35,000 TS; 200-220 Brin.

JOHNSTOWN JC GR V
Johnstown Corp.
C 3-3.5, Mn 0-1, Si 1.5-2.25, 0.50 Cr min, 0.50 Mo min, bal Fe.
Guides, dies, exhaust pipes, blow pipes. 40,000 TS; 200-240 Brin.

JOHNSTOWN JC GR VI
Johnstown Corp.
C 3.38-3.45, Mn 0-0.15, Si 2.5-4, Cr 0-0.1, Ni 0-0.1, Mo 0.2-1.5, Mg 0.05-1.08, bal Fe.
Coke oven jambs, furnace doors, grate bars, furnace stools and tracks, coke quenching car doors, grizzly bars, gears, bearings, diffusers. 65,000 TS; 45,000 YS; 12 El. ASTM A536.

JOHNSTOWN SS-1
Johnstown Corp.
C 0-1, Mn 0-1, Si 0-2, Cr 26-30, Ni 0-4.
Grate bars, end grate bars. Heat resistant properties to 1800°F. 250 Brin. ASTM A297 GR HC.

JOHNSTOWN SS-11
Johnstown Corp.
C 0-0.2, Mn 0-2, Si 0-2, Cr 19-23, Ni 9-12.
Sintering plant grizzly bars, breaker blades with a service temperature of 1200-1600°F. ASTM A297 GR HF.

JOHNSTOWN SS-6
Johnstown Corp.
C 0.2-0.5, Mn 0-2, Si 0-2, Cr 24-28, Ni 11-14.
Blast furnace blow pipes, heat treating furnace stools. 156 Brin. ASTM A297 GR HH.

JOHNSTOWN SS-7
Johnstown Corp.
C 2.5-2.7, Mn 0-1, Si 0-2, Cr 26-30, Ni 0-4.
Abrasion resistant chutes and liners. 500 Brin. ASTM A532.

JOHNSTOWN SS-8
Johnstown Corp.
C 0.2-0.6, Mn 0-2, Si 0-1, Cr 24-28, Ni 18-22.
High temperature applications to 2100°F. ASTM A297.

JOHNSTOWN VI H
Johnstown Corp.
C 3.38-3.45, Mn 0-0.15, Si 2.5-2.7, Cr 0-0.1, Ni 0-0.1, Mo 0-0.1, Mg 0.05-0.08, bal Fe.
Coke oven jambs, furnace doors, mill guides, heat shields. 65,000 TS; 45,000 YS; 12 El. ASTM A536.

JONAS & COLVER
Jonas & Culver Ltd.
C 1.36, Mn 0.41, bal Fe.
For tools, cutters; water hardened.

JORGENSEN 100
Earle M. Jorgensen
C 0.16-0.2, Mn 0.6-0.9, Si 0.2-0.3, Cr 0.5-0.75, Mo 0.4-0.6, Ni 1.1-1.4, V 0.06-0.12, bal Fe.
Plate, quenched and tempered: 115,000 psi TS; 100,000 psi YS; 16 El. For large, highly stressed structural equipment.

JORGENSEN J-20
Earle M. Jorgensen
C 0.2, Mn 1.25, P 0-0.04, S 0.25, Si 0.2, bal Fe.
Free machining steel. May be carburized and hardened for gears, cams, sprockets, molds. Similar to AISI 1119.

JORGENSEN J-45
Earle M. Jorgensen
C 0.45, Mn 1.25, P 0-0.04, S 0.25, Si 0.15, bal Fe.
Medium carbon free machining steel; may be hardened for use as cams, gears, trimming dies, molds. Similar to AISI 1144.

JORGENSEN NI-MO
Earle M. Jorgensen
C 0.28, Mn 0.15-0.45, Si 0.15-0.35, Mo 0.25-0.6, Ni 2.75-3.5, V 0.08, bal Fe.
Plate, quenched and tempered: 110,000 psi TS; 75,000 psi YS; 20 El. For stressed structural equipment.

JOSLYN STAINLESS AISI TYPE 304
Slater Steels
Stainless steel.
Now SLATER AISI TYPE 304.

JOSLYN STAINLESS TYPE 15-5 PH
Slater Steels
Stainless steel.
Now SLATER TYPE 15-5.

JOSLYN STAINLESS TYPE 17-4 PH
Slater Steels
Stainless steel.
Now SLATER TYPE 17-4.

JOSLYN STAINLESS TYPE 202
Joslyn Stainless Steels Co.
C 0-0.15, Mn 7.5-10, Si 0-1, Cr 17-19, Ni 4-6, N 0-0.25, bal Fe.
Annealed: 90,000 TS; 50,000 YS; 55 El; 170 Brin. Austenitic; hardenable only by work hardening, corrosion resistant; nonmagnetic. For antennas, bolts and fasteners, dairy equipment, spring wire, transportation equipment. AISI 202; ASTM A314; SAE 30202. *Obsolete*

JOSLYN STAINLESS TYPE 302 CU
Joslyn Stainless Steels Co.
C 0-0.1, Mn 0-2, Cr 17-19, Ni 8-10, Mo 0-0.5, Cu 3-4, bal Fe.
Annealed: 80,000 TS; 40,000 YS; 35 El; 163 Brin. For fasteners, automotive parts, aircraft and missile components, chemical equipment. Corrosion resistance, austenitic. *Obsolete*

JOSLYN STAINLESS TYPE 302B
Joslyn Stainless Steels Co.
C 0-0.15, Mn 0-2, Si 2-3, Cr 17-19, Ni 8-10, bal Fe.
Annealed: 90,000 TS; 40,000 YS; 50 El; 160 Brin. Austenitic: hardenable only by cold work; heat resistant; nonmagnetic. For heat treating fixtures, annealing boxes, furnace parts, fittings. AISI 302; ASTM A580; SAE 30302. *Obsolete*

JOSLYN STAINLESS TYPE 303
Slater Steels
Stainless steel.
Now SLATER TYPE 303 MAX MACHINABILITY.

JOSLYN STAINLESS TYPE 303 PB
Slater Steels
Stainless steel. C 0-0.15, Mn 0-2, S 0.12-0.3, Si 0-1, Cr 17-19, Ni 8-10, Mo 0-0.6, Pb 0.12-0.3, bal Fe.
Annealed: 90,000 TS; 35,000 YS; 50 El; 160 Brin. Austenitic, free machining, non-magnetic, hardenable only by cold work, corrosion resistant. For gears, shafts, studs, bolts, nuts, screw machine parts, Swiss automatic parts. QQ-S-764; AMS 5635; ASTM A581. *Obsolete*

JOSLYN STAINLESS TYPE 303 SE
Slater Steels
Stainless steel. C 0-0.15, Mn 0-2, S 0-0.06, Si 0-1, Cr 17-19, Ni 8-10, Mo 0-0.5, 0.15 Se min, bal Fe.
Annealed: 90,000 TS; 35,000 YS; 50 El; 160 Brin. Austenitic, free machining, non-magnetic, hardenable only by cold work, corrosion resistant. For gears, shafts, rivets, screw machine parts, bolts, nuts, studs, ordnance parts. AISI 303 Se; AMS 5640 Type 2; SAE 30303 Se. *Obsolete*

JOSLYN STAINLESS TYPE 304 N
Slater Steels
Stainless steel.
Now SLATER TYPE 304 N.

JOSLYN STAINLESS TYPE 304L
Slater Steels
Stainless steel.
Now SLATER TYPE 304l.

JOSLYN STAINLESS TYPE 305
Joslyn Stainless Steels Co.
C 0-0.12, Mn 0-2, Si 0-1, Cr 17-19, Ni 11-13, bal Fe.
Annealed: 85,000 TS; 35,000 YS; 55 El; 149 Brin. Austenitic, nonmagnetic, corrosion resistant. Less tendency to harden by cold work. For rivets, cold headed parts, extruded parts, cold formed parts. AISI 305; AMS 5685; SAE 30305. *Obsolete*

JOSLYN STAINLESS TYPE 305 MH
Joslyn Stainless Steels Co.
C 0-0.12, Mn 0-2, Cr 11.5-13, Ni 14-16, bal Fe.
Annealed: 78,000 TS; 40,000 YS; 55 El; 163 Brin. For Phillips head screws, nut forming, spinning, coining, extruded or swaged parts. Austenitic, corrosion resistant. *Obsolete*

JOSLYN STAINLESS TYPE 305H
Joslyn Stainless Steels Co.
C 0-0.12, Mn 0-2, Cr 16-17, Ni 17-18, bal Fe.
Annealed: 75,000 TS; 40,000 YS; 55 El; 156 Brin. For fasteners, cold headed bolts, screws, nuts. Low work hardening. Austenitic, corrosion resistant. *Obsolete*

JOSLYN STAINLESS TYPE 308
Slater Steels
Stainless steel. C 0-0.08, Mn 0-2, Si 0-1, Cr 19-21, Ni 10-12, bal Fe.
Annealed: 85,000 YS; 30,000 YS; 55 El; 149 Brin. Austenitic, non-magnetic, corrosion resistant. Less subject to work hardening, weldable. For weld rod, welded assemblies, heat treat equipment, furnace parts. AISI 308; ASTM A276; SAE 30308. *Obsolete*

JOSLYN STAINLESS TYPE 308 FM
Joslyn Stainless Steels Co.
C 0.15-0.25, Mn 1-2, S 0.12-0.25, Si 0.75-1.25, Cr 20-22, Ni 10.5-12, bal Fe.
Annealed: 100,000 TS; 35,000 YS; 40 El; 212 Brin. Austenitic, free-machining grade, good corrosion and oxidation resistance. For exhaust manifold valve parts, exhaust system shafts and bushings. *Obsolete*

JOSLYN STAINLESS TYPE 308L
Slater Steels
Stainless steel. C 0-0.03, Mn 0-2, Si 0-1, Cr 19-21, Ni 9-12, bal Fe.
Annealed: 85,000 TS; 35,000 YS; 55 El; 149 Brin. Austenitic, weldable, corrosion resistant. Main use is welding rod and parts to be welded. *Obsolete*

JOSLYN STAINLESS TYPE 309
Slater Steels
Stainless steel.
Now SLATER TYPE 309.

JOSLYN STAINLESS TYPE 309S
Slater Steels
Stainless steel.
Now SLATER TYPE 309S.

JOSLYN STAINLESS TYPE 310
Slater Steels
Stainless steel.
Now SLATER TYPE 310.

JOSLYN STAINLESS TYPE 310S
Slater Steels
Stainless steel.
Now SLATER TYPE 310S.

JOSLYN STAINLESS TYPE 316F
Slater Steels
Stainless steel.
Now SLATER TYPE 316F.

JOSLYN STAINLESS TYPE 316L
Slater Steels
Stainless steel.
Now SLATER TYPE 316L.

JOSLYN STAINLESS TYPE 317
Slater Steels
Stainless steel.
Now SLATER TYPE 317.

JOSLYN STAINLESS TYPE 317L
Slater Steels
Stainless steel.
Now SLATER TYPE 317L.

JOSLYN STAINLESS TYPE 321
Slater Steels
Stainless steel.
Now SLATER TYPE 321.

JOSLYN STAINLESS TYPE 330
Slater Steels
Stainless steel.
Now SLATER TYPE 330.

JOSLYN STAINLESS TYPE 347
Slater Steels
Stainless steel.
Now SLATER TYPE 347.

JOSLYN STAINLESS TYPE 347 FM (SE)
Slater Steels
Stainless steel. C 0-0.08, Mn 0-2, P 0.11-0.17, S 0-0.03, Si 0-1.5, Cr 17-19, Ni 9-12, Se 0.15-0.35, Mo 0-0.5, Cu 0-0.5, Cb + Ta = 10 x C min, bal Fe.
Annealed: 90,000 TS; 35,000 YS; 50 El; 160 Brin. Austenitic, non-magnetic; welding not recommended. For machined parts for assemblies operating at 800-1500°F. AMS 5642 Type 2. *Obsolete*

JOSLYN STAINLESS TYPE 348
Slater Steels
Stainless steel.
Now SLATER TYPE 348.

JOSLYN STAINLESS TYPE 403
Slater Steels
Stainless steel.
Now SLATER TYPE 403.

JOSLYN STAINLESS TYPE 405
Slater Steels
Stainless steel.
Now SLATER TYPE 405.

JOSLYN STAINLESS TYPE 410
Slater Steels
Stainless steel.
Now SLATER TYPE 410.

JOSLYN STAINLESS TYPE 410 CB
Slater Steels
Stainless steel.
Now SLATER TYPE 410 CB.

JOSLYN STAINLESS TYPE 410 LOW CARBON
Slater Steels
Stainless steel.
Now SLATER TYPE 410 LOW CARBON.

JOSLYN STAINLESS TYPE 416 SE
Slater Steels
Stainless steel. C 0-0.15, Mn 0-1.25, P 0-0.06, S 0-0.06, Si 0-1, Cr 12-14, 0.15 Se min, bal Fe.
Obsolete

JOSLYN STAINLESS TYPE 416 XS
Slater Steels
Stainless steel.
Now SLATER TYPE 416 XS.

JOSLYN STAINLESS TYPE 420
Slater Steels
Stainless steel.
Now SLATER TYPE 420.

JOSLYN STAINLESS TYPE 420 F
Slater Steels
Stainless steel.
Now SLATER TYPE 420 F.

JOSLYN STAINLESS TYPE 422
Slater Steels
Stainless steel.
Now SLATER TYPE 422.

JOSLYN STAINLESS TYPE 430
Slater Steels
Stainless steel.
Now SLATER TYPE 430.

JOSLYN STAINLESS TYPE 430 F
Slater Steels
Stainless steel.
Now SLATER TYPE 430 F.

JOSLYN STAINLESS TYPE 430 F SE
Joslyn Stainless Steels Co.
C 0-0.12, Mn 0-1.25, P 0-0.06, S 0-0.06, Si 0-1, Cr 14-18, Mo 0-0.6, 0.15% min Se, bal Fe.
Annealed: 80,000 TS; 45,000 YS; 25 El; 163 Brin. Ferritic, magnetic, free-machining, corrosion resistant, not hardenable, weldability-poor. For machined parts, bolts and nuts, shafts, fasteners, burner parts. AISI 430F Se; SAE 51430 F Se. *Obsolete*

JOSLYN STAINLESS TYPE 434
Joslyn Stainless Steels Co.
C 0-0.12, Mn 0-1, Si 0-1, Cr 16-18, Mo 0.75-1.25, bal Fe.
Annealed: 75,000 TS; 45,000 YS; 30 El; 156 Brin. Ferritic, magnetic, corrosion resistant. For automotive fasteners, miscellaneous cold headed parts. *Obsolete*

JOSLYN STAINLESS TYPE 434 FM
Joslyn Stainless Steels Co.
C 0-0.12, Mn 0-1, Si 0-1, Cr 16-18, Mo 0.75-1.25, 0.15% min S, bal Fe.
Annealed: 80,000 TS; 45,000 YS; 30 El; 156 Brin. Ferritic, magnetic, free-machining, corrosion resistant, not hardenable. For automotive fasteners, thermostat parts, machined parts. *Obsolete*

JOSLYN TYPE 303 FORGING QUALITY
Slater Steels
Stainless steel.
Now SLATER TYPE 303 FORGING QUALITY.

JOSLYN TYPE 304 MAX MACHINABILITY
Slater Steels
Stainless steel.
Now SLATER TYPE 304 max machinability.

JOSLYN TYPE 316 MAX MACHINABILITY
Slater Steels
Stainless steel.
Now SLATER TYPE 316 MAX MACHINABILITY.

JOSLYN TYPE 416 FORGING QUALITY
Slater Steels
Stainless steel.
Now SLATER TYPE 416 FORGING QUALITY.

JOSLYN TYPE 416 HIGH HARDENABILITY
Slater Steels
Stainless steel.
Now SLATER TYPE 416 HIGH HARDENABILITY.

JOSLYN TYPE 416 MACHINE SPECIAL
Slater Steels
Stainless steel.
Now SLATER TYPE 416 MAX MACHINABILITY.

JOSLYN TYPE 430 F SOLENOID QUALITY
Slater Steels
Stainless steel.
Now SLATER TYPE 430 F SOLENOID QUALITY.

JOSLYN TYPE 615 (GREEK ASCOLOY)
Slater Steels
Stainless steel.
Now SLATER TYPE 615 (GREEK ASCOLOY).

JS 700; JS 777
Now JESSOP JS 700 and JESSOP JS 777.

JS-700
Jessop Steel Co.
C 0.03, Mn 1.7, Si 0.5, Cr 21, Ni 25, Mo 4.5, Cb = 8 x C, bal Fe.
Annealed: 85,000 TS; 39,000 YS; 45 El; 50 RA; 170 Brin. Special engineering alloy for superior corrosion resistance. For agitators, piping, mixers, vessels, vaporizers, tanks and stills.

JUNKER 951
Otto Junker GmbH
C 0.4, Si 2, Cr 13, bal Fe.
Heat resistant casting. *Obsolete*

JUNKER A 1050
Otto Junker GmbH
C 0.15-0.35, Cr 17-19, Ni 8-10, bal Fe.
Cast: 64,000 TS min; 15 El min; 130-200 Brin. Heat resistant casting.

JUNKER A 1101
Otto Junker GmbH
C 0.3-0.5, Cr 21-23, Ni 9-11, bal Fe.
Cast: 64,000 TS min; 12 El min; 150-220 Brin. Heat resistant casting. Type HF.

JUNKER A 1102
Otto Junker GmbH
C 0.15-0.35, Cr 19-21, Ni 13-15, bal Fe.
Cast: 64,000 TS min; 12 El min; 150-220 Brin. Heat resistant casting.

JUNKER A 1151
Otto Junker GmbH
C 0.2-0.5, Cr 25-28, Ni 13-16, bal Fe.
Cast: 64,000 TS min; 8 El min; 150-220 Brin. Heat resistant casting. Type HI.

JUNKER A 1152
Otto Junker GmbH
C 0.3-0.5, Cr 17-19, Ni 36-39, bal Fe.
Cast: 58,000 TS min; 6 El min; 150-220 Brin. Heat resistant casting. Type HU.

JUNKER A 1153
Otto Junker GmbH
C 0.3-0.5, Cr 24-26, Ni 11-14, bal Fe.
Cast: 64,000 TS min; 8 El min; 130-200 Brin. Heat resistant casting. Similar to Type HH.

JUNKER A 12 MS
Otto Junker GmbH
C 0.06, Cr 17, Ni 13.5, Mo 4.5, bal Fe.
Corrosion resistant casting. 57,000 TS min; 28,500 YS min; 15 El min. *Obsolete*

JUNKER A 12 MSSN
Otto Junker GmbH
C 0-0.04, Cr 16.5-18.5, Ni 12.5-14.5, Mo 4-4.5, bal Fe.
Cast: 71,000 TS; 30,000 YS; 20 El; 150-200 Brin. Corrosion resistant steel casting.

JUNKER A 1201
Otto Junker GmbH
C 0.3-0.5, Cr 24-26, Ni 19-21, bal Fe.
Cast: 64,000 TS min; 8 El min; 150-220 Brin. Heat resistant casting. Type HK.

JUNKER A 1205
Otto Junker GmbH
C 0.35-0.55, Cr 27-30, Ni 47-50, W 4.5-5.5, bal Fe.
Cast: 58,000 psi TS min; 32,000 psi YS min; 3 El min; 150-220 Brin. Heat resistant casting.

JUNKER A 1234 NB
Otto Junker GmbH
C 0.05-0.15, Cr 19-21, Ni 31-33, Cb, bal Fe.
Cast: 64,000 psi TS min; 25,000 psi YS min; 20 El min. Heat resistant casting.

JUNKER A 1237
Otto Junker GmbH
C 0.3-0.5, Cr 24-26, Ni 34-36, bal Fe.
Cast: 64,000 psi TS min; 32,000 psi YS min; 8 El min. Heat resistant casting.

JUNKER A 13 MS
Otto Junker GmbH
C 0.06, Cr 17.5, Ni 12.5, Mo 2.8, bal Fe.
Corrosion resistant casting. 65,000 TS min; 30,000 YS min; 20 El min. *Obsolete*

JUNKER A 17 MK NB
Otto Junker GmbH
C 0.07, Cr 17.5, Ni 20, Mo 2.2, Cu 2, Cb, bal Fe.
Corrosion resistant casting. 65,000 min TS; 28,500 min YS; 15 min El. *Obsolete*

JUNKER A 17 MKN
Otto Junker GmbH
C 0-0.03, Cr 17-18.5, Ni 19-21, Mo 2.1-2.4, Cu 1.8-2.2, bal Fe.
Cast: 64,000 TS; 27,000 YS; 20 El; 130-200 Brin. Corrosion resisting steel casting.

JUNKER A 25 M NB
Otto Junker GmbH
C 0.07, Cr 25, Ni 25, Mo 2.2, Cb, bal Fe.
Corrosion resistant casting. 65,000 min TS; 31,500 min YS; 15 min El. *Obsolete*

JUNKER A 25 MK NB
Otto Junker GmbH
C 0.07, Cr 20, Ni 25, Mo 3, Cu 2, Cb, bal Fe.
Corrosion resistant casting. 65,000 min TS; 31,500 min YS; 15 min El. *Obsolete*

JUNKER A 25 MKN
Otto Junker GmbH
C 0-0.03, Cr 19-21, Ni 24-26, Mo 2.5-3.5, Cu 1.5-2, bal Fe.
Cast: 64,000 TS; 27,000 YS; 20 El; 130-200 Brin. Corrosion resistant steel casting.

JUNKER A 26 MKN
Otto Junker GmbH
C 0-0.03, Cr 19-21, Ni 24-26, Mo 4-5, Cu 1.5-2, bal Fe.
Cast: 64,000 TS; 27,000 YS; 20 El; 130-200 Brin. Corrosion resistant steel casting.

JUNKER A 27 MKN
Otto Junker GmbH
C 0-0.03, Cr 19-21, Ni 24-26, Mo 5.5-7, Cu 1-2, bal Fe.
68,000 psi YS min; 30,000 psi TS min; 30 El min. Corrosion resistant casting.

JUNKER A 28 MKN
Otto Junker GmbH
C 0-0.03, Cr 20-23, Ni 27-29, Mo 4-5, Cu 2-4, bal Fe.
Cast: 64,000 TS; 27,000 YS; 20 El; 130-200 Brin. Corrosion resistant steel casting.

JUNKER A 29 MKN
Otto Junker GmbH
C 0-0.03, Cr 19-21, Ni 28-30, Mo 2-3, Cu 3-4, bal Fe.
Cast: 64,000 TS; 27,000 YS; 20 El; 130-200 Brin. Corrosion resistant steel casting.

JUNKER A 31 MKN
Otto Junker GmbH
C 0-0.02, Cr 26-28, Ni 30-32, Mo 3-4, Cu 0.8-1.5, bal Fe.
71,000 psi TS min; 30,000 psi YS min; 30 El min. Corrosion resistant casting.

JUNKER A 35 MKN
Otto Junker GmbH
C 0-0.025, Cr 19-21, Ni 34-38, Mo 2-3, Cu 3-4, bal Fe.
Cast: 64,000 TS; 27,000 YS; 20 El; 130-200 Brin. Corrosion resistant casting.

JUNKER A 41 MKN
Otto Junker GmbH
C 0-0.02, Cr 19-21, Ni 40-42, Mo 4-6, Cu 2-3, bal Fe.
62,000 psi TS min; 37,000 psi YS min; 30 El min. Corrosion resistant casting.

JUNKER A 42 MKN
Otto Junker GmbH
C 0-0.02, Cr 19.5-23.5, Ni 38-46, Mo 2.5, Cu 1.5-3, bal Fe.
Cast: 64,000 TS; 27,000 YS; 20 El; 130-200 Brin. Corrosion resistant steel casting.

JUNKER A 56 M
Otto Junker GmbH
C 0-0.1, Cr 16.5, Ni 60, Mo 17, bal Fe.
Corrosion resistant casting. 78,000 min TS; 50,000 min YS; 5 min El. Type CW-12 M. *Obsolete*

JUNKER A 57 M
Otto Junker GmbH
C 0-0.03, Cr 15.5-17.5, Mo 16-18, bal Ni.
Cast: 72,000 TS; 33,000 YS; 15 El; 140-200 Brin. Corrosion resistant casting.

JUNKER A 65 M
Otto Junker GmbH
C 0-0.03, Mo 26-30, bal Ni.
Cast: 72,000 TS; 40,000 YS; 10 El; 160-220 Brin. Corrosion resistant casting.

JUNKER A 75
Otto Junker GmbH
C 0.05, Cr 15.5, Fe 8, bal Ni.
Cast: 58,000 psi TS; 19,000 psi YS; 20 El. Corrosion resistant casting. *Obsolete*

JUNKER A 8 M NB
Otto Junker GmbH
C 0-0.06, Cr 18-20, Ni 10.5-12.5, Mo 2-2.5, Cb, bal Fe.
Corrosion resistant casting. 65,000 TS min; 30,000 YS min; 20 El min.

JUNKER A 8 M S
Otto Junker GmbH
C 0-0.07, Cr 18-20, Ni 10-12, Mo 2-3, bal Fe.
Corrosion resistant casting. 65,000 TS min; 30,000 YS min; 20 El min. Type CF-8M.

JUNKER A 8 M SS
Otto Junker GmbH
C 0-0.03, Cr 17-20, Ni 9-13, Mo 2-3, bal Fe.
Corrosion resistant casting. 57,000 TS min; 27,000 YS min; 30 El min. Type CF-3M.

JUNKER A 8 NB
Otto Junker GmbH
C 0-0.06, Cr 18-20, Ni 9-11, Cb, bal Fe.
Cast: 64,000 TS; 25,000 YS; 20 El; 130-200 Brin. Corrosion resistant steel casting. Type CF-8C.

JUNKER A 8 NB TT
Otto Junker GmbH
C 0-0.04, Cr 18, Ni 10, Cb, bal Fe.
65,000 TS min; 28,500 YS min; 20 El min. For castings for low temperature applications to -195°C. *Obsolete*

JUNKER A 8 S
Otto Junker GmbH
C 0-0.07, Cr 17-20, Ni 9-11, bal Fe.
Corrosion resistant casting. 65,000 TS min; 28,500 YS min; 20 El min. Type CF-8.

JUNKER A 8 S TT
Otto Junker GmbH
C 0-0.07, Cr 18-20, Ni 9-11, bal Fe.
65,000 TS min; 28,500 YS min; 20 El min. For castings for low temperature applications to -195°C.

JUNKER A 8 SS
Otto Junker GmbH
C 0-0.03, Cr 17-20, Ni 8-12, bal Fe.
Corrosion resistant casting. 57,000 TS min; 25,500 YS min; 30 El min. Type CF-3.

JUNKER AF 1101
Otto Junker GmbH
C 0.3-0.5, Cr 25-28, Ni 3.5-5.5, bal Fe.
Cast: 71,000 TS min; 4 El min; 200-300 Brin. Heat resistant casting. Type HD.

JUNKER AF 1152
Otto Junker GmbH
C 0.25-0.4, Cr 27-29, Ni 9-11, bal Fe.
Cast: 71,000 TS min; 150-250 Brin. Heat resistant casting. Type HE.

JUNKER AF 21 NMN
Otto Junker GmbH
C 0-0.03, Cr 21-23, Ni 4.5-6.5, Mo 2.5-3.5, bal Fe.
83,000 psi TS min; 55 psi YS min; 25 El min. Corrosion resistant casting.

JUNKER AF 22 MN
Otto Junker GmbH
C 0-0.04, Cr 21.8-22.8, Ni 5.7-6.2, Mo 1.3-1.7, bal Fe.
830 psi TS min; 420 psi YS min; 25 El min. Corrosion resistant steel casting.

JUNKER AF 22 NMNB
Otto Junker GmbH
C 0-0.07, Cr 22-24, Ni 9-11, Mo 2-2.5, Cb, bal Fe.
Cast: 85,000 TS; 50,000 YS; 20 El; 180-230 Brin. Corrosion resistant steel casting.

JUNKER AF 25 N
Otto Junker GmbH
C 0-0.07, Cr 24-25, Ni 7.7-8.2, bal Fe.
Corrosion resistant casting. 85,000 TS min; 57,000 YS min; 20 El min. *Obsolete*

JUNKER AF 25 N M K
Otto Junker GmbH
C 0-0.03, Cr 24.5-26.5, Ni 5-7, Mo 2.5-3.5, Cu 2.75-3.5, bal Fe.
Cast: 920 psi TS min; 680 psi YS min; 22 El; 240 Brin min. Corrosion resistant steel casting.

JUNKER AF 26 N
Otto Junker GmbH
C 0-0.08, Cr 25-27, Ni 5.5-7.5, bal Fe.
83,000 psi TS min; 59,000 psi YS min; 20 El min. Corrosion resistant casting.

JUNKER AF 27 NC
Otto Junker GmbH
C 0.3-0.5, Cr 26-28, Ni 3.5-5.5, bal Fe.
Corrosion resistant casting. 71,000 TS min. Type CC-50.

JUNKER AF 27 NMN
Otto Junker GmbH
C 0-0.1, Cr 24-27, Ni 4.5-6, Mo 1.3-1.8, bal Fe.
Cast: 94,000 TS; 60,000 YS; 20 El; 200-250 Brin. Corrosion resistant steel casting.

JUNKER AUN 11
Otto Junker GmbH
C 0-0.07, Cr 16-18, Ni 10-12, bal Fe.
Cast: 64,000 TS min; 25,000 YS min; 20 El min; 150-190 Brin. Nonmagnetizable casting.

JUNKER AUN 14
Otto Junker GmbH
C 0-0.03, Cr 16.5-18.5, Ni 13-15, Mo 2.4-3, N_2, bal Fe.
69,000 psi TS min; 34,000 psi YS min; 30 El min. Nonmagnetizable casting.

JUNKER AUN 17
Otto Junker GmbH
C 0-0.03, Cr 20-21.5, Ni 15-17, Mo 3-3.5, Mn, N_2, Nb, bal Fe.
81,000 psi TS min; 44,000 psi YS min; 20 El min. Nonmagnetizable casting.

JUNKER AUN 9
Otto Junker GmbH
C 0-0.06, Cr 16-17, Ni 8.5-9.5, N_2, bal Fe.
For nonmagnetizable castings. 65,000 TS min; 28,500 min YS; 20 El min.

JUNKER F 1001
Otto Junker GmbH
C 0.3-0.5, Cr 16-18, bal Fe.
Cast: 71,000 TS min; 2 El min; 200-300 Brin. Heat resistant casting.

JUNKER F 1002 S
Otto Junker GmbH
C 1.4-1.8, Cr 17-19, bal Fe.
Cast: 300-400 Brin. Heat resistant casting.

JUNKER F 1051
Otto Junker GmbH
C 0.4, Cr 23, bal Fe.
Cast: 200-300 Brin. Heat resistant casting. *Obsolete*

JUNKER F 1101
Otto Junker GmbH
C 0.4, Cr 29, bal Fe.
Cast: 200-300 Brin. Heat resistant casting. *Obsolete*

JUNKER F 1102
Otto Junker GmbH
C 1.3, Cr 29, bal Fe.
Cast: 250-350 Brin. Heat resistant casting. *Obsolete*

JUNKER F 901 S
Otto Junker GmbH
C 0.2-0.4, Cr 6-8, bal Fe.
Cast: 71,000 psi TS min; 4 El min; 200-280 Brin. Heat resistant casting.

JUNKER G-NI 95
Otto Junker GmbH
C 0.3-1, bal Ni.
Cast: 45,000 psi TS; 17,000 psi YS; 12 El; 80 Brin min. Nickel casting.

JUNKER G-NI CU 30 NB
Otto Junker GmbH
C 0.05-0.15, Cu 27-33, Fe 1-2.5, Cb, bal Ni.
Cast: 57,000 TS; 25,000 YS; 25 El; 120 Brin min. Nickel-copper casting.

JUNKER IN 657
Otto Junker GmbH
C 0-0.1, Cr 48-52, Cb, bal Ni.
Cast: 78,000 psi TS min; 39,000 psi YS min; 8 El min. Heat resistant casting.

JUNKER M 13 N (GR. I)
Otto Junker GmbH
C 0-0.07, Cr 12-13.5, Ni 3.5-5, bal Fe.
Cast: 107,000 TS; 78,000 YS; 15 El; 240-300 Brin. Corrosion resistant cast steel.

JUNKER M 13 N (GR. II)
Otto Junker GmbH
C 0-0.07, Cr 12-13.5, Ni 3.5-5, bal Fe.
Cast: 128,000 TS; 121,000 YS; 12 El; 280-350 Brin. Corrosion resistant steel casting.

JUNKER M 14
Otto Junker GmbH
C 0.16-0.23, Cr 12.5-14.5, bal Fe.
Corrosion resistant casting. 85,000 TS min; 64,000 YS min; 12 El min. Type CA-40.

JUNKER M 14 S
Otto Junker GmbH
C 0.06-0.1, Cr 12-13.5, Ni 1-2, bal Fe.
Corrosion resistant casting. 85,000 TS min; 57,000 YS min; 15 El min. Type CA-15.

JUNKER M 16 N
Otto Junker GmbH
C 0-0.07, Cr 15-16.5, Ni 4.5-6, Mo 0.5-2, bal Fe.
108,000 psi TS min; 76,000 psi YS min; 15 El min. Corrosion resistant casting.

JUNKER M 18
Otto Junker GmbH
C 0.2-0.27, Cr 16-18, Ni 1-2, bal Fe.
Corrosion resistant casting. 113,000 TS min; 85,000 YS min; 4 El min.

JUNKER NI CR 50
Otto Junker GmbH
C 0.05-0.1, Cr 48-52, bal Ni.
Cast: 92,000 psi TS min; 57,000 psi YS min; 6 El min. Heat resistant casting.

JUNKER NI MO 59
Otto Junker GmbH
C 0-0.02, Cr 19-21, Mo 16-17.5, bal Ni.
Corrosion resistant casting. 71,000 TS min; 40,000 YS min; 30 El min.

JUNKER NI MO 69
Otto Junker GmbH
C 0-0.02, Mo 26-30, bal Ni.
Corrosion resistant casting. 71,000 TS min; 38,000 YS min; 20 El min; 140-200 Brin.

JUNKER NICRO 50
Now JUNKER NICR 50.

JUNKER THERMO 50
Otto Junker GmbH
C 0.05-0.25, Cr 27-30, Co 48-52, bal Fe.
Cast: 71,000 psi TS min; 35,000 psi YS min; 6 El min. Heat resistant casting. Resistant to thermal shock and high temperature corrosion.

JUNKER THERMO 51
Otto Junker GmbH
C 0.25-0.35, Cr 27-30, Co 48-52, Cb, bal Fe.
Cast: 78,000 psi TS min; 42,000 psi YS min; 3 El min. Heat resistant casting. Resistant to thermal shock and high temperature corrosion.

JUNKER X
English manufacture
Fe 6, Mg 0.5, bal Al.
For bearings, camshafts.

JUNKER Z
English manufacture
Ni 6.5, Ti 0.5, bal Al., ap k.
For bearings, camshafts.

K 059
Bergische Stahl Industrie
C 1.1-1.25, Si 0.2-0.4, Mn 0.2-0.4, Cr 1.2-1.5, bal Fe.
Cold work tool steel. W.-Nr. 1.2059.

K 833
Bergische Stahl Industrie
C 0.95-1.05, Si 0.15-0.25, Mn 0.15-0.3, V 0.05-0.15, bal Fe.
Cold work tool steel. W.-Nr. 1.2833.

K ALLOY
English manufacture
Cu 11.3, Mn 1.1, Mg 0.5, Si 0.79, bal Al.
For light alloy parts; leak-proof castings.

K F N 1
Kawasaki Steel Corp.
C 0.06, Mn 0.34, P 0.008, S 0.0125, B 0-0.007, bal Fe.
46,900 psi TS; 28,400 psi YS; 48 El. Low-alloy steel for deep drawing purposes.

K F N 2
Kawasaki Steel Corp.
c 0.005, Mn 0.1, P 0.01, S 0.004, B 0-0.007, bal Fe.
39,800 psi TS; 25,600 psi YS; 54 El. Low-alloy steel for extra deep drawing purposes.

K K K
Spencer Clark Metal Industries Ltd.
C 2.1, Mn 0.65, Si 0.75, Ni 0.85, Cr 13, Mo 0.65, bal Fe.
Chromium cold work tool steel, for press tools, molding dies, profilin rolls.

K O H
Kinite Corp.
C, alloy, bal Fe.
For tools; oil hardening.

K-390
Cytemp Specialty Steel Div.
C 0.25, Cr 2.75, W 10, Ni 2, bal Fe.
For brass forging and die casting dies, hot punches; hot work steel. *Obsolete*

K-4
German manufacture
Zn, Mg, bal Al.
For bearings.

K-42-B
Westinghouse Electric Corp.
Ni 42, Co 22, Fe 14, Cr 18, Ti 2, Al 0.6.
At 70°F: 165,000 psi TS; 19 El; 37 Brin. At 1110°F: 127,100 psi TS; 21 El; 22 Brin. Hardened: 185,000 psi TS; 90,000 psi YS; 30 El. For dies, valves, steam fitting and turbine blades, springs, bolts; retains strength at high temperature. *Obsolete*

K-4ONKHM
Russian manufacture
C 0.05-0.12, Co 40, Cr 20, Ni 15, Mn 2, 6 Mo + W, bal Fe.
Heat treated: 358,000 TS; 248,000 YS; 58 Rock C. For spiral springs in corrosive media and operating up to 400°C; high elastic properties imparted by cold work and subsequent tempering at 500°C

K-63
Mitsubishi Metals America Corp.
Cr 30, Ni 25, Co 2, bal Fe.
Heat resistant alloy; skid rails, riders and skid buttons of walking beam furnaces.

K-9 SPECIAL OIL HARDENING
A. Milne & Co.
C 0.9, W 0.5, Cr 0.7, Mn 1, Si 0.2, bal Fe.
For tools, cutters; oil hardening.

K-9 STEEL
Edgar Allen Balfour Ltd.
C 1, Mn 0.85, Cr 0.75, W 0.4, bal Fe.
Annealed: 100,000 TS; 54,000 YP; 21 El; 197 Brin. Heat treated: 220,000 TS; 155,000 YP; 10 El; 600 Brin. For dies, milling cutters, plugs, gages, circular cutters, master tools. Non-deforming, oil hardened, wear resistant.

K-MONEL
Now MONEL ALLOY K-500.

K-O
Kawasaki Steel Corp.
C 0.1-0.2, Mn 0.6-1, Si 0.15-0.3, B 0.002-0.006, Cr 0.4-0.8, Cu 0.15-0.5, Mo 0.4-0.6, Ni 0.7-1, V 0.03-0.1, bal Fe.
Plate: 114,000-135,000 TS; 100,000 min YS; 18 min El. For pressure vessels, bridges, mine cars, power shovels, cranes, trucks, trailers. Shock and wear resistant. *Obsolete*

K-O-1
Electronic Specialties Co.
Cu 4.8, Mg 0.23, Ti 0.27, Ag 0.64, bal Al.
Sand cast: 62,000-72,000 TS; 52,000-65,000 YS; 3-9 El. Permanent mold: 64,000-70,000 TS; 50,000-60,000 YS; 6-14 El. For landing gear struts, gear box housings, aircraft and auto/truck components, aerospace equipment. High strength casting alloy.

K-SPUN
Koppers Co. Inc.
TC 2.85-3.5, Si 0.95-1.45, Mn 0.4-0.8, bal Fe.
Heat treated: 70,000 TS; 285 Brin. For piston rings, cylinder liners; wear resistant cast iron.

K-TEN 60
Kobe Steel Ltd.
C 0.14, Si 0.38, Mn 1.26, P 0.015, S 0.01, Ni 0.2, Mo 0.05, V 0.04, bal Fe.
32 mm thick plate: 94,000 TS; 78,000 YS; 45 El. High strength low alloy steel for structural purposes.

K-TEN 70
Kobe Steel Ltd.
C 0.12, Si 0.31, Mn 1.03, P 0.013, S 0.01, Cu 0.21, Ni 0.62, Cr 0.31, Mo 0.33, V 0.24, bal Fe.
50 mm thick plate: 110,000 TS; 99,000 YS; 27 El. High strength low alloy steel for structural purposes.

K-TEN 80
Kobe Steel Ltd.
C 0.13, Si 0.32, Mn 0.86, P 0.009, S 0.001, Cu 0.24, Ni 0.92, Cr 0.45, Mo 0.4, V 0.03, B 0.002, bal Fe.
50 mm thick plate: 118,000 TS; 110,000 YS; 44 El. High strength low alloy steel for structural purposes.

K-V7
Russian manufacture
C 0.1-0.15, Cr 10.5-12.5, Mo 0.6-0.8, Ni 0-0.8, V 0.2-0.3, W 3.7-4.3, bal Fe.
Annealed: 85,000 TS; 45,000 YS; 20 El; 95 Rock B. For cutlery, surgical instruments, chemical plant equipment; corrosion resistant.

K.C.M. NICKEL BRONZE
Knowsley Cast Metal Co., Ltd.
Ni, Sn, bal Cu.
Cast: 33,600 TS; 10-15 El; 100-120 Brin. For slip rings for electrical motors; high conductivity.

K.D. NO. 10
Hidalgo Steel Co. Inc.
C, W, Co, bal Fe.
For high speed lathe tools, twist drills, reamers; high speed steel.

K.D. NO. 16
Hidalgo Steel Co. Inc.
C 0.7, Cr 4.5, Co 10, V 2.2, W 16, Mo 1.5, bal Fe.
For reamers, drills, broaches, taps, hobs; high speed steel, oil hardened.

K.D. NO. 6
Hidalgo Steel Co. Inc.
C, Co, W, bal Fe.
For high speed lathe tools, twist drills, reamers; high speed steel.

K.E. 1006
Sanderson Kayser Ltd.
C 1.05, Si 0.2, Mn 0.2, V 0.15, bal Fe.
Water hardening tool steel, soft center steel for cold heading dies. AISI W2.

K.E. 1029
Sanderson Kayser Ltd.
C, Si, Cr, bal Fe.
For valves; air hardened, heat resistant. *Obsolete*

K.E. 1036
Sanderson Kayser Ltd.
Cr 13, High C, bal Fe.
For sleeves, burners, atomizers, gas turbine parts; stainless, martensitic. *Obsolete*

K.E. 15
Sanderson Kayser Ltd.
C 0.07-0.1, Cr 13, bal Fe.
Annealed: 78,400 TS; 54,000 YS; 35 El; 70 RA; 152 Brin. Heat treated: 123,000 TS; 105,000 YS; 25 El; 65 RA; 255 Brin. For cutlery, stainless parts; stainless, martensitic.

K.E. 169
Sanderson Kayser Ltd.
C 0.14, Cr 0.85, Ni 3.4, Mo 0.15, bal Fe.
Nickel-chrome-moly case hardening steel. Core hardenable to 135,000-157,000 TS. For gears, spline shafts, axles, plastic mold dies.

K.E. 232
Sanderson Kayser Ltd.
C, Cr, bal Fe.
For valves; stainless, oxidation resistant. *Obsolete*

K.E. 25
Sanderson Kayser Ltd.
C 0.25, Cr 13, bal Fe.
Heat treated: 101,000-123,000 TS; 20 El; 201-255 Brin. For cutlery, stainless parts; stainless, martensitic. *Obsolete*

K.E. 332
Sanderson Kayser Ltd.
C, Cr, bal Fe.
Annealed: 98,500 TS; 56,000 YS; 20 El; 35 RA; 196 Brin. For valves; oxidation resistant. *Obsolete*

K.E. 339
Now SABEX.

K.E. 35
Sanderson Kayser Ltd.
C 0.35, Cr 13, bal Fe.
Heat treated: 99,000-274,000 TS; 68,000-242,000 YS; 9-30 El; 30-61 RA; 197-532 Brin. For cutlery, valves, surgical instruments; stainless, martensitic.

K.E. 354
Sanderson Kayser Ltd.
C, Cr, bal Fe.
Normalized: 112,000-135,000 TS; 67,000 YS; 20-30 El; 35-45 RA; 207-272 Brin. For valves, corrosion resistant parts; heat resistant. *Obsolete*

K.E. 355
Sanderson Kayser Ltd.
C 0.32, Cr 12, Ni 4.1, Mo 0.2, bal Fe.
For plastic mold dies. Oil or air hardening, tough and strong.

K.E. 37D

Sanderson Kayser Ltd.
C, Ni, Cr, Mo, bal Fe.
Heat treated: 157,000-168,000 TS; 141,000-153,100 YS;
18-20 El; 50-55 RA; 321-351 Brin. For shafts, gears, bolts; oil
or air hardened, tough. *Obsolete*

K.E. 396

Sanderson Kayser Ltd.
C 0.63, Cr 1, Ni 1.45, Mo 0.2, bal Fe.
Nickel-chrome-moly oil hardening tool steel. AISI L6. For
blanking and forming dies, shear blades.

K.E. 40 A

Sanderson Kayser Ltd.
C 0.1, Cr 12.5, Se, bal Fe.
Free cutting martensitic corrosion resistant steel, hardenable;
for pump shafts, hardware, fasteners, gears, axles.

K.E. 40A.M.

Sanderson Kayser Ltd.
C, Cr, S, bal Fe.
For valves, cutlery, surgical instruments; stainless, free-
cutting. *Obsolete*

K.E. 43

Sanderson Kayser Ltd.
C 0.15, Cr 13, bal Fe.
Heat treated: 123,000-145,000 TS; 15 El; 248-321 Brin. For
valves, cutlery, surgical instruments; stainless, hardenable.
Obsolete

K.E. 4L

Sanderson Kayser Ltd.
high C, high Cr, bal Fe.
For springs; oil hardened.

K.E. 581

Sanderson Kayser Ltd.
C, bal Fe.
Rolled: 67,000 TS. For screw machine products; free-cutting.
Obsolete

K.E. 595

Sanderson Kayser Ltd.
C 1.25, Mn 0.85, Cr 1.2, W 1.3, bal Fe.
Oil hardening alloy tool steel; for cold drawing and swaging
dies; oil or water hardening.

K.E. 672

Sanderson Kayser Ltd.
C 0.95, Mn 1.2, Cr 0.55, W 0.65, V 0.15, bal Fe.
Oil hardening alloy tool steel. AISI O1. For press and blanking
tools.

K.E. 805

Sanderson Kayser Ltd.
C 0.38, Cr 1, Ni 1.5, Mo 0.2, bal Fe.
Heat treated: 121,000-280,000 TS; 90,000-246,000 YS; 10-26
El; 35-68 RA; 250-515 Brin. For gears, machinery parts,
shafts, machinery tools, fasteners. Oil hardening, wear
resistant.

K.E. 839

Sanderson Kayser Ltd.
C 1.05, Mn 0.5, Cr 1.4, bal Fe.
Oil hardening alloy tool steel; for ball bearings, cutters,
gages, taps, swaging dies, shear blades. AISI L1.

K.E. 896

Sanderson Kayser Ltd.
C 0.5, Mn 0.65, Cr 1.1, V 0.2, bal Fe.
For arbors, axles, boring bars, connecting rods. Type L2 tool
steel, oil hardening, shock resistant.

K.E. 897

Sanderson Kayser Ltd.
C 0.3, Cr 1.25, Ni 4.25, Mo 0.2, bal Fe.
Heat treated: 224,000-270,000 TS; 10-14 El; 30-45 RA;
444-512 Brin. For gears, shafts, bolts, fasteners. Oil
hardening, shock resistant.

K.E. 954

Sanderson Kayser Ltd.
C, Co, Cr, bal Fe.
Air hardened: 600-627 Brin. For valves for air-cooled engines;
air-hardened. *Obsolete*

K.E. 960

Sanderson Kayser Ltd.
C 0.5, Si 0.65, Cr 1.5, W 2.25, V 0.2, bal Fe.
Extra tough alloy oil hardening tool steel. For shear blades,
pneumatic tools, stamping and riveting tools. AISI S1.

K.E. 965

Sanderson Kayser Ltd.
C 0.4, Cr 13, Ni 13, W 3, Si 1.5.
Heat treated: 152,000 TS; 27 El; 49 RA; 269 Brin. For valves,
supercharger buckets; non-distorting. *Obsolete*

K.E. 970

Sanderson Kayser Ltd.
C 2.1, Cr 13.5, W 0.5, bal Fe.
Chrome heavy duty tool steel; for plug and ring gages,
punches, lamination and blanking dies. AISI D3.

K.E. CARLISLE

Sanderson Kayser Ltd.
C 1, bal Fe.
For taps, drills, reamers, general tools; water hardened.
Obsolete

K.E. DIAMOND

Sanderson Kayser Ltd.
C 1.1, bal Fe.
For taps, drills, reamers, general tools; water hardened.

K.E. DIAMOND NO. 10

Sanderson Kayser Ltd.
C 1.5, Mn 0.4, Cr 0.6, W 5.75, bal Fe.
High carbon tungsten alloy tool steel. For reamer blades, roll
turning tools, form cutters; water hardenable.

K.E. EXTRA QUALITY

Sanderson Kayser Ltd.
C 1.1, bal Fe.
For taps, drills, reamers, general tools; water hardened.
Obsolete

K.E. H22

Sanderson Kayser Ltd.
C 0.1, Cr 18, Ni 8, bal Fe.
Rolled: 123,000 TS; 56,000 YS; 52 El; 57 RA; 228 Brin.
Annealed: 101,000 TS; 42,000 YS; 58 El; 65 RA; 163 Brin. For
chemical plant equipment; stainless, austenitic. *Obsolete*

K.E. KHRS

Sanderson Kayser Ltd.
C 0.4, bal Fe.
Rolled: 90,000 TS. For screw machine products; free-cutting.
Obsolete

K.E.127

Sanderson Kayser Ltd.
High C, high Cr, bal Fe.
For springs; oil hardened. *Obsolete*

K.E.144

Sanderson Kayser Ltd.
C, alloy, bal Fe.
For pneumatic chipping chisels; water hardened. *Obsolete*

K.E.156

Sanderson Kayser Ltd.
C, alloy, bal Fe.
For cold drawing and swaging dies; water hardened.
Obsolete

K.E.160

Sanderson Kayser Ltd.
C 1, bal Fe.
For taps, drills, reamers; water hardened. *Obsolete*

K.E.200

Sanderson Kayser Ltd.
C, Co, Cr, bal Fe.
For knurling and heading dies, press tools; air hardened.
Obsolete

K.E.226

Sanderson Kayser Ltd.
C 1, W 2, Cr 0.5, bal Fe.
For taps, drills, reamers, punches; water hardened. *Obsolete*

K.E.2301

Sanderson Kayser Ltd.
C 0.12, bal Fe.
Heat treated: 72,000-90,000 TS; 35-40 El; 55-65 RA; 156-170
Brin. For gears, pinions, shafts, worms; case hardened.
Obsolete

K.E.275

Sanderson Kayser Ltd.
C 0.4, W 10, bal Fe.
For die casting dies; hot work steel, oil hardened. *Obsolete*

K.E.287

Sanderson Kayser Ltd.
C 0.12, Ni 3, bal Fe.
Heat treated: 100,000-150,000 TS; 17-25 El; 45-55 RA;
235-286 Brin. For gears, pinions, worms, shafts; case
hardened, shock resistant. *Obsolete*

K.E.339

Sanderson Kayser Ltd.
C 0.4, W 10, bal Fe.
For die casting dies; hot work steel, oil hardened. *Obsolete*

K.E.40AM

Sanderson Kayser Ltd.
C 0.1, Cr 12-14, S, bal Fe.
Annealed: 75,000 TS; 40,000 YS; 35 El; 70 RA; 155 Brin. For
screw machine products, pinions, shafts; stainless, free-
cutting. *Obsolete*

K.E.41

Sanderson Kayser Ltd.
C 0.12, Ni 4.5, Cr, Mo, bal Fe.
Heat treated: 179,000-202,000 TS; 13-17 El; 40-50 RA;
375-390 Brin. For gears, pinions, shafts; case hardened,
shock resistant. *Obsolete*

K.E.484

Sanderson Kayser Ltd.
C 0.4, Ni, Cr, Mo, bal Fe.
For piercing and stamping dies, shears, die bolsters; hot work
steel, oil hardened. *Obsolete*

K.E.621

Sanderson Kayser Ltd.
C 1, bal Fe.
For press tools, drills, reamers, general tools; water
hardened. *Obsolete*

K.E.637

Sanderson Kayser Ltd.
C, alloy, bal Fe.
For gages, tools, taps, reamers, forming dies; non-deforming,
oil hardened. *Obsolete*

K.E.660

Sanderson Kayser Ltd.
C 0.12, Ni 5, Cr, bal Fe.
Heat treated: 139,000-162,000 TS; 14-18 El; 40-50 RA;
311-340 Brin. For gears, pinions, camshafts, cams; case
hardened, shock resistant. *Obsolete*

K.E.708

Sanderson Kayser Ltd.
C, alloy, bal Fe.
For cold drawing and swaging dies; water hardened.
Obsolete

K.E.753
Sanderson Kayser Ltd.
C 1, W 1.25, bal Fe.
For taps, drills, reamers, threading dies; water hardened.
Obsolete

K.E.795
Sanderson Kayser Ltd.
C 0.9, W, bal Fe.
For high temperature springs; oil or air hardened. *Obsolete*

K.E.933
Sanderson Kayser Ltd.
C 1.25, W 1, Cr 0.5, bal Fe.
For taps, drills, reamers, punches; water hardened. *Obsolete*

K.E.961
Sanderson Kayser Ltd.
C 1.5, Cr 12, bal Fe.
For plastic mold dies, hot and cold shears; oil or air hardened, resists abrasion. *Obsolete*

K.E.A. 108
Sanderson Kayser Ltd.
C 1, Mn 0.55, bal Fe, plus free cutting agent.
Free cutting water hardening carbon tool steel.

K.E.A. 138
Sanderson Kayser Ltd.
C 0.35, Cr 12, W 12, V 1, bal Fe.
Chrome-tungsten hot work tool steel. For pressure die casting dies, extrusion dies, punches. AISI H23.

K.E.A. 145
Sanderson Kayser Ltd.
C 0.38, Si 1.05, Cr 5.25, Mo 1.35, V 1, bal Fe.
5% chrome-vanadium-moly die casting steel. For die casting dies, extrusion dies, hot piercing and gripping dies. AISI H13.

K.E.A. 172
Sanderson Kayser Ltd.
C 0.7, Si 0.3, Mn 2, Cr 1, Mo 0.45, bal Fe.
Air hardening tool steel. AISI A6. *Obsolete*

K.E.A. 180
Sanderson Kayser Ltd.
C 1.55, Cr 12, Mo 0.85, V 0.3, bal Fe.
Heavy duty tool steel, air hardening. AISI D2.

K.E.A. 203
Sanderson Kayser Ltd.
C 0.85, Cr 17.5, Mo 0.55, bal Fe.
Heat treated: 280,000 TS; 270,000 YS; 3 El; 555 Brin. Annealed: 107,000 TS; 62,000 YS; 18 El; 220 Brin. For dental and surgical instruments, bearings, pivots, cutlery, valve parts. Type 440 B, hardenable, corrosion and wear resistant. *Obsolete*

K.E.A. 205
Sanderson Kayser Ltd.
C 1.4, Si 0.3, Mn 0.4, V 3.45, bal Fe.
Cold heading punch and die steel. For dies, punches, headers.

K.E.A. 207
Sanderson Kayser Ltd.
C 1.05, Cr 17, Mo 0.6, bal Fe.
Annealed: 110,000 TS; 70,000 YS; 15 El; B97 Rock. Heat treated: 285,000 TS; 275,000 YS; 2 El; 580 Brin. For bearings, surgical instruments, valve parts, nozzles, pivot pins. Type 440 C, hardenable, corrosion and wear resistant. *Obsolete*

K.E.A. 218
Sanderson Kayser Ltd.
C 0.32, Si 0.5, Mn 0.35, Cr 3, Mo 2.8, V 0.35, Co 3, bal Fe.
Hot work tool steel. B.S. 4659 Type BH10A.

K.E.A. 221
Sanderson Kayser Ltd.
C 0.09, Cr 18.5, Ni 10.5, Mo 2.7, bal Fe.
Annealed: 85,000 TS; 35,000 YS; 50 El; 165 Brin. Cold rolled: 150,000 TS; 135,000 YS; 6 El; 32 Rock C. For chemical and pharmaceutical equipment, digesters, valve trim tanks. AISI 316, stainless, austenitic, acid resistant.

K.E.A. 227
Sanderson Kayser Ltd.
C 0.5, Mn 1.2, Cr 0.65, Mo 0.2, bal Fe.
Easy machining bolster steel. For molds, press components.

K.E.A. 23
Sanderson Kayser Ltd.
C 0.07, Cr 18, Ni 9, Ti, bal Fe.
Weldable austenitic stainless; for welded structures and for operation at elevated temperatures.

K.E.A. 231
Sanderson Kayser Ltd.
C 0.7, Cr 18.25, Ni 8.5, bal Fe.
Annealed: 85,000 TS; 35,000 YS; 60 El; 150 Brin. Cold drawn: 180,000 TS; 125,000 YS; 10 El; C35 Rock. For architectural trim, welded structures, kitchen equipment, chemical and textile plant equipment. Type 304, austenitic, stainless. *Obsolete*

K.E.A. 28
Sanderson Kayser Ltd.
C 0.45, Cr 13.25, Ni 1, bal Fe.
Martensitic corrosion resistant steel, hardenable. For plastic molds.

K.E.A. 505
Sanderson Kayser Ltd.
C 0.17, Cr 17, Ni 1.75, Se, bal Fe.
Annealed: 125,000 TS; 95,000 YS; 20 El; 260 Brin. Cold drawn: 130,000 TS; 110,000 YS; 15 El; 270 Brin. For marine hardware, pump shafts, valve trim, aircraft structural components. AISI 431 Se, corrosion resistant, free cutting.

K.E.A. 507
Sanderson Kayser Ltd.
C 0.09, Cr 18, Ni 8.5, Se, bal Fe.
Free cutting austenitic stainless steel. For stainless parts requiring appreciable machining. AISI 303 Se.

K.E.A. 508
Sanderson Kayser Ltd.
C 0.2, Cr 13, Se, bal Fe.
Annealed: 95,000 TS; 50,000 YS; 25 El; 196 Brin. Cold drawn: 105,000 TS; 85,000 YS; 17 El; 215 Brin. For cutlery, marine hardware, valve trim, gears, ball bearings, gauges. AISI 420 Se, corrosion resistant. Heat treatable. Free-cutting. *Obsolete*

K.E.A. 521
Sanderson Kayser Ltd.
C 0.07, Cr 18, Ni 11.5, Mo 2.65, Se, bal Fe.
Free machining austenitic stainless steel with very good corrosion resistance. For machined parts in food, beverage, laundry plants. AISI 316 Se.

K.E.A.147
Sanderson Kayser Ltd.
C 1.5, Cr 12, V, Mo, bal Fe.
For blanking and forming dies, plug gages; oil or air hardened, non-deforming. *Obsolete*

K.E.X.369
Sanderson Kayser Ltd.
C 0.5, Cr 5.5, W, Mo, bal Fe.
For die casting dies; hot work steel, air hardened. *Obsolete*

K.L.S. 19 HOT WORK
Manufacturer not listed.
C 0.4, W 3, bal Fe.
For hot work tools. *Obsolete*

K.P. STYRIAN
English manufacture
C 1.15, Mn 0.24, Cr 0.04, W 4.67, bal Fe.
For tools, dies; oil hardened.

K.R. ALLOY
Carpenter Technology Corp.
C 0.5, Cr 3, bal Fe.
For hot work dies. *Obsolete*

K1022
Eagle & Globe Steel Ltd.
Carbon steel. C 0.18-0.23, Mn 0.7-1, P 0-0.05, S 0-0.05, Si 0.1-0.35, bal Fe.
478 MPa TS; 13 El. Fully killed steels are required and recommended for specialized work and heat treatment. BSS970 EN Series EN3D.

K1030
Eagle & Globe Steel Ltd.
Carbon steel. C 0.28-0.34, Mn 0.6-0.9, P 0-0.05, S 0-0.05, Si 0.1-0.35, bal Fe.
555 MPa TS; 9 El. Fully killed steels are required and recommended for specialized work and heat treatment. BSS970 EN Series EN6A.

K1040
Eagle & Globe Steel Ltd.
Carbon steel. C 0.37-0.44, Mn 0.6-0.9, P 0-0.05, S 0-0.05, Si 0.1-0.35, bal Fe.
620 MPa TS; 9 El. Fully killed carbon steel. Medium high tensile steel for shafts and parts. Responds to heat treatment. BSS970 EN Series EN8; BSS970 Part 1 1972 080M40.

K1045
Eagle & Globe Steel Ltd.
Carbon steel. C 0.45, Mn 0.75, Si 0.25, P 0-0.03, S 0-0.035, bal Fe.
For shafts, bolts and machine parts. AS1442 K1045; BS970 080A42/47; AISI 1045; BS970 EN8D/EN43D; Werkstoff 1.1191.

K1045
Eagle & Globe Steel Ltd.
Carbon steel. C 0.43-0.5, Mn 0.6-0.9, P 0-0.05, S 0-0.05, Si 0.1-0.35, bal Fe.
650 MPa TS; 8 El. Fully killed carbon steels are required and recommended for specialized work and heat treatment. BSS970 EN Series EN43B; BSS970 Part 1 1972 080A47.

K1050
Eagle & Globe Steel Ltd.
Carbon steel. C 0.48-0.55, Mn 0.6-0.9, P 0-0.05, S 0-0.05, Si 0.1-0.35, bal Fe.
680 MPa TS; 8 El. Fully killed carbon steels are required and recommended for specialized work and heat treatment. BSS970 EN Series EN43C.

K1050
Eagle & Globe Steel Ltd.
Carbon steel. C 0.48-0.55, Mn 0.6-0.9, Si 0.1-0.35, P 0-0.05, S 0-0.05, bal Fe.
For automotive forgings. AS1442 K1050; BS970 080A52 (EN43C); AISI 1050.

K1070S MA1100
Eagle & Globe Steel Ltd.
Carbon steel. C 0.65-0.75, P 0-0.05, S 0-0.05, Mn 0.6-0.9, Si 0.1-0.35, bal Fe.
Spring steel bars. 725 MPa TS; 212 Brin.

K1082S
Eagle & Globe Steel Ltd.
Carbon steel. C 0.78-0.9, P 0-0.05, S 0-0.05, Mn 0.6-0.9, Si 0.1-0.35, bal Fe.
Spring steel bars. 790 MPa TS; 235 Brin.

K1137
Eagle & Globe Steel Ltd.
Alloy steel. C 0.32-0.39, Mn 1.35-1.65, P 0-0.05, S 0.08-0.13, Si 0.1-0.35, bal Fe.
670 MPa TS; 9 El. Free cutting carbon steel for stub axles, pins and bolts requiring high strength and wear resistance. Resulfurized to improve machinability. BSS970 EN Series EN15AM; BSS970 Part 1 1972 216M36.

K1146
Eagle & Globe Steel Ltd.
Alloy steel. C 0.42-0.49, Mn 0.7-1, P 0-0.05, S 0.08-0.13, Si 0.1-0.35, bal Fe.
650 MPa TS; 7 El. Free cutting carbon steel for stub axles, pins and bolts requiring high strength and wear resistance. Resulfurized to improve machinability.

K12
Friedr. Lohmann GmbH
C 2.1, Cr 11.5, bal Fe.
For blanking and forming dies, punches. Oil hardened, nondeforming. Type D3. T30403

K1A
Fansteel/Wellman Dynamics
Magnesium. Zr 0.4-1, bal Mg.
Magnesium casting alloy. Cast: 24,000 psi TS; 6,000 psi YS; 14 El.

K4-20
Texas Instruments Inc./Materials Control
TI CAD K4 with fine silver backing to allow bonding and brazing.

K5-20
Texas Instruments Inc./Materials Control
TI CAD K5 with fine silver backing to allow bonding and brazing.

KA HO LOY
Kahl-Holt Co.
C 0.07-0.1, Cu 0.2-0.3, Mn 0.35-0.5, bal Fe.
32,000-37,000 TS; 18,000-23,000 YS. For roofs; resists atmosphere corrosion.

KA HO LOY
Apollo Steel Co.
C 0.07-0.1, Cu 0.2-0.3, Mn 0.35-0.5, bal Fe.
32,000-37,000 TS; 18,000-23,000 YS. For roofs; resists atmosphere corrosion.

KABEL OWS
Stahlwerke Kabel, C.
C 0.38, Si, Cr, V, bal Fe.
For bolts, springs, crankshafts; oil hardened, shock resistant.

KABEL SBN
Stahlwerke Kabel, C.
C 1.1, Si 0.2, Mn 0.3, Cr 0.4, bal Fe.
For springs, bearings, drills, cutters; water or oil hardened, wear resistant.

KAISALOY 42 CV
Kaiser Steel
C 0.2, Mn 0.9, Si 0.3, 0.01 Cb or V min, bal Fe.
Plate: 60,000 TS; 42,000 YS min; 24 El; 140 Brin. For bridges, booms, mine and railroad cars, derricks. High strength-low alloy steels. SAE J410C, Gr. 942X.

KAISALOY 45-CV
Kaiser Steel
C 0.22, Mn 1, Si 0.3, 0.01 Cb min, bal Fe.
Plate: 60,000 TS; 22 El; 80 Rock B. For bridges, booms, truck and bus bodies, derricks, structures. High strength-low alloy steel. SAE J410C, Gr. 945X.

KAISALOY 45-FG
Kaiser Steel
C 0.12, Mn 0.6, Si 0.5, Cu 0.3, Mo 0.1, Cr 0.25, Ni 0.6, 0.02 V min, 0.005 Ti min, bal Fe.
60,000 psi TS min; 45,000 YS min; 25 El min. High ductility and strength; for drawn frame members, formed channels. Easily welded. SAE J410C, Grade 945A.

KAISALOY 50 MM
Kaiser Steel
C 0.27, Mn 1.1-1.6, Si 0.3, 0.02 Cu min, bal Fe.
63,000-70,000 psi TS min; 42,000-50,000 YS min depending on thickness; 21 El min. For booms, riveted bridges, mixers, earth moving equipment. SAE J410C, Grade 950C.

KAISALOY 50 MV
Kaiser Steel
C 0.22, Mn 0.85-1.25, Si 0.3, 0.20 Cu min, 0.20 V min, bal Fe.
60,000-70,000 TS min; 40,000-50,000 psi YS min depending on thickness; 21 El min. Good weldability; for welded bridges and buildings. SAE J410C, Grade 950B.

KAISALOY 50-CR
Kaiser Steel
C 0.2, Mn 1.25, Si 0.25-0.75, Cu 0.2-0.35, Mo 0.15, Cr 0.1-0.25, Ni 0.3-0.6, V 0.02-0.1, Ti 0.005-0.03, bal Fe.
70,000 TS min; 50,000 psi YS min; 21 El min. Good weldability, abrasion and corrosion resistance. For material handling equipment, truck and trailer equipment.

KAISALOY 50-CV
Kaiser Steel
C 0.22, Mn 1.1, Si 0.3, 0.01 Cb min, bal Fe.
Plate: 65,000 TS; 50,000 YS; 22 El; 80 Rock B. For booms, bridges, bus and truck bodies, derricks. High strength-low alloy steel. SAE J410C, Gr. 950X.

KAISALOY 55-CV
Kaiser Steel
C 0.25, Mn 1.35, Si 0.3, 0.01 Cb min, bal Fe.
Plate: 70,000 TS; 55,000 YS; 20 El; 160 Brin. For bridges, structures, mine cars, booms, bus and truck bodies. High strength, low-alloy steel. SAE J410C, Gr. 955X.

KAISALOY 60 SG
Kaiser Steel
C 0.2, Mn 1.25, Si 0.35, Cu 0.8, Mo 0.25, Ni 0.9, 0.005 V min, bal Fe.
80,000 psi TS min; 60,000 psi YS min; 20 El min. Weldable; good wear and impact resistance. For dump trucks, freight cars, coal chutes, cranes, barges.

KAISALOY 60-CV
Kaiser Steel
C 0.26, Mn 1.35, Si 0.3, 0.01 Cb min, bal Fe.
Plate: 75,000 TS; 60,000 YS; 18 El; 85 Rock B. For mine cars, bridges, booms, truck and bus bodies, derricks. High strength, low alloy steel. SAE J410 C, Gr, 960x.

KAISALOY AR
Kaiser Steel
C 0.35-0.5, Mn 1.5-2, Si 0.15-0.35, bal Fe.
200-280 Brin, normal. Good abrasion resistance; for parts requiring good surface wear as chutes, road machinery.

KAISALOY NO. 1 STANDARD
Kaiser Steel
C 0-0.25, Mn 0-1.5, Cu 0-0.35, Cr 0-0.25, Mo 0-0.15, Ni 0-0.5, bal Fe.
Rolled: 70,000 TS; 50,000 YS. For bus and tractor bodies, railroad and mine cars; good weldability. Obsolete

KAISALOY NO. 2
Kaiser Steel
C 0-0.12, Mn 0-0.8, Cu 0-0.35, Ni 0-0.3, Cr 0-0.25, Mo 0-0.15, 0.02% min V, bal Fe.
Rolled: 70,000 TS; 50,000 YS; 23 El; 90 Brin. For auto bumpers, trucks, tractors, cars; high strength, low alloy steel, good weldability. Obsolete

KAISALOY NO. 3
Kaiser Steel
C 0-0.3, Mn 0-1.5, Cu 0-0.35, Ni 0-0.3, Cr 0-0.25, Mo 0-0.1, V, bal Fe.
Rolled: 70,000 TS; 50,000 YS; 23 El; 90 Brin. For auto bumpers, trucks, tractors, cars; high strength, low alloy steel, good weldability. Obsolete

KAISALOY SPECIAL FORMING
Now KAISALOY 45 FG.

KAISER 2 S
Kaiser Aluminum & Chemical Corp.
99% min Al.
S O-temper: 13,000 TS; 5,000 YS; 35 El. H 14-temper: 18,000 TS; 17,000 YS; 9 El. H 18-temper: 23,000 TS; 21,000 YS; 5 El. For gaskets, spun products. Obsolete

KAISER 24 S
Kaiser Aluminum & Chemical Corp.
Cu 4.5, Mn 0.6, Mg 1.5, bal Al.
O-temper: 26,000 TS; 11,000 YS; 19 El. T 4-temper: 64,000 TS; 44,000 YS; 18 El. T 6-temper: 69,000 TS; 54,000 YS; 11 El. For aircraft structures; heat treatable, high strength. Obsolete

KAISER 3 S
Kaiser Aluminum & Chemical Corp.
Mn 1-2, bal Al.
S O-temper: 17,000 TS; 7,000 YS; 34 El. S 14-temper: 23,000 TS; 22,000 YS; 8 El. S 18-temper: 29,000 TS; 27,000 YS; 4 El. For cooking utensils, building products; strain hardenable. Obsolete

KAISER 52 S
Kaiser Aluminum & Chemical Corp.
Mg 2.5, Cr 0.25, bal Al.
S O-temper: 28,000 TS; 12,000 YS; 25 El. S 14-temper: 37,000 TS; 31,000 YS; 10 El. S 18-temper: 41,000 TS; 36,000 YS; 7 El. For fan blades, pressure vessels; strain hardenable. Obsolete

KAISER 61 S
Kaiser Aluminum & Chemical Corp.
Mg 1, Cu 0.25, Si 1, Cr 0.25, bal Al.
O-temper: 18,000 TS; 8,000 YS; 22 El. T 4-temper: 35,000 TS; 22,000 YS; 20 El. T 6-temper: 47,000 TS; 41,000 YS; 12 El. For boats, containers, structures; heat treatable, good formability. Obsolete

KAISER 75 S
Kaiser Aluminum & Chemical Corp.
Zn 5.6, Mg 2.5, Cu 1.6, Cr 0.3, bal Al.
O-temper: 33,000 TS; 15,000 YS; 17 El. T-6 temper: 82,000 TS; 72,000 YS; 11 El. For aircraft structures; heat treatable, high strength. Obsolete

KAISER K2
Kaiser Aluminum & Chemical Corp.
Al-Mg-S.
Alloy aluminum casting. Develops integral color anodic coatings when properly anodized. Kalcolor trade name.

KAISER K3
Kaiser Aluminum & Chemical Corp.
Al-Mg-Si-Cr.
Alloy aluminum casting. Develops clear anodic coatings when properly anodized. Kalcolor trade name.

KAISER MO-B STEEL
Kaiser Steel
C 0-0.15, Mo 0.4-0.6, Mn 0-0.6, 0.001% min B, bal Fe.
Hot rolled: 75,000 TS; 65,000 YS; 15 El; 50 RA. For conveyor and crane buckets, jet engine components; low alloy, high strength. Obsolete

KAISER RP-1
Kaiser Aluminum & Chemical Corp.
Si 0.12-0.15, Fe 0.5-0.7, bal Al.
For rotors in large electric motors. 56% min electrical conductivity. Centrifugally cast. Obsolete

KAISER RP-S
Kaiser Aluminum & Chemical Corp.
Si 0.08-0.1, Fe 0-0.2, bal Al.
For rotors in small electric motors. 59% IACS min electrical conductivity. Centrifugally cast. *Obsolete*

KALCHOIDS
Manufacturer not listed.
Cu-Sn-Zn.
For bearings.

KALIF METAL
Hydril Co.
Cu 70, Pb 30, steel backed lead-copper mixture.
10,000 TS; 6 El; 25 Brin. For bearings, wrist pin bushings; steel backed bearings. *Obsolete*

KALIF METAL
Kalif Corp.
Cu 70, Pb 30, steel backed lead-copper mixture.
10,000 TS; 6 El; 25 Brin. For bearings, wrist pin bushings; steel backed bearings. *Obsolete*

KALKOS
Latrobe Steel Co.
C 0.3, Mn 0.35, Cr 12, W 12, V 0.9, Si 0.5, bal Fe.
For hot forming and working tools; hot work steel. *Obsolete*

KALLOY
Dunford Hadfields Ltd.
C 0.4, Cr 30, bal Fe.
For furnace parts, skids, retorts, carburizing and annealing boxes, baffle plates; heat resistant to 100 C. *Obsolete*

KALOR
Uddeholm Corp.
C 0.3, Cr 1.25, Ni 4.25, W 5.5, V 0.15, bal Fe.
For extrusions, dies, rams and liners; hot work steel, oil hardened. *Obsolete*

KALOR 2
Uddeholm Corp.
C 0.3, Si 1, Cr 1.1, V 0.18, W 3.75, bal Fe.
For extrusion press dies, liners and rams, punches; oil hardened, tough. *Obsolete*

KAMARSCH BEARING
Manufacturer not listed.
Sn 71, Sb 7.2, Cu 21.
For bearings; anti-friction.

KAMARSCH BEARING
Manufacturer not listed.
Sn 85, Sb 5, Cu 3.6, Bi 1.6, Zn 1.4.
For bearings; anti-friction.

KAMARSCH BEARING
Manufacturer not listed.
Sn 71, Sb 20, Cu 9.5.
For bearings; anti-friction.

KANTHAL A
Kanthal Corp.
C 0.05, Cr 22, Al 5, Co 0-1, bal Fe.
Annealed: 106,000-114,000 TS; 80,000 YS; 20 El; 65 RA; 230 Brin. For heating elements, furnaces; resists oxidation to 1330°C.

KANTHAL A
Kanthal A.B.
C 0.05, Cr 22, Al 5, Co 0-1, bal Fe.
Annealed: 106,000-114,000 TS; 80,000 YS; 20 El; 65 RA; 230 Brin. For heating elements, furnaces; resists oxidation to 1330°C.

KANTHAL A-1
Kanthal Corp.
C 0.05, Cr 22, Al 5.5, Co 0-1, bal Fe.
Annealed: 114,000 TS; 80,000 YS; 20 El; 65 RA; 230 Brin. For heating elements, furnaces; resists oxidation to 1375°C.

KANTHAL A-1
Kanthal A.B.
C 0.05, Cr 22, Al 5.5, Co 0-1, bal Fe.
Annealed: 114,000 TS; 80,000 YS; 20 El; 65 RA; 230 Brin. For heating elements, furnaces; resists oxidation to 1375°C.

KANTHAL DR
Kanthal A.B.
Cr 20, Al 4.5, Co 0.5, bal Fe.
Annealed: 100,000-140,000 TS; 3-12 El. For resistors, potentiometers; high electrical resistance. *Obsolete*

KANTHAL DR
Kanthal Corp.
Stainless Steel. Cr 20, Al 4.5, Co 0.5, bal Fe.
Annealed: 100,000-140,000 TS; 3-12 El. For resistors, potentiometers; high electrical resistance. *Obsolete*

KANTHAL DR
Kanthal Corp.
Cr 20, Al 4.5, Co 0.5, bal Fe.
Annealed: 100,000-140,000 TS; 3-12 El. For resistors, potentiometers; high electrical resistance. *Obsolete*

KANTHAL DR
Kanthal Corp.
Ni 29, Co 17, bal Fe.
For heating elements; high heat resistance. *Obsolete*

KANTHAL DSD
Kanthal Corp.
C 0.05, Cr 22, Al 4.5, Co 0-1, bal Fe.
Annealed: 114,000 TS; 80,000 YS; 20 El; 65 RA; 230 Brin. For heating elements, furnaces; resists oxidation to 1280°C.

KANTHAL DSD
Kanthal A.B.
C 0.05, Cr 22, Al 4.5, Co 0-1, bal Fe.
Annealed: 114,000 TS; 80,000 YS; 20 El; 65 RA; 230 Brin. For heating elements, furnaces; resists oxidation to 1280°C.

KANTHAL SUPER
Kanthal Corp.
Mo Si_2 + Si O_2.
For heating and resistance elements for service up to 1800°C.

KANTHAL SUPER
Kanthal A.B.
Mo Si_2 + Si O_2.
For heating and resistance elements for service up to 1800°C.

KANTHAL-DL
Kanthal Corp.
Al 4.5, Cr 22, Co 0.5, bal Fe.
For toaster heating elements. Corrosion and heat resistant. Max operating temperature 2200 F. *Obsolete*

KAPO
Manufacturer not listed.
C 2.2, Cr 12, V 1, bal Fe.
Hardened: C 60-66 Rock. For blanking and forming dies, burnishing tools and rolls, plug gauges, slitting cutters, drawing dies. Type D3 oil hardening cold work steel, deep hardening, abrasion resistant. *Obsolete*

KARABUK
Turkish manufacture
C 0.12-0.65, Mn 0.35-0.85, Si 0.15-1, bal Fe.
For structures, rails, gears, shafts, machinery parts. Constructional steel.

KARMA
Driver Harris Co.
Nickel. Cr 20, Al 3, Fe 3, bal Ni.
Heat treated: 150,000 TS; 35 El. For resistances, shunts, potentiometers; high specific resistance; vacuum melted. *Obsolete*

KARONI 10-18 MO
Stahlwerke Kabel, C.
C 0-0.1, Cr 18, Ni 9.5, Mo 2, Si 2.2, bal Fe.
Annealed: 85,000 TS; 35,000 YS; 50 El; 65 RA; 160 Brin. Cold drawn: 150,000 TS; 135,000 YS; 6 El; 300 Brin. For acid resistant chemical plant equipment; Type 316; stainless, austenitic.

KARONI 10-18 MO EXTRA
Stahlwerke Kabel, C.
C 0-0.12, Cr 18, Mo 2, Ni 10.5, Ti = 4 x C, bal Fe.
Annealed: 90,000 TS; 40,000 YS; 45 El; 60 RA; 180 Brin. For welded chemical plant equipment; Type 316 Ti; corrosion resistant.

KARONI 10-18 MO SU
Stahlwerke Kabel, C.
C 0-0.07, Cr 18, Ni 10.5, Mo 2, bal Fe.
Annealed: 85,000 TS; 35,000 YS; 50 El; 65 RA; 160 Brin. For acid resistant chemical plant equipment; Type 316; stainless, austenitic.

KARONI 10-18 MOCU EXTRA
Stahlwerke Kabel, C.
C 0-0.07, Si 0.4, Ni 17.5, Cr 17.5, Cu 2, Mo 2, Ti = 7 x C, bal Fe.
For valves, pumps, chemical plant equipment; corrosion and heat resistant.

KARONI 12-12 SU
Stahlwerke Kabel, C.
C 0-0.1, Cr 12.5, Ni 12, bal Fe.
For valves, pumps, turbine parts; corrosion and heat resistant.

KARONI 13-17 MO SU
Stahlwerke Kabel, C.
C 0-0.07, Cr 17, Ni 13, Mo 4.75, bal Fe.
Annealed: 85,000 TS; 35,000 YS; 50 El; 65 RA; 160 Brin. For acid resistant chemical plant equipment; Type 317; stainless, austenitic.

KARONI 17-17 MOCU EXTRA
Stahlwerke Kabel, C.
C 0-0.07, Cr 17.5, Ni 17.5, Mo 2, Cu 2, Ti = 7 x C, bal Fe.
For valves, pumps, turbine parts; corrosion resistant.

KARONI 20-13
Stahlwerke Kabel, C.
C 0-0.12, Cr 13, bal Fe.
Annealed: 75,000 TS; 40,000 YS; 35 El; 70 RA; 155 Brin. Cold drawn: 100,000 TS; 85,000 YS; 17 El; 60 RA; 205 Brin. For pumps, valves, turbine and jet parts; Type 310; corrosion resistant.

KARONI 20-13S
Stahlwerke Kabel, C.
C 0-0.12, Cr 13, S 0.2, bal Fe.
Annealed: 75,000 TS; 40,000 YS; 35 El; 70 RA; 155 Brin. For screw machine products, shafts, cutlery; Type 410F; corrosion resistant.

KARONI 20-16
Stahlwerke Kabel, C.
C 0.08, Cr 17, bal Fe.
Annealed: 95,000 TS; 50,000 YS; 25 El; 55 RA; 196 Brin. Cold drawn: 105,000 TS; 85,000 YS; 17 El; 50 RA; 215 Brin. For furnace parts, heat treating boxes, baffles; Type 430; corrosion resistant.

KARONI 20-16 MOS
Stahlwerke Kabel, C.
C 0.12, Cr 16.5, Mo 0.25, S 0.2, bal Fe.
Annealed: 95,000 TS; 50,000 YS; 25 El; 55 RA; 195 Brin. For screw machine products, shafts; Type 430F; corrosion resistant.

KARONI 210-15 MO
Stahlwerke Kabel, O.
C 1.1, Cr, Mo, S, bal Fe.
For bearings, liners, sleeves; water hardened.

KARONI 40-13
Stahlwerke Kabel, C.
C 0.2, Si 0.4, Cr 13, bal Fe.
Annealed: 95,000 TS; 50,000 YS; 25 El; 55 RA; 195 Brin. Cold drawn: 105,000 TS; 85,000 YS; 17 El; 50 RA; 215 Brin. For turbine blades, cutlery, valves, knives; Type 420; corrosion resistant.

KARONI 40-17
Stahlwerke Kabel, C.
C 0.22, Cr 17, Ni 1.5, bal Fe.
Annealed: 125,000 TS; 95,000 YS; 20 El; 55 RA; 260 Brin. Cold drawn: 130,000 TS; 110,000 YS; 15 El; 35 RA; 270 Brin. For marine hardware, valves, pumps; Type 431; corrosion resistant.

KARONI 40.13 MO
Stahlwerke Kabel, C.
C 0.2, Cr 13, Mo 1.2, bal Fe.
Annealed: 95,000 TS; 40,000 YS; 25 El; 55 RA; 92 Rock B. Heat treated: 240,000 TS; 200,000 YS; 9 El; 26 RA; 50 Rock C. For cutlery, knives, surgical instruments, oil refinery and chemical plant equipment. Corrosion resistant, hardenable.

KARONI 70-16 MO
Stahlwerke Kabel, C.
C 0.35, Si 0.4, Cr 16.5, Mo 1.15, bal Fe.
For chemical plant equipment, valves, pumps; corrosion resistant.

KARONI 8-18
Stahlwerke Kabel, C.
C 0-0.15, Cr 18, Ni 8.5, Si 0.4, bal Fe.
Annealed: 80,000 TS; 35,000 YS; 55 El; 75 RA; 150 Brin. Cold drawn: 180,000 TS; 150,000 YS; 10 El; 250 Brin. For chemical plant equipment, tanks, mixers, filters; stainless, austenitic; Type 302.

KARONI 8-18 EXTRA
Stahlwerke Kabel, C.
C 0-0.12, Cr 18, Ni 19.5, Ti = 4 x C, bal Fe.
For welded chemical plant equipment; corrosion and heat resistant.

KARONI 8-18 SU
Stahlwerke Kabel, C.
C 0-0.07, Si 0.4, Cr 18, Ni 9.5, bal Fe.
Annealed: 85,000 TS; 35,000 YS; 60 El; 70 RA; 150 Brin. Cold drawn: 180,000 TS; 125,000 YS; 10 El; 330 Brin. For welded chemical plant equipment; Type 304; stainless, austenitic.

KARONI 80-13
Stahlwerke Kabel, C.
C 0.4, Si 0.4, Cr 13, bal Fe.
Annealed: 95,000 TS; 50,000 YS; 25 El; 55 RA; 195 Brin. For turbine blades, cutlery, valves; Type 420; stainless.

KARONI 80-13 EXTRA
Stahlwerke Kabel, C.
C 0.85, Cr, V, bal Fe.
For bearings, liners; water hardened.

KASLE KA-2
Kasle Steel Co.
C 1, Mn 0.7, Si 0.3, Cr 5.25, V 0.3, Mo 1.1, bal Fe.
Type A2 air hardening tool steel (also available in free-machining grade). *Obsolete*

KASLE KA-6
Kasle Steel Co.
C 0.7, Mn 2.1, Si 0.3, Cr 1, Mo 1.3, bal Fe.
Type A6 air hardening tool steel. *Obsolete*

KASLE KA-9
Kasle Steel Co.
C 0.5, Mn 0.35, Si 0.3, Cr 5, Ni 1.5, V 1, Mo 1.4, bal Fe.
Type A9 air hardening tool steel. *Obsolete*

KASLE KA8-2H
Kasle Steel Co.
C 0.5, Mn 0.3, Si 1, Cr 5, V 1.4, Mo 1.35, W (optional), bal Fe.
Type A8 air hardening tool steel. *Obsolete*

KASLE KD-2
Kasle Steel Co.
C 1.55, Mn 0.3, Si 0.3, Cr 12, V 0.9, Mo 0.8, bal Fe.
Air or oil hardening high carbon, high chromium cold work tool steel. AISI D2. *Obsolete*

KASLE KD-3
Kasle Steel Co.
C 2, Mn 0.3, Si 0.4, Cr 12, bal Fe.
Air or oil hardening high carbon, high chromium cold work tool steel. AISI D3. *Obsolete*

KASLE KD-5
Kasle Steel Co.
C 1.5, Mn 0.4, Si 0.4, Cr 12.5, Ni 0.5, V 0.5, Mo 0.85, Co 3.3, bal Fe.
Air or oil hardening high carbon, high chromium cold work tool steel. AISI D5. *Obsolete*

KASLE KD-6
Kasle Steel Co.
C 2.1, Cr 12.5, V 0.15, W 1.3, bal Fe.
Air or oil hardening high carbon, high chromium cold work tool steel; AISI D6. *Obsolete*

KASLE KD-7
Kasle Steel Co.
C 2.4, Mn 0.4, Si 0.4, Cr 12.5, V 4, Mo 1.1, bal Fe.
Air or oil hardening high carbon, high chromium cold work tool steel. AISI D7. *Obsolete*

KASLE KF-2
Kasle Steel Co.
C 1.3, Mn 0.28, W 3.5, bal Fe.
Special purpose tool steel. AISI F2. *Obsolete*

KASLE KH-11
Kasle Steel Co.
C 0.35, Mn 0.4, Si 1, Cr 5, V 0.4, Mo 1.5, bal Fe.
Air or oil hardening hot work tool steel, chromium type. AISI H11. *Obsolete*

KASLE KH-12
Kasle Steel Co.
C 0.36, Mn 0.4, Si 1.1, Cr 5, V 0.35, W 1.3, Mo 1.5, bal Fe.
Air or oil hardening hot work tool steel, chromium type. AISI H12. *Obsolete*

KASLE KH-13
Kasle Steel Co.
C 0.4, Mn 0.4, Si 1.1, Cr 5, V 1.1, Mo 1.3, bal Fe.
Air or oil hardening hot work tool steel, chromium type. AISI H13. *Obsolete*

KASLE KH-19
Kasle Steel Co.
C 0.4, Mn 0.3, Si 0.3, Cr 4.25, V 2.2, W 4.25, Mo 0.4, Co 4.25, bal Fe.
Air or oil hardening hot work tool steel, chromium type (Cr-V-W-Co). AISI H19. *Obsolete*

KASLE KH-21
Kasle Steel Co.
C 0.3, Mn 0.3, Si 0.4, Cr 3.5, V 0.35, W 9, bal Fe.
Air or oil hardening hot work tool steel, tungsten type. AISI H21. *Obsolete*

KASLE KH-26
Kasle Steel Co.
C 0.55, Mn 0.3, Si 0.25, Cr 4, V 1.1, W 18, bal Fe.
Air or oil hardening hot work tool steel, tungsten type. AISI H26. *Obsolete*

KASLE KHD
Kasle Steel Co.
Mn 0.25, Si 0.25, P 0-0.015, S 0-0.015, C as desired, bal Fe.
Type W1 water hardening tool steel. *Obsolete*

KASLE KM-1
Kasle Steel Co.
C 0.85, Cr 4, V 1.35, W 1.9, Mo 9, bal Fe.
High speed tool steel, molybdenum type. AISI M1. *Obsolete*

KASLE KM-10
Kasle Steel Co.
C 0.87, Cr 4, V 2, Mo 8, bal Fe.
High speed tool steel, molybdenum type. AISI M10. *Obsolete*

KASLE KM-2
Kasle Steel Co.
C 0.85, Cr 4, V 2, W 6.5, Mo 5, bal Fe.
High speed tool steel, Mo-W type. AISI M2. *Obsolete*

KASLE KM-2C
Kasle Steel Co.
C 1, Cr 4, V 2, W 6.5, Mo 5, bal Fe.
High speed tool steel, Mo-W type. AISI M2. *Obsolete*

KASLE KM-3
Kasle Steel Co.
C 1.05, Cr 4, V 2.6, W 6, Mo 5.75, bal Fe.
High speed tool steel, Mo-W type. AISI M3 (Class 1). *Obsolete*

KASLE KM-35
Kasle Steel Co.
C 0.83, Cr 4.25, V 1.9, W 6.4, Mo 5, Co 5, bal Fe.
High speed tool steel, molybdenum type, (Mo-Cr-W-Co). AISI M35. *Obsolete*

KASLE KM-3II
Kasle Steel Co.
C 1.2, Cr 4, V 3.25, W 6.25, Mo 6.25, bal Fe.
High speed tool steel, Mo-W type. AISI M3 (Class 2). *Obsolete*

KASLE KM-4
Kasle Steel Co.
C 1.3, Cr 4, V 4, W 5.5, Mo 4.5, bal Fe.
High speed tool steel, molybdenum type. (Mo-Cr-V-W). AISI M4. *Obsolete*

KASLE KM-42
Kasle Steel Co.
C 1.08, Cr 4, V 1.15, W 1.55, Mo 9.7, Co 8, bal Fe.
High speed tool steel, Mo-Co type. AISI M42. *Obsolete*

KASLE KM-50
Kasle Steel Co.
C 0.8, Cr 4, V 1, Mo 4.5, bal Fe.
High speed tool steel, molybdenum type. AISI M50. *Obsolete*

KASLE KM-7
Kasle Steel Co.
C 1, Cr 4, V 2, W 1.75, Mo 8.75, bal Fe.
High speed tool steel, molybdenum type. AISI M7. *Obsolete*

KASLE KNL
Kasle Steel Co.
C 1.7, Cr 12, V 1, W 0.5, Mo 0.6, bal Fe.
Air or oil hardening high carbon, high chromium tool steel. *Obsolete*

KASLE KO-1
Kasle Steel Co.
C 0.95, Mn 1.2, Si 0.3, Cr 0.5, V 0.2, W 0.5, bal Fe.
Type O1 oil hardening tool steel. *Obsolete*

KASLE KO-2
Kasle Steel Co.
C 0.9, Mn 1.7, Si 0.25, Cr 0.4, V 0.15, bal Fe.
Type O2 oil hardening tool steel. *Obsolete*

KASLE KO-6
Kasle Steel Co.
C 1.45, Mn 0.75, Si 1.15, Cr 0-0.25, Mo 0.25, bal Fe.
Type O6 oil hardening tool steel. *Obsolete*

KASLE KS-1
Kasle Steel Co.
C 0.5-0.6, Mn 0.35, Si 0.8, Cr 1.35, V 0.25, W 2.5, bal Fe.
Type S1 shock resisting tool steel. *Obsolete*

KASLE KT-1
Kasle Steel Co.
C 0.72, Cr 4.3, V 1.1, W 18, bal Fe.
High speed tool steel, tungsten type. AISI T1. *Obsolete*

KASLE KT-15
Kasle Steel Co.
C 1.5, Cr 4, V 5, W 12.5, Co 5, bal Fe.
High speed tool steel, tungsten type, (W-Cr-V-Co). AISI T5. *Obsolete*

KASLE KT-5
Kasle Steel Co.
C 0.76, Cr 4.3, V 1.6, W 18, Mo 0.9, Co 9.5, bal Fe.
High speed tool steel, tungsten-cobalt type. AISI T5. *Obsolete*

KASLE KW-1
Kasle Steel Co.
Mn 0.2, Si 0.2, P 0-0.02, S 0-0.02, C as desired, bal Fe.
Type W1 water hardening tool steel. *Obsolete*

KASLE KW-2
Kasle Steel Co.
C 1.03, Mn 0.25, Si 0.23, V 0.22, bal Fe.
Type W2 water hardening tool steel. *Obsolete*

KASLE KW-5
Kasle Steel Co.
C 1.1, Mn 0.3, Si 0.25, Cr 0.5, bal Fe.
Type W5 water hardening tool steel. *Obsolete*

KATADYN SILVER
German manufacture
Ag, Cu.
For jewelry; corrosion resistant.

KAWASAKI NTK-M7
Japan Metal Industry Co. Ltd.
C 0-0.06, Cr 9-11, Ni 16-18, Mo 6-8, bal Fe.
Hot rolled: 105,000 TS; 86,000 YS; 34 El; 44 RA. Annealed: 78,000 TS; 35,000 YS; 59 El; 61 RA. For valve and pump parts, impellers. Corrosion and heat resistant. *Obsolete*

KAWASAKI-HTP52W
Kawasaki Steel Corp.
C 0-0.18, Mn 0.8-1.4, Si 0.3-0.5, Cu 0-0.3, bal Fe.
Rolled: 74,000 min TS; 47,000 min YP; 16 min El. For railroad and mine cars, derricks, agricultural equipment, structures. Constructional steel, tough. *Obsolete*

KAWASAKI-KO
Kawasaki Steel Corp.
C 0.1-0.2, Mn 0.6-1, Cu 0.15-0.5, Ni 0.7-1, Cr 0.4-0.8, Mo 0.4-0.6, V 0.03-0.1, B 0.002-0.006, bal Fe.
Heat treated: 114,000-135,000 TS; 100,000 min YP; 18 min El. For railroad and mine cars, derricks, agricultural equipment. High-strength, low alloy construction steel, good fabricability and weldability. *Obsolete*

KAWASAKI-QT60A
Kawasaki Steel Corp.
C 0-0.12, Mn 0.6-1, Cu 0-0.4, Ni 0.4-0.7, Cr 0.4-0.7, Mo 0-0.2, V 0.3-0.6, bal Fe.
Heat treated: 85,000 min TS; 68,000 min YP; 16 min El. For railroad and mine cars, derricks, agricultural equipment. High-strength low alloy constructional steel. *Obsolete*

KAWASAKI-QT60B
Kawasaki Steel Corp.
C 0-0.12, Mn 1.1-1.5, Si 0.35-0.55, Cu 0-0.3, bal Fe.
Heat treated: 85,000 min TS; 65,000 min YP; 16 min El. For railroad and mine cars, derricks, agricultural equipment, structures, buildings. Constructional steel, tough. *Obsolete*

KAY-BRUNNER NIRESIST
Kay-Brunner Steel Products Inc.
TC 2.3, Ni 12-15, Cu 5-7, Cr, Mn, Si, bal Fe.
For heat and corrosion resistant parts; heat and corrosion resistant. *Obsolete*

KAYEM
Pasminco Europe (Mazak) Ltd.
Al 3.9-4.3, Cu 2.5-3.2, Mg 0.04-0.06, bal Zn.
360 MPa TS (as cast); 7 El (as cast). For gravity or pressure die casting.

KAYEM-1
French manufacture
Al, Cu, bal Zn.
For die castings.

KAYEM-2
French manufacture
Al, Cu, bal Zn.
For die castings.

KB 90
Sintered Products Ltd.
Infiltrated, sintered iron. 43 kg/mm^2 TS; 30 Rock B min. Not recommended for hardening. Meets SAE 870; ASTM B303-58T Class-A.

KB KELLEY BRONZE
Kelly Foundry & Machine Co.
Nickel aluminum bronze for glass molds.

KB90/ 3/4
Sintered Products Ltd.
Infiltrated, sintered iron. 45.5 kg/mm^2 TS; 240 Vickers. Hardenable to 35 Rock C approx. For small hand tools, high duty gears. Meets SAE 872; ASTM B303-58T Class C.

KBI 40 ALLOY
Haynes International, Inc.
Refractory. Ta 60, Cb 40.
Solution alloy for chemical processes. Density 0.437 lb/In.^{3}; 40,000 psi TS; 28,000 psi YS (0.2% offset); 25 El.

KBI 41 ALLOY
Haynes International, Inc.
Refractory. Cb 37.5, W 2.5, Mo 2, bal Ta.
At 300°C: 47,400 psi YS; at RT: 50,000 psi YS. Welded tube bundles for heat exchangers, patches for glass-lined tank tubing.

KBI-1
Haynes International, Inc.
Refractory. Zr 0.8-1.2, bal Cb.
Aerospace applications; sodium vapor lamps; 35,000 psi TS; 20,000 psi YS; 20 El.

KBI-10
Haynes International, Inc.
Refractory. W 9-11, bal Ta.
Aerospace and chemical process applications; 70,000 psi TS; 55,000 psi YS; 20 El.

KBI-3
Haynes International, Inc.
Refractory. Hf 9-11, Ti 0.7 1.3, bal Cb.
Aerospace applications.

KBI-40
Haynes International, Inc.
Refractory. Cb 35-40, bal Ta.
Interchangeable with Ta in chemical process equipment; 40,000 psi TS; 28,000 psi YS; 25 El.

KBI-6
Haynes International, Inc.
Refractory. W 2-3, Cb 0-0.5, bal Ta.
For chemical process industry tubing, piping, heat exchangers; 40,000 psi TS; 28,000 psi YS; 20 El.

KBV1
Sintered Products Ltd.
7.2 g/cm^3 density; 55 kg/mm^2 TS; 250 Vickers min. Good conductivity. For exhaust valve seats of internal combustion engines. Copper infiltrated sintered Fe.

KBV2
Sintered Products Ltd.
7.3 g/cm^3 density; 61.5 kg/mm^2 TS; 250 Vickers min; good conductivity. For exhaust valve seats. Copper infiltrated sintered Fe. Higher copper than KBV1.

KE-2258
Sanderson Kayser Ltd.
C 0.7, Si 0.5, Mn 0.4, Cr 8, bal Fe.
For valves for internal combustion engines; heat resistant. *Obsolete*

KE-241
Sanderson Kayser Ltd.
C 0.55, Si 3, Cr 8.5, W 2, bal Fe.
For valves for water cooled engines; resists scaling to 1000 C. *Obsolete*

KE798
Sanderson Kayser Ltd.
C 1, Mn 0.6, Cr 1.5, W 0.5, bal Fe.
For tools, dies, cutters; oil hardened. *Obsolete*

KE881
Sanderson Kayser Ltd.
C 0.7, W 18, Cr 4, V 1, bal Fe.
For lathe and planer tools, drills, taps, hobs, reamers; high speed steel; Type T1. *Obsolete*

KEEN-KUT
Associated Steel Corp.
C, alloy, bal Fe.
For hollow drills; non-tempering.

KEENE'S ALLOY
Manufacturer not listed.
Cu 75, Ni 16, Sn 2.8, Zn 2.3, Co 2, Fe 1.5, Al 0.5.
For corrosion resistant and ornamental parts; corrosion resistant.

KEEWATIN
Atlas Specialty Steels
C 0.9, Mn 1, Cr 0.5, W 0.5, bal Fe.
For dies, cutters, punches, reamers, broaches; oil hardened, non-deforming, Type O1.

KEINMAYER'S AMALGAM
Manufacturer not listed.
Hg 50, Sn 25, Zn 25.
For amalgam.

KEL-CAST
Ohio Stainless & Commercial Steel Co.
C 3.6, Si 1.25-1.5, Mn 0.7, Cr 0.3, bal Fe.
Cast: 45,000 TS; 200-220 Brin. For bearings, gears, valve seats; cast iron.

KEL-CAST
Kelly Foundry & Machine Co.
C 3.6, Si 1.25-1.5, Mn 0.7, Cr 0.3, bal Fe.
Cast: 45,000 TS; 200-220 Brin. For bearings, gears, valve seats; cast iron.

KEL-CAST MM
Ohio Stainless & Commercial Steel Co.
Cast Iron. C 3.25, Si 2.5, Mn 0.7, Cr 0.3, Mo 0.4, bal Fe.
Cast: 45,000 TS; 200-220 Brin. For bushings, cams, gears, and valves. Cast iron.

KELCALOY
M.W. Kellog
C alloy, bal Fe.
For oil refinery equipment; corrosion resistant.

KELCAST 5MM
Kelly Foundry & Machine Co.
Alloy gray iron castings, with Cr, Mo, V. For I.S. plungers, neck rings, bottom plates, baffles.

KELCAST NO. 10 DUCTILE IRON
Kelly Foundry & Machine Co.
Ductile iron casting. Used as press mold plunger for forming boro-silicate ware and container molds.

KELCAST NO. 4
Kelly Foundry & Machine Co.
Gray iron casting; Class 25. 25,000 psi TS min; 160-190 Brin. General purpose containers mold metal; for molds, baffles, blanks, and bottom plates. ASTM A48-56.

KELCAST NO. 5
Kelly Foundry & Machine Co.
Gray iron casting; Class 35. 35,000 psi TS min; 180-210 Brin. Good strength general purpose casting. ASTM A48-56.

KELCAST NO. 5 MMP
Kelly Foundry & Machine Co.
Gray iron casting; Class 45. 45,000 psi TS min; 190-235 Brin. Specially alloyed and processed for Hartford I.S. 28 paste mold blanks, plungers, half bars and ring stock. ASTM A48-56.

KELCAST NO. 6XXX
Kelly Foundry & Machine Co.
Alloy gray iron castings. Specially alloyed and processed for packers' ware, tumbler plungers.

KELCAST NO. 7 FIRE X
Kelly Foundry & Machine Co.
Alloy cast iron. Heat resistant iron used principally for paste-mold plungers and glazier burners.

KELCAST TYPE 3
Kelly Foundry & Machine Co.
Alloy gray iron castings with Mo, V. Long life container and press mold iron. For boro-silicate glass operation.

KELLY-IRON
Kelly Foundry & Machine Co.
C 3.2, Si 2, alloy, bal Fe.
For bearings, cams, gears, rolls, valves; cast iron. *Obsolete*

KELMET
Manufacturer not listed.
Cu 68-70.5, Pb 22.5-25.5, Sn 6.4-6.6, S 0.03, Zn 0.03, Ni 0.04.
For bearings; heavy duty.

KELNOD NO. 10
Kelly Foundry & Machine Co.
Ductile iron, Type 100-70-03. 100,000 psi TS; 70,000 psi YS; 3 El; 230-270 Brin.

KELNOD NO. 6
Kelly Foundry & Machine Co.
Ductile iron, Type 60-45-10. 60,000 psi TS; 45,000 psi YS; 10 El; 160-180 Brin. ASTM A339-55.

KELNOD NO. 8
Kelly Foundry & Machine Co.
Ductile iron, Type 80-60-03. 80,000 psi TS; 60,000 psi YS; 3 El; 200-250 Brin. ASTM A339-55.

KELOCK 1014
Sanderson Kayser Ltd.
C 0.7, Cr 4, W, Mo, V, bal Fe.
For lathe and planer tools, cutters; high speed steel.
Obsolete

KELOCK 1021
Sanderson Kayser Ltd.
C 0.7, W 18.2, Cr 4, Mo 0.75, V 1.45, Co 10, bal Fe.
Hardened: 64-67 Rock C. For lathe and planer tools, cutters, reamers, broaches, form cutters. High-speed steel, Type T5, high red-hardness.

KELOCK 237
Sanderson Kayser Ltd.
C 0.7, W 18, Cr 4, V 1, bal Fe.
Hardened: 64-67 Rock C. For die cores, lathe and planer tools, drills, chasers, reamers, broaches. Type T1 high-speed steel, high red-hardness.

KELOCK 788
Sanderson Kayser Ltd.
C 0.7, Cr 4, W 13, V 1, Mo 2, bal Fe.
For tools, dies, cutters, reamers, drills; high speed steel, oil hardened. *Obsolete*

KELOCK 795
Sanderson Kayser Ltd.
C 0.7, W 14, Cr 4, V 0.6, bal Fe.
Hardened: 64-67 Rock C. For lathe and planer tools, cutters, hot and cold shear blades, drills, reamers, punches. High-speed steel, high red-hardness.

KELOCK 873
Sanderson Kayser Ltd.
C 0.7, Cr 4.25, W 18.25, Mo 0.75, V 1, Co 5, bal Fe.
Hardened: 64-66 Rock C. For lathe and planer tools, cutters, reamers,broaches, drills, form tools. Type T4 high-speed steel, high red-hardness.

KELOCK A157
Sanderson Kayser Ltd.
C 0.8, W 6, Cr 4, V 2, Mo 5, bal Fe.
Hardened: 64-67 Rock C. For lathe and planer tools, drills, hobs, reamers, taps, milling cutters. High-speed steel, high red-hardness. AISI M2.

KELOCK A182
Sanderson Kayser Ltd.
C 0.8, Cr 3.9, Mo 8.5, W 1.5, V 1.15, bal Fe.
High speed tool steel. B.S. 4659 Type BM 1; AISI M1.

KELOCK A229
Sanderson Kayser Ltd.
C 1.05, Cr 3.75, W 1.5, Mo 9.5, V 1.15, Co 8.25, bal Fe.
High speed steel for cutting tools that require extra red hardness. AISI M42.

KELOCK A72
Sanderson Kayser Ltd.
C 0.87, Cr 4.25, W 18.75, Mo 0.5, V 1, Co 5.5, bal Fe.
Hardened: C64-66 Rock. For engraving tools, high duty turning tools, broaches, milling cutters, hobs. High speed steel, good red-hardness, wear and abrasion resistant.
Obsolete

KELVAN
Now ELECTRITE KELVAN.

KELVAN
Latrobe Steel Co.
C 0.88, W 1.7, Cr 3.75, V 1.15, Mo 9.6, Co 8.4, bal Fe.
Hardened: 63-67 Rock C. For lathe tools and other machine tools, very good red hardness and wear resistance. AISI Type M-33 high speed tool steel.

KEMLER
English manufacture
Zn 76, Cu 9, Al 15.
For ornamental parts.

KEMLET
Manufacturer not listed.
Zn 67, Al 15, Cu 9.
For bearings.

KEN A-7
K.C. Glader Co.
C 2.5, Mn 0.7, Cr 5, V 4.5, Mo 1.1, bal Fe.
AISI A7 air hardening tool steel. *Obsolete*

KEN AIR A-4
K.C. Glader Co.
C 0.95, Mn 2, Cr 2.2, 1.1% Mo (lead added), bal Fe.
AISI A6 air hardening tool steel. *Obsolete*

KEN AIR A-6
K.C. Glader Co.
C 0.7, Mn 2, Cr 1, Mo 1.3, bal Fe.
AISI A6 air hardening tool steel.

KEN CHROME D-2
K.C. Glader Co.
C 1.5, Cr 12, V 0.8, Mo 0.75, bal Fe.
Air or oil hardening cold work tool and die steel; AISI D2. High carbon, high chromium type.

KEN CHROME D-3
K.C. Glader Co.
C 2, Cr 12, bal Fe.
Air or oil hardening cold work tool and die steel; AISI D3. High carbon, high chromium type. *Obsolete*

KEN CHROME D-4
K.C. Glader Co.
C 2.25, Cr 12, V 0.2, Mo 1, bal Fe.
Air or oil hardening cold work tool and die steel; AISI D4. High carbon, high chromium type. *Obsolete*

KEN CHROME D-7
K.C. Glader Co.
C 2.4, Cr 12.25, V 3.75, Mo 1.1, bal Fe.
Oil or air hardening cold work tool and die steel; AISI D7. High carbon, high chromium type. *Obsolete*

KEN DIE A-2
K.C. Glader Co.
C 1, Mn 0.5, Si 0.25, Cr 5, V 0.3, Mo 1.25, bal Fe.
Air hardening and cold work tool steel; AISI A2.

KEN F-2
K.C. Glader Co.
C 1.3, Mn 0.28, W 3.5, bal Fe.
Carbon-tungsten special purpose tool steel; AISI F2.
Obsolete

KEN GRAPH OIL O6
K.C. Glader Co.
C 1.45, Mn 0.8, Si 1.15, Mo 0.25, bal Fe.
Graphitic type oil hardening cold work tool steel; AISI O6.

KEN H 13
K.C. Glader Co.
C 0.4, Mn 0.4, Si 1, Cr 5, V 1, Mo 1.5, bal Fe.
Air or oil hardening hot work tool and die steel, chromium type; AISI H13.

KEN H-11
K.C. Glader Co.
C 0.36, Mn 0.35, Cr 4.85, V 0.2, Mo 1.5, bal Fe.
Air or oil hardening hot work tool and die steel, chromium type; AISI H11. *Obsolete*

KEN H-12
K.C. Glader Co.
C 0.35, Mn 0.35, Si 1, Cr 5, V 0.2, W 1.15, Mo 1.5, bal Fe.
Air or oil hardening hot work tool and die steel, chromium type; AISI H12. *Obsolete*

KEN H-21
K.C. Glader Co.
C 0.3, Mn 0.3, Si 0.3, Cr 3, V 0.3, W 10, bal Fe.
Air or oil hardening hot work tool and die steel, tungsten type; AISI H21. *Obsolete*

KEN H-24
K.C. Glader Co.
C 0.45, Mn 0.3, Si 0.3, Cr 3, V 0.35, W 15, bal Fe.
Air or oil hardening tool and die steel, tungsten type; AISI H24. *Obsolete*

KEN H-26
K.C. Glader Co.
C 0.55, Cr 4, V 1.1, W 18, bal Fe.
Air or oil hardening tool and die steel, tungsten type; AISI H26. *Obsolete*

KEN H-43
K.C. Glader Co.
C 0.6, Cr 3.6, V 1.75, Mo 8.5, bal Fe.
Air or oil hardening tool and die steel, molybdenum type; AISI H43. *Obsolete*

KEN L-2
K.C. Glader Co.
C 0.45, Mn 0.55, Si 0.2, Cr 0.95, V 0.2, bal Fe.
Oil hardening special purpose tool steel, low alloy type; AISI L2. *Obsolete*

KEN L-6
K.C. Glader Co.
C 0.75, Mn 0.75, Cr 0.9, Ni 1.75, Mo 0.35, bal Fe.
Oil hardening special purpose tool steel, low alloy type; AISI L6. *Obsolete*

KEN LUSTRE MOLD
K.C. Glader Co.
C 0.55, Mn 1, Si 0.3, Cr 1.1, Mo 0.25, bal Fe.
Oil hardenable tool steel, mainly for molds.

KEN M-1
K.C. Glader Co.
C 0.78, Cr 3.8, V 1.15, W 1.5, Mo 8.7, bal Fe.
High speed tool steel, molybdenum type; AISI M1. *Obsolete*

KEN M-2
K.C. Glader Co.
C 0.84, Cr 4.2, V 1.95, W 6.35, Mo 5, bal Fe.
High speed tool steel, Mo-W type; AISI M2.

KEN M-3
K.C. Glader Co.
C 1.03, Cr 4, V 2.5, W 6, Mo 5.5, bal Fe.
High speed tool steel, Mo-W type; AISI M3 Class 1. *Obsolete*

KEN M-4
K.C. Glader Co.
C 1.28, Cr 4.5, V 4, W 5.5, Mo 4.5, bal Fe.
High speed tool steel, molybdenum type; (Mo-Cr-V-W); AISI M4. *Obsolete*

KEN M-41
K.C. Glader Co.
C 1.1, Cr 4, W 6.75, Mo 3.75, V 2, Co 5, bal Fe.
High speed tool steel, W-Mo-Co type; AISI M41. *Obsolete*

KEN OIL O-1
K.C. Glader Co.
C 0.09, Mn 1.2, Si 0.3, Cr 0.5, V 0.2, W 0.5, bal Fe.
Oil hardening tool steel; AISI O1.

KEN P-2
K.C. Glader Co.
C 0.06, Mn 0.3, Si 0.15, Cr 0.95, Mo 0.25, B, bal Fe.
Low carbon mold steel; to be carburized before hardening; AISI P2. *Obsolete*

KEN P-20
K.C. Glader Co.
C 0.3, Mn 0.8, Si 0.5, Cr 1.7, Mo 0.4, bal Fe.
Oil hardenable mold steel; AISI P20.

KEN P-4
K.C. Glader Co.
C 0.07, Mn 0.35, Si 0.2, Cr 4.5, Mo 0.5, bal Fe.
Low carbon mold steel; to be carburized before hardening; AISI P4. *Obsolete*

KEN P-6
K.C. Glader Co.
C 0.1, Mn 0.5, Si 0.25, Cr 1.5, Ni 3.5, bal Fe.
Low carbon mold steel; to be carburized before hardening; AISI P6. *Obsolete*

KEN PREHARD NO. 2
K.C. Glader Co.
C 0.55, Mn 1, Si 0.3, Cr 1.1, Mo 0.25, bal Fe.
Oil hardened steel mainly for molds. *Obsolete*

KEN S-1
K.C. Glader Co.
C 0.5, Mn 0.3, Cr 1.15, V 0.2, W 2.4, bal Fe.
Oil hardening, shock resisting type tool steel; AISI S1. *Obsolete*

KEN S-2
K.C. Glader Co.
C 0.5, Mn 0.5, Si 0.7, V 0.2, Mo 0.45, bal Fe.
Oil or water hardening, shock resisting type tool steel; AISI S2. *Obsolete*

KEN S-4
K.C. Glader Co.
C 0.5, Mn 0.9, Si 0.2, bal Fe.
Water or oil hardening, shock resisting type tool steel; AISI S4. *Obsolete*

KEN S-5
K.C. Glader Co.
C 0.52, Mn 0.95, Si 2, Cr 0.15, V 0.2, Mo 0.5, bal Fe.
Water or oil hardening, shock resisting type tool steel; AISI S5. *Obsolete*

KEN T-1
K.C. Glader Co.
C 0.72, Cr 4.3, V 1.1, W 18, bal Fe.
High speed tool steel, tungsten type; AISI T1.

KEN T-15
K.C. Glader Co.
C 1.55, Cr 4.75, V 5, W 12.5, Co 5, bal Fe.
High speed tool steel, tungsten type, (W-Cr-V-Co); AISI T15. *Obsolete*

KEN T-2
K.C. Glader Co.
C 0.85, Cr 4, V 2.1, W 18.5, Mo 0.75, bal Fe.
High speed tool steel, tungsten type. AISI T2. *Obsolete*

KEN T-4
K.C. Glader Co.
C 0.74, Cr 4.2, V 1.1, W 18.5, Mo 0-0.5, Co 5, bal Fe.
High speed tool steel, W-Co type; AISI T5. *Obsolete*

KEN T-5
K.C. Glader Co.
C 0.78, Cr 4.2, V 1.95, W 18.5, Mo 0.7, Co 7.9, bal Fe.
High speed tool steel, W-Co type; AISI T4. *Obsolete*

KEN T-6
K.C. Glader Co.
C 0.78, Cr 4.2, V 1.75, W 19.5, Mo 0.75, Co 11.5, bal Fe.
High speed tool steel, tungsten-cobalt type; AISI T6. *Obsolete*

KEN T-8
K.C. Glader Co.
C 0.79, Cr 4, V 2, W 14, Mo 0.75, Co 5, bal Fe.
High speed tool steel, W-Co type; AISI T8. *Obsolete*

KEN TOUGH S-7
K.C. Glader Co.
C 0.5, Mn 0.7, Cr 3.25, Mo 1.4, bal Fe.
Air or oil hardening, shock resisting tool steel; AISI S7.

KEN TOUGH S-7 MOLD QUALITY
K.C. Glader Co.
C 0.5, Mn 0.7, Cr 3.25, Mo 1.4, bal Fe.
Air or oil hardening tool and die steel, mainly for molds.

KEN W-1
K.C. Glader Co.
C 1-1.1, Mn 0.3, Si 0.25, bal Fe.
Water hardening tool steel; AISI W1.

KEN W-2
K.C. Glader Co.
C 1-1.1, V 0.15-0.25, Mn, Si, bal Fe.
Water hardening tool steel; AISI W2. *Obsolete*

KENFACE
Kennametal Inc.
Cemented tungsten carbide-cobalt particles used in hot-press matrices to increase abrasion resistance of bonding systems.

KENNAMETAL
Kennametal Inc.
A series of sintered hard carbide alloys with WC base and various binder metals; for metal cutting, metal forming, rock cutting, wear and abrasion resistant parts and structural parts.

KENNAMETAL 303-SERIES
Kennametal Inc.
WC/Co blend.
Cobalt-based matrix powders for use in hot pressing, esp. in manufacture of hot-pressed diamond-impregnated products.

KENNAMETAL B-SERIES
Kennametal Inc.
copper alloys, Fe, Co.
Soft, low-temperature copper-base-alloy matrix powders for hot pressing. Intended uses are for sawing and drilling non-abrasive to very lightly abrasive material.

KENNAMETAL C-1-ABF
Kennametal Inc.
W, bronze alloy.
14-18 Rock C. Matrix powder used to hot press impregnated-diamond segments for sawing and drilling stones, concretes, reinforced concretes, asphalts, refractories, and glasses.

KENNAMETAL C-1-BF
Kennametal Inc.
W, bronze alloy.
22-26 Rock C. Matrix powder used to hot press impregnated-diamond segments for sawing and/or drilling stones, concretes, reinforced concretes, asphalts, and refractories.

KENNAMETAL F-70
Kennametal Inc.
WC, KF-110, Fe, copper alloy.
53-63 Rock C. Hot-pressed matrix used in saw segments for extremely abrasive material such as green concretes and asphalts.

KENNAMETAL F-SERIES
Kennametal Inc.

This series of iron-based hot-pressed matrices contains iron, a copper-based alloy, and fine microcrystalline WC in increasing amounts from F-40 to F-60. Major uses are for sawing abrasive to extremely abrasive materials.

KENNAMETAL GR. 420
Kennametal Inc.
WC, TIC, TaC, Co.
Sintered: 91.3 Rock A. For metal cutting of steels.

KENNAMETAL GR. K1
Kennametal Inc.
Co 11.5, WC.
Sintered: 89.7 Rock A. For metal cutting, wear parts.

KENNAMETAL GR. K11
Kennametal Inc.
Co 2.8, WC.
Sintered: 93.0 Rock A. For metal cutting.

KENNAMETAL GR. K21
Kennametal Inc.
WC, TiC, TaC, Co.
Sintered: 91.0 Rock A. For metal cutting of steels.

KENNAMETAL GR. K2884
Kennametal Inc.
WC, TiC, TaC, Co.
Sintered: 92.0 Rock A. For metal cutting (milling) of steels.

KENNAMETAL GR. K2S
Kennametal Inc.
WC, TiC, TaC, Co.
Sintered: 91.5 Rock. For metal cutting of steels.

KENNAMETAL GR. K3047
Kennametal Inc.
Co 10.5, WC.
Sintered: 88.5 Rock A. For wear-shock resistant parts.

KENNAMETAL GR. K3109
Kennametal Inc.
Co 12.2, WC.
Sintered: 88.0 Rock A. For wear-shock resistant parts.

KENNAMETAL GR. K3404
Kennametal Inc.
Co 5.5, WC.
Sintered: 90.5 Rock A. For wear-shock resistant parts.

KENNAMETAL GR. K3406
Kennametal Inc.
Co 7.8, WC.
Sintered: 89.5 Rock A. For wear-shock resistant parts.

KENNAMETAL GR. K3411
Kennametal Inc.
Co 9.5, WC.
Sintered: 88.7 Rock A. For wear-shock resistant parts.

KENNAMETAL GR. K45
Kennametal Inc.
WC, TiC, TaC, Co.
Sintered: 92.5 Rock A. For metal cutting of steels.

KENNAMETAL GR. K4H
Kennametal Inc.
WC, TiC, TaC, Co.
Sintered: 92.0 Rock A. For metal cutting of steels.

KENNAMETAL GR. K5H
Kennametal Inc.
WC, TiC, TaC, Co.
Sintered: 93.0 Rock A. For metal cutting of steels.

KENNAMETAL GR. K6
Kennametal Inc.
Co 5.5.
Sintered: 92.0 Rock A. For metal cutting, wear parts.

KENNAMETAL GR. K602
Kennametal Inc.
WC, TaC, under 1.5000 binder.
Sintered: 94.3 Rock A. For wear and corrosion resistant parts.

KENNAMETAL GR. K68
Kennametal Inc.
Co 5.8, WC.
Sintered: 92.6 Rock A. For metal cutting.

KENNAMETAL GR. K701
Kennametal Inc.
WC, Cr, Co.
Sintered: 92.0 Rock A. For severe abrasion-corrosion resistant parts.

KENNAMETAL GR. K703
Kennametal Inc.
WC, Cr, Co.
Sintered; 91.5 Rock A. For abrasion-corrosion resistant parts.

KENNAMETAL GR. K714
Kennametal Inc.
WC, TiC, Cr, Co.
Sintered: 92.5 Rock A. For abrasion-corrosion resistant parts.

KENNAMETAL GR. K7H
Kennametal Inc.
WC, TiC, TaC, Co.
Sintered: 93.5 Rock A. For metal cutting of steels.

KENNAMETAL GR. K8
Kennametal Inc.
Co 3.8, WC.
Sintered: Rock A 92.9. For metal cutting.

KENNAMETAL GR. K801
Kennametal Inc.
WC, Ni.
Sintered: 90.0 Rock A. For wear parts exposed to radiation or mineralized water.

KENNAMETAL GR. K803
Kennametal Inc.
WC, Ni.
Sintered: 91.0 Rock A. For wear parts which must be non-magnetic.

KENNAMETAL GR. K82
Kennametal Inc.
WC, TiC, TaC, Co.
Sintered: 90.0 Rock A. For metal forming dies and punches.

KENNAMETAL GR. K84
Kennametal Inc.
WC, TiC, TaC, Co.
Sintered: 91.0 Rock A. For metal forming dies and punches.

KENNAMETAL GR. K86
Kennametal Inc.
WC, TiC, TaC, Co.
Sintered: 91.7 Rock A. For metal forming dies and punches.

KENNAMETAL GR. K8735
Kennametal Inc.
WC, TiC, Co.
Sintered: 92.0 Rock A. For metal cutting of alloy irons.

KENNAMETAL GR. K9
Kennametal Inc.
Co 8.8, WC.
Sintered: 89.0 Rock A. For structural parts, high modulus and strength.

KENNAMETAL GR. K90
Kennametal Inc.
Co 25, WC.
Sintered: 85.0 Rock A. For impact forming of metals.

KENNAMETAL GR. K91
Kennametal Inc.
Co 19.5, WC.
Sintered: 86.8 Rock A. For impact forming of metals.

KENNAMETAL GR. K92
Kennametal Inc.
Co 15.5, WC.
Sintered: 88.5 Rock A. For punches and dies blanking metals.

KENNAMETAL GR. K94
Kennametal Inc.
Co 11.5, WC.
Sintered: 89.7 Rock A. For punches and dies blanking metals.

KENNAMETAL GR. K95
Kennametal Inc.
Co 9, WC.
Sintered: 90.2 Rock A. For wear resistant parts.

KENNAMETAL GR. K96
Kennametal Inc.
Co 5.5, WC.
Sintered: 92.0 Rock A. For wear resistant parts.

KENNAMETAL GR. KC210
Kennametal Inc.
WC, Co plus multi coating.
Sintered and coated. For metal cutting of alloy irons.

KENNAMETAL GR. KC810
Kennametal Inc.
WC, TiC, TaC, Co, plus multi coating.
Sintered and coated. For metal cutting of steels.

KENNAMETAL K151A
Kennametal Inc.
Ni 20, bal TiC, TaC, CbC.
Sintered: Rockwell A89. For brass extrusion dies; sintered; C.S. 150,000. *Obsolete*

KENNAMETAL K3H
Kennametal Inc.
Co 8, 76% WC, 4% TaC, 12% TiC.
Sintered: Rockwell A92. For cutting tools, rock bits, liners; sintered carbides; Tr.S.250,000. *Obsolete*

KENNAMETAL KE 7
Kennametal Inc.
WC & Co.
For wear resistant parts; sintered, extruded shapes only; Rockwell 91A. *Obsolete*

KENNAMETAL KE7
Kennametal Inc.
WC, Co.
For extrusion dies; sintered carbide, wear resistant. *Obsolete*

KENNAMETAL KF-110F
Kennametal Inc.
TC 3.6-4.1, Fe 0.5-0.8, WC.
Chill-cast eutectic tungsten carbide matrix wear-rate modifier used in hot-pressed bond systems.

KENNAMETAL KF-110MA
Kennametal Inc.
TC 3.8-4, Fe 0.1-0.35, Ni 0.1-0.35, Co 0.1-0.45, WC/W_2C.
Vacuum-fused eutectic tungsten carbide used as a component in matrices for hot pressing to regulate wear resistance.

KENNAMETAL KM
Kennametal Inc.
Co 10, 82% WC, 8% TiC.
Sintered: Rockwell A85. For cutting tools, die inserts, rock bits; sintered carbides; Tr.S.300,000. *Obsolete*

KENNAMETAL KWH
Kennametal Inc.
WTiC2 + cobalt.
For cutting tools, lathe centers; cemented carbides. *Obsolete*

KENNAMETAL MF-50BPF
Kennametal Inc.
W, WC, nickel-brass alloy.
26-32 Rock C. Matrix powder used in hot pressing of impregnated-diamond segments for sawing and drilling stones, rocks, concretes, reinforced concretes, asphalts, and refractories.

KENNAMETAL MF-75BPF
Kennametal Inc.
W, WC, copper alloy.
18-22 Rock C. Matrix powder used in hot pressing of
impregnated-diamond tools for drilling and sawing stones,
concretes, and asphalts.

KENNAMETAL N-SERIES
Kennametal Inc.
nickel alloy, Kenface, iron or a Cobamet alloy, other,
ingredients in small amounts.
This family of hot-press matrix powders is used to hot press
segments for asphalts, green concretes, and other extremely
abrasive materials.

KENNERTIUM
Kennametal Inc.
A series of sintered heavy tungsten alloys for use as radiation
shielding, inertial device components; machinable.

KENNERTIUM W-10
Kennametal Inc.
W 90, 7.5 Ni plus Cu.
Sintered: density 17.0. For parts of inertial devices, radiation
shielding, weights and counterbalances.

KENNERTIUM W-2
Kennametal Inc.
W 97.5, 2.5 Ni plus Cu.
Sintered: density 18.5. For parts of inertial devices, balancing
slugs, weights and counterbalance.

KENT
Time Steel Service Inc.
C 2.15, Mn 0.25, Si 0.25, Cr 12, V 0.8, bal Fe.
High carbon, high chromium tool and die steel; air or oil
hardened AISI D3.

KENTANIUM
Kennametal Inc.
A series of sintered hard carbide alloys with TiC base and Ni-
Mo binders: for metal cutting, metal forming, wear and
abrasion resistant parts.

KENTANIUM K151A
Kennametal Inc.
TiC, Ni, Mo.
Sintered: 90.0 Rock A. For high temperature abrasion and
wear resisting parts.

KENTANIUM K162B
Kennametal Inc.
TiC, Ni, Mo.
Sintered: 89.5 Rock A. For wear and abrasion resistant parts
combined with light weight.

KENTANIUM K165
Kennametal Inc.
TiC, Ni, Mo.
Sintered: 93.5 Rock A. For metal cutting and wear resistant
parts.

KENTUCKY
Newport Steel Corp.
C, Cu, bal Fe.
For construction steel.

KEOKUK ELECTRO-SILVERY
Keokuk Electro-Metals Co.
Si 15, C 6, bal Fe.
For iron and steel making; addition agent.

KERAU
Sanderson Kayser Ltd.
C 0.7, W 22, Cr 4, V 1, bal Fe.
For drills, milling cutters, saws, knives; high speed steel, oil
hardened. *Obsolete*

KERAU WUNDA
Sanderson Kayser Ltd.
W 18, Co 5, Cr 4, V 1.3, C, bal Fe.
For tools, dies, cutters; high speed steel. *Obsolete*

KERCHSTEEL
Manufacturer not listed.
As 0.12-0.18, C, bal Fe.
For construction steel.

KERN'S HYDRAULIC BRONZE
Manufacturer not listed.
Cu 78, Sn 12, Zn 10.
For vessels, pressure castings, valves, fittings; high strength.

KERROFESTAL
German manufacture
Si 1, Fe 0.3, Mn 0.7, Mg 0.65, bal Al.
For architecture, window frames, chemical industry; light
alloy.

KERUS
Sanderson Kayser Ltd.
C 0.7, W 14, bal Fe.
For punches, drills, taps, lathe and boring tools, nail dies;
high speed steel. *Obsolete*

KETOS
Crucible Materials Corp.
Tool material. C 0.9, Si 0.25, Mn 1.3, Cr 0.5, W 0.5, bal Fe.
For tools, dies, taps, reamers, broaches, gauges, hobs;
nondeforming, oil hardened.

KEY ALLOY-B
Manufacturer not listed.
Cu 65.2, Zn 20.15, Ni 12.7, Fe 0.4, Pb 1.1, Sn 0.15.
For hardware keys; free-cutting.

KEY STOCK-A
Manufacturer not listed.
Cu 60-65, Zn 22-26, Ni 12, Pb 1-2.
For hardware, keys; free-cutting.

KEYSTONE
Abex Corp.
Sn, Sb, bal Cu.
For bearings; heavy loads. *Obsolete*

KEYSTONE
Kidd Drawn Steel
C 0.9, Mn 1, Cr 0.5, W 0.5, bal Fe.
For blanking and forming dies, punches, gauges; Type O1;
oil hardened, non-deforming.

KEYSTONE ALLOY CHISEL
Disston Inc.
C 0.5, Cr 1.1, W 2, V 0.2, bal Fe.
For chisels, punches, shears; tough. *Obsolete*

KEYSTONE B
CCS Braeburn Alloy Steel
C 1.2, Mn 0.2, Cr 0.5, V 0.2, W 1.6, bal Fe.
For tools, cutters; fast finishing steel. *Obsolete*

KEYSTONE COPPER STEEL
U.S. Steel Corp.
C 0-0.1, Mn 0-0.5, P 0.01-0.1, S 0-0.05, 0.2% Cu min, bal Fe.
35,000-55,000 TS; 15,000-30,000 YS; 25-40 El. For roofing,
siding, railroad car construction; atmosphere corrosion
resistance. *Obsolete*

KEYSTONE DRILL ROD
Precision-Kidd Steel Co.
C 0.88-0.93, Mn 1.1-1.3, Si 0.2-0.35, Cr 0.15-0.2, V 0.15-0.2,
W 0.4-0.6, P 0-0.02, S 0-0.02, bal Fe.
Oil hardening drill rod; AISI O1.

KEYSTONE-A
Disston Inc.
C 0.5, W 2, Cr 1, V 0.2, bal Fe.
For chisels, die blocks; tough, hot work. *Obsolete*

KEYSTONE-B
CCS Braeburn Alloy Steel
C 1.2, Cr 0.5, V 0.2, W 1.7, bal Fe.
For form tools, dies, taps, chasers. *Obsolete*

KF15
Thyssen Edelstahlwerke AG
C 0.4, Cr 13, bal Fe.
For bakelite molds; corrosion resistant. *Obsolete*

KH 15N25
Russian manufacture
C 0.2, Cr 15, Ni 25, bal Fe.
For furnace parts, heat treating boxes; heat and corrosion
resistant.

KH 28
Russian manufacture
C 1.7-2.5, Si 1.3-1.7, Cr 28-32, bal Fe.
For chemical plant equipment; corrosion and heat resistant,
cast.

KH 31 (STANDARD ALLOY)
English manufacture
Th 2.5-4, Zr 0.5-0.7, bal Mg.
H24-temper: 38,900 TS; 29,800 YS; 6 El. T6-temper: 37,100
TS; 19,900 YS; 17 El. For airframes, rockets, missile
components; high creep resistance to 700°F.

KH34
Russian manufacture
C 1.5-2.2, Si 1.3-1.7, Cr 32-36, bal Fe.
For cast rabbles for furnaces, ore roasters; corrosion and
heat resistant, cast.

KHG
Russian manufacture
C 1.44, Mn 0.62, Mn 1.29, Si 0.22, bal Fe.
For cutters, tools, punches; oil hardened.

KHN 80
Russian manufacture
C 0-0.25, Mn 0-1.5, Si 0-1.2, Cr 18-22, bal Ni.
For electrical resistance and heating elements; application to
1000°C.

KICK PLATE BRASS
English manufacture
Cu 84, Zn 15, Pb 1.
For hardware, water pipe, fittings; free-cutting.

KIN-SHIBU-ICHI
Japanese manufacture
one part "Shaku do," two parts "Shibu ichi.".
For jewelry, ornaments; corrosion resistant.

KIND KW
Kind & Co. Edelstahlwerk
C 0.85, Cr 1.75, Mn 0.35, Si 0.3, bal Fe.
For bearings, cutters, dies, tools. Oil or water hardened, wear
resistant. *Obsolete*

KIND KWS
Kind & Co. Edelstahlwerk
C 0.85, Cr 1.75, Mn 0.35, bal Fe.
For bearings, piercing dies, cutters. Oil or water hardened,
wear resistant. *Obsolete*

KIND-EXTRA HART
Kind & Co. Edelstahlwerk
Tool material. C 1.25, Si 0.25, Mn 0.25, bal Fe.
For forming and blanking dies, engravers' tools. Type W1;
water hardened. W. Nr. 1663.

KIND-EXTRA MH
Kind & Co. Edelstahlwerk
Now KIND-EXTRA MITTELHART.

KIND-EXTRA MITTELHART
Kind & Co. Edelstahlwerk
C 1.05, Si 0.25, Mn 0.25, bal Fe.
For springs, tools, drills, taps, reamers. Type W1; water hardened. W. Nr. 1645.

KIND-EXTRA SCHWEISSBAR
Kind & Co. Edelstahlwerk
C 0.55, Si 0.1-0.4, Mn 0.5-0.7, bal Fe.
Heat treated: 150,000 TS; 110,000 YS; 15 El; 45 RA; 320 Brin. For gears, bolts, machine tool parts. Water hardened. *Obsolete*

KIND-EXTRA SEHR ZAH
Kind & Co. Edelstahlwerk
C 0.7, Si 0.25, Mn 0.25, bal Fe.
Heat treated: 175,000 TS; 128,000 YS; 12 El; 37 RA; 355 Brin. For springs, rails, punches, crimpers. Type W1; water hardened. W. Nr. 1620.

KIND-EXTRA ZAH
Kind & Co. Edelstahlwerk
C 0.8, Si 0.25, Mn 0.25, bal Fe.
Heat treated: 190,000 TS; 145,000 YS; 10 El; 30 RA; 400 Brin. For springs, tools, cutters, taps, drills, hobs. Type W1; water hardened.

KIND-EXTRA ZH
Kind & Co. Edelstahlwerk
C 1, Si 0-0.25, Mn 0-0.25, bal Fe.
Annealed: 100,000 TS; 53,000 YS; 21 El; 42 RA; 210 Brin. For springs, tools, drills, taps. Type W1; water hardened. *Obsolete*

KIND-PRIMA MH
Kind & Co. Edelstahlwerk
C 0.9, Si 0.25-0.5, Mn 0.3-0.8, bal Fe.
Heat treated: 180,000 TS; 115,000 YS; 10 El; 30 RA; 370 Brin. For springs, taps, reamers, broaches, drills. Type W1; water hardened. *Obsolete*

KIND-PRIMA SEHR ZAH
Kind & Co. Edelstahlwerk
C 0.6, Si 0.35, Mn 0.7, bal Fe.
Hot rolled: 98,000 TS; 60,000 YS; 24 El; 54 RA; 212 Brin. For axles, gears, bolts, shafts, crankpins. Water hardened. W. Nr. 1740.

KIND-PRIMA WEICH
Kind & Co. Edelstahlwerk
C 0.45, Si 0.35, Mn 0.7, bal Fe.
Hot rolled: 85,000 TS; 54,000 YS; 30 El; 53 RA; 183 Brin. For gears, plates, shafts, bolts, axles. Water hardened. W. Nr. 1730.

KIND-PRIMA ZAH
Kind & Co. Edelstahlwerk
C 0.75, Si 0.35, Mn 0.7, bal Fe.
Heat treated: 115,000-160,000 TS; 77,000-113,000 YS; 12-23 El; 40-54 RA; 230-320 Brin. For wheels, die blocks, girders, rails. Water hardened. W. Nr. 1750.

KIND-PRIMA ZH
Kind & Co. Edelstahlwerk
C 0.75, Si 0.25-0.5, Mn 0.3-0.8, bal Fe.
Heat treated: 180,000 TS; 140,000 YS; 12 El; 35 RA; 375 Brin. For springs, punches, clutch discs. Type W1; water hardened. *Obsolete*

KIND-SPEZIAL MH
Kind & Co. Edelstahlwerk
C 1.1, Si 0-0.25, Mn 0-0.25, bal Fe.
Annealed: 100,000 TS; 53,000 YS; 21 El; 42 RA; 200 Brin. For springs, taps, reamers, hobs. Type W1; water hardened. *Obsolete*

KIND-SPEZIAL SEHR ZAH
Kind & Co. Edelstahlwerk
C 0.7, Si 0-0.25, Mn 0-0.25, bal Fe.
Heat treated: 175,000 TS; 130,000 YS; 12 El; 37 RA; 355 Brin. For wheels, die blocks, rails, girders, springs. Type W1; water hardened. *Obsolete*

KIND-SPEZIAL ZAH
Kind & Co. Edelstahlwerk
C 0.8, Si 0.2, Mn 0.2, bal Fe.
Heat treated: 188,000 TS; 143,000 YS; 12 El; 35 RA; 390 Brin. For springs, taps, drills, hobs, cutters. Type W1; water hardened. W. Nr. 1525.

KIND-SPEZIAL ZAHHART
Kind & Co. Edelstahlwerk
C 1.05, Si 0.2, Mn 0.2, bal Fe.
Annealed: 100,000 TS; 53,000 YS; 21 El; 42 RA; 200 Brin. For springs, taps, reamers, drills, hobs. Type W1; water hardened. W. Nr. 1545.

KIND-SPEZIAL ZH
Kind & Co. Edelstahlwerk
Now KIND-SPEZIAL ZAHHART.

KING COBALT
Jessop Steel Co.
C 0.8, Cr 4.5, V 1.5, W 20, Co 12, bal Fe.
For lathe and planer tools, reamers, hobs, taps, drills, broaches; high speed steel. *Obsolete*

KINGHORN METAL
English manufacture
Cu 58.5, Zn 39.3, Sn 0.96, Fe 1.14.
For screws, nuts, hardware; high strength.

KINGSTON BRONZE
IMI Knoch Ltd.
Cu 83, Sn 4, Fe 0.5, bal Zn.
Drawn: 94,500 TS; 35,000 YS; 10.5 El; 183 Brin. For condenser tubes; high strength.

KINITE
Kinite Corp.
Si 0.55, C 1.5, Cr 12.5-14.5, Mo 1.1, Co 0.7, Mn 0.5, Ni 0.4, bal Fe, trace to 0.5 B.
Cast: 100,000 TS. Rolled: 200,000 TS. For dies, anvils, cutters, mandrels, press tools, blanking, forming and drawing dies, shear blades; abrasion and compression resistant.

KINNALLOY
Kinney Iron Works
C 3.3, Si 2.5, Ni 1.5, Cr 0.8, bal Fe.
40,000 TS; 500 Brin. For castings. Cast iron.

KINNITE
Kinney Iron Works
C 2.5, Cr 1.5, bal Fe.
Cast: 40,000 TS; 500 Brin. For abrasion resisting castings. Wear resistant.

KINSALLOY
International Nickel Inc.
Nickel. Mo 21-24, Al 7-8, C 0-0.1, bal Ni.
Cast: 100,000-130,000 TS. For high temperature applications; precision cast, heat resistant. *Obsolete*

KINSALLOY B
International Nickel Inc.
Nickel. Ni 70, Mo 22, Al 8.
For jet engine components, turbine rotor blades; heat resistant. *Obsolete*

KIRKALLOY
NL Industries
Al 3, Cu 8, Be 0.04, Pb 0-0.007, Sn 0-0.003, Cd 0-0.003, Fe 0-0.1, bal Zn.
Sand cast: 42,500 TS; 2.0 El; 7.6 impact strength; 124 Brin. *Obsolete*

KIRKSITE
Now HI-PHY KIRKSITE.

KIRKSITE A
NL Industries
Cu 3.5, Al 4, Mg 0.04, bal Zn.
Cast: 38,000 TS; 5 El; 107 Brin. Rolled: 62,000 TS; 3 El. For stamping dies, forming dies; MP 717°F. *Obsolete*

KISKI
CCS Braeburn Alloy Steel
C 0.9-1, Mn 1-1.2, Cr 0.43-0.57, V 0.15-0.25, W 0.45-0.75, bal Fe.
For tools, dies, blanking and forming dies, thread gauges. Non-deforming. *Obsolete*

KISSOCK STEEL
Bonney-Floyd Co.
C 0.65-0.75, Ni 1.9-2, Cr 1.3-1.6, Mo 0.35-0.45, bal Fe.
Cast: 301-550 Brin. For castings, wear plates; wear resisting. *Obsolete*

KL 33 A
Kawasaki Steel Corp.
C 0.06, Mn 1.32, P 0.011, S 0.013, Si 0.21, Ni 0.25, bal Fe.
64,000 psi TS; 49,800 psi YS; 43 El. Low-alloy normalized steel for structural and pressure vessels for low temperature service.

KL 33 B
Kawasaki Steel Corp.
C 0.06, Mn 1.3, P 0.011, S 0.014, Si 0.21, Ni 0.25, bal Fe.
65,400 psi TS; 54,000 psi YS; 41 El. Low-alloy quenched and tempered steel for low temperature pressure vessels and structures.

KL 36 B
Kawasaki Steel Corp.
C 0.05, Mn 1.44, P 0.008, S 0.009, Si 0.4, Ni 0.21, Cr 0.1, bal Fe.
71,300 psi TS; 57,200 psi YS; 51 El. Low-alloy quenched and tempered steel for low temperature pressure vessels and structures.

KL 85
Bergische Stahl Industrie
C 0.75-0.85, Si 0.25-0.4, Mn 0.3-0.5, Cr 0.4-0.7, V 0.15-0.25, bal Fe.
Cold work tool steel. W.-Nr. 1.2235.

KL1
Friedr. Lohmann GmbH
C 1, Cr 1, Mn 0.35, Si 0.25, W 1.1, bal Fe.
For bearings, cutters, forming dies. Oil or water hardened, wear resistant.

KLN 3A
Kawasaki Steel Corp.
C 0.09, Mn 0.66, P 0.008, S 0.006, Si 0.21, Ni 3.64, bal Fe.
78,700 psi TS; 64,900 psi YS; 38 El. Normalized 3.5% Ni steel for low temperature use.

KLN 3B
Kawasaki Steel Corp.
C 0.08, Mn 0.68, P 0.006, S 0.009, Si 0.21, Ni 3.65, bal Fe.
82,600 psi TS; 73,700 psi YS; 35 El. Quenched and tempered 3.5% Ni steel for low temperature use.

KLN 9
Kawasaki Steel Corp.
C 0.05, Mn 0.49, P 0.007, S 0.006, Si 0.3, Ni 9.16, bal Fe.
110,260 psi TS; 107,200 YS; 32 El. Quenched and tempered 9% Ni steel for low temperature use.

KLOSTER D-C-66
Manufacturer not listed.
C 0.35, W 8.7-9.7, Cr 3-3.5, V 0.4, bal Fe.
Hardened: 243,000 TS; 215,000 YS; 12 El; 37 RA. At 1000 F: 210,000 TS; 163,000 YS; 9 El; 32 RA. For brass extrusion dies, die casting dies, bolt threading dies, punches, shear blades, hot headers. Type H21 hot work steel, high toughness *Obsolete*

KLOSTER HOLLOW DRILL
Manufacturer not listed.
C 1, bal Fe.
For rock drills. *Obsolete*

KLOSTER HOT WORK STEEL
Manufacturer not listed.
C 0.5, W 10, Cr 4, V 1, bal Fe.
For dies, tools; hot work steel. *Obsolete*

KLOSTER KLS-44
Manufacturer not listed.
C, Cr, Ni, Mo, bal Fe.
For shear blades, hammer dies, die block; cold work steel, oil hardened. *Obsolete*

KLOSTER SOLID DRILL
Manufacturer not listed.
C 1.2, bal Fe.
For drills, taps. *Obsolete*

KLOSTER V-76
Manufacturer not listed.
C 0.5, Mn 0.6, Si 1.4, bal Fe.
For tools, dies, punches, upsetters; shock resistant, oil hardened. *Obsolete*

KLOSTER V-76
Manufacturer not listed.
C 0.9, Mn 1.2, Si 2, bal Fe.

KLS
Bergische Stahl Industrie
C 0.95-1.05, Si 0.15-0.3, Mn 0.25-0.4, Cr 1.4-1.7, bal Fe.
Cold work tool steel. W.-Nr. 1.2067.

KLS
W. Ossenberg & Cie Edelstahlwerke
C 1.25, Mn 0.7, Si 1.2, Cr 1.2, bal Fe.
For planing tools, threading tools, counter bores, grooving tools. W.-Nr. 1.2109.

KM 609
Bergische Stahl Industrie
C 1.55-1.75, Si 0.25-0.4, Mn 0.2-0.4, Cr 11-12, Mo 0.5-0.7, W 0.4-0.6, V 1.1-1.3, bal Fe.
Cold work tool steel. W.-Nr. 1.2609.

KM 80
German manufacture
C 0.2, Cr 14, Mo 2, V 1, bal Fe.
For turbine rotors; heat resistant.

KM12
Friedr. Lohmann GmbH
C 1.65, Cr 11.5, V 0.1, bal Fe.
For blanking and forming dies, punches. Air hardened, nondeforming.

KMV
George Cook & Co., Ltd.
C 1.5, Cr 12, Mo 0.85, V 0.5, bal Fe.
Cold work tool steel. For thread rolling, blanking and forming dies. BS 4659 BD2; AISI D2.

KN
Koerver & Nehring GmbH
Mi 1.59428e+025-0.5, C 0.32-0.4, Mn 0.6, Ni 2.25-2.75, Cr 0.55-0.95, bal Fe.
Low-alloy steel castings. W.-Nr. 5736.

KN 6 G
Koerver & Nehring GmbH
C 0.6-0.12, Si 0.4, Mn 0.75, Ni 1.3-1.8, bal Fe.
Low-alloy steel castings. W.-Nr. 5621.

KN GS 10NI6
Koerver & Nehring GmbH
C 0.06-0.12, Si 0.4, Mn 0.5-0.8, Ni 1.3-1.8, bal Fe.
Low-carbon, nickel-alloy steel castings. W.-Nr. 5621.

KN GS 38
Koerver & Nehring GmbH
C 0-0.25, Si 0.2-0.6, Mn 0.2-0.5, bal Fe.
Carbon steel castings. W.-Nr. 0416.

KN GS 45
Koerver & Nehring GmbH
C 0-0.25, Si 0-0.6, Mn 0.35, N 0-0.007, bal Fe.
Low-carbon steel castings. W.-Nr. 0443.

KN GS 52
Koerver & Nehring GmbH
C 0.3, Si 0.45, Mn 0.4, bal Fe.
Carbon steel castings. W.-Nr. 0551.

KN GS 60
Koerver & Nehring GmbH
C 0.4, Si 0.45, Mn 0.4, bal Fe.
Carbon steel castings. W.-Nr. 0553.

KN GS 70
Koerver & Nehring GmbH
C 0.5, Si 0.45, Mn 0.4, bal Fe.
Carbon steel castings. W.-Nr. 0554.

KNA
Koerver & Nehring GmbH
C 0.3-0.38, Mn 0.35, Ni 3.5-4, Cr 1.5-1.9, bal Fe.
Alloy steel castings. W.-Nr. 5952.

KNC 4 E
Koerver & Nehring GmbH
C 0.1-0.17, Mn 0.4, Ni 3.3-3.6, Cr 0.65-0.85, bal Fe.
Austenitic steel castings, for high temperature operations. W.-Nr. 2735.

KNC 5 E
Koerver & Nehring GmbH
C 0.1-0.17, Mn 0.4, Ni 4.2-4.7, Cr 0.9-1.2, bal Fe.
Austenitic steel castings, for high temperature operations. W.-Nr. 2745.

KNC-3
Cooper Alloy Corp.
Stainless steel. C 0.2-0.4, Cr 24-28, Ni 19-22, Si 1-3, Mn 1-2, bal Fe.
Cast: 65,000-85,000 psi TS; 35,000-50,000 psi YS; 10-30 El; 10-30 RA; 170-195 Brin. For heat and corrosion resistant parts; furnace parts.

KNCO 4 E
Koerver & Nehring GmbH
C 0.16-0.22, Mn 0.4, Ni 3.8-4.3, Cr 1.1-1.4, Mo 0.15-0.25, bal Fe.
Low-alloy steel castings. W.-Nr. 2764.

KNEISS METAL
Manufacturer not listed.
Zn 40-50, Pb 25-42, Sn 25-15, Cu 0-3.
For bearings; anti-friction.

KNIFE BOLSTERS
English manufacture
Cu 68, Zn 16.5, Ni 15, P 0.5.
For knife bolsters; corrosion resistant.

KNIFE BOLSTERS
English manufacture
Cu 56, Zn 28, Ni 16.
For ornamental parts; nickel silver.

KO-1 ALLOY
Olin Metals Research Labs
Cu 4.8, Ag 0.5, Mg 0.25, bal Al.
Heat treatable cast alloy. T6: 58,000-68,000 psi TS; 47,000-58,000 psi YS; 4-10 El. For high strength castings, aerospace housings, transmission line fittings, cylinder heads and pistons, turbine impellers. AA X201.0; AMS 4228 (T6) Tentative.

KOBALT
Ambo-Stahl-Gesellschaft
C 0.79, Co 4.7, Cr 4.3, Mo 0.75, V 1.55, W 18, bal Fe.
For lathe and planer tools, reamers, hobs, taps; high speed steel. *Obsolete*

KOBALT 10
Friedr. Lohmann GmbH
C 0.72, Co 10, Cr 4.2, Mo 0.8, V 1.5, W 18, bal Fe.
For lathe and planer tools, hobs, milling cutters. Type T5; high speed steel. T12005

KOBALT 10
W. Ossenberg & Cie Edelstahlwerke
C 0.8, Cr 4.5, Mo 1, W 18, V 1.6, Co 10, bal Fe.
High speed steel; for lathe and planing tools for tough jobs. W.-Nr. 1.3265.

KOBALT 10 SPEZIAL
W. Ossenberg & Cie Edelstahlwerke
C 1.25, Cr 4, Mo 3.8, W 10.5, V 3.2, Co 10.5, bal Fe.
High speed steel; automatic lathe tools. W.-Nr. 1.3207.

KOBALT 100
Ambo-Stahl-Gesellschaft
C 0.76, Co 10, Cr 4.2, Mo 0.8, V 1.8, W 18, bal Fe.
For lathe and planer tools, reamers, hobs, taps: high speed steel.

KOBALT 125
Thyssen Edelstahlwerke AG
C, Co, Cr, bal Fe.
For magnet steel. *Obsolete*

KOBALT 150
Ambo-Stahl-Gesellschaft
C 0.8, W, Co, Cr, V, Mo, bal Fe.
For lathe and planer tools, reamers, broaches, taps; high speed steel.

KOBALT 160
Thyssen Edelstahlwerke AG
Co, C, Cr, bal Fe.
For magnet steel. *Obsolete*

KOBALT 18
W. Ossenberg & Cie Edelstahlwerke
C 0.7, Cr 4.5, Mo 1, W 18, V 1.5, Co 18, bal Fe.
High cobalt high speed steel. For special difficult machining work.

KOBALT 200
Thyssen Edelstahlwerke AG
C, Co, Cr, bal Fe.
For magnet steel. *Obsolete*

KOBALT 3, ETC.
Now CORONA KOBALT 3, ETC.

KOBALT 30
SWB Stahlformguss Gesellschaft mbH
C 0.86, Co 2.8, Cr 4.3, Mo 0.85, V 2.1, W 12, bal Fe.
For lathe and planer tools, hobs, reamers; high speed steel. *Obsolete*

KOBALT 300
Thyssen Edelstahlwerke AG
C, Co, Cr, bal Fe.
For magnet steel. *Obsolete*

KOBALT 32V

Thyssen Edelstahlwerke AG
C 0.86, Co 2.8, Cr 4.3, Mo 0.8, V 2.1, W 12, bal Fe.
For lathe and planer tools, reamers, broaches, hobs; high speed steel. *Obsolete*

KOBALT 5

Friedr. Lohmann GmbH
C 0.79, Co 4.7, Cr 4.3, Mo 0.75, V 1.5, W 18, bal Fe.
For lathe and planer tools, reamers, taps, drills. Type T4; high speed steel. T12004

KOBALT 5

W. Ossenberg & Cie Edelstahlwerke
C 1.3, Cr 4.5, Mo 1, W 12, V 3.6, Co 5, bal Fe.
High speed steel. Finishing tools; cutting tools for higher speed operation and greater wear resistance. W.-Nr. 1.3202.

KOBALT 5 W

W. Ossenberg & Cie Edelstahlwerke
C 0.8, Cr 4.5, Mo 1, W 18, V 1.7, Co 5, bal Fe.
High speed steel; lathe and milling cutters for tough work.
W.-Nr. 1.3255; similar to AISI T4.

KOBALT 50

SWB Stahlformguss Gesellschaft mbH
C 0.8, Cr 4.3, Co, V, W, bal Fe.
For lathe and planer tools, reamers, drills, taps, hobs; high speed steel. *Obsolete*

KOBALT 98M

Thyssen Edelstahlwerke AG
C 0.9, Cr 4, Mo 8.75, V 2, W 1.7, bal Fe.
For drills and lathe tools operating at high speed; high speed steel with good red hardness, for roughing or finishing. Also used for milling cutters. *Obsolete*

KOBALT III N

SWB Stahlformguss Gesellschaft mbH
C 0.8, Cr 4.3, Co, W, V, Mo, bal Fe.
For lathe and planer tools, reamers, broaches, taps; high speed steel. *Obsolete*

KOBALT V33

Ambo-Stahl-Gesellschaft
C 0.86, Co 2.8, Cr 4.3, Mo 0.85, V 2.1, W 12, bal Fe.
For lathe and planer tools, reamers, taps, drills, hobs; high speed steel.

KOBALT V65

Ambo-Stahl-Gesellschaft
C 1.35, W, Cr, Co, V, bal Fe.
For engravers' tools, forming and blanking dies; high speed steel.

KOBALT-1

Marathon Specialty Steels Inc.
Tool material. C 0.85, W 18, Cr 4, V 2, Co 12, Mo 0.8, bal Fe.
Hardened: 64-66 Rock C. For lathe and planer tools, hobs, broaches, reamers, milling cutters. Type T-6 tool steel, high red-hardness, abrasion resistant.

KOBALT-II

Marathon Specialty Steels Inc.
C 0.7-0.74, W 18-19, Cr 4-4.25, V 1-1.2, Co 4.5-5.5, Mo 0.5, bal Fe.
Hardened: 64-66 Rock C. For boring and forming tools, gear cutters, lathe and planer cutters, taps, drills, reamers. Type T4 high speed steel, high red-hardness.

KOBITALIUM

English manufacture
Cu 1-5, Ni 0.2-2, Mn 0.25-2, Fe 1-2, Si 0.5-2, Mg 0.4-2, Ti 0.08-0.12, bal Al.
For light alloy castings; also rolled or forged.

KOCH WHITE GOLD

English manufacture
Ni 3.3-24.75, Mn 0.25-5, 75 min Au.
For jewelry, ornaments; corrosion resistant.

KOCHLINS BEARING

Manufacturer not listed.
Cu 90, Sn 10.
For bearings; Bronze.

KOERFLEX-20

Krupp Stahl AG
iron. Co 30, Cr 15, bal Fe.
For electromechanical devices, motors, generators.
Permanent magnet, high permeability. *Obsolete*

KOERFLEX-200

Krupp Stahl AG
react and refract. Co 52, V 10, bal Fe.
For hysteresis motors, automation devices, electromechanical equipment. Permanent magnet, similar to Vicalloy I. High permeability. *Obsolete*

KOERFLEX-30

Krupp Stahl AG
iron. Co 30, Cr 15, bal Fe.
For electrical and magnetic equipment, generators, motors.
Permanent magnet, high permeability. *Obsolete*

KOERFLEX-300

Krupp Stahl AG
react and refract. Co 52, Cr 8, V 4, bal Fe.
1000 Br, 2.8 (BH) max. For electromechanical devices, speedometers, recorders, hysteresis motors, computers.
Precipitation hardening, permanent magnet, similar to Vicalloy II. *Obsolete*

KOERVER-UMCO50

Koerver & Nehring GmbH
C 0.05-0.12, Ti 0.18, Cb 0.6, Cr 26-30, Co 47-52, bal Fe.
Cast: 135,000 TS; 48,000 YS; 10 El; 10 RA; 250 Brin.
Wrought: 132,000 TS; 61,000 YS; 7 El; 6 RA; 350 Brin. For furnace baffles, burner tips, sintering grates, quench baskets.
Corrosion, heat, thermal shock resistant.

KOERZIT 250S

Krupp Stahl AG
superalloy. Al 8, Ni 19, Co 24, Ti 5, bal Fe.
For electrical and magnetic equipment. Permanent magnet.
High permeability. *Obsolete*

KOERZIT 350

Krupp Stahl AG
superalloy. Ni 15, Co 30, Cu 4, Al 7, Ti 5, bal Fe.
8500 Br, 1150 H_c. For permanent magnets in electrical and magnetic equipment. High permeability. *Obsolete*

KOERZIT 400K

Krupp Stahl AG
superalloy. Co 25, Cb 2.5, bal Fe.
For permanent magnets in electrical and magnetic equipment. Alnico type, high permeability. *Obsolete*

KOERZIT 450

Krupp Stahl AG
superalloy. Co 34, Ti 5, bal Fe.
For permanent magnets in electrical and magnetic equipment. Alnico type magnet. High permeability. *Obsolete*

KOERZIT-H2

Krupp Stahl AG
react and refract. Co 52, Fe 38, V 10.
For hysteresis motors, automation devices, electro and magnetic devices. Requires cold rolling and tempering to develop the magnetic properties. Permanent magnet. Similar to Vicalloy I. *Obsolete*

KOERZIT-HI

Krupp Stahl AG
react and refract. Co 52, Cr 7, V 3, bal Fe.
For electric and magnetic equipment, hysteresis motors.
Magnetically soft; high permeability. *Obsolete*

KOERZIT-T

Krupp Stahl AG
react and refract. Co 52, Cr 8, V 4, bal Fe.
10,000 Br, 2.8 (BH) max. For electromechanical devices, speedometers, recorders, hysteresis motors, digital computers. Precipitation hardening. Permanent magnet.
Similar to Vicalloy II. *Obsolete*

KOERZIT-VS55

Krupp Stahl AG
superalloy. Co 38, Ti 7, bal Fe, Al, Ni, Cu.
For permanent magnets in electrical and magnetic equipment. Alnico type magnet, high permeability. *Obsolete*

KOLASSAL

Hoffman & Co. KG
C 2.1, Cr 11.5, Mn 0.3, bal Fe.
For forming and blanking dies, punches; oil hardening, nondeforming.

KOLASSAL EXTRA

Hoffman & Co. KG
C 2.1, Cr 11.5, W 0.7, bal Fe.
For forming and blanking dies, punches; oil hardened, nondeforming.

KOLASSAL SUPRA

Hoffman & Co. KG
C 1.65, Cr 11.5, V 0.1, bal Fe.
For blanking and forming dies, punches; air hardened, nondeforming.

KOLTCHOUGALUMIN

English manufacture
Al 91.7-93.9, Cu 4-6, Mn 0.6, Mg 0.4, Fe 0.8, Si 0.2, Cr 0.11.
For light alloy parts, aircraft structures; similar to "Dural."

KOLTSCHUG

Russian manufacture
Cu 4, Ni 0.3, bal Al.
For light alloy parts; age-hardenable.

KOMALP 3 HERZ

Now VEW S200.

KOMALP 3 HERZ EXTRA M

Now VEW S205.

KOMALP 3 HERZ M

Vereinigte Edelstahlwerke
C 0.95, Cr, W, Mo, V, bal Fe.
For form cutters, broaches, taps; high speed steel. *Obsolete*

KOMALP 300

Now VEW S610.

KOMALP 400

Vereinigte Edelstahlwerke
C 1.3, Cr 4.3, Mo 0.85, V 3.8, W 12, bal Fe.
For engravers' tools, blanking and forming dies; high speed steel. *Obsolete*

KOMALP EXTRA SPEZIAL

Vereinigte Edelstahlwerke
C 0.86, Cr 4.1, Mo 0.85, V 2.5, W 12, bal Fe.
For lathe and planer tools, reamers, broaches; high speed steel. *Obsolete*

KOMALP WM

Now VEW S600.

KOMET FRS27

Fakirstahl Hoffmanns GmbH & Co. KG
C 0.86, Cr 4.1, Mo 0.85, V 2.5, W 12, bal Fe.
For lathe and planer tools, reamers, hobs, taps, drills; high speed steel.

KOMO 205

Marathon Specialty Steels Inc.
C 0.8, Cr 4, Mo 5, W 6, V 2, Co 5, bal Fe.
Hardened: 63-66 Rock C. For reamers, hobs, lathe and planer tools, drills, broaches. Type M35, high speed steel, good red-hardness. Wear resistant.

KOMO 310

Thyssen Edelstahlwerke AG
C 1.25, W 9.5, Cr 4.2, V 3.2, Mo 3.15, Co 10, bal Fe.
Hardened: 64-68 Rock C. For reamers, broaches, chasers, hobs, lathe and planer tools, milling cutters. High-speed steel, high red-hardness.

KOMO 310

Marathon Specialty Steels Inc.
C 1.25, W 9.5, Cr 4.2, V 3.2, Mo 3.15, Co 10, bal Fe.
Hardened: 64-68 Rock C. For reamers, broaches, chasers, hobs, lathe and planer tools, milling cutters. High-speed steel, high red-hardness.

KONALLOY C-COCR20NI20W

Koerver & Nehring GmbH
C 0.35-0.45, Si 0-1, Mn 0-1.5, Cr 19-21, Ni 19-21, Mo 3.5-4.5, W 3.5-4.5, Nb 3.5-4.5, Fe 0-5, bal Co.
Components for gas turbines, combustion chambers, steel castings. W.-Nr. 4989.

KONALLOY C-COCR20W15NI

Koerver & Nehring GmbH
C 0.05-0.15, Si 0-1, Mn 1-2, Cr 19-21, Ni 9-11, W 14-16, Fe 0-3, bal Co.
High strength and corrosion resistance at elevated temperature; steel castings. W.-Nr. 4967.

KONALLOY COCR20W15N

Koerver & Nehring GmbH
C 0.5-0.15, Si 0-1, Mn 1-2, Cr 19-21, Ni 9-11, W 14-16, Fe 0-3, bal Co.
High-temperature alloy; for valves, gas turbine parts; steel castings. W.-Nr. 4964.

KONALLOY COCR25NIW

Koerver & Nehring GmbH
C 0.45-0.6, Si 0-1.2, Mn 0-1, Cr 22-28, Ni 9-12, W 9-12, Fe 0-2, bal Co.
High strength and corrosion resistance at elevated temperature; steel castings. W.-Nr. 4966.

KONALLOY COCR28FE

Koerver & Nehring GmbH
C 0-0.1, Si 0-1, Mn 0-1, Co 45-50, Cr 26-30, bal Fe.
High strength and corrosion resistance at elevated temperature; steel castings. W.-Nr. 4778.

KONALLOY COCR28MO

Koerver & Nehring GmbH
C 0.25-0.35, Si 0-1, Mn 0-1, Cr 27-29, Mo 5-6, Ni 1.5-3, bal Co.
High strength and corrosion resistance at elevated temperatures; steel castings. W.-Nr. 4979.

KONALLOY CRNI25 20

Koerver & Nehring GmbH
Stainless steel. C 0-0.2, Si 1.5-2.5, Mn 0-2, Cr 22-25, Ni 19-22, bal Fe.
Austenitic stainless steel for high temperature operation; steel castings. Similar to AISI 310; W.-Nr. 4843.

KONALLOY NI36

Koerver & Nehring GmbH
C 0-0.1, Mn 0-0.5, Ni 35-37, bal Fe.
High-temperature steel castings. W.-Nr. 3912.

KONALLOY NI38

Koerver & Nehring GmbH
C 0-0.1, Mn 0-1, Ni 36-40, bal Fe.
High-temperature steel castings. W.-Nr. 3913.

KONALLOY NI42

Koerver & Nehring GmbH
C 0-0.05, Mn 0-1, Ni 41-43, bal Fe.
High-temperature steel castings. W.-Nr. 3917.

KONALLOY NI43

Koerver & Nehring GmbH
C 0-0.1, Mn 0-1, Cr 0-1, Ni 42-44, bal Fe.
High-temperature steel castings. W.-Nr. 3918.

KONALLOY NI46

Koerver & Nehring GmbH
C 0-0.1, Mn 0-1, Cr 0-1, Ni 45-47, bal Fe.
High-temperature steel castings. W.-Nr 3920.

KONALLOY NI48

Koerver & Nehring GmbH
C 0-0.05, Mn 0-0.5, Ni 46-50, bal Fe.
High-temperature steel castings. W.-Nr. 3926.

KONALLOY NI49

Koerver & Nehring GmbH
C 0-0.05, Mn 0-1, Cr 0.7-1, Ni 48-50, bal Fe.
High-temperature steel castings. W.-Nr. 3921.

KONCOR

Latrobe Steel Co.
Si 1, Cr 5.25, V 4, Mo 1.1, C 1.1, Mn 0.35, bal Fe.
Hardened: 64-67 Rock C. For upsetter and forging dies, extrusion and pointing dies, chipping chisels, punches, die casting inserts. Air hardening, non-deforming, abrasion resistant, tough, heat resistant.

KONDUR 21

Koerver & Nehring GmbH
C 0.3-0.5, Si 0-1, Mn 0-1, Cr 26-30, Mo 4-5, bal Co.
Wear-resistant steel castings. W.-Nr. 4520.

KONDUR 6

Koerver & Nehring GmbH
C 0.3-0.35, Si 0-0.6, Mn 0-0.6, Cr 26-28, W 4.5-5, bal Co.
Wear-resistant steel castings. W.-Nr. 4519.

KONEL

Westinghouse Electric Corp.
Ni 73.07, Co 17.16, Ti 8.8, Si 0.55, Al 0.26, Mn 0.16.
100,000 psi TS; 40,000 psi YS; 35 El; 55 RA; 140 Brin. For lamp filaments, valves, valve stems; substitute for platinum; heat and corrosion resistant; hot short above 1250°C. *Obsolete*

KONIK

Continental Steel Corp.
C 0.15-0.25, Cu 0.2, Ni 0.4, Cr 0.3, bal Fe.
Cold drawn: 81,000 TS; 79,000 YS; 3 El; 5 RA. For cross chains, culvert sheets; case hardening steel.

KONIT B

Koerver & Nehring GmbH
Stainless steel. C 0-0.1, Si 0-1, Mn 0-1, Cr 0-1, Mo 26-30, Fe 4-7, 62.0 Ni min.
Stainless and heat-resistant steel castings. W.-Nr. 4600.

KONIT C

Koerver & Nehring GmbH
Stainless steel. C 0-0.1, Si 0-1, Mn 0-1, Cr 14-18, W 3-5, Fe 0-7, Mo 15-18, 52.0 Ni min.
Stainless and heat-resistant steel castings. W.-Nr. 4537.

KONIT D

Koerver & Nehring GmbH
Stainless steel. C 0.04-0.08, Si 10, Mn 1, Cu 0-3.5, Fe 0-2, bal Ni.
Stainless and heat-resistant steel castings. W.-Nr. 4566.

KONIT F

Koerver & Nehring GmbH
Stainless steel. C 0-0.1, Mn 0-1, Si 0-1, Cr 20-23, Fe 17-20, Mo 8-10, W 0.2-10, bal Ni.
Stainless and heat-resistant steel castings. W.-Nr. 4013.

KONIT X

Koerver & Nehring GmbH
Stainless steel. C 0-0.08, Si 0-1, Mn 1-2, Cr 21-23, Mo 5.5-7.5, Nb 1.75-2.5, Ni 44-47, bal Fe.
Stainless and heat-resistant castings. W.-Nr. 4557.

KONOLOY 800

Koerver & Nehring GmbH
C 0-0.2, Si 3-4, Mn 0-1.5, Cr 20-22, Ni 28-31, bal Fe.
Heat-resistant steel casting. W.-Nr. 4860.

KONOLOY 825

Koerver & Nehring GmbH
Stainless steel. C 0.15, Si 0-1, Mn 0-1, Cr 20.5-23, Co 0.5-2.5, Mo 6-10, W 0.2-1, Fe 17-20, bal Ni.
Austenitic stainless steel castings for high temperature operations. W.-Nr. 4972.

KONOMAG KN 3944

Koerver & Nehring GmbH
Stainless steel. C 0-0.07, Si 0-1, Mn 0-2, Cr 18-10, Ni 10-12, bal Fe.
Austenitic stainless steel castings. Similar to AISI 302.

KONONEL 600

Koerver & Nehring GmbH
Stainless steel. C 0-0.15, Si 0-0.5, Mn 0-1, Cr 14-17, Fe 6-10, 72.0 Ni min.
Austenitic stainless steel castings for high temperature operation. W.-Nr. 4816.

KONOX 4008

Koerver & Nehring GmbH
Stainless steel. C 0.08-0.15, Si 0-1, Mn 0-1, Cr 12-14, Mo 0.5-1.5, bal Fe.
Martensitic stainless steel castings. Similar to ACI CA-15.

KONSTANT

GWD Stahlformguss Gesellschaft mbH
C 0.9, Mn 1.9, V 0.1, Si 0.25, bal Fe.
For dies, punches, upsetters, crimpers; oil hardened, non-deforming. *Obsolete*

KONSTANT 15

Koerver & Nehring GmbH
C 1.4-1.6, Mn 0.6, Cr 1.3-1.6, bal Fe.
Wear-resistant steel casting. W.-Nr. 2063.

KONSTANTAN

VDM Nickel-Technologie AG
Cu 54, Ni 45, Mn 1.
Annealed: 83,000 TS; 30 El. Hard: 120,000 TS; 1 El. For precision resistances, reducing rheostats, shunts, electrical instruments; maximum operating temperature 600°C; MP 1276°C. *Obsolete*

KONTHERM KN 4710

Koerver & Nehring GmbH
C 0.2-0.4, Si 1-2.5, Mn 0-1, Cr 6-8, bal Fe.
Heat-resistant steel castings.

KONTRAB 2062

Koerver & Nehring GmbH
C 0.5-0.6, Si 0.45, Mn 0.45, Cr 1.3-1.6, bal Fe.
Wear-resistant steel castings.

KOPPER K-6

Koppers Co. Inc.
C 3.7, Si 2.9, Mn 1, Cr 0.5, Ni 1.2, bal Fe.
Cast: 30,000 TS; 180-220 Brin. For piston rings, parts to resist corrosive gases and liquids, packing; alloy cast iron. *Obsolete*

KOPPERS B-18

Koppers Co. Inc.
Cu 78-82, Sn 18-19.5, Pb 0-0.5, Fe 0-0.5, Zn 0-0.5.
Cast: 45,000-55,000 TS; 165-195 Brin. For oil sealing rings, piston and pump rings; corrosion and acid resisting.

KOPPERS B-19
Koppers Co. Inc.
Cu 78-82, Sn 12-14, Pb 4-6, Ni 0.75-1.25.
Cast: 30,000 TS; 100 Brin. For locomotive main cylinder packing and segmental rings; pressure tight.

KOPPERS B-23
Koppers Co. Inc.
Cu 78-82, Sn 15-17, Pb 4-6, Fe 0-0.5, Zn 0-0.5.
Cast: 30,000 TS; 130-156 Brin. For hydraulic sealing rings, piston rings; pressure tight.

KOPPERS B-48
Koppers Co. Inc.
Cu 64, Al 6, Mn 4, Fe 3, bal Zn.
Cast: 110,000 TS; 60,000 YS; 12 El; 225 Brin. For propellers; strong and corrosion resistant.

KOPPERS F-17
Koppers Co. Inc.
TC 3.5, Si 2.5, Ni 2.25, Mo 1, bal Fe.
Cast: 46,000 TS; 200 Brin. For piston rings; diesel rings; castings.

KOPPERS F-37
Koppers Co. Inc.
TC 3.5, Si 2.8, Mn 0.6, Ni 0.9, Mo 1.1, Cr 0.2, bal Fe.
Heat treated: 65,000-75,000 TS; 250-270 Brin. For piston rings; alloy cast iron, wear resistant. *Obsolete*

KOPPERS F-88
Koppers Co. Inc.
C 3.2, Mo 0.5, Si 1.4, Mo 0.5, Cu 0.5, bal Fe.
Heat treated: 88,000 TS; 265 Brin. For piston rings; high strength cast iron.

KOPPERS F-95
Koppers Co. Inc.
TC 3.2, Si 1.6, Mn 0.7, Ni 1, Mo 0.5, Cr 0.25, bal Fe.
Heat treated: 108,000 TS; 330 Brin. For piston rings; alloy cast iron, wear resistant. *Obsolete*

KOPPERS K-10
Koppers Co. Inc.
C 2.6, Si 2, Mn 1.2, Ni 15, Cu 6.5, Cr 2.1, bal Fe.
Cast: 32,000 TS; 137 Brin. For corrosion resistant castings; Type 1, Ni resist, ASM 5392.

KOPPERS K-13
Koppers Co. Inc.
C 1.75, Si 1.75, Ni 0.9, Cr 14, Mo 0.4, bal Fe.
Annealed: 90,000 TS; 270 Brin. For heat resistant castings; heat and abrasion resistant.

KOPPERS K-14
Koppers Co. Inc.
C 3.6, Si 2.3, Mn 0.6, Ni 1.1, Mo 1.1, Cr 0.3, bal Fe.
Heat treated: 45,000 TS; 270 Brin. For heavy duty piston rings; cast iron.

KOPPERS K-15
Koppers Co. Inc.
C 3.4, Si 2.75, Mn 0.5, bal Fe.
Annealed: 60,000 TS; 45,000 YS; 10 El; 140 Brin. For general castings, gears, housings; ductile iron, ferritic.

KOPPERS K-16
Koppers Co. Inc.
C 3.4, Si 2.2, Mn 0.5, bal Fe.
Heat treated: 90,000 TS; 65,000 YS; 2 El; 240 Brin. For gears, shafts, housings; pearlitic ductile iron.

KOPPERS K-25
Koppers Co. Inc.
C 3.25, Si 2.25, Mn 0.5, Ni 0.75, bal Fe.
Annealed: 60,000 TS; 40,000 YS; 15 El; 185 Brin. For gears, shafts, housings, ferritic ductile iron.

KOPPERS K-26
Koppers Co. Inc.
C 3.1, Si 1.3, Mn 0.75, Mo 0.5, Cu 0.5, bal Fe.
Heat treated: 95,000 TS; 400 Brin. For aircraft piston rings; cast iron.

KOPPERS K-27
Koppers Co. Inc.
C 3.3, Si 2.2, Mn 0.5, Mo 0.5, Cu 0.5, bal Fe.
Heat treated: 110,000 TS; 80,000 YS; 1 El; 265 Brin. For cylindrical shapes; alloy ductile iron.

KOPPERS K-28
Koppers Co. Inc.
C 3.3, Si 2.2, Mn 0.5, Mo 0.5, Cu 0.5, bal Fe.
Heat treated: 120,000 TS; 100,000 YS; 1 El; 400 Brin. For cylindrical shapes; alloy ductile iron.

KOPPERS K-35
Koppers Co. Inc.
C 2.3, Si 1.2, Mn 0.5, Ni 30, Cr 3, bal Fe.
Cast: 55,000 TS; 33,000 YS; 7 El; 140 Brin. For corrosion and abrasion resistant castings; Ni resist, Type 3.

KOPPERS K-37
Koppers Co. Inc.
C 2.7, Si 2.6, Mn 1.2, Ni 20, Cr 2.1, bal Fe.
55,000 psi TS; 170 Brin. Centrifugally or static cast; austenitic ductile iron. Used at temperatures up to 1400°F on applications such as engine exhaust seal rings, valve guides, turbo charger and gas turbine parts.

KOPPERS K-46
Koppers Co. Inc.
C 3, Si 3.15, Mn 7.75, Ni 5, bal Fe.
35,000 psi TS; 155 Brin. Coefficient of expansion 10.5×10^{-6} in./in.·°F. Centrifugally cast, austenitic Mn-Ni alloy cast iron. Used for aluminum piston inserts where a high coefficient of expansion is required.

KOPPERS K-53
Koppers Co. Inc.
C 3, Si 2.1, Mn 0.7, Ni 0.6, Mo 0.5, Cr 0.5, Cu 0.75, bal Fe.
Cast: 40,000 TS; 229 Brin. For cylinder liners; centrifugally cast gray iron.

KOPPERS K-60
Koppers Co. Inc.
C 0-0.07, Cr 16.5, Ni 4, Cb 0.45, Cu 4, N 0.05, bal Fe.
Heat treated: 150,000 TS; 140,000 YS; 6 El; 350 Brin. Corrosion resistant, hard castings; Armco 17-4PH, stainless, age hardened.

KOPPERS K-61
Koppers Co. Inc.
C 0-0.15, Cr 12.5, Ni 0.75, Mo 0.5, Cu 0.5, bal Fe.
Cast: 125,000 TS; 110,000 YS; 10 El; 310 Brin. For corrosion resistant castings; Type 410; stainless.

KOPPERS K-66
Koppers Co. Inc.
C 0.5, Cr 25.5, Ni 10.5, Co 54.5, Fe 0-2.
Cast: 110,000 TS; 95,000 YS; 2 El; 370 Brin. For chemical plant equipment, high temperature castings; Haynes Stellite No. 31; heat and corrosion resistant.

KOPPERS K-6B
Koppers Co. Inc.
C 2.8, Si 2, Mn 1.3, Ni 15, Cu 6.5, Cr 1.9, bal Fe.
Annealed: 25,000 TS; 145 Brin. For cylindrical shapes; austenitic, centrifugally cast iron.

KOPPERS K-6E
Koppers Co. Inc.
C 3.8, Si 2.9, Mn 0.6, Cr 0.3, Mg 0.45, bal Fe.
Cast: 30,000 TS; 193 Brin. For piston rings; centrifugally cast gray iron.

KOPPERS K-76
Koppers Co. Inc.
C 3.15, Si 4.75, bal Fe.
85,000 psi TS; 250 Brin. Centrifugally cast ductile iron. Used for exhaust manifold seal rings and other applications where resistance to oxidation and growth is necessary.

KOPPERS K-8
Koppers Co. Inc.
C 3.3, Si 2.3, Mn 0.65, Ni 0-1.5, Cr 0-0.4, bal Fe.
Cast: 27,000 TS; 180 Brin. For pistons ring inserts; centrifugally cast gray iron.

KOPPERS K-IRON
Koppers Co. Inc.
C 3.7, Si 2.85, Mn 0.6, bal Fe.
Cast: 25,000 TS; 180 Brin. For piston rings; cast iron.

KOPPERS XLS
Koppers Co. Inc.
C 3.5, Si 1.75, Mn 0.7, Cr 0.35, bal Fe.
Cast: 20,000 TS; 150 Brin. For large engine piston rings; gray cast iron.

KORA
Sanderson Kayser Ltd.
C 0.15-0.18, Si 0.1-0.15, Mn 0.7-0.9, bal Fe.
Hot rolled: 83,000 TS; 67,000 YS; 37 El; 55 RA; 163 Brin. Heat treated: 95,000 TS; 67,000 YS; 28 El; 60 RA; 180 Brin. For gears, camshafts, machinery parts; case hardening steel.

KORA NO. 2
Sanderson Kayser Ltd.
C 0.16, Cr 0.12, S 0-0.06, P 0-0.05, bal Fe.
Case hardening steel.

KORROFESTAL
Aluminium-Werke Wutoeschingen GmbH
Aluminum. Mg 0.6-1.4, Si 0.6-1.6, Mn 0.6-1, Cr 0-0.3, bal Al.
Annealed: 21,000 psi TS; 8000 psi YS; 24 El. For window frames, gutters, fan blades, boats; good forming and welding properties. *Obsolete*

KORRONIT 1212 T
Otto Wolff Handelsgesellschaft
C 0-0.1, Cr 12.5, Ni 12, bal Fe.
For valves, pumps, oil refinery equipment; corrosion and heat resistant.

KORRONIT 17
Otto Wolff Handelsgesellschaft
C 0.08, Cr 17, bal Fe.
Annealed: 80,000 TS; 50,000 YS; 25 El; 50 RA; 150 Brin. For oil refinery and food processsing equipment, sinks. Type 430; corrosion resistant, ferritic.

KORRONIT 1717
Otto Wolff Handelsgesellschaft
C 0-0.07, Cr 17.5, Mo 2, Ni 17.5, Cu 2, Ti = 7 x C, bal Fe.
For valves, pumps, chemical plant equipment; corrosion resistant.

KORRONIT 17E
Otto Wolff Handelsgesellschaft
C 0-0.1, Cr 17.5, Ti = 7 x C, bal Fe.
Annealed: 85,000 TS; 55,000 YS; 20 El; 4 RA; 180 Brin. For oil refinery and food processing equipment, weldments. Type 430 Ti; corrosion resistant.

KORRONIT 17S
Otto Wolff Handelsgesellschaft
C 0.12, Cr 16.5, Mo 0.25, S 0.2, bal Fe.
Annealed: 80,000 TS; 50,000 YS; 25 El; 50 RA; 160 Brin. For screw machine products, fasteners; Type 430F; stainless, free-cutting.

KORRONIT 188

Otto Wolff Handelgesellschaft
C 0-0.15, Cr 18, Ni 8.5, bal Fe.
Annealed: 80,000 TS; 35,000 YS; 55 El; 75 RA; 150 Brin. Cold drawn: 180,000 TS; 150,000 YS; 10 El; 250 Brin. For chemical plant equipment, tanks. Type 302; stainless, austenitic.

KORRONIT 188E

Otto Wolff Handelgesellschaft
C 0-0.12, Cr 18, Ni 8.5, Ti = 4 x C, bal Fe.
Annealed: 85,000 TS; 35,000 YS; 55 El; 65 RA; 150 Brin. Cold drawn: 95,000 TS; 60,000 YS; 40 El; 60 RA; 185 Brin. For welded chemical plant equipment, tanks, mixers. Type 321; stainless, austenitic.

KORRONIT 188P

Otto Wolff Handelgesellschaft
C 0-0.07, Cr 18, Ni 9.5, bal Fe.
Annealed: 85,000 TS; 35,000 YS; 60 El; 70 RA; 150 Brin. Cold drawn: 180,000 TS; 125,000 YS; 10 El; 330 Brin. For chemical plant equipment, tanks, mixers. Type 304; stainless, austenitic.

KORRONIT 189

Otto Wolff Handelgesellschaft
C 0-0.1, Si 2.2, Cr 18, Mo 2, Ni 9.5, bal Fe.
Annealed: 85,000 TS; 35,000 YS; 50 El; 65 RA; 160 Brin. Cold drawn: 150,000 TS; 135,000 YS; 6 El; 300 Brin. For acid resistant chemical plant equipment. Type 316; stainless austenitic.

KORRONIT 189E

Otto Wolff Handelgesellschaft
C 0-0.12, Cr 18, Mo 2, Ni 10.5, Ti = 4 x C, bal Fe.
Annealed: 85,000 TS; 40,000 YS; 50 El; 65 RA; 160 Brin. For welded chemical plant equipment, tanks. Type 316 Ti; stainless, austenitic.

KORRONIT 189P

Otto Wolff Handelgesellschaft
C 0-0.07, Cr 18, Mo 2, Ni 10.5, bal Fe.
Annealed: 85,000 TS; 35,000 YS; 50 El; 65 RA; 160 Brin. For acid resistant chemical plant equipment, tanks. Type 316; stainless, austenitic.

KORRONIT 18M

Otto Wolff Handelgesellschaft
C 0.15, Cr 18, Si 0.4, bal Fe.
Annealed: 80,000 TS; 50,000 YS; 25 El; 50 RA; 160 Brin. For oil refinery equipment, food processing equipment. Type 430; stainless, ferritic.

KORRONIT-C

Otto Wolff Handelgesellschaft
C 0.9, Cr 18, Mo 1.15, V 1, bal Fe.
Annealed: 105,000 TS; 60,000 YS; 18 El; 35 RA; 220 Brin. Heat treated: 280,000 TS; 270,000 YS; 3 El; 15 RA; 555 Brin. For cutlery, valves, ball bearings, surgical instruments. Type 440 B; corrosion resistant.

KORRONIT-H

Otto Wolff Handelgesellschaft
C 0.4, Si 0.4, Cr 13, bal Fe.
Annealed: 95,000 TS; 50,000 YS; 25 El; 55 RA; 195 Brin. For cutlery, valves, surgical instruments. Type 420; stainless, hardenable.

KORRONIT-M

Otto Wolff Handelgesellschaft
C 0.2, Si 0.4, Cr 13, bal Fe.
Annealed: 95,000 TS; 50,000 YS; 25 El; 55 RA; 195 Brin. For valves, cutlery, oil refinery equipment. Type 420; corrosion resistant.

KORRONIT-S

Otto Wolff Handelgesellschaft
C 0.22, Si 0.4, Cr 17, Ni 1.5, bal Fe.
Annealed: 125,000 TS; 95,000 YS; 20 El; 55 RA; 260 Brin. For marine hardware, pumps, valves. Type 431; corrosion and heat resistant.

KORRONIT-W

Otto Wolff Handelgesellschaft
C 0-0.12, Si 0.4, Cr 13, bal Fe.
For turbine blades, valves, cutlery, surgical instruments; Type 410; stainless.

KORROSIL 1

Georgsmarienwerke Selesiastahl GmbH
C 0-0.12, Si 0.4, Cr 13, bal Fe.
Annealed: 75,000 TS; 40,000 YS; 35 El; 70 RA; 155 Brin. Cold drawn: 100,000 TS; 85,000 YS; 17 El; 60 RA; 205 Brin. For turbine blades, surgical instruments; Type 410; stainless.

KORROSIL 2

Georgsmarienwerke Selesiastahl GmbH
C 0.2, Si 0.4, Cr 13, bal Fe.
Annealed: 95,000 TS; 50,000 YS; 25 El; 55 RA; 195 Brin. Cold drawn: 100,000 TS; 85,000 YS; 17 El; 50 RA; 215 Brin. For cutlery, valves, turbine blades, knives; Type 420; stainless.

KORROSIL 4

Georgsmarienwerke Selesiastahl GmbH
C 0.4, Si 0.4, Cr 13, bal Fe.
Annealed: 100,000 TS; 50,000 YS; 23 El; 52 RA; 200 Brin. For cutlery, valves, surgical and dental instruments; Type 420; stainless.

KORROSIL 60

Georgsmarienwerke Selesiastahl GmbH
C 0.22, Cr 17, Ni 1.5, bal Fe.
Annealed: 125,000 TS; 95,000 YS; 20 El; 55 RA; 260 Brin. For marine hardware, pumps, valves; Type 431.

KORROSIL 6A

Georgsmarienwerke Selesiastahl GmbH
C 0.12, Cr 16.5, Mo 0.25, S 0.2, bal Fe.
Annealed: 80,000 TS; 50,000 YS; 25 El; 50 RA; 150 Brin. For screw machine products, shafts, fasteners; Type 430F; stainless, free cutting.

KORROSIL 6U

Georgsmarienwerke Selesiastahl GmbH
C 0-0.1, Cr 17.5, Ti = 7 x C, bal Fe.
Annealed: 80,000 TS; 50,000 YS; 25 El; 50 RA; 150 Brin. For welded furnace parts and heat treating boxes; Type 430 Ti; corrosion resistant.

KORROSIL-2M

Klockner-Werke AG
C 0.2, Cr 13, Mo 1.2, bal Fe.
Annealed: 95,000 TS; 40,000 YS; 25 El; 55 RA; 92 Rock B. Heat treated: 240,000 TS; 210,000 YS; 8 El; 25 RA; 50 Rock C. For cutlery, knives, surgical instruments, shafts, gears. Corrosion resistant, hardenable.

KOSMOS

Rochling Burbach GmbH
C 0.82-0.9, Cr 4.5, W 12, Mo 0.7, V 2.5, bal Fe.
For tools, dies, cutters; high speed steel. *Obsolete*

KOSSIL

Sanderson Kayser Ltd.
C 0.9, Si 1.3-1.6, Cr 0.6-0.9, bal Fe.
Heat treated: 224,000 TS. For springs; tough, resilient. *Obsolete*

KOULTCHOOG-ALUMINUM (KAULTCHAAG)

Manufacturer not listed.
Cu 4-6, Mn, Mg, Si, bal Al.
For light alloy parts; similar to Duralumin, age-hardenable.

KOVAL

Stupakoff Laboratories Inc.
Ni 72, Co 17, Ti 2.2, Fe 6.25.
For internal combustion valves, molds and machine parts; high temperature resistant.

KOVAR

Now CARPENTER VACUMET KOVAR.

KOVAR

Stupakoff Laboratories Inc.
Ni 29, Co 17, Mn 0.2, bal Fe.
Annealed: 77,500 TS; 59,500 YS; 25 El. For electronic tubes, radio and X-ray tubes; same coefficient of expansion as glass; melting point 100°C.

KOVAR A

Westinghouse Electric Corp.
Ni 29, Co 17, Mn 0.3, bal Fe.
Annealed: 77,000 TS; 59,000 YS; 25 El; 882 Brin. For low expansion alloy for hard glass to metal seals; mean expansivity (25-400 C) 0.0000045. *Obsolete*

KP

W. Ossenberg & Cie Edelstahlwerke
C 1, Mn 0.2, Si 0.2, V 0.1, bal Fe.
Water hardening tool steel. W.-Nr. 1.2833; similar to AISI W1.

KP EXTRA

W. Ossenberg & Cie Edelstahlwerke
C 0.95, Mn 0.4, Si 0.3, V 0.4, bal Fe.
Water hardening tool steel. W.-Nr. 1.2835; similar to AISI W2.

KP-1

Russian manufacture
C 0.13, Mn 0.7, Si 0.2, Cr 12.2, Mo 0.7, bal Fe.
Annealed: 75,000 TS; 35,000 YS; 25 El; 92 Rock B. For springs, table flatware, oil refinery equipment; corrosion resistant.

KRAFFTS ALLOY

Swedish manufacture
Sn 8, Pb 26, Bi 66.
Melting point 100°C.

KRAMER "5" NICKEL SILVER

Illingworth Steel Co.
Zn 36, Ni 15, Sn 1, Pb 1, bal Cu.
Cast: 78,000 TS; 46,000 YS; 8 El; 8 RA; 180 Brin. For ornamental parts, fittings; white color. *Obsolete*

KRAMER "B" MANGANESE BRONZE

Illingworth Steel Co.
Mn, Fe, Zn, Sn, Al, bal Cu.
Cast: 75,000-82,000 TS; 34,000-38,000 YS; 20-28 El; 20-30 RA; 135-155 Brin. For marine propellers, fittings; corrosion resistant. *Obsolete*

KRAMER "K-MAG"

Illingworth Steel Co.
Mg 30, Cu 70.
For production of nodular gray iron castings; inoculator for Mg. *Obsolete*

KRAMER 12% NICKEL SILVER

Illingworth Steel Co.
Ni 12, Zn 22, Pb 10, Sn 3, Cr 53.
Cast: 38,000 TS; 20,000 YS; 18 El; 65 Brin. Corrosion resistant; for hardware, plumbing, fixtures, valves, ornamental castings.

KRAMER 15% NICKEL DIE CAST ALLOY

Illingworth Steel Co.
Ni 15, Zn, bal Cu.
Sand Cast: 80,000 TS; 45,000 YS; 8 El; 175 Brin. For ornamental castings, railroad car fittings.

KRAMER 15% PHOSPHOR COPPER

Illingworth Steel Co.
P 15, bal Cu.
Phosphor additive to copper alloys and brazing alloys.

KRAMER 68-1-3-28 ALLOY

Illingworth Steel Co.
Cu 68, Sn 1, Pb 3, Zn 28.
Cast: 34,000 TS; 12,500 YS; 40 El; 55 Brin. General purpose yellow brass, good machinability; for furniture, hardware, ornamental castings.

KRAMER 7-1/2% PHOSPHOR COPPER
Illingworth Steel Co.
P 7.5, bal Cu.
Phosphor additive to copper alloys and brazing alloys.

KRAMER 70/30 HARDENER
J.M. Ney Co.
Cu, Mn.
Obsolete

KRAMER 76-2.5-6.5-15 ALLOY
Illingworth Steel Co.
Cu 76, Sn 2.5, Pb 6.5, Zn 15.
Cast: 36,000 TS; 14,500 YS; 36 El; 58 Brin. For plumbing fixtures, faucets; good machinability.

KRAMER 78-7-15 ALLOY
Illingworth Steel Co.
Cu 78, Sn 7, Pb 15.
Cast: 34,000 TS; 17,500 YS; 30 El; 61 Brin. Anti-acid metal corrosion resistant for chemical pumps, mining equipment, bearings, and bushings.

KRAMER 80-10-10 ALLOY
Illingworth Steel Co.
Cu 80, Sn 10, Pb 1, bal Fe.
Cast: 41,000 YS; 18,500 YS; 35 El; 65 Brin. For bushings, bearings, pump bodies, and impellers, heavy duty; corrosion resistant.

KRAMER 81-2-2-15
Illingworth Steel Co.
Cu 81, Sn 2, Pb 2, Zn 15.
Cast: 37,000 TS; 15,500 YS; 37 El; 50 Brin. For ornamental castings; good machinability. *Obsolete*

KRAMER 81-2-2-15 ALLOY
Illingworth Steel Co.
Cu 81, Sn 2, Pb 2, Zn 15.
Cast: 35,000 TS; 15,000 YS; 35 El; 30 RA. For ornamental castings; free-cutting. *Obsolete*

KRAMER 81-3-7-9 ALLOY
Illingworth Steel Co.
Cu 81, Sn 3, Pb 7, Zn 9.
Cast: 34,000 TS; 17,000 YS; 27 El; 55 Brin. For low pressure valves and fittings; good machinability.

KRAMER 83-3-3-11 ALLOY
Illingworth Steel Co.
Cu 83, Sn 3, Pb 3, Zn 11.
Cast: 37,000 TS; 15,500 YS; 38 El; 53 Brin. Expendable fire hose couplings and fittings; good machinability. *Obsolete*

KRAMER 83-7-7-3 ALLOY
Illingworth Steel Co.
Cu 83, Sn 7, Pb 7, Zn 3.
Cast: 40,000 TS; 18,000 YS; 39 El; 67 Brin. General purpose bearing alloy, corrosion resistant, for bushings, bearing, heavy duty auto fittings.

KRAMER 85-5-5-5 ALLOY
Illingworth Steel Co.
Cu 85, Sn 5, Pb 5, Zn 5.
Cast: 37,000 TS; 16,500 YS; 34 El; 65 Brin. Red brass, good machinability; for low pressure valves and fittings.

KRAMER 85/15 HARDENER
Illingworth Steel Co.
Cu-Si.

KRAMER 86-6-1.5-4.5 ALLOY
Illingworth Steel Co.
Cu 88, Sn 6, Pb 1.5, Zn 4.5.
Cast: 42,000 TS; 18,000 YS; 43 El; 70 Brin. Corrosion resistant; for valves, fittings, flanges, gears; for high temperatures and pressures.

KRAMER 87-1-3-9 ALLOY
Illingworth Steel Co.
Cu 87, Sn 1, Pb 3, Zn 9.
Cast: 33,000 TS; 14,000 YS; 28 El; 30 RA. For ornamental castings, plaques; free-cutting. *Obsolete*

KRAMER 87-8-1-4 ALLOY
Illingworth Steel Co.
Sn 8, Pb 1, Zn 4, bal Cu.
Cast: 45,000 TS; 19,000 YS; 40 El,; 70 Brin. Corrosion resistant; for valves, steam pressure castings; resists high pressure.

KRAMER 88-12 ALUMINUM BRONZE
Illingworth Steel Co.
Al, bal Cu.
Cast: 55,000 TS; 40,000 YS; 1 El; 230 Brin. For welding fixtures; 20% IACS.

KRAMER 88-3-2-7 ALLOY
Illingworth Steel Co.
Cu 88, Sn 3, Pb 2, Zn 7.
Cast: 38,000 TS; 16,000 YS; 37 El; 60 Brin. For ornamental and statuary castings. *Obsolete*

KRAMER 88-6-1.5-4.5 ALLOY
Illingworth Steel Co.
Cu 88, Sn 6, Pb 1.5, Zn 4.5.
Cast: 42,000 TS; 18,000 YS; 43 El; 70 Brin. Corrosion resistant; for valves, fittings, flanges, gears; for high temperatures and pressures.

KRAMER 88-8-0-4 ALLOY
Illingworth Steel Co.
Cu 88, Sn 8, Zn 4.
Cast: 47,000 TS; 20,000 YS; 39 El; 70 Brin. Corrosion resistant; for valves, pumps.

KRAMER A MANGANESE BRONZE
Illingworth Steel Co.
Cu 58, Fe 1, Mn 0.5, Al 1, bal Zn.
Cast: 70,000-75,000 TS; 28,000-35,000 YS; 22-35 El; 120-140 Brin. For piston rods, valve stems, worm gears, propellers, structural castings, marine fittings. Corrosion resistant; substitute for steel and malleable iron where corrosion resistance is required.

KRAMER ALLOY NO. 66
Illingworth Steel Co.
Mg, Mn, Ca, P, bal Cu.
For use as a deoxidizer and degassifier for Cu-Ni alloys.

KRAMER ALLOY NO. 77
Illingworth Steel Co.
Mg, Mn, Ca, Si, Ti, Al, bal Cu.
For use as an aluminum bronze degassifier and to improve humidity.

KRAMER ANTI-ACID METAL
Illingworth Steel Co.
Cu 75, Sn 10, Pb 15.
For chemical apparatus; acid resistant.

KRAMER AX MANGANESE BRONZE
Illingworth Steel Co.
Cu 58, Mn 1, Fe 1.5, Al 1.5, bal Zn.
Cast: 80,000-88,000 TS; 38,000-40,000 YS; 18-24 El; 160-165 Brin. Corrosion resistant, for structural castings, valve stems, propellers, marine castings.

KRAMER C ALUMINUM BRONZE
Illingworth Steel Co.
Al 10, Fe 4, Ni 2, bal Cu.
Cast: 88,000-96,000 TS; 33,000-36,000 YS; 10-25 El; 7-30 RA; 150-175 Brin. Heat treated: 105,000-125,000 TS; 45,000-75,000 YS; 2-22 El; 3-20 RA; 180-300 Brin. For gears, worm wheels, bushings; corrosion resistant.

KRAMER CONTACT METAL
Illingworth Steel Co.
Sn, Pb, Zn, bal Cu.
Cast: 32,000 TS; 9500 YS; 30 El; 50 Brin. For electrical fittings; 32% IACS.

KRAMER DIE CAST YELLOW BRASS
Illingworth Steel Co.
Sn 1, Pb 1, Zn 35, Al, bal Cu.
Cast: 55,000 TS; 30,000 YS; 15-20 El; 120-130 Brin. For plumbing fixtures, ornamental castings; permanent mold castings.

KRAMER DIE MOLD BRONZE
Illingworth Steel Co.
Ni, Al, Zn, bal Cu.
Sand cast: 90,000 TS; 60,000 YS; 2.5 El; 200 Brin. For glass molds and high temperature applications; sand cast.

KRAMER EVERDUR
Illingworth Steel Co.
Cu, Si, Mn.
Corrosion resistant silicon bronze.

KRAMER G ALUMINUM BRONZE
Illingworth Steel Co.
Al, Fe, bal Cu.
Cast: 69,000 TS; 49,000 YS; 1 El; 265 Brin. For wear plates and guides.

KRAMER GLASS MOLD ALLOY
Illingworth Steel Co.
Cu, Zn, Ni, Al.
Used for making glass objects.

KRAMER I ALUMINUM BRONZE
Illingworth Steel Co.
Fe 1, Al 10, bal Cu.
As cast: 67,000-77,000 TS; 26,000-34,000 YS; 12-26 El; 10-24 RA; 130-150 Brin. Heat treated: 90,000-95,000 TS; 40,000-55,000 YS; 10-25 El; 160-220 Brin. Heat treatable aluminum bronze, corrosion resistant; for structural castings, gears, machine parts, steel mill nuts.

KRAMER L ALUMINUM BRONZE
Illingworth Steel Co.
Fe 0-4, Al 0-10, Mn 0-0.5, bal Cu.
As cast: 82,000-93,000 TS; 29,000-38,000 YS; 12-32 El; 140-165 Brin. Heat treated: 100,000-110,000 TS; 35,000-55,000 YS; 10-30 El; 160-210 Brin. Heat treatable aluminum bronze, corrosion resistant; for structural castings, gears, worm wheels.

KRAMER M ALUMINUM BRONZE
Illingworth Steel Co.
Fe 3-4, Al 9-9.5, bal Cu.
Cast: 75,000-85,000 TS; 23,000-33,000 YS; 25-45 El; 26-47 RA; 125-140 Brin. General utility aluminum bronze, corrosion resistant; for structural castings, valves, gears, nuts, pumps.

KRAMER N ALUMINUM BRONZE
Illingworth Steel Co.
Fe 4-5, Al 10-11, Ni 4-5, Mn 0-1, bal Cu.
As cast: 100,000-108,000 TS; 40,000-50,000 YS; 5-18 El; 6-20 RA; 180-210 Brin. Heat treated: 110,000-125,000 TS; 60,000-80,000 YS; 5-12 El; 220-260 Brin. Heat treatable aluminum bronze, corrosion resistant; for structural castings, pump and valve parts, gears, aircraft parts, marine casting.

KRAMER NICKEL ALUMINUM BRONZE
Illingworth Steel Co.
Fe, Al, Ni, Mn, bal Cu.
Cast: 90,000 TS; 40,000 YS; 20 El; 160 Brin. Corrosion resistant; for marine propellers.

KRAMER NICKEL SILVER I
Illingworth Steel Co.
Ni 20, Zn 8, Pb 4, Sn 5, Cu 63.
Cast: 46,000 TS; 22,000 YS; 20 El; 18 RA; 80 Brin. For hardware, plumbing fixtures, valves; corrosion resistant
Obsolete

KRAMER O ALUMINUM BRONZE
Illingworth Steel Co.
Fe 6-7, Al 11.5-12.5, Ni 3-4.5, Mn 0-0.5, bal Cu.
Cast: 90,000 TS; 70,000 YS; 1-3 El; 265-300 Brin. For castings; Al bronze. *Obsolete*

KRAMER R ALUMINUM BRONZE
Illingworth Steel Co.
Fe 4, Al 10, bal Cu.
As cast: 85,000 TS; 35,000 YS; 15 El; 160 Brin. Heat treated: 105,000 TS; 50,000 YS; 10 El; 200 Brin. Heat treatable aluminum bronze; corrosion resistant; for structural castings, valves, gears, bushings.

KRAMER S MANGANESE BRONZE
Illingworth Steel Co.
Cu 60, Fe 1, Mn 0.5, Al 0.75, Sn 1, Pb 1, bal Zn.
Cast: 60,000-65,000 TS; 25,000-30,000 YS; 15-20 El; 100-120 Brin. Corrosion resistant, free machining. For piston rods, valve stems, worm gears, propellers, valve bodies, fittings, structural castings, marine fittings.

KRAMER SILICON BRONZE
Illingworth Steel Co.
Si, Zn or Mn, bal Cu.
Cast: 60,000 TS; 25,000 YS; 40 El; 88 Brin. Corrosion resistant; for valves, pumps.

KRAMER SILICON-ALUMINUM BRONZE
Illingworth Steel Co.
Al, Si, bal Cu.
Sand cast: 70,000 TS; 30,000 YS; 25 El; 130 Brin. For valve stems, structural castings; corrosion resistant.

KRAMER SPECIAL "M" NICKEL SILVER
Illingworth Steel Co.
Ni 25, Sn 5, Zn 1, bal Cu.
Cast: 52,500 TS; 33,000 YS; 9 El; 12 RA; 135 Brin. For valve seats, discs; nongalling. *Obsolete*

KRAMER SPECIAL 20% NICKEL SILVER
Illingworth Steel Co.
Ni 20, Zn, bal Cu.
Sand cast: 50,000 TS; 25,000 YS; 20 El; 80 Brin. For valves, dairy and laundry castings, ornamental parts; corrosion resistant.

KRAMER SPECIAL 25% NICKEL SILVER
Illingworth Steel Co.
Ni 25, Zn, bal Cu.
Sand cast: 60,000 TS; 36,000 YS; 14 El; 135 Brin. For valve seats; corrosion resistant.

KRAMER SX MANGANESE BRONZE
Illingworth Steel Co.
Zn, Mn, Al, Fe, bal Cu.
Cast: 95,000-104,000 TS; 45,000-55,000 YS; 18-22 El; 18-22 RA; 180-195 Brin. For marine and aircraft castings; corrosion resistant.

KRAMER X MANGANESE BRONZE
Illingworth Steel Co.
Cu 60, Fe 1.5, Mn 1.5, Al 3, bal Zn.
Cast: 85,000-95,000 TS; 45,000 YS; 18-20 El; 175 Brin. Corrosion resistant; for hubcaps, piston rods, valve stems, propellers, marine fittings, structural castings.

KRAMER XX MANGANESE BRONZE
Illingworth Steel Co.
Cu 63, Fe 2, Mn 2.5, Al 5, bal Zn.
Cast: 110,000-115,000 TS; 60,000-72,000 YS; 13-18 El; 225 Brin. Heavy duty high strength alloy. For gears, cams, bridge bearings, screw down nuts, structural castings.

KREIDLER MO 60 R-NIET
Kreidler Werke G.m.b.H.
Copper. Cu 60.5, Pb 1.5, Zn 38.
Ductile, free-machining brass; for rivets.

KREIDLER MS 57/2
Kreidler Werke G.m.b.H.
Copper. Cu 58, Pb 2, bal Zn.
Good machinability and hot formability. Cu Zn 40 Pb 2.

KREIDLER MS 58
Kreidler Werke G.m.b.H.
Copper. Cu 58, Pb 3, bal Zn.
Good machinability and hot formability. Cu Zn 39 Pb 3.

KREIDLER MS 58A
Kreidler Werke G.m.b.H.
Copper. Cu 58, Pb 2, bal Zn.
For automatic machining, good machinability; good hot forming. Cu Zn 40 Pb 2.

KREIDLER MS 59/2
Kreidler Werke G.m.b.H.
Copper. Cu 59, Pb 2, bal Zn.
Free machining brass. Cu Zn 39 Pb 2.

KREIDLER MS 60 R
Kreidler Werke G.m.b.H.
Copper. Cu 60.5, Pb 1.5, Zn 38.
Free machining brass; good hot and cold forming. Cu Zn 38 Pb 1.5.

KREIDLER MS 63 D
Kreidler Werke G.m.b.H.
Copper. Cu 63, Zn 37.
For cold forming operations. Cu Zn 37.

KREIDLER SU 10
Kreidler Werke G.m.b.H.
Copper. Cu 58, Al 2, Zn 40.
55-65 kp/mm^2 TS; 24-32 kp/mm^2 YS; 10-18 El. High strength aluminum brass. Cu Zn 40 Al 2.

KREIDLER SU 4
Kreidler Werke G.m.b.H.
Copper. Cu 57-59, Mn 1-2.5, bal Zn.
45-50 kp/mm^2 TS; 18-28 kp/mm^2 YS; 18-20 El. Brass for appliance parts; good solderability. Cu Zn 40 Mn.

KRIEDLER SU 81
Kreidler Werke G.m.b.H.
Copper. Cu 58-61, Ni 2-3, bal Zn.
45-55 kp/mm^2 TS; 20-40 kp/mm^2 YS; 12-20 El. Medium strength brass for marine applications. Cu Zn 35 Ni.

KROKOLOY
Detroit Alloy Steel Co.
C-Cr-Co, bal Fe.
For dies for forming, blanking, drawing and coining, shear blades; cast to shape, air hardening.

KROM-ALOY
Chrome Alloys Mfg. Co.
Cr, Sn, Pb, Al, Mn, Ni, Zn, bal Cu.
77,210 TS; 54,600 YS; 10 El; 9 RA; 153 Brin. For marine hardware, propellers, pump valves and seats, trimmings; corrosion resistant.

KROMAIR
Now REPUBLIC A2.

KROMAL
Jessop Steel Co.
Mo 9, C 0.7, Cr 4, V 1.25, bal Fe.
Annealed: 228-241 Brin. For high speed tools and cutters; high speed steel. *Obsolete*

KROMAL NO. 1
Allied Steel & Tractor Products Inc.
C 0.9, Mn 1.2, Cr 0.5, bal Fe.
For punches, taps, dies, crimpers; nondeforming; oil hardening.

KROMAL NO. 2
Allied Steel & Tractor Products Inc.
C 0.9, Mn 1.5, Cr 0.4, V 0.3, bal Fe.
For tools, punches; nondeforming.

KROMAL NO. 3
Allied Steel & Tractor Products Inc.
C 0.5, Mn 0.9, Cr 0.9, Mo 0.2, bal Fe.
For maintenance and repair; oil hardening.

KROMAL NO. 4
Allied Steel & Tractor Products Inc.
C 0.48, Cr 1.85, Mo 0.4, Ni 1.25, bal Fe.
For maintenance and repair; oil hardening.

KROMAR D.70
British Steel Corp.
C 0-0.03, Si 0.1, Mn 0.1, Ni 4-5, Cr 11.5-12.5, Mo 4-5, Co 12-14, Ti 0.2-0.5, Al 0.05-0.15, bal Fe.
Precipitation hardening stainless; hardened: 180,000 TS; 170,000 YS, 16.5 El; 37-39 IZ. For aircraft components, pressure vessels, turbine blading. *Obsolete*

KROMARC 55
Westinghouse Electric Corp.
C 0.04, Cr 16, Ni 20, Mn 9.5, Mo 2, bal Fe.
For cast nozzle chambers, welding electrodes; austenitic, stainless, good weldability, resists hot tearing. *Obsolete*

KROMARC-58
Westinghouse Electric Corp.
Cr 15.5, Ni 21, Mn 10, Mo 2.25, Si 0.12, N 0.17, V 0.25, B 0.008, Zr 0.015, C 0.03, bal Fe.
At 400 F: 100,000 TS; 65,000 YS; 36 El; 65 RA. At 1200 F: 80,000 TS; 45,000 YS; 25 El; 50 RA. At 2000 F: 15,000 TS; 15,000 YS; 25 El; 25 RA. For welded components in steam turbines. High temperature tensile, creep and rupture strength; austenitic, stainless, and heat resistant. *Obsolete*

KROMAX
Manufacturer not listed.
Ni 80, Cr 20.
For resistance alloy; heat resistant.

KROMITE 14% MANGANESE PLATE
Associated Steel Corp.
C 1.3, Mn 13.5, Si 0.2, Mo 0.2, bal Fe.
Austenitic, manganese steel. As shipped: 190 Brin, work hardened to 580 Brin. Weldable with 14% Mn electrodes.

KROMITE BRAKE DIE
Associated Steel Corp.
C, alloy, bal Fe.
For brake dies, bead-forming dies; preheat treated, tough and wear resistant.

KROMITE NO. 2
Associated Steel Corp.
C, alloy, bal Fe.
For chain links, mandrels springs, tongs; shock resistant.

KROMITE NO. 3
Associated Steel Corp.
C, alloy, bal Fe.
Heat treated: 156,000 TS; 130,000 YS; 17 El; 56 RA; 270 Brin. For shafts, gears, bolts, tie rods; preheat treated.

KROMITE NO. 4
Associated Steel Corp.
C, alloy, bal Fe.
Heat treated: 211,000-304,000 TS; 450-560 Brin. For punches, picks, swages, vice jaws, bolts; tough, shock resistant.

KROMITE WEAR PLATE
Associated Steel Corp.
C 0.39, Mn 2.3, Si 0.28, Cr 0.9, Mo 0.34, Cu 0.5, bal Fe.
As delivered: 164,000 psi TS; 129,000 psi YS; 380-400 Brin. For chute liners, stamping dies, mixing blades, scraper blades.

KROMORE
Driver Harris Co.
Ni 85, Cr 15.
For heating elements, electrical resistances; heat resistant.
Obsolete

KROMOX
Capitol Castings Inc.
C 2.6, Mn 0.5, Cr 15.2, 1.25 Mo (typical), bal Fe.
High abrasion resistance for slurry pump parts, mill liners.

KRONA
Time Steel Service Inc.
C 0.9, Mn 1.2, Cr 0.5, V 0.2, W 0.5, bal Fe.
Oil hardened tool steel. AISI O1.

KRONOS
Peter A. Frasse & Co.
C 0.8, Mn 0.4, bal Fe.
For general tools. *Obsolete*

KROPP 23
Kropp Forge Co.
C 0.3, Mn 0.7, Si 0.35, Cr 0.96, Mo 0.2, bal Fe.
For plastic mold dies; water hardened.

KROPP 23C
Kropp Forge Co.
C 0.32, Mn 0.75, Si 0.35, Cr 0.9, Mo 0.2, bal Fe.
For zinc die casting dies; water hardened.

KROPP 56
Kropp Forge Co.
C 0.6, Mn 0.65, Ni 1.8, Cr 1.6, Mo 0.35, bal Fe.
For forging dies, shear blades, high pressure cylinders; oil hardening.

KROPP 61
Kropp Forge Co.
C 0.1, Mn 0.5, Ni 3.5, Cr 1.5, bal Fe.
For gears, pinions, spindels, cams, bearing rolls; tough.

KROPP 93
Kropp Forge Co.
C 0.35, Si 1, Cr 5, Mo 1, V 0.15, W 1, bal Fe.
For shear blades, extrusion and upsetting dies, mandrels; hot work steel.

KROPUNCH
Allied Steel & Tractor Products Inc.
C 0.4, W 2.5, V 0.3, Mo 2, Si 1.1, Cr, bal Fe.
For shear blades, punches, trim dies, mandrels; hot and cold work steel, air hardening.

KROSIL
Colt Industries
C 0.5, Mn 0.9, Si 1.5, Mo 0.35, bal Fe.
For rivet shearing tools, cold tie plate punches; shock and fatigue resistant. *Obsolete*

KROSIL CHISEL
Hawkridge Bros. Co.
C 0.5, Mn 0.9, Mo 0.35, Si 1.5, bal Fe.
For cold chisels, upsetters, dies, crimpers; oil hardened, shock resistant. *Obsolete*

KROTUNG
Allied Steel & Tractor Products Inc.
C 0.3, Cr 5, Mo 1.5, W 1.3, V 0.3, bal Fe.
For hot forging dies and tools, nut crowners; high red hardness; tough.

KROVAN
Colt Industries
C 0.95, Cr 1, V 0.2, bal Fe.
Water hardened. For jewelers' dies, heavy blanking dies, rolls; deep hardening, resists wear and impact; tough. *Obsolete*

KROVAN
Hawkridge Bros. Co.
C 0.9, Cr 1, V 0.2, bal Fe.
For rolls; water hardening. *Obsolete*

KROVAN
Crucible Specialty Metals
C 0.95, Cr 1, V 0.2, bal Fe.
Water hardened. For jewelers' dies, heavy blanking dies, rolls; deep hardening, resists wear and impact; tough. *Obsolete*

KRUPP 1.2334
Krupp Stahl AG
C 0.6-0.65, Si 0.2-0.4, Mn 0.4-0.6, P 0-0.03, S 0-0.03, Cr 1.3-1.5, Mo 0.25-0.3, V 0.07-0.12, bal Fe.
SEL 90; W. Nr. 1.2334.

KRUPP 10 CRMO 11
Krupp Stahl AG
C 0.08-0.12, Si 0.15-0.35, Mn 0.3-0.5, P 0-0.035, S 0-0.035, Cr 2.7-3, Mo 0.2-0.3, bal Fe.
SEW 590/61; W. Nr. 1.7276.

KRUPP 10 CRMO 9 10
Krupp Stahl AG
C 0.06-0.15, Si 0-0.5, Mn 0.4-0.7, P 0-0.035, S 0-0.03, Cr 2-2.5, Mo 0.9-1.1, bal Fe.
DIN 17155/83; W. Nr. 1.7380.

KRUPP 10 CRSIMOV 7
Krupp Stahl AG
C 0-0.12, Si 0.9-1.2, Mn 0.35-0.75, P 0-0.035, S 0-0.035, Cr 1.6-2, Mo 0.25-0.35, V 0.25-0.35, bal Fe.
SEL 90; W. Nr. 1.8075.

KRUPP 10 MNSI 5
Krupp Stahl AG
C 0.07-0.11, Si 0.55-0.75, Mn 1.03-1.27, P 0-0.02, S 0-0.02, Cr 0-0.12, Mo 0-0.12, Ni 0-0.12, Al 0-0.02, Cu 0-0.17, Ti 0-0.07, Zr 0-0.07, bal Fe.
DIN 17145/80; W. Nr. 1.5112.

KRUPP 10 MNSI 7
Krupp Stahl AG
C 0.08-0.13, Si 0.85-1.13, Mn 1.63-1.87, P 0-0.02, S 0-0.02, Cr 0-0.12, Mo 0-0.12, Ni 0-0.12, Al 0-0.02, Cu 0-0.17, Ti 0-0.07, Zr 0-0.06, bal Fe.
DIN 17145/80; W. Nr. 1.5130.

KRUPP 10 NI 14
Krupp Stahl AG
C 0-0.15, Si 0-0.35, Mn 0.3-0.8, P 0-0.025, S 0-0.02, Ni 3.25-3.75, V 0-0.05, bal Fe.
DIN 17280/85; W. Nr. 1.5637.

KRUPP 10 NICR 6 4
Krupp Stahl AG
C 0.07-0.12, Si 0.1-0.4, Mn 0.6-0.9, P 0-0.035, S 0-0.035, Cr 0.85-1.15, Ni 1.2-1.6, bal Fe.
SEL 90; W. Nr. 1.5805.

KRUPP 10 NIMNMO 6 5
Krupp Stahl AG
C 0.08-0.12, Si 0.1-0.25, Mn 1.3-1.5, P 0-0.02, S 0-0.02, Mo 0.35-0.5, Ni 1.5-1.7, bal Fe.
SEL 90; W. Nr. 1.6312.

KRUPP 10 NIMNMOCR 6 6
Krupp Stahl AG
C 0.07-0.13, Si 0.1-0.25, Mn 1.35-1.65, P 0-0.02, S 0-0.025, Cr 0.1-0.3, Mo 0.4-0.6, Ni 1.5-1.7, bal Fe.
SEL 90; W. Nr. 1.6313.

KRUPP 10 S 20
Krupp Stahl AG
C 0.07-0.13, Si 0.1-0.3, Mn 0.7-1.1, P 0-0.06, S 0.18-0.25, bal Fe.
DIN 1651/88; W. Nr. 1.0721.

KRUPP 10 SPB 20
Krupp Stahl AG
C 0.07-0.13, Si 0.1-0.3, Mn 0.7-1.1, P 0-0.06, S 0.18-0.25, Pb 0.15-0.35, bal Fe.
DIN 1651/88; W. Nr. 1.0722.

KRUPP 100 CR 2
Krupp Stahl AG
C 0.9-1.05, Si 0.15-0.35, Mn 0.25-0.4, P 0-0.03, S 0-0.025, Cr 0.4-0.6, Ni 0-0.3, Cu 0-0.3, bal Fe.
DIN 17230/80; W. Nr. 1.3501.

KRUPP 100 CR 6
Krupp Stahl AG
C 0.9-1.05, Si 0.15-0.35, Mn 0.25-0.45, P 0-0.03, S 0-0.025, Cr 1.35-1.65, Ni 0-0.3, Cu 0-0.3, bal Fe.
DIN 17230/80; W. Nr. 1.3505.

KRUPP 100 CRMN 6
Krupp Stahl AG
C 0.9-1.05, Si 0.5-0.7, Mn 1-1.2, P 0-0.03, S 0-0.025, Cr 1.4-1.65, Ni 0-0.3, Cu 0-0.3, bal Fe.
DIN 17230/80; W. Nr. 1.3520.

KRUPP 100 CRMNMO 8
Krupp Stahl AG
C 0.9-1.05, Si 0.4-0.6, Mn 0.8-1.1, P 0-0.03, S 0-0.025, Cr 1.8-2.05, Mo 0.5-0.6, Ni 0-0.3, Cu 0-0.3, bal Fe.
DIN 17230/80; W. Nr. 1.3539.

KRUPP 100 CRMO 7 3
Krupp Stahl AG
C 0.9-1.05, Si 0.2-0.4, Mn 0.6-0.8, P 0-0.03, S 0-0.025, Cr 1.65-1.95, Mo 0.2-0.35, Ni 0-0.3, Cu 0-0.3, bal Fe.
DIN 17230/80; W. Nr. 1.3536.

KRUPP 100 CRMO 7 4
Krupp Stahl AG
C 0.9-1.05, Si 0.2-0.4, Mn 0.6-0.8, P 0-0.03, S 0-0.025, Cr 1.65-1.95, Mo 0.4-0.5, Ni 0-0.3, Cu 0-0.3, bal Fe.
SEL 90; W. Nr. 1.3538.

KRUPP 105 CR 4
Krupp Stahl AG
C 1-1.1, Si 0.15-0.35, Mn 0.25-0.4, P 0-0.03, S 0-0.025, Cr 0.9-1.15, bal Fe.
SEL 90; W. Nr. 1.3503.

KRUPP 11 MN 4 AL
Krupp Stahl AG
C 0.08-0.14, Si 0-0.12, Mn 0.83-1.15, P 0-0.025, S 0-0.025, Cr 0-0.12, Ni 0-0.12, Al 0-0.04, Cu 0-0.17, bal Fe.
DIN 17145/80; W. Nr. 1.0494.

KRUPP 11 MN 4 SI
Krupp Stahl AG
C 0.08-0.14, Si 0.18-0.37, Mn 0.83-1.15, P 0-0.025, S 0-0.025, Cr 0-0.12, Ni 0-0.12, Al 0-0.03, Cu 0-0.17, bal Fe.
DIN 17145/80; W. Nr. 1.0492.

KRUPP 11 MNMO 4 5
Krupp Stahl AG
C 0.09-0.14, Si 0.08-0.22, Mn 0.85-1.15, P 0-0.02, S 0-0.02, Cr 0-0.12, Mo 0.48-0.62, Ni 0-0.12, Al 0-0.03, Cu 0-0.17, bal Fe.
DIN 17145/80; W. Nr. 1.5425.

KRUPP 11 MNNI 5 3
Krupp Stahl AG
C 0-0.14, Si 0-0.5, Mn 0.7-1.5, P 0-0.03, S 0-0.025, Ni 0.3-0.8, Nb 0-0.05, V 0-0.05, 0.02 Al min, bal Fe.
DIN 17280/85; W. Nr. 1.6212.

KRUPP 11 MNSI 6
Krupp Stahl AG
C 0.08-0.13, Si 0.75-0.95, Mn 1.33-1.57, P 0-0.02, S 0-0.02, Cr 0-0.12, Mo 0-0.12, Ni 0-0.12, Al 0-0.02, Cu 0-0.17, Ti 0-0.07, Zr 0-0.06, bal Fe.
DIN 17145/80; W. Nr. 1.5125.

KRUPP 11 NIMNCRMO 5 5
Krupp Stahl AG
C 0-0.15, Si 0.25-0.5, Mn 1-1.4, P 0-0.035, S 0-0.035, Cr 0.4-0.6, Mo 0.2-0.4, Ni 1-1.4, V 0.06-0.12, 0.02 Al min, bal Fe.
SEL 90; W. Nr. 1.6919.

KRUPP 12 CRMO 19 5
Krupp Stahl AG
C 0-0.15, Si 0.3-0.5, Mn 0.3-0.6, P 0-0.035, S 0-0.035, Cr 4.5-5.5, Mo 0.45-0.65, bal Fe.
SEL 90; W. Nr. 1.7362.

KRUPP 12 MN 6
Krupp Stahl AG
C 0.08-0.14, Si 0.08-0.22, Mn 1.35-1.65, P 0-0.025, S 0-0.025, Cr 0-0.12, Ni 0-0.12, Al 0-0.03, Cu 0-0.17, bal Fe.
DIN 17145/80; W. Nr. 1.0496.

KRUPP 12 MNNI 6 3
Krupp Stahl AG
C 0-0.15, Si 0.15-0.4, Mn 0.8-1.6, P 0-0.03, S 0-0.03, Ni 0.5-0.8, 0.025 Al min, bal Fe.
SEL 90; W. Nr. 1.6213.

KRUPP 12 MNNIMO 5 5
Krupp Stahl AG
C 0-0.15, Si 0.2-0.5, Mn 1.1-1.5, P 0-0.035, S 0-0.035, Cr 0-0.3, Mo 0.2-0.5, Ni 0.8-1.6, V 0-0.05, bal Fe.
SEL 90; W. Nr. 1.6343.

KRUPP 12 NI 19
Krupp Stahl AG
C 0-0.15, Si 0-0.35, Mn 0.3-0.8, P 0-0.025, S 0-0.02, Ni 4.5-5.3, V 0-0.05, bal Fe.
DIN 17280/85; W. Nr. 1.5680.

KRUPP 13 CR 2
Krupp Stahl AG
C 0.1-0.16, Si 0.15-0.35, Mn 0.4-0.6, P 0-0.035, S 0-0.035, Cr 0.3-0.5, bal Fe.
SEL 90; W. Nr. 1.7012.

KRUPP 13 CRMO 4 4
Krupp Stahl AG
C 0.08-0.18, Si 0.1-0.35, Mn 0.4-1, P 0-0.035, S 0-0.03, Cr 0.7-1.1, Mo 0.4-0.6, bal Fe.
DIN 17155/83; W. Nr. 1.7335.

KRUPP 13 MN 6
Krupp Stahl AG
C 0.08-0.14, Si 0.3-0.45, Mn 1.35-1.65, P 0-0.025, S 0-0.025, Cr 0-0.12, Ni 0-0.12, Cu 0-0.17, bal Fe.
DIN 17145/80; W. Nr. 1.0479.

KRUPP 13 MNMO 12 5
Krupp Stahl AG
C 0.08-0.17, Si 0.15-0.3, Mn 2.8-3.2, P 0-0.03, S 0-0.03, Mo 0.45-0.6, bal Fe.
SEL 90; W. Nr. 1.5428.

KRUPP 13 MNMO 6 5
Krupp Stahl AG
C 0.09-0.14, Si 0.08-0.22, Mn 1.35-1.65, P 0-0.02, S 0-0.02, Cr 0-0.12, Mo 0.48-0.62, Ni 0-0.12, Al 0-0.03, Cu 0-0.17, bal Fe.
DIN 17145/80; W. Nr. 1.5426.

KRUPP 13 MNMO 8 5
Krupp Stahl AG
C 0.08-0.14, Si 0.08-0.22, Mn 1.8-2.2, P 0-0.02, S 0-0.02, Cr 0-0.12, Mo 0.48-0.62, Ni 0-0.12, Al 0-0.03, Cu 0-0.17, bal Fe.
DIN 17145/80; W. Nr. 1.5427.

KRUPP 13 MNNI 6 3
Krupp Stahl AG
C 0-0.16, Si 0-0.5, Mn 0.85-1.65, P 0-0.03, S 0-0.025, Ni 0.3-0.85, Nb 0-0.05, V 0-0.05, 0.02 Al min, bal Fe.
DIN 17280/85; W. Nr. 1.6217.

KRUPP 13 NICR 6
Krupp Stahl AG
C 0.1-0.17, Si 0.15-0.35, Mn 0.3-0.5, P 0-0.035, S 0-0.035, Cr 0.65-0.85, Ni 1.35-1.5, bal Fe.
SEL 90; W. Nr. 1.5713.

KRUPP 14 CRMOV 6 9
Krupp Stahl AG
C 0.11-0.17, Si 0-0.25, Mn 0.8-1, P 0-0.02, S 0-0.015, Cr 1.25-1.5, Mo 0.8-1, V 0.2-0.3, bal Fe.
SEL 90; W. Nr. 1.7735.

KRUPP 14 MOV 6 3
Krupp Stahl AG
C 0.1-0.18, Si 0.1-0.35, Mn 0.4-0.7, P 0-0.035, S 0-0.035, Cr 0.3-0.6, Mo 0.5-0.7, V 0.22-0.32, bal Fe.
DIN 17175/79; W. Nr. 1.7715.

KRUPP 14 NI 6
Krupp Stahl AG
C 0-0.18, Si 0.1-0.35, Mn 0.3-0.8, P 0-0.035, S 0-0.035, Ni 1.3-1.6, bal Fe.
SEL 90; W. Nr. 1.5622.

KRUPP 14 NICR 10
Krupp Stahl AG
C 0.1-0.17, Si 0.15-0.35, Mn 0.4-0.7, P 0-0.035, S 0-0.035, Cr 0.55-0.95, Ni 2.25-2.75, bal Fe.
SEL 90; W. Nr. 1.5732.

KRUPP 14 NICR 14
Krupp Stahl AG
C 0.1-0.17, Si 0.15-0.35, Mn 0.4-0.7, P 0-0.035, S 0-0.035, Cr 0.55-0.95, Ni 3.25-3.75, bal Fe.
SEL 90; W. Nr. 1.5752.

KRUPP 14 NICR 18
Krupp Stahl AG
C 0.1-0.17, Si 0.15-0.35, Mn 0.4-0.7, P 0-0.035, S 0-0.035, Cr 0.9-1.3, Ni 4.25-4.75, bal Fe.
SEL 90; W. Nr. 1.5860.

KRUPP 14 NICRMO 13 4
Krupp Stahl AG
C 0.12-0.17, Si 0.15-0.4, Mn 0.3-0.6, P 0-0.025, S 0-0.02, Cr 0.8-1.1, Mo 0.2-0.3, Ni 3-3.5, bal Fe.
SEL 90; W. Nr. 1.6657.

KRUPP 15 CR 3
Krupp Stahl AG
C 0.12-0.18, Si 0.15-0.4, Mn 0.4-0.6, P 0-0.035, S 0-0.035, Cr 0.4-0.7, bal Fe.
DIN 17210/69; W. Nr. 1.7015.

KRUPP 15 CRMO 5
Krupp Stahl AG
C 0.13-0.17, Si 0.15-0.35, Mn 0.8-1.1, P 0-0.035, S 0-0.035, Cr 1-1.3, Mo 0.2-0.3, bal Fe.
SEL 90; W. Nr. 1.7262.

KRUPP 15 CRMOV 5 9
Krupp Stahl AG
C 0.13-0.18, Si 0-0.4, Mn 0.8-1.1, P 0-0.025, S 0-0.03, Cr 1.2-1.5, Mo 0.8-1.1, V 0.2-0.3, bal Fe.
DIN 17211/87; W. Nr. 1.8521.

KRUPP 15 CRNI 6
Krupp Stahl AG
C 0.14-0.19, Si 0-0.4, Mn 0.4-0.6, P 0-0.035, S 0-0.035, Cr 1.4-1.7, Ni 1.4-1.7, bal Fe.
DIN 17210/09.86; W. Nr. 1.5919.

KRUPP 15 M
Krupp Stahl AG
alloy steel. C 1.05, Mn 14, bal Fe.
Austenitic manganese steel, for heavy wear resistance.
Obsolete

KRUPP 15 MN 3
Krupp Stahl AG
C 0.12-0.18, Si 0.1-0.2, Mn 0.7-0.9, P 0-0.04, S 0-0.04, bal Fe.
SEL 90; W. Nr. 1.0467.

KRUPP 15 MN 3 AL
Krupp Stahl AG
C 0.12-0.18, Si 0-0.2, Mn 0.7-0.9, P 0-0.035, S 0-0.035, Al 0.02-0.05, Cu 0-0.25, N 0-0.012, bal Fe.
DIN 17115/87; W. Nr. 1.0468.

KRUPP 15 MNNI 6 3
Krupp Stahl AG
C 0.12-0.18, Si 0.15-0.35, Mn 1.2-1.65, P 0-0.015, S 0-0.005, Cr 0-0.15, Mo 0-0.05, Ni 0.5-0.85, Al 0.02-0.055, Cu 0-0.06, N 0-0.015, Nb 0-0.004, Ti 0-0.02, V 0-0.02, As 0-0.015, Sn 0-0.01, bal Fe.
SEL 90; W. Nr. 1.6210.

KRUPP 15 MNV 5
Krupp Stahl AG
C 0.12-0.18, Si 0.3-0.6, Mn 1.1-1.4, P 0-0.035, S 0-0.035, V 0.1-0.2, 0.03 Al min, bal Fe.
SEL 90; W. Nr. 1.5213.

KRUPP 15 MO 3
Krupp Stahl AG
C 0.12-0.2, Si 0.1-0.35, Mn 0.4-0.9, P 0-0.035, S 0-0.03, Mo 0.25-0.35, bal Fe.
DIN 17155/83; W. Nr. 1.5415.

KRUPP 15 NICR 6 4
Krupp Stahl AG
C 0.12-0.17, Si 0.1-0.4, Mn 0.6-0.9, P 0-0.035, S 0-0.035, Cr 0.85-1.15, Ni 1.2-1.6, bal Fe.
SEL 90; W. Nr. 1.5807.

KRUPP 15 NICRMO 10 6
Krupp Stahl AG
C 0.12-0.18, Si 0.15-0.35, Mn 0.1-0.4, P 0-0.025, S 0-0.025, Cr 1-1.8, Mo 0.2-0.6, Ni 2-3.25, Cu 0-0.25, Ti 0-0.02, V 0-0.03, bal Fe.
WW 1.6780/76; W. Nr. 1.6780.

KRUPP 15 NICRMO 16 5
Krupp Stahl AG
C 0.13-0.18, Si 0.15-0.4, Mn 0.25-0.55, P 0-0.025, S 0-0.02, Cr 1-1.4, Mo 0.2-0.3, Ni 3.8-4.3, bal Fe.
SEL 90; W. Nr. 1.6723.

KRUPP 15 NIMOCRB 4 5
Krupp Stahl AG
C 0.1-0.2, Si 0-0.5, Mn 0.6-1.3, P 0-0.00, S 0-0.025, Cr 0.4-1.1, Mo 0.4-0.6, Ni 0.7-1, V 0.03-0.08, 0.0005 B min, bal Fe.
SEL 90; W. Nr. 1.3421.

KRUPP 15 S 10
Krupp Stahl AG
C 0.12-0.18, Si 0.1-0.3, Mn 0.7-1.1, P 0-0.06, S 0.08-0.13, bal Fe.
DIN 1651/88; W. Nr. 1.0710.

KRUPP 15 S 20
Krupp Stahl AG
C 0.12-0.18, Si 0.1-0.4, Mn 0.5-0.9, P 0-0.07, S 0.18-0.26, bal Fe.
SEL 90; W. Nr. 1.0723.

KRUPP 15 SPB 20
Krupp Stahl AG
C 0.12-0.18, Si 0.1-0.4, Mn 0.5-0.9, P 0-0.07, S 0.18-0.26, Pb 0.15-0.3, bal Fe.
SEL 90; W. Nr. 1.0753.

KRUPP 150
Krupp Stahl AG
alloy steel. C 0.2, Si 0.35, Mn 1, Ti 0.2, bal Fe.
For hard facing electrodes; tough. *Obsolete*

KRUPP 1520
Krupp Stahl AG
C 0.65-0.74, Si 0.1-0.25, Mn 0.1-0.25, P 0-0.02, S 0-0.02, bal Fe.
SEL 90; W. Nr. 1.1520.

KRUPP 1525
Krupp Stahl AG
C 0.75-0.85, Si 0.1-0.25, Mn 0.1-0.25, P 0-0.02, S 0-0.02, bal Fe.
DIN 17350/80; W. Nr. 1.1525.

KRUPP 1545
Krupp Stahl AG
C 1.05, Mn 0.2, bal Fe.
DIN C105W1; W. Nr. 1.1545. Soft annealed, max 213 Brin. For shell hardened cold working tools.

KRUPP 16 CRMO 4
Krupp Stahl AG
C 0.13-0.2, Si 0.15-0.35, Mn 0.5-0.8, P 0-0.035, S 0-0.035, Cr 0.9-1.2, Mo 0.2-0.3, Ni 0-0.4, bal Fe.
SEI 90; W. Nr. 1.7242.

KRUPP 16 CRMO 4 4
Krupp Stahl AG
C 0.13-0.2, Si 0.15-0.35, Mn 0.5-0.8, P 0-0.035, S 0-0.035, Cr 0.9-1.2, Mo 0.4-0.5, Ni 0-0.4, bal Fe.
SEL 90; W. Nr. 1.7337.

KRUPP 16 CRMO 9 3
Krupp Stahl AG
C 0.12-0.2, Si 0.15-0.35, Mn 0.3-0.5, P 0-0.035, S 0-0.035, Cr 2-2.5, Mo 0.3-0.4, bal Fe.
SEW 590/61; W. Nr. 1.7281.

KRUPP 16 CRMOV 4
Krupp Stahl AG
C 0.13-0.2, Si 0.15-0.35, Mn 0.5-0.8, P 0-0.015, S 0-0.013, Cr 0.9-1.2, Mo 0.2-0.3, V 0.05-0.1, bal Fe.
SEL 90; W. Nr. 1.7728.

KRUPP 16 CRNIMO 6
Krupp Stahl AG
C 0.15-0.2, Si 0-0.4, Mn 0.4-0.6, P 0-0.035, S 0-0.035, Cr 1.5-1.8, Mo 0.25-0.35, Ni 1.4-1.7, Cu 0-0.3, bal Fe.
DIN 17230/80; W. Nr. 1.3531.

KRUPP 16 MNCR 5
Krupp Stahl AG
C 0.14-0.19, Si 0-0.4, Mn 1-1.3, P 0-0.035, S 0-0.035, Cr 0.8-1.1, bal Fe.
DIN 17210/09.86; W. Nr. 1.7131.

KRUPP 16 MNCRB 5
Krupp Stahl AG
C 0.14-0.18, Si 0.15-0.35, Mn 1-1.3, P 0-0.035, S 0.015-0.035, Cr 0.9-1.2, 0.0008 B min, bal Fe.
SEL 90; W. Nr. 1.7160.

KRUPP 16 MNCRS 5
Krupp Stahl AG
C 0.14-0.19, Si 0-0.4, Mn 1-1.3, P 0-0.035, S 0.02-0.035, Cr 0.8-1.1, bal Fe.
DIN 17210/09.86; W. Nr. 1.7139.

KRUPP 16 MNNI 6 3
Krupp Stahl AG
C 0-0.19, Si 0.15-0.5, Mn 1.2-1.65, P 0-0.025, S 0-0.015, Ni 0.5-0.85, N 0-0.02, Nb 0-0.004, Ti 0-0.02, V 0-0.02, 0.02 Al min, bal Fe.
SEL 90; W. Nr. 1.6211.

KRUPP 16 MO 5
Krupp Stahl AG
C 0.12-0.2, Si 0.15-0.5, Mn 0.5-0.8, P 0-0.04, S 0-0.04, Mo 0.45-0.65, bal Fe.
SEL 90; W. Nr. 1.5423.

KRUPP 16 NICRMO 12 6
Krupp Stahl AG
C 0.12-0.2, Si 0.15-0.35, Mn 0.1-0.4, P 0-0.025, S 0-0.025, Cr 1-1.8, Mo 0.2-0.6, Ni 2.25-3.5, Cu 0-0.25, Ti 0-0.02, V 0-0.03, bal Fe.
WW 1.6782/76; W. Nr. 1.6782.

KRUPP 1620
Krupp Stahl AG
C 0.65-0.74, Si 0.1-0.3, Mn 0.1-0.35, P 0-0.03, S 0-0.03, bal Fe.
DIN 17350/80; W. Nr. 1.1620.

KRUPP 1625
Krupp Stahl AG
C 0.75-0.85, Si 0.1-0.3, Mn 0.1-0.35, P 0-0.03, S 0-0.03, bal Fe.
SEL 90; W. Nr. 1.1625.

KRUPP 1645
Krupp Stahl AG
C 1-1.1, Si 0.1-0.3, Mn 0.1-0.35, P 0-0.03, S 0-0.03, bal Fe.
SEL 90; W. Nr. 1.1645.

KRUPP 1663
Krupp Stahl AG
C 1.2-1.35, Si 0.1-0.3, Mn 0.1-0.35, P 0-0.03, S 0-0.03, bal Fe.
SEL 90; W. Nr. 1.1663.

KRUPP 1673
Krupp Stahl AG
C 1.3-1.45, Si 0.1-0.3, Mn 0.1-0.35, P 0-0.03, S 0-0.03, bal Fe.
SEL 90; W. Nr. 1.1673.

KRUPP 17 CR 3
Krupp Stahl AG
C 0.14-0.2, Si 0-0.4, Mn 0.4-0.7, P 0-0.035, S 0-0.035, Cr 0.6-0.9, bal Fe.
DIN 17210/09.86; W. Nr. 1.7016.

KRUPP 17 CRMO 3 5
Krupp Stahl AG
C 0-0.21, Si 0.15-0.3, Mn 0.55-0.8, P 0-0.035, S 0-0.035, Cr 0.5-0.8, Mo 0.45-0.6, bal Fe.
SEL 90; W. Nr. 1.7332.

KRUPP 17 CRMOV 10
Krupp Stahl AG
C 0.15-0.2, Si 0.15-0.35, Mn 0.3-0.5, P 0-0.035, S 0-0.035, Cr 2.7-3, Mo 0.2-0.3, V 0.1-0.2, bal Fe.
SEW 590/61; W. Nr. 1.7766.

KRUPP 17 CRNI 6 6
Krupp Stahl AG
C 0.15-0.2, Si 0.15-0.35, Mn 0.4-0.6, P 0-0.035, S 0-0.035, Cr 1.5-1.7, Ni 1.5-1.7, bal Fe.
SEL 90; W. Nr. 1.5918.

KRUPP 17 CRNIMO 6
Krupp Stahl AG
C 0.15-0.2, Si 0-0.4, Mn 0.4-0.6, P 0-0.035, S 0-0.035, Cr 1.5-1.8, Mo 0.25-0.35, Ni 1.4-1.7, bal Fe.
DIN 17210/09.86; W. Nr. 1.6587.

KRUPP 17 MN 4
Krupp Stahl AG
C 0.14-0.2, Si 0-0.4, Mn 0.9-1.4, P 0-0.035, S 0-0.03, Cr 0-0.25, Mo 0-0.1, Ni 0-0.3, Cu 0-0.3, 0.02 Al min, bal Fe.
DIN 17155/83; W. Nr. 1.0481.

KRUPP 17 MNB 3
Krupp Stahl AG
C 0.15-0.2, Si 0-0.4, Mn 0.7-1, P 0-0.03, S 0-0.035, B 0.0008-0.005, 0.02 Al min, bal Fe.
SEL 90; W. Nr. 1.5506.

KRUPP 17 MNCR 5
Krupp Stahl AG
C 0.14-0.19, Si 0-0.4, Mn 1-1.3, P 0-0.035, S 0-0.035, Cr 0.8-1.1, Cu 0-0.3, bal Fe.
DIN 17230/80; W. Nr. 1.3521.

KRUPP 17 MNCRB 5 5
Krupp Stahl AG
C 0.13-0.2, Si 0-0.4, Mn 1-1.3, P 0-0.035, S 0-0.035, Cr 1-1.3, B 0.0008-0.005, bal Fe.
SEL 90; W. Nr. 1.7162.

KRUPP 17 MNMOV 6 4
Krupp Stahl AG
C 0-0.19, Si 0.2-0.5, Mn 1.4-1.7, P 0-0.035, S 0-0.035, Mo 0.2-0.5, Ni 0-1, Al 0-0.02, N 0-0.02, V 0-0.19, bal Fe.
SEL 90; W. Nr. 1.5403.

KRUPP 17 MNNI 4
Krupp Stahl AG
C 0.15-0.23, Si 0.13-0.32, Mn 0.83-1.15, P 0-0.02, S 0-0.02, Cr 0-0.15, Ni 0.68-0.87, Al 0-0.03, Cu 0-0.17, bal Fe.
DIN 17145/80; W. Nr. 1.6216.

KRUPP 17 MNV 7
Krupp Stahl AG
C 0-0.2, Si 0-0.55, Mn 1.4-1.9, P 0-0.035, S 0-0.035, N 0-0.025, Nb 0-0.08, V 0-0.2, 0.02 Al min, bal Fe.
SEL 90; W. Nr. 1.0870.

KRUPP 17 MOV 8 4
Krupp Stahl AG
C 0.14-0.22, Si 0.15-0.35, Mn 0.5-0.8, P 0-0.035, S 0-0.035, Cr 0.2-0.4, Mo 0.8-1, Ni 0-0.3, V 0.3-0.4, bal Fe.
SEL 90; W. Nr. 1.5406.

KRUPP 17 NICRMO 14
Krupp Stahl AG
C 0.15-0.2, Si 0-0.4, Mn 0.4-0.7, P 0-0.035, S 0-0.035, Cr 1.3-1.6, Mo 0.15-0.25, Ni 3.25-3.75, Cu 0-0.3, bal Fe.
DIN 17230/80; W. Nr. 1.3533.

KRUPP 1730
Krupp Stahl AG
C 0.45, Mn 0.7, bal Fe.
DIN C45W; W. Nr. 1.1730. As rolled, approx. 192 Brin, 650 N/mm^2 TS. Tools for hand use, shank material, base plates, build-up materials for tools.

KRUPP 1740
Krupp Stahl AG
C 0.6, Mn 0.7, bal Fe.
DIN C60W; W. Nr. 1.1740. As rolled, approx. 238 Brin, 800 N/mm^2 TS. Tools for hand use, shank material, base plates.

KRUPP 1744
Krupp Stahl AG
C 0.64-0.72, Si 0.15-0.4, Mn 0.6-0.8, P 0-0.035, S 0-0.035, bal Fe.
SEL 90; W. Nr. 1.1744.

KRUPP 1750
Krupp Stahl AG
C 0.72-0.82, Si 0.15-0.4, Mn 0.6-0.8, P 0-0.035, S 0-0.035, bal Fe.
SEL 90; W. Nr. 1.1750.

KRUPP 18 CRNI 8
Krupp Stahl AG
C 0.15-0.2, Si 0.15-0.4, Mn 0.4-0.6, P 0-0.035, S 0-0.035, Cr 1.8-2.1, Ni 1.8-2.1, bal Fe.
DIN 17210/69; W. Nr. 1.5920.

KRUPP 18 MNCRB 5
Krupp Stahl AG
C 0.16-0.2, Si 0.15-0.35, Mn 1-1.3, P 0-0.035, S 0.015-0.035, Cr 0.9-1.2, 0.0008 B min, bal Fe.
SEL 90; W. Nr. 1.7168.

KRUPP 18 NICR 6 4
Krupp Stahl AG
C 0.16-0.21, Si 0.1-0.4, Mn 0.6-0.9, P 0-0.035, S 0-0.035, Cr 0.85-1.15, Ni 1.2-1.6, bal Fe.
SEL 90; W. Nr. 1.5810.

KRUPP 1819
Krupp Stahl AG
C 0.85-0.95, Si 0.25-0.5, Mn 0.9-1.1, P 0-0.035, S 0-0.035, bal Fe.
SEL 90; W. Nr. 1.1819.

KRUPP 1820
Krupp Stahl AG
C 0.5-0.58, Si 0-0.15, Mn 0.3-0.5, P 0-0.03, S 0-0.03, bal Fe.
SEL 90; W. Nr. 1.1820.

KRUPP 1830
Krupp Stahl AG
C 0.8-0.9, Si 0.25-0.4, Mn 0.5-0.7, P 0-0.025, S 0-0.02, bal Fe.
DIN 17350/80; W. Nr. 1.1830.

KRUPP 19 MN 5
Krupp Stahl AG
C 0.17-0.22, Si 0.3-0.6, Mn 1-1.3, P 0-0.045, S 0-0.045, Cr 0-0.3, bal Fe.
SEL 90; W. Nr. 1.0482.

KRUPP 19 MN 6
Krupp Stahl AG
C 0.15-0.22, Si 0.3-0.6, Mn 1-1.6, P 0-0.035, S 0-0.03, Cr 0-0.25, Mo 0-0.1, Ni 0-0.3, Cu 0-0.3, 0.02 Al min, bal Fe.
DIN 17155/83; W. Nr. 1.0473.

KRUPP 19 MNB 4
Krupp Stahl AG
C 0.17-0.23, Si 0-0.4, Mn 0.8-1.15, P 0-0.035, S 0-0.035, B 0.0008-0.005, bal Fe.
DIN 1654 T.4/89; W. Nr. 1.5523.

KRUPP 19 MNCR 5
Krupp Stahl AG
C 0.17-0.22, Si 0-0.4, Mn 1.1-1.4, P 0-0.035, S 0-0.035, Cr 1-1.3, Cu 0-0.3, bal Fe.
DIN 17230/80; W. Nr. 1.3523.

KRUPP 2 MOV 5 3
Krupp Stahl AG
C 0.17-0.25, Si 0.15-0.35, Mn 0.5-0.8, P 0-0.035, S 0-0.035, Cr 0.2-0.4, Mo 0.45-0.55, Ni 0-0.3, V 0.25-0.35, bal Fe.
SEL 90; W. Nr. 1.5404.

KRUPP 20 CR 4
Krupp Stahl AG
C 0.17-0.23, Si 0-0.4, Mn 0.6-0.9, P 0-0.035, S 0-0.035, Cr 0.9-1.2, bal Fe.
DIN 17210/09.86; W. Nr. 1.7027.

KRUPP 20 CRMNS 3 3
Krupp Stahl AG
C 0.17-0.23, Si 0.2-0.35, Mn 0.6-1, P 0-0.04, Cr 0.6-1, 0.02 S min, bal Fe.
SEL 90; W. Nr. 1.7121.

KRUPP 20 CRMO 2
Krupp Stahl AG
C 0.18-0.23, Si 0.15-0.35, Mn 0.6-0.8, P 0-0.035, S 0.02-0.04, Cr 0.5-0.7, Mo 0.3-0.4, bal Fe.
SEL 90; W. Nr. 1.7311.

KRUPP 20 CRMO 5
Krupp Stahl AG
C 0.18-0.23, Si 0.15-0.35, Mn 0.9-1.2, P 0-0.035, S 0-0.035, Cr 1.1-1.4, Mo 0.2-0.3, bal Fe.
SEL 90; W. Nr. 1.7264.

KRUPP 20 CRMONIV 4 7
Krupp Stahl AG
C 0.17-0.25, Si 0-0.3, Mn 0.3-0.8, P 0-0.016, S 0-0.018, Cr 1.1-1.4, Mo 0.8-1, Ni 0.5-0.75, V 0.25-0.35, bal Fe.
SEW 555/84; W. Nr. 1.6979.

KRUPP 20 CRMOV 13 5
Krupp Stahl AG
C 0.17-0.23, Si 0.15-0.35, Mn 0.3-0.5, P 0-0.035, S 0-0.035, Cr 3-3.3, Mo 0.5-0.6, V 0.45-0.55, bal Fe.
SEW 590/61; W. Nr. 1.7779.

KRUPP 20 CRS 4
Krupp Stahl AG
C 0.17-0.23, Si 0-0.4, Mn 0.6-0.9, P 0-0.035, S 0.02-0.035, Cr 0.9-1.2, bal Fe.
DIN 17210/09.86; W. Nr. 1.7028.

KRUPP 20 MN 5
Krupp Stahl AG
C 0.17-0.23, Si 0.3-0.6, Mn 1-1.3, P 0-0.035, S 0-0.035, bal Fe.
SEW 550/76; W. Nr. 1.1133.

KRUPP 20 MN 6
Krupp Stahl AG
C 0.17-0.23, Si 0.3-0.6, Mn 1.3-1.6, P 0-0.035, S 0-0.035, bal Fe.
SEL 90; W. Nr. 1.1169.

KRUPP 20 MNCR 5
Krupp Stahl AG
C 0.17-0.22, Si 0-0.4, Mn 1.1-1.4, P 0-0.035, S 0-0.035, Cr 1-1.3, bal Fe.
DIN 17210/09.86; W. Nr. 1.7147.

KRUPP 20 MNCRS 5
Krupp Stahl AG
C 0.17-0.22, Si 0-0.4, Mn 1.1-1.4, P 0-0.035, S 0.02-0.035, Cr 1-1.3, bal Fe.
DIN 17210/09.86; W. Nr. 1.7149.

KRUPP 20 MNMO 3 5
Krupp Stahl AG
C 0.15-0.25, Si 0.15-0.25, Mn 0.7-1, P 0-0.025, S 0-0.025, Mo 0.4-0.6, bal Fe.
SEL 90; W. Nr. 1.5421.

KRUPP 20 MNMONI 4 5
Krupp Stahl AG
C 0.17-0.23, Si 0.1-0.4, Mn 1-1.5, P 0-0.035, S 0-0.035, Cr 0-0.5, Mo 0.45-0.6, Ni 0.4-0.8, bal Fe.
SEW 550/76; W. Nr. 1.6311.

KRUPP 20 MNMONI 5 5
Krupp Stahl AG
C 0.17-0.23, Si 0.15-0.3, Mn 1.2-1.5, P 0-0.012, S 0-0.008, Cr 0-0.2, Mo 0.4-0.55, Ni 0.5-0.8, Al 0.01-0.04, Cu 0-0.12, N 0-0.013, V 0-0.02, As 0-0.025, Sn 0-0.011, bal Fe.
SEL 90; W. Nr. 1.6310.

KRUPP 20 MOCR 4
Krupp Stahl AG
C 0.17-0.22, Si 0-0.4, Mn 0.7-1, P 0-0.035, S 0-0.035, Cr 0.3-0.6, Mo 0.4-0.5, bal Fe.
DIN 17210/09.86; W. Nr. 1.7321.

KRUPP 20 MOCRS 4
Krupp Stahl AG
C 0.17-0.22, Si 0-0.4, Mn 0.7-1, P 0-0.035, S 0.02-0.035, Cr 0.3-0.6, Mo 0.4-0.5, bal Fe.
DIN 17210/09.86; W. Nr. 1.7323.

KRUPP 20 NICRMO 14 6
Krupp Stahl AG
C 0.17-0.25, Si 0-0.35, Mn 0.2-0.4, P 0-0.02, S 0-0.02, Cr 1.5-2, Mo 0.4-0.6, Ni 3-3.8, Al 0-0.05, V 0-0.03, bal Fe.
SEL 90; W. Nr. 1.6742.

KRUPP 20 NICRMO 2
Krupp Stahl AG
C 0.17-0.23, Si 0-0.25, Mn 0.6-0.9, P 0-0.02, S 0-0.02, Cr 0.35-0.65, Mo 0.15-0.25, Ni 0.4-0.7, Al 0.02-0.05, Cu 0-0.25, N 0-0.012, bal Fe.
DIN 17115/87; W. Nr. 1.6522.

KRUPP 20 NICRMO 3
Krupp Stahl AG
C 0.17-0.23, Si 0-0.25, Mn 0.6-0.9, P 0-0.02, S 0-0.02, Cr 0.35-0.65, Mo 0.15-0.25, Ni 0.7-0.9, Al 0.02-0.05, Cu 0-0.25, N 0-0.012, bal Fe.
DIN 17115/87; W. Nr. 1.6527.

KRUPP 20 NIMOCR 6 5
Krupp Stahl AG
C 0.17-0.23, Si 0.15-0.4, Mn 0.6-0.9, P 0-0.035, S 0.02-0.035, Cr 0.3-0.5, Mo 0.4-0.5, Ni 1.4-1.8, bal Fe.
SEL 90; W. Nr. 1.6757.

KRUPP 20 NIMOV 14 5
Krupp Stahl AG
C 0.17-0.25, Si 0-0.35, Mn 0.2-0.4, P 0-0.02, S 0-0.02, Cr 0-0.3, Mo 0.4-0.6, Ni 3-3.8, Al 0-0.05, V 0.03-0.1, bal Fe.
SEL 90; W. Nr. 1.6348.

KRUPP 2002
Krupp Stahl AG
C 1.2-1.3, Si 0.15-0.3, Mn 0.25-0.4, P 0-0.03, S 0-0.03, Cr 0.3-0.4, bal Fe.
SEL 90; W. Nr. 1.2002.

KRUPP 2003
Krupp Stahl AG
C 0.7-0.8, Si 0.25-0.5, Mn 0.6-0.8, P 0-0.03, S 0-0.03, Cr 0.3-0.4, bal Fe.
DIN 17350/80; W. Nr. 1.2003.

KRUPP 2004
Krupp Stahl AG
C 0.8-0.9, Si 0.3-0.5, Mn 0.5-0.7, P 0-0.035, S 0-0.035, Cr 0.3-0.45, bal Fe.
SEL 90; W. Nr. 1.2004.

KRUPP 2007
Krupp Stahl AG
C 0.65-0.7, Si 0.2-0.3, Mn 0.75-0.9, P 0-0.03, S 0-0.03, Cr 0.55-0.7, bal Fe.
SEL 90; W. Nr. 1.2007.

KRUPP 2008
Krupp Stahl AG
C 1.35-1.5, Si 0.15-0.3, Mn 0.25-0.4, P 0-0.035, S 0-0.035, Cr 0.4-0.7, bal Fe.
SEL 90; W. Nr. 1.2008.

KRUPP 2018
Krupp Stahl AG
C 0.9-1, Si 0.15-0.3, Mn 0.2-0.4, P 0-0.025, S 0-0.025, Cr 0.3-0.4, bal Fe.
SEL 90; W. Nr. 1.2018.

KRUPP 204
Krupp Stahl AG
alloy steel. C 1.05, Cr 1, Mn 0.3, bal Fe.
For bearings, liners, sleeves, bushings; water hardened, wear resistant. *Obsolete*

KRUPP 2056
Krupp Stahl AG
C 0.85-0.95, Si 0.15-0.3, Mn 0.2-0.4, P 0-0.03, S 0-0.03, Cr 0.7-0.9, bal Fe.
SEL 90; W. Nr. 1.2056.

KRUPP 2057
Krupp Stahl AG
C 1-1.1, Si 0.15-0.35, Mn 0.25-0.4, P 0-0.03, S 0-0.03, Cr 0.9-1.1, bal Fe.
SEL 90; W. Nr. 1.2057.

KRUPP 206
Krupp Stahl AG
alloy steel. C 1.05, Mn 0.35, Cr 1.2, bal Fe.
For bearings, liners, sleeves, bushings; water hardened, wear resistant. *Obsolete*

KRUPP 2060
Krupp Stahl AG
C 1-1.1, Si 0.2-0.4, Mn 0-0.2, P 0-0.03, S 0-0.03, Cr 1.2-1.5, bal Fe.
SEL 77; W. Nr. 1.2060.

KRUPP 2063
Krupp Stahl AG
C 1.4-1.6, Si 0.15-0.3, Mn 0.5-0.7, P 0-0.035, S 0-0.035, Cr 1.3-1.5, bal Fe.
SEL 90; W. Nr. 1.2063.

KRUPP 2064
Krupp Stahl AG
C 0.8-0.9, Si 0.15-0.35, Mn 0.2-0.4, P 0-0.035, S 0-0.035, Cr 1.6-1.9, bal Fe.
SEL 81; W. Nr. 1.2064.

KRUPP 2067
Krupp Stahl AG
C 1, Cr 1.5, bal Fe.
DIN 100Cr6; W. Nr. 1.2067. Soft annealed, max 223 Brin. For small blanking and stamping tools, shear cutters, small rolls.

KRUPP 2080
Krupp Stahl AG
C 2.1, Cr 12, bal Fe.
DIN X210Cr12; W. Nr. 1.2080. Soft annealed, max 248 Brin. Tools for cutting sheets, cold forming and wear resistance.

KRUPP 2082
Krupp Stahl AG
C 0.17-0.22, Si 0.3-0.5, Mn 0.2-0.4, P 0-0.035, S 0-0.035, Cr 12.5-13.5, bal Fe.
SEL 90; W. Nr. 1.2082.

KRUPP 2083
Krupp Stahl AG
C 0.42, Cr 13, bal Fe.
DIN X42Cr13; W. Nr. 1.2083. Soft annealed, max 231 Brin. Moulds for corrosively acting plastics.

KRUPP 21 CRMONIV 4 7
Krupp Stahl AG
C 0.17-0.25, Si 0.15-0.35, Mn 0.35-0.85, P 0-0.03, S 0-0.035, Cr 0.9-1.2, Mo 0.65-0.8, Ni 0.2-0.8, V 0.25-0.35, bal Fe.
SEL 90; W. Nr. 1.6981.

KRUPP 21 CRMOV 5 11
Krupp Stahl AG
C 0.17-0.25, Si 0.3-0.6, Mn 0.3-0.6, P 0-0.035, S 0-0.035, Cr 1.2-1.5, Mo 1-1.2, Ni 0-0.6, V 0.25-0.35, bal Fe.
SEL 90; W. Nr. 1.8070.

KRUPP 21 CRMOV 5 7
Krupp Stahl AG
C 0.17-0.25, Si 0.15-0.35, Mn 0.35-0.85, P 0-0.03, S 0-0.035, Cr 1.2-1.5, Mo 0.65-0.8, V 0.25-0.35, bal Fe.
DIN 17240/76; W. Nr. 1.7709.

KRUPP 21 MN 4
Krupp Stahl AG
C 0.16-0.24, Si 0.1-0.25, Mn 0.8-1.1, P 0-0.035, S 0-0.035, bal Fe.
SEL 90; W. Nr. 1.0469.

KRUPP 21 MN 4 AL
Krupp Stahl AG
C 0.18-0.24, Si 0-0.25, Mn 0.8-1.1, P 0-0.035, S 0-0.035, Al 0.02-0.05, Cu 0-0.25, N 0-0.012, bal Fe.
DIN 17115/87; W. Nr. 1.0470.

KRUPP 21 MN 5
Krupp Stahl AG
C 0.18-0.24, Si 0-0.25, Mn 1.1-1.6, P 0-0.035, S 0-0.035, Al 0.02-0.05, Cu 0-0.25, N 0-0.012, bal Fe.
DIN 17115/87; W. Nr. 1.0495.

KRUPP 21 MNB 5
Krupp Stahl AG
C 0.17-0.24, Si 0-0.4, Mn 1.1-1.4, P 0-0.035, S 0-0.035, B 0.0008-0.005, bal Fe.
SEL 90; W. Nr. 1.5530.

KRUPP 21 MNCR 6 5
Krupp Stahl AG
C 0-0.25, Si 0-0.7, Mn 0-1.9, P 0-0.035, S 0-0.035, Cr 0.8-1.6, Mo 0-0.3, Ni 0-0.6, Cu 0-0.6, bal Fe.
SEL 90; W. Nr. 1.8705.

KRUPP 21 MNCRB 5 4
Krupp Stahl AG
C 0.18-0.24, Si 0-0.4, Mn 1.2-1.5, P 0-0.035, S 0-0.035, Cr 0.9-1.2, B 0.0008-0.005, bal Fe.
SEL 90; W. Nr. 1.7186.

KRUPP 21 MNSI 5
Krupp Stahl AG
C 0.18-0.24, Si 0.3-0.55, Mn 1.1-1.6, P 0-0.04, S 0-0.04, Al 0.02-0.04, bal Fe.
SEL 81; W. Nr. 1.0471.

KRUPP 21 NICRMO 2
Krupp Stahl AG
C 0.17-0.23, Si 0-0.4, Mn 0.65-0.95, P 0-0.035, S 0-0.035, Cr 0.4-0.7, Mo 0.15-0.25, Ni 0.4-0.7, bal Fe.
DIN 17210/09.86; W. Nr. 1.6523.

KRUPP 21 NICRMO 2 2
Krupp Stahl AG
C 0.18-0.23, Si 0.2-0.35, Mn 0.7-0.9, P 0-0.035, S 0-0.035, Cr 0.4-0.6, Mo 0.2-0.3, Ni 0.4-0.7, bal Fe.
SEL 90; W. Nr. 1.6543.

KRUPP 21 NICRMOS 2
Krupp Stahl AG
C 0.17-0.23, Si 0-0.4, Mn 0.65-0.95, P 0-0.035, S 0.02-0.035, Cr 0.4-0.7, Mo 0.15-0.25, Ni 0.4-0.7, bal Fe.
DIN 17210/09.86; W. Nr. 1.6526.

KRUPP 2101
Krupp Stahl AG
C 0.58-0.66, Si 0.9-1.2, Mn 0.9-1.2, P 0-0.03, S 0-0.03, Cr 0.4-0.7, bal Fe.
DIN 17350/80; W. Nr. 1.2101.

KRUPP 2103
Krupp Stahl AG
C 0.55-0.63, Si 1.7-2, Mn 0.6-0.9, P 0-0.035, S 0-0.035, Cr 0.35-0.45, bal Fe.
SEL 90; W. Nr. 1.2103.

KRUPP 2108
Krupp Stahl AG
C 0.85-0.95, Si 1.05-1.25, Mn 0.6-0.8, P 0-0.035, S 0-0.035, Cr 1.1-1.13, bal Fe.
SEL 90; W. Nr. 1.2108.

KRUPP 2109
Krupp Stahl AG
C 1.2-1.3, Si 1.05-1.25, Mn 0.6-0.8, P 0-0.035, S 0-0.035, Cr 1.1-1.3, bal Fe.
SEL 90; W. Nr. 1.2109.

KRUPP 2127
Krupp Stahl AG
C 1-1.1, Si 0.15-0.3, Mn 1-1.2, P 0-0.035, S 0-0.035, Cr 0.7-1, bal Fe.
SEL 90; W. Nr. 1.2127.

KRUPP 2129
Krupp Stahl AG
C 1.9-2.1, Si 0.15-0.3, Mn 0.8-1.1, P 0-0.035, S 0-0.035, Cr 1.9-2.2, bal Fe.
SEL 90; W. Nr. 1.2129.

KRUPP 2162
Krupp Stahl AG
C 0.21, Mn 1.3, Cr 1.2, bal Fe.
DIN 21MnCr5; W. Nr. 1.2162. Soft annealed, max 213 Brin. Case hardening steel for plastic moulds.

KRUPP 22 B 2
Krupp Stahl AG
C 0.19-0.25, Si 0-0.4, Mn 0.5-0.8, P 0-0.035, S 0-0.035, B 0.0008-0.005, bal Fe.
DIN 1654 T.4/89; W. Nr. 1.5508.

KRUPP 22 CRMO 4 4
Krupp Stahl AG
C 0.19-0.26, Si 0.15-0.4, Mn 0.5-0.8, P 0-0.035, S 0-0.035, Cr 0.9-1.2, Mo 0.4-0.5, Ni 0-0.6, bal Fe.
SEL 90; W. Nr. 1.7350.

KRUPP 22 CRMOS 3 5
Krupp Stahl AG
C 0.19-0.24, Si 0-0.4, Mn 0.7-1, P 0-0.035, S 0.02-0.035, Cr 0.7-1, Mo 0.4-0.5, bal Fe.
DIN 17210/86; W. Nr. 1.7333.

KRUPP 22 CRV 3
Krupp Stahl AG
C 0.2-0.24, Si 0-0.35, Mn 0.6-0.75, P 0-0.035, S 0-0.035, Cr 0.65-0.8, V 0.1-0.15, bal Fe.
SEL 90; W. Nr. 1.7511.

KRUPP 22 MN 6
Krupp Stahl AG
C 0.18-0.25, Si 0.15-0.3, Mn 1.3-1.7, P 0-0.035, S 0-0.035, bal Fe.
SEL 90; W. Nr. 1.1160.

KRUPP 22 MO 4
Krupp Stahl AG
C 0.18-0.25, Si 0.2-0.4, Mn 0.4-0.7, P 0-0.03, S 0-0.03, Cr 0-0.3, Mo 0.3-0.4, bal Fe.
SEL 90; W. Nr. 1.5419.

KRUPP 22 NIMOCR 3 7
Krupp Stahl AG
C 0.17-0.25, Si 0-0.35, Mn 0.5-1, P 0-0.02, S 0-0.02, Cr 0.3-0.5, Mo 0.5-0.8, Ni 0.6-1.2, Al 0-0.05, Cu 0-0.18, V 0-0.03, bal Fe.
SEL 90; W. Nr. 1.6751.

KRUPP 22 NIMOCR 4 7
Krupp Stahl AG
C 0.17-0.27, Si 0-0.4, Mn 0.5-1, P 0-0.035, S 0-0.035, Cr 0.3-0.5, Mo 0.5-0.8, Ni 0.6-1.2, bal Fe.
SEW 550/76; W. Nr. 1.6755.

KRUPP 22 S 20
Krupp Stahl AG
C 0.18-0.25, Si 0.1-0.4, Mn 0.5-0.9, P 0-0.07, S 0.15-0.25, bal Fe.
SEL 90; W. Nr. 1.0724.

KRUPP 22 SPB 20
Krupp Stahl AG
C 0.18-0.25, Si 0.1-0.4, Mn 0.5-0.9, P 0-0.07, S 0.15-0.25, Pb 0.15-0.3, bal Fe.
SEL 81; W. Nr. 1.0754.

KRUPP 2201
Krupp Stahl AG
C 1.55-1.75, Si 0.25-0.4, Mn 0.2-0.4, P 0-0.035, S 0-0.035, Cr 11-12, V 0.07-0.12, bal Fe.
SEL 90; W. Nr. 1.2201.

KRUPP 2206
Krupp Stahl AG
C 1.35-1.45, Si 0.15-0.35, Mn 0.25-0.4, P 0-0.025, S 0-0.025, Cr 0.2-0.4, V 0.1-0.15, bal Fe.
SEL 90; W. Nr. 1.2206.

KRUPP 2208
Krupp Stahl AG
C 0.31, Cr 0.55, V 0.1, bal Fe.
DIN 31CrV3; W. Nr. 1.2208. Untreated (as rolled and drawn). For spanners.

KRUPP 2210
Krupp Stahl AG
C 1.15, Cr 0.7, V 0.1, bal Fe.
DIN 115CrV3; W. Nr. 1.2210. Soft annealed, max 223 Brin (according to DIN 17350 for cold drawn products hardness can be up to 20 Brin higher). Silver steel for guiding pins, drills and punches.

KRUPP 2235
Krupp Stahl AG
C 0.75-0.85, Si 0.25-0.4, Mn 0.3-0.5, P 0-0.03, S 0-0.03, Cr 0.4-0.7, V 0.15-0.25, bal Fe.
DIN 17350/80; W. Nr. 1.2235.

KRUPP 2241
Krupp Stahl AG
C 0.51, Cr 1, V 0.1, bal Fe.
DIN 51CrV4; W. Nr. 1.2241. Soft annealed, max 231 Brin (according to DIN 17350 for cold drawn products hardness can be up to 20 Brin higher). Screw drivers, spanners, shank material for carbide tipped tools.

KRUPP 2242
Krupp Stahl AG
C 0.55-0.62, Si 0.15-0.35, Mn 0.8-1.1, P 0-0.035, S 0-0.035, Cr 0.9-1.2, V 0.07-0.12, bal Fe.
SEL 90; W. Nr. 1.2242.

KRUPP 2243
Krupp Stahl AG
C 0.61, Si 0.9, Cr 1.2, V 0.1, bal Fe.
DIN 61CrSiV5; W. Nr. 1.2243. Soft annealed, max 223 Brin (according to DIN 17350 for cold drawn products hardness can be up to 20 Brin higher). For chisels, tongs, pliers; screw drivers; shear blades.

KRUPP 2248
Krupp Stahl AG
C 0.35-0.42, Si 1.3-1.6, Mn 0.3-0.5, P 0 0.035, S 0-0.035, Cr 1.3-1.6, V 0.07-0.12, bal Fe.
SEL 90; W. Nr. 1.2248.

KRUPP 2249
Krupp Stahl AG
C 0.45, Si 1.5, Cr 1.5, V 0.1, bal Fe.
DIN 45SiCrV6; W. Nr. 1.2249. Soft annealed, max 223 Brin. For punches, chisels

KRUPP 23 CRMOB 3 3
Krupp Stahl AG
C 0.2-0.25, Si 0.15-0.35, Mn 0.7-0.9, P 0-0.035, S 0-0.035, Cr 0.7-0.9, Mo 0.3-0.4, 0.0008 B min, bal Fe.
SEL 90; W. Nr. 1.7271.

KRUPP 23 CRMOB 4
Krupp Stahl AG
C 0.2-0.25, Si 0.15-0.35, Mn 0.5-0.8, P 0-0.035, S 0-0.035, Cr 0.9-1.2, Mo 0.1-0.2, 0.0008 B min, bal Fe.
SEL 90; W. Nr. 1.7211.

KRUPP 23 CRNIMO 7 4 7
Krupp Stahl AG
C 0.2-0.26, Si 0-0.3, Mn 0.5-0.8, P 0-0.015, S 0-0.018, Cr 1.7-2, Mo 0.6-0.8, Ni 0.9-1.2, bal Fe.
SEL 90; W. Nr. 1.6749.

KRUPP 23 MNNICRMO 5 2
Krupp Stahl AG
C 0.2-0.26, Si 0-0.25, Mn 1.1-1.4, P 0-0.02, S 0-0.02, Cr 0.4-0.6, Mo 0.2-0.3, Ni 0.4-0.7, Al 0.02-0.05, Cu 0-0.25, N 0-0.012, bal Fe.
DIN 17115/87; W. Nr. 1.6541.

KRUPP 23 MNNICRMO 5 3
Krupp Stahl AG
C 0.2-0.26, Si 0-0.25, Mn 1.1-1.4, P 0-0.02, S 0-0.02, Cr 0.4-0.6, Mo 0.2-0.3, Ni 0.7-0.9, Al 0.02-0.05, Cu 0-0.25, N 0-0.012, bal Fe.
DIN 17115/87; W. Nr. 1.6540.

KRUPP 23 MNNICRMO 6 4
Krupp Stahl AG
C 0.2-0.26, Si 0.15-0.35, Mn 1.4-1.7, P 0-0.02, S 0-0.02, Cr 0.2-0.4, Mo 0.4-0.55, Ni 0.9-1.1, Al 0.02-0.05, bal Fe.
SEL 90; W. Nr. 1.6753.

KRUPP 23 MNNIMOCR 5 4
Krupp Stahl AG
C 0.2-0.26, Si 0-0.25, Mn 1.1-1.4, P 0-0.02, S 0-0.02, Cr 0.4-0.6, Mo 0.5-0.6, Ni 0.9-1.1, Al 0.02-0.05, Cu 0-0.25, N 0-0.012, bal Fe.
DIN 17115/87; W. Nr. 1.6758.

KRUPP 2303
Krupp Stahl AG
C 0.9-1.1, Si 0.15-0.3, Mn 0.2-0.4, P 0-0.035, S 0-0.035, Cr 1.1-1.3, Mo 0.2-0.4, bal Fe.
SEL 90; W. Nr. 1.2303.

KRUPP 2304
Krupp Stahl AG
C 0.8-0.9, Si 0.2-0.4, Mn 0.2-0.4, P 0-0.03, S 0-0.03, Cr 1.6-1.9, Mo 0.2-0.35, bal Fe.
SEL 90; W. Nr. 1.2304.

KRUPP 2305
Krupp Stahl AG
C 0.98-1.05, Si 0.5-0.65, Mn 1-1.2, P 0-0.02, S 0-0.02, Cr 1.4-1.6, Mo 0.12-0.17, bal Fe.
SEL 90; W. Nr. 1.2305.

KRUPP 2307
Krupp Stahl AG
C 0.26-0.34, Si 0.15-0.35, Mn 0.4-0.7, P 0-0.035, S 0-0.035, Cr 2.3-2.7, Mo 0.15-0.25, V 0.1-0.2, bal Fe.
SEL 90; W. Nr. 1.2307.

KRUPP 2309
Krupp Stahl AG
C 0.6-0.68, Si 0.3-0.5, Mn 1-1.2, P 0-0.035, S 0-0.035, Cr 0.6-0.8, Mo 0.2-0.3, bal Fe.
SEL 90; W. Nr. 1.2309.

KRUPP 2311
Krupp Stahl AG
C 0.4, Mn 1.45, Cr 2, Mo 0.2, bal Fe.
DIN 40CrMnMo7; W. Nr. 1.2311. Heat treated, approx. 295 Brin, 1000 N/mm² TS. Plastic moulds, quenched and tempered (or nitrided).

KRUPP 2312
Krupp Stahl AG
C 0.4, Mn 0.15, Cr 1.9, Mo 0.2, S 0.07, bal Fe.
DIN 40CrMnMoS86; W. Nr. 1.2312. Heat treated, approx 325 Brin, 1100 N/mm² TS. Plastic moulds, esp. frames.

KRUPP 2313
Krupp Stahl AG
C 0.16-0.23, Si 0.2-0.4, Mn 0.2-0.4, P 0-0.025, S 0-0.025, Cr 2.3-2.6, Mo 0.3-0.4, bal Fe.
SEL 90; W. Nr. 1.2313.

KRUPP 2314
Krupp Stahl AG
C 0.45-0.5, Si 0.3-0.5, Mn 0.3-0.5, P 0-0.035, S 0-0.035, Cr 14-15, Mo 0.2-0.3, bal Fe.
SEL 90; W. Nr. 1.2314.

KRUPP 2316
Krupp Stahl AG
C 0.36, Cr 17, Mo 1.1, Ni 0.5, bal Fe.
DIN X36CrMo17; W. Nr. 1.2316. Heat treated, 800 divided by 950 N/mm² TS. Moulds and extruding tools for corrosively acting plastics.

KRUPP 2318
Krupp Stahl AG
C 1.8 2, Si 0.25-0.4, Mn 0.4-0.6, P 0-0.03, S 0-0.03, Cr 11.5-12.5, Mo 0.5-0.7, bal Fe.
SEL 90; W. Nr. 1.2318.

KRUPP 2319
Krupp Stahl AG
C 0.6-0.7, Si 0.3-0.5, Mn 0.4-0.6, P 0-0.025, S 0-0.025, Cr 13.5-14.5, Mo 0.5-0.7, bal Fe.
SEL 90; W. Nr. 1.2319.

KRUPP 2323
Krupp Stahl AG
C 0.4-0.6, Si 0.15-0.35, Mn 0.6-0.9, P 0-0.03, S 0-0.03, Cr 1.3-1.6, Mo 0.65-0.85, V 0.25-0.35, bal Fe.
DIN 17350/80; W. Nr. 1.2323.

KRUPP 2327
Krupp Stahl AG
C 0.83-0.9, Si 0.15-0.35, Mn 0.3-0.45, P 0-0.03, S 0-0.03, Cr 1.6-1.9, Mo 0.2-0.35, V 0.05-0.15, bal Fe.
SEL 90; W. Nr. 1.2327.

KRUPP 2330
Krupp Stahl AG
C 0.32-0.37, Si 0.2-0.4, Mn 0.6-0.8, Cr 0.9-1.1, Mo 0.2-0.25, bal Fe.
SEL 90; W. Nr. 1.2330.

KRUPP 2332
Krupp Stahl AG
C 0.43-0.5, Si 0.15-0.35, Mn 0.6-0.8, P 0-0.025, S 0-0.025, Cr 0.9-1.2, Mo 0.25-0.4, bal Fe.
SEL 90; W. Nr. 1.2332.

KRUPP 2341
Krupp Stahl AG
C 0-0.07, Si 0-0.2, Mn 0-0.2, P 0-0.03, S 0-0.03, Cr 3.5-4, Mo 0.3-0.6, bal Fe.
SEL 90; W. Nr. 1.2341.

KRUPP 2343
Krupp Stahl AG
C 0.38, Si 1, Cr 5.1, Mo 1.3, V 0.4, bal Fe.
DIN X38CrMoV51; W. Nr. 1.2343. Soft annealed, max 229 Brin. Hot working tools in die forging presses for steels and light metals, in forging machines and extrusion presses; pressure casting dies for light metals; plastic moulds.

KRUPP 2344
Krupp Stahl AG
C 0.4, Si 1, Cr 5.1, Mo 1.3, V 1, bal Fe.
DIN X40CrMoV51; W. Nr. 1.2344. Soft annealed, max 229 Brin. Hot working tools in die forging presses for steels and light metals, in forging machines and extrusion presses; pressure casting dies for light metals.

KRUPP 2345
Krupp Stahl AG
C 0.48-0.53, Si 0.8-1.1, Mn 0.2-0.4, P 0-0.03, S 0-0.03, Cr 4.8-5.2, Mo 1.25-1.45, V 0.8-1, bal Fe.
SEL 90; W. Nr. 1.2345.

KRUPP 2353
Krupp Stahl AG
C 0.24-0.3, Si 0.4-0.6, Mn 0.3-0.7, P 0-0.03, S 0-0.03, Cr 1.3-1.5, Mo 1.1-1.4, V 0.35-0.45, bal Fe.
SEL 90; W. Nr. 1.2353.

KRUPP 2357
Krupp Stahl AG
C 0.45-0.55, Si 0.2-0.5, Mn 0.5-0.8, P 0-0.03, S 0-0.03, Cr 3-3.6, Mo 1.2-1.6, V 0.05-0.25, bal Fe.
SEL 90; W. Nr. 1.2357.

KRUPP 2360
Krupp Stahl AG
C 0.45-0.5, Si 0.7 0.0, Mn 0.05-0.45, P 0-0.02, S 0-0.005, Cr 7.3-7.8, Mo 1.3-1.5, V 1.3-1.5, bal Fe.
SEL 90; W. Nr. 1.2360.

KRUPP 2361
Krupp Stahl AG
C 0.85-0.95, Si 0-1, Mn 0-1, P 0-0.045, S 0-0.03, Cr 17-19, Mo 0.9-1.3, Ni 0-0.3, Cu 0-0.3, V 0.07-0.12, bal Fe.
SEL 90; W. Nr. 1.2361.

KRUPP 2362
Krupp Stahl AG
C 0.6-0.65, Si 1-1.2, Mn 0.3-0.5, P 0-0.035, S 0-0.035, Cr 5-5.5, Mo 1-1.3, V 0.25-0.35, bal Fe.
SEL 90; W. Nr. 1.2362.

KRUPP 2363
Krupp Stahl AG
C 0.9-1.05, Si 0.2-0.4, Mn 0.4-0.7, P 0-0.035, S 0-0.035, Cr 4.8-5.5, Mo 0.9-1.2, V 0.1-0.3, bal Fe.
SEL 90; W. Nr. 1.2363.

KRUPP 2365
Krupp Stahl AG
C 0.32, Cr 3, Mo 2.8, V 0.5, bal Fe.
DIN X32CrMoV33; W. Nr. 1.2365. Soft annealed, max 229 Brin. Hot working tools in die forging presses for steels and non ferrous metals, in forging machines and extrusion presses; pressure casting tools for heavy and light metals.

KRUPP 2367
Krupp Stahl AG
C 0.38, Cr 5, Mo 3, V 0.7, bal Fe.
DIN X38CrMoV53; W. Nr. 1.2367. Soft annealed, max 238 Brin. Large hot working tools in die forging presses.

KRUPP 2369
Krupp Stahl AG
C 0.77-0.85, Si 0-0.25, Mn 0-0.35, Cr 3.75-4.25, Mo 4-4.5, V 0.9-1.1, bal Fe.
SEL 90; W. Nr. 1.2369.

KRUPP 2376
Krupp Stahl AG
C 0.92-1, Si 0.2-0.4, Mn 0.2-0.4, P 0-0.03, S 0-0.03, Cr 11-12, Mo 0.8-1, V 0.8-1, bal Fe.
DIN 17350/80; W. Nr. 1.2376.

KRUPP 2378
Krupp Stahl AG
C 2.15-2.3, Si 0.15-0.3, Mn 0.25-0.4, P 0-0.035, S 0-0.035, Cr 12-13, Mo 0.8-1, V 2-2.3, bal Fe.
SEL 90; W. Nr. 1.2378.

KRUPP 2379
Krupp Stahl AG
C 1.55, Cr 12, Mo 0.7, V 1, bal Fe.
DIN X155CrVMo121; W. Nr. 1.2379. Soft annealed, max 255 Brin. Blanking and stamping dies, cold extrusion tools, circular shear blades, - suitable for nitriding.

KRUPP 2381
Krupp Stahl AG
C 0.7-0.77, Si 1-1.3, Mn 0.4-0.6, P 0-0.025, S 0-0.025, Mo 0.45-0.65, V 0.15-0.25, bal Fe.
SEL 90; W. Nr. 1.2381.

KRUPP 2382
Krupp Stahl AG
C 0.36-0.46, Si 0.8-1, Mn 0.85-1.1, P 0-0.035, S 0-0.035, Mo 0.1-0.25, bal Fe.
SEL 90; W. Nr. 1.2382.

KRUPP 24 CRMO 10
Krupp Stahl AG
C 0.2-0.28, Si 0.15-0.35, Mn 0.5-0.8, P 0-0.035, S 0-0.035, Cr 2.3-2.6, Mo 0.2-0.3, Ni 0-0.8, bal Fe.
SEW 590/61; W. Nr. 1.7273.

KRUPP 24 CRMO 5
Krupp Stahl AG
C 0.2-0.28, Si 0.15-0.35, Mn 0.5-0.8, P 0-0.03, S 0-0.035, Cr 0.9-1.2, Mo 0.2-0.35, bal Fe.
DIN 17240/76; W. Nr. 1.7258.

KRUPP 24 CRMOV 5 5
Krupp Stahl AG
C 0.2-0.28, Si 0.15-0.35, Mn 0.3-0.6, P 0-0.035, S 0-0.035, Cr 1.2-1.5, Mo 0.5-0.6, Ni 0-0.6, V 0.15-0.25, bal Fe.
SEL 90; W. Nr. 1.7733.

KRUPP 24 MNNICRMO 6 2
Krupp Stahl AG
C 0.21-0.27, Si 0.15-0.35, Mn 1.4-1.7, P 0-0.025, S 0-0.025, Cr 0.2-0.4, Mo 0.2-0.3, Ni 0.4-0.7, Al 0.02-0.05, bal Fe.
SEL 90; W. Nr. 1.6542.

KRUPP 24 NI 4
Krupp Stahl AG
C 0.2-0.28, Si 0.15-0.35, Mn 0.6-0.8, P 0-0.035, S 0-0.035, Cr 0-0.3, Ni 1-1.3, bal Fe.
SEL 90; W. Nr. 1.5613.

KRUPP 24 NI 8
Krupp Stahl AG
C 0.2-0.28, Si 0.15-0.35, Mn 0.6-0.8, P 0-0.035, S 0-0.035, Cr 0-0.3, Ni 1.9-2.2, bal Fe.
SEL 90; W. Nr. 1.5633.

KRUPP 24 NICRMOV 14 6
Krupp Stahl AG
C 0.2-0.28, Si 0.15-0.4, Mn 0.3-0.6, P 0-0.035, S 0-0.035, Cr 1.2-1.8, Mo 0.35-0.55, Ni 3-3.8, V 0.07-0.12, bal Fe.
SEL 90; W. Nr. 1.6952.

KRUPP 2414
Krupp Stahl AG
C 1.15-1.25, Si 0.15-0.3, Mn 0.2-0.35, P 0-0.035, S 0-0.035, Cr 0.15-0.25, W 0.9-1.1, bal Fe.
SEL 90; W. Nr. 1.2414.

KRUPP 2419
Krupp Stahl AG
C 1-1.1, Si 0.1-0.4, Mn 0.8-1.1, P 0-0.03, S 0-0.03, Cr 0.9-1.1, W 1-1.3, bal Fe.
DIN 17350/80; W. Nr. 1.2419.

KRUPP 2436
Krupp Stahl AG
C 2.1, Cr 12, W 0.7, bal Fe.
DIN X210CrW12; W. Nr. 1.2436. Soft annealed, max 255 Brin. Blanking and cold extrusion tools.

KRUPP 2442
Krupp Stahl AG
C 1.1-1.2, Si 0.15-0.3, Mn 0.2-0.4, P 0-0.035, S 0-0.035, Cr 0.15-0.25, W 1.8-2.1, bal Fe.
SEL 90; W. Nr. 1.2442.

KRUPP 2453
Krupp Stahl AG
C 1.25-1.35, Si 0.2-0.3, Mn 0.2-0.4, P 0-0.035, S 0-0.035, Cr 0-0.2, W 4.7-5.2, bal Fe.
SEL 90; W. Nr. 1.2453.

KRUPP 25 CRMO 4
Krupp Stahl AG
C 0.22-0.29, Si 0-0.4, Mn 0.6-0.9, P 0-0.035, S 0-0.03, Cr 0.9-1.2, Mo 0.15-0.3, bal Fe.
DIN 17200/87; W. Nr. 1.7218.

KRUPP 25 CRMOS 4
Krupp Stahl AG
C 0.22-0.29, Si 0-0.4, Mn 0.6-0.9, P 0-0.035, S 0.02-0.035, Cr 0.9-1.2, Mo 0.15-0.3, bal Fe.
DIN 17200/87; W. Nr. 1.7213.

KRUPP 25 MOCR 4
Krupp Stahl AG
C 0.23-0.29, Si 0.15-0.4, Mn 0.6-0.9, P 0-0.035, S 0-0.035, Cr 0.4-0.6, Mo 0.4-0.5, bal Fe.
DIN 17210/69; W. Nr. 1.7325.

KRUPP 25 MOCRS 4
Krupp Stahl AG
C 0.23-0.29, Si 0.15-0.4, Mn 0.6-0.9, P 0-0.035, S 0.02-0.035, Cr 0.4-0.6, Mo 0.4-0.5, bal Fe.
DIN 17210/69; W. Nr. 1.7326.

KRUPP 250
Krupp Stahl AG
alloy steel. C 0.25, Si 0.35, Mn 1.6, Ti 0.2, bal Fe.
For hard facing electrodes; tough. *Obsolete*

KRUPP 2510
Krupp Stahl AG
C 0.9-1.05, Si 0.15-0.35, Mn 1-1.2, P 0-0.035, S 0-0.035, Cr 0.5-0.7, V 0.05-0.15, W 0.5-0.7, bal Fe.
SEL 81; W. Nr. 1.2510.

KRUPP 2511
Krupp Stahl AG
C 0.75-0.85, Si 0.15-0.4, Mn 0.3-0.5, P 0-0.035, S 0-0.035, Cr 0.4-0.6, V 0.2-0.3, W 0.6-0.8, bal Fe.
SEL 90; W. Nr. 1.2511.

KRUPP 2515
Krupp Stahl AG
C 0.95-1.1, Si 0.1-0.25, Mn 0.15-0.3, P 0-0.035, S 0-0.035, Cr 0.1-0.25, V 0.1-0.2, W 0.9-1.2, bal Fe.
SEL 81; W. Nr. 1.2515.

KRUPP 2516
Krupp Stahl AG
C 1.15-1.25, Si 0.15-0.3, Mn 0.2-0.35, P 0-0.035, S 0-0.035, Cr 0.15-0.25, V 0.07-0.12, W 0.9-1.1, bal Fe.
SEL 90; W. Nr. 1.2516.

KRUPP 2519
Krupp Stahl AG
C 1.05-1.15, Si 0.15-0.3, Mn 0.2-0.4, P 0-0.03, S 0-0.03, Cr 1.1-1.3, V 0.15-0.25, W 1.2-1.4, bal Fe.
DIN 17350/80; W. Nr. 1.2519.

KRUPP 2521
Krupp Stahl AG
C 1.05-1.15, Si 0.15-0.3, Mn 0.3-0.4, P 0-0.035, S 0-0.035, Cr 1.1-1.3, V 0.15-0.25, W 1.8-2, bal Fe.
SEL 75; W. Nr. 1.2521.

KRUPP 253MA
Krupp Stahl AG
C 0-0.1, Si 1.5-2, Cr 20.5-21.5, Ni 10.5-11.5, Approx. 0.17 N and 0.05 Ce, bal Fe.
Austenitic. DIN X8CrNiSiN2111; W. Nr. 1.4893. 310 N/mm^2 YS, 650-850 N/mm^2 TS, 35 El.

KRUPP 2542
Krupp Stahl AG
C 0.4-0.5, Si 0.8-1.1, Mn 0.2-0.4, P 0-0.035, S 0-0.035, Cr 0.9-1.2, V 0.15-0.2, W 1.8-2.1, bal Fe.
SEL 90; W. Nr. 1.2542.

KRUPP 2550
Krupp Stahl AG
C 0.6, Cr 1.1, V 0.15, W 2, bal Fe.
DIN 60WCrV7; W. Nr. 1.2550. Soft annealed, max 231 Brin. Blanking dies, cold forming tools, shear blades for cold and hot work.

KRUPP 2552
Krupp Stahl AG
C 0.75-0.85, Si 0.4-0.6, Mn 0.3-0.5, P 0-0.035, S 0-0.035, Cr 1-1.2, V 0.25-0.35, W 1.8-2.1, bal Fe.
SEL 90; W. Nr. 1.2552.

KRUPP 2562
Krupp Stahl AG
C 1.35-1.5, Si 0.15-0.3, Mn 0.2-0.4, P 0-0.035, S 0-0.035, Cr 0.2-0.5, V 0.2-0.3, W 2.8-3.3, bal Fe.
SEL 90; W. Nr. 1.2562.

KRUPP 2564
Krupp Stahl AG
C 0.25-0.35, Si 0.8-1.1, Mn 0.3-0.5, P 0-0.035, S 0-0.035, Cr 0.9-1.2, V 0.15-0.2, W 3.5-4, bal Fe.
SEL 90; W. Nr. 1.2564.

KRUPP 2567
Krupp Stahl AG
C 0.25-0.35, Si 0.15-0.3, Mn 0.2-0.4, P 0-0.035, S 0-0.035, Cr 2.2-2.5, V 0.5-0.7, W 4-4.5, bal Fe.
SEL 90; W. Nr. 1.2567.

KRUPP 2581
Krupp Stahl AG
C 0.25-0.35, Si 0.15-0.3, Mn 0.2-0.4, P 0-0.035, S 0-0.035, Cr 2.5-2.8, V 0.3-0.4, W 8-9, bal Fe.
SEL 90; W. Nr. 1.2581.

KRUPP 26 CRMO 4
Krupp Stahl AG
C 0.22-0.29, Si 0-0.35, Mn 0.5-0.8, P 0-0.03, S 0-0.025, Cr 0.9-1.2, Mo 0.15-0.3, bal Fe.
DIN 17280/85; W. Nr. 1.7219.

KRUPP 26 MN 5
Krupp Stahl AG
C 0.22-0.29, Si 0.15-0.3, Mn 1.2-1.5, P 0-0.035, S 0-0.035, bal Fe.
SEL 90; W. Nr. 1161.

KRUPP 26 MNMOB 6 4
Krupp Stahl AG
C 0.25-0.3, Si 0.3-0.6, Mn 1.25-1.5, P 0-0.025, S 0-0.025, Cr 0.25-0.5, Mo 0.3-0.45, Ni 0.25-0.5, B 0.001-0.004, Cu 0-0.25, V 0-0.05, bal Fe.
SEL 90; W. Nr. 1.5465.

KRUPP 26 NICRMO 14 5
Krupp Stahl AG
C 0.22-0.3, Si 0-0.4, Mn 0.2-0.5, P 0-0.035, S 0-0.035, Cr 1-1.7, Mo 0.3-0.6, Ni 3.2-4, V 0-0.15, bal Fe.
SEL 90; W. Nr. 1.6953.

KRUPP 26 NICRMO 14 6
Krupp Stahl AG
C 0.25-0.3, Si 0.15-0.3, Mn 0.3-0.5, P 0-0.02, S 0-0.01, Cr 1.2-1.7, Mo 0.35-0.5, Ni 3.4-3.8, Al 0.02-0.05, V 0-0.12, bal Fe.
SEL 90; W. Nr. 1.6958.

KRUPP 26 NICRMOV 11 5
Krupp Stahl AG
C 0.22-0.32, Si 0-0.3, Mn 0.15-0.4, P 0-0.015, S 0-0.018, Cr 1.2-1.8, Mo 0.25-0.45, Ni 2.4-3.1, V 0.05-0.15, bal Fe.
SEL 90; W. Nr. 1.6948.

KRUPP 2601
Krupp Stahl AG
C 1.6, Cr 12, Mo 0.6, V 0.15, W 0.5, bal Fe.
DIN X165CrMoV12; W. Nr. 1.2601. Soft annealed, max 255 Brin. Blanking tools for lage sizes, cold forming tools, shear blades, plastic moulds.

KRUPP 2603
Krupp Stahl AG
C 0.4-0.5, Si 0.5-0.7, Mn 0.3-0.5, P 0-0.035, S 0-0.035, Cr 1.3-1.6, Mo 0.4-0.6, V 0.75-0.9, W 0.4-0.6, bal Fe.
SEL 90; W. Nr. 1.2603.

KRUPP 2604
Krupp Stahl AG
C 0.68-0.78, Si 0.2-0.4, Mn 0.4-0.6, P 0-0.035, S 0-0.035, Cr 0.4-0.6, Mo 0.25-0.4, V 0.15-0.3, W 0.4-0.7, bal Fe.
SEL 90; W. Nr. 1.2604.

KRUPP 2606
Krupp Stahl AG
C 0.37, Si 1, Cr 5.3, Mo 1.4, V 0.3, W 1.3, bal Fe.
DIN X37CrMoW51; W. Nr. 1.2606. Soft annealed, max 255 Brin. Hot forming tools in die forging presses for steel and metals, in forging machines and extrusion presses.

KRUPP 2609
Krupp Stahl AG
C 1.55-1.75, Si 0.25-0.4, Mn 0.2-0.4, P 0-0.035, S 0-0.035, Cr 11-12, Mo 0.5-0.7, V 1.1-1.3, W 0.4-0.6, bal Fe.
SEL 81; W. Nr. 1.2606.

KRUPP 2631
Krupp Stahl AG
C 0.45-0.55, Si 0.8-1, Mn 0.4-0.6, P 0-0.035, S 0-0.035, Cr 8-9, Mo 1.1-1.3, W 1.1-1.3, bal Fe.
SEL 90; W. Nr. 1.2631.

KRUPP 2662
Krupp Stahl AG
C 0.27-0.32, Si 0.15-0.3, Mn 0.2-0.4, P 0-0.035, S 0-0.035, Cr 2.2-2.5, Co 1.8-2.3, V 0.2-0.3, W 8-9, bal Fe.
SEL 90; W. Nr. 1.2662.

KRUPP 2678
Krupp Stahl AG
C 0.4-0.5, Si 0.3-0.5, Mn 0.3-0.5, P 0-0.025, S 0-0.025, Cr 4-5, Mo 0.4-0.6, Co 4-5, V 1.8-2.1, W 4-5, bal Fe.
SEL 90; W. Nr. 1.2678.

KRUPP 27 MNCRB 5 2
Krupp Stahl AG
C 0.24-0.3, Si 0-0.4, Mn 1.1-1.4, P 0-0.035, S 0-0.035, Cr 0.3-0.6, B 0.0008-0.005, bal Fe.
SEL 90; W. Nr. 1.7182.

KRUPP 27 MNCRV 4
Krupp Stahl AG
C 0.24-0.3, Si 0.15-0.35, Mn 1-1.3, P 0-0.035, S 0-0.035, Cr 0.6-0.9, V 0.07-0.12, bal Fe.
SEL 90; W. Nr. 1.8162.

KRUPP 27 MNSI 5
Krupp Stahl AG
C 0.24-0.3, Si 0.25-0.45, Mn 1.1-1.6, P 0-0.035, S 0-0.035, Al 0.02-0.05, Cu 0-0.25, N 0-0.012, bal Fe.
DIN 17115/87; W. Nr. 1.0412.

KRUPP 27 MNSIVS 6
Krupp Stahl AG
C 0.25-0.3, Si 0.5-0.8, Mn 1.3-1.6, P 0-0.035, S 0.03-0.05, V 0.08-0.13, bal Fe.
SEW 101/88; W. Nr. 1.5232.

KRUPP 2703
Krupp Stahl AG
C 0.7-0.78, Si 0.1-0.25, Mn 0.35-0.45, P 0-0.035, S 0-0.035, Cr 0.2-0.3, Ni 0.5-0.6, bal Fe.
SEL 90; W. Nr. 1.2703.

KRUPP 2705
Krupp Stahl AG
C 0.75-0.85, Si 0-0.4, Mn 0.25-0.6, P 0-0.025, S 0-0.025, Cr 0.2-0.5, Ni 2.4-2.9, bal Fe.
SEL 90; W. Nr. 1.2705.

KRUPP 2706
Krupp Stahl AG
C 0-0.03, Si 0-0.1, Mn 0-0.1, P 0-0.01, S 0-0.01, Cr 0-0.25, Mo 4.5-5.2, Ni 17-19, Co 7-9, Ti 0.35-0.55, bal Fe.
SEL 81; W. Nr. 1.2706.

KRUPP 2709
Krupp Stahl AG
C 0-0.03, Si 0-0.1, Mn 0-0.15, P 0-0.01, S 0-0.01, Cr 0-0.25, Mo 4.5-5.2, Ni 17-19, Co 8.5-10, Ti 0.8-1.2, bal Fe.
SEL 90; W. Nr. 1.2709.

KRUPP 2710
Krupp Stahl AG
C 0.4-0.5, Si 0.15-0.35, Mn 0.5-0.8, P 0-0.035, S 0-0.035, Cr 1.2-1.5, Ni 1.5-1.8, bal Fe.
SEL 90; W. Nr. 1.2710.

KRUPP 2711
Krupp Stahl AG
C 0.5-0.6, Si 0.15-0.35, Mn 0.5-0.8, P 0-0.025, S 0-0.025, Cr 0.6-0.8, Mo 0.25-0.35, Ni 1.5-1.8, V 0.07-0.12, bal Fe.
SEL 90; W. Nr. 1.2711.

KRUPP 2713
Krupp Stahl AG
C 0.55, Cr 0.7, Mo 0.3, Ni 1.7, V 0.1, bal Fe.
DIN 55NiCrMoV6; W. Nr. 1.2713. Soft annealed, max 248 Brin. or heat treated, approx. 368-412 Brin, 1250-1400 N/mm^2 TS. Dies for hammer forging of low and medium hardness; shank material, die holders.

KRUPP 2714
Krupp Stahl AG
C 0.55, Cr 1.1, Mo 0.5, Ni 1.7, V 0.1, bal Fe.
DIN 56NiCrMoV7; W. Nr. 1.2714. Soft annealed, max 248 Brin or heat treated, 368-412 Brin, 1250-1400 N/mm^2 TS. Large dies for hammer forging with higher hardness, extrusion rams.

KRUPP 2718
Krupp Stahl AG
C 0.5-0.57, Si 0.15-0.3, Mn 0.4-0.5, P 0-0.035, S 0-0.035, Cr 0.5-0.7, Ni 2.5-3, bal Fe.
SEL 90; W. Nr. 1.2718.

KRUPP 2721
Krupp Stahl AG
C 0.5, Cr 1.1, Ni 3.3, bal Fe.
DIN 50NiCr13; W. Nr. 1.2721. Soft annealed, max 248 Brin. Tools for cold forming and heading, shear blades.

KRUPP 2722
Krupp Stahl AG
C 0.15-0.2, Si 0.15-0.35, Mn 0.4-0.6, P 0-0.03, S 0-0.03, Cr 1.8-2.1, Ni 1.8-2.1, bal Fe.
SEL 90; W. Nr. 1.2722.

KRUPP 2726
Krupp Stahl AG
C 0.22-0.3, Si 0.3-0.5, Mn 0.2-0.4, P 0-0.03, S 0-0.03, Cr 0.6-0.9, Mo 0.2-0.4, Ni 1.3-1.6, V 0.15-0.2, bal Fe.
SEL 90; W. Nr. 1.2726.

KRUPP 2731
Krupp Stahl AG
C 0.45-0.55, Si 1.2-1.5, Mn 0.6-0.8, P 0-0.035, S 0-0.035, Cr 12-14, Ni 12.5-13.5, V 0.3-1, W 1.5-2.8, bal Fe.
SEL 90; W. Nr. 1.2731.

KRUPP 2735
Krupp Stahl AG
C 0.1-0.17, Si 0.2-0.35, Mn 0.3-0.5, P 0-0.03, S 0-0.03, Cr 0.65-0.85, Ni 3.3-3.6, bal Fe.
SEL 90; W. Nr. 1.2735.

KRUPP 2737
Krupp Stahl AG
C 0.24-0.32, Si 0.3-0.6, Mn 0.2-0.4, P 0-0.03, S 0-0.03, Cr 0.6-0.9, Ni 1-1.3, V 0.15-0.2, bal Fe.
SEL 90; W. Nr. 1.2737.

KRUPP 2738
Krupp Stahl AG
C 0.35-0.45, Si 0.2-0.4, Mn 1.3-1.6, P 0-0.035, S 0-0.035, Cr 1.8-2.1, Mo 0.15-0.25, Ni 0.9-1.2, bal Fe.
SEL 90; W. Nr. 1.2738.

KRUPP 2740
Krupp Stahl AG
C 0.24-0.32, Si 0.3-0.5, Mn 0.2-0.4, P 0-0.03, S 0-0.03, Cr 0.6-0.9, Mo 0.5-0.7, Ni 2.3-2.6, V 0.25-0.32, bal Fe.
SEL 90; W. Nr. 1.2740.

KRUPP 2744
Krupp Stahl AG
C 0.5-0.6, Si 0.15-0.35, Mn 0.6-0.8, P 0-0.035, S 0-0.035, Cr 0.9-1.2, Mo 0.7-0.9, Ni 1.5-1.8, V 0.07-0.12, bal Fe.
SEL 90; W. Nr. 1.2744.

KRUPP 2745
Krupp Stahl AG
C 0.1-0.17, Si 0.2-0.3, Mn 0.3-0.5, P 0-0.03, S 0-0.03, Cr 0.9-1.0, Ni 4.5-4.7, bal Fe.
SEL 90; W. Nr. 1.2745.

KRUPP 2746
Krupp Stahl AG
C 0.41-0.49, Si 0.15-0.35, Mn 0.6-0.8, P 0-0.025, S 0-0.02, Cr 1.4-1.6, Mo 0.73-0.85, Ni 3.8-4.2, V 0.45-0.55, bal Fe.
SEL 90; W. Nr. 1.2746.

KRUPP 2747
Krupp Stahl AG
C 0.24-0.31, Si 0.15-0.35, Mn 0.2-0.4, P 0-0.03, S 0-0.03, Cr 0.3-0.5, Mo 1.15-1.25, Ni 4.2-4.7, V 0.15-0.2, bal Fe.
SEL 90; W. Nr. 1.2747.

KRUPP 2758
Krupp Stahl AG
C 0.48-0.53, Si 1.3-1.5, Mn 0.5-0.7, P 0-0.025, S 0-0.025, Cr 3.8-4.2, Mo 0.6-0.8, Ni 11-12, Co 1.5-1.8, V 1-1.2, W 12-13, bal Fe.
SEL 90; W. Nr. 1.2758.

KRUPP 2762
Krupp Stahl AG
C 0.7-0.8, Si 0.15-0.3, Mn 0.15-0.35, P 0-0.025, S 0-0.035, Cr 1.4-1.6, Mo 0.6-0.8, Ni 0.4-0.6, W 0.2-0.4, bal Fe.
SEL 90; W. Nr. 1.2762.

KRUPP 2764
Krupp Stahl AG
C 0.19, Cr 1.3, Mo 0.2, Ni 4, bal Fe.
DIN X19NiCrMo4; W. Nr. 1.2764. Soft annealed, max 255 Brin. Case hardening steel for plastic moulds with very good toughness and polishability.

KRUPP 2765
Krupp Stahl AG
C 0.5-0.57, Si 0.15-0.3, Mn 0.2-0.4, P 0-0.025, S 0-0.025, Cr 1-1.2, Mo 0.2-0.4, Ni 3.7-4.2, W 0.4-0.6, bal Fe.
SEL 90; W. Nr. 1.2765.

KRUPP 2766
Krupp Stahl AG
C 0.32-0.38, Si 0.15-0.3, Mn 0.4-0.6, P 0-0.035, S 0-0.035, Cr 1.2-1.5, Mo 0.2-0.4, Ni 3.8-4.3, bal Fe.
SEL 90; W. Nr. 1.2766.

KRUPP 2767
Krupp Stahl AG
C 0.45, Cr 1.4, Mo 0.25, Ni 4, bal Fe.
DIN X45NiCrMo4; W. Nr. 1.2767. Soft annealed, max 262 Brin. Blanking tools for large sizes, cold forming tools, shear blades, plastic moulds.

KRUPP 2779
Krupp Stahl AG
C 0-0.08, Si 0-1, Mn 1-2, P 0-0.03, S 0-0.03, Cr 13.5-16, Mo 1-1.5, Ni 24-27, B 0.003-0.01, Ti 1.9-2.3, V 0.1-0.5, bal Fe.
SEL 90; W. Nr. 1.2779.

KRUPP 2780
Krupp Stahl AG
C 0-0.2, Si 1.8-2.3, Mn 0-2, P 0-0.035, S 0-0.035, Cr 19-21, Ni 11-13, bal Fe.
SEL 90; W. Nr. 1.2780.

KRUPP 2782
Krupp Stahl AG
C 0-0.2, Si 1.8-2.3, Mn 0-2, P 0-0.035, S 0-0.035, Cr 24-26, Ni 19-21, bal Fe.
SEL 90; W. Nr. 1.2782.

KRUPP 2786
Krupp Stahl AG
C 0-0.15, Si 1.5-2, Mn 0-2, P 0-0.035, S 0-0.035, CR 15-17, Ni 34-37, bal Fe.
SEL 90; W. Nr. 1.2786.

KRUPP 2787
Krupp Stahl AG
C 0.1-0.25, Si 0-1, Mn 0-1, P 0-0.035, S 0-0.035, Cr 15.5-18, Ni 1-2.5, bal Fe.
SEL 90; W. Nr. 1.2787.

KRUPP 28 B 2
Krupp Stahl AG
C 0.25-0.32, Si 0-0.4, Mn 0.5-0.8, P 0-0.035, S 0-0.035, B 0.0008-0.005, bal Fe.
DIN 1654 T.4/89; W. Nr. 1.5510.

KRUPP 28 CR 4
Krupp Stahl AG
C 0.24-0.31, Si 0-0.4, Mn 0.6-0.9, P 0-0.035, S 0-0.03, Cr 0.9-1.2, bal Fe.
DIN 17200/87; W. Nr. 1.7030.

KRUPP 28 CRMONIV 4 9
Krupp Stahl AG
C 0.25-0.32, Si 0-0.3, Mn 0.3-0.8, P 0-0.015, S 0-0.018, Cr 1.1-1.4, Mo 0.8-1, Ni 0.5-0.75, V 0.25-0.35, bal Fe.
SEL 90; W. Nr. 1.6985.

KRUPP 28 CRS 4
Krupp Stahl AG
C 0.24-0.31, Si 0-0.4, Mn 0.6-0.9, P 0-0.035, S 0.02-0.03, Cr 0.9-1.2, bal Fe.
DIN 17200/87; W. Nr. 1.7036.

KRUPP 28 MN 4
Krupp Stahl AG
C 0.24-0.32, Si 0.15-0.35, Mn 0.9-1.2, P 0-0.05, S 0-0.05, N 0-0.007, bal Fe.
SEL 90; W. Nr. 1.0560.

KRUPP 28 MN 6
Krupp Stahl AG
C 0.25-0.32, Si 0-0.4, Mn 1.3-1.65, P 0-0.035, S 0-0.03, bal Fe.
DIN 17200/87; W. Nr. 1.1170.

KRUPP 28 NICRMO 4
Krupp Stahl AG
C 0.24-0.34, Si 0.15-0.4, Mn 0.3-0.6, P 0-0.035, S 0-0.035, Cr 1-1.3, Mo 0.2-0.3, Ni 1-1.3, bal Fe.
SEL 90; W. Nr. 1.6513.

KRUPP 28 NICRMO 5 5
Krupp Stahl AG
C 0.26-0.32, Si 0-0.3, Mn 0.15-0.4, P 0-0.015, S 0-0.018, Cr 1-1.3, Mo 0.25-0.45, Ni 1-1.3, V 0-0.15, bal Fe.
SEL 90; W. Nr. 1.6732.

KRUPP 28 NICRMOV 8 5
Krupp Stahl AG
C 0.24-0.32, Si 0-0.4, Mn 0.3-0.6, P 0-0.035, S 0-0.035, Cr 1-1.5, Mo 0.35-0.55, Ni 1.8-2.1, V 0-0.15, bal Fe.
SEW 550/76; W. Nr. 1.6932.

KRUPP 2823
Krupp Stahl AG
C 0.65-0.75, Si 1.5-1.8, Mn 0.6-0.8, P 0-0.03, S 0-0.03, bal Fe.
SEL 90; W. Nr. 1.2823.

KRUPP 2825
Krupp Stahl AG
C 0.5-0.57, Si 0.8-1, Mn 0.8-1.2, P 0-0.035, S 0-0.035, bal Fe.
SEL 90; W. Nr. 1.2825.

KRUPP 2826
Krupp Stahl AG
C 0.58-0.65, Si 0.8-1, Mn 0.8-1.2, P 0-0.03, S 0-0.03, Cr 0.2-0.4, bal Fe.
DIN 17350/80; W. Nr. 1.2826.

KRUPP 2833
Krupp Stahl AG
C 1, V 0.1, bal Fe.
DIN 100V1; W. Nr. 1.2833. Soft annealed, suitable for hobbing, max 180 Brin. Shell hardening steel for cold heading dies.

KRUPP 2838
Krupp Stahl AG
C 1.4-1.5, Si 0.2-0.35, Mn 0.3-0.5, P 0-0.03, S 0-0.03, V 3-3.5, bal Fe.
DIN 17350/80; W. Nr. 1.2838.

KRUPP 2842
Krupp Stahl AG
C 0.9, Mn 2, Cr 0.3, V 0.1, bal Fe.
DIN 90MnCrV8; W. Nr. 1.2842. Soft annealed, max 231 Brin. Standard steel for small and middle large blanking and stamping tools.

KRUPP 2851
Krupp Stahl AG
C 0.3-0.37, Si 0.15-0.35, Mn 0.6-0.9, P 0-0.035, S 0-0.035, Cr 1.2-1.5, Al 0.8-1.1, bal Fe.
SEL 81; W. Nr. 1.2851.

KRUPP 2880
Krupp Stahl AG
C 1.55-1.75, Si 0.25-0.4, Mn 0.2-0.4, P 0-0.035, S 0-0.035, Cr 11-12, Mo 0.5-0.6, Co 1.2-1.4, bal Fe.
SEL 90; W. Nr. 1.2880.

KRUPP 2883
Krupp Stahl AG
C 0.75-0.85, Si 0.3-0.6, Mn 0.2-0.5, P 0-0.025, S 0-0.025, Cr 12.5-14.5, Co 0.9-1.3, bal Fe.
SEL 90; W. Nr. 1.2883.

KRUPP 2884
Krupp Stahl AG
C 2-2.3, Si 0.2-0.4, Mn 0.2-0.4, P 0-0.035, S 0-0.035, Cr 11.5-12.5, Mo 0.3-0.5, Co 0.8-1.1, W 0.6-0.8, bal Fe.
SEL 90; W. Nr. 1.2884.

KRUPP 2885
Krupp Stahl AG
C 0.28-0.35, Si 0.1-0.4, Mn 0.15-0.45, P 0-0.03, S 0-0.03, Cr 2.7-3.2, Mo 2.6-3, Co 2.5-3, V 0.4-0.7, bal Fe.
SEL 90; W. Nr. 1.2885.

KRUPP 2888
Krupp Stahl AG
C 0.17-0.23, Si 0.15-0.35, Mn 0.4-0.6, P 0-0.035, S 0-0.035, Cr 9-10, Mo 1.8-2.2, Co 9.5-10.5, W 5-6, bal Fe.
SEL 90; W. Nr. 1.2888.

KRUPP 2889
Krupp Stahl AG
C 0.4-0.5, Si 0.3-0.5, Mn 0.3-0.5, P 0-0.025, S 0-0.025, Cr 4-5, Mo 2.8-3.3, Co 4-5, V 1.8-2.1, bal Fe.
SEL 90; W. Nr. 1.2889.

KRUPP 30 CRMONIV 5 11
Krupp Stahl AG
C 0.28-0.34, Si 0-0.3, Mn 0.3-0.8, P 0-0.015, S 0-0.018, Cr 1.1-1.4, Mo 1-1.2, Ni 0.5-0.75, V 0.25-0.35, bal Fe.
SEL 90; W. Nr. 1.6946.

KRUPP 30 CRMOV 9
Krupp Stahl AG
C 0.26-0.34, Si 0-0.4, Mn 0.4-0.7, P 0-0.035, S 0-0.03, Cr 2.3-2.7, Mo 0.15-0.25, V 0.1-0.2, bal Fe.
DIN 17200/87; W. Nr. 1.7707.

KRUPP 30 CRNIMO 8
Krupp Stahl AG
C 0.26-0.34, Si 0-0.4, Mn 0.3-0.6, P 0-0.035, S 0-0.03, Cr 1.8-2.2, Mo 0.3-0.5, Ni 1.8-2.2, bal Fe.
DIN 17200/87; W. Nr. 1.6580.

KRUPP 30 MN 4
Krupp Stahl AG
C 0.26-0.34, Si 0.15-0.35, Mn 0.9-1.2, P 0-0.035, S 0-0.035, bal Fe.
SEL 90; W. Nr. 1.1146.

KRUPP 30 MN 5
Krupp Stahl AG
C 0.27-0.34, Si 0.15-0.4, Mn 1.2-1.5, P 0-0.035, S 0-0.035, Cr 0-0.3, bal Fe.
SEL 90; W. Nr. 1.1165.

KRUPP 30 MNB 5
Krupp Stahl AG
C 0.27-0.33, Si 0-0.4, Mn 1.15-1.45, P 0-0.035, S 0-0.035, B 0.0008-0.005, bal Fe.
SEL 90; W. Nr. 1.5531.

KRUPP 30 NICR 11
Krupp Stahl AG
C 0.27-0.34, Si 0.1-0.4, Mn 0.35-0.6, P 0-0.035, S 0-0.035, Cr 0.6-0.9, Ni 2.5-3, bal Fe.
SEL 90; W. Nr. 1.5737.

KRUPP 30 NICRMO 16 6
Krupp Stahl AG
C 0.27-0.33, Si 0.15-0.35, Mn 0.4-0.6, P 0-0.035, S 0-0.035, Cr 1.3-1.5, Mo 0.4-0.5, Ni 3.8-4.2, bal Fe.
SEL 90; W. Nr. 1.6747.

KRUPP 30 NICRMO 2 2
Krupp Stahl AG
C 0.27-0.34, Si 0.15-0.4, Mn 0.7-1, P 0-0.035, S 0-0.035, Cr 0.4-0.6, Mo 0.15-0.3, Ni 0.4-0.7, bal Fe.
SEL 90; W. Nr. 1.6545.

KRUPP 303
Krupp Stahl AG
tool material. C 1.4, V 0.1, Cr 0.3, bal Fe.
For engravers' tools, cutters, bearings, form dies; water hardened, wear resistant. *Obsolete*

KRUPP 31 CRMO 12
Krupp Stahl AG
C 0.28-0.35, Si 0-0.4, Mn 0.4-0.7, P 0-0.025, S 0-0.03, Cr 2.8-3.3, Mo 0.3-0.5, Ni 0-0.3, bal Fe.
DIN 17211/87; W. Nr. 1.8515.

KRUPP 31 CRMOV 9
Krupp Stahl AG
C 0.26-0.34, Si 0-0.4, Mn 0.4-0.7, P 0-0.025, S 0-0.03, Cr 2.3-2.7, Mo 0.15-0.25, V 0.1-0.2, bal Fe.
DIN 17211/87; W. Nr. 1.8519.

KRUPP 31 MN 4
Krupp Stahl AG
C 0.28-0.36, Si 0.2-0.5, Mn 0.8-1.1, P 0-0.045, S 0-0.045, 0.02 Al min, bal Fe.
DIN 21544/85; W. Nr. 1.0520.

KRUPP 31 NICR 14
Krupp Stahl AG
C 0.27-0.35, Si 0.15-0.35, Mn 0.4-0.8, P 0-0.035, S 0-0.035, Cr 0.55-0.95, Ni 3.25-3.75, bal Fe.
SEL 90; W. Nr. 1.5755.

KRUPP 32 CR 2
Krupp Stahl AG
C 0.28-0.35, Si 0-0.4, Mn 0.5-0.8, P 0-0.035, S 0-0.03, Cr 0.4-0.6, bal Fe.
DIN 17200/87; W. Nr. 1.7020.

KRUPP 32 CRB 4
Krupp Stahl AG
C 0.29-0.36, Si 0-0.4, Mn 0.6-0.9, P 0-0.035, S 0-0.035, Cr 0.9-1.2, B 0.0008-0.005, bal Fe.
SEL 90; W. Nr. 1.7076.

KRUPP 32 CRMO 12
Krupp Stahl AG
C 0.28-0.35, Si 0-0.4, Mn 0.4-0.7, P 0-0.035, S 0-0.035, Cr 2.8-3.3, Mo 0.3-0.5, bal Fe.
SEL 90; W. Nr. 1.7361.

KRUPP 32 CRMOV 12 10
Krupp Stahl AG
C 0.3-0.35, Si 0-0.35, Mn 0-0.6, P 0-0.025, S 0-0.01, Cr 2.8-3.2, Mo 0.8-1.2, V 0.25-0.35, bal Fe.
WW 1.7765/75; W. Nr. 1.7765.

KRUPP 32 CRS 2
Krupp Stahl AG
C 0.28-0.35, Si 0-0.4, Mn 0.5-0.8, P 0-0.035, S 0.02-0.035, Cr 0.4-0.6, bal Fe.
DIN 17200/87; W. Nr. 1.7021.

KRUPP 32 NICRMO 10 4
Krupp Stahl AG
C 0.28-0.36, Si 0.15-0.35, Mn 0.3-0.6, P 0-0.035, S 0-0.035, Cr 0.9-1.2, Mo 0.35-0.55, Ni 2.4-2.7, bal Fe.
SEL 90; W. Nr. 1.6743.

KRUPP 32 NICRMO 12 5
Krupp Stahl AG
C 0.29-0.36, Si 0.15-0.35, Mn 0.25-0.5, P 0-0.025, S 0-0.025, Cr 1.2-1.5, Mo 0.3-0.4, Ni 2.8-3, bal Fe.
WW 1.6655/61; W. Nr. 1.6655.

KRUPP 32 NICRMO 14 5
Krupp Stahl AG
C 0.28-0.36, Si 0.15-0.35, Mn 0.3-0.6, P 0-0.035, S 0-0.035, Cr 1-1.5, Mo 0.35-0.55, Ni 3-3.8, bal Fe.
SEL 90; W. Nr. 1.6746.

KRUPP 32 NICRMO 8 5
Krupp Stahl AG
C 0.28-0.36, Si 0-0.4, Mn 0.3-0.6, P 0-0.035, S 0-0.035, Cr 1-1.5, Mo 0.35-0.55, Ni 1.8-2.1, V 0-0.15, bal Fe.
SEL 90; W. Nr. 1.6581.

KRUPP 32 NICRMOV 14 5
Krupp Stahl AG
C 0.28-0.36, Si 0.15-0.35, Mn 0.3-0.6, P 0-0.035, S 0-0.035, Cr 1-1.5, Mo 0.35-0.55, Ni 3-3.8, V 0.07-0.12, bal Fe.
SEL 90; W. Nr. 1.6951.

KRUPP 3202
Krupp Stahl AG
C 1.3-1.45, Si 0-0.45, Mn 0-0.4, P 0-0.03, S 0-0.03, Cr 3.8-4.5, Mo 0.7-1, Co 4.5-5, V 3.5-4, W 11.5-12.5, bal Fe.
DIN 17350/80; W. Nr. 1.3202.

KRUPP 3207
Krupp Stahl AG
C 1.2-1.35, Si 0-0.45, Mn 0-0.4, P 0-0.03, S 0-0.03, Cr 3.8-4.5, Mo 3.2-3.9, Co 9.5-10.5, V 3-3.5, W 9-10, bal Fe.
DIN 17350/80; W. Nr. 1.3207.

KRUPP 3243
Krupp Stahl AG
C 0.92, Co 4.8, Cr 4, Mo 5, V 1.9, W 6.3, bal Fe.
DIN S6-5-2-5; W. Nr. 1.3243. Soft annealed, 240-300 Brin. High speed steel for heavy duty milling, planing and drilling.

KRUPP 3245
Krupp Stahl AG
C 0.88-0.96, Si 0-0.45, Mn 0-0.4, P 0-0.03, S 0.06-0.15, Cr 3.8-4.5, Mo 4.7-5.2, Co 4.5-5, V 1.7-2, W 6-6.7, bal Fe.
DIN 17350/80; W. Nr. 1.3245.

KRUPP 3246
Krupp Stahl AG
C 1.05-1.15, Si 0-0.45, Mn 0-0.4, P 0-0.03, S 0-0.03, Cr 3.8-4.5, Mo 3.6-4, Co 4.8-5.2, V 1.7-1.9, W 6.6-7.1, bal Fe.
DIN 17350/80; W. Nr. 1.3246.

KRUPP 3247
Krupp Stahl AG
C 1.05-1.12, Si 0-0.45, Mn 0-0.4, P 0-0.03, S 0-0.03, Cr 3.6-4.4, Mo 9-10, Co 7.5-8.5, V 1-1.3, W 1.2-1.8, bal Fe.
DIN 17350/80; W. Nr. 1.3247.

KRUPP 3255
Krupp Stahl AG
C 0.75-0.83, Si 0-0.45, Mn 0-0.4, P 0-0.03, S 0-0.03, Cr 3.8-4.5, Mo 0.5-0.8, Co 4.5-5, V 1.4-1.7, W 17.5-18.5, bal Fe.
DIN 17350/80; W. Nr. 1.3255.

KRUPP 33 MNCRB 5 2
Krupp Stahl AG
C 0.3-0.37, Si 0-0.4, Mn 1.2-1.5, P 0-0.035, S 0-0.035, Cr 0.3-0.6, B 0.0008-0.005, bal Fe.
SEL 90; W. Nr. 1.7185.

KRUPP 33 NICRMO 6
Krupp Stahl AG
C 0.3-0.37, Si 0.1-0.4, Mn 0.6-0.9, P 0-0.035, S 0-0.035, Cr 0.85-1.15, Mo 0.15-0.3, Ni 1.2-1.6, bal Fe.
SEL 90; W. Nr. 1.6567.

KRUPP 3302
Krupp Stahl AG
C 1.25, Cr 4, Mo 0.9, V 3.8, W 12, bal Fe.
DIN S12-1-4; W. Nr. 1.3302. Soft annealed, 240-300 Brin. High speed steel of extreme wear resistance.

KRUPP 3318
Krupp Stahl AG
C 0.9-1, Si 0-0.45, Mn 0-0.4, P 0-0.03, S 0-0.03, Cr 3.8-4.5, Mo 0.7-1, V 2.3-2.6, W 11.5-12.5, bal Fe.
SEL 90; W. Nr. 1.3318.

KRUPP 3333
Krupp Stahl AG
C 0.95-1.03, Si 0-0.45, Mn 0-0.4, P 0-0.03, S 0-0.03, Cr 3.8-4.5, Mo 2.5-2.8, V 2.2-2.5, W 2.7-3, bal Fe.
DIN 17350/80; W. Nr. 1.3333.

KRUPP 3339
Krupp Stahl AG
C 0.86-0.94, Si 0-0.45, Mn 0-0.4, P 0-0.03, S 0-0.03, Cr 3.8-4.5, Mo 2.8-3.2, V 1.9-2.2, W 5.7-6.3, bal Fe.
SEL 90; W. Nr. 1.3339.

KRUPP 3341
Krupp Stahl AG
C 0.86-0.94, Si 0-0.45, Mn 0-0.4, P 0-0.03, S 0.06-0.15, Cr 3.8-4.5, Mo 4.7-5.2, V 1.7-2, W 6-6.7, bal Fe.
DIN 17350/80; W. Nr. 1.3341.

KRUPP 3342
Krupp Stahl AG
C 0.95-1.05, Si 0-0.45, Mn 0-0.4, P 0-0.03, S 0-0.03, Cr 3.8-4.5, Mo 4.7-5.2, V 1.7-2, W 6-6.7, bal Fe.
DIN 17350/80; W. Nr. 1.3342.

KRUPP 3343
Krupp Stahl AG
C 0.9, Cr 4, Mo 5, V 1.9, W 6.4, bal Fe.
DIN S6-5-2; W. Nr. 1.3343. Soft annealed, 240-300 Brin. High speed steel of general applicability - also for cold and hot forming tools.

KRUPP 3344
Krupp Stahl AG
C 1.22, Cr 4, Mo 5, V 2.9, W 6.4, bal Fe.
DIN S6-5-3; W. Nr. 1.3344. Soft annealed, 240-300 Brin. High speed steel with very high wear resistance and good toughness for high duty tools.

KRUPP 3346
Krupp Stahl AG
C 0.78-0.86, Si 0-0.45, Mn 0-0.4, P 0-0.03, S 0-0.03, Cr 3.5-4.2, Mo 8-9.2, V 1-1.3, W 1.5-2, bal Fe.
SEL 90; W. Nr. 1.3346.

KRUPP 3348
Krupp Stahl AG
C 0.97-1.07, Si 0-0.45, Mn 0-0.4, P 0-0.03, S 0-0.03, Cr 3.5-4.2, Mo 8-9.2, V 1.8-2.2, W 1.5-2, bal Fe.
DIN 17350/80; W. Nr. 1.3348.

KRUPP 3355
Krupp Stahl AG
C 0.7-0.78, Si 0-0.45, Mn 0-0.4, P 0-0.03, S 0-0.03, Cr 3.8-4.5,
V 1-1.2, W 17.5-18.5, bal Fe.
SEL 90; W. Nr. 1.3355.

KRUPP 34 CR 4
Krupp Stahl AG
C 0.3-0.37, Si 0-0.4, Mn 0.6-0.9, P 0-0.035, S 0-0.03, Cr
0.9-1.2, bal Fe.
DIN 17200/87; W. Nr. 1.7033.

KRUPP 34 CRAL 6
Krupp Stahl AG
C 0.3-0.37, Si 0.15-0.35, Mn 0.6-0.9, P 0-0.035, S 0-0.035, Cr
1.2-1.5, Al 0.8-1.1, bal Fe.
SEL 90; W. Nr. 1.8504.

KRUPP 34 CRALMO 5
Krupp Stahl AG
C 0.3-0.37, Si 0-0.4, Mn 0.5-0.8, P 0-0.025, S 0-0.03, Cr 1-1.3,
Mo 0.15-0.25, Al 0.8-1.2, bal Fe.
DIN 17211/87; W. Nr. 1.8507.

KRUPP 34 CRALNI 7
Krupp Stahl AG
C 0.3-0.37, Si 0-0.4, Mn 0.4-0.7, P 0-0.025, S 0-0.03, Cr
1.5-1.8, Mo 0.15-0.25, Ni 0.85-1.15, Al 0.8-1.2, bal Fe.
DIN 17211/87; W. Nr. 1.8550.

KRUPP 34 CRALS 5
Krupp Stahl AG
C 0.3-0.37, Si 0.15-0.4, Mn 0.6-0.9, P 0-0.1, S 0.07-0.11, Cr
1-1.3, Al 0.8-1.2, bal Fe.
SEL 90; W. Nr. 1.8506.

KRUPP 34 CRMO 4
Krupp Stahl AG
C 0.3-0.37, Si 0-0.4, Mn 0.6-0.9, P 0-0.035, S 0-0.03, Cr
0.9-1.2, Mo 0.15-0.3, bal Fe.
DIN 17200/87; W. Nr. 1.7220.

KRUPP 34 CRMOS 4
Krupp Stahl AG
C 0.3-0.37, Si 0-0.4, Mn 0.6-0.9, P 0-0.035, S 0.02-0.035, Cr
0.9-1.2, Mo 0.15-0.3, bal Fe.
DIN 17200/87; W. Nr. 1.7226.

KRUPP 34 CRNIMO 6
Krupp Stahl AG
C 0.3-0.38, Si 0-0.4, Mn 0.4-0.7, P 0-0.035, S 0-0.03, Cr
1.4-1.7, Mo 0.15-0.3, Ni 1.4-1.7, bal Fe.
DIN 17200/87; W. Nr. 1.6582.

KRUPP 34 CRS 4
Krupp Stahl AG
C 0.3-0.37, Si 0-0.4, Mn 0.6-0.9, P 0-0.035, S 0.02-0.035, Cr
0.9-1.2, bal Fe.
DIN 17200/87; W. Nr. 1.7037.

KRUPP 34 MN 5
Krupp Stahl AG
C 0.3-0.37, Si 0.15-0.3, Mn 1.2-1.5, P 0-0.035, S 0-0.035, bal
Fe.
SEL 90; W. Nr. 1.1166.

KRUPP 34 NI 5
Krupp Stahl AG
C 0.3-0.38, Si 0.15-0.35, Mn 0.3-0.5, P 0-0.035, S 0-0.035, Cr
0-0.6, Ni 1.2-1.5, bal Fe.
SEL 90; W. Nr. 1.5620.

KRUPP 3401
Krupp Stahl AG
C 1.1-1.3, Si 0.3-0.5, Mn 12-13, P 0-0.1, S 0-0.04, Cr 0-1.5, bal
Fe.
SEL 90; W. Nr. 1.3401.

KRUPP 3402
Krupp Stahl AG
C 1-1.25, Si 0.35-0.7, Mn 13.5-14.5, P 0-0.08, S 0-0.02, bal
Fe.
DIN 17145/80; W. Nr. 1.3402.

KRUPP 35 B 2
Krupp Stahl AG
C 0.32-0.4, Si 0-0.4, Mn 0.5-0.8, P 0-0.035, S 0-0.035, B
0.0008-0.005, bal Fe.
DIN 1654 T.4/89; W. Nr. 1.5511.

KRUPP 35 CRNIMOV 8
Krupp Stahl AG
C 0.3-0.4, Si 0.15-0.35, Mn 0.3-0.6, P 0-0.025, S 0-0.025, Cr
1.8-2.1, Mo 0.25-0.35, Ni 1.8-2.1, V 0.07-0.12, bal Fe.
SEL 90; W. Nr. 1.6935.

KRUPP 35 NICR 18
Krupp Stahl AG
C 0.3-0.4, Si 0.15-0.35, Mn 0.4-0.8, P 0-0.035, S 0-0.035, Cr
1.1-1.5, Ni 4.25-4.75, bal Fe.
SEL 90; W. Nr. 1.5864.

KRUPP 35 NICRMOV 11 5
Krupp Stahl AG
C 0.3-0.4, Si 0.15-0.35, Mn 0.4-0.7, P 0-0.015, S 0-0.015, Cr
1-1.4, Mo 0.35-0.6, Ni 2.5-3, Al 0-0.02, V 0.08-0.2, bal Fe.
SEL 90; W. Nr. 1.6949.

KRUPP 35 NICRMOV 12 5
Krupp Stahl AG
C 0.3-0.4, Si 0.15-0.35, Mn 0.4-0.7, P 0-0.015, S 0-0.015, Cr
1-1.4, Mo 0.35-0.6, Ni 2.5-3.5, Al 0-0.015, V 0.08-0.2, bal Fe.
WW 1.6959/79; W. Nr. 1.6959.

KRUPP 35 S 20
Krupp Stahl AG
C 0.32-0.39, Si 0.1-0.3, Mn 0.7-1.1, P 0-0.06, S 0.18-0.25, bal
Fe.
DIN 1651/88; W. Nr. 1.0726.

KRUPP 35 SPB 20
Krupp Stahl AG
C 0.32-0.39, Si 0.1-0.3, Mn 0.7-1.1, P 0-0.06, S 0.18-0.25, Pb
0.15-0.35, bal Fe.
DIN 1651/88; W. Nr. 1.0756.

KRUPP 3537
Krupp Stahl AG
C 0.95-1.05, Si 0.2-0.4, Mn 0.25-0.4, P 0-0.03, S 0-0.025, Cr
1.65-1.95, Mo 0.15-0.25, Ni 0-0.3, Cu 0-0.3, bal Fe.
DIN 17230/80; W. Nr. 1.3537.

KRUPP 3541
Krupp Stahl AG
C 0.42-0.5, Si 0-1, Mn 0-1, P 0-0.04, S 0-0.03, Cr 12.5-14.5, Ni
0-1, Cu 0-0.3, bal Fe.
DIN 17230/80; W. Nr. 1.3541.

KRUPP 3543
Krupp Stahl AG
C 0.95-1.1, Si 0-1, Mn 0-1, P 0-0.04, S 0-0.03, Cr 16-18, Mo
0.35-0.75, Ni 0-0.5, Cu 0-0.3, bal Fe.
DIN 17230/80; W. Nr. 1.3543.

KRUPP 3549
Krupp Stahl AG
C 0.85-0.95, Si 0-1, Mn 0-1, P 0-0.045, S 0-0.03, Cr 17-19, Mo
0.9-1.3, Cu 0-0.3, V 0.07-0.12, bal Fe.
DIN 17230/80; W. Nr. 1.3549.

KRUPP 3551
Krupp Stahl AG
C 0.77-0.85, Si 0-0.25, Mn 0-0.35, P 0-0.015, S 0-0.015, Cr
3.75-4.25, Mo 4-4.5, V 0.9-1.1, bal Fe.
DIN 17230/80; W. Nr. 1.3551.

KRUPP 3553
Krupp Stahl AG
C 0.78-0.86, Si 0-0.4, Mn 0-0.4, P 0-0.03, S 0-0.03, Cr 3.8-4.5,
Mo 4.7-5.2, V 1.7-2, W 6-6.7, bal Fe.
DIN 17230/80; W. Nr. 1.3553.

KRUPP 3558
Krupp Stahl AG
C 0.7-0.78, Si 0-0.45, Mn 0-0.4, P 0-0.03, S 0-0.03, Cr 3.8-4.5,
Mo 0-0.6, V 1-1.2, W 17.5-18.5, bal Fe.
DIN 17230/80; W. Nr. 1.3558.

KRUPP 36 CRB 2
Krupp Stahl AG
C 0.32-0.4, Si 0-0.4, Mn 0.6-0.9, P 0-0.035, S 0-0.035, Cr
0.3-0.6, B 0.0008-0.005, bal Fe.
SEL 90; W. Nr. 1.7072.

KRUPP 36 CRB 4
Krupp Stahl AG
C 0.33-0.39, Si 0-0.4, Mn 0.7-1, P 0-0.035, S 0-0.035, Cr
0.9-1.2, B 0.0008-0.005, bal Fe.
SEL 90; W. Nr. 1.7077.

KRUPP 36 CRNIMO 4
Krupp Stahl AG
C 0.32-0.4, Si 0-0.4, Mn 0.5-0.8, P 0-0.035, S 0-0.03, Cr
0.9-1.2, Mo 0.15-0.3, Ni 0.9-1.2, bal Fe.
DIN 17200/87; W. Nr. 1.6511.

KRUPP 36 MN 4
Krupp Stahl AG
C 0.32-0.4, Si 0.25-0.5, Mn 0.9-1.2, P 0-0.05, S 0-0.05, N
0-0.007, bal Fe.
SEL 81; W. Nr. 1.0561.

KRUPP 36 MN 5
Krupp Stahl AG
C 0.32-0.4, Si 0-0.4, Mn 1.2-1.5, P 0-0.035, S 0-0.035, bal Fe.
SEL 90; W. Nr. 1.1167.

KRUPP 36 MN 6
Krupp Stahl AG
C 0.34-0.42, Si 0.15-0.35, Mn 1.4-1.65, P 0-0.035, S 0-0.035,
bal Fe.
SEL 90; W. Nr. 1.1127.

KRUPP 36 MN 7
Krupp Stahl AG
C 0.32-0.4, Si 0.3-0.45, Mn 1.6-1.9, P 0-0.03, S 0-0.03, bal Fe.
SEL 90; W. Nr. 1.5069.

KRUPP 36 NICR 10
Krupp Stahl AG
C 0.32-0.4, Si 0.15-0.35, Mn 0.4-0.8, P 0-0.035, S 0-0.035, Cr
0.55-0.95, Ni 2.25-2.75, bal Fe.
SEL 90; W. Nr. 1.5736.

KRUPP 36 NICR 6
Krupp Stahl AG
C 0.32-0.4, Si 0.15-0.35, Mn 0.4-0.8, P 0-0.035, S 0-0.035, Cr
0.3-0.7, Ni 1.25-1.75, bal Fe.
SEL 90; W. Nr. 1.5710.

KRUPP 36 NIR 6 4
Krupp Stahl AG
C 0.33-0.39, Si 0.1-0.4, Mn 0.6-0.9, P 0-0.035, S 0-0.035, Cr
0.85-1.15, Ni 1.2-1.6, bal Fe.
SEL 90; W. Nr. 1.5815.

KRUPP 37 CR 4
Krupp Stahl AG
C 0.34-0.41, Si 0-0.4, Mn 0.6-0.9, P 0-0.035, S 0-0.03, Cr
0.9-1.2, bal Fe.
DIN 17200/87; W. Nr. 1.7034.

KRUPP 37 CRB 1
Krupp Stahl AG
C 0.35-0.4, Si 0-0.4, Mn 0.5-0.8, P 0-0.035, S 0-0.035, Cr
0.3-0.4, 0.0008 B min, bal Fe.
SEL 90; W. Nr. 1.7007.

KRUPP 37 CRB 4
Krupp Stahl AG
C 0.34-0.4, Si 0-0.4, Mn 0.7-1, P 0-0.035, S 0-0.035, Cr
0.9-1.2, B 0.0008-0.005, bal Fe.
VDEH 88; W. Nr. 1.5545.

KRUPP 37 CRMO 3
Krupp Stahl AG
C 0.33-0.41, Si 0.15-0.35, Mn 0.6-0.9, P 0-0.035, S 0.02-0.04,
Cr 0.6-0.9, Mo 0.3-0.4, bal Fe.
SEL 90; W. Nr. 1.7315.

KRUPP 37 CRS 4
Krupp Stahl AG
C 0.34-0.41, Si 0-0.4, Mn 0.6-0.9, P 0-0.035, S 0.02-0.035, Cr
0.9-1.2, bal Fe.
DIN 17200/87; W. Nr. 1.7038.

KRUPP 37 MNSI 5
Krupp Stahl AG
C 0.33-0.41, Si 1.1-1.4, Mn 1.1-1.4, P 0-0.035, S 0-0.035, bal
Fe.
SEL 90; W. Nr. 1.5122.

KRUPP 38 CR 1
Krupp Stahl AG
C 0.34-0.41, Si 0.15-0.4, Mn 0.5-0.8, P 0-0.035, S 0-0.035, Cr
0.3-0.4, bal Fe.
SEL 90; W. Nr. 1.7001.

KRUPP 38 CR 2
Krupp Stahl AG
C 0.35-0.42, Si 0-0.4, Mn 0.5-0.8, P 0-0.035, S 0-0.03, Cr
0.4-0.6, bal Fe.
DIN 17200/87; W. Nr. 1.7003.

KRUPP 38 CR 4
Krupp Stahl AG
C 0.34-0.4, Si 0.15-0.4, Mn 0.6-0.9, P 0-0.025, S 0-0.035, Cr
0.9-1.2, bal Fe.
DIN 17212/72; W. Nr. 1.7043.

KRUPP 38 CRS 2
Krupp Stahl AG
C 0.35-0.42, Si 0-0.4, Mn 0.5-0.8, P 0-0.035, S 0.02-0.035, Cr
0.4-0.6, bal Fe.
DIN 17200/87; W. Nr. 1.7023.

KRUPP 38 MNB 5
Krupp Stahl AG
C 0.35-0.42, Si 0-0.4, Mn 1.15-1.45, P 0-0.035, S 0-0.035, B
0.0008-0.005, bal Fe.
SEL 90; W. Nr. 1.5532.

KRUPP 38 MNSI 4
Krupp Stahl AG
C 0.34-0.42, Si 0.7-0.9, Mn 0.9-1.2, P 0-0.035, S 0-0.035, bal
Fe.
SEL 90; W. Nr. 1.5120.

KRUPP 38 MNSIVS 5
Krupp Stahl AG
C 0.35-0.4, Si 0.5-0.8, Mn 1.2-1.5, P 0-0.035, S 0.03-0.065, V
0.08-0.13, bal Fe.
SEW 101/88; W. Nr. 1.5231.

KRUPP 38 SI 6
Krupp Stahl AG
C 0.35-0.42, Si 1.4-1.6, Mn 0.5-0.8, P 0-0.05, S 0-0.05, N
0-0.007, bal Fe.
SEL 90; W. Nr. 1.5022.

KRUPP 38 SI 7
Krupp Stahl AG
C 0.35-0.42, Si 1.5-1.8, P 0-0.03, S 0-0.03, bal Fe.
DIN 17221/88; W. Nr. 1.5023.

KRUPP 39 CRMOV 13 9
Krupp Stahl AG
C 0.35-0.42, Si 0.15-0.4, Mn 0.4-0.7, P 0-0.03, S 0-0.035, Cr
3-3.5, Mo 0.8-1.1, V 0.15-0.25, bal Fe.
SEL 90; W. Nr. 1.8523.

KRUPP 39 MNCRB 6 2
Krupp Stahl AG
C 0.36-0.42, Si 0-0.4, Mn 1.4-1.7, P 0-0.035, S 0-0.035, Cr
0.3-0.6, B 0.0008-0.005, bal Fe.
SEL 90; W. Nr. 1.7189.

KRUPP 40 CRMOV 4 7
Krupp Stahl AG
C 0.36-0.44, Si 0.15-0.35, Mn 0.35-0.85, P 0-0.03, S 0-0.035,
Cr 0.9-1.2, Mo 0.6-0.75, V 0.25-0.35, bal Fe.
DIN 17240/76; W. Nr. 1.7711.

KRUPP 40 MN 4
Krupp Stahl AG
C 0.36-0.44, Si 0.25-0.5, Mn 0.8-1.1, P 0-0.035, S 0-0.035, bal
Fe.
SEL 90; W. Nr. 1.1157.

KRUPP 40 MNB 4
Krupp Stahl AG
C 0.37-0.44, Si 0-0.4, Mn 0.8-1.1, P 0-0.035, S 0-0.035, B
0.0008-0.005, bal Fe.
SEL 90; W. Nr. 1.5527.

KRUPP 40 NICR 6
Krupp Stahl AG
C 0.38-0.43, Si 0.15-0.35, Mn 0.7-0.9, P 0-0.035, S 0-0.035, Cr
0.55-0.75, Ni 1.1-1.4, bal Fe.
SEL 90; W. Nr. 1.5711.

KRUPP 40 NICRMO 2 2
Krupp Stahl AG
C 0.37-0.44, Si 0.15-0.4, Mn 0.7-1, P 0-0.035, S 0-0.035, Cr
0.4-0.6, Mo 0.15-0.3, Ni 0.4-0.7, bal Fe.
SEL 90; W. Nr. 1.6546.

KRUPP 40 NICRMO 6
Krupp Stahl AG
C 0.35-0.45, Si 0.15-0.35, Mn 0.5-0.7, P 0-0.035, S 0-0.035, Cr
0.9-1.4, Mo 0.2-0.3, Ni 1.4-1.7, bal Fe.
SEL 90; W. Nr. 1.6565.

KRUPP 40 NICRMO 8 4
Krupp Stahl AG
C 0.37-0.44, Si 0.2-0.35, Mn 0.7-0.9, P 0-0.02, S 0-0.015, Cr
0.7-0.95, Mo 0.3-0.4, Ni 1.65-2, Al 0.005-0.05, bal Fe.
SEL 90; W. Nr. 1.6562.

KRUPP 40 NIMOCR 10 5
Krupp Stahl AG
C 0.37-0.43, Si 0.15-0.35, Mn 0.5-0.7, P 0-0.035, S 0-0.035, Cr
0.6-0.8, Mo 0.4-0.6, Ni 2.4-2.7, bal Fe.
SEL 90; W. Nr. 1.6745.

KRUPP 41 CR 4
Krupp Stahl AG
C 0.38-0.45, Si 0-0.4, Mn 0.6-0.9, P 0-0.035, S 0-0.03, Cr
0.9-1.2, bal Fe.
DIN 17200/87; W. Nr. 1.7035.

KRUPP 41 CRALMO 7
Krupp Stahl AG
C 0.38-0.45, Si 0-0.4, Mn 0.5-0.8, P 0-0.03, S 0-0.035, Cr
1.5-1.8, Mo 0.25-0.4, Al 0.8-1.2, bal Fe.
SEL 90; W. Nr. 1.8509.

KRUPP 41 CRMO 4
Krupp Stahl AG
C 0.38-0.44, Si 0.15-0.4, Mn 0.5-0.8, P 0-0.025, S 0-0.035, Cr
0.9-1.2, Mo 0.15-0.3, bal Fe.
DIN 17212/72; W. Nr. 1.7223.

KRUPP 41 CRS 4
Krupp Stahl AG
C 0.35-0.4, Si 0-0.4, Mn 0.6-0.9, P 0-0.035, S 0.02-0.035, Cr
0.9-1.2, bal Fe.
DIN 17200/87; W. Nr. 1.7039.

KRUPP 41 MNV 5
Krupp Stahl AG
C 0.38-0.44, Si 0.1-0.4, Mn 1.1-1.3, P 0-0.035, S 0-0.035, V
0.1-0.15, bal Fe.
SEL 90; W. Nr. 1.5219.

KRUPP 42 CR 4
Krupp Stahl AG
C 0.38-0.44, Si 0.15-0.4, Mn 0.5-0.8, P 0-0.025, S 0-0.035, Cr
0.9-1.2, bal Fe.
DIN 17212/72; W. Nr. 1.7045.

KRUPP 42 CRMO 4
Krupp Stahl AG
C 0.38-0.45, Si 0-0.4, Mn 0.6-0.9, P 0-0.035, S 0-0.03, Cr
0.9-1.2, Mo 0.15-0.3, bal Fe.
DIN 17200/87; W. Nr. 1.7225.

KRUPP 42 CRMOS 4
Krupp Stahl AG
C 0.38-0.45, Si 0-0.4, Mn 0.6-0.9, P 0-0.035, S 0.02-0.035, Cr
0.9-1.2, Mo 0.15-0.3, bal Fe.
DIN 17200/87; W. Nr. 1.7227.

KRUPP 42 CRMOV 7 3
Krupp Stahl AG
C 0.38-0.45, Si 0.15-0.35, Mn 0.5-0.8, P 0-0.035, S 0-0.035, Cr
1.6-1.9, Mo 0.3-0.4, V 0.07-0.12, bal Fe.
SEL 90; W. Nr. 1.7741.

KRUPP 42 CRV 6
Krupp Stahl AG
C 0.38-0.46, Si 0.15-0.35, Mn 0.5-0.8, P 0-0.035, S 0-0.035, Cr
1.4-1.7, V 0.07-0.12, bal Fe.
SEL 90; W. Nr. 1.7561.

KRUPP 42 MNMO 7
Krupp Stahl AG
C 0.38-0.45, Si 0.2-0.35, Mn 1.55-1.85, P 0-0.04, S 0-0.04, Mo
0.15-0.25, bal Fe.
SEL 90; W. Nr. 1.5432.

KRUPP 42 MNV 7
Krupp Stahl AG
C 0.38-0.45, Si 0.15-0.35, Mn 1.6-1.9, P 0-0.035, S 0-0.035, V
0.07-0.12, bal Fe.
SEL 90; W. Nr. 1.5223.

KRUPP 43 CRMO 4
Krupp Stahl AG
C 0.4-0.46, Si 0-0.4, Mn 0.6-0.9, P 0-0.025, S 0-0.035, Cr
0.9-1.2, Mo 0.15-0.3, Cu 0-0.3, bal Fe.
DIN 17230/80; W. Nr. 1.3563.

KRUPP 44 CR 2
Krupp Stahl AG
C 0.42-0.48, Si 0-0.4, Mn 0.5-0.8, P 0-0.025, S 0-0.035, Cr
0.4-0.6, Cu 0-0.3, bal Fe.
DIN 17230/80; W. Nr. 1.3561.

KRUPP 44 MNSIVS 6
Krupp Stahl AG
C 0.42-0.47, Si 0.5-0.8, Mn 1.3-1.6, P 0-0.035, S 0.02-0.035, V
0.1-0.15, bal Fe.
SEW 101/88; W. Nr. 1.5233.

KRUPP 45 B 2
Krupp Stahl AG
C 0.42-0.5, Si 0-0.4, Mn 0.5-0.8, P 0-0.035, S 0-0.035, B
0.0008-0.005, bal Fe.
SEL 90; W. Nr. 1.5513.

KRUPP 45 CR 2
Krupp Stahl AG
C 0.42-0.48, Si 0.15-0.4, Mn 0.5-0.8, P 0-0.025, S 0-0.035, Cr 0.4-0.6, bal Fe.
DIN 17212/72; W. Nr. 1.7005.

KRUPP 45 CRMOV 6 7
Krupp Stahl AG
C 0.4-0.5, Si 0.15-0.35, Mn 0.6-0.8, P 0-0.035, S 0-0.035, Cr 1.3-1.5, Mo 0.65-0.75, V 0.25-0.35, bal Fe.
SEL 90; W. Nr. 1.7737.

KRUPP 45 S 20
Krupp Stahl AG
C 0.42-0.5, Si 0.1-0.3, Mn 0.7-1.1, P 0-0.06, S 0.18-0.25, bal Fe.
DIN 1651/88; W. Nr. 1.0727.

KRUPP 45 SPB 20
Krupp Stahl AG
C 0.42-0.5, Si 0.1-0.3, Mn 0.7-1.1, P 0-0.06, S 0.18-0.25, Pb 0.15-0.35, bal Fe.
DIN 1651/88; W. Nr. 1.0757.

KRUPP 46 CR 1
Krupp Stahl AG
C 0.42-0.5, Si 0.15-0.4, Mn 0.5-0.8, P 0-0.035, S 0-0.035, Cr 0.3-0.4, bal Fe.
SEL 90; W. Nr. 1.7002.

KRUPP 46 CR 2
Krupp Stahl AG
C 0.42-0.5, Si 0-0.4, Mn 0.5-0.8, P 0-0.035, S 0-0.03, Cr 0.4-0.6, bal Fe.
DIN 17200/87; W. Nr. 1.7006.

KRUPP 46 CRB 2
Krupp Stahl AG
C 0.42-0.5, Si 0-0.4, Mn 0.6-0.9, P 0-0.035, S 0-0.035, Cr 0.3-0.6, B 0.0008-0.005, bal Fe.
SEL 90; W. Nr. 1.7075.

KRUPP 46 CRS 2
Krupp Stahl AG
C 0.42-0.5, Si 0-0.4, Mn 0.5-0.8, P 0-0.035, S 0.02-0.035, Cr 0.4-0.6, bal Fe.
DIN 17200/87; W. Nr. 1.7025.

KRUPP 46 MN 5
Krupp Stahl AG
C 0.42-0.48, Si 0.25-0.45, Mn 1.15-1.35, P 0-0.035, S 0-0.035, bal Fe.
SEL 90; W. Nr. 1.1128.

KRUPP 46 MN 7
Krupp Stahl AG
C 0.42-0.5, Si 0.15-0.35, Mn 1.6-1.9, P 0-0.05, S 0-0.05, N 0-0.007, bal Fe.
SEL 90; W. Nr. 1.0912.

KRUPP 46 MNSI 4
Krupp Stahl AG
C 0.42-0.5, Si 0.7-0.9, Mn 0.9-1.2, P 0-0.035, S 0-0.035, bal Fe.
SEL 90; W. Nr. 1.5121.

KRUPP 46 SI 7
Krupp Stahl AG
C 0.42-0.5, Si 1.5-1.8, Mn 0.5-0.8, P 0-0.05, S 0-0.05, N 0-0.007, bal Fe.
SEL 90; W. Nr. 1.5024.

KRUPP 46 SICRMO 6 3
Krupp Stahl AG
C 0.42-0.5, Si 1.3-1.7, Mn 0.5-0.8, P 0-0.03, S 0-0.025, Cr 0.5-0.75, Mo 0.15-0.3, bal Fe.
SEL 90; W. Nr. 1.8062.

KRUPP 48 CRMO 4
Krupp Stahl AG
C 0.46-0.52, Si 0-0.4, Mn 0.5-0.8, P 0-0.025, S 0-0.035, Cr 0.9-1.2, Mo 0.15-0.3, Cu 0-0.3, bal Fe.
DIN 17230/80; W. Nr. 1.3565.

KRUPP 49 CRMO 4
Krupp Stahl AG
C 0.46-0.52, Si 0.15-0.4, Mn 0.5-0.8, P 0-0.025, S 0-0.035, Cr 0.9-1.2, Mo 0.15-0.3, bal Fe.
DIN 17212/72; W. Nr. 1.7238.

KRUPP 49 MNVS 3
Krupp Stahl AG
C 0.44-0.5, Si 0-0.5, Mn 0.7-1, P 0-0.035, S 0.03-0.065, V 0.08-0.13, bal Fe.
SEW 101/88; W. Nr. 1.1199.

KRUPP 50 CRMO 4
Krupp Stahl AG
C 0.46-0.54, Si 0-0.4, Mn 0.5-0.8, P 0-0.035, S 0-0.03, Cr 0.9-1.2, Mo 0.15-0.3, bal Fe.
DIN 17200/87; W. Nr. 1.7228.

KRUPP 50 CRV 4
Krupp Stahl AG
C 0.47-0.55, Si 0-0.4, Mn 0.7-1.1, P 0-0.035, S 0-0.03, Cr 0.9-1.2, V 0.1-0.2, bal Fe.
DIN 17200/87; W. Nr. 1.8159.

KRUPP 50 MN 7
Krupp Stahl AG
C 0.45-0.55, Si 0-0.4, Mn 1.6-2, P 0-0.04, S 0-0.04, N 0-0.007, bal Fe.
SEL 90; W. Nr. 1.0913.

KRUPP 50 MNSI 4
Krupp Stahl AG
C 0.45-0.53, Si 0.7-1, Mn 0.9-1.2, P 0-0.035, S 0-0.035, bal Fe.
SEL 90; W. Nr. 1.5131.

KRUPP 51 CRMOV 4
Krupp Stahl AG
C 0.48-0.56, Si 0.15-0.4, Mn 0.7-1.1, P 0-0.03, S 0-0.03, Cr 0.9-1.2, Mo 0.15-0.25, V 0.08-0.15, bal Fe.
DIN 17221/88; W. Nr. 1.7701.

KRUPP 51 MNV 7
Krupp Stahl AG
C 0.48-0.55, Si 0.15-0.35, Mn 1.6-1.9, P 0-0.035, S 0-0.035, V 0.07-0.12, bal Fe.
SEL 90; W. Nr. 1.5225.

KRUPP 51 SI 7
Krupp Stahl AG
C 0.47-0.55, Si 1.5-1.8, Mn 0.5-0.8, P 0-0.045, S 0-0.045, N 0-0.007, bal Fe.
SEL 90; W. Nr. 1.5025.

KRUPP 52 MN 5
Krupp Stahl AG
C 0.47-0.55, Si 0.15-0.3, Mn 1.2-1.5, P 0-0.035, S 0-0.035, bal Fe.
SEL 90; W. Nr. 1.1226.

KRUPP 52 MNCRB 3
Krupp Stahl AG
C 0.48-0.55, Si 0.15-0.35, Mn 0.75-1, P 0-0.035, S 0-0.035, Cr 0.4-0.6, 0.0005 B min, bal Fe.
SEL 90; W. Nr. 1.7138.

KRUPP 53 MNSI 4
Krupp Stahl AG
C 0.5-0.57, Si 0.8-1, Mn 0.8-1.2, P 0-0.035, S 0-0.035, bal Fe.
SEL 90; W. Nr. 1.5141.

KRUPP 54 SICR 6
Krupp Stahl AG
C 0.51-0.59, Si 1.2-1.6, Mn 0.5-0.8, P 0-0.03, S 0-0.03, Cr 0.5-0.8, bal Fe.
DIN 17221/88; W. Nr. 1.7102.

KRUPP 55 CR 3
Krupp Stahl AG
C 0.52-0.59, Si 0.25-0.5, Mn 0.7-1.1, P 0-0.03, S 0-0.03, Cr 0.7-1, bal Fe.
DIN 17221/88; W. Nr. 1.7176.

KRUPP 55 SI 7
Krupp Stahl AG
C 0.52-0.6, Si 1.5-1.8, Mn 0.7-1, P 0-0.045, S 0-0.045, bal Fe.
DIN 17222/79; W. Nr. 1.5026.

KRUPP 55 SICR 6 3
Krupp Stahl AG
C 0.5-0.6, Si 1.2-1.6, Mn 0.5-0.9, P 0-0.03, S 0-0.025, Cr 0.5-0.8, Cu 0-0.12, bal Fe.
SEL 90; W. Nr. 1.7104.

KRUPP 55 SICR 7
Krupp Stahl AG
C 0.52-0.6, Si 1.5-1.8, Mn 0.7-1, P 0-0.05, S 0-0.05, Cr 0.2-0.4, N 0-0.007, bal Fe.
SEL 90; W. Nr. 1.7106.

KRUPP 58 CRV 4
Krupp Stahl AG
C 0.55-0.62, Si 0.15-0.4, Mn 0.7-1.1, P 0-0.035, S 0-0.035, Cr 0.9-1.2, V 0.1-0.2, bal Fe.
SEL 90; W. Nr. 1.8161.

KRUPP 6 P 15
Krupp Stahl AG
C 0-0.09, Si 0-0.01, Mn 0.2-0.45, P 0.12-0.2, S 0-0.06, bal Fe.
SEL 81; W. Nr. 1.0745.

KRUPP 60 MN 3
Krupp Stahl AG
C 0.57-0.65, Si 0.2-0.4, Mn 0.7-0.9, P 0-0.05, S 0-0.05, N 0-0.007, bal Fe.
SEL 90; W. Nr. 1.0642.

KRUPP 60 S 20
Krupp Stahl AG
C 0.57-0.65, Si 0.1-0.3, Mn 0.7-1.1, P 0-0.06, S 0.18-0.25, bal Fe.
DIN 1651/88; W. Nr. 1.0728.

KRUPP 60 SI 7
Krupp Stahl AG
C 0.56-0.64, Si 1.5-1.8, Mn 0.7-1, P 0-0.045, S 0-0.045, bal Fe.
SEL 90; W. Nr. 1.5027.

KRUPP 60 SICR 7
Krupp Stahl AG
C 0.57-0.65, Si 1.5-1.8, Mn 0.7-1, P 0-0.03, S 0-0.03, Cr 0.2-0.4, bal Fe.
DIN 17221/88; W. Nr. 1.7108.

KRUPP 60 SIMN 5
Krupp Stahl AG
C 0.55-0.65, Si 1-1.3, Mn 0.9-1.1, P 0-0.05, S 0-0.05, N 0-0.007, bal Fe.
SEL 90; W. Nr. 1.5142.

KRUPP 60 SPB 20
Krupp Stahl AG
C 0.57-0.65, Si 0.1-0.3, Mn 0.7-1.1, P 0-0.06, S 0.18-0.25, Pb 0.15-0.35, bal Fe.
DIN 1651/88; W. Nr. 1.0758.

KRUPP 61 CRMO 4
Krupp Stahl AG
C 0.57-0.65, Si 0.15-0.35, Mn 0.4-0.6, P 0-0.035, S 0-0.035, Cr 0.9-1.2, Mo 0.15-0.25, bal Fe.
SEL 90; W. Nr. 1.7229.

KRUPP 64 MN 3
Krupp Stahl AG
C 0.6-0.68, Si 0.2-0.4, Mn 0.5-0.8, P 0-0.05, S 0-0.05, N 0-0.007, bal Fe.
SEL 90; W. Nr. 1.0640.

KRUPP 65 CR 3
Krupp Stahl AG
C 0.6-0.7, Si 0.2-0.4, Mn 0.5-0.8, P 0-0.035, S 0-0.035, Cr 0.3-0.6, N 0-0.007, bal Fe.
SEL 90; W. Nr. 1.7017.

KRUPP 65 MN 4
Krupp Stahl AG
C 0.6-0.7, Si 0.25-0.5, Mn 0.9-1.2, P 0-0.035, S 0-0.035, bal Fe.
SEL 90; W. Nr. 1.1240.

KRUPP 65 SI 7
Krupp Stahl AG
C 0.6-0.7, Si 1.5-1.8, Mn 0.7-1, P 0-0.035, S 0-0.035, bal Fe.
SEL 90; W. Nr. 1.5028.

KRUPP 65 SICR 7
Krupp Stahl AG
C 0.6-0.8, Si 1.5-1.8, Mn 0.7-1, P 0-0.05, S 0-0.05, Cr 0.2-0.4, N 0-0.007, bal Fe.
SEL 90; W. Nr. 1.7107.

KRUPP 66 MN 4
Krupp Stahl AG
C 0.6-0.71, Si 0.15-0.3, Mn 0.85-1.15, P 0-0.035, S 0-0.035, bal Fe.
SEL 90; W. Nr. 1.1260.

KRUPP 67 SICR 5
Krupp Stahl AG
C 0.62-0.72, Si 1.2-1.4, Mn 0.4-0.6, P 0-0.035, S 0-0.035, Cr 0.4-0.6, bal Fe.
DIN 17222/79; W. Nr. 1.7103.

KRUPP 67 SICR 5 2
Krupp Stahl AG
C 0.62-0.72, Si 1.2-1.4, Mn 0.4-0.6, P 0-0.03, S 0-0.025, Cr 0.4-0.6, Cu 0-0.12, bal Fe.
SEL 90; W. Nr. 1.7105.

KRUPP 6907
Krupp Stahl AG
C 0-0.04, Si 0-1, Mn 0-2, P 0-0.045, S 0-0.03, Cr 17-19, Ni 9-11.5, N 0.1-0.18, bal Fe.
SEL 90; W. Nr. 1.6907.

KRUPP 6909
Krupp Stahl AG
C 0-0.07, Si 0-1, Mn 7.5-9.5, P 0-0.045, S 0-0.03, Cr 17-19, Ni 4.5-6.5, N 0.2-0.3, bal Fe.
SEL 90; W. Nr. 1.6909.

KRUPP 70 MN 3
Krupp Stahl AG
C 0.65-0.75, Si 0.2-0.4, Mn 0.6-0.9, P 0-0.05, S 0-0.05, N 0-0.007, bal Fe.
SEL 90; W. Nr. 1.0643.

KRUPP 70 S 20
Krupp Stahl AG
C 0.6-0.72, Si 0.15-0.25, Mn 0.5-0.7, P 0-0.07, S 0.15-0.25, bal Fe.
SEL 81; W. Nr. 1.0729.

KRUPP 70 SPB 20
Krupp Stahl AG
C 0.65-0.75, Si 0.1-0.4, Mn 0.5-0.9, P 0-0.07, S 0.15-0.25, Pb 0.15-0.3, bal Fe.
SEL 90; W. Nr. 1.0759.

KRUPP 71 SI 7
Krupp Stahl AG
C 0.68-0.75, Si 1.5-1.8, Mn 0.6-0.8, P 0-0.035, S 0-0.035, bal Fe.
DIN 17222/79; W. Nr. 1.5029.

KRUPP 75-2
Krupp Stahl AG
C 0.73-0.78, Si 0.1-0.3, Mn 0.3-0.7, P 0-0.04, S 0-0.04, bal Fe.
DIN 17140 T.1/83; W. Nr. 1.0614.

KRUPP 76 MN 3
Krupp Stahl AG
C 0.7-0.8, Si 0.2-0.4, Mn 0.6-0.9, P 0-0.05, S 0-0.05, N 0-0.007, bal Fe.
SEL 90; W. Nr. 1.0645.

KRUPP 8 MN 4
Krupp Stahl AG
C 0.05-0.1, Si 0.2-0.4, Mn 0.9-1.2, P 0-0.02, S 0-0.02, bal Fe.
SEL 90; W. Nr. 1.1117.

KRUPP 8 NI 9
Krupp Stahl AG
C 0-0.1, Si 0-0.35, Mn 0.3-0.8, P 0-0.025, S 0-0.02, Mo 0-0.1, Ni 8-10, V 0-0.05, bal Fe.
DIN 17280/85; W. Nr. 1.5662.

KRUPP 80 MN 4
Krupp Stahl AG
C 0.75-0.85, Si 0.25-0.5, Mn 0.9-1.2, P 0-0.035, S 0-0.035, bal Fe.
SEL 75; W. Nr. 1.1259.

KRUPP 85 MN 3
Krupp Stahl AG
C 0.8-0.9, Si 0.15-0.35, Mn 0.7-0.9, P 0-0.05, S 0-0.05, N 0-0.007, bal Fe.
SEL 90; W. Nr. 1.0647.

KRUPP 9 S 20
Krupp Stahl AG
C 0-0.13, Si 0-0.05, Mn 0.6-1.2, P 0-0.1, S 0.18-0.25, bal Fe.
SEL 81; W. Nr. 1.0711.

KRUPP 9 SMN 28
Krupp Stahl AG
C 0-0.14, Si 0-0.05, Mn 0.9-1.3, P 0-0.1, S 0.27-0.33, bal Fe.
DIN 1651/88; W. Nr. 1.0715.

KRUPP 9 SMN 36
Krupp Stahl AG
C 0-0.15, Si 0-0.05, Mn 1.1-1.5, P 0-0.1, S 0.34-0.4, bal Fe.
DIN 1651/88; W. Nr. 1.0736.

KRUPP 9 SMNPB 28
Krupp Stahl AG
C 0-0.14, Si 0-0.05, Mn 0.9-1.3, P 0-0.1, S 0.27-0.33, Pb 0.15-0.35, bal Fe.
DIN 1651/88; W. Nr. 1.0718.

KRUPP 9 SMNPB 36
Krupp Stahl AG
C 0-0.15, Si 0-0.05, Mn 1.1-1.5, P 0-0.1, S 0.34-0.4, Pb 0.15-0.35, bal Fe.
DIN 1651/88; W. Nr. 1.0737.

KRUPP 90 MN 4
Krupp Stahl AG
C 0.85-0.95, Si 0.25-0.5, Mn 0.9-1.1, P 0-0.035, S 0-0.035, bal Fe.
SEL 90; W. Nr. 1.1273.

KRUPP A 163
Krupp Stahl AG
stainless steel. C 0.13, Cr 16, Ni 5, Mo 3, bal Fe.
Stainless steel. *Obsolete*

KRUPP A 18 AZ
Krupp Stahl AG
stainless steel. C 0.1, Cr 18, Ni 10, Ti, S, bal Fe.
Austenitic stainless steel. *Obsolete*

KRUPP A11K
Krupp Stahl AG
alloy steel. C 0.52-0.58, Mn 0.5, bal Fe.
For forging die blocks; water hardened. *Obsolete*

KRUPP A12-14P
Krupp Stahl AG
tool material.
See KRUPP CF70

KRUPP A12.0
Krupp Stahl AG
tool material. C 0.6, Si 0.4, Mn 0.6, bal Fe.
Heat treated: 160,000 TS; 113,000 YS; 12 El; 40 RA; 321 Brin.
For punches, rails, crimpers; water hardened. *Obsolete*

KRUPP A14K
Krupp Stahl AG
tool material. C 0.6, Si 0.4, Mn 0.6, bal Fe.
Heat treated: 160,000-115,000 TS; 113,000-77,000 YS; 12-13 El; 40-54 RA; 321-230 Brin. For punches, rails, crimpers; water hardened. *Obsolete*

KRUPP A15-17P
Krupp Stahl AG
tool material. C 0.85, Si 0.25, Mn 0.25, bal Fe.
Heat treated: 190,000 TS; 145,000 YS; 10 El; 30 RA; 400 Brin.
For lathe and planer tools, drills, taps, hobs; Type W1; water hardened. *Obsolete*

KRUPP A16K
Krupp Stahl AG
tool material. C 0.75-0.83, Mn 0.5, bal Fe.
For forging die blocks; water hardened. *Obsolete*

KRUPP A17 G
Krupp Stahl AG
tool material. C 0.8-0.9, Si 0.25-0.4, Mn 0.45-0.6, bal Fe.
For tools, drills, springs; water hardening. *Obsolete*

KRUPP A18-20P
Krupp Stahl AG
tool material. C 1, Mn 0.25, Si 0.25, bal Fe.
Annealed: 100,000 TS; 53,000 YS; 21 El; 42 RA; 200 Brin. For drills, taps, reamers, tools, dies; Type W1; water hardened. *Obsolete*

KRUPP A22P/A23P
Krupp Stahl AG
tool material. C 1.15, Si 0-0.25, Mn 0-0.25, bal Fe.
Annealed: 110,000 TS; 60,000 YS; 18 El; 38 RA; 210 Brin. For reamers, drills, taps, broaches, hobs, tools; Type W1; water hardened. *Obsolete*

KRUPP A24P
Krupp Stahl AG
tool material. C 1.3, Si 0-0.25, Mn 0-0.25, bal Fe.
For engravers' tools, form cutters, milling cutters; Type W1; water hardened. *Obsolete*

KRUPP A26P
Krupp Stahl AG
tool material. C 1.3, Si 0-0.25, Mn 0-0.25, bal Fe.
For engravers' tools, form and milling cutters; Type W1; water hardened. *Obsolete*

KRUPP A30
Krupp Stahl AG
alloy steel. C 0.15, Si 0.25, Mn 0.4, bal Fe.
Cold drawn: 72,000 TS; 60,000 YS; 22 El; 58 RA; 145 Brin.
For gears, cams, camshafts, fasteners; carburizing steel.
Obsolete

KRUPP A3P
Krupp Stahl AG
alloy steel. C 0.15, Mn 0.4, Si 0.25, bal Fe.
Annealed: 70,000 TS; 40,000 YS; 25 El; 60 RA; 145 Brin. For gears, cams, camshafts, bolts, fasteners; case hardening steel. *Obsolete*

KRUPP A3PO
Krupp Stahl AG
alloy steel. C 0.15, Si 0.25, Mn 0.35, bal Fe.
Annealed: 70,000 TS; 40,000 YS; 25 El; 60 RA; 145 Brin. For gears, cams, camshafts, machine tool parts; case hardening steel. *Obsolete*

KRUPP A50
Krupp Stahl AG
alloy steel. C 0.22, Si 0.25, Mn 0.45, bal Fe.
Annealed: 73,000 TS; 41,000 YS; 22 El; 58 RA; 150 Brin. For gears, cams, bolts, fasteners, camshafts; case hardening steel. *Obsolete*

KRUPP A70
Krupp Stahl AG
alloy steel. C 0.35, Si 0.25, Mn 0.45, bal Fe.
Hot rolled: 85,000 TS; 54,000 YS; 30 El; 53 RA; 185 Brin. For gears, pinions, shafts, bolts, fasteners; water hardened. *Obsolete*

KRUPP A90
Krupp Stahl AG
alloy steel. C 0.45, Si 0.25, Mn 0.45, bal Fe.
Hot rolled: 98,000 TS; 59,000 YS; 24 El; 45 RA; 212 Brin. For gears, pinions, shafts, bolts, fasteners; water hardened. *Obsolete*

KRUPP ALC 184 H
Krupp Stahl AG
stainless steel. C 0.02, Cr 19, Ni 17, Mo, N, bal Fe.
Austenitic stainless steel. *Obsolete*

KRUPP ALC 187 H
Krupp Stahl AG
stainless steel. C 0.02, Ni 20, Cr 18, Mo, N, bal Fe.
Austenitic stainless steel. *Obsolete*

KRUPP ALC 250
Krupp Stahl AG
stainless steel. C 0.01, Cr 25, Ni 21, bal Fe.
Stainless steel for high temperature operation. *Obsolete*

KRUPP AM 17
Krupp Stahl AG
stainless steel. C 0.1, Cr 17, Mn 7, Ni, bal Fe.
Cr-Mn stainless steel; austenitic. *Obsolete*

KRUPP AM 18
Krupp Stahl AG
stainless steel. C 0.08, Cr 18, Mn 9, Ni, bal Fe.
Cr-Mn stainless steel. *Obsolete*

KRUPP AM 18 H
Krupp Stahl AG
stainless steel. C 0.05, Cr 18, Mn 9, Ni, N, bal Fe.
Cr-Mn stainless steel. *Obsolete*

KRUPP AM 213
Krupp Stahl AG
stainless steel. C 0.03, Cr 21, Ni 15, Mn 7, Mo 3, bal Fe.
Stainless steel. *Obsolete*

KRUPP ANALYSIS
Krupp Huttenwerke
C 0.08-0.16, Cr 1.2-1.7, Ni 3.8-4.3, bal Fe.
For armor plate, gears; tough, case-hardening alloy steel.

KRUPP ANALYSIS
International Nickel Inc.
C 0.08-0.16, Cr 1.2-1.7, Ni 3.8-4.3, bal Fe.
For armor plate, gears; tough, case-hardening alloy steel.

KRUPP AP 17 7
Krupp Stahl AG
stainless steel.
See NIROSTA 4568

KRUPP AP 173
Krupp Stahl AG
stainless steel. C 0.09, Cr 17, Ni 5, Mo 3, N, bal Fe.
Stainless steel. *Obsolete*

KRUPP AS 183 A
Krupp Stahl AG
stainless steel. C 0.03, Cr 17, Ni 11, Mo, S bal Fe.
Stainless steel. *Obsolete*

KRUPP AST 35
Krupp Stahl AG
C 0-0.17, Si 0-0.35, P 0-0.045, S 0-0.045, Cr 0-0.3, 0.40 Mn min, 0.025 Al min, bal Fe.
DIN 17135/64; W. Nr. 1.0346.

KRUPP AST 41
Krupp Stahl AG
C 0-0.2, Si 0-0.35, P 0-0.045, S 0-0.045, Cr 0-0.3, 0.45 Mn min, 0.025 Al min, bal Fe.
DIN 17135/64; W. Nr. 1.0426.

KRUPP AST 45
Krupp Stahl AG
C 0-0.22, Si 0-0.35, P 0-0.045, S 0-0.045, Cr 0-0.3, 0.45 Mn min, 0.025 Al min, bal Fe.
DIN 17135/64; W. Nr. 1.0436.

KRUPP AST 52
Krupp Stahl AG
C 0-0.2, Si 0-0.55, Mn 0-1.5, P 0-0.045, S 0-0.045, Cr 0-0.3, 0.025 Al min, bal Fe.
DIN 17135/64; W. Nr. 1.0577.

KRUPP B126 P
Krupp Stahl AG
tool material. C 0.62-0.68, Si 1.7, Mn 0.6, bal Fe.
For tools, springs, punches; shock resistant. *Obsolete*

KRUPP B136M
Krupp Stahl AG
tool material. C 0.7, Si 1.7, Mn 0.7, bal Fe.
For punches, crimpers, upsetters, springs; oil hardened, shock resistant. *Obsolete*

KRUPP BC 1044
Krupp Stahl AG
tool material. C 0.53, Si 0.8, Mn 1.05, bal Fe.
For springs, punches, upsetters; oil hardened. *Obsolete*

KRUPP BC1244
Krupp Stahl AG
tool material. C 0.57-0.63, Si 0.9-1.1, Mn 0.9-1.1, bal Fe.
For tools, springs, punches; shock resistant. *Obsolete*

KRUPP BC40
Krupp Stahl AG
tool material. C 0.37, Si 1.25, Mn 1.25, bal Fe.
For punches, crimpers, upsetters; oil hardened. *Obsolete*

KRUPP BDF SPEZIAL
Krupp Stahl AG
tool material. C 0.54-0.6, Si 0.8-1.1, W 1.7-2, bal Fe.
For tools, hot work dies; hot work steel. *Obsolete*

KRUPP BDF30
Krupp Stahl AG
tool material. C 0.3-0.36, Si 0.8-1, Mn 0.3, Cr 0.9-1.1, V 0.2, W 1.7-2, bal Fe.
For tools, dies, hot work tools; hot work steel. *Obsolete*

KRUPP BDF50
Krupp Stahl AG
tool material. C 0.4-0.46, Si 0.8-1, Cr 0.9-1.1, W 1.7-2, bal Fe.
For tools, hot work dies; hot work steel. *Obsolete*

KRUPP BDFD
Krupp Stahl AG
tool material. C 0.5, Cr 1.5, W 2.25, V 0.2, bal Fe.
For tools, dies, punches; oil hardened. *Obsolete*

KRUPP BEARING
Krupp Stahl AG
aluminum. Al 87, Cu 8, Sn 5.
For bearings. *Obsolete*

KRUPP BF1552
Krupp Stahl AG
tool material. C 0.67, Si 1.3, Cr 0.5, Mn 0.5, bal Fe.
For springs; oil hardened, tough. *Obsolete*

KRUPP BF2555
Krupp Stahl AG
tool material. C 1.2-1.3, Si 1.05-1.2, Mn 0.7, Cr 1.1-1.3, bal Fe.
For tools, dies, bearings; oil hardening. *Obsolete*

KRUPP BFD-50G
Krupp Stahl AG
tool material. C, alloy, bal Fe.
For tools. *Obsolete*

KRUPP BFG
Krupp Stahl AG
tool material. C 0.95, Mn 1.2, Cr 0.5, W 0.5, bal Fe.
For tools, dies; non-deforming. *Obsolete*

KRUPP BFM SPEZIAL
Krupp Stahl AG
tool material.
See KRUPP 2249

KRUPP BFM755
Krupp Stahl AG
tool material. C 0.3-0.38, Si 1.4-1.6, Cr 1.2-1.5, V 0.1-0.2, bal Fe.
For tools, punches, dies; shock resistant. *Obsolete*

KRUPP BFV
Krupp Stahl AG
tool material. C 0.65-0.7, Si 0.15-0.25, Mn 0.5-0.6, bal Fe.
For tools, springs, punches; water hardening. *Obsolete*

KRUPP BMH
Krupp Stahl AG
tool material. C 0.42-0.48, Si 1.5-1.7, Mn 0.6, bal Fe.
For tools, springs, punches; shock resistant. *Obsolete*

KRUPP BS 45 W
Krupp Stahl AG
tool material. C 0.34, Ni 4, Cr, W, bal Fe.
Hot work tool or mold steel. *Obsolete*

KRUPP BS1
Krupp Stahl AG
tool material. C 0.95, bal Fe.
For tools, drills, taps; water hardened. *Obsolete*

KRUPP BS2
Krupp Stahl AG
tool material. C 0.95, bal Fe.
For tools, drills, taps; water hardened. *Obsolete*

KRUPP BS3
Krupp Stahl AG
tool material. C 1.1, Si 0.25, Mn 0.25, bal Fe.
For drills, taps, reamers, cutters; Type W1; water hardened. *Obsolete*

KRUPP BS4
Krupp Stahl AG
tool material. C 1, Si 0.25, Mn 0.25, bal Fe.
For springs, drills, taps, punches, reamers; Type W1; water hardened. *Obsolete*

KRUPP BS5
Krupp Stahl AG
tool material. C 0.85, Si 0-0.25, Mn 0-0.25, bal Fe.
Heat treated: 190,000 TS; 145,000 YS; 10 El; 30 RA; 400 Brin.
For drills, taps, reamers, punches, cutters; Type W1; water hardened. *Obsolete*

KRUPP BS6
Krupp Stahl AG
tool material. C 0.7, Si 0-0.25, Mn 0-0.25, bal Fe.
Heat treated: 175,000 TS; 128,000 YS; 12 El; 37 RA; 355 Brin.
For rails, punches, hammers, axes; Type W1; water hardened. *Obsolete*

KRUPP BVR 1
Krupp Stahl AG
alloy steel. C 0.28, Ni 2, Cr 1.3, Mo 0.45, bal Fe.
For deep hardening shafts and gears. W. Nr. 1.6932.
Obsolete

KRUPP BVR 1 SS
Krupp Stahl AG
alloy steel. C 0.26, Ni 3.5, Cr 1.5, Mo 0.45, bal Fe.
For very deep hardening structural parts. *Obsolete*

KRUPP BVR 1S
Krupp Stahl AG
alloy steel. C 0.26, Ni 2.75, Cr 1.5, Mo 0.45, bal Fe.
For deep hardening structural parts. W.Nr. 1.6948. *Obsolete*

KRUPP C 10
Krupp Stahl AG
C 0.07-0.13, Si 0-0.4, Mn 0.3-0.6, P 0-0.045, S 0-0.045, bal Fe.
DIN 17210/09.86; W. Nr. 1.0301.

KRUPP C 10 PB
Krupp Stahl AG
C 0.07-0.13, Si 0-0.4, Mn 0.3-0.6, P 0-0.045, S 0-0.045, Pb 0.15-0.3, bal Fe.
DIN 17210/09.86; W. Nr. 1.0302.

KRUPP C 15
Krupp Stahl AG
C 0.12-0.18, Si 0-0.4, Mn 0.3-0.6, P 0-0.045, S 0-0.045, bal Fe.
DIN 17210/09.86; W. Nr. 1.0401.

KRUPP C 15 PB
Krupp Stahl AG
C 0.12-0.18, Si 0-0.4, Mn 0.3-0.6, P 0-0.045, S 0-0.045, Pb 0.15-0.3, bal Fe.
DIN 17210/09.86; W. Nr. 1.0403.

KRUPP C 21
Krupp Stahl AG
C 0.18-0.23, Si 0.15-0.35, Mn 0.6-1.05, P 0-0.04, S 0-0.05, bal Fe.
SEL 90; W. Nr. 1.0432.

KRUPP C 22
Krupp Stahl AG
C 0.17-0.24, Si 0-0.4, Mn 0.3-0.6, P 0-0.045, S 0-0.045, bal Fe.
DIN 17200/87; W. Nr. 1.0402.

KRUPP C 22 PB
Krupp Stahl AG
C 0.17-0.24, Si 0-0.4, Mn 0.3-0.6, P 0-0.045, S 0-0.045, Pb 0.15-0.3, bal Fe.
SEL 90; W. Nr. 1.0404.

KRUPP C 22.3
Krupp Stahl AG
C 0.18-0.23, Si 0.15-0.35, Mn 0.4-0.9, P 0-0.035, S 0-0.03, Cr 0-0.3, 0.015 Al min, bal Fe.
VDTUEV-BLATT 304/83, W. Nr. 1.0427.

KRUPP C 22.8
Krupp Stahl AG
C 0.18-0.23, Si 0.15-0.35, Mn 0.4-0.9, P 0-0.035, S 0-0.03, Cr 0-0.3, 0.015 Al min, bal Fe.
VDTUEV-BLATT 350-1/82; W. Nr. 1.0460.

KRUPP C 22.8 S 1
Krupp Stahl AG
C 0.18-0.23, Si 0.15-0.35, Mn 0.4-0.9, P 0-0.016, S 0-0.01, Cr 0-0.3, Al 0.015-0.05, Cu 0-0.15, V 0-0.02, bal Fe.
VDTUEV-BLATT 453/84. W. Nr. 1.1338.

KRUPP C 25
Krupp Stahl AG
C 0.22-0.29, Si 0-0.4, Mn 0.4-0.7, P 0-0.045, S 0-0.045, bal Fe.
DIN 17200/87; W. Nr. 1.0406.

KRUPP C 25 PB
Krupp Stahl AG
C 0.22-0.29, Si 0-0.4, Mn 0.4-0.7, P 0-0.045, S 0-0.045, Pb 0.15-0.3, bal Fe.
SEL 90; W. Nr. 1.0411.

KRUPP C 35
Krupp Stahl AG
C 0.32-0.39, Si 0-0.4, Mn 0.5-0.8, P 0-0.045, S 0-0.045, bal Fe.
DIN 17200/87; W. Nr. 1.0501.

KRUPP C 35 PB
Krupp Stahl AG
C 0.32-0.39, Si 0-0.4, Mn 0.5-0.8, P 0-0.045, S 0-0.045, Pb 0.15-0.3, bal Fe.
SEL 90; W. Nr. 1.0502.

KRUPP C 40
Krupp Stahl AG
C 0.37-0.44, Si 0-0.4, Mn 0.5-0.8, P 0-0.045, S 0-0.045, bal Fe.
DIN 17200/87; W. Nr. 1.0511.

KRUPP C 40 PB
Krupp Stahl AG
C 0.37-0.45, Si 0.15-0.35, Mn 0.5-0.8, P 0-0.045, S 0-0.045, Pb 0.15-0.3, bal Fe.
SEL 90; W. Nr. 1.0512.

KRUPP C 45
Krupp Stahl AG
C 0.42-0.5, Si 0-0.4, Mn 0.5-0.8, P 0-0.045, S 0-0.045, bal Fe.
DIN 17200/87; W. Nr. 1.0503.

KRUPP C 45 PB
Krupp Stahl AG
C 0.42-0.5, Si 0-0.4, Mn 0.5-0.8, P 0-0.045, S 0-0.045, Pb 0.15-0.3, bal Fe.
SEL 90; W. Nr. 1.0504.

KRUPP C 55
Krupp Stahl AG
C 0.52-0.6, Si 0-0.4, Mn 0.6-0.9, P 0-0.045, S 0-0.045, bal Fe.
DIN 17200/87; W. Nr. 1.0535.

KRUPP C 60
Krupp Stahl AG
C 0.57-0.65, Si 0-0.4, Mn 0.6-0.9, P 0-0.045, S 0-0.045, bal Fe.
DIN 17200/87; W. Nr. 1.0601.

KRUPP C 60 PB
Krupp Stahl AG
C 0.57-0.65, Si 0-0.4, Mn 0.6-0.9, P 0-0.045, S 0-0.045, Pb 0.15-0.3, bal Fe.
SEL 90; W. Nr. 1.0602.

KRUPP C 67
Krupp Stahl AG
C 0.65-0.72, Si 0.15-0.35, Mn 0.6-0.9, P 0-0.045, S 0-0.045, bal Fe.
DIN 17222/79; W. Nr. 1.0603.

KRUPP C 68
Krupp Stahl AG
C 0.65-0.72, Si 0.25-0.5, Mn 0.6-0.8, P 0-0.045, S 0-0.045, bal Fe.
SEL 90; W. Nr. 1.0627.

KRUPP C 75
Krupp Stahl AG
C 0.7-0.8, Si 0.15-0.35, Mn 0.6-0.8, P 0-0.045, S 0-0.045, bal Fe.
DIN 17222/79; W. Nr. 1.0605.

KRUPP C66
Krupp Stahl AG
alloy steel. C 0.3, Mn 1.35, Si 0.25, bal Fe.
For gears, shafts, machine tool parts; water hardened, tough. *Obsolete*

KRUPP C7E
Koch Light Alloys Ltd.
Aluminum. C 0.15, Cr 0.65, Mn 0.5, Si 0.25, bal Fe.
For gears, pinions, machine tool parts; cast hardening steel.

KRUPP CA 1220 SPEZIAL
Krupp Stahl AG
tool material. C 2, Cr 12, V 1, Mo, bal Fe.
High chromium cold work tool steel. *Obsolete*

KRUPP CARBON
Thomas Prosser & Sons
C 0.7-1, bal Fe.
For tools, drills, taps; water hardened.

KRUPP CF 35
Krupp Stahl AG
C 0.33-0.39, Si 0.15-0.35, Mn 0.5-0.8, P 0-0.025, S 0-0.035, bal Fe.
DIN 17212/72; W. Nr. 1.1183.

KRUPP CF 45
Krupp Stahl AG
C 0.43-0.49, Si 0.15-0.35, Mn 0.5-0.8, P 0-0.025, S 0-0.035, bal Fe.
DIN 17212/72; W. Nr. 1.1193.

KRUPP CF 53
Krupp Stahl AG
C 0.5-0.57, Si 0.15-0.35, Mn 0.4-0.7, P 0-0.025, S 0-0.035, bal Fe.
DIN 17212/72; W. Nr. 1.1213.

KRUPP CF 54
Krupp Stahl AG
C 0.5-0.57, Si 0-0.4, Mn 0.4-0.7, P 0-0.025, S 0-0.035, Cu 0-0.3, bal Fe.
DIN 17230/80; W. Nr. 1.1219.

KRUPP CF 70
Krupp Stahl AG
C 0.68-0.75, Si 0.15-0.35, Mn 0.2-0.35, P 0-0.025, S 0-0.035, bal Fe.
DIN 17212/72; W. Nr. 1.1249.

KRUPP CK 10
Krupp Stahl AG
C 0.07-0.13, Si 0-0.4, Mn 0.3-0.6, P 0-0.035, S 0-0.035, bal Fe.
DIN 17210/09.86; W. Nr. 1.1121.

KRUPP CK 101
Krupp Stahl AG
C 0.95-1.05, Si 0.15-0.35, Mn 0.4-0.6, P 0-0.035, S 0-0.035, bal Fe.
DIN 17222/79; W. Nr. 1.1274.

KRUPP CK 15
Krupp Stahl AG
C 0.12-0.18, Si 0-0.4, Mn 0.3-0.6, P 0-0.035, S 0-0.035, bal Fe.
DIN 17210/09.86; W. Nr. 1.1141.

KRUPP CK 15 AL
Krupp Stahl AG
C 0.12-0.18, Si 0.15-0.35, Mn 0.25-0.5, P 0-0.035, S 0-0.035,
N 0-0.007, 0.02 Al min, bal Fe.
SEL 90; W. Nr. 1.1148.

KRUPP CK 16 AL
Krupp Stahl AG
C 0.13-0.18, Si 0-0.15, Mn 0.4-0.6, P 0-0.025, S 0-0.025, Ni
0-0.1, Al 0.03-0.08, Cu 0-0.15, N 0-0.007, bal Fe.
SEL 90; W. Nr. 1.1135.

KRUPP CK 19
Krupp Stahl AG
C 0.15-0.23, Si 0-0.15, Mn 0.4-0.6, P 0-0.03, S 0-0.025, Cr
0-0.05, Ni 0-0.1, Al 0.03-0.08, Cu 0-0.15, bal Fe.
WW 1.1134/79; W. Nr. 1.1134.

KRUPP CK 22
Krupp Stahl AG
C 0.17-0.24, Si 0-0.4, Mn 0.3-0.6, P 0-0.035, S 0-0.03, bal Fe.
DIN 17200/87; W. Nr. 1.1151.

KRUPP CK 25
Krupp Stahl AG
C 0.22-0.29, Si 0-0.4, Mn 0.4-0.7, P 0-0.035, S 0-0.03, bal Fe.
DIN 17200/87; W. Nr. 1.1158.

KRUPP CK 30
Krupp Stahl AG
C 0.27-0.34, Si 0-0.4, Mn 0.5-0.8, P 0-0.035, S 0-0.03, bal Fe.
DIN 17200/87; W. Nr. 1.1178.

KRUPP CK 35
Krupp Stahl AG
C 0.32-0.39, Si 0-0.4, Mn 0.5-0.8, P 0-0.035, S 0-0.03, bal Fe.
DIN 17200/87; W. Nr. 1.1181.

KRUPP CK 4
Krupp Stahl AG
C 0-0.04, Si 0-0.01, Mn 0.12-0.2, P 0-0.02, S 0-0.02, Cr 0-0.03,
Cu 0-0.15, 0.02 Al min, bal Fe.
SEL 90; W. Nr. 1.1005.

KRUPP CK 40
Krupp Stahl AG
C 0.37-0.44, Si 0-0.4, Mn 0.5-0.8, P 0-0.035, S 0-0.03, bal Fe.
DIN 17200/87; W. Nr. 1.1186.

KRUPP CK 42 AL
Krupp Stahl AG
C 0.39-0.44, Si 0.25-0.4, Mn 0.75-0.9, P 0-0.035, S 0-0.035, N
0-0.007, 0.02 Al min, bal Fe.
SEL 90; W. Nr. 1.1190.

KRUPP CK 45
Krupp Stahl AG
C 0.42-0.5, Si 0-0.4, Mn 0.5-0.8, P 0-0.035, S 0-0.03, bal Fe.
DIN 17200/87; W. Nr. 1.1191.

KRUPP CK 45 PB
Krupp Stahl AG
C 0.42-0.5, Si 0-0.4, Mn 0.5-0.8, P 0-0.035, S 0-0.035, Pb
0.15-0.3, bal Fe.
SEL 90; W. Nr. 1.1195.

KRUPP CK 48 MN
Krupp Stahl AG
C 0.45-0.51, Si 0.1-0.4, Mn 0.5-0.8, P 0-0.03, S 0-0.035, Cr
0-0.4, Mo 0-0.1, Ni 0-0.4, bal Fe.
SEL 90; W. Nr. 1.198.

KRUPP CK 50
Krupp Stahl AG
C 0.47-0.55, Si 0-0.4, Mn 0.6-0.9, P 0-0.035, S 0-0.03, bal Fe.
DIN 17200/87; W. Nr. 1.1206.

KRUPP CK 53
Krupp Stahl AG
C 0.5-0.57, Si 0-0.4, Mn 0.6-0.9, P 0-0.035, S 0-0.035, bal Fe.
SEL 90; W. Nr. 1.1210.

KRUPP CK 55
Krupp Stahl AG
C 0.52-0.6, Si 0-0.4, Mn 0.6-0.9, P 0-0.035, S 0-0.03, bal Fe.
DIN 17200/87; W. Nr. 1.1203.

KRUPP CK 60
Krupp Stahl AG
C 0.57-0.65, Si 0-0.4, Mn 0.6-0.9, P 0-0.035, S 0-0.03, bal Fe.
DIN 17200/87; W. Nr. 1.1221.

KRUPP CK 67
Krupp Stahl AG
C 0.65-0.72, Si 0.15-0.35, Mn 0.6-0.9, P 0-0.035, S 0-0.035,
bal Fe.
DIN 17222/79; W. Nr. 1.1231.

KRUPP CK 68
Krupp Stahl AG
C 0.6-0.7, Si 0.2-0.65, Mn 0.7-1.2, P 0-0.035, S 0-0.035, bal
Fe.
SEL 90; W. Nr. 1.1233.

KRUPP CK 7
Krupp Stahl AG
C 0-0.08, Si 0-0.15, Mn 0-0.5, P 0-0.035, S 0-0.035, bal Fe.
SEL 90; W. Nr. 1.1009.

KRUPP CK 75
Krupp Stahl AG
C 0.7-0.8, Si 0.15-0.3, Mn 0.6-0.8, P 0-0.035, S 0-0.035, bal
Fe.
DIN 17222/79; W. Nr. 1.1248.

KRUPP CK 85
Krupp Stahl AG
C 0.8-0.9, Si 0.15-0.35, Mn 0.45-0.65, P 0-0.035, S 0-0.035,
bal Fe.
DIN 17222/79; W. Nr. 1.1269.

KRUPP CM 15
Krupp Stahl AG
C 0.12-0.18, Si 0-0.4, Mn 0.3-0.6, P 0-0.035, S 0.02-0.035, bal
Fe.
DIN 17210/09.86; W. Nr. 1.1140.

KRUPP CM 22
Krupp Stahl AG
C 0.17-0.24, Si 0-0.4, Mn 0.3-0.6, P 0-0.035, S 0.02-0.035, bal
Fe.
DIN 17200/87; W. Nr. 1.1149.

KRUPP CM 35
Krupp Stahl AG
C 0.32-0.39, Si 0-0.4, Mn 0.5-0.8, P 0-0.035, S 0.02-0.035, bal
Fe.
DIN 17200/87; W. Nr. 1.1180.

KRUPP CM 45
Krupp Stahl AG
C 0.42-0.5, Si 0-0.4, Mn 0.5-0.8, P 0-0.035, S 0.02-0.035, bal
Fe.
DIN 17200/87; W. Nr. 1.1201.

KRUPP CM 53
Krupp Stahl AG
C 0.5-0.55, Si 0-0.4, Mn 0.6-0.9, P 0-0.035, S 0.035, Cr 0-0.2,
bal Fe.
SEL 90; W. Nr. 1.1205.

KRUPP CM 55
Krupp Stahl AG
C 0.52-0.6, Si 0-0.4, Mn 0.6-0.9, P 0-0.035, S 0.02-0.035, bal
Fe.
DIN 17200/87; W. Nr. 1.1209.

KRUPP CM 60
Krupp Stahl AG
C 0.57-0.65, Si 0-0.4, Mn 0.6-0.9, P 0-0.035, S 0.02-0.035, bal
Fe.
DIN 17200/87; W. Nr. 1.1223.

KRUPP CMV SPECIAL
Krupp Stahl AG
stainless steel. Cr 12.5, V 1, Ni 0.2, C 18 Mn, bal Fe.
For blading for gas turbine on jet engine; corrosion and heat
resistant. *Obsolete*

KRUPP CQ 10
Krupp Stahl AG
C 0.07-0.13, Si 0.15-0.35, Mn 0.25-0.3, P 0-0.035, S 0-0.035,
bal Fe.
SEL 90; W. Nr. 1.1122.

KRUPP CQ 15
Krupp Stahl AG
C 0.12-0.18, Si 0-0.4, Mn 0.3-0.6, P 0-0.035, S 0-0.035, bal
Fe.
DIN 1654 T.3/89; W. Nr. 1.1132.

KRUPP CQ 22
Krupp Stahl AG
C 0.17-0.24, Si 0-0.4, Mn 0.3-0.6, P 0-0.035, S 0-0.035, bal
Fe.
DIN 1654 T.4/89; W. Nr. 1.1152.

KRUPP CQ 35
Krupp Stahl AG
C 0.32-0.39, Si 0-0.4, Mn 0.5-0.8, P 0-0.035, S 0-0.035, bal
Fe.
DIN 1654 T.4/89; W. Nr. 1.1172.

KRUPP CQ 45
Krupp Stahl AG
C 0.42-0.5, Si 0-0.4, Mn 0.5-0.8, P 0-0.035, S 0-0.035, bal Fe.
DIN 1654 T.4/89; W. Nr. 1.1192.

KRUPP D 12-2
Krupp Stahl AG
C 0-0.12, Si 0-0.01, Mn 0-0.5, P 0-0.05, S 0-0.05, N 0-0.007,
bal Fe.
SEL 75; W. Nr. 1.0012.

KRUPP D 15-2
Krupp Stahl AG
C 0.13-0.18, Si 0.1-0.3, Mn 0.3-0.6, P 0-0.04, S 0-0.04, bal Fe.
DIN 17140 T.1/83; W. Nr. 1.0413.

KRUPP D 20-2
Krupp Stahl AG
C 0.18-0.23, Si 0.1-0.3, Mn 0.3-0.6, P 0-0.04, S 0-0.04, bal Fe.
DIN 17140 T.1/83; W. Nr. 1.0414.

KRUPP D 25-2
Krupp Stahl AG
C 0.23-0.28, Si 0.1-0.3, Mn 0.3-0.6, P 0-0.04, S 0-0.04, bal Fe.
DIN 17140 T.1/83; W. Nr. 1.0415.

KRUPP D 30-2
Krupp Stahl AG
C 0.28-0.33, Si 0.1-0.3, Mn 0.3-0.6, P 0-0.04, S 0-0.04, bal Fe.
DIN 17140 T.1/83; W. Nr. 1.0530.

KRUPP D 35-2
Krupp Stahl AG
C 0.33-0.38, Si 0.1-0.3, Mn 0.3-0.6, P 0-0.04, S 0-0.04, bal Fe.
DIN 17140 T.1/83; W. Nr. 1.0516.

KRUPP D 45-2
Krupp Stahl AG
C 0.43-0.48, Si 0.1-0.3, Mn 0.3-0.7, P 0-0.04, S 0-0.04, bal Fe.
DIN 17140 T.1/83; W. Nr. 1.0517.

KRUPP D 55-2
Krupp Stahl AG
C 0.53-0.58, Si 0.1-0.3, Mn 0.3-0.7, P 0-0.04, S 0-0.04, bal Fe.
DIN 17140 T.1/83; W. Nr. 1.0518.

KRUPP D 6-2
Krupp Stahl AG
C 0-0.06, Mn 0-0.4, P 0-0.04, S 0-0.04, bal Fe.
DIN 17140 T.1/83; W. Nr. 1.0314.

KRUPP D 65-2
Krupp Stahl AG
C 0.63-0.68, Si 0.1-0.3, Mn 0.3-0.7, P 0-0.04, S 0-0.04, bal Fe.
DIN 17140 T.1/83; W. Nr. 1.0612.

KRUPP D 8-2
Krupp Stahl AG
C 0-0.08, Mn 0-0.45, P 0-0.04, S 0-0.04, bal Fe.
DIN 17140 T.1/83; W. Nr. 1.0313.

KRUPP D 85-2
Krupp Stahl AG
C 0.83-0.88, Si 0.1-0.3, Mn 0.3-0.7, P 0-0.04, S 0-0.04, bal Fe.
DIN 17140 T.1/83; W. Nr. 1.0616.

KRUPP D 9
Krupp Stahl AG
C 0-0.1, Si 0-0.3, Mn 0-0.5, P 0-0.07, S 0-0.06, bal Fe.
DIN 17140 T.1/83; W. Nr. 1.0010.

KRUPP D 95-2
Krupp Stahl AG
C 0.9-0.99, Si 0.1-0.3, Mn 0.3-0.7, P 0-0.04, S 0-0.04, bal Fe.
DIN 17140 T.1/83; W. Nr. 1.0618.

KRUPP DF108C
Krupp Stahl AG
tool material. C 1.5, Cr 11.5, Mo 0.75, bal Fe.
For tools, dies; non-deforming. *Obsolete*

KRUPP DF109CN
Krupp Stahl AG
tool material. C 0.8, W 9.5, Cr 2.5, V 0.1, bal Fe.
For hot work tools and dies; hot work steel. *Obsolete*

KRUPP DF109CW
Krupp Stahl AG
alloy steel.
See KRUPP 2581

KRUPP DF168
Krupp Stahl AG
tool material.
See KRUPP 2436

KRUPP DF35
Krupp Stahl AG
tool material. 0.9 C 0.45 Cr, bal Fe.
For tools, drills, taps; water hardening. *Obsolete*

KRUPP DF51
Krupp Stahl AG
tool material. C 1.05, Mn 0.9, Cr 1, W 1.15, bal Fe.
For cutters, bearings, dies, tools; water hardened. *Obsolete*

KRUPP DFM
Krupp Stahl AG
tool material. C 0.7, W 18, Co 5, Cr 4, V 1.3, bal Fe.
For tools, dies, cutters; high speed steel. *Obsolete*

KRUPP DFM EXTRA SPECIAL
Krupp Stahl AG
tool material. C 0.7, W 18, Co 9, Mo 1, Cr 4, V 2, bal Fe.
For tools, dies, cutters; high speed steel. *Obsolete*

KRUPP DFM SUPRA
Krupp Stahl AG
tool material. C 0.79, Co 4.75, Cr 4.3, Mo 0.75, V 1.5, W 18, bal Fe.
For lathe and planer tools, reamers, hobs, taps, drills; high speed steel, heavy duty. *Obsolete*

KRUPP DFM SUPRA SPEZIAL
Krupp Stahl AG
tool material. C 0.76, Co 10, Cr 4.2, Mo 0.8, V 1.8, W 18, bal Fe.
For lathe and planer tools, reamers, broaches; high speed steel. *Obsolete*

KRUPP DFM ZWEIKARRO
Krupp Stahl AG
tool material.
See KRUPP 3558

KRUPP DFMC
Krupp Stahl AG
tool material. C 0.7, W 18, Cr 4, V 1, bal Fe.
For tools, dies, cutters; high speed steel. *Obsolete*

KRUPP DFMN E-SPEZIAL
Krupp Stahl AG
tool material. C 0.86, Cr 4.1, Mo 0.85, V 2.5, W 12, bal Fe.
For lathe and planer tools, reamers, broaches, taps; high speed steel. *Obsolete*

KRUPP DFMN SUPRA
Krupp Stahl AG
tool material. C 0.86, Mo 0.85, V 2.1, W 12, Co 2.8, Cr 4.3, bal Fe.
For lathe and planer tools, reamers, broaches; high speed steel. *Obsolete*

KRUPP DFMN ZWEIKARRO
Krupp Stahl AG
tool material. C 0.82, Cr 4.1, Mo 0.85, V 1.6, W 8.7, bal Fe.
For lathe and planer tools, broaches; high speed steel. *Obsolete*

KRUPP DFMV 5
Krupp Stahl AG
tool material. Co 1.35, Si 0.3, Mn 0.3, Cr 4.25, W 10.5, V 4.35, bal Fe.
For hard facing electrodes; wear resistant. *Obsolete*

KRUPP DFMV3
Krupp Stahl AG
tool material. C 0.86, Cr 4.1, Mo, W, V, bal Fe.
For lathe and planer tools, reamers, drills, hobs, high speed steel. *Obsolete*

KRUPP E33E
Krupp Stahl AG
alloy steel. C 0.2, Ni 1.5, bal Fe.
For gears, pinions, machine tool parts; case hardening steel. *Obsolete*

KRUPP E67
Krupp Stahl AG
alloy steel.
See KRUPP 18CRNIP

KRUPP EB7
Krupp Stahl AG
alloy steel. C 0.55, Mn 0.5-0.7, Si 0.1-0.4, bal Fe.
Heat treated: 160,000 TS; 113,000 YS; 12 El; 40 RA; 320 Brin.
For gears, bolts, machine tool parts; water hardened. *Obsolete*

KRUPP ED 3 R
Krupp Stahl AG
C 0-0.004, N 0-0.007, 0.02 Al min, bal Fe.
DIN 1623 T.3/83; W. Nr. 1.0393.

KRUPP ED 3 U
Krupp Stahl AG
C 0-0.004, N 0-0.007, bal Fe.
DIN 1623 T.3/83; W. Nr. 1.0393.

KRUPP ED 4
Krupp Stahl AG
C 0-0.004, 0.02 Al min, bal Fe.
DIN 1623 T.3/83; W. Nr. 1.0394.

KRUPP EF1514
Krupp Stahl AG
tool material. C 0.8, Mn 0.7, Cr 1-1.2, Ni 0.5, bal Fe.
For tools, dies; tough. *Obsolete*

KRUPP EF23V
Krupp Stahl AG
alloy steel. C 0.28, Ni 1.5, Cr 0.5, bal Fe.
For bolts, gears, machine tool parts; oil hardened, shock resistant. *Obsolete*

KRUPP EF24V
Krupp Stahl AG
alloy steel. C 0.35, Cr 0.5, Ni 1.5, bal Fe.
For bolts, gears, machine tool parts; oil hardened, shock resistant. *Obsolete*

KRUPP EF29V
Krupp Stahl AG
alloy steel. C 0.28, Cr 0.7, Ni 2.5, bal Fe.
For bolts, gears, machine tool parts; oil hardened, shock resistant. *Obsolete*

KRUPP EF30V
Krupp Stahl AG
alloy steel. C 0.35, Cr 0.7, Ni 2.5, bal Fe.
For gears, bolts, crankshafts; oil hardened, shock resistant. *Obsolete*

KRUPP EF35E
Krupp Stahl AG
alloy steel. C 0.13, Cr 0.7, Ni 2.5, bal Fe.
For gears, cams, camshafts, fasteners; case hardening steel, shock resistant. *Obsolete*

KRUPP EF48V
Krupp Stahl AG
alloy steel. C 0.22, Cr 0.7, Ni 3.5, Mn 0.6, bal Fe.
For gears, cams, camshafts, fasteners; case hardening steel, shock resistant. *Obsolete*

KRUPP EF49V
Krupp Stahl AG
alloy steel. C 0.3, Cr 0.7, Ni 3.5, bal Fe.
For gears, bolts, crankshafts, fasteners; oil hardened, shock resistant. *Obsolete*

KRUPP EF58E
Krupp Stahl AG
alloy steel.
See KRUPP 14NICR14

KRUPP EF59E
Krupp Stahl AG
alloy steel.
See KRUPP 14NICR18

KRUPP EF62V
Krupp Stahl AG
alloy steel. C 0.35, Cr 1.3, Ni 4.5, bal Fe.
For gears, bolts, crankshafts, machine tool parts; oil hardened, shock resistant. *Obsolete*

KRUPP EFS
Krupp Stahl AG
alloy steel. C 0.45, Cr 0.7, Ni 2.5, bal Fe.
For gears, bolts, crankshafts, fasteners; oil hardened, shock resistant. *Obsolete*

KRUPP EFS SPECIAL S
Krupp Stahl AG
tool material. C 0.45-0.5, Mn 0.5-0.8, Cr 1.2-1.5, Ni 1.1-1.4, bal Fe.
For tools, gears, shafts; oil hardening. *Obsolete*

KRUPP EFWP
Krupp Stahl AG
tool material. C 0.5-0.55, Cr 0.9-1.2, Ni 3-3.5, bal Fe.
For tools, gears, shafts; oil hardening. *Obsolete*

KRUPP EK 2 R
Krupp Stahl AG
C 0-0.08, N 0-0.007, 0.02 Al min, bal Fe.
DIN 1623 T.3/83; W. Nr. 1.0391.

KRUPP EK 2 U
Krupp Stahl AG
C 0-0.08, N 0-0.007, bal Fe.
DIN 1623 T.3/83; W. Nr. 1.0391.

KRUPP EK 4
Krupp Stahl AG
C 0-0.08, 0.02 Al min, bal Fe.
DIN 1623 T.3/83; W. Nr. 1.0392.

KRUPP ESTE 255
Krupp Stahl AG
C 0-0.16, Si 0-0.4, Mn 0.5-1.3, P 0-0.025, S 0-0.015, Cr 0-0.3, Mo 0-0.08, Ni 0-0.3, Cu 0-0.2, N 0-0.02, Nb 0-0.03, 0.02 Al min, bal Fe.
DIN 17102/83; W. Nr. 1.1103.

KRUPP ESTE 285
Krupp Stahl AG
C 0-0.16, Si 0-0.4, Mn 0.6-1.4, P 0-0.025, S 0-0.015, Cr 0-0.3, Mo 0-0.08, Ni 0-0.3, Cu 0-0.2, N 0-0.02, Nb 0-0.03, 0.02 Al min, bal Fe.
DIN 17102/83; W. Nr. 1.1104.

KRUPP ESTE 315
Krupp Stahl AG
C 0-0.16, Si 0-0.45, Mn 0.7-1.5, P 0-0.025, S 0-0.015, Cr 0-0.3, Mo 0-0.08, Ni 0-0.3, Cu 0-0.2, N 0-0.02, Nb 0-0.03, 0.02 Al min, bal Fe.
DIN 17102/83; W. Nr. 1.1105.

KRUPP ESTE 355
Krupp Stahl AG
C 0-0.18, Si 0.1-0.5, Mn 0.9-1.65, P 0-0.025, S 0-0.015, Cr 0-0.3, Mo 0-0.08, Ni 0-0.3, Cu 0-0.2, N 0-0.02, Nb 0-0.05, V 0-0.1, 0.02 Al min, bal Fe.
DIN 17102/83; W. Nr. 1.1106.

KRUPP ESTE 380
Krupp Stahl AG
C 0-0.2, Si 0.1-0.6, Mn 1-1.7, P 0-0.025, S 0-0.015, Cr 0-0.3, Mo 0-0.08, Ni 0-1, Cu 0-0.2, N 0-0.02, Nb 0-0.05, V 0-0.2, 0.02 Al min, bal Fe.
DIN 17102/83; W. Nr. 1.8911.

KRUPP ESTE 420
Krupp Stahl AG
C 0-0.2, Si 0.1-0.6, Mn 1-1.7, P 0-0.025, S 0-0.015, Cr 0-0.3, Mo 0-0.1, Ni 0-1, Cu 0-0.2, N 0-0.02, Nb 0-0.05, V 0-0.2, 0.02 Al min, bal Fe.
DIN 17102/83; W. Nr. 1.8913.

KRUPP ESTE 460
Krupp Stahl AG
C 0-0.2, Si 0.1-0.6, Mn 1-1.7, P 0-0.025, S 0-0.015, Cr 0-0.3, Mo 0-0.1, Ni 0-1, Cu 0-0.2, N 0-0.02, Nb 0-0.05, V 0-0.2, 0.02 Al min, bal Fe.
DIN 17102/83; W. Nr. 1.8918.

KRUPP ESTE 500
Krupp Stahl AG
C 0-0.21, Si 0.1-0.6, Mn 1-1.7, P 0-0.025, S 0-0.015, Cr 0-0.3, Mo 0-0.1, Ni 0-1, Cu 0-0.2, N 0-0.02, Nb 0-0.05, V 0-0.22, 0.02 Al min, bal Fe.
DIN 17102/83; W. Nr. 1.8919.

KRUPP EV420
Krupp Stahl AG
alloy steel. C 0.08, Mn 1-1.25, bal Fe.
Welded: 55,000 TS; 45,000 YS; 33-40 El; 60-80 RA. For welding rod. *Obsolete*

KRUPP EXTRA
Krupp Stahl AG
tool material. C 0.35, Cr 2.5, V 0.65, W 4.5, bal Fe.
For tools, hot work dies; hot work steel. *Obsolete*

KRUPP EXTRA SPEZIAL P
Krupp Stahl AG
tool material. C 0.5, Mn 0.85, Cr 1.5, Mo 0.65, V 0.3, bal Fe.
For tools, gears, shafts; oil hardening. *Obsolete*

KRUPP F 16
Krupp Stahl AG
stainless steel. C 0.07, Cr 16, bal Fe.
Ferritic type stainless steel. *Obsolete*

KRUPP F 161
Krupp Stahl AG
stainless steel. C 0.06, Cr 16, Mo, bal Fe.
Ferritic type stainless steel. *Obsolete*

KRUPP F 17 A ZN
Krupp Stahl AG
stainless steel. C 0.04, Cr 17, Nb, Mo, Si, bal Fe.
For high temperature operation. *Obsolete*

KRUPP F 17 B
Krupp Stahl AG
stainless steel.
See NIROSTA 4016

KRUPP F1248 C
Krupp Stahl AG
tool material. C 1.65, Cr 11.5, V 0.1, bal Fe.
For blanking and forming dies, punches; air hardened, non-deforming. *Obsolete*

KRUPP F13
Krupp Stahl AG
stainless steel.
See FERROTHERM 4722

KRUPP F1448C
Krupp Stahl AG
tool material. C 1.65, Cr 11.5, V 0.1, bal Fe.
For blanking and forming dies, punches; air hardened, non-deforming. *Obsolete*

KRUPP F153
Krupp Stahl AG
tool material. C 1.4, Cr 0.3, Mn 0.3, Si 0.25, V 0.1, bal Fe.
For bearings, cutters, tools, dies; water hardened, wear resistant. *Obsolete*

KRUPP F168
Krupp Stahl AG
tool material. C 87, Cr 1.7-1.9, 0.82 0, bal Fe.
For tools, bearings; water hardening. *Obsolete*

KRUPP F182
Krupp Stahl AG
tool material. C 0.9, Cr 0.45, bal Fe.
For tools, drills, taps; water hardening. *Obsolete*

KRUPP F193
Krupp Stahl AG
tool material. C 0.9-1, Cr 0.7-0.9, bal Fe.
For tools, drills, taps; water hardening. *Obsolete*

KRUPP F202
Krupp Stahl AG
tool material. C 1.1, Mn 0.3, Cr 0.4, Si 0.2, bal Fe.
For cutters, tools, bearings, drills; water hardened, wear resistant. *Obsolete*

KRUPP F2048
Krupp Stahl AG
tool material. C 2.25, Cr 13, bal Fe.
For tools, dies; non-deforming. *Obsolete*

KRUPP F206 G
Krupp Stahl AG
tool material. C 0.95-1.05, Cr 1.25-1.5, bal Fe.
For tools, bearings; wear resistant. *Obsolete*

KRUPP F20S
Krupp Stahl AG
tool material. C 1.6, Si 0.9, Mn 0.2, Cr 20, bal Fe.
For hard facing electrodes; corrosion resistant. *Obsolete*

KRUPP F20SH
Krupp Stahl AG
cast iron. C 2.5, Si 0.9, Mn 0.2, Cr 20, bal Fe.
For hard facing electrodes; corrosion resistant. *Obsolete*

KRUPP F22
Krupp Stahl AG
alloy steel. C 0.13, Cr 0.5, Mn 0.3, Si 0.2, bal Fe.
For gears, bolts, machine tool parts; case hardened. *Obsolete*

KRUPP F261
Krupp Stahl AG
tool material. C 1.26-1.32, Cr 0.2-0.3, S 0.03, bal Fe.
For tools, cutters, drills; water hardening. *Obsolete*

KRUPP F261 P
Krupp Stahl AG
tool material. C 1.26-1.32, S 0.025, P 0.025, Cr 0.2-0.3, bal Fe.
For tools, cutters, drills; water hardening. *Obsolete*

KRUPP F283
Krupp Stahl AG
tool material. C 1.35-1.45, Cr 0.6-0.8, bal Fe.
For tools, cutters, hobs; water hardening. *Obsolete*

KRUPP F28S
Krupp Stahl AG
tool material. C 1.6, Si 0.9, Mn 0.55, Cr 28, bal Fe.
For hard facing electrodes; corrosion resistant. *Obsolete*

KRUPP F28SH
Krupp Stahl AG
cast iron. C 2.1, Si 0.9, Mn 0.55, Cr 28, bal Fe.
For hard facing electrodes; corrosion resistant. *Obsolete*

KRUPP F3014-LCS
Krupp Stahl AG
tool material. C 1.4-1.5, Si 1.4-1.6, Cr 3.5, bal Fe.
For tools, dies; tough. *Obsolete*

KRUPP F306
Krupp Stahl AG
tool material. C 1.4-1.5, Cr 1.3-1.5, bal Fe.
For tools, bearings; wear resistant. *Obsolete*

KRUPP F306 G
Krupp Stahl AG
tool material. C 1.4-1.5, Cr 1.3-1.5, bal Fe.
For tools, bearings; wear resistant. *Obsolete*

KRUPP F33
Krupp Stahl AG
alloy steel. C 0.15, Cr 0.65, Si 0.25, Mn 0.5, bal Fe.
For gears, bolts, machine tool parts; case hardened. *Obsolete*

KRUPP F4048
Krupp Stahl AG
tool material. C 2-2.2, Cr 11-12, bal Fe.
For tools, dies; non-deforming. *Obsolete*

KRUPP F74
Krupp Stahl AG
alloy steel.
See KRUPP 41CR4

KRUPP FALC 223
Krupp Stahl AG
C 0-0.03, Si 0-1, Mn 0-2, P 0-0.03, S 0-0.02, Cr 21-23, Mo
2.5-3.5, Ni 4.5-6.5, N 0.08-0.2, bal Fe.
SEW 400/88; W. Nr. 1.4462.

KRUPP FC29
Krupp Stahl AG
alloy steel.
See KRUPP 20MNCR5

KRUPP FEDERSTAHLDRAHT A
Krupp Stahl AG
alloy steel. Mn 0.3-0.7, Si 0.2-0.3, C, bal Fe.
Spring steel wire. W. Nr. 1.0500. *Obsolete*

KRUPP FEDERSTAHLDRAHT B
Krupp Stahl AG
alloy steel. Mn 0.3-0.7, Si 0.1-0.3, C, bal Fe.
Spring steel wire. W. Nr. 1.0600. *Obsolete*

KRUPP FF112
Krupp Stahl AG
alloy steel. C 0.3, Si 2.2, Cr 6, bal Fe.
For oil refinery equipment; heat and creep resistant.
Obsolete

KRUPP FF120
Krupp Stahl AG
stainless steel. C 0.6, Si 1.5, Cr 22, bal Fe.
For furnace grates, wear plates, rabble arms; heat and creep
resistant. *Obsolete*

KRUPP FF128
Krupp Stahl AG
stainless steel. C 0.6, Si 1.5, Cr 29, bal Fe.
For frunace grates, wear plates, rabble arms; heat resistant.
Obsolete

KRUPP FF13
Krupp Stahl AG
stainless steel.
See FERROTHERM 4722

KRUPP FF18
Krupp Stahl AG
stainless steel.
See FERROTHERM 4741

KRUPP FF228
Krupp Stahl AG
tool material. C 1.3, Si 1.5, Cr 29, bal Fe.
For wear blades, scrapers, crushers, rollers; heat and
abrasion resistant. *Obsolete*

KRUPP FF25
Krupp Stahl AG
stainless steel. C 0.35, Cr 23-27, bal Fe.
Annealed: 90,000 TS; 60,000 YS; 20 El; 45 RA; 180 Brin. For
furnace parts, preheaters, heat treating boxes; Type 446;
corrosion resistant. *Obsolete*

KRUPP FF30
Krupp Stahl AG
stainless steel. C 0.35, Cr 23-27, bal Fe.
Annealed: 90,000 TS; 60,000 YS; 20 El; 45 RA; 180 Brin. For
furnace parts, preheaters, heat treating boxes; Type 446;
corrosion resistant. *Obsolete*

KRUPP FF6
Krupp Stahl AG
alloy steel. C 0-0.12, Si 2.3, Cr 6, bal Fe.
For oil refinery equipment; creep and heat resistant.
Obsolete

KRUPP FF6N
Krupp Stahl AG
alloy steel. C 0.3, Si 2.2, Cr 6, bal Fe.
For oil refinery equipment; creep and heat resistant.
Obsolete

KRUPP FFE25
Krupp Stahl AG
stainless steel. C 0.2, Si 1.2, Cr 25, Ni 4, bal Fe.
Cast: 90,000 TS; 65,000 YS; 2 El; 212 Brin. For cylinder liners,
bushings, valves, corroson and heat resistant. *Obsolete*

KRUPP FFE28
Krupp Stahl AG
stainless steel. C 0.2, Si 1.2, Cr 25, Ni 4, bal Fe.
Cast: 90,000 TS; 65,000 YS; 2 El; 212 Brin. For cylinder liners,
bushings, valve seats and bodies; corrosion and heat
resistant. *Obsolete*

KRUPP FFP13
Krupp Stahl AG
stainless steel.
See FERROTHERM 4724

KRUPP FFP18
Krupp Stahl AG
alloy steel. C 0 0.12, Al 1, Cr 1, bal Fe.
Annealed: 120,000 TS; 90,000 YS; 20 El; 55 RA; 250 Brin. For
oil refinery equipment; heat and creep resistant. *Obsolete*

KRUPP FFP24
Krupp Stahl AG
stainless steel.
See FERROTHERM 4762

KRUPP FFP6
Krupp Stahl AG
alloy steel. C 0.12, Si 0.8, Cr 0.5, Al 0.8, bal Fe.
For oil refinery equipment; heat and creep resistant.
Obsolete

KRUPP FK15
Krupp Stahl AG
alloy steel. C 0.15, Cr, Mo, bal Fe.
For gears, shafts, machine tool parts; case hardening steel,
tough. *Obsolete*

KRUPP FK24
Krupp Stahl AG
alloy steel. C 0.2, Cr, Mo, bal Fe.
For machine tool parts, gears, shafts; case hardening steel,
tough. *Obsolete*

KRUPP FK30
Krupp Stahl AG
alloy steel.
See KRUPP 25CRMO4

KRUPP FK34
Krupp Stahl AG
alloy steel.
See KRUPP 34CRMO4

KRUPP FK44
Krupp Stahl AG
alloy steel.
See KRUPP 42CRMO4

KRUPP FKM45
Krupp Stahl AG
alloy steel. C 0.42, Cr, Mo, bal Fe.
For gears, bolts, fasteners, shafts; oil hardened, tough.
Obsolete

KRUPP FKM54
Krupp Stahl AG
alloy steel. C 0.3, Cr, Mo, V, bal Fe.
For gears, bolts, machine tool parts; oil hardened, tough.
Obsolete

KRUPP FM1041
Krupp Stahl AG
alloy steel.
See KRUPP 50CRV4

KRUPP FM1251
Krupp Stahl AG
tool material. C 0.6-0.68, Si 0.7-1, Mn 0.7-1, Cr 1.1-1.3, V
0.1-0.18, bal Fe.
For tools, gears, springs; tough. *Obsolete*

KRUPP FM21
Krupp Stahl AG
alloy steel. C 0.5, Cr 1, V 0.09, bal Fe.
For springs, gears, bolts, shafts; oil hardened, shock
resistant. *Obsolete*

KRUPP FM2432
Krupp Stahl AG
tool material. C 1.15-1.25, Cr 0.6-0.8, V 0.15, bal Fe.
For tools, bearings; water hardening. *Obsolete*

KRUPP FM881
Krupp Stahl AG
alloy steel. C 0.34-0.39, Cr 1.6-1.9, V 0.15, bal Fe.
For gears, springs; tough. *Obsolete*

KRUPP FMC 24
Krupp Stahl AG
alloy steel. C 0.27, Cr, V, Mn, bal Fe.
For gears, bolts, machine tool parts; oil hardened, tough.
Obsolete

KRUPP FP12
Krupp Stahl AG
alloy steel. C 0.34, Al 1.1, Cr 1.4, bal Fe.
For oil refinery equipment; creep and heat resistant.
Obsolete

KRUPP FP15
Krupp Stahl AG
alloy steel. C 0.27, Al 1.1, Cr 1.4, bal Fe.
For oil refinery equipment; creep and heat resistant.
Obsolete

KRUPP FPE23
Krupp Stahl AG
alloy steel.
See KRUPP 34CRALNI7

KRUPP FPK13
Krupp Stahl AG
alloy steel.
See KRUPP 34CRALMO5

KRUPP GA 350
Krupp Stahl AG
alloy steel. C 0.45, Si 0.35, Mn 2.1, Cr 1.15, Ti 0.2, bal Fe.
For hard facing electrodes; wear resistant. *Obsolete*

KRUPP GA 500
Krupp Stahl AG
alloy steel. C 0.9, Si 0.35, Mn 2.1, Cr 1.3, Ti 0.2, bal Fe.
For hard facing electrodes; wear resistant. *Obsolete*

KRUPP GB 7 C
Krupp Stahl AG
tool material. C 0.75, Mn 0.6, Cr 0.5, bal Fe.
For hand tools. *Obsolete*

KRUPP GL-A
Krupp Stahl AG
C 0-0.28, P 0-0.04, S 0-0.04, 2.50 Mn min, bal Fe.
SEL 90; W. Nr. 1.0441.

KRUPP GL-D
Krupp Stahl AG
C 0-0.21, Si 0-0.35, P 0-0.04, S 0-0.04, 0.60 Mn min, 0.015 Al
min, bal Fe.
SEL 90; W. Nr. 1.0474.

KRUPP H I
Krupp Stahl AG
C 0-0.16, Si 0-0.35, Mn 0.4-1.2, P 0-0.035, S 0-0.03, Cr 0-0.25, Mo 0-0.1, Ni 0-0.3, Cu 0-0.3, 0.02 Al min, bal Fe.
DIN 17155/83; W. Nr. 1.0345.

KRUPP H II
Krupp Stahl AG
C 0-0.2, Si 0-0.35, Mn 0.5-1.3, P 0-0.035, S 0-0.03, 0.02 Al min, bal Fe.
DIN 17155/83; W. Nr. 1.0425.

KRUPP H III
Krupp Stahl AG
C 0-0.22, Si 0-0.35, P 0-0.05, S 0-0.05, Cr 0-0.3, N 0-0.007, 0.55 Mn min, bal Fe.
SEL 90; W. Nr. 1.0435.

KRUPP H IV
Krupp Stahl AG
C 0-0.26, Si 0-0.35, P 0-0.05, S 0-0.05, Cr 0-0.3, N 0-0.007, 0.60 Mn min, bal Fe.
SEL 90; W. Nr. 1.0445.

KRUPP HCT1
Krupp Stahl AG
stainless steel. C 0-0.2, Cr 22-24, Ni 12-15, Mn 0-2, Si 0-3.5, bal Fe.
Annealed: 85,000-95,000 TS; 40,000-50,000 YS; 45-55 El; 150-185 Brin. For heat treating boxes, oil refinery and chemical plant equipment; Type 309; austenitic, heat resistant. *Obsolete*

KRUPP HCT3
Krupp Stahl AG
stainless steel. C 0-0.25, Cr 24-26, Ni 19-22, bal Fe.
Annealed: 95,000 TS; 45,000 YS; 50 El; 65 RA; 180 Brin. At 1200°F: 57,000 TS; 22,000 YS; 32 El; 45 RA. For furnace parts and equipment, heat treating boxes; Type 310; austenitic, heat resistant. *Obsolete*

KRUPP HFX 6356
Krupp Stahl AG
C 0-0.03, Si 0-0.1, Mn 0-0.1, P 0-0.01, S 0-0.01, Mo 3-4.5, Ni 17-18.5, Al 0-0.2, Co 11.5-13.5, Ti 1.5-2, bal Fe.
SEL 90; W. Nr. 1.6356.

KRUPP HFX 6358
Krupp Stahl AG
C 0-0.03, Si 0-0.1, Mn 0-0.1, P 0-0.01, S 0-0.01, Mo 4.5-5.5, Ni 17-19, Al 0.05-0.15, B 0-0.005, Co 8-10, Ti 0.5-0.8, bal Fe.
SEL 90; W. Nr. 1.6358.

KRUPP HFX 6926
Krupp Stahl AG
C 0.35-0.4, Si 0.15-0.4, Mn 0.5-0.8, P 0-0.015, S 0-0.01, Cr 0.65-0.95, Mo 0.3-0.4, Ni 1.6-2, V 0.08-0.15, bal Fe.
SEL 90; W. Nr. 1.6926.

KRUPP HFX 6928
Krupp Stahl AG
C 0.38-0.45, Si 1.4-1.9, Mn 0.4-0.7, P 0-0.02, S 0-0.015, Cr 0.6-1, Mo 0.3-0.5, Ni 1.2-1.7, V 0.05-0.15, bal Fe.
SEL 90; W. Nr. 1.6928.

KRUPP HFX 7783
Krupp Stahl AG
C 0.38-0.43, Si 0.8-1, Mn 0.2-0.4, P 0-0.015, S 0-0.01, Cr 4.75-5.25, Mo 1.2-1.4, V 0.4-0.6, bal Fe.
SEL 90; W. Nr. 1.7783.

KRUPP HGS
Krupp Stahl AG
tool material. C 0.55-0.6, Cr 0.6-0.8, Ni 1.5-1.8, Mo 0.15-0.2, bal Fe.
For forging die blocks; oil hardened, tough. *Obsolete*

KRUPP HGS SPEZIAL
Krupp Stahl AG
tool material. C 0.34, Cr 0.68, Mo 0.15, bal Fe.
For tools, gears, shafts; oil hardening. *Obsolete*

KRUPP HGSE
Krupp Stahl AG
tool material. C 0.55, Mn 0.7, Cr 0.8, Ni 1.8, Mo 0.5, bal Fe.
For forging die blocks; oil hardening. *Obsolete*

KRUPP HIGH SPEED
Thomas Prosser & Sons
C, W, alloy, bal Fe.
For high speed tools and cutters; high speed steel.

KRUPP HM1
Krupp Stahl AG
tool material. C, bal Fe.
For tools. *Obsolete*

KRUPP HM2
Krupp Stahl AG
tool material. C, bal Fe.
For tools. *Obsolete*

KRUPP HM3
Krupp Stahl AG
tool material. C 1.1, bal Fe.
For tools, drills, taps; water hardened. *Obsolete*

KRUPP HM4
Krupp Stahl AG
tool material. C 1.1, bl Fe.
For tools, drills, taps; water hardened. *Obsolete*

KRUPP HM4 G
Krupp Stahl AG
tool material. C 0.9-1, Cr 0.5, bal Fe.
For tools, bearings, drills; water hardening. *Obsolete*

KRUPP HSB 8-SO
Krupp Stahl AG
tool material.
See KRUPP 2705

KRUPP HSK
Krupp Stahl AG
tool material. C 0.9, Si 0.25-0.5, Mn 0.3-0.8, bal Fe.
Heat treated: 190,000 TS; 145,000 YS; 10 El; 30 RA; 400 Brin.
For drills, taps, springs, tools, reamers; Type W1; water hardened. *Obsolete*

KRUPP IF 18
Krupp Stahl AG
C 0-0.02, Ti 0-0.3, bal Fe.
SEW 095/87; W. Nr. 1.0873.

KRUPP K.A.-2
Krupp Stahl AG
stainless steel. Ni 7-10, Cr 16-20, Si 0.75, C 0.15, bal Fe.
For heat and corrosion resisting parts; for high temperature use. *Obsolete*

KRUPP KFM
Krupp Stahl AG
alloy steel. C 0.95, Si 0.3, Mn 0.3, Cr 3.75, W 1.3, Ti 0.12, V 2.8, Mo 2.35, bal Fe.
For hard facing electrodes; wear resistant. *Obsolete*

KRUPP KFM EXTRA
Krupp Stahl AG
tool material. C 0.85, W, Mo, bal Fe.
For cutters, dies; oil or water hardened. *Obsolete*

KRUPP KFM ZWEISTERN
Krupp Stahl AG
tool material. C 0.95, W, Mo, bal Fe.
For cutters, dies; oil or water hardened. *Obsolete*

KRUPP KST 37-2
Krupp Stahl AG
C 0-0.17, P 0-0.05, S 0-0.05, N 0-0.009, bal Fe.
DIN 17100/80; W. Nr. 1.0113.

KRUPP KST 37-3
Krupp Stahl AG
C 0-0.17, P 0-0.04, S 0-0.04, 0.02 Al min, bal Fe.
DIN 17100/80; W. Nr. 1.0127.

KRUPP KST 44-2
Krupp Stahl AG
C 0-0.21, P 0-0.05, S 0-0.05, N 0-0.009, bal Fe.
DIN 17100/80; W. Nr. 1.0148.

KRUPP KST 44-3
Krupp Stahl AG
C 0-0.2, P 0-0.04, S 0-0.04, 0.02 Al min, bal Fe.
DIN 17100/80; W. Nr. 1.0137.

KRUPP KST 52-3
Krupp Stahl AG
C 0-0.2, Si 0-0.55, Mn 0-1.6, P 0-0.04, S 0-0.04, 0.02 Al min, bal Fe.
DIN 17100/80; W. Nr. 1.0575.

KRUPP LST 45.8
Krupp Stahl AG
C 0-0.22, Si 0.2-0.35, P 0-0.05, S 0-0.05, N 0-0.007, 0.45 Mn min, bal Fe.
SEL 81; W. Nr. 1.0407.

KRUPP LSTE 36
Krupp Stahl AG
C 0.16-0.22, Si 0-0.55, Mn 0-1.5, P 0-0.045, S 0-0.045, Cr 0-0.3, N 0-0.009, bal Fe.
SEL 81; W. Nr. 1.0569.

KRUPP M 13 H
Krupp Stahl AG
tool material. C 0.9, Cr 13, bal Fe.
Martensitic stainless steel; for tools, stainless ball bearings; punches. *Obsolete*

KRUPP M 131 CO
Krupp Stahl AG
tool material. C 0.65, Cr 14, Co 1, Mo, V, bal Fe.
For surgical scissors and similar tools. *Obsolete*

KRUPP M 131 V
Krupp Stahl AG
tool material. C 0.79, Cr 13, V 2, Mo, bal Fe.
Cold work tool steel; for stamping and forming dies. *Obsolete*

KRUPP M 171 W
Krupp Stahl AG
tool material. C 0.68, Cr 17.
Martensitic type stainless steel; hardenable to above 52 Rock C. COMP:bal Fe. *Obsolete*

KRUPP M 2
Krupp Stahl AG
C 0-0.03, Si 0-0.01, Mn 0-0.03, P 0-0.01, S 0-0.035, Al 0-0.001, Cu 0-0.03, N 0-0.005, bal Fe.
SEL 90; W. Nr. 1.0340.

KRUPP M202
Krupp Stahl AG
tool material. C 0.9-1, V 0.3, bal Fe.
For tools, drills, taps; water hardening. *Obsolete*

KRUPP MA 16
Krupp Stahl AG
C 0.12-0.18, Si 0.15-0.3, Mn 0.5-0.7, P 0-0.03, S 0-0.04, N 0-0.007, bal Fe.
SEL 75; W. Nr. 1.0423.

KRUPP MINIMUM R
Krupp Stahl AG
iron. C 0-0.04, Mn 0-0.2, Al 0.05-0.15, bal Fe.
For magnet cores. W. Nr. 1.1004. *Obsolete*

KRUPP MINIMUM RA
Krupp Stahl AG
iron. C 0-0.04, Mn 0.12-0.2, bal Fe.
For magnet cores. W. Nr. 1.1005. *Obsolete*

KRUPP MK 3
Krupp Stahl AG
C 0-0.04, Si 0-0.01, Mn 0-0.2, P 0-0.035, S 0-0.035, Al
0.05-0.15, bal Fe.
SEL 90; W. Nr. 1.1004.

KRUPP MOG 310
Krupp Stahl AG
tool material.
See KRUPP 2365

KRUPP N48
Krupp Stahl AG
alloy steel. C 0.3, Cr, V, bal Fe.
For gears, bolts, machine tool parts; oil hardened, tough.
Obsolete

KRUPP N54
Krupp Stahl AG
alloy steel. C 0.31, Cr 2.35, Mo 0.18, V 0.13, bal Fe.
For gears, bolts, machine tool parts; oil hardened, tough.
Obsolete

KRUPP NCT1A
Krupp Stahl AG
stainless steel. C 0.15, Si 2, Cr 19.5, Ni 9.5, bal Fe.
Annealed: 80,000 TS; 35,000 YS; 55 El; 75 RA; 150 Brin. For
chemical plant equipment, tanks, mixers; Type 302; stainless,
austenitic. *Obsolete*

KRUPP NCT1A GUSS
Krupp Stahl AG
stainless steel. C 0.4, Cr 22, Ni 9.5, bal Fe.
Cast: 85,000 TS; 45,000 YS; 35 El; 165 Brin. For heat treat
boxes, furnace parts, conveyors; Type HF; corrosion and
heat resistant. *Obsolete*

KRUPP NCT2 GUSS
Krupp Stahl AG
C 0.4, Cr, Ni, Si, bal Fe.
For furnace parts, heat treat boxes; corrosion and heat
resistant. *Obsolete*

KRUPP NCT3
Krupp Stahl AG
stainless steel. C 0.15, Cr 24, Ni 19, bal Fe.
Annealed: 100,000 TS; 45,000 YS; 50 El; 65 RA; 185 Brin. For
valves, pumps, turbine and jet parts; Type 310; stainless,
austenitic. *Obsolete*

KRUPP NCT3 GUSS
Krupp Stahl AG
superalloy. C 0.4, Cr 25, Ni 19, bal Fe.
Cast: 75,000 TS; 50,000 YS; 17 El; 170 Brin. Aged: 85,000 TS;
50,000 YS; 10 El; 190 Brin. For heat treat boxes, burner tips,
hearth plates; Type HK; austenitic, stainless. *Obsolete*

KRUPP NCT3/133
Krupp Stahl AG
superalloy. Cr 20, Ni 25, bal Fe.
For welding electrodes; stainless, austenitic. *Obsolete*

KRUPP NCT30
Krupp Stahl AG
stainless steel. C 0.15, Ni, Cr, bal Fe.
For chemical plant equipment; stainless. *Obsolete*

KRUPP NCT3A GUSS
Krupp Stahl AG
superalloy. C 0.4, Cr 26, Ni 14, bal Fe.
Cast: 75,000 TS; 50,000 YS; 15 El; 170 Brin. For heat treat
boxes, furnace parts, hearth plates; Type HI; austenitic, heat
resistant. *Obsolete*

KRUPP NCT6
Krupp Stahl AG
stainless steel. C 0.15, Ni, Cr, bal Fe.
For chemical plant equipment; stainless. *Obsolete*

KRUPP NCT8
Krupp Stahl AG
stainless steel. C 0.15, Ni, Cr, bal Fe.
For chemical plant equipment; stainless. *Obsolete*

KRUPP NFKC
Krupp Stahl AG
alloy steel. C 1.4, Si 0.2, Mn 0.3, Cr 0.45, W 3.25, V 0.2, bal
Fe.
For hard facing electrodes; wear resistant. *Obsolete*

KRUPP NI 36
Krupp Stahl AG
iron. C 0-0.1, Ni 36, bal Fe.
Low co-efficient of thermal expansion. *Obsolete*

KRUPP NI 48
Krupp Stahl AG
iron. C 0-0.05, Ni 48, bal Fe.
For glass to metal seals, or soft magnetic parts as magnet
cores. *Obsolete*

KRUPP NIMN 20 6
Krupp Stahl AG
C 0-0.2, Si 0-1, Mn 6-7, Ni 19-21, bal Fe.
SEW 385/57; W. Nr. 1.3932.

KRUPP NIROSTAGUSS 1
Krupp Stahl AG
cast iron. C 1-2, 24.Cr min, bal Fe.
Cast: 58,000 TS; 250 Brin. For chemical equipment, HNO_3
and pulp industries; corrosion resistant. *Obsolete*

KRUPP NIROSTAGUSS 2
Krupp Stahl AG
cast iron. C 1-2, 24. Cr min, bal Fe.
Cast: 65,000 TS; 300 Brin. For chemical equipment, HNO_3
and pulp industries; corrosion resistant. *Obsolete*

KRUPP NIROSTAGUSS 3
Krupp Stahl AG
cast iron. C 1-2, 24. Cr min, bal Fe.
Cast: 72,000-114,000 TS; 400-500 Brin. For paper drying
cylinders, chemical equipment; corrosion resistant. *Obsolete*

KRUPP NITROPLAT
Krupp Stahl AG
composite sheet Krupp V2A on C steel.
For corrosion resistant parts. *Obsolete*

KRUPP NON-DEFORMING
Thomas Prosser & Sons
C, alloy, bal Fe.
For tools, dies; non-deforming.

KRUPP NOVONIT 8405
Krupp Stahl AG
C 0.65-0.75, Si 0.15-0.35, Mn 1.8-2.2, P 0-0.025, S 0-0.025, Cr
0.9-1.2, Al 0-0.1, Ti 0.1-0.25, bal Fe.
Weld filler metal. SEL 90; W. Nr. 1.8405.

KRUPP P141
Krupp Stahl AG
tool material. C 0.85, W, Mo, Cr, V, bal Fe.
For tools, dies, cutters; oil hardened. *Obsolete*

KRUPP P193
Krupp Stahl AG
superalloy. C, Ni, bal Fe.
For jet engine components, turbine rotor blades; austenitic,
heat resistant. *Obsolete*

KRUPP P427
Krupp Stahl AG
tool material. C 1.65, Cr, Co, bal Fe.
For blanking and forming dies; oil hardened, wear resistant.
Obsolete

KRUPP P77
Krupp Stahl AG
tool material. C 0.9, Mn 1.9, V 0.1, bal Fe.
For punches, dies, shears, upsetters; oil hardened, non-
deforming. *Obsolete*

KRUPP PFM
Krupp Stahl AG
tool material. C 0.95-1.05, Cr 0.3, V 0.15, bal Fe.
For tools, drills, taps; water hardening. *Obsolete*

KRUPP PRESSGUETE 260
Krupp Stahl AG
carbon steel. C 0-0.16, Si 0-0.5, Mn 0-1.2.
For beams, frame construction. W.Nr. 1.8941; QSTE 260 [A-
Z]C *Obsolete*

KRUPP PRESSGUETE 340
Krupp Stahl AG
carbon steel. C 0-0.16, Si 0-0.5, Mn 0-1.5, Ti or Cb or V, bal
Fe.
For beams, frame construction. W. Nr. 1.8945; QSTE 340 N.
Obsolete

KRUPP PRESSGUETE 380
Krupp Stahl AG
hsla steel. C 0-0.18, Si 0-0.5, Mn 0-1.6, Ti 0.12-0.2, bal Fe.
For beams, frame construction. W. Nr. 1.8950; QSTE 380 N.
Obsolete

KRUPP PRESSGUETE 420
Krupp Stahl AG
hsla steel. C 0-0.2, Si 0-0.5, Mn 0-1.6, Ti 0.12-0.2, bal Fe.
For beams, frame construction. W.Nr. 1.8952; QSTE 420 N.
Obsolete

KRUPP PRESSGUETE 460
Krupp Stahl AG
hsla steel. C 0-0.21, Si 0-0.5, Mn 0-1.7, Ti 0.12-0.2, bal Fe.
For beams, frame construction. W.Nr. 1.8955; QSTE 460 N.
Obsolete

KRUPP PRESSGUETE 500
Krupp Stahl AG
hsla steel. C 0-0.22, Si 0-0.5, Mn 0-1.7, Ti 0.12-0.2, bal Fe.
For beams, frame construction. W.Nr. 1.8957; QSTE 500 N.
Obsolete

KRUPP PST 37-3
Krupp Stahl AG
C 0-0.17, P 0-0.04, S 0-0.04, 0.02 Al min, bal Fe.
DIN 17100/80; W. Nr. 1.0176.

KRUPP PST 44-2
Krupp Stahl AG
C 0-0.21, P 0-0.05, S 0-0.05, N 0-0.009, bal Fe.
DIN 17100/80; W. Nr. 1.0146.

KRUPP PST 44-3
Krupp Stahl AG
C 0-0.2, P 0-0.04, S 0-0.04, 0.02 Al min, bal Fe.
DIN 17100/80; W. Nr. 1.0135.

KRUPP PST 52-3
Krupp Stahl AG
C 0-0.2, Si 0-0.55, Mn 0-1.6, P 0-0.04, S 0-0.04, 0.02 Al min,
bal Fe.
DIN 17100/80; W. Nr. 1.0572.

KRUPP QST 32-3
Krupp Stahl AG
C 0-0.06, Si 0-0.1, Mn 0.2-0.4, P 0-0.04, S 0-0.04, 0.02 Al min,
bal Fe.
DIN 1654 T.2/89; W. Nr. 1.0303.

KRUPP QST 34-3
Krupp Stahl AG
C 0.05-0.1, Si 0-0.1, Mn 0.2-0.4, P 0-0.04, S 0-0.04, 0.02 Al min, bal Fe.
DIN 1654 T.2/89; W. Nr. 1 .0213.

KRUPP QST 36-3
Krupp Stahl AG
C 0.06-0.13, Si 0-0.1, Mn 0.25-0.45, P 0-0.04, S 0-0.04, 0.02 Al min, bal Fe.
DIN 1654 T.2/89; W. Nr. 1.0214.

KRUPP QST 37-3
Krupp Stahl AG
C 0-0.17, P 0-0.04, S 0-0.04, 0.02 Al min, bal Fe.
DIN 17100/80; W. Nr. 1.0123.

KRUPP QST 38-3
Krupp Stahl AG
C 0.1-0.18, Si 0-0.1, Mn 0.25-0.45, P 0-0.04, S 0-0.04, 0.02 Al min, bal Fe.
DIN 1654 T.2/89; W. Nr. 1.0234

KRUPP QST 42-3
Krupp Stahl AG
C 0-0.23, Si 0.03-0.3, Mn 0.2-0.5, P 0-0.045, S 0-0.045, bal Fe.
SEL 75; W. Nr. 1.0143.

KRUPP QST 44-2
Krupp Stahl AG
C 0-0.21, P 0-0.05, S 0-0.05, N 0-0.009, bal Fe.
DIN 17100/80; W. Nr. 1.0128.

KRUPP QST 44-3
Krupp Stahl AG
C 0-0.2, P 0-0.04, S 0-0.04, 0.02 Al min, bal Fe.
DIN 17100/80; W. Nr. 1.0133.

KRUPP QST 52-3
Krupp Stahl AG
C 0-0.2, Si 0-0.55, Mn 0-1.6, P 0-0.04, S 0-0.04, 0.02 Al min, bal Fe.
DIN 17100/80; W. Nr. 1.0573.

KRUPP QST 52-3 CU 3
Krupp Stahl AG
C 0-0.2, Si 0-0.55, Mn 0-1.6, P 0-0.04, S 0-0.04, Cu 0.25-0.35, 0.02 Al min, bal Fe.
DIN 17100/80; W. Nr. 1.0587.

KRUPP R 10 S 10
Krupp Stahl AG
C 0-0.15, Si 0-0.4, Mn 0.3-0.8, P 0-0.05, S 0.08-0.12, bal Fe.
DIN 17111/80; W. Nr. 1.0703.

KRUPP REZ EXTRA
Krupp Stahl AG
tool material. C 0.35, Cr 0.8, Mo 0.55, bal Fe.
For tools, gears, shafts; oil hardening. *Obsolete*

KRUPP RFE 100
Krupp Stahl AG
C 0-0.05, Si 0-0.1, Mn 0.2-0.35, P 0-0.03, S 0-0.035, Al 0.04-0.1, bal Fe.
DIN 17405/79; W. Nr. 1.1013.

KRUPP RFE 120
Krupp Stahl AG
C 0-0.05, Si 0-0.1, Mn 0.2-0.35, P 0-0.03, S 0-0.035, Al 0.04-0.1, bal Fe.
DIN 17405/79; W. Nr. 1.1012.

KRUPP RFE 160
Krupp Stahl AG
C 0-0.1, Si 0-0.1, Mn 0.5-0.9, P 0-0.08, S 0.18-0.27, Al 0.04-0.1, bal Fe.
DIN 17405/79; W. Nr. 1.1011.

KRUPP RFE 60
Krupp Stahl AG
C 0-0.03, Si 0-0.05, Mn 0-0.2, P 0-0.025, S 0-0.015, Al 0.04-0.1, bal Fe.
DIN 17405/79; W. Nr. 1.1015.

KRUPP RFE 80
Krupp Stahl AG
C 0-0.05, Si 0-0.1, Mn 0.2-0.35, P 0-0.03, S 0-0.035, Al 0.04-0.1, bal Fe.
DIN 17405/79; W. Nr. 1.1014.

KRUPP RKST 37-2
Krupp Stahl AG
C 0-0.17, P 0-0.05, S 0-0.05, N 0-0.009, bal Fe.
DIN 17100/80; W. Nr. 1.0125.

KRUPP RNI 24
Krupp Stahl AG
iron. C 0-0.15, Mn 0-1, Si 0-1, Ni 38, bal Fe.
For high temperature operation, and for glass to metal seals.
W.Nr. 1.3911. *Obsolete*

KRUPP ROST 2
Krupp Stahl AG
C 0-0.1, Si 0-0.01, Mn 0.3-0.6, P 0-0.045, S 0-0.045, N 0-0.007, bal Fe.
SEL 90; W. Nr. 1.0331.

KRUPP ROST 37-3
Krupp Stahl AG
C 0-0.17, P 0-0.04, S 0-0.04, 0.02 Al min, bal Fe.
DIN 17100/80; W. Nr. 1.0175.

KRUPP ROST 4
Krupp Stahl AG
C 0-0.1, Si 0-0.1, Mn 0.3-0.6, P 0-0.03, S 0-0.035, N 0-0.007, bal Fe.
SEL 90; W. Nr. 1.0337.

KRUPP ROST 44-2
Krupp Stahl AG
C 0-0.21, P 0-0.05, S 0-0.05, N 0-0.009, bal Fe.
DIN 17100/80; W. Nr. 1.0149.

KRUPP ROST 44-3
Krupp Stahl AG
C 0-0.2, P 0-0.04, S 0-0.04, 0.02 Al min, bal Fe.
DIN 17100/80; W. Nr. 1.0138.

KRUPP ROST 52-3
Krupp Stahl AG
C 0-0.2, Si 0-0.55, Mn 0-1.6, P 0-0.04, S 0-0.04, 0.02 Al min, bal Fe.
DIN 17100/80; W. Nr. 1.0576.

KRUPP RPST 37-2
Krupp Stahl AG
C 0-0.17, P 0-0.05, S 0-0.05, N 0-0.009, bal Fe.
DIN 17100/80; W. Nr. 1.0172.

KRUPP RPST 42-2
Krupp Stahl AG
C 0-0.23, Si 0.03-0.3, Mn 0.2-0.5, P 0-0.05, S 0-0.05, N 0-0.007, bal Fe.
SEL 75; W. Nr. 1.0191.

KRUPP RQST 34-2
Krupp Stahl AG
C 0-0.15, Si 0.03-0.3, Mn 0.2-0.5, P 0-0.05, S 0-0.05, N 0-0.007, bal Fe.
DIN 17100/66; W. Nr. 1.0109.

KRUPP RQST 37-2
Krupp Stahl AG
C 0-0.17, P 0-0.05, S 0-0.05, N 0-0.009, bal Fe.
DIN 17100/80; W. Nr. 1.0122.

KRUPP RQST 37-2 CU 3
Krupp Stahl AG
C 0-0.17, P 0-0.05, S 0-0.05, Cu 0.25-0.35, N 0-0.009, bal Fe.
DIN 17100/80; W. Nr. 1.0170.

KRUPP RQST 42-2
Krupp Stahl AG
C 0-0.23, Si 0.03-0.3, Mn 0.2-0.5, P 0-0.05, S 0-0.05, N 0-0.007, bal Fe.
SEL 75; W. Nr. 1.0142

KRUPP RROST 37-2
Krupp Stahl AG
C 0-0.17, P 0-0.05, S 0-0.05, N 0-0.009, bal Fe.
DIN 17100/80; W. Nr. 1.0174.

KRUPP RRSD 10
Krupp Stahl AG
C 0.07-0.11, Si 0-0.12, Mn 0.38-0.57, P 0-0.025, S 0-0.025, Cr 0-0.12, Ni 0-0.12, Al 0-0.04, Cu 0-0.17, bal Fe.
DIN 17145/80; W. Nr. 1.0351.

KRUPP RRSD 8
Krupp Stahl AG
C 0.05-0.1, Si 0.05-0.15, Mn 0.35-0.55, P 0-0.03, S 0-0.03, Al 0-0.03, N 0-0.007, bal Fe.
SEL 90; W. Nr. 1.0325.

KRUPP RRST 13
Krupp Stahl AG
C 0-0.1, 0.02 Al min, bal Fe.
DIN 1623 T.1/83; W. Nr. 1.0347.

KRUPP RRST 13-ZE
Krupp Stahl AG
C 0-0.1, 0.02 Al min, bal Fe.
DIN 17163/88; W. Nr. 1.0347.

KRUPP RRST 23
Krupp Stahl AG
C 0-0.1, Mn 0-0.45, P 0-0.03, S 0-0.03, 0.02 Al min, bal Fe.
DIN 1614 T.1/86; W. Nr. 1.0359.

KRUPP RRST 3
Krupp Stahl AG
C 0-0.1, 0.02 Al min, bal Fe.
DIN 1624/87; W. Nr. 1.0347.

KRUPP RRSTE 210.7
Krupp Stahl AG
C 0-0.17, Si 0-0.45, P 0-0.04, S 0-0.035, 0.35 Mn min, 0.02 Al min, bal Fe.
DIN 17172/78; W. Nr. 1.0319.

KRUPP RRSTE 240.7
Krupp Stahl AG
C 0-0.17, Si 0-0.45, P 0-0.04, S 0-0.035, 0.40 Mn min, bal Fe.
DIN 17172/78; W. Nr. 1.0459.

KRUPP RRSTW 23
Krupp Stahl AG
C 0-0.1, 0.02 Al min, bal Fe.
DIN 1614 T.2/86; W. Nr. 1.0398.

KRUPP RSD 10 SI
Krupp Stahl AG
C 0.07-0.11, Si 0.18-0.45, Mn 0.38-0.57, P 0-0.025, S 0-0.025, Cr 0-0.12, Ni 0-0.12, Al 0-0.01, Cu 0-0.17, bal Fe.
DIN 17145/80; W. Nr. 1.0339.

KRUPP RSD 4
Krupp Stahl AG
C 0.03-0.07, Si 0.07-0.17, Mn 0.5-0.7, P 0-0.025, S 0-0.025, Cu 0-0.2, bal Fe.
SEL 81; W. Nr. 1.0317.

KRUPP RSD 7
Krupp Stahl AG
C 0.05-0.09, Si 0.05-0.17, Mn 0.38-0.62, P 0-0.025, S 0-0.02, Cr 0-0.12, Ni 0-0.12, Al 0-0.01, Cu 0-0.17, bal Fe.
SEL 90; W. Nr. 1.0324.

KRUPP RST 34-2
Krupp Stahl AG
C 0-0.15, Si 0.03-0.3, Mn 0.2-0.5, P 0-0.05, S 0-0.05, N 0-0.007, bal Fe.
SEL 90; W. Nr. 1.0034.

KRUPP RST 35-2
Krupp Stahl AG
C 0.06-0.12, Si 0-0.25, Mn 0.4-0.6, P 0-0.035, S 0-0.035, Cu 0-0.25, N 0-0.012, bal Fe.
DIN 17115/87; W. Nr. 1.0208.

KRUPP RST 36
Krupp Stahl AG
C 0-0.14, Si 0-0.3, Mn 0.25-0.5, P 0-0.05, S 0-0.05, bal Fe.
DIN 17111/80; W. Nr. 1.0205.

KRUPP RST 37-2
Krupp Stahl AG
C 0-0.17, P 0-0.05, S 0-0.05, N 0-0.009, bal Fe.
DIN 17100/80; W. Nr. 1.0038.

KRUPP RST 37-2 CU 3
Krupp Stahl AG
C 0-0.17, P 0-0.05, S 0-0.05, Cu 0.25-0.35, N 0-0.009, bal Fe.
DIN 17100/80; W. Nr. 1.0167.

KRUPP RST 38
Krupp Stahl AG
C 0-0.19, Si 0-0.3, Mn 0.25-0.5, P 0-0.05, S 0-0.05, bal Fe.
DIN 17111/80; W. Nr. 1.0223.

KRUPP RZST 37-2
Krupp Stahl AG
C 0-0.17, P 0-0.05, S 0-0.05, N 0-0.009, bal Fe.
DIN 17100/80; W. Nr. 1.0165.

KRUPP SC2448
Krupp Stahl AG
tool material. C 1.3, Mn 12, bal Fe.
For tools, wear resistant parts; non-deforming, austenitic.
Obsolete

KRUPP SK 135
Krupp Stahl AG
tool material. C 1.35, W 1.2, Cr, Mo, V, bal Fe.
Cold work tool steel. *Obsolete*

KRUPP SKV 73
Krupp Stahl AG
tool material. C 0.77, W 3, Cr, V, bal Fe.
Tool steel. *Obsolete*

KRUPP SKV 81
Krupp Stahl AG
tool material. C 0.8, W 0.7, V, bal Fe.
Cold work tool steel. *Obsolete*

KRUPP SONDERSTAHL
Krupp Stahl AG
hsla steel. C 0-0.12, Mn 0.4-1, Ti 0.1-0.16, bal Fe.
For special applications. W.Nr. 1.8882. *Obsolete*

KRUPP SPECIAL
Thomas Prosser & Sons
C, alloy, bal Fe.
For tools, oil hardened.

KRUPP SPEZIAL S
Krupp Stahl AG
tool material. C 0.45, Cr, Ni, bal Fe.
For machine tool parts; oil hardened. *Obsolete*

KRUPP ST 12
Krupp Stahl AG
C 0-0.1, N 0-0.007, bal Fe.
DIN 1623 T.1/83; W. Nr. 1.0330.

KRUPP ST 12 CU 3
Krupp Stahl AG
C 0-0.1, Si 0-0.01, Mn 0.2-0.45, P 0-0.05, S 0-0.05, Cu 0.25-0.35, N 0-0.008, bal Fe.
DIN 5512 T.2/75; W. Nr. 1.0344.

KRUPP ST 12-ZE
Krupp Stahl AG
C 0-0.1, N 0-0.007, bal Fe.
DIN 17163/88; W. Nr. 1.0330.

KRUPP ST 13 CU 3
Krupp Stahl AG
C 0-0.1, Si 0-0.01, Mn 0.2-0.4, P 0-0.025, S 0-0.025, Cu 0.25-0.35, N 0-0.007, bal Fe.
SEL 90; W. Nr. 1.0353.

KRUPP ST 14
Krupp Stahl AG
C 0-0.08, 0.02 Al min, bal Fe.
DIN 1623 T.1/83; W. Nr. 1.0338.

KRUPP ST 14 CU 3
Krupp Stahl AG
C 0-0.08, Si 0.03-0.1, Mn 0-0.4, P 0-0.025, S 0-0.025, Cu 0.25-0.35, N 0-0.007, 0.02 Al min, bal Fe.
DIN 5512 T.2/75; W. Nr. 1.0354.

KRUPP ST 14-ZE
Krupp Stahl AG
C 0-0.08, 0.02 Al min, bal Fe.
DIN 17163/88; W. Nr. 1.0338.

KRUPP ST 15
Krupp Stahl AG
C 0-0.06, Mn 0-0.35, P 0-0.025, S 0-0.025, bal Fe.
SEL 90; W. Nr. 1.0312.

KRUPP ST 22
Krupp Stahl AG
C 0-0.1, Mn 0-0.45, P 0-0.035, S 0-0.035, N 0-0.007, bal Fe.
DIN 1614 T.1/86; W. Nr. 1.0320.

KRUPP ST 24
Krupp Stahl AG
C 0-0.08, Mn 0-0.4, P 0-0.025, S 0-0.025, 0.02 Al min, bal Fe.
DIN 1614 T.1/86; W. Nr. 1.0327.

KRUPP ST 33
Krupp Stahl AG
C 0-0.3, Si 0-0.3, Mn 0.2-0.5, P 0-0.06, S 0-0.05, bal Fe.
DIN 17100/80; W. Nr. 1.0035.

KRUPP ST 34-2
Krupp Stahl AG
C 0-0.15, Si 0-0.3, Mn 0.2-0.5, P 0-0.05, S 0-0.05, N 0-0.007, bal Fe.
SEL 90; W. Nr. 1.0032.

KRUPP ST 35
Krupp Stahl AG
C 0-0.18, Si 0-0.35, P 0-0.05, S 0-0.05, N 0-0.007, 0.40 Mn min, bal Fe.
SEL 90; W. Nr. 1.0308.

KRUPP ST 37-2
Krupp Stahl AG
C 0-0.17, P 0-0.05, S 0-0.05, N 0-0.009, bal Fe.
DIN 17100/80; W. Nr. 1.0037.

KRUPP ST 37-2 G
Krupp Stahl AG
C 0-0.17, P 0-0.04, S 0-0.035, N 0-0.009, bal Fe.
DIN 1623 T.2/86; W. Nr. 1.0037.

KRUPP ST 37-3
Krupp Stahl AG
C 0-0.17, P 0-0.04, S 0-0.04, 0.02 Al min, bal Fe.
DIN 17100/80; W. Nr. 1.0116.

KRUPP ST 37-3 CU 3
Krupp Stahl AG
C 0-0.17, P 0-0.04, S 0-0.04, Cu 0.25-0.35, 0.02 Al min, bal Fe.
DIN 17100/80; W. Nr. 1.0166.

KRUPP ST 37-3 G
Krupp Stahl AG
C 0-0.17, P 0-0.04, S 0-0.035, bal Fe.
DIN 1623 T.2/86; W. Nr. 1.0116.

KRUPP ST 37-3 G-ZE
Krupp Stahl AG
C 0-0.17, P 0-0.04, S 0-0.035, bal Fe.
DIN 17163/88; W. Nr. 1.0116.

KRUPP ST 37.8
Krupp Stahl AG
C 0-0.17, Si 0.1-0.35, Mn 0.4-0.8, P 0-0.04, S 0-0.04, bal Fe.
DIN 17177/79; W. Nr. 1.0315.

KRUPP ST 42-3
Krupp Stahl AG
C 0-0.23, Si 0.03-0.3, Mn 0.2-0.5, P 0-0.045, S 0-0.045, N 0-0.009, bal Fe.
SEL 75; W. Nr. 1.0136.

KRUPP ST 42.8
Krupp Stahl AG
C 0-0.21, Si 0.1-0.35, Mn 0.4-1.2, P 0-0.04, S 0-0.04, bal Fe.
DIN 17177/79; W. Nr. 1.0498.

KRUPP ST 44-2
Krupp Stahl AG
C 0-0.21, P 0-0.05, S 0-0.05, N 0-0.009, bal Fe.
DIN 17100/80; W. Nr. 1.0044.

KRUPP ST 44-3
Krupp Stahl AG
C 0-0.2, P 0-0.04, S 0-0.04, 0.02 Al min, bal Fe.
DIN 17100/80; W. Nr. 1.0144.

KRUPP ST 44-3 G
Krupp Stahl AG
C 0-0.2, P 0-0.04, S 0-0.035, 0.02 Al min, bal Fe.
DIN 1623 T.2/86; W. Nr. 1.0144.

KRUPP ST 45.8
Krupp Stahl AG
C 0-0.21, Si 0.1-0.35, Mn 0.4-1.2, P 0-0.04, S 0-0.04, bal Fe.
DIN 17175/79; W. Nr. 1.0405.

KRUPP ST 50-2 G
Krupp Stahl AG
C 0-0.4, P 0-0.05, S 0-0.05, N 0-0.009, bal Fe.
DIN 1623 T.2/86; W. Nr. 1.0050.

KRUPP ST 52-3
Krupp Stahl AG
C 0-0.2, Si 0-0.55, Mn 0-1.6, P 0-0.04, S 0-0.04, bal Fe.
DIN 17100/80; W. Nr. 1.0570.

KRUPP ST 52-3 CU 3
Krupp Stahl AG
C 0-0.2, Si 0-0.55, Mn 0-1.6, P 0-0.04, S 0-0.04, Cu 0.25-0.35, 0.02 Al min, bal Fe.
DIN 17100/80; W. Nr. 1.0585.

KRUPP ST 60-2 G
Krupp Stahl AG
C 0-0.5, P 0-0.05, S 0-0.05, N 0-0.009, bal Fe.
DIN 1623 T.2/86; W. Nr. 1.0060.

KRUPP ST 70-2 G
Krupp Stahl AG
C 0-0.65, P 0-0.05, S 0-0.05, N 0-0.009, bal Fe.
DIN 1623 T.2/86; W. Nr. 1.0070.

KRUPP STE 210.7
Krupp Stahl AG
C 0-0.17, Si 0-0.45, P 0-0.04, S 0-0.035, 0.35 Mn min, bal Fe.
DIN 17172/78; W. Nr. 1.0307.

KRUPP STE 240.7
Krupp Stahl AG
C 0-0.17, Si 0-0.45, P 0-0.04, S 0-0.035, 0.40 Mn min, bal Fe.
DIN 17172/78; W. Nr. 1.0457.

KRUPP STE 250 Z
Krupp Stahl AG
C 0-0.16, Mn 0-0.6, P 0-0.04, S 0-0.04, bal Fe.
DIN 17162 T.2/88; W. Nr. 1.0242.

KRUPP STE 250-2 Z
Krupp Stahl AG
C 0-0.16, Mn 0-0.6, P 0-0.04, S 0-0.04, bal Fe.
DIN 17162 T.2/80; W. Nr. 1.0242.

KRUPP STE 255
Krupp Stahl AG
C 0-0.18, Si 0-0.4, Mn 0.5-1.3, P 0-0.035, S 0-0.03, Cr 0-0.3, Mo 0-0.08, Ni 0-0.3, Cu 0-0.2, N 0-0.02, Nb 0-0.03, 0.02 Al min, bal Fe.
DIN 17102/83; W. Nr. 1.0461.

KRUPP STE 280 Z
Krupp Stahl AG
C 0-0.2, Mn 0-0.8, P 0-0.04, S 0-0.04, bal Fe.
DIN 17162 T.2/88; W. Nr. 1.0244.

KRUPP STE 280-2 Z
Krupp Stahl AG
C 0-0.2, Mn 0-0.8, P 0-0.04, S 0-0.04, bal Fe.
DIN 17162 T.2/80; W. Nr. 1.0244.

KRUPP STE 280-3 Z
Krupp Stahl AG
C 0-0.2, Mn 0-0.8, P 0-0.04, S 0-0.04, 0.02 Al min, bal Fe.
DIN 17162 T.2/80; W. Nr. 1.0246.

KRUPP STE 285
Krupp Stahl AG
C 0-0.18, Si 0-0.4, Mn 0.6-1.4, P 0-0.035, S 0-0.03, Cr 0-0.3, Mo 0-0.08, Ni 0-0.3, Cu 0-0.2, N 0-0.02, Nb 0-0.03, 0.02 Al min, bal Fe.
DIN 17102/83; W. Nr. 1.0486.

KRUPP STE 290.7
Krupp Stahl AG
C 0-0.22, Si 0-0.45, Mn 0.5-1.1, P 0-0.04, S 0-0.035, 0.02 Al min, bal Fe.
DIN 17172/78; W. Nr. 1.0484.

KRUPP STE 290.7 TM
Krupp Stahl AG
C 0.04-0.12, Si 0-0.4, Mn 0.5-1.5, P 0-0.035, S 0-0.025, V 0-0.12, bal Fe.
DIN 17172/78; W. Nr. 1.0429.

KRUPP STE 315
Krupp Stahl AG
C 0-0.18, Si 0-0.45, Mn 0.7-1.5, P 0-0.035, S 0-0.03, Cr 0-0.3, Mo 0-0.08, Ni 0-0.3, Cu 0-0.2, N 0-0.02, Nb 0-0.03, 0.02 Al min, bal Fe.
DIN 17102/83; W. Nr. 1.0505.

KRUPP STE 320 Z
Krupp Stahl AG
C 0-0.23, Mn 0-1, P 0-0.04, S 0-0.04, 0.02 Al min, bal Fe.
DIN 17162 T.2/88; W. Nr. 1.0250.

KRUPP STE 320-3 Z
Krupp Stahl AG
C 0-0.23, Mn 0-1, P 0-0.04, S 0-0.04, 0.02 Al min, bal Fe.
DIN 17162T.2/80; W. Nr. 1.0250.

KRUPP STE 320.7
Krupp Stahl AG
C 0-0.22, Si 0-0.45, Mn 0.7-1.3, P 0-0.04, S 0-0.035, bal Fe.
DIN 17172/78; W. Nr. 1.0409.

KRUPP STE 320.7 TM
Krupp Stahl AG
C 0.04-0.12, Si 0-0.4, Mn 0.7-1.5, P 0-0.035, S 0-0.025, V 0-0.12, bal Fe.
DIN 17172/78; W. Nr. 1.0430.

KRUPP STE 350 Z
Krupp Stahl AG
C 0-0.25, Mn 0-1.5, P 0-0.04, S 0-0.04, 0.02 min Al, bal Fe.
DIN 17162 T.2/88; W. Nr. 1.0529.

KRUPP STE 350-3 Z
Krupp Stahl AG
C 0-0.25, Mn 0-1.5, P 0-0.04, S 0-0.04, 0.02 Al min, bal Fe.
DIN 17162 T.2/80; W. Nr. 1.0529.

KRUPP STE 355
Krupp Stahl AG
C 0-0.2, Si 0.1-0.5, Mn 0.9-1.65, P 0-0.035, S 0-0.03, Cr 0-0.3, Mo 0-0.08, Ni 0-0.3, Cu 0-0.2, N 0-0.02, Nb 0-0.05, V 0-0.1, 0.02 Al min, bal Fe.
DIN 17102/83; W. Nr. 1.0562.

KRUPP STE 360.7
Krupp Stahl AG
C 0-0.22, Si 0-0.55, Mn 0.9-1.5, P 0-0.04, S 0-0.035, V 0-0.12, 0.02 Al min, bal Fe.
DIN 17172/78; W. Nr. 1.0582.

KRUPP STE 360.7 TM
Krupp Stahl AG
C 0.04-0.12, Si 0-0.45, Mn 0.9-1.5, P 0-0.035, S 0-0.025, V 0-0.12, bal Fe.
DIN 17172/78; W. Nr. 1.0578.

KRUPP STE 380
Krupp Stahl AG
C 0-0.2, Si 0.1-0.6, Mn 1-1.7, P 0-0.035, S 0-0.03, Cr 0-0.3, Mo 0-0.08, Ni 0-1, Cu 0-0.2, N 0-0.02, Nb 0-0.05, V 0-0.2, 0.02 Al min, bal Fe.
DIN 17102/83; W. Nr. 1.8900.

KRUPP STE 385.7
Krupp Stahl AG
C 0-0.23, Si 0-0.55, Mn 1-1.5, P 0-0.04, S 0-0.035, V 0-0.12, 0.02 Al min, bal Fe.
DIN 17172/78; W. Nr. 1.8970.

KRUPP STE 385.7 TM
Krupp Stahl AG
C 0.04-0.14, Si 0-0.45, Mn 1-1.6, P 0-0.035, S 0-0.025, V 0-0.12, 0.02 Al min, bal Fe.
DIN 17172/78; W. Nr. 1.8971.

KRUPP STE 415.7
Krupp Stahl AG
C 0-0.23, Si 0-0.55, Mn 1-1.5, P 0-0.04, S 0-0.035, V 0-0.12, 0.02 Al min, bal Fe.
DIN 17172/78; W. Nr. 1.8972.

KRUPP STE 415.7 TM
Krupp Stahl AG
C 0.04-0.14, Si 0-0.45, Mn 1-1.6, P 0-0.035, S 0-0.025, V 0-0.12, 0.02 Al min, bal Fe.
DIN 17172/78; W. Nr. 1.8973.

KRUPP STE 420
Krupp Stahl AG
C 0-0.2, Si 0.1-0.6, Mn 1-1.7, P 0-0.035, S 0-0.03, Cr 0-0.3, Mo 0-0.1, Ni 0-1, Cu 0-0.2, N 0-0.02, Nb 0-0.05, V 0-0.2, 0.02 Al min, bal Fe.
DIN 17102/83; W. Nr. 1.8902.

KRUPP STE 445.7 TM
Krupp Stahl AG
C 0.04-0.16, Si 0-0.55, Mn 1-1.6, P 0-0.035, S 0-0.025, V 0-0.12, 0.02 Al min, bal Fe.
DIN 17172/78; W. Nr. 1.8975.

KRUPP STE 460
Krupp Stahl AG
C 0-0.2, Si 0.1-0.6, Mn 1-1.7, P 0-0.035, S 0-0.03, Cr 0-0.3, Mo 0-0.1, Ni 0-1, Cu 0-0.2, N 0-0.02, Nb 0-0.05, V 0-0.2, 0.02 Al min, bal Fe.
DIN 17102/83; W. Nr. 1.8905.

KRUPP STE 47
Krupp Stahl AG
alloy steel.
See STE460

KRUPP STE 480.7 TM
Krupp Stahl AG
C 0.04-0.16, Si 0-0.55, Mn 1.1-1.7, P 0-0.035, S 0-0.025, V 0-0.12, 0.02 Al min, bal Fe.
DIN 17172/78; W. Nr. 1.8977.

KRUPP STE 500
Krupp Stahl AG
C 0-0.21, Si 0.1-0.6, Mn 1-1.7, P 0-0.035, S 0-0.03, Cr 0-0.3, Mo 0-0.1, Ni 0-1, Cu 0-0.2, N 0-0.02, Nb 0-0.05, V 0-0.22, 0.02 Al min, bal Fe.
DIN 17102/83; W. Nr. 1.8907.

KRUPP STE 690 V
Krupp Stahl AG
alloy steel. C 0.15, Mn 0.8, Ni 0.85, Cr 0.55, Mo 0.5, V, Cu, B, bal Fe.
Steel for case hardening. W.Nr. 1.8920. *Obsolete*

KRUPP STSCH 600
Krupp Stahl AG
C 0.3-0.5, Si 0.05-0.35, Mn 0.7-1.2, P 0-0.04, S 0-0.04, bal Fe.
SEL 90; W. Nr. 1.0544.

KRUPP STSCH 700
Krupp Stahl AG
C 0.4-0.6, Si 0.05-0.35, Mn 0.8-1.25, P 0-0.05, S 0-0.05, bal Fe.
UIC 860 V/86; W. Nr. 1.0521.

KRUPP STSCH 800
Krupp Stahl AG
C 0.45-0.65, Si 0.1-0.5, Mn 0.8-1.3, P 0-0.04, S 0-0.04, bal Fe.
SEL 90; W. Nr. 1.0524.

KRUPP STSCH 900 A
Krupp Stahl AG
C 0.6-0.8, Si 0.1-0.5, Mn 0.8-1.3, P 0-0.04, S 0-0.04, bal Fe.
UIC 860 V/86; W. Nr. 1.0623.

KRUPP STSCH 900 B
Krupp Stahl AG
C 0.55-0.75, Si 0.1-0.5, Mn 1.3-1.7, P 0-0.04, S 0-0.04, bal Fe.
UIC 86 V/86; W. Nr. 1.0624.

KRUPP STSP 37
Krupp Stahl AG
C 0-0.2, Si 0-0.3, Mn 0.2-0.5, P 0-0.08, S 0-0.05, bal Fe.
SEL 90; W. Nr. 1.0021.

KRUPP STSP 45
Krupp Stahl AG
C 0-0.25, Si 0-0.3, Mn 0.25-0.6, P 0-0.08, S 0-0.05, bal Fe.
SEL 90; W. Nr. 1.0023.

KRUPP STSP S
Krupp Stahl AG
C 0-0.22, Si 0-0.6, Mn 0-1.5, P 0-0.06, S 0-0.05, bal Fe.
SEL 90; W. Nr. 1.0083.

KRUPP STW 22
Krupp Stahl AG
C 0-0.1, N 0-0.007, bal Fe.
DIN 1614 T.2/86; W. Nr. 1.0332.

KRUPP STW 24
Krupp Stahl AG
C 0-0.08, 0.02 Al min, bal Fe.
DIN 1614 T.2/86; W. Nr. 1.0335.

KRUPP T 13 A
Krupp Stahl AG
stainless steel. C 0.22, Cr 14, S, bal Fe.
Martensitic type stainless steel; hardenable to above 150,000
psi TS. *Obsolete*

KRUPP TNC 18 + MO
Krupp Stahl AG
alloy steel. C 0.35, Ni 4.5, Cr 1.2, Mo, bal Fe.
For axles, shafts, truck and tractor parts. *Obsolete*

KRUPP TSTE 255
Krupp Stahl AG
C 0-0.16, Si 0-0.4, Mn 0.5-1.3, P 0-0.03, S 0-0.025, Cr 0-0.3,
Mo 0-0.08, Ni 0-0.3, Cu 0-0.2, N 0-0.02, Nb 0-0.03, 0.02 Al
min, bal Fe.
DIN 17102/83; W. Nr. 1.0463.

KRUPP TSTE 285
Krupp Stahl AG
C 0-0.16, Si 0-0.4, Mn 0.6-1.4, P 0-0.03, S 0-0.025, Cr 0-0.3,
Mo 0-0.08, Ni 0-0.3, Cu 0-0.2, N 0-0.02, Nb 0-0.03, 0.02 Al
min, bal Fe.
DIN 17102/83; W. Nr. 1.0488.

KRUPP TSTE 315
Krupp Stahl AG
C 0-0.16, Si 0-0.45, Mn 0.7-1.5, P 0-0.03, S 0-0.025, Cr 0-0.3,
Mo 0-0.08, Ni 0-0.3, Cu 0-0.2, N 0-0.02, Nb 0-0.03, 0.02 Al
min, bal Fe.
DIN 17102/83; W. Nr. 1.0508.

KRUPP TSTE 355
Krupp Stahl AG
C 0-0.18, Si 0.1-0.3, Mn 0.9-1.65, P 0-0.03, S 0-0.025, Cr
0-0.3, Mo 0-0.08, Ni 0-0.3, Cu 0-0.2, N 0-0.02, Nb 0-0.05, V
0-0.1, 0.02 Al min, bal Fe.
DIN 17102/83; W. Nr. 1.0566.

KRUPP TSTE 380
Krupp Stahl AG
C 0-0.2, Si 0.1-0.6, Mn 1-1.7, P 0-0.03, S 0-0.025, Cr 0-0.3,
Mo 0-0.08, Ni 0-1, Cu 0-0.2, N 0-0.02, Nb 0-0.05, V 0-0.2, 0.02
Al min, bal Fe.
DIN 17102/83; W. Nr. 1.8910.

KRUPP TSTE 420
Krupp Stahl AG
C 0-0.2, Si 0.1-0.6, Mn 1-1.7, P 0-0.03, S 0-0.025, Cr 0-0.3,
Mo 0-0.1, Ni 0-1, Cu 0-0.2, N 0-0.02, Nb 0-0.05, V 0-0.2, 0.02
Al min, bal Fe.
DIN 17102/83; W. Nr. 1.8912.

KRUPP TSTE 460
Krupp Stahl AG
C 0-0.2, Si 0.1-0.6, Mn 1-1.7, P 0-0.03, S 0-0.025, Cr 0-0.3,
Mo 0-0.1, Ni 0-1, Cu 0-0.2, N 0-0.02, Nb 0-0.05, V 0-0.2, 0.02
Al min, bal Fe.
DIN 17102/83; W. Nr. 1.8915.

KRUPP TSTE 500
Krupp Stahl AG
C 0-0.21, Si 0.1-0.6, Mn 1-1.7, P 0-0.03, S 0-0.025, Cr 0-0.3,
Mo 0-0.1, Ni 0-1, Cu 0-0.2, N 0-0.02, Nb 0-0.05, V 0-0.22, 0.02
Al min, bal Fe.
DIN 17102/83; W. Nr. 1.8917.

KRUPP TSTE 690 V
Krupp Stahl AG
C 0-0.2, Si 0.15-0.8, Mn 0-1.7, P 0-0.025, S 0-0.025, Cr 0-1,
Mo 0-0.6, Ni 0-1.5, B 0-0.005, Cu 0-0.5, Nb 0-0.05, V 0-0.12,
bal Fe.
SEL 90; W. Nr. 1.8928.

KRUPP TSTE 690 V A
Krupp Stahl AG
C 0.1-0.2, Si 0.15-0.35, Mn 0.6-1, P 0-0.025, S 0-0.025, Cr
0.4-0.65, Mo 0.4-0.6, Ni 0.7-1, Cu 0.15-0.5, V 0.03-0.08, 0.02
Al min, 0.0005 B min, bal Fe.
SEL 90; W. Nr. 1.8920.

KRUPP TSTE 690 V B
Krupp Stahl AG
C 0.12-0.21, Si 0.2-0.35, Mn 0.7-1, P 0-0.025, S 0-0.025, Cr
0.4-0.65, Mo 0.15-0.25, Cu 0.2-0.4, Ti 0.01-0.03, V 0.03-0.08,
0.02 Al min, 0.0005 B min, bal Fe.
SEL 90; W. Nr. 1.8921.

KRUPP TSTE 690 V C
Krupp Stahl AG
C 0.12-0.21, Si 0.2-0.35, Mn 0.95-1.3, P 0-0.025, S 0-0.025, Cr
0.4-0.65, Mo 0.2-0.3, Ni 0.3-0.7, V 0.03-0.08, 0.02 Al min,
0.0005 B min, bal Fe.
SEL 90; W. Nr. 1.8922.

KRUPP TT 17
Krupp Stahl AG
stainless steel. C 0.16, Cr 17, Ni, bal Fe.
Chromium stainless steel. *Obsolete*

KRUPP TT 171 N
Krupp Stahl AG
stainless steel. C 0.03, Cr 16, Ni 6, Mo, bal Fe.
Austenitic stainless steel; for cold drawn springs. *Obsolete*

KRUPP TTST 35
Krupp Stahl AG
C 0-0.17, Si 0-0.35, P 0-0.03, S 0-0.025, 0.40 Mn min, 0.02
min, bal Fe.
SEL 90; W. Nr. 1.0356.

KRUPP TTST 41
Krupp Stahl AG
C 0-0.2, Si 0-0.35, P 0-0.045, S 0-0.045, 0.45 Mn min, bal Fe.
SEL 81; W. Nr. 1.0437.

KRUPP TTST 45
Krupp Stahl AG
C 0-0.22, Si 0-0.35, P 0-0.045, S 0-0.045, N 0-0.009, 0.45 Mn
min, bal Fe.
SEL 90; W. Nr. 1.0456.

KRUPP TTSTE 47
Krupp Stahl AG
alloy steel. C 0.2, Mn 1.4, Ni 0.5, Cu 0.4, V 0.15, bal Fe.
Fine grain structural steel. W. Nr. 1.8915. *Obsolete*

KRUPP TYPE METAL
German manufacture
Pb 60, Sb 18, Sn 12, Cu 4.7, Ni 4.7, Bi 1.
For type metal.

KRUPP U 10 S 10
Krupp Stahl AG
C 0-0.15, Si 0-0.01, Mn 0.3-0.6, P 0-0.05, S 0.08-0.12, bal Fe.
DIN 17111/80; W. Nr. 1.0702.

KRUPP U 10 S 6
Krupp Stahl AG
C 0-0.15, Si 0-0.01, Mn 0.3-0.6, P 0-0.08, S 0.03-0.08, bal Fe.
SEL 75; W. Nr. 1.0706.

KRUPP U 7 S 10
Krupp Stahl AG
C 0-0.1, Si 0-0.01, Mn 0.4-0.7, P 0-0.08, S 0.08-0.12, bal Fe.
SEL 75; W. Nr. 1.0700.

KRUPP UH I
Krupp Stahl AG
C 0-0.14, Mn 0.2-0.8, P 0-0.035, S 0-0.03, bal Fe.
DIN 17155/83; W. Nr. 1.0348.

KRUPP UK17F EXTRA
Krupp Stahl AG
stainless steel. C 0-0.1, Cr 17, Mo 1.8, Ti = 7 x C, bal Fe.
For welded oil refinery and chemical plant equipment;
corrosion resistant. *Obsolete*

KRUPP UKST 37-2
Krupp Stahl AG
C 0-0.17, P 0-0.05, S 0-0.05, N 0-0.007, bal Fe.
DIN 17100/80; W. Nr. 1.0124.

KRUPP UPST 34-2
Krupp Stahl AG
C 0-0.15, Mn 0.2-0.5, P 0-0.05, S 0-0.05, N 0-0.007, bal Fe.
SEL 75; W. Nr. 1.0177.

KRUPP UPST 37-2
Krupp Stahl AG
C 0-0.17, P 0-0.05, S 0-0.05, N 0-0.007, bal Fe.
DIN 17100/80; W. Nr. 1.0160.

KRUPP UQST 34-2
Krupp Stahl AG
C 0-0.15, Si 0-0.01, Mn 0.2-0.5, P 0-0.05, S 0-0.05, N 0-0.007,
bal Fe.
SEL 75; W. Nr. 1.0104.

KRUPP UQST 36
Krupp Stahl AG
C 0-0.14, Si 0-0.01, Mn 0.25-0.5, P 0-0.04, S 0-0.04, bal Fe.
DIN 17111/80; W. Nr. 1.0204.

KRUPP UQST 37-2
Krupp Stahl AG
C 0-0.17, P 0-0.05, S 0-0.05, N 0-0.007, bal Fe.
DIN 17100/80; W. Nr. 1.0121.

KRUPP UQST 37-2 CU 3
Krupp Stahl AG
C 0-0.17, P 0-0.05, S 0-0.05, Cu 0.25-0.35, N 0-0.007, bal Fe.
DIN 17100/80; W. Nr. 1.0164.

KRUPP UQST 38
Krupp Stahl AG
C 0-0.19, Si 0-0.01, Mn 0.25-0.5, P 0-0.04, S 0-0.04, bal Fe.
DIN 17111/80; W. Nr. 1.0224

KRUPP UROST 37-2
Krupp Stahl AG
C 0-0.17, P 0-0.05, S 0-0.05, N 0-0.007, bal Fe.
DIN 17100/80; W. Nr. 1.0173.

KRUPP USD 5
Krupp Stahl AG
C 0.06-0.1, Si 0-0.01, Mn 0.45-0.65, P 0-0.01, S 0-0.01, Cr
0-0.12, Ni 0-0.12, Cu 0-0.17, bal Fe.
DIN 17145/80; W. Nr. 1.1112.

KRUPP USD 6
Krupp Stahl AG
C 0.06-0.1, Si 0-0.01, Mn 0.45-0.65, P 0-0.02, S 0-0.02, Cr
0-0.12, Ni 0-0.12, Cu 0-0.17, bal Fe.
DIN 17145/80; W. Nr. 1.1116.

KRUPP USD 7
Krupp Stahl AG
C 0.06-0.1, Si 0-0.01, Mn 0.45-0.65, P 0-0.025, S 0-0.025, Cr
0-0.12, Ni 0-0.12, Cu 0-0.17, bal Fe.
DIN 17145/80; W. Nr. 1.0323.

KRUPP UST 13
Krupp Stahl AG
C 0-0.1, N 0-0.007, bal Fe.
DIN 1623 T.1/83; W. Nr. 1.0333.

KRUPP UST 23
Krupp Stahl AG
C 0-0.08, Mn 0.2-0.4, P 0-0.025, S 0-0.025, N 0-0.007, bal Fe.
DIN 1614 T.1/86; W. Nr. 1.0321.

KRUPP UST 34-2
Krupp Stahl AG
C 0-0.15, Si 0-0.01, Mn 0.2-0.5, P 0-0.05, S 0-0.05, N 0-0.007, bal Fe.
SEL 90; W. Nr. 1.0028.

KRUPP UST 35-2
Krupp Stahl AG
C 0.06-0.14, Si 0-0.01, Mn 0.4-0.6, P 0-0.035, S 0-0.035, Cu 0-0.25, N 0-0.012, bal Fe.
DIN 17115/87; W. Nr. 1.0207.

KRUPP UST 36
Krupp Stahl AG
C 0-0.14, Si 0-0.01, Mn 0.25-0.5, P 0-0.05, S 0-0.05, bal Fe.
DIN 17111/80; W. Nr. 1.0203.

KRUPP UST 37-2
Krupp Stahl AG
C 0-0.17, P 0-0.05, S 0-0.05, N 0-0.007, bal Fe.
DIN 17100/80; W. Nr. 1.0036.

KRUPP UST 37-2 CU 3
Krupp Stahl AG
C 0-0.18, Si 0-0.01, Mn 0.2-0.5, P 0-0.05, S 0-0.05, Cu 0.25-0.35, bal Fe.
SEL 75; W. Nr. 1.0162.

KRUPP UST 37-2 G
Krupp Stahl AG
C 0-0.17, P 0-0.04, S 0-0.035, N 0-0.007, bal.
DIN 1623T.2/86; W. Nr. 1.0036.

KRUPP UST 38
Krupp Stahl AG
C 0-0.19, Si 0-0.01, Mn 0.25-0.5, P 0-0.05, S 0-0.05, bal Fe.
DIN 17111/80; W. Nr. 1.0217.

KRUPP UST 4
Krupp Stahl AG
C 0-0.09, Si 0-0.01, Mn 0.25-0.5, P 0-0.03, S 0-0.03, N 0-0.007, bal Fe.
SEL 90; W. Nr. 1.0336.

KRUPP USTW 23
Krupp Stahl AG
C 0-0.1, N 0-0.007, bal Fe.
DIN 1614 T.2/86; W. Nr. 1.0334.

KRUPP UZST 34-2
Krupp Stahl AG
C 0-0.15, Si 0-0.01, Mn 0.2-0.5, P 0-0.05, S 0-0.05, bal Fe.
SEL 75; W. Nr. 1.0151.

KRUPP UZST 37-2
Krupp Stahl AG
C 0-0.17, P 0-0.05, S 0-0.05, N 0-0.007, bal Fe.
DIN 17100/80; W. Nr. 1.0161.

KRUPP V M STEEL
Krupp Stahl AG
stainless steel. Cr 11-15, low Ni, bal Fe.
For chemical equipment; corrosion resistant. *Obsolete*

KRUPP V10A
Krupp Stahl AG
stainless steel. C 0.12, Mn 8, Si 0.8, Cr 19.5, Ni 9, bal Fe.
Welded: 85,000 TS; 45,000 YS; 45 El; 45 RA. For welding rod; stainless, austenitic. *Obsolete*

KRUPP V12A NORMAL
Krupp Stahl AG
stainless steel. C 0-0.1, Cr 12.5, Ni 12, bal Fe.
For valves, cutlery; corrosion resistant. *Obsolete*

KRUPP V12A SUPRA
Krupp Stahl AG
stainless steel. C 0.08, Cr 18, Ni 8, bal Fe.
Annealed: 78,000 TS; 31,000 YS; 50 El; 60 RA; 140 Brin. For architectural trim, tanks, vessels, stainless, austenitic. *Obsolete*

KRUPP V13F
Krupp Stahl AG
stainless steel. C 0.12, Cr 13, Ni 0.2, bal Fe.
Heat treated: 150,000 TS; 20 El; 250 Brin. For tableware, instruments; Type 403; corrosion resistant. *Obsolete*

KRUPP V13FA1
Krupp Stahl AG
stainless steel. C 0.05, Cr 11.5-14.5, Al 0.1-0.3, bal Fe.
Annealed: 70,000 TS; 40,000 YS; 30 El; 160 Brin. For annealing boxes, quenching racks, furnace parts, oil refinery equipment. Type 405 stainless steel, heat and corrosion resistant. *Obsolete*

KRUPP V14A SUPRA
Krupp Stahl AG
stainless steel. C 0-0.1, Cr 18-20, Ni 10-14, Mo 2-3, bal Fe.
Annealed: 85,000 TS; 31,000 YS; 50 El; 60 RA; 140 Brin. For hypochlorite plant equipment; stainless, austenitic; Type 316. *Obsolete*

KRUPP V14AS
Krupp Stahl AG
stainless steel. C 0-0.1, Cr 18-20, Ni 11-14, Mo 3-4, bal Fe.
Annealed: 85,000-95,000 TS; 35,000-45,000 YS; 60-50 El; 75-60 RA; 150-190 Brin. For acid resistant chemical plant equipment; Type 317; stainless, austenitic. *Obsolete*

KRUPP V15F
Krupp Stahl AG
C 0.08, Cr, bal Fe.
For corrosion resistant parts. *Obsolete*

KRUPP V16 SUPRA
Krupp Stahl AG
stainless steel. C 0.15, Cr 18, Ni 8, bal Fe.
Rolled: 78,000-101,000 TS; 55-65 El; 140-175 Brin. For stainless parts, chemical plant equipment; austenitic, stainless. *Obsolete*

KRUPP V16A EXTRA
Krupp Stahl AG
stainless steel. C 0-0.07, Cr 17.5, Ni 17.5, Mo 2, Ti = 7 x C, bal Fe.
For valves, acid resistant equipment, pumps; stainless, austenitic. *Obsolete*

KRUPP V16A SUPRA
Krupp Stahl AG
stainless steel. C 0-0.1, Cr 18-20, Ni 10-14, Mo 3-4, bal Fe.
Annealed: 90,000 TS; 35,000 YS; 50 El; 60 RA; 160 Brin. For chemical plant equipment; Type 317; stainless, austenitic. *Obsolete*

KRUPP V17F
Krupp Stahl AG
stainless steel. C 0.1, Cr 17, Ni 0.2, Ti 0.2, bal Fe.
Annealed: 78,000-93,000 TS; 50,000-55,000 YS; 20-18 El; 137-150 Brin. For chemical plant equipment; Type 430; corrosion resistant. *Obsolete*

KRUPP V17F EXTRA
Krupp Stahl AG
stainless steel.
See NIROSTA 4510

KRUPP V17FS
Krupp Stahl AG
stainless steel. C 0-0.12, Cr 14-18, 0.07 min S or Se, bal Fe.
Annealed: 75,000-85,000 TS; 40,000-55,000 YS; 30-25 El; 137-150 Brin. For chemical plant equipment, screw machine products; Type 430F; stainless, free-cutting. *Obsolete*

KRUPP V17FX EXTRA
Krupp Stahl AG
stainless steel. C 0.08, Cr 17, Ni 8, Cb 0.6, bal Fe.
Annealed: 85,000 TS; 40,000 YS; 50 El; B 82 Brin. For welded structures, agitators, evaporators, tanks, digesters, chemical plant equipment. Type 17-8 Cb stainless steel, austenitic, stabilized. *Obsolete*

KRUPP V1M
Krupp Stahl AG
stainless steel. C 0.15, Ni 2, Cr 15, Mn 0.4, Si 0.5, bal Fe.
Annealed: 85,000 TS; 55,000 YS; 30 El; 150 Brin. Heat treated: 213,000 TS; 160,000 YS; 450 Brin. For furnace equipment, cutlery, turbine parts; Type 431; corrosion resistant. *Obsolete*

KRUPP V2A
Krupp Stahl AG
stainless steel. Cr 23, Ni 9.5, C 0.4, bal Fe.
85,000-105,000 TS; 35,000 YS; 50 El; 150-200 Brin. For heat and corrosion resisting parts, chemical equipment; heat and corrosion resistant; austenitic. *Obsolete*

KRUPP V2A EXTRA
Krupp Stahl AG
stainless steel. C 0.1, Cr 18, Ni 8-9.5, Ti = 4 x C, bal Fe.
Annealed: 83,000 TS; 29,000 YS; 45 El; 55 RA. For crucibles, autoclaves, chemical plant equipment; Type 321; stainless, Austenitic. *Obsolete*

KRUPP V2A NIROSTA
Krupp Stahl AG
stainless steel. C 0.2, Cr 20, Ni 7, bal Fe.
Annealed: 85,000 TS; 35,000 YS; 60 El; 70 RA; 150 Brin. Rolled: 125,000 TS; 95,000 YS; 25 El; 55 RA; 277 Brin. For oil refinery and chemical plant equipment; Type 302; stainless, austenitic. *Obsolete*

KRUPP V2A NORMAL
Krupp Stahl AG
stainless steel. C 0.11-0.16, Cr 18, Ni 8, bal Fe.
For chemical plant equipment, mixers, Type 302; stainless, austenitic. *Obsolete*

KRUPP V2A SPECIAL
Krupp Stahl AG
stainless steel. C 0.08, Cr 18, Ni 8, bal Fe.
Annealed: 90,000 TS; 40,000 YS; 40 El; 60 RA; 160 Brin. For chemical plant equipment, architectural trim; Type 304; stainless, austenitic. *Obsolete*

KRUPP V2A SUPRA
Krupp Stahl AG
stainless steel.
See NIROSTA 4307

KRUPP V2AB SUPRA
Krupp Stahl AG
stainless steel.
See NIROSTA 4300

KRUPP V2AE
Krupp Stahl AG
stainless steel.
See NIROSTA 4541

KRUPP V2AS
Krupp Stahl AG
stainless steel. C 0-0.08, Cr 18-20, Ni 8-11, Mn 0-2, bal Fe.
Annealed: 80,000 TS; 45,000 YS; 60 El; 135 Brin. Cold drawn: 180,000 TS; 150,000 YS; 10 El; 330 Brin. For chemical plant equipment, welded structures; Type 304; stainless, austenitic. *Obsolete*

KRUPP V2AX EXTRA
Krupp Stahl AG
stainless steel.
See NIROSTA 4550

KRUPP V3M

Krupp Stahl AG
stainless steel. C 0.4, Cr 14, Ni 0.5, bal Fe.
Heat treated: 256,000 TS; 190,000 YS; 6 El; 10 RA; 540 Brin.
For cutlery, razors, knives; Type 420; corrosion resistant.
Obsolete

KRUPP V3M EXTRA

Krupp Stahl AG
tool material. C 1, Cr 12-18, bal Fe.
Annealed: 210 Brin. Heat treated: 620 Brin. For ball and roller
bearings, disks for meat grinders; corrosion resistant.
Obsolete

KRUPP V3ME

Krupp Stahl AG
stainless steel. C 0.75-0.95, Cr 16-18, Mo 0-0.75, bal Fe.
Annealed: 107,000 TS; 18 El; 35 RA; 220 Brin. Heat treated:
280,000 TS; 270,000 YS; 3 El; 15 RA; 555 Brin. For cutlery,
valves, bearings, instruments; Type 440B; corrosion resistant.
Obsolete

KRUPP V3MS

Krupp Stahl AG
stainless steel. C 0-0.15, Cr 12-14, Ni 1.25-2.5, .07 min P, S,
or Se, bal Fe.
Annealed: 75,000 TS; 40,000 YS; 30 El; 60 RA; 155 Brin. Cold
drawn: 100,000 TS; 85,000 YS; 13 El; 50 RA; 205 Brin. For
screw machine products, valve trim, pump shafts; Type 416;
free-cutting, stainless. *Obsolete*

KRUPP V4 AE-TI

Krupp Stahl AG
stainless steel. C 0-0.1, Cr 16-18, Ni 10-14, Mo 1.7-2.7, Cb =
10 x C, bal Fe.
Annealed: 85,000-95,000 TS; 35,000-45,000 YS; 60-50 El;
75-60 RA; 150-190 Brin. For acid resistant chemical plant
equipment; Type 316 Cb; stainless, austenitic. *Obsolete*

KRUPP V11A SUPRA

Krupp Stahl AG
stainless steel. C 0.06, Cr 17, Ni 12, Mo 2-3, bal Fe.
Annealed: 85,000 TS; 35,000 YP; 50 El; 80 Rock B. For
chemical plant equipment, tanks, evaporators, digesters.
Type 316 stainless steel, austenitic.

KRUPP V4A

Krupp Stahl AG
stainless steel. C 0-0.07, Ni 10.5, Cr 18, Mo 2.5, bal Fe.
Rolled: 85,000-108,000 TS; 36,000-70,000 YS; 50-45 El;
150-185 Brin. For chemical and textile plant equipment; Type
316; stainless, austenitic. *Obsolete*

KRUPP V4A EXTRA

Krupp Stahl AG
stainless steel. C 0-0.12, Cr 18, Ni 10, Mo 2, Ti = 4 x C, bal
Fe.
Annealed: 85,000 TS; 39,000 YS; 50 El; 60 RA; 160 Brin. For
sulphide, paper and pulp and textile industries; stainless,
austenitic; Type 316 Ti. *Obsolete*

KRUPP V4A SUPRA

Krupp Stahl AG
stainless steel. C 0-0.12, Cr 18, Ni 8, Mo 2.5, bal Fe.
Annealed: 78,000 TS; 32,000 YS; 50 El; 60 RA; 140 Brin. For
sulphide, paper and pulp industries, textile equipment; Type
316; stainless, austenitic. *Obsolete*

KRUPP V4AB SUPRA

Krupp Stahl AG
stainless steel. C 0.1, Cr 18, Ni 9.5, Mo 2, bal Fe.
Annealed: 85,000 TS; 35,000 YS; 50 El; 65 RA; 160 Brin. For
acid resistant chemical plant equipment; Type 316; stainless,
austenitic. *Obsolete*

KRUPP V4AS

Krupp Stahl AG
stainless steel.
See NIROSTA 4401

KRUPP V4AX EXTRA

Krupp Stahl AG
stainless steel. C 0-0.12, Cr 18, Mo 2, Ni 10.5, Cb = 8 x C, bal
Fe.
Annealed: 85,000 TS; 35,000 YS; 45 El; 60 RA; 165 Brin. For
welded acid resistant chemical plant equipment; Type 318;
stainless, austenitic. *Obsolete*

KRUPP V5M

Krupp Stahl AG
stainless steel.
See NIROSTA 4021

KRUPP V6A STEEL

Krupp Stahl AG
stainless steel. Cr 20, Ni 7, C 0.2, Cu 2.5, bal Fe.
For chemical equipment; corrosion resistant; resists NH_4Cl.
Obsolete

KRUPP V8A EXTRA

Krupp Stahl AG
stainless steel. C 0-0.12, Cr 18, Ni 10.5, Mo 2, Ti = 4 x C, bal
Fe.
Annealed: 85,000 TS; 35,000 YS; 45 El; 60 RA; 165 Brin. For
welded acid resistant chemical plant equipment; Type 316 Ti;
stainless, austenitic. *Obsolete*

KRUPP VCM 21-10

Krupp Stahl AG
stainless steel. C 0.6, Cr 21, Mn 10, Nb, Mo, V, bal Fe.
For Exhaust valves. W.Nr. 1.4785

KRUPP VCMO 18 3

Krupp Stahl AG
C 0.8-0.9, Si 0-1, Mn 0-1.5, P 0-0.04, S 0-0.03, Cr 16.5-18.5,
Mo 2-2.5, V 0.3-0.6, bal Fe.
SEL 90; W. Nr. 1.4748.

KRUPP VCMO 10 3

Krupp Stahl AG
stainless steel. C 0.85, Cr 17.5, Mo 2.25, V 0.5, bal Fe.
For valves. W. Nr. 1.4748.

KRUPP VCN 188

Krupp Stahl AG
C 0.4-0.5, Si 2-3, Mn 0.8-1.5, P 0-0.045, S 0-0.03, Cr 17-19, Ni
8-10, W 0.8-1.2, bal Fe.
SEL 81; W. Nr. 1.4873.

KRUPP VCN 21-2

Krupp Stahl AG
stainless steel. C 0.55, Mn 8, Cr 21, Ni 2, N 0.3, bal Fe.
For automotive exhaust valves. W.Nr. 1.4785.

KRUPP VCN 21-4

Krupp Stahl AG
stainless steel. C 0.53, Cr 21, Ni 4, Mn 9, N 0.5, bal Fe.
For exhaust valves. W.Nr. 1.4871.

KRUPP VCN 21-4 ZNW

Krupp Stahl AG
C 0.45-0.55, Si 0-0.45, Mn 8-10, P 0-0.05, S 0-0.03, Cr 20-22,
Ni 3.5-5, N 0.4-0.6, Nb 1.8-2.5, W 0.8-1.5, bal Fe.
DIN 17480/84; W. Nr. 1.4882.

KRUPP VCN 21-7

Krupp Stahl AG
stainless steel. C 0.7, Mn 6.5, Cr 21, Ni 1.7, 0.23 N bal Fe.
For exhaust valves. W.Nr. 1.4881. *Obsolete*

KRUPP VCS 10

Krupp Stahl AG
C 0.35-0.45, Si 2-3, Mn 0-0.8, P 0-0.04, S 0-0.03, Cr 9-11, Mo
0.8-1.3, bal Fe.
DIN 17480/84; W. Nr. 1.4731.

KRUPP VCS 15

Krupp Stahl AG
stainless steel. C 0.8, Cr 15, Mo 1, Ni 0.75, W 1, bal Fe.
For automotive valves. W.Nr. 1.4732.

KRUPP VCS 2

Krupp Stahl AG
alloy steel. C 0.45, Si 4, Cr 3, bal Fe.
For valves. W.Nr. 1.4704. *Obsolete*

KRUPP VCS 20

Krupp Stahl AG
stainless steel. C 0.8, Si 2.25, Cr 20, Ni 1.5, bal Fe.
For automotive exhaust valves. W.Nr. 1.4747. *Obsolete*

KRUPP VCS 9

Krupp Stahl AG
alloy steel. C 0.45, Si 3.25, Cr 9, bal Fe.
For exhaust valves. W.Nr. 1.4718

KRUPP VK3M EXTRA

Krupp Stahl AG
tool material. C 0.9, Mo, Cr, V, W, bal Fe.
For lathe and planer tools, reamers; high speed steel.
Obsolete

KRUPP VK5M

Krupp Stahl AG
stainless steel.
See NIROSTA 4120

KRUPP W 53

Krupp Stahl AG
tool material. C 0.53, Si 0.3, Mn 0.7, bal Fe.
Tool steel for hand tools as hammers, shears, screwdrivers.
Obsolete

KRUPP W 6

Krupp Stahl AG
tool material. C 1, C 1.75, Mo 1.25, bal Fe.
For shears, chisels, knives. *Obsolete*

KRUPP W 93

Krupp Stahl AG
tool material. C 0.9, Mn 0.7, bal Fe.
For knives, hand saws, shears. *Obsolete*

KRUPP WA

Krupp Stahl AG
tool material. C 0.4, Cr 1.2, Mo 0.4, V 0.75, W 0.4, bal Fe.
For tools, gears, shafts; water hardening. *Obsolete*

KRUPP WA 342

Krupp Stahl AG
tool material. C 0.45, Mn 1, Cr 1.7-2, Mo 0.2, bal Fe.
For forging die blocks; oil hardened. *Obsolete*

KRUPP WA100

Krupp Stahl AG
stainless steel. C 0.43, Si 2.3, Mn 1, Cr 17.5, Ni 8.5, Ti 0.9, bal
Fe.
For stainless parts; corrosion resistant, austenitic. *Obsolete*

KRUPP WA402

Krupp Stahl AG
tool material. C 0.45-0.5, Mn 0.9-1.1, Cr 1.1-1.3, Mo 0.2, bal
Fe.
For forging die blocks; oil hardened. *Obsolete*

KRUPP WA5 M

Krupp Stahl AG
stainless steel.
See KRUPP 2082

KRUPP WA594

Krupp Stahl AG
tool material. C 0.32-0.38, Si 0.8-1, Cr 2.2-2.5, V 0.4, bal Fe.
For dies, punches; shock resistant. *Obsolete*

KRUPP WA650

Krupp Stahl AG
tool material. C 0.3-0.38, Si 1.4-1.6, Cr 1.2-1.5, V 0.1-0.2, bal
Fe.
For tools, punches, dies; shock resistant. *Obsolete*

KRUPP WA904
Krupp Stahl AG
tool material. C 0.25-0.3, Cr 1-1.2, V 0.2, W 3.5-4, bal Fe.
For tools, hot work dies; hot work steel. *Obsolete*

KRUPP WA904 EXTRA
Krupp Stahl AG
tool material.
See KRUPP 2567

KRUPP WA930 N
Krupp Stahl AG
tool material. C 0.26, W 9, Cr 3, Ni, V, bal Fe.
For die casting molds. *Obsolete*

KRUPP WAGS
Krupp Stahl AG
tool material.
See Krupp 2343

KRUPP WAM03
Krupp Stahl AG
tool material.
See KRUPP 2365

KRUPP WB136 M
Krupp Stahl AG
tool material.
See KRUPP 65SI7

KRUPP WB97 M
Krupp Stahl AG
tool material. C 0.42-0.48, Si 1-1.3, Mn 0.5-0.7, bal Fe.
For tools, springs, punches; shock resistant. *Obsolete*

KRUPP WBC1255
Krupp Stahl AG
tool material. C 0.6-0.7, Si 1-1.3, Mn 0.9-1.2, bal Fe.
For tools, springs, punches; shock resistant. *Obsolete*

KRUPP WBF1552
Krupp Stahl AG
tool material. C 0.67-0.73, Si 1.2-1.4, Cr 0.4-0.55, bal Fe.
For tools, punches, dies; oil hardening. *Obsolete*

KRUPP WC 128
Krupp Stahl AG
tool material. C 0.7, Si 1.9, S 0.07, bal Fe.
For tools, chisels, dies; tough. *Obsolete*

KRUPP WCV 3
Krupp Stahl AG
tool material. C 0.48, Cr 0.9, V, bal Fe.
Cold work tool steel. Similar to AISI 6150. *Obsolete*

KRUPP WCV 4 SPEZIAL
Krupp Stahl AG
tool material. C 0.56, Cr 1, Mo, V, bal Fe.
For axes, chisels, hand tools. *Obsolete*

KRUPP WF 3
Krupp Stahl AG
tool material.
See KRUPP 2825-SO

KRUPP WF100
Krupp Stahl AG
tool material. C 0.45, Cr, Ni, W, bal Fe.
For forging and heading dies, punches; oil hardened, tough. *Obsolete*

KRUPP WF100 D
Krupp Stahl AG
stainless steel. C 0.38, Mn 0.52, Si 1.84, Cr 14.8, Ni 12.9, Mo 0.23, W 2.5, bal Fe.
For valves; high heat and corrosion resistance. *Obsolete*

KRUPP WF202
Krupp Stahl AG
tool material. C 1-1.15, Cr 0.3-0.5, bal Fe.
For tools, drills, taps; water hardening. *Obsolete*

KRUPP WF204
Krupp Stahl AG
tool material. C 1-1.1, Cr 0.8-1, bal Fe.
For tools, drills, taps; water hardening. *Obsolete*

KRUPP WF206
Krupp Stahl AG
tool material. C 0.95-1.05, Cr 1.25-1.5, bal Fe.
For tools, bearings; wear resistant. *Obsolete*

KRUPP WF33
Krupp Stahl AG
alloy steel.
See KRUPP 15CR3

KRUPP WF50
Krupp Stahl AG
tool material. C 0.55, Ni, Cr, W, bal Fe.
For forging and heading dies; shear blades; oil hardened, tough. *Obsolete*

KRUPP WFC27
Krupp Stahl AG
tool material.
See KRUPP 16MNCR5

KRUPP WFC29
Krupp Stahl AG
tool material.
See KRUPP 20MNCR5 *Obsolete*

KRUPP WFC31
Krupp Stahl AG
tool material. C 0.2-0.25, Mn 1.3-1.6, Cr 1.2-1.5, bal Fe.
For tools, bearings; case hardening. *Obsolete*

KRUPP WFF24
Krupp Stahl AG
stainless steel. C 0.15, Si 1-1.3, Cr 23-25, Al 2, bal Fe.
For corrosion resistant parts. *Obsolete*

KRUPP WFM1041
Krupp Stahl AG
alloy steel.
See KRUPP 50CRV4

KRUPP WFM1141
Krupp Stahl AG
tool material.
See KRUPP 58CRV4

KRUPP WFM48
Krupp Stahl AG
tool material.
See KRUPP 30CRMOV9

KRUPP WFM961
Krupp Stahl AG
alloy steel. C 0.43-0.48, Cr 1.3-1.5, V 0.2, bal Fe.
For tools, gears, springs; tough. *Obsolete*

KRUPP WFP12
Krupp Stahl AG
tool material. C 0.3-0.35, Cr 1.3-1.5, Al 1-1.2, bal Fe.
For tools, gears, shafts; oil hardening. *Obsolete*

KRUPP WK 143 C
Krupp Stahl AG
tool material. C 0.75, W 1.35, Ni, Mo, bal Fe.
Tool or mold steel. *Obsolete*

KRUPP WL 1.1144
Krupp Stahl AG
C 0.12-0.18, Si 0.15-0.3, Mn 0.3-0.6, P 0-0.035, S 0-0.035, bal Fe.
Aircraft material. WL 1.1144/78; W. Nr. 1.1144.

KRUPP WL 1.1174
Krupp Stahl AG
C 0.32-0.39, Si 0.15-0.35, Mn 0.5-0.8, P 0-0.035, S 0-0.035, bal Fe.
Aircraft material. WL 1.1174/78; W. Nr. 1.1174.

KRUPP WL 1.1194
Krupp Stahl AG
C 0.42-0.5, Si 0.15-0.35, Mn 0.5-0.8, P 0-0.035, S 0-0.035, bal Fe.
Aircraft material. WL 1.1194/78; W. Nr. 1.1194.

KRUPP WL 1.1654
Krupp Stahl AG
C 1-1.1, Si 0.1-0.3, Mn 0.1-0.35, P 0-0.03, S 0-0.03, bal Fe.
Aircraft material. WL 1.1654/79; W. Nr. 1.1654.

KRUPP WL 1.3544
Krupp Stahl AG
C 0.95-1.1, Si 0-1, Mn 0-1, P 0-0.025, S 0-0.015, Cr 16-18, Mo 0.35-0.75, Ni 0-0.5, bal Fe.
Aircraft material. WL 1.3544/79; W. Nr. 1.3544.

KRUPP WL 1.3554
Krupp Stahl AG
C 0.78-0.86, Si 0-0.4, Mn 0-0.4, P 0-0.015, S 0-0.015, Cr 3.8-4.5, Mo 4.7-5.2, V 1.7-2, W 6-6.7, bal Fe.
Aircraft material. WL 1.3554/79; W. Nr. 1.3554.

KRUPP WL 1.4014
Krupp Stahl AG
C 0.17-0.22, Si 0-1, Mn 0-1, P 0-0.035, S 0-0.03, Cr 12-14, Ni 0-1, bal Fe.
Aircraft material. WL 1.4014/81; W. Nr. 1.4014.

KRUPP WL 1.4044
Krupp Stahl AG
C 0.12-0.2, Si 0-1, Mn 0-1, P 0-0.035, S 0-0.025, Cr 15-18, Ni 2-3, bal Fe.
Aircraft material. WL 1.4044/81; W. Nr. 1.4044.

KRUPP WL 1.4314
Krupp Stahl AG
C 0-0.07, Si 0-1, Mn 0-2, P 0-0.045, S 0-0.03, Cr 17-19, Ni 8.5-10.5, bal Fe.
Aircraft material. WL 1.4314/77; W. Nr. 1.4314.

KRUPP WL 1.4324
Krupp Stahl AG
C 0-0.12, Si 0-1, Mn 0-2, P 0-0.045, S 0-0.03, Cr 16-18, Ni 7-9, bal Fe.
Aircraft material. WL 1.4324/76; W. Nr. 1.4324.

KRUPP WL 1.4544
Krupp Stahl AG
C 0-0.08, Si 0-1, Mn 0-2, P 0-0.035, S 0-0.025, Cr 17-19, Ni 9-11.5, Ti 0-0.6, bal Fe.
Aircraft material. WL 1.4544/82; W. Nr. 1.4544.

KRUPP WL 1.4545
Krupp Stahl AG
C 0-0.07, Si 0-1, Mn 0-1, P 0-0.03, S 0-0.015, Cr 14-15.5, Mo 0-0.5, Ni 3.5-5.5, Cu 2.5-4.5, Nb 0-0.45, bal Fe.
Aircraft material. WL 1.4545/89; W. Nr. 1.4545.

KRUPP WL 1.4546
Krupp Stahl AG
C 0-0.08, Si 0-1, Mn 0-2, P 0-0.045, S 0-0.03, Cr 17-19, Ni 9-11.5, Nb 0-1, bal Fe.
Aircraft material. WL 1.4546/82; W. Nr. 1.4546.

KRUPP WL 1.4548
Krupp Stahl AG
C 0-0.07, Si 0-1, Mn 0-1, P 0-0.025, S 0-0.025, Cr 15-17.5, Ni 3-5, Cu 3-5, Nb 0.15-0.45, bal Fe.
Aircraft material. WL 1.4548/89; W. Nr. 1.4548.

KRUPP WL 1.4564
Krupp Stahl AG
C 0-0.09, Si 0-1, Mn 0-1, P 0-0.04, S 0-0.03, Cr 16-18, Ni 6.5-7.75, Al 0.75-1.5, Cu 0-0.5, bal Fe.
Aircraft material. WL 1.4564/77; W. Nr. 1.4564.

KRUPP WL 1.4574
Krupp Stahl AG
C 0-0.09, Si 0-1, Mn 0-1, P 0-0.04, S 0-0.03, Cr 14-16, Mo 2-3, Ni 6.5-7.75, Al 0.75-1.5, bal Fe.
Aircraft material. WL 1.4574/77; W. Nr. 1.4574.

KRUPP WL 1.4911
Krupp Stahl AG
C 0.05-0.12, Si 0.1-0.8, Mn 0.2-1.35, P 0-0.025, S 0-0.02, Cr 9.8-11.5, Mo 0.5-1.1, Ni 0.2-1.2, B 0.005-0.015, Co 5-7, N 0-0.035, Nb 0.2-0.6, V 0.1-0.6, bal Fe.
Aircraft material. WL 1.4911/81; W. Nr. 1.4911.

KRUPP WL 1.4914
Krupp Stahl AG
C 0.11-0.19, Si 0.15-0.65, Mn 0.2-1.25, P 0-0.03, S 0-0.025, Cr 10-12, Mo 0.4-1, Ni 0.5-1.2, N 0.03-0.09, Nb 0.1-0.6, V 0.1-0.7, bal Fe.
Aircraft material. WL 1.4914/81; W. Nr. 1.4914.

KRUPP WL 1.4924
Krupp Stahl AG
C 0.2-0.25, Si 0.2-0.5, Mn 0.4-0.7, P 0-0.035, S 0-0.035, Cr 11-12, Mo 0.9-1.2, Ni 0.4-0.8, V 0.25-0.35, bal Fe.
Aircraft material. WL 1.4924/82; W. Nr. 1.4924.

KRUPP WL 1.4933
Krupp Stahl AG
C 0.08-0.13, Si 0-0.35, Mn 0.5-0.9, P 0-0.03, S 0-0.025, Cr 11-12.5, Mo 1.5-2, Ni 2-3, N 0.02-0.04, V 0.25-0.4, bal Fe.
Aircraft material. WL 1.4933/81; W. Nr. 1.4933.

KRUPP WL 1.4934
Krupp Stahl AG
C 0.18-0.26, Si 0.15-0.4, Mn 0.4-0.7, P 0-0.035, S 0-0.035, Cr 10.5-12.5, Mo 0.9-1.2, Ni 0.3-0.8, V 0.25-0.35, bal Fe.
Aircraft material. WL 1.4934/82; W. Nr. 1.4934.

KRUPP WL 1.4943
Krupp Stahl AG
C 0-0.06, Si 0-1, Mn 0-2, P 0-0.025, S 0-0.015, Cr 13.5-16, Mo 1-1.5, Ni 24-27, Al 0-0.35, B 0.003-0.01, Ti 1.7-2, V 0.1-0.5, bal Fe.
Aircraft material. WL 1.4943/81; W. Nr. 1.4943.

KRUPP WL 1.4944
Krupp Stahl AG
C 0-0.08, Si 0-1, Mn 0-2, P 0-0.025, S 0-0.015, Cr 13.5-16, Mo 1-1.5, Ni 24-27, Al 0-0.35, B 0.003-0.01, Ti 1.9-2.3, V 0.1-0.5, bal Fe.
Aircraft material. WL 1.4944/84; W. Nr. 1.4944.

KRUPP WL 1.4974
Krupp Stahl AG
C 0.08-0.16, Si 0-1, Mn 1-2, P 0-0.04, S 0-0.03, Cr 20-22.5, Mo 2.5-3.5, Ni 19-21, Co 18.5-21, N 0.1-0.2, Nb 0.75-1.25, W 2-3, bal Fe.
Aircraft material. WL 1.4974/82; W. Nr. 1.4974.

KRUPP WL 1.4984
Krupp Stahl AG
C 0.04-0.1, Si 0.3-0.6, Mn 0-1.5, P 0-0.045, S 0-0.03, Cr 15.5-17.5, Mo 1.6-2, Ni 15.5-17.5, Nb 0-1.2, bal Fe.
Aircraft material. WL 1.4984/78; W. Nr. 1.4984.

KRUPP WL 1.5924
Krupp Stahl AG
C 0.14-0.19, Si 0-0.04, Mn 0.4-0.6, P 0-0.035, S 0-0.035, Cr 1.4-1.7, Ni 1.4-1.7, bal Fe.
Aircraft material. WL 1.5924/E08.86; W. Nr. 1.5924.

KRUPP WL 1.5934
Krupp Stahl AG
C 0.15-0.2, Si 0.15-0.4, Mn 0.4-0.6, P 0-0.035, S 0-0.035, Cr 1.8-2.1, Ni 1.8-2.1, bal Fe.
WL 1.5934/76; W. Nr. 1.5934.

KRUPP WL 1.6354
Krupp Stahl AG
C 0-0.03, Si 0-0.1, Mn 0-0.1, P 0-0.01, S 0-0.01, Mo 4.6-5.2, Ni 17-19, Al 0.05-0.15, Co 8-9.5, Ti 0.6-0.9, bal Fe.
Aircraft material. WL 1.6354/81; W. Nr. 1.6354.

KRUPP WL 1.6359
Krupp Stahl AG
C 0-0.00, Si 0-0.1, Mn 0-0.1, P 0-0.01, S 0-0.01, Mo 4.6-5.2, Ni 17-19, Al 0.05-0.15, Co 7-8.5, Ti 0.3-0.6, bal Fe.
Aircraft material. WL 1.6359/81; W. Nr. 1.6359.

KRUPP WL 1.6604.
Krupp Stahl AG
C 0.26-0.34, Si 0.15-0.4, Mn 0.3-0.6, P 0-0.025, S 0-0.025, Cr 1.8-2.3, Mo 0.3-0.5, Ni 1.8-2.0, bal Fe.
Aircraft material. WL 1.6604/82; W. Nr. 1.6604.

KRUPP WL 1.6722
Krupp Stahl AG
C 0.13-0.18, Si 0.15-0.4, Mn 0.25-0.55, P 0-0.015, S 0-0.01, Cr 1-1.4, Mo 0.2-0.3, Ni 3.8-4.3, bal Fe.
Aircraft material. WL 1.6722/79; W. Nr. 1.6722.

KRUPP WL 1.6944
Krupp Stahl AG
C 0.35-0.4, Si 0.15-0.35, Mn 0.5-0.8, P 0-0.015, S 0-0.01, Cr 0.65-0.9, Mo 0.3-0.4, Ni 1.65-2, V 0.08-0.15, bal Fe.
Aircraft material. WL 1.6944/84; W. Nr. 1.6944.

KRUPP WL 1.6974
Krupp Stahl AG
C 0.29-0.34, Si 0-0.1, Mn 0.1-0.35, P 0-0.01, S 0-0.01, Cr 0.9-1.1, Mo 0.9-1.1, Ni 7-8, Co 4.25-4.75, V 0.06-0.12, bal Fe.
Aircraft material. WL 1.6974/77; W. Nr. 1.6974.

KRUPP WL 1.7214
Krupp Stahl AG
C 0.22-0.29, Si 0.15-0.35, Mn 0.5-0.8, P 0-0.02, S 0-0.015, Cr 0.9-1.2, Mo 0.15-0.25, Ni 0-0.3, bal Fe.
Aircraft material. WL 1.7214/82; W. Nr. 1.7214.

KRUPP WL 1.7334
Krupp Stahl AG
C 0.17-0.22, Si 0.15-0.4, Mn 0.6-0.9, P 0-0.035, S 0-0.035, Cr 0.3-0.5, Mo 0.4-0.5, bal Fe.
WL 1.7334/78; W. Nr. 1.7334.

KRUPP WL 1.7736
Krupp Stahl AG
C 0.12-0.18, Si 0-0.2, Mn 0.8-1.1, P 0-0.02, S 0-0.005, Cr 1.25-1.5, Mo 0.8-1, V 0.2-0.3, bal Fe.
Aircraft material. WL 1.7736/E86; W. Nr. 1.7736.

KRUPP WL 1.7784
Krupp Stahl AG
C 0.38-0.43, Si 0.8-1, Mn 0.2-0.4, P 0-0.015, S 0-0.01, Cr 4.75-5.25, Mo 1.2-1.4, V 0.4-0.6, bal Fe.
Aircraft material. WL 1.7784/76; W. Nr. 1.7784.

KRUPP WL 1.8154
Krupp Stahl AG
C 0.47-0.55, Si 0.15-0.4, Mn 0.7-1.1, P 0-0.035, S 0-0.035, Cr 0.9-1.2, V 0.1-0.2, bal Fe.
Aircraft material. WL 1.8154/81; W. Nr. 1.8154.

KRUPP WL 1.8514
Krupp Stahl AG
C 0.26-0.34, Si 0-0.4, Mn 0.4-0.7, P 0-0.025, S 0-0.025, Cr 2.3-2.7, Mo 0.15-0.25, V 0.1-0.2, bal Fe.
Aircraft material. WL 1.8514/81; W. Nr. 1.8514.

KRUPP WL 1.8544
Krupp Stahl AG
C 0.3-0.37, Si 0.2-0.5, Mn 0.5-0.8, P 0-0.025, S 0-0.025, Cr 1-1.3, Mo 0.15-0.25, Al 0.8-1.2, bal Fe.
Aircraft material. WL 1.8544/76; W. Nr. 1.8544.

KRUPP WL 1.8564
Krupp Stahl AG
C 0.28-0.35, Si 0.1-0.4, Mn 0.4-0.7, P 0-0.025, S 0-0.02, Cr 2.8-3.3, Mo 0.3-0.5, Ni 0-0.3, bal Fe.
Aircraft material. WL 1.8564/80; W. Nr. 1.8564.

KRUPP WL 17734
Krupp Stahl AG
C 0.12-0.18, Si 0-0.2, Mn 0.8-1.1, P 0-0.02, S 0-0.015, Cr 1.25-1.5, Mo 0.8-1, V 0.2-0.3, bal Fe.
Aircraft material. WL 1.7734/82; W. Nr. 1.7734.

KRUPP WM 131 V
Krupp Stahl AG
tool material. C 0.79, Cr 13-14, V 1.7-2, Mo, bal Fe.
Cold work tool steel, or mold steel. Corrosion resistant knives.

KRUPP WN48
Krupp Stahl AG
tool material. C 0.28-0.33, Cr 2.2-2.5, V 0.2, bal Fe.
For tools, dies. *Obsolete*

KRUPP WQL1.4939
Krupp Stahl AG
C 0.08-0.13, Si 0-0.35, Mn 0.5-0.9, P 0-0.025, S 0-0.02, Cr 11-12.5, Mo 1.5-2, Ni 2-3, N 0.02-0.04, V 0.26-0.4, bal Fe.
Aircraft material. WL 1.4939/81; W. Nr. 1.4939.

KRUPP WSTE 255
Krupp Stahl AG
C 0-0.18, Si 0-0.4, Mn 0.5-1.3, P 0-0.035, S 0-0.03, Cr 0-0.3, Mo 0-0.08, Ni 0-0.3, Cu 0-0.2, N 0-0.02, Nb 0-0.03, 0.02 Al min, bal Fe.
DIN 17102/83; W. Nr. 1.0462.

KRUPP WSTE 285
Krupp Stahl AG
C 0-0.18, Si 0-0.4, Mn 0.6-1.4, P 0-0.035, S 0-0.03, Cr 0-0.3, Mo 0-0.08, Ni 0-0.3, Cu 0-0.2, N 0-0.02, Nb 0-0.03, 0.02 Al min, bal Fe.
DIN 17102/83; W. Nr. 1.0487.

KRUPP WSTE 315
Krupp Stahl AG
C 0-0.18, Si 0-0.45, Mn 0.7-1.5, P 0-0.035, S 0-0.03, Cr 0-0.3, Mo 0-0.08, Ni 0-0.3, Cu 0-0.2, N 0-0.02, Nb 0-0.03, 0.02 Al min, bal Fe.
DIN 17102/83; W. Nr. 1.0506.

KRUPP WSTE 355
Krupp Stahl AG
C 0-0.2, Si 0.1-0.5, Mn 0.9-1.65, P 0-0.035, S 0-0.03, Cr 0-0.3, Mo 0-0.08, Ni 0-0.3, Cu 0-0.2, N 0-0.02, Nb 0-0.05, V 0-0.1, 0.02 Al min, bal Fe.
DIN 17102/83; W. Nr. 1.0565.

KRUPP WSTE 380
Krupp Stahl AG
C 0-0.2, Si 0.1-0.6, Mn 1-1.7, P 0-0.035, S 0-0.03, Cr 0-0.3, Mo 0-0.08, Ni 0-1, Cu 0-0.2, N 0-0.02, Nb 0-0.05, V 0-0.2, 0.02 Al min, bal Fe.
DIN 17102/83; W. Nr. 1.8930.

KRUPP WSTE 420
Krupp Stahl AG
C 0-0.2, Si 0.1-0.6, Mn 1-1.7, P 0-0.035, S 0-0.03, Cr 0-0.3, Mo 0-0.1, Ni 0-1, Cu 0-0.2, N 0-0.02, Nb 0-0.05, V 0-0.2, 0.02 Al min, bal Fe.
DIN 17102/83; W. Nr. 1.8932.

KRUPP WSTE 460
Krupp Stahl AG
C 0-0.2, Si 0.1-0.6, Mn 1-1.7, P 0-0.035, S 0-0.03, Cr 0-0.3, Mo 0-0.1, Ni 0-1, Cu 0-0.2, N 0-0.02, Nb 0-0.05, V 0-0.2, 0.02 Al min, bal Fe.
DIN 17102/83; W. Nr. 1.8935.

KRUPP WSTE 47
Krupp Stahl AG
alloy steel.
See WSTE460

KRUPP WSTE 500
Krupp Stahl AG
C 0-0.21, Si 0.1-0.6, Mn 1-1.7, P 0-0.035, S 0-0.03, Cr 0-0.3, Mo 0-0.1, Ni 0-1, Cu 0-0.2, N 0-0.02, Nb 0-0.05, V 0-0.22, 0.02 Al min, bal Fe.
DIN 17102/83; W. Nr. 1.8937.

KRUPP WT10
Krupp Stahl AG
alloy steel. C, Si, Mn, bal Fe.
Obsolete

KRUPP WTK 52
Krupp Stahl AG
C 0-0.12, Si 0.25-0.75, Mn 0.2-0.5, P 0.07-0.15, S 0-0.035, Cr 0.5-1.25, Ni 0-0.65, Cu 0.25-0.55, bal Fe.
SEL 90; W. Nr. 1.8962.

KRUPP WTST 37-2
Krupp Stahl AG
C 0-0.13, Si 0.1-0.4, Mn 0.2-0.5, P 0-0.05, S 0-0.035, Cr 0.5-0.8, Cu 0.3-0.5, N 0-0.009, bal Fe.
SEW 087/81; W. Nr. 1.8960.

KRUPP WTST 37-3
Krupp Stahl AG
C 0-0.13, Si 0.1-0.4, Mn 0.2-0.5, P 0-0.045, S 0-0.035, Cr 0.5-0.8, Cu 0.3-0.5, bal Fe.
SEW 087/81; W. Nr. 1.8961.

KRUPP WTST 52-3
Krupp Stahl AG
C 0-0.15, Si 0.1-0.5, Mn 0.9-1.3, P 0-0.045, S 0-0.035, Cr 0.5-0.8, Cu 0.3-0.5, V 0.02-0.1, bal Fe.
SEW 087/81; W. Nr. 1.8963.

KRUPP WW
Krupp Stahl AG
tool material. C 0.4, Si 0.01, Mn 0.14, Cr 0.05, Ni 0.15, bal Fe.
For tools, hubbing dies; case hardened. *Obsolete*

KRUPP WW 1.6659
Krupp Stahl AG
C 0.28-0.35, Si 0.15-0.4, Mn 0.4-0.7, P 0-0.025, S 0-0.025, Cr 0.9-1.2, Mo 0.2-0.3, Ni 3-3.5, bal Fe.
WW 1.6659/79; W. Nr. 1.6659.

KRUPP WX 25
Krupp Stahl AG
tool material. C 0.25, Ni 4, Cr 1.25, Mo, bal Fe.
Cold work tool steel.

KRUPP WX 70
Krupp Stahl AG
tool material. C 0.72, Si 1, Ni, Mo, V, bal Fe.
Tool steel, shock resisting. *Obsolete*

KRUPP WX69
Krupp Stahl AG
C 0.72, Si 1.2, Cr 0.6, Mo 0.6, Ni 1.2, V 0.2, bal Fe.
W. Nr. 1.2790. Soft annealed, max 265 Brin, (according to DIN 17350 for cold drawn products hardness can be up to 20 Brin higher). For screw drivers, bits.

KRUPP X 12 CRMO 9 1
Krupp Stahl AG
C 0-0.15, Si 0.25-1, Mn 0.3-0.6, P 0-0.03, S 0-0.03, Cr 8-10, Mo 0.9-1.1, bal Fe.
SEL 90; W. Nr. 1.7386.

KRUPP X 7 CRMO 9 1
Krupp Stahl AG
C 0.04-0.09, Si 0.45-0.75, Mn 0.43-0.72, P 0-0.015, S 0-0.015, Cr 8.6-9.9, Mo 0.9-1.1, bal Fe.
DIN 17145/80; W. Nr. 1.7388.

KRUPP ZF
Krupp Stahl AG
tool material. C 0.85, Si 0-0.25, Mn 0-25, bal Fe.
Heat treated: 190,000 TS; 145,000 YS; 10 El; 30 RA; 400 Brin. For springs, taps, hobs, reamers, broaches; Type W1; water hardened. *Obsolete*

KRUPP ZST 37-2
Krupp Stahl AG
C 0-0.17, P 0-0.05, S 0-0.05, N 0-0.009, bal Fe.
DIN 17100/80; W. Nr. 1.0159.

KRUPP ZST 37-3
Krupp Stahl AG
C 0-0.17, P 0-0.04, S 0-0.04, 0.02 Al min, bal Fe.
DIN 17100/80; W. Nr. 1.0168.

KRUPP ZST 44-2
Krupp Stahl AG
C 0-0.21, P 0-0.05, S 0-0.05, N 0-0.009, bal Fe.
DIN 17100/80; W. Nr. 1.0129.

KRUPP ZST 44-3
Krupp Stahl AG
C 0-0.2, P 0-0.04, S 0-0.04, 0.02 Al min, bal Fe.
DIN 17100/80; W. Nr. 1.0153.

KRUPP ZST 52-3
Krupp Stahl AG
C 0-0.2, Si 0-0.55, Mn 0-1.6, P 0-0.04, S 0-0.04, 0.02 Al min, bal Fe.
DIN 17100/80; W. Nr. 1.0597.

KRUPP ZSTE 180 BH
Krupp Stahl AG
C 0-0.04, Si 0-0.5, Mn 0-0.7, P 0-0.06, S 0-0.03, 0.02 Al min, bal Fe.
SEW 094/87; W. Nr. 1.0395.

KRUPP ZSTE 220 BH
Krupp Stahl AG
C 0-0.06, Si 0-0.5, Mn 0-0.7, P 0-0.08, S 0-0.03, 0.02 Al min, bal Fe.
SEW 094/87; W. Nr. 1.0396.

KRUPP ZSTE 220 P
Krupp Stahl AG
C 0-0.06, Si 0-0.5, Mn 0-0.7, P 0-0.08, S 0-0.03, 0.02 Al min, bal Fe.
SEW 094/87; W. Nr. 1.0397.

KRUPP ZSTE 260 P
Krupp Stahl AG
C 0-0.08, Si 0-0.5, Mn 0-0.7, P 0-0.1, S 0-0.03, 0.02 Al min, bal Fe.
SEW 0904/87; W. Nr. 1.0417.

KRUPP ZSTE 260-ZE
Krupp Stahl AG
C 0-0.1, Si 0-0.5, Mn 0-0.6, P 0-0.03, S 0-0.03, Nb 0-0.09, Ti 0-0.22, 0.015 Al min, bal Fe.
DIN 17163/88; W. Nr. 1.0480.

KRUPP ZSTE 300 BH
Krupp Stahl AG
C 0-0.1, Si 0-0.5, Mn 0-0.7, P 0-0.12, S 0-0.03, 0.02 Al min, bal Fe.
SEW 094/87; W. Nr. 1.0444.

KRUPP ZSTE 300 P
Krupp Stahl AG
C 0-0.1, Si 0-0.5, Mn 0-0.7, P 0-0.12, S 0-0.03, 0.02 Al min, bal Fe.
SEW 094/87; W. Nr. 1.0448.

KRUPP ZSTE 300-ZE
Krupp Stahl AG
C 0-0.1, Si 0-0.5, Mn 0-0.8, P 0-0.03, S 0-0.03, Nb 0-0.09, Ti 0-0.22, 0.015 Al min, bal Fe.
DIN 17163/88; W. Nr. 1.0489.

KRUPP ZSTE 340-ZE
Krupp Stahl AG
C 0-0.1, Si 0-0.5, Mn 0-1, P 0-0.03, S 0-0.03, Nb 0-0.09, Ti 0-0.22, 0.015 Al min, bal Fe.
DIN 17163/88; W. Nr. 1.0548.

KRUPP ZSTW 260 BH
Krupp Stahl AG
C 0-0.08, Si 0-0.5, Mn 0-0.7, P 0-0.1, S 0-0.03, 0.02 Al min, bal Fe.
SEW 094/87; W. Nr. 1.0400.

KRUPP-V2AFH
Krupp Stahl AG
stainless steel. C 0.1, Cr 16-18, Ni 6-8, bal Fe.
Annealed: 110,000 TS; 40,000 YS; 60 El; 85 Rock B. For aircraft structural members, railroad cars, household utensils, diaphragms, springs. Type 301 stainless steel, austenitic, non-hardenable. *Obsolete*

KRUPPIN
Krupp Stahl AG
iron. Ni 28, C, bal Fe.
For resistance alloy; heat resistant. *Obsolete*

KS "Y" ALLOY
Karl Schmidt Co.
Aluminum. Cu 4, Si 0.4, Ni 2, Fe 0.4, Mg 1.5, bal Al.
35,000 TS; 28,500 YS; 1.0 El; 90 Brin. For pistons for steam engines; coefficient of expansion 0.0000245.

KS 1
Saarstahl AG
C 0.08, Si 0.12, Mn 0.5, bal Fe.
For welded chains. W.-Nr. 0208. *Obsolete*

KS 1 R
Saarstahl AG
C 0.08, Si 0.12, Mn 0.5, bal Fe.
For welded chains. W.-Nr. 0209. *Obsolete*

KS 10
Saarstahl AG
Now SAARSTAHL 23 MNNIMOCR 64.

KS 1275 KOLBENLEG
English manufacture
Mg 1, bal Al.
For light alloy parts; corrosion resistant.

KS 13
English manufacture
Sb 6-8, bal Al.
For bearings.

KS 2
Saarstahl AG
C 0.16, Si 0.15, Mn 0.8, bal Fe.
For welded chains. W.-Nr. 0847. *Obsolete*

KS 3
Saarstahl AG
C 0.21, Si 0.15, Mn 1, bal Fe.
For welded chains. W.-Nr. 0848. *Obsolete*

KS 4
Saarstahl AG
Now SAARSTAHL 21 MNSI 5.

KS 5
Saarstahl AG
Now SAARSTAHL 27 MNSI 5.

KS 9

Saarstahl AG

C 0.23, Si 0.2, Mn 1.5, Cr 0.35, Mo 0.27, Ni 0.62, bal Fe.
Heat treated: 1080 min N/mm^2 TS. For strong welded chains.
W.-Nr. 6542. *Obsolete*

KS ALUSIL

Karl Schmidt Co.

Aluminum. Si 20, Cu 1.5-2, Ni 0.5, bal Al.
23,000 TS; 21,000 YS; 0.25 El; 100 Brin. For pistons for
general use; low temperature coefficient; coefficient of
expansion 0.000018.

KS MAGNET STEEL

Japanese manufacture

Co 30-40, Cr 1.5-3, W 5-9, Mn 0.35, Si 0.12, C 0.4-0.8, bal Fe.
Hardened: 444-652 Brin. For tools, magnets; corrosion and
heat resistant; Br 10,000-10,500; Hc 200-240.

KS MAGNET STEEL

Japanese manufacture

C 0.7-1, W 6-8, Co 1-2, bal Fe.
For magnets; magnetos; high permeability.

KS NO. 1275

Karl Schmidt Co.

Aluminum. Si, Cu, Ni, bal Al.
For light alloy parts; corrosion resistant.

KS NO. 245

Karl Schmidt Co.

Aluminum. Si 14, Cu 4.5, Ni 1.5, Mg 0.7, Mn 1, bal Al.
29,000 TS; 25,000 YS; 0.2 El; 130 Brin. For pistons for small
diesel motors and motorcycles; high wear resistance;
coefficient of expansion 0.00002.

KS NO. 280

Karl Schmidt Co.

Aluminum. Si 21, Cu 1.5, Mn 0.7, Ni 1.5, Mg 0.5, Co 1.2, bal
Al.
27,000 TS; 26,000 YS; 0.25 El; 120 Brin. For pistons for air-
cooled motors; coefficient of expansion 0.000018.

KS NO. 280B

Karl Schmidt Co.

Aluminum. Si, bal Al.
For light alloy parts; corrosion resistant.

KS NO. 282

Karl Schmidt Co.

Aluminum. Si 23-25, Ni 0.8-1.3, Ti 0.2, Mn 0.2, Co 0.3-0.5, Fe
0.7, Zn 0.2, Mg 0.8-1.3, Cu 0.81, bal Al.
For pistons in internal combustion engines. Medium thermal
stressing, corrosion resistant.

KS NO. 283

Karl Schmidt Co.

Aluminum. Si 17-19, Cu 4.7-5.3, Ni 3.8-4.2, Ti 0.1, Mn 0.6-0.8,
Co 0.5-0.7, Fe 0.7, Zn 0.2, Mg 0.4-0.6, bal Al.
For pistons in internal combustion engines. High thermal
stressing, corrosion resistant.

KS NO. 411B

Karl Schmidt Co.

Aluminum. Si, bal Al.
For light alloy parts; corrosion resistant.

KS NO. 837

Karl Schmidt Co.

Aluminum. Si, bal Al.
For light alloy parts; corrosion resistant.

KS NO. 83A

Karl Schmidt Co

Aluminum. Al alloy.
For light alloy parts.

KS RED

English manufacture

Cu 16.5, Ni 0.8, Fe 0.5, Si 0.5, bal Al.
For pistons; cast.

KS SEEWASSER

Vereinigte Leichtmetallwerke, G.m.b.H.

Mn 1.3, Mg 2.2, Sb 0.2, Si 0.7, bal Al.
Sand cast: 26,000 TS; 12,800 YS; 2.5 El; 5 RA; 60 Brin. Die
cast: 32,720 TS; 12,800 YS; 1.6 El; 81 Brin. Rolled: 45,500 TS;
36,000 YS; 2 El; 35 RA; 83 Brin. For furniture, interior light
fixtures, wire castings, ship parts, chemical industry; light
alloy; resists sea water corrosion.

KS SEEWASSER

Karl Schmidt Co.

Mn 1.3, Mg 2.2, Sb 0.2, Si 0.7, bal Al.
Sand cast: 26,000 TS; 12,800 YS; 2.5 El; 5 RA; 60 Brin. Die
cast: 32,720 TS; 12,800 YS; 1.6 El; 81 Brin. Rolled: 45,500 TS;
36,000 YS; 2 El; 35 RA; 83 Brin. For furniture, interior light
fixtures, wire castings, ship parts, chemical industry; light
alloy; resists sea water corrosion.

KS SEEWASSER JUB

Japanese manufacture

Al alloy.
For light alloy parts.

KS SPECIAL PISTON ALLOY

Karl Schmidt Co.

Aluminum. Cu 15, Si 0.6, Fe 0.6, Ni 0.6, Mg 0.3, bal Al.
26,000 TS; 0.2 El; 120 Brin. For pistons for Zeppelin motors
and diesel engines; coefficient of expansion 0.000024.

KS15V

Friedr. Lohmann GmbH

C 1.45, Cr 1.4, Mn 0.6, bal Fe.
For bearings, liners, bushings. Oil hardened, wear resistant.

KSS SUPRA

Thyssen Edelstahlwerke AG

C 0.95-1.05, Si 0.2-0.35, Mn 0.1-0.3, bal Fe.
Water hardened: 215,000 TS; 187,000 YS; 9 El; 24 RA; 400
Brin. Annealed: 88,000 TS; 53,000 YS; 19 El; 187 Brin. For
cold heading and forming dies, pneumatic hammers, drills,
reamers. Type W3, water hardening. *Obsolete*

KU A-10

Kulite Tungsten Corp.

Tungsten. W 90, bal Ni-Fe.
Density 17; 125,000 TS; 90,000 YS; 15 El; 27 Rock C. High
density alloy for counterweights, balancing and shielding
applications. See Kulite K1700, K1701. *Obsolete*

KU A-5

Kulite Tungsten Corp.

Tungsten. W 95, bal Ni-Fe.
Density 18; 120,000 TS; 92,000 YS; 12 El; 30 Rock C. High
density alloy for counterweights, balancing and shielding
applications. See Kulite K1800, K1801. *Obsolete*

KU-112

Kulite Tungsten Corp.

Tungsten. W 90, bal Ni-Cu.
Density 17; 110,000 TS; 70,000 YS; 5 El; 25 Rock C. High
density alloy for counterweights, balancing and shielding
applications. See Kulite K1700, K1701. *Obsolete*

KU-112-18

Kulite Tungsten Corp.

Tungsten. W 95, bal Ni-Cu.
Density 18; 105,000 TS; 90,000 YS; 3 El; 28 Rock C. High
density alloy for counterweights, balancing and shielding
applications. See Kulite K1800, K1801. *Obsolete*

KUBAX

Rochling Burbach GmbH

C 1.2-1.3, Cr 0.1-0.2, V 0.15-0.25, bal Fe.
For tools, dies, drills; oil hardening. *Obsolete*

KUFIL

IMI Knoch Ltd.

Cu 99, Ag 1.
Welded: 29,000 TS. For welding rod for Cu; free flowing.

KUHBIER A12

C. Kuhbier & Sohn

C 0-0.1, Cr 12.5, Ni 12, bal Fe.
For chemical plant equipment, valves; heat and corrosion
resistant.

KUHBIER A18

C. Kuhbier & Sohn

C 0-0.15, Cr 18, Ni 9, bal Fe.
Annealed: 80,000 TS; 35,000 YS; 55 El; 75 RA; 150 Brin. For
chemical plant equipment, tanks; Type 302; stainless,
austenitic.

KUHBIER F13

C. Kuhbier & Sohn

C 0-0.12, Si 0.4, Cr 13, bal Fe.
Annealed: 75,000 TS; 40,000 YS; 35 El; 70 RA; 155 Brin. For
valves, cutlery, surgical instruments; Type 410; stainless.

KUHBIER F17

C. Kuhbier & Sohn

C 0.8, Cr 17, bal Fe.
Annealed: 107,000 TS; 62,000 YS; 18 El; 35 RA; 220 Brin. For
bearings, instrument pivots, liners; corrosion resistant,
hardenable.

KUHBIER F17A

C. Kuhbier & Sohn

C 0.12, Cr 16.5, Mo 0.25, S 0.2, bal Fe.
Annealed: 80,000 TS; 50,000 YS; 20 El; 40 RA; 150 Brin. For
screw machine products, shafts; free-cutting; Type 430 Mo;
corrosion resistant.

KUHBIER F17T

C. Kuhbier & Sohn

C 0-0.1, Cr 17.5, Ti = 7 x C, bal Fe.
Annealed: 80,000 TS; 50,000 YS; 25 El; 50 RA; 150 Brin. For
oil refinery equipment, welded structures; corrosion and
creep resistant.

KUHBIER F19T

C. Kuhbier & Sohn

C 0-0.1, Cr 17, Mo 1.8, Ti = 7 x C, bal Fe.
Annealed: 125,000 TS; 95,000 YS; 20 El; 55 RA; 260 Brin. For
oil refinery equipment, welded structures; corrosion and
creep resistant.

KUHBIER I

C. Kuhbier & Sohn

C 0.15, Cr 24, Ni 19, bal Fe.
Annealed: 100,000 TS; 45,000 YS; 50 El; 65 RA; 185 Brin. For
furnace parts, heat treat boxes; heat resistant, austenitic.

KUHBIER IA

C. Kuhbier & Sohn

C 0.15, Cr 19.5, Ni 9.5, bal Fe.
Annealed: 80,000 TS; 35,000 YS; 55 El; 75 RA; 150 Brin. For
chemical plant equipment, tanks, vessels; Type 302;
stainless, austenitic.

KUHBIER M13

C. Kuhbier & Sohn

C 0.2, Si 0.4, Cr 13, bal Fe.
Annealed: 95,000 TS; 50,000 YS; 25 El; 55 RA; 195 Brin. For
turbine blades, valves, cutlery, surgical instruments; Type
420; stainless.

KUHBIER M13.4

C. Kuhbier & Sohn

C 0.4, Si 0.4, Cr 13, bal Fe.
Annealed: 100,000 TS; 55,000 YS; 20 El; 50 RA; 210 Brin. For
surgical instruments, valves, cutlery, knives; Type 420;
stainless.

KUHBIER M15

C. Kuhbier & Sohn

C 0.2, Si 0.4, Cr 13, Mo 1.15, bal Fe.
Annealed: 100,000 TS; 55,000 YS; 20 El; 50 RA; 210 Brin. For
turbine blades, valves, cutlery, knives; Type 420 Mo; stainless.

KUHBIER M18

C. Kuhbier & Sohn
C 0.22, Si 0.4, Cr 17, Ni 1.5, bal Fe.
Annealed: 125,000 TS; 95,000 YS; 20 El; 55 RA; 260 Brin. For pumps, marine hardware, valves, gauges; Type 431; corrosion and heat resistant.

KUHBIER M18.9

C. Kuhbier & Sohn
C 0.9, Cr 18, Mo 1.15, V 1, bal Fe.
Annealed: 110,000 TS; 65,000 YS; 17 El; 32 RA; 230 Brin. For cutlery, valves, bearings, surgical instruments; corrosion and wear resistant.

KUHBIER M19

C. Kuhbier & Sohn
C 0.35, Cr 16.5, Mo 1.15, bal Fe.
For chemical plant and oil refinery equipment; corrosion and heat resistant.

KUHNE PHOSPHOR BRONZE

English manufacture
Cu 78, Sn 11, Pb 10, Ni 0.3, P 0.6.
For hard bearings; heavy duty.

KUKI EXTRA MH

Carl Urbach & Co., Stahlwerk KG
C 1.1, Si 0-0.25, Mn 0-0.25, bal Fe.
Annealed: 100,000 YS; 53,000 YS; 21 El; 42 RA; 200 Brin. For drills, hobs, taps, springs; Type W1; water hardened.

KUKI EXTRA ZAH

Carl Urbach & Co., Stahlwerk KG
C 0.85, Mn 0-0.25, Si 0-0.25, bal Fe.
Heat treated: 190,000 TS; 145,000 YS; 12 El; 35 RA; 320 Brin. For drills, punches, cutters, springs; Type W1; water hardened.

KUKI EXTRA ZH

Carl Urbach & Co., Stahlwerk KG
C 1, Si 0-0.25, Mn 0-0.25, bal Fe.
Annealed: 100,000 TS; 53,000 YS; 21 El; 42 RA; 200 Brin. For drills, taps, reamers, cutting tools; Type W1; water hardened.

KUKI PRIMA H

Carl Urbach & Co., Stahlwerk KG
C 1.3, Si 0-0.25, Mn 0-0.25, bal Fe.
For cutters, reamers, broaches; Type W1; water hardened.

KUKI PRIMA MH

Carl Urbach & Co., Stahlwerk KG
C 1.15, Mn 0-0.25, Si 0-0.25, bal Fe.
Annealed: 110,000 TS; 55,000 YS; 20 El; 40 RA; 210 Brin. For cutters, hobs, reamers, broaches; Type W1; water hardened.

KUKI PRIMA ZAH

Carl Urbach & Co., Stahlwerk KG
C 1, Si 0-0.25, Mn 0-0.25, bal Fe.
Annealed: 100,000 TS; 53,000 YS; 21 El; 42 RA; 200 Brin. For cutters, hobs, reamers, broaches; Type W1; water hardened.

KUKI PRIMA ZH

Carl Urbach & Co., Stahlwerk KG
C 0.85, Si 0-0.25, Mn 0-0.25, bal Fe.
Heat treated: 190,000 TS; 145,000 YS; 12 El; 45 RA; 330 Brin. For cutters, tools, springs, punches, drills; Type W1; water hardened.

KUKI PRIMA ZW

Carl Urbach & Co., Stahlwerk KG
C 0.7, Si 0-0.25, Mn 0-0.25, bal Fe.
Heat treated: 175,000 TS; 128,000 YS; 12 El; 37 RA; 355 Brin. For punches, crimpers, springs; Type W1; water hardened.

KUKI SPEZIAL H

Carl Urbach & Co., Stahlwerk KG
C 0.9, Si 0.25-0.5, Mn 0.3-0.8, bal Fe.
Heat treated: 190,000 TS; 145,00 YS; 10 El; 30 RA; 400 Brin. For punches, crimpers, drills, springs; Type W1; water hardened.

KUKI SPEZIAL MH

Carl Urbach & Co., Stahlwerk KG
C 0.75, Si 0.4, Mn 0.6, bal Fe.
Heat treated: 175,000 TS; 130,000 YS; 12 El; 36 RA; 355 Brin. For punches, drills, springs, crimpers; Type W1; water hardened.

KUKI SPEZIAL ZAH

Carl Urbach & Co., Stahlwerk KG
C 0.67, Si 0.4, Mn 0.6, bal Fe.
Heat treated: 170,000 TS; 125,000 YS; 15 El; 38 RA; 350 Brin. For punches, springs, crimpers; Type W1; water hardened.

KUKI SPEZIAL ZH

Carl Urbach & Co., Stahlwerk KG
C 0.6, Si 0.3-0.5, Mn 0.5-0.8, bal Fe.
Heat treated: 160,000 TS; 113,000 YS; 12 El; 40 RA; 325 Brin. For punches, crimpers; water hardened.

KUKI SPEZIAL ZW

Carl Urbach & Co., Stahlwerk KG
C 0.45, Si 0.4, Mn 0.6, bal Fe.
Hot rolled: 98,000 TS; 59,000 YS; 24 El; 45 RA; 212 Brin. For gears, shafts, crankshafts; water hardened.

KULGRID "C"

GTE Sylvania
Ni, clad Cu.
For lead-in wires for vacuum tubes; resists high temperatures.

KULGRID "C"

Eisler Electric Co.
Ni, clad Cu.
For lead-in wires for vacuum tubes; resists high temperatures.

KULITE K1700

Kulite Tungsten Corp.
Tungsten. W 90, bal Ni-Cu-Fe.
Density 17 g/cm^3; 125,000 psi TS; 85,000 psi YS; 23 Rock C. High density alloy for aircraft counterweights, radiation shielding, darts, sinker weights and boring bars. ASTM B777 Class 1; MIL-T-21014(D) Class 1.

KULITE K1701

Kulite Tungsten Corp.
Tungsten. W 90, bal Ni-Cu-Fe.
Density 17 g/cm^3; 110,000 psi TS; 80,000 psi YS; 22 Rock C. High density alloy for aircraft counterweights, radiation shielding, darts, sinker weights and boring bars. ASTM B777 Class 1; MIL-T-21014(D) Class 1.

KULITE K1750

Kulite Tungsten Corp.
Tungsten. W 92.5, bal Ni-Cu-Fe.
Density 17.5 g/cm^3; 125,000 psi TS; 90,000 psi YS; 24 Rock C. High density alloy for aircraft counterweights, radiation shielding, darts, sinker weights and boring bars. ASTM B777 Class 2; MIL-T-21014(D) Class 2.

KULITE K1800

Kulite Tungsten Corp.
Tungsten. W 95, bal Ni-Cu-Fe.
Density 18 g/cm^3; 125,000 psi TS; 90,000 psi YS; 25 Rock C. High density alloy for aircraft counterweights, radiation shielding, darts, sinker weights and boring bars. ASTM B777 Class 3; MIL-T-21014(D) Class 3.

KULITE K1801

Kulite Tungsten Corp.
Tungsten. W 95, bal Ni-Cu-Fe.
Density 18 g/cm^3; 110,000 psi TS; 85,000 psi YS; 24 Rock C. High density alloy for aircraft counterweights, radiation shielding, darts, sinker weights and boring bars. ASTM B777 Class 3; MIL-T-21014(D) Class 3.

KULITE K1850

Kulite Tungsten Corp.
Tungsten. W 97, bal Ni-Cu-Fe.
Density 18.5 g/cm^3; 120,000 psi TS; 95,000 psi YS; 26 Rock C. High density alloy for aircraft counterweights, radiation shielding, darts, sinker weights and boring bars. ASTM B777 Class 4; MIL-T-21014(D) Class 4.

KULITE KLT-115

Kulite Tungsten Corp.
Tungsten. Pb, W.
Sintered: 18,000-20,000 TS; 200-280 Brin. For shielding for nuclear materials. High density. *Obsolete*

KUMANAL

IMI Knoch Ltd.
Cu 88, Mn 10, Al 2.
Annealed: 50,000 TS; 30 El; 73 Brin. For resistance wire, instrument shunts; low temperature coefficient.

KUMANIC

IMI Knoch Ltd.
Cu 60, Mn 20, Ni 20.
Annealed: 70,000 TS; 21,000 YS; 35 El; 97 Brin. Heat treated: 155,000 TS; 130,000 YS; 2 El; 320 Brin. For springs, contacts; corrosion resistant.

KUMIUM

IMI Knoch Ltd.
Cu 99.5, Cr 0.5.
Annealed: 33.600 TS; 16,700 YS; 60 El; 59 Brin. Heat treated: 70,000 TS; 60,000 YS; 14 El; 127 Brin. For welding tips, electrical contacts; hardenable, high conductivity.

KUNHEIM METAL

German manufacture
Ce 36, La 49, Di 10, Mg 1, Hydrides of Misch metal.
For cigarette lighters; pyrophoric.

KUNHEIM METAL

German manufacture
Mg 12, Al 2, 86 rare earths.
For cigarette lighters.

KUNIAL ALLOYS

IMI Knoch Ltd.
Cu-Zn alloys.
For engineering applications; can be heat treated.

KUNIAL BRASS

IMI Knoch Ltd.
Cu 72.5, Ni 6, Al 1.5, bal Zn.
Heat treated: 108,000 TS; 81,000 YS; 11 El; 240 Brin. For nuts, bolts, valves, primers, fuse bodies, keys, springs; hardened and strengthened by heat treatment.

KUNIAL BRONZE

IMI Knoch Ltd.
Sn, bal Cu.
For hardware, plumbing, pipes; tough.

KUNIAL COPPER

IMI Knoch Ltd.
Cu 90, Ni, Al.
For general use, tube, wire, plate, sheet; temper hardening.

KUNIAL NICKEL SILVER

IMI Knoch Ltd.
Ni, Zn, bal Cu.
For hardware, utensils; corrosion resistant.

KUNIFER 10

IMI Rod & Wire
Ni 10, Cu 90.
Annealed: 45,000 TS; 22,400 YS; 55 El; 90 Brin. 1/2 hard: 62,000 TS; 45,000 YS; 25 El; 110 Brin. Hard: 83,000 TS; 81,000 YS; 15 El; 190 Brin. For condenser tubes; corrosion resistant, no season cracking.

KUNIFER 10 (COPPER-NICKEL)
IMI Yorkshire Alloys Ltd.
Ni 10-11.5, Fe 1.5-1.8, Mn 0.5-1.8, bal Cu.
Annealed: 52,200 psi TS; 23,200 psi YS; 48 El; 95 DPN (88 Brin). Tubes for condensers, hydraulic and sea water pipelines. Corrosion resistant.

KUNIFER 30
IMI Rod & Wire
Ni 30, Mn 1, Fe 1, bal Cu.
Annealed: 63,000 TS; 25,000 YS; 55 El; 100 Brin. 1/2 hard: 75,000 TS; 42,000 YS; 25 El; 150 Brin. Hard: 100,000 TS; 94,000 YS; 10 El; 210 Brin. For condenser tubes; corrosion resistant, no season cracking.

KUNIFER 30 (COPPER-NICKEL)
IMI Yorkshire Alloys Ltd.
Ni 30-32, Fe 0.7-1, Mn 0.5-1, bal Cu.
Annealed: 56,500 psi TS; 26,100 psi YS; 48 El; 100 DPN (97 Brin). Tubes for condensers and hydraulic lines. Erosion and corrosion resistant.

KUNIFER 30A
IMI Rod & Wire
Cu 66, Ni 30, Mn 2, Fe 2.
For condenser tubes; corrosion resistant.

KUNIFER 5
IMI Rod & Wire
Ni 5, Fe 1, bal Cu.
Annealed: 40,000-45,000 TS; 20,000-25,000 YS; 40-50 El; 68-80 Brin. For marine equipment, hardware, fasteners; corrosion resistant.

KUPFER-NICKEL 54/45
VDM Nickel-Technologie AG
Cu 54, Ni 45.
For electrical resistances; constantan. *Obsolete*

KUPFER-NICKEL 67/30/3
VDM Nickel-Technologie AG
Cu 67, Ni 30, Mn 3.
For electrical resistances. *Obsolete*

KUPFER-SILUMIN
Metallgesellschaft Reuterweg
Si 12, Cu 0.8, Mn 0.3, bal Al.
Sand cast: 25,000 TS; 14,000 YS; 2-4 El; 60 Brin. For wheels, rolls, engine blocks, motor housing; high fluidity.

KUPFERNICKEL 75/25
VDM Nickel-Technologie AG
Ni 25, Cu 75.
Annealed: 45,000 TS; 18,500 YS; 42 El; 62 Brin. For coinage; corrosion resistant. *Obsolete*

KUPFERNICKEL 80/20
VDM Nickel-Technologie AG
Cu 80, Ni 20.
Rolled: 62,500 TS; 59,700 YS; 12 El; 120 Brin. Annealed: 45,500 TS; 17,100 YS; 41 El; 64 Brin. For condenser tubes; corrosion resistant. *Obsolete*

KUPRODUR
German manufacture
Si 0.5, Ni 0.7, bal Cu.
For staybolts, propeller parts; age-hardenable, corrosion resistant.

KUROMI
Japanese manufacture
Cu + Sn + Co. Japanese alloy.

KUT KOST
General Tool & Die Co.
C 0.7, B 1.5, W 18, Cr 4, V 1, bal Fe.
For tools, dies, tipped tools; high speed steel.

KUT KOST GRADE V
General Tool & Die Co.
B 1.5, C, W-Co, bal Fe.
For tools, cutters; for heavy cuts.

KUT KOST GRADE X
General Tool & Die Co.
B 1.5, C, W-Co, bal Fe.
For tools, cutters; centrifugally cast.

KUT KOST GRADE XV
General Tool & Die Co.
B 1.5, C, W, Co, bal Fe.
For tools, cutters; centrifugally cast.

KUT KOST GRADE XX
General Tool & Die Co.
B 1.5, W, Co, bal Fe.
For tools, cutters; light cuts.

KUTERN
IMI Knoch Ltd.
Cu 99.5, Te 0.5.
Annealed: 33.600 TS; 5500 YS; 50 El; 48 Brin. For machined parts; free-cutting.

KUTHERM
Now IMI 575 KUTHERM.

KWIK-KUT
North American Steel Corp.
C 1.1, Cr 0.5, bal Fe.
For hollow drills; fatigue resistant.

KWS45
Friedr. Lohmann GmbH
C 0.45, Cr 1.05, V 0.18, W 1.85, Si 0.9, bal Fe.
For header dies, upsetters, shears, punches. Type S2; oil hardened, tough. T41902

KWS55
Friedr. Lohmann GmbH
C 0.6, Si 0.9, Cr 1.05, V 0.18, W 1.85, bal Fe.
For header dies, punches, tools, shears. Type S1; oil hardened, tough. T41901

KXA STEEL
Rex Buckeye Co.
C, Cr, Mo, bal Fe.
For die casting machine parts, plungers; oil hardened.

KYNAL C 67
IMI Knoch Ltd.
Cu 4, Mn, Si, Mg, bal Al.
Heat treated, tempered: 63,840 TS; 39,200 YS; 15 El; 138 Brin. For high strength applications; hardenable.

KYNAL C65
IMI Knoch Ltd.
Cu 4, Mn 0.6, Fe 0.7, Si 0.5, bal Al.
Heat treated: 56,000-60,500 TS; 40,500-49,000 YS; 6-10 El; 100 Brin. For aircraft construction; age hardened, high strength.

KYNAL C66
IMI Knoch Ltd.
Cu 4, Si 0.85, Mg 0.85, Mn 1.2, bal Al.
Heat treated: 56,000-62,700 TS; 33,600-51,500 YS; 8-15 El; 120-150 Brin. For aircraft construction; age hardened, high strength.

KYNAL C69
IMI Knoch Ltd.
Cu 1-2, Mg 0.5-1.25, Si 1, Mn 1, bal Al.
Heat treated: 38,100-52,000 TS; 29,100-41,000 YS; 6-15 El; 105 Brin. For aircraft construction; age hardened, high strength.

KYNAL C70
IMI Knoch Ltd.
Cu 2.5-4, Mg 0.25-0.75, Sb 0.3-1, bal Al.
Rolled: 36,000 TS; 15,700 YS; 10 El. For fuel caps, structures; free-cutting.

KYNAL C71
IMI Knoch Ltd.
Cu 1.3-3, Mg 0.2-0.5, Sn 0.1-0.5, bal Al.
Heat treated: 38,000 TS. For rivets, structural members; heat treatable.

KYNAL M 35/1
IMI Knoch Ltd.
Mg 2, Mn 0.5, bal Al.
Annealed: 24,650 TS; 13,400 YS; 18 El; 45 Brin. Hard: 33,600 TS; 26,880 YS; 5 El; 67 Brin. For medium strength applications; good ductility and corrosion resistance.

KYNAL M 35/2
IMI Knoch Ltd.
Mg 3, Mn 1, bal Al.
Annealed: 31,360 TS; 15,680 YS; 18 El; 50 Brin. Hard: 40,320 TS; 33,600 YS; 5 El; 75 Brin. For medium strength applications; good ductility and corrosion resistance.

KYNAL M 35/3
IMI Knoch Ltd.
Mg 3.75-4.25, Si 0-0.6, Fe 0-0.7, Ti 0-0.2, bal Al.
Soft: 38,000 TS; 15,700 YS. For marine applications; corrosion resistant.

KYNAL M33
IMI Knoch Ltd.
Mg 0.9-1.1, Cu 0-0.05, Mn 0-0.05, bal Al.
Soft: 17,000 TS; 7500 YS; 22 El. 1/2 H-temper: 25,000 TS; 20,000 YS; 4 El. H-temper: 29,200 TS; 29,200 YS; 2 El. For medium strength structures; good ductility and drawability.

KYNAL M36
IMI Knoch Ltd.
Mg 5, Mn 1, bal Al.
Annealed: 38,000 TS; 17,920 YS; 18 El; 55 Brin. Hard: 44,800 TS; 38,080 YS; 5 El; 80 Brin. For marine parts; good corrosion resistance.

KYNAL M37
IMI Knoch Ltd.
Mg 7, Mn 1, bal Al.
Annealed: 49,300 TS; 20,200 YS; 20 El; 60 Brin. For marine parts; corrosion resistant.

KYNAL M39
IMI Knoch Ltd.
Si 0.7-1, Mg 0.5-1, bal Al.
Soft: 16,000 TS; 25 El. Heat treated: 38,000 TS; 31,000 YS; 10 El. For brazing alloy; heat treatable.

KYNAL M39/1
IMI Knoch Ltd.
Mg 0.4-1.5, Si 0.3-0.7, bal Al.
Heat treated: 33,600-38,100 TS; 20,400-31,600 YS; 14-22 El; 65-80 Brin. For aircraft construction; corrosion resistant, age hardened.

KYNAL M39/2
IMI Knoch Ltd.
Mg 0.5-1, Si 0.7-1.3, Mn 1, Fe 0.6, Ti 0.2, bal Al.
Heat treated: 33,600-44,800 TS; 20,200-41,500 YS; 10-22 El; 64-95 Brin. For aircraft construction; corrosion resistant, age hardened.

KYNAL M40
IMI Knoch Ltd.
Cu 0.15-0.4, Mn 0.2-0.8, Mg 0.8-1.2, Si 0.4-0.8, Fe 0-0.7, bal Al.
Heat treated: 38,000-41,000 TS; 30,000-34,000 YS; 7-14 El. For structural members; heat treatable, corrosion resistant.

KYNAL M41

IMI Knoch Ltd.
Mg 0.4-1.5, Si 0.6-1.3, Ti 0-0.2, Cr 0-0.5, bal Al.
Heat treated: 36,000-41,000 TS; 30,000-34,000 YS; 7-14 El.
For light alloy parts; heat treatable, corrosion resistant.

KYNAL P1

IMI Knoch Ltd.
Al 99.99.
Soft: 8950 TS; 45 El. 1/2 H-temper: 13,000 TS; 12 El. H-temper: 14,500 TS; 6 El. For heat and light reflectors; corrosion resistant and ductile.

KYNAL P10

IMI Knoch Ltd.
Cu 0.1, Si 0.5, Fe 0.7, Mn 0.1, bal Al.
Soft: 12,300 TS; 7500 YS; 35 El; 21 Brin. Hard: 20,100 TS; 16,000 YS; 5 El; 38 Brin. For trim, architectural applications; good ductility.

KYNAL P3

IMI Knoch Ltd.
Al 99.8, Cu 0-0.02, Si 0-0.15, Fe 0-0.15.
Soft: 10,000 TS. 1/2 H-temper: 16,000 TS. H-temper: 18,000 TS. For heat and light reflectors; corrosion resistant and ductile.

KYNAL P5

IMI Knoch Ltd.
99.6 Al min.
Annealed: 11,200 TS; 40 El; 21 Brin. Hard: 19,000 TS; 7 El; 38 Brin. For chemical plant equipment; high resistance to corrosion.

KYNAL PA 19

IMI Knoch Ltd.
Mn 1.25, bal Al.
Annealed: 15,680 TS; 25 El; 25 Brin. Hard: 24,700 TS; 5 El; 47 Brin. For cold formed and welded parts; corrosion- resistant parts.

KYNAL PA16

IMI Knoch Ltd.
Si 4-5.5, Cu 0.1, Fe 0.6, Mn 0.5, bal Al.
Sand cast: 18,000 TS; 9,000 YS; 6 El; 40 Brin. Permanent mold: 24,000 TS; 9,000 YS; 9 El; 45 Brin. For instrument housings, intricate castings; corrosion resistant, weldable. *Obsolete*

KYNAL PA17

IMI Knoch Ltd.
Si 7-8, Fe 0-0.6, bal Al.
For welding rod; corrosion resistant. *Obsolete*

KYNAL PA20

IMI Knoch Ltd.
Zn 0.9-1.1, bal Al.
For cladding sheet for Z93A alloy.

KYNAL S57

IMI Knoch Ltd.
Cu 0.7-1.3, Mn 0.8-1.5, Si 10.5-13.5, Ni 0.7-1.3, bal Al.
Heat treated: 40,000 TS; 3 El. For medium strength applications; heat treatable, corrosion resistant.

KYNAL Y88

IMI Knoch Ltd.
Cu 1.5-4, Mg 0.3-1.5, Si 0.5-1.3, Fe 0.6-1.5, Ni 2, Ce 0.3, Ti 0.3, bal Al.
Heat treated: 60,500 TS; 47,100 YS; 10 El; 130 Brin. For aircraft construction, pistons, cylinder heads; age hardenable.

KYNAL Y89

IMI Knoch Ltd.
Cu 1.5-3, Mg 1.2-1.8, Si 0.55-1.25, Ni 0.5-1.5, Ti 0.2, bal Al.
WP-temper: 58,000-60,500 TS; 44,500-47,500 YS; 8-10 El.
Forgings for high temperature applications; age hardened, high strength.

KYNAL-PA15

IMI Knoch Ltd.
Si 9-14, Cu 0.1, Fe 0.6, Mn 0.5, bal Al.
O-Temper: 18,000 TS; 30 El. H-Temper: 27,000 TS; 7 El. For low and medium strength parts. Good formability and weldability. *Obsolete*

KYNALCORE C65A

IMI Knoch Ltd.
Cu 3.5-5, Mg 0.4-1.2, Mn 0.4-1.2, Ti 0.3, Fe 0.7, bal Al.
Heat treated: 54,000 TS; 31,500 YS; 15 El. For structural members; heat treatable, Al-clad.

KYNALCORE C66A

IMI Knoch Ltd.
Cu 3.5-4.8, Mg 0.8, Si 0.9, Fe 1, Mn 1.2, Ti 0.3, bal Al.
W-temper: 53,800 TS; 31,400 YS; 15 El. WP-temper: 58,200 TS; 44,800 YS; 8 El. For structural members; age hardened, Al-clad.

KYNALCORE C68A

IMI Knoch Ltd.
Cu 4.4, Mg 0.6, Si 0.7, Mn 0.6, bal Al.
T4-temper: 56,000 TS; 33,000 YS; 15 El. T6-temper: 60,100 TS; 47,000 YS; 8 El. For aircraft structures; Al-coated.

KYNALCORE Z93A

IMI Knoch Ltd.
Cu 0-1.5, Mg 2-3.5, Zn 4.5-6.5, Mn 0.8, bal Al.
Heat treated: 74,000 TS; 62,700 YS; 9 El; 170 Brin. For structural members; heat treatable, Al-clad.

KYNALCORE Z93C

IMI Knoch Ltd.
Cu 1.5, Mg 2-3.5, Si 0.5, Fe 0.5, Zn 4.5-6.5, Ti 0.2, bal Al.
WP-temper: 72,000 TS; 60,000 YS; 8 El. For structural members; age hardened, high strength.

KYRO-KAY 316L

Teledyne McKay
Stainless steel. C 0.025, Mn 2.1, Si 0.3, Cr 17.7, Ni 13.6, Mo 2.1, bal Fe.
Covered welding electrode. AWS E316L.

L 11

Birmingham Aluminium Casting Co.
Aluminum. Cu 6-8, Sn 0-1, Fe 0-0.8, Ti 0-0.2, bal Al.
Cast: 22,000 TS; 16,000 YS; 4 El; 60 Brin. For gear boxes, cylinder heads; sand or permanent mold castings.

L 14

Bergische Stahl Industrie
C 0.95-1.05, Si 0.1-0.25, Mn 0.1-0.25, bal Fe.
Carbon tool steel. W.-Nr. 1.1540.

L 23

Bergische Stahl Industrie
C 1.05-1.15, Si 0.1-0.3, Mn 0.1-0.35, bal Fe.
Carbon tool steel. W.-Nr. 1.1650.

L 24

Bergische Stahl Industrie
C 0.95-1.05, Si 0.1-0.3, Mn 0.1-0.35, bal Fe.
Carbon tool steel. W.-Nr. 1.1640.

L 25

Bergische Stahl Industrie
C 0.8-0.9, Si 0.1-0.3, Mn 0.1-0.35, bal Fe.
Carbon tool steel. W.-Nr. 1.1630.

L 5

Birmingham Aluminium Casting Co.
Aluminum. Zn 12.5-14.5, Cu 2.5-3, Fe 0-0.8, Si 0-0.7, Ti 0-0.2, bal Al.
Cast: 20,000 TS; 11,000 YS; 4 El; 65 Brin. For crankcases, gear boxes, fans, brackets; sand or permanent mold castings.

L 8

Birmingham Aluminium Casting Co.
Aluminum. Cu 11-13, Fe 0-0.8, Si 0-0.7, Ti 0-0.2, bal Al.
Cast: 22,000 TS; 16,000 YS; 1.5 El; 80 Brin. For carburetors, automobile pistons; high pressure castings.

L C H S CHISEL

Jessop-Saville Ltd.
C, alloy, bal Fe.
For tools, chisels, punches; tough.

L C N-155

Cytemp Specialty Steel Div.
C 0.08-0.16, Mn 1-2, Cr 20-22.5, Ni 19-21, Co 18.5-21, W 2-3, Mo 2.5-3.5, 0.75-1.25 Cb + Ta, 0.10-0.20 N_2, bal Fe.
Heat treated: 80,000-150,000 TS; 40,000-110,000 YS; 50-15 El; 55-20 RA. For rotors, blades, bolts, buckets for gas engine and jet engines; high oxidation and heat resistance.

L T A H

Sanderson Kayser Ltd.
C 0.7, Si 0.3, Mn 2, Cr 1, Mo 1.45, bal Fe.
Cold work tool steel. B.S. 4659 Type BA6; AISI A6.

L-10 ALLOY

British Aluminium Co., Ltd.
Al 89, Cu 10, Sn 1, Fe, Si.
For light alloy parts. *Obsolete*

L-35

Darwin & Milner Inc.
Tool material. C 0.9, Mn 0.4, V 0.35, bal Fe.
For cold heading dies; water hardened.

L-605

Driver Harris Co.
Reactive/refractory. Ni 10, Cr 20, W 15, bal Co.
Wire and rod for welding wire and fastener stock. *Obsolete*

L-605

Now UDIMET L-605; ALLOY L-605; and UNITEMP L-605.

L-IV

Vereinigte Leichtmetallwerke, G.m.b.H.
Aluminum. Cu 4, Si 2, bal Al.
Heat treated: 43,000 TS; 36,000 YS; 4 El. For light alloy parts; same as "Lautal."

L-NICKEL

International Nickel Inc.
Ni 99.4, Cu 0.1, Fe 0.15, Mn 0.2, Si 0.05, C 0-0.02.
Annealed: 55,000-75,000 TS; 15,000-25,000 YS; 35-55 El; 60-75 RA; 80-100 Brin. For immersion heaters; corrosion and heat resisting. *Obsolete*

L. C. ELECTRITE DOUBLE SIX M-2

Latrobe Steel Co.
C 0.68, Cr 4.1, W 6.4, Mo 5, V 1.9, bal Fe.
For cold forming and working tools; oil hardening, high speed steel. *Obsolete*

L.C. IRON

Power Products Inc.
TC 3.5, Alloy, bal Fe.
Cast: 40,000 TS; 255 Brin. For castings for high temperature service; Tr.S. 2800; Tr.D. 210. *Obsolete*

L.F.M. 4-6 CR-MO

LFM Mfg. Co.
C 0.13-0.2, Cr 4-7, Mo, bal Fe.
For corrosion resistant parts; corrosion resistant.

L.L.C.T.

Colt Industries
C 0.2, Cr 4, V 0.5, W 15, bal Fe.
For hot work dies and tools; shear blades, crowning tools; tough and heat resistant. *Obsolete*

L.L.C.T.

Crucible Specialty Metals
C 0.2, Cr 4, V 0.5, W 15, bal Fe.
For hot work dies and tools; shear blades, crowning tools; tough and heat resistant. *Obsolete*

L.L.L. EXTRA

S.K.F. Industries Inc.
C, bal Fe.
For tools. *Obsolete*

L.L.L. NO. 44

S.K.F. Industries Inc.
C, alloy, bal Fe.
For tools. *Obsolete*

L.L.L. NO. 82

S.K.F. Industries Inc.
C, alloy, bal Fe.
For tools. *Obsolete*

L.L.L. NO. 83

S.K.F. Industries Inc.
C, alloy, bal Fe.
For tools. *Obsolete*

L.L.L. STANDARD

S.K.F. Industries Inc.
C, bal Fe.
For tools. *Obsolete*

L.M.

Alais Forges et Camargue
Si 0.75, Cu 4.75, Mn 0.75, bal Al.
58,000-65,000 TS; 18-20 El. For light alloy parts; age-hardenable.

L.M. STEEL

Republic Steel Corp.
Ni 2.8-3.3, Cr 0.7-0.9, Mo 0.3-0.5, C 0.2-0.4, bal Fe.
For gears, shafts, machinery parts. *Obsolete*

L.M.-0

Aluminiumwerke Maulbronn
Si 4, Cu 1.5, bal Al.
23,000-28,000 TS; 3-4 El. For light alloys, food handling machinery; age hardenable.

L.M.S

English manufacture
Pb Sn-Sb.
For bearings; anti-friction.

L.S.D.

Edgar Allen Balfour Ltd.
C 0.75, bal Fe.
For tools, springs, punches; water hardened. *Obsolete*

L.T.C.

Sigmund Cohn Corp.
Au 64-66, bal Ni-Cr.
Precision potentiometer resistance wire, bridgewire for electroexplosive devices.

L13

Sheffield Smelting Co. Ltd.
Ag 20, Cu, Zn.
Melting range: 770-810°C. Maximum stress: 50.4 kgf/mm^2; 4 El. For silver brazing; general use.

L13S

Sheffield Smelting Co. Ltd.
Ag 20, Cu, Zn, Si.
Melting range: 690-810°C. Maximum stress: 52.3 kgf/mm^2; 30 El. For silver brazing; general use.

L161

Friedr. Lohmann GmbH
C 1.65, Cr 11.2, V 0.4, Mo 0.4, Si 0.8, bal Fe.
For blanking and forming dies, punches. Air hardened, nondeforming.

L17

Sheffield Smelting Co. Ltd.
Ag 24, Cu, Zn.
Melting range: 740-780°C. Maximum stress: 35.3 kgf/mm^2; 13 El. For silver brazing; general use.

L2 ALLOY

Russian manufacture
Co 8, 0.8 Cr_2O_3, 2.0 TaC, 89.2 WC.
Sintered: 165,000 Transverse Strength; 90 Rock A. For cutting tools to machine cast iron, high alloy steels, hard plastics; sintered carbide, wear resistant.

L3B

Acciaierie Valbruna s.p.a.
Tool Material. C 1.12, Si 0-0.35, Mn 0-0.35, bal Fe.
Cold work tool steel. AISI W1-10; W. Nr. 1.1654.

L4B

Acciaierie Valbruna s.p.a.
Tool Material. C 1.25, Mn 0-0.3, Cr 0.35, Si 0-0.3, bal Fe.
Cold work tool steel. W. Nr. 1.2002.

L4BL

Acciaierie Valbruna s.p.a.
Tool Material. C 1.17, Si 0-0.3, Mn 0-0.4, Cr 0.65, V 0.1, bal Fe.
Cold work tool steel. W. Nr. 1.2210.

LA 21

Swiss manufacture
6 Cu + Ni, Mn, Mg, Si, bal Al.
Cast: 60,000-89,000 TS; 56,000-80,000 YS; 3-1 El; 100-120 Brin. Wrought: 63,000-76,000 TS; 44,000-60,000 YS; 4-2 El; 80-100 Brin. For bearings.

LA 31
Swiss manufacture
Sn 8, Cu, Ni, Mg, bal Al.
Cast: 44,000-60,000 TS; 22,000-31,000 YS; 8-3 El; 40-55 Brin.
Wrought: 50,000-63,000 TS; 28,000-41,000 YS; 12-6 El; 45-60 Brin. For bearings.

LA BELLE "C" & "H"
Colt Industries
C 0.5-0.8, Ni 0.4, Cr 0.2, V 0.1, bal Fe.
For chipper knives; water hardening. *Obsolete*

LA BELLE 1089
Colt Industries
C 0.3, Ni 2.25, Cr 3, W 10, bal Fe.
For hot work tools, gripper dies; hot work steel. *Obsolete*

LA BELLE 2-70
Crucible Materials Corp.
Tool material. C 0.6, Si 2, Mn 0.8, bal Fe.
For punches, chisels, stamps; tough.

LA BELLE CCV
Colt Industries
C 1, V 0.2, Cr 0.2, bal Fe.
For dies for plow bolts, square shoulder bolts; cold heading dies. *Obsolete*

LA BELLE CHROMONIC
Colt Industries
C 0.5, Mn 1.25, Cr 7, Ni 1.5, Mo 0.75, bal Fe.
For flying shears, billet shears; oil hardening. *Obsolete*

LA BELLE COLD HEADER DIE
Crucible Materials Corp.
Tool material. C 0.95, bal Fe.
For cold header dies; water hardening.

LA BELLE COLD STRIKING DIE
Crucible Materials Corp.
Tool material. C 0.95, bal Fe.
For cold forging dies; water hardening.

LA BELLE CSD
Colt Industries
C 0.95, bal Fe.
For dies for wheel bolts and single extrusion applications; cold heading dies. *Obsolete*

LA BELLE EAGLE
Colt Industries
C, bal Fe.
For tools, cutters, drills, dies, hammers; water hardening. *Obsolete*

LA BELLE EXTRA
Crucible Materials Corp.
Tool material. C 0.95, bal Fe.
For cold work dies, cold forming dies; water hardened.

LA BELLE EXTRA 2-1/2
Colt Industries
C 0.95, bal Fe.
For dies for small rivets and non-ferrous specialties; cold heading dies. *Obsolete*

LA BELLE EXTRA 3-1/2
Colt Industries
C 0.95, bal Fe.
For dies for bolts; cold heading dies. *Obsolete*

LA BELLE EXTRA 3A
Colt Industries
C 0.95, bal Fe.
For small screw and bolt dies; cold heading dies. *Obsolete*

LA BELLE HOT HEADER
Colt Industries
C, alloy, bal Fe.
For tools, hot header dies. *Obsolete*

LA BELLE N-183 MIX
Colt Industries
C, Ni, V, bal Fe.
For keen edge dies, machine knives. *Obsolete*

LA BELLE NICKEL MOLY
Colt Industries
C 1.1, Ni 0.5, Mo 1.25, bal Fe.
For tools, dies; oil hardening. *Obsolete*

LA BELLE NO. 211 DIE
Colt Industries
C 0.5, Ni 2.2, Cr 0.7, Mo 0.5, bal Fe.
For tools, hot work dies; oil hardened. *Obsolete*

LA BELLE NO. 24 DIE
Colt Industries
C 1.1, Cr 12, Mo 1, V 0.25, bal Fe.
For forming and crimping rolls, perforating dies, plug gages. *Obsolete*

LA BELLE NO. 89 DIE
Colt Industries
C 0.5, Cr 4, V 1, Mo 0.5, bal Fe.
For forging mandrels, bolt and rivet dies; hot work steel. *Obsolete*

LA BELLE SELFTEM
Colt Industries
C 0.35, Mn 0.8, Cr 0.9, Mo 0.5, bal Fe.
For chisels, shock tools; oil hardening. *Obsolete*

LA BELLE SEMI-HOT
Colt Industries
C, Cr, W, V, bal Fe.
For hot work tools and dies. *Obsolete*

LA BELLE SILICON 2
Crucible Materials Corp.
Tool material. C 0.6, Mn 0.8, Si 1.9, Mo 0.3, Cr 0.25, bal Fe.
For punches, chisels, rivet sets; tough.

LA BELLE SILICON PUNCH NO. 1
Colt Industries
C, Cr, Si, Mn, bal Fe.
For punches, chisels. *Obsolete*

LA BELLE SPECIAL HOT DIE
Colt Industries
C 0.4, Cr 2.25, Ni 2.5, bal Fe.
For hot heading dies; hot work steel. *Obsolete*

LA BELLE STAMP
Colt Industries
C 0.75, Mn 0.95, Mo 0.35, bal Fe.
For steel stamps; water hardening. *Obsolete*

LA BELLE STRIKING DIE
Colt Industries
C 0.95, bal Fe.
For cold forging dies; water hardening. *Obsolete*

LA BELLE TNU
Colt Industries
C 1.15, Ni 0.65, W 2, bal Fe.
For tools; oil hardening. *Obsolete*

LA BELLE TYPE 3
Colt Industries
C, bal Fe.
For dies for bolts, cold heading dies. *Obsolete*

LA BOUR R-55
LaBour Pump Co.
Nickel. Si 4, Cr 23, Cu 6, Mo 4, W 2, C 0.2-0.3, Ni 52, Fe 8.
For pump parts; modification of "R-50."

LA SALLE 1541 A,B
Now COLFORM E.T.D. 1541 A,B.

LA-1 ALLOY
Russian manufacture
C 0.16, Cr 15, Ni 15, Co 3, Mo 2, W 1, bal Fe.
Heat resisting alloy.

LA-4 ALLOY
Russian manufacture
C 0.12, Cr 15, Ni 15, Co 3, Mo 2, W 1, bal Fe.
Heat resisting alloy.

LA-5 ALLOY
Russian manufacture
C 0.16, Cr 15, Ni 16, Co 3, Mo 2, W 1, Cb 1, bal Fe.
Heat resisting alloy.

LA-685
General Communications Co.
Ni, bal Fe.
Rolled: 70,000 TS; 24,000 YS; 36 El; 68 RA. For wave meter cavities, gauges, instruments, optical structures; low temperature coefficient. *Obsolete*

LA-BOUR G-60
LaBour Pump Co.
Ni 63-65, C 0.06, Cr 23-25, Cu 5-6, Mo 2-4, W 2-4, Si 0.8, Fe 1, Mn 0.2.
For valves, pump parts; corrosion resisting. *Obsolete*

LA-BOUR R-50
LaBour Pump Co.
Ni 55, Cr 23-25, Cu 5-6, W 4, Mo 2, Fe 10.
For valves, pump parts; corrosion and heat resistant. *Obsolete*

LA-BOUR R-50 (MODIFIED)
LaBour Pump Co.
Ni 8-9, Cr 23, Cu 6, Mo 4, W 2, Si 0.75, Fe 53-54, C.
For valves, pump parts. *Obsolete*

LA-LED
LaSalle Steel Co.
C 0.08-0.09, Mn 0.85-1.15, S 0.26-0.35, Pb 0.15-0.35, P 0.04-0.07, bal Fe.
Drawn: 70,000 TS; 60,000 YS; 15 El; 45 RA; 140 Brin. For screw machine products; free-cutting, carburizing steel.

LA-LED X
LaSalle Steel Co.
C 0.09, Mn 0.85-1.15, P 0.04-0.09, S 0.26-0.35, Pb 0.15-0.35, Te, bal Fe.
Bar: 70,000 TS; 60,000 YS; 15 El; 45 RA; 140 Brin. For screw machine products, fasteners, hardware. Free machining.

LA-SALLE E.T.D. 180
Now "E.T.D" 180.

LABELLE HT
Crucible Materials Corp.
Tool material. C 0.43, Mn 1.3, Si 2.25, Cr 1.3, V 0.3, Mo 0.4, bal Fe.
Heat treated: 314,000 TS; 255,000 YS; 9 El; 29 RA; 560 Brin.
For shear blades, chisels, punches; oil hardened, shock resistant.

LABORATORY 33
J.F. Jelenko & Co.
Precious metal. Au 60, Pd 4, Ag 27.
Quenched: 59,000 psi TS; 29,500 psi YS; 34 El; 103 Brin.
Hardened: 86,000 psi TS; 52,000 psi YS; 11 El; 175 Brin.
Type III hard dental alloy. For inlays, crowns, and fixed bridgework.

LABORATORY 44
J.F. Jelenko & Co.
Precious metal. Ag 56, Pd 4, Ag 25.
Quenched: 73,100 psi TS; 54,000 psi YS; 38 El; 169 Brin.
Hardened: 108,000 psi TS; 104,500 psi YS; 2.5 El; 231 Brin.
Type IV extra-hard dental alloy. For hard inlays, thin crowns, fixed bridgework, and partial dentures.

LAC 10

Birmingham Aluminium Casting Co.
Aluminum. Cu 9-10.5, Mg 0.15-0.35, Fe 0.3-1, Si 0-0.6, bal Al.
30,000-44,000 TS; 28,000-40,000 YS; 110-130 Brin. For light alloy parts for high temperature service; hardenable.

LAC 10

British manufacture
Mg 0.15-0.35, Si 0.6, Cu 9-10.5, Mn 0.6, Fe 0.3-1, bal Al.
For pistons; non-hardenable.

LAC 112

British manufacture
Mg 0.3, Si 7-13, Cu 2-3, Mn 0.5, Ni 1.5, Zn 1.2, bal Al.
For die castings; high fluidity.

LAC 112A

Birmingham Aluminium Casting Co.
Aluminum. Cu 0.75-2.5, Si 9-11.5, bal Al.
For die castings.

LAC 113A

British manufacture
Si 1.3, Cu 2.5-4.5, Fe 1, Zn 9-13, bal Al.
For sand castings; non-hardenable.

LACOLITE

Lakeside Malleable Castings Co.
Cast iron. C 3, Si, Mn, alloy, bal Fe.
For gears, shafts, machine tool parts.

LACOMO

Latrobe Steel Co.
C 0.8, Cr 4, Co 5, V 1.25, W 1.5, Mo 8.5, bal Fe.
For hobs, saw teeth, lathe tools, reamers; high speed steel.
AISI M30. *Obsolete*

LACTOVAC

Fonderie de Precision S.A.
C 0.2-0.5, bal Fe.
For machinery parts; water hardening.

LADISH D-11

Ladish Co., Inc. (developer)
C 0.42-0.5, Mn 0.6-0.9, Si 0.25-0.45, Ni 0.4-0.7, V 0.45-0.6, Cr 0.9-1.2, Mo 1.9-2.15, bal Fe.
Air or oil hardened; good high temperature strength. For hot work tool and die applications.

LADISH D-6

Colt Industries
C 0.42-0.48, Cr 0.9-1.2, Ni 0.4-0.7, Mo 0.9-1.1, V 0.05-0.1, bal Fe.
For solid fuel rocket motor cases; tough, shock resistant.
Obsolete

LADISH D6A

Ladish Co., Inc. (developer)
Same composition as LADISH D6AC, but normally not vacuum melted. For stone crushers, dies and thrust bearings.

LADISH D6AC

Ladish Co., Inc. (developer)
C 0.42-0.5, Mn 0.6-0.9, Si 0.15-0.3, Ni 0.4-0.7, Cr 0.9-1, Mo 0.9-1.1, V 0.05-0.15, bal Fe.
Vacuum melted, hardenable to 300,000 psi TS. Used in aerospace industry for airframe components and solid rocket motor boosters.

LAFOND'S AXLE BEARING

Manufacturer not listed.
Cu 80, Sn 18, Zn 2.
For axle bearings; heavy duty.

LAFOND'S HEAVY BEARING

Manufacturer not listed.
Cu 83, Sn 15, Zn 1.5, Pb 0.5.
For heavy duty bearings; high strength.

LAFONDS MALLEABLE BRONZE

French manufacture
Sn 2, bal Cu.
For electrical equipment; corrosion resistant.

LAFONDS PUMP BRONZE

French manufacture
Sn 10, Zn 2, bal Cu.
For pump parts, gears, worm wheels, shafts; tough, corrosion resistant.

LAGAL

German manufacture
Al alloy.
For bearings; Zn-free.

LAGERBRONZE

Eisenwerke Neubrandenburg G.m.b.H.
Sn 4-5, Zn 3.5-5, Pb 3.5-4.5, bal Cu.
Cast: 50,000-58,000 TS; 50 El; 70-85 Brin. For machine and engine parts; corrosion resistant bronze.

LAITON YELLOW BRASS

English manufacture
Cu 60-70, Zn 27-40, Pb 5.3, Sn 0-1.
For plumbing, hardware, bolts, nuts; free-cutting.

LAKE'S METAL

Manufacturer not listed.
Pb 87, Sn 6, Sb 7.
For bearings for heavy loads; anti-friction.

LALL

Swiss manufacture
Mg, Zn, bal Al.
Cast: 44,000-56,000 TS; 19,000-34,000 YS; 12-4 El; 35-45 Brin. Wrought: 44,000-60,000 TS; 19,000-28,000 YS; 22-14 El; 35 RA; 50 Brin. For bearings.

LAMITE GRADE 1

Teledyne Vasco
Co-Cr-Mo-W.
For hard facing electrodes; cast. *Obsolete*

LAMITE GRADE 3

Teledyne Vasco
Co-Cr-Mo-W.
For hard facing electrodes; cast, corrosion resistant.
Obsolete

LAMITE GRADE 6

Teledyne Vasco
Co-Cr-Mo-W.
For hard facing electrodes; cast, corrosion resistant.
Obsolete

LAN-CER-AMP

American Metallurgical Products Co.
Ce 45-50, 30.0 La min, 20-24 Nd + Pr.
For alloying ferrous and nonferrous alloys.

LANARK

Latrobe Steel Co.
C 0.6, Cr 0.18, Mn 0.9, V 0.28, Mo 1.28, Si 1.9, bal Fe.
For chisels, punches, shear blades, stamps; shock resistant.

LANCASTER SILEXITE 55

Lancaster Steel Co. Inc.
C, Cr, Ni, bal Fe.
For pump shafts, valve parts; heat and corrosion resistant.
Obsolete

LANDERIG'S SPECULUM

Manufacturer not listed.
Cu 70, Sn 30.
For mirrors, reflectors; bronze.

LANGALLOY 10V

Langley Alloys Ltd.
C 0-0.12, Ni 10, Cr 18, bal Fe.
For chemical plant equipment at low temperature operation; stainless. *Obsolete*

LANGALLOY 12V

Langley Alloys Ltd.
C 0-0.08, Ni 12, Cr 18, Mo 3.5, S, bal Fe.
For chemical plant equipment; acid resistant, free-cutting.

LANGALLOY 13V

Langley Alloys Ltd.
C 0-0.08, Ni 12, Cr 18, Mo 3.5, bal Fe.
For chemical plant equipment; acid resistant.

LANGALLOY 14V

Langley Alloys Ltd.
Cr 18, Ni 10, Mo 3, Nb 1, bal Fe.
Cast: 400 N/mm^2 TS; 240 N/mm^2 YS; 12 El. Niobium stabilized austenitic for chemical equipment. Free machining casting.

LANGALLOY 15V

Langley Alloys Ltd.
Cr 18, Ni 8, bal Fe.
Austenitic stainless steel casting. Similar to ASTM A296 CF-8.

LANGALLOY 16V

Langley Alloys Ltd.
C 0-0.08, Ni 8, Cr 18, Mo 3, bal Fe.
For chemical plant equipment; acid resistant.

LANGALLOY 18V

Langley Alloys Ltd.
C 0-0.08, Ni 8, Cr 18, Mo 3, S, bal Fe.
For chemical plant equipment; acid resistant, free-cutting.

LANGALLOY 1R

Langley Alloys Ltd.
C 0-0.5, Cr 14, Ni 75, Fe 8, Si 0-2.5, Mn 0-0.7.
For furnace equipment, salt pots, heat treating boxes; high heat resistance, good strength.

LANGALLOY 1V

Langley Alloys Ltd.
C 0.2, Cr 17.5-19.5, Ni 10-12, Cb = 10 x C, bal Fe.
Sand cast: 70,000 TS; 30,000 YS; 20 El; 200 Brin. For valves, pump parts, welded structures; stainless, austenitic, stabilized.

LANGALLOY 20V

Langley Alloys Ltd.
C 0-0.07, Cr 20, Ni 29, Mo 3, Cb 1, bal Fe.
For sulfuric acid equipment; stainless casting for H_2SO_4.

LANGALLOY 22V

Langley Alloys Ltd.
Cr 18, Ni 10, Mo 3, Ni 1, bal Fe.
Stabilized austenitic stainless steel casting. For weldable chemical equipment. BS 3100 318C17.

LANGALLOY 25V

Langley Alloys Ltd.
C 0-0.5, Ni 12, Cr 25, Si 0-2, Mn 0-2, bal Fe.
For superheaters in oil refineries, furnace parts, heat treating equipment. Oxidation resistant to 1100°C, good hot strength.

LANGALLOY 26V

Langley Alloys Ltd.
C 0-0.35, Ni 12, Cr 25, Si 0-1.75, Mn 0-2, bal Fe.
For superheaters in oil refineries, furnace parts, heat treating equipment; oxidation resistant to 1100 C, good hot strength.
Obsolete

LANGALLOY 2V

Langley Alloys Ltd.
C 0-0.12, Ni 10-12, Cr 17.5-19, Mo 2.75-4, S 0.15-0.35, bal Fe.
For chemical equipment, welded acid resistant units; stainless, free cutting. *Obsolete*

LANGALLOY 33V
Langley Alloys Ltd.
Cr 18, Ni 10, Mo 3, bal Fe.
Austenitic stainless steel casting. For chemical equipment. Similar to ASTM A296 CF-8M.

LANGALLOY 3V
Langley Alloys Ltd.
Cr 17.5-19, Ni 10-12, Mo 2.5-3, C 0.2, bal Fe.
Sand cast: 70,000 TS; 30,000 YS; 15 El; 200 Brin. For valves, pump parts, chemical plant equipment; Type 317, stainless, austenitic.

LANGALLOY 4R
Langley Alloys Ltd.
C 0-0.12, Fe 5, Mn 0.75, Si 0.75, Mo 25-35, bal Ni.
Cast: 72,000-80,000 TS; 45,000-56,000 YS; 5-10 El; 200-250 Brin. For chemical plant equipment, pumps, valves; corrosion resistant. *Obsolete*

LANGALLOY 4V
Langley Alloys Ltd.
C 0.2-0.3, Ni 10.5-12.5, Cr 21-24, W 2-3, bal Fe.
Cast: 78,500 TS; 33,000 YS; 20 El; 165-185 Brin. For oil refinery cracking plant, gas turbine parts; stainless steel casting, heat and creep resistant.

LANGALLOY 5R
Langley Alloys Ltd.
Fe 5, Mn 0.75, Si 0.75, Cr 15-18, Mo 17, W 5, bal Ni.
Cast: 72,000-80,000 TS; 45,000-56,000 YS; 8-12 El; 180-220 Brin. For chemical plant equipment, pumps, valves; corrosion resistant. *Obsolete*

LANGALLOY 5V
Langley Alloys Ltd.
C 0.12, Cr 18, Ni 15, Mo 3, bal Fe.
Cast. For specialized instrumentation applications; stainless, nonmagnetic.

LANGALLOY 6R
Langley Alloys Ltd.
Si 10, Cu 3, bal Ni.
Sand cast: 56,000-78,000 TS; 56,000-78,000 YS; 0-3 El; 350-400 Brin. For valve cones and seats, pump parts; resists corrosion and erosion. *Obsolete*

LANGALLOY 6V
Langley Alloys Ltd.
C 0-0.12, Cr 13, bal Fe.
For pump parts, valve trim; heat treatable. *Obsolete*

LANGALLOY 7R
Langley Alloys Ltd.
Cr 23, Mo 6, Cu 6, Fe 5, W 2, bal Ni.
Sand cast: 56,000 TS; 40,000 YS; 6 El; 180 Brin. For valves, valve seats, pump parts; resists H_2SO_4 acids.

LANGALLOY 7V
Langley Alloys Ltd.
C 0-0.15, Ni 8, Cr 18, Mo 3, bal Fe.
For chemical plant equipment; acid resistant.

LANGALLOY 8R
Langley Alloys Ltd.
C 0-0.75, Ni 60, Cr 15, Si 0-2.
For heat treating equipment, salt pots; high thermal fatigue and hot gas resistance.

LANGALLOY 8V
Langley Alloys Ltd.
C 0-0.15, Ni 8, Cr 18, Mo 3, Mn 0-2, S, bal Fe.
For chemical plant equipment; acid resistant, free-cutting.

LANGALLOY 9R
Langley Alloys Ltd.
C 0-0.5, Ni 37, Cr 18, Si 0-3, Mn 0-2, bal Fe.
For heat treating equipment, furnace parts; good for cyclic heating up to 1100°C.

LANTHANUM
Atomergic Chemetals Corp.
La.
Purities: 99.9%, 99.6%, 99+%. Forms: ingot, lump, rod, sheet, foil, wire, powder.

LANTHANUM METAL
Ronson Metals Corp.
Comparatively pure La, for alloying.

LANZ CAST IRON
Heinrich Lanz A.G.
C 3.4, Si 2.8, Mn 0.7, bal Fe.
56,500 TS; 26 El. For housings, frames, gears, castings, high strength castings.

LAPCO
Light Alloys Products Co. Ltd.
Al alloy.
For light alloy parts.

LAPELLOY
Universal Cyclops
C 0.25-0.35, Cr 11-12, Mo 2.5-3, V 0.3, Ni 0-0.5, bal Fe.
At 20°F: 155,000 TS; 140,000 YS; 17 El; 35 RA. At 1000°F: 60,000 TS; 50,000 YS; 35 El; 85 RA. For high temperature bolts, valve stems, turbine buckets and blades; scale and oxidation resistant to 1400°F.

LAPELLOY
Carpenter Technology Corp.
C 0.25-0.35, Cr 11-12, Mo 2.5-3, V 0.3, Ni 0-0.5, bal Fe.
At 20°F: 155,000 TS; 140,000 YS; 17 El; 35 RA. At 1000°F: 60,000 TS; 50,000 YS; 35 El; 85 RA. For high temperature bolts, valve stems, turbine buckets and blades; scale and oxidation resistant to 1400°F.

LAPELLOY C
Carpenter Technology Corp.
C 0.2-0.25, Cr 11-12, Mo 2.5-3, Cu 1.75-2.25, N 0.1, bal Fe.
Heat treated: 135,000-203,000 TS; 105,000-170,000 YS; 17-18 El; 47-55 RA. For turbine shafts, compression wheels and buckets. Heat and oxidation resistant to 1400°F.

LARPORT
Osborn Steels Ltd.
C 0.3, Cr 1.25, Mo 0.3, Ni 4.25, bal Fe.
For plastic mold dies; oil or air hardened, tough. *Obsolete*

LASALLE 1018
LaSalle Steel Co.
C 0.15-0.2, Mn 0.6-0.9, P 0-0.04, S 0-0.05, bal Fe.
Carbon, cold finished steel bar. 64,000 psi TS; 54,000 psi YS; 15 El (in 2 in.); 40 RA; 126 Brin. For bending and cold forming.

LASALLE 1020-90
LaSalle Steel Co.
C 0.18-0.23, Mn 0.7-1, bal Fe.
Rolled: 57,000-80,000 TS; 40,000-72,000 YS; 17-38 El; 60-65 RA; 125-156 Brin. For machinery parts, piston pin shafts, pinions, gears; Modified X-1020. *Obsolete*

LASALLE 1045
LaSalle Steel Co.
C 0.43-0.5, Mn 0.6-0.9, P 0-0.04, S 0-0.05, bal Fe.
Carbon, cold finished steel bar. 91,000 psi TS; 77,000 psi YS; 12 El (in 2 in.); 35 RA; 179 Brin. For induction hardening applications.

LASALLE 1117
LaSalle Steel Co.
C 0.14-0.2, Mn 1-1.3, P 0-0.04, S 0.08-0.13, bal Fe.
Carbon cold finished steel bar. 69,000 psi TS; 58,000 psi YS; 15 El (in 2 in.); 40 RA; 137 Brin. Resulfurized.

LASALLE 1144
LaSalle Steel Co.
C 0.4-0.48, Mn 1.35-1.65, P 0-0.04, S 0.24-0.33, bal Fe.
Cold finished steel bar. 108,000 psi TS; 90,000 psi YS; 10 El (in 2 in.); 30 RA; 217 Brin. Resulfurized.

LASALLE 1212
LaSalle Steel Co.
C 0-0.13, Mn 0.7-1, P 0.07-0.12, S 0.16-0.23, bal Fe.
Carbon, cold finished steel bar. 78,000 psi TS; 60,000 psi YS; 10 El (in 2 in.); 35 RA; 167 Brin. Basic screw machine steel.

LASALLE 1215/1213
LaSalle Steel Co.
C 0.09-0.13, Mn 0.7-1.05, P 0.04-0.12, S 0.24-0.35, bal Fe.
Carbon, cold finished steel bar. 78,000 psi TS; 60,000 psi YS; 10 El (in 2 in.); 35 RA; 167 Brin. Used where mechanical properties can be sacrificed for machinability.

LASALLE 4140
LaSalle Steel Co.
C 0.38-0.43, Mn 0.75-1, P 0-0.035, S 0-0.04, Si 0.15-0.35, Cr 0.8-1.1, Mo 0.15-0.25.
Annealed alloy. Cold finished steel bar.

LASALLE 4140 LEADED
LaSalle Steel Co.
C 0.38-0.43, Mn 0.75-1, P 0-0.035, S 0-0.04, Si 0.15-0.35, Cr 0.8-1.1, Mo 0.15-0.25, Pb 0.15-0.35.
Annealed alloy. Cold finished steel bar.

LASALLE 8620
LaSalle Steel Co.
C 0.18-0.23, Mn 0.7-0.9, P 0-0.035, S 0-0.04, Si 0.15-0.35, Ni 0.4-0.7, Cr 0.4-0.6, Mo 0.15-0.25, bal Fe.
Alloy. Cold finished steel bar.

LASALLE 8620 LEADED
LaSalle Steel Co.
C 0.18-0.23, Mn 0.7-0.9, P 0-0.035, S 0-0.04, Si 0.15-0.35, Ni 0.4-0.7, Cr 0.4-0.6, Mo 0.15-0.25, Pb 0.15-0.35, bal Fe.
Leaded alloy. Cold finished steel bar.

LASALLE SERIES 83-420
LaSalle Steel Co.
C 0-0.09, Mn 0.85-1.15, P 0.04-0.09, S 0.26-0.35, Bi added, bal Fe.
Cold finished steel bar. 70,000-90,000 psi TS; 60,000-80,000 psi YS; 15-20 El; 35-55 RA; 75-90 Rock B; 137-183 Brin. 250% of 1212 machining characteristics.

LATROBE
Teledyne Vasco
C 1, Mn 0.3, Si 0.25, bal Fe.
For drills, taps, punches, cutters; Type W1. *Obsolete*

LATROBE
Teledyne Vasco
C 0.6-1.4, bal Fe.
For tools, general purposes. *Obsolete*

LATROBE "EXTRA"
Latrobe Steel Co.
Mn 0.2-0.35, bal Fe.
For taps, reamers, twist drills, shear blades, springs, tools; water hardening. *Obsolete*

LATROBE "SPECIAL"
Latrobe Steel Co.
Mn 0.2-0.35, C, bal Fe.
For lathe and planer tools, taps, threading dies, reamers, punch. *Obsolete*

LATROBE "STANDARD"
Latrobe Steel Co.
Mn 0.2-0.35, C, bal Fe.
For blacksmith tools, cold chisels. *Obsolete*

LATROBE AGT
Latrobe Steel Co.
C 0.08-0.13, Ni 3-3.5, Cr 1-1.4, Mo 0.08-0.15, bal Fe.
Rolled: 131,000 TS; 88,000 YS; 19 El; 61 RA; 269 Brin. Normalized: 131,000 TS; 83,000 YS; 19 El; 58 RA; 269 Brin. For gears, pinions, camshafts, crankshafts; case hardening, tough, shock resistant. *Obsolete*

LATROBE BR-3
Latrobe Steel Co.
C 2.8, Cr 5.2, V 4.5, bal Fe.
For cold working dies; for cold working applications.
Obsolete

LATROBE BR-4
Latrobe Steel Co.
C 2.3, Si 0.4, Mn 0.4, Cr 12.5, V 4, Mo 1.1.
Die steel. For applications involving extreme abrasive wear.
AISI Type D7.

LATROBE BR-4
Latrobe Steel Co.
C 2.4, Cr 12.75, Mn 0.4, V 4, Mo 1.1, bal Fe.
For shaving, stamping and deep draw dies; abrasion
resistant.

LATROBE BR-4FM
Latrobe Steel Co.
C 2.4, Cr 12.75, Mo 1.1, V 4, Mn 0.4, S 0.15, bal Fe.
For stamping and deep drawing dies; oil hardened, non-
deforming. *Obsolete*

LATROBE CARBON
Latrobe Steel Co.
C 1.05, Si 0.2, Mn 0.2, Cr 0.08, bal Fe.
For tools, drills, taps; water hardening.

LATROBE CC
Latrobe Steel Co.
C, alloy, bal Fe.
For tools. *Obsolete*

LATROBE CLW NO. 1
Latrobe Steel Co.
Now CLW NO. 1.

LATROBE CM 52
Latrobe Steel Co.
C 0.89, Cr 4, V 1.85, Mo 4.5.
Intermediate high speed steel. For cutting; commercial twist
drills and woodworking tools.

LATROBE CM-50
Latrobe Steel Co.
C 0.84, Cr 4.1, V 1, Mo 4.25.
Intermediate type high speed steel; for moderate cutting;
woodworking tools.

LATROBE GSN-FM
Latrobe Steel Co.
C 2.2, Cr 13, S, bal Fe.
For stamping and forming dies; oil hardened, non-deforming.
Obsolete

LATROBE HW 108
Latrobe Steel Co.
C 0.38, Si 1, Mn 0.3, Cr 3.75, Mo 3.6, V 0.5, Co 2.
Hot work steel. For forging dies and punches, extrusion dies
and die casting tooling

LATROBE MCH
Latrobe Steel Co.
C 0.5, W 1, Mo 6.25, Cr 3.75, V 0.75, bal Fe.
For hot shears and punches, extrusion dies and rams; hot
work steel, oil hardened. *Obsolete*

LATROBE MCL
Latrobe Steel Co.
C 0.3, W 1, Mo 6.25, Cr 3.75, V 0.75, bal Fe.
For extrusion dies and mandrels, hot punches, shears; hot
work steel, oil hardened. *Obsolete*

LATROBE MP159
Latrobe Steel Co.
Ni 25.5, Co 35.7, Cr 19, Fe 9, Mo 7, Ti 3, Cb 0.6, Al 0.2.
Annealed: 58,000 YS; 123,000 TS; 60 El; 69 RA. For
fasteners, jet engines, drive components, prosthetic devices,
marine and petroleum industry applications.

LATROBE MP35N
Latrobe Steel Co.
Ni 35, Co 35-0, Cr 20, Mo 10.
Annealed: 55,000 YS; 130,000 TS; 65 El; 75 RA; 90 Rock B.
Nickel-cobalt base produced by vacuum induction.

LATROBE NO. 426
Latrobe Steel Co.
C, W, bal Fe.
For tools, dies. *Obsolete*

LATROBE NO. 5
Latrobe Steel Co.
Tool material. C 0.6, W 18, Cr 4, V 1, bal Fe.
For bolt trimmers, header inserts, punches; high speed steel.
Obsolete

LATROBE NO. 7
Latrobe Steel Co.
C 0.65, W 6.5, Mo 5, Cr 4, V 2.
For extrusion dies, header inserts; high speed steel.
Obsolete

LATROBE O.K.V.
Latrobe Steel Co.
C, alloy, bal Fe.
For tools. *Obsolete*

LATROBE VDC-MF
Latrobe Steel Co.
C 0.4, Cr 5, Mo 1, Si 1, V 1, S, bal Fe
For hot work tools, die casting dies; Type H13; resists heat
checking. *Obsolete*

LATTEN (LAITON)-1
English manufacture
Cu 60-70, Zn 27-40, Pb 0-5.3, Sn 0-1.
For hardware, fixtures; yellow brass.

LATTEN (LAITON)-2
English manufacture
Cu 71, Zn 28.53, Pb 0.25, Sn 0.02, Fe 0.2.
Annealed: 56,000 TS; 45,000 YS; 35 El; 65 RA. For hardware,
fixtures; yellow brass.

LAUTAL-1
Vereinigte Leichtmetallwerke, G.m.b.H.
Aluminum. Cu 4.5-5.5, Si 0.2-0.5, bal Al.
Wrought: 35,000-63,000 TS; 18,000-40,000 YS; 20-25 El;
65-110 Brin. For aircraft construction; age-hardening.

LAUTAL-2
Vereinigte Leichtmetallwerke, G.m.b.H.
Aluminum. Cu 4.5, Si 0.75, Mn 0.75, bal Al.
Annealed: 23,000-35,000 TS; 7,000-12,000 YS; 12-20 El;
45-55 Brin. For electric cables; light alloy.

LAVELSSIERE BRONZE
Manufacturer not listed.
Cu 61, Zn 38, Sn 1.
For condenser tubing, marine parts; corrosion resistant.

LAVIN 10A NICKEL SILVER
R. Lavin & Sons, Inc.
Ni 12, Zn 20, Pb 9, Sn 2, Cu 57.
Cast: 30,000 TS; 15,000 YS; 8 El; 50-60 Brin. For hardware
fittings, valves, plumbing. Corrosion resistant. CDA 973.

LAVIN 10B NICKEL SILVER
R. Lavin & Sons, Inc.
Ni 16, Zn 16, Pb 5, Sn 3, Cu 60.
Cast: 35,000 TS; 17,000 YS; 15 El; 65-80 Brin. For plumbers'
fittings, statuary, bolts. Corrosion resistant. CDA 974.

LAVIN 11A NICKEL SILVER
R. Lavin & Sons, Inc.
Ni 20, Zn 8, Pb 4, Sn 4, Cu 64.
Cast: 30,000 TS; 17,000 YS; 8 El; 76-120 Brin. For hardware,
fittings, plumbing, valves. Corrosion resistant. CDA 976.

LAVIN 11B NICKEL SILVER
R. Lavin & Sons, Inc.
Ni 25, Zn 2, Pb 1.5, Sn 5, Cu 66.5.
Cast: 45,000 TS; 22,000 YS; 15 El; 120-150 Brin. For
hardware and plumbing fixtures. Corrosion resistant. CDA
978.

LAVIN 7A MANGANESE BRONZE
R. Lavin & Sons, Inc.
Cu 59, Sn 0.75, Pb 0.75, Fe 1, Mn 0.3, Al 1, bal Zn.
Cast: 60,000 TS; 20,000 YS; 15 El; 80-95 Brin. For valve
stems, propellers, worm gears. Corrosion resistant. CDA 864.

LAVIN 8A MANGANESE BRONZE
R. Lavin & Sons, Inc.
Cu 57, Fe 1, Al 1, Mn 0.25, bal Zn.
Cast: 65,000 TS; 25,000 YS; 20 El; 90-120 Brin. For valves,
propellers, worm gears, valve stems. Corrosion resistant. CDA
865.

LAVIN 8C MANGANESE BRONZE
R. Lavin & Sons, Inc.
Cu 64, Fe 3, Al 5, Mn 4, bal Zn.
Cast: 110,000 TS; 60,000 YS; 12 El; 190-235 Brin. For
screwdown nuts, gears, bridge parts. Corrosion resistant.
CDA 863.

LAVIN 9A ALUMINUM BRONZE
R. Lavin & Sons, Inc.
Cu 87.5, Fe 3.5, Al 9.
Cast: 65,000 TS; 25,000 YS; 20 El; 110-140 Brin. For worm
wheels, bearings, bushings. Corrosion resistant. CDA 952.

LAVIN 9B ALUMINUM BRONZE
R. Lavin & Sons, Inc.
Cu 89, Fe 1, Al 10.
Cast: 65,000 TS; 25,000 YS; 20 El; 110 Brin. Heat treated:
80,000 TS; 40,000 YS; 12 El; 160 Brin. For valve seats,
stripper nuts, gears. Corrosion resistant, heat treatable. CDA
953.

LAVIN 9C ALUMINUM BRONZE
R. Lavin & Sons, Inc.
Cu 85, Fe 4, Al 11.
Cast: 75,000 TS; 30,000 YS; 12 El; 150 Brin. Heat treated:
90,000 TS; 45,000 YS; 6 El; 190 Brin. For worm wheels, pump
parts, bushings. Corrosion resistant, heat treatable. CDA 954.

LAVIN 9D ALUMINUM BRONZE
R. Lavin & Sons, Inc.
Cu 81, Ni 4, Fe 4, Al 11.
Cast: 90,000 TS; 40,000 YS; 6 El; 190 Brin. Heat treated:
110,000 TS; 60,000 YS; 5 El; 200 Brin. For worm wheels,
pump parts, bushings. Corrosion resistant, heat treatable.
CDA 955.

LAVIN CF NO. 4
R. Lavin & Sons, Inc.
Cu alloy.
Deoxidizer for copper alloys; densifier.

LAVIN NDZ BRONZE
R. Lavin & Sons, Inc.
Zn 0-5, Ni 0-5.5, Fe 0-0.5, Al 0-2, Si 0-2, Pb 0-0.25, bal Cu.
Cast: 65,000-69,000 TS; 30,000-35,000 YS; 20-34 El; 17-31
RA; 123-134 Brin. Heat treated: 78,900 TS; 63,400 YS; 5 El;
179 Brin. For valve stems and bodies, valve bonnets,
propeller wheels, marine outboard gears. High corrosion
resistance with good strength. CDA 994.

LAVIN NDZ-S BRONZE

R. Lavin & Sons, Inc.
Zn 0-5, Ni 0-5.5, Fe 0-5.5, Al 0-2, Si 0-2, Pb 0-0.25, bal Cu.
Cast: 69,000-75,000 TS; 40,000-45,000 YS; 13-23 El; 10-20 RA; 143-149 Brin. Heat treated: 86,400 TS; 62,400 YS; 8 El; 196 Brin. For valve stems and bodies, valve bonnets, propeller wheels, marine outboard gears. High corrosion and dezincification resistance. High yield strength. CDA 995.

LAVIN NO. 230

R. Lavin & Sons, Inc.
Cu 55, Mn, bal Zn.
Cast: 65,000 TS; 40 El; 112 Brin. For castings; Mn bronze. Obsolete

LAVIN NO. 300

R. Lavin & Sons, Inc.
Cu 55, Zn, Mn.
Cast: 70,000 TS; 38 El; 131 Brin. For castings; Mn bronze. Obsolete

LAVIN NO. 310

R. Lavin & Sons, Inc.
Cu 55, Zn, Mn.
Cast: 60,000 TS; 45 El; 103 Brin. For castings; Mn bronze. Obsolete

LAVIN NO. 320

R. Lavin & Sons, Inc.
Cu 55, Zn, Mn.
Cast: 80,000 TS; 30 El; 156 Brin. For castings; Mn bronze. Obsolete

LAVIN NO. 384

R. Lavin & Sons, Inc.
Cu 60, Zn, Mn.
Cast: 90,000 TS; 20 El; 178 Brin. For castings, propellers; Mn bronze. Obsolete

LAVIN NO. 420

R. Lavin & Sons, Inc.
Cu 55, Zn, Mn.
Cast: 70,000 TS; 35 El; 137 Brin. For castings, propellers; Mn bronze. Obsolete

LAVIN NO. 483

R. Lavin & Sons, Inc.
Cu 60, Zn, Mn.
Cast: 80,000 TS; 25 El; 170 Brin. For castings, propellers; Mn bronze. Obsolete

LAVIN NO. 598

R. Lavin & Sons, Inc.
Cu 60, Zn, Mn.
Cast: 110,000 TS; 15 El; 228 Brin. For castings, propellers; Mn bronze. Obsolete

LAVIN NO. 599

R. Lavin & Sons, Inc.
Cu 60, Zn, Mn.
Cast: 115,000 TS; 12 El; 235 Brin. For castings, propellers; Mn bronze. Obsolete

LAVIN NO. 685

R. Lavin & Sons, Inc.
Cu 60, Zn, Mn.
Cast: 100,000 TS; 15 El; 207 Brin. For castings, propellers; Mn bronze. Obsolete

LAVIN NSF NO. 5

R. Lavin & Sons, Inc.
Degasifier for nickel-silver castings; densifier.

LAVIN SPECIAL

R. Lavin & Sons, Inc.
Cu, Ni, Al, bal Zn.
Cast: 70,000-102,000 TS; 40,000-68,000 YS; 1-6 El; 165-227 Brin. For neck rings and plungers for glass molding machines. High thermal conductivity.

LAWN MOWER

Disston Inc.
C 1, Cr 1.15, bal Fe.
For lawn mower blades; oil hardening. Obsolete

LAWS PHOSPHOR BRONZE

English manufacture
Sn 9.5-11, P 0.7-1, bal Cu.
For bearings, gears, bushings; heavy duty.

LC 406

Capitol Castings Inc.
C 0.5-0.6, Cr 1.9-2.3, Mo 0.28-0.38, bal Fe.
Mill liners.

LC LOW CARBON

Teledyne Pittsburgh Tool Steel
C 0.2-0.25, Mn, bal Fe.
AISI C1020; precision ground flats and squares.

LC SPEZIAL

Friedr. Lohmann GmbH
C 2.1, Cr 11.5, W 0.7, bal Fe.
For blanking and forming dies. Type D6; oil hardened, nondeforming.

LC-SILVERIN 400

Vereinigte Deutsche Nickel-Werke AG
Nickel. Fe 1.5, C 0-0.04, Mn 1, Si 0.05, 65 Ni + Co, bal Cu.
Strip, wire, bar, forged and turned parts for electronics, controls, optical parts, fasteners, apparatus, welding, drawing and stamping parts, automotive parts.

LCN-1

Union Carbide Corp.
Cu 55, Ni 41, In 4.
Metallic spray coating for D-Gun. High strength and anti-galling; 300 DPH.

LCT

Colt Industries
C 0.45, Cr 3, V 0.4, W 15, bal Fe.
For hot swaging dies, extrusion dies, shear blades; hot work steel. Obsolete

LCT NO. 2

Colt Industries
C 0.45, Cr 3, V 0.4, W 11.5, bal Fe.
For brass forging dies, punches, forming rolls; hot work steel; Peerless LCT 2. Obsolete

LCU-2

Union Carbide Corp.
Cu 90, Al 10.
Metallic spray coating for plasma torch. Machinable; 175 DPH. For gear bushings.

LE MATS METAL

Manufacturer not listed.
Cu 80, Zn 5, Ni 16, Fe 5, Sn 2, Co 1.
For jewelry; v."Lutecin."

LE MOYNE ALCOCAN

Le Moyne Steel Co.
C 0.7, Ni, Cr, bal Fe.
For hand tools, chisels; oil or air hardened.

LE MOYNE BEST

Le Moyne Steel Co.
C 1.2, Cr 0.3, W 1.2, bal Fe.
For broaches, reamers, dies, cutters; water hardened.

LE MOYNE EXTRA

Le Moyne Steel Co.
C 0.7-1, bal Fe.
For tools, drills, punches; water hardened.

LE MOYNE FINISHING

Le Moyne Steel Co.
C 1.3, W 4, Mo 4, bal Fe.
For finishing tools and cutters; water or oil hardened.

LE MOYNE HAK

Le Moyne Steel Co.
C 0.9-1.05, Cr 3.7-4.1, bal Fe.
Annealed: 90,000 TS; 78,000 YS; 28 El; 57 RA; 195 Brin. Heat treated: 205,000 TS; 168,000 YS; 2 El; 12 RA; 412 Brin. For riveters, gripper dies, compression dies; wear and heat resistant.

LE MOYNE HS

Le Moyne Steel Co.
C 0.56-0.75, W 18, Cr 4, V 1, bal Fe.
Annealed: 105,000 TS; 75,000 YS; 14 El; 17 RA; 255 Brin. For chasers, taps, reamers, drills, broaches; high speed steel, oil hardened.

LE MOYNE K760

Le Moyne Steel Co.
C 0.72, W 17.5, Cr 4, Co 4.5, V 1, Mo 0.5, bal Fe.
For reamers, taps, hobs, drills, lathe tools; high speed steel, oil hardened.

LE MOYNE NON-SHRINK

Le Moyne Steel Co.
C 0.9, Mn 1.2, Cr 0.5, W 0.5, bal Fe.
For punches, tools, crimpers, cutters; oil hardened, nondeforming.

LE MOYNE PYRO

Le Moyne Steel Co.
C 0.33, W 10.2, Cr 3.5, V 0.45, bal Fe.
For punches, shears, hot dies; hot work steel, oil hardened.

LE MOYNE SPECIAL

Le Moyne Steel Co.
C 0.8-1.1, bal Fe.
For tools, drills, punches, cutters; water hardened.

LE MOYNE SUPREME

Le Moyne Steel Co.
C 0.7, W, Co, Mo, bal Fe.
For tools, dies, cutters; high speed steel.

LE PROVINOX

French manufacture
C, alloy, bal Fe.
50,000-57,000 TS; 35,000-45,000 YS; 36 El. For general engineering applications; semi-non-oxidizing steel. Oil quenched and tempered, 900°F: 45 Rock C. For gas turbine compressor wheels and structural members.

LEAD

Atomergic Chemetals Corp.
Pb.
Purities, zone refined: 99.9999%, 99.9995%, 99.999%, 99.99%. Forms: rod, bar, shot, powder, wire, sheet, foil, single crystals.

LEAD ALLOY

Manufacturer not listed.
Sn 75, Sb 20, Pb 5.
For bearings; anti-friction.

LEAD BRONZE NO. 1

Knowsley Cast Metal Co., Ltd.
Sn 9, Pb 15, bal Cu.
Cast: 22,400 TS; 4 El; 50-70 Brin. For soft shaft bearings; for poor lubrication.

LEAD BRONZE NO. 2

Knowsley Cast Metal Co., Ltd.
Sn 10, Pb 10, bal Cu.
Cast: 24,600 TS; 4 El; 60-75 Brin. For bearings for medium hard shafts and heavy loads; good anti-friction properties.

LEAD BRONZE NO. 3
Knowsley Cast Metal Co., Ltd.
Sn 10, Pb 5, bal Cu.
Cast: 26,900 TS; 5 El; 65-75 Brin. For bearings; good corrosion resistance, high strength.

LEAD FOIL
Manufacturer not listed.
Pb 86, Fe 6.9, Al 5.5, Sn 1.9.
For lead foil.

LEAD FOIL (CALIN)
Manufacturer not listed.
Pb 86, Sn 13, Cu 1.
For lead foil; corrosion resistant.

LEAD NO. 1 HARD
NL Industries
Pb 90, Sb 10.
8220 TS; 17 El; 17 Brin. For type metal; MP 486°F. *Obsolete*

LEAD NO. 1 HARD
Climax Performance Materials Corp.
Pb 99.25, Cd 0.25, Sb 0.5.
For valves, cocks, cable sheathing.

LEAD NO. 2 HARD
NL Industries
Pb 85, Sb 15.
9000 TS; 11.7 El; 17 Brin. For type metal; MP 476°F. *Obsolete*

LEAD NO. 2 HARD
Climax Performance Materials Corp.
Pb 98.25, Cd 0.25, Sn 1.5.
For valves, cocks, tank lining.

LEAD SHOT
Manufacturer not listed.
Pb 99.8, As 0.2.
For lead shot.

LEAD TAPE
Manufacturer not listed.
Pb 95, Sb 4.5, Sn 0.5.
For lead tape; hard.

LEAD-CALCIUM
Manufacturer not listed.
Pb 97, Ca 0.79, Ba 0.66.
For bearings; anti-friction.

LEADBEATER BR
Leadbeater & Scott Ltd.
High C, medium Cr, bal Fe.
For dies; oil hardened.

LEADBEATER L.S. 11
Leadbeater & Scott Ltd.
C 2, Cr 12, bal Fe.
For blanking and wire drawing dies, press tools; oil hardened, non-deforming.

LEADBEATER L.S. 25
Leadbeater & Scott Ltd.
C, W, bal Fe.
For hot heading, swaging and piercing dies; hot die steel, shock and abrasion resistant.

LEADBEATER L.S. 3000
Leadbeater & Scott Ltd.
W 7, C, bal Fe.
For drawing dies, oil hardened.

LEADBEATER L.S. 54
Leadbeater & Scott Ltd.
high C, Cr, Mn, bal Fe.
For taps, reamers, chasers, dies; oil hardened.

LEADBEATER N.S.O.H.
Leadbeater & Scott Ltd.
High C, Mn, bal Fe.
For punches, dies; oil hardened, non-deforming.

LEADBEATER P.B.
Leadbeater & Scott Ltd.
C 0.6, Cr, bal Fe.
For taps, reamers, threading dies, punches, chasers; oil hardened.

LEADBEATER S.P.C.
Leadbeater & Scott Ltd.
High C, alloy, bal Fe.
For chisels, punches, marking dies; shock resistant, oil hardened.

LEADED BRASS 244
Anaconda Co.
Copper. Zn 37, Pb 1, bal Cu.
Hard: 70,000 TS; 60,000 YS; 8 El; 150 Brin. Soft: 45,000 TS; 17,000 YS; 50 El; 60 Brin. For hardware, screw machine parts. Free cutting.

LEADED BRONZE
American manufacture
Zn 10-35, Pb 1.5-2.5, bal Cu.
For hardware, screws, fittings, bolts; free-cutting.

LEADED COMMERCIAL BRONZE
Chase Brass & Copper Co., Inc.
Copper. Cu 89, Zn 9, Pb 2.
Drawn: 52,000 psi TS; 45,000 psi YS; 18 El; 60 Brin. For hardware, screw machine parts; free cutting.

LEADED COMMERCIAL BRONZE 202
Anaconda Co.
Copper. Cu 88.5, Zn 9.25, Pb 2.25.
Hard: 54,000 TS; 45,000 YS; 15 El; 58 Rock B. Soft: 37,000 TS; 12,000 YS; 40 El; 1 Rock B. For screw machine parts, hardware. Free cutting.

LEADED COMMERCIAL BRONZE-201
Anaconda Co.
Copper. Zn 9.5, Pb 0.5, bal Cu.
Soft: 37,000 TS; 12,000 YS; 40 El. Hard: 62,000 TS; 47,000 YS; 6 El. For percussion caps, gilding. Good workability.

LEADED COPPER 126
Anaconda Co.
Pb 1, bal Cu.
Soft: 30,000 TS; 10,000 YS; 38 El. Hard: 50,000 TS; 40,000 YS; 10 El. For screw machine parts, gas welding tips; free machining. *Obsolete*

LEADED COPPER-187
Anaconda Co.
Copper. Cu 99, Pb 1.
Hard: 48,000 TS; 40,000 YS; 12 El; 50 Rock B. Soft: 32,000 TS; 10,000 YS; 45 El; 45 Rock B. For current carrying studs, nuts, bolts, fasteners. Free cutting, 98% electrical conductivity.

LEADED FLANGING BRASS
Chase Brass & Copper Co., Inc.
Copper. Cu 62.5, Zn 35.6, Pb 1.9.
For screw machine products; flange grade.

LEADED GUN METAL
English manufacture
gun metal + 1.0 Pb.
For corrosion resistant parts; improved machinability.

LEADED HIGH BRASS
Anaconda Co.
Copper. Cu 65-78, Pb 0.3-0.8, bal Zn.
For tanks, vessels, containers. Good ductility and strength.

LEADED MONEL METAL
English manufacture
Ni 60, Cu 32, Pb 2.2, Fe 2.2, Mn 2, Si 0.9, C 0.2.
For corrosion resistant parts; improved machinability.

LEADED MUNTZ METAL
Revere Copper Products Inc.
Cu 58-61, Pb 0.25-0.7, bal Zn.
As hot rolled: 54,000 TS; 20,000 YS; 45 El. For heat exchangers, tube plates and baffles. C36500

LEADED MUNTZ METAL-365
Anaconda Co.
Copper. Cu 60, Zn 39.35, Pb 0.65.
Soft: 54,000 TS; 20,000 YS; 45 El; 45 Rock B. For industrial condensers. Strong, stiff, elastic, free cutting.

LEADED NAVAL BRASS
Now NAVAL BRASS, HIGH LEADED.

LEADED NAVAL BRASS 29
Olin Brass, Indianapolis
Copper. Cu 60, Pb 1.75, Sn 0.75, Zn 37.5.
Hard: 75,000 psi TS; 53,000 psi YS; 15 El. Soft: 57,000 psi TS; 25,000 psi YS; 40 El. For hardware; free-cutting; corrosion resistant.

LEADED NAVAL BRASS 482
Anaconda Co.
Copper. Zn 38.55, Sn 0.75, Pb 0.7, bal Cu.
Hard: 63,000 TS; 35,000 YS; 28 El; 116 Brin. Soft: 56,000 TS; 25,000 YS; 38 El; 83 Brin. For marine hardware, screw machine products. Free cutting, forgeable.

LEADED NAVAL BRASS-485
Anaconda Co.
Copper. Cu 60, Zn 37.5, Pb 1.75, Sn 0.75.
Hard: 63,000 TS; 35,000 YS; 25 El; 65 Rock B. Soft: 56,000 TS; 25,000 YS; 35 El; 50 Rock B. For marine hardware, valve stems, screw machine products. Free machining, high strength.

LEADED NICKEL COMMERCIAL BRONZE
Chase Brass & Copper Co.
Cu 90.25, Zn 6.9, Ni 1, P 0.1, Pb 1.75.
Copper Alloy No. 316. Hard: 70,000 TS; 60,000 YS; 10 El; 78 Rock B. For bolts, fasteners, nuts, pole line hardware. ASTM B140 Alloy B. *Obsolete*

LEADED NICKEL COPPER 7021
Anaconda Co.
Copper. Cu 97.8, Pb 1, Ni 1, P 0.2.
Rod: 85,000 TS; 75,000 YS; 5 El; 55% electrical conductivity; 80% machinability. For electrical contacts, connectors, control elements for power tubes.

LEADED NICKEL COPPER-831
Anaconda Co.
Copper. Pb 1, Ni 1, P 0.2, bal Cu.
Heat treated: 80,000 TS; 70,000 YS; 7 El. For screw machine products, fasteners. Free cutting, corrosion resistant.

LEADED NICKEL SILVER 10%-796
Anaconda Co.
Copper. Zn 42, Pb 1, Ni 10, Mn 2, bal Cu.
Hard: 70,000 TS; 40,000 YS; 15 El; 137 Brin. Soft: 60,000 TS; 20 El. For screw machine products, architectural parts. Free cutting, corrosion resistant.

LEADED NICKEL SILVER 10%-823
Anaconda Co.
Copper. Pb 2.75, Ni 10, Zn 40.5, Mn 0.15, bal Cu.
For architectural trim, forgings. Extruded.

LEADED NICKEL SILVER 12%-796
Anaconda Co.
Pb 1, Ni 12, Zn 22, bal Cu.
Hard: 68,000 TS; 60,000 YS; 15 El. For hardware, screw machine products; free cutting. *Obsolete*

LEADED NICKEL SILVER 18%-789

Anaconda Co.
Ni 18, Pb, Zn, bal Cu.
For hardware, cutlery; corrosion resistant. *Obsolete*

LEADED NICKEL SILVER, 12%

Chase Brass & Copper Co.
Cu 64, Zn 23, Ni 12, Pb 1.
Annealed: 56,000 TS; 35 El. Hardened: 78,000 TS; 5 El. For hardware, screw machine products; corrosion resistant, free-cutting. *Obsolete*

LEADED PHOSPHOR BRONZE

Frontier Bronze Corp.
Copper. Cu 80, Sn 10, Pb 10.
Cast: 30,000-37,000 TS; 16,000-23,000 YS; 5-18 El; 6-16 RA; 55-60 Brin. Chilled: 30,000-33,000 TS; 19,000-21,000 YS; 4-7 El; 7-10 RA. For bearings for high speeds and heavy pressures; resists shock and vibrations.

LEADED PHOSPHOR BRONZE (B) 379

Anaconda Co.
Copper. Sn 5, Pb 1, P 0.1, bal Cu.
Hard: 65,000 TS; 55,000 YS; 25 El. For tubes, bushings, perforated sheet, clutch plates. Resists fatigue and corrosion.

LEADED PHOSPHOR BRONZE GR B

Now PHOSPHOR BRONZE B-1.

LEADED RED BRASS

Chase Brass & Copper Co.
Cu 85, Pb 1.75, bal Zn.
1/2 Hard temper: 60,000 TS; 50,000 YS; 15 El. For hardware, bolts, screws; free-cutting. *Obsolete*

LEADED TS4140 MODIFIED

LaSalle Steel Co.
C 0.4, Cr 1, Mo 0.2, Pb 0.2, bal Fe.
For screw machine products; free-cutting. *Obsolete*

LEADED TUBE BRASS 3301

Anaconda Co.
Copper. Zn 33.25, Pb 0.25, bal Cu.
Hard: 73,000 TS; 60,000 YS; 10 El; 150 Brin. Soft: 45,000 TS; 17,000 YS; 45 El; 60 Brin. For plumbing goods, tubes. Yellow brass.

LEADED TUBE BRASS 331

Anaconda Co.
Copper. Zn 33, Pb 1, bal Cu.
Hard: 73,000 TS; 60,000 YS; 10 El; 150 Brin. Soft: 45,000 TS; 17,000 YS; 55 El; 60 Brin. For screw machine parts. Free cutting tubes.

LEAKPRUF

Alpha Metals Inc.
Pb 40-60, bal Sn.
For soft solder for Ni, Monel and stainless steels; acid flux filled.

LEANTIN

Lumen Bearing Co.
Pb 82, Sb 15, Sn 35.
Cast. For machine bearings; low pressure bearings. *Obsolete*

LEBANON 1005

Now LEBANON 1010.

LEBANON 1010

Lebanon Foundry and Machine Co.
C 0-0.13, Si 0-0.7, Mn 0-0.5, bal Fe.
For electromagnet components.

LEBANON 1040

Lebanon Foundry and Machine Co.
C 0.35-0.45, Si 0-0.6, Mn 0.6-0.8, bal Fe.
Medium carbon steel casting. Normalized and tempered: 80,000 TS; 40,000 YS; 18 El. ASTM A148 Grade 80-40.

LEBANON 15-5

Lebanon Foundry and Machine Co.
C 0-0.06, Cr 14-15.5, Ni 4.5-5.5, Cb 0.15-0.25, Cu 2.5-3.2, bal Fe.
Precipitation hardenable stainless steel casting. Heat treated: 180,000 TS; 150,000 YS; 6 El. For pumps, impellers, fuel controls. ASTM A747 Grade CB7Cu-2.

LEBANON 17-22

Lebanon Foundry and Machine Co.
C 0.25-0.35, Si 0.3-0.6, Mn 0.45-0.7, Cr 1-1.5, Ni 0.4-0.6, bal Fe.
Low carbon, low alloy steel casting. Normalized and tempered: 125,000 TS; 100,000 YS; 12 El; 262-306 Brin. For aircraft brake components. Reference: Timken 17-22 A. *Obsolete*

LEBANON 17-4

Lebanon Foundry and Machine Co.
C 0-0.06, Cr 15.5-16.7, Ni 3.6-4.6, Cb 0.1-0.35, Cu 2.5-3.2, bal Fe.
Precipitation hardenable stainless steel casting. Heat treated: 180,000 TS; 150,000 YS; 6 El. For pumps, impellers, fuel controls. ASTM A747 Grade CB7Cu-1.

LEBANON 33

Lebanon Foundry and Machine Co.
C 0-0.07, Cr 18-20, Ni 22-25, Mo 2.5-3, Cu 1.5-2, bal Fe.
Non-magnetic stainless steel casting. Not hardenable by heat treatment. For paper pulp machinery, food machinery. ASTM A743 Grade CN7MS, ASTM A744 Grade CN7MS.

LEBANON 4140

Lebanon Foundry and Machine Co.
C 0.35-0.45, Si 0-0.6, Mn 0.6-0.8, Cr 0.8-1.2, Mo 0.2-0.3, bal Fe.
Medium carbon low alloy steel casting. Normalized and tempered: 100,000 TS; 80,000 YS; 15 El. Can be oil hardened to about 200,000 psi TS. Structural parts; AISI 4140.

LEBANON 431

Lebanon Foundry and Machine Co.
C 0.16-0.22, Cr 15-16.5, Ni 1.5-2.5, bal Fe.
Corrosion resistant steel casting; oil or air hardenable. For special purpose hardenable components as impellers, pumps, fuel controls. AISI 431; SAE 51431.

LEBANON 4330

Lebanon Foundry and Machine Co.
C 0.25-0.35, Si 0-0.6, Mn 0.6-0.8, Cr 0.6-1.25, Ni 1.25-2.5, Mo 0.25-0.5, bal Fe.
Medium carbon low alloy steel casting. Oil hardenable to 140,000-180,000 TS. For structural parts, gears, dies, wheels, housings. AISI 4330.

LEBANON 4335

Lebanon Foundry and Machine Co.
C 0.3-0.4, Si 0-0.6, Mn 0.6-0.8, Cr 0.6-1.25, Ni 1.25-2.5, Mo 0.25-0.5, bal Fe.
Medium carbon low alloy steel casting. Oil hardenable to 180,000-240,000 psi TS. For structural parts requiring good strength and toughness. AISI 4335.

LEBANON 4340

Lebanon Foundry and Machine Co.
C 0.35-0.45, Si 0-0.6, Mn 0.6-0.8, Cr 0.6-1.25, Ni 1.25-2.5, Mo 0.25-0.5, bal Fe.
Medium carbon low alloy steel casting. Oil hardenable in fairly heavy sections to 200,000-260,000 psi TS. For heavily loaded structural parts. AISI 4340.

LEBANON 440 C

Lebanon Foundry and Machine Co.
Mo 0-0.75, C 0.95-1.2, Cr 16-18, bal Fe.
Corrosion resistant steel casting. Oil or air hardenable to above 55 Rock C. For stainless ball bearings, races. AISI 440C; SAE 51440C. *Obsolete*

LEBANON 442

Lebanon Steel Foundry
C 0.2-0.3, Cr 18-22, Ni 0-2, bal Fe.
Corrosion resistant steel casting. For pumps, chemical equipment, valves; good resistance to nitric acid, alkalis, organics. AISI 442; ACI CB-30. *Obsolete*

LEBANON 8615

Lebanon Foundry and Machine Co.
C 0.1-0.2, Si 0-0.6, Mn 0.6-0.9, Cr 0.4-0.9, Ni 0.4-1.1, Mo 0.15-0.25, bal Fe.
Low carbon low alloy steel casting. For structural and carburized parts. AISI 8615. ASTM A148 Grade 80-50.

LEBANON 8630

Lebanon Foundry and Machine Co.
C 0.25-0.35, Si 0-0.6, Mn 0.6-0.95, Cr 0.4-0.9, Ni 0.4-1.1, Mo 0.15-0.25, bal Fe.
Medium carbon low alloy steel casting. Normalized and tempered: 90,000 TS; 60,000 YS; 20 El. Weldable; water or oil quench. Structural parts. ASTM A148 Grade 90-60.

LEBANON 8630/1

Lebanon Foundry and Machine Co.
C 0.25-0.35, Si 0-0.6, Mn 0.6-0.95, Cr 0.4-0.9, Ni 0.4-1.1, Mo 0.15-0.25, bal Fe.
Medium carbon low alloy steel casting. Quenched and tempered: 105,000 TS; 85,000 YS; 17 El; 220-227 Brin. Structural parts; can be reheat treated; water or oil quenched. ASTM A148 Grade 105-85.

LEBANON 8630/2

Lebanon Foundry and Machine Co.
C 0.25-0.35, Si 0-0.6, Mn 0.6-0.95, Cr 0.4-0.9, Ni 0.4-1.1, Mo 0.15-0.25, bal Fe.
Medium carbon low alloy steel castings. Quenched and tempered: 120,000 TS; 95,000 YS; 14 El; 250-311 Brin. Structural parts; can be reheat treated; water or oil quenched. ASTM A148 Grade 120-95.

LEBANON 8630/3

Lebanon Foundry and Machine Co.
C 0.25-0.35, Si 0-0.6, Mn 0.6-0.95, Cr 0.4-0.9, Ni 0.4-1.1, Mo 0.15-0.25, bal Fe.
Medium carbon low alloy steel castings. Quenched and tempered: 150,000 TS; 125,000 YS; 9 El; 310-375 Brin. For structural parts, gears, rollers, sprockets, rollers, housings. ASTM A148 Grade 150-125.

LEBANON C12

Lebanon Foundry and Machine Co.
C 0-0.2, Si 0-1, Mn 0.35-0.65, Cr 8-10, Mo 0.9-1.2, bal Fe.
Low carbon, chromium alloy steel casting. Normalized and tempered: 90,000 TS; 60,000 YS; 18 El. Air hardenable. For valves, fittings for moderate corrosion. ASTM A217 Grade C12; ASME SA217 Grade C12.

LEBANON C5

Lebanon Foundry and Machine Co.
C 0-0.2, Si 0-0.75, Mn 0.4-0.7, Cr 4-6.5, Mo 0.45-0.65, bal Fe.
Low carbon alloy steel casting. Normalized and tempered: 90,000 TS; 60,000 YS; 18 El. Air hardenable. For valves, bonnets, fittings, yokes, flanges, to 1200°F. ASTM A217 Grade C5; ASME SA217 Grade C5.

LEBANON CA15

Lebanon Foundry and Machine Co.
C 0-0.15, Cr 11.5-14, Ni 0-1, bal Fe.
Corrosion resistant steel casting. Oil or air hardenable, weldable. For valves, pump parts, oil refinery equipment. ASTM A217 Grade CA15; AISI 410; ASME SA217 Grade CA15.

LEBANON CA15M

Lebanon Foundry and Machine Co.
C 0-0.15, Cr 11.5-14, Ni 0-1, Mo 0.4-0.8, Se 0.2-0.35, bal Fe.
Free machining, corrosion resistant steel casting. Oil or air hardenable. For valve and pump parts requiring considerable machining. Similar to AISI 416SE. *Obsolete*

LEBANON CA40

Lebanon Foundry and Machine Co.
C 0.2-0.3, Cr 11.5-14, Ni 0-1, bal Fe.
Corrosion resistant steel castings. Oil or air hardenable to 180,000-240,000 psi TS. For cylinder liners, valves, cutter blades. ASTM A743 Grade CA40; AISI 420.

LEBANON CA40A

Now LEBANON CA40B.

LEBANON CA40B

Lebanon Foundry and Machine Co.
C 0.3-0.4, Cr 12-14, Mo 1, bal Fe.
Corrosion resistant steel castings. Oil or air hardenable to above 48 Rock C. For glass molds and plungers of mirror quality finish.

LEBANON CA6NM

Lebanon Foundry and Machine Co.
C 0-0.06, Si 0-1, Cr 11.5-14, Ni 3.5-4.5, Mo 0.4-1, bal Fe.
For compressors, pump impellers, valves, and turbine castings. ASTM A743 Grade CA6NM; ASME SA487 Grade CA6NM.

LEBANON CC50

Lebanon Foundry and Machine Co.
C 0-0.5, Cr 26-30, Ni 0-4, bal Fe.
Corrosion resistant steel casting. Good resistance to dilute sulfuric acid, in mine waters, and other dilute acids. For mine pumps, chemical plants. ASTM A743 Grade CC50; AISI 446.

LEBANON CD

Lebanon Foundry and Machine Co.
Now LEBANON CD4MCU.

LEBANON CD4MCU

Lebanon Foundry and Machine Co.
C 0-0.04, Cr 24.5-26.5, Ni 4.75-6, Mo 1.75-2.25, Cu 2.75-3.25, bal Fe.
Corrosion resistant steel casting. Precipitation hardening alloy; very good corrosion resistance. ASTM A351, A743, A744 Grade CD4MCu.

LEBANON CE30

Lebanon Foundry and Machine Co.
C 0-0.3, Cr 26-30, Ni 8-11, bal Fe.
Stainless steel casting; resistant to sulfurous acid. For pulp and paper mill equipment, mine equipment, pumps and valves. ASTM A743 Grade CE30.

LEBANON CF 20

Lebanon Foundry and Machine Co.
C 0-0.2, Cr 18-21, Ni 8-11, bal Fe.
Non-magnetic stainless steel casting. Non-hardenable by heat treatment. General purpose stainless cast components. ASTM A743 Grade CF20; AISI 302.

LEBANON CF3

Lebanon Foundry and Machine Co.
C 0-0.03, Cr 17-21, Ni 8-11, bal Fe.
Non-magnetic stainless steel casting. Weldable, not hardenable by heat treatment. For valves, pumps, headers for chemical plants. ASTM A743, A744 Grade CF3; AISI 304L; ASME SA351 Grade CF3.

LEBANON CF3A

Lebanon Foundry and Machine Co.
C 0-0.03, Cr 17-21, Ni 8-11, bal Fe.
Non-magnetic stainless steel casting; solution treated. Weldable, non-hardenable by heat treatment. Good resistance to stress corrosion cracking. For valves, pumps in chemical plants. ASTM A351 Grade CF-3A; ASME SA351 Grade CF3A.

LEBANON CF3M

Lebanon Foundry and Machine Co.
C 0-0.03, Cr 18-21, Ni 9-12, Mo 2-3, bal Fe.
Non-magnetic stainless steel casting. Weldable, not hardenable by heat treating. Exceptional corrosion resistance; pumps, valves, impellers, fittings. ASTM A743, A744 Grade CF3M. AISI 316L; ASME SA351 Grade CF3M.

LEBANON CF8

Lebanon Foundry and Machine Co.
C 0-0.08, Cr 18-21, Ni 8-11, bal Fe.
Non-magnetic stainless steel casting; quenched. Non-hardenable by heat treatment. General purpose stainless casting. ASTM A743, A744 Grade CF8; AISI 304; ASME SA351 Grade CF8.

LEBANON CF8A

Lebanon Foundry and Machine Co.
C 0-0.08, Cr 18-21, Ni 8-11, bal Fe.
Non-magnetic stainless steel casting; solution treated. Non-hardenable by heat treatment. Solution treatment increases ferrite content to improve resistance to stress corrosion cracking. ASTM A351 Grade CF-8A.

LEBANON CF8C

Lebanon Foundry and Machine Co.
C 0-0.08, Cr 18-21, Ni 9-12, Cb + Ta = 8 x C min, bal Fe.
Non-magnetic stainless steel casting; stabilized; weldable without reheat treatment. For welded stainless assemblies. AISI 347; ASTM A743, A744 Grade CF8C; ASME SA351 Grade CF8C.

LEBANON CF8F

Lebanon Foundry and Machine Co.
C 0-0.08, Cr 18-21, Ni 9-12, Mo 0-1.5, Se 0.2-0.35, P 0-0.15, bal Fe.
Non-magnetic stainless steel casting, free machining; non-hardenable heat treatment; not recommended for welding. For castings requiring much machining; AISI 303 Se. *Obsolete*

LEBANON CF8M

Lebanon Foundry and Machine Co.
C 0-0.08, Cr 18-21, Ni 9-12, Mo 2-3, bal Fe.
Non-magnetic stainless steel casting. Non-hardenable by heat treatment. Good resistance to corrosion. For chemical plant equipment. ASTM A743, A744 Grade CF8M; AISI 316; ASME SA351 Grade CF8M.

LEBANON CG8M

Lebanon Foundry and Machine Co.
C 0-0.08, Cr 18-21, Ni 9-12, Mo 3-3.5, bal Fe.
Non-magnetic stainless steel casting. Non-hardenable by heat treatment. Good corrosion resistance, especially in reducing environment. AISI 317. ASTM A351, A743, A744 Grade CG8M.

LEBANON CH20

Lebanon Foundry and Machine Co.
C 0-0.2, Cr 22-26, Ni 12-15, bal Fe.
Non-magnetic stainless steel casting. Non-hardenable by heat treatment. Improved corrosion resistance in hot dilute sulfuric acid. For digester parts, pumps, strainers. ASTM A743 Grade CH20; AISI 309; ASME SA351 Grade CH20.

LEBANON CHW

Lebanon Foundry and Machine Co.
C 0.15-0.3, Cr 22-25, Ni 11-14, W 2.5-3.5, bal Fe.
Non-magnetic stainless steel casting. Non-hardenable by heat treatment. For flanged rings and tail cone rings requiring both corrosion and heat resistance. *Obsolete*

LEBANON CK20

Lebanon Foundry and Machine Co.
C 0-0.25, Cr 23-27, Ni 19-22, bal Fe.
Non-magnetic stainless steel casting. Non-hardenable by heat treatment. Improved resistance to corrosion and high temperatures; weldable. For jet engine parts, pumps, valves. ASTM A743 Grade CK20; AISI 310; ASME SA351 Grade CK20.

LEBANON CK45

Lebanon Foundry and Machine Co.
C 0.35-0.45, Cr 23-27, Ni 19-22, bal Fe.
Non-magnetic stainless steel casting. Non-hardenable by heat treatment. Heat resistant alloy for use in valves and fittings for chemical and refinery parts.

LEBANON CN37

Lebanon Foundry and Machine Co.
C 0-0.03, Cr 19-22, Ni 27.5-30.5, Mo 2-3, Cu 3-4, bal Fe.
Non-magnetic stainless steel casting. Weldable; not hardenable by heat treatment. Good resistance to hot sulfuric acid. For paper pulp machinery, food machinery. ASTM A743 Grade CN3M.

LEBANON CN3M

Lebanon Foundry and Machine Co.
Now LEBANON CN37.

LEBANON CN7M

Lebanon Foundry and Machine Co.
C 0-0.07, Cr 19-22, Ni 27.5-30.5, Mo 2-3, Cu 3-4, bal Fe.
Non-magnetic stainless steel casting. Non-hardenable by heat treatment. Good resistance to hot sulfuric acid. For paper pulp machinery, food machinery. ASTM A743, A744 Grade CN7N; ASME SA 351 Grade CN7M.

LEBANON CU-NI

Lebanon Foundry and Machine Co.
C 0-0.15, Ni 28-32, Fe 0.25-1, Cb 0-1.5, bal Cu.
Corrosion resistant alloy casting for marine pumps, valves, fittings ASTM B369.

LEBANON H810 (INCOLOY)

Lebanon Foundry and Machine Co.
Cr 19-21, Ni 31-33, bal Fe.
Heat resistant steel casting. Maximum operating temperature 2000°F (1093°C).

LEBANON HAB

Lebanon Foundry and Machine Co.
C 0-0.12, Mo 26-30, Fe 4-6, V 0.2-0.6, Co 0-1, Cr 0-1, bal Ni.
Corrosion resistant alloy casting; for valves, pump parts, fittings in hydrochloric acid environment. Similar to Hastelloy B; ASTM A404 Grade N12MV.

LEBANON HAC

Lebanon Foundry and Machine Co.
C 0-0.12, Cr 15.5-17.5, Mo 16-18, Fe 4.5-7, W 3.75-5.25, V 0.2-0.4, Co 0-1, bal Ni.
Corrosion resistant alloy casting for use with wet chlorine gas, hypochloric acid and other acids as valves and fittings. ASTM A494 Grade CW12MW. Similar to Hastelloy C.

LEBANON HE

Lebanon Foundry and Machine Co.
Cr 26-30, Ni 8-11, bal Fe.
Heat resistant steel casting. Maximum operating temperature 2000°F (1093°C). ACI HE; SAE 70312; ASTM 297 Grade HE.

LEBANON HH

Lebanon Foundry and Machine Co.
Cr 24-28, Ni 11-14, bal Fe.
Heat resistant steel casting. Maximum operating temperature 2000°F (1093°C). ACI HH: SAE 70309; ASTM A297 Grade HH.

LEBANON HK

Lebanon Foundry and Machine Co.
Cr 24-28, Ni 18-22, bal Fe.
Heat resistant steel casting. Maximum operating temperature 2100°F. ACI HK; SAE 70310; ASTM A297 Grade HK.

LEBANON HP

Lebanon Foundry and Machine Co.
C 0.5, Cr 24-28, Ni 33-37, bal Fe.
Heat resisting steel casting. Service temperature 2000°F (1093°C). ASTM A297 Grade HP.

LEBANON HR

Lebanon Foundry and Machine Co.
Cr 21-25, Ni 10-13, bal Fe.
Heat resistant steel casting.

LEBANON HR-A

Lebanon Steel Foundry
Cr 21, Ni 14, bal Fe.
Heat resistant steel casting. *Obsolete*

LEBANON HT

Lebanon Foundry and Machine Co.
Cr 15-19, Ni 33-37, bal Fe.
Heat resistant steel casting. Maximum operating temperature 2000°F (1093°C). ACI HT; SAE 70330; ASTM A297 Grade HT.

LEBANON HX

Lebanon Foundry and Machine Co.
C 0.4, Cr 15-19, Ni 64-68, bal Fe.
Heat resistant steel casting for use in furnace parts and environments of highly corrosive nature. ASTM A296 Grade HX.

LEBANON HY-100

Lebanon Foundry and Machine Co.
C 0-0.22, Si 0-0.5, Mn 0.55-0.75, Cr 1.35-1.85, Ni 2.75-3.5, Mo 0.3-0.6, bal Fe.
Low carbon alloy steel casting; 100,000-120,000 YS. Oil hardenable. For valves, fittings; good strength with good impact values; low notch sensitivity. MIL-S-23008 B HY-100.

LEBANON HY-80

Lebanon Foundry and Machine Co.
C 0-0.2, Si 0-0.5, Mn 0.55-0.75, Cr 1.35-1.65, Ni 2.5-3.25, Mo 0.3-0.6, bal Fe.
Low carbon alloy steel casting. Oil hardenable to 100,000-160,000 TS. For valves, fittings; good strength with high impact energy. MIL-S-23008 B HY-80.

LEBANON INC

Lebanon Foundry and Machine Co.
C 0-0.1, Cr 14-17, Fe 6-10, 1.0-2.0 Cb + Ta, bal Ni.
Corrosion resistant alloy casting, resistant to alkalies, oxidizing and organic acids. For food processing and textile processing equipment, high temperature service. ASTM A494 Grade CY40 (Inconel).

LEBANON LC2

Lebanon Foundry and Machine Co.
C 0-0.25, Si 0-0.6, Mn 0.5-0.8, Ni 2-3, bal Fe.
Low carbon, nickel alloy steel casting. Normalized, quenched and tempered: 65,000 TS; 40,000 YS; 24 El. For valves for cryogenic service to -100°F (-73°C). ASTM A352 Grade LC2; ASME SA352 Grade LC2.

LEBANON LC3

Lebanon Foundry and Machine Co.
C 0-0.15, Si 0-0.6, Mn 0.5-0.8, Ni 3-4, bal Fe.
Low carbon, nickel alloy steel casting. Normalized, quenched and tempered: 65,000 TS; 40,000 YS; 24 El. Valves for cryogenic service to -150°F (-101°C). ASTM A352-76 Grade LC3; ASME SA352 Grade LC3.

LEBANON LCB

Lebanon Foundry and Machine Co.
C 0-0.3, Si 0-0.6, Mn 0-1, bal Fe.
Low carbon steel casting. Normalized, quenched and tempered: 65,000 TS; 35,000 YS; 24 El. General purpose cast housings. ASTM A352 Grade LCB; ASME SA352 Grade LCB.

LEBANON M-30C

Lebanon Foundry and Machine Co.
C 0-0.3, Cu 26-33, Fe 1.5-3.5, 60.0 Ni min, 1.0-3.0 Cb + Ta.
Corrosion resistant alloy casting, resistant to mineral and organic acids, sea water, caustic solutions. For valves and fittings for marine service. Fed: QQ-N-288-Comp E (Monel). ASTM A494 Grade M-30C.

LEBANON M-35-1

Lebanon Foundry and Machine Co.
C 0-0.3, Cu 26-33, Fe 1.5-3.5, 60.0 Ni min.
Corrosion resistant alloy casting, resistant to mineral and organic acids, sea water, caustic solutions. For valves and fittings for marine service. Fed: QQ-N-288-Comp A (Monel). ASTM A494 Grade M-35-1.

LEBANON ME

Lebanon Foundry and Machine Co.
Now LEBANON M-30C.

LEBANON NI

Lebanon Foundry and Machine Co.
C 0-1, Cu 0-1.25, Fe 0-1.25, bal Ni.
Nickel casting. As cast: 45,000 TS; 20,000 YS; 15 El. Resistant to many industrial chemicals. For equipment for food processing, pharmaceuticals. ASTM A494 Grade CZ100.

LEBANON WC1

Lebanon Foundry and Machine Co.
C 0-0.25, Si 0-0.6, Mn 0.5-0.8, Mo 0.45-0.65, bal Fe.
Low carbon, low alloy steel casting. For valves, fittings, turbines and other pressure castings to 1100°F. ASTM A217-75 Grade WC1: A352 Grade LC1; ASME SA352 Grade LC.

LEBANON WC6

Lebanon Foundry and Machine Co.
C 0-0.2, Si 0-0.6, Mn 0.5-0.8, Cr 1-1.5, Mo 0.45-0.65, bal Fe.
Low carbon, low alloy steel casting. Normalized and tempered: 70,000 TS; 40,000 YS; 20 El. For valves, fittings, turbines and other pressure castings for service to 1200°F (649°C). ASTM A217 Grade WC6; ASME SA217 Grade WC6.

LEBANON WC6A

Lebanon Foundry and Machine Co.
C 0-0.25, Si 0-0.6, Mn 0-0.7, Cr 0.4-0.7, Mo 0.4-0.6, bal Fe.
Low carbon, low alloy steel casting. Normalized and tempered: 70,000 TS; 40,000 YS; 22 El. For valves, fittings, turbines and other pressure castings for service below 1200°F (649°C). ASTM A356 Grade 5.

LEBANON WC6B

Lebanon Foundry and Machine Co.
C 0-0.3, Si 0-0.75, Mn 0-1, Cr 1.5-2.25, Mo 0.45-0.65, bal Fe.
Medium carbon low alloy steel casting. Normalized, quenched and tempered: 105,000 TS; 85,000 YS; 15 El. For valves, fittings, turbines and other pressure castings for service below 1200°F (649°C).

LEBANON WC9

Lebanon Foundry and Machine Co.
C 0-0.18, Si 0-0.6, Mn 0.4-0.7, Cr 2-2.75, Mo 0.9-1.2, bal Fe.
Low carbon alloy steel casting. Normalized and tempered: 70,000 TS; 40,000 YS; 20 El. For valves, fittings, turbines, and other pressure castings for service below 1200°F (649°C). ASTM A217-75 Grade WC9; ASME SA217 Grade WC9.

LEBANON WCA

Lebanon Foundry and Machine Co.
C 0-0.25, Si 0-0.6, Mn 0-0.7, bal Fe.
Low carbon steel casting. Normalized and tempered: 60,000 TS; 30,000 YS; 24 El. ASTM A216-75 Grade WCA; AISI 1020; ASME SA216 Grade WCA.

LEBANON WCB

Lebanon Foundry and Machine Co.
C 0-0.3, Si 0-0.6, Mn 0-1, bal Fe.
Low carbon steel casting. 70,000-95,000 psi (485-655 MPa) TS; 36,000 psi (250 MPa) YS. ASTM A216 Grade WCB; ASME SA216 Grade WCB.

LEBANON WCB/2

Now LEBANON WCB.

LEBANON WCC

Lebanon Foundry and Machine Co.
C 0-0.25, Mn 0-1.2, bal Fe.
Low carbon steel casting. 70,000-95,000 psi (485-655 MPa) TS; 40,000 psi (275 MPa) YS. For valves, fittings, turbines for service to 1100°F (593°C). ASTM A216 Grade WCC; ASME SA216 Grade WCC.

LECHESNE

Manufacturer not listed.
Cu 60-90, Ni 10-40, Al 0.05-0.2.
For chemical engineering equipment; corrosion resistant.

LECO

Lehigh Steel Corp.
C 0.33, Cr 0.75, Ni 0.3, Mo 0.75, Cu 0.75, Ti 0.15, bal Fe.
For chisels, dies, punches, pneumatic tools; oil hardened, shock resistant.

LECO EXTRA

Lehigh Steel Corp.
C 0.44, Cr 0.75, Ni 0.35, Mn 0.4, Mo 0.75, Si 0.65, Cu 0.75, Ti 0.15, bal Fe.
For dies, rivet sets, chisels, punches, cutters; fatigue and abrasion resistant.

LECO NON-TEMPERING

Teledyne Firth Sterling
Steel. C, bal Fe.
For dies, tools; water hardened, nontempering. *Obsolete*

LECTRI-LED 4140

Lukens Steel
C 0.41, Mn 0.9, Si 0.28, Cr 0.93, Mo 0.19, Pb 0.15, bal Fe.
Heat treated: 176,000 TS; 158,000 YP; 16 El; 45 RA; C 38 Rock. For rubber molds, plastic molds, steam platens, gears, machine tool parts, high pressure valve bodies. Free machining, tough. *Obsolete*

LED-O-LOY

Bridgeport Rolling Mills Co.
Cu 87, Sn 2, Pb 2, bal Zn.
Half hard: 58,000 TS; 49,000 YS; 14 El; Rock B 70. Hard: 76,000 TS; 70,000 YS; 10 El; Rock B 80. Extra hard: 86,000 TS; 75,000 YS; 3 El; Rock B 86. For meter and gauge components, bearings, thrust washers, valve parts. Free-cutting, corrosion resistant. *Obsolete*

LEDA

Firth Brown Ltd.
C 0.7, Cr, W, V, bal Fe.
For lathe and planer tools, reamers, taps, hobs; high speed steel. *Obsolete*

LEDALOYL

Johnson Bronze Co.
Sn 10, Pb 5, 1% graphite, bal Cu.
For self-lubricating bearings; sintered type, oil impregnated. *Obsolete*

LEDDEL ALLOY

Manufacturer not listed.
Zn 90, Cu 5, Al 5.
For die castings.

LEDDEL BEARING

Manufacturer not listed.
Zn 87.5, Cu 6.26, Al 6.25.
For bearings; anti-friction.

LEDEBUR'S BEARING-1

Manufacturer not listed.
Zn 85, Sb 10, Cu 5.
For bearings; anti-friction.

LEDEBUR'S BEARING-2

Manufacturer not listed.
Zn 77, Cu 5.5, Sn 18.
For bearings; anti-friction.

LEDLOY

Joseph T. Ryerson & Son Inc.
C 0.2, Pb 0.2, Mn 1, S 0.25, bal Fe.
For screw machine products, fasteners, shafts, gears, cams, camshafts. Free-machining. *Obsolete*

LEDLOY "A"

Copperweld Steel Co.
C, Pb, bal Fe.
For valve spools; free-cutting. *Obsolete*

LEDLOY 1018
Inland Steel Co.
Alloy steel. C 0.15-0.2, Mn 0.6-0.9, Pb 0.15-0.35, bal Fe.
Free-machining steel.

LEDLOY 1117
Inland Steel Co.
Alloy steel. C 0.14-0.2, Mn 1-1.3, P 0-0.04, S 0.08-0.13, Pb 0.15-0.35, bal Fe.
Free machining steel.

LEDLOY 1137
Inland Steel Co.
Alloy steel. C 0.32-0.39, Mn 1.35-1.65, P 0-0.04, S 0.08-0.13, Pb 0.15-0.35, bal Fe.
Free machining medium carbon steel.

LEDLOY 1144
Inland Steel Co.
Alloy steel. C 0.4-0.48, Mn 1.35-1.65, P 0-0.04, S 0.24-0.33, Pb 0.15-0.35, bal Fe.
Free machining medium carbon steel.

LEDLOY 1215
Inland Steel Co.
Alloy steel. C 0-0.09, Mn 0.75-1.05, P 0.04-0.09, S 0.26-0.35, Pb 0.15-0.35, bal Fe.
Free machining low carbon steel.

LEDLOY 170
Joseph T. Ryerson & Son Inc.
Pb 0.15-0.35, C, bal Fe.
For machine tool parts, gears; free-cutting. *Obsolete*

LEDLOY 300
Joseph T. Ryerson & Son Inc.
C 0.12, Mn 1, P 0.07, S 0.3, Pb 0.15-0.35, bal Fe.
79 ksi TS; 71 ksi YS; 16 El; 52 RA. Free-cutting; for screw machine products. AISI 12L14.

LEDLOY 375
Joseph T. Ryerson & Son Inc.
C 0.09, Mn 1, P 0.04-0.09, S 0.26-0.35, Pb 0.15-0.35, Te 0.03, bal Fe.
79 ksi TS; 71 ksi YS; 16 El; 52 RA. Free-machining steel, particularly for threaded parts, gears.

LEDLOY 4140
Inland Steel Co.
Alloy steel. C 0.39, Mn 0.8, S 0.03, Si 0.24, Cr 0.84, Ni 0.09, Mo 0.18, Pb 0.15-0.35, bal Fe.
Heat treated: 158,000 TS; 144,000 YS; 16 El; 56 RA; 321 Brin. For fasteners, gears, camshafts, housings, machine tools, crankshafts. Free-cutting. Oil hardened.

LEDLOY 5120
Copperweld Steel Co.
C 0.17-0.23, Mn 0.6-1, Cr 0.6-1, Pb 0.15-0.25, bal Fe.
For fuel valve burner tubes; free cutting. *Obsolete*

LEDLOY 8620
Inland Steel Co.
C 0.18-0.23, Mn 0.7-0.9, Pb 0.15-0.25, Si 0.25-0.35, Ni 0.4-0.7, Cr 0.4-0.6, Mo 0.15-0.25, bal Fe.
Cold drawn: 110,000 TS; 101,000 YS; 15 El; 55 RA; 223 Brin. Heat treated: 192,000 TS; 150,000 YS; 12 El; 49 RA; 338 Brin. For gears, cams, camshafts, fasteners, chuck jaws, pneumatic chucks. Free-cutting. Case hardening. *Obsolete*

LEDLOY A
Inland Steel Co.
Alloy steel. C 0-0.09, Mn 0.85-1.15, P 0.04-0.09, S 0.26-0.35, Pb 0.15-0.35, bal Fe.
Free machining low carbon steel.

LEDLOY C-1045
Inland Steel Co.
Alloy steel. C 0.4-0.5, Pb 0.15-0.35, bal Fe.
Rolled: 98,000 TS; 65,000 YS; 28 El; 52 RA. For machinery parts; water hardened; free cutting.

LEDLOY C-1120
Inland Steel Co.
C 0.15-0.25, Pb 0.15-0.35, bal Fe.
Rolled: 62,300 TS; 44,000 YS; 36 El; 63 RA. For machinery parts; free-cutting. *Obsolete*

LEDLOY-AX
Inland Steel Co.
C 0.08, Mn 1.07, P 0.075, S 0.32, Si 0.02, Ni 0.003, Pb 0.21, Te 0.048, bal Fe.
Bar: 75,600 TS; 71,900 YS; 13 El; 37 RA; 91 Rock B. For screw machine products, fasteners, screws, bolts, spark wheels. Free machining.

LEDLOY-AX
Bliss & Laughlin Steel Co.
C 0.08, Mn 1.07, P 0.075, S 0.32, Si 0.02, Ni 0.003, Pb 0.21, Te 0.048, bal Fe.
Bar: 75,600 TS; 71,900 YS; 13 El; 37 RA; 91 Rock B. For screw machine products, fasteners, screws, bolts, spark wheels. Free machining.

LEDO
Colt Industries
C 0.4, Mo 0.2, bal Fe.
Heat treated: 118,000-134,000 TS; 83,000-94,000 YS; 15-20 El; 43-53 RA; 248-320 Brin. For crankshafts, gears, bolts, spindles, shafts; oil or water hardened. *Obsolete*

LEDRITE 2
Olin Brass, Indianapolis
Copper. Cu 61, Pb 1.8, Zn 35.2.
Hard: 55,000 psi TS; 42,000 psi YS; 30 El. For screw machine parts and hardware; free-cutting.

LEDRITE 6
Olin Brass, Indianapolis
Copper. Cu 61, Pb 3.4, Zn 35.6.
Hard: 58,000 psi TS; 45,000 psi YS; 25 El. For screw machine parts and hardware; free-cutting.

LEDRITE BRASS
Olin Brass, Indianapolis
Copper. Cu 61, Pb 3.4, Zn 35.6.
Soft: 45,000 psi TS; 35 El. Hard: 80,000 psi TS; 10 El. For screw machine parts; free cutting.

LEDUCT 370-17
Ley's Malleable Castings Co. Ltd.
C 3.5-3.8, Si 2-2.6, Mn 0.1-0.6, S 0.005-0.01, P 0-0.06, Mg 0.03-0.055, bal Fe.
Annealed: 370 N/mm^2 TS; 230 N/mm^2 YS; 115-179 Brin. Spheroidal graphitic iron for automotive, truck and tractor industries.

LEDUCT 420/12
Ley's Malleable Castings Co. Ltd.
C 3.5-3.8, Si 2-2.6, Mn 0.1-0.6, S 0.005-0.01, P 0-0.06, Mg 0.03-0.055, bal Fe.
As cast: 420 N/mm^2 TS; 250 N/mm^2 YS; 12 El; 149-201 Brin. Spheroidal graphitic iron for automotive, truck and tractor industries.

LEDUCT 500-7
Ley's Malleable Castings Co. Ltd.
C 3.5-3.8, Si 2-2.6, Mn 0.1-0.6, Si 0.005-0.01, P 0-0.06, Mg 0.03-0.055, bal Fe.
As cast: 500 N/mm^2 TS; 310 N/mm^2 YS; 7 El; 170-241 Brin. Spheroidal graphitic iron for automotive, truck and tractor industries.

LEDUCT 600-3
Ley's Malleable Castings Co. Ltd.
C 3.5-3.8, Si 2-2.6, Mn 0.1-0.6, S 0.005-0.01, P 0-0.06, Mg 0.03-0.055, bal Fe.
As cast: 600 N/mm^2 TS; 350 N/mm^2 YS; 3 El; 192-269 Brin. Spheroidal graphitic iron for automotive, truck and tractor industries.

LEDUCT 700-2
Ley's Malleable Castings Co. Ltd.
C 3.5-3.8, Si 2-2.6, Mn 0.1-0.6, S 0.005-0.01, P 0-0.06, Mg 0.03-0.055, bal Fe.
As cast: 700 N/mm^2 TS; 400 N/mm^2 YS; 2 El; 229-302 Brin. Spheroidal graphitic iron for automotive, truck and tractor industries.

LEDUCT 800-2
Ley's Malleable Castings Co. Ltd.
C 3.5-3.8, Si 2-2.6, Mn 0.1-0.6, S 0.005-0.01, P 0-0.06, Mg 0.03-0.055, bal Fe.
As cast: 800 N/mm^2 TS; 460 N/mm^2 YS; 2 El; 248-352 Brin. Spheroidal graphitic iron for automotive, truck and tractor industries.

LEDURIT Z IV
Thyssen Edelstahlwerke AG
C 2.5, Mn 0.8, Cr 4.3, bal Fe.
For hard facing electrode; wear resistant. *Obsolete*

LEGA-Y
Soc. Alluminio Veneto per Azioni
Aluminum. Cu 3.8-4.2, Mg 1.3-1.7, Ni 1.8-2.3, bal Al.
Heat treated: 36,000-47,000 TS; 32,000-43,000 YS; 0.3-0.5 El; 95-115 Brin. For light alloy parts; age hardened.

LEGACY
J.F. Jelenko & Co.
Precious metal. Au 2, Pd 86, Ag 1.
125,000 psi TS; 95,500 psi YS; 20 El. Dental alloy for fusing porcelain to metal.

LEGAL
Siemens-Schukert AG.
Aluminum. Mg 0.5-2, Si 0.3-1.5, Mn 0-15, bal Al.
For light alloy parts; corrosion resistant, hardenable.

LEGIERUNG 600
Kreidler Werke G.m.b.H.
Mg 1.5-3, Mn 0.5-1.5, Cr 0-0.3, bal Al.
Soft: 28,000 TS; 13,000 YS; 30 El; 47 Brin. Hard: 42,000 TS; 37,000 YS; 8 El; 77 Brin. For marine products, aircraft structures; resists sea water corrosion. *Obsolete*

LEGIERUNG 800
Kreidler Werke G.m.b.H.
Mn 1-1.5, Cr 0-0.3, bal Al.
Soft: 16,000 TS; 6,000 YS; 40 El; 28 Brin. Hard: 29,000 TS; 27,000 YS; 10 El; 55 Brin. For aircraft and engineering structures; good formability and weldability. *Obsolete*

LEHIGH DD1
Lehigh Steel Corp.
C, alloy, bal Fe.
For machine parts, gears, shafts, mandrels; oil hardened.

LEHIGH DIE AND TOOL
Bethlehem Steel Corp.
Cr 12, C 1.65, V 0.3, Mo 0.75, bal Fe.
For dies, tools, thread rolling dies, blanking dies, punches; non-warping; non-shrinking. *Obsolete*

LEHIGH H
Bethlehem Steel Corp.
C 1.6, Cr 11.5, Mo 0.8, V 0.4, bal Fe.
For gages, punches, master hobs; Type D2; oil hardened. *Obsolete*

LEHIGH L
Bethlehem Steel Corp.
C 0.85, Cr 11.5, Ni 1.05, V 0.3, Mo 0.45, bal Fe.
For dies, shear blades, punches; abrasion resistant, non-deforming. *Obsolete*

LEHIGH N.C.
Lehigh Steel Corp.
C 0.7, Cr 1, Ni 1.6, Cu 0.35, Mn 0.4, Si 0.2, bal Fe.
For shear blades, rivet sets, punches, dies; oil hardened, shock resistant.

LEHIGH S
Bethlehem Steel Corp.
C 2, Cr 11.7, V 0.6, bal Fe.
For dies and shear blades, gauges; abrasion resistant, non-deforming. *Obsolete*

LEHIGH SPECIAL NO. 3
Lehigh Steel Corp.
C 0.7-1.2, bal Fe.
For tools and dies; general use.

LEHIGH SPECIAL NO. 5
Lehigh Steel Corp.
C 0.7-1.4, bal Fe.
For tools and dies; abrasion resistant.

LEHIGH SS
Lehigh Steel Corp.
C 0.75, Cr 4.5, Co 12, V 2, W 20, bal Fe.
For lathe and planer tools, reamers, taps, broaches, hobs; high speed steel, oil hardened.

LEHIGH XXX
Lehigh Steel Corp.
C 0.7, Cr 4, V 1, W 18, bal Fe.
For lathe and planer tools, drills, taps, reamers; high speed steel; Type T2.

LEICHMETALLWERKE M115
Vereinigte Leichtmetallwerke, G.m.b.H.
Aluminum. Mn 1-1.5, Cr 0.3, bal Al.
Soft: 16,000 TS; 6000 YS; 40 El. Hard: 29,000 TS; 27,000 YS; 10 El. For cooking utensils, tanks, furniture; good forming and welding properties.

LEICHTMETALL MN 20
Durener Metallwerke
Aluminum. Mn 1-2, bal Al.
Soft: 13,000 TS; 6000 YS; 35 El; 35 Brin. Hard: 43,000 TS; 36,000 YS; 4 El; 70 Brin. For light alloy parts; corrosion resistant.

LEICHTMETALLWERKE MZB
Vereinigte Leichtmetallwerke, G.m.b.H.
Aluminum. Cu 2.5-5, Mg 0.2-1.8, Mn 0.3-1.5, Pb 0.5-2.5, Sn, Cd, Bi, bal Al.
For screw machine products; free-cutting.

LEITPANTAL
VAW Vereinigte Aluminium-Werke AG
Aluminum.
Aluminum alloy AlMgSi0.5.

LEKTROCAST
Detroit Alloy Steel Co.
C 3-3.4, Si 2, Ni 1.5-2.5, bal Fe.
Cast: 50,000 TS; 225-280 Brin. For castings, draw dies for metal stamping; wear resistant.

LEMAG
Now "LEDUCT".

LEMARQUANDS ALLOY
Manufacturer not listed.
Cu 39, Ni 7, Zn 37, Sn 9, Co 8.
For jewelry; corrosion resistant.

LEMAX HEAT TREATED PEARLITIC 38/30/4
Ley's Malleable Castings Co. Ltd.
C 2.3-2.6, Si 1.3-1.55, Mn 0.4-0.57, S 0.15-0.25, P 0-0.08, bal Fe.
Heat treated pearlitic malleable iron castings. 85,100 TS; 67,200 YS; 4 El; 197-255 Brin. For automotive, truck and tractor industries.

LEMAX HEAT TREATED PEARLITIC 45/40/2
Ley's Malleable Castings Co. Ltd.
C 2.3-2.6, Si 1.3-1.55, Mn 0.4-0.57, S 0.15-0.25, P 0-0.08, bal Fe.
Heat treated pearlitic malleable iron castings. 100,800 TS; 89,600 YS; 2 El; 241-269 Brin. For automotive, truck and tractor industries.

LEMS 1
Thyssen Edelstahlwerke AG
C 1.1, Cr 1.2, V 0.2, W 1.3, bal Fe.
For cold work punching tools; oil or water hardened. *Obsolete*

LENIN-T58
Skoda Works National Corp.
C 0.16-0.2, Cr 11.5-12.5, Mo 0.5, Ni 0.5-1, V 0.15-0.25, W 2-2.5, bal Fe.
Annealed: 80,000 TS; 40,000 YS; 24 El; 95 Rock B. Hardened: 230,000 TS; 195,000 YS; 12 El; 48 Rock C. For valves, bearings, cutlery, surgical instruments, gears; corrosion resistant, hardenable.

LENNALMAGMAN
Reynolds Aluminiumwerke GmbH
Aluminum.
Aluminum alloy AlMgMn(AlMg2Mn0.8).

LENNALMAN
Reynolds Aluminiumwerke GmbH
Aluminum.
Aluminum alloy AlMn.

LENNALSIL
Westfalische Leichtmetallwerke GmbH
Mg 0.6-1.4, Si 0.6-1.6, Mn 0.6-1, Cr 0-0.3, bal Al.
Annealed: 21,000 TS; 8000 YS; 24 El. For window frames, fan blades, gutters, boats; good forming and welding properties.

LENNEDUR
Westfalische Leichtmetallwerke GmbH
Cu 2.5-5, Mg 0.2-1.8, Mn 0.3-1.5, bal Al.
Annealed: 27,000 TS; 11,000 YS; 22 El; 47 Brin. Heat treated: 72,000 TS; 57,000 YS; 130 Brin. For aircraft structures and fittings, fasteners; age hardenable.

LENOX
Manufacturer not listed.
C 0.9, Mn 1.5, Mo 0.3, Si 0.3, bal Fe.
Annealed: 200 Brin. For tools, dies, punches, knives; water hardened.

LEONARD
Jessop-Saville Ltd.
C 0.9-1, bal Fe.
For tools, drills, punches; water hardened.

LEPAZ PEARLITIC 30/22/8
Ley's Malleable Castings Co. Ltd.
C 2.3-2.6, Si 1.3-1.55, Mn 0.4-0.57, S 0.15-0.25, P 0-0.08, bal Fe.
Pearlitic malleable iron castings. 67,200 psi TS; 49,280 psi YS; 8 El; 163-207 Brin. For automotive, truck and tractor industries.

LEPAZ PEARLITIC 36/24/7
Ley's Malleable Castings Co. Ltd.
C 2.3-2.6, Si 1.3-1.55, Mn 0.4-0.57, S 0.15-0.25, P 0-0.08, bal Fe.
Pearlitic malleable iron castings. 80,600 TS; 53,800 YS; 7 El; 179-217 Brin. For automotive, truck and tractor industries.

LEPPE HWF 48
Chr. Hover & Sohn Edelstahlwerk
C 0.06-0.0600004, Cr 18, Ni 11, bal Fe.
Stainless steel for high temperature pipelines, pressure vessels.

LEPPESTAHL
Hover, Gebruder, Edelstahlwerk
C 2.1, Cr 11.5, bal Fe.
For blanking and forming dies, punches; oil hardened, nondeforming.

LEPPESTAHL EXTRA 6
Hover, Gebruder, Edelstahlwerk
C 2.1, Cr 11.5, W 0.7, bal Fe.
For blanking and forming dies, punches; nondeforming, oil hardened.

LEPPESTAHL X
Hover, Gebruder, Edelstahlwerk
C 1.65, Cr 11.5, V 0.1, bal Fe.
For blanking and forming dies, punches; air hardened, nondeforming.

LESCALLOY 13-8 MO VIM-VAR
Latrobe Steel Co.
C 0.04, Cr 12.6, Ni 8.3, Mo 2.15, Al 1.
Solution treated: 363 Brin max. Martensitic, precipitation hardening stainless steel.

LESCALLOY 15-5 VAC-ARC
Latrobe Steel Co.
C 0.04, Si 0.35, Mn 0.7, Cr 14.75, Ni 4.75, Cu 3.5, 0.30 Cb + Ta.
Solution treated: 363 Brin. Martensitic, precipitation hardening stainless steel.

LESCALLOY 300 M
Latrobe Steel Co.
Now LESCALLOY 300 M VAC ARC.

LESCALLOY 300 M VAC ARC
Latrobe Steel Co.
S 0.42, Mn 0.75, Si 1.65, Ni 1.8, Cr 0.8, Mo 0.4, V 0.07, bal Fe.
Heat treated: 285,000-290,000 TS; 238,000-243,000 YS; 10-13 El; 32-47 RA. For landing gears, air frames, missile cases, pressure vessels. High strength structural steel, tough, shock resistant.

LESCALLOY 35NCD16 VAN-ARC
Latrobe Steel Co.
C 0.4, Mn 0.45, Si 0.3, Cr 1.8, Ni 4, Mo 0.45, V 0.03.
Tempered: 225,000 YS; 285,000 TS; 10 El; 40 RA; 285-321 Brin. Vacuum arc remelted, high strength alloy steel. For aerospace structural applications.

LESCALLOY 422 STAINLESS
Latrobe Steel Co.
C 0.22, Mn 0.75, Si 0.3, Cr 12, W 1, Mo 1, Ni 0.75, V 0.25, bal Fe.
Oil quenched: C 50-55 Rock. Tempered at 1150 F: 150,000 TS; 130,000 YS; 14-17 El. Magnetic, good strength, corrosion resistant. For steam turbine blades, and components, valves, gas turbine and jet engine parts. *Obsolete*

LESCALLOY 4330 +V VAC ARC
Latrobe Steel Co.
C 0.3, Mn 0.85, Si 0.3, Cr 0.35, Ni 1.8, Mo 0.4, V 0.07, bal Fe.
Vacuum arc structural steel. Hardened: up to 235,000 TS; 195,000 YS; 11 El; 47-50 Rock C. For highly stressed structural parts.

LESCALLOY 4330 VAC ARC
Latrobe Steel Co.
Now LESCALLOY 4330 +V VAC ARC.

LESCALLOY 4335 +V VAC ARC
Latrobe Steel Co.
C 0.35, Mn 0.75, Si 0.5, Cr 0.8, Ni 1.85, Mo 0.35, V 0.2, bal Fe.
Vacuum arc melted structural steel. Hardened: up to 275,000 TS; 240,000 YS; 10 El. Deep hardening; for highly stressed structural parts.

LESCALLOY 4335 VAC ARC
Latrobe Steel Co.
Now LESCALLOY 4335 +V VAC ARC.

LESCALLOY 4340 VAC ARC
Latrobe Steel Co.
C 0.4, Mn 0.75, Si 0.3, Cr 0.8, Ni 1.8, V 0.2, Mo 0.25, bal Fe.
Vacuum arc melted structural steel. Hardened: up to 276,000 TS; 222,000 YS; 11 El in 3-1/2 inch round sections. Very deep hardening; for highly stressed structural parts.

LESCALLOY 52100 VAC ARC
Latrobe Steel Co.
C 1, Mn 0.35, Si 0.3, Cr 1.5, bal Fe.
Vacuum arc melted; wrought. Hardenable to 60-66 Rock C; oil or water quench. For aircraft engine bearings and instrument bearings for use below 400-450°F. AMS 6444.

LESCALLOY 5616
Latrobe Steel Co.
C 0.18, Mn 0.35, Si 0.35, Cr 13, W 3, Ni 2, bal Fe.
Heat treated: 145,000 TS; 114,000 YS; 17 El; 53 RA; 302 Brin. For jet engine and gas turbine compressor wheels, blades and auxiliary parts. Heat resistant, creep strength. *Obsolete*

LESCALLOY 600
Latrobe Steel Co.
C 0.07, Mn 0.3, Si 0.2, Cr 15, Fe 7, bal Ni.
Heat treated: 112,000 TS; 49,000 YS; 41 El; 180 Vickers. For nuclear reactor parts, turbine components. High creep and stress-rupture strength. Resists corrosion at elevated temperatures and stress-corrosion cracking. *Obsolete*

LESCALLOY 6304 VAC ARC
Latrobe Steel Co.
Now LESCALLOY 6305 VAC ARC.

LESCALLOY 6305 VAC ARC
Latrobe Steel Co.
C 0.45, Si 0.25, Mn 0.6, Cr 1, V 0.25, Mo 0.5, bal Fe.
Quenched and tempered to 100,000-150,000 psi TS. For high temperature bolts, jet engine shafts, fittings and high temperature lances.

LESCALLOY 718
Latrobe Steel Co.
C 0.03-0.1, Cr 17-21, Ni 50-55, Co 0-1, Ti 2.8-3.3, Al 0-0.8, Mo 3, B 0-0.006, Cu 0-0.1, 5.0-5.5% Cb + Ta, bal Fe.
Aged: 215,000 TS; 180,000 YS; 18 El. For aircraft structural parts, space ship components, missiles, gas turbine and jet engine parts. Age hardenable. High tensile, creep and rupture strength. *Obsolete*

LESCALLOY 8620 VAC ARC
Latrobe Steel Co.
C 0.18-0.23, Ni 0.4-0.7, Cr 0.4-0.6, Mo 0.15-0.25, bal Fe.
Low alloy steel for case hardening for gears, cam shafts. AISI 8620.

LESCALLOY 901
Latrobe Steel Co.
C 0.05, Cr 12.8, Mo 5.7, Ti 2.4, Fe 35, Co 1, Cu 0-0.5, Al 0-0.35, B 0.01-0.02, bal Ni.
At 70 F: 180,000 TS; 125,000 YS; 15 El; 20 RA. At 1200 F: 148,000 TS; 106,000 YS; 17 El; 24 RA. For aircraft and industrial gas turbines, rotors, and compressor discs, structural parts. High creep and rupture strength to 1400 F. Precipitation hardening, high strength alloy. *Obsolete*

LESCALLOY 9310 (AGT)
Latrobe Steel Co.
Now LESCALLOY 9310 (AGT) VAC ARC.

LESCALLOY 9310 (AGT) VAC ARC
Latrobe Steel Co.
C 0.1, Mn 0.5, Si 0.25, Ni 3.25, Cr 1.2, Mo 0.12, bal Fe.
Vacuum arc melted; wrought. Heat treated: 331-363 Brin. Carburized for aircraft engine gears and pinions; oil quench. AMS 6265 A.

LESCALLOY A-286
Latrobe Steel Co.
C 0.05, Mn 1.5, Si 0.75, Cr 15, V 0.3, Ni 26, Mo 1.25, Ti 2.2, Al 0.2, bal Fe.
Heat treated: 160,000 TS; 110,000 YS; 24 El; 43 RA; 302 Brin. For jet engine and supercharger component parts, turbine wheels and blades, nozzles. High strength and good corrosion at elevated temperatures. Age hardenable. *Obsolete*

LESCALLOY AF1410 VIM-VAR
Latrobe Steel Co.
C 0.16, Ni 10, Co 14, Cr 2, Mo 1, bal Fe.
High strength, tough. For aerospace structural applications; aircraft landing gear.

LESCALLOY BG 42 VAC ARC
Latrobe Steel Co.
Now LESCALLOY BG 42 VIM VAR.

LESCALLOY BG 42 VIM VAR
Latrobe Steel Co.
C 1.15, Cr 14.5, Mo 4, V 1, bal Fe.
For bearings operating in corrosive environment and at high temperature; stainless, retains hot hardness.

LESCALLOY BG66
Latrobe Steel Co.
C 0.85, Cr 4.15, W 6.3, V 1.85, Mo 5, bal Fe.
This is Type M2 high speed tool steel, specially melted and processed for very critical high temperature bearings. *Obsolete*

LESCALLOY D-979
Latrobe Steel Co.
C 0.04, Cr 15, Ni 45, Mo 4, W 4, Ti 3, Al 1, B 0.01, bal Fe.
Bar heat treated: 204,000 TS; 146,000 YS; 14 El; 23 RA. For jet engine components, turbine discs. Precipitation hardening. Good strength, oxidation and creep resistant. *Obsolete*

LESCALLOY D6AC
Latrobe Steel Co.
Now LESCALLOY D6AC VAC ARC.

LESCALLOY D6AC VAC ARC
Latrobe Steel Co.
C 0.46, Mn 0.75, Si 0.25, Cr 1.1, Ni 0.6, Mo 1, V 0.12, bal Fe.
Heat treated: 228,000-280,000 TS; 195,000-250,000 YS; 7 El; 23-25 RA; 46-53 Rock C. For solid fuel rocket motor cases, gears, crankshafts. High strength structural steel, tough, shock resistant.

LESCALLOY EXPANDAL
Latrobe Steel Co.
C 0.6, Mn 5.75, Si 0.2, Ni 10, bal Fe.
High thermal expansion alloy; coefficient of thermal expansion at 72-600°F: 11.5-12.5 x 10⁻⁶in./in.·°F. Austenitic, cold drawn; 140,000 TS. *Obsolete*

LESCALLOY HP 9-4-20 VAC ARC
Latrobe Steel Co.
C 0.2, Si 0-0.1, Mn 0.3, Cr 0.75, V 0.1, Ni 9, Mo 1, bal Fe.
180,000 psi YS (typical); good weldability. For pressure vessels, rocket motor cases, hulls for deep submersible vessels, aircraft components.

LESCALLOY HP 9-4-30 VAC ARC
Latrobe Steel Co.
C 0.3, Mn 0.3, Si 0.05, Cr 1, V 0.1, Mo 1, Ni 7.5, Co 4.5, bal Fe.
Low alloy, tough, high strength steel, hardenable to 50-52 Rock C; resists softening at elevated temperature; weldable. For armor plate, underwater pressure vessels, rocket motor cases, aircraft structural parts.

LESCALLOY JETHETE M-152
Latrobe Steel Co.
C 0.12, Mn 0.7, Si 0.12, Cr 11.25, Ni 2.9, Mo 1.6, V 0.3, N 0.035, bal Fe.
Nitrogen bearing, vacuum melted, hardenable stainless. Oil quenched and tempered: 900°F: 45 Rock C. For gas turbine compressor wheels and structural members. *Obsolete*

LESCALLOY LINCO
Latrobe Steel Co.
C 0.3, Mn 1.1, Si 0.35, Cr 11.5, Mo 2.75, V 0.25, Ni 0.35, bal Fe.
Oil quenched: C 50-56 Rock. Tempered at 1150 F: 160,000 TS; 130,000 YS; 15 El. Magnetic, good strength, corrosion resistant. For jet engine and gas turbine compressor wheels and auxiliary parts. *Obsolete*

LESCALLOY M50 VAC ARC
Latrobe Steel Co.
C 0.85, Si 0.2, Cr 4.1, Mo 4.52, V 1, bal Fe.
Hardenable to 63-64 Rock C. For aircraft engine bearings; withstands elevated temperatures. AISI M50.

LESCALLOY MARVAC 250
Latrobe Steel Co.
C 0.01, Ni 18.25, Co 8, Mo 5, Ti 0.4, Al 0.1, bal Fe.
Vacuum induction melt plus vacuum consumable electrode, maraging steel. Annealed: 290 Brin. Solution treated and aged: 250,000-265,000 TS; 240,000-255,000 YS; 9-13 El. For high strength missile and aircraft components; high fatigue properties; weldable. *Obsolete*

LESCALLOY MARVAC 300
Latrobe Steel Co.
C 0.01, Ni 18.25, Co 9, Mo 5, Ti 0.65, Al 0.1, bal Fe.
Double vacuum melted maraging steel. Annealed: 300 Brin. Solution treated and aged: 290,000-300,000 TS; 280,000-290,000 YS; 6-9 El. For high strength missile and aircraft components; high fatigue properties weldable. *Obsolete*

LESCALLOY MP35N
Latrobe Steel Co.
Now MP35N ALLOY.

LESCALLOY NITRALLOY 135 M VAC ARC
Latrobe Steel Co.
C 0.4, Mn 0.6, Si 0.3, Cr 1.6, Al 1.2, Mo 0.2, bal Fe.
Vacuum arc melted nitriding steel. Oil quenched and tempered, 1200°F: 135,000 TS; 100,000 YS; 16 El; 280-340 Brin. High case hardness develops when nitrided. For aircraft gears and shafts.

LESCALLOY NITRALLOY N VAC ARC
Latrobe Steel Co.
C 0.24, Si 0.3, Mn 0.6, Cr 1.15, Al 1.25, Ni 3.5, Mo 0.25, bal Fe.
Vacuum arc melted, age hardenable, nitriding steel. Oil quenched and tempered, 1200°F: 29 Rock C. Nitrided and aged, 975°F: Case: 60 Rock C min; core: 38-43 Rock C. For aircraft gears and shafts.

LESCALLOY SUPER NITRALLOY
Latrobe Steel Co.
C 0.23, Si 0.25, Mn 0.35, Cr 0.5, Al 2, Ni 5, Mo 0.25, V 0.1, bal Fe.
Vacuum arc melted, age hardenable, nitriding steel. Oil quenched and tempered 1275 F: C 30-35 Rock. Nitride (and age) 1050 F: Case: above C 60 Rock. Core: 200,000-212,000 TS; 180,000-190,000 YS. For aircraft gears and shafts. *Obsolete*

LESCALLOY V-57
Latrobe Steel Co.
C 0-0.08, Si 0-0.75, Mn 0-0.35, Cr 13.5-16, Ni 25.5-28.5, Mo 1-1.5, Ti 2.7-3.2, V 0-0.5, Al 0.1-0.35, B 0.005-0.025, bal Fe.
At room temperature: 178,000 TS; 128,000 YS; 21 El; 35 RA. At 1200 F 140,000 TS; 110,000 YS; 20 El; 30 RA. For jet engine and missile components, gas turbines, fuel nozzles. High temperature alloy steel, precipitation hardening, austenitic. *Obsolete*

LESCALLOY WASPALLOY
Latrobe Steel Co.
C 0.06, Cr 19.5, Mo 4, Co 13.5, Al 1.3, Ti 3, Fe 0-2, bal Ni. Vacuum arc melted high temperature alloy. Solution treated and aged: room temperature: 190,000-200,000 TS; 140,000-145,000 YS; 18-20 El. Test at 1000 F: 180,000-185,000 TS; 125,000-130,000 YS; 17 El. Weldable; for structural parts to be used up to 1000 F. *Obsolete*

LESCALLOY X-750
Latrobe Steel Co.
C 0.04, Cr 15.5, Ni 73, Ti 2.5, Al 0.8, Fe 7, 0.9% Cb + Ta. Age hardened: 170,000 TS; 115,000 YS; 18 El; 25 RA. For turbine wheels, springs, bolts, vanes. Precipitation hardening, creep resistant to 1500 F. *Obsolete*

LESCALLOY-UT18 VAC ARC
Latrobe Steel Co.
C 0.4, Mn 0.6, Si 0.25, Mo 0.95, Ni 0.2, V 0.2, Cr 3.25, bal Fe. Heat treated: 209,000 TS; 173,000 YS; 16 El; 60 Rock A; 388-429 Brin. For turbine and compressor rotor shafts. High strength and good impact resistance up to 1000°F. Heat resistant.

LESCALLOY-UT19 VAC ARC
Latrobe Steel Co.
C 0.16, Mn 0.45, Si 0.23, Cr 1.25, Ni 4.25, Mo 0.25, bal Fe. Heat treated: 212,000 TS; 145,000 YS; 18 El; 67 Rock A; 388-444 Brin (core). For gears, pinions, shafts, cams, aircraft components. Carburizing steel, tough and shock resistant.

LESCO "18-8"
Latrobe Steel Co.
Cr 17-20, Ni 7-10, C 0.07-0.25, bal Fe. Annealed: 87,000 TS; 33,000 YS; 58 El; 70 RA; 143 Brin. Heat treated: 100,000 TS; 44,000 YS; 48 El; 62 RA; 179 Brin. For stainless steel parts; corrosion and heat resisting. *Obsolete*

LESCO A-6
Latrobe Steel Co.
C 0.7, Si 0.3, Mn 2, Cr 1, Mo 1.25. Medium alloy, air hardening die steel. For die and mold applications.

LESCO BG41
Latrobe Steel Co.
C 1.05, Cr 14.5, Mo 4, V 0.12, S 0.3, bal Fe. For highly stressed bearings for high temperature operation; stainless, retains hot hardness. *Obsolete*

LESCO BRAKE DIE
Latrobe Steel Co.
C 0.42, Mn 0.9, Cr 1, Mo 0.2. Quenched, tempered and stress relieved medium carbon alloy; 262-311 Brin. For tooling and maintenance applications and machine component parts. AISI Type 4140/42h.

LESCO E.H.W.
Latrobe Steel Co.
C 0.25, W 15, Cr 4, V 0.5, bal Fe. For header, gripper and extrusion dies; hot work steel, oil hardened. *Obsolete*

LESCO E.S.A.
Latrobe Steel Co.
C, W, Cr, bal Fe. For cutting tools, taps, threading dies, reamers; fast machining, will not resist shock. *Obsolete*

LESCO EXTRA
Latrobe Steel Co.
Mn 0.2-0.35, C, bal Fe. For taps, reamers, twist drills, shear blades; springs; tools; water hardening steel. *Obsolete*

LESCO HW 108
Latrobe Steel Co.
C 0.38, Si 1, Mn 0.3, Cr 3.75, Mo 3.6, V 0.5, Co 2, bal Fe. Hot work tool steel; for forging dies, hot working tools.

LESCO M-4 PM
Latrobe Steel Co.
C 1.37, Si 0.4, Mn 0.35, S 0.07, W 5.5, Cr 4.25, V 4, Mo 4.5. High speed steel. For cutting tools; to machine abrasive alloys, castings and heat treated materials. For broaches, end mills, form tools, cutters, punches and pins, threading tools, cold extrusion dies, die inserts, and special drills, reamers and taps.

LESCO NINETEEN
Latrobe Steel Co.
C 0.4, Si 0.3, Mn 0.3, Cr 4.25, W 4.25, V 2, Co 4.25, bal Fe. Hot work die steel; retains 50 Rock C hardness to 1100°F. For forging dies, brass extrusion tooling, hot punches, brass die casting dies. AISI Type H19. Hot work steel.

LESCO S-7
Latrobe Steel Co.
C 0.5, Si 0.25, Mn 0.75, Cr 3.25, Mo 1.4. Air or oil hardening die steel. For tooling; shear blades, swaging dies, gripper dies, chisels and punches.

LESCO SPECIAL
Latrobe Steel Co.
Mn 0.2-0.35, C, bal Fe. For lathe and planer tools, threading dies, taps, reamers, punches; water hardening. *Obsolete*

LESCO STANDARD
Latrobe Steel Co.
Mn 0.2-0.35, C, bal Fe. For blacksmith tools, cold chisels; cheap grade water hardening tool steel. *Obsolete*

LESCO T-15 PM
Latrobe Steel Co.
C 1.57, Si 0.3, Mn 0.3, S 0.07, W 12.25, Cr 4, V 4.9, Co 5. High speed steel produced by powder metallurgy. For cutting tools; to machine abrasive alloys.

LESJOFORS GLA
Lesjofors Aktiebolag
C 0.45, Si 1, Mn 1, bal Fe. For springs; oil hardened.

LESJOFORS LB11
Lesjofors Aktiebolag
C 0.55, Mn 0.5, bal Fe. For wire ropes; water or oil hardened.

LESJOFORS LB12
Lesjofors Aktiebolag
C 0.6, Mn 0.5, bal Fe. For wire ropes; water or oil hardened.

LESJOFORS LB13 "LEWI"
Lesjofors Aktiebolag
C 0.65, Mn 0.6, bal Fe. Heat treated: 160,000 TS; 114,000 YS; 12 El; 40 RA; 325 Brin. For springs; water or oil hardened.

LESJOFORS LB13MN
Lesjofors Aktiebolag
C 0.65, Mn 1.5, bal Fe. For springs; oil hardened, tough.

LESJOFORS LB14
Lesjofors Aktiebolag
C 0.7, Mn 0.6, bal Fe. Heat treated: 175,000 TS; 128,000 YS; 12 El; 37 RA; 355 Brin. For springs; oil hardened.

LESJOFORS LB145
Lesjofors Aktiebolag
C 0.65, Si 1.3, Mn 0.5, Cr 0.5, bal Fe. For springs, punches; oil hardened, tough.

LESJOFORS LB15
Lesjofors Aktiebolag
C 0.75, Mn 0.5, bal Fe. Heat treated: 180,000 TS; 130,000 YS; 10 El; 35 RA; 375 Brin. For springs; oil hardened.

LESJOFORS LB16
Lesjofors Aktiebolag
C 0.8, Mn 0.4, bal Fe. Heat treated: 188,000 TS; 145,000 YS; 12 El; 35 RA; 390 Brin. For tools, drills, taps, hobs, reamers; Type W1; water hardened.

LESJOFORS LB50
Lesjofors Aktiebolag
C 0.5, Mn 0.8, V 0.1, Cr, bal Fe. For springs; oil hardened, tough.

LESJOFORS LB85
Lesjofors Aktiebolag
C 0.4, Mn 0.8, Cr 0.8, Ni 1.25, bal Fe. For gears, shafts, machine tool parts; oil hardened.

LESJOFORS LB8MN
Lesjofors Aktiebolag
C 0.4, Mn 1.5, bal Fe. For springs; water hardened, tough.

LESJOFORS LERO
Lesjofors Aktiebolag
C 0.1, Cr 18, Ni 8, bal Fe. For springs; Type 302; stainless.

LESJOFORS LSMN
Lesjofors Aktiebolag
C 0.55, Si 1.75, Ni 1.5, Mn 0.85, bal Fe. For springs; oil hardened, tough.

LESJOFORS LSMO
Lesjofors Aktiebolag
C 0.55, Si 1.75, Mn 0.8, Cr 0.2, bal Fe. For springs; oil hardened, tough.

LESTEM
Lehigh Steel Corp.
C, alloy, bal Fe. For dies, tools; water hardened, non-tempering.

LEWIS IRON
Joseph T. Ryerson & Son Inc.
C 0.02, Si 0.02, 3% slag, bal Fe. Rolled: 45,000 TS; 28,000 YS; 30 El; 44 RA; 95 Brin. For chains, hooks, staybolts, mine car parts; wrought iron. *Obsolete*

LEWIS SPECIAL IRON
Joseph T. Ryerson & Son Inc.
C 0.02, Si 0.01, 2.5% slag, bal Fe. Rolled: 48,000 TS; 30,000 YS; 30 El; 48 RA; 95 Brin. For locomotive staybolts; wrought iron. *Obsolete*

LFM GR.A
LFM Mfg. Co.
C 0.15-0.22, Mn 0.65-0.85, Si 0.35-0.5, bal Fe. Annealed: 73,000 TS; 41,000 YS; 22 El; 58 RA; 140 Brin. For nails, rivets, gears, fasteners, fan blades; Type 1020; case hardened.

LFM GR.B
LFM Mfg. Co.
C 0.2-0.26, Mn 0.65-0.85, Si 0.35-0.5, bal Fe. Annealed: 75,000 TS; 45,000 YS; 20 El; 55 RA; 160 Brin. For rivets, gears, fan blades, fasteners; Type 1022; water hardened.

LFM GR.C
LFM Mfg. Co.
C 0.25-0.35, Mn 0.65-0.8, Si 0.35-0.5, bal Fe. Hot rolled: 80,000 TS; 50,000 YS; 30 El; 56 RA; 165 Brin. For gears, bolts, crankshafts, brackets; Type 1030; water hardened.

LFM GR.D

LFM Mfg. Co.

C 0.3-0.4, Mn 0.65-0.8, Si 0.35-0.5, bal Fe.

Hot rolled: 85,000 TS; 72,000 YS; 26 El; 51 RA; 180 Brin. For armature shafts, gears, axles, bolts; screws; Type 1035; water hardened.

LFM GR.E

LFM Mfg. Co.

C 0.4-0.5, Mn 0.65-0.8, Si 0.35-0.5, bal Fe.

Hot rolled: 98,000 TS; 59,000 YS; 24 El; 45 RA; 212 Brin. For axles, gears, bolts, tie rods, rails; Type 1045; water hardened.

LFM GR.G

LFM Mfg. Co.

C 0.26-0.32, Mn 1.1-1.3, Si 0.35-0.5, Mo 0.2-0.3, bal Fe.

For gears, bolts, crankshafts; oil hardened, tough.

LFM GR.H

LFM Mfg. Co.

C 0.4-0.5, Mn 0.9-1.2, Mo 0.3-0.4, bal Fe.

For gears, bolts, crankshafts; oil hardened, tough.

LFM GR.I

LFM Mfg. Co.

C 0.2-0.3, Mn 0.65-0.8, Ni 2-2.25, bal Fe.

For gears, bolts, shafts, machine tool parts; oil hardened, tough.

LFM GR.J

LFM Mfg. Co.

C 0.22-0.33, Mo 0.15-0.2, Mn 0.65-0.8, bal Fe.

For gears, bolts, machine tool parts; Type 4027; water hardened.

LFM GR.K

LFM Mfg. Co.

C 0.13-0.2, Mn 0.4-0.7, Cr 4-6.5, Mo 0.45-0.65, bal Fe.

For oil refinery equipment; Type 502; creep resistant.

LFM GR.L

LFM Mfg. Co.

C 0.28-0.33, Cr 1-1.5, Mo 0.15-0.25, bal Fe.

For gears, bolts, machine tool parts; Type 4130; oil hardened.

LFM GR.M

LFM Mfg. Co.

C 0.35-0.45, Cr 1-1.5, Mo 0.15-0.25, bal Fe.

For gears, bolts, crankshafts, machinery parts; Type 4140; oil hardened.

LFM GR.N

LFM Mfg. Co.

C 0.13-0.2, Cr 1-1.5, Mo 0.45-0.65, bal Fe.

For gears, bolts, camshafts, cams; Type 4115; case hardened.

LFM GR.O

LFM Mfg. Co.

C 0.15-0.2, Cr 1.25-1.75, Ni 2.25-2.75, Mo 0.2-0.4, bal Fe.

For gears, bolts, camshafts, cams; case hardened, tough.

LFM GR.P

LFM Mfg. Co.

C 0.25-0.35, Cr 1.2-1.7, Ni 2.2-2.7, Mo 0.2-0.4, bal Fe.

For gears, bolts, crankshafts; oil hardened, tough.

LFM GR.R

LFM Mfg. Co.

C 0.55-0.65, Cr 2-2.5, Mo 0.35-0.4, bal Fe.

For dies, crimpers, upsetters; oil hardened, tough.

LFM GR.S

LFM Mfg. Co.

C 0.28-0.35, Cr 2.75-3.25, Mo 0.2-0.3, bal Fe.

For oil refinery equipment; creep resistant.

LFM GR.T

LFM Mfg. Co.

C 0.16-0.25, Mo 0.45-0.65, Mn 0.65-0.8, bal Fe.

For housings, machine tool parts; Type WC1; water hardened.

LG

J.F. Jelenko & Co.

Au 55, Pt 27, Pd 14.

100,000 psi TS; 75,000 psi YS; 10 El; 300 Brin. Dental alloy for partial dentures.

LIBERTY

J.F. Jelenko & Co.

Precious metal. Au 2, Pd 76.5.

166,000 psi TS; 115,500 psi YS; 20 El; 321 Brin. Dental allo for fusing porcelain to metal.

LIBERTY

Crucible Specialty Metals

C 1.2, Cr 0.3, W 1.2, bal Fe.

For broaches, reamers, dies; water hardening; tough. *Obsolete*

LIBERTY

Hawkridge Bros. Co.

C 1.2, Cr 0.3, W 1.25, bal Fe.

For chaser and threading tools; water hardening. *Obsolete*

LIBERTY

Colt Industries

C 1.2, Cr 0.3, W 1.2, bal Fe.

For broaches, reamers, dies; water hardening; tough. *Obsolete*

LIBERTY PISTONS

Manufacturer not listed.

Al 77, Zn 1.1, Fe 0.5.

For pistons; non-heat treatable.

LICHTENBERG FUSIBLE ALLOY

German manufacture

Bi 50, Sn 20, Pb 30.

For boiler safety plugs, fire extinguishers; melting point 75°C.

LICO

Logan Iron & Steel Co.

C 0.3, bal Fe.

For castings; water hardening.

LIDDELS ALLOY

Manufacturer not listed.

Zn 88-90, Cu 5-6.5, Al 6.5.

For die castings.

LIDEOX

Wolverine Tube, Inc.

Copper. Hg, Be, S, Li, bal Cu.

For condensers; antibiofouling. *Obsolete*

LIEBKNECHT WHITE GOLD

Manufacturer not listed.

Au 80, Ni 13.9, Cu 1, Pd 0.1.

For jewelry; corrosion resistant.

LIGHT DUTY BEARINGS-1

Manufacturer not listed.

Pb 80-90, Sb 10-17, Sn 0-10, Cu 0-1.

For bearings; light duty.

LIGHT DUTY BEARINGS-2

Manufacturer not listed.

Pb 89.3, Sb 9.9, Sn 0.9.

For light duty bearings; anti-friction.

LIGHT LEADED BRASS

Chase Brass & Copper Co.

Cu 65, Zn 34.75, Pb 0.25.

Annealed: 47,000 TS; 15,000 YS; 62 El. Drawn: 74,000 TS; 60,000 YS; 8 El. For drawn and stamped parts; high ductility. *Obsolete*

LIGMALLOY

Kling Bros. Engineering Works

For molded bearings.

LILY BRAND NO. 1

Charles Carr Ltd.

Sn 10, Zn 2, Pb 0-0.5, Ni 0-1, bal Cu.

Cast: 44,800 TS; 20,200 YS; 40 El; 75 Brin. For bearings, hardware, machine tool parts; Gun metal, BS1400G1C.

LILY BRAND NO. 10

Charles Carr Ltd.

Sn, P, Ni, bal Cu.

Cast: 42,600 TS; 25,700 YS; 8 El; 115 Brin. For bearings, hardware; P-Bronze.

LILY BRAND NO. 12

Charles Carr Ltd.

Sn 10, Pb 10, Ni 0-1.5, P 0-0.3, bal Cu.

Cast: 35,900 TS; 13,500 YS; 8 El; 85 Brin. For bearings, hardware; free-cutting. Leaded bronze, BS1400LB2C.

LILY BRAND NO. 13

Charles Carr Ltd.

Sn 10, Pb 10, Ni 0-1.5, P 0-0.3, bal Cu.

Cast: 40,400 TS; 4 El; 100 Brin. For bearings, hardware. Leaded bronze, BS1400LB1C.

LILY BRAND NO. 15

Charles Carr Ltd.

Fe 4, Mn 0-1.5, Al 10, Ni 5, bal Cu.

Cast: 94,100 TS; 38,100 YS; 15 El; 170 Brin. For bearings, propellers, hardware, machine tool parts; Al-Bronze, BS1400AB2C.

LILY BRAND NO. 16

Charles Carr Ltd.

Al, bal Cu.

Cast: 92,000 TS; 74,000 YS; 6 El; 300 Brin. For non-sparking tools; Al-Bronze, corrosion resistant.

LILY BRAND NO. 17

Charles Carr Ltd.

Sn 0-1.5, Pb 0-0.5, Al 0-2.5, Fe 1.2, Mn 0-3, 55.0 Cu min, bal Zn.

Cast: 76,200 TS; 36,000 YS; 24 El; 130 Brin. For propellers, hardware, marine parts; Mn-bronze, BS1400HTB1C.

LILY BRAND NO. 18

Charles Carr Ltd.

Sn 0-1, Pb 1-5, Cu 66-73, bal Zn.

Cast: 33,600 TS; 35 El. For hardware; commercial brass, BS1400B2.

LILY BRAND NO. 19

Charles Carr Ltd.

Zn, bal Cu.

Cast: 33,600 TS; 35 El. For hardware; commercial brass, BS1400B3.

LILY BRAND NO. 2

Charles Carr Ltd.

10.0 Sn min, 0.05 Zn min, 0.50 P min, bal Cu.

Cast: 40,400 TS; 22,400 YS; 14 El; 95 Brin. For bearings, hardware; Bronze, BS1400PB1C.

LILY BRAND NO. 3

Charles Carr Ltd.

Sn 8, Zn 4, Pb 0-0.5, Ni 0-1, bal Cu.

Cast: 44,800 TS; 20,200 YS; 38 El; 70 Brin. For bearings, hardware, machine tool parts; Gun metal, BS1400G2C.

LILY BRAND NO. 4
Charles Carr Ltd.
Sn 5, Zn 5, Pb 5, Ni 0-1, bal Cu.
Cast: 36,000 TS; 15,700 YS; 25 El; 60 Brin. For hardware, machine tool parts; Gun metal, BS1400LG2C.

LILY BRAND NO. 5
Charles Carr Ltd.
Sn 7, Zn 5, Pb 2, Ni 0-1, bal Cu.
Cast: 40,400 TS; 18,000 YS; 35 El; 60 Brin. For hardware, plumbing; Gun metal, BS1400LG3C.

LILY BRAND NO. 7
Charles Carr Ltd.
10.0 Sn min, 0.05 Zn min, 0.50 P min, bal Cu.
Cast: 49,300 TS; 26,900 YS; 3 El; 110 Brin. For bearings, hardware; Bronze, BS1400PB1C.

LILY BRAND NO. 8
Charles Carr Ltd.
Sn 12, Zn 0-0.3, Pb 0-0.5, Ni 0-0.5, 0.15 P min, bal Cu.
Cast: 40,400 TS; 20,200 YS; 10 El; 100 Brin. For bearings, hardware; Bronze, BS1400PB2C.

LILY BRAND NO. 9
Charles Carr Ltd.
Sn 12, Zn 0-0.3, Pb 0-0.5, 0.15 P min, bal Cu.
Cast: 44,800 TS; 22,400 YS; 4 El; 120 Brin. For bearings, hardware; Bronze, BS1400PB2C.

LILY PHOSPHOR BRONZE
Non-Ferrous Castings Co. Ltd.
Sn 5, P 0.05, bal Cu.
For bushings, bearings. Chill cast.

LINCO
Now LESCALLOY LINCO.

LINCOLNWELD L-50
Lincoln Electric Co.
C, bal Fe.
Submerged arc welding electrode. AWS Class EM13K.

LINCOLNWELD L-60
Lincoln Electric Co.
C, bal Fe.
Submerged arc welding electrode. AWS Class EL12.

LINCOLNWELD L-61
Lincoln Electric Co.
C, bal Fe.
Submerged arc welding electrode. AWS Class EM12K.

LINCOLNWELD L-70
Lincoln Electric Co.
C, bal Fe.
Submerged arc welding electrode.

LINDENBERG 55SWC
Bergische Stahl Industrie
C 0.55, Si 0.9, Cr 1.05, V 0.18, W 1.85, bal Fe.
For cold work tools and dies; oil hardened. *Obsolete*

LINDENBERG A18V150CO
Bergische Stahl Industrie
C 0.79, Co 4.7, Cr 4.3, Mo 0.75, V 1.5, W 18, bal Fe.
For lathe and planer tools, reamers, hobs, taps; high speed steel. *Obsolete*

LINDENBERG A18V80
Bergische Stahl Industrie
C 0.74, Cr 4, V 1.1, W 18.5, bal Fe.
For lathe and planer tools, reamers, hobs, taps; high speed steel. *Obsolete*

LINDENBERG A300
Bergische Stahl Industrie
C 1.65, Cr, Co, bal Fe.
For blanking and forming dies; air hardened, non-deforming. *Obsolete*

LINDENBERG A3V
Bergische Stahl Industrie
C 1.4, W, V, bal Fe.
For fast finishing cutters, tools; water hardened. *Obsolete*

LINDENBERG A600
Bergische Stahl Industrie
C 1.45, Cr 1.4, Mn 0.6, bal Fe.
For bearings, liners, sleeves; water hardened. *Obsolete*

LINDENBERG BFH
Bergische Stahl Industrie
C 0.55, Cr 1.05, V 0.18, W 1.85, bal Fe.
For cold work tools, header dies, punches; oil hardened, tough. *Obsolete*

LINDENBERG BLA702
Bergische Stahl Industrie
C 0.5, Cr 1.05, Ni 3.25, bal Fe.
For gears, pinions, crankshafts, bolts, fasteners; oil hardened, shock resistant. *Obsolete*

LINDENBERG C615
Bergische Stahl Industrie
C 1, Cr 1.1, Mn 0.07, bal Fe.
For bearings, liners, sleeves, cutters; water hardened, wear resistant. *Obsolete*

LINDENBERG CCN
Bergische Stahl Industrie
C 2.1, Cr 11.5, Mn 0.3, bal Fe.
For forming and blanking dies, punches; oil hardened, nondeforming. *Obsolete*

LINDENBERG CCNW
Bergische Stahl Industrie
C 2.1, Cr 11.5, W 0.7, bal Fe.
For forming and blanking dies, punches; oil hardened, nondeforming. *Obsolete*

LINDENBERG EXTRA 2
Bergische Stahl Industrie
C 1.3, Si 0-0.25, Mn 0-0.25, bal Fe.
For engravers' tools, form cutters, reamers; Type W1; water hardened. *Obsolete*

LINDENBERG EXTRA 3
Bergische Stahl Industrie
C 1.15, Si 0-0.25, Mn 0-0.25, bal Fe.
Annealed: 110,000 TS; 55,000 YS; 20 El; 40 RA; 210 Brin. For drills, taps, reamers, milling cutters; Type W1; water hardened. *Obsolete*

LINDENBERG EXTRA 4
Bergische Stahl Industrie
C 1, Si 0-0.25, Mn 0-0.25, bal Fe.
Annealed: 100,000 TS; 53,000 YS; 21 El; 42 RA; 200 Brin. For drills, taps, reamers, springs, hobs; Type W1; water hardened. *Obsolete*

LINDENBERG EXTRA 5
Bergische Stahl Industrie
C 0.85, Si 0-0.25, Mn 0-0.25, bal Fe.
Heat treated: 190,000 TS; 145,000 YS; 10 El; 30 RA; 400 Brin. For drills, springs, tools, cutters, taps, hobs; Type W1; water hardened. *Obsolete*

LINDENBERG EXTRA 6
Bergische Stahl Industrie
C 0.7, Si 0-0.25, Mn 0-0.25, bal Fe.
Heat treated: 175,000 TS; 130,000 YS; 12 El; 37 RA; 355 Brin. For springs, rails, hammers, tools; Type W1; water hardened. *Obsolete*

LINDENBERG HSO
Bergische Stahl Industrie
C 1, Cr 1.1, Mn 0.07, bal Fe.
For bearings, bushings, liners; water hardened, wear resistant. *Obsolete*

LINDENBERG HW3MH
Bergische Stahl Industrie
C 1.1, Si 0-0.25, Mn 0-0.25, bal Fe.
Annealed: 110,000 TS; 58,000 YS; 18 El; 40 RA; 210 Brin. For springs, taps, cutters, drills, reamers; Type W1; water hardened. *Obsolete*

LINDENBERG HW4ZH
Bergische Stahl Industrie
C 1, Si 0-0.25, Mn 0-0.25, bal Fe.
Heat treated: 130,000-185,000 TS; 80,000-120,000 YS; 10-20 El; 30-45 RA; 270-380 Brin. For springs, taps, drills, reamers, hobs; Type W1; water hardened. *Obsolete*

LINDENBERG HW5Z
Bergische Stahl Industrie
C 0.85, Si 0-0.25, Mn 0-0.25, bal Fe.
Heat treated: 130,000-190,000 TS; 88,000-145,000 YS; 12-20 El; 35-50 RA; 255-390 Brin. For drills, taps, reamers, springs; Type W1; water hardened. *Obsolete*

LINDENBERG IL15
Bergische Stahl Industrie
C 0.8, Cr 4.1, Mo 0.85, V 1.5, W 8.7, bal Fe.
For lathe and planer tools, reamers, broaches, taps; high speed steel. *Obsolete*

LINDENBERG IL232
Bergische Stahl Industrie
C 0.95, W, Mo, Cr, V, bal Fe.
For reamers, hobs, broaches, taps; high speed steel. *Obsolete*

LINDENBERG IL25
Bergische Stahl Industrie
C 0.86, Cr 4.1, Mo 0.85, V 2.5, W 12, bal Fe.
For lathe and planer tools, drills, taps, hobs; high speed steel. *Obsolete*

LINDENBERG IL35
Bergische Stahl Industrie
C 0.86, Co 2.8, Cr 4.3, Mo 0.85, V 2.1, W 12, bal Fe.
For lathe and planer tools, reamers, broaches, hobs; high speed steel. *Obsolete*

LINDENBERG IL45
Bergische Stahl Industrie
C 1.3, Cr 4.3, Mo 0.85, V 3.8, W 12, bal Fe.
For engravers tools, blanking and forming dies; high speed steel. *Obsolete*

LINDENBERG IL45CO
Bergische Stahl Industrie
C 1.35, Cr 4, Co, V, Mo, W, bal Fe.
For forming and blanking dies, engravers tools; high speed steel. *Obsolete*

LINDENBERG K
Bergische Stahl Industrie
C 0.55, Si 0.9, Mn 0.3, Cr 1.05, V 0.18, W 1.85, bal Fe.
For heading dies, punches, upsetters; oil hardened, tough. *Obsolete*

LINDENBERG NWNH2
Bergische Stahl Industrie
C 1.65, Cr, Co, bal Fe.
For blanking and punching dies; oil hardened, tough. *Obsolete*

LINDENBERG RS
Bergische Stahl Industrie
C 1.65, Cr 11.5, V 0.1, bal Fe.
For blanking and forming dies, punches; air hardened, nondeforming. *Obsolete*

LINDENBERG T18V
Bergische Stahl Industrie
C 1.05, W 1.15, Cr 1, Mn 0.9, bal Fe.
For cold work tools, heading dies; oil hardened, tough. *Obsolete*

LINDENBERG WLH
Bergische Stahl Industrie
C 0.56, Ni, Cr, Mo, V, bal Fe.
For upsetting and forging dies; oil hardened, tough.
Obsolete

LINEAR METAL
Sheepbridge Engineering Ltd.
C 3.3, Si 1, P 0.2, bal Fe.
Cast: 180-220 Brin. For castings, cylinder liners; cast iron.

LINEAR METAL
Sheepbridge Alloy Castings Ltd.
C 3.3, Si 1, P 0.2, bal Fe.
Cast: 180-220 Brin. For castings, cylinder liners; cast iron.

LINOTYPE
Manufacturer not listed.
Pb 85, Sb 11, Sn 3.5.
11,700 TS; 9 El; 21 Brin. For type metal; M.P. 476°F.

LION
Edgar T. Ward's Sons Co.
C 0.7-0.9, bal Fe.
For tools, drills, taps; water hardening.

LION
Burys & Co. Ltd.
C 0.7, W 14, Cr 3.5, V 0.6, bal Fe.
For tools, reamers, hobs, drills, broaches; high speed steel.

LION (AISI W-1 GR3)
Jessop Steel Co.
C 0.6-1.4, bal Fe.
For tools, general tools, broaches, chisels, dies, knives, files; water hardened.

LION 22.0
Burys & Co. Ltd.
C 0.75, W 22, Cr 4.5, V 1.2, bal Fe.
For lathe and planer tools, hobs, reamers, drills; high speed steel.

LION BRAND ANTIFRICTION METAL
Blackwell's Metallurgical Works
Sn 5-15, Sb 10-15, bal Pb.
For bearings for rolling mills and stone crushers; for heavy pressures and high speeds.

LION BRAND PLASTIC WHITE METAL
Blackwell's Metallurgical Works
Sb 3-10, Cu 1-3, Pb 10-17, bal Sn.
For main bearings for marine engines. Babbitt.

LION EXTRA (AISI W1 GR 2)
Jessop Steel Co.
C 0.6-1.3, Si 0.25, Mn 0.3, bal Fe.
For drills, hobs, reamers, taps, broaches; Type W1; water hardened.

LION G.19.F.
Burys & Co. Ltd.
C 0.9, Mn 2, bal Fe.
For taps, dies, reamers, forming dies; oil hardened, nondeforming.

LION G.19.H.
Burys & Co. Ltd.
C 0.9, Si 0.45, Mn 1.25, W 0.5, Cr 0.5, bal Fe.
For taps, dies, reamers, forming dies; oil hardened, nondeforming.

LION H.18.V.
Burys & Co. Ltd.
C 1, Cr 1.5, V 0.2, bal Fe.
For taps, dies, forming rolls, broaches, reamers; oil hardened, wear resistant.

LION K.15
Burys & Co. Ltd.
C 0.3, Cr 13, S, bal Fe.
Annealed: 95,000 TS; 50,000 YS; 25 El; 55 RA; 196 Brin. For stainless screw machine products; free-cutting, corrosion resistant.

LION P.16
Burys & Co. Ltd.
C 1, Mn 1, W 0.75, Cr 0.75, bal Fe.
For blanking and forming dies, reamers, broaches; oil hardened, nondeforming.

LION PEG STEEL 5.5.0.
Burys & Co. Ltd.
C 0.5, W 14, Cr 3, V 0.5, bal Fe.
For hot punches, core and die inserts; hot work steel, oil hardened.

LION R.19
Burys & Co. Ltd.
C 0.25, Cr 13, bal Fe.
Annealed: 95,000 TS; 50,000 YS; 25 El; 55 RA; 196 Brin. For dies, cutlery, valves; Type 420; corrosion resistant.

LION REGENT
Burys & Co. Ltd.
C 0.8, W 20, Cr 5, V 1.25, Co 18, Mo 0.75, bal Fe.
For cutting tools, broaches, milling cutters; high speed steel.

LION SPECIAL
Burys & Co. Ltd.
C 0.75, W 19, Cr 4.25, V 1.25, Co 8, Mo 0.75, bal Fe.
For lathe and planer tools, hobs, drills; high speed steel.

LION VANADIUM
Jessop Steel Co.
C 0.6 1, V 0.25, bal Fe.
AISI W2 Gr. 1.

LIPCOR
British Steel plc
Mild steel. For noncritical machine applications.

LISCO
Logan Iron & Steel Co.
C 0.35, bal Fe.
For castings; water hardening.

LITHALOYS
Lithaloys Corp.
Li, Cu.
Master alloy for Cu castings; for sound castings.

LITHIUM
Atomergic Chemetals Corp.
Li.
Purities: 99.99%, reactor grade 99.9+%, 99.8%. Forms: Ingot, rod, sheet, foil, wire, shot, powder, billets, ribbon.

LITHIUM-CALCIUM
Maywood Chemical Works
Li 50, Ca 50.
For deoxidizer, metallurgical applications.

LITHIUM-CONDUCTIVITY BRONZE
Maywood Chemical Works
Copper. Cd 0-2, 98 Cu min, Sn, Si.
For high strength conductors; bronze treated with Li with increased conductivity.

LITHIUM-COPPER
Maywood Chemical Works
Li, bal Cu.
Annealed: 31,500-36,500 TS; 60-72 El. For electrical conductors; high conductivity, oxygen free.

LITHOBRAZE 720
Handy & Harmon
Precious metal. Ag 71.3-72.3, Li 0.15-0.3, bal Cu.
MP: 1435°F. For brazing alloy for stainless steels; eutectic alloy. AWS BAg-8a.

LITHOBRAZE 925
Handy & Harmon
Precious metal. Ag 92.5, Cu 7.3, Li 0.2.
MP: 1435°F; FP: 1635°F. For high temperature brazing, short time operation to 900°F, continuous to 500°F. AWS BAg-19.

LITINUM BRONZE
English manufacture
Cu 90, Sn 4.7, Zn 3.86, Pb 1.6.
For nuts, bushings, bearings, screw machine parts; free-cutting.

LITTITE
Little Bros. Foundry Co.
C 3, Si, Mn, alloy, bal Fe.
For gears, shafts, machine tool housings; cast iron.

LITTLES SPECULUM
Manufacturer not listed.
Al 65, Sn 31, Zn 2.3, As 1.9.
For mirrors, reflectors; high polish.

LK4
Russian manufacture
C 0.2, Cr 26.5, Mo 5, Ni 3.3, bal Co.
Cast cobalt-base superalloy. For nozzle guide vanes.

LK4YA
Russian manufacture
C 0.26, Cr 26.5, W 5, B 0.02, Ni 0-3, bal Co.
Cast cobalt base superalloy.

LK66YA
Russian manufacture
C 0.3, Cr 22.5, W 9.5, B 0.02, Cb 1.75, Ni 0-2, bal Co.
Cast cobalt base superalloy.

LM 0
English manufacture
Al 99.5, Fe 0-0.4.
Aluminium casting. AFNOR A5; BS 1490 LM0.

LM 1
British Aluminium Co., Ltd.
Cu 7, Si 3, Zn 3, bal Al.
Sand cast: 19,000 TS; 12,300 YS; 73 Brin. Permanent mold: 24,600 TS; 15,700 YS; 1 El; 80 Brin. For general purpose castings; Brit. DTD428.

LM 10
Sterling Metals Ltd.
Mg 9.5-11, Si 0.25, Fe 0.35, bal Al.
Sand cast, solution treated: 280 MPa TS; 170 MPa YS; 8 El; 70 Brin. Special purpose cast aluminum alloy; high proof stress, resistant to shock and corrosion. BS 1490 LM10; AFNOR A-G 10.

LM 10
British Aluminium Co., Ltd.
Mg 9.5-11, Si 0.25, Fe 0.35, bal Al.
Sand cast, solution treated: 280 MPa TS; 170 MPa YS; 8 El; 70 Brin. Special purpose cast aluminum alloy; high proof stress, resistant to shock and corrosion. BS 1490 LM10; AFNOR A-G 10.

LM 10A
Johnson Matthey plc
Ag, Cu, Sn.
Soft solder; 214-275°C MP.

LM 11
English manufacture
Cu 4-5, Si 0-0.25, Fe 0-0.25, Ti 0.05-0.3, bal Al.
Cast aluminium alloy. BS 1490 LM11.

LM 12

British Aluminium Co., Ltd.
Cu 9-11, Si 0-2.5, Zn 0-0.8, Fe 0-1, Mn 0-0.6, Ni 0-0.5, Mg 0.2-0.4, bal Al.
Cast aluminum alloy. BS 1490 LM12; AFNOR A-S 12.

LM 13

British Aluminium Co., Ltd.
Cu 0.7-1.5, Si 10-12, Zn 0-0.5, Fe 0-1, Mn 0-0.5, Ni 0-1.5, Mg 0.8-1.5, bal Al.
Cast aluminum alloy. BS 1490 LM13; AFNOR A-S 12UN.

LM 14

English manufacture
Cu 3.5-4.5, Si 0-0.6, Fe 0-0.6, Mn 0-0.6, Ni 1.8-2.3, Mg 1.2-1.7, bal Al.
Cast aluminium alloy. BS 1490 LM14; AFNOR A-U 4 NT.

LM 15

Johnson Matthey plc
Zn, Ag, Cd.
Zinc base soft solder; 280-320°C MP.

LM 15

English manufacture
Cu 1.3-3, Si 0.6-2, Fe 0.8-1.4, Ni 0.5-2, Mg 0.5-1.7, Ti 0.05-0.3, bal Al.
Cast aluminium alloy. BS 1490 LM15.

LM 16

Sterling Metals Ltd.
Cu 1-1.5, Mg 0.4-0.6, Si 1.5-5.5, Ni 0.25, Fe 0.6, Mn 0.5, bal Al.
Sand cast, solution treated & aged: 230 MPa TS; 220 MPa YS; 90 Brin. General purpose alloy; good pressure tightness and castability.

LM 16

British Aluminium Co., Ltd.
Cu 1-1.5, Mg 0.4-0.6, Si 1.5-5.5, Ni 0.25, Fe 0.6, Mn 0.5, bal Al.
Sand cast, solution treated & aged: 230 MPa TS; 220 MPa YS; 90 Brin. General purpose alloy; good pressure tightness and castability.

LM 18

British Aluminium Co., Ltd.
Si 4.5-6, Fe 0-0.6, Mn 0-0.5, bal Al.
Cast aluminum alloy. BS 1490 LM18.

LM 2 TYPE 1

British Aluminium Co., Ltd.
Cu 1.1, Si 10.3, bal Al.
Sand cast: 21,300 TS; 9,800 YS; 2 El; 70 Brin. For pressure tight castings; corrosion resistant.

LM 2 TYPE 2

British Aluminium Co., Ltd.
Cu 2.1, Si 10.3, Zn 0.9, bal Al.
Permanent mold: 25,500 TS; 12,000 YS; 2 El; 78 Brin. For pressure tight castings; corrosion resistant.

LM 20

British Aluminium Co., Ltd.
Si 11.5, bal Al.
Sand cast: 25,200 TS; 8900 YS; 5.0 El; 55 Brin. Die cast: 31,400 TS; 9500 YS; 7.5 El; 60 Brin. For pressure tight castings; Brit. LM20M, corrosion resistant.

LM 21

British Aluminium Co., Ltd.
Cu 3, Si 5, Mn 0.5, bal Al.
Sand cast: 20,200 TS; 1.5 El. Permanent mold: 22,400 TS; 1.5 El. For general engineering castings; Brit. LM21M, corrosion resistant.

LM 22

Sterling Metals Ltd.
Cu 2.8-3.8, Si 4-6, Fe 0.6, Mn 0.2-0.6, bal Al.
Chill cast, solution treated: 245 MPa TS; 110 MPa YS; 8 El; 70 Brin. Permanent mould casting only; for heavy duty applications.

LM 22

British Aluminium Co., Ltd.
Cu 2.8-3.8, Si 4-6, Fe 0.6, Mn 0.2-0.6, bal Al.
Chill cast, solution treated: 245 MPa TS; 110 MPa YS; 8 El; 70 Brin. Permanent mould casting only; for heavy duty applications.

LM 24

British Aluminium Co., Ltd.
Cu 3.5, Si 8.5, bal Al.
Permanent mold: 25,700 TS; 1.5 El. For general engineering castings; Brit. LM24M, corrosion resistant.

LM 24 TYPE A

English manufacture
Cu 3-4, Si 7.5-9.5, Zn 0-1.5, Fe 0-1.3, Mn 0-0.5, Ni 0-0.5, Mg 0-0.3, Pb 0-0.3, Sn 0-0.2, Ti 0-0.2, bal Al.
Cast aluminium alloy. BS 1490 LM24-A; AFNOR A-S 9U 3/A.

LM 24 TYPE B

English manufacture
Cu 3-4, Si 7.5-9.5, Zn 0-3, Fe 0-1.3, Mn 0-0.5, Ni 0-0.5, Mg 0-0.3, Pb 0-0.3, Sn 0-0.2, Ti 0-0.2, bal Al.
Cast aluminium alloy. BS 1490 LM24-A; AFNOR A-S 9U 3/B.

LM 25

Sterling Metals Ltd.
Mg 0.2-0.45, Si 6.5-7.5, bal Al.
Sand cast, solution treated and aged: 230 MPa TS; 215 MPa YS 80 Brin. General purpose alloy with good casting qualities.

LM 25

British Aluminium Co., Ltd.
Mg 0.2-0.45, Si 6.5-7.5, bal Al.
Sand cast, solution treated and aged: 230 MPa TS; 215 MPa YS 80 Brin. General purpose alloy with good casting qualities.

LM 26

British Aluminium Co., Ltd.
Cu 2-4, Si 8.5-10.5, Zn 0-1, Fe 0-1.2, Mn 0-0.5, Ni 0-1, Mg 0.5-1.5, Pb 0-0.2, Ti 0-0.2, bal Al.
Cast aluminum alloy. BS 1490 LM26.

LM 27

Sterling Metals Ltd.
Cu 1.5-2.5, Mg 0.3, Si 6-8, Fe 0.8, Mn 0.2-0.6, Ni 0.3, Zn 1, bal Al.
Sand cast: 140 MPa TS; 90 MPa YS; 1 El; 75 Brin. Good castability; pressure tight.

LM 27

British Aluminium Co., Ltd.
Cu 1.5-2.5, Mg 0.3, Si 6-8, Fe 0.8, Mn 0.2-0.6, Ni 0.3, Zn 1, bal Al.
Sand cast: 140 MPa TS; 90 MPa YS; 1 El; 75 Brin. Good castability; pressure tight.

LM 28

British Aluminium Co., Ltd.
Cu 1.3-1.8, Si 17-20, Zn 0-0.2, Fe 0-0.7, Mn 0-0.6, Ni 0.8-1.5, Mg 0.8-1.5, Cr 0-0.6, Co 0-0.5, bal Al.
Cast aluminum alloy. BS 1490 LM28; AFNOR A-S 20U.

LM 29

British Aluminium Co., Ltd.
Cu 0.8-1.3, Si 22-25, Zn 0-0.2, Fe 0-0.7, Mn 0-0.6, Ni 0.8-1.3, Mg 0.8-1.3, Cr 0-0.6, Co 0-0.5, bal Al.
Cast aluminum alloy. BS 1490 LM29; AFNOR A-S 22 UNK.

LM 3

English manufacture
Cu 2.5-4.5, Si 0-1.3, Zn 9-13, Fe 0-1, Mn 0-0.5, Ni 0-0.5, Pb 0-0.3, bal Al.
Cast aluminium alloy. BS 1490 LM3.

LM 30

British Aluminium Co., Ltd.
Cu 4-5, Si 16-18, Zn 0-0.2, Fe 0-1.1, Mn 0-0.3, Mg 0.4-0.7, Ti 0-0.2, bal Al.
Cast aluminum alloy. BS 1490 LM30.

LM 4

British Aluminium Co., Ltd.
Cu 3, Si 5, Mn 0.5, bal Al.
Sand cast: 22,400 TS; 12,300 YS; 2.5 El; 65 Brin. Permanent mold: 23,500 TS; 12,300 YS; 2.5 El; 70 Brin. For pressure tight castings; Brit. DTD424, age hardenable.

LM 5

Sterling Metals Ltd.
Mg 3-6, Si 0.3, Fe 0.6, Mn 0.3-0.7, bal Al.
Sand cast: 140 MPa TS; 90 MPa YS; 3 El; 55 Brin. Special purpose cast aluminum alloy; high corrosion resistance-for marine applications. BS 1490 LM5; AFNOR A-G 3T.

LM 5

British Aluminium Co., Ltd.
Mg 3-6, Si 0.3, Fe 0.6, Mn 0.3-0.7, bal Al.
Sand cast: 140 MPa TS; 90 MPa YS; 3 El; 55 Brin. Special purpose cast aluminum alloy; high corrosion resistance-for marine applications. BS 1490 LM5; AFNOR A-G 3T.

LM 6

British Aluminium Co., Ltd.
Si 11.3, bal Al.
Sand cast: 25,700 TS; 9000 YS; 7 El; 55 Brin. Die cast: 34,000 TS; 10,200 YS; 10 El; 60 Brin. For pressure tight thin wall castings; Brit. BS3L33, corrosion resistant.

LM 630

Light Metals Inc.
Al 6, Zn 3, Mn 0.2, bal Mg.
Cast: 27,000 TS; 5 El; 50 Brin. Heat treated: 38,000 TS; 5 El; 73 Brin. For light alloy castings; heat treatable.

LM 7

English manufacture
Cu 1-2.5, Si 1.5-3.5, Fe 0.3-1.4, Ni 0.5-1.7, Mg 0.05-0.2, Ti 0.05-0.3, bal Al.
Cast aluminium alloy. BS 1490 LM7.

LM 8

English manufacture
Si 3.5-6, Fe 0-0.6, Mg 0.3-0.8, Mn 0-0.5, bal Al.
Cast aluminium alloy. BS 1490 LM8; AFNOR A-S 4G.

LM 9

Sterling Metals Ltd.
Mg 0.2-0.6, Si 10-13, Fe 0.6, Mn 0.3-0.7, bal Al.
Sand cast, precipitation treated: 170 MPa TS; 120 MPa YS; 1.5 El; 60 Brin. Special purpose cast aluminum alloy; high strength, rigid, good corrosion resistance. BS 1490 LM9; AFNOR A-S 9G.

LM 9

British Aluminium Co., Ltd.
Mg 0.2-0.6, Si 10-13, Fe 0.6, Mn 0.3-0.7, bal Al.
Sand cast, precipitation treated: 170 MPa TS; 120 MPa YS; 1.5 El; 60 Brin. Special purpose cast aluminum alloy; high strength, rigid, good corrosion resistance. BS 1490 LM9; AFNOR A-S 9G.

LM 920

Light Metals Inc.
Al 9, Zn 2, Mn 0.1, bal Mg.
Cast: 24,000 TS; 2 El; 65 Brin. Heat treated: 40,000 TS; 2 El; 87 Brin. For light alloy castings; heat treatable.

LM6M, ETC.

Now MILLS LM6M, ETC.

LMC

Pechiney Electrometallurgie
Miscellaneous nonferrous. Al 0.8, Ba 0.8, Ca 1.7, Si 66, bal Fe.
Cast iron inoculant.

LMI 108 ALLOY

Europa Metalli-LMI S.p.A.
Copper. Mg 0.22, P 0.2, Sn 0.1, Ca 0.01, bal Cu.
Age hardened copper alloy for use in the electronic industry production of lead frames, relay springs, wire wrap, terminals, fuse clips, insulation displacement connectors, electrical springs, bus plates and as a substitute for Cu-Cd alloys in electrical cables. Two types of LMI 108 are produced by varying the work-cycle; type A has higher electrical conductivity, type B has better mechanical characteristics.

LN-2

Union Carbide Corp.
99 Ni min.
Metallic spray coating for plasma torch. For build-up of worn parts; 200 DPH.

LO-AIR

Cytemp Specialty Steel Div.
C 0.7, Mn 2.25, Cr 1, Mo 1.35, bal Fe.
Heat treated: 293,000 TS; 264,000 YS; 560 Brin. For blanking, piercing and embossing dies. Type A6. Air hardening, wear resistant.

LO-CRO TYPE 501

Colt Industries
Cr 4-6, 0.10% C min, bal Fe.
For oil refineries; corrosion resistant. *Obsolete*

LO-CRO-TYPE 502

Colt Industries
C 0-0.1, Cr 4-6, bal Fe.
For oil refineries; corrosion resistant. *Obsolete*

LO-EX

Metallgesellschaft Reuterweg
Cu 0.5-1.3, Mn 0.8-1.5, Si 11-13, Ni 2-3, bal Al.
Sand cast: 37,000 TS; 35,900 YS; 0 El; 125 Brin. Permanent mold: 44,800 TS; 42,600 YS; 0 El; 130 Brin. For pistons; age hardenable.

LO-EX

Alluminio SA
Cu 0.5-1, Si 12.8, Mg 0.8, Ni 2.2, bal Al.
Cast: 36,000-47,000 TS; 36,000-46,000 YS; 0.2-0.5 El; 120-135 Brin. For pistons, sleeves, bushings; low thermal expansion.

LO-EX

Soc. Alluminio Veneto per Azioni
Cu 0.5-1, Si 12.8, Mg 0.8, Ni 2.2, bal Al.
Cast: 36,000-47,000 TS; 36,000-46,000 YS; 0.2-0.5 El; 120-135 Brin. For pistons, sleeves, bushings; low thermal expansion.

LO-EX ALLOY

Birmingham Aluminium Casting Co.
Aluminum. Si 11-13, Ni 1-2.5, Mg 1, Cu 0.7, bal Al.
Cast: 26,000 TS; 23,000 YS; 0.5 El; 75 Brin. Heat treated: 38,000 TS; 33,000 YS; 0.5 El; 140 Brin. For pistons; low thermal expansion.

LO-FLO NO. S-0

Fusion Inc.
For silver solder; MP 1500 F. *Obsolete*

LO-FLO NO. S-1

Fusion Inc.
For silver solder; MP 1300-1400 F. *Obsolete*

LO-FLO NO. S-4

Fusion Inc.
For silver solder; MP 1150 F. *Obsolete*

LO-LUMINIUM

English manufacture
Sn 55, Zn 33, Al 11, Cu 1.
For aluminum solder.

LO-TEMP

John Hewson Co.
Zn 40, bal Cu.
For repairing castings; brass.

LO-TIN SILVER FUSE

Fusion Inc.
For solder. *Obsolete*

LOADED CENTRICAST

Sheepbridge Engineering Ltd.
C 3.2, Si 2.8, Cr 0.8, bal Fe.
Cast: 36,000 Ts: 240-300 Brin. For cylinder liners; centrifugal.

LOADED CENTRICAST

Sheepbridge Alloy Castings Ltd.
C 3.2, Si 2.8, Cr 0.8, bal Fe.
Cast: 36,000 Ts: 240-300 Brin. For cylinder liners; centrifugal.

LOADED IRON

Sheepbridge Engineering Ltd.
C 3.3, Si 2.2, Cr 1, P 0.4, bal Fe.
320 Brin. For cylinder liners; cast iron.

LOADED IRON

Sheepbridge Alloy Castings Ltd.
C 3.3, Si 2.2, Cr 1, P 0.4, bal Fe.
320 Brin. For cylinder liners; cast iron.

LOCKALLOY

NGK Metals Corp.
Al 38, Be 62.
Sheet: 56,000 TS; 44,000 YS; 8 El. At 400°F: 44,000 TS; 40,000 YS; 11 El. For nuclear fuel canning, gyro cages, aircraft brakes, computer memory drums and discs, spacecraft parts, missiles, satellite launch vehicles. High modulus, low density alloy.

LOCKPORT SPECIAL

Wallace Murray Corp.
C 0.7, W 18, Cr 4, V 2, bal Fe.
For tools, cutters; high speed steel.

LOCOMOTIVE TUBE

English manufacture
Cu 97, Ni 3.
For tubes for locomotives; corrosion resistant.

LODEX

Manufacturer not listed.
Fe, Co single domain particle.
Permanent magnet material. Anisotropic or isotropic. Pressed or extruded shapes. Uniform magnetic and physical properties. *Obsolete*

LOFLEX

Engelhard Corp.
Bimetal. For thermostatic bimetals; maximum temperature 800°F. *Obsolete*

LOGAN

Logan Iron & Steel Co.
C 0.4, bal Fe.
For castings; water hardening.

LOGAN (W2 GRADE 4)

Time Steel Service Inc.
C 1, Mn 0.25, Si 0.25, bal Fe.
Water hardened tool steel; AISI W2.

LOGAN FORGING BRASS

Cerro Metal Products Co.
Cu 58.5, Pb 2, bal Zn.
Extruding: 61,000 TS; 30,000 YS; 40 El; 104 Brin. For brass forgings and hot pressings. CA 377.

LOHM

Driver Harris Co.
Copper. Ni 7-7.5, bal Cu.
Wrought: 50,000-100,000 TS. For electrical equipment, radio rheostats. Low electrical resistance, maximum operating temperature 200°C. *Obsolete*

LOHMANIT

Manufacturer not listed.
W + Mo carbides (chiefly W_2C.) For hard cutting tools and dies; cast alloy.

LOHMANN 120

Friedr. Lohmann GmbH
C 1.2, Si 0-0.25, Mn 0-0.25, bal Fe.
Annealed: 115,000 TS; 60,000 YS; 18 El; 38 RA; 225 Brin. For springs, taps, drills, cutters, broaches. Type W1; water hardened. *Obsolete*

LOHMANN 440C

Friedr. Lohmann GmbH
C 1.1, Cr 17, Mo 0.6, bal Fe.
For disc knives, surgical and dental instruments. Type 440C; corrosion resistant. S44004

LOHMANN A2

Friedr. Lohmann GmbH
C 1, Cr 5, Mo 1, V 0.2, bal Fe.
For blanking dies, rolls, shear blades, cold coining dies. Minimal change in size with heat treatment. Type A2. T30102

LOHMANN A37

Friedr. Lohmann GmbH
C 0.85, Cr, bal Fe.
For tools, springs, bearings, dies. Water hardened. *Obsolete*

LOHMANN A7

Friedr. Lohmann GmbH
C 2.2, Cr 5.2, Mo 1.1, V 4.8, W 1.1, bal Fe.
For loopers and tufting knives for the carpet industry, press tools. Type A7. T30107

LOHMANN C112/50

Friedr. Lohmann GmbH
C 1.05, Cr, bal Fe.
For bearings, liners, cutters. Water hardened, wear resistant. *Obsolete*

LOHMANN CNI 40

Friedr. Lohmann GmbH
C 0.45, Cr 1.3, Mo 0.2, Ni 4, bal Fe.
For cutlery dies, blanking dies, billet shears, bending tools, plastic molds. Type 6F7.

LOHMANN CO 10

Friedr. Lohmann GmbH
C 0.76, Co 10, Cr 4.2, Mo 0.8, V 1.8, W 18, bal Fe.
For lathe and planer tools, reamers, drills, cutters. High speed steel. *Obsolete*

LOHMANN CO 3

Friedr. Lohmann GmbH
C 0.86, Co 2.8, Cr 4.3, Mo 0.85, V 2.1, W 12, bal Fe.
For lathe and planer tools, reamers, broaches. High speed steel. *Obsolete*

LOHMANN CO 5

Friedr. Lohmann GmbH
C 0.79, Co 4.7, Cr 4.3, Mo 0.75, V 1.5, W 18, bal Fe.
For lathe and planer tools, reamers, broaches, drills, taps. High speed steel. *Obsolete*

LOHMANN CRMO

Friedr. Lohmann GmbH
C 0.9, Cr 18, V 0.1, Mo 1.1, bal Fe.
For bearings, liners, valves, disc knives. Type 440B; corrosion resistant. S44003

LOHMANN E120
Friedr. Lohmann GmbH
C 1.1, Si 0-0.25, Mn 0-0.25, bal Fe.
Annealed: 110,000 TS; 55,000 YS; 20 El; 40 RA; 210 Brin. For springs, taps, drills, hobs, reamers. Type W1; water hardened. *Obsolete*

LOHMANN E85
Friedr. Lohmann GmbH
C 0.85, Si 0-0.25, Mn 0-0.25, bal Fe.
Heat treated: 190,000 TS; 145,000 YS; 10 El; 30 RA; 400 Brin. For springs, tools, punches, drills, taps. Type W1; water hardened. *Obsolete*

LOHMANN E95
Friedr. Lohmann GmbH
C 0.95, Si 0-0.25, Mn 0-0.25, bal Fe.
Heat treated: 195,000 TS; 150,000 YS; 8 El; 28 RA; 420 Brin. For drills, taps, reamers, broaches. Type W1; water hardened. *Obsolete*

LOHMANN GP35
Friedr. Lohmann GmbH
C 1.5, Cr, Si, bal Fe.
For engravers' tools, forming and blanking dies. Water or oil hardened. *Obsolete*

LOHMANN GS125
Friedr. Lohmann GmbH
C 1.2, Cr 0.2, V 0.1, W 1, bal Fe.
For fast finishing cutters, bushings. Water or oil hardened. *Obsolete*

LOHMANN HCS
Friedr. Lohmann GmbH
C 1.25, Si 1.15, Mn 0.7, Cr 1.2, bal Fe.
For header dies, punches, bushings. Water or oil hardened. *Obsolete*

LOHMANN HKL
Friedr. Lohmann GmbH
C 0.58, Mn 0.95, Cr 1, V 0.09, bal Fe.
For gears, springs, crankshafts. Oil hardened, shock resistant. *Obsolete*

LOHMANN HKM
Friedr. Lohmann GmbH
Now E100.

LOHMANN HPW
Friedr. Lohmann GmbH
C 0.38, Si, Cr, V, bal Fe.
For machine tool parts. Oil hardened, tough. *Obsolete*

LOHMANN HPW EXTRA
Friedr. Lohmann GmbH
C 0.45, Si, Cr, V, bal Fe.
For springs, gears, crankshafts. Oil hardened, tough. *Obsolete*

LOHMANN HPWH
Friedr. Lohmann GmbH
C 0.55, Si 0.9, Cr 1.05, V 0.18, W 1.85, bal Fe.
For header dies, shears, grippers. Oil hardened, tough. *Obsolete*

LOHMANN HPWW
Friedr. Lohmann GmbH
C 0.35, Si 0.9, Cr 1.05, V 0.18, W 1.85, bal Fe.
For header dies, shears, grippers. Oil hardened, tough. *Obsolete*

LOHMANN HSL
Friedr. Lohmann GmbH
C 2.1, Cr 11.5, Mn 0.3, bal Fe.
For blanking and forming dies, punches. Oil hardened, nondeforming. *Obsolete*

LOHMANN HSL EXTRA
Friedr. Lohmann GmbH
C 1.65, Cr 11.5, V 0.1, Mn 0.3, bal Fe.
For blanking and forming dies, punches. Air hardened, nondeforming. *Obsolete*

LOHMANN HSL SPEZIAL
Friedr. Lohmann GmbH
C 2.1, Cr 11.5, Mn 0.3, bal Fe.
For blanking and forming dies, punches. Oil hardened, nondeforming. *Obsolete*

LOHMANN HSO EXTRA
Friedr. Lohmann GmbH
C 1.15, V 0.1, Cr 0.65, Mn 0.3, bal Fe.
For dies, cutters, punches. Oil hardened, tough. *Obsolete*

LOHMANN HSO SPEZIAL
Friedr. Lohmann GmbH
C 0.9, Mn 1.9, V 0.1, bal Fe.
For punches, dies, shears, crimpers. Oil hardened, nondeforming. *Obsolete*

LOHMANN HSO-111
Friedr. Lohmann GmbH
C 1.45, Cr 1.4, Mn 0.6, bal Fe.
For bearings, punches, bushings. Water hardened, wear resistant. *Obsolete*

LOHMANN HSO-121
Friedr. Lohmann GmbH
C 1.05, Cr 1, W 1.15, Mn 0.9, bal Fe.
For header dies, cutters, upsetters. Oil or water hardened, tough. *Obsolete*

LOHMANN HWR
Friedr. Lohmann GmbH
C 1.42, W, V, bal Fe.
For header dies, cutters. Oil or water hardened. *Obsolete*

LOHMANN HWS
Friedr. Lohmann GmbH
C 0.3, Cr 1.1, V 0.18, W 3.75, bal Fe.
For extrusion rams and liners, punches, riveters. Oil hardened, tough. *Obsolete*

LOHMANN HWS EXTRA
Friedr. Lohmann GmbH
C 0.3, Cr 2.35, V 0.6, W 4.25, bal Fe.
For extrusion rams and liners, punches. Oil hardened, tough. *Obsolete*

LOHMANN HWS SPEZIAL
Friedr. Lohmann GmbH
C 0.3, Cr 2.65, V 0.35, W 8.5, bal Fe.
For extrusion rams and liners, punches. Oil hardened, tough. *Obsolete*

LOHMANN KS85
Friedr. Lohmann GmbH
C 1.15, Cr 0.65, V 0.1, bal Fe.
For cutters, header dies, blanking dies. Oil hardened, tough. *Obsolete*

LOHMANN KST
Friedr. Lohmann GmbH
C 0.58, Cr 1, V 0.1, bal Fe.
For springs, gears, forging dies, crankshafts. Oil hardened, shock resistant. *Obsolete*

LOHMANN KV1
Friedr. Lohmann GmbH
C 1.55, Cr 11.5, Mo 0.7, V 1, bal Fe.
For thread rolling dies, cold extrusion tools, blanking and stamping tools, cold piercers, circular shear blades, deep drawing tools, pressure pads. Type D2. T30402

LOHMANN KWS75
Friedr. Lohmann GmbH
C 0.35, Cr 1.05, W 1.85, V 0.18, bal Fe.
For heading dies, upsetters, piercers, tools. Oil hardened, tough. *Obsolete*

LOHMANN KZV30
Friedr. Lohmann GmbH
C 0.8, W, Cr, V, bal Fe.
For tools, cutters. Oil hardened. *Obsolete*

LOHMANN L170
Friedr. Lohmann GmbH
C 1.65, Cr 11.5, Mo 0.6, V 0.3, W 0.5, bal Fe.
For thread rolling dies, cold extrusion tools, blanking and stamping tools, shear blades. Type D2. T30402

LOHMANN LMS125
Friedr. Lohmann GmbH
C 1.2, W, bal Fe.
For cutters, forming rolls, header dies. Oil hardened. *Obsolete*

LOHMANN LMS200
Friedr. Lohmann GmbH
C 1.15, W, bal Fe.
For header dies, forming rolls, cutters. Oil hardened. *Obsolete*

LOHMANN LO44
Friedr. Lohmann GmbH
C 0.4, Si 1, Cr 5, Mo 1.3, V 1, bal Fe.
Hot work steel. For pressure casting dies and metal extrusion tools, forging dies, molds, hot shear blades. Type H13. T20813

LOHMANN MC 100
Friedr. Lohmann GmbH
C 1.05, Mn, Cr, bal Fe.
For bearings, liners, bushings. Water hardened, wear resistant. *Obsolete*

LOHMANN MO1
Friedr. Lohmann GmbH
C 0.82, Cr 3.8, Mo 8.5, V 1.1, Fe 1.6, bal Fe.
For reamers, broaches, milling cutters. Type M1; high speed steel. T11301

LOHMANN MO42
Friedr. Lohmann GmbH
C 1.1, Cr 4, Mo 9.5, V 1.1, W 1.5, Co 8, bal Fe.
For reamers, broaches, milling cutters. Type M42; high speed steel. T11342

LOHMANN MO55
Friedr. Lohmann GmbH
C 0.092-0.92, Cr 4.1, Mo 5, V 1.8, W 6.2, Co 4.7, bal Fe.
For reamers, broaches, milling cutters. Type M35; high speed steel.

LOHMANN MSV4
Friedr. Lohmann GmbH
C 0.7, Si 0-0.25, Mn 0-0.25, bal Fe.
Heat treated: 175,000 TS; 128,000 YS; 12 El; 37 RA; 355 Brin. For springs, rails, tools, hammers, axes. Type W1; water hardened. *Obsolete*

LOHMANN MSVE
Friedr. Lohmann GmbH
C 1, Si 0-0.25, Mn 0-0.25, bal Fe.
Annealed: 100,000 TS; 53,000 YS; 21 El; 42 RA; 200 Brin. For springs, tools, cutters, drills, taps, reamers. Type W1; water hardened. *Obsolete*

LOHMANN N2
Friedr. Lohmann GmbH
C 0.82, Cr 4.1, Mo 0.85, V 1.6, W 8.7, bal Fe.
For lathe and planer tools, drills, hobs, broaches. High speed steel. *Obsolete*

LOHMANN N3
Friedr. Lohmann GmbH
C 0.95, Cr 4.1, Mo, W, V, bal Fe.
For lathe and planer tools, reamers, broaches, hobs. High speed steel. *Obsolete*

LOHMANN N4
Friedr. Lohmann GmbH
C 0.85, Cr 4.1, Mo, W, V, bal Fe.
For lathe and planer tools, reamers. High speed steel. *Obsolete*

LOHMANN NE
Friedr. Lohmann GmbH
C 0.74, Cr 4.1, V 1.1, W 18.5, bal Fe.
For lathe and planer tools, reamers, drills, taps, hobs, broaches. High speed steel. *Obsolete*

LOHMANN NR
Friedr. Lohmann GmbH
C 1.3, Cr 4.3, Mo 0.85, V 3.8, W 12, bal Fe.
For blanking and forming dies, cutters, reamers. High speed steel. *Obsolete*

LOHMANN NR5
Friedr. Lohmann GmbH
C 1.35, W, Co, Mo, V, Cr, bal Fe.
For engravers' tools, blanking and forming dies. High speed steel. *Obsolete*

LOHMANN NSP
Friedr. Lohmann GmbH
C 0.86, Cr 4.1, Mo 0.85, V 2.5, W 12, bal Fe.
For lathe and planer tools, reamers, broaches, taps. High speed steel. *Obsolete*

LOHMANN P85
Friedr. Lohmann GmbH
C 0.85, Mn 0-0.25, bal Fe.
Heat treated: 190,000 TS; 145,000 YS; 10 El; 30 RA; 400 Brin. For springs, tools, cutters, drills, taps. Type W1; water hardened. *Obsolete*

LOHMANN P95
Friedr. Lohmann GmbH
C 1, Si 0-0.25, Mn 0-0.25, bal Fe.
Annealed: 100,000 TS; 53,000 YS; 21 El; 42 RA; 200 Brin. For springs, taps, drills, hobs, cutters. Type W1; water hardened. *Obsolete*

LOHMANN PM90
Friedr. Lohmann GmbH
C 0.9, Mn 1.9, V 0.1, Si 0.25, bal Fe.
For punches, dies, crimpers, upsetters, cutters. Oil hardened, nondeforming. *Obsolete*

LOHMANN PV15
Friedr. Lohmann GmbH
C 0.61, Cr 1.18, V 0.1, Mn 0.75, Si 0.85, bal Fe.
For springs, gears, bolts, crimpers, punches. Oil hardened, shock resistant. *Obsolete*

LOHMANN PZ105
Friedr. Lohmann GmbH
C 1.05, Cr 1, Mn 0.3, bal Fe.
For bearings, sleeves, liners, bushings. Water hardened, wear resistant. *Obsolete*

LOHMANN RFW
Friedr. Lohmann GmbH
Stainless steel. C 0.2, Cr 13, Mn 0.3, bal Fe.
Annealed: 95,000 TS; 50,000 YS; 25 El; 55 RA; 195 Brin. For valves, cutlery, turbine blades. Type 420 stainless. *Obsolete*

LOHMANN S39
Friedr. Lohmann GmbH
C 0.85, Si 0.1-0.4, Mn 0.5-0.7, bal Fe.
Heat treated: 190,000 TS; 145,000 YS; 10 El; 30 RA; 400 Brin.
For springs, tools, cutters, drills, taps. Type W1; water hardened. *Obsolete*

LOHMANN SCV14
Friedr. Lohmann GmbH
C 0.38, Cr, V, Si, bal Fe.
For gears, bolts, machine tool parts. Oil hardened, tough. *Obsolete*

LOHMANN SM4
Friedr. Lohmann GmbH
C 0.35, Si 0.25-0.5, Mn 0.3-0.8, bal Fe.
Hot rolled: 85,000 TS; 54,000 YS; 30 El; 53 RA; 183 Brin. For gears, bolts, machine tool parts. Water hardened. *Obsolete*

LOHMANN SM5
Friedr. Lohmann GmbH
C 0.6, Si 0.25-0.5, Mn 0.3-0.8, bal Fe.
Heat treated: 160,000 YS; 113,000 YS; 12 El; 40 RA; 325 Brin.
For machine tool parts, punches, springs. Water hardened. *Obsolete*

LOHMANN SM7
Friedr. Lohmann GmbH
Now SM7CR.

LOHMANN SM9
Friedr. Lohmann GmbH
Now SM9CR.

LOHMANN SNI
Friedr. Lohmann GmbH
C 0.45, Cr, Ni, bal Fe.
For gears, bolts, machine tool parts. Oil hardened, tough. *Obsolete*

LOHMANN SS1030
Friedr. Lohmann GmbH
C 1.25, Cr 4.1, Mo 3.5, V 3.2, W 9.5, Co 10, bal Fe.
For reamers, broaches, milling cutters. Type T42; high speed steel.

LOHMANN SW50
Friedr. Lohmann GmbH
C 1, Mn 1.1, Cr 0.6, V 0.1, W 0.6, bal Fe.
For blanking and stamping dies, threading tools, drills, broaches, measuring tools, shear blades. Type O1. T31501

LOHMANN T130
Friedr. Lohmann GmbH
C 1.3, Si 0-0.25, Mn 0-0.25, bal Fe.
For engravers' tools, cutters, taps, reamers. Type W1; water hardened. *Obsolete*

LOHMANN WCMO
Friedr. Lohmann GmbH
C 0.4, Si 1, Cr 5.1, Mo 1.2, V 0.3, bal Fe.
For pressure casting dies, metal extrusion tools, forging dies, molds, hot shear blades. Type H11.

LOHMANN WM4
Friedr. Lohmann GmbH
C 0.3, Si 1, Cr 1.1, V 0.18, W 3.75, bal Fe.
For header dies, upsetters, crimpers. Oil hardened, tough. *Obsolete*

LOHMANN WM5
Friedr. Lohmann GmbH
C 0.3, Cr 2.35, V 0.6, W 4.25, bal Fe.
For extrusion press rams and liners, upsetters, punches. Oil hardened, tough. *Obsolete*

LOHMANN WMCO3
Friedr. Lohmann GmbH
C 0.3, Co 2, Cr 2.4, V 0.25, W 8.5, bal Fe.
For extrusion press rams and liners, punches, shears. Hot work steel, oil hardened. *Obsolete*

LOHMANN WMO
Friedr. Lohmann GmbH
C 0.3, Cr 3, Mo 2.8, V 0.5, bal Fe.
For press and piercing mandrels, die inserts, heavy metal pressure casting tools. Type H10. T20810

LOHMANN WORI EXTRA
Friedr. Lohmann GmbH
C 1.42, W, V, bal Fe.
For cutters, engravers' tools. Water hardened. *Obsolete*

LOHMANN WSCR60
Friedr. Lohmann GmbH
C 0.45, Si 1.15, Cr 1.2, V 0.1, bal Fe.
For springs, gears, bolts. Oil hardened. *Obsolete*

LOHMANN WSP
Friedr. Lohmann GmbH
C 1.25, Si 1.15, Cr 1.2, Mn 0.7, bal Fe.
For liners, bushings. Oil hardened. *Obsolete*

LOHMANN WWMV
Friedr. Lohmann GmbH
C 0.45, Cr 1.35, Mo 0.45, V 0.8, W 0.45, bal Fe.
For punches, crimpers. Oil hardened. *Obsolete*

LOHMANN ZCR
Friedr. Lohmann GmbH
C 1.4, Cr, V, bal Fe.
For blanking and forming dies, cutters, bearers. Oil or water hardened, abrasion resistant. *Obsolete*

LOHMANN-RFK
Friedr. Lohmann GmbH
C 0.2, Cr 13, Mo 1, bal Fe.
Annealed: 95,000 TS; 50,000 YS; 25 El; 92 Rock B. Heat treated: 250,000 TS; 215,000 YS; 8 El; 52 Rock C. For cutlery, surgical instruments, gears, shafts, needle valves, gauges. Corrosion resistant, hardenable. *Obsolete*

LOIRE A.S.
Compagnie Ateliers et Forges de la Loire
C 0.5, bal Fe.
Heat treated: 120,000 TS; 77,200 YS; 22 El; 245 Brin. For cutting tools, chisels, hammers, gears, punches, shafts, axles. Water hardening.

LOIRE CMY22
Compagnie Ateliers et Forges de la Loire
C 0.3, Cr 3, Mo 0.5, bal Fe.
Heat treated: 135,000 TS; 121,000 YS; 15 El. For alternator and turbine parts, rotors. Resists oxidation to 600°C.

LOIRE CMY6
Compagnie Ateliers et Forges de la Loire
C 0.15, Cr 0.5, Mo 0.5, bal Fe.
Heat treated: 71,000 TS; 42,700 YS; 20 El. For furnace tubes, oil refineries, boilers. Resists heat to 600°C.

LOIRE ICN 164T
Compagnie Ateliers et Forges de la Loire
Stainless steel. C 0-0.07, Cr 17, Ni 11.5, Mo 2.8, Ti 0.4, bal Fe.
Annealed: 85,300 TS; 35,600 YS; 40 El. For chemical plant equipment, acid mixing tanks, vessels, agitators. Austenitic, stainless, welding grade.

LOIRE S I C
Compagnie Ateliers et Forges de la Loire
C 0.98, Cr 0.5, bal Fe.
Heat treated: 200,000 TS; 180,000 YS; 3 El; 390 Brin. For bearings, rollers, bushings, gages, mandrels. Cold heading tool steel, oil or water hardening.

LOIRE-ALC6
Compagnie Ateliers et Forges de la Loire
C 0-0.08, Cr 23, Al 5.2, bal Fe.
For heating elements, industrial furnaces, parabolic radiators. High heat resistance to 1100°C.

LOIRE-AMCR
Compagnie Ateliers et Forges de la Loire
C 0.3, Cr 10, Mn 18, Ni 1, bal Fe.
Forged: 109,700 TS; 46,600 YS; 45 El. Cast: 92,500 TS; 42,700 YS; 28 El. For marine hardware, diesel engine motor frames. Nonmagnetic, corrosion-resistant steel.

LOIRE-AMCRN

Compagnie Ateliers et Forges de la Loire
C 0.65, Ni 8, Mn 8, Cr 4, bal Fe.
Rolled: 154,000 TS; 114,000 YS; 20 El. For electrical construction. Non-magnetic, corrosion-resistant steel.

LOIRE-CMY16

Compagnie Ateliers et Forges de la Loire
C 0.12, Cr 5.5, Mo 0.6, bal Fe.
Heat treated: 78,000-107,000 TS; 35,600-92,500 YS; 14-23 El. For oil refining and hot hydrogenization equipment. Corrosion resistant to sour crude oils.

LOIRE-CMY17

Compagnie Ateliers et Forges de la Loire
C 0.13, Cr 2.3, Mo 1, bal Fe.
Heat treated: 71,000-135,000 TS; 34,000-121,000 YS; 15-25 El. For oil refining equipment, hydrogenization equipment, gas turbine housings. Maximum operating temperature 600°C.

LOIRE-CMY18

Compagnie Ateliers et Forges de la Loire
C 0.13, Cr 1, Mo 0.5, bal Fe.
Heat treated: 85,000-128,000 TS; 57,000-92,500 YS; 12-20 El. For rotors, blades and arbors in steam and gas turbines. Resists oxidation to 550°C.

LOIRE-CMYV

Compagnie Ateliers et Forges de la Loire
C 0.15, Cr 1, Mo 1, V 0.2, bal Fe.
Heat treated: 88,000 TS; 58,000 YS; 16 El. For steam and gas turbine rotors and blades, superheated steam units. Resists oxidation to 600°C.

LOIRE-COQ2

Compagnie Ateliers et Forges de la Loire
C 1, bal Fe.
Annealed: 100,000 TS; 54,000 YP; 20 El; 190 Brin. Heat treated: 216,000 TS; 152,000 YP; 11 El; 601 Brin. For chisels, rivet sets, fixtures, bearings. Water hardening, wear resistant.

LOIRE-HKMV

Compagnie Ateliers et Forges de la Loire
C 0.14, Cr 0.55, Mo 0.55, V 0.12, bal Fe.
Annealed: 78,200 TS; 57,000 YS; 22 El. For flanges for condensers, steam turbine discs and arbors, superheater tubes. Free from temper embrittlement.

LOIRE-HR200

Compagnie Ateliers et Forges de la Loire
C 0.38, Cr 5, Mo 1.3, Si 1, V 0.5, bal Fe.
Heat treated: 300,000 TS; 250,000 YS; 6 El; 55 Rock C. Annealed: 102,000 TS; 66,000 YS; 28 El; 94 Rock B. For aircraft landing gears, rocket motors, bolts, missile structures, cylinder liners, dies. Martensitic steel. Hardenable.

LOIRE-ICN 001

Compagnie Ateliers et Forges de la Loire
C 0.1, Cr 18, Ni 9, bal Fe.
Annealed: 92,500 TS; 31,300 YS; 50 El. Rolled: 170,000 TS; 149,000 YS; 20 El. For decorations, trim, cooking utensils, chemical plant and food processing equipment. Corrosion resistant. Austenitic, nonhardenable by heat treatment.

LOIRE-ICN 162T

Compagnie Ateliers et Forges de la Loire
Stainless steel. C 0-0.08, Cr 17, Ni 11.5, Mo 2.2, Ti 0.4, bal Fe.
Annealed: 85,300 TS; 35,600 YS; 40 El. For chemical plant equipment, acid mixing tanks, vessels, agitators. Austenitic, stainless, welding grade.

LOIRE-ICN 164

Compagnie Ateliers et Forges de la Loire
Stainless steel. C 0-0.07, Cr 17, Ni 11.5, Mo 2.2, bal Fe.
Annealed: 192,500 TS; 35,600 YS; 45 El. For chemical plant equipment, acid mixing tanks, agitators, vessels. Austenitic stainless, Type 316.

LOIRE-ICN 164BC

Compagnie Ateliers et Forges de la Loire
Stainless steel. C 0-0.03, Cr 16.5, Ni 13.5, Mo 2.2, bal Fe.
Annealed: 85,300 TS; 28,400 YS; 50 El. For chemical plant equipment, acid mixing tanks, agitators, vessels. Austenitic, stainless, non-heat-treatable.

LOIRE-ICN 164BCN

Compagnie Ateliers et Forges de la Loire
Stainless steel. C 0-0.03, Cr 17.5, Ni 13.5, Mo 2.8, bal Fe.
Annealed: 78,200 TS; 31,300 YS; 40 El. For chemical plant equipment, acid mixing tanks, agitators, vessels. Austenitic, stainless, non-heat-treatable.

LOIRE-ICN 164BCS

Compagnie Ateliers et Forges de la Loire
Stainless steel. C 0-0.03, Cr 18.5, Ni 14, Mo 3.3, bal Fe.
Annealed: 78,200 TS; 32,700 YS; 38 El. For chemical plant equipment, acid mixing tanks, agitators, vessels. Austenitic, stainless, Type 317.

LOIRE-ICN 164K

Compagnie Ateliers et Forges de la Loire
Stainless steel. C 0-0.04, Cr 17, Ni 11.5, Mo 2.2, bal Fe.
Annealed: 85,300 TS; 31,300 YS; 45 El. For chemical plant equipment, acid mixing tanks, agitators. Austenitic, stainless, Type 316.

LOIRE-ICN 212

Compagnie Ateliers et Forges de la Loire
C 0-0.1, Cr 12.5, Ni 12, bal Fe.
Rolled, 36%: 135,100 TS; 128,000 YS; 15 El. Annealed: 78,200 TS; 35,600 YS; 50 El. For trim, watch cases, instrument housings, jewelry, decorations. Corrosion resistant.

LOIRE-ICN 472

Compagnie Ateliers et Forges de la Loire
Stainless steel. C 0-0.07, Cr 18, Ni 9.5, bal Fe.
Annealed: 82,500 TS; 35,600 YS; 48 El; 80 Rock B. Cold drawn: 150,000 TS; 100,000 YS; 25 El; 25 Rock C. For chemical plant equipment, tanks, vessels, mixers, agitators. Austenitic, stainless, Type 304.

LOIRE-ICN 472BC

Compagnie Ateliers et Forges de la Loire
Stainless steel. C 0-0.03, Cr 18.5, Ni 11.5, bal Fe.
Annealed: 78,200 TS; 29,900 YS; 48 El; 72 Rock B. Cold drawn: 125,000 TS; 80,000 YS; 30 El; 20 Rock C. For chemical plant equipment, tanks, agitators, mixers, welded structures. austenitic, stainless, Type 304L.

LOIRE-ICN 472K

Compagnie Ateliers et Forges de la Loire
Stainless steel. C 0-0.04, Cr 18, Ni 10.5, bal Fe.
Annealed: 78,200 TS; 29,900 YS; 48 El. Cold drawn: 130,000 TS; 85,000 YS; 28 El; 20 Rock C. For chemical plant equipment, tanks, vessels, welded structures. Austenitic, stainless, Type 304L.

LOIRE-ICN 472NB

Compagnie Ateliers et Forges de la Loire
Stainless steel. C 0-0.06, Cr 17.5, Ni 12.5, Cb 1, bal Fe.
Annealed: 82,500 TS; 35,600 YS; 48 El; 80 Rock B. At 1000°F: 55,000 TS; 35,000 YS; 36 El; 68 RA. For chemical plant equipment, tanks, agitators, mixers, vessels, welded structures. Austenitic, stainless, Type 347. Welding grade. Stabilized.

LOIRE-ICN 472T

Compagnie Ateliers et Forges de la Loire
C 0-0.07, Cr 18, Ni 10.5, Ti 0.4, bal Fe.
Annealed: 78,200 TS; 29,900 YS; 48 El; 80 Rock B. Cold drawn: 95,000 TS; 60,000 YS; 40 El; 185 Brin. For chemical plant equipment, cold headed parts, tanks, vessels, agitators. Corrosion resistant. Welding grade. Type 321. Austenitic.

LOIRE-IMC 201

Compagnie Ateliers et Forges de la Loire
C 0.1, Cr 17, Mn 8, Ni 4, 0.2 N_2, bal Fe.
Annealed: 102,000 TS; 54,000 YS; 45 El. For food and chemical processing equipment, trim, decorations. Stainless, austenitic. Low work hardenability.

LOIRE-IMC 202

Compagnie Ateliers et Forges de la Loire
Stainless steel. C 0.05, Cr 18, Mn 9, Ni 5, 0.2 N_2, bal Fe.
Annealed: 78,200 TS; 45,500 YS; 45 El. For food and chemical processing equipment, trim, decorations. Austenitic, stainless. Weldable grade.

LOIRE-NYS

Compagnie Ateliers et Forges de la Loire
C 0-0.1, Ni 78, Cr 20, Fe 0-1.
Annealed: 106,000 TS; 50,000 YS. For heating elements, furnace parts, combustion chambers of gas and turbine reactors. High heat resistance to 1200°C.

LOIRE-NYSR

Compagnie Ateliers et Forges de la Loire
C 0-0.1, Ni 45, Cr 23, Si 1.2, Mn 1.2, bal Fe.
For radiators, industrial furnaces, heating elements. High heat resistance to 1050°C.

LOIRE-RBL

Compagnie Ateliers et Forges de la Loire
C 1.4, Cr 0.7, bal Fe.
Hardened: 62-64 Rock C. For files. Water hardening, wear and abrasion resistant.

LOIRE-RM12

Compagnie Ateliers et Forges de la Loire
C 0.3, Cr 25, Ni 12, Si 1.5, Mn 1.2, bal Fe.
Heat treated: 110,000 TS; 60,000 YS; 50 El; 90 Rock B. For furnaces, heat treat boxes, rails, furnace hearths, salt and lead pots. Austenitic, heat and corrosion resistant.

LOIRE-RM20

Compagnie Ateliers et Forges de la Loire
C 0.3, Cr 25, Ni 20, Si 1.8, Mn 1.3, bal Fe.
Annealed: 95,000 TS; 45,000 YS; 50 El; 90 Rock B. For furnace parts, heat treat boxes, salt and lead pots, glass industry molds. Austenitic, heat and corrosion resistant.

LOIRE-RM3

Compagnie Ateliers et Forges de la Loire
C 0.2, Cr 27, Ni 3, bal Fe.
Annealed: 85,300 TS; 64,000 YS; 12 El. For furnace parts, heat treat boxes, retorts, salt pots, lead pots, burners. Cast, heat-resistant steel.

LOIRE-RM35

Compagnie Ateliers et Forges de la Loire
C 0.45, Ni 35, Cr 15, Si 2, Mn 1.2, bal Fe.
Cast: 70,000 TS; 40,000 YS; 10 El; 180 Brin. At 1600°F: 18,800 TS; 15,000 YS; 26 El. For furnace parts, salt and lead pots, heat treat boxes, pyrometer tubes, grids, glass making equipment. Austenitic, corrosion and heat resistant to 1150°C. ACI Type HT.

LOIRE-RM6

Compagnie Ateliers et Forges de la Loire
C 0.32, Cr 26, Ni 5.5, Si 1.8, bal Fe.
Annealed: 99,600 TS; 78,200 YS; 18 El. For furnace parts, salt and lead pots, oil burners. Austenitic, ferritic cast steel. Corrosion and heat resistant to 1100°C.

LOIRE-RM60

Compagnie Ateliers et Forges de la Loire
C 0.45, Ni 60, Cr 12, Si 2.2, Mn 1.2, bal Fe.
Annealed: 80,000 TS; 35,600 YS; 4 El. For cyanide and salt baths, furnace parts, heat treat boxes, glass furnace parts. Corrosion and heat resistant.

LOIRE-SCH4

Compagnie Ateliers et Forges de la Loire
C 0.4, Cr 9.5, Mo 0.8, Si 2.4, bal Fe.
Heat treated: 114,000-135,000 TS; 71,000-99,600 YS; 16 El.
For inlet and exhaust valves for internal combustion engines.
Martensitic, heat resistant to 900°C.

LOIRE-SP20

Compagnie Ateliers et Forges de la Loire
C 0.77, Cr 19.5, Ni 1.5, Si 2, bal Fe.
Heat treated: 121,000-143,000 TS; 71,100-114,000 YS; 12-14
El. For exhaust valves, jet engine components. Martensitic,
corrosion and heat resistant to 1000°C.

LOMINIUM STEEL

Edgar Allen Balfour Ltd.
C 0.47, Mn 0.6, Cr 1.75, V 0.2, bal Fe.
For die casting die steel; oil hardened.

LOMU

Carpenter Technology Corp.
Ni 6-10, Mn 8-12, Cr 0-1, bal Fe.
Rolled: 94,000 TS; 36,000 YS; 58 El; 62 RA. For springs;
non-magnetic. *Obsolete*

LONG TERNE

United States Steel Corp.
Steel sheet; coated with lead-tin alloy for roofing, siding.

LOPHOS

Ziv Steel & Wire Co.
C, V, bal Fe.
For cold heading dies.

LORRAINE A48S

Vallourec S.A.
C 0-0.25, Mn 0.64-1.06, bal Fe.
Annealed: 54,000 TS; 30,000 YS; 35 El; 163 Brin. For low
temperature tubing; Afnor A48S, minimum operating
temperature -45 C. *Obsolete*

LOSIL

British Steel plc
Silicon bearing electrical steel. For rotating machine
applications.

LOSTA

Friedr. Lohmann GmbH
C 0.9, Mn 1.9, V 0.1, Cr 0.4, bal Fe.
For punches, dies, upsetters, cutters. Type O2; oil hardened,
nondeforming. T31502

LOTMESSING 60

VDM Nickel-Technologie AG
Zn 38, Si 2, bal Cu.
Soft: 52,000 TS; 48 El; 75 Brin. 1/2 H-temper: 65,000 TS; 30
El; 105 Brin. For hardware, bolts, fasteners; corrosion
resistant. *Obsolete*

LOTUN

Disston Inc.
C, alloy, bal Fe.
For tools, dies; oil hardening. *Obsolete*

LOTUNG

Ziv Steel & Wire Co.
C 0.3, Cr 3, W 9, V 0.45, bal Fe.
For Al die casting dies, brass extrusion dies; hot work steel,
oil hardened.

LOTUS

Lumen Bearing Co.
Sn 10, Sb 15, Pb 75.
For bearings; Babbitt; medium bearing loads.

LOW BRASS

Chase Brass & Copper Co.
Cu 80, Zn 20.
Annealed: 44,000 TS; 10,000 YS; 60 El. Rolled: 65,000 TS;
45,000 YS; 12 El. For formed and drawn parts requiring high
finish; high ductility.

LOW BRASS

Anaconda Co.
Cu 80, Zn 20.
Annealed: 44,000 TS; 10,000 YS; 60 El. Rolled: 65,000 TS;
45,000 YS; 12 El. For formed and drawn parts requiring high
finish; high ductility.

LOW BRASS 5

Olin Brass, Indianapolis
Copper. Cu 80, Zn 20.
Hard: 74,000 psi TS; 59,000 psi YS; 7 El; 156 Brin. Soft:
44,000 psi TS; 14,000 psi YS; 50 El; 56 Brin. For hardware
and diaphragms; corrosion resistant.

LOW BRASS-32

Anaconda Co.
Copper. Cu 80, Zn 20.
Soft: 43,000 TS; 16,000 YS; 50 El. Hard: 73,000 TS; 60,000
YS; 8 El. For formed and drawn parts. High ductility.

LOW CARBON GFS

Teledyne Pittsburgh Tool Steel
C 0.2, bal Fe.
Low carbon ground flat stock steel. For jigs and fixtures.
Obsolete

LOW CARBON NICKEL

Now NICKEL 201.

LOW CARBON ONTARIO

Allegheny Ludlum Steel
C 0.87, Mn 0.5, Si 0.5, Cr 12, Ni 0.35, V 0.15, Mo 0.8, bal Fe.
Air or oil hardening cold work tool steel, chromium type; AISI
D1. *Obsolete*

LOW CARBON TATMO

Latrobe Steel Co.
C 0.74, W 1.5, Cr 3.75, V 1.2, Mo 8.25, Si 0.3, Mn 0.25, bal
Fe.
High speed steel for punches, thread rolling dies, pipe taps,
hot and cold heading tools. Improved toughness.

LOW CHROME-NICKEL

Johnson Matthey plc
C 0.35-0.45, Cr 0.25-0.5, Mn 0.8-1, Ni 0.5-0.75, Si 0.3-0.4, bal
Fe.
Annealed: 100,000 TS; 60,000 YS; 20 El; 18 RA; 200 Brin.
Heat treated: 120,000-200,000 TS; 100,000-175,000 YS;
10-35 El; 5-20 RA. For gears, shafts, pinions; tough.
Obsolete

LOW LEADED BRASS

Chase Brass & Copper Co., Inc.
Copper. Cu 66.5, Zn 33, Pb 0.5.
For drain tubes, plumbing goods, pump liners; free-cutting.

LOW LEADED BRASS-226

Anaconda Co.
Copper. Zn 35, Pb 0.5, bal Cu.
Soft: 45,000 TS; 17,000 YS; 55 El. Hard: 73,000 TS; 60,000
YS; 9 El. For hardware. Good formability.

LOW LEADED TUBE BRASS

Chase Brass & Copper Co.
Cu 66.5, Zn 33, Pb 0.5.
Annealed: 52,000 TS; 20,000 YS; 50 El. For pump liners,
plumbing. *Obsolete*

LOW LEADED TUBE BRASS-330

Chase Brass & Copper Co.
Cu 66.5, Zn 33, Pb 0.5.
Hard: 73,000 TS; 60,000 YS; 10 El; 80 Rock B. Soft: 45,000
TS; 17,000 YS; 55 El; 15 Rock B. For pump and power
cylinders and liners, plumbing parts, munitions. Free-cutting
yellow brass.

LOW LEADED TUBE BRASS-330

Anaconda Co.
Cu 66.5, Zn 33, Pb 0.5.
Hard: 73,000 TS; 60,000 YS; 10 El; 80 Rock B. Soft: 45,000
TS; 17,000 YS; 55 El; 15 Rock B. For pump and power
cylinders and liners, plumbing parts, munitions. Free-cutting
yellow brass.

LOW PHOSPHORUS COPPER

Now COPPER, DEOXIDIZED, DLP.

LOW PHOSPHORUS COPPER

Chase Brass & Copper Co., Inc.
Copper. P 0.007, 99.9 Cu min.
For electrical conductors, wave guides; high conductivity,
deoxidized copper.

LOW RESISTANCE 180 ALLOY

Carpenter Technology Corp.
Cu 77, Ni 23.
Annealed: 345 MPa TS; 25 El. Cold worked: 689 MPa TS; 25
El. Resistivity 180 ohms/circular mil-ft. Maximum operating
temperature 538°C. For lead wires for resistors, heater wires
for electric blankets, fuse wires, support wires and wire used
in high resistance magnetic coils. ASTM B-267 Class 7.

LOW RESISTANCE 30 ALLOY

Carpenter Technology Corp.
Cu 98, Ni 2.
Annealed: 207 MPa TS; 25 El. Cold worked: 414 MPa TS; 25
El. Resistivity 30 ohms/circular mil-ft. Maximum operating
temperature 316°C. For lead wires for resistors, heater wires
for electric blankets, fuse wires, support wires and wire used
in high resistance magnetic coils. ASTM B-267 Class 11.

LOW RESISTANCE 60 ALLOY

Carpenter Technology Corp.
Cu 94, Ni 6.
Annealed: 241 MPa TS; 25 El. Cold worked: 483 MPa TS; 25
El. Resistivity 60 ohms/circular mil-ft. Maximum operating
temperature 316°C. For lead wires for resistors, heater wires
for electric blankets, fuse wires, support wires and wire used
in high resistance magnetic coils. ASTM B-267 Class 10.

LOW RESISTANCE 90 ALLOY

Carpenter Technology Corp.
Cu 90, Ni 10.
Annealed: 241 MPa TS; 25 El. Cold worked: 517 MPa TS; 25
El. Resistivity 90 ohms/circular mil-ft. Maximum operating
temperature 427°C. For lead wires for resistors, heater wires
for electric blankets, fuse wires, support wires and wire used
in high resistance magnetic coils. ASTM B-267 Class 9.

LOW SILICON BRONZE

Chase Brass & Copper Co., Inc.
Copper. Cu 98.5.
Annealed: 40,000 psi TS; 15,000 psi YS; 50 El. Drawn (50%):
90,000 psi TS; 67,000 psi YS; 12 El. For bolts, marine and
pole line hardware.

LOW TEMPERATURE AIR HARDENING

DoAll Co.
C 0.7, Mn 2.1, Si 0.3, Cr 1, Mo 1.3, S 0.12, bal Fe.
Free machining grade. AISI Type A6 air hardening tool steel.

LOW-TIN COMMERCIAL BRONZE

Chase Brass & Copper Co.
Cu 90, Zn 9.5, Sn 0.5.
1/2 Hard temper: 53,000 TS; 46,000 YS; 10 El; 108 Brin. For
bearings, bushings, thrust washers; bushing bronze.
Obsolete

LOWMOOR IRON

Lowmoor Best Yorkshire Iron Ltd.
C 3.2, Si 2.6, bal Fe.
For castings; cast iron.

LOWROFF PHOSPHOR BRONZE

English manufacture
Pb 5-16, Sn 4-13, P 0.5-1, bal Cu.
For bearings, bushings; heavy duty.

LOWSCORE
Dunford Hadfields Ltd.
C 0-0.12, Cr 18-22, bal Fe.
Annealed: 68,000-80,000 TS; 30 El; 130-153 Brin. For heat resisting equipment, furnace parts, pots; Type 442; corrosion resistant. *Obsolete*

LOXLEY
Sanderson Kayser Ltd.
C 0.7-0.8, bal Fe.
For tools, taps, punches; water hardened. *Obsolete*

LOYCON N
British Steel plc
Now QT.445 (1).

LOYCON QT
British Steel plc
Now QT.445 (2).

LPD
Latrobe Steel Co.
C 0.35, Cr 5, W 1.3, Mo 1.6, Si 1, Mn 0.3, V 0.3, bal Fe.
For hot forging dies, punches, mandrels; hot work steel.

LR 80
Sigmund Cohn Corp.
Au 78-82, Cu, Ag, bal Pt.
Resistance wire for potentiometer.

LS
Thyssen Edelstahlwerke AG
C, Mn, bal Fe.
For rock press plates. *Obsolete*

LS 33, ETC.
Now MILLS LS 33, ETC.

LS-31
Union Carbide Corp.
Cr 25, Ni 10, W 7, bal Co.
Metallic spray coating for plasma torch. Good wear resistance; 350 DPH. For jet engine ducts.

LSC
Allegheny Ludlum Steel
C, alloy, bal Fe.
For hot work tools and dies; hot work steel. *Obsolete*

LT-75
Now LUKENS LT-75.

LTA-60
Youngstown Steel
C 0.07, Mn 0.4, P 0.01, S 0.025, bal Fe.
60,000 YS. For high strength cold rolled steel sheet applications.

LTA-70
Youngstown Steel
C 0.07, Mn 0.4, P 0.01, S 0.025, bal Fe.
70,000 YS. For high strength cold rolled steel sheet applications.

LUBECO
Lumen Bearing Co.
Sn 40, Pb 47, Sb 12, Cu 1.
Cast. For Babbitt metal. *Obsolete*

LUBRAL
Istituto Sperimentali Metalli Leggeri
Sn 6.5, Cu 1.07, Ni 1, Si 0.14, Fe 0.19, bal Al.
Cast: 14,000 TS; 6.8 El; 37 Brin. For bearings.

LUBRAL SN6
Montecatini Settore Alluminio
Cu 1, Si 1.2, Ti 0.12, Ni 1, Sn 6, bal Al.
Rolled: 21,000-27,000 TS; 9000-12,000 YS; 6-12 El; 45-55 Brin. For light alloy parts.

LUBRI-DIE
Ziv Steel & Wire Co.
C 1.45, Mn 0.8, Si 1.15, Mo 0.25, bal Fe.
Annealed: 84,500 TS; 49,500 YS; 25 El; 197 Brin. Heat treated: 164,000 TS; 136,000 YS; 13 El; 302 Brin. Heat treated: 218,000 TS; 177,000 YS; 8 El; 388 Brin. For blanking and forming dies, punches, gauges, hobs, cams, taps, pneumatic hammers, piercing dies. Type O6; oil hardening, cold work tool steel.

LUBRICO NO. 1
Buckeye Brass & Mfg. Co.
Copper. Cu 70-75, Pb 20-22, Sn 5-10.
For bearings and bushings; heavy duty.

LUBRICO NO. 2
Buckeye Brass & Mfg. Co.
Copper. Cu 70-75, Pb 20-22, Sn 5-10, bal Ni.
For bearings and bushings; heavy duty.

LUBRICO NO. 3
Buckeye Brass & Mfg. Co.
Copper. Cu 70-75, Pb 20-22, Sn 5-10, bal Ni.
For bearings and bushings; heavy duty.

LUBROTEC 19985
Eutectic Corp.

Alloy powder for metal spraying final coat; machinable but tough and wear resistant.

LUCALOX
American manufacture
Al_2O_3.
For cutting tools; sintered.

LUCAS NIFAL
Manufacturer not listed.
Ni 25, Al 12, Fe 63.
For permanent magnets.

LUCERNO
Manufacturer not listed.
Ni 65-68, Cu 27-30, Fe 0-2.4, Mn 2.2-5.
For resistance alloy; heat resistant.

LUCERO
Driver Harris Co.
Ni 70, Cu 30.
140,000 TS. For springs, electrical resistance wire, bolts, valves; resists heat to 500 C. *Obsolete*

LUCKY 7
Plew Tool Co.
C, V, bal Fe.
Water hardened. For tools.

LUDENSCHEIDT BUTTON METAL
German manufacture
Cu 20, bal Zn.
For ornamental parts; die castings.

LUDLOY
Ludlow Steel Corp.
C, alloy bal Fe.
Heat treated: 265,000 TS; 13 El; 53 Rock C. For chisels, sledges, arbors, dies, punches; shock resistant, water hardening.

LUDLUM
Manufacturer not listed.
Cr 13-17, Si 1, Mo 1, C 0.4, bal Fe.
For corrosion resisting parts, cutlery; corrosion resistant.

LUDLUM 602
AL Tech Specialty Steel Corp.
C 0.5, Mo 0.4, Si 1.5, V 0.12, bal Fe.
Hardened: 330,000 TS; 270,000 YS; 9 El; 28 RA; 578 Brin. For chisels, punches; shock resistant. *Obsolete*

LUDLUM LMW-EZ
AL Tech Specialty Steel Corp.
C 0.7, Cr 3.25-4.25, W 1.2-2, V 0.75-1, Mo 7.5-9.5, bal Fe.
For reamers, lathe tools, drills, broaches; high speed steel, oil hardened. *Obsolete*

LUDLUM NO. 545
Allegheny Ludlum Steel
C, alloy, bal Fe.
For tools, dies. *Obsolete*

LUDLUM S.S.V.
Allegheny Ludlum Steel
C 0.7-1.2, bal Fe.
For tools, jugs, fixtures. *Obsolete*

LUDLUM VLM-EZ
AL Tech Specialty Steel Corp.
C 0.85, Mo 8, Cr 4, V 2, bal Fe.
For reamers, drills, hobs, taps, lathe cutters; high speed steel, oil hardened. *Obsolete*

LUDLUM XCM
AL Tech Specialty Steel Corp.
C 1.2, Cr 1.4, Mo 0.45, bal Fe.
For tools, taps, gauges, lathe centers; oil hardened. *Obsolete*

LUETECIN
French manufacture
Cu 80, Ni 6-16, Zn 5, Fe 5, Sn 2, Co 1.
For cheap jewelry; corrosion resistant.

LUKENS 10L45
Lukens Steel
C 0.42-0.5, Mn 0.6-0.9, Si 0.1-0.35, Pb 0.1-0.35, bal Fe.
Plate: 93,000-103,000 TS; 53,000-56,000 YP; 20-23 El; 39-42 RA; 190-216 Brin. For rubber and fiberglass molds, dies, platens, sprockets, gears, cams, wheels. Tough, wear resistant, free-cutting. *Obsolete*

LUKENS 45
Lukens Steel
C 0.2-0.25, Mn 1.2-1.35, Si 0.15-0.3, bal Fe.
Plate: 65,000 TS; 45,000 YS; 24 El min. For bridges, barges, field erected tanks, heavy industrial equipment. High-strength carbon steel plate.

LUKENS 50
Lukens Steel
C 0.2-0.25, Mn 1.35, Si 0.2, bal Fe.
Plate: 70,000 TS min; 50,000 YS min; 24 El min. For bridges, barges, field erected tanks, heavy industrial equipment. High-strength carbon steel plate.

LUKENS 55
Lukens Steel
C 0.23, Si 0.25, Mn 1.3-1.6, bal Fe.
Plate: 75,000 TS min; 55,000 YS min; 23 El min. For bridges, barges, field erected tanks, heavy industrial equipment. High-strength carbon steel plate.

LUKENS 60
Lukens Steel
C 0.23, Mn 1.6, Si 0.25, bal Fe.
Plate: 80,000 TS min; 60,000 YS min; 23 El min. For bridges, barges, field erected tanks, heavy industrial equipment. High-strength carbon steel plate.

LUKENS 80
Lukens Steel
C 0.2-0.25, Mn 1.1-1.6, Si 0.2-0.6, bal Fe.
Heat treated: 95,000-120,000 TS; 75,000-80,000 YS; 19 El min. For welded structures, heavy industrial equipment, bridges. Good low temperature notch toughness.

LUKENS A440
Lukens Steel
C 0.28, Mn 1.1-1.6, 0.20 Cu min, bal Fe.
Plate: 70,000 TS; 50,000 YS; 18 El. For derricks, booms, bridges, mine cars, bus and truck bodies. High-strength low-alloy steel.

LUKENS A441
Lukens Steel
C 0.22, Mn 1.25, 0.20 Cu min, 0.02 V min, bal Fe.
Plate: 70,000 TS; 50,000 YS; 22 El. For bridges, booms, derricks, mine cars, bus and truck bodies. High-strength low-alloy steel.

LUKENS AAR-M128 GR.A
Lukens Steel
C 0-0.25, Mn 1.35-1.5, P 0-0.04, S 0-0.05, Si 0.3-0.5, V 0-0.02, bal Fe.
Plate: 81,000-101,000 TS; 50,000 YP; 19 El. For rail tank cars, pressure vessels, bridges, mine cars, derricks. High-strength, low-alloy structural steel. *Obsolete*

LUKENS AAR-M128 GR.B
Lukens Steel
C 0-0.25, Mn 1.35-1.5, P 0-0.04, S 0-0.05, Si 0.3-0.5, Mo 0-0.08, Cr 0-0.25, Ni 0-0.25, Cu 0-0.35, bal Fe.
Plate: 81,000-101,000 TS; 50,000 YP; 19 El. For rail tank cars, pressure vessels, bridges, mine cars, derricks. High-strength low alloy construction and structural steel. *Obsolete*

LUKENS AR-300
Lukens Steel
C 0.28, Mn 1.4, Si 0.2-0.5, Cu 0.2, bal Fe.
Hardened: 285-321 Brin; 137,000-158,000 TS; 123,000-148,000 YS; 10-11 El. For materials handling equipment, reactor vessels, floor wear plates. Quenched and tempered carbon steel plate. Wear and abrasion resistant.

LUKENS AR-350
Lukens Steel
C 0.3, Mn 1-1.6, Si 0.4, Cu 0.2, bal Fe.
Heat treated: 171,000 TS; 165,000 YS; 13 El; 320-380 Brin. For materials handling equipment, reactor vessels, floor wear plates. Quenched and tempered carbon steel plate. Wear and abrasion resistant.

LUKENS CARBON-MOLYBDENUM STEEL
Lukens Steel
C 0-0.25, Mn 0-0.9, Mo 0.25-0.6, bal Fe.
Rolled: 65,000-87,000 TS; 33,000-41,000 YS. For shell and vessel work, general fabrication; good creep values at elevated temperature. *Obsolete*

LUKENS CHROME-NICKEL STEEL
Lukens Steel
C 0.35-0.45, Mn 0.3-0.6, Ni 3.25-3.75, Cr 1.25-1.75, bal Fe.
Heat treated: 200-300 Brin. For hot rolled steel plates, wear and abrasion resistant parts; resists abrasion. *Obsolete*

LUKENS CROMANSIL (GRADE A)
Lukens Steel
C 0-0.17, Mn 1.05-1.4, Cr 0.3-0.6, Si 0.6-0.9, bal Fe.
Rolled: 75,000-90,000 TS; 45,000-54,000 YS. For high-tensile plate work; tough.

LUKENS CROMANSIL STEEL (GRADE B)
Lukens Steel
C 0-0.25, Mn 1.05-1.4, Cr 0.3-0.6, Si 0.6-0.9, bal Fe.
Rolled: 85,000-100,000 TS; 47,000-55,000 YS. For high-tensile plate work; tough.

LUKENS FROSTLINE (CLASS 1)
Lukens Steel
C 0-0.22, Mn 1-1.6, Si 0.15-0.3, Cb 0.01-0.05, bal Fe.
70,000-90,000 psi TS; 50,000 psi YS min; 21 El. For LNG ship components, coal handling equipment.

LUKENS FROSTLINE (CLASS 2)
Lukens Steel
C 0-0.22, Mn 1-1.6, Si 0.15-0.3, Cb 0.01-0.05, bal Fe.
80,000-115,000 psi TS; 65,000 psi YS min; 21 El. For pipeline fittings, offroad construction equipment.

LUKENS HP
Lukens Steel
C 0.06, Mn 0.35, Si 0.05, S 0.025, P 0.025, bal Fe.
Plate: 35,000 TS; 25,000 YS; 28 El. For magnet cores in linear accelerators, cyclotrons, synchrotrons, bubble chambers. Soft magnet steel; high permeability, high flux density, low hysteresis loss.

LUKENS INCONEL-CLAD STEEL
Lukens Steel
Inconel bonded to steel base.
55,000 TS; 27,000 YS. For corrosion resistant parts; 10% Inconel of total plate thickness. *Obsolete*

LUKENS L-XX
Lukens Steel
C 0.2-0.25, Mn 1.2-1.6, Si 0.15-0.3, bal Fe.
Plate: 65,000 min TS; 45,000 min YS; 23 min El. For bridges, barges, booms, derricks, heavy industrial equipment. High-strength low alloy steel plate. *Obsolete*

LUKENS LT-75
Lukens Steel
C 0-0.24, Mn 0.7-1.35, Si 0.15-0.3, bal Fe.
LT-75 N: 65,000-90,000 TS; 46,000-50,000 YS; 23-24 El. For low temperature applications, cold storage tanks, rail tank cars. Low temperature steel, good notch toughness.

LUKENS LT-75HS
Lukens Steel
C 0-0.22, Mn 1.1-1.6, Si 0.2-0.6, bal Fe.
Plate: 75,000 YS min; 95,000-115,000 TS. For heavy construction equipment, offshore drilling platforms, rail tank cars, cold storage tanks, bridges. Good notch toughness at -75°F. Quenched and tempered steel plate.

LUKENS LT-75QT
Lukens Steel
C 0-0.24, Mn 0.7-1.35, Si 0.15-0.3, bal Fe.
Plate: 75,000-100,000 TS; 56,000-60,000 YS; 23-24 El. For low temperature applications, cold storage tanks, rail tank cars. Good low temperature notch toughness.

LUKENS M-131
Lukens Steel
C 0-0.25, Mn 0.56-0.94, Si 0-0.3, bal Fe.
Grades A, B, C: 64,000-84,000 TS; 37,000 min YP. For bridges, buildings, hoppers, shafts, deck and hull plates, crushers, hatch covers. ASTM-A131 Gr. A, B, C. *Obsolete*

LUKENS M113
Lukens Steel
C 0.15, Mn 0.6, Si 0.2, S 0-0.05, P 0-0.06, 0.20% min Cu, bal Fe.
Gr. A: 66,000-86,000 TS; 38,000 min YP. Gr. B: 55,000-75,000 TS; 31,000 min YP. Gr. C: 53,000-73,000 TS; 30,000 min YP. For bridges, buildings, hoppers, shafts, deck plates, hatch covers, crushers, locomotives and cars. ASTM-A113 Gr. A, B, C; high strength. *Obsolete*

LUKENS M283
Lukens Steel
C 0.2, Mn 0.8, Si 0.2, 0.2% min Cu, bal Fe.
Gr. A: 50,000-70,000 TS; 28,000 min YP. Gr. B: 55,000-75,000 TS; 32,000 min YP. Gr. C: 61,000-81,000 TS; 35,000 min YP. Gr. D: 66,000-86,000 TS; 38,000 min YP. For bridges, buildings, hoppers, shafts, deck and hull plates, hatch covers, crushers. ASTM-A283 Gr. A, B, C, D. *Obsolete*

LUKENS M284
Lukens Steel
C 0-0.42, Mn 0-0.94, Si 0.1-0.3, bal Fe.
Gr. A: 55,000 min TS; 29,000 min YP. Gr. B: 61,000 min TS; 32,000 min YP. Gr. C: 66,000 min TS; 35,000 min YP. Gr. D: 66,000 min TS; 38,000 min YP. For bridges, buildings, hoppers, shafts, ship deck and hull plates, hatch covers, crushers. Good fabricability and weldability. ASTM-A284 Gr. A, B, C, D. *Obsolete*

LUKENS M36
Lukens Steel
C 0.25, Mn 0.8-1.2, Si 0.15, bal Fe.
Plate: 60,000-80,000 TS; 42,000 min YP; 23 El. For bridges, buildings, hoppers, shafts, deck and hull plates, hatch covers, crushers, backing plates. ASTM-A36. *Obsolete*

LUKENS M573 GR.65
Lukens Steel
C 0.24-0.26, Mn 0.85-1.2, P 0-0.04, S 0-0.05, Si 0.15-0.3, bal Fe.
Plate: 65,000-77,000 TS; 35,000 min YP; 20 El (in 8 in.). For mine cars, derricks, booms, bridges, general structures. Improved toughness. *Obsolete*

LUKENS M573 GR.70
Lukens Steel
C 0.27, Mn 0.85-1.2, P 0-0.04, S 0-0.05, Si 0.15-0.3, bal Fe.
Plate: 70,000-85,000 TS; 38,000 min YS; 18 El (in 8 in.). For mine cars, bus bodies, booms, derricks, bridges. Improved toughness. *Obsolete*

LUKENS M7
Lukens Steel
C 0.15, Mn 0.6, Si 0.15, S 0-0.05, 0.20% min Cu, bal Fe.
Plate: 66,000-86,000 TS; 38,000 min YP. For bridges, buildings, hoppers, shafts, deck and hull plates, hatch covers, crushers. ASTM-A7. *Obsolete*

LUKENS MANGANESE-VANADIUM STEEL
Lukens Steel
C 0-0.18, Mn 1-1.45, V 0.08-0.18, bal Fe.
Rolled: 80,000-95,000 TS; 50,000 YS. For structural purposes; welds satisfactorily.

LUKENS MP
Lukens Steel
C 0.1, Mn 0.5, Si 0.2, S 0.03, P 0.03, bal Fe.
Plate: 30,000 YS; 45,000 TS; 28 El. For magnet cores in linear accelerators, cyclotrons, synchrotrons, bubble chambers. Soft magnet steel, medium permeability, high flux density, low hysteresis loss. *Obsolete*

LUKENS NICKEL STEEL
Lukens Steel
C 0-0.22, Mn 0.4-0.8, Ni 2-2.5, Si 0.15-0.25, bal Fe.
Rolled: 70,000 TS; 35,000 YS. For vessels, boiler shells, dewaxing equipment, oil refinery; resists low temperature embrittlement. *Obsolete*

LUKENS NINE NICKEL
Lukens Steel
C 0-0.13, Ni 9, Mn 0-0.8, bal Fe.
Normalized: 90,000 TS; 60,000 YS; 22 El. For storage and process vessels for handling liquid N_2 and O_2 for low temperature service. ASTM A203.

LUKENS SP-40
Lukens Steel
C 0.24-0.31, Mn 1-1.4, Si 0.15-0.3, bal Fe.
Normalized: 40,000 psi YS. For press bases.

LUKENS SPA-90
Lukens Steel
C 0.14-0.21, Mn 0.95-1.3, Si 0.15-0.35, Cr 1-1.5, Ni 1.2-1.5, Mo 0.4-0.55, V 0.03-0.08, bal Fe.
Plate, quenched and tempered: 105,000-135,000 psi TS; 90,000 psi YS min, 14-16 El. For earthmoving and transport equipment, booms, bridges, tower and building members.

LUKENS T-1
Lukens Steel
C 0.1-0.2, Mn 0.6-1, Si 0.15-0.35, Cr 0.4-0.65, Ni 0.7-1, Mo 0.4-0.6, Cu 0.15-0.5, V 0.03-0.08, B 0.002-0.006, bal Fe. Plate, quenched and tempered: 115,000-135,000 psi TS; 100,000 psi YS min. For bridge, tower members, pressure vessels. ASTM A514; Type F; A517 Grade F.

LUKENS T-1 TYPE B
Lukens Steel
C 0.12-0.21, Mn 0.95-1.3, Si 0.2-0.35, Cr 0.4-0.65, Ni 0.3-0.7, Mo 0.2-0.3, Cu 0.2-0.4, V 0.03-0.08, 0.0005 B min, bal Fe. Plate, quenched and tempered: 115,000-135,000 psi TS; 100,000 psi YS min. For bridges, buildings, mining equipment. ASTM A514; Type H; A517 Grade H.

LUKENS T-1A
Lukens Steel
C 0.15-0.21, Mn 0.7-1, Si 0.2-0.35, Cr 0.4-0.65, Mo 0.15-0.25, V 0.03-0.08, Ti 0.01-0.03, B 0.0005-0.005, bal Fe. Plate, quenched and tempered: 115,000-135,000 psi TS; 100,000 psi YS min. For earth moving equipment, truck frames and bodies, storage tanks, oil field rigs. ASTM A514; Type B; A517 Grade B.

LUKENS TRANSLINE BCV 42, ETC.
Now BCV-42, ETC.

LUKENS TRANSLINE UCV-60
Now UCV-60.

LUMDIE
Latrobe Steel Co.
C 0.4, Si 1, Mn 0.25, W 4.75, Cr 5.25, bal Fe. For dies for Al die casting, header gripper dies on bolt machines; hot work steel.

LUMEN 00-A
Lumen Bearing Co.
Cu 80, Sn 20.
Cast: 33,000 TS; 21,000 YS; 0.5 El; 0.4 RA; 143 Brin. For bells and bearings; alloy deoxidized with P.

LUMEN 00-C
Lumen Bearing Co.
Cu 84, Sn 16.
Cast: 32,000 TS; 27,000 YS; 1-1.5 El; 0.8-1.2 RA; 70-80 Brin. For bearing metal, washers, dies, trunnions; alloy deoxidized with P.

LUMEN 1
Lumen Bearing Co.
Cu 88, Sn 8-10, Zn 2-4.
Cast: 40,000 TS; 20,000 YS; 15-25 El; 14-23 RA; 57-74 Brin. For machine parts, spur and bevel gears, air valves, pumps; alloy deoxidized with Zn.

LUMEN 11 C
Lumen Bearing Co.
Cu 89, Fe 1, Al 10.
Cast: 75,000 TS; 30,000 YS; 15-25 El; 15-25 RA; 114-140 Brin. For gears, stripper nuts; shock resisting Al Bronze; Izod 30-36.

LUMEN 14
Lumen Bearing Co.
Cu 90, Sn 6.5, Zn 2, Pb 1.5.
Cast: 40,000 TS; 18,000 YS; 15-33 El; 14-33 RA; 44-48 Brin. For valve bodies, carburetors; oil pump; known as Valve or Steam Bronze; Izod 11-15.

LUMEN 15
Lumen Bearing Co.
Cu 88.75, Sn 11, Pb 0.25.
Sand cast: 40,000 TS; 23,000 YS; 5-10 El; 5-10 RA; 63-70 Brin. For worm wheels, heavy duty bearings, spur gears; deoxidized with P.

LUMEN 15A
Lumen Bearing Co.
Sn 11, Pb 1.5, bal Cu.
Sand cast: 36,000 TS; 20,400 YS; 12 El; 10 RA; 54 Brin. Permanent mold: 45,000 TS; 23,000 YS; 10 El; 8 RA; 72 Brin. For gears, worms, bearings; P-deoxidizer.

LUMEN 19
Lumen Bearing Co.
Cu 62.5, Zn 28, Fe 2.5, Mn 3, Al 4.
Cast: 90,000 TS; 45,000 YS; 18 El; 17 RA; 179 Brin. For propellers, shafts, gears; Mn-bronze.

LUMEN 2
Lumen Bearing Co.
Cu 86, Sn 9.5, Pb 2.5, Zn 2.
Cast: 40,000 TS; 19,000 YS; 15-25 El; 12-23 RA; 52-65 Brin. For bearing bronze, worm gears; tough.

LUMEN 20
Lumen Bearing Co.
Cu 63.75, Zn 23, Fe 2.75, Mn 3.75, Al 6.75.
Cast: 120,000 TS; 75,000 YS; 5-10 El; 4-8 RA; 240-270 Brin. For spur gears, gibs, cams; known as Super Mn Bronze; Izod 7-11.

LUMEN 31
Lumen Bearing Co.
Cu 70, Sn 9, Pb 21.
Cast: 27,000-29,000 TS; 17,000-19,000 YS; 14-17 El; 59-63 Brin. For bearings; heavy duty.

LUMEN 33
Lumen Bearing Co.
Cu 70, Sn 4, Pb 26.
Cast: 18,000-23,000 TS; 12,000-16,000 YS; 12-20 El; 41-45 Brin. For bearings; heavy duty.

LUMEN 4
Lumen Bearing Co.
Cu 80, Sn 10, Pb 10.
Cast: 36,000 TS; 20,000 YS; 6-8 El; 5-7.5 RA; 46-70 Brin. For bearings, bushings, feed nuts; for high speeds.

LUMEN 43
Lumen Bearing Co.
Sn 10.5, Ni 1.5, bal Cu.
Cast: 50,000 TS; 27,000 YS; 15 El; 13 RA; 74 Brin. For bearings, worm gears, nuts; corrosion resistant.

LUMEN 48
Lumen Bearing Co.
Cu 84, Sn 10, Pb 2.5, Ni 3.5.
Chill cast: 45,000 TS; 26,000 YS; 7 El; 8 RA; 83 Brin. For bearings, worms, gears, nuts; deoxidized with P.

LUMEN 5
Lumen Bearing Co.
Cu 85, Sn 5, Pb 5, Zn 5.
Cast: 37,000 TS; 18,600 YS; 15-20 El; 15-20 RA; 45-50 Brin. For medium and low pressure valve bodies, carburetors, general brass castings; Izod 6-7.

LUMEN 54
Lumen Bearing Co.
Cu 85, Sn 10, Pb 5.
Sand cast: 34,000 TS; 22,000 YS; 12 El; 10 RA; 50 Brin. Permanent mold: 40,000 TS; 20,000 YS; 8 El; 7 RA; 70 Brin. For elevator worm gears, feed nuts, tool bearings; Izod 4-7.

LUMEN 6
Lumen Bearing Co.
Cu 78, Sn 8, Pb 14.
Chill cast: 32,000 TS; 18,000 YS; 10-15 El; 8-15 RA; 48-65 Brin. For bushings, pumps, mine and acid machines; high speed heavy duty bushings; Izod 4-5.

LUMEN 9
Lumen Bearing Co.
Cu 57.5, Sn 0.77, Zn 40, Fe 1, Al 1, Mn 0.25.
Cast: 65,000 TS; 25,000 YS; 10-35 El; 9-30 RA; 109-120 Brin. For propeller blades and hubs, valves, pump bodies; known as Manganese Bronze; Izod 20-40.

LUMEN 90
Lumen Bearing Co.
Cu 81, Al 14, Fe 5.
Cast: 302 Brin. For die blocks; Al-bronze, wear resistant.

LUMEN 91
Lumen Bearing Co.
Cu 80, Al 15, Fe 5.
Cast: 340 Brin. For die blocks; Al-bronze, wear resistant.

LUMEN 96
Lumen Bearing Co.
Al 8.5, Fe 3.5, bal Cu.
Cast: 73,000 TS; 25,000 YS; 120 Brin. For gears, bolts, hardware, bearings; Al-bronze, corrosion resistant.

LUMEN 97
Lumen Bearing Co.
Cu 85, Al 11, Fe 4.
Cast: 80,000 TS; 35,000 YS; 8 El; 6 RA; 163 Brin. Heat treated: 90,000 TS; 45,000 YS; 3 El; 3 RA; 200 Brin. For gears, feed nuts, shifter forks; Al-bronze, hardenable.

LUMEN 98
Lumen Bearing Co.
Cu 84, Al 12, Fe 4.
Cast: 80,000 TS; 37,000 YS; 2 El; 1 RA; 212 Brin. For feed shoes, gibs, cams, gears, forming dies; Al bronze. Wear resistant.

LUMEN ALLOY NO. 3
Lumen Bearing Co.
Cu 86, Sn 12, Pb 0-0.5, Zn 1.5.
Cast: 43,000 TS; 22,000 YS; 10-16 El; 9-13 RA; 57-65 Brin. For mine pumps and mining machinery, bearings; resists corrosion. *Obsolete*

LUMEN ALLOY NO. 4 A
Lumen Bearing Co.
Cu 79.5, Sn 12, Pb 7.5, P 1.
Cast: 32,000 TS; 24,000 YS; 1-3 El; 1-3 RA; 57-65 Brin. For bearings, cam shaft, grinding machine bearings. *Obsolete*

LUMEN ALLOY NO. 42
Lumen Bearing Co.
Cu 87.75, Sn 10, Ni 1.75, Pb 0.25, P 0.25.
Cast: 57,000-58,000 TS; 30,500-33,750 YS; 17-26 El; 90-115 Brin. For gears; for severe service. *Obsolete*

LUMEN ALLOY NO. 7
Lumen Bearing Co.
Cu 93.5, Sn 6.5.
Cast: 42,100 TS; 19,600 YS; 20-34 El; 24-32 RA; 52-57 Brin. For bearings, springs, clips, trolley wheels. *Obsolete*

LUMEN BRONZE
Lumen Bearing Co.
Cu 10, Zn 86, Al 4.
Cast: 36,000 TS; 36,000 YS; 119 Brin. For bearings for electric motors and lathes; will not resist excessive heat or live steam.

LUMEN METAL-1
Lumen Bearing Co.
Sn 85, Cu 10, Al 5.
Cast: 40,000 TS; 4 El. For bearings for electric motors. *Obsolete*

LUMEN METAL-2
Lumen Bearing Co.
Sn 88, Cu 4, Al 8.
Heat treated: 45,000 TS; 30 El. For bearings. *Obsolete*

LUMEN NO. 27

Lumen Bearing Co.
Cu 79.75, Fe 6, Mn 3.5, Al 10.75.
Cast: 96,000 TS; 38,000 YS; 10-20 El; 9-17 RA; 163-170 Brin.
For valve seat rings; resists heat to 800 C. *Obsolete*

LUMEN-1A

Lumen Bearing Co.
Cu 88, Sn 8, Zn 4.
Cast: 40,000 TS; 18,000 YS; 20 El; 18 RA; 57 Brin. For
bearings, bushings, liners. Corrosion resistant.

LUMINARC 2-S

Universal Power Corp.
99.0 Al min.
For Al welding rod.

LUNAR

J.M. Ney Co.
Pd 60, Ag 30.
White noble porcelain alloy for restorations subject to stress;
hard. 102,000 psi TS; 62,000 psi YS (0.1% offset); 28 El; 244
HV.

LUNDENHEIMER A-7-S

Lunkenheimer Co.
C 0-0.08, Cr 18-20, Ni 8-10, bal Fe.
Annealed: 80,000-95,000 TS; 30,000-45,000 YS; 55-60 El;
60-70 RA; 160-180 Brin. For valves and valve parts; Type 304,
stainless, austenitic. *Obsolete*

LUNKENHEIMER H-1

Lunkenheimer Co.
Sn, alloy, bal Fe.
For bonnets, bushings, valves; valve bronze.

LUNKENHEIMER HB-1

Lunkenheimer Co.
Sn, alloy, bal Fe.
For key cocks, valves; valve bronze.

LUNKENHEIMER ML-3

Lunkenheimer Co.
Fe 0-3.5, 60.0 Ni min, 23.0 Cu min, 2.0 C + Si max.
Cast: 65,000-90,000 TS; 32,000-40,000 YS; 25-45 El; 125-150
Brin. For valve parts, discs, stems, seat rings; corrosion
resistant.

LUNKENHEIMER ML-5

Lunkenheimer Co.
Fe 0-3.5, Al 0-0.5, 60.0 Ni min, 23.0 Cu min, 2.0 C + Si max.
Cast: 65,000-90,000 TS; 32,000-40,000 YS; 25-45 El. For
valve parts, seat rings, discs, stems; corrosion resistant, cast
Monel.

LUNKENHEIMER N-31

Lunkenheimer Co.
Cu 63, Ni 28.
Cast: 40,000 psi TS; 70 Brin (500). For valve components;
corrosion resistant, ductile, wear resistant, withstands high
temperature for short exposures.

LUNKENHEIMER N3

Lunkenheimer Co.
Cu 67, Ni 30.
Cast: 45,000 psi TS; 70 Brin (500). For pressure containing
valve components; corrosion resistant, wear resistant, ductile.

LUNKENHEIMER NO. 25 IRON

Lunkenheimer Co.
C 2.6, Ni 1-1.5, bal Fe.
Cast: 29,000 TS. For valves, flanges, pipe fittings; cast iron.

LUNKENHEIMER NO. 28 IRON

Lunkenheimer Co.
C 2.6, Ni 1-1.5, bal Fe.
Cast: 39,000 TS. For valves, flanges, pipe fittings; cast iron.

LUNKENHEIMER NO. 30 IRON

Lunkenheimer Co.
C 2.6, Ni 1-1.5, bal Fe.
Cast: 43,000 TS. For valves, flanges, pipe fittings; cast iron.

LUNKENHEIMER NO. 50 IRON

Lunkenheimer Co.
C 2.6, Ni 1.5, Cr 0.3, Mo 0.7, bal Fe.
Cast: 56,000 TS. For valves, flanges, pipe fittings; alloy cast
iron.

LUNKENHEIMER NS-5

Lunkenheimer Co.
Ni 47, Cu 44.
Cast: 110,000 TS; 301 Brin. For valve components such as
seat rings and discs; good hardness and corrosion
resistance.

LUNKENHEIMER NT-4

Lunkenheimer Co.
Cu 66, Ni 27.
Cast: 65,000 psi TS; 97 Rock B. For valve seats and discs;
good resistance to wear and corrosion.

LUNKENHEIMER NT-7

Lunkenheimer Co.
Ni 50, Cu 37.
Cast: 60,000 psi TS; 201 Brin. For valve components such as
seat rings; good wearing and non-galling in high temperature
steam service.

LUNKENHEIMER S-1

Lunkenheimer Co.
Sn 6, Pb 1.7, Zn 3.7, Ni 0.47, bal Cu.
Cast: 42,500 TS; 18,900 YS; 50 El. For valves; corrosion
resistant, free-cutting.

LUNKENHEIMER T-1

Lunkenheimer Co.
Sn 5, Pb 4.6, Zn 4.6, Ni 0.75, bal Cu.
Cast: 38,200 TS; 18,300 YS; 35 El. For valves; corrosion
resistant, free-cutting.

LUNKENHEIMER WC 5

Lunkenheimer Co.
C 0.14, Mo 0.93, Cr 0.79, Ni 0.88, bal Fe.
Cast: 88,000 TS; 61,000 YS; 27 El; 54 RA. For castings,
valves; resists graphitization.

LUNKENHEIMER WC1

Lunkenheimer Co.
C 0.19, Si 0.47, Mn 0.66, Mo 0.56, bal Fe.
At 70°F: 77,800 TS; 52,200 YS; 29.9 El; 48.5 RA. At 950°F:
63,000 TS; 37,000 YS; 30 El; 65 RA. For castings, body
bonnets.

LUNKENHEIMER WC4

Lunkenheimer Co.
C 0.15, Mo 0.5, Cr 0.7, Ni 0.9, bal Fe.
Cast: 86,000 TS; 60,000 YS; 29 El; 54 RA. For valves,
castings; resists graphitization.

LUNKENHEIMER WCB

Lunkenheimer Co.
C 0.27, Mn 0.68, Si 0.43, bal Fe.
At 70°F: 75,600 TS; 49,100 YS; 30 El; 47.2 RA. At 750°F:
64,000 TS; 27,000 YS; 35 El; 54 RA. For castings; good
weldability.

LUNORIUM

Manufacturer not listed.
Cr 14.9, Co 0.9, Ni 55.6, Mo 18.5, W 4, Fe 5, Mn 0.2, Si 0.4, C
0.2.
Cast: 69,000 TS; 32,000 YS; 1.4 El.

LUNZ IRON

E. Lunn & Co.
C 1.5, Cr 13, bal Fe.
For glass molds. Non-warping.

LURGIMETALL

Metallgesellschaft Reuterweg
Pb 96.5, Ba 2.8, Ca 0.4, Na 0.3.
For bearing metals; anti-friction.

LURIUM 107B

Fromson Co. Inc.
Aluminum. Mg 0.5-1, Si 0.2-1, bal Al.
Heat treated: 28,500-42,700 TS; 21,400-35,600 YS; 5-15 El;
80-100 Brin. For jewelry, reflectors, auto trim, and lighting
fixtures. Takes good polishing and bright anodizing.

LURIUM L

Fromson Co. Inc.
Aluminum. Al 99.99.
O: 6000 TS; 2100 YS; 60 El; 13 Brin. 1/2 H: 13,000 TS; 10,000
YS; 8 El; 25 Brin. H: 20,000 TS; 18,500 YS; 4 El; 35 Brin. For
reflectors and domestic appliances. High reflectivity.

LURIUM L10

Fromson Co. Inc.
Aluminum. Mg 0.08-1.2, bal Al.
Soft: 14,200 TS; 4200 YS; 20 El; 30 Brin. 1/2 H-temper:
24,200 TS; 21,400 YS; 8 El; 50 Brin. H-temper: 35,500 TS;
30,000 YS; 2 El; 60 Brin. For reflectors, ornamental trim, and
domestic appliances. High reflectivity.

LURIUM L20

Fromson Co. Inc.
Aluminum. Mg 1.5-2.5, bal Al.
Soft: 21,400 TS; 7000 YS; 35 El; 40 Brin. 1/2 H-temper:
35,000 TS; 32,600 YS; 6 El; 75 Brin. H-temper: 50,000 TS;
45,500 YS; 2 El; 80 Brin. For reflectors, ornamental trim, and
domestic appliances. High reflectivity.

LURIUM L5

Fromson Co. Inc.
Aluminum. Mg 0.4-0.6, bal Al.
Soft: 10,000 TS; 2800 YS; 40 El; 22 Brin. 1/2 H-temper:
21,400 TS; 20,000 YS; 6 El; 45 Brin. H-temper: 27,000 TS;
25,800 YS; 2 El; 50 Brin. For reflectors, ordnance and
domestic appliances. High reflectivity.

LUSCO

Ludlow Steel Corp.
C, alloy, bal Fe.
For pneumatic tools, wear and shock resisting tools and dies;
oil hardened.

LUSTER

J.F. Ratcliff Metals Ltd.
C, Cr, Ni, bal Fe.
For stainless parts; corrosion resistant.

LUSTERITE 440C

Latrobe Steel Co.
C 1.1, Cr 17, Mo 0.5, bal Fe.
Annealed: 110,000 TS; 65,000 YS; 15 El; 30 RA; 210 Brin.
Heat treated: 280,000 TS; 270,000 YS; 3 El; 15 RA; 555 Brin.
For bearings, knives, cutlery; Type 440C; stainless,
hardenable. *Obsolete*

LUSTERITE-440 B

Latrobe Steel Co.
C 0.8, Cr 16.5-18, Mo 0.5, bal Fe.
For surgical and dental instruments, knives; stainless,
capable of extreme hardness. *Obsolete*

LUSTRE-DIE

Peninsular Steel Co.
C 0.5, Mn 1, Si 0.3, Mo 0.25, Cr 1.1, bal Fe.
302-352 Brin; for plastic mold dies; good weldability.

LUSTRE-DIE

Bethlehem Steel Corp.
C 0.5, Mn 1, Si 0.3, Mo 0.25, Cr 1.1, bal Fe.
302-352 Brin; for plastic mold dies; good weldability.

LUTETIUM

Atomergic Chemetals Corp.

Lu.

Purities: 99.9% special distilled grade, 99.5+%, low tantalum contained. Forms: ingot, lump, turnings, sheet, wire, sponge.

LUXAL 63/25

Vereinigte Leichtmetallwerke, G.m.b.H.

Aluminum. Mg 2-4, Mn 0-0.4, Cr 0-0.3, bal Al.

Soft: 28,000 TS; 13,000 YS; 30 El; 47 Brin. Hard: 40,000 TS; 35,000 YS; 10 El; 73 Brin. For aircraft tanks and fittings, fuel lines, marine parts; resists sea water corrosion.

LUXALLOY NON GAMMA 2

Degussa AG

Silver.

Silver alloy for mixing amalgams for dentistry and dental engineering.

LW 2325

German manufacture

C 0.1, Si 0.42, Mn 0.26, Cr 15.68, W 5.25, Mo 1.38, Al 2.1, Ti 1.58, Fe 1.68, bal Ni.

For hot extrusion dies; good high temperature properties and wear resistance.

LW 2326

German manufacture

C 0.18, Si 0.34, Mn 0.39, Cr 14.91, W 1.39, Al 4.35, Fe 1.6, bal Ni.

For hot extrusion dies; good high temperature properties and wear resistance.

LW-1

Union Carbide Corp.

Co 9, 91 WC.

Cemented carbide spray coating for D-Gun. Extreme wear resistance; 1300 DPH.

LW-11 B

Union Carbide Corp.

Co 12, 81 WC.

Cemented carbide spray coating for plasma torch. Good wear resistance. 750 DPH.

LW-15

Union Carbide Corp.

Co 10, Cr 4, 86 WC.

Cemented carbide spray coating for D-Gun. Wear and corrosion resistant. 1100 DPH.

LW-1N40

Union Carbide Corp.

Co 15, 85 WC.

Cemented carbide spray coating for D-Gun. Wear resistance; better impact resistance than LW-1; 1050 DPH.

LX 8

Sheffield Smelting Co. Ltd.

Ag 15, Cu, Zn, Cd.

Melting range: 700-780°C. Maximum stress: 39.4 kgf/mm^2; 10 El. For silver brazing; general use.

LXB

Acciaierie Valbruna s.p.a.

Tool Material. C 0.95, Si 0-0.3, Mn 0-0.3, Cr 0.45, bal Fe.

Cold work tool steel. AISI W1C; W. Nr. 1.2004.

LXX HIL

Atlas Steels Ltd.

W 19, Cr 4, V 2.25, C, bal Fe.

For tools, dies, cutters; high speed steel. *Obsolete*

LXX-5T

Allegheny Ludlum Steel

C 0.4, W 18, bal Fe.

For hot work, tools and dies; hot work steel. *Obsolete*

LYNCAST

Lynchburg Foundry Co.

High C, Mn, Si, bal Fe.

For chemical engineering equipment; cast iron. *Obsolete*

LYNITE

Aluminum Company of America

Al 89, Cu 11, Mg 0.5.

For pistons. *Obsolete*

LYNUX BRONZE

English manufacture

Cu 89, Fe 7.2, Al 3.8.

For corrosion resisting hardware and fittings.

LYON'S GOLD

English manufacture

Zn 27, Cu 72.

For cartridge cases, condenser tubes, brazing; "Tombac."

LYONORE

Lyon, Conklin & Co.

C 0.2, Cu 0.2, Cr 0.5, Ni 0.8, bal Fe.

For structural parts; copper open hearth steel. *Obsolete*

M 921

George Cook & Co., Ltd.
C 0.8, Cr 4, Mo 9, V 1.1, W 1.5, bal Fe.
Molybdenum high speed steel. For twist drills, taps, lathe tools. BS 4659 BT1; AISI M1.

M AND A ALLOY NO. 13

Silver Creek Precision Co.
Cu 80, Sn 8, Pb 10, Ni 1, bal Zn.
Cast: 28,000 TS; 19,000 YS; 8 El; 40-50 Brin. For bearings; free machining. *Obsolete*

M AND A ALLOY NO. 24

Silver Creek Precision Co.
Cu 50-64, Sn 2-6, Pb 4-6, Ni 20-30, Zn 6-12.
Cast: 40,000 TS; 15 El. For valves, coffee urns; nickel silver. *Obsolete*

M AND A ALLOY NO. 27

Silver Creek Precision Co.
Cu 50-64, Sn 2-6, Pb 4-6, Ni 20-30, Zn 6-12.
Cast: 40,000 TS; 15 El. For valves, trim; nickel silver. *Obsolete*

M AND A ALLOY NO. 32

Silver Creek Precision Co.
Si 5, Fe 0-1, bal Al.
Cast: 17,000 TS; 8,000 YS; 1-3 El; 40 Brin. For light alloy castings; corrosion resistant. *Obsolete*

M AND A ALLOY NO. 33

Silver Creek Precision Co.
Cu 7.5, Fe 1.5, Zn 2, Si 3, bal Al.
Cast: 19,000 TS; 14,000 YS; 1.5 El; 60 Brin. For castings; good machinability. *Obsolete*

M AND A NO. 1

Silver Creek Precision Co.
Cu 88.5, Sn 6, Pb 1.5, Zn 4.
Cast: 38,000 TS; 20,000 YS; 40 El; 35 Brin. For pressure valves, pump housing; pressure-tight. *Obsolete*

M AND A NO. 10

Silver Creek Precision Co.
Cu 90, Sn 10.
Cast: 40,000 TS; 20,000 YS; 10 El; 86 Brin. For worm gears; tough. *Obsolete*

M AND A NO. 11

Silver Creek Precision Co.
Cu 85, Zn 15.
Cast. For castings to be brazed; red brass. *Obsolete*

M AND A NO. 14

Silver Creek Precision Co.
Cu 60, Zn 30, Mn, Fe, bal Al.
Cast: 110,000 TS; 75,000 YS; 5 El; 220 Brin. For gears, shafts; wear and corrosion resistant. *Obsolete*

M AND A NO. 15

Silver Creek Precision Co.
Cu 55, Zn 40, bal Al, Mn, Fe.
Cast: 95,000 TS; 50,000 YS; 15 El; 160 Brin. For cams, gears, cylinders; wear and corrosion resistant. *Obsolete*

M AND A NO. 16

Silver Creek Precision Co.
Cu 57, Al 1, Fe 1, Mn 0.5, Zn 40.5.
Cast: 65,000 TS; 35,000 YS; 20 El; 110 Brin. For worms, gears, valve stems, propellers; corrosion and wear resistant. *Obsolete*

M AND A NO. 17

Silver Creek Precision Co.
Cu 57, Sn 0.5, Pb 0.5, Fe 1, Al 0.5, Mn 0.25, bal Zn.
Cast: 60,000 TS; 30,000 YS; 15 El; 85 Brin. For valve stems, gears; free-machining. *Obsolete*

M AND A NO. 18

Silver Creek Precision Co.
Cu 62, Sn 1, Pb 1, Al 0.15, bal Zn.
Cast: 40,000 TS; 14,000 YS; 30 El; 50 Brin. For fittings, ship trimmings, commercial castings; free-cutting. *Obsolete*

M AND A NO. 19

Silver Creek Precision Co.
Cu 66, Sn 1, Pb 3, bal Zn.
Cast: 30,000 TS; 10,000 YS; 40 El; 45 Brin. For fittings, ship trimmings, commercial castings; free-cutting. *Obsolete*

M AND A NO. 2

Silver Creek Precision Co.
Cu 85, Sn 5, Pb 5, Zn 5.
Cast: 34,000 TS; 19,000 YS; 20 El; 43-50 Brin. For valves, fittings; "Red Brass." *Obsolete*

M AND A NO. 20

Silver Creek Precision Co.
Cu 60-70, bal Zn.
Cast: 22,000 TS. For castings; yellow brass. *Obsolete*

M AND A NO. 21

Silver Creek Precision Co.
Cu 88.5, Fe 3.5, Al 8.
Cast: 75,000 TS; 35,000 YS; 30 El; 110 Brin. For worm gears, propellers, cams, pump parts; corrosion resistant, tough. *Obsolete*

M AND A NO. 22

Silver Creek Precision Co.
Cu 89, Fe 1, Al 10.
Cast: 80,000 TS; 30,000 YS; 35 El; 130 Brin. For worm gears, propellers, cams, pump parts; heat treatable, corrosion resistant. *Obsolete*

M AND A NO. 23

Silver Creek Precision Co.
Cu 96, Si 4, Mn 1.
Cast: 45,000 TS; 30 El. For castings; non-sparking, corrosion resistant. *Obsolete*

M AND A NO. 25

Silver Creek Precision Co.
Cu 50-64, Sn 2-6, Pb 4-6, Ni 20-30, Zn 6-12.
Cast: 40,000 TS; 15 El. For valves, trim; nickel silver. *Obsolete*

M AND A NO. 26

Silver Creek Precision Co.
Cu 50-64, Sn 2-6, Pb 4-6, Ni 20-30, Zn 6-12.
Cast: 40,000 TS; 15 El. For valves, trim; nickel silver. *Obsolete*

M AND A NO. 28

Silver Creek Precision Co.
Cu 23, Fe 3.5, Ni 60, bal C, Si, Mn.
Cast: 65,000 TS; 32,500 YS; 25 El. For trim, dairy equipment; Monel. *Obsolete*

M AND A NO. 3

Silver Creek Precision Co.
Cu 80, Sn 10, Pb 10.
Cast: 28,000 TS; 19,000 YS; 8 El; 40-52 Brin. For bearings; for poor lubrication. *Obsolete*

M AND A NO. 32

Silver Creek Precision Co.
Si 5, Fe 0-1, bal Al.
Cast: 17,000 TS; 8,000 YS; 1-3 El; 40 Brin. For light alloy castings; corrosion resistant. *Obsolete*

M AND A NO. 34

Silver Creek Precision Co.
Cu 10, Fe 1.2, Mg 0.2, bal Al.
Cast: 23,000 TS; 21,000 YS; 0 El; 80 Brin. For pistons, cylinder heads; non-hardenable. *Obsolete*

M AND A NO. 35

Silver Creek Precision Co.
Si 5, bal Al.
Cast: 17,000 TS; 9,000 YS; 3 El; 40 Brin. For castings; corrosion resistant. *Obsolete*

M AND A NO. 36

Silver Creek Precision Co.
Cu 1, Fe 0.7, Mn 1, bal Al.
Cast: 16,000 TS; 9,000 YS; 10 El; 35 Brin. For castings; non-hardenable. *Obsolete*

M AND A NO. 37

Silver Creek Precision Co.
Cu 8, Fe 1, Si 1.2, bal Al.
Cast: 19,000 TS; 14,000 YS; 2 El; 65 Brin. For castings; non-hardenable. *Obsolete*

M AND A NO. 38

Silver Creek Precision Co.
Cu 4.5, Si 0.75, Fe 0.5, Zn 0.2, bal Al.
T4-temper: 32,000 TS; 16,500 YS; 8.5 El; 60 Brin. T6-temper: 36,000 TS; 24,000 YS; 0.5 El; 75 Brin. For crankcases, general purpose; age-hardened. *Obsolete*

M AND A NO. 4

Silver Creek Precision Co.
Cu 88, Sn 10, Zn 2.
Cast: 40,000 TS; 22,000 YS; 20 El; 65-75 Brin. For pressure castings; "Gun Metal." *Obsolete*

M AND A NO. 40

Silver Creek Precision Co.
Si 5, Cu 1.3, Mg 0.5, bal Al.
T7-temper: 38,000 TS; 36,000 YS; 0.5 El; 85 Brin. T6-temper: 35,000 TS; 25,000 YS; 2.5 El; 80 Brin. For liquid cooled engine parts; age-hardened. *Obsolete*

M AND A NO. 5

Silver Creek Precision Co.
Cu 88, Sn 8, Zn 4.
Cast: 40,000 TS; 21,000 YS; 20 El; 55-65 Brin. For gears, pump impellers, machine parts; resists salt water corrosion. *Obsolete*

M AND A NO. 6

Silver Creek Precision Co.
Cu 87, Sn 8, Pb 1, Zn 4.
Cast: 40,000 TS; 20,000 YS; 30 El; 50-63 Brin. For gears, pump impellers, machine parts; resists salt water corrosion. *Obsolete*

M AND A NO. 7

Silver Creek Precision Co.
Cu 83.5, Sn 8, Pb 8, P 0.5.
Cast: 25,000 TS; 20,000 YS; 8 El; 50-60 Brin. For bearings; heavy duty. *Obsolete*

M AND A NO. 8

Silver Creek Precision Co.
Cu 78, Sn 6, Pb 16.
Cast: 20,000 TS; 15,000 YS; 10 El; 50-90 Brin. For bearings; heavy duty. *Obsolete*

M C V

Sanderson Kayser Ltd.
C 0.46, Cr 0.6, Ni 0.55, Mo 0.25, Mn 1.2, B, Ti, Zr, V, bal Fe.
Oil hardening structural steel.

M H ALLOY NO. 00C

McCallum-Hatch Bronze Co. Inc.
Copper. Cu 84, Sn 16.
Cast: 25,000-35,000 TS; 20,000-28,000 YS; 1-2 El; 80-100 Brin. For bearings for turn tables and movable bridges; compression (0.1%) 18,000 psi.

M H ALLOY NO. 1

McCallum-Hatch Bronze Co. Inc.
Copper. Cu 88, Sn 8-10, Zn 2-4.
Cast: 40,000-50,000 TS; 18,000-22,000 YS; 20-40 El; 20-30 RA; 65-74 Brin. For steam and hydraulic castings, air valves, small gears; U.S.N. "G" Bronze; SAE No.62; "Gun Metal".

M H ALLOY NO. 11

McCallum-Hatch Bronze Co. Inc.
Copper. Cu 89, Al 10, Fe 1.
Cast: 65,000-80,000 TS; 23,000-27,000 YS; 20-27 El; 20-27 RA; 93-100 Brin. Heat treated: 75,000-95,000 TS; 55,000-65,000 YS; 3-15 El; 140-200 Brin. For gears, feed nuts, bearings; resists wear, repeated shock and corrosion.

M H ALLOY NO. 14

McCallum-Hatch Bronze Co. Inc.
Copper. Cu 90, Sn 6.5, Pb 1.5, Zn 2.
Cast: 34,000-40,000 TS; 16,000-19,000 YS; 23-35 El; 23-53 RA; 50-60 Brin. For plain bearings, backs of Babbit-lined shells, low coefficient of friction and expansion.

M H ALLOY NO. 15

McCallum-Hatch Bronze Co. Inc.
Copper. Cu 88.5, Sn 11, Pb 0.25, P 0.25.
Sand cast: 33,000-40,000 TS; 21,000-24,000 YS; 10-15 El; 8-15 RA; 7-80 Brin. Chill cast: 50,000 TS; 30,000 YS; 6 El; 100 Brin. For worm gears mating with hardened worms; SAE-65; compression (0.1%) 17,000 psi.

M H ALLOY NO. 20

McCallum-Hatch Bronze Co. Inc.
Copper. Cu 65, Zn 23, Fe 2, Mn 3, Al 7.
Cast: 90,000-100,000 TS; 50,000-60,000 YS; 5-20 El; 180-200 Brin. For general castings; compression (0.1%) 65,000 psi.

M H ALLOY NO. 25 B

McCallum-Hatch Bronze Co. Inc.
Copper. Cu 69, Sn 5, Pb 25, Ni 1.
20,000-24,000 TS; 14,000-16,000 YS; 12-16 El. For general castings, bearings; acid resisting.

M H ALLOY NO. 4

McCallum-Hatch Bronze Co. Inc.
Copper. Cu 80, Sn 10, Pb 10.
Cast: 30,000-50,000 TS; 19,000-21,000 YS; 10-20 El; 10-20 RA; 55-65 Brin. For high speed bearings; for high pressures.

M H ALLOY NO. 48

McCallum-Hatch Bronze Co. Inc.
Copper. Cu 84, Sn 10, Pb 2.5, Ni 3.5.
Sand cast: 40,000-50,000 TS; 25,000-28,000 YS; 15-25 El; 10-20 RA; 80-93 Brin. For gears; compression (0.1%) 20,000-24,000 psi.

M H ALLOY NO. 5

McCallum-Hatch Bronze Co. Inc.
Copper. Cu 85, Sn 5, Pb 5, Zn 5.
Cast: 30,000-38,000 TS; 15,000-19,000 YS; 15-20 El; 15-20 RA; 50-59 Brin. For bearings, pumps, steam valves; carburetors; "Ounce Metal."

M H ALLOY NO. 9

McCallum-Hatch Bronze Co. Inc.
Copper. Cu 58, Zn 40, Fe 1.5, Al 0.5-1.5, Mn 0-0.25.
Grade A: 70,000-80,000 TS; 30,000-34,000 YS; 25-40 El; 100 Brin. Grade B: 80,000-90,000 TS; 40,000-45,000 YS; 15-25 El; 120 Brin. For propeller blades and hubs, valve stems, engine framing; compression (0.1%) 20,000 psi.

M H ALLOYS

Now BARR ALLOYS.

M S M ALLOY

A. Milne & Co.
C 0.5, Mn 0.7, Si 1.8-2, Mo 0.25, bal Fe.
For shear blades, chisels, punches; tough.

M-10 AIR HARDENING

McInnes Steel Co.
C 1, Cr 5, Mo 1, V 0.25, bal Fe.
For gauges, punches, cold work dies; air hardening. *Obsolete*

M-11

Detroit Tap & Tool Co.
C 0.7, W 18, Cr 4, V 1, Co 5, bal Fe.
For taps, threading tools; high speed steel.

M-2 DREADNAUGHT

Hawkridge Bros. Co.
C 0.8, Cr 4, V 2, W 6, Mo 5, bal Fe.
For lathe and planer tools, reamers, broaches; high speed steel; Type M2

M-2 STEEL

Accurate Brass Co.
C 0.3, Mn 1.15, Ni 1.25-1.5, bal Fe.
Cast: 90,000-110,000 TS; 55,000-65,000 YS; 20-30 El; 40-50 RA; 200-250 Brin. For lifting forks for trucks; oil hardening, shock resistant.

M-203

General Electric Co.
C 0.07, Cr 1.95, Ni 24.5, Co 36.5, W 12, Cb 1.5, Ti 2.15, Al 0.75, Fe 1.6.
High temperature alloys, heat and corrosion resistant.

M-204

General Electric Co.
C 0.07, Cr 18.5, Ni 24.5, Co 40.5, W 12, Cb 1.2, Fe 1.6, B. 0.22.
High temperature alloy; heat and corrosion resistant.

M-205

General Electric Co.
C 0.07, Cr 18.5, Co 37.5, W 12, Cb 1.2, Al 2.75, Fe 1.6, B. 0.22, 24.5 Ni min.
High temperature alloy; heat and corrosion resistant.

M-250

Bethlehem Steel Corp.
C 0.01, Mn 0.07, Si 0.07, Ni 18, Co 8, Mo 4.75, Ti 0.4, Al 0.1, bal Fe.
High temperature alloy; heat and corrosion resistant. *Obsolete*

M-252

Driver Harris Co.
Nickel. Cr 19, Co 10, Mo 10, Ti 2, Al 1, bal Ni.
Wire and rod for welding wire and fastener stock. *Obsolete*

M-252

Cannon-Muskegon Corp.
C 0.15, Mn 0.5, Si 0.5, Cr 20, Co 10, Mo 10, B 0.005, Ti 2.6, Al 1, bal Ni.
Bar: 180,000 TS; 122,000 YS; 16 El. At 1400°F: 137,000 TS; 104,000 YS; 10 El. For high temperature service, bolts, jet engine and gas turbine parts. Heat and corrosion resistant.

M-252

Universal Cyclops
C 0.15, Mn 0.5, Si 0.5, Cr 20, Co 10, Mo 10, B 0.005, Ti 2.6, Al 1, bal Ni.
Bar: 180,000 TS; 122,000 YS; 16 El. At 1400°F: 137,000 TS; 104,000 YS; 10 El. For high temperature service, bolts, jet engine and gas turbine parts. Heat and corrosion resistant.

M-252

Now UDIMET M-252; ALLOY M-252; ALLVAC M-252; CARPENTER PYROMET M-252; CRUCIBLE M-252; and UNITEMP M-252.

M-2A STEEL

Accurate Brass Co.
C 0.25-0.3, Ni 1.15-1.25, bal Fe.
Cast: 80,000-95,000 TS; 53,000-58,000 YS; 24-30 El; 35-50 RA; 175-200 Brin. For sheave wheels, gears; oil hardened, shock resistant.

M-3 DREADNAUGHT

Hawkridge Bros. Co.
C 1, Cr 4, V 2.7, W 6, Mo 5, bal Fe.
For lathe and planer tools, reamers, broaches, taps; Type M3; high speed steel. *Obsolete*

M-3 SILECTRON

Allegheny Ludlum Steel
Si 3, bal Fe.
Core loss 0.49 watts/lb. max. For transformers, laminations. Electrical steel, high permeability. *Obsolete*

M-600

General Electric Co.
C 0.08, Cr 19, Mo 7, Ti 2.3, Al 1.1, Fe 13, 55.5 Ni min.
For high temperature structural parts.

M-813

General Electric Co.
C 0.08, Cr 18, Ni 35, Mo 4, Ti 2.25, Al 1.4, bal Fe.
For aircraft gas turbine parts, high temperature bolting.

M-841

George Cook & Co., Ltd.
C 1.05, Cr 4, Mo 9.5, V 1.2, W 1.5, Co 8, bal Fe.
Molybdenum-cobalt high speed steel. For drills, reamers, lathe tools. AISI M-42; BS 4659 BM 42.

M-A, VM

Teledyne Vasco
Now MATRIX I.

M-ALLOY

George Cook & Co., Ltd.
C 0.58, Mn 0.35, Cr 0.95, Si 0.95, Mo 0.2, bal Fe.
For plastic mold dies, hot and cold punches, form tools; oil hardened, tough. *Obsolete*

M-ALLOY

Pettibone Corp.
C, alloy, bal Fe.
For foundry sling liners; wear resistant.

M-ALLOY

Thomas Bolton Ltd.
C, alloy, bal Fe.
For aircraft forgings and parts; corrosion resistant. *Obsolete*

M-CHROME

Latrobe Steel Co.
C 1, Cr 1.5, bal Fe.
For tools, dies, gauges; deep hardening. *Obsolete*

M-H COMPOSITE

Metallgesellschaft Reuterweg
C, alloy, bal Fe.
For heavy duty rollers; in sheet mills.

M-H SPECIAL ALLOY STEEL

Gulf & Western Mfg. Co.
C, alloy, bal Fe.
For rolls for slabbing mills; water hardened.

M-K-9

Major Engineering Co.
Cu 4, bal Al.
Rolled: 60,000 TS; 30 El; 70 Brin. For light alloy parts; age-hardenable.

M-M-0011 ALLOY

Martin-Marietta Corp.
C 0.15, Cr 8.25, Co 10, W 10, Mo 0.7, Ta 3, Al 5.5, Ti 1, Hf 1.5, B 0.015, Zr 0.5, bal Ni.
Cast high temperature alloy; for turbine wheels and blades. Same as MAR-M-247.

M-M-1
A. Milne & Co.
C 0.65-0.85, Mo 7.5-9.5, W 1.25-2.25, Cr 3.5-4.5, V 0.9-1.5, bal Fe.
For tools, cutters; high speed steel.

M. W. METAL
English manufacture
Mg 90-95, Al 3-7, Zn 2-5, Mn 0.5.
36,000-39,000 TS; 33,000 TS; 16-13 El. For light alloy parts; age-hardenable.

M.C. MOLD AND CAVITY STEEL
Teledyne Vasco
C, alloy, bal Fe.
For dies and molds; deep hardening. *Obsolete*

M.C.C
English manufacture
Cu 10, Mg 1.3, Ni 2, Fe 0.5, Si 0.5, Cr 2.5, bal Al.
For pistons; self-aging.

M.H. ALLOY NO. 6
McCallum-Hatch Bronze Co. Inc.
Copper. Cu 77.5, Sn 7, Pb 14, Ni 1.5.
30,000-36,000 TS; 17,000-19,000 YS; 20-28 El; 21-25 RA; 64 Brin. For machinery bearings for high speed and medium pressures; "Bar Alloy No. 6;" acid resistant.

M.K. STEEL
Japanese manufacture
Ni 10-40, Al 1-20, bal Fe.
For magnets; not forged readily.

M.K. STEEL
Japanese manufacture
Ni 25, Al 10, bal Fe.
For magnets; must be cast to shape.

M.M.M
Dresser Industries
Sn 0-9.25, 57.5 min Ni, 26.0 min Cu, bal Fe, Si, Mn.
Cast: 70,000 TS; 45,000 YS; 17.5 El. For valves for superheated steam; modified "Monel" Metal.

M.M.M
Consolidated Ashcroft Hancock Co. Inc.
Sn 0-9.25, 57.5 Ni min, 26.0 Cu min, bal Fe, Si, Mn.
Cast: 70,000 TS; 45,000 YS; 17.5 El. For valves for superheated steam; modified "Monel" Metal.

M.O. 81 STEEL
Acieries de Champagnole
C 0.87, Cr 4, V 2, Mo 8.5, W trace, bal Fe.
Air or oil hardens to 61-65 Rock C. For twist drills, "hand" taps, dies, and extra long drills. AFNOR: Z 90 DV 09.02; AISI M10.

M.O. 82 STEEL
Acieries de Champagnole
C 0.84, Cr 4, W 1.4, Mo 8, V 1.5, bal Fe.
Air or oil hardens to 62-66 Rock C. For twist drills, end mills, reamers, taps, counter bores, and hobbing cutters. AFNOR: Z 85 DVW 08.02.01; AISI M1; Germany: S 2.9.1 (B Mo 9); W.Nr 1.3346.

M.O. 83 STEEL
Acieries de Champagnole
C 1, Cr 4, W 1.4, Mo 8.5, V 1.9, bal Fe.
Air or oil hardens to 62-66 Rock C. For drills (deep hole), taps (blind holes), counter bores, and end mills. AFNOR: Z 100 DVW 08.02.01; AISI M7; Germany: S 2.9.2 (B Mo 9V); W.Nr 1.3348.

M.O. 88 STEEL
Acieries de Champagnole
C 1.1, Cr 3.8, W 1.5, Mo 9.5, V 1.2, Co 8, bal Fe.
Air or oil hardens to 63-70 Rock C. For turning, drilling, milling, and planing extra hard or tough materials. AFNOR: Z 110 DKWV 10.08.02.01; AISI M 42.

M.P.M.-3C-STEEL
Colt Industries
C 0.3-0.4, Mn 1.6-1.9, bal Fe.
Annealed: 100,000 TS; 60,000 YS; 19 El; 50 RA; 250 Brin. As rolled: 113,000 TS; 77,000 YS; 22 El; 57 RA; 228 Brin. For gears, wheels, castings, axles, drive shafts, steering arms; tough, pearlitic. *Obsolete*

M.S. BRONZE
Associated Spring Co.
Cu 96, Sn 3.5, P 0.25.
For springs; for watches.

M.S. STEEL
AMAX Corp.
Cr 0.8-1.1, Mo 0.3-0.4, Mn 0.6-0.9, C 0.4-0.6, bal Fe.
For gears, pinions, shafts, axles, tools, forgings. *Obsolete*

M.V.C. ALLOY
Metropolitan-Vickers Electrical Co. Ltd.
Si 7.5-13, bal Al.
Sand cast: 18,000-27,000 TS; 10,000-11,000 YS; 5-15 El; 50 Brin. Chill cast: 22,000-30,000 TS; 10 El. For airplane and automobile parts, marine parts, tubes; great resistance to atmospheric and sea water corrosion; modified Al-Si alloy.

M/S VENUS
Paul Bergsoe & Son
Sn 83.5, Sb 8, Cu 6.5, Pb 2.
Cast: 17,100 TS; 31 Brin. For diesel engine bearings. MP: 355-710°F. Wear resistant.

M1
Creusot-Loire
C 0.15, Mn 0.5, Si 0-0.35, bal Fe.
Carburizing steel; for cams, small machine parts. Italie: UNI 16.

M1-CARBIDE
German manufacture
Co 0, 10 TiO + TaO, 04 WC.
For hard tools, cutters, dies; sintered carbides, wear and abrasion resistant.

M115
Aluminium-Zentral e.V.
Aluminum. Mn 1.5, bal Al.
Half hard: 22,000 TS; 16,000 YS; 7-14 El; 40-45 Brin. Hard: 35,000 TS; 31,000 YS; 3-6 El; 60 Brin. For roofing, structures, commercial vehicles. Corrosion resistant, nonhardenable. *Obsolete*

M13
Acciaierie Valbruna s.p.a.
C 1.2, Mn 12, Cr 0-0.5, Si 0-0.5, P 0.1, S 0.03, bal Fe.
Austenitic manganese steel. W. Nr. 1.3802.

M2-CARBIDE
German manufacture
Co 7.5, 10.5 TiC + TaC, 82 WC.
For hard tools, cutters, dies; sintered carbides, wear and abrasion resistant.

M200
Bergische Stahl Industrie
C 0.85-0.95, Si 0.15-0.3, Mn 1.9-2.1, V 0.05-0.15, 0.2-0.5 Cr (optional), bal Fe.
Cold work tool steel. W.-Nr. 1.2842.

M21
Atrax Cemented Carbide

Sintered carbide. 325,000 transverse strength; 14.75-15.0 g/cm^3 density; 90.5-91.5 RA. Industry code C-1-9; ISO K30.

M22
International Nickel Inc.
C 0.13, Cr 5.7, Mo 2, W 11, Al 6.3, Zr 0.6, Ta 3, bal Ni.
High temperature alloy.

M22
Huntington Alloys Inc.
C 0.13, Cr 5.7, Mo 2, W 11, Al 6.3, Zr 0.6, Ta 3, bal Ni.
High temperature alloy.

M22VC
Mond Nickel Co. Ltd.
C 0.08-0.16, Cr 5-6.5, Al 5.9-6.6, Mo 1.5-2.5, W 10.5-11.5, Ta 2.6-3.4, Zr 0.4-0.8, Fe 0-1, bal Ni.
Cast: 106,700 TS; 99,000 YS; 5.6 El; 14 RA. At 600°C: 117,000 TS; 109,000 YS; 4.5 El; 13.8 RA. For aircraft gas turbines, spinners for glass fiber production. High temperature vacuum casting alloy; creep and oxidation resistant.

M25T
Sheffield Smelting Co. Ltd.
Ag 55, Cu, Zn, Sn.
Melting range: 630-660°C. Maximum stress: 44.7 kgf/mm^2; 24 El. For silver brazing; excellent fluidity.

M2A
Westinghouse Electric Corp.
Mo alloy.
For spot welding tips. *Obsolete*

M3
Delta Enfield Metals Ltd.
Now ERM CCS (M3).

M308
Plykrome Corp.
C 0.08, Cr 14, Ni 33, Mo 4, W 6.5, Ti 2, Al 0.25, Zr 0.25, bal Fe.
High temperature alloy; heat and corrosion resistant.

M31
Atrax Cemented Carbide
Sintered carbide. 350,000 transverse strength; 14.15-14.45 g/cm^3 density; 88.3-89.3 RA. Industry code: C-11; ISO K40.

M328
Delta Enfield Metals Ltd.
Now ERM CCS/Z (M328).

M331
A. Milne & Co.
C 0.32, Cr 3.4, V 0.4, Mo 2.5, bal Fe.
Air or oil hardening; hot work tool steel.

M333
A. Milne & Co.
C 0.3, Cr 3, V 0.6, Mo 3, Co 2.25, bal Fe.
Air or oil hardening; hot work tool steel.

M40
Atrax Cemented Carbide
Sintered carbide. 400,000 transverse strength; 14.15-14.35 g/cm^3 density; 85.5-86.5 RA. Industry code: C-16.

M41
Atrax Cemented Carbide
Sintered carbide. 400,000 transverse strength; 14.15-14.45 g/cm^3 density; 86.8-87.8 RA. Industry code: C-16.

M45
Creusot-Loire
C 0.46, Mn 0.7, Si 0-0.4, bal Fe.
Carbon structural steel. AISI-SAE 1049.

M5
Atrax Cemented Carbide
Sintered carbide. 400,000 transverse strength; 14.4-14.6 g/cm^3 density; 87.5-88.5 RA. Industry code: C-16.

M50
Now VASCO M50 CVM; CRUCIBLE M50 VAR; and CARPENTER CONSUMET M-50.

M56

Atrax Cemented Carbide
Sintered carbide. 400,000 transverse strength; 14.4-14.6 g/cm^3 density; 87.7-88.7 RA. Industry code: C-16.

M5T

Sheffield Smelting Co. Ltd.
Ag 35, Cu, Zn, Sn.
Melting range: 645-735°C. Maximum stress: 47.5 kgf/mm^2; 10 El. For silver brazing; general engineering purposes.

M6

Atrax Cemented Carbide
Sintered carbide. 400,000 transverse strength; 14.4-14.6 g/cm^3 density; 88.3-89.3 RA. Industry code: C-16.

M7

Atrax Cemented Carbide
Sintered carbide. 360,000 transverse strength; 14.5-14.7 g/cm^3 density; 89.0-90.0 RA. Industry code: C-16.

M9T

Sheffield Smelting Co. Ltd.
Ag 39, Cu, Zn, Sn.
Melting range: 635-710°C. Maximum stress: 49.0 kgf/mm^2; 10 El. For silver brazing; general engineering purposes.

MA-18NICOMO

International Nickel Inc.
Superalloy. Ni 18, C, Co, Mo, bal Fe.
Heat treated: 220,000-305,000 TS; 215,000-300,000 YS; 60-64 RA. Ductile and weldable.

MA-754

Cannon-Muskegon Corp.
C 0.04, Cr 20, Ti 0.5, Al 0.3, 0.6 Y$_2$O$_3$, bal Ni.
Mechanically alloyed powder for blades and vanes.

MAB (MITER-AL-BRAZE) NO. 2

Belmont Metals Inc.
Cu 3, Al 6, bal Zn.
33,000 psi TS. For brazing aluminum; melting point 730°F

MAB (MITER-AL-BRAZE) REGULAR

Belmont Metals Inc.
Cu 3, Al 4, bal Zn.
37,800 psi TS. For aluminum brazing; melting point 720°F

MAC HEMPITE GRADE A

Gulf & Western Mfg. Co.
Ni 1.5-3.5, Mn 0.7-4, Cr 0-1.25, Mo 0-0.75, C 0.4-3, bal Fe.
Soft: 100,000 TS; 65,000 YS; 25 El; 45 RA. Hard: 600 Brin. For jaws, rollers, tires, balls, liners, impellers; heavy crusher grade; resists wear and spalling.

MAC HEMPITE GRADE B

Gulf & Western Mfg. Co.
Ni 1.5-3.5, Mn 0.7-4, Mo 0-0.75, C 0.4-3, bal Fe.
Soft: 90,000 TS; 65,000 YS; 25 El; 45 RA; 200 Brin. For heavy duty gears, pinions; medium hardened grade; shock resistant.

MAC HEMPITE GRADE C

Gulf & Western Mfg. Co.
Ni 1.5-3.5, Mn 0.7-4, Cr 0-1.25, Mo 0-0.75, C 0.4-3, bal Fe.
For gears, pinions, hammers, beaters, conveyor parts, brake drums; hardened grade.

MAC-IT

Strong, Carlisle & Hammond Co.
C 0.4, Cr 0.8, Ni 1.5, bal Fe.
For screws, threaded fasteners; oil hardening.

MACALLOY

McCalls Special Products
C 0.6, Cr 0.75, bal Fe.
1000 N/mm^2 TS. Steel bar for prestressed concrete.

MACALLOY

Teledyne Vasco
C 0.35-0.65, Ni 1.25, Cr 0.75, Mo 0.2, bal Fe.
Rolled. For gears, shafts. *Obsolete*

MACCO 33

P.F. McDonald & Co.
C 0.4, Cr 5.5, Mo 1.4, V 1, Si 1, bal Fe.
For die casting dies; air hardening.

MACCO 35

P.F. McDonald & Co.
C, alloy, bal Fe.
For tools, air-hardened.

MACCO 35 AIR HARD

P.F. McDonald & Co.
Mn 0.6, Cr 5.25, Mo 1.15, V 0.25, C 1, bal Fe.
For punch and cold work dies; air hardening.

MACCO 99

P.F. McDonald & Co.
C 0.35, Mn 0.8, Cr 0.85, Mo 0.35, Si 0.6, bal Fe.
For die casting dies; water or oil hardening.

MACCO ALLOY RAZOR BLADE

P.F. McDonald & Co.
C 0.7, Cr 1.2, bal Fe.
For razor blades; water hardened.

MACCO B-29

P.F. McDonald & Co.
C 0.98, Mn 0.26, V 0.2, bal Fe.
For cold heading and forming dies; water hardening.

MACCO BELT KNIFE

P.F. McDonald & Co.
C, Cr, Ni, V, Si, Mn, bal Fe.
For knives, leather splitting belt knives; oil hardened.

MACCO BRAKE DIE

P.F. McDonald & Co.
C 0.5, Mn 0.9, Cr 1.02, Mo 0.22, bal Fe.
For brake dies; oil hardened.

MACCO BRAND SWEDISH

P.F. McDonald & Co.
C, bal Fe.
For tools; special grade of tool steel.

MACCO CARBON

P.F. McDonald & Co.
C 1, Mn 0.2, Si 0.16, bal Fe.
Annealed: 100,000 TS; 53,000 YS; 21 El; 42 RA; 200 Brin. For drills, taps, chasers, lathe and planer tools; Type W1; water hardened.

MACCO CERTIFIED

P.F. McDonald & Co.
C, bal Fe.
For general purpose tools; water hardening.

MACCO ENORMOUS

P.F. McDonald & Co.
C 0.8, Cr 5, W 17, V 1.5, Co 5, bal Fe.
For taps, dies, twist drills, cutters; high speed steel.

MACCO FOOLPROOF O.H.

P.F. McDonald & Co.
C 0.55, Cr 1.25, W 2.75, V 0.2, bal Fe.
Heat treated: 231,000-315,000 TS; 209,000-275,000 YS; 9-11 El; 27-42 RA; 48-56 Rock C. For header and swaging dies, chipping chisels, rivet busters, track tools. Type S1 shock resisting tool steel, good fatigue resistance.

MACCO HARDTUF

P.F. McDonald & Co.
C 0.6, Mo 0.45, V 0.2, Si 1.85, bal Fe.
For plastic mold dies; oil hardened.

MACCO HOBOMOLD A

P.F. McDonald & Co.
C 0.07, Cr 4.5, Mo 0.45, Mn 0.4, bal Fe.
For plastic mold dies; Type P4; air hardened, hobbing steel.

MACCO HOBOMOLD B

P.F. McDonald & Co.
C 0.06, Mn 0.03, Cr 1, Si 0.2, bal Fe.
For plastic mold dies; Type P5; oil hardened, hobbing steel.

MACCO HOBOMOLD C

P.F. McDonald & Co.
C 0.04, Mn 0.2, Si 0.16, bal Fe.
For plastic mold dies; case-hardening.

MACCO HOLLOW-DRILL

P.F. McDonald & Co.
C, bal Fe.
For hollow drills, tools, rock drills; water hardened.

MACCO KROMAX

P.F. McDonald & Co.
C 2.5, Si 0.6, Ni 0.5, Cr 12-14, bal Fe.
For dies, tools; does not possess red hardness.

MACCO KROMAX 1

P.F. McDonald & Co.
C 1.5, Cr 11.9, V 0.7, bal Fe.
For blanking, forming and punching dies; air hardened, non-deforming.

MACCO KROMAX 2

P.F. McDonald & Co.
C 1.4, Cr 12.5, Mo 0.85, Co 3.25, Si 0.4, bal Fe.
For blanking, punching, forming dies; air hardening, cold forming.

MACCO KROMAX SPECIAL

P.F. McDonald & Co.
C 1.6, Cr 12.4, Mo 0.72, bal Fe.
For tools, dies; non-deforming.

MACCO LENS MOLD

P.F. McDonald & Co.
C 0.4, Cr 5.25, W 4.65, Si 1, bal Fe.
For die casting lens mold; oil hardening.

MACCO M.L.V.

P.F. McDonald & Co.
C 0.35, Si 1.05, W 1.55, Mo 1.65, Cr 5.15, bal Fe.
Hardened: 216,000 TS; 185,000 YS; 14 El; 53 RA. At 1000°F: 144,000 TS; 110,000 YS; 19 El; 64 RA. For die casting dies, hot shear blades, forging and heading dies, punches. Type H12 hot work tool steel, tough and shock resistant.

MACCO ML

P.F. McDonald & Co.
C 0.35, Cr 5, W 1.5, Mo 1.65, Si 1, bal Fe.
For hot work dies; hot work steels.

MACCO NLS

P.F. McDonald & Co.
C 0.9, Mn 0.6, Cr 5.2, Mo 1.2, V 0.3, bal Fe.
For plastic molding dies; oil hardening.

MACCO NON-TEMP

P.F. McDonald & Co.
C 0.35, Cu 0.3, Si 0.45, Mo 0.7, Cr 0.8, Mo 0.3, bal Fe.
For general purpose tools, dies, punches; oil or water hardened, shock resistant.

MACCO P-125

P.F. McDonald & Co.
C 0.25, Cr 4.2, W 14.5, V 0.5, bal Fe.
For hot work dies; hot work steels.

MACCO P-150

P.F. McDonald & Co.
C 0.5, Cr 2.9, W 15.3, V 0.6, bal Fe.
For hot work dies; high speed steel.

MACCO P-175
P.F. McDonald & Co.
C 0.3, Cr 3.3, W 9.5, V 0.5, Si 0.4, bal Fe.
For hot work dies; hot work steel.

MACCO RADIO
P.F. McDonald & Co.
C 0.8, Cr 4, W 6, Mo 5.5, V 1.8, bal Fe.
For tools, cutters; high speed steel.

MACCO ROYAL CROWN
P.F. McDonald & Co.
C 1, Cr 1, W 0.5, Mn 0.5, bal Fe.
For dies, punches, taps, drills, reamers, drawing dies; non-deforming steel.

MACCO SILICON MANGANESE
P.F. McDonald & Co.
C 0.63, Mn 0.78, Si 2, bal Fe.
For punches, chisels, crimpers; Type S4; oil hardened, shock resistant.

MACCO SILVER DIE
P.F. McDonald & Co.
C 0.52, Mn 1, Cr 1.1, Mo 0.25, bal Fe.
For dies for plastic molds; oil hardened.

MACCO SPECIAL
P.F. McDonald & Co.
C 1-1.1, bal Fe.
For tools, drills, taps; water hardening.

MACCO SUPER MOLY
P.F. McDonald & Co.
C 0.8, Cr 3.9, W 1.7, Mo 8.8, V 1.1, bal Fe.
For cutting tools, reamers; high speed steel.

MACCO SUPERIOR HIGH SPEED
P.F. McDonald & Co.
C 0.72, Cr 4, V 1, W 18.5, bal Fe.
For tools, cutters, drills; high speed steel.

MACCO SUPERIOR SWEDISH
P.F. McDonald & Co.
C, bal Fe.
For tools, special punches, dies; oil hardened.

MACCO SWEDISH CHROME TUNGSTEN
P.F. McDonald & Co.
C, Cr, W, bal Fe.
For tools, special taps, dies; water hardened.

MACCO SWEDISH MAGNETIC IRON
P.F. McDonald & Co.
C 0.2, Si 0.02, bal Fe.
For magnetic parts for all kinds of electrical instruments; high permeability.

MACCO WJF
P.F. McDonald & Co.
C 1.4, W 4, V 0.35, Cr 0.6, bal Fe.
For cutters, taps, reamers; water hardened, fast finishing.

MACCOMAX
P.F. McDonald & Co.
C 0.7, W 18, Cr 4, V 1, bal Fe.
For tools, cutters; high speed steel.

MACH 5
United States Steel Corp.
C 0-0.09, Mn 0.85-1.15, P 0.04-0.09, S 0.26-0.35, Pb 0.15-0.35, Bi 0.05-0.1, bal Fe.
Cold drawn: 73 ksi TS; 68 ksi YS; 15.8 El. Free machining steel.

MACHINE BRONZE
Lumen Bearing Co.
Cu 50, Ni 25, Sn 25.
For bearings; wear resisting.

MACHINERY 30
Atlas Specialty Steels
C 0.3, Mn 0.8, Si 0.2, bal Fe.
For machinery parts; construction steel. *Obsolete*

MACHINERY 40
Atlas Specialty Steels
C 0.4, Mn 0.8, Si 0.2, bal Fe.
For machinery parts; construction steel. *Obsolete*

MACHINERY BRASS
English manufacture
Zn 16, Sn 1, bal Cu.
For machinery parts; corrosion resistant.

MACHS ALLOY
Manufacturer not listed.
Al 90-98, Mg 2-10.
For light alloy parts.

MACHS SPECULUM
Manufacturer not listed.
Al 69, Mg 31.
For light alloy parts.

MACHTS METAL
Manufacturer not listed.
Cu 57, Zn 43.
For brazing metal, hardware.

MACHTS YELLOW METAL
Manufacturer not listed.
Cu 57, Zn 43.
For hardware, bolts, nuts; high strength.

MACKENITE METAL
Mackenzie's Sons Co. Inc.
C 0.2, Cr 20, Ni 12, bal Fe.
For annealing pots, furnace pots; heat resistant.

MACKENZIE METAL
Manufacturer not listed.
Pb 68-70, Sb 16-17, Sn 13-16.
For bearings; anti-friction.

MACLOY B
Firth-Vickers Stainless Steels Ltd.
C 0.27, Si 0.3, Mn 1.25, Cr 11, Ni 36, bal Fe.
Heat treated: 90,000 TS; 64,000 YS; 30 El; 55 RA; 200 Brin.
For corrosion and heat resistant parts; austenitic, stainless.

MACLOY G
Firth-Vickers Stainless Steels Ltd.
C 0.5, Si 1.8, Mn 0.6, Cr 17, Ni 37, bal Fe.
94,000 TS; 54,000 YS; 30 El; 40 RA; 200 Brin. For heat and corrosion resistant parts; austenitic, heat and corrosion resistant.

MACOLOY TYPE E
McCauley Alloy Sales Co.
P, bal Cu.
For brazing; melting point 1150°F.

MACOLOY TYPE F
McCauley Alloy Sales Co.
P, bal Cu.
For brazing; melting point 1180°F.

MACOLOY TYPE G
McCauley Alloy Sales Co.
P, bal Cu.
For brazing; melting point 1600-1900°F.

MACRO 40F
Kennametal Inc.
WC, copper-base alloy, others.
32-36 Rock C. Diamond-matrix powder for hot-pressed impregnated products, especially core bits used for drilling extremely abrasive materials.

MACROCRYSTALLINE WC
Kennametal Inc.
TC 0-6.2, Fe 0-0.15, Mo 0-0.1, Ti 0-0.15, Si 0-0.02, Ta 0-0.07, Nb 0-0.05, 0.03 free C max.
Hard powder used as a wear-rate modifier, hard additive, and bond hardener for hot-pressed diamond-product matrices.

MACROFIL 49
Kennametal Inc.
Cu 49, Ni 10, Zn 41.
Melting point 930°C (1706°F). Infiltration alloy for use in the manufacture of surface-set diamond tools.

MACROFIL 56
Kennametal Inc.
Cu 56, Zn 43, Sn 1.
Melting point 870°C (1598°F). Infiltration alloy used in the manufacture of surface-set diamond tools.

MACROFIL 65
Kennametal Inc.
Cu 65, Ni 15, Zn 20.
Infiltration alloy used in the manufacture of surface-set diamond tools.

MACROMAL
Rheinische Rohrenwerke Aktiengesellschaf
C 0.2, Mn 12-13, bal Fe.
Rolled: 100,000-135,000 TS; 35,600 YS; 30 El. For compass housings, deck superstructures, transformer shields; austenitic, nonmagnetic.

MACROMAL S
Rheinische Rohrenwerke Aktiengesellschaf
C 0.2, Mn 17,18, bal Fe.
Rolled: 100,000-135,000 TS; 35,600 YS; 30 El. For compass housings, deck superstructures, transformer shields; austenitic, non-magnetic.

MACROSIL
VDM Nickel-Technologie AG
C 0-0.15, Cr 11-13, Ni 1.5-2.5, Mo 0.3-0.6, Mn 17-19, bal Fe.
For chemical plant equipment; corrosion resistant. *Obsolete*

MADISON-KIPP NO. 400
Madison-Kipp Corp.
Si 8, bal Al.
For die castings; corrosion resistant.

MAERKER IRRUBIGO 1
Schmidt & Clemens Edelstahlwerke
C 0.1, Cr 13, bal Fe.
Air or oil hardenable; corrosion resistant. Heat treated: 85,000-107,000 TS; 18 min El; 140-210 Brin. Yield strength at 750 F: 44,100 psi min. For jet engine parts, steam turbine parts; parts subject to corrosion by water and steam. DIN X10Cr13; AISI 410; VDEh spec 4006. *Obsolete*

MAERKER IRRUBIGO 1 MO
Schmidt & Clemens Edelstahlwerke
C 0.15, Cr 13, Mo 1.15, bal Fe.
Air or oil hardenable; corrosion resistant. Heat treated: 107,000-128,000 TS; 14 min El; 220-260 Brin. Yield strength at 750 F: 51,000 psi. For fittings, brackets, hardware for increased strength and oxidation resistance at elevated temperatures. DIN XCrMo13; VDEh Spec 4119. *Obsolete*

MAERKER IRRUBIGO 12
Schmidt & Clemens Edelstahlwerke
C 0.05, Cr 17.5, Mo 4.75, Ni 13.5, bal Fe.
Quenched: 71,000-107,000 TS; 35,500 YS; 45 min El; 130-190 Brin. Resistant to corrosion by halogen salt solutions. Indifferent to localized corrosion. DIN X5CrNiMo1713; VDEh Spec 4449; similar to AISI 317. *Obsolete*

MAERKER IRRUBIGO 17H
Schmidt & Clemens Edelstahlwerke
C 0.22, Cr 17, Ni 2, bal Fe.
Air or oil hardenable; corrosion resistant. Heat treated: 114,000-135,000 TS; 14 min El; 225-275 Brin. Yield strength at 750 F: 54,000 psi. For machine parts subject to high mechanical stress with increased resistance to weak acids, salt solutions; resistant to sea water. DIN X22CrNi17; VDEh Spec 4057. *Obsolete*

MAERKER IRRUBIGO 1818
Schmidt & Clemens Edelstahlwerke
C 0.05, Cr 17.5, Mo 2.25, Ni 20, Cu 2, Nb = 8 x C, bal Fe.
Quenched: 71,000-107,000 TS; 31,300 YS; 40 min El; 130-190 Brin. Resistant to sulphuric acid. DIN X5CrNiMoCuNb1818; VDEh Spec 4505. *Obsolete*

MAERKER IRRUBIGO 188
Schmidt & Clemens Edelstahlwerke
C 0.1, Cr 18, Ni 9, bal Fe.
Quenched: 71,000-100,000 TS; 31,300 YS; 50 min El; 130-180 Brin. Basic type of corrosion resistant austenitic stainless steel; general use. DIN X12CrNi188; VDEh Spec 4300; AISI 302 *Obsolete*

MAERKER IRRUBIGO 188 EL
Schmidt & Clemens Edelstahlwerke
C 0.02, Cr 18.5, Ni 11.5, bal Fe.
Quenched: 64,000-100,000 TS; 25,600 YS; 50 min El; 130-180 Brin. Welding grade austenitic stainless; not necessary to heat treat after welding. DIN X2CrNi189; VDEh Spec 4306; AISI 304L. *Obsolete*

MAERKER IRRUBIGO 188 S
Schmidt & Clemens Edelstahlwerke
C 0.08, Cr 18, Ni 10.5, Ti = 5 x %C, bal Fe.
Quenched: 71,000-107,000 TS; 29,800 YS; 40 min El; 130-190 Brin. Welding grade austenitic stainless; also used for parts operating at elevated temperatures. DIN X10CrNiTi189; VDEh Spec 4541; AISI 321. *Obsolete*

MAERKER IRRUBIGO 188 SS
Schmidt & Clemens Edelstahlwerke
C 0.05, Cr 18.5, Ni 11.5, bal Fe.
Quenched: 71,000-100,000 TS; 27,000 YS; 50 min El; 130-180 Brin. Improved quality austenitic stainless; weldable without following heat treatment. DIN X5CrNi189; VDEh Spec 4301; AISI 304. *Obsolete*

MAERKER IRRUBIGO 188A
Schmidt & Clemens Edelstahlwerke
C 0.12, Cr 18, Ni 9, S 0.15, bal Fe.
Quenched: 71,000-100,000 TS; 31,300 YS; 50 min El; 130-180 Brin. Good machinability; for stainless parts requiring much machining. DIN X12CrNiS188; VDEh Spec 4305; AISI 303. *Obsolete*

MAERKER IRRUBIGO 188N
Schmidt & Clemens Edelstahlwerke
C 0.08, Cr 18, Ni 10.5, Nb = 8 x %C, bal Fe.
Quenched: 71,000-107,000 TS; 29,800 YS; 40 min El; 130-190 Brin. Welding grade austenitic stainless; also used for parts operating at elevated temperatures. DIN X10CrNiNb189; VDEh Spec 4550; AISI 347. *Obsolete*

MAERKER IRRUBIGO 199
Schmidt & Clemens Edelstahlwerke
C 0.1, Cr 18, Mo 2.25, Ni 10, bal Fe.
Quenched: 71,000-107,000 TS; 31,300 YS; 40 min El; 130-190 Brin. Excellent corrosion resistance; resistant to sulphites, phosphoric acid, organic acids, etc. DIN X12CrNiMo1810. *Obsolete*

MAERKER IRRUBIGO 199 EL
Schmidt & Clemens Edelstahlwerke
C 0.02, Cr 17.5, Mo 2.25, Ni 12, bal Fe.
Quenched: 64,000-100,000 TS; 28,400 YS; 45 min El; 130-180 Brin. Excellent corrosion resistance; resistant to sulphites, phosphoric acid, organic acids, etc.; weldable without subsequent heat treat. DIN X2CrNiMo1810; VDEh Spec 4404; AISI 316 L. *Obsolete*

MAERKER IRRUBIGO 199 N
Schmidt & Clemens Edelstahlwerke
C 0.08, Cr 17.5, Mo 2.25, Ni 12, Nb = 8 x C, bal Fe.
Quenched: 71,000-107,000 TS; 32,700 YS; 40 min El; 130-190 Brin. Excellent corrosion resistance; resistant to sulphites, phosphoric acid, organic acids, etc.; weldable without subsequent heat treat. DIN X10CrNiMoNb1810; VDEh Spec 4580. *Obsolete*

MAERKER IRRUBIGO 199 S
Schmidt & Clemens Edelstahlwerke
C 0.08, Cr 17.5, Mo 2.25, Ni 12, Ti = 5 x C, bal Fe.
Quenched: 71,000-107,000 TS; 32,700 YS; 40 min El; 130-190 Brin. Excellent corrosion resistance; resistant to sulphites, phosphoric acid, organic acids, etc.; weldable without subsequent heat treat. DIN 10CrNiMoTi1810; VDEh Spec 4571. *Obsolete*

MAERKER IRRUBIGO 199 SS
Schmidt & Clemens Edelstahlwerke
C 0.05, Cr 17.6, Mo 2.25, Ni 12, bal Fe.
Quenched: 71,000-100,000 TS; 29,900 YS; 45 min El; 130-180 Brin. Excellent corrosion resistance; resistant to sulphites, phosphoric acid, organic acids, etc.; weldable without subsequent heat treat. DIN 5CrNiMo1810; VDEh Spec 4401; AISI 316. *Obsolete*

MAERKER IRRUBIGO 1S
Schmidt & Clemens Edelstahlwerke
C 0.13, Cr 13, Mo 0.35, Ni 1.6, bal Fe.
Oil or air hardenable; corrosion resistant. Heat treated: 114,000-135,000 TS; 14 min El; 225-275 Brin. For highly stressed machine parts subject to corrosion by water and steam. DIN X13CrNiMo13. *Obsolete*

MAERKER IRRUBIGO 2
Schmidt & Clemens Edelstahlwerke
C 0.2, Cr 13, bal Fe.
Air or oil hardenable. Heat treated: 114,000-135,000 TS; 14 min El; 225-275 Brin. For higher stressed parts subject to corrosion by water and steam. DIN X20Cr13; AISI 420; VDEh Spec 4021. *Obsolete*

MAERKER IRRUBIGO 2 MO
Schmidt & Clemens Edelstahlwerke
C 0.2, Cr 13, Mo 1.15, Ni 0-1, bal Fe.
Air or oil hardenable; corrosion resistant. Heat treated: 107,000-128,000 TS; 14 min El; 220-260 Brin. Yield strength at 750 F: 51,000 psi. For fittings, brackets, hardware for increased strength and oxidation resistance at elevated temperature. DIN X20CrMo13; VDEh Spec 4120. *Obsolete*

MAERKER IRRUBIGO 200 EL
Schmidt & Clemens Edelstahlwerke
C 0.02, Cr 17.5, Mo 2.75, Ni 13.5, bal Fe.
Quenched: 64,000-100,000 TS; 28,400 YS; 45 min El; 130-180 Brin. Increased resistance to sulphites and organic acids; weldable without subsequent heat treat. DIN X2NiCrMo1812; VDEh Spec 4435; AISI 316L. *Obsolete*

MAERKER IRRUBIGO 200 N
Schmidt & Clemens Edelstahlwerke
C 0.08, Cr 17.5, Mo 2.75, Ni 13.5, Nb = 8 x C, bal Fe.
Quenched: 71,000-107,000 TS; 32,700 YS; 40 min El; 140-190 Brin. Increased resistance to sulphites and organic acids, weldable without subsequent heat treat. DIN X10CrNiMoNb1812; VDEh Spec 4583; AISI 316 Cb. *Obsolete*

MAERKER IRRUBIGO 200 SS
Schmidt & Clemens Edelstahlwerke
C 0.05, Cr 17.5, Mo 2.75, Ni 13.5, bal Fe.
Quenched: 71,000-100,000 TS; 29,800 YS; 45 min El; 140-180 Brin. Increased resistance to sulphites and organic acids; weldable without subsequent heat treat. DIN X5CrNiMo1812; VDEh Spec 4436; AISI 316. *Obsolete*

MAERKER IRRUBIGO 25
Schmidt & Clemens Edelstahlwerke
C 0.05, Cr 25, Mo 2.25, Ni 25, Nb = 8 x %C, bal Fe.
Quenched: 71,000-107,000 TS; 35,500 YS; 35 min El; 130-190 Brin. Good resistance to non-oxidizing acids; for use in the manufacture of viscose products, coking plants, chemical industry. DIN X5CrNiMoNb2525; VDEh Spec 4578. *Obsolete*

MAERKER IRRUBIGO 3
Schmidt & Clemens Edelstahlwerke
C 0.45, Cr 13, bal Fe.
Air or oil hardenable to about C 54 Rock. Annealed: 92,000-114,000 TS; 180-225 Brin. For corrosion resistant cutting appliances; parts subject to wear and tear, shears, cutlery. DIN X40Cr13; VDEh Spec. 4034. *Obsolete*

MAERKER IRRUBIGO 42
Schmidt & Clemens Edelstahlwerke
C 0.06, Cr 20, Mo 5, Ni 42, Cu 2.25, Nb = 10 x %C, bal Fe.
Quenched: 78,000-114,000 TS; 28,400 YS; 30 min El; 140-200 Brin. Resistant to sulphuric acid, phosphoric acid, organic acids and their salt solutions. DIN X5CrNiMoCuNb2042. *Obsolete*

MAERKER IRRUBIGO 6M
Schmidt & Clemens Edelstahlwerke
C 0.05, Cr 25, Mo 1.55, Ni 7, Nb = 10 x C, bal Fe.
Quenched: 92,000-114,000 TS; 25 min El; 190-230 Brin. For machine parts highly stressed, but which must resist rust and corrosion. DIN X5CrNiMo257; VDEh Spec 4582. *Obsolete*

MAERKER IRRUBIGO EXTRA M
Schmidt & Clemens Edelstahlwerke
C 0.38, Cr 16.5, Mo 1.15, Ni 0-1, bal Fe.
Air or oil hardenable; corrosion resistant. Heat treated: 114,000-135,000 TS; 14 min El; 225-275 Brin. Yield strength at 750 F: 56,000 psi. For shafts, valves, spindles, fittings resistant to corrosion by sea water, organic acids, halogen salts; good at elevated temperature. DIN X35CrMo17; VDEh Spec 4122. *Obsolete*

MAERKER IRRUBIGO MC 18 W
Schmidt & Clemens Edelstahlwerke
C 0.9, Cr 18, Mo 1.15, V 0.1, bal Fe.
Air or oil hardenable; corrosion resistant. Annealed: 128,000 max TS; 260 max Brin. For cutting appliances and parts with a high resistance to wear and tear. DIN X90CrMoV18; VDEh Spec 4112; similar to AISI 440B. *Obsolete*

MAESTRO
J.F. Jelenko & Co.
Precious metal. Au 3, Pd 30, Ag 50.
Quenched: 88,000 psi TS; 62,000 psi YS; 20 El; 150 Brin. Hardened: 125,000 psi TS; 105,000 psi YS; 4.5 El; 230 Brin. Type IV extra-hard dental alloy. For hard inlays, thin crowns, fixed bridgework, and partial dentures.

MAFERITE 14
Usines de A. Manoir, Pitres (Eure)
Cr 14, C, bal Fe.
200 Brin. For oil refinery equipment; scale resistant to 850°C.

MAFERITE 20
Usines de A. Manoir, Pitres (Eure)
Cr 20, C, bal Fe.
230 Brin. For oil refinery equipment, furnace parts; scale resistant to 1000°C.

MAFERITE 25
Usines de A. Manoir, Pitres (Eure)
Cr 25, C, bal Fe.
230 Brin. For furnace parts; scale resistant to 1050°C.

MAFERITE 28N
Usines de A. Manoir, Pitres (Eure)
Cr 28, Ni 4, C, bal Fe.
For furnace parts; scale resistant to 1150°C.

MAFERITE 30

Usines de A. Manoir, Pitres (Eure)
Cr 30, C, bal Fe.
250 Brin. For furnace parts; scale resistant to 1150°C.

MAFERITE 30P

Usines de A. Manoir, Pitres (Eure)
Cr 30, C, Mo, bal Fe.
260 Brin. For furnace parts; scale resistant to 1150°C.

MAFERITE 6

Usines de A. Manoir, Pitres (Eure)
C 0.1, Cr 6, bal Fe.
Rolled: 200 Brin. For oil refinery equipment; scale resistant to 750°C.

MAG 1

Sterling International Technology Ltd.
Al 7.5-9, Zn 0.3-1, Mn 0.15-0.4, bal Mg.
Sand cast: 140 MPa TS; 85 MPa YS; 2 El; 55 Brin. General purpose cast magnesium alloy.

MAG 2

Sterling International Technology Ltd.
Al 7.5-9, Zn 0.3-1, Mn 0.15-0.7, bal Mg.
Sand cast: 140 MPa TS; 85 MPa YS; 2 El; 55 Brin. Special purpose cast magnesium alloy. High intrinsic corrosion resistance.

MAG 3

Sterling International Technology Ltd.
Al 9-10.5, Zn 0.3-1, Mn 0.15-0.4, bal Mg.
Sand cast: 125 MPa TS; 95 MPa YS; 55 Brin. General purpose cast magnesium alloy. For intricate castings.

MAG 4

Sterling International Technology Ltd.
Zn 3.5-5.5, Mn 0.15, Zr 0.4-1, bal Mg.
Sand cast, stress relieved: 230 MPa TS; 145 MPa YS; 7 El; 65 Brin. General purpose cast magnesium alloy. High strength for use up to 150°C.

MAG 5

Sterling International Technology Ltd.
Zn 3.5-5, Mn 0.15, Zr 0.4-1, 0.75-1.75 rare earth metals, bal Mg.
Sand cast, precipitation treated: 200 MPa TS; 135 MPa YS; 3 El, 65 Brin. Special purpose cast magnesium alloy. For thin, narrow sectioned castings.

MAG 6

Sterling International Technology Ltd.
Zn 0.8-3, Mn 0.15, Zr 0.4-1, 2.5-4.0 rare earth metals, bal Mg.
Sand cast, precipitation treated: 155 MPa TS; 110 MPa YS; 3 El; 50 Brin. Special purpose cast magnesium alloy. Good creep resistance up to 250°C.

MAG 7

Sterling International Technology Ltd.
Al 7.5-9.5, Zn 0.3-1.5, Mn 0.15-0.8, Cu 0.35, Si 0.4, bal Mg.
Sand cast, solution treated, aged: 185 MPa TS; 110 MPa YS; 70 Brin. General purpose cast magnesium alloy.

MAG 8

Sterling International Technology Ltd.
Zn 1.7-2.5, Mn 0.15, Zr 0.4-1, Th 2.5-4, 0.10 rare earth metals, bal Mg.
Sand cast, precipitation treated: 185 MPa TS; 85 MPa YS; 5 El; 50 Brin. Special purpose cast magnesium alloy. Good creep resistance up to 350°C.

MAG 9

Sterling International Technology Ltd.
Zn 5-6, Mn 0.15, Zr 0.4-1, Th 1.5-2.3, 0.20 rare earth metals, bal Mg.
Sand cast, precipitation treated: 255 MPa TS; 155 MPa YS; 5 El; 65 Brin. Special purpose cast magnesium alloy. For heavy duty structural use.

MAG METAL

J.L. Snowbar
Cu, Ni.
Rolled: 55 Brin. For jewelry; platinum substitute.

MAGALLOY PX-4

Bohn Aluminium & Brass Corp.
Al 2, Mn 0.2, Cd 2, Cu 4, bal Mg.
21,000 TS; 13,000 YS; 3 El; 45 Brin. For pistons, permanent mold castings; high thermal conductivity. *Obsolete*

MAGALLOY RX-1

Bohn Aluminium & Brass Corp.
Al 2.5-3.5, Zn 0.7-1.3, bal Mg.
Extruded: 34,000 TS; 20,000 YS; 10 El; 50 Brin. For extrusions. *Obsolete*

MAGALLOY RX-13

Bohn Aluminium & Brass Corp.
Al 7.8-9.2, Zn 0.2-0.8, 0.12% Mn min, bal Mg.
Forged: 42,000 TS; 24,000 YS; 5 El; 75 Brin. Aged: 42,000 TS; 28,000 YS; 2 El; 80 Brin. For highly stressed parts; heat treatable. *Obsolete*

MAGALLOY RX-2

Bohn Aluminium & Brass Corp.
Al 3-4, Sn 4-6, bal Mg.
Forged: 36,000 TS; 22,000 YS; 7 El; 52 Brin. For forgings. *Obsolete*

MAGALLOY RX-3

Bohn Aluminium & Brass Corp.
Mn 1.2-2, bal Mg.
Extruded: 39,000-45,000 TS; 25,000-29,000 YS; 4-9 El; 39-46 Brin. For extruded parts; resists sea water. *Obsolete*

MAGALLOY RX-9

Bohn Aluminium & Brass Corp.
Al 6-7, Zn 0.75, Mn 0.15-0.2, bal Mg.
Wrought: 45,000-48,000 TS; 30,000-34,000 YS; 11-13 El; 55-60 Brin. For light alloy parts, screw machine products. *Obsolete*

MAGALLOY X-10

Bohn Aluminium & Brass Corp.
Al 9-11, Mn 0.1, bal Mg.
Cast: 18,000 TS; 10,000 YS; 1 El; 52 Brin. Heat treated: 30,000 TS; 17,000 YS; 1 El; 65 Brin. For leak proof castings; general casting alloy. *Obsolete*

MAGALLOY X-11

Bohn Aluminium & Brass Corp.
Al 11.5-12.5, Mn 0.1, bal Mg.
Cast: 20,000 TS; 16,000 YS; 1 El; 70 Brin. Heat treated: 32,000 TS; 22,000 YS; 0.5 El; 90 Brin. For sand and permanent mold castings; age-hardenable. *Obsolete*

MAGALLOY X-12

Bohn Aluminium & Brass Corp.
Al 8.3-9.7, Zn 1.7-2.3, 0.1% Mn min, bal Mg.
Cast: 20,000 TS; 10,000 YS; 1 El; 60 Brin. T4-temper: 30,000 TS; 10,000 YS; 6 El; 59 Brin. T6-temper: 32,000 TS; 17,000 YS; 1 El; 77 Brin. For high pressure parts; sand castings, heat treatable. *Obsolete*

MAGALLOY X-5

Bohn Aluminium & Brass Corp.
Al 7.8-9.2, Mn 0.15-0.2, bal Mg.
Cast: 28,000 TS; 13,000 YS; 7 El; 50 Brin. Heat treated: 36,000 TS; 13,000 YS; 8 El; 50 Brin. For light alloy castings; high impact strength. *Obsolete*

MAGALLOY X-6

Bohn Aluminium & Brass Corp.
Mn 1.2-2, bal Mg.
Sand cast: 13,000 TS; 4,500 YS; 5 El; 35 Brin. Die cast: 26,000 TS; 10,000 YS; 7 El; 37 Brin. For gasoline tank fittings, high pressure die castings; corrosion resistant. *Obsolete*

MAGALLOY X-7

Bohn Aluminium & Brass Corp.
Al 5.3-6.7, Zn 2.5-3, Mn 0.15-0.2, bal Mg.
Cast: 28,000 TS; 12,000 YS; 7 El; 50 Brin. Heat treated: 37,000 TS; 19,000 YS; 5 El; 71 Brin. For aircraft castings; age-hardenable. *Obsolete*

MAGALLOY X-8

Bohn Aluminium & Brass Corp.
Ce 2.4-4.5, Zr 0.1-0.4, bal Mg.
Cast: 29,000 TS; 11,000 YS; 8 El; 50 Brin. For light alloy sand castings; high temperature applications. *Obsolete*

MAGALLOY Z-104

Bohn Aluminium & Brass Corp.
Mg 3.5-4.5, bal Al.
22,000-25,000 TS; 12,000 YS; 6-9 El; 46-54 Brin. For sand castings, casings; corrosion resistant. *Obsolete*

MAGALLOY Z-107

Bohn Aluminium & Brass Corp.
Mg 6.5-7.5, bal Al.
25,000-27,000 TS; 15,000 YS; 4-6 El; 55-65 Brin. For sand castings, casings; corrosion resistant. *Obsolete*

MAGALLOY Z-110

Bohn Aluminium & Brass Corp.
Mg 9-11, bal Al.
Heat treated: 42,000-45,000 TS; 22,000-25,000 YS; 12-14 El; 68-80 Brin. For sand castings, gasoline instruments, meters; heat treatable. *Obsolete*

MAGALOY

Aetna Standard Engineering Co.
C 3.2, Si 2, Mg 0.5, Mn 0.7, bal Fe.
For rolls; ductile cast iron.

MAGALUMA

Aluminium Belge, S.A.
Mg 3.4, Mn 0.15, bal Al.
Soft: 29,000-39,000 TS; 10,000-14,500 YS; 15-20 El; 50-60 Brin. Hard: 42,500-50,000 TS; 36,000-42,500 YS; 2-6 El; 85-100 Brin. For light alloy parts; heat treatable, corrosion resistant.

MAGAN-NICKEL ALNI

Vereinigte Deutsche Nickel-Werke AG
Al 4, 94% min Ni.
For spark plug electrodes; corrosion and heat resistant. *Obsolete*

MAGAN-NICKEL NIC 1.5

Vereinigte Deutsche Nickel-Werke AG
Mn 1.5-2, 97.5% min Ni.
For spark plug electrodes; corrosion and heat resistant. *Obsolete*

MAGAN-NICKEL NIC 4

Vereinigte Deutsche Nickel-Werke AG
Mn 4, 95% min Ni.
For spark plug electrodes; corrosion and heat resistant. *Obsolete*

MAGAN-NICKEL NIC 5

Vereinigte Deutsche Nickel-Werke AG
Mn 5, 94% min Ni.
For spark plug electrodes; corrosion and heat resistant. *Obsolete*

MAGAN-NICKEL ZKNI

Vereinigte Deutsche Nickel-Werke AG
Mn 2, Cu 2, 95% min Ni.
For spark plug electrodes; corrosion and heat resistant. *Obsolete*

MAGDAL

Aluminium Francais
Mg 1, Mn 1.25, bal Al.
Soft: 21,000 TS; 16 El. Hard rolled: 34,000 TS; 8 El. For light alloy parts; wrought.

MAGEX A

Pyramid Steel Company
C, alloy, bal Fe.
Heat treated: 155,000 TS. For pump shafts, gears, drive shafts; preheat treated. *Obsolete*

MAGEX B

Pyramid Steel Company
C, alloy, bal Fe.
For cams, chucks, reamers; oil hardened. *Obsolete*

MAGIC

Jessop Steel Co.
C 0.5, Si 2, Mn 0.9, Mo 1, bal Fe.
Annealed: 207 Brin. For chisels, track tools, punches, cement breakers; Type S5; shock resistant. *Obsolete*

MAGNA METAL

Manufacturer not listed.
Mg alloy.
For light alloy parts; same as "Electron."

MAGNADUR-3

English manufacture
$BaFe_{12}O_{19}$.
For permanent magnets in metering devices, magnetic and electrical equipment; high permeability.

MAGNADURE

N.American Phillips Co./Ferroxcube Div.
$BaO + Fe_2O_3$.
For TV ring magnets; permanent magnets.

MAGNALITE

Walker M. Levett
Aluminum. Al 94.2, Cu 2.5, Ni 1.5, Zn 0.5, Mg 1.3.
26,000 TS; 2.5 El. For airplane construction and engine pistons; casting alloy.

MAGNALIUM

English manufacture
Cu 0-2.5, Mg 1-5.5, Ni 0-1.2, Al 85-95, Sn 0-3, Si 0.2-0.6, Fe 0-0.9, Mn 0-0.3.
41,800 TS; 34 El; 29 RA. 14,900 TS; 31 El; 69 RA. For light castings, ornamental, commercial shapes; very brittle; melting point 1110-1280°F.

MAGNALIUM (CAST X)

English manufacture
Al 95, Cu 1.8, Mg 1.6, Ni 1.2.
For light alloy castings.

MAGNALIUM (CAST Y)

English manufacture
Al 97, Cu 1.8, Mg 1.5, Sn + Pb.
For light alloy castings.

MAGNALIUM (CAST Z)

English manufacture
Al 95, Sn 3.2, Mg 1.6, Cu 0.2, Pb 0.7.
For light alloy castings.

MAGNALIUM (CAST)

English manufacture
Al 85, Mg 15.
For light alloy castings.

MAGNALIUM (ORIGINAL)

English manufacture
Al 70-95, Mg 5-30.
For light alloy castings.

MAGNALIUM (SHEET)

English manufacture
Al 95, Mg 5.
For pistons.

MAGNALIUM ALLOY NO 1

English manufacture
Al 95.5, Mg 1.75, Cu 1.75, Ni 1.
For light alloy parts; cast alloy.

MAGNALIUM ALLOY NO 2

English manufacture
Al 95.5, Mg 1.75, Cu 1.75, Pb 1.
For light alloy parts; cast alloy.

MAGNALLOY 1151

Magnacast Corp.
Cu 85, Sn 5, Pb 5, Zn 5.
Cast: 45,000 TS; 21,400 YS; 28 El; 72 Brin. Leaded red brass. ASTM B145 (4A); B271 (4A); SAE 40; CDA 836; AMS 4855.

MAGNALLOY 1152

Magnacast Corp.
Cu 85, Sn 5, Pb 5, Zn 5.
Cast: 55,000 TS; 24,200 YS; 13.2 El.

MAGNALLOY 1232

Magnacast Corp.
Cu 80, Sn 3, Pb 7, Zn 9, Ni 1.
Cast: 34,000 TS; 18,000 YS; 22 El; 62 Brin. Leaded semi-red brass. ASTM B145 (5A); B271 (5A); CDA 844.

MAGNALLOY 2051

Magnacast Corp.
Cu 89, Sn 11.
Cast: 51,500 TS; 29,000 YS; 18 El; 95 Brin. Tin bronze; for bearings, bushings. SAE 65; CDA 907.

MAGNALLOY 2101

Magnacast Corp.
Cu 88, Sn 10, Zn 2.
Cast: 51,500 TS; 29,000 YS; 18 El; 92 Brin. Tin bronze, formerly called gun metal. SAE 62; ASTM B22(D), B143(1A). AMS 4845; CDA 905.

MAGNALLOY 2251

Magnacast Corp.
Cu 88, Sn 8, Zn 4.
Cast: 49,000 TS; 23,000 YS; 18 El; 77 Brin. ASTM B143(1B); SAE 620; CDA 903.

MAGNALLOY 2451

Magnacast Corp.
Cu 88, Sn 6, Pb 1, Zn 4, Ni 1.
Cast: 45,500 TS; 23,000 YS; 35 El; 76 Brin. ASTM B143(2A), B271(2A); SAE 622, CDA 922.

MAGNALLOY 3051

Magnacast Corp.
Cu 80, Sn 10, Pb 10.
Cast: 41,000 TS; 24,000 YS; 10 El; 80 Brin. High leaded tin bronze. ASTM B144(3A), B271(3A); SAE 64. AMS 4842; CDA 937.

MAGNALLOY 3111

Magnacast Corp.
Cu 83, Sn 8, Pb 8, Ni 1.
Cast: 37,400 TS; 22,000 YS; 19.8 El. High leaded tin bronze. Fed. Std. 00153:E8; QQ-B-1005(8); CDA 934.

MAGNALLOY 3151

Magnacast Corp.
Cu 83, Sn 7, Pb 7, Zn 3.
Cast: 44,000 TS; 27,000 YS; 16 El; 72 Brin. ASTM B144(3B); SAE 660; CDA 932.

MAGNALLOY 3191

Magnacast Corp.
Cu 78, Sn 6, Pb 16.
Cast: 34,200 TS; 23,000 YS; 12 El; 62 Brin. High leaded tin bronze. SAE 67; CDA 941 (Similar specs).

MAGNALLOY 3221

Magnacast Corp.
Cu 70, Sn 5, Pb 25.
Cast: 23,100 TS; 13,200 YS; 7.7 El. High leaded tin bronze. ASTM B144(3E); CDA 943.

MAGNALLOY 3242

Magnacast Corp.
Cu 68, Sn 2, Pb 30.
Cast: 19,800 TS; 11,000 YS; 5.5 El.

MAGNALLOY 3261

Magnacast Corp.
Cu 84, Sn 5, Pb 9, Zn 9.
Cast: 38,000 TS; 21,000 YS; 20 El; 66 Brin.

MAGNALLOY 3271

Magnacast Corp.
Cu 75, Sn 5, Pb 19.
Cast: 28,700 TS; 22,800 YS; 8.0 El; 57 Brin. High leaded tin bronze. SAE 67; CDA 941.

MAGNALLOY 4151

Magnacast Corp.
Cu 88, Fe 3, Al 9.
Cast: 71,500 TS; 27,500 YS; 22 El. Aluminum bronze. ASTM B148(9A); SAE 68A; CDA 952.

MAGNALLOY 4152

Magnacast Corp.
Cu 88, Fe 1, Al 11.
Cast: 71,500 TS; 27,500 YS; 22 El. Aluminum bronze. ASTM B148(9B); SAE 68B; CDA 953.

MAGNALLOY 4153

Magnacast Corp.
Cu 86, Fe 4, Al 10.
Cast: 102,700 TS; 45,800 YS; 18.8 El; 190 Brin. Aluminum bronze. ASTM B148(9C); CDA 954 (Similar).

MAGNALLOY 4211

Magnacast Corp.
Cu 58, Sn 1, Zn 38, Fe 1, Al 1, Mn 1.
Cast: 71,500 TS; 27,500 YS; 22 El. ASTM B147(8C); AMS 4860; CDA 865.

MAGNALLOY 4231

Magnacast Corp.
Cu 64, Zn 26, Fe 2, Al 4, Mn 4.
Cast: 99,000 TS; 49,500 YS; 22 El. Manganese bronze. ASTM B147(8B); CDA 862.

MAGNALLOY 4241

Magnacast Corp.
Cu 64, Zn 24, Fe 2, Al 6, Mn 4.
Cast: 121,000 TS; 66,000 YS; 13.0 El. Manganese-aluminum bronze. ASTM B147 8 B or 8 C; CDA 862, 863. AMS 4862.

MAGNALLOY 4401

Magnacast Corp.
Cu 88, Sn 5, Zn 2, Ni 5.
Cast: 80,000 TS; 55,000 YS; 10 El. Nickel-tin bronze. ASTM B292-A; CDA 947.

MAGNE 1

Uddeholm Corp.
C 0.9, Cr 3.25, bal Fe.
For permanent magnets. *Obsolete*

MAGNE 2

Uddeholm Corp.
C 0.9, Cr 6.25, bal Fe.
For permanent magnets. *Obsolete*

MAGNEL

Capitol Castings Inc.
C 3, Si 1.5, bal Fe.
Annealed: 52,000 TS; 33,000 YS; 13 El; 120 Brin. For electrical and electromagnetic instruments, magnet cores; high permeability, low hysteresis loss. *Obsolete*

MAGNEQUENCH MQ1-A

Magnequench
Iron.
Iron-neodymium-boron alloy. Annealed magnet alloy
powder; energy product 8.75 MGOe max.

MAGNEQUENCH MQ1-B

Magnequench
Iron.
Iron-neodymium-boron alloy. Annealed magnet alloy
powder; energy product 9.5 MGOe max.

MAGNEQUENCH MQ2

Magnequench
Iron.
Iron-neodymium-boron alloy. Isotropic magnet alloy powder;
energy product 13.7 MGOe max.

MAGNEQUENCH MQ3

Magnequench
Iron.
Iron-neodymium-boron alloy. Anisotropic magnet alloy
powder; energy product 28-45 MGOe max.

MAGNEQUENCH MQA

Magnequench
Iron.
Iron-neodymium-boron alloy. Anisotropic powder; energy
product 20-40 MGOe max.

MAGNEQUENCH MQP-A

Magnequench
Iron.
Iron-neodymium-boron alloy. Isotropic magnet alloy powder;
energy product 12 MGOe max.

MAGNEQUENCH MQP-B

Magnequench
Iron.
Iron-neodymium-boron alloy. Isotropic magnet alloy powder;
energy product 11 MGOe max.

MAGNEQUENCH MQP-C

Magnequench
Iron.
Iron-neodymium-boron alloy. Isotropic magnet alloy powder;
energy product 11 MGOe max.

MAGNEQUENCH MQP-D

Magnequench
Iron.
Iron-neodymium-boron alloy. Isotropic magnet alloy powder;
energy product > 12.5 MGOe max.

MAGNESIL

Spang Specialty Metals
Si 3.25, bal Fe.
Alloy strip, soft magnetically. Often used for magnetic
shielding. *Obsolete*

MAGNESIL N

Spang Specialty Metals
Si 3.25, bal Fe.
Insulated soft magnetic iron strip; non-oriented. *Obsolete*

MAGNESIL-O

Spang Specialty Metals
Si 3.25, bal Fe.
Insulated soft magnetic strip; with pronounced directional
magnetic properties in the direction of rolling. *Obsolete*

MAGNESIUM

Atomergic Chemetals Corp.
Mg.
Purities: special distilled 99.99%, 99.9+% (nuclear grade),
99.5% (commercial). Forms: ingot, rod, powder, wire, sheet,
foil, granule, single crystals.

MAGNESIUM -ZE63A

Magnesium Elektron Ltd.
Magnesium. Zn 5.5-6, 0.4 Zr min, 2.0-3.0 rare earths, bal Mg.
T6 temper: 42,000-45,000 TS; 29,000-32,000 YS; 5-10 El. At
257°F: 30,000 TS; 19,000 YS; 34 El. At 347°F: 21,000 TS;
16,000 YS; 32 El. For helicopter structural and transmission
castings, gearboxes, aircraft landing wheels. Castings,
operating up to 300°F. High fatigue strength.

MAGNESIUM A10

English manufacture
Al 10, Mn 0.1, bal Mg.
Cast: 12,000 TS; 12,000 YS; 2 El; 53 Brin. T4-temper: 40,000
TS; 13,000 YS; 10 El; 52 Brin. T6-temper: 40,000 TS; 19,000
YS; 1 El; 69 Brin. For high-strength castings; age-hardened.

MAGNESIUM ALLOY NO. 23 (6% AL)

Sifbronze
Magnesium.
Now SIF MAGNESIUM NO. 23.

MAGNESIUM AZ31X

Dominion Magnesium Ltd.
Al 2.5-3.5, Zn 0.6-1.4, 0.20 Mn min, bal Mg.
Rolled: 32,000-40,000 TS; 16,000-29,000 YS; 4-12 El. For
light alloy parts.

MAGNESIUM AZ51

English manufacture
Al 4.8, Zn 0.8, Mn 0.25, bal Mg.
Soft: 40,000 TS; 21,000 YS; 19 El; 57 Brin. Hard: 45,000 TS;
34,000 YS; 10 El; 71 Brin. For light alloy structures; sheet and
plate.

MAGNESIUM AZ53

Dow Chemical Co.
Mg alloy.
For light alloy parts. *Obsolete*

MAGNESIUM AZ80

English manufacture
Al 8.5, Zn 0.5, Mn 0.15, bal Mg.
Forged: 46,000 TS; 31,000 YS; 8 El; 69 Brin. Aged: 50,000
TS; 34,000 YS; 6 El; 72 Brin. For high-strength forgings,
aircraft parts; age-hardened.

MAGNESIUM AZ81A

Dow Chemical Co.
Magnesium. Al 7-8.1, Zn 0.4-1, 0.13 Mn min, bal Mg.
F-temper: 28,000 psi TS; 13,000 psi YS; 6 El. T4-temper:
40,000 psi TS; 30,000 psi YS; 12 El; 55 Brin. For aircraft parts,
structural members; heat treatable. *Obsolete*

MAGNESIUM AZ916

English manufacture
Mg alloy.
T6-temper: 19,000 TS. For aircraft and general purpose
castings; age-hardenable.

MAGNESIUM AZ91A

Dow Chemical Co.
Magnesium. Al 0, Zn 0.7, Mn 0.2, bal Mg.
Die cast: 33,000 psi TS; 22,000 psi YS; 3 El; 60 Brin. For
instrument housings and portable tools; general die castings.
Obsolete

MAGNESIUM AZ91A

Dow Chemical Co.
Al 9, Zn 0.5, bal Mg.
Sand cast: 24,000 TS; 14,000 YS; 2 El; 52 Brin. T4-temper:
40,000 TS; 14,000 YS; 11 El; 53 Brin. T6-temper: 40,000 TS;
19,000 YS; 4 El; 66 Brin. For portable tools, instrument
housings; formerly Dowmetal R; age-hardenable. *Obsolete*

MAGNESIUM AZ91A

Dow Chemical Co.
Al 9, Zn 0.7, Mn 0.2, bal Mg.
Die cast: 33,000 TS; 22,000 YS; 3 El; 60 Brin. For instrument
housings, portable tools, general die castings. *Obsolete*

MAGNESIUM AZ91C

Dow Chemical Co.
Cu 8.7, Zn 0.7, Mn 0.15, bal Mg.
F-temper: 24,000 TS; 14,000 YS; 2 El; 52 Brin. T4-temper:
40,000 TS; 14,000 YS; 11 El; 53 Brin. T6-temper: 40,000 TS;
19,000 YS; 4 El; 66 Brin. For aircraft engine parts, brake
castings, pressure tight castings; age-hardenable. *Obsolete*

MAGNESIUM AZ92

English manufacture
Al 9, Zn 2, Mn 0.1, bal Mg.
Cast: 25,000 TS; 14,000 YS; 2 El; 65 Brin. T5-temper: 25,000
TS; 17,000 YS; 1 El; 69 Brin. T4-temper: 40,000 TS; 14,000
YS; 10 El; 63 Brin. T6-temper: 40,000 TS; 22,000 YS; 3 El; 81
Brin. For high-strength casting; age-hardened.

MAGNESIUM EK-30

Dow Chemical Co.
Zn 0.2, 3% misch-metal, bal Mg.
At 70 F: 22,900 TS; 16,500 YS; 2.4 El. At 400 F: 20,000 TS;
14,000 YS; 13.1 El. At 600 F: 11,800 TS; 7,800 YS; 69.6 El.
For high temperature applications, aircraft components; heat
treatable. *Obsolete*

MAGNESIUM EK30A

Dow Chemical Co.
Zn 0-0.3, 2.5-4.0% rare earths, 0.2% min Zr, bal Mg.
T6-temper: 23,000 TS; 16,000 YS; 3 El. For aircraft engine
components; age hardenable, used up to 500 F. *Obsolete*

MAGNESIUM EK31A

Dow Chemical Co.
Zr 0.55, 3.0% misch-metal, bal Mg.
T5-temper: 22,000 TS; 17,000 YS; 2 El. T6-temper: 25,000 TS;
19,000 YS; 4 El. For aircraft engine components; age
hardenable, high temperature service. *Obsolete*

MAGNESIUM EK31D

Dow Chemical Co.
Zr 0.55, Di 2, bal Mg.
T6-temper: 38,000 TS; 21,000 YS; 7 El. For aircraft engine
components; age hardenable, high temperature service.
Obsolete

MAGNESIUM EK41A

Dow Chemical Co.
Zr 0.4-1, Zn 0-0.3, 3-5% rare earth metals, bal Mg.
T5-temper: 23,000 TS; 16,000 YS; 2 El. T6-temper: 25,000 TS;
20,000 YS; 3 El. For aircraft engine components, missiles,
high temperature parts. Age hardenable. Useful to 500 F.
Good creep resistance. *Obsolete*

MAGNESIUM EM22

German manufacture
Mn 1.5-2, 2-6 rare earths, bal Mg.
For aircraft engine components; high temp. use.

MAGNESIUM EM42

German manufacture
Mn 1.5-2, 2-6 rare earths, bal Mg.
For aircraft engine components; high temp. use.

MAGNESIUM EX31A

Dow Chemical Co.
Zr 0.4-1, 2.5-4.0% rare earths; bal Mg.
T6-temper: 35,000 TS; 22,000 YS; 5 El; 64 Brin. At 600 F:
15,000 TS; 13,000 YS. For aircraft and missile components
up to 550 F; airframe and engine components for 1000 h at
450 F. Sand castings. Age hardenable. *Obsolete*

MAGNESIUM HM11A

Dow Chemical Co.
Th 1, Mn 1, bal Mg.
At 70 F: 30,200 TS. At 400 F: 16,800 TS. At 800 F: 7,800 TS.
For aircraft and missile components; die castings for high
temperature use. *Obsolete*

MAGNESIUM INGOT (PRIMARY), GRADE MG-1

Dow Chemical Co.

Magnesium. Mn 0-0.1, Fe 0-0.05, Cu 0-0.02, Ni 0-0.001, Pb 0-0.01, Sn 0-0.01, Ca 0-0.0015, Na 0-0.003, 99.8 Mg min, 0.005 others max.

For remelting. ASTM B92. Gr. 9980 A.

MAGNESIUM INGOT (PRIMARY), GRADE MG-2

Dow Chemical Co.

Magnesium. Mn 0-0.01, Fe 0-0.05, Cu 0-0.02, Ni 0-0.001, Pb 0-0.01, Sn 0-0.01, Ca 0-0.0015, Na 0-0.003, 99.9 Mg min, 0.05 others max.

For remelting.

MAGNESIUM INGOT (PRIMARY), GRADE MG-4

Dow Chemical Co.

Magnesium. Mn 0-0.006, Fe 0-0.04, Al 0-0.003, Si 0-0.005, Ca 0-0.0015, Ni 0-0.001, Pb 0-0.005, Sn 0-0.005, Cu 0-0.004, Na 0-0.003, 99.9 Mg min, 0.00007 B max, 0.01 others max.

For remelting.

MAGNESIUM K1A

Dow Chemical Co.

Magnesium. Zr 0.7, bal Mg.

Cast: 26,000 psi TS; 80,000 psi YS; 19 El. Cast: 26,000 psi TS; 80,000 psi YS; 19 El. For light alloy parts, missiles; fine grained. *Obsolete*

MAGNESIUM KIXI

Dow Chemical Co.

Zr 0.4-1, bal Mg.

For missile and aircraft components; good damping capacity, weldable. *Obsolete*

MAGNESIUM LA141

Brooks & Perkins Inc.

Li 14.1, Al 1.5, bal Mg.

Rolled: 21,000 TS; 17,000 YS; 25-30 El; 58 Brin. For satellite and spacecraft construction; light weight. *Obsolete*

MAGNESIUM LA141A

Brooks & Perkins Inc.

Li 13-15, Al 0.75-1.25, Mn 0-0.05, Na 0-0.003, Fe 0-0.005, bal Mg.

At room temperature: 21,000 TS; 18,000 YS; 28 El. Forged: 31,500 TS; 30,000 YS; 12 El; 25 RA. For space vehicles, satellites, pressure vessels, bulkheads, micrometeorite shielding. Good formability and weldability. *Obsolete*

MAGNESIUM LA142

Brooks & Perkins Inc.

Li 13-15, Al 1-2, Mn 0.08-0.15, bal Mg.

Rolled sheet: 28,800 TS; 27,000 YS; 18 El. At 200 F: 15,300 TS; 13,800 YS; 16 El. For armoured ordnance vehicles. Good formability and weldability. *Obsolete*

MAGNESIUM LA91

Brooks & Perkins Inc.

Li 9, Al 1, bal Mg.

Rolled: 22,000 TS; 16,500 YS; 30-35 El; 52 Brin. For satellite and spacecraft construction; lightweight. *Obsolete*

MAGNESIUM LA91A

Brooks & Perkins Inc.

Li 9, Al 1, bal Mg.

Sheet: 22,000 TS; 16,500 YS; 30-35 El. Forged: 23,500 TS; 20,800 YS; 39 El; 62 RA. For space craft construction, pressure vessels. Good formability and weldability. *Obsolete*

MAGNESIUM LAZ933A

Brooks & Perkins Inc.

Li 9, Al 3, Zn 3, bal Mg.

Sheet: 30,000 TS; 21,000 YS; 30 El; E 79 Rock. For aerospace equipment operating at cryogenic temperatures. Good low temperature properties. *Obsolete*

MAGNESIUM M-1

English manufacture

Mn 1.5, bal Mg.

Soft: 33,000 TS; 18,000 YS; 17 El. Hard: 35,000 TS; 26,000 YS; 7 El. Extruded: 38,000 TS; 28,000 YS; 10 El. For structural members, aircraft parts; corrosion resistant.

MAGNESIUM MSR

Manufacturer not listed.

Ag 2.5, Zr 0.7, 2 rare earths, bal Mg.

Heat treated: 40,000 TS; 30,000 YS; 4 El; 70-85 Brin. For high temperature applications, missile components; good creep strength, heat treatable.

MAGNESIUM MSR-A

Magnesium Elektron Ltd.

Magnesium. Ag 2.5, Zr 0.7, 1.7 rare earths, bal Mg.

Heat treated: 78,400 TS; 50,400 YS; 4 El; 70-90 Brin. For high strength castings for high temperature use; age-hardenable, cast alloy.

MAGNESIUM MSR-B

Magnesium Elektron Ltd.

Magnesium. Ag 2.5, Zr 0.7, 2.5 rare earths, bal Mg.

Heat treated: 78,400 TS; 54,900 YS; 2 El; 70-90 Brin. For high strength castings for high temperature use; age-hardenable, cast alloy.

MAGNESIUM MTZ

English manufacture

Th 3, Zr 0.7, bal Mg.

Heat treated: 65,000 TS; 24,600 YS; 5 El; 50-60 Brin. For aircraft parts; high creep resistance.

MAGNESIUM ZE10A

Dow Chemical Co.

Zn 1-1.5, Ce 0.2, bal Mg.

H24-temper: 38,000 TS; 28,000 YS; 12 El. For missile shipping and storage containers; requires no stress relief after welding. *Obsolete*

MAGNESIUM ZK60A

Aluminum Company of America

Zn 4.8-6.2, 0.45 Zr min, bal Mg.

F-temper: 49,000 TS; 38,000 YS; 12 El. TS-temper: 51,000 TS; 42,000 YS; 10 El. T6-temper: 53,000 TS; 48,000 YS. For brake housings, fuel meter bodies, bulkheads. Heat treatable; extrusions; high strength and tough.

MAGNESIUM ZK60B

Dow Chemical Co.

Zn 4.6-6.8, 0.45% min Zr, bal Mg.

T5-temper: 45,000 TS; 35,000 YS; 4 El. For aircraft loading ramps; heat treatable, extrusions. *Obsolete*

MAGNESIUM ZK61

Dominion Magnesium Ltd.

Zn 6, Zr 0.8, bal Mg.

Cast: 21,700 TS; 10,900 YS; 12 El. Heat treated: 228,000 TS; 14,100 YS; 8 El. Aged: 255,000 TS; 16,400 YS; 10 El. For engine components; age hardenable.

MAGNESIUM ZK62

English manufacture

Zn 6, Zr 0.5, 2 Misch Metal, bal Mg.

Extruded: 50,000 TS; 45,000 YS; 12 El. For aircraft components.

MAGNESIUM ZM41

Dow Chemical Co.

Zn 4, Mn, 0.7% rare earths, bal Mg.

For light alloy parts; sheets. *Obsolete*

MAGNESIUM-AZ61A

Magnesium Elektron Ltd.

Magnesium. Al 6.5, Zn 1, Mn 0.2, bal Mg.

Tube: 40,000 TS; 22,000 YS; 15 El; 55 Brin. Bar: 45,000 TS; 32,000 YS; 15 El; 56 Brin. For light alloy structures, extrusion and forgings.

MAGNESIUM-AZ63A

Magnesium Elektron Ltd.

Magnesium. Al 6, Zn 3, Mn 0.2, bal Mg.

Cast: 29,000 TS; 14,000 YS; 6 El; 50 Brin. T6-Temper: 40,000 TS; 19,000 YS; 5 El; 73 Brin. For airplane wheels and brakes, oil pumps, crankcases. Age hardenable casting alloy.

MAGNESIUM-I14

Manufacturer not listed.

Li 7, Zr 1, Th 3, Zn 6, Cd 5, Ag 6, bal Mg.

Rolled: 45,000 min TS; 35,000 min YS; 15 min El. At -452°F: 8 min El. For cryogenic applications. Good low temperature properties to -423°F.

MAGNESIUM-IA6

Manufacturer not listed.

Li 9, Th 3, Zn 2, Al 4, Ag 4, Mn 1, bal Mg.

Heat treated: 45,000 min TS; 35,000 min YS; 15 min El. For cryogenic applications. Good low temperature properties to -423°F.

MAGNESIUM-ZE63B

Magnesium Elektron Ltd.

Magnesium. Zn 5.5-6, Ag 0.75-1.25, 0.4 Zr min, 2.0-3.0 rare earths, bal Mg.

T6 temper: 43,000-46,000 TS; 30,000-34,000 YS; 4-8 El. For helicopter structural and transmission castings, gearboxes, aircraft landing wheels. Cast alloy. High fatigue strength.

MAGNESIUM-ZK61A

Magnesium Elektron Ltd.

Magnesium. Zn 6, Zr 0.6, bal Mg.

T6 temper: 45,000 TS; 28,000 YS; 8 El; 75 Brin. T5 temper: 40,000 TS; 26,000 YS; 8 El; 65 Brin. For ordnance vehicles, missiles, aircraft components. Heat treatable castings.

MAGNESIUM-ZLH972

Manufacturer not listed.

Li 7, Zn 9, Th 2, bal Mg.

For cryogenic applications. Good low temperature properties to -423°F.

MAGNESIUM-ZTY

Magnesium Elektron Ltd.

Magnesium. Zn 0.5, Th 0.75, Zr 0.6, bal Mg.

Forging: 43,000 TS; 26,000 YS; 18 El. Sheet: 38,000 TS; 24,000 YS; 12 El. For compressor cases, valve covers, pistons. Good creep and thermal properties. Hot formable, weldable.

MAGNET C

Thyssen Edelstahlwerke AG

C 1, Cr 3.3, bal Fe.

For magnet steel. *Obsolete*

MAGNET CM

Thyssen Edelstahlwerke AG

C 0.95, Mn 1.1, Cr 5, bal Fe.

For permanent magnets. *Obsolete*

MAGNET W

Thyssen Edelstahlwerke AG

C, W, bal Fe.

For magnet steel. *Obsolete*

MAGNET WH

Thyssen Edelstahlwerke AG

C, alloy, bal Fe.

For magnets; high permeability. *Obsolete*

MAGNETHERM

Pechiney/SOFREM

Mg 99.8-99.85, Other Mg alloys.

For sand and shell casting.

MAGNETICS ROUND PERMALLOY 80

Spang Specialty Metals

Now PERMALLOY 80 (1).

MAGNETOFLEX 20
Vacuum Metals Corp.
Fe 20, Ni 20, Cu 60.
Wire: 100,000-120,000 TS; 200 Brin. For magnets in electrical and electronic equipment; age-hardenable, permanent magnet; same as Cunife.

MAGNETOFLEX 35
Vacuumschmelze GmbH
Fe, Co, V.
Deformable permanent magnet alloy for rings of hysteresis motors.

MAGNETOFLEX 40
Vacuumschmelze GmbH
Fe, Co, V.
Deformable permanent magnet alloy for rings of hysteresis motors.

MAGNETOFLEX-12
German manufacture
Cu 68, Ni 20, bal Co.
For permanent magnets in electrical and magnetic equipment; age hardenable, high permeability.

MAGNEWIN
Wintershall, A.G.
Al, bal Mg.
For light alloy parts.

MAGNEWIN 3515
German manufacture
Al 7.3, Zn 0.75, Mn 0.12, bal Mg.
For light alloy parts; heat treatable.

MAGNICO
American manufacture
Co 12, Cu 6, Al 10, Ni 18, bal Fe.
For permanent magnets.

MAGNICO
Russian manufacture
Fe 50, Co 24, Ni 14, Al 9, Cu 3.
For electrical and magnetic equipment; magnet.

MAGNIKO
Russian manufacture
Co 25.3, Ni 14.6, Cu 2.8, Al 8.5, bal Fe.
For electrical and magnetic equipment; permanent magnet, high magnetic permeability.

MAGNIL
American Silver Co.
C 0.1, Mn 15.5, Ni 0-0.75, Cr 18, bal Fe.
Cold rolled: 219,400 TS; 187,400 YS; 6 El. Heat treated. 296,000 TS; 268,000 YS. For electronics, computers, instrument and control industries, high modulus diaphragms, bellows, springs, wear plates. Austenitic, nonmagnetic, stainless.

MAGNO
Midland-Ross Corp.
C 0.05, bal Fe.
Annealed: 60,000 TS; 22 El. For magnet cores; high permeability, low hysteresis loss. *Obsolete*

MAGNO
Capitol Castings Inc.
C 0.05, bal Fe.
Annealed: 60,000 TS; 22 El. For magnet cores; high permeability, low hysteresis loss. *Obsolete*

MAGNO (ELALCO)
English manufacture
Ni 95, Mn 5.
Hard drawn: 140,000 TS. Soft: 65,000 TS; 22 El. For electrical resistance wire, electromagnetic uses.

MAGNO-522
Forjas Alavesas S.A.
C 0.95, Mn 1.1, Si 0-0.4, Cr 0.5, W 0.5, V 0.15, bal Fe.
Oil hardening tool steel for dies, gages. IHA F-522; DIN 105 MnCr4; AISI O1.

MAGNO-90
Forjas Alavesas S.A.
C 0.9, Mn 1.9, Si 0-0.4, Cr 0.4, V 0.15, bal Fe.
Cold work tool steel; for small tools as reamers, punches, hand stamps. DIN 90Mn8; AFNOR 80M8; UNI U85MV8.

MAGNO-NICKEL
International Nickel Inc.
Mn 2-6, bal Ni.
50,900-69,600 TS; 28,400-34,100 YS; 39-50 El. For tools. *Obsolete*

MAGNOLIA 120 BRONZE
Magnolia Metal Corp.
Cu 75, Pb 20, Sn 5.
Cast: 25,500-28,000 TS; 12,000-18,000 YS; 9-10 El; 55-67 Brin. For bearings for light loads; resists many acid solutions.

MAGNOLIA AA BRONZE
Magnolia Metal Corp.
Cu 88, Sn 10, Zn 2.
Cast: 34,000-40,000 TS; 26,000-30,000 YS; 8 El; 85 Brin. For bearings for underate loads; resists brine solutions.

MAGNOLIA ANTIFRICTION METAL
Magnolia Metal Corp.
Sb 10-15, Sn 2-7, bal Pb.
Cast: 18,000 TS; 8400 YS; 23 Brin. For high-speed and heavy-pressure bearings; antifriction metal; 825-900°F.

MAGNOLIA BEARING BRONZE
Magnolia Metal Corp.
Cu 80, Sn 10, Pb 10.
Cast: 26,000 TS; 12,000 YS; 8 El; 60 Brin. For bearings, bushings; high speed. *Obsolete*

MAGNOLIA CADMIUM NICKEL
Magnolia Anti-Friction Metal Co.
Cd, bal Ni.
For bearings, liners, high loads.

MAGNOLIA CONTINUOUS CAST BRONZE
Magnolia Metal Corp.
Cu 80, Sn 6-8, Pb 11-13, Zn 0-1.
As cast: 44,000 TS; 26,000 YS; 17 El; 80 Brin. For bearings, bushings.

MAGNOLIA ISOTROPIC BRONZE
Magnolia Metal Corp.
Sn 10, Pb 10, Cu 80.
Cast: 31,250 TS; 26,000 YS; 8.5 El; 70 Brin. For die cast bronze bearings; die cast.

MAGNOLIA METAL
Magnolia Metal Corp.
Pb 70-84, Sb 15-16, Fe 0.03, Sn 0-7.
Cast: 15,000 TS; 3400 YS; 21 Brin. For antifriction bearing metal; for bearings subjected to moderate loads.

MAGNOLIA MODIFIED SAE 64 BRONZE
Magnolia Metal Corp.
Cu 78-82, Sn 6-9, Pb 11-13, Zn 0-1.
31,250 psi TS; 20,000 psi YS; 8.5 El; 70 Brin. Good machinability bearing bronze with improved bearing characteristics.

MAGNOWIN
Wintershall, A.G.
Mg alloy.
For light alloy parts.

MAGNOX ZA
Birmetals Ltd.
Zr 0.6, bal Mg.
For light alloy parts; good creep ductility. *Obsolete*

MAGNOX-A12
Birmetals Ltd.
Al 0.7-0.9, Be 0.002-0.03, Ca 0-0.008, Fe 0-0.006, bal Mg.
For nuclear reactors; low creep ductility at 200 C. *Obsolete*

MAGNOX-C
Magnesium Elektron Ltd.
Magnesium. Al 0.7-0.9, Be 0.005, bal Mg.
12,800 TS; 8 El (at 212°F); 3000 TS; 53 El (at 575°F); 1000 TS; 88 El (at 750°F). For light alloy parts, structural members.

MAGNUMINIUM 127
High Duty Alloys Ltd.
Al 9, Zn 1, Mn 0.2, bal Mg.
Sand cast: 20,200 TS; 9000 YS; 2 El. Permanent mold: 26,900 TS; 9000 YS; 4 El. For aircraft parts, crankcases, pressure tight castings and housings; heat treatable. *Obsolete*

MAGNUMINIUM 133
Magnesium Castings & Products Ltd.
Al 0.2, Zn 0.2, Mn 2.5, Si 0.4, Cu 0.2, bal Mg.
30,000 TS; 16,000 YS; 10 El; 50 Brin. For light alloy parts, fuel tanks, structures; weldable.

MAGNUMINIUM 155
Magnesium Castings & Products Ltd.
Al 0-9, Zn 0-1.5, Mn 0-1, Cu 0-0.3, bal Mg.
40,000 TS; 20,000 YS; 10 El; 55 Brin. For light alloy parts; non-hardenable.

MAGNUMINIUM 166
Magnesium Castings & Products Ltd.
Al 5-11, Zn 0-1.5, Mn 0-1, bal Mg.
40,000-45,000 TS; 16,000-28,000 YS; 10-14 El; 50-60 Brin. For light alloy forgings and stampings; heat treatable.

MAGNUMINIUM 177
Magnesium Castings & Products Ltd.
Al 0-8.5, Zn 3.5, bal Mg.
Sand cast: 20,000 TS; 9000 YS; 2-8 El; 50 Brin. For die castings; MP 610°C.

MAGNUMINIUM 181
Magnesium Castings & Products Ltd.
Al 0-8.5, Zn 0-3.5, Mn 0-0.5, bal Mg.
Cast: 23,000 TS; 10,000 YS; 8 El; 45 Brin. Heat treated: 32,000 TS; 10,000 YS; 12 El; 55 Brin. For light castings; heat treatable.

MAGNUMINIUM 199
Magnesium Castings & Products Ltd.
Al 8.5, Zn 3.5, Si 0.4, Mn 0.4, bal Mg.
Cast: 25,000-30,000 TS; 4-8 El; 45-55 Brin. For sand and die light alloy castings; heat treatable.

MAGNUMINIUM 220
Magnesium Castings & Products Ltd.
Al 9-11, Zn 0-3.5, Mn 0-0.5, up to 1.5 impurities, bal Mg.
Sand cast: 24,000 TS; 12,000 YS; 2-4 El; 50-60 Brin. Die cast: 26,000 TS; 2-5 El; 50-60 Brin. For sand and die light alloy castings; casting alloy, heat treatable.

MAGNUMINIUM 266
Magnesium Castings & Products Ltd.
Al 0-11, Zn 0-1.5, Mn 0-1, bal Mg.
45,000-50,000 TS; 20,000-30,000 YS; 10-14 El; 50-60 Brin. For light alloy extruded bars and shapes; for structures.

MAGNUMINIUM 288 A
Magnesium Castings & Products Ltd.
Al 0-11, Zn 0-2, Mn 0-1, bal Mg.
35,000-48,000 TS; 20,000-33,000 YS; 5-12 El; 80-90 Brin. For aircraft parts; heat treatable.

MAGNUMINIUM 299
Magnesium Castings & Products Ltd.
Al 2-6, Sn 3-10, Ag 0.25-4, bal Mg.
Cast: 18,000-26,000 TS; 10,000-12,000 YS; 4-5 El; 40-50 Brin. For light castings, meter parts, heat treatable.

MAGNUMINIUM 299
Birmingham Aluminium Casting Co.
Al 2.5-6, Sn 4.5-7.5, Ag 0.2-2, Mn 0.1-0.4, bal Mg.
Cast: 22,000 TS; 10,000 YS; 5-9 El; 45 Brin. Aged: 31,400 TS; 10,000 YS; 4-14 El; 50 Brin. For airplane wheels, motor and instrument housings; pressure tight castings.

MAGNUMINIUM 299
Magnesium Castings & Products Ltd.
Al 2.5-6, Sn 4.5-7.5, Ag 0.2-2, Mn 0.1-0.4, bal Mg.
Cast: 22,000 TS; 10,000 YS; 5-9 El; 45 Brin. Aged: 31,400 TS; 10,000 YS; 4-14 El; 50 Brin. For airplane wheels, motor and instrument housings; pressure tight castings.

MAGOTTEAUX-UM CO50
Magotteaux SA
C 0.05-0.12, Ti 0.18, Cb 0.6, Cr 26-30, Co 47-52, bal Fe.
Cast: 135,000 TS; 48,000 YS; 10 El; 10 RA; 250 Brin. Wrought: 132,000 TS; 61,000 YS; 7 El; 6 RA; 350 Brin. For furnace baffles, burner tips, sintering grates, quench baskets. Corrosion, heat, thermal shock resistant. *Obsolete*

MAHLE 124
Mahle GmbH
Si 11-13, Cu 0.8-1.5, Ni 1.3, Mg 0.8-1.3, Fe 0.7, bal Al.
Cast: 28,000-35,000 psi TS; 90-125 Brin. Forged: 41,000-51,000 psi TS; 90-125 Brin. For pistons, cylinders.

MAHLE 136
Mahle GmbH
Si, Cu, Ni, bal Al.
For light alloy parts, engine components; high temperature use. *Obsolete*

MAHLE 138
Mahle GmbH
Si 17-19, Cu 0.8-1.5, Ni 1.3, Mg 0.8-1.3, Fe 0.7, bal Fe.
Cast: 25,000-31,000 psi TS; 90-125 Brin. Forged: 31,000-41,000 psi TS; 90-125 Brin. For pistons, cylinders.

MAHLE 244
Mahle GmbH
Si 23-26, Cu 0.8-1.5, Ni 1.3, Mg 0.8-1.3, Fe 0.7, Cr 0.6, bal Al.
Cast: 24,000-29,000 psi TS; 90-125 Brin. For pistons, cylinders.

MAHLE Y (1)
Mahle GmbH
Cu 3.5-4.5, Ni 1.75-2.25, Mg 1.25-1.75, Fe 0.6, Si 0.5, bal Al.
Cast: 31,000-38,000 psi TS; 95-125 Brin. Forged: 48,000-58,000 psi TS; 95-125 Brin. For pistons, cylinders.

MAHLE Y (2)
Mahle GmbH
Cu 3.8-4.2, Ni 1.7-2.2, Mg 1.5, Si 0.3, Fe 0.7, bal Al.
For pistons, cylinder heads; high temperature use. *Obsolete*

MAILLECHORT
Manufacturer not listed.
Cu 65-67, Zn 13.5, Ni 13-19, Fe 0.5-3.2, Sn + Pb.
For ornamental white metal parts, nickel silver.

MAILLECHORT, GERMAN
German manufacture
Cu 65.4, Zn 13.4, Ni 16.8, Fe 3.4.
For hardware, fittings; corrosion resistant.

MAILLECHORT, PARIS
French manufacture
Cu 66.24, Zn 13.42, Ni 16.42, Fe 3.2.
For hardware, fittings; corrosion resistant.

MAILLECHORT, VIENNA
Austrian manufacture
Cu 66.6, Zn 13.6, Ni 19.3, Fe 0.48.
For hardware, fittings; corrosion resistant.

MAIN METAL
Now INTAL N 89.

MAINTENAL
Pyramid Steel Company
C, alloy, bal Fe.
Heat treated: 164,000 TS; 141,000 YS; 17 El; 57 RA; 280 Brin. For wrenches, drills, gears, pump shafts, bolts; preheat treated, fatigue resistant.

MAJOR
A. Milne & Co.
C 0.7, W 21, Cr 4, V 1.5, Co 13, Mo 0.5, bal Fe.
For tools, cutters; high speed steel.

MAJOR METAL
Manufacturer not listed.
Cu 3, Fe 2, Zn 0.4, Ni 0.4, bal Al.
For light alloy parts; hardenable.

MAJORITY
J.M. Ney Co.
Au 50, Cu 31.3, Ag 10.
Nonporcelain gold alloy for inlays, partial and full coverage crowns, pontics, and multiple unit restorations. Age hardened: 93,000 psi TS; 71,000 psi YS (0.1% offset); 7 El; 220 HV.

MAL COLLOY
Japanese manufacture
Al 15, Ni 19.8, bal Co.
Coercive force 1500; Residual induction 3200; Max energy product 1,450,000. For electrical and magnet equipment and instruments; permanent magnet after heat treatment.

MAL-ARC
CMW Inc.
Cr-Mo-Co.
For hard facing welding rod; abrasion resistant. *Obsolete*

MAL-DIE
Allied Steel & Tractor Products Inc.
C 1, Cr 5, Mo 1.1, bal Fe.
For blanking and forming dies; air-hardening.

MALAX A
Allied Steel & Tractor Products Inc.
C 0.83, Mo 5, Cr 4, V 1.9, W 6.4, bal Fe.
For drills, taps, lathe and planer tools; oil hardening.

MALAX AA
Allied Steel & Tractor Products Inc.
C 0.75, Cr 4.25, W 19, V 1, bal Fe.
For tools, cutters, reamers; high speed steel.

MALAX AAA
Allied Steel & Tractor Products Inc.
C 0.75, Cr 4.25, W 19, V 2, Co 5, bal Fe.
For tools, cutters, lathe and planer tools; high speed steel.

MALAX AAAA
Allied Steel & Tractor Products Inc.
C 0.8, Cr 4.2, W 19, V 2, Mo 1, Co 7.5, bal Fe.
For cutters, forming tools; high speed steel.

MALCROME
Allied Steel & Tractor Products Inc.
C, Cr, Mo, W, bal Fe.
For tools, dies; oil hardening.

MALGA
Allied Steel & Tractor Products Inc.
C, Cr, Mo, W, bal Fe.
For tools, dies; oil hardening.

MALGA ELEKTRO SPECIAL
Allied Steel & Tractor Products Inc.
C, Cr, W, bal Fe.
For tools, dies; water or oil hardening.

MALGA MRO SPECIAL
Allied Steel & Tractor Products Inc.
C, Mo, W, Cr, bal Fe.
For tools, dies; wear and shock resistant.

MALGA SPECIAL NON-TEMPERING
Allied Steel & Tractor Products Inc.
C, W, Mo, Cr, bal Fe.
For tools, dies; no tempering after quenching.

MALGALOY
Allied Steel & Tractor Products Inc.
Si, Cr, Mo, V, W, C, bal Fe.
For shear blades; shock resistant.

MALLEABLE CAST IRON "BLACK HEART"
English manufacture
Mn 0.1-0.35, S 0.1, P 0.2, Si 0.5-1.2, C 1.5-2.5, bal Fe.
35,000-45,000 TS; 10-5 El. For railway and automotive castings; machines easily.

MALLEABLE CAST IRON "WHITE HEART"
English manufacture
Mn 0.1-0.4, S 0.25, P 0.1, Si 0.5-1.1, CC 0-0.4, TC 0-0.8, bal Fe.
40,000-55,000 TS; 5-3 El. For railway and automotive castings; machines easily.

MALLEABLE IRON 32510
Alloy Foundries
TC 2.5, Si 1, Mn 0.4, bal Fe.
Cast: 50,000 TS; 32,500 YS; 10 El; 115-135 Brin. For hardware, agriculture and auto parts; tough, shock resistant.

MALLEABLE IRON 35018
Alloy Foundries
TC 2.2, Si 1, Mn 0.4, bal Fe.
Cast: 53,000 TS; 35,000 YS; 18 El; 115-135 Brin. For hardware, agriculture and auto parts; tough, shock resistant.

MALLEABLE IRON GM 11M
General Motors Corp./Central Foundry Div
50,000 psi TS; 32,500 psi YS; 10 El; 156 Brin max. For less highly stressed parts; steering gear housing.

MALLEABLE IRON GM11M
Gimo-Osterby Bruks, A.B.
C 2.55, Si 1.4, Mn 0.45, S 0.12, P 0.05, bal Fe.
Malleable iron. 50,000 TS; 32,500 YS; 10 El; 156 Brin max. For lightly stressed parts; e.g., steering gear housings. SAE and ASTM Grade 32510.

MALLET ALLOY
Manufacturer not listed.
Zn 75, Cu 25.
For bearings.

MALLIX
Capitol Castings Inc.
C 1.8, Si 1, Mn 0.5, bal Fe.
Cast: 78,000 TS; 55,000 YS; 5 El; 180 Brin. For grate bars, elevator buckets; pearlitic, malleable iron. *Obsolete*

MALLORY 1000 GYROMET
CMW Inc.
Cu, Ni, bal W.
For gyroscopes, counter weights; sintered heavy metal. *Obsolete*

MALLORY 1000, ETC.
Now CMW 1000, ETC.

MALLORY 125
CMW Inc.
Co, Ni, Be, bal Cu.
Cold drawn: 110,000 TS; 12 El; 240 Brin. For spot and seam welding stainless steel; 48% conductivity. *Obsolete*

MALLORY 22
CMW Inc.
Cd, Zr, bal Cu.
Wrought: 78,000 TS; 72,000 YS; 10 El; 50 RA; 144 Brin. For resistance welding electrodes; high conductivity. *Obsolete*

MALLORY 2960
CMW Inc.
W 96.1, Ni 2.8, Cu 1.1.
Sintered: 110,000 min TS; 29 Rock C. Density: 18.3. For counterweights, gyroscopic rotors. High density, high strength. *Obsolete*

MALLORY 2985
CMW Inc.
W 98.5, Ni 1.12, Cu 0.38.
Sintered: 80,000 min TS. Density: 18.9. High density, high strength. *Obsolete*

MALLORY 333
CMW Inc.
Cr, Li, bal Cu.
Cast: 53,000 TS; 20 El. Wrought: 75,000 TS; 15 El. For welding electrodes, springs, collector rings; 85% electrical conductivity. *Obsolete*

MALLORY 333
CMW Inc.
Li, Cr, bal Cu.
Cast: 50,000 TS; 27,000 YS; 20 El; 125 Brin. Rolled: 75,000 TS; 35,000 YS; 15 El; 150 Brin. For welding dies, brush holders, terminal studs; high electrical conductivity and strength. *Obsolete*

MALLORY 53
CMW Inc.
Cu alloy.
Wrought: 100,000 TS; 10 El; 200 Brin. Cast: 70,000 TS; 200 Brin. For flash welding dies, bushings; fusion welding rod; 45% conductivity. *Obsolete*

MALLORY 53-Z
CMW Inc.
Ni, Zr, Si, bal Cu.
Cast: 60,000 TS; 8 El; 165 Brin. For electrode holders; high conductivity 40%. *Obsolete*

MALLORY 84
CMW Inc.
Cu alloy.
Wrought: 90,000 TS; 15 El; 180 Brin. For welding electrodes, dies, electrical contacts; 80% electrical conductivity. *Obsolete*

MALLORY ALLOYS
Now CMW ALLOYS.

MALLORY D-154X
CMW Inc.
Ag 86.7, 13.3% CdO.
Annealed: 29,000 TS; 48 Rock F. Cold worked: 84 Rock F. Electrical conductivity: 80%. For electrical contacts. *Obsolete*

MALLORY D-157
CMW Inc.
Fe 50, Cu 25, Ag 25.
Annealed: 84 Rock F; cold rolled: 94 Rock F. Electrical conductivity: 21%. For electrical contacts. *Obsolete*

MALLORY D-258F
CMW Inc.
Ag 98, C 2.
Electrical conductivity: 77%. For electrical contacts. *Obsolete*

MALLORY D-51
CMW Inc.
Ag-Ni.
For electric contacts; sintered; 43% conductive. *Obsolete*

MALLORY D-511
CMW Inc.
Ag 40, Ni 60.
Annealed: 42 Rock F; cold worked: 97 Rock F. Electrical conductivity: 25%. For electrical contacts. *Obsolete*

MALLORY D-52
CMW Inc.
Ag-CdO.
Sintered: 24,000 TS. For electrical contacts; 90% conductivity. *Obsolete*

MALLORY D-53
CMW Inc.
Ag-CdO.
Sintered: 21,000 TS. For electrical contacts; 85% conductivity. *Obsolete*

MALLORY D-54
CMW Inc.
Now CMW D54.

MALLORY D-558F
CMW Inc.
Ag 98.5, C 1.5.
Annealed: 22,000 TS; 33 Rock F. Cold worked: 33,500 TS; 66 Rock F. Electrical conductivity: 97%. For electrical contacts. *Obsolete*

MALLORY D-58F
CMW Inc.
Ag 95, C 5.
Cold worked: 40 Rock F. Electrical conductivity: 75%. For electrical contacts. *Obsolete*

MALLORY D-59
CMW Inc.
Ag-Ni-graphite.
For electrical contacts; electrical conductivity 70%. *Obsolete*

MALLORY D53X
CMW Inc.
Ag, CdO.
Annealed: 26,000 TS; 20 El; 30 Brin. For electrical contacts; 90% electrical conductivity. *Obsolete*

MALLORY D581
CMW Inc.
Ag, graphite.
Sintered: 1 El; 58 Brin. For electrical contacts; 90% electrical conductivity. *Obsolete*

MALLORY D582
CMW Inc.
Ag, graphite.
Drawn: 42,000 TS; 1 El. For electrical contacts; sintered, 98% electrical conductivity. *Obsolete*

MALLORY ELKON BRONZE
CMW Inc.
Cu alloy.
Cast: 55,000 TS; 37,500 YS; 10 El. For resistance welding electrodes; 40% electrical conductivity. *Obsolete*

MALLORY HA TUNGSTEN
CMW Inc.

W-alloy. Wrought: 120,000 TS. For electrical contacts; 32% electrical conductivity. *Obsolete*

MALLORY HB METAL
CMW Inc.
W-alloy.
For electrical contacts. *Obsolete*

MALLORY L-2748
Manufacturer not listed.
Al 5, Cr 5, bal Ti.
Rolled: 165,000 TS; 153,000 YS; 10 El; 71 Rock A. For high temperature applications; corrosion and heat resistant.

MALLORY L-2749
Manufacturer not listed.
Al, Cr, bal Ti.
Rolled: 90,000 TS; 75,000 YS; 18 El; 62 Rock A. For high temperature applications; corrosion and heat resistant.

MALLORY MANGANESE ALUMINUM BRONZE
CMW Inc.
Cu 63-68, Fe 3, Mn 2.5-5, Al 3-6, bal Zn.
Rolled: 60,000-125,000 TS; 30,000-70,000 YS; 8-30 El; 120-275 Brin. For propeller blades, worm gears, valve stems; corrosion resistant. *Obsolete*

MALLORY MANGANESE BRONZE
CMW Inc.
Cu 57-60, Sn 1, Mn 0.5, bal Zn.
Forged: 60,000-125,000 TS; 30,000-70,000 YS; 8-30 El; 120-275 Brin. For propeller blades, gears, valve stems; corrosion resistant. *Obsolete*

MALLORY MK-TUNGSTEN
CMW Inc.
Wrought: 300,000 TS. For magneto contacts; 32% conductivity. *Obsolete*

MALLOY
George Cook & Co., Ltd.
C 0.6, Mn 0.45, Si 1.1, Cr 1.1, Mo 0.25, bal Fe.
Shock resisting tool steel; for punches and dies, shear blades, collets.

MALLOYDIUM
Manufacturer not listed.
Cu 60, Ni 23, Zn 13, Fe 0.9.
For tableware; acid resisting.

MALTA GR
Jessop Steel Co.
WC, Co.
For cutting tools; sintered. *Obsolete*

MALTA JC GR. CR.
Jessop Steel Co.
WC, Co.
For tipped tools, cutters; carbides, sintered. *Obsolete*

MALTA JC GR. MF.
Jessop Steel Co.
WC, Co.
For tipped cutting tools; sintered carbide. *Obsolete*

MALTA JC GR. SF.
Jessop Steel Co.
WC, Co.
For tipped cutting tools; sintered carbide. *Obsolete*

MALTA SR
Jessop Steel Co.
WC, Co.
For cutting tools; sintered. *Obsolete*

MALTA SS
Jessop Steel Co.
WC, Co.
For cutting tools; sintered. *Obsolete*

MALUMINUM
Manufacturer not listed.
Al 87, Cu 6.4, Zn 4.8, Fe 1.4, Si 0.2, Mn 0.1, Pb 0.2.
21,000 TS; 1.5 El. For light alloy parts.

MAMMUT
Styria-Stahl Steirische Gusstahlwerke AG
C 0.7, W 18, Co 5, Cr 4, V 1.3, bal Fe.
For lathe and planer tools, drills, taps, reamers, hobs; high speed steel, oil hardened. *Obsolete*

MAMMUT
Vereinigte Edelstahlwerke
C 0.74, Cr 4.1, V 1.1, W 18.5, bal Fe.
For lathe and planer tools, drills, reamers, hobs; high speed steel. *Obsolete*

MAMMUT SPECIAL
Now VEW S200.

MAMMUT SPECIAL KN
Now VEW S205.

MAMMUT SPEZIAL XX
Vereinigte Edelstahlwerke
C 0.7, Cr 4, Mo, V, W, bal Fe.
For lathe and planer tools, broaches, reamers; high speed steel. *Obsolete*

MAN
Bohler Gesellschaft M.B.H.
C 0.3, Si 0.25, Mn 1.35, bal Fe.
For gears, shafts, machine tool parts; water hardened.

MAN-VAN
American manufacture
C 0.95, Mo 0.15, V 0.1, bal Fe.
For cutting tools, dies; water hardening.

MANAURITE 10
Usines de A. Manoir, Pitres (Eure)
C 0.2, Ni 10, Cr 22, bal Fe.
Annealed: 72,000 psi TS; 37,000 psi YS; 18 El. For chemical plant and oil refinery equipment; resists scaling to 1050°C, stainless.

MANAURITE 12
Usines de A. Manoir, Pitres (Eure)
Ni 12, Cr 25, C, bal Fe.
Annealed: 77,000 psi TS; 40,000 psi YS; 20 El. For furnace parts, heat treating equipment; resists scaling to 1080°C, heat resistant.

MANAURITE 12
Societe Nouvelle des Acieries de Pompey
C 0.3, Cr 25, Ni 12, bal Fe.
For furnace parts, heat treat boxes; austenitic, corrosion and heat resistant. *Obsolete*

MANAURITE 20
Usines de A. Manoir, Pitres (Eure)
Ni 20, Cr 25, C, bal Fe.
Annealed: 80,000 psi TS; 36,000 psi YS; 22 El. For furnace parts, heat treating equipment; resists scaling to 1150°C, heat resistant.

MANAURITE 35
Usines de A. Manoir, Pitres (Eure)
Ni 35, Cr 15, C, bal Fe.
Annealed: 72,000 psi TS; 39,000 psi YS; 22 El. For heat treating boxes, salt pots; resists scaling to 1200°C, heat resistant.

MANAURITE 50W
Societe Nouvelle des Acieries de Pompey
C 0.5, Ni 50, Cr 27, W 5.
Cast: 64,500 TS; 3.5 El; 2.7 RA. For gas generator retorts, hearth plates, furnace parts. High heat resistance to 2200 F. OMPOSITION al Fe. *Obsolete*

MANAURITE 60
Usines de A. Manoir, Pitres (Eure)
Ni 60, Cr 15, C, bal Fe.
Annealed: 64,000 psi TS; 39,000 psi YS; 22 El. For salt pots, heat treating equipment; resists scaling to 1200°C, heat resistant.

MANAURITE 8S
Usines de A. Manoir, Pitres (Eure)
C 0.2, Ni 8, Cr 18, bal Fe.
Annealed: 72,000 psi TS; 39,000 psi YS; 20 El. For chemical plant equipment; resists scaling to 900°C, stainless austenitic.

MANCRO
Uddeholm Corp.
C 0.15, Si 0.25, Mn 0.5, Cr 0.75, bal Fe.
For gears, shafts; case hardened. *Obsolete*

MANCRO 32
Uddeholm Corp.
C 0.15, Mn 1.2, Cr 1, bal Fe.
For gears, shafts; case hardened. *Obsolete*

MANCRO 4
Uddeholm Corp.
C 0.2, Cr 1.35, Mn 1.35, bal Fe.
For gears, shafts; case hardened. *Obsolete*

MANCRO 71
Uddeholm Corp.
C 0.34, Mn 1.05, Cr 1.05, bal Fe.
For gears, shafts; oil hardened. *Obsolete*

MANCRO 72
Uddeholm Corp.
C 0.4, Cr, Mn, Mo, bal Fe.
For gears, shafts, machine tool parts; water hardened. *Obsolete*

MANCRO 8
Uddeholm Corp.
C 0.38, Si 0.65, Mn 1.15, Cr 1.15, bal Fe.
For gears, shafts; tough. *Obsolete*

MANDUR
Aluminium-Zentral e.V.
Aluminum. Cu 3.5-5.5, Si 0.3-1, Mn 0.5-1, Mg 0.5-1.2, bal Al.
Heat treated: 78,000-85,000 TS; 64,000-71,000 YS; 10-15 El; 130-150 Brin. For oil pans, crankcases, housings, engine cylinder heads. Age hardenable. High strength. *Obsolete*

MANDURA
KM-kabelmetal AG
Copper. Mn 1, Si 3, bal Cu.
Components for chemical industry and pressure vessels. Product forms: strip, sheet, plate, finished parts. Cold rolled: 60-100 ksi TS; 30-70 ksi YS; 5-40 El.

MANELEC
Empire Sheet & Tin Plate Co.
Si 1, bal Fe.
For electrical equipment, motors; high permeability.

MANG-ROD NO. 250
Marquette Corp.
C 0.9-1.05, Cr 0.9-1.2, Si 0.4, Mn 14, bal Fe.
Welded: 250 Brin. For hard facing electrode; for Mn steels.

MANG-TRODE
Ampco Pittsburgh Corp.
Zn, Mn, bal Cu.
For welding electrode for Mn-bronze; arc welding. *Obsolete*

MANGA-KOTE
Grand Northern Products Ltd.
C 0.8, Si 0.8, Mn 14, Cr 4.7, Mo 3.24, Ni 3.5, B 0.1, bal Fe.
Coated hardfacing electrode. For overlay or buildup of bucket teeth, grading buckets, tractor rollers, crusher parts.

MANGA-TONE
Resisto-Loy Company, Inc.
C 1.7, Ni 10, Mn 30, V 2, bal Fe.
Welded: 70,000 TS; 500 Brin. For hard facing electrode; wear resistant. *Obsolete*

MANGA-TONE N.M.
Grand Northern Products Ltd.
C 2, Si 0.8, Mn 26.2, Ni 8.3, bal Fe.
Filler rod for repair or buildup of austenitic manganese steel parts as shovel teeth, railroad frogs, crossovers; pump impellers, tractor bottom rollers, crusher plates.

MANGABRAZE
Baldwin Steel Co.
C 0.35, Mn 1.92, Mo 0.32, Si 0.28, Cu 0.42, bal Fe.
Heat treated: 157,000 TS; 146,000 YS; 16 El; 58 RA; 360-385 Brin. For chutes, hoppers, conveyors, screens, liners, scrapers, buckets; wear and impact resistant, mill heat treated.

MANGAL
Vereinigte Leichtmetallwerke, G.m.b.H.
Aluminum. Mn 1.5, bal Al.
Annealed: 17,000 TS; 20 El. For light alloy parts; corrosion resistant.

MANGAL
VAW Vereinigte Aluminium-Werke AG
Aluminum. Mn 1.5, bal Al.
Soft: 14,000 TS; 20-30 El; 20-25 Brin. Hard: 36,000 TS; 28,000 YS; 2 El; 60 Brin. For commercial vehicles, roofing, structures. Nonhardening, corrosion resistant.

MANGALAL
Now ALCAN GB-3S.

MANGALOY
Bergstrom Alloys Corp.
C, Mn, Ni, Cr, bal Fe.
For hard surfacing electrodes; for manganese steel.

MANGALOY
English manufacture
Ni, Mn, bal Fe.
For electrical resistors; heat resistant.

MANGAN
Krupp Stahl AG
alloy steel.
See KRUPP 3401

MANGAN-NEUSILBER
German manufacture
Cu 59-73, Ni 10-18, Mn 2.4-20, Zn 5-20.
For white metal parts; corrosion resistant.

MANGANAL
Stulz Sickles Steel Co.
C 0.6-0.9, Mn 11-14, Ni 2.5-3.5, bal Fe.
140,000-155,000 TS; 55,000-60,000 YS; 55-72 El; 35-54 RA; 180-500 Brin. For welding rod, resurfacing broken and worn high Mn steel parts; abrasion resistant.

MANGANEND 1 M
Arcos Alloys
C, Mo, bal Fe.
Welded: 90,000-100,000 TS; 70,000-85,000 YS; 20-30 El; 40-70 RA. For welding electrodes; for steel, low H$_2$. *Obsolete*

MANGANEND 13
Arcos Alloys
Ni 12-14, C, Mn, bal Fe.
For welding electrodes. *Obsolete*

MANGANEND 13A
Arcos Alloys
C 0.85, Si 0.8, Mn 13.5, Mo 1, bal Fe.
Welded: 16 Rock C; work hardened: 48 Rock C. For hardfacing electrodes; austenitic 13.5% manganese steel. *Obsolete*

MANGANEND 2M
Arcos Alloys
C 0.1, Mn 1.6, Mo 0.35, bal Fe.
Welded: 125,000 TS; 105,000 YS; 10 El; 15 RA. For welding electrodes for steel; low H$_2$. *Obsolete*

MANGANESE

Atomergic Chemetals Corp.

Mn.

Purities: 99.99+%, 99.9%, dehydrogenated. Forms: flake, powder, foil, vacuum melted lump.

MANGANESE ANTIFRICTION

United American Metals Corp.

Mn, Sb, Sn, Pb.

5,670 psi TS. For machinery bearings; self-lubricating Babbitt metal. Meets C.S. 19450.

MANGANESE BORON

Foote Mineral Co.

B 20, Mn 80.

For use in manufacture of non-ferrous alloys. *Obsolete*

MANGANESE BRASS 510

Anaconda Co.

Copper. Zn 29, Mn 1, bal Cu.

Hard: 76,000 TS; 62,000 YS; 10 El; 160 Brin. Soft: 47,000 TS; 16,000 YS; 65 El; 60 Brin. For strip for resistance spot and seam welded products.

MANGANESE BRASS 667

Olin Brass

Cu 70, Zn 28.8, Mn 1.2.

Annealed: 47,000-67,000 TS; 13,000-37,000 YS; 30-60 El. Cold rolled: 51,000-105,000 TS; 2-40 El. Seam and spot welding brass; for communication equipment, welded assemblies. *Obsolete*

MANGANESE BRASS 73

Olin Brass, Indianapolis

Copper. Cu 70, Mn 1.25, Zn 28.75.

Annealed: 50,000 psi TS; 55 El. Drawn: 90,000 psi TS; 6 El. For resistance seam and spot welding.

MANGANESE BRONZE

American Manganese Bronze Co.

Copper. Cu 56-60, Fe 0.4-1.5, Al 0.5-1, Pb 0-0.4, bal Zn. Cast: 65,000 TS; 30,000 YS; 28 El; 25 RA; 100 Brin. For propellers, hubs, valves, pump bodies; corrosion resistant.

MANGANESE BRONZE

Manufacturer not listed.

Cu 82-86, Sn 6-17, Mn 0.2-2.7, Zn 0-5, Pb 0.3.

For gears, general castings; tough.

MANGANESE BRONZE (A) COPPER ALLOY NO. 67

Chase Brass & Copper Co., Inc.

Copper. Cu 58.5, Zn 39.25, Sn 1, Fe 1, Mn 0.25.

Half hard: 72,000 psi TS; 40,000 psi YS; 35 El; 70 Rock B. For balls, forgings, valve stems, welding rod. ASTM B124 Alloy 4; CDA 675.

MANGANESE BRONZE 19

Olin Brass, Indianapolis

Copper. Cu 58.5, Mn 0.3, Sn 0.7, Fe 1, bal Zn.

Hard: 83,000 psi TS; 55,000 psi YS; 25 El. Soft: 72,000 psi TS; 30,000 psi YS; 45 El. For bolts, valve parts, and tie rods; corrosion resistant.

MANGANESE BRONZE 937

Anaconda Co.

Copper. Cu 57-62, Zn 36-40, Sn 0.5-1.5, Fe 0.5-1, Mn 0.5.

Soft: 60,000 TS; 30,000 YS; 30 El; 30 RA. Hard: 80,000 TS; 45,000 YS; 20 El; 15 RA. For valve stem forgings, slotted and perforated screens. High strength and toughness; resists action of salt water.

MANGANESE BRONZE 984

Anaconda Co.

Copper. Zn 00, Sn 1, Fe 0.5, Mn 0.5, bal Cu.

For welding rod for steel, cast iron, copper and nickel alloys. 1598°F MP.

MANGANESE BRONZE A 675

Anaconda Co.

Copper. Cu 58.5, Zn 39.25, Sn 1, Mn 0.25, Fe 1.

Hard rod: 75,000 TS; 45,000 YS; 20 El; 85 Rock B. Soft rod: 65,000 TS; 30,000 YS; 30 El; 65 Rock B. For clutch discs, pump rods, shafts, balls, valve stems and bodies. Readily machined and welded.

MANGANESE BRONZE CAST

English manufacture

Cu 58, Mn 2, Zn 40.

For marine propellers; corrosion resistant.

MANGANESE BRONZE E-77

Accurate Brass Co.

Cu 57-60, Sn 0.5-1.5, Al 0-0.25, Mn 0-0.5, bal Zn.

Forged: 75,000-85,000 TS; 40,000-45,000 YS; 15-20 El; 120-142 Brin. For marine parts, hardware, propellers; corrosion resistant, high strength.

MANGANESE BRONZE ROLLED

English manufacture

Cu 59, Sn 1, Mn 0.3, Zn 31.

For marine parts, nuts, bolts; corrosion resistant.

MANGANESE CARBON ALLOY

LaSalle Steel Co.

C 0.35-0.45, Mn 1.2, bal Fe.

Heat treated: 125,000 TS; 105,000 YS; 16 El; 50 RA; 260-320 Brin. For drive shafts, bolts, studs, spindles, pins. *Obsolete*

MANGANESE CASTING BRASS

General Motors Corp./Central Foundry

Cu 58.5, Zn 40, Sn 0.85, Al 0.5, Pb 0-1.5, Mn 0.15.

70,000 psi TS; 20 El. A substitute for malleable iron; high strength.

MANGANESE COPPER "A"

English manufacture

Cu 29.2, Mn 51.65, Fe 9.68, Al 6.25, C 3.23.

For resistances, heat and corrosion resistant parts.

MANGANESE COPPER "B"

English manufacture

Cu 56.3, Mn 40.9, Fe 1.5, Si 1.1.

For resistances, heat and corrosion resistant parts.

MANGANESE COPPER "C"

English manufacture

Cu 75, Mn 25.

For resistances, heat and corrosion resistant parts.

MANGANESE COPPER "D"

English manufacture

Cu 75.3, Mn 22.4, Fe 2.15.

For resistances, heat and corrosion resistant parts.

MANGANESE COPPER "E"

English manufacture

Cu 85, Mn 10.92, Fe 1.83, Zn 2.

For resistances, heat and corrosion resistant parts.

MANGANESE COPPER "F"

English manufacture

Cu 89.7, Mn 8.72, Fe 1.54.

For resistances, heat resisting parts.

MANGANESE COPPER "G"

English manufacture

Cu 85.55, Mn 10.66, Fe 2.66, Sn 0.39, Pb 0.45.

For resistances; heat resistant.

MANGANESE COPPER "H"

English manufacture

Cu 84.33, Mn 10.61, Fe 2.31, Zn 2.1, Sn 0.4, Pb 0.3.

For resistances; heat resistant.

MANGANESE GRADE A

United States Steel Corp.

C 0.8, Mn 13, Si 0.5-0.8, Ni 3-3.5, bal Fe.

Work hardened: 450 Brin. Weldable; for severe impact, crushers, wear plates; nonmagnetic.

MANGANESE GRADE B

United States Steel Corp.

C 0.8, Mn 12, Si 0.2, Cr 0-0.5, bal Fe.

Work hardened: 500 Brin. For severe impact, crushers, wear plates.

MANGANESE GRADE C

United States Steel Corp.

C 0.8, Mn 13, Si 0.65, Ni 2, Mo 0.5, Cr 0-0.5, bal Fe.

Work hardened: 450 Brin. Weldable; for severe impact, hammers, crushers.

MANGANESE M1

Sight Feed Generator Co.

C, Mn, Ni, bal Fe.

Welded: 350 Brin. For arc welding electrodes, hard facing; forgeable, wear resistant.

MANGANESE METAL

Foote Mineral Co.

Mn 97, approximately 0.2 C.

Constituent of alloys. *Obsolete*

MANGANESE MOLYBDENUM STEEL

Lukens Steel

C 0-0.25, Mn 0-1.65, Mo 0-0.75, bal Fe.

Rolled: 95,000 TS; 60,000 YS; 25 El; 200 Brin. For shaft gears; tough. *Obsolete*

MANGANESE NICKEL

International Nickel Inc.

Ni 95-98.5, Mn 1.5-5.

52,100-59,100 TS; 16,500-23,100 YS; 36-51 El. For fittings; corrosion resistant. *Obsolete*

MANGANESE NICKEL

English manufacture

Cu 52-85, Mn 14-31, Ni 3-16.

For heat resistant parts.

MANGANESE NICKEL 5%

Wilbur B. Driver Co.

Nickel. Mn 5, Ni 95.

Annealed: 75,000 TS; 35,000 YS; 40 El; 140 Brin. For lead wires for electrical appliances; heat resistant.

MANGANESE NICKEL NO. 484 ALLOY

Hoskins Mfg. Co.

Ni 98.5, Mn 1.5.

For spark plug electrodes; heat resistant. *Obsolete*

MANGANESE NICKEL SILVER

German manufacture

Cu 60-73, Mn 2.4-20, Ni 10-17, Sn 0-10, Zn 0-8.8.

For white metal parts; corrosion resistant.

MANGANESE NICKEL-2%

Wilbur B. Driver Co.

Nickel. Mn 3, bal Ni.

Annealed: 75,000 TS; 35,000 YS; 40 El; 140 Brin. For lead wires for electrical appliances; heat resistant.

MANGANESE RED BRASS 507

Anaconda Co.

Copper. Zn 14, Mn 1, bal Cu.

Hard: 69,000 TS; 55,000 YS; 7 El; 140 Brin. Soft: 40,000 TS; 15,000 YS; 45 El; 55 Brin. For strip for resistance spot and seam welded products.

MANGANESE S3

Sight Feed Generator Co.

C, Mn, Ni, bal Fe.

Welded: 180 Brin. Work hardened: 450 Brin. For hard facing electrodes; wear resistant.

MANGANESE SCREW STOCK
LaSalle Steel Co.
C 0.15-0.25, Mn 0.9-1.2, bal Fe.
Rolled: 65,000-80,000 TS; 44,000-75,000 YS; 16-35 El; 55-60 RA; 130-156 Brin. For case hardened parts, gears, shafts; SAE-X-1314. *Obsolete*

MANGANESE STEEL MEDIUM
Bethlehem Steel Corp.
C 0.1, Mn 1.2, bal Fe.
Rolled: 50,000 TS. For structures, bridges; high tensile steel. *Obsolete*

MANGANESE STEEL PEARLITIC
English manufacture
Mn 1.4-3.5, Si 0.2-0.3, C 0.2-0.6, bal Fe.
For gears, shafts; British Patent 131980.

MANGANESE-NICKEL (2%)
Henry Wiggin & Co. Ltd.
Mn 2, bal Ni.
Annealed: 81,000 TS; 24,000 YS; 56 El; 76 RA; 139 Brin. For grid and support wires in radio valve lamps; heat resistant. *Obsolete*

MANGANESE-NICKEL (5%)
Henry Wiggin & Co. Ltd.
Mn 5, bal Ni.
Annealed: 81,000 TS; 24,000 YS; 56 El; 76 RA; 139 Brin. For grid and support wires in radio valve lamps; heat resistant. *Obsolete*

MANGANIN
Harrison Alloys Inc.
Copper.
Now HAI-MANGANIN 13.

MANGANIN
Isabellenhuette
Copper. Cu 86, Ni 2, Mn 12.
Annealed: 390 N/mm^2 TS. For electrical equipment and instruments. Resistance alloy. Maximum working temperature to 140°C.

MANGANIN
Molecu-Wire Corp.
Ni 4, Mn 12-14, bal Cu.
Annealed: 80,000 TS; 45,000 YS; 50 El. For resistors, thermocouples; low temperature coefficient of resistance.

MANGANIN
Wilbur B. Driver Co.
Ni 4, Mn 12-14, bal Cu.
Annealed: 80,000 TS; 45,000 YS; 50 El. For resistors, thermocouples; low temperature coefficient of resistance.

MANGANIN
Gilby-Fodor S.A.
Ni 4, Mn 12-14, bal Cu.
Annealed: 80,000 TS; 45,000 YS; 50 El. For resistors, thermocouples; low temperature coefficient of resistance.

MANGANIN ALLOY 130
Carpenter Technology Corp.
Cu 84, Mn 12, Ni 4.
Annealed: 276 MPa TS. Cold worked: 621 MPa TS. For use in precision built electrical apparatus such as Wheatstone bridges, decade boxes, voltage dividers, potentiometers and resistance standards. Maximum operating temperature in air 93°C.

MANGANINGOT
Now SPECIALLOY 5025 MANGANINGOT TM.

MANGANNICKEL 1
VDM Nickel-Technologie AG
Mn 1.5, C 0.25, bal Ni.
Annealed: 74,000 TS; 36 El; 95 Brin. Drawn: 132,000 TS; 1 El; 250 Brin. For German silver, candlesticks, water meter parts; corrosion resistant. *Obsolete*

MANGANNICKEL 2
VDM Nickel-Technologie AG
Mn 1.5, bal Ni.
Rolled: 135,000 TS; 1 El; 270 Brin. Annealed: 76,800 TS; 35 El; 100 Brin. For thermocouples; corrosion resistant. *Obsolete*

MANGANNICKEL 4
VDM Nickel-Technologie AG
Mn 4, bal Ni.
Rolled: 142,200 TS; 1 El; 280 Brin. Annealed: 81,000 TS; 34 El; 110 Brin. For thermocouples, spark plugs; corrosion resistant. *Obsolete*

MANGANNICKEL 5
VDM Nickel-Technologie AG
Mn 4.5-5.5, bal Ni.
For spark plug electrodes; heat resistant. *Obsolete*

MANGANO
Latrobe Steel Co.
C 0.95, Mn 1.65, Si 0.25, bal Fe.
For dies, gages, tools, blanking dies; non-deforming, oil hardening. *Obsolete*

MANGANO SPECIAL
Latrobe Steel Co.
C 0.95, Cr 0.5, Mn 1.2, W 0.5, bal Fe.
For reamers, dies, threading taps, trimming dies; nondeforming. *Obsolete*

MANGANOID
Dresser Industries
C 1, Mn 1.2, bal Fe.
For grinding balls for crushers, grinders and pulverizers. *Obsolete*

MANGANWELD A
Lincoln Electric Co.
C, Cr, Ni, high Mn, bal Fe.
For welding electrodes, hard facing electrodes; shielded arc. *Obsolete*

MANGANWELD B
Lincoln Electric Co.
C 0.4, Mn 13, bal Fe.
For welding electrodes for manganese steel; hard facing. *Obsolete*

MANGCRAFT A
Manufacturer not listed.
C 0.5-0.8, Mn 11-14, bal Fe.
For welding rod for 14% Mn steel; austenitic, wear resistant.

MANGCRAFT B
Manufacturer not listed.
C 0.5-0.8, Mn 11-14, Ni 4.5-5.5, bal Fe.
For welding rod for 14% Mn-Ni steel; austenitic, abrasion resistant.

MANGDIE
British Steel Corp.
C 0.95, Mn 1.25, Cr 0.5, V 0.2, W 0.5, bal Fe.
Oil hardening non-distorting alloy tool steel; for brake-press tools, taps, threading dies, plug gages. *Obsolete*

MANGJET
Lincoln Electric Co.
C 0.65, Mn 14.5, Si 0.14, Mo 1.15, bal Fe.
Hard surfacing, arc welding electrodes; resistance to severe impact.

MANGO-PLATE
Pyramid Steel Company
C, alloy, bal Fe.
Heat treated: 153,000 TS; 140,000 YS; 14 El; 54 RA; 390 Brin. For coal chutes, conveyor lines, scraper blades; work hardened, abrasion resistant.

MANGONIC 2
Henry Wiggin & Co. Ltd.
Mn 2, bal Ni.
For electrode support wires in radio valves and tungsten filament lamps; grid supports and windings. *Obsolete*

MANGONIC 3
Henry Wiggin & Co. Ltd.
Mn 3, bal Ni.
For radio valve support wires, lamps; heat resistant. *Obsolete*

MANGONIC 5
Henry Wiggin & Co. Ltd.
Mn 5, bal Ni.
For radio valve support wires, lamps; heat resistant. *Obsolete*

MANGRID D
Wilbur B. Driver Co.
Mn 4.5, bal Ni.
Annealed: 75,000 TS; 35,000 YS; 40 El; 140 Brin. For vacuum tube grid wire; magnetic. *Obsolete*

MANGRID E
Wilbur B. Driver Co.
Mn 20, bal Ni.
Annealed: 79,000 TS. For furnace lead wire, grid wire; magnetic. *Obsolete*

MANHARDT'S ALLOY
English manufacture
Al 83, Sn 10, Cu 6.2, Mg 0.1, P.
For light alloy parts; non-hardenable.

MANIFLEX
Now CARPENTER MANIFLEX.

MANIOR APS10
Usines de A. Manoir, Pitres (Eure)
C 0.12, Cr 2, Al 0.4-0.9, Mo, bal Fe.
Normalized: 69,000-92,000 psi TS; 42,000-54,000 psi YS; 12-16 El. For oil refinery equipment; corrosion resistant.

MANMO
Teledyne Vasco
C 0.55-0.75, Mn 0.8, Cr 1, Mo 0.45, bal Fe.
For brake dies, plastic molds, cams, chisels, punches, oil hardening, tough. *Obsolete*

MANNESMANNSTAHL-F12
Mannesmann-Huttenwerk AG
C 0.2, Cr 12, Mo 1, Ni 0.4, V 0.3, bal Fe.
Annealed: 90,000 TS; 38,000 YS; 26 El; 92 Rock B. Hardened: 230,000 TS; 190,000 YS; 11 El; 46 Rock C. For valves, bearings, cutlery, surgical instruments. Corrosion resistant, hardenable.

MANNHEIM GOLD
American Smelting & Refining Co.
Cu 80-89, Zn 7-20, Sn 0.9.
25,200 TS; 25 El; 69 Brin. For inexpensive jewelry; moderately corrosion resistant.

MANOFORT 160
Usines de A. Manoir, Pitres (Eure)
C 0.3, Ni 2, Cr 1, Mo 0.4, bal Fe.
Heat treated: 212,000-242,000 psi TS; 163,000-200,000 psi YS; 7-9 El. For gears, shafts, crankshafts; oil hardened, shock resistant.

MANOFORT 180
Usines de A. Manoir, Pitres (Eure)
C 0.4, Ni 3.5, Cr 1.4, Mo 0.4, bal Fe.
Heat treated: 228,000-275,000 psi TS; 185,000-214,000 psi YS; 4-6 El. For gears, shafts, countershafts; oil hardened, shock resistant.

MANOIR ABRADUR 220
Usines de A. Manoir, Pitres (Eure)
C 0.7, Mn 1, bal Fe.
Heat treated: 100,000-115,000 psi TS; 65,000-78,000 psi YS; 6-8 El; 220-260 Brin. For rails, tools, hammers; water or oil hardened.

MANOIR ABRADUR 240
Usines de A. Manoir, Pitres (Eure)
C 0.5, Mn 1, Cr 1, bal Fe.
Heat treated: 100,000-114,000 psi TS; 65,000-78,000 psi YS; 6-8 El; 220-260 Brin. For camshafts, countershafts, gears, shafts; wear resistant, oil hardened.

MANOIR ABRADUR 400
Usines de A. Manoir, Pitres (Eure)
C 0.3, Ni 1.8, Cr 1, Mo 0.35, bal Fe.
Heat treated: 186,000-228,000 psi TS; 156,000-186,000 psi YS; 4-6 El; 400-480 Brin. For gears, shafts, countershafts; oil hardened, tough, shock resistant.

MANOIR ABRADUR 500
Usines de A. Manoir, Pitres (Eure)
C 0.4, Ni 3.5, Cr 1.4, Mo 0.4, bal Fe.
Heat treated: 144,000-156,000 psi TS; 92,000-114,000 psi YS; 10-12 El; 280-32 Brin. For gears, shafts, countershafts, mining equipment; oil hardened, tough, shock resistant.

MANOIR ABRADUR 600
Usines de A. Manoir, Pitres (Eure)
C, Cr, W, V, bal Fe.
600 Brin. For abrasion resistant parts, sand blasting nozzles; wear and abrasion resistant.

MANOIR ABRADUR M14
Usines de A. Manoir, Pitres (Eure)
C 1.2, Mn 13, bal Fe.
Rolled: 120,000-157,000 psi TS; 50,000-65,000 psi YS; 30-35 El. For shovels, dippers, frogs, cross tracks; wear resistant.

MANOIR ABRADUR M14K
Usines de A. Manoir, Pitres (Eure)
C 1.2, Mn 13, Cr, bal Fe.
For dippers, shovels, cross tracks; wear and abrasion resistant.

MANOIR APS20
Usines de A. Manoir, Pitres (Eure)
C 0.12, Cr 4, Al 0.4-0.9, Mo, bal Fe.
Normalized: 69,000-92,000 psi TS; 42,000-54,000 psi YS; 12-16 El. For oil refinery equipment: corrosion resistant.

MANOIR APS25
Usines de A. Manoir, Pitres (Eure)
C 0.12, Cr 4, Al 0.9, Ni 0.9, bal Fe.
Normalized: 114,000-105,000 psi TS; 85,000-107,000 psi YS; 7-11 El. For oil refinery equipment; corrosion resistant.

MANOIR EL38M
Usines de A. Manoir, Pitres (Eure)
Low C, Mn, bal Fe.
Annealed: 55,000 psi TS; 26,000 psi YS; 22 El; 120 Brin. For motors, electrical equipment, high magnetic permeability.

MANOIR EL40
Usines de A. Manoir, Pitres (Eure)
Low C, Mn, bal Fe.
Annealed: 58,000 psi TS; 30,000 psi YS; 22 El; 120 Brin.

MANOIR EL45
Usines de A. Manoir, Pitres (Eure)
C, Mn, bal Fe.
Annealed: 65,000 psi TS; 32,000 psi YS; 20 El; 140 Brin.

MANOIR EL50
Usines de A. Manoir, Pitres (Eure)
C, Mn, bal Fe.
Annealed: 72,000 psi TS; 36,000 psi YS; 18 El; 155 Brin.

MANOIR EL55
Usines de A. Manoir, Pitres (Eure)
C, Mn, bal Fe.
Annealed: 79,000 psi TS; 40,000 psi YS; 15 El; 170 Brin. For bed plates.

MANOIR EL65
Usines de A. Manoir, Pitres (Eure)
C, Mn, bal Fe.
Annealed: 92,000 psi TS; 50,000 psi YS; 10 El; 210 Brin.

MANOIR P17
Usines de A. Manoir, Pitres (Eure)
C 0.12, Mn, Mo, V, bal Fe.
Rolled: 76,000-92,000 psi TS; 58,000-65,000 psi YS; 18-22 El. For gears, cams, camshafts; case hardened, tough.

MANOIR PF-0
Usines de A. Manoir, Pitres (Eure)
C 0-0.15, Mo 0.5, bal Fe.
Annealed: 65,000 psi TS; 37,000 psi YS; 24 El. For oil refinery equipment.

MANOIR PF-1
Usines de A. Manoir, Pitres (Eure)
C 0-0.15, Cr 0.5, Mo 0.5, bal Fe.
Annealed: 78,000 psi TS; 40,000 psi YS; 18 El. For oil refinery equipment.

MANOIR PF-15
Usines de A. Manoir, Pitres (Eure)
C 0-0.2, Cr 0.8, Mo 0.5, bal Fe.
Annealed: 92,000 psi TS; 58,000 psi YS; 16 El. For oil refinery equipment.

MANOIR PF-2
Usines de A. Manoir, Pitres (Eure)
C 0-0.2, Cr 1, Mo 0.5, V 0.15, bal Fe.
Annealed: 78,000 psi TS; 46,000 psi YS; 16 El. For oil refinery equipment.

MANOIR PF-5
Usines de A. Manoir, Pitres (Eure)
C 0-0.15, Cr 5, Mo 0.5, bal Fe.
Annealed: 78,000 psi TS; 36,000 psi YS; 18 El. For oil refinery equipment, stills; corrosion resistant.

MANOIR PF-6
Usines de A. Manoir, Pitres (Eure)
C 0-0.15, Cr 2.2, Mo 1, bal Fe.
Annealed: 85,000 psi TS; 36,000 psi YS; 15 El. For oil refinery equipment; corrosion resistant.

MANOIR PFV-55
Usines de A. Manoir, Pitres (Eure)
C 0-0.15, Cr 2, Mo 0.35, V, Al, bal Fe.
Annealed: 78,000 psi TS; 36,000 psi YS; 15 El. For oil refinery equipment; corrosion resistant.

MANOIR PM35
Usines de A. Manoir, Pitres (Eure)
C 0.17, Mn, Cr, Mo, V, bal Fe.
Normalized: 143,000-170,000 psi TS; 107,000-121,000 psi YS; 7-10 El. For gears, cams, camshafts; case hardened, tough.

MANOIR RS1
Usines de A. Manoir, Pitres (Eure)
C 0.45, Mn 0.7, Si 1.8, bal Fe.
Heat treated: 177,000-200,000 psi TS; 156,000-177,000 psi YS; 4 El. For punches, air hammers, chisels; shock resistant, oil hardened.

MANSILOY
Union Carbide Corp.
Mn 60-63, Si 28-31, C 0-0.07, P 0-0.05.
For production of stainless steels; reduces metal oxides from the slag.

MAPLE LEAF
Atlas Specialty Steels
C 0.8-1.2, Si 0.25, Mn 0.25, bal Fe.
Water hardened: 165,000-215,000 TS; 110,000-150,000 YS; 11-15 El; 32-37 RA; 330-600 Brin. For stamps, knurls, drills, taps, mandrels, reamers, cutters; water hardening, Type W1. *Obsolete*

MAPLE LEAF 8
Atlas Specialty Steels
C 0.8, bal Fe.
For blacksmith tools; water hardened. *Obsolete*

MAR-CON 660
Bunting Bearings Corp.
Sn 7, Pb 7, Zn 3, bal Cu.
Cast: 44,000 TS; 27,000 YS; 16 El; 14 RA; 73 Brin. For bearings, bushings, liners. Continuous cast bronze. *Obsolete*

MAR-M 905
Martin-Marietta Corp.
C 0.05, Cr 20, Ni 20, Ti 0.5, Zr 0.1, Ta 7.5, bal Co.
Sheet alloy for use to 1400°F.

MAR-M ALLOY 200
Martin-Marietta Corp.
C 0.12-0.17, Cr 8-10, W 11.5-13.5, Co 9-11, Zr 0.05, Cb 0.75-1.25, Al 4.75-5.25, Ti 1.75-2.25, B 0.015, Fe 0-1.5, bal Ni.
As Cast: 135,000 TS; 120,000 YS; 7 El. At 1400°F: 135,000 TS; 122,500 YS; 3.5 El. For turbine blades and vanes in aircraft gas turbine engines. High strength and oxidation resistant to 1900°F.

MAR-M ALLOY 200
Cannon-Muskegon Corp.
C 0.12-0.17, Cr 8-10, W 11.5-13.5, Co 9-11, Zr 0.05, Cb 0.75-1.25, Al 4.75-5.25, Ti 1.75-2.25, B 0.015, Fe 0-1.5, bal Ni.
As Cast: 135,000 TS; 120,000 YS; 7 El. At 1400°F: 135,000 TS; 122,500 YS; 3.5 El. For turbine blades and vanes in aircraft gas turbine engines. High strength and oxidation resistant to 1900°F.

MAR-M ALLOY 211
Martin-Marietta Corp.
C 0.15, Cr 9, Co 10, Mo 2.5, W 5.5, Cb 2.7, B 0.015, Zn 0.05, Ti 2, Al 5, bal Ni.
For integrally cast turbine wheels and blades. Cast alloy. High stress-rupture strength. Corrosion and heat resistant.

MAR-M ALLOY 211
Cannon-Muskegon Corp.
C 0.15, Cr 9, Co 10, Mo 2.5, W 5.5, Cb 2.7, B 0.015, Zn 0.05, Ti 2, Al 5, bal Ni.
For integrally cast turbine wheels and blades. Cast alloy. High stress-rupture strength. Corrosion and heat resistant.

MAR-M ALLOY 246
Martin-Marietta Corp.
C 0.15, Cr 9, Co 10, W 10, Mo 2.5, Ta 1.5, Ti 1.5, Al 5.5, B 0.015, Zr 0.05, bal Ni.
Cast: 139,000 TS; 122,000 YS; 4.5 El. At 1600°F: 132,000 TS; 100,000 YS; 4.5 El. For turbine vanes, nozzles, jet engine components. Precipitation hardening. For service up to 1900°F. High oxidation resistant.

MAR-M ALLOY 246
Cannon-Muskegon Corp.
C 0.15, Cr 9, Co 10, W 10, Mo 2.5, Ta 1.5, Ti 1.5, Al 5.5, B 0.015, Zr 0.05, bal Ni.
Cast: 139,000 TS; 122,000 YS; 4.5 El. At 1600°F: 132,000 TS; 100,000 YS; 4.5 El. For turbine vanes, nozzles, jet engine components. Precipitation hardening. For service up to 1900°F. High oxidation resistant.

MAR-M ALLOY 302

Martin-Marietta Corp.

C 0.78-0.93, Fe 0-1.5, B 0-0.01, Cr 20-23, W 9-11, Ta 8-10, Zr 0.1-0.3, bal Co.

Cast: 140,000 TS; 100,000 YS; 2 El; 40 Rock C. At 1600°F: 56,000-78,000 TS; 40,000-50,000 YS; 7-14 El. For turbine vanes, nozzle guide vanes and buckets in gas turbines. High oxidation and thermal shock resistance. For service to 2100°F.

MAR-M ALLOY 302

Cannon-Muskegon Corp.

C 0.78-0.93, Fe 0-1.5, B 0-0.01, Cr 20-23, W 9-11, Ta 8-10, Zr 0.1-0.3, bal Co.

Cast: 140,000 TS; 100,000 YS; 2 El; 40 Rock C. At 1600°F: 56,000-78,000 TS; 40,000-50,000 YS; 7-14 El. For turbine vanes, nozzle guide vanes and buckets in gas turbines. High oxidation and thermal shock resistance. For service to 2100°F.

MAR-M ALLOY 322

Martin-Marietta Corp.

C 0.9-1.1, Cr 20-23, W 8-10, Ta 4-5, Zr 2-2.5, Ti 0.65-0.85, Fe 0-1.5, bal Co.

Cast: 121,000 TS; 91,000 YS; 3 El; 4 RA; 36 Rock C. At 1500°F: 95,000 TS; 55,000 YS; 10 El; 10 RA. For turbine vanes and blades, jet engine components. High temperature strength and ductility. Oxidation resistant to 2000°F.

MAR-M ALLOY 322

Cannon-Muskegon Corp.

C 0.9-1.1, Cr 20-23, W 8-10, Ta 4-5, Zr 2-2.5, Ti 0.65-0.85, Fe 0-1.5, bal Co.

Cast: 121,000 TS; 91,000 YS; 3 El; 4 RA; 36 Rock C. At 1500°F: 95,000 TS; 55,000 YS; 10 El; 10 RA. For turbine vanes and blades, jet engine components. High temperature strength and ductility. Oxidation resistant to 2000°F.

MAR-M ALLOY 421

Martin-Marietta Corp.

C 0.15, Cr 15.5, Mo 1.75, Co 10, W 3.5, Cb 1.75, Ti 1.75, Al 4.25, Zr 0.05, B 0.015, Fe 0-1, bal Ni.

As Cast: 132,000 TS; 115,000 YS; 6 El; 10 RA. Heat treated: 150,000 TS; 132,000 YS; 4.4 El; 6.5 RA. Wrought: 198,000 TS; 136,000 YS; 20 El; 26 RA. For turbine rotors and discs, jet engine components. Precipitation hardening, high strength and sulfidation resistant.

MAR-M ALLOY 421

Cannon-Muskegon Corp.

C 0.15, Cr 15.5, Mo 1.75, Co 10, W 3.5, Cb 1.75, Ti 1.75, Al 4.25, Zr 0.05, B 0.015, Fe 0-1, bal Ni.

As Cast: 132,000 TS; 115,000 YS; 6 El; 10 RA. Heat treated: 150,000 TS; 132,000 YS; 4.4 El; 6.5 RA. Wrought: 198,000 TS; 136,000 YS; 20 El; 26 RA. For turbine rotors and discs, jet engine components. Precipitation hardening, high strength and sulfidation resistant.

MAR-M ALLOY 432

Martin-Marietta Corp.

Co 20, Cr 15.5, Ti 4.3, Al 2.8, W 3, Ta 2, Cb 2, B 0.015, Zr 0.05, bal Ni.

Cast: 180,000 TS; 150,000 YS; 5 El; 6.5 RA. For integrally cast turbine wheels, jet engine components, turbine blades. Sulphidation resistant. Resists creep rupture and hot corrosion.

MAR-M ALLOY 432

Cannon-Muskegon Corp.

Co 20, Cr 15.5, Ti 4.3, Al 2.8, W 3, Ta 2, Cb 2, B 0.015, Zr 0.05, bal Ni.

Cast: 180,000 TS; 150,000 YS; 5 El; 6.5 RA. For integrally cast turbine wheels, jet engine components, turbine blades. Sulphidation resistant. Resists creep rupture and hot corrosion.

MAR-M ALLOY 509

Martin-Marietta Corp.

C 0.6, Cr 23.5, Ni 10, W 7, Ta 3.5, Ti 0.2, Zr 0.5, bal Co.

At 70°F: 113,000 TS; 85,000 YS; 3.5 El; 5.8 RA. At 1600°F: 68,000 TS; 45,000 YS; 7 El; 13 RA. For aircraft and industrial turbines. Cast superalloy, high strength, low creep rate. Resists shock and oxidation.

MAR-M ALLOY 509

Cannon-Muskegon Corp.

C 0.6, Cr 23.5, Ni 10, W 7, Ta 3.5, Ti 0.2, Zr 0.5, bal Co.

At 70°F: 113,000 TS; 85,000 YS; 3.5 El; 5.8 RA. At 1600°F: 68,000 TS; 45,000 YS; 7 El; 13 RA. For aircraft and industrial turbines. Cast superalloy, high strength, low creep rate. Resists shock and oxidation.

MAR-M ALLOY 918

Martin-Marietta Corp.

Ni 20, Cr 20, Ta 7.5, Zr 0.1, C 0.05, bal Co.

Bar: 130,000 TS; 56,000 YS; 48 El. For burner cans and afterburner liners in gas turbine engines. High heat, oxidation, and corrosion resistance.

MAR-M ALLOY 918

Cannon-Muskegon Corp.

Ni 20, Cr 20, Ta 7.5, Zr 0.1, C 0.05, bal Co.

Bar: 130,000 TS; 56,000 YS; 48 El. For burner cans and afterburner liners in gas turbine engines. High heat, oxidation, and corrosion resistance.

MAR-M-247

Martin-Marietta Corp.

C 0.15, Cr 8.25, Co 10, W 10, Mo 0.7, Ta 3, Al 5.5, Ti 1, Hf 1.5, B 0.015, Zr 0.05, bal Ni.

Cast high temperature alloy; for turbine wheels and blades. Same as M-M-0011 ALLOY.

MARADAMIT

Manufacturer not listed.

Al alloy.

For light alloy parts.

MARATHON

Colt Industries

C 2.2, Cr 12, bal Fe.

For dies, swedges, punches, drawing, shearing and riveting dies; resists wear and abrasion. *Obsolete*

MARATHON

Crucible Specialty Metals

C 2.2, Cr 12, bal Fe.

For dies, swedges, punches, drawing, shearing and riveting dies; resists wear and abrasion. *Obsolete*

MARATHON

Duraloy Blaw-Knox/Union Steel Casting

C, alloy, bal Fe.

For machine tool parts; oil hardened. *Obsolete*

MARATHON CRM SPEZIAL

Marathon Specialty Steels Inc.

Tool material. C 1, Mn 1, Cr 1, Mo 0.25, bal Fe.

Heat treated: 288,000 TS; 278,000 YS; 540 Brin. For precision and plug gauges, bearings, liners, arbors. High toughness and wear resistance. Type L-5 tool steel.

MARATHON E612

Marathon Specialty Steels Inc.

C 0.4, Si 1.15, Cr 5.25, W 4.25, bal Fe.

Hardened: 269,000 TS; 207,000 YS; 5 El; 9 RA. At 900°F: 217,000 TS; 172,000 YS; 10 El; 31 RA. For die casting dies, forging and extrusion tools, upsetters, dummy blocks. Type H14 hot work steel, deep hardening.

MARATHON MO10

Marathon Specialty Steels Inc.

C 0.8, Cr 4, Mo 8, W 1.5, V 1, bal Fe.

Hardened: 64-66 Rock C. For lathe and planer tools, hobs, drills, reamer, broaches, form cutters. Type M1 high speed steel, high red-hardness, wear resistant.

MARATHON MO19

Marathon Specialty Steels Inc.

C 1, Cr 4, Mo 8.75, W 1.75, V 2, bal Fe.

Hardened: 64-67 Rock C. For reamers, hobs, chasers, lathe and planer tools. Type M7 high-speed steel, high red-hardness. Abrasion resistant.

MARATHON MO20

Marathon Specialty Steels Inc.

C 0.85, Mo 5, W 6.5, Cr 4, V 2, bal Fe.

Hardened: 64-66 Rock C. For lathe and planer tools, drills, taps, chasers, drawing dies, punches, hobs, form cutters. Type M2 high speed steel, high red-hardness, tough.

MARATHON MO20S

Marathon Specialty Steels Inc.

C 0.62-0.68, Cr 3.8-4.4, V 1.8-2.1, W 6.5-6.7, Mo 4.7-5.2, bal Fe.

Hardened: 58-62 Rock C. For hot extrusion dies, punches, shear blades, forging mandrels. Type H42 high speed steel for hot working.

MARATHON PW16

Marathon Specialty Steels Inc.

C 0.5, Cr 4, W 18, V 1, bal Fe.

Hardened: 60-65 Rock C. Heat treated: 130,000 TS; 95,000 YS; 4 El; 52 Rock C. For punches, hot work tools and dies, extrusion and die casting dies. Type H26 hot work steel, tough, good high temperature strength.

MARATHON SA200

Marathon Specialty Steels Inc.

C 0.75, Cr 4, W 14, V 2, bal Fe.

Hardened: 64-66 Rock C. For drills, reamers, taps, punches, lathe and planer tools. Type T7 high speed steel, high red-hardness.

MARATHON SA900

Marathon Specialty Steels Inc.

C 1.5, W 13, Cr 4.5, V 5, Co 5, bal Fe.

Hardened: 64-66 Rock C. For cutting tools, broaches, drills, hobs, lathe and planer cutters, reamers. Type T15 high speed steel, high-red-hardness, abrasion resistant.

MARATHON SS4

Saarstahl AG

C 1.05-1.15, bal Fe.

For tools, taps, drills; water hardened.

MARATHON-000 EXTRA

Marathon Specialty Steels Inc.

Tool material. C 0.8, Cr 4, W 18, V 2, bal Fe.

Hardened: 64-66 Rock C. For lathe and planer tools, drills, hobs, reamers, chasers, form cutters. Type T-2 tool steel, high red-hardness.

MARATHON-CRS

Marathon Specialty Steels Inc.

Tool material. C 1.05, Cr 1.4, bal Fe.

Heat treated: 200,000 TS; 185,000 YS; 3 El; 390 Brin. For gauges, bushings, knurls, taps, dies, arbors, rolls, bearings. Oil hardening, tough and wear resistant. Type L-1 tool steel.

MARATHON-E38 MO

Marathon Specialty Steels Inc.

Tool material. C 0.35, Si 1, Cr 5, V 1, Mo 1.5, bal Fe.

Heat treated: 290,000 TS; 227,000 YS; 3 El; 55 Rock C. For forging and drawing dies, die casting dies, hot piercing and forming dies, swaging and gripping dies. Type H13 tool steel, air hardening. High resistance to heat checking.

MARATHON-E38V

Marathon Specialty Steels Inc.

Tool material. C 0.35, Cr 5, V 1, Mo 1.2, bal Fe.

Heat treated: 290,000 TS; 225,000 YS; 3 El; 55 Rock C. For die casting dies, swaging and gripping dies, forging and extrusion dies. Type H-13 tool steel, air hardening. High resistance to heat checking.

MARATHON-E38W

Marathon Specialty Steels Inc.
Tool material. C 0.36, Cr 5, Mo 1.4, W 1.4, V 0.5, bal Fe.
Heat treated: 290,000 TS; 235,000 YS; 8 El; 54 Rock C. For die casting dies, hot punches and heading dies, shear blades. Type H12 tool steel, hot work steel, high toughness and wear resistance.

MARATHON-SA

Marathon Specialty Steels Inc.
Tool material. C 0.55, Mn 0.9, Si 2, Cr 0.3, V 0.2, bal Fe.
Hardened: 338,000 TS; 281,000 YS; 5 El; 600 Brin. For punches, shear blades, pneumatic tools, rivet busters, knurling tools. Type S-4 tool steel, deep hardening, shock and wear resistant.

MARINE 490

NKK Corp.
Mn 0-1.5, Cr 0.5-0.8, Al 0.15-0.55, Cu 0.2-0.35, Ni 0-0.4, 0.10 C max, Cb or V, bal Fe.
High tensile, low alloy steel for sea-water equipment.

MARINE 50

NKK Corp.
Now MARINE 490.

MARINE ALLOY

American manufacture
Cu 2, Sn 40, Pb 48, Sb 10.
For bearings: submerged bearings.

MARINE BABBITT

Manufacturer not listed.
Pb 72, Sn 21, Sb 7.
For marine bearings; anti-friction.

MARINE BRONZE

American manufacture
Cu 57.5, Ni 0.8, Sn 0.15, Al 0.5, bal Zn.
83,500 TS; 15 El.For marine parts.

MARINE BRONZE NO. 8

Manufacturer not listed.
Copper. Mn, Zn, Sn, Al, bal Cu.
Cast: 65,000 TS; 20-25 El; 29 RA; 150 Brin. For marine propellers, ship parts, etc., subjected to salt water corrosion; resists corrosion. *Obsolete*

MARINE GLYCO

Joseph T. Ryerson & Son Inc.
Sn, Sb, bal Pb.
For marine bearings; Babbitt.

MARINE NICKEL

Lewin Metals Corp.
Pb, Ni, bal Sn.
For bearings. Babbitt metal.

MARINER

United States Steel Corp.
C 0.22, Mn 0.8, P 0.11, bal Fe.
Sheet. as rolled: 70 ksi TS: 50 ksi YS; 18 El. Sheet piling for use in marine environment. ASTM A690.

MARK 12KH14A

Russian manufacture
C 0.12, Cr 14, bal Fe.
Annealed: 75,000 TS; 40,000 YS; 35 El; 70 RA; 155 Brin. For valves, cutlery, valve turbines; type 410; stainless.

MARK 18KH14A

Russian manufacture
C 0.18, Cr 14, bal Fe.
Annealed: 95,000 TS; 50,000 YS; 25 El; 55 RA; 195 Brin. For valves, cutlery, surgical and dental instruments; Type 420; stainless.

MARK 2, ETC.

Now CENTRICAST MARK 2, ETC.

MARK 50

Russian manufacture
C 0.5, bal Fe.
Annealed: 96,000 TS; 52,000 YS; 16 El; 23 RA; 170 Brin. For gears, bolts, fasteners; water hardened.

MARK E169

Russian manufacture
Cr 14, Ni 14, W 2, bal Fe.
For valves, cutlery; corrosion resistant.

MARK EYA-1T

Russian manufacture
C 0.12, Cr 18, Ni 8, Ti = 7 x C, bal Fe.
Annealed: 85,000 TS; 35,000 YS; 55 El; 65 RA; 150 Brin. For welded chemical plant equipment; Type 321; stainless, austenitic.

MARK I

Sheepbridge Engineering Ltd.
C 3.3, Si 1.4, P 0.2, bal Fe.
Cast: 180-250 Brin. For medium section castings; cast iron.

MARK I

Sheepbridge Alloy Castings Ltd.
C 3.3, Si 1.4, P 0.2, bal Fe.
Cast: 180-250 Brin. For medium section castings; cast iron.

MARK III

Sheepbridge Engineering Ltd.
C 3.2, Si 1.9, Ni 1.4, Cr 0.4, bal Fe.
Cast: 220-280 Brin. For heat treated castings, gears; cast iron.

MARK III

Sheepbridge Alloy Castings Ltd.
C 3.2, Si 1.9, Ni 1.4, Cr 0.4, bal Fe.
Cast. 220-280 Brin. For heat treated castings, gears; cast iron.

MARK U7

Russian manufacture
C 0.7, Mn 0-0.25, Si 0-0.25, bal Fe.
Heat treated: 174,000 TS; 128,000 YS; 12 El; 37 RA; 352 Brin. For springs, rails, clutch discs; water hardened.

MARK U8

Russian manufacture
C 0.8, Si 0-0.25, Mn 0-0.25, bal Fe.
Heat treated: 188,000 TS; 143,000 YS; 12 El; 35 RA; 388 Brin. For springs, taps, reamers, drills, hobs; water hardened.

MARK VII

Sheepbridge Engineering Ltd.
C 3.4, Si 1.4, P 0.2, bal Fe.
Cast: 170-220 Brin. For piston rings; cast iron.

MARK VII

Sheepbridge Alloy Castings Ltd.
C 3.4, Si 1.4, P 0.2, bal Fe.
Cast: 170-220 Brin. For piston rings; cast iron.

MARKANA METAL

Vereinigte Deutsche Nickel-Werke AG
Cu 58-60, Zn 38-40, 1.5-2.0% Mn + Fe.
For sheets for spinning, lavatories, wash basins, sinks; corrosion resistant; MP 880 C. *Obsolete*

MARKER 10AX

Schmidt & Clemens Edelstahlwerke
C 0.26, Cr 20, Ni 11, bal Fe.
Cast: 71,000-100,000 TS; 35,000 YS; 15 El; 200 Brin. Austenitic, corrosion resistant, weldable; good strength at elevated temperature. For parts operating at 550-1000 C. DIN G-X25CrNiSi2014. *Obsolete*

MARKER 11AXS

Schmidt & Clemens Edelstahlwerke
C 0.2-0.5, Si 1-2.5, Mn 0-1.5, Cr 24-26, Mo 0-0.5, Ni 11-14, bal Fe.
Cast, annealed: 64,000-92,000 TS; 10 min El; 130-200 Brin. For furnace parts; useable in air to 2000 F. Austenitic stainless. DIN G-X35CrNiSi2512; VDEh Spec 4837; ACI HH; AISI 309 cast. *Obsolete*

MARKER 11F

Schmidt & Clemens Edelstahlwerke
C 0.4, Cr 22, Ni 1.8, bal Fe.
Cast: 57,000-85,000 TS; 50,000 YS; 1 El; 280 Brin. For furnace castings for use at 1650-1920 F. DIN G-X40CrSi22. *Obsolete*

MARKER 12 AC 15

Schmidt & Clemens Edelstahlwerke
C 0.1-0.2, Si 1-2, Mn 0-1.5, Cr 24-27, Mo 0-0.5, Ni 19-21, bal Fe.
Cast, annealed: 64,000-92,000 TS; 25 min El, 130-200 Brin. Austenitic stainless casting; for parts operating at temperatures up to 1920 F. DIN G-X15CrNiSi2520; VDEh Spec 4849; ACI CK-20; AISI 310 cast. *Obsolete*

MARKER 12 AXN

Schmidt & Clemens Edelstahlwerke
C 0.15-0.3, Si 0.5-1.5, Mn 0-1.5, Cr 16-19, Mo 0-0.5, Ni 34-36, Nb 1-1.5, bal Fe.
Cast, annealed: 57,000-85,000 TS; 8 min El; 150-220 Brin. Austenitic stainless steel casting; good in most atmospheres at elevated temperatures. DIN G-X20NiCrNb3515. *Obsolete*

MARKER 12 AXS

Schmidt & Clemens Edelstahlwerke
C 0.2-0.4, Si 1-2, Mn 0-1.5, Cr 13-16, Mo 0.5, Ni 34-36, bal Fe.
Cast, annealed: 57,000-85,000 TS; 12 min El; 130-200 Brin. Austenitic stainless steel casting; for parts in carburizing atmosphere at temperatures up to 1920 F. DIN G-X25NiCrSi3515; ACI HT. *Obsolete*

MARKER 12A30

Schmidt & Clemens Edelstahlwerke
C 0.2-0.4, Si 1-2, Mn 0-1.5, Cr 29-31, Mo 0-0.5, Ni 19-21, bal Fe.
Cast, annealed: 64,000-92,000 TS; 6 min El; 150-220 Brin. Austenitic stainless casting; useable in air to 2160 F. DIN G-X30CRNiSi3020; ACI HL. *Obsolete*

MARKER 210 M

Schmidt & Clemens Edelstahlwerke
C 0.35, Si 0.9, Mn 0.3, Cr 1.05, V 0.2, W 1.85, bal Fe.
Cold work tool and die steel; oil or water hardenable to 53-56 Rock C. For pneumatic tools, cold shear blades, header chisels. Werkstoff Nr. 1.2541. *Obsolete*

MARKER 215 M

Schmidt & Clemens Edelstahlwerke
C 0.45, Si 0.9, Mn 0.3, Cr 1.05, V 0.2, W 1.85, bal Fe.
Cold work tool and die steel; oil hardenable to 54-58 Rock C. For pneumatic hammers, chisels, riveting tools. Werkstoff Nr. 1.2542.

MARKER 220 M

Schmidt & Clemens Edelstahlwerke
C 0.6, Si 0.9, Mn 0.3, Cr 1.05, V 0.2, W 1.85, bal Fe.
Cold work tool and die steel; oil or water hardenable to 57-61 Rock C. Punching and trimming dies, shear knives to 9 mm thick, wood working tools. Werkstoff Nr. 1.2550.

MARKER 2X COLUMBUS

Schmidt & Clemens Edelstahlwerke
C 0.85, bal Fe.
Water hardening carbon tool steel; for hand and pneumatic chisels, stone splitting hammers and wedges. Werkstoff Nr. 1.1631. *Obsolete*

MARKER 2X COLUMBUS-HART
Schmidt & Clemens Edelstahlwerke
C 1.15, bal Fe.
Water hardening carbon tool steel; special tools largely for stone industry. *Obsolete*

MARKER 3X ROSE
Schmidt & Clemens Edelstahlwerke
C 0.55, Si 0.08, Mn 0.4, bal Fe.
Water hardening carbon tool steel; for pliers, hatchets, axes, anvils, wood and stone working tools. Werkstoff Nr. 1.1820. *Obsolete*

MARKER 430 M
Schmidt & Clemens Edelstahlwerke
C 1, Cr 0.6, W 0.6, V 0.1, Mn, bal Fe.
Tool steel for hand punches, chisels. Werkstoff Nr. 1.2510.

MARKER 465 M
Schmidt & Clemens Edelstahlwerke
C 0.9, Mn 1.9, V 0.1, bal Fe.
Cold work tool and die steel; oil hardenable to 63-64 Rock C. For cutting dies of complicated shapes for thin metal to 3 mm thick, threading dies. Werkstoff Nr. 1.2842.

MARKER 4712, ETC.
Now WERKSTOFF 1.4712, ETC.

MARKER 476 M EXTRA
Schmidt & Clemens Edelstahlwerke
C 1.65, Cr 12, Mo 0.6, V 0.1, W 0.5, bal Fe.
Cold work tool and die steel; air or oil hardenable to 63-65 Rock C. For cold forming and stamping laminations, saw blades for precision parts as gauges; some corrosion resistance. Werkstoff Nr. 1.2601.

MARKER 476 M EXTRA B
Schmidt & Clemens Edelstahlwerke
C 1.65, Cr 12, Mo 0.65, V 0.5, W 0.5, bal Fe.
Cold work tool and die steel; air or oil hardenable to 63-65 Rock C. For cold forming and stamping dies, shear blades. Werkstoff Nr. 1.2602. *Obsolete*

MARKER 476 M SPEZIAL
Schmidt & Clemens Edelstahlwerke
C 2, Cr 12, Mo 0.5, V 0.5, W 1, bal Fe.
Cold work tool and die steel; air or oil hardenable to 63-65 Rock C. For cold forming and stamping dies, broaches, thread rolling dies, gages. *Obsolete*

MARKER 476 MEL
Schmidt & Clemens Edelstahlwerke
C 0.96, Cr 12, Mo 0.9, V 2.2, bal Fe.
High chromium cold work tool steel. Werkstoff Nr. 1.2376.

MARKER 477 M
Schmidt & Clemens Edelstahlwerke
C 1.05, Mn 0.9, Cr 1, W 1.15, bal Fe.
Cold work tool and die steel; oil hardenable to 63-66 Rock C. For punching and trimming dies for thin sheet and strip up to 4 mm thick, shears. Werkstoff Nr. 1.2419.

MARKER 480 M
Schmidt & Clemens Edelstahlwerke
C 1.05, Cr 0.9, Mn, bal Fe.
Tool steel for punches, cutting knives. Werkstoff Nr. 1.2127.

MARKER 4922
Schmidt & Clemens Edelstahlwerke
C 0.2, Cr 11.5, Mo 1, Ni 0.6, V 0.3, bal Fe.
Temperature resisting and corrosion resisting steel. Martensitic type. Werkstoff Nr. 1.4922.

MARKER 4948
Schmidt & Clemens Edelstahlwerke
C 0.06, Cr 18, Ni 11, Mo 0.3, bal Fe.
Austenitic stainless steel. Werkstoff Nr. 1.4948.

MARKER 4961
Schmidt & Clemens Edelstahlwerke
C 0.08, Cr 16, Ni 13, Nb 1, bal Fe.
Stabilized austenitic stainless steel. Werkstoff Nr. 1.4961.

MARKER 4980
Schmidt & Clemens Edelstahlwerke
C 0.08, Cr 15, Ni 26, Mo 1.25, V 0.3, Ti 2.1, Al, B, bal Fe.
Special alloy for high temperature operation. Werkstoff Nr. 1.4980.

MARKER 4981
Schmidt & Clemens Edelstahlwerke
C 0.08, Cr 16.5, Ni 16.5, Mo 1.8, Nb 1, bal Fe.
Werkstoff Nr. 1.4981.

MARKER 4988
Schmidt & Clemens Edelstahlwerke
C 0.08, Cr 16.5, Ni 13.5, Mo 1.3, V 0.75, Nb 1, bal Fe.
Werkstoff Nr. 1.4988.

MARKER 8A
Schmidt & Clemens Edelstahlwerke
C 0.15-0.35, Si 1-2.5, Mn 0-1.5, Cr 17-19, Mo 0-0.5, Ni 8-10, bal Fe.
Cast, annealed: 64,000-92,000 TS; 15 min El; 130-200 Brin. Austenitic stainless casting; for use in air up to 1550 F. DIN G-X25CrNiSi189; VDEh 4525. *Obsolete*

MARKER AW50
Schmidt & Clemens Edelstahlwerke
C 0.45, Cr 1.65, Mo 0.5, V 0.9, W 0.6, bal Fe.
Hot work tool and die steel; oil or water hardenable to 250,000-270,000 TS. Pressing dies for hot forming of screws, hot shears. Werkstoff Nr. 1.2603.

MARKER B 12
Schmidt & Clemens Edelstahlwerke
C 0.2, Cr 13, bal Fe.
Cold work tool steel. Werkstoff Nr. 1.2082.

MARKER C 3
Schmidt & Clemens Edelstahlwerke
C 1.4, Si 0.25, Mn 0.3, Cr 0.3, V 0.1, bal Fe.
Water hardening tool steel, high surface hardness possible; for drawing dies and tools. Werkstoff Nr. 1.2206.

MARKER C 73
Schmidt & Clemens Edelstahlwerke
C 1.45, Si 0.25, Mn 0.6, Cr 1.4, bal Fe.
Oil hardening tool steel; for woodworking tools, files, rubber cutting knives. Werkstoff Nr. 1.2063.

MARKER CDD
Schmidt & Clemens Edelstahlwerke
C 0.9, Si 0.25, Mn 0.3, Cr 0.8, bal Fe.
Water or oil hardening tool steel; for drawing dies, embossing tools, cold forming press and bend rollers. Werkstoff Nr. 1.2056.

MARKER CVC
Schmidt & Clemens Edelstahlwerke
C 0.9, Si 0.3, Mn 0.3, Cr 3.25, Mo 0.25, V 0.15, bal Fe.
Oil hardening tool steel; also used for large roller bearings and races; hardenable to 64-66 Rock C. *Obsolete*

MARKER D 606 G
Schmidt & Clemens Edelstahlwerke
C 0-0.1, Si 0-1, Mn 0-1.5, Cr 15-17, Ni 12-14, Nb = 10 x %C min, bal Fe.
Austenitic stainless casting. Good elevated temperature strength and creep properties to about 1100 F. DIN G-X8CrNiNb1613; VDEh Spec 4961. *Obsolete*

MARKER D 607 G
Schmidt & Clemens Edelstahlwerke
C 0-0.1, Si 0-1, Mn 0-1.5, Cr 15.5-17.5, Mo 1.5-2, Ni 15.5-17.5, Nb = 10 x %C min, bal Fe.
Austenitic stainless casting. Good elevated temperature strength and creep properties to about 1100 F. DIN G-8CrNiMoNb1616; VDEh Spec 4981. *Obsolete*

MARKER D690
Schmidt & Clemens Edelstahlwerke
C 0.06, Cr 15, Mo 1.25, Ni 26, V 0.3, Ti, B, Al, bal Fe.
Steel for parts operated at high temperature. Werkstoff Nr. 1.2779.

MARKER DC 10
Schmidt & Clemens Edelstahlwerke
C 0.23, Si 1.25, Cr 10, Mo 1.2, Ni 0.75, V 1, W 0.45, N 0.1, bal Fe.
Hot work tool and die steel; oil or water hardenable to 250,000-265,000 TS. For hot working of zinc, aluminum and other light metals. *Obsolete*

MARKER DCM
Schmidt & Clemens Edelstahlwerke
C 0.35, Si 1, Cr 5.15, Mo 1.45, V 0.5, bal Fe.
Hot work tool and die steel; oil, water or air hardenable to 256,000-274,000 TS. For forging dies, pressure casting molds. Werkstoff Nr. 1.2343; AISI H 11.

MARKER DCMC
Schmidt & Clemens Edelstahlwerke
C 0.36, Si 1, Mn 0.4, Co 2, Cr 5.15, Mo 1.45, V 0.5, bal Fe.
Hot work tool and die steel; oil or air hardenable to 250,000-270,000 TS. For forging dies, pressure casting molds. *Obsolete*

MARKER DCMX
Schmidt & Clemens Edelstahlwerke
C 0.27, Si 1, Cr 5.15, Mo 1.45, V 0.5, bal Fe.
Hot work tool and die steel; air or oil hardenable to 240,000-270,000 TS. For pressure casting molds, forging dies; better toughness than Marker DCM.

MARKER DCS
Schmidt & Clemens Edelstahlwerke
C 0.4, Si 1, Mn 0.4, Cr 5.15, Mo 1.2, V 1.05, S 0.15, bal Fe.
Hot work tool and die steel. Free machining grade similar to Marker DCV.

MARKER DCV
Schmidt & Clemens Edelstahlwerke
C 0.4, Si 1, Cr 5.15, Mo 1.35, V 1, bal Fe.
Hot work tool and die steel; oil or water hardenable to 256,000-276,000 TS. For hot pressing dies for light metals, pressure casting molds, forging dies. Werkstoff Nr. 1.2344; AISI H 13.

MARKER DCW
Schmidt & Clemens Edelstahlwerke
C 0.37, Si 1, Cr 5.15, Mo 1.45, V 0.3, W 1.5, bal Fe.
Hot work tool and die steel; oil or water hardenable to 280,000-300,000 TS. For hot shears, punches, forming tools for screws, rivets; hot tools for welding machines, forging dies. Werkstoff Nr. 1.2606; AISI H 12.

MARKER ED 12 G
Schmidt & Clemens Edelstahlwerke
C 0.15-0.23, Si 0-1, Mn 0.4-0.8, Cr 10-12, Mo 0.9-1.2, Ni 0.6-0.9, V 0.25-0.35, bal Fe.
Casting, air or oil hardenable. Good elevated temperature yield strength and creep strength up to about 1022 F. DIN G-X20CrMoV121; VDEh Spec 4937. *Obsolete*

MARKER ED 5 SG
Schmidt & Clemens Edelstahlwerke
C 0-0.15, Si 0-0.8, Mn 0.4-0.7, Cr 4-6.5, Mo 0.45-0.65, bal Fe.
Casting, air or oil hardenable. Good elevated temperature yield strength and creep strength up to about 932 F. DIN GS-12CrMo195; VDEh Spec 7362; AISI 501 cast. *Obsolete*

MARKER EUZONIT 60
Schmidt & Clemens Edelstahlwerke
C 0-0.05, Cr 15.5-18, Mo 16-19, Fe 0-1, bal Ni.
Cast, quenched: 71,000-100,000 TS; 42,600 YS; 6 min El; 170-230 Brin. Yield strength at 752 F: 28,400 psi. Resistant to chemical attack; good oxidizing resistance at elevated temperature. DIN G-X3NiMoCr6418; VDEh Spec 4472. *Obsolete*

MARKER EUZONIT 60.S
Schmidt & Clemens Edelstahlwerke
C 0.02, Cr 20, Mo 15.5, Fe, bal Ni.
Nickel-chromium-molybdenum stainless alloy. Werkstoff Nr.
2.4811.

MARKER EUZONIT 70
Schmidt & Clemens Edelstahlwerke
C 0-0.05, Mo 26-29, Fe 0-1, bal Ni.
Cast, quenched: 71,000-100,000 TS; 42,600 YS; 6 min El;
170-230 Brin. Cast stainless nickel base alloy; resistant to
boiling hydrochloric acid, humid hydrochloric gas, sulfuric
and phosphoric acids. DIN G-X3NiMo7028; VDEh Spec 4482.
Obsolete

MARKER EUZONIT 70.5
Schmidt & Clemens Edelstahlwerke
C 0.02, Mo 28, Fe, bal Ni.
Nickel-molybdenum stainless alloy. Werkstoff Nr. 2.4810.

MARKER EUZONIT G-4810
Schmidt & Clemens Edelstahlwerke
C 0.05, Mo 20, Fe, bal Ni.
Stainless nickel-molybdenum casting. Werkstoff Nr. 2.4810.

MARKER EUZONIT G-60
Schmidt & Clemens Edelstahlwerke
C 0.05, Cr 17, Mo 18, Fe, bal Ni.
Stainless nickel base steel casting. Werkstoff Nr. 2.4537.

MARKER EXTRA-HART
Schmidt & Clemens Edelstahlwerke
C 1.3, bal Fe.
Water hardening carbon tool steel; for engraving and
watchmaking tools, sculpturing tools. Werkstoff Nr. 1.1660.
Obsolete

MARKER EXTRA-MITTELHART
Schmidt & Clemens Edelstahlwerke
C 1.15, bal Fe.
Water hardening carbon tool steel; for drills, knives, chisels,
knife sharpening steels. Werkstoff Nr. 1.1650. *Obsolete*

MARKER EXTRA-SEHR ZAH
Schmidt & Clemens Edelstahlwerke
C 0.7, bal Fe.
Water hardening carbon tool steel; for blacksmith tools; very
tough. Werkstoff Nr. 1.1620. *Obsolete*

MARKER EXTRA-ZAH
Schmidt & Clemens Edelstahlwerke
C 0.85, bal Fe.
Water hardening carbon tool steel; for chisels, punches,
forming tools, hammers, anvils, rollers for mechanical
handling; tough. Werkstoff Nr. 1.1630. *Obsolete*

MARKER EXTRA-ZAHHART
Schmidt & Clemens Edelstahlwerke
C 1, bal Fe.
Water hardening carbon tool steel; for scythes, sickles, hand
stamping dies. Werkstoff Nr. 1.1640. *Obsolete*

MARKER F9
Schmidt & Clemens Edelstahlwerke
C 0.7, Si 1.65, Mn 0.7, bal Fe.
Water or oil hardening steel; for wrenches, pliers, screw
drivers. Werkstoff Nr. 1.2823.

MARKER G-4008
Schmidt & Clemens Edelstahlwerke
C 0.1, Cr 13.5, bal Fe.
Rustproof chromium steel casting. ACI CA 15; Werkstoff Nr.
1.4008.

MARKER G-4027
Schmidt & Clemens Edelstahlwerke
C 0.2, Cr 14, bal Fe.
Rustproof chromium steel casting. ACI CA 15; Werkstoff. Nr.
1.4027.

MARKER G-4034
Schmidt & Clemens Edelstahlwerke
C 0.45, Cr 15, bal Fe.
Rustproof chromium steel casting. ACI CA 40.

MARKER G-4059
Schmidt & Clemens Edelstahlwerke
C 0.25, Cr 17, Ni 1.5, bal Fe.
Rustproof chromium steel casting. ACI CB 30.

MARKER G-4085
Schmidt & Clemens Edelstahlwerke
C 0.7, Cr 29, bal Fe.
Rustproof and acid resisting steel casting. Werkstoff Nr.
1.4085.

MARKER G-4086
Schmidt & Clemens Edelstahlwerke
C 1.1, Cr 29, bal Fe.
Rustproof and acid resisting steel casting. Werkstoff Nr.
1.4086.

MARKER G-4088
Schmidt & Clemens Edelstahlwerke
C 1.7, Cr 18, bal Fe.
Rust resistant and wear resistant steel casting.

MARKER G-4122
Schmidt & Clemens Edelstahlwerke
C 0.4, Cr 17, Ni 1, Mo 1, bal Fe.
Rustproof chromium steel casting. Werkstoff Nr. 1.4122.

MARKER G-4136
Schmidt & Clemens Edelstahlwerke
C 0.7, Cr 29, Mo 2.2, bal Fe.
Rustproof and acid resisting steel casting. Werkstoff Nr.
1.4136.

MARKER G-4138
Schmidt & Clemens Edelstahlwerke
C 1.1, Cr 29, Mo 2.2, bal Fe.
Rustproof and acid resisting steel casting. ACI CC 50;
Werkstoff Nr. 1.4138.

MARKER G-4306
Schmidt & Clemens Edelstahlwerke
C 0.03, Cr 19, Ni 10, bal Fe.
Stainless chromium-nickel steel casting. ACI CF 3; Werkstoff
Nr. 1.4306.

MARKER G-4308
Schmidt & Clemens Edelstahlwerke
C 0.07, Cr 20, Ni 10, bal Fe.
Stainless chromium-nickel steel casting. ACI CF 8; Werkstoff
Nr. 1.4308.

MARKER G-4312
Schmidt & Clemens Edelstahlwerke
C 0.1, Cr 18, Ni 9, bal Fe.
Stainless chromium-nickel steel casting. ACI CF 12; Werkstoff
Nr. 1.4312.

MARKER G-4313
Schmidt & Clemens Edelstahlwerke
C 0.08, Cr 13, Ni 3.5, Mo 0.5, bal Fe.
Rustproof chromium steel casting. Werkstoff Nr. 1.4313.

MARKER G-4340
Schmidt & Clemens Edelstahlwerke
C 0.4, Cr 27, Ni 4, bal Fe.
Rustproof and acid resisting steel casting. ACI CC 50;
Werkstoff Nr. 1.4340.

MARKER G-4347
Schmidt & Clemens Edelstahlwerke
C 0.08, Cr 25, Ni 7, bal Fe.
Stainless chromium-nickel steel casting. Werkstoff Nr. 1.4347.

MARKER G-4404
Schmidt & Clemens Edelstahlwerke
C 0.03, Cr 18, Ni 11, Mo 2.2, bal Fe.
Stainless chromium-nickel steel casting. ACI CF 3M; AISI
316L; Werkstoff Nr. 1.4404.

MARKER G-4405
Schmidt & Clemens Edelstahlwerke
C 0.07, Cr 16, Ni 4.5, Mo 1, bal Fe.
Rustproof chromium steel casting. Werkstoff Nr 1.4405.

MARKER G-4408
Schmidt & Clemens Edelstahlwerke
C 0.7, Cr 20, Ni 10, Mo 2.2, bal Fe.
Stainless chromium-nickel steel casting. ACI CF 8 M; AISI
316; Werkstoff Nr. 1.4408.

MARKER G-4410
Schmidt & Clemens Edelstahlwerke
C 0.1, Cr 18, Ni 10, Mo 2.2, bal Fe.
Stainless chromium-nickel steel casting. Werkstoff Nr. 1.4410.

MARKER G-4457
Schmidt & Clemens Edelstahlwerke
C 0.25, Cr 25, Ni 9, Mo 2.2, bal Fe.
Stainless chromium-nickel steel casting.

MARKER G-4460
Schmidt & Clemens Edelstahlwerke
C 0.1, Cr 25, Ni 7, Mo 1.5, bal Fe.
Stainless chromium-nickel steel casting.

MARKER G-4464
Schmidt & Clemens Edelstahlwerke
C 0.35, Cr 26, Ni 5, Mo 2.5, bal Fe.
Rustproof and acid resisting steel casting. Werkstoff Nr.
1.4464.

MARKER G-4500
Schmidt & Clemens Edelstahlwerke
C 0.08, Cr 20, Ni 25, Mo 3, Nb, Cu, bal Fe.
Stainless chromium-nickel steel casting. ACI CN 7 M;
Werkstoff Nr. 1.4500.

MARKER G-4552
Schmidt & Clemens Edelstahlwerke
C 0.08, Cr 18, Ni 10, Nb, bal Fe.
Stainless chromium-nickel steel casting. ACI CF 8 C;
Werkstoff Nr. 1.4552.

MARKER G-4559
Schmidt & Clemens Edelstahlwerke
C 0.08, Cr 20, Ni 42, Mo 5, Nb, Cu, bal Fe.
Stainless chromium-nickel steel casting. Werkstoff Nr. 2.4557.

MARKER G-4579
Schmidt & Clemens Edelstahlwerke
C 0.08, Cr 17.5, Ni 13.5, Mo 4.5, Nb, bal Fe.
Stainless chromium-nickel steel casting. ACI CG 8 M;
Werkstoff Nr. 1.4579.

MARKER G-4581
Schmidt & Clemens Edelstahlwerke
C 0.08, Cr 18, Ni 11, Mo 2.2, Nb, bal Fe.
Stainless chromium-nickel steel casting. Werkstoff Nr. 1.4581.

MARKER G-4585
Schmidt & Clemens Edelstahlwerke
C 0.08, Cr 18, Ni 20, Mo 2.2, Nb, Cu, bal Fe.
Stainless chromium-nickel steel casting. Werkstoff Nr. 1.4585.

MARKER H18F
Schmidt & Clemens Edelstahlwerke
C 0-0.12, Si 2, Cr 18, bal Fe.
Annealed: 80,000 TS; 50,000 YS; 25 El; 50 RA; 150 Brin. For
furnace parts, heat treat boxes; Type 430; corrosion resistant.
Obsolete

MARKER H24 FS
Schmidt & Clemens Edelstahlwerke
C 0.1, Cr 24, Ni 1.5, bal Fe.
Wrought, annealed: 71,000-92,000 TS; 43,000 YS; 10 El; 200 Brin. Ferritic, corrosion resistant, good high temperature strength; for parts operating at 950-1200 C. *Obsolete*

MARKER H40A
Schmidt & Clemens Edelstahlwerke
C 0.12, Cr 16, Ni 36, bal Fe.
Wrought, annealed: 78,000-106,000 TS; 38,000 YS; 60 El; 180 Brin. Austenitic, for carburizing and annealing boxes, furnace parts operating at 550-1150 C. *Obsolete*

MARKER H8A
Schmidt & Clemens Edelstahlwerke
C 0.15, Cr 18, Ni 10, bal Fe.
Wrought, annealed: 78,000-106,000 TS; 52,500 YS; 40 El; 160 Brin. Austenitic general purpose stainless and heat resisting steel; for parts operating at 550-850 C. *Obsolete*

MARKER HM 3
Schmidt & Clemens Edelstahlwerke
C 0.31, Si 0.3, Mn 0.6, Cr 2.5, Mo 0.2, V 0.15, bal Fe.
For case hardening by nitriding. Case hardness expected: 750 VDH. Core strength: 128,000-156,000 psi TS. *Obsolete*

MARKER HOV D
Schmidt & Clemens Edelstahlwerke
C 0.35, Si 0.25, Mn 0.6, Cr 1.6, Mo 0.2, Ni 1, Al 0.9, bal Fe.
For case hardening by nitriding. Case hardness expected: 900 VDH. Core strength: 112,000-142,000 psi TS. *Obsolete*

MARKER IRR 4006, ETC.
Now WERKSTOFF NR. 1.4006, ETC.

MARKER K 6 R
Schmidt & Clemens Edelstahlwerke
C 0.6, Si 0.85, Mn 0.75, Cr 1.2, V 0.1, bal Fe.
Cold work tool steel; oil hardenable to 59-62 Rock C. For hole punching tools in heavy metal up to 4 mm thick, wood working tools as planes, saws. Werkstoff Nr. 1.2243.

MARKER K 8 R
Schmidt & Clemens Edelstahlwerke
C 0.85, Si 1.15, Mn 0.7, Cr 1.2, bal Fe.
Cold work tool steel; oil hardenable to 63-65 Rock C. For punches, chisels, woodworking tools, milling cutters. Werkstoff Nr. 1.2108.

MARKER K 97
Schmidt & Clemens Edelstahlwerke
C 0.06, Si 0.2, Mn 0.25, Cr 3.75, Mo 0.5, V 0.1, bal Fe.
Case hardening steel; case hardness: 62-65 Rock C. Core strength: 112,000-142,000 psi TS.

MARKER KEW
Schmidt & Clemens Edelstahlwerke
C 0.6, Cr 3.75, Mo 0.85, V 0.7, W 9, bal Fe.
Hot work tool steel, oil or air hardenable to 60-62 Rock C. For hot punching, stamping and cutting, tube pressing mandrels. Werkstoff Nr. 1.2622.

MARKER KL 9
Schmidt & Clemens Edelstahlwerke
C 1, Si 0.25, Mn 0.4, Cr 1.55, bal Fe.
Oil hardening tool steel; used for cutters for meat grinders, for ball and roller bearings and races. Werkstoff Nr. 1.2067.

MARKER KO 109
Schmidt & Clemens Edelstahlwerke
C 0.8, Co 4.75, Cr 4.15, Mo 0.65, V 1.55, W 18, bal Fe.
High speed steel for turning and punching tools for steel and cast iron; for deep hole drills. AISI T4; Werkstoff Nr. 1.3255.

MARKER KO 11
Schmidt & Clemens Edelstahlwerke
C 0.78, Co 9.5, Cr 4.15, Mo 0.65, V 1.55, W 18, bal Fe.
High speed steel for turning, planing, and drilling steel and cast iron (AISI T5); Werkstoff Nr. 1.3265.

MARKER KO 12
Schmidt & Clemens Edelstahlwerke
C 1.32, Co 10.5, Cr 4.5, Mo 4, V 3.25, W 10, bal Fe.
High speed steel for finishing tools, for inserted cutters, for cold flow press tools. Werkstoff Nr. 1.3207.

MARKER KO 13
Schmidt & Clemens Edelstahlwerke
C 1.1, Co 8, Cr 3.8, Mo 9.5, V 1.2, W 1.5, bal Fe.
High speed steel for lathe tools, form cutters, planing tools, drills, reamers. Good high temperature hardness. AISI M42; Werkstoff Nr. 1.3247.

MARKER KO 15
Schmidt & Clemens Edelstahlwerke
C 1.5, Co 4.75, Cr 4.15, Mo 3.75, V 5, W 6.6, bal Fe.
High speed steel for lathe tools, gear cutters, counterbores, drills, taps. AISI M15. *Obsolete*

MARKER KO 55
Schmidt & Clemens Edelstahlwerke
C 1.32, Co 4.75, Cr 4.15, Mo 0.85, V 3.75, W 12, bal Fe.
High speed steel for finishing tools for turning, reaming, form cutting, for inserted cutters. Werkstoff Nr. 1.3202.

MARKER KSP
Schmidt & Clemens Edelstahlwerke
C 0.5, Si 0.3, Mn 0.65, Cr 0.8, Mo 0.3, Ni 1.8, V 0.1, bal Fe.
Oil hardening steel, to 56-58 Rock C. For forming tools, can also be nitrided to 600-650 Vickers; smooth finish possible.

MARKER KSPE
Schmidt & Clemens Edelstahlwerke
C 0.5, Cr 1.1, Mo 0.5, Ni 1.7, V 0.1, bal Fe.

MARKER LW 10
Schmidt & Clemens Edelstahlwerke
C 1.2, Si 0.25, Mn 0.3, Cr 0.2, V 0.1, W 1, bal Fe.
Water hardening tool steel, high surface hardness possible; for drills, drawing tools, knives for pipe cutting. Werkstoff Nr. 1.2516.

MARKER MAT
Schmidt & Clemens Edelstahlwerke
C 0.03, Ni 18, Co 9, Mo 4.9, Ti, bal Fe.
Werkstoff Nr. 1.2709.

MARKER MEWE
Schmidt & Clemens Edelstahlwerke
C 0.3, Cr 2.35, V 0.6, W 4.25, bal Fe.
Hot work tool and die steel; air or oil hardenable to 240,000-265,000 TS. For hot forming and thread rolling of screws, nuts, rivets; for hot shears for pressure casting molds. Werkstoff Nr. 1.2567.

MARKER MH
Schmidt & Clemens Edelstahlwerke
C 0.35, Cr 1.4, Mo 0.3, Ni 4.1, bal Fe.
Hot work tool steel as hot pressing dies. Werkstoff Nr. 1.2766.

MARKER MHM
Schmidt & Clemens Edelstahlwerke
C 0.35, Si 0.3, Mn 0.5, Cr 0.9, Mo 1.15, Ni 3.5, V 0.12, bal Fe.
Hot work tool steel; oil hardenable for 228,000-242,000 TS; for deep impression forging dies. *Obsolete*

MARKER MNSK
Schmidt & Clemens Edelstahlwerke
C 0.62, Cr 0.6, Mn, Si, bal Fe.
Tool steel for punches, shear knives. Werkstoff Nr. 1.2101.

MARKER MO 5
Schmidt & Clemens Edelstahlwerke
C 0.85, Cr 4.15, Mo 5.1, V 1.85, W 6.35, bal Fe.
High speed steel for tools for general purpose cutting, lathe turning, milling, drilling of steel and nonferrous metals; tough. Werkstoff Nr. 1.3343; AISI M2.

MARKER MO 5 CO
Schmidt & Clemens Edelstahlwerke
C 0.85, Co 4.75, Cr 4.15, Mo 5.1, V 1.85, W 6.35, bal Fe.
High speed steel for turning tools, threading tools, deep hole drills, cutting tough iron and steel. Werkstoff Nr. 1.3243; AISI M35.

MARKER MO 5 CO H
Schmidt & Clemens Edelstahlwerke
C 1.1, Co 5, Cr 4.25, Mo 3.75, V 2, W 6.75, bal Fe.
High speed steel for fast finishing tools for turning, planing, deep hole drilling. Werkstoff Nr. 1.3246; AISI M41.

MARKER MO 5 H
Schmidt & Clemens Edelstahlwerke
C 1, Cr 4.15, Mo 5.1, V 1.85, W 6.35, bal Fe.
High speed steel for finish cutting lathe, milling and drilling tools. Werkstoff Nr. 1.3342

MARKER MO 9 CO 8
Schmidt & Clemens Edelstahlwerke
C 0.94, Co 8, Cr 4, Mo 8.5, V 2, W 2, bal Fe.
High speed steel for turning and planing steel and cast iron under difficult conditions. AISI M34; Werkstoff Nr. 1.3249.

MARKER MO 9 V
Schmidt & Clemens Edelstahlwerke
C 1.05, Cr 3.8, Mo 9, V 2, W 1.75, bal Fe.
High speed steel for turning, milling, drilling steel and nonferrous metals. Werkstoff Nr. 1.3348; AISI M7.

MARKER MOV 4
Schmidt & Clemens Edelstahlwerke
C 1.23, Cr 4.15, Mo 5.1, V 2.9, W 6.35, bal Fe.
High speed steel for turning, reaming, gear cutting of steel and nonferrous metals; high hardness. Werkstoff Nr. 1.3344; AISI M3(2).

MARKER MSV
Schmidt & Clemens Edelstahlwerke
C 1, Si 0.2, Mn 0.25, V 0.15, bal Fe.
Water hardening tool steel; for small cold forming dies for making screws, rivets, coins; for drawing dies.

MARKER MW 3
Schmidt & Clemens Edelstahlwerke
C 1.42, Si 0.25, Mn 0.3, Cr 0.3, V 0.25, W 3.25, bal Fe.
Water hardening tool steel; for lathe and planer tools, precision tools for watchmaking, engraving tools. Werkstoff Nr. 1.2562. 7
   Use *Obsolete*

MARKER MW 5
Schmidt & Clemens Edelstahlwerke
C 1.3, Si 0.25, Mn 0.3, Cr 0.3, W 4.75, bal Fe.
Water hardening tool steel. For drawing dies, mandrels for drawing pipe. Werkstoff Nr. 1.2453. *Obsolete*

MARKER P 42 W
Schmidt & Clemens Edelstahlwerke
C 0.85, Cr 26, W 15, Fe, Nb, bal Co.
Special alloy for high temperature operation.

MARKER P 63
Schmidt & Clemens Edelstahlwerke
C 0.3, Cr 28, Mo 5.5, Ni 2.5, Fe, bal Co.
Special alloy for high temperature operation. Werkstoff Nr. 2.4979

MARKER PHM
Schmidt & Clemens Edelstahlwerke
C 0.45, Cr 1.4, Mo 0.7, V 0.3, bal Fe.
Hot work tool steel; oil or water hardenable to 300,000 TS. For centrifugal casting molds for zinc alloys, hot forming tools. Werkstoff Nr. 1.2323.

MARKER POLYTEN
Schmidt & Clemens Edelstahlwerke
C 1, Si 0.3, Mn 0.5, Cr 5.2, Mo 1, V 0.2, bal Fe.
Air or oil hardening tool steel; for cold forming dies, shears, knives, slotting saws, nail making tools. Werkstoff Nr. 1.2363; AISI A2. *Obsolete*

MARKER PSH
Schmidt & Clemens Edelstahlwerke
C 0.5, Si 0.25, Mn 0.5, Cr 1.05, Ni 3.5, bal Fe.
Oil hardening tool steel, to 55-58 Rock C. For dies for forming emblems, plaques, buttons, commemorative coins, cutlery. Werkstoff Nr. 1.2767.

MARKER PSHB
Schmidt & Clemens Edelstahlwerke
C 0.45, Si 0.25, Mn 0.5, Cr 1.25, Ni 3.75, 0.20 Mo or 0.50 W, bal Fe.
Oil hardening tool steel, to 53-56 Rock C. For pressing dies requiring deep impressions, thread rolling dies, embossing tools. Werkstoff Nr. 1.2767.

MARKER PT
Schmidt & Clemens Edelstahlwerke
C 1, Cr 5.2, Mo 1.1, V 0.2, bal Fe.
Air hardening tool steel for thread rolling dies, shear punches, trimming tools. Werkstoff Nr. 1.2363.

MARKER PT 13
Schmidt & Clemens Edelstahlwerke
C 0.8, Cr 13.5, Co 1.1, bal Fe.
Werkstoff Nr. 1.2883.

MARKER PT 8
Schmidt & Clemens Edelstahlwerke
C 0.5, Cr 9, Mo 1.2, W 1.2, bal Fe.
Cold work tool steel for shear knives and cutting dies. Werkstoff Nr. 1.2631.

MARKER PTE
Schmidt & Clemens Edelstahlwerke
C 0.63, Cr 5.3, Mo 1.2, V 0.3, bal Fe.
Air hardening tool steel. Werkstoff Nr. 1.2362.

MARKER PTM
Schmidt & Clemens Edelstahlwerke
C 0.81, Cr 4, Mo 43, V 1, bal Fe.
Air hardening cold work tool steel. Werkstoff Nr. 1.2369.

MARKER PWC
Schmidt & Clemens Edelstahlwerke
C 0.45, Cr 4.25, Mo 0.5, V 2.25, W 4.25, Co 4.25, bal Fe.
Hot work tool steel, air or oil hardenable to 240,000-265,000 TS. For hot punching and stamping operations. Werkstoff Nr. 1.2678; AISI H 19.

MARKER PWC 2
Schmidt & Clemens Edelstahlwerke
C 0.2, Cr 9.5, Mo 2, W 5.5, Co 10, bal Fe.
Hot work tool steel. Werkstoff Nr. 1.2888.

MARKER R 17
Schmidt & Clemens Edelstahlwerke
C 0.38, Si 0.4, Mn 0.8, Cr 17, Mo 1.25, Ni 1, bal Fe.
Air or oil hardenable to 50-55 Rock C. For surgical and dental tools, cutlery, kitchen tools; corrosion resistant.

MARKER R 18
Schmidt & Clemens Edelstahlwerke
C 1.1, Si 0.4, Mn 0.5, Cr 17.5, Mo 1, V 0.1, bal Fe.
Air or oil hardenable to 58-60 Rock C. Corrosion resisting; for discs for meat grinders, surgical and dental tools. *Obsolete*

MARKER REM
Schmidt & Clemens Edelstahlwerke
C 0.45, Cr 13.5, Ni 13, V 0.5, W 2.5, bal Fe.
Austenitic, hot work tool steel. For dies and forming tools for working copper-nickel and nickel alloy that are difficult to work. Werkstoff Nr. 1.2731.

MARKER REM SPEZIAL
Schmidt & Clemens Edelstahlwerke
C 0.5, Cr 4, Mo 0.7, Ni 11.5, V 1.1, W 12.5, Co 1.7, bal Fe.
Hot work tool steel. Werkstoff Nr. 1.2758.

MARKER RH
Schmidt & Clemens Edelstahlwerke
C 0.4, Si 0.4, Mn 0.3, Cr 13, bal Fe.
Air or oil hardening high chromium steel; for synthetic resin molds, cold stamping and punching dies, surgical tools; corrosion resistant. Werkstoff Nr. 1.2083.

MARKER S & C S 14
Schmidt & Clemens Edelstahlwerke
C 0.15-0.25, Cr 12.5-14.5, Ni 0-0.1, bal Fe.
Cast, heat treated: 85,000-114,000 TS; 56,900 YS; 12 min El; 180-240 Brin. Ductile, weldable, cast corrosion resistant steel; favorable mechanical properties; pumping parts, mechanical engineering. DIN G-X25Cr14; VDEh Spec 4021. *Obsolete*

MARKER S & C S 14 S
Schmidt & Clemens Edelstahlwerke
C 0.08-0.15, Cr 12.5-14, Mo 0-0.5, Ni 0.5-1.5, bal Fe.
Cast, heat treated: 85,000-114,000 TS; 56,900 YS; 14 min El; 180-240 Brin. Ductile, weldable, cast corrosion resistant steel; turbine and pump construction. DIN G-X12Cr14; ACI CA-15; VDEh Spec 4006. *Obsolete*

MARKER S & C S 15
Schmidt & Clemens Edelstahlwerke
C 0.4-0.5, Cr 14-15.5, bal Fe.
Cast, annealed: 240 max Brin. Corrosion resistant, hardenable; good sliding properties, for pump construction. DIN G-X45Cr15; VDEh Spec 4034. *Obsolete*

MARKER S & C S 17 M
Schmidt & Clemens Edelstahlwerke
C 0.33-0.43, Cr 15.5-17.5, Mo 1-1.3, Ni 0-1, bal Fe.
Cast, heat treated: 107,000-135,000 TS; 65,400 YS; 5 min El; 220-300 Brin. Yield strength at 752 F: 49,800 psi. Hardenable, corrosion resistant, good high temperature strength, resistant to organic acids. DIN G-X35CrMo17; VDEh Spec 4122. *Obsolete*

MARKER S & C S 18
Schmidt & Clemens Edelstahlwerke
C 0.15-0.25, Cr 15.5-17.5, Ni 1-2, bal Fe.
Cast, heat treated: 100,000-128,000 TS; 64,000 YS; 6 min El; 220-290 Brin. Hardenable, increased corrosion resistance, ship building and marine work. DIN G-X22CrNi17; VDEh Spec 4059; similar to ACI Cb-30 and AISI 431. *Obsolete*

MARKER S & C S 30
Schmidt & Clemens Edelstahlwerke
C 0.5-0.9, Cr 27-30, bal Fe.
Cast, annealed: 210-280 Brin. Good corrosion resistant casting, resistant to oxidizing acids and salts. DIN G-X70Cr29; VDEh Spec 4085. *Obsolete*

MARKER S & C S 30 H
Schmidt & Clemens Edelstahlwerke
C 0.9-1.3, Cr 27-30, bal Fe.
Cast, annealed: 280-330 Brin. Corrosion resistant casting, good wear and abrasion resistance. DIN G-X120Cr29; VDEh Spec 4086. *Obsolete*

MARKER S & C S 30 M
Schmidt & Clemens Edelstahlwerke
C 0.5-0.9, Cr 27-30, Mo 2-2.5, bal Fe.
Cast, annealed: 210-280 Brin. Corrosion resistant casting; for potash mining equipment, and chemical industry. DIN G-X70CrMo292; VDEh Spec 4136. *Obsolete*

MARKER S & C S 30 MH
Schmidt & Clemens Edelstahlwerke
C 0.9-1.3, Cr 27-30, Ni 2-2.5, bal Fe.
Cast, annealed: 260-330 Brin. Corrosion resistant casting; good hardness and very good wear resistance. DIN G-X120CrMo292; VDEh Spec 4138. *Obsolete*

MARKER S & C SG 25
Schmidt & Clemens Edelstahlwerke
C 1.5-1.8, Cr 24-26, bal Fe.
Cast, annealed: 350 max Brin. Hardenable chromium corrosion resistant casting; good wear resistance. DIN G-X170Cr25. *Obsolete*

MARKER S & C SG 25 M
Schmidt & Clemens Edelstahlwerke
C 1.5-1.8, Cr 24-26, Mo 2-2.5, bal Fe.
Cast, annealed: 400 max Brin. Hardenable chromium corrosion resistant steel casting; good resistance to salt solutions, diluted organic acids; wear resistant. DIN G-X170CrMo252. *Obsolete*

MARKER S & C SN 10
Schmidt & Clemens Edelstahlwerke
C 0.2-0.3, Cr 24-26, Ni 8-10, bal Fe.
Cast, quenched: 71,000-107,000 TS; 89,800 YS; 10 min El; 180-240 Brin. Corrosion resistant steel casting; good ductility; for pump components. DIN G-X25CrNi259; similar to ACI CE-30. *Obsolete*

MARKER S & C SN 10 M
Schmidt & Clemens Edelstahlwerke
C 0.2-0.3, Cr 23.5-25, Mo 2-2.5, Ni 8-10, bal Fe.
Cast, quenched: 78,000-107,000 TS; 89,800 YS; 10 min El; 180-240 Brin. Corrosion resistant steel casting; improved resistance to sulfites and chlorides. DIN G-X25CrNiMo259. *Obsolete*

MARKER S & C SN 12
Schmidt & Clemens Edelstahlwerke
C 0-0.08, Cr 16-18, Mo 4-5, Ni 12.5-14.5, Nb = 8 x %C, bal Fe.
Cast, quenched: 57,000-92,000 TS; 28,400 YS; 15 min El; 140-190 Brin. Austenitic corrosion resistant steel casting; weldable, good at elevated temperatures, indifferent to localized corrosion. DIN G-X5CrNiMoNb1713; VDEh Spec 4579; similar to ACI CG-8M, AISI 317 (cast). *Obsolete*

MARKER S & C SN 18
Schmidt & Clemens Edelstahlwerke
C 0-0.08, Cr 16.5-18.5, Mo 2-2.5, Ni 19-21, Cu 1.8-2.4, Nb = 8 x %C, bal Fe.
Cast, quenched: 64,000-92,000 TS; 25,600 YS; 15 min El; 130-180 Brin. Austenitic corrosion resistant steel casting; resistant to sulfuric acid, phosphoric acid, strong organic acids. DIN G-X7CrNiMoCu1818; VDEh Spec 4585. *Obsolete*

MARKER S & C SN 25
Schmidt & Clemens Edelstahlwerke
C 0-0.08, Cr 19-21, Mo 2.5-3.5, Ni 24-26, Cu 1.5-2, Nb = 10 x %C, bal Fe.
Cast, quenched: 64,000-92,000 TS; 28,400 YS; 14 min El; 130-180 Brin. Austenitic corrosion resistant steel casting; resistant to sulfuric acid, phosphoric acid, strong organic acids. DIN G-X7NiCrMoCu2520; VDEh Spec 4500; ACI CN-7M. *Obsolete*

MARKER S & C SN 26 M
Schmidt & Clemens Edelstahlwerke
C 0-0.08, Cr 24-26, Mo 2-2.5, Ni 24-26, Nb = 8 x %C, bal Fe.
Cast, quenched: 64,000-92,000 TS; 28,400 YS; 15 min El; 130-180 Brin. Austenitic corrosion resistant steel casting; resistant to sulfuric acid, phosphoric acid, strong organic acids. DIN G-X7CrNiMoNb2525; VDEh Spec 4588. *Obsolete*

MARKER S & C SN 42
Schmidt & Clemens Edelstahlwerke
C 0-0.08, Cr 20-22, Mo 5-6, Ni 42-44, Nb = 12 x %C, bal Fe.
Cast, quenched: 71,000-100,000 TS; 31,300 YS; 14 min El; 150-200 Brin. Austenitic stainless alloy; resistant to sulfuric acid, phosphoric acid, strong organic acids, many pickling and galvanizing baths. DIN G-X5NiCrMoNb4422; VDEh Spec 4557. *Obsolete*

MARKER S & C SN 5 M

Schmidt & Clemens Edelstahlwerke
C 0.3-0.5, Cr 26-28, Mo 2-2.5, Ni 4-6, bal Fe.
Cast, annealed: 240-300 Brin. Corrosion resistant steel casting; for intricate casting. DIN G-X40CrNiMo275; VDEh Spec 4464. *Obsolete*

MARKER S & C SN 6 M

Schmidt & Clemens Edelstahlwerke
C 0-0.1, Cr 24-28, Mo 1.3-1.8, Ni 5-7, bal Fe.
Cast, annealed: 85,000-107,000 TS; 59,700 YS; 15 min El; 170-220 Brin. Corrosion resistant steel casting; good mechanical properties and good resistance to oxidizing acids and salts. DIN G-X8CrNiMo275. *Obsolete*

MARKER S & C SN 8

Schmidt & Clemens Edelstahlwerke
C 0-0.12, Cr 17-19, Ni 8-10, bal Fe.
Cast, quenched: 64,000-92,000 TS; 28,400 YS; 20 min El; 140-190 Brin. Basic grade of austenitic stainless steel casting; excellent ductility; universal application. DIN X10CrNi189; VDEh Spec 4312; similar to ACI CF-20. *Obsolete*

MARKER S & C SN 8 EL

Schmidt & Clemens Edelstahlwerke
C 0-0.03, Cr 17-19.5, Ni 10-12, bal Fe.
Cast, quenched: 64,000-92,000 TS; 24,200 YS; 25 min El; 130-180 Brin. Lowest carbon grade of basic 18-8 austenitic stainless steel casting; for welding without subsequent heat treat. DIN G-X2CrNi1811; VDEh Spec 4306; ACI Cf-3. *Obsolete*

MARKER S & C SN 8 N

Schmidt & Clemens Edelstahlwerke
C 0-0.08, Cr 17.5-20, Ni 9-11, Nb = 8 x %C, bal Fe.
Cast, quenched: 64,000-92,000 TS; 28,400 YS; 20 min El; 140-190 Brin. Niobium stabilized 18-8 casting, weldable without subsequent heat treat; also used at temperatures above 600 F. DIN G-X8CrNiNb1810; VDEh Spec 4552; ACI CF-8C. *Obsolete*

MARKER S & C SN 8 S

Schmidt & Clemens Edelstahlwerke
C 0-0.07, Cr 17.5-20, Ni 9-11, bal Fe.
Cast, quenched: 60,000-88,000 TS; 28,400 YS; 20 min El; 140-190 Brin. Lower carbon grade of basic 18-8 austenitic cast steel; weldable without subsequent heat treat. DIN G-X6CrNi1810; VDEh Spec 4308; ACI CF-8. *Obsolete*

MARKER S & C SN 9

Schmidt & Clemens Edelstahlwerke
C 0-0.12, Cr 17-19, Mo 2-2.5, Ni 9-11, bal Fe.
Cast, quenched: 64,000-92,000 TS; 29,800 YS; 20 min El; 140-190 Brin. Austenitic stainless casting; better corrosion resistance than ordinary 18-8 type. DIN G-X10CrNiMo1810; VDEh Spec 4410. *Obsolete*

MARKER S & C SN 9 EL

Schmidt & Clemens Edelstahlwerke
C 0-0.03, Cr 16.5-18.5, Mo 2-2.5, Ni 11-13, bal Fe.
Cast, quenched: 60,000-88,000 TS; 24,200 YS; 25 min El; 130-180 Brin. Low carbon, improved corrosion resistant austenitic stainless casting; weldable without subsequent heat treat. DIN G-X2CrNiMo1812; VDEh Spec 4404; ACI CF-3M; AISI 316L (cast). *Obsolete*

MARKER S & C SN 9 N

Schmidt & Clemens Edelstahlwerke
C 0-0.08, Cr 16.5-18.5, Mo 2-2.5, Ni 10.5-12, Nb = 8 x %C, bal Fe.
Cast, quenched: 64,000-92,000 TS; 29,800 YS; 20 min El; 140-190 Brin. Niobium stabilized austenitic casting; very good corrosion resistance, weldable without subsequent heat treat; usable above 600 F. DIN G-X7CrNiMoNb1811; VDEh Spec 4581. *Obsolete*

MARKER S & C SN 9 S

Schmidt & Clemens Edelstahlwerke
C 0-0.07, Cr 17-19.5, Mo 2-2.5, Ni 10-12, bal Fe.
Cast, quenched: 64,000-92,000 TS; 29,800 YS; 20 min El; 140-190 Brin. Austenitic stainless casting; very good corrosion resistance, weldable without subsequent heat treat. DIN G-X6CrNiMo1811; VDEh Spec 4408; ACI CF-8M; AISI 316 (cast). *Obsolete*

MARKER S & C SN 99 N

Schmidt & Clemens Edelstahlwerke
C 0-0.08, Cr 18-20, Mo 2.8-3.5, Ni 12-14, Nb = 8 x C, bal Fe.
Cast, quenched: 64,000-92,000 TS; 29,800 YS; 20 El min; 140-190 Brin. Yield strength at 752 F: 18,500 psi. Austenitic stainless casting; good corrosion resistance; for welded assemblies above 600 F. DIN G-X8CrNiMoNb1812; VDEh Spec 4583. *Obsolete*

MARKER S & G SN 5

Schmidt & Clemens Edelstahlwerke
C 0.3-0.5, Cr 26-28, Ni 3.5-5.5, bal Fe.
Cast, annealed: 71,000-114,000 TS; 230-280 Brin. Chromium corrosion resistant casting; good fluidity for intricate castings. DIN G-X40CrNi274; VDEh Spec 4340; ACI CC-50. *Obsolete*

MARKER S 7 A

Schmidt & Clemens Edelstahlwerke
C 0.2, Si 0.25, Mn 1.25, Cr 1.15, bal Fe.
Case hardening steel. Case hardness: 62-64 Rock C. Core strength: 142,000-182,000 psi TS. Werkstoff Nr. 1.2162.

MARKER S 7 N

Schmidt & Clemens Edelstahlwerke
C 0.3, Si 0.3, Mn 0.6, Cr 1.35, Mo 0.25, Ni 0.5, V 0.12, bal Fe.
Case hardening steel. Case hardness expected: 62-64 Rock C. Core strength: 200,000-240,000 psi TS. *Obsolete*

MARKER S.N.18

Mannesmannrohren-Werke AG
C 0.08, Cr 18, Ni 18, Mo 2, Cu 2, bal Fe.
For chemical plant equipment, pump and valve parts; corrosion resistant, austenitic. *Obsolete*

MARKER S.N.25

Mannesmannrohren-Werke AG
C 0.1, Cr 18, Ni 25, Mo 4, Cu 2, bal Fe.
For chemical plant equipment, pump and valve parts; corrosion resistant, austenitic. *Obsolete*

MARKER S.N.42

Mannesmannrohren-Werke AG
C 0.1, Cr 18, Ni 42, Mo 5, Cu 2, bal Fe.
For chemical plant equipment; corrosion and heat resistant. *Obsolete*

MARKER SL 10

Schmidt & Clemens Edelstahlwerke
C 0.06, Cr 20, Co 18, Fe, Ti, Al, bal Ni.
Special alloy for high temperature operation. Werkstoff Nr. 2.4969. Similar to Nimonic 90.

MARKER SL 15

Schmidt & Clemens Edelstahlwerke
C 0.08, Cr 19, Mo 10, Co 11, Fe, Ti, Al, bal Ni.
Special alloy for high temperature operation. Werkstoff Nr. 2.4973.

MARKER SL 8

Schmidt & Clemens Edelstahlwerke
C 0.05, Cr 20, Fe 2, Al 1.2, Ti 2.2, bal Ni.
Special alloy for high temperature operation. Werkstoff Nr. 2.4952. Similar to Nimonic 80 A.

MARKER SPEZIAL M EXTRA

Schmidt & Clemens Edelstahlwerke
C 1, V 0.1, bal Fe.
Water hardening carbon tool steel; for small cold working dies, as for rivets, screws, coins, medals. *Obsolete*

MARKER SPEZIAL M-MITTELHART

Schmidt & Clemens Edelstahlwerke
C 1.1, bal Fe.
Water hardening carbon tool steel; for drills, taps, woodworking tools, knife sharpening or whetting steel. Werkstoff Nr. 1.1550. *Obsolete*

MARKER SPEZIAL M-SEHR ZAH

Schmidt & Clemens Edelstahlwerke
C 0.7, bal Fe.
Water hardening carbon tool steel; for shears, woodworking tools, riveting hammers; very tough. Werkstoff Nr. 1.1520. *Obsolete*

MARKER SPEZIAL M-ZAH

Schmidt & Clemens Edelstahlwerke
C 0.85, bal Fe.
Water hardening tool steel; for punches and dies, for chisels, shears, cutlery. Werkstoff Nr. 1.1530. *Obsolete*

MARKER SPEZIAL M-ZAHHART

Schmidt & Clemens Edelstahlwerke
C 1, bal Fe.
Water hardening carbon tool steel; for workshop tools as drills, chisels, hand lettering and stamping dies. Werkstoff Nr. 1.1540. *Obsolete*

MARKER SPM

Schmidt & Clemens Edelstahlwerke
C 0.3, Cr 2.35, V 0.25, W 9, Co 2.05, bal Fe.
Hot work tool steel, oil or air hardenable to 240,000-265,000 TS max. For dies for hot forming bronze, brass and other copper alloys. Werkstoff Nr. 1.2662.

MARKER SRS

Schmidt & Clemens Edelstahlwerke
C 0.55, Cr 0.7, Mo 0.3, Ni 1.65, V 0.1, bal Fe.
Hot work tool and die steel; oil hardenable to 300,000 TS. For drop forge dies, jaws for welding machines, shears. Werkstoff Nr. 1.2713.

MARKER SRS SPEZIAL

Schmidt & Clemens Edelstahlwerke
C 0.57, Cr 1.1, Mo 0.8, Ni 1.7, V 0.1, bal Fe.
Hot work tool steel. Werkstoff Nr. 1.2744.

MARKER SRSE

Schmidt & Clemens Edelstahlwerke
C 0.55, Cr 1.1, Mo 0.5, Ni 1.65, V 0.1, bal Fe.
Hot work tool and die steel; air, oil or water hardenable to 270,000-300,000 TS. For casting molds for zinc, tin and lead; hot forming tools. Werkstoff Nr. 1.2714.

MARKER SS 11

Schmidt & Clemens Edelstahlwerke
C 0.88, Cr 4.15, Mo 0.85, V 2.45, W 12, bal Fe.
High speed steel for turning, milling, drilling and gear cutting steel. Werkstoff Nr. 1.3318. *Obsolete*

MARKER SS 13

Schmidt & Clemens Edelstahlwerke
C 1.25, Cr 4.15, Mo 0.85, V 3.75, W 12, bal Fe.
High speed steel, high hardness for finishing tools as reamers, finish turning and precision boring. Werkstoff Nr. 1.3302. *Obsolete*

MARKER SS 19

Schmidt & Clemens Edelstahlwerke
C 0.74, Cr 4.15, V 1.1, W 18, bal Fe.
High speed steel for general purpose tools for turning, planing, milling and drilling steel and nonferrous metals. Werkstoff Nr. 1.3355; AISI T1.

MARKER TAM

Schmidt & Clemens Edelstahlwerke
C 0.85, Si 0.25, Mn 0.25, Mo 0.15, Ni 0.2, V 0.3, bal Fe.
Water hardening tool steel; for cold forming screws, nuts, bolts, balls; good resistance to cracking under pressure. *Obsolete*

MARKER VG 508

Schmidt & Clemens Edelstahlwerke
C 0.2-0.4, Si 1-2, Mn 0-1.5, Cr 24-26, Mo 0.5, Ni 34-36, bal Fe.
Cast, annealed: 64,000-92,000 TS; 8 min El; 150-220 Brin.
Austenitic stainless casting; for use in parts operating at
temperatures up to 2100 F. DIN G-X40NiCrSi3525. *Obsolete*

MARKER VK-25

Schmidt & Clemens Edelstahlwerke
C 1.7, Cr 25, bal Fe.
Rust resistant and wear resistant steel casting.

MARKER W 11

Schmidt & Clemens Edelstahlwerke
C 0.3, Cr 2.65, V 0.35, W 9, bal Fe.
Hot work tool steel, air or oil hardenable to 240,000-265,000
TS. For hot stamping dies, punches, shears, hot extrusion
dies. Werkstoff Nr. 1.2581.

MARKER W 18 K

Schmidt & Clemens Edelstahlwerke
C 0.58, Cr 4.15, Mo 0.3, V 1, W 18, bal Fe.
Cold work tool and die steel; oil or air hardenable to 57-60
Rock C. For thread rolling dies for screws, bolts. Werkstoff Nr.
1.2587. *Obsolete*

MARKER W5/0

Schmidt & Clemens Edelstahlwerke
C 0.3, Si 0.85, Cr 1.05, V 0.2, W 3.75, bal Fe.
Hot work tool and die steel; oil or water hardenable to
210,000-250,000 TS. For hot forming rivets; thread rolling
screws; mandrels, jaws for electric welders. Werkstoff Nr.
1.2564.

MARKER WAGT

Schmidt & Clemens Edelstahlwerke
C 0.4, Mn 1.25, Cr 1.85, Mo 0.2, bal Fe.
Hot work tool and die steel; oil hardenable to 270,000 TS. For
drop hammer forging dies, hot punches, frames for casting
molds. Werkstoff Nr. 1.2311.

MARKER WAGT EXTRA

Schmidt & Clemens Edelstahlwerke
C 0.43, Mn 1.4, Cr 2.3, Mo 0.25, Ni 0.5, V 0.1, bal Fe.
Hot work tool and die steel; oil or air hardenable to
234,000-284,000 TS. For forging hammers. *Obsolete*

MARKER WAGT S

Schmidt & Clemens Edelstahlwerke
C 0.4, Si 0.3, Mo 0.2, S 0.1, Mn 1.25, Cr 1.85, bal Fe.
Hot work tool steel; free machining grade of MARKER WAGT.

MARKER WL 4

Schmidt & Clemens Edelstahlwerke
C 1, Si 0.5, Mn 1.1, Cr 1.55, bal Fe.
Oil hardening tool steel; used for meat grinders and larger
ball and roller bearings. *Obsolete*

MARKER WM 25

Schmidt & Clemens Edelstahlwerke
C 0.22, Cr 1.4, Mo 1.1, V 0.3, bal Fe.
Hot work tool steel.

MARKER WM 28

Schmidt & Clemens Edelstahlwerke
C 0.32, Cr 3, Mo 2.8, V 0.6, bal Fe.
Hot work tool and die steel; oil or water hardenable to
235,000-250,000 TS. For hot forming heavy and light metals,
hot shears, thread rolling, forming railroad spikes, pressure
casting molds. Werkstoff Nr. 1.2365; AISI H 10.

MARKER WM 28 S

Schmidt & Clemens Edelstahlwerke
C 0.25, Cr 2.85, Mo 2.55, V 0.45, bal Fe.
Hot work tool steel.

MARKER WM 30

Schmidt & Clemens Edelstahlwerke
C 0.4, Cr 5, Mo 3, V 0.9, bal Fe.
Hot work tool steel. Werkstoff Nr. 1.2367.

MARKER WMC

Schmidt & Clemens Edelstahlwerke
C 0.33, Cr 2.5, Mo 2.5, V 0.5, Co 2.5, bal Fe.
Hot work tool and die steel; oil or water hardenable to
235,000-250,000 TS. For threading dies for screws, hot
forming rivets, valve cones. *Obsolete*

MARKER WP 0

Schmidt & Clemens Edelstahlwerke
C 0.14, Si 0.25, Mn 0.4, Cr 0.75, Ni 3.5, bal Fe.
Case hardening steel for heavy sections. Case hardness:
60-64 Rock C. Core strength: 112,000-170,000 psi TS. For
large gears and shafts requiring good strength and high
surface hardness. Werkstoff Nr. 1.2735.

MARKER WP 1

Schmidt & Clemens Edelstahlwerke
C 0.19, Si 0.25, Mn 0.4, Cr 1.25, Ni 3.75, 0.20 Mo or W, bal
Fe.
Case hardening steel for heavy sections. Case hardness:
60-64 Rock C. Core strength: 155,000-200,000 psi TS. For
case hardening heavy drive gears, tractor transmission gears,
aircraft gears and shafts. Werkstoff Nr. 1.2764.

MARKER ZES

Schmidt & Clemens Edelstahlwerke
C 2.1, Cr 12, bal Fe.
Cold work tool and die steel; air or oil hardenable to 63-65
Rock C. For stamping and forming tools on thin metal sheet
and strip; for rolls and presses for processing rubber,
graphite, artificial resins and iron powder. Werkstoff Nr.
1.2080.

MARKER ZES SPEZIAL

Schmidt & Clemens Edelstahlwerke
C 2.1, Cr 12, Mo 0.35, V 0.15, Co 1.6, bal Fe.
Cold work tool and die steel; air or oil hardenable to 63-65
Rock C. Similar to Marker Zes but with improved toughness.
Obsolete

MARKER ZESE

Schmidt & Clemens Edelstahlwerke
C 1.65, Cr 12, V 0.1, bal Fe.
Cold work tool and die steel; air or oil hardenable to 63-65
Rock C. For dies and shears for cutting thin sheet and strip
up to 5 mm thick, for broaches. Werkstoff Nr. 1.2201.

MARKER ZESEK

Schmidt & Clemens Edelstahlwerke
C 1.55, Cr 12, Mo 0.7, V 1, bal Fe.
High carbon, high chrome cold work tool steel for cutting and
punching dies. Werkstoff Nr. 1.2379.

MARKER ZESV

Schmidt & Clemens Edelstahlwerke
C 2.2, Cr 12.5, Mo 0.9, V 2.2, bal.
High carbon, high chrome cold work tool steel; for punching
and stamping dies. Werkstoff Nr. 1.2378.

MARKER-ED12

Mannesmannrohren-Werke AG
C 0.2, Cr 11.5, Mo 1, V 0.3, bal Fe.
Annealed: 90,000 TS; 40,000 YS; 25 El; 92 Rock B.
Hardened: 240,000 TS; 200,000 YS; 10 El; 50 Rock C. For
valves, bearings, cutlery, surgical instruments. Corrosion
resistant, hardenable. *Obsolete*

MARKER-ED12G

Mannesmannrohren-Werke AG
C 0-0.18, Cr 11, Mo 1.05, Ni 0.75, V 0.3, bal Fe.
Annealed: 90,000 TS; 38,000 YS; 26 El; 92 Rock B.
Hardened: 235,000 TS; 200,000 YS; 10 El; 48 Rock C. For
valves, bearings, cutlery, surgical instruments. Corrosion
resistant, hardenable. *Obsolete*

MARKER-ED12W

Mannesmannrohren-Werke AG
C 0.2, Cr 12, Mo 1, V 0.3, W 0.5, bal Fe.
Annealed: 90,000 TS; 40,000 YS; 25 El; 92 Rock B.
Hardened: 240,000 TS; 205,000 YS; 9 El; 50 Rock C. For
cutlery, valves, bearings, surgical instruments. Corrosion
resistant, hardenable. *Obsolete*

MARKERITE

German manufacture
carbide.
For tool tips, cutters; hard metal.

MARKET BRASS

American manufacture
Cu 65, Zn 35.
For condenser tubes, hardware; corrosion resistant.

MARKEY BRONZE ALLOY NO. M-1, ETC.

Now MARKEY M-55.

MARKEY M-1

Bunting Bearings Corp.
83.0 Cu min, 7.0 Sn min, 3.0 Zn min, 7.0 Pb min.
Cast leaded bronze; SAE 660. Copper alloy No. 932. ASTM
B505-89-C93200 and ASTM B271-89-C93200.

MARKEY M-11

Markey Bronze Corp.
Cu 83, Al 11, Fe 4.
Cast aluminum bronze; SAE 68 A-B; Copper alloy No. 954.
Obsolete

MARKEY M-12

Markey Bronze Corp.
Cu 77, Sn 8, Pb 15.
Cast bearing bronze. *Obsolete*

MARKEY M-13

Markey Bronze Corp.
Cu 67.5, bal Zn.
Cast yellow brass; SAE 41. *Obsolete*

MARKEY M-14

Markey Bronze Corp.
Cu 81, Sn 8, Pb 11.
Cast high leaded bronze. *Obsolete*

MARKEY M-15

Bunting Bearings Corp.
85.0 Cu min, 14.0 Sn min, 1.0 Zn min.
Cast tin bronze. Copper alloy No. 910. ASTM B505-89-
C91000 and ASTM B271-89-C91000.

MARKEY M-16

Bunting Bearings Corp.
78.0 Cu min, 6.0 Sn min, 16.0 Pb min.
Cast high lead tin bronze. Copper alloy No. 939. ASTM
B505-89-C93900 and ASTM B271-89-C93900.

MARKEY M-17

Bunting Bearings Corp.
87.0 Cu min, 11.0 Sn min, 1.0 Pb min, 1.0 Ni min.
Cast leaded tin bronze; SAE 640. Copper alloy No. 925.
ASTM B505-89-C92500 and ASTM B271-89-C92500.

MARKEY M-18

Markey Bronze Corp.
Cu 87, Sn 8, Pb 1, Zn 4.
Cast leaded tin bronze; Copper alloy No. 923; ASTM
B143-2B. *Obsolete*

MARKEY M-19

Markey Bronze Corp.
Cu 70, Sn 9, Pb 21.
Cast high leaded tin bronze. *Obsolete*

MARKEY M-2

Now MARKEY M-55.

MARKEY M-20
Markey Bronze Corp.
Cu 72, Sn 3, Pb 25.
Cast high leaded tin bronze. *Obsolete*

MARKEY M-21
Markey Bronze Corp.
Cu 70, Sn 5, Pb 25.
Cast high leaded tin bronze; Copper alloy No. 943. *Obsolete*

MARKEY M-22
Markey Bronze Corp.
Manganese bronze. SAE 43. *Obsolete*

MARKEY M-23
Markey Bronze Corp.
Cu 66, Zn 34.
Cast yellow brass. *Obsolete*

MARKEY M-24
Markey Bronze Corp.
Cu 83, Sn 10, Pb 5, Zn 2.
Cast leaded bronze. *Obsolete*

MARKEY M-25
Markey Bronze Corp.
Cu 75, Sn 5, Pb 20.
Cast high leaded tin bronze. *Obsolete*

MARKEY M-29
Bunting Bearings Corp.
81.0 Cu min, 3.0 Sn min, 7.0 Pb min, 9.0 Zn min.
Cast leaded semi-red brass. Copper alloy No. 844. ASTM B505-89-C84400 and ASTM B271-89-C84400.

MARKEY M-3
Bunting Bearings Corp.
85.0 Cu min, 5.0 Sn min, 5.0 Zn min, 5.0 Pb min.
Cast leaded red brass; SAE 40. Copper alloy No. 836. ASTM B505-89-C83600 and ASTM B271-89-C83600.

MARKEY M-31
Bunting Bearings Corp.
Cu 72-79, Sn 4.5-6.5, Pb 18-22, Zn 0-1.
Cast high leaded tin bronze. ASTM B505-89-C94100 and ASTM B271-89-C94100.

MARKEY M-32
Bunting Bearings Corp.
Cu 82.5-86, Sn 9-11, Pb 2-3.2, Ni 2.8-4, P 0-0.25.
Leaded tin bronze casting. ASTM B505-89-C92900 and ASTM B271-89-C92900.

MARKEY M-33
Bunting Bearings Corp.
Cu 86-89, Sn 9.7-10.8, Pb 0-0.25, Ni 1.2-2, P 0-1.5.
Tin bronze casting. ASTM B505-89-C91600 and ASTM B271-89-C91600.

MARKEY M-34
Markey Bronze Corp.
Cu 75, Sn 5, Pb 20, Ni 0-1.
Cast high leaded tin bronze. *Obsolete*

MARKEY M-35
Bunting Bearings Corp.
P 0-1.5, Pb 0-0.5, Zn 0-0.5, 89.0 Cu min, 11.0 Sn min.
Cast phosphor gear bronze; SAE 65. Copper alloy No. 907. ASTM B505-89-C90700 and ASTM B271-89-C90700.

MARKEY M-36
Bunting Bearings Corp.
Cu 70, Sn 3.5-5, Pb 19-23, P 0-0.5, 1.0 others max.
Cast high leaded tin bronze. *Obsolete*

MARKEY M-37
Bunting Bearings Corp.
Sn 1.5-2.5, Pb 28-29, Zn 0-0.5, Ni 0.25-0.75, bal Cu.
Cast copper lead alloy.

MARKEY M-38
Bunting Bearings Corp.
Cu 82-85, Sn 7-9, Pb 7-9, Zn 0-0.8, Ni 0-1, P 0-1.5.
Leaded tin bronze casting. ASTM B505-89-C93400 and ASTM B271-89-C93400.

MARKEY M-4
Bunting Bearings Corp.
85.0 Cu min, 5.0 Sn min, 9.0 Pb min, 1.0 Zn min.
Cast leaded bronze; SAE 66. Copper alloy No. 935. ASTM B505-89-C93500 and ASTM B271-89-C93500.

MARKEY M-40
Bunting Bearings Corp.
Cu 72-76, Sn 5-6, Pb 18-20, Zn 0-0.5, Ni 0-0.5, P 0-1.5.
Leaded tin bronze casting.

MARKEY M-41
Bunting Bearings Corp.
Cu 68.5-73.5, Pb 23-27, Sn 4.5-6, Zn 0-0.8.
Cast high leaded bronze. Copper alloy No. 943. ASTM B505-89-C94300 and ASTM B271-89-C94300.

MARKEY M-42
Bunting Bearings Corp.
Cu 76-78, Sn 2.35-4, Pb 9-11, Ni 2-2.5, Zn 6-9.
Cast high leaded bronze.

MARKEY M-43
Bunting Bearings Corp.
Cu 86-89, Sn 9-11, Pb 0-0.3, Zn 1-3.
Cast tin bronze; SAE 62. Copper alloy No. 905.

MARKEY M-44
Bunting Bearings Corp.
Cu 84-89, Sn 9-13.5, Pb 0-1, Zn 1-3.
Cast tin bronze.

MARKEY M-45
Bunting Bearings Corp.
Cu 86-89, Sn 7.5-9, Pb 0-0.3, Zn 3-5, Ni 1.
Cast in bronze. Copper alloy No. 903. ASTM B505-89-C90300 and ASTM B271-89-C90300.

MARKEY M-46
Bunting Bearings Corp.
Cu 86-89, Sn 5.5-6.5, Pb 1-2, Zn 3-5, Ni 0-1.
Cast leaded tin bronze. Copper alloy No. 922. ASTM B271-89-C92200 and ASTM B505-89-C92200.

MARKEY M-47
Bunting Bearings Corp.
Cu 78-82, Sn 2-3, Pb 9-11, Sn 6.5-8.5, P 0-0.04.
Cast leaded brass.

MARKEY M-50
Bunting Bearings Corp.
Cu 67-73.5, Sn 9-11, Pb 18-20, Zn 0-0.5.
Cast high leaded tin bronze; 48-78 Brin.

MARKEY M-53
Bunting Bearings Corp.
Cu 79-83, Sn 6-8, Pb 11-13, Zn 0-1, P 0-1.5.
Leaded tin bronze casting. ASTM B505-89-C93600 and ASTM B271-89-C93600.

MARKEY M-55
Bunting Bearings Corp.
Cu 78-82, Sn 9-11, Pb 8-11, Zn 0-0.8, Ni 0-1, P 0-1.5.
Leaded tin bronze casting. UNS C93700. ASTM B505-89-C93700 and ASTM B271-89-C93700.

MARKEY M-56
Bunting Bearings Corp.
Cu 77-81, Sn 10-11, Pb 8-9.5, Zn 0-0.75, Ni 0-0.5, P 0-1.5.
Leaded tin bronze casting.

MARKEY M-57
Bunting Bearings Corp.
Cu 78-82, Sn 4-6, Pb 12-14, Zn 0-3, Ni 0-1.
Leaded tin bronze casting; high lead content.

MARKEY M-58
Bunting Bearings Corp.
Cu 75-79, Sn 6.3-7.5, Pb 13-16, Zn 0-0.8, Ni 0-1, P 0-1.5.
Leaded tin bronze casting; high lead content. ASTM B505-89-C93800 and ASTM B271-89-C93800.

MARKEY M-59
Bunting Bearings Corp.
Cu 73-76.5, Sn 4.75-6, Pb 18.25-20, Ni 0.5-1, P 0-0.05.
Leaded tin bronze casting; high lead content.

MARKEY M-6
Bunting Bearings Corp.
88.0 Cu min, 10.0 Sn min, 2.0 Zn min.
Cast bronze; SAE 62. Copper alloy No. 905. ASTM B505-89-C90500 and ASTM B271-89-C90500.

MARKEY M-61
Bunting Bearings Corp.
Cu 66-74, Sn 6.5-9.5, Pb 19.5-24.5, Zn 0-0.8, Ni 0-0.25, P 0-0.05.
Leaded tin bronze casting; high lead content.

MARKEY M-62
Bunting Bearings Corp.
Cu 78-80, Sn 20-22, Pb 0-0.3, Zn 0-0.3.
High tin bronze.

MARKEY M-63
Bunting Bearings Corp.
Cu 78.5-81.5, Sn 2.15-2.85, Pb 9.5-10.5, Zn 6.25-8.25, Ni 0.25-0.75.

MARKEY M-64
Bunting Bearings Corp.
Cu 82.25-86.25, Sn 9-11, Pb 4-6, Zn 0-0.5, Ni 0-0.15.

MARKEY M-65
Bunting Bearings Corp.
Cu 79-81, Sn 7-9, Pb 9-11, Zn 1.5-3, Ni 0.5-0.75.

MARKEY M-66
Bunting Bearings Corp.
Cu 67-71, Sn 10.5-11.5, Pb 18-20, Zn 0-0.5, Ni 0-0.5.

MARKEY M-67
Bunting Bearings Corp.
Pb 28-32, Zn 0-1, Ni 0-0.3, 4.0 Sn min, bal Cu.

MARKEY M-7
Bunting Bearings Corp.
88.0 Cu min, 10.0 Sn min, 2.0 Pb min.
Cast low lead bronze; SAE 63. Copper alloy No. 927. ASTM B505-89-C92700 and ASTM B271-89-C92700.

MARKEY M-8
Markey Bronze Corp.
Cu 76, Sn 6, Pb 18.
Cast bearing bronze; Copper alloy No. 941. *Obsolete*

MARKEY M-9
Markey Bronze Corp.
Sn 9, Pb 1.2, Ni 5.2, bal Cu.
Cast nickel bronze. *Obsolete*

MARKUS ALLOY
English manufacture
Cu, Ni, Zn.
For decorative parts; corrosion resistant.

MAROG
Teledyne Vasco
C 0.7, W 18, Cr 4, V 1, bal Fe.
For tools, drills, taps, lathe cutters; high speed steel. *Obsolete*

MARQUES AGPV
Now AGPV 11.

MARQUES D2S
Societe Nouvelle des Acieries de Pompey
C 0.2, S 0.12-0.18, bal Fe.
For screw machine products; free-cutting. *Obsolete*

MARQUES D2SS
Societe Nouvelle des Acieries de Pompey
C 0.2, Mn 1, S 0.12-0.18, bal Fe.
For screw machine products; free-cutting. *Obsolete*

MARQUES D3S
Societe Nouvelle des Acieries de Pompey
S 0.08-0.13, C, bal Fe.
For screw machine products; free-cutting. *Obsolete*

MARQUES D3SS
Societe Nouvelle des Acieries de Pompey
C, Mn, S, bal Fe.
For screw machine products; free-cutting. *Obsolete*

MARQUES DIS
Societe Nouvelle des Acieries de Pompey
C, S, bal Fe.
For screw machine products; free-cutting. *Obsolete*

MARQUES DISS
Societe Nouvelle des Acieries de Pompey
C, S, bal Fe.
For screw machine products; free-cutting. *Obsolete*

MARREL 50SS
Marrel Freres, S.A.
C 0.18-0.25, Mn 0.8-1, bal Fe.
Normalized: 74,000-83,000 TS; 53,000 YS; 24 El; 58 RA;
149-166 Brin. For machinery and welded parts; parts
workable at low temperature.

MARREL 5NS
Marrel Freres, S.A.
C 0.12-0.16, Cr 0.2, Ni 5-5.5, bal Fe.
Hardened: 170,000-200,000 TS; 108,000 YS; 6 El; 40 RA;
350-420 Brin. For motor valves, boilers, gears; case
hardening steel, shock resistant.

MARREL AMMO
Marrel Freres, S.A.
C 0-0.2, Mn 0.9-1.2, Cr 0.2, Mo 0.4-0.6, bal Fe.
Normalized: 84,000-90,000 TS; 50,000 YS; 18 El; 50 RA;
150-180 Brin. For sheet iron parts for superchargers; for use
at temperatures up to 500°C.

MARREL ARM
Marrel Freres, S.A.
C 1-1.15, Cr 1.4-1.6, bal Fe.
Hardened: 630-740 Brin. For cutlery, stamping dies, cams,
tools; oil hardening.

MARREL ASK1
Marrel Freres, S.A.
C 0.15-0.21, Mn 0.8-0.9, Cr 0.6-1, bal Fe.
Hardened: 135,000-165,000 TS; 86,000 YS; 8 El; 33 RA;
280-340 Brin. For gears, shafts; case hardening steel.

MARREL ASK2
Marrel Freres, S.A.
C 0.2-0.25, Mn 0.6-0.9, Cr 0.85-1, bal Fe.
Hardened: 162,000-191,000 TS; 103,000 YS; 5 El; 28 RA. For
gears, shafts; case hardening steel.

MARREL ASK3
Marrel Freres, S.A.
C 0.3-0.4, Mn 0.6-0.9, Cr 0.85-1, bal Fe.
Hardened: 114,000-128,000 TS; 102,000 YS; 15 El; 62 RA;
230-272 Brin. For mechanical parts, gears, shafting; oil
hardening.

MARREL ASK4
Marrel Freres, S.A.
C 0.4-0.48, Mn 0.6-0.9, Cr 0.85-1, bal Fe.
Hardened: 128,000-142,000 TS; 115,000 YS; 14 El; 58 RA;
260-310 Brin. For forged parts, gears; oil hardening.

MARREL ASK5
Marrel Freres, S.A.
C 0.48-0.58, Mn 0.6-0.9, Cr 0.85-1.15, bal Fe.
Hardened: 140,000-155,000 TS; 128,000 YS; 13 El; 50 RA;
290-330 Brin. For wear rings, grinding parts; wear resistant.

MARREL ASM3
Marrel Freres, S.A.
C 0.29-0.36, Mn 0.45-0.55, Cr 0.4-0.6, Ni 0.4-0.6, bal Fe.
Hardened: 114,000-128,000 TS; 100,000 YS; 13 El; 60 RA;
235-265 Brin. For connecting rods, center bits, cylinders; oil
hardening.

MARREL ASM4
Marrel Freres, S.A.
C 0.36-0.44, Mn 0.5-0.7, Cr 0.4-0.6, Ni 0.4-0.6, bal Fe.
Hardened: 128,000-141,000 TS; 114,000 YS; 12 El; 58 RA;
260-295 Brin. For mechanical parts, gears, shafts; oil
hardening.

MARREL ASM5
Marrel Freres, S.A.
C 0.44-0.54, Mn 0.5-0.7, Ni 0.4-0.6, Cr 0.4-0.6, bal Fe.
Hardened: 142,000 TS; 125,000 YS; 11 El; 50 RA; 290-330
Brin. For gears, shafts; oil hardening.

MARREL ASMY
Marrel Freres, S.A.
C 0.35-0.4, Mn 0.4-0.6, Cr 0.4-0.6, Ni 0.4-0.6, Mo 0.15-0.2, bal
Fe.
Hardened: 121,000-135,000 TS; 114,000 YS; 15 El; 61 RA;
260-298 Brin. For compressed gas bottles and cylinders; oil
hardening.

MARREL A3T
Marrel Freres, S.A.
C 0.32-0.38, Mn 0.5-0.7, Cr 1.6-2, Ni 3.6-4, bal Fe.
Hardened: 240,000-320,000 TS; 158,000 YS; 6 El; 10 RA;
470-560 Brin. For rollers, gears, connecting rods; oil
hardening, tough.

MARREL C14A
Marrel Freres, S.A.
C 0.1-0.15, 13.0 Cr min, bal Fe.
Heat treated: 114,000-128,000 TS; 100,000 YS; 14 El; 50 RA;
230-270 Brin. For pump shafts, steam turbine parts; corrosion
resistant.

MARREL C14B
Marrel Freres, S.A.
C 0.15-0.23, 13.0 Cr min, bal Fe.
Hardened: 136,000-150,000 TS; 121,000 YS; 12 El; 48 RA;
270-320 Brin. For motor valves, steam turbine parts; corrosion
resistant.

MARREL C14C
Marrel Freres, S.A.
C 0.23-0.32, 13.0 Cr min, bal Fe.
Hardened: 140,000-170,000 TS; 128,000 YS; 10 El; 42 RA;
300-360 Brin. For motor valves, pump parts; corrosion
resistant.

MARREL C14M
Marrel Freres, S.A.
C 0.32-0.4, 13.0 Cr min, bal Fe.
Hardened: 165,000-200,000 TS; 142,000 YS; 8 El; 36 RA;
360-420 Brin. For cutlery, surgical instruments; corrosion
resistant, hardenable.

MARREL C2N
Marrel Freres, S.A.
C 0.045-0.095, Mn 0.3-0.5, Ni 1.7-2, bal Fe.
Rolled: 78,000-95,000 TS; 57,000 YS; 15 El; 62 RA; 150-210
Brin. For axles, connecting rods; case hardening steel.

MARREL CH1
Marrel Freres, S.A.
C 0.08-0.11, Cr 0.6-0.8, Ni 3.3-3.7, bal Fe.
Hardened: 135,000-156,000 TS; 108,000 YS; 12 El; 48 RA;
270-340 Brin. For gears, shafts, pinions; case hardening
steel.

MARREL CH2
Marrel Freres, S.A.
C 0.11-0.15, Cr 0.6-0.8, Ni 3.3-3.7, bal Fe.
Hardened: 156,000-185,000 TS; 111,000 YS; 9 El; 33 RA;
340-430 Brin. For gears, shafts, pinions; case hardening
steel.

MARREL CH2D
Marrel Freres, S.A.
C 0.15-0.2, Cr 0.6-0.8, Ni 3.3-3.7, bal Fe.
Hardened: 185,000-210,000 TS; 128,000 YS; 8 El; 30 RA;
390-460 Brin. For gears, pinions, shafts; case hardening
steel.

MARREL CH3
Marrel Freres, S.A.
C 0.21-0.29, Cr 0.7-0.9, Ni 3.3-3.7, bal Fe.
Hardened: 114,000-129,000 TS; 110,000 YS; 16 El; 60 RA;
230-270 Brin. For spindles, shafts, gears; oil hardening.

MARREL CH4
Marrel Freres, S.A.
C 0.29-0.38, Cr 0.7-0.9, Ni 3.3-3.7, bal Fe.
Hardened: 114,000-141,000 TS; 121,000 YS; 15 El; 58 RA;
260-300 Brin. For spindles, shafts, gears; oil hardening.

MARREL CNC1
Marrel Freres, S.A.
C 0.1-0.13, Cr 0.6-0.8, Ni 2.6-3, bal Fe.
Hardened: 135,000-165,000 TS; 90,000 YS; 10 El; 45 RA;
270-350 Brin. For automotive and aircraft parts, gears, shafts;
case hardening steel.

MARREL CNC2
Marrel Freres, S.A.
C 0.13-0.17, Cr 0.6-0.8, Ni 2.6-3, bal Fe.
Hardened: 162,000-185,000 TS; 107,000 YS; 9 El; 33 RA;
340-430 Brin. For automotive and aircraft parts, gears, shafts;
case hardening steel.

MARREL CNC2D
Marrel Freres, S.A.
C 0.17-0.21, Cr 0.6-0.8, Ni 2.6-3, bal Fe.
Hardened: 185,000-210,000 TS; 114,000 YS; 7 El; 30 RA;
390-460 Brin. For gears, shafts; case hardening steel.

MARREL CNC3
Marrel Freres, S.A.
C 0.22-0.3, Cr 0.6-0.8, Ni 2.6-3, bal Fe.
Hardened: 114,000-128,000 TS; 109,000 YS; 17 El; 66 RA;
230-270 Brin. For gears, shafts, axles; oil hardening.

MARREL CNC4
Marrel Freres, S.A.
C 0.3-0.38, Cr 0.5-0.7, Mn 0.5-0.7, Ni 2.6-3, bal Fe.
Hardened: 128,000-141,000 TS; 121,000 YS; 15 El; 60 RA;
260-300 Brin. For gears, shafts, axles; oil hardening.

MARREL CNC5
Marrel Freres, S.A.
C 0.38-0.46, Mn 0.5-0.7, Cr 0.5-0.7, Ni 2.6-3, bal Fe.
Hardened: 142,000-157,000 TS; 128,000 YS; 13 El; 48 RA;
300-350 Brin. For gears, shafts, axles; oil hardening.

MARREL CNCO
Marrel Freres, S.A.
C 0.06-0.1, Cr 0.5-0.6, Ni 2.6-3, bal Fe.
Hardened: 107,000-135,000 TS; 83,000 YS; 13 El; 50 RA;
210-280 Brin. For gears, axles, camshafts, connecting rods;
case hardening steel.

MARREL CNPO
Marrel Freres, S.A.
C 0.3-0.35, Mn 0.5-0.7, Cr 0.6-0.8, Ni 2.5-2.8, Mo 0.25-0.35, bal Fe.
Hardened: 135,000-150,000 TS; 128,000 YS; 14 El; 60 RA; 270-320 Brin. For shafts, cranks, gears; tough, oil hardening.

MARREL CNY1
Marrel Freres, S.A.
C 0.13-0.17, Mn 0.6-0.9, Cr 0.5-0.7, Ni 1.1-1.3, Cr 0.16-0.2, bal Fe.
Hardened: 135,000-165,000 TS; 92,000 YS; 9 El; 38 RA; 270-340 Brin. For gears, shafts; case hardening steel.

MARREL CNY2
Marrel Freres, S.A.
C 0.17-0.2, Mn 0.6-0.9, Ni 1.1-1.3, Cr 0.5-0.7, Mo 0.16-0.2, bal Fe.
Hardened: 155,000-185,000 TS; 107,000 YS; 7 El; 30 RA; 340-390 Brin. For gears, crankshafts; case hardening steel.

MARREL CNY2D
Marrel Freres, S.A.
C 0.2-0.24, Mn 0.6-0.9, Cr 0.5-0.7, Ni 1.1-1.3, Mo 0.16-0.2, bal Fe.
Hardened: 185,000-225,000 TS; 110,000 YS; 6 El; 28 RA; 390-460 Brin. For gears, crankshafts; case hardening steel.

MARREL CNY3
Marrel Freres, S.A.
C 0.22-0.28, Mn 0.6-0.9, Cr 0.5-0.7, Ni 1.1-1.3, Mo 0.16-0.2, bal Fe.
Hardened: 113,000-128,000 TS; 110,000 YS; 16 El; 60 RA; 230-270 Brin. For center bits, shafts, gears, propellers; oil hardening.

MARREL CNY4
Marrel Freres, S.A.
C 0.28-0.36, Mn 0.6-0.9, Cr 0.5-0.7, Ni 1.1-1.3, Mo 0.16-0.2, bal Fe.
Hardened: 128,000-142,000 TS; 120,000 YS; 14 El; 63 RA; 260-300 Brin. For shafts, gears, propellers; oil hardening.

MARREL CNY5
Marrel Freres, S.A.
C 0.36-0.46, Mn 0.6-0.9, Cr 0.5-0.7, Ni 1.1-1.3, Mo 0.16-0.2, bal Fe.
Hardened: 141,000-155,000 TS; 124,000 YS; 13 El; 58 RA; 300-350 Brin. For shafts, axles, gears; oil hardening.

MARREL CTD
Marrel Freres, S.A.
C 0.05-0.1, Mn 0.4, bal Fe.
Normalized: 72,000-93,000 TS; 57,000 YS; 28 El; 70 RA; 140-187 Brin. For gears, cams, crankshafts; case hardening steel.

MARREL M1
Marrel Freres, S.A.
C 0.65-0.75, Mn 0.45-0.65, bal Fe.
Normalized: 104,000 TS; 64,000 YS; 14 El; 29 RA; 207-229 Brin. For springs, hammers, files, chisels; oil or water hardening.

MARREL M2
Marrel Freres, S.A.
C 0.55-0.65, Mn 0.5-0.7, bal Fe.
Normalized: 107,000-121,000 TS; 66,000 YS; 14 El; 36 RA; 207-258 Brin. For hammers, springs, punches; water hardening.

MARREL M3
Marrel Freres, S.A.
C 0.45-0.55, Mn 0.5-0.7, bal Fe.
Normalized: 93,000-107,000 TS; 60,000 YS; 17 El; 45 RA; 180-224 Brin. For forgings, axes, shafts; water hardening.

MARREL M3W
Marrel Freres, S.A.
C 0.48-0.55, Mn 0.45, bal Fe.
Normalized: 93,000-107,000 TS; 60,000 YS; 17 El; 45 RA; 180-224 Brin. For cutlery, table flatware; water hardening.

MARREL M4
Marrel Freres, S.A.
C 0.32-0.38, Mn 0.5-0.7, bal Fe.
Normalized: 78,000-93,000 TS; 55,000 YS; 20 El; 51 RA; 156-190 Brin. For gears, pinions, shafts; water hardening.

MARREL M5
Marrel Freres, S.A.
C 0.18-0.23, Mn 0.5-0.7, bal Fe.
Normalized: 72,000 TS; 48,500 YS; 23 El; 57 RA; 140-160 Brin. For axles, cotter pins, bolts, forged parts; water hardening.

MARREL M6
Marrel Freres, S.A.
C 0.1-0.175, Mn 0.5-0.7, bal Fe.
Normalized: 64,000-72,000 TS; 43,000 YS; 25 El; 65 RA; 120-146 Brin. For gears, pinions, shafts; case hardening steel.

MARREL M7
Marrel Freres, S.A.
C 0.09-0.125, Mn 0.5-0.7, bal Fe.
Normalized: 59,000-64,000 TS; 40,000 YS; 27 El; 72 RA; 112-128 Brin. For bolts, rivets, stampings; water hardening.

MARREL MO
Marrel Freres, S.A.
C 0.95-1.05, Mn 0.3, bal Fe.
Hardened: 114,000-160,000 TS; 72,000-100,000 YS; 10 El; 25 RA; 265-290 Brin. For tools, dies, files, punches, hammers; water hardening.

MARREL MSG
Marrel Freres, S.A.
C 0.45-0.5, Mn 0.4-0.8, Si 1.8-2, bal Fe.
Hardened: 200,000-220,000 TS; 185,000 YS; 6 El; 28 RA; 420-500 Brin. For springs; oil hardening.

MARREL NK1
Marrel Freres, S.A.
C 0.07-0.11, Mn 0.6-0.9, Cr 0.9-1.25, Ni 1.2-1.6, bal Fe.
Hardened: 142,000-180,000 TS; 120,000 YS; 8 El; 42 RA; 290-390 Brin. For gears, pinions, shafts, camshafts; case hardening steel.

MARREL NK2
Marrel Freres, S.A.
C 0.12-0.18, Mn 0.6-0.9, Cr 0.9-1.25, Ni 1.2-1.6, bal Fe.
Hardened: 180,000-210,000 TS; 142,000 YS; 7 El; 40 RA; 370-450 Brin. For aircraft gears, pinions, shafts; case hardening steel.

MARREL NK2D
Marrel Freres, S.A.
C 0.16-0.22, Mn 0.6-0.9, Cr 0.9-1.25, Ni 1.2-1.6, bal Fe.
Hardened: 185,000-230,000 TS; 155,000 YS; 13 El; 36 RA; 390-500 Brin. For aircraft gears, pinions, shafts; case hardening steel.

MARREL NK3
Marrel Freres, S.A.
C 0.22-0.29, Mn 0.6-0.9, Cr 0.75-1.1, Ni 1.2-1.6, bal Fe.
Hardened: 195,000-240,000 TS; 170,000 YS; 6 El; 24 RA; 420-540 Brin. For gears, shafts, machinery parts; oil hardening.

MARREL NK4
Marrel Freres, S.A.
C 0.3-0.38, Mn 0.6-0.9, Cr 0.75-1.1, Ni 1.2-1.6, bal Fe.
Hardened: 225,000-280,000 TS; 200,000 YS; 5 El; 22 RA; 480-630 Brin. For gears, pinions, axles; oil hardening.

MARREL SKM 2
Marrel Freres, S.A.
C 0.15-0.19, Mn 0.65-0.8, Cr 0.85-1.15, Mo 0.2-0.3, bal Fe.
Hardened: 160,000-185,000 TS; 107,000 YS; 7 El; 31 RA; 340-400 Brin. For bolts, cylinder liners, gears; case hardening steel.

MARREL SKM 2 D
Marrel Freres, S.A.
C 0.19-0.23, Mn 0.65-0.8, Cr 0.85-1.15, Mo 0.2-0.3, bal Fe.
Hardened: 182,000-210,000 TS; 111,000 YS; 6 El; 28 RA; 390-460 Brin. For gears, shafts, camshafts, bolts; case hardening steel.

MARREL SKM 3
Marrel Freres, S.A.
C 0.23-0.29, Mn 0.65-0.8, Cr 0.85-1.15, Mo 0.2-0.3, bal Fe.
Hardened: 114,000-128,000 TS; 100,000 YS; 15 El; 63 RA; 230-270 Brin. For shafts, axles, gears; oil hardening.

MARREL SKM 3 S
Marrel Freres, S.A.
C 0.21-0.28, Mn 0.65-0.8, Cr 0.85-1.15, Mo 0.2-0.3, bal Fe.
Hardened: 128,000-156,000 TS; 99,500 YS; 10 El; 50 RA; 260-340 Brin. For gears, aircraft parts; good weldability.

MARREL SKM 4
Marrel Freres, S.A.
C 0.32-0.38, Mn 0.65-0.8, Cr 0.85-1.15, Mo 0.2-0.3, bal Fe.
Hardened: 128,000-141,000 TS; 118,000 YS; 14 El; 60 RA; 260-300 Brin. For shafts, pneumatic hammer pistons, torsion bars; oil hardening.

MARREL SKM 5
Marrel Freres, S.A.
C 0.38-0.48, Mn 0.65-0.8, Cr 0.85-1.15, Mo 0.2-0.3, bal Fe.
Hardened: 141,000-156,000 TS; 131,000 YS; 13 El; 53 RA; 300-340 Brin. For shafts, torsion bars, pneumatic hammer pistons; oil hardening.

MARREL SKM 5 B
Marrel Freres, S.A.
C 0.4-0.47, Mn 0.7-0.8, Cr 1-1.3, Mo 0.4-0.5, bal Fe.
Hardened: 141,000-156,000 TS; 132,000 YS; 13 El; 53 RA; 300-340 Brin. For cutting pliers, tools, dies; oil hardening, tough.

MARREL SKM1
Marrel Freres, S.A.
C 0.11-0.15, Mn 0.65-0.8, Cr 0.85-1.15, Mo 0.2-0.3, bal Fe.
Hardened: 135,000-170,000 TS; 86,000 YS; 10 El; 35 RA; 220-350 Brin. For gears, pinions, cams, shafts; case hardening steel.

MARREL SKMC
Marrel Freres, S.A.
C 0.1-0.15, Mn 0.65-0.9, Cr 0.6-0.9, Mo 0.45-0.6, bal Fe.
Rolled: 68,000-83,000 TS; 43,000 YS; 20 El; 70 RA; 130-170 Brin. For tubes for high temperature and steam pipes; good weldability.

MARREL SKMO
Marrel Freres, S.A.
C 0.08-0.11, Mn 0.65-0.8, Cr 0.85-1.15, Mo 0.2-0.3, bal Fe.
Hardened: 114,000-135,000 TS; 83,000 YS; 10 El; 40 RA; 235-285 Brin. For gears, pinions, cams, shafts; case hardening steel.

MARS SUPERIEUR
Ugine Aciers
C 0.7, W 18, Cr 4, V 1, bal Fe.
For tools, dies, cutters; high speed steel.

MARSH CH5
Marsh Bros. & Co. Ltd.
C 0.34-0.4, Si 0.9, Cr 5, Mo 1.25-1.5, V 0.5-0.7, bal Fe.
For hot forging dies, punches; hot work steel, oil hardened.

MARSH CMN
Marsh Bros. & Co. Ltd.
C 0.5-0.55, Si 0.15-0.2, Cr 1.5, bal Fe.
For drop stamping dies; oil hardened.

MARSH CTC 1
Marsh Bros. & Co. Ltd.
C 0.45, Cr 1.25, W 2, Si 0.75, bal Fe.
For pneumatic chisels, boiler making tools; oil hardened, shock resistant.

MARSH CTH
Marsh Bros. & Co. Ltd.
C 0.43, Cr 0.25, W 0.5, bal Fe.
For hot forging dies and punches; hot work steel, oil hardened.

MARSH CTHV
Marsh Bros. & Co. Ltd.
C 0.3, Cr 3.4, V 0.34, W 8.4, bal Fe.
For hot forging dies, punches; hot work steel, oil hardened.

MARSH CYC
Marsh Bros. & Co. Ltd.
C 0.7, W 18, Cr 4, V 1, bal Fe.
For tools, cutters, reamers, hobs, drills; high speed steel.

MARSH CYC 5% COBALT
Marsh Bros. & Co. Ltd.
C 0.7, Cr 4.5, Mo 1, V 1.5, W 18, Co 5, bal Fe.
For lathe and boring tools, drills, shapers, cutters; high speed steel.

MARSH CYC EXTRA
Marsh Bros. & Co. Ltd.
C 0.7, W 18, Cr 4, V 1, Co 5, bal Fe.
For drills, hobs, reamers, broaches, lathe and planer cutters; high speed steel; Type T4.

MARSH CYC SPECIAL 14%
Marsh Bros. & Co. Ltd.
C 0.7, W 14, Cr 4, V 2, bal Fe.
For lathe and planer tools, drills, reamers, taps; high speed steel.

MARSH CYC275
Marsh Bros. & Co. Ltd.
C 0.8, W 20, Cr 4.5, Mo 1.25, V 1.2, Co 10, bal Fe.
For lathe and planer tools, hobs, reamers, milling cutters.

MARSH H.F.C.
Marsh Bros. & Co. Ltd.
C 2.15, Cr 12.5, Si 0.6, Mn 0.3, bal Fe.
For forming and drawing dies, punches, shears; oil hardened, nondeforming.

MARSH M.N.
Marsh Bros. & Co. Ltd.
C 0.9, Mn 1.7, Si 0.2, Cr 0.25, bal Fe.
For dies, gages, precision tools, taps, drills; oil hardened, nondeforming.

MARSH N3
Marsh Bros. & Co. Ltd.
C 0.4, Ni 3-3.5, Si 0.4, Mn 0.6, bal Fe.
For chisels, gears, punches, crimpers; oil hardened, shock resistant.

MARSH NC
Marsh Bros. & Co. Ltd
C 0.45, Ni 3.75, Cr 1.1, Mn 0.5, Si 0.35, bal Fe.
For chisels, snaps, gears, crimpers, punches; oil hardened, shock resistant.

MARSH NC1
Marsh Bros. & Co. Ltd.
C 0.4, Ni 1.5, Cr 1, bal Fe.
For spring collets; oil hardened, tough.

MARSH PATENT
Manufacturer not listed.
Ni 75, Cr 25.
For electrical resistances, heating elements; heat resistant.

MARSH SPECIAL
Marsh Bros. & Co. Ltd.
C 1.2, Mn 0.3, bal Fe.
For wood working tools, drills; water hardened; Type W1.

MARSH WTS (0.70C)
Marsh Bros. & Co. Ltd.
C 0.7-0.8, Si 0.1-0.2, Mn 0.3-0.4, bal Fe.
Heat treated: 185,000 TS; 140,000 YS; 13 El; 37 RA; 375 Brin.
For lathe centers, vice jaws, press tools; Type W1; water hardened.

MARSH WTS (0.95C)
Marsh Bros. & Co. Ltd.
C 0.95-1.05, Mn 0.3-0.4, Si 0.1-0.2, bal Fe.
Annealed: 100,000 TS; 53,000 YS; 21 El; 42 RA; 200 Brin. For blanking tools, punches, shears, press tools; Type W1; water hardened.

MARSHALL CRAT
Marshall Steel Co.
C 0.18, Mn 0.5, Si 0.2, bal Fe.
Annealed: 72,000 TS; 40,000 YS; 22 El; 58 RA; 140 Brin. For jigs, fixtures, patterns, machinery parts; Type 1018 steel, killed.

MARTIES ALLOY
Manufacturer not listed.
Zn 18, Ni 35, Fe 10, Sn 10, bal Cu.
For resistances, heat and corrosion resistant parts; heat and corrosion resistant.

MARTIES' NON-OXIDIZABLE ALLOY
Manufacturer not listed.
Ni 35, Fe 10, Zn 10, Sn 20, Br 25.
For heat and corrosion resistant parts; heat and corrosion resistant.

MARTIES' NON-OXIDIZABLE ALLOY
Manufacturer not listed.
Ni 35, Zn 18, Cu 17, Fe 10, Sn 10.
For heat and corrosion resistant parts; also called "Marties."

MARTIN
Detroit Alloy Steel Co.
C 1.4-1.6, Cr 12-14, Co 0.6-0.8, Mo 0.8-0.9, V 0.35-0.4, bal Fe.
For dies, rotary shears, gages, punches, shear blades, liners for dust mills; castings; cast to shape; air hardening.

MARTIN STEEL
English manufacture
C 0.73, Si 0.4, bal Fe.
For blacksmith tools, dies, hammers; water hardening.

MARTINEL
Martinel Steel Co., Ltd.
Mn 0.6-0.9, C, 2.75 Ni min, bal Fe.
85,000 TS; 50,000 YS; 20 El; 40 RA. For shipbuilding purposes in strength members, flat plate keel, bilge, deck and side plates; special method of rolling.

MARTINSITE M130
Inland Steel Co.
Carbon steel. C 0.04-0.22, Mn 0.2-0.6, bal Fe.
130,000 psi TS (900 MPa). For fasteners, tubing, and miscellaneous parts.

MARTINSITE M160
Inland Steel Co.
Carbon steel. C 0.04-0.22, Mn 0.2-0.6, bal Fe.
160,000 psi TS (1100 MPa). For fasteners, various automotive and appliance parts.

MARTINSITE M190
Inland Steel Co.
Carbon steel. C 0.04-0.22, Mn 0.2-0.6, bal Fe.
190,000 psi TS (1300 MPa). For miscellaneous parts.

MARTINSITE M220
Inland Steel Co.
Carbon steel. Mn 0.2-0.6, C 0.04-0.22, bal Fe.
220,000 psi TS (1500 MPa). Fasteners and miscellaneous parts for appliances and automotive.

MARVAC 250, ETC.
Now LESCALLOY MARVAC 250.

MARVAL
Now AUBERT & DUVAL MARVAL.

MARVAL 18
Aubert & Duval
C 0.02, Ni 18, Co 8, Mo 5, Ti 0.5, bal Fe.
Maraged: 170 kg/mm^2 TS; 155 kg/mm^2 YS; 8 El. Maraging steel. AFNOR Z 2 NKD 18.

MARVAL 18 H
Aubert & Duval
Alloy steel. C 0.02, Ni 18, Co 9, Mo 5, Ti 0.6, bal Fe.
Annealed: 1900 MPa TS; 6 El; 302 Brin. For ball or roller bearings, valve seats or guiding rings, in contact with steam.

MARVAL X 12
Aubert & Duval
Alloy steel. C 0.02, Cr 12, Ni 9, Mo 2, Ti, Al, bal Fe.
Annealed: 1450 MPa TS; 11 El; 293 Brin. For aviation and aerospace industries.

MARVEL
Teledyne Vasco
C 0.33, W 10, Cr 3.5, V 0.45, bal Fe.
For cutting tools, forming dies, punches, shears; hot die steel.

MARWE 126E
Mannesmannrohren-Werke AG
C 0.15, Cr 0.75, Mo 0.15, bal Fe.
For gears, cams, machine tool parts; case hardening, tough.
Obsolete

MARWE 127M
Mannesmannrohren-Werke AG
C 0.14, Mn 1, Si 0.3, bal Fe.
For gears, cams, camshafts; case hardening steel. *Obsolete*

MARWE 134A
Mannesmannrohren-Werke AG
C, Si, Mn, bal Fe.
For gears, machine tool parts; case hardening steel.
Obsolete

MARWE 13P
Mannesmannrohren-Werke AG
C 0.15, Mn 0.6, Si 0.25, Mo 0.3, bal Fe.
Annealed: 70,000 TS; 40,000 YS; 25 El; 60 RA; 145 Brin. For gears, cams, camshafts, fasteners; case hardening steel.
Obsolete

MARWE 14D
Mannesmannrohren-Werke AG
C 0.14, Si 0.25, Mn 0.6, Mo 0.3, bal Fe.
Annealed: 70,000 TS; 40,000 YS; 25 El; 60 RA; 145 Brin. For gears, cams, machine tool parts; case hardening steel.
Obsolete

MARWE 176M
Mannesmannrohren-Werke AG
C 0.36, Si 0.3, Mn 1, bal Fe.
For gears, machine tool parts; water hardened. *Obsolete*

MARWE 17L

Mannesmannrohren-Werke AG
C 0.13, Cr 0.85, Mo 0.45, bal Fe.
For gears, cams, camshafts, machine tool parts; case hardening, tough. *Obsolete*

MARWE 213ESV

Mannesmannrohren-Werke AG
C 0.1, Si 1.1, Cr 1.8, Mo 0.3, V 0.3, bal Fe.
For dies, cams, crankshafts; case hardening steel. *Obsolete*

MARWE 215E

Mannesmannrohren-Werke AG
C 0.1, Cr 2.25, Mo 1.1, bal Fe.
For oil refinery equipment; creep and heat resistant. *Obsolete*

MARWE 215ESV

Mannesmannrohren-Werke AG
C 0.08, Si 1.25, Cr 2.1, Mo 0.45, bal Fe.
For oil refinery equipment; creep and heat resistant. *Obsolete*

MARWE 220EV

Mannesmannrohren-Werke AG
C 0.13, Cr 1.05, Mo 0.25, V 0.2, bal Fe.
For gears, cams, camshafts; case hardening steel. *Obsolete*

MARWE 220MH

Mannesmannrohren-Werke AG
C, Si, Mn, bal Fe.
For machine tool parts, fasteners; case hardening steel. *Obsolete*

MARWE 221M

Mannesmannrohren-Werke AG
C 0.16, Mn 1, Si, bal Fe.
For machine tool parts; case hardening steel. *Obsolete*

MARWE 228E

Mannesmannrohren-Werke AG
C 0.12, Cr 5, Mo 0.55, bal Fe.
For oil refinery equipment; creep and heat resistant. *Obsolete*

MARWE 251CBH

Mannesmannrohren-Werke AG
C 0.3, Al 0.9, Cr 1.2, bal Fe.
For oil refinery equipment; creep and heat resistant. *Obsolete*

MARWE 251DV

Mannesmannrohren-Werke AG
C 0.27, Cr, V, bal Fe.
For gears, cams, camshafts; case hardening steel. *Obsolete*

MARWE 253DV

Mannesmannrohren-Werke AG
C 0.32, Cr, V, bal Fe.
For gears, bolts, crankshafts; oil hardened, shock resistant. *Obsolete*

MARWE 262EB

Mannesmannrohren-Werke AG
C 0.32, Cr 1.1, Al 1.1, Mo 0.18, bal Fe.
For oil refinery equipment; creep and heat resistant. *Obsolete*

MARWE 291DV

Mannesmannrohren-Werke AG
C 0.5, Mn 0.95, Cr 1.05, V 0.1, bal Fe.
For gears, bolts, springs, shafts; oil hardened, shock resistant. *Obsolete*

MARWE 310CW

Mannesmannrohren-Werke AG
C, Si, Mn, bal Fe.
For machine tool parts, gears; water hardened. *Obsolete*

MARWE 320ES

Mannesmannrohren-Werke AG
C, Si, Mn, bal Fe.
For machine tool parts; water hardened. *Obsolete*

MARWE 321ES

Mannesmannrohren-Werke AG
C, Si, Mn, bal Fe.
For machine tool parts; water hardened. *Obsolete*

MARWEDUR A61

Mannesmannrohren-Werke AG
C 0.15, Cr 19.5, Ni 9.5, bal Fe.
Annealed: 80,000 TS; 35,000 YS; 55 El; 75 RA; 150 Brin. Cold drawn: 180,000 TS; 150,000 YS; 10 El; 250 Brin. For chemical plant equipment, tanks, mixers, filters; Type 302; stainless, austenitic. *Obsolete*

MARWEDUR A63

Mannesmannrohren-Werke AG
C 0.15, Cr 24, Ni 19, bal Fe.
Annealed: 100,000 TS; 45,000 YS; 50 El; 65 RA; 185 Brin. For furnace parts, valves, pumps, retorts; Type 310; stainless. *Obsolete*

MARWEDUR AN11

Mannesmannrohren-Werke AG
C 0.08, Cr 17, Ni 13, Cb, bal Fe.
Annealed: 90,000 TS; 45,000 YS; 50 El; 65 RA; 160 Brin. For welded chemical plant equipment; Type 347; stainless, austenitic. *Obsolete*

MARWEDUR AN21

Mannesmannrohren-Werke AG
C 0.08, Cr 17, Mo 2, Ni 13, bal Fe.
Annealed: 85,000 TS; 35,000 YS; 50 El; 65 RA; 160 Brin. Cold drawn: 150,000 TS; 135,000 YS; 6 El; 300 Brin. For acid resistant chemical plant equipment; Type 316; stainless, austenitic. *Obsolete*

MARWEDUR-F-12

Mannesmann-Huttenwerk AG
C 0.2, Cr 12, Mo 1, Ni 0.4, V 0.3, bal Fe.
Annealed: 95,000 TS; 40,000 YS; 25 El; 92 Rock B. Hardened: 240,000 TS; 195,000 YS; 10 El; 50 Rock C. For valves, cutlery, bearings, surgical instruments. Corrosion resistant, hardenable.

MARWEDUR-F11

Mannesmann-Huttenwerk AG
C 0.16-0.23, Cr 12, Ni 0.5, Mo 1, W 0.5, V 0.3, Mn 0.6, Si 0.3, bal Fe.
For furnace parts, oil refinery equipment, cutlery. Corrosion and heat resistant, hardenable.

MARWENIT 12M

Mannesmannrohren-Werke AG
C 0.22, Cr 17, Ni 1.5, bal Fe.
Annealed: 125,000 TS; 95,000 YS; 20 El; 55 RA; 260 Brin. Cold drawn: 130,000 TS; 110,000 YS; 15 El; 35 RA; 270 Brin. For marine hardware, pumps, valves; Type 431; heat resistant. *Obsolete*

MARWENIT A10S

Mannesmannrohren-Werke AG
C 0-0.12, Cr 18, Ni 9, bal Fe.
Annealed: 80,000 TS; 35,000 YS; 55 El; 75 RA; 150 Brin. Cold drawn: 180,000 TS; 150,000 YS; 10 El; 250 Brin. For chemical plant equipment, tanks, mixers, filters; Type 302; stainless, austenitic. *Obsolete*

MARWENIT A20S

Mannesmannrohren-Werke AG
C 0-0.12, Cr 18, Ni 9.5, bal Fe.
Annealed: 80,000 TS; 35,000 YS; 55 El; 75 RA; 150 Brin. Cold drawn: 180,000 TS; 150,000 YS; 10 El; 250 Brin. For chemical plant equipment, tanks, mixers, filters; Type 302; stainless, austenitic. *Obsolete*

MARWENIT AN10

Mannesmannrohren-Werke AG
C 0-0.12, Cr 18, Ni 9.5, Cb = 8 x C, bal Fe.
Annealed: 90,000 TS; 45,000 YS; 56 El; 65 RA; 160 Brin. Cold drawn: 100,000 TS; 65,000 YS; 45 El; 60 RA; 205 Brin. For welded chemical plant equipment, tanks, mixers; Type 347; stainless, austenitic. *Obsolete*

MARWENIT AN20

Mannesmannrohren-Werke AG
C 0-0.12, Cr 18, Mo 2, Ni 10.5, Cb = 8 x C, bal Fe.
Annealed: 90,000 TS; 40,000 YS; 40 El; 60 RA; 170 Brin. For welded acid resistant chemical plant equipment; Type 316 Cb; stainless, austenitic. *Obsolete*

MARWENIT AT10

Mannesmannrohren-Werke AG
C 0-0.12, Cr 18, Ni 9.5, Ti = 4 x C, bal Fe.
Annealed: 85,000 TS; 35,000 YS; 55 El; 65 RA; 150 Brin. For welded chemical plant equipment, tanks, mixers; Type 321; stainless, austenitic. *Obsolete*

MARWENIT AT20

Mannesmannrohren-Werke AG
C 0-0.12, Cr 18, Mo 2, Ni 10.5, Ti = 4 x C, bal Fe.
Annealed: 90,000 TS; 40,000 YS; 40 El; 60 RA; 170 Brin. For welded acid resistant equipment; Type 316 Ti; stainless, austenitic. *Obsolete*

MARWENIT AX10

Mannesmannrohren-Werke AG
C 0-0.07, Cr 18, Ni 9.5, bal Fe.
Annealed: 85,000 TS; 35,000 YS; 60 El; 70 RA; 150 Brin. Cold drawn: 180,000 TS; 125,000 YS; 10 El; 330 Brin. For chemical plant equipment, tanks, mixers; Type 304; stainless, austenitic. *Obsolete*

MARWENIT AX20

Mannesmannrohren-Werke AG
C 0-0.07, Cr 18, Mo 2, Ni 10.5, bal Fe.
Annealed: 85,000 TS; 35,000 YS; 50 El; 65 RA; 160 Brin. Cold drawn: 150,000 TS; 135,000 YS; 6 El; 300 Brin. For acid resistant chemical plant equipment; Type 316; stainless, austenitic. *Obsolete*

MARWENIT F10

Mannesmannrohren-Werke AG
C 0-0.12, Cr 13, Si 0.4, bal Fe.
Annealed: 75,000 TS; 40,000 YS; 35 El; 70 RA; 155 Brin. Cold drawn: 100,000 TS; 85,000 YS; 17 El; 60 RA; 205 Brin. For turbine blades, cutlery, valves, surgical instruments; Type 410; corrosion resistant. *Obsolete*

MARWENIT F15

Mannesmannrohren-Werke AG
C 0-0.12, Cr 16.5, Mo 0.25, S 0.2, bal Fe.
Annealed: 80,000 TS; 50,000 YS; 25 El; 50 RA; 160 Brin. For screw machine products, dairy equipment; Type 430F; corrosion resistant. *Obsolete*

MARWENIT F20

Mannesmannrohren-Werke AG
C 0.1, Cr 13, Mo 1.15, bal Fe.
Annealed: 75,000 TS; 40,000 YS; 35 El; 70 RA; 155 Brin. For turbine blades, valves, pumps; corrosion resistant. *Obsolete*

MARWENIT FT13

Mannesmannrohren-Werke AG
C 0-0.1, Cr 17.5, Ti = 7 x C, bal Fe.
Annealed: 80,000 TS; 50,000 YS; 25 El; 50 RA; 160 Brin. For welded oil refinery equipment; Type 430 Ti; stainless. *Obsolete*

MARWENIT FT23

Mannesmannrohren-Werke AG
C 0-0.1, Cr 17, Mo 1.8, Ti = 7 x C, bal Fe.
Annealed: 80,000 TS; 50,000 YS; 25 El; 50 RA; 160 Brin. For welded oil refinery equipment; corrosion resistant. *Obsolete*

MARWENIT M10

Mannesmannrohren-Werke AG
C 0.4, Cr 13, Si 0.4, bal Fe.
Annealed: 95,000 TS; 50,000 YS; 25 El; 50 RA; 160 Brin. For cutlery, valves, surgical instruments; Type 420; stainless. *Obsolete*

MARWENIT M11

Mannesmannrohren-Werke AG
C 0.2, Si 0.4, Cr 13, bal Fe.
Annealed: 95,000 TS; 50,000 YS; 25 El; 55 RA; 195 Brin. For cutlery, valves, turbine blades, surgical instruments; Type 420; stainless. *Obsolete*

MARWENIT M12

Mannesmannrohren-Werke AG
C 0.22, Si 0.4, Cr 17, Ni 1.5, bal Fe.
Annealed: 125,000 TS; 95,000 YS; 20 El; 55 RA; 260 Brin. Cold drawn: 130,000 TS; 110,000 YS; 15 El; 35 RA; 270 Brin. For marine hardware, pumps, valves; Type 431; corrosion and heat resistant. *Obsolete*

MARWETHERM A61

Mannesmannrohren-Werke AG
C 0.15, Cr 19.5, Ni 9.5, bal Fe.
Annealed: 80,000 TS; 35,000 YS; 55 El; 75 RA; 150 Brin. For chemical plant equipment, tanks, mixers, filters; Type 302, stainless, austenitic. *Obsolete*

MARWETHERM A63

Mannesmannrohren-Werke AG
C 0.15, Cr 24, Ni 19, bal Fe.
Annealed: 100,000 TS; 45,000 YS; 50 El; 65 RA; 185 Brin. For furnace parts, valves, pumps, turbine blades; Type 310; stainless, austenitic. *Obsolete*

MARWETHERM F105

Mannesmannrohren-Werke AG
C 0-0.12, Si 2, Cr 18, bal Fe.
Annealed: 80,000 TS; 50,000 YS; 25 El; 50 RA; 150 Brin. For furnace parts, heat treat boxes; Type 430; corrosion resistant. *Obsolete*

MARWETHERM F120

Mannesmannrohren-Werke AG
C 0-0.12, Al 1.5, Cr 24, bal Fe.
For oil refinery equipment; heat and creep resistant. *Obsolete*

MARWETHERM F90

Mannesmannrohren-Werke AG
C 0-0.12, Si 2.3, Cr 6, bal Fe.
Annealed: 70,000 TS; 30,000 YS; 28 El; 65 RA; 160 Brin. For oil refinery equipment, heat exchangers; Type 502; creep and heat resistant. *Obsolete*

MARWETHERM F95

Mannesmannrohren-Werke AG
C 0-0.12, Si 2.2, Cr 13, bal Fe.
Annealed: 75,000 TS; 40,000 YS; 35 El; 70 RA; 155 Brin. For turbine blades, oil refinery equipment; corrosion and heat resistant. *Obsolete*

MARWIN

W. Martin Winn Ltd.
C 0.25, Si 0.3, bal Fe.
For machinery parts; free machining.

MARWIN SUPASTUFF

W. Martin Winn Ltd.
C 0.3, bal Fe.
For machinery parts, bolts; water hardened.

MAS

Thyssen Edelstahlwerke AG
C 0.55, Si 0.9, Mn 1.2, bal Fe.
For tools, chisels, punches; oil hardening, tough. *Obsolete*

MASILOY I TM 669

Olin Corp.
Cu 63.5, Zn 24.5, Mn 12.
Annealed: 54,000-64,000 TS; 17,000-30,000 YS; 38-46 El. Cold rolled: 58,000-108,000 TS; 27,000-105,000 YS; 1-40 El. Low cost, white colored, brass alloy. For flatware, hollow ware, hardware. *Obsolete*

MASILOY II TM 672

Olin Corp.
Cu 60, Zn 28, Mn 7, Ni 5.
Annealed: 57,000-66,000 TS; 18,000-32,000 YS; 48-56 El. Cold rolled: 65,000-114,000 TS; 30,000-106,000 YS; 1-45 El. A low nickel white brass; for flatware, hollow ware, hardware. *Obsolete*

MASSALLOY

Massillon Steel Casting Co.
C, Cr, Ni, Mo, bal Fe.
Heat treated: 135,000 TS; 110,000 YS; 17 El; 30 RA; 286 Brin. For steel castings. Oil hardened.

MASSILLON

Massillon Steel Casting Co.
C, Mn, bal Fe.
Normalized: 100,000 TS; 60,000 YS; 20 El; 35 RA; 207 Brin. For stokerworms. Cast steel.

MASTALLOY 1011

Magnacast Corp.
Cu 80, Sn 5, Pb 3, Zn 12.
Cast: 32,000 TS; 16,000 YS; 13 El. Leaded semi-red brass. CDA 842. *Obsolete*

MASTALLOY 1151, ETC.

Now MAGNALLOY 1151, ETC.

MASTALLOY 1201

Magnacast Corp.
Cu 82, Sn 4, Pb 6, Zn 7, Ni 1.
Cast: 33,000 TS; 17,600 YS; 17.6 El. Leaded red brass. ASTM B-145 (4B), B-271 (4B); CDA 838. *Obsolete*

MASTALLOY 1941

Magnacast Corp.
Cu 81, Sn 19.
Cast: 50,600 TS; 46,000 YS; 1.1 El; 160 Brin. Tin bronze castings. ASTM B-22 (A); AMS 7322; CDA 913. *Obsolete*

MASTALLOY 1971

Magnacast Corp.
Cu 85, Sn 14, Zn 1.
Cast: 35,200 TS; 34,600 YS; 2.0 El. Tin bronze; for bearings, bushings. Fed. Standard 00153-D2; CDA 910. *Obsolete*

MASTALLOY 2001

Magnacast Corp.
Cu 87, Sn 11, Pb 1, Ni 1.
Cast: 40,000 TS; 24,000 YS; 10 El; 67 Brin. Leaded tin bronze casting. SAE 640; CDA 925. *Obsolete*

MASTALLOY 2061

Magnacast Corp.
Cu 88, Sn 10, Pb 2.
Cast: 49,000 TS; 25,000 YS; 18 El; 86 Brin. Leaded tin bronze casting. SAE 63; CDA 927. *Obsolete*

MASTALLOY 2062

Magnacast Corp.
Cu 84, Sn 10, Pb 2, Ni 4.
Cast: 53,000 TS; 31,000 YS; 15 El; 92 Brin. Leaded nickel tin bronze casting. *Obsolete*

MASTALLOY 2351

Magnacast Corp.
Cu 86, Sn 8, Pb 1, Zn 4, Ni 1.
Cast: 49,000 TS; 23,000 YS; 18 El; 77 Brin. ASTM B-143 (2B), B-271 (2B); SAE 621; CDA 923. *Obsolete*

MASTALLOY 2961

Magnacast Corp.
Cu 71, Sn 13, Pb 15, Ni 1.
Cast: 26,000 TS; 16,000 YS; 8.5 El; 80 Brin. High leaded tin bronze. Fed. Standard 00153:E2; CDA 940. QQ-B-1005(13); QQ-C-525(13). *Obsolete*

MASTER

Duke Steel Co. Inc.
C, Mn, bal Fe.
For tools and dies; water hardening.

MASTER METAL

United American Metals Corp.
Sn, Cu, Pb.
For machinery bearings; Babbitt.

MATADOR 122

Dohlen-Stahl Gusstahl-Handels GmbH
C 0.86, Cr 4.1, Mo 0.85, V 2.5, W 12, bal Fe.
For lathe and planer tools, drills, taps, hobs; high speed steel.

MATADOR 181

Dohlen-Stahl Gusstahl-Handels GmbH
C 0.74, Cr 4.1, V 1.1, W 18.5, bal Fe.
For lathe and planer tools, reamers, drills, broaches, taps; high speed steel.

MATADOR 185

Dohlen-Stahl Gusstahl-Handels GmbH
C 0.79, Co 4.7, Cr 4.3, Mo 0.75, V 1.5, W 18, bal Fe.
For drills, taps, reamers, hobs, broaches; high speed steel.

MATADOR 333

Dohlen-Stahl Gusstahl-Handels GmbH
C 0.95, Cr 4, V, W, Mo, bal Fe.
For cutters, taps, broaches, reamers, drills; high speed steel.

MATADOR 91

Dohlen-Stahl Gusstahl-Handels GmbH
C 0.82, Cr 4.1, Mo 0.85, C 1.6, W 8.7, bal Fe.
For lathe and planer tools, reamers, hobs, drills; high speed steel.

MATHER & PLATT NO. 7 ALLOY

Mather & Platt Ltd.
C 3, Cu 4-5, Ni 12-14, Mn 1-1.2, Cr 4-6, Si 1.2-1.5, bal Fe.
25,000-30,000 TS; 200-220 Brin. For valves, centrifugal pumps. Austenitic cast iron. Abrasion and corrosion resistant.

MATHESIUS METAL

Manufacturer not listed.
Ca 3, Pb, 1-2 alkali earth metals.
For bearings; anti-friction.

MATIDENT B

Johnson Matthey plc
Au 11, Ag 54, Pd 20, Cu 12, Zn, Pt.
Heat-treatable, general white dental casting alloy; Type 4.

MATOBAR

Reinforcement Steel Services
Hard drawn steel wire or bar. See Nizec. *Obsolete*

MATRELOY

Materials Research Corporation
Cr 39, Mo 4, Ti 2, Al 1, bal Ni.
Bar: 375,000 TS; 200,000-275,000 YS; 8 El. At 1400 F: 120,000 YS. For furnace parts, springs, chemical processing equipment, electronic components. Corrosion resistant, heat treatable, nonmagnetic. *Obsolete*

MATRIX

Manufacturer not listed.
Bi 48, Sn 14.5, Pb 28.5, Sb 9.
19,000 TS; 1 El; 19 Brin. For fusible alloy for die mounting, dental models; slight expansion in cooling. *Obsolete*

MATRIX BRASS
Manufacturer not listed.
Cu 62, Zn 37, Pb 1.5.
For engraving, hardware; free-cutting.

MATRIX I
Teledyne Vasco
C 0.51, W 2, Mo 2.75, Cr 4.5, V 1, Mo, Cr, V, bal Fe.
Hardened: 361,000 TS; 292,000 YS; 20 RA; 6 El; 130,000 fatigue strength. For airframes, fasteners, pressure vessels, gears, engine mounts. Tough and fatigue resistant.

MATRIX-35
Forjas Alavesas S.A.
C 0.33, Mn 0.35, Si 1, Cr 1.2, W 4, Mo 0.25, bal Fe.
Hot work tool steel; dies, molds. IHA F-527; DIN 30 WCrV15.

MATRIX-54
Forjas Alavesas S.A.
C 0.55, Si 0.4, Mn 0.6, Cr 1.1, Mo 0.5, V 0.15, bal Fe.
Special purpose steel for hot work operation; drop hammer dies for deep engraving. IHA F-528; DIN 56 NiCrMoV7.

MATRIX-II
Teledyne Vasco
C 0.54, W 1, Mo 5, Cr 4, V 1, Co 8, 12.0 W-Mo-Cr-V, bal Fe.
Annealed: 95,000 TS; 48,000 YS; 25 El; 55 RA; 90 Rock B.
Heat treated: 350,000 TS; 290,000 YS; 8 El; 33 RA; 60 Rock C. Heat treated: 404,000 TS; 363,000 YS; 18 RA; 6 El. For rocket motor cases, high speed rotors, gears, pressure vessels, engine mounts, nuclear applications, molds. High temperature strength and heat resistance.

MATRIX-W
Forjas Alavesas S.A.
C 0.3, Mn 0.4, Si 0-0.5, Cr 2.5, W 9, V 0.25, bal Fe.
Hot work tool steel; for hot blanking dies, punches, mandrels, forming dies. IHA F-526; similar to AISI H21.

MATTHEY 100
Johnson Matthey plc
Cu, Be, Co.
Beryllium copper casting; 45% IACS conductivity. *Obsolete*

MATTHEY 1000
Johnson Matthey plc
W, Cu.
Sintered alloy for spot welding electrodes; 14% IACS conductivity.

MATTHEY 20S
Johnson Matthey plc
Ag 27.5, W 72.5.
Sintered; for spot welding electrodes. 43% IACS conductivity.

MATTHEY 3
Johnson Matthey plc
Cu, Cr.
Cast. For resistance welding electrodes for mild steel; 80% IACS conductivity. *Obsolete*

MATTHEY 328
Johnson Matthey plc
Cu, Cr, Zr.
For resistance welding electrodes; 80% IACS conductivity. *Obsolete*

MATTHEY 35S
Johnson Matthey plc
Ag 35, W 64.
Sintered; for electrical contacts. 52% IACS conductivity.

MATTHEY 50S
Johnson Matthey plc
Ag 49, W 51.
Sintered; for electrical contacts. 61% IACS conductivity.

MATTHEY 53
Johnson Matthey plc
Cu, Si, Ni.
For welding electrodes. *Obsolete*

MATTHEY 73
Johnson Matthey plc
Be 1.8, Co 0.1, bal Cu.
Beryllium copper castings; 20% IACS conductivity. *Obsolete*

MATTHEY A 25
Johnson Matthey plc
Pb, Ag.
Lead base soft solder; 304°C MP.

MATTHEY A 5
Johnson Matthey plc
Pb, Ag.
Lead base soft solder; 304-370°C MP.

MATTHEY D510
Johnson Matthey plc
Ag, Ni.
For electrical contacts; 87% IACS conductivity.

MATTHEY D520
Johnson Matthey plc
Ag, Ni.
For electrical contacts; 57% IACS conductivity.

MATTHEY D54
Johnson Matthey plc
Ag, Cd.
For electrical contacts; 82% IACS conductivity.

MATTHEY D54L
Johnson Matthey plc
Ag, Cd.
For electrical contacts; 82% IACS conductivity. *Obsolete*

MATTHEY D54X
Johnson Matthey plc
Ag, Cd.
For electrical contacts; 82% IACS conductivity.

MATTHEY D55X
Johnson Matthey plc
Ag, Cd.
For electrical contacts; 75% IACS conductivity.

MATTHEY D56
Johnson Matthey plc
Ag, Ni.
For electrical contacts; 72% IACS conductivity.

MATTHEY D58/1
Johnson Matthey plc
1.0 graphite, bal Ag.
For sliding contacts; 96% IACS conductivity.

MATTHEY D58/2
Johnson Matthey plc
2.0 graphite, bal Ag.
For sliding contacts; 87% IACS conductivity.

MATTIBEL B
Johnson Matthey plc
Au 10, Ag 67, Cu 21.
White dental casting alloy.

MATTIBEL G
Johnson Matthey plc
Au 78, Ag 15, Cu 6.
Rich dental casting alloy; Class 2.

MATTIBEL R
Johnson Matthey plc
Au 80, Ag 9, Cu 8, bal Pt.
Red dental casting alloy; Type 2.

MATTIBRAZE 12
Johnson Matthey plc
Ag 12, Cu 50, Zn 31, Cd 7.
Cadmium bearing silver brazing alloy; 620-825°C MP; DIN 8513 L-Ag12Cd.

MATTIBRAZE 34
Johnson Matthey plc
Ag, Cu, Zn.
Brazing alloy; 612-668°C MP.

MATTIBRAZE 45
Johnson Matthey plc
Ag 45, Cu 17, Zn 18, Cd 20.
Cadmium bearing silver brazing alloy; 620-635°C MP; DIN 8513 L-Ag45Cd.

MATTICAST R
Johnson Matthey plc
Au 74.5, Ag 13, Cu 7, Pt, Pd, Zn.
Pink, yellow dental casting alloy; Class 3.

MATTICRAFT 45
Johnson Matthey plc
Au 45, Pd 45, Sn 10.
High-strength dental bonding alloy.

MATTICRAFT 80
Johnson Matthey plc
Pd 80, Ga 8, Cu 9, bal Au.
White dental bonding alloy.

MATTICRAFT B
Johnson Matthey plc
Ag 28, Pd 60, Sn 7, Au, In.
Low cost, general purpose dental bonding alloy.

MATTICRAFT C
Johnson Matthey plc
Pd 79, Sn 11, Cu, Ga.
High-strength, silver-free dental bonding alloy.

MATTICRAFT E
Johnson Matthey plc
Au 78, Pt 10, Pd 9, Ag, In.
Self-hardening ivory colored dental bonding alloy.

MATTICRAFT G
Johnson Matthey plc
Au 86, Pt 10, Ag, Pd, In, Rh.
Yellow dental bonding alloy for short span bridges.

MATTICRAFT JMP
Johnson Matthey plc
Au 86, Pt 10, Ag, Pd, Cu, Zn, In.
Yellow dental bonding alloy.

MATTICRAFT M
Johnson Matthey plc
Au 52, Pd 38, In, Ga.
Dental bonding alloy.

MATTICRAFT S
Johnson Matthey plc
Au 82, Pt 11, Ag, Pd, In.
General purpose yellow dental bonding alloy.

MATTIDENT 4L
Johnson Matthey plc
Au 50, Ag 15, Cu 25, Pt, Pd, Zn.
Yellow dental casting alloy; Class 4.

MATTIDENT 60
Johnson Matthey plc
Au 66, Ag 16, Cu 19, Pt, Pd, Zn.
Yellow dental casting alloy; Class 4.

MATTIDENT AE
Johnson Matthey plc
Au 99.99.
Assay enhancing dental casting alloy.

MATTIDENT ALPHA
Johnson Matthey plc
Au 86, Pt 8, Pd, Cu, Fe.
Rich, yellow dental bonding alloy.

MATTIDENT E
Johnson Matthey plc
Au 55, Ag 24, Cu 11, Pd 8, bal Zn.
Yellow dental casting alloy; Class 4.

MATTIDENT EC
Johnson Matthey plc
Au 49, Ag 32, Cu 10, Pt, Pd, Zn.
Class 4; yellow.

MATTIDENT EF
Johnson Matthey plc
Au 55, Ag 27, Pd 7.
Yellow dental casting alloy; Class 4.

MATTIDENT G
Johnson Matthey plc
Au 68.5, Ag 11, Cu 12, Pt, Pd, Zn.
Yellow dental casting alloy; Class 4.

MATTIDENT L
Johnson Matthey plc
Au 59, Ag 22, Cu 13, Pt, Pd, Zn.
Yellow dental casting allow; Class 4.

MATTIDENT P
Johnson Matthey plc
Au 46, Pd 34.
Economy white dental casting alloy; Class 4.

MATTIDENT R
Johnson Matthey plc
Au 72, Ag 11, Cu 12.
Pink, yellow dental casting alloy; Class 4.

MATTILOY FINE 68
Johnson Matthey plc
Ag 68.
Conventional dental amalgam alloy.

MATTILOY PLUS 43
Johnson Matthey plc
Ag 43.
Non-gramme-2 dental amalgam alloy.

MATTINAX GA
Johnson Matthey plc
Au 92, Ag 5, Pt, Cu.
Rich yellow dental casting alloy; Class 1.

MATTINAX R
Johnson Matthey plc
Au 90, Ag 6, bal Pd.
Pink yellow dental casting alloy; Class 1.

MAUSTINOX A
Usines de A. Manoir, Pitres (Eure)
C 0-0.12, Cr 18, Ni 8, bal Fe.
Annealed: 70,000-85,000 psi TS; 30,000-37,000 psi YS; 30 El; 140-190 Brin. For chemical and textile plant equipment; stainless, austenitic.

MAUSTINOX B
Usines de A. Manoir, Pitres (Eure)
C 0-0.12, Cr 17, Ni 9, Mo 2.5, bal Fe.
Annealed: 71,000-92,000 psi TS; 30,000-37,000 psi YS; 30 El; 150-200 Brin. For acid resistant and chemical plant equipment; stainless, austenitic.

MAUSTINOX C
Usines de A. Manoir, Pitres (Eure)
C 0-0.12, Cr 20, Ni 8, Mo 2.5, bal Fe.
Annealed: 71,000-85,000 psi TS; 40,000-50,000 psi YS; 25 El; 150-200 Brin. For acid resistant chemical plant equipment; stainless, austenitic.

MAUSTINOX D
Usines de A. Manoir, Pitres (Eure)
C 0-0.2, Cr 25, Ni 5, bal Fe.
Annealed: 92,000 psi TS; 65,000 psi YS; 16 El; 190-200 Brin. For chemical plant equipment, furnace parts; corrosion and heat resistant.

MAUSTINOX F
Usines de A. Manoir, Pitres (Eure)
C 0-0.2, Cr 24, Ni 13, bal Fe.
Annealed: 78,000-100,000 psi TS; 30,000-37,000 psi YS; 30 El; 150-200 Brin. For furnace parts, salt pots, heat treat boxes; heat and corrosion resistant, austenitic.

MAUSTINOX H
Usines de A. Manoir, Pitres (Eure)
C 0-0.2, Cr 25, Ni 20, bal Fe.
Annealed: 65,000-92,000 psi TS; 30,000-37,000 psi YS; 30 El; 140-180 Brin. For oil refinery equipment, furnaces, salt pots, heat treat boxes; austenitic, heat and corrosion resistant.

MAUSTINOX SA
Usines de A. Manoir, Pitres (Eure)
C 0-0.12, Cr 18, Ni 8, Ti, bal Fe.
Annealed: 75,000-85,000 psi TS; 30,000-37,000 psi YS; 30 El; 140-190 Brin. For welded chemical and textile plant equipment, tanks, agitators; stainless, austenitic, stabilized.

MAUSTINOX SB
Usines de A. Manoir, Pitres (Eure)
C 0-0.12, Cr 17, Ni 9, Mo 2.5, Ti, bal Fe.
Annealed: 71,000-92,000 psi TS; 30,000-37,000 psi YS; 30 El; 150-200 Brin. For welded acid resistant and chemical plant equipment; stainless, austenitic, stabilized.

MAUSTINOX X
Usines de A. Manoir, Pitres (Eure)
C 0-0.12, Cr 15, Ni 30, Mo, bal Fe.
Annealed: 65,000-74,000 psi TS; 29,000-36,000 psi YS; 35 El. For heat and corrosion resistant parts; austenitic.

MAUSTINOX Y
Usines de A. Manoir, Pitres (Eure)
C 0-0.12, Cr 20, Ni 20, Mo 4.5, bal Fe.
Annealed: 65,000-78,000 psi TS; 30,000-40,000 psi YS; 32 El. For heat and corrosion resistant parts; austenitic.

MAX-EL 1-B
Crucible Materials Corp.
Alloy steel. C 0.2, Mn 1, Mo 0.2, bal Fe.
Rolled: 70,000 TS; 50,000 YS; 30 El; 55 RA; 156 Brin. Heat treated: 135,300 TS; 114,000 YS; 18 El; 54 RA; 332 Brin. For shafts, gears, pinions, bolts, studs; case hardening.

MAX-EL 2-B
Crucible Materials Corp.
Alloy steel. C 0.4, Mn 1, Mo 0.2, bal Fe.
Rolled: 90,000 TS; 55,000 YS; 20 El; 45 RA. Heat treated: 119,500 TS; 85,000 YS; 20 El; 53 RA; 251 Brin. For shafts, spindles, worms, racks, gears, pinions; oil hardening.

MAX-EL 3-1/2
Crucible Materials Corp.
Alloy steel. C 0.5, Mn 1.25, Cr 0.6, Mo 0.15, bal Fe.
Heat treated: 230,000 TS; 225,000 YS; 9.5 El; 36 RA; 514 Brin. For screws, racks, shafts, gears, pinions, spindles; oil hardening.

MAX-EL BRAKE DIE
Crucible Materials Corp.
Tool material. C 0.5, Mn 1.5, Mo 0.18, Cr 0.65, S 0.08, bal Fe. For press brake dies, vee dies, bending and flanging dies; free-cutting, wear and impact resistant.

MAX-WEAR
Colt Industries
C 0.8, Mn 13, Si 0.75, Ni 1.5, Mo 0.5, bal Fe.
Heat treated: 550 Brin. For dredge buckets, rock crushers, chute liners; work hardens, abrasion resistant. *Obsolete*

MAX. 4
Edgar Allen Balfour Ltd.
C 0.15, Ni 20, Cr 25, Mn 0.6, Si 1, bal Fe.
Annealed: 95,000 TS; 45,000 YS; 50 El; 90 Rock B. For furnace and heat treating equipment, furnace linings, heat exchangers, retorts. Type 310; stainless, austenitic, corrosion and heat resistant.

MAXAL CO
Italian manufacture
Co 24, Ni 14, Al 8, Cu 3.
For permanent magnet, magnetic and electrical equipment; high permeability.

MAXCHIP NO. 1
Edgar Allen Balfour Ltd.
C, Ni, Cr, bal Fe.
For hand tools, chisels, punches; air hardened, tough, shock resistant.

MAXCHIP NO. 2
Edgar Allen Balfour Ltd.
C, Cr, bal Fe.
For chisels, punches; oil hardened, tough, shock resistant.

MAXEL 1
Colt Industries
C 0.2, Mn 1-1.2, bal Fe.
Rolled: 75,000-80,000 TS; 45,000-55,000 YS; 30-40 El; 55-65 RA; 156 Brin. Heat treated: 185,000 TS; 150,000 YS; 7 El; 24 RA; 392 Brin. For shafts, gears, pinions, bolts, studs; water hardening. *Obsolete*

MAXEL 2
Colt Industries
C 0.4, Mn 1-1.2, bal Fe.
Rolled: 90,000-110,000 TS; 50,000-65,000 YS; 23-35 El; 50-60 RA. For shafts, gears; tough. *Obsolete*

MAXEL 3
Colt Industries
C 0.45, Mn 1-1.2, Cr 0.4-0.6, bal Fe.
Rolled: 130,000 TS; 100,000 YS; 22 El; 62 RA; 260 Brin. Heat treated: 280,000 TS; 190,000 YS; 7.5 El; 29 RA; 550 Brin. For screws, racks, shafts, gears; tough. *Obsolete*

MAXEL 400 FM
Crucible Materials Corp.
Stainless steel. C 0.32, Cr 16.75, S 0.12, Mn 1.25, Mo 0.25.
Hardness: 32-37 Rock C. Corrosion resistant; pre-hardened, free machining steel for holder or backing applications in plastic molding.

MAXEL 7
Colt Industries
C 0.6-0.7, Mn 0.75-1, Mo 0.7-0.9, bal Fe.
For machine tool parts, water hardening. *Obsolete*

MAXEL HOLDER BLOCK STEEL
Eagle & Globe Steel Ltd.
Alloy steel. C 0.5, Mn 1.25, S 0.08, Cr 0.65, Mo 0.18, bal Fe.
Heat treated: 269-302 Brin. Free machining high tensile low alloy steel. For structural members of die mounting frames and supports, frames for plastic molds and die casting dies, holders and backers for forging dies, tools for brake presses. AS1239 L101A-S.

MAXEL NO. 4 COLLET
Colt Industries
C 0.75, Cr, Mn, bal Fe.
For collets, feed fingers, spindles; oil hardened. *Obsolete*

MAXEL SHANK STEEL
Colt Industries
C 0.5, Cr, Mn, bal Fe.
For shanks for tipped tools; oil hardened. *Obsolete*

MAXELOY
Colt Industries
C 0-0.15, Mn 0.9-1.1, Ni 0.3-0.5, Cr 0-0.25, 0.20% min Cu, bal Fe.
Rolled: 70,000 TS; 50,000 YS; 22 El; 140 Brin. For truck frames, mine and railroad cars, pumps, dredges, coal chutes, ships; high strength low alloy steel. *Obsolete*

MAXELOY
Colt Industries
C 0-0.15, Mn 1.2, P 0.7, Cu 0.3, Ni 0.5, Si 0.7, V, bal Fe.
Rolled: 67,000-70,000 TS; 47,000-50,000 YS; 22 El; 18-19 RA. For structures, railroad and mine cars; low carbon-high strength steel. *Obsolete*

MAXHETE NO. 1
Edgar Allen Balfour Ltd.
C 0.35, Cr 18.5, Ni 8.5, W 2.5, bal Fe.
Cast: 137,000 TS; 35 El. At 900°C: 38,000 TS; 47 El. For burner tubes, pipe unions; austenitic, heat resistant to 1000°C.

MAXHETE NO. 1A
Edgar Allen Balfour Ltd.
C 0-0.4, Cr 13.5, Ni 13.5, W 2.2, Si 1.5, Mn 0.6, bal Fe.
At 20°C: 106,000 TS; 27 El; 44 RA. At 90°C: 61,000 TS; 45 El; 54 RA. For furnace parts, heat treating boxes, nozzles; heat resistant, austenitic.

MAXHETE NO. 2
Edgar Allen Balfour Ltd.
C 0-0.25, Cr 25, Ni 12, Mn 0.6, Si 0.85, bal Fe.
For furnace parts, heat treat equipment, conveyors; resists heat and scaling to 1100°C.

MAXHETE NO. 4
Edgar Allen Balfour Ltd.
C 0-0.1, Cr 25, Ni 33.5, Mn 0.4, Si 0.75, bal Fe.
For furnace parts, hearth plates, grids; austenitic, high temperature strength.

MAXHETE NO. 5
Edgar Allen Balfour Ltd.
C 0-0.2, Ni 65, Cr 15, Mn 0.85, bal Fe.
For carburizing boxes, retorts, oil burner parts; high corrosion and cyclic heat resistance.

MAXHETE NO. 7
Edgar Allen Balfour Ltd.
C 0.12, Cr 20, Si 0.5, bal Fe.
For furnace parts and equipment, heat treating boxes. Heat resistant. Maximum operating temperature 1150°C.

MAXHETE NO. 8
Edgar Allen Balfour Ltd.
C 0-0.1, Ni 79, Cr 19.5, Mn 0.85, bal Fe.
For electrical resistances; high heat resistance.

MAXHETE-3
Edgar Allen Balfour Ltd.
C, high Cr, bal Fe.
For rabble arms and blades in roasting furnaces; heat resistant to 1150°C.

MAXHETE-3
Edelstahlwerk Rochling AG
C, high Cr, bal Fe.
For rabble arms and blades in roasting furnaces; heat resistant to 1150°C.

MAXHETE-6
Edgar Allen Balfour Ltd.
C, Ni, Cr, bal Fe.
For furnace skids, superheater supports; heat resistant, austenitic.

MAXHETE-6
Edelstahlwerk Rochling AG
C, Ni, Cr, bal Fe.
For furnace skids, superheater supports; heat resistant, austenitic.

MAXI-FORM 50
Gulf States Steel, Inc.
HSLA steel. C 0-0.09, Mn 0-0.9, P 0-0.015, S 0-0.02, 0.02 Al min, 0.01 Cb min, bal Fe.
High strength low alloy steel with impact toughness and ductility. 50,000 YS min; 60,000 TS min; 25 El min.

MAXI-FORM 60
Gulf States Steel, Inc.
HSLA steel. C 0-0.09, Mn 0-0.9, P 0-0.015, S 0-0.02, 0.02 Al min, 0.01 Cb min, bal Fe.
High strength low alloy steel with impact toughness and ductility. 60,000 YS min; 70,000 TS min; 22 El min.

MAXILVRY
Edgar Allen Balfour Ltd.
C 0.12, Cr 18.5, Ni 8.5, bal Fe.
Annealed: 83,000 TS; 38,000 YS; 48 El; 60 RA; 143 Brin. Cold drawn: 140,000 TS; 100,000 YS; 21 El; 45 RA; 190 Brin. For fittings, cutlery, valve plates, chemical retorts; Type 302; stainless, austenitic.

MAXILVRY "A.W."
Edgar Allen Balfour Ltd.
C, Cr, Ni, W, Cu, bal Fe.
For welded stainless steel parts; stainless. *Obsolete*

MAXILVRY ADS
Edgar Allen Balfour Ltd.
Stainless steel. C 0.1, Cr 12, Ni 12, Cu 0.5, bal Fe.
Annealed: 90,000 TS; 38,000 YS; 40 El; 45 RA; 160 Brin. For deep drawn parts, chemical plant equipment; stainless, austenitic.

MAXILVRY AM
Edgar Allen Balfour Ltd.
Stainless steel. C 0.07, Cr 18, Ni 13, Cu 0.2, Mo 3, bal Fe.
Annealed: 100,000 TS; 40,000 YS; 40 El; 45 RA; 190 Brin. For chemical equipment to resist H_2SO_4, mixers, tanks; Type 316; stainless, austenitic.

MAXILVRY AT
Edgar Allen Balfour Ltd.
Stainless steel. C 0.2, Cr 18, Ni 8, bal Fe.
Annealed: 100,000 TS; 40,000 YS; 40 El; 45 RA; 190 Brin. For chemical plant equipment; Type 302; stainless, austenitic.

MAXILVRY AWP
Edgar Allen Balfour Ltd.
Stainless steel. C 0-0.1, Cr 18, Ni 8.5, Ti 0.5, bal Fe.
Annealed: 100,000 TS; 40,000 YS; 40 El; 45 RA; 190 Brin. For chemical plant equipment; Type 321; stainless, austenitic.

MAXILVRY C.B.
Edgar Allen Balfour Ltd.
Stainless steel. C 0.1, Cr 18.5, Ni 8.5, Cb 1, bal Fe.
Annealed: 78,000-100,000 TS; 33,000-40,000 YS; 40-60 El; 45-65 RA; 160-195 Brin. For impellers, tanks, chemical plant equipment; Type 347; stainless, austenitic.

MAXILVRY M.B.T.
Edgar Allen Balfour Ltd.
Stainless steel. C 0.04, Cr 17.5, Ni 12, Mo 2.75, Cb 0.75, bal Fe.
Annealed: 78,000-100,000 TS; 45,000-67,000 YS; 40-60 El; 60-75 RA; 170-200 Brin. For chemical plant equipment, welded structures; stainless, austenitic.

MAXILVRY ML
Edgar Allen Balfour Ltd.
C, Ni, Cr, bal Fe.
For cutlery, valves; stainless.

MAXILVRY ML
Edelstahlwerk Rochling AG
C, Ni, Cr, bal Fe.
For cutlery, valves; stainless.

MAXILVRY SPECIAL
Edgar Allen Balfour Ltd.
Stainless steel. C 0.06, Cr 18.5, Ni 8.5, bal Fe.
Annealed: 78,000-100,000 TS; 33,000-40,000 YS; 40-60 El; 45-65 RA; 160-190 Brin. For chemical plant equipment, mixers, agitators, tanks; austenitic, stainless; Type 304.

MAXIMOLD
Ziv Steel & Wire Co.
C 0.4, Cr 5.2, Si 1, V 1, Mo 1.2, bal Fe.
For Al and Zn die casting dies; resists heat checks, air hardened.

MAXIMUM
Westa-Westdeutsche
C 0.74, Cr 4.1, V 1.1, W 18.5, bal Fe.
For lathe and planer tools, drills, reamers, taps; high speed steel.

MAXIMUM SPEZIAL 30
Westa-Westdeutsche
C 0.86, Co 2.8, Cr 4.3, Mo 0.85, V 2.1, W 12, bal Fe.
For lathe and planer tools, drills, reamers, taps; high speed steel.

MAXIMUM SPEZIAL 55
Westa-Westdeutsche
C 0.76, Co 10, Cr 4.2, Mo 0.8, V 1.8, W 18, bal Fe.
For lathe and planer tools, taps, broaches, drills; high speed steel.

MAXIMUM SPEZIAL 55G
Robert Zapp Werkstofftechnik GmbH
C 0.8, Co, Cr, W, Mo, V, bal Fe.
For lathe and planer tools, reamers, broaches; high speed steel.

MAXIMUM SPEZIAL G
Westa-Westdeutsche
C 0.82, Cr 4.1, Mo 0.85, V 1.6, W 8.7, bal Fe.
For lathe and planer tools, reamers, broaches, taps; high speed steel.

MAXIMUM SPEZIAL G EXTRA
Westa-Westdeutsche
C 0.86, Cr 4.1, Mo 0.85, V 2.5, W 12, bal Fe.
For lathe and planer tools, drills, taps, hobs; high speed steel.

MAXIMUM SPEZIAL MO
Westa-Westdeutsche
C 0.95, Cr, W, Mo, V, bal Fe.
For lathe and planer tools, drills; high speed steel.

MAXIMUM STEEL
Peter A. Frasse & Co.
C, W, bal Fe.
For cutting tools. *Obsolete*

MAXINIUM STEEL
Edgar Allen Balfour Ltd.
C 0.32, Si 1.2, Cr 5.25, W 5, Mo 0.5, bal Fe.
For die casting dies, molds. Oil hardened.

MAXITE
Columbia Tool Steel Co.
Tool material. C 0.8, W 14, Cr 4, V 2, Co 5.2, Mo 0.6, bal Fe.
Annealed: 228 Brin. For cutting tools, drills, reamers, boring and shaping tools. Type T8; high-speed steel; high red hardness.

MAXITE 15
Columbia Tool Steel Co.
Tool material. C 1.57, W 12.65, Mo 0.65, Cr 4.75, V 5, Co 5, bal Fe.
For milling cutters, lathe tools, shaper tools, broaches. Type T15; high speed tool steel.

MAXMITH
Edgar Allen Balfour Ltd.
C 0.4, Mn 0-0.6, Ni 3.25, bal Fe.
For chisels, riveting tools, caulking and beading tools; requires no tempering, tough.

MAXNAP
Edgar Allen Balfour Ltd.
C 0.3, Mn 0.6, Cr 1.05, V 0.2, bal Fe.
For pneumatic hammers, riveters, nut piercers; tough and fatigue resistant.

MAXTACK
Edgar Allen Balfour Ltd.
C, Cr, W, Mn, bal Fe.
For nail and tack dies, woodworking tools; air hardened.

MAXTACK
A. Milne & Co.
C 2.25, W 10, Cr 2, Mn 2.5, Si 1, bal Fe.
For tools, cutters; oil or air hardening.

MAXTENSILE
Farrell Co.
Cast Iron. C 3.3, Si 2.4, Ni 1.5, Cr 0.8, bal Fe.
For hydraulic castings, sliding parts, spindles, couplings, sprockets, mill rolls; alloy cast iron. *Obsolete*

MAXTENSILE
USM Corp. (Farrell Co.)
C 3.3, Si 2.4, Ni 1.5, Cr 0.8, bal Fe.
For hydraulic castings, sliding parts, spindles, couplings, sprockets, mill rolls; alloy cast iron. *Obsolete*

MAXTUFF
Ziv Steel & Wire Co.
Tool Material. C 0.5, Mn 0.3, Si 0.75, Cr 1.15, V 0.2, W 2.5, bal Fe.
Type S1 shock resisting tool steel.

MAYARI A
Bethlehem Steel Corp.
Ni 1.25, Cr 0.6, C, bal Fe.
Rolled: 130,000 TS; 110,000 YS; 20 El; 61 RA. For oil well and refinery equipment, auto parts; ductile, tough. *Obsolete*

MAYARI B
Bethlehem Steel Corp.
C 0.34, Mn 0.6, Ni 0.8, Cr 0.3, bal Fe.
For bolts and studs for reactor chambers and superheaters; for service up to 800 F. *Obsolete*

MAYARI IRON
Bethlehem Steel Corp.
C 4-4.5, Mn 1, Si 0.5-3, Ni 2, Cr 2, bal Fe.
For heat, acid, wear, or hard castings; pig iron. *Obsolete*

MAYARI PIG IRON
Bethlehem Steel Corp.
C 4-4.5, Ni 2, Cr 2, Si 0.5-3, Mn 1, Ti 0.15, V 0.2, bal Fe.
For heat and wear resistant castings; heat and wear resistant. *Obsolete*

MAYARI R
Bethlehem Steel Corp.
C 0-0.12, Mn 0.5-1, Cu 0.5-0.7, Ni 0.25-0.75, Cr 0.2-1, P 0.08-0.12, bal Fe.
Rolled: 70,000 TS; 50,000 YS; 22 El; 150 Brin. For transportation equipment, structures; good weldability

MAYARI R-50
Bethlehem Steel Corp.
C 0.1-0.2, Mn 0.75-1.25, Cu 0.2-0.4, Cr 0.4-0.7, Ni 0.25-0.5, V 0.01-0.1, bal Fe.
Plate: 50,000 YS min; 70,000 TS min; 21 El min. For structural railroad and general manufacturing applications. Low-alloy high-tensile structural steel.

MAYARI R-60
Bethlehem Steel Corp.
C 0.1-0.2, Mn 0.75-1.35, Cu 0.2-0.4, Cr 0.4-0.7, Ni 0.25-0.5, V 0.01-0.1, bal Fe.
Plate: 60,000 YS min; 80,000 TS min; 16 El min. For structural, architectural, railroad and general manufacturing applications. Low-alloy high-strength structural steel.

MAYARI SUPERHEATER BOLT STEEL
Bethlehem Steel Corp.
C 0.25-0.35, Ni 1.25-1.75, Cr 0.6-0.9, bal Fe.
Water quenched: 115,000 TS; 100,000 YS; 19 El; 55 RA; 229 Brin. For engine and superheater bolts; heat treated.
Obsolete

MAYOR
Vereinigte Edelstahlwerke
C 0.75, bal Fe.
For tools, drills, punches, springs; water hardened. *Obsolete*

MAZAK 2
National Alloys Ltd
Al 4.1, Cu 2.7, Mg 0.03, bal Zn.
Die cast: 47,300 TS; 83 Brin, 15 IS (Izod), 93,100 Compressive Ultimate Strength. For die castings free from hot shortness.

MAZAK 2
Morris Ashby Ltd.
Al 4.1, Cu 2.7, Mg 0.03, bal Zn.
Die cast: 47,300 TS; 83 Brin, 15 IS (Izod), 93,100 Compressive Ultimate Strength. For die castings free from hot shortness.

MAZAK 3
Birmingham Aluminium Casting Co.
Zinc. Al 3.9-4.3, Mg 0.03-0.06, bal Zn.
Die cast: 42,000 TS; 15 El; 80 Brin. For instrument cases, housings, ornamental grills; impact strength and dimensional stability.

MAZAK 3
National Alloys Ltd.
Al 4, Mg 0.04, bal Zn.
For gears, frames, hardware, die castings.

MAZAK 3, MAZAK 5
Pasminco Europe (Mazak) Ltd.
Al 4.1, Mg 0.05, bal Zn.
283 MPa TS (as cast); 10 El (as cast). For automotive engineering applications, builder's hardware, domestic appliances, business equipment, locks, toys, and giftware.

MAZAK 5
Birmingham Aluminium Casting Co.
Zinc. Al 3.9-4.3, Cu 0.75-1.25, Mg 0.03-0.06, bal Zn.
Die cast: 42,000 TS; 3 El; 65 Brin For motor frames, gears, instrument cases; corrosion resistant to atmosphere.

MAZAK 5
National Alloys Ltd.
Al 4, Cu 1, Mg 0.3, bal Zn.
Die cast: 42,000 TS; 3 El; 65 Brin. For gears, die castings.

MAZAK 6
National Alloys Ltd.
Al 4, Cu 1.25, bal Zn.
Die cast: 42,000 TS; 5 El; 65 Brin. For die castings. Maximum fluidity.

MAZAK 8
Pasminco Europe (Mazak) Ltd.
Al 8.2-8.8, Cu 0.9-1.3, Mg 0.02-0.03, bal Zn.
375 MPa TS (as cast); 6-10 El (as cast). For high-temperature automotive applications, electrical components, and pneumatic equipment.

MAZAK ZA12
Pasminco Europe (Mazak) Ltd.
Al 10.8-11.5, Cu 0.5-1.25, Mg 0.02-0.03, bal Zn.
404 MPa TS (as cast); 4-7 El (as cast). Used for sand, gravity die, and graphite mold casting.

MAZAK ZA27
Pasminco Europe (Mazak) Ltd.
Al 25.5-28, Cu 2-2.5, Mg 0.015-0.02, bal Zn.
426 MPa TS (as cast); 2-3 El (as cast). For sand casting and cold chamber pressure die casting.

MB1
Delta (Manganese Bronze) Ltd.
Copper. Cu 57, Mn 1.25, Pb 2.25, bal Zn.
General-purpose manganese brass suitable for architectural purposes. Good machinability and can be hot stamped. Available forms: rod, bar, section, hollow rod. Extruded: 340-385 N/mm^2 TS; 25-40 El. Drawn: 430-500 N/mm^2 TS; 25-35 El.

MB5
Delta (Manganese Bronze) Ltd.
Copper. Cu 58.5, Mn 0.35, Fe 1.25, Sn 1, bal Zn.
American-type manganese brass, with tin and iron added to give strength and corrosion resistance superior to that of naval brasses. Forms available: rod, bar, coil, section, hollow rod. Extruded: 430-460 N/mm^2 TS; 30-40 El. Drawn: 500-570 N/mm^2 TS; 20-30 El.

MBA2
Mitsubishi Metals America Corp.
Copper. Cu 62, Al 3, Mn 3, Si 1, Cr 0-1, Ni 0-1, bal Zn.
Bearing bronze; for synchronizer rings of automobile manual transmissions. Hydraulic parts; bearing and worm wheel.

MBA5
Mitsubishi Metals America Corp.
Copper. Cu 60, Al 5, Ni 2.5, Ti 1.5, bal Zn.
Bearing bronze; for synchronizer rings of automobile manual transmissions.

MC 100
Thyssen Edelstahlwerke AG
C, Cr, Ni, Mo, bal Fe.
16 El. For machinery parts; oil hardened. *Obsolete*

MC 102
Mond Nickel Co. Ltd.
C 0.02-0.06, Si 0.1-0.4, Mn 0.1-0.5, Fe 0-4, Cr 19-20.5, Co 0-5, Mo 5.5-6.5, Cb 6.2-6.7, W 2-3, bal Ni.
Heat treated: 103,000 TS; 85,000 YS; 10 El; 340 Vickers. For gas turbine stator blades, diesel hot plugs, turbine rotors. Good oxidation resistance and strength to 900°C. High temperature casting alloy. Age-hardenable.

MC 40
Thyssen Edelstahlwerke AG
C, Cr, Mo, bal Fe.
16 El. For machinery parts; oil hardened. *Obsolete*

MC 406
Capitol Castings Inc.
C 0.8-0.9, Cr 2-2.35, Mo 0.28-0.38, bal Fe.
Mill liners.

MC ALLOY
Mitsubishi Metals America Corp.
Cr 50, Mo 0-5, Fe 0-25, bal Ni.
Excellent corrosion resistance to nitric-hydrofluoric, phosphoric and sulfuric acid. Catching tray for wafers; electrodes in galvanizing tanks.

MC GILL METAL
McGill Mfg. Co. Inc.
Cu 89, Al 9, Fe 2.
70,000-80,000 TS; 30,000-40,000 YS; 10-20 El; 180 Brin. For gears, bushings, bearings; resists corrosion and wear.
Obsolete

MC III V

Thyssen Edelstahlwerke AG
C 0.2, Cr 2.5-2.7, Mo 0.4, V 8, W 0.4, bal Fe.
For case hardened parts; carburizing steel. *Obsolete*

MC KAY 17-4 PH

Now MC KAY 630.

MC KAY 18

Teledyne McKay
C 0.06-0.1, Mn 0.2-0.4, bal Fe.
Welded: 67,000 TS; 55,000 YS; 26 El; 46 RA; 156 Brin. For welding electrodes for flat position fillet; shielded arc coating. *Obsolete*

MC KAY TOOL & DIE

Teledyne McKay
C 0.48, Cr 3.7, Mn 0.4, Mo 9, W 1.6, V 1.1, bal Fe.
550-625 Brin. For welding rod, hard surfacing; shielded arc. *Obsolete*

MC KENNA K4

Kennametal Inc.
WC+Co.
For tools, cutters; for cutting cast iron, glass, porcelain. *Obsolete*

MCADAMITE

Manufacturer not listed.
Zn 12-18, Cu 3.1, Mg 0.2, bal Al.
For strong light alloy parts; non-hardenable.

MCADAMS ALLOY "A"

Manufacturer not listed.
Al 60, Cu 11-55, Cr 10-43, Zn 20.
For light alloy parts; non-hardenable.

MCADAMS ALLOY "B"

Manufacturer not listed.
Al 69, Cu 7.7, Ni 0.6, Zn 23.
For light alloy parts; non-hardenable.

MCADAMS ALLOY "C"

Manufacturer not listed.
Al 70, Cu 3, Zn 22, Sb 5.
For light alloy parts; non-hardenable.

MCADAMS ALLOY "D"

Manufacturer not listed.
Al 80, Cd 8, Ag 4, Sn 8.
For light alloy parts; non-hardenable.

MCADAMS ALLOY "E"

Manufacturer not listed.
Al 82, Cu 12, Cd 5, Ag 1.
For light alloy parts; non-hardenable.

MCADAMS ALLOYS

Manufacturer not listed.
Al 60-82, Cu 0-55, Cd 0-8, Ag 0-4, Sn 0-8, Cr 0-43, Zn 0-23, Sb 0-5.
For light alloy parts.

MCCINNES SPECIAL CR-NI STEEL

McInnes Steel Co.
Cr 0.5, Ni 1.5, Mn 0.6, C 0.7, bal Fe.
For hot work tools, bolt heading and forming dies; gripper dies, shear blades, hot punching dies, heat resisting, high strength and toughness. *Obsolete*

MCFARLAND & HARDER ALLOY "A"

Manufacturer not listed.
Cr 10, Ni 48, Cu 43.
For heat and corrosion resistant parts; stainless and corrosion resistant.

MCFARLAND & HARDER ALLOY "B"

Manufacturer not listed.
Cr 16, Ni 29, Cu 55.
For heat and corrosion resistant parts; stainless and corrosion resistant.

MCFARLAND & HARDER ALLOY "C"

Manufacturer not listed.
Cr 30, Ni 59, Cu 11.
For heat and corrosion resistant parts; stainless and corrosion resistant.

MCFARLAND & HARDER ALLOY "D"

Manufacturer not listed.
Cr 43, Ni 46, Cu 11.
For heat and corrosion resistant parts; stainless and corrosion resistant.

MCFARLAND & HARDER ALLOYS

Manufacturer not listed.
Ni 29-59, Cu 11-55, Cr 10-43.
For corrosion and heat resistant parts; stainless and corrosion resistant.

MCGILL BRASS PRESSURE CASTING ALLOY

McGill Mfg. Co. Inc.
Cu 57-62, Zn 38-41, Sn 0.5, Pb 1.5.
60,000 TS; 30,000 YS; 20 El. For pressure die castings; structural parts; tough. *Obsolete*

MCGILL NO. 2

McGill Mfg. Co. Inc.
Si, bal Fe.
95,000 TS; 160 Brin. For machinery parts; corrosion and abrasion resistant. *Obsolete*

MCGILL NO. 3

McGill Mfg. Co. Inc.
Sn 10, Zn 0.5, Pb 2, bal Cu.
For hydraulic pressure castings. *Obsolete*

MCGILL NO. 4

McGill Mfg. Co. Inc.
Cu 62-67, Zn 38-41, Pb 1.5, Sn 0.8.
For hardware; free cutting. *Obsolete*

MCH

Latrobe Steel Co.
C 0.5, Mo 6.2, W 1, Cr 3.7, V 0.75, bal Fe.
For forging dies, punches, shear blades; hot work steel, resists heat checking. *Obsolete*

MCINNES "V"

McInnes Steel Co.
C 0.7, W 18, Cr 4, V 1, bal Fe.
For turning and boring tools, special dies, taps, reamers, milling cutters; high speed steel. *Obsolete*

MCINNES FOLDIER DIE

McInnes Steel Co.
C 0.9-1.05, Mn 0.25-0.35, bal Fe.
Heat treated: 190,000 TS; 120,000 YS; 10 El; 30 RA; 380 Brin. For brake dies, forming dies; water or oil hardened. *Obsolete*

MCINNES HC-HC

McInnes Steel Co.
C 2, Cr 13, V 1, bal Fe.
For dies, punches, reamers, gauges; very tough. *Obsolete*

MCINNES HIGH CARBON HIGH CHROME

McInnes Steel Co.
C 2.1, Cr 13, V 1, Ni 0.5, bal Fe.
For broaches, drawing dies; oil or air hardening. *Obsolete*

MCINNES MACHINE

McInnes Steel Co.
C 0.35-0.45, Mn 0.4, Si 0.3, bal Fe.
Hot rolled: 90,000 TS; 58,000 YS; 27 El; 50 RA; 200 Brin. For spindles, lead screws, piston rods, gears; water hardened. *Obsolete*

MCINNES RECORD A

McInnes Steel Co.
C 0.3-0.5, W 12-14, Cr 2-4, V 1, bal Fe.
For piercers, punches, heading and trimming dies; hot work steel, oil hardened. *Obsolete*

MCINNES SPECIAL TOOL STEEL

McInnes Steel Co.
C 0.95-1.1, bal Fe.
For hard wearing surface and extreme depth hardness; deep hardening. *Obsolete*

MCINNES STANDARD VA HIGH SPEED

McInnes Steel Co.
W 18, Cr 4, V 1.5, C 0.5-0.9, bal Fe.
For turning and boring tools, cutters, special dies, reamers. *Obsolete*

MCINNES VANADIUM CRUCIBLE

McInnes Steel Co.
C 0.75-1, Mn 0.3, V 0.2, bal Fe.
For cold forming dies, punches, chisels, shear blades, dies, rivet sets; water hardened. *Obsolete*

MCK-ALLOY

Teledyne McKay
C 0.2, Ni 0.5, Cr 0.5, Mo 0.2, bal Fe.
125,000 TS; 15 El; 260 Brin. For chains. *Obsolete*

MCKAY "C-S"

Teledyne McKay
C 0.03, Mn 0.6, Si 0.6, Cr 14.5, Mo 15, W 3.3, Fe 4, bal Ni.
Wire for submerged arc welding. Deposit: 200-225 Brin. Work hardens to 36-40 Rock C. Excellent corrosion resistance.

MCKAY 023

Teledyne McKay
Stainless steel. C 0.04, Mn 2.25, Si 0.25, Cr 19.7, Ni 32.9, Mo 2.15, Cu 3.1, bal Fe.
Covered welding electrode. 76,000 psi TS; 48,000 psi YS; 40 El.

MCKAY 1-C

Teledyne McKay
C 2, Cr 30, Co 50, W 12.
Coated electrode. For hard surfacing. 52 Rock C; corrosion and wear resistant; retains hardness above 1500°F. For rocker arms, cams, wire drawing blocks, slag ladles. Meets AWS A5.13; Class E CoCr-C.

MCKAY 10

Teledyne McKay
C 0.06-0.1, Mn 0.3-5, bal Fe.
Welded: 70,000 TS; 62,000 YS; 20.5 El; 28 RA; 154 Brin. For welding rods; AWS E6013; shielded arc. *Obsolete*

MCKAY 10016 G

Teledyne McKay
C 0.09, Mn 0.9, Si 0.45, Ni 1.75, Mo 0.3, V 0.08, bal Fe.
Welded: 108,000 TS; 99,000 YS; 23 El. Low carbon alloy. For welding low alloy, high strength steels. 100,000 psi TS; AWS E10016-G. *Obsolete*

MCKAY 10018-D2 XLM

Teledyne McKay
C 0.11, Mn 1.85, Si 0.45, Mo 0.35, Ni 0.75, bal Fe.
Low alloy low hydrogen covered welding electrode. Welded: 110,000 psi TS; 98,000 psi YS; 25 El. For low-alloy high-strength steels in the 100,000 psi TS range. AWS E10018-D2.

MCKAY 10018-M XLM
Teledyne McKay
C 0.06, Mn 1.25, Si 0.4, Cr 0.1, Ni 1.55, Mo 0.3, bal Fe.
Low alloy low hydrogen covered welding electrode. Welded: 104,000 psi TS; 95,000 psi YS; 24 El. For welding low-alloy, high-strength steels in the 100,000 psi TS range. AWS E10018-M.

MCKAY 11016
Teledyne McKay
C 0.08, Mn 1.25, Si 0.45, Cr 1.2, Ni 2, Mo 0.3, Mo, bal Fe.
Welded: 117,000 TS; 106,000 YS; 23 El. Low carbon alloy steel weld rod. For welding low alloy, high strength steels. 110,000 psi TS; AWS E11016-G. *Obsolete*

MCKAY 11018-M XLM
Teledyne McKay
C 0.07, Mn 1.5, Si 0.4, Ni 1.6, Mo 0.4, bal Fe.
Low alloy low hydrogen covered welding electrode. Welded: 115,000 psi TS; 104,000 psi YS; 24 El. For welding low-alloy high-strength steels in the 110,000 psi TS range. MIL 11018; AWS E11018-M.

MCKAY 116 HV
Teledyne McKay
C 0.06-0.1, Mn 0.3-0.5, bal Fe.
Welded: 79,000 TS; 69,000 YS; 19 El; 37 RA; 162 Brin. For welding rods; shielded arc. *Obsolete*

MCKAY 117
Teledyne McKay
C 0.06-0.1, Mn 0.3-0.5, bal Fe.
Welded: 65,000 TS; 57,000 YS; 19 El; 35 RA; 120 Brin. For welding rods; shielded arc. *Obsolete*

MCKAY 12018-M XLM
Teledyne McKay
C 0.07, Mn 1.5, Si 0.4, Cr 0.45, Ni 2, Mo 0.4, bal Fe.
Low alloy low hydrogen covered welding electrode. Welded: 128,000 psi TS; 114,000 psi YS; 24 El. For low-alloy high-strength steels in the 120,000 psi TS range. MIL 12018; AWS E12018-M.

MCKAY 13018
Teledyne McKay
C 0.085, Mn 1.63, Si 0.42, Cr 0.73, Ni 1.64, Mo 0.24.
Typical deposit analysis. Low carbon alloy steel weld rod. Welded: 134,000 TS; 124,000 YS; 21 El. For welding high yield strength steels. *Obsolete*

MCKAY 14
Teledyne McKay
C 0.06-0.1, Mn 0.3-0.5, bal Fe.
Welded: 70,000 TS; 62,000 YS; 23 El; 30 RA; 150 Brin. For welding rods; AWS E6013; shielded arc. *Obsolete*

MCKAY 14018
Teledyne McKay
C 0.08, Mn 1, Si 0.4, Cr 0.5, Ni 3.5, Mo 0.75, bal Fe.
Welded: 147,000 TS; 141,000 YS; 18 El. Low carbon alloy steel weld rod. For welding high yield strength steels.

MCKAY 14018 HT
Teledyne McKay
C 0.085, Mn 0.6, Si 0.45, Cr 0.4, Ni 8.2, Mo 0.5, V 0.08, bal Fe.
Low carbon alloy steel weld rod. Welded: 165,000 TS; 155,000 YS; 16 El. For welding HY-130 steel prior to heat treatment. *Obsolete*

MCKAY 15-60
Teledyne McKay
C 0.2, Mn 1.6, Si 0.35, Cr 14.3, Ni 61, bal Fe.
Bare wire for welding. *Obsolete*

MCKAY 16-25-6
Teledyne McKay
C 0-0.12, Cr 15-17, Ni 24-27, Mo 5.5-7, bal Fe.
Welded: 95,000 TS; 65,000 YS; 43 El. For welding electrodes; shielded arc, high heat resistant. *Obsolete*

MCKAY 16-8-2 (1)
Teledyne McKay
C 0.06, Mn 0.25, Si 0.25, Cr 15.8, Ni 8.2, Mo 1.6, bal Fe.
Bare wire for welding. *Obsolete*

MCKAY 16-8-2 (2)
Teledyne McKay
C 0.08, Mn 2.25, Si 0.35, Cr 15.3, Ni 8.2, Mo 0.5.
Welded: 94,000 TS; 70,500 YS; 40 El. Coated stainless steel weld rod. For welding Types 316, 317, and 347 stainless steels for high temperature operation. AWS A5-4; ASTM A298. *Obsolete*

MCKAY 17-14 CU MO
Teledyne McKay
C 0.12, Mn 1.4, Si 0.4, Cr 16.1, Ni 14.4, Mo 2.4, Cb 0.75, Cu 3, bal Fe.
Bare wire for welding. *Obsolete*

MCKAY 17-4 PH
Teledyne McKay
Now MCKAY 630, WIRE.

MCKAY 17-7 PH
Teledyne McKay
C 0.07, Mn 0.8, Si 0.7, Cr 17, Ni 7.2, Al 1, bal Fe.
Bare wire for welding. *Obsolete*

MCKAY 18-8 MO-ELC
Teledyne McKay
C 0-0.038, Cr 17-19, Ni 12-14, Mo 2.5, bal Fe.
Weld: 80,000 TS; 55,000 YS; 40 El. For welding rod; shielded arc. *Obsolete*

MCKAY 18-8 CB ELC
Teledyne McKay
C 0-0.038, Cr 18-20.5, Ni 9-11, Cb = 10 x C, bal Fe.
Welded: 90,000 TS; 70,000 YS; 35 El. For welding electrode; shielded arc. *Obsolete*

MCKAY 19
Teledyne McKay
C 0.08-0.12, Mn 0.3-0.5, bal Fe.
Welded: 55,000 TS; 46,000 YS; 30 El; 50 RA; 110 Brin. For welding rods; shielded arc. *Obsolete*

MCKAY 20
Teledyne McKay
C 0.08-0.12, Mn 0.2-0.4, bal Fe.
Welded: 70,000 TS; 63,000 YS; 27 El; 46 RA; 140 Brin. For welding rod; AWS-E6020, shielded arc. *Obsolete*

MCKAY 20 (320)
Teledyne McKay
C 0.05, Mn 1.25, Si 0.35, Cr 20.5, Ni 33.5, Mo 2.5, Cb 0.75, Cu 3.4, bal Fe.
Bare wire for welding. *Obsolete*

MCKAY 20-80
Teledyne McKay
C 0-0.1, Cr 18-19.5, Si 0-1.3, bal Ni.
For welding electrodes; Nichrome V; shielded arc, heat resistant. *Obsolete*

MCKAY 20-H
Teledyne McKay
C 0.08-0.12, Mn 0.2-0.4, bal Fe.
Welded: 69,000 TS; 60,000 YS; 26.5 El; 44 RA; 140 Brin. For welding rod; shielded arc. *Obsolete*

MCKAY 20-SP
Teledyne McKay
C 0.08-0.12, Mn 0.2-0.4, bal Fe.
Welded: 70,000 TS; 62,000 YS; 26 El; 45 RA; 140 Brin. For welding rod; shielded arc. *Obsolete*

MCKAY 21
Teledyne McKay
C 0.25, Cr 28, Co 56, Mo 5.75.
Coated electrode. For build-up requiring corrosion resistance and good strength at elevated temperatures up to 1500°F. For hot shears, hot trim dies, hot extrusion dies, and pressure valves. *Obsolete*

MCKAY 2209
Teledyne McKay
Stainless steel. C 0.03, Mn 0.95, Si 0.45, Cr 23, Ni 9.7, Mo 3, bal Fe.
Covered welding electrode. 115,000 psi TS; 90,000 psi YS; 27 El. For welding 22Cr-5Ni-3Mo (2205 Type) duplex stainless steel.

MCKAY 236-S
Teledyne McKay
C 0.12-0.24, Mn 1.2-1.8, Si 0.45-0.7, Ni 5.3, Mo 5.3, P 0.01-0.02, S 0.01-0.015, bal Fe.
Submerged arc surfacing wire. 41 Rock C.

MCKAY 242-S
Teledyne McKay
C 0.14-0.16, Mn 1.6-1.8, Si 0.7-0.8, Cr 1.8-2.2, Mo 0.6, V 0.21-0.31, P 0.013-0.016, S 0.013-0.014, bal Fe.
Submerged arc surfacing wire. 36-37 Rock C.

MCKAY 25
Teledyne McKay
C 0.05, Cr 20.5, Co 45, W 15.5, Ni 10.
Coated electrode. For build-up requiring good wear and corrosion resistance and strength at temperatures to and above 1200°F. Machinable with carbide tools. *Obsolete*

MCKAY 250-S
Teledyne McKay
C 0.19-0.28, Mn 0.68-2.18, Si 0.36-0.97, Cr 9.19-11.3, P 0.007-0.045, S 0.016-0.018, bal Fe.
Submerged arc surfacing wire. 48 Rock C.

MCKAY 252-S
Teledyne McKay
C 0.17-0.2, Mn 1.8-2.2, Si 0.45-0.6, Cr 3.5-3.8, P 0.014-0.03, S 0.012-0.016, bal Fe.
Submerged arc surfacing wire. 45 Rock C.

MCKAY 308
Teledyne McKay
Stainless steel. C 0.04, Mn 1.9, Si 0.45, Cr 20.5, Ni 9.8, bal Fe.
Welding wire. AWS ER308.

MCKAY 308 HC
Teledyne McKay
C 0.09, Mn 1.8, Si 0.5, Cr 19.5, Ni 10.2, bal Fe.
Welded: 93,500 TS; 68,000 YS; 40 El. Coated stainless steel weld rod. For welding or build-up on austenitic 18-8 stainless steel; ferrite.

MCKAY 308, 308H
Teledyne McKay
C 0.06, Mn 1, Si 0.4, Cr 20.2, Ni 9.6, bal Fe.
General welding electrode: 86,000 TS; 65,000 YS; 45 El. For use on Types 301, 302, 304, 305, and 308 base metals. AWS E308, 308H.

MCKAY 308, WIRE
Teledyne McKay
C 0.04, Mn 1.9, Si 0.45, Cr 20.5, Ni 9.8, bal Fe.
Bare wire for welding. AWS ER308.

MCKAY 308L
Teledyne McKay
Stainless steel. C 0.017, Mn 1.8, Si 0.5, Cr 20.3, Ni 9.7, bal Fe.
Welding wire. AWS ER308L.

MCKAY 308L HI SIL
Teledyne McKay

C 0.017, Mn 1.8, Si 0.8, Cr 20.3, Ni 9.7, bal Fe.
Bare wire for welding. AWS ER308L-Si.

MCKAY 308L HI SIL
Teledyne McKay

Stainless steel. C 0.017, Mn 1.8, Si 0.8, Cr 20.3, Ni 9.7, bal Fe.

Welding wire. AWS ER308L-Si.

MCKAY 308L, ELECTRODE
Teledyne McKay

Stainless steel. C 0.03, Mn 1, Si 0.4, Cr 20.2, Ni 9.8, bal Fe.
Welded: 83,000 TS; 64,000 YS; 37 El. Covered welding electrode for Type 308L stainless steel. AWS E308L.

MCKAY 308L, WIRE
Teledyne McKay

C 0.017, Mn 1.8, Si 0.5, Cr 20.3, Ni 9.7, bal Fe.
Bare wire for welding. AWS ER308L.

MCKAY 309
Teledyne McKay

Stainless steel. C 0.06, Mn 1.75, Si 0.5, Cr 23.7, Ni 12.7, bal Fe.
Welding wire. AWS ER309.

MCKAY 309, ELECTRODE
Teledyne McKay

C 0.07, Mn 1, Si 0.35, Cr 23.4, Ni 12.5, bal Fe.
Covered welding electrode: 88,000 TS; 67,000 YS; 37 El. For welding and build-up on type 309 stainless. AWS E309.

MCKAY 309, WIRE
Teledyne McKay

Stainless steel. C 0.06, Mn 1.75, Si 0.5, Cr 23.7, Ni 12.7, bal Fe.
Bare wire for welding. AWS ER309.

MCKAY 309CB
Teledyne McKay

C 0.06, Mn 1.6, Si 0.4, Cr 24.5, Ni 13, Cb 0.6, bal Fe.
Bare wire for welding. *Obsolete*

MCKAY 309CB, ELECTRODE
Teledyne McKay

C 0.07, Mn 1, Si 0.5, Cr 23, Ni 13, Cb 0.85.
Coated stainless steel weld rod. Welded: 100,000 TS; 80,000 YS; 34 El. Covered welding electrode for 347 and 321 clad steels.

MCKAY 309L
Teledyne McKay

Stainless steel. C 0.016, Mn 2, Si 0.45, Cr 23.7, Ni 13.2, bal Fe.
Welding wire. AWS ER309L.

MCKAY 309L, ELECTRODE
Teledyne McKay

C 0.035, Mn 1, Si 0.5, Cr 23, Ni 13.2, bal Fe.
Covered welding electrode for weld overlay or welding stainless to mild or low alloy steels. 79,000 TS; 64,000 YS; 41 El. AWS 309L.

MCKAY 309L, WIRE
Teledyne McKay

Stainless steel. C 0.016, Mn 2, Si 0.45, Cr 23.7, Ni 13.2, bal Fe.
Bare wire for welding. AWS ER309L.

MCKAY 309MO
Teledyne McKay

C 0.07, Mn 1, Si 0.4, Cr 22.3, Ni 13, Mo 2.3, bal Fe.
Covered welding electrode for 316 clad steels and joining Mo-containing stainless to carbon steels. 94,000 TS; 74,000 YS; 35 El. AWS 309Mo.

MCKAY 309MOL
Teledyne McKay

Stainless steel. C 0.03, Mn 1, Si 0.45, Cr 23, Ni 13.5, Mo 2.3, bal Fe.
Covered welding electrode. 90,000 psi TS; 70,000 psi YS; 35 El. AWS E309Mo.

MCKAY 310
Teledyne McKay

Stainless steel. C 0.13, Mn 2.1, Si 0.5, Cr 26.2, Ni 21, bal Fe.
Covered welding electrode. 86,000 psi TS; 63,000 psi YS; 40 El. AWS E310. For welding base metal of similar composition.

MCKAY 310, WIRE
Teledyne McKay

Stainless steel. C 0.1, Mn 1.8, Si 0.5, Cr 27, Ni 21.1, bal Fe.
Welding wire. AWS ER310.

MCKAY 310CB
Teledyne McKay

Stainless steel. C 0.1, Mn 2.1, Si 0.45, Cr 26, Ni 21, bal Fe.
Covered welding electrode. 90,000 psi TS; 65,000 psi YS; 35 El. AWS E310Cb. For 347 clad steels.

MCKAY 310H
Teledyne McKay

Stainless steel. C 0.4, Mn 2.25, Si 0.4, Cr 26.2, Ni 21.4, bal Fe.
Covered welding electrode. 118,000 psi TS; 90,000 psi YS; 28 El. AWS E310H. For HK casting repair.

MCKAY 310MO
Teledyne McKay

Stainless steel. C 0.1, Mn 2.1, Si 0.45, Cr 26, Ni 21, Mo 2.25, bal Fe.
Covered welding electrode. 90,000 psi TS; 65,000 psi YS; 39 El. AWS E310Mo. Similar to Type 310.

MCKAY 312
Teledyne McKay

Stainless steel. C 0.1, Mn 1.3, Si 0.6, Cr 29, Ni 9, bal Fe.
Covered welding electrode. 115,000 psi TS; 95,000 psi YS; 25 El. AWS E312. For joining dissimilar metals.

MCKAY 312MO
Teledyne McKay

Stainless steel. C 0.1, Mn 1.7, Si 0.6, Cr 29, Ni 9, Mo 2, bal Fe.
Covered welding electrode. 120,000 psi TS; 95,000 psi YS; 29 El. For joining dissimilar metals.

MCKAY 316, 316H
Teledyne McKay

Stainless steel. C 0.06, Mn 1.85, Si 0.35, Cr 18, Ni 13, Mo 2.15, bal Fe.
Covered welding electrode. 85,000 psi TS; 68,000 psi YS; 45 El. AWS E316/316H. For welding Type 316 steel.

MCKAY 316, WIRE
Teledyne McKay

Stainless steel. C 0.04, Mn 1.7, Si 0.48, Cr 18.7, Ni 12.7, Mo 2.3, bal Fe.
Welding wire. AWS ER316.

MCKAY 316/316H/HF
Teledyne McKay

Stainless steel. C 0.04, Mn 1.2, Si 0.35, Cr 19.3, Ni 11.5, Mo 2.35, bal Fe.
Covered welding electrode. 85,000 psi TS; 65,000 psi YS; 40 El. AWS E316/316H.

MCKAY 316L
Teledyne McKay

Stainless steel. C 0.019, Mn 1.7, Si 0.45, Cr 18.9, Ni 12.6, Mo 2.2, bal Fe.
Welding wire. AWS ER316L.

MCKAY 316L
Teledyne McKay

Stainless steel. C 0.03, Mn 1.6, Si 0.35, Cr 18, Ni 13.2, Mo 2.25, bal Fe.
Covered welding electrode. 82,000 psi TS; 61,000 psi YS; 42 El. AWS E316L. For welding Type 316L material.

MCKAY 316L HF
Teledyne McKay

Stainless steel. C 0.03, Mn 1, Si 0.35, Cr 19.5, Ni 11.6, Mo 2.25, bal Fe.
Covered welding electrode. 85,000 psi TS; 66,000 psi YS; 41 El. AWS E316L.

MCKAY 316L HI SIL
Teledyne McKay

Stainless steel. C 0.015, Mn 1.9, Si 0.8, Cr 19, Ni 12, Mo 2.2, bal Fe.
Welding wire. AWS ER316L-Si.

MCKAY 317
Teledyne McKay

Stainless steel. C 0.06, Mn 1.5, Si 0.45, Cr 18.4, Ni 13.6, Mo 3.3, bal Fe.
Covered welding electrode. 95,000 psi TS; 70,000 psi YS; 33 El. AWS E317.

MCKAY 317, WIRE
Teledyne McKay

C 0.02, Mn 1.75, Si 0.45, Cr 19, Ni 13.3, Mo 3.2, bal Fe.
Bare wire for welding. AWS ER317L.

MCKAY 317L
Teledyne McKay

Stainless steel. C 0.02, Mn 1.75, Si 0.45, Cr 19, Ni 13.3, Mo 3.2, bal Fe.
Welding wire. AWS ER317L.

MCKAY 317L
Teledyne McKay

Stainless steel. C 0.03, Mn 1.5, Si 0.45, Cr 18.4, Ni 13.6, Mo 3.2, bal Fe.
Covered welding electrode. 92,000 psi TS; 69,000 psi YS; 35 El. AWS E317L. Similar to Type 316L; high temperature creep resistance.

MCKAY 318
Teledyne McKay

Stainless steel. C 0.05, Mn 1.75, Si 0.4, Cr 19.5, Ni 12.5, Mo 2.3, Cb 0.55, bal Fe.
Covered welding electrode. 95,000 psi TS; 75,000 psi YS; 35 El. AWS E318.

MCKAY 320
Teledyne McKay

Stainless steel. C 0.02, Mn 0.45, Si 0.25, Cr 19.7, Ni 33.4, Mo 2.1, Cu 3.2, 0.40 Cb + Ta, bal Fe.
Welding wire. AWS ER320.

MCKAY 320
Teledyne McKay

Stainless steel. C 0.04, Mn 2.25, Si 0.25, Cr 19.7, Ni 32.9, Mo 2.15, Cb 0.5, Cu 3.1, bal Fe.
Covered welding electrode. 84,000 psi TS; 54,000 psi YS; 39 El. AWS E320. For fabricating Carpenter 20 and 20 Cb-3 stainless steels.

MCKAY 330
Teledyne McKay

Stainless steel. C 0.2, Mn 1.75, Si 0.35, Cr 16, Ni 35, bal Fe.
Welding wire. AWS ER330.

MCKAY 330
Teledyne McKay

Stainless steel. C 0.2, Mn 2.25, Si 0.5, Cr 14.5, Ni 34, bal Fe.
Covered welding electrode. 86,000 psi TS; 58,000 psi YS; 40 El. AWS E330. For welding Type 330 base metal.

MCKAY 330, WIRE
Teledyne McKay
Stainless steel. C 0.2, Mn 1.75, Si 0.35, Cr 16, Ni 35, bal Fe.
Bare wire for welding. AWS ER330.

MCKAY 347
Teledyne McKay
Stainless steel. C 0.045, Mn 1.95, Si 0.45, Cr 19.7, Ni 9.3, 0.75 Cb + Ta, bal Fe.
Welding wire. AWS ER347.

MCKAY 347
Teledyne McKay
Stainless steel. C 0.05, Mn 1.2, Si 0.5, Cr 19.6, Ni 9.8, Cb 0.65, bal Fe.
Covered welding electrode. 96,000 psi TS; 64,000 psi YS; 36 El. AWS E347. For welding Type 347 and 321 steel.

MCKAY 347, WIRE
Teledyne McKay
Stainless steel. C 0.045, Mn 1.95, Si 0.45, Cr 19.7, Ni 9.3, Cb 0.75, bal Fe.
Bare wire for welding. AWS ER347.

MCKAY 349
Teledyne McKay
C 0.09, Mn 1.5, Si 0.35, Cr 20.5, Ni 8.5, Mo 0.5, Cb 1.25, W 1.5, Ti 0.2, bal Fe.
Bare wire for welding; ASM 5782. *Obsolete*

MCKAY 349
Teledyne McKay
Stainless steel. C 0.1, Mn 1.2, Si 0.6, Cr 18.9, Ni 8.7, Cb 0.95, Mo 0.5, W 1.4, Ti 0.07, bal Fe.
Covered welding electrode. 110,000 psi TS; 84,000 psi YS; 30 El. AWS E349. For high temperature applications; aircraft industry.

MCKAY 350
Teledyne McKay
C 0.1, Mn 0.8, Si 0.35, Cr 17.25-17, Ni 4.5, Mo 2.9, N 0.08, bal Fe.
Bare wire for welding. *Obsolete*

MCKAY 360
Teledyne McKay
C 0.2, Mn 1.25, Si 0.5, Cr 15, Ni 61.5.
Typical deposit analysis. Coated stainless steel alloy weld rod. For welding Nichrome type alloys for high temperature service. MIL-E-17496. *Obsolete*

MCKAY 363
Teledyne McKay
C 0.03, Mn 0.3, Si 0.24, Cr 11.5, Ni 4.2, Ti 0.4, bal Fe.
Bare wire for welding. *Obsolete*

MCKAY 380
Teledyne McKay
C 0.12, Mn 0.4, Si 0.5, Cr 19, Ni 79.
Typical deposit analysis. Coated stainless alloy weld rod. For welding 80 Ni-20 Cr type alloys for high temperature applications. (Nichrome V). MIL-E-8844A. *Obsolete*

MCKAY 4-6 CR
Teledyne McKay
C 0-0.1, Cr 5, Mo 0.5, bal Fe.
For welding electrodes; welds type 501 and 502 steels of 5% Cr content. *Obsolete*

MCKAY 410
Teledyne McKay
Stainless steel. C 0.08, Mn 0.5, Si 0.4, Cr 12.9, bal Fe.
Welding wire. AWS ER410.

MCKAY 410
Teledyne McKay
Stainless steel. C 0.09, Mn 0.5, Si 0.4, Cr 11.8, bal Fe.
Covered welding electrode. Stress relieved 2 h at 1575°F: 80,000 psi TS; 44,000 psi YS; 28 El. AWS E410. Air hardening stainless for welding 12% chromium material; requires pre- and post-heat treatment.

MCKAY 410NIMO
Teledyne McKay
Stainless steel. C 0.05, Mn 0.75, Si 0.4, Cr 11.7, Ni 4.5, Mo 0.5, bal Fe.
Covered welding electrode. Stress relieved 1 h at 1125°F: 134,000 psi TS; 123,000 psi YS; 18 El. AWS E410NiMo. For welding ASTM CA6NM castings; 410, 410S, and 405 base metals.

MCKAY 430
Teledyne McKay
C 0.08, Mn 0.5, Si 0.6, Cr 15.5.
Typical deposit analysis. Coated stainless steel weld rod. Welded, stress relieve 1400 F: 85,000 TS; 60,000 YS; 28 El. For welding Type 430 and other 16% Cr materials. *Obsolete*

MCKAY 442
Teledyne McKay
C 0.08, Mn 0.6, Si 0.4, Cr 19.5.
Typical deposit analysis. Coated stainless steel weld rod. Welded, stress relieve 1400 F: 79,000 TS; 57,000 YS; 27 El. For welding 18% chromium materials. *Obsolete*

MCKAY 446
Teledyne McKay
C 0.1, Mn 0.75, Si 0.6, Cr 27.
Typical deposit analysis. Coated stainless steel weld rod. Welded: 90,000 TS; 65,000 YS; 8 El. For welding 28% chromium materials. *Obsolete*

MCKAY 502
Teledyne McKay
C 0.07, Mn 0.55, Si 0.38, Cr 5.25, Mo 0.5, bal Fe.
Welded, stress relieved at 1350°F: 75,000 TS; 50,000 YS; 30 El. Coated stainless steel weld rod. For welded assemblies subject to high temperature conditions; AWS E502. *Obsolete*

MCKAY 502, WIRE
Teledyne McKay
C 0.035, Mn 0.46, Si 0.42, Cr 5.6, Ni 0.18, Mo 0.55, bal Fe.
Bare wire for welding. AWS ER502. *Obsolete*

MCKAY 502-18
Teledyne McKay
C 0.06, Mn 0.8, Si 0.4, Cr 5, Mo 0.5, bal Fe.
Low alloy low hydrogen covered welding electrode. Stress relieved 2 h at 1575°F: 67,000 psi TS; 31,000 psi YS; 38 El. For welding chromium molybdenum steels for service conditions too severe for McKay 9018-B3. AWS A5.4.

MCKAY 505-18
Teledyne McKay
C 0.06, Mn 0.8, Si 0.45, Cr 9.25, Mo 1, bal Fe.
Low alloy low hydrogen covered welding electrode. Stress relieved 2 h at 1575°F: 132,000 psi TS; 112,000 psi YS; 17 El. Composite iron powder for welding chromium molybdenum steels for service conditions too severe for McKay 9018-B3 or 502.18.

MCKAY 6-C
Teledyne McKay
C 1, Cr 30, Co 61, W 4.5.
Coated electrode. For hard surfacing. 42 Rock C; corrosion and metal-to-metal wear resistant, retains hardness above 1200°F. For exhaust valves, pistons, steam valves, and hot punches.

MCKAY 6010
Teledyne McKay
C 0.1, Mn 0.3, Si 0.25, bal Fe.
Mild steel covered electrode. 70,000 psi TS; 62,000 psi YS; 30 El. AWS E6010. X-ray quality weld deposits.

MCKAY 6010 IP
Teledyne McKay
Low C, bal Fe.
Iron powder modified E6010 electrode. For DC welding. Welded: 69,000 TS; 58,000 YS; 30 El. For mild steel welding. AWS E6010. *Obsolete*

MCKAY 6011
Teledyne McKay
C 0.08, Mn 0.35, Si 0.25, bal Fe.
Mild steel covered electrode. 67,000 psi TS; 60,000 psi YS; 28 El. AWS E6011. X-ray quality weld metal.

MCKAY 6011 IP
Teledyne McKay
Low C, bal Fe.
Iron powder modified E6011 electrode. For AC welding. Welded: 69,000 TS; 61,000 YS; 30 El. For mild steel welding; AWS E6011. *Obsolete*

MCKAY 6013
Teledyne McKay
C 0.09, Mn 0.5, Si 0.25, bal Fe.
Mild steel covered electrode. 74,000 psi TS; 64,000 psi YS; 30 El. For use with low-voltage AC welders. AWS E6013.

MCKAY 6020
Teledyne McKay
Low C, bal Fe.
Weld rod. Welded: 70,000 TS; 62,000 YS; 27 El. AC or DC welding of heavy plate. AWS E6020. *Obsolete*

MCKAY 630 ELECTRODE
Teledyne McKay
Stainless steel. C 0.035, Mn 0.45, Si 0.4, Cr 16.35, Ni 4.75, Cb 0.2, Cu 3.3, bal Fe.
Covered welding electrode. AWS E630. For welding ASTM A564 Type 630 (17-4 PH) precipitation hardening steel.

MCKAY 630, WIRE
Teledyne McKay
Stainless steel. C 0.03, Mn 0.6, Si 0.45, Cr 16.5, Ni 5, Mo 0.2, Cu 3.4, 0.21 Cb + Ta, bal Fe.
Welding wire. AWS ER630.

MCKAY 7014
Teledyne McKay
C 0.08, Mn 0.7, Si 0.45, bal Fe.
Mild steel covered electrode. 81,000 psi TS; 71,000 psi YS; 26 El. Iron powder electrode. AWS E7014.

MCKAY 7016
Teledyne McKay
C 0.08, Mn 0.8, Si 0.3, bal Fe.
Mild steel low hydrogen covered welding electrode. 78,000 psi TS; 71,000 psi YS; 32 El. For mild and free-machining steels. AWS E7016.

MCKAY 7018-A1 XLM
Teledyne McKay
C 0.06, Mn 0.6, Si 0.45, Mo 0.5, bal Fe.
Low alloy low hydrogen covered welding electrode. Welded: 84,000 psi TS; 73,000 psi YS; 27 El. For welding 50% Mo steels. AWS E7018-A1.

MCKAY 7018-C2L XLM
Teledyne McKay
C 0.04, Mn 0.65, Si 0.4, Ni 3.3, bal Fe.
Low alloy low hydrogen covered welding electrode. Welded: 78,000 psi TS; 65,000 psi YS; 32 El. For welding 3% nickel steels. AWS E7018-C2L.

MCKAY 7018XLM (1)
Teledyne McKay
C 0.07, Mn 1.1, Si 0.5, bal Fe.
Mild steel low hydrogen covered welding electrode. 78,000 psi TS; 68,000 psi YS; 31 El. For mild and free-machining steels and joining mild to low alloy steels. AWS E7018.

MCKAY 7018XLM (2)
Teledyne McKay
C 0.07, Mn 1.35, Si 0.5, bal Fe.
Mild steel low hydrogen covered welding electrode. 79,000 psi TS; 69,000 psi YS; 20 El. Supplied to 20 ft·lb Charpy min. AWS E7018.

MCKAY 7024
Teledyne McKay
C 0.07, Mn 0.9, Si 0.45, bal Fe.
Mild steel covered electrode. 74,000 psi TS; 64,000 psi YS; 26 El. High speed electrode; iron powder or contact type. AWS E7024.

MCKAY 711
Teledyne McKay
C 0.06-0.1, Mn 0.3-0.5, Mo 0.4-0.6, bal Fe.
Welded: 74,000 TS; 66,000 YS; 25 El; 42 RA; 170 Brin. For welding rod; AWS-E-7011. *Obsolete*

MCKAY 714
Teledyne McKay
C 0.06-0.1, Mn 0.3-0.5, Mo 0.4-0.6, bal Fe.
Welded: 72,000 TS; 58,000 YS; 22 El; 35 RA; 160 Brin. For welding rod; AWS-E7013; shielded arc. *Obsolete*

MCKAY 715
Teledyne McKay
C 0.06-0.1, Mn 0.5-1.3, Mo 0.4-0.6, bal Fe.
Welded: 75,000 TS; 67,000 YS; 24 El; 40 RA; 172 Brin. For welding rod; AWS-E7010; shielded arc. *Obsolete*

MCKAY 720
Teledyne McKay
C 0.08-1.12, Mn 0.3-0.5, Mo 0.4-0.6, bal Fe.
Welded: 77,000 TS; 69,000 YS; 28 El; 43 RA; 170 Brin. For welding rod; AWS E-7020; shielded arc. *Obsolete*

MCKAY 720-H
Teledyne McKay
C 0.08-0.12, Mn 0.3-0.5, Mo 0.4-0.6, bal Fe.
Welded: 76,000 TS; 62,000 YS; 23 El; 45 RA; 170 Brin. For welding rod; shielded arc. *Obsolete*

MCKAY 724
Teledyne McKay
C 0.08-0.12, Mn 0.3-0.5, Mo 0.4-0.6, bal Fe.
Welded: 86,000 TS; 77,000 YS; 20 El; 35 RA; 176 Brin. For welding rod; shielded arc; AWS-E 7013. *Obsolete*

MCKAY 8016-C3
Teledyne McKay
C 0.09, Mn 0.85, Si 0.4, Ni 0.95, bal Fe.
Welded: 91,000 TS; 77,000 YS; 27 El. For welding 1% nickel alloy steels for low temperature service. Low carbon; low alloy weld rod. AWS E8 16-C3. *Obsolete*

MCKAY 8018-B2L XLM
Teledyne McKay
C 0.04, Mn 0.8, Si 0.35, Cr 1.25, Mo 0.5, bal Fe.
Low alloy low hydrogen covered welding electrode. Welded: 95,000 psi TS; 78,000 psi YS; 26 El. For welding 1-1/4 and 1/2 Mo steels. AWS E8018-B2L

MCKAY 8018-C1 XLM
Teledyne McKay
C 0.06, Mn 0.9, Si 0.45, Ni 2.3, bal Fe.
Low alloy low hydrogen covered welding electrode. Welded: 86,000 psi TS; 73,000 psi YS; 32 El. For welding 2% nickel steel. AWS E8018-C1.

MCKAY 8018-C2 XLM
Teledyne McKay
C 0.06, Mn 0.85, Si 0.4, Ni 3.3, bal Fe.
Low alloy low hydrogen covered welding electrode. Welded: 92,000 psi TS; 77,000 psi YS; 32 El. For welding 3% nickel steels. AWS E8018-C2.

MCKAY 8018-C3 XLM
Teledyne McKay
C 0.05, Mn 0.9, Si 0.45, Ni 0.95, bal Fe.
Low alloy low hydrogen covered welding electrode. Welded: 84,000 psi TS; 75,000 psi YS; 30 El. For welding low-alloy high-strength steels requiring sub zero temperatures. For MIL 8018; AWS E8018-C3.

MCKAY 8018-G XLM
Teledyne McKay
C 0.08, Mn 1.3, Si 0.45, Mo 0.2, bal Fe.
Low alloy low hydrogen covered welding electrode. Welded: 86,000 psi TS; 73,000 psi YS; 32 El. For welding low alloy steels in the 80,000-85,000 TS range. AWS E8018-G.

MCKAY 8018-W XLM
Teledyne McKay
C 0.07, Mn 0.8, Si 0.45, Cr 0.6, Ni 0.5, Cu 0.5, bal Fe.
Low alloy low hydrogen covered welding electrode. Welded: 86,000 psi TS; 73,000 psi YS; 26 El. For welding weathering steels. AWS E8018-W.

MCKAY 9018-B3 XLM
Teledyne McKay
C 0.07, Mn 0.8, Si 0.35, Cr 2.25, Mo 1.05, bal Fe.
Low alloy low hydrogen covered welding electrode. Welded: 130,000 psi TS; 110,000 psi YS; 21 El. For welding 2-1/4 Cr, 1 Mo steel when heat treating, long term stress relieving, or high elevated temperature strength is required. AWS E9018-B3.

MCKAY 9018-B3L XLM
Teledyne McKay
C 0.04, Mn 0.8, Si 0.35, Cr 2.25, Mo 1.05, bal Fe.
Low alloy low hydrogen covered welding electrode. Welded: 112,000 psi TS; 96,000 psi YS; 21 El. For 2-1/4 Cr, 1 Mo steel. AWS E9018-B3L.

MCKAY 9018-G XLM
Teledyne McKay
C 0.07, Mn 1.5, Si 0.45, Mo 0.55, bal Fe.
Low alloy low hydrogen covered welding electrode. Welded: 103,000 psi TS; 95,000 psi YS; 26 El. AWS E9018G.

MCKAY 9018-M XLM
Teledyne McKay
C 0.08, Mn 1, Si 0.4, Ni 1.6, Mo 0.2, bal Fe.
Low alloy low hydrogen covered welding electrode. Welded: 97,000 psi TS; 84,000 psi YS; 28 El. For welding low-alloy high-strength steel in 90,000 psi TS range. MIL 9018; AWS E9018-M.

MCKAY AM 363
Teledyne McKay
C 0.075, Mn 0.25, Si 0.4, Cr 11.8, Ni 4.4, Ti 0.5.
Typical deposit analysis. Coated stainless steel weld rod. Welded: 135,000 TS; 130,000 YS; 10 El. For welding AM 363 material and for repair of large Type 410 castings. *Obsolete*

MCKAY AM 363
Teledyne McKay
C 0.03, Mn 0.3, Si 0.24, Cr 11.5, Ni 4.2, Al 0.4, Co 0.05, bal Fe.
Bare wire for welding. *Obsolete*

MCKAY BARE STAINLESS STEEL WELD WIRE
Teledyne McKay
Non-AISI grades have individual listing. Twelve grades conforming to AISI: 308, 309, 310, 316, 316L, 32, 347-348, 410, 430, 442, 446 and 502. *Obsolete*

MCKAY BU-S
Teledyne McKay
C 0.13-0.15, Mn 1.5-1.8, Si 0.6-0.83, Cr 0.68-1.1, P 0.009-0.025, S 0.006-0.018, bal Fe.
Submerged arc surfacing wire. 25 Rock C.

MCKAY C
Teledyne McKay
C 0.03, Mn 0.6, Si 0.4, Cr 15.5, Mo 16, W 3.8, Fe 3.5, bal Ni.
Special maintenance covered welding electrode. 85-95 Rock B. For hot forging dies, twist guide rolls, hot shear blades, hot punches, hot extrusion dies, hot trimming dies, and mill guides.

MCKAY C-G
Teledyne McKay
Special maintenance covered welding electrode. Use with CO_2; slag coverage not as complete as C-T1.

MCKAY C-T1
Teledyne McKay
Special maintenance covered welding electrode. Use with CO_2; full slag coverage.

MCKAY CAST ALLOY
Teledyne McKay
C 1.1, Mn 0.4, Si 2.7, Cu 1.4, Fe 5.5, bal Ni.
Covered welding electrode for cast iron. For machinable welds.

MCKAY CAST ALLOY 60
Now MCKAY NICKALLOY 60.

MCKAY CAST ALLOY 60
Teledyne McKay
C 1.3, Mn 0.5, Si 0.6, Ni 49, bal Fe.
Covered welding electrode for cast iron. For joining or surfacing.

MCKAY CAST ALLOY T-60
Teledyne McKay
C 1.3, Mn 0.2, Si 0.8, Ni 46, bal Fe.
Covered welding electrode for cast iron.

MCKAY GP
Teledyne McKay
C 0.06, Mn 1, Si 0.5, Ni 9, Cr 26.5, bal Fe.
Special maintenance covered welding electrode. 120,000 psi TS; 90,000 psi YS; 27 El. For joining dissimilar metals and hard-to-weld steels; joining wedge bars to bucket teeth, welding attachments to manganese castings, welding T-1 steel lips to manganese buckets, and welding grouser bars to grousers.

MCKAY HARDALLOY 1
Teledyne McKay
Non-ferrous Co-Cr-W hard facing electrode. For deposits to resist heat and corrosion as turbine valves, steam valves. *Obsolete*

MCKAY HARDALLOY 44
Teledyne McKay
Hard surfacing electrode, coated rod. For overlaying carbon steel and austenitic manganese steel for wear resistant surfaces. Work hardens to 40-55 Rock C; for crusher screens, dredge pump impellers. *Obsolete*

MCKAY HARDALLOY 58 TIC
Teledyne McKay
C 2, Mn 1.75, Si 1.25, Cr 5, Mo 0.5, V 0.4, Ti 2.6, Cb 2.1, bal Fe.
Hard surfacing electrode. For overlaying carbon steel with hard surface, 53-61 Rock C, for resistance to severe abrasion and impact. For cement conveyor screws, augers, scraper blades.

MCKAY HARDALLOY 6
Teledyne McKay
Cobalt base hard facing electrode. For build-up; deposits have high resistance to thermal shock and impact; for gasoline exhaust valves, shear blades, needle valves. *Obsolete*

MCKAY HW-T
Teledyne McKay
C 4, Mn 1, Si 0.6, Cr 5, Mo 1.5, V 0.4, W 1.3, bal Fe.
Special maintenance covered welding electrode. Welded: 56 Rock C.

MCKAY MON-ALLOY
Teledyne McKay
C 0.12, Mn 1.8, Si 0.8, Ni 65, Fe 1.8, bal Cu.
Monel base weld rod. Welded: 96,000 TS; 49,000 YS; 37 El; 150 Brin. For welding Monel and Ni-Cu alloys to steel, or for surfacing steel. AWS ENiCu-1. *Obsolete*

MCKAY N-155
Teledyne McKay
C 0.1-0.2, Cr 20-22, Ni 19-21, Cb 0.75-1.25, Mo 2.75-3.75, Co 18.5-21, W 2-3, N 0.1-0.2.
Welded: 106,000 TS; 75,000 YS; 24 El. For welding electrodes for high temperature service; shielded arc; high heat resistant. *Obsolete*

MCKAY NICKALLOY
Teledyne McKay
Pure nickel weld rod. Welded: 45,000 TS; 2 El. For machinable welds on cast iron. AWS E Ni-C1. *Obsolete*

MCKAY NICKALLOY 60
Teledyne McKay
Ni 59, Fe 41.
Nickel iron weld rod. Welded: 77,000 TS; 61,000 YS; 8 El. For machinable welds on cast iron. AWS E NiFe-C1. *Obsolete*

MCKAY NICKEL MANGANESE
Teledyne McKay
Hard facing electrode; DC reverse. For build-up on carbon steels or manganese austenitic steels where severe impact is expected. For railroad car castings, crusher screens, railroad frogs and switches. *Obsolete*

MCKAY PLURALLOY 70 AC
Teledyne McKay
C 0.08, Mn 0.8, Si 0.3, bal Fe. Welded: 78,000 TS; 71,000 YS; 32 El. Mild steel weld rod for welding mild steel and free machining steel. AWS E7016. *Obsolete*

MCKAY S-3
Teledyne McKay
C 0.09, Mn 1.1, Si 0.6, Cu 0.3, bal Fe.
Mild steel solid welding wire. 78,300 psi TS; 65,700 psi YS; 27 El. Copper coated.

MCKAY S-6
Teledyne McKay
C 0.08, Mn 1.5, Si 0.9, Cu 0.3, bal Fe.
Mild steel solid welding wire. 76,100 psi TS; 64,600 psi YS; 25 El. Similar to S-3 with higher deoxidizers. AWS ER70S-6

MCKAY S7-T
Teledyne McKay
C 0.5, Mn 0.8, Si 0.5, Cr 3.25, Mo 1.45, bal Fe.
Welding wire for build-up of shock-resisting tools. Heat treatable.

MCKAY STAINLESS 20
Now MCKAY 320.

MCKAY TOOL FORGE 36
Teledyne McKay
C 0.095, Mn 1.1, Si 0.6, Cr 5.4, Ni 1.15, Mo 1.48, W 0.4, V 0.4, bal Fe.
Deposit: 192,000 TS; 163,000 YS; 14 El; 38-42 Rock C. Weld wire for GMA welding (CO_2 shielding). For build-up and repair gears, shafts, forging die blocks.

MCKAY TOOL-AGE 400
Teledyne McKay
C 0.045, Mn 0.5, Si 0.3, Cr 5, Mo 12.5, Co 19, bal Fe.
Coated AC-DC electrode, all position. Deposit: 44-51 Rock C. Aged: 56-64 Rock C. Age hardenable deposit; good impact, erosion, thermal shock resistant. For build-up of hot punches, forging dies, extrusion dies. *Obsolete*

MCKAY TOOL-AGE 400-S
Teledyne McKay
Similar to TOOL-AGE 400, but is 1/8 in. wire for submerged arc. *Obsolete*

MCKAY TOOL-ALLOY "822"
Teledyne McKay
Coated Ni-C alloy steel electrode, AC or DC. For build-up or repair of forming and drawing dies, large drive gears. *Obsolete*

MCKAY TOOL-ALLOY "A"
Teledyne McKay
C 0.85, Mn 1, Si 0.5, Cr 5.2, Mo 1.5, V 0.35.
Air hardening weld rod and wire (deposit from wire is slightly different). For repair and build-up of air hardening tools and dies. *Obsolete*

MCKAY TOOL-ALLOY "FH-30"
Teledyne McKay
Coated steel electrodes, AC or DC. As welded: 34 Rock C; flame hardenable to 53 Rock C. For repair of plastic molds, SAE 4130, 6145 steel casting and for build-up of soft material. *Obsolete*

MCKAY TOOL-ALLOY "FH-45"
Teledyne McKay
Coated steel electrodes, AC or DC. As welded: 44 Rock C; flame hardenable to 60 Rock C. For repair of plastic molds, build-up and surface hardener of softer material. *Obsolete*

MCKAY TOOL-ALLOY "HS"
Teledyne McKay
C 0.65, Mn 0.8, Si 0.45, Cr 3.95, Mo 8.75, V 0.9, W 1.8.
High speed steel rod and wire (deposit from wire is slightly different). For repair and build-up of parts made of AISI M1 tool steel or to deposit such material on carbon steel parts. *Obsolete*

MCKAY TOOL-ALLOY "HW"
Teledyne McKay
C 0.3, Mn 0.5, Si 0.5, Cr 4.75, Mo 1.75, V 0.25, W 1.3.
Alloy steel welding rod and wire (deposit from wire is slightly different). For repair and build-up of AISI H-12 and similar hot work tool and die materials. *Obsolete*

MCKAY TOOL-ALLOY "HW-2"
Teledyne McKay
Alloy steel coated electrode for DC reverse; bare wire for repair or build-up of various hot-work tool steel dies or tools. *Obsolete*

MCKAY TOOL-ALLOY "HW-2FC"
Teledyne McKay
Flux cored open arc wire. For repair or build-up of various hot work tool steel dies or tools. *Obsolete*

MCKAY TOOL-ALLOY "O"
Teledyne McKay
C 0.65, Mn 1.12, Si 0.4, Cr 1.35, Mo 1.12, W 0.5, V 0.35.
Oil hardening weld rod and wire (wire is slightly different composition). For repair and build-up of oil-hardening tools and dies. *Obsolete*

MCKAY TOOL-ALLOY "W"
Teledyne McKay
C 0.8, Mn 0.6, Si 0.45, Cr 0.4, Mo 0.4.
Water hardening weld rod and wire. For repair and build-up of water-hardening tools and dies. *Obsolete*

MCKAY TOOL-ALLOY C
Teledyne McKay
C 0.03, Mn 0.6, Si 0.6, Cr 15.5, Mo 16, W 3.75, Fe 6, bal Ni.
Low hydrogen, AC-DC coated electrode. As welded: 200-225 Brin; work hardens to 36-40 Rock C. For build-up and repair of forging equipment. *Obsolete*

MCKAY TOOL-ALLOY C-O
Teledyne McKay
Now MCKAY C-T1.

MCKAY TOOL-FAB
Teledyne McKay
Low alloy coated electrode. As welded: 96,000 TS; 85,500 YS; 30 El. For repairing broken or cracked tool steel dies and for build-up. *Obsolete*

MCKAY TOOL-FORGE
Teledyne McKay
Cr Mo V steel flux cored wire. For repairing sow blocks, rams, etc. As welded: 35-45 Rock C. *Obsolete*

MCKAY TOOL-FORGE 29
Teledyne McKay
C 0.08, Mn 1.2, Si 0.6, Cr 0.85, Ni 2, Mo 0.6, bal Fe.
Deposit: 129,000 TS; 118,000 YS; 19 El; 23-27 Rock C. Weld wire for GMA welding (CO_2 shielding). For build-up and repair forge dies, sow blocks, shafts; used also for undercoating.

MCKAY TOOL-MAR "300"
Teledyne McKay
Ni Co-Mo-Fe bare wire for TIG welding or build-up of maraging steel, or of H11, H12, H13 hot work tool steels. *Obsolete*

MCKAY TUBE-ALLOY 204-0
Teledyne McKay
Low carbon, 2.5% alloy steel wire. Deposit: 90,000 TS; 70,000 YS; 23 El. Crack resistant, machinable build-up on steel shafts, gears, crane wheels, back-up collets. *Obsolete*

MCKAY TUBE-ALLOY 218-0
Teledyne McKay
Low phosphorus austenitic manganese 19.5% alloy steel wire. For work-hardenable build-up and surfacing of shovel teeth, crusher jaws, breaker bars, railroad frogs.

MCKAY TUBE-ALLOY 218-S
Teledyne McKay
Low phosphorus austenitic manganese 19.5% alloy steel wire. For work-hardenable build-up and surfacing of shovel buckets, crusher rolls, hammer mill hammers, railroad frogs and switches. *Obsolete*

MCKAY TUBE-ALLOY 230-0
Teledyne McKay
High carbon, Cr-Mn alloy (21%) weld wire. Deposit: 28-37 Rock C; work hardens to 50 Rock C. For build-up on bucket teeth, impactor bars. *Obsolete*

MCKAY TUBE-ALLOY 240 TIC-0
Teledyne McKay
C 4, Mn 1.5, Ti 7-9, Cr 2.1, Mo 10.5, TiC, bal Fe.
High chromium cast iron alloy wire for hard facing hammer mill hammers, crusher rolls.

MCKAY TUBE-ALLOY 252-0
Teledyne McKay
6.5% alloy steel hard-facing wire. For surfacing tractor track rollers and idlers, shovel rollers, brake drum ditcher rolls, mine car wheels. *Obsolete*

MCKAY TUBE-ALLOY 828-S
Teledyne McKay
Medium carbon 3.5% alloy steel wire. For medium hardness deposits on gears, sheaves, steel shafting, conveyor rolls. *Obsolete*

MCKECHNIE 115
McKechnie Metals Ltd.
Cu 56, Pb 3, bal Zn.
Extruded: 380 N/mm^2 TS; 180 N/mm^2 YS; 20 El; 90-120 Vickers. For high speed machining.

MCKECHNIE 160
McKechnie Metals Ltd.
Cu 90.5, Al 9.5.
Drawn: 540 N/mm^2 TS; 250 N/mm^2 YS; 20 El; 150 Vickers. Moderate hot working, fair machinability. Corrosion resistant aluminum bronze.

MCKECHNIE 164
McKechnie Metals Ltd.
Cu 87.7, Fe 1.5, Al 9.3, Ni 1.5.
Drawn: 590 N/mm^2 TS; 310 N/mm^2 YS; 15 El; 170 Vickers. Moderate hot working, fair machinability. Medium strength aluminum bronze.

MCKECHNIE 197

McKechnie Metals Ltd.
Cu 81, Fe 4.5, Al 10, Ni 4.5.
Drawn: 770 N/mm^2 TS; 430 N/mm^2 YS; 15 El; 220 Vickers.
Good hot working, fair machinability. High strength, corrosion resistant aluminum bronze.

MCKECHNIE 20

McKechnie Metals Ltd.
Cu 61.5, Pb 2, bal Zn.
Soft: 320 N/mm^2 TS; 140 N/mm^2 YS; 30 El; 80-110 Vickers.
Drawn: 380 N/mm^2 TS; 210 N/mm^2 YS; 25 El; 110-130 Vickers. Moderate hot working, moderate machinability. BS 2874 CZ 119.

MCKECHNIE 21

McKechnie Metals Ltd.
Cu 65, bal Zn.
Soft: 320 N/mm^2 TS; 100 N/mm^2 YS; 40 El; 80-100 Vickers.
Drawn: 350 N/mm^2 TS; 140 N/mm^2 YS; 35 El; 100-120 Vickers. Fair hot working, excellent cold working. BS 2874 CZ107.

MCKECHNIE 210

McKechnie Metals Ltd.
Cu 57, Pb 2.5, bal Zn.
Extruded: 380 N/mm^2 TS; 180 N/mm^2 YS; 25 El; 90-120 Vickers. Drawn: 410 N/mm^2 TS; 200 N/mm^2 YS; 20 El; 120-150 Vickers. Excellent hot working and good machinability. Similar to BS 2874 CZ122.

MCKECHNIE 226

McKechnie Metals Ltd.
Cu 59.5, Pb 2, bal Zn.
Extruded: 360 N/mm^2 TS; 140 N/mm^2 YS; 30 El; 90-120 Vickers. Drawn: 380 N/mm^2 TS; 200 N/mm^2 YS; 25 El; 110-140 Vickers. Good hot working, good machinability. ASTM B124 Alloy 2 (UNS C 37700).

MCKECHNIE 310

McKechnie Metals Ltd.
Cu 80, Pb 0.2, bal Zn.
Soft: 290 N/mm^2 TS; 100 N/mm^2 YS; 40 El; 80-100 Vickers.
Drawn: 320 N/mm^2 TS; 140 N/mm^2 YS; 30 El; 90-120 Vickers. Fair hot working, excellent cold working. Gilding brass. BS 2874 CZ104.

MCKECHNIE 311

McKechnie Metals Ltd.
Cu 70, bal Zn.
Soft: 290 N/mm^2 TS; 100 N/mm^2 YS; 50 El; 80-100 Vickers.
Drawn: 350 N/mm^2 TS; 140 N/mm^2 YS; 40 El; 100-120 Vickers. Fair hot working, excellent cold working. For severe bending and cold forming. BS 2874 CZ106.

MCKECHNIE 312

McKechnie Metals Ltd.
Cu 60, bal Zn.
Soft: 320 N/mm^2 TS; 140 N/mm^2 YS; 35 El; 90-110 Vickers.
Drawn: 380 N/mm^2 TS; 180 N/mm^2 YS; 30 El; 110-140 Vickers. Good hot working, moderate cold working. "Muntz" metal. BS 2874 CZ109.

MCKECHNIE 4

McKechnie Metals Ltd.
Cu 58, Pb 3, bal Zn.
Drawn: 410 N/mm^2 TS; 200 N/mm^2 YS; 20 El; 130-150 Vickers. Rod for high speed machining. BS 2874 CZ121.

MCKECHNIE 4 STAR

McKechnie Metals Ltd.
Cu 58, Pb 4.25, bal Zn.
Drawn: 430 N/mm^2 TS; 210 N/mm^2 YS; 15 El; 140-170 Vickers. Excellent machinability. BS 2874 CZ121.

MCKECHNIE 540

McKechnie Metals Ltd.
Cu 57.5, Sn 0.75, Pb 1.25, Fe 0.75, Mn 1.25, bal Zn.
Extruded: 460 N/mm^2 TS; 185 N/mm^2 YS; 20 El; 100 Vickers. Drawn: 500 N/mm^2 TS; 215 N/mm^2 YS; 15 El; 130 Vickers. Good hot working, good machinability. Manganese bronze. BS 2874 CZ115.

MCKECHNIE 65

McKechnie Metals Ltd.
Cu 66.5, Fe 0.75, Al 4.75, Mn 0.75, bal Zn.
Extruded: 610 N/mm^2 TS; 300 N/mm^2 YS; 15 El; 130 Vickers. Good hot working, fair machinability. Super high strength bronze. BS 2874 CZ114.

MCKECHNIE 651

McKechnie Metals Ltd.
Cu 89.2, Fe 2, Al 8.8.
Drawn: 540 N/mm^2 TS; 240 N/mm^2 YS; 35 El; 150 Vickers. Moderate hot working, fair machinability. Tough, corrosion resistant aluminum bronze.

MCKECHNIE 75

McKechnie Metals Ltd.
Cu 77.5, Fe 5, Al 11.5, Mn 1, Ni 5.
Extruded: 900 N/mm^2 TS; 460 N/mm^2 YS; 5 El; 250 Vickers. Good hot working, fair machinability. Super high strength aluminum bronze. Very good wear resistance.

MCKECHNIE 830

McKechnie Metals Ltd.
Cu 60, Sn 0.4, Si 0.4, bal Zn.
Melting range: 890-900°C. Low fuming, free flowing brazing rod. BS 1453/C2.

MCKECHNIE 850

McKechnie Metals Ltd.
Cu 59, Si 0.2, bal Zn.
Melting range: 890-900°C. Tin free, low fuming welding or brazing rod. Similar to BS 1845/10.

MCKECHNIE AB

McKechnie Metals Ltd.
Cu 88.7, Pb 1.5, Al 9.8.
Drawn: 560 N/mm^2 TS; 280 N/mm^2 YS; 15 El; 200 Vickers. Moderate hot working, good machinability. Free machining aluminum bronze.

MCKECHNIE CON

McKechnie Metals Ltd.
Cu 58, Sn 0.5, Pb 0.5, Fe 1, Al 1.5, Mn 2.75, Ni 1.75, bal Zn.
Extruded: 520 N/mm^2 TS; 230 N/mm^2 YS; 20 El; 120 Vickers. Drawn: 560 N/mm^2 TS; 280 N/mm^2 YS; 15 El; 150 Vickers. Corrosion resistant pump and valve stock.

MCKECHNIE CSC STAR

McKechnie Metals Ltd.
Cu 62.5, bal Zn.
Soft: 320 N/mm^2 TS; 100 N/mm^2 YS; 40 El; 80-100 Vickers.
Drawn: 350 N/mm^2 TS; 140 N/mm^2 YS; 35 El; 100-130 Vickers. Moderate hot working, excellent cold working. For rivets and pin wire. BS 2874 CZ108.

MCKECHNIE DO

McKechnie Metals Ltd.
Cu 99.95, P 0.03.
Electrical conductivity: 85.0% IACS. Deoxidized, non-arsenical copper; suitable for welding. BS 2874 C106.

MCKECHNIE E102

McKechnie Metals Ltd.
Cu 59, Pb 0.7, Al 1, Mn 2, Si 0.5, bal Zn.
Extruded: 530 N/mm^2 TS; 250 N/mm^2 YS; 15 El; 130 Vickers. Good hot working, moderate machinability. Hard, wear resistant, forgeable bearing bronze.

MCKECHNIE E78

McKechnie Metals Ltd.
Cu 90, bal Zn.
Soft: 250 N/mm^2 TS; 90 N/mm^2 YS; 40 El; 80-100 Vickers.
Drawn: 320 N/mm^2 TS; 140 N/mm^2 YS; 30 El; 90-120 Vickers. Fair hot working, good cold working. Gilding brass. BS 2874 CZ101.

MCKECHNIE E86

McKechnie Metals Ltd.
Cu 91, Al 7, Si 2.
Drawn: 590 N/mm^2 TS; 310 N/mm^2 YS; 30 El; 150 Vickers. Moderate hot working, moderate machinability. Corrosion resistant, tough, aluminum bronze.

MCKECHNIE EC

McKechnie Metals Ltd.
Cu 99.99.
Half hard: 250 N/mm^2 TS; 150 N/mm^2 YS; 30 El; 90 Vickers.
Hard: 300 N/mm^2 TS; 240 N/mm^2 YS; 15 El; 105 Vickers.
Electrolytic tough pitch copper. BS 2874 C101.

MCKECHNIE ETS

McKechnie Metals Ltd.
Cu 60, Sn 0.75, Pb 0.1, bal Zn.
Drawn: 380 N/mm^2 TS; 200 N/mm^2 YS; 20 El; 100-130 Vickers. Good hot working, fair machinability. American type naval brass. ASTM B21 Alloy A (CDA 464).

MCKECHNIE F58

McKechnie Metals Ltd.
Cu 58.5, Sn 0.9, Pb 0.4, bal Zn.
Drawn: 380 N/mm^2 TS; 210 N/mm^2 YS; 25 El; 100-130 Vickers. Good hot working, moderate machinability. Naval brass, for stamping. BS 2874 CZ113.

MCKECHNIE F7S

McKechnie Metals Ltd.
Cu 61.5, Sn 1.2, Pb 0.4, bal Zn.
Drawn: 380 N/mm^2 TS; 140 N/mm^2 YS; 25 El; 100-130 Vickers. Good hot working, moderate machinability. Standard admiralty naval brass. BS 2874 CZ112.

MCKECHNIE FX

McKechnie Metals Ltd.
Cu 60, Si 0.75, Pb 1.75, bal Zn.
Drawn: 380 N/mm^2 TS; 200 N/mm^2 YS; 20 El; 100-130 Vickers. Good hot working, good machinability. Naval brass. ASTM B21 Alloy C (CDA 485).

MCKECHNIE GB

McKechnie Metals Ltd.
Cu 58, Pb 3.75, bal Zn.
Extruded: 380 N/mm^2 TS; 180 N/mm^2 YS; 20 El; 90-120 Vickers. Excellent machinability; "Gill" brass for deep drilling operations. BS 2874 CZ121.

MCKECHNIE HC

McKechnie Metals Ltd.
Cu 99.95, O 0.04.
Electrical conductivity: 98.0% IACS. Fire refined, tough pitch, high conductivity copper. BS 2874 C102.

MCKECHNIE K

McKechnie Metals Ltd.
Cu 57.5, Sn 0.75, Pb 1.25, Fe 0.75, Al 0.25, Mn 1.25, bal Zn.
Extruded: 460 N/mm^2 TS; 215 N/mm^2 YS; 20 El; 110 Vickers. Drawn: 500 N/mm^2 TS; 250 N/mm^2 YS; 15 El; 140 Vickers. Standard free-machining manganese bronze. BS 2874 CZ114.

MCKECHNIE KP

McKechnie Metals Ltd.
Cu 59, Sn 0.75, Pb 0.75, Fe 0.25, Al 0.25, Mn 1.25, bal Zn.
Drawn: 460 N/mm^2 TS; 215 N/mm^2 YS; 20 El; 130 Vickers. Good hot working, moderate machinability. For propeller shafting.

MCKECHNIE KS

McKechnie Metals Ltd.
Cu 57.5, Pb 0.75, Fe 0.25, Mn 1.75, bal Zn.
Extruded: 430 N/mm^2 TS; 185 N/mm^2 YS; 25 El; 100
Vickers. Good hot working, moderate machinability.
Architectural (manganese) bronze.

MCKECHNIE KW

McKechnie Metals Ltd.
Cu 58, Pb 2, Mn 1, bal Zn.
Extruded: 430 N/mm^2 TS; 185 N/mm^2 YS; 20 El; 100
Vickers. Good hot working, good machinability. Free
machining manganese bronze.

MCKECHNIE KZ

McKechnie Metals Ltd.
Cu 58, Sn 0.5, Pb 0.75, Fe 0.75, Al 0.75, Mn 0.75, bal Zn.
Extruded: 500 N/mm^2 TS; 220 N/mm^2 YS; 20 El; 110
Vickers. Drawn: 530 N/mm^2 TS; 280 N/mm^2 YS; 15 El; 140
Vickers. High strength forging and machining stock. BS 2874
CZ114.

MCKECHNIE NS

McKechnie Metals Ltd.
Cu 46, Mn 0.1, Ni 10, Si 0.2, bal Zn.
Melting range: 905-915°C. Low fuming nickel bronze welding
rod. BS 1453/C5.

MCKECHNIE PW

McKechnie Metals Ltd.
Cu 58.5, Sn 0.75, Fe 0.3, Mn 0.04, Ni 0.2, Si 0.1, bal Zn.
Melting range: 885-895°C. Low fuming manganese bronze
filler rod. Similar to BS 1453/C4.

MCKECHNIE S

McKechnie Metals Ltd.
Cu 57, Pb 3, Al 0.3, bal Zn.
Extruded: 410 N/mm^2 TS; 200 N/mm^2 YS; 20 El; 100-130
Vickers. For high speed machining. Similar to BS 2874
CZ121.

MCKECHNIE SC

McKechnie Metals Ltd.
Cu 99.7, S 0.3.
Electrical conductivity: 95.0% IACS. Good machinability.

MCKECHNIE SIB

McKechnie Metals Ltd.
Cu 59, Al 1.75, Mn 3, Si 1, bal Zn.
Extruded: 580 N/mm^2 TS; 280 N/mm^2 YS; 15 El; 130
Vickers. Good hot working, fair machinability. Hard, wear
resistant, forgeable bearing bronze. CDA 674.

MCKECHNIE SS

McKechnie Metals Ltd.
Cu 59, Pb 1.2, bal Zn.
Extruded: 360 N/mm^2 TS; 140 N/mm^2 YS; 30 El; 100-120
Vickers. Drawn: 380 N/mm^2 TS; 200 N/mm^2 YS; 25 El;
120-150 Vickers. Good hot working, moderate machinability.
Similar to BS 2874 CZ122.

MCKECHNIE SS SPECIAL

McKechnie Metals Ltd.
Cu 60.5, Pb 0.5, bal Zn.
Soft: 320 N/mm^2 TS; 140 N/mm^2 YS; 35 El; 90-110 Vickers.
Drawn: 380 N/mm^2 TS; 180 N/mm^2 YS; 30 El; 110-140
Vickers. Good hot working, fair machinability. BS 2874
CZ123.

MCKECHNIE SS1

McKechnie Metals Ltd.
Cu 59, Pb 0.5, bal Zn.
Soft: 320 N/mm^2 TS; 140 N/mm^2 YS; 35 El; 90-110 Vickers.
Drawn: 380 N/mm^2 TS; 180 N/mm^2 YS; 30 El; 110-140
Vickers. Good hot working, fair machinability. BS 2874
CZ123.

MCKECHNIE SWM

McKechnie Metals Ltd.
Cu 45, Pb 1.5, Mn 1, Ni 8.5, bal Zn.
Extruded: 530 N/mm^2 TS; 250 N/mm^2 YS; 15 El; 130
Vickers. Good hot working, good machinability. For
architectural purposes.

MCKECHNIE TC

McKechnie Metals Ltd.
Cu 99.5, Te 0.5, trace P.
Drawn: 300 N/mm^2 TS; 240 N/mm^2 YS; 15 El; 100 Vickers.
Electrical conductivity: 90.0% IACS. Excellent machinability.
BS 2874 C109.

MCKECHNIE W

McKechnie Metals Ltd.
Cu 58, Pb 2, bal Zn.
Extruded: 380 N/mm^2 TS; 200 N/mm^2 YS; 25 El; 90-120
Vickers. Drawn: 410 N/mm^2 TS; 210 N/mm^2 YS; 20 El;
120-150 Vickers. Excellent hot working and good
machinability. BS 2874 CZ122.

MCKECHNIE WM

McKechnie Metals Ltd.
Cu 46, Pb 2, Mn 0.3, Ni 9.5, bal Zn.
Drawn: 530 N/mm^2 TS; 280 N/mm^2 YS; 15 El; 150 Vickers.
Good hot working, good machinability. Standard free
machining nickel brass. BS 2874 NS101.

MCKECHNIE WMW

McKechnie Metals Ltd.
Cu 47, Pb 1.5, Mn 0.5, Ni 8, bal Zn.
Drawn: 530 N/mm^2 TS; 280 N/mm^2 YS; 15 El; 150 Vickers.
Moderate hot working, good machinability. Nickel brass for
decorative hardware.

MCKECHNIE XX

McKechnie Metals Ltd.
Cu 50, bal Zn.
Melting range: 875-885°C. Hard brazing rod. BS 1845/8.

MCKECHNIE YR

McKechnie Metals Ltd.
Cu 61.5, Pb 1.5, bal Zn.
Soft: 350 N/mm^2 TS; 140 N/mm^2 YS; 30 El; 80-110 Vickers.
Drawn: 380 N/mm^2 TS; 180 N/mm^2 YS; 25 El; 110-140
Vickers. Moderate hot working, moderate machinability. BS
2874 CZ119.

MCKECHNIE YS

McKechnie Metals Ltd.
Cu 61, Pb 3, bal Zn.
Soft: 350 N/mm^2 TS; 150 N/mm^2 YS; 25 El; 80-110 Vickers.
Drawn: 400 N/mm^2 TS; 180 N/mm^2 YS; 20 El; 110-140
Vickers. Standard grade free-machining brass. BS 2874
CZ124; ASTM B16.

MCKECHNIES BRONZE

English manufacture
Cu 57, Zn 41, Sn 1, Pb 0.5, Fe 1.
For rods, nuts, bolts; high strength.

MCKINNEY ALLOYS

English manufacture
Al 95-97, Cu 2-3, Mn 1-2.
For light alloy parts exposed to sea water; resists sea water
corrosion.

MCL

Latrobe Steel Co.
C 0.3, Mo 6.2, W 1, Cr 3.7, V 0.75, bal Fe.
For forging dies, dummy blocks, extrusion dies; hot work
steel, oil hardened. *Obsolete*

MCLOUTH ML-50

McLouth Steel Corp.
C 0.13, Mn 0.48, P 0.01, S 0.015, Si 0.17, Cb 0.01, Cr 0.4, Ni
0.61, Cu 0.22, Mo 0.01, bal Fe.
Bar: 73,000 TS min; 50,000 YS min; 22 El. For chutes, crane
booms, derricks, dump bodies, marine parts, pressure tanks.
Good formability and weldability. Low alloy high-strength
steel.

MCLOUTH ML-60

McLouth Steel Corp.
C 0.13, Mn 0.9, P 0.01, S 0.016, Si 0.2, Cu 0.29, Ni 0.68, Cb
0.014, Mo 0.01, Cr 0.51, bal Fe.
Bar: 75,000 TS min; 60,000 YS min; 22 El min. For chutes,
crane booms, derricks, dump bodies, marine parts, pressure
tanks. Good formability and weldability. Low alloy high-
strength steel.

MCLOUTH ML-70

McLouth Steel Corp.
C 0.16, Mn 0.95, P 0.01, S 0.021, Si 0.17, Cu 0.3, Mo 0.01, Cb
0.029, bal Fe.
Bar: 85,000 TS min; 70,000 YS min; 20 El min. For chutes,
pressure tanks, crane booms, derricks, dump bodies, marine
parts. Good formability and weldability. Low-alloy high-
strength steel.

MCLOUTH ML-F

McLouth Steel Corp.
C 0-0.22, Mn 0-1.25, Cb 0.012, 0.2 Cu min, 0.02 V min, bal
Fe.
Plate: 70,000 TS; 50,000 YS; 22 El. For automobile parts,
buckets, chutes, dump bodies, truck frames, wheels,
transmission towers, derricks. Good formability and
weldability. Low alloy-high strength steel.

MCLOUTH-MLX

McLouth Steel Corp.
C 0-0.26, Mn 0-1.5, 0.005 Cb min, 0.02 V min, bal Fe.
Gr. 45: 60,000 TS; 45,000 YS; 22 El. Gr. 50: 65,000 TS; 50,000
YS; 22 El. Gr. 55: 70,000 TS; 55,000 YS; 22 El. Gr. 60: 75,000
TS; 60,000 YS; 20 El. For derricks, transmission towers,
chutes, crane booms, buckets, dump bodies. Good
formability and weldability. High strength-low alloy steel.

MCLURE ALLOY

English manufacture
Al 85, Cu 8.2, Sn 5-6, Fe 0.9, Si 0.3, Mn 0.2.
For light alloy parts; non-hardenable.

MD

Now FINKL MD.

MD-22

Alcan Metal Powders Division
Al 96.7, Cu 2, Mg 1, Si 0.3.
Powder to be compacted and sintered to make powder metal
parts.

MD-24

Alcan Metal Powders Division
Aluminum. Al 93.8, Cu 4.4, Mg 0.5, Si 0.9, Mn 0.4.
Powder to be compacted and sintered to make powder metal
parts. Equivalent to AA 2014.

MD-69

Alcan Metal Powders Division
Aluminum. Al 98.05, Cu 0.25, Mg 1, Si 0.6, Cr 0.1.
Powder to be compacted and sintered to make powder metal
parts. Equivalent to AA 6061.

MD-76

Alcan Metal Powders Division
Aluminum. Al 90.1, Cu 1.6, Mg 2.5, Cr 0.2, Zn 5.6.
Powder to be compacted and sintered to make powder metal
parts. Equivalent to AA 7075.

MD-9824

Alcan Metal Powders Division
Ag 45, Cu 30, Zn 25.
100 mesh powder for making special solders.

MECO
Manufacturer not listed.
Cu 50, Ni 25, Zn 20.
For chemical equipment; corrosion resistant.

MEDAL BRONZE-1
Manufacturer not listed.
Cu 92, Sn 8.
For medals, ornaments; corrosion resistant.

MEDAL BRONZE-2
Manufacturer not listed.
Cu 95, Sn 4, Zn 1.
For medals, ornaments; corrosion resistant.

MEDAL BRONZE-3
Manufacturer not listed.
Cu 97, Sn 1, Zn 2.
For medals, ornaments; corrosion resistant.

MEDAL METAL
Manufacturer not listed.
Cu 84, Zn 16.
For medals, ornaments; corrosion resistant.

MEDIUM
Engelhard Corp.
Ag 70, Cu 20, Zn 10.
Silversmithing solder for silver; 1335-1390°F MP. *Obsolete*

MEDIUM HARD
Electro-Steel Co.
C 0.9, bal Fe.
For tools, dies, fixtures; water hardened.

MEDIUM LEADED BRASS
Chase Brass & Copper Co.
Cu 65, Zn 34, Pb 1.
Annealed: 47,000 TS; 15,000 YS; 60 El. Rolled: 74,000 TS; 60,000 YS; 7 El. For screws, rivets, nuts, free cutting. *Obsolete*

MEDIUM LEADED BRASS 229
Anaconda Co.
Copper. Cu 64, Zn 35, Pb 1.
Soft: 45,000 TS; 17,000 YS; 57 El. Hard: 73,000 TS; 60,000 YS; 8 El. For hardware, bolts. Moderate cold working, free cutting.

MEDIUM LEADED BRASS-340
Anaconda Co.
Copper. Cu 64.5, Zn 34.5, Pb 1.
Hard rod: 65,000 TS; 50,000 YS; 15 El; 75 Rock B. Soft rod: 46,000 TS; 17,000 YS; 60 El; 15 Rock B. For plaques, hinges, gears, wheels, ratchets, pinions, valve stems, rivets, channel plates. Corrosion resistant. Electrical conductivity: 26.

MEDIUM LEADED NAVAL BRASS
Chase Brass & Copper Co.
Cu 60.5, Sn 0.75, Pb 0.7, bal Zn.
Hard temper: 75,000 TS; 56,000 YS; 28 El. For marine hardware, screw machine products; corrosion resistant. *Obsolete*

MEDIUM SILVER SOLDER
Manufacturer not listed.
Ag 70-75, Cu 20-23, Zn 5-7.5.
For silver solder; corrosion resistant.

MEDUSA
Gulf Steel Corp.
C 0.5, Cr 1.5, W 2.5, bal Fe.
For rivet sets, punches, pneumatic tools; Type S1; shock resistant.

MEECHITE
Meech Foundry Inc.
C 0.2, Cr 29, bal Fe.
For melting pots; nonferrous metals, corrosion resistant.

MEEHANITE A
Hamilton Allied Corp.
C 3, Si 1, bal Fe.
Cast: 50,000 TS; 17,000 YS; 196 Brin. Heat treated: 75,000 TS; 600 Brin. For castings, gears, shafts; superseded by Meehanite GA. *Obsolete*

MEEHANITE A
Hamilton Allied Corp.
C 3, Si 1, bal Fe.
Cast: 50,000 TS; 17,000 YS; 196 Brin. Heat treated: 75,000 TS; 600 Brin. For castings, gears, shafts; superseded by Meehanite GA. *Obsolete*

MEEHANITE A
Meehanite Metal Corp.
C 3, Si 1, bal Fe.
Cast: 50,000 TS; 17,000 YS; 196 Brin. Heat treated: 75,000 TS; 600 Brin. For castings, gears, shafts; superseded by Meehanite GA. *Obsolete*

MEEHANITE A
Meehanite Metal Corp.
C 3, Si 1, bal Fe.
Cast: 50,000 TS; 17,000 YS; 196 Brin. Heat treated: 75,000 TS; 600 Brin. For castings, gears, shafts; superseded by Meehanite GA. *Obsolete*

MEEHANITE ALMANITE TYPE W
Meehanite Metal Corp.
Austenitic-martensitic white cast iron. W_1: 500-600 Brin; pearlitic matrix. W_2: 500-600 Brin; martensitic matrix. W_4: 400-700 Brin; austenitic as cast, but can be converted to martensitic. For parts requiring severe abrasive wear.

MEEHANITE ALMANITE TYPE W5
Meehanite Metal Corp.
Martensitic cast iron with nodular graphite. 415-552 N/mm^2 (60,000-80,000 psi) TS; 2-4 El; good impact strength. For crusher jaws, pulverizers, hammers.

MEEHANITE ALMANITE TYPE WSH
Meehanite Metal Corp.
Austenitic cast iron with nodular graphite. 690 N/mm^2 (100,000 psi) TS; 350-500 Brin; 4-10 El. Work hardens readily; for crusher liners, dredge buckets, dipper teeth.

MEEHANITE B
Hamilton Allied Corp.
C 3, Si 1, bal Fe.
Cast: 45,000 TS; 17,000 YS; 196 Brin. Heat treated: 60,000 TS; 500 Brin. For castings, gears, shafts; cast iron superseded by Meehanite GB. *Obsolete*

MEEHANITE B
Meehanite Metal Corp.
C 3, Si 1, bal Fe.
Cast: 45,000 TS; 17,000 YS; 196 Brin. Heat treated: 60,000 TS; 500 Brin. For castings, gears, shafts; cast iron superseded by Meehanite GB. *Obsolete*

MEEHANITE C
Hamilton Allied Corp.
C 3, Si 1, bal Fe.
Cast: 40,000 TS; 32,000 YS; 196-450 Brin. For cast iron castings, gears; superseded by Meehanite GC. *Obsolete*

MEEHANITE C
Meehanite Metal Corp.
C 3, Si 1, bal Fe.
Cast: 40,000 TS; 32,000 YS; 196-450 Brin. For cast iron castings, gears; superseded by Meehanite GC. *Obsolete*

MEEHANITE CB
Hamilton Allied Corp.
C, alloy, bal Fe.
Cast: 45,000 TS; 187 Brin. For acid pans, kettles, pumps, valves, fittings; acid resistant castings. *Obsolete*

MEEHANITE CB
Meehanite Metal Corp.
C, alloy, bal Fe.
Cast: 45,000 TS; 187 Brin. For acid pans, kettles, pumps, valves, fittings; acid resistant castings. *Obsolete*

MEEHANITE CB3
Hamilton Allied Corp.
C, alloy, bal Fe.
Cast: 48,000 TS; 187 Brin. For pumps, valves, fittings; acid resistant castings. *Obsolete*

MEEHANITE CB3
Meehanite Metal Corp.
C, alloy, bal Fe.
Cast: 48,000 TS; 187 Brin. For pumps, valves, fittings; acid resistant castings. *Obsolete*

MEEHANITE D
Hamilton Allied Corp.
C, alloy, bal Fe.
Cast: 35,000 TS; 175-400 Brin. For cast iron castings, gears, housings, casings, superseded by Meehanite GD. *Obsolete*

MEEHANITE D
Meehanite Metal Corp.
C, alloy, bal Fe.
Cast: 35,000 TS; 175-400 Brin. For cast iron castings, gears, housings, casings, superseded by Meehanite GD. *Obsolete*

MEEHANITE E
Hamilton Allied Corp.
C 3, Alloy, bal Fe.
Cast: 30,000 TS; 27,000 YS; 140-450 Brin. For cast iron castings, gears, shafts, casings; superseded by Meehanite GE. *Obsolete*

MEEHANITE E
Meehanite Metal Corp.
C 3, Alloy, bal Fe.
Cast: 30,000 TS; 27,000 YS; 140-450 Brin. For cast iron castings, gears, shafts, casings; superseded by Meehanite GE. *Obsolete*

MEEHANITE GA
Hamilton Allied Corp.
Over 3.0% C, over 1.0% Si, bal Fe.
Cast: 50,000 TS; 36,000 YS; 5 El; 220 Brin. Heat treated: 77,000 TS; 380 Brin. For machine tool castings, gears, shafts, housings; close grain iron. *Obsolete*

MEEHANITE GA
Meehanite Metal Corp.
Over 3.0% C, over 1.0% Si, bal Fe.
Cast: 50,000 TS; 36,000 YS; 5 El; 220 Brin. Heat treated: 77,000 TS; 380 Brin. For machine tool castings, gears, shafts, housings; close grain iron. *Obsolete*

MEEHANITE GAH
Hamilton Allied Corp.
C 3, Alloy, bal Fe.
Cast: 70,000 TS; 250-600 Brin. For cast iron castings, gears; strong. *Obsolete*

MEEHANITE GAH
Meehanite Metal Corp.
C 3, Alloy, bal Fe.
Cast: 70,000 TS; 250-600 Brin. For cast iron castings, gears; strong. *Obsolete*

MEEHANITE GB
Hamilton Allied Corp.
Over 3.0% C, over 1.0% Si, bal Fe.
Cast: 45,000; 196 Brin. For machine tool castings, gears, shafts, housings; close grain iron. *Obsolete*

MEEHANITE GB
Meehanite Metal Corp.
Over 3.0% C, over 1.0% Si, bal Fe.
Cast: 45,000; 196 Brin. For machine tool castings, gears, shafts, housings; close grain iron. *Obsolete*

MEEHANITE GC
Hamilton Allied Corp.
Over 3.0% C, over 1.0% Si, bal Fe.
Cast: 40,000 TS; 192 Brin. For pistons, engine castings, gears, shafts, housings; close grain iron. *Obsolete*

MEEHANITE GC
Meehanite Metal Corp.
Over 3.0% C, over 1.0% Si, bal Fe.
Cast: 40,000 TS; 192 Brin. For pistons, engine castings, gears, shafts, housings; close grain iron. *Obsolete*

MEEHANITE GD
Hamilton Allied Corp.
C 3, Alloy, bal Fe.
Cast: 33,000-38,000 TS; 193 Brin. For pressure castings, gears; cast iron. *Obsolete*

MEEHANITE GD
Meehanite Metal Corp.
C 3, Alloy, bal Fe.
Cast: 33,000-38,000 TS; 193 Brin. For pressure castings, gears; cast iron. *Obsolete*

MEEHANITE GE
Hamilton Allied Corp.
Over 3.0% C, over 1.0% Si, bal Fe.
Cast: 30,000 TS; 27,000 YS; 174 Brin. For machine tool castings, gears, shafts, housings, lathe beds; close grain iron. *Obsolete*

MEEHANITE GE
Meehanite Metal Corp.
Over 3.0% C, over 1.0% Si, bal Fe.
Cast: 30,000 TS; 27,000 YS; 174 Brin. For machine tool castings, gears, shafts, housings, lathe beds; close grain iron. *Obsolete*

MEEHANITE GM
Hamilton Allied Corp.
C 3, Alloy, bal Fe.
Cast: 50,000-60,000 TS; 17,000 YS; 241 Brin. For castings, gears, shafts; high strength. *Obsolete*

MEEHANITE GM
Meehanite Metal Corp.
C 3, Alloy, bal Fe.
Cast: 50,000-60,000 TS; 17,000 YS; 241 Brin. For castings, gears, shafts; high strength. *Obsolete*

MEEHANITE HA
Hamilton Allied Corp.
Over 3.0% C, over 1.0% Si, alloy, bal Fe.
Cast: 50,000 TS; 220 Brin. For diesel engines components, valves, liners, superheaters; up to 900 F operating temperature, alloy iron. *Obsolete*

MEEHANITE HA
Meehanite Metal Corp.
Over 3.0% C, over 1.0% Si, alloy, bal Fe.
Cast: 50,000 TS; 220 Brin. For diesel engines components, valves, liners, superheaters; up to 900 F operating temperature, alloy iron. *Obsolete*

MEEHANITE HB
Hamilton Allied Corp.
Over 3.0% C, over 1.0% Si, Cu, Cr, Mo, bal Fe.
Cast: 38,000 TS; 300 Brin. For stoker plates, furnace rolls, skid castings; heat resistant to 1400 F, alloy iron. *Obsolete*

MEEHANITE HB
Meehanite Metal Corp.
Over 3.0% C, over 1.0% Si, Cu, Cr, Mo, bal Fe.
Cast: 38,000 TS; 300 Brin. For stoker plates, furnace rolls, skid castings; heat resistant to 1400 F, alloy iron. *Obsolete*

MEEHANITE HC
Hamilton Allied Corp.
C, alloy, bal Fe.
Cast: 38,000 TS; 229 Brin. For cast iron castings; heat resistant. *Obsolete*

MEEHANITE HC
Meehanite Metal Corp.
C, alloy, bal Fe.
Cast: 38,000 TS; 229 Brin. For cast iron castings; heat resistant. *Obsolete*

MEEHANITE HD
Hamilton Allied Corp.
C, alloy, bal Fe.
Cast: 33,000 TS; 202 Brin. For heat resisting cast iron castings; heat resistant. *Obsolete*

MEEHANITE HD
Meehanite Metal Corp.
C, alloy, bal Fe.
Cast: 33,000 TS; 202 Brin. For heat resisting cast iron castings; heat resistant. *Obsolete*

MEEHANITE HR
Meehanite Metal Corp.
Si 0-1, 3.15 C min, Cu, Cr, bal Fe.
Cast: 40,000 TS; 300 Brin. For glass molds, furnace and burner parts, rolls; heat resistant to 1550°F, alloy iron.

MEEHANITE HR
Hamilton Allied Corp.
Si 0-1, 3.15 C min, Cu, Cr, bal Fe.
Cast: 40,000 TS; 300 Brin. For glass molds, furnace and burner parts, rolls; heat resistant to 1550°F, alloy iron.

MEEHANITE HS
Meehanite Metal Corp.
C 0-2.8, 5.0 Si min, Mn, others, bal Fe.
Cast: 60,000 TS; 196 Brin. For oil refinery supports, blast furnace parts, trays, dampers; heat resisting cast iron up to 1700°F.

MEEHANITE HS-100
Meehanite Metal Corp.
C 2.5, Si 5, Mn 0.3, Special alloys, bal Fe.
Cast: 60,000 TS; 196 Brin. For oil refinery tube supports, hot gas valves; up to 1700 F heat resistant. *Obsolete*

MEEHANITE KC
Hamilton Allied Corp.
C, alloy, bal Fe.
Cast. For pumps, valves, fittings, evaporators; alkali resistant. *Obsolete*

MEEHANITE KC
Meehanite Metal Corp.
C, alloy, bal Fe.
Cast. For pumps, valves, fittings, evaporators; alkali resistant. *Obsolete*

MEEHANITE SC
Hamilton Allied Corp.
Si 4.7, 2.5% min C, alloy, bal Fe.
Cast: 27,000 TS; 300 Brin. For annealing and carburizing boxes, furnace parts; heat resistant to 1650 F, alloy iron. *Obsolete*

MEEHANITE SC
Meehanite Metal Corp.
Si 4.7, 2.5% min C, alloy, bal Fe.
Cast: 27,000 TS; 300 Brin. For annealing and carburizing boxes, furnace parts; heat resistant to 1650 F, alloy iron. *Obsolete*

MEEHANITE SF-60
Meehanite Metal Corp.
C 3.5, Si 2.7, Mn 0.5, Addition agents, bal Fe.
Cast: 60,000 TS; 161 Brin. For gears, frames, housings, shafts; cast iron, replacement for malleable iron. *Obsolete*

MEEHANITE SH-100
Meehanite Metal Corp.
C 3.5, Si 2.5, Mn 0.5, Addition agents, bal Fe.
Cast: 90,000 TS; 227 Brin. For heavy duty gears, cams, tires, crankshafts; cast iron. *Obsolete*

MEEHANITE SP-80
Meehanite Metal Corp.
Si 2.5, Mn 0.4, 3.5% min C, addition agents, bal Fe.
Cast: 80,000 TS; 207 Brin. For gears, housings, frames, shafts; cast iron for high pressures. *Obsolete*

MEEHANITE TYPE AQ
Meehanite Metal Corp.
Gray iron casting type; hardenable. As cast: N/mm² 345 (50,000 psi) TS; 280 Brin. Heat treated: N/mm² 448 (65,000 psi) TS; up to 500 Brin. For cams, dies, punches, rollers.

MEEHANITE TYPE AQS
Meehanite Metal Corp.
Hardenable cast iron, as cast. As cast: N/mm² 550 TS; 225 Brin. Hardened: N/mm² 480 TS; up to 500 Brin. For wear and abrasion resisting parts.

MEEHANITE TYPE CC
Meehanite Metal Corp.
Moderately corrosion resistant cast iron. 276 N/mm² (40,000 psi) TS; 200 Brin. For pumps, valves, evaporators, filter presses.

MEEHANITE TYPE CR
Meehanite Metal Corp.
Austenitic cast iron with flake graphite. Good corrosion resistance; for handling acid and alkali solutions to 700°C. 25,000 psi TS; 131-183 Brin. ASTM A436-72a.

MEEHANITE TYPE CRS
Meehanite Metal Corp.
Austenitic cast iron with nodular graphite. 58,000 psi TS min; 8.0 El min. Good corrosion resistance, for handling acids and alkali solutions to 700°C.

MEEHANITE TYPE GA-350
Meehanite Metal Corp.
Gray iron casting type. N/mm² 350 (50,000 psi) TS; Brin 220. General purpose cast iron, including diesel engine cylinders and liners. ASTM A48-74; QQ-I-652C.

MEEHANITE TYPE GC-275
Meehanite Metal Corp.
Gray iron casting type. N/mm² 275 (40,000 psi) TS; 190 Brin. For small and medium size castings. ASTM A48-74; Federal QQ-I-652C.

MEEHANITE TYPE GE-200
Meehanite Metal Corp.
Gray iron casting type. N/mm² 200 (30,000 psi) TS; 180 Brin. Replaces ordinary cast iron. ASTM A48-74; Federal QQ-I-652c.

MEEHANITE TYPE GF-150
Meehanite Metal Corp.
Gray iron casting type. N/mm² 150 (120,000 psi) TS; 160 Brin. For lightly loaded castings requiring much machining.

MEEHANITE TYPE GM-400
Meehanite Metal Corp.
Gray iron casting type. N/mm² 400 (55,000 psi) TS; Brin 230. General purpose cast iron, including pressing, blanking and header dies. ASTM A48-74; Federal QQ-I-652c.

MEEHANITE TYPE H5V
Meehanite Metal Corp.
Heat resisting type cast iron. 670/828 N/mm² (100,000-120,000 psi) TS; 2-10 El; 200 Brin. For hot forming dies, turbo and supercharger castings, furnace parts.

MEEHANITE TYPE SF-400
Meehanite Metal Corp.
Cast ductile iron, essentially ferritic. N/mm² 400 TS min (60,000 psi TS); N/mm² 310 YS min (45,000 psi YS); 15-20 El; Brin. For steel weldments; replaces malleable iron. ASTM A-395-74; AMS 5315.

MEEHANITE TYPE SH-700 (SH 100)
Meehanite Metal Corp.
Cast ductile iron: 100,000 psi TS; 65,000 YS; 1-5 El; 240 Brin. Hardenable, often surface hardened for cams, dies, brake drums. ASTM A-536-72; Mil-I-11466 B (MR).

MEEHANITE TYPE SH-800 (SH-100)
Meehanite Metal Corp.
Cast ductile iron, heat treated. 100,000-170,000 psi TS (N/mm^2 700-1190 TS); 263-600 Brin.

MEEHANITE TYPE SP-600 (SP 80)
Meehanite Metal Corp.
Cast ductile iron: 80,000-100,000 psi TS; 60,000-75,000 psi YS; 3-10 El; 200 Brin approx. Pearlitic. For automotive connecting rods, crankshafts, gears, cams, car journal boxes. ASTM A536-72; ASM 5316.

MEEHANITE WA
Hamilton Allied Corp.
C, alloy, bal Fe.
Cast: 50,000 TS; 196-321 Brin. For gears, brake drums, cams, dies, pinions; wear resistant. *Obsolete*

MEEHANITE WA
Meehanite Metal Corp.
C, alloy, bal Fe.
Cast: 50,000 TS; 196-321 Brin. For gears, brake drums, cams, dies, pinions; wear resistant. *Obsolete*

MEEHANITE WAH
Hamilton Allied Corp.
C, alloy, bal Fe.
Cast: 70,000 TS; 200-600 Brin. For cast iron castings; wear resistant. *Obsolete*

MEEHANITE WAH
Meehanite Metal Corp.
C, alloy, bal Fe.
Cast: 70,000 TS; 200-600 Brin. For cast iron castings; wear resistant. *Obsolete*

MEEHANITE WB
Hamilton Allied Corp.
C, alloy, bal Fe.
Cast: 38,000 TS; 350-550 Brin. For mill liners, crushers, pump impellers; wear resistant. *Obsolete*

MEEHANITE WB
Meehanite Metal Corp.
C, alloy, bal Fe.
Cast: 38,000 TS; 350-550 Brin. For mill liners, crushers, pump impellers; wear resistant. *Obsolete*

MEEHANITE WBC
Hamilton Allied Corp.
C, alloy, bal Fe.
Chilled: 45,000 TS; 550 Brin. For cement mill gears, rolls, rollers, cams; hard exterior. *Obsolete*

MEEHANITE WBC
Meehanite Metal Corp.
C, alloy, bal Fe.
Chilled: 45,000 TS; 550 Brin. For cement mill gears, rolls, rollers, cams; hard exterior. *Obsolete*

MEEHANITE WEC
Meehanite Metal Corp.
C 3.5, Si 1, Mn 0.6, Cr 0.75, bal Fe.
Cast: 30,000 TS; 444 Brin. For chill cast castings, gears, housing; wear resistant cast iron. *Obsolete*

MEEHANITE WH
Meehanite Metal Corp.
C 3.3, Si 0.75, Mn 0.4, Cr 0.2-0.3, bal Fe.
Cast: 30,000 TS; 575 Brin. For wear resistant castings for mining equipment; wear resistant cast iron. *Obsolete*

MEEHANITE-HE
Meehanite Metal Corp.
C 0-3.55, 1.9 Si min, Cu, Cr, Mn, bal Fe.
Cast: 30,000 TS; 220 min Brin. For slag pots, furnace castings, ingot molds, sinter grates. Heat resistant cast iron.

MEEHANITE-HE
Hamilton Allied Corp.
C 0-3.55, 1.9 Si min, Cu, Cr, Mn, bal Fe.
Cast: 30,000 TS; 220 min Brin. For slag pots, furnace castings, ingot molds, sinter grates. Heat resistant cast iron.

MEGAPERM 40 L
Vacuumschmelze GmbH
Soft magnetic 35 to 40% nickel alloy for relay parts, core material, electromechanical transducers. 75,000 microns permeability max.

MEGAPERM 4510
Vacuumschmelze GmbH
Ni 45, Mn 10, bal Fe.
For magnets; high permeability. *Obsolete*

MEGAPERM 6510
Vacuumschmelze GmbH
Ni 65, Mn 10, bal Fe.
For magnets; high permeability. *Obsolete*

MEGAPYR II
German manufacture
Cr 30, Fe 65, Al 5.
For cast magnets, electrical resistances; heat resistant to 1300 C. *Obsolete*

MEIGH METAL
Meigh Castings Co. Ltd.
Al 9, Fe 4, Ni 5, bal Cu.
Cast: 101,000 TS; 43,000 YS; 14 El. Non-magnetic, non-sparking corrosion resistant bronze. BS ABCD; DTD 412.

MEL-TROL
Now CARPENTER MEL-TROL.

MEL-TROL HAMDEN
Carpenter Technology Corp.
C 2.1, Mn 0.35, Si 0.25, Cr 12.5, Ni 0.5, bal Fe.
Hardened: 63-65 Rock C. For spindles, hubs, cold rolls, slitting cutters, master tools, blanking and forming dies, lamination dies. Type D3. Oil hardening, nondeforming. *Obsolete*

MELCHOIR WIRE-1
Manufacturer not listed.
Cu 57.3, Ni 41.6, Mn 1.1.
For white metal wire; corrosion resistant.

MELCHOIR WIRE-2
Manufacturer not listed.
Cu 62.7, Zn 25.9, Ni 10.8, Mn 0.6.
For white metal wire; corrosion resistant.

MELCHOR WIRE
Manufacturer not listed.
Cu 57.3, Ni 41.6, Mn 1.1.
For wire; corrosion resistant.

MELCLIF A5
Magnesium Elektron Ltd.
Magnesium. Al 5, Zn 0.5, Mn 0.4, bal Mg.
Cast: 29,700 TS; 12,000 YS; 11 El; 50 Brin. For light alloy parts; wrought.

MELKHIOR
Russian manufacture
Ni, bal Cu.

MELLOID A
Bull's Metal & Marine Ltd.
Sn 4.25, P 0.15, bal Cu.
For marine parts, hardware, condenser tubes; corrosion resistant; malleable bronze. *Obsolete*

MELLOID AA
Bull's Metal & Marine Ltd.
Sn, P, bal Cu.
For bearings, worm wheels; P-Bronze. *Obsolete*

MELLOID AAA
Bull's Metal & Marine Ltd.
Sn, P, bal Cu.
For side valves, bearings; P-Bronze. *Obsolete*

MELLOID B
Bull's Metal & Marine Ltd.
Sn, P, bal Cu.
For worm wheels, bushings; P-Bronze. *Obsolete*

MELLOID C
Bull's Metal & Marine Ltd.
Sn, P, bal Cu.
For pumps and valve bodies; P-Bronze. *Obsolete*

MELLOID D
Bull's Metal & Marine Ltd.
Sn, P, bal Cu.
For bearings, valve bodies; P-Bronze. *Obsolete*

MELLOID E
Bull's Metal & Marine Ltd.
Sn, P, bal Cu.
For worm wheels, bearings, bushings; P-Bronze. *Obsolete*

MELMAG 75 BATTERY PLATE
Magnesium Elektron Ltd.
Magnesium. T1 6.6-7.6, Al 4.6-5.6, Mn 0-0.25, bal Mg.
For electrochemical applications.

MELMAG AP 65 BATTERY PLATE
Magnesium Elektron Ltd.
Magnesium. Al 6-7, Pb 4.5-5, Zn 0.4-1.5, Mn 0.15-0.3, bal Mg.
High voltage alloy for electrochemical applications.

MELMAG AZ61 BATTERY PLATE
Magnesium Elektron Ltd.
Magnesium. Al 5.8-7.2, Zn 0.4-1.5, Mn 0.15-0.25, bal Mg.
General purpose alloy for electrochemical applications.

MELOTTE FUSIBLE ALLOY
English manufacture
Bi 50, Sn 31, Pb 19.
For fire extinguishers; melting point 99.5°C.

MELTRITE
Pickands Mather Sales Co., Inc.
Ferrous. C 4-4.4, bal Fe.
For making steel and cast iron. Pig iron.

MELTROL HAMDEN
Now CARPENTER HAMPDEN.

MELTRON A8
Magnesium Elektron Ltd.
Magnesium. Al 7.5-9, Zn 0.3-1, Mn 0.15-0.4, bal Mg.
Cast: 25,000 TS; 13,500 YS; 2 El; 60 Brin. Heat treated: 38,000 TS; 14,000 YS; 6 El; 70 Brin.

MELTRON AZ31
Magnesium Elektron Ltd.
Magnesium. Al 3, Zn 1, Mn 0.3, bal Mg.
O-temper: 37,000 TS; 22,000 YS; 21 El; 56 Brin. H24-temper: 42,000 TS; 32,000 YS; 16 El; 73 Brin. F-temper: 37,000 TS; 22,000 YS; 21 El. For truck bodies, aircraft cowling and frames; good formability.

MELTRON AZ855
Magnesium Elektron Ltd.
Magnesium. Al 5.5-8.5, Zn 0-1.5, Mn 0.15-0.4, bal Mg.
Forged: 44,800 TS; 29,200 YS; 10 El; 70 Brin. For bearing
housings, cylinder heads, control levers; age hardenable,
forgings and extrusions.

MELTRON ZTX
Darwins Alloy Castings
Th 2.5, Zn 1, Zr 0.6, bal Mg.
Sheet: 16,200 TS; 8000 YS; 18 El. At 300°C: 5200 TS; 4200
YS; 43 El. For aircraft and missile components; creep
resistant and good properties to 600°F. *Obsolete*

MERAL
Swiss manufacture
Cu 3.2, Mg 0.8, Mn 0.3, Ni 1, bal Al.
56,000 TS; 36,000 YS; 18 El; 110 Brin. For light alloy parts;
age-hardenable.

MERCOLOY
Merco Nordstrom Valve Co.
Cu alloy.
For valves. Corrosion resistant.

MERCURY
Atomergic Chemetals Corp.
Hg.
Purities: UHP (99.99999%), 99.9999%, 99.999%, triple
distilled (99.99%). Packed in vacuum sealed containers,
flasks, polybottles.

MERCURY
Universal Cyclops
C 0.7-1.2, bal Fe.
For tools, dies. *Obsolete*

MERICO TOOL STEEL
Meridian Steel Co. Inc.
C 0.35, Mn 0.4, Si 0.64, Cr 0.77, Ni 0.5, Mo 0.76, Cu 0.73, bal
Fe.
159,000 psi TS; 98,000 psi YS; 22 El; 34 Rock C. Hot work
tool steel.

MERICO-1
Meridian Steel Co. Inc.
C 0.35, Cr 0.77, Co 0.5, Mo 0.75, Cu 0.73, bal Fe.
For blacksmith tools, chisels, punches; water hardened, non-
tempering.

MERICO-2
Meridian Steel Co. Inc.
C 0.4, Cr 0.77, Mo 0.75, Cu 0.75, Si 0.6, Mn 0.4, bal Fe.
For shear blades, rivet sets, pneumatic tools; oil hardened,
non-tempering.

MERICROME ALLOY STEEL
Meridian Steel Co. Inc.
C 0.4, Mn 1.7, Mo 0.33, 0.25 Cr + Ni max, bal Fe.
Heat treated: 155,000 psi TS; 135,000 psi YS; 21 El; 315 Brin.

MERIDIAN ABRASION RESISTANT STEEL
Meridian Steel Co. Inc.
C 0.45-0.5, Mn 0.9-1, Cr 0.6-0.7, Mo 0.25-0.35, Si 0.3-0.4, Ni
0.7-0.8, Ti 0.05, Zr 0.1, bal Fe.
200,000-210,000 psi TS; 180,000-186,000 YS; 16 El; 45 RA;
42 Rock C. For parts to resist wear and abrasion.

MERIDIAN AR STEEL
Meridian Steel Co. Inc.
C 0.6-1.4, V 0.25, bal Fe.
For drills, reamers, lathe and planer tools, hobs; Type W2;
water hardened.

MERIDIAN CARBIDE HIGH SPEED
Meridian Steel Co. Inc.
C 0.65, Cr 4.5, W 18, V 2, Mo 0.75, Co 12, bal Fe.
For cutting tools, drills, reamers, hobs; high speed steel.

MERIDIAN DIE "O" STEEL
Meridian Steel Co. Inc.
C 0.9, Mn 1, W 0.5, bal Fe.
For cold work tools, header and blanking dies; Type O1; oil
hardened, non-deforming.

MERIDIAN DIE STEEL
Meridian Steel Co. Inc.
C 1, Cr 5, Mo 1, bal Fe.
For dies, rolls, spindles, punches; Type A2; oil hardened,
non-deforming.

MERIDIAN HS STEEL
Meridian Steel Co. Inc.
C 0.8, Cr 4, V 2, W 18, bal Fe.
For drills, taps, hobs, reamers, lathe cutters; Type T2; high
speed steel.

MERIDIAN SUPER C HIGH SPEED TOOL BITS
Meridian Steel Co. Inc.
C 1.25, Cr 4.1, Mo 3.1, V 3.1, W 9, Co 12, bal Fe.
67 Rock C. High speed steel.

MERIT METAL
American Smelting & Refining Co.
Sb, Sn, bal Pb.
For bearings; Babbitt.

MERMAID
Spear & Jackson (Industrial) Ltd.
C 0.75, Cr 4.25, W 18, V 1.1, bal Fe.
High speed steel for cutting tools. AISI T1.

MERTEN
Meridian Steel Co. Inc.
C, alloy, bal Fe.
For machine tool parts; fatigue and wear resistant.

MERTEN ALLOY STEEL
Meridian Steel Co. Inc.
C 0.4, Mn 0.85, Si 0.35, Cr 1, Ni 1, Mo 0.45, bal Fe.
165,000 psi TS; 142,000 psi YS; 18 El; 325 Brin. Hot work tool
steel.

MESMERIC
Firth Brown Ltd.
C 0.2, Si 0.3, Mn 0.4-1, bal Fe.
72,000 TS; 39,600 YS; 25 El; 50 RA. For general engineering
purposes; case-hardening steel; "Firth Brown F-100", "Atlas
C.C.H." *Obsolete*

MESOLOY
Molecu-Wire Corp.
Fe 72, Cr 23, Al 5.
Wire: 105,000-180,000 TS. For resistors; high heat resistance.
Obsolete

MESSING 63
VDM Nickel-Technologie AG
Zn 37, bal Cu.
Soft: 48,000 TS; 55 El; 70 Brin. 1/2 H-temper: 58,000 TS; 32
El; 100 Brin. Hard: 69,000 TS; 20 El; 135 Brin. For hardware;
corrosion-resistant brass. *Obsolete*

MESSING 63PB
VDM Nickel-Technologie AG
Zn 37, Pb 2, bal Cu.
Soft: 48,000 TS; 55 El; 70 Brin. 1/2 hard: 58,000 TS; 32 El;
100 Brin. Hard: 69,000 TS; 20 El; 136 Brin. For hardware,
screw machine products; free-cutting, leaded brass.
Obsolete

MESSING 72
VDM Nickel-Technologie AG
Zn 28, bal Cu.
Soft: 47,000 TS; 54 El; 70 Brin. 1/2 H-temper: 58,000 TS; 30
El; 100 Brin. Hard: 65,000 TS; 16 El; 130 Brin. For hardware;
corrosion-resistant brass. *Obsolete*

MESSING 75
VDM Nickel-Technologie AG
Zn 25, bal Cu.
Soft: 47,000 TS; 54 El; 70 Brin. 1/2 H-temper: 58,000 TS; 29
El; 100 Brin. Hard: 66,000 TS; 15 El; 125 Brin. For hardware;
corrosion-resistant brass. *Obsolete*

MESSING 80
VDM Nickel-Technologie AG
Zn 20, bal Cu.
Soft: 44,000 TS; 63 El; 65 Brin. 1/2 H-temper: 52,000 TS; 28
El; 95 Brin. Hard: 62,000 TS; 15 El; 120 Brin. For hardware,
ornamental parts; corrosion-resistant brass. *Obsolete*

MESSING 85
VDM Nickel-Technologie AG
Zn 15, bal Cu.
Soft: 42,000 TS; 50 El; 65 Brin. 1/2 H-temper: 52,000 TS; 26
El; 95 Brin. Hard: 61,000 TS; 12 El; 115 Brin. For ornamental
and decorative parts; corrosion-resistant brass. *Obsolete*

MESSING 90
VDM Nickel-Technologie AG
Zn 10, bal Cu.
Soft: 40,000 TS; 48 El; 60 Brin. 1/2 H-temper: 48,000 TS; 24
El; 90 Brin. Hard: 56,000 TS; 10 El; 110 Brin. For ornamental
and decorative parts; corrosion-resistant brass. *Obsolete*

MESTA SPECIAL
Mesta Machine Co.
C 0.4, Ni 1.5, Cr 0.8, bal Fe.
For rolls; cast steel.

METAL BAND SAW
Colt Industries
C 1.35, Cr 0.3, bal Fe.
For band saws; oil hardening. *Obsolete*

METALFLO
Johnson Matthey plc
Ag 25, Cu 33, Zn 25, Cd 16.8, Si 0.2.
Cadmium bearing silver brazing alloy; 606-720°C MP; DIN
8513.

METALINE
R.W. Rhodes Metaline Co.
Sn 10, bal Cu.
For oil-less bronze bearings; sintered.

METALJOINER
Colonial Alloys Co.
Pb-Sn-Cd.
For solder; M.P. 650°F.

METALLIC PACKING
American manufacture
Pb 82.25, Sn 4.75, Sb 13.
For metallic packing.

METALLINE
American manufacture
Cu 35, Fe 10, Cu 30, Al 25.
For tools; heat and corrosion resistant.

METALOY NO. I
Silver Creek Precision Co.
Cu 70, Sn 2, Pb 28.
Cast: 20,000 TS; 15 El; 42 Brin. For bearings, bushings;
heavy duty. *Obsolete*

METALOY NO. II
Silver Creek Precision Co.
Cu 70, Sn 4, Pb 26.
Cast: 22,000 TS; 16 El; 45 Brin. For bearings, bushings;
heavy duty. *Obsolete*

METALOY NO. III
Silver Creek Precision Co.
Cu 70, Sn 10, Pb 20.
Cast: 32,000 TS; 14 El; 72 Brin. For bearings; heavy duty.
Obsolete

METALSIL
Johnson Matthey plc
Ag 20, Cu 40, Zn 25, Cd 15.
Cadmium bearing silver brazing alloy; 605-765°C MP; DIN 8513 L-Ag20Cd.

METAMIC 247
Morgan Crucible Co. Ltd.
Ni 75, 25.0 mullite.
Sintered: 200-300 Brin. For bearings for 500-900°C operating temperature; cermet.

METAMIC LT-1
Morgan Crucible Co. Ltd.
Cr 70, 30 Al$_2$O$_3$.
Sintered: 350 Brin. For gas turbine blades; sintered, refractory.

METAMOLD
Ziv Steel & Wire Co.
C, alloy, bal Fe.
For Zn and Al die casting dies; oil hardened. *Obsolete*

METARSAL
Pechiney/Trefimetaux
Cu 76-79, Al 2, As 0.2-1, bal Zn.
Heat exchanger tubing. Cu Zn 22 Al 2 As.

METARSIC
Pechiney/Trefimetaux
Cu 7, As 0.5, bal Zn.
Tube for heat exchanger operations. Cu Zn 30 As.

METARSTAN
Pechiney/Trefimetaux
Cu 70, Sn 1, As 0.2-1, bal Zn.
Heat exchanger tubing. Cu Zn 29 Sn 1 As; ASTM B111, Alloy 443.

METCO 12C
METCO Inc.
Fe 2.5, Cr 10, C 0.15, Si 2.5, B 2.5, bal Ni.
Corrosion resistant metallic powders for spraying hard coating.

METCO 14E-14F
METCO Inc.
Fe 4, Cr 14, C 0.6, Si 3.5, B 2.75, bal Ni.
Corrosion resistant metallic powders for spraying hard coating.

METCO 15E-15F
METCO Inc.
Fe 4, Cr 17, C 1, Si 4, B 3.5, bal Ni.
Corrosion resistant metallic powder for spraying hard coating.

METCO 16C
METCO Inc.
Fe 2.5, Cr 16, C 0.5, Si 4, B 4, Cu 3, Mo 3, bal Ni.
Corrosion resistant metallic powders for spraying hard coating.

METCO 18C
METCO Inc.
Fe 2.5, Cr 18, C 0.2, Si 3.5, B 3, Mo 6, Co 40, bal Ni.
Corrosion resistant metallic powders for spraying hard coating.

METCO 19E
METCO Inc.

Self-fluxing, hard facing alloy powder for flame spraying. 55-60 Rock C. Excellent wear resistance; for plug gages, fuel rod mandrels, cam followers.

METCO 31C
METCO Inc.
Fe 2.5, Ni 46, Cr 11, C 0.5, Si 2.5, B 2.5, 35 WC-Co aggregate.
Metallic powder for spraying; gives extreme wear-resistant coating.

METCO 32C
METCO Inc.
Fe 0.8, Ni 14, Cr 3.5, C 0.1, Si 0.8, B 0.8, 80.0 WC-Co aggregate.
Metal powder for spraying; gives extreme wear resistant coating.

METCO 34F-34FP
METCO Inc.
Fe 3.5, Ni 33, Cr 9, C 0.5, Si 2, B 2, 50.0 WC-Co aggregate.
Metal powder for spraying; gives extreme wear resistant coating.

METCO 404
METCO Inc.
Ni 80, Al 20.
Metal powder for spraying; self-bonding.

METCO 41C
METCO Inc.
Ni 12, Cr 17, C 0.1, Si 1, Mo 2.5, bal Fe.
Corrosion resistant alloy for metal spraying.

METCO 42C
METCO Inc.
Ni 2, Cr 16, C 0.2, bal Fe.
Corrosion resistant alloy for metal spraying.

METCO 439
METCO Inc.
Fe 1.5, Cr 6, C 0.5, Si 1.5, B 1, Al 3, 50.0 WC-Co aggregate, bal Ni.
Metal powder for spraying; self-bonding and extreme wear resistance.

METCO 43C, 443CNS, 43F, 43F-NS
METCO Inc.
Ni 80, Cr 80.
Corrosion resistant alloy for metal spraying.

METCO 450
METCO Inc.
Al 4.5, bal Ni.
Metal powder for spraying; self-bonding.

METCO 451
METCO Inc.
Cr 9.5, Si 2.5, B 1.5, Al 0.5, bal Ni.
Metal powder for spraying; self-bonding.

METCO 54
METCO Inc.
99.0+ Al.
For metal spraying.

METCO 55
METCO Inc.
99.0+ Cu.
For metal spraying.

METCO 56F-NS
METCO Inc.
Ni 99.3.
For metal spraying.

METCO 63 NS
METCO Inc.
99.0+ Mo.
For metal spraying.

METCO 70C-NS
METCO Inc.
99.0+ chromium carbide.
For metallic spraying.

METCO 71-NS
METCO Inc.
Fe 1, C 4, Co 12, bal WC-Co aggregate.
For metal spraying.

METCO 72F-NS
METCO Inc.
Tungsten carbide/cobalt powder. For spraying guillotine knives, can-making seaming chucks and rails, jet engine parts.

METCO 80-NS
METCO Inc.
Ni 12, Cr 3, 85.0 chromium carbide.
For metal spraying.

METCO 81-NS
METCO Inc.
Ni 20, Cr 5, 75.0 chromium carbide.
For metal spraying.

METCO MONEL METAL
Monarch Alloy Co.
Ni 67, Cu 23.
Monel metal wire for metal spraying.

METCO-15F
METCO Inc.
Cr 17, Fe 4, Si 4, B 3.5, C 1, bal Ni.
Cast: 60 Rock C. For hard facing application. Self fluxing metal spray. Melting point 1875°F; wear resistant.

METCOLOY NO. 1
METCO Inc.
C, Cr, Ni, bal Fe.
For metal spraying; stainless steel wire.

METCOLOY NO. 2
METCO Inc.
C, Cr, Ni, bal Fe.
For metal spraying; stainless steel wire.

METEOR
Teledyne Firth Sterling
Steel. C 1.25, Cr 0.25, V 0.15, W 1.5, bal Fe.
For taps, punches, dental burrs; oil hardening. *Obsolete*

METILLURE
French manufacture
Si 14-15, Mn 0.7, bal Fe.
For drains, anodes, evaporators; acid resistant, brittle.

METITE
General Electric Co.
Cu 79.5, Sn 9, Pb 7.5, 4.0% graphite.
Sintered: 5000-8000 TS. For brush and current collector on electric machines; high conductivity. *Obsolete*

METONAL 10P
Pechiney/Trefimetaux
Al 9, Ni 2, Fe 1, bal Cu.
Cu-Al condenser tube plates.

METONAL 11
Pechiney/Trefimetaux
Al 6, bal Cu.
Cu-Al tubing for heat exchanger applications. Cu Al 6; ASTM B111, Alloy 608.

METONAL 15
Pechiney/Trefimetaux
Al 9, Ni 5, Fe 3, bal Cu.
Cu-Al-Ni alloy for condenser tube plates. Cu Al 9 Ni 5 Fe 3; BSS 2875, CA 105.

METONAL 20
Pechiney/Trefimetaux
Al 7, Fe 2, bal Cu.
Cu-Al tubing for heat exchanger applications. Cu Al 7 Fe 2; BSS 2871, Alloy CA 102.

METONIC 10
Pechiney/Trefimetaux
Ni 9-11, Fe 1-1.8, Mn 0-1, bal Cu.
Tubing for heat-exchanger applications. Cu Ni 10 Fe 1 Mn;
ASTM B111, Alloy 706.

METONIC 30
Pechiney/Trefimetaux
Ni 30, Mn 1, Fe 0-1, bal Cu.
Tubing for heat exchanger applications. Cu Ni 3 Mn 1 Fe;
ASTM B111, Alloy 715.

METORITE
Manufacturer not listed.
Al 94-98, P 1-4, Zn 1-2.
For light alloy parts; non-hardenable.

METROL NO. 610-FM
Now CARPENTER NO. 610-FM.

METSPEC 117
Metal Specialties Inc.
Bismuth base low melting alloy. MP 117°F (47.5°C); 5400 psi
TS; 15 Brin.

METSPEC 136
Metal Specialties Inc.
Bismuth base low melting alloy. MP 136°F (58°C); 5700 psi
TS; 15 Brin.

METSPEC 158
Metal Specialties Inc.
Bismuth base low melting alloy. MP 158°F (70°C); 3500 psi
TS; 13.5 Brin.

METSPEC 158/190
Metal Specialties Inc.
Bismuth base low melting alloy. Melting range 159-190°F
(70-88°C); 4200 psi TS; 13.5 Brin.

METSPEC 255
Metal Specialties Inc.
Bismuth base low melting alloy. MP 255°F (124°C); 6300 psi
TS; 14.5 Brin.

METSPEC 281
Metal Specialties Inc.
Bismuth base low melting alloy. MP 281°F (138.5°C); 8800
psi TS; 23 Brin.

METSPEC 281/338
Metal Specialties Inc.
Bismuth base low melting alloy. Melting range: 281-338°F,
(138.5-170°C); 8700 psi TS; 23.5 Brin.

MEZZO STEEL
Cleveland Twist Drill Co.
C 0.7, W 1.5, Mo 9, Cr 4, V 1, bal Fe.
For tools, cutters; high speed steel. *Obsolete*

MFA (LEAD)
Now SJMFA.

MFRS
Saarstahl AG
Now SAARSTAHL 1.2312.

MG FUSEKOTE 1010
MG Industries
Hot spray powder for joining and buildup of cast iron.

MG FUSEKOTE 1020
MG Industries
Hot spray powder for nickel base for stainless steels, carbon
steels, and cast iron. 18-22 Rock C.

MG FUSEKOTE 1030
MG Industries
High nickel chromium hot spray powder. 37-40 Rock C.

MG FUSEKOTE 1040
MG Industries
Self-wetting copper hot spray powder for joining and
repairing worn copper parts.

MG FUSEKOTE 1060
MG Industries
High nickel chromium hot spray powder for severe abrasion
and functional wear. 58-62 Rock C.

MG FUSEKOTE 1070
MG Industries
Tungsten carbide bearing hot spray powder for high abrasion
applications. Matrix: 63 Rock C. Tungsten carbide: 70 Rock
C.

MG FUSEKOTE 1075
MG Industries
Coarse particle tungsten carbide bearing hot spray powder.
Matrix: 64 Rock C. Tungsten carbide: 70 Rock C.

MG KOOLBOND 900
MG Industries
Bondcoat for cold spray powder process.

MG KOOLKOTE 910
MG Industries
Low alloy steel cold spray powder.

MG KOOLKOTE 920
MG Industries
Austenitic chromium nickel molybdenum cold spray powder.

MG KOOLKOTE 930
MG Industries
High nickel alloy cold spray powder.

MG KOOLKOTE 931
MG Industries
One-step high nickel alloy cold spray powder.

MG KOOLKOTE 935
MG Industries
Aluminum bronze cold spray powder.

MG KOOLKOTE 936
MG Industries
One-step aluminum bronze cold spray powder.

MG KOOLKOTE 940
MG Industries
Nickel-aluminum-chromium alloy cold spray powder. 30-40
Rock C.

MG SLIK STIK 6308
MG Industries
Stainless steel welding electrode for type 304 stainless steel.
Meets AWS A5.4 E308-16.

MG SLIK STIK 6308L
MG Industries
Stainless steel welding electrode for type 304L stainless steel.
Meets AWS A5.4 E308L-16.

MG SLIK STIK 6309L
MG Industries
Stainless steel welding electrode for type 309L stainless steel
and dissimilar metals. Meets AWS A5.4 E309L-16.

MG SLIK STIK 6316
MG Industries
Stainless steel welding electrode for type 316 stainless steel.
Meets AWS A5.4 E316-16.

MG SLIK STIK 6316L
MG Industries
Stainless steel welding electrode for type 316L stainless steel.
Meets AWS A5.4 E316L-16.

MG SLIK STIK 6347
MG Industries
Stainless steel welding electrode for types 347, 304L, 321
stainless steel. Meets AWS A5.4 E347-16.

MG SLIK-SIL 102
MG Industries
Flux coated silver braze type rod for joining ferrous and
nonferrous metals. 72,000 psi TS max.

MG SLIK-SIL 104
MG Industries
Flux coated silver braze type rod (cadmium free) for joining
ferrous, nonferrous, and dissimilar metals. 78,000 psi TS
max.

MG SLIK-SIL 106
MG Industries
Flux coated silver braze type rod (cadmium free) for food,
pharmaceutical and general industrial use. 76,000 psi TS
max.

MG-120
MG Industries
Tin silver soft solder; lead, zinc, antimony, and cadmium free.
Melts at 430°F; 15,000 psi TS.

MG-120A
MG Industries
Flux cored, tin silver solder; lead, zinc, antimony, and
cadmium free. Melts at 430°F; 15,000 psi TS.

MG-130
MG Industries
Flux coated or bare nickel silver, multi-temperature brazing
rod for general repair. At 1400°F: for building up gear teeth
and overlay. At 1750°F: thin flowing for close fitting joints and
attaching cutting tools.

MG-200
MG Industries
High nickel alloy for welding cast iron. Pulsed arc on DC,
straight polarity. Also DC reverse and AC. Good
machinability. 50,000 psi TS.

MG-210
MG Industries
Nickel-iron alloy for welding cast iron. AC or DC reverse. 350
Brin; 60,000 psi TS.

MG-240
MG Industries
Bare cast iron welding rod for oxyfuel welding of cast iron.
Machinable deposits match gray cast iron in color. 200 Brin;
40,000 psi TS.

MG-250
MG Industries
High nickel electrode for welding cast iron. AC-DC, reverse or
straight polarity. 50,000 psi TS. Meets AWS A5.13 E NiCl.

MG 260
MG Industries
Nickel iron alloy electrode for welding cast iron. High strength
with good machinability. AC-DC, reverse. 70,000 psi TS.
Meets AWS A5.13 E NiFeCl.

MG-289
MG Industries
Maximum strength nickel iron electrode for welding dirty or
greasy cast iron. AC or DC, reverse polarity. 210 Brin; 75,000
psi TS.

MG-300
MG Industries
Ni, Mn.
Aluminum bronze welding electrode for joining and
surfacing. Corrosion, cavitation, and erosion resistant. DC,
reverse polarity. 200 Brin; 87,000 psi TS.

MG-310
MG Industries
Phosphor bronze welding electrode for use with AC welding current and arc brazing. AC only. 80 Brin; 45,000 psi TS.

MG-320
MG Industries
Phosphor bronze welding electrode for surfacing and joining. DC, reverse. 120 Brin; 60,000 psi TS.

MG-340
MG Industries
Flux cored bronze brazing rod. Similar to AWS A5.8 RBCuZnA.

MG-350
MG Industries
Bare and flux coated low fuming bronze brazing rod for general fabrication and maintenance. 100 Brin.

MG-360
MG Industries
hosphor core copper brazing; self-fluxing on copper. AWS A5.

MG-370
MG Industries
Phosphor copper silver brazing alloy; self-fluxing on copper. AWS A5.8 BCuP-3.

MG-380
MG Industries
Phosphor copper silver brazing alloy. AWS A5.8 BCuP-5.

MG-382
MG Industries
Bronze brazing rod for automotive repair of unibody and sheet metal parts. 90 Brin; 45,000 psi TS.

MG-390
MG Industries
Copper welding electrode for welding electrolytic and deoxidized copper. DC, reverse polarity. 31,000 psi TS. Meets AWS A5.6 ECu.

MG-400
MG Industries
Aluminum arc or oxyfuel aluminum welding electrode for welding most common types of aluminum. DC, reverse only. 30,000 psi TS. Meets AWS A5.3 E4043.

MG-410
MG Industries
Thin flowing aluminum brazing rod for brazing most common grades of aluminum. 30,000 psi TS.

MG-420
MG Industries
Flux cored aluminum brazing rod for brazing most common grades of aluminum.

MG-460
MG Industries
Low temperature solder for aluminum and dissimilar applications. 18,000 psi TS.

MG-470
MG Industries
Self-fluxing solder for joining and build up of aluminum, zinc die castings and kirksite. 31,000 psi TS.

MG-500
MG Industries
All position mild steel welding electrode for general repair work. Non-sticking, AC-DC, either polarity. 78,000 psi TS; 24 El.

MG-505
MG Industries
General purpose all-position mild steel welding electrode for deep penetration for dirty steels. AC-DC, either polarity. 80,000 psi TS.

MG-506
MG Industries
General purpose all-position mild steel welding electrode for deep penetration for dirty, greasy, painted steels. AC-DC, either polarity. 82,000 psi TS.

MG-510
MG Industries
Mild steel welding electrode for poor fit up and in heavier weldments. AC-DC, either polarity. 72,000 psi TS.

MG-518
MG Industries
General purpose low hydrogen welding electrode for general fabrication and repair. AC-DC, reverse polarity. 75,000 psi TS. AWS A5.1 E7018.

MG-530
MG Industries
High strength low alloy welding electrode for quenched and tempered high strength low alloy steels. AC-DC, reverse polarity. 110,000 psi TS.

MG-540
MG Industries
All-position low hydrogen welding electrode for steels where underbead cracking must be avoided. AC-DC, reverse polarity. 80,000 psi TS.

MG-560
MG Industries
High speed cutting electrode to cut all steels. AC-DC, straight polarity.

MG-570
MG Industries
High speed chamfering electrode to gouge and chamfer all metals. AC-DC, straight polarity.

MG-600
MG Industries
High strength welding electrode for joining most steels. AC-DC, reverse polarity. As welded: 120,000 psi TS. Work hardened: 180,000 psi TS; 28 El.

MG-610
MG Industries
Heat and corrosion resistant welding electrode for most grades of stainless steels. AC-DC, reverse polarity. 89,000 psi TS.

MG-650
MG Industries
General purpose low carbon welding electrode for welding types 302, 304, 347, and other nonmolybdenum bearing grades stainless steels. AC-DC, reverse polarity. 74,000 psi TS.

MG-660
MG Industries
Low carbon welding electrode for welding molybdenum bearing stainless steels. AC-DC, reverse polarity. 91,000 psi TS.

MG-670
MG Industries
High deposition stainless steel welding electrode for overlay and joining of types 302, 304, and 316L stainless steels. AC-DC, reverse polarity. 92,000 psi TS.

MG-690
MG Industries
Universal welding electrode for high heat or cryogenic applications. DC, reverse polarity. 85,000 psi TS.

MG-700
MG Industries
High speed tool welding electrode. AC-DC, either polarity. As welded: 58-62 Rock C. Heat treated: 63-65 Rock C. Hot hardness: 56 Rock C at 1100°F.

MG-710
MG Industries
Air hardening tool steel welding electrode. AC-DC, either polarity. 55-60 Rock C.

MG-740
MG Industries
Surfacing welding electrode for build up of parts on construction equipment, steel mill equipment, railroad rails. AC-DC, either polarity. As welded: 33-41 Rock C.

MG-745
MG Industries
Chromium, manganese, nickel welding electrode for joining and rebuilding manganese steel parts. AC-DC, reverse polarity. As welded: 19 Rock C. Work hardened: 45 Rock C max.

MG-750
MG Industries
Chromium, manganese, nickel welding electrode for joining and hardsurfacing steels to withstand high impact. AC-DC, reverse polarity. 90,000 psi TS. As welded: 15 Rock C. Work hardened: 45 Rock C max.

MG-755
MG Industries
Hardsurfacing welding electrode to impact medium hardness and impact resistance. AC-DC, reverse polarity. As welded: 48-53 Rock C.

MG-760
MG Industries
Hardsurfacing welding electrode for parts subjected to abrasion wear and impact. AC-DC, either polarity. As welded: 55-60 Rock C.

MG-765
MG Industries
Hardsurfacing welding electrode for severe abrasion and mild impact. For agricultural tillage tools; deposits smooth bead. AC-DC, reverse polarity. As welded: 57-61 Rock C.

MG-770
MG Industries
Hardsurfacing welding electrode for severe abrasion and high impact. AC-DC, reverse polarity. 56-60 Rock C.

MG-788
MG Industries
Composite rod made of nickel silver brazing alloy with sharp tungsten carbide chips. Applications include drill bits, horseshoes, or any part requiring a rough, hard surface.

MG-790
MG Industries
Cr, W, Mo, Co.
High deposition, high hardness hardsurfacing electrode. Retains hardness and impact strength up to 1600°F. AC-DC, reverse polarity. 63-65 Rock C.

MG-799
MG Industries
Cr, Ni, B.
Self-fluxing oxyfuel hardsurfacing rod for hardness and metal to metal wear. 56-62 Rock C. Bonding temperature: 1800°F.

MG-9Y-1 ZN-0.5 ZR
Dow Chemical Co.
Y 9, Zn 1, Zr 0.5-50000, bal Mg.
Heat treated: 59,000-61,000 TS; 46,000-52,000 YS; 6-8 El; 48,000-52,000 CYS; 21,000 Fatigue Strength. For aircraft and missile components. Heat treatable. Good elevated temperature properties. *Obsolete*

MGN

Creusot-Loire
W 5, C, Si, Mn, bal Fe.
For permanent magnets, electrical equipment; retains magnetization. *Obsolete*

MGR

Latrobe Steel Co.
C 0.55, Si 0.95, Mn 0.3, W 1.25, Cr 5, Mo 1.25, bal Fe.
Air hardening tool steel, tough and wear resistant. For punches, rivet sets, tools and dies. AISI A8.

MIAMI

AL Tech Specialty Steel Corp.
C 0.38, Mn 0.8, P 0-0.03, S 0-0.008, Si 0.7, Cr 13, bal Fe. Annealed: 95 ksi TS; 50 psi YS; 25 El; 55 RA; 197 Brin. For injection molds, impression molds, glass molds, transfer molds, lens quality molds.

MIAMI FAST CUT

American manufacture
C 1.2-1.3, bal Fe.
For cutters, broaches; fast finishing.

MIAMI NO CHARGE

American manufacture
C 0.9, Mn 1.2, bal Fe.
For tools, dies, broaches, punches, shears; non-deforming.

MIARMI IRON

Dayton Malleable Iron Co.
C 3, Si 2, Mn 0.7, bal Fe.
For high strength castings, gears, shafts; cast iron. *Obsolete*

MICHALLOY

American manufacture
C, Ni, Cr, bal Fe.
For grinding balls, mill liners, pumps; high abrasion resistance; Ni-Hard.

MICHIANA 111

Michiana Products Corp.
Stainless steel. Cr 25, Ni 12, Mo 3.5, C 0.2, bal Fe.
For chemical engineering equipment; stainless, heat resistant.

MICHIANA 49

Michiana Products Corp.
Stainless steel. Cr 18, Ni 8, C 0.2, bal Fe.
For chemical engineering equipment; stainless.

MICHIANA 55

Michiana Products Corp.
C 0.5, Cr 30, Ni 2, bal Fe.
Cast: 45,000 TS; 30,000 YS; 1 El; 2 RA; 250 Brin. For parts in contact with brass and copper at rolling temperature; heat and corrosion resistant.

MICHIANA NO. 100

Michiana Products Corp.
Cr 25, Ni 12, C 0-0.5, bal Fe.
Cast: 63,000-65,000 TS; 40,000-45,000 YS; 2.5-3.5 El; 3-5 RA; 210 Brin. For grids, furnace parts; corrosion and heat resisting.

MICHIANA NO. 100S

Michiana Products Corp.
C 0.21-0.35, Cr 24-30, S 0.25, Ni 12-15, bal Fe.
For heat and corrosion resistant parts; heat and corrosion resistant.

MICHIANA NO. 119

Michiana Products Corp.
Cr 25, Ni 20, Si 1.25, Mn 1.5, C 0.2, bal Fe.
Cast. For heat and corrosion resisting parts; heat and corrosion resisting.

MICHIANA NO. 122

Michiana Products Corp.
C 0.2, Cr 18, Ni 8, Mo 3.5, bal Fe.
For use in sulfite paper industry; corrosion and heat resistant.

MICHIANA NO. 147

Michiana Products Corp.
C 0-0.15, Mn 0-1, Si 0-1.5, Cr 11-14, Ni 0-1, bal Fe.
Cast: 90,000 TS; 65,000 YS; 18 El. For castings; corrosion resistant.

MICHIANA NO. 233

Michiana Products Corp.
Ni 20, Cr 3, Si 1.5, Mn 1.5, C 2.2-2.4, bal Fe.
For chemical handling equipment; corrosion resisting.

MICHIANA NO. 241

Michiana Products Corp.
C 2.6-2.8, Cr 28, Si 1.25, Mn 1, bal Fe.
For coal-coke handling equipment; abrasion resisting.

MICHIANA NO. 48

Michiana Products Corp.
Cr 28, Ni 8, C 0-0.5, bal Fe.
Cast: 75,000 TS; 55,000 YS; 1 El; 1 RA; 220 Brin. For resistance to sulfurous gases at elevated temperatures; corrosion and heat resisting.

MICHIANA NO. 49 MO

Michiana Products Corp.
Stainless steel. C 0.08, Mn 0-1.5, Si 0-1.5, Cr 18-21, Ni 9-12, Mo 2-3, bal Fe.
Cast: 70,000 TS; 30,000 YS; 30 El. For castings; stainless.

MICHIANA NO. 49 A

Michiana Products Corp.
Stainless steel. C 0.2, Mn 0-1.5, Si 0-2, Cr 18-21, Ni 8-11, bal Fe.
Cast: 70,000 TS; 30,000 YS; 30 El. For castings; stainless.

MICHIANA NO. 49 CB

Michiana Products Corp.
Stainless steel. C 0.08, Mn 0-1.5, Si 0-2, Cr 18-21, Ni 9-12, C = 8 x Cb, bal Fe.
Cast: 70,000 TS; 30,000 YS; 35 El. For welded castings; stainless, stabilized.

MICHIANA NO. 63

Michiana Products Corp.
Cr 28, Ni 15, C, bal Fe.
Cast. For resistance to sulfurous gases at high temperatures and also molten salts; heat and corrosion resistant.

MICHIGAN APEX

Michigan Smelting & Refining Co.
Cu 86.5, Fe 3.5, Al 10.
Cast: 85,000 TS; 35,000 YS; 20 El; 160 Brin. For castings, gears; Al bronze.

MICHIGAN GRADE A

Michigan Smelting & Refining Co.
Cu 56, 3.0 hardener, bal Zn.
Cast: 70,000-80,000 TS; 30,000-35,000 YS; 25-40 El; 25-34 RA; 114-140 Brin. For castings; Mn bronze.

MICHIGAN GRADE AX

Michigan Smelting & Refining Co.
Cu 57, 4.0 hardener, bal Zn.
Cast: 80,000-90,000 TS; 40,000-46,000 YS; 25-40 El; 20-30 RA; 120-130 Brin. For castings; Mn bronze.

MICHIGAN GRADE C

Michigan Smelting & Refining Co.
Cu 56, 2.5 hardener, bal Zn.
Cast: 60,000-70,000 TS; 25,000-40,000 YS; 15-25 El; 30-40 RA; 80-95 Brin. For castings; Mn bronze.

MICHIGAN GRADE X

Michigan Smelting & Refining Co.
Cu 56, 6.0 hardener, bal Zn.
Cast: 85,000-95,000 TS; 40,000-45,000 YS; 20-30 El; 20-25 RA; 105-155 Brin. For castings; Mn bronze.

MICHIGAN GRADE XX

Michigan Smelting & Refining Co.
Cu 57, 9.0 hardener, bal Zn.
Cast: 95,000-110,000 TS; 45,000-55,000 YS; 20-30 El; 18-25 RA; 150-175 Brin. For castings; Mn bronze.

MICHIGAN GRADE XXX

Michigan Smelting & Refining Co.
Cu 62, 13.0 hardener, bal Zn.
Cast: 110,000-125,000 TS; 75,000-90,000 YS; 12-18 El; 12-17 RA; 200-240 Brin. For castings; Mn bronze.

MICHIGAN NO. 90

Michigan Smelting & Refining Co.
Cu 90, Al 10.
Cast: 77,000 TS; 30,000 YS; 30 El; 120 Brin. Heat treated: 90,000 TS; 55,000 YS; 5 El; 160 Brin. For castings, gears; Al bronze.

MICHIGAN NO. 90H

Michigan Smelting & Refining Co.
Cu 84.5, Fe 3.5, Al 12.
Cast: 85,000 TS; 40,000 YS; 10 El; 160 Brin. Heat treated: 95,000 TS; 50,000 YS; 5 El; 200 Brin. For castings; Al bronze.

MICHIGAN NO. 90M

Michigan Smelting & Refining Co.
Cu 85.5, Fe 3.5, Al 11.
Cast: 80,000 TS; 40,000 YS; 10 El; 166 Brin. Heat treated: 90,000 TS; 50,000 YS; 7 El; 200 Brin. For castings; Al bronze.

MICHIGAN NO. 90S

Michigan Smelting & Refining Co.
Cu 88, Fe 3.5, Al 8.5.
Cast: 65,000-85,000 TS; 22,000-30,000 YS; 20-30 El; 115-135 Brin. For castings, gears; Al bronze.

MICHIGAN NO. 90V

Michigan Smelting & Refining Co.
Cu 79, Fe 5, Ni 5, Al 11.
Cast: 90,000-105,000 TS; 45,000-60,000 YS; 3-7 El; 180-215 Brin. For castings; Al bronze.

MICRO

IMI Knoch Ltd.
Cu 98, Ni 2.
Cold worked: 37,500 TS; 23,500 YS; 39 El; 80 Brin. Extruded: 33,600 TS; 6000 YS; 55 El; 59 Brin. For locomotive boiler tubes and plates; corrosion resistant.

MICRO MACH

Manufacturer not listed.
C 0.08-0.12, Cr 17.3, Ni 6.2, bal Fe.
Longitudinal: 200,000 TS; 180,000 YS; 12 El. Transverse: 209,000 TS; 175,000 YS; 13 El. For aircraft and missile wing and skin surfaces; austenitic, stainless.

MICRO-STAR

J.F. Jelenko & Co.
Precious metal. Au 2, Pd 79.
100,000 psi TS; 80,000 psi YS; 20 El; 240 Brin. Dental alloy for fusing porcelain to metal.

MICROFLAT

Washington Steel Corp.
Cold rolled, stretcher leveled sheet and coils.

MICROFLEX

Washington Steel Corp.
Soft tempered austenitic stainless sheet for roofing, flashing and architectural work.

MICROROLD

Washington Steel Corp.
Sendzimir mill precision rolled sheet and strip steel products; all sheet and strip products.

MICROSIL
Magnetic Metals Co.
Fe 97, Si 2.9-3.3.
Grain oriented. Generally used for high power, relatively low frequency applications in high performance power transformers, saturable reactors, inverter transformers, magnetic amplifiers.

MID-MAX
Midvale-Heppenstall Co.
C 0.74, Mn 0.2, Cr 3.75, W 1.65, Mo 8.78, V 1.15, Si 0.3, bal Fe.
For tools, cutters; high speed steel. *Obsolete*

MIDAS
J.F. Jelenko & Co.
Precious metal. Au 46, Pd 6, Ag 39.5.
Quenched: 65,000 psi TS; 35,000 psi YS; 30 El; 125 Brin.
Hardened: 100,000 psi TS; 85,000 psi YS; 13 El; 210 Brin.
Type III hard dental alloy. For inlays, crowns, and fixed bridgework.

MIDCYL
Midland Motor Cylinder Co.
C 3.2, Si 2.5, bal Fe.
Cast: 30,000 TS. For cylinder liners; cast iron. *Obsolete*

MIDFLEX
Engelhard Corp.
Bimetal. For thermo-metal; bimetal element. *Obsolete*

MIDLING HARD 115
Raven Steel & Tool Co.
C 1.1, Si 0.25, bal Fe.
For tools; drill rod.

MIDOHM
Driver Harris Co.
Copper. Ni 22-23, bal Cu.
Wrought: 100,000-50,000 TS. For resistances, rheostats; load banks; heat resistant; maximum operating temperature 200°C. *Obsolete*

MIDVAC-422
Midvale-Heppenstall Co.
C 0.25, Cr 12, Mo 0.9, Ni 0.9, V 0.2, W 0.9, bal Fe.
Annealed: 85,000 TS; 42,000 YS; 22 El; 95 Rock B.
Hardened: 240,000 TS; 205,000 YS; 9 El; 50 Rock C. For cutlery, surgical instruments, hardware, bearings, valves. Corrosion resistant, hardenable.

MIDVALE EXTRA GRADE
Midvale-Heppenstall Co.
C 0.6-1.2, bal Fe.
For tools, punches, drills; water hardened. *Obsolete*

MIDVALE EXTRA TOOL
Midvale-Heppenstall Co.
C 0.6-1.4, bal Fe.
For milling cutters, reamers, dies, taps, mill picks, cold heading dies. *Obsolete*

MIDVALE N.D.
Midvale-Heppenstall Co.
C 0.9, Mn 1.5, bal Fe.
For tools, dies; non-deforming steel. *Obsolete*

MIDVALE NO. 44
Midvale-Heppenstall Co.
C 0.85, Mn 0.9, Cr 11, W 0.85, V 0.2, Si 1.15, bal Fe.
For knife and shear blades; air hardening. *Obsolete*

MIDVALE NON-TEMPERING
Midvale-Heppenstall Co.
C, alloy, bal Fe.
For tools, dies; shock resistant. *Obsolete*

MIDVALE SPECIAL
Midvale-Heppenstall Co.
high C, bal Fe.
For tools, dies, machine parts, cutting tools, taps, mill picks, reamers; made in all tempers. *Obsolete*

MIDVALE STAINLESS IRON
Midvale-Heppenstall Co.
Cr 10-14, Mn 0-0.5, C 0-0.13, bal Fe.
For stainless articles; corrosion resistant. *Obsolete*

MIDVALOY "A"
Midvale-Heppenstall Co.
Cr 19, Ni 35, Low C, bal Fe.
For corrosion and heat resistant parts, furnaces. *Obsolete*

MIDVALOY "A.M.F." ALLOY
Midvale-Heppenstall Co.
C 0.15, Ni 48, Mn 1.5, bal Fe.
Wrought: 87,000-149,000 TS; 42,000-152,000 YS; 10-40 El; 50-70 RA; 175-310 Brin. Untreated: 87,500 TS; 52,000 YS; 20 El; 25 RA; 175 Brin. For parts of refrigeration machines operating at low temperatures; corrosion resisting. *Obsolete*

MIDVALOY "B"
Midvale-Heppenstall Co.
C 1, Cr 23, Ni 11, bal Fe.
For corrosion and heat resistant parts; grids. *Obsolete*

MIDVALOY "HR-1"
Midvale-Heppenstall Co.
Cr 20, Ni 7, W 4, C 0.35, bal Fe.
For recuperators, chain grate stockers, gas turbine rotors; high temperature work. *Obsolete*

MIDVALOY "HY-X"
Midvale-Heppenstall Co.
C 0.5, Cr 8, Ni 22, Cu 1, bal Fe.
Wrought: 100,000-120,000 TS; 60,000-85,000 YS; 18-34 El; 30-58 RA; 200-215 Brin. For plugs in oil still heater tubes; resists corrosion by sulfuric acid. *Obsolete*

MIDVALOY "K.A. 2 MO"
Midvale-Heppenstall Co.
N 0-0.16, Cr 18-20, Ni 8-9, Mo 3, bal Fe.
Cast: 90,000 TS; 50,000 YS; 25 El; 25 RA. Air cooled: 92,500 TS; 50,000 YS; 25 El; 25 RA. For stainless parts, paper industry; resists sulfurous acid. *Obsolete*

MIDVALOY "K.A. 2 S"
Midvale-Heppenstall Co.
C 0-0.07, Cr 18-20, Ni 8-9, bal Fe.
Annealed: 75,000 TS; 35,000 YS; 40 El; 50 RA; 160 Brin. For stainless parts, chemical plant equipment; corrosion resistant. *Obsolete*

MIDVALOY "K.A. 2"
Midvale-Heppenstall Co.
C 0-0.16, Cr 18-19, Ni 8-9, bal Fe.
Cast: 70,000-80,000 TS; 35,000-40,000 YS; 35-60 El; 30-60 RA; 130-150 Brin. Air cooled: 75,500 TS; 37,000 YS; 47 El; 41 RA; 130 Brin. For stainless parts, chemical plant equipment; corrosion resistant. *Obsolete*

MIDVALOY "N" METAL
Midvale-Heppenstall Co.
C 0.45, Cr 19, Ni 60, bal Fe.
Cast: 60,000-75,000 TS; 34,000-40,000 YS; 5-11 El; 5-15 RA; 156 Brin. Air cooled: 73,000 TS; 37,000 YS; 11 El; 14 RA. For castings for chemical industries and parts subjected to rapid changes of temperature; corrosion resistant. *Obsolete*

MIDVALOY 0500
Midvale-Heppenstall Co.
C 0.3, Cr 5, Mo 0.3-0.5, bal Fe.
For oil refinery equipment; heat resistant. *Obsolete*

MIDVALOY 1225
Midvale-Heppenstall Co.
Cr 12, Ni 25, C, bal Fe.
Cast: 55,000-65,000 TS; 40,000-45,000 YS; 5-12 El; 10-20 RA; 150 Brin. For furnace and heat resisting parts, carburizing and annealing boxes; corrosion and abrasion resisting. *Obsolete*

MIDVALOY 13
Midvale-Heppenstall Co.
C 0.12, Cr 12-14, Ni 0-0.5, bal Fe.
Rolled: 80,000-200,000 TS; 35,000-170,000 YS; 7-30 El; 140-400 Brin. For corrosion resistant parts, oil refinery equipment; corrosion resistant. *Obsolete*

MIDVALOY 1300
Midvale-Heppenstall Co.
C 0.07-0.4, Cr 13, bal Fe.
Wrought: 70,000-200,000 TS; 40,000-165,000 YS; 13-28 El; 53-70 RA; 130-400 Brin. For corrosion resisting parts; engineering construction; corrosion and abrasion resisting.

MIDVALOY 1300-15
Midvale-Heppenstall Co.
C 0.14, Cr 13, Ni 0.4, bal Fe.
Wrought: 90,000 TS; 55,000 YS; 25 El; 64 RA; 230 Brin. For cutlery, machine parts; corrosion resistant. *Obsolete*

MIDVALOY 1300-40
Midvale-Heppenstall Co.
C 0.4, Cr 12.5, bal Fe.
Heat treated: 400-500 Brin. For valve trim, valve parts; corrosion resistant. *Obsolete*

MIDVALOY 1700
Midvale-Heppenstall Co.
C 0.07-0.4, Cr 17, bal Fe.
Wrought: 80,000 TS; 50,000 YS; 28 El; 55 RA; 160 Brin. For corrosion resistant parts; corrosion resistant. *Obsolete*

MIDVALOY 1700-30
Midvale-Heppenstall Co.
C 0.27, Cr 18, Ni 0.5, bal Fe.
Annealed: 90,500 TS; 59,000 YS; 21 El; 28.5 RA; 197 Brin. For chemical equipment; resists nitric acid; corrosion and heat resistant. *Obsolete*

MIDVALOY 1735
Midvale-Heppenstall Co.
C 0.21-0.5, Cr 17, Ni 35, bal Fe.
For heat and corrosion resistant parts, furnace parts; heat and corrosion resistant. *Obsolete*

MIDVALOY 1760
Midvale-Heppenstall Co.
Cr 17, Ni 60, C, bal Fe.
For high temperature applications; heat and corrosion resistant. *Obsolete*

MIDVALOY 1808
Midvale-Heppenstall Co.
C 0.07-0.2, Cr 18, Ni 8, bal Fe.
For stainless parts, chemical plant equipment; stainless, austenitic.

MIDVALOY 1808 C
Midvale-Heppenstall Co.
C 0.07, Cr 18, Ni 8, Cb = 10 x C, bal Fe.
For stainless parts; stabilized. *Obsolete*

MIDVALOY 1808 M
Midvale-Heppenstall Co.
C 0.07, Cr 18, Ni 8, Mo 2.1-3.5, bal Fe.
Annealed: 70,000 TS; 35,000 YS; 40 El; 60 RA; 160 Brin. For corrosion resistant parts, chemical plant equipment; stainless, austenitic.

MIDVALOY 1808 SE
Midvale-Heppenstall Co.
C 0.07, Se 0.2-0.3, Cr 18, Ni 8, bal Fe.
Wrought: 95,000 TS; 40,000 YS; 48 El; 60 RA; 160 Brin. Cast: 77,000 TS; 40,000 YS; 45 El; 50 RA. For stainless parts, machine parts; stainless, free cutting. *Obsolete*

MIDVALOY 1808-20
Midvale-Heppenstall Co.
C 0.17, Cr 18.5, Ni 9.8, bal Fe.
Annealed: 75,000 TS; 41,000 YS; 49 El; 68 RA; 171 Brin. For castings; stainless. *Obsolete*

MIDVALOY 1808-7
Midvale-Heppenstall Co.
C 0.07, Cr 18.5, Ni 9, bal Fe.
Annealed: 75,500 TS; 41,000 YS; 46 El; 59 RA; 150 Brin. For castings; corrosion resistant, stainless. *Obsolete*

MIDVALOY 1808-7 CB
Midvale-Heppenstall Co.
C 0.07, Cr 19, Ni 9, Cb, bal Fe.
Annealed: 78,000 TS; 42,000 YS; 43 El; 58 RA; 155 Brin. For castings; stainless. *Obsolete*

MIDVALOY 1808-7 MO
Midvale-Heppenstall Co.
C 0.07, Cr 19, Ni 9.6, Mo, bal Fe.
Annealed: 89,500 TS; 55,000 YS; 37 El; 65 RA; 175 Brin. For castings; stainless. *Obsolete*

MIDVALOY 1808-7S
Midvale-Heppenstall Co.
C 0.06, Cr 19.9, Ni 9, Se, bal Fe.
Annealed: 76,000 TS; 39,000 YS; 47 El; 62 RA; 150 Brin. For castings; stainless, free machining. *Obsolete*

MIDVALOY 20
Midvale-Heppenstall Co.
C 0-0.07, Ni 29, Cr 20, Mo 2-3, Cu 4, bal Fe.
Cast: 65,000-75,000 TS; 28,000-38,000 YS; 35-50 El; 40-50 RA; 120-150 Brin. For chemical plant equipment, tanks; resists mixed acids, austenitic.

MIDVALOY 2025
Midvale-Heppenstall Co.
Cr 20, Ni 25, C, bal Fe.
For high temperature applications; heat and corrosion resistant. *Obsolete*

MIDVALOY 2060
Midvale-Heppenstall Co.
C 0.36-0.5, Cr 20, Ni 60, bal Fe.
For heat and corrosion resistant parts, chemical industries; heat and corrosion resistant. *Obsolete*

MIDVALOY 2100
Midvale-Heppenstall Co.
C 0-0.12, Cr 21, bal Fe.
Wrought: 63,000 TS; 45,000 YS; 30 El; 40 RA. For corrosion resistant parts, furnace parts; heat and corrosion resistant. *Obsolete*

MIDVALOY 2512 B
Midvale-Heppenstall Co.
C 1, Cr 25, Ni 12, bal Fe.
Cast: 70,000 TS; 50,000 YS; 8 El; 10 RA. For furnace parts, heat treating boxes, retorts; corrosion and abrasion resisting. *Obsolete*

MIDVALOY 2512-10
Midvale-Heppenstall Co.
C 0.1, Ni 12, Cr 25, bal Fe.
Forged: 75,000-115,000 TS; 45,000-65,000 YS; 25-45 El; 30-50 RA; 150-240 Brin. Cast: 65,000-85,000 TS; 40,000-50,000 YS; 30-35 El; 30-40 RA; 130-195 Brin. Annealed: 113,000 TS; 62,700 YS; 43 El; 49 RA; 196 Brin. For apparatus to resist nitric and acetic acids, sulfite liquors and sulfurous acids; corrosion and heat resisting.

MIDVALOY 2512-10 M
Midvale-Heppenstall Co.
C 0-0.1, Cr 25, Ni 12, Mo 2.5-3.5, bal Fe.
For chemical plant equipment; heat and corrosion resistant. *Obsolete*

MIDVALOY 2512-10 MC
Midvale-Heppenstall Co.
C 0-0.1, Cr 25, Ni 12, Mo 2.5-3.5, Cb = 10 x C, bal Fe.
For heat resistant welded structures and equipment; heat and corrosion resistant. *Obsolete*

MIDVALOY 2512-16
Midvale-Heppenstall Co.
C 0.16, Cr 25, Ni 12, bal Fe.
For automatic furnaces, skid rails, walking beams; heat resistant to 1850°F.

MIDVALOY 2512-16 M
Midvale-Heppenstall Co.
C 0-0.16, Cr 25, Ni 12, Mo 2.5-3.5, bal Fe.
Cast: 82,000 TS; 41,000 YS; 40 El; 40 RA; 165 Brin. *Obsolete*

MIDVALOY 2512-16 SE
Midvale-Heppenstall Co.
C 0.16, Cr 25, Ni 12, Se 0.2-0.35, bal Fe.
For heat resistant parts, furnace equipment; heat and corrosion resistant; free-cutting. *Obsolete*

MIDVALOY 2512-20
Midvale-Heppenstall Co.
C 0.22, Cr 25.4, Ni 12.8, bal Fe.
Annealed: 80,000 TS; 40,000 YS; 405 El; 39.4 RA; 165 Brin. For castings; corrosion and heat resistant. *Obsolete*

MIDVALOY 2512-7
Midvale-Heppenstall Co.
C 0.07, Cr 23-26, Ni 11-13, bal Fe.
Rolled: 70,000-115,000 TS; 40,000-60,000 YS; 9-45 El; 230-240 Brin. For heat and corrosion resisting parts; heat and corrosion resistant.

MIDVALOY 2512-7 M
Midvale-Heppenstall Co.
C 0-0.07, Cr 25, Ni 12, Mo 2.5-3.5, bal Fe.
For chemical plant equipment; heat and corrosion resistant. *Obsolete*

MIDVALOY 2520
Midvale-Heppenstall Co.
Cr 25, C 0-0.25, Ni 20, Mo 0.18, bal Fe.
Cast: 70,000-80,000 TS; 30,000-36,000 YS; 25-50 El; 20-50 RA; 145 Brin. Wrought: 75,000-165,000 TS; 30,000-120,000 YS; 5-48 El; 30-55 RA; 160 Brin. For corrosion resisting apparatus; corrosion resisting, austenitic.

MIDVALOY 2602
Midvale-Heppenstall Co.
Cr 26.5, C 0.25, Ni 1.5, bal Fe.
Cast: 75,000-100,000 TS; 60,000-80,000 YS; 20-30 El; 40-60 RA; 165-200 Brin. For furnace parts; heat and corrosion resisting. *Obsolete*

MIDVALOY 2700
Midvale-Heppenstall Co.
C 0.35-1, Cr 27-30, Ni 0-3, bal Fe.
For stainless parts; stainless iron; austenitic. *Obsolete*

MIDVALOY 2802
Midvale-Heppenstall Co.
C 0.2, Cr 28, Ni 0-3, bal Fe.
For ore roaster parts in mining industry; corrosion and heat resistant. *Obsolete*

MIDVALOY 2803
Midvale-Heppenstall Co.
C 0.79, Cr 29.3, Ni 2.5, bal Fe.
68,000 TS; 200 Brin. For pump bodies, pump parts, impellers; hardenable, heat resistant, resists sulfur dioxide. *Obsolete*

MIDVALOY 3030
Midvale-Heppenstall Co.
Cr 27, Ni 30, C 0.5, bal Fe.
Cast: 42,500-75,000 YS. For heat resisting and furnace parts; corrosion and abrasion resisting. *Obsolete*

MIDVALOY 77 F
Midvale-Heppenstall Co.
Cr 1.2, Mo 0.3, C, bal Fe.
For dies, taps, ball races, cams, balls and bearings; oil hardening. *Obsolete*

MIDVALOY 77W
Midvale-Heppenstall Co.
C 1.2, Mn 0.55, Cr 1.65, Mo 0.45, bal Fe.
For tools, cutters, dies, taps, oil cams; oil hardening. *Obsolete*

MIDVALOY 976
Midvale-Heppenstall Co.
Cr 9.7, Ni 1.5, Al 2.3, C, bal Fe.
For corrosion resistant parts, exhaust valves; corrosion resistant. *Obsolete*

MIDVALOY AERO-VALVE
Midvale-Heppenstall Co.
C 0.6, Cr 12.5, Ni 20, W 2, bal Fe.
Wrought: 130,000-160,000 TS; 120,000-130,000 YS; 14-20 El; 25-35 RA; 260-340 Brin. For poppet valves for aeronautic engines, heavy trucks, etc.; austenitic. *Obsolete*

MIDVALOY ATV-1
Midvale-Heppenstall Co.
Ni 36, Cr 11, Mn 1.4, C 0.25-0.35, bal Fe.
Cast: 65,000 TS; 40,000 YS; 10 El; 10 RA; Wrought: 90,000 TS; 50,000 YS; 28 El; 35 RA; 170 Brin. For turbine blades, valve stems; parts for superheated steam; austenitic; heat resisting. *Obsolete*

MIDVALOY ATV-3
Midvale-Heppenstall Co.
Ni 26, Cr 14.5, Mn 1.2, C 0.45, W 4, bal Fe.
Cast: 80,000 TS; 43,000 YS; 20 El; 20 RA. Wrought: 100,000 TS; 50,000 YS; 33 El; 45 RA; 185 Brin. For exhaust valves, gas turbine rotors; austenitic; heat resisting.

MIDVALOY BTG
Midvale-Heppenstall Co.
Ni 60, Cr 12, Mn 1.2, C 0.3, W 3, bal Fe.
Cast: 70,000 TS; 27,000 YS; 25 El; 24 RA; 140 Brin. Wrought: 106,000 TS; 50,000 YS; 37 El; 50 RA. For autovalves, cast vessels for chemical industries; austenitic; heat resisting. *Obsolete*

MIDVALOY EME
Midvale-Heppenstall Co.
C 0-0.12, Cr 18-20, Ni 10-12, W 3, Cb 1.5, 0.1% nitrogen gas, bal Fe.
Worked: 130,000 TS; 90,000 YS; 14 El; 37 RA; 275 Brin. For discs for gas turbines, high temperature bolts; high heat resistance. *Obsolete*

MIDVALOY EXTRA HIGH SPEED
Midvale-Heppenstall Co.
C 0.66, Mn 0.27, Cr 4.32, W 13.3, V 1.94, bal Fe.
For tools, high speed cutters; high speed steel. *Obsolete*

MIDVALOY FINISHING
Midvale-Heppenstall Co.
C 1.26, Mn 0.36, W 4, Cr 1, bal Fe.
For tools, finishing tools; oil hardening. *Obsolete*

MIDVALOY GTA
Midvale-Heppenstall Co.
C 0.19, Cr 20, Ni 32, Co 30, W 5.5, Cb 4, bal Fe.
For gas turbine parts; high temperature alloy, heat resistant. *Obsolete*

MIDVALOY H.C.
Midvale-Heppenstall Co.
C 0.2-0.25, Cr 26, Ni 1.5, bal Fe.
Wrought: 55,000-85,000 TS; 30,000-60,000 YS; 10-30 El;
20-55 RA. Cast: 55,000-75,000 TS; 40,000-60,000 YS; 1-5 El;
1-5 RA. For furnace parts, chemical plant equipment; heat
and corrosion resistant. *Obsolete*

MIDVALOY NO. 11
Midvale-Heppenstall Co.
C 0.35-0.5, Cr 0.6-0.9, Ni 3, bal Fe.
Heat treated: 135,000-275,000 TS; 110,000-245,000 YS;
10-25 El; 38-58 RA; 275-500 Brin. For airplane crankshafts,
gears, splines, transmission units; shock resistant;

MIDVALOY NO. 11-MO
Midvale-Heppenstall Co.
C 0.4, Cr 0.6-0.9, Ni 2.6-3.2, Mo 0.2-0.3, bal Fe.
Heat treated: 125,000-160,000 TS; 100,000-140,000 YS;
15-22 El; 45-55 RA; 250-325 Brin. For shafts, gears; tough.

MIDVALOY NO. 26
Midvale-Heppenstall Co.
C 0.15-0.45, Ni 1.5, Cr 0.5, bal Fe.
Heat treated: 118,000-248,000 TS; 98,000-242,000 YS; 6-20
El; 22-61 RA; 225-400 Brin. For gears, pinions, shafts, bolts,
nuts, axles; shock resistant. *Obsolete*

MIDVALOY NO. 3-3-4
Midvale-Heppenstall Co.
C 0.5-0.65, Cr 24-30, Ni 24-30, bal Fe.
For heat and corrosion resistant parts; heat and corrosion
resistant. *Obsolete*

MIDVALOY NO.21-00 CU
Midvale-Heppenstall Co.
C 0.2-0.35, Cr 21, Cu 1, bal Fe.
For heat and corrosion resistant parts, chemical industries;
heat and corrosion resistant. *Obsolete*

MIDVALOY NUT PIERCER
Midvale-Heppenstall Co.
C 0.5, Cr 4, W 13, bal Fe.
For tools, piercers. *Obsolete*

MIDVALOY SPECIAL FINISHING
Midvale-Heppenstall Co.
C 1.31, Mn 0.43, Cr 0.75, W 6.8, bal Fe.
For tools, finishing tools; oil hardening. *Obsolete*

MIDVALOY STAINLESS FC
Midvale-Heppenstall Co.
C 0-0.12, Cr 12-15, S, bal Fe.
Wrought. For corrosion resistant parts; corrosion resistant,
stainless, free cutting. *Obsolete*

MIDVALOY STAINLESS IRON, C-1
Midvale-Heppenstall Co.
C 0-0.12, Cr 0-15, bal Fe.
Wrought: 65,000 TS; 37,000 YS; 35 El; 65 RA; 135 Brin.
Treated: 111,500-200,000 TS; 88,500-185,000 YS; 15-22.7 El;
45-68.5 RA; 229-425 Brin. For material for engineering
construction; corrosion resisting. *Obsolete*

MIDVALOY STAINLESS IRON, C-2
Midvale-Heppenstall Co.
C 0-0.12, Cr 15-18, bal Fe.
Treated: 69,000 TS; 48,400 YS; 32 El; 62 RA; 166 Brin. For
storage structures for nitric acid; corrosion resisting.
Obsolete

MIDVALOY STAINLESS IRON, C-3
Midvale-Heppenstall Co.
C 0-0.12, Cr 18-23, bal Fe.
Wrought: 65,000 TS; 45,000 YS; 40 El; 60 RA. Treated: 85,000
TS; 55,000 YS; 20 El; 25 RA. For structures for storage of
nitric acid; corrosion resisting. *Obsolete*

MIDVALOY STAINLESS IRON, C-4
Midvale-Heppenstall Co.
C 0-0.12, Cr 23-30, bal Fe.
Cast: 90,000 TS; 55,000 YS; 3 El; 3 RA. Wrought: 100,000 TS;
80,000 YS; 15 El; 40 RA; 200 Brin. For burner parts, chemical
plant equipment; corrosion resisting. *Obsolete*

MIDVALOY STAINLESS STEEL "B"
Midvale-Heppenstall Co.
C 0.65, Cr 17.5, bal Fe.
Wrought: 70,000 TS; 35,000 YS; 30 El; 50 RA; 180 Brin. Heat
treated: 78,400-230,000 TS; 40,100-200,000 YS; 8-31 El;
20-57 RA; 550 Brin. For cutlery and similar objects; corrosion
resisting, stainless. *Obsolete*

MIDVALOY STAINLESS STEEL 7
Midvale-Heppenstall Co.
C 0.25, Cr 20, Cu 1, bal Fe.
For stainless steel castings. *Obsolete*

MIDVALOY STAINLESS STEEL, A
Midvale-Heppenstall Co.
Cr 9-16, 0.12% min C, bal Fe.
Wrought: 86,000 TS; 42,000 YS; 30 El; 65 RA; 175 Brin. Heat
treated: 116,000-230,000 TS; 94,700-200,000 YS; 9-21 El;
20-44 RA; 200-241 Brin. For cutlery and similar objects;
stainless. *Obsolete*

MIDVALOY V-2-A
Midvale-Heppenstall Co.
Ni 8, Cr 18, Mn 0.5, C 0.1, bal Fe.
Rolled: 88,500-107,000 TS; 34,500-55,000 YS; 45-59 El; 62-74
RA. For stainless steel parts. *Obsolete*

MIG 20
Sifbronze
Stainless steel. C 0.04, Si 0.4, Mn 1.5, Ni 10, Cr 20, Nb 0.6,
bal Fe.
Stainless steel filler wire for aircraft fabrication, stainless steel
pipelines, tanks, fittings, and hospital equipment. 650
N/mm^2 TS; 180 Brin. BS 2901 347 S96.

MIG 21
Sifbronze
Stainless steel. C 0.02, Si 0.4, Mn 1.5, Ni 12, Cr 19, Mo 2, bal
Fe.
Stainless steel filler wire with low carbon content for nuclear
and chemical engineering industries. 650 N/mm^2 TS; 180
Brin. BS 2901 316 S92.

MIG 32
Sifbronze
Copper. Al 10, Fe 1, bal Cu.
Aluminum bronze wire suitable for welding materials of
similar composition; corrosion resistant. 550 N/mm^2 TS; 150
Brin. BS 2901 C13.

MIG 33
Sifbronze
Stainless steel. C 0.02, Si 0.4, Mn 1.5, Ni 10, Cr 21, bal Fe.
Stainless steel filler wire suitable for welding 18/8 (304)
austenitic stainless steels; corrosion and wear resistant. 650
N/mm^2 TS; 180 Brin. BS 2901 308 S92.

MIG 34
Sifbronze
Stainless steel. C 0.1, Si 0.4, Mn 1.5, Ni 13, Cr 24, bal Fe.
Stainless steel wire suitable for joining material of similar
composition and also dissimilar stainless steels. 650 N/mm^2
TS; 180 Brin. BS 2901 309 S94.

MIG 35
Sifbronze
Stainless steel. C 0.1, Si 0.4, Mn 1.5, Ni 21, Cr 26, bal Fe.
25/20 stainless steel wire suitable for heat resistant,
austenitic stainless steels. 650 N/mm^2 TS; 180 Brin. BS 2901
310 S94.

MIG 8
Sifbronze
Copper. Sn 7, bal Cu.
Phosphor bronze wire suitable for fusion welding of phosphor
bronze castings and copper alloys. 500 N/mm^2 TS; 120 Brin.
BS 2901 C11.

MIG 968
Sifbronze
Copper. Mn 1, Si 3, bal Cu.
Copper wire suitable for fusion welding materials of similar
composition. 350 N/mm^2 TS; 90 Brin. BS 2901 C9.

MIGRA IRON NO. 1
Friedrich Wilhelms-Hutte
Si 1-3, Mn 0.4-0.9, C 3.8-4.1, bal Fe.
Cast. For pistons and cylinders; cast iron.

MIGRA IRON NO. 2
Friedrich Wilhelms-Hutte
Si 1-3, Mn 0.4-0.9, C 3.8-4.1, bal Fe.
For pistons and cylinders; cast iron.

MIGRA IRON NO. 3
Manufacturer not listed.
Si 1-3, C 3.8-4.1, Ni 1-2, Cr 0.5-1, Mn 1-1.5, bal Fe.
For pistons and cylinders; cast iron.

MIKADO
Hover, Gebruder, Edelstahlwerk
C 1.45, Cr 1.4, Mn 0.6, Si 0.25, bal Fe.
For bearings, bushings, liners, sleeves; water hardened, wear
resistant.

MIL 48
A. Milne & Co.
C 0.55, Si 2, Cr 0.25, V 0.25, bal Fe.
Water hardening; tool steel, shock resistant; AISI S4.

MILBRITE
A. Milne & Co.
C 0.5, Mn 1, Si 0.3, Cr 1.1, Mo 0.25, bal Fe.
Oil hardening; designed for molds; similar to AISI 4150.

MILCO 9
A. Milne & Co.
C 0.73, Mn 0.28, Cr 4, V 1.75, W 18.25, Co 8.5, bal Fe.
For tools, cutters, reamers; high speed steel.

MILD SELF HARDENING
Colt Industries
C 1.7, Cr 3.35, W 5.25, bal Fe.
For dies; oil hardening. *Obsolete*

MILL BRASS MIX
E.A. Williams & Sons
Sn, Pb, bal Cu.
For bearings, bushings; tough.

MILLALOY
Doelger & Kirsten Inc.
C 0.4, Ni 4, Cr 1.5, bal Fe.
Heat treated: 312,000 TS; 272,000 YS; 11 El, 35 RA; 532 Brin.
For shear blades; oil hardened.

MILLARD
Anaconda Co.
Cu 90.48, Zn 0.1, Sn 0.07, Fe 0.55, Al 8.8.
97,100 TS; 44,900 YS; 29 El; 35 RA; 156 Brin. For valve seats,
spark plug bushings in aircraft engines. *Obsolete*

MILLENITE
Lake & Elliot Ltd.
C 0.3, Ni 1-5, Cr 0.8, bal Fe.
For high duty castings; oil hardened.

MILLER 200 PLUS GR. C

Miller Co.
Sn 7.82, P 0.18, Cu 92.
Spring: 110,000 TS; 80,000 YS; 24 El; 235 Brin. For diaphragms, bellows, springs, clips, fasteners; high ductility, corrosion resistant.

MILLER 200PLUS

Miller Co.
Sn 4.8-9.8, P 0.15, bal Cu.
Rolled: 96,000-118,000 TS; 70,000-85,00 YS; 20-24 El; 210-240 Brin. For clips, springs, bellows, diaphragms; phosphor bronze.

MILLER 200PLUS GR. A

Miller Co.
Sn 4.82, P 0.18, Cu 95.
Spring: 96,000 TS; 70,000 YS; 22 El; 210 Brin. For diaphragms, bellows, springs, clips, fasteners; high ductility, corrosion resistant.

MILLER 200PLUS GR. D

Miller Co.
Sn 9.85, P 0.15, Cu 90.
Spring: 118,000 TS; 85,000 YS; 20 El; 240 Brin. For diaphragms, bellows, springs, clips, fasteners; high ductility, corrosion resistant.

MILLING

English manufacture
Cu 54-56, Zn 27.5-31, Ni 15-18, Pb 0.5-1.
For white metal parts; easy to machine.

MILLING SILVER-1

English manufacture
Cu 56, Zn 31, Ni 12, Pb 1.
For white metal parts; easy to machine.

MILLING SILVER-2

English manufacture
Cu 56, Zn 27.5, Ni 16, Pb 0.5.
For white metal parts; easy to machine.

MILLS DTD 361B

William Mills & Co. Ltd.
Aluminum. Cu 4.5, bal Al.
Aluminum alloy casting; chill cast, solution treated and aged. BS Aerospace DTD 361B.

MILLS DTD 5008A

William Mills & Co. Ltd.
Aluminum. Mn 0.6, Zn 5, Ti 0.2, Cr 0.5, bal Al.
Aluminum alloy casting; chill cast and aged. BS Aerospace DTD 5008A; AA D712.

MILLS DTD 5018

William Mills & Co. Ltd.
Aluminum. Si 7.6, Mn 0.2, Zn 1.2, bal Al.
Aluminum alloy sand casting. BS Aerospace DTD 5018.

MILLS DTD 5028

William Mills & Co. Ltd.
Aluminum. Si 7, Mg 0.3, bal Al.
Aluminum alloy sand casting. BS Aerospace DTD 5028; AA A356

MILLS DTD 716A

William Mills & Co. Ltd.
Aluminum. Mg 0.5, Si 5, bal Al.
Aluminum alloy casting; chill cast. BS Aerospace DTD 716A.

MILLS DTD 722A

William Mills & Co. Ltd.
Aluminum. Mg 0.5, Si 5, bal Al.
Aluminum alloy casting; chill cast and aged. BS Aerospace DTD 722A.

MILLS DTD 727A

William Mills & Co. Ltd.
Aluminum. Mg 0.5, Si 5, bal Al.
Aluminum alloy casting; chill cast and solution treated. BS Aerospace DTD 727A.

MILLS DTD 735A

William Mills & Co. Ltd.
Aluminum. Mg 0.5, Si 5, bal Al.
Aluminum alloy casting; chill cast, solution treated and aged. BS Aerospace DTD 735A.

MILLS DTD 741A

William Mills & Co. Ltd.
Aluminum. Cu 4.2, Mg 2, Co 0.7, Nb 0.2, bal Al.
Aluminum alloy casting; chill cast, solution treated and aged. BS Aerospace DTD 741A.

MILLS L33

William Mills & Co. Ltd.
Aluminum. Si 11, bal Al.
Aluminum alloy, chill cast, BS 1490 LM-6M.

MILLS L35

William Mills & Co. Ltd.
Aluminum. Cu 4, Mg 1.5, Ni 2, Ti 0.2, bal Al.
Aluminum alloy casting (Y Alloy). BS Aerospace L35; AA 242.

MILLS L51

William Mills & Co. Ltd.
Aluminum. Cu 1, Si 2.5, Ni 0.9, Fe 1, Ti 0.2, bal Al.
Aluminum alloy casting; chill cast and aged. BS Aerospace L51.

MILLS L52

William Mills & Co. Ltd.
Aluminum. Cu 2.5, Mg 1, Ni, Si, Fe, bal Al.
Aluminum alloy casting; chill cast.

MILLS L53

William Mills & Co. Ltd.
Aluminum. Mg 10, bal Al.
Aluminum alloy casting; chill cast, solution treated. BS 1490 LM 10TB; AA 520.

MILLS L78

William Mills & Co. Ltd.
Aluminum. Cu 1.25, Mg 0.5, Si 5, bal Al.
Aluminum alloy casting; chill cast. BS 1490 LM-16TF; AA 355.

MILLS L91

William Mills & Co. Ltd.
Aluminum. Cu 4.5, bal Al.
Aluminum alloy sand casting. BS Aerospace L91; AA 295.

MILLS L92

William Mills & Co. Ltd.
Aluminum. Cu 4.5, bal Al.
Aluminum alloy sand casting. BS Aerospace L92; AA 295.

MILLS L99

William Mills & Co. Ltd.
Aluminum. Mg 0.3, Si 7, bal Al.
Aluminum alloy sand casting. BS Aerospace L99; AA A356.

MILLS LM10TB

William Mills & Co. Ltd.
Aluminum. Mg 10, bal Al.
Magnesium-aluminum sand or die casting. Solution treated. BS 1490 LM10W.

MILLS LM12M

William Mills & Co. Ltd.
Aluminum. Cu 10, Mg 0.3, bal Al.
Aluminum alloy casting; as chill cast. AA 222.

MILLS LM13TE

William Mills & Co. Ltd.
Aluminum. Cu 1, Mg 1, Si 12, Ni 2, bal Al.
Aluminum alloy sand or die casting; precipitation treated. AA A332.

MILLS LM13TF

William Mills & Co. Ltd.
Aluminum. Cu 1, Mg 1, Si 12, Ni 2, bal Al.
Aluminum alloy sand or die casting; chill cast, solution treated, and aged. BS 1490 LM13WP.

MILLS LM13TF7

William Mills & Co. Ltd.
Aluminum. Cu 1, Mg 1, Si 12, Ni 2, bal Al.
Aluminum alloy sand or die casting; full heat treatment plus stabilization.

MILLS LM16TB

William Mills & Co. Ltd.
Aluminum. Cu 1, Mg 0.5, Si 5, bal Al.
Aluminum alloy sand or die casting; chill cast and solution treated. AA 355.

MILLS LM16TF

William Mills & Co. Ltd.
Aluminum. C 1, Mg 0.5, Si 5, bal Al.
Aluminum alloy sand or die casting; chill cast, solution treated and aged. BS 1490 LM16WP.

MILLS LM18M

William Mills & Co. Ltd.
Aluminum. Si 5, bal Al.
Aluminum sand or die casting alloy; as chill cast. BS 1490 LM18M.

MILLS LM20M

William Mills & Co. Ltd.
Aluminum. Si 12, bal Al.
Aluminum die casting alloy; as chill cast. BS 1490 LM20M.

MILLS LM21M

William Mills & Co. Ltd.
Aluminum. Cu 4, Si 6, Mn 0.5, bal Al.
Aluminum alloy sand or die casting; as chill cast. BS 1490 21M; AA 319.

MILLS LM22TB

William Mills & Co. Ltd.
Aluminum. Cu 3, Si 5, Mn 0.5, bal Al.
Aluminum alloy casting; solution treated. BS 1490 LM22W.

MILLS LM24M

William Mills & Co. Ltd.
Aluminum. Cu 3.5, Si 8, bal Al.
Aluminum alloy die casting; as cast. BS 1490 LM24M.

MILLS LM25M

William Mills & Co. Ltd.
Aluminum. Mg 0.3, Si 7, bal Al.
Aluminum alloy sand casting; as cast. BS 1490 LM25; AA 356.

MILLS LM25TB7

William Mills & Co. Ltd.
Aluminum. Mg 0.3, Si 7, bal Al.
Aluminum alloy sand casting; full heat treatment plus stabilization.

MILLS LM25TE

William Mills & Co. Ltd.
Mg 0.3, Si 7, bal Fe.
Aluminum alloy sand casting; precipitation treated.

MILLS LM25TF

William Mills & Co. Ltd.
Aluminum. Mg 0.3, Si 7, bal Al.
Aluminum alloy sand casting; solution treated and precipitation treated.

MILLS LM26TE
William Mills & Co. Ltd.
Aluminum. Cu 3, Mg 1, Si 9.5, bal Al.
Aluminum alloy casting; precipitation treated. AA 332.

MILLS LM27M
William Mills & Co. Ltd.
Aluminum. Cu 2, Si 7, Mn 0.4, bal Al.
Aluminum alloy casting; as chill cast.

MILLS LM28TE
William Mills & Co. Ltd.
Aluminum. Cu 1.5, Mg 1.2, Si 19, Ni 1.2, bal Al.
Aluminum alloy casting; precipitation treated.

MILLS LM28TF
William Mills & Co. Ltd.
Aluminum. Cu 1.5, Mg 1.2, Si 19, Ni 1.2, bal Al.
Aluminum alloy casting; solution treated and precipitation treated.

MILLS LM29TE
William Mills & Co. Ltd.
Aluminum. Cu 1.1, Mg 1.1, Si 23, Ni 1.1, bal Al.
Aluminum alloy casting; precipitation treated.

MILLS LM29TF
William Mills & Co. Ltd.
Aluminum. Cu 1.1, Mg 1.1, Si 23, Ni 1.1, bal Al.
Aluminum alloy casting; solution treated and precipitation treated.

MILLS LM2M
William Mills & Co. Ltd.
Aluminum. Cu 1.5, Si 10, bal Al.
Silicon-aluminum die casting, as cast. BS 1490 LM2M.

MILLS LM30M
William Mills & Co. Ltd.
Aluminum. Cu 4.5, Mg 0.5, Si 17, bal Al.
Aluminum alloy casting; as chill cast.

MILLS LM30TS
William Mills & Co. Ltd.
Aluminum. Cu 4.5, Mg 0.5, Si 17, bal Al.
Aluminum alloy casting; chill cast and stress relieved.

MILLS LM4M
William Mills & Co. Ltd.
Aluminum. Cu 3, Si 5, Mn 0.5, bal Al.
Aluminum die casting, as cast. BS 1490. LM4M; AA 319.

MILLS LM4TF
William Mills & Co. Ltd.
Aluminum. Cu 3, Si 5, Mn 0.5, bal Al.
Aluminum die casting alloy; solution treated and precipitation treated. BS 1490 LM4WP.

MILLS LM5M
William Mills & Co. Ltd.
Aluminum. Mg 5, Mn 0.5, bal Al.
Sand or die cast aluminum alloy, as cast. BS 1490 LM5M.

MILLS LM6M
William Mills & Co. Ltd.
Aluminum. Si 11, bal Al.
Aluminum sand or die casting alloy. As chill cast. BS 1490 LM 6M.

MILLS LM9M
William Mills & Co. Ltd.
Aluminum. Mg 0.4, Si 12, Mn 0.5, bal Al.
Aluminum sand or die casting; as chill cast. BS 1490 LM9.

MILLS LM9TE
William Mills & Co. Ltd.
Aluminum. Mg 0.4, Si 12, Mn 0.5, bal Al.
Aluminum alloy casting; precipitation treated.

MILLS LM9TF
William Mills & Co. Ltd.
Aluminum. Mg 0.4, Si 12, Mn 0.5, bal Al.
Aluminum alloy casting; solution treated and precipitation treated. BS 1490 LM9WP.

MILLS LMOM
William Mills & Co. Ltd.
Aluminum. Al 99.5.
Aluminum casting. BS 1490 LM 0M.

MILLS-LM10W
William Mills & Co. Ltd.
Si 9.5-11, Ti 0.1, Cu 0-0.1, bal Al.
Cast: 40,500 TS; 23,500 YS; 8 El; 75 Brin. For high stressed castings; shock and corrosion resistant. *Obsolete*

MILLS-LM11W
William Mills & Co. Ltd.
Cu 4-5, Ti 0.2, bal Al.
Cast: 31,500 TS; 18,000 YS; 7 El; 65 Brin. Aged: 40,500 TS; 27,000 YS; 4 El; 100 Brin. For high stressed castings; age-hardenable. *Obsolete*

MILLS-LM1M
William Mills & Co. Ltd.
Cu 6-8, Mg 0.15, Si 2-4, Zn 2-4, bal Al.
Cast: 18,000 TS; 11,000 YS; 80 Brin. For die castings, housings; not subject to shock. *Obsolete*

MILLS-LM7P
William Mills & Co. Ltd.
Cu 0.8-2, Mg 0.05-0.2, Si 1.5-2.8, Fe 0.8-1.4, Ni 0.8-1.7, Ti 0.2, bal Al.
Cast: 22,400 TS; 15,700 YS; 2 El; 65 Brin. For general engineering castings; high ductility. *Obsolete*

MILMOLD
A. Milne & Co.
C 0.3, Mn 0.8, Si 0.5, Cr 1.7, Mo 0.4, bal Fe.
Oil hardening; tool steel designed for molds; AISI P20.

MILNAIR
A. Milne & Co.
C 0.95, Mn 2, Cr 2, Mo 1, bal Fe.
For tools; abrasion resistant.

MILNAIR 4
A. Milne & Co.
C 0.95, Mn 2, Si 0.35, Cr 2.2, Mo 1.1, bal Fe.
Air or oil hardening; cold work tool steel; AISI A4.

MILNAIR 5
A. Milne & Co.
C 1, Cr 5.25, V 0.25, Mo 1.1, bal Fe.
For tools, dies; air hardening; nondeforming.

MILNE 3074
A. Milne & Co.
C 0.35, V 0.5, Cr 4, W 9.25, bal Fe.
For punches, shears, extrusion dies; hot work steel, oil hardening.

MILNE CMV
A. Milne & Co.
C 0.38, Cr 5.25, Mo 1.25, V 1.05, bal Fe.
For hot work tools, punches; hot work steel, oil hardening.

MILNE CMW
A. Milne & Co.
C 0.37, Cr 5, Mo 1.5, V 0.3, W 1.3, bal Fe.
For hot work tools, punches; hot work steel, oil hardening.

MILNE DOUBLE SIX
A. Milne & Co.
C 2.25, Cr 12, bal Fe.
For blanking and forming dies; Type D3; oil hardening.

MILNE HOLLOW DIE STEEL
A. Milne & Co.
C 0.8-1, bal Fe.
For hollow dies; water hardening.

MILNE M-330
A. Milne & Co.
C 0.3, Cr 3, Mo 3, V 0.6, bal Fe.
For punches, crimpers, upsetters; hot work steel, resists heat checking.

MILNE M-331
A. Milne & Co.
C 0.4, Cr 3.3, Mo 2.25, V 0.4, bal Fe.
For punches, upsetters, dies, shears; hot work steel, resists softening.

MILNE M-333
A. Milne & Co.
C 0.3, Cr 3, Mo 3, V 0.6, Co 2.25, bal Fe.
For heavy duty hot work tools, punches, dies; hot work steel, oil hardening.

MILNE MM6+6
A. Milne & Co.
C 0.85, Cr 4.15, W 6.4, Mo 5, V 1.9, bal Fe.
For tools, cutters, dies; high speed steel.

MILNE MMCO
A. Milne & Co.
C 0.7, W 5.8, Mo 5.2, Cr 4, V 2, Co 9, bal Fe.
For lathe and planer tools, reamers, hobs, cutters; high speed steel.

MILNE MT-9
A. Milne & Co.
C 1.5, Cr 12.5, bal Fe.
For blanking and drawing dies; oil hardening; nondeforming.

MILNE MX-15
A. Milne & Co.
C 0.5, W 12, Cr 4, Ni 12, V 1, bal Fe.
For extrusion dies and mandrels, punches; hot work steel, oil hardening.

MILNE ORANGE LABEL
A. Milne & Co.
C 1.01-1.05, bal Fe.
For tools, drills, taps; water hardening.

MILNE RED LABEL
A. Milne & Co.
C 0.95-1.1, Mn 0.3, Si 0.25, bal Fe.
For tools, drills, hobs, taps, punches; Type W1; water hardening.

MILNE WHITE LABEL
A. Milne & Co.
C 0.9-1, bal Fe.
For tools, drills, springs; water hardening.

MILO
Hidalgo Steel Co. Inc.
C 0.4, Mn 1.1, bal Fe.
For flogging tools, scarifer and fire tools; water hardened.

MILO 35
Hidalgo Steel Co. Inc.
C 0.7-1.1, bal Fe.
For pneumatic tools; water hardening.

MILO 38
Hidalgo Steel Co. Inc.
C 0.7-0.9, bal Fe.
For pneumatic tools, chisels, punches; water hardened; Type W1.

MILRITE
NL Industries
Sn, Sb, Cu, bal Pb.
For bearings; Babbitt. *Obsolete*

MILTUFF
A. Milne & Co.
C 0.5, Mn 0.7, Si 0.25, Cr 3.25, Mo 1.4, bal Fe.
Air or oil hardening; tool steel, shock resistant; AISI S7.

MILVAN
A. Milne & Co.
C 0.7, W 19, Cr 4.25, V 2.25, bal Fe.
For tools, cutters; high speed steel.

MILVAN NO. 1
A. Milne & Co.
W 19, Cr 4, V 2.25, C, bal Fe.
For tools, cutters; high speed steel.

MILWALOY 1 1/4 MN
Milwaukee Steel Foundry Co.
C 0.3-0.4, Mn 1.1-1.4, bal Fe.
Cast: 75,000-95,000 TS; 45,000-60,000 YS; 23-30 El; 30-55 RA; 160-196 Brin. For tractor parts, road machinery, sprockets; water hardened.

MILWALOY 13
Manufacturer not listed.
C 0.08-0.12, Cr 12-14, bal Fe.
For stainless parts; stainless.

MILWALOY 18
Milwaukee Steel Foundry Co.
C 0.3-0.4, Ni 13-15, Mn 3.9-4.5, bal Fe.
For circuit breakers, switches; austenitic, non-magnetic.

MILWALOY 18-8
Milwaukee Steel Foundry Co.
C 0.08-0.15, Ni 8-10, Cr 17-20, bal Fe.
Cast: 70,000 TS; 25,000 YS; 40 El; 35 RA; 160 Brin. Heat treated: 80,000 TS; 40,000 YS; 70 El; 60 RA; 187 Brin. For food machinery; stainless.

MILWALOY 26
Milwaukee Steel Foundry Co.
C 0.08-0.12, Cr 18-20, Ni 8-10, bal Fe.
For corrosion resistant parts; A Krupp-Nirosta Steel; 18-8.

MILWALOY 29-9
Milwaukee Steel Foundry Co.
C 0.1-0.2, Ni 8-10, Cr 27-31, bal Fe.
Cast: 80,000 TS; 40,000 YS; 35 El; 35 RA; 187 Brin. Heat treated: 90,000 TS; 50,000 YS; 50 El; 60 RA; 207 Brin. For paper mill machinery; heat resistant.

MILWALOY 35-15
Milwaukee Steel Foundry Co.
C 0.2-0.4, Ni 32-38, Cr 13-17, bal Fe.
Cast. For heat resisting castings; heat resistant.

MILWALOY 38
Milwaukee Steel Foundry Co.
C 0.12-0.15, Cr 28-30, Ni 8-10, bal Fe.
For corrosion resistant parts; A Krupp-Nirosta 29-9 steel.

MILWALOY 50
Milwaukee Steel Foundry Co.
C 0.3-0.4, Cr 15-18, Ni 32-36, bal Fe.
For furnace parts to resist high temperatures; heat and corrosion resistant.

MILWALOY 7
Milwaukee Steel Foundry Co.
C 0.3-0.4, Cr 1.5-1.75, V 0.6-0.7, bal Fe.
For nitriding; nitralloy steel.

MILWALOY CC
Milwaukee Steel Foundry Co.
C 0.4-0.5, Mn 0.5-0.8, Cr 0.9-1.1, bal Fe.
Cast: 85,000-100,000 TS; 50,000-70,000 YS; 17-22 El; 25-45 RA; 179-210 Brin. For crusher machinery, screen plates, wear segments; wear resistant.

MILWALOY COMMERCIAL
Milwaukee Steel Foundry Co.
C 0.25-0.35, Mn 0.5-0.8, bal Fe.
Cast: 65,000-80,000 TS; 38,000-55,000 YS; 25-35 El; 40-55 RA; 140-165 Brin. For structural and machine castings; water hardened.

MILWALOY CR NI
Milwaukee Steel Foundry Co.
C 0.35-0.45, Mn 0.6-0.9, Ni 1.25-1.75, Cr 0.6-0.9, bal Fe.
Cast: 90,000-110,000 TS; 55,000-75,000 YS; 18-25 El; 30-50 RA; 180-220 Brin. For gears, cams, rollers; wear resistant.

MILWALOY CR NI MO
Milwaukee Steel Foundry Co.
C 0.35-0.45, Mn 0.5-0.8, Ni 1.25-1.75, Cr 0.6-0.9, Mo 0.3-0.4, bal Fe.
Cast: 95,000-155,000 TS; 65,000-135,000 YS; 12-24 El; 25-50 RA; 189-255 Brin. For castings; wear resistant.

MILWALOY DYNAMO
Milwaukee Steel Foundry Co.
C 0.05-0.15, Mn 0-0.2, bal Fe.
Cast: 50,000-65,000 TS; 25,000-37,000 YS; 30-37 El; 50-70 RA; 130-150 Brin. For magnet bodies, armature frames, electrical machinery; high magnetic permeability.

MILWALOY HI-CARBON
Milwaukee Steel Foundry Co.
C 0.4-0.5, Mn 0.5-0.8, bal Fe.
Cast: 80,000-95,000 TS; 47,000-65,000 YS; 20-30 El; 30-50 RA; 170-196 Brin. For gears, racks, sprockets; water hardened.

MILWALOY KA2SMO
Milwaukee Steel Foundry Co.
C 0.4-0.6, Ni 8-10, Cr 18-20, Mo 2-4, bal Fe.
Cast: 70,000 TS; 30,000 YS; 35 El; 45 RA; 178 Brin. Heat treated: 80,000 TS; 40,000 YS; 50 El; 70 RA; 196 Brin. For corrosion resisting castings; corrosion resistant.

MILWALOY MN BO
Milwaukee Steel Foundry Co.
C 0.3-4, Mn 1.1-1.4, B 0.003, bal Fe.
Cast: 80,000-95,000 TS; 50,000-65,000 YS; 20-30 El; 35-55 RA; 174-196 Brin. For tractors, road machinery; wear resisting.

MILWALOY MN MO BO
Milwaukee Steel Foundry Co.
C 0.3-0.4, Mn 1.1-1.4, Mi 0.15-0.2, B 0.003, bal Fe.
Cast: 90,000-110,000 TS; 70,000-80,000 YS; 20-30 El; 35-55 RA; 179-217 Brin. For tractor parts, road machinery; wear resisting.

MILWALOY NIT
Milwaukee Steel Foundry Co.
C 0.2-0.3, Mn 0.5-0.9, Cr 2.5-3, Mo 0.35-0.45, V 0.2, bal Fe.
Cast: 80,000-95,000 TS; 70,000-80,000 YS; 20-30 El; 30-55 RA; 170-200 Brin. For pistons, cylinders, valve parts; nitriding alloy.

MIN-OX-GRADE 51 C
Binney Castings Co.
C 2.5, Si 3, Al 3, V 0.2, Cr 2.5, Mn 0.15, bal Fe.
Natural: 45,000 TS; 320 Brin. Heat treated: 600 Brin. For glass molds, dies for tile pressing, oven enameling racks. Heat and wear resistant.

MIN-OX-GRADE A
Binney Castings Co.
Cu 80, Al 7, Zn 5, Fe 3, Mn 3.
55,000 TS; 13 El; 12 RA; 96 Brin. For glass molds, slides, bearings; heat and wear resistant. *Obsolete*

MIN-OX-GRADE D-V
Binney Castings Co.
Copper. Cu 62, Al 7, Zn 7, Ni 22.
80,000 TS; 0.5-2.0 El; 0 RA. For glass molds, cams, machine parts. Heat and wear resistant.

MINARGENT
Manufacturer not listed.
Cu 46-57, Ni 32-40, W 0-28, Al 0.2-0.5.
For silver solder, substitute for silver in silverware.

MINE TALABOT SPECIAL
Societe Nouvelle du Saut-du-Tarn
C 0.75, bal Fe.
For tools, springs, punches; water hardened.

MINEOR
Darwin & Milner Inc.
Tool material. C 1, Cr 5, Mo 1, bal Fe.
For punches, mandrels, rolls, dies; Type A2; air hardened; nondeforming.

MINEOR FM
Darwin & Milner Inc.
Tool material. C 1, Cr 5, Mo 1, S 0.2, bal Fe.
For punches, mandrels, rolls; air hardened; free cutting.

MINERVA CHISEL
A. Milne & Co.
C 0.5, Cr 1.75, W 1.9-2, V 0.2, bal Fe.
For tools, dies; oil hardening.

MINERVA H.C.
Edgar Allen Balfour Ltd.
C 0.53, Cr 1.8, W 1.9, V 0.2, bal Fe.
For chisels, shear blades, extrusion dies; hot-work steel, oil hardened.

MINERVA L.C.
Edgar Allen Balfour Ltd.
C 0.43, Cr 1.8, W 1.9, V 0.2, bal Fe.
For chisels, bits, screwdrivers, scarfing tools; oil hardened, shock resistant.

MINERVA SPECIAL
Edgar Allen Balfour Ltd.
C 0.5, Cr 1.8, W 2.2, V 0.25, bal Fe.
For pneumatic chisels, cold chisels, rivet busters; shock resistant.

MINIMAX NO. 178
Graham Chemical Corp.
Ag-Hg.
For dental amalgams; corrosion resistant.

MINIMUM
Thyssen Edelstahlwerke AG
C 0.04, bal Fe.
For electrical and magnetic equipment; magnetically soft. *Obsolete*

MINIMUM EXTRA
Thyssen Edelstahlwerke AG
C 0.04, Al, bal Fe.
For electrical and magnetic equipment; magnetically soft. *Obsolete*

MINIMUM V SPECIAL
Thyssen Edelstahlwerke AG
C 0.04, bal Fe.
For vacuum equipment; magnetically soft. *Obsolete*

MINOFOR
English manufacture
Sn 66-69, Sb 18-20, Zn 9-10, Cu 3-4, Fe 0-1.
For bearings; anti-friction.

MINOVAR
International Nickel Inc.
Cast iron. TC 0-2.4, Si 1-2, Ni 34-36, Cr 0-0.1, Mn 0-0.5, bal Fe.
Cast: 20,000-25,000 TS; 100-125 Brin. For electrical equipment; low coefficient of expansion. Formerly INVAR CAST IRON. *Obsolete*

MINOX 10
Usines de A. Manoir, Pitres (Eure)
C 0.1, Cr 13, bal Fe.
Annealed: 69,000-85,000 psi TS; 40,000-48,000 psi YS; 18 El; 140-160 Brin. For turbine blades, surgical instruments, cutlery; corrosion resistant, hardenable.

MINOX 15
Usines de A. Manoir, Pitres (Eure)
C 0.15, Cr 13, bal Fe.
Annealed: 74,000-92,000 psi TS; 42,000-52,000 psi YS; 14 El; 150-180 Brin. For turbine blades, surgical instruments, cutlery; corrosion resistant, hardenable.

MINOX 1820
Usines de A. Manoir, Pitres (Eure)
C 0.2, Cr 18, Ni 2, bal Fe.
Annealed: 85,000-106,000 psi TS; 58,000-92,000 psi YS; 12 El; 200-240 Brin. For chemical plant equipment; resists nitric acid.

MINOX 20
Usines de A. Manoir, Pitres (Eure)
C 0.2, Cr 14, bal Fe.
Annealed: 78,000-107,000 psi TS; 46,000-54,000 psi YS; 12 El; 160-200 Brin. For turbine blades, surgical instruments, cutlery; corrosion resistant, hardenable.

MINOX 30
Usines de A. Manoir, Pitres (Eure)
C 0.3, Cr 14, bal Fe.
Annealed: 85,000-114,000 psi TS; 52,000-68,000 psi YS; 10 El; 160-200 Brin. For surgical instruments, cutlery; corrosion resistant, hardenable.

MINOX-15
Societe Nouvelle des Acieries de Pompey
C 0.2, Cr 13, Mo 1, bal Fe.
Cast: 90,000 psi TS; 45,000 psi YS; 15 El; 95 Rock B. Hardened: 200,000 psi TS; 180,000 psi YS; 10 El; 45 Rock C. For valves, cutlery, oil and chemical plant equipment. Corrosion resistant castings, hardenable.

MINT DIE STEEL
Saville & Co. Ltd., J.J.
C, bal Fe.
For coining, cold-heading and embossing dies; water hardening.

MINT DIE STEEL
Jessop-Saville Ltd.
C, bal Fe.
For coining, cold-heading and embossing dies; water hardening.

MIR-O-COL
Mir-o-Col Alloy Co.
Cr, Mo, W, Co.
For hard surfacing welding electrode; wear resistant.

MIR-O-COL BR
Mir-o-Col Alloy Co.
C, Cr, Ni, Mo, bal Co.
For hard facing rod; abrasion and impact resistant.

MIR-O-COL NO. 1
Mir-o-Col Alloy Co.
C 3-4, Cr 11-13, Mn 1-1.5, Si 2-3, Ni 4-6, Mo 8-10, bal Fe.
Welded: 600-650 Brin. For welding rod for plowshares, shovels, root cutters; austenitic, abrasion resistant.

MIR-O-COL NO. 11
Mir-o-Col Alloy Co.
C, Co, Cr, W.
Cast: 500-520 Brin. For hard facing electrode; corrosion and abrasion resistant.

MIR-O-COL NO. 3
Mir-o-Col Alloy Co.
C 1.75-2.2, Mn 0.5-1, Si 0.6-0.9, Cr 8-9, bal Fe.
Cast: 45,000 TS; 0 El; 480 Brin. For hard facing welding rod; impact and abrasion resistant, tough, hard.

MIR-O-COL NO. 4
Mir-o-Col Alloy Co.
C 1-1.5, Mn 12-14, Si 0.8, Mo 0.5-1, bal Fe.
Welded: 60,000 TS; 22 El; 250 Brin. For hard facing rod; martensitic; impact and abrasion resistant.

MIR-O-COL NO. 4T-S
Mir-o-Col Alloy Co.
Cr, Mo, W, Co.
For hard facing electrode; ductile.

MIR-O-COL NO. 5
Mir-o-Col Alloy Co.
C 0.7-1, Mn 0.75-1.25, Ni 3-3.5, Mo 6.5-8, Co 0.4, B 0.8, Si 1.2, Cr 3.5-4.5, bal Fe.
Welded: 500 Brin. For welding rod for hot cutting dies, shears and tools; corrosion, heat and abrasion resistant.

MIR-O-COL NO. 6
Mir-o-Col Alloy Co.
C 0.9-1.2, Cr 26-30, Ni 0-0.5, Mo 0-0.5, Fe 0-3, W 3.5-5, bal Co.
Welded: 400-440 Brin. For welding rod for diesel engine valves; corrosion, abrasion and heat resistant.

MIRA METAL
American manufacture
Cu 75, Pb 16.3, Sb 6.8, Ni 0.24, Fe 0.43, Zn 0.62, Sn 0.91.
For valves, pipes; acid and corrosion resisting.

MIRACAST
J.M. Ney Co.
Au 41, Pd 4, Pt 1.
Gold color nonporcelain alloy for inlays, partial coverage and full coverage crowns, and multiple unit fixed restorations. 69,000 psi TS; 35,000 psi YS (0.1% offset); 46 El; 128 HV.

MIRACULOY
Sivyer Steel Corp.
C 0.35, Mn 1.25, Si 0.4, Cr 0.65, Ni 1.5, Mo 0.3, bal Fe.
Normalized: 115,000 TS; 85,000 YS; 18 El; 40 RA; 275 Brin. Heat treated: 125,000-245,000 TS; 90,000-190,000 YS; 8-20 El; 10-50 RA; 250-650 Brin. For heavy duty castings; high strength. *Obsolete*

MIRAMINT
Manufacturer not listed.
W, C + Co.
For hard cutting tools and dies; sintered.

MIRRALOY
Associated Steel Corp.
C, bal Fe.
For journals, shafting; turned, ground and polished.

MIRROMOLD
Now CARPENTER MIRROMOLD.

MIRYCAL
Great Western Steel Co.
C 0.5, Mn 0.25, Cr 0.95, W 1, Mo 0.2, bal Fe.
For tools, dies, punches, shear blades, boiler maker's tools; shock resistant.

MIRYCAL CHISEL
Ryer Inc. Ltd.
C 0.45, W 1-1.25, Cr 0.85-0.95, Mo 0.15-0.2, Mn 0.2-0.3, bal Fe.
For battering tools, chipping chisels, beading tools, rivet busters, cold sets, swages, track chisels, hot dies. Tough, shock resistant.

MISCHMETAL
Now CERALLOY MISCHMETAL.

MISCHMETAL
Uddeholm Corp.
Ce 50, 45 La + Nd.
For pyrophoric alloy in cigarette lighters; removes gas from radio tubes. *Obsolete*

MISCO "A"
McInnes Steel Co.
C 1, bal Fe.
For mining tools, chisels, quarry and track tools; water hardening. *Obsolete*

MISCO "C"
Michigan Steel Casting Co.
C 0.2-0.3, Cr 28-30, Ni 8-10, Si 0.6-0.8, bal Fe.
Cast: 90,000-95,000 TS; 60,000-65,000 YS; 23-28 El; 30-36 RA; 197-212 Brin. For use in sulfite industry, castings, valves; corrosion resistant.

MISCO "H.N."
Michigan Steel Casting Co.
C 0.6-0.7, Cr 15-18, Ni 60-65, bal Fe.
For heat treating boxes, furnace parts, valves, stills; resists H_2SO_4; heat, wear and corrosion resistant.

MISCO 16
Michigan Steel Casting Co.
C 0-0.2, Cr 16, bal Fe.
For heat and corrosion resistant parts; heat and corrosion resistant.

MISCO 16-8.16C
Michigan Steel Casting Co.
C 0-0.16, Ni 8-10, Cr 18-20, Mn 0-1, Si 0-2, bal Fe.
Annealed: 80,000 TS; 40,000 YS; 50 El; 60 RA; 170 Brin. For stainless castings; corrosion resistant.

MISCO 18-8
Michigan Steel Casting Co.
C 0.16, Cr 18, Ni 8, bal Fe.
85,000 TS; 45,500 YS; 40 El; 50 RA. For stainless parts, chemical engineering equipment; resists sulfurous acid.

MISCO 18-8 MO
Michigan Steel Casting Co.
C 0.1, Cr 18, Ni 8, Mo 3, bal Fe.
Annealed: 85,000 TS; 45,000 YS; 50 El; 60 RA; 179 Brin. For stainless parts; stainless.

MISCO 18-8 SE
Michigan Steel Casting Co.
C 0-0.16, Ni 8-10, Cr 18-20, Se 0.2-0.3, bal Fe.
Annealed: 80,000 TS; 40,000 YS; 50 El; 50 RA; 183 Brin. For corrosion resistant parts, castings; corrosion resistant.

MISCO 18-8 TI
Michigan Steel Casting Co.
C 0-0.1, Ni 8-10, Cr 18-20, Ti = 4 x C, bal Fe.
Annealed: 70,000 TS; 35,000 YS; 40 El; 40 RA. For stainless castings; stainless.

MISCO 18-8.07C
Michigan Steel Casting Co.
C 0-0.07, Ni 8-10, Cr 18-20, Mn 0-1, Si 0-2, bal Fe.
Annealed: 75,000 TS; 42,000 YS; 55 El; 65 RA; 156 Brin. For stainless castings; corrosion resistant.

MISCO 18-8.10C
Michigan Steel Casting Co.
C 0-0.1, Mn 0-1, Si 0-2, Ni 8-10, Cr 18-20, bal Fe.
Annealed: 75,000 TS; 42,000 YS; 55 El; 65 RA; 156 Brin. For stainless castings; corrosion resistant.

MISCO 18-8CB
Michigan Steel Casting Co.
C 0-0.1, Ni 8-10, Cr 18-20, Cb = 8 x C, bal Fe.
Cast: 70,000 TS; 35,000 YS; 40 El; 40 RA; 190 Brin.
Annealed: 82,000 TS; 42,000 YS; 40 El; 40 RA; 179 Brin. For chemical and plastic plant equipment; Type 347; stainless, austenitic.

MISCO 20
Michigan Steel Casting Co.
C 0-0.07, Ni 27.5-30.5, Cr 19-22, Mo 2-3, Cu 3.5-4.5, Mn 0.7, Si 0-1.5, bal Fe.
Annealed: 70,000 TS; 34,000 YS; 38 El; 45 RA; 143 Brin. For corrosion resistant castings; resists H_2SO_4.

MISCO 22
Michigan Steel Casting Co.
C 0.18-0.25, Mn 0.6-0.7, Si 0.4-0.5, bal Fe.
Cast: 60,000 TS; 30,000 YS; 26 El; 38 RA; 120-149 Brin. For gears, pinions, shafts, housings; water hardened.

MISCO 25
Michigan Steel Casting Co.
C 0.22-0.28, Mn 0.7, Si 0.5, bal Fe.
Cast: 70,000 TS; 38,000 YS; 24 El; 36 RA; 143-179 Brin. For gears, shafts, housings, machinery parts; water hardened.

MISCO 25-20
Michigan Steel Casting Co.
C 0.3-0.4, Ni 19-21, Cr 24-26, bal Fe.
Cast: 229 Brin. For heat resistant castings; corrosion and heat resistant.

MISCO 27
Michigan Steel Casting Co.
C 0.24-0.3, Mn 0.8, Si 0.5, bal Fe.
Cast: 75,000 TS; 43,000 YS; 22 El; 30 RA; 149-187 Brin. For machine tool castings, gears, housings, shafts; water hardened.

MISCO 28
Michigan Steel Casting Co.
C 0-0.2, Cr 28, bal Fe.
For heat and corrosion resistant parts; heat and corrosion resistant.

MISCO 28-15
Michigan Steel Casting Co.
C 0-0.4, Ni 14.5-16.5, Cr 27-30, bal Fe.
Cast: 70,000 TS; 40,000 YS; 12 El; 15 RA. For heat resistant castings; corrosion and heat resistant.

MISCO 28-20
Michigan Steel Casting Co.
C 0-0.4, Ni 19-21, Cr 27-30, bal Fe.
Cast. For heat resistant castings; corrosion and heat resistant.

MISCO 30-30
Michigan Steel Casting Co.
Cr 30, Ni 30, C, bal Fe.
For chemical engineering; heat and corrosion resistant.

MISCO 430
Michigan Steel Casting Co.
C 0.2, Cr 17, bal Fe.
For corrosion resistant castings; corrosion resistant; Type 430.

MISCO B
McInnes Steel Co.
C 0.75, Si 0.3, Mn 0.3, bal Fe.
For mining tools, chisels, hammers; water hardened.
Obsolete

MISCO B
Michigan Steel Casting Co.
C 0.25, Cr 25, Ni 13, bal Fe.
For corrosion and heat resisting parts, chemical engineering equipment, corrosion and heat resistant.

MISCO B-1
Michigan Steel Casting Co.
C 0.5, Cr 25, Ni 13, bal Fe.
For heat and corrosion resisting parts; heat and corrosion resistant.

MISCO C-1
Michigan Steel Casting Co.
C 0.5, Cr 29, Ni 9, bal Fe.
For heat and corrosion resisting parts; heat and corrosion resistant.

MISCO C-MO
Michigan Steel Casting Co.
C 0.18-0.25, Mn 0.7, Si 0.5, Mo 0.5, bal Fe.
Cast: 70,000 TS; 45,000 YS; 24 El; 35 RA; 143-179 Brin. For machine tool castings, gears, housings, shafts; water hardened.

MISCO CROMO
Michigan Steel Casting Co.
C 0.27-0.33, Mn 1.4, Si 0.5, Cr 0.9, Mo 0.3, bal Fe.
Cast: 100,000 TS; 70,000 YS; 18 El; 30 RA; 202-235 Brin. For machinery castings, gears, shafts, housings; oil hardened.

MISCO GRADE A
Michigan Steel Casting Co.
C 0.5-0.7, Ni 35-37, Cr 15-17, bal Fe.
For furnace parts, retorts, carburizing boxes; resists heat to 1950°F.

MISCO HN-1
Michigan Steel Casting Co.
C 0.5-0.8, Cr 16-20, Ni 66-70, bal Fe.
Cast: 60,000 TS; 36,000 YS; 3 El; 217 Brin. For stainless steel castings, furnace equipment; acid resistant.

MISCO HN-2
Michigan Steel Casting Co.
C 0.7, Cr 18, Ni 65, bal Fe.
For chemical engineering equipment; resists H_2SO_4.

MISCO K
Michigan Steel Casting Co.
C 0.2, Cr 25, Ni 20, bal Fe.
For stainless and heat resistant castings; Type 310; corrosion and heat resistant.

MISCO METAL
Michigan Steel Casting Co.
C 0.5, Ni 38, Cr 18, Si 1.5, bal Fe.
Cast: 68,000 TS; 40,000 YS; 10 El; 10 RA; 187 Brin. For furnace parts, annealing boxes, retorts; heat and corrosion resistant.

MISCO MS
Michigan Steel Casting Co.
C 0.45, Cr 18, Ni 38, bal Fe.
Cast: 70,000 TS; 45,000 YS; 5 El; 6 RA; 190 Brin. For castings subject to cyclic heating; resists thermal fatigue.

MISCO N
Michigan Steel Casting Co.
C 0.4, Cr 9, Ni 21, bal Fe.
75,000 TS; 45,000 YS; 30 El; 40 RA. For chemical engineering equipment; resists sulfuric acid.

MISCO N-5
Michigan Steel Casting Co.
C 0.4, Ni 30, Si 4, bal Fe.
For chemical engineering equipment; resists H_2SO_4.

MISCROME 1
Michigan Steel Casting Co.
C 0.25-0.35, Cr 13-15, bal Fe.
Cast: 10,000 TS; 72,000 YS; 17 El; 25 RA; 200 Brin. Heat treated: 123,000 TS; 98,000 YS; 20 El; 44 RA. For chemical engineering equipment; resists HNO_3.

MISCROME 2
Michigan Steel Casting Co.
C 0.25, Cr 21, bal Fe.
For chemical engineering equipment; resists HNO_3.

MISCROME 3
Michigan Steel Casting Co.
C 0.25, Cr 28, bal Fe.
For chemical engineering equipment; resists HNO_3.

MISCROME 5
Michigan Steel Casting Co.
C 1.4-1.6, Cr 12-14, Mo 0.9-1.1, Ni 0-0.5, Mn 0.5-0.7, bal Fe.
Annealed: 269 Brin. Heat treated: 131,000 TS; 352 Brin. For wear and corrosion resistant castings; castings, hard, wear resistant.

MISCROME CR
Michigan Steel Casting Co.
C 2.5, Cr 16, bal Fe.
For heat and corrosion resisting parts, grids; heat and corrosion resistant.

MISCROME KR
Michigan Steel Casting Co.
Cr 24-30, 1.6 C min, bal Fe.
For heat and corrosion resistant parts; heat and corrosion resistant.

MISCROME NO. 4
Michigan Steel Casting Co.
C 0.15, Cr 13, bal Fe.
Heat treated: 185,000 TS; 138,000 YS; 1 El; 4 RA; 328 Brin. For corrosion resisting parts; corrosion resistant.

MISHIMA STEEL
Japanese manufacture
Al 10, Ni 25, bal Fe.
For permanent magnets; high coercive force, sintered.

MITCHALLOY A
Robert Mitchell Co. Inc.
TC 2.9-3.1, Ni 2-2.5, Cr 0.3-0.6, bal Fe.
45,000-50,000 TS; 240-280 Brin. For brake drums. Wear resistant.

MITCHALLOY B
Robert Mitchell Co. Inc.
C 2.8-3.5, Ni 4-5, Cr 1.5-2, bal Fe.
Sand cast: 40,000-50,000 TS; 550-650 Brin. For pulverizers, grinders, mixers. Wear and abrasion resistant.

MITCHALLOY C
Robert Mitchell Co. Inc.
TC 2.7-3, Ni 12-13, Cu 5-7, Cr 1.5-4, bal Fe.
Cast: 20,000-35,000 TS; 140-190 Brin. For heat and corrosion resistant parts. Austenitic.

MITCHALLOY D
Robert Mitchell Co. Inc.
C 3.3, Si 2.6, Ni 1.5, Cr 0.8, bal Fe.
For gratebars. Alloy cast iron.

MITIA GR. A
Firth Brown Ltd.
WC, Co.
For tipped cutters, for cast iron, tough, intermittent cutting.
Obsolete

MITIA GR. B
Firth Brown Ltd.
WC, Co.
For tipped cutters; for general purpose. *Obsolete*

MITIA GR. C
Firth Brown Ltd.
WC, Co.
For tipped cutters; hard grade for cutting chilled cast iron.
Obsolete

MITIA GR. TA
Firth Brown Ltd.
WC, Co.
For tipped cutters; general purpose cutting. *Obsolete*

MITIA GR. TA5
Firth Brown Ltd.
WC, Co.
For tipped cutters; for rough cutting steel. *Obsolete*

MITIA GR. TE
Firth Brown Ltd.
WC, Co.
For tipped cutters; hard and wear resistant grade for finish turning. *Obsolete*

MITIA GR. TEIO
Firth Brown Ltd.
WC, Co.
For tipped cutters; for mild and low alloy steels. *Obsolete*

MITIFINE
H. Kramer & Co. (Ajax Metal Div.)
Babbitt containing no Pb and more Sn.
For heaviest marine engine bearings, crossheads, heavy machine Babbitts; free flowing; great toughness, slight shrinkage. *Obsolete*

MITIS IRON
English manufacture
Al 0.06-0.27, C, bal Fe.
For pipes, fittings; wrought iron deoxidized with Al.

MITSUBISHI MKA
Mitsubishi Metals America Corp.
Ag 0.08, bal oxygen-free Cu.
Equivalent to C 10700. For bus bars, wire, commutator segments.

MITSUBISHI MKC
Mitsubishi Metals America Corp.
Cr 0.8, bal oxygen-free Cu.
Equivalent to C 18400. For wire, resistance welding electrodes.

MITSUBISHI MKT
Mitsubishi Metals America Corp.
Te 0.5, P 0.08, bal oxygen-free Cu.
Equivalent to C 14500. For wire, rod.

MITSUBISHI MKZ
Mitsubishi Metals America Corp.
Zr 0.08, bal oxygen-free Cu.
Equivalent to C 15100. For wire, lead pins, lead frames.

MITSUBISHI OMCL
Mitsubishi Metals America Corp.
Cr 0.3, Zr 1, Mg 0.05, Si 0.02.
Thermal conductivity: 0.072 cal/cm·s·$^\circ$C. Coefficient of thermal expansion: 17.1 x 10^{-6}·$^\circ$C 20 to ~200°C. Modulus of elasticity: 14000 kgf/mm^2; electrical conductivity: 80% IACS. For lead frames, wire.

MITSUBISHI OOFC
Mitsubishi Metals America Corp.
Equivalent to C 10100. For connectors, wire, rod, electronic devices.

MITSUBISHI SH-IS-54
Mitsubishi Metals America Corp.
Now SH-IS-54.

MITSUBISHI SH-IS-60
Mitsubishi Metals America Corp.
Now SH-IS-60.

MIXEND
Arcos Alloys
67% min Ni, 23% min Cu.
For weld metal for cast iron. *Obsolete*

MIXEND 60
Arcos Alloys
Ni 60, Fe 40.
Welded: 262 Brin. For welding rod; machinable welds on high- P cast iron. *Obsolete*

MIXEND 99
Arcos Alloys
Ni 99.
Welded: 150 Brin. For welding rod; machinable welds on cast iron. *Obsolete*

MK9-AS
Major Engineering Co.
Si 4, Cu 5, bal Al.
Cast: 35,000 TS. Wrought: 47,500 TS. For light alloy parts; age-hardenable.

ML
Italian manufacture
C 0.8, W 18, Cr 4, V 2, Mo 0.75, bal Fe.
Hardened: 64-66 Rock C. For lathe tools, reamers, hobs, broaches, milling cutters. Type T2 high speed steel.

ML-ALLOY
American manufacture
Cu 4, Ni 2, Mg 2, Mn 0.3, V 0.05, Ti 0.1, bal Al.
Cast: 33,000 TS. At 600°F: 17,000 TS. For aircraft castings, cylinder heads, pistons; age-hardenable, high temperature use.

ML-ALUMINUM
American manufacture
Cu 4, Ni 2, Mg 2, Ti 0.1, Cr 0.3, V 0.1, Mn 0.3, bal Al.
Cast: 36,000 TS; 31,000 YS; 1.2 El. At 600°F: 17,000 TS; 14,500 YS; 9.0 El. For aircraft parts, cylinder heads, pistons; good high temperature properties.

ML1700
Manufacturer not listed.
C 0.2, Cr 25, W 15, B 0.4, bal Co.
Corrosion resistant, high temperature alloy.

ML1700
General Electric Co.
C 0.2, Cr 25, W 15, B 0.4, bal Co.
Corrosion resistant, high temperature alloy.

ML25
Acciaierie Valbruna s.p.a.
C 0.22-0.28, Mn 0.5-0.8, Cr 0.8-1.1, Mo 0.15-0.25, bal Fe.
Cr-Mo structural steel. 25 Cr Mo 4. *Obsolete*

ML30
Acciaierie Valbruna s.p.a.
C 0.27-0.33, Mn 0.4-0.7, Cr 0.8-1.1, Mo 0.15-0.25, bal Fe.
Cr-Mo structural steel. 30 Cr Mo 4. AISI 4130. *Obsolete*

ML35
Acciaierie Valbruna s.p.a.
C 0.32-0.38, Mn 0.6-0.9, Cr 0.8-1.1, Mo 0.15-0.25, bal Fe.
Cr-Mo structural steel. 35 Cr Mo 4. AISI 4135. *Obsolete*

ML40
Acciaierie Valbruna s.p.a.
C 0.37-0.44, Mn 0.7-1, Cr 0.9-1.2, Mo 0.15-0.25, bal Fe.
Cr-Mo structural steel. 40 Cr Mo 4. Similar to AISI 4140. *Obsolete*

ML50
Acciaierie Valbruna s.p.a.
C 0.47-0.55, Mn 0.6-0.9, Cr 0.8-1.1, Mo 0.15-0.25, bal Fe.
Cr-Mo structural steel. 50 Cr Mo 4; AISI 4150. *Obsolete*

MLV
Acciaierie Valbruna s.p.a.
C 0.45, Cr 1, Mo 0.35, V 0.25, bal Fe.
Low alloy steel; good for high temperature bolts. *Obsolete*

MM
Cooper Alloy Corp.
Nickel. C 0.25, Ni 63, Si 1.5, Fe 2.5, bal Cu.
Cast: 65,000 psi TS; 33,000 psi YS; 25 El; 28 RA; 150 Brin.
For chemical plant equipment. Corrosion resistant.

MM 6 & 6
A. Milne & Co.
C 0.83, Cr 4.15, V 1.9, W 6.35, Mo 5, bal Fe.
High speed steel for cutting tools; Mo-W type; AISI M2.

MM BOLT STEEL
Eagle & Globe Steel Ltd.
Alloy steel. C 0.33, Mn 1.55, Mo 0.28, bal Fe.
Heat treated for bolts, nuts and studs. 770-930 MPa TS. For bolts up to SAE Grade 5; crankshafts, connecting rods, armature shafts, general automotive forgings, and axles. ASG EN16B; BS970 EN16B.

MM DREADNOUGHT
Colt Industries
C, alloy, bal Fe.
For tools, oil hardening. *Obsolete*

MM DREADNOUGHT
Crucible Specialty Metals
C, alloy, bal Fe.
For tools, oil hardening. *Obsolete*

MM-007
Cannon-Muskegon Corp.
C 0.1, Cr 8, Co 10, Mo 6, W 0-0.1, Ti 1, Al 6, B 0.015, Zr 0.075, Ta 4.25, Hf 1.3, Cb 0-0.1, bal Ni.
Hafnium modified B-1900; improved ductility for turbine parts.

MM-008
Cannon-Muskegon Corp.
C 0.07, Cr 14.6, Co 15.2, Mo 4.4, Ti 3.35, Al 4.3, B 0.015, Zr 0.03, Hf 1.3, bal Ni.
Hafnium containing alloy; improved ductility for turbine parts.

MM-H
Cooper Alloy Corp.
Nickel. C 0.25, Ni 63, Si 3, Fe 2, bal Cu.
Cast: 80,000 psi TS; 50,000 psi YS; 10 El; 14 RA; 225 Brin.
For valve trim, valve parts, pump and turbine parts. Corrosion and abrasion resistant.

MM-S
Cooper Alloy Corp.
Nickel. C 0.25, Ni 63, Si 4, Fe 2, bal Cu.
Aged hardened: 130,000 psi TS; 100,000 psi YS; 2 El; 2 RA; 350 Brin. For valve trim, pump and turbine parts. Age-hardenable; corrosion resistant.

MM001
Martin-Marietta Corp.
C 0.12, Cr 10, Co 10, Mo 3, V 0.5, Ti 3.8, Al 5.5, Hf 1.4, B 0-0.015, Zr 0-0.05, bal Ni.
High temperature alloy.

MM002
Martin-Marietta Corp.
C 0.15, Cr 9, Co 10, W 10, Ta 2.5, Ti 1.5, Al 5.5, Hf 1.5, Zr 0.05, B 0.015, bal Ni.
Cast Aged (1600°F): 145,000 psi TS; 125,000 psi YS; 8 El.
High temperature alloy; for turbine wheels and blades.

MM002

Cannon-Muskegon Corp.
C 0.15, Cr 9, Co 10, W 10, Ta 2.5, Ti 1.5, Al 5.5, Hf 1.5, Zr 0.05, B 0.015, bal Ni.
Cast Aged (1600°F): 145,000 psi TS; 125,000 psi YS; 8 El.
High temperature alloy; for turbine wheels and blades.

MM004

Martin-Marietta Corp.
C 0.05, Cr 12, Mo 4.5, Cb 2, Al 5.9, Ti 0.6, Zr 0.1, Hf 1.3, B 0.01, bal Ni.
High temperature alloy; for turbine wheels and blades.

MM004

Cannon-Muskegon Corp.
C 0.05, Cr 12, Mo 4.5, Cb 2, Al 5.9, Ti 0.6, Zr 0.1, Hf 1.3, B 0.01, bal Ni.
High temperature alloy; for turbine wheels and blades.

MMCB

Cooper Alloy Corp.
Nickel. C 0.3, Si 1-2, Mn 1.5, Cu 26-33, Fe 3.5, Cb 1-3, bal Ni.
Corrosion resistant casting. FED QQ-N-288-F

MMM DREADNOUGHT

Colt Industries
C, alloy, bal Fe.
For tools; oil hardening. *Obsolete*

MMM DREADNOUGHT

Crucible Specialty Metals
C, alloy, bal Fe.
For tools; oil hardening. *Obsolete*

MMV

A. Milne & Co.
C 1.15, Cr 4, V 2.8, W 6, Mo 5.25, bal Fe.
High speed steel for cutting tools; molybdenum-tungsten type; AISI M3 Class 2.

MNC

W. Ossenberg & Cie Edelstahlwerke
C 1.05, Mn 1.1, Si 0.3, Cr 0.9, bal Fe.
For punching and stamping dies for light work. W.-Nr. 1.2127.

MNMO 18

Walsingham Steel Co., Ltd.
C 0.16-0.2, Si 0.3-0.6, Mn 1.3-1.7, Ni 0-0.4, Cr 0-0.25, Mo 0.25-0.35, Cu 0-0.3, bal Fe.

MNMO 38

Walsingham Steel Co., Ltd.
C 0.36-0.4, Si 0.3-0.6, Mn 1.35-1.55, Ni 0-0.4, Cr 0-0.25, Mo 0.3-0.35, Cu 0-0.3, bal Fe.
Meets BS 1458 Gr. A, B.

MNS NICKEL

Henry Wiggin & Co. Ltd.
Mn 1.5-2.25, High C, Si, 96% min Ni.
Rolled: 80,000 TS; 24,000 YS; 56 El; 76 RA; 139 Brin. For water meter components, small machined parts; corrosion resi *Obsolete*

MNV3

American manufacture
C 0.45, Cr 0, Si 3, bal Fe.
Valve steel.

MO

Sheffield Smelting Co. Ltd.
Ag 30, Cu, Zn.
Melting range: 680-770°C. Maximum stress: 49 kgf/mm^2; 5 El. For silver brazing; general use.

MO 18

Saarstahl AG
Now SAARSTAHL 16 CRMO 4.

MO RAPID EXTRA 3

Now VEW S610.

MO RAPID EXTRA 3A

Vereinigte Edelstahlwerke
C 1, W 1.35, Cr 3.8, Mo 2.35, V 2.35, bal Fe.
For tools, dies, cutters; high speed steel. *Obsolete*

MO TAP

Latrobe Steel Co.
C 1.2, Mo 0.6, Cr 0.5, Mn 0.8, bal Fe.
For taps, drills, reamers; water or oil hardening. *Obsolete*

MO-0.5 TI

Climax Performance Materials Corp.
C 0.03, Ti 0.5, bal Mo.
High Young's modulus for boring bar shanks.

MO-1

Japanese manufacture
Mo 23, Cr 2.5, bal Fe.
For permanent magnets in magnetic and electrical equipment; high permeability, precipitation hardening.

MO-1225 X

Thyssen Edelstahlwerke AG
C 0.9, Cr 4, Mo 2.5, V 2.5-3, W 1, Co 1.05, bal Fe.
For tools, dies, cutters; oil hardening. *Obsolete*

MO-20

Thyssen Edelstahlwerke AG
C 0.85, Cr 4.2, Mo 5, V 2, W 6.5, bal Fe.
For lathe and planer tools, hobs, drills, reamers; high speed steel, oil hardened. *Obsolete*

MO-30

Thyssen Edelstahlwerke AG
C 1.2, Cr 4.2, Mo 5, V 3.2, W 6.2, bal Fe.
For lathe and planer tools, milling cutters, hobs; high speed steel. *Obsolete*

MO-325

Thyssen Edelstahlwerke AG
C 0.95, Cr 4, Mo 2.5, V 2.5, W 2.8, bal Fe.
For milling cutters, hobs, drills, broaches; high speed steel, oil hardened. *Obsolete*

MO-325 X

Thyssen Edelstahlwerke AG
C 0.9, Cr 4, Mo 2.5, V 2.5-3, W 2.2, Co 2.5, bal Fe.
For tools, dies, cutters; oil hardening. *Obsolete*

MO-CHIP

Teledyne Firth Sterling
Steel. C 0.7, Mo 8, Co 2.5, Cr 0.5, V 1, bal Fe.
For tools, cutters, lathe and planer tools; high speed steel, oil hardened. *Obsolete*

MO-LO

American Art Alloys Inc.
C 3, Mn, Si, bal Fe.
For piston rings, cylinder liners, bushings; cast iron.

MO-MANG

Abex Corp.
C 0.7-0.9, Mn 12-14, bal Fe.
For welding rod; for Mn steel.

MO-MANG

Allis-Chalmers Mfg. Co.
C 0.8, Mn 14, Mo 0.2, bal Fe.
Cast: 200 Brin. For welding electrodes; work hardens to 550 Brin.

MO-MAX

Cleveland Twist Drill Co.
C 0.64-0.84, Cr 3.25-4.25, W 1.25-2, V 0.75-1.25, Mo 7.5-9.5, bal Fe.
For high speed tools and cutters, hot work dies, drills; high speed steel. *Obsolete*

MO-PERMALLOY 4-79 ALLOY

Wilbur B. Driver Co.
Nickel. Ni 79, Mo 4.3, bal Fe.
For laminated cores for communication inductors, transformers, electro-magnetic shields. Soft magnet, high permeability.

MO-RE 1, ETC.

Now DBK MO-RE 1, ETC.

MO-STAR

Midvale-Heppenstall Co.
C, alloy, bal Fe.
For tools, oil hardening. *Obsolete*

MO-STEEL NO. 1

AMAX Corp.
Cr 0.7-1, Mo 0.25-1, C 0.15-0.23, Mn 0.4-0.7, bal Fe.
Hardened: 200,000 TS; 182,000 YS; 25 El; 65 RA; 360 Brin.
For gears, shafts; case-hardening steel; tough, shock resistant. *Obsolete*

MO-STEEL NO. 2

AMAX Corp.
Cr 0.8-1.1, Mo 0.25-0.4, C 0.23-0.3, Mn 0.5-0.8, bal Fe.
Heat treated: 144,000-230,000 TS; 133,000-218,000 YS; 12-18 El; 48-61 RA; 300-450 Brin. For crankshafts, connecting rods, steering knuckles, front axles, propeller shafts, bolts; water hardening. *Obsolete*

MO-STEEL NO. 3

AMAX Corp.
Cr 0.8-1.1, Mo 0.25-0.4, C 0.3-0.4, Mn 0.5-0.8, bal Fe.
Water hardened: 145,000-225,000 TS; 140,000-200,000 YS; 11-20 El; 47-64 RA; 275-400 Brin. Oil hardened: 130,000-145,000 TS; 112,000-135,000 YS; 15-22 El; 55-64 RA; 255-300 Brin. For gears, heavy crankshafts, driving axles, connecting rods, crank pins, piston rods; tough, shock resistant. *Obsolete*

MO-TECHNICKILL

Gulf & Western Mfg. Co.
C, Mo, bal Fe.
For heavy duty rollers; wear resistant.

MO-TIGER

Bethlehem Steel Corp.
C 0.7, Mo 9, W 1.5, Cr 4, V 1, bal Fe.
For tools, cutters, drills; high speed steel. *Obsolete*

MO-TUNG

Cytemp Specialty Steel Div.
C 0.65-0.85, Mn 0.2-0.35, Mo 7.5-8.5, W 1.5, Cr 3.75-4.5, V 1-1.25, bal Fe.
For high speed steel cutting tools, reamers, broaches, drills, lathe tools; high speed steel.

MO-TUNG 54

Universal Cyclops
C 0.7, Cr 4, W 5, Mo 5, V 1, bal Fe.
For tools, cutters, taps, drills, reamers; high speed steel. *Obsolete*

MO-TUNG 652

Cytemp Specialty Steel Div.
C 0.8, Cr 4, V 2, Mo 5, W 6.5, bal Fe.
For lathe and planer tools, drills, form cutters, hot punches; high speed steel.

MO-VAN

Colt Industries
C 0.85, Cr 4, V 1.75, Mo 7.5, bal Fe.
For tools, cutters, reamers; high speed steel. *Obsolete*

MO10

Thyssen Edelstahlwerke AG
C 0.8, Mo 9, Cr 3.8, W 1.75, V 1.2, bal Fe.
For lathe and planer tools, reamers, broaches, taps; high speed steel. *Obsolete*

MO18
Walsingham Steel Co., Ltd.
C 0.16-0.2, Si 0.2-0.5, Mn 0.5-1, Ni 0-0.4, Cr 0-0.25, Mo 0.4-0.7, Cu 0-0.3, bal Fe.
Carbon-molybdenum steel for use up to 450°C. Meets BS 1398 Gr. A.

MO26
Saarstahl AG
C 0.25, Si 0.25, Mn 0.75, Cr 0.5, Mo 0.45, S 0.02-0.035, bal Fe.
Alloy carburizing steel. For gears, pinions, spline couplings. W.-Nr. 7325, 7326. *Obsolete*

MO330
Saarstahl AG
Now SAARSTAHL 32 CRMOV 1210.

MOCAR
Hewitt Metals Corp.
Sn 10, Sb 15, bal Pb.
For bearings for connecting rods and camshafts. Babbitt.

MOCARB
CCS Braeburn Alloy Steel
C 1, Mo 5, W 6.5, Cr 4.2, V 1.9, bal Fe.
Hardened: C 64-67 Rock. For lathe and planer tools, broaches, rolls, hobs, reamers, end mills, milling cutters. High speed steel. High hot-hardness and wear resistance. *Obsolete*

MOCASCO
Motor Castings Co.
Iron. C 3.2, Si 2.5, Cr 1, bal Fe.
For castings; cast iron. *Obsolete*

MOCASCO 30 GRAY CAST IRON
Motor Castings Co.
Iron. TC 3.3-3.6, Si 1.9-2.6, Ni, Cu, Mo, Cr residuals, bal Fe.
Cast: 30,000 TS; 179-200 Brin. For air compressor cylinders, motor blocks, and cylinder heads.

MOCASCO 35 GRAY CAST IRON
Motor Castings Co.
Iron. TC 3.2-3.5, Si 1.8-2.5, Cr 0.2-0.4, Ni, Cu, Mo residuals, bal Fe.
Cast: 187-241 Brin. For air compressor cylinders, motor blocks, and cylinder heads.

MOCASCO 40 COMPACTED GRAPHITE IRON
Motor Castings Co.
Iron. TC 3.5-3.8, Si 2-2.8, Ni, Cu, Mo, Cr residuals, bal Fe.
37,500-45,000 psi TS; 27,500-33,000 YS; 3-6 El; 130-179 Brin. Pearlite 10% max, CG 80% min.

MOCASCO 40 GRAY CAST IRON
Motor Castings Co.
Iron. TC 3.2-3.5, Si 1.8-2.4, Cr 0.2-0.4, Ni, Cu, Mo residuals, bal Fe.
40,000 TS; 187-255 Brin. For cylinder heads, liners, pistons, and valves.

MOCASCO 45 GRAY CAST IRON
Motor Castings Co.
Iron. TC 3.2-3.5, Si 1.8-2.4, Cr 0.2-0.4, Cu 0.3-0.5, Ni, Mo residuals, bal Fe.
197-255 Brin. For cylinder heads, pistons, and hydraulic valve bodies.

MOCASCO 50 COMPACTED GRAPHITE IRON
Motor Castings Co.
Iron. TC 3.5-3.7, Si 2.1-2.5, Ni 0.4-0.6, Cu 0.4-0.6, Mo 0.05-0.2, Cr 0.15-0.3, bal Fe.
45,000-60,000 psi TS; 35,000-45,000 YS; 2-3 El; 163-241 Brin. Pearlite 10-85%, CG 80% min.

MOCASCO 50 GRAY CAST IRON
Motor Castings Co.
Iron. TC 3.2-3.5, Si 1.8-2.5, Ni 0.5-1, Cu 0.2-1, Mo 0.4-0.75, Cr 0.2-0.45, bal Fe.
Cast: 50,000 TS; 217-262 Brin. For cylinder heads, pistons, and hydraulic valve bodies.

MOCASCO 60
Motor Castings Co.
Iron. Ni 1.5, Mo 0.9, C, Cr, bal Fe.
Cast: 60,000 TS; 228-260 Brin. For gears, brake drums, pump and hydraulic castings; cast iron. *Obsolete*

MOCASCO 60 COMPACTED GRAPHITE IRON
Motor Castings Co.
Iron. TC 3.4-3.8, Si 1.8-2.2, Cu 0.4-1, Mo 0.45-0.7, Cr 0-0.2, Ni residual, bal Fe.
Cast: 60,000-75,000 psi TS; 45,000-60,000 YS; 1-3 El; 207-270 Brin. Pearlite 85% min, CG 80% min.

MOCASCO CG GRADE 60
Motor Castings Co.
Iron.
Special cast iron. Cast: 60,000 TS; 40,000 YS; 3 El; 180 Brin. *Obsolete*

MOCASCO CG GRADE 68
Motor Castings Co.
Iron.
Special cast iron. Cast; 68,000 TS; 45,000 YS; 2.5 El; 220 Brin. *Obsolete*

MOCASCO CG GRADE 70
Motor Castings Co.
Iron.
Special cast iron. Cast: 70,000 TS; 50,000 YS; 2 El; 260 Brin. *Obsolete*

MOCASCO DUCTILE IRON 60-40-18
Motor Castings Co.
Iron. TC 3.3-3.7, Si 2.2-2.6, bal Fe.
156-217 Brin, ferritic ductile as cast.

MOCASCO DUCTILE IRON 65-45-12
Motor Castings Co.
Iron. TC 3.3-3.6, Si 2.2-2.5, bal Fe.
156-217 Brin, 50% ferritic/pearlitic as cast.

MOCASCO DUCTILE IRON 80-55-06
Motor Castings Co.
Iron. TC 3.3-3.6, Si 2.2-2.5, bal Fe.
187-255 Brin, pearlitic as cast.

MOCK GOLD-1
English manufacture
Cu 67-80, Pt 20-29, Zn 0-4.
For ornaments; corrosion resistant.

MOCK GOLD-2
English manufacture
Ni 6, Pt 1, Ag 1, 1.0 brass.
For ornaments; corrosion resistant.

MOCK SILVER
English manufacture
Al 84, Sn 10, Cu 5.5, P.
For instruments, fittings, ornamental; corrosion resistant.

MOCUT
CCS Braeburn Alloy Steel
C 0.6-0.85, Cr 3.5-4, V 0.9-1.3, W 1.3-1.8, Mo 7.75-9, bal Fe.
For twist drills, lathe and planer cutters, reamers, hobs. Type M1 high speed steel. *Obsolete*

MOD-1
United States Steel Corp.
C 0.9-1.05, Mn 0.95-1.25, Cr 0.9-1.15, Si 0.5-0.7, bal Fe.
High hardness after heat treatment. For tools or wear resistant parts.

MOD-2
United States Steel Corp.
C 0.85-1, Mn 1.4-1.7, Cr 1.4-1.7, Si 0.6-0.8, bal Fe.
High hardness after heat treatment. For tools or wear resistant parts.

MODERN
Becker Stahlwerk A.G.
C 0.7, W 18, Cr 4, V 1, bal Fe.
For cutters, tools; high speed steel.

MODIFIED "OUNCE" METAL
Belmont Metals Inc.
Cu 84, Sn 5, Pb 5, Zn 5, Ni 1.
For fittings, valves, hardware castings; pressure tight.

MODIFIED MONEL
Manning, Maxwell & Moore Co.
Ni 60-65, Cu 24-27, Sn 9-11, Fe 1-3.
Cast: 70,000 TS; 45,000 YS; 17 El. For valves for super heated steam; corrosion resistant.

MODIFIED MONEL METAL
Manning, Maxwell & Moore Co.
Ni 60-65, Cu 24-27, Sn 9-11, Fe 1-3, Mn, Si.
For valves for superheated steam.

MODULAY
J.F. Jelenko & Co.
Precious metal. Au 77, Pd 9, Ag 14.
Quenched: 58,000 psi TS; 27,000 psi YS; 38 El; 92 Brin.
Dental alloy, Type II; for medium hard inlays and crowns.

MODULVAR
Creusot-Loire
Ni 36, bal Fe.
Annealed: 70,000 TS; 24,000 YS; 36 El; 68 RA; 143 Brin. Cold drawn: 90,000 TS; 70,000 YS; 20 El; 60 RA; 185 Brin. For instruments, chronometers; high positive thermoelastic coefficient. *Obsolete*

MOGUL (AISI M2)
Jessop Steel Co.
C 0.78, Cr 4, W 1.5, V 1.15, Mo 8.7, bal Fe.
Annealed: 228-241 Brin. For reamers, drills, lathe and planer tools, hobs; Type M1; high speed steel.

MOGUL ALLOY BABBITT
Federal-Mogul Corp.
Cu 4-5, Sb 6-8, bal Sn.
For engine bearings, bushings; Babbitt. *Obsolete*

MOHAWK HOT DIE
AL Tech Specialty Steel Corp.
C 0.45, W 14, Cr 3.5, V 0.6, bal Fe.
Oil hardened: 310,000-268,000 TS; 1-23 El; 5-79 RA; 600-321 Brin. Heavy duty tungsten hot work tool steel suitable for hot shear blades, hot punches, long run hot forging and extrusion dies. AISI H-24.

MOHDI HOT DIE STEEL
Eagle & Globe Steel Ltd.
Tool material. C 0.4, Cr 4.4, W 2, Mo 5.5, V 1.65, Co 1.2, bal Fe.
Annealed: 240 Brin max. High wear resistance, high strength and resistance to tempering at elevated temperatures. Hot forging dies and punches for nuts and bolts. Warm forging dies and punches for similar products. Extrusion tools for copper and copper based alloys. Die casting dies for brass. Forging and pressing dies for hot forging brass.

MOHECO
Motor Castings Co.
TC 2.85-3.05, Si 1.75-2.1, Mn 0.6-0.8, Ni 1.75, Cr 0.9, bal Fe.
Heat treated: 525 Brin. Cast: 220 Brin. For dies, drawing dies; wear resistant. *Obsolete*

MOHICAN
Atlas Specialty Steels
C 0.8, W 1.5, Cr 4, V 1, Mo 8.5, bal Fe.
For tools, cutters, broaches, milling cutters; high speed steel.
Obsolete

MOHICAN 6
Atlas Specialty Steels
C 0.62, W 1.7, Cr 3.75, Mo 8.7, V 1, bal Fe.
At 400°F: 243,000 TS; 3.6 El; 8.1 RA; 477 Brin. For hot punches, shear blades, forming dies; shock and wear resistant, high red hardness.

MOHICAN 8
Atlas Specialty Steels
C 0.8, W 1.5, Cr 4, Mo 9, V 1.2, bal Fe.
For twist drills, hobs, mills, hot punches; tough, high speed steel. *Obsolete*

MOIL POINT
Colt Industries
C 0.8, Mn 0.5, Si 0.35, bal Fe.
For battering tools, moil points; tough, oil hardening.
Obsolete

MOIL POINT
Midvale-Heppenstall Co.
C 0.75, Mo 0.25, bal Fe.
For shock resistant tools; oil hardening. *Obsolete*

MOLDALOY
Trethaway Assoc.
Bi, Pb, Sn, Sb.
Cast: 11,500 TS; 22 Brin. For molds, forming dies, forging dies, chuck jaws; Melting point 430°F; molding alloy.

MOLDIE
Boyd-Wagner Co.
C 0.9, Mn 2.25, V 0.1, bal Fe.
For Bakelite and rubber molds; steam resistant.

MOLDINOC 65
Pechiney Electrometallurgie
Miscellaneous nonferrous. Al 0.95, Ca 1, Si 66, bal Fe.
Cast iron inoculant.

MOLDINOC 75
Pechiney Electrometallurgie
Miscellaneous nonferrous. Al 3, Ca 0.4, Si 77, bal Fe.
Cast iron inoculant.

MOLDTEM
Heppenstall Co.
Tool material. C, Cr, Mo, V, bal Fe.
Heat treated: 340 Brin. For zinc die casting dies, plastic mold dies; prehardened, hot work steel.

MOLECULOY
Molecu-Wire Corp.
Cr 20, Al 3, Co 0.2, bal Ni.
Resistance wire. Electrical resistivity: 800 ohms/circular mil ft. Annealed: 130,000 psi TS (average). Hard drawn: 250,000 psi TS. For winding precision resistors and potentiometers; maximum operating temperature: 250°C.

MOLECULOY III
Molecu-Wire Corp.
Cr 20, Al 4, Mn 5, Si 1, bal Ni.
Resistance wire. Electrical resistivity: 835 ohms/circular mil ft. Annealed: 130,000 psi TS. Hard drawn: 260,000 psi TS. Maximum operating temperature: 250°C.

MOLEGRAIN
Manufacturer not listed.
C 3.3, alloy, Si, Mn, bal Fe.
For gears, shafts, housings; cast iron.

MOLEL
Delsteel Inc.
C, alloy, bal Fe.
For tools, cutters, punches; oil hardened.

MOLEX NO. 5
Associated Steel Corp.
C, alloy, bal Fe.
For crushers, punches, wrenches, wedges; shock and abrasion resistant.

MOLEX NO. 6
Associated Steel Corp.
C, alloy, bal Fe.
For gages, chuck jaws, knives, reamers; oil hardened, shock and wear resistant.

MOLEX NO. 7
Associated Steel Corp.
C, alloy, bal Fe.
For bushings, cam rollers, shear blades, forming dies; oil hardened, wear resistant.

MOLEX NO. 8
Associated Steel Corp.
C 1.1, Mn 0.75, Si 0.5, Cr 5.5, V 0.3, Mo 1.3, P 0.015, S 0.015, bal Fe.
Air-hardened tool steel for cold-working tools.

MOLEX NO. 8 MODIFIED
Associated Steel Corp.
C 0.55, Mn 0.9, Si 0.25, Cr 3.5, Ni 0.25, V 0.25, Mo 1.4, bal Fe.
Air-hardened tool steel.

MOLEX-A
Associated Steel Corp.
C, alloy, bal Fe.
For dies, gauges, broaches, shears, reamers; wear resistant, oil hardened.

MOLEX-O
Associated Steel Corp.
C, alloy, bal Fe.
Annealed: 202 Brin. For blanking and drawing dies, shear blades; oil hardened, non-deforming.

MOLFOR-35
Forjas Alavesas S.A.
C 0.35, Mn 0.4, Si 1, Cr 5, Mo 1.3, V 0.4, bal Fe.
Hot work tool steel; die casting dies, forging dies and inserts. IHA F-537; similar to AISI H13.

MOLFOR-40
Forjas Alavesas S.A.
C 0.3, Mn 0.5, Si 0-1, Cr 3, Mo 2.8, V 0.5, bal Fe.
Hot work tool steel; pressure casting molds, punches, extrusion cylinders. DIN XC 32 CrMo 33.

MOLIBLOC
Creusot-Loire
C 0.35, Mn 0.8, Si 0.5, Cr 1.6, Mo 0.4, bal Fe.
Oil hardening tool steel for molds. AFNOR 34 CD 7; similar to AISI P20.

MOLIN METAL
American Art Alloys Inc
Al, bal Cu.
For molds, dies; Al-bronze.

MOLIN METAL
Dirilyte Co. of America
Al, bal Cu.
Cast. For gears, shafts, bearings; aluminum bronze.

MOLITE
Columbia Tool Steel Co.
C 0.8, Cr 4, Mo 5, W 6.5, V 1.9, bal Fe.
For tools, cutters, taps, reamers, broaches; high speed steel.
Obsolete

MOLITE 2
Columbia Tool Steel Co.
Tool material. C 0.86, W 6.25, Mo 5, Cr 4.15, V 1.9, bal Fe.
For milling cutters, twist drills, broaches, reamers; AISI M2; high speed steel.

MOLITE 2 SMOOTHCUT
Columbia Tool Steel Co.
Tool material. C 0.86, W 6.25, Mo 5, Cr 4.15, V 1.9, bal Fe, plus free machining additions.
For gear cutters, twist drills, reamers, broaches; AISI M2; high speed tool steel.

MOLITE 3
Columbia Tool Steel Co.
Tool material. C 1.03, Mo 6.25, W 6.25, Cr 4, V 2.5, bal Fe.
Mo-W high speed tool steel; for broaches, taps, form tools, twist drills, blanking dies. AISI M3 Class 1.

MOLITE 3 SMOOTHCUT TYPE 1
Columbia Tool Steel Co.
Tool material. C 1.03, W 6.25, Mo 6.25, Cr 4, V 2.5, sulfides, bal Fe.
Hardened: 64-66 Rock C. For broaches, taps, dies, reamers; form tools, hobs, slitting saws. High speed steel; high red hardness. Good machinability rating. AISI M3 Class 1.

MOLITE 3 TYPE 2
Columbia Tool Steel Co.
Tool material. C 1.2, W 6.25, Mo 6.25, Cr 4, V 3, bal Fe.
Hardened: 64-66 Rock C. For broaches, form tools, taps, dies, milling cutters, cutoff tools. High speed steel; high red hardness. AISI Type M3 Class 2.

MOLITE 4
Columbia Tool Steel Co.
Tool material. C 1.28, W 5.5, Mo 4.5, Cr 4.5, V 4, bal Fe.
Hardened: 64-66 Rock C. For broaches, counter bores, drills, reamers, milling cutters. High speed steel; high red hardness. AISI Type M4.

MOLITE 42
Columbia Tool Steel Co.
Tool material. C 1.06, W 1.6, Mo 9.5, Cr 3.75, V 1.15, Co 8, bal Fe.
Mo-Co high speed tool steel. For broaches, shaving tools, lathe tools. AISI M42.

MOLITE 5
Columbia Tool Steel Co.
C, alloy, bal Fe.
For tools, dies; oil hardening. *Obsolete*

MOLITE 8
Columbia Tool Steel Co.
C, alloy, bal Fe.
For tools, dies; oil hardening. *Obsolete*

MOLITE 9
Columbia Tool Steel Co.
C, alloy, bal Fe.
For tools, dies; oil hardening. *Obsolete*

MOLITE HW10
Columbia Tool Steel Co.
C 0.85, Cr 4, V 2, Mo 8, bal Fe.
For tools, cutters, reamers, drills; high speed steel, oil hardened. *Obsolete*

MOLITE HW60
Columbia Tool Steel Co.
C 0.63, Mo 8.25, Cr 4, V 1.9, bal Fe.
Heat treated: 202,000 TS; 107,000 YS; 2 El; O 50 Rock. For hot upsetting and forming dies, hot dummy blocks, hot or cold header dies. Hot work steel Type H 43, shock resistant. *Obsolete*

MOLITE M-1
Columbia Tool Steel Co.
C, Cr, Mo, V, bal Fe.
For cutting tools; high speed steel. *Obsolete*

MOLITE M2
Columbia Tool Steel Co.
C 0.7, Cr 4, Mo 5, W 6.5, bal Fe.
For lathe and planer tools, drills, reamers, taps; high speed steel. *Obsolete*

MOLITE SMOOTHCUT

Columbia Tool Steel Co.
C 0.8, Cr 4, V 2, W 6, Mo 5, bal Fe.
For lathe and planer tools, drills, taps, hobs, reamers; Type M2; high speed steel. *Obsolete*

MOLMANG

Stulz Sickles Steel Co.
Mn 11-13.5, C, bal Fe.
Welded: 500-600 Brin. For welding electrodes for build-up work; austenitic, work hardening.

MOLTROP

Manufacturer not listed.
C 0.3-0.5, bal Fe.
For gears, shafts; water hardened.

MOLVA C

Wallace Murray Corp.
C 1.02, Cr 3.75, W 1.75, Mo 8.5, V 1.9, bal Fe.
Molybdenum high speed steel for drills, reamers, lathe tools. AISI M7.

MOLVA-T

Wallace Murray Corp.
C 0.8, Cr 4, V 2, Mo 5, W 6, bal Fe.
For lathe and planer tools, drills, form cutters, hot punches; high speed steel.

MOLY 8

Pennsylvania Steel Corp.
C 0.6, Cr 3.6, V 1.75, Mo 8.5, bal Fe.
For hot work tools, shears, punches; Type H12; hot work steel.

MOLY ARK

Jessop-Saville Ltd.
C 0.7, W 5.5, Mo 4, Cr 4, V 1.5, bal Fe.
For tools, cutters; high speed steel.

MOLY ASCOLOY

Usines de A. Manoir, Pitres (Eure)
C 0.08, Cr 13, Mo 2, bal Fe.
For jet engine parts, afterburners, nozzles, bolts; high strength and high heat resistance.

MOLY ASTROLOY

Now CARPENTER MOLY ASTROLOY.

MOLY B 100

Plansee, Metallwerk Gesellschaft
Mo 50, W 50.
For high vacuum tubes. Sintered alloy.

MOLY HIGH SPEED

McInnes Steel Co.
C 0.8, Cr 4, Mo 4.2, V 1.5, W 5.5, bal Fe.
For lathe and planer tools, hobs, broaches; Type M1; high speed steel. *Obsolete*

MOLY PERMALLOY

Allegheny Ludlum Steel
Ni 79, Mo 4, Fe 17.
Annealed: 64,000 TS; 18,500 YS; 27 El; 116 Brin. Rolled: 160,000 TS; 150,000 YS; 1 El; 280 Brin. For toroid, tape and lamination cores for transformers and relays; magnetically soft, high permeability. *Obsolete*

MOLY PERMINVAR-3

Bell Telephone Laboratories
Fe 33.7, Ni 34, Co 29, Mo 3, Mn 0.3.
For magnetic amplifiers.

MOLY TELASTIC

Falk Corp.
C 0.3-0.4, Mn 0.7-1, Si 0-0.6, Cr 0.4-0.6, Mo 0.15-0.25, bal Fe.
Low alloy cast steel; hardenable to 300-340 Brin in 1 in. diameter size. Weldable, tough, for gears, construction machinery, pressure vessels.

MOLY-IRON

Weatherly Casting & Machine Co.

ow DIAMITE MO.

MOLY-MANG

Westinghouse Electric Corp.
Mn 11-13, C, bal Fe.
For welding rod for high manganese steel; austenitic, wear and abrasion resistant. *Obsolete*

MOLY-NICKEL

Chicago Hardware Foundry Co.
C, Mo, Ni, bal Fe.
For welding rod.

MOLYBDENITE

Continental Foundry & Machine Co.
C, Cr, Mo, bal Fe.
For mill pinions, guides, rolls.

MOLYBDENUM

Atomergic Chemetals Corp.
Mo.
Purities: zone and chemical refined 99.995% 99.95%. Forms: powder, rod, wire, sheet, tubing, sintered, wrought bar, foil, arc castings, single crystals.

MOLYBDENUM CHISEL

A. Milne & Co.
C 0.6, Mo 0.5, bal Fe.
For chisels, tools; oil hardening.

MOLYBDENUM METAL

Foote Mineral Co.
100% and 97% Mo.
For special metallurgical applications. *Obsolete*

MOLYBDENUM PERMALLOY

Magnetic Metals Co.
Ni 80, Fe 16, Mo 4.
Soft magnetic alloy.

MOLYBDENUM PERMALLOY 2-81

ITT Components Group Europe
Nickel. Mo 2, Ni 81, Fe 17.
For inductance and loading coils, low magnetic core loss, high permeability.

MOLYBDENUM PERMALLOY 2-81

Bell Telephone Laboratories
Mo 2, Ni 81, bal Fe.
For inductance and loading coils; low magnetic core loss.

MOLYBDENUM PERMALLOY 2-81

Western Electric Co.
Mo 2, Ni 81, bal Fe.
For inductance and loading coils; low magnetic core loss.

MOLYBDENUM PERMALLOY, 4-79

Western Electric Co.
Mo 4, Ni 79, Fe 17.
For high frequency and electrical apparatus; high permeability and resistivity.

MOLYBDENUM PERMALLOY, 4-79

Bell Telephone Laboratories
Mo 4, Ni 79, Fe 17.
For high frequency and electrical apparatus; high permeability and resistivity.

MOLYBDENUM SELF-HARDENING

Universal Cyclops
C, alloy, bal Fe.
For tools, dies. *Obsolete*

MOLYBDENUM SILICON

Textron Inc.
C, Mo, Si, bal Fe.
Heat treated: 100,000-110,000 TS; 197-228 Brin. For pistons, brake drums; wear resistant.

MOLYBDENUM STEEL

English manufacture
Mo 0.15-0.4, C 0.1-0.45, Mn 0.6-1.3, Si 0.3-0.5, bal Fe.
For machinery parts, gears, shafts; tough.

MOLYBDENUM-0.05 ZR

AMAX Corp.
Zr 0.05, bal Mo.
Rolled rod: 127,000 TS; 108,400 YS; 20 El; 60 RA.
Recrystallized sheet: 78,800 TS; 63,500 YS; 40 El. Stress relieved sheet: 126,000 TS; 115,800 YS; 13 El. For high temperature applications, heat engines, nuclear reactors, glass molding dies. High heat and corrosion resistance. *Obsolete*

MOLYBDENUM-50W

Plansee, Metallwerk Gesellschaft
Refractory. C 0.008, W 49.3, bal Mo.
Stress Relieved: 144,000 TS; 133,800 YS; 14 El.
Recrystallized: 97,700 TS; 0 El. For high temperature applications and heat engines. High heat and corrosion resistance.

MOLYBDIE

A. Finkl & Sons Co.
C 0.3-0.6, Mn 0.5-0.8, Cr 0.8-1.1, Ni 1.2-1.7, Mo 0.2-0.9, bal Fe.
106,500 TS; 86,500 YS; 21 El; 64 RA; 208 Brin. For gears, dies, heavy forged parts, crankshafts, piston rods; tough. *Obsolete*

MOLYBDIE GRADE FG

A. Finkl & Sons Co.
High C, alloy, bal Fe.
For coining and straightening dies; high resistance to abrasion. *Obsolete*

MOLYBDIE GRADE FM

A. Finkl & Sons Co.
C, Cr, Ni, Mo, bal Fe.
For dies adapted for the limited production of all types of drop forgings; resistant to heat abrasion. *Obsolete*

MOLYBDIE GRADE FS

A. Finkl & Sons Co.
C 0.55, Mn 0.7, Cr 0.9, Ni 1.5, Mo 0.3, bal Fe.
For die blocks for the production of large drop forgings of heavy sections; wear resistant. *Obsolete*

MOLYBDIE GRADE FX

A. Finkl & Sons Co.
C 0.55, Mn 0.7, Cr 0.9, Ni 1.5, Mo 0.3, bal Fe.
269-429 Brin. For tempered die blocks, inserts, sow blocks, forging machine dies, half roll dies, dies for drop forgings; resistant to heat. *Obsolete*

MOLYBDIE GRADE WG

A. Finkl & Sons Co.
C, Cr, Ni, Mo, bal Fe.
For cutlery, dies, dies for shallow impression forgings; will not distort on hardening. *Obsolete*

MOLYBDIE NI-MO

A. Finkl & Sons Co.
C, Cr, Ni, Mo, bal Fe.
For piston rods and heads; high impact and fatigue resistance. *Obsolete*

MOLYBDIE TYPE C

A. Finkl & Sons Co.
C 0.4, Mn 0.6, Cr 0.85, Ni 1.5, Mo 0.3, bal Fe.
Heat treated: 130,000 TS; 100,000 YS; 18 El; 50 RA; 285 Brin.
For machine parts, piston rods, rams; shock resistant. *Obsolete*

MOLYBDIE TYPE R
A. Finkl & Sons Co.
C 0.31, Cr 0.75, Ni 1.5, Mo 0.3, bal Fe.
Heat treated: 130,000 TS; 100,000 YS; 18 El; 50 RA; 285 Brin.
For machine parts, shafting, gears; shock resistant. *Obsolete*

MOLYBESCO 1
Vallourec S.A.
C 0.16, Cr 0.3, Mo 0.3, bal Fe.
Good tensile strength, good creep properties up to 550°C.
For boilers, super heaters.

MOLYBESCO 2
Vallourec S.A.
C 0.1, Cr 0.15, Mo 0.55, bal Fe.
High tensile strength; for heat exchangers, condensers, furnace tubes.

MOLYBESCO 4
Vallourec S.A.
C 0.1, Cr 0.15, Mo 0.55, bal Fe.
High tensile strength; for heat exchangers, condensers, furnace tubes.

MOLYBESCO 5
Vallourec S.A.
C 0.12, Cr 0.15, Mo 0.55, bal Fe.
High tensile strength; for heat exchangers, condensers, furnace tubes.

MOLYCUT 562
Firth Brown Ltd.
C 0.82, Cr 4.1, Mo 5, V 1.9, W 6.4, bal Fe.
For lathe and planer tools, drills, broaches, taps; high speed steel. *Obsolete*

MOLYITE
Columbia Tool Steel Co.
Mo 9, Cr 4, V 2, bal Fe.
For tools, cutters; high speed steel. *Obsolete*

MOLYMET
Wilbur B. Driver Co.
Ni 45, Fe 45, Mo 10.
For radio grid wire. *Obsolete*

MOLYNEAUX FUSIBLE ALLOY
French manufacture
Bi 41.5, Sn 16.7, Pb 25, Cd 16.7.
For fuses, fire extinguishers; melting point 60°C.

MOMARC
Teledyne Vasco
Now MOMARC VM.

MOMARC VM
Teledyne Vasco
C 1, Cr 1.4, Mo 1, bal Fe.
Normalized: 185,000 TS; 139,000 YS; 13 El; 360 Brin. For ball bearings, special rolls, precision gauges, arbors, dies. Wear and abrasion resistant, oil or water hardening.

MONACA
Denman & Davis Co.
C 0.5, bal Fe.
For gears and shafts. Water hardening. *Obsolete*

MONACA DRILL ROD
Teledyne Pittsburgh Tool Steel
C 1.18-1.28, bal Fe.
For small tools and parts, drills; water hardening. *Obsolete*

MONACA SPECIAL
Teledyne Pittsburgh Tool Steel
C 0.7-0.9, V 0.2, bal Fe.
For tools, drills, taps, reamers, hobs, broaches; Type W2; water hardened. *Obsolete*

MONAR 816
Arcos Alloys
Now ARCOS 816.

MONARCH
Hardenite Steel Co. Ltd.
C, alloy, bal Fe.
For tools, dies. Oil hardening.

MONARCH 2
Atlas Steels Ltd.
C 0.5, Mn 0.7, Si 1.8-2, Mo 0.25, bal Fe.
For tools, chisels; tough. *Obsolete*

MONARCH METAL
Monarch Alloy Co.
Pb 20, Sn 5, bal Cu.
For bearings; leaded.

MONARCH NO. 10
Monarch Alloy Co.
Pb, bal Cu.
Cast: 30,000 TS; 22,000 YS; 60 Brin. For bearings for pumps, electric motors and machine tools; for high speed and moderate load.

MONARCH NO. 12
Monarch Alloy Co.
Pb, bal Cu.
Cast: 30,900 TS; 25,100 YS; 70 Brin. For bearings for large presses, nail and wire machines; for high speed and moderate, heavy loads.

MONARCH NO. 16
Monarch Alloy Co.
Pb, bal Cu.
Cast: 37,000 TS; 29,800 YS; 75 Brin. For bearings for large presses, cranes; heavy duty at low speeds.

MONARK
Atlas Specialty Steels
C 0.6, Si 2, Cr 0.3, Mo 0.2, bal Fe.
For tools, chisels; shock resisting.

MONARK-1
Atlas Specialty Steels
C 0.5, Mn 0.4, Si 1.2, Mo 0.5, bal Fe.
For chisels, pneumatic tools; water hardened.

MONARK-2
Atlas Specialty Steels
C 0.6, Mn 0.75, Si 2, Cr 0.3, Mo 0.2, bal Fe.
For chisels, pneumatic tools; water hardened.

MOND
Mond Nickel Co. Ltd.
Cu 26, Ni 70, Mn 4.
For turbine blades, pump rods, valve stems; corrosion resistant.

MOND "70"
Mond Nickel Co. Ltd.
Cu 26, Ni 70, Mn 4.
85,000-94,000 TS; 36,000-40,400 YS; 45-65 El. For turbine blades, pumps, rods, valve stems, valve seats. *Obsolete*

MOND NICKEL
Mond Nickel Co. Ltd.
Ni 70, Mn 4, Cu 26.
For resistance wires, chemical and mining machinery. *Obsolete*

MONEL (1)
Ancast, Inc.
Nickel. C 0-0.35, Mn 0-1.5, Si 0-1.25, Fe 0-3.5, Cu 26-33, P 0-0.03, S 0-0.03, bal Ni.
ASTM A-494 GR M-35-1(a), ASTM A-743 GR M-35-1.

MONEL (2)
Ancast, Inc.
Nickel. C 0-0.35, Mn 0-1.5, Si 0-2, Fe 0-3.5, Cu 26-33, P 0-0.03, S 0-0.03, bal Ni.
ASTM A-494 GR M-35-2.

MONEL 130
Gilby-Fodor S.A.
Cu 28-34, bal Ni.
For corrosion resistant parts. Corrosion resistant.

MONEL 187
Inco Alloys International Inc.
Now MONEL ALLOY 187. *Obsolete*

MONEL 190
Inco Alloys International Inc.
Now MONEL ALLOY 190. *Obsolete*

MONEL 40
Inco Alloys International Inc.
Now MONEL FILLER METAL 40. *Obsolete*

MONEL 400
Criterion Metals, Inc.
Nickel. Ni 66, C 0.12, Mn 0.9, Fe 1.35, S 0.005, Si 0.15, Cu 31.5.
Thin gauge sheet, various tempers. Soft: 70-85 ksi TS; 25-45 ksi YS. Hard: 100 ksi TS min; 90 ksi YS min. ASTM B-127.

MONEL 400
Enpar Sonderwerkstoffe GmbH
Nickel.
Alternate manufacturer.

MONEL 401
Criterion Metals, Inc.
Nickel. Ni 44, C 0.03, Mn 1.7, Fe 0.2, S 0.005, Si 0.01, Cu 53, Co 0.5.
Thin gauge sheet, various tempers. Soft: 70-85 ksi TS; 25-45 ksi YS. Hard: 100 ksi TS min; 90 ksi YS min. ASTM B-127.

MONEL 402
Huntington Alloys Inc.
C 0.12, Mn 0.9, Fe 1.2, Cu 39.8, bal Ni.
Hot rolled: 90,000 TS; 65,000 YS; 35 El; 175 Brin. Cold drawn: 95,000 TS; 85,000 YS; 25 El; 205 Brin. For corrosion resistant parts, chains, hooks, grates for H 2 pickling; corrosion resistant, resists hydrogen embrittlement. *Obsolete*

MONEL 403
Huntington Alloys Inc.
Ni 55-60, Fe 0-2.5, Mn 0-2.2, Si 0-0.75, bal Cu.
Annealed: 65,000-85,000 TS; 23,000-45,000 YS; 35-50 El; 130-140 Brin. Hot rolled: 70,000-90,000 TS; 35,000-60,000 YS; 20-40 El; 130-170 Brin. For pickling equipment for H 2 SO 4, cable sheathing, high corrosion resistance. *Obsolete*

MONEL 404
Criterion Metals, Inc.
Nickel. Ni 55, C 0.06, Mn 0.01, Fe 0.05, S 0.005, Si 0.02, Cu 44, Al 0.02.
Thin gauge sheet, various tempers. Soft: 70-85 ksi TS; 25-45 ksi YS. Hard: 100 ksi TS min; 90 ksi YS min.

MONEL 60
Inco Alloys International Inc.
Now MONEL FILLER METAL 60. *Obsolete*

MONEL ALLOY 187
Inco Alloys International Inc.
C 0.02, Ni 32, Mn 2, Si 0.15, Cu 65.
Welding electrode for shielded metal-arc welding 70/30, 80/20 and 90/10 Cu-Ni alloys and the clad side of Cu Ni clad steels. AWS A5.6 (Class E Cu Ni); ASME SFB 5.6. *Obsolete*

MONEL ALLOY 190

Inco Alloys International Inc.
C 0.01, Ni 65, Mn 3.1, Fe 0.3, Si 0.75, S 0.007, Cu 30.5, Al 0.15, Ti 0.55.
Welding electrode for shielded metal-arc welding Monel alloy 400 to itself and to steel; for overlaying on steel. AWS A5.11 (Class E Cu Ni-2); ASME SFB 5.11. *Obsolete*

MONEL ALLOY 400

Inco Alloys International Inc.
C 0.12, Ni 65, Cu 32, Fe 1.5, Mn 1.
As cast: 550/mm^2 TS; 200 N/mm^2 YS; 40 El; 12-90 VHN.
Good corrosion resistance for valves, pumps, propeller shafts and marine hardware. BS 3072-76; NA13; ASTM B127; ASME SB 127; ASM 4544, etc.

MONEL ALLOY 400

Teledyne Rodney Metals
Ni 66.5, C 0.15, Fe 1.25, S 0.012, Si 0.25, Cu 31.5.
Annealed: 70,000-90,000 TS; 25,000-50,000 YS; 60-35 El; 110-149 Brin. Cold drawn, stress relieved: 84,000-120,000 TS; 55,000-100,000 YS; 40-22 El 160-225 Brin. For valves and pumps, marine fixtures, tanks, piping, heat exchangers; corrosion resistant.

MONEL ALLOY 400

Huntington Alloys Inc.
Ni 66.5, C 0.15, Mn 1, Fe 1.25, S 0.012, Si 0.25, Cu 31.5.
Annealed: 70,000-90,000 TS; 25,000-50,000 YS; 60-35 El; 110-149 Brin Cold drawn, stress relieved: 84,000-120,000 TS; 55,000-100,000 YS; 40-22 El 160-225 Brin. For valves and pumps, marine fixtures, tanks, piping, heat exchangers; corrosion resistant.

MONEL ALLOY 400

Titanium Metals Corp.
Ni 66.5, C 0.15, Mn 1, Fe 1.25, S 0.012, Si 0.25, Cu 31.5.
Annealed: 70,000-90,000 TS; 25,000-50,000 YS; 60-35 El; 110-149 Brin Cold drawn, stress relieved: 84,000-120,000 TS; 55,000-100,000 YS; 40-22 El 160-225 Brin. For valves and pumps, marine fixtures, tanks, piping, heat exchangers; corrosion resistant.

MONEL ALLOY 401

Inco Alloys International Inc.
Copper. Ni 40-45, Mn 0-2.25, Fe 0-0.75, Si 0-0.25, C 0-0.1, S 0-0.015, bal Cu.
Annealed: 64,000 TS; 19,500 YS; 51 El. Copper-nickel alloy designed for specialized electrical and electronic applications. 4401

MONEL ALLOY 401

Henry Wiggin & Co. Ltd.
Ni 42.5, C 0.05, Mn 1.13, Fe 0.38, S 0.008, Si 0.13, bal Cu.
For specialized electrical and electronic applications, wire wound resistors; bimetal contacts.

MONEL ALLOY 401

Huntington Alloys Inc.
Ni 42.5, C 0.05, Mn 1.13, Fe 0.38, S 0.008, Si 0.13, bal Cu.
For specialized electrical and electronic applications, wire wound resistors; bimetal contacts.

MONEL ALLOY 404

Inco Alloys International Inc.
Ni 54.5, C 0.08, Mn 0.05, Fe 0.25, S 0.012, Si 0.05, Cu 44, Al 0.03.
For wave guides, metal to ceramic seals, transistor capsules, power tubes. Low magnetic permeability; excellent brazing characteristics.

MONEL ALLOY 450

Inco Alloys International Inc.
Copper. Ni 29-33, Fe 0.4-1, Mn 0-1, Zn 0-1, Pb 0-0.05, P 0-0.02, S 0-0.02, bal Cu.
Annealed: 56,000 TS; 24,000 YS; 46 El. Copper-nickel alloy; for seawater condensers, condenser plates, distiller tubes, evaporator and heat exchanger tubes, and salt-water piping. C71500

MONEL ALLOY 474

International Nickel Inc.
Ni 54, Cu 46, C 0.01, Fe 0.01, S 0.001.
At room temperature: 60,000 TS; 25,000 YS. At 800 F: 35,000 TS; 16,000 YS. For wave guides, transistor capsules, electronic power tubes, metal-to-ceramic seals. Corrosion and heat resistant. *Obsolete*

MONEL ALLOY 502

Inco Alloys International Inc.
Ni 66.5, C 0.05, Mn 0.75, Fe 1, S 0.005, Si 0.25, Cu 28, Al 3, Ti 0.25.
Cold drawn: 87,000 TS; 55,000 YS; 42 El; 158 Brin. Cold drawn and aged: 140,000 TS; 94,000 YS; 25 El; 255 Brin. For fasteners, pump and propeller shafts, valve stems; machinable; age hardenable. *Obsolete*

MONEL ALLOY K-500

Inco Alloys International Inc.
Nickel. Cu 27-33, Al 2.3-3.15, Ti 0.35-0.85, Fe 0-2, C 0-0.25, Mn 0-1.5, S 0-0.01, Si 0-0.5, 63.0 Ni min.
Hardened: 160,000 TS; 115,000 YS; 20 El. Low permeability; non-magnetic. For pump shafts, oil-well tools and instruments, doctor blades and scrapers, springs, valve trim, fasteners, and marine propeller shafts. 5500

MONEL ALLOY K-500

Henry Wiggin & Co. Ltd.
Ni 66.5, Cu 0.13, Mn 0.75, Fe 1, S 0.005, Cu 29.5, Al 3, Ti 0.63.
Annealed: 90,000-110,000 TS; 40,000-60,000 YS; 45-25 El; 140-185 Brin. Annealed and aged: 130,000-165,000 TS; 85,000-120,000 YS; 35-20 El; 250-315 Brin. For pump shafts and impellers, oil well drill collars and instruments, electronic components; corrosion resistant.

MONEL ALLOY K-500

Huntington Alloys Inc.
Ni 66.5, Cu 0.13, Mn 0.75, Fe 1, S 0.005, Cu 29.5, Al 3, Ti 0.63.
Annealed: 90,000-110,000 TS; 40,000-60,000 YS; 45-25 El; 140-185 Brin. Annealed and aged: 130,000-165,000 TS; 85,000-120,000 YS; 35-20 El; 250-315 Brin. For pump shafts and impellers, oil well drill collars and instruments, electronic components; corrosion resistant.

MONEL ALLOY R-405

Inco Alloys International Inc.
Nickel. Cu 28-34, Fe 0-2.5, S 0.025-0.06, Mn 0-2, C 0-0.3, Si 0-0.5, 63.0 Ni min.
Annealed: 80,000 TS; 35,000 YS; 40 El. Free-machining version of MONEL alloy 400. Used for meter and valve parts, fasteners, and screw-machine products. 4405

MONEL ALLOY R-405

Huntington Alloys Inc.
Ni 66.5, C 0.15, Mn 1, Fe 1.25, S 0.043, Si 0.25, Cu 31.5.
Annealed: 70,000-85,000 TS; 25,000-40,000 YS; 50-35 El; 110-140 Brin. Cold drawn: 85,000-115,000 TS; 50,000-105,000 YS; 35-15 El; 160-245 Brin. For water meter parts, screw machine products, valve seat inserts, fasteners for nuclear applications. Corrosion resistant, free machining.

MONEL ALLOY R-405

Henry Wiggin & Co. Ltd.
Ni 66.5, C 0.15, Mn 1, Fe 1.25, S 0.043, Si 0.25, Cu 31.5.
Annealed: 70,000-85,000 TS; 25,000-40,000 YS; 50-35 El; 110-140 Brin. Cold drawn: 85,000-115,000 TS; 50,000-105,000 YS; 35-15 El; 160-245 Brin. For water meter parts, screw machine products, valve seat inserts, fasteners for nuclear applications. Corrosion resistant, free machining.

MONEL C

Henry Wiggin & Co. Ltd.
C 0.2, Fe 0-2.5, Mn 0-2, Ni 63-70, bal Cu.
Cold drawn: 100,000 TS; 75,000 YS; 25 El; 60 RA; 200 Brin.
For water meter components, machined parts; corrosion resistant. *Obsolete*

MONEL FILLER METAL 40

Inco Alloys International Inc.
Ni 66, C 0.1, Mn 0.9, Fe 1.35, S 0.005, Si 0.15, Cu 31.5.
For oxyacetylene welding of Monel alloys 400 and 404. AWS A 5.14 Class RNiCu-5; ASME SFB 5.14. *Obsolete*

MONEL FILLER METAL 60

Inco Alloys International Inc.
Ni 65, C 0.03, Mn 3.5, Fe 0.2, S 0.005, Si 1, Cu 27, Ti 2.2.
For gas-shielded arc welding of Monel alloys 400 and 404.
AWS A 5.14 Class ERNiCu-7; ASME SFB 5.14.

MONEL FILLER METAL 64

Huntington Alloys Inc.
Ni 65, C 0.15, Mn 0.6, Fe 1, S 0.005, Si 0.15, Cu 30, Al 2.8, Ti 0.5.
For oxyacetylene and gas-shielded arc welding of Monel alloy K-500. AWS A 5.14 Class ERNiCu-8; MIL-E-21562. *Obsolete*

MONEL FILLER METAL 67

Inco Alloys International Inc.
Ni 31, C 0.02, Mn 0.75, Fe 0.5, S 0.005, Si 0.1, Cu 67.5, Ti 0.3.
For oxyacetylene and gas-shielded arc welding of 70/30, 80/20, 90/10, copper-nickel alloys and clad side of copper-nickel clad steel. AWS A 5.6 Class ECuNi; ASME SFB 5.6.

MONEL G

Henry Wiggin & Co. Ltd.
C 0.3, Fe 0-2.5, Mn 0-2, Ni 63-70, bal Cu.
Cold drawn: 100,000 TS; 75,000 YS; 25 El; 60 RA; 200 Brin.
For water meter components, machines parts; good corrosion resistance. *Obsolete*

MONEL K

Gilby-Fodor S.A.
Cu 30, Al 2, bal Ni.
For heat resistant parts. Corrosion and heat resistant.

MONEL K 500

Enpar Sonderwerkstoffe GmbH
Nickel.
Alternate manufacturer.

MONEL K-500

Criterion Metals, Inc.
Nickel. Ni 65, C 0.15, Mn 0.6, Fe 1, S 0.005, Si 0.15, Cu 29.5, Al 2.8, Ti 0.5.
Thin gauge sheet, various tempers. Soft: 75-85 Rock B. Hard: 15-32 Rock C.

MONEL K-502

Techalloy Co. Inc.
C 0.1, Ni 65, Mn 1, Fe 1, Si 0.5, Cu 29.5, Al 2.8, Ti 0.5.
Age hardenable, corrosion resistant alloy for gyroscope components, small machined parts. QQ-N-286A.

MONEL K44

Gilby-Fodor S.A.
Cu 30, Mn 1, Al 2, Fe 2, bal Ni.
For corrosion and heat resistant parts. Corrosion and heat resistant.

MONEL K64

Gilby-Fodor S.A.
Cu 30, Mn 1, Al 2, Fe 2, bal Ni.
For corrosion and heat resistant parts. Corrosion and heat resistant.

MONEL R405

Driver Harris Co.
Nickel. Cu 30, bal Ni.
Machinable grade wire and rod. *Obsolete*

MONEL WELDING ELECTRODE 134

Huntington Alloys Inc.
Ni 64, C 0.2, Mn 2.5, Fe 1, S 0.005, Si 0.3, Cu 30, Al 1.8, Ti 0.75.
Electrode for shielded metal-arc welding of Monel alloy K-500; overlaying on steel. AWS A 5.11 Class ENiCuAl-1; ASME SFB 5.11. *Obsolete*

MONEL WELDING ELECTRODE 187

Inco Alloys International Inc.
Ni 32, C 0.02, Mn 2, Fe 0.6, S 0.01, Si 0.15, Cu 65.
Electrode for shielded metal-arc welding of 70/30, 80/20, 90/10 copper-nickel alloys and clad side copper-nickel clad steel. AWS A 5.6 Class E CuNi; MIL-E-22200/4.

MONEL WELDING ELECTRODE 190

Inco Alloys International Inc.
Ni 65, C 0.01, Mn 3.1, Fe 0.3, S 0.007, Si 0.75, Ti 0.55, Cu 30.5, Al 0.15.
Electrode for shielded metal-arc welding of Monel alloys 400 and 404; overlaying on steel. AWS A 5.11 Class E NiCu-2; ASME SFB 5.11.

MONEL-140

International Nickel Inc.
Fe 0-3.5, C 0-13, 23% Cu min, 60% Ni min.
For welding electrodes, Monel overlay; corrosion resistant. *Obsolete*

MONEND

Now MONEND 806.

MONEND 806

Arcos Alloys
Now ARCOS 9N10.

MONEVAL 806

Arcos Alloys
Ni 64-68, Cu 27-30, Fe 0-1.
Welded: 75,500 TS; 42,500 YS; 42 El; 60 RA. For welding electrodes for Monel; Monel type. *Obsolete*

MONIK-30

Forjas Alavesas S.A.
C 0.32, Mn 0.4, Si 0-0.5, Ni 4.2, Cr 0.5, Mo 1.1, V 0.15, bal Fe.
Hot work tool steel; mandrels for hot fabrication of tube. DIN 28 Ni Mo 179

MONIKROM

Midland Motor Cylinder Co.
C 3.1-3.4, Si 2-2.4, Mn 0.7, Cr 0.8-1, Ni 0.15-0.25, Mo 0.15-0.25, bal Fe.
Cast: 43,000 TS; 240-280 Brin. For camshafts, cam sleeves; alloy cast iron; 80,000 transverse strength.

MONIMAX

Allegheny Ludlum Steel
Ni 47, Mo 3, bal Fe.
Magnetically soft; saturation inductance 14,500. For electrical equipment, special transformers, tape recording heads, vibrator cores, small motors. *Obsolete*

MONIX

Saarstahl AG
C 0.16, Si 0.25, Mn 0.75, Cr 0.6, Mo 0.15, Ni 1.15, bal Fe.
Alloy carburizing steel. For highly stressed case hardened parts. *Obsolete*

MONIX 10

Saarstahl AG
C 0.37, Si 0.25, Mn 0.65, Cr 1.05, Mo 0.2, Ni 1.05, bal Fe.
Alloy steel hardenable to 900-1050 N/mm² TS. For automotive parts. W.-Nr. 6511. *Obsolete*

MONIX 15

Saarstahl AG
Now SAARSTAHL 34 CRNIMO 6.

MONIX 2

Saarstahl AG
Now SAARSTAHL 30 CRNIMO 8.

MONIX 20

Rochling Burbach GmbH
C 0.3, Cr 2, Mo 0.3, Ni 2, bal Fe.
For gears, bolts, oil refinery equipment; oil hardened, tough. *Obsolete*

MONIX 3

Saarstahl AG
C 0.35, Si 0.25, Mn 0.4, Cr 1.35, Mo 0.25, Ni 3.55, bal Fe.
Deep hardening alloy steel hardenable to 1300-1500 N/mm² TS. For highly stressed heavy parts. *Obsolete*

MONIX 3 K

Saarstahl AG
Now SAARSTAHL 20 NICRMO 145.

MONIX 30

Saarstahl AG
Now SAARSTAHL 30 NICRMO 22.

MONIX 4

Saarstahl AG
Now SAARSTAHL 40 NICRMO 186.

MONIX 40

Saarstahl AG
C 0.4, Si 0.3, Mn 0.9, Cr 0.5, Mo 0.2, Ni 0.55, bal Fe.
Alloy steel hardenable to 950-1150 N/mm² TS for automotive parts and equipment. W.-Nr. 6546; AISI 8640. *Obsolete*

MONIX E

Saarstahl AG
Now SAARSTAHL 21 NICRMO 2.

MONIX F

Saarstahl AG
Now SAARSTAHL 17 CRNIMO 6.

MONIX H

Saarstahl AG
C 0.17, Si 0.25, Mn 0.8, Cr 0.9, Mo 0.2, Ni 1.45, bal Fe.
Alloy carburizing steel for highly stressed case hardened parts. W.-Nr. 6566; similar to AISI 4320. *Obsolete*

MONO-LOY

Grand Northern Products Ltd.
C 1.4, Si 0.9, Mn 31, Cr 0.5, Mo 0.5, Ni 5.5, bal Fe.
Hard facing electrode. For overlaying or buildup on crusher rolls, dredge buckets, tractor sprockets, hammers. Good shock resistance.

MONOTYPE STANDARD

English manufacture
Pb 74, Sb 18, Sn 8.
For type metal; slightly expansive on solidifying.

MONOWELD

Now ALL-STATE MONOWELD.

MONOX

George W. Prentiss & Co.
Cu 23, bal Ni.
For instruments. *Obsolete*

MONTAN

SWB Stahlformguss Gesellschaft mbH
C 0.7, Si 0-0.25, Mn 0-0.35, bal Fe.
For rails, tools, springs, axes; water hardened. *Obsolete*

MONTAN 70

SWB Stahlformguss Gesellschaft mbH
C 0.7, Si 0-0.25, Mn 0-0.25, bal Fe.
Heat treated: 175,000 TS; 130,000 YS; 12 El; 37 RA; 355 Brin. For rails, punches, springs, axles, crimpers; water hardened. *Obsolete*

MONTANIUM

English manufacture
Cu 2.5-3.5, Mg 0.5, bal Al.
51,000 TS; 16 El; 50 RA. For light alloy parts; similar to "Duralumin."

MONTEGAL

Metallgesellschaft Reuterweg
Mg 0.95, Si 0.8, Ca 0.2, bal Al.
50,000 TS; 6 El; 100 Brin. For light alloy parts; non-hardenable.

MOONESTONE BRONZE

Thomas Bolton Ltd.
Cu alloy.
For instrument terminals; free cutting. *Obsolete*

MOP 115 VITAC

Creusot-Loire
C 0.34, Mn 1, Cr 1.8, Mo 0.4, bal Fe.
Oil hardening for dies for plastic molds. AFNOR 34 CD 7; similar to AISI P20.

MOP 16

Creusot-Loire
C 0.38, Ni 4, Cr 1.7, Mo 0.5, bal Fe.
Air or oil hardening mold steel; shock resistant. AFNOR Y 35NCD 16.

MOP 82

Creusot-Loire
C 0.38, Si 1, Cr 5.25, Mo 1.35, V 0.5, bal Fe.
Air hardening hot work tool steel for die casting or pressure casting molds. AFNOR Z38 CDV 5; AISI H11.

MOPERMALLOY 4-79

Wilbur B. Driver Co.
Nickel. Ni 79, Fe 17.
Saturation inductance 8700. For laminated cores for communication inductors, transformers, electro-magnetic shields. Soft magnet, high permeability.

MORAINE

Moraine Mfg., Inc.
Sn, bal Cu.
For bearings; sintered.

MOREX

Darwins Alloy Castings
C, Mo, W, bal Fe.
For tools, cutters; high-speed steel. *Obsolete*

MOREX

Darwin Tools Ltd.
C 0.7, W 0.6, Mo 4-6, Cr, bal Fe.
For tools, cutters; high speed steel.

MORFLEX

Engelhard Corp.
Cu 18, Mn 22, bal Fe.
High expanding side: 10.0% nickel. Low expanding side: 36.0% nickel. For thermostatic bimetals; maximum temperature 500°F. *Obsolete*

MORINS CHINESE BRONZE

English manufacture
Cu 83, Pb 10, Sn 5, Zn 2.
For ornaments.

MORRISON'S DUCTILE BRONZE

Wm. Gallimore & Sons Ltd.
Cu 91, Sn 9.
For hardware, pump rods, valves, gears. *Obsolete*

MOSAIC GOLD

English manufacture
Cu 63, Zn 37.
46,000 TS. For brass ornaments; yellow brass.

MOSIL

Teledyne Vasco
C 0.57, Si 1.9, Mn 0.8, Cr 0.25, Mo 0.33, bal Fe.
For punches, chisels; shock resistant. Type S5.

MOSTAR

Societe Nouvelle des Acieries de Pompey
C 0.32, Cr 5, Mo 1.15, Si 1, V 0.35, bal Fe.
For upsetters, punches, extrusion dies; oil or air hardened, hot work steel. *Obsolete*

MOTAL

Aluminiumwerke Maulbronn
Si 5, bal Al.
21,500-27,000 TS; 3-6 El. For light alloys, automobile parts, motor housings; sand-cast alloy.

MOTELEC

Empire Sheet & Tin Plate Co.
Si 2.5, bal Fe.

MOTEMP

CCS Braeburn Alloy Steel
C 0.88, Cr 4, Mo 8, V 2, bal Fe.
Hardened: 64-66 Rock C. For circular saws, taps, drills, broaches, counterbores, milling cutters, hobs, chasers, form tools. Type M-10 high speed steel. High red-hardness. *Obsolete*

MOTEMP RSP

CCS Braeburn Alloy Steel
C 0.68, Mo 8, Cr 4, V 2, bal Fe.
Molybdenum hot work tool steel. For dies, hot forming tools. AISI H43. *Obsolete*

MOTOR 03A

Richard W. Carr & Co., Ltd.
C 1.5, Cr 0.5, V 0.35, W 4.5, Mo 0.35, bal Fe.
For drawing dies, turning tools; water hardened, abrasion resistant. *Obsolete*

MOTOR 05S

Richard W. Carr & Co., Ltd.
C 1.4, Mn 1.5, Cr 12, V 1, bal Fe.
For rim rolling dies, hot shears, punches; oil or water hardened, non-deforming. *Obsolete*

MOTOR 07S

Richard W. Carr & Co., Ltd.
C 1.85, Cr 11, Ni 0.1, V 0.25, bal Fe.
For gages, guide rolls, impact extrusion dies; air or oil hardened, non-deforming. *Obsolete*

MOTOR 21SW

Richard W. Carr & Co., Ltd.
C 1.15, Mn 1, Cr 1, V 0.25, W 1.35, bal Fe.
For blanking and drawing dies, shears, hobs; oil hardened, non-deforming. *Obsolete*

MOTOR 25B

Richard W. Carr & Co., Ltd.
C 1, Mn 3, Cr 1, Mo 1, bal Fe.
For punches, shears, blanking and drawing dies; air hardened, abrasion resistant. *Obsolete*

MOTOR 25S

Richard W. Carr & Co., Ltd.
C 0.27, Ni 4.25, W 6.75, Mo 0.2, Si 0.3, bal Fe.
For extrusion mandrels and dies; hot work steel, air or oil hardened. *Obsolete*

MOTOR MAGNUS

Richard W. Carr & Co., Ltd.
C 0.8, Cr 4, V 2, W 6, Mo 5.5, bal Fe.
For drills, reamers, hobs, broaches, shear blades; high speed steel. AISI M2.

MOTOR MAXIMUM

Richard W. Carr & Co., Ltd.
C 0.75, Cr 4.5, V 1.2, W 18, bal Fe.
For die casting dies, bushings, ejectors; high speed steel. AISI T1.

MOTOR O6S, ETC.

Now CARR'S QUALITY 06S, ETC.

MOTOR SPECIAL

Richard W. Carr & Co., Ltd.
C 0.7, Cr 4, W 14, bal Fe.
For drills, slitters, shear blades, chisels; high speed steel.

MOTORK

Magnetic Metals Co.
Si 0-0.6, bal Fe.
Low corrosion, low silicon alloy for motor laminations.

MOTUF

CCS Braeburn Alloy Steel
C 1, Cr 3.75, V 2.1, W 1.75, Mo 8.75, bal Fe.
Hardened: 65-66 Rock C. For broaches, burnishing tools, reamers, milling cutters, counterbores, taps, end mills, roll turning tools. Type M-7 high-speed steel. High wear resistance and high red-hardness. *Obsolete*

MOTUNG P & D

Cytemp Specialty Steel Div.
C 0.74, Mo 8.7, W 1.65, V 1.15, Cr 4, bal Fe.
Annealed: 228 Brin. For trim dies; cavity punches; high speed steel, oil hardened. *Obsolete*

MOTUNG-CV

Cytemp Specialty Steel Div.
C 1, Mo 8.75, W 1.75, Cr 4, V 2, bal Fe.
Hardened: 64-66 Rock C. For cutters, drills, reamers, end mills, slotting saws, routers, woodworking tools, milling cutters. Good wear resistance. AISI M7. High speed steel.

MOULIMPHY

Creusot-Loire
C, alloy, bal Fe.
For glass rolls and molds. *Obsolete*

MOULTREX

Saarstahl AG
Now SAARSTAHL 1.2738X1.

MOUREY WHITE GOLD

English manufacture
Au 50, Ag 35, Pd 15.
For jewelry; corrosion resistant.

MOUSSETS SILVER

English manufacture
Ag 27.5, Cu 59.5, Zn 9.5, Ni 3.5.
For instruments and cheap jewelry; corrosion resistant.

MOVA

Firth Brown Ltd.
C 0.2, Si 0.25, Mn 0.5, Mo 0.7, V 0.25, bal Fe.
For jet engine components, turbine rotor blades; heat resistant. *Obsolete*

MOW

Thyssen Edelstahlwerke AG
C, Ni, Mo, bal Fe.
Water treated: 101,000-121,000 TS; 85,000-107,000 YS; 17-20 El. For gears, shafts; oil or water hardening. *Obsolete*

MP 210-MULTIPHASE ALLOY MP210N

SPS Technologies
Co 36, Cr 19.5, Mo 7.5, Ti 3.8, Cb 1.1, bal Ni.
Heat treated: 240,000 TS; 225,000 YS; 10 El. For high strength bolts and fasteners. High fatigue strength; corrosion resistant.

MP-159 ALLOY

Latrobe Steel Co.
Ni 25.5, Co 35.5, Cr 19, Mo 7, Fe 9, Ti 3, Cb 0.6, Al 0.2.
R.T.: 260,000 psi TS. 1100°F: 200,000 psi TS. High temperature superalloy.

MP159-MULTIPHASE ALLOY MP159

SPS Technologies
Ni 25, Co 36, Cr 19, Mo 7, Ti 2.9, Fe 9, Cb 0.5, Al 0.2.
Ultra high strength: 260 ksi TS at 1000°F. Good high temperature properties, including creep rupture. For high temperature fasteners. Meets AMS 5841, 5842, 5843. R30159

MP35N

Now MULTIPHASE MP35N.

MP35N

Standard Pressed Steel Co.
Ni 35, Co 35, Cr 20, Mo 10.
Heat treated: 286,300 TS; 225,000 YS; 11 El; 39 RA.
Annealed 146,000 TS; 61,000 YS; 70 El; B 100 Rock. For high strength fasteners, bolts, aircraft and spacecraft components, steam turbine parts. Corrosion resistant, high fatigue strength.

MP35N

Cannon-Muskegon Corp.
Ni 35, Co 35, Cr 20, Mo 10.
Heat treated: 286,300 TS; 225,000 YS; 11 El; 39 RA.
Annealed 146,000 TS; 61,000 YS; 70 El; B 100 Rock. For high strength fasteners, bolts, aircraft and spacecraft components, steam turbine parts. Corrosion resistant, high fatigue strength.

MP35N

Latrobe Steel Co.
Ni 35, Co 35, Cr 20, Mo 10.
Heat treated: 286,300 TS; 225,000 YS; 11 El; 39 RA.
Annealed 146,000 TS; 61,000 YS; 70 El; B 100 Rock. For high strength fasteners, bolts, aircraft and spacecraft components, steam turbine parts. Corrosion resistant, high fatigue strength.

MP35N

DuPont de Nemours & Co., E.I.
Ni 35, Co 35, Cr 20, Mo 10.
Heat treated: 286,300 TS; 225,000 YS; 11 El; 39 RA.
Annealed 146,000 TS; 61,000 YS; 70 El; B 100 Rock. For high strength fasteners, bolts, aircraft and spacecraft components, steam turbine parts. Corrosion resistant, high fatigue strength.

MP35N ALLOY

Latrobe Steel Co.
Ni 35, Co 35, Cr 20, Mo 10.
Annealed: 132,000 TS. Work strengthened and aged: 200,000-300,000 TS; 160,000-290,000 YS; 9-18 El; 17-95 CPY. High temperature strength and corrosion resistance.

MP35N-MULTIPHASE ALLOY MP35N

SPS Technologies
Ni 35, Co 35, Cr 20, Mo 10.
Ultra high strength, 260 ksi TS. Tough, ductile, corrosion resistant, high fatigue strength. For springs, fasteners, tubing; to 700°F. Meets AMS 5758, 5844, 5845. R30035

MRL-A11

Olin Metals Research Labs
Si 0.6-0.9, Fe 0-0.7, Mg 0.4-0.6, Zn 0-0.25, Cu 0-0.2, Ti 0-0.15, Mn 0.05-0.15, Cr 0.05-0.15, Zr 0.05-0.15, bal Al.
T5: 38,000 psi TS; 35,000 psi YS; 8 El. Good impact strength: 20-60 Charpy V-notch (above 30 ft·lb). For highway guard rails, semi-hollow type. Hardenable.

MRL-A8
Olin Metals Research Labs
Mg 6-8, Si 0-0.5, Fe 0-0.5, Mn 0-0.35, Cu 0-0.25, Cr 0.05-0.3, Zn 0-0.2, B 0.001-0.05, Be 0-0.02, Ti 0-0.015, bal Al.
50,000 psi TS; 22,000 psi YS; 25 El. H38: 69,000 psi TS; 54,000 YS; 10 El. Good forming properties; especially good corrosion resistance, including marine atmosphere. AA-X5090.

MS ALLOY
Japanese manufacture
Ni 30-60, Cr 1-18, bal Fe.
For magnetic shunts; magnetically soft.

MS-15
Swiss manufacture
C 1, Co 15, Cr 9, Mo 1.5, bal Fe.
Hardened: 60-64 Rock C. For permanent magnets in magnetic and electrical equipment; hardened, high permeability.

MS-250
Precision Castparts Corp.
C 0-0.03, Ni 16-17.5, Co 9.5-11, Mo 4.4-4.9, Al 0.05-0.15, Ti 0.15-0.45, Zr 0-0.03, B 0-0.005, bal Fe.
Heat treated: 275,000 TS; 263,000 YS; 6 El; 16 RA. For compressor rotors, high impact pressure devices. High strength and ductility. Maraging cast alloy steel.

MS-35
Swiss manufacture
C 0.9, Co 35, Cr 6, W 5, bal Fe.
For permanent magnets in magnetic and electrical equipment; hardened, high permeability.

MS-6
Swiss manufacture
C 1, Cr 9, Co 6, Mo 1.5, bal Fe.
Hardened: 60-64 Rock C. For permanent magnets, in magnetic and electrical equipment; hardened, high permeability.

MS4
Acciaierie Valbruna s.p.a.
C 0.48, Si 1.75, Mn 0.6, P 0.035, S 0.035, bal Fe.
Spring steel, W. Nr. 1.0902.

MS4-DE
Acciaierie Valbruna s.p.a.
C 0.46, Si 1.65, Mn 0.85, P 0-0.03, S 0-0.05, bal Fe.
Spring steel. W. Nr. 1.0902.

MS5
Acciaierie Valbruna s.p.a.
C 0.55, Si 1.75, Mn 0.85, Cr 0.3, P 0.035, S 0.035, bal Fe.
Silico-manganese spring steel. W. Nr. 1.0904; AISI 9255.

MS5-DE
Acciaierie Valbruna s.p.a.
C 0.6, Si 1.15, Mn 1, P 0-0.035, S 0-0.035, Spring steel. W. Nr. 1.0908., bal Fe.

MS5E
Acciaierie Valbruna s.p.a.
C 0.6, Si 1.75, Mn 0.8, P 0.035, S 0.035, bal Fe.
Silico-manganese spring steel. W. Nr. 1.0906; AISI 9260.

MS6
Acciaierie Valbruna s.p.a.
C 0.49, Si 1.65, Mn 0.8, P 0-0.035, S 0-0.035, bal Fe.
Spring steel. W. Nr. 1.0903.

MS7
Acciaierie Valbruna s.p.a.
C 0.6, Si 2, Mn 0.9, Cr 0.3, P 0.035, S 0.035, bal Fe.
Silico-manganese spring steel.

MS7-DE
Acciaierie Valbruna s.p.a.
C 0.6, Si 1.65, Mn 0.9, P 0-0.045, S 0-0.045, Cr 0.3, bal Fe.
Spring steel. W. Nr. 1.0906.

MS8Y
Acciaierie Valbruna s.p.a.
C 0.42-0.55, Si 1.5-2, Mn 0.4-0.8, bal Fe.
Silicon shock resisting steel. 50 Si 7. *Obsolete*

MSA
Thyssen Edelstahlwerke AG
C 0.8, Mn 0.3, W 0.6, bal Fe.
For dies, drills, taps; water hardening. *Obsolete*

MSM 3AL-2 1\2V
RMI Company
Al 3, V 2.5, bal Ti.
Heat treated: 120,000 TS; 105,000 YS; 15 El; 47 RA; 140 Brin. Annealed: 56,000 TS; 48,000 YS; 16 El. For aircraft and missile components; age-hardenable, good weldability. *Obsolete*

MSM185
RMI Company
Ti alloy.
Heat treated: 260,000 TS. For aircraft and missile components; age-hardenable, good weldability. *Obsolete*

MSO
Thyssen Edelstahlwerke AG
C 0.9, Mn 2, V 0.1, bal Fe.
For blanking dies, profile rollers; nondeforming, oil hardened. *Obsolete*

MSR-A
Sterling Metals Ltd.
Ag 2-3, Zr 0.4-1, 1.2-2.0% Rare Earth, bal Mg.
Heat treated: 34,800-41,500 TS; 22,400-28,000 YS; 4-8 El; 65-80 Brin. High strength magnesium alloy. Aircraft DTD 5025 (England). *Obsolete*

MSR-B
Sterling Metals Ltd.
Ag 2-3, Zr 0.4-1, 2.0-3.0% Rare Earth, bal Mg.
Heat treated: 34,800-41,500 TS; 24,700-30,200 YS; 2-4 El; 78-85 Brin. High strength magnesium alloy. Aircraft DTD 5035 (England). *Obsolete*

MST 2AL-2FE
RMI Company
C 0.5, Al 2, Fe 2, bal Ti.
Annealed: 145,000 TS; 135,000 YS; 12 El. Hard: 180,000 TS; 160,000 YS; 5 El. For jet engine components; corrosion resistant. *Obsolete*

MST 3MN COMPLEX
RMI Company
Mn 3, Fe 1, Cr 1, V 1, Mo 1, bal Ti.
Heat treated: 151,000-196,000 TS; 147,000-190,000 YS; 5-16 El; 16-44 RA. For jet engine components; heat treatable. *Obsolete*

MST ALLOYS
Now RMI ALLOYS.

MT-17
English manufacture
C 0.07, Mn 1.5, Cr 21, Ni 30, Co 21, Mo 3, W 2.2, Ti 1.6, bal Fe.
For high temperature applications; heat resistant.

MT-6
Now DARWIN MT6.

MT-STEEL
Olin Corp.
Al 8, C 2, Si 0-0.2, Co 0.6, bal Fe.
Heat treated: 5000 Gausses residual induction; 200 Oersted coercive force. For electrical and magnetic equipment. Permanent magnet. Quenched and tempered for maximum magnetic properties.

MTR
Marrel Freres, S.A.
C 0.13, Mn 0.65-0.95, Cr 0.4-0.7, Cu 0.25, Mo 0.4-0.6, Ni 1.1-1, V 0.04-0.1, bal Fe.
Plate: 107,000-135,000 TS; 78,000 YS min; 13 El min. For pressure vessels, bridges, mine cars, power shovels, cranes, trucks, trailers. Shock and wear resistant.

MTS4-CR
Thyssen Edelstahlwerke AG
C 0.2, Cr 12.5, Mo 1.1, V 0.25, W 0.5, bal Fe.
For surgical and dental instruments; corrosion resistant. *Obsolete*

MU-GUARD 48
Spang Specialty Metals
Ni 48, bal Fe.
Soft magnetic alloy strip for magnetic shielding; medium effectiveness. *Obsolete*

MU-GUARD 80
Spang Specialty Metals
Ni 80, Mo 4.5, bal Fe.
Soft magnetic alloy strip. Very high magnetic shielding value. *Obsolete*

MU-HOLE PUNCH
Midvale-Heppenstall Co.
C, alloy, bal Fe.
For tools; oil hardened. *Obsolete*

MUELLER 1020
Mueller Brass Co.
Cu 99.95, 0.04 O_2.
Half hard: 53,000 TS; 46,000 YS; 5 El; 90 Rock F. Soft: 32,000 TS; 10,000 YS; 40 El; 35 Rock F. For high conductivity tubing and forgings. CDA 102; formerly Mueller 2140F. Oxygen-free high conductivity copper. *Obsolete*

MUELLER 1100
Mueller Brass Co.
Cu 99.9, O 0.04.
Hard rod: 48,000 TS; 44,000 YS; 16 El; 87 Rock F. For trolley wire, cables, conductors; connectors, terminals, fittings, pipes. Tough pitch copper. CDA 110; formerly Mueller 214 ETP.

MUELLER 1220
Mueller Brass Co.
Cu 99.9, P 0.015-0.04.
Annealed: 32,000 TS; 10,000 YS; 45 El; 40 Rock F. Hard drawn: 55,000 TS; 50,000 YS; 8 El; 60 Rock B. For air conditioners, refrigerators, kettles, oil coolers, rotating bands. Phosphorus deoxidized copper. CDA 122; formerly Mueller 214 DHP.

MUELLER 1450
Mueller Brass Co.
Te 0.5, P 0.01, Cu 99.49.
Tellurium copper rod and screw machine products; free machining. CDA 145; was Mueller 799W. *Obsolete*

MUELLER 1470
Mueller Brass Co.
S 0.3, bal Cu.
Sulfurized copper. Half hard: 43,000 TS; 40,000 YS; 18 El; 42 Rock B. Hard: 48,000 TS; 46,000 YS; 13 El; 50 Rock B. 96% conductivity; 85% machinability. Free machining copper; for screw machined parts requiring properties similar to copper. CDA 147; was Mueller 771W. *Obsolete*

MUELLER 1476
Mueller Brass Co.
S 0.2-0.4, 99.9 Cu min + Ag + S.
Sulfur bearing copper, free cutting; CDA 14720. *Obsolete*

MUELLER 1477
Mueller Brass Co.
S 0-0.2, 99.90 Cu min + Ag + S.
Sulfur bearing copper, free machining; CDA 14710. *Obsolete*

MUELLER 14S
Now MUELLER 2014.

MUELLER 17S
Now MUELLER 2017.

MUELLER 1820
Mueller Brass Co.
Cr 1, Si 0.02, bal Cu.
Heat treated: 78,000 TS; 70,000 YS; 13 El; 80 Rock B. For resistance welding tips, holders and wheels. CDA 182 and 184; formerly Mueller 902. Heat treatable; high strength. *Obsolete*

MUELLER 200
Now MUELLER 3770.

MUELLER 201
Now MUELLER 3600.

MUELLER 2014
Mueller Brass Co.
Cu 4.4, Mn 0.8, Mg 0.4, Si 0.8, bal Al.
Heat treated: 70,000 TS; 60,000 YS; 14 El; 140 Brin. For hardware; forging alloy.

MUELLER 2017
Mueller Brass Co.
Cu 4, Mn 0.5, Mg 0.5, Si 0.75, bal Al.
Heat treated: 60,000 TS; 37,500 YS; 22 El; 105 Brin. For hardware; forging alloy.

MUELLER 203
Now MUELLER 4640.

MUELLER 204
Now MUELLER 4851.

MUELLER 207
Now MUELLER 4850.

MUELLER 210
Now MUELLER 4643.

MUELLER 211
Now MUELLER 3771.

MUELLER 212
Now MUELLER 3450.

MUELLER 213
Now MUELLER 3801.

MUELLER 214
Now MUELLER 1100.

MUELLER 214A
Now MUELLER 1220.

MUELLER 224C
Now MUELLER 6230.

MUELLER 224E
Now MUELLER 6180.

MUELLER 224H
Now MUELLER 6240.

MUELLER 224K
Now MUELLER 6300.

MUELLER 2300
Mueller Brass Co.
Cu 84-86, Pb 0-0.05, bal Zn.
Red brass tubing and pipe. CDA 230.

MUELLER 241A
Now MUELLER 6750.

MUELLER 241W
Now MUELLER 6780.

MUELLER 245
Now MUELLER 4820.

MUELLER 246
Now MUELLER 3602.

MUELLER 248
Now MUELLER 3530.

MUELLER 2600
Mueller Brass Co.
Cu 63-68.5, Pb 0-0.15, bal Zn.
Yellow brass tubing. CDA 270.

MUELLER 2740
Mueller Brass Co.
Cu 61-64, Pb 0-0.1, bal Zn.
Yellow brass tubing. CDA 274.

MUELLER 3300
Mueller Brass Co.
Cu 65-68, Pb 0.2-0.8, bal Zn.
Low leaded brass tube. CDA 330.

MUELLER 3320
Mueller Brass Co.
Cu 65-68, Pb 1.3-2, bal Zn.
High leaded brass tube. CDA 332. *Obsolete*

MUELLER 3400
Mueller Brass Co.
Pb 1, Cu 65, bal Zn.
HH temper: 58,000 TS; 45,000 YS; 30 El; 70 Rock B. For cold headed fasteners, hardware, marine hardware. CDA 340; formerly Mueller 314. Leaded brass; good machinability. *Obsolete*

MUELLER 3406
Now MUELLER 3400.

MUELLER 3450
Mueller Brass Co.
Pb 2, Cu 62.5, bal Zn.
1/4 Hard: 53,000 TS; 35,000 YS; 35 El; 68 Rock B. For thread rolled fasteners, marine hardware. CDA 345; formerly Mueller 212. Leaded brass; good cold workability.

MUELLER 3470
Mueller Brass Co.
Cu 63, Pb 1.5, bal Zn.
Medium leaded brass. 3/8 Hard: 55,000 TS; 35,000 YS; 30 El; 68-70 Rock B. Thread rolling brass; good machining and cold working characteristics. 85% machinability. CDA No. 347; formerly Mueller 312. *Obsolete*

MUELLER 3500
Mueller Brass Co.
Cu 61.7, Pb 1, bal Zn.
Thread rolling brass. 1/4 Hard: 53,000 TS; 35,000 YS; 35 El; 70 Rock B. For roll threaded brass parts. CDA No. 350; formerly Mueller 251.

MUELLER 3530
Mueller Brass Co.
Zn 35.25, Pb 2.75, bal Cu.
Cast. For screw machine products; commercial brass.

MUELLER 3600
Mueller Brass Co.
Pb 3, Cu 61.5, bal Zn.
HH temper: 60,000 TS; 45,000 YS; 25 El; 72 Rock B. Hard temper: 75,000 TS; 55,000 YS; 10 El; 78 Rock B. For screw machine products, fasteners, bearings, bushings. CDA 360; formerly Mueller 201. Leaded brass; free-cutting.

MUELLER 3602
Mueller Brass Co.
Pb 2.75, Cu 61.5, bal Zn.
3/8 Hard: 55,000 TS; 40,000 YS; 25 El; 68 Rock B. For fasteners, marine hardware. CDA 360; formerly Mueller 246. Leaded brass; free-cutting.

MUELLER 3770
Mueller Brass Co.
Pb 1.5-2.5, Cu 58-61, Fe 0-0.3, bal Zn.
Forgings: 50,000-60,000 TS; 20,000-25,000 YS; 40-50 El; 40-60 Rock B. Extruded: 58,000 TS; 23,000 YS; 42 El; 48 Rock B. For hardware, valve bodies, automobile parts, plumbing fittings. Brass forging alloy; formerly Mueller 200.

MUELLER 3771
Mueller Brass Co.
Pb 2, Cu 60, bal Zn.
Forged: 58,000 TS; 23,000 YS; 42 El; 48 Rock B. For forged hardware, shafts, fasteners. CDA 377; formerly Mueller 211. Leaded forging brass.

MUELLER 3801
Mueller Brass Co.
Pb 2, Cu 56, bal Zn.
Forged: 65,000 TS; 30,000 YS; 15 El; 60 Rock B. For forged hardware, fasteners. Formerly Mueller 213; leaded forging brass. *Obsolete*

MUELLER 4430
Mueller Brass Co.
Cu 70-73, Pb 0-0.07, Sn 0.8-1.2, As 0.02-0.1, bal Zn.
Admiralty arsenical copper tubing. CDA 443. *Obsolete*

MUELLER 4450
Mueller Brass Co.
Cu 70-73, Pb 0-0.07, Sn 0.8-1.2, P 0.02-0.1, bal Zn.
Admiralty phosphorized copper tubing. CDA 445. *Obsolete*

MUELLER 4620
Mueller Brass Co.
Cu 62-65, Sn 0.5-1, bal Zn.
HH temper: 65,000 TS; 45,000 YS; 30 El; 72 Rock B. For fasteners, bolts, marine hardware, rivets, clamps, connectors. Corrosion resistant. Naval brass; CDA 462. Formerly Mueller 308. *Obsolete*

MUELLER 4640
Mueller Brass Co.
Sn 0.75, Cu 60.5, bal Zn.
Forged: 64,000 TS; 26,000 YS; 40 El; 55 Rock B. Hard: 75,000 TS; 60,000 YS; 20 El; 80 Rock B. For marine products, gears, valve stems, light bearings, screw machine products. Naval brass; CDA 464. Formerly Mueller 203. *Obsolete*

MUELLER 4643
Mueller Brass Co.
Cu 60, Pb 0-0.3, Sn 0.75, bal Zn.
1/2 H-temper: 68,000 TS; 45,000 YS; 22 El; 150 Brin. Soft: 54,000 TS; 22,000 YS; 40 El; 83 Brin. For propeller shafts, gears, bolts, screw machine products; Tobin bronze. *Obsolete*

MUELLER 4700
Mueller Brass Co.
Cu 58.5, Sn 0.9, bal Zn.
Naval brass welding rod. Melting point, liquidus: 1640 F. For welding bronzes, brasses, and for braze welding steel and cast iron. CDA No. 470. Formerly Mueller 228. *Obsolete*

MUELLER 4701
Mueller Brass Co.
Cu 59.6, Sn 0.75, bal Zn.
Naval brass welding rod. Melting point, liquidus: 1640 F. For welding bronzes, brasses and for braze welding steel and cast iron. CDA No. 470. Formerly Mueller 234. *Obsolete*

MUELLER 4820
Mueller Brass Co.
Pb 0.75, Sn 0.75, Cu 60.5, bal Zn.
Forged: 60,000 TS; 30,000 YS; 40 El; 55 Rock B. Hard: 70,000 TS; 48,000 YS; 30 El; 78 Rock B. For marine parts, gears, screw machine products. Leaded Naval brass, free-turning; CDA 482. Formerly Mueller 245. *Obsolete*

MUELLER 4850
Mueller Brass Co.
Pb 2, Sn 0.75, Cu 60.5, bal Zn.
HH temper: 65,000 TS; 48,000 YS; 15 El; 72 Rock B. For marine products, gears, valve stems, light bearings, screw machine products. Leaded Naval brass; CDA 485. Formerly Mueller 207.

MUELLER 4851
Mueller Brass Co.
Pb 2, Sn 0.75, Cu 59.5, bal Zn.
Forged: 62,000 TS; 24,000 YS; 40 El; 55 Rock B. For light bearings, forged hardware. CDA 485. Formerly Mueller 204. Leaded Naval brass, free-cutting. *Obsolete*

MUELLER 600
Now MUELLER 6741.

MUELLER 601
Now MUELLER 6680.

MUELLER 602
Now MUELLER 6730.

MUELLER 603
Now MUELLER 6731.

MUELLER 604
Now MUELLER 6732.

MUELLER 605
Now MUELLER 6733.

MUELLER 6140
Mueller Brass Co.
Cu 90.5, Fe 2, Al 7.5.
Aluminum bronze. Drawn: 88,000 TS; 52,000 YS; 35 El; 85 Rb. High strength, good ductility bronze for shafts, and various load carrying sliding parts. Non-heat treatable. CDA No. 614. Formerly Mueller 224-7. *Obsolete*

MUELLER 6162
Now MUELLER 6181.

MUELLER 6180
Mueller Brass Co.
Al 10, Fe 1, Cu 89.
HH temper: 100,000 TS; 65,000 YS; 16 El; 208 Brin. For gears, valve parts, worm wheels, marine hardware. Aluminum bronze, corrosion resistant. CDA 617. Formerly Mueller 224 E-75. *Obsolete*

MUELLER 6181 TUF-STUF ALUMINUM BRONZE
Mueller Brass Co.
Cu 90, Fe 1, Al 9.
1/2 Hard: 82,000 TS; 50,000 YS; 30 El; 175 Brin (1000 kg). Forged: 75,000 TS; 36,000 YS; 20 El; 140 Brin (1000 kg). High strength, heat treatable bronze for shafts and other sliding parts. CDA No. 617. Formerly Mueller 224 E-30. *Obsolete*

MUELLER 6230
Mueller Brass Co.
Cu 88, Fe 3, Al 9.
HH temper: 95,000 TS; 62,000 YS; 9 El; 185 Brin. Forged: 84,000 TS; 40,000 YS; 40 El; 135 Brin. For gears, worm wheels, marine hardware, shafts. Aluminum bronze. Not heat treatable. CDA 616. Formerly 224C. *Obsolete*

MUELLER 6236
Mueller Brass Co.
Al 8-10, Fe 2-4, Si 0-0.25, Sn 0-0.2, bal Cu.
Aluminum bronze rod. CDA 623. *Obsolete*

MUELLER 6246
Mueller Brass Co.
Cu 85.5, Fe 3.6, Al 10.6.
Aluminum bronze, heat treatable. Extruded: 100,000 TS; 50,000 YS; 7 El; 195 Brin (3000 kg). Moderately high strength aluminum bronze, possessing good corrosion and high temperature properties. CDA 616. Formerly Mueller 224-11. *Obsolete*

MUELLER 6250
Mueller Brass Co.
Cu 82.8, Fe 4.3, Al 12.9.
Aluminum bronze. Extruded: 105,000 TS; 65,000 YS; 1.5 El; 285 Brin (3000 kg). Heat treatable, good strength, very good wear resistance; for gears, worm wheels, wear plates. Formerly Mueller 224-13. *Obsolete*

MUELLER 6300
Mueller Brass Co.
Ni 5, Mn 1, Al 10, Fe 3, Cu 81.
Annealed: 120,000 TS; 75,000 YS; 15 El; 241 Brin. Heat treated: 114,000 TS; 68,000 YS; 15 El; 223 Brin. For gears, aviation landing gears, shafts, marine hardware. Aluminum bronze, heat treatable. CDA 628. Formerly Mueller 224K. *Obsolete*

MUELLER 6390
Now MUELLER 6420.

MUELLER 6391
Now MUELLER 6241.

MUELLER 6420
Mueller Brass Co.
Al 7, Si 2, Cu 91.
Hard: 98,000 TS; 65,000 YS; 25 El; 95 Rock B. Forged: 75,000 TS; 35,000 YS; 45 El; 85 Rock B. For nuts, bolts, pole line hardware. Aluminum bronze, corrosion resistant. CDA 639. Formerly Mueller 802. *Obsolete*

MUELLER 6421
Mueller Brass Co.
Cu 91, Al 6.7, Si 1.9.
Aluminum silicon bronze. Hard: 85,000 TS; 45,000 YS; 40 El; 84 Rock B. As forged: 75,000 TS; 35,000 YS; 45 El; 80 Rock B. 50% machinability. Free turning, good strength, non-magnetic, corrosion resistant; for shafts, nuts, bolts, pole line hardware. CDA 639. Formerly Mueller 808. *Obsolete*

MUELLER 6550
Mueller Brass Co.
Si 2.8-3.8, Pb 0-0.05, Zn 0-1.5, Ni 0-0.6, Mn 0-1.5, bal Cu.
Bar: 60,000 TS; 20,000 YS; 60 El; B 30 Rock. For marine hardware, fasteners, screws, bolts, diaphragms. CDA 655. Formerly Mueller 800. Silicon Bronze, corrosion resistant. *Obsolete*

MUELLER 6680
Mueller Brass Co.
Cu 60.5, Si 1, Mn 2.5, bal Zn.
HH temper: 70,000-85,000 TS; 40,000-65,000 YS; 15-25 El; 70-87 Rock B. For bearings, gears, connecting rods, marine hardware. Bearing bronze; ductile. Formerly Mueller 601.

MUELLER 6700
Fe 3, Al 5, Mn 4, Cu 64, bal Zn.
Soft: 90,000 TS; 50,000 YS; 20 El; 84 Rock B. Hard: 124,000 TS; 72,000 YS; 18 El; 95 Rock B. For rollers, rotors, valve stems. Manganese bronze; shock resistant. CDA 670. Formerly Mueller 721B. *Obsolete*

MUELLER 6701
Mueller Brass Co.
Fe 3, Al 5, Mn 4, Cu 64, bal Zn.
HH temper: 112,000 TS; 67,000 YS; 13 El; 94 Rock B. For rollers, rotors, valve stems. Manganese bronze; shock resistant. CDA 670. Formerly Mueller 721E. *Obsolete*

MUELLER 6702
Mueller Brass Co.
Cu 64, Fe 3, Al 5, Mn 4, bal Zn.
Forged: 115,000 TS; 67,000 YS; 18 El; 95 Rock B. For rotors, rollers, valve stems. Manganese bronze; shock resistant. CDA 670. Formerly Mueller 721x. *Obsolete*

MUELLER 6730
Mueller Brass Co.
Pb 1, Mn 2.5, Si 1, Cu 60.5, bal Zn.
HH-temper: 70,000-85,000 TS; 40,000-65,000 YS; 15-25 El; 70-87 Rock B. For gears, sleeve and thrust bearings, bushings, cams, marine hardware. Good bearing characteristics. Formerly Mueller 602.

MUELLER 6731
Mueller Brass Co.
Cu 59.75, Pb 1, Si 1, Mn 2.5, CO 8.27544e+008-0.501271, He 5.92392e+020-2, ge 3.45959e-012-62.5, Be 1.80621e+028-0.00109864, 60 0.00195588-0.006, CO 8.90405e+011-3.37774e+009, ba 1.50001-1.5.

MUELLER 6732
Mueller Brass Co.
Cu 61.5, Pb 2.5, Si 1, Mn 2.5, bal Zn.
HH temper: 65,000-80,000 TS; 40,000-60,000 YS; 10-20 El; 75-86 Rock B. For bearings used against soft or hard mating surfaces. Bearing bronze. Can be soldered. CDA 673. Formerly Mueller 604. *Obsolete*

MUELLER 6733
Mueller Brass Co.
Cu 62, Pb 0.6, Si 1, Mn 2.5, bal Zn.
HH temper: 70,000-75,000 TS; 45,000-55,000 YS; 18-25 El; 75-82 Rock B. For bearing mating with soft or hard members. CDA 673. Formerly Mueller 605. *Obsolete*

MUELLER 6741
Mueller Brass Co.
Cu 58, Si 1, Al 1.5, Mn 2.5, bal Zn.
HH temper: 68,000-100,000 TS; 34,000-65,000 YS; 12-18 El; 78-88 Rock B. Forged: 68,000 TS; 34,000 YS; 18 El; 78 Rock B. For bearings against hardened surfaces, bushings, connecting rods, hardware. Bearing bronze; corrosion resistant. CDA 674. Formerly Mueller 600.

MUELLER 6750
Mueller Brass Co.
Cu 59, Sn 1, Fe 1, Mn 0.25, bal Zn.
Hard: 83,000 TS; 56,000 YS; 20 El; 80 Rock B. Forged: 68,000 TS; 35,000 YS; 27 El; 85 Rock B. For screw machine products, valve stems, airplane parts, balls. Manganese bronze; corrosion resistant. CDA 675. Formerly Mueller 241A. *Obsolete*

MUELLER 6780
Mueller Brass Co.
Cu 57, Fe 1, Al 1, Mn 0.25, bal Zn.
Forged: 80,000 TS; 29,000 YS; 35 El; 130 Brin. For counterweights for airplane engine crankshafts. Manganese bronze. Formerly Mueller 241W. *Obsolete*

MUELLER 6800
Mueller Brass Co.
Cu 58.5, Fe 0.4, Sn 0.9, Mn 0.3, Ni 0.5, bal Zn.
Welding rod, low fuming. Melting point, liquidus: 1650 F. For braze welding of steel and cast iron and for surfacing. CDA No. 680. Formerly Mueller 227. *Obsolete*

MUELLER 6801
Mueller Brass Co.
Cu 59, Fe 0.4, Sn 0.75, Mn 0.2, Ni 0.3, bal Zn.
Welding rod, low fuming. Melting point, liquidus: 1650 F. For braze welding of steel and cast iron and for surfacing. CDA No. 680. Formerly Mueller 235. *Obsolete*

MUELLER 6810
Mueller Brass Co.
Cu 58, Fe 0.6, Sn 0.9, bal Zn.
Welding rod, low fuming. Melting point, liquidus: 1650 F. For welding brass and bronze, braze welding steel and cast iron, and surfacing. CDA No. 681. Formerly Mueller 236. *Obsolete*

MUELLER 6940
Mueller Brass Co.
Cu 80-83, Pb 0-0.3, Si 3.5-4.5, bal Zn.
Rod: 90,000 TS; 57,000 YS; 38 El. For valve stems. Silicon red brass; corrosion resistant. CDA 694. *Obsolete*

MUELLER 6970
Mueller Brass Co.
Cu 77, Pb 1, Si 3, bal Zn.
Silicon brass. 1/4 hard: 80,000 TS; 46,000 YS; 42 El; 82 Rock B. For shafts, bolts, nuts, cap screws, screw machine parts; good machinability and corrosion resistance; valve stem alloy. CDA No 697. Formerly Mueller 804. *Obsolete*

MUELLER 721
Now MUELLER 6700.

MUELLER 7986
Mueller Brass Co.
Cu 46, Ni 10, Pb 2.5, Mn 0.25, bal Zn.
Forged: 70,000 TS; 42,000 YS; 30 El; 120 Brin. For hardware, faucets, fixtures; forging alloy, nickel silver; "Niag 515." *Obsolete*

MUELLER 799
Now MUELLER 1450.

MUELLER 802
Now MUELLER 6420.

MUELLER 803
Mueller Brass Co.
Si 2.4-4, Fe 1.5-4, Zn 1.5-4, bal Cu.
Annealed: 56,000-68,000 TS; 20,000-32,000 YS; 30-40 El; 89-120 Brin. For jet engine parts, anti-friction roller bearing cages; non-galling, corrosion resistant. *Obsolete*

MUELLER 8360
Mueller Brass Co.
Cu 83, Pb 6, Sn 4, Zn 7.
Cast: 30,000 TS; 15,000 YS; 20 El; 55 Brin. For water pump impellers, fittings for gasoline and oil lines, bushings, miscellaneous castings. Formerly Mueller 218 cast alloy. *Obsolete*

MUELLER 8386
Mueller Brass Co.
Cu 85, Pb 5, Zn 5, Sn 5.
Cast: 35,000 TS; 15,000 YS; 23 El; 62 Brin. For water pump impellers, fittings for gasoline and oil lines, bushings, miscellaneous castings. Red Brass, good castability and machinability. *Obsolete*

MUELLER 8420
Mueller Brass Co.
Cu 80, Zn 10, Sn 5, Pb 3, Cu alloy.
For screw pipe fittings, hardware; free-cutting, cast. *Obsolete*

MUELLER 8440
Mueller Brass Co.
Cu 80, Pb 7, Sn 3, Ni 0.5, Zn 9.5.
Cast: 34,000 TS; 15,000 YS; 24 El; 55 Brin. For bronze backing of lined bearings, bushings. Good embeddability, corrosion resistant. Was Mueller 272. *Obsolete*

MUELLER 902
Now MUELLER 1920.

MUELLER 9030
Mueller Brass Co.
Cu 88, Zn 4, Sn 8.
Cast: 38,000 TS; 18,000 YS; 15 El; 75 Brin. For bearings, bolts, nuts, gears; modified "G-Metal"; wear resistant. *Obsolete*

MUELLER 9050
Mueller Brass Co.
Cu 88, Zn 2, Sn 10.
Cast: 38,000 TS; 18,000 YS; 15 El; 75 Brin. For bearings, bolts, nuts, gears, pump pistons; "G Metal" or gun metal, wear resistant. *Obsolete*

MUELLER 9220
Mueller Brass Co.
Cu 88.75, Pb 1.5, Zn 3.75, Sn 6.
Cast: 34,000 TS; 18,000 YS; 25 El; 67 Brin. For steam pressure castings, gears, bushings, oil pumps; "Steam or Valve Bronze," "M"-Bronze. *Obsolete*

MUELLER 9230
Mueller Brass Co.
Sn 8, Pb 1, bal Cu.
Cast. For bearings, bushing; bronze, free-cutting. *Obsolete*

MUELLER 9340
Mueller Brass Co.
Sn 8, Pb 8, bal Cu.
For bushings, bearings; cast. *Obsolete*

MUELLER 9370
Mueller Brass Co.
Cu 80, Pb 10, Sn 10.
Cast: 30,000 TS; 10,000 YS; 10 El; 68 Brin. For bushings, bearings, castings; acid resistant. *Obsolete*

MUELLER 9380
Mueller Brass Co.
Cu 78, Pb 15, Sn 7.
Soft: 25,000 TS; 17,000 YS; 14 El; 100 Brin. For bearings, mine water pump parts; for moderate pressures, bearing bronze. *Obsolete*

MUELLER TUF-STUF 6240
Mueller Brass Co.
Cu 86, Al 11, Fe 3.
Forged: 80,000 TS; 45,000 YS; 10 El; 170 Brin. Heat treated: 95,000 TS; 70,000 YS; 3 El; 200 Brin. For valve seat inserts, gears, worm wheels; forging alloy, heat treatable; "Tuf-Stuf H." *Obsolete*

MUFLEX
Engelhard Corp.
Iron invar. For thermostatic bimetals; high-iron, low-invar. *Obsolete*

MULTI-ALLOY
English manufacture
C 0.25, Ni 46.5, Cr 20.5, Mo 2.7, Co 3.3, Ti 1.2, Cb 2.9, W 3.5, bal Fe.
For gas turbine parts; heat resistant.

MULTICORE 95A
Multicore
Lead. Sn 95, Sb 5.
Lead free, creep resistant solder. MP 457°F. FP 469°F.

MULTICORE HMP
Multicore
Lead. Pb 93.5, Sn 5, Ag 1.5.
Solder for high melting point applications. MP 565°F. FP 574°F.

MULTICORE PURE TIN
Multicore
Tin. Sn 100.
Lead free solder. MP 450°F. FP 450°F.

MULTICORE SN15
Multicore
Lead. Pb 85, Sn 15.
Solder for lamps. MP 437°F. FP 554°F.

MULTICORE SN20
Multicore
Lead. Pb 80, Sn 20.
Solder for lamps. MP 361°F. FP 527°F.

MULTICORE SN30
Multicore
Lead. Pb 70, Sn 30.
Solder for radiators, motors. MP 361°F. FP 491°F.

MULTICORE SN40
Multicore
Lead. Pb 60, Sn 40.
General purpose solder. MP 361°F. FP 453°F.

MULTICORE SN50
Multicore
Pb 50, Sn 50.
General purpose solder. MP 361°F. FP 414°F.

MULTICORE SN60
Multicore
Pb 40, Sn 60.
Solder for electrical applications and electronics. MP 361°F. FP 370°F.

MULTICORE SN62
Multicore
Pb 36, Sn 62, Ag 2.
Solder for silver and gold surfaces. MP 354°F. FP 354°F.

MULTICORE SN63
Multicore
Pb 37, Sn 63.
Solder for electronics. MP 361°F. FP 361°F.

MULTICORE SN96
Multicore
Tin. Sn 96, Ag 4.
Bright, nontoxic solder. MP 430°F. FP 430°F.

MULTICORE TLC
Multicore
Pb 32, Sn 50, Cd 18.
Low melting point solder for heat sensitive joints. MP 293°F. FP 293°F.

MULTIMET
Enpar Sonderwerkstoffe GmbH
Superalloy.
Alternate manufacturer.

MULTIMET ALLOY
Haynes International, Inc.
Alloy steel. C 0.08-0.16, Cr 20-22.5, Ni 19-21, Mo 2.5-3.5, W 2-3, Co 18.5-21, Cb 0.75-1.5, N 0.1-0.2, Mn 1-2, bal Fe.
Cast: 98,000 TS; 55,000 YS; 23 El; 24 RA. Forged: 110,000-140,000 TS; 15-30 El; 45-48 RA. For turbine blading, jet and combustion chambers, welding rod; high heat resistant.

MULTIMOLD
Bethlehem Steel Corp.
C 0.35, Mn 0.7, Cr 0.8, Mo 0.3, bal Fe.
For plastic mold dies, die casting dies; oil hardened.
Obsolete

MULTIPASS
Now HAYNES MULTIPASS.

MULTIPHASE - MP20N
SPS Technologies
Now MP 210-MULTIPHASE ALLOY MP 210N.

MULTIPHASE MP-159
Latrobe Steel Co.
Now MP-159 ALLOY.

MULTOLE
Boyd-Wagner Co.
C 0.6, Mn 0.7, Si 1.9, bal Fe.
For punches, shear blades, knives; keen edge tools.

MULTOLE
Midvale-Heppenstall Co.
C 0.57, Mn 0.7, Si 1.95, bal Fe.
For tools, punches; retains keen cutting edge. *Obsolete*

MUMETAL
Spang Specialty Metals
Ni 75-78, Mn 0-1.5, Si 0-0.5, C 0-0.05, Cu 4-6, Cr 2, bal Fe.
Soft magnetic alloys in sheet, strip, bar and rod forms.
Conforms to ASTM A753 and MIC N 14411C with regards to
composition, temper and magnetic properties.

MUMETAL
Telcon Metals Ltd.
Ni 77, Fe 14, Cu 5, Mo 4.
Soft magnetic alloy, high permeability at low field strengths
for transformers, chokes, etc.

MUMETAL 40
Telcon Metals Ltd.
Cu 5, Mo 4, Fe 14, bal Ni.
Soft magnetic alloy; high permeability.

MUMETAL PLUS
Telcon Metals Ltd.
Soft magnetic alloy, high permeability, low losses for current
transformer, etc.

MUNGOOSE
Barker & Allen Ltd.
Ni 12-15, Cu, Zn.
For domestic utensils, ornaments; nickel silver.

MUNTZ METAL
P.H. Muntz & Co. Ltd.
Copper. Cu 60, Zn 40.
Cast: 48,200 TS; 54 El; 52 RA; 93 Brin. Hard sheet: 80,000 TS;
9.5 El. For bolts, pins, spindles, wire, condenser tubes; high
corrosion resistance.

MUNTZ METAL
Revere Copper Products Inc.
Cu 59-63, bal Zn.
Eighth hard: 60,000 TS; 35,000 YS; 30 El. As hot rolled:
54,000 TS; 21,000 YS; 45 El. For architectural panels and
sheets. C28000

MUNTZ METAL
Chase Brass & Copper Co., Inc.
Copper. Cu 60, Zn 40.
Annealed: 54,000 psi TS; 21,000 psi YS; 45 El. Rolled: 80,000
psi TS; 60,000 psi YS; 5 El. For trim, brazing rod, valve stems,
condenser tubes; high strength.

MUNTZ METAL
Now IMI 452.

MUNTZ METAL NO. 3
Cerro Metal Products Co.
Cu 61, bal Zn.
For forgings; 58,000 TS; 25,000 YS; 45 El; 55 Rock B. For
forgings and hardware. Yellow brass; high strength; CA 280.

MUNTZ METAL-280
Anaconda Co.
Copper. Cu 60, Zn 40.
Soft: 54,000 TS; 20,000 YS; 45 El. For architecture, trim,
perforated sheets. High strength.

MUNTZ PATENTS
English manufacture
Cu 56-63, Zn 37-42, Pb 0-4.
For tubes, hardware; free-cutting.

MUNZBRONZE
VDM Nickel-Technologie AG
Sn 4, Zn 1, bal Cu.
Soft: 48,000 TS; 44 El; 68 Brin. Hard: 82,000 TS; 5 El; 156
Brin. For chemical plant equipment; corrosion resistant.
Obsolete

MURALOY
LaClede Brass (or Steel) Works
Sn 55, Zn 5, Pb 5, bal Cu.
For castings. *Obsolete*

MUREX 4-6 CHROME
Metal & Thermit Corp.
Cr 3.21, Mo 0.5, C, bal Fe.
83,000 TS; 62,000 YS; 26 El; 45 RA. For welding electrodes;
corrosion resistant.

MUREX ALTERNEX-A
Metal & Thermit Corp.
C 0.1, bal Fe.
Welded: 76,400 TS; 66,200 YS; 24 El. For welding electrodes;
for mild steel; Type E-6013.

MUREX CARBON-MOLY
Metal & Thermit Corp.
C 0.12, Si 0.08, Mn 0.64, Mo 0.48, bal Fe.
70,000 TS; 54,000 YS; 34 El; 66.5 RA. For welding electrodes;
creep strength at high temperature. *Obsolete*

MUREX CHROMANSIL
Metal & Thermit Corp.
C 0.12, Cr 0.18, Mo 0.5, Mn 0.45-0.55, bal Fe.
81,800 TS; 76,800 YS; 25 El; 61.5 RA. For welding electrodes;
high tensile strength. *Obsolete*

MUREX CHROME-COPPER
Metal & Thermit Corp.
Cr 0.4, Cu 0.2, Si 0.08, C 0.08, bal Fe.
84,500 TS; 63,000 YS; 26 El; 46.5 RA. For welding electrodes;
corrosion resistant. *Obsolete*

MUREX CHROME-MOLY
Metal & Thermit Corp.
C 0.08, Cr 0.32, Mo 0.67, bal Fe.
80,000 TS; 60,000 YS; 25 El; 62 RA. For welding electrodes;
creep strength. *Obsolete*

MUREX CRESTA
Metal & Thermit Corp.
C 0.15, bal Fe.
For welding electrodes; for mild steel. *Obsolete*

MUREX FILLEX
Metal & Thermit Corp.
C 0.2, bal Fe.
Welded: 66,000 TS; 53,000 YS; 28 El. For welding electrodes;
for mild steel; Type E-6020.

MUREX GENEX-M
Metal & Thermit Corp.
C 0.2, bal Fe.
Welded: 68,000 TS; 54,000 YS; 28 El. For welding electrodes;
Type E-6012.

MUREX HARDEX 20
Metal & Thermit Corp.
C, Cr, Mn, bal Fe.
For hard surfacing electrodes; for moderate abrasion and
high impact. *Obsolete*

MUREX HARDEX 25
Metal & Thermit Corp.
C 0.2, Mn 1.2, Cr 0.7, Si 0.9, bal Fe.
For hard facing electrodes; for moderate abrasion and high
impact.

MUREX HARDEX 45
Metal & Thermit Corp.
C 0.6, Mn 1.4, Cr 2.7, Si 0.8, bal Fe.
For hard facing electrodes; for high abrasion and moderate
impact.

MUREX HARDEX 60
Metal & Thermit Corp.
C, Cr, Mn, bal Fe.
For hard facing electrode; for high abrasion and moderate
impact. *Obsolete*

MUREX MANGANESE STEEL
Metal & Thermit Corp.
C 1, Mn 13, Ni 0.85, bal Fe.
For welding electrodes; wear resistant. *Obsolete*

MUREX MOLEX
Metal & Thermit Corp.
C 0.15, bal Fe.
Welded: 75,000 TS; 62,000 YS; 25 El. For welding electrodes;
for C-Mo steel; Type E-7010 Al.

MUREX NICKEL STEEL
Metal & Thermit Corp.
Mo 0.29, Mn 0.54, Si 0.08, C 0.08, Ni 2-3.5, bal Fe.
86,000 TS; 72,000 YS; 25 El; 64 RA. For electrodes for
welding; high impact. *Obsolete*

MUREX NO. 14 CHROME
Metal & Thermit Corp.
Cr 14, C, bal Fe.
For welding electrodes; for 12-14 Cr steel. *Obsolete*

MUREX NO. 5 CHROME
Metal & Thermit Corp.
Cr 4-6, C, bal Fe.
For welding electrodes; for 4-6 Cr steel. *Obsolete*

MUREX ROLEX
Metal & Thermit Corp.
C 0.18, bal Fe.
For hard surfacing electrodes; for carbon steels. *Obsolete*

MUREX SPECIAL A
Metal & Thermit Corp.
C 0.08, Si 0.08, Mn 0.46, Ni 0.85, bal Fe.
73,000 TS; 59,000 YS; 31 El; 63.5 RA. For welding electrodes;
high tensile strength. *Obsolete*

MUREX STAINLESS
Metal & Thermit Corp.
C 0.08, Cr 18, Ni 8, bal Fe.
For stainless steel welding electrodes. *Obsolete*

MUREX STAINLESS NO. 15
Metal & Thermit Corp.
C 0.15, Cr 18, Ni 8, bal Fe.
For welding electrodes; stainless. *Obsolete*

MUREX STAINLESS NO. 7
Metal & Thermit Corp.
C 0.15, Cr 18, Ni 8, bal Fe.
For welding electrodes; stainless. *Obsolete*

MUREX SURFACING
Metal & Thermit Corp.
C, alloy, bal Fe.
For welding rod; for high speed surfacing. *Obsolete*

MUREX TYPE F
Metal & Thermit Corp.
C, bal Fe.
For welding electrodes; for mild steel. *Obsolete*

MUREX TYPE FHP
Metal & Thermit Corp.
C 0.08-0.13, Mn 0.2-0.4, bal Fe.
Welded: 67,000 TS; 56,000 YS; 28 El. For welding, electrodes; shielded arc; Type E-6020.

MUREX VERTEX
Metal & Thermit Corp.
C, bal Fe.
For welding electrodes; for mild steel. *Obsolete*

MURMANS ALLOY-1
English manufacture
Al 72, Zn 15, Mg 14.
For light alloy parts; non-hardenable.

MURMANS ALLOY-2
English manufacture
Al 92, Zn 4.4, Mg 3.6.
For light alloy parts; non-hardenable.

MUSHET STEEL
Japanese manufacture
C 1.5-2, W 4-6, Cr 0.25-0.3, Mn 0.3-0.5, bal Fe.
For tools, dies; cannot be cold worked.

MUSIC SPRING
Wickwire Spencer Steel Co.
C 0.8, Mn 0.25, Si 0.25, bal Fe.
For springs; Swedish steel.

MUSTANG
Jessop Steel Co.
C 0.84, Cr 4.3, V 1.9, W 6.4, Mo 5, bal Fe.
For cutters, taps, reamers, hobs; high speed steel. *Obsolete*

MUSTANG M-2 (AISI M2)
Jessop Steel Co.
C 0.84, Cr 4.2, W 6.5, V 2, Mo 5, bal Fe.
For cutting tools, broaches, chasers, drills, hobs, reamers; high speed steel.

MUSTANG SPECIAL
Jessop Steel Co.
C 0.8, W 6, Mo 5, Cr 4, V 1, bal Fe.
For cutters, taps, tools, dies, reamers, drills; high speed steel. *Obsolete*

MUSTANG-LC (AISI H-42)
Jessop Steel Co.
C 0.65, Cr 4, V 2, W 6.5, Mo 5, bal Fe.
As quenched: 58-59 Rock C. Tempered 1000°F: 60-61 Rock C. For hot forming and swaging dies, nut piercers, hot punches, extrusion dies, forging mandrels, chipper dies. Good resistance to high temperature softening. High speed steel. Type H42.

MUTEMP
British Steel Corp.
Fe 70, Ni 30.
For compensating shunts for electrical equipment; thermo-sensitive, magnetic. *Obsolete*

MUVAR
Hamilton Technology Inc.
Mn 0.3, Mo 4, Fe 16.7, Ni 79.
For electrical and magnetic equipment; magnetically soft. *Obsolete*

MV 350
Bergische Stahl Industrie
C 1.4-1.5, Si 0.2-0.35, Mn 0.3-0.5, V 3-3.5, bal Fe.
Cold work tool steel. W.-Nr 1.2838.

MV-1
Latrobe Steel Co.
C 0.8, Cr 4.1, Mo 4.25, V 1.1, bal Fe.
For cutting tools, small taps, thread rolls; low alloy high speed steel. *Obsolete*

MV-2
Latrobe Steel Co.
C 0.88, Cr 4.1, Mo 4.25, V 2, bal Fe.
For cutting tools, pipe taps, thread chasers; low alloy high speed steel. *Obsolete*

MVD
Acciaierie Valbruna s.p.a.
C 0.4, Si 0-0.35, Mn 0.6, P 0.035, S 0.035, Cr 0.95, Mo 0.6, V 0.3, bal Fe.
Steel for high temperature bolting; ASTM A193-B16.

MVD-DE
Acciaierie Valbruna s.p.a.
C 0.45, Si 0.22, Mn 0.75, P 0-0.03, S 0-0.03, Cr 1.45, Mo 0.75, V 0.3, bal Fe.
Heat resisting structural steel. W. Nr. 1.2323.

MX-2
American manufacture
C 0.39, Cr 1.1, Co 1, Si 1, Mn 0.7, V 0.15, Mo 0.25, bal Fe.
Modified AISI 4037 (4037 + Co).

MX10
Sheffield Smelting Co. Ltd.
Ag 40, Cu, Zn, Cd.
Melting range: 605-635°C. Maximum stress: 44.9 kgf/mm^2; 15 El. For silver brazing; good fluidity.

MX18N
Sheffield Smelting Co. Ltd.
Ag 48, Cu, Zn, Cd, Ni.
Melting range; 640-660°C. Maximum stress: 47.2 kgf/mm^2; 35 El. For silver brazing; WC tipped tools.

MX20
Sheffield Smelting Co. Ltd.
Ag 50, Cu, Zn, Cd.
Melting range: 620-640°C. Maximum stress: 45.7 kgf/mm^2; 35 El. For silver brazing; optimum fluidity.

MX20N
Sheffield Smelting Co. Ltd.
Ag 50, Cu, Zn, Cd, Ni.
Melting range: 47.2 kgf/mm^2; 35 El. For silver brazing; WC tipped tools.

MX8N
Sheffield Smelting Co. Ltd.
Ag 38, Cu, Zn, Cd, Ni.
Melting range: 640-670°C. Maximum stress; 45.7 kgf/mm^2; 30 El. For silver brazing; WC tipped tools.

MY NO. 3
Vereinigte Edelstahlwerke
C 0.5, Mn 0.7, Si 1.8-2, Mo 0.2, bal Fe.
For tools, chisels; tough. *Obsolete*

MY-A-CHROME
Houghton & Richards Inc.
C 0.7, Cr 1, bal Fe.
For tools, drills, punches; oil or water hardening.

MYA
Houghton & Richards Inc.
C 0.48, Mn 0.3, Si 1.45, Cr 1.45, V 0.25, bal Fe.
Oil hardening tool steel; shock resistant.

MYSTIC METAL
Manufacturer not listed.
Sn 11, Bi 0.1, Pb 89.
For solder.

N 0721
VILLARES
C 0.09, Mn 0.7, P 0-0.07, S 0.15-0.25, bal Fe.
Free cutting steel, similar to Werkstoff Nr. 1.0721.

N 0726
VILLARES
C 0.36, Mn 0.7, P 0-0.06, S 0.15-0.25, bal Fe.
Free cutting steel. Similar to Werkstoff Nr. 1.0726.

N 1, ETC.
Now EASTERN N 1, ETC.

N 12
Walsingham Steel Co., Ltd.
C 0.1-0.14, Si 0.3-0.6, Mn 0.6-0.8, Ni 3.4-3.6, Cr 0-0.25, Mo 0-0.15, Cu 0-0.3, bal Fe.
Low temperature, impact resistant. Meets BS 1504-503; ASTM A352-66 Gr. LC3.

N 52 (NEW BIDE)
Newcomer Products Inc.
WC 72, Co 11, 8.0 TiC, 10.0 TaC.
Sintered: 375,000 Transverse strength, 91.0 RA. For tools.

N-153 ALLOY
American manufacture
C 0.35-0.45, Mn 1.6, Ni 15-16, Cr 15-16, W 1.25-1.75, Mo 4, Cb 0.8-1.2, bal Fe.
Rolled: 176,000 TS; 146,000 YS; 15 El; 30 RA. For jet engine parts; heat resistant.

N-155
Driver Harris Co.
Superalloy. Ni 10, Cr 23, Co 20, W 2, Mo 3, Cb 1, bal Fe.
Wire, rod, ribbon for welding wire and fastener stock.
Obsolete

N-155
Cannon-Muskegon Corp.
C 0.1-0.2, Ni 18-22, Cr 18-22, Mo 2.5-3.5, W 2-3, Cb 0.7-1.2, Co 18-22, 0.1-0.2 N_2.
Solution treated: 30,000-50,000 TS. For jet engine parts, gas turbine blades, ship propulsion blades; heat resistant, similar to "Multimet."

N-155
Stellite Division
C 0.1-0.2, Ni 18-22, Cr 18-22, Mo 2.5-3.5, W 2-3, Cb 0.7-1.2, Co 18-22, 0.1-0.2 N_2.
Solution treated: 30,000-50,000 TS. For jet engine parts, gas turbine blades, ship propulsion blades; heat resistant, similar to "Multimet."

N-155
Wallace Murray Corp.
C 0.1-0.2, Ni 18-22, Cr 18-22, Mo 2.5-3.5, W 2-3, Cb 0.7-1.2, Co 18-22, 0.1-0.2 N_2.
Solution treated: 30,000-50,000 TS. For jet engine parts, gas turbine blades, ship propulsion blades; heat resistant, similar to "Multimet."

N-155
Now UDIMET N-155; ALLOY N-155; CARPENTER PYR N-155.

N-155
Ancast, Inc.
Nickel. C 0-0.2, Mn 1-2, Si 0-1, Cr 20-22.5, Ni 19-21, Mo 2.5-3.5, W 2-3, Co 18.5-21, P 0-0.04, S 0-0.03, 0.75-1.25 Cb + Ta, 0.10-0.20 N_2, bal Fe.
AMS 5378D; ASTM A-567 GR3.

N-2
Now ALOYCO N-2.

N-238
English manufacture
Ni, Cr.
For high temperature applications; heat and corrosion resistant.

N-3
Now ALOYCO N-3.

N-3 ALLOY
American manufacture
C 0.4-0.7, Si 0-2, Mn 0-1.5, Cr 20-26, bal Fe.
Cast: 80,000 TS; 40,000 YS; 0 El; 200 Brin. For furnace parts, grids, combustion chambers. High heat and corrosion resistance.

N-3 METAL
Lunkenheimer Co.
Ni 32, Pb 2.75, Si 9.5, Mn 0.75, bal Cu.
For hardware; free-cutting, corrosion resistant.

N-A-X
Sharon Steel Corp.
C 0.12, Mn 0.8, Si 0.7, Cr 0.55, Zr 0.08, bal Fe.
HSLA hot rolled strip or sheet. 70,000 psi TS min; 50,000 psi YS min; 22 El min. For automotive and structural applications. SAE 950A.

N-A-X 80
National Intergroup Inc.
C 0.12, Mn 1, Cu 0.8, Cb, bal Fe.
Sheet: 90,000 psi TS min; 80,000 psi YS min. Zirconium treated. For automotive, truck, agricultural and railroad applications. ASTM A-715; SAE J410c, Grade 980 XK.

N-A-X AC9111
National Steel Corp.
C 0.1, Mn 0.6, Si 0.7, Cr 0.6, Zr 0.1, bal Fe.
Rolled: 75,000 TS; 50,000 YS; 40 El; 75 RA. For structures, bridges; good weldability. *Obsolete*

N-A-X AC9115
National Steel Corp.
C 0.1-0.17, Cr 0.6, Mo 0-0.15, Zr 0.05-0.15, bal Fe.
Hot rolled: 76,000 TS; 52,000 YS; 40 El; 74 RA; 150 Brin. For gas turbine structures; good weldability. *Obsolete*

N-A-XTRA 100
National Intergroup Inc.
C 0-0.21, Mn 0.6-1.1, Si 0.4-0.9, Cr 0.5-0.8, Mo 0-0.3, Zr 0.05-0.15, B 0-0.0025, bal Fe.
Hardened (min): 115,000 TS; 100,000 YS; 18 El. For structural parts.

N-A-XTRA 110
National Intergroup Inc.
C 0-0.21, Mn 0.6-1.1, Si 0.4-0.9, Cr 0.5-0.8, Mo 0-0.3, Zr 0.05-0.15, B 0-0.0025, bal Fe.
Hardened (min): 125,000 TS; 110,000 YS; 18 El. For structural parts.

N-A-XTRA 125
National Steel Corp.
C 0.18-0.22, Mn 0.8, Si 0.8, Cr 0.6, Mo 0.16, Zr 0.12, bal Fe.
Heat treated: 139,000 TS; 126,800 YS; 20 El; 55 RA; 293 Brin. For excavator buckets, mine cars and equipment, road machinery; structural steel, good formability. *Obsolete*

N-A-XTRA 150
National Steel Corp.
C 0.26-0.3, Mn 0.8, Si 0.8, Cr 0.6, Mo 0.14-0.18, Zr 0.08-0.15, bal Fe.
Heat treated: 165,000 TS; 158,600 YS; 13 El; 47 RA; 352 Brin. For excavator buckets, mine cars and equipment, road machinery; structural steel, good formability. *Obsolete*

N-A-XTRA 55
National Steel Corp.
C 0.18, Si 0.7, Mn 0.9, P 0.018, S 0.021, Cr 0.89, Mo 0.32, Zr 0.09, bal Fe.
Plate: 92,000-114,000 TS; 78,000 minimum YP; 18 minimum El. For bridges, tanks, railroad cars, containers, ships, aircraft. Good formability and weldability. *Obsolete*

N-A-XTRA 60
National Steel Corp.
C 0.18, Si 0.7, Mn 0.9, P 0.018, S 0.021, Cr 0.9, Mo 0.3, Zr 0.09, bal Fe.
Plate: 100,000-120,000 TS; 86,000 minimum YP; 18 minimum El. For tanks, containers, machine parts, ships, bridges, aircraft, railroad cars. Good formability and weldability. *Obsolete*

N-A-XTRA 65
National Steel Corp.
C 0.18, Si 0.7, Mn 0.9, P 0.018, S 0.021, Cr 0.9, Mo 0.32, Zr 0.09, bal Fe.
Plate: 107,000-127,000 TS; 92,000 minimum YP; 17 minimum El. For tanks, containers, machine parts, bridges, railroad cars, aircraft. Good formability and weldability. *Obsolete*

N-A-XTRA 70
National Steel Corp.
C 0.18, Si 0.7, Mn 0.9, P 0.018, S 0.021, Cr 0.9, Mo 0.3, Zr 0.09, bal Fe.
Plate: 114,000-135,000 TS; 100,000 minimum YP; 16 minimum El. For containers, tanks, machine parts, ships, bridges, aircraft, railroad cars. Good weldability, and formability. *Obsolete*

N-A-XTRA 80
National Intergroup Inc.
C 0-0.21, Mn 0.6-1.1, Si 0.4-0.9, Cr 0.5-0.8, Mo 0-0.3, Zr 0.05-0.15, B 0-0.0025, bal Fe.
Hardened (min): 95,000 TS; 80,000 YS; 20 El. For structural parts.

N-A-XTRA 90
National Intergroup Inc.
C 0-0.21, Mn 0.6-1.1, Si 0.4-0.9, Cr 0.5-0.8, Mo 0-0.3, Zr 0.05-0.15, B 0-0.0025, bal Fe.
Hardened (min): 105,000 TS; 90,000 YS; 18 El. For structural parts.

N-S 1000
Duraloy
C 0.37-0.43, Cr 4.75-5.25, Mo 1.2-1.4, bal Fe.
Drawn: 200,000 TS; 1-2 El. For dies, tools. Oil or air hardened. *Obsolete*

N-S 25
Duraloy
Now DURALOY JPT.

N-S STAINLESS STEEL WELDING WIRE
National Standard Co..
Welding wire of 22 stainless compositions for arc welding stainless steels, bare wire.

N-TUF-CR 196
Nippon Steel USA Inc.
C 0-0.13, Mn 0.9-1.5, Ni 5-6, Mo 0.1-0.3, Cr 0.6, bal Fe.
For welded pressure vessels.

N.A.-4A
Duraloy Blaw-Knox/Union Steel Casting
C 0.2, Cr 20, Ni 10, bal Fe.
For chemical plant equipment; austenitic, stainless. *Obsolete*

N.B.M. SILVER BABBITT
Abex Corp.
Sb, Ag, Sn, bal Pb.
For bearings; Babbitt. *Obsolete*

N.C. ALLOY
Lehigh Steel Corp.
C, alloy, bal Fe.
For tools, dies; oil hardening.

N.C. HEAT RESISTING
Crucible Steel Castings Co.
C 0.25, Cr 28, Ni 8, bal Fe.
For heat and corrosion resistant parts; heat and corrosion resistant. *Obsolete*

N.C.A. (NON-CORRODIBLE ALUMINUM)

Haywoods NCA Metal Ltd.
Ni 2.2, Cu 3, Mg 0.2, N 0.2, bal Al.
Rolled: 45,000 TS. Sand cast: 24,000-31,000 TS;
12,000-15,000 YS; 7-5 El; 63 Brin. For light alloy parts.
Resists seawater corrosion.

N.C.C. ALLOY

Henry Wiggin & Co. Ltd.
C 0-1.5, Si 0-1.75, 56% Ni min, 23% Cu min, 7.5% Cr min.
For production of nickel-resistant castings; MP 2300 F.
Obsolete

N.C.C. PIG

International Nickel Inc.
Nickel. C 0-1.5, Si 0-1.75, 56.0 Ni min, 23.0 Cu min, 7.5 Cr min.
For foundry alloy for NI-RESIST IRON; MP 2300°F. *Obsolete*

N.C.M.

British Steel Corp.
C 0.1-0.2, Ni 4-4.5, Cr 1.25-1.5, Mo 0.15-0.3, bal Fe.
Heat treated: 200,000 TS; 165,000 YS; 15 El; 47 RA; 418 Brin.
For case hardened parts for abrasion resistance; case hardening steel; Iz-26. *Obsolete*

N.C.M.

Joseph T. Ryerson & Son Inc.
C, Ni, Cr, Mo, bal Fe.
Heat treated: 125,000 TS; 105,000 YS; 16 El; 50 RA; 270-310 Brin. For gears, shafts; oil hardened, tough. *Obsolete*

N.C.O.H.

British Steel Corp.
C, alloy, bal Fe.
Annealed: 94,000 TS; 52,000 YS; 28 El; 50 RA; 196 Brin. Heat treated: 230,000 TS; 210,000 YS; 8 El; 25 RA; 477 Brin. For gears, axles, crankshafts. *Obsolete*

N.D. FORGING STEEL

Teledyne Firth Sterling
C 0.3, W 15, Cr 4, V 0.5, bal Fe.
For forging dies, extrusion rams and liners; hot work steel, oil hardened. *Obsolete*

N.D.S.

Latrobe Steel Co.
C 0.75, Cr 1, Ni 1.75, bal Fe.
For tools, shear blades; shock resisting.

N.G.F. ALLOY

American manufacture
Cu 1-1.5, Mn 0.75-2, bal Al.
25,000 TS; 18,000 YS; 4 El. For light alloy castings.

N.K.L. METAL

J. Shaw, Son & Greenhalgh Ltd.
For superheated steam.

N.M.C

English manufacture
Ni 12, Mn 5, Cr 3.5, C 0.5, Si 0.5, bal Fe.
For valve inserts; heat resistant.

N.O.H.

British Steel Corp.
Ni 3, C, bal Fe.
Annealed: 90,000 TS; 60,000 YS; 20 El; 50 RA. Heat treated: 110,000 TS; 90,000 YS; 27 El; 55 RA. For gears, shafts; tough. *Obsolete*

N.P.L. "Y" ALLOY

Alcan-Booth Industries, Ltd.
Al 93, Ni 2, Mg 1.5, Cu 4.
Chill cast: 29,000 TS; 25,000 YS; 1 El; 80 Brin. Heat treated: 43,000 TS; 34,000 YS; 3 El; 105 Brin. For airplane castings, pistons, crankcases, cylinder heads; corrosion resistant.

N.P.L. "Y" ALLOY

National Physical Laboratory
Al 93, Ni 2, Mg 1.5, Cu 4.
Chill cast: 29,000 TS; 25,000 YS; 1 El; 80 Brin. Heat treated: 43,000 TS; 34,000 YS; 3 El; 105 Brin. For airplane castings, pistons, crankcases, cylinder heads; corrosion resistant.

N.W.S

Andrews Toledo Ltd.
C 0.9, Mn 1.2, bal Fe.
For tools, dies; non-shrinkable.

N.W.S

Balfour Darwins Ltd.
C 0.9, Mn 1.2, bal Fe.
For tools, dies; non-shrinkable.

N10 (NEW BIDE)

Newcomer Products Inc.
Sintered carbide tool material.
For heavy roughing cuts on cast iron, non-ferrous metals and non-metallics.

N18K8M3T

Russian manufacture
C 0.05, Ni 18.3, Co 8.5, Mo 3.5, Ti 0.34, Al 0.14, bal Fe.
Heat treated: 300,000 TS; 250,000 YP. For jet engine and spacecraft components, landing gears, missiles, crimping tools; maraging steel, tough, shock resistant.

N20 (NEW BIDE)

Newcomer Products Inc.
Sintered carbide tool material.
For general purpose machining and milling of cast iron and non-ferrous metals.

N22 (NEW BIDE)

Newcomer Products Inc.
Sintered carbide tool material.
For machining of high temperature alloys, non-ferrous and non-metallics.

N30 (NEW BIDE)

Newcomer Products Inc.
Sintered carbide tool material.
For semi-finishing and finishing of cast iron non-ferrous and non-metallic materials, and high temperature alloys.

N35K6

Russian manufacture
C 0.01, Ni 35, Co 5, Ti 2.3, Cu 0.2, Si 0.4, Mn 0.6, bal Fe.
Hardened: 142,000 TS. For electrical equipment, instruments, pumps; precipitation hardenable, low expansion alloy.

N40 (NEW BIDE)

Newcomer Products Inc.
Sintered carbide tool material.
For high speed finishing and precision boring of cast iron, non-ferrous metals and non-metallics.

N50 (NEW BIDE)

Newcomer Products Inc.
Sintered carbide tool material.
For heavy duty rough machining and milling of carbon and alloy steels.

N60 (NEW BIDE)

Newcomer Products Inc.
Sintered carbide tool material.
For general purpose turning, planning and milling of carbon and alloy steel.

N70 (NEW BIDE)

Newcomer Products Inc.
Sintered carbide tool material.
For light roughing and finishing cuts on carbon and alloy steel.

N7131

VILLARES
C 0.16, Si 0.25, Mn 1.15, Cr 0.95, bal Fe.
Alloy carburizing steel. DIN 16 Mn Cr 5.

N7147

VILLARES
C 0.2, Si 0.25, Mn 1.25, Cr 1.15, bal Fe.
Alloy carburizing steel. DIN 20 Mn Cr 5.

N72 (NEW BIDE)

Newcomer Products Inc.
Sintered carbide tool material.
For moderate roughing and finishing of carbon and alloy steels.

N80

Societe Nouvelle des Acieries de Pompey
C 0.12, Mn 1.5, Mo 0.2, bal Fe.
Tube: 100,000 psi TS min; 80,000 psi YS min. For gas well piping and tubing. Resists sulfide stress corrosion.

N80 (NEW BIDE)

Newcomer Products Inc.
Sintered carbide tool material.
For fast finishing and precision boring of carbon and alloy steels.

N8161

VILLARES
C 0.58, Si 0.25, Mn 0.95, Cr 1.05, V 0.1, bal Fe.
Spring steel. DIN 58 Cr V 4.

N93 (NEW MET)

Newcomer Products Inc.
Sintered carbide tool material.
For normal to high velocity machining of carbon and alloy steel where extra toughness is required.

N95 (NEW MET)

Newcomer Products Inc.
Sintered carbide tool material.
For normal to high velocity machining of carbon and alloy steel to close tolerance finishes.

NA 3, ETC.

Now DBK NA-3, ETC.

NA-1

Duraloy
C 0.1-0.5, Cr 28, Ni 10, bal Fe.
For heat and corrosion resistant parts, furnace parts. Heat and corrosion resistant to sulfur atmospheres. *Obsolete*

NA-12

Duraloy
C 0.15-0.4, Cr 11.5-14, Ni 0-1, bal Fe.
Annealed: 95,000 TS; 65,000 YS; 25 El; 190 Brin. For glass mold dies, steam turbine parts, valves, pumps. Type CA40; corrosion resistant. *Obsolete*

NA-18

Duraloy
Cr 17, C 0-0.3, bal Fe.
For corrosion resisting parts. Corrosion resistant. *Obsolete*

NA-19

Duraloy
Cr 19-26, C 0.5-0.7, Ni 3, bal Fe.
Cast: 65,000 TS; 50,000 YS; 2 El; 2 RA; 175 Brin. For stainless parts, heat resisting parts, Mg pots, pipe. Corrosion resistant, heat resistant. ACI-CCHC. *Obsolete*

NA-2

Duraloy
C 0.35-0.7, Cr 15, Ni 35, bal Fe.
Cast: 70,000 TS; 35,000 YS; 10 El; 10 RA; 175 Brin. For furnace parts, retorts, heat treating boxes. Type HT; corrosion and heat resistant. *Obsolete*

NA-20
Duraloy
C 0-0.5, Cr 26-30, Ni 4-7, Mo 0-0.5, bal Fe.
Cast: 85,000 TS; 48,000 YS; 16 El; 190 Brin. For furnace blowers, copper melting equipment. Type HD; resists flue gases. *Obsolete*

NA-26
Duraloy
Special alloy limited to chutes (or snouts) in continuous galvanizing lines.

NA-2T
Duraloy
C 0.4-0.6, Ni 36-39, Cr 16-19, Si 1.5, Mn 1.2, bal Fe.
Cast: 68,000 TS; 40,000 YS; 10 El; 10 RA; 187 Brin. For high temperature chains and conveyors, heat treating boxes. Type HU. Good thermal fatigue. *Obsolete*

NA-34
Duraloy
C 0-0.1, Cr 18-20, Ni 8-10, Mo 2-4, bal Fe.
Heat treated: 75,000 TS; 35,000 YS; 50 El; 50 RA; 200 Brin. For castings, pipe, paper mill equipment. Corrosion resistant. *Obsolete*

NA-4
Duraloy Blaw-Knox/Union Steel Casting
C 0-0.1, Cr 18-20, Ni 8-10, bal Fe.
Annealed: 75,000 TS; 35,000 YS; 50 El; 50 RA; 150 Brin. For cast pipe, chemical plant equipment; Type CF; corrosion resistant. *Obsolete*

NA-4CB
Duraloy Blaw-Knox/Union Steel Casting
C 0-0.1, Cr 18-20, Ni 8-10, Cb = 10 x C, bal Fe.
Annealed: 80,000 TS; 40,000 YS; 45 El; 45 RA; 15 Brin. For welded structures, chemical plant equipment; Type CF-8C; stainless, austenitic. *Obsolete*

NA-60
Duraloy
Now DURALOY HW.

NA-65
Duraloy
Cr 18-20, Ni 65-70, C 0-0.3, bal Fe.
Cast: 75,000 TS; 45,000 YS; 15 El; 15 RA; 200 Brin. For high temperature service castings, pipe, retorts. Heat resistant to 2100°F. ACI-HX. *Obsolete*

NA-7
Duraloy
C 0.3-0.6, Cr 24-30, Ni 18-22, bal Fe.
Cast: 75,000 TS; 50,000 YS; 15 RA; 175 Brin. For furnace parts. ACI-HK. Nonmagnetic, high creep resistance up to 2000°F. *Obsolete*

NA226
Birmingham Aluminum Casting Co.
Aluminum. Cu 4-5, Ti 0-0.2, bal Al.
32,000-42,000 TS; 14,000-28,000 YS; 5-12 El; 65-100 Brin. For light alloy parts; hardenable.

NA22H
Duraloy
C 0.44, Ni 46, Cr 26.3, W 5.3, Mn 1.4, Si 1, bal Fe.
At 70°F: 64,500 TS; 3.5 El; 2.7 RA. At 1800°F: 18,000 TS; 32.0 El; 48.0 RA. For radiant tubes for furnaces, gas generator retorts. High heat resistance to 2200°F.

NA350
Birmingham Aluminum Casting Co.
Aluminum. Cu 9.5-10.5, Cu 0-0.15, Si 0-0.25, bal Al.
40,000 TS; 21,000 YS; 10 El; 75 Brin. For light alloy parts; hardenable.

NA6LC ALLOY
Duraloy
C, Ni, Cr, bal Fe.
For pickling hooks, hangers, frames. Corrosion resistant.
Obsolete

NACO
Capitol Castings Inc.
Cast steel. C, Cr, Mn, bal Fe.
For mill balls, linings; corrosion and abrasion resistant.

NACO 123
Nordisk Aluminium Industry, A/S
Aluminum. Si 5, Ti 0-0.2, bal Al.
Cast: 19,000-24,000 TS; 9000-19,000 YS; 10-6 El; 40-50 Brin. For instrument housings, cases; corrosion resistant, good castability.

NACO 123
Norsk Aluminium Co. A/S
Aluminum. Si 5, Ti 0-0.2, bal Al.
Cast: 19,000-24,000 TS; 9000-19,000 YS; 6-10 El; 40-50 Brin. For instrument housings, cases; corrosion resistant, good castability.

NACO 160
Nordisk Aluminium Industry, A/S
Aluminum. Si 12, bal Al.
Die cast: 37,000 TS; 18,000 YS; 2 El. For instrument housings, cases; corrosion resistant; for thin wall sections.

NACO 160
Norsk Aluminium Co. A/S
Aluminum. Si 12, bal Al.
Die cast: 37,000 TS; 18,000 YS; 2 El. For instrument housings, cases; corrosion resistant; for thin wall sections.

NACO 162
Nordisk Aluminium Industry, A/S
Aluminum. Cu 1, Mg 1, Si 12, bal Al.
Cast: 36,000 TS; 28,000 YS; 0.5 El; 105 Brin. For pistons; low coefficient of expansion.

NACO 162
Norsk Aluminium Co. A/S
Aluminum. Cu 1, Mg 1, Si 12, bal Al.
Cast: 36,000 TS; 28,000 YS; 0.5 El; 105 Brin. For pistons; low coefficient of expansion.

NACO 1S
Nordisk Aluminium Industry, A/S
Aluminum. 99.5 Al min.
O-temper: 12,000 TS; 4500 YS; 37 El. 1/2 Hard: 16,000 TS; 14,000 YS; 10 El. Hard: 20,500 TS; 19,500 YS; 9 El. For light weight structures; corrosion resistant.

NACO 1S
Norsk Aluminium Co. A/S
Aluminum. 99.5 Al min.
O-temper: 12,000 TS; 4500 YS; 37 El. 1/2 Hard: 16,000 TS; 14,000 YS; 10 El. Hard: 20,500 TS; 19,500 YS; 9 El. For light weight structure; corrosion resistant.

NACO 2S
Nordisk Aluminium Industry, A/S
Aluminum. 1 Fe + Si max, 99.0 Al min.
O-temper: 13,500 TS; 5500 YS; 35 El; 23 Brin. Hard temper: 24,000 TS; 23,000 YS; 5 El; 44 Brin. For light weight structures, tanks, containers; corrosion resistant.

NACO 2S
Norsk Aluminium Co. A/S
1 Fe + Si max, 99.0 Al min.
O-temper: 13,500 TS; 5500 YS; 35 El; 23 Brin. Hard temper: 24,000 TS; 23,000 YS; 5 El; 44 Brin. For light weight structures, tanks, containers; corrosion resistant.

NACO 320
Nordisk Aluminium Industry, A/S
Aluminum. Mg 4, Si 0.5, Ti 0-0.2, bal Al.
Cast: 25,000 TS; 12,000 YS; 9 El; 50 Brin. For marine parts, architectural trim; corrosion resistant.

NACO 320
Norsk Aluminium Co. A/S
Aluminum. Mg 4, Si 0.5, Ti 0-0.2, bal Al.
Cast: 25,000 TS; 12,000 YS; 9 El; 50 Brin. For marine parts, architectural trim; corrosion resistant.

NACO 350
Nordisk Aluminium Industry, A/S
Aluminum. Mg 10, bal Al.
Heat treated: 45,000 TS; 25,000 YS; 14 El; 75 Brin. For gasoline fuel meters, aircraft and marine castings; heat treatable, corrosion resistant.

NACO 350
Norsk Aluminium Co. A/S
Aluminum. Mg 10, bal Al.
Heat treated: 45,000 TS; 25,000 YS; 14 El; 75 Brin. For gasoline fuel meters, aircraft and marine castings; heat treatable, corrosion resistant.

NACO 3S
Nordisk Aluminium Industry, A/S
Aluminum. Mn 1.2, bal Al.
O-temper: 17,000 TS; 7500 YS; 30 El; 28 Brin. Hard temper: 29,000 TS; 27,000 YS; 4 El; 55 Brin. For structures, tank cars, containers; resists reducing and oxidizing atmospheres.

NACO 3S
Norsk Aluminium Co. A/S
Aluminum. Mn 1.2, bal Al.
O-temper: 17,000 TS; 7500 YS; 30 El; 28 Brin. Hard temper: 29,000 TS; 27,000 YS; 4 El; 55 Brin. For structures, tank cars, containers; resists reducing and oxidizing atmospheres.

NACO 50S
Nordisk Aluminium Industry, A/S
Aluminum. Mg 0.7, Si 0.4, bal Al.
O-temper: 16,000 TS; 9000 YS; 26 El. T-temper: 33,000 TS; 29,000 YS; 18 El. For architectural shapes, trim, molding; heat treatable, corrosion resistant.

NACO 50S
Norsk Aluminium Co. A/S
Aluminum. Mg 0.7, Si 0.4, bal Al.
O-temper: 16,000 TS; 9000 YS; 26 El. T-temper: 33,000 TS; 29,000 YS; 18 El. For architectural shapes, trim, molding; heat treatable, corrosion resistant.

NACO 54S
Nordisk Aluminium Industry, A/S
Aluminum. bal Al.
Aluminum alloy for light weight parts.

NACO 54S
Norsk Aluminium Co. A/S
Aluminum. bal Al.
Aluminum alloy for light weight parts.

NACO 57S
Nordisk Aluminium Industry, A/S
Aluminum. Mg 2.5, Cr 0.3, bal Al.
O-temper: 29,000 TS; 14,000 YS; 25 El; 45 Brin. Hard Temper: 43,000 TS; 38,000 YS; 7 El; 85 Brin. For marine construction and hardware; resists seawater corrosion.

NACO 57S
Norsk Aluminium Co. A/S
Aluminum. Mg 2.5, Cr 0.3, bal Al.
O-temper: 29,000 TS; 14,000 YS; 25 El; 45 Brin. Hard temper: 43,000 TS; 38,000 YS; 7 El; 85 Brin. For marine construction and hardware; resists seawater corrosion.

NACO STEEL
Capitol Castings Inc.
C 0.3-0.4, Mn 1.5, bal Fe.
110,000 TS; 70,000 YS; 8 El; 15 RA; 250 Brin. For car coupler knuckles, car wheels, anchor chains; water hardened. *Obsolete*

NADA C
Nassau Smelting & Refining Co.
Si 2.2, bal Cu.
Cast: 45,000-55,000 TS; 60-70 El. For bronze castings; corrosion resistant. *Obsolete*

NADA NT
Nassau Smelting & Refining Co.
Sn 3.5-3.75, Pb 0.5-1, Ni 3.5-3.75, bal Cu.
Cast: 40,000-45,000 TS; 20,000 YS; 30-40 El; 70-75 Brin. For pressure and structural castings; corrosion and erosion resisting. *Obsolete*

NADA NT-2
Nassau Smelting & Refining Co.
Sn 4.5-5, Pb 0.5-1, Ni 3-3.5, bal Cu.
Cast: 42,000-48,000 TS; 20,000-22,000 YS; 30-40 El; 72-80 Brin. For pressure and structural castings; corrosion and erosion resisting. *Obsolete*

NADA NT-3
Nassau Smelting & Refining Co.
Cu 86-88, Sn 4.5-5, Pb 0.5-1, Ni 2-2.5.
Cast: 40,000-43,000 TS; 18,000-20,000 YS; 30-40 El; 70-75 Brin. For pressure and structural castings; corrosion and erosion resisting. *Obsolete*

NAK 55
U.S. Metalsource
C 0.15, Mn 1.5, S 0.1, Si 0.3, Ni 3, Al 1, Cu 1, bal Fe.
For plastic and rubber molds, o-rings in home appliances, seals in automobiles, computers, cameras, communications.

NALOY
National Broach & Machine
C 0.7, W 18, Cr 4, V 1, bal Fe.
For broaches, form tools; high speed steel.

NAPAC
National Intergroup Inc.
C 0.07, Mn 0.32, Cb 0.003, Al 0.07, bal Fe.
Rolled: 52,000-60,000 TS; 38,000-45,000 YS; 35 El; 62 Rock B (as ordered). For structural parts such as automobile bumpers.

NAPAC-35
National Intergroup Inc.
C 0.1, Mn 0.35, Cb, bal Fe.
Aluminum-killed. Hot rolled sheet: 35,000 psi YS min. Weldable; for automotive equipment.

NAPAC-40
National Intergroup Inc.
C 0.1, Mn 0.55, Cb, bal Fe.
Aluminum killed. Hot rolled sheet: 40,000 psi YS min. Weldable; for automotive equipment.

NAPAC-45
National Intergroup Inc.
C 0.1, Mn 0.55, Cb, bal Fe.
Aluminum killed. Hot rolled sheet: 45,000 psi YS min. Weldable; for automotive equipment.

NAPAC-50
National Intergroup Inc.
C 0.1, Mn 0.55, Cb, bal Fe.
Aluminum killed. Hot rolled sheet: 50,000 psi YS min. Weldable; for automotive equipment.

NAPAC-F-45
National Intergroup Inc.
C 0.12, Mn 0.75, Cb, Zr, bal Fe.
Aluminum killed. Hot rolled sheet: 55,000 TS min; 45,000 YS min. Weldable; for automotive equipment.

NAPAC-F-50
National Intergroup Inc.
C 0.12, Mn 0.75, Cb, Zr, bal Fe.
Aluminum killed. Hot rolled sheet: 60,000 TS min; 50,000 YS min. Weldable; for automotive equipment.

NAPAC-F-55
National Intergroup Inc.
C 0.12, Mn 0.75, Cb, Zr, bal Fe.
Aluminum killed. Hot rolled sheet: 65,000 TS min; 55,000 YS min. Weldable; for automotive equipment.

NAPAC-F-60
National Intergroup Inc.
C 0.12, Mn 0.75, Cb, Zr, bal Fe.
Aluminum killed. Hot rolled sheet: 70,000 TS min; 60,000 YS min. Weldable; for automotive equipment.

NAPAC-S-45
National Intergroup Inc.
C 0.15, Mn 0.75, Cb, bal Fe.
Zirconium killed. Hot rolled sheet: 60,000 TS min; 45,000 YS min. Weldable; for automotive equipment.

NAPAC-S-50
National Intergroup Inc.
C 0.15, Mn 0.75, Cb, bal Fe.
Aluminum killed. Hot rolled sheet: 65,000 TS min; 50,000 YS min. Weldable; for automotive equipment.

NAPRALOY
Napraloy Co.
C 4.5, Cr 24, bal Fe.
Cast: 600 Brin. For welding rod, hard facing electrodes; wear resistant.

NARCOLOY
N.C. Ashton Ltd.
Al 7, Sn, Co, bal Cu.
Cold forging aluminum bronze rod.

NARITE
N.C. Ashton Ltd.
Al 14, Ni 1, Fe 4.5, bal Cu.
Casting for deep draw dies; superior in compression.

NARITE HT
N.C. Ashton Ltd.
Al 11.25, Ni 5, Fe 5, Mn 1, bal Cu.
Heat treatable to 400 Brin. For nonmagnetic, non-sparking tools.

NARLOY
American manufacture
Ag 3, bal Cu.
Brazing alloy.

NARRMAC HNA
N.C. Ashton Ltd.
Al 10.5, Ni 4.8, Fe 4, Mn 0.5, bal Cu.
50 tsi TS; 12 El. Wrought aluminum bronze.

NARRMAC II
N.C. Ashton Ltd.
Al 9.5, bal Cu.
35 tsi TS; 25 El. Wrought aluminum bronze. DTD 160; BS 2032 CA 103.

NARRMAC III
N.C. Ashton Ltd.
Al 9.75, Fe 2.5, bal Cu.
32 tsi TS; 20 El. Wrought aluminum bronze. DTD 197.

NARRMAC V
N.C. Ashton Ltd.
Al 9.5, Ni 4.5, Fe 4, Mn 0.5, bal Cu.
45 tsi TS; 15 El. Wrought aluminum bronze. DTD 197; BS 2033 CA 104.

NARTRODE E
N.C. Ashton Ltd.
Al 9.25, Ni 4.5, Fe 3.25, bal Cu.
Weld metal: 40 tsi TS; 15 El. Welding wire for automatic processes.

NARTRODE S
N.C. Ashton Ltd.
Al 9.5, Fe 1, bal Cu.
Weld metal: 38 tsi TS; 25 El Welding wire for automatic processes.

NARVE
Uddeholm Corp.
C 0.65, Si 1, Mn 0.5, Cr 1.1, Mo 0.6, bal Fe.
For dies, rams, shafts, tools; cold work steel. *Obsolete*

NASA CO-W-RE
Cannon-Muskegon Corp.
C 0.4, Co 3, W 25, Ti 1, Zr 1, Re 2, bal Co.
Cast alloy for high temperature space applications.

NASA-TRW VI A
Now TRW VI A.

NASCO
North American Steel Corp.
C 0.5, Mn 0.8, Cr 1, Mo 0.4, bal Fe.
Hardened: 263,000-274,000 TS; 209,000-220,000 YS; 550-600 Brin. For tools, chisels, punches, wrenches; non-tempering, shock resistant, tough.

NASCOLOY O
North American Steel Corp.
C 0.5, Mn 0.8, Si 1.8, bal Fe.
For pneumatic tools, chisels, rivet sets; wear resistant, tough, oil hardened, requires tempering.

NASCOLOY W
North American Steel Corp.
C 0.4, Mn 0.7, Si 0.2, Cr 0.8, Mo 0.3, Cu 0.3, bal Fe.
For pneumatic tools, chisels, rivet sets; wear resistant, tough, water hardened, no tempering.

NAT
Eagle & Globe Steel Ltd.
Tool material. C 0.4, Si 1, Cr 5, V 0.4, Mo 1.3, bal Fe.
Annealed: 235 Brin max. Water cooled, high resistance to heat checking. Die casting dies and inserts for aluminum and magnesium alloys, extrusion mandrels, hot blanking tools, punches, aluminum extrusion tools, tools for press forging of steel and shear blades. AS1239 H11A; AISI H11; BS4659 BH11; Werkstoff 1.2343.

NATALLOY NO. 2
National Cleveland Corp.
C 0.45, Mn 0.8, Cr 0.95, Mo 0.06, V 0.15, bal Fe.
Heat treated: 108,000 TS; 88,000 YS; 16 El; 34 RA; 225-300 Brin. For plastic and die casting dies; oil hardened, AISI 6145.

NATIONAL
Delsteel Inc.
W 18, C, Cr, V, bal Fe.
For high speed steel cutters and tools; high speed steel.

NATIONAL 217
Hollup Corp.
C, alloy, bal Fe.
For hard facing electrodes; wear resistant.

NATIONAL 459
Hollup Corp.
C, alloy, bal Fe.
For hard facing electrodes; wear resistant.

NATIONAL BEARING SILVER BABBITT NO. 397

Abex Corp.
Ag 1.5-2.6, Sn 2.5-4, Sb 9-11, bal Pb.
Cast: 9500-10,000 psi TS; 5 El; 15.3 Brin. For bearings; 8000 psi crushing strength; Babbitt.

NATIONAL ECONOMY

Hollup Corp.
C, 10 alloy, bal Fe.
For hard facing electrodes; wear resistant.

NATIONAL GRAPHITIC STEEL

Capitol Castings Inc.
C 0.3, 0.50 graphite, bal Fe.
Annealed: 75,000 TS; 200 Brin. For steel castings; abrasion resistant. *Obsolete*

NATIONAL HTM

Capitol Castings Inc.
C 3, Mn, Si, bal Fe.
Cast: 70,000-110,000 TS; 48,000-85,000 YS; 7-12 El; 163-302 Brin. For gears, shafts, housings, staybolts; pearlitic, malleable iron. *Obsolete*

NATIONAL NO. 1

Hollup Corp.
C 0.06, bal Fe.
For welding electrodes for low carbon steel; copper coated.

NATIONAL NO. 6

Hollup Corp.
C 0.1, alloy, bal Fe.
For welding electrodes; for high pressure piping.

NATIONAL TOOLFACE

Hollup Corp.
C, alloy, bal Fe.
For hard facing electrodes; wear resistant.

NATIONAL TUBE 1 1/4 CR-1/2 MO

U.S. Steel Corp.
C 0-0.15, Si 0-1, Cr 1-1.5, Mo 0.45-0.65, bal Fe.
Annealed: 60,000 TS; 25,000 YS; 30 El; 163 Brin. For steam pipes, oil refinery still tubing. *Obsolete*

NATIONAL TUBE 1 CR-1/2 MO

U.S. Steel Corp.
C 0-0.15, Cr 0.8-1.1, Mo 0.45-0.65, bal Fe.
Annealed: 60,000 TS; 25,000 YS; 30 El; 163 Brin. For steam pipes, boiler and superheating tubing. *Obsolete*

NATIONAL TUBE 2 1/2 CR-1/2 MO-3/4 SI

U.S. Steel Corp.
C 0-0.15, Si 0-1, Cr 2.25-2.75, Mo 0.45-0.65, bal Fe.
Annealed: 60,000 TS; 25,000 YS; 30 El; 163 Brin. For refinery still tubing; corrosion resistant. *Obsolete*

NATIONAL TUBE 2 1/4 CR-1 MO

U.S. Steel Corp.
C 0-0.15, Cr 2-2.5, Mo 0.9-1.1, bal Fe.
Annealed: 60,000 TS; 25,000 YS; 30 El; 163 Brin. For refinery still tubing; corrosion resistant. *Obsolete*

NATIONAL TUBE 2 CR-1/2 MO

U.S. Steel Corp.
C 0-0.15, Cr 1.75-2.25, Mo 0.45-0.65, bal Fe.
Annealed: 60,000 TS; 25,000 YS; 30 El; 163 Brin. For steam pipes, superheater tubing. *Obsolete*

NATIONAL TUBE 2-1/2 NICKEL STEEL

U.S. Steel Corp.
C 0.1-0.2, Mn 0.3-0.6, Si 0.1-0.2, Ni 2.25-2.75, bal Fe.
Normalized: 65,000 TS; 45,000 YS; 30 El. For piping for low temperature service (sub-zero), stationary boiler tubes; high impact value. *Obsolete*

NATIONAL TUBE 3 1/2 NI

U.S. Steel Corp.
C 0-0.2, Ni 2.25-3.75, bal Fe.
Normalized: 60,000 TS; 30,000 YS; 25 El; 190 Brin. For caustic evaporators, low temperature piping; high impact value. *Obsolete*

NATIONAL TUBE 3 CR-1 MO

U.S. Steel Corp.
C 0-0.15, Cr 2.75-3.25, Mo 0.8-1, bal Fe.
Annealed: 60,000 TS; 25,000 YS; 30 El; 163 Brin. For superheater and refinery still tubing; corrosion resistant. *Obsolete*

NATIONAL TUBE 5 CR-1/2 MO-1 1/2 SI

U.S. Steel Corp.
C 0-0.15, Si 1-2, Cr 4-6, Mo 0.45-0.65, bal Fe.
Annealed: 60,000 TS; 25,000 YS; 30 El; 163 Brin. For heat exchanger tubing; high temperature use. *Obsolete*

NATIONAL TUBE 5 NI

U.S. Steel Corp.
C 0-0.17, Ni 4.75-5.25, bal Fe.
Normalized: 60,000 TS; 30,000 YS; 25 El; 207 Brin. For caustic evaporators, low temperature piping; high impact value. *Obsolete*

NATIONAL TUBE 7 CR-1/2 MO

U.S. Steel Corp.
C 0-0.15, Si 0.5-1, Cr 6-8, Mo 0.45-0.65, bal Fe.
Annealed: 60,000 TS; 25,000 YS; 30 El; 179 Brin. For heat exchanger tubing; higher temperature use. *Obsolete*

NATIONAL TUBE 8 CR-1 MO

U.S. Steel Corp.
C 0-0.15, Cr 7-9, Mo 0.9-1.1, bal Fe.
Annealed: 60,000 TS; 25,000 YS; 30 El; 179 Brin. For heat exchanger tubing; higher heat resistance. *Obsolete*

NATIONAL TUBE 9 CR-1 MO

U.S. Steel Corp.
C 0-0.15, Si 0.5-1, Cr 8-10, Mo 0.9-1.1, bal Fe.
Annealed: 60,000 TS; 25,000 YS; 30 El; 179 Brin. For heat exchanger tubing; oxidation and corrosion resistant. *Obsolete*

NATIONAL TUBE 9 NI

U.S. Steel Corp.
C 0-0.12, Ni 8-10, bal Fe.
Normalized: 80,000 TS; 50,000 YS; 20 El. For oil well pump tubing, caustic evaporators, low temperature tubing; high impact value. *Obsolete*

NATIONAL TUBE C-1118

U.S. Steel Corp.
C 0.14-0.2, Mn 1.3-1.6, S 0.08-0.13, bal Fe.
For machined tubular parts; free-cutting. *Obsolete*

NATIONAL TUBE DIAMOND METAL

Hollup Corp.
WC.
For hard facing electrodes; wear resistant.

NATIONALLOY 1

Now AISI 303.

NATIONALLOY 14

National Forge Co.
C 0.28-0.38, Mn 0.4-0.7, P 0-0.015, S 0-0.015, Mn 0-0.4, Ni 3-4, Cr 0.8-1.2, Mo 0.4-0.8, V 0.05-0.2, bal Fe.
Deep hardening low alloy, low carbon steel.

NATIONALLOY 22

National Forge Co.
P 0-0.015, S 0-0.015, Si 0-0.4, Ni 1.5-2, Cr 0.8-1.2, Mo 0.2-0.3, C 0.28-0.36, Mn 0.6-0.9, 0.05 V, bal Fe.
Modified AISI 4330 steel.

NATIONALLOY 3

Now AISI 410.

NATIONALLOY 7

National Forge Co.
C 0.3-0.4, Mn 0.6-0.9, P 0-0.015, S 0-0.015, Si 0-0.4, Ni 2.25-3, Cr 0.8-1.2, Mo 0.3-0.5, V 0-0.2, bal Fe.
Medium hardening alloy steel. Modified AISI 4335.

NAUBUC

English manufacture
Cu 58, Zn 16.25, Ni 25, Fe 0.75.
For knives; corrosion resistant.

NAUTAL

Hungarian manufacture
Mn 4.5, Mg 0.4, bal Al.
For light alloy parts; corrosion resistant.

NAVAHO

AL Tech Specialty Steel Corp.
C 0.55, Mn 0.3, Si 0.9, Cr 5, W 1.25, Mo 1.3, bal Fe.
Usual working hardness: 57-59 Rock C. Punch and die steel for hot and cold work applications. Used for shear blades, chipper knives, forming and forging dies, backup rolls, and plastic molds. AISI A-8.

NAVAL

Manufacturer not listed
Cu 61, Sn 1-1.5, bal Zn.
Drawn: 52,000 TS; 30,000 YS; 35 El; 100 Brin. For extrusions; corrosion resistant.

NAVAL ALUMINUM

American manufacture
Cu 1.5, Mn 0.9, Ni 0.4, Fe + Si, bal Al.
For instruments and fittings; resists corrosion.

NAVAL ALUMINUM BRONZE

Manufacturer not listed
Cu 87-85, Al 7-9, Fe 2.5-4.5.
For marine parts, propellers, pumps; tough.

NAVAL BRASS 24

Olin Brass, Indianapolis
Copper. Cu 60, Sn 0.75, Zn 39.25.
Hard: 80,000 psi TS; 52,000 psi YS; 15 El. Soft: 55,000 psi TS; 25,000 psi YS; 50 El. For hardware, marine parts, and valve parts; resists sea water corrosion.

NAVAL BRASS 28

Olin Brass, Indianapolis
Copper. Cu 60, Pb 0.6, Sn 0.75, Zn 38.65.
Hard: 75,000 psi TS; 53,000 psi YS; 20 El. Soft: 57,000 psi TS; 25,000 psi YS; 47 El. For hardware; strong.

NAVAL BRASS E-24

Accurate Brass Co.
Cu 58.5-60, Sn 0.5-1, bal Zn.
Forged: 60,000-65,000 TS; 25,000-30,000 YS; 3-45 El; 83-94 Brin. For hardware, bolts, machinery parts; corrosion resistant.

NAVAL BRASS NO. 63

Accurate Brass Co.
Cu 57.5-60.5, Sn 1.25-1.75, Pb 0.25-0.75, bal Zn.
Forged: 60,000-65,000 TS; 26,000-30,000 YS; 20-25 El; 95-110 Brin. For hardware, bolts, machinery parts, fasteners; free-cutting, leaded bronze.

NAVAL BRASS, 64%

Chase Brass & Copper Co., Inc.
Copper. Cu 64, Sn 0.75, bal Zn.

NAVAL BRASS, HIGH LEADED

Chase Brass & Copper Co., Inc.
Copper. Cu 60, Zn 37.5, Pb 1.8, Sn 0.7.
Annealed: 57,000 psi TS; 25,000 psi YS; 40 El. Drawn (20%): 75,000 psi TS; 53,000 psi YS; 15 El. For marine hardware, valve stems, screw machine products.

NAVAL BRASS, UNINHIBITED
Chase Brass & Copper Co., Inc.
Copper. Cu 60, Sn 0.75, bal Zn.
Hot rolled: 55,000 psi TS; 25,000 psi YS; 50 El; 30 RA. Cold rolled: 70,000 psi TS; 58,000 psi YS; 17 El; 40 RA. For condenser tubes, bolts, spindles; resists sea water corrosion.

NAVAL BRASS-462
Anaconda Co.
Copper. Cu 60, Zn 39.25, Sn 0.75.
Soft: 56,000 TS; 22,000 YS. For condenser plates, marine hardware. Corrosion resistant.

NAVAL BRASS-464
Anaconda Co.
Copper. Cu 60, Zn 39.25, Sn 0.75.
Hard: 63,000 TS; 35,000 YS; 30 El; 65 Rock B. Soft: 56,000 TS; 25,000 YS; 40 El; 50 Rock B. For nuts, bolts, fasteners, rivets, valve stems, pump shafts, marine hardware. Corrosion resistant to seawater.

NAVAL BRONZE
Anaconda Co.
Copper. Cu 88, Sn 8, bal Zn.
For steam and structural parts, expansion joints, gears, valves. Tough bronze.

NAVAL BRONZE NO. 4
Manufacturer not listed
Pb 44, Sn 36, Sb 16, Cu 4.
For bearings; anti-friction.

NAVAL GUN METAL "G"
American manufacture
Cu 88, Sn 10, Zn 2.
For gears, pistons, bearings, bushings; high strength.

NAVAL JOURNAL BEARING, SPEC "HX"
American manufacture
Cu 83, Sn 14, Pb 3.5.
For bearings; heavy duty.

NAVAL JOURNAL BEARING, SPEC. "H"
American manufacture
Cu 83, Sn 14, Pb 3.5.
50,000 TS; 9.5 El. Bearings; heavy duty.

NAVAL NO. 6
Cerro Metal Products Co.
Cu 60, Sn 0.75, bal Zn.
Soft: 58,000 TS; 28,000 YS; 42 El. 1/2 H-temper: 65,000 TS; 38,000 YS; 37 El. H-temper: 75,000 TS; 52,000 YS; 25 El. For marine hardware, bolts, rivets, and propeller shafts. Corrosion resistant; CA 464.

NAVAL NO. 95
Cerro Metal Products Co.
Cu 63, Sn 0.7, bal Zn.
For marine hardware, bolts, and propeller shafts. Corrosion resistant; CA 462.

NAVAL PHOSPHOR BRONZE (P-C) CAST
American manufacture
Cu 88, Sn 8, Zn 4, Pb 0.5.
57,000-35,000 TS; 31,000 YS; 18 El. For bearings, gears, marine parts; resists sea water corrosion; U.S.N.-46 B5f.

NAVAL PHOSPHOR BRONZE (P-R) ROLLED
American manufacture
Cu 94, Sn 3.5, P 0.5.
80,000 TS; 60,000 YS; 12 El. For pump parts, valve stems, bolts; resists sea water corrosion; U.S.N.-46 B14d.

NAVAL VALVE BRONZE
American manufacture
Cu 88, Sn 6.5, Zn 4, Pb 1.5.
58,000-32,000 TS; 29,000 YS; 9-17 El; 80 Brin. For valve stems, valve seats, valve bodies; "Composition M"; U.S.N.-46 B8d.

NAVALIUM
English manufacture
Al 96.6, Cu 0.1, Sn 0.7, Fe 0.6, Mn 2.3, Si 0.24.
For light alloy parts.

NAVALOY
English manufacture
Sn, Sb, bal Pb.
For bearings; Babbitt metal.

NAVAN
George Cook & Co., Ltd.
C, bal Fe.
For gears, shafts, wear resistant parts; case hardened; B.S.I. 5005/101.

NAVIBRONZE
Le Bronze Industriel
Sn 8, P 0-0.25, bal Cu.
Bronze, wrought.

NAVY
Paul Bergsoe & Son
Sn 74, Sb 9, Cu 4, Pb 13.
Cast: 14,200 TS; 28 Brin. MP: 355-665°F. For steam turbine bearings. Wear resistant.

NAVY "N" ALLOY
American manufacture
Mn 3, Cu 6, bal Al.
20,000 TS; 8 El. For general castings; corrosion resisting.

NAVY ALUMINUM ALLOY
American manufacture
Cu 1.5, Mn 0.9, Ni 0.4, Fe 0.4, Si 0.3, bal Al.
For instruments, fittings, light alloy parts; non-hardenable.

NAVY ANTIFRICTION METAL GRADE 1
Puget Sound Metal Works
Sn 90-93, Sb 3.5-5, Pb 0.5, Cu 3.5-5.
10,770 TS; 18-30 Brin. For antifriction metal, bearings; Babbitt metal, aircraft engine bearings; Babbitt.

NAVY ANTIFRICTION METAL GRADE 2
Puget Sound Metal Works
Sn 87.5-89.5, Sb 7-8, Pb 0.35, Cu 3.5-4.5, As 0.1.
For antifriction metal, genuine Babbitt, automotive engine bearings; for moderately severe service.

NAVY ANTIFRICTION METAL GRADE 3
Puget Sound Metal Works
Sn 83-85, Sb 7.5-8.5, Pb 0.35, Cu 7.5-8.5, As 0.1.
For hard Babbitt, bearings; for moderately heavy pressures.

NAVY ANTIFRICTION METAL GRADE 4
Puget Sound Metal Works
Sn 80.5-82.5, Sb 12-14, Pb 0.25, Cu 5-6, As 0.1.
For hard bearings; for heavy pressure and high speeds.

NAVY ANTIFRICTION METAL GRADE 5
Puget Sound Metal Works
Sn 61-63, Sb 9.5-10.5, Pb 24-26, Cu 2.5-3.5, As 0.15.
For electric motor bearings; for low pressures and high speeds.

NAVY ANTIFRICTION METAL GRADE 6
Puget Sound Metal Works
Sn 4.5-5.5, Sb 14-16, Pb 79-81, Cu 0.5, As 0.2.
For inexpensive Babbitt bearings; for light service.

NAVY ANTIFRICTION METAL GRADE 7
Puget Sound Metal Works
Sn 9-10, Sb 14-16, Pb 74-76, Cu 0.5, As 0.2.
For inexpensive Babbitt bearings; for light service.

NAVY BEARING
American manufacture
Sn 80-91, Sb 4.5-15, Cu 3.7-5.
For bearings, bushings; anti-friction.

NAVY BEARING, HARD
American manufacture
Sn 80, Sb 15, Cu 5.
For bearings; anti-friction.

NAVY COMPOSITION "W"
American manufacture
Sn 89, Sb 7.3, Cu 3.7.
For bearings; anti-friction.

NAVY GEAR BRONZE
American manufacture
Cu 84-86, Sn 13-15, Zn 1.5, P 0.5.
30,000 TS; 1-8 El. For gears and worm wheels; tough.

NAVY NO. 4 ALLOY
American manufacture
Al 96, Cu 4.
15,000 TS; 5 El. For light alloy parts, boxes, covers, face plates.

NAVY TOMBASIL
Illingworth Steel Co.
Si 5, Zn 6, bal Cu.
Cast: 60,000 TS; 35,000 YS; 15 El; 110 Brin. Corrosion resistant; for valve stems.

NAX 9112
National Intergroup Inc.
C 0.1-0.15, Mn 0-1.1, Si 0.5-0.9, Cr 0.5-0.8, Zr 0.05-0.15, bal Fe.
Water or oil hardenable; for structural parts.

NAX 9115
National Intergroup Inc.
C 0.13-0.18, Mn 0-1.1, Si 0.5-0.9, Cr 0.5-0.8, Zr 0.05-0.15, bal Fe.
Water or oil hardenable; for structural parts.

NAX 9120
National Intergroup Inc.
C 0.18-0.23, Mn 0-1.1, Si 0.5-0.9, Cr 0.5-0.8, Zr 0.05-0.15, bal Fe.
Water or oil hardenable; for structural parts.

NAX 9120 MOD
National Intergroup Inc.
C 0.15-0.2, Mn 0-1.1, Si 0.5-0.9, Cr 0.5-0.8, Zr 0.05-0.15, bal Fe.
Water or oil hardenable; for structural parts.

NAX 9130
National Intergroup Inc.
C 0.28-0.33, Mn 0-1.1, Si 0.5-0.9, Cr 0.5-0.8, Zr 0.05-0.15, bal Fe.
Water or oil hardenable; for structural parts.

NAX FINE GRAIN
National Intergroup Inc.
C 0-0.22, Mn 1.1, Si 0.4-0.9, Zr 0.03-0.15, bal Fe.
Rolled (min): 70,000 TS; 50,000 YS; 22 El; 140-160 Brin. For structural parts; weldable.

NAX HIGH TENSILE
National Intergroup Inc.
C 0-0.18, Mn 0.5-0.9, Si 0.6-0.9, Cr 0.4-0.8, Zr 0.03-0.12, bal Fe.
Rolled (min): 70,000 TS; 50,000 YS; 22 El; 140-160 Brin. For structural parts; improved corrosion resistance over most carbon steels; weldable.

NAX X9115
National Intergroup Inc.
C 0.13-0.18, Mn 0-1.1, Si 0.5-0.9, Cr 0.5-0.8, Mo 0.1-0.2, Zr 0.05-0.15, bal Fe.
Water or oil hardenable; for structural parts.

NAX X9120

National Intergroup Inc.
C 0.18-0.23, Mn 0-1.1, Si 0.5-0.9, Cr 0.5-0.8, Mo 0.1-0.2, Zr 0.05-0.15, bal Fe.
Water or oil hardenable; for structural parts.

NAX X9120 MOD

National Intergroup Inc.
C 0.15-0.2, Mn 0-1.1, Si 0.5-0.9, Cr 0.5-0.8, Mo 0.1-0.2, Zr 0.05-0.15, bal Fe.
Water or oil hardenable; for structural parts.

NAX X9130

National Intergroup Inc.
C 0.28-0.33, Mn 0-1.1, Si 0.5-0.9, Cr 0.5-0.8, Mo 0.1-0.2, Zr 0.05-0.15, bal Fe.
Water or oil hardenable; for structural parts.

NB

Japan Metal Industry Co. Ltd.
C 0.04, Ni 65.32, Mo 28.2, Fe 5.18.
For corrosion and heat resistant parts; chemical plant equipment, pump and valve parts. Corrosion and heat resistant.

NBD "A" GRADE

Abex Corp.
Pb, Sb, bal Sn.
For low grade bearings; Babbitt metal.

NBD ALUMINUM BABBITT

Abex Corp.
Al, bal Sn.
Cast: 12,100 psi TS; 8.8 El; 6.6 RA; 28.6 Brin. For heavy pressure bearings; Babbitt metal; 23200 psi crushing strength.

NBD ARCTIC BRONZE

Abex Corp.
Sb, Sn, bal Cu.
For locomotive bearings.

NBD ARMATURE BABBITT

Abex Corp.
Sn, Cu, Ni, Sb.
Cast: 13,600 psi TS; 6.1 El; 7.0 RA; 33.8 Brin. For armature bearings and marine turbine bearings; Babbitt metal; 26700 psi crushing strength.

NBD BETA CRUSHER BABBITT

Abex Corp.
Sb, Sn, Pb, Cu.
Cast: 11,300 psi TS; 4.9 El; 5.3 RA; 27.5 Brin. For bearings for cement mills and mines; Babbitt metal; resists heavy loads.

NBD CRESCENT BABBITT

Abex Corp.
Cu, Sn, Sb.
Cast: 8,500 psi TS; 1.4 El; 2.1 RA; 25.1 Brin. For heavy duty bearings for rolling mills, paper mills; Babbitt metal; 18300 psi crushing strength.

NBD CUPRO NICKEL

Abex Corp.
Cu 65-67, Ni 30-31.5, Fe 0.6-0.8, Cb 0.7-1, Si 0.25-0.5, Mn 1-1.25, Pb 0-0.01.
60,000 psi TS; 32,000 psi YS; 20 El.

NBD ENGINE BABBITT

Abex Corp.
Sb 9-11, Cu 4-6, bal Sn.
Cast: 10,000-15,000 psi TS; 5 El. For bearings, bushings, liners; Babbitt; 10600 psi crushing strength.

NBD EXTRA COPPER-HARDENED BABBITT

Abex Corp.
Cu, Sn, Sb.
Cast: 8,500 psi TS; 5.0 El; 5.2 RA; 27.1 Brin. For bearings; Babbitt metal; 7400 psi crushing strength.

NBD GENUINE BABBITT (ORIGINAL)

Abex Corp.
Sn 88.9, Cu 3.7, Sb 7.4.
Cast: 13,000 psi TS; 10 El; 14 RA; 28.7 Brin. For high temperature resistant bearings; Babbitt metal; 21200 psi crushing strength.

NBD GENUINE BABBITT (SPECIAL)

Abex Corp.
Cu, Sb, bal Sn.
Cast: 10,000 psi TS; 5.0 El; 12.9 RA; 28.9 Brin. For heavy duty bearings; Babbitt metal; 22200 psi crushing strength.

NBD HOO-HOO

Abex Corp.
Sn, Cu, Ni, Sb.
Cast: 12,800 psi TS; 11.1 El; 16 RA; 28 Brin. For bearings; Babbitt metal; 25600 psi crushing strength.

NBD IMPROVED BABBITT

Abex Corp.
Cu, Sn, Sb.
Cast: 9,000 psi TS; 2.2 El; 1.5 RA; 24.1 Brin. For general utility bearings; Babbitt metal; 19500 psi crushing strength.

NBD NICKEL BABBITT

Abex Corp.
Sn, Cu, Ni.
Cast: 11,900 psi TS; 5.0 El; 15.8 RA; 28.9 Brin. For bearings; high speed and high pressures; 10600 psi crushing strength.

NBD NO. 1 STANDARD BABBITT

Abex Corp.
Sn, Sb, bal Pb.
For bearings; Babbitt metal.

NBD NO. 10A

Abex Corp.
Cu 60, Al 1, Mn 3, Fe 1, bal Zn.
Cast: 60,000 psi TS; 20,000 psi YS; 15 El; 110 Brin. For high strength castings.

NBD NO. 10B

Abex Corp.
Cu 64, Al 6, Mn 4, Fe 3, bal Zn.
Cast: 90,000 psi TS; 45,000 psi YS; 20 El; 180 Brin. For high strength castings; 64000 psi compressive yield point.

NBD NO. 10C

Abex Corp.
Al 1.2, Mn 1, Fe 1.2, Cu 58, bal Zn.
Cast: 65,000 psi TS; 25,000 psi YS; 20 El; 100 Brin. For bearings, bushings, sleeves; SAE 43; manganese bronze.

NBD NO. 10D

Abex Corp.
Cu 64, Al 6, Mn 3, Fe 3, bal Zn.
Cast: 110,000 psi TS; 60,000 psi YS; 12 El; 200 Brin. For bearings, worm wheels, gears, and valves; SAE 430B; manganese bronze.

NBD NO. 11

Abex Corp.
Sn, Pb, bal Cu.
Cast: 33,000 psi TS; 16,000 psi YS; 12 El; 58 Brin. For bearings and bushings.

NBD NO. 11 C

Abex Corp.
Cu 86, Sn 8, Pb 2, Zn 4.
Cast: 34,000 psi TS; 16,000 psi YS; 20 El; 65 Brin. For bearings; heavy duty.

NBD NO. 11A

Abex Corp.
Cu 88, Sn 8, Pb 4.
Cast: 28,000 psi TS; 16,000 psi YS; 12 El; 60 Brin. For bearings; heavy duty.

NBD NO. 11D

Abex Corp.
Sn 5, Zn 4, Ni 3, bal Cu.
Cast: 40,000 psi TS; 17,000 psi YS; 25 El; 65 Brin. For bearings, bushings, and sleeves; heavy duty nickel bronze.

NBD NO. 12

Abex Corp.
Sn, Pb, bal Cu.
Cast: 40,000 psi TS; 21,000 psi YS; 19 El; 18 RA; 81 Brin. For bearings, and bushings; 16600 psi compressive yield point.

NBD NO. 13

Abex Corp.
Sn, Pb, Zn, bal Cu.
Cast: 30,000 psi TS; 19,000 psi YS; 13 El; 18 RA; 69 Brin. For bearings and bushings; for steam and water pressure.

NBD NO. 13 A

Abex Corp.
Cu 80, Sn 8, Pb 12.
Cast: 25,000 psi TS; 12,000 psi YS; 8 El; 60 Brin. For bearings; heavy duty.

NBD NO. 15A

Abex Corp.
Sn 5, Pb 20, Zn 3, bal Cu.
Cast: 21,000 psi TS; 10,000 psi YS; 7 El; 45 Brin. For bearings, bushings, and sleeves; heavy duty leaded bronze.

NBD NO. 197

Abex Corp.
Cu 95, Si 4, Mn 1.
Cast: 45,000 psi TS; 18,000 psi YS; 20 El; 80 Brin. For gears, bolts, shafts, and fasteners; silicon bronze.

NBD NO. 198

Abex Corp.
Cu 91, Sn 1, Zn 5, Si 3.
Cast: 45,000 psi TS; 18,000 psi YS; 20 El. For hardware, bolts, and shafts; silicon bronze.

NBD NO. 199

Abex Corp.
Cu 82, Zn 14, Si 4.
Cast: 60,000 psi TS; 24,000 psi YS; 16 El. For hardware and valves; silicon bronze.

NBD NO. 2

Abex Corp.
Sn, Pb, Ni, bal Cu.
Cast: 33,000 psi TS; 21,000 psi YS; 7 El; 7 RA; 75 Brin. For bearings, engine and machinery castings; 14000 psi compressive yield point.

NBD NO. 2 STANDARD BABBITT

Abex Corp.
Pb, Sn, Sb.
For large, slow running bearings; Babbitt metal.

NBD NO. 20 A

Abex Corp.
Cu 97, Sn 2, Zn 1.
For bearings, copper, and electrodes.

NBD NO. 20-C

Abex Corp.
Cu 99.9.
Cast: 25,000 psi TS; 7000 psi YS; 45 El; 35 Brin. For electrode holders and water cooled linings; 80% electrical conductivity.

NBD NO. 20-H

Abex Corp.
Cu 99.3, Cr 0.7.
Heat treated: 55,000 psi TS; 35,000 psi YS; 25 El; 100 Brin. For castings, electrode holders; heat treatable, corrosion resistant.

NBD NO. 20-K

Abex Corp.
Cu 98.2, Ni 1.5, Be 0.3.
Heat treated: 75,000 psi TS; 50,000 psi YS; 3 El; 200 Brin. For castings; age-hardenable, corrosion resistant.

NBD NO. 20B

Abex Corp.
Cu 99.7.
For bearings; blast furnace copper.

NBD NO. 20L

Abex Corp.
Cu 99.9.
Cast: 25,000 psi TS; 8500 psi YS; 45 El; 35 Brin. For conductors; 90% electrical conductivity.

NBD NO. 21 B

Abex Corp.
Cu 63, Sn 1, Pb 2, Zn 34.
Cast: 40,000 psi TS; 14,000 psi YS; 15 El; 40 Brin. For bearings; heavy duty.

NBD NO. 21 C

Abex Corp.
Cu 72, Sn 1, Pb 2, Zn 25.
Cast: 35,000 psi TS; 12,000 psi YS; 25 El; 40 Brin. For bearings; heavy duty.

NBD NO. 22

Abex Corp.
Cu 88, Sn 2.
Cast: 25,00 psi TS; 7000 psi YS; 40 El; 45 Brin. For bells; welding bell copper.

NBD NO. 22 B

Abex Corp.
Cu 93, Sn 7.
Cast: 35,000 psi TS; 20,000 psi YS; 10 El; 85 Brin. For bearings; heavy duty.

NBD NO. 22 D

Abex Corp.
Cu 84, Sn 16.
Cast: 18,000 psi TS. For bearings; bridge bronze.

NBD NO. 22 E

Abex Corp.
Cu 80, Sn 20.
Cast: 24,000 psi TS. For bearings; bridge bronze.

NBD NO. 22 F

Abex Corp.
Cu 91, Sn 5.5, Zn 3.5.
Cast: 37,000 psi TS; 14,000 psi YS; 30 El; 40 Brin. For bearings and trolley wheels; heavy duty.

NBD NO. 22-C

Abex Corp.
Cu 91, Sn 9.
Cast. For plumbing and hardware; acid bronze.

NBD NO. 22A

Abex Corp.
Cu 90, Sn 10.
Cast: 35,000 psi TS; 21,000 psi YS; 10 El; 90 Brin. For bearings and gears; SAE 65.

NBD NO. 295

Abex Corp.
Cu 91, Zn 9.
Cast. For ornamental accessories; gilding bronze.

NBD NO. 3

Abex Corp.
Sn, Pb, bal Cu.
Cast: 32,000 psi TS; 21,000 psi YS; 10 El; 10 RA; 74 Brin. For bushings, gearings; 14000 psi compressive yield point.

NBD NO. 3 A

Abex Corp.
Cu 80, Sn 10, Pb 10.
Cast: 25,000 psi TS; 12,000 psi YS; 8 El; 60 Brin. For bearings; SAE 64.

NBD NO. 3 B

Abex Corp.
Cu 83, Sn 7, Pb 7, Zn 3.
Cast: 30,000 psi TS; 14,000 psi YS; 12 El; 60 Brin. For bearings; SAE 660.

NBD NO. 3 D

Abex Corp.
Cu 79.75, Sn 9, Pb 10, P 0.75, Ni 0.25.
Cast: 25,000 psi TS; 12,000 psi YS; 8 El; 60 Brin. For bearings; heavy duty.

NBD NO. 3 STANDARD BABBITT

Abex Corp.
Pb, Sn, Sb.
For bearings; Babbitt metal.

NBD NO. 36 BRONZE

Abex Corp.
Sn, Sb, bal Cu.
For car journal bearings. *Obsolete*

NBD NO. 4

Abex Corp.
Sn, Zn, bal Cu.
Cast: 42,000 psi TS; 23,000 psi YS; 21 El; 22 RA; 88 Brin. For gears, pinions, worm wheels, and nuts; 16000 psi compressive yield point.

NBD NO. 4 F

Abex Corp.
Cu 83, Sn 11, Pb 3, Zn 3.
Cast: 35,000 psi TS; 16,000 psi YS; 16 El; 60 Brin. For bearings; heavy duty.

NBD NO. 4 H

Abex Corp.
Cu 87, Sn 7, Pb 3, Zn 3.
Cast: 30,000 psi TS; 15,000 psi YS; 15 El; 60 Brin. For bearings; heavy duty.

NBD NO. 4 I

Abex Corp.
Cu 85, Sn 5, Pb 5.
Cast: 30,000 psi TS; 14,000 psi YS; 20 El; 60 Brin. For bearings; SAE 40.

NBD NO. 4 J

Abex Corp.
Cu 87, Sn 9, Pb 2, Zn 2.
Cast: 30,000 psi TS; 15,000 psi YS; 15 El; 60 Brin. For bearings; modified Gun Bronze.

NBD NO. 4 K

Abex Corp.
Cu 88, Sn 10, Zn 2.
Cast: 40,000 psi TS; 18,000 psi YS; 20 El; 70 Brin. For bearings; SAE 62; Gun Bronze.

NBD NO. 4 L

Abex Corp.
Cu 88, Sn 8, Zn 4.
Cast: 40,000 psi TS; 18,000 psi YS; 20 El; 70 Brin. For bearings; SAE 620; Navy "G".

NBD NO. 4 STANDARD BABBITT

Abex Corp.
Pb, Sn, Sb.
For severe service bearings; Babbitt metal.

NBD NO. 46

Abex Corp.
Cu 84, Sn 8, Pb 8.
Cast: 25,000 psi TS; 12,000 psi YS; 8 El. For bearings.

NBD NO. 5

Abex Corp.
Sn, bal Cu.
Cast: 37,000 psi TS; 21,000 psi YS; 10 El. For gears; acid resistant.

NBD NO. 6

Abex Corp.
Sn, Pb, bal Cu.
Cast: 27,500 psi TS; 17,000 psi YS; 16 El; 15 RA; 47 Brin. For bearings with poor lubrication; 9700 psi compressive yeild point.

NBD NO. 6 A

Abex Corp.
Cu 77, Sn 7, Pb 15.
Cast: 25,000 psi TS; 14,000 psi YS; 10 El; 50 Brin. For bearings; hard bronze.

NBD NO. 6 G

Abex Corp.
Cu 78, Sn 6, Pb 16.
Cast: 30,000 psi TS; 16,000 psi YS; 12 El; 55 Brin. For bearings; heavy duty.

NBD NO. 6 I

Abex Corp.
Cu 78, Sn 6, Pb 16.
Cast: 25,000 psi TS; 14,000 psi YS; 10 El; 50 Brin. For bearings; heavy duty.

NBD NO. 6-H

Abex Corp.
Cu 74, Sn 6, Pb 20.
Cast: 28,000 psi TS; 14,000 psi YS; 10 El; 50 Brin. For bearings, bushings; leaded bronze.

NBD NO. 622

Abex Corp.
Cu 88, Sn 6, Pb 1.5, Zn 4.
Cast: 34,000 psi TS; 16,000 psi YS; 22 El; 50 Brin. For valves and plumbing; valve bronze; SAE 622.

NBD NO. 63

Abex Corp.
Cu 88, Sn 10, Pb 2.
Cast: 35,000 psi TS; 10 El. For plumbing, hardware, and bearings; SAE 63; leaded tin bronze.

NBD NO. 64

Abex Corp.
Cu 80, Sn 10, Pb 10.
Cast: 25,000 psi TS; 12,000 psi YS; 8 El; 50 Brin. For bearings, liners, and sleeves; leaded bronze; SAE 64.

NBD NO. 65

Abex Corp.
Cu 88, Sn 12.
Cast: 35,000 psi TS; 21,000 psi YS; 10 El; 90 Brin. For gears, worm wheels, and bearings; SAE 65; gear bronze.

NBD NO. 65-N

Abex Corp.
Cu 87.5, Sn 11, Ni 1.5.
Cast: 35,000 psi TS; 21,000 psi YS; 10 El; 90 Brin. For gears; nickel gear bronze.

NBD NO. 66

Abex Corp.
Cu 85, Sn 5, Pb 9, Zn 1.
Cast: 25,000 psi TS; 12,000 psi YS; 8 El. For bearings; SAE 66.

NBD NO. 660

Abex Corp.
Cu 83, Sn 7, Pb 7, Zn 3.
Cast: 30,000 psi TS; 14,000 psi YS; 12 El; 50 Brin. For plumbing and hardware; SAE 660; modified red brass.

NBD NO. 6M
Abex Corp.
Cu 75, Sn 7, Pb 18.
For bearings, bushings, and sleeves; heavy duty.

NBD NO. 6S
Abex Corp.
Sn 5, Pb 24, bal Cu.
Cast: 21,000 psi TS; 7 El. For bearings, bushings, and liners; heavy duty.

NBD NO. 7
Abex Corp.
Sn, Pb, Ni, bal Cu.
Cast: 40,000 psi TS; 27,000 psi YS; 12 El; 15 RA; 90 Brin. For bearings; for heavy loads.

NBD NO. 7A
Abex Corp.
Sn 10, Pb 2, Ni 3, bal Cu.
Cast: 42,000 psi TS; 18,000 psi YS; 15 El; 80 Brin. For railway bearings; leaded nickel bronze.

NBD NO. 9-AF
Abex Corp.
Cu 86, Al 10.5, Fe 3.5.
Cast: 75,000 psi TS; 30,000 psi YS; 12 El; 150 Brin. Heat treated: 90,000 psi TS; 45,000 psi YS; 6 El; 190 Brin. For bearing segments, wearing plates, and screwdown nuts; Al-bronze, corrosion and wear resistant.

NBD NO. 9A
Abex Corp.
Al 10, Fe 1, bal Cu.
Cast, heat treated: 65,000 psi TS; 25,000 psi YS; 20 El; 11 RA; 120 Brin. For bearings and wear plates; 15000 psi compressive yield point; SAE-68B.

NBD NO. 9B
Abex Corp.
Al, Fe, bal Cu.
Cast: 64,000 psi TS; 28,000 psi YS; 11 El; 12 RA; 144 Brin. For acid resisting parts, crates, and pickling racks; 24000 psi compressive yield point.

NBD NO. 9C
Abex Corp.
Al, Fe, bal Cu.
Cast: 84,000 psi TS; 33,000 psi YS; 25 El; 24 RA; 137 Brin. For bushings, and bearings; 24000 psi compressive yield point.

NBD NO. 9D
Abex Corp.
Al, Fe, bal Cu.
Cast: 88,000 psi TS; 43,000 psi YS; 13 El; 13 RA; 149 Brin. For bushings, and bearings; 30000 psi compressive yield point.

NBD NO. 9E
Abex Corp.
Al, Fe, bal Cu.
Cast: 90,000 psi TS; 43,000 psi YS; 12 El; 12 RA; 170 Brin. For gears, pinions, worms, and worm wheels; 35000 psi compressive yield point.

NBD NO. 9F
Abex Corp.
Al 9, Fe 3, bal Cu.
Cast, heat treated: 65,000 psi TS: 25,000 psi YS; 20 El; 130 Brin. For slides, gibs, and gears; 40000 psi compressive yield point; SAE-68A.

NBD NO. 9G
Abex Corp.
Al, Fe, bal Cu.
Cast: 55,500 psi TS; 55,500 psi YS; 0 El; 0 RA; 300 Brin. For forming or drawing dies; 61000 psi compressive yield point.

NBD NO. 9H
Abex Corp.
Al, Fe, bal Cu.
Cast: 120,000 psi TS; 101,000 psi YS; 3 El; 5.5 RA; 235 Brin. For propellers and high strength castings; heat treatable.

NBD NO. 9K
Abex Corp.
Al 10, Fe 4, Ni 2, bal Cu.
Heat treated: 75,000-90,000 psi TS; 30,000-45,000 psi YS; 8-12 El; 150-190 Brin. For propellers, bearings, gears; Al-bronze; heat treatable.

NBD NO. 9L
Abex Corp.
Al 10, Fe 5, Ni 5, bal Cu.
Heat treated: 90,000-110,000 psi TS; 40,000-60,000 psi YS; 5-6 El; 109-200 Brin. For bearings, propellers, gears; Al-bronze; heat treatable.

NBD PHOSPHOR BRONZE BABBITT
Abex Corp.
P, Cu, bal Sn.
Cast: 11,150 psi TS; 5.6 El; 7.1 RA; 28.1 Brin. For marine and stationary engine bearings; Babbitt metal; 18100 psi crushing strength.

NBD REGENT BABBITT
Abex Corp.
Sb 14, Sn 5, bal Pb.
Cast: 9,000 psi TS; 2.0 El; 20 Brin. For relining railroad bearings; Babbitt metal.

NBD REX BABBITT
Abex Corp.
Sn 10, Sb 15, bal Cu.
Cast: 10,000 psi TS; 2.5 El; 1.8 RA; 24.8 Brin. For medium high speed and heavy pressure bearings; Babbitt metal; 10100 psi crushing strength.

NBD SPECIAL MOTOR BABBITT
Abex Corp.
Sb, Sn.
Cast: 9,100 psi TS; 2.2 El; 3.6 RA; 26.5 Brin. For motor bearings; Babbitt metal; 18000 psi crushing strength.

NBD SPECIAL NO. 1 BABBITT
Abex Corp.
Sn alloy.
Cast: 7,500 psi TS; 1.0 El; 1.7 RA; 24.2 Brin. For bearings for rolling and paper mills; Babbitt metal; 14200 psi crushing strength.

NBD TIGER BRONZE
Abex Corp.
Pb, Sn, bal Cu.
Cast: 33,000 psi TS; 18,000 psi YS; 19 El; 15 RA; 53 Brin. For engine castings, bearings; shock resistant.

NBD UNIVERSAL BABBITT
Abex Corp.
Sn, Pb, bal Cu.
Cast: 14,900 TS; 1.0 El; 1.0 RA; 40 Brin. For bearings to resist heat; tough Babbitt metal. M. P. 1100 F. *Obsolete*

NBD. NO. 21 A
Abex Corp.
Cu 68, Pb 2, Zn 30.
Cast: 30,000 psi TS; 11,000 psi YS; 20 El; 40 Brin. For bearings; heavy duty.

NC
Japan Metal Industry Co. Ltd.
C 0.05, Cr 15.32, Ni 58.07, Mo 13.73, Fe 5.68, W 4.24.
For chemical plant equipment, pump and valve parts. Corrosion and heat resistant.

NC 80/20
Inco Alloys International Inc.
Now NC 80/20 FILLER METAL.

NC 80/20 FILLER METAL
Inco Alloys International Inc.
C 0.26, Mn 1.2, Fe 0.5, Si 0.5, Cu 0.2, Cr 19.6, bal Ni.
Filler metal for gas-shielded arc welding BRIGHTRAY alloys, INCONEL alloy 600, INCOLOY ALLOYS DS and NIMONIC alloy 75. For high temperature applications. BS 2901-NA-34; DIN S-NiCr 20.

NC-4
French manufacture
Ni 24, Cr 3, bal Fe.
For thermostatic bimetal.

NCM
Walsingham Steel Co., Ltd.
C 0.25-0.35, Si 0.35-0.6, Mn 0.7-1, Ni 0.6-1.1, Cr 0.6-0.9, Mo 0.3-0.5, bal Fe.
Low alloy structural steel.

NCM
George Cook & Co., Ltd.
C 0.4, Mn 0.5, Si 0.3, Cr 1.25, Mo 0.2, Ni 1.5, bal Fe.
Oil hardening tool steel; for shear blades, collets, plastic molds.

NCM 30
Walsingham Steel Co., Ltd.
C 0.25-0.35, Si 0.3-0.6, Mn 0.6-0.8, Ni 1.5-1.8, Cr 0.9-1.2, Mo 0.3-0.4, Cu 0-0.3, bal Fe.
High tensile steel with high abrasion and impact resistance. Meets BS 1458 Gr. A, B; ASTM A148-65 Cr, 105-85; SAE 0105. Similar to AISI 4330.

NCM 35
Walsingham Steel Co., Ltd.
C 0.33-0.37, Si 0.25-0.6, Mn 0.6-0.8, Ni 1.65-2, Cr 0.7-0.8, Mo 0.2-0.3, bal Fe.
250-320 Brin. Similar to AISI 4335.

NCM 70
Walsingham Steel Co., Ltd.
C 0.65-0.75, Si 0.3-0.7, S 0-0.025, P 0-0.025, Mn 0.6-0.8, Ni 0.65-0.85, Cr 1.3-1.7, Mo 0.25-0.35, bal Fe.
Oil hardenable, low alloy steel.

NCM 75
Walsingham Steel Co., Ltd.
C 0.7-0.8, Si 0.3-0.6, Mn 0.7-0.9, Ni 0.8-1.2, Cr 1.5-2, Mo 0.25-0.35, bal Fe.

NCMV
British Steel Corp.
C 0.4-0.45, Si 0.6-0.8, Mn 0.3-0.5, Ni 1.7-2, Cr 1.4-1.6, Mo 0.9-1.1, V 0.2-0.3, bal Fe.
Hardened: 210,000-310,000 TS; 194,000-257,000 YS; 6-12 El; 45-59 Rock C. For high strength parts as airframe, undercarriage components and pressure vessels. Can be nitrided. *Obsolete*

NCMV
Walsingham Steel Co., Ltd.
C 0.25-0.3, Si 0.2-0.5, Mn 0.6-0.9, Ni 1.45-1.8, Cr 1.3-1.7, Mo 0.3-0.4, V 0.15-0.25, bal Fe.
40 tsi YS min; 65 tsi TS min; 300 Brin min.

NCR-238
Colt Industries
C 0.4, Si 1.4, Cr 8.25, Ni 2.25, Cu 1.25, bal Fe.
For stainless parts, pump and furnace parts; stainless. *Obsolete*

NDHTC-OILITE
Chrysler Corp.
Fe alloy.
Sintered: 80,000 TS. For bearings; ferrous, porous.

NDZ BRONZE

R. Lavin & Sons, Inc.
Pb 0-0.25, Zn 0-5, Ni 0-5.5, Fe 0-5.5, Al 0-2, Si 0-2, bal Cu.
As sand cast: 65,000-69,000 TS; 30,000-35,000 YS; 20-34 El;
123-134 Brin. For valve stems, propeller wheels, gears for
marine and outboard industry, marine parts requiring
resistance to dezincification and for dealuminization, water
works equipment. Heat treatable to slightly higher properties.

NDZ-S BRONZE

R. Lavin & Sons, Inc.
Pb 0-0.25, Zn 0-2, Ni 0-5.5, Fe 0-5.5, Al 0-2, Si 0-2, bal Cu.
As sand cast: 69,000-75,000 TS; 40,000-45,000 YS; 13-23 El;
143-149 Brin. For high strength valve stems, propeller wheels,
gears, marine hardware requiring resistance to dezincification
and/or dealuminization. Heat treatable to slightly higher
properties. CDA 995.

NEACID

Degussa AG
Pickling agent for the jewelry industry.

NEALLOY

Cambridge Wire Cloth Co.
Cr 0.6, Ni 0-1, C 0-0.12, Si 0.6, Cu 0-0.5, bal Fe.
For woven wire conveyor belts, wire cloth and slings.
Obsolete

NEASCO

Belmont Metals Inc.
Si, Fe, bal Cu.
For master alloy; for silicon bronze.

NEATRO

Teledyne Vasco
C 1.27, W 55, Cr 4.5, V 4, Mo 4.5, bal Fe.
For cutting tools, broaches, chasers, hobs; high speed steel.
Type M4.

NEATRO FM

Teledyne Vasco
C 1.25, W 5.5, Cr 4.5, V 4, Mo 4.5, bal Fe.
For cutting tools, broaches, chasers, hobs; high speed steel,
free-machining. *Obsolete*

NEB-BRONZE

Now 1% NEB-BRONZE.

NEBALOY

New England Brass Co.
Cu 62-65, Pb 0-0.07, Fe 0-0.05, bal Zn.
Electrical conductivity: 26.5% IACS at 68°F. Good cold
working properties. For electrical terminals and connectors,
hardware, jewelry, washers, shells, stampings. Copper alloy
No. 272.

NECOMICLE

Japanese manufacture
C, Ni, Cr, Mo, bal Fe.
For turbine blades, chemical apparatus; stainless.

NECRONI STEEL

National Erie Corp.
C 0.3-0.35, Si 0-0.35, Ni 0.5, Cr 1, bal Fe.
Oil treated: 110,000-220,000 TS; 80,000-185,000 YS; 6-18 El;
10-46 RA; 200-555 Brin. For gears, crankshafts. *Obsolete*

NEEDLE METAL

American manufacture
Cu 85, Sn 8, Zn 5.3, Pb 1.7.
For needles, valves, fittings; free-cutting.

NEEDLE WIRE

Colt Industries
C 0.7, bal Fe.
For spring beard, latch, sewing machine needles; tough.
Obsolete

NEELIUM

General Thermoelectric Corp.
Bi, Te, Se, Sb.
For thermoelectric cooling; semi-conductor.

NELOY

National Erie Corp.
C 0.3-0.4, Mn 1-1.25, bal Fe.
Heat treated: 190,000-220,000 TS; 175,000-195,000 YS; 8-14
El; 25-45 RA; 364-477 Brin. For gears, crankshafts; tough;
hard after heat treatment.

NELOY MOLYBDENUM

National Erie Corp.
C 0.3, Ni 0.6, Cr 0.6, Mo 0.2, bal Fe.
Annealed: 85,000-95,000 TS; 55,000-65,000 YS; 20-25 El;
30-40 RA; 170-197 Brin. For steel castings, gears; tough.

NELOY MOLYBDENUM NO. 2A

National Erie Corp.
C 0.3, Ni 0.6, Cr 0.6, Mo 0.2, bal Fe.
100,000-115,000 TS; 70,000-90,000 YS; 15-20 El; 35-45 RA;
190-220 Brin. For steel castings, gears; tough.

NELOY MOLYBDENUM NO. 3A

National Erie Corp.
C 0.3, Ni 0.6, Cr 0.6, bal Fe.
Heat treated: 125,000-140,000 TS; 115,000-130,000 YS;
12-15 El; 30-40 RA; 250-280 Brin. For steel castings, gears;
tough.

NELOY MOLYBDENUM NO. 5A

National Erie Corp.
C 0.3, Ni 0.6, Cr 0.6, bal Fe.
Heat treated: 140,000-160,000 TS; 130,000-145,000 YS;
10-13 El; 20-30 RA; 302-331 Brin. For steel castings, gears;
tough.

NELOY MOLYBDENUM NO. 6A

National Erie Corp.
C 0.3, Ni 0.6, Cr 0.6, bal Fe.
Heat treated: 160,000-180,000 TS; 145,000-160,000 YS;
7.5-10 El; 15-25 RA; 341-375 Brin. For steel castings, gears;
tough.

NELOY NO. 1

National Erie Corp.
C, alloy, bal Fe.
Cast: 85,000-95,000 TS; 23-30 El; 50-60 RA; 163-170 Brin. For
steel castings, gears; tough.

NELOY NO. 2

National Erie Corp.
C 0.4, Si 0.4, Mn 0.75, bal Fe.
Drawn: 90,000-93,000 TS; 65,000-75,000 YS; 23-28 El; 55-65
RA; 174-198 Brin. For steel castings, gears; tough.

NELOY NO. 3

National Erie Corp.
C 0.4, Si 0.4, Mn 0.75, bal Fe.
Heat treated: 100,000-110,000 TS; 80,000-90,000 YS; 20-26
El; 50-60 RA; 202-240 Brin. For steel castings, gears; tough.

NELOY NO. 4

National Erie Corp.
C 0.4, Si 0.4, Mn 0.75, bal Fe.
Heat treated: 115,000-130,000 TS; 95,000-110,000 YS; 15-20
El; 40-45 RA; 240-268 Brin. For steel castings, gears; tough.

NELOY NO. 5

National Erie Corp.
C, alloy, bal Fe.
Heat treated: 125,000-135,000 TS; 110,000-125,000 YS;
12-20 El; 35-45 RA; 268-288 Brin. For steel castings, gears;
tough.

NELOY NO. 6

National Erie Corp.
C, alloy, bal Fe.
Heat treated: 135,000-150,000 TS; 120,000-135,000 YS; 9-18
El; 30-40 RA; 286-302 Brin. For steel castings, gears; tough.

NELSON-BOHNALITE

Karl Schmidt Co.
Aluminum. Cu 10, Si 0.2, Mg 0.3, Ni 0.3, bal Al.
25,500 TS; 23,400 YS; 0.2 El; 110 Brin. For pistons for motor
vehicles; coefficient of expansion 0.000024.

NELSON-BOHNALITE

Aluminium-Zentral e.V.
Aluminum. Cu 9-11, Si 0.2, Mg 0.3, bal Al.
For general castings, fittings, hardware. Good machinability,
good strength. *Obsolete*

NEMICLE

Japanese manufacture
C, Ni, Cr, Mo, bal Fe.
For turbine blades, chemical apparatus; stainless.

NEMICLE C

Japanese manufacture
C, Ni, Cr, Mo, bal Fe.
For turbine blades, chemical apparatus; stainless.

NEMICLE F

Japanese manufacture
C, Ni, Cr, Mo, bal Fe.
For turbine blades, chemical apparatus; stainless.

NEO 2001 A

Stackpole Magnet Division
Permanent magnet. Residual flux density: 4600 Gauss.
Coercive force: 4100 Oersted.

NEO 2001 B

Stackpole Magnet Division
Permanent magnet. Residual flux density: 5200 Gauss.
Coercive force: 4300 Oersted.

NEO-BAROS

Creusot-Loire
Ni 90, Cr 10.
For weights, pen points; similar to "Baros," non-magnetic.
Obsolete

NEOCHRAN

Skoda Works National Corp.
Cr 22-30, Si 0.5-2.5, C 1-1.5, bal Fe.
For stainless castings; corrosion resistant.

NEODYMIUM

Atomergic Chemetals Corp.
Nd.
Purities: 99.9%, 99.5+%. Forms: ingot, lump, sheet, wire,
turnings, foil, rod.

NEOGEN

American manufacture
Cu 58, Zn 27, Ni 12, Sn 2, Al 0.5, Bi 0.5.
For ornamental and structural parts; corrosion resistant.

NEOMAGNAL A

German manufacture
Mg, Zn, bal Al.
For bearings.

NEONALIUM

Strasser Co.
Aluminum. Al 86-94, Cu 6-14, 0.4-1.0 other elements.
Heat treated: 22,800-34,000 TS; 11,000-16,000 YS; 1.0 RA;
80-120 Brin. For light alloy parts; good heat resistance.

NEONALIUM

Aluminium-Zentral e.V.
Aluminum. Cu 6-14, bal Al.
Sand cast: 26,000 TS; 18,000 YS; 0.8 El; 90 Brin. Chill cast:
32,000 TS; 26,000 YS; 0.3 El; 110 Brin. For housings,
casings, general castings. Good machinability. *Obsolete*

NEOR

Darwin & Milner Inc.
C 2.3, Cr 13, Si 0.6, Ni 0.6, Mn 0.4, bal Fe.
For press tools, punches, dies, reamers, broaches, gages; remarkable resistance to abrasion.

NEOR

Ziv Steel & Wire Co.
C 2.3, Cr 13, Si 0.6, Ni 0.6, Mn 0.4, bal Fe.
For press tools, punches, dies, reamers, broaches, gages; remarkable resistance to abrasion.

NEOREX

Creusot-Loire
C 2, Cr 13, bal Fe.
Cold work tool steel; air or oil hardening. For blanking and coining dies. Italy: UNI X210 Cr 13 KU.

NEOSPANAL

Kreidler Werke G.m.b.H.
Aluminum.
Aluminum alloy AlCuBiPb. *Obsolete*

NEOXYDIN

W. Seibel AG
Aluminum.
Aluminum alloy AlMg3 (Cu).

NEPTUNE

Pyramid Steel Company
C 0.3, Mn 0.75, Si 0.5, Cr 1.7, Mo 0.4, bal Fe.
Oil hardened steel designed for molds. AISI P20. *Obsolete*

NERGANDIN

IMI Knoch Ltd.
Cu 70, Zn 28, Pb 2.
89,000 TS; 10 El; 135 Brin. For condenser tubes; resistant to sea water corrosion.

NERGANDIN

P.H. Muntz & Co., Ltd.
Cu 70, Zn 28, Pb 2.
89,000 TS; 10 El; 135 Brin. For condenser tubes; resistant to sea water corrosion.

NERO 3

U.S. Spring & Bumper Co.
C 1.05, Cr 1, bal Fe.
For bearings, liners, sleeves, punches, cutters; water hardened, wear resistant.

NERO EXTRA

Hoffman & Co. KG
C 1, Cr 1.1, Mn 0.07, Si 0.25, bal Fe.
For bearings, liners, sleeves, punches, cutters; water hardened, wear resistant.

NERO EXTRA SPEZIAL

Hoffman & Co. KG
C 1.45, Cr 1.4, Mn 0.6, Si 0.25, bal Fe.
For bearings, liners, dies; water hardened, wear resistant.

NERO HBK

Hoffman & Co. KG
C 1.05, Cr 1, W 1.15, Mn 0.9, Si 0.25, bal Fe.
For bearings, forming and blanking dies; oil hardened, abrasion resistant.

NERO HFG

Hoffman & Co. KG
Cu 1.25, Si 1.15, Mn 0.7, Cr 1.2, bal Fe.
For bearings, liners, punches, dies; water or oil hardened, wear resistant.

NERO HWF1

Hoffman & Co. KG
C 1.2, V 0.1, W 1, Mn 0.28, Si 0.25, bal Fe.
For wear plates, punches, tools, dies; oil hardened.

NERO KST

Hoffman & Co. KG
C 0.9, Mn 1.9, Si 0.25, V 0.1, bal Fe.
For punches, shears, crimpers, dies, upsetters; oil hardened, nondeforming.

NERO SPEZIAL

Hoffman & Co. KG
C 1.05, Cr, bal Fe.
For bearings, liners, tools, dies; water hardened, wear resistant.

NES-3

Nippon Stainless Steel Co. Ltd.
Cr 15, Mo 17, Fe 5, W 4, bal Ni.
For valves, pumps, pharmaceutical and chemical plant equipment. Corrosion and heat resistant.

NESALOY

Nesaloy Products Inc.
Li alloy.
To disperse Pb in Cu; alloying.

NETIC S3-5

Polymer Corp. Ltd.
Ni, bal Fe.
Sheet: 46,100 TS; 28,200 YS; 25 El; 45 Rock B. For magnetic shielding. Low magnetic retentivity.

NETIC S3-6

Polymer Corp. Ltd.
Ni, bal Fe.
Sheet: 40,000-45,000 TS; 28,000 YP; 45 Rock B. For magnetic shields. Low magnetic retentivity.

NETIC-S3

Polymer Corp. Ltd.
Ni, bal Fe.
Sheet: 67,100 TS; 29,500 YP; 7.5 El; 68-71 Rock B. For magnetic shields. Low magnetic retentivity.

NEU-TEC-TRONIC 157 BN

Eutectic Corp.
Neutral flux core modifications of Eutec Rod 157. *Obsolete*

NEUF ECLAIRS

Creusot-Loire
C 0.85, W 19, Co 9, V 1.8, Mo 0.6, bal Fe.
For cutting tools, lathe and planer cutters; oil hardened. *Obsolete*

NEUMAL BD

Neumayer Kabel u. Metallwerke AG
Cu 2.5-5, Mg 0.2-1.8, Mn 0.3-1.5, Pb 0.5-2.5, Sn, Cd, Bi, bal Al.
For screw machine products; free-cutting.

NEUMAL D3

Neumayer Kabel u. Metallwerke AG
Cu 2.5-5, Mg 0.2-1.8, Mn 0.3-1.5, bal Al.
Annealed: 27,000 TS; 11,000 YS; 22 El; 47 Brin. Heat treated: 72,000 TS; 57,000 YS; 130 Brin. For aircraft structures and fittings, fasteners; age-hardenable.

NEUMAL-S

Neumayer Kabel u. Metallwerke AG
Mg 0.6-1.4, Si 0.6-1.6, Mn 0.6-1, Cr 0-0.3, bal Al.
Annealed: 21,000 TS; 8000 YS; 24 El. For window frames, gutters, fan blades, boats. Good forming and welding properties.

NEUSILBER 47-11 PB

VDM Nickel-Technologie AG
Cu 47, Ni 11, Pb 1.5, bal Zn.
Soft: 58,000 TS; 25 El; 100 Brin. Hard: 82,000 TS; 0.5 El; 185 Brin. For optical and camera parts, hardware, tableware; leaded nickel silver, corrosion resistant. *Obsolete*

NEUSILBER 4D

Vereinigte Deutsche Nickel-Werke AG
Ni 13-14, Cu 63-64, Pb 1, bal Zn.
For hardware, cutlery; leaded nickel silver, corrosion resistant. *Obsolete*

NEUSILBER 57-12 PB

VDM Nickel-Technologie AG
Cu 57, Ni 12, Pb 1.7, bal Zn.
Soft: 58,000 TS; 35 El; 80 Brin. Hard: 82,000 TS; 8 El; 160 Brin. For optical and camera parts, tableware; leaded nickel silver, corrosion resistant. *Obsolete*

NEUSILBER 60-25

VDM Nickel-Technologie AG
Cu 60, Ni 25, bal Zn.
Soft: 65,000 TS; 30 El; 90 Brin. Hard: 90,000 TS; 7 El; 170 Brin. Spring: 114,000 TS; 0.5 El; 200 Brin. For optical and camera parts, tableware, jewelry; nickel silver, corrosion resistant. *Obsolete*

NEUSILBER 62-18

VDM Nickel-Technologie AG
Cu 62, Ni 18, bal Zn.
Soft: 61,000 TS; 40 El; 85 Brin. Hard: 85,000 TS; 7 El; 165 Brin. Spring: 108,000 TS; 1 El; 195 Brin. For optical and camera parts, tableware, jewelry; nickel silver, corrosion resistant. *Obsolete*

NEUSILBER 62-18 PB

VDM Nickel-Technologie AG
Cu 62, Ni 18, Pb 2, bal Zn.
Soft: 58,000 TS; 35 El; 85 Brin. Hard: 82,000 TS; 8 El; 160 Brin. For optical and camera parts, hardware, tableware; leaded nickel silver, corrosion resistant. *Obsolete*

NEUSILBER 65-12

VDM Nickel-Technologie AG
Cu 65, Ni 12, bal Zn.
Soft: 58,000 TS; 40 El; 85 Brin. Hard: 80,000 TS; 10 El; 160 Brin. Spring: 103,000 TS; 1.5 El; 190 Brin. For tableware, jewelry, camera parts; nickel silver, corrosion resistant. *Obsolete*

NEUSILBER 71-7

VDM Nickel-Technologie AG
Cu 71, Ni 8, bal Zn.
Soft: 58,000 TS; 35 El; 80 Brin. Hard: 80,000 TS; 8 El; 150 Brin. Spring: 103,000 TS; 3 El; 180 Brin. For tableware, costume jewelry, hardware; nickel silver, corrosion resistant. *Obsolete*

NEUSILBER B. AND D. I

VDM Nickel-Technologie AG
Cu 57, Ni 12, Pb 2, bal Zn.
Rolled: 65,400 TS; 8 El; 120 Brin. Annealed: 46,800 TS; 35 El; 70 Brin. For instruments, optical frames; corrosion resistant, free-cutting. *Obsolete*

NEUSILBER B. AND D. II

VDM Nickel-Technologie AG
Cu 54, Ni 8, Pb 1, bal Zn.
Rolled: 79,500 TS; 4 El; 180 Brin. Annealed: 56,800 TS; 48 El; 85 Brin. For instruments, optical frames; free cutting, corrosion resistant. *Obsolete*

NEUSILBER B. AND D. III

VDM Nickel-Technologie AG
Cu 62, Ni 15, Pb 1, bal Zn.
Rolled: 81,000 TS; 5 El; 170 Brin. Annealed: 58,300 TS; 37 El; 95 Brin. For instruments, optical frames; free cutting, corrosion resistant. *Obsolete*

NEUSILBER B. AND D. IV

VDM Nickel-Technologie AG
Cu 47, Ni 11, Pb 1, bal Zn.
Rolled: 120,800 TS; 2 El; 210 Brin. Annealed: 85,200 TS; 30 El; 140 Brin. For instruments, optical frames; free-cutting, corrosion resistant. *Obsolete*

NEUSILBER EXCELSIOR
VDM Nickel-Technologie AG
Ni 25, Zn 15, Cu 60.
Annealed: 57,000 TS; 32 El; 85 Brin. For food handling equipment; German silver. *Obsolete*

NEUSILBER PDS
Vereinigte Deutsche Nickel-Werke AG
Ni 13.5, Cu 55, Pb 1.5, bal Zn.
For hardware, cutlery; leaded nickel silver, corrosion resistant. *Obsolete*

NEUSILBER, ENAMEL QUALITY
VDM Nickel-Technologie AG
Ni 20, Zn 5, bal Cu.
For bullet jackets, condenser tubes; German silver. *Obsolete*

NEUSILBER, NICKELIN
VDM Nickel-Technologie AG
Ni 22, Zn 20, Cu 56.
For hardware, electrical resistances; German silver. *Obsolete*

NEUSILBER, PRIMA
VDM Nickel-Technologie AG
Ni 12, Zn 23, Cu 65.
Annealed: 50,000 TS; 35 El; 75 Brin. For ornamental parts; German silver. *Obsolete*

NEUSILBER, PRIMA-PRIMA
VDM Nickel-Technologie AG
Ni 18, Zn 20, Cu 62.
Annealed: 50,000 TS; 35 El; 75 Brin. For white metal parts; German silver. *Obsolete*

NEUSILBER, QUARTA
VDM Nickel-Technologie AG
Ni 7, Zn 22, Cu 71.
Annealed: 51,200 TS; 45 El; 70 Brin. For ornaments; German silver. *Obsolete*

NEUSILBER, SEKUNDA
VDM Nickel-Technologie AG
Ni 10, Zn 24, Cu 66.
For ornamental parts; German silver. *Obsolete*

NEUSILBER, TERTIA
VDM Nickel-Technologie AG
Ni 9, Zn 25, Cu 66.
For ornaments; German silver. *Obsolete*

NEUSILBER, TUBES
VDM Nickel-Technologie AG
Ni 16, Zn 19.5.
For German silver tubes; German silver. *Obsolete*

NEUSTADT
German manufacture
Cu 71.5, Zn 28.5.
For condenser tubing; deep drawn.

NEUTRALEISEN
German manufacture
Si 14-15, Mn 0.7, bal Fe.
For insoluble anodes, crucibles, condensers, evaporators; acid resistant, brittle.

NEUTRALLOY
Bethlehem Foundry & Machine Co.
Ni 73, Cr 14, Si 2, Fe 10.
For castings; corrosion and heat resistant. *Obsolete*

NEUTRALLOY
F.B. Oldham & Co.
Ni 75, Cr 15, Fe 10, Si, Ti, Mn.
For castings; corrosion and acid resistant.

NEUTROLOY
Molecu-Wire Corp.
Cu 55, Ni 45.
Resistance wire. Electrical resistivity: 300 ohms/circular mil ft. Low temperature coefficient of resistance. Annealed: 50,000 psi TS. Hard drawn: 140,000 psi TS. Maximum operating temperature: 500°C.

NEUTRON FLUX TI-CU ALLOY
Chicago Development Corp.
Cu 1, bal Ti.
For neutron flux density measurement.

NEUTROSORB AND NEUTROSORB PLUS
Carpenter Technology Corp.
Stainless steel. C 0-0.08, Mn 0-2, P 0-0.045, S 0-0.03, Si 0-0.75, Cr 18-20, Ni 12-15, B 0-2, N 0-0.1, bal Fe.
Similar to conventional Type 304, with the addition of boron to provide higher thermal neutron absorption cross section, higher hardness, tensile strength and yield strength. For use in the nuclear industry in spent fuel storage racks and cask baskets, control rods, burnable poison and shielding.

NEVADA
Specialty Steel Co. of America
C 1.1, Mn 0.75, Si 0.5, Cr 5.5, V 0.3, Mo 1.3, P 0.015, S 0.015, bal Fe.
Air hardened tool steel for cold work tools.

NEVADA MODIFIED
Specialty Steel Co. of America
C 0.55, Mn 0.9, Si 0.25, Cr 3.5, Ni 0.25, V 0.25, Mo 1.4, bal Fe.
Air hardened tool steel.

NEVADA SILVER
American manufacture
Cu-Ni.
For ornaments, electrical resistances; nickel silver.

NEVASTAIN "D"
Associated Steel Corp.
C 0.15, Si 1, Cu 20-21, bal Fe.
At 70°F: 65,000-75,000 TS; 40,000-45,000 YS; 32 El; 63 RA. At 1500°F: 10,000 TS; 8,000 YS; 90 El; 98 RA. For furnace linings, conveyors, heat treating boxes; heat resistant; see Silcrome 21.

NEVASTAIN "H"
Associated Steel Corp.
C 1-1.1, Cr 16-18, Si 0.6, bal Fe.
Annealed: 115,000 TS; 40,000-50,000 YS; 12 El; 17-18 RA; 195-240 Brin. Air cooled: 140,000-240,000 TS; 140,000-220,000 YS; 0-2.8 El; 0-13.5 RA. For furnace parts, heat treating equipment; see Silcrome H-17.

NEVASTAIN "K.N.C.-3"
Associated Steel Corp.
Cr 24-26, Ni 19-21, C 0-0.15, bal Fe.
For furnace parts; heat and corrosion resistant; see Silcrome 25-20.

NEVASTAIN "S"
Associated Steel Corp.
Cr 11.5-13, C 0-0.12, bal Fe.
Rolled: 125,000-150,000 TS; 8-10 El; 25-35 RA; 280-320 Brin. Heat treated: 130,000 TS; 21 El; 68 RA; 237 Brin. Annealed: 50,000 TS; 35 El; 60 RA; 140 Brin. For turbine blading, pump rods, machine parts, valves, spoons, and forks; high resistance to shock and impact; see Silcrome 12.

NEVASTAIN C. "A"
Associated Steel Corp.
Cr 12.5-13.5, C 0.3-0.35, bal Fe.
Annealed: 90,000-105,000 TS; 20-25 El; 40-50 RA: 195-210 Brin. Heat treated: 145,000-225,000 TS; 9-14 El; 25-47 RA; 270-420 Brin. For cutlery, turbine blades; corrosion resistant; see Silcrome L-12.

NEVASTAIN C. "B"
Associated Steel Corp.
Cr 16.5-17.5, C 0.65-0.7, bal Fe.
For cutlery, knives; corrosion resistant; see Silcrome M-17.

NEVASTAIN E. "Z"
Associated Steel Corp.
Cr 13.5-15, C 0-0.12, bal Fe.
For corrosion resistant parts; corrosion resistant; see Silcrome 12-EZ.

NEVASTAIN K.A. "2"
Associated Steel Corp.
Cr 17-18.5, Ni 8.25-10, C 0.08-0.16, bal Fe.
Heat treated: 86,000-112,000 TS; 37,000-90,000 YS; 41-75 El; 64-82 RA. For stainless parts, chemical plant equipment; corrosion resistant; see Silcrome KA2.

NEVASTAIN K.A. 2 "S"
Associated Steel Corp.
Cr 17-18.5, Ni 8.25-10, C 0-0.07, bal Fe.
Heat treated: 132,000 TS; 42,000 YS; 50 El; 76 RA. At 1600°F: 20,000 TS; 7,000 YS; 28 El; 26 RA. For stainless parts, chemical plant equipment; corrosion resistant; see Silcrome KA2S.

NEVASTAIN RA
Cincinnati Steel Castings Co.
C 0.1, Cr 16, Cu 1, Si 1, bal Fe.
For rustless and stainless parts. Stainless.

NEVASTAIN, GRADE A
Associated Steel Corp.
C 0.3-0.35, Cr 12.5-13.5, bal Fe.
Annealed: 90,000-105,000 TS; 40,000-50,000 YS; 20-25 El; 40-50 RA; 195-210 Brin. Hardened: 225,000 TS; 185,000 YS; 9 El; 25 RA; 420 Brin. For stainless cutlery, tanks; corrosion resistant; see Silcrome L-12.

NEVASTAIN, GRADE B
Associated Steel Corp.
C 0.65-0.7, Cr 16.5-17.5, bal Fe.
Annealed: 90,000-95,000 TS; 40,000-45,000 YS; 26 El; 45-50 RA; 180-210 Brin. For stainless cutlery; corrosion resistant; see Silcrome M-17.

NEVASTAIN, GRADE RA
Associated Steel Corp.
Stainless steel. C 0-0.1, Cr 16, Cu 1, Si 1, Mn 0.4, bal Fe.
Annealed: 75,000 TS; 40,000 YS; 40 El; 75 RA; 150 Brin. Oil treated: 90,000-103,000 TS; 50,000-98,000 YS; 25-30 El; 60-63 RA; 170-217 Brin. For general fabricated stainless products; stainless; see Silcrome R.A.

NEVEROIL-21
American manufacture
Ni 32.5, Cu 64.5.
For corrosion resistant parts.

NEVYANSKITE
Russian manufacture
Ir 44-58, Os 27-49, Pt 0-10, Ru 0-6, Rh 1.5-3, Pd + Fe + Cu.
For fountain pen points; mined by U.S.S.R.

NEW BIDE
Latrobe Steel Co.
Carbides.
For dies, cutting tools; cemented carbides. *Obsolete*

NEW CAPITAL
Styria-Stahl Steirische Gusstahlwerke AG
C 0.7, Cr 3.75, W 14, V 0.5, bal Fe.
For lathe and planer tools, reamers, drills, hobs; high speed steel. *Obsolete*

NEW CAPITAL STEEL
English manufacture
W 14, Cr 3.7, V 0.1, C 0.6, bal Fe.
For high speed tools, reamers, cutters, punches, gages; high speed steel.

NEW K.S.
Colt Industries
Co 27, Ni 18, Ti 7, Al 3.7, bal Fe.
Bar: 7150 Br, 785 H c , 4300 Bo, 2,030,000 (BH) maximum.
For electrical and magnetic equipment. Permanent magnet.
High permeability. *Obsolete*

NEW KATHODE
Lincoln Electric Co.
C, bal Fe.
For welding electrodes; shielded arc for mild steel. *Obsolete*

NEW LIGHTWELD
Lincoln Electric Co.
C 0.2, bal Fe.
For welding electrode; coated. *Obsolete*

NEW MET
Newcomer Products Inc.
Mo, Ti, carbides.
For cutters; sintered carbides.

NEW MET N93
Newcomer Products Inc.
Ni 8, 61.0 TiC, 31.0 MoC.
Sintered: 175,000 Transverse strength; 93.5 Rock A. For tools, bearings and seals. Resists heat, oxidation and wear.

NEW MET N95
Newcomer Products Inc.
Ni 2, 64.0 TiC, 34.0 MoC.
Sintered: 150,000 Transverse strength; 95 Rock A. For tools, bearings and seals. Resists heat, oxidation and wear.

NEW PROCESS COLD HEADER
Jessop Steel Co.
C 1, bal Fe.
Annealed: 100,000 TS; 53,000 YS; 21 El; 42 RA; 200 Brin. For cutters, drills, taps, reamers, broaches; Type W1; water hardened.

NEW PROCESS COLD HEADER DIE STEEL
Jessop Steel Co.
C 1, Mn 0.25, Si 0.18, bal Fe.
For cold header punches and dies; water hardened.

NEW RAPID
Houghton & Richards Inc.
C, W, bal Fe.
For punches, dies, cutting tools; high speed steel.

NEW RYCUT 50
Now RYCUT 50.

NEW TOOL STEEL CAST
English manufacture
Ni 58, Zn 20, Al 12, Si 10.
For cuttings tools and dies; corrosion resistant.

NEW-BIDE
Now N20, ETC.

NEWCO
New England High Carbon Wire Co.
C 0.1-0.3, Cr 13, bal Fe.
For corrosion resistant parts. *Obsolete*

NEWCOMER C-2
Newcomer Products Inc.
Carbide.
Rockwell A83. For cutting tools for high speeds; sintered, Tr.S. 220,000. *Obsolete*

NEWCOMER C-3
Newcomer Products Inc.
Carbide.
Rockwell A92. For cutting tools for roughing and finishing; sintered, Tr.S. 260,000. *Obsolete*

NEWCOMER C-4
Newcomer Products Inc.
Carbide.
Rockwell A91. For cutting tools for heavy duty; sintered, Tr.S. 300,000. *Obsolete*

NEWCOMER C-5
Newcomer Products Inc.
Carbide.
Rockwell A89. For cutting tools for heavy duty; sintered, Tr.S. 375,000. *Obsolete*

NEWCOMER S-2
Newcomer Products Inc.
Carbide.
Rockwell A92. For cutting tools for light roughing; sintered, Tr.S. 220,000. *Obsolete*

NEWCOMER S-4
Newcomer Products Inc.
Carbide.
Rockwell A91. For cutting tools for roughing; sintered; Tr.S. 275,000. *Obsolete*

NEWCOMER S-6
Newcomer Products Inc.
Carbide.
Rockwell A90. For heavy duty cutting tools; sintered, Tr.S. 340,000. *Obsolete*

NEWCOR
British Steel plc
Non-silicon bearing electrical steel. For fractional horsepower motor laminations.

NEWHALL
Sanderson Kayser Ltd.
C 0.95, Mn 1.2, Cr 0.55, W 0.55, V 0.2, bal Fe.
Cold work tool steel; oil hardening. B.S. 4659 Type BO1; AISI O1.

NEWLOY
Harrison, Fischer & Co. Ltd.
Copper. Cu 64, Ni 35, Sn 1.
For base metal for tableware, resistance wire; good corrosion and acid resistance.

NEWMAX
Ziv Steel & Wire Co.
C 0.45-0.53, Mn 0.75-1, Cr 0.8-1.1, Mo 0.15-0.25, bal Fe.
Oil hardening steel, designed for molds. Similar to AISI 4150.

NEWPORT ARMATURE
Newport Steel Corp.
Si, bal Fe.
For electrical generators; high permeability.

NEWPORT ELECTRICAL "A"
Newport Steel Corp.
Si, bal Fe.
For electric generators; high permeability.

NEWPORT ELECTRICAL "B"
Newport Steel Corp.
Si, bal Fe.
For motors, armatures; high permeability.

NEWPORT ELECTRICAL "C"
Newport Steel Corp.
Si, bal Fe.
For motors, armatures; high permeability.

NEWPORT FIELD
Newport Steel Corp.
Si, bal Fe.
For fields, armatures, electrical equipment; high permeability.

NEWPORT TRANSFORMER
Newport Steel Corp.
Si, bal Fe.
For transformers; high permeability.

NEWPORT TRANSFORMER EXTRA SPECIAL
Newport Steel Corp.
Si, bal Fe.
For transformers; high permeability.

NEWPORT TRANSFORMER SPECIAL
Newport Steel Corp.
Si, bal Fe.
For transformers; high permeability.

NEWTON FUSIBLE ALLOY
English manufacture
Bi 50, Sn 18.75, Pb 31.25.
For fire and signal alarms, fire extinguisher plugs; melting point 95°C.

NEY .490 FINE SOLDER
J.M. Ney Co.
Gold color solder. Melting range: 1405-1440°F.

NEY .585 FINE SOLDER
J.M. Ney Co.
Gold color solder. Melting range: 1395-1455°F.

NEY .615 FINE SOLDER
J.M. Ney Co.
For post soldering of porcelain fused-to-metal restorations. Melting range: 1395-1505°F.

NEY .650 FINE SOLDER
J.M. Ney Co.
For post soldering of porcelain fused-to-metal restorations. Melting range: 1455-1505°F.

NEY 14K JEWELRY ALLOY
J.M. Ney Co.
Au 58, Ag 3.
Gold color alloy for casting jewelry and artwork. Melting range: 1620-1648°F.

NEY 24K
J.M. Ney Co.
Melting temperature: 1945°F.

NEY 76
J.M. Ney Co.
Pd 25, Ag 59.
Dental casting alloy. Melting range: 1675-1810°F. Casting temperature: 1950°F. Cast: 100,000 psi TS; 74,000 psi YS (0.1% offset); 110 DWT/in.^{3} density. 200 Brin can be reduced to 140 Brin by heat treatment. For white crown and bridge; hard.

NEY 90
J.M. Ney Co.

Now NEYDIUM 90.

NEY ORO "A"
J.M. Ney Co.
93% Au-Pt, bal Cu, Ag.
31,300 TS; 7,800 YS; 25 El; 46 Brin. For dental inlays; soft. *Obsolete*

NEY ORO "A-W"
J.M. Ney Co.
95 Au-Pt, bal Zn.
33,800 TS; 14,500 YS; 12 El; 70 Brin. For dental inlays; soft. *Obsolete*

NEY ORO "B"
J.M. Ney Co.
85% Au-Pt, bal Cu, Ag, Zn.
Soft: 52,000 TS; 25,000 YS; 20 El; 105 Brin. Heat treated: 60,000 TS; 30,500 YS; 12 El; 115 Brin. For dental inlays; hard. *Obsolete*

NEY ORO "B-W"
J.M. Ney Co.
81% Au-Pt, bal Cu, Ag, Zn.
Soft: 57,000 TS; 27,000 YS; 17 El; 102 Brin. Heat treated: 60,700 TS; 31,000 YS; 15 El; 113 Brin. For dental inlays; hard. *Obsolete*

NEY ORO "E"
J.M. Ney Co.
79% Au-Pt, bal Cu, Ag, Zn.
Soft: 75,000 TS; 45,500 YS; 6 El; 150 Brin. Heat treated: 112,000 TS; 82,500 YS; 1 El; 235 Brin. For dental inlays, partial dentures; extra hard. *Obsolete*

NEY ORO "G-W"
J.M. Ney Co.
74% Au-Pt, bal Cu, Ag, Zn.
Soft: 67,500 TS; 41,000 YS; 13 El; 132 Brin. Heat treated: 117,000 TS; 76,500 YS; 3 El; 228 Brin. For dental inlays, partial dentures; extra hard. *Obsolete*

NEY WHITE PRE-SOLDER
J.M. Ney Co.
Solder for porcelain fused-to-metal alloys (pre-porcelain). Melting range: 1935-2020°F.

NEY'S NO. 125
J.M. Ney Co.
Hg, Ag.
For dental amalgam. *Obsolete*

NEY-ORO "A-1"
J.M. Ney Co.
Au 77.5, Ag 12.5, Pd 2.
Cast: 62,000 TS (nominal); 28,000 psi YS (at 0.1% offset, nominal); 168 DWT/in.^{3} density. Melting range 1690-1795°F; casting temperature 1900°F; 105 HV. For dental multiple surface inlays, medium hard, gold color.

NEY-ORO "B-2"
J.M. Ney Co.
Au 74, Ag 11.5, Cu 9.5, Pd 4.
Cast: 74,000 psi TS (nominal); 39,000 psi YS (0.1% offset, nominal); 135 HV; 164 DWT/in.^{3} density. For dental inlays, full and partial coverage crowns, and multiple unit fixed restorations; gold color; hard.

NEY-ORO "G"
J.M. Ney Co.
Pt 8.5, Ag 4.5, Cu 14.5, Zn 1, Au 71.5.
Heat treated: 185,000 TS (nominal); 165,000 YS (nominal); 1 El (nominal); 280 Brin. Electronic alloy for pivot in instruments bearings, slip rings, commutator bars, make and break contacts. Age hardenable, corrosion and wear resistant.

NEY-ORO "G-3"
J.M. Ney Co.
Cast: 123,500 psi TS (nominal); 90,000 YS (0.1% offset, nominal); 250 HV; 159 DWT/in.^{3} density. Melting range 1645-1743°F; casting temperature 1900°F. For dental inlays, partial dentures; gold color; extra hard.

NEY-ORO 28
J.M. Ney Co.
Ag 25, Au 75.
Electronic alloy. Cold worked: 60,000 TS (nominal); 35,000 PL (nominal); 100 Brin (nominal). For contact brushes, used against coin silver slip rings. Low contact resistance.

NEY-ORO 28A
J.M. Ney Co.
Au 75, Ag 22, Ni 3.
Electronic alloy. Rolled: 90,000 TS (nominal); 50,000 PL (nominal); 140 Brin (nominal). Annealed: 51,000 TS (nominal); 84 Brin (nominal). For sliding contacts, electrical brush contacts. 72-75 ohms/circular mill feet electrical resistivity.

NEY-ORO 28B
J.M. Ney Co.
Au 75, Ag 23.5, Ni 1.5.
Electronic alloy. Rolled: 75,000 TS (nominal); 40,000 PL (nominal); 130 Brin (nominal). For sliding contacts on Constantan, make and break contacts. 71-75 ohms/circular mill feet electrical resistivity.

NEY-ORO 41
J.M. Ney Co.
Au 20, Pd 21, Ag 38.5.
Gold color nonporcelain alloy for inlays, partial and full coverage crowns, pontics, and multiple unit restorations. 80,000 psi TS; 45,000 psi YS (0.1% offset); 6 El; 180 HV.

NEY-ORO 5
J.M. Ney Co.
Now NEY-ORO NO. 5.

NEY-ORO 6
J.M. Ney Co.
Dental alloy. Melting range: 1550-1635°F; 235 Brin. For partial dentures. *Obsolete*

NEY-ORO 60
J.M. Ney Co.
Au 56, Pd 4.
Gold alloy for inlays, partial and full coverage crowns, pontics and multiple unit restorations. 121,000 psi TS; 106,000 psi YS (0.1% offset); 7.3 El; 265 HV.

NEY-ORO 65 CF
J.M. Ney Co.
Au 55, Pd 10, Ag 25.5.
Gold color copper-free nonporcelain alloy for onlays, partial and full coverage crowns, pontics and multiple unit restorations. Cast: 100,000 psi TS max; 70,000 psi YS (0.1% offset); 5 El; 215 HV.

NEY-ORO 69
J.M. Ney Co.
Au 69, Ag 25, Pt 6.
Electronic alloy. Annealed: 40,000 TS (nominal); 18,000 YP (nominal); 85 Brin (nominal). Work hardened: 70,000 TS (nominal); 30,000 YP (nominal); 120 Brin (nominal). For make and break contacts, telephone relays, slip rings. Corrosion resistant.

NEY-ORO A-A
J.M. Ney Co.
Au 81, Pd 4.
Dental alloy for single surface inlays; casting gold. Melting range: 1865-1980°F; casting temperature 2100°F. Cast: 46,000 psi TS (nominal); 16,000 psi YS (0.1% offset, nominal); 175 DWT/in.^{3} density; 70 HV.

NEY-ORO B-20
J.M. Ney Co.
Au 62, Ag 26, Pd 3.
Melting range: 1636-1740°F; casting temperature 1850°F. Cast: 115,000 psi TS (nominal); 75,000 psi YS (0.1% offset, nominal); 190 HV; 152 DWT/in.^{3} density. Dental casting alloy for inlay and bridge work. For casting bridge retainers and pontics, thin inlays, 3/4 crowns; hard.

NEY-ORO CB
J.M. Ney Co.
Au 59, Ag 22.5, Pd 4, Zn, bal Cu.
Dental casting alloy; gold color. High noble non-porcelain alloy for full and partial coverage crowns, inlays, and multiple unit fixed restorations; hard. Melting range: 1550-1640°F; casting temperature: 1750°F. Cast: 136,000 psi TS (nominal); 109,000 psi YS (0.1% offset, nominal); 275 HV; 146.6 DWT/in.^{3} density.

NEY-ORO ELASTIC
J.M. Ney Co.
79.5% Au-Pt, Ag, Zn, bal Cu.
Age hardened: 173,000 TS; 131,000 YP; 270 Brin. For wire clasps, pinlays in dentures. Corrosion resistant. *Obsolete*

NEY-ORO ELASTIC NO. 2
J.M. Ney Co.
78 Au-Pt, Ag, Zn, bal Cu.
Soft: 100,000 TS; 60,000 YS; 22 El. Heat treated: 152,000 TS; 113,500 YS; 9 El. For dental alloy applications, wire clasps and bars; M.P. 1835 F. *Obsolete*

NEY-ORO ELASTIC NO. 3
J.M. Ney Co.
73 Au-Pt, bal Cu, Ag, Zn.
Soft: 92,500 TS; 62,000 YS; 26 El. Heat treated: 147,500 TS; 113,000 YS; 4 El. For wire clasps and bars, dental alloy applications; M.P. 1700 F. *Obsolete*

NEY-ORO NO. 5
J.M. Ney Co.
Au 63, Pd 5.
Dental alloy. Gold color. Melting range 1650-1730°F; 220 Brin; 260 HV. For removable partial dentures and long span fixed restoration. Casting temperature: 1875°F. Cast: 134 ksi TS nominal; 100 ksi YS (0.1% offset); 152 DWT/in.^{3} density.

NEYCAST
J.M. Ney Co.
Dental solder; gold color. Flows at 1425°F; color matched to NEYCAST III casting gold. *Obsolete*

NEYCAST III
J.M. Ney Co.
Dental casting alloy; gold color. Melting range: 1600-1790°F; casting temperature: 1950°F. 175 Brin. *Obsolete*

NEYDIUM 90
J.M. Ney Co.
Cu 10, Ag 90.
Electronic alloy for sliding contacts, slip rings, commutator segments, rivet head contacts. Low resistivity, high electrical conductivity, and high corrosion resistance. 82-88 Rock 15T (over 0.015 in. thick wall thickness); 60,000-80,000 psi TS (under 0.015 in. thick).

NEYDIUM GOLD CERAMIC
J.M. Ney Co.
Au 49, Ag 32.
Dental alloy; white gold color. Melting range: 2165-2310°F; 230 Brin, 260 HV. Cast: 109 ksi TS (nominal); 66 ksi YS (0.1% offset, nominal); 143 DWT/in.^{3} density; casting temperature: 2425°F.

NEYDIUM N-PRE
J.M. Ney Co.
Base metal. Melting range: 1950-2045°F.

NEYDIUM NO. 8
J.M. Ney Co.
72 Au-Pt, Ag, Zn, Sn, bal Cu.
Soft: 121,000 TS; 91,000 YS; 9 El; 195 Brin. Heat treated: 154,000 TS; 122,000 YS; 4 El; 255 Brin. For dental applications; M.P. 1860 F. *Obsolete*

NEYDIUM NON PRECIOUS
J.M. Ney Co.
Dental alloy. Melting range: 2220-2430°F; 195 Brin. For porcelain fused to metal restorations. *Obsolete*

NEYDIUM NON-PRECIOUS SLDR.
J.M. Ney Co.
Dental solder (non-precious). Melting range: 1950-2045°F; white color; for use with NEYDIUM non-precious alloy.

NEYDIUM NP
J.M. Ney Co.
Now NEYDIUM NON-PRECIOUS SLDR.

NEYLASTIC H. F
J.M. Ney Co.
Gold color wire for orthodontic use. Fusing temperature: 1830°F.

NGSA
Saarstahl AG
C 0.43, Cr 1.85, Mo 0.5, V 0.8, W 0.4, bal Fe.
Hot work tool steel. For metal upsetting tools, shear knives, pressing punches. W.-Nr. 2603; AFNOR 45 CVD8. *Obsolete*

NH 11
Saarstahl AG
Now SAARSTAHL 1.4020.

NH 22
Saarstahl AG
Now SAARSTAHL 1.2782.

NH 40
Saarstahl AG
Now SAARSTAHL 1.2786X2.

NH PIG
International Nickel Inc.
Nickel. Ni 44-47, Cr 15-17, bal Fe.
For foundry alloys; MP 2250°F. *Obsolete*

NH18V
Friedr. Lohmann GmbH
C 0.74, Cr 4.1, V 1.1, W 18.5, bal Fe.
For lathe and planer tools, reamers, woodcutting knives. Type T1; high speed steel. T12001

NH260
Friedr. Lohmann GmbH
C 0.86, Cr 4.1, Mo 0.85, V 2.5, W 12, bal Fe.
For lathe and planer tools, reamers, broaches, drills, taps. High speed steel.

NHMO
Friedr. Lohmann GmbH
C 0.9, Cr 4.1, Mo 5, V 1.8, W 6.4, Mo, W, V, bal Fe.
For lathe and planer tools, reamers, hobs, drills. Type M2; high speed steel. 11302

NI 14
TRW Inc.
WC 88, Ni 12.
300,000 transverse strength; 87.5 Rock A; density 14.40. Sintered carbide tool material.

NI 99 ELECTRODE
J.W. Harris Co., Inc.
Cast iron machinable electrode; AC DC, straight or reverse polarity. For repairing cast iron cylinder heads, motor blocks.

NI C 1.5
Vereinigte Deutsche Nickel-Werke AG
Mn 1.5, C, bal Ni.
For chemical equipment. *Obsolete*

NI C 4
Vereinigte Deutsche Nickel-Werke AG
Mn 4, C, bal Ni.
For spark plug electrodes. *Obsolete*

NI C 5
Vereinigte Deutsche Nickel-Werke AG
Mn 5, C, bal Ni.
For spark plug electrodes. *Obsolete*

NI CHILLITE
Gulf & Western Mfg. Co.
Ni, Cr, Mo, C, bal Fe.
For rolls; chilled iron.

NI CHILLITE NO. 2
Gulf & Western Mfg. Co.
C, alloy, bal Fe.
For rod mill rods; heavy duty.

NI CLAD TO LOW CARBON STEEL CLAD TO CU
Texas Instruments Inc./Materials Control
1008 low carbon steel clad with 201 nickel on one side and CDA 101 copper on the other. For button battery anode caps.

NI CR HOBBING
Bethlehem Steel Corp.
C 0.09, Mn 0.5, Cr 0.55, Si 0.1, Ni 1.25, bal Fe.
For plastic mold dies; oil hardening. *Obsolete*

NI RW 4 ZR
German manufacture
W 4.2, Zr 0.1, bal Ni.
For thermionic valves; heat and corrosion resistant.

NI SPAN ALLOY C-902
Inco Alloys International Inc.
Now NI-SPAN-C ALLOY 902.

NI STAINLESS STEEL
Latrobe Steel Co.
C 0.3-0.4, Cr 13-15, Ni 1.5-2, bal Fe.
For pump rods, pistons, valves, cutlery; corrosion resistant; resists heat up to 1000 F. *Obsolete*

NI TENSILIRON
International Nickel Inc.
Cast iron. C 2.5-3.1, Mn 0.5-0.9, Si 1.2-2.75, Ni 1-4, Mo 0-1, Cr 0-0.5, bal Fe.
Cast: 40,000 100,000 TS; 220-350 Brin. For gears, turbines, casings and rotors, valves, bushings; Nickel "Tensile;" high strength cast iron. *Obsolete*

NI TENSYLE
Sheepbridge Alloy Castings Ltd.
C 2.8, Ni 1.5-2, Si 1.25-1.75, bal Fe.
Cast: 56,000 TS; 200 Brin. Heat treated: 66,000 TS; 280 Brin. For flywheels; cast iron.

NI TENSYLE
Sheepbridge Engineering Ltd.
C 2.8, Ni 1.5-2, Si 1.25-1.75, bal Fe.
Cast: 56,000 TS; 200 Brin. Heat treated: 66,000 TS; 280 Brin. For flywheels; cast iron.

NI-20 CR-2THO₂
Now TD NICR.

NI-BAR IRON
Ingersoll-Rand Co.
TC 3.3, Mn 0.6, Si 1.5, Cr 0.6, Ni 1.5, bal Fe.
36,000 TS; 210 Brin. For grate bars, stoker links.

NI-BRAI
Ampco Metal
Ni 5, Al 10, Fe 5, Mn 1.5, bal Cu.
Cast: 95,000 TS; 43,000 YS; 18 El; 15 RA; 170 Brin. For pumps, valves, bearings; Al-bronze, corrosion resistant. *Obsolete*

NI-CHRO-ZINK
German manufacture
Ni, Cr, Zn.
For die casting alloy.

NI-COPPER
Manufacturer not listed
Ni-Cu.
For wire for motor winding; heat resistant. *Obsolete*

NI-CR ALLOY, 50CR-50NI
Ancast, Inc.
Nickel. C 0-0.1, Mn 0-0.3, Si 1, Cr 48-52, Fe 0-1, P 0-0.02, S 0-0.02, Al 0-0.25, Ti 0-0.5, 0.30 N₂ max, bal Ni.
ASTM A-560.

NI-CR ALLOY, 60CR-40NI
Ancast, Inc.
Nickel. C 0-0.1, Mn 0-0.3, Si 0-1, Cr 58-62, Fe 0-1, P 0-0.02, S 0-0.02, Al 0-0.25, Ti 0-0.5, 0.30 N₂ max, bal Ni.
ASTM A-560.

NI-CR-MO ABRASION RESISTANT STEEL
United States Steel Corp.
C 0.3, Mn 0.3, Si 0.25, Ni 3, Cr 1, Mo 0.3, bal Fe.
Water quenched and tempered: 400 Brin. Will take mild forming and welding. For heavy impact and abrasion; chutes, wear plates, dump trucks.

NI-CU ALLOY, QQN 288 (1)
Ancast, Inc.
Nickel. C 0-0.35, Mn 0-1.5, Si 0-2, Ni 62-68, Fe 0-2.5, Cu 28-33, Al 0-0.5.
Comp. A (Alloy 410).

NI-CU ALLOY, QQN 288 (2)
Ancast, Inc.
Nickel. C 0-0.3, Mn 0-1.5, Si 1-2, Fe 0-3.5, P 0-0.05, S 0-0.03, bal Ni.
Comp. E (Alloy 411).

NI-FE 30
Edgar Allen Balfour Ltd.
Ni 30, bal Fe.
For shunts in electrical equipment. Temperature compensating alloy. Low coefficient of expansion, high permeability.

NI-FLEX 76A
Materials Development Corp.
Nickel. Cr 13, Fe 4, Si 4.5, B 3.2, Ni 75.3.
Brazing alloy. MP 1790°F. FP 1990°F. Meets AMS 4776A.

NI-FLEX 77
Materials Development Corp.
Nickel. Cr 7, Fe 3, Si 4.5, B 3, Ni 82.5.
Brazing alloy. MP 1780°F. FP 1850°F. Meets AMS 4777.

NI-FLEX 78
Materials Development Corp.
Nickel. Si 4.5, B 3, Ni 92.5.
Brazing alloy. MP 1800°F. FP 1900°F. Meets AMS 4778.

NI-FLEX 79
Materials Development Corp.
Nickel. Si 3.5, B 2, Ni 94.5.
Brazing alloy. MP 1800°F. FP 1950°F. Meets AMS 4779.

NI-FLEX 95
Materials Development Corp.
Nickel. Cr 15, B 3.5, Ni 81.5.
Brazing alloy. MP 1930°F. FP 1950°F

NI-HARD
Robins Engineers & Constructors Inc.
Ni 4.5, Cr 1.5, Mn 1.5, C 2.75-3.75, bal Fe.
Chilled: 600-750 Brin. For abrasion resisting applications, crushers, rolls, feeder vanes; cast iron, abrasion resistant. Formerly SUPER MANGA IRON.

NI-HARD
Sheepbridge Alloy Castings Ltd.
C 3, Ni 3.5-4.5, Cr 1.5, bal Fe.
Cast: 40,000 TS; 500-650 Brin. For castings; abrasion resisting.

NI-HARD
Sheepbridge Engineering Ltd.
C 3, Ni 3.5-4.5, Cr 1.5, bal Fe.
Cast: 40,000 TS; 500-650 Brin. For castings; abrasion resisting.

NI-HARD

Follsain-Wycliffe Foundries, Ltd.
C 3.2-3.8, Si 0.3-0.8, Mn 0.3-0.8, Ni 3.5-5.5, Cr 1.5-2.5, bal Fe.
Abrasion resistant casting; 500 Brin min.

NI-HARD TYPE 2

International Nickel Inc.
Cast iron. Mn 0.4-0.6, Ni 4.25-4.75, Cr 1.4-2.5, C 0-2.9, Si 0.5-0.8, bal Fe.
Cast: 45,000-55,000 TS; 525-565 Brin. For ball mill liners, crushers, pump parts; abrasion and corrosion resistant. *Obsolete*

NI-HARD TYPE 3

International Nickel Inc.
Cast iron. C 1-1.6, Si 0.4-0.7, Mn 0.4-0.7, Ni 4-4.75, Cr 1.4-1.6, bal Fe.
Sand cast: 75,000-125,000 TS; 350-500 Brin. Chill cast: 90,000-140,000 TS; 300-600 Brin. For jaw crushers, grinding balls, mill liners, slurry pumps. Wear resistant, tough. *Obsolete*

NI-HARD TYPE 4

Thomas Foundries, Inc.
C 2.5-3.6, Mn 0-1.3, Si 1-2.2, Ni 5-7, Cr 7-11, Mo 0-1, P 0-0.1, S 0-0.15, bal Fe.
Sand cast: 550 Brin min. Chill cast: 600 Brin min. For large section cast iron with greater abrasion resistance. ASTM A 532 Cl. I Type D.

NI-HARD TYPE 4

International Nickel Inc.
Cast iron. TC 3-3.6, GC 0-0.1, Si 1.5-2, Mn 0.4-0.7, Ni 5.5-6.5, Cr 7-10, bal Fe.
Sand cast: 75,000-85,000 TS; 5000-6000 transverse strength, 0.08-0.11 def., 580 Vickers min. For jaw crushers, hammer mills, grinding balls, mill liners. Wear resistant, martensitic white iron resistant to fracture under impact. *Obsolete*

NI-HARD TYPE A

Thomas Foundries, Inc.
C 3-3.6, Mn 1.3, Si 0-0.8, Ni 3.3-3.5, Cr 1.4-4, Mo 0-1, P 0-0.3, S 0-0.15, bal Fe.
Sand cast: 550 Brin min. Chill cast: 600 Brin min. Abrasion resistant cast iron for medium size sections. ASTM A 532 Cl. I Type A.

NI-HARD TYPE D

Thomas Foundries, Inc.
C 2.5-3.6, Mn 0-1.3, Si 1-2.2, Ni 5-7, Cr 7-11, Mo 0-1, P 0-0.1, S 0-0.15, bal Fe.
Sand cast: 550 Brin min. Chill cast: 600 Brin min. Abrasion resistant cast iron for tough large section size castings. ASTM A 532 Cl. I Type D.

NI-HARD TYPE N

Thomas Foundries, Inc.
C 3-3.6, Mn 1.3, Si 0-0.8, Ni 3.3-3.5, Cr 1.4-4, Mo 0-1, P 0-0.3, S 0-0.15, bal Fe.
Sand cast: 550 Brin min. Chill cast: 600 Brin min. Abrasion resistant cast iron for larger size sections. ASTM A 532 Cl. I Type A.

NI-HARD, TYPE 1

International Nickel Inc.
Cast iron. Ni 4.2-4.7, C 3-3.6, Cr 1.4-2.5, Si 0.5, Mn 0.3-0.7, bal Fe.
Cast: 40,000-50,000 TS; 550-650 Brin. Chilled: 80,000 TS; 700 Brin. For die casting pots, pump plungers, roller bearing races, chilled rolls and liners; tough, corrosion and wear resistant. *Obsolete*

NI-HTC

Gilby-Fodor S.A.
Fe 0.04, Ni 99.9.
For heating elements. Useful operating temperature up to 100°C.

NI-MOC

Societe Nouvelle du Saut-du-Tarn
Cr 16, Mo 17, Fe 7, W 4, bal Ni.
For valves, pumps, pharmaceutical and chemical plant equipment. Heat and corrosion resistant. *Obsolete*

NI-RESIST

Sheepbridge Alloy Castings Ltd.
C 2.8, Ni 14, Cu 7, Cr 2, bal Fe.
Cast: 28,000 TS; 140-200 Brin. For castings; heat and corrosion resisting.

NI-RESIST

Sheepbridge Engineering Ltd.
C 2.8, Ni 14, Cu 7, Cr 2, bal Fe.
Cast: 28,000 TS; 140-200 Brin. For castings; heat and corrosion resisting.

NI-RESIST G

Symington-Gould Corp.
C 2.7, Ni, Cu, Cr, bal Fe.
Annealed: 30,000 TS; 150 Brin. For chemical plant equipment; resists acids, oxidation.

NI-RESIST I

Usines Emile Henricott, SA
C 2-3, Ni 14, Cr 2, Cu 6, bal Fe.
Cast: 25,000-30,000 TS; 120-180 Brin. For corrosion and heat resistant castings; stainless cast iron, austenitic.

NI-RESIST II

Usines Emile Henricott, SA
C 3, Ni 20, Cr 2, Si 2, bal Fe.
Cast: 25,000-30,000 TS; 120-180 Brin. For corrosion and heat resistant castings; stainless cast iron, austenitic.

NI-RESIST III

Usines Emile Henricott, SA
C 2.7, Ni 30, Si 2, bal Fe.
Cast: 25,000-35,000 TS; 110-170 Brin. For corrosion and heat resistant castings; stainless cast iron, austenitic.

NI-RESIST IV

Usines Emile Henricott, SA
Ni 30, Si 3, C, bal Fe.
Cast: 25,000-35,000 TS; 130-200 Brin. For food industry equipment; stainless cast iron.

NI-RESIST LOW EXPANSION

International Nickel Inc.
Ni 30-36, Cr 1-4, Si 1-2, Mn 0.8-1.2, TC 2-2.8, bal Fe.
Cast: 20,000-35,000 TS; 100-166 Brin. For pumps, valves, pistons; corrosion resistant. *Obsolete*

NI-RESIST N

Symington-Gould Corp.
C 2.7, Ni, Cr, bal Fe.
Annealed: 30,000 TS; 150 Brin. For chemical plant equipment; resists caustic, ammonia.

NI-RESIST TYPE 1

Dominion Wheel & Foundry Co.
C 2.5-3.5, Si 1.2, Mn 0.75-1.2, Ni 12-15, Cr 1-3, Cu 5-7, bal Fe.
Cast: 25,000-30,000 TS; 130-160 Brin. For corrosion and heat resistant castings; resists corrosion, heat and wear.

NI-RESIST TYPE 1

International Nickel Inc.
Cast iron. Cr 1.7-2.5, Si 1-2.5, Mn 1-1.5, Ni 13-17, Cu 5-7, TC 0-3, bal Fe.
Cast: 25,000-30,000 TS; 2 El; 0 RA; 130-160 Brin. For pipes, valves, pumps, propellers, hydraulic turbines, oil burners, automotive engine pistons and sleeves. Austenitic, nonmagnetic cast iron; corrosion and heat resistant. *Obsolete*

NI-RESIST TYPE 1-B

International Nickel Inc.
Cast iron. TC 0-3, Si 1-2.8, Mn 1-1.5, Ni 13.5-17.5, Cr 2.5-4, bal Fe.
Cast: 25,000 TS; 1 El; 140-190 Brin. For pumps, valves, filter presses, impellers, nozzles; nonmagnetic, erosion and corrosion resistant cast iron. *Obsolete*

NI-RESIST TYPE 2

Fahralloy Co.
C 0-3, Cr 1.7-2.5, Ni 18-22, Si 1-2.5, Mn 0.8-1.5, bal Fe.
Cast: 25,000-30,000 TS; 130-160 Brin. For castings. Corrosion and heat resistant.

NI-RESIST TYPE 2

Dominion Wheel & Foundry Co.
C 0-3, Cr 1.7-2.5, Ni 18-22, Si 1-2.5, Mn 0.8-1.5, bal Fe.
Cast: 25,000-30,000 TS; 130-160 Brin. For corrosion and heat resistant castings; resists heat, corrosion and wear.

NI-RESIST TYPE 2-B

International Nickel Inc.
Cast iron. TC 0-3, Si 1-2.8, Mn 0.8-1.5, Ni 18-22, Cr 2.5-4, bal Fe.
Cast: 25,000 TS; 1 El; 130-190 Brin. For turbocharger casings, manifolds, steam turbine nozzles; nonmagnetic, heat resistant cast iron. *Obsolete*

NI-RESIST TYPE 2A

Fahralloy Co.
C 0-2.8, Cr 1.7-2.5, Ni 18-22, Si 1.5-2.7, Mn 0.8-1.5, bal Fe.
Cast: 30,000-50,000 TS; 145-190 Brin. For castings. Corrosion and heat resistant.

NI-RESIST TYPE 2B

Fahralloy Co.
C 0-3, Cr 3-6, Ni 18-22, Si 1.2, Mn 0.8-1.5, bal Fe.
Cast: 25,000-45,000 TS; 170-250 Brin. For castings. Corrosion and heat resistant.

NI-RESIST TYPE 3

Dominion Wheel & Foundry Co.
C 0-2.75, Cr 2.5-3.5, Ni 28-32, Si 1-2, Mn 0.4-0.8, bal Fe.
Cast: 25,000-35,000 TS; 120-150 Brin. For corrosion and heat resistant castings; resists heat, corrosion and wear.

NI-RESIST TYPE 3

International Nickel Inc.
C 0-2.75, Cr 2.5-3.5, Ni 28-32, Si 1-2, Mn 0.4-0.8, bal Fe.
Cast: 25,000-35,000 TS; 120-150 Brin. For heat and corrosion resistant castings; corrosion and heat resistant.

NI-RESIST TYPE 3

Fahralloy Co.
C 0-2.75, Cr 2.5-3.5, Ni 28-32, Si 1-2, Mn 0.4-0.8, bal Fe.
Cast: 25,000-35,000 TS; 120-150 Brin. For heat and corrosion resistant castings; corrosion and heat resistant.

NI-RESIST TYPE 4

Dominion Wheel & Foundry Co.
C 0-2.6, Cr 4.5-5.5, Ni 29-32, Si 5-6, Mn 0.4-0.8, bal Fe.
Cast: 25,000-35,000 TS; 150-180 Brin. For corrosion and heat resistant castings; resists heat, corrosion and wear.

NI-RESIST TYPE 4

International Nickel Inc.
C 0-2.6, Cr 4.5-5.5, Ni 29-32, Si 5-6, Mn 0.4-0.8, bal Fe.
Cast: 25,000-35,000 TS; 150-180 Brin. For heat resistant castings, stove stops, cookware; corrosion and heat resistant.

NI-RESIST TYPE 4

Fahralloy Co.
C 0-2.6, Cr 4.5-5.5, Ni 29-32, Si 5-6, Mn 0.4-0.8, bal Fe.
Cast: 25,000-35,000 TS; 150-180 Brin. For heat resistant castings, stove stops, cookware; corrosion and heat resistant.

NI-RESIST TYPE 5

Fahralloy Co.
C 0-2.4, Cr 0-3, Ni 34-36, Si 1-2, Mn 0.4-0.8, bal Fe.
Cast: 20,000-25,000 TS; 100-140 Brin. For heat resistant castings. Corrosion and heat resistant.

NI-RESIST TYPE 5
International Nickel Inc.
Cast iron. TC 0-2.4, Si 1-2, Mn 0.4-0.8, Ni 34-36, Cu 0-0.5, Cr 0-0.1, bal Fe.
Cast: 20,000-25,000 TS; 100-125 Brin. For gages, glass molds, paper dies, chemical equipment; low thermal expansion. *Obsolete*

NI-RESIST TYPE D-2
International Nickel Inc.
Cast iron. TC 0-3, Si 1.5-3, Mn 0.7-1.2, Ni 18-22, Cr 1.75-2.75, bal Fe.
Cast: 60,000 TS; 30,000 YS; 8-20 El; 140-200 Brin. For pumps, valves, pipe fittings, paper rolls; corrosion and heat resistant cast iron. *Obsolete*

NI-RESIST TYPE D-2B
International Nickel Inc.
Cast iron. TC 0-3, Si 1.5-3, Mn 0.7-1.2, Ni 18-22, Cr 2.7-4, bal Fe.
Cast: 60,000 TS; 30,000 YS; 7-15 El; 150-210 Brin. For impellers, pumps, valves, engine parts; corrosion and heat resistant cast iron. *Obsolete*

NI-RESIST TYPE D-2C
International Nickel Inc.
Cast iron. Mn 1.8-2.4, Ni 28-32, C 0-2.9, Si 1.3, bal Fe.
Cast: 60,000 TS; 30,000 YS; 20-40 El; 120-170 Brin. For switch gears, pumps, valves, bearings, seals; corrosion and heat resistant cast iron. *Obsolete*

NI-RESIST TYPE D-3
International Nickel Inc.
Cast iron. TC 2.6, Si 1-2.8, Ni 28-32, Cr 2.5-3.5, bal Fe.
Cast: 55,000 TS; 30,000 YS; 6-20 El; 140-200 Brin. For steam turbines, engines, liners, valves, kettles; corrosion and heat resistant cast iron. *Obsolete*

NI-RESIST TYPE D-3A
International Nickel Inc.
Cast iron. TC 2.6, Si 1-2.8, Ni 28-32, Cr 1.5, bal Fe.
Cast: 55,000 TS; 30,000 YS; 10-20 El; 130-190 Brin. For liners, bearings, valve guides; corrosion and heat resistant cast iron. *Obsolete*

NI-RESIST TYPE D-4
International Nickel Inc.
Cast iron. TC 2.6, Si 5-6, Ni 28-32, Cr 4.5-5.5, bal Fe.
Cast: 60,000 TS; 200-270 Brin. For cookware, range tops; nonmagnetic, corrosion resistant cast iron. *Obsolete*

NI-RESIST TYPE D-5
International Nickel Inc.
Cast iron. TC 0-2.4, Si 1-2.8, Ni 34-36, bal Fe.
Cast: 55,000 TS; 30,000 YS; 20-40 El; 130-280 Brin. For dies, glass molds, ingot molds; corrosion and heat resistant cast iron. *Obsolete*

NI-RESIST TYPE D-5B
International Nickel Inc.
Cast iron. TC 0-2.4, Si 1-2.8, Ni 34-36, Cr 2-3, bal Fe.
Cast: 55,000 TS; 30,000 YS; 6-15 El; 140-190 Brin. For dies, glass molds, ingot molds; corrosion and heat resistant cast iron. *Obsolete*

NI-RESIST V
Usines Emile Henricott, SA
C 2.4, Ni 36, Si 1-2, bal Fe.
Cast: 20,000-25,000 TS; 110-160 Brin. For heat resistant castings; stainless cast iron, low coefficient of expansion.

NI-RESIST, TYPE 2
International Nickel Inc.
Cast iron. Ni 18-22, Cr 2-4, C 2.2-3, Mn 1-1.5, Si 1-2.5, bal Fe.
Cast: 25,000-30,000 TS; 2 El; 0 RA; 120-170 Brin. For heat resistant parts, grids, furnace parts, pumps, pipe; corrosion and heat resistant. *Obsolete*

NI-ROD
Now NI-ROD WELDING ELECTRODE.

NI-ROD 44 WELDING ELECTRODE
Inco Alloys International Inc.
For cast irons, especially for high strength and ductility.

NI-ROD 55
International Nickel Inc.
Ni 60, Fe 40.
For welding electrodes for cast iron; machinable. *Obsolete*

NI-ROD 55 WELDING ELECTRODE
Inco Alloys International Inc.
Ni 53, C 1.5, Mn 0.3, Fe 45, S 0.005, Si 0.5, Cu 0.1.
Electrode for shielded metal-arc welding of cast and ductile irons; cast irons to wrought alloys. AWS A 5.15 Class E NiFe-Cl.

NI-ROD 55X WELDING ELECTRODE
Inco Alloys International Inc.
For cast irons, especially for out-of-position welding and high phosphorus irons.

NI-ROD 99X WELDING ELECTRODE
Inco Alloys International Inc.
For cast irons, especially for out-of-position welding, thin sections, and machinability.

NI-ROD FC 55 CORED WIRE
Inco Alloys International Inc.
Ni 50, Fe 44, C 1, Mn 4.2, Si 0.6.
For automatic and semi-automatic welding of gray, malleable and ductile cast irons.

NI-ROD FILLER METAL 44
Inco Alloys International Inc.
For cast irons, especially robotic and automatic welding.

NI-ROD WELDING ELECTRODE
Inco Alloys International Inc.
Ni 95, C 1, Mn 0.2, Fe 3, S 0.005, Si 0.7, Cu 0.1.
Electrode for shielded metal-arc welding of cast iron. AWS A 5.15 Class E Ni-Cl.

NI-SPAN C
Engelhard Corp.
Ni 41-43, Ti 2.4, Cr 5.1-5.7, C 0-0.6, Al 0.6, Si 0.8, bal Fe.
Annealed: 90,000 psi TS; 35,000 psi YS; 40 El; 145 Brin.
Aged: 200,000 psi TS; 180,000 psi YS; 7 El; 395 Brin. For instruments, springs, diaphragms; age-hardenable, constant modulus. *Obsolete*

NI-SPAN HI
International Nickel Inc.
Superalloy. Ni 28-30, Ti 2.4, Cr 8-9, C 0-0.06, Si 0.5, Mn 0.4, Al 0.6, bal Fe.
Aged: 140,000 TS; 90,000 YS; 20 El. For thermostats, bimetals; high coefficient of expansion, age hardenable. *Obsolete*

NI-SPAN LO 42
International Nickel Inc.
Nickel. Ni 40.5-42.5, Ti 2.2-2.6, C 0-0.06, Mn 0.4, Si 0.5, Al 0.6, bal Fe.
Annealed: 80,000 TS; 40,000 YS; 32 El; 330 Brin. Hardened: 165,000 TS; 120,000 YS; 14 El; 330 Brin. For thermostats, bimetals; age hardenable, low coefficient of expansion. *Obsolete*

NI-SPAN LO 45
International Nickel Inc.
Nickel. Ni 44.4-46.5, Ti 2.4, C 0-0.06, Mn 0.4, Si 0.5, Al 0.6, bal Fe.
For thermostats, bimetals; low coefficient of expansion, age hardenable. *Obsolete*

NI-SPAN LO 52
International Nickel Inc.
Nickel. Ni 51-53, Ti 2.4, C 0-0.06, Mn 0.4, Si 0.5, Al 0.6, bal Fe.
Annealed: 85,000 TS; 35,000 YS; 27 El; 125 Brin. Hardened: 120,000 TS; 95,000 YS; 17 El; 305 Brin. For thermostats, bimetals; low coefficient of expansion, age hardenable. *Obsolete*

NI-SPAN-C ALLOY 902
Inco Alloys International Inc.
C 0.03, Ni 42.5, Fe 49, Cr 5.3, Ti 2.4, Al 0.5.
Low thermal expansion alloy. Thermal expansion at 20-100°C: 6.2×10^{-6}/K.

NI-SPAN-C ALLOY 902
Ulbrich Stainless & Spec.Metals Inc.
Ni 42.25, C 0.03, Mn 0.4, Fe 48.5, S 0.02, Cr 5.33, Al 0.55, Ti 2.58, Si 0.5, Cu 0.05.
Hot rolled and aged: 175,000 TS; 110,000 YS; 25 El. For tuning forks and other mechanical resonators, electromechanical filter watch and clock hairsprings; age-hardenable; controllable thermo-elastic coefficient.

NI-SPAN-C ALLOY 902
Huntington Alloys Inc.
Ni 42.25, C 0.03, Mn 0.4, Fe 48.5, S 0.02, Cr 5.33, Al 0.55, Ti 2.58, Si 0.5, Cu 0.05.
Hot rolled and aged: 175,000 TS; 110,000 YS; 25 El. For tuning forks and other mechanical resonators, electromechanical filter watch and clock hairsprings; age-hardenable; controllable thermo-elastic coefficient.

NI-SPAN-D
Henry Wiggin & Co. Ltd.
Ni 42.2, Cr 6.5, Ti 2.8, Al 0.45, bal Fe.
For diaphragms, capsules, tuning forks, Bourdon tubes, transducer load cells. Maximum service temperature of 85 C. *Obsolete*

NI-VEE
Olds Alloys Co.
Ni 5, Sn 5, Zn 2, bal Cu.
Aged: 80,000 TS; 50,000 YS; 10 El; 150 Brin. For gears, valves, construction castings; age-hardenable, corrosion resistant.

NI-VEE 1
A.W. Cadman Mfg. Co.
Cu 88, Ni 5, Sn 5, Zn 2.
Cast: 50,000 TS; 22,000 YS; 40 El; 85 Brin. Heat treated: 65,000-85,000 TS; 40,000-55,000 YS; 8-10 El; 130-180 Brin. For gears, cams, valves; fine grain, heat treatable.

NI-VEE 2
A.W. Cadman Mfg. Co.
Cu 80, Ni 5, Sn 5, Pb 5, Zn 5.
Cast: 40,000 TS; 20,000 YS; 20 El; 80 Brin. Heat treated: 50,000 TS; 30,000 YS; 5 El; 130 Brin. For pressure castings, valves, fittings, plumbing parts; fine grain, pressure tight.

NI-VEE 3
A.W. Cadman Mfg. Co.
Cu 80, Ni 5, Sn 5, Pb 10.
Cast: 35,000 TS; 20,000 YS; 10 El; 80 Brin. Heat treated: 40,000 TS; 25,000 YS; 5 El; 110 Brin. For bearings, bushings, acid resistant castings; pressure tight, fine grain.

NI-VEE 4
A.W. Cadman Mfg. Co.
Cu 70, Ni 5, Sn 5, Pb 20.
Cast: 25,000 TS; 18,000 YS; 10 El; 70 Brin. Heat treated: 30,000 TS; 22,000 YS; 5 El; 80 Brin. For bearings, bushings, acid resistant castings; pressure tight, fine grain.

NI-VEE L15
Olds Alloys Co.
Ni 5, Sn 5, Pb 15, Zn 0-0.5, bal Cu.
Cast: 30,000 TS; 21,000 YS; 12 El; 78 Brin. Heat treated: 35,000 TS; 24,000 YS; 5 El; 88 Brin. For bearings, bushings, acid resistant castings; fine grain, heat treatable, good castability.

NI-VEE L2
Olds Alloys Co.
Ni 5, Sn 5, Pb 2.5, Zn 0-0.5, bal Cu.
Cast: 44,000 TS; 23,000 YS; 30 El; 92 Brin. Heat treated: 55,000 TS; 40,000 YS; 10 El; 135 Brin. For gears, bushings, bearings, construction castings; fine grain, heat treatable, good castability.

NI-VEE L2-Z2
Olds Alloys Co.
Ni 5, Sn 5, Pb 2.5, Zn 0-2, bal Cu.
Cast: 42,000 TS; 24,000 YS; 25 El; 90 Brin. Heat treated: 49,000 TS; 35,000 YS; 8 El; 120 Brin. For gears, bearings, bushings, construction castings, fine grain, heat treatable, good castability.

NI-VEE L5
Olds Alloys Co.
Ni 5, Sn 5, Pb 5, Zn 0-0.5, bal Cu.
Cast: 41,000 TS; 20,000 YS; 22 El; 92 Brin. Heat treated: 47,000 TS; 35,000 YS; 7 El; 130 Brin. For bearings, bushings, acid resistant castings; fine grain, heat treatable, good castability.

NI-VEE L5-Z5
Olds Alloys Co.
Ni 5, Sn 55, Pb 5, Zn 5, P 0.05, bal Cu.
Cast: 40,000 TS; 20,000 YS; 20 El; 80 Brin. Heat treated: 50,000 TS; 30,000 YS; 5 El; 120 Brin. For pumps, valves, fittings; corrosion resistant.

NI-VEE TYPE A
International Nickel Inc.
Copper. Cu 88, Ni 5, Sn 5, Zn 2.
Cast: 50,000 TS; 22,000 YS; 40 El; 50 RA; 85 Brin. Tempered: 65,000 TS; 40,000 YS; 10 El; 130 Brin. Heat treated: 85,000 TS; 55,000 YS; 10 El; 26 RA; 180 Brin. For machine tools, cams, rollers, guides, gears; age hardenable, corrosion resistant. *Obsolete*

NI-VEE TYPE B
International Nickel Inc.
Copper. Cu 87, Ni 5, Sn 5, Pb 1, Zn 2.
Cast: 45,000 TS; 20,000 YS; 30 El; 80 Brin. Tempered: 60,000 TS; 30,000 YS; 8 El; 120 Brin. For machine tools, cylinder, cams, gears, rollers; age hardenable, corrosion resistant. *Obsolete*

NI-VEE TYPE C
International Nickel Inc.
Copper. Cu 80, Ni 5, Pb 5, Zn 5, Sn 5.
Cast: 40,000 TS; 20,000 YS; 15 El; 15 RA; 80 Brin. Tempered: 50,000 TS; 30,000 YS; 5 El; 2.5 RA; 130 Brin. For pressure castings, pumps, valves, fittings; age hardenable, corrosion resistant. *Obsolete*

NI-VEE TYPE D
International Nickel Inc.
Copper. Cu 80, Ni 5, Sn 5, Pb 10.
Cast: 35,000 TS; 20,000 YS; 10 El; 10 RA; 80 Brin. Tempered: 40,000 TS; 25,000 YS; 2 El; 3 RA; 110 Brin. For bearings, bushings, liners; age hardenable, corrosion resistant. *Obsolete*

NI-VEE TYPE E
International Nickel Inc.
Copper. Cu 70, Ni 5, Sn 5, Pb 20.
Cast: 25,000 TS; 18,000 YS; 5 El; 5 RA; 70 Brin. Tempered: 30,000 TS; 22,000 YS; 2 El; 2 RA; 80 Brin. For bearings, bushings, liners; age hardenable, corrosion resistant. *Obsolete*

NI-VEE-L10
Olds Alloys Co.
Cu 80, Sn 5, Pb 10, Ni 5.
Cast: 35,000 TS; 10 El; 80 Brin. Heat treated: 40,000 TS; 5 El; 110 Brin. For valves, fittings; pressure tight.

NI-VEE-L20
Olds Alloys Co.
Cu 70, Ni 5, Sn 5, Pb 20.
Cast: 25,000 TS; 10 El; 70 Brin. Heat treated: 30,000 TS; 5 El; 80 Brin. For pressure castings, bearings; free-cutting.

NI-WELD
Crucible Materials Corp.
Nickel. Ni 100.
For welding rod for cast iron; machinable welds.

NIAG
Mueller Brass Co.
Cu 46, Pb 1, Ni, bal Zn.
70,000 TS; 42,000 YS; 30 El; 120 Brin. For hardware, plated parts; wear and corrosion resistant. *Obsolete*

NIAG
American manufacture
Cu 46.7, Zn 40.7, Ni 9.1, Pb 2.8, Mn 0.3.
For white metal parts; corrosion resistant.

NIAGARA
Colt Industries
C 0.45-0.55, bal Fe.
Rolled: 80,000 TS; 50,000 YS. For machine tool parts, gears, shafts; water hardened. *Obsolete*

NIAGRA
Wallace Murray Corp.
C 0.7, W 14, Cr 4, V 2, bal Fe.
For tools, cutters, taps, hobs; high speed steel. *Obsolete*

NIAGRA BRAND FERRO-CHROME
Pittsburgh Metallurgical Co. Inc.
Cr 66-70, bal Fe.
For metallurgical applications in steel; Cr-additions.

NIAGRA BRAND FERRO-SILICON
Pittsburgh Metallurgical Co. Inc.
Si 15-90, bal Fe.
For metallurgical applications in steel; Si-additions.

NIAGRA BRAND SILICO MANGANESE
Pittsburgh Metallurgical Co. Inc.
Si 12-20, Mn 65-70, bal Fe.
For steel metallurgical applications; Mn-additions.

NIAL
Edgar Allen Balfour Ltd.
Ni 24, Al 13, Cu 4, bal Fe.
For loud speakers, lighting and ignition equipment. Permanent magnet. High permeability.

NIAL I
Wilbur B. Driver Co.
Nickel. Mn 2.5, Al 2, Si 1, bal Ni.
For negative thermoelement of standard Type K thermocouple.

NIALCO
Ugine Aciers
Ni 19, Co 12, Al 10, Cu 6, bal Fe.
For permanent magnets, electrical and magnetic equipment. High magnetic permeability.

NIALCO 200
Austrian manufacture
Al 10, Ni 20, Co 15, Cu 3, bal Fe.
For permanent magnets.

NIALCO 400
Austrian manufacture
Al 8, Ni 14, Co 24, Cu 3, bal Fe.
For permanent magnets.

NIALCO I
Ugine Aciers
Ni 12, Al 10, Co 4-20, Cu 2, bal Fe.
Residual induction 6500 gauss; coercive force 530 oersted; energy product 1.4 max; 45 Rock C. For magnets in Wattmeters. Permanent magnet. High permeability.

NIALCO II
Ugine Aciers
Ni 12, Al 10, Co 4-20, Cu 2, bal Fe.
Residual induction 6300 gauss; coercive strength 650 oersted; energy product 1.5 max; 45 Rock C. For magnets in electrical relays. Permanent magnet. High permeability.

NIALCO III
Ugine Aciers
Ni 12, Al 10, Co 4-20, Cu 2, bal Fe.
Residual induction 7000 gauss; coercive force 690 oersted; energy product 1.7 max; 45 Rock C. For magnets in magnetos. Permanent magnet, high permeability.

NIALCO IV
Ugine Aciers
Ni 12, Al 10, Co 4-20, Cu 2, bal Fe.
Residual induction 5700 gauss; coercive force 1000 oersted; energy product 1.9 max; 58 Rock C. For magnets in electrical equipment. Permanent magnet, high permeability.

NIALITE
Baldwin-Lima-Hamilton Corp.
Copper. Al 10, Ni 5, Fe 5, Mn 1.5, bal Cu.
Cast: 85,700 TS; 39,000 YS; 24 El; 220 Brin. For pumps, valves, propellers; corrosion and cavitation resistant.

NIBORIUM B
Niborium Industries Inc.
Electrical contact alloy. Annealed: 65,000 psi TS; 40,000 psi YS; 40 El. Spring temper: 135,000 psi TS; 133,000 psi YS; 1 El. Solderable, weldable, machinable, corrosion resistant electrical contact material.

NIBSI
Western Gold & Platinum Co.
Si 3.5, B 1.8, bal Ni.
Brazing powder; melt range: 1800-1950°F. For brazing stainless and high temperature alloys. AMS 4779.

NIC ALLOY
Swedish Crucible Steel Co.
C, Ni, bal Fe.
For heat and corrosion resistant parts. *Obsolete*

NIC-MOTAL
J.T. Wing & Co.
Pb, Sb, Ni, bal Sn.
Babbitt metal for bearings.

NICA-0
Breda Co.
C 0-0.08, Cr 18, Ni 8, bal Fe.
Annealed: 85,000 TS; 35,000 YS; 60 El; 70 RA; 150 Brin. For chemical plant equipment, tanks, mixers; Type 304; stainless, austenitic.

NICA-00
Breda Co.
C 0-0.05, Cr 18, Ni 8, bal Fe.
Annealed: 85,000 TS; 35,000 YS; 60 El; 70 RA; 150 Brin. For chemical plant equipment, tanks, mixers; Type 304; stainless, austenitic.

NICA-1
Breda Co.
C 0.11-0.16, Cr 18, Ni 8, bal Fe.
Annealed: 80,000 TS; 35,000 YS; 55 El; 75 RA; 150 Brin. For chemical plant equipment, tanks, mixers, filters; Type 302; stainless austenitic.

NICA-2
Breda Co.
C 0.17-0.25, Cr 18, Ni 8, bal Fe.
Annealed: 85,000 TS; 40,000 YS; 50 El; 70 RA; 160 Brin. For chemical plant equipment, tanks, mixers, filters; Type 301 and 302; stainless, austenitic.

NICALLOY
English manufacture
Ni 47, Fe 53.
For electrical equipment; magnetically soft, high permeability.

NICALOI
Allegheny Ludlum Steel
Ni 47, bal Fe.
For armature punchings, electrical machinery and motors; high magnetic permeability. *Obsolete*

NICALOY
Manufacturer not listed
Ni 49, Fe 51.
For electrical equipment and apparatus; resistance alloy; high magnetic permeability. *Obsolete*

NICALUN
American Abrasive Metals Co.
Ni, Cu, alloy base with embedded abrasive grains.
For elevator door sills, stair treads; castings; wear resistant.

NICAR
Arcos Alloys
C 3.5-4, Si 1, Cr 12-18, Ni 65-75, Co 0.2, B 2.5-4.5, Fe 4.
For hardfacing bare rod; class RNICr-C. *Obsolete*

NICAST
Chemetron Corp.
Si 1, Fe 1.96, bal Ni.
For welding rod; for cast iron. *Obsolete*

NICHRANEL-C
Le Bronze Industriel
Ni 60, Cr 15, Mo 15, Fe 5, W 5.
For pumps, valves, pharmaceutical and chemical plant equipment. Heat and corrosion resistant. *Obsolete*

NICHRO-ZINK
NL Industries
Zn, Cu.
16,000 TS; 2 El; 40 Brin. Easily soldered. *Obsolete*

NICHROFRY 152
Societe des AFY
C 0-0.25, Cr 24-26, Ni 19-22, bal Fe.
Annealed: 95,000 TS; 45,000 YS; 50 El; 65 RA; 180 Brin. At 1200°F: 57,000 TS; 22,000 YS; 32 El; 45 RA. For furnace parts, valves, pumps, heat treating boxes; Type 310; austenitic, heat resistant.

NICHROFY 152
Creusot-Loire
C 0-0.25, Cr 25, Ni 20, bal Fe.
Annealed: 100,000 TS; 45,000 YS; 50 El; 65 RA; 185 Brin. For furnace parts, heat treating boxes; Type 310; heat resistant. *Obsolete*

NICHROFY 345
Creusot-Loire
C 0-0.2, Cr 24, Ni 12, bal Fe.
Annealed: 90,000 TS; 40,000 YS; 50 El; 65 RA; 170 Brin. For furnace parts, heat treating boxes; Type 309; heat resistant. *Obsolete*

NICHROFY 345SP
Creusot-Loire
C 0-0.08, Cr 20, Ni 11, bal Fe.
For furnace parts, heat treating boxes; Type 308; corrosion and heat resistant. *Obsolete*

NICHROFY 424
Creusot-Loire
C 0.13-0.25, Cr 26, Ni 4, bal Fe.
Cast: 90,000 TS; 65,000 YS; 2 El; 212 Brin. For cylinder liners, valve seats and bodies, bushings; Type 327; heat resistant. *Obsolete*

NICHROLLOY I
Walker Metal Products Ltd.
C 0.3, Ni 23, Cr 20, Mn 1, V 1, Al 0.5, bal Fe.
Annealed: 60,000 psi TS. For heat treating boxes, resistance wire; high heat resistance.

NICHROLOY 37
Walker Metal Products Ltd.
C 0.2, Cr 25, Ni 12, bal Fe.
Cast; 70,000 psi TS; 30,000 psi YS; 35 El; 60 RA; 150 Brin. For furnace parts, cast heat treating boxes; heat resistant to 2000°F.

NICHROLOY 45
Walker Metal Products Ltd.
C 0.4, Cr 25, Ni 20, bal Fe.
Cast: 75,000 psi TS; 50,000 psi YS; 17 El; 170 Brin. For furnace parts, retorts, stills; heat resistant to 2100°F.

NICHROLOY 50
Walker Metal Products Ltd.
C 0.4, Cr 15, Ni 35, bal Fe.
Cast: 70,000 psi TS; 40,000 psi YS; 10 El; 12 RA; 170 Brin. For salt pots, furnace parts, heat treating boxes; heat resistant to 2100°F.

NICHROLOY 72
Walker Metal Products Ltd.
C 0.4, Cr 12, Ni 60, bal Fe.
Cast: 70,000 psi TS; 40,000 psi YS; 6 El. For lead and cyanide pots, furnace equipment; heat resistant.

NICHROLOY A
Baldwin Steel Co.
C 0.51, Cr 1.05, Mo 0.25, Ni 0.53, V 0.21, Mn 0.97, bal Fe.
Heat treated: 165,000 TS; 150,000 YS; 20 El; 59 RA; 280-310 Brin. For arbors, axles, bolts, cams, gears, hubs, die liners; shock and fatigue resistant, preheat treated at mill.

NICHROLOY II
Walker Metal Products Ltd.
Ni 40, Cr 7, Mn 3, bal Fe.
For resistance wire, heat treating boxes; heat and corrosion resistant.

NICHROLOY III
Walker Metal Products Ltd.
Ni 75, Cr 16, Mn 3, bal Fe.
105,000-112,000 psi TS; 60,000-80,000 psi YS; 45-25 El; 65-59 RA; 180-210 Brin. For resistance wire, heat treating boxes; heat and corrosion resistant.

NICHROLOY L
Baldwin Steel Co.
C 0-0.5, Cr 1.05, Mo 0.25, Ni 0.53, V 0.21, Mn 0.97, bal Fe.
For arbors, axles, bolts, cams, gears, hubs, die liners; preheat treated at mill, shock and fatigue resistant.

NICHROME
Harrison Alloys Inc.
Nickel.
Now HAI-NICR 60.

NICHROME 62-16
Now WIRESPRAY NICHROME 62-16.

NICHROME II
Driver Harris Co.
Ni 66, Cr 22, bal Fe.
Cast: 100,000 TS. For electric resisting units, annealing pots; heat and corrosion resisting. *Obsolete*

NICHROME III
Driver Harris Co.
Ni 85, Cr 15.
110,000 TS. For electrical resistance units, annealing pots, heating elements; heat and corrosion resisting. *Obsolete*

NICHROME I
Driver Harris Co.
Ni 65, Cr 11.18, Fe 22.36, Mn 0.7, Si 0.26.
100,000 TS; 157-187 Brin. For electrical resistance units, annealing pots, heating elements; heat and corrosion resisti *Obsolete*

NICHROME II
British Driver-Harris Co. Ltd.
Ni 65-67, Cr 20-22, Fe 12-14, Mn 1.5-2.
For heavy ribbon for large heating furnaces; maximum working temperature 1100 C. *Obsolete*

NICHROME II
Driver Harris Co.
Ni 66, Cr 22, bal Fe.
100,000 TS. For electrical resistance units, annealing pots, heating elements; heat and corrosion resisting. *Obsolete*

NICHROME IV
Driver Harris Co.
Ni 82.36, Cr 15.9, Fe 0.84, Mn 0.48, Si 0.16.
120,000 TS; 60,000 YS; 25 El; 35 RA; 183 Brin. For electrical resistance units, annealing pots, heating elements; heat and corrosion resisting. *Obsolete*

NICHROME IV
Driver Harris Co.
Ni 80, Cr 20.
120,000 TS; 60,000 YS; 25 El; 35 RA; 183 Brin. For heating elements, heat treating boxes, electrical resistances; heat resistant. *Obsolete*

NICHROME S
Driver Harris Co.
Ni 25, Cr 20, bal Fe.
For heating elements, annealing and carburizing boxes; corrosion and heat resistant. *Obsolete*

NICHROME TYPE A
Driver Harris Co.
Ni 62, Cr 15, bal Fe.
For rheostats, potentiometers, heating elements; heat resistant. *Obsolete*

NICHROME TYPE B
Driver Harris Co.
Cr-Ni.
For addition agent to cast iron. *Obsolete*

NICHROME V
Harrison Alloys Inc.
Nickel.
New HAI NICR 80.

NICHROME V-242
Now D-H NO 242.

NICHROME V-245
Now D-H NO 245.

NICHROTHERM NCT-1
Krupp Stahl AG
stainless steel. C 0.15, Ni 15, Cr 20, bal Fe.
Untreated: 100,000-114,000 TS; 78,000-93,000 YS; 35-25 El; 55-45 Brin. For furnace parts, crucibles, autoclaves, recuperators; heat and corrosion resistant to 1050°C. *Obsolete*

NICHROTHERM NCT-3

Krupp Stahl AG

stainless steel. C 0.15, Ni 20, Cr 25, bal Fe.
Untreated: 100,000-114,000 TS; 65,000-78,000 YS; 35-25 El; 55-45 RA. For furnace parts, crucibles, autoclaves, recuperators; heat and corrosion resistant to 1200°C. *Obsolete*

NICHROTHERM NCT-6

Krupp Stahl AG

nickel. C 0.15, Ni 60, Cr 15, bal Fe.
Untreated: 85,000-100,000 TS; 35,000-50,000 YS; 35-25 El; 55-45 RA. For furnace parts, crucibles, autoclaves, recuperators; heat and corrosion resistant to 1150°C. *Obsolete*

NICHROTHERM NCT-6A

Krupp Stahl AG

nickel. Fe, Cr, Ni.
For furnace parts, crucibles, autoclaves, recuperators; heat and corrosion resistant to 1150°C. *Obsolete*

NICHROTHERM NCT-8

Krupp Stahl AG

nickel. C 0.15, Ni 80, Cr 17, bal Fe.
Untreated: 85,000-100,000 TS; 35,000-50,000 YS; 25-35 El; 45-55 RA. For furnace parts, crucibles, autoclaves, recuperators; heat and corrosion resistant; austenitic, max temperature 1300°C. *Obsolete*

NICHROTHERM-NCT-1A

Krupp Stahl AG

stainless steel. C 0.15, Si 2.5, Mn 1.5, Cr 18, Ni 10, bal Fe.
Bar: 90,000 TS; 40,000 YS; 50 El; B 85 Rock. For heat treating fixtures, annealing boxes, tube supports, furnace parts. Type 302B stainless steel, austenitic. *Obsolete*

NICHROTHERMSTEEL

Westinghouse Electric Corp.

Ni, bal Fe.
Low coefficient of expansion. *Obsolete*

NICK SOLDER

J.W. Harris Co., Inc.

Ag 0-2, Cu 4, Ni 0-1, bal Sn.
Working temperature: 440-600°F. Lead-free, nickel/silver bearing solder for use in potable water systems. Awaiting approval from BOCA, IATMO, SPCCI.

NICKAHL IRON

Aluminium Industrie Aktiengesellschaft

C 1.5, Cr 0.8, Si 2.2, bal Fe.
For dies to stamp fenders, door panels and seat sides for automobiles; tough and wear resistant cast iron.

NICKAHL STEEL

Jessop Steel Co.

Ni 35-42, bal Fe.
For valves, thermostatic elements low coefficient of expansion. *Obsolete*

NICKEL

Atomergic Chemetals Corp.

Ni.
Purities, zone and chemical refined: 99.999%, 99.99%, 99.95%. Forms: powder, rod, pellet, sheet, wire, foil, platelets, cathodes, single crystals.

NICKEL "C"

Vereinigte Deutsche Nickel-Werke AG

Mn 1.5, bal Ni.
69,000 TS; 45 El. For chemical apparatus, thermocouple elements; heat and corrosion resistant to salt solutions and caustics. *Obsolete*

NICKEL "GFA"

Henry Wiggin & Co. Ltd.

Ni alloy.
For electronic equipment; corrosion and heat resistant. *Obsolete*

NICKEL "HPA"

Henry Wiggin & Co. Ltd.

Ni alloy.
For electronic equipment; corrosion and heat resistant. *Obsolete*

NICKEL "HPB"

Henry Wiggin & Co. Ltd.

Ni alloy.
For electronic equipment; corrosion and heat resistant. *Obsolete*

NICKEL "O"

Henry Wiggin & Co. Ltd.

Ni alloy.
For electronic equipment; corrosion and heat resistant. *Obsolete*

NICKEL 141

Gilby-Fodor S.A.

C 0.03, Ni 96, Mn 0.3, Fe 0.05, Si 0.6, Ti 2.5, Al 0.25, S 0.005.
Welding electrode for shielded metal-arc Welding Nickel 200 and 201, and overlaying on steel, and joining nickel to steel. AWS A5.11 (class ENi-1); ASME SFB 5.11.

NICKEL 141

Henry Wiggin & Co. Ltd.

C 0.03, Ni 96, Mn 0.3, Fe 0.05, Si 0.6, Ti 2.5, Al 0.25, S 0.005.
Welding electrode for shielded metal-arc Welding Nickel 200 and 201, and overlaying on steel, and joining nickel to steel. AWS A5.11 (class ENi-1); ASME SFB 5.11.

NICKEL 200

Criterion Metals, Inc.

Nickel. Ni 99.5, C 0.06, Mn 0.25, Fe 0.15, S 0.005, Si 0.05, Cu 0.05, contains small amounts of Co.
Thin gauge sheet, various tempers: 55-90 ksi TS; 15-70 ksi YS; ASTM B-162.

NICKEL 200

Inco Alloys International Inc.

Ni 99.5, C 0.08, Mn 0.18, Fe 0.2, S 0.005, Si 0.18, Cu 0.13.
Annealed: 55,000-80,000 TS; 15,000-30,000 YS; 40-55 El; 90-120 Brin. Cold drawn: 65,000-110,000 TS; 40,000-100,000 YS; 10-35 El; 140-230 Brin. Hot finished: 60,000-85,000 TS; 15,000-45,000 YS; 35-55 El; 90-150 Brin. Commercially pure nickel; for chemical handling, food processing, electronic equipment; corrosion resistant.

NICKEL 201

Criterion Metals, Inc.

Nickel. Ni 99.5, C 0.01, Mn 0.2, Fe 0.15, S 0.005, Si 0.05, Cu 0.05, contains small amounts of Co.
Thin gauge sheet, various tempers: 50 ksi TS min; 12 ksi YS; ASTM B-162.

NICKEL 201

Enpar Sonderwerkstoffe GmbH

Nickel.
Alternate manufacturer.

NICKEL 201

Henry Wiggin & Co. Ltd.

Ni 99.5, C 0.01, Mn 0.18, Fe 0.2, S 0.005, Si 0.18, Cu 0.13.
Annealed: 50,000-60,000 TS; 10,000-25,000 YS; 60-40 El; 75-100 Brin. Cold drawn: 60,000-100,000 TS; 39,000-90,000 YS; 35-10 El; 125-200 Brin. Lower carbon than Nickel 200; for caustic evaporators, plater bars, combustion boats; preferred for application above 600°F.

NICKEL 201

Huntington Alloys Inc.

Ni 99.5, C 0.01, Mn 0.18, Fe 0.2, S 0.005, Si 0.18, Cu 0.13.
Annealed: 50,000-60,000 TS; 10,000-25,000 YS; 60-40 El; 75-100 Brin. Cold drawn: 60,000-100,000 TS; 39,000-90,000 YS; 35-10 El; 125-200 Brin. Lower carbon than Nickel 200; for caustic evaporators, plater bars, combustion boats; preferred for application above 600°F.

NICKEL 205

Criterion Metals, Inc.

Nickel. Ni 99.5, C 0.06, Mn 0.2, Fe 0.1, S 0.005, Si 0.05, Cu 0.05, Ti 0.02, Mg 0.04, contains small amounts of Co.
Thin gauge sheet, various tempers: 55-90 ksi TS; 15-70 ksi YS (0.2% offset).

NICKEL 205

Inco Alloys International Inc.

Nickel. Mg 0.01-0.08, Ti 0.01-0.05, Cu 0-0.15, Fe 0-0.2, C 0-0.15, Si 0-0.15, S 0-0.008, Mn 0-0.35, 99.0 Ni min.
Annealed: 50,000 TS; 13,000 YS; 45 El; for the anodes and grids of electronic valves. 2205

NICKEL 205

Henry Wiggin & Co. Ltd.

Ni 99.5, C 0.08, Mn 0.18, Fe 0.1, S 0.004, Si 0.08, Cu 0.08, Ti 0.03, Mg 0.05.
Annealed: 50,000 TS; 13,000 YS; 45-40 El; 77 Brin. Cold rolled: 95,000 TS; 90,000 YS; 3 El; 210 Brin. For electrical and electronic applications.

NICKEL 205

Huntington Alloys Inc.

Ni 99.5, C 0.08, Mn 0.18, Fe 0.1, S 0.004, Si 0.08, Cu 0.08, Ti 0.03, Mg 0.05.
Annealed: 50,000 TS; 13,000 YS; 45-40 El; 77 Brin. Cold rolled: 95,000 TS; 90,000 YS; 3 El; 210 Brin. For electrical and electronic applications.

NICKEL 211

Criterion Metals, Inc.

Nickel. Ni 95, C 0.1, Mn 4.75, Fe 0.05, S 0.005, Si 0.05, Cu 0.03, contains small amounts of Co.
Thin gauge sheet, various tempers: 55-90 ksi TS; 15-70 ksi YS (0.2% offset).

NICKEL 211

Inco Alloys International Inc.

Ni 96.85, C 0.1, Mn 4.75, Fe 0.38, S 0.008, Si 0.08, Cu 0.13.
For electronic applications; formerly "D" Nickel. *Obsolete*

NICKEL 212

Inco Alloys International Inc.

Ni 98, C 0.1, Mn 2, Fe 0.05, S 0.005, Si 0.05, Cu 0.03.
Electron tube supports; Formerly "E" Nickel.

NICKEL 220

Enpar Sonderwerkstoffe GmbH

Nickel.
Alternate manufacturer.

NICKEL 220

Inco Alloys International Inc.

Ni 99.5, C 0.04, Mn 0.1, Fe 0.05, S 0.004, Si 0.03, Cu 0.05, Ti 0.03, Mg 0.05.
Annealed: 70,000 TS; 20,000 YS; 40 El; 100 Brin. For electronic receiving tube cathodes. *Obsolete*

NICKEL 222

Inco Alloys International Inc.

C 0.01, Ni 99.8, Mg 0.05.
Cathode nickel; also used for sleeves of indirectly heated oxide coated cathodes in radio valves. ASTM F239; BS 3504.

NICKEL 225

Huntington Alloys Inc.

Ni 99.5, C 0.06, Mn 0.13, Fe 0.05, S 0.005, Si 0.2, Cu 0.03, Ti 0.02, Mg 0.04.
Annealed: 70,000 TS; 20,000 YS; 40 El; 100 Brin. Formerly "225" Nickel. *Obsolete*

NICKEL 230

Inco Alloys International Inc.

Ni 99.5, C 0.05, Mn 0.08, Fe 0.05, S 0.004, Si 0.02, Cu 0.05, Ti 0.003, Mg 0.06.
Annealed: 70,000 TS; 20,000 YS; 40 El; 100 Brin. Electron tube applications. *Obsolete*

NICKEL 233
Criterion Metals, Inc.
Nickel. Ni 99.5, C 0.09, Mn 0.18, Fe 0.05, S 0.005, Si 0.03, Cu 0.03, Ti 0.003, Mg 0.07, contains small amounts of Co.
Thin gauge sheet, various tempers: 55-90 ksi TS; 15-70 ksi YS; ASTM F-239.

NICKEL 233
Inco Alloys International Inc.
Ni 99.5, C 0.09, Mn 0.18, Fe 0.05, Si 0.005, Si 0.03, Cu 0.03, Ti 0.003, Mg 0.07. *Obsolete*

NICKEL 240
Inco Alloys International Inc.
Ni 95, Cr 1.7, Ti 0.3, Mn 2, Si 0.45, Zr 0.15.
Special nickel grade for spark plug centers and earth electrodes. *Obsolete*

NICKEL 241
Inco Alloys International Inc.
Ni 90, Cr 5, Mn 3, Si 1.7.
Special nickel grade for spark plug centers and earth electrodes. *Obsolete*

NICKEL 270
Criterion Metals, Inc.
Nickel. Ni 99.97, C 0.02, Mn 0.001, Fe 0.005, S 0.001, Si 0.001, Cu 0.001, Ti 0.001, Mg 0.001, Cb 0.001, contains small amounts of Co.
Thin gauge sheet, various tempers: 55-90 ksi TS; 15-70 ksi YS (0.2% offset).

NICKEL 270
Huntington Alloys Inc.
Ni 99.98, C 0.01, Mn 0.003.
High purity nickel; for electronic applications; heat exchangers.

NICKEL 270
Henry Wiggin & Co. Ltd.
Ni 99.98, C 0.01, Mn 0.003.
High purity nickel; for electronic applications; heat exchangers.

NICKEL 270
Inco Alloys International Inc.
Nickel. Cu 0-0.01, Fe 0-0.05, Mn 0-0.003, C 0-0.02, S 0-0.003, Ti 0-0.005, Mg 0-0.005, Si 0-0.005, 99.9 Ni min.
Annealed: 50,000 TS; 16,000 YS; 50 El. High purity grade. For components of hydrogen; also used for electrical resistance thermometers. 2270

NICKEL 400
Allegheny Ludlum Steel
C 0-0.3, Cu 25-32, Ni 63.
Ni-Cu alloy with good high strength, excellent corrosion resistance and good weldabiltiy. N04400

NICKEL 61
Inco Alloys International Inc.
C 0.06, Ni 96, Mn 0.3, Fe 0.1, Si 0.4, Cu 0.02, Ti 3, S 0.005.
Filler metal for gas-shielded arc welding Nickel 200 and 201, and overlaying on steel. AWS A5.14 (class ERNi-3); BS 2901-NA-32. *Obsolete*

NICKEL 99.6/99.6 K
VDM Nickel-Technologie AG
99.6 Ni min.
For manufacturing and processing of mineral products, especially caustic alkalis. *Obsolete*

NICKEL ALLOY, GRADE A
Chase Brass & Copper Co.
Cu 65, Zn 17, Ni 18.
Plates: 58,000-95,000 TS; 2-32 El; 77-158 Brin. For silver plated ware, plumbing pipes, resistance wire; high resistance. *Obsolete*

NICKEL ALUMINUM BRONZE
American Manganese Bronze Co.
Copper. Cu 78-0, Al 9-11.5, Ni 3-5.5, Fe 3-5, Mn 0-1.5.
Cast: 80,000 TS; 35,000 YS; 15 El; 15 RA; 140 Brin. For ship propellers, pump and turbine parts; resists cavitation and corrosion, tough.

NICKEL ALUMINUM BRONZE
American manufacture
Cu 10, Ni 40, Al 30, Sn 20.
For ornaments; corrosion resistant.

NICKEL ALUMINUM BRONZE
American manufacture
Cu 88, Ni 10, Al 2, Sn.
For heat and corrosion resisting parts; high strength.

NICKEL BEARING
English manufacture
Cu 50, Ni 25, Sn 25.
For bearings; heavy duty.

NICKEL BORON STEEL
Manufacturer not listed
Ni 2.8-3.6, B 0.1-0.5, C 0.15-0.7, bal Fe.
For dynamically stressed parts; water or oil hardened.

NICKEL BRASS
American Nickeloid Co.
Ni coated brass.
For fabricated parts; easily stamped, formed, drawn.

NICKEL BRASS
American manufacture
Cu 50-54, Zn 35-44, Ni 1.5, Fe 0.5, Al.
For condenser tubes, hardware; corrosion resistant.

NICKEL BRONZE
Belmont Metals Inc.
Cu 82, Sn 8, Zn 2, Ni 8.
For superheated steam parts; corrosion resistant.

NICKEL CAST IRON NO. 1
International Nickel Inc.
Cast iron. Ni 3, Si 1.5, Cr 0.8-1, C, bal Fe.
50,000 TS; 385 Brin. For cast gears; wear resisting castings. *Obsolete*

NICKEL CAST IRON NO. 2
International Nickel Inc.
Ni 1.1, Si 1.5, Cr 0.5, 3.30 T.C., bal Fe.
34,400-71,000 TS; 10 El; 223 Brin. For cast gears; wear resistant. *Obsolete*

NICKEL CAST IRON NO. 4
International Nickel Inc.
Ni 3, Si 1.3, Cr 0.57, 3.0 T.C., bal Fe.
41,550 TS; 286 Brin. For wear resistant castings; wear resistant. *Obsolete*

NICKEL CERIUM STEEL
English manufacture
Ni 2.2-3, Ce 0.1-0.9, C 0.4-0.75, bal Fe.
For machinery parts; oil hardening.

NICKEL CHROME ALUMINUM
English manufacture
Ni 88, Al 12, Cr 8.
For corrosion and heat resistant parts.

NICKEL CHROME CAST IRON
Manufacturer not listed
Ni 1.5-1.75, Cr 0.6-0.8, C 3-3.4, Si 0.9-1.75, Mn 0.5-0.7, bal Fe.
Cast: 35,000-48,000 TS; 0 El; 0 RA; 170-230 Brin. For grids; corrosion and abrasion resisting.

NICKEL CHROME PEERLESS
English manufacture
Cr 16.5, Fe 3, Mn 2, C 0.1, bal Ni.
For heat resisting parts.

NICKEL CHROME PREMIER
English manufacture
Fe 25, Cr 11, Mn 3, bal Ni.
For heat resisting parts.

NICKEL CHROME SUPERIOR
English manufacture
Cr 19.5, Fe 0.5, Mn 2, C 0.2, bal Ni.
For heat resisting parts.

NICKEL CLAD 304 SS - THREE LAYER
Texas Instruments Inc./Materials Control
201 nickel clad to both sides of type 304 stainless steel. For button cell battery cans.

NICKEL CLAD 304 SS - TWO LAYER
Texas Instruments Inc./Materials Control
201 Nickel clad to one side of type 304 stainless steel. For button cell batteries; ratio of nickel to steel can be specified from 5-50%.

NICKEL CLAD 430 SS
Texas Instruments Inc./Materials Control

Type 430 stainless steel clad with 201 nickel on one side. For button cell battery cans.

NICKEL CLAD TO BOTH SIDES OF 430 SS
Texas Instruments Inc./Materials Control

Type 430 stainless steel clad with 201 nickel on both sides. For button cell battery cans.

NICKEL COBALT 9-H
Hanson-Van Winkle-Munning Co.
Ni-Co.
For anodes; electroplating.

NICKEL COINAGE U.S.A
American manufacture
Cu 75, Ni 25.
For coinage; corrosion resistant.

NICKEL COPPER
American Nickeloid Co.
Ni coated Cu.
For fabricated parts; easily stamped, formed, drawn.

NICKEL COPPER STEEL
American manufacture
Ni 1-25, Cu 0.4-10, C 0.15-0.8, bal Fe.
For structures, machinery parts, fences, gates; resists soil corrosion.

NICKEL CZ100
Alloy Foundries
C 0.1, Mn 0.15, Si 0.2, Mo 0.125, Fe 0.3, bal Ni.
Alloy castings; corrosion and temperature resistant. Meets ACI CZ 100; ASTM A-296 CZ-100.

NICKEL FILLER METAL 61
Inco Alloys International Inc.
Ni 96, C 0.06, Mn 0.3, Fe 0.1, S 0.005, Si 0.4, Cu 0.02, Ti 3.
For gas shielded arc welding of Nickel 200 and Nickel 201; overlaying on steel. AWS A 5.14 Class ERNi-3; ASME SFB 5.14.

NICKEL HP
Criterion Metals, Inc.
Nickel. Ni 99.5, C 0.02, Mn 0.2, Fe 0.01, S 0.003, Si 0.01, Cu 0.02, Al 0.003, Mg 0.002, Zr 0.001, Cb 0.001, Pb 0.001, contains small amounts of Co.
Thin gauge sheet, various tempers: 50 ksi TS min; 8 ksi YS (0.2% offset); ASTM B-162.

NICKEL LEADED COMMERCIAL BRONZE
Chase Brass & Copper Co.
Cu 90.25, Zn 6.9, Ni 1, P 0.1, Pb 1.75.
For bolts, fasteners, pole line hardware; free-cutting.
Obsolete

NICKEL MALLEABLE
International Nickel Inc.
Nickel. Ni 99.4, Mn 0.15, Cu 0.1, C 0.05, Fe 0.15.
For coinage, constituent of alloys; corrosion resistant.
Obsolete

NICKEL MANGANESE BRONZE
American manufacture
Cu 53, Zn 39, Sn 2.6, In 2.5, Mn 1.7, Al, Pb.
For corrosion resistant strong castings.

NICKEL MOLYBDENUM STEEL
American manufacture
Ni 1.5-3, Mo 0.1-0.7, C 0.4, Mn, bal Fe.
For gears, shafts; oil hardened.

NICKEL N. 100
English manufacture
W 2, Al 1, C 0.2, Mg 0.2, bal Ni.
For tube cathodes and valve components; corrosion and heat resistant.

NICKEL N. 93
English manufacture
W 2, Al 1, C 0.2, bal Ni.
For tube cathodes and valve components; corrosion and heat resistant.

NICKEL OREIDE
Vereinigte Deutsche Nickel-Werke AG
Cu 87, Zn 6.7, Ni 6.7.
For hardware, ornamental parts; corrosion resistant.
Obsolete

NICKEL SILICON STEEL
American manufacture
Ni 2.8-3.3, Si 0.5-2.2, Cu 0.35-0.5, bal Fe.
For machinery parts, gears, shafts, axles; oil hardening.

NICKEL SILVER
INSILCO Corp.
Ni 18, Cu 65, Zn 18.
59,900 TS; 31,800 YS; 34 El; 65 RA. For silver plated table ware; Rockwell "B"-50.

NICKEL SILVER 10%
Manufacturer not listed
Cu 56-65, Zn 25-34, Ni 10.
For ornaments, plated ware; corrosion resistant.

NICKEL SILVER 10% 745
Anaconda Co.
Copper. Cu 65, Zn 24.75, Ni 10, Mn 0.25.
Hard sheet: 87,000 TS; 70,000 YS; 6 El; 87 Rock B. Soft sheet: 55,000 TS; 20,000 YS; 42 El; 30 Rock B. For optical goods, costume jewelry, holloware, radio dials, nameplates, camera parts. Corrosion resistant, high strength.

NICKEL SILVER 12%-766
Anaconda Co.
Copper. Zn 31.25, Ni 12, Mn 0.25, bal Cu.
Hard: 97,000 TS; 65,000 YS; 5 El; 176 Brin. Soft: 60,000 TS; 24,000 YS; 40 El; 80 Brin. For springs. Corrosion resistant.

NICKEL SILVER 13%-776
Anaconda Co.
Copper. Cu 43.25, Zn 43.6, Ni 13, Mn 0.15.
Extruded: 60,000 TS; 20,000 YS; 20 El. For architectural decoration. Warm silver color.

NICKEL SILVER 14%
Manufacturer not listed
Cu 56-60, Zn 26-28, Ni 14.
For ornaments, plated ware; corrosion resistant.

NICKEL SILVER 15%
Manufacturer not listed
Cu 57-64, Zn 21-28, Ni 15.
For ornaments, plated ware; corrosion resistant.

NICKEL SILVER 15% 767
Anaconda Co.
Copper. Cu 56.5, Zn 28.25, Ni 15, Mn 0.25.
Hard sheet: 100,000 TS; 75,000 YS; 4 El; 93 Rock B. Soft sheet: 60,000 TS; 24,000 YS; 40 El; 50 Rock B. For architectural panel and trim, nameplates, dials, musical instruments, watch cases. Corrosion resistant, high strength.

NICKEL SILVER 18%
Manufacturer not listed
Cu 55-65, Zn 17-27, Ni 18.
For ornaments, plated ware; corrosion resistant.

NICKEL SILVER 18%
Chase Brass & Copper Co.
Cu 65, Ni 18, Zn 17.
Plates: 58,000-95,000 TS; 2-35 El; 70-160 Brin. For spinning, stamping; corrosion resisting. *Obsolete*

NICKEL SILVER 18% (A) COPPER ALLOY NO. 7
Chase Brass & Copper Co., Inc.
Copper. Cu 65, Ni 18, Zn 17.
Annealed, strip: 58,000 psi TS; 25,000 psi YS; 40 El; 45 Rock B. Hard, strip: 85,000 psi TS; 74,000 psi YS; 3 El; 87 Rock B. For tableware, hollow ware, base for silver plated parts, welding rod. ASTM: B122 Alloy 2, B151 Alloy A, B206 Alloy A; CDA 752.

NICKEL SILVER 18% 735
Anaconda Co.
Cu 72, Ni 18, Zn 10.
Ann: 48,000-56,000 TS; 14,000-24,000 YS; 34-44 El. Cold rolled: 57,000-90,000 TS; 28,000-80,000 YS; 1-25 El. Good deep drawing white brass; for ferrules, ink cartridges, condenser cans, instruments.

NICKEL SILVER 18% 735
Olin Corp.
Cu 72, Ni 18, Zn 10.
Ann: 48,000-56,000 TS; 14,000-24,000 YS; 34-44 El. Cold rolled: 57,000-90,000 TS; 28,000-80,000 YS; 1-25 El. Good deep drawing white brass; for ferrules, ink cartridges, condenser cans, instruments.

NICKEL SILVER 18% 752
Anaconda Co.
Cu 65, Ni 18, Zn 17.
Ann: 51,000-61,000 TS; 16,000-29,000 YS; 35-42 El. Cold rolled: 59,000-104,00 TS; 22,000-96,000 YS; 1-35 El. Silver color, good formability. For flatware, holloware, musical instruments, coins.

NICKEL SILVER 18% 752
Olin Corp.
Cu 65, Ni 18, Zn 17.
Ann: 51,000-61,000 TS; 16,000-29,000 YS; 35-42 El. Cold rolled: 59,000-104,00 TS; 22,000-96,000 YS; 1-35 El. Silver color, good formability. For flatware, holloware, musical instruments, coins.

NICKEL SILVER 18% B
Chase Brass & Copper Co.
Cu 60-62, Ni 18, Zn 30-32.
Sheet, wire: 60,000-113,000 TS; 25,000-70,000 YS; 2-45 El. For stamping, drawing and forming; corrosion resisting.
Obsolete

NICKEL SILVER 18%-719
Anaconda Co.
Copper. Zn 17, Ni 18, Mn 0.25, bal Cu.
Soft: 58,000 TS; 40 El. Hard: 70,000 TS. For hardware, marine trim. Corrosion resistant.

NICKEL SILVER 18%-7641
Anaconda Co.
Copper. Zn 20.25, Ni 18, Mn 0.25, bal Cu.
Hard: 80,000 TS; 6,000 YS; 8 El; 137 Brin. Soft: 60,000 TS; 25,000 YS; 35 El; 83 Brin. For tubes. Corrosion resistant.

NICKEL SILVER 18%-770
Anaconda Co.
Copper. Zn 27, Ni 18, Mn 0.25, bal Cu.
Soft: 60,000 TS; 22,000 YS; 45 El. Hard: 80,000 TS; 60,000 YS; 20 El. For springs, hardware. Corrosion resistant.

NICKEL SILVER 20%
Manufacturer not listed
Cu 53-64, Zn 16-27, Ni 20.
For ornaments, plated ware; corrosion resistant.

NICKEL SILVER 25%
Manufacturer not listed
Cu 55, Ni 25, Zn 20.
For ornaments, plated ware; corrosion resistant.

NICKEL SILVER 30%
Manufacturer not listed
Cu 47-65, Ni 30, Zn 5-23.
For ornaments, plated ware; corrosion resistant.

NICKEL SILVER 548
Olin Brass, Indianapolis
Copper. Cu 48, Ni 9.5, bal Zn.
For braze welding rods; corrosion resistant.

NICKEL SILVER 55-18
Anaconda Co.
Cu 53.5-56.5, Zn 27, Ni 16.5-19.5.
Soft: 60,000 TS; 27,000 YS; 40 El. Hard: 100,000 TS; 85,000 YS; 3 El. For optical goods, springs, resistance wire; corrosion resistant, good workability.

NICKEL SILVER 55-18
Seymour Products Co.
Cu 53.5-56.5, Zn 27, Ni 16.5-19.5.
Soft: 60,000 TS; 27,000 YS; 40 El. Hard: 100,000 TS; 85,000 YS; 3 El. For optical goods, springs, resistance wire; corrosion resistant, good workability.

NICKEL SILVER 55-18
Chase Brass & Copper Co.
Cu 53.5-56.5, Zn 27, Ni 16.5-19.5.
Soft: 60,000 TS; 27,000 YS; 40 El. Hard: 100,000 TS; 85,000 YS; 3 El. For optical goods, springs, resistance wire; corrosion resistant, good workability.

NICKEL SILVER 565-18% (A)
Olin Brass, Indianapolis
Copper. Cu 65, Ni 18, bal Zn.
Annealed: 60,000 psi TS; 32 El. Drawn: 85,000 psi TS; 3 El. For hardware, hollowware, and jewelry; corrosion resistant.

NICKEL SILVER 567-10%
Olin Brass, Indianapolis
Copper. Cu 65, Ni 10, bal Zn.
Annealed: 50,000 psi TS; 45 El. Drawn: 90,000 psi TS; 3 El. For hardware, hollowware, and jewelry; corrosion resistant.

NICKEL SILVER 65-10
Now CDA C78800.

NICKEL SILVER 65-10
Seymour Products Co.
Cu 63.5-68.5, Zn 25, Ni 9-11.
1/4-H temper: 65,000 TS; 45,000 YS; 25 El. 1/2-H temper: 73,000 TS; 60,000 YS; 12 El. H temper: 86,000 TS; 75,000 YS; 4 El. For hardware, optical parts, holloware; corrosion resistant, good formability.

NICKEL SILVER 65-10
Chase Brass & Copper Co.
Cu 63.5-68.5, Zn 25, Ni 9-11.
1/4-H temper: 65,000 TS; 45,000 YS; 25 El. 1/2-H temper:
73,000 TS; 60,000 YS; 12 El. H temper: 86,000 TS; 75,000 YS;
4 El. For hardware, optical parts, holloware; corrosion
resistant, good formability.

NICKEL SILVER 65-10
Anaconda Co.
Cu 63.5-68.5, Zn 25, Ni 9-11.
1/4-H temper: 65,000 TS; 45,000 YS; 25 El. 1/2-H temper:
73,000 TS; 60,000 YS; 12 El. H temper: 86,000 TS; 75,000 YS;
4 El. For hardware, optical parts, holloware; corrosion
resistant, good formability.

NICKEL SILVER 65-12
Now CDA C75700.

NICKEL SILVER 65-12
Seymour Products Co.
Cu 63.5-66.5, Zn 23, Ni 11-13.
1/4 H-temper: 65,000 TS; 45,000 YS; 23 El. 1/2 H-temper:
73,000 TS; 60,000 YS; 11 El. H-temper: 85,000 TS; 75,000 YS;
4 El. For fasteners, hardware, optical parts; corrosion
resistant, good formability.

NICKEL SILVER 65-12
Chase Brass & Copper Co.
Cu 63.5-66.5, Zn 23, Ni 11-13.
1/4 H-temper: 65,000 TS; 45,000 YS; 23 El. 1/2 H-temper:
73,000 TS; 60,000 YS; 11 El. H-temper: 85,000 TS; 75,000 YS;
4 El. For fasteners, hardware, optical parts; corrosion
resistant, good formability.

NICKEL SILVER 65-12
Anaconda Co.
Cu 63.5-66.5, Zn 23, Ni 11-13.
1/4 H-temper: 65,000 TS; 45,000 YS; 23 El. 1/2 H-temper:
73,000 TS; 60,000 YS; 11 El. H-temper: 85,000 TS; 75,000 YS;
4 El. For fasteners, hardware, optical parts; corrosion
resistant, good formability.

NICKEL SILVER 65-15
Now CDA C75400.

NICKEL SILVER 65-15
Seymour Products Co.
Cu 63.5-66.5, Ni 14-16, Zn 20.
1/4 H-temper: 65,000 TS; 49,000 YS; 21 El. 1/2 H-temper:
74,000 TS; 62,000 YS; 10 El. H-temper: 85,000 TS; 75,000 YS;
3 El. For optical goods, jewelry, hardware; corrosion resistant,
good workability.

NICKEL SILVER 65-15
Chase Brass & Copper Co.
Cu 63.5-66.5, Ni 14-16, Zn 20.
1/4 H-temper: 65,000 TS; 49,000 YS; 21 El. 1/2 H-temper:
74,000 TS; 62,000 YS; 10 El. H-temper: 85,000 TS; 75,000 YS;
3 El. For optical goods, jewelry, hardware; corrosion resistant,
good workability.

NICKEL SILVER 65-15
Anaconda Co.
Cu 63.5-66.5, Ni 14-16, Zn 20.
1/4 H-temper: 65,000 TS; 49,000 YS; 21 El. 1/2 H-temper:
74,000 TS; 62,000 YS; 10 El. H-temper: 85,000 TS; 75,000 YS;
3 El. For optical goods, jewelry, hardware; corrosion resistant,
good workability.

NICKEL SILVER 65-18
Now CDA C75200.

NICKEL SILVER 65-18
Seymour Products Co.
Cu 63-66.5, Zn 17, Ni 16.5-19.5.
Soft: 58,000 TS; 25,000 YS; 40 El. 1/2 H-temper: 74,000 TS;
62,000 YS; 8 El. Hard: 85,000 TS; 74,000 YS; 3 El. For
hardware, optical goods, holloware; corrosion resistant,
good workability.

NICKEL SILVER 65-18
Chase Brass & Copper Co.
Cu 63-66.5, Zn 17, Ni 16.5-19.5.
Soft: 58,000 TS; 25,000 YS; 40 El. 1/2 H-temper: 74,000 TS;
62,000 YS; 8 El. Hard: 85,000 TS; 74,000 YS; 3 El. For
hardware, optical goods, holloware; corrosion resistant,
good workability.

NICKEL SILVER 65-18
Anaconda Co.
Cu 63-66.5, Zn 17, Ni 16.5-19.5.
Soft: 58,000 TS; 25,000 YS; 40 El. 1/2 H-temper: 74,000 TS;
62,000 YS; 8 El. Hard: 85,000 TS; 74,000 YS; 3 El. For
hardware, optical goods, holloware; corrosion resistant,
good workability.

NICKEL SILVER 8.0
Waterbury Rolling Mills Inc.
Ni 7-9, Cu 63-66, bal Zn.
CDA 741,743.

NICKEL SILVER 828
Anaconda Co.
Copper, Zn 40, Ni 10.2, Si 0.15, P 0.02, bal Cu.
For welding rod. Low fuming nickel silver. 1690°F MP.

NICKEL SILVER BARE
J.W. Harris Co., Inc.

Nickel silver brazing rod. For repair and buildup of gears,
valve seats, shafts. *Obsolete*

NICKEL SILVER BARE & FLUX COATED
J.W. Harris Co., Inc.
Nickel silver brazing rod.
For maintenance repair and build up of steel tubing, bicycles,
furniture, buildings.

NICKEL SILVER CASTING
Manufacturer not listed
Cu 56-70, Ni 13-20, Zn 5.6-24, Sn 0-4, Pb 0-3.5.
For plumbing fixtures, fittings; corrosion resistant.

NICKEL SILVER GRADE A
Riverside Metals Corp.
Cu 65, Ni 18, Zn 17.
55,000-110,000 TS; 2-50 El; 50-175 Brin. For cutlery, cheap
jewelry, resistance wire, auto fittings, plumbing; corrosion
resisting. *Obsolete*

NICKEL SILVER GRADE B
Riverside Metals Corp.
Cu 55, Ni 18, Zn 27.
Rolled: 40 El; 175 Brin. For cutlery, jewelry, plumbing;
corrosion resisting. *Obsolete*

NICKEL SILVER GRADE B
Riverside Metals Corp.
Cu 55, Ni 18, Zn 27.
1-40 El; 50-175 Brin. For cutlery, cheap jewelry, resistance
wire, auto fittings, plumbing; corrosion resisting. *Obsolete*

NICKEL SILVER WELDING ROD
Century Brass Products Inc.
Cu 48, Ni 10, Zn 42.
Rod for welding nickel silver alloys. CDA 773.

NICKEL SILVER, ROLLING
Manufacturer not listed
Cu 49, Zn 39, Ni 12.
For ornamental sheets and rolled parts; corrosion resistant.

NICKEL SILVER, TURNING
Manufacturer not listed
Cu 59-66, Zn 22-29, Ni 12, Pb 0-5.
For screws, bolts; machining grade.

NICKEL SPECIAL
Isabellenhuette
Ni 99.4, Fe 0.2-0.6.
For electrical equipment and instruments. Resistance alloy.
Maximum working temperature to 150°C. *Obsolete*

NICKEL STAYBOLT
Atlas Specialty Steels
C 0.08, Mn 0.25, Ni 2.25, bal Fe.
For boiler staybolts; tough. *Obsolete*

NICKEL STEEL
American Nickeloid Co.
Ni bonded to steel.
For floor plates, reflectors, hardware, stampings; resists heat
to 500°F; Ni bonded to steel.

NICKEL TIN
American Nickeloid Co.
Ni coated Sn.
For fabricated parts; easily stamped, formed, drawn.
Obsolete

NICKEL URANIUM STEEL
American manufacture
Ni 0.3-0.4, U 0.2-0.4, C 0.2-0.8, bal Fe.
183,000-209,000 TS; 159,000-175,000 YS; 8.5-9.0 El; 34-31
RA. For general engineering applications; corrosion resistant.

NICKEL VANADIUM STEEL
Manufacturer not listed
Ni 3, V 0.1-0.45, C 0.36, Si 0.2-0.4, bal Fe.
For machinery parts, gears, pinions, shafts, crankshafts;
tough, shock resistant.

NICKEL WELD FILLER
Telcon Metals Ltd.
Cu 70, Ni 30.
For brazing.

NICKEL WELDING ELECTRODE 141
Inco Alloys International Inc.
Ni 96, C 0.03, Mn 0.3, Fe 0.05, S 0.005, Si 0.6, Cu 0.03, Al
0.25, Ti 2.5.
Electrode for shielded metal-arc welding of Nickel 200 and
Nickel 201, welding clad side of nickel-clad steel, joining
nickel to steel, overlaying on steel. AWS A 5.11 Class ENi-1;
ASME SFB 5.11.

NICKEL ZIRCONIUM STEEL
Manufacturer not listed
Ni 3, Zr 0.24, C 0.4, Si 2.4, bal Fe.
For crankshafts, axles, machinery parts; oil hardening.

NICKEL, GRADE E
International Nickel Inc.
Ni 97.65, Co 97.65, Cu 0.1, Fe 0.2, Mn 1.9, C 0.1.
For foundry applications, furnace lead-in wires. *Obsolete*

NICKEL, GRADE T
International Nickel Inc.
Cu 0.07, Fe 1.5, Mn 0.05, Si 0.5, C 0.12, Ni 98.92.
For foundry applications. *Obsolete*

NICKEL-131
Gilby-Fodor S.A.
C 0-0.15, bal Ni.
For welding rod.

NICKEL-ARC
Chemetron Corp.
high C, Ni, bal Fe.
For welding electrodes for cast iron; machinable cast iron.

NICKEL-ARC 55
Chemetron Corp.
Ni 55, Fe 45.
As welded: 49,000 psi TS (approx.) For welding cast iron;
machinable weld metal. AWS class ENiFe-C1.

NICKEL-ARC 99

Chemetron Corp.
99 + Nickel.
As welded: 32,800 psi (approx.) For welding and overlay on cast iron. AWS class ENi-C1.

NICKEL-AT

Gilby-Fodor S.A.
C 0-0.15, bal Ni.
For welding rod.

NICKEL-AT

Henry Wiggin & Co. Ltd.
Cu 0-0.25, Fe 0-0.4, Mn 0-0.35, Si 0-0.15, C 0-0.15, S 0-0.01, 99.0% min Ni.
For chemical plant and other corrosion resisting applications. Good welding qualities, corrosion and heat resistant. *Obsolete*

NICKEL-BRONZE

Baldwin-Lima-Hamilton Corp.
Copper. Cu 85-90, Sn 3-6, Ni 3-7, Zn 1-3.
Heat treated: 50,000-80,000 TS; 25,000-60,000 YS; 15-25 El; 15-25 RA; 95-170 Brin. For gears, screw down nuts; castings.

NICKEL-CHROME-COPPER

English manufacture
Ni 80-85, Cr 20-25, Cu 15-20.
For electric irons, percolators.

NICKEL-CHROMIUM-IRON ALLOY 37/18

Henry Wiggin & Co. Ltd.
C 0.2, Ni 37, Cr 18, bal Fe.
Annealed: 105,000 TS; 54,000 YS; 39 El; 50 RA; 183 Brin. For furnace belts, heat treating equipment, furnace parts; heat resistant. *Obsolete*

NICKEL-CLAD STEEL

Lukens Steel
Pure nickel on steel base 10% Ni, bal steel.
Plate: 55,000-65,000 TS; 40,000-50,000 YS; 29-41 El; 52-62 RA. For vessels, tanks, tank cars, digesters, retorts, piping, stills; corrosion resisting. *Obsolete*

NICKEL-COPPER-TITANIUM

LTV Steel
C 0-0.15, Mn 0-1, Si 0-0.5, Ni 0-0.7, Ti 0-0.05, 0.3 Cu min, bal Fe.
50,000 psi YS min. Good cold forming, weldability, atmosphere corrosion resistance. For guard rails and automobile bumpers.

NICKEL-MANGANESE

Gilby-Fodor S.A.
Mn 2-5, bal Ni.
Annealed: 78,000 TS; 50,000 YS; 30 El. For grid wires and electronic tubes. Corrosion and heat resistant.

NICKEL-NI-C

Gilby-Fodor S.A.
W 2, C, bal Ni.
For electron tube elements. Heat resistant.

NICKEL-SILC-ON

Silicum Pistons Ltd.
Ni, Si.
Cold rolled: 70,000-85,000 TS; 60,000-70,000 YS; 25-15 El; 45-55 RA; 149-170 Brin. For pistons.

NICKEL-TUNGSTEN

Manufacturer not listed
W 50-75, Ni 25-50.
For chemical apparatus; resistant to acids.

NICKEL-ZIRCONIUM

Manufacturer not listed
Ni 86, Si 6, Zr 1.5, C 0.1.
For chemical apparatus; corrosion resistant.

NICKELALLOY

Teledyne McKay
Ni 100.
Welded: 45,000 TS; 3 El. For welding electrodes; shielded arc. *Obsolete*

NICKELCAST

Hobart Welding Products
C 2, Si 4, Mn 1, Fe 0-8, Ni 0-85, Cu 0-2.5, 1.0 others max.
Weld metal. As welded: 40,100 psi TS; 38,200 psi YS; 3.6 El. For welding ductile iron and cast iron. AWS A5.15-69 (E Ni-Cl). *Obsolete*

NICKELCAST 55

Hobart Welding Products
C 2, Si 4, Mn 1, Ni 45-60, Cu 2.5, 1.0 others, bal Fe.
Weld metal. As welded: 57,000-84,000 psi TS; 43,000-63,000 psi YS; 6-13 El. For welding ductile iron and cast iron. AWS A5.15-69 (E NiFeCl). *Obsolete*

NICKELCHROMEIGHT

Westinghouse Electric Corp.
C, Cr, Ni, bal Fe.
For welding electrodes; for type 18/8 stainless steel. *Obsolete*

NICKELCHROMTWELVE

Westinghouse Electric Corp.
C, Ni, Cr, bal Fe.
For welding electrodes; for type 25/12 stainless steel. *Obsolete*

NICKELCHROMWELD

Lincoln Electric Co.
Ni, Cr, bal Fe.
For welding electrodes; shielded arc. *Obsolete*

NICKELCOLUMBIUM-REGULAR

Reading Alloys, Inc.
Reactive/refractive. C 0.1, Cb 55-65, Fe 3, Mn 0.5, Ni 35-45, Si 2.5.
Master alloy.

NICKELCOLUMBIUM-VACUUM

Reading Alloys, Inc.
Reactive/refractive. C 0.1, Cb 60-65, Fe 1, Mn 0.05, Ni 33-38, Si 0.2.
Master alloy.

NICKELDUR

Janney Cylinder Co.
Cu 80, Pb 10, Ni 5, Sn 5.
Cast: 19,000 TS; 85 Brin. For centrifugal cast liner; nickel bronze, corrosion resistant.

NICKELEISEN 36K

VDM Nickel-Technologie AG
Ni 36-38, bal Fe.
For magnetic and electrical equipment, motors; soft magnet, high permeability. *Obsolete*

NICKELEISEN 36W

VDM Nickel-Technologie AG
Ni 36-38, bal Fe.
For magnetic and electrical equipment, motors; soft magnet, high permeability. *Obsolete*

NICKELEND

Arcos Alloys
Fe 0.5, bal Ni.
For Ni weld rod. *Obsolete*

NICKELENE-1

English manufacture
Ni 5-30, Cu 52-80, Zn 10-35.
For electrical instruments, resistance wires, thermocouples; German silver.

NICKELENE-2

English manufacture
Ni 13, Cu 55, Zn 21, Sn 2, Pb 10.
For resistance wires; heat resistant.

NICKELIN

VDM Nickel-Technologie AG
Cu 58, Ni 22, Fe 20.
For regulating and checking electrical resistances; German silver. *Obsolete*

NICKELIN "A"

English manufacture
Ni 18, Cu 62, Zn 20.
For resistance alloy; heat resistant.

NICKELIN "B"

English manufacture
Ni 32, Cu 68.
For resistance alloy; heat resistant.

NICKELIN "C"

English manufacture
Ni 31.5, Cu 55.3, Zn 13.1.
For resistance alloy; heat resistant.

NICKELIN I

Vereinigte Deutsche Nickel-Werke AG
Ni 33.3, bal Cu.
For chemical equipment, fruit presses, boilers, kettles; corrosion resistant. *Obsolete*

NICKELIN I

VDM Nickel-Technologie AG
Cu 54, Ni 26, Zn 20.
Hard: 121,000 TS; 1.5 El. Annealed: 85,000 TS; 30 El. For starting rheostats for motors, field regulators, loading resistances; maximum operating temperature 500°C. *Obsolete*

NICKELIN II

Vereinigte Deutsche Nickel-Werke AG
Ni 25, bal Cu.
For fruit presses, coinage; corrosion resistant. *Obsolete*

NICKELIN II

VDM Nickel-Technologie AG
Cu 67, Ni 30, Mn 3.
Hard: 100,000 TS; 2 El; 170 Brin. Annealed: 70,000 TS; 44 El; 92 Brin. For starting rheostats, field regulators, loading resistances; maximum operating temperature 500°C. *Obsolete*

NICKELIN III

Vereinigte Deutsche Nickel-Werke AG
Ni 20, bal Cu.
For medals, ornaments; corrosion resistant. *Obsolete*

NICKELIN III

VDM Nickel-Technologie AG
Cu 58, Ni 22, Zn 20.
Hard: 118,700 TS; 1 El. Annealed: 72,000 TS; 34 El. For regulating and control resistances; maximum operating temperature 500°C. *Obsolete*

NICKELIN IV

Vereinigte Deutsche Nickel-Werke AG
Cu 85, Ni 15.
45,000 TS; 48 El. For projectiles or bullets to be nickel-jacketed; tough. *Obsolete*

NICKELIN RESISTANCE

VDM Nickel-Technologie AG
Cu 54, Zn 20, Ni 26.
For motor starters, electrical equipment; nickel silver. *Obsolete*

NICKELIN V

Vereinigte Deutsche Nickel-Werke AG
Ni 10, Cu 90.
For armature wires, electrical equipment; corrosion resistant.
Obsolete

NICKELIN-40

Isabellenhuette
Cu 67, Ni 30, Mn 3.
Hard: 115,000-128,000 TS. Soft: 50,000-60,000 TS.
Temperature coefficient resistance per °C: +0.00014. For electrical equipment and instruments. Resistance alloy. Maximum operating temperature 400°C. *Obsolete*

NICKELIN-W

Isabellenhuette
Copper. Cu 67, Ni 30, Mn 3.
Annealed: 400 N/mm^2 TS. For electrical equipment and instruments. Resistance alloy. Maximum working temperature to 500°C.

NICKELINE

Manufacturer not listed
Cu 55-75, Ni 18-32, Zn 0-20, Fe 0.2-0.45.
For electrical resistance; heat resistant.

NICKELINE

NL Industries
Pb, Sb, bal Sn.
24 Brin. For Babbitt, bearings; Babbitt metal. *Obsolete*

NICKELMANG

Westinghouse Electric Corp.
C, alloy, bal Fe.
For welding electrodes for manganese steel; bare and coated. *Obsolete*

NICKELOID

American Nickeloid Co.
Ni coated Zn.
For construction; corrosion resistant. *Obsolete*

NICKELOID

Wm. McPhail & Sons
Sn, Cu, bal Ni.
Cast: 36,000-50,000 TS; 180-200 Brin. For valves for severe steam services, valve seats; corrosion resistant. *Obsolete*

NICKELOID

Pacific Metal Co.
Pb, Sb, Ni, bal Sn.
For bearings. Babbitt metal.

NICKELOID

Barker & Allen Ltd.
Ni 40-45, bal Cu.
For non-rusting and corrosion resistant parts; corrosion resistant.

NICKELOY

English manufacture
Al 93.8, Cu 4.15, Ni 1.41.
20,000 TS; 5 RA. For automotive engine parts; age-hardenable.

NICKELOY-ROD NO. 2512

Marquette Corp.
C 0.1-0.15, Mn 0.35-0.55, Si 0.25-0.35, Ni 4.75-5.25, bal Fe.
Welded: 80,000-90,000 TS; 65,000-75,000 YS; 25-30 El; 50-55 RA. For all-position welding rod; low temperature toughness.

NICKELVAC 600

Teledyne Allvac
C 0.07, Cr 15.5, Ti 0.3, Al 0.2, Fe 9, Mn 0.5, bal Ni.
For high temperature equipment.

NICKELVAC 625

Teledyne Allvac
C 0.05, Cr 21.5, Mo 9, Ti 0.1, Al 0.1, 3.8 Cb + Ta, bal Ni.

NICKELVAC 90

Now NICKELVAC N90.

NICKELVAC A-286

Teledyne Allvac
C 0.06, Cr 14.75, Mo 1.25, Ni 25.5, Ti 2.1, Al 0.2, Mn 1.5, bal Fe.
At room temperature: 146,000 TS; 105,00 YS; 25 El. At 1400°F: 64,000 TS; 62,000 YS; 19 El. For jet engine and supercharger parts, turbine wheels and blades, afterburners, bolting. Austenitic, heat and corrosion resistant.

NICKELVAC B

Now NICKELVAC H-B.

NICKELVAC C-263

Teledyne Allvac
C 0.07, Cr 20, Co 20, Mo 5.8, Ti 2.1, Al 0.5, B 0.005, Si 0.3, Mn 0.4, Zr 0.06, bal Ni.
For gas turbine and heat engine components.

NICKELVAC H-B

Teledyne Allvac
C 0.1, Mn 0.8, Si 0.7, Cr 0.6, Co 2.5, Mo 28, Fe 5, V 0.3, bal Ni.
Annealed: 127,000 TS; 56,000 YS; 52 El. At 1200°F: 75,000 TS; 42,000 YS; 13 El. For chemical and oil refinery equipment, valves, pumps, bolts, gas turbine components. Acid resistant, austenitic, tough. Good creep and high temperature properties.

NICKELVAC H-B2

Teledyne Allvac
C 0.01, Mo 26.5, Ti 0.1, Al 0.1, Mn 0.75, Fe 1, bal Ni.
Corrosion resistant alloy.

NICKELVAC H-C

Teledyne Allvac
C 0.04, Cr 15.5, Mo 15.5, Ti 0.1, Al 0.1, Fe 6.5, W 3.5, Mn 0, Si 0.8, V 0.2, bal Ni.
Corrosion and heat resistant; for chemical and oil refinery equipment.

NICKELVAC H-C276

Teledyne Allvac
C 0.01, Cr 15.5, Mo 15.5, Ti 0.1, Al 0.1, Fe 6.5, W 3.8, Mn 0, V 0.2, bal Ni.
For chemical and oil refinery equipment.

NICKELVAC H-W

Teledyne Allvac
C 0.02, Cr 5, Mo 24.2, Ti 0.1, Al 0.1, B 0.003, Fe 5.8, V 0.3, bal Ni.
For gas turbine and heat engine components.

NICKELVAC H-X

Teledyne Allvac
C 0.1, Cr 21.5, Co 1.5, Mo 9, Ti 0.1, Al 0.1, Fe 19, bal Ni.
For gas turbines and heat engine components.

NICKELVAC HN

Teledyne Allvac
C 0.06, Cr 7, Mo 16.5, Ti 0.1, Al 0.1, Fe 2.5, bal Ni.
For chemical plant equipment.

NICKELVAC L-605

Teledyne Allvac
C 0.05-0.15, Mn 1-2, Cr 19-21, W 14-16, Ni 9-11, Fe 0-3, bal Co.
At 70 F: 161,000 TS; 86,000 YS; 47 El; Rock C 24. At 1500 F: 55,000 TS; 45,000 YS; 16 El. For afterburners, nozzle diaphragm valves, high temperature valves and springs, buckets. High strength above 1600 F, oxidation and corrosion resistant. *Obsolete*

NICKELVAC N

Now NICKELVAC HN.

NICKELVAC N-155

Teledyne Allvac
C 0.15, Mn 1.5, Si 0.5, Cr 20, Ni 20, Co 20, Mo 3.5, W 2.5, Cb 1, N 0.15, bal Fe.
Sheet: 118,000 TS; 58,000 YS; 49 El. At 1000 F: 94,000 TS; 40,000 YS; 50 El. For aircraft cabin heaters, afterburners, rocket chambers, combustion chambers and manifolds. High heat and corrosion resistance, good high temperature strength. *Obsolete*

NICKELVAC N-80A

Teledyne Allvac
C 0.05, Cr 20, Ti 2.4, Al 1.3, bal Ni.
High temperature alloy.

NICKELVAC N-90

Teledyne Allvac
C 0.07, Cr 20, Co 16, Ti 2.5, Al 1.5, bal Ni.
High temperature alloy.

NICKELVAC SUPER-X

Teledyne Allvac
C 0.045-0.09, Cr 15-16, Fe 6.25-7.25, Al 0.55-0.9, Ti 2.35-2.7, Cu 0-0.1, Mn 0-0.25, 0.80-1.1% Cb + Ta, 72.0% min Ni + Co.
Heat treated: 188,000 TS; 129,000 YS; 23.6 El; 33.8 RA. At 600 F: 178,000 TS; 117,000 YS; 21 El; 36.7 RA. For applications requiring oxidation and corrosion resistance coupled with high strength to 1200-1500 F. *Obsolete*

NICKELVAC W-722

Teledyne Allvac
C 0.06, Cr 15.5, Ti 2.4, Al 0.7, Fe 8, bal Ni.

NICKELVAC X-750

Teledyne Allvac
C 0.06, Cr 15.5, Ti 2.55, Al 0.75, Fe 8, Cb 1, bal Ni.
At room temperature: 162,000 TS; 92,000 YS; 24 El. At 1400°F: 70,000 TS; 62,000 YS; 9 El. For gas turbine blades and wheels, springs, turbochargers, afterburners, fasteners. Age hardenable. High heat and corrosion resistant.

NICKELVAC X-751

Teledyne Allvac
C 0.06, Cr 16, Ti 2.5, Al 1.2, Fe 8.5, Cb 1, B 0.007, bal Ni.
For exhaust valves.

NICKELVAC-700

Now ALLVAC 700.

NICKELVAC-751

Now NICKELVAC X751.

NICKELVAC-901

Teledyne Allvac
C 0.06, Cr 12.5, Mo 6.1, Ni 43, Ti 3, B 0.015, bal Fe.
At room temperature: 175,000 TS; 130,000 YS; 14 El. At 1400°F: 105,000 TS; 92,000 YS; 19 El. For aircraft and gas turbine components, rotors and compressor discs, fasteners, bolts. Heat and corrosion resistant. High creep and rupture strength.

NICKELVAC-C

Now NICKELVAC H-C.

NICKELVAC-F

Teledyne Allvac
C 0.05, Mn 1.5, Si 1, Cr 22, Co 2.5, Mo 6.5, W 1, Cb 2, Ni 45.5, bal Fe.
Annealed: 108,000-116,000 TS; 48,000-53,000 YS; 40-46 El; 47-54 RA; B 89-93 Rock. At 1000 F: 79,000 TS; 32,000 YS; 50 El. For paper and pulp digesters, chemical and petroleum processing equipment. High heat and corrosion resistance. Good high temperature properties. *Obsolete*

NICKELVAC-INOR-8
Teledyne Allvac
C 0.06, Mn 0.8, Si 0.5, Cr 7, Co 0.5, Mo 16.5, Ti 3, Fe 5, B 0.01, bal Ni.
At RT: 100,000 TS; 45,000 YS; 50 El. At 1200 F: 70,000 TS; 30,000 YS; 45 El. For high temperature applications, jet engine and gas turbine components. High heat and corrosion resistance. *Obsolete*

NICKELVAC-N
Now NICKELVAC HN.

NICKELVAC-W
Now NICKELVAC HW.

NICKELVAC-X
Now NICKELVAC HX.

NICKEND 2
Arcos Alloys
Ni 2.25, C, bal Fe.
Welded: 80,000-90,000 TS; 65,000-75,000 YS; 20-30 El; 45-60 RA. For welding electrodes; for steel, low H_2. *Obsolete*

NICKEND 3
Arcos Alloys
C 0.06, Mn 0.8, Si 0.3, Ni 3.4, bal Fe.
For welding electrodes; class E8016-C2, for welding 3-1/4 nickel steel. *Obsolete*

NICKIMPHY
Creusot-Loire
C 0.02, Mn 0.1, Cu 0.1, bal Ni.
For radio tube anodes; heat and corrosion resist. *Obsolete*

NICKKAL STEEL
Jessop Steel Co.
Ni 39, C, bal Fe.
For valves, thermostats; controlled expansion. *Obsolete*

NICKOLITE
Manufacturer not listed
Cu 60, Sn 3, Pb 6, Ni 20, Zn 11.
For typewriter parts, door knobs; nickel silver.

NICKONOMY
H. Boker & Co.
C 0.2-0.4.
bal Fe. *Obsolete*

NICKRALEX K5
Le Bronze Industriel
Al 3, Ni 14, bal Cu.
Nickel-aluminum-copper alloy; wrought or cast: 205-275 Brin.
AFNOR UN 14 A2.

NICKRALEX KC1
Le Bronze Industriel
Al 3, Ni 16, Cr 2, bal Cu.
Copper-nickel alloy; wrought: 250-310 Brin.

NICKREL
Le Bronze Industriel
Ni 67, Fe 1.5, Mn 1.5, bal Cu.
Nickel-copper alloy; wrought; work-hardenable: 100-200 Brin.
ASTM B164-70.

NICKRELK
Le Bronze Industriel
Al 3, Ni 66, Fe 1.5, Mn 1.5, bal Cu.
Nickel-copper alloy; wrought: 240-290 Brin.

NICKROFRY 345
Societe des AFY
C 0-0.2, Cr 22-24, Ni 12-15, Mn 0-2, Si 0-1, bal Fe.
Annealed: 85,000-95,000 TS; 40,000-50,000 YS; 45-55 El; 150-185 Brin. For heat treating boxes, oil refinery and chemical plant equipment; Type 309; austenitic, heat resistant.

NICKROTHERM
Krupp Stahl AG
stainless steel. C, Cr, Ni, Si, bal Fe.
For chemical plant equipment; corrosion and heat resistant. *Obsolete*

NICKROTHERM "F.F"
Krupp Stahl AG
stainless steel. C 0.2, Si 1.5, Cr 18, bal Fe.
For heating muffles, autoclaves, roasting furnaces, rabbles, boilers, grates, recuperators, dampers; heat and corrosion resistant. *Obsolete*

NICKROTHERM "N.C.T"
Krupp Stahl AG
stainless steel. C 0.2, Cr 19, Ni 9, bal Fe.
For heating muffles, autoclaves, roasting furnaces, rabbles, boilers, grates, recuperators, dampers; heat and corrosion resistant. *Obsolete*

NICLOY 5
Babcock & Wilcox Co.
C 0-0.14, Ni 4.75-5.25, Mn 0.4, Si 0.2, bal Fe.
Rolled: 100,000 TS; 75,000 YS; 30 El; 70 RA. At 1000 F: 36,000 TS; 45 El; 84 RA. For heat exchangers, refrigerators, liquid air equipment; impact resistant and tough at low temperatures. *Obsolete*

NICLOY-36
Babcock & Wilcox
Ni 36, bal Fe.
Annealed: 65,000 TS; 40,000 YS; 35 El; 70 Rock B. Cold Worked: 105,000 TS; 95,000 YS; 8 El; 217 Brin. For hairsprings, time pieces, precision instruments, bourdon tubes, bimetals. Low coefficient of expansion. Constant modulus. Corrosion resistant. *Obsolete*

NICO
Northfield Iron Co.
C 0.2, Cu 0.5, Ni 0.8, bal Fe.
For corrugated culverts. Rust resistant.

NICO
C.E. Phillips & Co.
C 0.1, bal Fe.
For cast iron welding electrodes; ductile.

NICO NO. 1
English manufacture
Sn 4-5, Ni 2-3, Sb 0-23, bal Pb.
36 Brin. For bearings; White Metal.

NICO NO. 2
English manufacture
Sb 10, Sn 10, Ni 1, bal Pb.
29 Brin. For bearings; White Metal.

NICOL
Columbia Bronze Corp.
Al 10, Ni 5, Fe 5, Mn 1.5, bal Cu.
Cast: 93,000 TS; 40,000 YS; 20 El; 137 Brin. For ship propellers; heat treatable, aluminum bronze, tough.

NICOL-ROD NO. 44
Marquette Corp.
Cu, Ni.
Welded: 221 Brin. For welding rod; for malleable iron.

NICOL-ROD NO. 99
Marquette Corp.
Ni.
For welding rod for cast iron; special iron coating.

NICOLOY 3-1/2, ETC.
Now B & W NICOLOY 3-1/2, ETC.

NICORO
GTE Products Corp./Wesgo Div.
Au 35, Cu 62, Ni 3.
Brazing alloy available in foil, flexibraze, wire, powder, extrudable paste and preform. Liquidus 1886°F. Solidus 1832°F. Meets BAu-3.

NICORO
Western Gold & Platinum Co.
Au 35, Cu 62, Ni 3.
For brazing Kovar, copper, nickel and steel. Melting point: 1832-1886°F. Corrosion resistant.

NICORO 80
GTE Products Corp./Wesgo Div.
Au 81.5, Cu 16.5, Ni 2.
Brazing alloy available in foil, flexibraze, wire, powder, extrudable paste and preform. Liquidus 1697°F. Solidus 1670°F.

NICORROS AL-ALLOY K-500
VDM Nickel-Technologie AG
Nickel. Cu 27-33, Al 2.3-3.15, Mn 0-1.5, Fe 0.5-2, Ti 0.35-0.85, C 0-0.2, 63.0 Ni min.
Minimum values at 20°C: 965 N/mm² TS; 690 N/mm² YS; 20 El. For marine technology, oil and gas production. Hot finished, age hardened. Material No. 2.4375. N05500

NICORROS B 6530-FM 60
VDM Nickel-Technologie AG
Fe 0-2.5, Mn 3-4, Ti 1.5-3, Si 0-1, C 0-0.1, S 0-0.02, Cu 28-34, Al 0-1, 62.0 Ni min.
Submerged-arc and electroslag overlay welding on carbon steel or pressure vessel steel for use in the chemical and petrochemical industries. Material No. 2.4377. N04060

NICORROS LC
VDM Nickel-Technologie AG
Fe 0.5-2.5, Mn 0-2, C 0-0.04, 63.0 Ni min, bal Cu.
For nuclear technology, manufacture and processing of mineral products. *Obsolete*

NICORROS S 6530-FM 60
VDM Nickel-Technologie AG
Fe 0.5-2.5, Cu 28-34, Mn 2-4, Ti 1.5-3, Si 0-1, C 0-0.15, S 0-0.02, Al 0-1, 62.0 Ni min.
Filler metal. Minimum values at RT: 460 N/mm² TS; 200 N/mm² YS; 25 El. Material No. 2.4377. N04060

NICORROS-ALLOY 400
VDM Nickel-Technologie AG
Nickel. Cu 28-34, Al 0-0.5, Mn 0-1.25, Fe 1-2.5, Ti 0-0.2, C 0-0.15, 63.0 Ni min.
Minimum values at 20°C: 450 N/mm² TS; 180 N/mm² YS; 35 El. For marine technology and petroleum refining. Hot finished, age hardened. Material No. 2.4360. N04400

NICOSEAL
Now CARPENTER KOVAR.

NICOSEL
Firth Brown Ltd.
C 0.04, Si 0.15, Mn 0.3, Ni 29, Co 17.1, bal Fe.
For instruments; low coefficient of expansion. *Obsolete*

NICOVAR
Carpenter Technology Corp.
Ni 29, Co 17.5, bal Fe.
For glass and ceramic seals. Controlled expansion. *Obsolete*

NICRAL "A"
Nicralium Co.
Ni 0.1, Cr 0.5, Cu 0.5, Mg 0.5, bal Al.
Annealed: 20,000 TS; 10,000 YS; 24 El; 40 Brin. Heat treated: 46,000 TS; 41,000 YS; 8 El; 120 Brin. For light Al alloy parts; high strength and workability. *Obsolete*

NICRAL "B"
Nicralium Co.
Ni 0.5, Cr 0.25, Cu 0.25, Mg 0.25, bal Al.
Annealed: 17,000 TS; 6,000 YS; 27 El. Heat treated: 38,000 TS; 32,000 YS; 12 El. For light Al alloy parts; for stamping and forming. *Obsolete*

NICRAL "X"
Nicralium Co.
Ni 1, Cr 0.5, Cu 1, Mg 0.5, bal Al.
Extruded: 35,000 TS; 24,000 YS; 16 El. Heat treated: 53,000 TS; 43,000 YS; 15.5 El. For light Al alloy parts, extruded structural shapes; formerly known as Super-Hyblum. *Obsolete*

NICRAL B
Creusot-Loire
C 0-0.35, Cr 20, bal Fe.
Rolled: 128,000 TS; 101,000 YS; 10 El. For domestic utensils, electric irons, boiler parts; Type 442; corrosion resistant. *Obsolete*

NICRAL C
Creusot-Loire
C 0.07, Ni 33.5, Cr 21, Ti 0.35, Al 0.3, bal Fe.
High temperature corrosion resistant alloy. For heat exchangers, furnace parts. AFNOR 5 NC 35 20; W.Nr. 2.4856; Similar to INCOLOY 800.

NICRAL C (SIRIUS 35)
Creusot-Loire
C 0-0.2, Cr 18, Ni 37, bal Fe.
Wrought, annealed: 590 N/mm^2 psi TS min. Austenitic stainless for high temperature operation, as furnace parts. AFNOR Z12NC37-18; AISI 330.

NICRAL D (SIRIUS 3)
Creusot-Loire
C 0.25, Cr 25, Ni 20, Si 2, bal Fe.
Wrought, annealed: 540 N/mm^2 psi TS min. Austenitic stainless for high temperature operation. AFNOR Z12CNS 25-20; AISI 310.

NICRAL DC
Creusot-Loire
C 0.15-0.5, Cr 23-25, Ni 18-20, Si, bal Fe.
Annealed: 100,000 TS; 45,000 YS; 50 El; 65 RA; 185 Brin. For furnace parts, valves, pumps, turbine parts; Type 310; stainless, austenitic. *Obsolete*

NICRAL DS
Creusot-Loire
C 0.15-0.5, Cr 23-25, Ni 18-20, Si, bal Fe.
For furnace parts, heat treat boxes; austenitic; Type 310; stainless. *Obsolete*

NICRAL E
Creusot-Loire
C 0.35, Cr 30, bal Fe.
Rolled: 85,000 psi TS; 58,000 psi YS; 20 El. For carburizing boxes, furnace parts; Type 416; resists sulfur atmospheres.

NICRAL F
Creusot-Loire
C 0-0.5, Ni 65, Cr 18, Si 1.5, Mn 1, bal Fe.
Cast: 73,000 TS; 34,000 YS; 6.5 El; 200 Brin. For furnace parts, gas turbine and gas engine parts; corrosion and oxidation resistant. *Obsolete*

NICRAL H (SIRIUS 345)
Creusot-Loire
C 0-0.2, Cr 23, Ni 13.5, bal Fe.
Wrought, annealed: 540 N/mm^2 psi TS min. Austenitic stainless for high temperature operation. AFNOR Z15CN24.13; AISI 309.

NICRAL H9
Creusot-Loire
C 0.9, Cr 25, Ni 12, Si, bal Fe.
For furnace parts, valves; austenitic; corrosion and heat resistant. *Obsolete*

NICRAL J
Creusot-Loire
C 0-0.5, Cr 24, Ni 3, bal Fe.
For furnace parts and equipment; Type 327; corrosion and heat resistant. *Obsolete*

NICRAL K 25
Creusot-Loire
C 0.03, Ni 41, Cr 21, Mo 3, Cu 2, Ti 0.9, bal Fe.
Corrosion resistant alloy; for use with phosphoric acid and other chemicals. AFNOR NC 21 Fe DU; INCOLOY 825.

NICRAL KM
Creusot-Loire
C 0.9, Cr 20, Ni 42, Si, bal Fe.
For heat resistant parts; corrosion and heat resistant. *Obsolete*

NICRAL M
Creusot-Loire
C 0-0.15, Cr 13, Si 2, bal Fe.
For valves, cutlery; corrosion resistant. *Obsolete*

NICRAL P
Creusot-Loire
C 2.5, Cr 26, bal Fe.
For petroleum and oil refinery equipment; corrosion and heat resistant. *Obsolete*

NICRAL S
Creusot-Loire
Si 2, Cr 24, C, bal Fe.
For heat and corrosion resistant parts; corrosion and heat resistant. *Obsolete*

NICRAL S-CU
Creusot-Loire
C 0-0.3, Cr 22, Cu, bal Fe.
For heat and corrosion resistant parts; heat and corrosion resistant. *Obsolete*

NICRAL T
Creusot-Loire
C 0.12, Cr 17, Ni 13, W 3, Ti, bal Fe.
Water quenched: 540 N/mm^2 psi TS min; 225 N/mm^2 min YS; 40 El. AFNOR Z 12 CNWT 17-13 B.

NICRAL V
Creusot-Loire
C 0.55, Cr 22, Ni 36, Co, Mo, bal Fe.
For furnace parts and equipment; austenitic, corrosion and heat resistant. *Obsolete*

NICRAL W
Creusot-Loire
C 0-0.3, Cr 22, Ni 12, W, bal Fe.
For turbo reactors, jet engine components; corrosion and heat resistant. *Obsolete*

NICRAL X
Creusot-Loire
C 0-0.1, Cr 17, Ni 13, Mo, bal Fe.
Annealed: 85,000 TS; 35,000 YS; 50 El; 65 RA; 160 Brin. For acid resistant equipment; Type 316; austenitic, stainless. *Obsolete*

NICRAL Z
Creusot-Loire
C 0.04, Cr 15.5, Fe 0-10, bal Ni.
High temperature alloy. For furnace muffles, heat exchanger tubing, jet engine parts. AFNOR NC 15 Fe; W.Nr. 2.4640; similar to INCONEL 600.

NICRAL-O
Creusot-Loire
C 0.2, Si 2.5, Cr 25, Ni 21, bal Fe.
Bar: 100,000 TS; 50,000 YS; 45 El; B 89 Rock. For furnace parts, annealing boxes, heat treating fixtures. Type 314 Stainless Steel, high heat resistant. *Obsolete*

NICRALLOY-A
Osborn Steels Ltd.
Ni 80, Cr 20.
Cold drawn: 131,000 TS; 80,000 YS; 35 El; 61 RA. For electrical resistors, heating elements; operates up to 1150 C, heat resistant. *Obsolete*

NICRALLOY-B
Osborn Steels Ltd.
Ni 65, Cr 15, bal Fe.
Cold drawn: 105,000 TS; 59,000 YS; 40 El; 66 RA. For electrical resistors, heating elements; operates up to 950 C, heat resistant. *Obsolete*

NICRANOR
Acieries et Forges d'Anor
C 0.15-0.25, Si 2-2.5, Mn 2-2.5, Cr 20-22, Ni 45-47, Nb 1-1.4, bal Fe.
Heat resisting nickel alloy.

NICRFE
American manufacture
C 0.08, Mn 0.65, P 0-0.015, Si 0.65, Cr 17, Cb 0-2, Cu 0, Fe 8, bal Ni.
Similar to Inconel 604. *Obsolete*

NICRO 31
Uddeholm Corp.
C 0.15, Mn 0.55, Cr 0.65, Ni 1.25, bal Fe.
For gears, crankshafts, camshafts; case hardened. *Obsolete*

NICRO 33
Uddeholm Corp.
C 0.12, Mn 0.55, Cr 0.75, Ni 3, bal Fe.
For gears, crankshafts, camshafts; case hardened. *Obsolete*

NICRO 34
Uddeholm Corp.
C 0.12, Mn 0.55, Cr 1.25, Ni 4.5, bal Fe.
For gears, crankshafts, camshafts; case hardened. *Obsolete*

NICRO 63
Uddeholm Corp.
C 0.3, Ni 3.5, Mn 0.6, Cr 0.75, bal Fe.
For gears, shafts, bolts; shock resistant. *Obsolete*

NICRO 632
Uddeholm Corp.
C 0.3, Mn 0.65, Cr 1, Ni 3.25, Mo 0.25, bal Fe.
For gears, shafts, bolts; shock resistant. *Obsolete*

NICRO 64
Uddeholm Corp.
C 0.3, N 0.65, Cr 1.25, Ni 4.25, bal Fe.
For gears, shafts, bolts; shock resistant. *Obsolete*

NICRO 642
Uddeholm Corp.
C 0.3, Mn 0.5, Cr 1.25, Ni 4.25, Mo 0.25, bal Fe.
For gears, shafts, bolts; shock resistant. *Obsolete*

NICRO 71
Uddeholm Corp.
C 0.37, Mn 0.6, Cr 0.7, Ni 1.5, bal Fe.
For gears, shafts; tough. *Obsolete*

NICRO 81
Uddeholm Corp.
C 0.38, Mn 0.8, Cr 0.8, Ni 1.25, bal Fe.
For gears, shafts; tough. *Obsolete*

NICRO 82
Uddeholm Corp.
C 0.38, Mn 0.55, Cr 1.15, Ni 2.6, bal Fe.
For gears, shafts; oil hardened. *Obsolete*

NICROBRAZ 120
Wall Colmonoy Corp.
Cr 13.5, B 3.5, Si 4.5, Fe 4.5, C 0.8, bal Ni.
For heat resistant brazing alloy; 1760-1875°F MP. *Obsolete*

NICROBRAZ 1351

Wall Colmonoy Corp.
Si 3, B 1.5, bal Ni.
Brazing temperature: 2000-2150°F. Modified Nicrobraz 135, for wide gap brazing.

NICROBRAZ 160

Wall Colmonoy Corp.
Cr 9-12, B 1.75-2.75, Si 3-4, Fe 2.5-4, C 0.4-0.6, bal Ni.
1780-2120°F MP. For wide clearance joints where heavier fillets or greater joint ductility and machinability are desired.

NICROBRAZ 180

Wall Colmonoy Corp.
C 0.25, Cr 5, Si 3, Fe 1, B 1.8, bal Ni.
Melt: 1780-2160 F; Braze: 2150-2250 F. For torch brazing wide gap joints in stainless steels and nickel base alloys. Machinable; corrosion and temperature resistant. *Obsolete*

NICROBRAZ 200

Wall Colmonoy Corp.
Cr 6-8, Si 4-5, W 5-7, Fe 2-4, B 2.75-3.5, bal Ni.
Melting temperature: 1790-1900°F. Brazing temperature: 1950-2150°F. Recommended for components where hardenable base metals are used. High creep and stress rupture strength.

NICROBRAZ 213

Wall Colmonoy Corp.
C 0.4, Cr 19.25, Ni 17, Si 10, W 4.25, bal Co.
Similar to Nicrobraz 210, but with no boron and more silicon.

NICROBRAZ 220

Wall Colmonoy Corp.
Cr 4, Mn 45, B 0.8, bal Ni.
Melt: 1852-1975 F; Braze: 2000-2150 F. For atmosphere brazing of high temperature alloys, for rocket thrust chambers and complex heat exchangers. Ductile, usable with any base high melting metal. *Obsolete*

NICROBRAZ 230

Wall Colmonoy Corp.
Cr 3.5, Si 2.5, Fe 1, Mn 35, B 0.9, bal Ni.
Melt: 1800-1950 F; Braze: 1950-2050 F. For atmosphere brazing high temperature heat exchangers; uses from cryogenic to over 1500 F. *Obsolete*

NICROBRAZ 300

Wall Colmonoy Corp.
C 0.6-1, Cr 18-21, Si 2.5-3.5, Ni 16-20, W 8.5-11.5, B 3-4, bal Co.
Melting temperature: 1900-2050°F. Brazing temperature: 2150-2250°F. For atmosphere brazing of T joints or wide gap joints where maximum strength is needed.

NICROBRAZ 3001

Wall Colmonoy Corp.
Cr 11.5, Si 6, bal Ni.
Brazing temperature: 2225-2250°F. Heat and corrosion resistant brazing alloy; J-8101. Meets brazing process specification P50T9D. *Obsolete*

NICROBRAZ 3002

Wall Colmonoy Corp.
Cr 15, Si 8, bal Ni.
Brazing temperature: 2150-2200°F. For brazing thin gauge honeycomb. B50TF143 (J-8102).

NICROBRAZ 3003

Wall Colmonoy Corp.
Cr 17, Si 9.5, B 0.1, bal Ni.
Brazing temperature: 2100-2150°F. Similar to Nicrobraz 30, but with improved flow. B50TF142 (J-8103).

NICROBRAZ 3004

Wall Colmonoy Corp.
Cr 11.5, Si 7, B 0.4, bal Ni.
Brazing temperature: 2125-2150°F. Modified Nicrobraz 30; J-8104. Meets brazing process spec. P50T9K. *Obsolete*

NICROBRAZ 3005

Wall Colmonoy Corp.
Cr 13, Si 8, B 0.35, bal Ni.
Brazing temperature: 2125-2150°F. Modified Nicrobraz 30; J-8105. Meets brazing process spec. P50T9L. *Obsolete*

NICROBRAZ 35

Wall Colmonoy Corp.
Cr 18.5-20.5, Mn 9-10, Si 9.25-10, bal Ni.
Brazing alloy for heat resistant joints; 1975-2025°F MP. Similar to Nicrobraz 30, but contains manganese. Has lower brazing temperature with minimum erosion effects and no loss of flowability. B50T50; B50TF99.

NICROBRAZ 5007

Wall Colmonoy Corp.
Cr 11.2, Co 0-0.06, P 8, bal Ni.
Brazing temperature: 1850-2050°F. Modified Nicrobraz 50 for wide gap brazing.

NICROBRAZ 5025

Wall Colmonoy Corp.
Cr 7, Cu 50, P 5, bal Ni.
Brazing temperature: 1950-2100°F. For brazing thin walled stainless tubing, primarily in automotive applications.

NICROBRAZ 5027

Wall Colmonoy Corp.
Cr 4.9, Cu 65, P 3.5, bal Ni.
Brazing temperature: 1950-2100°F. For brazing thin walled stainless tubing, primarily in automotive applications.

NICROBRAZ 5060

Wall Colmonoy Corp.
Cr 8, Si 1.4, Fe 0.4, P 6, B 0.8, bal Ni.
Brazing temperature: 1850-2050°F. For brazing thin walled and delicate structures where heavy and ductile fillets are desired.

NICROBRAZ 5075

Wall Colmonoy Corp.
Cr 10, Si 0.9, Fe 0.25, P 7.5, B 0.5, bal Fe.
Brazing temperature: 1800-2000°F. For brazing thin walled and delicate structures where heavy and ductile fillets are desired.

NICROBRAZ 51

Wall Colmonoy Corp.
Cr 24-26, P 9-11, bal Ni.
Brazing temperature: 1800-2000°F. Similar to Nicrobraz 50, except for better strength, heat, and corrosion resistance.

NICROBRAZ 65

Wall Colmonoy Corp.
Mn 21-24.5, Cu 4-5, Si 6-8, bal Ni.
Melting temperature: 1800-1850°F. Brazing temperature: 1850-2000°F. Low melting, free flowing for vacuum brazing of ferrous, nickel and cobalt alloys. BNi-8.

NICROBRAZ LM01

Wall Colmonoy Corp.
Cr 3, Si 3, Fe 1.3, B 1.9, bal Ni.
Brazing temperature: 2000-2110°F. Modified Nicrobraz LM; J-8201. Meets brazing process spec. P50T9A. *Obsolete*

NICROBRAZ LM02

Wall Colmonoy Corp.
Cr 2.25, Si 2.75, Fe 1, B 1.7, bal Ni.
Brazing temperature: 2000-2110°F. Modified Nicrobraz LM; J-8202. Meets brazing process spec. P50T9B. *Obsolete*

NICROBRAZ LM03

Wall Colmonoy Corp.
Cr 2.5, Si 3, Fe 1.1, B 1.8, bal Ni.
Brazing temperature: 2000-2110°F. Modified Nicrobraz LM; J-8203. Meets brazing process spec. P50T9C. *Obsolete*

NICROBRAZ LM04

Wall Colmonoy Corp.
Cr 2.3, Si 3, Fe 1.2, B 1.8, bal Ni.
Brazing temperature: 2000-2120°F. Modified Nicrobraz LM; J-8204. Meets brazing process spec. P50T9E. *Obsolete*

NICROBRAZ LM05

Wall Colmonoy Corp.
Cr 2.7, Si 3, Fe 1.2, B 1.8, bal Ni.
Brazing temperature: 2000-2120°F. Modified Nicrobraz LM; J-8205. Meets brazing process spec. P50T9F. *Obsolete*

NICROBRAZ LM06

Wall Colmonoy Corp.
Cr 2.3, Si 2.4, Fe 1, B 1.5, bal Ni.
Brazing temperature: 2000-2125°F. Modified Nicrobraz LM; J-8206. Meets brazing process spec. P50T9G. *Obsolete*

NICROBRAZ LM07

Wall Colmonoy Corp.
Cr 5.6, Si 3.6, Fe 2.4, B 2.5, bal Ni.
Brazing temperature: 2050-2120°F. Modified Nicrobraz LM; J-8207. Good for wide gap brazing. Meets brazing process spec. P50T9N.

NICROBRAZ LM08

Wall Colmonoy Corp.
Cr 3.8, Si 3.5, Fe 1.7, B 2.2, bal Ni.
Brazing temperature: 2000-2110°F. Modified Nicrobraz LM; J-8208. *Obsolete*

NICROBRAZ STANDARD

Wall Colmonoy Corp.
Cr 13.5, B 3.5, Si 4.5, Fe 4.5, C 0.8, bal Ni.
For brazing stainless steel and nickel base alloys; 1790-1900°F MP. *Obsolete*

NICROBRAZ-60

Wall Colmonoy Corp.
Si 8, Mn 17, C 0-0.15, bal Ni.
For heat resistant brazing alloy; M.P. 1850-1890 F. *Obsolete*

NICROCOAT 1

Wall Colmonoy Corp.
Cr 13.5, Ni 73.2, 13.3 others.
Fusing temperature: 2000-2150°F. For build-up on Inconel 600, Hastelloy X, 300 series stainless steel.

NICROCOAT 2

Wall Colmonoy Corp.
Si 4.5, Ni 92.35, 3.15 others.
Fusing temperature: 1900-1950°F. For build-up of thin walled structures of 300 series stainless.

NICROCOAT 3

Wall Colmonoy Corp.
Cr 13, Ni 76.85, 10.15 others.
Fusing temperature: 1800-1900°F. For build-up on mild steel and low carbon steel as mufflers, heat exchangers.

NICROCOAT 4

Wall Colmonoy Corp.
Cr 15, Ni 81.5, 3.5 others.
Fusing temperature: 1950-2050°F. For build-up on jet engine combustion chambers, etc., made of Hastelloy X, Inconel 600, 304 stainless.

NICROCOAT 5

Wall Colmonoy Corp.
Cr 7, Fe 3, Ni 82.5, 7.5 others.
Fusing temperature: 1900-2000°F. For build-up on 300 series stainless for operation up to 1500°F.

NICROCOAT 6

Wall Colmonoy Corp.
Cr 7, W 6, Ni 77, 10 others.
Fusing temperature: up to 1950°F. For build-up on jet engine components containing W and Mo.

NICROCOAT 610

Wall Colmonoy Corp.
Cr 13.5, Ni 76.4, 10.1 TiB_2.
Fusing treatment: 15-30 min at 2000-2050°F, vacuum. For build-up on high temperature alloys to resist heat and erosion.

NICROCOAT 620

Wall Colmonoy Corp.
Cr 17.9, Si 4.9, Ni 68.6, 8.6 $TiSi_2$.
Fusing treatment: 15-30 min at 2080-2100°F, vacuum. For build-up on high temperature alloys to resist heat and erosion.

NICROCOAT 630

Wall Colmonoy Corp.
Cr 16.9, Si 5.3, Ni 64.4, 6.7 TiN, 6.7 $TiSi_2$.
Fusing treatment: 15-30 min at 2080-2100°F, vacuum. For build-up on high temperature alloys for maximum resistance to oxidation and erosion.

NICROCOAT 7

Wall Colmonoy Corp.
Si 3.5, Ni 94.5, 2 others.
Fusing temperature: 1950°F and up. For build-up on 300 series stainless and low alloy steel for petrochemical equipment, glass molds.

NICROCOAT 700

Wall Colmonoy Corp.
Cr 13, Ni 72, 10.0 TiB_2, 5.0 Al_2O_3.
Fusing treatment: 15-30 min at 2100-2150°F, vacuum. For build-up on high temperature alloys for maximum resistance to sulfidation.

NICROCOAT 8

Wall Colmonoy Corp.
Cr 10, Ni 83, 7 others.
Fusing temperature: 2000°F and up. For build-up on 300 series stainless for operation up to 1500°F.

NICROCOAT 9

Wall Colmonoy Corp.
Cr 19, Si 10, Ni 66, 5 others.
Fusing temperature: 2050°F and up. For build-up on turbine blades, vanes and other jet engine parts.

NICRODIE

Columbia Tool Steel Co.
Tool material. C 0.72, Mn 0.6, Si 0.35, Mo 0.25, Cr 0.9, Ni 1.5, bal Fe.
Oil hardened; cold work tool steel. For forming dies, shear blades, brake dies. AISI L6.

NICROEX

Columbia Tool Steel Co.
C 0.7, Cr 0.8, Ni 1.25, Mo 0.35, bal Fe.
For brake dies; oil hardening. *Obsolete*

NICROFER 3127HMO-ALLOY 31

VDM Nickel-Technologie AG
Ni 30-32, Cr 26-28, Mo 6-7, Cu 1-1.4, C 0-0.015, N 0.15-0.25, bal Fe.
At 20°C: 690 N/mm^2 TS; 320 N/mm^2 YS; 50 El. For manufacturing of acids, seawater piping, pulp and paper industry, sour gas production. Material No. 1.4562. N08031

NICROFER 3127LC-ALLOY 28

VDM Nickel-Technologie AG
Ni 30-32, Cr 26-28, Mo 3-4, Cu 1-1.4, C 0-0.015, N 0.04-0.07, bal Fe.
Minimum values at 20°C: 500 N/mm^2 TS; 220 N/mm^2 YS; 35 El. For heat exchangers, oil and gas production tubing, sour water strippers. Material No. 1.4563. N08028

NICROFER 3220

VDM Nickel-Technologie AG
Ni 30-32, Cr 19-21.5, Si 0.2-0.6, Al 0.2-0.4, Ti 0.2-0.5, C 0.04-0.08, 0.7 Al + Ti max, bal Fe.
Minimum values at 20°C: 520 N/mm^2 TS; 170 N/mm^2 YS; 30 El. For heat exchangers and piping systems in chemical and petrochemical industry, furnace parts and tubular sheaths for heating elements. Material No. 1.4958. N08800

NICROFER 3220-ALLOY 800

VDM Nickel-Technologie AG
Ni 30-32, Cr 19-21.5, Si 0.2-0.6, Al 0.2-0.4, Ti 0.2-0.5, C 0.04-0.08, 0.7 Al + Ti max, bal Fe.
Minimum values at 20°C: 520 N/mm^2 TS; 210 N/mm^2 YS; 30 El. For heat exchangers and piping systems in chemical and petrochemical industry, furnace parts and tubular sheaths for heating elements. Material No. 1.4876. N08800

NICROFER 3220H-ALLOY 800H

VDM Nickel-Technologie AG
Ni 30-32, Cr 19-22, Si 0.2-0.6, Al 0.2-0.4, Ti 0.2-0.5, C 0.06-0.08, 0.7 Al + Ti max, bal Fe.
Minimum values at 20°C: 500 N/mm^2 TS; 170 N/mm^2 YS; 35 El. For chemical and petrochemical equipment, pigtails and furnaces, convection and radiant tubing in ethylene pyrolysis, steam generator tubing in helium-cooled, high-temperature reactor systems. Material No. 1.4958. N08810

NICROFER 3220HT-ALLOY 800HP

VDM Nickel-Technologie AG
Ni 30-32, Cr 19-22, Si 0.2-0.6, Al 0.3-0.6, Ti 0.3-0.6, C 0.06-0.1, 1.2 Al + Ti max, bal Fe.
Minimum values at 20°C: 500 N/mm^2 TS; 170 N/mm^2 YS; 35 El. For chemical and petrochemical equipment, pigtails and furnaces, convection and radiant tubing in ethylene pyrolysis, steam generator tubing in helium-cooled, high-temperature reactor systems. Material No. 1.4959. N08811

NICROFER 3220LC-ALLOY 800L

VDM Nickel-Technologie AG
Ni 32-34, Cr 20-22, C 0-0.025, Ti 0.35-0.6, Al 0.15-0.4, 1.0 Al + Ti max, bal Fe.
Minimum values at 20°C: 450 N/mm^2 TS; 180 N/mm^2 YS; 35 El. For steam generators and feed water heaters. Material No. 1.4558. N08800

NICROFER 3228NBCE-ALLOY AC66

VDM Nickel-Technologie AG
Ni 31-33, Cr 26-28, Si 0-0.3, Al 0-0.025, C 0.04-0.08, Nb 0.6-1, Ce 0.05-0.1, bal Fe.
Minimum values at 20°C: 500 N/mm^2 TS; 185 N/mm^2 YS; 35 El. For components in waste incinerators and other aggressive process gases at high temperatures. Material No. 1.4877.

NICROFER 3620NB-ALLOY 20

VDM Nickel-Technologie AG
Ni 36-38, Cr 19-20, Mo 2-3, Cu 3-4, C 0-0.025, Nb 0.1-0.3, bal Fe.
Minimum values at 20°C: 550 N/mm^2 TS; 240 N/mm^2 YS; 30 El. For heat exchangers and piping systems in acid plants. Material No. 2.4660. N08020

NICROFER 3718-(ALLOY 330)

VDM Nickel-Technologie AG
Ni 34-37, Cr 15-17, Si 1-2, C 0-0.15, Mn 0-2, bal Fe.
Minimum values at 20°C: 550 N/mm^2 TS; 230 N/mm^2 YS; 30 El. For structural parts in furnaces for temperatures up to 1000°C (1830°F) in air. Material No. 1.4864. N08330

NICROFER 3718SO-ALLOY DS

VDM Nickel-Technologie AG
Ni 35-39, Cr 17-19, Si 1.9-2.5, C 0-0.1, bal Fe.
Minimum values at 20°C: 550 N/mm^2 TS; 230 N/mm^2 YS; 30 El. For structural parts in furnaces for temperatures up to 1000°C (1830°F) in air. Material No. 1.4862.

NICROFER 4221-ALLOY 825

VDM Nickel-Technologie AG
Ni 38-46, Cr 19.5-23.5, Mo 2.5-3.5, Cu 1.5-3, C 0-0.025, Ti 0.6-1.2, bal Fe.
Minimum values at 20°C: 550 N/mm^2 TS; 240 N/mm^2 YS; 30 El. For nuclear waste reprocessing, chemical plant, oil and gas production. Material No. 2.4858. N08825

NICROFER 4626MOW-ALLOY 333

VDM Nickel-Technologie AG
Ni 44-47, Cr 24-26, Mo 2.5-3.5, C 0.03-0.06, Si 0.8-1.2, Co 2.5-3.5, W 2.5-3.5, Mn 1.2-2, bal Fe.
Minimum values at 20°C: 690 N/mm^2 TS; 330 N/mm^2 YS; 30 El. For high temperature components in gas turbines, annealing furnaces and petrochemical engineering. Material No. 2.4608. N06333

NICROFER 4722CO-ALLOY X

VDM Nickel-Technologie AG
Cr 20.5-22.5, Fe 17-20, Mo 8.5-9.5, C 0.05-0.09, Al 0-0.2, Si 0.2-0.8, Co 0.5-1.5, W 0.2-1, bal Ni.
Minimum values at 20°C: 720 N/mm^2 TS; 310 N/mm^2 YS; 35 El. For casings and combustion chambers for gas turbines and rocket engines, components for heat treatment furnaces, catalyst support grids in nitric acid production. Material No. 2.4665. N06002

NICROFER 4823NMO-ALLOY G-3

VDM Nickel-Technologie AG
Cr 21.5-23.5, Fe 18-21, Mo 6.5-8, Cu 1.5-2.5, C 0-0.015, Nb 0.2-0.5, W 0-1.5, Co 0-5, bal Ni.
Minimum values at 20°C: 620 N/mm^2 TS; 240 N/mm^2 YS; 45 El. For components for the chemical and petrochemical industry handling acids. Material No. 2.4619. N06985

NICROFER 4823TI-ALLOY 48

VDM Nickel-Technologie AG
Cr 21-23.5, Fe 18-21, Mo 5-7, Cu 1.5-2.2, C 0-0.015, Ti 1.5-2, Al 0.4-0.9, bal Ni.
At 20°C: 1100 N/mm^2 TS; 850 N/mm^2 YS. For equipment and components for the pulp and paper industry and seawater applications. Material No. 2.4661.

NICROFER 5120COTI-ALLOY C-263

VDM Nickel-Technologie AG
Cr 19-21, Fe 0-0.7, Mo 5.6-6.1, C 0.04-0.08, Ti 1.9-2.4, Al 0.3-0.6, Co 19-21, bal Ni.
Minimum values at 20°C: 1000 N/mm^2 TS; 590 N/mm^2 YS; 35 El. For rings and forgings for gas turbines. Material No. 2.4650. N07263

NICROFER 5219NB-ALLOY 718

VDM Nickel-Technologie AG
Ni 50-55, Cr 17-21, Mo 2.8-3.3, C 0.02-0.08, Nb 4.8-5.5, Ti 0.7-1.15, Al 0.3-0.7, B 0.002-0.006, bal Fe.
Minimum values at 20°C: 1240 N/mm^2 TS; 1040 N/mm^2 YS; 12 El. For equipment in sour gas fields, marine industry. Material No. 2.4668. N07718

NICROFER 5219NB-ALLOY 718

VDM Nickel-Technologie AG
Ni 50-55, Cr 17-21, Mo 2.8-3.3, C 0.02-0.08, Ti 0.7-1.15, Al 0.3-0.7, Nb 4.8-5.5, B 0.002-0.006, bal Fe.
Minimum values at 20°C: 1240 N/mm^2 TS; 1040 N/mm^2 YS; 12 El. For components for gas turbines and rocket engines, high strength structural parts and fixtures in nuclear power plants. Material No. 2.4668. N07718

NICROFER 5520COSO-ALLOY 617

VDM Nickel-Technologie AG
Cr 20-23, Fe 0-2, Mo 8-10, C 0.05-1, Ti 0.2-0.6, Al 0.6-1.5, Co 10-13, Zr 0.015-0.045, bal Ni.
Minimum values at 20°C: 750 N/mm^2 TS; 350 N/mm^2 YS; 35 El. For gas turbines, aircraft and space applications, furnace components, radiant tubes, forged parts for helium/helium intermediate heat exchanger of the nuclear process heat prototype reactor. Material No. 2.4663. N06617

NICROFER 5621HMOW-ALLOY 22

VDM Nickel-Technologie AG
Cr 20-22.5, Fe 2-6, Mo 12.5-14.5, C 0-0.01, W 2.5-3.5, V 0-0.35, Co 0-2.5, Si 0-0.08, bal Ni.
Minimum values at 20°C: 690 N/mm² TS; 310 N/mm² YS; 45 El. For equipment handling strongly oxidized acid, pulp and paper industry. Material No. 2.4602. N06022

NICROFER 5716HMOW-ALLOY C-276

VDM Nickel-Technologie AG
Cr 15-16.5, Fe 4-7, Mo 15-17, C 0-0.01, W 3-4.5, V 0.1-0.3, Co 0-2.5, Si 0-0.08, bal Ni.
Minimum values at 20°C: 750 N/mm² TS; 310 N/mm² YS; 30 El. For chemical and petrochemical processes, pulp and paper. Material No. 2.4819. N10276

NICROFER 5923HMO-ALLOY 59

VDM Nickel-Technologie AG
Cr 22-24, Fe 0-1.5, Mo 15-16.5, C 0-0.01, Co 0-0.3, Al 0.1-0.4, Si 0-0.1, bal Ni.
Minimum values at 20°C: 710 N/mm² TS; 350 N/mm² YS; 45 El. For chemical and petrochemical processing, pulp and paper. Material No. 2.4605. N06059

NICROFER 6020HMO-ALLOY 625

VDM Nickel-Technologie AG
Cr 21-23, Fe 0-3, Mo 8-10, C 0-0.025, Nb 3.2-3.8, bal Ni.
Minimum values at 20°C: 830 N/mm² TS; 415 N/mm² YS; 35 El. For flue gas scrubbers, acid production. Material No. 2.4856. N06625

NICROFER 6022HMO-ALLOY 625H

VDM Nickel-Technologie AG
Cr 21-23, Fe 0-3, Mo 8-10, C 0.04-0.06, Ti 0-0.4, Al 0-0.4, Nb 3.5-4.3, bal Ni.
Minimum values at 20°C: 690 N/mm² TS; 275 N/mm² YS; 25 El. For high temperature sections of heavy water plants, flare stacks on offshore platforms. Material No. 2.4856. N06625

NICROFER 6023-ALLOY 601

VDM Nickel-Technologie AG
Cr 22-24, Fe 13-15, Si 0-0.5, Al 1.1-1.6, Ti 0.1-0.4, C 0.03-0.08, 59.0 Ni min.
Minimum values at 20°C: 600 N/mm² TS; 270 N/mm² YS; 30 El. For radiant tubes for industrial furnaces, exhaust gas detoxizer systems, gas turbines, petrochemical plants. Material No. 2.4851. N06601

NICROFER 6023H-ALLOY 601H

VDM Nickel-Technologie AG
Cr 22-24, Fe 13-15, Si 0-0.5, Al 1.1-1.6, Ti 0.3-0.6, C 0-0.1, Zr 0.01-0.03, 59.0 Ni min.
Minimum values at 20°C: 600 N/mm² TS; 240 N/mm² YS; 30 El. For components in heat treatment and furnaces, exhaust gas systems, combustion chambers in solid waste incinerators, titanium dioxide production. Material No. 2.4851. N06601

NICROFER 6030-ALLOY 690

VDM Nickel-Technologie AG
Ni 61-64, Cr 27-30, Fe 8-10, C 0-0.02.
Minimum values at 20°C: 600 N/mm² TS; 300 N/mm² YS; 45 El. For oxidizing sulfur containing atmospheres, where oil ash corrosion is a problem, against corrosive attack by alkaline oxides. Material No. 2.4642. N06690

NICROFER 6616HMO-ALLOY C-4

VDM Nickel-Technologie AG
Cr 14.5-17.5, Fe 0-3, Mo 14-17, C 0-0.009, Ti 0-0.7, Co 0-2, Si 0-0.05, bal Ni.
Minimum values at 20°C: 700 N/mm² TS; 305 N/mm² YS; 40 El. For chemical process industry, FGD systems. Material No. 2.4610. N06455

NICROFER 7016TIAL-ALLOY 751

VDM Nickel-Technologie AG
Cr 15-17, Fe 5-9, C 0-0.08, Ti 2.1-2.6, Al 1-1.6, Nb 0.8-1.2, 70.0 Ni min.
Minimum values at 20°C: 1150 N/mm² TS; 700 N/mm² YS; 20 El. For structural parts and fasteners in gas turbines, nuclear power plants and petrochemical engineering, exhaust valves for combustion engines. Material No. 2.4694. N07751

NICROFER 7016TINB-ALLOY X-750

VDM Nickel-Technologie AG
Cr 14-17, Fe 5-9, C 0-0.08, Ti 2.25-2.75, Al 0.4-1, Nb 0.7-1.2, 70.0 Ni min.
Minimum values at 20°C: 1190 N/mm² TS; 800 N/mm² YS; 30 El. For structural parts and fasteners in gas turbines, nuclear power plants and petrochemical engineering. Material No. 2.4669. N07750

NICROFER 7216-ALLOY 600

VDM Nickel-Technologie AG
Cr 14-17, Fe 6-10, Si 0-0.5, Al 0-0.3, Ti 0-0.3, C 0.05-0.08, 72.0 Ni min, 0.5 others max.
Minimum values at 20°C: 550 N/mm² TS; 240 N/mm² YS; 30 El. For low temperature chemical processes, such as production of vinyl chloride, construction of industrial furnace components. Material No. 2.4816. N06600

NICROFER 7216H-ALLOY 600H

VDM Nickel-Technologie AG
Cr 14-17, Fe 6-10, Si 0-0.5, Al 0-0.3, Ti 0-0.3, C 0.05-0.08, Cu 0-0.5, 72.0 Ni min.
Minimum values at 20°C: 500 N/mm² TS; 180 N/mm² YS; 35 El. For industrial furnace components, steam generators in nuclear power stations. Material No. 2.4816.

NICROFER 7216LC-ALLOY 600L

VDM Nickel-Technologie AG
Cr 14-17, Fe 6-10, C 0-0.025, Ti 0-0.3, 72.0 Ni min.
Minimum values at 20°C: 550 N/mm² TS; 180 N/mm² YS; 30 El. For temperatures below 450°F. Material No. 2.4817. N06600

NICROFER 7520-ALLOY 75

VDM Nickel-Technologie AG
Cr 19-21, Fe 0-5, Si 0.3-0.7, Al 0-0.3, Ti 0.2-0.6, Cu 0-0.5, C 0.08-0.13, bal Ni.
Minimum values at 20°C: 650 N/mm² TS; 240 N/mm² YS; 25 El. For gas turbine casings and components in heat-treatment furnaces. Material No. 2.4951. N06075

NICROFER 7520TI-ALLOY 80A

VDM Nickel-Technologie AG
Cr 19-21, Fe 0-1, C 0.04-0.09, Ti 2-2.6, Al 1.1-1.7, bal Ni.
Minimum values at 20°C: 1000 N/mm² TS; 620 N/mm² YS; 20 El. For bolts and fasteners for steam generators and power plants, gas turbine components. Material No. 2.4952. N07080

NICROFER B 5621-FM 22

VDM Nickel-Technologie AG
Cr 20-22.5, Fe 2-6, Mo 12.5-14.5, Si 0-0.08, C 0-0.01, Co 0-2.5, W 2.5-3.5, V 0-0.35, bal Ni.
Submerged-arc and electroslag overlay welding on carbon steel or pressure vessel steel for use in the chemical and petrochemical industries. Material No. 2.4635. N06022

NICROFER B 5716-FM C-276

VDM Nickel-Technologie AG
Cr 15-16.5, Fe 4-7, Mo 15-17, Mn 0-1, Si 0-0.08, C 0-0.01, S 0-0.015, W 3-4.5, V 0.1-0.3, bal Ni.
Submerged-arc and electroslag overlay welding on carbon steel or pressure vessel steel for use in the chemical and petrochemical industries. Material No. 2.4886. N10276

NICROFER B 6020-FM 625

VDM Nickel-Technologie AG
Cr 20-23, Fe 0-2, Mo 8-10, Mn 0-0.5, Ti 0-0.4, Nb 3.2-4.1, Si 0-0.5, C 0-0.05, S 0-0.015, Al 0-0.3, 60.0 Ni min.
Submerged-arc and electroslag overlay welding on carbon steel or pressure vessel steel for use in the chemical and petrochemical industries, for marine applications. Material No. 2.4831. N06625

NICROFER B 6616-FM C-4

VDM Nickel-Technologie AG
Cr 14-18, Fe 0-3, Mo 14-17, Mn 0-0.5, Ti 0-0.7, Si 0-0.05, C 0-0.01, S 0-0.015, Co 0-2, bal Ni.
Submerged-arc and electroslag overlay welding on carbon steel or pressure vessel steel for use in the chemical and petrochemical industries. Material No. 2.4611. N06455

NICROFER B 7020-FM 82

VDM Nickel-Technologie AG
Cr 18-22, Fe 0-3, Cu 0-0.5, Mn 2.5-3.5, Ti 0-0.75, Nb 2-3, Si 0-0.5, C 0-0.05, S 0-0.015, Co 0-0.1, 67.0 Ni min.
Submerged-arc and electroslag overlay welding on carbon steel or pressure vessel steel for use in the chemical and petrochemical industries, power stations. Material No. 2.4806. N06082

NICROFER S 4225-(FM 65)

VDM Nickel-Technologie AG
Ni 37-42, Cr 23-27, Fe 0-30, Mo 3.5-7.5, Cu 1.5-3, Mn 1-3, Ti 0-1, Si 0-0.5, C 0-0.02, S 0-0.015, Al 0-0.2.
Filler metal. Minimum values at RT: 550 N/mm² TS; 240 N/mm² YS; 25 El. Material No. 2.4655.

NICROFER S 4722-FM X

VDM Nickel-Technologie AG
Cr 20.5-22.5, Fe 17-20, Mo 8.5-9.5, Mn 0-1, Si 0.2-0.8, C 0.05-0.09, Co 0.5-1.5, W 0.2-1, bal Ni.
Filler metal. Minimum values at RT: 660 N/mm² TS; 400 N/mm² YS; 20 El. Material No. 2.4613. N06002

NICROFER S 5520-FM 617

VDM Nickel-Technologie AG
Cr 20-24, Fe 0-1, Mo 8-10, Mn 0-1, Ti 0-0.6, Si 0-0.5, C 0-0.1, S 0-0.015, Co 10-14, Al 0.8-1.5, Cu 0-0.5, Zr 0.015-0.045, 50.0 Ni min.
Filler metal. Minimum values at RT: 700 N/mm² TS; 400 N/mm² YS; 25 El. Material No. 2.4627.

NICROFER S 5621-FM 22

VDM Nickel-Technologie AG
Cr 20-22.5, Fe 2-6, Mo 12.5-14.5, Si 0-0.08, C 0-0.01, Co 0-2.5, W 2.5-3.5, V 0-0.35, bal Ni.
Filler metal. Minimum values at RT: 700 N/mm² TS; 400 N/mm² YS; 25 El. Material No. 2.4635. N06022

NICROFER S 5716-FM C-276

VDM Nickel-Technologie AG
Cr 14.5-16.5, Fe 4-7, Mo 15-17, Mn 0-1, Si 0-0.08, C 0-0.01, S 0-0.015, W 3-4.5, V 0-0.35, bal Ni.
Filler metal. Minimum values at RT: 700 N/mm² TS; 400 N/mm² YS; 25 El. Material No. 2.4886. N10276

NICROFER S 5923-FM 59

VDM Nickel-Technologie AG
Cr 22-24, Fe 0-1.5, Mo 15-16.5, Mn 0-0.5, Si 0-0.1, C 0-0.01, Co 0-0.3, Al 0.1-0.4, bal Ni.
Filler metal. Material No. 2.4607. N06059

NICROFER S 6020-FM 625

VDM Nickel-Technologie AG
Cr 20-23, Fe 0-2, Mo 8-10, Mn 0-0.5, Ti 0-0.4, Nb 3-4, Si 0-0.5, C 0-0.05, S 0-0.015, Al 0-0.3, 60.0 Ni min.
Filler metal. Minimum values at RT: 700 N/mm² TS; 420 N/mm² YS; 25 El. Material No. 2.4831. N06625

NICROFER S 6616-FM C-4

VDM Nickel-Technologie AG
Cr 14.4-17.5, Fe 0-3, Mo 14-17, Mn 0-1, Ti 0.2-0.7, Si 0-0.08, C 0-0.015, S 0-0.015, bal Ni.
Filler metal. Minimum values at RT: 650 N/mm² TS; 300 N/mm² YS; 25 El. Material No. 2.4611. N06455

NICROFER S 7020-FM 82

VDM Nickel-Technologie AG
Cr 18-22, Fe 0-3, Mn 2.5-3.5, Ti 0-0.75, Nb 2-3, Si 0-0.5, C 0-0.05, S 0-0.015, Cu 0-0.5, Co 0-0.1, 67.0 Ni min.
Filler metal. Minimum values at RT: 600 N/mm^2 TS; 360 N/mm^2 YS; 25 El. Material No. 2.4806. N06082

NICROGAP 106

Wall Colmonoy Corp.
Si 0.35, B 0.2, bal Ni.
For use with copper or silver alloys or Nicrobraz 50 to fill wide joints in mild and low alloy steels.

NICROGAP 108

Wall Colmonoy Corp.
Fe 7, Cr 15, Si 0.75, B 0.2, bal Ni.
Used to fill wide joints in stainless and heat resistant alloys.

NICROGAP 112

Wall Colmonoy Corp.
Ni 8, Cr 18, Si 1.25, B 0.2, bal Fe.
Used to fill wide joints in 300 series stainless.

NICROGAP 114

Wall Colmonoy Corp.
Cr 12, Si 1.25, B 0.2, bal Fe.
Used to fill wide joints in 400 series stainless.

NICROMA 120

Ambo-Stahl-Gesellschaft
C 0-0.1, Cr 12.5, Ni 12, bal Fe.
For valves, pumps, oil refinery equipment; corrosion and heat resistant. *Obsolete*

NICROMA 130

Ambo-Stahl-Gesellschaft
C 0-0.07, Cr 17, Mo 4.7, Ni 13, bal Fe.
Annealed: 90,000 TS; 40,000 YS; 45 El; 60 RA; 180 Brin. For acid resistant equipment; Type 317; stainless, austenitic. *Obsolete*

NICROMA 150

Ambo-Stahl-Gesellschaft
C 0-0.07, Cr 17.5, Mo 2, Ni 17.5, Cu 2, Ti = 7 x C, bal Fe.
For valves, pump parts, chemical plant equipment; corrosion and heat resistant. *Obsolete*

NICROMA 17

Ambo-Stahl-Gesellschaft
C 0.22, Cr 17, Ni 1.5, bal Fe.
Annealed: 85,000 TS; 50,000 YS; 30 El; 55 RA; 180 Brin. For furnace parts, grids, conveyors; corrosion and heat resistant.

NICROMAN

Disston Inc.
C 0.7, Mn 0.4, Ni 1.65, Cr 1, Cu 0.35, bal Fe.
For dies, hobs, taps, shear blades; tough. *Obsolete*

NICROMANG

Now AMSCO NICROMANG.

NICROMAZ C1

Creusot-Loire
C 0-0.04, Mn 0-2, Si 0-1, Ni 24-27, Cr 19-22, Mo 4-4.8, Cu 2-3, bal Fe.
Cast, annealed: 450 N/mm^2 psi TS min. For chemical industries; phosphoric and sulfuric acids, petrochemicals. AFNOR Z3 NCDU 25.20 M; similar to ACI CN-7M.

NICROMAZ CM

Creusot-Loire
C 0-0.08, Mn 0-1.5, Si 0-1.5, Ni 24-27, Cr 19-22, Mo 4-4.8, Cu 1.5-2.5, Nb 0.5-0.8, bal Fe.
Cast, annealed: 460 N/mm^2 psi TS min. For chemical industry; phosphoric and sulfuric acids, petrochemicals. AFNOR Z6 NCDU Nb 25.20 M; similar to ACI CN-7M.

NICROMAZ SP B

Creusot-Loire
C 0-0.2, Mn 0-2, Si 4.5-5.5, Ni 39-42, Cr 13-15, Mo 4-6, bal Fe.
Cast, annealed: 180-228 Brin. Resists corrosion by concentrated sulfuric acid up to 150°C. AFNOR Z8 NCDS 40.14.5.5 M.

NICROMAZ-B

Creusot-Loire
C, alloy, bal Fe.
For furnace parts, heat treating boxes; high resistance to carburizing atmosphere. *Obsolete*

NICROMINA-2

TradeARBED Inc.
Stainless steel. C 0-0.12, Mn 0-2, Si 0-1, Cr 17-19, Ni 10.5-13, bal Fe.
Austenitic stainless steel. AISI 305. *Obsolete*

NICROMINA-2 + MO

TradeARBED Inc.
Stainless steel. C 0-0.08, Mn 0-2, Si 0-1, Cr 16-18, Ni 10-14, Mo 2-3, bal Fe.
Austenitic stainless steel, for chemical equipment. AISI 316. *Obsolete*

NICROMINA-2 + NB

TradeARBED Inc.
Stainless steel. C 0-0.08, Mn 0-2, Si 0-1, Cr 17-19, Ni 9-13, Nb/Ta = 10 x C min, bal Fe.
Stabilized austenitic stainless steel. AISI 347. *Obsolete*

NICROMINA-2 + TI

TradeARBED Inc.
Stainless steel. C 0-0.08, Mn 0-2, Si 0-1, Cr 17-19, Ni 9-12, Ti = 5 x C min, bal Fe.
Stabilized austenitic stainless steel. AISI 321. *Obsolete*

NICROSIL

Russian manufacture
C 1.7-2, Cr 1.8-3, Ni 16-20, Mn 0.8-1.3, bal Fe.
For furnace equipment; corrosion resistant, cast iron.

NICROSILAL

Sheepbridge Alloy Castings Ltd.
C 1.8, Si 5, Ni 18, Cr 2, bal Fe.
Cast: 36,000 TS; 140-200 Brin. For furnace grids, heat treating boxes, corrosion resistant, non-magnetic.

NICROSILAL

Mond Nickel Co. Ltd.
C 1.8, Si 6, Ni 18, Cr 2, N 1, bal Fe.
36,000 TS; 3 El; 110-250 Brin. For furnace grids, generator boxes, annealing pots; heat resistant; does not glow at high temperature; austenitic.

NICROSILAL

Sheepbridge Engineering Ltd.
C 1.8, Si 5, Ni 18, Cr 2, bal Fe.
Cast: 36,000 TS; 140-200 Brin. For furnace grids, heat treating boxes, corrosion resistant, non-magnetic.

NICROSILAL 5 CR

English manufacture
C 2.1, Si 4.9, Mn 0.9, Ni 22.8, Cr 4.6, P 0.05, bal Fe.
For grids; heat resistant, cast iron.

NICROTUNG

Sankey & Sons Ltd.
C 0.08-0.13, B 0.02-0.08, Zr 0.02-0.08, Cr 11-13, Co 9-11, W 7-8.5, Al 3.75-4.75, Ti 3.75-4.75, bal Ni.
Cast: 130,000 TS; 120,000 YS; 5 El; 9 RA; 380 Brin. At 1800°F: 67,000 TS; 52,000 YS; 6 El. For missile and rocket engine components, gas turbines; heat resistant to 2000°F.

NICROTUNG

Westinghouse Electric Corp.
C 0.08-0.13, B 0.02-0.08, Zr 0.02-0.08, Cr 11-13, Co 9-11, W 7-8.5, Al 3.75-4.75, Ti 3.75-4.75, bal Ni.
Cast: 130,000 TS; 120,000 YS; 5 El; 9 RA; 380 Brin. At 1800°F: 67,000 TS; 52,000 YS; 6 El. For missile and rocket engine components, gas turbines; heat resistant to 2000°F.

NICU STEEL

English manufacture
Ni 2.2, Mn 0.6, Cu 0.5, C 0.3, bal Fe.
For machinery parts, structural work; oil hardening.

NICUAR 813

Arcos Alloys
Now ARCOS 813.

NICUEND

Arcos Alloys
Now ARCOS 803/

NICUITE

A.W. Cadman Mfg. Co.
Sn 10, Ni 3.5, Zn 2.5, bal Cu.
Cast: 50,000 TS; 25,000 YS; 24 El; 70 Brin. Heat treated: 95,000 TS; 75,000 YS; 6 El; 160 Brin. For bearings, worm gears, slippers, slides; heat treatable, heavy loads, slow speeds.

NICULOY

Belmont Metals Inc.
Cu, Ni, Fe.
For master alloy; for making alloys.

NICUMAN 23

GTE Products Corp./Wesgo Div.
Cu 67.5, Mn 23.5, Ni 9.
Brazing alloy available in foil, flexibraze, wire, powder, extrudable paste and preform. Liquidus 1751°F. Solidus 1697°F.

NICUMAN 37

GTE Products Corp./Wesgo Div.
Cu 52.5, Mn 38, Ni 9.5.
Brazing alloy available in foil, flexibraze, wire, powder, extrudable paste and preform. Liquidus 1697°F. Solidus 1616°F. Meets AMS-4764.

NICUSIL 3

GTE Products Corp./Wesgo Div.
Ag 71.15, Cu 28.1, Ni 0.75.
Brazing alloy available in foil, flexibraze, wire, powder, extrudable paste and preform. Liquidus 1463°F. Solidus 1436°F.

NICUSIL 8

GTE Products Corp./Wesgo Div.
Ag 56, Cu 42, Ni 2.
Brazing alloy available in foil, flexibraze, wire, powder, extrudable paste and preform. Liquidus 1639°F. Solidus 1420°F. Meets AMS-4765.

NICUSIL-3

Western Gold & Platinum Co.
Ag 71.15, Cu 28.1, Ni 0.75.
Melting point: 1436-1463°F. For brazing. Good wetting and filleting.

NICUSILTIN 6

GTE Products Corp./Wesgo Div.
Ag 62.5, Cu 29, Ni 2.5, Sn 6.
Brazing alloy available in foil, flexibraze, wire, powder, extrudable paste and preform. Liquidus 1476°F. Solidus 1275°F. Meets AMS-4774A.

NIFER

Texas Instruments Inc./Materials Control
Low C steel clad on both sides with Ni. For electron tubes. *Obsolete*

NIGY
Creusot-Loire
C 0.1, Mn 0.5-1, Si 2-3, bal Ni.
For spark plugs; heat and corrosion resistant. *Obsolete*

NIKA
Japanese manufacture
Pb, Sn, bal Cu.
For bearings; tough and non-corrosive.

NIKALIUM
Delta Metal (BW) Ltd.
Al, bal Cu.
Rolled: 94,000 TS; 36,000 YS; 10 El; 180 Brin. For gears, ship propellers; corrosion and wear resistant, tough. *Obsolete*

NIKE-3
Breda Co.
C 0-0.1, Cr 18, Ni 8, bal Fe.
Annealed: 80,000 TS; 35,000 YS; 55 El; 75 RA; 150 Brin. For chemical plant equipment, tanks, mixers, filters; Type 302; stainless, austenitic.

NIKE-B
Breda Co.
C 0-0.08, Cr 18, Ni 8, Ti, bal Fe.
Annealed: 85,000 TS; 35,000 YS; 55 El; 65 RA; 150 Brin. For welded chemical plant equipment, tanks, vessels; Type 321; stainless, austenitic.

NIKE-M
Breda Co.
C 0-0.12, Cr 18, Ni 10, Mo 2.5, bal Fe.
Annealed: 85,000 TS; 35,000 YS; 50 El; 65 RA; 160 Brin. For acid resistant chemical plant equipment, mixers, tanks; Type 316; stainless, austenitic.

NIKON
Massey-Harris Ltd.
Cast Iron. C 3.3, Mn 0.7, Si 2.2, bal Fe.
For machinery castings. Cast iron.

NIKRO M
Teledyne Vasco
C 0.55-0.75, Ni 1.4, Cr 0.85, Mo 0.4, bal Fe.
Heat treated: 290,000 TS. For collets, races, arbors, gears.
Type L6.

NIKROME
Joseph T. Ryerson & Son Inc.
C 0.4, Cr 0.8, Ni 1.5, bal Fe.
Rolled: 125,000 TS; 105,000 TS; 105,000 YS; 16 El; 50 RA; 300 Brin. For drive shafts, axles, spindles, roll mandrels; oil hardened. *Obsolete*

NIKROME 285
Joseph T. Ryerson & Son Inc.
C 0.41, Mn 0.75, Ni 1.9, Cr 0.82, Mo 0.26, bal Fe.
Heat treated: 130 ksi TS; 110 ksi YS; 15 El; 45 RA; 285-341 Brin. Preheat treated steel; for shafts, axles.

NIKROME 302
Joseph T. Ryerson & Son Inc.
C 0.41, Mn 0.75, Ni 1.9, Cr 0.82, Mo 0.26, bal Fe.
Heat treated: 140 ksi TS; 120 ksi YS; 14 El; 42 RA; 302-363 Brin. Preheat treated steel; for shafts, axles.

NIKROME M
Now NIKROME 285.

NIKROTHAL
Now NIKROTHAL 40.

NIKROTHAL 20
Kanthal Corp.
Superalloy. Cr 25, Fe 55, Ni 20.
Annealed: 95,000-114,000 psi TS; 50,000 psi YS. For resistors, heating elements; max operating temp 1050 °C. *Obsolete*

NIKROTHAL-2
Kanthal A.B.
Ni 20, Cr 25, bal Fe.
Resistivity 570 ohms/cmf. For unsupported heating elements. Maximum operating temperature 1920 F. *Obsolete*

NIKROTHAL-2
Kanthal Corp.
Ni 20, Cr 25, bal Fe.
Resistivity 570 ohms/cmf. For unsupported heating elements. Maximum operating temperature 1920 F. *Obsolete*

NIKROTHAL-8
Kanthal A.B.
Cr 20, bal Ni.
Annealed: 95,000-200,000 TS; 40,000-50,000 YS; 22-33 El. For resistors, heating elements; high electrical resistance, max operating temperature 2100°F.

NIKROTHAL-8
Kanthal Corp.
Cr 20, bal Ni.
Annealed: 95,000-200,000 TS; 40,000-50,000 YS; 22-33 El. For resistors, heating elements; high electrical resistance, max operating temperature 2100°F.

NIKROTHAL-LX
Kanthal A.B.
Cr 20, Ni 75, plus Al, Si, Mn, Cu.
Wire: 155,000-200,00 TS; 5-25 El. Elect. resistivity 800 ohms/cmf. For rheostats, resistors, potentiometers, electric appliances, precision and vitreous enamel resistors. High electrical resistivity, low temperature coefficient. Max. operating temperature 572°F.

NIKROTHAL-LX
Kanthal Corp.
Cr 20, Ni 75, plus Al, Si, Mn, Cu.
Wire: 155,000-200,00 TS; 5-25 El. Elect. resistivity 800 ohms/cmf. For rheostats, resistors, potentiometers, electric appliances, precision and vitreous enamel resistors. High electrical resistivity, low temperature coefficient. Max. operating temperature 572°F.

NILCOR
National Standard Co. (Athenia Steel)
Ni alloy.
For springs; non-magnetic. *Obsolete*

NILGRO 36
Darwins Alloy Castings
Ni 36, bal Fe.
For thermostats; controlled expansion. *Obsolete*

NILGRO 42
Darwins Alloy Castings
Ni 42, bal Fe.
For thermostats; controlled expansion. *Obsolete*

NILO 36
Huntington Alloys Inc.
Ni 36, bal Fe.
Annealed: 70,000 TS; 24,000 YS; 36 El; 68 RA; 143 Brin. For thermostats, glass-to-metal seals; controlled expansion.

NILO 36
Henry Wiggin & Co. Ltd.
Ni 36, bal Fe.
Annealed: 70,000 TS; 24,000 YS; 36 El; 68 RA; 143 Brin. For thermostats, glass-to-metal seals; controlled expansion.

NILO 40
Mond Nickel Co. Ltd.
Ni 40, Fe 60.
For glass to metal seals, thermostats; controlled coefficient of expansion.

NILO 42
Huntington Alloys Inc.
Ni 42, bal Fe.
Annealed: 78,000 TS; 41,000 YS; 45 El; 73 RA; 143 Brin. For thermostats, bimetals; controlled low expansion.

NILO 42
Henry Wiggin & Co. Ltd.
Ni 42, bal Fe.
Annealed: 78,000 TS; 41,000 YS; 45 El; 73 RA; 143 Brin. For thermostats, bimetals; controlled low expansion.

NILO 475
Mond Nickel Co. Ltd.
Ni 47, Cr 5, bal Fe.
For glass to metal seals, thermostats; controlled coefficient of expansion.

NILO 475
Henry Wiggin & Co. Ltd.
Ni 47, Cr 5, bal Fe.
For glass to metal seals, thermostats; controlled coefficient of expansion.

NILO 48
Inco Alloys International Inc.
Now NILO ALLOY 48.

NILO 50
Mond Nickel Co. Ltd.
Ni 50, Fe 50.
For glass to metal seals, thermostats; controlled coefficient of expansion.

NILO 501
Henry Wiggin & Co. Ltd.
Ni 50, bal Fe.
For glass to metal seals; controlled expansion. *Obsolete*

NILO 51
Inco Alloys International Inc.
Now NILO ALLOY 51. *Obsolete*

NILO ALLOY 36
Inco Alloys International Inc.
Ni 35-38, C 0-0.1, Mn 0-0.6, P 0-0.025, S 0-0.025, Si 0-0.35, Cr 0-0.5, Mo 0-0.5, Co 0-1, bal Fe.
Annealed: 71,000 TS; 35,000 YS; 42 El. Nickel-iron low-expansion alloy.

NILO ALLOY 42
Inco Alloys International Inc.
Ni 42, C 0-0.05, Mn 0-0.8, P 0-0.025, S 0-0.025, Si 0-0.3, Cr 0-0.25, Al 0-0.15, Co 0-1, bal Fe.
Annealed: 71,000 TS; 36,000 YS; 43 El. Nickel-iron controlled-expansion alloy. For semiconductor lead frames in integrated circuits, bi-metal thermostat strips, thermostat rods, ceramic-to-metal seals with alumina ceramics. K94100

NILO ALLOY 45
Inco Alloys International Inc.
C 0.05, Ni 45, Fe 54.
Magnetic alloy with high saturation flux density. Initial permeability at 20°C: 6000-10,000. For rocking armature telephone receivers. *Obsolete*

NILO ALLOY 475
Inco Alloys International Inc.
Ni 47, Fe 48, Cr 4.8.
Annealed: 75,000 TS; 26,000 YS; 37 El. Nickel-iron-chromium controlled-expansion alloy. For glass-to-metal seals and for anode-cavity caps in television tubes.

NILO ALLOY 48
Inco Alloys International Inc.
Ni 48, bal Fe.
For thermostats, glass to metal seals, instrument components. Controlled and low coefficient of expansion.

NILO ALLOY 51

Inco Alloys International Inc.
C 0.05, Ni 51, Fe 48.
Controlled expansion alloy; for sealing to soft glasses, e.g., in reed relay switch blades. AFNOR Fe-N 505. *Obsolete*

NILO ALLOY K

Inco Alloys International Inc.
Ni 29, Co 17, bal Fe.
Annealed: 77,200 TS; 54,000 YS; 41 El; 72 RA; 170 Brin. For glass to metal seals; expansion of medium hard glass.

NILO K

Inco Alloys International Inc.
Now NILO ALLOY K.

NILOMAG 36

Inco Alloys International Inc.
Ni 47, Mo 3, bal Fe.
For transformers, alternators; soft magnet for low power losses. *Obsolete*

NILOMAG 45

Inco Alloys International Inc.
Ni 64.3, Fe 34.7, Mo 1.
For magnetic amplifiers, pulse transformers; soft magnet, domain oriented. *Obsolete*

NILOMAG 77

Inco Alloys International Inc.
Now NILOMAG ALLOY 77.

NILOMAG 772

Henry Wiggin & Co. Ltd.
Ni 77, Fe 14, Cu 5.5, Mo 3.5.
For cores for telephone transformers, magnetic amplifiers; soft magnet, high permeability. *Obsolete*

NILOMAG 800

Henry Wiggin & Co. Ltd.
Ni 80.5, Fe 15.5, Mo 4.
For cores in logical type circuits; fast switching alloy. *Obsolete*

NILOMAG 801

Henry Wiggin & Co. Ltd.
Ni 80, Fe 8, Mo 12.
For computer work for storage applications; fast switching alloy, soft magnet. *Obsolete*

NILOMAG ALLOY 77

Inco Alloys International Inc.
Ni 77, Fe 14, Cu 5, Mo 4.
For cores for telephone transformers, magnetic amplifiers; soft magnet, high permeability.

NILOY 48

Manufacturer not listed
Ni 48, bal Fe.
Annealed: 50,000 TS; 33,000 YS; 46 El; 67 RA; 143 Brin. For thermostats, bimetals; controlled low expansion.

NILSTAIN 302

Wilbur B. Driver Co.
Stainless steel. Cr 18-20, Ni 8-10, C 0.2, Mn 2, bal Fe.
Rolled: 85,000 TS; 35,000 YS; 5 El; 9 RA; 140 Brin. For stainless parts, springs; corrosion resistant.

NILSTAIN 304

Wilbur B. Driver Co.
Stainless steel. Ni 8-10, Cr 18-20, C 0-0.08, Mn 0-2, bal Fe.
Rolled: 85,000-250,000 TS; 35,000-200,000 YS; 5-60 El; 55-70 RA; 140-455 Brin. For springs, screws, bolts; stainless wire.

NILSTAIN 305

Wilbur B. Driver Co.
Stainless steel. C 0-0.12, Cr 17-19, Ni 10-13, bal Fe.
Annealed: 85,000 TS; 38,000 YS; 50 El; 150 Brin. For cold headed parts; corrosion resistant, austenitic.

NILSTAIN 308

Wilbur B. Driver Co.
Ni 10-12, Cr 19-22, C 0.08, bal Fe.
Rolled: 85,000-350,000 TS; 35,000-300,000 YS; 55-60 El; 55-70 RA; 140-400 Brin. For weaving wire, bolts, nuts, screws; stainless. *Obsolete*

NILSTAIN 309

Wilbur B. Driver Co.
Ni 12-14, Cr 22-26, C 0-0.2, bal Fe.
Rolled: 100,000-225,000 TS; 52,000-150,000 YS; 5-75 El; 50-60 RA; 160-400 Brin. For weaving wire, heat resisting baskets; heat resistant. *Obsolete*

NILSTAIN 309 CB

Wilbur B. Driver Co.
Ni 13-15, Cr 23-25, Mn 1.5-2, C 0-0.07, Cb = 15 x C, bal Fe.
For welding wire; stainless. *Obsolete*

NILSTAIN 310

Wilbur B. Driver Co.
Ni 19-22, Cr 24, C 0-0.25, bal Fe.
Cold drawn: 105,000 TS; 50,000 YS; 50 El; 55 RA; 165 Brin. For welding wire, furnace parts; heat resistant. *Obsolete*

NILSTAIN 314

Wilbur B. Driver Co.
Cr 25, Ni 20, Si 1.5-3, bal Fe.
Annealed: 85,000 TS; 40,000 YS; 40 El. For wire belts; resists carburizing atmosphere. *Obsolete*

NILSTAIN 316

Wilbur B. Driver Co.
Ni 0-14, Cr 16-18, Mo 2-3, C 0-0.1, bal Fe.
Cold drawn: 95,000 TS; 45,000 YS; 55 El; 75 RA; 185 Brin. For welding wire, weaving wire; stainless. *Obsolete*

NILSTAIN 317

Wilbur B. Driver Co.
Ni 0-14, Cr 18-20, Mo 3-4, C 0-0.1, bal Fe.
Cold drawn: 95,000 TS; 45,000 YS; 55 El; 75 RA; 185-300 Brin. For weaving wire; heat resistant. *Obsolete*

NILSTAIN 321

Wilbur B. Driver Co.
Ni 7-10, Cr 17-20, C 0-0.1, Ti = 4 x C, bal Fe.
Cold drawn: 85,000 TS; 35,000 YS; 55 El; 70 RA; 130-300 Brin. For welding wire; stainless. *Obsolete*

NILSTAIN 325

Wilbur B. Driver Co.
Ni 19-23, Cr 7-10, Cu 1-1.5, C 0-0.25, bal Fe.
Cold drawn: 90,000 TS; 50,000 YS; 30-40 El; 45-55 RA; 160-190 Brin. For corrosion resisting parts; corrosion resistant. *Obsolete*

NILSTAIN 330

Wilbur B. Driver Co.
Stainless steel. Ni 33-36, Cr 14-16, C 0-0.25, bal Fe.
Cold drawn: 75,000 TS; 15-25 El. For heat resistant parts; heat resistant.

NILSTAIN 347

Wilbur B. Driver Co.
Cr 17-20, Ni 8-12, C 0-0.1, Cb = 10 x C, bal Fe.
Cold drawn: 85,000 TS; 35,000 YS; 55-60 El; 65-75 RA; 130-300 Brin. For welding wire; corrosion resistant. *Obsolete*

NILSTAIN 416

Wilbur B. Driver Co.
Cr 23-30, C 0-0.35, bal Fe.
Cold drawn: 85,000 TS; 60,000 YS; 20-30 El; 40-50 RA; 160-200 Brin. For glass sealing wire; heat resistant. *Obsolete*

NILSTAIN 430

Wilbur B. Driver Co.
Stainless steel. Cr 14-18, C 0-0.12, bal Fe.
Cold drawn: 75,000-100,000 TS; 50,000-90,000 YS; 10-35 El; 60-75 RA; 140-250 Brin. For instruments; corrosion resistant.

NILSTAIN C-20

Wilbur B. Driver Co.
C 0-0.07, Cr 20, Ni 29, Mo 2, Cu 3, bal Fe.
Annealed: 85,000 TS; 35,000 YS; 35-50 El; 50-70 RA; 150-180 Brin. For chemical plant equipment; stainless, resists sulfuric acid. *Obsolete*

NILSTAIN X

Wilbur B. Driver Co.
Ni 7-8, Cr 17-19, Cu 1.5-3, Mn 4-6, C, bal Fe.
Rolled: 80,000 TS; 50 El. For lock washers, cold heading parts; non-magnetic; stainless. *Obsolete*

NILVAR

Harrison Alloys Inc.
Nickel.
Now HAI-36 ALLOY.

NIMAG 100

Spang Specialty Metals
Ni 99.5, C 0.06, Fe 0.15, Mn 0.25, 0.105. others.
Good mechanical properties; corrosion resistant. *Obsolete*

NIMAG 101

Spang Specialty Metals
Ni 99.5, C 0.01, Fe 0.15, Mn 0.2, 0.15 others.
Good mechanical properties. Corrosion resistant. *Obsolete*

NIMAG 104

Spang Specialty Metals
Ni 95.2, Co 4.5, 0.30 others.
Good magnetostrictive properties. *Obsolete*

NIMAG 105

Spang Specialty Metals
Ni 99.5, C 0.06, Mn 0.2, Fe 0.1, Mg 0.04, Ti 0.02, 0.08 others.
Easily formed or drawn. Low oxidation at high temperatures. High Curie temperature. *Obsolete*

NIMAG 111

Alcan Canada Products Ltd.
Ni 95, Mn 4.75, 0.25 others.
Good strength and base hardness. Resistant to sulfur compounds at high temperature.

NIMAG 120

Spang Specialty Metals
Ni 99.5, C 0.06, Mn 0.12, Mg 0.04, Ti 0.02. *Obsolete*

NIMAG 130

Spang Specialty Metals
Ni 99.5, C 0.09, Mn 0.1, Mg 0.06, Fe 0.05, Cu 0.01, Si 0.03, Ti 0.003. *Obsolete*

NIMAG 133

Spang Specialty Metals
Ni 99.5, C 0.09, Mn 0.18, Mg 0.07, Ti 0.003, Fe 0.05, Cu 0.03, Si 0.03.
For vacuum tube anodes, oxide coated and cold cathodes, structural tube parts. *Obsolete*

NIMAG 60

Spang Specialty Metals
Ni 76, Cu 0.1, Fe 7.2, Mn 0.2, Si 0.2, Cr 15.8.
Resists oxidation at high temperatures; resists corrosion. Non-magnetic. *Obsolete*

NIMAR 110

British Steel Corp.
C 0.02, Si 0.1, Mn 0.1, Ni 17-19, Co 7-8.5, Mo 4.6-5.1, Ti 0.4-0.6, Al 0.05-0.15, bal Fe.
Maraging steel, 1 1/2 in. bar, as hardened: 266,000 TS; 256,000 YS; 6-12 El. For parts requiring high strength. *Obsolete*

NIMAR 125

British Steel Corp.
C 0.02, Si 0.1, Mn 0.1, Ni 17-19, Co 8.5-9.5, Mo 4.7-5.2, Ti 0.6-0.8, Al 0.05-0.15, bal Fe.
Maraging steel, 3 in. bar, hardened: 294,000 TS; 282,000 YS; 5-12 El. For parts requiring very high strength. *Obsolete*

NIMAR 700

Arcos Alloys

C 0.02, Mn 0.06, Si 0.06, Ni 18.2, Mo 4.8, Ti 0.6, Al 0.1, Co 9.8, bal Fe.

For bare welding wire; maraging steel grade 250, for gas metal arc welding. *Obsolete*

NIMAR 701

Arcos Alloys

C 0.02, Mn 0.06, Si 0.06, Ni 18.2, Mo 4.8, Ti 1, Al 0.1, Co 9.8, bal Fe.

For bare welding wire; maraging steel grade 250, for submerged arc welding. *Obsolete*

NIMARK

Now CARPENTER NIMARK.

NIMARK ALLOY 250

Carpenter Technology Corp.

C 0-0.03, Mn 0-0.1, Si 0-0.1, P 0-0.01, S 0-0.01, Ni 18-19, Mo 4.7-5, Co 7-8, Ti 0.3-0.5, Al 0.05-0.15, Zr 0-0.03, B 0-0.003, Ca 0-0.05, bal Fe.

Heat treated, longitudinal, RT: 1758 MPa TS; 1724 MPa YS; 12 El; 62 RA; 49 Rock C. Maraging nickel steel with good ductility at high strength.

NIMARK-II

Carpenter Technology Corp.

C 0.03, Ni 20, Ti 1.45, Al 0.25, B 0.003, Zr 0.01, Cb 0.5, bal Fe.

Solution annealed: 162,000 TS; 123,000 YS; 16 El; 74 RA; Rock C 31. Aged: 280,000 TS; 267,000 YS; 13 El; 58 RA; Rock C 52. For missile and rocket components. Vacuum melted, maraging steel. Good ductility at high strength levels. *Obsolete*

NIMBUS DC 02

Horbach & Schmitz GmbH

C 1.35, Cr 4.2, W 12, V 3.75, Mo 0.8, Co 5, bal Fe.

High speed steel; for finishing; good abrasion resistance. W. Nr. 1.3202.

NIMBUS DC 07

Horbach & Schmitz GmbH

C 1.25, Cr 4.2, W 10.5, V 3.25, Mo 3.75, Co 10.5, bal Fe.

High speed steel; for lathe tools; good red hardness and abrasion resistance. W. Nr. 1.3207.

NIMBUS DC 55

Horbach & Schmitz GmbH

C 0.8, Cr 4.2, W 18, V 1.6, Mo 0.7, Co 5, bal Fe.

High speed steel; for lathe and planing tools for severe service. W. Nr. 1.3255; similar to AISI T4.

NIMBUS DC 65

Horbach & Schmitz GmbH

C 0.76, Cr 4.2, W 18, V 1.6, Mo 0.7, Co 9, bal Fe.

High speed steel; for heavy duty lathe tools. W. Nr. 1.3265; similar to AISI T5.

NIMBUS DD 16

Horbach & Schmitz GmbH

C 0.82, Cr 4.2, W 9, V 1.6, Mo 0.8, bal Fe.

High speed steel; for milling cutters, twist drills, end mills. W. Nr. 1.3316.

NIMBUS DD 18

Horbach & Schmitz GmbH

C 0.95, Cr 4.2, W 12, V 2.5, Mo 0.9, bal Fe.

High speed steel; for lathe tools, taps, threading cutters. W. Nr. 1.3318.

NIMBUS DD 33

Horbach & Schmitz GmbH

C 1, Cr 4.2, Mo 2.7, W 2.8, V 2.4, bal Fe.

High speed steel; for twist drills, milling cutters, broaches, reamers. W. Nr. 1.3333.

NIMBUS DD 43

Horbach & Schmitz GmbH

C 0.88, Cr 4.2, Mo 5, W 6.3, V 1.8, bal Fe.

High speed steel; for milling cutters, rough cutting lathe tools, broaches. W. Nr. 1.3343; similar to AISI M2.

NIMBUS DD 55

Horbach & Schmitz GmbH

C 0.75, Cr 4.2, W 18, V 1.1, bal Fe.

High steed steel; for lathe and planer tools, taps, broaches, twist drills. W. Nr. 1.3355; AISI T1.

NIMEND

Arcos Alloys

C 0.1-0.15, Cr 14-18, W 4-5, Mo 15-18, Fe 5.3, bal Ni.

Welded: 20-25 Rock C; aged 50 h at 1475°F: 38-40 Rock C.

For hardfacing electrodes; similar to Hastelloy C. *Obsolete*

NIMO

Darwin & Milner Inc.

Tool material. C 0.5-0.6, Cr 0.3-0.5, V 0.1-0.15, Mo 0.4-0.5, Ni 2.5-2.8, bal Fe.

For tools, dies; oil hardened.

NIMOCAST 235D

Mond Nickel Co. Ltd.

Cr 14-17, Mo 4.5-6, Fe 3.5-5, Al 3.25-4, Ti 2-3, Si 0-0.3, C 0.1-0.2, Mn 0-0.1, B 0.05-0.1, 5.6-6.5 Al + Ti, bal Ni.

At 1200°F: 83,400 fatigue strength. At 1650°F: 65,000 fatigue strength. For jet engine and aerospace equipment and parts. High heat and corrosion resistance.

NIMOCAST 242

Inco Alloys International Inc.

C 0.3, Co 10, Cr 20, Ti 0-3, Al 0-0.2, Mo 10, bal Ni.

Cast: 70,000 TS; 43,000 YS; 8 El; 220 Brin. For high temperature applications; high resistance to thermal shock. *Obsolete*

NIMOCAST 257

Sheepbridge Alloy Castings Ltd.

C 0.8, Co 16, Cr 20, Fe 0-2, Ti 1.6, Al 0.9, bal Ni.

Cast. For gas turbine engine rings; good properties to 650°C.

NIMOCAST 258

Henry Wiggin & Co. Ltd.

C 0.2, Cr 10, Co 20, Fe 0-2, Ti 3.7, Al 4.8, Mo 5, bal Ni.

Cast: 122,000 TS; 114,000 YS; 3.4 El; 375 Brin. Heat treated: 117,400 TS; 110,000 YS; 4.5 El; 383 Brin. For high temperature applications; creep resistance to 1000 C. *Obsolete*

NIMOCAST 713

Mond Nickel Co. Ltd.

Cr 14, Mo 4.5, Ti 1, Al 6, Fe 0-1.5, 2 Cb + Ta, bal Ni.

For gas turbine stator and rotor blades. Creep resistant for service to 1000°C.

NIMOCAST 713

Henry Wiggin & Co. Ltd.

Cr 14, Mo 4.5, Ti 1, Al 6, Fe 0-1.5, 2 Cb + Ta, bal Ni.

For gas turbine stator and rotor blades. Creep resistant for service to 1000°C.

NIMOCAST 75

Sheepbridge Alloy Castings Ltd.

C 0.1, Cr 20, Fe 0-5, Ti 0.4, Al 0.2, bal Ni.

Cast: 75,000 TS; 29,400 YS; 34.8 El; 164 Brin. Heat treated: 81,500 TS; 27,500 YS; 40.5 El; 172 Brin. For high temperature applications; high oxidation resistance.

NIMOCAST 80

Inco Alloys International Inc.

C 0.05, Cr 20, Fe 0-5, Ti 2.4, Al 1.2, bal Ni.

Cast: 72,000 TS; 66,700 YS; 3 El; 253 Brin. Heat treated: 111,600 TS; 78,000 YS; 14 El; 270 Brin. For high temperature applications; high creep resistance to 750°C. *Obsolete*

NIMOCAST 90

Sheepbridge Alloy Castings Ltd.

C 0.1, Co 16, Fe 0-5, Ti 2.4, Al 1.2, Cr 20, bal Ni.

Cast: 92,800 TS; 72,700 YS; 8.1 El; 280 Brin. Heat treated: 106,000 TS; 79,600 YS; 12.6 El; 291 Brin. For high temperature applications; high creep resistance to 870°C.

NIMOCAST ALLOY 263

Inco Alloys International Inc.

C 0.06, Ni 51, Cr 20, Co 20, Mo 5.9, Ti 2.2, Al 0.5.

Vacuum melted cast alloy, similar to wrought alloy NIMONK alloy 263. *Obsolete*

NIMOCAST ALLOY 713LC

Inco Alloys International Inc.

C 0.05, Ni 74, Cr 12, Ti 0.7, Al 6.

At 700°C: 1010 N/mm^2 TS; 775 N/mm^2 YS; 13 El. Low carbon grade of NIMOCAST alloy 713. For gas turbine rotor blades, turbine and turbocharger rotors. *Obsolete*

NIMOCAST ALLOY 738

Inco Alloys International Inc.

C 0.18, Ni 61, Cr 16, Co 8.5, Mo 1.8, Ti 3.4, Al 3.4, Ta 1.6, W 2.5.

At 1000°C: 450 N/mm^2 TS; 325 N/mm^2 YS; 16 El. Good high temperature creep strength and hot corrosion resistance. Similar to IN-738. *Obsolete*

NIMOCAST ALLOY 738 LC

Inco Alloys International Inc.

C 0.11, Ni 61, C 16, Co 8.5, Mo 1.8, Ti 3.4, Al 3.4, Ta 1.6, W 2.5.

At 700°F: 1010 N/mm^2 TS; 500 N/mm^2 YS; 4 El. Low carbon version of NIMOCAST alloy 738. *Obsolete*

NIMOCAST ALLOY 739

Inco Alloys International Inc.

C 0.15, Ni 48, Cr 22.4, Co 19, Ti 3.7, Al 1.9, Ta 1.4, W 2, Nb 1.

At 700°F: 980 N/mm^2 TS; 590 N/mm^2 YS; 7 El. Vacuum melted, cast alloy; good high temperature strength and corrosion resistance; for blades and vanes in marine and industrial gas turbines. Similar to IN-939. *Obsolete*

NIMOCAST ALLOY PD21

Inco Alloys International Inc.

C 0.1, Ni 73, Cr 6, Mo 2, Al 6, W 10.5.

At 1000°C: 575 N/mm^2 TS; 375 N/mm^2 YS; 4 El. High stress rupture properties up to 1050°C. For stator blades. Similar to IN-M-21. *Obsolete*

NIMOCAST PE 10

Mond Nickel Co. Ltd.

Cr 20, Mo 6, W 2.5, 6.5 Cb + Ta, bal Ni.

For turbo-charger rotors, gas turbine components, diesel engine pre-combustion chambers. Creep resistant for service to 870°C, oxidation and corrosion resistant.

NIMOCAST PE 10

Henry Wiggin & Co. Ltd.

Cr 20, Mo 6, W 2.5, 6.5 Cb + Ta, bal Ni.

For turbo-charger rotors, gas turbine components, diesel engine pre-combustion chambers. Creep resistant for service to 870°C, oxidation and corrosion resistant.

NIMOCAST PK 24

Mond Nickel Co. Ltd.

C 0.18, Co 15, Cr 10, Mo 3, Ti 5.2, Al 5.6, B, Zr, bal Ni.

For gas turbines, engines, turbine-stator blades, turbocharger rotors. High temperature characteristics for max. service about 1040°C.

NIMOCAST PK 24

Henry Wiggin & Co. Ltd.

C 0.18, Co 15, Cr 10, Mo 3, Ti 5.2, Al 5.6, B, Zr, bal Ni.

For gas turbines, engines, turbine-stator blades, turbocharger rotors. High temperature characteristics for max. service about 1040°C.

NIMOCAST PK 36
Henry Wiggin & Co. Ltd.
Cr 10, Co 10, Ti 5, Al 5, Mo 4, Zr 0.12, C 0.1, B 0.015, bal Ni.
For jet engine and gas turbine components. Heat and corrosion resistant. High temperature casting alloy. *Obsolete*

NIMOCAST-C 242
Mond Nickel Co. Ltd.
C 0.3, Si 0.3, N 0.3, Cr 20, Co 10, Fe 0-1, Ti 0-0.3, Al 0-0.2, Mo 10, bal Ni.
Cast: 69,700 TS; 43,000 YS; 8 El; 220 Vickers. For high temperature cast components, turbine nozzle guide vanes. High resistance to thermal shock. High corrosion and heat resistant castings.

NIMOCAST-MC 57
Mond Nickel Co. Ltd.
C 0.08, Si 0.4, Mn 0.3, Cr 20, Co 16, Fe 0-2, Ti 1.6, Al 0.9, bal Ni.
Heat treated: 100,000 TS; 65,000 YS; 16 El. For high temperature cast components, gas turbine engine rings, jet engine parts. Cast high temperature alloy. Good combination of yield strength and ductility at 650°C. Heat treatable.

NIMOCAST-MC 58
Mond Nickel Co. Ltd.
C 0.2, Mn 0.4, Si 0.4, Cr 10, Co 20, Fe 0-2, Ti 3.7, Al 4.8, Mo 5, bal Ni.
Heat treated: 117,500 TS; 110,000 YS; 4.5 El; 383 Vickers.
Cast: 122,000 TS; 114,000 YS; 3.4 El; 375 Vickers. For high temperature cast components, turbine and jet engine components. Good creep resistance up to 1000°C. Casting alloy, good castability. Heat treatable.

NIMOCAST-PD 16
Henry Wiggin & Co. Ltd.
W 11, Cr 6, Al 6, Mo 2, Cb 1.5, Zr 0.12, C 0.12, B 0.02, bal Ni.
For investment cast rotors and stators, gas turbine blades. High temperature casting alloy, heat and corrosion resistant. *Obsolete*

NIMOFER 6928-ALLOY B-2
VDM Nickel-Technologie AG
Cr 0-1, Fe 0-2, Mo 26-30, C 0-0.01, Co 0-1, Si 0-0.08, bal Ni.
Minimum values at 20°C: 760 N/mm^2 TS; 350 N/mm^2 YS; 40 El. For chemical process industry involving acids. Material No. 2.4617. N10665

NIMOFER S 6928-FM B-2
VDM Nickel-Technologie AG
Cr 0-1, Fe 0-2, Mo 26-30, Mn 0-1, Si 0-0.08, C 0-0.01, S 0-0.015, bal Ni.
Filler metal. Minimum values at RT: 720 N/mm^2 TS; 350 N/mm^2 YS; 25 El. Material No. 2.4615. N10665

NIMOL
Mond Nickel Co. Ltd.
Ni 14, Cu 6, Cr 0-14, 75 cast iron.
20,000-27,000 TS; 1.5-2.5 El; 120-240 Brin. For retorts for fuel carbonization, pump and engine liners, pans for fusing caustic alkalis; superseded by Ni-Resist.

NIMOLOY ALLOY PK37
Inco Alloys International Inc.
C 0.12, Ni 60, Cr 18.5, Co 17, Ti 2.2, Al 1.2
At 700°C: 970 N/mm^2 TS; 680 N/mm^2 YS; 12 El. High strength alloy with good abrasion and shock resistance up to 850°C. For forging press anvils, hot shear blades, other hot working tools. *Obsolete*

NIMONIC
Mond Nickel Co. Ltd.
Cr 20, Ti 2, Al 2, bal Ni.
For high temperature applications; heat resistant.

NIMONIC 100
Mond Nickel Co. Ltd.
Cr 11, Mo 5, Co 20, Ti 1.5, Al 5, bal Ni.
Heat treated: 320-400 Brin. For gas turbine rotor blades; creep resistant to 1000°C.

NIMONIC 105
Enpar Sonderwerkstoffe GmbH
Nickel.
Alternate manufacturer.

NIMONIC 105
Inco Alloys International Inc.
Now NIMONIC ALLOY 105.

NIMONIC 110
Mond Nickel Co. Ltd.
C 0.15, Cr 15, Co 20, Ti 1.75, Al 5.75, Mo 5, bal Ni.
Hot rolled: 180,000 TS; 124,000 YS; 18 El; 17 RA. At 800°F: 113,000 TS; 84,000 YS; 37 El; 35 RA. For high temperature applications, aircraft gas turbines, high temperature fasteners, gas turbine blades. Creep and high oxidation resistance.

NIMONIC 115
Enpar Sonderwerkstoffe GmbH
Nickel.
Alternate manufacturer.

NIMONIC 115
Inco Alloys International Inc.
Now NIMONIC ALLOY 115.

NIMONIC 118
Mond Nickel Co. Ltd.
C 0.16, Cr 15, Ti 3.85, Al 4.8, Co 14.9, Mo 3.5, Zr 0.045, B 0.016, Fe 0-0.7, Mn 0-0.5, bal Ni.
Heat treated: 170,000 TS; 125,000 YS; 29 El; 45 RA. At 600°C: 150,000 TS; 110,000 YS; 25 El; 34 RA. For gas turbine and jet engine components, missiles, aircraft. Age-hardenable, heat and oxidation resistant.

NIMONIC 58
Mond Nickel Co. Ltd.
Cr 11, Mo 5, Co 20, Al 5, Ti 2, bal Ni.
Cast: 123,000 TS; 114,000 YS; 3.4 El; 375 Vickers. For high temperature castings, jet engine and gas turbine parts. Heat and corrosion resistant. High creep and rupture strength.

NIMONIC 75
Enpar Sonderwerkstoffe GmbH
Nickel.
Alternate manufacturer.

NIMONIC 75
Mond Nickel Co. Ltd.
C 0.08-0.15, Ti 0.2-0.6, Cr 18-21, Si 0-1, Fe 5-11, 1.0 Mn max, 0.5 Cu max, bal Ni + Co.
Solution treated: 103,000-96,000 TS; 60,000 YS; 50 El; 50 RA. For combustion chambers, turbines; age-hardenable, good resistance to creep and oxidation.

NIMONIC 75
Henry Wiggin & Co. Ltd.
C 0.08-0.15, Ti 0.2-0.6, Cr 18-21, Si 0-1, Fe 5-11, 1.0 Mn max, 0.5 Cu max, bal Ni + Co.
Solution treated: 103,000-96,000 TS; 60,000 YS; 50 El; 50 RA. For combustion chambers, turbines; age-hardenable, good resistance to creep and oxidation.

NIMONIC 75 Г
Mond Nickel Co. Ltd.
C 0.08-0.15, Ti 0.2-0.6, Cr 18-21, Si 0-1, Mn 0-1, Cu 0-0.5, Fe 5-11, bal Ni + Co.
For turbines, combustion chambers; heat resistant, age-hardenable, creep and oxidation resistant.

NIMONIC 80
Mond Nickel Co. Ltd.
C 0.04, Si 0.47, Ni 75, Cr 21, Mn 0.56, Ti 2.45, Al 0.63.
Heat treated: 147,000 TS; 41 El; 35 RA; 250 Brin. At 20°C: 132,000 TS; 80,000 YS; 45 El; 36 RA. At 800°C: 63,000 TS; 53,000 YS; 8 El; 10 RA. For rotor blades, gas turbine engines; high creep resistance, age-hardenable.

NIMONIC 80A
Enpar Sonderwerkstoffe GmbH
Nickel.
Alternate manufacturer.

NIMONIC 80A
Henry Wiggin & Co. Ltd.
C 0.04, Cr 21, Mn 0.6, Ti 2.5, Al 0.7, bal Ni.
At 20°C: 132,000 TS; 80,000 YS; 45 El; 36 RA. At 800°C: 62,000 TS; 53,000 YS; 8 El; 10 RA. For gas turbine blades, and valves; high creep resistant.

NIMONIC 80A
Mond Nickel Co. Ltd.
C 0.04, Cr 21, Mn 0.6, Ti 2.5, Al 0.7, bal Ni.
At 20°C: 132,000 TS; 80,000 YS; 45 El; 36 RA. At 800°C: 62,000 TS; 53,000 YS; 8 El; 10 RA. For gas turbine blades, and valves; high creep resistant.

NIMONIC 90
Enpar Sonderwerkstoffe GmbH
Nickel.
Alternate manufacturer.

NIMONIC 90
Henry Wiggin & Co. Ltd.
C 0-0.1, Al 0.8-1.8, Ti 1.8-3, Fe 0-5, Co 15-21, Cr, bal Ni.
At 20°C: 166,000 TS; 101,000 YS; 39 El; 20 RA. At 600°C: 132,000 TS; 90,000 YS; 26 El; 21 RA. at 800°C 85,000 TS; 63,000 YS; 7 El; 4 RA. For turbine blades, combustion chambers, rotor dies, and valves; age hardened, super heat and creep resistance.

NIMONIC 90
Mond Nickel Co. Ltd.
C 0-0.1, Al 0.8-1.8, Ti 1.8-3, Fe 0-5, Co 15-21, Cr, bal Ni.
At 20°C: 166,000 TS; 101,000 YS; 39 El; 20 RA. At 600°C: 132,000 TS; 90,000 YS; 26 El; 21 RA. at 800°C 85,000 TS; 63,000 YS; 7 El; 4 RA. For turbine blades, combustion chambers, rotor dies, and valves; age hardened, super heat and creep resistance.

NIMONIC 93
Mond Nickel Co. Ltd.
C 0.1, Cr 19.5, Ti 2.75, Al 1.5, Co 18, Mo 0-0.3, Zr 0.08, B 0.008, bal Ni.
Heat treated: 185,000 TS; 120,000 YS; 28 El; 41 RA. At 600°C: 160,000 TS; 105,000 YS; 24 El; 23 RA. For jet engine and gas turbine components. Age-hardenable, heat and oxidation resistant.

NIMONIC 95
Henry Wiggin & Co. Ltd.
Cr 18-21, Co 15-21, C 0-0.15, Al 1.4-2.5, Fe 5, Ti 4, bal Ni.
At 20 C: 184,000 TS; 25 El; 24 RA; 290 Brin. At 800 C: 89,600 TS; 5 El; 6 RA; 360 Brin. For gas turbine rotor blades; age hardened, creep resistant. *Obsolete*

NIMONIC ALLOY 105
Inco Alloys International Inc.
Cr 15, Co 20, Mo 5, Al 4.5, Ti 1.4, C 0.15, bal Ni.
At 20°C: 144,000 TS; 116,000 YS; 7 El; 7 RA. At 800°C: 107,000 TS; 78,500 YS; 16 El; 17 RA. For gas turbine components, jet engines, combustion chambers. Corrosion and heat resistance.

NIMONIC ALLOY 115
Inco Alloys International Inc.
C 0.15, Cr 15, Co 20, Mo 5, Al 4, bal Ni.
For gas turbine engine blades; high creep and rupture strength at high temperature.

NIMONIC ALLOY 263
Inco Alloys International Inc.
C 0.06, Ni 51, Cr 20, Co 20, Mo 5.9, Ti 2, Al 0.5.
At 700°C: 750 N/mm^2 TS; 460 N/mm^2 YS; 23 El. Precipitation hardening, creep-resisting alloy for gas turbine rings and sheet components for use to 850°C. BS HR10; ANOR NCK 20D.

NIMONIC ALLOY 75

Inco Alloys International Inc.
Nickel. Cr 18-21, Ti 0.2-0.6, C 0.08-0.15, Si 0-1, Cu 0-5, Mn 0-1, bal Ni.
Annealed: 6,000 rupture strength (1000 h) at 1400°F. Nickel-chromium alloy. For sheet-metal fabrications in gas-turbine engines, for components of industrial furnaces, for heat-treating equipment and fixtures, and in nuclear engineering. 6075

NIMONIC ALLOY 80A

Inco Alloys International Inc.
Nickel. Cr 18-21, Ti 1.8-2.7, Al 1-1.8, C 0-0.1, Si 0-1, Cu 0-0.2, Fe 0-3, Mn 0-1, Co 0-2, B 0-0.008, Zr 0-0.15, S 0-0.015, bal Ni.
Precipitation hardened: 94,000 rupture strength (1000 h) at 1100°F. Nickel-chromium alloy; precipitation hardenable by additions of aluminum and titanium. For gas turbine components, bolts, tube supports in nuclear steam generators, die-casting inserts and cores, and exhaust valves. 7080

NIMONIC ALLOY 81

Inco Alloys International Inc.
C 0.03, Ni 66, Cr 30, Ti 1.8, Al 1.4.
At 700°C: 790 N/mm^2 TS; 500 N/mm^2 YS; 25 El.
Precipitation hardening alloy; for exhaust valves in internal combustion engines.

NIMONIC ALLOY 86

Inco Alloys International Inc.
C 0.05, Ni 64.5, Cr 25, Mo 10, Mn 0.15, Ce 0.03.
At 700°C: 500 N/mm^2 TS; 260 N/mm^2 YS; 74 El. Alloy sheet with good ductility, high creep strength, and resistance to cyclic oxidation up to 1050°C. For gas turbine components and heat treat equipment.

NIMONIC ALLOY 90

Inco Alloys International Inc.
Nickel. Cr 18-21, Co 15-21, Ti 2-3, Al 1-2, C 0-0.13, Si 0-1, Cu 0-0.2, Fe 0-1.5, Mn 0-1, B 0-0.02, S 0-0.015, Zr 0-0.15, bal Ni.
Precipitation hardened: 52,000 rupture strength (1000 h) at 1300 °F. Nickel-chromium-cobalt alloy. For blades and discs in gas turbines, hot-working tools, and springs. 7090

NIMONIC ALLOY 901

Inco Alloys International Inc.
C 0.05, Cr 12.8, Ni 43, Mo 5.7, Ti 2.4, Fe 35.
Heat treated: 168,000 TS; 110,000 YS; 23 El. At 1200°F: 125,000 TS; 95,000 YS; 5 El. For aircraft gas turbine blades, turbine discs; age hardenable, up to 1400°F service.

NIMONIC ALLOY 91

Inco Alloys International Inc.
C 0.08, Ni 47.5, Cr 28.5, Co 20, Ti 2.3, Al 1.2.
At 700°C: 945 N/mm^2 TS; 580 N/mm^2 YS; 28 El.
Precipitation hardening alloy; improved corrosion resistance to salt and sulfur environments. For gas turbine blades in engines burning impure fuels. *Obsolete*

NIMONIC ALLOY AP1

Inco Alloys International Inc.
Nickel. Ni 55.5, Co 17, Cr 15, Mo 5, Al 4, Ti 3.5, C 0.02, B 0.0025.
Precipitation hardened and forged: 110,000 rupture strength (80 h) at 1300°F. Nickel-cobalt-chromium alloy produced by powder metallurgy. For discs in gas turbines.

NIMONIC ALLOY C263

Mond Nickel Co. Ltd.
C 0.03, Cr 20, Ti 2.15, Al 0.45, Co 20, Mo 5.9, Zr 0-0.02, S 0-0.007, B 0-0.001, bal Ni.
Heat treated: 146,000 TS; 90,000 YS; 45 El; 42 RA. At 600°C: 120,000 TS; 70,000 YS; 44 El; 50 RA. For jet engine and gas turbine components, missiles. Age-hardenable, heat and oxidation resistant.

NIMONIC ALLOY PE11

Inco Alloys International Inc.
Nickel. Ni 37-41, Cr 17-19, Mo 4.75-5.75, Ti 2.2-2.5, Al 0.7-1, C 0.03-0.08, Si 0-0.5, Cu 0-0.5, Mn 0-0.2, Co 0-1, B 0-0.001, Zr 0.02-0.05, S 0-0.015, bal Fe.
Precipitation hardened: 49,000 rupture strength (1000 h) at 1200°F. Nickel-iron-chromium alloy. For components of gas turbines.

NIMONIC ALLOY PE16

Inco Alloys International Inc.
Nickel. Ni 42-45, Cr 15.5-17.5, Mo 2.8-3.8, Ti 1.1-1.3, Al 1.1-1.3, C 0.04-0.08, Si 0-0.5, Cu 0-0.5, Mn 0-0.2, Co 0-2, B 0-0.005, Zr 0.02-0.04, S 0-0.015, bal Fe.
Precipitation hardened: 52,600 rupture strength (1000 h) at 1200°F. Nickel-iron-chromium alloy. For gasoline components and in nuclear reactors.

NIMONIC ALLOY PK33

Inco Alloys International Inc.
Nickel. Cr 16-20, Co 12-16, Mo 5-9, Ti 1.5-3, Al 1.7-2.5, C 0-0.07, Si 0-0.5, Cu 0-0.2, Fe 0-1, Mn 0-0.5, S 0-0.015, B 0-0.005, Zr 0-0.06, bal Ni.
Precipitation hardened: 87,000 rupture strength (1000 h) at 1200°F. Nickel-chromium-cobalt alloy. For flame tubes in gas turbines and other components.

NIMONIC B

Mond Nickel Co. Ltd.
C 0-0.1, Ti 1.8-2.7, Al 0.8-1.8, Fe 0-5, Co 15-21, Cr 18-21, bal Ni.
Rolled: 160,000 TS; 90,000 YS; 39 El; 20 RA. At 1500°F: 76,000 TS; 56,000 YS; 7 El; 4 RA. For turbine gland springs; heat resistant, heat treatable.

NIMONIC C

Mond Nickel Co. Ltd.
Al 0.5-1.8, Ti 1.8-2.7, Cr 18-21, bal Ni.
At 20°C: 134,000 TS; 80,000 YS; 45 El; 36 RA. At 800°C: 64,000 TS; 54,000 YS; 8 El; 10 RA. For heat exchangers, valves, valve inserts, gas turbine parts; heat and corrosion resistant.

NIMONIC C-263

Cytemp Specialty Steel Div.
Nickel. C 0.06, Cr 20, Co 20, Fe 0.5, Mo 5.9, Ti 2.1, Al 0.45, bal Ni.
Nickel base alloy.

NIMONIC C.C.

Mond Nickel Co. Ltd.
C 0-0.1, Al 0.5-1.8, Fe 0-5, Ti 1.8-2.7, Cr 18-21, Co 0-2, bal Ni.
Heat treated: 112,000 TS; 80,000 YS; 14 El; 270 Vickers. For gas engine and jet engine and gas turbine components. High creep resistance to 900°C. Cast alloy, oxidation resistant.

NIMONIC C242

Mond Nickel Co. Ltd.
C 0.3, Cr 20, Co 10, Mo 10, Fe 0-1, bal Ni.
Cast: 70,000 TS; 44,000 YP; 8 El; 220 Vickers. For high temperature castings, nozzle guide vanes, jet engine and gas turbine parts. Heat and corrosion resistant. High resistance to creep and thermal shock.

NIMONIC C75

Mond Nickel Co. Ltd.
C 0.08-0.15, Fe 0-5, Ti 0.2-0.6, Cr 18-21, Al 0.2, Cu 0-0.5, bal Ni.
Cast: 74,500 TS; 30,000 YS; 35 El; 164 Vickers. Heat treated: 82,000 TS; 28,000 YS; 41 El; 172 Vickers. For furnace trays, gas turbine parts, combustion chambers, jet engine components. Resists scaling and thermal shock. Heat and corrosion resistant.

NIMONIC CF

Mond Nickel Co. Ltd.
C 0.08-0.15, Ti 0.2-0.6, Fe 5-11, Cr 8-21, bal Ni.
For jet engine components; cast alloy.

NIMONIC D

Mond Nickel Co. Ltd.
Ni 37, Cr 18, Si 2, bal Fe.
At 70°F: 94,000 TS; 48,000 YS; 38 El; 50 RA. At 2000°F: 10,000 TS; 124 El; 83 RA. For gas turbine parts, impulse blades; heat resistant.

NIMONIC DS

Henry Wiggin & Co. Ltd.
C 0-0.15, Cr 17-19, Si 2-2.5, Mn 1.2, Ni 36-39, bal Fe.
At 20 C: 105,300 TS; 38 El; 50 RA. At 800 C: 23,500 TS; 83 El; 71 RA. At 1000 C: 12,400 TS; 124 El; 83 RA. For woven furnace belts, radiant tubes, heat treat equipment; oxidation resistant to 950 C. *Obsolete*

NIMONIC DT

Henry Wiggin & Co. Ltd.
Ni alloy.
For high temperature applications; heat resistant. *Obsolete*

NIMONIC F

Mond Nickel Co. Ltd.
C 0.08-0.15, Ti 0.2-0.6, Fe 5-11, Cr 18-21, bal Ni.
For gas turbine flame tubes; oxidation and heat resistant.

NIMONIC M14V

Henry Wiggin & Co. Ltd.
Nickel alloy.
For high temperature applications; corrosion and heat resistant. *Obsolete*

NIMONIC M15V

Henry Wiggin & Co. Ltd.
Nickel alloy.
For high temperature applications; corrosion and heat resistant. *Obsolete*

NIMONIC M17V

Henry Wiggin & Co. Ltd.
Nickel alloy.
For high temperature applications; corrosion and heat resistant. *Obsolete*

NIMONIC M4VC

Henry Wiggin & Co. Ltd.
Nickel alloy.
For high temperature applications; high temperature resistance. *Obsolete*

NIMONIC M6VC

Henry Wiggin & Co. Ltd.
Nickel alloy.
For high temperature applications; high temperature resistance. *Obsolete*

NIMONIC MC 57

Mond Nickel Co. Ltd.
C 0.08, Co 16, Cr 20, Al 0.9, Ti 1.6, bal Ni.
Heat treated: 180,000 TS; 65,000 YS; 16 El. For high temperature castings, gas turbine components. Heat and corrosion resistant. Good properties to 650°C.

NIMONIC PE 11

Mond Nickel Co. Ltd.
Cr 18, Mo 5.2, Ti 2.3, Al 0.8, Ni 38, bal Fe.
Heat treated: 160,000 TS; 100,000 YS; 21 El; 35 RA. At 600°C: 140,000 TS; 100,000 YS; 23 El; 34 RA. For gas turbine thrust reversers, noise suppressors, jet pipes. Creep resistant. Heat and corrosion resistant; age-hardenable.

NIMONIC PE 13

Mond Nickel Co. Ltd.
C 0.05-0.15, Cr 20.5-23, Fe 17-20, Co 0.5-2.5, Mo 8-10, W 0.2-1, bal Ni.
Annealed: 116,500 TS; 51,500 YS; 43 El; 200 Vickers. At 1000°C; 20,200 TS; 13,500 YS; 44 El. For elevated temperature applications, missile and jet engine components, furnace parts. High temperature, heat resistant.

NIMONIC PE 13
Henry Wiggin & Co. Ltd.
C 0.05-0.15, Cr 20.5-23, Fe 17-20, Co 0.5-2.5, Mo 8-10, W 0.2-1, bal Ni.
Annealed: 116,500 TS; 51,500 YS; 43 El; 200 VPN. At 1000°C; 20,200 TS; 13,500 YS; 44 El. For elevated temperature applications, missile and jet engine components, furnace parts. High temperature, heat resistant.

NIMONIC PE 16
Mond Nickel Co. Ltd.
Cr 16.5, Co 0-2, Mo 3.2, Ti 1.2, Al 1.2, Ni 43.5, bal Fe.
Heat Treated: 121,000 TS; 65,000 YS; 29 El. At 600°C: 99,000 TS; 54,000 YS; 30 El. For gas turbine compressor delivery casings, turbine casings. Creep and oxidation resistant for service up to 750°C., age-hardenable.

NIMONIC PE 16
Henry Wiggin & Co. Ltd.
Cr 16.5, Co 0-2, Mo 3.2, Ti 1.2, Al 1.2, Ni 43.5, bal Fe.
Heat Treated: 121,000 TS; 65,000 YS; 29 El. At 600°C: 99,000 TS; 54,000 YS; 30 El. For gas turbine compressor delivery casings, turbine casings. Creep and oxidation resistant for service up to 750°C., age-hardenable.

NIMONIC PE 7
Mond Nickel Co. Ltd.
C 0.1, Cr 18, Mo 5, Ni 37, Co 0-2, Ti 1.2, Al 1.2, bal Fe.
For power generating equipment, gas turbines, steam plant, casings and compressors. Creep and high temperature resistance. For service up to 580°C.

NIMONIC PE13 (FILLER METAL)
Inco Alloys International Inc.
C 0.1, Mn 1, Fe 17.2, Si 1, S 0.02, Cr 22, Mo 8.1, W 0.6, Pb 0.002, bal Ni.
Filler metal for gas-shielded arc welding NIMONIC alloy PE13. BS 2901-NA-40. *Obsolete*

NIMONIC PK 25
Enpar Sonderwerkstoffe GmbH
Nickel.
Alternate manufacturer.

NIMONIC PK 25
Henry Wiggin & Co. Ltd.
C 19, Co 18, Mo 4.2, Ti 3, Al 2.7, bal Ni.
Heat Treated: 197,000 TS; 18 El; 22 RA. At 1400°F: 155,000 TS; 20 El; 28 RA. For gas turbine blades and other components, bolts, valves, jet engine components. Creep and heat resistant to 940°C.

NIMONIC PK 25
Mond Nickel Co. Ltd.
C 19, Co 18, Mo 4.2, Ti 3, Al 2.7, bal Ni.
Heat Treated: 197,000 TS; 18 El; 22 RA. At 1400°F: 155,000 TS; 20 El; 28 RA. For gas turbine blades and other components, bolts, valves, jet engine components. Creep and heat resistant to 940°C.

NIMONIC PK 31
Mond Nickel Co. Ltd.
Co 14, C 0.06, Cr 20, Mo 4.5, Ti 2.3, Al 0.6, Cb 5, bal Ni.
Heat treated: 190,000 TS; 140,000 YS; 25 El; 34 RA. At 600°C: 160,000 TS; 120,000 YS; 25 El; 34 RA. For rotor discs in gas turbines, jet engine and missile components. High tensile strength and creep resistant. High oxidation and corrosion resistance.

NIMONIC PK 33
Mond Nickel Co. Ltd.
Co 14, C 0.05, Cr 19, Mo 7.5, Ti 2, Al 2, bal Ni.
Heat Treated: 165,000 TS; 95,000 YS; 33 El; 41 RA. At 600°C: 140,000 TS; 88,000 YS; 32 El; 40 RA. For engine rotor and sheet components, gas turbine parts, jet pipes for service up to 950°C. High oxidation and corrosion resistance. Age-hardenable.

NIMONIC PK 33
Henry Wiggin & Co. Ltd.
Co 14, C 0.05, Cr 19, Mo 7.5, Ti 2, Al 2, bal Ni.
Heat Treated: 165,000 TS; 95,000 YS; 33 El; 41 RA. At 600°C: 140,000 TS; 88,000 YS; 32 El; 40 RA. For engine rotor and sheet components, gas turbine parts, jet pipes for service up to 950°C. High oxidation and corrosion resistance. Age-hardenable.

NIMONIC-CB
Mond Nickel Co. Ltd.
Ti 1.8-2.7, Al 0.8-1.8, Fe 0-5, Co 15-21, Cr 18-21, bal Ni.
Heat treated: 106,000 TS; 80,000 YS; 13 El; 290 Vickers. For gas turbine rotor blades. High creep and oxidation resistant. Cast alloy.

NIMOPLY
Henry Wiggin & Co. Ltd.
Clad Nimonic 75.
For gas turbine construction; copper rolled between sheets of Nimonic 75. *Obsolete*

NIO-O-NEL ALLOY 825
Now INCOLOY ALLOY 825.

NIOLOY
Teledyne Ohiocast
C 3, Ni, bal Fe.
For rolls for steel mills; chilled iron. *Obsolete*

NIORO
GTE Products Corp./Wesgo Div.
Au 82, Ni 18.
Brazing alloy available in foil, flexibraze, wire, powder, extrudable paste and preform. Liquidus 1742°F. Solidus 1742°F. Meets AMS-4787; BAu-4.

NIORO
Western Gold & Platinum Co.
Au 82, Ni 18.
For brazing W, Mo, Cu, Ni, and Kovar. Melting point: 950°C. Excellent flow and wetting properties.

NIORONI
GTE Products Corp./Wesgo Div.
Au 73.8, Ni 26.2.
Brazing alloy available in foil, flexibraze, wire, powder, extrudable paste and preform. Liquidus 1850°F. Solidus 1796°F.

NIPERM 50
Vereinigte Edelstahlwerke
C 0.1, Ni 48, Cr, bal Fe.
For soft magnet; high permeability. *Obsolete*

NIPERMAG
Cinaudagraph Corp.
Al 12, Ni 32, Ti 0.4, bal Fe.
Cast: 10,500 TS; 23,000 traverse strength; 45 Rock C; 560 Brin, 660 Hc; 3400 Bo. For loud speakers, motors, generators. Permanent magnet, high permeability.

NIPIGON
Atlas Specialty Steels
C 0.8, W 19, V 2, Cr 4, Mo 1, Co 9, bal Fe.
For tools, dies, cutting tools; high speed steel. *Obsolete*

NIPPERT ALLOY N 4
Now ZIRCONIUM COPPER N-4.

NIPPON EVERSHINING STEEL
Nihon Jyokiko Seikosho Goshi
C 0.15, Cr 18, Ni 8, bal Fe.
For stainless parts; stainless.

NIPPON NST-M1
Nippon Stainless Steel Co. Ltd.
C 0.07-0.12, Mn 0.3-0.6, Si 0-0.5, Cr 12-13.5, Mo 0.4-0.6, Ni 0-0.5, bal Fe.
Annealed: 75,000 TS; 35,000 YS; 25 El; 92 Rock B. For table flatware, springs, oil refinery equipment. Corrosion resistant.

NIPPON NST-M2
Nippon Stainless Steel Co. Ltd.
C 0.1-0.2, Mn 0.2-0.8, Si 0-0.5, Cr 11.5-13, Mo 0.6-1.5, Ni 0-0.5, bal Fe.
Annealed: 75,000 TS; 40,000 YS; 25 El; 92 Rock B. For springs, table flatware, oil refinery equipment, surgical instruments, valves. Corrosion resistant, hardenable.

NIPURE
Wilbur B. Driver Co.
Nickel. Si 0-0.01, Fe 0-0.05, Mn 0-0.02, Mg 0-0.01, Al 0-0.01, Ti 0-0.01, Cr 0-0.05, Pb 0-0.01, Cu 0-0.04, C 0-0.02, bal Ni.
For electronic components. Vacuum melted high purity nickel.

NIRANIUM
Manufacturer not listed
Cr 28.8, Co 64.2, Ni 4.3, W 2, Si 0.1, C 0.2, Al 0.7.
At 1000°C: 82,000 TS; 43,000 YS; 0.7 El. At 700°C: 85,000 TS; 47,000 YS; 1.3 El.

NIRESIST
Russian manufacture
C 2.7, Cr 3-3.4, Ni 12-15, Cu 5-8, Si 1.5, bal Fe.
Cast: 20,000-35,000 TS; 125-200 Brin. For heat and corrosion resistant parts; cast iron.

NIRESULT
Michigan Steel Casting Co.
Cr 24-30, Ni 12-15, Cu 5-7, 1.1 C min, bal Fe.
For heat and corrosion resistant parts; heat and corrosion resistant.

NIREX
Harrison Alloys Inc.
Nickel.
Now HAI-66 ALLOY.

NIROMET 36
Wilbur B. Driver Co.
Nickel. Ni 36, Fe 64.
Wire. Annealed: 70,000 TS. Cold worked: 150,000 TS. For bimetals, precision springs, time devices. Low expansion.

NIROMET 42
Wilbur B. Driver Co.
Nickel. Ni 42, bal Fe.
Bar: 70,000-150,000 TS; 50,000 YS; 35 El. For glass to metal hermetic sealing, leads and terminals for resistors. Matches Corning 1075 glass. Controlled expansion.

NIROMET 426
Wilbur B. Driver Co.
Nickel. Ni 42, Cr 6, bal Fe.
Wire: 80,000-150,00 TS. For glass to metal seals (Corning 0120 glass). Controlled expansion.

NIROMET 44
Wilbur B. Driver Co.
Nickel. Ni 44, Fe 56.
Wire, annealed: 70,000 TS; cold worked: 150,00 TS. For special glass sealing and fiber optics. Controlled expansion.

NIROMET 46
Wilbur B. Driver Co.
Nickel. Ni 46, bal Fe.
Bar: 70,000-150,000 TS; 50,000 YS; 35 El. For terminal bands in vitreous enameled resistors, cores and armatures for relays, motors, transformers; controlled expansion alloy.

NIROMET 48
Wilbur B. Driver Co.
Nickel. Ni 48, Fe 52.
Wire: 70,000-150,000 TS. For glass to metal seals. Controlled expansion.

NIRON 46
Wilbur B. Driver Co.
Ni 46, Fe 54
Annealed: 70,000 TS; 50,000 YS; 35 El. For terminal bands for vitreous enamel resistors; controlled expansion. *Obsolete*

NIRON 52

Wilbur B. Driver Co.
Nickel. Ni 51, Fe 49.
Annealed: 70,000 TS; 50,000 YS; 35 El. For glass to metal seals; controlled expansion.

NIRONITE

Gulf & Western Mfg. Co.
C, alloy, bal Fe.
For rolls for strip mills; heavy duty.

NIRONITE B

Gulf & Western Mfg. Co.
C 3.2, Si 2, Cr 1, bal Fe.
For rolling mill casting; special iron alloy.

NIRONITE C

Gulf & Western Mfg. Co.
C, alloy, bal Fe.
For rolls for bar and billet mills; heavy duty.

NIRONITE D

Gulf & Western Mfg. Co.
C, alloy, bal Fe.
For rolls for strip mills; heavy duty.

NIRONZE 635

Olin Brass, Indianapolis
Ni 1.9, Si 0.6, bal Cu.
Annealed: 40,000 psi TS; 12,000 psi YS; 50 El; 90 RA; 56 Brin. Aged: 88,000 psi TS; 70,000 psi YS; 12 El; 20 RA; 170 Brin. For cold headed bolts, fasteners, switch gear, and marine hardware; age-hardenable, corrosion resistant.
Obsolete

NIROSAD

Metalltechnik Schmidt GmbH & Co.
Cr 18, Ni 10.5, Mn 1.1, Si 1, Mo 0.3, C 0.1, bal Fe.
Blasting shot; austenitic with controlled amounts of martensite.

NIROSTA

Now B & W NIROSTA.

NIROSTA

Krupp Stahl AG
C 0-0.08, Si 0-1, Mn 0-2, P 0-0.045, S 0-0.03, Cr 17-19, Ni 9-12, Ti 0-0.8, bal Fe.
DIN 17440/85; W. Nr. 1.4541.

NIROSTA

Krupp Stahl AG
C 0-0.03, Si 0-1, Mn 0-2, P 0-0.045, S 0-0.03, Cr 16.5-18.5, Ni 6-8, N 0.1-0.15, bal Fe.
SEW 400/88; W. Nr. 1.4318.

NIROSTA 18-8

Michiana Products Corp.
Si 0.6-1.25, Ni 8-10, Cr 16-19, C 0-0.35, bal Fe.
Corrosion resisting articles; now Michiana No. 49.

NIROSTA 19-9-4

Michiana Products Corp.
Si 0.6-1.25, Ni 8-12, Cr 18-20, C 0.35, Mo 2-4, bal Fe.
For hot sulfuric acid tanks; corrosion resistant and hot H_2SO_4 resistant, under pressure.

NIROSTA 25-20

Michiana Products Corp.
Si 0-1.5, Ni 18-22, Cr 24-27, C 0-0.35, bal Fe.
Applicable to temperature and corrosion work where temperature exceeds 1400°F; heat and corrosion resistant up to 1400°F.

NIROSTA 4000

Krupp Stahl AG
Stainless Steel. C 0-0.08, Cr 12-14, bal Fe.
Ferritic. DIN X6Cr13; W. Nr. 1.4000. Softened: 250 N/mm² YS, 400-600 N/mm² TS, 20 El. Hardened/tempered: 420 N/mm² YS, 600-800 N/mm² TS, 18 El.

NIROSTA 4001

Krupp Stahl AG
C 0-0.08, Si 0-1, Mn 0-1, P 0-0.045, S 0-0.03, Cr 13-15, bal Fe.
SEL 90; W. Nr. 1.4001.

NIROSTA 4002

Krupp Stahl AG
Stainless Steel. C 0-0.08, Cr 12-14, Al 0.1-0.3, bal Fe.
Ferritic. DIN X6CrAl13; W. Nr.1.4002. Softened: 250 N/mm² YS, 400-600 N/mm² TS, 20 El. Hardened/tempered: 400 N/mm² YS, 550-700 N/mm² TS, 18 El.

NIROSTA 4003

Krupp Stahl AG
C 0-0.03, Si 0-1, Mn 0.5-1.5, P 0-0.045, S 0-0.03, Cr 10.5-12.5, Ni 0.3-1, N 0-0.03, bal Fe.
SEW 400/88; W. Nr. 1.4003.

NIROSTA 4005

Krupp Stahl AG
Stainless Steel. C 0-0.15, Cr 12-13, S 0.15-0.35, bal Fe.
Martensitic. DIN X12CrS13; W. Nr. 1.4005. Softened: 0-720 N/mm² TS. Hardened/tempered: 440 N/mm² YS, 620-820 N/mm² TS, 12 El.

NIROSTA 4006

Krupp Stahl AG
Stainless Steel. C 0.08-0.12, Cr 12-14, bal Fe.
Martensitic. DIN X10Cr13; W. Nr. 1.4006. Softened: 250 N/mm² YS, 450-650 N/mm² TS, 20 El. Hardened/tempered: 420 N/mm² YS, 600-800 N/mm² TS, 15 El.

NIROSTA 4013

Krupp Stahl AG
C 0-0.1, Si 0-0.5, Mn 0-0.5, P 0-0.045, S 0-0.03, Cr 15-16, Ni 0-0.04, bal Fe.
SEL 90; W. Nr. 1.4013.

NIROSTA 4016

Krupp Stahl AG
Stainless Steel. C 0-0.08, Cr 15.5-17.5, bal Fe.
Ferritic. DIN X6Cr17; W. Nr. 1.4016. Softened: 270 N/mm² YS, 450-600 N/mm² TS, 20 El.

NIROSTA 4021

Krupp Stahl AG
Stainless Steel. C 0.17-0.25, Cr 12-14, bal Fe.
Martensitic. DIN X20Cr13; W. Nr. 1.4021. Softened: 0-740 N/mm² TS. Hardened/tempered in oil at 1000 + 700 C: 450 N/mm² YS, 650-800 N/mm² TS, 14 El. Hardened/tempered in oil at 1000 + 650 C: 550 N/mm² YS, 750-900 N/mm² TS, 12 El.

NIROSTA 4024

Krupp Stahl AG
Stainless Steel. C 0.12-0.17, Cr 12-14, bal Fe.
Martensitic. DIN X15Cr13; W. Nr. 1.4024. Softened: 0-720 N/mm² TS. Hardened/tempered: 450 N/mm² YS, 650-800 TS, 14 El.

NIROSTA 4028

Krupp Stahl AG
Stainless Steel. C 0.28-0.35, Cr 12-14, bal Fe.
Martensitic. DIN X30Cr13; W. Nr. 1.4028. Softened: 0-780 N/mm² TS. Hardened/tempered: 600 N/mm² YS, 800-1000 N/mm² TS, 11 El.

NIROSTA 4028

Krupp Stahl AG
C 0.58-0.7, Si 0-1, Mn 0-1, P 0-0.045, S 0-0.03, Cr 12.5-14.5, bal Fe.
SEW 400/88; W. Nr. 1.4037.

NIROSTA 4031

Krupp Stahl AG
Stainless Steel. C 0.35-0.42, Cr 12.5-14.5, bal Fe.
Martensitic. DIN X38Cr13; W. Nr. 1.4031. Softened: 0-800 N/mm² TS.

NIROSTA 4034

Krupp Stahl AG
Stainless Steel. C 0.42-0.5, Cr 12.5-14.5, bal Fe.
Martensitic. DIN X46Cr13; W. Nr. 1.4034. Softened: 0-840 N/mm² TS. As ball bearing steel W. Nr. 1.3541.

NIROSTA 4057

Krupp Stahl AG
Stainless Steel. C 0.14-0.23, Cr 15.5-17.5, Ni 1.5-2.5, bal Fe.
Martensitic. DIN X20CrNi172; W. Nr. 1.4057. Softened: 0-950 N/mm² TS. Hardened/tempered: 550 N/mm² YS, 750-950 N/mm² TS, 12 El.

NIROSTA 4104

Krupp Stahl AG
Stainless Steel. C 0.1-0.17, Cr 15.5-17.5, Mo 0.2-0.6, S 0.15-0.35, bal Fe.
Martensitic. DIN X12CrMoS17; W. Nr. 1.4104. Softened: 540-740 N/mm² TS, 16 El. Hardened/tempered: 450 N/mm² YS, 640-840 N/mm² TS, 11 El.

NIROSTA 4105

Krupp Stahl AG
Stainless Steel. C 0-0.06, Cr 16.5-18.5, Mo 0.2-0.6, S 0.15-0.35, bal Fe.
Ferritic. DIN X4CrMoS18; W. Nr. 1.4105. Softened: 270 N/mm² YS, 450-650 N/mm² TS, 20 El.

NIROSTA 4106

Krupp Stahl AG
C 0.08-0.13, Si 0-1, Mn 0-1, P 0-0.045, S 0-0.03, Cr 11.5-13.5, Mo 0.4-0.6, Ni 0.5-1, bal Fe.
SEL 81; W. Nr. 1.4106.

NIROSTA 4108

Krupp Stahl AG
C 1-1.1, Si 0-1, Mn 0-1, P 0-0.045, S 0-0.03, Cr 12-14, Mo 0.4-0.6, bal Fe.
SEL 81; W. Nr. 1.4108.

NIROSTA 4109

Krupp Stahl AG
C 0.6-0.75, Si 0-1, Mn 0-1, P 0-0.045, S 0-0.03, Cr 13-15, Mo 0.5-0.6, bal Fe.
SEL 90; W. Nr. 1.4109.

NIROSTA 4110

Krupp Stahl AG
C 0.48-0.6, Si 0-1, Mn 0-1, P 0-0.045, S 0-0.03, Cr 13-15, Mo 0.5-0.8, V 0-0.15, bal Fe.
SEW 400/88; W. Nr. 1.4110.

NIROSTA 4112

Krupp Stahl AG
Stainless Steel. C 0.85-0.95, Cr 17-19, Mo 0.9-1.3, V 0.07-0.12, bal Fe.
Martensitic. DIN X90CrMoV18; W. Nr. 1.4112.

NIROSTA 4113

Krupp Stahl AG
C 0-0.08, Si 0-1, Mn 0-1, P 0-0.045, S 0-0.03, Cr 16-18, Mo 0.9-1.3, bal Fe.
DIN 17441/85; W. Nr. 1.4113.

NIROSTA 4116

Krupp Stahl AG
Stainless Steel. C 0.42-0.5, Cr 13.8-15, Mo 0.45-0.6, V 0.1-0.15, bal Fe.
Martensitic. DIN X45CrMoV15; W. Nr. 1.4116. Softened: 0-900 N/mm² TS.

NIROSTA 4117

Krupp Stahl AG
Stainless Steel. C 0.35-0.4, Cr 14-15, Mo 0.4-0.6, V 0.1-0.15, bal Fe.
Martensitic. DIN X38CrMoV15; W. Nr. 1.4117. Softened: 0-900 N/mm² TS.

NIROSTA 4120

Krupp Stahl AG

Stainless Steel. C 0.17-0.22, Cr 12-14, Mo 0.9-1.3, Ni 0-0.8, bal Fe.

Martensitic. DIN X20CrMo13; W. Nr. 1.4120. Softened: 0-770 N/mm^2 TS. Hardened/tempered: 550 N/mm^2 YS, 750-900 N/mm^2 TS, 14 El.

NIROSTA 4122

Krupp Stahl AG

Stainless Steel. C 0.33-0.45, Cr 15.5-17.5, Mo 0.8-1.3, Ni 0-0.8, bal Fe.

Martensitic. DIN X35CrMo17; W. Nr. 1.4122. Softened: 0-900 N/mm^2 TS. Hardened/tempered: 550 N/mm^2 YS, 750-950 N/mm^2 TS, 12 El.

NIROSTA 4125

Krupp Stahl AG

Stainless Steel. C 0.95-1.2, Cr 16-18, Mo 0.4-0.8, bal Fe.

Martensitic. DIN 105CrMo17; W. Nr. 1.4125.

NIROSTA 4126

Krupp Stahl AG

C 1.05-1.15, Si 0-1, Mn 0-1, P 0-0.045, S 0-0.03, Cr 17-18, Mo 0.8-1, bal Fe.

SEL 81; W. Nr. 1.4126.

NIROSTA 4300

Krupp Stahl AG

C 0-0.12, Si 0-1, Mn 0-2, P 0-0.045, S 0-0.03, Cr 17-19, Ni 8-10, bal Fe.

SEL 75; W. Nr. 1.4300.

NIROSTA 4301

Krupp Stahl AG

Stainless Steel. C 0-0.07, Cr 17-19, Ni 8.5-10.5, bal Fe.

Austenitic. DIN X5CrNi1810; W. Nr. 1.4301. 195 N/mm^2 YS, 500-700 N/mm^2 TS, 45 El. For cold drawn bars with diameter 4.0 and 20 mm max, a max TS of 850 N/mm^2 and A$_5$ 20% max is permissible.

NIROSTA 4303

Krupp Stahl AG

Stainless Steel. C 0-0.07, Cr 17-19, Ni 11-13, bal Fe.

Austenitic. DIN X5CrNi1812; W. Nr. 1.4303. 185 N/mm^2 YS, 490-690 N/mm^2 TS, 45 El.

NIROSTA 4305

Krupp Stahl AG

Stainless Steel. C 0-0.12, Cr 17-19, Ni 8-10, S 0.15-0.35, bal Fe.

Austenitic. DIN X10CrNiS189; W. Nr. 1.4305. 195 N/mm^2 YS, 500-700 N/mm^2 TS, 35 El.

NIROSTA 4306

Krupp Stahl AG

Stainless Steel. C 0-0.03, Cr 18-20, Ni 10-12.5, bal Fe.

Austenitic. DIN X2CrNi1911; W. Nr. 1.4306. 180 N/mm^2 YS, 460-680 N/mm^2 TS, 45 El.

NIROSTA 4310

Krupp Stahl AG

Stainless Steel. C 0-0.12, Cr 16-18, Mo 0-0.8, Ni 6-9, bal Fe.

Austenitic. DIN X12CrNi177; W. Nr. 1.4310. 260 N/mm^2 YS, 600-950 N/mm^2 TS, 35 El.

NIROSTA 4311

Krupp Stahl AG

Stainless Steel. C 0-0.03, Cr 17-19, Ni 8.5-11.5, N 0.12-0.22, bal Fe.

Austenitic. DIN X2CrNiN1810; W. Nr. 1.4311. 270 N/mm^2 YS, 550-760 N/mm^2 TS, 40 El.

NIROSTA 4313

Krupp Stahl AG

Stainless Steel. C 0-0.05, Cr 12.5-14, Mo 0.4-0.7, Ni 3.5-4.5, 0.02 N min, bal Fe.

Martensitic. DIN X4CrNi134; W. Nr. 1.4313. 550-850 N/mm^2 YS, 760-1200 N/mm^2 TS, 14-17 El.

NIROSTA 4319

Krupp Stahl AG

C 0-0.07, Si 0-1, Mn 0-2, P 0-0.045, S 0-0.03, Cr 16-18, Ni 7-8, bal Fe.

SEL 90; W. Nr. 1.4319.

NIROSTA 4335

Krupp Stahl AG

C 0-0.02, Si 0-0.15, Mn 0-2, P 0-0.025, S 0-0.005, Cr 24-26, Mo 0-0.1, Ni 20-22, bal Fe.

SEW 400/88; W. Nr. 1.4335.

NIROSTA 4401

Krupp Stahl AG

Stainless Steel. C 0-0.07, Cr 16.5-18.5, Mo 2-2.5, Ni 10.5-13.5, bal Fe.

Austenitic. DIN X5CrNiMo17122; W. Nr. 1.4401. 205 N/mm^2 YS, 510-710 N/mm^2 TS, 40 El. For cold drawn bars with diameter less than or equal to 4.0 and 20 mm, a max tensile strength of 850 N/mm^2 and A$_5$ less than or equal to 20% is permissible.

NIROSTA 4404

Krupp Stahl AG

Stainless Steel. C 0-0.03, Cr 16.5-18.5, Mo 2-2.5, Ni 11-14, bal Fe.

Austenitic. DIN X2CrNiMo17132; W. Nr. 1.4404. 190 N/mm^2 YS, 490-690 N/mm^2 TS, 40 El.

NIROSTA 4406

Krupp Stahl AG

Stainless Steel. C 0-0.03, Cr 16.5-18.5, Mo 2-2.5, Ni 10.5-13.5, N 0.12-0.22, bal Fe.

Austenitic. DIN X2CrNiMoN17122; W. Nr. 1.4406. 280 N/mm^2 YS, 580-800 N/mm^2 TS, 40 El.

NIROSTA 4418

Krupp Stahl AG

Stainless Steel. C 0-0.05, Cr 15-16.5, Mo 0.8-1.5, Ni 5-6, 0.02 N min, bal Fe.

Martensitic. DIN X4CrNiMo165; W. Nr. 1.4418. 550-850 N/mm^2 YS, 830-1200 N/mm^2 TS, 14-16 El.

NIROSTA 4420

Krupp Stahl AG

C 0-0.07, Si 0-1, Mn 0-2, P 0-0.045, S 0-0.03, Cr 16.5-18.5, Mo 1.2-1.7, Ni 9-12, bal Fe.

SEL 90; W. Nr. 1.4420.

NIROSTA 4427

Krupp Stahl AG

C 0-0.12, Si 0-1, Mn 0-2, P 0-0.06, S 0.15-0.35, Cr 16.5-18.5, Mo 2-2.5, Ni 10.5-13.5, bal Fe.

SEL 90; W. Nr. 1.4427.

NIROSTA 4428

Krupp Stahl AG

C 0-0.03, Si 0-1, Mn 0-2, P 0-0.025, S 0-0.01, Cr 17-18.5, Mo 2.7-3.2, Ni 13-14.5, N 0.14-0.22, bal Fe.

DIN 17443/86; W. Nr. 1.4428.

NIROSTA 4429

Krupp Stahl AG

Stainless Steel. C 0-0.03, Cr 16.5-18.5, Mo 2.5-3, Ni 11.5-14.5, N 0.14-0.22, S 0-0.025, bal Fe.

Austenitic. DIN X2CrNiMoN17133; W. Nr. 1.4429. 295 N/mm^2 YS, 580-800 N/mm^2 TS, 40 El.

NIROSTA 4435

Krupp Stahl AG

Stainless Steel. C 0-0.03, Cr 17-18.5, Mo 2.5-3, Ni 12.5-15, S 0-0.25, bal Fe.

Austenitic. DIN X2CrNiMo18143; W. Nr. 1.4435. 190 N/mm^2 YS, 490-690 N/mm^2 TS, 35 El.

NIROSTA 4436

Krupp Stahl AG

Stainless Steel. C 0-0.07, Cr 16.5-18.5, Mo 2.5-3, Ni 11-14, S 0-0.025, bal Fe.

Austenitic. DIN X5CrNiMo17133; W. Nr. 1.4436. 205 N/mm^2 YS, 510-710 N/mm^2 TS, 40 El.

NIROSTA 4438

Krupp Stahl AG

C 0-0.03, Si 0-1, Mn 0-2, P 0-0.045, S 0-0.025, Cr 17.5-19.5, Mo 3-4, Ni 14-17, bal Fe.

DIN 17440/85; W. Nr. 1.4438.

NIROSTA 4439

Krupp Stahl AG

Stainless Steel. C 0-0.03, Cr 16.5-18.5, Mo 4-5, Ni 12.5-14.5, N 0.12-0.22, S 0-0.025, bal Fe.

Austenitic. DIN X2CrNiMoN17135; W. Nr. 1.4439. 285 N/mm^2 YS, 580-800 N/mm^2 TS, 35 El.

NIROSTA 4441

Krupp Stahl AG

C 0-0.03, Si 0-1, Mn 0-2, P 0-0.025, S 0-0.01, Cr 17-18.5, Mo 2.7-3.2, Ni 13.5-15.5, N 0-0.1, bal Fe.

DIN 17443/86; W. Nr. 1.4441.

NIROSTA 4442

Krupp Stahl AG

C 0-0.03, Si 0-1, Mn 0-2, P 0-0.025, S 0-0.01, Cr 17-18.5, Mo 3.7-4.2, Ni 14-16, N 0.1-0.2, bal Fe.

DIN 17443/86; W. Nr. 1.4442.

NIROSTA 4449

Krupp Stahl AG

C 0-0.07, Si 0-1, Mn 0-2, P 0-0.045, S 0-0.03, Cr 16-18, Mo 4-5, Ni 12.5-14.5, bal Fe.

SEL 81; W. Nr. 1.4449.

NIROSTA 4460

Krupp Stahl AG

Stainless Steel. C 0-0.05, Cr 25-28, Mo 1.3-2, Ni 4.5-6, N 0.05-0.2, bal Fe.

Ferritic-austenitic. DIN X4CrNiMoN2752; W. Nr. 1.4460. 450 N/mm^2 YS, 600-800 N/mm^2 TS, 20 El.

NIROSTA 4461

Krupp Stahl AG

C 0-0.03, Si 0-0.75, Mn 5.5-7.5, P 0-0.025, S 0-0.01, Cr 21-23, Mo 2.7-3.7, Ni 10-16, N 0.35-0.5, Nb 0.1-0.25, bal Fe.

DIN 17443/86; W. Nr. 1.4461.

NIROSTA 4465

Krupp Stahl AG

C 0-0.02, Si 0-0.7, Mn 0-2, P 0-0.02, S 0-0.015, Cr 24-26, Mo 2-2.5, Ni 22-25, N 0.08-0.16, bal Fe.

SEW 400/88; W. Nr. 1.4465.

NIROSTA 4466

Krupp Stahl AG

C 0-0.02, Si 0-0.4, Mn 1.5-2, P 0-0.02, S 0-0.015, Cr 24.5-25.5, Mo 2-2.3, Ni 21.5-22.5, N 0.1-0.14, bal Fe.

SEW 400/84; W. Nr. 1.4466.

NIROSTA 4467

Krupp Stahl AG

C 0-0.03, Si 0-0.8, Mn 4-6, P 0-0.03, S 0-0.015, Cr 24.5-26.5, Mo 2-3, Ni 3.5-4.5, N 0.3-0.45, W 0-1.5, bal Fe.

SEL 90; W. Nr. 1.4467.

NIROSTA 4505 (1)

Krupp Stahl AG

C 0-0.04, Si 0-0.75, Mn 0-0.75, P 0-0.03, S 0-0.015, Cr 22-24, Mo 2.5-3, Ni 26-28, Cu 2.5-3.5, Ti 0.4-0.7, bal Fe.

SEL 90; W. Nr. 1.4503.

NIROSTA 4505 (2)

Krupp Stahl AG

C 0-0.05, Si 0-1, Mn 0-2, P 0-0.045, S 0-0.015, Cr 16.5-18.5, Mo 2-2.5, Ni 19-21, Cu 1.8-2.2, bal Fe.

SEW 400/88; W. Nr. 1.4505.

NIROSTA 4506

Krupp Stahl AG

C 0-0.07, Si 0-1, Mn 0-2, P 0-0.045, S 0-0.03, Cr 16.5-18.5, Mo 2-2.5, Ni 19-21, Cu 1.8-2.2, bal Fe.

SEL 90; W. Nr. 1.4506.

NIROSTA 4509
Krupp Stahl AG
C 0-0.08, Si 0-1, Mn 0-1, P 0-0.045, S 0-0.02, Cr 17-19, Nb 0.3-1.2, Ti 0.1-0.8, bal Fe.
SEL 90; W. Nr. 1.4509.

NIROSTA 4510
Krupp Stahl AG
C 0-0.08, Si 0-1, Mn 0-1, P 0-0.045, S 0-0.03, Cr 16-18, Ti 0-1.2, bal Fe.
DIN 17440/85; W. Nr. 1.4510.

NIROSTA 4511
Krupp Stahl AG
C 0-0.08, Si 0-1, Mn 0-1, P 0-0.045, S 0-0.03, Cr 16-18, Nb 0-1.2, bal Fe.
DIN 17441/85; W. Nr. 1.4511.

NIROSTA 4512
Krupp Stahl AG
C 0-0.08, Si 0-1, Mn 0-1, P 0-0.045, S 0-0.03, Cr 10.5-12.5, Ti 0-1, bal Fe.
DIN 17441/85; W. Nr. 1.4512.

NIROSTA 4520
Krupp Stahl AG
C 0-0.015, Si 0-0.5, Mn 0-0.5, P 0-0.025, S 0-0.02, Cr 14-16, N 0-0.015, Ti 0.25-0.4, bal Fe.
SEW 400/88; W. Nr. 1.4520.

NIROSTA 4521
Krupp Stahl AG
Stainless Steel. C 0-0.025, Cr 17-19, Mo 1.8-2.3, Ti 0-0.8, 7 X (C + N) less than or equal Ti, bal Fe.
Ferritic. DIN X2CrMoTi182; W. Nr. 1.4521. Softened: 320 N/mm^2 YS, 450-650 N/mm^2 TS, 20 El.

NIROSTA 4522
Krupp Stahl AG
C 0-0.025, Si 0-1, Mn 0-1, P 0-0.045, S 0-0.03, Cr 17-19, Mo 1.8-2.3, Ni 0-0.25, Nb 0.6-1.2, bal Fe.
SEW 400/88; W. Nr. 1.4522.

NIROSTA 4527
Krupp Stahl AG
C 0-0.04, Si 0-1, Mn 0-2, P 0-0.035, S 0-0.01, Cr 26-28, Mo 1-2, Ni 5-6, N 0.15-0.3, Nb 0.15-0.25, V 0.15-0.25, bal Fe.
SEL 90; W. Nr. 1.4527.

NIROSTA 4529
Krupp Stahl AG
C 0-0.02, Si 0-1, Mn 0-2, P 0-0.03, S 0-0.015, Cr 19-21, Mo 6-7, Ni 24-26, Cu 0.5-1.5, N 0.1-0.25, bal Fe.
SEW 400/88; W. Nr. 1.4529.

NIROSTA 4530
Krupp Stahl AG
C 0-0.03, Si 0-1, Mn 1-3, P 0-0.035, S 0-0.01, Cr 23.5-26.5, Mo 3-4, Ni 5-6, N 0.15-0.3, Nb 0.15-0.25, V 0.15-0.25, bal Fe.
SEL 90; W. Nr. 1.4530.

NIROSTA 4532
Krupp Stahl AG
C 0-0.09, Si 0-1, Mn 0-1, P 0-0.03, S 0-0.03, Cr 14-16, Mo 2-2.5, Ni 6.5-7.8, Al 0.75-1.5, bal Fe.
SEL 90; W. Nr. 1.4532.

NIROSTA 4534
Krupp Stahl AG
Stainless Steel. C 0-0.05, Si 0-0.1, Mn 0-0.1, P 0-0.01, S 0-0.008, Cr 12.25-13.25, Ni 7.5-8.5, Mo 2-2.5, Al 0.9-1.35, N 0-0.01, bal Fe.
Precipitation-hardenable; W. Nr. 1.4534. 363 max Brin. S13800

NIROSTA 4535
Krupp Stahl AG
C 0.85-0.95, Si 0-1, Mn 0-1, P 0-0.045, S 0-0.03, Cr 15.5-17.5, Mo 0.4-0.6, Co 1.2-1.8, V 0.2-0.3, bal Fe.
SEL 90; W. Nr. 1.4535.

NIROSTA 4539
Krupp Stahl AG
Stainless Steel. C 0-0.02, Cr 19-21, Mo 4-5, Ni 24-26, Cu 1-2, N 0.04-0.15, bal Fe.
Austenitic special. DIN X1NiCrMoCuN25205; W. Nr. 1.4539. 220 N/mm^2 YS, 520-720 N/mm^2 TS, 40 El.

NIROSTA 4541
Krupp Stahl AG
Stainless Steel. C 0-0.08, Cr 17-19, Ni 9-12, Ti 0-0.8, 5 X C less than or equal to Ti, bal Fe.
Austenitic. DIN X6CrNiTi1810; W. Nr. 1.4541. 200 N/mm^2 YS, 500-730 N/mm^2 TS, 40 El. For cold drawn bars with diameter 4.0 and 20 mm max, a max tensile strength of 850 N/mm^2 and A_5 20% max is permissible.

NIROSTA 4542
Krupp Stahl AG
C 0-0.07, Si 0-1, Mn 0-1, P 0-0.045, S 0-0.03, Cr 15-17, Ni 3-5, Cu 3-5, Nb 0.15-0.45, bal Fe.
SEL 90; W. Nr. 1.4542.

NIROSTA 4542 OR 4548
Krupp Stahl AG
Stainless Steel. C 0-0.07, Si 0-1, Mn 0-1, P 0-0.035, S 0-0.025, Cr 15-16.5, Ni 3-5, Cu 3-5, Nb 0.15-0.45, bal Fe.
Precipitation-hardenable; W. Nr. 1.4542 or 1.4548. 363 max Brin. S17400

NIROSTA 4545
Krupp Stahl AG
Stainless Steel. C 0-0.07, Si 0-1, Mn 0-1, P 0-0.03, S 0-0.015, Cr 14-15.5, Ni 3.5-5.5, Mo 0-0.5, Cu 2.5-4.5, 5 X C to 0.45 Nb, bal Fe.
Precipitation-hardenable; W. Nr. 1.4545. 363 max Brin. S15500

NIROSTA 4550
Krupp Stahl AG
Stainless Steel. C 0-0.08, Cr 17-19, Ni 9-12, Nb 0-1, 10 X C less than or equal to Nb, bal Fe.
Austenitic. DIN X6CrNiNb1810; W. Nr. 1.4550. 205 N/mm^2 YS, 510-740 N/mm^2 TS, 40 El.

NIROSTA 4553
Krupp Stahl AG
C 0-0.04, Si 0-1, Mn 0-2, P 0-0.035, S 0-0.02, Cr 17-19, Ni 9-12, Co 0-0.2, Nb 0-0.65, bal Fe.
SEL 90; W. Nr. 1.4553.

NIROSTA 4558
Krupp Stahl AG
C 0-0.03, Si 0-0.7, Mn 0-1, P 0-0.02, S 0-0.015, Cr 20-23, Ni 32-35, Al 0.15-0.45, N 0-0.015, Ti 0-0.6, bal Fe.
SEW 400/88; W. Nr. 1.4558.

NIROSTA 4561
Krupp Stahl AG
C 0-0.02, Si 0-0.5, Mn 0-2, P 0-0.035, S 0-0.015, Cr 17-18.5, Mo 2-2.5, Ni 11.5-13.5, Ti 0.4-0.6, bal Fe.
SEL 90; W. Nr. 1.4561.

NIROSTA 4563
Krupp Stahl AG
Stainless Steel. C 0-0.02, Cr 26-28, Mo 3-4, Ni 30-32, Cu 0.8-1.5, N 0.04-0.15, bal Fe.
Austenitic special. DIN X1NiCrMoCuN31274; W. Nr. 1.4563. 215 N/mm^2 YS, 500-750 N/mm^2 TS, 40 El.

NIROSTA 4565
Krupp Stahl AG
Stainless Steel. C 0-0.03, Cr 21-25, Mo 3-4.5, Ni 15-18, Mn 4.5-6.5, Nb 0-0.3, N 0.3-0.5, bal Fe.
Austenitic special. DIN X2CrNiMnMoNNb231753; W. Nr. 1.4565. 420 N/mm^2 YS, 800-1000 N/mm^2 TS, 35 El.

NIROSTA 4567
Krupp Stahl AG
Stainless Steel. C 0-0.05, Cr 17-19, Ni 8-10, Cu 3-4, bal Fe.
Austenitic. DIN X3CrNiCu189; W. Nr. 1.4567. 195 N/mm^2 YS, 470-670 N/mm^2 TS, 45 El.

NIROSTA 4568
Krupp Stahl AG
Stainless Steel. C 0-0.09, Si 0-1, Mn 0-1, P 0-0.035, S 0-0.015, Cr 16.5-17.5, Ni 6.75-7.75, Al 0.75-1.35, bal Fe.
Precipitation-hardenable; W. Nr. 1.4568, 229 max Brin. S17700

NIROSTA 4571
Krupp Stahl AG
Stainless Steel. C 0-0.08, Cr 16.5-18.5, Mo 2-2.5, Ni 10.5-13.5, Ti 0-0.8, 5 x C less than or equal to Ti, bal Fe.
Austenitic. DIN X6CrNiMoTi17122; W. Nr. 1.4571. 210 N/mm^2 YS, 500-730 N/mm^2 TS, 35 El. For cold drawn bars with a diameter less than or equal to 4.0 and less than or equal to 20 mm, a max TS of 850 N/mm^2 and A_5 less than or equal to 20% is permissible.

NIROSTA 4573
Krupp Stahl AG
C 0-0.1, Si 0-1, Mn 0-2, P 0-0.045, S 0-0.03, Cr 16.5-18.5, Mo 2.5-3, Ni 12-14.5, bal Fe.
SEL 81; W. Nr. 1.4573.

NIROSTA 4577
Krupp Stahl AG
C 0-0.04, Si 0-0.5, Mn 0-2, P 0-0.03, S 0-0.015, Cr 24-26, Mo 2-2.5, Ni 24-26, Ti 0-0.6, bal Fe.
SEW 400/88; W. Nr. 1.4577.

NIROSTA 4580
Krupp Stahl AG
C 0-0.08, Si 0-1, Mn 0-2, P 0-0.045, S 0-0.03, Cr 16.5-18.5, Mo 2-2.5, Ni 10.5-13.5, Nb 0-1, bal Fe.
DIN 17440/85; W. Nr. 1.4580.

NIROSTA 4582
Krupp Stahl AG
C 0-0.06, Si 0-1, Mn 0-2, P 0-0.045, S 0-0.03, Cr 24-26, Mo 1.3-2, Ni 6.5-7.5, bal Fe.
SEL 90; W. Nr. 1.4582.

NIROSTA 4586
Krupp Stahl AG
C 0-0.07, Si 0-1, Mn 0-2, P 0-0.045, S 0-0.03, Cr 16.5-18.5, Mo 3-3.5, Ni 21.5-23.5, Cu 1.5-2, bal Fe.
SEL 90; W. Nr. 1.4586.

NIROSTA 4589
Krupp Stahl AG
C 0-0.08, Si 0-1, Mn 0-1, P 0-0.045, S 0-0.03, Cr 13.5-15.5, Mo 0.2-1.2, Ni 1-2.5, Ti 0.3-0.5, bal Fe.
SEL 90; W. Nr. 1.4589.

NIROSTA 4594
Krupp Stahl AG
Stainless Steel. C 0-0.07, Si 0-0.7, Mn 0-1, P 0-0.035, S 0-0.025, Cr 13.2-14.7, Ni 5-6, Mo 1.2-2, Cu 1.2-2, Nb 0.2-0.5, bal Fe.
Precipitation-hardenable; W. Nr. 1.4594. AISI XM-25. 380 max Vickers.

NIROSTA KA2
Associated Steel Corp.
C 0-0.16, Mn 0-0.5, Si 0.5, Cr 17-20, Ni 7-10, bal Fe. 85,000-95,000 TS; 30,000-40,000 YS; 60-55 El; 75-50 RA; 135-150 Brin. For cutlery, stainless articles; non-magnetic, austenitic; scaling point 1700°F.

NIROSTA KA2
Krupp Huttenwerke
C 0-0.16, Mn 0-0.5, Si 0.5, Cr 17-20, Ni 7-10, bal Fe. 85,000-95,000 TS; 30,000-40,000 YS; 60-55 El; 75-50 RA; 135-150 Brin. For cutlery, stainless articles; non-magnetic, austenitic; scaling point 1700°F.

NIROSTA KA4
Michigan Standard Alloy Inc.
C 0.25, Mn 0.5, Si 0.5-0.75, Cr 17-20, Ni 8-10, Mo 2-4, bal Fe.
At 70°F: 105,000 TS; 49,000 YS; 55 El; 62 RA. At 1110°F:
70,000 TS; 22,000 YS; 30 El. At 1830°F: 9000 TS; 4500 YS;
47 El. For pipe fittings, pump castings, impellers, shafts; high
corrosion resistance; austenitic; same as "Standard alloy
KA4".

NIROSTA VK5M
Krupp Stahl AG
stainless steel. C 0.2, Cr 13, Mo 1.2, bal Fe.
Annealed: 95,000 TS; 40,000 YS; 25 El; 55 RA; 92 Rock B.
Cold Drawn: 105,000 TS; 85,000 YS; 17 El; 50 RA; 95 Rock B.
Heat Treated: 250,000 TS; 215,000 YS; 8 El; 25 RA; 50 Rock
C. For cutlery, surgical instruments, oil refinery equipment,
valves. Corrosion resistant, hardenable. *Obsolete*

NIROSTA VK7M
Krupp Stahl AG
stainless steel. C 0.15, Cr 13, Mo 1.2, bal Fe.
Annealed: 80,000 TS; 40,000 YS; 25 El; 93 Rock B. Heat
Treated: 200,000 TS; 160,000 YS; 12 El; 45 Rock C. For
surgical instruments, cutlery, oil refining equipment, table
flatware. Corrosion resistant, hardenable. *Obsolete*

NIRUS TYPE 1
Spencer Clark Metal Industries Ltd.
C 0.25, Mn 0.65, Si 0.25, Cr 3.25, Ni 23.5, bal Fe.
Nonmagnetic steel. 35 tsi TS min; 18 tsi YS min. For bus
bars, studs, nuts, bolts.

NIRUS TYPE 3
Spencer Clark Metal Industries Ltd.
C 0.25, Mn 1.25, Si 0.25, Cr 0.15, Ni 25, bal Fe.
Nonmagnetic steel, similar to NIRIUS TYPE 1, but is readily
weldable.

NISILOY
International Nickel Inc.
Nickel. Ni 60, Si 30, Fe 10.
For iron inoculant; for gray iron castings; MP 1800°F.
Obsolete

NISIMAZ
Creusot-Loire
Ni 21, Cr 2.5, Si 7, C, bal Fe.
For chemical plant equipment; resists HC1 and H 2 SO 4 .
Obsolete

NISPAN C 902
Criterion Metals, Inc.
Nickel. Ni 42, C 0.02, Mn 0.4, Fe 48.5, S 0.008, Si 0.5, Cu
0.05, Cr 0.54, Ti 2.4, Al 0.65.
Thin gauge sheet, various tempers.

NIT CA1
Ambo-Stahl-Gesellschaft
C 0.27, Al 1.1, Cr 1.4, bal Fe.
For oil refinery equipment; creep resistant. *Obsolete*

NIT CA2
Ambo-Stahl-Gesellschaft
C 0.34, Al 1.1, Cr 1.4, bal Fe.
For oil refinery equipment; creep resistant. *Obsolete*

NIT CMA
Ambo-Stahl-Gesellschaft
C 0.32, Al 1.1, Cr 1.1, Mo 0.18, bal Fe.
For oil refinery equipment; creep resistant. *Obsolete*

NIT CMA
Ambo-Stahl-Gesellschaft
C 0.31, Cr 2.35, Mo 0.18, V 0.13, bal Fe.
For gears, shafts, crankshafts, bolts, studs, oil hardened,
shock resistant. *Obsolete*

NIT CMNA
Ambo-Stahl-Gesellschaft
C 0.33, Al 1.1, Cr 1.7, Ni 1, bal Fe.
For oil refinery equipment; creep resistant. *Obsolete*

NITAL
CCS Braeburn Alloy Steel
C 0.27, Cr 3, V 0.2, W 10.5, Mo 0.2, Ni 1.5, bal Fe.
For brass extrusion dies; hot work steel, oil hardened.
Obsolete

NITEC 10224
Eutectic Corp.
Nickel base alloy powder for cast iron and steel brazing or
build-up.

NITECTIC 222
Eutectic Corp.
Now NUCLEOTEC 2222. *Obsolete*

NITINOL
American manufacture
Ti 40-45, Ni 55-60.
54.5 Ni: 110,000-124,000 TS; 40,000-55,000 YS; 15.5 El; 16
RA; 42-52 Rock A. For nonmagnetic tools, sensing devices,
chemical plant equipment, space components.
Nonmagnetic, corrosion resistant, hardenable, alloy with a
memory.

NITINOL-55
American manufacture
Ni 55, bal Ti.
Bar: 120,000 TS; 50,000 YS; 15 El; 16 RA; 50 Rock A. For
non-magnetic tools, sensing devices, chemical plant
equipment. Alloy with a memory. Non-magnetic, corrosion
resistant.

NITRALLOY "I"
Republic Steel Corp.
C 0.1-0.2, Mn 0.4-0.7, Si 0.5, Cr 0.9-1.4, Al 0.9-1.4, Mo
0.15-0.3, bal Fe.
For nitrided parts; nitriding steel. *Obsolete*

NITRALLOY "L"
Teledyne Vasco
C 0.7, Mn 0.5, Al 1, Cr 1.25, Mo 0.75, V 0.15, W 1, bal Fe.
For hot working dies; nitriding steel; nitride at 950-975 F.
Obsolete

NITRALLOY 125, TYPE H
Allegheny Ludlum Steel
C 0.23, Mn 0.7-4, Si 0.2-0.4, Al 0.85-1.2, Cr 1.58, Mo 0.2, bal
Fe.
Heat treated; for nitrided gears, shafts, cams, clutches,
rollers; nitriding steel. *Obsolete*

NITRALLOY 135 TYPE G
AL Tech Specialty Steel Corp.
C 0.36, Mn 0.4-0.7, Si 0.2-0.4, Al 0.85-1.2, Cr 1.49, Mo 0.18,
bal Fe.
Heat treated: 224,000-104,000 TS; 180,000-85,000 YS; 11-18
El; 36-59 445-200 Brin. For nitrided gears, shafts, cams,
clutches, rollers; nitriding steel.

NITRALLOY 230
Nitralloy Corp.
C 0.25-0.35, Mn 0.4-0.6, Si 0.2-0.3, Al 1-1.5, Mo 0.6-1, bal Fe.
For nitrided parts, gears, shafts, pinions; nitriding steel;
Nitralloy N.

NITRALLOY 640
Nitralloy Corp.
C 0.35-0.45, Mn 0.4-0.6, Si 0.2-0.3, Cr 1.4-1.6, 0.45 V min, bal
Fe.
For nitrided parts, gears, shafts, pinions; nitriding steel, hard
case.

NITRALLOY ALAMO
Allegheny Ludlum Steel
C 0.25-0.35, Al 1-1.4, Mo 0.6-1, bal Fe.
For gears, shafts, camshafts, connecting rods; nitriding steel.
Obsolete

NITRALLOY CM
Firth Brown Ltd.
C 0.2-0.4, Si 3, Mn 6, Cr 1, Ni 0.6, Mo 1.2, bal Fe.
For gears, shafts, cams, crankshafts; nitriding steel. *Obsolete*

NITRALLOY EZ
Allegheny Ludlum Steel
C 0.3-0.4, Cr 1.25, Al 1.1, Mo 0.2, 0.15-0.25 Se or S, bal Fe.
Heat treated: 155,000-200,000 TS; 135,000-175,000 YS;
18-16 El; 55-50 RA; 300-400 Brin. For nitrided parts, gears,
shafts; nitriding steel.

NITRALLOY EZ
Bethlehem Steel Corp.
C 0.3-0.4, Cr 1.25, Al 1.1, Mo 0.2, 0.15-0.25 Se or S, bal Fe.
Heat treated: 155,000-200,000 TS; 135,000-175,000 YS;
18-16 El; 55-50 RA; 300-400 Brin. For nitrided parts, gears,
shafts; nitriding steel.

NITRALLOY EZ
Joseph T. Ryerson & Son Inc.
C 0.3-0.4, Cr 1.25, Al 1.1, Mo 0.2, 0.15-0.25 Se or S, bal Fe.
Heat treated: 155,000-200,000 TS; 135,000-175,000 YS;
18-16 El; 55-50 RA; 300-400 Brin. For nitrided parts, gears,
shafts; nitriding steel.

NITRALLOY EZ 505
Allegheny Ludlum Steel
Al 0.85-1.2, Si 0.2-0.4, C, Cr, Mo, bal Fe.
For nitrided parts, gears, shafts; nitriding steel, free-cutting.
Obsolete

NITRALLOY G (135)
Colt Industries
C 0.3-0.4, Mn 0.4-0.7, Cr 0.9-1.4, Mo 0.15-0.25, Al 0.8-1.2, bal
Fe.
Heat treated: 105,000-225,000 TS; 80,000-177,000 YS; 11-28
El; 35-60 RA; 200-445 Brin. For cylinder liners, gears, cams,
camshafts; nitriding steel. *Obsolete*

NITRALLOY G GR. 69
Allegheny Ludlum Steel
C 0.38-0.45, Mn 0.4-0.7, Cr 1.4-1.8, Mo 0.3-0.45, Al 0.95-1.35,
bal Fe.
For nitrided parts, gears, shafts; nitriding steel. *Obsolete*

NITRALLOY GK3
Firth Brown Ltd.
C 0.35, Mn 0.5, Cr 2, Mo 0.25, V 0.15, bal Fe.
For gears, shafts; nitriding steel. *Obsolete*

NITRALLOY GK5
Firth Brown Ltd.
C 0.25, Mn 0.5, Cr 2, Mo 0.25, V 0.15, bal Fe.
For gears, shafts; nitriding steel.

NITRALLOY GK7
Firth Brown Ltd.
C 0.18, Mn 0.5, Cr 2, Mo 0.25, V 0.15, bal Fe.
For gears, shafts; nitriding steel. *Obsolete*

NITRALLOY GR
Nitralloy Corp.
C 1.25-1.5, Mn 0.4-0.6, Si 1.25-1.5, Cr 0.2-0.4, Mo 0.2-0.3, Al
1-1.5, bal Fe.
Rolled: 108,500 TS; 84,000 YS; 17.5 El; 19.4 RA; 363 Brin.
For seal rings, cylinder liners; self lubricating, free-cutting.

NITRALLOY GR. 3
Firth Brown Ltd.
C 0.38, Al 1.1, Mo 0.2, Cr 1.5, bal Fe.
Heat treated: 123,000-145,000 TS. For nitrided rolls,
crankshafts, camshafts, cams, brick press plates, gears.
Nitriding steel, wear resistant. *Obsolete*

NITRALLOY GR. 5
Firth Brown Ltd.
C 0.26-0.35, Cr 1.5, Mo 0.2, Al 1.1, bal Fe.
Heat treated: 90,000-125,000 TS. For pump rods, camshafts,
cams, crankshafts, gears. Nitriding steel. Wear resistant.
Obsolete

NITRALLOY H (125)

Colt Industries
C 0.2-0.3, Cr 0.9-1.4, Mo 0.15-0.25, Al 0.85-1.2, bal Fe.
Heat treated: 102,000-178,000 TS; 85,000-155,000 YS; 12-26 El; 46-70 RA; 225-400 Brin. For cylinder liners, bushings, shafts, gears; nitriding steel. *Obsolete*

NITRALLOY H70

Allegheny Ludlum Steel
C 0.25, Cr 1.2, Al 1.1, Mo 0.2, bal Fe.
For nitrided parts; nitriding steel. *Obsolete*

NITRALLOY HCM3

Firth Brown Ltd.
C 0.4, Mn 0.5, Ni 0.3, Cr 3, Mo 1, V 0.25, bal Fe.
For gears, shafts; nitriding steel.

NITRALLOY HCM5

Firth Brown Ltd.
C 0.3, Mn 0.45, Ni 0.5, Cr 3, Mo 0.4, bal Fe.
For gears, shafts; nitriding steel.

NITRALLOY HCM7

Firth Brown Ltd.
C 0.2, Mn 0.45, Ni 0.5, Cr 3, Mo 0.4, bal Fe.
For gears, shafts; nitriding steel.

NITRALLOY LK1

Firth Brown Ltd.
C 0.5, Mn 0.65, Cr 1.6, Al 1.1, Mo 0.2, bal Fe.
For gears, shafts; nitriding steel. *Obsolete*

NITRALLOY LK3

Firth Brown Ltd.
C 0.4, Mn 0.65, Cr 1.6, Mo 0.2, Al 1.1, bal Fe.
For gears, shafts; nitriding steel.

NITRALLOY LK5

Firth Brown Ltd.
C 0.3, Mn 0.65, Cr 1.6, Mo 0.2, Al 1.1, bal Fe.
For gears, shafts; nitriding steel.

NITRALLOY LK7

Firth Brown Ltd.
C 0.2, Mn 0.65, Cr 1.6, Mo 0.2, Al 1.1, bal Fe.
For gears, shafts; nitriding steel. *Obsolete*

NITRALLOY N

AL Tech Specialty Steel Corp.
C 0.2-0.27, Mn 0.4-0.7, Cr 1-1.3, Mo 0.2-0.3, Ni 3.25-3.75, Al 1.1-1.4, bal Fe.
For nitrided parts; nitriding steel.

NITRALLOY N

Bethlehem Steel Corp.
C 0.2-0.27, Mn 0.4-0.7, Cr 1-1.3, Mo 0.2-0.3, Ni 3.25-3.75, Al 1.1-1.4, bal Fe.
For nitrided parts; nitriding steel.

NITRALLOY NO. 115

Nitralloy Corp.
C 0.11-19, Cr 1.3-1.5, Mo 0.15-0.25, Al 0.85-1.2, bal Fe.
Water quenched and drawn: 90,000-142,000 TS; 65,000-130,000 YS; 18-32 El; 63-77 RA; 175-300 Brin. For nitrided parts, gears, shafts. *Obsolete*

NITRALLOY NO. 115

Teledyne Vasco
C 0.11-19, Cr 1.3-1.5, Mo 0.15-0.25, Al 0.85-1.2, bal Fe.
Water quenched and drawn: 90,000-142,000 TS; 65,000-130,000 YS; 18-32 El; 63-77 RA; 175-300 Brin. For nitrided parts, gears, shafts. *Obsolete*

NITRALLOY NO. 135 CVM

Nitralloy Corp.
C 0.3-0.4, Cr 0.9-1.4, Mo 0.15-0.25, Al 1-1.4, bal Fe.
Oil quenched and drawn: 105,000-225,000 TS; 90,000-180,000 YS; 10-27 El; 37-62 RA; 175-440 Brin. For nitrided parts, gears, shafts; Nitralloy G.

NITRALLOY SPECIAL SULFUR

Republic Steel Corp.
C 0.3-0.4, Mn 0.4-0.7, S 0.08-0.18, Cr 0.9-1.4, Mo 0.15-0.25, Al 0.9-1.4, bal Fe.
For nitrided parts; nitriding steel. *Obsolete*

NITRALLOY, HI-CARBON

Republic Steel Corp.
C 0.55-0.65, Cr, Mo, Al, bal Fe.
For nitrided parts; nitriding steel. *Obsolete*

NITREX

Now CARPENTER NITREX.

NITREX-I

Carpenter Technology Corp.
C 0.38-0.45, Mn 0.6, Si 0.3, Cr 1.4-1.8, Al 0.85-1.2, Mo 0.3-0.45, bal Fe.
Tempered 1100°F: 181,000 TS; 165,000 YS; 15 El; 54 RA; 41 Rock C. Tempered 1000°F: 206,000 TS; 182,000 YS; 13 El. For cylinder liners, gears, cams, thread guides, camshafts. Nitriding steel, wear and abrasion resistant case.

NITREX-II

Carpenter Technology Corp.
C 0.2-0.27, Mn 0.6, Si 0.3, Cr 1-1.5, Al 0.8-1.2, Mo 0.2-0.3, Ni 3.25-3.75, bal Fe.
Tempered 1000°F: 198,000 TS; 191,000 YP; 14 El; 45 RA; 38 Rock C. Tempered 1100°F: 145,000 TS; 130,000 YP; 21 El; 58 RA; 30 Rock C. After nitriding: 190,000 TS; 180,000 YP; 6-15 El; 43 RA; core Rock 41 C; case 94 Rock 15-N. For cylinder liners, bushings, shafts, gears, cams, camshafts. Precipitation hardening nitriding steel.

NITRICAST

Certified Alloy Products Inc.
C 2.5, Si 2.5, Cr 1, Mo 0.2, Al 1, bal Fe.
As nitrided: 50,000 TS; 277 Brin (core) file hard case. For oil well tooling, sleeves, liners, cams. *Obsolete*

NITRICASTIRON

Nitralloy Corp.
TC 2.75, GC 1.89, CC 0.86, Si 2.58, Cr 1.22, V 0.16, Mo 0.24, Al 1.01, bal Fe.
Annealed: 56,000-67,000 TS; 302 Brin. For nitrided cast iron, cams, cylinders, valves; wear resistant.

NITRIDING NTR

Creusot-Loire
C 0.38, Mn 0.6, Si 0-0.4, Al 1, Cr 1.65, Mo 0.32, bal Fe.
For nitrided parts. Similar to Nitralloy 135.

NITRIDING STEEL 125 TYPE H

Colt Industries
C 0.25, Cr 1.15, Mo 0.2, Al 1.15, bal Fe.
Heat treated: 122,000 TS; 103,000 YS; 61 El; 67 RA; 255 Brin.
For nitrided parts, gears, shafts; Nitralloy H. *Obsolete*

NITRIDING STEEL 135 MODIFIED

Colt Industries
Al 1, Cr 1.75, Mo 0.4, 0.4 C, 0.5 Mn, bal Fe.
Heat treated: 158,700 TS; 141,000 YS; 17 El; 56 RA; 320 Brin.
For cylinder barrels, nitrided parts; nitriding steel.

NITRIDING STEEL 135 MODIFIED

Republic Steel Corp.
Al 1, Cr 1.75, Mo 0.4, 0.4 C, 0.5 Mn, bal Fe.
Heat treated: 158,700 TS; 141,000 YS; 17 El; 56 RA; 320 Brin.
For cylinder barrels, nitrided parts; nitriding steel.

NITRIDING STEEL 135 TYPE 6

Colt Industries
C 0.35, Cr 1.15, Mo 0.2, Al 1.15, bal Fe.
Heat treated: 138,000 TS; 120.000 YS; 20 El; 58 RA; 280 Brin.
For nitrided parts, gears, shafts; Nitralloy C. *Obsolete*

NITRIDING STEEL 230

Colt Industries
C 0.25-0.35, Mn 0.4-0.6, Si 0.2-0.3, Al 1-1.5, Mo 0.6-1, bal Fe.
For nitrided parts; nitriding steel. *Obsolete*

NITRIDING STEEL BM

Colt Industries
C 0.3, Cr 1, Mo 0.4, Al 0.8, bal Fe.
For gears, shafts, pinions, cams, camshafts; nitriding steel. *Obsolete*

NITRIDING STEEL N

Colt Industries
C 0.2-0.27, Mn 0.4-0.7, Cr 1-1.3, Mo 0.2-0.3, Ni 3.5, Al 1.1-1.4, bal Fe.
Heat treated: 198,000 TS; 191,000 YS; 14 El; 45 RA; 390 Brin.
For nitrided parts, gears, crankshafts; nitriding steel. *Obsolete*

NITRIX 1470

Saarstahl AG
C 0.31, Cr 2.35, Mo 0.18, V 0.13, bal Fe.
For gears, shafts, crankshafts, bolts, fasteners; oil hardened, shock resistant.

NITRIX 1471

Saarstahl AG
C 0.32, Al 1.1, Cr 1.1, Mo 0.18, bal Fe.
For oil refinery equipment; creep and heat resistant.

NITRIX 1472

Saarstahl AG
C 0.34, Al 1.1, Cr 1.1, bal Fe.
For oil refinery equipment; creep and heat resistant.

NITRIX 1473

Saarstahl AG
C 0.3, Mn 0.6, Cr, V, bal Fe.
For gears, shafts, bolts, fasteners; oil hardened, shock resistant.

NITRIX 65

Saarstahl AG
C 0.27, Al 1.1, Cr 1.4, bal Fe.
For oil refinery equipment; creep and heat resistant.

NITRIX 65A

Saarstahl AG
C 0.3, Al 0.9, Cr 1.2, bal Fe.
For oil refinery equipment, bolts; creep and heat resistant.

NITRIX 71

Saarstahl AG
C 0.32, Al 1.1, Cr 1.1, Mo 0.18, bal Fe.
For oil refinery equipment, bolts, fasteners; creep and heat resistant.

NITRIX 72

Saarstahl AG
C 0.34, Al 1.1, Cr 1.4, bal Fe.
For oil refinery equipment; heat and creep resistant.

NITRIX 73

Saarstahl AG
C 0.31, Cr 2.35, Mo 0.18, V 0.13, bal Fe.
For gears, shafts, crankshafts, bolts, studs; oil hardened, shock resistant.

NITRIX 80

Saarstahl AG
C 0.32, Al 1.1, Mo 0.18, Cr 1.1, bal Fe.
For oil refinery equipment; heat and creep resistant.

NITRO

Carpenter Technology Corp.
C 1, Mn 0.35, Si 0.25, V 0.2, bal Fe.
For drills, taps, springs, cutters. Type W2.

NITROCHROME DG

SMC (Shieldalloy Metallurgical Corp.)
Cr 65, C 0-0.1, Si 0-1.5, 3.0 N min, bal Fe.
Master alloy for ladle additions.

NITROCHROME GS

SMC (Shieldalloy Metallurgical Corp.)
Cr 64, C 0-0.1, Si 0-1.5, 6.0 N min, bal Fe.
Master alloy for ladle additions.

NITRODUR 65

Thyssen Edelstahlwerke AG
C 0.27, Cr 1.4, Al 1.1, bal Fe.
For gears, shafts, machine tool parts; nitriding steel.
Obsolete

NITRODUR 65A

Thyssen Edelstahlwerke AG
C 0.3, Cr 1.2, Al 0.9, P 0.09, S 0.09, bal Fe.
For gears, shafts, machine tool parts; nitriding steel, free-cutting. *Obsolete*

NITRODUR 80

Thyssen Edelstahlwerke AG
C 0.32, Cr 1.1, Mo 0.2, Al 1.1, bal Fe.
For gears, shafts, machine tool parts; nitriding steel.
Obsolete

NITRODUR 81

Thyssen Edelstahlwerke AG
C 0.34, Cr 1.4, Al 1.1, bal Fe.
For gears, shafts, machine tool parts; nitriding steel.
Obsolete

NITRODUR 85

Thyssen Edelstahlwerke AG
C 0.33, Cr 1.7, Ni 1, Al 1.1, bal Fe.
For gears, shafts, machine tool parts; nitriding steel.
Obsolete

NITRODUR 90

Thyssen Edelstahlwerke AG
C 0.3, Cr 2.3, Mo 0.15, V 0.1, bal Fe.
For gears, shafts, machine tool parts; nitriding steel.
Obsolete

NITRODUR 91

Thyssen Edelstahlwerke AG
C 0.3, Cr, V, bal Fe.
For gears, shafts, bolts, pinions, fasteners; oil hardened, shock resistant. *Obsolete*

NITROFIL

IMI Rod & Wire
Al 0.2, Ti 0.2, bal Cu.
For filler rod and wire for nitrogen arc and inert gas shielded metal arc welding of copper.

NITRONIC

Now ARMCO NITRONIC.

NITRONIC 20

Armco
Stainless steel. Cr 22-24, Ni 7-9, C 0-0.38, Mn 1.5-3.5, N 0.28-0.4, bal Fe.
Bar, rod and wire, billets. For elevated temperature use where high mechanical strength and resistance to oxidation and sulfidation are important. Austenitic, age-hardenable by heat treatment.

NITRONIC 30

Armco
Stainless steel. Cr 15-17, Ni 1.5-3, C 0-0.1, Mn 7-9, N 0.15-0.3, bal Fe.
Bar, rod and wire, sheet, strip, billets. Good aqueous corrosion resistance coupled with good resistance to abrasive and metal-to-metal wear. Austenitic, non-hardenable by heat treatment.

NITRONIC 33

Now ARMCO NITRONIC 33.

NITRONIC 50 STAINLESS STEEL

G.O. Carlson Inc.
Stainless steel. C 0-0.06, Mn 4-6, P 0-0.04, S 0-0.03, Si 0-1, Cr 20.5-23.5, Ni 11.5-13.5, Mo 1.5-3, Ni 0.2-0.4, Cb 0.1-0.3, V 0.1-0.3, bal Fe.
Annealed and water quenched (plate), RT: 834 MPa TS; 414 MPa YS (0.2%); 50 El (in 2 in.); 70 RA; 98 Rock B. For use in the petroleum, petrochemical, chemical, pulp and paper, textile, food industries and marine industries.

NITROSA STEEL

Teledyne Firth Sterling
Tool steel. Cr 18, Ni 8, Mn 0.35, C 0.35, bal Fe.
Cold worked: 200,000 TS; 100,000 YS; 13 El. For stainless parts, chemical apparatus; resists corrosion of acids and gases; resists high temperatures to 1650°F. *Obsolete*

NITTANY

Cerro Wire & Cable Company
Zn 24, Pb 2, bal Cu.
For screws, nuts, bolts; free-cutting. *Obsolete*

NITTANY (FREE CUTTING) BRASS

Cerro Metal Products Co.
Cu 61, Pb 3, bal Zn.
Cold drawn: 66,000 TS; 50,000 YS; 29 El; 46 RA; 126 Brin.
For screw machine parts, bolts, and screws. Free-turning yellow brass; CA 360.

NITTANY NO. 2

Cerro Metal Products Co.
Cu 61.5-62, Pb 2.2-2.5, bal Zn.
51,000-68,000 TS; 23,000-37,000 YS; 29-54 El; 46-50 RA; 58-110 Brin. For screw machine parts. Free cutting. *Obsolete*

NITTANY NO. 30

Cerro Metal Products Co.
Cu 63, Pb 1.2, bal Zn.
Rolled: 67-79 Brin. For screw machine products and hardware. Free cutting; CA 350.

NITTANY NO. 31

Cerro Metal Products Co.
Cu 63, Pb 1.75, bal Zn.
Rolled: 67-79 Brin. For screw machine products and hardware. Free cutting; CA 353.

NITTANY NO. 35

Cerro Metal Products Co.
Cu 64, Pb 0.85, bal Zn.
Rolled: 100-107 Brin. For hardware and screw machine products. For spinning and swaging parts. CA 340.

NITTANY NO. 38

Cerro Metal Products Co.
Cu 62, Pb 2.3, bal Zn.
Rolled: 107-125 Brin. For hardware and screw machine products. Free cutting; CA 356.

NITTANY NO. 4

Cerro Metal Products Co.
Cu 62-62.5, Pb 3-3.25, bal Zn.
50,000-65,000 TS; 27,000-37,000 YS; 29-54 El; 46-51 RA; 63-110 Brin. For screw machine parts. Free cutting; CA 360.

NITTANY NO. 49

Cerro Metal Products Co.
Cu 61, Pb 3, bal Zn.
Soft: 50,000 TS; 22,000 YS; 32 El; 83 Brin. 1/2 H-temper: 60,000 TS; 42,000 YS; 22 El; 125 Brin. H-temper: 85,000 TS; 56,000 YS; 7 El; 160 Brin. For screw machine products, screws, and bolts. Free-cutting; CA 360.

NITTANY NO. 7

Cerro Metal Products Co.
Cu 62-62.5, Pb 1.25-1.5, bal Zn.
48,000-60,000 TS; 19,000-36,000 YS; 26-44 El; 62-75 RA; 63-90 Brin. For screw machine parts. Free cutting. *Obsolete*

NITTANY NO. 77

Cerro Metal Products Co.
Cu 62.5, Pb 3, bal Zn.
For screw machine products, hardware, and bolts. Free-cutting; CA 360.

NITUF

CCS Braeburn Alloy Steel
C 0.5, Mn 0.4, Si 1, Cr 5.25, Ni 1.5, V 1, Mo 1.35, bal Fe.
Air hardening tool and die steel, AISI A9. *Obsolete*

NITUNG

Pennsylvania Steel Corp.
C 0.3, Cr 2.75, Mo 0.3, W 9.5, Ni 1.6, bal Fe.
For extrusion rams and liners, hot work tools; hot work steel, oil hardened.

NITUTEC 10020

Eutectic Corp.
Nickel-copper base powder for thin overlays on other metals.
Obsolete

NIVAC

Colt Industries
Ni 99.92.
For equipment handling fluoride at high temperatures of 500-600°C. Corrosion and heat resistant.

NIVAC

Crucible Specialty Steel
Ni 99.92.
For equipment handling fluoride at high temperatures of 500-600°C. Corrosion and heat resistant.

NIVAC 50 NI

Crucible Materials Corp.
Nickel. Ni 50, bal Fe.
Annealed: 58,000 TS; 19,000 YP; 27 El. Coercive force 0.06 for 10,000 Gauss. For magnetic temperature compensators, speedometers, tachometers, voltage regulators, watt-hour meters. High permeability, magnetically soft.

NIVAC 79 NI

Crucible Materials Corp.
Nickel. Ni 79, bal Fe.
For telephone loading coils, magnetic shielding, sensitive relays, pulse transformers. High permeability.

NIVAC 80 NI

Crucible Materials Corp.
Nickel. Ni 80, bal Fe.
For telephone loading coils, magnetic shielding, sensitive relays, pulse transformers. High permeability.

NIVAC P

Crucible Materials Corp.
C 0.007, Si 0.005, P 0.002, Co 0.13, Cu 0.005, Fe 0.01, 0.005 O_2, bal Ni.
For vacuum tubes, diaphragms, resistance wire; high purity nickel. *Obsolete*

NIVAC-77 NI

Crucible Materials Corp.
Nickel. Ni 77, bal Fe.
For telephone loading coils, magnetic shielding, sensitive relays, pulse transformers. High permeability, soft magnet.

NIVAC-78 NI

Crucible Materials Corp.
Nickel. Ni 78, bal Fe.
For telephone loading coils, magnetic shielding, sensitive relays, pulse transformers. High permeability, soft magnet.

NIVAFLEX

Vacuumschmelze GmbH
C 0.03, Cr 18, Ni 21, Co 45, W 4, Mo 4, Ti 1, Be 0.3, bal Fe.
For high temperature springs. Corrosion and heat resistant.
Obsolete

NIVAN

Disston Inc.
Ni 1.4, Cr 0.35, V 0.2, C, bal Fe.
For cutlery; tough.

NIVAROX

Vacuumschmelze GmbH
Ni 28, W, Mo, Be, bal Fe.
For instrument springs, diaphragms; constant modulus.
Obsolete

NIVAROX CT

Vacuumschmelze GmbH
C 0-0.1, Ni 37, Cr 8, Fe 54, Mn 0.8, Be 0.9, Ti 1, Si 0.2.
For watch springs; corrosion resistant. *Obsolete*

NIVAROX M

Vacuumschmelze GmbH
C 0-0.1, Ni 31.5, Fe 60, Mn 0.7, Be 0.9, Mo 6.5, Si 0.1.
For watch springs; corrosion resistant. *Obsolete*

NIVAROX N

Vacuumschmelze GmbH
C 0-0.1, Ni 36, Fe 60, Mn 0.8, Be 0.9, Si 0.1, Ti 2.
For watch springs; corrosion resistant. *Obsolete*

NIVAROX W

Vacuumschmelze GmbH
C 0-0.1, Ni 30, Fe 61, Mn 0.7, Be 0.9, W 7.5, Si 0.1, Ti 0.1.
For watch springs; corrosion resistant. *Obsolete*

NIVCO

Westinghouse Electric Corp.
C 0.02, Mn 0.35, Si 0.15, Ni 22.5, Zr 1.1, Ti 1.8, Al 0.22, Fe 1, bal Co.
At room temperature: 165,000 psi TS; 110,000 psi YS; 25 El. At 1200°F: 105,000 psi TS; 75,000 psi YS; 20 El. For steam turbine blading; high heat strength to 1200°F. *Obsolete*

NIVCO-10

Westinghouse Electric Corp.
C 0.02, Zr 1.1, Ti 1.8, Ni 22.5, bal Co.
At 70°F: 165,000 psi TS; 110,000 psi YS; 25 El. At 1200°F: 105,000 psi TS; 75,000 psi YS; 20 El. At 1300°F: 85,000 psi TS; 65,000 psi YS; 26 El. For steam turbine blades; used where vibratory stresses are critical. *Obsolete*

NIZEC

British Steel plc
Electrolytically zinc-nickel alloy coated steel. Mild steel.

NK STS-50

NKK Corp.
C 0-0.15, Si 0.2-0.5, Mn 1-1.5, P 0-0.02, S 0-0.02, bal Fe.
High strength plate to meet shipbuilding code. *Obsolete*

NK-AC 90

NKK Corp.
C 0.18, Mn 0.6, Cr 0.9, Mo 0.18, bal Fe.
Heat treated: 90,000-105,000 psi YS. Seamless tube and pipe. Good impact resistance at low temperature; resistant to hydrogen sulfide corrosion cracking.

NK-CMV12

NKK Corp.
C 0.2, Mn 0.55, Ni 0.55, Cr 12, Mo 1, V 0.3, bal Fe.
Heat and oxidation resisting seamless pipe and tube; for operation at 500-550°C.

NK-HITEN 100

NKK Corp.
Now NK-HITEN 980.

NK-HITEN 590

NKK Corp.
C 0-0.18, Si 0-0.55, Mn 0-1.5, Cu, Cr, Mo, Ti as required, bal Fe.
Plates, quenched and tempered; 590-710 N/mm² TS; 450 N/mm² YS; 20 El min. 6-100 mm thick plate for structural purposes.

NK-HITEN 590C

NKK Corp.
C 0-0.18, Si 0-0.7, Mn 0-1.6, Cu 0-0.8, Ni 0-0.8, 0.09 V or Nb, bal Fe.
Plate, as rolled or normalized: 590-710 N/mm² TS; 450 N/mm² YS; 20-28 El. 6-50 mm thick plate for structural purposes.

NK-HITEN 590U

NKK Corp.
C 0.05-0.12, Si 0.15-0.4, Mn 0.9-1.4, Cu 0-0.3, Cr 0-0.3, Mo 0-0.2, V 0-0.08, B 0-0.003, bal Fe.
Quenched and tempered, 6-50 mm thick plate: 590-710 N/mm² TS; 450 N/mm² YS; 20-28 El. Plate for structural purposes.

NK-HITEN 60

NKK Corp.
Now NK-HITEN 590.

NK-HITEN 60

Japan Steel & Tube Co.Ltd.
C 0-0.16, Mn 0-1.35, Ni 0-0.6, Mo 0-0.3, V 0-0.15, bal Fe.
Normalized: 85,000 min TS; 65,000 min YP; 16 min El. Heat Treated: 114,000 TS; 100,000 YS; 18-20 El. For agricultural equipment, mine and railroad cars. High strength low-alloy constructional steel. Tough, shock resistant.

NK-HITEN 60

American Carbide Corp.
C 0-0.16, Mn 0-1.35, Ni 0-0.6, Mo 0-0.3, V 0-0.15, bal Fe.
Normalized: 85,000 TS min; 65,000 YP min; 16 El min. Heat treated: 114,000 TS; 100,000 YS; 18-20 El. For agricultural equipment, mine and railroad cars. High strength low-alloy constructional steel. Tough, shock resistant.

NK-HITEN 60C

NKK Corp.
Now NK-HITEN 590C.

NK-HITEN 60U

NKK Corp.
Now NK-HITEN 590U.

NK-HITEN 610

NKK Corp.
C 0-0.18, Si 0-0.55, Mn 0-1.5, Cu, Cr, Mo, Ti as required, bal Fe.
Plate, quenched and tempered; 610-730 N/mm² TS; 490 N/mm² YS; 19-29 El. 6-100 mm thick plate for structural purposes.

NK-HITEN 610C

NKK Corp.
C 0-0.18, Si 0-0.7, Mn 0-1.6, Cu 0-0.8, Ni 0-0.8, 0.09 V or Nb, bal Fe.
Plate, as rolled or normalized: 610-730 N/mm² TS; 490 N/mm² YS; 19-29 El. 6-40 mm thick plate for structural purposes.

NK-HITEN 610U

NKK Corp.
C 0.05-0.12, Si 0.15-0.4, Mn 0.9-1.4, Cu 0-0.3, Cr 0-0.3, Mo 0-0.2, V 0-0.08, B 0-0.003, bal Fe.
Quenched and tempered, 6-50 mm thick plate: 610-730 N/mm² TS; 490 N/mm² YS; 19-29 El. Plate for structural purposes.

NK-HITEN 62

NKK Corp.
Now NK-HITEN 610.

NK-HITEN 62C

NKK Corp.
Now NK-HITEN 610C.

NK-HITEN 62U

NKK Corp.
Now NK-HITEN 610U.

NK-HITEN 670

NKK Corp.
C 0-0.16, Si 0-0.35, Mn 0-1.2, Cu 0-0.3, Ni 0-0.8, Cr 0-0.7, Mo 0-0.4, V 0-0.07, B 0-0.003, bal Fe.
Plate, quenched and tempered: 670-800 N/mm² TS; 550 N/mm² YS; 18-26 El. 6-100 mm thick plate for structural purposes.

NK-HITEN 68

NKK Corp.
Now NK-HITEN 670.

NK-HITEN 690B

NKK Corp.
C 0-0.14, Si 0-0.35, Mn 0-1, Cu 0-0.3, Mo 0-0.4, V 0-0.07, B 0-0.003, Ni 0-1.3, Cr 0-0.7, bal Fe.
Plate, quenched and tempered: 690-830 N/mm² TS; 620 N/mm² YS; 17-25 El. 6-100 mm thick plate for structural purposes.

NK-HITEN 70B

NKK Corp.
Now NK-HITEN 690B.

NK-HITEN 710

NKK Corp.
C 0-0.16, Si 0-0.35, Mn 0-1.2, Cu 0-0.3, Ni 0-0.8, Cr 0-0.7, Mo 0-0.4, V 0-0.07, B 0-0.003, bal Fe.
Plate, quenched and tempered: 710-840 N/mm² TS; 620 N/mm² YS; 17-25 El. 6-100 mm thick plate for structural purposes.

NK-HITEN 72

NKK Corp.
Now NK-HITEN 710.

NK-HITEN 780

NKK Corp.
C 0-0.18, Si 0-0.35, Mn 0-1, Cu 0.15-0.5, Ni 0-1, Cr 0-0.8, Mo 0-0.6, V 0-0.1, B 0.006, bal Fe.
Plate, quenched and tempered: 780-930 N/mm² TS; 690 N/mm² YS; 16-24 El. 6-100 mm thick plate for structural purposes.

NK-HITEN 780A

NKK Corp.
C 0-0.18, Si 0-0.6, Mn 0-1, Cu 0.15-0.5, Cr 0-1.2, Mo 0-0.6, V 0-0.1, B 0-0.006, bal Fe.
Plate, quenched and tempered: 780-930 N/mm² TS; 690 N/mm² YS; 16-24 El. 6-100 mm thick plate for structural purposes.

NK-HITEN 780B

NKK Corp.
C 0-0.14, Si 0-0.35, Mn 0-1.2, Cu 0-0.3, Ni 0-1.5, Cr 0-0.7, Mo 0-0.4, V 0-0.1, B 0-0.003, bal Fe.
Plate, quenched and tempered: 780-930 N/mm² TS; 690 N/mm² YS; 16-24 El. 6-100 mm thick plate for structural purposes.

NK-HITEN 80

NKK Corp.
Now NK-HITEN 780.

NK-HITEN 80A

NKK Corp.
Now NK-HITEN 780A.

NK-HITEN 80B

NKK Corp.
Now NK-HITEN 780B.

NK-HITEN 980

NKK Corp.
C 0-0.18, Si 0-0.55, Mn 0-1.2, Cu 0.15-0.5, Ni 0-1.2, Cr 0-0.8, Mo 0-0.7, V 0-0.15, Zr 0-0.15, B 0-0.006, bal Fe.
Plate, quenched and tempered: 950-1130 N/mm² TS; 880 N/mm² YS; 12-19 El. 6-26 mm thick plate for structural purposes.

NK-MARINE 490F

NKK Corp.
C 0-0.15, Si 0-0.55, Mn 0-1.5, Cu 0.2-0.5, Nb 0-0.1, Al 0.15-0.55, Cr 0.5-0.8, bal Fe.
Hot rolled strip and sheet, as rolled, 1.4-13.0 mm thick: 490-610 N/mm^2 TS; 360 N/mm^2 YS; 25 El. Seawater resistant steel.

NK-MARINE 50F

NKK Corp.
Now NK-MARINE 490F.

NK-T 95

NKK Corp.
C 0-0.24, bal Fe.
Quenched, tempered and straightened seamless pipe and tube: 110,000 psi TS min; 95,000-125,000 psi YS. High collapse value.

NKBH 390

NKK Corp.
Low carbon, low alloy strip and sheet. Annealed, cold rolled: 390 N/mm^2 TS; 235 N/mm^2 YS; 30 El min. 0.6-1.2 mm thick, good formability.

NKBH 40

NKK Corp.
Now NKBH 390.

NKBH 440

NKK Corp.
Low carbon, low alloy strip and sheet. Annealed, cold rolled: 440 N/mm^2 TS; 275 N/mm^2 YS; 26 El min. 0.6-1.2 mm thick, good formability.

NKBH 45

NKK Corp.
Now NKBH 440.

NKBH 490

NKK Corp.
Low carbon, low alloy strip and sheet. Annealed, cold rolled: 490 N/mm^2 TS; 315 N/mm^2 YS; 23 El min. 0.6-1.2 mm thick, good formability.

NKBH 50

NKK Corp.
Now NKBH 490.

NKBH 540

NKK Corp.
Low carbon, low alloy strip and sheet. Annealed, cold rolled: 540 N/mm^2 TS; 355 N/mm^2 YS; 20 El min. 0.6-1.2 mm thick, good strength and formability.

NKBH 55

NKK Corp.
Now NKBH 540.

NKBH 590

NKK Corp.
Low carbon, low alloy strip and sheet. Annealed, cold rolled: 590 N/mm^2 TS; 390 N/mm^2 YS; 17 El min. 0.6-1.2 mm thick, good strength and formability.

NKBH 60

NKK Corp.
Now NKBH 590.

NKCA 390

NKK Corp.
Low carbon, low alloy strip and sheet. Annealed, cold rolled: 390 N/mm^2 TS; 235 N/mm^2 YS; 30 El min. 0.6-2.0 mm thick, good formability.

NKCA 40

NKK Corp.
Now NKCA 390.

NKCA 440

NKK Corp.
Low carbon, low alloy strip and sheet. Annealed, cold rolled: 440 N/mm^2 TS; 275 N/mm^2 YS; 26 El min. 0.6-2.0 mm thick, good formability.

NKCA 45

NKK Corp.
Now NKCA 440.

NKCA 490

NKK Corp.
Low carbon, low alloy strip and sheet. Annealed, cold rolled: 490 N/mm^2 TS; 315 N/mm^2 YS; 23 El min. 0.6-2.0 mm thick, good formability.

NKCA 50

NKK Corp.
Now NKCA 490.

NKCA 540

NKK Corp.
Low carbon, low alloy strip and sheet. Annealed, cold rolled: 540 N/mm^2 TS; 355 N/mm^2 YS; 20 El min. 0.6-2.0 mm thick, good strength and formability.

NKCA 55

NKK Corp.
Now NKCA 540.

NKCA 590

NKK Corp.
Low carbon, low alloy strip and sheet. Annealed, cold rolled: 590 N/mm^2 TS; 390 N/mm^2 YS; 17 El min. 0.6-2.0 mm thick, good strength and formability.

NKCA 60

NKK Corp.
Now NKCA 590.

NKHA 490

NKK Corp.
Hot rolled strip and sheet; composition by agreement. As rolled, 1.6-3.2 mm thick: 490 N/mm^2 TS; 345 N/mm^2 YS; 22-24 El. For structural purposes; good strength and formability.

NKHA 50

NKK Corp.
Now NKHA 490.

NKHA 540

NKK Corp.
Hot rolled strip and sheet; composition by agreement. As rolled, 1.6-3.2 mm thick: 540 N/mm^2 TS; 375 N/mm^2 YS; 21-23 El. For structural purposes; good strength and formability.

NKHA 55

NKK Corp.
Now NKHA 540.

NKHA 590

NKK Corp.
Hot rolled strip and sheet; composition by agreement. As rolled, 1.6-3.2 mm thick: 590 N/mm^2 TS; 440 N/mm^2 YS; 19-21 El. For structural purposes; good strength and formability.

NKHA 60

NKK Corp.
Now NKHA 590.

NKHF 490

NKK Corp.
C 0-0.18, Si 0-0.55, Mn 0-1.5, Nb 0-0.1, Ti 0-0.1, bal Fe.
Hot rolled strip and sheet, as rolled, 3.2-8.0 mm thick: 490 N/mm^2 TS; 355 N/mm^2 YS; 24 El. For structural applications, good formability.

NKHF 50

NKK Corp.
Now NKHF 490.

NKHF 540

NKK Corp.
C 0-0.18, Si 0-0.55, Mn 0-1.5, Nb 0-0.1, bal Fe.
Hot rolled strip and sheet, as rolled, 3.2-8.0 mm thick: 540 N/mm^2 TS; 390 N/mm^2 YS; 24 El. For structural applications, good strength and formability.

NKHF 55

NKK Corp.
Now NKHF 540.

NKHF 590

NKK Corp.
C 0-0.18, Si 0-0.55, Mn 0-1.5, Nb 0-0.1, Zr 0-0.2, bal Fe.
Hot rolled strip and sheet, as rolled, 3.2-8.0 mm thick: 590 N/mm^2 TS; 450 N/mm^2 YS; 20-24 El. For structural applications, good strength and formability.

NKHF 60

NKK Corp.
Now NKHF 590.

NKHF 690

NKK Corp.
C 0-0.15, Si 0-0.7, Mn 1-1.5, Cu 0-0.5, Ni 0-0.5, Cr 0-0.5, V 0-0.15, Nb 0-0.1, Zr 0-0.2, bal Fe.
Hot rolled strip and sheet, as rolled, 3.2-8.0 mm thick: 690 N/mm^2 TS; 550 N/mm^2 YS; 14-18 El. For structural applications; good strength and formability.

NKHF 70

NKK Corp.
Now NKHF 690.

NKHF 780

NKK Corp.
C 0-0.1, Si 0-0.7, Mn 1.5-2.2, Cu 0-0.8, Ni 0-0.8, Cr 0-0.5, V 0-0.15, Nb 0-0.1, Zr 0-0.2, Ti 0-0.15, bal Fe.
Hot rolled strip and sheet, as rolled, 3.2-8.0 mm thick: 780 N/mm^2 TS; 685 N/mm^2 YS; 12-16 El. For structural applications; good strength and formability.

NKHF 80

NKK Corp.
Now NKHF 780.

NKS

Japanese manufacture
Al 3.7, Ni 17.7, Co 27.2, Ti 6.7, bal Fe.
For permanent magnets, electrical and magnetic equipment; high permeability.

NKS MAGNET

Japanese manufacture
Al 8, Ni 14, Co 24, Cu 3.
For permanent magnets; high coercive force.

NKS-1

Japanese manufacture
Al 8, Ni 15, Cu 3, Ti 1.25, Co 24, bal Fe.
Cast: 20,000 TS; For electrical and magnetic equipment; similar to Alnico VI, permanent magnet.

NM STEEL

AMAX Corp.
Ni 3-5, Mo 0.3-0.7, C 0.2-0.4, Mn 0.3-0.5, bal Fe.
250,000 TS; 210,000 YS; 8 El; 25 RA; 51 Brin. For aircraft and automobile parts. *Obsolete*

NM STEEL

British Steel Corp.
Ni 3-5, Mo 0.3-0.7, C 0.2-0.4, Mn 0.3-0.5, bal Fe.
250,000 TS; 210,000 YS; 8 El; 25 RA; 51 Brin. For aircraft and automobile parts. *Obsolete*

NM-100
Nuclear Metals Inc.
C 1.25, Cr 17.5, W 10.5, Co 9.5, V 0.75, bal Fe.
Hardened, room temperature: 290,000 TS; 245,000 YS; 1000°F: 240,000 TS; 175,000 YS. For aircraft bearings and other bearings; parts subject to dry friction at elevated temperatures; resistant to galling, corrosion and high temperature softening and oxidation.

NM-100
Whittaker Corp.
C 1.25, Cr 17.5, W 10.5, Co 9.5, V 0.75, bal Fe.
Hardened, RT: 290,000 TS; 245,000 YS. At 1000°F: 240,000 TS; 175,000 YS. For aircraft bearings and other bearings and parts subject to dry friction at elevated temperatures; resistant to galling, corrosion and high temperature softening and oxidation.

NMHG
Creusot-Loire
Ni 29, Cr 3, C, bal Fe.

NO 1002
American manufacture
Cr 22, Ni 16, W 7, Fe 1.5, Al 0.3, Ti 0.2, Zr 0.3, C 0.6, Si 0.5, M 0.7, La 0.05, bal Co.
Good stress rupture at elevated temp. and resistant to oxidation.

NO-CHAT
CMW Inc.
W 90, Ni 6, Cu 4.
Sintered: 112,000 TS; 6 El; 250 Brin. Density: 16.96. Modulus of rigidity: 19.2 x 10^6 psi. For boring bars, grinding quills or arbors, cut-off tools.

NO-CO-RO
Republic Steel Corp.
C 0-0.04, Mn 0.1-0.15, 0.20% min Cu, bal Fe.
Rolled: 40,000-43,000 TS; 25,000-28,000 YS; 35-38 El; 45-50 RA. For culverts. *Obsolete*

NO-DU-MAG
Ferranti Ltd.
Cast iron. C 3.3, Ni 10-11, Mn 5.6, Mg 0.17, Si 2.5, bal Fe.
Cast: 53,700-62,700 TS; 40,300-49,300 YS; 8-12 El; 13-18 RA; 260 BHN. For switch gears, resistant grids, magnetic chucks. Austenitic, ductile cast iron.

NO-KOR-O 12-F
Atlas Steels Ltd.
C 0.1, Cr 12, S 0.25, bal Fe.
Annealed: 100,000 TS; 50,000 YS; 40 El; 60 RA; 183 Brin. For golf heads, pumps, valves; free machining. *Obsolete*

NO-KOR-O 25-20
Atlas Steels Ltd.
C 0.2, Cr 25, Ni 20, bal Fe.
Annealed: 100,000 TS; 50,000 YS; 40 El; 60 RA; 183 Brin. For heat and corrosion resistant parts, exhaust valves; heat *Obsolete*

NO-KOR-O-12
Atlas Steels Ltd.
C 0.1, Cr 12, bal Fe.
Heat treated: 95,000-180,000 TS; 75,000-160,000 YS; 16-27 El; 62-75 RA; 196-364 Brin. For golf club heads, valve parts, pump rods; formerly "S". *Obsolete*

NO-KOR-O-14
Atlas Steels Ltd.
C 0.35, Cr 14, bal Fe.
Heat treated: 255,000 TS; 225,000 YS; 6 El; 12 RA; 504 Brin. For cutlery and dental instruments; formerly "Cutlery". *Obsolete*

NO-KOR-O-18
Atlas Steels Ltd.
C 0.1, Cr 18, bal Fe.
For golf club heads, valve parts, pump rods; formerly "A". *Obsolete*

NO-KOR-O-18-2
Atlas Steels Ltd.
C 0.15, Cr 18, Ni 2, bal Fe.
Hardened: 137,000-260,000 TS; 117,000-178,000 YS; 16-27 El; 38-55 RA; 225-402 Brin. For heat and corrosion resisting parts, valve seats; formerly "RA". *Obsolete*

NO-KOR-O-18-8
Atlas Steels Ltd.
C 0.1, Cr 18, Ni 9, bal Fe.
Annealed: 85,000 TS; 35,000 YS; 60 El; 65 RA; 143 Brin. For general stainless applications; formerly "KA2". *Obsolete*

NO-KOR-O-18-8F
Atlas Steels Ltd.
C 0.1, Mn 0.4, Si 0.4, Cr 18, Ni 9, Se 0.25, bal Fe.
Annealed: 85,000 TS; 35,000 YS; 60 El; 65 RA; 143 Brin. For stainless parts; corrosion resistant. *Obsolete*

NO-KOR-O-18-8M
Atlas Steels Ltd.
C 0.1, Cr 18, Ni 12, Mo 2, Mn 1.3, bal Fe.
Annealed: 100,000 TS; 50,000 YS; 50 El; 60 RA; 183 Brin. For equipment and apparatus in paper and pulp industries; formerly "KA4". *Obsolete*

NO-KOR-O-18-8S
Atlas Steels Ltd.
C 0.07, Cr 18, Ni 9, bal Fe.
Annealed: 85,000 TS; 35,000 YS; 60 El; 65 RA; 143 Brin. For tubes, stainless parts; formerly "KA2S". *Obsolete*

NO-KOR-O-18-H-60
Atlas Steels Ltd.
C 0.65, Mn 0.5, Si 0.4, Cr 17, bal Fe.
207 Brin. For corrosion resisting parts; corrosion resistant. *Obsolete*

NO-KOR-O-H
Atlas Steels Ltd.
C 0.21-0.35, Cr 12-15, bal Fe.
For cutlery, instruments; corrosion resistant. *Obsolete*

NO-KOR-O-HR
Atlas Steels Ltd.
C 0.25, Cr 25, bal Fe.
For heat and corrosion resisting parts. *Obsolete*

NO-KOR-O-KA2-HS
Atlas Steels Ltd.
C 0.1, Cr 18, Ni 8, High Si, bal Fe.
For stainless parts. *Obsolete*

NO-OX
Boiler Development Corp.
Al 6, C, Si, Mn, bal Fe.
Bar, annealed: 68,000 TS; 50,000 YS; 30 El; 78.8 Rock B. For pack carburizing containers, baffles and flame deflectors, burners, furnace parts, retorts, muffles. High temperature alloy. Aluminized. Oxidation resistant to 2200°F.

NO-TIN SILVER FUSE
Fusion Inc.
For solder. *Obsolete*

NO-WEAR
GTE Sylvania
WC, Cr, Co, Fe.
For hard facing welding electrodes; gas welding. *Obsolete*

NO. 000
Uddeholm Corp.
C 0.7-0.9, bal Fe.
Annealed: 180 Brin. For dies and punches, taps, shears, broaches. *Obsolete*

NO. 000 DOUBLE EXTRA CARBON
Adams & Osgood Steel Co.
C 1.1-1.3, bal Fe.
For intricate tools, broaches, drawing and threading dies; water hardening.

NO. 1 STANDARD
Delta Metal (BW) Ltd.
Cu 55-60, Pb 0-0.2, Al 0-1.5, Mn 0-3.5, Sn 0-1.5, bal Zn.
Extruded: 76,000 TS; 37,000 YS; 28 El; 20 RA; 135 Brin.
Drawn: 86,000 TS; 41,000 YS; 25 El; 18 RA; 140 Brin. For pump valves, propellers, condenser tubes; Parson's manganese bronze. *Obsolete*

NO. 1 TA QUALITY
Delta Metal (BW) Ltd.
Cu 55-60, Pb 0-0.2, Mn 2-3, bal Zn.
Rolled: 76,000 TS; 36,000 YS; 25 El; 25 RA; 125 Brin. For hydraulic tubes; Parson's manganese bronze. *Obsolete*

NO. 1015-65
LaSalle Steel Co.
C 0.1-0.2, bal Fe.
Rolled: 55,000-65,000 TS; 18-38 El; 60-65 RA; 120-156 Brin. For case hardened parts; Modified X-1015. *Obsolete*

NO. 1020-90 STEEL
Republic Steel Corp.
C 0.15-0.25, Mn 0.7-1.05, bal Fe.
Cold drawn: 75,000-90,00 TS; 55,000-75,000 YS; 15-25 El; 50-55 RA; 170-187 Brin. For carburized parts; case-hardening steel. *Obsolete*

NO. 1040 ALLOY
German manufacture
Cu 15, Mn 1, Mo 3, Ni 71, Fe 10.
For electrical equipment; magnetic alloy.

NO. 11V
Carpenter Technology Corp.
C, V, bal Fe.
For tools; shock resistant. *Obsolete*

NO. 155 PREMIUM SILVER BRAZING ROD
Now ALL-STATE NO. 155.

NO. 155 PREMIUM SILVER BRAZING ROD
Chemetron Corp.
Ag 55, bal Cu.
Cast: 50,000 TS. For silver brazing; 1155°F working temperature.

NO. 2 B
Crucible Specialty Metals
C, W, bal Fe.
For hot die work. *Obsolete*

NO. 2 DIE METAL
Henning Bros. & Smith Inc.
Zinc. Al 3.5-4.5, Cu 2.5-3.5, Mg 0.02-1, bal Zn.
Sand cast: 37,000 TS; 3 El. For stamping and drop hammer dies; subject to intergranular attack.

NO. 3 C
Jessop Steel Co.
C, Cr, Co, bal Fe.
For cold work dies. *Obsolete*

NO. 3 MINE TALABOT SPECIAL D
Societe Nouvelle du Saut-du-Tarn
C 0.95, bal Fe.
For tools, springs, taps, drills; water hardened.

NO. 3074 HOT WORK
A. Milne & Co.
C 0.4-0.45, W 8-10, Cr 2.5, V 0.1-0.15, bal Fe.
For hot work tools and dies, forging mandrels; hot work steel.

NO. 35 MONEL
International Nickel Inc.
Cu 31.25, Ni 65.5, Fe 1.
Cold rolled: 78,000-85,000 TS; 45,000-65,000 YS; 20-40 El.
For table tops; corrosion resistant. *Obsolete*

NO. 4 VASCO FINISH

Teledyne Vasco
C 1.3, W 3.5, bal Fe.
For rifling tools and reamers, forming rolls. *Obsolete*

NO. 426 ALLOY

General Electric Co.
Co 35, Ni 18, Al 6, Ti 8, bal Fe.
For permanent magnets; high coercive force. *Obsolete*

NO. 444 ALLOY

Now CLEVITE S 56.

NO. 446 ALLOY

English manufacture
Cr 27-30, bal Fe.
For glass-to-metal seals.

NO. 50 SOLDER ALLOY

Soldering Specialties Co.
In, Pb.
600°F MP. For solder.

NO. 500 ALLOY

American manufacture
Al alloy.
Cast: 27,000 TS; 20,000 YS; 2 El. For casting; light weight.

NO. 6 MINE TULIPE SPECIAL D

Societe Nouvelle du Saut-du-Tarn
C 0.75, bal Fe.
For tools, springs, punches; water hardened.

NO. 712 ALLOY

Engelhard Corp.
Os 29, Ir 40-50, bal Pt.
For contact points; corrosion resistant. *Obsolete*

NO. 812

Disston Inc.
C 1.8, Cr 12.25, bal Fe.
For blanking and coining dies; taps, reamers; non-deforming.

NO. 812

Teledyne Firth Sterling
C 1.8, Cr 12.5, bal Fe.
For dies; non-deforming. *Obsolete*

NO. 999

Hobson, Houghton & Co.
C 0.7, W 18, Cr 4, V 1, Co 5, bal Fe.
For hogging cuts for fast speeds, cutters; high speed steel.

NO. MR-100 BRONZE

Michigan Powdered Metal Products Inc.
Cu composite.
For bearings; sintered; 25-30% porosity. *Obsolete*

NO. PS-10 SINTERED IRON

Michigan Powdered Metal Products Inc.
Fe.
For pole pieces, stators and rotors in motors; for electromagnetic uses, sintered. *Obsolete*

NO. R-35 SINTERED IRON

Michigan Powdered Metal Products Inc.
Fe.
For bearings; sintered; 18% porosity. *Obsolete*

NO. X-10 COMPOSITE METAL

Michigan Powdered Metal Products Inc.
Fe composite.
For machine parts; sintered. *Obsolete*

NO. X-20 COMPOSITE METAL

Michigan Powdered Metal Products Inc.
Fe composite.
Hardened: 500 Brin. For machine parts; sintered, hardenable. *Obsolete*

NO.55 ALLOY

English manufacture
Cr 27-30, bal Fe.
For glass-to-metal seals; same coefficient of expansion as glass.

NOBELOY

Manufacturer not listed
Au alloy.
For jewelry; corrosion resistant.

NOBILIUM

American manufacture
Co 65, Cr 28, Ni 0.1, Mo 5, C 0.4, V 1, Fe 0.5, Mn 0.1, Si 0.05.
Cast: 168,000 TS; 88,000 YS; 4 El. For dentures; corrosion resistant.

NODULITE 100-70-03

Hamilton Foundry
C 3.3, Si 2.5, Mn 0.7, Mg 0.05, bal Fe.
Cast: 100,000-120,000 TS; 70,000-90,000 YS; 3-10 El; 241-302 Brin. For gears, crankshafts, pistons, camshafts, track shoes, brake drums. Ductile iron; high strength and wear resistant, tough. *Obsolete*

NODULITE 120-90-02

Hamilton Foundry
C 3.3, Si 2.5, Mn 0.7, Mg 0.05, bal Fe.
Cast: 120,000-175,000 TS; 90,000-150,000 YS; 2-7 El; 269-388 Brin. For pinions, gears, cams, machine guides, dies, pumps. Wear-resistant ductile iron. *Obsolete*

NODULITE 60-40-18

Hamilton Foundry
C 3.3, Si 2.5, Mn 0.7, Mg 0.05, bal Fe.
Cast: 60,000-80,000 TS; 40,000-60,000 YS; 18-25 El; 137-192 Brin. For gears, pinions, cams, dies, idlers, track rollers, pumps. Ductile iron; resists thermal shock; tough. *Obsolete*

NODULITE 65-45-12

Hamilton Foundry
C 3.3, Si 2.5, Mn 0.7, Mg 0.05, bal Fe.
Cast: 65,000-85,000 TS; 45,000-60,000 YS; 12-20 El; 143-207 Brin. For pressure castings, pipe fittings, valves, cylinders, pump bodies. Ductile iron; tough, high strength. *Obsolete*

NODULITE 80-55-06

Hamilton Foundry
C 3.3, Si 2.5, Mn 0.7, Mg 0.05, bal Fe.
Cast: 80,000-100,000 TS; 55,000-75,000 YS; 6-12 El; 179-269 Brin. For gears, cams, bearings, pistons, crankshafts, sheaves, sprockets. Ductile iron; high strength, wear resistant, tough. *Obsolete*

NODULOY 3

Cyprus Foote Mineral Co.
Mg 2.8-3.2, Si 44-48, Ca 0.8-1.3, Al 0-1.2, bal Fe.
Magnesium-ferrosilicon additive for nodularization of ductile iron. Also known as NODULOR type. *Obsolete*

NODULOY 3R

Cyprus Foote Mineral Co.
Mg 2.8-3.3, Si 44-48, Ca 0.8-1.3, Al 0-1.2, 0.70-1.0 rare earths, bal Fe.
Magnesium-ferrosilicon with rare earths (NODULOY R group) for nodularization of ductile iron. *Obsolete*

NODULOY 5

Cyprus Foote Mineral Co.
Mg 5-6, Si 44-48, Ca 0.8-1.3, Al 0-1.2, bal Fe.
Magnesium-ferrosilicon additive for nodularization of ductile iron. Also known as NODULOY. *Obsolete*

NODULOY 5-1C

Cyprus Foote Mineral Co.
Mg 5-6, Si 44-48, Ce 0.9-1.1, Ca 0.8-1.3, Al 0-1.2, bal Fe.
Magnesium-ferrosilicon with cerium (NODULOY C group) for nodularization of ductile iron. *Obsolete*

NODULOY 5C

Cyprus Foote Mineral Co.
Mg 5-6, Si 44-48, Ce 0.5-0.75, Ca 0.8-1.3, Al 0-1.2, bal Fe.
Magnesium-ferrosilicon with cerium (NODULOY C group) for nodularization of ductile iron. *Obsolete*

NODULOY 5LC

Cyprus Foote Mineral Co.
Mg 5-6, Si 44-48, Ce 0.3-0.4, Ca 0.8-1.3, Al 0-1.2, bal Fe.
Magnesium-ferrosilicon with cerium (NODULOY C group) for nodularization of ductile iron. *Obsolete*

NODULOY 5R-1

Cyprus Foote Mineral Co.
Mg 5-6, Si 44-48, Ca 0.8-1.3, Al 0-12, 0.60-0.80 rare earths, bal Fe.
Magnesium-ferrosilicon with rare earths (NODULOY R group) for nodularization of ductile iron. *Obsolete*

NODULOY 5R-2

Cyprus Foote Mineral Co.
Mg 5-6, Si 44-48, Ca 0.8-1.3, Al 0-1.2, 0.90-1.20 rare earths, bal Fe.
Magnesium-ferrosilicon with rare earths (NODULOY R group) for nodularization of ductile iron. *Obsolete*

NODULOY 5R-3

Cyprus Foote Mineral Co.
Mg 5-6, Si 44-48, Ca 0.8-1.3, Al 0-1.2, 1.7-2.0 rare earths, bal Fe.
Magnesium-ferrosilicon with rare earths (NODULOY R group) for nodularization of ductile iron. *Obsolete*

NODULOY 9

Cyprus Foote Mineral Co.
Mg 8.5-10, Si 44-48, Ca 1-1.5, Al 0-1.2, bal Fe.
Magnesium-ferrosilicon additive for nodularization of ductile iron. Also known as NODULOY type. *Obsolete*

NODULOY 9C

Cyprus Foote Mineral Co.
Mg 8.5-10, Si 44-48, Ce 0.5-0.7, Ca 1-1.5, Al 0-1.2, bal Fe.
Magnesium-ferrosilicon with cerium (NODULOY C group) for nodularization of ductile iron. *Obsolete*

NODULOY 9LC

Cyprus Foote Mineral Co.
Mg 8.5-10, Si 44-48, Ce 0.3-0.4, Ca 1-1.5, Al 0-1.2, bal Fe.
Magnesium-ferrosilicon with cerium (NODULOY C group) for nodularization of ductile iron. *Obsolete*

NODULOY 9R

Cyprus Foote Mineral Co.
Mg 8.5-10, Si 44-48, Ca 0.8-1.3, Al 0-1.2, 0.70-1.0 rare earths, bal Fe.
Magnesium-ferrosilicon with rare earths (NODULOY R group) for nodularization of ductile iron. *Obsolete*

NODULOY TYPE 18 C

Foote Mineral Co.
Mg 16-20, Si 60-65, Ca 1-3, 0.5% min Ce, bal Fe.
For inoculating cast iron to nodularize the graphite making ductile iron. Cast iron inoculant. *Obsolete*

NODULOY TYPE 7 C

Foote Mineral Co.
Mg 8-9.5, Si 43-47, Ce 0.5-0.6, Fe 4.5-6.5.
For inoculating cast iron to nodularize the graphite making ductile iron. Cast iron inoculant. *Obsolete*

NODULOY TYPE 8 C

Foote Mineral Co.
Mg 8-9.5, Si 44-48, Ca 1-3, 0.5% min Ce, bal Fe.
For inoculating cast iron to nodularize the graphite making ductile iron. Cast iron inoculant. *Obsolete*

NOGROTH

Q. & C. Co.
TC 1.85, Si 1.05, P 0.16, Ni 1.5, Cr 0.33, Mn 0.35, bal Fe.
Annealed: 102,000 TS; 85,000 YS; 9.5 El; 250 Brin. For pump liners, pistons, valves; cast iron.

NOGROTH
Empire Steel Castings Co.
Cr, Ni, white cast iron.
86,000-89,000 TS; 50,000-56,000 YS; 10 El; 200 Brin. For bearings, bushings, pump liner, piston, valves; abrasive resisting. *Obsolete*

NOHEET
Ardal Ltd.
Pb 98, Na 1.4, Sb 0.11, Sn 0.1.
For bearings, metal or die cast. Anti-friction metal.

NOIL
Baker, Perkins & Co. Ltd.
Copper. Sn 20, Cu 80.
158 Brin. For piston rings.

NOMAG
Ferranti Ltd.
C 3, Mn 6, Si 2.3, Ni 10, bal Fe.
Cast: 22,000-26,000 TS; 158-160 Brin. For switch covers, resistance grids; austenitic, cast iron, non-magnetic.

NOMAG
Mond Nickel Co. Ltd.
C 3, Mn 6, Si 2.3, Ni 10, bal Fe.
Cast: 22,000-26,000 TS; 158-160 Brin. For switch covers, resistance grids; austenitic, cast iron, non-magnetic.

NON MAGNETIC
Jessop Steel Co.
C 0.35, Mn 12.4, Ni 3.25, Cr 4.15, Mo 0.48, bal Fe.
For wear resistant parts; work hardened. *Obsolete*

NON-CERTIFIED OXYGEN FREE
Criterion Metals, Inc.
Copper. Cu 99.95.
Thin gauge sheet, various tempers: 23-52 ksi TS min; 5-51 ksi YS min. C10200

NON-CORRODITE NO. 25
Millbury Steel Foundry Co.
Stainless steel. Cr 22.2, Ni 0.4, Si 1.57, Mn 1, Cu 0.15, C 0.34, bal Fe.
For castings, corrosion resisting parts; non-corrosive; stainless.

NON-GRAN BRONZE
American Non-Gran Bronze Co.
Copper. Sn, bal Cu.
Sintered: 40,000 TS; 22,000 YS; 15 El; 11 RA; 80 Brin. For bolt headers, pulleys, and machinery. Nongranular structure.

NON-OXIDIZABLE
Manufacturer not listed
Fe 62, Cr 25, Mn 10, C 1.1, Si 0.95.
For corrosion and heat resisting parts, abrasion resistant parts; U.S. Patent 1,333,151; corrosion, heat and abrasion resistant.

NON-PAREIL
Theodore Hiertz Metal Co.
Pb 78, Sb 17, Sn 5.
24 Brin. For bearings, solders. Antifriction metal; MP 300°C.

NON-SCALING-1
Manufacturer not listed
Ni 30-40, Cr 15-20, Si 3.5, Cu 1.25, bal Fe.
For furnace parts; heat resisting.

NON-SCALING-2
Manufacturer not listed
Ni 24, Cr 24, Si 3, bal Fe.
For furnace parts; heat resisting.

NON-SCUFF
Coulter Steel & Forge Company
C 0.6, Cr 3.7, V 1.8, Mo 8.5, bal Fe.
Heat treated: 290,000 TS; 238,000 YS; 0.5 El; 3 RA. At 1000 F: 216,000 TS; 150,000 YS; 3 El; 7 RA. For hot punches, rivet sets, spike cutters, hot shear blades, headers. Type H43 hot work tool steel, tough, high wear resistance. *Obsolete*

NON-SHRINK
Bisset Steel Co.
C 1.5, Cr 13, W 1.5, bal Fe.
For dies and tools, production dies; non-shrinking, oil-hardening. *Obsolete*

NON-SHRINK
Osborn Steels Ltd.
C 0.95, Mn 1.25, Cr 0.5, W 0.5, bal Fe.
For tools, dies; non-deforming. *Obsolete*

NON-SHRINK DIE
Atlantic Steel Corp.
C 0.9, Mn 1.2, bal Fe.
For dies; nondeforming. *Obsolete*

NON-SHRINKABLE
Teledyne Vasco
C 0.9, Mn 1.2, Cr 0.5, W 0.5, bal Fe.
For general tools, broaches; non-deforming. Type O1.

NON-SHRINKABLE DRILL ROD
Teledyne Allvac
C 0.95, Mn 1.2, Cr 0.5, W 0.5, V 0.2, bal Fe.
For tools, punches; nondeforming.

NON-SHRINKING PATENT
Manufacturer not listed
Pb 87, Sb 6, Sn 6, Cd 1.3.
For impressions, type.

NON-STRETCH STEEL
LaSalle Steel Co.
C 0.9, Mn 1.2, bal Fe.
125,000 YS. For dies; nondeforming. *Obsolete*

NON-TARNISHABLE
Manufacturer not listed
Cu 63.6, Zn 31, Sn 3.25, Pb 2.
For corrosion and tarnish resisting parts; corrosion resisting.

NON-TEMP. CHISEL
Midvale-Heppenstall Co.
C 0.5, Mn 0.6, Cr 0.65, Mo 0.25, bal Fe.
For shock and impact resistant tools; oil hardening. *Obsolete*

NON-TEMPERING
Bethlehem Steel Corp.
C 0.35, Mn 0.7, Cr 0.8, Mo 0.3, Cu 0.3, bal Fe.
For chisels, punches, blacksmith tools; shock resistant.

NON-TEMPERING
Allied Steel & Tractor Products Inc.
C 0.35, Mn 0.75, Si 0.45, Cr 0.8, W 0.25, Mo 0.5, bal Fe.
Water or oil hardening; shock resistant tool steel.

NON-WAIR
British Steel Corp.
C 2.2, Cr 13, bal Fe.
Air or oil hardening non-distorting alloy tool steel; for brick molds, tire molds, thread rolling dies, hobs, gauges. *Obsolete*

NON-WARP STEEL
LaSalle Steel Co.
C 0.9, Mn 1.4, bal Fe.
100,000 YS. For dies; nondeforming. *Obsolete*

NONERODE
Coulter Steel & Forge Company
C 0.35, Cr 5, V 0.4, Mo 1.5, bal Fe.
For die casting dies, punches; Type H11; oil hardened, hot work steel. *Obsolete*

NONGRAM
Manufacturer not listed
Cu 87, Sn 11, Zn 2-3.
35,000 TS; 16 El; 62 Brin. For bushings, bearings, valves.

NONSHOCK TUNGSTEN
Coulter Steel & Forge Company
C 0.55, Cr 1.25, W 2.75, V 0.2, bal Fe.
Heat treated: 231,000-315,000 TS; 209,000-275,000 YS; C 48-56 Rock; 9-11 El; 27-42 RA. For header and swaging dies, chipping chisels, rivet busters, track tools. Type S1 shock resisting tool steel. Good fatigue resistance. *Obsolete*

NONSULITE
Michiana Products Corp.
C 0.36-0.5, Cr 28, Ni 8, bal Fe.
For heat and corrosion resistant parts; now Michiana No. 48.

NONVAR
Firth Brown Ltd.
C 0.92, Mn 1.75, Si 0.3, bal Fe.
For tools, dies; oil hardening, non-deforming.

NORAL
Alcan-Booth Industries, Ltd.
Cu 2-4, Si 4-6, Mn 0.3-0.7, Ni 0.35, Zn 0.3, Ti 0.2, bal Al.
Sand cast: 23,500 TS; 11,200 YS; 3.5 El; 75 Brin. Permanent mold: 26,900 TS; 14,500 YS; 5.0 El; 85 Brin. For general purpose castings; good castability; age-hardenable. *Obsolete*

NORAL 10 S
Alcan-Booth Industries, Ltd.
Mn 0.5, Mg 0.5, bal Al.
Rolled: 12,000-13,000 TS; 25-33 El; 25-28 Brin. For light alloy parts; extruded. *Obsolete*

NORAL 100
Alcan-Booth Industries, Ltd.
99.5% min Al.
Cast: 10,000 TS; 3,000 YS; 24 Brin. For cables; commercially pure aluminum. *Obsolete*

NORAL 11 2
Alcan-Booth Industries, Ltd.
Cu 5.5, Pb 0.5, Bi 0.5, bal Al.
Rolled: 24,000-26,000 TS; 23-27 El; 45-55 Brin. For light alloy parts, screws; free machining. *Obsolete*

NORAL 111
Alcan-Booth Industries, Ltd.
Si, bal Al.
For light alloy castings; high fluidity. *Obsolete*

NORAL 115
Alcan-Booth Industries, Ltd.
Si, bal Al.
For light alloy castings; high fluidity. *Obsolete*

NORAL 123
Alcan-Booth Industries, Ltd.
Si 5, bal Al.
Cast: 22,000 TS; 10,000 YS; 6 El; 50 Brin. For light alloy castings; corrosion resistant. *Obsolete*

NORAL 124
Alcan-Booth Industries, Ltd.
Si, bal Al.
For light alloy castings; high fluidity. *Obsolete*

NORAL 125
Alcan-Booth Industries, Ltd.
Cu 1-1.5, Mg 0.5, Si 5, Ti 0-0.2, bal Al.
Cast: 30,000 TS; 20,000 YS; 3 El; 80 Brin. Aged: 42,000 TS; 35,000 YS; 2 El; 110 Brin. For cylinder heads, valve bodies, hose couplings; age-hardenable. *Obsolete*

NORAL 126
Alcan-Booth Industries, Ltd.
Si, bal Al.
For light alloy castings; corrosion resistant. *Obsolete*

NORAL 127
Alcan-Booth Industries, Ltd.
Si, bal Al.
For light alloy castings; corrosion resistant. *Obsolete*

NORAL 13 S
Alcan-Booth Industries, Ltd.
Si 12, bal Al.
Cast; 36,000-38,000 TS; 25-29 El; 70-75 Brin. For light alloy parts; die casting. *Obsolete*

NORAL 135
Alcan-Booth Industries, Ltd.
Si, bal Al.
For light alloy castings; corrosion resistant. *Obsolete*

NORAL 140
Alcan-Booth Industries, Ltd.
Si, bal Al.
For light alloy castings; corrosion resistant. *Obsolete*

NORAL 15 S
Alcan-Booth Industries, Ltd.
Zn 12.5, Cu 1.5, bal Al.
Heat treated: 44,000-46,000 TS; 30-31 El; 90-93 Brin. For architectural and marine parts; heat treatable. *Obsolete*

NORAL 150
Alcan-Booth Industries, Ltd.
Si 10, bal Al.
For light alloy castings; high fluidity. *Obsolete*

NORAL 155
Alcan-Booth Industries, Ltd.
Si, bal Al.
For light alloy castings. *Obsolete*

NORAL 156
Alcan-Booth Industries, Ltd.
Si, bal Al.
For light alloy castings. *Obsolete*

NORAL 158
Alcan-Booth Industries, Ltd.
Si 11, Ni 3, bal Al.
Sand cast: 26,900 TS; 10,100 YS; 2 El. Permanent mold: 35,900 TS; 13,500 YS; 3 El. For aircraft castings; corrosion resistant. *Obsolete*

NORAL 160
Alcan-Booth Industries, Ltd.
Si 12, Mn 0.5, Fe 0.6, bal Al.
Cast: 31,000 TS; 11,000 YS; 10 El; 10 RA; 60 Brin. For general marine castings; high strength. *Obsolete*

NORAL 161
Alcan-Booth Industries, Ltd.
Mg 0.5, Mn 0.5, Si 12.5, bal Al.
Aged: 27,000 TS; 16,000 YS; 1.5 El; 75 Brin. T-temper: 45,000 TS; 35,000 YS; 1 El; 115 Brin. For light castings; heat treatable. *Obsolete*

NORAL 162
Alcan-Booth Industries, Ltd.
Si 12, Mg 1, Cu 1, Ni 2.5, bal Al.
Aged: 46,000 TS; 38,000 YS; 5 El; 125 Brin. For pistons, cylinder heads; age-hardenable, high temperature use. *Obsolete*

NORAL 165
Alcan-Booth Industries, Ltd.
Si, bal Al.
For light alloy castings. *Obsolete*

NORAL 16S
Alcan-Booth Industries, Ltd.
Cu 2.5, Mg 0.3, bal Al.
T6-temper: 40,000 TS; 20,000 YS; 20 El; 70 Brin. For rivets, light alloy parts; age-hardenable. *Obsolete*

NORAL 17 S
Alcan-Booth Industries, Ltd.
Cu 3.5-5, Mg 0.4-1.2, Mn 0.6, bal Al.
Heat treated: 50,000-54,000 TS; 42,600 YS; 15-18 El; 95-104 Brin. For structural and aircraft parts, levers, brackets; wrought, heat treatable. *Obsolete*

NORAL 172
Alcan-Booth Industries, Ltd.
Cu 8, Si 2.5, bal Al.
Cast: 22,000 TS; 15,000 YS; 1 El; 65 Brin. For light alloy castings. *Obsolete*

NORAL 19S
Alcan-Booth Industries, Ltd.
Cu 4, Mg 1.25, Ni 2, bal Fe.
Heat treated: 58,200 TS; 39,200 YS; 10 El; 130 Brin. For pistons, cylinder heads; age-hardenable, high temperature application. *Obsolete*

NORAL 1S
Alcan-Booth Industries, Ltd.
Al 99.5.
For impact extruded containers, food and chemical equipment; high corrosion resistance and high ductility. *Obsolete*

NORAL 1SC
Alcan-Booth Industries, Ltd.
Al 99.5.
For electrical conductors, busbars, overhead lines; high electrical conductivity. *Obsolete*

NORAL 211
Alcan-Booth Industries, Ltd.
Cu, Al.
For light alloy castings; good machinability. *Obsolete*

NORAL 218
Alcan-Booth Industries, Ltd.
Cu 4, Mg 1.5, Ni 2.5, bal Al.
Aged: 45,000 TS; 35,000 YS; 1 El; 110 Brin. For pistons, cylinder heads; age-hardenable, high temperature use. *Obsolete*

NORAL 22 S
Alcan-Booth Industries, Ltd.
Cu-Mg-Mn-Si, bal Al.
Heat treated: 56,000-62,000 TS; 8-13 El; 120-128 Brin. For aircraft parts; wrought, heat treatable. *Obsolete*

NORAL 222
Alcan-Booth Industries, Ltd.
Cu 4-4.6, Si 0-0.9, Fe 0-0.7, Ti 0-0.25, bal Al.
36,000 TS; For light alloy castings; age hardened. *Obsolete*

NORAL 224
Alcan-Booth Industries, Ltd.
Cu, bal Al.
For castings; light alloy. *Obsolete*

NORAL 225
Alcan-Booth Industries, Ltd.
Cu, bal Al.
For castings; light alloy. *Obsolete*

NORAL 226
Alcan-Booth Industries, Ltd.
Si 0-0.9, Fe 0-0.7, Ti 0-0.25, Cu 4-4.6, bal Al.
W: 34,000 TS; 18,000 YS; 10 El; 90 Brin. T: 61,000 TS; 54,000 YS; 5 El; 130 Brin. For castings, aircraft parts, gear boxes, levers, age hardened, high strength. *Obsolete*

NORAL 232
Alcan-Booth Industries, Ltd.
Cu, bal Al.
For castings; light alloy. *Obsolete*

NORAL 234
Alcan-Booth Industries, Ltd.
Cu, bal Al.
For castings; light alloy. *Obsolete*

NORAL 235
Alcan-Booth Industries, Ltd.
Cu, bal Al.
For castings; light alloy. *Obsolete*

NORAL 236
Alcan-Booth Industries, Ltd.
Cu, bal Al.
For castings; light alloy. *Obsolete*

NORAL 238
Alcan-Booth Industries, Ltd.
Cu, bal Al.
For castings; light alloy. *Obsolete*

NORAL 24 S
Alcan-Booth Industries, Ltd.
Cu 4.5, Mg 1.6, Mn 0.6, bal Al.
Heat treated: 56,000-58,000 TS; 10-17 El; 100-110 Brin. For aircraft parts; wrought, heat treatable. *Obsolete*

NORAL 240
Alcan-Booth Industries, Ltd.
Cu, bal Al.
For castings; light alloy. *Obsolete*

NORAL 242
Alcan-Booth Industries, Ltd.
Cu, bal Al.
For castings; light alloy. *Obsolete*

NORAL 244
Alcan-Booth Industries, Ltd.
Cu, bal Al.
For castings; light alloy. *Obsolete*

NORAL 25 S
Alcan-Booth Industries, Ltd.
Mn 0.8, Cu 4.5, bal Al.
Heat treated: 28,000-58,000 TS; 37,000 YS; 19 El; 110 Brin. For aircraft parts, propellers; wrought, heat treatable. *Obsolete*

NORAL 250
Alcan-Booth Industries, Ltd.
Cu, bal Al.
For castings; light alloy. *Obsolete*

NORAL 252
Alcan-Booth Industries, Ltd.
Cu 10, Fe 0.75, Mg 0.25, bal Al.
T-temper: 31,400-44,000 TS; 28,000-36,000 YS; 1 El; 125-130 Brin. For pistons, castings; heat treatable. *Obsolete*

NORAL 26 S
Alcan-Booth Industries, Ltd.
Cu 3-4.5, Si 0-1, Mn 0-1.2, Mg 0-1, bal Al.
O-temper: 60,000 TS; 12,000 YS; 20 El; 80 Brin. W-temper: 63,000 TS; 38,000 YS; 18 El; 115 Brin. T-temper: 69,500 TS; 58,000 YS; 9 El; 135 Brin. For aircraft parts, structures, gears, levers, brackets; heat treatable, high strength. *Obsolete*

NORAL 260
Alcan-Booth Industries, Ltd.
Cu 12, bal Al.
For castings; light alloy. *Obsolete*

NORAL 28S
Alcan-Booth Industries, Ltd.
Al alloy.
Rolled: 49,300 TS; 14 El. For machinery parts; free-cutting. *Obsolete*

NORAL 2S
Alcan-Booth Industries, Ltd.
Cu 0.1, Si 0.5, Fe 0.7, bal Al.
Soft: 11,000 TS; 7,000 YS; 35 El; 20 Brin. Hard: 23,000 TS; 14,500 YS; 4 El; 44 Brin. For chemical and brewing equipment, tanks, paneling; ductile and corrosion resistant. *Obsolete*

NORAL 3 S
Alcan-Booth Industries, Ltd.
Mn 1-1.5, Cu 0.15, Si 0.6, bal Al.
Soft: 16,000 TS; 7,000 YS; 39 El; 29 Brin. Hard: 29,000 TS; 25,000 YS; 4 El; 30 Brin. For panels, tanks, architecture, trim, food containers; wrought. *Obsolete*

NORAL 305
Alcan-Booth Industries, Ltd.
Mg 1.5, Si 1, bal Al.
T: 34,000 TS; 22,000 YS; 3 El; 85 Brin. For castings; light alloy, age-hardenable. *Obsolete*

NORAL 307
Alcan-Booth Industries, Ltd.
Mg, bal Al.
For castings; light alloy. *Obsolete*

NORAL 31 S
Alcan-Booth Industries, Ltd.
Si 9-14, bal Al.
20,000 TS; 30 El; 38 Brin. For marine parts, paneling; extruded. *Obsolete*

NORAL 320
Alcan-Booth Industries, Ltd.
Cu 0.1, Mg 3-6, Mn 0.4, bal Al.
Cast: 20,200 TS; 3 El. For gasoline flow meters; corrosion resistant. *Obsolete*

NORAL 322
Alcan-Booth Industries, Ltd.
Mg 3.7, Zn 2, bal Al.
Cast: 26,000 TS; 17,000 YS; 5 El; 65 Brin. For castings; corrosion resistant. *Obsolete*

NORAL 330
Alcan-Booth Industries, Ltd.
Mg, bal Al.
For castings; light alloy. *Obsolete*

NORAL 33S
Alcan-Booth Industries, Ltd.
Si 5, bal Al.
M-temper: 19,000 TS; 10,000 YS; 20 El; 33 Brin. For architectural trim, welding wire; corrosion resistant. *Obsolete*

NORAL 350
Alcan-Booth Industries, Ltd.
Mg 9.5-10.5, Si 0.35, Ti 0.25, Fe 0.35, bal Al.
Sand cast: 40,000 TS; 25,000 YS; 13 El; 95 Brin. Permanent mold: 47,000 TS; 29,000 YS; 18 El; 105 Brin. For levers, brackets, marine and bus parts, hardware; age-hardenable, corrosion and shock resistant. *Obsolete*

NORAL 38 S
Alcan-Booth Industries, Ltd.
Si 11.5, Cu 1, Mg 1, Ni 1, bal Al.
T6-temper: 54,000 TS; 42,000 YS; 4 El; 100 Brin. For light alloy parts, pistons; heat treatable. *Obsolete*

NORAL 42S
Alcan-Booth Industries, Ltd.
Cu 2.2, Si 0.85, Ni 1.2, Mg 1.5, Fe 1, Ti 0.2, bal Al.
Heat treated: 58,000 TS; 38,100 YS; 6 El; 125 Brin. For aircraft and general engineering structural parts; age hardenable, extrusions. *Obsolete*

NORAL 450
Alcan-Booth Industries, Ltd.
Zn, bal Al.
For castings; light alloy. *Obsolete*

NORAL 465
Alcan-Booth Industries, Ltd.
Zn, Cu, bal Al.
For castings; light alloy. *Obsolete*

NORAL 4S
Alcan-Booth Industries, Ltd.
Mn 1.25, Mg 1, bal Al.
Rolled: 21,000-22,000 TS; 18-27 El; 37-42 Brin. For architectural panels, roofing, floors; wrought. *Obsolete*

NORAL 50 S
Alcan-Booth Industries, Ltd.
Mg 0.6, Si 0.5, bal Al.
Extruded: 22,000 TS; 16,000 YS; 20 El; 40 Brin. T-temper: 36,000 TS; 27,000 YS; 25 El; 45 Brin. For light alloy parts, auto bodies, panels, trim; heat treatable. *Obsolete*

NORAL 51 S
Alcan-Booth Industries, Ltd.
Mg 0.6, Si 1, Cu 0.15, Fe 0.6, Mn 1, bal Al.
Heat treated: 44,000 TS; 40,000 YS; 13 El; 90 Brin. For structural, architectural and marine parts; heat treatable. *Obsolete*

NORAL 510
Alcan-Booth Industries, Ltd.
Mn, bal Al.
For castings; light alloy. *Obsolete*

NORAL 54S
Alcan-Booth Industries, Ltd.
Mg 3-4, Cu 0.15, Fe 0.75, Cr 0.5, Mn 1, Ti 0.2, bal Al.
O-temper: 33,500 TS; 14,500 YS; 24 El; 55 Brin. 1/4 H-temper: 48,600 TS; 34,200 YS; 9 El; 75 Brin. For marine parts, car bodies, shipbuilding; sea water corrosion resistant. *Obsolete*

NORAL 55 S
Alcan-Booth Industries, Ltd.
Cu, bal Al.
Heat treated: 33,000 TS; 26 El; 76 Brin. For architectural and marine parts; age-hardenable. *Obsolete*

NORAL 57 S
Alcan-Booth Industries, Ltd.
Mg 0.2, Mn 0.25, bal Al.
Extruded: 24,000 TS; 11,000 YS; 30 El; 45 Brin. For architectural and marine parts; work hardens. *Obsolete*

NORAL 58S
Alcan-Booth Industries, Ltd.
Mg 6.5-7.5, Mn 1, Ti 0.2, Cu 0.15, Si 0.6, bal Al.
O-temper: 44,800 TS; 20,200 YS; 18 El; 50 Brin. 1/4 H-temper: 51,500 TS; 36,000 YS. 1/2 H-temper: 56,000 TS; 5 El; 70 Brin. Extruded: 49,300 TS; 35 El; 85 Brin. For light weight structures, marine parts; good formability and corrosion resistance. *Obsolete*

NORAL 62S
Alcan-Booth Industries, Ltd.
Cu 1.5, Mg 1, Mn 0.75, Si 1, bal Al.
S-W temper: 53,700 TS; 33,600 YS; 17 El; 90 Brin. S-T temper: 62,700 TS; 58,200 YS; 12 El; 120 Brin. For aircraft structures, tubular furniture; heat treatable, wrought. *Obsolete*

NORAL 65S
Alcan-Booth Industries, Ltd.
Cu 0.25, Mg 1, Si 0.5, Cr 0.25, bal Al.
S-O temper: 19,000 TS; 7,840 YS; 24 El; S-W temper: 32,500 TS; 19,000 YS; 20 El; 60 Brin. S-T temper: 43,700 TS; 38,000 YS; 12 El; 90 Brin. For body panels; wrought, heat treatable. *Obsolete*

NORAL 80 S
Alcan-Booth Industries, Ltd.
Al alloy.
38,000 TS; 24 El; 77 Brin. For light alloy parts; free machining. *Obsolete*

NORAL 8000 S
Alcan-Booth Industries, Ltd.
Al alloy.
For conductors in overhead transmission lines; high electrical conductivity. *Obsolete*

NORAL A56S
Alcan-Booth Industries, Ltd.
Mg 5, Mn 1, Cu 0.15, Si 0.6, Fe 0.75, Cr 0.5, Ti 0.2, bal Al.
O-temper: 40,100 TS; 18,000 YS; 28 El; 70 Brin. 1/4 H-temper: 52,600 TS; 31,400 YS; 8-11 El; 100 Brin. For marine parts, shipbuilding, rivets; sea water corrosion resistant. *Obsolete*

NORAL B116
Alcan-Booth Industries, Ltd.
Cu 0.1, Mg 0.5, Si 3.5-5.5, Fe 0.5, Mn 0.5, Ti 0.2, bal Al.
Sand cast: 22,200 TS; 12,300 YS; 3.0 El; 65 Brin. Permanent mold: 25,800 TS; 14,500 YS; 5.5 El; 70 Brin. For aircraft and auto parts; age-hardened, good castability. *Obsolete*

NORAL B26S
Alcan-Booth Industries, Ltd.
Cu 3.5-4.8, Mg 0.8, Si 0.9, Fe 1, Mn 1.2, Ti 0.3, bal Al.
W-temper: 53,800 TS; 31,400 YS; 15 El; 115 Brin. For aircraft parts; high strength, fair ductility. *Obsolete*

NORAL B320
Alcan-Booth Industries, Ltd.
Mg 5, Mn 0.5, bal Al.
Sand cast: 22,400 TS; 13,500 YS; 4 El; 70 Brin. Permanent mold: 31,400 TS; 13,500 YS; 12 El; 70 Brin. For architectural and ornamental hardware; corrosion resistant. *Obsolete*

NORAL B51S
Alcan-Booth Industries, Ltd.
Si 1, Mg 0.6, Mn 0.5, bal Al.
O-temper: 20,200 TS; 10,000 YS; 26 El. W-temper: 33,600 TS; 24,600 YS; 20 El; 60 Brin. WP-temper: 49,300 TS; 41,500 YS; 13 El; 95 Brin. For road and rail transport vehicles, bridges, cranes; heat treatable, corrosion resistant. *Obsolete*

NORAL B54S
Alcan-Booth Industries, Ltd.
Mg 4.25, Mn 0.75, bal Al.
Hot rolled: 43,500 TS; 29,000 YS; 20 El; 70 Brin. O-temper: 40,400 TS; 18,000 YS; 26 El; 65 Brin. For ship building, marine parts; good formability, corrosion resistant. *Obsolete*

NORAL B75S
Alcan-Booth Industries, Ltd.
Zn 6, Mg 2.5, Cu 0.6, Mn 0.25, Cr 0.2, bal Al.
WP-temper: 90,000 TS; 85,000 YS; 9.5 El; 170 Brin. For \ aircraft and engineering structures; heat treatable, high strength. *Obsolete*

NORAL C77 S
Alcan-Booth Industries, Ltd.
Cu 1.75, Mg 2, Zn 7, Cr 0.13, bal Al.
S-T temper: 94,000 TS; 85,100 YS; 11 El; 175 Brin. For aircraft structures, spar booms; heat treatable. *Obsolete*

NORAL D50S
Alcan-Booth Industries, Ltd.
Mg 0.5, Si 0.5, bal Al.
WP-temper: 29,200 TS; 23,500 YS; 12 El. For busbars; good strength and conductivity. *Obsolete*

NORAL D57S

Alcan-Booth Industries, Ltd.
Mg 1.3, bal Al.
1/4-H temper: 19,100 TS; 14,600 YS; 22 El; 40 Brin. 1/2-H temper: 28,000 TS; 24,600 YS; 10 El; 63 Brin. H-temper: 37,000 TS; 31,400 YS; 5 El; 70 Brin. For auto trim, domestic appliances; high reflectivity. *Obsolete*

NORAL H10

Alcan-Booth Industries, Ltd.
Mg 0.4-1.5, Si 0.75-1.3, bal Al.
W-temper: 33,600 TS; 20,200 YS; 20 El; 60 Brin. WP-temper: 44,800 TS; 40,400 YS; 13 El; 90 Brin. For aircraft structural members; age-hardened. *Obsolete*

NORAL M57S

Alcan-Booth Industries, Ltd.
Mg 2, Cu 0.15, Si 0.6, Fe 0.75, Mn 0.5, Ti 0.2, Cr 0.5, bal Al.
O-temper: 24,600 TS; 11,200 YS; 24 El; 48 Brin. 1/4 H-temper: 31,000 TS; 25,000 YS; 10 El; 63 Brin. 1/2 H-temper: 35,000 TS; 30,900 YS; 7 El; 70 Brin. For shipbuilding, motor car bodies, marine parts, corrosion resistant, work hardens readily. *Obsolete*

NORAL M75 S

Alcan-Booth Industries, Ltd.
Cu 1.33, Mg 2.5, Mn 0.25, Zn 5.75, Cr 0.12, bal Al.
S-O temper: 28,000 TS; 12,300 YS; 20 El. S-T temper: 76,000 TS; 65,999 YS; 12 El. For aircraft structures; heat treatable; wrought. *Obsolete*

NORAL NA17S ALCLAD

Alcan-Booth Industries, Ltd.
Cu 3.5-4.5, Mg 0.5, Mn 0.7, bal Al.
For structures, aircraft; heat treatable. *Obsolete*

NORAL NA22S ALCLAD

Alcan-Booth Industries, Ltd.
Cu, Mg, Si, Mn, bal Al.
For structures, aircraft; heat treatable. *Obsolete*

NORAL NA22S ALCLAD

Alcan-Booth Industries, Ltd.
Cu, Mg, Si, Mn, bal Al.
For structures, aircraft; heat treatable. *Obsolete*

NORAL NA24S ALCLAD

Alcan-Booth Industries, Ltd.
Cu 4.4, Mn 1.7, Mg 1.5, bal Al.
For structures, aircraft; heat treatable. *Obsolete*

NORAL NA26ST ALCLAD

Alcan-Booth Industries, Ltd.
Cu 4.2, Mg 1.5, Mn 0.7, bal Al.
Rolled: 50,000 TS; 8 El. For structures, aircraft; clad. *Obsolete*

NORANDA 1100

Noranda Metal Industries Ltd.
99.9 Cu min.
Electrolytic tough pitch copper; for bus bars, conductors, wave guides, copper to brass seals. CDA and UNS C11000; was NORANDA 102.

NORANDA 1140

Noranda Metal Industries Ltd.
Cu 99.9, 0.034 Ag (10 oz/ton) min.
Tough pitch copper with silver. CDA and UNS C11400.

NORANDA 1150

Noranda Metal Industries Ltd.
99.9 Cu min, 0.054 Ag (16 oz/ton) min.
Tough pitch copper with silver; resists softening by heat. CDA and UNS C11500; was NORANDA 3111.

NORANDA 1160

Noranda Metal Industries Ltd.
Cu 99.9, 0.085 Ag (25 oz/ton) min.
Tough pitch copper with silver; resists softening by heat; for commutator segments. CDA and UNS C11600.

NORANDA 1200

Noranda Metal Industries Ltd.
P 0-0.012, 99.9 Cu min.
Phosphorus deoxidized copper. For radiators, commutators, switches. CDA and UNS C12000; was NORANDA 106.

NORANDA 1220

Noranda Metal Industries Ltd.
P 0.02, 99.9 Cu min.
Phosphorus deoxidized copper. For air conditioners, gas lines, hydraulic and oil lines. CDA and UNS C12200; was NORANDA 110.

NORANDA 1450

Noranda Metal Industries Ltd.
Te 0.4-0.6, P 0-0.012, 99.9 Cu min.
Tellurium bearing phosphorus deoxidized copper. For improved machining. CDA and UNS C14500.

NORANDA 2100

Noranda Metal Industries Ltd.
Cu 94-96, Zn 4-6.
Gilding brass (5%); for jewelry, tokens. CDA and UNS C21000; was NORANDA 26.

NORANDA 2200

Noranda Metal Industries Ltd.
Cu 90, Zn 10.
Commercial bronze; for costume jewelry, ornamental trim, weather stripping. CDA and UNS C22000; was NORANDA 25.

NORANDA 2300

Noranda Metal Industries Ltd.
Cu 85, Zn 15.
Red brass (85%); for radiator cores, conduit, pump lines, trim. CDA and UNS C23000; was NORANDA 85.

NORANDA 2400

Noranda Metal Industries Ltd.
Cu 80, Zn 20.
Low brass (80%); for ornamental metal work, medallions, pump liners. CDA and UNS C24000; was NORANDA 5.

NORANDA 2600

Noranda Metal Industries Ltd.
Zn 30.5, Cu 69.5.
Cartridge brass; good strength and ductility. For grillwork, lamp fixtures, cartridge cases. CDA and UNS C26000; was NORANDA 69.

NORANDA 2680

Noranda Metal Industries Ltd.
Cu 66, Zn 34.
Yellow brass; good strength, particularly after cold work. For hardware, reflectors, plumbing parts. CDA and UNS C26800; was NORANDA 1.

NORANDA 3140

Noranda Metal Industries Ltd.
Pb 2, Zn 8.5, bal Cu.
Leaded commercial bronze; free-machining for screw machine parts. CDA and UNS C31400; was NORANDA 80.

NORANDA 3200

Noranda Metal Industries Ltd.
Pb 2, Cu 85, Zn 13.
Leaded red brass; free-machining, for screw machine operations. CDA and UNS C32000.

NORANDA 3300

Noranda Metal Industries Ltd.
Pb 0.5, Zn 32.5, Cu 67.
Low-leaded brass; for tubing, plumbing, pumps, power cylinders. Good machining. CDA and UNS C33000; was NORANDA 18.

NORANDA 3320

Noranda Metal Industries Ltd.
Pb 1.75, Cu 67, bal Zn.
High-leaded brass; good machinability. For screw machine parts. CDA and UNS C33200; was NORANDA 04.

NORANDA 3400

Noranda Metal Industries Ltd.
Pb 1.1, Cu 64.5, bal Zn.
Medium-leaded brass; good machinability. For hardware, clock plates, fasteners. CDA and UNS C34000; was NORANDA 63.

NORANDA 3530

Noranda Metal Industries Ltd.
Pb 2, Cu 62, bal Zn.
High-leaded brass (62%); for hardware, watch cases, clock and watch parts. CDA and UNS C35300; was NORANDA 62.

NORANDA 3600

Noranda Metal Industries Ltd.
Pb 3.4, Cu 61.5, bal Zn.
Free-cutting brass; for automatic screw machine parts; clock gears. CDA and UNS C36000; was NORANDA 6.

NORANDA 3770

Noranda Metal Industries Ltd.
Pb 2, Cu 59.5, bal Zn.
Forging brass; for forged marine hardware, valve stems, fasteners, bolts, plumbing. CDA and UNS C37700.

NORANDA 4250

Noranda Metal Industries Ltd.
Sn 1.9, Zn 9.1, bal Cu.
Tin commercial bronze; for heat exchangers. CDA and UNS C42500; was NORANDA 92.

NORANDA 4430

Noranda Metal Industries Ltd.
Sn 1, As 0.02-0.1, Cu 71, bal Zn.
Arsenical admiralty grade; for marine hardware. CDA and UNS C44300; was NORANDA 30.

NORANDA 4620

Noranda Metal Industries Ltd.
Sn 0.5-1, Cu 64, bal Zn.
Naval brass; for hardware, fixtures. CDA and UNS C46200; was NORANDA 46.

NORANDA 4640

Noranda Metal Industries Ltd.
Sn 0.65, Pb 0.2, Cu 60.5, bal Zn.
Naval brass, uninhibited; for marine hardware, valve stems, shafts. CDA and UNS C46400; was NORANDA 24.

NORANDA 4820

Noranda Metal Industries Ltd.
Sn 0.75, Pb 0.7, Cu 60.5, bal Zn.
Naval brass, medium leaded; for marine hardware, screw machine products. CDA and UNS C48200; was NORANDA 28.

NORANDA 4850

Noranda Metal Industries Ltd.
Sn 0.75, Pb 1.75, Cu 60.5, bal Zn.
Naval brass, high leaded; free machining, for screw machine products. CDA and UNS C48500; was NORANDA 29.

NORANDA 5050

Noranda Metal Industries Ltd.
Sn 1.0, Zn 0.3, P 0-0.35, bal Cu.
Phosphor bronze (1.25% E); for electrical contacts, flexible hose, pole-line hardware. CDA and UNS C50500; was NORANDA 32.

NORANDA 5100

Noranda Metal Industries Ltd.
Sn 5, Zn 0.3, P 0-0.35, bal Cu.
Phosphor bronze (5% A); for hardware. CDA and UNS C51000.

NORANDA 5190

Noranda Metal Industries Ltd.
Sn 6, P 0-0.35, Zn 0.3, bal Cu.
Phosphor bronze; for springs, clutch discs, hardware. CDA and UNS C51900; was NORANDA 1502.

NORANDA 5210
Noranda Metal Industries Ltd.
Sn 8, P 0-0.35, Zn 0.2, bal Cu.
Phosphor bronze (8% C); high strength. For springs, switch parts, wire brushes. CDA and UNS C52100; was NORANDA 1532.

NORANDA 5211
Noranda Metal Industries Ltd.
Sn 8, P 0.1, bal Cu.
Phosphor bronze; for Bourdon tubing, springs, clutch discs, sleeve bushings. Was NORANDA 35.

NORANDA 6080
Noranda Metal Industries Ltd.
Al 5.5, As 0.25, bal Cu.
Aluminum bronze; for fasteners, structural components, condensers. CDA and UNS C60800; was NORANDA 53.

NORANDA 6371
Noranda Metal Industries Ltd.
Al 7.15, Si 2, bal Cu.
Aluminum-silicon bronze; good strength. For bushings, fasteners, marine fittings. Was NORANDA 707.

NORANDA 6510
Noranda Metal Industries Ltd.
Si 1.9, bal Cu.
Low-silicon bronze; for hydraulic pressure lines, bolts, pole line hardware. CDA and UNS C65100; was NORANDA 609.

NORANDA 6670
Noranda Metal Industries Ltd.
Mn 1.25, Cu 70, bal Zn.
Manganese brass; for shafting, valve stems, pump rods, hardware. CDA and UNS C66700; was NORANDA 73.

NORANDA 6750
Noranda Metal Industries Ltd.
Mn 0.3, Fe 1, Sn 1, Cu 58, bal Zn.
Manganese bronze A; for automotive clutch discs, shafting, hardware. CDA and UNS C67500; was NORANDA 19.

NORANDA 6810
Noranda Metal Industries Ltd.
Sn 1, Fe 1, Mn 0-0.5, Cu 58, bal Zn.
Bronze, low fuming. CDA and UNS C68100.

NORANDA 6870
Noranda Metal Industries Ltd.
Al 2, As 0-0.1, Cu 78, bal Zn.
Aluminum brass, arsenical; for condenser tubes, heat exchangers. CDA and UNS C68700; was NORANDA 54.

NORANDA 7060
Noranda Metal Industries Ltd.
Mn 0.75, Fe 1, Ni 10, bal Cu.
Copper nickel, 10%; for fasteners, decorative trim. CDA and UNS C70600.

NORANDA 7100
Noranda Metal Industries Ltd.
Mn 0.75, Fe 0.4, Ni 20, bal Cu.
Copper nickel 20%; for valves, condenser plates, evaporators, fasteners. CDA and UNS C71000; was NORANDA 520.

NORANDA 7150
Noranda Metal Industries Ltd.
Mn 0.75, Fe 0.4, Ni 30, bal Cu.
Copper nickel, 30%; for valves, pumps, tanks, fasteners. CDA and UNS C71500; was NORANDA 531.

NORANDA 7450
Noranda Metal Industries Ltd.
Ni 10, Cu 65, Mn 0.15, bal Zn.
Nickel silver, 10%; for costume jewelry, camera parts, tableware. CDA and UNS C74500; was NORANDA 567.

NORANDA 7520
Noranda Metal Industries Ltd.
Ni 18, Mn 0.15, Cu 65, bal Zn.
Nickel silver 18%; for diaphragms, springs, slide fasteners, jewelry. CDA and UNS C75200; was NORANDA 565.

NORANDA 7620
Noranda Metal Industries Ltd.
Ni 12, Cu 60, bal Zn.
Nickel silver; for decorative hardware. CDA and UNS C76200.

NORANDA 7700
Noranda Metal Industries Ltd.
Ni 18, Mn 0.15, Cu 55, bal Zn.
Nickel silver, 55-18; for tableware, springs, fixtures, hardware. CDA and UNS C77000; was NORANDA 555.

NORANDA 7930
Noranda Metal Industries Ltd.
Ni 12, Pb 1.5, Zn 25, Mn 0.15, bal Cu.
Leaded nickel silver; good machinability. For costume jewelry, tableware, hardware. CDA and UNS C79300; was NORANDA 580.

NORANDA-1
Now NORANDA 2680.

NORANDA-102
Now NORANDA 1100.

NORANDA-106
Now NORANDA 1200.

NORANDA-110
Now NORANDA 1220.

NORANDA-113
Noranda Metal Industries Ltd.
P 0.02, As 0.3, 99.4% Cu min.
Hard: 40,000 TS; 32,000 YS; 25 El; F 77 Rock. For gas and oil lines, air conditioning. Arsenical copper. Deoxidized. *Obsolete*

NORANDA-136
Now NORANDA 3770.

NORANDA-1502
Now NORANDA 5190.

NORANDA-1532
Now NORANDA 5210.

NORANDA-18
Now NORANDA 3300.

NORANDA-19
Now NORANDA 6750.

NORANDA-20
Now NORANDA 4430.

NORANDA-24
Now NORANDA 4640.

NORANDA-25
Now NORANDA 2200.

NORANDA-26
Now NORANDA 2100.

NORANDA-28
Now NORANDA 4820.

NORANDA-29
Now NORANDA 4850.

NORANDA-3
Now NORANDA 3300.

NORANDA-30
Now NORANDA 4430.

NORANDA-3111
Now NORANDA 1150.

NORANDA-32
Now NORANDA 5050.

NORANDA-35
Now NORANDA 5210.

NORANDA-3770
Noranda Metal Industries Ltd.
Cu 58-61, Pb 1.5-2.5, Sn 0.15-0.25, bal Zn.
Hard drawn: 72,000 TS; 50,000 YS; 15 El; B 78 Rock. Soft: 52,000 TS; 20,000 YS; 45 El; F 78 Rock. For forged marine hardware, valve stems, bolts, fasteners, plumbing. Forging brass. *Obsolete*

NORANDA-46
Now NORANDA 4620.

NORANDA-5
Now NORANDA 2400.

NORANDA-520
Now NORANDA 7100.

NORANDA-53
Now NORANDA 6080.

NORANDA-531
Now NORANDA 7150.

NORANDA-54
Now NORANDA 6870.

NORANDA-555
Now NORANDA 7700.

NORANDA-565
Now NORANDA 7520.

NORANDA-567
Now NORANDA 7450.

NORANDA-580
Now NORANDA 7930.

NORANDA-6
Now NORANDA 3600.

NORANDA-609
Now NORANDA 6510.

NORANDA-62
Now NORANDA 3530.

NORANDA-63
Now NORANDA 3400.

NORANDA-64
Now NORANDA 3320.

NORANDA-69
Now NORANDA 2600.

NORANDA-707
Now NORANDA 6371.

NORANDA-73
Now NORANDA 6670.

NORANDA-820
Noranda Metal Industries Ltd.
Sn 5.5, P 0.15, bal Cu.
Hard: 81,000 TS; 75,000 YS; 10 El; B 87 Rock. Soft: 50,000 TS; 21,000 YS; 52 El; F 77 Rock. For bridge bearing plates, bellows, Bourdon tubing, clutch discs, sleeve bushings, springs. Tough, high strength, corrosion resistant. Phosphor bronze Gr. A. *Obsolete*

NORANDA-85
Now NORANDA 2300.

NORANDA-89
Now NORANDA 3140.

NORANDA-92
Now NORANDA 4250.

NORBIDE
Norton Co.
B 78, Fe 0.14, C 21.
For wire drawing dies; drilling WC die nibs, nozzles for abrasive blasting; cemented boron-carbide.

NORDIC
Reading Iron Co.
C 0.02, Si 0.03, 3% slag, bal Fe.
Rolled: 45,000 TS; 28,000 YS; 30 El; 44 RA; 95 Brin. For pipes; wrought iron. *Obsolete*

NORDIC 360
British Steel plc
C 0-0.16, Si 0-0.4, Mn 0-1.4, P 0-0.02, S 0-0.008, Al 0.02-0.06, Nb 0-0.04, V 0-0.05, N 0-0.012, 0.41 CEV max, bal Fe.
Seamless hollow sections: 490-630 N/mm^2 TS; 360 N/mm^2 YS. High strength; weldable.

NORELCO CONTACT 15
North American Philips Co.Inc.
C 0.05-0.1, Mn 0.5-1, Si 0.1-0.3, bal Fe.
Welded: 75,000 TS; 65,000 YS; 35 El. For welding electrodes; lime coated.

NORELCO CONTACT 18
North American Philips Co.Inc.
C 0.08-0.15, Mn 0.5-0.7, Si 0.1-0.2, bal Fe.
For contact welding electrodes; arc welding, organic coating.

NORELCO CONTACT 20
North American Philips Co.Inc.
C 0.08-0.15, Mn 0.5-0.7, Si 0.1-0.2, bal Fe.
Welded: 67,000 TS; 53,000 YS; 25 El. For welding electrodes; iron oxide coating.

NORGRIP
Pechiney/Eurotungstene
WC plus alloy.
For anti-skid tire studs and studded straps; for snow and ice chains.

NORIS CME
Hoffman & Co. KG
C 0.2, Mn 1.25, Cr 1.15, bal Fe.
For gears, cams, camshafts, mandrels; case hardening steel.

NORMAG A6
Norsk Hydro
Magnesium. Al 5.5-6.5, Zn 0-0.1, Mn 0.1-0.4, Si 0-0.1, Cu 0-0.05, 0.15 others, bal Mg.
Ductile magnesium alloy for pressure die casting; tough and cold workability. For car wheels. Pressure die cast: 190-230 N/mm^2 TS; 120-150 N/mm^2 YS; 4-8 El; 55-70 Brin. Sand cast: 180-250 N/mm^2 TS; 80-110 N/mm^2 YS; 8-15 El; 50-65 Brin.

NORMAG AS 21
Norsk Hydro
Magnesium. Al 1.9-2.5, Zn 0.15-0.2, Si 0.7-1.2, Cu 0-0.04, Ni 0-0.002, 0.35 Mn min, 0.30 others, bal Mg.
Magnesium alloy; creep resistant up to 150°C. Pressure die cast: 170 N/mm^2 TS; 110 N/mm^2 YS; 4 El; 63 Brin.

NORMAG AS 41
Norsk Hydro
Magnesium. Al 4-5, Zn 0-0.1, Mn 0.2-0.5, Si 0.4-1, 0.15 others, bal Mg.
Magnesium alloy; creep resistant up to 150°C. Pressure die cast: 200-250 N/mm^2 TS; 120-150 N/mm^2 YS; 3-6 El; 60-90 Brin.

NORMAG AZ 61
Norsk Hydro
Magnesium. Al 5.5-6.5, Zn 0.2-1, Mn 0.1-0.4, 0.15 others, bal Mg.
Cold worked magnesium alloy. Pressure die cast: 200-240 N/mm^2 TS; 130-160 N/mm^2 YS; 3-6 El; 55-70 Brin.

NORMAG AZ 63
Norsk Hydro
Magnesium. Al 5.5-6.5, Zn 2.4-3.5, Si 0-0.05, Cu 0-0.02, Ni 0-0.003, 0.15 Mg min, 0.30 others, bal Mg.
Magnesium anode alloy.

NORMAG AZ 71
Norsk Hydro
Magnesium. Al 6.5-7.4, Zn 0.25-1, Mn 0.1-0.3, Si 0-0.15, Cu 0-0.05, 0.30 others, bal Mg.
Workable magnesium alloy for pressure die casting. Good strength, ductility, and toughness in as cast and heat treated conditions. Weldable when sand or gravity die cast. Pressure die cast: 200 N/mm^2 TS; 140 N/mm^2 YS; 3 El; 60 Brin.

NORMAG AZ 81
Norsk Hydro
Magnesium. Al 7-8.5, Zn 0.3-1, Mn 0.1-0.3, Si 0-0.1, Cu 0-0.05, 0.15 others, bal Mg.
Magnesium alloy for pressure die casting. Good strength, ductility, and toughness in as cast and heat treated conditions. Weldable when sand or gravity die cast. Sand cast: 160-280 N/mm^2 TS; 90-120 N/mm^2 YS; 2-12 El; 50-65 Brin. Gravity die cast: 160-280 N/mm^2 TS; 90-120 N/mm^2 YS; 8-12 El; 50-65 Brin. Weldable when sand or die gravity cast.

NORMAG AZ 91
Norsk Hydro
Magnesium. Al 8-9.5, Zn 0.3-1, Mn 0.1-0.3, Si 0.1-0.2, Cu 0.05-0.08, Ni 0-0.01, 0.30 others, bal Mg.
Magnesium alloy for pressure die casting. Sand cast: 160-300 N/mm^2 TS; 90-190 N/mm^2 YS; 2-12 El; 50-90 Brin. Gravity die cast: 160-300 N/mm^2 TS; 110-190 N/mm^2 YS; 2-10 El; 55-90 Brin. Weldable when sand or gravity die cast.

NORMALLOY
Foote Mineral Co.
C 0.4, Mn 0.8, Cr 1.5, V 0.2, bal Fe.
For crankshafts; "Vancoram" normalized alloy steel. *Obsolete*

NORMALLOY
U.S. Steel Corp.
C 0.4-0.5, Mn 0.9-1.3, Cr 0.3-0.6, 0.05% V min, bal Fe.
Normalized: 110,000 TS; 70,000 YS; 15 El; 40 RA. For sprockets, gear hubs; for heavy duty automotive machinery. *Obsolete*

NORMANNA
Swedish American Steel Corp.
C, alloy, bal Fe.
For tools; oil hardening.

NORMAR
Ziv Steel & Wire Co.
C 0.9, Cr 0.5, Mn 1.5, Mo 0.25, bal Fe.
For dies, punches, crimpers; non-deforming, shock resistant.

NORO 30
Hoffman & Co. KG
C 0.95, W, Co, Cr, V, bal Fe.
For lathe and planer tools, reamers, broaches; high speed steel.

NORO 40
Hoffman & Co. KG
C 0.82, Cr 4, Mo 0.85, V 1.6, W 8.7, bal Fe.
For lathe and planer tools, hobs, reamers, taps; high speed steel.

NORO 50
Hoffman & Co. KG
C 0.85, Cr, V, W, Mo, bal Fe.
For lathe and planer tools, drills, hobs, taps; high speed steel.

NORO 60
Hoffman & Co. KG
C 0.86, Cr 4.1, V 2.5, Mo 0.85, W 12, bal Fe.
For lathe and planer tools, hobs, broaches; high speed steel.

NORO 60 EXTRA
Hoffman & Co. KG
C 1.3, Cr 4.3, Mo 0.85, V 3.8, W 12, bal Fe.
For engraving tools, form dies, taps; high speed steel.

NORO 60 SPEZIAL
Hoffman & Co. KG
C 0.74, Cr 4.1, V 1.1, W 18.5, bal Fe.
For lathe and planer tools, reamers, hobs, taps, drills; high speed steel.

NORO EXTRA D
Hoffman & Co. KG
C 0.35, Cr 1.05, V 0.18, W 1.85, Si 0.9, bal Fe.
For cold work tools, upsetters, headers, dies; oil hardened, tough.

NORO EXTRA L
Hoffman & Co. KG
C 0.55, Cr 1.05, V 0.18, W 1.8, Si 0.9, bal Fe.
For cold work tools, upsetters, headers, dies; oil hardened, tough.

NORO MKW
Hoffman & Co. KG
C 0.38, Si, Cr, V, bal Fe.
For gears, pinions, shafts, bolts, crankshafts; oil hardened, shock resistant.

NORO MRO
Hoffman & Co. KG
C 0.45, Si, Cr, V, bal Fe.
For gears, springs, shafts, arbors, crankshafts; oil hardened, shock resistant.

NORO REKORD 110
U.S. Spring & Bumper Co.
C 0.76, Co 10, Cr 4.2, Mo 0.8, V 1.8, W 18, bal Fe.
For lathe and planer tools, cutters, reamers, hobs; high speed steel.

NORO REKORD 30
Hoffman & Co. KG
C 0.86, Co 2.8, Cr 4.3, Mo 0.85, V 2.1, W 12, bal Fe.
For lathe and planer tools, reamers, hobs, taps; high speed steel.

NORO REKORD 50
U.S. Spring & Bumper Co.
C 0.8, Co, Cr, V, Mo, W, bal Fe.
For lathe and planer tools, drills, taps, reamers; high speed steel.

NORO REKORD 53
U.S. Spring & Bumper Co.
C 0.79, Co 4.75, Cr 4.3, Mo 0.75, V 1.5, W 18, bal Fe.
For lathe and planer tools, hobs, reamers; high speed steel.

NORO REKORD 53V
Hoffman & Co. KG
C 1.3, W, Co, V, Cr, Mo, bal Fe.
For engraving tools, forming dies; reamers; high speed steel.

NORSEC 360
British Steel plc
C 0-0.16, Si 0-0.45, Mn 0-1.6, P 0-0.025, S 0-0.01, 0.02 Al min, 0.12 Nb + V max, bal Fe.
490-630 N/mm^2 TS; 360 N/mm^2 YS min.

NORTH STAR
Swedish American Steel Corp.
C, W, bal Fe.
For dies; non-deforming.

NORTON 90-10
NRC, Inc.
Now NRC 90-10.

NORTON BORON MASTER ALLOY
Norton Co.
B 8.5, Si 2.7-3.3, Fe 81-84, C 1-1.7, Al, bal Ti.
For steel inoculant; increase depth of hardening of steel.

NORTON T-111
NRC, Inc.
Ta 90, W 8, Hf 2.
Weldable, ductile; very high strength to weight ratio at high temperatures. For missile hardware, supersonic air and spacecraft, liquid reactors.

NOVALITE
English manufacture
Cu 12.5, Mg 0.3, Ni 1.4, Fe 0.5, Si 0.5, bal Al.
For pistons; cast.

NOVANTIOX
Chiers-Chatillon
C 0.06, Cr 19, Ni 10, bal Fe.
Annealed: 85,000 TS; 35,000 YS; 60 El; 150 Brin. For architectural molding and trim, kitchen equipment, chemical plant apparatus and processing equipment. Type 304 stainless steel, austenitic. *Obsolete*

NOVAR B
Stahlwerke Sudwestfalen
C 0-0.1, Cr 17, Mo 1.8, Ti = 7 x C, bal Fe.
For welded chemical plant and oil refinery equipment; stainless, stabilized. *Obsolete*

NOVITE
American Marsh Pumps Inc.
Ni 1.5, Cr 0.5, C 3, bal Fe.
Cast: 38,000 TS; 220 Brin. For impellers, cylinder heads, water pumps; cast iron. *Obsolete*

NOVO 2
H. Boker & Co.
C 0.55, W 18, Cr 4, V 2, bal Fe.
For lathe and planer tools, reamers, broaches; Type T2; high speed steel. *Obsolete*

NOVO HIGH SPEED
H. Boker & Co.
C 0.5, Cr 2.7, W 15, V 0.6, bal Fe.
For tools, cutters, punches, dies; high speed steel, oil hardened. *Obsolete*

NOVO SPECIAL
H. Boker & Co.
C 0.7, Cr 3-4, W 18, V 0.9, bal Fe.
For lathe tools; high speed steel. *Obsolete*

NOVO STEEL
H. Boker & Co.
C 0.7, W 19, Cr 3, Co 0.6, bal Fe.
Heat treated: 130,000 TS; 5 El. For lathe and planer tools, hobs, reamers, drills, chasers; high speed steel. *Obsolete*

NOVO SUPERIOR
H. Boker & Co.
C 0.8, Mn 0.17, Cr 2.6, W 18.63, V 1.01, bal Fe.
For tools, lathe and cutting tools, taps, drills, reamers; high speed steel. *Obsolete*

NOVO SUPERIOR VANADIUM
H. Boker & Co.
C 1.05, Cr 4, V 3, W 18, bal Fe.
For reamers, taps, broaches, drills; Type T3; high speed steel. *Obsolete*

NOVOKONSTANT
VDM Nickel-Technologie AG
Cu 82.5, Al 3, Mn 13.5, Fe 1.
Rolled: 142,200 TS; 0.5 El. Annealed: 78,200 TS; 25 El. For electrical resistances; heat resistant. *Obsolete*

NOVONIT 309 MO R
Krupp Stahl AG
C 0-0.12, Si 0-1.5, Mn 0-2.5, P 0-0.03, S 0-0.025, Cr 22-25, Mo 2-3, Ni 11-15, bal Fe.
Weld filler metal. DIN 8556 T.1/86; W. Nr. 1.4459.

NOVONIT 3954
Krupp Stahl AG
C 0-0.03, Si 0-1, Mn 7-9, P 0-0.025, S 0-0.015, Cr 21-23, Mo 3.4-4, Ni 16-19, N 0.15-0.35, bal Fe.
Weld filler metal. WW 1.3954/87; W. Nr. 1.3954.

NOVONIT 3986
Krupp Stahl AG
C 0-0.02, Si 0.4-0.8, Mn 6.5-7.5, P 0-0.025, S 0-0.015, Cr 22-23, Mo 2.6-2.9, Ni 16.5-17.5, N 0.3-0.4, V 0.1-0.18, bal Fe.
Weld filler metal. SEL 90; W. Nr. 1.3986.

NOVONIT 4009
Krupp Stahl AG
C 0-0.1, Si 0-1, Mn 0-1.5, P 0-0.03, S 0-0.025, Cr 12-15, bal Fe.
Weld filler metal. DIN 8556 T.1/86; W. Nr. 1.4009.

NOVONIT 4015
Krupp Stahl AG
C 0-0.1, Si 0-1, Mn 0-1.5, P 0-0.03, S 0-0.025, Cr 16-19, bal Fe.
Weld filler metal. DIN 8556 T.1/86; W. Nr. 1.4015.

NOVONIT 4115
Krupp Stahl AG
C 0-0.25, Si 0-1, Mn 0-1.5, P 0-0.03, S 0-0.025, Cr 15.5-18.5, Mo 0.5-1.5, Ni 0-1, bal Fe.
Weld filler metal. DIN 8556 T.1/86; W. Nr. 1.4115.

NOVONIT 4302
Krupp Stahl AG
C 0-0.06, Si 0-1.5, Mn 0-2, P 0-0.03, S 0-0.025, Cr 18.5-21, Ni 9-11, bal Fe.
Weld filler metal. DIN 8556 T.1/86; W. Nr. 1.4302.

NOVONIT 4302 B
Krupp Stahl AG
C 0-0.07, Si 0-1.5, Mn 0-2, P 0-0.03, S 0-0.025, Cr 18-21, Ni 8-11, bal Fe.
Weld filler metal. DIN 8556 T.1/86; W. Nr. 1.4302.

NOVONIT 4302 R
Krupp Stahl AG
C 0-0.07, Si 0-1.5, Mn 0-2, P 0-0.03, S 0-0.025, Cr 18-21, Ni 8-11, bal Fe.
Weld filler metal. DIN 8556 T.1/86; W. Nr. 1.4302.

NOVONIT 4316
Krupp Stahl AG
C 0-0.025, Si 0-1.5, Mn 0-2, P 0-0.03, S 0-0.025, Cr 18.5-21, Ni 9-11, bal Fe.
Weld filler metal. DIN 8556 T.1/86; W. Nr. 1.4316.

NOVONIT 4316 B
Krupp Stahl AG
C 0-0.04, Si 0-1.5, Mn 0-2, P 0-0.03, S 0-0.025, Cr 18-21, Ni 8-11, bal Fe.
Weld filler metal. DIN 8556 T.1/86; W. Nr. 1.4316.

NOVONIT 4316 R
Krupp Stahl AG
C 0-0.04, Si 0-1.5, Mn 0-2, P 0-0.03, S 0-0.025, Cr 18-21, Ni 8-11, bal Fe.
Weld filler metal. DIN 8556 T.1/86; W. Nr. 1.4316.

NOVONIT 4332
Krupp Stahl AG
C 0-0.025, Si 0-1.5, Mn 0-2.5, P 0-0.03, S 0-0.025, Cr 22-25, Ni 11-15, bal Fe.
Weld filler metal. DIN 8556 T.1/86; W. Nr. 1.4332.

NOVONIT 4332 R
Krupp Stahl AG
C 0-0.04, Si 0-1.5, Mn 0-2.5, P 0-0.03, S 0-0.025, Cr 22-25, Ni 11-15, bal Fe.
Weld filler metal. DIN 8556 T.1/86; W. Nr. 1.4332.

NOVONIT 4337
Krupp Stahl AG
C 0-0.15, Si 0-1.5, Mn 0-2.5, P 0-0.035, S 0-0.025, Cr 27-31, Ni 8-12, bal Fe.
Weld filler metal. DIN 8556 T.1/86; W. Nr. 1.4337.

NOVONIT 4337 R
Krupp Stahl AG
C 0-0.15, Si 0-1.5, Mn 0-2.5, P 0-0.035, S 0-0.025, Cr 27-31, Ni 8-12, bal Fe.
Weld filler metal. DIN 8556 T.1/86; W. Nr. 1.4337.

NOVONIT 4370
Krupp Stahl AG
C 0-0.2, Si 0-1.5, Mn 5-8, P 0-0.035, S 0-0.025, Cr 17-20, Ni 7-10, bal Fe.
Weld filler metal. DIN 8556 T.1/86; W. Nr. 1.4370.

NOVONIT 4403
Krupp Stahl AG
C 0-0.06, Si 0-1.5, Mn 0-2, P 0-0.03, S 0-0.025, Cr 18.5-21, Mo 2.5-3, Ni 10-13, bal Fe.
Weld filler metal. DIN 8556 T.1/86; W. Nr. 1.4403.

NOVONIT 4430
Krupp Stahl AG
C 0-0.025, Si 0-1.5, Mn 0-2, P 0-0.03, S 0-0.025, Cr 18-20, Mo 2.5-3, Ni 10-13, bal Fe.
Weld filler metal. DIN 8556 T.1/86; W. Nr. 1.4430.

NOVONIT 4430 B
Krupp Stahl AG
C 0-0.04, Si 0-1.5, Mn 0-2, P 0-0.03, S 0-0.025, Cr 17-20, Mo 2.5-3, Ni 10-13, bal Fe.
Weld filler metal. DIN 8556 T.1/86; W. Nr. 1.4430.

NOVONIT 4440
Krupp Stahl AG
C 0-0.025, Si 0-1.5, Mn 2.5-5, P 0-0.035, S 0-0.025, Cr 17-20, Mo 4-5, Ni 16-19, bal Fe.
Weld filler metal. DIN 8556 T.1/86; W. Nr. 1.4440.

NOVONIT 4440 B
Krupp Stahl AG
C 0-0.04, Si 0-1.5, Mn 1-4, P 0-0.035, S 0-0.025, Cr 17-20, Mo 4-5, Ni 16-19, bal Fe.
Weld filler metal. DIN 8556 T.1/86; W. Nr. 1.4440.

NOVONIT 4455
Krupp Stahl AG
C 0-0.025, Si 0-1.5, Mn 5-9, P 0-0.035, S 0-0.025, Cr 19-22, Mo 2.5-3.5, Ni 15-18, N 0.15-0.2, bal Fe.
Weld filler metal. DIN 8556 T.1/86; W. Nr. 1.4455.

NOVONIT 4502
Krupp Stahl AG
C 0-0.1, Si 0-1, Mn 0-1.5, P 0-0.03, S 0-0.025, Cr 16-19, Ti 0.3-0.7, bal Fe.
Weld filler metal. DIN 8556 T.1/86; W. Nr. 1.4502.

NOVONIT 4519
Krupp Stahl AG
C 0-0.025, Si 0-1.5, Mn 2-5, P 0-0.03, S 0-0.025, Cr 19-22, Mo 4-6, Ni 24-27, Cu 1-2, bal Fe.
Weld filler metal. DIN 8556 T.1/86; W. Nr. 1.4519.

NOVONIT 4519 B
Krupp Stahl AG
C 0-0.04, Si 0-1.5, Mn 1-4, P 0-0.03, S 0-0.025, Cr 19-22, Mo 4-6, Ni 23-26, Cu 1-2, bal Fe.
Weld filler metal. DIN 8556 T.1/86; W. Nr. 1.4519.

NOVONIT 4551
Krupp Stahl AG
C 0-0.07, Si 0-1.5, Mn 0-2, P 0-0.03, S 0-0.025, Cr 18.5-21, Ni 8.5-10.5, Nb 0-1.1, bal Fe.
Aircraft material. DIN 8556 T.1/86; W. Nr. 1.4551.

NOVONIT 4551 B
Krupp Stahl AG
C 0-0.08, Si 0-1.5, Mn 0-2, P 0-0.03, S 0-0.025, Cr 18-21, Ni 8-11, Nb 0-1.1, bal Fe.
Weld filler metal. DIN 8556 T.1/86; W. Nr. 1.4551.

NOVONIT 4556
Krupp Stahl AG
C 0-0.025, Si 0-1.5, Mn 0-2.5, P 0-0.03, S 0-0.025, Cr 22-25, Ni 11-15, Nb 0-1.1, bal Fe.
Weld filler metal. DIN 8556 T.1/86; W. Nr. 1.4556.

NOVONIT 4576
Krupp Stahl AG
C 0-0.07, Si 0-1.5, Mn 0-2, P 0-0.03, S 0-0.025, Cr 18.5-21, Mo 2.5-3, Ni 10-13, Nb 0-1.1, bal Fe.
Weld filler metal. DIN 8556 T.1/86; W. Nr. 1.4576.

NOVONIT 4576 B
Krupp Stahl AG
C 0-0.08, Si 0-1.5, Mn 0-2, P 0-0.03, S 0-0.025, Cr 17-20, Mo 2.5-3, Ni 10-13, Nb 0-1.1, bal Fe.
Weld filler metal. DIN 8556 T.1/86; W. Nr. 1.4576.

NOVONIT 4773
Krupp Stahl AG
stainless steel. C 0-0.1, Cr 30, Mo 0-2, bal Fe.
Stainless for high temperature operation. *Obsolete*

NOVONIT 4773 B
Krupp Stahl AG
C 0-0.1, Si 0-2, Mn 0-2, P 0-0.035, S 0-0.025, Cr 27-30, bal Fe.
Weld filler metal. DIN 8556 T.1/86; W. Nr. 1.4773.

NOVONIT 4820
Krupp Stahl AG
C 0-0.15, Si 0-2, Mn 0-2, P 0-0.03, S 0-0.025, Cr 24.5-27.5, Ni 4-6, bal Fe.
Weld filler metal. DIN 8556 T.1/86; W. Nr. 1.4820.

NOVONIT 4829
Krupp Stahl AG
C 0-0.15, Si 0-2, Mn 0-2, P 0-0.03, S 0-0.025, Cr 20.5-23.5, Ni 10-13, bal Fe.
Weld filler metal. DIN 8556 T.1/86; W. Nr. 1.4829.

NOVONIT 4842
Krupp Stahl AG
C 0-0.15, Si 0-2, Mn 2-5, P 0-0.03, S 0-0.025, Cr 24-27, Ni 19-22, bal Fe.
Weld filler metal. DIN 8556 T.1/86; W. Nr. 1.4842.

NOVONIT 4842 B
Krupp Stahl AG
C 0-0.15, Si 0-2, Mn 2-5, P 0-0.03, S 0-0.025, Cr 23-27, Ni 18-22, bal Fe.
Weld filler metal. DIN 8556 T.1/86; W. Nr. 1.4842.

NOVONIT 4850
Krupp Stahl AG
C 0.1-0.2, Si 0.5-1.5, Mn 1.5-2, P 0-0.025, S 0-0.02, Cr 20-22, Ni 31-34, Nb 1-2.5, bal Fe.
Weld filler metal. SEL 90; W. Nr. 1.4850.

NOVONIT 4853
Krupp Stahl AG
C 0.35-0.45, Si 0.5-1.5, Mn 1.5-2.5, P 0-0.025, S 0-0.02, Cr 24-26, Ni 34-36, Nb 1-2, bal Fe.
Weld filler metal. SEL 90; W. Nr. 1.4853.

NOVONIT 4863
Krupp Stahl AG
C 0-0.2, Si 0-2, Mn 0-2, P 0-0.025, S 0-0.02, Cr 17-19, Ni 36-40, bal Fe.
Weld filler metal. SEL 90; W. Nr. 1.4863.

NOVONIT FALC 233
Krupp Stahl AG
stainless steel. C 0.02, Cr 23, Ni 6, Mo, bal Fe.
Stainless steel welding metal.

NOVONOX-A17
Stahlwerke Sudwestfalen
C 0.1, Cr 17, Ni 7, bal Fe.
Annealed: 110,000 TS; 40,000 YS; 60 El; B 85 Rock. Hard: 185,000 TS; 140,000 YS; 8 El; C 41 Rock. For aircraft structural members, diaphragms, household utensils. Type 301 stainless steel, good ductility. Work hardens. *Obsolete*

NOVONOX-A18
Stahlwerke Sudwestfalen
C 0.12, Cr 18, Ni 9, bal Fe.
Annealed: 90,000 TS; 40,000 YS; 50 El; B 85 Rock. For chemical and textile plant equipment, food processing apparatus, tanks, vessels, agitators. Type 302 Stainless steel, austenitic. *Obsolete*

NOVONOX-A18Z
Stahlwerke Sudwestfalen
C 0.06, Cr 18, Ni 11, Cb 0.4, bal Fe.
Annealed: 85,000 TS; 30,000 YS; 55 El; B 80 Rock. For welded structures, chemical and pharmaceutical plant equipment. Type 321 stainless steel, stabilized, austenitic. *Obsolete*

NOVONOX-A18ZN
Stahlwerke Sudwestfalen
C 0.06, Ni 12, Cr 18, Cb 0.5, bal Fe.
Annealed: 85,000 TS; 35,000 YS; 60 El; B 80 Rock. For welded structures, chemical and pharmaceutical plant equipment. Type 347 stainless steel, stabilized, austenitic. *Obsolete*

NOVONOX-ALC 18
Stahlwerke Sudwestfalen
C 0-0.03, Cr 19, Ni 10, bal Fe.
Annealed: 77,000 TS; 30,000 YS; 60 El; 140 Brin. For architectural trim, chemical plant equipment, food processing equipment. Type 304L stainless steel, austenitic, non-hardenable. *Obsolete*

NOVONOX-AS 182
Stahlwerke Sudwestfalen
C 0.05, Cr 17, Ni 11, Mo 3, bal Fe.
Annealed: 85,000 TS; 35,000 YS; 50 El; B 80 Rock. For chemical plant equipment, kettles, agitators, evaporators, tanks, valve trim. Type 316 stainless steel, austenitic, acid resistant. *Obsolete*

NOVONOX-AS 183
Stahlwerke Sudwestfalen
C 0.06, Cr 17, Ni 12, Mo 2.5, bal Fe.
Annealed: 80,000-90,000 TS; 30,000-40,000 YS; 40-60 El; 70-80 RA; B 78-85 Rock. For chemical plant equipment, agitators, evaporators, tanks, valve trim. Type 316-319 stainless steels, austenitic, acid resistant. *Obsolete*

NOVONOX-AS18
Stahlwerke Sudwestfalen
C 0-0.08, Cr 19, Ni 10, bal Fe.
Annealed: 85,000 TS; 35,000 YS; 60 El; 150 Brin. For architectural trim, chemical plant equipment, tanks, food processing equipment. Type 304 stainless steel, austenitic. Non-hardenable. *Obsolete*

NOVONOX-F 13 A1
Stahlwerke Sudwestfalen
C 0.06, Cr 13, Al 0.2, bal Fe.
Annealed: 70,000 TS; 40,000 YS; 30 El; 160 Brin. For annealing boxes, oil refining equipment, quenching racks. Type 405 stainless steel, ferritic. *Obsolete*

NOVONOX-F17
Stahlwerke Sudwestfalen
C 0.1, Cr 16, bal Fe.
Annealed: 70,000 TS; 40,000 YS; 30 El; 140 Brin. For automotive trim, hardware, oil burners, fasteners. Type 430 stainless steel, high heat and corrosion resistance. *Obsolete*

NOVONOX-FA26
Stahlwerke Sudwestfalen
C 0.12, Cr 28, Ni 5, Mo 1, bal Fe.
Annealed: 105,000 TS; 80,000 YS; 25 El; 230 Brin. Ht. Tr.: C 45-50 Rock. For valves, valve fittings, pumps. Type 329 stainless steel, precipitation hardening, corrosion and heat resistant. *Obsolete*

NOVONOX-T13
Stahlwerke Sudwestfalen
C 0.2, Cr 13, bal Fe.
Annealed: 95,000 TS; 50,000 YS; 25 El; B 92 Rock. Cold drawn: 105,000 TS; 85,000 YS; 17 El; B 95 Rock. For cutlery, surgical and dental equipment, gears, shafts. Type 420 stainless steel, hardenable. *Obsolete*

NOVONOX-T131
Stahlwerke Sudwestfalen
C 0.2, Cr 13, Mo 1.2, bal Fe.
Annealed: 95,000 TS; 40,000 YS; 25 El; 55 RA; B 92 Rock. Heat Treated: 250,000 TS; 215,000 YS; 8 El; 25 RA; C 50 Rock. For cutlery, surgical instruments, ball bearings, valves, hardware. Corrosion resistant, hardenable. *Obsolete*

NOVONOX-TT131
Stahlwerke Sudwestfalen
C 0.15, Cr 13, Mo 1.2, bal Fe.
Annealed: 80,000 TS; 40,000 YS; 25 El; B 93 Rock. Heat Treated: 200,000 TS; 160,000 YS; 12 El; C 45 Rock. For surgical instruments, cutlery, table flatware, oil refinery equipment. Corrosion resistant, hardenable. *Obsolete*

NOVOSTON
Stone Manganese - J. Stone & Co. Ltd.
Ni 1.5-5, Al 7-9, Mn 11-13, Fe 2-4, bal Cu.
Cast. For ship propellers; Aluminum bronze, corrosion resistant.

NOVOTHERM 10 FZ
Krupp Stahl AG
stainless steel. C 0.12, Cr 21, Ti, bal Fe.
Stainless for high temperature operation, and oil refinery equipment. *Obsolete*

NOVOTHERM 100
Stahlwerke Sudwestfalen
C 0-0.12, Si 1, Al 1, Cr 18, bal Fe.
Annealed: 80,000 TS; 50,000 YS; 25 El; 50 RA; 150 Brin. For oil refinery equipment; heat and creep resistant. *Obsolete*

NOVOTHERM 105
Stahlwerke Sudwestfalen
C 0.15, Si 2, Cr 19.5, Ni 9.5, bal Fe.
Annealed: 80,000 TS; 35,000 YS; 55 El; 75 RA; 150 Brin. For chemical plant equipment, tanks, mixers, filters; Type 302; stainless, austenitic. *Obsolete*

NOVOTHERM 10A
Robert Zapp Werkstofftechnik GmbH
C 0-0.15, Si 2, Cr 20, Ni 12, bal Fe.
Annealed: 85,000-108,000 TS; 43,000 YS; 40 El; 145-190 Brin. For heat treating boxes, furnace parts; AISI 309; austenitic, resists scaling to 1050°C.

NOVOTHERM 10F
Robert Zapp Werkstofftechnik GmbH
C 0-0.1, Si 1, Al 1, Cr 18, bal Fe.
Annealed: 72,000-92,000 TS; 43,000 YS; 12 El; 165-210 Brin. For oil refinery equipment; AISI 405; resists scaling to 1050°C.

NOVOTHERM 110
Stahlwerke Sudwestfalen
C 0-0.12, Si 1.5, Al 1.5, Cr 24, bal Fe.
For furnace parts, oil refinery equipment; heat and creep resistant. *Obsolete*

NOVOTHERM 11FA
Robert Zapp Werkstofftechnik GmbH
C 0-0.2, Si 1, Cr 25, Ni 4, bal Fe.
Annealed: 85,000-107,000 TS; 58,000 YS; 25 El; 175-200 Brin. For heat treating boxes, furnace parts, oil refinery equipment; AISI 446; resists scaling to 1100°C.

NOVOTHERM 120 A
Krupp Stahl AG
superalloy. Co 32, Cr 25, Ni, bal Fe.
High temperature alloy. *Obsolete*

NOVOTHERM 12A
Robert Zapp Werkstofftechnik GmbH
C 0-0.15, Si 2, Cr 25, Ni 20, bal Fe.
Annealed: 85,000-108,000 TS; 43,000 YS; 40 El; 175-190 Brin. For furnace parts and equipment; AISI 310; austenitic, resists scaling to 1200°C.

NOVOTHERM 12F
Robert Zapp Werkstofftechnik GmbH
C 0-0.1, Si 1.5, Al 1.5, Cr 24, bal Fe.
Annealed: 72,000-92,000 TS; 50,000 YS; 10 El; 170-215 Brin. For oil refinery equipment; resists scaling to 1200°C.

NOVOTHERM 130
Stahlwerke Sudwestfalen
C 0.15, Si 2, Cr 24, Ni 19, bal Fe.
Annealed: 100,000 TS; 45,000 YS; 50 El; 65 RA; 185 Brin. For furnace parts, valves, pumps, turbine parts; Type 310; stainless, austenitic. *Obsolete*

NOVOTHERM 70
Stahlwerke Sudwestfalen
C 0-0.12, Al 0.8, Cr 6.5, bal Fe.
For oil refinery equipment; heat and creep resistant. *Obsolete*

NOVOTHERM 7A
Krupp Stahl AG
stainless steel. C 0.1, Mn 18, Cr 12, bal Fe.
For special oil refinery equipment. *Obsolete*

NOVOTHERM 85
Stahlwerke Sudwestfalen
C 0-0.12, Si 1.2, Al 1, Cr 13, bal Fe.
Annealed: 75,000 TS; 40,000 YS; 35 El; 70 RA; 155 Brin. For oil refinery equipment; heat and creep resistant. *Obsolete*

NOVOTHERM 85 F
Krupp Stahl AG
stainless steel. C 0.08, Cr 17, bal Fe.
Ferritic stainless steel, for oil refinery equipment. *Obsolete*

NOVOTHERM 8A
Robert Zapp Werkstofftechnik GmbH
C 0-0.12, Si 0.6, Cr 18, Ni 10, Ti, bal Fe.
Annealed: 78,000-108,000 TS; 38,000 YS; 40 El; 140-190 Brin. For welded chemical plant equipment; austenitic, resists scaling to 800°C.

NOVOTHERM 8F
Robert Zapp Werkstofftechnik GmbH
C 0-0.1, Si 0.8, Al 0.8, Cr 6.5, bal Fe.
Annealed: 64,000-85,000 TS; 36,000 YS; 20 El; 140-185 Brin. For oil refinery equipment; AISI 501 and 502; resists scaling to 800°C.

NOVOTHERM 9F
Robert Zapp Werkstofftechnik GmbH
C 0-0.1, Si 1.2, Al 1, Cr 13, bal Fe.
Annealed: 72,000-92,000 TS; 43,000 YS; 15 El; 160-205 Brin. For oil refinery equipment; resists scaling to 950°C.

NOVOTHERM NC36
Robert Zapp Werkstofftechnik GmbH
C 0-0.12, Si 1.8, Cr 16, Ni 36, bal Fe.
Annealed: 78,000-108,000 TS; 38,000 YS; 40 El; 140-185 Brin. For furnace parts and equipment; austenitic, resists scaling to 1100°C.

NOXIDA-C
Krupp Stahl AG
nickel. Ni 63, Cr 16, Mo 16, Fe 0-6.
For pharmaceutical and chemical plant equipment, valves, pumps. Heat and corrosion resistant. *Obsolete*

NOXIS 1
Aubert & Duval
Alloy steel. C 0.45, Si 1.4, Ni 13, Cr 13, W 2.7, bal Fe.
Austenitic, nitrided steel. Oil quenched: 800 kg/mm^2 TS; 30 El. For motorcar, lorry and diesel engine valves.

NOXIS 4
Aubert & Duval
C 0.42, Ni 14, Cr 14, W 3, Co 11, Nb 4, bal Fe.
Wrought iron-nickel.

NP-464
Armco
C 0.35, Mn 2.3, Si 0.2, Cr 25, Ni 5.5, N 0.4, bal Fe.
For diesel exhaust valves. *Obsolete*

NR-106
Now SYLVANIA NR-106.

NR-203, ETC.
Now INNERSHIELD NR-203, ETC.

NR-NICKEL 99.6
Vereinigte Deutsche Nickel-Werke AG
Nickel. Cu 0.02, Fe 0.05, C 0-0.02, Mn 0.15, Si 0.03, 99.6 Ni + Co.
Strip, wire, bar, forged and turned parts for electronics, controls, optical parts, fasteners, apparatus, drawing and stamping parts.

NRC 76
NRC, Inc.
Ta 97.5, W 2.5.
Increased strength and equivalent corrosion resistance of pure Ta. High formability for chemical process equipment.

NRC 90-10
NRC, Inc.
Ta 90, W 10.
High temperature strength and corrosion resistance For missile components, electronics, high temperature devices.

NRC T111
NRC, Inc.
Ta 90, W 8, Hf 2.
Weldable, ductile; very high strength to weight ratio at high temperatures. For missile hardware, supersonic air and spacecraft, liquid reactors.

NS
Delta Enfield Metals Ltd.
Now ERM N.S. (M100).

NS 190
French manufacture
C, alloy, bal Fe.
For heat, creep and oxidation resistant.

NS-1 1/4 CR-1/2 MO GMA
National Standard Co.
C 0.08-0.13, Mn 0.55-0.8, Si 0.45-0.7, Cr 1.2-1.35, Mo 0.5-0.7, bal Fe.
Welding wire for gas metal arc welding of low alloy steels; designed for assemblies to be annealed or heat treated later. *Obsolete*

NS-1 1/4 CR-1/2 MO SA
National Standard Co.
C 0.08-0.13, Mn 0.55-0.8, Si 0-0.25, Cr 1.3-1.65, Mo 0.5-0.7, bal Fe.
Welding wire for submerged arc welding of low alloy steels; designed for assemblies to be annealed or heat treated later. *Obsolete*

NS-101
National Standard Co.. C 0.1, Mn 1, Si 0.55, bal Fe.
Welding wire: as welded: 76,000 TS; 60,000 YS; 28 El; 130 Brin. For inert gas welding mild steel.

NS-102
National Standard Co.. C 0.08, Mn 1.95, Si 0.6, Mo 0.5, bal Fe.
Welding wire; as welded: 102,000 TS; 84,000 YS; 21 El; 163 Brin. For inert gas welding 4130 and some high strength steels.

NS-103
National Standard Co.. C 0.05, Mn 1.25, Si 0.5, Al 0.1, Zr 0.09, Ti 0.1, bal Fe.
Welding wire; as welded: 83,000 TS; 71,000 YS; 27.5 El; 140 Brin. For improved quality welds in inert gas welding mild steel.

NS-107
National Standard Co.. C 0.15, Mn 1.2, Si 0.55, Al 0.65, bal Fe.
Welding wire; as welded: 80,000 TS; 64,000 YS; 25 El; 130 Brin. For inert gas welding oily and rusty mild steel without porosity.

NS-115
National Standard Co.. C 0.1, Mn 1.7, Si 1, bal Fe.
Welding wire; as welded: 90,000 TS; 68,000 YS; 28 El; 160 Brin. For producing neat appearing welds when inert gas welding mild steel.

NS-116
National Standard Co.. C 0.1, Cr 1.95, Mn 0.55, bal Fe.
Welding wire; as welded: 87,000 TS; 71,000 YS; 26 El; 42 ft. lbs. Charpy V notch. For inert gas welding of mild steel.

NS-18-2
National Standard Co.. C 0.1, Mn 12, Si 0.5, Cr 18, Ni 1.6, N 0.34, bal Fe.
Low nickel austenitic stainless steel wire. Drawn (0.018" dia): 320,000 TS; 3.2 El. Non-magnetic after severe cold work. Hardens by cold work. For non-magnetic stainless steel springs.

NS-2 1/4 CR-1 MO GMA
National Standard Co.
C 0.08-0.13, Mn 0.55-0.8, Si 0.45-0.7, Cr 2.2-2.75, Mo 1-1.2, bal Fe.
Welding wire for gas metal arc welding of low alloy steels; designed for assemblies to be annealed or heat treated later. *Obsolete*

NS-2 1/4 CR-1 MO SA
National Standard Co.
C 0.08-0.13, Mn 0.55-0.8, Si 0-0.25, Cr 2.4-2.75, Mo 1-1.2, bal Fe.
Welding wire for submerged arc welding of low alloy steels; designed for assemblies to be annealed or heat treated later *Obsolete*

NS-2 1/4 NICKEL

National Standard Co.
C 0-0.03, Mn 0.3-0.6, Ni 2-2.9, bal Fe.
Welding wire for submerged arc and TIG welding of 3-1/2% nickel fabricated pipe for cryogenic applications. 25 ft, lbs. U-notch impact at -150 F. *Obsolete*

NS-22

National Standard Co.. C 0-0.06, Mn 4-6, Si 0-1, Cr 20.5-23.5, Ni 11.5-13.5, Mo 1.5-3, N 0.2-0.4, Cb 0.1-0.3, V 0.1-0.3, bal Fe.
Wire for springs; non-magnetic after cold work; excellent chloride resistance. Spring temper (0.062 inch): 255,000-285,000 psi TS.

NS-25

National Standard Co.
C 0.05-0.15, Mn 1-2, Cr 19-21, Ni 9-11, W 14-16, Fe 0-3, bal Co.
Drawn: 178,000 TS. Heat treated: 191,000 TS. For springs; L605 alloy, high temperature applications. *Obsolete*

NS-3-1/2 NICKEL

National Standard Co.
C 0-0.03, Mn 0.3-0.6, Ni 2.9-3.55, bal Fe.
Welding wire for submerged arc and TIG welding of 3-1/2% nickel fabricated pipe for cryogenic applications. 25 ft, lbs. U-notch impact at -150 F. *Obsolete*

NS-308-HISI

National Standard Co.. C 0.05, Mn 1.7, Si 0.85, P 0.015, S 0.01, Cr 21, Ni 10.25, bal Fe.
Bare stainless steel wire for welding 308 and similar steels.

NS-308L HISI

National Standard Co.. C 0.022, Mn 1.7, Si 0.85, P 0.015, S 0.01, Cr 21, Ni 10.25, bal Fe.
Bare stainless steel wire for welding 308L and similar steels.

NS-309 HISI

National Standard Co.. C 0.05, Mn 1.7, Si 0.85, P 0.015, S 0.01, Cr 24.5, Ni 13.5, bal Fe.
Bare stainless steel wire for welding 309 and similar steels.

NS-316L HISI

National Standard Co.. C 0.022, Mn 1.7, Si 0.85, P 0.015, S 0.01, Cr 19, Ni 13, Mo 2.25, bal Fe.
Bare stainless steel wire for welding 316L and similar steels.

NS-355

National Standard Co.
Cr 15.65, Ni 4.38, Mo 2.68, Mn 1, C 0.14, P 0.03, S 0.01, Si 0.32, Cu 0.12, bal Fe.
Spring: 290,000-500,000 TS. For springs; high ultimate strength, corrosion resistant. *Obsolete*

NS-5

Lunkenheimer Co.
Ni 50, Cu 46, Si 2.2, Mn 1.9.
100,000 TS; 82,500 YS; 2 El; 300 Brin. For valve seats and disks; wear, abrasion and corrosion resistant.

NS-60M

National Standard Co.
C 0-0.15, Mn 0-4, Fe 0-2.5, Si 0-1.25, Al 0-1.25, Ti 1.5-3, 62-69% Ni-Co, bal Cu.
Welding wire; as welded: 76,600 TS; 37.5 El. For welding Monel pipe. AWS 5.14 ASTM B304 C1. ER NiCu-7. *Obsolete*

NS-61N

National Standard Co.
C 0-0.15, Mn 0-1, Fe 0-1, Si 0-0.75, Cu 0-0.25, Al 0-1.5, Ti 2-3.5, 93% min Ni-Co.
Welding wire; as welded: 73,700 TS; 32.5 El. For welding nickel, overlaying steel, welding cast iron. AWS 5.14 ASTM B304 C1. ER Ni-3. *Obsolete*

NS-62I

National Standard Co.
C 0-0.08, Mn 0-1, Fe 6-8, Si 0-0.35, Cu 0-0.5, Cr 14-17, 70% min Ni-Co, 1.5-3.0% Cb-Ta.
Welding wire; as welded: 87,900 TS; 45 El. For welding Inconel. AWS 5.14 ASTM B304 C1. ER NiCr Fe5. *Obsolete*

NS-67C

National Standard Co.
C 0-0.04, Mn 0-1, Fe 0.4-0.7, Si 0-0.15, Ti 0.2-0.5, 29-32% Ni-Co, bal Cu.
Welding wire; as welded: 54,200 TS; 37.5 El. For welding 70/30, 80/20, 90/10 copper-nickel. AWS A5.5 ASTM B225 C1. E CuNi. *Obsolete*

NS-750X

National Standard Co.. Co 0-1, Cr 14-17, Ni 69-0, Fe 5-9, C 0-0.08, Al 0.4-1, Ti 2-2.5.
Drawn (15% red.) and heat treated wire has: 190,000-205,000 TS. Corrosion resistant. For springs operating at 550-1100°F.

NS-82I

National Standard Co.
C 0-0.08, Mn 2.5-3.5, Fe 0-1, Si 0-0.5, Cu 0-0.5, Ti 0-0.75, Cr 18-22, 67% min Ni-Co, 2-3% Cb-Ta.
Welding wire; as welded: 97,100 TS; 50.5 El. For welding Inconel, 9% Ni steel, and for overlaying steel. AWS 5.14 ASTM B304 C1. ER NiCr-3. *Obsolete*

NS-92I

National Standard Co.
C 0-0.08, Mn 2-2.75, Fe 0-8, Si 0-0.35, Cu 0-0.5, Ti 2.5-3.5, Cr 14-17, 67% min Ni-Co.
Welding wire; for welding dissimilar metals; for welding 9% nickel steel. AWS 5.14 ASTM B304 C1. ER NiCr Fe-6; AMS 5675. *Obsolete*

NS-A286

National Standard Co.. Cr 13.5-16, Ni 24-27, Mo 1-1.75, Al 0-0.35, Ti 1.75-2.25, V 0.1-0.5, bal Fe.
Drawn (68% red.) and heat treated wire has: 210,000-238,000 TS. Austenitic, corrosion resistant. For springs operating at 600-1000°F.

NS-L605

National Standard Co.. Co 50, Cr 19-21, Ni 9-11, Fe 0-3, C 0.05-0.15, W 14-16.
Drawn (28% red.) and heat treated wire has: 205,000-225,000 TS. Corrosion resistant. For springs operating at 1100-1500°F range; for wire holding parts together for dip brazing at 1400°F.

NSCD

Vallourec S.A.
C 0.03, Cr 17.5, Ni 16, Mo 5, Co 3, bal Fe.
Resistant to pitting. For seawater treatment.

NSK

W. Ossenberg & Cie Edelstahlwerke
C 1, Mn 0.3, Si 0.25, Cr 1.5, bal Fe.
Cold work tool steel for punches, stamping dies, trimming dies. W.-Nr. 1.2067

NSS 6

Eagle & Globe Steel Ltd.
Tool material. C 0.7, Si 0.3, Mn 1.9, Cr 1, Mo 1.35, bal Fe.
Annealed: 230 Brin max. Air hardening die steel. For large blanking and forming dies, trimming dies, heavy duty punches, rolls, master hobs, shear blades, bending tools. AS1239 A6A; AISI A6; BS4659 BA6.

NSZ

W. Ossenberg & Cie Edelstahlwerke
C 1.45, Mn 0.6, Si 0.2, Cr 1.5, bal Fe.
Cold work tool steel for reamers, threading cutters, broaches. W.-Nr. 1.2063.

NT-2

Newcomer Products Inc.
Titanium carbide coated N-2 carbide tool material.

NT-5

Newcomer Products Inc.
Titanium carbide coated N-52 carbide tool material.

NT-6

Newcomer Products Inc.
Titanium carbide coated N-60 carbide tool material.

NTK-M7

Japan Metal Industry Co. Ltd.
Cr 9-11, Ni 16-18, Mo 6-8, C 0-0.06, Si 0-1, Mn 0-2, 0.03 S + P max, bal Fe.
Rolled: 106,000 TS; 86,000 YS; 34 El; 43 RA. Annealed: 79,000 TS; 35,000 YS; 60 El; 62 RA. For chemical and food processing equipment, pumps, impellers. Stainless. Resists organic chlorination media. Resists moderate concentrations of hot HCl.

NTK-T1

Japan Metal Industry Co. Ltd.
Mo 1-3, bal TiC.
Sintered: 120,000-140,000 TS; 93-94 Rock A. For tools, bearings, seals. Resists heat and wear.

NTK-T3

Japan Metal Industry Co. Ltd.
Mo 13-16, bal TiC.
Sintered: 155,000-185,000 TS; 92-93 Rock A. For tools, bearings, seals. Resists heat and wear.

NU-BRAZE III

Sherman & Co.
Ag, P, Cu.
For brazing of non-ferrous metals; brazing point 1300°F.

NU-BRAZE SUPER FLO

Sherman & Co.
Ag, bal Cu.
For silver brazing; melting point 1076°F.

NU-BRAZE VI

Sherman & Co.
Ag, Cu, Zn, Cd.
For brazing of non-ferrous metals; melting point 1170 °F.

NU-BRONZE

English manufacture
Cu 95.4, Ni 3.25, Mn 0.25, Si 0.73.
For corrosion resistant parts.

NU-DIE

Crucible Materials Corp.
Tool material. C 0.4, Si 1.2, Cr 5.2, Mo 1.5, bal Fe.
For die casting dies, plastic mold dies; oil hardening.

NU-DIE

Flockton, Tompkin & Co., Ltd.
C 0.4, Si 1.2, Cr 5.2, Mo 1.5, bal Fe.
For plastic molds, die casting dies. Oil or air hardening.

NU-DIE CASTING

Hawkridge Bros. Co.
C 0.35, Cr 5, V 0.4, Mo 1.5, bal Fe.
For die casting dies, punches; Type H11; hot work steel.

NU-DIE DENSIFIED HOT WORK STEEL

Crucible Materials Corp.
Tool material. C 0.4, Mn 0.4, Si 1, Cr 5, V 0.55, Mo 1.35, bal Fe.
Air hardened: 40-55 Rock C. Designed for extrusion dies for aluminum and magnesium. AISI H11. Hot work tool steel.

NU-DIE V

Slater Steels Corp.
Tool material. C 0.4, Cr 5.25, Mn 0.35, V 1.05, Si 1.05, Mo 1.35, bal Fe.
Die casting steel for aluminum and magnesium die casting dies and die inserts. Typical applications include die casting dies, die inserts, cores, ejector pins, plungers, sleeves, slides, aluminum and magnesium extrusion dies, bolsters, dummy blocks, mandrels.

NU-DIE V

Crucible Materials Corp.
Tool material. C 0.3-0.4, Cr 5, Mo 1.4, V 1.1, Si 1, bal Fe.
For die casting dies for Al-base alloys; air hardening.

NU-GILD (SCOVILL ALLOY 125)

Century Brass Products Inc.
Cu 87.5, bal Zn.
Annealed: 40,000 TS; 44 El. Hard: 65,000 TS; 5 El. For
fasteners, jewelry; deep drawing. *Obsolete*

NU-GILD (SCOVILL ALLOY 226)

Century Brass Products Inc.
Cu 87.5, bal Zn.
Annealed: 40,000 TS; 44 El. Hard: 65,000 TS; 5 El. For
fasteners, jewelry; deep drawing. *Obsolete*

NU-GOLD

English manufacture
Cu 87.73, Zn 12.22.
For jewelry, ornaments; red brass.

NU-MOL

E.C. Atkins & Co.
C 0.9, Alloy, bal Fe.
For hacksaws; tough.

NU-PYR-LOY

Pyramid Steel Company
C, alloy, bal Fe.
Heat treated: 309,000-342,000 TS; 264,000-282,000 YS;
570-600 Brin. For pneumatic and shock tools, chisels, shear
blades; tough and shock resistant, oil hardened.

NUALL

Flockton, Tompkin & Co., Ltd.
C, alloy, bal Fe.
For chuck jaws, reamers, drills; oil hardened.

NUCALLOY 41

Stoody Company
Cr 12, B 2, Si 4, C 0.5, bal Ni.
Centrifugal castings, as cast: 44 Rock C. For high
temperature and corrosion applications and good wear.

NUCALLOY 45

Stoody Company
Cr 14, B 3, Si 4.5, C 0.65, bal Ni.
Centrifugal castings, as cast: 53 Rock C. For high
temperature and corrosion applications and good wear.

NUCAST

Apothecaries Hall Co.
99.0 + Ni.
For nickel anodes.

NUCLEOTEC 2222

Eutectic Corp.
Electrode for DC welding of nickel alloys and dissimilar
alloys. 100,000 psi TS.

NUCUT

Wallace Murray Corp.
C, alloy, bal Fe.
For dies; oil hardened.

NUE-DIE XL

Slater Steels Corp.
Tool material. C 0.4, Cr 5.2, Mn 0.35, S 0-0.005, V 0.95, Si 1,
Mo 1.3, bal Fe.
H13 die steel. Typical applications include long run
aluminum die casting dies, die inserts, precision forging dies,
cores, plungers, sleeves, slides, aluminum and magnesium
extrusion dies.

NUERAL

Aluminiumwerke Nurnberg G.m.b.H.
Aluminum. Si, Cu, Ni, bal Al.
For cylinder heads; heat resistant.

NUERAL 132

Aluminiumwerke Nurnberg G.m.b.H.
Aluminum. Cu 0.8-1, Si 13-14, Ni 1.9-2.4, bal Al.
Sand cast: 32,000 TS; 0.3 El. For cylinder heads; heat
treatable.

NUERENBERGER GOLD

German manufacture
Al 2-7.5, Au 0.2-2.5, bal Cu.
For gold substitute; corrosion resistant.

NUGILD

Seymour Products Co.
Zn 13, bal Cu.
Soft: 43,000 TS; 56 Brin. 1/2 Hard: 54,000 TS; 116 Brin.

NULOY

Sivyer Steel Corp.
C 0.3, Ni 1.5, Cr 0.8, bal Fe.
For steel castings; tough. *Obsolete*

NUMETAL

Allegheny Ludlum Steel
Ni 77, Cu 4.5, Cr 1.5, bal Fe.
For audio transformers, sensitive relays; high permeability.

NURAL

Aluminium-Zentral e.V.
Aluminum. Si 0.2-1, Mn 0.2-0.5, Mg 3-12, bal Al.
Chill cast: 32,000-37,000 TS; 5-10 El; 78-80 Brin. For aircraft
fittings, car frames, marine parts, lever brackets, hardware,
general housings. Permanent mold and die cast, corrosion
resistant. *Obsolete*

NURAL 122

Aluminiumwerke Nurnberg G.m.b.H.
Aluminum. Cu 9.5-10.2, Mg 0.15-0.35, Fe 0.8-1.5, bal Al.
Sand cast: 22,000-30,000 TS; 1.5-3.3 El. For light alloy parts;
nonhardenable.

NURAL 122

Aluminium-Zentral e.V.
Aluminum. Cu 9.5-10.2, Mg 0.15-0.35, Fe 0.8-1.15, bal Al.
Chill cast: 26,000-37,000 TS; 1 El; 70-90 Brin. For sole plates
for electric hand irons, general castings. Permanent mold
cast, good machinability. *Obsolete*

NURAL 132

Aluminium-Zentral e.V.
Aluminum. Cu 0.8-2, Si 12.5-14.2, Mg 0.8-1, Fe 0.5, Ni
0.8-2.4, bal Al.
Chill cast: 28,000-37,000 TS; 0.5-1.2 El. For meter cases,
switches boxes, manifolds, fittings, aircraft parts. Permanent
mold cast, good castability and corrosion resistant. *Obsolete*

NURAL 132 A

Aluminiumwerke Nurnberg G.m.b.H.
Aluminum. Cu 0.8-1, Si 12.2-12.8, Mg 0.8-1, Fe 0.5, Ni
1.9-2.4, bal Al.
Chill cast: 29,000-37,000 TS; 0.5-1.2 El. For light alloy parts;
corrosion resistant.

NURAL 132 B

Aluminiumwerke Nurnberg G.m.b.H.
Aluminum. Cu 2, Si 12.5, Ni 0.8-1, Mg 0.8-1, Fe 0.5, bal Al.
Chill cast: 29,000-37,000 TS; 0.5-1.2 El. For light alloy parts;
corrosion resistant.

NURAL 142

Aluminiumwerke Nurnberg G.m.b.H.
Aluminum. Cu 4-4.5, Mg 1.3-1.8, Ni 1.8-2.2, bal Al.
Sand cast: 21,000-28,000 TS; 1-4 El. For light alloy parts; age
hardenable.

NURAL 142

Aluminium-Zentral e.V.
Aluminum. Cu 4-4.5, Mg 1.3-1.8, Ni 1.8-2.2, bal Al.
Chill cast: 23,000-32,000 TS; 1-5 El; 70-85 Brin. Heat treated:
26,000-35,000 TS; 1-6 El; 80-110 Brin. For cylinder heads,
pistons, generator housings, fittings. Permanent mold
castings, heat treatable, for high temperature service.
Obsolete

NURAL 1761

Aluminiumwerke Nurnberg G.m.b.H.
Aluminum. Cu 0.8-1.2, Si 16.4-17.5, Ni 3.2-3.6, Mn 0.5, Mg
0.7-1.2, Cr 0.5, bal Al.
Heat treated: 28,000-36,000 TS; 0.3-0.5 El. For auto engine
pistons, permanent mold castings; low density.

NURAL 1761P

Aluminiumwerke Nurnberg G.m.b.H.
Aluminum. Cu 0.8-1.2, Si 16.4-17.5, Ni 3.2-3.6, Mn 0.4-0.6,
Mg 0.7-1.2, Cr 0.5, bal Al.
Permanent mold: 29,000-33,000 TS; 28,000-30,000 YS;
0.3-0.8 El; 14-90 Brin. For pistons; low density.

NURAL 195

Aluminiumwerke Nurnberg G.m.b.H.
Aluminum. Cu 4-5.5, Si 0.5-1.5, Fe 0.4-1, bal Al.
Sand cast: 23,000-36,000 TS; 2-8 El. For light alloy parts; age
hardenable.

NURAL 200

Aluminiumwerke Nurnberg G.m.b.H.
Aluminum. Cu 14-16.8, Fe 0.5-1, Mg 0.15-0.35, bal Al.
Sand cast: 20,000-28,000 TS; 0.3-0.5 El. For light alloy parts;
heat treatable.

NURAL 2361

Aluminiumwerke Nurnberg G.m.b.H.
Aluminum. Cu 0.8-1.2, Si 22-25, Ni 0.8-1, Mg 0.7-1.2, Cr 0.5,
bal Al.
Permanent mold: 29,000-33,000 TS; 28,000-30,000 YS;
0.1-0.3 El; 12-90 Brin. For pistons; low density.

NURAL 25

Aluminiumwerke Nurnberg G.m.b.H.
Aluminum. Mg 1-2, Si 5.6, Mn 0.1-0.4, Fe 0.7, Cu 0.5, Ti 0.2,
Zn 0.2, bal Al.
Sand: 19,000 TS; 14,000 YS; 3 El. Cast: 26,000 TS; 19,000
YS; 1 El. Hardened: 40,000 TS; 36,000 YS; 1 El. For sand and
permanent mold castings; hardenable.

NURAL 30

Aluminiumwerke Nurnberg G.m.b.H.
Aluminum. Cu 6-7.5, Si 0.5-2, bal Al.
Sand cast: 18,000-28,000 TS; 0.8-2.0 El. For light alloy parts;
nonhardenable.

NURAL 3210

Aluminiumwerke Nurnberg G.m.b.H.
Aluminum. Cu 0.8-1.5, Si 11-13, Ni 0.8-1.3, Mg 0.8-1.3, bal Al.
Permanent mold: 31,000-35,000 TS; 26,000-30,000 YS; 0.3-1
El; 90-125 Brin. For light alloy parts; corrosion resistant.

NURAL 43

Aluminiumwerke Nurnberg G.m.b.H.
Aluminum. Si 4.5-6, bal Al.
Sand cast: 18,000-24,000 TS; 2-5 El. For light alloy parts;
corrosion resistant.

NURAL 43

Aluminium-Zentral e.V.
Aluminum. Si 4.5-6, Fe 1.2, bal Al.
Sand cast: 17,000-24,000 TS; 2-5 El. Die cast:
20,000-28,000 TS; 3-5 El; 50-60 Brin. For marine castings,
manifolds, meter housings, carburetors, food handling
equipment. Sand, permanent mold and die cast. *Obsolete*

NURAL 511

Aluminiumwerke Nurnberg G.m.b.H.
Aluminum. Mg 4.5-5.5, Mn 0.1-0.5, Si 0.6-1.5, Ti 0.2, bal Al.
Sand cast: 26,000 TS; 14,000 YS; 3 El; 65 Brin. For light alloy
castings; corrosion resistant.

NURAL 77

Aluminiumwerke Nurnberg G.m.b.H.
Aluminum. Cr 4-7, Zn 4-6, Si 1-3, Fe 0-1.5, bal Al.
Sand cast: 18,000-25,000 TS; 0.5-2 El. For light alloy parts; heat treatable.

NURAL 85

Aluminiumwerke Numberg G.m.b.H.
Aluminum. Cu 2-4, Si 5-7, Mn 0.4-0.6, bal Al.
Chill cast: 28,000-35,000 TS; 1.0-4.5 El. For light alloy parts; heat treatable.

NURAL 85 S

Aluminiumwerke Nurnberg G.m.b.H.
Aluminum. Cu 2-3, Si 4-6, Fe 0-2, Mn 0-0.5, bal Al.
Sand cast: 20,000-27,000 TS; 1-4 El. For light alloy parts; heat treatable.

NURAL 93

Aluminiumwerke Nurnberg G.m.b.H.
Aluminum. Cu 4-4.5, Si 1.5-2.5, Ni 3-4.5, bal Al.
Sand cast: 20,000-26,000 TS; 0.5-1.2 El. For light alloy parts; heat treatable.

NURAL AL-MG-MN

Aluminiumwerke Nurnberg G.m.b.H.
Aluminum. Si 0.2-1.3, Mn 0.6-1.5, Mg 1.5-3, bal Al.
Chill cast: 32,000-36,000 TS; 5-10 El. For light alloy parts; age hardenable.

NURAL DZNAL4CUL

Aluminiumwerke Nurnberg G.m.b.H.
Cu 0.6-1, Al 3.7-4.3, Mg 0.02-0.05, bal Zn.
Die cast: 36,000-47,000 TS; 2-5 El; 80-100 Brin. For housings, cases, machinery casting; Zamak 5.

NURAL GALCU6SI3

Aluminiumwerke Nurnberg G.m.b.H.
Aluminum. Cu 4-7, Si 2-4, Zn 2.5, Fe 1.1, bal Al.
Sand cast: 29,000 TS; 23,000 YS; 0.5 El; 100 Brin. Permanent mold: 32,000 TS; 28,000 YS; 0.2 El; 110 Brin. For light alloy castings; age hardenable.

NURAL GALSI12

Aluminiumwerke Nurnberg G.m.b.H.
Aluminum. Si 11-13.5, Mn 0.3-0.5, bal Al.
Sand cast: 31,000 TS; 13,000 YS; 4 El; 60 Brin. For light alloy castings; corrosion resistant.

NUREMBERG GOLD

Manufacturer not listed
Cu 90, Al 7.5, Au 2.5.
Jewelry; gold color.

NUREX

Capitol Castings Inc.
Now NACO.

NUROX

Capitol Castings Inc.
C 3.3, Si, Mn, bal Fe.
For gears, shafts, housings; cast iron. *Obsolete*

NUSHANK

Atlas Specialty Steels
C 0.45, Mn 0.6, Cr 0.4, Ni 3, Mo 0.25, bal Fe.
For tools, shanks; wear resistant. *Obsolete*

NUSITE

Nusite Steel Process Co.
C 0.9, Co 13, bal Fe.
For permanent magnets; heat treated.

NUTHERM

Atlas Specialty Steels
C 0.7, Mn 2, Cr 1, Mo 1.35, Si 0.3, bal Fe.
For cold work tools and dies; low temperature air hardening. *Obsolete*

NW 1

Thyssen Edelstahlwerke AG
C, Ni, bal Fe.
Water treated: 78,000-100,800 TS; 50,000-72,000 YS; 22-29 El. For gears, shafts; water hardening. *Obsolete*

NW 3

Thyssen Edelstahlwerke AG
C, Ni, bal Fe.
Oil treated: 85,000-114,000 TS; 65,000-85,000 YS; 18-24 El. For gears, shafts; oil hardening, tough. *Obsolete*

NW 5

Thyssen Edelstahlwerke AG
C, Ni, bal Fe.
Oil treated: 101,000-127,000 TS; 72,000-101,000 YS; 20-26 El. For gears, shafts; oil hardening, tough. *Obsolete*

NWA 99.7

Northwest Alloys, Inc.
Magnesium. Cu 0-0.02, Pb 0-0.01, Mn 0-0.1, Ni 0-0.001, Si 0-0.15, Sn 0-0.01, 99.70 Mg min, 0.05 other impurities max. Ingots for alloying with aluminum and other ferrous and nonferrous metals.

NWA 99.8

Northwest Alloys, Inc.
Magnesium. Cu 0-0.02, Pb 0-0.01, Mn 0-0.1, Ni 0-0.001, Sn 0-0.01, 99.80 Mg min, 0.05 other impurities max. Ingots for alloying with aluminum and other ferrous and nonferrous metals. Conforms to ASTM B92.

NX-188

Cannon-Muskegon Corp.
C 0.04, Mo 18, Al 8, bal Ni.
Cast alloy for jet engine vanes. Good high temperature strength. Directionally solidified.

NY

French manufacture
C 0.07-0.2, Mn 0.35, Ni 4-7, bal Fe.
Heat treated: 78,000-190,000 TS; 58,000-156,000 YS; 25-8 El. For construction work, mine car wheels.

NYBLADE

Sanderson Kayser Ltd.
C 0.48, Si 1, C 0.65, Ni 3, bal Fe.
Cold work tool steel for chisels, punches, shear blades.

NYBY 0908 MO

Nyby, Granges, AB
C 0-0.08, Cr 9, Mo 1, bal Fe.
Annealed: 64,000 TS; 36,000 YS; 20 El; 50 RA; 190 Brin. For oil refinery equipment, heat exchangers. Good oxidation resistance to 1400 F. Resists sulphide corrosion. *Obsolete*

NYBY 12-12

Nyby, Granges, AB
C 0.08, Cr 12, Ni 12, bal Fe.
Annealed: 110,000 TS; 44,000 YS; 62 El; 65 RA; 140 Brin. For valves. Austenitic stainless steel, heat and corrosion resistant. *Obsolete*

NYBY 12-12 EL

Nyby, Granges, AB
C 0-0.05, Cr 12, Ni 13, bal Fe.
Annealed: 72,000 TS; 29,000 YS; 55 El; 50 RA; 160 Brin. For pump and valve bodies. Austenitic, corrosion resistant. *Obsolete*

NYBY 14-12

Nyby, Granges, AB
C 0.08-0.16, Cr 12, Ni 12, bal Fe.
For valves, corrosion resistant. *Obsolete*

NYBY 1408 AL

Nyby, Granges, AB
C 0-0.08, Cr 13, Al 0.2, bal Fe.
Annealed: 58,000 TS; 31,000 YS; 20 El; 55 RA; 190 Brin. For oil refinery equipment, stills. Corrosion resistant. Type 405 stainless steel. *Obsolete*

NYBY 1420

Nyby, Granges, AB
C 0.2, Cr 14, bal Fe.
Annealed: 145,000 TS; 97,000 YS; 27 El; 60 RA; 225 Brin. Heat treated: 220,000 TS; 8 El; 510 Brin. For turbine blades, knives, surgical instruments, cutlery. Martensitic stainless steel, Type 420. *Obsolete*

NYBY 16-13 LNB

Nyby, Granges, AB
C 0-0.08, Cr 16, Ni 13, Cb = 10 x C, bal Fe.
Annealed: 78,000 TS; 32,000 YS; 35 El; 50 RA; 190 Brin. For welded chemical plant equipment, tanks, vessels, mixers. Austenitic stainless steel. *Obsolete*

NYBY 16-14 LMONBV

Nyby, Granges, AB
C 0-0.08, Cr 16, Ni 14, Mo 1.3, N 0.15, Cb = 10 x C, bal Fe.
Annealed: 78,000 TS; 39,000 YS; 30 El; 50 RA; 190 Brin. For chemical plant equipment, acid mixers, agitators, retorts. Austenitic stainless steel. *Obsolete*

NYBY 16-16 LMONB

Nyby, Granges, AB
C 0-0.08, Cr 16, Ni 16, Mo 1.8, Cb = 10 x C, bal Fe.
Annealed: 78,000 TS; 36,000 YS; 35 El; 50 RA; 190 Brin. For welded structures, chemical plant equipment tanks, acid containers. Austenitic stainless steel. *Obsolete*

NYBY 17-12 UL

Nyby, Granges, AB
C 0-0.03, Cr 17, Ni 12, bal Fe.
Annealed: 70,000 TS; 28,000 YS; 40 El; 50 RA; 160 Brin. For chemical plant equipment, stiolls, mixers, agitators, tanks. Austenitic stainless steel. Not hardenable. *Obsolete*

NYBY 17-7

Nyby, Granges, AB
C 0-0.12, Cr 17, Ni 7, bal Fe.
Annealed: 78,000 TS; 34,000 YS; 50 El; 55 RA; 180 Brin. For trailer and truck bodies, household utensils, diaphragms. Austenitic stainless steel type 301. *Obsolete*

NYBY 1706

Nyby, Granges, AB
C 0-0.06, Cr 17, bal Fe.
Annealed: 63,000 TS; 40,000 YS; 20 El; 50 RA; 180 Brin. For chemical plant and oil refinery equipment. Corrosion resistant. Type 430 stainless steel. Ferritic. *Obsolete*

NYBY 1708

Nyby, Granges, AB
C 0-0.08, Cr 17, bal Fe.
Annealed: 63,000 TS; 43,000 YS; 20 El; 50 RA; 200 Brin. For furnace parts, chemical plant equipment. Corrosion resistant. Type 430 stainless steel. Ferritic. *Obsolete*

NYBY 1708 MO

Nyby, Granges, AB
C 0-0.08, Cr 17, Mo 1, bal Fe.
Annealed: 65,000 TS; 36,000 YS; 20 El; 50 RA; 200 Brin. For chemical plant and oil refinery equipment. Ferritic, corrosion resistant, not hardenable. *Obsolete*

NYBY 1708 MO NB

Nyby, Granges, AB
C 0-0.08, Cr 17, Mo 1, Cb = 12 x C, bal Fe.
Annealed: 65,000 TS; 36,000 YS; 20 El; 50 RA; 200 Brin. For welded furnace parts and heat treating boxes. Ferritic, corrosion resistant, stabilized, welding grade. *Obsolete*

NYBY 1708 MOT

Nyby, Granges, AB
C 0-0.08, Cr 17, Mo 1, Ti = 7 x C, bal Fe.
Annealed: 65,000 TS; 36,000 YS; 20 El; 50 RA; 200 Brin. For welded furnace parts, heat treatment boxes. Ferritic, corrosion resistant, weldable. *Obsolete*

NYBY 1708 NB
Nyby, Granges, AB
C 0-0.08, Cr 17, Cb = 12 x C, bal Fe.
Annealed: 63,000 TS; 36,000 YS; 20 El; 50 RA; 180 Brin. For welded oil refinery equipment. Corrosion resistant. Type 430 Cb stainless steel. Ferritic. Welding grade. Stabilized. *Obsolete*

NYBY 1708 T
Nyby, Granges, AB
C 0-0.08, Cr 17, bal Fe.
Annealed: 64,000 TS; 36,000 YS; 20 El; 50 RA; 180 Brin. For auto trim, kitchen sinks, septic tanks, builder's hardware. Ferritic stainless steel. Not hardenable by heat treatment. *Obsolete*

NYBY 1710 SI
Nyby, Granges, AB
C 0-0.12, Cr 17, Si 2.2, bal Fe.
Annealed: 79,000 TS; 50,000 YS; 15 El; 230 Brin. For retorts, salt pots, furnace parts. Ferritic, corrosion resistant Type 442. Not hardenable. *Obsolete*

NYBY 18-10 E MO
Nyby, Granges, AB
C 0-0.06, Cr 17, Ni 11, Mo 2.3, bal Fe.
Annealed: 78,000 TS; 29,000 YS; 50 El; 55 RA; 180 Brin. For chemical plant equipment, vats, filters, agitators, digestors, tanks. Austenitic stainless steel, acid resistant, not hardenable.

NYBY 18-10 LMOT
Nyby, Granges, AB
C 0.08, Cr 17, Ni 12, Mo 2.3, Ti = 5 x C, bal Fe.
Annealed: 78,000 TS; 32,000 YS; 40 El; 50 RA; 190 Brin. For welded chemical plant equipment, tanks, mixers, agitators, vessels. Austenitic stainless steel, stabilized, welding grade, not hardenable.

NYBY 18-10 UMO
Nyby, Granges, AB
C 0-0.03, Cr 17, Ni 11, Mo 2.2, bal Fe.
Annealed: 78,000 TS; 29,000 YS; 50 El; 160 Brin. For chemical plant equipment. Austenitic stainless steel; Type 316L.

NYBY 18-10L MONB
Nyby, Granges, AB
C 0.08, Cr 17, Ni 12, Mo 2.3, Cb = 10 x C, bal Fe.
Annealed: 78,000 TS; 34,000 YS; 40 El; 50 RA; 190 Brin. For welded chemical plant equipment, mixers, tanks, vessels, retorts. Type 318 stainless steel, austenitic, stabilized, acid resistant.

NYBY 18-12 EMO
Nyby, Granges, AB
C 0.06, Cr 17, Ni 12, Mo 2.8, bal Fe.
Annealed: 78,000 TS; 30,000 YS; 50 El; 55 RA; 180 Brin. For acid resistant chemical plant equipment, tanks, mixers, vats. Type 316 stainless steel, austenitic, acid resistant.

NYBY 18-12 LMO
Nyby, Granges, AB
C 0-0.08, Cr 17, Ni 11, Mo 2.8, bal Fe.
Annealed: 78,000 TS; 30,000 YS; 50 El; 55 Ra; 180 Brin. For acid resistant chemical plant equipment, tanks, mixers, retorts. Type 316 and Type 319 stainless steel. Austenitic, acid resistant. *Obsolete*

NYBY 18-12 LMONB
Nyby, Granges, AB
C 0-0.08, Cr 17, Ni 13, Mo 2.8, Cb = 10 x C, bal Fe.
Annealed: 78,000 TS; 30,000 YS; 40 El; 50 RA; 190 Brin. For welded chamical plant equipment, tanks, vessels, mixers, agitators. Type 318 stainless steel, austenitic, stabilized, acid resistant. *Obsolete*

NYBY 18-12 LMOT
Nyby, Granges, AB
C 0-0.08, Cr 17, Ni 13, Mo 2.8, Ti = 5 x C, bal Fe.
Annealed: 78,000 TS; 30,000 YS; 40 El; 50 RA; 190 Brin. For welded chemical plant equipment, tanks, vessels, agitators. Austenitic stainless steel, stabilized, welding grade.

NYBY 18-12 UMO
Nyby, Granges, AB
C 0-0.03, Cr 17, Ni 12, Mo 2.8, bal Fe.
Annealed: 78,000 TS; 30,000 YS; 50 El; 55 RA; 180 Brin. For acid resistant chemical plant equipment, tanks, mixers, vats. Type 316 L, stainless steel, austenitic, acid resistant.

NYBY 18-12 UMON
Nyby, Granges, AB
C 0-0.03, Cr 17, Ni 13, Mo 2.7, N 0.15, bal Fe.
Annealed: 100,000 TS; 45,000 YS; 50 El; 190 Brin. For low temperature purposes in marine atmosphere. Austenitic, corrosion resistant stainless steel, Type 316 LN, with high yield strength and good low temperature properties.

NYBY 18-12MO
Nyby, Granges, AB
C 0-0.1, Cr 19, Ni 12, Mo 3.5, bal Fe.
Annealed: 90,000 TS; 40,000 YS; 45 El; 60 RA; 180 Brin. For acid resistant chemical plant equipment; Type 317; austenitic, stainless. *Obsolete*

NYBY 18-14 EMO
Nyby, Granges, AB
C 0-0.06, Cr 19, Ni 14, Mo 3.5, bal Fe.
Annealed: 78,000 TS; 29,000 YS; 45 El; 45 RA; 190 Brin. For chemical plant equipment, vats, tanks. Austenitic, stainless steel, acid resistant. Type 317 stainless steel, acid resistant.

NYBY 18-14 LMO
Nyby, Granges, AB
C 0-0.08, Cr 18.5, Ni 13.5, Mo 3.5, bal Fe.
Annealed: 128,000 TS; 44,000 YS; 62 El; 65 RA; 160 Brin. For acid resistant chemical plant equipment, mixers, tanks, retorts. Austenitic stainless steel, Type 317, not hardenable. *Obsolete*

NYBY 18-14 UMO
Nyby, Granges, AB
C 0-0.03, Cr 18, Ni 14, Mo 3.5, bal Fe.
Annealed: 80,000 TS; 30,000 YS; 45 El; 165 Brin. For chemical plant equipment. Austenitic stainless steel; Type 317L.

NYBY 18-15 EMO
Nyby, Granges, AB
C 0.06, Cr 17, Ni 15, Mo 4.5, bal Fe.
Annealed: 78,000 TS; 29,000 YS; 45 El; 45 RA; 190 Brin. For chemical plant equipment, vats, tanks. Austenitic, stainless, acid resistant. Type 317 stainless steel, not hardenable.

NYBY 18-15 UMO
Nyby, Granges, AB
C 0-0.03, Cr 17, Ni 15, Mo 4.5, bal Fe.
Annealed: 80,000 TS; 32,000 YS; 45 El; 165 Brin. For chemical plant equipment. Austenitic stainless steel with high resistance to general corrosion and pitting.

NYBY 18-5-9
Nyby, Granges, AB
C 0-0.1, Cr 18, Ni 5, Mn 9, N 0.15, bal Fe.
Annealed: 85,000 TS; 47,000 YS; 45 El; 40 RA; 230 Brin. For chemical plant equipment, auto and truck bodies, bus bodies. Austenitic stainless steel Type 204. Not hardenable. *Obsolete*

NYBY 18-8
Nyby, Granges, AB
C 0.1, Cr 18, Ni 8, bal Fe.
Annealed: 80,000 TS; 35,000 YS; 55 El; 75 RA; 150 Brin. For chemical plant and dairy equipment, tanks; Type 302; stainless, austenitic. *Obsolete*

NYBY 18-8 LT
Nyby, Granges, AB
C 0-0.08, Cr 18, Ni 10, Ti = 5 x C, bal Fe.
Annealed: 78,000 TS; 31,000 YS; 40 El; 50 RA; 190 Brin. For welded chemical plant equipment, tanks, vessels, agitators, mixers. Austenitic stainless steel Type 321, stabilized, welding grade.

NYBY 18-8 UL
Nyby, Granges, AB
C 0-0.03, Cr 19, Ni 11, bal Fe.
Annealed: 78,000 TS; 29,000 YS; 50 El; 55 RA; 180 Brin. For chemical plant equipment, tanks, vessels, mixers, filters, agitators. Austenitic stainless steel. Type 304 L, welding grade.

NYBY 18-8 ULN
Nyby, Granges, AB
C 0-0.03, Cr 18, Ni 9, N 0.15, bal Fe.
Annealed: 95,000 TS; 44,000 YS; 50 El; 190 Brin. For LNG tanks and other low temperature equipment. Austenitic stainless steel, Type 304 LN, with high yield strength and good low temperature properties.

NYBY 18-8 UMO
Nyby, Granges, AB
C 0-0.03, Cr 17, Ni 11, Mo 1.6, bal Fe.
Annealed: 78,000 TS; 29,000 YS; 50 El; 55 RA; 180 Brin. For chemical plant and pickling equipment, agitators, mixers, retorts. Austenitic stainless steel, acid resistant. *Obsolete*

NYBY 18-8-LNB
Nyby, Granges, AB
C 0-0.08, Cr 18, Ni 10, Cb = 10 x C, bal Fe.
Annealed: 78,000 TS; 31,000 YS; 40 El; 50 RA; 190 Brin. For welded chemical plant equipment, tanks, mixers, agitators, stills. Austenitic Type 347 stainless steel, stabilized, welding grade.

NYBY 18-8EL
Nyby, Granges, AB
C 0-0.05, Cr 18, Ni 8, bal Fe.
Annealed: 85,000 TS; 35,000 YS; 60 El; 70 RA; 150 Brin. For chemical plant equipment, tanks, mixers; Type 304; stainless, austenitic.

NYBY 18-8EMO
Nyby, Granges, AB
C 0-0.12, Cr 18, Ni 10, Mo 2.5, bal Fe.
Annealed: 85,000 TS; 35,000 YS; 50 El; 65 RA; 160 Brin. For acid resistant chemical plant equipment; Type 316; stainless, austenitic. *Obsolete*

NYBY 18-8H
Nyby, Granges, AB
C 0.11-0.16, Cr 18, Ni 8-0.8, bal Fe.
Annealed: 80,000 TS; 35,000 YS; 55 El; 75 RA; 150 Brin. For chemical plant and dairy equipment, tanks; Type 302; stainless, austenitic. *Obsolete*

NYBY 18-8L
Nyby, Granges, AB
C 0.06-0.08, Cr 18, Ni 8, bal Fe.
Annealed: 85,000 TS; 35,000 YS; 60 El; 70 RA; 150 Brin. For chemical plant equipment, tanks, mixers; Type 304; stainless, austenitic.

NYBY 18-8LMO
Nyby, Granges, AB
C 0.07-0.12, Cr 18, Ni 9.5, Mo 1.5, bal Fe.
Annealed: 85,000 TS; 35,000 YS; 50 El; 65 RA; 160 Brin. For acid resistant chemical plant equipment; Type 316; stainless, austenitic. *Obsolete*

NYBY 18-8LMONB
Nyby, Granges, AB
C 0-0.08, Cr 17, Ni 10, Mo 1.6, Ti = 5 x C, bal Fe.
Annealed: 78,000 TS; 32,000 YS; 40 El; 50 RA; 190 Brin. For welded chemical plant equipment, tanks, vessels, agitators. Austenitic stainless steel, stabilized, welding grade. *Obsolete*

NYBY 18-8T

Nyby, Granges, AB
C 0-0.08, Cr 18, Ni 8, Ti, bal Fe.
Annealed: 85,000 TS; 35,000 YS; 55 El; 65 RA; 150 Brin. For welded chemical plant equipment; Type 321; stainless, austenitic. *Obsolete*

NYBY 1803 MOT

Nyby, Granges, AB
C 0-0.025, Cr 18, Mo 2, Ti 0.4, N 0-0.025, bal Fe.
Annealed: 79,000 TS; 52,000 YS; 30 El; 180 Brin. For water heaters, hot water pipes, heat exchangers. Ferritic stainless steel with good weldability, immune to stress corrosion cracking and good resistance to pitting and general corrosion.

NYBY 1803 T

Nyby, Granges, AB
C 0-0.025, Cr 18, Ti 0.4, N 0-0.025, bal Fe.
Annealed: 73,000 TS; 40,000 YS; 35 El; 160 Brin. For water heaters, hot water pipes, coolers. Ferritic stainless steel with good weldabiltiy and immune to stress corrosion cracking.

NYBY 20-12 L

Nyby, Granges, AB
C 0.08, Cr 20, Ni 11, bal Fe.
Annealed: 78,000 TS; 32,000 YS; 50 El; 55 RA; 180 Brin. For heat treating furnaces, salt pots. Heat and corrosion resistant. Austenitic, Type 308 stainless steel. *Obsolete*

NYBY 20-12 SI

Nyby, Granges, AB
C 0-0.2, Cr 20, Ni 12, Si 2, bal Fe.
Annealed: 85,000 TS; 36,000 YS; 45 El; 45 RA; 190 Brin. For heat treating furnaces and boxes, retorts, salt pots. Austenitic. Heat and corrosion resistant. Type 309B stainless steel.

NYBY 20-20

Nyby, Granges, AB
C 0-0.12, Cr 20, Ni 20, bal Fe.
For valves, furnace parts and equipment; corrosion and heat resistant. *Obsolete*

NYBY 20-25 UMOCU

Nyby, Granges, AB
C 0-0.025, Cr 20, Ni 25, Mo 4.5, Cu 1.5, bal Fe.
Annealed: 85,000 TS; 33,000 YS; 40 El; 165 Brin. For use in high-corrosive environments such as H_2SO_4. Austenitic stainless steel with very good resistance to general corrosion and pitting.

NYBY 23-14

Nyby, Granges, AB
C 0.15, Cr 23, Ni 14, bal Fe.
Annealed: 121,000 TS; 48,500 YS; 50 El; 55 RA; 170 Brin. For furnace parts, slat pots, oil refinery equipment. Austenitic stainless steel. Type 309. Heat resistant. *Obsolete*

NYBY 23-14 L NB

Nyby, Granges, AB
C 0-0.08, Cr 23, Ni 14, Cb = 10 x C, bal Fe.
Annealed: 78,000 TS; 31,000 YS; 35 El; 50 RA; 200 Brin. For furnace parts and equipment, welded structures and heat treating boxes, retorts. Type 309 Cb stainless steel. Austenitic. Weldable grade, heat resistant. *Obsolete*

NYBY 23-14L

Nyby, Granges, AB
C 0-0.08, Cr 23, Ni 13, bal Fe.
Annealed: 78,000 TS; 31,000 YS; 35 El; 50 RA; 190 Brin. For heat treating boxes, furnace parts and equipment, retorts, valves. Type 309S stainless steel. Austenitic, heat resistant.

NYBY 24-6 UMO

Nyby, Granges, AB
C 0-0.03, Cr 24, Ni 6, Mo 1.5, N 0.1, bal Fe.
Annealed: 100,000 TS; 68,000 YS; 25 El; 240 Brin. For chemical plant equipment as heat-exchangers, tanks, mixers. Ferritic-austenitic stainless steel with high yield strength and good resistance to stress corrosion and pitting.

NYBY 25-21

Nyby, Granges, AB
C 0.15, Cr 25, Ni 20, bal Fe.
Annealed: 132,000 TS; 52,500 YS; 45 El; 55 RA; 170 Brin. For furnace parts, salt pots, heat treating boxes. Austenitic stainless steel. Type 310, heat resistant. *Obsolete*

NYBY 25-21 L

Nyby, Granges, AB
C 0-0.08, Cr 25, Ni 20, bal Fe.
Annealed: 78,000 TS; 31,000 YS; 45 El; 45 RA; 190 Brin. For furnace parts and equipment, valves, pumps, gas turbine components, retorts. Type 310 S stainless steel, austenitic. Resists corrosion and oxidation.

NYBY 25-21 SI

Nyby, Granges, AB
C 0.2, Cr 25, Ni 21, Si 2, bal Fe.
Annealed: 85,000 TS; 36,000 YS; 30 El; 35 RA; 200 Brin. For furnace parts, heat treating boxes, retorts. Type 314 stainless steel, austenitic. Resists oxidation and carburization.

NYBY 25-22 UMON

Nyby, Granges, AB
C 0-0.02, Cr 25, Ni 22, Mo 2, N 0.12, bal Fe.
Annealed: 78,000 TS; 38,000 YS; 35 El; 180 Brin. For chemical plant equipment, particularly urea fabrication. Austenitic stainless steel with high resistance to general corrosion and intercrystalline corrosion.

NYBY 25-24

Nyby, Granges, AB
C 0-0.25, Cr 25, Ni 20, Si 1.5-3, bal Fe.
For furnace parts and equipment, salt pots; Type 314; corrosion and heat resistant. *Obsolete*

NYBY 25-24 EMO

Nyby, Granges, AB
C 0.06, Cr 25, Ni 25, Mo 2.3, bal Fe.
Annealed: 78,000 TS; 36,000 YS; 35 El; 45 RA; 180 Brin. For furnaces, retorts, heat treating equipment. Austenitic stainless steel. Heat and acid resistant. *Obsolete*

NYBY 2520

Nyby, Granges, AB
C 0-0.2, Cr 25, N 0.15, bal Fe.
Annealed: 78,000 TS; 44,000 YS; 15 El; 50 RA; 230 Brin. For furnace parts, heat treating boxes. Ferritic Type 446 stainless steel, heat resistant. *Obsolete*

NYBY 27-4

Nyby, Granges, AB
C 0-0.12, Cr 26, Ni 4, bal Fe.
Cast: 90,000 TS; 65,000 YS; 2 El; 212 Brin. For cylinder liners, bushings, valve seats and bodies; Type CC50; heat resistant. *Obsolete*

NYBY 27-5MO

Nyby, Granges, AB
C 0-0.1, Cr 26, Ni 5, Mo 1.5, bal Fe.
For furnace parts and equipment, heat treating boxes; Type 329; heat resistant. *Obsolete*

NYBY-1408

Nyby, Granges, AB
C 0-0.15, Cr 11.5-13, bal Fe.
Annealed: 75,000 TS; 40,000 YS; 35 El; 70 RA; 155 Brin. For turbine blades, dental and surgical instruments, valves; Type 410; corrosion resistant. *Obsolete*

NYBY-1410

Nyby, Granges, AB
C 0-0.15, Cr 11.5-13.5, Si 0-1, bal Fe.
Annealed: 75,000 TS; 40,000 YS; 35 El; 70 RA; 155 Brin. For turbine blades, dental and surgical instruments, valves; Type 410; corrosion resistant. *Obsolete*

NYBY-1410MO

Nyby, Granges, AB
C 0.08, Cr 13, Mo 1, bal Fe.
Annealed: 75,000 TS; 40,000 YS; 35 El; 70 RA; 155 Brin. For oil refinery and chemical plant equipment; corrosion resistant. *Obsolete*

NYBY-1415

Nyby, Granges, AB
C 0.13-0.18, Cr 13, bal Fe.
Annealed: 80,000 TS; 42,000 YS; 33 El; 68 RA; 170 Brin. For turbine blades, valves, cutlery, surgical instruments; Type 420; corrosion resistant. *Obsolete*

NYBY-1415MO

Nyby, Granges, AB
C 0.14, Cr 13, Mo 1, bal Fe.
Annealed: 75,000 TS; 40,000 YS; 35 El; 70 RA; 155 Brin. For oil refinery and chemical plant equipment; corrosion resistant. *Obsolete*

NYBY-1425

Nyby, Granges, AB
C 0.19-0.25, Cr 13, bal Fe.
Annealed: 95,000 TS; 50,000 YS; 25 El; 55 RA; 195 Brin. For valves, cutlery, surgical instruments, knives; Type 420; corrosion resistant. *Obsolete*

NYBY-1435

Nyby, Granges, AB
C 0.27-0.35, Cr 13, bal Fe.
Annealed: 95,000 TS; 50,000 YS; 25 El; 55 RA; 195 Brin. For valves, cutlery, surgical instruments, knives; Type 420; corrosion resistant. *Obsolete*

NYBY-1710

Nyby, Granges, AB
C 0.07-0.12, Cr 17, bal Fe.
Annealed: 80,000 TS; 50,000 YS; 25 El; 50 RA; 150 Brin. For oil refinery equipment, oil burners and heaters; Type 430; corrosion resistant. *Obsolete*

NYBYLOID-1

British Oxygen Co., Ltd.
Fe 10, bal Ni.
Weld: 96,000-101,000 TS. For welding 9% nickel steel. Low hydrogen welding electrode. Core wire of Ni and a covering containing 13 Cr, 10 Fe, Mn, Cb, Mo

NYMPHE

Creusot-Loire
C, bal Fe.
For tools. *Obsolete*

O-6
Bethlehem Steel Corp.
C 1.45, Mn 0.8, Si 1.05, Mo 0.25, bal Fe.
For cold working, forming, shaping and drawing dies. AISI Type O6; oil-hardened tool steel. *Obsolete*

O-NICKEL
English manufacture
99.5 min Ni + Co, controlled Mg.
For valve components, tube cathodes; corrosion and heat resistant.

O.B. ALLOY NO. 17
Ohio Brass Co.
Cu 85.77, Sn 4.48, Pb 3.43, Zn 6.31.
Cast: 36,000 TS; 16,600 YS; 38 El; 63 Brin. For high grade valves; IZ-13.2. *Obsolete*

O.B. ALLOY NO. 4 B
Ohio Brass Co.
Cu 82.17, Sn 2.87, Pb 7.05, Zn 7.81.
Cast: 31,200 TS; 13,500 YS; 29 El; 53 Brin. For valves. *Obsolete*

O.B. ALLOY NO. 5
Ohio Brass Co.
Cu 84.42, Sn 2.75, Pb 2.98, Zn 9.84.
Cast: 34,300 TS; 13,300 YS; 37 El; 55 Brin. For electric railway appliances; IZ-15. *Obsolete*

O.B. ALLOY NO. 5 A
Ohio Brass Co.
Cu 88.38, Sn 5.09, Pb 1.94, Zn 4.58.
Cast: 36,800 TS; 16,300 YS; 37 El; 58 Brin. For electric railway appliances; IZ-14.8. *Obsolete*

O.H. 38
Oscar W. Hedstrom Corp.
Zn 15, Cu 2-3, Si 2, bal Al.
Cast: 30,000 TS; 17,500 YS; 2.5 El; 65 Brin. For light alloy parts, gear guards.

O.P.
Japanese manufacture
For permanent magnets; high coercive force.

OA 3010
Eutectic Corp.
Continuous electrode for AC-DC build-up on low and medium carbon steel for joining and filling. 25 Rock C.

OA 3205
Eutectic Corp.
Continuous electrode for AC-DC build-up and coating of manganese steel; 25 Rock C; work hardens to 50-55 Rock C.

OA 4601
Eutectic Corp.
Continuous electrode for AC-DC coating on all steels. 55-60 Rock C; maintains hardness at moderate temperatures.

OA 4625
Eutectic Corp.
Continuous electrode for AC-DC build-up on all steels; resists abrasion, impact, compression. 40-45 Rock C.

OA 690
Eutectic Corp.
Continuous electrode for high strength steel joining and tough, wear resistant cladding; DC; 90,000 psi TS.

OAKES TRODALOY 1
Oakes Bronze & Aluminum Co.
Be 0.4, Co 2.6, bal Cu.
Cast: 90,000 TS; 50,000 YS; 10-15 El. Wire: 125,000 TS; 60,000 YS; 8-15 El. For springs; age hardenable.

OB ALLOY NO. 29
Ohio Brass Co.
Cu 57.5, Sn 0.5, Pb 0.25, Zn 39.25, Al 1, Fe 1.
Cast: 77,000 TS; 31,500 YS; 26 El; 120 Brin. For splicers, fair electrical conductivity. *Obsolete*

OCTANIUM
Parker Pen Co.
Co 40, Ni 15.5, Cr 20, Mo 7, Mn 2, Fe 15, Be 0.03, C 0.15.
Heat treated: 368,000 TS. For fountain pen nibs; noncorrosive, non-magnetic. *Obsolete*

ODESSA
Manufacturer not listed
Ag 33.25, Cu 42.5, Zn 15.75, Ni 8.5.
For silver solder; corrosion resistant.

OERDERLIN FMB
Oederlin & Co. Ltd.
C 0.1, Cr 18, Ni 8, Mo 2.5-3, bal Fe.
Rolled: 58,000-78,000 TS; 29,000-32,000 YS; 12-18 El; 170-200 Brin. For chemical plant equipment; stainless, acid resistant, austenitic.

OERDERLIN FST
Oederlin & Co. Ltd.
C 0.1, Ni 8, Cr, Si, Mn, W, bal Fe.
Annealed: 65,000-85,000 TS; 31,000-40,000 YS; 15-20 El; 160-200 Brin. For chemical plant equipment; stainless, austenitic.

OERSTIT 1000
Thyssen Edelstahlwerke AG
Ni 17.5, Al 6-7.5, Cu 3, Ti 7-8, 19 or 35% Co, bal Fe.
For magnetic and electrical equipment. Permanent magnet. *Obsolete*

OERSTIT 120
Thyssen Edelstahlwerke AG
C 0.04, Ni 27, Cu 2, Al 13, Ti 0.5, bal Fe.
For permanent magnets; cast. *Obsolete*

OERSTIT 120 CU
Thyssen Edelstahlwerke AG
C 0.04, Ni 25, Al 13, Cu 4, bal Fe.
For magnets; cast. *Obsolete*

OERSTIT 120 K
Thyssen Edelstahlwerke AG
Al 13, Ni 26, Co 4, Cu 3, Ti 1, bal Fe.
For electrical and magnetic equipment. Permanent magnet. Isotropic. *Obsolete*

OERSTIT 130
Thyssen Edelstahlwerke AG
C 0.04, Ni 26, Co 4.5, Al 12, Cu 4.5, bal Fe.
For magnets; cast. *Obsolete*

OERSTIT 160
Thyssen Edelstahlwerke AG
C 0.04, Ni 24, Co 9.5, Al 11, Cu 4, bal Fe.
For magnets; cast. *Obsolete*

OERSTIT 160K
Marathon Specialty Steels Inc.
Ni 22, Co 17, Al 10, Cu 3, Ti 1, bal Fe.
For permanent magnets.

OERSTIT 160R
Marathon Specialty Steels Inc.
Ni 22, Co 14, Al 10, Cu 3, Ti 1, bal Fe.
For permanent magnets.

OERSTIT 190
Marathon Specialty Steels Inc.
Ni 20, Co 16, Al 9, Cu 3, Ti 1, bal Fe.
For permanent magnets.

OERSTIT 190 K
Thyssen Edelstahlwerke AG
Al 10, Ni 21, Co 17, Cu 3, Ti 1, bal Fe.
For electrical and magnetic equipment. Permanent magnet. Anisotropic. *Obsolete*

OERSTIT 250
Thyssen Edelstahlwerke AG
C 0.04, Si 1, Ni 18, Co 25, Al 7, Cu 3, Ti 5.5, bal Fe.
For permanent magnets; cast. *Obsolete*

OERSTIT 260
Thyssen Edelstahlwerke AG
Al 7, Ni 15, Co 28, Cu 5, Ti 8, bal Fe.
For electrical and magnetic equipment. Permanent magnet. Isotropic. *Obsolete*

OERSTIT 30
Thyssen Edelstahlwerke AG
C 1, Cr 3, bal Fe.
For magnets; hardened for max permeability. *Obsolete*

OERSTIT 300
Marathon Specialty Steels Inc.
Ni 17, Co 2, Al 7.5, Cu 3, Ti 2.5, bal Fe.
For permanent magnets.

OERSTIT 35
Marathon Specialty Steels Inc.
C 1, Cr 4.5, Mn 1, bal Fe.
For permanent magnets; workable, maximum service temperature 175°F.

OERSTIT 350
Marathon Specialty Steels Inc.
Ni 15, Co 30, Al 7.5, Cu 4, Ti 5, bal Fe.
For permanent magnets.

OERSTIT 360
Thyssen Edelstahlwerke AG
C 0.04, Ni 15, Co 23.5, Al 8, Cu 3, Ti 1, bal Fe.
For magnets; cast. *Obsolete*

OERSTIT 40
Thyssen Edelstahlwerke AG
C 1, Cr 4, W 0.5, Co 2, bal Fe.
For magnets; hardened for max permeability. *Obsolete*

OERSTIT 400
Thyssen Edelstahlwerke AG
C 0.04, Ni 15, Co 22, Al 8, Cu 4, bal Fe.
For magnets; cast. *Obsolete*

OERSTIT 400K
Marathon Specialty Steels Inc.
Ni 14, Co 24, Al 8, Cu 3, Ti 0.5, bal Fe.
For permanent magnets.

OERSTIT 400R
Marathon Specialty Steels Inc.
Ni 14, Co 24, Al 8, Cu 3, Ti 0.5, bal Fe.
For permanent magnets.

OERSTIT 450
Thyssen Edelstahlwerke AG
Al 7, Ni 15, Co 30, Cu 4, Ti 5, bal Fe.
For electrical and magnetic equipment. Permanent magnet. Anisotropic. *Obsolete*

OERSTIT 450 K
Thyssen Edelstahlwerke AG
Al 7, Ni 14, Co 42, Cu 4, Ti 8, bal Fe.
For electrical and magnetic equipment. Permanent magnet. Anisotropic. *Obsolete*

OERSTIT 50
Thyssen Edelstahlwerke AG
C 1, Cr 8, Mo 1.2, Co 6, bal Fe.
For magnets; hardened for max permeability. *Obsolete*

OERSTIT 500

Marathon Specialty Steels Inc.
Ni 14, Co 24, Al 8, Cu 3, Ti 0.5, bal Fe.
For permanent magnets; high permeability.

OERSTIT 60

Thyssen Edelstahlwerke AG
C 1, Cr 8.5, Mo 1.5, Co 10.5, bal Fe.
For magnets; hardened for max permeability. *Obsolete*

OERSTIT 600

Marathon Specialty Steels Inc.
Ni 14, Co 24, Al 8, Cu 3, Ti 0.5, bal Fe.
For magnets; maximum service temperature 660°F, cast and
sintered alloy.

OERSTIT 70

Thyssen Edelstahlwerke AG
C 1, Cr 8.5, Mo 1.5, Co 15.5, bal Fe.
For magnets; hardened for max permeability. *Obsolete*

OERSTIT 700

Thyssen Edelstahlwerke AG
Ni 21, Co 12, Al 10, Cu 6, bal Fe.
For permanent magnets; high permeability. *Obsolete*

OERSTIT 800

Thyssen Edelstahlwerke AG
Ni 18, Co 19, Al 9, Cu 4, Ti 4, bal Fe.
For permanent magnets; high permeability. *Obsolete*

OERSTIT 900 CP

Thyssen Edelstahlwerke AG
Pt 78, bal Co.
For electrical and magnetic equipment. Permanent magnet.
Isotropic. *Obsolete*

OERSTIT 90G

Marathon Specialty Steels Inc.
C 0.04, Ni 22, Cu 2, Al 11.5, Ti 0.5, bal Fe.
For permanent magnets; cast.

OERSTIT 90W

Thyssen Edelstahlwerke AG
C 0.9, Cr 4.5, Mo 0.3, W 4.5, Co 30, bal Fe.
For magnets. *Obsolete*

OERSTIT SPEZIAL

Thyssen Edelstahlwerke AG
C, Co, Al, Cu.
For permanent magnets; high permeability. *Obsolete*

OFHC COPPER

Climax Performance Materials Corp.
Cu 99.98.
For electrical equipment; oxygen free, high conductivity.

OFHC COPPER 101

Anaconda Co.
Copper. 99.9 Cu min.
Hard: 48,000 TS; 40,000 YS; 15 El; 83 Brin. Soft: 32,000 TS;
10,000 YS; 50 El; 42 Brin. For electrical apparatus. High
conductivity, oxygen free.

OFHC COPPER 102

Anaconda Co.
Copper. 99.95 Cu min.
Hard: 55,000 TS; 50,000 YS; 10 El; 60 Rock B. Soft: 32,000
TS; 10,000 YS; 55 El; 40 Rock F. For radar and electronic
components, waveguide tubes, Dumet wire, connectors.
Electrical conductivity: 101. Corrosion resistant.

OFHC SULFUR COPPER

Anaconda Co.
Copper. S 0.2-0.6, 99.90 Cu + S min.
1/2 H temper: 42,000 TS; 36,000 YS; 25 El; 73 Brin. Hard:
48,000 TS; 42,000 YS; 15 El; 83 Brin. For contact pins and
inserts, screw machine products. High thermal and electrical
conductivity, free cutting, corrosion resistant.

OHIO 0-10

Teledyne Ohiocast
C 0-0.3, Mn 0.4-0.7, Si 0.3-0.7, Mo 0.5, Cr 1-1.7, bal Fe.
Cast: 85,000 TS; 55,000 YS; 20 El; 35 RA; 170-229 Brin. For
castings, gears, shafts, tough. *Obsolete*

OHIO 0-11A

Teledyne Ohiocast
C 0-0.15, Mn 0-1, Si 0-1, Mo 1-1.5, Cr 8-10, bal Fe.
Cast: 70,000 TS; 40,000 YS; 24 El; 40 RA; 163-217 Brin. For
castings, gears, shafts; tough. *Obsolete*

OHIO 0-14

Teledyne Ohiocast
C 0-0.35, Mn 0.9-1.2, Si 0.3-0.6, Mo 0.3-0.4, bal Fe.
Cast: 105,000 TS; 85,000 YS; 15 El; 30 RA; 229-262 Brin. For
castings, gears, shafts; tough. *Obsolete*

OHIO 0-15E

Teledyne Ohiocast
C 0-0.35, Mn 0.9-1.2, Si 0.4, Mo 0.1-0.2, Ni 0.4, Cr 0.4, bal Fe.
Cast: 90,000 TS; 70,000 YS; 20 El; 30 RA; 187-217 Brin. For
castings, gears, shafts; tough. *Obsolete*

OHIO 0-17

Teledyne Ohiocast
C 0-0.3, Mn 0-1, Si 0-1, Mo 0.5-0.7, Cr 6-8, bal Fe.
Normalized: 90,000 TS; 60,000 YS; 18 El; 30 RA; 197-241
Brin. For castings, oil refinery equipment; corrosion resistant.
Obsolete

OHIO 0-19

Teledyne Ohiocast
C 0-0.35, Mn 1.4-1.5, Mo 0.4-0.6, V 0.1-0.18, bal Fe.
Water quenched: 140,000 TS; 130,000 YS; 14 El; 30 RA;
293-341 Brin. For castings, gears, shafts; high strength.
Obsolete

OHIO 0-1A

Teledyne Ohiocast
C 0.4-0.5, Mn 0.6-0.9, Si 0.3-0.6, bal Fe.
Cast: 80,000 TS; 45,000 YS; 17 El; 25 RA; 163-217 Brin. For
castings, gears, shafts; general purpose. *Obsolete*

OHIO 0-1W

Teledyne Ohiocast
C 0-0.25, Mn 0-0.7, Si 0-0.6, bal Fe.
Cast: 60,000 TS; 30,000 YS; 24 El; 35 RA; 137-187 Brin. For
castings, gears, shafts; low strength. *Obsolete*

OHIO 0-20

Teledyne Ohiocast
C 0-0.35, Mn 1.2-1.4, Si 0.3-0.6, Mo 0.4-0.6, V 0.1-0.18, bal
Fe.
Water quenched: 125,000 TS; 115,000 YS; 35 RA; 262-302
Brin. For castings, gears, shafts; high strength. *Obsolete*

OHIO 0-2W

Teledyne Ohiocast
C 0-0.25, Mn 0-0.7, Si 0-0.6, Mo 0.4-0.6, bal Fe.
Cast: 65,000 TS; 35,000 YS; 24 El; 35 RA; 143-187 Brin. For
castings, gears, shafts; tough. *Obsolete*

OHIO 0-9

Teledyne Ohiocast
C 0.15-0.3, Mn 0.7-1, Si 0.3-0.6, Ni 1.75-2.25, bal Fe.
Normalized: 80,000 TS; 50,000 YS; 25 El; 45 RA; 163-217
Brin. For castings, gears, shafts. *Obsolete*

OHIO 8-16 M

Teledyne Ohiocast
C 0-0.16, Mn 0-1.2, Mo 2.5-3.5, 18% min Cr, 9% min Ni, bal
Fe.
For stainless castings; austenitic, acid resistant. *Obsolete*

OHIO C-12

Teledyne Ohiocast
C 0-0.15, Mn 0-1, Si 0-1.25, Ni 0-1, Cr 11-14, bal Fe.
For stainless castings; stainless. *Obsolete*

OHIO C-12P

Teledyne Ohiocast
C 0-0.2, Mn 1.25, Ni 0-1.25, Cr 10-14, bal Fe.
For castings; heat and corrosion resistant. *Obsolete*

OHIO C-2

Teledyne Ohiocast
C 0-0.25, Mn 1.25, Si 1.5, Ni 34-37, Cr 8-11, bal Fe.
For castings; heat and corrosion resistant. *Obsolete*

OHIO C-8A

Teledyne Ohiocast
C 0-0.07, Mn 0-1.25, Si 0-1.5, 8% min Ni, 18% min Cr, bal Fe.
For stainless castings; stainless. *Obsolete*

OHIO C-8C

Teledyne Ohiocast
C 0-0.15, Mn 1.25, Si 1.5, Ni 8, Cb = 10 x C, bal Fe.
For castings; heat and corrosion resistant. *Obsolete*

OHIO C-8M

Teledyne Ohiocast
C 0-0.15, Mn 1.25, Si 1.5, Ni 8, Cr 18, Mo 3, bal Fe.
For castings; heat and corrosion resistant. *Obsolete*

OHIO DIE FM

Teledyne Vasco
C 1.55, Cr 12, V 0.8, Co 0.4, Mo 0.8, bal Fe.
Annealed: 105,000 TS; 60,000 YS; 13 El; 18 RA; 212 Brin. For
roll threading dies, punches, gages, shears; air hardening,
free-cutting. *Obsolete*

OHIO DIE STEEL

Teledyne Vasco
C 1.55, Cr 12, V 0.85, Mo 0.8, bal Fe.
105,000 TS; 60,000 YS; 13 El; 18 RA; 212 Brin. For dies, roller
threading dies, punches, gages, shear blades; non-warping
wear resisting.

OHIO N-3A

Teledyne Ohiocast
C 0-0.35, Mn 1.5, Si 1.75, Ni 11-14, Cr 24-27, bal Fe.
For castings; heat and corrosion resistant. *Obsolete*

OHIO N-3K

Teledyne Ohiocast
C 0-0.4, Mn 1.5, Si 1.75, Ni 10-13, Cr 24-27, bal Fe.
For castings; heat and corrosion resistant. *Obsolete*

OHIO N-3M

Teledyne Ohiocast
C 0-0.4, Mn 1.5, Si 1.75, Ni 14-17, Cr 25-29, bal Fe.
For castings; heat and corrosion resistant. *Obsolete*

OHIO N-3S

Teledyne Ohiocast
C 0-0.35, Mn 1.5, Si 1.75, Ni 10-12, Cr 24-26, bal Fe.
For castings; heat and corrosion resistant. *Obsolete*

OHIO O-11

Teledyne Ohiocast
C 0-0.3, Mn 0-1, Cr 8-10, Mo 1.2, bal Fe.
Normalized: 90,000 TS; 60,000 YS; 18 El; 30 RA; 221 Brin.
For castings, fittings; corrosion resistant. *Obsolete*

OHIO O-12

Teledyne Ohiocast
C 0.3-0.45, Mn 0.8, Cr 0.8-1.1, Mo 0.15-0.25, bal Fe.
Normalized: 100,000 TS; 65,000 YS; 18 El; 30 RA; 221 Brin.
For castings, fittings; oil hardening. *Obsolete*

OHIO O-12-B

Teledyne Ohiocast
C 0.3-0.45, Mn 0.6-0.9, Mo 0.15-0.25, Cr 0.8-1.1, bal Fe.
Normalized: 100,000 TS; 65,000 YS; 18 El; 30 RA; 221 Brin.
Heat treated: 110,000 TS; 85,000 YS; 16 El; 30 RA; 248 Brin.
For castings, fittings; general. *Obsolete*

OHIO O-2
Teledyne Ohiocast
C 0-0.35, Mn 0-1, Mo 0.4, bal Fe.
Normalized: 70,000 TS; 45,000 YS; 22 El; 35 RA; 143 Brin.
For valves, fittings; castings. *Obsolete*

OHIO O-21
Teledyne Ohiocast
C 0-0.3, Mn 0-0.7, Mo 0.4-0.6, Cr 0.4-0.7, bal Fe.
Normalized: 70,000 TS; 45,000 YS; 22 El; 35 RA; 163 Brin.
For castings, fittings; general. *Obsolete*

OHIO O-21-20
Teledyne Ohiocast
C 0-0.18, Mn 0-0.7, Mo 0.4-0.6, Cr 0.5-0.8, bal Fe.
Normalized: 70,000 TS; 45,000 YS; 22 El; 35 RA; 163 Brin.
For castings, fittings; general. *Obsolete*

OHIO O-22
Teledyne Ohiocast
C 0-0.35, Mn 0-0.85, Mo 0.2-0.3, Ni 1-1.25, bal Fe.
Normalized: 85,000 TS; 53,000 YS; 22 El; 35 RA; 197 Brin.
For castings; general. *Obsolete*

OHIO O-23
Teledyne Ohiocast
C 0-0.3, Mn 0-0.7, Mo 0.45-0.65, Cr 1-1.5, bal Fe.
Normalized: 85,000 TS; 55,000 YS; 20 El; 35 RA; 197 Brin.
For castings, fittings; general. *Obsolete*

OHIO O-24
Teledyne Ohiocast
C 0-0.18, Mn 0-0.7, Mo 0.4-0.6, Cr 0.8-1.1, bal Fe.
Normalized: 70,000 TS; 45,000 YS; 22 El; 35 RA; 163 Brin.
For castings; general. *Obsolete*

OHIO O-27
Teledyne Ohiocast
C 0.25-0.35, Mn 1.45-1.65, Ti, bal Fe.
Normalized: 90,000 TS; 60,000 YS; 22 El; 45 RA; 190 Brin.
For truck (R.R.) side frames, bolsters; tough. *Obsolete*

OHIO O-28
Teledyne Ohiocast
C 0-0.15, Si 0-0.6, bal Fe.
Normalized: 60,000 TS; 30,000 YS; 30 El; 50 RA; 120 Brin.
For lifting magnets, housings; castings. *Obsolete*

OHIO O-29
Teledyne Ohiocast
C 0.25-0.35, Cr 0.9-1.1, Ni 2.3-2.7, Mo 0.3-0.4, V 0.9-0.15, Mn 0.7-0.9, bal Fe.
Water quenched: 135,000 TS; 120,000 YS; 15 El; 35 RA; 300 Brin. For high tensile heavy castings; water hardening. *Obsolete*

OHIO O-4
Teledyne Ohiocast
C 0-0.3, Mn 0-1, Cr 4-6, Mo 0.5, bal Fe.
Normalized: 90,000 TS; 65,000 YS; 18 El; 30 RA; 221 Brin.
For oil refinery fittings and valves; castings. *Obsolete*

OHIO O-5-20
Teledyne Ohiocast
C 0-0.2, Mn 0-1, Si 0-0.7, Mo 0.45-0.65, Cr 1.5-2.25, bal Fe.
Normalized: 75,000 TS; 50,000 YS; 20 El; 35 RA; 170 Brin.
For valves, fittings; castings. *Obsolete*

OHIO O-6
Teledyne Ohiocast
C 0.6-0.7, Mn 0.5-0.8, Cr 0-1.5, V 0-0.2, bal Fe.
For castings; abrasion resisting. *Obsolete*

OHIOLOY 1045
Teledyne Ohiocast
C 0.4-0.5, Mn 0.6-0.9, Si 0-0.6, bal Fe.
Carbon steel casting, 3 strength ranges; 80,000 to 100,000 min TS; 40,000-70,000 min YS; 10-18 El, 163-269 Brin. Weldable; for industrial machinery and tools as pinions, sheaves, sprockets, rollers, gears. AISI 1045; SAE 0050 A and 0050 B. Formerly Ohio 0-1-50. *Obsolete*

OHIOLOY 1670
Teledyne Ohiocast
C 0.6-0.75, Mn 0.75-1, Si 0.3-0.7, Cr 1.3-1.7, Mo 0.35-0.65, bal Fe.
Alloy steel casting; hardened to 375 min Brin. For abrasive service in cement mills and similar operations. Formerly Ohio 0-16. *Obsolete*

OHIOLOY 4130
Teledyne Ohiocast
C 0.2-0.3, Mn 0.75-1, Si 0.3-0.6, Cr 0.8-1, Mo 0.2-0.4, bal Fe.
Low alloy steel casting; 80,000 min TS, 50,000 min YS; 22 min El; 183-224 Brin. Weldable, hardenable; for industrial machinery, ordnance, mining equipment as gears, knuckles. ASTM A148-60 80-50; AISI 4130; SAE 080. Formerly OHIO 0-41-30. *Obsolete*

OHIOLOY 4140
Teledyne Ohiocast
C 0.35-0.45, Mn 0.75-1, Si 0.3-0.6, Cr 0.8-1.1, Mo 0.2-0.4, bal Fe.
Low alloy steel casting: 105,000 min TS; 85,000 min YS; 17 min El; 229-269 Brin. Weldable with care, hardenable; for strong parts in industrial as gears, spindles, housings, in construction, road building, ordnance. ASTM A148-60 105-85; AISI 4140; SAE 0105. Formerly OHIO 0-41-40. *Obsolete*

OHIOLOY 4330
Teledyne Ohiocast
C 0.26-0.32, Mn 0.65-0.85, Si 0-0.6, Ni 1.65-2, Cr 0.7-0.9, Mo 0.2-0.3, bal Fe.
Alloy steel casting: 100,000 min TS; 65,000 min YS; 17 min El; 202-248 Brin. Weldable, hardenable; for high quality castings for industrial equipment. AISI 4330. *Obsolete*

OHIOLOY 8035
Teledyne Ohiocast
C 0.3-0.4, Mn 1.2-1.4, Si 0-0.6, Mo 0.2-0.3, bal Fe.
Low alloy steel casting; 90,000 min TS; 60,000 min YS; 18 min El; 192-241 Brin. Weldable, hardenable; small casting to be hardened, or large castings not hardened, for construction equipment. Formerly OHIO 0-34A. *Obsolete*

OHIOLOY 8730
Teledyne Ohiocast
C 0.25-0.35, Mn 0.75-1, Si 0.3-0.6, Ni 0.4-0.7, Cr 0.4-0.6, Mo 0.2-0.3, bal Fe.
Low alloy steel casting: 80,000 min TS; 50,000 min YS; 22 min El; 183-229 Brin. Weldable, hardenable; for small to medium size castings to be hardened for industrial machinery, construction equipment, ordnance. QQ-S-681d 80-50; AISI 8730; SAE 080. Formerly OHIO 0-15. *Obsolete*

OHIOLOY 8740
Teledyne Ohiocast
C 0.35-0.45, Mn 0.75-1, Si 0.3-0.6, Ni 0.4-0.7, Cr 0.4-0.6, Mo 0.2-0.3, bal Fe.
Low alloy steel casting: 100,000 min TS; 65,000 min YS; 17 min El; 202-248 Brin. Weldable with care, hardenable; for medium size casting, usually to be hardened, for ordnance, construction equipment, mining, railroad. AISI 8740; formerly OHIO 0-15-45-C. *Obsolete*

OHIOLOY C-16
Teledyne Ohiocast
C 0.3-0.35, Mn 0.75-1.25, Si 0.8-1.2, Cr 16-18, Ni 1.2-1.7, bal Fe.
Annealed: 125,000 TS; 95,000 YS; 20 El; C 24 Rock. Hardenable to 180,000-200,000 TS. For cutter bars, hammers, impellers, pump shafts, marine hardware. Abrasion and corrosion resistance. AISI 431; formerly FAHRITE C-16. *Obsolete*

OHIOLOY C-355
Teledyne Ohiocast
C 0.08-0.12, Mn 0.75-1.1, Si 0.45-0.75, Cr 14.5-15.5, Ni 4-4.5, Mo 2-2.6, N 0.07-0.11, bal Fe.
Age hardenable, stainless. Cast: 150,000 TS; 95,000 YS; 10 El. For compressor blades, aircraft castings. Reference AM-355; formerly C-355. *Obsolete*

OHIOLOY C-411
Teledyne Ohiocast
C 0.3, Mn 1.5, Si 1-2, Ni 62-66, Cb 1-3, Cu 26-33.
Corrosion resistant casting. Cast: 65,000 TS; 32,500 YS; 25 El. For marine hardware, pulp and paper equipment, valves, fasteners: QQ-N-288 E; Reference: Monel. Formerly FAHRITE C-95E. *Obsolete*

OHIOLOY C12
Teledyne Ohiocast
C 0-0.2, Mn 0.35-0.65, Si 0-1, Cr 8-10, Mo 0.9-1.1, bal Fe.
Cr-Mo alloy steel casting; 90,000 min TS; 60,000 min YS; 18 min El; 192-241 Brin. Weldable, hardenable; for industrial castings of larger size and better corrosion resistance; for use to 1200 F. ASTM A217-60T C12; formerly OHIO 0-11-20. *Obsolete*

OHIOLOY C5
Teledyne Ohiocast
C 0-0.2, Mn 0.4-0.7, Si 0-0.75, Cr 4-6.5, Mo 0.45-0.65, bal Fe.
Cr-Mo alloy steel casting; 90,000 min TS; 60,000 min YS; 18 min El; 192-241 Brin. Weldable, hardenable; for industrial castings of larger size; pressure castings for service temperatures below 1200 F. ASTM A217-60T C5; AISI 501. Formerly OHIO 0-4-20. *Obsolete*

OHIOLOY CE-30
Teledyne Ohiocast
C 0.3, Mn 1.5, Si 2, Cr 26-30, Ni 8-11, bal Fe.
Corrosion resistant casting. Annealed: 80,000 TS; 40,000 YS; 10 El. For digestor fittings, fractionating towers, pump bodies and casings, valve bodies. ACI CE-30; AISI 312; SAE 60312. Formerly FAHRITE C-73. *Obsolete*

OHIOLOY LC3
Teledyne Ohiocast
C 0-0.15, Mn 0.5-0.8, Si 0-0.6, Ni 3-4, bal Fe.
Nickel alloy steel casting: 65,000 min TS; 40,000 min YS; 24 min El; 137-187 Brin. Weldable; for pressure castings for sub-zero service to -150 F. ASTM A352-60T LC3; formerly OHIO 0-25. *Obsolete*

OHIOLOY NI
Teledyne Ohiocast
C 0-0.5, Mn 0-1.75, Si 0-1.5, Fe 0-2, Cu 0-0.8, bal Ni.
Corrosion resistant nickel casting. Cast: 50,000 TS; 25,000 YS; 10 El. For equipment in chemical industry to avoid metal contamination. Formerly FAHRITE Ni. *Obsolete*

OHIOLOY WC1 (OR LC1)
Teledyne Ohiocast
C 0-0.25, Mn 0.5-0.8, Si 0-0.6, Mo 0.45-0.65, bal Fe.
Carbon-Moly steel casting: 65,000 min TS; 35,000 min YS; 24 min El; 137-187 Brin. Weldable; for pressure castings, service to 1100 F. ASTM A217-60T WC1; ASTM A352-60T LC1. Formerly OHIO 0-2-25. *Obsolete*

OHIOLOY WC6
Teledyne Ohiocast
C 0-0.2, Mn 0.5-0.8, Si 0-0.6, Cr 1-1.5, Mo 0.45-0.65, bal Fe.
Cr-Mo low alloy steel casting; 70,000 min TS; 40,000 min YS; 20 min El; 137-187 Brin. Weldable; for yokes, valves, fittings, flanges, bonnets; for service up to 1200 F. ASTM A217-60T WC6; formerly OHIO 0-23-20. *Obsolete*

OHIOLOY WC9
Teledyne Ohiocast
C 0-0.18, Mn 0.4-0.7, Si 0-0.6, Cr 2-2.75, Mo 0.9-1.2, bal Fe.
Cr-Mo alloy steel casting: 70,000 min TS; 40,000 min YS; 20 min El; 143-187 Brin. Weldable; for pressure castings with service temperatures not exceeding 1200 F. ASTM A217-60T WC9; formerly OHIO 0-5M-20. *Obsolete*

OHIOLOY WCA
Teledyne Ohiocast
C 0-0.25, Mn 0-0.7, Si 0-0.6, bal Fe.
Carbon steel casting: 60,000 min TS; 30,000 min YS; 25 El; 137-187 Brin. Weldable; for light loaded parts and assemblies operating below 1000 F. ASTM A27-60 60-30; AISI 1020; SAE 0022. Formerly OHIO 0-1-25. *Obsolete*

OHIOLOY WCB
Teledyne Ohiocast
C 0-0.3, Mn 0-0.7, Si 0-0.6, bal Fe.
Carbon steel casting: 65,000 min TS; 35,000 min YS; 22 min El; 137-217 Brin. Weldable; for light loaded parts and assemblies operating below 1000 F. AISI 1030; SAE 0030; formerly OHIO 0-1. *Obsolete*

OHMAL
Manufacturer not listed
Cu 87.5, Mn 9, Ni 3.5.
For electrical resistors; heat resistant.

OHMALLOY
Gilby-Fodor S.A.
Cr 15, Al 5, Fe 80.
Annealed: 100,000 TS; 70,000 YS; 25 El. For cathodes and filaments in electronic tubes. Heat resistant.

OHMALON
Wilbur B. Driver Co.
Cr 12, Al 14, bal Fe.
For heat resistant parts; heat resistant. *Obsolete*

OHMALOY
Now FECRALOY.

OHMALOY
Wilbur B. Driver Co.
Cr 13, C 0.6, Al 4, bal Fe.
Cold rolled: 128,000 TS; 103,800 YS; 5 El. Annealed: 85,700 TS; 67,640 YS; 33 El. For resistance wire, rheostats, resistors; brittle above 1600 F. *Obsolete*

OHMALOY
Allegheny Ludlum Steel
Cr 12-14, Ni 0-0.5, C 0-0.15, Al 3-4.5, bal Fe.
Annealed: 80,000 TS; 50,000 YS; 20 El; 60 RA; 240 Brin. For chemical engineering equipment; furnace resistors; corrosion resistant; Type 406. *Obsolete*

OHMAX
Driver Harris Co.
Cr 20, Al 8.5, bal Fe.
Rolled: 100,000-130,000 TS. For electrical resistances; heat resistant to 1200 C. *Obsolete*

OHMSNIT 110
Vereinigte Edelstahlwerke
C 0.1, Cr 20, Ni 80.
For heating elements; heat resistant. *Obsolete*

OIL CUPS
Manufacturer not listed
Cu 88, Sn 5, Zn 7.
For oil cups, fittings.

OIL DIE SMOOTHCUT
Columbia Tool Steel Co.
Tool material. C 1.05, Mn 0.8, W 0.5, Cr 1.6, bal Fe.
Oil hardened; cold work tool steel. For blanking and forming dies, punches, hobs. Type O3.

OIL ENGINE BABBITT
Hoyt Metal Co. of London Ltd.
Sn, Pb, Sb.
At 70 F: 34.4 Brin. At 212 F: 15.7 Brin. For bearings for oil engines; M.P. 462 F. *Obsolete*

OIL HARDENING
DoAll Co.
C 0.9, Mn 1.25, Si 0.3, Cr 0.5, V 0.2, W 0.5, bal Fe.
Oil hardening tool steel; AISI O1.

OIL HARDENING GAUGE PLATE
T. Inman & Co. Ltd.
C 0.85-0.9, Mn 1.6-1.7, Cr 0.5, V 0.25, bal Fe.
Supplied precision ground; for gauges.

OIL PUMP
English manufacture
Cu 85, Sn 3, Zn 9, P 3.
For oil pump parts.

OILCRAT
Marshall Steel Co.
C 0.95, Mn 1.25, V 0.15, Cr 0.5, W 0.5, bal Fe.
For taps, punches, spindles, crimpers; oil hardened, nondeforming.

OILGRAPH
AL Tech Specialty Steel Corp.
C 1.45, Mn 0.8, Si 1.15, Cr 0.2, Mo 0.25, bal Fe.
Usual heat treated hardness: 58/62 Rock C. Graphitic oil hardening tool steel of outstanding machinability. For blanking and forming dies, punches, gages, fixtures and machine parts. AISI O-6.

OILGRAPH EZ
Allegheny Specialty Steel
C 1.45, Si 1.15, Cr 0.2, Mo 0.25, S 0.1, 0.2-0.4% graphite, bal Fe.
For tools, dies; free-cutting, non-galling graphitic steel. *Obsolete*

OILITE 148 HARDENED
Chrysler Corp.
C 0.5, Ni 2, Fe 97.
Sintered, hardened, tempered at 500°F: 110,000 psi TS; 103,000 psi YS; 32 Rc. MPIF FN-0205-S; ASTM B-484 Gr. 1, Class B, Type II.

OILITE 160 HARDENED
Chrysler Corp.
C 0.5, Ni 2, Mo 0.5, Fe 97.
Sintered, hardened, tempered at 500°F; 110,000 psi TS; 103,000 psi YS; 32 Rc.

OILITE 182 HARDENED
Chrysler Corp.
C 0.5, Ni 0.5, Mo 0.5, Mn 0.4, Fe 98.
Sintered, hardened, tempered at 500°F: 110,000 psi TS; 103,000 psi YS; 32 Rc.

OILITE 304
Chrysler Corp.
Cr 19, Ni 10, Fe 71.
Sintered: 35,000 TS; 32,000 YS; 1.0 El; Density 61-64. Austenitic stainless powder metal. MPIF SS-304-P.

OILITE 316
Chrysler Corp.
Cr 17, Ni 12, Mo 2, Fe 69.
Sintered: 38,000 TS; 3,000 YS; 1.0 El; Density 6.1-6.4. Austenitic stainless powder metal. MPIF SS-316-P.

OILITE 410
Chrysler Corp.
Cr 12, Fe 88.
Sintered: 55,000 TS; 54,000 YS; 1.0 El; Density 6.1-6.4. Chromium type stainless steel. MPIF SS-410-P.

OILITE 993
Chrysler Corp.
Cu 1-3, C 0.25-0.6, Ni 3-5, bal Fe.
Sintered and heat treated: 120,000 TS; 6.8 min density; 40,000 fatigue.

OILITE ALUMINUM
Chrysler Corp.
Cu 4-5, Pb 3-4, Sn 3-4, Al 85-87, ot 0-3.
Sintered: 15,000 TS; 2.2-2.4 density. Oilite bearing material.

OILITE BRASS
Chrysler Corp.
Cu 77-80, Fe 0-0.25, Pb 1-2, Sn 0-0.1, bal Zn.
Sinter: 20,000 TS; 9 El; 7.2-7.7 density, 25,000 shear; 8500 fatigue. ASTM B282-60 Class A; SAE 890; PMPMA BZ-0218-T.

OILITE COPPER
Chrysler Corp.
Cu 98-0, Fe 0-0.1.
Sintered: 20,000 TS; 8 El; 8.0 min density. MIL-B-20296.

OILITE IM-20
Chrysler Corp.
C 0.8, Cu 20, Fe 79.
Sintered: 75,000 YS; 1.0 El; 80 Rb; Density 7.2-7.6. Infiltered type powder metal. MPIF FX-2008-T; SAE 872; ASTM B-303 Class C.

OILITE IN-10
Chrysler Corp.
C 0.8, Cu 10, Fe 89.
Sintered: 90,000 TS; 75,000 YS; 2.5 El; 80 Rb; Density 7.2-7.6. Infiltered type powder metal. MPIF FX-1008-T.

OILITE LEAD BRONZE
Chrysler Corp.
Cu 76-78.5, Fe 0-1, Pb 14-16, C 0-1, Sn 6.5-8.5.
Sintered: 8000 TS; 6.8-7.2 density. Oilite bearing material.

OILITE NICKEL SILVER
Chrysler Corp.
Cu 63-66.5, Ni 16.5-19.5, Zn 16.5-19.5.
Sintered: 30,000-37,500 TS; 10-15 El; 7.5-8.80 density. ASTM B458-67 Gr 1, Type 1; PMPMA BZN-1818-V.

OILITE R-112 FRICTION MATERIAL
Chrysler Corp.
Sintered: 2500 TS; Density 5.8-6.2; Hardness 55 RL; 1.0 El. Coefficient of friction 0.30-0.35.

OILITE STAINLESS STEEL
Chrysler Corp.
C 0.08-0.2, Cr 17-19, Ni 8-9, Fe 70.5-73.
Sintered: 35,000 TS; 6.0-6.4 density. 20,000 shear 26,000 fatigue. PMPMA SS-304-L-P.

OILITE STESSITE BRONZE
Chrysler Corp.
Cu 93-96, Sn 4-6, 2.5 others.
Sintered 20,000 TS; 10 El; 7.2-7.8 density. 10,000 fatigue stength.

OILTEMP
Bethlehem Steel Corp.
C 0.9, Mn 1.1, Cr 0.5, W 0.5, V 0.2, bal Fe.
For tools, dies, cold work dies; oil hardening. *Obsolete*

OILWAY
H. Boker & Co.
C 0.94, Cr 0.44, W 0.5, Mn 1.2, bal Fe.
For tools, dies; oil hardening. *Obsolete*

OJIBWAY
Atlas Specialty Steels
C 0.95, Mn 0.3, Si 0.3, bal Fe.
For open headed dies; water hardened. *Obsolete*

OKADUR 10
KM-kabelmetal AG
Cu 2.5-5, Mg 0.2-1.8, Mn 0.3-1.5, bal Al.
For light alloy structural parts; age hardenable. *Obsolete*

OKADUR 58
KM-kabelmetal AG
Cu 2.5-5, Mg 0.2-1.8, Mn 0.3-1.5, Pb 0.5-2.5, Sn, Cd, Bi.
For screw machine products; free-cutting. *Obsolete*

OKADUR 6
KM-kabelmetal AG
Cu 2.5-5, Mg 0.2-1.8, Mn 0.3-1.5, bal Al.
For light alloy structural parts; age hardenable. *Obsolete*

OKADURPLAT
KM-kabelmetal AG
Cu 2.5-5, Mg 0.2-1.8, Mn 0.3-1.5, bal Al.
For aircraft structures; clad, age hardenable. *Obsolete*

OKER BRASS
English manufacture
Zn 44.5, Sn 0.5, bal Cu.
For architectural structures and trim; high strength, corrosion resistant.

OKER BRASS
German manufacture
Cu 64.25, Zn 35.25, Sn 0.39, Bp 0.12.
For hardware bolts, fittings; high strength.

OKER I BRASS
English manufacture
Zn 30, Pb 1, bal Cu.
For fittings, hardware; good workability, free-cutting.

OKER-1
English manufacture
Cu 69, Zn 30, Pb 0.97.
Rolled: 40,000 TS; 35 El; 37 Brin. For tubes, sheets.

OKER-2
English manufacture
Cu 55, Zn 45, Sn 0.5.
Rolled 50,000 TS. For sheets, architectural purposes; corrosion resistant.

OKER-CAST
English manufacture
Cu 72, Zn 24, Pb 1.1, Fe 2.3.
Cast: 60,000 TS; 33 El; 30.6 RA; 86 Brin. For turbine parts; free-cutting.

OLD GENUINE BABBITT
Lumen Bearing Co.
Sn 89, Sb 7.5, Cu 3.5.
For bearings; heavy duty.

OLD HYCC
Colt Industries
C 1.5, Cr 12, bal Fe.
For dies; oil hardening, non-deforming. *Obsolete*

OLDS BEARING BRONZE
Olds Alloys Co.
Cu 60-65, Pb 17-19, Zn 9-11, Ni 4.5-6, Sn 4.5-6.5.
Cast: 22,000 TS; 16,000 YS; 3 El; 56 Brin. For formed parts, bearings, bushings; corrosion resistant.

OLDS HI TENSIL BRONZE NO. 11515
Olds Alloys Co.
Cu alloy.
Rolled: 118,000 TS; 80,000 YS; 15 El; 215 Brin. For hardware gears; tough.

OLDS HI TENSIL BRONZE NO. 7025
Olds Alloys Co.
Cu alloy.
Rolled: 75,000 TS; 32,000 YS; 31.5 El; 140 Brin. For hardware, gears; tough.

OLDS HI TENSIL BRONZE NO. 9025
Olds Alloys Co.
Cu alloy.
Rolled: 92,000 TS; 42,000 YS; 28 El; 180 Brin. For hardware, gears; tough.

OLDS LEADED BRONZE, OA-10
Olds Alloys Co.
Cu 70, Sn 10, Pb 20.
As cast: 54-58 Brin (500 kg). For bearings. QQ-B691 Comp 7; SAE 794.

OLDSMOLOY
Olds Alloys Co.
Ni 15, Zn 35, Cu 45, Cr, Mn, Sn.
Rolled: 70,000-76,000 TS; 15-20 El; 21 RA; 150-160 Brin. For bearings, gears, hardware, food machinery; resists salt water and food acids.

OLIN 15100
Olin Brass
Zr 0.05-0.15, bal Cu.
Annealed: 37,000-42,000 TS; 9000-17,000 YS; 35-41 El. Cold rolled: 42,000-71,000 TS; 28,000-69,000 YS; 1-33 El. For lead frames, electronic connectors. C15100

OLIN ALCOLOY 688
Olin Brass
Cu 73.5, Zn 22.7, Al 3.4, Co 0.4.
Annealed: 77,000-87,000 TS; 47,000-62,000 YS; 30-40 El. Cold rolled: 87,000-140,000 TS; 63,000-120,000 YS; 2-29 El. Good corrosion resistance, high strength. For relay springs, electric terminals, shells, connectors. C68800

OLIN ALLOY 263-HIGH BRASS
Olin Brass
Cu 68.5, Zn 31.5.
Annealed: 45-61 ksi TS; 10-33 ksi YS; 52-62 El. Rolled: 49-102 ksi TS; 21-93 ksi YS. For belt buckles, cabinet handles, electrical connectors, strike plates, table lamps, weatherstrip. *Obsolete*

OLIN ALLOY 425-LUBALOY X
Now LUBALOY X 425.

OLIN ALLOY 638, CORONZE
Now CORONZE.

OLIN ALLOY 664
Now COBRON 664.

OLIN ALLOY 688 ALCALOY
Now ALCALOY 688.

OLIN ALLOY NO. 411
Now LUBALOY 411.

OLIN ALLOY NO. 422
Now LUBRONZE 422.

OLIN ARSENIC INHIBITED CARTRIDGE BRASS
Olin Brass
Zn 30, As 0.05, bal Cu.
For bullet jackets, fasteners, sockets, radiators; easily formed, good corrosion resistance. C26130

OLIN BERYLLIUM COPPER
Olin Brass
Be 1.6-1.79, 0.60 Co + Fe + Ni, bal Cu.
Annealed: 60,000-78,000 TS; 30,000-55,000 YS; 35 El. Cold rolled: 75,000-210,000 TS; 60,000-180,000 YS; 1-3 El. For contact springs, switches, relays, connectors. Excellent elevated temperature strength. C17000

OLIN BERYLLIUM COPPER
Olin Brass
Be 1.8-2, 0.60 Co + Fe + Ni, bal Cu.
Annealed: 100,000-110,000 TS; 70,000-95,000 YS; 18 El. Cold rolled: 100,000-220,000 TS; 80,000-205,000 YS; 1-33 El. For contact springs, switches, relays, connectors. Excellent elevated temperature strength. C17200

OLIN BRAZING BRASS 282
Olin Brass
Cu 59, Zn 38.8, P 0.2.
Brazing sheet, annealed. Low melting temperature. C28200

OLIN C10100
Olin Brass
Annealed: 26,000-36,000 TS; 6000-12,000 YS; 20-44 El. Cold rolled: 34,000-60,000 TS; 30,000-50,000 YS; 1-33 El. For electrical assemblies and severely formed parts; resists hydrogen embrittlement. C10100

OLIN C11300
Olin Brass
0.027 Ag min.
Annealed: 26,000-32,000 TS; 6000-12,000 YS; 20-42 El. Cold rolled: 34,000-57,000 TS; 30,000-57,000 YS; 1-33 El. For gaskets, electrical connectors, terminals, contacts. C11300

OLIN C11400
Olin Brass
0.034 Ag min.
Annealed: 26,000-32,000 TS; 6000-12,000 YS; 20-42 El. Cold rolled: 34,000-57,000 TS; 30,000-57,000 YS; 1-33 El. For gaskets, electrical connectors, terminals, contacts, good high temperature softening resistance. C11400

OLIN C11600
Olin Brass
0.085 Ag min.
Annealed: 26,000-32,000 TS; 6000-12,000 YS; 20-42 El. Cold rolled: 34,000-57,000 TS; 30,000-57,000 YS; 1-33 El. For gaskets, electrical connectors, terminals, contacts, good high temperature softening resistance. C11600

OLIN C12000
Olin Brass
P 0.004-0.012, bal Cu.
Annealed: 34,000-35,000 TS; 6000-8000 YS; 47-48 El. Cold rolled: 34,000-57,000 TS; 30,000-57,000 YS; 1-33 El. For electrical applications, tubular bus bars; good welding and brazing. C12000

OLIN C19700
Olin Brass
Fe 0.75, P 0.25, Mg 0.11, Zn 0-0.2, Sn 0-0.2, bal Cu.
Annealed: 43,000-53,000 TS; 16,000-30,000 YS; 20 El. Cold rolled: 53,000-80,000 TS; 36,000-78,000 YS; 1-29 El. For lead frames, electronic connectors and automotive connectors. C19700

OLIN C41300
Olin Brass
Zn 9, Sn 1, Pb 0-0.1, bal Cu.
Annealed: 40,000-48,000 TS; Cold rolled: 43,000-79,000 TS; 2-41 El. For electrical springs, jewelry. C41300

OLIN C65400
Olin Brass
Sn 1.5, Si 3, Cr 0.06, Zn 0-0.5, bal Cu.
Cold rolled: 75,000-137,000 TS; 45,000-127,000 YS; 1-45 El. For contact springs, connectors; stress relaxation resistance. C65400

OLIN C70250
Olin Brass
Ni 3, Si 0.65, Mg 0.15, bal Cu.
Annealed: 88,000 TS min; 80,000 YS min; 6 El. Cold rolled: 106,000 TS min; 100,000 YS min; 20 El. For lead frames, connectors, springs, contacts; stress relaxation resistance. C70250

OLIN CARTRIDGE BRASS
Olin Brass
Zn 30, bal Cu.
Annealed: 45,000-61,000 TS; 10,000-33,000 YS; 40-67 El. Cold rolled: 49,000-104,000 TS; 21,000-93,000 YS; 1-59 El. For bullet jackets, fasteners, sockets, radiators; easily formed, good corrosion resistance. C26000

OLIN COBRAZE 6991
Now COBRAZE 6991.

OLIN COBRAZE 6991
Olin Brass
Mn 30, Zn 4, Fe 1.2, Al 0.5, bal Cu.
Annealed: 85,000 TS max. Hard: 100,000 TS min. Brazing filler metal; brazing temperature 885-900°C. *Obsolete*

OLIN COBRON 664
Now COBRON 664.

OLIN COBRON 664
Olin Brass
Zn 11, Fe 1.5, Co 0.5, bal Cu.
Annealed: 63,000 TS; 45,000 YS; 25 El. Spring temper: 100,000 TS; 94,000 YS; 3 El. For fuse clips, electrical terminals, connectors, springs. Good formability and strength. 30% electrical conductivity. C66400

OLIN COMMERCIAL BRONZE
Olin Brass
Zn 10, bal Cu.
Annealed: 36,000-42,000 TS; 8000-17,000 YS; 46-49 El. Cold rolled: 40,000-80,000 TS; 19,000-74,000 YS; 1-33 El. For coins, terminals, marine hardware; easily formed, good corrosion resistance. C22000

OLIN COPPER 102 OF
Olin Brass
99.95 Cu min.
Annealed: 26,000-36,000 TS; 6000-12,000 YS; 20-44 El. Cold rolled: 34,000-60,000 TS; 30,000-58,000 YS; 1-33 El. For electrical assemblies and severely formed parts; resists hydrogen embrittlement. C10200

OLIN COPPER 104 OFS
Olin Brass
99.95 Cu + Ag min, 0.027 Ag min.
Annealed: 26,000-36,000 TS; 6000-12,000 YS; 20-44 El. Cold rolled: 34,000-60,000 TS; 30,000-58,000 YS; 1-33 El. For electrical assemblies and severely formed parts; resistant to hydrogen embrittlement and elevated temperature softening. C10400

OLIN COPPER 110 ETP
Olin Brass
99.9 Cu min.
Annealed: 26,000-36,000 TS; 6,000-12,000 YS; 20-44 El. Cold rolled: 34,000-60,000 TS; 30,000-58,000 YS; 1-33 El. Drawn or formed electrical shells and parts; General purpose copper for electrical parts. C11000

OLIN COPPER 1102, LOW OXYGEN ETP
Olin Brass
O 0-0.02, 99.9 Cu min.
Annealed: 26,000-36,000 TS; 6,000-12,000 YS; 20-44 El. Cold rolled: 34,000-60,000 TS; 30,000-58,000 YS; 1-33 El. Formed and deep drawn parts; good ductility (soft); good tear resistance. *Obsolete*

OLIN COPPER 1142 LOW OXYGEN STP
Olin Brass
O 0-0.02, 99.9 Cu min, 0.044 Ag min.
Annealed: 26,000-36,000 TS; 6000-12,000 YS; 20-44 El. Cold rolled: 34,000-60,000 TS; 30,000-58,000 YS; 1-33 El. For commutator and collector rings, severely formed parts; resists softening at elevated temperatures. *Obsolete*

OLIN COPPER 122 DHP
Olin Brass
99.9 Cu min, 0.015 P min.
Annealed: 26,000-36,000 TS; 6000-12,000 YS; 20-44 El. For deep drawn copper shells, brazed or welded assemblies. Cold rolled: 34,000-60,000 TS; 30,000-58,000 YS; 1-33 El. For formed copper parts. C12200

OLIN COPPER 129 FRSTP
Olin Brass
99.9 Cu min, 0.054 Ag min.
Annealed: 26,000-36,000 TS; 6000-12,000 YS; 20-44 El. Cold rolled: 34,000-60,000 TS; 30,000-58,000 YS; 1-33 El. Good solderability and high softening temperature. For radiator air channels, radiator fins. *Obsolete*

OLIN COPPER NO. 1092
Olin Brass
O 0-0.02, 99.9 Cu (incl Ag) min.
Annealed (soft): 30,000 TS; 10,000 YS; 30 El. Rolled: 34-60 ksi TS; 26-58 ksi YS; 1-33 El. For electrical parts requiring good formability. Low oxygen electrolytic copper. C10920

OLIN COPPER NO. 1093
Olin Brass
O 0-0.02, 99.90 Cu (incl Ag) min, 0.044 Ag min.
Annealed (soft): 30,000 TS; 10,000 YS; 30 El. Rolled: 34-60 ksi TS; 26-58 ksi YS; 1-33 El. For electronics, lead frames, commutator rings. Low oxygen electrolytic copper with silver. C10930

OLIN COPPER NO. 1094
Olin Brass
O 0-0.02, 99.90 Cu (incl Ag) min, 0.085 Ag min.
Annealed (soft): 30 ksi TS; 10 ksi YS; 30 El. Rolled: 34-60 ksi TS; 26-58 ksi YS; 1-33 El. For electronics, lead frames. Low oxygen electrolytic copper with silver. C10940

OLIN COPPER, LOW OXYGEN ETP 1102
Olin Brass
O 0-0.02, 99.9 Cu min.
Annealed: 26,000-36,000 TS; 6,000-12,000 YS; 20-44 El. Cold rolled: 34,000-60,000 TS; 30,000-58,000 YS; 1-33 El. For deep drawn parts. *Obsolete*

OLIN COPPER, LOW OXYGEN STP 1142
Olin Brass
O 0-0.02, 99.9 Cu min, 0.044 Ag min (13 oz/ton).
Annealed: 26,000-36,000 TS; 6000-12,000 YS; 20-44 El. Cold rolled: 34,000-60,000 TS; 30,000-58,000 YS; 1-33 El. For commutator and collector rings. Resists softening at elevated temperatures. *Obsolete*

OLIN COPPER, LOW OXYGEN STP 1162
Olin Brass
O 0.02, 99.9 Cu min, 0.085 Ag min (25 oz/ton).
Annealed: 26,000-30,000 TS; 6000-12,000 YS; 20-44 El. Cold rolled: 34,000-60,000 TS; 30,000-58,000 YS; 1-3 El. For lead frames, heat sinks, electronic components. *Obsolete*

OLIN COPPER, LOW OXYGEN STP 1162
Olin Brass
O 0-0.02, 99.9 Cu min, 0.085 Ag min (25 oz/ton).
Annealed: 26,000-36,000 TS; 6000-12,000 YS; 20-44 El. Cold rolled: 34,000-60,000 TS; 30,000-58,000 YS; 1-33 El. For commutator bars, electronic components. Resists softening at elevated temperatures. *Obsolete*

OLIN COPPER-NICKEL-TIN ALLOY 725
Olin Brass
Cu 88.2, Ni 9.5, Sn 2.3.
Annealed: 55,000 TS; 22,000 YS; 35 El. Spring temper: 91,000 TS; 90,000 YS; 1 El. For springs in connectors, relays and switches. C72500

OLIN CORONZE 638
Olin Brass
Cu 95, Al 2.8, Si 1.8, Co 0.4.
Annealed: 77,000-87,000 TS; 41,000-67,000 YS; 29-42 El. Cold rolled: 90,000-130,000 TS; 70,000-114,000 YS; 2-30 El. Soft for cans, bases, electronic parts; cold rolled for springs, terminals circuit frames. Stress corrosion resistant, oxidation resistant, good strength. C63800

OLIN CUPRO-NICKEL 10% 706
Olin Brass
Cu 88, Ni 10, Fe 1.4.
Annealed: 40,000-50,000 TS; 10,000-25,000 YS; 33-40 El. Cold rolled: 54,000-88,000 TS; 48,000-84,000 YS; 1-30 El. For heat exchanger tubes and plates. Good salt water corrosion resistance. C70600

OLIN CUPRO-NICKEL 25% 713
Olin Brass
Cu 75, Ni 25.
Annealed: 48,000-58,000 TS; 14,000-24,000 YS; 35-45 El. Cold rolled: 54,000-94,000 TS; 16,000-89,000 YS; 2-38 El. For heat exchanger tubes and plates. Good salt water corrosion resistance. C71300

OLIN CUPRO-NICKEL 30% 715
Olin Brass
Cu 68.5, Ni 31, Fe 0.5.
Annealed: 54,000-60,000 TS; 20,000-22,000 YS; 40-45 El. Cold rolled: 58,000-94,000 TS. For heat exchanger tubes and sheet. Good corrosion resistance to high velocity salt water. C71500

OLIN CUPRO-NICKEL 5% 704
Olin Brass
Cu 93, Ni 5.5, Fe 1.5.
Available annealed or cold rolled. For welded heat exchanger tubes, corrosion resistant. C70400

OLIN CUPRO-NICKEL 7% 705
Olin Brass
Cu 93, Ni 7.
Annealed: 36,000-42,000 TS; 35-42 El. Cold rolled: 39,000-70,000 TS; 1-32 El. For resistance strips for fuses. Constant resistivity and melting temperature. C70500

OLIN GILDING METAL 210
Olin Brass
Cu 95, Zn 5.
Annealed: 34,000-38,000 TS; 5000-11,000 YS; 45-48 El. Cold rolled: 37,000-69,000 TS; 14,000-59,000 YS; 2-45 El. For coins, medallions, bullet jackets, terminals; red color, easily formed, good corrosion resistance. C21000

OLIN HIGH BRASS 262
Olin Brass
Cu 68.5, Zn 31.5.
Annealed: 44,000-61,000 TS; 13,000-34,000 YS; 35-65 El. Cold rolled: 49,000-110,000 TS; 20,000-100,000 YS; 1-52 El. A general purpose yellow brass. For formed parts, hardware, electrical assemblies. C26200

OLIN HIGH CONDUCTIVITY BRONZE 405
Olin Brass
Cu 95, Zn 4, Sn 1.
Annealed: 38,000-45,000 TS; 9000-17,000 YS; 40-50 El. Cold rolled: 41,000-82,000 TS; 15,000-74,000 YS; 20-45 El. Moderate strength high-conductivity bronze for fuse clips, meter clips, electrical assemblies. C40500

OLIN HIGH LEADED BRASS 353
Olin Brass
Cu 61, Zn 37, Pb 2.
Annealed: 45,000-54,000 TS; 12,000-24,000 YS; 45-60 El. Cold rolled: 49,000-99,000 TS; 20,000-80,000 YS; 1-50 El. High strength yellow brass, good machinability. For clock and watch plates, and gears, brass keys. C35300

OLIN HSM COPPER 194
Olin Brass
Cu 97.5, Fe 2.35, Zn 0.12, P 0.03.
Annealed: 40,000-60,000 TS; 24,000-45,000 YS; 20-35 El. Cold rolled: 53,000-85,000 TS; 48,000-78,000 YS; 1-13 El. For electrical connectors and terminals, springs, tubular products; high-conductivity strength, corrosion resistance. C19400

OLIN INCRAMUTE 699
Now INCRAMUTE 699.

OLIN INCRAMUTE 699
Olin Brass
Mn 44, Al 1.9, bal Cu.
Heat treated: 81,000 TS; 44,000 YS; 27 El. Vibration damping; noise damping; good strength. *Obsolete*

OLIN JEWELRY BRONZE 226
Olin Brass
Cu 87, Zn 13.
Annealed: 37,000-45,000 TS; 10,000-21,000 YS; 35-48 El. Cold rolled: 42,000-86,000 TS; 16,000-75,000 YS; 2-40 El. For jewelry, buckles, slide fasteners. Gold color, easily buffed and finished. C22600

OLIN KO-1
Electronic Specialties Co.(H.& S.Metals)
Cu 4.8, Mg 0.23, Ti 0.27, Ag 0.64, bal Al.
Sand cast: 68,000 TS; 60,000 YS; 6 El. Permanent mold: 67,000 TS; 55,000 YS; 10 El. For landing gear struts, gear box housings, aircraft and auto truck components, aerospace equipment. High strength casting alloy. *Obsolete*

OLIN KO-1
Olin Corp.
Cu 4.8, Mg 0.23, Ti 0.27, Ag 0.64, bal Al.
Sand cast: 68,000 TS; 60,000 YS; 6 El. Permanent mold: 67,000 TS; 55,000 YS; 10 El. For landing gear struts, gear box housings, aircraft and auto truck components, aerospace equipment. High strength casting alloy. *Obsolete*

OLIN LEADED BEARING BRONZE 544
Olin Brass
Cu 89, Sn 4, Pb 4, Zn 3.
Annealed: 42,000-52,000 TS; 14,000-27,000 YS; 40-50 El. Cold rolled: 47,000-103,000 TS; 20,000-92,000 YS; 2-40 El. For bearings, bushing, thrust washers. Good bearing properties. C54400

OLIN LOW BRASS 240
Olin Brass
Cu 80, Zn 20.
Annealed: 44,000-54,000 TS; 12,000-29,000 YS; 42-58 El. Cold rolled: 48,000-97,000 TS; 18,000-84,000 YS; 1-35 El. Low zinc brass with good corrosion resistance; for bellows, flexible hose, water meters. C24000

OLIN LUBALLOY X ALLOY 80
Olin Brass
Sn 1.75, Zn 8.25, bal Cu.
Hard: 70,000 TS; 4 El. Soft: 45,000 TS; 50 El. For contact springs, radio tube sockets, fuse clips; corrosion resistant. *Obsolete*

OLIN LUBALOY 411
Olin Brass
Cu 90, Zn 9.5, Sn 0.5.
Annealed: 38,000-44,000 TS; 9000-18,000 YS; 40-44 El. Cold rolled: 42,000-87,000 TS; 17,000-78,000 YS; 1-40 El. For bushings, washers, clutch plates, fuse clips; wear resistant and good corrosion resistance. C41100

OLIN LUBALOY X425
Olin Brass
Cu 88, Zn 10, Sn 2.
Annealed: 40,000-50,000 TS; 12,000-24,000 YS; 44-55 El. Cold rolled: 45,000-106,000 TS; 20,000-88,000 YS; 1-40 El. For electrical springs, connectors, terminals. High strength, fatigue-corrosion resistant. C42500

OLIN LUBRONZE 422
Olin Brass
Fu 87, Zn 12, Sn 1.
Annealed: 40,000-50,000 TS; 10,000-20,000 YS; 44-50 El. Cold rolled: 47,000-94,000 TS; 20,000-83,000 YS; 1-40 El. For electrical terminals, connectors, clips, chains; strength, stress-corrosion resistance. C42200

OLIN MEDIUM LEADED BRASS 350
Olin Brass
Cu 61, Zn 38, Pb 1.
Annealed: 45,000-55,000 TS; 12,000-26,000 YS; 45-60 El. Cold rolled: 49,000-99,000 TS; 20,000-79,000 YS; 1-44 El. Good machinability and formability. For hose coupling nuts, sink strainers, bearing cages. C35000

OLIN NICKEL SILVER 10% 740
Olin Brass
Cu 71, Ni 10, Zn 19.
Annealed: 50,000-60,000 TS; 15,000-33,000 YS; 33-43 El. Cold rolled: 56,000-95,000 TS; 30,000-86,000 YS; 1-34 El. A high strength copper nickel silver, good corrosion resistance; for musical instruments, electronic parts. C74000

OLIN NICKEL SILVER 12% 738
Olin Brass
Cu 70, Ni 12, Zn 18.
Annealed: 48,000-58,000 TS; 12,000-30,000 YS; 35-45 El. Silver color; for coinage. *Obsolete*

OLIN NICKEL SILVER 12% 762
Olin Brass
Cu 59, Ni 12, Zn 29.
Annealed: 57,000-75,000 TS; 21,000-51,000 YS; 35-50 El. Cold rolled: 65,000-125,000 TS; 36,000-115,000 YS; 1-40 El. High strength corrosion fatigue resistant, for relay springs, contacts, connectors, clips. C76200

OLIN NICKEL SILVER 18% 770
Olin Brass
Cu 55, Ni 18, Zn 27.
Annealed: 61,000-75,000 TS; 22,000-40,000 YS; 35-48 El. Cold rolled: 72,000-126,000 TS; 40,000-116,000 YS; 2-35 El. High strength spring nickel silver. For relay springs, contact springs, diaphragms. C77000

OLIN NICKEL SILVER 65-18
Olin Brass
Ni 18, Zn 17.
Annealed: 53,000-63,000 TS; 18,000-32,000 YS; 29-42 El. Cold rolled: 58,000-96,000 TS; 26,000-64,000 YS; 1-35 El. For rivets, flatware, core parts, terminals, holloware; good corrosion resistance, silver color. C75200

OLIN NICKEL SILVER 8% 65-8 743
Olin Brass
Cu 65, Ni 8, Zn 27.
Annealed: 53,000 TS; 23,000 YS; 40 El. Hard: 85,000 TS; 73,000 YS; 4 El. For hollow ware, flatware, optical goods. C74300

OLIN NICKEL SILVER, 18%
Olin Brass
Ni 18, Zn 10, bal Cu.
Annealed: 50,000-68,000 TS; 15,000-60,000 YS; 11-36 El. Cold rolled: 56,000-88,000 TS; 28,000 YS; 10-25 El. For pen and pencil bodies, other drawn parts. C73500

OLIN NO. 605
Olin Corp.
Fe 2.3, P 0.03, bal Cu.
Annealed: 48,000 TS; 20,000 YS; 28 El; 95 DPH. 1/2 hard: 61,000 TS; 56,000 YS; 6.5 El; 131 DPH. Spring: 75,000 TS; 69,000 YS; 1.5 El; 149 DPH. For springs, diaphragms, hollow ware, gaskets, clips, terminals, contacts. 65% electrical conductivity, corrosion resistant. *Obsolete*

OLIN PHOSPHOR BRONZE 1.25% E, 505
Olin Brass
Cu 98.7, Sn 1.25, P 0.04.
Annealed: 37,000-42,000 TS; 8000-17,000 YS; 40-47 El. Cold rolled: 41,000-84,000 TS; 16,000-77,000 YS; 1-45 El. For flexible metal hose, pole line hardware. Corrosion resistant and corrosion fatigue resistant. C50500

OLIN PHOSPHOR BRONZE 4%, 511
Olin Brass
Sn 4.2, Pb 0-0.1, Fe 0-0.3, bal Cu.
40,000-52,000 TS; 12,000-24,000 YS; 45-58 El. Cold rolled: 45,000-109,000 TS; 18,000-96,000 YS; 1-44 El. High conductivity spring phosphor bronze. For electric terminals, clips, connectors. C51100

OLIN PHOSPHOR BRONZE, 5%
Olin Brass
Sn 5, P 0.2, bal Cu.
Annealed: 46,000-56,000 TS; 19,000-29,000 YS; 48-62 El. Cold rolled: 49,000-114,000 TS; 22,000-105,000 YS; 1-50 El. For bellows, fasteners, springs, brushes, welding rod. C51000

OLIN PHOSPHOR BRONZE, 8%
Olin Brass
Sn 8, P 0.1, bal Cu.
Annealed: 56,000-65,000 TS; 23,000-35,000 YS; 60-69 El. Cold rolled: 63,000-122,000 TS; 35,000-116,000 YS; 2-60 El. For bellows, fasteners, springs, brushes, welding rod. C52100

OLIN RED BRASS
Olin Brass
Zn 15, bal Cu.
Annealed: 39,000-47,000 TS; 8000-19,000 YS; 43-48 El. Cold rolled: 44,000-90,000 TS; 22,000-80,000 YS; 2-39 El. For coins, ornamental, terminals, bellows; easily formed, good corrosion resistance. C23000

OLIN STRESCON TM 195
Olin Brass
Cu 97, Fe 1.5, Co 0.8, Sn 0.6, P 0.1.
Annealed: 50,000-65,000 TS; 22,000-45,000 YS; 22-35 El. Cold rolled: 60,000-100,000 TS; 35,000-87,000 YS; 3-24 El. For electrical terminals, connectors, contact springs; high strength, high conductivity, corrosion resistant. C19500

OLIN YELLOW BRASS 268
Olin Brass
Cu 66, Zn 34.
Annealed: 44,000-61,000 TS; 13,000-34,000 YS; 35-65 El. Cold rolled: 49,000-99,000 TS; 24,000-76,000 YS; 3-50 El. General purpose yellow brass. For formed parts, electrical hardware, hinges. C26800

OLIN-0629
Olin Corp.
Al 9-10, Fe 3.5-4.5, Zn 0-0.8, bal Cu.
Annealed: 85,000-100,000 TS; 45,000-65,000 YS; 30-35 El; B 70 Rock. Hard: 112,000-130,000 TS; 90,000-110,000 YS; 10 El; B 96-101 Rock. For fuses, clips, springs, fasteners. Aluminum bronze, corrosion resistant. *Obsolete*

OLIN-0629M
Olin Corp.
Al 9.4-10, Fe 3.5-4.5, Zn 0-0.8, bal Cu.
Annealed: 85,000-100,000 TS; 45,000-65,000 YS; 30-35 El; B 80-90 Rock. Heat treated: 150,000-170,000 TS; 100,000-120,000 YS; 3 El. For bushings, worm gears, bearings, shafts. Aluminum bronze, heat treatable. *Obsolete*

OLITE BRONZE (WAS OILITE)
Chrysler Corp.
Cu 87.5-90.5, Fe 0-1, C 0-1.75, Sn 9.5-10.5.
Sintered: 14,000 TS; 11,000 YS; 6.4-6.8 density. Oilite bearing material. ASTM B438-67 Gr 1 Type 11; SAE 841. PMPMA BT-0010-R.

OLYMPIA
J.F. Jelenko & Co.
Precious metal. Au 51.5, Pd 38.5.
115,000 psi TS; 83,000 psi YS; 20 El; 200 Brin. Dental alloy for fusing porcelain to metal.

OLYMPIA II
J.F. Jelenko & Co.
Precious metal. Au 35, Pd 57.
118,000 psi TS; 81,000 psi YS; 20 El; 220 Brin. Dental alloy for fusing porcelain to metal.

OLYMPIC
William Oxley & Co.
C 0.7, W 18, Cr 4, Cb 1, bal Fe.
For high speed tools and cutters. High speed steel.

OLYMPIC
Latrobe Steel Co.
C 1.5, Cr 12, V 1, Mo 0.7, bal Fe.
For dies, tools, lamination dies, punches, forming dies; air hardening, abrasion resistant, non-deforming.

OLYMPIC
Latrobe Steel Co.
C 1.5, Si 0.3, Mn 0.5, Cr 12, V 0.9, Mo 0.75.
Air hardening high carbon, high chromium die steel. For long run tool and die applications. AISI Type D2.

OLYMPIC BRONZE TYPE A
Chase Brass & Copper Co.
Cu 96, Zn 1, Si 3.
Annealed: 60,000 TS; 25,000 YS; 60 El; 60 RA; 50 Brin. Hard rolled: 94,000 TS; 58,000 YS; 8 El; 14 RA; 140 Brin. For sheet, rod, wire, tubes, bolts, nuts, hardware, cable, welding rod; corrosion resisting. *Obsolete*

OLYMPIC BRONZE TYPE C
Chase Brass & Copper Co.
Cu 94.75, Zn 1, Si 4.25.
Sand cast: 50,000 TS; 15,000 YS; 13 RA. Wrought: 115,000 TS. For high strength sand castings, bolts, nuts, screws, welding rods, hardware, cable; resists "season cracking". *Obsolete*

OLYMPIC BRONZE TYPE D
Chase Brass & Copper Co.
Cu 95.5, P 0.5, Si 3, Zn 1.
Rolled: 55,000-150,000 TS; 70-200 Brin. For screw machine parts; corrosion resistant. *Obsolete*

OLYMPIC BRONZE TYPE G
Chase Brass & Copper Co.
Cu 77, Si 1, Zn 22.
For hardware, springs, hinges, sheathing; corrosion resistant. *Obsolete*

OLYMPIC FM
Latrobe Steel Co.
C 1.5, Cr 12, V 1, Mo 0.7, bal Fe.
For lamination and blanking dies, punches; air hardening, free-cutting, non-deforming. *Obsolete*

OM METAL
American Smelting & Refining Co.
Cu 63, Pb 2, Ni 0.5, Mn 0.5, Al 0.3, Si 1, Sn 0.5, Sb 0.5, bal Zn.
53,000 psi TS; 22,300 psi YS; 25 El. Yellow brass; resists dezincification.

OMAN
Woodworkers Tool Works
Pb, Cu.
For bearings;: self-lubricating.

OMAN METAL
Oman Non-Friction Metal Co.
Pb 25, bal Cu.
For bearings; heavy duty.

OMANITBRONZE
Ostermann GmbH & Co.
Copper. Cu 81, Al 10, Fe, bal Ni.
10 El. For valve gauge fittings; corrosion resistant.

OMC
Mitsubishi Metal America Corp.
Copper. Cr 1, Zr 0.1, bal Cu.
High electrical conductivity and high mechanical strength at room and elevated temperatures. Suitable for the mold of steel continuous casting.

OMC 105 COMP I GRADE 2 (CAST)
Manufacturer not listed.
Ti unalloyed.
Cast: 50,000 min TS; 40,000 min YS; 20.0 min El. For aircraf equipment, marine and chemical processing equipment.

OMC 105 COMP. I GRAD 3 (CAST)
Manufacturer not listed.
Ti unalloyed. Cast: 65,000 min TS; 55,000 min YS; 15.0 min El. For heat exchangers, valve trim, anodizing racks, chemical plant equipment.

OMC 105 COMP. I GRADE 4 (CAST)
Manufacturer not listed.
Ti unalloyed. Cast 80,000 min TS; 70,000 min YS; 12.0 min El. For aircraft equipment, marine equipment, brackets, housings, anodizing racks.

OMC 105 COMP. II (CAST)
Manufacturer not listed.
Pd 0.15, bal Ti.
Cast: 65,000 min TS; 55,000 min YS; 15.0 min El. For chemica process equipment for oxidizing and mildly reducing environments.

OMC 105 COMP. IV (CAST)
Manufacturer not listed.
Al 5, Sn 2.5, bal Ti.
Cast: 115,000 min TS; 105,000 min YS; 10.0 min El. Compresso case housings, stiffeners, cryogenic tankage.

OMC 163 (CAST)
Manufacturer not listed.
Al 6, V 4, bal Ti.
Cast: 130,000 min TS; 120,000 min YS; 6 min El. For compressor wheels, blades, rocket cases, cryogenic equipment, marine equipment.

OMC 164
Now OMC GRADE 163 (CAST).

OMC 302 (CAST)
Manufacturer not listed.
Zr, unalloyed.
Cast: 55,000 min TS; 40,000 min YS; 12 min El. Chemical process equipment, dry chlorine environments.

OMC 3A1-2.5V
Manufacturer not listed.
Al 3, V 2.5, bal Ti.
Wrought, annealed: 125,000 min TS; 105,000 min YS; 10.0 min El. Weldable, suitable for fabricating into welded titanium tubing or seamless tubing. AMS 4943.

OMC 5AL-2.5 SN
Manufacturer not listed.
Al 5, Sn 2.5, bal Ti.
Wrought annealed: 115,000 min TS; 110,000 min YS; 10.0 min El; 25.0 min RA. For aircraft tailcones, compressor caps, housings, stiffeners, cryogenic tankage. Weldable, forgeable good creep resistance to 900°F. AMS 4910, 4926, 4953, 4966. MIL-T-9046, 9047, 81556.

OMC 6AL-2SN-4ZR-2MO
Manufacturer not listed.
Al 6, Sn 2, Zr 4, Mo 2, bal Ti.
Wrought, annealed: 130,000 min TS; 120,000 min YS; 10.0 min El; 25.0 min RA. Compressor blades and wheels, good strength and toughness to 800-900°F heat treatable. AMS 4975, 4976; MIL-T-9046, 9047.

OMC 6AL-2SN-4ZR-6MO
Manufacturer not listed.
Al 6, Sn 2, Zr 4, Mo 6, bal Ti.
Wrought, solution treated and aged: 170,000 min TS; 160,000 min YS; 12.0 min El; 30.0 min RA. Engine discs and compresso blades, deep hardenable. PWA 1216; AMS 4981.

OMC 6AL-4V
Manufacturer not listed.
Al 6, V 4, bal Ti.
Wrought, annealed: 130,000 min TS; 120,000 min YS; 10.0 min El, 25.0 min RA. For aircraft and engine forgings, compresso wheels and blades, spacers, cryogenic and marine equipment; good response to heat treatment in thinner sections. AMS 4928; MIL-T-9046, 9047, 81556.

OMC 6AL-6V-2SN
Manufacturer not listed.
Al 6, V 6, Sn 2, bal Ti.
Wrought, annealed: 140,000 min TS; 130,000 min YS; 8.0 min El; 20.0 min RA. For pressure vessels, rocket motor cases, airframes, ordance equipment; heat treatable. AMS 4971, 4978; MIL-T-9046, 9047, 81556.

OMC 85W-15MO (CAST)
Manufacturer not listed.
Mo 15, bal W.
For rocket throat inserts, rocket engine vanes exposed to solid fuel exhausts.

OMC 8AL-1MO, 1V
Manufacturer not listed.
Al 8, Mo 1, V 1, bal Ti.
Wrought, annealed: 130,000 min TS; 120,000 min YS; 10.0 min El, 20.0 min RA. High temperature jet engine forging alloy. For airframes, turbine parts, blades, vanes and discs. AMS 4915, 4972, 4973. MIL-T-9046, 9047, 81556.

OMC GRADE I
Manufacturer not listed.
Ti, unalloyed.
Wrought, annealed: 35,000 min TS; 25,000 min YS; 24.0 min El; 30.0 RA. For heat exchangers, aircraft ducting, good corrosion resistance. ASTM B265, B348.

OMC GRADE II
Manufacturer not listed.
Ti unalloyed.
Wrought, annealed: 50,000 min TS; 40,000 min YS; 20 min El; 30.0 min RA. For heat exchangers, aircraft ducting, good corrosion resistance. ASTM B265, B348.

OMC GRADE III
Manufacturer not listed.
Ti unalloyed.
Wrought, annealed: 65,000 min TS; 55,000 min YS; 18.0 min El 30.0 min RA. For heat exchangers, valve trim, welding rods, anodizing racks, and chemical plant equipment. AMS 4900.

OMC GRADE IV
Manufacturer not listed.
Ti unalloyed.
Wrought, annealed: 80,000 min TS; 70,000 min YS; 15.0 min El; 25.0 min RA. For aircraft equipment, ammunition boxes, anodizing racks, shroud spacers. AMS 4901, 4921.

OMC PD
Now OMC TI-PD.

OMC TI-17
Manufacturer not listed.
Al 5, Zr 2, Sn 2, Mo 4, Cr 2, bal Ti.
Wrought, solution treated and aged: 160,000 min TS; 150,000 min YS; 5 min El; 10.0 min RA. For engine discs, heat treatable. GE C50TF44, TF 57, TF 62.

OMC TI-PD
Manufacturer not listed.
Pd 0.15, bal Ti.
Wrought, annealed: 50,000 min TS; 40,000 min YS; 20.0 min El; 35.0 min RA. For chemical processing equipment, oxidizin and mildly reducing environments.

OMC-166 A (CAST)
Manufacturer not listed
Al 5, Sn 2.5, bal Ti.
Cast, min: 120,000 TS; 115,000 YS; 10 El. For aircraft tailcones, compressor case housings, stiffeners, cryogenic tankage. Weldable, forgeable; good creep resistance to 900 F. *Obsolete*

OMC-40
Manufacturer not listed
Ti, unalloyed.
Wrought, min: 40,000 TS; 30,000 YS; 25 El; 35 RA. For valves, heat exchangers, aircraft ducting, good corrosion resistance. MIL-T-12117 Class 40. *Obsolete*

OMC-55
Now OMC GRADE III.

OMC-7 AL-4 MO
Manufacturer not listed
Al 7, Mo 4, bal Ti.
Wrought, HT, min: 145,000 TS; 135,000 YS; 10 El. Heat treatable forging alloy; for aircraft turbines, compressor wheels, and blades, spacer rings, airframes. *Obsolete*

OMC-70
Now OMC GRADE IV.

OMEGA
Bethlehem Steel Corp.
C 0.6, Mn 0.7, Si 1.85, V 0.25, Mo 0.5, bal Fe.
Heat treated: 320,000-340,000 TS; 280,000-310,000 YS; 5-6 El; 11-21 RA. For pneumatic chisels, rivet sets, punches, blacksmith tools; 7-15 Izod; resists shock and fatigue. *Obsolete*

OMEGA BRAND BE CU ALLOY NO. 32
Riverside Metals Corp.
Be 1.5-2.25, bal Cu.
Annealed: 60,000 TS. Heat treated: 180,000 TS. For contact clips, springs, diaphragms, radio socket tubes, valve seats; age-hardening, corrosion resistant. *Obsolete*

OMEGA BRAND NICKEL SILVER NO. 6
Riverside Metals Corp.
Cu 70.5-73.5, Ni 16.5-19.5, Zn 8.5-11.5, Fe 0-0.35.
Hard sheet: 50,000-88,000 TS; 1-35 El. For deep drawing purposes; corrosion resistant. *Obsolete*

OMEGA BRAND NICKEL SILVER, ALLOY NO. 3
Riverside Metals Corp.
Cu 65, Ni 18, Zn 17.
Sheet: 55,000-90,000 TS; 1-40 El. For plated silverware, cutlery, hardware; corrosion resisting. *Obsolete*

OMEGA BRAND NICKEL SILVER, ALLOY NO. 7
Riverside Metals Corp.
Cu 55, Ni 18, Zn 27.
Wire: 60,000-160,000 TS; 1-50 El. For springs for electrical, radio and telephone equipment, resistance wire; corrosion resisting. *Obsolete*

OMEGA BRAND PHOSPHOR BRONZE NO. 30
Riverside Metals Corp.
Cu 95.5, Sn 4.3, P 0.2.
Wire: 45,000-100,000 TS; 0.5-60 El. For springs and contacts in radio and electrical equipment, screen plates for paper manufacturing; EIRES 11.6; M.E.-16,000,000. *Obsolete*

OMEGA BRAND PHOSPHOR BRONZE NO. 47
Riverside Metals Corp.
Cu 91.6 0, Sn 8, P 0.15.
Sheet: 51,000-130,000 TS; 1.0-75.0 El. For paper mill equipment; EIRES 14.8. *Obsolete*

OMEGA EXTRA
Remystahl
C 0.55, Cr 0.7, Mo 0.25-0.35, Ni 1.6, V 0.1, bal Fe.
For gears, machine tool parts, crankshafts; oil hardened, shock resistant. W.-Nr. 1.2713.

OMEGA EXTRA MO
Remystahl
C 0.55, Cr 1, Ni 1.65, Mo 0.8, V 0.1, bal Fe.
Hot work tool steel. W.-Nr. 1.2744.

OMEGA EXTRA V
Remystahl
C 0.56, Cr 1.1, Mo 0.45-0.55, Ni 1.65, V 0.1, bal Fe.
For gears, machine tool parts, crankshafts; oil hardened, shock resistant. W.-Nr. 1.2714.

OMMET "A" ALLOY
Molybdenum Co.,N.O.
Mo Ni-Fe.
280 Brin. For machinery parts; sintered alloy, heat resistant.

OMMET A
Plansee, Metallwerk Gesellschaft
Nickel. Ni 50, Mo 20, Fe 22.
For amplifiers and transmitter tubes. Corrosion and heat resistant.

OMMET IRON
Molybdenum Co.,N.O.
C 0.001, Si 0.01, P 0.01, Mn 0.01, Fe 99.98.
For vacuum tubes; sintered iron.

OMMET-B
Plansee, Metallwerk Gesellschaft
Ni-Mo-Fe.
For amplifiers and transmitter tubes; chemical plant equipment. Sintered; corrosion resistant.

OMMET-FE
Plansee, Metallwerk Gesellschaft
Fe.
For structural parts for amplifiers, transmitter tubes, rectifiers, and valves. Sintered; highest purity.

OMMET-SIVAR 48
Molybdenum Co.,N.O.
Fe Co-Ni.
For glass to metal seals, vacuum tubes; sintered alloy.

OMMET-SIVAR 60
Molybdenum Co.,N.O.
Fe Co-Ni.
For glass to metal seals, vacuum tubes; sintered alloy.

OMMET-SIVAR 90
Molybdenum Co.,N.O.
Fe Co-Ni.
For glass to metal seals; sintered alloy.

ON
Manufacturer not listed.
C 0.7, W 18, Cr 4, V 1, bal Fe.
For tools, high speed cutters; high speed steel.

ON-PLUS
Peninsular Steel Co.
C 0.7, Mo 9.5, W 1.5, Cr 4, V 1, bal Fe.
For tools, high speed cutters. *Obsolete*

ONE FIVE ONE
George Cook & Co., Ltd.
C 1, Mn 0.5, Cr 5, Mo 1, V 0.3, bal Fe.
Air hardening cold work tool steel. For punching dies, slitting cutters, knurling tools. BS 4659 BA2; AISI A2.

ONE TON BRASS
English manufacture
Cu 61, Zn 38, Sn 1.
For marine parts, condenser tubes; corrosion resistant.

ONEIDA
Associated Steel Corp.
C 0.9, Mn 1.2, Cr 0.5, V 0.2, W 0.5, bal Fe.
Oil-hardened tool steel; AISI O1.

ONERAL
English manufacture
C 0.9, Mo 10, Ti 0.03, Ni 6, Zr 0.03, Fe 4, Cr 28, bal Co.
For cast turbine blades, high temperature components. High heat resistance.

ONION FUSIBLE ALLOY
English manufacture
Bi 50, Sn 2, Pb 30.
For fuses, safety plugs; melting point 92°C.

ONPLUS
Manufacturer not listed.
C 0.7, W 18, Cr 4, V 2, bal Fe.
For cutting tools; high speed steel.

ONTARIO
AL Tech Specialty Steel Corp.
C 1.5, Cr 12, V 0.9, Mo 1, bal Fe.
Typical heat treated hardness: 58-61 Rock C. High performance cold work die steel; high carbon-high chromium type for long run blanking and forming dies, punches, thread rolling dies. AISI D-2.

ONTARIO-EZ
AL Tech Specialty Steel Corp.
C 1.5, Cr 12, V 0.9, Mo 1, S 0.12, bal Fe.
Same as ONTARIO except free-machining grade.

ONTOP
Manufacturer not listed.
C 0.7, W 18, Cr 4, V 1, Co 5, bal Fe.
For tools, high speed cutters; high speed steel.

ONYX SPRING STEEL
Crucible Materials Corp.
Carbon steel. C 0.7, bal Fe.
For springs; water hardening.

OO
Eagle & Globe Steel Ltd.
Tool material. C 0.62, Mn 0.7, Cr 1.1, Mo 0.25, W 1.9, V 0.15, Si 0.7, bal Fe.
Shock resisting steel. Tough, wear resistant. For punches, shear blades, dies. AS1239 S1A-6; AISI S1; Werkstoff 1.2550.

OPTAL 3
Now WIELAND A13.

OPTAL 5
Now WIELAND A13.

OPTAL 7
Wieland-Werke AG Metallwerke
Mg 6.8, Mn 0.15, Cr 0.3, bal Al.
1/2 H-temper: 50,000 TS; 31,000 YS; 9 El; 90 Brin. *Obsolete*

OPTICAL WIRE ALLOY
American manufacture
Cu 54, Zn 28, Ni 18.
Annealed: 33,000 TS; 20,000 YS; 33 El; 38 RA; 50 Brin. For optical instruments; German Silver.

OPTIMUM
Otto Wolff Handelgesellschaft
C 2.1, Cr 11.5, bal Fe.
For forming and blanking dies, punches; oil or air hardened, nondeforming.

OPTIMUM CO20
Otto Wolff Handelgesellschaft
C 1.65, Cr 12, Co, bal Fe.
For forming and blanking dies, punches; air hardened, nondeforming.

OPTIMUM W
Otto Wolff Handelgesellschaft
C 1.65, Cr 11.5, V 0.1, bal Fe.
For forming and blanking dies, punches; air hardened, nondeforming.

OPTIMUM Z

Otto Wolff Handelgesellschaft
C 2.1, Cr 11.5, W 0.7, bal Fe.
For forming and blanking dies, punches; oil or air hardened, nondeforming.

OPTION

J.M. Ney Co.
Pd 79, Au 2.
Silver-free white gold color alloy. Melting range: 2012-2174°F; casting temperature 2300°F; 330 Brin; 425 HV.

OPTION POST SOLDER

J.M. Ney Co.
Gold color solder. Melting range: 1515-1560°F.

OPTION PRE-SOLDER

J.M. Ney Co.
Solder for porcelain fused-to-metal alloys (pre-porcelain).
Melting range: 1866-1947°F.

ORA-FUTURA II

Certified Alloy Products Inc.
C 0.1, Si 1.2, Cr 14, Mo 1.1, Cb 0.5, Al 1.7, Ti 2.3, 0.20 Ce and La, bal Ni.
As cast: 101,000 psi TS; 79,000 psi YS; 7 El; 16 RA; 262 Brin.
Dental alloy for bridges and restorations either polished or porcelainized. *Obsolete*

ORALLOY

Precious Metals Research Works Inc.
Au alloy.
For dentures.

ORANGE LABEL

A. Milne & Co.
C 1-1.05, bal Fe.
For tools; precision cast.

ORANGE LABEL

Wallace Murray Corp.
C 0.45-0.55, Cr 1.25-1.75, W 2-3, V 0.2-0.3, bal Fe.
For chisels, punches, rivet sets; "Special Alloy Tool."

ORANGE LABEL SPECIAL ALLOY TOOL STEEL

Wallace Murray Corp.
C, alloy, bal Fe.
For chisels, punches, heading tools, rivet sets; fatigue resistant.

ORANIUM BRONZE

Olin Brass, Indianapolis
Cu 90, Al 10.
Cast: 65,000 psi TS; 20 El. Rolled: 80,000 psi TS. For castings and gears; aluminum bronze. *Obsolete*

ORANIUM BRONZE "M"

Anaconda Co.
Cu 95, Al 5.
100,000 TS; 10 El. For hardware; medium. *Obsolete*

ORANIUM BRONZE "M.H"

Manufacturer not listed
Cu 92, Al 8.
130,000 TS; 4 El. For hardware; medium hard-soft.

ORANIUM BRONZE "S"

Manufacturer not listed
Cu 97, Al 3.
30,000 TS; 60 El. For hardware; soft-soft.

ORANIUM BRONZE, "H.H"

Manufacturer not listed
Cu 89, Al 11.
For gears, propellers; Al-bronze, tough.

ORANIUM BRONZE, "H.X"

Manufacturer not listed
Cu 88.5, Al 11.5.
For gears, bearings; Al-bronze, tough.

ORBIS

Uddeholm Corp.
C 0.95, Si 1.5, Mn 0.75, Cr 1, V 0.1, bal Fe.
For dies; for cold work. *Obsolete*

ORBIT

Crucible Materials Corp.
Tool material. C 0.7, Mn 2, S 0.15, Cr 1, Mo 1.35, bal Fe.
For dies for blanking and forming, rim rolls, master hubs; tough, air hardened, free machining.

ORDIX

Saarstahl AG
Now SAARSTAHL 1.2103.

ORDIX EXTRA

Rochling Burbach GmbH
C 0.67, Si 1.3, Mn 0.5, Cr 0.5, bal Fe.
For springs, punches; oil hardened, shock resistant.
Obsolete

ORDIX SPECIAL

Saarstahl AG
Now SAARSTAHL 1.2242.

OREIDE "A"

English manufacture
Cu 68, Zn 32, Sn 0.5.
60,000 TS; 30 El. For hardware; Tobin Bronze.

OREIDE "B"

English manufacture
Cu 81-90, Zn 10-15, Sn 0-4.
French gold, carriage and harness hardware, ornamental work.

OREIDE .5%

Century Brass Products Inc.
Cu 89-92, Sn 0.3-0.7, Pb 0-1, bal Zn.
Annealed: 42,000 TS; 12,000 YS; 43 El; F 60 Rock. Hard: 68,000 TS; 57,000 YS; 4 El; B 76 Rock. For jewelry and weather stripping. Good formability, weldability and corrosion resistance. *Obsolete*

OREIDE SCOVILL ALLOY 321

Century Brass Products Inc.
Cu 88-91, Sn 0.35-0.65, bal Zn.
Annealed: 40,000 TS; 40 El. Hard: 75,000 TS; 4 El. For bearings, jewelry; corrosion resistant. *Obsolete*

OREIDE, BRUNSWICK

English manufacture
Cu 68, Zn 32, Sn 0.5.
For condenser tubing; corrosion resistant.

OREIDE, FRENCH GOLD

Manufacturer not listed
Cu 81-90, Zn 10-15, Sn 0-9, Pb 1-2.
For plumbing, hardware; free-cutting.

ORELLOY 100

Oregon Steel Mills
C 0.1-0.2, Mn 1.1-1.5, Si 0.15-0.3, Mo 0.2-0.3, B 0.001-0.005, bal Fe.
Quenched and tempered: 321 Brin. Abrasion resistant steel.

ORELLOY 100A

Oregon Steel Mills
C 0.1-0.2, Mn 1.1-1.5, Si 0.15-0.3, Mo 0.2-0.3, B 0.001-0.005, bal Fe.
Quenched and tempered: 110-130 ksi TS; 100 ksi YS; 18 El.
HSLA steel, extra high strength. ASTM A514.

ORELLOY 100B

Oregon Steel Mills
C 0.1-0.2, Mn 1.1-1.5, Si 0.15-0.3, Mo 0.2-0.3, B 0.001-0.005, bal Fe.
Quenched and tempered: 115-135 ksi TS; 100 ksi YS; 16 El.
HSLA steel, extra high strength. ASTM A517.

ORELLOY 242

Oregon Steel Mills
C 0.1, Mn 0.5, P 0.09, S 0.03, Si 0.48, Cr 0.86, Cu 0.43.
Rolled: 70,000 TS; 50,000 YS; 22 El. For structural members, bridges, booms, derricks, tanks. High strength low-alloy steel.

ORELLOY 242 1

Oregon Steel Mills
C 0.1, Mn 0.5, P 0.09, S 0.03, Si 0.48, Cu 0.43, Ni 0.43, Cr, bal Fe.
Wrought: 70 ksi TS; 50 ksi YS; 22 El. HSLA; improved atmosphere corrosion resistance. ASTM A242.

ORELLOY 242 2

Oregon Steel Mills
C 0.17, Mn 1.05, S 0.2, Cu 0.25, Cr 0.5, Ni 0.06, bal Fe.
Wrought: 70 ksi TS; 50 ksi YS; 18 El. HSLA steel; improved atmosphere corrosion resistance. ASTM A242.

ORELLOY 440

Oregon Steel Mills
C 0.25, Mn 1.1-1.6, Si 0.3, 0.2 Cu min, bal Fe.
Wrought: 70 ksi TS; 50 ksi YS; 21 El. HSLA steel; improved atmosphere corrosion resistance. ASTM A440.

ORELLOY 441

Oregon Steel Mills
C 0.22, Mn 1.25, Si 0.3, 0.02 Cu min, 0.02 V min, bal Fe.
Wrought: 70 ksi TS; 50 ksi YS; 21 El. HSLA steel; improved atmosphere corrosion resistance. ASTM A441.

ORELLOY 45

Oregon Steel Mills
C 0.22, Mn 1.35, Si 0.3, 0.005 Cb min, bal Fe.
Wrought: 60 ksi TS; 45 ksi YS; 22 El. HSLA steel. ASTM A572.

ORELLOY 50

Oregon Steel Mills
C 0.23, Mn 1.35, Si 0.3, 0.005 Cb min, bal Fe.
Wrought: 65 ksi TS; 50 ksi YS; 21 El. HSLA steel. ASTM A572.

ORELLOY 50 LT

Oregon Steel Mills
C 0.2, Mn 1.5, P 0.025, S 0.025, Si 0.3, 0.001 C min, bal Fe.
Wrought: 70 ksi TS; 50 ksi YS; 25 El. High strength low alloy steel.

ORELLOY 55

Oregon Steel Mills
C 0.25, Mn 1.35, Si 0.3, 0.005 Cb min, bal Fe.
Wrought: 70 ksi TS; 55 ksi YS; 20 El. HSLA steel. ASTM A572.

ORELLOY 588

Oregon Steel Mills
C 0.19, Mn 1.25, Si 0.3, Cu 0.25-0.4, Cr 0.4-0.65, bal Fe.
Wrought: 70 ksi TS; 50 ksi YS; 21 El. HSLA steel; improved atmosphere corrosion resistance. ASTM A588.

ORELLOY 60

Oregon Steel Mills
C 0.26, Mn 1.35, Si 0.3, 0.005 Cb min, bal Fe.
Wrought: 75 ksi TS; 60 ksi YS; 18 El. HSLA steel. ASTM A572.

ORELLOY 65

Oregon Steel Mills
C 0.26, Mn 1.35, Si 0.3, 0.005 Cb min, bal Fe.
Wrought: 80 ksi TS; 65 ksi YS; 17 El. HSLA steel. ASTM A572.

ORELLOY 70

Oregon Steel Mills
C 0.26, Mn 1.6, Si 0.3, 0.005 Cb min, bal Fe.
Wrought: 85 ksi TS; 70 ksi YS; 16 El. HSLA steel. ASTM A572.

ORELLOY 70 FG
Oregon Steel Mills
C 0.15, Mn 1.6, P 0.025, S 0.025, Si 0.3, 0.001 Cb min, bal Fe.
Wrought: 80 ksi TS; 70 ksi YS; 20 El. High strength low alloy steel.

ORELLOY AR
Oregon Steel Mills
C 0.43, Mn 1.5, Si 0.29, Cu 0.08, bal Fe.
Hot rolled: 235 Brin. Abrasion resistant steel.

ORELLOY AR 320
Oregon Steel Mills
C 0.31, Mn 1.65, Si 0.15-0.3, Mo 0.2-0.3, B 0.001-0.005, bal Fe.
Quenched and tempered: 320 Brin. Abrasion resistant steel.

ORELLOY AR 340
Oregon Steel Mills
C 0.31, Mn 1.65, Si 0.15-0.3, Mo 0.2-0.3, B 0.001-0.005, bal Fe.
Quenched and tempered: 340 Brin. Abrasion resistant steel.

ORELLOY AR 360
Oregon Steel Mills
C 0.31, Mn 1.65, Si 0.15-0.3, Mo 0.2-0.3, B 0.001-0.005, bal Fe.
Quenched and tempered: 360 Brin. Abrasion resistant steel.

ORELLOY AR 400
Oregon Steel Mills
C 0.31, Mn 1.65, Si 0.15-0.3, Mo 0.2-0.3, B 0.001-0.005, bal Fe.
Quenched and tempered: 400 Brin. Abrasion resistant steel.

ORELLOY-42
Oregon Steel Mills
C 0.21, Mn 1.35, Si 0.0, 0.005 Cb min, bal Fe.
Wrought: 60 ksi TS; 42 ksi YS; 24 El. HSLA steel. ASTM A572.

ORIENTED M-6W
Armco
Si, bal Fe.
For transformers; formerly Tran-Cor 4W-O; grain oriented. *Obsolete*

ORIENTED M-7W
Armco
Si, bal Fe.
For transformers; formerly Tran-Cor 3W-O; grain oriented. *Obsolete*

ORIENTED M-7X
Armco
Si, bal Fe.
For transformers; formerly Tran-Cor 3X-O; grain oriented. *Obsolete*

ORIENTED M-8X
Armco
Si, bal Fe.
For transformers; formerly Tran-Cor 2X-O; grain oriented. *Obsolete*

ORIENTED T
Armco
Si, bal Fe.
For armatures, electric generators; high permeability; formerly Tran-Cor T-O. *Obsolete*

ORIENTED T-S
Armco
Si, bal Fe.
For wound type transformers and reactors; high permeability; formerly Tran-Cor T-O-S. *Obsolete*

ORIGINAL MONEL METAL
Eisenwerke Neubrandenburg G.m.b.H.
Nickel. Cu 23, Si 2.5, bal Ni.
Sand cast: 56,000-72,000 TS; 28,000-43,000 YS; 20-30 El; 120-140 Brin. For pump parts, chemical industries; corrosion resistant.

ORION
National Alloys Ltd.
C 0.5, Cr 1, V 0.2, bal Fe.
Heat treated: 115,000-240,000 TS; 80,000-195,000 YS; 5-23 El; 440-262 Brin. For machine parts, tools, automobile springs and gears, forging dies, shear blades. Resists heavy vibrations.

ORION
J.M. Ney Co.
Au 51, Pd 39.
Porcelain alloy for single and multiple unit fixed restorations. 122,000 psi TS; 72,000 psi YS (0.1% offset); 28 El; 222 HV.

ORION 26-1
Creusot-Loire
C 0-0.01, Cr 26, Mo 1, N 0-0.015, bal Fe.
Wrought, annealed: 400 N/mm^2 psi TS min. Stainless, ferritic grade; good with organic acids; resistant to stress corrosion. AFNOR Z-01CD26-1; W.-Nr. 1.4131.

ORION-100 CAST
Compagnie Ateliers et Forges de la Loire
Stainless steel. C 0.35, Cr 27, Si 1.2, bal Fe.
Cast: 95,000 TS; 65,000 YS; 2 El; 212 Brin. For furnace parts, heat treating boxes, salt pots, oil burners, exhaust manifolds. Ferritic stainless steel, heat and oxidation resistant.

ORION-100 FORGE
Compagnie Ateliers et Forges de la Loire
Stainless steel. C 0.15, Cr 27, bal Fe.
Annealed: 92,500 TS; 57,000 YS; 25 El. For furnace parts, heat treating boxes, salt pots. Ferritic stainless steel. Heat and oxidation resistant.

ORION-120
Compagnie Ateliers et Forges de la Loire
Stainless steel. C 0-0.02, Cr 27, bal Fe.
Annealed: 80,000 TS; 50,000 YS; 25 El; 82 Rock B. For furnace parts, heat treating boxes, salt pots, oil burners. Ferritic stainless steel. Heat resistant to 1150°C in oxidizing atmosphere.

ORKAN
Hufnagel GmbH
C 1.3-1.4, Cr 4.2, Mo 1, V 4, W 12, Co 5, bal Fe.
Co-W high speed steel, for finishing tools.

ORKAN-S
Hufnagel GmbH
C 1.2, Cr 4, Mo 4, W 10, V 3, Co 10, bal Fe.
Co-W high speed steel; lathe tools for finish machining.

ORLEANS
Wallace Murray Corp.
C 0.6, Mn 0.8, Si 2, Cr 0.25, V 0.2, Mo 0.45, bal Fe.
For pneumatic chisels, punches, concrete breakers, cutters; Type S5; oil hardening, shock resistant.

ORMULU
Lumen Bearing Co.
Cu 58, Sn 16.7, Zn 25.3.
For bushings, hardware. *Obsolete*

ORMULU, LARGE
Lumen Bearing Co.
Cu 91, Sn 6.5, Zn 3.
30,000-40,000 TS; 15-25 El; 14-23 RA. For instruments, utensils, bearings, hardware. *Obsolete*

ORMULU, SMALL
Lumen Bearing Co.
Cu 94, Sn 5.9.
33,000-42,100 TS; 19,600-19,700 YS; 20-34 El; 24-32 RA. For switches, contacts, electrical parts, springs. *Obsolete*

ORNAL 0.5
Kreidler Werke G.m.b.H.
Aluminum. Mg 0.5, Si 0.5, bal Al.
13-27 kp/mm^2 TS; 7-20 kp/mm^2 YS; 10-15 El. Parts for instruments and optical equipment. Al Mg Si 0.5.

ORNAL 0.8
Kreidler Werke G.m.b.H.
Aluminum. Mg 0.8, Si 0.8, bal Al.
13-28 kp/mm^2 TS; 7-20 kp/mm^2 YS; 12-16 El. Parts for instruments and optical equipment. Al Mg Si 0.8.

ORO-FUTURA I
Certified Alloy Products Inc.
C 0.1, Si 1.2, Cr 15, Mo 0.5, Al 1.5, Ti 2, 0.20 Ce and La, bal Ni.
As cast: 99,000 psi TS; 72,000 psi YS; 11 El; 17 RA; 229 Brin. For dental use; porcelain coating restorations, crowns. *Obsolete*

OROBRAZE 1040
Johnson Matthey plc
Ag 30, bal Au.
Brazing alloy; 1030-1040°C MP.

OROBRAZE 910
Johnson Matthey plc
Au 80, Cu, Fe.
Brazing alloy; 908-910°C MP.

OROBRAZE 940
Johnson Matthey plc
Cu, bal Au.
Brazing alloy; 930-940°C MP.

OROBRAZE 950
Johnson Matthey plc
Ni 17.5, Au 82.5, bal Cu.
Brazing alloy; 950°C MP.

OROBRAZE 990
Johnson Matthey plc
Ni 25, bal Au.
Brazing alloy; 950-990°C MP.

ORTHODUR
Vereinigte Metall. Ranshofen-Berndorf
Aluminum.
Aluminum alloy AlCuMg1.

ORTHOMETAL
Telcon Metals Ltd.
Soft magnetic alloy, square hysteresis loop for low level magnetic amplifers, etc.

ORTHOMUMETAL
Telcon Metals Ltd.
Soft magnetic alloy.

ORTHONAL, ROUND; ORTHONAL, SQUARE
Spang Specialty Metals
Ni 50, Fe 50.
Grain oriented thin strip soft magnetic alloy. For magnetic amplifiers, flux counters, flux switching, etc. *Obsolete*

ORTHONIK
Armco
Ni 50, Fe 53.
For magnetic and electrical equipment, magnetic tape and foil. Magnetically soft, high permeability. *Obsolete*

ORTHOSIL
Thomas & Skinner Inc.
Si, bal Fe.
For laminations for transformers, chokes, filters, reactors; high magnetic saturation. *Obsolete*

ORVAR 1
Uddeholm Corp.
C 0.3, Si 0.9, Mn 0.5, Cr 5.25, Mo 1.4, V 0.1, bal Fe.
For punches, crimpers, upsetters, riveters; hot work steel, oil hardened. AISI H13.

ORVAR 2 MICRODIZED
Uddeholm Corp.
C 0.38, Si 1, Mn 0.4, Cr 5.3, Mo 1.3, V 0.9, bal Fe.
263,000 psi TS; 220,000 psi YP; 10 El; 45 RA; 52 Rock C. Chromium-molybdenum-vanadium alloyed steel. For tools for extrusion, plastic molding applications, cold punching, hot shearing, shrink rings and wear resistant parts. AISI H13; W.-Nr. 1.2344.

ORVAR SUPREME
Uddeholm Corp.
C 0.38, Si 1, Mn 0.4, Cr 5.3, Mo 1.3, V 0.9, bal Fe.
263,000 psi TS; 220,000 psi YP; 10 El; 45 RA; 52 Rock C. Chromium-molybdenum-vanadium alloyed steel. For tools for die casting and extrusion. AISI H13; W.-Nr. 1.2344.

ORVAR-1
Uddeholm Corp.
C 0.38, Cr 5.2, Mo 1.3, V 1, bal Fe.
Heat treated: 300,000 TS; 227,000 YS; 9 El; 55 Rock C. For hot dies, fasteners, extrusion and die casting dies, punches. Type H11; hot work tool steel, high toughness and wear resistance. *Obsolete*

ORVAR-2
Uddeholm Corp.
C 0.35-0.4, Si 0.8-1.2, Cr 5-5.5, Mo 1.2-1.5, V 0.9-1.1, bal Fe.
For die casting dies, punches, hot shears, forging dies; Type H13; hot work steel.

OSBORN 303
Osborn Steels Ltd.
Stainless steel. C 0-0.08, Cr 18, Ni 10, S 0.2, bal Fe.
Free cutting, Type 18/8; stainless steel. BS 970 303SZ1; EN 58AM; similar to AISI 303.

OSBORN 304
Osborn Steels Ltd.
Stainless steel. C 0-0.06, Cr 18, Ni 10, bal Fe.
Austenitic, Type 18/8; stainless steel, for food and dairy equipment. BS 970 304S15; EN 58E; similar to AISI 304.

OSBORN 304 L
Osborn Steels Ltd.
Stainless steel. C 0-0.03, Cr 18, Ni 11, bal Fe.
Low carbon, Type 18/8; austenitic, stainless steel. Preferred for welded assemblies. BS 970 304S12; similar to AISI 304L.

OSBORN 309
Osborn Steels Ltd.
Stainless steel. C 0-0.12, Cr 24, Ni 15, bal Fe.
Heat resistant, stainless steel. BS 970 309S24; similar to AISI 309.

OSBORN 310
Osborn Steels Ltd.
Stainless steel. C 0-0.12, Cr 25, Ni 21, bal Fe.
Heat resistant, stainless steel. For mildly carburizing atmospheres up to 1100°C. BS 970 310S24; similar to AISI 310.

OSBORN 316
Osborn Steels Ltd.
Stainless steel. C 0.07, Cr 18, Ni 11, Mo 2.5, bal Fe.
Acid resistant, stainless steel for severe conditions. BS 970 PT4 316S16; EN 58J; similar to AISI 316.

OSBORN 316 L
Osborn Steels Ltd.
Stainless steel. C 0-0.03, Cr 18, Ni 13, Mo 2.5, bal Fe.
Extra low carbon grade of 316 for welded assemblies. BS 970 PT4 316S12; similar to AISI 316L.

OSBORN 317
Osborn Steels Ltd.
Stainless steel. C 0-0.06, Cr 19, Ni 14, Mo 3.5, bal Fe.
Austenitic, stainless steel for highly corrosive conditions. BS 970 317S16; similar to AISI 317.

OSBORN 320
Osborn Steels Ltd.
C 0-0.08, Cr 18, Ni 11, Mo 2.5, Ti, bal Fe.
Titanium stabilized weldable version of 316. BS 970 PT4 320S17; EN 58J.

OSBORN 321
Osborn Steels Ltd.
Stainless steel. C 0-0.08, Cr 18, Ni 11, Ti, bal Fe.
Titanium stabilized weldable grade of 18/8 stainless steel. BS 970 PT4 321S12; EN 58B/C; similar to AISI 321.

OSBORN 325
Osborn Steels Ltd.
C 0.08, Cr 18, Ni 10, S 0.2, Ti, bal Fe.
Free machining version of 321. BS 970 PT4 325S21; EN 58M.

OSBORN 330
Osborn Steels Ltd.
C 0-0.1, Cr 19, Ni 38, Si 2, bal Fe.
Heat resistant steel for furnace equipment.

OSBORN 347
Osborn Steels Ltd.
Stainless steel. C 0-0.08, Cr 18, Ni 10, Nb, bal Fe.
Niobium stabilized weldable grade of 18/8 stainless steel. BS 970 PT4 347S17; EN 58F/G; similar to AISI 347.

OSBORN 403
Osborn Steels Ltd.
Stainless steel. C 0-0.08, Cr 13, bal Fe.
Ferritic, martensitic chromium stainless steel for lightly stressed engineering fittings. BS 970 PT4 403S17; similar to AISI 403.

OSBORN 405
Osborn Steels Ltd.
Stainless steel. C 0-0.08, Cr 13, Al, bal Fe.
Non-hardened, weldable, ferritic, stainless steel. BS 970 PT4 405S17; similar to AISI 405.

OSBORN 410
Osborn Steels Ltd.
Stainless steel. C 0-0.12, Cr 12.5, bal Fe.
Martensitic, stainless steel for mildly corrosive conditions. BS 970 PT4 410S21; EN 56A; similar to AISI 410.

OSBORN 416A
Osborn Steels Ltd.
Stainless steel. C 0.12, Cr 12.5, S 0.25, bal Fe.
Free machining, martensitic, stainless steel. BS 970 416S21; EN 56AM; similar to AISI 416.

OSBORN 416C
Osborn Steels Ltd.
Stainless steel. C 0.25, C 13, S 0.25, bal Fe.
High carbon, free machining, martensitic, stainless steel. BS 970 416S37; EN 56CM; similar to AISI 420F.

OSBORN 420 C
Osborn Steels Ltd.
Stainless steel. C 0.23, Cr 12.5, bal Fe.
Martensitic, stainless steel. BS 970 420S37; EN 56C; similar to AISI 420.

OSBORN 420 D
Osborn Steels Ltd.
Stainless steel. C 0.32, Cr 12.5, bal Fe.
Martensitic, stainless steel; hardened to 550 HV. For cutlery and edge tools. BS 970 420S45; EN 56D; similar to AISI 420.

OSBORN 430
Osborn Steels Ltd.
Stainless steel. C 0-0.1, Cr 17, bal Fe.
Ferritic, stainless, non-hardenable. BS 970 PT4 430S15; EN 60; similar to AISI 430.

OSBORN 431
Osborn Steels Ltd.
Stainless steel. C 0-0.16, Cr 16, Ni 2.3, bal Fe.
Martensitic, stainless, hardenable; improved corrosion resistance. BS 970 PT4 431S29; EN 57; similar to AISI 431.

OSBORN A2
Osborn Steels Ltd.
C 1, Cr 5, Mo 1.05, V 0.3, bal Fe.
Air hardened, cold work tool steel. BS 4659 BA2; similar to AISI A2.

OSBORN A7
Osborn Steels Ltd.
C 2.3, Cr 5.3, W 1.05, Mo 1.05, V 4.7, bal Fe.
Air hardened, cold work tool steel. BS 4659 BA2; similar to AISI A7.

OSBORN D2
Osborn Steels Ltd.
C 1.5, Cr 12, Mo 0.9, V 0.9, bal Fe.
Air or oil hardened, cold work tool steel. BS 4659 BD2; similar to AISI D2.

OSBORN D3
Osborn Steels Ltd.
C 2.2, Cr 12.5, V 0.25, bal Fe.
Air or oil hardened, cold work tool steel. BS 4659 BD3; similar to AISI D3.

OSBORN D4
Osborn Steels Ltd.
C 2.1, Cr 12.25, Mo 0.8, V 0.65, bal Fe.
Air or oil hardened, cold work tool steel. Similar to AISI D4.

OSBORN H10A
Osborn Steels Ltd.
C 0.37, Cr 3, Mo 2.75, V 0.45, Co 3, bal Fe.
Hot work tool steel, chromium type. BS 4659 BH10A; similar to AISI H10.

OSBORN H12
Osborn Steels Ltd.
C 0.35, Cr 5, W 1.4, Mo 1.6, V 0.25, bal Fe.
Hot work tool steel, chromium type. BS 4659 BH12; similar to AISI H12.

OSBORN H13
Osborn Steels Ltd.
C 0.39, Cr 5, Mo 1.4, V 1, bal Fe.
Hot work tool steel, chromium type. BS 4659 BH13; similar to AISI H13.

OSBORN H21
Osborn Steels Ltd.
C 0.3, Cr 3, W 8.8, V 0.3, bal Fe.
Hot work tool steel, tungsten type. BS 4659 BH21; similar to AISI H21.

OSBORN H21 N
Osborn Steels Ltd.
C 0.25, Cr 3, W 8.8, V 0.3, Ni 2.3, bal Fe.
Hot work tool steel, tungsten type. BS 4659 BH21A.

OSBORN LV10N
Osborn Steels Ltd.
C 0.4, Si 3, Ni 8, Cr 19, 0.20 N_2, bal Fe.
Austenitic, valve steel. For exhaust valves for diesel engines. US Sil10N.

OSBORN LV20
Osborn Steels Ltd.
C 0.8, Si 2, Ni 1.4, Cr 19.5, bal Fe.
Martensitic, steel. For exhaust valves. BS 443S65; US XB.

OSBORN LV21/12N
Osborn Steels Ltd.
C 0.2, Ni 11, Cr 21, 0.20 N_2, bal Fe.
Austenitic, valve steel. For exhaust valves for diesel engines.
BS 970 381S34; US 21/12N.

OSBORN LV21/2N
Osborn Steels Ltd.
C 0.55, Mn 8.5, Ni 2, Cr 20, 0.30 N^2, bal Fe.
Austenitic, valve steel. US 21/2N.

OSBORN LV21/42
Osborn Steels Ltd.
C 0.05, N 9, Ni 4, Cr 21, Nb 2, 0.45 N_2, bal Fe.
Austenitic, valve steel. For exhaust valves; improved stress-rupture properties. BS 970 3523529.

OSBORN LV21/43
Osborn Steels Ltd.
C 0.05, Mn 9, Ni 4, Cr 21, Nb 2, W 1, 0.45 N_2, bal Fe.
Austenitic, valve steel.

OSBORN LV21/4N
Osborn Steels Ltd.
C 0.5, Mn 9, Ni 4, Cr 21, 0.45 N^2, bal Fe.
Austenitic, valve steel. For exhaust valves in petroleum engines. BS 970 349S52; US 21/4N.

OSBORN LV21/4NS
Osborn Steels Ltd.
C 0.05, Mn 9, Ni 4, Cr 21, S 0.05, 0.45 N^2, bal Fe.
Austenitic, valve steel. For exhaust valves in petroleum engines. BS 970 349S54.

OSBORN LV52
Osborn Steels Ltd.
C 0.45, Si 3.4, Cr 8.2, bal Fe.
Martensitic, steel. For inlet valves in petroleum engines. BS 970 401S45; US Sil1.

OSBORN LV54A
Osborn Steels Ltd.
C 0.4, Si 1.5, Ni 14, Cr 13.5, W 2.7, Mo 0.5, bal Fe.
Austenitic, steel. For making hollow valves; easily hard faced. BS 970 331S42; AMS 5700.

OSBORN LV55
Osborn Steels Ltd.
C 0.45, Si 2.3, Mn 1.2, Ni 9, Cr 17.2, W 1, bal Fe.
Austenitic, steel. For hollow valves; to be hard faced. W.-Nr. 1.4873.

OSBORN M1
Osborn Steels Ltd.
C 0.8, Cr 4, W 1.8, Mo 8.75, V 1.1, bal Fe.
Molybdenum high speed steel. BS 4659 BM1; similar to AISI M1.

OSBORN M15
Osborn Steels Ltd.
C 1.55, Cr 4.7, W 6.5, Mo 3, V 5, Co 5, bal Fe.
Mo-W-Co high speed steel. BS 4659 BM15.

OSBORN M2
Osborn Steels Ltd.
C 0.85, Cr 4, W 6.3, Mo 5, V 1.9, bal Fe.
Mo-W high speed steel. BS 4659 BM2; similar to AISI M2.

OSBORN M2 S
Osborn Steels Ltd.
C 0.85, Cr 4, W 6.3, Mo 5, V 1.9, S 0.1, bal Fe.
Mo-W high speed steel; free-machining.

OSBORN M35
Osborn Steels Ltd.
C 0.9, Cr 4, W 6.3, Mo 5, V 1.9, Co 5, bal Fe.
Mo-W-Co high speed steel. W.-Nr. 1.3243.

OSBORN M42
Osborn Steels Ltd.
C 1.07, Cr 3.8, W 1.5, Mo 9.7, V 1.1, Co 8.3, bal Fe.
Mo-Co high speed steel. BS 4659 BM42; similar to AISI M42.

OSBORN MN
Osborn Steels Ltd.
C 1.2, Mn 12, bal Fe.
12% manganese austenitic, steel. Good wear resistance for wearing plates, shutes, screens and security applications.

OSBORN O1
Osborn Steels Ltd.
C 0.9, Mn 1.25, Cr 0.5, W 0.5, V 0.2, bal Fe.
Oil hardened, cold work tool steel. BS 4659 BO1; similar to AISI O1.

OSBORN QJ
Osborn Steels Ltd.
C 1.1, Cr 1.3, bal Fe.
Cold work tool steel; hardenable to 60-64 Rock C.

OSBORN RAB1
Osborn Steels Ltd.
C 0.31, Ni 4.1, Cr 1.3, Mo 0.3, bal Fe.
Supplied at 285 Brin max; can be rehardened to 26-53 Rock C. For molds. W.-Nr. 1.2766.

OSBORN RAB20
Osborn Steels Ltd.
C 0.4, N 1.5, Cr 2, Mo 0.2, bal Fe.
Supplied prehardened to 30 Rock C; can be rehardened to 26-53 Rock C. For molds.

OSBORN S1
Osborn Steels Ltd.
C 0.52, Cr 1.4, W 2.2, V 0.2, bal Fe.
Cold work tool steel, shock resistant. BS 4659 BS1; similar to AISI S1.

OSBORN S510
Osborn Steels Ltd.
C 0.21, bal Fe.
Carbon steel with 21 tsi TS; aircraft grade. BS 2S510.

OSBORN S511
Osborn Steels Ltd.
C 0.1, bal Fe.
Deep drawing carbon steel; aircraft grade. BS 2S511.

OSBORN S514/5
Osborn Steels Ltd.
C 0.21, Mn 1.5, bal Fe.
Carbon-manganese steel with 30-50 tsi TS; aircraft grade. BS 2S514/5.

OSBORN S516/7
Osborn Steels Ltd.
C 0.46, Mn 1.5, bal Fe.
Carbon-manganese steel with 60-75 tsi TS; aircraft grade. BS 2S516/7.

OSBORN S524/5
Osborn Steels Ltd.
C 0.08, Cr 18, Ni 10, Ti, Nb, bal Fe.
Stabilized corrosion resistant steel, cold rolled to 52 tsi TS; aircraft grade. BS 2S524/5.

OSBORN S526/7
Osborn Steels Ltd.
C 0.08, Cr 18, Ni 10, Ti, Nb, bal Fe.
Stabilized corrosion resistant steel, cold rolled to 35 tsi TS; aircraft grade. BS 2S526/7

OSBORN S530/1
Osborn Steels Ltd.
C 0.12, Cr 24.5, Ni 17.5, Ti, Nb, bal Fe.
Stabilized heat resistant steel; aircraft grade. BS 2S530/1.

OSBORN S532/3
Osborn Steels Ltd.
C 0.05, Cr 15.6, Ni 5.4, Mo 1.7, Ti, Cu, bal Fe.
Corrosion resistant steel, precipitation hardened to 64-76 tsi BS 2S532/3.

OSBORN S534/5
Osborn Steels Ltd.
C 0.25, Cr 1, Mo 0.2, bal Fe.
Chrome-molybdenum steel with 57-74 tsi TS; suitable for welding. BS 2S534/5.

OSBORN T1
Osborn Steels Ltd.
C 0.75, Cr 4, W 18, V 1.1, bal Fe.
Tungsten high speed steel. BS 4659 BT1; similar to AISI T1.

OSBORN T4
Osborn Steels Ltd.
C 0.78, Cr 4, W 18, Mo 0.8, V 1.1, Co 5, bal Fe.
Tungsten-cobalt high speed steel. BS 4659 BT4; similar to AISI T4.

OSBORN T6
Osborn Steels Ltd.
C 0.8, Cr 4, W 20.5, Mo 0.8, Co 1.5, bal Fe.
Tungsten-cobalt high speed steel. BS 4659 BT6; similar to AISI T6.

OSBORN WJ
Osborn Steels Ltd.
C 0.04, Cr 21, Ni 25, Mo 4.5, Ti, bal Fe.
Austenitic, stainless. For severe conditions.

OSBORN-EWC
Osborn Steels Ltd.
C 0.03, Cr 18.25, Ni 13.5, Mo 3.75, bal Fe.
Annealed: 108,000 TS; 45,000 YP; 47 El; 57 RA. For scrubber liners, chemical plant equipment, digesters. Acid resistant; Type 317, austenitic. *Obsolete*

OSCILLUMIN
German manufacture
Cu 0.8, Fe 0.4-0.6, Si 12.6-13.2, bal Al.
28,000 TS; 4 El; 65 Brin. For light alloy castings; high fluidity.

OSEMUND
Stahlwerke R. & H. Plate
C 1.4, Cr 0.3, V 0.1, bal Fe.
For cutters, bearings, liners; oil hardened, wear resistant. *Obsolete*

OSMAGAL
KM-kabelmetal AG
Mn 1.8, bal Al.
For sheet; rolled. *Obsolete*

OSMAGAL
Kabel-und Metallwerke AG
Aluminum. Mn 1.8, bal Al.
Soft: 15,000 TS; 6400 YS; 30 El; 25 Brin. Hard: 35,000 TS; 28,000 YS; 2 El; 60 Brin. For heat exchangers, truck panels, fixtures, duct work. Nonheat treatable.

OSMAGAL S
KM-kabelmetal AG
Mg 1.5-3, Mn 0.5-1.5, Cr 0-0.3, bal Al.
For aircraft and structural parts; resists sea water corrosion. *Obsolete*

OSMIRIDIUM (NATURAL) NEVYANSKITE
Manufacturer not listed
Ir 44-58, Os 27-49, Pt 0-10, Ru 0-6, Rh 1.5-3.
For pen points; corrosion resistant.

OSMIRIDIUM (NATURAL) SISERSKITE
Manufacturer not listed
Os 57, Ru 8, 34 Rh + Ir.
For pen points; corrosion resistant.

OSMIRIDIUM SISEROKITE
Manufacturer not listed
Os 57, Ru 8, 34 Rh + Ir.
For pen points.

OSMIUM
Atomergic Chemetals Corp.
Os.
Purities: 99.999%, 99.99%, 99.9%. Forms: sponge, powder, arc melted buttons.

OSNA CU58PB
KM-kabelmetal AG
Copper. Pb 1, P 0.005, bal Cu.
Free-machining copper with very good cold forming properties. Product forms: rods, profiles, tubes, wires. Annealed: 32 ksi TS min; 10 ksi YS min; 35 El min. Cold drawn: 40 ksi TS min; 36 ksi YS min; 7 El min.

OSNA CU58S
KM-kabelmetal AG
Copper. S 0.3, P 0.005, bal Cu.
Free-machining copper: current carrying components, bolts, nuts, studs, welding nozzles. Product forms: rods, profiles, tubes, wires. Annealed: 32 ksi TS min; 10 ksi YS min; 35 El min. Cold drawn: 40 ksi TS min; 36 ksi YS min; 7 El min.

OSNA CU58TE
KM-kabelmetal AG
Copper. Te 0.5, P 0.005, bal Cu.
Free-machining copper: current carrying components, bolts, nuts, studs, welding nozzles. Product forms: rods, profiles, tubes, wires. Annealed: 32 ksi TS min; 10 ksi YS min; 35 El min.Cold drawn: 40 ksi TS min; 36 ksi YS min; 7 El min.

OSNALIUM 3
KM-kabelmetal AG
Mg 2-4, Mn 0-0.4, Cr 0-0.3, bal Al.
Soft: 28,000 TS; 13,000 YS; 30 El; 47 Brin. Hard: 42,000 TS; 37,000 YS; 8 El; 77 Brin. For light alloy parts, marine hardware; resists sea water corrosion. *Obsolete*

OSNALIUM 5
KM-kabelmetal AG
Mg 4-5.5, Mn 0-0.8, Cr 0-0.3, bal Al.
Soft: 42,000 TS; 22,000 YS; 35 El; 65 Brin. Hard: 60,000 TS; 50,000 YS; 15 El; 100 Brin. For aircraft structures, light alloy parts; corrosion resistant. *Obsolete*

OSNALIUM 7
KM-kabelmetal AG
Mg 5.5-7.5, Mn 0-0.8, Cr 0-0.3, bal Al.
For light alloy parts; corrosion resistant. *Obsolete*

OSNISIL 1
KM-kabelmetal AG
Copper. Ni 1.3, Si 0.5, Cr 0.2, bal Cu.
Connection elements for overhead transmission lines, fasteners, mechanical components. Product forms: rods, profiles, tubes, wires. Heat treated: 85 ksi TS; 78 ksi YS; 10 El.

OSNISIL 2
KM-kabelmetal AG
Copper. Ni 1.9, Si 0.6, Cr 0.3, bal Cu.
Connection elements for overhead transmission lines, fasteners, bearings, mechanical components. Product forms: rods, profiles, tubes, wires, finished parts, strip. Heat treated: 93 ksi TS; 85 ksi YS; 10 El.

OSNISIL 3
KM-kabelmetal AG
Copper. Ni 3.5, Si 1, Cr 0.2, bal Cu.
Connection elements for overhead transmission lines, fasteners, bearings, mechanical components. Product forms: rods, profiles, tubes, wires, finished parts. Heat treated: 114 ksi TS; 107 ksi YS; 10 El.

OSNISIL D
KM-kabelmetal AG
Copper. Ni 2.5, Si 0.6, Cr 0.3, bal Cu.
Pistons for die casting machines, welding electrodes. Product forms: rods, profiles, finished parts. Heat treated: 96 ksi TS; 83 ksi YS; 10 El.

OSNISIL G
KM-kabelmetal AG
Copper. Ni 2.1, Si 0.7, Cr 0.3, bal Cu.
Dam and filler blocks for wire casting machines. Product forms: finished parts. Heat treated: 94 ksi TS; 76 ksi TS; 15 El.

OSSENBERG ESS, ETC.
Now ESS ETC.

OSTERMANN CAST BRONZE NO. 10
Ostermann GmbH & Co.
Copper. Cu 90, Sn 10.
29,000 TS; 15 El. For general construction of machinery fittings, apparatus; corrosion resistant.

OSTERMANN CAST BRONZE NO. 14
Ostermann GmbH & Co.
Copper. Cu 86, Sn 14.
29,000 TS; 3 El. For bearings, gears, high pressure hydraulic apparatus; wear resistant.

OSTERMANN CAST BRONZE NO. 20
Ostermann GmbH & Co.
Copper. Cu 80, Sn 20.
21,000 TS. For step-bearings, slide valves, wearing plates, bells; for parts with high friction loading.

OSTERMANN LEAD BRONZE NO. 10
Ostermann GmbH & Co.
Copper. Cu 86, Sn 10, Pb 4.
28,000 TS; 15 El. For bearings for hot rolling works and electrical machinery; heavy duty.

OSTERMANN LEAD TIN BRONZE NO. 8
Ostermann GmbH & Co.
Copper. Cu 80, Sn 8, Pb 12.
23,000 TS; 8 El. For bearings with high compressive loading and cold rolling mills; heavy duty.

OSTERMANN RED BRASS NO. 10
Ostermann GmbH & Co.
Copper. Cu 86, Sn 10, Zn 4.
29,000 TS; 10 El. For pipe lines and fittings, general machinery construction; also called "Machine Bronze."

OSTERMANN RED BRASS NO. 4
Ostermann GmbH & Co.
Copper. Cu 93, Sn 4, Zn 2, Pb 1.
29,100 TS; 25 El. For pipe flanges; known as "Flange Bronze."

OSTERMANN RED BRASS NO. 5
Ostermann GmbH & Co.
Copper. Cu 85, Sn 5, Zn 7, Pb 3.
21,000 TS; 10 El. For railroad machinery fittings to be bright finished or polished; free-cutting.

OSTERMANN RED BRASS NO. 8
Ostermann GmbH & Co.
Copper. Cu 82, Sn 8, Zn 7, Pb 3.
21,000 TS; 6 El. For machine fittings to be bright finished or polished; free-cutting.

OSTERMANN RED BRASS NO. 9
Ostermann GmbH & Co.
Copper. Cu 85, Sn 9, Zn 6.
29,000 TS; 12 El. For railroad bearings, fittings; heavy duty.

OSTERMANN WROUGHT BRONZE NO. 6
Ostermann GmbH & Co.
Copper. Cu 94, Sn 6.
For wire, sheet strip; tin bronze.

OT4
Russian manufacture
Al 3, Mn 1.5, bal Ti.
Alpha titanium alloy.

OT4-0
Russian manufacture
Al 1, Mn 1.5, bal Ti.
Alpha titanium alloy.

OT4-1
Russian manufacture
Al 2, Mn 1.5, bal Ti.
Alpha titanium alloy.

OT4-2
Russian manufacture
Al 6, Mn 1.5, bal Ti.
Titanium alloy.

OTISCOLOY
LTV Steel
C 0.08-0.12, Mn 0.9-1.25, Cu 0.35, Ni 0.1, Cr 0.05, bal Fe.
Drawn: 70,000 psi TS; 50,000 psi YS; 25 El; 55 Brin. For deep drawn parts.

OTISEL K-4 MOLYBDENUM
Otis Elevator Co.
C 0.13-0.35, Cr 5-7, bal Fe, Mo.
For corrosion resistant parts; corrosion resistant.

OTISEL K-5 TUNGSTEN
Otis Elevator Co.
C 0.13-0.35, Cr 5-7, W, bal Fe.
For corrosion resistant parts; corrosion resistant.

OTISEL K2
Otis Elevator Co.
C, bal Fe.
For machinery parts. *Obsolete*

OTISEL K3
Otis Elevator Co.
C, bal Fe.
For machinery parts. *Obsolete*

OTISEL K7
Otis Elevator Co.
C, bal Fe.
For machinery parts. *Obsolete*

OTISEL K8
Otis Elevator Co.
C, bal Fe.
For machinery parts. *Obsolete*

OTISEL K9
Otis Elevator Co.
C, bal Fe.
For machinery parts. *Obsolete*

OTISEL NO. 1
Otis Elevator Co.
Cr 18, Ni 8, bal Fe.
For heat and corrosion resistant parts; heat and corrosion resistant alloy.

OTISEL NO. 2
Otis Elevator Co.
C 0.1-0.15, Cr 12-16, bal Fe.
For corrosion resistant parts; corrosion resistant.

OTISEL NO. 3
Otis Elevator Co.
C 0.1-0.15, Cr 15-18, bal Fe.
For corrosion resistant parts; corrosion resistant.

OTISEL NO. 4
Otis Elevator Co.
C 0-0.2, Cr 20-30, Ni 12-20, bal Fe.
For heat and corrosion resistant parts; heat and corrosion resistant.

OTISEL NO. 5
Otis Elevator Co.
C 0-0.25, Cr 26-30, bal Fe.
For heat and corrosion resistant parts; heat and corrosion resistant.

OTISEL O-10
Otis Elevator Co.
C, alloy, bal Fe.
For castings. *Obsolete*

OTISEL O-12
Otis Elevator Co.
C 0.12-0.35, Cr 11-14, bal Fe.
Cast: 75,000-130,000 TS; 39,000-90,000 YS; 10-25 El; 140-150 Brin. For acid resistant castings; corrosion resistant.

OTISEL O-16
Otis Elevator Co.
C 0.12, Cr 14-18, bal Fe.
Cast: 70,000-80,000 TS; 39,000-43,000 YS; 15-20 El; 140-175 Brin. For acid resistant castings; corrosion resistant.

OTISEL O-18
Otis Elevator Co.
Stainless steel. C 0.08-0.2, Cr 18-20, Ni 8-10, bal Fe.
Cast: 78,000-85,000 TS; 39,000-42,000 YS; 60-65 El; 135-160 Brin. For acid resistant castings; stainless.

OTISEL O-2
Otis Elevator Co.
C, alloy, bal Fe.
For castings. *Obsolete*

OTISEL O-24
Otis Elevator Co.
C 0-0.2, Cr 22-25, Ni 10-13, bal Fe.
Cast: 80,000-85,000 TS; 40,000-43,000 YS; 45-50 El; 140-150 Brin. For heat resistant castings; heat resistant.

OTISEL O-3
Otis Elevator Co.
C, alloy, bal Fe.
For castings. *Obsolete*

OTISEL O-30
Otis Elevator Co.
C 0-0.25, Cr 25-30, Mn 0-1.3, bal Fe.
Cast: 45,000-99,000 TS; 35,000-60,000 YS; 2-3 El; 170-190 Brin. For heat resistant castings; resists heat to 2150°F.

OTISEL O-4
Otis Elevator Co.
C 0.15-0.35, Mo 0.25-2, Cr 4-8, bal Fe.
Cast: 100,000-110,000 TS; 65,000-80,000 YS; 18 El; 190-230 Brin. For castings in petroleum refineries; stainless. *Obsolete*

OTISEL O-5
Otis Elevator Co.
C, alloy, bal Fe.
For castings. *Obsolete*

OTISEL O-6
Otis Elevator Co.
C, alloy, bal Fe.
For castings. *Obsolete*

OTISEL O-7
Otis Elevator Co.
C, alloy, bal Fe.
For castings. *Obsolete*

OTISEL O-8
Otis Elevator Co.
C, alloy, bal Fe.
For castings. *Obsolete*

OTISEL O-9
Otis Elevator Co.
C, alloy, bal Fe.
For castings. *Obsolete*

OTISEL O-D
Otis Elevator Co.
C, alloy, bal Fe.
For castings. *Obsolete*

OTISEL O-H
Otis Elevator Co.
C, alloy, bal Fe.
For castings. *Obsolete*

OTISEL O-M
Otis Elevator Co.
C, alloy, bal Fe.
For castings. *Obsolete*

OTISEL O-S
Otis Elevator Co.
C, alloy, bal Fe.
For castings.

OTOTANI-ALLOY GRADE 2
Schulte Eisenhandlung GmbH
Si 35-50, Ca 20-30, Mn 5-15, Fe 10-20, Al 0-1.5, P 0-0.08, S 0-0.05.
For melting or ladle additions to produce cast steel with nodular graphite to increase the wear resistance, as on steel rolls for rolling mills.

OTOTANI-ALLOY GRADE 3
Schulte Eisenhandlung GmbH
Si 35-50, Ca 20-25, Fe 10-25, Mn 5-15, Al 4-10, C 0-0.3, S 0-0.05.
For melting or ladle additions to produce fine grain steel.

OTTAWA
Atlas Specialty Steels
C 0.8, Mn 0.2, Si 0.15, bal Fe.
For mining drills and rods; hollow drill steel.

OTTAWA 60
Allegheny Ludlum Steel
C 3.25, Cr 1, Mo 1, V 12, bal Fe.
For drawing rings for stainless steel; heat treatable, galling resistant. *Obsolete*

OTTO'S SPECULUM
English manufacture
Cu 69, Sn 31.
40,000 TS; 10,000 YS; 38 El. For telescope reflectors, mirrors.

OUNCE METAL
Belmont Metals Inc.
Cu 85, Sn 5, Zn 5, Pb 5.
27,000-33,000 psi TS; 15-20 El; 15-20 RA; 50-59 Brin. For bearing metal, casting valves, carburetor and pump parts; Composition Brass, also called "Std. Red Composition."

OUTOKUMPU AP105
Outokumpu Metals (USA) Inc.
Copper. Cu 93, Al 5, Ni 2.
Aluminum bronze for jewelry, coins, token blanks, medals. Density 8.20 kg/dm^3; electrical conductivity 15% IACS; thermal conductivity 80 W/m·K.

OUTOKUMPU HK003
Outokumpu Metals (USA) Inc.
Copper. O 0-0.001, 99.98 Cu min, 0.025-0.050 AgO.
Silver bearing high conductivity oxygen free copper for busbars, hollow conductors, windings, commutators, contacts, and switches, transistor and rectifier bases, induction coils, magnet tubes, thyristors. Density 8.94 kg/dm^3; electrical conductivity 100% IACS; thermal conductivity 395 W/m·K. C10400

OUTOKUMPU HK01
Outokumpu Metals (USA) Inc.
Copper. O 0-0.001, 99.98 Cu min, 0.035-0.050 AgO.
Silver bearing high conductivity oxygen free copper for busbars, hollow conductors, windings, commutators, contacts, and switches, transistor and rectifier bases, induction coils, magnet tubes, thyristors. Density 8.94 kg/dm^3; electrical conductivity 100% IACS; thermal conductivity 395 W/m·K. C10500

OUTOKUMPU HK015
Outokumpu Metals (USA) Inc.
Copper. O 0-0.001, 99.98 Cu min, 0.09-0.12 AgO.
Silver bearing high conductivity oxygen free copper for busbars, hollow conductors, windings, commutators, contacts, and switches, transistor and rectifier bases, induction coils, magnet tubes, thyristors. Annealed: 26,000-36,000 TS; 6000-12,000 YS; 20-44 El. Cold rolled: 34,000-60,000 TS; 30,000-58,000 YS; 1-33 El. Density 8.94 kg/dm^3; electrical conductivity 100% IACS; thermal conductivity 395 W/m·K. C10700

OUTOKUMPU HK025
Outokumpu Metals (USA) Inc.
Copper. O 0-0.001, 99.98 Cu min, 0.15-0.25 AgO.
Silver bearing high conductivity oxygen free copper for busbars, hollow conductors, windings, commutators, contacts, and switches, transistor and rectifier bases, induction coils, magnet tubes, thyristors. Density 8.94 kg/dm^3; electrical conductivity 99% IACS; thermal conductivity 390 W/m·K.

OUTOKUMPU K1
Outokumpu Metals (USA) Inc.
Copper. P 0.005-0.012, 99.90 Cu min.
Common copper, deep drawing copper for stills, storage tanks, cisterns, cylinders, air conditioners, kitchenware, automotive radiators, roofing, gutters, flashings, claddings, facades. Annealed: 26,000-36,000 TS; 6000-12,000 YS; 20-44 El. Cold rolled: 34,000-60,000 TS; 30,000-58,000 YS; 1-33 El. Density 8.94 kg/dm^3; electrical conductivity 96% IACS; thermal conductivity 380 W/m·K. C12000

OUTOKUMPU K1E
Outokumpu Metals (USA) Inc.
Copper. O 0-0.001, 99.98 Cu min.
High conductivity oxygen free copper for electrical assemblies and severely formed parts, electron beam welding, resists hydrogen embrittlement. Annealed: 26,000-36,000 TS; 6000-12,000 YS; 20-44 El. Cold rolled: 34,000-60,000 TS; 30,000-58,000 YS; 1-33 El. Density 8.94 kg/dm^3; electrical conductivity 100% IACS min; thermal conductivity 395 W/m·K. C10200

OUTOKUMPU K1S
Outokumpu Metals (USA) Inc.
Copper. O 0-0.0005, 99.99 Cu min.
High conductivity oxygen free copper for electrical assemblies and severely formed parts, high vacuum applications, electron beam welding, resists hydrogen embrittlement. Annealed: 26,000-36,000 TS; 6000-12,000 YS; 20-44 El. Cold rolled: 34,000-60,000 TS; 30,000-58,000 YS; 1-33 El. Density 8.94 kg/dm^3; electrical conductivity 101% IACS min; thermal conductivity 399 W/m·K. C10100

OUTOKUMPU K2

Outokumpu Metals (USA) Inc.
Copper. P 0.015-0.04, 99.90 Cu min.
For deep drawn copper shells, brazed or welded assemblies, tubes for air conditioning refrigeration, water, gas and sanitation. Annealed: 26,000-36,000 TS; 6000-12,000 YS; 20-44 El. Cold rolled: 34,000-60,000 TS; 30,000-58,000 YS; 1-33 El. Density 8.94 kg/dm^3; electrical conductivity 85% IACS; thermal conductivity 335 W/m·K. C12200

OUTOKUMPU K4

Outokumpu Metals (USA) Inc.
Copper. O 0.02-0.06, 99.90 Cu min.
High conductivity copper for electrical applications. Density 8.90 kg/dm^3; electrical conductivity 100% IACS min; thermal conductivity 395 W/m·K. C11000

OUTOKUMPU KRK101

Outokumpu Metals (USA) Inc.
Copper. Cu 99, Cr 0.9, Zr 0.1.
Chrome-zirconium copper for resistance welding electrodes, heat treatable, high strength and conductivity at high temperatures. Aged: 72,000 TS; 65,000 YS; 15 El. Density 8.94 kg/dm^3; electrical conductivity 75% IACS min; thermal conductivity 320 W/m·K. C18100

OUTOKUMPU MS63

Outokumpu Metals (USA) Inc.
Copper. Cu 63, Zn 37.
Brass for cold pressworking and heading, pins, rivets, nuts, springs. Density 8.45 kg/dm^3; electrical conductivity 26% IACS; thermal conductivity 115 W/m·K. C27000

OUTOKUMPU MS70

Outokumpu Metals (USA) Inc.
Copper. Cu 70, Zn 30.
Deep drawing brass, castridge brass for components, cartridge cases and primer caps. Density 8.55 kg/dm^3; electrical conductivity 28% IACS; thermal conductivity 120 W/m·K. C26000

OUTOKUMPU MS80

Outokumpu Metals (USA) Inc.
Copper. Cu 80, Zn 20.
Brass for facades, outdoor claddings, panels, formed angles and channels, wire cloths, screens, zip fasteners, pins, rivets and screws. Density 8.65 kg/dm^3; electrical conductivity 32% IACS; thermal conductivity 140 W/m·K. C24000

OUTOKUMPU MS85

Outokumpu Metals (USA) Inc.
Copper. Cu 85, Zn 15.
Red brass resistant to dezincification and stress corrosion for decorative items, jewelry, pins, rivets and screws, fasteners, screens, zip fasteners. Density 8.75 kg/dm^3; electrical conductivity 37% IACS; thermal conductivity 160 W/m·K. C23000

OUTOKUMPU MS90

Outokumpu Metals (USA) Inc.
Copper. Cu 90, Zn 10.
Brass resistant to dezincification and stress corrosion for decorative items, coins, jewelry, medals, cartridge cases and primer caps, bullet jackets. Density 8.80 kg/dm^3; electrical conductivity 44% IACS; thermal conductivity 190 W/m·K. C22000

OUTOKUMPU NK110

Outokumpu Metals (USA) Inc.
Copper. Ni 10, Fe 1.5, Mn 0.7, bal Cu.
Copper nickel for seawater tubes, PHE plates, brake tubes, wire meshes, welding rods. Density 8.90 kg/dm^3; electrical conductivity 9% IACS; thermal conductivity 45 W/m·K. C70600

OUTOKUMPU NK25

Outokumpu Metals (USA) Inc.
Copper. Ni 25, Mn 0.3, bal Cu.
Copper nickel for coins, token blanks, medals. Density 8.90 kg/dm^3; electrical conductivity 5% IACS; thermal conductivity 25 W/m·K. C71300

OUTOKUMPU NO. 2

Outokumpu Metals (USA) Inc.
Zinc. Al 3.9-4.3, Cu 2.5-3.5, Mg 0.03-0.06, Fe 0-0.05, Pb 0-0.003, Cd 0-0.003, Sn 0-0.001, bal Zn.
Zinc die casting alloy equivalent to BS 1004; ASTM B-86.

OUTOKUMPU NO. 3

Outokumpu Metals (USA) Inc.
Zinc. Al 3.9-4.3, Mg 0.03-0.06, Fe 0-0.03, Pb 0-0.003, Cd 0-0.003, Cu 0-0.03, Sn 0-0.001, bal Zn.
Zinc die casting alloy equivalent to BS 1004; ASTM B-86.

OUTOKUMPU NO. 5

Outokumpu Metals (USA) Inc.
Zinc. Al 3.9-4.3, Cu 0.5-1.25, Mg 0.03-0.06, Fe 0-0.03, Pb 0-0.003, Cd 0-0.003, Sn 0-0.001, bal Zn.
Zinc die casting alloy equivalent to BS 1004; ASTM B-86.

OUTOKUMPU PP102

Outokumpu Metals (USA) Inc.
Copper. Si 2, Mn 0.3, bal Cu.
Weldable silicon bronze for bolts, nails, rivets, screws and fasteners especially for outdoor and underwater construction, welding rods. Density 8.70 kg/dm^3; electrical conductivity 12% IACS; thermal conductivity 70 W/m·K. C65100

OUTOKUMPU PP103

Outokumpu Metals (USA) Inc.
Copper. Si 3, Mn 1, bal Cu.
Weldable silicon bronze for bolts, nails, rivets, screws and fasteners especially for outdoor and underwater construction, welding rods. Density 8.50 kg/dm^3; electrical conductivity 7% IACS; thermal conductivity 35 W/m·K. C65500

OUTOKUMPU TEK06

Outokumpu Metals (USA) Inc.
Copper. Cu 99.4, Te 0.4-0.7, P 0.01.
Tellerium copper for screw machine products, welding tips and nozzles, high conductivity, free machining, resists hydrogen embrittlement. Annealed rod: 32,000 TS; 10,000 YS; 45 El. Hard rod: 48,000 TS; 40,000 YS; 12 El. Density 8.90 kg/dm^3; electrical conductivity 90% IACS; thermal conductivity 370 W/m·K. C14500

OUTOKUMPU TP105

Outokumpu Metals (USA) Inc.
Copper. Cu 95, Sn 5.
Tin bronze for electrical contacts, springs, fasteners, cold headed screws, rivets and bolts, welding rods, wire cloths. Density 8.80 kg/dm^3; electrical conductivity 15% IACS; thermal conductivity 80 W/m·K. C51000

OUTOKUMPU TP107

Outokumpu Metals (USA) Inc.
Copper. Cu 94, Sn 6.
Tin bronze for electrical contacts, springs, fasteners, cold headed screws, rivets and bolts, welding rods, wire cloths. Density 8.80 kg/dm^3; electrical conductivity 12% IACS; thermal conductivity 70 W/m·K. C51900

OUTOKUMPU TP108

Outokumpu Metals (USA) Inc.
Copper. Cu 92, Sn 8.
Tin bronze for electrical contacts, springs, fasteners, cold headed screws, rivets and bolts, welding rods, wire cloths. Density 8.80 kg/dm^3; electrical conductivity 12% IACS; thermal conductivity 70 W/m·K. C52100

OUTOKUMPU UH112

Outokumpu Metals (USA) Inc.
Copper. Cu 64, Ni 12, Mn 0.3, bal Zn.
Nickel silver for springs, clips, rivets, instrument and camera parts, spectacle frames, optical parts, jewelry, zip fasteners. Density 8.65 kg/dm^3; electrical conductivity 8% IACS; thermal conductivity 40 W/m·K. C75700

OUTOKUMPU UH118

Outokumpu Metals (USA) Inc.
Copper. Cu 62, Ni 18, Mn 0.4, bal Zn.
Nickel silver for springs, clips, rivets, instrument and camera parts, spectacle frames, optical parts, jewelry, zip fasteners. Density 8.75 kg/dm^3; electrical conductivity 6% IACS; thermal conductivity 35 W/m·K. C75900

OUTOKUMPU ZRK015

Outokumpu Metals (USA) Inc.
Copper. Cu 99.8, Zr 0.15.
Zirconium copper for resistance welding electrodes, heat treatable, high strength and conductivity at high temperatures. Annealed: 30,000 TS; 10,000 YS; 45 El. Aged: 60,000 TS; 55,000 YS; 15 El. Density 8.94 kg/dm^3; electrical conductivity 92% IACS; thermal conductivity 365 W/m·K. C15000

OV

W. Ossenberg & Cie Edelstahlwerke
C 0.5, Mn 1, Si 0.25, Cr 1, V 0.1, bal Fe.
Cold work tool steel for shear knives, hand tools, shanks for carbide tipped lathe tools. W.-Nr. 1.2241.

OVAKO 047

Ovako Steel Hellefors AB
Quenched and tempered steel. C 0.42-0.5, Si 0.1-0.4, Mn 0.5-0.8, bal Fe.
Oil or water quenched. Flame or induction hardened.

OVAKO 145

Ovako Steel Hellefors AB
Case hardened steel. C 0.13-0.18, Si 0.15-0.4, Mn 0.7-1.1, Cr 0.6-1, Ni 0.8-1.2, bal Fe.
Oil hardened. For transmission components.

OVAKO 146

Ovako Steel Hellefors AB
Case hardened steel. C 0.15-0.23, Si 0.15-0.4, Mn 0.7-1.1, Cr 0.6-1, Ni 0.8-1.2, bal Fe.
Oil hardened. For transmission components.

OVAKO 152

Ovako Steel Hellefors AB
Case hardened steel. C 0.17-0.23, Si 0.15-0.4, Mn 0.6-0.95, Cr 0.35-0.65, Ni 0.35-0.75, Mo 0.15-0.25, bal Fe.
For transmission components.

OVAKO 214

Ovako Steel Hellefors AB
Structural steel. C 0-0.2, Si 0-0.5, Mn 1-1.6, bal Fe.
Fine-grained treated steel with good weldability.

OVAKO 234

Ovako Steel Hellefors AB
Case hardened steel. C 0.16-0.2, Si 0.15-0.25, Mn 1-1.3, Cr 0.8-1.1, bal Fe.
High resistance to wear and high core toughness.

OVAKO 253

Ovako Steel Hellefors AB
Case hardened steel. C 0.17-0.21, Si 0.2-0.3, Mn 0.3-0.5, Cr 1.2-1.4, Ni 2.5-2.9, Mo 0.2-0.26, bal Fe.
High-hardenability case-hardened steel. For high demands.

OVAKO 264

Ovako Steel Hellefors AB
Chain steel. C 0.18-0.23, Si 0.1-0.3, Mn 0.9-1.1, Cr 0.4-0.5, Ni 0.4-0.5, bal Fe.
Boron steel. For chains; grade 8.

OVAKO 280

Ovako Steel Hellefors AB
Structural steel. C 0.16-0.2, Si 0.25-0.45, Mn 1.4-1.6, V 0.07-0.1, bal Fe.
High-strength micro-alloyed steel. For shafts and pistons.

OVAKO 327

Ovako Steel Hellefors AB
Quenched and tempered steel. C 0.38-0.45, Si 0.1-0.4, Mn
0.6-0.9, Cr 0.9-1.2, Mo 0.15-0.3, bal Fe.
Oil quenched, flame or induction hardened. For shafts and
gears.

OVAKO 356

Ovako Steel Hellefors AB
Quenched and tempered steel. C 0.32-0.39, Si 0.1-0.4, Mn
0.5-0.8, Cr 1.3-1.7, Ni 1.3-1.7, Mo 0.15-0.3, bal Fe.
Oil quenched, flame or induction hardened. For heavy shafts
and gears.

OVAKO 382

Ovako Steel Hellefors AB
Structural steel. C 0.42-0.46, Si 0.15-0.4, Mn 0.7-1, V 0.07-0.1,
bal Fe.
High-strength micro-alloyed steel. For forged components.

OVAKO 482

Ovako Steel Hellefors AB
Structural steel. C 0.36-0.42, Si 0.3-0.5, Mn 1.1-1.4, V
0.11-0.16, bal Fe.
High-strength micro-alloyed steel. For automotive
components, shafts and oil drilling equipment.

OVAKO 483

Ovako Steel Hellefors AB
Chain steel. C 0.27-0.31, Si 0.15-0.35, Mn 1.65-1.9, Cr
0.1-0.3, Ni 0.05-0.25, V 0.06-0.12, bal Fe.
For chains; grade 3.

OVAKO 528

Ovako Steel Hellefors AB
Quenched and tempered steel. C 0.5-0.55, Si 0.15-0.4, Mn
0.6-0.9, Cr 0.9-1.2, Mo 0.15-0.25, bal Fe.
Oil quenched, induction hardened. For heavy driveshafts.

OVAKO 803

Ovako Steel Hellefors AB
Rolling bearing steel. C 0.98-1.1, Si 0.2-0.35, Mn 0.25-0.4, Cr
1.35-1.65, bal Fe.
Rolling bearings; SKF 3.

OVAKO 824

Ovako Steel Hellefors AB
Rolling bearing steel. C 0.92-1.02, Si 0.25-0.4, Mn 0.25-0.4,
Cr 1.65-1.95, Mo 0.15-0.25, bal Fe.
SKF 24.

OVAKO 825

Ovako Steel Hellefors AB
Rolling bearing steel. C 0.92-1.02, Si 0.25-0.4, Mn 0.25-0.4,
Cr 1.65-1.95, Mo 0.3-0.4, bal Fe.
SKF 25.

OVAKO 831

Ovako Steel Hellefors AB
Rolling bearing steel. C 0.92-1.02, Si 0.5-0.7, Mn 1-1.2, Cr
0.95-1.15, bal Fe.
ASTM A485-1.

OVAKO 832

Ovako Steel Hellefors AB
Rolling bearing steel. C 0.87-0.97, Si 0.6-0.8, Mn 1.5-1.7, Cr
1.4-1.7, bal Fe.
ASTM A485 2.

OVAKO/SKF 214

Eagle & Globe Steel Ltd.
Alloy steel. C 0.18, Si 0.3, Mn 1.5, 0.47 C equivalent (approx),
bal Fe.
Cold rolled OVAKO/SKF 214 (SKF 75), hot rolled
OVAKO/SKF 214: 330-740 MPa YS; 490-760 MPa TS; 10-30
El. BS150M19 EN14A; SS 2172; DIN St52-3; AFNOR 20M5;
SAE 1518H; API H40.

OVAKO/SKF 280

Eagle & Globe Steel Ltd.
Alloy steel. C 0.18, Si 0.3, Mn 1.5, V 0.08, 0.49 C equivalent
(approx), bal Fe.
Hot rolled OVAKO/SKF 280: 400-470 MPa YS; 580-650 MPa
TS; 20-25 El. SS 2142; DIN StE460; AFNOR 20MV6; API
J+K55.

OVH

W. Ossenberg & Cie Edelstahlwerke
C 0.58, Mn 1, Si 0.25, Cr 1, V 0.1, bal Fe.
Cold work stamps, punches, split chucks. W.-Nr. 1.2242.

OXALLOY-28

GTE Products Corp.
99.92 Cu min.
For high temperature wire. Heat resistant to 400°F.

OXWELD NO. 1 HT

Union Carbide Corp.
O, alloy, bal Fe.
Welded: 70,000 TS. For welding rod for alloy steel.

OXWELD NO. 19 CUPRO

Union Carbide Corp.
Cu, alloy.
Welded: 35,000 TS. For bronze welding rod; phosphor
bronze.

OXWELD NO. 2

Union Carbide Corp.(Linde Div.)
C, bal Fe.
For welding rod for high carbon steel. *Obsolete*

OXWELD NO. 21 H.S.

Union Carbide Corp.(Linde Div.)
Cu alloy.
For bronze welding rod. *Obsolete*

OXWELD NO. 23

Union Carbide Corp.
Si 5, bal Al.
For Al welding rod.

OXWELD NO. 25 M

Union Carbide Corp.
Sn, Zn, bal Cu.
Welded: 50,000 TS; 96 Brin. For bronze welding rod.

OXWELD NO. 26

Union Carbide Corp.
Si 2.5, Si-Cu alloy, bal Cu.
For bronze welding rod.

OXWELD NO. 28

Union Carbide Corp.
C 0.08, Cr 18, Ni 8, Cb, bal Fe.
For stainless steel welding rod.

OXWELD NO. 31 T

Union Carbide Corp.
Sn, Zn, bal Cu.
For bronze welding rod.

OXWELD NO. 32 CMS

Union Carbide Corp.
C, bal Fe.
Welded: 98,000 TS. For welding rod.

OXWELD NO. 38

Union Carbide Corp.
Cu alloy.
For Cu welding rod.

OXWELD NO. 7

Union Carbide Corp.
C, bal Fe.
Welded: 45,000 TS. For welding rod for steel; copper coated.

OXWELD NO. 9

Union Carbide Corp.
C 3.2, Si 2.2, bal Fe.
For cast iron welding rod; for gray cast iron.

OXYGEN FREE COPPER

Chase Brass & Copper Co., Inc.
Copper. Cu 99.98.
Annealed: 32,000 psi TS; 10,000 psi YS; 45 El. Rolled: 50,000
psi TS; 45,000 psi YS; 6 El. For electrical conductors; high
conductivity.

P & A ENAMELING
Empire Sheet & Tin Plate Co.
C 0.2, bal Fe.
For appliances. *Obsolete*

P & H 70 LA-2
Chemetron Corp.
C 0-0.12, Mn 0.85-1.2, bal Fe.
Welded: 79,000 TS; 69,000 YS; 31 El; 65 RA. For welding rods for high C and alloy steel; low H 2 type. *Obsolete*

P & H 75 LP
Chemetron Corp.
C 0-0.12, Ni 2.5, bal Fe.
Welded: 85,000 TS; 75,000 YS; 26 El; 43 RA. For welding rods for low temperature applications; high impact at low temperature. *Obsolete*

P & H 80 LE
Chemetron Corp.
C 0-0.1, Cr 0.8-1.1, Mo 0.4-0.6, bal Fe.
Relieved: 80,000 TS; 67,000 YS; 27 El. Welded: 86,000 TS; 77,000 YS; 24 El; 41 RA. For welding rods for Cr-Mo steel; high temperature-pressure power piping. *Obsolete*

P & H AW-2B
Chemetron Corp.
C 0.4, bal Fe.
Welded: 90,000-100,000 TS; 22-28 El; 220-270 Brin. For welding electrodes. *Obsolete*

P & H NO. 21
Chemetron Corp.
C 0.14-0.2, Cr 1-1.3, Ni 1.75, Mn 0.75-1, Mo 0.8, V 0.2, bal Fe.
Welded: 125,000-143,000 TS; 80,000-125,000 YS; 17-22 El; 29 RA. For welding rods for aircraft steels; for heat treating applications. *Obsolete*

P ALLOY
English manufacture
Fe 0.11, Si 0.11, bal Al.
8,000 TS; 34 El. For light alloy parts.

P H S
Edgar Allen Balfour Ltd.
C 0.1, Si 0.1, Mn 0.4, bal Fe.
For plastic mold dies; hobbing steel.

P H VAN
Pennsylvania Steel Corp.
C 0.35, Si 1, Cr 5, V 1, Mo 1.5, bal Fe.
Annealed: 98,000 TS; 74,000 YS; 28 El; 210 Brin. Hardened: 135,000-290,000 TS; 100,000-228,000 YS; 3-16 El; 7-48 RA; 27-55 Rock C. For forging and heading dies, compression tools, casting dies, hot piercing and forming punches, bolt dies, swaging dies. Type H13 hot work steel, red-tough, shock and impact resistant.

P N BRONZE
Langley Alloys Ltd.
Ni 3-4, P 0.05-0.3, Sn 11.5-13, bal Cu.
Cast: 31,000-45,000 TS; 5-20 El; 70-100 Brin. For pump and valve components. Good for high pressure work.

P-2
GTE Valenite Corp.
Multi-phase coating. For cast iron and non-ferrous machining. General purpose. Code C-2/C-3/C-4.

P-47
GTE Valenite Corp.
Multi-phase coating. For medium roughing to high speed finishing applications on alloyed steels where crater and wear resistance is required. Code C-6/C-7/C-8.

P-5
GTE Valenite Corp.
Multi-phase coating. General purpose steel cutting grade. Used in a wide variety of applications. Code C-5/C-6/C-7.

P-50 NICKEL
International Nickel Inc.
Nickel. Cu 0-0.04, Mn 0-0.02, Fe 0-0.05, C 0-0.05, Mg 0-0.01, max Si, bal Ni + Co.
For cathodes in electron tubes; high purity nickel.

P-52
GTE Valenite Corp.
Multi-phase coating. General purpose steel cutting grade. Code C-5/C-6/C-7.

P-54
GTE Valenite Corp.
Multi-phase coating. For medium to heavy roughing of steel forgings and castings. Code C-5/C-6/C-7.

P-57
GTE Valenite Corp.
Multi-phase coating. General purpose coated grade for heavy stock removal of alloyed steel rolls and parts where long cuts generate a lot heat in the cutting edge. Code C-5/C-6/C-7.

P-D TUNGSTEN
Delsteel Inc.
W 9.5, C, alloy, bal Fe.
For hot work and punch dies; high speed steel for maximum red-hardness.

P-TIN ALLOY
Capper Pass & Son Ltd.
Sn 55.5, Pb 41.1, Sb 3.4.
For soldering. Soft solder.

P.1000
Now CARR'S QUALITY P.1000.

P.A.C.
Eagle & Globe Steel Ltd.
Tool material. C 0.43, Cr 4.25, Mo 0.35, W 4.25, V 2.2, Co 4.25, bal Fe.
Hot work steel for brass pressing and die casting; die cavities, pegs, cores. AS1239 H19A; AISI H19.

P.B.
Patriarche & Bell
C 0.7-0.9, bal Fe.
For tools; water hardening.

P.E.R 1
Aubert & Duval
Nickel. C 0.1, Si 0-0.3, Mn 0-0.3, Cr 20, Co 0-5, Ti 0.3, Cu 0-0.5, Fe 0-5, bal Ni.
Wrought: 740 MPa TS; 280 MPa YS. Equivalent to Nimonic 75.

P.E.R 10
Aubert & Duval
Nickel. C 0.05, Si 0-0.5, Mn 0-0.5, Cr 20, Mo 6, W 2.8, Co 0-2, Fe 0-2, Zr 0-0.2, 6.50 Nb + Ta, bal Ni.
Cast. Equivalent to P.E. 10.

P.E.R 13 B.C
Aubert & Duval
Nickel. C 0.05, Si 0-0.2, Mn 0-0.2, Cr 12.5, Mo 4.2, Al 6, Ti 0.9, Fe 0-2, Zr 0.1, 2.40 Nb + Ta, bal Ni.
Cast. Equivalent to Inconel 713 L.C.

P.E.R 15
Aubert & Duval
Nickel. C 0.02, Si 0-0.2, Mn 0-0.2, Cr 12, Mo 1.8, W 4.5, Al 3.5, Co 8.5, Ti 4, Fe 0-1, B 0.15, Zr 0.1, 4.0 Nb + Ta, bal Ni.
Cast. Equivalent to B 1925.

P.E.R 16
Aubert & Duval
Nickel. C 0.12, Si 0-0.2, Mn 0-0.2, Cr 6, Mo 2, W 11, Al 6, Co 0-1, Fe 0-1, Zr 0.13, 1.60 Nb + Ta, bal Ni.
Cast. Equivalent to E.P.D. 16.

P.E.R 263
Aubert & Duval
Nickel. C 0.05, Si 0-0.4, Mn 0-0.5, Cr 19.5, Mo 5.8, Al 0.4, Co 20, Ti 2.2, Cu 0-0.2, Fe 0-0.7, 2.60 Ti + Al, Zr, bal Ni.
Wrought: 970 MPa TS; 580 MPa YS. Equivalent to Alloy C 263.

P.E.R 263 F
Aubert & Duval
Nickel. C 0.06, Si 0-0.3, Mn 0-0.2, Cr 20, Mo 5.8, Al 0.5, Co 20, Ti 2.2, Fe 0-0.5, Zr, bal Ni.
Cast. Equivalent to Alloy C 263.

P.E.R 59
Aubert & Duval
Nickel. C 0.08, Si 0-0.2, Mn 0-0.2, Cr 15, Mo 4.5, Al 4.5, Co 22, Ti 5, Cu 0-0.1, Fe 0-1, bal Ni.
Cast.

P.E.R 7
Aubert & Duval
Nickel. C 0.06, Si 0-0.2, Mn 0-0.2, Cr 14.5, Mo 5, Al 4, Co 19, Ti 3, Cu 0-0.2, Fe 0-1, Zr 0.05, bal Ni.
Wrought: 1410 MPa TS; 970 MPa YS. Equivalent to Udimet 700.

P.E.R 7 H
Aubert & Duval
Nickel. C 0.07, Si 0-0.2, Mn 0-0.2, Cr 18, Mo 3, W 1.5, Al 2.5, Co 15, Ti 5, Cu 0-0.2, Fe 0-1, Zr 0.05, bal Ni.
Wrought: 1180 MPa TS; 910 MPa YS. Equivalent to Udimet 710.

P.E.R 72
Aubert & Duval
Nickel. C 0.04, Si 0-0.2, Mn 0-0.2, Cr 18, Mo 3, W 1.2, Al 2.5, Co 15, Ti 5, Cu 0-0.2, Fe 0-0.7, Zr 0.03, bal Ni.
Wrought: 1530 MPa TS; 1220 MPa YS. Equivalent to Udimet 720.

P.E.R C 1023
Aubert & Duval
Nickel. C 0.15, Si 0-0.2, Mn 0-0.2, Cr 15.5, Mo 8.5, Al 4.1, Co 10, Ti 3.5, Fe 0-0.5, bal Ni.
Cast. Equivalent to C 1023.

P.E.R C 130
Aubert & Duval
Nickel. C 0.08, Si 0-0.6, Mn 0-0.6, Cr 21.5, Mo 10, Al 0.8, Co 0-1, Ti 2.6, Fe 0-0.5, bal Ni.
Cast. Equivalent to C 130.

P.H.W.
Pennsylvania Steel Corp.
C 0.35, Si 1.05, W 1.55, Mo 1.65, Cr 5.15, bal Fe.
Hardened: 216,000 TS; 185,000 YS; 14 El; 53 RA. At 1000°F: 144,000 TS; 110,000 YS; 19 El; 64 RA. For die casting dies, hot shear blades, forging and heading dies, punches. Type H12 hot work tool steel, tough and shock resistant.

P.M.G. ALLOY
British Steel Corp.
Cu 95.5, Si 3, Fe 1.5.
Sand cast: 53,000 TS; 22,000 YS; 30 El; 100 Brin. Chill cast: 53,000 TS; 22,000 YS; 30 El; 150 Brin. For pump parts, gears, shafts, trolley wheels, marine fittings; substitute for Admiralty Gun Metal. *Obsolete*

P.M.G. ALLOY
Phelps Dodge Industries
Cu 91.5, Si 3.5, Fe 1.5, Zn 4.
Forged: 70,000 TS; 28,000 YS; 35 El; 140 Brin. For spindles, stems, gears, shafts; wear resistant. *Obsolete*

P.M.G. ALLOY
Phelps Dodge Industries
Cu 95.5, Si 3, Fe 1.5-1.8, bal Zn.
Cold drawn: 95,000 TS; 45,000 YS; 30 El; 150 Brin. For nuts, bolts, screws; valve stems, pump spindles, shafting; corrosion resistant. *Obsolete*

P.M.G. ALLOY

Phelps Dodge Industries
Cu 95.5, Si 3, Fe 1.5.
Sand cast: 53,000 TS; 22,000 YS; 30 El; 100 Brin. Chill cast: 53,000 TS; 22,000 YS; 30 El; 150 Brin. For pump parts, gears, shafts, trolley wheels, marine fittings; substitute for Admiralty Gun Metal. *Obsolete*

P.M.G. NO. 1

Phelps Dodge Industries
Si 1-3, bal Cu.
Cast: 53,000 TS; 18,000 YS; 15 El; 115 Brin. For tanks, pressure vessels, fire extinguishers; corrosion resistant. *Obsolete*

P.M.G. NO. 13

Phelps Dodge Industries
Si 1-3, bal Cu.
Cast: 50,000 TS; 23,000 YS; 10 El; 116 Brin. For pressure vessels, tanks, marine construction; corrosion resistant. *Obsolete*

P.M.G. NO. 14

Phelps Dodge Industries
Si 1-3, bal Cu.
Cast: 45,000 TS; 40,000 YS; 1 El; 150 Brin. For marine hardware, pressure vessels, fasteners; corrosion resistant. *Obsolete*

P.M.G. NO. 4

Phelps Dodge Industries
Si 1-3, bal Cu.
Cast: 50,000 TS; 16,000 YS; 30 El; 90 Brin. For marine hardware, tanks, pressure vessels; corrosion resistant. *Obsolete*

P.M.G. NO. 6

Phelps Dodge Industries
Si 1-3, bal Cu.
Cast: 40,000 TS; 13,000 YS; 50 El; 80 Brin. For fire extinguishers, marine hardware; corrosion resistant. *Obsolete*

P.M.G., GRADE 30

Phelps Dodge Industries
Cu 62, Si 0.8, Fe 0.4, Zn 37.8.
Cold drawn: 75,000-95,000 TS; 30,000-65,000 YS; 20-30 El; 25-50 RA; 140-200 Brin. For valve stems, pump spindles, nuts, bolts, screws, shafts. *Obsolete*

P.M.G., GRADE 77

Phelps Dodge Industries
Cu 77, Si 2.8, Fe 0.4, Zn 19.8.
Cold drawn: 80,000-85,000 TS; 40,000-45,000 YS; 30-45 El; 45-50 RA; 150-160 Brin. For valve stems, pump spindles, shafting. *Obsolete*

P.M.G., GRADE 8

Phelps Dodge Industries
Cu 95.5, Si 3, Fe 1.5.
Cast: 50,000-55,000 TS; 28,000-32,000 YS; 20-30 El; 15-20 RA; 95-105 Brin. For marine parts, valves, pump bodies and liners, tail shaft liners. *Obsolete*

P.R.K. (POTTS)

Horace T. Potts Co.
C 1.5, Cr 13, bal Fe.
For tools, dies, shears. *Obsolete*

P.X.D.

Latrobe Steel Co.
C, W, Ni, Mo, bal Fe.
For piercing points, hot forging dies; for hot working. *Obsolete*

P07

Keystone Carbon Co.
Fe 50, Ni 50.
6.8 density; coercive force 1.2-2.5 oersteds. Remanence: 5160-8500 Gausses. Soft magnetic material for corrosion resistant cores and solenoids.

P10

Pelton Costeel, Inc.
Alloy steel. C 0.25-0.35, Mn 0-1, Si 0.3-0.6, P 0-0.05, S 0-0.05, Cr 0.4-0.7, Ni 0.4-0.7, Mo 0.15-0.25, bal Fe.
Weldable chromium, nickel, molybdenum steel for lightweight high strength applications such as gear blanks, spindles, planetary gears, housings, and ground engaging tools. Normalized and tempered: 60,000 YS; 90,000 TS; 18 El; 179-223 Brin. Quenched and tempered: 85,000-125,000 YS; 105,000-150,000 TS; 9-17 El; 212-363 Brin.

P11

Pelton Costeel, Inc.
Alloy steel. C 0.31-0.41, Mn 0-1, Si 0.3-0.6, P 0-0.05, S 0-0.05, Cr 0.4-0.7, Ni 0.4-0.7, Mo 0.15-0.25, bal Fe.
High strength carbon, chromium, nickel, molybdenum steel for applications requiring toughness and flame or induction hardening. Quenched and tempered: 95,000-145,000 YS; 120,000-175,000 TS; 6-14 El; 241-375 Brin.

P12

Pelton Costeel, Inc.
Carbon steel. C 0-0.12, Mn 0-0.2, Si 0-0.6, P 0-0.04, S 0-0.05, bal Fe.
Casting alloy for rotors, stators and magnet bodies. Normalized: 27,000 YS min; 48,000 TS min; 27 El min; 163 Brin max. Low electrical resistivity, high magnetic saturation, and low magnetic hysterisis loss.

P15

Pelton Costeel, Inc.
Carbon steel. C 0.1-0.15, Mn 0.6-0.9, Si 0.7-1, P 0-0.05, S 0-0.05, 0.20 Al min, bal Fe.
Casting alloy for electrical brake mechanisms, electrical switch gear, and magnet bodies. Normalized: 35,000 YS min; 56,000 TS min; 30 El min; 131-163 Brin. Low electrical resistivity, high magnetic properties, and weldable.

P2

Styria-Stahl Steirische Gusstahlwerke AG
C 0.8-1, bal Fe.
For tools, dies, fixtures; water hardened. *Obsolete*

P2 ALLOY

Birmingham Aluminum Casting Co.
Aluminum. Cu 3-4.5, Si 4-5, Fe 2-4, Ni 1.75-2.5, bal Al.
Cast: 25,000 TS; 23,000 YS; 2 El; 101 Brin. For carburetors, brake shoes, cameras, electric meters; pressure die casting.

P20

Pelton Costeel, Inc.
Carbon steel. C 0.1-0.2, Mn 0-0.75, Si 0.3-0.6, P 0-0.05, S 0-0.05, bal Fe.
Ductile, weldable casting alloy for construction equipment components, and valves. Normalized: 30,000 YS min; 60,000 TS min; 24 El min; 131-170 Brin.

P25

Pelton Costeel, Inc.
Carbon steel. C 0.21-0.28, Mn 0.6-0.9, Si 0.3-0.6, P 0-0.05, S 0-0.05, Cr (residual), Ni (residual), and Mo (residual), bal Fe.
Ductile, weldable casting alloy for levers, gear boxes, brackets, and hydraulic cylinders. Normalized: 36,000 YS min; 70,000 TS min; 22 El min; 140-179 Brin.

P3

Styria-Stahl Steirische Gusstahlwerke AG
C 1.05-1.15, bal Fe.
For tools, drills, taps; water hardened. *Obsolete*

P35

Pelton Costeel, Inc.
Carbon steel. C 0.31-0.38, Mn 0.7-1, Si 0.3-0.6, P 0-0.05, S 0-0.05, Cr (residual), Ni (residual), and Mo (residual), bal Fe.
Ductile, weldable casting alloy for gears and sheaves. Normalized: 40,000 YS; 80,000 TS; 18 El; 163-187 Brin; 47 Rock C. Quenched and tempered: 50,000-60,000 YS; 80,000-90,000 TS; 35-40 El; 163-235 Brin.

P4

Styria-Stahl Steirische Gusstahlwerke AG
C 0.9-1, bal Fe.
For tools, taps, reamers; water hardened. *Obsolete*

P45

Pelton Costeel, Inc.
Carbon steel. C 0.4-0.5, Mn 0.6-0.9, Si 0.3-0.6, P 0-0.05, S 0-0.05, Cr (residual), Ni (residual), and Mo (residual), bal Fe.
High strength, wear resistant steel. Normalized and tempered: 45,000 YS; 85,000 TS; 16 El; 170-229 Brin; 54 Rock C. Quenched and tempered: 60,000-70,000 YS; 90,000-100,000 TS; 10-14 El; 187-255 Brin.

P45

Gulf States Steel, Inc.
Carbon steel. C 0-0.25, Mn 0.9, bal Fe.
High strength carbon steel for structural applications that do not involve repetitive shock loading or lower temperature environments such as freight car flooring, purlins, and girts in pre-engineered steel buildings. 45,000 YS min; 60,000 TS min; 17 El.

P5

Styria-Stahl Steirische Gusstahlwerke AG
C 0.7-0.8, bal Fe.
For tools; water hardened. *Obsolete*

P50

Gulf States Steel, Inc.
Carbon steel. C 0-0.25, Mn 0-1.35, bal Fe.
High strength carbon steel for structural applications that do not involve repetitive shock loading or lower temperature environments such as freight car flooring, purlins, and girts in pre-engineered steel buildings. 50,000 YS min; 65,000 TS min; 15 El.

P55

Gulf States Steel, Inc.
Carbon steel. C 0-0.25, Mn 0-1.35, bal Fe.
High strength carbon steel for structural applications that do not involve repetitive shock loading or lower temperature environments such as freight car flooring, purlins, and girts in pre-engineered steel buildings. 55,000 YS min; 70,000 TS min.

P6

Pelton Costeel, Inc.
Alloy steel. C 0.18-0.26, Mn 1.2-1.5, Si 0.3-0.6, P 0-0.05, S 0-0.05, Cr (residual), Ni (residual), and Mo (residual), bal Fe.
Weldable manganese steel for lightweight transportation and machinery components. Normalized: 50,000 YS; 80,000 TS; 22 El; 207 Brin max. Quenched and tempered: 60,000-85,000 YS; 90,000-105,000 TS; 17-20 El; 187-277 Brin.

P6

Styria-Stahl Steirische Gusstahlwerke AG
C 0.7-0.8, bal Fe.
For tools; water hardened. *Obsolete*

P6 ALLOY

Wilbur B. Driver Co.
React and refract. V 4.8, Ni 6.1, Co 45, bal Fe.
For hysteresis motors. Saturation induction 16,000 Gausses. coercive force 60 oersteds, magnetically semihard.

P75

Friedr. Lohmann GmbH
Cr 0.8, Si 0-0.25, Mn 0.25, bal Fe.
Heat treated: 174,000 TS; 128,000 YS; 12 El; 37 RA; 355 Brin. For springs, rails, clutch discs. Type W1; water hardened.

P9

Pelton Costeel, Inc.
Alloy steel. C 0.14-0.2, Mn 0-1, Si 0.3-0.6, P 0-0.05, S 0-0.05, Cr 0.5-0.8, Ni 0.4-0.6, Mo 0.15-0.25, bal Fe.
Weldable chromium, nickel, molybdenum steel for lightweight high strength applications such as brackets for transportation industry, cylinder ends, boom connectors, and crawler parts. Normalized and tempered: 50,000 YS; 80,000 TS; 22 El; 163-201 Brin. Quenched and tempered: 60,000-95,000 YS; 90,000-120,000 TS; 14-20 El; 179-302 Brin.

PA 22 NICKEL

Henry Wiggin & Co. Ltd.
Mg 0.06-0.09, 99.5% min Ni.
For sleeves of indirectly heated, oxide coated cathodes in radio valves. Cathode nickel. *Obsolete*

PA 23 NICKEL

Henry Wiggin & Co. Ltd.
Mg 0.035-0.065, 99.5% min Ni.
For sleeves of indirectly heated, oxide-coated cathodes in radio valves. Cathode nickel. *Obsolete*

PAC 10

Powder Alloy Corporation
Copper.
Pure copper powder for electrical conductivity and radiofrequency shielding. 60-85 Rock H; 1200 psi strength min. Meets RR 9507/11.

PAC 1002

Powder Alloy Corporation
Nickel.
Nickel-silicon-chromium brazing filler for joints with high strength and oxidation resistance. Applications include joining turbine blades and other jet parts. Meets AMS 4782; AWS BNi-5.

PAC 1004

Powder Alloy Corporation
Precious metal.
Gold-palladium-nickel filler for joining corrosion and heat resistant steels and alloys. Meets AMS 4784.

PAC 1100

Powder Alloy Corporation
Aluminum.
Aluminum 12% silicon brazing filler for furnace and dip brazing of aluminum. Meets AMS 4185; AWS BAlSi-5.

PAC 1102

Powder Alloy Corporation
Cobalt.
Cobalt base metallic powder containing silicon, chromium, nickel, tungsten, and boron for joining cobalt alloys. Corrosion and oxidation resistant at high temperatures. Meets AMS 4783; AWS BCo-1.

PAC 118

Powder Alloy Corporation
Refractory.
Molybdenum powder for hard, wear resistant self-bonding coatings. 30-40 Rock C; 3000 psi strength min.

PAC 118FNS

Powder Alloy Corporation
Refractory.
Molybdenum powder for hard, wear resistant self-bonding thin coatings. 30-40 Rock C, 3000 psi strength min. Meets PWA 1338C.

PAC 118NS

Powder Alloy Corporation
Refractory.
Molybdenum powder for hard, wear resistant self-bonding coatings. 30-40 Rock C; 3000 psi strength min. Meets PWA 1313; USAF AF04(694)-916; RR 9507/19.

PAC 12

Powder Alloy Corporation
Copper.
Copper-nickel alloy for fretting wear applications. 65-75 Rock B; 6000 psi lap shear strength min.

PAC 1200

Powder Alloy Corporation
Silver. Cu 40, Zn 5, Ni 2, bal Ag.
Silver base alloy for joining of ferrous alloys. Meets AWS BAg-13; AMS 4765.

PAC 1202

Powder Alloy Corporation
Silver. Mn 15, bal Ag.
Silver base filler for joining iron base alloys. Meets AMS 4766.

PAC 1204

Powder Alloy Corporation
Silver.
Silver base filler for applications where precipitation hardening operations can be used during brazing cycle. Meets AMS 4767; AWS BAg-19.

PAC 1206

Powder Alloy Corporation
Silver.
Alloy of silver-copper-zinc-cadmium for general purpose applications. Meets AMS 4768; AWS BAg-2.

PAC 1208

Powder Alloy Corporation
Silver.
Low flow temperature silver filler material used for capillary joints. Meets AMS 4769; AWS BAg-1.

PAC 1210

Powder Alloy Corporation
Silver.
Low flow temperature silver filler material used for capillary joints. Meets AMS 4770; AWS BAg-1a.

PAC 1212

Powder Alloy Corporation
Silver.
Silver filler material used in applications requiring marine and caustic corrosion resistance. Meets AMS 4771; AWS BAg-3.

PAC 1214

Powder Alloy Corporation
Silver.
Silver filler material used for joining ferrous materials especially austenitic steels. Meets AMS 4772; AWS BAg-13.

PAC 1216

Powder Alloy Corporation
Silver.
Silver-copper filler material used for pure dry atmosphere brazing applications joining stainless or precipitation hardening steels. Meets AMS 4773; AWS BAg-18.

PAC 1218

Powder Alloy Corporation
Silver.
Silver filler for joining of corrosion resistant steels or alloys. Meets AMS 4774; AWS BAg-21.

PAC 1220

Powder Alloy Corporation
Precious metal.
Gold-copper-nickel filler for brazing of iron, nickel and cobalt base materials that require oxidation and corrosion resistance. Meets AMS 4785; AWS BAu-5.

PAC 1222

Powder Alloy Corporation
Precious metal.
Gold alloy containing nickel and palladium for joining corrosion and heat resistant steels and alloys. Meets AMS 4786.

PAC 1224

Powder Alloy Corporation
Precious metal.
Gold-nickel alloy to braze high temperature iron and nickel base alloys. Meets AMS 4787; AWS BAu-4.

PAC 1226

Powder Alloy Corporation
Copper.
Copper-phosphorus brazing filler used for joint preplacing. Meets AWS BCuP-1.

PAC 1228

Powder Alloy Corporation
Copper.
Copper-phosphorus brazing filler fluid used to penetrate tight joints. Meets AWS BCuP-2.

PAC 123

Powder Alloy Corporation
Chromium carbide for fretting wear applications above 1200°F. 35-40 Rock C; 1500 psi strength min. Meets PWA 1304.

PAC 1230

Powder Alloy Corporation
Copper.
Copper-phosphorus brazing filler for applications where tight tolerances and fits cannot be maintained. Meets AWS BCuP-3.

PAC 1232

Powder Alloy Corporation
Copper.
Copper-phosphorus brazing filler that contains 6% silver for a lower brazing range. Meets AWS BCuP-4.

PAC 1234

Powder Alloy Corporation
Copper
Copper-phosphorus brazing filler that contains 15% silver for applications that require low and high temperature strength. Meets AWS BCuP-5.

PAC 1236

Powder Alloy Corporation
Copper.
Copper-phosphorus brazing filler for joining copper and copper base alloys. Meets AWS BCuP-6.

PAC 1238

Powder Alloy Corporation
Copper.
Copper-phosphorus brazing filler with self-fluxing properties when joining copper. Meets AWS BCuP-7.

PAC 124

Powder Alloy Corporation
Chromium carbide for fretting wear applications above 1200°F. 35-40 Rock C; 1500 psi strength min. Meets PWA 1306.

PAC 1240

Powder Alloy Corporation
Precious metal.
Gold-copper alloy for joining iron, nickel, and cobalt base metals. Meets AWS BAu-1.

PAC 1242

Powder Alloy Corporation
Precious metal.
Gold-copper alloy for joining iron, nickel, and cobalt base metals. Meets AWS BAu-2.

PAC 1244

Powder Alloy Corporation
Precious metal.
Gold-copper-nickel alloy for joining iron, nickel, and cobalt base metals. Meets AWS BAu-3.

PAC 1246
Powder Alloy Corporation
Silver. Ag 99.5.
Silver for coating, brazing, powder metallurgy.

PAC 125
Powder Alloy Corporation
Crushed tungsten carbide/12% cobalt composite powder for hard, dense coatings resistant to fretting wear below 1000°F. 40-45 Rock C; 5000 psi strength min. Meets PWA 1302.

PAC 126
Powder Alloy Corporation
Crushed tungsten carbide/12% cobalt composite powder for thin, smooth as-sprayed coatings. 50-55 Rock C; 5000 psi strength min. Meets PWA 1301; AMS 7879-15.

PAC 127
Powder Alloy Corporation
Sintered tungsten carbide/12% cobalt aggregate powder for coatings resistant to abrasive and sliding wear below 1000°F. 55-60 Rock C; 5500 psi strength min.

PAC 128F
Powder Alloy Corporation
Crushed tungsten carbide/12% cobalt aggregate powder for plasma arc sprayed coatings resistant to abrasive and fretting wear below 1000°F. 47-55 Rock C; 5500 psi strength min.

PAC 128SF
Powder Alloy Corporation
Crushed tungsten carbide/12% cobalt aggregate powder for plasma arc sprayed thin dense coatings resistant to abrasive and fretting wear below 1000°F. 50-55 Rock C; 5500 psi strength min.

PAC 129
Powder Alloy Corporation
Blend of 85% chromium carbide/15% nickel chromium alloy powders for wear resistance at high temperatures. 30-40 Rock C; 2500 psi strength min. Meets PWA 1308.

PAC 12CO
Powder Alloy Corporation
Cobalt.
Cobalt base chromium tungsten alloy for plasma transferred arc coatings for applications requiring high temperature corrosion, abrasion, and oxidation resistance. 37-44 Rock C.

PAC 130
Powder Alloy Corporation
Blend of 75% chromium carbide/25% nickel chromium alloy powders for abrasion and fretting wear resistance from 1000 to 1500°F. 35-40 Rock C; 3000 psi strength min. Meets PWA 1307; RR 9507/2.

PAC 1300
Powder Alloy Corporation
Copper.
Water thinning brazing copper paste for joining ferrous and nonferrous metals. Meets AMS 3430; AWS BCu-1.

PAC 1302
Powder Alloy Corporation
Copper. 99.9 Cu min.
Copper powder for joining ferrous and nonferrous metals. Meets AMS 4740; AWS BCu-1A.

PAC 131
Powder Alloy Corporation
Blend of 75% chromium carbide/25% nickel chromium alloy powders for thin, smooth coatings with abrasion and fretting wear resistance from 1000 to 1500°F. 50-55 Rock C; 3000 psi strength min. Meets PWA 1305; RR 9507/17.

PAC 131-1
Powder Alloy Corporation
Blend of 93% chromium carbide/7% nickel chromium alloy powders for thin, smooth coatings with wear resistance at high temperatures. 60-65 Rock C; 3000 psi strength min. Meets PWA 1364.

PAC 132
Powder Alloy Corporation
Carbon steel.
Low carbon steel thermal spray and plasma transferred arc powder. Wear resistant for salvage, general machine work. 90-100 Rock B; 2200 psi strength min.

PAC 133
Powder Alloy Corporation
Self-fusing blend of PAC 60, 124, and 900 powders for thin, smooth, hard coatings. 45-50 Rock C; 3000 psi strength min. Meets PWA 1309; RR 9507/34.

PAC 134
Powder Alloy Corporation
Nickel.
Nickel-chromium-aluminum composite thermal spray powder. Wear and cavitation resistant. 25-35 Rock B; strength on roughened surfaces 5000 psi min.

PAC 135
Powder Alloy Corporation
Refractory.
Self-fusing molybdenum/nickel-chromium-boron-silicon powder blend for hard, dense wear resistant coatings. 50-55 Rock C; 2200 psi strength approx.

PAC 136
Powder Alloy Corporation
Refractory.
Titanium alloy for hard, thin dense coatings that self-bond to titanium substrates. 35-40 Rock C; 4500 psi strength min.

PAC 1400
Powder Alloy Corporation
Nickel.
Nickel-silicon-chromium-manganese alloy powder.

PAC 1402
Powder Alloy Corporation
Nickel.
Nickel-silicon-chromium alloy powder for high temperature brazing.

PAC 1406
Powder Alloy Corporation
Cobalt.
Cobalt base alloy powder containing silicon, chromium, nickel, and tungsten. Meets PWA 713.

PAC 1408
Powder Alloy Corporation
Copper.
Copper-nickel-indium brazing alloy produced by inert gas atomization.

PAC 1410
Powder Alloy Corporation
Copper.
Copper-manganese-nickel brazing alloy for joining corrosion and heat resistant alloys. Meets AMS 4764.

PAC 1412
Powder Alloy Corporation
Nickel.
Blended mixture of nickel-silicon-chromium alloy (80%) and pure nickel (20%).

PAC 1414
Powder Alloy Corporation
Copper.
Copper-manganese-nickel high temperature brazing alloy produced by inert gas atomization.

PAC 1416
Powder Alloy Corporation
Nickel.
Nickel-chromium (19%)/silicon (10%) alloy for joining superalloys at high brazing temperatures.

PAC 1418
Powder Alloy Corporation
Nickel.
Nickel-silicon-boron filler alloy for heat resistant joining of highly stressed components. Meets AMS 4778; AWS BNi-3.

PAC 1424
Powder Alloy Corporation
Nickel.
Nickel-chromium-silicon powder (90%) with a nickel-silicon powder containing boron for vacuum brazing operations.

PAC 1426
Powder Alloy Corporation
Nickel.
Nickel-chromium-silicon alloy blended with pure nickel (20%) for high temperature brazing applications.

PAC 1428
Powder Alloy Corporation
Nickel.
Nickel base alloy for hydrogen atmosphere brazing applications at 2100°F.

PAC 1430
Powder Alloy Corporation
Nickel.
Nickel base alloy for hydrogen atmosphere brazing applications at 2110°F.

PAC 1432
Powder Alloy Corporation
Nickel.
Nickel base alloy for hydrogen atmosphere brazing applications at 2125°F.

PAC 1434
Powder Alloy Corporation
Nickel.
Nickel base alloy for wide gap brazing applications at 2145°F.

PAC 1436
Powder Alloy Corporation
Nickel.
Nickel base alloy for wide gap brazing applications at 2140°F.

PAC 1440
Powder Alloy Corporation
Nickel.
Nickel-chromium-silicon-boron brazing filler for heat resistant high strength joints. Applications include joining turbine components in jet engines, highly stressed sheet metal parts and other highly stressed structures. Meets AMS 4775; AWS BNi-1.

PAC 1442
Powder Alloy Corporation
Nickel.
Nickel-chromium-silicon-boron brazing filler for heat resistant high strength joints. Applications include joining turbine components in jet engines, highly stressed sheet metal parts and other highly stressed structures. Meets AMS 4776; AWS BNi-1A.

PAC 1444
Powder Alloy Corporation
Nickel.
Nickel-chromium-silicon-boron-iron filler for heat resistant high strength joints at reduced temperatures. Meets AMS 4777; AWS BNi-2.

PAC 1446
Powder Alloy Corporation
Nickel.
Nickel base silicon-boron alloy for wide ductile fillet joints. Meets AMS 4779; AWS BNi-4.

PAC 1448
Powder Alloy Corporation
Manganese-nickel-cobalt-boron powdered alloy for joining corrosion and heat resistant steels and alloys. Meets AMS 4780.

PAC 1450
Powder Alloy Corporation
Nickel.
High phosphorus nickel base filler for brazing applications in marginal atmospheres and for joining low chromium bearing steels in exothermic atmospheres. Meets AWS BNi-6.

PAC 1452
Powder Alloy Corporation
Nickel.
Nickel base alloy containing manganese for brazing properties on nickel. Applications include honeycomb brazements and other stainless and corrosion resistant metals. Meets AWS BNi-8.

PAC 156CO
Powder Alloy Corporation
Cobalt.
Cobalt base chromium tungsten alloy for plasma transferred arc coatings for applications requiring high temperature corrosion, impact, and oxidation resistance. 40-45 Rock C.

PAC 157CO
Powder Alloy Corporation
Cobalt.
Cobalt base chromium tungsten alloy with boron for plasma transferred arc coatings for applications requiring high temperature corrosion, impact, and oxidation resistance. 48-54 Rock C.

PAC 158CO
Powder Alloy Corporation
Cobalt.
Cobalt base chromium tungsten alloy for plasma transferred arc coatings for applications requiring high temperature corrosion, impact, and oxidation resistance. 40-46 Rock C.

PAC 16
Powder Alloy Corporation
Copper.
Aluminum bronze alloy powder for fretting cavitational wear applications. 50-80 Rock H; 1200 psi strength min. Meets RR 9507/24.

PAC 16F
Powder Alloy Corporation
Copper.
Aluminum bronze alloy powder for as-sprayed coatings. 80-90 Rock H; 3500 psi strength min. Meets RR 9507/29.

PAC 17
Powder Alloy Corporation
Refractory.
Pure tantalum for hard, dense self bonding coatings. Corrosion and oxidation resistant for petrochemical applications. 30-40 Rock C; 3000 psi strength min.

PAC 19
Powder Alloy Corporation
Aluminum.
Aluminum powder for dense corrosion and oxidation coatings with electrical conductivity and radiofrequency shielding applications. 25-55 Rock H; 1200 psi strength min.

PAC 19NS
Powder Alloy Corporation
Aluminum.
Aluminum powder for dense corrosion and oxidation coatings with electrical conductivity and radiofrequency shielding applications. 25-55 Rock H; 1200 psi strength min. Meets PWA 1320; USAF 67A60753; RR 9507/13.

PAC 1CO
Powder Alloy Corporation
Cobalt.
Cobalt base chromium tungsten alloy for plasma transferred arc coatings for applications requiring high temperature corrosion, abrasion, and oxidation resistance. 45-53 Rock C.

PAC 200S
Powder Alloy Corporation
Agglomerated 16% cobalt composite powder for thin, dense coatings resistant to abrasion and fretting wear below 1000°F. 55-65 Rock C; 9000 psi strength min. Meets RR 9507/1.

PAC 200SF
Powder Alloy Corporation
Agglomerated 16% cobalt composite powder for thin, dense coatings resistant to abrasion and fretting wear below 1000°F. 55-65 Rock C; 8500 psi strength min. Meets GE B50TF167, Class B.

PAC 21CO
Powder Alloy Corporation
Cobalt.
Cobalt base chromium molybdenum alloy for plasma transferred arc coatings for applications requiring high temperature corrosion, impact, and oxidation resistance. 26-31 Rock C.

PAC 600
Powder Alloy Corporation
Nickel.
Self-fluxing nickel base alloy containing chromium, boron, silicon, molybdenum, and copper. Wear resistant to fretting, erosion cavitation, and abrasion. 58-62 Rock C.

PAC 60E
Powder Alloy Corporation
Nickel.
Self-fluxing nickel-chromium-boron-silicon alloy for applications requiring resistance to wear by abrasive grains, particle erosion and cavitation. 60-65 Rock C. Meets AMS 4775; MIL-R-17131A; API Standard 610.

PAC 60F
Powder Alloy Corporation
Nickel.
Self-fluxing nickel-chromium-boron-silicon alloy for applications requiring high resistance to wear by abrasive grains, particle erosion and caviation. 60-65 Rock C. Meets AMS 4775A.

PAC 63
Powder Alloy Corporation
Cobalt.
Self-fluxing cobalt base alloy containing nickel, chromium, boron, silicon, and molybdenum for coating martensitic steels. Wear resistant to fretting, erosion cavitation, and abrasion. 50-55 Rock C.

PAC 63
Powder Alloy Corporation
Blend containing 70% PAC 63 self-fluxing cobalt base alloy and tungsten carbide-cobalt composite. Wear resistant to fretting and abrasion. Matrix: 50-55 Rock C. Carbides: 70-75 Rock C.

PAC 64
Powder Alloy Corporation
Nickel.
Self-fluxing nickel-chromium-boron alloy for applications requiring resistance to abrasion, fretting wear, and erosion caviation. 54-61 Rock C.

PAC 65
Powder Alloy Corporation
Nickel.
Self-fluxing nickel-chromium-boron alloy that produces machinable fused coatings. Corrosion and wear resistant. 30-35 Rock C.

PAC 658C
Powder Alloy Corporation
Copper.
Copper-nickel-indium alloy; fretting wear resistant. 65-75 Rock B; 6000 psi lap shear strength min. Meets RR 9507/31.

PAC 658F
Powder Alloy Corporation
Copper.
Copper-nickel-indium alloy for thin, smooth coatings; fretting wear resistant. 65-75 Rock B; 6000 psi lap shear strength min.

PAC 661
Powder Alloy Corporation
Refractory.
Pure tungsten for heat resistant, dense coatings for rocket engine throat liners, tail cones, refractory crucibles, and electrical conductors on quartz substrates. 40-50 Rock A; 2000 psi strength min. Meets USN OS 10682D.

PAC 661F
Powder Alloy Corporation
Refractory.
Pure tungsten for heat resistant, thin dense coatings. 40-50 Rock A; 2000 psi strength min. Meets USN OS 10682D.

PAC 69E
Powder Alloy Corporation
Nickel.
Self-fluxing nickel-chromium-boron-silicon alloy for applications on hardened steel substrates. 50-55 Rock C.

PAC 69F
Powder Alloy Corporation
Nickel.
Self-fluxing nickel-chromium-boron-silicon alloy for thin smooth coatings on hardened steel substrates. 50-55 Rock C.

PAC 6CO
Powder Alloy Corporation
Cobalt.
Cobalt base chromium tungsten alloy for plasma transferred arc coatings for applications requiring high temperature corrosion, abrasion, and oxidation resistance. 35-40 Rock C. Meets AMS 5788.

PAC 86
Powder Alloy Corporation
Blend containing 65% PAC 60F and 35% tungsten carbide-cobalt composite. Sliding wear resistant. Matrix: 60-65 Rock C. Carbides: 70-75 Rock C.

PAC 87
Powder Alloy Corporation
Blend containing 20% PAC 60F and 80% tungsten carbide. Resistant to abrasion, fretting and erosion. Matrix: 60-65 Rock C. Carbides: 70-75 Rock C.

PAC 89F
Powder Alloy Corporation
Blend containing 50% PAC 60F and 50% tungsten carbide for combustion gas and plasma arc spraying. Resistant to abrasion, fretting and erosion. Matrix: 60-65 Rock C. Carbides: 70-75 Rock C. Plasma coatings: 55-60 Rock C.

PAC 89FP
Powder Alloy Corporation
Blend containing 50% PAC 60F and 50% tungsten carbide for thin coatings plasma arc sprayed. Matrix: 60-65 Rock C. Carbides: 70-75 Rock C. Plasma coatings: 55-60 Rock C.

PAC 900
Powder Alloy Corporation
Nickel.
Nickel thermal spray and plasma transferred arc powder. 50-70 Rock B; 2600 psi strength min. Meets PWA 1324; USAF 67A60653.

PAC 901
Powder Alloy Corporation
Aluminum.
Aluminum powder alloyed with 12% silicon for hard dense coatings. 85-95 Rock H; 2000 psi strength min. Meets PWA 1335.

PAC 902
Powder Alloy Corporation
Refractory.
75% molybdenum/nickel-chromium-boron-silicon powder blend for plasma arc sprayed scuff resistant coatings. 35-40 Rock C; 3000 psi strength min.

PAC 903
Powder Alloy Corporation
Self-fusing blend of PAC 128, 124, and 900 powders for hard, dense, wear resistant coatings. 50-55 Rock C; 5500 psi strength min.

PAC 906
Powder Alloy Corporation
Nickel.
Aluminum clad nickel composite thermal spray powder that reacts exothermically to self-bond onto metallic surfaces during spraying. Oxidation and high temperature resistant to 2100°F. 60-80 Rock B; strength on roughened surfaces 5000 psi min; on smooth surfaces 3000 psi min. Meets PWA 1337; RR 9507/5.

PAC 907
Powder Alloy Corporation
Nickel.
Aluminum clad nickel-chromium composite powder. Oxidation and high temperature corrosion resistant. 85-92 Rock B; strength on roughened surfaces 5000 psi min; on smooth surfaces 3000 psi min.

PAC 908C
Powder Alloy Corporation
Nickel.
Aluminum clad nickel-chromium composite powder. Oxidation and high temperature corrosion resistant. 85-92 Rock B; strength on roughened surfaces 5000 psi min; on smooth surfaces 3000 psi min. Meets PWA 1347; RR 9507/14; AVCO M 3956C.

PAC 909
Powder Alloy Corporation
Nickel.
Nickel clad aluminum composite powder that reacts exothermically during spraying to form bonds. Resistant to temperatures to 1800°F, thermal shock and erosion. 70-80 Rock B; strength on roughened surfaces 5000 psi min; on smooth surfaces 2500 psi min. Meets PWA 1321; RR 9507/4; USAF 67A60753.

PAC 90C
Powder Alloy Corporation
Cobalt.
Cobalt base alloy thermal spray and plasma transferred arc powder. Abrasion and high temperature resistant. 30-35 Rock C; 3000 psi strength min. Meets PWA 1318; RR 9507/3.

PAC 90VF
Powder Alloy Corporation
Cobalt.
Cobalt base alloy thermal spray and plasma transferred arc powder. Abrasion and high temperature resistant; suited for thin coatings. 30-35 Rock C; 2500 psi strength min. Meets PWA 1316; RR 9507/23.

PAC 911
Powder Alloy Corporation
Self-fusing blend of PAC 128, 124, and 900 powders for hard, dense, wear resistant coatings. 35-50 Rock C; 7000 psi strength approx. Meets PWA 1322.

PAC 912
Powder Alloy Corporation
Nickel.
Self-bonding molybdenum-nickel-aluminum composite powder for stainless steel. Wear and corrosion resistant. 70-80 Rock B; 7000 psi strength.

PAC 91C
Powder Alloy Corporation
Cobalt.
Cobalt aluminum prealloyed powder used in coating processes that increases oxidation, erosion, sulfidation, and thermal shock resistance of turbine blades and vanes. Meets PWA 1353-1.

PAC 91F
Powder Alloy Corporation
Cobalt.
Cobalt aluminum prealloyed powder used in coating processes that increases oxidation, erosion, sulfidation, and thermal shock resistance of turbine blades and vanes. Meets PWA 1353.

PAC 920
Powder Alloy Corporation
Stainless steel.
Self-bonding nickel-chromium-molybdenum-aluminum composite powder for stainless steel. Wear and corrosion resistant. 28-32 Rock C; 4500 psi strength.

PAC 922
Powder Alloy Corporation
Stainless steel.
Self-bonding aluminum clad nickel-chrome-iron composite powder for stainless steel. Wear and corrosion resistant. 78-88 Rock B; 5500 psi strength.

PAC 930
Powder Alloy Corporation
Carbon steel.
Carbon steel type composite powder for hard bearing and wear resistant applications. Abrasion and fretting wear resistant. 82-87 Rock B; 4000 psi strength.

PAC 932
Powder Alloy Corporation
Carbon steel.
Carbon steel type composite powder for hard bearing and wear resistant applications. Abrasion and fretting wear resistant. 28-32 Rock C; 5500 psi strength.

PAC 940
Powder Alloy Corporation
Copper.
Aluminum clad aluminum bronze composite powder for soft bearing and wear resistant applications. Abrasion and fretting wear resistant. 50-60 Rock B; 2500 psi strength.

PAC 96C
Powder Alloy Corporation
Stainless steel.
Type 316 stainless steel thermal spray and plasma transferred arc powder. Corrosion resistant; may be applied in thick layer. 85-95 Rock B; 2500 psi strength min. Meets RR 9507/26.

PAC 96F
Powder Alloy Corporation
Stainless steel.
Type 316 stainless steel thermal spray and plasma transferred arc powder. Corrosion resistant; coating 0.015 in. max. 85-95 Rock B; 2300 psi strength min. Meets RR 9507/22.

PAC 96PTA
Powder Alloy Corporation
Stainless steel.
Type 316 stainless steel for plasma transferred arc coatings for applications requiring corrosion resistance. 85-95 Rock B.

PAC 97
Powder Alloy Corporation
Stainless steel.
High chromium 400 stainless steel thermal spray and plasma transferred arc powder. Wear resistant; low shrinkage. 35-40 Rock C; 3200 psi strength min.

PAC 97PTA
Powder Alloy Corporation
Stainless steel.
High chromium 400 stainless steel for plasma transferred arc coatings for applications requiring wear resistance. 35-40 Rock C.

PAC 98C
Powder Alloy Corporation
Nickel.
Nickel base thermal spray and plasma transferred arc powder. Oxidation and corrosion resistant. 90-100 Rock B; 3000 psi strength min.

PAC 98C-1
Powder Alloy Corporation
Nickel.
Nickel base thermal spray and plasma transferred arc powder. Oxidation and corrosion resistant. 90-100 Rock B; 3000 psi strength min. Meets US Naval Specifications OS 8293; OS 10602.

PAC 98F
Powder Alloy Corporation
Nickel.
Nickel base thermal spray and plasma transferred arc powder. Oxidation and corrosive gas resistant up to 1800°F. 90-95 Rock B; 2500 psi strength min. Meets PWA 1317; RR 9507/27.

PAC 98F-1
Powder Alloy Corporation
Nickel.
Nickel base thermal spray and plasma transferred arc powder. Oxidation and corrosive gas resistant up to 1800°F. 90-95 Rock B; 2500 psi strength min. Meets PWA 1319.

PAC 99
Powder Alloy Corporation
Nickel.
Nickel base chromium alloy thermal spray and plasma transferred arc powder. Wear and corrosion resistant for salvage. 75-85 Rock B; 2500 psi strength min.

PACKING
Manufacturer not listed
Pb 82, Sn 4.8, Sb 13.
20 Brin. For metallic packing; pouring temperature 324°C.

PACKING (VALVE)
Manufacturer not listed
Sn 71, Sb 24, Cu 5.
For valve packing: anti-friction.

PACKING METAL
Manufacturer not listed
Cu 51.8, Zn 14.3, Ni 15, Pb 17, Sn 1.8.
For metallic packing.

PACKING PISTON
Compagnie de Orleans
Lead. Pb 73-76, Sn 12-14, Sb 10-15.
11,000 TS; 26 Brin. For metallic packing.

PACKING RINGS, FRENCH
French manufacture
Cu 51.8, Zn 14.3, Ni 15, Pb 17, Sn 1.8.
For packing rings.

PACKING RUSSIAN
Russian manufacture
Zn 99, Sn 0.9, Pb 0.3, Fe 0.2.
For packing metal.

PAGE 3-1/2 NI
American Chain & Cable
Ni 3.5, C, bal Fe.
For gas welding rod. Corrosion resistant.

PAGE A-S-10
American Chain & Cable
C, bal Fe.
For welding wire for submerged arc welding. Bare.

PAGE A-S-110
American Chain & Cable
C, bal Fe.
For welding wire. Bare.

PAGE A-S-15
American Chain & Cable
C, bal Fe.
For welding wire for submerged arc welding. Bare.

PAGE A-S-15-MO
American Chain & Cable
C, alloy, bal Fe.
For welding wire for submerged arc welding. Bare.

PAGE A-S-20
American Chain & Cable
C, bal Fe.
For welding wire; bare.

PAGE A-S-3-1/2N
American Chain & Cable
C 0.15-0.2, Mn 0.3-0.6, Si 0.15-0.75, Ni 3.25-3.75, bal Fe.
For welding wire for Ni and alloy steels. Bare.

PAGE A-S-6
American Chain & Cable
C, bal Fe.
For welding wire for submerged arc welding. Bare.

PAGE A-S-65
American Chain & Cable
C, bal Fe.
For welding wire. Bare.

PAGE ALLEGHENY 4-6 CHROMIUM MOLY
American Chain & Cable
C 0-0.1, Cr 4, Mo, bal Fe.
Annealed: 70,000 TS; 30,000 YS; 37 El; 70 Rock B. For welding electrodes. Corrosion resistant.

PAGE B
American Chain & Cable
C 0.13-0.18, bal Fe.
Welded: 55,000 TS; 7 El. For welding electrodes. Rust and lime coated.

PAGE C
American Chain & Cable
C 0-0.06, Mn 0-0.15, bal Fe.
Welded: 52,000-60,000 TS; 23-27 El. For gas welding rod and radio tube leads. AWS-GA-50; gas welding.

PAGE CE
American Chain & Cable
C 0.06, bal Fe.
Welded: 50,000 TS; 8 El. For welding electrodes. Rust and lime coated.

PAGE DENTAL ALLOY
Manufacturer not listed
Al 93, Cu 2.25, Au 4.75.
For dental applications.

PAGE E
American Chain & Cable
C 0.13-0.18, bal Fe.
Welded: 55,000 TS; 5 El. For welding electrodes. Rust and lime coated.

PAGE HARD FACING
American Chain & Cable
C 0.6, Mn 12-14, Ni 2, bal Fe.
For hard facing electrodes. High manganese steel.

PAGE HC
American Chain & Cable
C 0.9, Mn 1.4, bal Fe.
For welding electrodes. Abrasion resistant.

PAGE HI-TENSILE C
American Chain & Cable
C 0.07, bal Fe.
Welded: 75,000 TS; 20 El. For welding electrodes. Shielded arc.

PAGE HI-TENSILE F
American Chain & Cable
C 0.06, bal Fe.
Welded: 80,000 TS; 17 El. For welding electrodes. Shielded arc.

PAGE HI-TENSILE G
American Chain & Cable
C 0.11, bal Fe.
Welded: 85,000 TS; 17 El. For welding electrodes. Shielded arc.

PAGE HI-TENSILE M
American Chain & Cable
C 0.06-0.08, Mn 0.32-0.42, bal Fe.
Welded: 70,000-85,000 TS; 55,000 YS; 25-32 El. For welding electrodes for low alloy steels. E-6015.

PAGE HI-TENSILE SHIELDED ARC
American Chain & Cable
C 0.07, bal Fe.
Welded: 80,000 TS; 25 El. For welding electrodes. Shielded arc.

PAGE HIGH CARBON
American Chain & Cable
C 0.9-1.1, Mn 0.6, bal Fe.
For hard facing electrodes. Wear resistant.

PAGE HT
American Chain & Cable
C 0.2, bal Fe.
For welding electrodes.

PAGE MANGANESE BRONZE
American Chain & Cable
Mn 0.3, Zn 40, P 0.75-0.85, bal Cu.
Welded: 60,000-65,000 TS; 20-25 El. For gas welding wire and cast iron welding.

PAGE MANGANESE NICKEL
American Chain & Cable
C 0.8-1, Mn 12-14, Ni 3-3.5, Mo 0.2, Si 0.8-1.1, bal Fe.
Welded: 150 Brin. For welding electrode and hard facing electrode. Shielded arc, austenitic.

PAGE MC
American Chain & Cable
C 0.36, Mn 1, Si 1.2, bal Fe.
For welding electrodes. Mild; wear resistant.

PAGE MEDIUM CARBON
American Chain & Cable
C 0.32-0.45, Mn 0.9-1.1, bal Fe.
For hard facing electrodes. Tough.

PAGE NAVAL BRONZE
American Chain & Cable
Cu 59-61, P 0.5-1, bal Zn.
Welded: 49,000-55,000 TS; 25-35 El. For gas welding wire and cast iron welding.

PAGE-ALLEGHENY "B" SP
American Chain & Cable
C 0-0.11, Cr 18-20, Ni 8-10, bal Fe.
For welding electrodes. Stainless.

PAGE-ALLEGHENY 12% CR
American Chain & Cable
Cr 12, C 0-0.12, bal Fe.
Annealed: 70,000 TS; 35,000 YS; 25 El; 80 Rock B. For welding electrodes. Corrosion resistant.

PAGE-ALLEGHENY 12-14 CR
American Chain & Cable
Cr 12-15, 0.12 C min, bal Fe.
Annealed: 85,000 TS; 40,000 YS; 30 El; 95 Rock B. For combustion and steam engine parts and fans. Heat resistant to 1500°F.

PAGE-ALLEGHENY 15-35
American Chain & Cable
C 0-0.25, Cr 15, Ni 35, bal Fe.
For welding electrodes for stainless steel. Coated; stainless.

PAGE-ALLEGHENY 16% CR
American Chain & Cable
C 0-0.12, 16.0 Cr min, bal Fe.
For welding electrodes. Corrosion resistant.

PAGE-ALLEGHENY 17-7
American Chain & Cable
C 0.1-0.2, Cr 16-18, Ni 7-8.5, bal Fe.
Annealed: 110,000 TS; 40,000 YS; 35 El; 88 Rock B. Drawn: 270,000 TS; 240,000 YS; 1 El; 47 Rock C. For trim and household articles. Type 301X; stainless; austenitic.

PAGE-ALLEGHENY 18% CR
American Chain & Cable
C 0-0.35, 18.0 Cr min, bal Fe.
For welding electrodes. Corrosion resistant.

PAGE-ALLEGHENY 18-8
American Chain & Cable
Cr 18, Ni 8, C 0.07, Cb 1.3, bal Fe.
Annealed: 90,000 TS; 35,000 YS; 55 El; 80 Rock B. Drawn: 350,000 TS; 175,000 YS; 45 Rock C. For welding electrodes, strand springs, and rope. Stainless steel.

PAGE-ALLEGHENY 18-8 EZ
American Chain & Cable
C 0-0.15, Cr 17-19, Ni 8-10, 0.7 min P or S or Se, bal Fe.
Annealed: 90,000-120,000 TS; 35,000 YS; 35-50 El; 130-150 Brin. For screw machine products and hardware. Stainless; austenitic; Type 303; free-cutting.

PAGE-ALLEGHENY 18-8 MO
American Chain & Cable
C 0-0.07, Cr 18-20, Ni 12-15, Mo 2-2.5, bal Fe.
Annealed: 90,000 TS; 30,000 YS; 50 El. Drawn: 210,000 TS; 190,000 YS; 0.5 El. For welding electrodes. Austenitic; stainless.

PAGE-ALLEGHENY 18-8 TYPE 307
American Chain & Cable
C 0.07-0.15, Mn 4, Si 0.5, Cr 19.5-22, Ni 9-10.5, bal Fe.
Annealed: 80,000 TS; 30,000 YS; 50 El; 80 RA; 180 Brin. For stainless parts. Type 307; stainless, austenitic.

PAGE-ALLEGHENY 18-85
American Chain & Cable
C 0.03, Cr 19-22, Ni 10-12, bal Fe.
For welding electrodes. Stainless steel.

PAGE-ALLEGHENY 18-8S
American Chain & Cable
C 0.03, Cr 19-22, Ni 10-12, bal Fe.
For welding electrodes. Type 304; stainless.

PAGE-ALLEGHENY 19-9
American Chain & Cable
Cr 19, Ni 9, C 0.07, bal Fe.
Annealed: 80,000 TS; 30,000 YS; 50 El; 80 Rock B. For welding electrodes. Stainless steel.

PAGE-ALLEGHENY 23% CR
American Chain & Cable
C 0-0.16, Cr 23, bal Fe.
Annealed: 85,000 TS; 25 El; 84 Rock B. For welding electrodes. Corrosion resistant.

PAGE-ALLEGHENY 25-20
American Chain & Cable
C 0-0.25, Cr 27, Ni 21, bal Fe.
Annealed: 90,000 TS; 40,000 YS; 50 El; 80 Rock B. Drawn: 200,000 TS; 150,000 YS; 36 Brin C. For welding electrodes. Stainless steel.

PAGE-ALLEGHENY 25-20 CB
American Chain & Cable
C 0.06-0.1, Ni 20-22, Cb 1.2-1.4, 27.0 Cr min, bal Fe.
Annealed: 75,000 TS; 30,000 YS; 40 El; 50 RA; 180 Brin. For welding electrodes. Stainless; austenitic; stabilized.

PAGE-ALLEGHENY 25-20 MO
American Chain & Cable
C 0.06-0.1, Mn 1.5-2, Ni 20-22, Mo 2, 27.5 Cr min, bal Fe.
For welding electrodes. Heat and corrosion resistant.

PAGE-ALLEGHENY 33 GRADE C1
American Chain & Cable
C 0-0.12, Cr 12-14, bal Fe.
For fans, blowers, condensers, and engine parts. Heat resistant to 1500°F.

PAGE-ALLEGHENY 33 N.H.
American Chain & Cable
C 0-0.08, Cr 11.5-13.5, Al 0.1-0.2, bal Fe.
For welding electrodes. Stainless.

PAGE-ALLEGHENY 33 T.Q.
American Chain & Cable
C 0-0.12, Cr 11.5-13, bal Fe.
For welding electrodes. Stainless.

PAGE-ALLEGHENY 34
American Chain & Cable
C 0-0.12, Cr 14-16, bal Fe.
For fans, blowers. Corrosion resistant.

PAGE-ALLEGHENY 430F
American Chain & Cable
C 0-0.12, Cr 14-18, 0.07 P min, S, Se, or 0.6 Mo max, Zn, bal Fe.
For welding rod. Free-cutting, stainless.

PAGE-ALLEGHENY 44
American Chain & Cable
C 0-0.2, Cr 22-26, Ni 11-13, bal Fe.
For furnace parts, boiler baffles, and pumps. Heat resistant to 2100°F.

PAGE-ALLEGHENY 46
American Chain & Cable
Cr 4-6, 0.1 C min, bal Fe.
For welding electrodes and 4-6 Cr steel.

PAGE-ALLEGHENY 55
American Chain & Cable
C 0-0.35, Cr 23-30, bal Fe.
For furnace parts and boiler baffles. Heat resistant to 2100°F.

PAGE-ALLEGHENY 66 GRADE B
American Chain & Cable
Cr 15-18, 0.12 C min, bal Fe.
For fans, blowers, condensers, and evaporators. Heat resistant to 1600°F.

PAGE-ALLEGHENY 66 GRADE C2
American Chain & Cable
C 0-0.12, Cr 16-18, bal Fe.
For corrosion resisting parts. Corrosion resistant.

PAGE-ALLEGHENY 66W
American Chain & Cable
C 0-0.12, Cr 16-18, W 2.5-3.5, bal Fe.
For welding electrodes. Corrosion resistant.

PAGE-ALLEGHENY 67
American Chain & Cable
C 0-0.35, Cr 18-23, bal Fe.
For welding electrodes. Corrosion resistant.

PAGE-ALLEGHENY 8-10 CR-MO
American Chain & Cable
C 0-0.08, Cr 8.5-10.5, Mo 1.25-1.75, bal Fe.
For welding electrodes. Martensitic stainless steel.

PAGE-ALLEGHENY A-TI
American Chain & Cable
C 0-0.1, Cr 17-20, Ni 7-10, Ti = 6 x C, bal Fe.
For welding electrodes. Stainless.

PAGE-ALLEGHENY ALLOY 22 CR
American Chain & Cable
C 0.08-0.2, Cr 19-22, Ni 9-12, bal Fe.
For digesters, pipes, tanks, agitators, and strainers. Corrosion resistant.

PAGE-ALLEGHENY AMO
American Chain & Cable
C 0-0.1, Cr 16-19, Ni 0-14, Mo 2-4, bal Fe.
For welding electrodes. Stainless.

PAGE-ALLEGHENY METAL A
American Chain & Cable
C 0-0.11, Cr 17-19, Ni 7-9, bal Fe.
For welding electrodes. Stainless.

PAGE-ALLEGHENY METAL B
American Chain & Cable
C 0.08-0.2, Cr 18-20, Ni 8-10, bal Fe.
For welding electrodes. Stainless.

PAGE-ALLEGHENY METAL C
American Chain & Cable
C 0.08-0.2, Cr 17-19, Ni 7-9, bal Fe.
For stainless parts. Type 302.

PAGE-ALLEGHENY METAL CB
American Chain & Cable
C 0-0.1, Cr 17-20, Ni 8-12, Cb = 10 X C, bal Fe.
For welding electrodes. Stainless.

PAGE-ALLEGHENY METAL FM
American Chain & Cable
C 0-0.2, Ni 7-9.5, Cr 17-19, S or Se or Mo, bal Fe.
For welding electrodes. Stainless.

PAGE-ALLEGHENY OHMALOY
American Chain & Cable
C 0-0.12, Cr 12-14, Al 4-4.5, bal Fe.
For welding electrodes. Stainless.

PAGE-ARMCO
American Chain & Cable
C 0-0.03, bal Fe.
Weled: 55,000 TS; 8 El. For welding electrodes; bare and coated.

PAGE-AUTO
American Chain & Cable
C 0.06, bal Fe.
For welding electrodes.

PAINI
GTE Products Corp./Wesgo Div.
Pd 60, Ni 40.
Brazing alloy available in foil, flexibraze, wire, powder, extrudable paste and preform. Liquidus 2260°F. Solidus 2260°F.

PAINICUSIL
GTE Products Corp./Wesgo Div.
Pd 22.5, Ni 10, Cu 18.9, Ag 48.6.
Brazing alloy available in foil, flexibraze, wire, powder, extrudable paste and preform. Liquidus 2155°F. Solidus 1670°F.

PAINTGRIP
Armco
See ARMCO GAINEX and FORMABLE series.

PAINTWELL
United States Steel Corp.
Low C, Mn, bal Fe.
Zinc coated steel; chemically treated for painting.

PAITUNG WHITE COPPER
Manufacturer not listed
Cu 26-40, Zn 1-32, Ni 16-37, Fe 0-2.6.
For ornamental white metal parts; corrosion resistant.

PAKTONG (PACKFONG)
English manufacture
Ni 32-41, Cu 26-40, Zn 16-37, Fe 0-2.6.
130,000 TS; 2 El. For tableware and ornamental uses; Nickel Silver.

PAKTONG, COOKSON
Manufacturer not listed
Cu 40.9, Zn 45, Ni 11.1, Fe 2.5, Co 0.16.
For domestic utensils; corrosion resistant.

PAKTONG, FYFE
Manufacturer not listed
Cu 40.4, Zn 25.4, Ni 31.6, Fe 2.6.
For tableware, domestic utensils; corrosion resistant.

PAKTONG, KEFERSTEIN
Manufacturer not listed
Cu 26.3, Zn 36.8, Ni 36.8.
For tableware; corrosion resistant.

PAKTONG, PEAT
Manufacturer not listed
Cu 57.9, Zn 32.2, Ni 7.7, Fe 2.5.
For ornamental ware, domestic utensils; corrosion resistant.

PAKTONG, THURSTON
Manufacturer not listed
Cu 43.8, Zn 40.6, Ni 15.6.
For tableware; corrosion resistant.

PAKTONG, TUTENAG
Manufacturer not listed
Cu 44-45.7, Zn 16-39.6, Ni 17.4-40.
For domestic ware, ornamental parts; corrosion resistant.

PALAU-A
English manufacture
Ni 60, Pt 20, Pd 10, V 10.
For jewelry, chemical apparatus; Pt substitute, white gold.

PALAU-B
English manufacture
Pd 20, Au 80.
For jewelry, chemical apparatus; Pt substitute, white gold.

PALAURAL
Johnson Matthey plc
Au 46, Pd 30, Cu 19, Au, Zn.
Economy white dental casting alloy for posts and pins; Class 4.

PALCO
GTE Products Corp./Wesgo Div.
Pd 65, Co 35.
Brazing alloy available in foil, flexibraze, wire, powder, extrudable paste and preform. Liquidus 2255°F. Solidus 2246°F.

PALCO
Western Gold & Platinum Co.
Pd 65, Co 35.
For brazing molybdenum and tungsten. For cathode structures. Melting point: 2244-2250°F.

PALCRONIRY T-49
Western Gold & Platinum Co.
Au 41, Pd 27, Ni 22, Cr 10.
Melting point: 2150-2200°F. For high temperature brazing. Good flowability and wettability.

PALCUSIL 10
GTE Products Corp./Wesgo Div.
Ag 58.5, Cu 31.8, Pd 9.7.
Brazing alloy available in foil, flexibraze, wire, powder, extrudable paste and preform. Liquidus 1566°F. Solidus 1515°F.

PALCUSIL 15
GTE Products Corp./Wesgo Div.
Ag 65, Cu 20.3, Pd 14.7.
Brazing alloy available in foil, flexibraze, wire, powder, extrudable paste and preform. Liquidus 1652°F. Solidus 1562°F.

PALCUSIL 25
GTE Products Corp./Wesgo Div.
Ag 54, Cu 21.3, Pd 24.7.
Brazing alloy available in foil, flexibraze, wire, powder, extrudable paste and preform. Liquidus 1742°F. Solidus 1652°F.

PALCUSIL 5
GTE Products Corp./Wesgo Div.
Ag 68.5, Cu 26.8, Pd 4.7.
Brazing alloy available in foil, flexibraze, wire, powder, extrudable paste and preform. Liquidus 1490°F. Solidus 1485°F.

PALCUSIL-10
Western Gold & Platinum Co.
Ag 58, Cu 32, Pd 10.
Melting point: 1515-1566°F. For high temperature brazing of Kovar to ceramic seals. Good wettability.

PALCUSIL-15
Western Gold & Platinum Co.
Ag 65, Cu 20, Pd 15.
Melting point: 1562-1652°F. For high temperature brazing for Kovar and ceramic seals. Good flowability and wettability.

PALCUSIL-25
Western Gold & Platinum Co.
Ag 54, Cu 21, Pd 25.
Melting point: 1652-1742°F. For high temperature brazing for Kovar. High vapor pressure. Does not embrittle Kovar.

PALCUSIL-5
Western Gold & Platinum Co.
Ag 68, Cu 27, Pd 5.
Melting point: 1485-1490°F. For brazing Kovar to ceramic seals. Good wettability.

PALE YELLOW GOLD
Manufacturer not listed
Au 92, Ag 0.8-3, Fe 0-8.3.
For jewelry; corrosion resistant.

PALID
German manufacture
Sb 11, As 7, bal Pb.
For bearings; anti-friction.

PALINEY 9
J.M. Ney Co.
Pd 35, Ag 30, Pt 10, Au 10.
Stress relieved: 155,000 TS (nominal); 120,000 psi PL (nominal); 300 Knoop (nominal). Electronic alloy for telephone spring wire relays.

PALINEY CB
J.M. Ney Co.
Au 15, Pd 23, Pt 1.
Melting range: 1715-1845°F; casting temperature: 1900°F. Cast: 120,000 psi TS (nominal); 86,000 psi YS (0.1% offset, nominal); 280 HV; 119 DWT/in.^{3} density. Dental alloy for all fixed restorations, white gold color.

PALINEY M
J.M. Ney Co.
Au 20, Ag 30, Pt 5, Pd 45.
Wire: 60,000 TS (nominal); 20,000 PL (nominal); 23 El (nominal); 95 Brin (nominal). Electronic alloy for sliding contacts in potentiometers; corrosion and tarnish resistant, resistance wire.

PALINEY MEDIUM FUSING SOLDER
J.M. Ney Co.
Au 57, Ag 26.5, Pd 5.
Dental white gold solder. Melting range: 1395-1615°F.

PALINEY NO. 4
J.M. Ney Co.
For electric welding; melting range: 1735-1885°F. Platinum color white, triple thick.

PALINEY NO. 6
J.M. Ney Co.
Pd 44, Ag 38, Pt 1, Cu 16, Ni 1.
Heat treated: 165,000 TS (nominal); 110,000 PL (nominal); 15 El (nominal); 260 Brin (nominal). Annealed: 110,000 TS (nominal); 55,000 psi PL (nominal); 24 El (nominal); 150 Brin (nominal). Electronic alloy for pivots, springs, bearings, electrical contacts. Heat treatable, nonmagnetic, corrosion resistant.

PALINEY NO. 7
J.M. Ney Co.
Pd 35, Pt 10, Ag 30, Cu 14, Au 10, Zn 1.
Heat treated: 185,000 TS (nominal); 145,000 psi PL (nominal); 10 El (nominal); 280 Brin (nominal). Annealed: 120,000 TS (nominal); 90,000 psi PL (nominal); 24 El (nominal); 180 Brin (nominal). Electronic alloy for springs, pivots, bearings, contacts, potentiometers. Nonmagnetic, corrosion resistant.

PALINEY NO. 8
J.M. Ney Co.
Pt 1, Ag 38, Pd 44.
Wire, annealed: 100,000-125,000 TS; 20 min El. Sheet, heat treated: 170,000-220,000 psi TS; 0.5 min El; 155,000 psi PL (nominal); 390 Knoop. Electronic alloy for sliding contact in potentiometer windings of Nichrome. Corrosion and wear resistant.

PALIUM A
French manufacture
Sn 1, Pb 4, Cu 4, Mg 0.8, Mn 0.3, Zn 22, bal Al.
For bearings; antifriction.

PALIUM Z
French manufacture
Sn 2.6, Pb 4, Cu 4.5, Mg 0.6, Mn 0.3, Zn 0.3, bal Al.
For bearings; antifriction.

PALLABRAZE -1090
Johnson Matthey plc
Pd 18, bal Cu.
For brazing thermionic valves, magnetrons, klystrons. 1080-1090°C MP. High temperature brazing alloy.

PALLABRAZE-1010
Johnson Matthey plc
Pd 5, bal Ag.
For brazing thermionic valves, magnetrons, klystrons. 970-1010°C MP. High temperature brazing alloy.

PALLABRAZE-1225
Johnson Matthey plc
Pd 30, bal Ag.
For brazing thermionic valves, magnetrons, klystrons. 1150-1225°C MP. High temperature brazing alloy.

PALLABRAZE-1237
Johnson Matthey plc
Pd 60, bal Ni.
For brazing thermionic valves, magnetrons, klystrons. 1237°C MP. High temperature brazing alloy.

PALLABRAZE-810
Johnson Matthey plc
Pd 5, bal Ag + Cu.
For brazing thermionic valves, magnetrons, klystrons. 870-810°C MP. High temperature brazing alloy.

PALLABRAZE-840
Johnson Matthey plc
Pd 10, Ag, bal Cu.
For brazing thermionic valves, magnetrons, klystrons. 830-840°C MP. High temperature brazing alloy.

PALLABRAZE-850
Johnson Matthey plc
Pd 10, Cu, bal Ag.
For brazing thermionic valves, magnetrons, klystrons. 824-850°C MP. High temperature brazing alloy.

PALLABRAZE-880
Johnson Matthey plc
Pd 15, Cu, bal Ag.
For brazing thermionic valves, magnetrons, klystrons. 856-880°C MP. High temperature brazing alloy.

PALLABRAZE-900
Johnson Matthey plc
Pd 20, Cu, bal Ag.
For brazing thermionic valves, magnetrons, klystrons. 876-900°C MP. High temperature brazing alloy.

PALLABRAZE-950
Johnson Matthey plc
Pd 25, Cu, bal Ag.
For brazing thermionic valves, magnetrons, klystrons. 901-950°C MP. High temperature brazing alloy.

PALLADENT
Manufacturer not listed
Pd 40, bal Al.

PALLADIUM
Fansteel Metals
Precious metal. Pd 99.9.
Annealed: 66 Rock-15T; electrical conductivity 10% IACS. For electrical contacts. *Obsolete*

PALLADIUM
Atomergic Chemetals Corp.
Pd.
Purities: 99.999%, 99.99%, 99.9%. Forms: sponge, rod, wire, ingot, sheet, foil, single crystals.

PALLADIUM ALLOY NO. 312
Sigmund Cohn Corp.
Pd 56, Pt 37.5, Mo 6.5.
High resistance, high tensile strength.

PALLADIUM ALLOY-1
Manufacturer not listed
Rh 10, Pd 90.
For jewelry; corrosion resistant.

PALLADIUM ALLOY-2
Manufacturer not listed
Pd 67, Ag 33.
For jewelry; corrosion resistant.

PALLADIUM GOLD-1
Manufacturer not listed
Au 90, Pd 10.
For jewelry; white gold, Pt substitute.

PALLADIUM GOLD-2
Manufacturer not listed
Cu 40, Au 31, Ag 10, Pd 10.
For jewelry; white gold, Pt substitute.

PALLAS
Hover, Gebruder, Edelstahlwerk
C 1.05, Cr, bal Fe.
For bearings, cutters, liners, sleeves; water hardened, wear resistant.

PALLIAG
Degussa AG
Precious metal.
Precious metal alloy for dentistry and dental engineering.

PALMANSIL 5
GTE Products Corp./Wesgo Div.
Ag 75, Pd 20, Mn 5.
Brazing alloy available in foil, flexibraze, wire, powder, extrudable paste and preform. Liquidus 1962°F. Solidus 1846°F.

PALMANSIL-5
Western Gold & Platinum Co.
Ag 75, Pd 20, Mn 5.
Melting point: 1846-1962°F. For high temperature brazing. Good flowability and weldability.

PALNI
Western Gold & Platinum Co.
Pd 60, Ni 40.
Melting point: 2260°F. For high temperature brazing. Good flowability and wettability.

PALNIRO 1
GTE Products Corp./Wesgo Div.
Au 50, Pd 25, Ni 25.
Brazing alloy available in foil, flexibraze, wire, powder, extrudable paste and preform. Liquidus 2050°F. Solidus 2016°F. Meets AMS 4784.

PALNIRO 4
GTE Products Corp./Wesgo Div.
Au 30, Pd 34, Ni 36.
Brazing alloy available in foil, flexibraze, wire, powder, extrudable paste and preform. Liquidus 2136°F. Solidus 2075°F. Meets AMS 4785.

PALNIRO 7
GTE Products Corp./Wesgo Div.
Au 70, Pd 8, Ni 22.
Brazing alloy available in foil, flexibraze, wire, powder, extrudable paste and preform. Liquidus 1899°F. Solidus 1841°F. Meets AMS-4786.

PALNIRO-1
Western Gold & Platinum Co.
Au 50, Pd 25, Ni 25.
Melting point: 2016-2050°F. For high temperature brazing. Wets and flows well on Mo-W and stainless steel.

PALNIRO-4
Western Gold & Platinum Co.
Au 30, Pd 34, Ni 36.
Melting point: 2075-2136°F. For high temperature brazing. Wets and flows well on Mo-W and stainless steel.

PALNIRO-7
Western Gold & Platinum Co.
Au 70, Pd 8, Ni 22.
Melting point: 1841-1899°F. For high temperature brazing. Wets W-Mo and stainless steel.

PALORO
GTE Products Corp./Wesgo Div.
Precious metal. Au 92, Pd 8.
Brazing alloy available in foil, flexibraze, wire, powder, extrudable paste and preform. Liquidus 2264°F. Solidus 2192°F.

PALORO
Western Gold & Platinum Co.
Au 92, Pd 8.
For brazing on molybdenum and tungsten. Melting point: 2192-2264°F.

PALSIL 10
GTE Products Corp./Wesgo Div.
Ag 90, Pd 10.
Brazing alloy available in foil, flexibraze, wire, powder, extrudable paste and preform. Liquidus 1949°F. Solidus 1836°F.

PALSIL-10
Western Gold & Platinum Co.
Ag 90, Pd 10.
Melting point: 1835-1950°F. For high temperature brazing. Good wettability and flowability.

PAN-444
Machinery & Machine Supplies Co. Inc.
Cu 83.75, Sn 4, Ni 4, Pd 4, P 0.25.
Cast: 27,000-36,000 TS; 16,000-20,000 YS; 8-18 El; 65-80 Brin. For rocker bracket bushings in aircraft engines; leaded bronze, high ductility.

PAN-7
Machinery & Machine Supplies Co. Inc.
Cu 81.9, Sn 10, Ni 1, P 7, P 0.1.
Cast: 22,500-29,000 TS; 13,500-20,000 YS; 5-11 El; 65-75 Brin. For connecting rod bearings, liners, sleeves; leaded bronze, anti-friction.

PAN-B
Machinery & Machine Supplies Co. Inc.
Cu 72.45, Sn 9.5, Ni 3, Pd 15, P 0.05.
Cast: 22,500-38,000 TS; 13,500-22,500 YS; 6-15 El; 60-75 Brin. For bearings, bushings, liners, sleeves; leaded phosphor bronze, corrosion resistant.

PAN-BS
Machinery & Machine Supplies Co. Inc.
Pd 37, Sn 0.5, P 0.05, Ni 0.5, bal Cu.
Cast. For main bearings, locomotive slippers; leaded bronze.

PAN-H
Machinery & Machine Supplies Co. Inc.
Cu 66.85, Sn 10, Ni 3, Pd 20, P 0.15.
Cast: 20,000-27,000 TS; 11,500-18,000 YS; 6-13 El; 60-70 Brin. For rotors, sleeves, bearings, bushings; leaded bronze, corrosion resistant.

PANDALOY
Allegheny Ludlum Steel
C, Cr, Ni, bal Fe.
For paper making equipment, feed screws; stainless.

PANDEX
Latrobe Steel Co.
C 0-0.08, Cr 13-16, Ni 24-28, Mo 1-1.5, Ti 1.75-2.25, V 0.1-0.5, bal Fe.
Rolled: 135,000-160,000 TS; 85,000-115,000 YS; 20-30 El; 30-55 RA. At 1400 F: 64,000 TS; 18.5 El; 23.5 RA. For jet engine parts, superchargers, turbine wheels and blades; austenitic, heat resistant, age-hardenable. *Obsolete*

PANSERI
English manufacture
Cu 1, Mg 0.4, Ni 4.5, Fe 0.5, Si 11.5, bal Al.
For pistons; high temperature resistance.

PANTAL
Vereinigte Leichtmetallwerke G.m.b.H.
Aluminum. Mg 0.5-1, Mn 0.4-1.4, Si 0.8-2, Ti 0.3, bal Al.
Annealed: 16,000 TS; 14,000 YS; 25 El; 50 Brin. Aged: 60,000 TS; 50,000 YS; 10 El; 100 Brin. For sheathing, body construction, architectural use; corrosion resistant, age-hardenable.

PANTAL
VAW Vereinigte Aluminium-Werke AG
Aluminum. Si 0.7, Mg 1.4, Mg 0.9, T 0-0.2, bal Al.
Heat treated: 43,000-50,000 TS; 26,000-36,000 YS; 12-15 El; 70-95 Brin. For light alloy parts.

PANTAL
VDM Aluminium GmbH
Aluminum. Si 0.5-1, Mn 0.4-1.4, Mg 0.8-2, bal Al.
Half hard: 22,000 TS; 17,000 YS; 8 El; 45 Brin. Hard: 36,000 TS; 28,000 YS; 4 El; 60 Brin. For general structures, scaffolds, booms, transmission towers. Corrosion resistant.

PANTAL 5
Metallgesellschaft Reuterweg
Si 9-13, Mg 0.25-0.4, Mn 0.3-0.5, bal Al.
For light alloy parts; high corrosion resistance.

PANTAL-5
VDM Aluminium GmbH
Aluminum. Si 5, Mn 0.7, Mg 0.7, bal Al.
Hard: 26,000-37,000 TS; 21,000-33,000 YS; 4-8 El; 55-70 Brin. For general purposes, welding rods. Corrosion resistant.

PANTANAX
Thyssen Edelstahlwerke AG
C 1.2, Si 0.4, Mn 12.5, bal Fe.
For earth moving equipment, crushers, wear plates; wear and abrasion resistant. *Obsolete*

PANTANAX 1273, ETC.
Thyssen Edelstahlwerke AG
See Werkstoff Nr. 1.1273, etc.
Wear resistant steels.

PANTANAX M 14
Thyssen Edelstahlwerke AG
C 1, Si 1, Mn 14.5, Cr 3, Ni 0.5, bal Fe.
Flux cored wire for excavator teeth and blades. Meets DIN 8555.

PANTANAX M 14 CR
Thyssen Edelstahlwerke AG
C 0.5, Si 1, Mn 15, Cr 14.5, bal Fe.
Flux cored wire for excavator teeth and blades. Meets DIN 8555.

PANTHER 12
Allegheny Ludlum Steel
C, alloy, bal Fe.
For tools, dies. *Obsolete*

PANTHER 5
AL Tech Specialty Steel Corp.
C 1.5, Cr 4.7, W 12.5, V 5, Co 5, bal Fe.
Usual heat treated hardness: 66-68 Rock C. For heavy duty cutting tools requiring extra red hardness and abrasion resistance. W-C-Co high speed steel. AISI T-15.

PANTHER 5N
Styria-Stahl Steirische Gusstahlwerke AG
C 0.8, Cr 4.3, Mo 0.85, V 2.1, Co, W, bal Fe.
For lathe and planer tools, broaches, reamers, taps; high speed steel. *Obsolete*

PANTHER 5N

Vereinigte Edelstahlwerke
C 0.8, Cr 4.3, Mo 0.85, V 2.1, Co, W, bal Fe.
For lathe and planer tools, broaches, reamers, taps; high speed steel. *Obsolete*

PANTHER 750 SPEZIAL

Styria-Stahl Steirische Gusstahlwerke AG
C 1.3, Cr 4.3, Mo 0.85, V 3.8, W 12, bal Fe.
For blanking and forming dies, broaches, taps; high speed steel. *Obsolete*

PANTHER 750 SPEZIAL

Vereinigte Edelstahlwerke
C 1.3, Cr 4.3, Mo 0.85, V 3.8, W 12, bal Fe.
For blanking and forming dies, broaches, taps; high speed steel. *Obsolete*

PANTHER EXTRA

Allegheny Ludlum Steel
C 0.8, W 14, Cr 4, V 2, Mo 0.68, Co 5, bal Fe.
For tools, cutters; high speed steel. *Obsolete*

PANTHER EXTRA 655

Now VEW S705.

PANTHER N

Styria-Stahl Steirische Gusstahlwerke AG
C 0.86, Co 2.8, Cr 4.3, Mo 0.85, V 2.1, W 12, bal Fe.
For lathe and planer tools, reamers, drills, taps; high speed steel. *Obsolete*

PANTHER N

Vereinigte Edelstahlwerke
C 0.86, Co 2.8, Cr 4.3, Mo 0.85, V 2.1, W 12, bal Fe.
For lathe and planer tools, reamers, drills, taps; high speed steel. *Obsolete*

PANTHER SPECIAL

Styria-Stahl Steirische Gusstahlwerke AG
C 0.7, W 18, Co 9, Mo 1, Cr 4, V 2, bal Fe.
For lathe and planer tools, milling cutters, hobs, taps; Type T4; high speed steel. *Obsolete*

PANTHER SPECIAL

AL Tech Specialty Steel Corp.
C 0.75, W 19, Cr 4, V 1, Co 5, bal Fe.
18-4-1+5 Co high speed steel for heavy duty single point cutting applications; lathe and planer tools. AISI T-4. *Obsolete*

PANTHER ULTRA

Now VEW S300.

PANTHER XX

Allegheny Ludlum Steel
C 0.7, Cr, Ni, bal Fe.
For tools, cutters. *Obsolete*

PANZER

Rudolf Schmidt Stahlwerke
C 0.7, W 19, Cr 4, V 2, bal Fe.
For cutters, tools; high speed steel.

PAR 1

Crucible Steel Castings Co.
Cr, W, Co, Fe.
For heat and wear resisting parts. Resists wear at high temperature up to 2000°F. *Obsolete*

PAR 10

Crucible Steel Castings Co.
C 0.8, Cr 15, Mo 0.45, W 0.5, Al 0.04, bal Fe.
For guides for hot work. Corrosion and heat resistant. *Obsolete*

PAR 10A

Farrell Co.
Cast Iron. C, Ni, Cr, bal Fe.
For heat and corrosion resisting castings; heat, wear and corrosion resistant.

PAR 11

Crucible Steel Castings Co.
C 0.55, Cr 0.75, Mo 0.3-1.5, bal Fe.
For forging dies. Oil hardened. *Obsolete*

PAR 11A

Crucible Steel Castings Co.
C, Ni, Cr, bal Fe.
For heat and corrosion resisting castings. Heat, wear and corrosion resistant. *Obsolete*

PAR 2

Crucible Steel Castings Co.
C 0.3, Cr 1.5, Ni 2.5, Mo 0.3, bal Fe.
For gears, pinions, shafts. Wear and shock resistant. *Obsolete*

PAR 2C

Crucible Steel Castings Co.
C, Cr, Ni, bal Fe.
For heat and corrosion resisting castings. Heat, wear and corrosion resistant. *Obsolete*

PAR 3

Crucible Steel Castings Co.
C 0.35, Cr 3, bal Fe.
For machinery parts. Wear and abrasion resistant. *Obsolete*

PAR 4

Crucible Steel Castings Co.
C 0.4, Cr 25, Ni 2, bal Fe.
For heat and corrosion resisting parts. Heat and acid resistant. *Obsolete*

PAR 5

Crucible Steel Castings Co.
C 0.3, Cr 18, Ni 8, bal Fe.
For heat and corrosion resisting parts. Heat and acid resistant. *Obsolete*

PAR 6

Crucible Steel Castings Co.
C 0.4, Cr 28, Ni 10, bal Fe.
For heat and corrosion resisting parts. Heat and acid resistant. *Obsolete*

PAR 7

Crucible Steel Castings Co.
C 0.4, Cr 16, Ni 35, bal Fe.
For heat and corrosion resisting parts. Heat and acid resistant. *Obsolete*

PAR 8

Crucible Steel Castings Co.
C 0.25, Cr 18, Ni 65, bal Fe.
For heat and corrosion resisting parts. Heat and acid resistant. *Obsolete*

PAR 9

Crucible Steel Castings Co.
C 0.4, Cr 30, bal Fe.
For heat and corrosion resisting parts. Heat and acid resistant. *Obsolete*

PAR 9 A

Crucible Steel Castings Co.
C, Ni, Cr, bal Fe.
For heat and corrosion resisting castings. Heat, wear and corrosion resistant. *Obsolete*

PAR-EXC

Teledyne Vasco
C 0.53, W 2, Cr 1.65, V 0.25, bal Fe.
Annealed: 70,000 TS; 30 El; 64 RA; 165 Brin. Heat treated: 245,000 TS; 5 El; 18 RA; 525 Brin. For punches, dies, chisels, shear blades, cutters, pneumatic tools, hot working dies; extremely tough. Type S1.

PAR-TEN

United States Steel Corp.
C 0.12, Mn 0.75, Si 0.1, V 0.04, bal Fe.
Rolled: 65,000 TS; 45,000 YS; 28 El. For railroad and agricultural equipment, mine cars, auto bodies. High strength, low alloy construction steel.

PARA

Cytemp Specialty Steel Div.
C 1.25, Cr 0.4, W 1.6, V 0.2, bal Fe.
For cutters, dies, taps, drills, shear blades, punches; dense deep hardening steel. *Obsolete*

PARAGON

Colt Industries
C 0.95, Mn 1.7, Cr 0.27, V 0.25, bal Fe.
For stamping dies, taps, reamers; non-deforming. *Obsolete*

PARAGON A

Colt Industries
C 0.98, Mn 1.68, V 0.26, Cu 0.11, Cr 0.63, bal Fe.
For tools, dies; non-deforming. *Obsolete*

PARAGON OIL HARDENING

Colt Industries
C, alloy, bal Fe.
For dies, rollers, punches; air hardened, non-deforming. *Obsolete*

PARALLOY

Youngstown Foundry & Machine Co.
C 3, Mn 1, Si 2, Mo 0.2-0.25, Ni 2.5-3, Cr 0.5-0.75, bal Fe.
Cast: 50,000 psi TS; 250 Brin. For castings, cylinders, dies; cast iron.

PARALOY NO. 2

Youngstown Foundry & Machine Co.
C 3, Si 1.5, Ni 2-4, Cr 0.75-1, bal Fe.
Cast: 40,000-50,000 psi TS; 250-300 Brin. For cast dies for sheet metal stamping and drawing; cast iron.

PARFORCE SPECIAL 3

Now VEW M310.

PARISON ALLOY

English manufacture
Cu 69, Ni 19.5, Zn 6.5, Cd 5.
For cheap jewelry; gilt finish.

PARK A HOLLO DRILL

Colt Industries
C 0.3, Mn 0.95, Ni 2.25, Cr 0.7, Mo 0.3, bal Fe.
For hollow drill rod for detachable bits; oil hardening. *Obsolete*

PARK ALLOY CLIPPER BLADE

Colt Industries
C 1.1, Ni 0.35, Cr 0.4, V 0.3, bal Fe.
For hair clippers; oil hardening. *Obsolete*

PARK GATE

Hemmings & Co.
C 0.7, W 18, Cr 4, V 1, bal Fe.
For high speed tools and cutters. High speed steel.

PARK ORTHOPEDIC STEEL

Colt Industries
C, alloy, bal Fe.
For orthopedic braces. *Obsolete*

PARK SILVER
Colt Industries
C 0.7-1.2, bal Fe.
For cutters, drills, broaches, dies; water hardening. *Obsolete*

PARK SPECIAL
Crucible Materials Corp.
Tool material. C, alloy, bal Fe.
For tools.

PARKALOY
Parker-Kalon Corp.
C 0.3, Ni 1.2, Cr 7, bal Fe.
For screws, bolts; cold forging alloy.

PARKERS CHROME ALLOY
Manufacturer not listed
Cu 60, Zn 20, Ni 10, Cr 10.
For fountain pen points; corrosion resistant.

PARR
English manufacture
Ni 80, Cr 15, Cu 5.
73,000 TS; 15 RA; 137 Brin. For chemical machinery parts; stainless.

PARR
English manufacture
Ni 66.6, Cr 18, Cu 8.5, W 3.3, Al 2, Mn 1, Ti 0.2, B 0.2.
For chemical machinery parts; corrosion resistant.

PARSON'S ALLOY
American Manganese Bronze Co.
Copper. Cu 56, Zn 41.5, Fe 1.2, Sn 0.7, Mn 0.1, Al 0.46.
70,000 TS; 25 El; 119 Brin. For propeller blades, valve stems, engine frames, machine parts; tough.

PARSON'S MANGANESE BRONZE
Delta Metal (BW) Ltd.
Cu 55-60, Mn 0-3.5, Sn, Al, bal Zn.
Forged: 64,000 TS; 28,000 YS; 35 El; 120 Brin. Rolled: 70,000 TS; 32,000 YS; 34 El; 130 Brin. Extruded: 70,000 TS; 38,000 YS; 28 El. For pump and sluice valve spindles, propellers, marine engine parts, condenser tubes, valves; non-corrosive, resists mild acids. *Obsolete*

PARSON'S WHITE BRASS
Baldwin-Lima-Hamilton Corp.
Tin. Sn 74, Cu 5, Sd 7, Pb 14.
Cast: 12,250 TS; 108,700 YS; 3 El. For marine and automobile bearings; Babbitt.

PARSON'S WHITE BRASS
Delta Metal (BW) Ltd.
Sn 62, Zn 35, Cu 3.
Cast: 12,250 TS. For marine and automobile bearings; hard, tough alloy. *Obsolete*

PARSONS 2 S.A.
Delta Metal (BW) Ltd.
Sn base alloy.
Cast. For bearings for internal combustion engines and steam turbines; antifriction white metal. *Obsolete*

PARSONS STAR
Delta Metal (BW) Ltd.
Sn base alloy.
Cast. For marine main bearings, thrust blocks, steam pumps, and reciprocating engines; antifriction white metal. *Obsolete*

PARSONS WHITE BRASS DA
Baldwin-Lima-Hamilton Corp.
Tin. Cu 3.7, Sb 7.5, bal Sn.
Cast: 10,500 TS; 9400 YS; 3.5 El; 3.9 RA; 29 Brin. For bearings for diesel turbine and gasoline engines; Babbitt.

PARSTEEL
Paragon Steel Co.
C 0.5, Ni 3, Cr 1, bal Fe.
For gears, shafts; tough.

PARTINIUM-1
English manufacture
Al 89, Cr 7.4, Zn 1.7, Fe 1.3, Si 1.1.
19,000 TS; 12,000 YS; 2.5 El; 50 Brin. For light aircraft and automobile parts; similar to Alcoa No. 113.

PARTINIUM-2
English manufacture
Al 96, Sb 2.4, W 0.8, Cu 0.6, Sn 0.2.
65.0 Brin. For light aircraft and automobile parts.

PATENT COBALT-CHROME STEEL
Darwin & Milner Inc.
C 1.27-1.43, Cr 11.75-13.75, Co 2.7-3.3, Ni 0-0.6, Mo 0.55-0.85, Mn 0.35, Si 0-0.6.
For air cooled aeronautic engine valves, blanking, trimming, forming and shearing tools; air hardening, non-deforming; resists wear.

PATHER EXTRA SPEZIAL
Now VEW S305.

PATINA STEEL
Vereinigte Stahlwerke
Cu 0.2-0.3, C, bal Fe.
Rolled: 52,000-64,000 TS; 35,000-50,000 YS; 725 El; 750 RA; 100 Brin. For sheet metal structures, roofing, fences, pipes, bridges, nails; rust resisting. *Obsolete*

PATINA STEEL BEIZEREI
Vereinigte Stahlwerke
Cu 0.4, Mn, Si, P, bal Fe.
Rolled: 52,000-64,000 TS; 35,000-50,000 YS; > 25 El; > 50 RA; > 100 Brin. For pickling installations and structures; high acid resistance. *Obsolete*

PATINA STEEL RAUCHGAS
Vereinigte Stahlwerke
Cu 0.4, Mn, Si, P, bal Fe.
Rolled: 52,000-64,000 TS; 35,000-50,000 YS; > 25 El; > 50 RA; > 100 Brin. For structures exposed to smoke and acid fumes; high smoke resistance. *Obsolete*

PATRIUS CNM
Welded Carbide Co. Inc.
C 0.56, Cr 0.8, Mo 0.2, Ni 1.75, V 0.1, bal Fe.
For forging and heading dies; oil hardened, tough.

PATRIUS V
Welded Carbide Co. Inc.
C 0.55, Cr 0.7, Mi 0.18, Ni 1.65, V 0.1, bal Fe.
For forging and heading dies; oil hardened, tough.

PATTERN ALLOY-1
English manufacture
Al 90, Cu 8, Sn 2.
20,000-29,000 TS; 6-4 El. For pistons; similar to Alcoa No. 12.

PATTERN ALLOY-2
English manufacture
Pb 87, Sb 13.
18,00 TS; 11 El; 10 RA. For bearings; hard.

PATTERN METAL
Manufacturer not listed
Zn 30-40, Sn 15-40, Pb 20-42, Cu 0-3.
For bearings; anti-friction.

PAULITE
Simpson Bros. Machine Works
C 3.2, Si 2.5, Ni 1.5, Cr 0.8, bal Fe.

PAX NO. 2
Sanderson Kayser Ltd.
C 0.5, Si 0.8, Cr 1.5, W 2.25, V 0.2, bal Fe.
Shock resisting tool steel. B.S. 4659 Type BS1; AISI S1.

PAX NON BREAK
Sanderson Kayser Ltd.
C 0.5, Cr 1.5, W 2.25, V 0.25, bal Fe.
For hot work dies; hot work steel.

PAX NON-BREAK
Sanderson Kayser Ltd.
C 0.4, Cr 1.5, W 2.2, V 0.2, bal Fe.
For chisels and other shock tools.

PCLW-BPX ALLOY
Carpenter Technology Corp.
At 20°C: 15,000 psi YS; 40,000 psi TS; 37 El. Copper base positive extension wire used with copper in Type B thermocouples.

PCLW-SNX ALLOY
Carpenter Technology Corp.
At 20°C: 18,000 psi YS; 38,000 TS; 20 El. Copper base negative extension wire used in place of platinum in Type S or R thermocouples.

PD 135
Phelps Dodge Industries
Cr-Cd, bal Cu.
Hardened plus cold work: 65,000 psi TS; 60,000 psi YS; 12 El; good conductivity. For springs.

PD 135 FM
Phelps Dodge Industries
Free machining grade of PD 135.

PD-135
International Wire Products
Copper base precipitation hardened alloy. Solution treated, cold work, plus aging gives high strength, spring properties, and good flex life.

PD-135 FM
International Wire Products
Free machining grade of PD-135.

PD1
Teledyne Firth Sterling
Tool material.
Polycrystalline diamond for machining aluminum alloys, cast iron, copper alloys, plastics and other nonmetallics.

PDCP
Phelps Dodge Industries
Cu.
For electric generators and motors; electrolytic pure copper.
Obsolete

PDOF NO. 101 AND 102
Phelps Dodge Industries
99.95 Cu min, 10 ppm Cu max.
High conductivity.

PDOF NO. 104
Phelps Dodge Industries
99.95 Cu min, 8 oz/ton Ag min.
Oxygen free. High conductivity; resists softening with heat.

PDOF NO. 105
Phelps Dodge Industries
99.95 Cu min, 10 oz/ton Ag min.
Oxygen free. Resists softening with heat; high conductivity.

PDOF NO. 107
Phelps Dodge Industries
99.95 Cu min, 25 oz/ton Ag min.
Oxygen free. Resists softening with heat; good conductivity. Commutator segments.

SECTION I: ALLOY DATA / 929

PDRL 162
Cannon-Muskegon Corp.
C 0.12, Cr 10, Mo 4, W 2, Cb 1, Ti 1, Al 6.5, B 0.02, Zr 0.1, Ta 2, bal Ni.
Cast: 146,000 TS; 118,000 YS; 7 El. At 1400°F: 146,000 TS; 123,000 YS; 5.5 El. For high temperature service jet engine and turbine components. Heat and corrosion resistant. Casting alloy.

PDRL 163
Cannon-Muskegon Corp.
Cr 16.7, Al 6.3, Ti 0.1, Mo 1.6, W 2, Cb 1, C 0.05, Ta 2, B 0.02, Zr 0.1, Fe 0-0.3, bal Ni.
For jet engine and gas turbine components, high temperature parts. High heat and oxidation resistance. Cast alloy.

PE 50
Italian manufacture
Mg 5, bal Al.
Annealed: 42,000 TS; 22,000 YS; 35 El; 65 Brin. For rivets; non-heat treatable.

PE 8
Russian manufacture
WC 92, Co 8.
For cutting tools; cemented.

PECHINEY 412
Pechiney Electrometallurgie
Miscellaneous nonferrous. Mg 3.5-4.5, Ca 1, Si 45-50, Al 0-1, 1.3 RE, bal Fe.
Cast iron nodularizer.

PECHINEY 522
Pechiney Electrometallurgie
Miscellaneous nonferrous. Mg 4.7-5.3, Ca 2, Si 45-50, Al 0-1, 2.0 RE, bal Fe.
Cast iron nodularizer.

PECHINEY 610
Pechiney Electrometallurgie
Miscellaneous nonferrous. Mg 5.5-6.5, Ca 1, Si 45-50, Al 0-1, bal Fe.
Cast iron nodularizer.

PECHINEY 611
Pechiney Electrometallurgie
Miscellaneous nonferrous. Mg 5.5-6.5, Ca 1, Si 45-50, Al 0-1, 0.5 RE, bal Fe.
Cast iron nodularizer.

PECHINEY 611 A
Pechiney Electrometallurgie
Miscellaneous nonferrous. Mg 5.5-6.5, Ca 1, Si 45-50, Al 0-1, 1.0 RE, bal Fe.
Cast iron nodularizer.

PECHINEY 731
Pechiney Electrometallurgie
Miscellaneous nonferrous. Mg 6-7, Ca 3, Si 45-50, Al 0-1, 0.6 RE, bal Fe.
Cast iron nodularizer.

PECHINEY 910
Pechiney Electrometallurgie
Miscellaneous nonferrous. Mg 8.5-10, Ca 1, Si 45-50, Al 0-1, bal Fe.
Cast iron nodularizer.

PECHINEY 911
Pechiney Electrometallurgie
Miscellaneous nonferrous. Mg 8.5-10, Ca 1, Si 45-50, Al 0-1, 1.0 RE, bal Fe.
Cast iron nodularizer.

PECHINEY 931
Pechiney Electrometallurgie
Miscellaneous nonferrous. Mg 8.5-10, Ca 3, Si 45-50, Al 0-1, 1.0 RE, bal Fe.
Cast iron nodularizer.

PECHINEY STRONTIUM
Pechiney Electrometallurgie
Miscellaneous nonferrous. Ba 0.4, Ca 0.1, Na 0.1, Mg 0.05, N 0.05, Fe 0.01, P 0.0005, 99 Sr min.
For modification of eutectic structure of aluminum-silicon alloys to improve mechanical properties and machinability.

PECHKO WHITE GOLD
Manufacturer not listed
Au 60, Pd 30, Pt 10, Ir 0.1-2.
For jewelry, ornaments; corrosion resistant.

PECKRITE
Peckovers, Ltd.
TC 2.8-3, Si 1.6-1.8, Mn 0.75, Ni 1.2-1.5, bal Fe.
Cast: 50,000 TS. Heat treated: 70,000 TS; 450 Brin. For bushings, gears, piston rings, cams; cast iron.

PECO
Now VEW K244.

PEERLESS
Manufacturer not listed
Cr 16.5, Fe 3, Mn 2, bal Ni.
For heating elements; resistance alloy.

PEERLESS "A"
Crucible Materials Corp.
Tool material. C 0.3, W 9, Cr 3.25, V 0.25, bal Fe.
For hot heading and forging dies, gripping dies, swedges; hot work steel.

PEERLESS 56
Colt Industries
C 0.4, Cr 3.25, V 0.33, Mo 2.5, bal Fe.
Heat treated: 307,000 TS; 243,000 YS; 8 El; 36 RA; 600 Brin. For extrusion mandrels; hot work steel, oil hardened. *Obsolete*

PEERLESS 750
Crucible Materials Corp.
Tool material. C 0.45, Mn 0.25, Si 0.95, Cr 7.5, W 1, Mo 1, bal Fe.
Hot work tool steel; forging dies.

PEERLESS B
Colt Industries
C 0.45, W 15, Cr 2.75, V 0.4, bal Fe.
For extrusion dies, brass forging dies; hot work steel. *Obsolete*

PEERLESS C
Colt Industries
C 0.35, W 15, Cr 2.75, V 0.4, bal Fe.
For extrusion dies, die casting dies; hot steel work. *Obsolete*

PEERLESS D
Colt Industries
C 0.45, W 11, Cr 2, V 0.2, bal Fe.
For spike dies, brass forging dies, permanent molds; hot work steel. *Obsolete*

PEERLESS EXTRA
Colt Industries
C 0.7-1.2, bal Fe.
For tools and punches, cutters; water hardening. *Obsolete*

PEERLESS J
Colt Industries
C 0.53, Ni 0.11, Cr 1.37, V 1.3, W 2.8, Mo 1.5, bal Fe.
Annealed: 88,000 TS; 60,000 YS; 29 El; 56 RA. Heat treated: 209,000-256,000 TS; 188,000-232,000 YS; 7-8 El; 25-27 RA.
For hot work dies, aircraft structural members, hot work steel, oil hardened. *Obsolete*

PEERLESS LCT-2
Colt Industries
C 0.4, Mn 0.3, Cr 2, V 0.35, W 11.5, bal Fe.
Heat treated: 190,000 TS; 160,000 YS; 9 El; C 42 Rock. For dummy blocks, die holders, brass extrusion and forging dies, punches, die casting dies. Abrasion and heat resistant. Type H22 hot work steel. *Obsolete*

PEERLESS LLCT
Crucible Materials Corp.
Tool material. C 0.25, Mn 0.3, Si 0.3, Cr 4, W 15, V 0.5, bal Fe.
Hot work tool steel.

PEERLESS SPECIAL
Wallace Murray Corp.
C, Co, bal Fe.
For tools, dies; oil hardening.

PEERLESS W CO
Colt Industries
C 0.3, Cr 2.5, W 12, Co 0.5, bal Fe.
For extrusion dies and tools; hot work steel. *Obsolete*

PEGASE
Societe Nouvelle des Acieries de Pompey
Free machining medium carbon steel.

PELCOLOY
Molecu-Wire Corp.
Ni 70, Fe 30.
Resistance wire. Electrical resistivity: 120 ohms/circular mil ft. Annealed: 70,000 psi TS. Hard drawn: 150,000 psi TS. High temperature coefficient of resistance: 4500 ppm between 0 and 100°C. Maximum operating temperature: 590°C.

PEN HOB
Manufacturer not listed
C 0-0.1, Cr 0.6, Ni 1.25, bal Fe.
For hobbed cavity molds; case hardened.

PEN METAL-1
English manufacture
Au 67, Cu 25, Ag 8.
145,000 TS. For pen points; corrosion resistant.

PEN METAL-2
English manufacture
Cu 85, Zn 13, Sn 2.
For pen points; corrosion resistant.

PEN-O-FOUR
Peninsular Steel Co.
C 0.75, Mn 0.7, Cr 0.9, Ni 1.75, Mo 0.35, bal Fe.
For tools, dies; oil hardening.

PEN-VAN NO. 12
Manufacturer not listed
C 0.55, Cr 1, V 0.15, bal Fe.
For tools, chisels, punches; oil hardening.

PENAIR 5
Peninsular Steel Co.
C 1, Mn 0.6, Cr 5.25, Mo 1.1, V 0.25, bal Fe.
For dies, rollers, punches; air hardened, non-deforming. AISI A2.

PENCO
Manufacturer not listed
C 0.95-1.05, bal Fe.
For tools, drills, taps; water hardened.

PENCO 70
Peninsular Steel Co.
C 0.7, Mn 0.7, Si 0.25, Cr 0.7, Ni 1.5, Mo 0.3, bal Fe.
Oil hardening; for forming rolls, punches, dies, clutch parts, knuckle pins, spindles. Reference AISI-SAE 4370. *Obsolete*

PENCO ACS
Peninsular Steel Co.
C 0-0.07, Mn 0.4, Si 0.25, Cr 4.5, bal Fe.
For hubbed cavity plastic molds; oil hardening.

PENCO WH
Peninsular Steel Co.
C 0.95-1.1, bal Fe.
For tools, dies; water hardening. *Obsolete*

PENCO-OCS
Peninsular Steel Co.
C 0.06, Mn 0.3, Cr 1, Mo 0.25, B, bal Fe.
For hobbed cavity molds; case hardened.

PENCOYD
U.S. Steel Corp.
C 0.3, Ni 1, bal Fe.
For bridge construction. *Obsolete*

PENN-AIR
Pennsylvania Steel Corp.
C 1, Mn 2, Cr 1, Mo 1, bal Fe.
For punches, dies, blanking and forming tools; Type A2; air hardened.

PENN-CUT
Pennsylvania Steel Corp.
C 0.7, Cr 4, V 1, W 18, bal Fe.
For taps, hobs, reamers, lathe cutters; Type T1; high speed steel.

PENN-CUT 5
Pennsylvania Steel Corp.
C 0.8, Cr 4, V 1, W 18, Co 5, bal Fe.
For lathe and planer tools, form cutters, hobs; Type T5; high speed steel.

PENN-CUT MOLY
Pennsylvania Steel Corp.
C 0.8, Cr 4, V 2, W 6.4, Mo 5, bal Fe.
For lathe and planer tools, form cutters, hobs; Type M2; high speed steel.

PENN-FLEX
Pennsylvania Steel Corp.
C 0.33, Mn 0.72, Si 0.25, Cr 0.85, W 0.42, Mo 0.45, bal Fe.
Oil or water hardening tool steel, designed for plastic molds.

PENNALOY "A"
Pennsylvania Steel Foundry & Machine Co.
C 0.4, Ni 1.5, Cr 0.8, Mo 0.2, bal Fe.
Medium: 225-250 Brin. For castings for abrasion resistance, bucket lips and teeth; abrasion resistant. *Obsolete*

PENNALOY "B"
Pennsylvania Steel Foundry & Machine Co.
C, alloy, bal Fe.
Hardened: 500-600 Brin. For crusher jaws, cement mills, sand pumps; wear resistant. *Obsolete*

PENNALOY HEAT RESISTANT
Pennsylvania Steel Foundry & Machine Co.
C 0.13-0.2, Cr 5-7, bal Fe.
For oil refineries; corrosion resistant. *Obsolete*

PENNALOY HEAT RESISTANT
Pennsylvania Steel Foundry & Machine Co.
C 0.21-0.25, Cr 7-11, bal Fe.
For oil refineries; corrosion resistant. *Obsolete*

PENNANT
Delsteel Inc.
C 0.9, Mn 1.6, V 0.2, bal Fe.
For tools, blanking and forming dies, broaches, taps, reamers; non-shrinkable.

PENNROLD-10
Brush Wellman Corp.
Be 0.4-0.6, 2.35-2.60 Co or Ni, bal Cu.
Annealed: 40,000 TS; 25,000 YS; 30 El; 40 RA. Heat treated: 150,000 TS; 140,000 YS; 1 El. For current carrying springs, switch parts, circuit breaker parts. Age hardenable, fatigue and corrosion resistant. *Obsolete*

PENNROLD-165
Brush Wellman Corp.
Be 1.6-1.8, 0.20 Co or Ni min, 0.60 Co + Ni max, bal Cu.
Annealed: 60,000 TS; 25,000 YS; 60 El. Hard: 120,000 TS; 110,000 YS; 2 El. For current carrying and mechanical springs. Age hardenable; fatigue and corrosion resistant. *Obsolete*

PENNROLD-25
Brush Wellman Corp.
Be 1.8-2.05, 0.20 Co or Ni min, 0.60 Co + Ni max, bal Cu.
Annealed: 60,000 TS; 28,000 YS; 60 El. Hard: 120,000 TS; 112,000 YS; 2 El. Heat treated: 215,000 TS; 205,000 YS; 1 El. For current carrying springs, diaphragms, switch blades, contacts, bellows. Age hardenable; fatigue and corrosion resistant. *Obsolete*

PENNSYLVANIA L.C.D.
Pennsylvania Steel Corp.
C 0.4, Cr 4.5, Mo 1, W 0.8, Mn 0.4, bal Fe.
For shears, punches, blanking and trimming dies; oil hardened, non-deforming.

PENNSYLVANIA L.T.A.
Pennsylvania Steel Corp.
C 1, Mn 2, Cr 0.9, Mo 0.9, bal Fe.
For chisels, punches, dies, mandrels, crimpers; air hardened, non-deforming.

PENNSYLVANIA P.B. DRILL ROD
Patriarche & Bell
C 1-1.2, bal Fe.
Annealed: 100,000 TS; 53,000 YS; 21 El; 42 RA; 200 Brin. For drills, taps, reamers; Type W1; water hardened.

PENNSYLVANIA P.H. 14
Pennsylvania Steel Corp.
C 0.42, Cr 3.5, V 0.3, W 14, bal Fe.
For extrusion dies, liners and rams, hot punches; hot work steel, oil hardened.

PENNSYLVANIA P.H. 9
Pennsylvania Steel Corp.
C 0.35, Cr 2.75, V 0.3, W 9, bal Fe.
For extrusion rams, liners, die casting dies; hot work steel, oil hardened.

PENNSYLVANIA P.H. VAN
Pennsylvania Steel Corp.
C 0.4, Si 1.1, Cr 5.25, V 0.9, Mo 1.2, bal Fe.
For hot work tools and dies; hot work steel, oil hardened.

PENNSYLVANIA P.H.W.
Pennsylvania Steel Corp.
C 0.35, Mn 1.5, Cr 5, Si 1, W 1.25, Mo 1.25, V 0.2, bal Fe.
For hot work tools and dies; hot work steel, oil hardened.

PENNTEMP 165
Brush Wellman Corp.
Be 1.6-1.8, Co 0.2-0.35, bal Cu.
AM temper: 100,000-110,000 TS; 75,000-90,000 YS; 18-22 El. XHM temper: 160,000-175,000 YS; 135,000-150,000 YS; 3-7 El. For contacts, switch members, clips, springs. Age hardenable. *Obsolete*

PENSTOCK A
Lukens Steel
C 0-0.25, C 0.9-1.35, Si 0.15-0.3, bal Fe.
As rolled: 63,000-70,000 psi TS; 42,000-50,000 psi YS; 18-22 El. Used for penstock.

PENSTOCK B
Lukens Steel
C 0.25, Mn 0.9-1.35, Si 0.15-0.3, bal Fe.
Normalized, fine grain: 63,000-70,000 psi TS; 42,000-50,000 psi YS; 18-22 El. Fine grain grade, for penstocks.

PENSTOCK-B
Lukens Steel
C 0-0.5, Mn 0.9-1.35, Si 0.15-0.3, bal Fe.
Normalized: 70,000 TS min; 42,00 YS min; 23 El min. For welded structures, hydroelectric penstocks; shock resistant.

PENTANAX S
Thyssen Edelstahlwerke AG
C 0.85, Si 1, Mn 17.5, bal Fe.
For wear and abrasion resistant parts; wear and abrasion resistant. *Obsolete*

PEQUOT
Allegheny Ludlum Steel
C, alloy, bal Fe.
For tools, dies. *Obsolete*

PER 2
French manufacture
C 0.25, Cr 22, Ti 2, Al 2, bal Ni.
For gas turbine components; heat and oxidation resistant.

PERALUMAN 1
Aluminium Industrie Aktiengesellschaft
Mg 0.5-1.5, Mn 0.5-1.5, bal Al.
Soft: 25,000 TS; 10,000 YS; 20 El; 45 Brin. Hard: 41,000 TS; 38,000 YS; 8 El; 75 Brin. Extruded: 26,000 TS; 14,000 YS; 20 El; 45 Brin. For paneling, roofing, containers; corrosion resistant, non-hardenable.

PERALUMAN 1
Lavorazione Leghe Leggere SpA
Mg 0.5-1.5, Mn 0.5-1.5, bal Al.
Soft: 25,000 TS; 10,000 YS; 20 El; 45 Brin. Hard: 41,000 TS; 38,000 YS; 8 El; 75 Brin. Extruded: 26,000 TS; 14,000 YS; 20 El; 45 Brin. For paneling, roofing, containers; corrosion resistant, non-hardenable.

PERALUMAN 10
Soc. Alluminio Veneto per Azioni
Aluminum. Mg 10, Mn 0.4, bal Al.
Heat treated: 50,000-56,000 TS; 21,000-26,000 YS; 6-13 El; 75-85 Brin. For marine and aircraft parts; heat treatable, resists seawater corrosion.

PERALUMAN 15
Alluminio SA
Mn 0-0.3, Mg 1.3-1.8, bal Al.
Soft: 21,000 TS; 8,000 YS; 28 El; 40 Brin. Hard: 32,000 TS; 30,000 YS; 6 El; 65 Brin. For panels, containers. Corrosion resistant.

PERALUMAN 15
Aluminium Industrie Aktiengesellschaft
Mn 0-0.3, Mg 1.3-1.8, bal Al.
Soft: 21,000 TS; 8,000 YS; 28 El; 40 Brin. Hard: 32,000 TS; 30,000 YS; 6 El; 65 Brin. For panels, containers. Corrosion resistant.

PERALUMAN 2
Multiple manufacturers
Aluminum. Mn 1.3-1.5, Mg 2-2.3, bal Al.
Soft: 31,000 TS; 14,000 YS; 20 El; 50 Brin. Hard: 60,000 TS; 57,000 YS; 2 El; 105 Brin. For vessels, storage tanks, welded structures. Nonhardenable, corrosion resistant.

PERALUMAN 2
Lavorazione Leghe Leggere SpA
Mg 2.15, Mn 1.4, bal Al.
Soft: 31,000-36,000 TS; 15,000-21,000 YS; 20-16 El. Hard: 50,000-65,000 YS; 45,000-58,000 YS; 5-2 El. For paneling in architecture, ship building, automotive parts; corrosion resistant.

PERALUMAN 2

Aluminium Industrie Aktiengesellschaft
Mg 2.15, Mn 1.4, bal Al.
Soft: 31,000-36,000 TS; 15,000-21,000 YS; 20-16 El. Hard: 50,000-65,000 YS; 45,000-58,000 YS; 5-2 El. For paneling in architecture, ship building, automotive parts; corrosion resistant.

PERALUMAN 25

Alluminio SA
Mg 2.2-2.8, Fe 0-0.4, Si 0-0.3, 0.15-0.45 Mn + Cr, bal Al.
Annealed: 22,000-31,300 TS; 11,400-17,000 YS; 22-30 El; 45-45 Brin. Hard: 37,000-41,200 TS; 34,000-40,000 YS; 5-9 El; 70-85 Brin. For boat hulls, guided missile containers. Corrosion resistant.

PERALUMAN 25

Lavorazione Leghe Leggere SpA
Mg 2.2-2.8, Fe 0-0.4, Si 0-0.3, 0.15-0.45 Mn + Cr, bal Al.
Annealed: 22,000-31,300 TS; 11,400-17,000 YS; 22-30 El; 45-45 Brin. Hard: 37,000-41,200 TS; 34,000-40,000 YS; 5-9 El; 70-85 Brin. For boat hulls, guided missile containers. Corrosion resistant.

PERALUMAN 3

Alluminio SA
Aluminum. Mg 3-3.3, Mn 0.3, bal Al.
Cast: 21,400-28,500 TS; 9000-12,000 YS; 6-10 El; 45-50 Brin. For light alloy castings; corrosion resistant.

PERALUMAN 3

Soc. Alluminio Veneto per Azioni
Aluminum. Mg 3, Mn 0.3, bal Al.
Cast: 21,000-29,000 TS; 8,000-12,000 YS; 6-10 El; 45-55 Brin. For light alloy parts; corrosion resistant.

PERALUMAN 3

Oederlin & Co. Ltd.
Mg 2.0, Mn 0 0.4, bal Fe.
Sand cast: 22,800 TS; 11,400 YS; 5 El; 50 Brin. Permanent mold: 29,000 TS; 11,400 YS; 10 El; 55 Brin. For aircraft and marine equipment; corrosion resistant to seawater.

PERALUMAN 3

Aluminium Industrie Aktiengesellschaft
Mg 2-3, Mn 0-0.7, bal Al.
Soft: 32,000 TS; 16,000 YS; 25 El; 60 Brin. Hard: 45,000 TS; 40,000 YS; 7 El; 90 Brin. For architectural trim, containers, ship building; corrosion resistant.

PERALUMAN 3

Aluminium Walzwerke Singen GmbH
Mg 2-3, Mn 0-0.7, bal Al.
Soft: 32,000 TS; 16,000 YS; 25 El; 60 Brin. Hard: 45,000 TS; 40,000 YS; 7 El; 90 Brin. For architectural trim, containers, ship building; corrosion resistant.

PERALUMAN 30G

Aluminium Industries, AG
Mg 2-4, Mn 0.25-0.35, bal Al.
Annealed: 21,400-28,500 TS; 9,000-12,000 YS; 20-6 El; 45-50 Brin. For architecture, shipbuilding; corrosion resistant.

PERALUMAN 30G

Aluminium Industrie Aktiengesellschaft
Mg 2-4, Mn 0.25-0.35, bal Al.
Annealed: 21,400-28,500 TS; 9,000-12,000 YS; 20-6 El; 45-50 Brin. For architecture, shipbuilding; corrosion resistant.

PERALUMAN 35

Alluminio SA
Mg 3.2-3.8, Mn 0.2-0.4, Ti 0-0.1, Fe 0-0.4, Cu 0-0.05, bal Al.
Annealed: 30,000-37,000 TS; 12,800-20,000 YS; 20-30 El; 55-70 Brin. Hard: 41,200-47,000 TS; 35,600-44,100 YS; 4-9 El; 85-100 Brin. For boat hulls, guided missile containers. Corrosion resistant.

PERALUMAN 35

Lavorazione Leghe Leggere SpA
Mg 3.2-3.8, Mn 0.2-0.4, Ti 0-0.1, Fe 0-0.4, Cu 0-0.05, bal Al.
Annealed: 30,000-37,000 TS; 12,800-20,000 YS; 20-30 El; 55-70 Brin. Hard: 41,200-47,000 TS; 35,600-44,100 YS; 4-9 El; 85-100 Brin. For boat hulls, guided missile containers. Corrosion resistant.

PERALUMAN 3G

Swiss Aluminium Ltd.
Aluminum. Si 0.2-0.4, Mn 0.25-0.35, Mg 3-3.3, Ti 0.08-0.15, bal Al.
As cast: 27,000 TS; 14,000 YS; 3-8 El; 45-60 Brin. Heat treated: 33,000 TS; 20,000 YS; 2-8 El; 50-70 Brin. For sand and die castings, marine hardware, crankcases, oil pans, housings. Fair castability, high corrosion resistance. Resists seawater. Heat treatable. *Obsolete*

PERALUMAN 5

Soc. Alluminio Veneto per Azioni
Aluminum. Mg 5, Mn 0.4, bal Al.
Heat treated: 28,000-33,000 TS; 13,000-17,000 YS; 8-12 El; 60-80 Brin. For light alloy parts; corrosion resistant.

PERALUMAN 5

Soc. Alluminio Veneto per Azioni
Mg 5, Mn 0.5, bal Al.
Soft: 41,000 TS 21,000 YS; 24 El; 70 RA; 160 Brin. For architectural trim; corrosion resisant.

PERALUMAN 5

Aluminium Industrie Aktiengesellschaft
Mg 5, Mn 0.5, bal Al.
Soft: 41,000 TS 21,000 YS; 24 El; 70 RA; 160 Brin. For architectural trim; corrosion resisant.

PERALUMAN 5

Aluminium Walzwerke Singen GmbH
Mg 5, Mn 0.5, bal Al.
Soft: 41,000 TS 21,000 YS; 24 El; 70 RA; 160 Brin. For architectural trim; corrosion resisant.

PERALUMAN 50

Alluminio SA
Mn 0.2-0.4, Mg 4.5-5.6, bal Al.
1/2 H-temper: 45,5000 TS; 25,000 YS; 30 El; 80 Brin. H-25 temper: 57,000 TS; 48,000 YS; 8-13 El; 110-95 Brin. For chemical and food processing equipment; resists sea water corrosion.

PERALUMAN 50

Aluminium Industrie Aktiengesellschaft
Mn 0.2-0.4, Mg 4.5-5.6, bal Al.
1/2 H-temper: 45,5000 TS; 25,000 YS; 30 El; 80 Brin. H-25 temper: 57,000 TS; 48,000 YS; 8-13 El; 110-95 Brin. For chemical and food processing equipment; resists sea water corrosion.

PERALUMAN 50G

Aluminium Industries, AG
Mg 4-6, Mn 0.3, Si 0.7-1, bal Al.
Annealed: 41,000 TS; 21,000 YS; 24 El; 70 Brin. For architectural trim, corrosion resistant.

PERALUMAN 50G

Aluminium Industrie Aktiengesellschaft
Mg 4-6, Mn 0.3, Si 0.7-1, bal Al.
Annealed: 41,000 TS; 21,000 YS; 24 El; 70 Brin. For architectural trim, corrosion resistant.

PERALUMAN 5G

Swiss Aluminium Ltd.
Aluminum. Si 0.8-1, Mn 0.25-0.35, Mg 5-5.5, Ti 0.08-0.15, bal Al.
Sand cast: 28,500 TS; 15,500 YS; 2-5 El; 55-70 Brin. Die cast: 36,000 TS; 16,000 YS; 3-8 El; 60-80 Brin. For marine hardware, crankcases, instrument cases, fuel gages. Fair castability, high strength and corrosion resistant to seawater. *Obsolete*

PERALUMAN 7

Multiple manufacturers
Aluminum. Si 0.3-0.5, Mg 7, bal Al.
Soft: 42,000 TS; 21,000 YS; 24 El; 75 Brin. Hard: 64,000 TS; 51,000 YS; 4 El; 130 Brin. For rivets. Nonheat treatable.

PERALUMAN 7

Lavorazione Leghe Leggere SpA
Mg 7, Mn 0.5, bal Al.
Half hard: 60,000 TS; 43,000 YS; 15-20 El; 90-95 Brin. Annealed: 43,000 TS; 29,000 YS; 26-30 El; 80-90 Brin. For light alloy parts, aircraft parts; corrosion resistant.

PERALUMAN 7

Aluminium Industrie Aktiengesellschaft
Mg 7, Mn 0.5, bal Al.
Half hard: 60,000 TS; 43,000 YS; 15-20 El; 90-95 Brin. Annealed: 43,000 TS; 29,000 YS; 26-30 El; 80-90 Brin. For light alloy parts, aircraft parts; corrosion resistant.

PERALUMAN 7

Soc. Alluminio Veneto per Azioni
Aluminum. Mg 7, Mn 0.4, bal Al.
Heat treated: 37,000-43,000 TS; 19,000-22,000 YS; 5-11 El; 70-80 Brin. For marine hardware; resists seawater corrosion, age hardenable.

PERALUMAN 9

Alluminio SA
Aluminum. Mg 8-10, Mn 0.3, bal Al.
Cast: 35,600-40,000 TS; 21,400-27,000 YS; 6-13 El; 75-80 Brin. For fuel meters; corrosion resistant.

PERALUMAN 9

Soc. Alluminio Veneto per Azioni
Aluminum. Mg 9.3, Mn 0.3, Be 0.02, bal Al.
Heat treated: 28,000-35,000 YS; 14,000-19,000 YS; 1.5-2.5 El; 70-80 Brin. For marine hardware, resists seawater corrosion, age hardenable.

PERALUMAN 9G

Swiss Aluminium Ltd.
Aluminum. Mn 0.25-0.35, Mg 9-9.5, Be 0.03, bal Al.
Die cast: 38,000 TS; 23,000 YS; 1-3 El; 60-80 Brin. For general die castings, marine hardware. Fair castability, good strength, high corrosion resistance. Heat treatable. *Obsolete*

PERALUMAN-100

Aluminium Walzwerke Singen GmbH
Aluminum. Mg 1, bal Al.
Annealed: 105 MPa TS; 27 El. Hard: 210 MPa TS; 3 El. Decorative plate.

PERALUMAN-100

Swiss Aluminium Ltd.
Aluminum. Si 0.3, Fe 0.45, Cu 0.05, Mn 0.15, Mg 0.7-1.1, Cr 0.1, Zn 0.2, Ti 0.05, bal Al.
Extrusion alloy in billet form; not age hardenable. 100-140 N/mm^2 TS; 40-90 N/mm^2 YS; 15-17 El; 30 Brin. Also available as wrought alloy in rolling slab form.

PERALUMAN-150

Swiss Aluminium Ltd.
Aluminum. Si 0.3, Fe 0.45, Cu 0.05, Mn 0.15, Mg 1.4-1.7, Cr 0.1, Zn 0.2, Ti 0.05, bal Al.
Extrusion alloy in billet form; not age hardenable. 145-180 N/mm^2 TS; 50-100 N/mm^2 YS; 13-15 El; 40 Brin. Also available as wrought alloy in rolling slab form.

PERALUMAN-200

Swiss Aluminium Ltd.
Aluminum. Si 0.4, Fe 0.5, Cu 0.15, Mn 0.1-0.5, Mg 1.7-2.4, Cr 0.15, Zn 0.15, Ti 0.15, bal Al.
Wrought alloy in rolling slab form.

PERALUMAN-210

Swiss Aluminium Ltd.
Aluminum. Si 0.4, Fe 0.55, Cu 0.1, Mn 0.5-1.1, Mg 1.6-2.5, Cr 0.3, Zn 0.2, Ti 0.1, bal Al.
Wrought alloy in rolling slab form.

PERALUMAN-212
Swiss Aluminium Ltd.
Aluminum. Si 0.4, Fe 0.55, Cu 0.1, Mn 0.4-1, Mg 1.7-2.4, Cr 0.15, Zn 0.15, Ti 0.15, bal Al.
Extrusion alloy in billet form; not age hardenable. 200-290 N/mm^2 TS; 100-170 N/mm^2 YS; 11-14 El; 55 Brin.

PERALUMAN-260
Swiss Aluminium Ltd.
Aluminum. Si 0.25, Fe 0.4, Cu 0.1, Mn 0.4-1, Mg 2.4-3, Cr 0.05-0.2, Zn 0.25, Ti 0.2, bal Al.
Extrusion alloy in billet form; not age hardenable. 220-290 N/mm^2 TS; 100-180 N/mm^2 YS; 11-13 El; 60 Brin.

PERALUMAN-30
Swiss Aluminium Ltd.
Aluminum. Si 0.5, Fe 0.15, Cu 0.01, Mn 0.01-0.4, Mg 2.7-3.5, Zn 0.1, Ti 0.01-0.2, bal Al.
Primary foundry alloy in ingot form. 170-210 N/mm^2 TS; 70-100 N/mm^2 YS; 4-16 El; 50-60 Brin.

PERALUMAN-30
Aluminium Industrie Aktiengesellschaft
Mn 0-0.4, Mg 3-4, bal Al.
1/2 H-temper: 39,750 TS; 35,500 YS; 8 El; 80 Brin. H-temper: 49,500 TS. For chemical and food processing equipment, ship building; resists sea water corrosion.

PERALUMAN-300
Swiss Aluminium Ltd.
Aluminum. Si 0.4, Fe 0.4, Cu 0.1, Mn 0.15-0.5, Mg 2.6-3.2, Cr 0.1, Zn 0.2, Ti 0.15, bal Al.
Wrought alloy in rolling slab form.

PERALUMAN-302
Swiss Aluminium Ltd.
Aluminum. Si 0.4, Fe 0.4, Cu 0.1, Mn 0.15-0.5, Mg 2.9-3.4, Cr 0.1, Zn 0.2, Ti 0.15, bal Al.
Extrusion alloy in billet form; not age hardenable. 180-230 N/mm^2 TS; 80-150 N/mm^2 YS; 12-14 El; 50 Brin.

PERALUMAN-36
Swiss Aluminium Ltd.
Aluminum. Si 0.9-1.3, Fe 0.15, Cu 0.01, Mn 0.01-0.4, Mg 2.7-3.5, Zn 0.1, Ti 0.01-0.2, bal Al.
Primary foundry alloy in ingot form. 220-300 N/mm^2 TS; 160-220 N/mm^2 YS; 2-15 El; 70-90 Brin.

PERALUMAN-401
Aluminium Walzwerke Singen GmbH
Aluminum. Mg 4, bal Al.
Annealed: 240 MPa TS; 18 El. Hard: 330 MPa TS; 4 El.

PERALUMAN-412
Swiss Aluminium Ltd.
Aluminum. Si 0.4, Fe 0.4, Cu 0.1, Mn 0.2-0.7, Mg 3.5-4.5, Cr 0.05-0.2, Zn 0.25, Ti 0.15, bal Al.
Extrusion alloy in billet form; not age hardenable. 240-300 N/mm^2 TS; 100-180 N/mm^2 YS; 12-14 El; 65 Brin.

PERALUMAN-460
Swiss Aluminium Ltd.
Aluminum. Si 0.4, Fe 0.4, Cu 0.1, Mn 0.4-1, Mg 4-4.9, Cr 0.05-0.2, Zn 0.25, Ti 0.15, bal Al.
Wrought alloy in rolling slab form.

PERALUMAN-462
Swiss Aluminium Ltd.
Aluminum. Si 0.4, Fe 0.4, Cu 0.1, Mn 0.4-1, Mg 4-4.9, Cr 0.05-0.2, Zn 0.25, Ti 0.15, bal Al.
Extrusion alloy in billet form; not age hardenable. 270-350 N/mm^2 TS; 140-220 N/mm^2 YS; 10-12 El; 65 Brin.

PERALUMAN-50
Swiss Aluminium Ltd.
Aluminum. Si 0.3, Fe 0.15, Cu 0.01, Mn 0.01-0.4, Mg 4.8-5.5, Zn 0.1, Ti 0.01-0.2, bal Al.
Primary foundry alloy in ingot form. 160-250 N/mm^2 TS; 100-140 N/mm^2 YS; 10-25 El; 55-75 Brin.

PERALUMAN-502
Swiss Aluminium Ltd.
Aluminum. Si 0.4, Fe 0.4, Cu 0.1, Mn 0.2-0.5, Mg 4.9-5.6, Cr 0.1, Zn 0.25, Ti 0.15, bal Al.
Extrusion alloy in billet form; not age hardenable. 250-310 N/mm^2 TS; 110-180 N/mm^2 YS; 11-13 El; 60 Brin.

PERALUMAN-56
Swiss Aluminium Ltd.
Aluminum. Si 0.9-1.5, Fe 0.15, Cu 0.01, Mn 0.01-0.4, Mg 4.8-5.5, Zn 0.1, Ti 0.01-0.2, bal Al.
Primary foundry alloy in ingot form. 150-260 N/mm^2 TS; 110-160 N/mm^2 YS; 3-14 El; 60-85 Brin.

PERALUMAN-860
Swiss Aluminium Ltd.
Aluminum. Si 0.08, Fe 0.08, Cu 0.02, Mn 0.03, Mg 0.7-1.1, Cr 0.02, Zn 0.05, Ti 0.02, bal Al.
Extrusion alloy in billet form for electrical conductors. 100-130 N/mm^2 TS; 30-80 N/mm^2 YS; 13-15 El; 30 Brin.

PERALUMAN-98
Swiss Aluminium Ltd.
Aluminum. Si 0.25, Fe 0.3, Cu 0.01, Mn 0.1, Mg 9.5-11, Zn 0.1, Ti 0.01-0.2, bal Al.
Primary foundry alloy in ingot form. 280-330 N/mm^2 TS; 180-260 N/mm^2 YS; 8-15 El; 75-85 Brin.

PERAX
Marsh Bros. & Co. Ltd.
C, alloy, bal Fe.
For punches, dies, tools; oil hardened, tough.

PERCEIT
English manufacture
Co 55-80, Cr 20-35, W 0.1.
40,000-130,000 TS; 570 Brin. For high speed tools and dies; high speed steel.

PERCIT
Manufacturer not listed
For tools; similar to Stellite.

PERCIT EXTRA
Krupp Stahl AG
react and refract. C 1.1, Si 2.5, Mn 0.4, Cr 27, Co 60, W 4.3, bal Fe.
For hard facing electrodes; heat resistant. *Obsolete*

PERCIT SPECIAL
Krupp Stahl AG
superalloy. C 1.5, Si 1.45, Mn 0.4, Cr 27.5, Co 32.5, W 4.25, bal Fe.
For hard facing electrodes; heat resistant. *Obsolete*

PERCUSSION CAP BRASS
Chase Brass & Copper Co.
Cu 90, Zn 9.6, Pb 0.4.
Soft: 35,000 TS; 40 El; 52 Brin. Hard: 55,000 TS; 39,000 YS; 4 El; 135 Brin. For percussion caps, gilding, electrical parts; gilding metal. *Obsolete*

PERCY ALUMINUM
English manufacture
Cu 86-90, Al 7.5-13, Pb 0-2, Mn 0-1.5.
Rolled: 53,000 TS; 9 El. Cast: 12,000 TS; 30 El; 30 RA. For bearings, stripper nuts; heavy duty.

PERDONAL
Wieland-Werke AG Metallwerke
Mg 1.5-3, Mn 0.5-1.5, Cr 0-0.3, bal Al.
Soft: 26,000 TS; 10,000 YS; 20 El; 45 Brin. Hard: 41,000 TS; 36,000 YS; 5 El; 77 Brin. For roofing, hydraulic tubing, architectural trim; good forming and welding properties. *Obsolete*

PERDONAL
Wieland-Werke AG Metallwerke
Aluminum.
Aluminum alloy AlMg2Mn0.8.

PERDONAL
Wieland-Werke AG Metallwerke
Mg 2, Mn 0.5, bal Al.
Extruded: 29,000 TS; 20,000 YS: 15 El; 55 Brin. For ship building, car bodies, window frames; corrosion resistant. *Obsolete*

PERDURO
Dresser Industries
TC 1.6-1.8, CC 0.4-0.6, Mn 0.5-0.7, Si 1.5, Cu 0.9-1.1, bal Fe.
Heat treated: 80,000-90,000 TS; 60,000-70,000 YS; 6-8 El; 179-207 Brin. For cast chains, sprockets, chain links; resists heat to 1100°F; malleable iron.

PEREKS CC
Darwin & Milner Inc.
Nickel. C 0.66-0.8, Ni 49-57, Cr 24-30, Si, Mn, bal Fe.
For heat and corrosion resistant parts; heat and corrosion resistant.

PERFECTION ANTIFRICTION
NL Industries
Pb, Sn, Sb.
For bearings; babbitt. *Obsolete*

PERFORM 260 K
Krupp Stahl AG
C 0-0.1, Si 0-0.5, Mn 0-0.8, P 0-0.03, S 0-0.03, bal Fe.
W. Nr. 1.0480.

PERFORM 260 N
Krupp Stahl AG
C 0-0.16, Si 0-0.5, Mn 0-1.2, P 0-0.03, S 0-0.03, 0.015 Al min, bal Fe.
W. Nr. 1.0971.

PERFORM 260 N
Krupp Stahl AG
C 0-0.16, Si 0-0.5, Mn 0-1.2, P 0-0.03, S 0-0.03, 0.015 Al min, bal Fe.
W. Nr. 1.8941.

PERFORM 260 TM
Krupp Stahl AG
C 0-0.1, Mn 0-0.25, P 0-0.015, S 0-0.02, Al 0.03-0.07, Ti 0.12-0.2, bal Fe.
W. Nr. 1.0970.

PERFORM 260 TM
Krupp Stahl AG
C 0-0.1, Mn 0-0.25, P 0-0.015, S 0-0.02, Al 0.03-0.07, Ti 0.12-0.2, bal Fe.
W. Nr. 1.8940.

PERFORM 300 K
Krupp Stahl AG
carbon steel. C 0-0.1, Si 0-0.5, Mn 0-1, Al or Nb or Ti or V, bal Fe.
Structural steel, for plates, beams. W.Nr. 1.0489

PERFORM 300 TM
Krupp Stahl AG
C 0-0.12, Si 0-0.5, Mn 0-1.2, P 0-0.03, S 0-0.03, Nb 0-0.09, Ti 0-0.22, 0.015 Al min, bal Fe.
W. Nr. 1.0972.

PERFORM 300 TM
Krupp Stahl AG
C 0-0.12, Si 0-0.5, Mn 0-1.2, P 0-0.03, S 0-0.03, Nb 0-0.09, Ti 0-0.22, 0.015 Al min, bal Fe.
W. Nr. 1.8967.

PERFORM 340 K
Krupp Stahl AG
carbon steel. C 0-0.12, Si 0-0.5, Mn 0-1.5, Al, or Nb, or Ti or V, bal Fe.
Structural steel, for plates, girders, W. Nr. 1.0548.

PERFORM 340 N
Krupp Stahl AG
C 0-0.16, Si 0-0.5, Mn 0-1.5, P 0-0.03, S 0-0.03, 0.015 Al min, bal Fe.
W. Nr. 1.0975.

PERFORM 340 N
Krupp Stahl AG
C 0-0.16, Si 0-0.5, Mn 0-1.5, P 0-0.03, S 0-0.03, 0.015 Al min, bal Fe.
W. Nr. 1.8945.

PERFORM 340 TM
Krupp Stahl AG
C 0-0.12, Si 0-0.5, Mn 0-1.3, P 0-0.03, S 0-0.03, 0.015 Al min, bal Fe.
W. Nr. 1.0974.

PERFORM 340 TM
Krupp Stahl AG
C 0-0.12, Si 0-0.5, Mn 0-1.3, P 0-0.03, S 0-0.03, 0.015 Al min, bal Fe.
W. Nr. 1.8942.

PERFORM 340 W
Krupp Stahl AG
carbon steel. C 0-0.12, Si 0-0.5, Mn 0-1.5, bal Fe.
Structural steel, for plates, girders. W. Nr. 1.0534. *Obsolete*

PERFORM 360 TM
Krupp Stahl AG
C 0-0.1, Si 0-0.5, Mn 0-0.8, P 0-0.03, S 0-0.03, Ti 0.12-0.2, 0.015 Al min, bal Fe.
W. Nr. 1.0976.

PERFORM 360 TM
Krupp Stahl AG
C 0-0.1, Si 0-0.5, Mn 0-0.8, P 0-0.03, S 0-0.03, Ti 0.12-0.2, 0.015 Al min, bal Fe.
W. Nr. 1.8946.

PERFORM 380 K
Krupp Stahl AG
C 0-0.1, Si 0-0.5, Mn 0-1.3, P 0-0.03, S 0-0.03, bal Fe.
W. Nr. 1.0550.

PERFORM 380 N
Krupp Stahl AG
C 0-0.18, Si 0-0.5, Mn 0-1.6, P 0-0.03, S 0-0.03, 0.015 Al min, bal Fe.
W. Nr. 1.0979.

PERFORM 380 N
Krupp Stahl AG
C 0-0.18, Si 0-0.5, Mn 0-1.6, P 0-0.03, S 0-0.03, 0.015 Al min, bal Fe.
W. Nr. 1.8950.

PERFORM 380 TM
Krupp Stahl AG
C 0-0.12, Si 0-0.5, Mn 0-1.4, P 0-0.03, S 0-0.03, 0.015 Al min, bal Fe.
W. Nr. 1.0978.

PERFORM 380 TM
Krupp Stahl AG
C 0-0.12, Si 0-0.5, Mn 0-1.4, P 0-0.03, S 0-0.03, 0.015 Al min, bal Fe.
W. Nr. 1.8951.

PERFORM 420 K
Krupp Stahl AG
carbon steel. C 0-0.12, Si 0-0.5, Mn 0-1.6, Al or Nb or Ti or V, bal Fe.
Structural steel, for plates, girders. W. Nr. 1.0556.

PERFORM 420 K
Krupp Stahl AG
C 0-0.1, Si 0-0.5, Mn 0-1.4, P 0-0.03, S 0-0.03, bal Fe.
W. Nr. 1.0556.

PERFORM 420 N
Krupp Stahl AG
C 0-0.2, Si 0-0.5, Mn 0-1.6, P 0-0.03, S 0-0.03, 0.015 Al min, bal Fe.
W. Nr. 1.0981.

PERFORM 420 N
Krupp Stahl AG
C 0-0.2, Si 0-0.5, Mn 0-1.6, P 0-0.03, S 0-0.03, 0.015 Al min, bal Fe.
W. Nr. 1.8952.

PERFORM 420 TM
Krupp Stahl AG
C 0-0.12, Si 0-0.5, Mn 0-1.5, P 0-0.03, S 0-0.03, 0.015 Al min, bal Fe.
W. Nr. 1.0980.

PERFORM 420 TM
Krupp Stahl AG
C 0-0.12, Si 0-0.5, Mn 0-1.5, P 0-0.03, S 0-0.03, 0.015 Al min, bal Fe.
W. Nr. 1.8953.

PERFORM 420 W
Krupp Stahl AG
carbon steel. C 0-0.12, Si 0-0.5, Mn 0-1.6, bal Fe.
Structural steel, for beams, girders. W. Nr. 1.0590. *Obsolete*

PERFORM 460 K
Krupp Stahl AG
C 0-0.12, Si 0-0.5, Mn 0-1.6, P 0-0.03, S 0-0.03, bal Fe.
W. Nr. 1.0574.

PERFORM 460 N
Krupp Stahl AG
C 0-0.21, Si 0-0.5, Mn 0-1.7, P 0-0.03, S 0-0.03, 0.015 Al min, bal Fe.
W. Nr. 1.0983.

PERFORM 460 N
Krupp Stahl AG
C 0-0.21, Si 0-0.5, Mn 0-1.7, P 0-0.03, S 0-0.03, 0.015 Al min, bal Fe.
W. Nr. 1.8955.

PERFORM 460 TM
Krupp Stahl AG
C 0-0.12, Si 0-0.5, Mn 0-1.6, P 0-0.03, S 0-0.03, 0.015 Al min, bal Fe.
W. Nr. 1.0982.

PERFORM 460 TM
Krupp Stahl AG
C 0-0.12, Si 0-0.5, Mn 0-1.6, P 0-0.03, S 0-0.03, 0.015 Al min, bal Fe.
W. Nr. 1.8956.

PERFORM 460 W
Krupp Stahl AG
carbon steel. C 0-0.12, Si 0-0.5, Mn 0-1.7, bal Fe.
Structural steel, for beams, girders. W. Nr. 1.0592. *Obsolete*

PERFORM 500 K
Krupp Stahl AG
C 0-0.15, Si 0-0.5, Mn 0-1.7, P 0-0.03, S 0-0.03, bal Fe.
W. Nr. 1.0599.

PERFORM 500 N
Krupp Stahl AG
C 0-0.22, Si 0-0.5, Mn 0-1.7, P 0-0.03, S 0-0.03, 0.015 Al min, bal Fe.
W. Nr. 1.0985.

PERFORM 500 N
Krupp Stahl AG
C 0-0.22, Si 0-0.5, Mn 0-1.7, P 0-0.03, S 0-0.03, 0.015 Al min, bal Fe.
W. Nr. 1.8957.

PERFORM 500 TM
Krupp Stahl AG
C 0-0.12, Si 0-0.5, Mn 0-1.7, P 0-0.03, S 0-0.03, 0.015 Al min, bal Fe.
W. Nr. 1.0984.

PERFORM 500 TM
Krupp Stahl AG
C 0-0.12, Si 0-0.5, Mn 0-1.7, P 0-0.03, S 0-0.03, 0.015 Al min, bal Fe.
W. Nr. 1.8959.

PERFORM 500 W
Krupp Stahl AG
carbon steel. C 0-0.12, Si 0-0.5, Mn 0-1.7, bal Fe.
Structural steel, for plates, beams, girders. W. Nr. 1.0596.
Obsolete

PERFORM 550 TM
Krupp Stahl AG
C 0-0.12, Si 0-0.5, Mn 0-1.8, P 0-0.03, S 0-0.03, 0.015 Al min, bal Fe.
W. Nr. 1.0986.

PERFORM 550 TM
Krupp Stahl AG
C 0-0.12, Si 0-0.5, Mn 0-1.8, P 0-0.03, S 0-0.03, 0.015 Al min, bal Fe.
W. Nr. 1.8948.

PERFORM 690 TM
Krupp Stahl AG
C 0-0.15, Si 0.2-0.5, Mn 1.8-2.1, P 0-0.025, S 0-0.02, Nb 0-0.06, Ti 0.12-0.22, 0.015 Al min, bal Fe.
W. Nr. 1.8949.

PERKING BRASS
English manufacture
Cu 76.2-80, Sn 19.8-23.9, Zn 0-0.14.
For ornamental parts, reflectors.

PERLIT (PERLITGUSS)
Dursar Corp.
C 2.4-3.5, Si 0.5-1.5, P 0.2-0.6, Mn 0.6-1, Ni, Cr, bal Fe.
30,000 TS; 200 Brin. For automotive engine blocks, brake drums; nickel cast iron; wear, resistant, tough.

PERLIT (PERLITGUSS)
Heinrich Lanz A.G.
C 2.4-3.5, Si 0.5-1.5, P 0.2-0.6, Mn 0.6-1, Ni, Cr, bal Fe.
30,000 TS; 200 Brin. For automotive engine blocks, brake drums; nickel cast iron; wear, resistant, tough.

PERLIT NICKEL CAST IRON
Manufacturer not listed
C 1.7-3, <1 Si, Ni, Mn, bal Fe.
For casting, frames, housings; U.S. Patent 1564284; sufficient Ni to ppt. graphite.

PERMADOR
Degussa AG
Precious metal.
Precious metal alloy for dentistry and dental engineering.

PERMADUR H
Otto Wolff Handelgesellschaft
C 0.55, Cr 1.05, V 0.18, W 1.85, bal Fe.
For cold work tools, upsetters, crimpers, punches: oil hardened, tough.

PERMADUR W
Otto Wolff Handelgesellschaft
Cr 1.05, V 0.18, W 1.85, bal Fe.
For cold work tools, upsetters, punches, dies; oil hardened tough.

PERMADUR-ADS

J.C. Soding & Halbach
C 1.05, Si 0.5, Mn 0.5, Cr 5.25, Mo 1.15, V 0.3, bal Fe.
Heat treated: 215,000 TS; 195,000 YS; 5 El; C 50 Rock. For forming and blanking dies, shear blades, master hubs, punches. Wear resistant, Type A2, air hardening. *Obsolete*

PERMADUR-ADWS

J.C. Soding & Halbach
C 1.65, Si 0.32, Mn 0.3, Cr 11.5, Mo 0.6, W 0.5, V 0.09, bal Fe.
Heat treated: 278,000 TS; 214,000 YS; 1 El; C 56 Rock. For blanking and drawing dies, punches, thread rolling dies. Type D2 air hardening. Wear resistant, nondeforming. *Obsolete*

PERMAFLY

Creusot-Loire
Ni 80, Fe 20.
For magnets for electrical equipment; magnetically soft. *Obsolete*

PERMAG

Now ALLOYMET PERMAG.

PERMAL

Perry Barr Metal Co., Ltd.
Si 5, bal Al.
For die casting; good fluidity.

PERMALLOY

Western Electric Co.
Ni 39-81, Fe 19-61.
For loading material for signaling conductors, audio frequency transformers; greater magnetic qualities than iron.

PERMALLOY 12.5-80 MO

Western Electric Co.
Mo 12.5, Fe 7.5, Ni 80.
For magnets; soft.

PERMALLOY 3.8-78.5 CR

American manufacture
Cr 3.8, Mn 0.6, Fe 17.1, bal Ni.
For soft magnets; high permeability.

PERMALLOY 4-79 MO

Western Electric Co.
Mn 0.3, Mo 4, Fe 16.7, bal Ni.
85,000 TS; 24,000 YS. For laminated cores, inductors, transformers, magnetic field detectors.

PERMALLOY 45

Wilbur B. Driver Co.
Nickel. Ni 45, Fe 55.
Saturation induction 16,000 Gausses. For cores and armatures in relays, motors, inductors, transformers, electromagnetic shields. Soft magnet.

PERMALLOY 45

Western Electric Co.
Ni 45, Mn 0.3, bal Fe.
Sheet: 65,800 TS; 21,000 YS; 30 El; 10, Brin. Initial permeability 2500; max permeability 25,000. For audio frequency apparatus and output transformers; high permeability.

PERMALLOY 45

Bell Telephone Laboratories
Ni 45, Mn 0.3, bal Fe.
Sheet: 65,800 TS; 21,000 YS; 30 El; 10, Brin. Initial permeability 2500; max permeability 25,000. For audio frequency apparatus and output transformers; high permeability.

PERMALLOY 49

Wilbur B. Driver Co.
Nickel. Ni 49-51, bal Fe.
Saturation induction 16,000 Gausses. For leads and armatures for glass shielded switches, laminated cores for magnetic amplifiers, motor transformers, shielding. Soft magnet.

PERMALLOY 70

Wilbur B. Driver Co.
Ni 70, Fe 30.
Saturation inductance 12,500. For square loop laminated cores for magnetic amplifiers. Soft magnet. *Obsolete*

PERMALLOY 78

Bell Telephone Laboratories
Ni 78.5, Fe 21.1, Mn 0.3.
Sheet: 75,000 TS; 15,000 YS; 35 El; 115 Brin. Initial permeability 8000; max permeability 100,000; 0.05 Hc. For sensitive D.C. appaaratus, relays; high permeability.

PERMALLOY 78

Western Electric Co.
Ni 78.5, Fe 21.1, Mn 0.3.
Sheet: 75,000 TS; 15,000 YS; 35 El; 115 Brin. Initial permeability 8000; max permeability 100,000; 0.05 Hc. For sensitive D.C. appaaratus, relays; high permeability.

PERMALLOY 80

Now MAGNETICS ROUND PERMALLOY 80.

PERMALLOY 80 (1)

Spang Specialty Metals
Ni 79.5-80.6, Mo 3.8-4.8, Si 0-0.5, Mn 0-0.8, C 0-0.05.
Soft magnetic alloy in sheet, strip, bar and rod form. Conforms to ASTM A 753 and MILN 14411 with regards to composition, temper and magnetic properties.

PERMALLOY 80 (2)

Spang Specialty Metals
Ni 80, Mo 4.5, bal Fe.
Soft magnetic alloy strip, very high shielding value. *Obsolete*

PERMALLOY B

ITT Components Group Europe
Ni 46, Fe 54.
For telecommunications; low coefficient of expansion.

PERMALLOY D

ITT Components Group Europe
Ni 36, Fe 64.
For magnetic equipment; low coefficient of expansion.

PERMALLOY F

ITT Components Group Europe
Nickel. Ni 65, Fe 35.
Domain oriented alloy for electrical equipment; high magnetic permeability.

PERMALLOY G

ITT Components Group Europe
Ni 46, Fe 54.
Domain oriented alloy for electrical equipment; high magnetic permeability.

PERMALLOY STANDARD

Bell Telephone Laboratories
Ni 78.5, Fe 21.5, C 0.04, Co 0.37, Mn 0.022.
For loading submarine cables; high electrical resistance.

PERMALLOY SUPER C

ITT Components Group Europe
Nickel. Ni 77.4, Fe 13.3, Mo 3.7, Cu 5.
For transformers, chokes, magnetic shielding (0.001) 50,000 min. Soft magnet; high permeability.

PERMALLOY-68

Bell Telephone Laboratories
Ni 68, bal Fe.
For magnetic apparatus; high permeability.

PERMALLOY-78.5

Wilbur B. Driver Co.
Nickel. Ni 78.5, Fe 21.5.
Saturation induction 10,700 Gausses. For reeds in mercury wetted switches, relay armatures. Soft magnet.

PERMAN

SWB Stahlformguss Gesellschaft mbH
C 1.05, Cr 1, W 1.15, bal Fe.
For bearings, cutters, liners, sheaves; water hardened, wear resistant. *Obsolete*

PERMAN N

SWB Stahlformguss Gesellschaft mbH
C 1, Mn, W, Cr, V, bal Fe.
For bearings, cutters, sleeves, liners; water hardened, wear resistant. *Obsolete*

PERMANENT

Now SODING PERMANENT.

PERMANENT 10

J.C. Soding & Halbach
C 0.8, W, Co, Cr, V, Mo, bal Fe.
For lathe and planer tools, reamers, taps, hobs; high speed steel. *Obsolete*

PERMANENT 17

J.C. Soding & Halbach
C 0.86, Co 2.8, Cr 4.3, Mo 0.85, V 2.1, W 12, bal Fe.
For lathe and planer tools, reamers, broaches; high speed steel. *Obsolete*

PERMANENT 35

J.C. Soding & Halbach
C 0.8, Co 5, Cr 4.5, Mo 0.8, V 1.7, W 18.5, bal Fe.
High-speed steel, for lathe and planer tools, reamers, drills, hobs.

PERMANENT 40

J.C. Soding & Halbach
C 0.86, Cr 4, Mo 0.8, V 2.5, W 12, bal Fe.
For lathe and planer tools, hobs; high speed steel. *Obsolete*

PERMANENT 50

J.C. Soding & Halbach
C 1.3, Cr 4.3, Mo 0.85, V 3.8, W 12, bal Fe.
For engravers' tools, forming and drawing dies, cutters; high speed steel. *Obsolete*

PERMANENT 70

J.C. Soding & Halbach
C 1.4, Cr 4.5, Co 5, V 4, W 12.5, Mo 1, bal Fe.
High-speed steel; for lathe and planer tools, reamers, hobs, taps.

PERMANENT 80

J.C. Soding & Halbach
C 0.76, Co 10, Cr 4.2, Mo 0.8, V 1.8, W 18, bal Fe.
For lathe and planer tools, form cutters; high speed steel. *Obsolete*

PERMANENTE

Permanente Magnesium Inc.
Zn 2.89, Al 6, Mn 0.25, bal Mg.
Cast: 28,300 TS; 7 El; 52 Brin. For light alloy castings; heat treatable. *Obsolete*

PERMANICKEL

Now PERMANICKEL ALLOY 300.

PERMANICKEL 300

Driver Harris Co.
Nickel. Ni 98.5, C, Mn, Fe, Si, Cu, Ti, Mg.
For high temperature age hardenable springs. *Obsolete*

PERMANICKEL 300
Criterion Metals, Inc.
Nickel. Ni 98.6, C 0.25, Mn 0.1, Fe 0.1, S 0.005, Si 0.06, Cu 0.02, Ti 0.5, Mg 0.35.
Thin gauge sheet, various tempers. Annealed: 90-120 ksi TS; 35-60 ksi YS. Age hardened: 150-190 ksi TS. ASTM F-290.

PERMANICKEL ALLOY 300
Huntington Alloys Inc.
Ni 98.5, C 0.2, Mn 0.25, Fe 0.3, Cu 0.13, Ti 0.4, Mg 0.35, S 0.005, Si 0.18.
Hot-finished: 90,000-120,000 TS; 35,000-65,000 YS; 45-20 El; 140-230 Brin. Hot-finished and aged: 160,000-200,000 TS; 120,000-150,000 YS; 20-10 El; 285-360 Brin. For grid lateral winding wires; magnetostriction devices; thermostat contact arms, solid-state capacitors; grid slide rods, diaphragms, springs; clips, fuel cells.

PERMANICKEL ALLOY 300
Techalloy Co. Inc.
Ni 98.5, C 0.2, Mn 0.25, Fe 0.3, Cu 0.13, Ti 0.4, Mg 0.35, S 0.005, Si 0.18.
Hot-finished: 90,000-120,000 TS; 35,000-65,000 YS; 45-20 El; 140-230 Brin. Hot-finished and aged: 160,000-200,000 TS; 120,000-150,000 YS; 20-10 El; 285-360 Brin. For grid lateral winding wires; magnetostriction devices; thermostat contact arms, solid-state capacitors; grid slide rods, diaphragms, springs; clips, fuel cells.

PERMANIT 110
Vereinigte Edelstahlwerke
C 0.05, Ni 21, Al 12, bal Fe.
For cast permanent magnets; high permeability. *Obsolete*

PERMANIT 120
Now VEW P752.

PERMANIT 140
Vereinigte Edelstahlwerke
C 0.05, Ni 26, Co 8, Al 11, bal Fe.
For cast magnets; permanent magnet. *Obsolete*

PERMANIT 160
Now VEW P754.

PERMANIT 180K
Vereinigte Edelstahlwerke
C 0.04, Ni 20, Co 20, Al 7, Cu 7, Ti 6, bal Fe.
For magnets; permanent. *Obsolete*

PERMANIT 200
Vereinigte Edelstahlwerke
C 0.05, Ni 20, Co 15, Al 9.5, Cu 3.5, bal Fe.
For cast magnets; permanent magnet. *Obsolete*

PERMANIT 250
Vereinigte Edelstahlwerke
C 0.04, Ni 20, Co 24, Al 8, Cu 4, Ti 5, bal Fe.
For magnets; permanent. *Obsolete*

PERMANIT 30
Now VEW P702.

PERMANIT 35
Now VEW P702.

PERMANIT 38
Vereinigte Edelstahlwerke
C 1, Mn 1, Cr 5, bal Fe.
For magnets; permanent. *Obsolete*

PERMANIT 40
Now VEW P712.

PERMANIT 400
Now VEW P760.

PERMANIT 400K
Vereinigte Edelstahlwerke
C 0.01, Ni 14, Co 24, Al 8, Cu 3, Ti, bal Al.
For magnets; permanent. *Obsolete*

PERMANIT 50
Vereinigte Edelstahlwerke
C 1, Cr 8, Mo 1-2, Co 6.5, bal Fe.
For permanent magnets; high permeability. *Obsolete*

PERMANIT 500
Now VEW P758.

PERMANIT 600
Vereinigte Edelstahlwerke
C 0.05, Ni 14, Co 24, Al 8, Cu 3, bal Fe.
For magnets; permanent. *Obsolete*

PERMANIT 70
Vereinigte Edelstahlwerke
O 0.06, Cr 8.5, Mo 1, Co 15.5, bal Fe.
For permanent magnet; 9000 gausses residual. *Obsolete*

PERMANITE
American manufacture
Co, Cr, W, bal Fe.
For permanent magnet steel; high coercive force.

PERMANT
Firth Brown Ltd.
C 0.08, Mn 0.2, Si 0.05, Ni 36, bal Fe.
For scales, clocks, watches, instruments; low coefficient of expansion. *Obsolete*

PERMAS
Fisher Scientific Co.
C 0-0.25, Cr 24-26, Ni 19-22, bal Fe.
Rolled: 95,000 TS; 45,000 YS; 50 El; 65 RA; 185 Brin. For balance weights. Type 011; corrosion resistant.

PERMAT
Manufacturer not listed
Cu 45, Co 30, Ni 25.
For magnets for electrical equipment; magnetically soft.

PERMAX
Alloy Foundries
C, alloy, bal Fe.
Cast: 60,000 TS; 38,000 YS; 10 El; 150 Brin. For bell yokes, direct current magnets.

PERMAX F
Vacuumschmelze GmbH
Ni 65, bal Fe.
Soft magnetic alloy with flat hysteresis loop for power electronics; low remanence and low coercivity. Saturation flux density 1.25 T.

PERMAX M
Vacuumschmelze GmbH
Soft magnetic 54 to 68% nickel iron alloy for transducers, leakage current protection switches, transformers.

PERMAX Z
Vacuumschmelze GmbH
Ni 65, bal Fe.
Soft magnetic alloy with square hysteresis loop for magnetic amplifiers, converters, chokes and cores. Saturation flux density 1.25 T.

PERMENDUR
Western Electric Co.
Co 50, Fe 50.

PERMENDUR 2-V
Bell Telephone Laboratories
Co 48.8, Fe 48.8, Mn 0.4, V. 1.7.
Sheet: 87,000 TS; 60,000 YS. For receiver diaphragms.

PERMENDUR 2-V
Western Electric Co.
Co 48.8, Fe 48.8, Mn 0.4, V. 1.7.
Sheet: 87,000 TS; 60,000 YS. For receiver diaphragms.

PERMENDUR 24
Telcon Metals Ltd.
Co 24, bal Fe.
Soft magnetic alloy, high saturation and good mechanical properties.

PERMENDUR 49
Telcon Metals Ltd.
Fe 49, Co 49, V 2.
Soft magnetic alloy, high saturation, used for stators, etc.

PERMENDUR V
Spang Specialty Metals
Co 47.5-49.5, V 1.75-2.1, C 0-0.25, Mn 0-0.1, Si 0-0.1.
Soft magnetic alloy in sheet and strip form. Conforms to ASTM A801 and MIL A47182 with regards to composition, temper and magnetic properties.

PERMENDUR-50 KF
Russian manufacture
Co 49.95, V 1.4, Si 0.04, Mn 0.22, C 0.02, S 0.007, P, bal Fe.
For magnetic and electrical equipment; soft magnetic alloy, high permeability, high magnetostriction saturation.

PERMENOR-4801
Vacuumschmelze GmbH
Ni 48, bal Fe.
Cold rolled and annealed: 58,000 TS; 19,000 YS; 27 El. For high temperature low expansion applications, audio transformers, motor rotors. High permeability, controlled expansion. *Obsolete*

PERMENORM 3601
Vacuumschmelze GmbH
Soft magnetic 36 to 40% nickel-iron alloy for transformers, chokes, relay components; high permeability; K1, K2, K3.

PERMENORM 500 H2
Vacuumschmelze GmbH
Soft magnetic 40 to 50% nickel-iron alloy for transducers, leakage current protective switches, magnetic shields, systems for measuring instruments; high permeability.

PERMENORM 500 S2
Vacuumschmelze GmbH
Soft magnetic 45 to 50% nickel alloy for applications requiring high permeability and high flux density; transformers, measurement technology, and relays. 140,000 microns permeability max.

PERMENORM 5000 H3
Vacuumschmelze GmbH
Soft magnetic 40 to 50% nickel-iron alloy for magnetic amplifiers, counting and storage cores, chokes, pulse transformers; high permeability.

PERMENORM 5000 Z
Vacuumschmelze GmbH
Now PERMENORM 5000 H3.

PERMENORM 5000 Z
Vacuumschmelze GmbH
Ni 50, bal Fe.
Soft magnetic alloy with square hysteresis loop for magnetic amplifiers, converters, chokes and cores. Saturation flux density 1.55 T.

PERMENORM 5000 ZE
Vacuumschmelze GmbH
Ni 50, bal Fe.
Soft magnetic alloy with square hysteresis loop for magnetic amplifiers, converters, chokes and cores. Saturation flux density 1.60 T.

PERMENORM 5050 F

Vacuumschmelze GmbH
Ni 50, bal Fe.
Soft magnetic alloy with flat hysteresis loop for power electronics; low remanence and low coercivity. Saturation flux density 1.52 T.

PERMENORM 5050 H4

Vacuumschmelze GmbH
Soft magnetic 45 to 50% nickel alloy for applications requiring high permeability and high flux density; transformers, measurement technology, and relays. 90,000 microns permeability max.

PERMET PF-1

Colt Industries
Fe 100.
For magnets; Br 5700, Hc 470. *Obsolete*

PERMET PF-2

Crucible Materials Corp.
Reactive and refractory. Co 30, Fe 70.
For magnets; 1,520,000 BH max. 6000 Br; 3830 Bo; 625 Hc.

PERMINVAR 25-45

Bell Telephone Laboratories
Ni 45, Co 25, Fe 30.
1.2 Hc, 400 initial permeability, 2000 max permeability, 10,000 B. For electrical communication, circuits, magnetic circuits. High permeability, soft magnet.

PERMINVAR 25-45

ITT Components Group Europe
Ni 45, Co 25, Fe 30.
1.2 Hc, 400 initial permeability, 2000 max permeability, 10,000 B. For electrical communication, circuits, magnetic circuits. High permeability, soft magnet.

PERMINVAR 43-23

Western Electric Co.
Mn 0.3, Ni 43, Co 23, Fe 33.7.
For magnetic coils, transformers; soft.

PERMINVAR 7-70

Kanthal Corp.
Nickel. Co 7, Fe 23, Ni 70.
Permeability 850-4000; coercive force 0.6; saturation induction 12,500. For magnets for electrical and magnetic instruments. Curie temperature 650°C. Soft magnetic alloy.
Obsolete

PERMINVAR 7.5-45-25 MO

Western Electric Co.
Mn 0.6, Mo 7.5, Ni 45, Co 25, Fe 21.9.
Initial permeability 850; maximum permeability 4000; Hc = 0.6. Soft magnetic material.

PERMINVAR 7.5-70

Bell Telephone Laboratories
Co 7.5, Ni 70, Fe 22.7, Mn 0.3.
For communication equipment for modulus cores; high initial permeability.

PERMINVAR 7.5-70

Western Electric Co.
Co 7.5, Ni 70, Fe 22.7, Mn 0.3.
For communication equipment for modulus cores; high initial permeability.

PERMITE

Aluminum Industries Inc.
Si 0.5, Cu 10, Fe 1, Mg 0.4, bal Al.
40,000 TS; 38,000 YS; 0.5 El; 120 Brin. For internal combustion engine pistons; coefficient expansion 0.0000127.

PERMITE 1002 (2002)

Aluminum Industries Inc.
Cu 9-11, Si 0.4-0.6, Mg 0.25-0.4, Fe 1-1.5, bal Al.
Cast: 26,000-45,000 TS; 21,000-35,000 YS; 1.0 El; 60-150 Brin. For light alloy parts, pistons, permanent mold castings; non-hardenable.

PERMITE 1005

Aluminum Industries Inc.
Cu 7, Fe 1.2, Si 1.7, bal Al.
For permanent mold castings; similar to Alcoa 12.

PERMITE 1011 (2011)

Aluminum Industries Inc.
Fe 0.75, Si 4-5, Zn 0.5, Mg 0.1, bal Al.
Chill cast: 21,000 TS; 2.5 El; 50 Brin. Sand cast: 17,000 TS; 3.0 El; 45 Brin. For general castings; corrosion resistant.

PERMITE 1018

Aluminum Industries Inc.
Cu 1-1.5, Si 0-0.5, Fe 0-0.6, Mn 0.7-1.2, bal Al.
23,000-30,000 TS; 3500-10,000 YS; 8-12 El; 45-70 Brin. For light alloy parts; non-hardenable.

PERMITE 1020

Aluminum Industries Inc.
Si 7, Mg 0.3, bal Al.
Heat treated and T-6 temper: 38,000 TS; 5 El; 90 Brin. For permanent mold castings; similar to Alcoa 356.

PERMITE 1021

Aluminum Industries Inc.
Zn 1.8, Mg 3.8, bal Al.
Cast: 25,000 TS; 8 El; 60 Brin. For permanent mold castings; similar to Alcoa A214.

PERMITE 1024

Aluminum Industries Inc.
Cu 7, Si 5.5, Mg 0.3, bal Al.
T-551 temper: 31,000 TS; 1 El; 110 Brin. For permanent mold castings; similar to Alcoa 152.

PERMITE 1027

Aluminum Industries Inc.
Cu 4.5, Si 5.5, bal Al.
Cast: 28,000 TS; 4 El; 70 Brin. For permanent mold castings; similar to Alcoa A 108.

PERMITE 1029

Aluminum Industries Inc.
Cu 3.5, Si 6, bal Al.
Temper: 38,000 TS; 4 El; 95 Brin. For permanent mold castings; similar to Alcoa 319.

PERMITE 1031

Aluminum Industries Inc.
Cu 1, Ni 1, Sn 6.5, bal Al.
T-533 temper: 20,000 TS; 10 El; 100 Brin. For permanent mold castings; similar to Alcoa 750.

PERMITE 1034

Aluminum Industries Inc.
Cu 3, Si 9.5, Mg 1, Ni 1, bal Al.
Heat treated: 34,000 TS; 10 El; 105 Brin. For permanent mold castings; similar to Alcoa D132.

PERMITE 2001 (KANT SKORE PISTON ALLOY)

Aluminum Industries Inc.
Cu 9-11, Fe 0.5, Si 1.25, Zn 0.25, Mg 0.35, Mn 0.1, bal Al.
Heat treated: 20,000-38,000 TS; 0-2 El; 65-140 Brin. Cast: 25,000 TS; 0.5 El; 85 Brin. For pistons. *Obsolete*

PERMITE 2003

Aluminum Industries Inc.
Cu 0.8, Fe 0.8, Si 12, Mg 1, Ni 2.5, bal Al.
For sand castings; similar to Alcoa A132 alloy.

PERMITE 2004

Aluminum Industries Inc.
Cu 7, Fe 1.2, Zn 1.7, bal Al.
Cast: 23,000 TS; 3 El; 70 Brin. For sand castings; similar to Alcoa 112.

PERMITE 2010

Aluminum Industries Inc.
Cu 4-5, Fe 1, Si 1, Zn 0.25, Mg 0.2, bal Al.
Sand cast: 27,000 TS; 2.0 El; 50 Brin. Heat treated: 28,000-38,000 TS; 1.0-5.0 El; 60-100 Brin. For general castings; age hardened.

PERMITE 2025

Aluminum Industries Inc.
Cu 4, Si 3, bal Al.
Cast: 21,000 TS; 2.5 El; 55 Brin. For sand castings; similar to Alcoa 108.

PERMITE 3011

Aluminum Industries Inc.
Si 5, bal Al.
Cast: 30,000 TS; 14,000 YS; 7 El. For die castings; similar to Alcoa 43.

PERMITE 3012

Aluminum Industries Inc.
Si 12, bal Al.
Cast: 37,000 TS; 18,000 YS; 1.8 El. For die castings; similar to Alcoa 13.

PERMITE 3027

Aluminum Industries Inc.
Cu 4, Si 5, bal Al.
Cast: 40,000 TS; 22,000 YS; 3.5 El. For die castings; similar to Alcoa 85.

PERMITE 3032

Aluminum Industries Inc.
Mg 8, bal Al.
Cast: 42,000 TS; 23,000 YS; 7 El. For die castings; similar to Alcoa 218.

PERMITE 3033

Aluminum Industries Inc.
Si 9.5, Mg 0.5, bal Al.
Cast: 42,000 TS; 23,000 YS; 1.8 El. For die castings; similar to Alcoa 360.

PERMITE 3034

Aluminum Industries Inc.
Cu 3.5, Si 8.5, bal Al.
Cast: 45,000 TS; 25,000 YS; 2.0 El. For die castings; similar to Alcoa 380.

PERMITE A-2023 SPECIAL

Aluminum Industries Inc.
Si 0-0.2, Mg 9.5-10.5, Zn 0.4-0.6, Fe, bal Al.
27,000 TS; 18,000 YS; 4 El; 60-70 Brin. For light alloy parts, pistons, permanent mold castings, pumps; heat treatable.

PERMITE NO. 1003 (LOW EXPANSION ALLOY)

Aluminum Industries Inc.
Cu 1, Fe 1, Si 13.5, Zn 0.25, Mg 1, Mn 0.1, Ni 2.5, bal Al.
Heat treated: 28,000-35,000 TS; 0.5-1.0 El; 90-100 Brin. Cast: 25,000 TS; 1.0 El; 85 Brin. For low expansion alloy; coefficient expansion 0.0000110.

PERMITE NO. 1004 (2004)

Aluminum Industries Inc.
Cu 6-8, Fe 1.5, Si 2, Zn 2.5, Mg 0.2, Mn 0.2, bal Al.
Chill cast: 23,000 TS; 1.0 El; 80 Brin. Sand cast: 16,000-20,000 TS; 1.0-2.5 El; 60 Brin. For general castings; coefficient expansion 0.000013.

PERMITE NO. 1006 (2006)

Aluminum Industries Inc.
Cu 11-13.5, Fe 0.75, Si 0.5, Zn 0.25, bal Al.
Chill cast: 25,000 TS; 1.0 El; 100 Brin. Sand cast: 23,000 TS; 1.0 El; 90 Brin. For leak proof castings; high fluidity.

PERMITE NO. 1007 (2007)

Aluminum Industries Inc.
Cu 6-8, Fe 1, Si 1-2, Zn 0.25, Mg 0.2, Mn 0.2, bal Al.
Chill cast: 26,000 TS; 2.5 El; 75 Brin. Sand cast: 24,000 TS; 2.0 El; 70 Brin. For general castings; good machinability.

PERMITE NO. 1008 (2008)

Aluminum Industries Inc.
Cu 4-5, Fe 1.2, Si 2-3, Zn 0.25, bal Al.
Chill cast: 25,000 TS; 2.0 El; 70 Brin. Heat treated: 35,000-38,000 TS; 1.0-4.0 El; 70-100 Brin. For general castings; age hardened.

PERMITE NO. 1009 (2009)

Aluminum Industries Inc.
Cu 4-5, Fe 0.5, Si 0.5, bal Al.
Sand cast: 25,000 TS; 2 El. Heat treated: 28,000-36,000 TS; 1.0-6.0 El; 60-75 Brin. For general castings; age hardened.

PERMITE NO. 1010

Aluminum Industries Inc.
Cu 4, Si 1, bal Al.
For machine parts; age hardenable.

PERMITE NO. 1014 (2014) (Y-ALLOY)

Aluminum Industries Inc.
Cu 4, Fe 0.75, Si 0.75, Zn 0.2, Mg 1-1.5, Ni 2, bal Al.
Sand cast: 23,000 TS; 0.5 El; 90 Brin. Heat treated: 30,000 TS; 0 El; 110 Brin. For general castings; coefficient expansion 0.0000136.

PERMITE NO. 1015 (2015)

Aluminum Industries Inc.
Cu 2.5, Fe 0.75, Si 1, Zn 30-31, bal Al.
Sand cast: 25,000 TS; 3 El; 70 Brin. Chill cast: 30,000 TS; 3 El; 70 Brin. For general castings; non-hardenable.

PERMITE NO. 1016

Aluminum Industries Inc.
Cu 10, Fe 1, Si 3-4, Zn 0.25, Mg 0.2, bal Al.
Chill cast: 22,000 TS; 1.0 El; 85 Brin. For general castings; non-hardenable.

PERMITE NO. 1019 (2019)

Aluminum Industries Inc.
Cu 1.25, Fe 1, Si 4.5-5.5, Zn 0.25, Mg 0.5, bal Al.
Heat treated: 27,000-35,000 TS; 1.0-4.5 El; 60-90 Brin. For general castings; age hardened.

PERMITE NO. 2005

Aluminum Industries Inc.
Cu 6-8.5, Fe 0.75, Si 0.5, Zn 0.25, bal Al.
Sand cast: 20,000 TS; 2.0 El; 65 Brin. For general castings; non-hardenable.

PERMITE NO. 2012

Aluminum Industries Inc.
Fe 0.75, Si 9-11, Zn 0.5, Mg 0.1, bal Al.
Sand cast: 23,000 TS; 5.0 El; 50 Brin. For general castings; corrosion resistant.

PERMITE NO. 2013

Aluminum Industries Inc.
Fe 0.75, Si 13, Zn 0.5, Mg 0.1, bal Al.
Sand cast: 28,000 TS; 8.0 El; 50 Brin. For general castings; corrosion resistant.

PERMITE NO. 2017 (PATTERN ALLOY)

Aluminum Industries Inc.
Cu 6-8, Fe 1, Si 1, Zn 2.5, Sn 1, bal Al.
Sand cast: 18,000 TS; 2.0 El; 70 Brin. For general castings; good machinability.

PERMITE NO. 2018

Aluminum Industries Inc.
Cu 2.5, Mn 0.5-2, bal Al.
Sand cast: 20,000 TS; 8 El. Heat treated: 28,000-30,000 TS; 4-6 El. For general castings; high ductility.

PERMITE NO. 2020

Aluminum Industries Inc.
Fe 1, Si 6.5-7.5, Zn 0.25, Mg 0.35, bal Al.
Heat treated: 26,000-30,000 TS; 3-5 El; 55-70 Brin. For general castings; age hardened

PERMITE NO. 2021

Aluminum Industries Inc.
Cu 0.1, Fe 0.5, Si 0.5, Mg 3-4, bal Al.
Sand cast: 22,000 TS; 6 El; 50 Brin. For general castings; corrosion resistant.

PERMITE NO. 2022

Aluminum Industries Inc.
Cu 0.1, Fe 0.5, Si 0.5, Mg 5-6, bal Al.
Sand cast: 25,000 TS; 4 El; 60 Brin. For general castings; corrosion resistant.

PERMITE NO. 2023

Aluminum Industries Inc.
Cu 0.1, Fe 0.5, Si 0.5, Mg 9-10, bal Al.
Sand cast: 27,000 TS; 3 El; 60 Brin. Heat treated: 38,000 TS; 11 El; 75 Brin. For general castings; corrosion resistant.

PERMIUM NO. 205

Paraloy Co.
Os, Ru, Rh, Co, Ni.
For jewelry, instrument jewels; sintered.

PERMO

Permo Inc.
Os, Rh, Ru.
For instrument bearings, fountain pen tips; corrosion and wear resistant.

PERMOLD

Permold Inc.
Cu, Si, bal Al.
For washing machine and typewriter parts; permanent mold. *Obsolete*

PERMOLD 103.7N

Permold Inc.
Cu 3, Si 10, Mg 0.7, Ni 0.8, bal Al.
Sand cast-T65: 30,000 TS; 25,000 YS; 0.5 El; 100 Brin. Permanent mold T6: 36,000 TS; 28,000 YS; 1 El; 105 Brin For automotive pistons; Type D132; age-hardenable. *Obsolete*

PERMOLD 110.5

Permold Inc.
Cu 1.5, Si 10, Mn 0.7, Mg 0.5, bal Al.
Stabilized: 34,000 TS; 0 El; 90 Brin. For pistons; heat treatable. *Obsolete*

PERMOLD 113.5

Permold Inc.
Cu 1.5, Si 13, Mn 0.7, Mg 0.5, bal Al.
Stabilized: 34,000 TS; 0 El; 90 Brin. For automotive pistons; heat treatable. *Obsolete*

PERMOLD 120.5

Permold Inc.
Cu 1.5, Si 20, Mn 0.7, Mg 15, bal Al.
Stabilized: 28,000 TS; 0 El; 90 Brin. For pistons; heat treatable. *Obsolete*

PERMOLD 135W

Permold Inc.
Cu 3.75, Si 6.25, Mg 0.1, Fe 0-0.8, bal Al.
Sand cast-T6: 36,000 TS; 24,000 YS; 2 El; 80 Brin. Permanent mold-T6: 40,000 TS; 27,000 YS; 3 El; 95 Brin. For general purpose castings; age-hardenable. *Obsolete*

PERMOLD 15.3

Permold Inc.
Cu 1.3, Si 5, Mn 0.3, Mg 0.3, bal Al.
Heat treated: 32,000 TS; 2 El; 75 Brin. For aircraft castings; heat treatable, permanent mold cast. *Obsolete*

PERMOLD 15.5

Permold Inc.
Cu 1.5, Si 5, Mn 0.4, Mg 0.4, bal Al.
Heat treated and aged: 37,000 TS; 1.5 El; 75 Brin. For castings; heat treatable. *Obsolete*

PERMOLD 18.3

Permold Inc.
Cu 1.3, Si 8, Mn 0.3, Mg 0.3, bal Al.
Heat treated: 34,000 TS; 1 El; 80 Brin. For aircraft castings; heat treatable, permanent mold cast. *Obsolete*

PERMOLD 3-5

Permold Inc.
Cu 3.75, Si 6.25, Mg 0.1, bal Al.
Sand cast-T6: 31,000 TS; 24,000 YS; 1.5 El; 80 Brin. Permanent mold-T6: 38,000 TS; 26,000 YS; 1.5 El; 90 Brin. For aircraft castings; Type 319; age-hardenable. *Obsolete*

PERMOLD 40

Permold Inc.
Cu 4.5, bal Al.
Heat treated: 32,000 TS; 3 El; 65 Brin. For castings; heat treatable, permanent mold cast. *Obsolete*

PERMOLD 40E

Permold Inc.
Mg 0.6, Zn 5.5, Cr 0.5, Ti 0.15, bal Al.
T5-temper: 32,000 TS; 25,000 YS; 3 El; 75 Brin. For high strength castings; self-aging. *Obsolete*

PERMOLD 42N

Permold Inc.
Cu 4, Fe 1.1, Si 2, bal Al.
Heat treated and aged: 40,000 TS; 2 El; 75 Brin. For aircraft parts; heat treatable. *Obsolete*

PERMOLD 45

Permold Inc.
Cu 4, Si 5, bal Al.
Cast permanent mold: 23,000 TS; 1 El; 60 Brin. For light alloy castings; not heat treatable. *Obsolete*

PERMOLD 7.3

Permold Inc.
Si 7, Mg 0.3, bal Al.
Heat treated and aged: 33,000 TS; 3 El; 70 Brin. For castings; heat treatable, permanent mold cast. *Obsolete*

PERMOLD 75.3

Permold Inc.
Cu 7, Si 5, Mg 0.3, bal Al.
Stabilized: 35,000 TS; 0 El; 100 Brin. For pistons; heat treatable. *Obsolete*

PERMOLD 90-10

Permold Inc.
Cu 10, Si 4, Mg 0.3, bal Al.
Permanent mold: 30,000 TS; 24,000 YS; 1.5 El; 100 Brin. For sole plates for electric irons; Type 138. *Obsolete*

PERMOLD 90-10S

Permold Inc.
Cu 10, Fe 1.2, Si 4, Mg 0.5, bal Al.
Cast: 26,000 TS; 0 El; 90 Brin. For sole plates for electric irons; heat treatable, permanent mold cast. *Obsolete*

PERMOLD 95-5

Permold Inc.
Si 5.25, bal Al.
Sand cast: 19,000 TS; 9,000 YS; 6 El; 40 Brin. Permanent mold: 21,000 TS; 10,000 YS; 5 El; 50 Brin. For cooking utensils, architectural castings; corrosion resistant; Type 43. *Obsolete*

PERMOLD 95-5

Permold Inc.
Si 5, bal Al.
Cast permanent mold: 21,000 TS; 5 El; 35 Brin. For Instrument cases, carburetor parts, covers; not heat treatable. *Obsolete*

PERMOLD H.P. 15.5

Permold Inc.
Cu 1.3, Si 5, Mg 0.5, bal Al.
T61-temper: 41,000 TS; 31,000 YS; 3 El; 85 Brin. For aircraft castings; Type C355; age-hardenable. *Obsolete*

PERMOLD H.P. 40

Permold Inc.
Cu 4.5, Fe 0-1, bal Al.
T6-temper: 40,000 TS; 30,000 YS; 2 El; 95 Brin. For machinery and aircraft castings; age-hardenable; Type 195. *Obsolete*

PERMOLD H.P. 42

Permold Inc.
Cu 4.5, Fe 0-1, Si 2.5, bal Al.
T6-temper: 40,000 TS; 26,000 YS; 5 El; 90 Brin. For machinery and aircraft castings; age-hardenable; Type B195. *Obsolete*

PERMOLD H.P. 7.3

Permold Inc.
Fe 0-0.2, Si 0.7, Mg 0.3, bal Al.
Permanent mold-T61: 38,000 TS; 28,000 YS; 5 El; 85 Brin. For high strength castings; good strength and ductility, age-hardenable. *Obsolete*

PERMOLD X-12

Permold Inc.
Cu 7.5, Fe 1, Zn 2, bal Al.
Cast permanent mold: 26,000 TS; 1 El; 90 Brin. For washing machine and home appliance parts; not heat treatable. *Obsolete*

PERMOLD Y

Permold Inc.
Cu 4, Mg 1.5, Ni 2, bal Al.
Heat treated: 32,000 TS; 0 El; 90 Brin. For cylinder heads, pistons; heat treatable, permanent mold cast. *Obsolete*

PERMOMETAL NO. 11

Permo Inc.
Os, Ru, Ir.
For instrument pivots; wear and corrosion resistant.

PERMOMETAL NO. 115

Permo Inc.
Os, Ru, Ir.
For instrument pivots; corrosion resistant.

PERMOMETAL NO. 81

Permo Inc.
Os, Ru.
For instrument pivots, fountain pens, needles; corrosion resistant.

PERNIFER 36-ALLOY 36

VDM Nickel-Technologie AG
Ni 35-37, Co 0-0.5, Mn 0-0.5, Si 0-0.2, C 0-0.03, bal Fe.
Minimum values at 20°C: 450 N/mm^2 TS; 240 N/mm^2 YS; 30 El. For production, storage and transportation of liquefied gases. Material No. 1.3912. K93601

PERPLEX

SWB Stahlformguss Gesellschaft mbH
C 1.05, Cr 1, W 1.15, Mn 0.9, bal Fe.
For bearings, cutters, liners, sleeves; water hardened, wear resistant. *Obsolete*

PERUNAL

Aluminium Industries, AG
Zn 6, Mg 2, Cu 1.5, bal Al.
74,000-83,000 TS; 64,000-72,000 YS; 10-5 El; 155 Brin. For light alloy parts.

PERUNAL

Aluminium Industrie Aktiengesellschaft
Zn 6, Mg 2, Cu 1.5, bal Al.
74,000-83,000 TS; 64,000-72,000 YS; 10-5 El; 155 Brin. For light alloy parts.

PERUNAL

Aluminium Walzwerke Singen GmbH
Zn 6, Mg 2, Cu 1.5, bal Al.
74,000-83,000 TS; 64,000-72,000 YS; 10-5 El; 155 Brin. For light alloy parts.

PERUNAL 215

Aluminium Walzwerke Singen GmbH
Aluminum. Zn 5.5, Mg 2.5, Cu 1.5, bal Al.
Solution heat treated: 530 MPa TS; 7 El. For highly stressed construction of vehicles and machines.

PERUNAL-215

Swiss Aluminium Ltd.
Aluminum. Si 0.4, Fe 0.5, Cu 1.2-2, Mn 0.3, Mg 2.1-2.9, Cr 0.18-0.28, Zn 5.1-6.1, Ti 0.05, bal Al.
Extrusion age hardenable alloy in billet form. 530-670 N/mm^2 TS; 460-630 N/mm^2 YS; 6-7 El; 150 Brin.

PETERSON 52100

Peterson Steels Inc.
C 0.95-1.1, Mn 0.25-0.45, Cr 1.3-1.6, bal Fe.
Annealed: 90,000 TS; 51,000 YS; 31 El;'61 RA; 180 Brin. Hardened: 300,000 TS; 650 Brin. For tools, dies, bearings, bushings; oil hardening.

PETITE IRON

Crucible Steel Castings Co.
Cr, Ni, V, bal Fe.
For cams, forming and spinning rolls, machine slides, forming dies. Wear resisting. *Obsolete*

PEWTER

American manufacture
Sn 85-90, bal Pb.
For household utensils, dishes, ornamental articles; soft, ductile, easily worked.

PF

Bergische Stahl Industrie
C 0.5-0.6, Si 0.15-0.35, Mn 0.5-0.8, Cr 0.6-0.8, Mo 0.25-0.35, Ni 1.5-1.8, V 0.07-0.12, bal Fe.
Cold work tool steel. W.-Nr. 1.2713.

PF-EXTRA

Bergische Stahl Industrie
C 0.5-0.6, Si 0.15-0.35, Mn 0.6-0.8, Cr 1-1.2, Mo 0.45-0.55, Ni 1.5-1.8, V 0.07-0.12, bal Fe.
Cold work tool steel. W-Nr. 1.2714.

PF1

TradeARBED Inc.
C 0.35, Cr 5, Mo 1.5, V 0.4, bal Fe.
Hot work tool steel, for dies. AISI H11. *Obsolete*

PF1 + V

TradeARBED Inc.
C 0.35, Cr 5, Mo 1.5, V 1, bal Fe.
Hot work tool steel, for dies. AISI H13. *Obsolete*

PF1+W

TradeARBED Inc.
C 0.35, Cr 5, Mo 1.5, W 1.5, V 0.4, bal Fe.
Hot work tool steel, for dies. AISI H12. *Obsolete*

PFA

Chemetron Corp.
Mn 0.45, Si 0.28, weld metal: 0.09 C, bal Fe.
As welded: 77,500 psi TS; 65,500 psi YS; 22 El. Covered electrode, AC-DC, straight polarity, for form equipment, metal furniture and general welding.

PFIZER DUCTILE COBALT

Pfizer Inc.
Co 95, Fe 5.
High cobalt content with excellent ductility 80,000 TS; 50 El. Sheath for cobalt base composite welding rods.

PFIZER HP COBALT

Pfizer Inc.
99.9 Co + Ni.
High purity cobalt in strip form. 110,000 TS; 20 El. For electroplating anodes, high temperature magnetic applications, Co^{60} gamma radiation sources, sheath for cobalt base composite welding rods, x-ray tube targets.

PFIZER HP NICKEL

Pfizer Inc.
99.9 Ni min (6 grades of hardness).
Soft: 45,000-50,000 TS. Half hard: 65,000-75,000 TS. Full hard: 85,000-100,000 TS. High purity, ductile, corrosion resistant, gas free. For electronic tube components, deep drawing and stamping small parts, magnetostrictive transducers, heat exchange fins.

PFIZER NICKEL CLAD STEEL

Pfizer Inc.
Nickel on one or both sides of carbon steel base, (as per customer requirements).

PFIZER PM 36 ALLOY

Pfizer Inc.
Ni-Fe alloy. Controlled expansion alloy.

PFIZER PM 42 ALLOY

Pfizer Inc.
Ni 42, bal Fe.
80,000 TS; 34,000 YS. Controlled expansion. For glass to metal sealing on hard or soft glass.

PFIZER PM 46 ALLOY

Pfizer Inc.
Ni 46, bal Fe.
82,000 TS; 34,000 YS. Controlled expansion. For glass to metal seals, especially terminal caps and bonds on vitreous enameled resistors.

PFIZER PM 48 ALLOY

Pfizer Inc.
Ni-Fe alloy. Controlled expansion alloy.

PFIZER PM 52 ALLOY

Pfizer Inc.
Ni-Fe alloy. Controlled expansion alloy.

PFIZER PM NICKEL-COPPER ALLOY

Pfizer Inc.
Ni 65, Fe 1, bal Cu.
Corrosion resistant, especially in marine environments. For tubing, banding, small parts.

PFIZER SEAL VAR TM

Pfizer Inc.
Ni 30, Co 15, bal Fe.
Controlled thermal expansivity to match hard glasses. For hermetic glass to metal seals (hard glass) and ceramic to metal seals. ASTM F-15.

PFIZER STAINLESS CLAD CARBON STEEL

Pfizer Inc.
Stainless steel on one or both sides of carbon steel, base (as per customer requirements).

PFIZER STAINLESS STEEL CLAD ALUMINUM

Pfizer Inc.
Al or Al alloys clad on one or both sides with a, stainless steel (as per customer requirements).

PFIZER TITANIUM CLAD ALUMINUM

Pfizer Inc.
For chemical processing applications.

PGP

J.M. Ney Co.
Au 25, Pt 50, Pd 25.
Nonoxidizing wire for endodontic posts. Can be electric welded to cobalt-chromium removable partial dentures. Fusing temperature: 2790°F.

PH 15-7 MO

Now ARMCO PH 15-7 MO and REPUBLIC PH 15-7 MO.

PH 55B
Cooper Alloy Corp.
Stainless steel. C 0-0.08, Cr 20, Ni 9, Mo 5, Cu 4, bal Fe.
For corrosion resistant castings. Shock resistant. *Obsolete*

PH 55C
Cooper Alloy Corp.
Stainless steel. C 0-0.08, Si 3.5, Cr 20, Ni 9, Mo 4, Cu 3, bal Fe.
For corrosion resistant castings. Stainless; austenitic. *Obsolete*

PH NO. 2
Colt Industries
C, alloy, bal Fe.
387 Brin. For dies for Zn die castings; age-hardenable, die steel. *Obsolete*

PH-2
Russian manufacture
W 7-10, bal Cb.
Resists molten lithium.

PH-55 A
Cooper Alloy Corp.
Stainless steel. C 0.04, Mn 0.5, Si 3, Cr 20, Ni 8.8, Mo 4, bal Fe.
Corrosion resistant casting. Same as COOPER PH-55 A.

PH-55D
Cooper Alloy Corp.
Stainless steel. C 0-0.05, Cr 18-21, Ni 9-12, Si 3-5, Mo 3.75-4.25, Cb 0.75-1.25, N 0.1, bal Fe.
For pump parts. Precipitation hardening; corrosion and abrasion resistant. *Obsolete*

PH13-8 MO
Now CARPENTER PH 13-8 MO; REPUBLIC PH 13-8; and A.

PHENIX
International Nickel Inc.
Fe 75, Ni 25.
Rolled: 104,000 TS; 56,000 YS; 45 El; 68 RA. For resistance alloy; Sc-21. *Obsolete*

PHENIX-HD301
Vereinigte Edelstahlwerke
C 0.15, Cr 12.75, Mo 1.15, Ni 0.5, bal Fe.
Annealed: 75,000 TS; 40,000 YS; 35 El; 70 RA; 155 Brin. Cold drawn: 100,000 TS; 85,000 YS; 17 El; 60 RA; 205 Brin. For oil refinery and chemical plant equipment, table flatware, knives, cutlery. Corrosion resistant. *Obsolete*

PHENIX-HD32
Vereinigte Edelstahlwerke
C 0.2, Mn 0.3, Si 0.4, Cr 12.75, Mo 1.15, Ni 0.4, bal Fe.
Annealed: 95,000 TS; 50,000 YS; 25 El; 55 RA. Heat treated: 250,000 TS; 215,000 YS; 8 El; C 52 Rock. For cutlery, oil refinery and chemical plant equipment, surgical instruments. Corrosion resistant, hardenable. *Obsolete*

PHENIX-HD40
Vereinigte Edelstahlwerke
C 0.2, Cr 12, Mo 1.15, Ni 0.5, V 0.3, bal Fe.
Annealed: 90,000 TS; 38,000 YS; 26 El; B 93 Rock. Hardened: 235,000 TS; 195,000 YS; 10 El; C 48 Rock. For bearings, hardware, cutlery, valves, surgical instruments. Corrosion resistant, hardenable. *Obsolete*

PHENIX-HD50
Vereinigte Edelstahlwerke
C 0.2, Mn 0.3, Si 0.4, Cr 12, Mo 1.15, Ni 0.65, V 0.3, W 0.5, bal Fe.
Annealed: 95,000 TS; 40,000 YS; 25 El; B 92 Rock. Hardened: 240,000 TS; 210,000 YS; 10 El; C 50 Rock. For cutlery, bearings, valves, surgical instruments. Corrosion resistant, hardenable. *Obsolete*

PHI 84
Creusot-Loire
C 0.38, Si 1, Cr 5.25, Mo 1.3, V 1, bal Fe.
Air hardening hot work tool steel for die casting or pressure casting dies. AFNOR Z 38 CDV5; AISI H 13.

PHILADELPHIA BRONZE 137
Ampco Metal
Ni 5, Zn 2, Sn 5, bal Cu.
Cast: 45,000 TS; 22,000 YS; 25 El; 85 Brin. For valves, pump fittings, gears, cams; corrosion resistant. *Obsolete*

PHILADELPHIA BRONZE 137
CMW Inc.
Ni 5, Zn 2, Sn 5, bal Cu.
Cast: 45,000 TS; 22,000 YS; 25 El; 85 Brin. For valves, pump fittings, gears, cams; corrosion resistant. *Obsolete*

PHILADELPHIA BRONZE 137 T
Ampco Metal
Ni 5, Zn 2, Sn 5, bal Cu.
Heat treated: 70,000-90,000 TS; 50,000-75,000 YS; 10-20 El; 140-200 Brin. For bearings, pole line hardware, ratchets; shock resistant, high fatigue resistant. *Obsolete*

PHILADELPHIA BRONZE 40
Ampco Metal
Al 8, bal Cu.
Cast: 70,000 TS; 28,000 YS; 25 El; 110 Brin. For gun mountings, worm wheels, gears, valve seats; aluminum-bronze, corrosion resistant. *Obsolete*

PHILADELPHIA BRONZE 4010
Ampco Metal
Cu alloy.
Cast: 70,000 TS; 32,000 YS; 10 El; 170 Brin. For dies for flash welding, gears, valve components; corrosion resistant. *Obsolete*

PHILADELPHIA BRONZE 404
Ampco Metal
Al 10, Ni 5, Fe 5, bal Cu.
Cast: 90,000 TS; 45,000 YS; 15 El; 180 Brin. For impeller blades, valve seats, worm gears; aluminum bronze; heat treatable. *Obsolete*

PHILADELPHIA BRONZE 4040
Ampco Metal
Al 10, Ni 5, Fe 5, bal Cu.
Cast: 90,000 TS; 45,000 YS; 15 El; 180 Brin. For valve seats, discs, bearings, gears, bushings; aluminum bronze; heat treatable. *Obsolete*

PHILADELPHIA BRONZE 41
Ampco Metal
Al 9, bal Cu.
Rolled: 75,000 TS; 35,000 YS; 20 El. For valve stems, propeller blades, bolts, pumps; corrosion resistant, aluminum bronze. *Obsolete*

PHILADELPHIA BRONZE 77
Ampco Metal
Cu alloy.
Cast: 70,000-85,000 TS; 35,000-70,000 YS; 1-12 El; 200-350 Brin. For bushings, bearings, drawing dies; wear and erosion resistant. *Obsolete*

PHILADELPHIA BRONZE 85
Ampco Metal
Ni 12, Al 1.5, bal Cu.
Cast: 85,000-95,000 TS; 60,000-70,000 YS; 7-11 El; 200-220 Brin. For marine hardware, fittings, valves, pumps; resists sea water corrosion. *Obsolete*

PHILO BRAND FERROCHROME
Ohio Ferro Alloys Corp.
Cr 65-70, C 4-6, Under 2% Si, bal Fe.
For metallurgical applications; Cr-additions. *Obsolete*

PHILO BRAND FERROMANGANESE
SiMETCO
P 0.3, Mn 78-82, C 0.7, Si, bal Fe.
For metallurgical applications; Mn additions. *Obsolete*

PHILO BRAND FERROSILICON 50% GRADE
SiMETCO
Si 50, Fe 48.5, 1.5 others.
For metallurgical applications; Si additions. *Obsolete*

PHILO BRAND FERROSILICON 75% GRADE
SiMETCO
Si 75, Fe 22.5, 2.5 others.
For metallurgical applications; Si additions. *Obsolete*

PHILO BRAND FERROSILICON 85% GRADE
SiMETCO
Si 85, Fe 12.5, 2.5 others.
For metallurgical applications; Si additions. *Obsolete*

PHILO BRAND FERROSILICON 90% GRADE
SiMETCO
Si 91, Fe 6.5, 2.5 others.
For metallurgical applications; Si additions. *Obsolete*

PHOENIX
NL Industries
Sn, Sb, bal Cu.
Cast: 24 Brin. For bearings for crossheads; heavy loads and high speeds. *Obsolete*

PHOENIX
Vereinigte Edelstahlwerke
C, alloy, bal Fe.
For machine tool parts; oil hardened. *Obsolete*

PHOENIX 430A
Phoenix Steel Corp.
C 0.08, Mn 0.45, P 0.017, S 0.01, Si 0.38, Cr 15.68, bal Fe.
Plate: 73,200 TS; 48,600 YS; 31 El; 57 RA; 143 Brin. For chemical plant industries, agitators, furnace parts. Resists nitric acid and sulfur bearing gases to 1500 F. Tough and weldable. *Obsolete*

PHOENIX 430B
Phoenix Steel Corp.
C 0.08, Mn 0.42, P 0.017, S 0.013, Si 0.38, Cr 16.88, bal Fe.
Plate: 76,000 TS; 52,000 YS; 30 El; 56 RA; 143 Brin. For chemical plant industries, furnace parts, acid containers. Resists nitric acid and sulfur bearing gases to 1500 F. Tough and weldable. *Obsolete*

PHOENIX A
Duraloy
C, alloy, bal Fe.
For steel castings.

PHOENIX A SPECIAL
Duraloy
C, alloy, bal Fe.
For steel castings.

PHOENIX CM-70
British Steel Corp.
C 0.16-0.21, Mn 1.1-1.4, Si 0.15-0.35, bal Fe.
Heat treated: 156,000 TS. For general purpose structural components. Suitable for carburizing and cold extrusion. *Obsolete*

PHOENIX CM-70 (SPT 539)
British Steel Corp.
C 0.16-0.21, Mn 1.1-1.4, Si 0.15-0.35, bal Fe.
Hardened: 157,000 TS; 116,000 YS; 32 Izod. For case hardened gears, cams, sprockets, spline shafts. *Obsolete*

PHOENIX CM-80F (SPT 566)
British Steel Corp.
C 0.17-0.23, Mn 1.2-1.5, Si 0.15-0.35, B 0-0.004, bal Fe.
Hardened: 179,000 TS; 104,000 YS; 10 El. For case hardened gears, cams, sprockets, spline shafts. *Obsolete*

PHOENIX CM-90F (SPT 567)
British Steel Corp.
C 0.2-0.25, Mn 1.3-1.6, Si 0.15-0.35, B 0-0.004, bal Fe.
Hardened: 200,000 TS; 134,000 YS; 10 El. For case hardened gears, cams, sprockets, spline shafts. *Obsolete*

PHOENIX HANSA
Vereinigte Edelstahlwerke
C 0.7, W 18, Cr 4, V 1, bal Fe.
For cutters, tools; high speed steel. *Obsolete*

PHOENIX K
Duraloy
C 3.2, Si 1.8, Cr 1, bal Fe.
For castings. Cast iron.

PHOENIX MANGEAR
United Engineering Steels Ltd.
C 0.12, Si 0.2, Mn 1.5, bal Fe.
Used in manufacture of chains. *Obsolete*

PHOENIX METAL
NL Industries
Pb alloy.
For bearings; solidification temperature 234°C. *Obsolete*

PHOENIX METAL
Duraloy
C 3.2, Ni 1.5, Cr 1, Si 1.7, bal Fe.
Cast. For rolls. Hard cast iron.

PHOENIX MOLYBDENUM CHILL
Duraloy
C 3.2, Si 1.8, Mo 0.5, bal Fe.
For castings. Cast iron.

PHOENIX NICKEL CHILL
Duraloy
C 3.2, Si 1.8, Ni 1.5, bal Fe.
For castings, wear plates. Cast iron.

PHOENIX PIROCO
Duraloy
C, alloy, bal Fe.
For steel castings.

PHOENIX RAPID MACHINING STEEL
British Steel Corp.
C 0.15, S 0.3, bal Fe.
63,500 TS; 44,000 YS; 48 El; 61 RA. For screws, bolts, nuts; free cutting. *Obsolete*

PHOENIX-A387 GR.D
Phoenix Steel Corp.
C 0.1, Cr 2.25, Mo 1, bal Fe.
Normalized: 60,000-85,000 TS. For structures operating up to 800 F. Heat exchangers, high pressure boilers. High strength, high temperature steel. *Obsolete*

PHOENIXITE
German manufacture
For cutting tools; hard metal.

PHOENIXLOY
Duraloy Blaw-Knox/Union Steel Casting
5 Ni + Cr, bal Fe.
For mill rolls for heavy duty; super hard alloy cast iron; Sc 80-90.

PHOENIXLOY
International Nickel Inc.
5 Ni + Cr, bal Fe.
For mill rolls for heavy duty; super hard alloy cast iron; Sc 80-90.

PHONO BRONZE
Olin Brass, Indianapolis
Copper. Cu 98.74, Sn 1.25, Fe 0.008.
Hard: 90,000 psi TS; 65,000 psi YS; 4 El; 70 RA. For trolley wire and hardware; resists air corrosion.

PHONO CADMIUM 955
Olin Brass, Indianapolis
Cd 0.85, Sn 0.55, bal Cu.
Annealed: 40,000 psi TS; 50 El. Drawn: 85,000 psi TS; 10 El. For trolley wire and cable; 55% conductivity. *Obsolete*

PHONO ELECTRIC 865
Olin Brass, Indianapolis
Sn 0.4, bal Cu.
Annealed: 35,000 psi TS; 50 El. Drawn: 63,000 psi TS; 10 El. For trolley wire and marine hardware; 65% conductivity. *Obsolete*

PHONO ELECTRIC BRONZE 840
Olin Brass, Indianapolis
Copper. Sn 1.4, bal Cu.
Annealed: 40,000 psi TS; 50 El. Drawn: 85,000 psi TS; 10 El. For trolley wire and hardware; 40% minimum conductivity.

PHONO ELECTRIC WIRE
Manufacturer not listed
Cu 98.55, Sn 1.4, Si 0.05.
For trolley wire, electrical conductors; high electrical conductivity.

PHONO HI CONDUCTIVITY
Olin Brass, Indianapolis
Cd 0.9, bal Cu.
Annealed: 37,000 psi TS; 50 El. Drawn: 65,000 psi TS; 6 El. For trolley wire and hardware; 80% minimum conductivity. *Obsolete*

PHONO HI-CONDUCTIVITY 985
Olin Brass, Indianapolis
Cd 0.9, bal Cu.
Annealed:; 37,000 psi TS; 50 El. Drawn: 80,000 psi TS; 6 El. For cable and trolley wire; 85% conductivity. *Obsolete*

PHONO HI-STRENGTH 715
Olin Brass, Indianapolis
Copper. Al 2.7, Si 0.35, Cu 96.95.
Annealed: 50,000 psi TS; 50 El. Drawn: 125,000 psi TS; 5 El. For cable, hardware, and bolts; 15% conductivity.

PHONO TELEPHONE 830
Olin Brass, Indianapolis
Copper. Sn 1.7, Cu 98.3.
Annealed: 40,000 psi TS; 50 El. Drawn: 90,000 psi TS; 10 El. For cable and drop wire; 30% conductivity.

PHONO-HI-STRENGTH
Olin Brass, Indianapolis
Cu 97, Sn 2, Si 1. *Obsolete*

PHOS 15
Sheffield Smelting Co. Ltd.
Ag 14.5, Cu, P.
Melting range: 645-700°C. Maximum stress: 66.9 kgf/mm^2; 25 El. For silver brazing copper; self-fluxing.

PHOS 2
Sheffield Smelting Co. Ltd.
Ag 2, Cu, P.
Melting range: 645-740°C. Maximum stress: 44.1 kgf/mm^2; 5 El. For silver brazing copper; self-fluxing.

PHOS 5
Sheffield Smelting Co. Ltd.
Ag 5, Cu, P.
Melting range: 645-730°C. Maximum stress: 47.2 kgf/mm^2; 6 El. For silver brazing copper; self-fluxing.

PHOS BRONZE C ELECTRODE
J.W. Harris Co., Inc.
Phosphor bronze electrode; DC reverse polarity. AWS AS.6 ECuSn-C. For welding and buildup on bronze castings, for joining steel, cast iron to bronze.

PHOS O
Sheffield Smelting Co. Ltd.
Cu, P.
Melting range: 705-800°C. Maximum stress: 50.4 kgf/mm^2; 2 El. For brazing copper; fluid, self-fluxing.

PHOS SIL-0
American Brazing Alloys Co.
Cu 92.8, P 7.2.
Cast: 93,000 TS. For brazing alloy for copper to copper. BCuP-2 specification; melting point 1305°F. Corrosion resistant.

PHOS SIL-15
American Brazing Alloys Co.
Ag 15, Cu 80, P 5.
Cast: 86,000 TS. For brazing alloy for non-ferrous metals. BCuP-5 specification; melting point 1185°F. Rapid penetration.

PHOS SIL-2
American Brazing Alloys Co.
Cu 91, Ag 2, P 7.
Cast: 93,000 TS. For brazing alloy for copper to copper. BCuP specification; melting point 1190°F. Corrosion resistant.

PHOS SIL-6
American Brazing Alloys Co.
Cu 87.75, Ag 6, P 6.25.
Cast: 92,000 TS. For brazing alloy for joints with poor fit-up. BCuP-3 specification; melting point 1185°F. Very ductile.

PHOS SIL-6F
American Brazing Alloys Co.
Ag 6, Cu 86.75, P 7.25.
Cast: 90,000 TS. For brazing alloy for critical joints. BCuP-4 specification; melting point 1190°F. Good penetration.

PHOS-COPPER
Westinghouse Electric Corp.
P 8, bal Cu.
Welded: 85,000 psi TS; 17 El. For brazing alloy for copper and brass parts; MP 1305-1460°F. *Obsolete*

PHOS-COPPER-5
Westinghouse Electric Corp.
P 4.8-5.2, bal Cu.
For filler metal for electrical connections; brazing MP 1640-1310°F. Meets AWS C1.BCuP-1. *Obsolete*

PHOS-SILVER
Westinghouse Electric Corp.
Ag 6, P 7.8, bal Cu.
Cast: 85,000 psi TS. For brazing for copper alloys; self-fluxing, MP 1185-1230°F. *Obsolete*

PHOS-SILVER 18
Westinghouse Electric Corp.
Ag 17.5-18.5, P 7-7.5, bal Cu.
For brazing copper, brass, and bronze. Lowest melting point 1190°F. Brazing temperature 1200-1250°F. *Obsolete*

PHOS-SILVER-15
Westinghouse Electric Corp.
P 4.9-5.1, Ag 14.8-15.2, bal Cu.
For brazing alloy; MP 1190-1485°F. Meets AWS Cl.BCuP-5. *Obsolete*

PHOS-SILVER-2
Westinghouse Electric Corp.
P 6.9-7.1, Ag 1.8-2.2, bal Cu.
For brazing alloy; MP 1190-1145°F. *Obsolete*

PHOS-SILVER-6
Westinghouse Electric Corp.
P 7.2-7.35, Ag 5.8-6.2, bal Cu.
For brazing alloy; MP 1190-1330°F. For close fit-up work. Meets AWS Cl-BCuP-4. *Obsolete*

PHOS-SILVER-65
Westinghouse Electric Corp.
P 4.9-5.1, Ag 5.8-6.2, bal Cu.
For brazing alloy for lap joints; MP 1190-1595°F. *Obsolete*

PHOS-SILVER-6M
Westinghouse Electric Corp.
P 6.05-6.2, Ag 5.8-6.2, bal Cu.
For brazing alloy; MP 1190-1465°F; very ductile. Meets AWS
Cl.BCuP-3. *Obsolete*

PHOS-TRODE
Ampco Metal
Sn 7-9, P 0.35, 0.5 others max, bal Cu.
Cast: 55,000 TS; 29,000 YS; 35 El; 33 RA; 89 Brin. For
shielded arc welding rod. Grade C. Phosphor Bronze.
Obsolete

PHOSCO
United Wire & Supply Co.
P 7.5, Cu 92.5.
For brazing non-ferrous metals. Self-fluxing. MP
1320-1450°F.

PHOSNIC BRONZE
Chase Brass & Copper Co.
Ni 0.85-1.35, P 0.18-0.3, bal Cu.
Soft: 38,000 TS; 10,000 YS; 40 El. Age-hardened: 85,000 TS;
68,000 YS; 15 El. Spring: 110,000 TS; 100,000 YS; 4 El. For
marine hardware, spring clips, electrical conductors;
corrosion resistant, age-hardenable. *Obsolete*

PHOSON
J.W. Harris Co., Inc.
Ag 0-1, Cu 0-96, P 0-7.
Melting range: 1309°F solidus; 1455°F liquidus. For
connections in transformers, motors and other apparatus.

PHOSON 2
United Wire & Supply Co.
Ag 2, P 7, Cu 91.
Melting range: 1185-1450°F. Brazing: 1300-1500°F. 90,000
psi TS; 18 El; 96 Rock 15-T. For various silver brazing
applications.

PHOSON 5
United Wire & Supply Co.
Ag 5, P 6, Cu 89.
Melting range: 1190-1485°F. Brazing: 1300-1500°F. 92,000
psi TS; 24 El; 86 Rock 15-T. For various silver brazing
applications.

PHOSON 6
United Wire & Supply Co.
P 7, Ag 6, bal Cu.
Cast: 95,000 TS. For silver brazing alloy. MP 1190-1380°F;
self-fluxing.

PHOSON-0
United Wire & Supply Co.
P 7.2, bal Cu.
Cast: 93,000 TS; 18 El. For brazing alloy for copper and brass
parts for close joint tolerance. MP 1305-1485°F.

PHOSON-15
United Wire & Supply Co.
Ag 15, P 5, Cu 85.
Cast: 86,000 TS. For brazing alloy for copper and copper
alloys. MP 1185-1500°F; self fluxing.

PHOSPHATIZED (PAINTBOND)
United States Steel Corp.
Low C, Mn, bal Fe.
Zinc coated steel; chemically treated for painting.

PHOSPHOR 36 BRONZE GRADE A
Olin Brass, Indianapolis
Copper. Sn 5, P 0.15, Cu 94.05.
Annealed: 50,000 psi TS; 52 El. Drawn: 81,000 psi TS; 10 El.
For springs, diaphragms, and clutch discs; corrosion and
fatigue resistant.

PHOSPHOR BRONZE
Manufacturer not listed
Cu 80, Sn 8, Pb 10, 2 P Sn.
For railroad bearings; heavy duty.

PHOSPHOR BRONZE (A) 351
Anaconda Co.
Copper. Sn 5, P 0.25, bal Cu.
Hard drawn: 130,000 TS. Soft: 45,000 TS; 50 El. For tubes,
sheets, wire, general parts. Resists fatigue, corrosion and
abrasion.

PHOSPHOR BRONZE (A) 5090
Anaconda Co.
Copper. Sn 4, P 0.05, bal Cu.
Sheet. Hard: 80,000 TS; 65,000 YS; 8 El; 86 Rock B. Soft:
48,000 TS; 20,000 YS; 48 El; 28 Rock B. For drawing into
containers such as kettles, electrical terminals. Weldable and
solderable.

PHOSPHOR BRONZE (A)-305
Anaconda Co.
Copper. Sn 5, P 0.05, bal Cu.
Hard: 65,000 TS; 57,000 YS; 20 El; 150 Brin. Soft: 48,000 TS;
20,000 YS; 50 El; 67 Brin. For strips and tubes. Corrosion
resistant.

PHOSPHOR BRONZE (C) 353
Anaconda Co.
Copper. Sn 8, P 0.25, bal Cu.
Hard drawn: 130,000 TS. Hard sheet: 110,000 TS; 3 El. For
tubes, sheets, wire, general parts. Resists fatigue and
corrosion.

PHOSPHOR BRONZE (D) 354
Anaconda Co.
Copper. Sn 10, P 0.25, bal Cu.
Hard: 130,000 TS; 5 El. Soft: 60,000 TS; 65 El. For tubes,
sheets, wire, general parts. Resists fatigue and corrosion.

PHOSPHOR BRONZE 1.25% E
Chase Brass & Copper Co., Inc.
Copper. Cu 98.55, Sn 1.25, P 0.2.
Annealed: 40,000 psi TS; 14,000 psi YS; 48 El. Rolled (60%):
75,000 psi TS; 4 El. For electrical contacts, pole line
hardware, flexible hose. CDA 505.

PHOSPHOR BRONZE 1.5
Waterbury Rolling Mills Inc.
Sn 1.5, P 0.04, bal Cu.
UNS C50200; C50500.

PHOSPHOR BRONZE 10% D
Chase Brass & Copper Co., Inc.
Copper. Cu 90, Sn 9.85, P 0.15.
Annealed: 66,000 psi TS; 28,000 psi YS; 68 El. Hard rolled:
100,000 TS; 13 El. For springs, bridge bearing plates; tough.

PHOSPHOR BRONZE 301
Anaconda Co.
Copper. Sn 3, P 0.25, bal Cu.
Hard: 90,000 TS; 70,000 YS; 4 El. For Fourdrinier wire.

PHOSPHOR BRONZE 310
Anaconda Co.
Sn 10.5, P 0.3, bal Cu.
For welding rod; carbon arc. *Obsolete*

PHOSPHOR BRONZE 320
Anaconda Co.
Copper. Sn 6.5, P 0.3, bal Cu.
Hard: 120,000 TS; 2 El. Soft: 57,000 TS; 40 El. For Fourdrinier
wire.

PHOSPHOR BRONZE 4%-903 GRADE A
Anaconda Co.
Cu 95.75, Sn 4, P 0.25.
Soft: 48,000 TS; 20,000 YS; 50 El. Hard: 65,000 TS; 55,000
YS; 30 El. For cotter pins, diaphragms, welding rod; P-
Bronze. *Obsolete*

PHOSPHOR BRONZE 5% A 510
Chase Brass & Copper Co.
Cu 94.8, Sn 5, P 0.2.
Annealed: 46,000-56,000 TS; 19,000-29,000 YS; 48-62 El.
Cold rolled: 49,000-122,000 TS; 21,000-110,000 YS; 1-50 El.
High strength, fatigue and corrosion resistant. For bellows,
electric spring contacts, connectors, terminals.

PHOSPHOR BRONZE 5% A 510
Olin Corp.
Cu 94.8, Sn 5, P 0.2.
Annealed: 46,000-56,000 TS; 19,000-29,000 YS; 48-62 El.
Cold rolled: 49,000-122,000 TS; 21,000-110,000 YS; 1-50 El.
High strength, fatigue and corrosion resistant. For bellows,
electric spring contacts, connectors, terminals.

PHOSPHOR BRONZE 7.0
Waterbury Rolling Mills Inc.
Sn 7, P 0.04, bal Cu.

PHOSPHOR BRONZE 8% C 521
Chase Brass & Copper Co.
Cu 91.8, Sn 8, P 0.2.
Annealed: 56,000-65,000 TS; 23,000-34,000 YS; 60-67 El.
Cold rolled: 63,000-134,000 TS; 33,000-122,000 YS; 1-60 El.
Very high strength spring bronze. For diaphragms, springs,
bellows, pen clips.

PHOSPHOR BRONZE 8% C 521
Olin Corp.
Cu 91.8, Sn 8, P 0.2.
Annealed: 56,000-65,000 TS; 23,000-34,000 YS; 60-67 El.
Cold rolled: 63,000-134,000 TS; 33,000-122,000 YS; 1-60 El.
Very high strength spring bronze. For diaphragms, springs,
bellows, pen clips.

PHOSPHOR BRONZE A510
Anaconda Co.
Copper. Cu 94.8, Sn 5, P 0.2.
Hard sheet: 80,000 TS; 65,000 YS; 10 El; 86 Rock B. Soft
sheet: 48,000 TS; 20,000 YS; 50 El; 28 Rock B. For chemical
hardware, springs, bridge bearing plates, Bourdon tubes.
Good resilience and resistance to fatigue and corrosion.

PHOSPHOR BRONZE B-1
Chase Brass & Copper Co., Inc.
Copper. Cu 93.8, Sn 5, Pb 1, P 0.2.
Drawn (20%): 70,000 psi TS; 58,000 psi YS; 25 El. For
bearings, bushings, gears, spindles, CDA 534.

PHOSPHOR BRONZE B-2
Chase Brass & Copper Co., Inc.
Copper. Pb 4, Sn 4, Zn 4, Cu 88, P.
Annealed: 44,000 psi TS; 19,000 psi YS; 50 El. Drawn (35%):
75,000 psi TS; 63,000 psi YS; 15 El. For bearings, gears,
pinions, valve parts. CDA 544.

PHOSPHOR BRONZE BEARINGS-1
Manufacturer not listed
Cu 83, Sn 14, P 1, Zn 2.
For bearings, bushings; heavy duty.

PHOSPHOR BRONZE BEARINGS-2
Manufacturer not listed
Cu 88.1, Sn 8, P 0.15, Pb 4.7.
For bearings, bushings; heavy duty.

PHOSPHOR BRONZE BRIDGE-1
Manufacturer not listed
Cu 80, Sn 20, P 1-0.2.
For bearings, bushings; high strength.

PHOSPHOR BRONZE BRIDGE-2
Manufacturer not listed
Cu 85, Sn 15, P 1.
For bearings, bushings; high strength.

PHOSPHOR BRONZE ENGLISH
Manufacturer not listed
Cu 79.2, Sn 10.2, Pb 9.6, P 0.97.
For tubes, hardware; free-cutting.

PHOSPHOR BRONZE GEAR-1
Manufacturer not listed
Cu 88, Sn 10, Pb 2, P 0.1.
For gears; tough.

PHOSPHOR BRONZE GEAR-2
Manufacturer not listed
Cu 85, Sn 13, Zn 2, P 0.1.
For gears; tough.

PHOSPHOR BRONZE GRADE A
Century Brass Products Inc.
Sn 4-6, P 0-0.35, bal Cu.
Annealed: 50,000 TS; 50 El. Hard: 95,000 TS; 2 El. For springs, diaphragms; phosphorus bronze. *Obsolete*

PHOSPHOR BRONZE GRADE A
Chase Brass & Copper Co.
Cu 95, Sn 4.75, P 0.25.
Annealed: 49,000 TS; 20,000 YS; 58 El. Rolled: 81,000 TS; 75,000 YS; 10 El. For springs, diaphragms, wire rope; corrosion resisting. *Obsolete*

PHOSPHOR BRONZE GRADE C
Century Brass Products Inc.
Sn 7-9, P 0-0.35, Zn 0-0.2, bal Cu.
Annealed: 55,000 TS; 60 El. Hard: 110,000 TS; 2 El. For springs, bearings; phosphorus bronze. *Obsolete*

PHOSPHOR BRONZE GRADE C
Olin Brass, Indianapolis
Sn 8, P 0.1, Cu 91.9.
Annealed: 60,000 psi TS; 65 El. Drawn: 112,000 psi TS; 3 El. For springs and diaphragms; corrosion and fatigue resistant. *Obsolete*

PHOSPHOR BRONZE GRADE C
Chase Brass & Copper Co.
Cu 92, Sn 7.75, P 0.25.
Annealed: 60,000 TS; 24,000 YS; 65 El. Rolled: 93,000 TS; 72,000 YS; 10 El. For springs, welding rod; corrosion resisting. *Obsolete*

PHOSPHOR BRONZE GRADE E-1
Chase Brass & Copper Co.
Cu 98.25, Sn 1.75.
Annealed: 45,000 TS; 45 El. Drawn: 105,000 TS; 3 El. For welding rod, springs; corrosion resisting. *Obsolete*

PHOSPHOR BRONZE HAIRSPRINGS
Manufacturer not listed
Cu 93, Sn 6.6, P 0.12-0.2.
For hairsprings; high strength.

PHOSPHOR BRONZE NO. 1
Knowsley Cast Metal Co., Ltd.
Cu 90, 10.0 Sn min, 0.5 P min.
Cast: 27,000 TS; 1.5 El; 65-90 Brin. For bearings with hardened shafts, heavy loads and high speeds.

PHOSPHOR BRONZE NO. 2
Knowsley Cast Metal Co., Ltd.
Cu 88, Sn 12, P 0.15.
Cast: 31,500 TS; 7 El; 69 Brin. For gears, bushings, bearings with hardened shafts; heavy loads and high speeds.

PHOSPHOR BRONZE NO. 3
Knowsley Cast Metal Co., Ltd.
Sn 9-11, P 0.03-0.25, bal Cu.
Cast: 36,000 TS; 10 El. For pressure tight castings; corrosion resistant.

PHOSPHOR BRONZE NO. 30
Riverside Metals Corp.
Cu 94.8, Sn 5, P 0.2.
Wire: 105,000-135,000 TS. For hardware, springs, wire rope; corrosion resistant. *Obsolete*

PHOSPHOR BRONZE NO. 4
Knowsley Cast Metal Co., Ltd.
Sn 7.5, P 0.3, Pb 2-5, bal Cu.
Cast: 27,000 TS; 3 El; 60-80 Brin. For medium duty castings; good bearing qualities.

PHOSPHOR BRONZE PRR-B
Manufacturer not listed
Cu 76.8, Sn 8, Pb 15, P 0.2.
For railroad bearings; heavy duty.

PHOSPHOR BRONZE PRR-P
Manufacturer not listed
Cu 86.5, Sn 9.85, Zn 3.77, P 0.05.
For railroad bearings; heavy duty.

PHOSPHOR BRONZE S PA
American manufacture
Cu 79.7, Sn 10, Pb 9.5, P 0.8.
For heavy duty bearings; high strength.

PHOSPHOR BRONZE, AMERICAN NO. 1
Riverside Metals Corp.
Cu 70-0, Sn 13, Pb 16, P 1.
For fittings exposed to sea water gun fittings, small springs; high resistance to corrosion and fatigue. *Obsolete*

PHOSPHOR BRONZE, AMERICAN NO. 2
Riverside Metals Corp.
Cu 95, Sn 4.9, P 0.1.
For screw propellers, airplane parts, worm wheels, bearings, fuse clips; high resistance to corrosion and fatigue. *Obsolete*

PHOSPHOR BRONZE, ENGLISH NO. 1
Riverside Metals Corp.
Cu 79.5, Sn 10.2, Pb 9.6, P 0.7.
For fittings exposed to sea water, gun fittings, small springs, bearings; high resistance to corrosion and fatigue. *Obsolete*

PHOSPHOR BRONZE, ENGLISH NO. 2
Riverside Metals Corp.
Cu 76.8, Sn 8, Pb 15, P 0.2.
For screw propellers, airplane parts, worm wheels, fuse clips; high resistance to corrosion and fatigue. *Obsolete*

PHOSPHOR BRONZE, LOCOMOTIVE BEARING
Riverside Metals Corp.
Cu 90.34-94.7, Sn 4.4-8.9, P 0.35-0.76.
For screw propellers, fittings exposed to sea water, worm wheels, bearings; high resistance to corrosion and fatigue. *Obsolete*

PHOSPHOR BRONZE, RUSSIAN
Riverside Metals Corp.
Cu 93.7, Sn 5.8, P 0.16, Zn 0.34.
For screw propellers, fittings exposed to sea water, worm wheels;, bearings; high resistance to corrosion and fatigue. *Obsolete*

PHOSPHOR BRONZE-314
Anaconda Co.
Copper. Sn 4, Mn 0.25, P 0.8, bal Cu.
Soft: 48,000 TS; 20,000 YS; 50 El. Hard: 65,000 TS; 55,000 YS; 30 El. For hot work parts. Strong.

PHOSPHOR BRONZE-316
Anaconda Co.
Cu 88, Sn 10, Mn 1, Fe 1, P.
Soft: 65,000 TS; 65 El. Hard: 85,000 TS; 25 El. For hot work parts; strong. *Obsolete*

PHOSPHOR BRONZE-356
Anaconda Co.
Copper. Sn 1.25, P 0.05, bal Cu.
Soft: 40,000 TS; 14,000 YS; 48 El. Hard: 65,000 TS; 50,000 YS; 6 El. For electrical conductors, metal hose. Corrosion resistant.

PHOSPHOR BRONZE-507
Anaconda Co.
Copper. Sn 1.75, P 0.01, bal Cu.
Soft: 45,000 TS. Hard: 105,000 TS. For trolley and line wires. Strong.

PHOSPHOR BRONZE-D524
Anaconda Co.
Copper. Cu 89.75, Sn 10, P 0.25.
Hard sheet: 102,000 TS; 70,000 YS; 12 El; 97 Rock B. Soft sheet: 66,000 TS; 28,000 YS; 65 El; 55 Rock B. For clips, beater bars, chemical hardware, springs, condenser tubes. High strength, resilience and resistance to fatigue and corrosion.

PHOSPHOR COPPER GRADE A
General Motors Corp./Central Foundry
Fe 0-0.15, 0.14 P min, 99.75 P + Cu min.
25,000 psi TS; 8 El. For bearings and machine castings; corrosion resistant.

PHOSPHOR COPPER GRADE B
General Motors Corp./Central Foundry
P 0.1-0, Fe 0-0.15, 0.10 P min, 99.75 P + Cu min.
For bearings and machine castings; corrosion resistant.

PHOSPHORBRONZE SN BZ 4
VDM Nickel-Technologie AG
Sn 4, P 0.3, bal Cu.
Rolled: 82,000 TS; 5 El; 156 Brin. Annealed: 48,300 TS; 44 El; 70 Brin. For springs; tough. *Obsolete*

PHOSPHORBRONZE SN BZ 6
VDM Nickel-Technologie AG
Sn 6.5, P 0.3, bal Cu.
Rolled: 113,800 TS; 2 El; 210 Brin. Annealed: 59,700 TS; 70 El; 75 Brin. For springs, contacts; tough. *Obsolete*

PHOSPHORBRONZE SN BZ 8
VDM Nickel-Technologie AG
Sn 8, P 0.3, bal Cu.
Rolled: 156,200 TS; 1 El. Annealed: 71,100 TS; 55 El. For springs; contacts; tough. *Obsolete*

PHOSPHORIZED ADMIRALTY
Century Brass Products Inc.
Cu 71, Sn 1, Zn 28, P 0.03.
Tubing for naval equipment. CDA 445.

PHOSPHORIZED ADMIRALTY METAL
Century Brass Products Inc.
Cu 70-73, Sn 0.9-1.2, P 0-0.1, bal Zn.
Annealed: 52,000 TS; 65 El. Hard: 85,000 TS; 10 El. For condenser tubes, ferrules; corrosion resistant. *Obsolete*

PHOSPHORIZED ARSENICAL COPPER-142
Anaconda Co.
Copper. Cu 99.68, P 0.02, As 0.3.
Hard tube: 45,000 TS; 40,000 YS; 10 El; 50 Rock B. Soft tube: 33,000 TS; 10,000 YS; 45 El; 45 Rock F. For condensers and heat exchangers. Electrical conductivity: 45%. Pitting and corrosion resistant.

PHOSPHORIZED COPPER-122
Anaconda Co.
Copper. Cu 99.9, P 0.02.
Hard sheet: 48,000 TS; 40,000 YS; 6 El; 50 Rock B. Soft sheet: 33,000 TS; 10,000 YS; 45 El; 45 Rock F. For refrigeration and air conditioning units, plumbing and heating units, oil carriers, hydraulic and gas lines. Electrical conductivity: 85%. Hydrogen embrittlement.

PHV DIE
Colt Industries
C 0.27, Ni 2.8, Cr 1.15, V 0.4, Mo 0.25, Al 1.15, bal Fe.
Heat treated: 170,500 TS; 159,000 YS; 16 El; 42 RA; Rockwell C54-C62. For dies for plastic and injection molding inserts; age-hardenable. *Obsolete*

PIERROT METAL, BEUGNOT

Manufacturer not listed.
Zn 83, Cu 8.3, Sn 7.6-7.7, Sb 3.5, Pb 3.
For bearings; anti-friction.

PIERROTS B.M

English manufacture
Zn 83.3, Sn 7.6, Cu 2.3, Sb 3.8, Pb 3.
For bearings; will not resist heat or live steam.

PILADUC

Youngstown Steel
C 0.2, bal Fe.
Rolled. For normalized tin plate; general use. *Obsolete*

PINCHBECK

Manufacturer not listed
Cu 88-94, Zn 6-12.
For jewelry; red brass.

PINKUS BRASS

English manufacture
Cu 88.1, Zn 6.9, Sn 2.5, Pb 1.8, Ni 0.3, Sb 0.32.
For hardware, fittings; free-cutting.

PINKUS BRONZE

English manufacture
Sn 14.7, Zn 1.5, Pb 8.8, Sb 2.5, bal Cu.
For bearings, bushings; heavy duty.

PINSBAC

Aktiebolaget Svenska Metallverken
C 0.3-0.5, bal Fe.
For machinery parts, gears; water hardening.

PIONEER

Time Steel Service Inc.
C 0.55, Mn 0.3, Si 0.9, Cr 5.1, W 1.25, Mo 1.45, bal Fe.
Air hardened tool steel, AISI A8.

PIONEER 921-T

Pioneer Metals, Inc.
Si 2, Cu 3.5, Fe 0-0.5, Ti 0.08, bal Al.
Cast: 20,000 TS; 18,000 YS; 1 El; 70 Brin. T6-temper: 45,000
TS; 22,000 YS; 0.5 El; 104 Brin. For tooling plates; heat
treatable, dimensionally stable.

PIONEER HEAT RESISTING METAL

Pioneer Alloy Products Co.
Cr 20, Ni 20, Mo 1, C, bal Fe.
For heat treating equipment between 1700-2000°F; resists
heat up to 2000°F.

PIONEER METAL

Pioneer Alloy Products Co.
Nickel. Ni 65, Cr-Mo, bal Fe.
Cast: 74,000 TS; 36,500 YS; 42 El; 150 Brin. For castings,
fittings, pumps, valve parts; acid resistant.

PIONEER METAL

Pioneer Alloy Products Co.
Ni 35, Cr 25, Mo 0-5, C 0.2-0.5, bal Fe.
Pioneer "a" acid resisting. Cast: 74,000 TS; 36,500 YS; 42 El;
150 Brin. Heat treated: 65,000 TS; 45,000 YS; 20 El. For
valves, fittings, pump parts, castings; corrosion and acid
resistant.

PIREKS 12/25

Darwins Alloy Castings
Ni 12, Cr 25, C, bal Fe.
For furnace parts and equipment; heat resistant to 1100°C.
Obsolete

PIREKS 12/25

E.A. Balfour Steel
Heat resisting alloy; resists sulfurous atmospheres.

PIREKS 20/25

Darwins Alloy Castings
Ni 20, Cr 25, C, bal Fe.
For furnace parts and equipment; creep and heat resistant to
950°C. *Obsolete*

PIREKS 20/25

E.A. Balfour Steel
Heat resisting alloy; good creep resistance up to 1050°C.

PIREKS 228

Morris Ashby Ltd.
C, alloy, bal Fe.
For damper plates, pyrometer tubes; heat resistant to sulfur
atmosphere.

PIREKS 25/20

Darwins Alloy Castings
Ni 25, Cr 20, C, bal Fe.
For furnace parts and equipment; heat resistant to 950°C.
Obsolete

PIREKS 25/20

E.A. Balfour Steel
Heat resisting alloy; for continuous service to 1050°C.

PIREKS 35/15

Darwins Alloy Castings
Ni 35, Cr 15, C, bal Fe.
For furnace parts and equipment; heat resistant to 950°C.
Obsolete

PIREKS 35/15

E.A. Balfour Steel
Nickel-chromium heat resisting alloy for cycling use in
carburizing atmospheres.

PIREKS 37/18/2

E.A. Balfour Steel
Heat resisting nickel-chromium-niobium alloy for cycling use
in carburizing atmospheres, such as carburizing boxes.

PIREKS 529

E.A. Balfour Steel
Heat resisting alloy, ferritic type, good resistance to attack
from high sulfur.

PIREKS 60

Darwins Alloy Castings
C, alloy, bal Fe.
For quenching jigs, and fixtures; heat resistant to 1050°C.
Obsolete

PIREKS 60

E.A. Balfour Steel
Heat resisting alloy; for continuous use up to 1050°C.

PIREKS 60/13

Darwins Alloy Castings
C, alloy, bal Fe.
For furnace trays, containers, electrical resistors, high- heat
resistance. *Obsolete*

PIREKS 60/13

E.A. Balfour Steel
Heat resisting alloy; for cycle heating and cooling to about
1050 °C.

PIREKS METAL

Sybry, Searls & Co. Ltd.
C 0.25, Cr 18, Ni 8, bal Fe.
At 20°C: 80,500 TS; 250 Brin. At 1000°C: 34,000 TS. For
annealing and carburizing boxes, heat treating appliances.
Cast alloy; resists heat up to 1000°C.

PIREKS RCC

Darwin & Milner Inc.
Nickel. C 0.21-0.35, Ni 49-57, Cr 24-30, Si, Mn, bal Fe.
For heat and corrosion resistant parts; heat and corrosion
resistant.

PIREKS-REACTAL

Darwin & Milner Inc.
Nickel. C 0.6, Cr 20, Ni 65, Si 2, bal Fe.
For furnace parts, carburizing and annealing boxes; heat
resistant.

PIRO-R

Forjas Alavesas S.A.
C 0.1, Mn 1.1, Si 0-0.06, P 0-0.05, S 0.3, bal Fe.
Free-machining, low carbon steel. AISI 1215; BS EN1A; DIN
95 Mn z3.

PIRSCH'S GERMAN SILVER

Manufacturer not listed
Cr 71-80, Ni 16-17, Zn 1-7.5, Zn 1-2.5, Sb 1-2.8, Co 1-2, Fe
1-1.5, Al 0-0.5.
For ornaments, tableware; corrosion resistant.

PISTON

CCS Braeburn Alloy Steel
C, alloy, bal Fe.
For engineering parts. *Obsolete*

PISTON

Uddeholm Corp.
C 1.05, V 0.1, bal Fe.
For pistons, drills, taps; water hardened. *Obsolete*

PISTONS-1

Manufacturer not listed.
Cu 83, Zn 16, Sn 1.
For pistons; corrosion resistant.

PISTONS-2

Manufacturer not listed
Al 93.5, Cu 2.45-3.4, Mg 1.39, Si 0.4, Zn 0-0.28, Ni 0-1.47.
For pistons, cylinder heads; age-hardenable.

PITALOY NO. 100

Pittsburgh Steel Foundry Corp.
C 0.35-0.4, V 1, Mn 0.9, Mo 0.35, Si 0.35, bal Fe.
Cast: 100,000 TS; 70,000 YS; 22 El; 40 RA. For general
castings, locomotive frames; tough.

PITALOY NO. 90

Pittsburgh Steel Foundry Corp.
C 0.3, V 0.1, Mn 0.9, Si 0.35, Ni 1.6, bal Fe.
Cast: 90,000 TS; 60,000 YS; 25 El; 50 RA. For general
castings, locomotive frames, roll mill machinery, cross-heads;
tough.

PITHO

Sanderson Kayser Ltd.
C 0.95, Mn 1.2, Cr 0.5, W 0.5, bal Fe.
For reamers, taps, broaches, gauges; nonshrinking.
Obsolete

PITT-TEN "X"

Hansell-Elcock Co.
C 0.1-0.2, Mn 0.5-1, Cb 0.01-0, bal Fe.
Rolled: 60,000-75,000 TS; 45,000-60,000 YS; 18-24 El. For
agricultural equipment, bus and truck bodies. High strength
structural steel.

PITT-TEN A441

Wheeling-Pittsburgh Steel Corp.
C 0.22, Mn 1, Si 0.3, Cu 0.2, 0.02 V min, bal Fe.
Wrought: 70 ksi TS; 50 ksi YS; 18 El. HSLA steel. ASTM A441.
Obsolete

PITT-TEN NO. 1

Wheeling-Pittsburgh Steel Corp.
C 0-0.12, Si 0-0.1, Mn 0.7, Cr 0.8, Ni 0.7, P 0-0.07, S 0-0.05,
bal Fe.
For truck and bus bodies, mine cars; high strength structural
steel. *Obsolete*

PITT-TEN NO. 2

Metropolitan-Vickers Electrical Co. Ltd.
C 0-0.15, Mn 0-0.75, Si 0-0.1, P 0-0.07, S 0-0.05, Cr 0.05, bal Fe.
For trucks and bus bodies, mine cars; high strength structural steel.

PITT-TEN X45W

Wheeling-Pittsburgh Steel Corp.
C 0.2, Mn 1, Si 0.1, 0.01 Cb min, 0.01 V min, bal Fe.
Wrought: 60 ksi TS; 45 ksi YS; 24 El. HSLA steel. ASTM A572. *Obsolete*

PITT-TEN X50W

Wheeling-Pittsburgh Steel Corp.
C 0.2, Mn 1, Si 0.1, 0.01 Cb min, 0.01 V min, bal Fe.
Wrought: 65 ksi TS; 50 ksi YS; 22 El. HSLA steel. ASTM A572. *Obsolete*

PITT-TEN X55W

Wheeling-Pittsburgh Steel Corp.
C 0.2, Mn 1, Si 0.1, 0.01 Cb min, 0.01 V min, bal Fe.
Wrought: 70 ksi TS; 55 ksi YS; 20 El. HSLA steel. ASTM A572. *Obsolete*

PITT-TEN X60W

Wheeling-Pittsburgh Steel Corp.
C 0.2, Mn 1, Si 0.1, 0.01 Cb min, 0.01 V min, bal Fe.
Wrought: 75 ksi TS; 60 ksi YS; 18 El. HSLA steel. ASTM A572. *Obsolete*

PITTSBURG

Teledyne Pittsburgh Tool Steel
C 1, Cr 5.25, Mn 0.5, Mo 1.1, V 0.25, bal Fe.
Air hardenable to 64 Rock C. For tools, dies, jigs, precision parts. Type A2; air hardening tool steel.

PITTSBURGH

Wheeling-Pittsburgh Steel Corp.
C 0.1, bal Fe.
For welding rod. *Obsolete*

PIVOT DRILL ROD

Colt Industries
C 1.25, bal Fe.
For drills, pivots; water hardening. *Obsolete*

PLACET

English manufacture
Ni 60, Fe 20, Cr 15, Mn 5.
Cast: 50,000 TS; 1 El; 179 Brin. For resistance alloy; heat resistant.

PLACOVAR

Hamilton Technology Inc.
50 at.% Co, 50 at.% Pt.
Permanent magnet alloy, ductile. Coercive force: 4,300 Oersteds. Residual induction: 6,450 Gausses. Energy product: 9.5 x 10 6 Gauss-Oersteds. For use in instrument *Obsolete*

PLANCHER

Ziv Steel & Wire Co.
C 0.55-0.65, Mo 0.5, Mn 0.8, Si 2, bal Fe.
For chisels, punches, shear blades, plastic master hobs; oil hardened, shock resistant.

PLANET CHOICE

A.R. Purdy Co. Inc.
C 0.7-1.2, bal Fe.
For tools, drills, taps; water hardening.

PLANET COLD ROLLED

A.R. Purdy Co. Inc.
C 0.7-0.9, bal Fe.
For tools; water hardening.

PLANET DRILL ROD

A.R. Purdy Co. Inc.
C 1.2, bal Fe.
For drill rods, drills, tools; cold drawn.

PLANET EXTRA

A.R. Purdy Co. Inc.
C 0.3-1.2, bal Fe.
For tools, drills, taps; water hardening.

PLANET HIGH SPEED

A.R. Purdy Co. Inc.
C 0.7, W 18, Cr 4, V 1, bal Fe.
For cutting tools, taps; high speed steel.

PLANET REGULAR

A.R. Purdy Co. Inc.
C 0.8-1.2, bal Fe.
For tools, drills, taps; water hardening.

PLANET SHEFFOIL

A.R. Purdy Co. Inc.
C 0.7, W 18, Cr 4, V 1, bal Fe.
For high speed tools, cutters; high speed steel.

PLANET SPECIAL

A.R. Purdy Co. Inc.
C 0.7-1, V 0.2, bal Fe.
For dies and tools; water hardening.

PLANEWELD NO. 1

Lincoln Electric Co.
C 0.1, Cr 0.8, Mo 0.2, bal Fe.
For welding electrodes for Cr-Mo steel; shielded arc. *Obsolete*

PLANEWELD NO. 2

Lincoln Electric Co.
C 0.15, Cr 0.8, Mo 0.2, bal Fe.
For welding rod for Cr-Mo steel; shielded arc. *Obsolete*

PLANSEE WZ 12D

Plansee, Metallwerk Gesellschaft
Ni 39, Co 13, Cr 13, 35.0 TiC.
For jet engine components. High heat resistance.

PLASDIE

Columbia Tool Steel Co.
C 0.3, Cr 0.75, Mo 0.25, bal Fe.
For Zn die casting dies; Type H13; oil hardened. *Obsolete*

PLASMEX 1000

North American Steel Corp.
C 0.3, Mn 0.75, Si 0.5, Cr 1.7, Mo 0.4, bal Fe.
Oil hardened, tool steel; for molds.

PLASMOLD

Firth Brown Ltd.
C 0.35, Si 0.25, Mn 0.5, Ni 4.3, Cr 1.3, Mo 0.3, bal Fe.
For gears, bolts, crankshafts, forging dies; oil hardened, shock resistant. *Obsolete*

PLAST-IRON

National Radiator Co.
C 0.008, Mn 0.001, Si 0.005, Fe 99.98.
For magnets, radio cores; powdered metal.

PLASTALLOY

Disston Inc.
C 0.08, Mn 0.43, Ni 1.3, Cr 0.56, bal Fe.
For dies, plastic molds; water hardened. *Obsolete*

PLASTIC BRONZE

Belmont Metals Inc.
Cu 66, Sn 5, Pb 28, Ni 1.
For bearings; heavy duty.

PLASTIC CSM NO. 2

Colt Industries
C 0.3-0.35, Mn 0.7, Si 0.5, Cr 0.8, Mo 0.25, bal Fe.
For plastic mold dies, zinc die-casting dies; water hardening. *Obsolete*

PLASTIC HOBBING

Edgar Allen Balfour Ltd.
C 0-0.1, Si 1, Mn 0.4, bal Fe.
110 Brin. For hobbed dies and plastic molds.

PLASTIC METAL

English manufacture
Sn 81, Cr 9.5, Sb 8.6, Fe 1.4.
For bearings, bushings; Babbitt.

PLASTIFORM

British Steel Corp.
C 0.3, Ni 4, Cr 1.3, Mo 0.25, bal Fe.
Air or oil hardening alloy tool steel; for stamping dies, plastic molds, drop stamping dies. *Obsolete*

PLASTIFORM 1

3M Co.

Flexible permanent magnet. 1650 oersted coercive force; 2150 gauss residual inductance. Flexible and easily cut. For instruments and electrical systems and controls.

PLASTIFORM 1 H

3M Co.

Flexible permanent magnet. Coercive force 1940 oersted; residual inductance 2150 gauss. Flexible and easily cut. For instruments and electrical systems and controls.

PLASTIFORM 1.4H

3M Co.

Flexible permanent magnet. Coercive force 2200 oersted; residual inductance 2450 gauss. Flexible and easily cut. For instruments and electrical systems and controls.

PLASTIRON

Disston Inc.
C 0-0.12, bal Fe.
For plastic molds; carburized grade. *Obsolete*

PLASTO C

SWB Stahlformguss Gesellschaft mbH
C 0.15, Cr, bal Fe.
For gears, cams, camshaft, fasteners; case hardening steel. *Obsolete*

PLASTO CC

SWB Stahlformguss Gesellschaft mbH
C, Cr, bal Fe.
For gears, cams, camshafts, fasteners; case hardening steel. *Obsolete*

PLASTO MC

SWB Stahlformguss Gesellschaft mbH
C 0.2, Mn 1.25, Cr 1.15, bal Fe.
For gears, cams, camshafts, fasteners; case hardening steel. *Obsolete*

PLASTO MCW

SWB Stahlformguss Gesellschaft mbH
C 0.15, Cr 1, Mn 1.25, Si 0.25, bal Fe.
For gears, cams, camshafts, fasteners; case hardening steel. *Obsolete*

PLASTO NI

SWB Stahlformguss Gesellschaft mbH
C 0.19, Cr 1.75, Mo 0.2, Ni 3.75, bal Fe.
For gears, cams, camshafts, fasteners; case hardening steel, shock resistant. *Obsolete*

PLASTO RR

SWB Stahlformguss Gesellschaft mbH
C 0.4, Cr 13, Si 0.4, Mn 0.3, bal Fe.
Annealed: 100,000 TS; 55,000 YS; 22 El; 52 RA; 200 Brin. For cutlery, valves, oil refinery equipment; Type 420; stainless. *Obsolete*

PLASTO U

SWB Stahlformguss Gesellschaft mbH
C 0.15, Si 0.15-0.35, Mn 0.25-0.5, bal Fe.
Annealed: 70,000 TS; 40,000 YS; 25 El; 60 RA; 145 Brin. For gears, cams, fasteners, bolts; case hardening steel. *Obsolete*

PLASTO V

SWB Stahlformguss Gesellschaft mbH
C 0.5, Ni, Cr, V, bal Fe.
For gears, bolts, crankshafts; oil hardened, shock resistant. *Obsolete*

PLASTODUR

VAW Vereinigte Aluminium-Werke AG
Aluminum.
Aluminum alloy AlCuMg.

PLATA METAL NO. 5

Atkinson Co.
Sb, Sn, Pb.
For bearings. Babbitt metal.

PLATALARGAN

Manufacturer not listed
Pt, Al, Ag.
For pen points; corrosion resistant.

PLATE AJ 30/4

Stahlwerke R. & H. Plate
C 0.45, Si 1.5, Mn 0.6, Cr 1.4, V 0.1, bal Fe.
Oil hardening tool steel. For cold punches, mandrels, embossing tools, lettering and numbering dies. DIN 45SiCrV6; Werkstoff Nr. 1.2249. *Obsolete*

PLATE AJ 30/H

Stahlwerke R. & H. Plate
C 0.61, Si 0.9, Mn 0.8, Cr 1.2, V 0.1, bal Fe.
Oil hardening tool steel. For cold punches for heavy sheet, large embossing tools, staking tools. DIN 61CrSiV5; W. Nr. 1.2243.

PLATE AJ30N

Stahlwerke R. & H. Plate
C 0.61, Cr 1.18, V 0.1, bal Fe.
For springs, upsetters; oil hardened, shock resistant. *Obsolete*

PLATE AK14

Stahlwerke R. & H. Plate
C 0.7, Si 0.25, Mn 0.25, bal Fe.
Heat treated: 175,000 TS; 130,000 YS; 12 El; 36 RA; 355 Brin. For rails, springs, hammers, axes; water hardened. *Obsolete*

PLATE AK20

Stahlwerke R. & H. Plate
C 1, Si 0.2, Mn 0.2, bal Fe.
Annealed: 100,000 TS; 53,000 YS; 21 El; 42 RA; 200 Brin. For drills, hobs, reamers, taps; Type W1; water hardened. *Obsolete*

PLATE AR 40/1

Stahlwerke R. & H. Plate
C 0.45, Si 1, Mn 0.35, Cr 1.1, V 0.2, W 2, bal Fe.
Oil hardening tool steel. For pneumatic tools, chisels, staking tools, riveting hammers. DIN 45 WCrV7; W. Nr. 1.2542

PLATE AR40/2

Stahlwerke R. & H. Plate
C 0.35, Cr 1.05, V 0.2, W 1.85, bal Fe.
For cold work tools, upsetters, crimpers; oil hardened, shock resistant. *Obsolete*

PLATE AR404

Stahlwerke R. & H. Plate
C 0.55, Si 0.9, Mn 0.3, Cr 1.05, V 0.2, W 1.85, bal Fe.
For upsetters, punches, shears, cold work tools; oil hardened, tough. *Obsolete*

PLATE BM 260

Stahlwerke R. & H. Plate
C 0.4, Si 0.4, Mn 0.3, Cr 13.5, bal Fe.
Air or oil hardening tool steel. For plastic molds for corrosive plastics. DIN X40Cr13; W. Nr. 1.2083.

PLATE BP 16 E

Stahlwerke R. & H. Plate
C 0.15, Si 0.25, Mn 0.5, Cr 0.65, bal Fe.
For case hardening parts for roller bearings, measuring instruments, piston pins. DIN 15Cr3. *Obsolete*

PLATE BP 18 E

Stahlwerke R. & H. Plate
C 0.16, Si 0.25, Mn 1.2, Cr 1, bal Fe.
For case hardening small cog wheels, shafts, control levers. DIN 16MnCr5. *Obsolete*

PLATE BP 20 E

Stahlwerke R. & H. Plate
C 0.21, Si 0.3, Mn 1.2, Cr 1.2, bal Fe.
For case hardening gears, spline shafts, pinions, crank shafts for small engines. DIN 21 MnCr5.

PLATE BS 30 E

Stahlwerke R. & H. Plate
C 0.15, Si 0.25, Mn 0.5, Cr 1.55, Ni 1.55, bal Fe.
For case hardening cog wheels, chain drives, splined shafts, gears. DIN 15 CrNi6.

PLATE BS 70 E

Stahlwerke R. & H. Plate
C 0.14, Si 0.3, Mn 0.4, Cr 0.7, Ni 3.5, bal Fe.
For case hardening parts, such as gears, cam shafts, splined coupling of middle stress. DIN 14NiCr14; W9 Nr. 1.5752.

PLATE BS 90 E

Stahlwerke R. & H. Plate
C 0.14, Si 0.3, Mn 0.4, Cr 1.1, Ni 4.5, bal Fe.
For case hardening parts as crankshafts, highly stressed gears and shafts of large sections for trucks and ordnance. DIN 14Ni, Cr18; W. Nr. 1.5860.

PLATE BS 90 E MO

Stahlwerke R. & H. Plate
C 0.19, Si 0.2, Mn 0.4, Cr 1.25, Mo 0.25, Ni 4, bal Fe.
For case hardening parts such as truck and tractor gears and splined shafts for case hardened ordnance parts. DIN 19NiCrMo15; W.Nr. 1.6587.

PLATE BV36

Stahlwerke R. & H. Plate
C 0.5, Si 0.25, Cr 1.05, Ni 3.25, bal Fe.
For gears, pinions, crankshafts, bolts, studs; oil hardened, shock resistant. *Obsolete*

PLATE CLIMAX TZM

Stahlwerke R. & H. Plate
Mo 99.3, Ti 0.5, Zr 0.1.
For hot extrusion dies, mandrels, turbine components, casting dies and molds.

PLATE E 5 S

Stahlwerke R. & H. Plate
C 0-0.06, Si 0-0.2, Mn 0.2, Cr 4.5, Mo 0.5, bal Fe.
For case hardening heavy gears, axles, shafts, tractor, railroad and ordnance equipment. DIN X6CrMo5.

PLATE EXTRA EXTRA P10

Stahlwerke R. & H. Plate
C 0.45, Ni, Cr, bal Fe.
For gears, bolts, machine tool parts; oil hardened, tough. *Obsolete*

PLATE EXTRA SPEZIAL MH

Stahlwerke R. & H. Plate
C 1.1, Si 0-0.25, Mn 0-0.25, bal Fe.
Annealed: 110,000 TS; 58,000 YS; 20 El; 40 RA; 210 Brin. For springs, drills, hobs, reamers; Type W1; water hardened. *Obsolete*

PLATE EXTRA SPEZIAL ZAH

Stahlwerke R. & H. Plate
C 0.85, Si 0-0.25, Mn 0-0.25, bal Fe.
Heat treated: 190,000 TS; 145,000 YS; 10 El; 30 RA; 400 Brin. For springs, taps, reamers; Type W1; water hardened. *Obsolete*

PLATE EXTRA SPEZIAL ZH

Stahlwerke R. & H. Plate
C 1, Si 0-0.25, Mn 0-0.25, bal Fe.
Annealed: 100,000 TS; 53,000 YS; 21 El; 42 RA; 200 Brin. For springs, taps, reamers, broaches; Type W1; water hardened. *Obsolete*

PLATE K 14

Stahlwerke R. & H. Plate
C 0.7, Si 0-0.3, Mn 0-0.35, bal Fe.
Water hardening tool steel; for stone tools for soft stones, chisels, snap head dies. DIN C70W2; Werkstoff Nr. 1.1620. *Obsolete*

PLATE K 20

Stahlwerke R. & H. Plate
C 1, Si 0-0.3, Mn 0-0.35, bal Fe.
Water hardening tool steel; for stone tools, for hard stone, embossing tools, scythes. DIN C100 W2; W. Nr. 1.1640.

PLATE K18

Stahlwerke R. & H. Plate
C 0.85, Si 0-0.3, Mn 0-0.35, bal Fe.
Water hardening tool steel; for stone tools for middle hard stones, hammers, leather and spoon dies. DIN C85W2; W. Nr. 1.1630.

PLATE KM 20 V

Stahlwerke R. & H. Plate
C 1, Si 0.2, Mn 0.25, V 0.1, bal Fe.
Water hardening tool steel. For piercing dies, upsetting dies, small shear blades, shovels, hand tools. DIN 100 V1; W. Nr. 1.2833.

PLATE KMV SUPRA

Stahlwerke R. & H. Plate
C 1.45, Si 0.2-0.35, Mn 0.3-0.5, V 3-3.5, bal Fe.
Cold work tool steel. W. Nr. 1.2838.

PLATE KP260W

Stahlwerke R. & H. Plate
C 1.65, Cr 11.5, V 0.1, bal Fe.
For blanking and forming dies, punches; air hardened, nondeforming. *Obsolete*

PLATE KS 10

Stahlwerke R. & H. Plate
C 1, Si 0-0.25, Mn 0-0.25, bal Fe.
Water hardening tool steel; for cutting and punching dies, shear blades, hollow and massive embossing dies. DIN C100W1; W. Nr. 1.1540.

PLATE KS 12

Stahlwerke R. & H. Plate
C 1.1, Si 0-0.25, Mn 0-0.25, bal Fe.
Water hardening tool steel; for small cutting and punching dies, milling cutters, scrapers, cutting tools, twist drills. DIN C110W1; Werkstoff Nr. 1.1550. *Obsolete*

PLATE KS 66

Stahlwerke R. & H. Plate
C 0.5, Si 0.3, Mn 0.5, Cr 1.1, Mo 0.2, Ni 3.25, bal Fe.
Oil hardening tool steel. For cutlery dies, artificial resin molding dies, dies for tableware. DIN (similar to) 50 NiCr13; W. Nr. 1.2721.

PLATE KS 7

Stahlwerke R. & H. Plate
C 0.7, Si 0-0.25, Mn 0-0.25, bal Fe.
Water hardening carbon tool steel; for cold trimming dies, trimming punches, centering drifts little drop hammer dies. DIN C70W1, Werkstoff Nr. 1.1520. *Obsolete*

PLATE KS 9
Stahlwerke R. & H. Plate
C 0.85, Si 0-0.25, Mn 0-0.25, bal Fe.
Water hardening tool steel; for cold impact tools, cold cutting and punching dies, strainers, snap dies. DIN C85W1; W. Nr. 1.1530.

PLATE KW 83
Stahlwerke R. & H. Plate
C 0.45, Si 0.3, Mn 0.65, Cr 1.4, Mo 0.5, Ni 4, V 0.15, W 0.5, bal Fe.
Oil or air hardening tool steel. For large, tough embossing tools, cold upsetting tools, air hardening dies. DIN X45NiCrMo4; W. Nr. 1.2767.

PLATE KW 83 SPEZIAL
Stahlwerke R. & H. Plate
C 0.4-0.5, Cr 1.2-1.5, Mo 0.15-0.35, Ni 3.8-4.3, W 0.5.
Cold work tool steel. W. Nr. 1.2767.

PLATE LM22
Stahlwerke R. & H. Plate
C 0.9, Cr 0.8, Mn 0.3, bal Fe.
For bearings, liners, punches, bushings; water hardened, wear resistant. *Obsolete*

PLATE MS4
Stahlwerke R. & H. Plate
C 0.45, Si 0.1-0.4, Mn 0.5-0.7, bal Fe.
Hot rolled: 98,000 TS; 60,000 YS; 24 El; 45 RA; 212 Brin. For axles, gears, bolts, crankshafts; water hardened. *Obsolete*

PLATE MS7
Stahlwerke R. & H. Plate
C 0.6, Si 0.25-0.5, Mn 0.3-0.8, bal Fe.
Heat treated: 115,000-160,000 TS; 77,000-113,000 YS; 12-23 El; 40-54 RA; 230-320 Brin. For wheels, die blocks, girders, rails; water hardened. *Obsolete*

PLATE MS8
Stahlwerke R. & H. Plate
C 0.75, Si 0.25-0.5, Mn 0.3-0.8, bal Fe.
Heat treated: 122,000-174,000 TS; 82,000-128,000 YS; 12-22 El; 37-52 RA; 240-350 Brin. For springs, clutch discs, punches, die blocks; water hardened; Type W1. *Obsolete*

PLATE MSS
Stahlwerke R. & H. Plate
C 0.53, Si 0.9, Mn 0.9, bal Fe.
Oil hardenable to 54-60 Rock C. For axles, pins, staking tools, cold chisels. DIN 53MnSi4; W. Nr. 1.2825.

PLATE NM 110 MO
Stahlwerke R. & H. Plate
C 1, Si 0.3, Mn 0.5, Cr 5.2, Mo 1.15, V 0.3, bal Fe.
Air hardening tool steel. For thread rolling dies, trimming dies, shear punches. DIN X100CrMoV51; W. Nr. 1.2363.

PLATE NM 150 V
Stahlwerke R. & H. Plate
C 1.5-1.6, Cr 11.5-12.5, Mo 0.6-0.8, V 0.9-1.1, bal Fe.
Cold work tool steel. W. Nr. 1.2379; AISI D2.

PLATE NM 240
Stahlwerke R. & H. Plate
C 2.1, Si 0.3, Mn 0.3, Cr 12, bal Fe.
Oil or air hardening tool steel. For heavy duty stamping dies, broaches, cold shears, wood milling cutters. DIN X210Cr12; W. Nr. 1.2080.

PLATE NM 240 CO
Stahlwerke R. & H. Plate
C 2.1, Si 0.3, Mn 0.4, Cr 12, Mo 0.4, V 0.15, W 0.7, Co 1, bal Fe.
Air or oil hardening. For thread rolling dies, punching and forming dies, broaches. DIN X210CrCoW12; W. Nr. 1.2884.

PLATE NM 240 V
Stahlwerke R. & H. Plate
C 2.1, Si 0.3, Mn 0.3, Cr 12, W 0.8, bal Fe.
Oil or air hardening tool steel. For heavy duty stamping dies, punches, broaches, die rings, plastic molds. DIN X210CrW12; W. Nr. 1.2436.

PLATE NM 240 W
Stahlwerke R. & H. Plate
C 1.65, Si 0.3, Mn 0.3, Cr 12, V 0.1, bal Fe.
Air or oil hardening tool steel. For heavy duty and long wearing stamping and punching dies, broaches. DIN X165CrV12; W. Nr. 1.2201.

PLATE NM 240 WMO
Stahlwerke R. & H. Plate
C 1.65, Si 0.3, Mn 0.35, Cr 12, Mo 0.7, V 0.35, W 0.5, bal Fe.
Air or oil hardening. For heavy duty punching and stamping dies and die blocks, broaches, plastic molds for corrosive plastics. DIN X165CrMoV12; W. Nr. 1.2601.

PLATE NM 30
Stahlwerke R. & H. Plate
C 1.05, Si 0.3, Mn 0.2, Cr 1.4, bal Fe.
Water or oil hardening tool steel for small dies, punches, gages. DIN 105Cr5; W. Nr. 1.2060.

PLATE NN 20
Stahlwerke R. & H. Plate
C 1.05, Si 0.25, Mn 1.1, Cr 0.9, bal Fe.
Oil hardening tool steel. For threading tools, cutting and punching dies for medium duty. DIN 105MnCr4; W. Nr. 1.2127.

PLATE NN 40
Stahlwerke R. & H. Plate
C 0.9, Si 0.2, Mn 2, V 0.1, bal Fe.
Oil hardening tool steel. For difficult cutting dies and punches for sheet up to 3 mm thick. DIN 90MnV8; W. Nr. 1.2842.

PLATE NN 90
Stahlwerke R. & H. Plate
C 0.85-0.95, Si 1.05-1.25, Mn 0.6-0.9, Cr 1.1-1.3, bal Fe.
Cold work tool steel. W. Nr. 1.2108.

PLATE NN26M
Stahlwerke R. & H. Plate
C 1.25, Si 1.15, Mn 0.7, Cr 1.2, bal Fe.
For blanking and forming dies, bushings, liners; oil hardened, wear resistant. *Obsolete*

PLATE NR 40/D
Stahlwerke R. & H. Plate
C 0.45, Si 1, Mn 0.35, Cr 1.65, V 0.2, W 2, bal Fe.
Oil hardening tool steel. For cold shear blades, pneumatic tools, chisels, cold upset dies. DIN 45WCrV77; Werkstoff Nr. 1.2547. *Obsolete*

PLATE NR 40/H
Stahlwerke R. & H. Plate
C 0.6, Si 0.6, Mn 0.3, Cr 1.1, V 0.2, W 2, bal Fe.
Oil hardening tool steel. For highly stressed perforating dies, trimming dies, lower dies. DIN 60WCrV7; W. Nr. 1.2550.

PLATE ON 22
Stahlwerke R. & H. Plate
C 1.15, Si 0.2, Mn 0.35, Cr 0.7, V 0.1, bal Fe.
Water or oil hardening tool steel. For twist drills, reamers, punches, taps. DIN 115CrV3; W. Nr. 1.2210.

PLATE ON28
Stahlwerke R. & H. Plate
C 1.45, Cr 1.4, Mn 0.6, bal Fe.
For bearings, liners, blanking dies, cutters; water or oil hardened, wear resistant. *Obsolete*

PLATE OR 33 M
Stahlwerke R. & H. Plate
C 1.05, Si 0.25, Mn 1, Cr 1, W 1.2, bal Fe.
Oil hardening tool steel. For cutting and stamping dies for sheet steel up to 5 mm thick. DIN 105WCr6; W. Nr. 1.2419.

PLATE OW 90
Stahlwerke R. & H. Plate
C 1.42, Si 0.2, Mn 0.3, Cr 0.35, V 0.25, W 3.25, bal Fe.
Water or oil hardening tool steel. For lathe and planing tools, engraving needles, scrapers, serrating steels. DIN 142WV13; W. Nr. 1.2562.

PLATE OW20V
Stahlwerke R. & H. Plate
C 1.2, V 0.1, W 1, N 0.28, bal Fe.
For cutters, bearings, liners; water hardened.

PLATE PA 1810
Stahlwerke R. & H. Plate
C 0-0.1, Si 0-1, Mn 0-2, Cr 17-19, Ni 9-11.5, Nb = 8 x C, bal Fe.
Stabilized austenitic stainless steel. W. Nr. 1.4550; AISI 347.

PLATE PA 188
Stahlwerke R. & H. Plate
C 0-0.12, Cr 18, Ni 9, bal Fe.
Annealed: 71,000-100,000 TS; 31,000 minimum YS; 50 minimum El; 130-180 Brin. Austenitic, stainless, hardenable only by cold work. For stainless equipment in food, beverage, dairy, paper, chemical industries. DIN X12CrNi188; Werkstoff Nr. 1.4300. *Obsolete*

PLATE PA 188 S
Stahlwerke R. & H. Plate
C 0-0.15, Cr 18, Ni 9, S, bal Fe.
Annealed: 71,000-100,000 TS; 31,000 minimum YS; 50 minimum El; 130-180 Brin. Austenitic, stainless, free machining. For machined shafts, bolts, screws and similar parts in food and beverage plants. DIN X12CrNiS188; Werkstoff Nr. 1.4305. *Obsolete*

PLATE PA 188 TI
Stahlwerke R. & H. Plate
C 0-0.1, Cr 18, Ni 11, Ti, bal Fe.
Annealed: 71,000-106,000 TS; 35,000 YS min; 40 El min; 130-190 Brin. Austenitic, stainless, weldable. For parts and welded assemblies for use in food, beverage, dairy and chemical industries. DIN X10CrNiTi1810; W. Nr. 1.4541.

PLATE PA 188 W
Stahlwerke R. & H. Plate
C 0.07, Cr 18, Ni 10, bal Fe.
Annealed: 71,000-100,000 TS; 31,000 YS min; 50 El min; 130-180 Brin. Austenitic, stainless, weldable. For parts and welded assemblies in food, dairy, beverage, and chemical industries. DIN X5CrNi189; W. Nr. 1.4301.

PLATE PAO 188/2 TI
Stahlwerke R. & H. Plate
C 0-0.1, Cr 18, Ni 11, Mo 2.3, Ti, bal Fe.
Annealed: 71,000-106,000 TS; 35,000 YS min; 40 El min; 130-190 Brin. Austenitic, weldable very good stainless properties. For parts and welded assemblies in paper, textile and chemical industries. DIN X10CrNiMoTi810; W. Nr. 1.4571.

PLATE PAO 188/2 W
Stahlwerke R. & H. Plate
C 0-0.07, Cr 18, Ni 11, Mo 2.3, bal Fe.
Annealed: 71,000-100,000 TS; 28,000 YS min; 45 El min; 130-180 Brin. Austenitic, very good stainless properties. For parts and assemblies in paper, textile, and chemical industries. DIN X5CrNiMo1810; W. Nr. 1.4401.

PLATE PCL
Stahlwerke R. & H. Plate
C 1.05, Cr, bal Fe.
For bearings, cutters, liners; water hardened.

PLATE PFC 141

Stahlwerke R. & H. Plate
C 0-0.08, Cr 14, bal Fe.
Annealed: 71,000-92,000 TS; 43,000 YS; 20 El min; 140-180 Brin. Corrosion resistant steel, not hardenable by heat treatment, ferritic. For structural parts, building fittings. DIN X7Cr14; W. Nr. 1.4001.

PLATE PFC 171

Stahlwerke R. & H. Plate
C 0-0.1, Cr 17, bal Fe.
Ferritic corrosion resistant steel; not hardenable by heat treating. Annealed: 64,000-85,000 TS; 42,000 YS min; 20 El min; 130-170 Brin. For building fittings, corrosion resistant hardware, spoons. DIN X8Cr17; W. Nr. 1.4016.

PLATE PFC 171 S

Stahlwerke R. & H. Plate
C 0.12, Cr 17, Mo 0.25, S, bal Fe.
Oil quenched: 100,000-120,000 TS; 64,000 YS min; 12 El min; 190-235 Brin. Ferritic, corrosion resistant steel, free machining grade. For machined parts as corrosion resistant screws, bolts, studs, nuts. DIN X12CrMoS17; W. Nr. 1.4104.

PLATE PFMC 14

Stahlwerke R. & H. Plate
C 0.2, Cr 13, bal Fe.
Heat treated: 113,000-135,000 TS; 78,000 YS min; 5 El min; 225-275 Brin. Martensitic corrosion resistant steel, oil hardenable to 40-48 Rock C. High strength structural parts for pumps, impellers, turbine wheels. DIN X20Cr13; W. Nr. 1.4021.

PLATE PGS 1

Stahlwerke R. & H. Plate
C 0.15, Si 0.25, Mn 0.35, bal Fe.
For case hardening small parts of machines as levers, links, bolts, pins. DIN C15W3. *Obsolete*

PLATE PGS 4

Stahlwerke R. & H. Plate
C 0.45, Si 0.35, Mn 0.7, bal Fe.
Water hardening tool steel; for hammers, forks, axes, knives, shears, screw drivers. DIN C45W3; W. Nr. 1.1730.

PLATE PGS 6

Stahlwerke R. & H. Plate
C 0.6, Si 0.35, Mn 0.7, bal Fe.
Water or oil hardening tool steel; for shanks for tools, bars, needle beds, stone breakers, hammers. DIN C60W3; W. Nr. 1.1740.

PLATE PLATIT 40

Stahlwerke R. & H. Plate
C 1, Cr 31, W 14, Co 53.
For parts for high temperature operation.

PLATE PLATIT EXTRA

Stahlwerke R. & H. Plate
C 2.5, Cr 33, W 14.5, Co 48.
For parts for high temperature operation. *Obsolete*

PLATE PLATIT HH

Stahlwerke R. & H. Plate
C 0.25, Cr 27, Mo 6, Co 55, bal Fe.
For high temperature operations.

PLATE PM 1291

Stahlwerke R. & H. Plate
C 0.3, Si 0.5, Mn 0.3, Cr 12, Mo 1, V 0.5, W 10, Co 1.5, bal Fe.
Hot work tool steel; air or oil hardening. For extrusion dies, die cast dies. Werkstoff Nr. 1.2871. *Obsolete*

PLATE PM 512

Stahlwerke R. & H. Plate
C 0.37, Si 1, Mn 0.5, Cr 5.3, Mo 1.5, V 0.2, W 1.3, bal Fe.
Hot work tool steel; air or oil hardening. For forging and pressing dies, ferrous and nonferrous. DIN X37CrMoW51; W. Nr. 1.2606.

PLATE PM 512 CO

Stahlwerke R. & H. Plate
C 0.35, Si 0.7, Mn 0.4, Cr 5, Mo 1.6, V 0.5, W 2, bal Fe.
Hot work tool steel; air or oil hardening. *Obsolete*

PLATE PM 524

Stahlwerke R. & H. Plate
C 0.38, Si 1, N 0.4, Cr 5.3, Mo 1.4, V 0.6, bal Fe.
Hot work tool steel; air or oil hardening. For pressure casting molds for light alloys. DIN X40CrMoV51; W. Nr. 1.2344.

PLATE PMC 145

Stahlwerke R. & H. Plate
C 0.4, Cr 13, bal Fe.
Martensitic corrosion resistant steel. Air or oil hardenable to 45-55 Rock C. For shafts, spline shafts, impellers, turbine wheels, pump parts, cutlery. DIN X40Cr13; W. Nr. 1.4034.

PLATE PMC NI 162

Stahlwerke R. & H. Plate
C 0.22, Cr 17, Ni 2, bal Fe.
Heat treated: 115,000-135,000 TS; 85,000 YS min; 4 El min; 225-275 Brin. Corrosion resistant. For structural parts requiring good strength and non-rusting properties. DIN X22CrNi17; W. Nr. 1.4057.

PLATE PMCO 1310

Stahlwerke R. & H. Plate
C 1.05, Cr 13, Mo 0.5, bal Fe.
Martensitic, corrosion resistant steel. Air or oil hardenable to 52-60 Rock C. For stainless ball bearings, valves, balls for ball point pens, cutlery, knife blades. DIN X105CrMo13; Werkstoff Nr. 1.4108. *Obsolete*

PLATE PMCO 174

Stahlwerke R. & H. Plate
C 0.38, Cr 17, Mo 1.2, Ni 0.5, bal Fe.
Heat treated, RT: 115,000-135,000 TS; 85,000 YS min; 14 El min; 225-275 Brin. At 400°C: 71,000 YS min. Corrosion resistant, good temperature resistance. For arbors, shafts, spindels, bolts operating up to 400°C. DIN X35CrMo17; W. Nr. 1.4122.

PLATE PMCO 189

Stahlwerke R. & H. Plate
C 0.9, Cr 18, Mo 1.2, V, bal Fe.
Martensitic, corrosion resistant, air or oil hardenable to 50-60 Rock C. For valves, knives, cutlery, ball bearings and races, shafts, couplings, surgical instruments, dental tools. DIN X90CrMoV18; W. Nr. 1.4112.

PLATE PN 24

Stahlwerke R. & H. Plate
C 0.5, Si 0.25, Mn 0.9, Cr 1, V 0.1, bal Fe.
Oil hardening tool steel. For cold punches. DIN 50CrV4; W. Nr. 1.2241.

PLATE PP 21

Stahlwerke R. & H. Plate
C 0.19-0.24, Si 0.35-0.55, Mn 0.3-0.5, Cr 1.3-1.5, Mo 1-1.2, V 0.25-0.35, bal Fe.
Alloy steel, low-carbon, for carburizing. W. Nr. 1.2052.

PLATE PP 32

Stahlwerke R. & H. Plate
C 0.4, Si 0.3, Mn 1.5, Cr 2, Mo 0.2, bal Fe.
Hot work tool steel; oil hardening. For punches, centrifugal casting molds. DIN 48CrMoV67; W. Nr. 1.2323.

PLATE PP36J

Stahlwerke R. & H. Plate
C 0.45, Cr 1.4, Mo 0.7, V 0.3, bal Fe.
For gears, bolts, crankshafts, axles; oil hardened, tough. *Obsolete*

PLATE PR97K

Stahlwerke R. & H. Plate
C 0.3, Cr 1.1, V 0.18, W 3.75, Si 1, bal Fe.
For upsetters, riveters, punches, crimpers; oil hardened, tough. *Obsolete*

PLATE PS 300

Stahlwerke R. & H. Plate
C 0.5, Si 1.3, Mn 0.7, Cr 14, Ni 13, V 1.2, W 1.3, bal Fe.
Hot work tool steel; not heat treatable. For pressing dies of simple profile; austenitic. DIN X50NiCrWV1313; W. Nr. 1.2731.

PLATE PU 27

Stahlwerke R. & H. Plate
C 0.56, Si 0.3, Mn 0.7, Cr 1, Mo 0.5, Ni 1.7, V 0.1, bal Fe.
Oil hardening tool steel. For coining and stamping dies, large mandrels, cold headers, embossing tools. DIN 56NiCrMoV7; Werkstoff Nr. 1.2714. *Obsolete*

PLATE PU 37

Stahlwerke R. & H. Plate
C 0.55, Si 0.3, Mn 0.7, Cr 1, Mo 0.5, Ni 1.7, V 0.1, bal Fe.
Hot work tool steel; oil or air hardening. For drop forge hammer dies. DIN 56NiCrMoV7; W. Nr. 1.2714.

PLATE PU 83

Stahlwerke R. & H. Plate
C 0.45, Si 0.2, Mn 0.5, Cr 1.3, Mo 0.5, Ni 4.5, V 0.2, W 0.5, bal Fe.
Hot work tool steel; air or oil hardening. For hot pressing tools with difficult gravures. DIN X45NiCrMo4; Werkstoff Nr. 1.2767. *Obsolete*

PLATE PV30

Stahlwerke R. & H. Plate
C 0.56, Cr 0.7, Mo 0.18, Ni 1.65, V 0.1, bal Fe.
For gears, bolts, crankshafts; oil hardened, shock resistant. *Obsolete*

PLATE PV36

Stahlwerke R. & H. Plate
C 0.45, Cr 0.7, Ni 1.65, bal Fe.
For gears, bolts, crankshafts; oil hardened, shock resistant. *Obsolete*

PLATE PVX 1212

Stahlwerke R. & H. Plate
C 0.45, Si 0.6, Mn 0.6, Cr 4.5, Mo 0.8, Ni 12, V 1.2, W 12, Co 0.5, bal Fe.
Hot work tool steel; not heat treatable. For high temperature pressing using light loads.

PLATE PVX 2615

Stahlwerke R. & H. Plate
C 0.08, Si 0-1, Mn 1-2, Cr 13.5-16, Mo 1-1.5, Ni 24-27, Ti 1.9, V 0-0.5, Al 0-0.35, B, bal Fe.
For high temperature applications. W. Nr. 1.4980.

PLATE PW 100

Stahlwerke R. & H. Plate
C 0.3, Si 0.2, Mn 0.3, Cr 2.4, V 0.6, W 4.3, bal Fe.
Hot work tool steel; oil hardening. For pressure casting molds and dies for nonferrous processing. DIN X30WCrV53; W. Nr. 1.25679.

PLATE PW 1096

Stahlwerke R. & H. Plate
C 0.17-0.23, Co 9.5-10.5, Cr 9-10, Mo 1.8-2.2, W 5-6, bal Fe.
Steel for high temperature applications. W. Nr. 1.2888.

PLATE PW 190

Stahlwerke R. & H. Plate
C 0.3, Si 0.2, Mn 0.3, Cr 2.6, V 0.4, W 8.5, bal Fe.
Hot work tool steel; oil or air hardening. Hot extrusion dies, pressure casting molds and dies for nonferrous metals. DIN X30WCrV93; W. Nr. 1.2581.

PLATE PW 78

Stahlwerke R. & H. Plate
C 0.45, Si 0.6, N 0.7, Cr 1.7, Mo 0.6, V 0.8, W 0.8, bal Fe.
Hot work tool steel; oil hardening. Shear blades, pressing dies and punches. DIN 45CrVMoW58; W. Nr. 1.2603.

PLATE PW100 EXTRA
Stahlwerke R. & H. Plate
C 0.3, Cr 2.65, V 0.35, W 8.5, bal Fe.
For extrusion press rams and liners, punches; hot work steel, oil hardened. *Obsolete*

PLATE PW134
Stahlwerke R. & H. Plate
C 0.65, Cr 3.75, Mo 0.85, V 0.7, W 8.5, bal Fe.
For lathe and planer tools, drills, reamers, taps; high speed steel. *Obsolete*

PLATE PW32
Stahlwerke R. & H. Plate
C 0.4, Cr, Mn, Mo, bal Fe.
For gears, bolts, machine tool parts; oil hardened, tough. *Obsolete*

PLATE PW83
Stahlwerke R. & H. Plate
C 0.4, Ni, Cr, Mo, bal Fe.
For gears, bolts, machine tool parts; oil hardened, tough. *Obsolete*

PLATE PW97
Stahlwerke R. & H. Plate
C 0.3, Cr 1.1, V 0.18, W 3.75, bal Fe.
For header dies, upsetters, punches; oil hardened, tough. *Obsolete*

PLATE PZ 194
Stahlwerke R. & H. Plate
C 0.32, Si 0.6, Mn 0.5, Cr 2.3, Mo 3.2, V 0.5, W 6.5, Co 4.5, bal Fe.
Hot work tool steel; air or oil hardening. Pressure casting molds, extrusion dies. Werkstoff Nr. 1.2676. *Obsolete*

PLATE PZ190
Stahlwerke R. & H. Plate
C 0.3, Co 2, Cr 2.4, V 0.25, W 8.5, bal Fe.
For upsetters, die casting dies, punches; hot work steel, oil hardened. *Obsolete*

PLATE SVX 33
Stahlwerke R. & H. Plate
C 0.95, Cr 4, Mo 2.7, V 2.4, W 3, bal Fe.
For twist drills, milling cutters, broaches, band saws, hack saw blades. High speed steel. W. Nr. 1.3333; DIN S3-3-2.

PLATE SVX 34
Stahlwerke R. & H. Plate
C 0.85, Cr 4, Mo 0.9, V 1.6, W 9, bal Fe.
Milling cutters, twist drills and taps for working on soft materials. High speed steel. W. Nr. 1.3316; DIN S9-1-2.

PLATE SVX 360
Stahlwerke R. & H. Plate
C 0.75, Cr 4, Mo 0.5, V 1.1, W 18, bal Fe.
For twist drills, threading tools, milling cutters, gear shapers. High speed steel. W. Nr. 1.3355; DIN S18-0-1.

PLATE SVX 526
Stahlwerke R. & H. Plate
C 0.85, Cr 4.2, Mo 5, V 2, W 6.3, bal Fe.
For twist drills, broaches, milling cutters, segments for circular saws, lathe tools for roughing cuts. High speed steel. W. Nr. 1.3343; DIN S6-5-2.

PLATE SVX 536
Stahlwerke R. & H. Plate
C 1.2, Cr 4.2, Mo 5, V 3.3, W 6.3, bal Fe.
For heavy duty milling cutters, highly stressed broaches, reamers. High speed steel. W. Nr. 1.3344; DIN S6-5-3.

PLATE SVX 54
Stahlwerke R. & H. Plate
C 0.85, Cr 4, Mo 0.9, V 2.5, W 12, bal Fe.
For lathe tools, milling knives, relieved cutters for working on hard materials. High speed steel. W. Nr. 1.3318; DIN S12-1-2.

PLATE SVX 811
Stahlwerke R. & H. Plate
C 0.82, Cr 3.8, Mo 8.6, V 1.2, W 1.8, bal Fe.
For twist drills, reamers, milling cutters, threading tools and taps. High speed steel. W. Nr. 1.3346; DIN S2-9-1.

PLATE SVX 90
Stahlwerke R. & H. Plate
Mo 0.9, V 3.8, W 12, C 1.25, Cr 4.2, bal Fe.
For finishing and milling tools, pinion type cutters, steel shapes, broaches. High speed steel. W. Nr. 1.3302; DIN S12-1-4.

PLATE SVZ 203
Stahlwerke R. & H. Plate
C 0.83, Cr 4.2, Mo 0.9, V 1.9, W 12, Co 2.8, bal Fe.
For lathe tools and milling cutters, good retention of hardness. High speed steel. Werkstoff Nr. 1.3211; DIN S12-1-2-3. *Obsolete*

PLATE SVZ 245
Stahlwerke R. & H. Plate
C 1.3, Cr 4.2, Mo 0.9, V 3.8, W 12, Co 4.8, bal Fe.
For finishing and roughing cutting tools for best wear characteristics. High speed steel. W. Nr. 1.3202; DIN S12-1-4-5.

PLATE SVZ 365
Stahlwerke R. & H. Plate
C 0.8, Cr 4.2, Mo 0.7, V 1.6, W 18, Co 4.8, bal Fe.
Lathe and planing tools with eminent cutting strength and toughness. High speed steel. W. Nr. 1.3255; DIN S18-1-2-5.

PLATE SVZ 410
Stahlwerke R. & H. Plate
C 0.76, Cr 4.2, Mo 0.7, V 1.6, W 18, Co 9.5, bal Fe.
Lathe and planing tools for heavy work requiring good red hardness, and for machining austenitic steels. High speed steel. W. Nr. 1.3265; DIN S18-1-2-10.

PLATE SVZ 526 CO
Stahlwerke R. & H. Plate
C 0.82, Cr 4.2, Mo 5, V 2, W 6.3, Co 4.8, bal Fe.
For lathe tools, drills, milling cutters for tough austenitic steels. High speed steel, good red hardness. W. Nr. 1.3243; DIN S6-5-2-5.

PLATE TV27
Stahlwerke R. & H. Plate
C 0.31, Cr 2.35, Mo 0.18, V 0.13, bal Fe.
For die casting and plastic mold dies; oil hardened, tough. *Obsolete*

PLATE UP24
Stahlwerke R. & H. Plate
C 0.45, Si, Cr, V, bal Fe.
For springs, gears, bolts, punches, crankshafts; oil hardened, shock resistant. *Obsolete*

PLATE US MO
Stahlwerke R. & H. Plate
C 0.3, Si 0.3, N 0.3, Cr 3, Mo 2.8, V 0.5, bal Fe.
Hot work tool steel; oil hardening. Pressure casting molds for heavy metal alloy, pressing dies. DIN X32CrMoV33; W. Nr. 1.2365.

PLATE US MO CO
Stahlwerke R. & H. Plate
C 0.3, Si 0.3, Mn 0.4, Cr 2.8, Mo 2.7, V 0.5, Co 3, bal Fe.
Hot work tool steel; air or oil hardening. *Obsolete*

PLATE US MO K
Stahlwerke R. & H. Plate
C 0.3, Si 0.2, Mn 0.4, Cr 2.8, Mo 2.8, V 0.5, bal Fe.
Hot work tool steel; oil or air hardening. For pressure castings molds, pressing dies. DIN X3CrMoV33; W. Nr. 1.2365.

PLATE UW280
Stahlwerke R. & H. Plate
C 0.2, Cr 13, Mn 0.3, bal Fe.
Annealed: 95,000 TS; 50,000 YS; 25 El; 55 RA; 196 Brin. For turbine blades, valves, cutlery, surgical instruments; Type 420; stainless. *Obsolete*

PLATE WCN
Stahlwerke R. & H. Plate
C 0.5, Ni, Cr, bal Fe.
For gears, bolts, machine tool parts, shafts; oil hardened, shock resistant. *Obsolete*

PLATE WKL
Stahlwerke R. & H. Plate
C 0.56, Ni, Cr, Mo, V, bal Fe.
For forging and heading dies, punches; oil hardened, shock resistant. *Obsolete*

PLATE WL30
Stahlwerke R. & H. Plate
C 0.53, Si 0.9, Mn 0.9, bal Fe.
For chisels, pneumatic tools, punches; oil hardened, tough. *Obsolete*

PLATE WP32
Stahlwerke R. & H. Plate
C 0.4, Cr, Mn, Mo, bal Fe.
For gears, bolts, machine tool parts; oil hardened, tough. *Obsolete*

PLATE WV 27
Stahlwerke R. & H. Plate
C 0.55, Si 0.3, Mn 0.6, Cr 0.7, Mo 0.3, Ni 1.7, V 0.1, bal Fe.
Oil hardening tool steel, for hot work. For coining dies, embossing tools, counterbores, large mandrels. DIN 55NiCrMoV6; W. Nr. 1.2713.

PLATE WV30
Stahlwerke R. & H. Plate
C 0.55, Cr 0.7, Mo 0.18, Ni 1.65, V 0.1, bal Fe.
For die casting and plastic mold dies; oil hardened, tough. *Obsolete*

PLATE ZM 14
Stahlwerke R. & H. Plate
C 1.4, Si 0.3, Mn 0.3, Cr 0.35, V 0.1, bal Fe.
Water hardening tool steel. For punches, cold drawing dies, cutting tools that operate cold, bearings. DIN 140CrV1; W. Nr. 1.2206.

PLATE ZM13
Stahlwerke R. & H. Plate
C 0.9, Cr 0.8, Mn 0.3, Si 0.25, bal Fe.
For bearings, cutters, sleeves, liners; water hardened, wear resistant. *Obsolete*

PLATE ZW 100
Stahlwerke R. & H. Plate
C 1.3, Si 0.25, Mn 0.3, Cr 0.3, W 5, bal Fe.
Water or oil hardening tool steel. For drawing dies, highly stressed cold forming dies. DIN X130W5; W. Nr. 1.2453.

PLATE-LOY
Manufacturer not listed
Pb.
For hot lead plating.

PLATERGAL
Lavorazione Leghe Leggere SpA
Aluminum.
Aluminum alloy AlZn1.

PLATERS METAL
Century Brass Products Inc.
Cu 85-88, Sn 1.75-2.5, bal Zn.
Annealed: 50,000 TS; 55 El. Hard: 94,000 TS; 1 El. For springs, diaphragms; high strength. *Obsolete*

PLATIKUT
Disston Inc.
C 0.2, Mn 0.8, Ni 0.55, Cr 0.5, Mo 0.2, bal Fe.
150 Brin. For molds, dies; for plastic industry. *Obsolete*

PLATINA, BIRMINGHAM
English manufacture
Cu 46.6, Zn 53.15, Fe 0.25.
For hardware, ornaments; corrosion resistant.

PLATINA, BIRMINGHAM
English manufacture
Cu 20.25, Zn 79.4, Fe 0.33.
For hardware, ornaments.

PLATINA, PLATING
English manufacture
Cu 20.25, Zn 79.4, Fe 0.33.
For hardware ornaments.

PLATINAM
Hopkinsons Ltd.
Ni 50, Cu 35, Sn 10, Fe 5, Al 0.3.
For steam valves up to 480°C, valve disks and seats; heat resisting, abrasion and erosion resistant.

PLATINAX-II
Johnson Matthey plc
Co 23.3, bal Pt.
Remanance 6400 gauss; 9.2 Brin max. For permanent magnets, metering devices. Can be machined, rolled and drawn. Outstanding magnetic properties obtained by a two-stage heating treatment. Corrosion resistant to H_2SO_4 and HNO_3.

PLATINE
English manufacture
Zn 57, Cu 43.
For ornaments; weak and brittle.

PLATINE-AU-TITRE (PROPLATINUM)
French manufacture
Ag 83-65.
For substitute for platinum; corrosion resistant.

PLATINEL 1503
Engelhard Corp.
Au 65, Pd 35.
For high temperature thermocouples. Negative leg.

PLATINEL 1786
Engelhard Corp.
Au 3, Pd 83, bal Pt.
For high temperature thermocouples. Positive leg.

PLATINEL 1813
Engelhard Corp.
Pd 55, Pt 31, Au 14.
For high temperature thermocouples.

PLATINIRIDIUM
French manufacture
Pt 53, Ir 28, Rh 7, Cu 3, Fe 4, traces Pd + As.
For pen nibs, ornamental parts; mined in Russia.

PLATINIT
English manufacture
Ni 46, bal Fe.
For chemical apparatus; corrosion resistant.

PLATINITE
Creusot-Loire
C 0.15, Ni 49, bal Fe.
Annealed: 78,000-90,000 TS; 48,000-60,000 YS; 30-45 El. To replace platinum in electric light bulbs; heat resistant; coefficient expansion of glass. *Obsolete*

PLATINITE + CR
Creusot-Loire
C 0.15, Ni 49, Cr, bal Fe.
For glass-to-metal seals. Stepped expansion. *Obsolete*

PLATINOID
Manufacturer not listed
Cu 50-90, Ni 3-40, Al 0.1, Zn 0.4.
For chemical equipment and construction; corrosion resistant.

PLATINOID A
English manufacture
Cu 60, Zn 24, Ni 14, W 2.
57,000 TS; 33 El. High resistance alloy, heating elements, thermocouples; Tungsten, German Silver.

PLATINOID B
English manufacture
Cu 54, Zn 20, Ni 25, Fe 0.5, Mn 0.2, W.
72,000 TS; 30 El; 89 Brin. For high resistance alloys, heating elements, thermocouples, Tungsten, German Silver.

PLATINOR
Manufacturer not listed
Cu 45, Pt 18, Br 18, Ag 9, Ni 9.
For jewelry, ornaments; corrosion resistant.

PLATINUM
Fansteel Metals
Precious metal. Pt 99.9.
Annealed: 60 Rock-15T; electrical conductivity 15% IACS. For electrical contacts. *Obsolete*

PLATINUM
Atomergic Chemetals Corp.
Pt.
Purities: 99.999%, 99.99%, 99.9%. Forms: sponge, powder, rod, wire, sheet, foil, single crystals.

PLATINUM ALLOY
Manufacturer not listed
Pt 1, 2-5 parts Ag, 0-1 part Cu.
For ornaments, jewelry; corrosion resistant.

PLATINUM ALLOY NO. 417
Sigmund Cohn Corp.
Pt 62, Pd 33, Mo 5.
High resistance; high tensile strength.

PLATINUM ALLOY NO. 479
Sigmund Cohn Corp.
Pt 91-93, W 7-9.
Cold drawn: 300,000 TS. High tensile strength elements.

PLATINUM ALLOY NO. 851
Sigmund Cohn Corp.
Pt 79-81, Rh 14-16, Ru 4-6.
Cold drawn: 300,000 TS. For electroexplosive devices, variable resistors; high electrical and heat resistance.

PLATINUM GOLD-1
Manufacturer not listed
Au 70, Pt 30.
For ornaments, jewelry; white, corrosion resistant.

PLATINUM GOLD-2
Manufacturer not listed
Pt 58, Ag 25, Au 17.
For jewelry; corrosion resistant.

PLATINUM GOLD-3
Manufacturer not listed
Au 60, Pt 40.
For jewelry; corrosion resistant.

PLATINUM IRIDIUM-1
Manufacturer not listed
Ir 5, Pt 95.
170 Brin. For jewelry; corrosion resistant.

PLATINUM IRIDIUM-2
Manufacturer not listed
Ir 30, Pt 70.
400 Brin. For surgical instruments; corrosion resistant.

PLATINUM LEAD (BIRMINGHAM PLATINUM)
Manufacturer not listed
Zn 53-79, Cu 20-47, Fe 0.3.
For ornamental parts; castings.

PLATINUM RHODIUM
Manufacturer not listed
Pt 80-100, Rh 0-20.
For thermocouples; heat resistant.

PLATINUM SILVER
Manufacturer not listed
Ag 66.7, Pt 33.3.
For ornaments, jewelry; corrosion resistant.

PLATINUM SOLDER
Manufacturer not listed
Ag 73.
For solder for platinum alloys; corrosion resistant.

PLATINUM SUBSTITUTE
Manufacturer not listed
Al 23.6-24, Bi 3.7, Au 0.7, Ni 72.
For ornamental white metal parts; corrosion resistant.

PLATINUM SUBSTITUTE COOPERS-1
Manufacturer not listed
Ag 70, Pd 25, Co 5.
For jewelry; corrosion resistant.

PLATINUM SUBSTITUTE COOPERS-2
Manufacturer not listed
Ag 70, Pt 25, Ni 5.
For jewelry; corrosion resistant.

PLATINUM SUBSTITUTE ELECTRICAL-1
Manufacturer not listed
Au 70, Ag 25, 5 Ni or Pt.
For electrical contacts; heat resistant.

PLATINUM SUBSTITUTE ELECTRICAL-2
Manufacturer not listed
Ag 25, Au 68, Pt 7.5.
For electrical contacts; heat resistant.

PLATINUM/NICKEL ALLOY
Sigmund Cohn Corp.
Pt 89-91, Ni 9-11.
260,000 psi TS.

PLATNIK
Manufacturer not listed
Pt, Ni.
For ornaments, jewelry; corrosion resistant.

PLETTENBERG CMV
Plettenberger Gusstahlfabrik
C 0.45, Ni, Cr, W, bal Fe
For forging dies, upsetters, header dies; oil hardened, tough.

PLETTENBERG EXTRA MH
Plettenberger Gusstahlfabrik
C 1.1, Si 0-0.25, Mn 0-0.25, bal Fe.
Annealed: 110,000 TS; 58,000 YS; 18 El; 38 RA; 210 Brin. For springs, tools, drills, taps, reamers; Type W1; water hardened.

PLETTENBERG EXTRA WEICH
Plettenberger Gusstahlfabrik
C 0.7, Si 0-0.25, N 0-0.25, bal Fe.
Heat treated: 175,000 TS, 128,000 YS; 12 El; 37 RA; 055 Brin. For springs, rails, punches, axes, tools; Type W1; water hardened.

PLETTENBERG EXTRA ZAH
Plettenberger Gusstahlfabrik
C 0.85, Si 0-0.25, Mn 0-0.25, bal Fe.
Heat treated: 190,000 TS; 145,000 YS; 10 El; 30 RA; 400 Brin.
For springs, tools, cutters, taps, drills; Type W1; water hardened.

PLETTENBERG EXTRA ZH
Plettenberger Gusstahlfabrik
C 1, S¹ 0-0.25, Mn 0-0.25, bal Fe.
Annealed: 100,000 TS; 53,000 YS; 21 El; 42 RA; 200 Brin. For springs, drills, taps, hobs, reamers; Type W1; water hardened.

PLETTENBERG GC
Plettenberger Gusstahlfabrik
C 0.4, Cr, Mn, Mo, bal Fe.
For gears, bolts, machine tool parts; oil hardened, shock resistant.

PLETTENBERG GCN1
Plettenberger Gusstahlfabrik
C 0.55, Cr 0.7, Mo 0.18, Ni 1.65, V 0.1, bal Fe.
For springs, gears, crankshafts, bolts, studs; oil hardened, shock resistant.

PLETTENBERG GCN2
Plettenberger Gusstahlfabrik
C 0.56, Ni, Cr, Mo, V, bal Fe.
For gears, bolts, crankshafts; oil hardened, shock resistant.

PLETTENBERG GMS
Plettenberger Gusstahlfabrik
C 0.53, Si 0.9, Mn 0.9, bal Fe.
For upsetters, punches, crimpers; water hardened.

PLETTENBERG HW
Plettenberger Gusstahlfabrik
C 0.67, Si 1.3, N 0.5, Cr 0.5, bal Fe.
For upsetters, punches, die blocks; oil hardened, tough.

PLETTENBERG KCR13
Plettenberger Gusstahlfabrik
C 0.4, Cr 13, Mn 0.3, bal Fe.
Annealed: 100,000 TS; 55,000 YS; 20 El; 50 RA; 200 Brin. For valves, cutlery, surgical and dental instruments; Type 420; stainless.

PLETTENBERG KCR2
Plettenberger Gusstahlfabrik
C 0.2, Mn 1.25, Cr 1.15, bal Fe.
For cams, bolts, camshafts; case hardened.

PLETTENBERG KL
Plettenberger Gusstahlfabrik
C 0.55, Si 0.9, Cr 1.05, V 0.18, W 1.85, bal Fe.
For header dies, forging dies; oil hardened, tough.

PLETTENBERG KLV
Plettenberger Gusstahlfabrik
C 0.61, Si 0.85, Mn 0.75, Cr 1.18, V 0.1, bal Fe.
For crankshafts, header dies, punches; oil hardened.

PLETTENBERG KS
Plettenberger Gusstahlfabrik
C 0.45, W, Cr, V, bal Fe.
For forging and header dies, upsetters; oil hardened, tough.

PLETTENBERG LC13
Plettenberger Gusstahlfabrik
C 2.1, Cr 11.5, Mn 0.3, bal Fe.
For blanking and forming dies, punches; oil hardened, nondeforming.

PLETTENBERG LC13S
Plettenberger Gusstahlfabrik
C 1.65, Cr 11.5, V 0.1, Mn 0.3, bal Fe.
For blanking and forming dies, punches; air hardened, nondeforming.

PLETTENBERG LC13SK
Plettenberger Gusstahlfabrik
C 1.65, Cr 11.5, Mn 0.3, Co, bal Fe.
For blanking and forming dies, punches; air hardened, nondeforming.

PLETTENBERG LC13W
Plettenberger Gusstahlfabrik
C 2.1, Cr 11.5, W 0.7, Mn 0.3, bal Fe.
For blanking and piercing dies, punches; oil hardened, nondeforming.

PLETTENBERG MO10
Plettenberger Gusstahlfabrik
C 0.8, Cr, Mo, W, V, bal Fe.
For cutters, tools, dies; oil hardened.

PLETTENBERG MO6
Plettenberger Gusstahlfabrik
C 0.85, W, Mo, bal Fe.
For cutters, tools, dies; oil hardened.

PLETTENBERG PCR1
Plettenberger Gusstahlfabrik
C 0.45, Mo 0.7, V 0.3, Cr 1.4, bal Fe.
For forging dies, headers, upsetters; oil hardened, tough.

PLETTENBERG PCR2
Plettenberger Gusstahlfabrik
C 0.9, Cr 0.8, Mn 0.3, bal Fe.
For bearings, cutters, bushings, liners, races; water or oil hardened.

PLETTENBERG PD
Plettenberger Gusstahlfabrik
C 0.35, Si 0.9, Cr 1.05, V 0.18, W 1.85, bal Fe.
For heading and forging dies, punches, upsetters; oil hardened, tough.

PLETTENBERG PF
Plettenberger Gusstahlfabrik
C 0.7, Si 1.7, Mn 0.7, bal Fe.
For springs, punches; oil hardened, tough.

PLETTENBERG PG
Plettenberger Gusstahlfabrik
C 0.15, Si 0.35, Mn 0.3, bal Fe.
Annealed: 70,000 TS; 55,000 YS; 25 El; 60 RA; 145 Brin. For screws, bolts, gears, machine tool parts; case hardened.

PLETTENBERG PG100
Plettenberger Gusstahlfabrik
C 0.9, Si 0.25-0.5, Mn 0.3-0.8, bal Fe.
Heat treated: 130,000-180,000 TS; 80,000-120,000 YS; 10-20 El; 30-47 RA; 270-375 Brin. For drills, hobs, taps, springs, reamers; water hardened; Type W1.

PLETTENBERG PG65
Plettenberger Gusstahlfabrik
Mn 0.3-0.8, C 0.45, Si 0.25-0.5, bal Fe.
Hot rolled: 98,000 TS; 59,000 YS; 24 El; 45 RA; 212 Brin. For axles, gears, bolts, tie rods, crankshafts; water hardened.

PLETTENBERG PG85
Plettenberger Gusstahlfabrik
C 0.6, Si 0.25-0.5, Mn 0.3-0.8, bal Fe.
Heat treated: 115,000-160,000 TS; 77,000-113,000 YS; 12-23 El; 40-54 RA; 230-320 Brin. For wheels, die blocks, girders, tie rods, springs; water hardened.

PLETTENBERG PG90
Plettenberger Gusstahlfabrik
C 0.67, Si 0.25-0.5, Mn 0.3-0.8, bal Fe.
Heat treated: 122,000-174,000 TS; 82,000-128,000 YS; 12-22 El; 37-52 RA; 240-352 Brin. For springs, clutch discs, die blocks, girders, rails; water hardened.

PLETTENBERG PG95
Plettenberger Gusstahlfabrik
C 0.75, Si 0.25-0.5, Mn 0.3-0.8, bal Fe.
Heat treated: 125,000-180,000 TS; 85,000-130,000 YS; 10-20 El; 35-50 RA; 245-360 Brin. For springs, clutch discs, die blocks, rails; water hardened.

PLETTENBERG PN
Plettenberger Gusstahlfabrik
C 0.5, Cr 1.05, Ni 3.25, N 0.5, bal Fe.
For gears, bolts, crankshafts; oil hardened, shock resistant.

PLETTENBERG PRIMA WEICH
Plettenberger Gusstahlfabrik
C 0.7, Si 0-0.25, Mn 0-0.25, bal Fe.
Heat treated: 122,000-174,000 TS; 82,000-128,000 YS; 12-22 El; 37-52 RA; 240-350 Brin. For wheels, die blocks, girders, springers; Type W1; water hardened.

PLETTENBERG PRIMA ZAH
Plettenberger Gusstahlfabrik
C 0.85, Si 0-0.25, Mn 0-0.25, bal Fe.
Heat treated: 129,000-188,000 TS; 87,000-143,000 YS; 12-21 El; 35-50 RA; 235-388 Brin. For springs, taps, reamers, drills, cutters; Type W1; water hardened.

PLETTENBERG PRIMA-MH
Plettenberger Gusstahlfabrik
C 1.15, Si 0-0.25, Mn 0-0.25, bal Fe.
Annealed: 110,000 TS; 58,000 YS; 18 El; 38 RA; 210 Brin. For springs, taps, reamers, drills, hobs; Type W1; water hardened.

PLETTENBERG PRIMA-ZH
Plettenberger Gusstahlfabrik
C 1, Si 0-0.25, Mn 0-0.25, bal Fe.
Heat treated: 200,000 TS; 150,000 YS; 8 El; 28 RA; 410 Brin. For springs, taps, reamers, drills, hobs; Type W1; water hardened.

PLETTENBERG RAPID EXTRA III
Plettenberger Gusstahlfabrik
C 0.95, W, Mo, bal Fe.
For cutters, dies, tools; oil hardened.

PLETTENBERG S111
Plettenberger Gusstahlfabrik
C 1.05, Cr 1, W 1.15, Mn 0.9, bal Fe.
For heading dies, drawing and forming dies; oil hardened, tough.

PLETTENBERG S113
Plettenberger Gusstahlfabrik
C 1.45, Cr 1.4, Mn 0.6, bal Fe.
For blanking and forming dies, bearings; oil hardened, abrasion and wear resistant.

PLETTENBERG SK
Plettenberger Gusstahlfabrik
C 1, V 0.1, Mn 0.3, bal Fe.
For blanking and forming dies, reamers; Type W2; water hardened.

PLETTENBERG SP
Plettenberger Gusstahlfabrik
C 0.45, Si, Cr, V, bal Fe.
For gears, bolts, crankshafts; oil hardened, shock resistant.

PLETTENBERG SP35
Plettenberger Gusstahlfabrik
C 0.38, Si, Cr, V, bal Fe.
For gears, bolts, crankshafts; oil hardened, shock resistant.

PLETTENBERG SVM2
Plettenberger Gusstahlfabrik
C 0.9, Mn 1.9, V 0.1, bal Fe.
For punches, shears, blanking and forming dies; oil hardened, non-deforming.

PLETTENBERG WCR13

Plettenberger Gusstahlfabrik
C 0.2, Cr 13, Mn 0.3, bal Fe.
Annealed: 95,000 TS; 50,000 YS; 25 El; 55 RA; 195 Brin. For valves, cutlery, surgical and dental instruments; Type 420; stainless.

PLETTENBERG WP SPEZIAL EXTRA

Plettenberger Gusstahlfabrik
C 0.3, Cr 1.1, V 0.18, W 3.75, bal Fe.
For forging and heading dies, upsetters, shears; hot work steel, oil hardened.

PLETTENBERG WP5

Plettenberger Gusstahlfabrik
C 0.3, Cr 2.35, V 0.6, W 4.25, bal Fe.
For shear blades, upsetters, forging dies; hot work steel, oil hardened.

PLETTENBERG WP9

Plettenberger Gusstahlfabrik
C 0.3, Cr 2.8, V 0.35, W 8.5, bal Fe.
For extrusion rams and liners, shears, punches; hot work steel, oil hardened.

PLETTENBERG WPK

Plettenberger Gusstahlfabrik
C 0.3, Co 2, Cr 2.4, V 0.25, W 8.5, bal Fe.
For extrusion rams and liners, punches; hot work steel, oil hardened.

PLETTENBERG WPV

Plettenberger Gusstahlfabrik
C 0.65, Cr 3.75, Mo 0.85, V 0.7, W 8.5, bal Fe.
For shear blades, cutters, upsetters; oil hardened, hot work steel.

PLOW STEEL

Manufacturer not listed
C 0.6-0.9, bal Fe.
For plow shares, blacksmith tools; water hardening.

PLOWFACE

Champion Rivet Co.
C 4.5, Mn 6.5, Si 2, Cr 30, bal Fe.
550 Brin. For hard surfacing electrodes; abrasion and impact resistant.

PLUMBERS SOLDER

Manufacturer not listed
Pb 67, Sn 33.
For solder; soft solder.

PLUMBERS WHITE NO. 1

Manufacturer not listed
Cu 54, Zn 27, Ni 17, Pb 2.
For plumbing fixtures; free-cutting.

PLUMBERS WHITE NO. 2

Manufacturer not listed
Cu 54, Zn 25, Ni 13, Pb 7, Sn 1.
For faucets, cocks; free-cutting.

PLUMBERS WHITE NO. 3

Manufacturer not listed
Cu 58, Zn 25, Ni 15, Pb 1, Fe 1, Mn 0.3.
For faucets, cocks; corrosion resistant.

PLUMBIC BRONZE

English manufacture
Pb 26, Mn 1.7, Sn 1.5, Fe 1.2, bal Cu.
For bearings, bushings; heavy duty.

PLUMBITE

H. Kramer & Co. (Ajax Metal Div.)
Pb, Sn, Sb.
Used in slow running and light weight machines; free flowing, durable plastic. *Obsolete*

PLUMBSOL

Johnson Matthey plc
Ag, Sn.
Cast: 3350 TS; 60 El. For soft solder for plumbing installations; 221-225°C MP.

PLUMRITE

Olin Brass, Indianapolis
Copper. Cu 61, Zn 38.6, Pb 0.4.
For brass pipe for fresh water service and chemical plants; high strength.

PLUMRITE 85

Olin Brass, Indianapolis
Copper. Cu 85, Zn 15.
Annealed: 45,000 psi TS; 45 El. For water pipes and plumbing; red brass.

PLUMRITE COPPER

Olin Brass, Indianapolis
Copper. Cu 99.9.
Annealed: 30,000 psi TS; 40 El. Drawn: 48,000 psi TS; 4 El. For piping and water tube; corrosion resistant.

PLUMRITE NO. 67

Olin Brass, Indianapolis
Copper. Cu 67, Zn 32.5, Pb 0.5.
For water pipes and plumbing; free-cutting.

PLUMRITE STANDARD 85

Olin Brass, Indianapolis
Copper. Cu 85, Zn 15.
Hard: 70,000 psi TS; 57,000 psi YS; 5 El. Soft: 40,000 psi TS; 12,000 psi YS; 47 El. For flexible hose and deep drawn parts; corrosion resistant.

PLURALLOY 100

Teledyne McKay
C 0.06-0.1, Mn 0.6-0.8, Mo 0.7-0.9, bal Fe.
Welded: 95,000 TS; 83,000 YS; 27 El; 197 Brin. For welding electrodes; AWS-E9015; shielded arc. *Obsolete*

PLURALLOY 110

Teledyne McKay
C 0.06-0.1, Mn 0.6-0.8, Mo 0.2-0.4, Ni 1.25-1.75, V 0.05-10, bal Fe.
Welded: 104,000 TS; 97,000 YS; 25 El; 207 Brin. For welding electrodes; AWS-E-10015; shielded arc. *Obsolete*

PLURALLOY 120

Teledyne McKay
C 0.06-0.1, Mn 0.6-0.8, Mo 0.2-0.4, Ni 1.75-2.25, V 0.1-0.2, bal Fe.
Welded: 110,000 TS; 106,000 YS; 22 El; 220 Brin. For welding electrodes; shielded arc. *Obsolete*

PLURALLOY 3 1/2 NI

Teledyne McKay
C 0.06-0.1, Mn 0.6-0.8, Ni 3.25-3.75, bal Fe.
Welded: 83,000 TS; 71,000 YS; 29 El; ;70 RA; 163 Brin. For welding electrodes; shielded arc. *Obsolete*

PLURALLOY 70

Teledyne McKay
C 0.06-0.1, Mn 0.6-0.8, bal Fe.
Welded: 75,000 TS; 64,000 YS; 31 El; 137 Brin. For welding electrodes; AWS-E6015; shielded arc; low hydrogen. *Obsolete*

PLURALLOY 80

Teledyne McKay
C 0.06-0.1, Mn 0.6-0.8, Mo 0.3-0.5, bal Fe.
Welded: 82,000 TS; 69,000 YS; 29 El; 160 Brin. For welding electrodes; AWS-E7015; shielded arc. *Obsolete*

PLURALLOY 90

Teledyne McKay
C 0.06-0.1, Mn 0.6-0.8, Mo 0.55-0.75, bal Fe.
Welded: 87,600 TS; 76,000 YS; 29 El; 170 Brin. For welding electrode; AWS-E8015; shielded arc. *Obsolete*

PLUTAIR-675

Richard W. Carr & Co., Ltd.
C 0.5, Cr 7.5, Mo 0.5, Si 1.3, Mn 0.45, bal Fe.
For hot shears, die holders, hot dies; air hardened. *Obsolete*

PLUTEOUS 1000

Manufacturer not listed
Now CARR'S QUALITY P.1000.

PLUTEOUS 1000

Richard W. Carr & Co., Ltd.
C 0.1, Cr, bal Fe.
Heat treated: 78,500-101,000 TS; 67,000-89,600 YS; 25 El; 65 RA. For aircraft parts, fittings, pump rods, plastic mold dies; martensitic, corrosion and heat resistant. *Obsolete*

PLUTEOUS 1001

Richard W. Carr & Co., Ltd.
Medium C, Cr, bal Fe.
Heat treated: 103,000-116,500 TS; 78,500-101,000 YS; 20 El; 55 RA. For ball and roller bearings, needles, springs, valves; martensitic, stainless. *Obsolete*

PLUTEOUS 1001/31 C

Richard W. Carr & Co., Ltd.
Medium C, Cr, S, bal Fe.
For oil pump and valve parts; free-cutting, stainless, martensitic. *Obsolete*

PLUTEOUS 1003

Richard W. Carr & Co., Ltd.
Cr 18, Ni 8, C, bal Fe.
Annealed: 78,400-94,400 TS; 30-50 El; 60 RA. For chemical and textile equipment, marine parts; stainless, austenitic. *Obsolete*

PLUTEOUS 1009

Richard W. Carr & Co., Ltd.
Cr 16, Ni 2, C, bal Fe.
Heat treated: 123,000 TS; 96,300-114,200 YS; 20 El; 40-60 RA. For aircraft parts, fittings, pump shafts, valves; martensitic, stainless. *Obsolete*

PLUTO

Vereinigte Edelstahlwerke
C 0.7, W 19, Cr 4, V 2, bal Fe.
For cutters, tools; high speed steel. *Obsolete*

PLUTO G

Now VEW 5200.

PLUTO PARAMOUNT

Richard W. Carr & Co., Ltd.
C 0.78, Cr 4.5, W 18.75, Mo 0.7, V 1.25, Co 10, bal Fe.
Cobalt-tungsten high speed steel. For tableware; corrosion resistant.

PLUTO PEERLESS

Richard W. Carr & Co., Ltd.
C 1.4, Cr 4, Mo 3.2, W 9, V 3.1, Co 9.5, bal Fe.
Tungsten-molybdenum-cobalt high speed steel. AISI T42.

PLUTO PERFECTUM

Richard W. Carr & Co., Ltd.
C 0.85, Cr 5, W 10, Co 5.0, bal Fe.
For milling cutters, lathe and planer tools; high speed steel.

PLUTO PLUS

Richard W. Carr & Co., Ltd.
C 1.1, Cr 3.75, Mo 9.5, W 1.5, V 1.2, Co 8, bal Fe.
Molybdenum-cobalt high speed steel. AISI M42.

PLUTO PREMIUM

Richard W. Carr & Co., Ltd.
C 1.55, Cr 4.5, Mo 3.5, W 6.5, V 5, Co 5, bal Fe.
High speed steel.

PLUTOCRAT
Richard W. Carr & Co., Ltd.
C 0.8, Cr 4.5, V 1, W 22, bal Fe.
For milling cutters, drills, reamers, boring tools; high speed steel.

PLUTOIL 704
Now CARR'S QUALITY P.704.

PLUTOIL 708
Richard W. Carr & Co., Ltd.
C 0.9, Mn 1.5, Cr 0.5, bal Fe.
For forming, punching, and trimming dies; thread gages; oil hardened, non-deforming. *Obsolete*

PLUTOIL 723
Richard W. Carr & Co., Ltd.
C 1, Mn 0.5, Cr 1.5, bal Fe.
For collets, taps, drills, cams, broaches; oil hardened, abrasion resistant. *Obsolete*

PLUTONIC 151, ETC.
Now CARR'S QUALITY P.151, ETC.

PLUTONIC-12S
Richard W. Carr & Co., Ltd.
C 0.3, W 8.5, Cr 4, Mn 0.3, bal Fe.
For Al die casting dies; resists heat checking. *Obsolete*

PLUTONIC-157
Richard W. Carr & Co., Ltd.
C 0.15, Cr 1.2, Ni 4.25, bal Fe.
For gears, pinions, camshafts, cams; case hardened, shock resistant. *Obsolete*

PLUTONIC-163
Richard W. Carr & Co., Ltd.
C 0.15, Ni 0.8, Mo 0.1, Cr 0.6, Mn 0.8, bal Fe.
For gears, pinions, camshafts, cams; case hardened, shock resistant. *Obsolete*

PLUTONIC-164
Richard W. Carr & Co., Ltd.
C 0.2, Cr 0.8, Ni 1, Mo 0.1, bal Fe.
For gears, pinions, camshafts, cams, bolts; case hardened, shock resistant. *Obsolete*

PLUTONIC-165
Richard W. Carr & Co., Ltd.
C 0.2, Ni 1.25, Mo 0.12, Mn 0.75, bal Fe.
For gears, pinions, camshafts, cams; case hardened, shock resistant. *Obsolete*

PLUTONIC-167
Richard W. Carr & Co., Ltd.
C 0.2, Cr 1.6, Ni 2, Mo 0.2, Mn 0.55, bal Fe.
For gears, pinions, camshafts, cams; case hardened, shock resistant. *Obsolete*

PLUTONIC-451
Richard W. Carr & Co., Ltd.
C 0.3, Cr 0.25, Ni 3, Mn 0.55, bal Fe.
For die casting dies; oil hardened. *Obsolete*

PLUTONIC-601
Richard W. Carr & Co., Ltd.
C 0.4, Mn 0.65, Cr 1.25, V 0.15, bal Fe.
For die casting and permanent mold dies; oil hardened. *Obsolete*

PLUTONIC-606
Richard W. Carr & Co., Ltd.
C 1.25, Cr 0.5, W 1.5, bal Fe.
For draw bench dies, taps; water hardened, tough. *Obsolete*

PLYKROME
Krupp Huttenwerke
composite steel plus welded stainless surface high, Cr, high Ni Ferrous alloy.
For stainless steel parts, tanks, vessels; stainless steel sheet welded onto steel surface.

PLYKROME
Plykrome Corp.
composite steel plus welded stainless surface high, Cr, high Ni Ferrous alloy.
For stainless steel parts, tanks, vessels; stainless steel sheet welded onto steel surface.

PLYMITE
P.L. & M. Co.
W alloy.
For wear resistant parts.

PM 18
Walsingham Steel Co., Ltd.
C 0.16-0.2, Si 0.3-0.5, Mn 1.35-1.55, Ni 0-0.4, Cr 0-0.25, Mo 0-0.15, Cu 0-0.3, bal Fe.
Pearlitic manganese steel. Meets BS 2772.

PM 18/8/6
Pose-Marre Edelstahlwerk G.m.b.H.
C 0-0.2, Cr 18, Ni 8, Mn 6, bal Fe.
Wire or electrode: 45-65 kp/mm^2 TS; 18 kp/mm^2 YS min; 15 El min. For welding of comparable base metals. *Obsolete*

PM 20/70 NB
Pose-Marre Edelstahlwerk G.m.b.H.
C 0-0.05, Cr 20, Ni 70, Nb 2.5, Ti 0.5, bal Fe.
Wire or electrode: 45-65 kp/mm^2 TS; 22 kp/mm^2 YS min; 10 El min. For welding of high nickel alloys. *Obsolete*

PM 23
Walsingham Steel Co., Ltd.
C 0.18-0.23, Si 0-0.5, Mn 1.2-1.6, bal Fe.
Pearlitic manganese steel. Meets BS 1456 Gr. A, B.

PM 24/24 NB
Pose-Marre Edelstahlwerk G.m.b.H.
C 0.35, Cr 24, Ni 24, Nb 1.5, bal Fe.
Wire or electrode: 50-70 kp/mm^2 TS; 25 kp/mm^2 YS min; 8 El min. For welding of comparable base metals. *Obsolete*

PM 25/20 R
Pose-Marre Edelstahlwerk G.m.b.H.
C 0.4, Cr 25, Ni 20, bal Fe.
Wire or electrode: 50-70 kp/mm^2 TS; 25 kp/mm^2 YS min; 5 El min. For welding of comparable base metals. *Obsolete*

PM 25/35 NB
Pose-Marre Edelstahlwerk G.m.b.H.
C 0.4, Cr 25, Ni 35, Nb 1.5, bal Fe.
Wire or electrode: 50-70 kp/mm^2 TS; 25 kp/mm^2 YS min; 8 El min. For welding of comparable base metals. *Obsolete*

PM 25/35 R
Pose-Marre Edelstahlwerk G.m.b.H.
C 0.4, Cr 25, Ni 35, bal Fe.
Wire or electrode: 50-70 kp/mm^2 TS; 25 kp/mm^2 YS min; 5 El min. For welding of comparable base metals. *Obsolete*

PM 26/36 MO
Pose-Marre Edelstahlwerk G.m.b.H.
C 0.45, Cr 26, Ni 36, Mo 1.5, bal Fe.
Wire or electrode: 50-70 kp/mm^2 TS; 25 kp/mm^2 YS min; 5 El min. For welding of comparable base metals. *Obsolete*

PM 28/48/5
Pose-Marre Edelstahlwerk G.m.b.H.
C 0-0.1, Cr 28, Ni 48, W 5, bal Fe.
Wire or electrode: 45-65 kp/mm^2 TS; 18 kp/mm^2 YS min; 3 El min. For welding of comparable base metals. *Obsolete*

PM 30/30
Pose-Marre Edelstahlwerk G.m.b.H.
C 0.45, Cr 30, Ni 30, bal Fe.
Wire or electrode: 50-70 kp/mm^2 TS; 25 kp/mm^2 YS min; 4 El min. For welding of comparable base metals. *Obsolete*

PM IN-792
International Nickel Inc.
Nickel. C 0-0.1, Cr 12.5, Mo 2, Co 9.3, Al 3.2, Ti 4.3, W 4, B 0.015, Zr 0.1, bal Ni.
Heat treated. At room temperature: 1400 N/mm^2 TS; 1100 N/mm^2 YS; 8 El. At 1400°F: 1000 N/mm^2 TS; 950 N/mm^2 YS; 20 El. For special elevated temperature operations. *Obsolete*

PM UMCO 50
Pose-Marre Edelstahlwerk G.m.b.H.
C 0-0.1, Cr 28, Co 50, bal Fe.
Wire or electrode: 45-65 kp/mm^2 TS; 18 kp/mm^2 YS min; 3 El min. For welding of comparable base metals. *Obsolete*

PM-20
Societe Nouvelle des Acieries de Pompey
C 0.05, Mn 0.9, Ni 3, Cu 0.7, Mo 0.18, bal Fe.
Rolled: 85,000-95,000 TS; 57,000-68,000 YS; 18 El. For automotive and railway bodies; good workability. *Obsolete*

PM150
Friedr. Lohmann GmbH
C 1.05, Cr 1, W 1.15, Mn 0.9, bal Fe.
For cutters. Oil hardened.

PMB 20, ETC.
Now BOHNALLOY PMB 20, ETC.

PMD-45
Atlas Specialty Steels
C 0.4, Mn 0.45, Si 0.3, Cr 5, Mo 2.25, V 1, bal Fe.
Hot work tool steel. *Obsolete*

PMG NO. 10
Phelps Dodge Industries
Cu 95.6, Si 3.2, FW 1.2.
For tanks, vats, pump rods, pistons; corrosion resistant. *Obsolete*

PMG NO. 3
Phelps Dodge Industries
Cu 98.2, Si 1.2, Fe 0.6.
For screws, rivets, bolts, nails; corrosion resistant. *Obsolete*

PMG NO. 95
Phelps Dodge Industries
Cu 96.6, Pb 0.4, Si 2.5, Fe 0.5.
For screw machine products; tough, corrosion resistant. *Obsolete*

PMM NO. 96
Phelps Dodge Industries
Cu 97, Si 2.5, Fe 0.5.
For bolts, nuts, tubing; cold heading. *Obsolete*

PMP-Z-70
Powder Metal Products Co.
Cu alloy.
For cams, latches, keys, bushings; powder metals; sintered.

PNEU-DIE
Disston Inc.
C 0.48, Cr 2.15, Mo 0.3, bal Fe.
For chisels, punches; tough. *Obsolete*

PNEUMO
George Cook & Co., Ltd.
C 0.45, Cr 1.25, W 0.2, Mn 0.35, Si 0.75, bal Fe.
For upsetters, rivet sets; oil hardened, tough.

PNEUTOUGH

British Steel Corp.
C 0.5, Si 0.7, Cr 1.5, V 0.25, W 2.25, bal Fe.
Oil hardening shock resisting special alloy tool steel; for pneumatic chisels, coal picks, diecasting dies (zinc), pneumatic tools, guillotine blades. *Obsolete*

PNUSNAP OH

Firth Brown Ltd.
O 0.40, Si 1, Mn 0.5, Cr 1, W 1.8, bal Fe.
For tools, dies; hot work steel. *Obsolete*

PNUSNAP WH

Firth Brown Ltd.
C 0.35, Si 0.6, Mn 0.3, Cr 1.1, W 1.9, bal Fe.
For tools, dies; hot work steel. *Obsolete*

POBEDIT

Russian manufacture
Co 10, 90.0 WC.
For cutting tools; cemented.

POBEDIT-ALPHA

Russian manufacture
TiC, Co.
For cutting tools; cemented.

POFORS SR1855

Bofors AB
C 1, Si 1.5, Cr 1, bal Fe.
For cutting and threading tools; oil hardened. *Obsolete*

POLAR

Teledyne Allvac
C 0.2, bal Fe.
Annealed: 73,000 TS; 41,000 YS; 22 El; 58 RA; 140 Brin. For pole pieces for magnetos; high permeability. *Obsolete*

POLARIS

Disston Inc.
C 0.6, Ni 0.7, Cr 0.15, Mo 0.15, bal Fe.
For lawn mower blades, dies; tough, wear resistant. *Obsolete*

POLARIS 2

Time Steel Service Inc.
C 0.06, Mn 0.3, Si 0.15, Cr 0.95, Mo 0.25, bal Fe (boron added).
Steel for plastic molds. AISI P2.

POLARIS 20

Time Steel Service Inc.
C 0.37, Mn 1, Si 0.3, Cr 1.25, V 0.15, Mo 0.35, bal Fe.
Steel for plastic molds. AISI P20.

POLARIT 720

Outokumpu Metals (USA) Inc.
Stainless steel. C 0-0.03, Cr 17-19, Ni 9-12.
Corrosion resistant stainless steel equivalent to BS 304S11; ASTM 304L. 170 N/mm^2 YS min; 485 N/mm^2 TS; 40 El; 183 Brin.

POLARIT 721

Outokumpu Metals (USA) Inc.
Stainless steel. C 0.025, Si 0.5, Mn 1.5, Cr 18, Ni 9, N 0.14, bal Fe.
Corrosion resistant stainless steel equivalent to AISI 304LN; SS 2371; BS 304S62; AFNOR Z2 CN18-10 AZ. Annealed: 270 N/mm^2 YS min; 550-760 N/mm^2 TS; 35 El min; 210 Brin max.

POLARIT 725

Outokumpu Metals (USA) Inc.
Stainless steel. C 0-0.05, Cr 17-19, Ni 8-11, bal Fe.
Corrosion resistant stainless steel equivalent to BS 304S15; ASTM 304. 205 N/mm^2 YS min; 515 N/mm^2 TS; 40 El; 183 Brin.

POLARIT 731

Outokumpu Metals (USA) Inc.
Stainless steel. C 0-0.08, Cr 17-19, Ni 9-12, Ti = 5 x C up to 0.80, bal Fe.
Corrosion resistant stainless steel equivalent to BS 321S12; ASTM 321. 205 N/mm^2 YS min; 515 N/mm^2 TS; 40 El; 183 Brin.

POLARIT 750

Outokumpu Metals (USA) Inc.
Stainless steel. C 0-0.03, Cr 16-18.5, Ni 11-14, Mo 2-2.5, bal Fe.
Corrosion resistant stainless steel equivalent to BS 316S11; ASTM 316L. 170 N/mm^2 YS min; 485 N/mm^2 TS; 40 El; 217 Brin.

POLARIT 751

Outokumpu Metals (USA) Inc.
Stainless steel. C 0.025, Si 0.5, Mn 1.5, Cr 18, Ni 11, Mo 2.2, N 0.14, bal Fe.
Corrosion resistant stainless steel equivalent to AISI 316LN; AFNOR Z2 CND17-12 AZ. Annealed: 280 N/mm^2 YS min; 580-800 N/mm^2 TS; 35 El min; 210 Brin max.

POLARIT 752

Outokumpu Metals (USA) Inc.
Stainless steel. C 0-0.03, Cr 16-18.5, Ni 11.5-14.5, Mo 2.5-3, bal Fe.
Corrosion resistant stainless steel equivalent to BS 316S13; ASTM 316L. 170 N/mm^2 YS min; 485 N/mm^2 TS; 40 El; 217 Brin.

POLARIT 753

Outokumpu Metals (USA) Inc.
Stainless steel. C 0.025, Si 0.5, Mn 1.5, Cr 18, Ni 12, Mo 2.7, N 0.15, bal Fe.
Corrosion resistant stainless steel equivalent to AISI 316LN; SS 2375; BS 326S62; AFNOR Z2 CND17-13 AZ. Annealed: 295 N/mm^2 YS min; 580-800 N/mm^2 TS; 35 El min; 220 Brin max.

POLARIT 755

Outokumpu Metals (USA) Inc.
Stainless steel. C 0-0.05, Cr 16-18.5, Ni 10.5-14, Mo 2-2.5, bal Fe.
Corrosion resistant stainless steel equivalent to BS 316S31; ASTM 316. 205 N/mm^2 YS min; 515 N/mm^2 TS; 40 El; 217 Brin.

POLARIT 757

Outokumpu Metals (USA) Inc.
Stainless steel. C 0-0.05, Cr 16-18.5, Ni 10.5-14, Mo 2.5-3, bal Fe.
Corrosion resistant stainless steel equivalent to BS 316S33; ASTM 316. 205 N/mm^2 YS min; 515 N/mm^2 TS; 40 El; 217 Brin.

POLARIT 761

Outokumpu Metals (USA) Inc.
Stainless steel. C 0-0.08, Cr 16-18.5, Ni 10.5-14, Mo 2-2.5, Ti = 5 x C up to 0.80, bal Fe.
Corrosion resistant stainless steel equivalent to BS 320S31; ASTM 316Ti.

POLARIT 770

Outokumpu Metals (USA) Inc.
Stainless steel. C 0.025, Si 0.5, Mn 1.5, Cr 18, Ni 14, Mo 3.5, bal Fe.
Corrosion resistant stainless steel equivalent to SS 2367; AISI 317; AFNOR Z2 CND 19-15. Annealed: 220 N/mm^2 YS min; 490-690 N/mm^2 TS; 40 El min; 200 Brin max. S31703

POLARIT 772

Outokumpu Metals (USA) Inc.
Stainless steel. C 0.025, Si 0.5, Mn 1.5, Cr 17, Ni 13, Mo 4.5, N 0.15, bal Fe.
Corrosion resistant stainless steel equivalent to AISI 317LNM. Annealed: 285 N/mm^2 YS min; 590-780 N/mm^2 TS; 40 El min; 210 Brin max.

POLARIT 773

Outokumpu Metals (USA) Inc.
Stainless steel. C 0.025, Si 0.5, Mn 1.5, Cr 17, Ni 15, Mo 4.5, bal Fe.
Corrosion resistant stainless steel equivalent to AISI 317LM. Annealed: 205 N/mm^2 YS min; 540-740 N/mm^2 TS; 40 El min; 190 Brin max.

POLARIT 774

Outokumpu Metals (USA) Inc.
Stainless steel. C 0.02, Si 0.5, Mn 1.5, Cr 20, Ni 25, Mo 4.5, Cu 1.5, bal Fe.
Corrosion resistant stainless steel equivalent to SS 2562; AFNOR Z1 NCDU 25-20. Annealed: 220 N/mm^2 YS min; 500-750 N/mm^2 TS; 35 El min; 220 Brin max. N08904

POLARIT 778

Outokumpu Metals (USA) Inc.
Stainless steel. C 0.02, Si 0.5, Mn 0.5, Cr 20, Ni 18, Mo 6.2, N 0.2, bal Fe.
Corrosion resistant stainless steel. Annealed: 300 N/mm^2 YS min; 600-800 N/mm^2 TS; 35 El min; 220 Brin max. S31254

POLDI T5 HART

Poldi Steel Works
C 0.61, Mn 0.65, Si 0.25, bal Fe.
Hot rolled: 118,000 TS; 70,000 YS; 16 El; 35 RA; 229 Brin. For punches, crimpers, axles; water hardened.

POLDI T5 HART

Westa-Westdeutsche
C 0.61, Mn 0.65, Si 0.25, bal Fe.
Hot rolled: 118,000 TS; 70,000 YS; 16 El; 35 RA; 229 Brin. For punches, crimpers, axles; water hardened.

POLISH 2H18N9

Polish manufacture
C 0.2, Mn 0.7, Cr 18, Ni 8.3, Mo 0.9, bal Fe.
Annealed: 90,000 TS; 40,000 YS; 55 El; 155 Brin. For chemical plant equipment, agitators, digesters, tanks; stainless, austenitic, nonmagnetic.

POLITAL

Durener Metallwerke
Aluminum. Mg 0.5-2, Si 0.3-1.5, Mn 0-1.5, bal Al.
For light alloy parts; corrosion resistant, hardenable.

POLITAL

Durener Metallwerke
Aluminum. Si 0.5-1.5, Mn 0.4-1, Mg 0.4-1, bal Al.
Hardened: 40,000-50,000 TS; 26,000-36,000 YS; 10-20 El; 70-100 Brin. For general structures, scaffolds, booms, transmission towers. Heat treatable, high strength. *Obsolete*

POLITAL 38

Durener Metallwerke
Aluminum. Mg 2.5-4, bal Al.
Soft: 26,500 TS; 12,000 YS; 26 El; 60 Brin. Hard: 43,000 TS; 37,000 YS; 4 El; 85 Brin. For light alloy parts; corrosion resistant.

POLITIT

Hoffman & Co. KG
C 0.4, Cr 13, Mn 0.3, Si 0.4, bal Fe.
Annealed: 95,000 TS; 50,000 YS; 25 El; 55 RA; 196 Brin. For turbine blades, cutlery, valves, knives, fasteners; Type 420; stainless, hardenable.

POLITIT 13/1

Hoffman & Co. KG
C 0-0.12, Cr 13, Si 0.4, bal Fe.
Annealed: 75,000 TS; 40,000 YS; 35 El; 70 RA; 155 Brin. Cold drawn: 100,000 TS; 85,000 YS; 17 El; 60 RA; 205 Brin. For turbine blades, valve trim; Type 410; stainless.

POLITIT 13/2

Hoffman & Co. KG
C 0.2, Si 0.4, Cr 13, bal Fe.
Annealed: 95,000 TS; 50,000 YS; 25 El; 55 RA; 195 Brin. Cold drawn: 105,000 TS; 85,000 YS; 17 El; 50 RA; 215 Brin. For turbine blades, valve trim, cutlery, surgical instruments; Type 420; stainless, hardenable.

POLITIT 13/4

Hoffman & Co. KG
C 0.4, Cr 13, Si 0.4, bal Fe.
Annealed: 95,000 TS; 50,000 YS; 25 El; 55 RA; 195 Brin. For cutlery, knives, surgical instruments; Type 420; stainless, hardenable.

POLITIT 13M

Hoffman & Co. KG
C 0.2, Cr 13, Mo 1.15, bal Fe.
Annealed: 95,000 TS; 50,000 YS; 25 El; 55 RA; 200 Brin. For acid resistant and oil refinery equipment; corrosion resistant.

POLITIT 17M

Hoffman & Co. KG
C 0.35, Cr 16.5, Mo 1.15, bal Fe.
For chemical plant and oil refinery equipment; corrosion resistant.

POLITIT 17N

Hoffman & Co. KG
C 0.22, Cr 17, Ni 1.5, bal Fe.
Annealed: 125,000 TS; 95,000 YS; 20 El; 55 RA; 240 Brin. For furnace parts and accessories; corrosion resistant, hardenable.

POLITIT 18/10M

Hoffman & Co. KG
C 0-0.07, Cr 18, Mo 2, Ni 10.5, bal Fe.
Annealed: 85,000 TS; 35,000 YS; 50 El; 65 RA; 160 Brin. Cold drawn: 150,000 TS; 135,000 YS; 6 El; 300 Brin. For acid resistant chemical plant equipment; Type 316; stainless, austenitic.

POLITIT 18/10MT

Hoffman & Co. KG
C 0-0.12, Cr 18, Mo 2, Ni 10.5, Ti = 4 x C, bal Fe.
Annealed: 90,000 TS; 40,000 YS; 45 El; 60 RA; 180 Brin. For welded chemical plant equipment; Type 316 Ti, stainless, austenitic.

POLITIT 18/8

Hoffman & Co. KG
C 0-0.15, Cr 18, Ni 8.5, bal Fe.
Annealed: 80,000 TS; 35,000 YS; 55 El; 75 RA; 150 Brin. For chemical plant equipment, tanks, dryers, mixers; Type 302; stainless, austenitic.

POLITIT 18/8T

Hoffman & Co. KG
C 0-0.12, Cr 18, Ni 9.5, Ti = 4 x C, bal Fe.
Annealed: 85,000 TS; 35,000 YS; 55 El; 65 RA; 150 Brin. For welded chemical plant equipment, tanks, mixers; Type 321; stainless, austenitic.

POLITIT 18/9

Hoffman & Co. KG
C 0.07, Cr 18, Ni 9.5, bal Fe.
Annealed: 85,000 TS; 35,000 YS; 60 El; 70 RA; 150 Brin. Cold drawn: 180,000 TS; 125,000 YS; 10 El; 330 Brin. For welded chemical plant equipment; Type 304; stainless, austenitic.

POLITIT 18CM

Hoffman & Co. KG
C 0.9, Cr 18, V 1, Mo 1.15, bal Fe.
Annealed: 108,000 TS; 62,000 YS; 18 El; 35 RA; 220 Brin. Heat treated: 280,000 TS; 270,000 YS; 3 El; 15 RA; 555 Brin. For bearings, valves, gauges, sleeves; stainless, wear resistant.

POLIVIT

German manufacture
Al 97.4, Cu 0.6, Mn 1.8, Ag 0.2.
For light alloy parts.

POLYMET A-G

Polymer Corp. Ltd.
Ag + plastic lubricant.
For instrument bearings, valve seats, rings and brushes. Low friction. Withstands temperatures to 700°F.

POMET 111

Powder Metals Inc.
Sn 5-10, bal Cu.
For mechanical and electrical parts; powder metals.

POMET 117B

Powder Metals Inc.
Cu 94-96, Sn 4-6.
Sintered: 18 El. For precision parts; sintered.

POMET 117C

Powder Metals Inc.
Cu 89-92, Sn 8-11.
Sintered: 39,000 TS. For instrument parts; sintered, good wear resistance.

POMET 141

Powder Metals Inc.
C 0.2, Cr 18, Ni 8, bal Fe.
For mechanical parts; sintered, stainless.

POMET 300

Powder Metals Inc.
Fe.
For magnetic cores, pole shoes; powder metallurgy.

POMET 309

Powder Metals Inc.
C 0.2-0.3, bal Fe.
Sintered: 38,000 TS. For machine and instrument parts; sintered alloy.

POMET 309H

Powder Metals Inc.
C 0.3, bal Fe.
Sintered and heat treated: 400 Brin. For clutch parts, cams, gears; sintered.

POMET 389

Powder Metals Inc.
C 0.1, bal Fe.
For general parts; powdered metals.

POMET 560

Powder Metals Inc.
Cu 5, bal Al.
For light alloy parts; powder metals.

POMOLOY C-2

Fairbanks, Morse & Co.
C 3.2, Si 2.2, bal Fe.
Cast: 40,000 TS; 215 Brin. For pump casings; gray cast iron.

POMOLOY C-3

Fairbanks, Morse & Co.
C 3.2, Si 2, bal Fe.
Cast: 40,000 TS; 210 Brin. For pump casings; gray cast iron.

POMPEY 10 CNK1

Societe Nouvelle des Acieries de Pompey
C 0.09, Mn 0.75, Si 0.25, Ni 1.4, Cr 1.1, bal Fe.
For gears, bolts, cams, camshafts; AFNOR 10NC6, case hardening steel, tough.

POMPEY 10 CNK3

Societe Nouvelle des Acieries de Pompey
C 0.1, Ni 3, Cr 0.75, bal Fe.
For gears, cams, machine tool parts; Afnor 10NC12, case hardening steel, tough. *Obsolete*

POMPEY 10 M1

Societe Nouvelle des Acieries de Pompey
C 0.15, Mn 1, Si 0.25, bal Fe.
Carburizing steel. AFNOR A 42 FP 2.

POMPEY 10K2

Societe Nouvelle des Acieries de Pompey
C 1, Cr 1.5, bal Fe.
Annealed: 95,000 TS; 62,000 YS; 27 El; 62 RA; 179 Brin. Heat treated: 200,000 TS; 185,000 YS; 3 El; 30 RA; 390 Brin. For punches, bearings, bushings; water hardened, wear resistant. *Obsolete*

POMPEY 10WK2

Societe Nouvelle des Acieries de Pompey
C 1, W 2, Cr, bal Fe.
For metal saws; water hardened. *Obsolete*

POMPEY 11

Societe Nouvelle des Acieries de Pompey
C 1.1, bal Fe.
For files; water hardened. *Obsolete*

POMPEY 1144

Societe Nouvelle des Acieries de Pompey
C 0.45, Mn 1.5, Si 0.25, P 0.04, S 0.28, bal Fe.
Manganese structural steel. AFNOR 45 MF 6; Free machining; AISI 1144.

POMPEY 11WK1

Societe Nouvelle des Acieries de Pompey
C 1.1, W 1, Cr, bal Fe.
For metal saws; water hardened. *Obsolete*

POMPEY 12 CNK3

Societe Nouvelle des Acieries de Pompey
C 0.14, Ni 3, Cr 0.75, bal Fe.
For gears, cams, camshafts, machine tool parts; AFNOR 14NC12, case hardening steel, tough.

POMPEY 12KV

Societe Nouvelle des Acieries de Pompey
C 1.2, Cr 1, V, bal Fe.
For milling cutters, reamers, taps, threading dies; water or oil hardened. *Obsolete*

POMPEY 12W1

Societe Nouvelle des Acieries de Pompey
C 1.1, W 1, V, bal Fe.
For drills, taps, cutters, reamers; water hardened. *Obsolete*

POMPEY 14 CNK 1

Societe Nouvelle des Acieries de Pompey
C 0.14, Mn 0.5, Ni 3, Cr 0.75, bal Fe.
Alloy carburizing steel, deep hardening. AFNOR 14 NC 12.

POMPEY 14 CNK3

Societe Nouvelle des Acieries de Pompey
C 0.14, Mn 0.4, Si 0.2, Ni 3.45, Cr 0.75, bal Fe.
Heat treated: 157,000-193,000 TS; 128,000 YS. For gears, cams, camshafts, machine tool parts; DIN-ECN35, case hardeni steel. *Obsolete*

POMPEY 14 N3

Societe Nouvelle des Acieries de Pompey
C 0.14, Mn 0.45, Si 0.2, Ni 3.5, bal Fe.
Nickel carburizing steel; deep hardening. AFNOR 12 N 14 and 3.5% Ni.

POMPEY 14CNK3

Societe Nouvelle des Acieries de Pompey
C 0.14, Mn 0.4, Si 0.2, Ni 3.5, Cr 0.75, bal Fe.
Heat treated: 178,000 TS; 135,000 YS; 10 El; 55 RA. For gears, bolts, camshafts, cams; D1N14NiCr14; case hardened. *Obsolete*

POMPEY 1541

Societe Nouvelle des Acieries de Pompey
C 0.4, Mn 1.5, Si 0.25, bal Fe.
Manganese structural steel. AFNOR 40 M 6; AISI 1541.

POMPEY 15N3
Societe Nouvelle des Acieries de Pompey
C 0.15, Mn 0.5, Si 0.35, Ni 3, bal Fe.
Heat treated: 127,000-163,000 TS; 100,000 YS; 10 El. For gears, cams, machine tool parts, fasteners; case hardening, shock resistant. *Obsolete*

POMPEY 16 CNK1
Societe Nouvelle des Acieries de Pompey
C 0.15, Mn 0.75, Si 0.25, Ni 1.4, Cr 1.1, bal Fe.
For gears, bolts, cams, machine tool parts; AFNOR 16NC6, case hardening steel, tough.

POMPEY 16 NC 4
Societe Nouvelle des Acieries de Pompey
C 0.16, Mn 0.85, Si 0.25, Ni 1, Cr 1, bal Fe.
Low alloy carburizing steel. AFNOR 16 NC 4.

POMPEY 17 M 1
Societe Nouvelle des Acieries de Pompey
C 0.18, Mn 1.1, Si 0.25, bal Fe.
Carburizing steel. AFNOR A 48 Fp 2.

POMPEY 18 M 2
Societe Nouvelle des Acieries de Pompey
C 0.18, Mn 1.35, Si 0.25, bal Fe.
Manganese carburizing steel. AFNOR 52 Fp 2.

POMPEY 20 CD 2
Societe Nouvelle des Acieries de Pompey
C 0.2, Mn 1, Si 0.25, Cr 0.5, Mo 0.2, bal Fe.
Low alloy carburizing steel. AFNOR 20 CD 2.

POMPEY 20 MN 5
Societe Nouvelle des Acieries de Pompey
C 0.2, Mn 1.25, Si 0.25, bal Fe.
Manganese carburizing steel. AFNOR 20 Mn 5.

POMPEY 20 NC 4
Societe Nouvelle des Acieries de Pompey
C 0.2, Mn 0.9, Si 0.25, Ni 1, Cr 1, bal Fe.
Low alloy carburizing steel. AFNOR 20 Nc 4.

POMPEY 20 NK1
Societe Nouvelle des Acieries de Pompey
C 0.2, Ni 1.4, Cr 1.1, bal Fe.
For gears, bolts, cams, machine tool parts; AFNOR 20NC6, case hardening steel, tough.

POMPEY 20 NK3
Societe Nouvelle des Acieries de Pompey
C 0.22, Mn 0.5, Ni 2.8, Cr 0.75, bal Fe.
Heat treated: 128,000-157,000 TS; 114,000 YS. For gears, cams, camshafts, machine tool parts; Afnor 20NC11, case hardening steel. *Obsolete*

POMPEY 20N2
Societe Nouvelle des Acieries de Pompey
C 0.22, Mn 0.45, Si 0.2, Ni 2, bal Fe.
Heat treated: 143,000-177,000 TS; 121,000 YS; 7 El. For gears, cams, camshafts, fasteners; Afnor 20N8, case hardened, shock resistant. *Obsolete*

POMPEY 20ND2
Societe Nouvelle des Acieries de Pompey
C 0.2, Ni 1.8, Mo 0.25, bal Fe.
Heat treated: 143,000-177,000 TS; 121,000 YS; 7 El. For gears, machine tool parts, fasteners; Afnor 20ND8, case hardened, shock resistant. *Obsolete*

POMPEY 20NKD
Societe Nouvelle des Acieries de Pompey
C 0.2, Ni 0.6, Cr 0.5, Mo 0.2, bal Fe.
Hot rolled: 93,000 psi TS; 65,000 psi YS; 25 El; 63 RA; 162 Brin. For gears, bolts, machine tool parts; case hardening, shock resistant, AFNOR 20 NCD2.

POMPEY 21 CDV 5
Societe Nouvelle des Acieries de Pompey
C 0.21, Mn 0.4, Si 0.45, Cr 1.3, V 0.25, bal Fe.
Cr-V carburizing steel. AFNOR 20 CDV 5-08.

POMPEY 22 M 1
Societe Nouvelle des Acieries de Pompey
C 0.22, Mn 1.2, Si 0.25, bal Fe.
Manganese carburizing steel. AFNOR A.58.F.P.2.

POMPEY 23 D 5 E
Societe Nouvelle des Acieries de Pompey
C 0.23, Mn 0.65, Si 0.2, Mo 0.55, bal Fe.
Molybdenum carburizing steel. AFNOR 23 D 5.

POMPEY 28 CD 12
Societe Nouvelle des Acieries de Pompey
C 0.3, Mn 0.85, Si 0.25, Cr 3, Mo 0.4, bal Fe.
Cr-Mo medium carbon steel. AFNOR 30 CD 12.

POMPEY 29 NKD 2
Societe Nouvelle des Acieries de Pompey
C 0.28, Mn 0.5, Si 0.25, P 0-0.02, S 0-0.02, Ni 2.35, Cr 0.6, Mo 0.5, bal Fe.
Ni-Cr-Mo structural steel. AFNOR 25 NCD 9.

POMPEY 30NKD2
Societe Nouvelle des Acieries de Pompey
C 0.3, Ni 2, Cr 2, Mo 0.35, bal Fe.
Heat treated: 170,000-200,000 psi TS; 143,000 psi YS; 8 El. For bolts, gears, machine tool parts; oil hardened, shock resistant.

POMPEY 30NK3
Societe Nouvelle des Acieries de Pompey
C 0.29, Ni 2.8, Cr 0.75, bal Fe.
Heat treated: 143,000-175,000 psi TS; 128,000 psi YS. For gears, bolts, crankshafts, fasteners; AFNOR 30NC11, oil hardened, shock resistant.

POMPEY 30NKD3
Societe Nouvelle des Acieries de Pompey
C 0.3, Ni 3, Cr 0.6, Mo 0.35, bal Fe.
Heat treated: 135,000-185,000 psi TS; 128,000 psi YS; 12 El. For bolts, gears, crankshafts, fasteners; oil hardened, shock resistant.

POMPEY 32NK3
Societe Nouvelle des Acieries de Pompey
C 0.32, Ni 3.45, Cr 0.7, bal Fe.
Heat treated: 143,000-175,000 TS; 128,000 YS. For gears, bolts, crankshafts, fasteners; DIN-VCN35, oil hardened, shock resistant. *Obsolete*

POMPEY 35 NK1
Societe Nouvelle des Acieries de Pompey
C 0.34, Ni 1.45, Cr 1, bal Fe.
For gears, bolts, crankshafts, axles, studs; AFNOR 35NC6, oil hardened, shock resistant.

POMPEY 35 NKD 2
Societe Nouvelle des Acieries de Pompey
C 0.34, Mn 0.75, Si 0.25, Ni 1.4, Cr 1, Mo 0.25, bal Fe.
Ni-Cr-Mo structural steel. AFNOR 35 NCD 6.

POMPEY 35CP
Societe Nouvelle des Acieries de Pompey
C 0.35, Cr 0.95, Mo 0.25, bal Fe.
For cutters, dies; oil hardened. *Obsolete*

POMPEY 35NKD1
Societe Nouvelle des Acieries de Pompey
C 0.35, Ni 1.2, Cr 0.6, Mo 0.15, bal Fe.
Heat treated: 150,000-177,000 TS; 143,000 YS, 10 El. For gears, bolts, fasteners, crankshafts; Afnor 35NCD4. *Obsolete*

POMPEY 38 NKD
Societe Nouvelle des Acieries de Pompey
C 0.38, Mn 0.65, Si 0.25, Ni 0.85, Cr 0.8, Mo 0.25, bal Fe.
Ni-Cr-Mo structural steel. AFNOR 40 NCD 3.

POMPEY 4.330
Societe Nouvelle des Acieries de Pompey
C 0.3, Mn 0.65, Si 0.25, Ni 1.8, Cr 0.75, Mo 0.25, bal Fe.
Ni-Cr-Mo structural steel. AFNOR 30 NCD 7 (AISI 4330).

POMPEY 4.340
Societe Nouvelle des Acieries de Pompey
Mn 0.65, Si 0.25, Ni 1.8, Cr 0.75, Mo 0.25, C 0.4, bal Fe.
Ni-Cr-Mo structural steel. AFNOR 40 NCD 7; AISI 4340.

POMPEY 40 CD 2
Societe Nouvelle des Acieries de Pompey
C 0.4, Mn 1, Si 0.25, Cr 0.55, Mo 0.2, bal Fe.
Cr-Mo structural steel. AFNOR 40 CD 2.

POMPEY 40NKD
Societe Nouvelle des Acieries de Pompey
C 0.4, Ni 0.6, Cr 0.5, Mo 0.2, bal Fe.
Annealed: 107,000 TS; 90,000 YS; 14 El; 45 RA; 217 Brin. For bolts, gears, crankshafts, fasteners; oil hardened, shock resistant. *Obsolete*

POMPEY 4142
Societe Nouvelle des Acieries de Pompey
C 0.42, Mn 0.85, Si 0.25, Cr 1, Mo 0.2, bal Fe.
Cr-Mo structural steel. AFNOR (42 CD 4); AISI 4142.

POMPEY 45CP
Societe Nouvelle des Acieries de Pompey
C 0.45, Cr 0.95, Mo 0.25, bal Fe.
For cutters, dies, gears, shafts; oil hardened, shock resistant. *Obsolete*

POMPEY 48 MN 5
Societe Nouvelle des Acieries de Pompey
C 0.48, Mn 1.25, Si 0.25, bal Fe.
Manganese structural steel. AFNOR 48 M 5.

POMPEY 8CN2
Societe Nouvelle des Acieries de Pompey
C 0.08, Mn 0.4, Si 0.2, Ni 2, bal Fe.
Heat treated: 85,000-121,000 TS; 64,000 YS; 12 El. For gears, cams, camshafts, fasteners; case hardening, shock resistant, Afnor 10N8h. *Obsolete*

POMPEY 8NK
Societe Nouvelle des Acieries de Pompey
C 0.8, Ni 1.5, Cr, bal Fe.
For wood saws, circular saws; water hardened. *Obsolete*

POMPEY 8SRX
Societe Nouvelle des Acieries de Pompey
C 0.8, Ni, Cr, V, bal Fe.
For band saws, water or oil hardened. *Obsolete*

POMPEY 9KX
Societe Nouvelle des Acieries de Pompey
C 0.9, Cr 0.8, V, bal Fe.
For bearings, stamps, coining dies; water hardened. *Obsolete*

POMPEY ABRADUR M14
Societe Nouvelle des Acieries de Pompey
C 1.2, Mn 13, bal Fe.
For pulverizers, crushers, rock breakers; high wear and abrasion resistance.

POMPEY ANTICHOC-1
Societe Nouvelle des Acieries de Pompey
C 0.6, Si 1.8, Mn 0.9, Cr 0.4, bal Fe.
For punches, pneumatic tools, rivet sets; oil hardened, shock resistant. *Obsolete*

POMPEY ANTICHOC-2
Societe Nouvelle des Acieries de Pompey
C 0.5, Si + Cr, bal Fe.
For punches, chisels, pneumatic tools; oil hardened, shock resistant. *Obsolete*

POMPEY ANTICHOC-H
Societe Nouvelle des Acieries de Pompey
C 0.45, Cr 1.1, W 2, V 0.2, bal Fe.
For pneumatic chisels, crimpers, punches; shock resistant, hot work steel. *Obsolete*

POMPEY ANTICHOC-HD
Societe Nouvelle des Acieries de Pompey
C 0.55, Cr 1.1, W 2, V 0.2, bal Fe.
For punches, crimpers, pneumatic tools; impact resistant, hot work steel. *Obsolete*

POMPEY APS10
Societe Nouvelle des Acieries de Pompey
C 0.12, Cr 2-2.5, Al 0.35-0.9, Mo 0-0.35, bal Fe.
Normalized: 92,000 psi TS; 52,000 psi YS; 20 El. For evaporators, transportation tanks; corrosion resistant.

POMPEY APS1001
Societe Nouvelle des Acieries de Pompey
Cr 2, Al 0.9, C, bal Fe.
For agricultural equipment. *Obsolete*

POMPEY APS20
Societe Nouvelle des Acieries de Pompey
C 0.12, Cr 4, Al 0.9, Cr 0-0.2, Mo 0-0.12, bal Fe.
Normalized: 70,000-92,000 TS; 43,000-52,000 YS; 20-22 El. For chemical plant equipment; resists salt water and marine corrosion. *Obsolete*

POMPEY APS20M
Societe Nouvelle des Acieries de Pompey
Cr 4, Al 0.9, C, bal Fe.
For agricultural equipment. *Obsolete*

POMPEY APS25
Societe Nouvelle des Acieries de Pompey
C 0.12, Cr 4, Al 0.9, Ni 0.9, Cu 0.2, Mo 0.12, bal Fe.
Normalized: 121,000-157,000 psi TS; 85,000 psi YS; 14 El. For automobile bodies, railroad cars, diesel motor parts; resists marine and industrial atmospheres.

POMPEY APS5
Societe Nouvelle des Acieries de Pompey
C 0.07, Mn 0.4, Cu 0.4-0.55, W, Mo bal Fe.
Annealed: 50,000-56,000 TS; 36,000 YS; 30 El. For siphons, oil industry parts, heat exchangers; corrosion resistant. *Obsolete*

POMPEY APS6
Societe Nouvelle des Acieries de Pompey
C 0.08, Mn 0.4-0.55, Cu 0.4-0.55, Mo, bal Fe.
Annealed: 60,000 TS; 40,000 YS; 26 El. For siphons, oil industry parts, heat exchangers; corrosion resistant. *Obsolete*

POMPEY B 1021
Societe Nouvelle des Acieries de Pompey
C 0.21, Mn 0.75, Si 0.25, B, bal Fe.
Boron-carbon structural steel. AFNOR 21.B.3.

POMPEY B 1419
Societe Nouvelle des Acieries de Pompey
C 0.18, Mn 1.15, Si 0.25, B, bal Fe.
Boron-manganese carburizing steel. AFNOR 18 MB 4.

POMPEY B 1422
Societe Nouvelle des Acieries de Pompey
C 0.2, Mn 1.25, Si 0.25, B, bal Fe.
Boron-manganese carburizing steel. AFNOR 20 MB 5.

POMPEY B 1435
Societe Nouvelle des Acieries de Pompey
C 0.35, Mn 1, Si 0.25, B, bal Fe.
Boron-manganese structural steel. AFNOR 35 MB 4.

POMPEY B 1440
Permo Inc.
C 0.4, Mn 1, Si 0.25, B, bal Fe.
Boron-Mn structural steel. AFNOR 40 MB 4.

POMPEY B 8620
Societe Nouvelle des Acieries de Pompey
C 0.2, Mn 0.8, Si 0.25, Ni 0.55, Cr 0.5, Mo 0.2, B, bal Fe.
AFNOR 19 NCDB 2; AISI 8620+B.

POMPEY B1038
Societe Nouvelle des Acieries de Pompey
Si 0.25, C 0.37, M 0.75, B, bal Fe.
Boron-carbon structural steel. AFNOR 38 B 3.

POMPEY C 2 MK
Societe Nouvelle des Acieries de Pompey
C 0.2, Mn 1.15, Si 0.25, bal Fe.
Manganese carburizing steel. AFNOR 20 MC 5.

POMPEY C.1.MK
Societe Nouvelle des Acieries de Pompey
C 0.16, Mn 1.15, Si 0.25, bal Fe.
Manganese carburizing steel. AFNOR 16 MC 5.

POMPEY C1K
Societe Nouvelle des Acieries de Pompey
C 0.12, Mn 0.75, Cr 0.8, bal Fe.
Heat treated: 107,000-128,000 TS; 72,000 YS; 15 El; 40 RA. For gears, bolts, machine tool parts; Afnor 12C3h, case hardening steel. *Obsolete*

POMPEY C2K
Societe Nouvelle des Acieries de Pompey
C 0.18, Mn 0.75, Cr 0.8, bal Fe.
Heat treated: 135,000-172,000 TS; 120,000 YS; 8 El; 30 RA. For gears, bolts, machine tool parts; Afnor 18C3h, case hardening steel. *Obsolete*

POMPEY CKD1
Societe Nouvelle des Acieries de Pompey
C 0.12, Cr 1, Mo 0.25, bal Fe.
Heat treated: 150,000-186,000 TS; 121,000 YS. For gears, cams, camshafts, machine tool parts; AISI 4110, case hardening steel, tough. *Obsolete*

POMPEY CKD2
Societe Nouvelle des Acieries de Pompey
C 0.18, Cr 1, Mo 0.25, bal Fe.
Heat treated: 157,000-193,000 psi TS; 130,000 psi YS. For gears, bolts, machine tool parts; AISI 4120, case hardening steel, tough.

POMPEY CNKD
Societe Nouvelle des Acieries de Pompey
C 0.1, Ni 1.2, Cr 0.6, Mo 0.15, bal Fe.
Heat treated: 135,000-164,000 TS; 114,000 YS; 11 El. For gears, fasteners, cams, machine tool parts; Afnor 10NCD4, case hardening, shock resistant. *Obsolete*

POMPEY CNKD1
Societe Nouvelle des Acieries de Pompey
C 0.15, Ni 1.2, Cr 0.5, Mo 0.15, bal Fe.
Heat treated: 143,000-185,000 TS; 114,000 YS; 8 El. For gears, fasteners, cams, machine tool parts; Afnor 16NCD4, case hardening, shock resistant. *Obsolete*

POMPEY CNKD2
Societe Nouvelle des Acieries de Pompey
C 0.17, Ni 1.4, Cr 1, Mo 0.2, bal Fe.
Heat treated: 170,000-206,000 psi TS; 143,000 psi YS; 9 El. For gears, cams, camshafts, fasteners; AFNOR 18NCD6, case hardening, shock resisting.

POMPEY CNKD3
Societe Nouvelle des Acieries de Pompey
C 0.17, Ni 3.2, Cr 1, Mo 0.25, bal Fe.
Heat treated: 186,000-221,000 psi TS; 143,000 psi YS; 8 El. For gears, cams, camshafts, fasteners; AFNOR 16NCD13, case hardening, shock resistant.

POMPEY CROSTAR
Societe Nouvelle des Acieries de Pompey
C 2, Cr 13, V, bal Fe.
For forming and blanking dies, cutters; oil hardened, non-deforming. *Obsolete*

POMPEY D 1 S
Societe Nouvelle des Acieries de Pompey
C 0.11, Mn 0.75, Si 0.2, P 0.04, bal Fe.
Free machining low carbon steel. AFNOR. 10.F.2.

POMPEY D 15 SS
Societe Nouvelle des Acieries de Pompey
C 0.17, Mn 1.15, Si 0.25, P 0.04, S 0.11, bal Fe.
Free machining steel. AFNOR 17 MF 4; AISI 1117.

POMPEY D 37 SS
Societe Nouvelle des Acieries de Pompey
C 0.36, Mn 1.5, Si 0.25, P 0.04, S 0.11, bal Fe.
Free machining steel; hardenable. AFNOR 35 MF 6; AISI 1137.

POMPEY D 38 K
Societe Nouvelle des Acieries de Pompey
C 0.38, Mn 0.75, Si 0.25, Cr 0.45, bal Fe.
Low chromium structural steel. AFNOR 38 C 2.

POMPEY D 42 K
Societe Nouvelle des Acieries de Pompey
C 0.43, Mn 0.75, Si 0.25, Cr 0.45, bal Fe.
Low chromium structural steel. AFNOR 42.C.2.

POMPEY D 46 SS
Societe Nouvelle des Acieries de Pompey
C 0.45, Mn 1, Si 0.25, P 0.04, S 0.11, bal Fe.
Free machining structural steel. AFNOR 45 MF 4; AISI 1146.

POMPEY D 52 L
Societe Nouvelle des Acieries de Pompey
C 0.18, Mn 1.35, Si 0.25, bal Fe.
Low carbon structural steel. AISI 1518.

POMPEY D Z S
Societe Nouvelle des Acieries de Pompey
C 0.18, Mn 0.75, Si 0.2, P 0.04, S 0.18, bal Fe.
Free machining low carbon steel. AFNOR 20.F.2.

POMPEY D3
Societe Nouvelle des Acieries de Pompey
C 0.32, Mn 0.7, Si 0.3, bal Fe.
Normalized: 85,000 TS; 58,000 YS; 25 El. For gears, fasteners, shafts, bolts; water hardened. *Obsolete*

POMPEY D3SS
Societe Nouvelle des Acieries de Pompey
C 0.35, Mn 1.15, Si 0.25, S 0.15, bal Fe.
For screw machine products, bolts, fasteners; free cutting.

POMPEY D48
Societe Nouvelle des Acieries de Pompey
C 0.48, Mn 0.6, Si 0.3, bal Fe.
Normalized: 103,000 TS; 61,000 YS; 19 El. For gears, fasteners, axles, shafts, bolts; water hardened. *Obsolete*

POMPEY DA 42 L
Societe Nouvelle des Acieries de Pompey
C 0.15, Mn 1, Si 0.25, bal Fe.
Low carbon structural steel.

POMPEY DA 48 L
Societe Nouvelle des Acieries de Pompey
C 0.18, Mn 1.1, Si 0.25, bal Fe.
Low carbon structural steel.

POMPEY DC1
Societe Nouvelle des Acieries de Pompey
C 0.11, Mn 0.45, Si 0.2, P 0-0.03, S 0-0.025, bal Fe.
Normalized: 60,000 TS; 40,000 YS; 35 El. For gears, shafts, cams, fasteners; case hardening. *Obsolete*

POMPEY DK 35
Societe Nouvelle des Acieries de Pompey
C 0.35, Mn 0.8, Si 0.25, Cr 1, bal Fe.
Chromium structural steel; hardenable. AFNOR 35 C 4.

POMPEY DK3
Societe Nouvelle des Acieries de Pompey
C 0.32, Mn 0.75, Cr 1, bal Fe.
For gears, bolts, machine tool parts. AFNOR 32C4, oil hardened.

POMPEY DK4
Societe Nouvelle des Acieries de Pompey
C 0.4, Mn 0.75, Cr 1, bal Fe.
For gears, bolts, machine tool parts. AFNOR 38C4, oil hardened.

POMPEY DMS
Societe Nouvelle des Acieries de Pompey
C 0.37, Mn 1.25, Si 1.25, bal Fe.
Structural steel, hardenable. AFNOR 37 MS.5.

POMPEY E7
Societe Nouvelle des Acieries de Pompey
C 0.7, S 0-0.04, P, bal Fe.
Heat treated: 174,000 psi TS; 128,000 psi YS; 12 El; 37 RA; 352 Brin. For cams, forging tools, nippers, lifters; water hardened.

POMPEY E7FX
Societe Nouvelle des Acieries de Pompey
C 0.75, S 0.015, P 0.025, bal Fe.
Heat treated: 180,000 psi TS; 140,000 psi YS; 14 El; 38 RA; 370 Brin. For crimpers, punches, tools; water hardened.

POMPEY E8FX
Societe Nouvelle des Acieries de Pompey
C 0.8, S 0.015, P 0.025, bal Fe.
Heat treated: 188,000 psi TS; 143,000 psi YS; 12 El; 35 RA; 390 Brin. For tools, dies, drills, taps; water hardened.

POMPEY EF10
Societe Nouvelle des Acieries de Pompey
C 1, S 0-0.025, P 0-0.025, bal Fe.
Annealed: 100,000 psi TS; 53,000 psi YS; 21 El; 42 RA; 200 Brin. For cams, knives, cutters, drills, gages; water hardened; Type W1.

POMPEY EF10X
Societe Nouvelle des Acieries de Pompey
C 1, 0.025% max S and P, bal Fe.
For cutters, reamers, drills, punches; water hardened, tool steel. *Obsolete*

POMPEY EF12
Societe Nouvelle des Acieries de Pompey
C 1.2, 0.025% max P and S, bal Fe.
For cutters, reamers, drills, broaches; water hardened, tool steel. *Obsolete*

POMPEY EGALIT
Societe Nouvelle des Acieries de Pompey
C 0.85, Mn 2, V 0.2, bal Fe.
For drills, taps, cutting tools, stamps; water hardened.

POMPEY EP 7
Societe Nouvelle des Acieries de Pompey
C 0.7, Mn 0.6, P 0-0.03, S 0-0.025, bal Fe.
For radial tire mesh in tire industry.

POMPEY ETAD 1
Societe Nouvelle des Acieries de Pompey
C 0.3, Mn 0.4, Si 0.25, Ni 3.75, Cr 1.35, Mo 0.5, bal Fe.
Ni-Cr-Mo structural steel, deep hardening. AFNOR 30 NCD 16

POMPEY ETAD 3
Societe Nouvelle des Acieries de Pompey
C 0.34, Mn 0.45, Si 0.25, P 0-0.03, S 0-0.025, Ni 4, Cr 1.8, Mo 0.4, bal Fe.
Alloy structural steel; deep hardening. AFNOR 35 NCD 16.

POMPEY ETAN
Societe Nouvelle des Acieries de Pompey
C 0.38, Ni 4.3, Cr 1.75, bal Fe.
Heat treated: 260,000 TS; 200,000 YS; 5 El; 40 RA. For gears, pinions, crankshafts; Afnor 40NC17, oil hardened, shock resistant. *Obsolete*

POMPEY ETANO
Societe Nouvelle des Acieries de Pompey
C 0.35, Ni 3.75, Cr 1.7, bal Fe.
Heat treated: 250,000 psi TS; 192,000 psi YS; 6 El; 40 RA. For gears, pinions, crankshafts. AFNOR 35NC15, oil hardened, shock resistant.

POMPEY ETH
Societe Nouvelle des Acieries de Pompey
C 0.3, Cr 3.75, Mo 0.25, Ni 1, bal Fe.
Heat treated: 235,000-300,000 TS; 192,000-213,000 YS; 6 El; 40 RA. For gears, pinions, shafts; air or oil hardened, shock resistant. *Obsolete*

POMPEY ETH
Societe Nouvelle des Acieries de Pompey
C 0.4, Cr 1, Mo 0.25, bal Fe.
Heat treated: 270,000-310,000 TS; 235,000-270,000 YS; 5 El; 40 RA. For gears, pinions, crankshafts; Afnor 42CD4, oil hardened, shock resistant. *Obsolete*

POMPEY F8
Societe Nouvelle des Acieries de Pompey
C 0.8, 0.035% max S and P, bal Fe.
For cutting tools, springs, drills, hammers; water hardened. *Obsolete*

POMPEY FF 1
Societe Nouvelle des Acieries de Pompey
C 0.1, Mn 0.45, Si 0.2, Al, bal Fe.
AFNOR XC 10 FF.

POMPEY FF 15
Societe Nouvelle des Acieries de Pompey
C 0.15, Mn 0.7, Si 0.2, Al, bal Fe.
AFNOR XC 15 FF.

POMPEY FF 2
Societe Nouvelle des Acieries de Pompey
C 0.19, Mn 0.55, Si 0.2, Al, bal Fe.
AFNOR XC 20 FF.

POMPEY FF 2 H
Societe Nouvelle des Acieries de Pompey
C 0.2, Mn 0.85, si 0-0.2, Cr 0.25, Al, bal Fe.
AFNOR 20 C 1 FF.

POMPEY FF 3 H
Societe Nouvelle des Acieries de Pompey
C 0.38, Mn 0.7, Si 0.2, Cr 0.25, Al, bal Fe.
AFNOR 38 C 1 FF.

POMPEY FF 32
Societe Nouvelle des Acieries de Pompey
C 0.32, Mn 0.65, Si 0-0.2, Al, bal Fe.
AFNOR XC 32 FF.

POMPEY FF 35
Societe Nouvelle des Acieries de Pompey
C 0.35, Mn 0.65, Si 0.2, Al, bal Fe.
AFNOR X C 35 FF.

POMPEY FF 38
Societe Nouvelle des Acieries de Pompey
C 0.38, Mn 0.65, Si 0-0.2, Al, bal Fe.
AFNOR XC 38 FF.

POMPEY FF 42 H
Societe Nouvelle des Acieries de Pompey
C 0.42, Mn 0.7, Si 0.2, Cr 0.25, Al, bal Fe.
AFNOR 42 C 1 FF.

POMPEY FFC
Societe Nouvelle des Acieries de Pompey
C 0.06, Mn 0.4, Si 0-0.1, Al, bal Fe.
AFNOR XC 6 FF.

POMPEY FFC 1
Societe Nouvelle des Acieries de Pompey
C 0.1, Mn 0.45, Si 0-0.1, Al, bal Fe.
AFNOR XC 10 FF.

POMPEY FFC 15
Societe Nouvelle des Acieries de Pompey
C 0.15, Mn 0.7, Si 0-0.1, Al, bal Fe.
AFNOR XC 15 FF.

POMPEY FFC 2
Societe Nouvelle des Acieries de Pompey
C 0.18, Mn 0.55, Si 0-0.1, Al, bal Fe.
AFNOR XC 18 F.F.

POMPEY FFC 22
Societe Nouvelle des Acieries de Pompey
C 0.22, Mn 0.6, Si 0-0.1, Al, bal Fe.
AFNOR XC 22 FF.

POMPEY FFC 3
Societe Nouvelle des Acieries de Pompey
C 0.34, Mn 0.4, Si 0-0.2, Al, bal Fe.
AFNOR XC 35 FF.

POMPEY GK
Societe Nouvelle des Acieries de Pompey
C 1.4, V 0.15, Cr 1.5, bal Fe.
For bearings, taps, cutters; water hardened. *Obsolete*

POMPEY HARDINOX 3
Societe Nouvelle des Acieries de Pompey
C 0.35, Cr 13.5, bal Fe.
Annealed: 100,000 TS; 60,000 YS; 22 El; 52 RA; 200 Brin. For cutlery, surgical and dental instruments; AISI 420, corrosion resistant. *Obsolete*

POMPEY K74
Societe Nouvelle des Acieries de Pompey
C 0.75, Cr 0.5, bal Fe.
For wood saws; water hardened. *Obsolete*

POMPEY KD 16
Societe Nouvelle des Acieries de Pompey
C 0.16, Mn 0.75, Si 0.25, Cr 1.1, Mo 0.25, bal Fe.
Cr-Mo case hardening steel. AFNOR 18 CD 4.

POMPEY KD 20
Societe Nouvelle des Acieries de Pompey
C 0.19, Mn 0.75, Si 0.25, Cr 1.1, Mo 0.25, bal Fe.
Cr-Mo case hardening steel. AFNOR 20 CD 4; AISI 4120.

POMPEY KD 27
Societe Nouvelle des Acieries de Pompey
C 0.27, Mn 0.75, Si 0.25, Cr 1.1, Mo 0.25, bal Fe.
Cr-Mo structural steel. AFNOR 27 CD 4.

POMPEY KD 30
Societe Nouvelle des Acieries de Pompey
C 0.3, Mn 0.75, Si 0.25, Cr 1.1, Mo 0.25, bal Fe.
Cr-Mo structural steel. AFNOR 30 CD 4; AISI 4130.

POMPEY KD 33
Societe Nouvelle des Acieries de Pompey
C 0.33, Mn 0.75, Si 0.25, Cr 1.1, Mo 0.25, bal Fe.
Cr-Mo structural steel. AFNOR 33 CD 4; AISI 4130-4135.

POMPEY KD 39
Societe Nouvelle des Acieries de Pompey
C 0.39, Mn 0.75, Si 0.25, Cr 1, Mo 0.25, bal Fe.
Cr-Mo structural steel. AFNOR 39 CD 4; AISI 4140.

POMPEY KD2
Societe Nouvelle des Acieries de Pompey
C 0.26, Cr 1, Mo 0.25, bal Fe.
For gears, bolts, machine tool parts; Type 4125; oil or water hardened, tough. AFNOR 25 Cd4.

POMPEY KD2S
Societe Nouvelle des Acieries de Pompey
C 0.23, Cr 1, Mo 0.25, bal Fe.
For gears, bolts, machine tool parts; Type 4125; case
hardening steel, tough. AFNOR 25 CD 2 S.

POMPEY KD3
Societe Nouvelle des Acieries de Pompey .
C 0.35, Cr 1, Mo 0.25, bal Fe.
For gears, bolts, crankshafts, machine tool parts; Type 4135;
oil hardened, tough. AFNOR 35 CD 4.

POMPEY KD4
Societe Nouvelle des Acieries de Pompey
C 0.42, Cr 1, Mo 0.25, bal Fe.
Hardened; tough. AFNOR 42 CD 4.

POMPEY KR6
Societe Nouvelle des Acieries de Pompey
C 1, Cr 1.4, bal Fe.
Heat treated: 120,000-237,000 psi TS; 95,000-226,000 psi YS;
3-14 El; 47-30 RA; 227-444 Brin. For bearings, sleeves, liners;
SAE52100; AFNOR 100C6; wear resistant.

POMPEY L12
Societe Nouvelle des Acieries de Pompey
C 1.2, bal Fe.
For files; water hardened. *Obsolete*

POMPEY L13
Societe Nouvelle des Acieries de Pompey
C 1.3, bal Fe.
For files; water hardened. *Obsolete*

POMPEY L7
Societe Nouvelle des Acieries de Pompey
C 0.7, bal Fe.
Heat treated: 174,000 TS; 128,000 YS; 12 El; 37 RA; 352 Brin.
For files; water hardened. *Obsolete*

POMPEY LCKD
Societe Nouvelle des Acieries de Pompey
C 0.6, Mn 0.9, Si 1.75, Cr 0.4, Mo 0.25, bal Fe.
For wire cutters, punches, crimpers, upsetters; oil hardened,
non-deforming. *Obsolete*

POMPEY LK
Societe Nouvelle des Acieries de Pompey
C 1.4, Cr, bal Fe.
For files; water hardened. *Obsolete*

POMPEY LRT
Societe Nouvelle des Acieries de Pompey
C 0.5, bal Fe.
Annealed: 96,000 TS; 52,000 YS; 16 El; 23 RA; 170 Brin. For
files; water hardened. *Obsolete*

POMPEY MANGANESE-CHROME
Societe Nouvelle des Acieries de Pompey
C 0.12-0.2, Mn 1.2-1.3, Cr 0.25-0.4, Cu 0.25-0.4, bal Fe.
Rolled: 85,000 TS; 50,000 YS; 20 El. For bridges, auto and
railroad bodies; high tensile construction steel. *Obsolete*

POMPEY MAS
Societe Nouvelle des Acieries de Pompey
C 0.55, Ni 2, Cr 1.1, Mo 0.35, V 0.1, bal Fe.
For punches, shears, upsetters, forging dies; oil hardened,
hot work steel. *Obsolete*

POMPEY MC12
Societe Nouvelle des Acieries de Pompey
C 1.2, bal Fe.
For ore and rock crushers; water hardened. *Obsolete*

POMPEY MC7
Societe Nouvelle des Acieries de Pompey
C 0.75, bal Fe.
Heat treated: 185,000 TS; 140,000 YS; 13 El; 36 RA; 375 Brin.
For rock crushers; water hardened. *Obsolete*

POMPEY MC8
Societe Nouvelle des Acieries de Pompey
C 0.8, bal Fe.
For grate bars; water hardened. *Obsolete*

POMPEY MKND
Societe Nouvelle des Acieries de Pompey
C 0.4, Cr 4, 1 Ni + Mo, bal Fe.
For hot work tools, shears; oil hardened. *Obsolete*

POMPEY MKND-1
Societe Nouvelle des Acieries de Pompey
C 0.1, Cr 4, Ni 1, Mo, bal Fe.
For hot work tools; oil hardened. *Obsolete*

POMPEY MODAL-1
Societe Nouvelle des Acieries de Pompey
C 0.08, bal Fe.
For molds and dies for plastics; case hardened. *Obsolete*

POMPEY MODAL-2
Societe Nouvelle des Acieries de Pompey
C 0.1, Ni 2, bal Fe.
For molds and dies for plastics; case hardened. *Obsolete*

POMPEY MODAL-3
Societe Nouvelle des Acieries de Pompey
C 0.1, Ni 1.5, Cr, bal Fe.
For molds and dies for plastics; case hardened. *Obsolete*

POMPEY MODAL-4
Societe Nouvelle des Acieries de Pompey
C 0.1, Ni 4.5, Cr, bal Fe.
For molds and dies for plastics; case hardened. *Obsolete*

POMPEY MSV
Societe Nouvelle des Acieries de Pompey
C 0.06-0.11, Mn 0.3-0.5, Si 0.1, bal Fe.
Annealed: 65,000 psi TS; 40,000 psi YS; 28 El; 65 RA; 130
Brin. For nails, rivets, screws, bolts, bushings; Type 1008;
case hardened.

POMPEY MSV3
Societe Nouvelle des Acieries de Pompey
C 0.26-0.35, Mn 0.4-0.9, Si 0.12, bal Fe.
Hot rolled: 80,000 TS; 50,000 psi YS; 30 El; 56 RA; 165
Brin. For gears, bolts, armature shafts, keys, brackets; Type
1030; water hardened.

POMPEY MSVD
Societe Nouvelle des Acieries de Pompey
C 0.15-0.23, Mn 0.4-0.6, Si 0.12, bal Fe.
Annealed: 73,000 psi TS; 42,000 psi YS; 22 El; 58 RA; 140
Brin. For screws, bolts, gears, cams; Type 1020; case
hardened.

POMPEY NITRAL 3
Societe Nouvelle des Acieries de Pompey
C 0.4, Cr 1.7, Mo 0.2, Al 1, bal Fe.
Heat treated: 144,000 TS; 121,000 YS; 12 El; 50 RA. For
engine cylinders, pistons, pumps, arbors; heat and creep
resistant. *Obsolete*

POMPEY PF 2
Societe Nouvelle des Acieries de Pompey
C 0.12, Mn 0.5, Si 0.6, Cr 1.1, Mo 0.55, bal Fe.
Cr-Mo structural steel. For oil refinery equipment; creep
resistant. AFNOR 12 CD4-05.

POMPEY PF 2 A
Societe Nouvelle des Acieries de Pompey
C 0.13, Mn 0.6, Si 0.25, Cr 1, Mo 0.5, bal Fe.
Cr-Mo carburizing or structural steel. AFNOR 15 CD 4-05.

POMPEY PF1
Societe Nouvelle des Acieries de Pompey
C 0.15, Cr 0.55, Mo 0.55, bal Fe.
Normalized: 72,000 psi TS; 25,000 psi YS; 18 El. For oil
refinery equipment; creep resistant. AFNOR 13CD2.

POMPEY PF5
Societe Nouvelle des Acieries de Pompey
C 0.13, Cr 5, Mo 0.55, bal Fe.
Normalized: 85,000 psi TS; 25,000 psi YS; 25 El. For oil
refinery equipment; creep and heat resistant. AFNOR
Z15CD5-05.

POMPEY PF6
Societe Nouvelle des Acieries de Pompey
C 0.13, Cr 2.2, Mo 1, bal Fe.
Normalized: 82,000 psi TS; 30,000 psi YS; 21 El. For oil
refinery equipment; creep resistant. AFNOR 10CD9.

POMPEY PFOA
Societe Nouvelle des Acieries de Pompey
C 0.16, Mn 0.65, Si 0.25, Mo 0.55, bal Fe.
Molybdenum carburizing steel. AFNOR 15 D 6.

POMPEY PFOB
Societe Nouvelle des Acieries de Pompey
C 0.15, Mn 0.65, Si 0.2, Mo 0.3, bal Fe.
Molybdenum carburizing steel. AFNOR 15 D 3.

POMPEY PFV 55
Societe Nouvelle des Acieries de Pompey
C 0.12, Mn 0.4, Si 0.35, Cr 2.1, Mo 0.35, V 0.08, Al 0.45, bal
Fe.
Cr-Mo structural steel; for oil refinery equipment; creep
resistant. AFNOR 12 CADV 8.

POMPEY PM17
Societe Nouvelle des Acieries de Pompey
C 0.12, Mn 1.5, bal Fe.
Rolled: 77,000-92,000 TS; 58,000 YS; 20 El. For agricultural
and tractor parts, automobiles, scrapers, bulldozers.
Obsolete

POMPEY PM18
Societe Nouvelle des Acieries de Pompey
C 0.2, Mn 1.7, Mo 0.2, bal Fe.
Rolled: 72,000-117,000 TS; 50,000-65,000 YS; 15 El. For
machine tool parts; case hardened. *Obsolete*

POMPEY PM20
Societe Nouvelle des Acieries de Pompey
C 0.1, Mn 0.6-1, Ni 2.3-3.2, Mo 0.15-0.25, Cu 0.4-0.6, bal Fe.
Rolled: 82,000-97,000 TS; 58,000 YS; 20 El. For automobile
parts, gears, fasteners; case hardened. *Obsolete*

POMPEY PM35
Societe Nouvelle des Acieries de Pompey
C 0.18, Mn 1.5, Cr 1.5, Mo 0.12, bal Fe.
Heat treated: 143,000-186,000 TS; 103,000 YS; 11 El; 45 RA.
For agricultural and mining equipment. *Obsolete*

POMPEY PP
Societe Nouvelle des Acieries de Pompey
C 0.3, Mo 0.25, Cr 2.4, V 0.25, bal Fe.
For gears, shafts, crankshafts, bolts, studs; oil hardened,
tough, high temperature use. *Obsolete*

POMPEY PSK
Societe Nouvelle des Acieries de Pompey
C, Si, Cr, bal Fe.
For vises, ejectors, molds, mandrels, punches; oil hardened.
Obsolete

POMPEY RKV
Societe Nouvelle des Acieries de Pompey
C 0.5, Mn 0.8, Cr 1, V 0.15, bal Fe.
Heat treated: 170,000-206,000 psi TS; 8 El; 35 RA. For gears,
bolts, springs, crankshafts; oil hardened, shock resistant.
AFNOR 50CV4.

POMPEY RMK
Societe Nouvelle des Acieries de Pompey
C 0.48, Mn 0.8, Si 0.3, Cr 1, bal Fe.
Heat treated: 193,000-221,000 psi TS; 171,000 psi YS; 5 El;
30 RA. For springs, torsion bars; oil hardened. AFNOR 48C4.

POMPEY RMKR
Societe Nouvelle des Acieries de Pompey
C 0.8, Mn 0.95, Si 0.6, Cr 0.4, bal Fe.
Heat treated: 150,000-177,000 TS; 107,000 YS; 8 El; 40 RA.
For springs; oil hardened. *Obsolete*

POMPEY RP
Societe Nouvelle des Acieries de Pompey
C, Ni, Cr, V, bal Fe.
For cams, guides, rollers, machine tool parts; oil hardened.
Obsolete

POMPEY RS
Societe Nouvelle des Acieries de Pompey
C 0.45, Mn 0.65, Si 1.8, bal Fe.
Heat treated: 200,000-235,000 psi TS; 170,000 psi YS; 6 El;
30 RA. For springs; water hardened. AFNOR 45S7.

POMPEY RS 2
Societe Nouvelle des Acieries de Pompey
C 0.51, Mn 0.65, Si 1.8, bal Fe.
Silicon structural steel; for springs. AFNOR 51 S 7.

POMPEY RS 2 K
Societe Nouvelle des Acieries de Pompey
C 0.56, Mn 0.75, Si 1.8, Cr 0.3, bal Fe.
Silico-manganese steel for springs. AFNOR 56 SC 7; AISI
9255.

POMPEY RS 4
Societe Nouvelle des Acieries de Pompey
C 0.6, Mn 0.85, Si 1.8, bal Fe.
Silico-manganese steel for springs. AFNOR 60 S 7; AISI 9260.

POMPEY RS 4 K
Societe Nouvelle des Acieries de Pompey
C 0.6, Mn 0.75, Si 1.6, Cr 0.6, bal Fe.
Silico-manganese steel for springs. AFNOR 61 SC 7; similar
to AISI 9260.

POMPEY RS3
Societe Nouvelle des Acieries de Pompey
C 0.55, Mn 0.75, Si 1.7, bal Fe.
Heat treated: 200,000-235,000 psi TS; 177,000 psi YS; 5 El;
25 RA. For springs; oil hardened. AFNOR 55S7.

POMPEY RS3K
Societe Nouvelle des Acieries de Pompey
C 0.6, Mn 0.85, Si 1.8, Cr 0.4, bal Fe.
Heat treated: 213,000-242,000 psi TS; 185,000 psi YS; 5 El;
25 RA. For springs, coils; oil hardened. AFNOR 61SC7.

POMPEY RS4E
Societe Nouvelle des Acieries de Pompey
C 0.6, Mn 0.85, Si 1.9, bal Fe.
Heat treated: 213,000-242,000 TS; 185,000 YS; 4 El; 25 RA.
For springs; Afnor 55s7, Afnor 65s7, oil hardened. *Obsolete*

POMPEY RSKD
Societe Nouvelle des Acieries de Pompey
C 0.48, Mn 0.65, Si 1.5, Cr 0.6, Mo 0.25, bal Fe.
Special alloy steel for springs, torsion bars. AFNOR 45 SCD
6.

POMPEY RSOB
Societe Nouvelle des Acieries de Pompey
C 0.38, Mn 0.65, Si 1.5, bal Fe.
Silicon structural steel; for springs; refineries. AFNOR 38.S.6.

POMPEY S3
Societe Nouvelle des Acieries de Pompey
C 0.35, bal Fe.
Hot rolled: 85,000 psi TS; 54,000 psi YS; 30 El; 53 RA; 185
Brin. For gears, shafts, housings; water hardened.

POMPEY S3F
Societe Nouvelle des Acieries de Pompey
C 0.35, bal Fe.
Hot rolled: 85,000 psi TS; 54,000 psi YS; 30 El; 53 RA; 185
Brin. For gears, shafts, axles, mowers; water hardened.

POMPEY S4
Societe Nouvelle des Acieries de Pompey
C 0.45, bal Fe.
Hot rolled: 98,000 psi TS; 59,000 psi YS; 24 El; 45 RA; 212
Brin. For machine tool parts, gears, shafts; water hardened.

POMPEY S4F
Societe Nouvelle des Acieries de Pompey
C 0.4, bal Fe.
Hot rolled: 91,000 psi TS; 58,000 psi YS; 27 El; 50 RA; 200
Brin. For gears, shafts, axles, mowers; water hardened.

POMPEY S4M
Societe Nouvelle des Acieries de Pompey
C 0.9, bal Fe.
Heat treated: 190,000 psi TS; 145,000 psi YS; 10 El; 30 RA;
400 Brin. For cutters, dies; water hardened.

POMPEY S4S
Societe Nouvelle des Acieries de Pompey
C 0.4, S 0.035, P 0.035, bal Fe.
For rolled: 91,000 psi TS; 58,000 psi YS; 27 El; 50 RA; 200
Brin. For agricultural equipment, plows; water hardened.

POMPEY S5M
Societe Nouvelle des Acieries de Pompey
C 1.1, bal Fe.
Annealed: 110,000 psi TS; 56,000 psi YS; 20 El; 40 RA; 210
Brin. For cutters, dies; water hardened.

POMPEY S65
Societe Nouvelle des Acieries de Pompey
C 0.65, bal Fe.
Heat treated: 170,000 psi TS; 116,000 psi YS; 10 El; 38 RA;
340 Brin. For shafts, mandrels, dies, tools; water hardened.

POMPEY S6F
Societe Nouvelle des Acieries de Pompey
C 0.62, bal Fe.
Heat treated: 160,000 psi TS; 113,000 psi YS; 12 El; 40 RA;
325 Brin. For agricultural equipment, mowers; water
hardened.

POMPEY S7 TRIPLEX
Societe Nouvelle des Acieries de Pompey
C 0.7, bal Fe.
Heat treated: 174,000 psi TS; 128,000 psi YS; 12 El; 37 RA;
352 Brin. For agricultural equipment, plows; water hardened.

POMPEY S75
Societe Nouvelle des Acieries de Pompey
C 0.75, bal Fe.
Heat treated: 180,000 psi TS; 140,000 psi YS; 13 El; 36 RA;
370 Brin. For crimpers, punches, dies, tools; water hardened.

POMPEY S7F
Societe Nouvelle des Acieries de Pompey
C 0.68, bal Fe.
Heat treated: 174,000 psi TS; 128,000 psi YS; 12 El; 37 RA;
355 Brin. For agricultural equipment, mowers; water
hardened.

POMPEY S8
Societe Nouvelle des Acieries de Pompey
C 0.8, bal Fe.
Heat treated: 188,000 TS; 143,000 YS; 12 El; 35 RA; 385 Brin.
For tools, dies, mandrels; water hardened. *Obsolete*

POMPEY S8R
Societe Nouvelle des Acieries de Pompey
C 0.8, S 0.035, P 0.035, bal Fe.
Heat treated: 188,000 TS; 143,000 YS; 12 El; 35 RA; 390 Brin.
For agricultural equipment, rakes; water hardened. *Obsolete*

POMPEY S90
Societe Nouvelle des Acieries de Pompey
C 0.9, S 0.035, P 0.035, bal Fe.
Heat treated: 190,000 psi TS; 145,000 psi YS; 10 El; 30 RA;
400 Brin. For agricultural equipment, mowing machines;
water hardened.

POMPEY SH4
Societe Nouvelle des Acieries de Pompey
C 0.48, bal Fe.
Annealed: 96,000 psi TS; 52,000 psi YS; 16 El; 23 RA; 170
Brin. For machine tool parts, gears, shafts; water hardened.

POMPEY SH5
Societe Nouvelle des Acieries de Pompey
C 0.54, bal Fe.
Annealed: 100,000 psi TS; 55,000 psi YS; 15 El; 22 RA; 180
Brin. For shafts, axles, tools; water hardened.

POMPEY TR 42
Societe Nouvelle des Acieries de Pompey
Mn 0.65, Si 0.25, C 0.42, bal Fe.
Structural steel for shafts, bolts. AFNOR XC 42 psi TS; similar
to AISI 1040.

POMPEY TR4M
Societe Nouvelle des Acieries de Pompey
C 0.45, Cr 0.15, Mn 0.9, Si 0.25, bal Fe.
For machine tool parts, shafts, gears, induction or flame
hardened parts. AFNOR XC 45TS.

POMPEY TR5
Societe Nouvelle des Acieries de Pompey
C 0.52, Mn 0.7, Si 0.25, Cr 0.15, bal Fe.
For machine tool parts, shafts, gears; induction or flame
hardened parts. AFNOR XC 52 psi TS.

POMPEY TR5M
Societe Nouvelle des Acieries de Pompey
C 0.52, Mn 0.8, Si 0.25, Cr 0.15, bal Fe.
For machine tool parts, shafts, gears; induction or flame
hardened parts. AFNOR XC 52TS.

POMPEY TRJ
Societe Nouvelle des Acieries de Pompey
C 0.45, Cr 0.15, Mn 0.7, Si 0.25, bal Fe.
For machine tool parts, shafts, gears; induction or flame
hardened parts. AFNOR XC 45TS.

POMPEY TRK
Societe Nouvelle des Acieries de Pompey
C 0.5, Mn 0.9, Si 0.3, Cr 1, bal Fe.
Heat treated: 163,000-193,000 TS; 150,000 YS; 8 El; 30 RA;
610 Brin. For machine tool parts, shafts, gears; induction or
flame hardened parts. *Obsolete*

POMPEY TRKD
Societe Nouvelle des Acieries de Pompey
C 0.42, Mn 0.7, Si 0.3, Cr 1, Mo 0.25, bal Fe.
For machine tool parts, shafts, gears; induction or flame
hardened parts. AFNOR 42CD 4TS.

POMPEY TRNKD
Societe Nouvelle des Acieries de Pompey
C 0.4, Mn 0.75, Ni 0.85, Cr 0.75, Mo 0.25, bal Fe.
Ni-Cr-Mo structural steel; oil hardenable for highly stressed
parts. AFNOR 40 NCD 3 psi TS.

POMPEY TUREX 4
Societe Nouvelle des Acieries de Pompey
C 0.4, W 4, Cr 1.4, V 0.2, bal Fe.
For hot work tools; oil hardened. *Obsolete*

POMPEY TUREX 9
Societe Nouvelle des Acieries de Pompey
C 0.3, Cr 2.8, W 10, V 0.3, bal Fe.
For extrusion press rams and liners, punches, shears; hot
work steel, oil hardened. *Obsolete*

POMPEY XC-10, ETC.
Now AFNOR TABLES.

POMPEY-APS10A
Societe Nouvelle des Acieries de Pompey
C 0.12, Cr 2, Al 0.9, bal Fe.
Normalized: 69,000-85,000 TS; 44,000 YS. For evaporators,
dryers, graters, nitrate plants, chemical plant equipment.
Corrosion resistant. *Obsolete*

POMPEY-APS10C
Societe Nouvelle des Acieries de Pompey
C 0.12, Cr 2.5, Al 0.4, Cu 0.2, bal Fe.
Normalized: 76,000-92,000 TS; 52,000 YS. For marine
equipment, vats, hardware. Resists marine corrosion.
Obsolete

POMPEY-APS10M
Societe Nouvelle des Acieries de Pompey
C 0.12, Cr 2, Al 0.9, Mo 0.12, bal Fe.
Normalized: 72,000-88,000 psi TS; 46,000 psi YS. For
evaporators, dryers, graters, nitrate plants, chemical
equipment. Corrosion resistant to stress cracking due to
nitrates.

POMPEY-APS10M4
Societe Nouvelle des Acieries de Pompey
C 0.12, Cr 2, Al 0.35, Mo 0.35, bal Fe.
Normalized: 87,000 TS; 42,000 YS. For marine equipment,
vats, hardware. Corrosion resistant to marine atmospheres.
Obsolete

POMPEY-APS20A
Societe Nouvelle des Acieries de Pompey
C 0.12, Cr 4, Al 0.9, bal Fe.
Normalized: 69,000-85,000 psi TS; 43,000 psi YS. For oil
refinery and coking equipment. Corrosion resistant to
industrial and marine atmospheres.

POMPEY-APS20C
Societe Nouvelle des Acieries de Pompey
C 0.12, Cr 4, Al 0.4, Cu 0.2, bal Fe.
Normalized: 76,000-92,000 TS; 52,000 YS. For oil refinery
equipment, furnace parts. Corrosion resistant to marine and
industrial atmospheres. *Obsolete*

POMPTON
AL Tech Specialty Steel Corp.
C 0.7-1.4, Mn 0.25, Si 0.25, bal Fe.
Typical heat treated hardness: 56-60 Rock C. For threading
dies, shear knives, chisels, drills, punches, trimming and
blanking dies, cold heading dies, pneumatic tools. Water
hardening. AISI W-1.

POMPTON D.R.
Allegheny Ludlum Steel
C 1.2, bal Fe.
For drills, taps; drill rod, water hardening. *Obsolete*

POMPTON EXTRA
AL Tech Specialty Steel Corp.
C 0.7-1.4, Mn 0.25, Si 0.25, bal Fe.
Typical heat treated hardness: 56-60 Rock C. For threading
dies, shear knives, chisels, drills, punches, trimming and
blanking dies, cold heading dies, pneumatic tools. Water
hardening. AISI W-1. *Obsolete*

POMPTON SPECIAL
AL Tech Specialty Steel Corp.
C 0.7-1.4, Mn 0.25, Si 0.25, bal Fe.
Typical heat treated hardness: 56-60 Rock C. For threading
dies, shear knives, chisels, drills, punches, trimming and
blanking dies, cold heading dies, pneumatic tools. Water
hardening. AISI W-1. *Obsolete*

PONSARD'S HIGH MANGANESE BRASS
English manufacture
Cu 50-75, Mn 20-25, Zn 2-15, Fe 0-16.
For corrosion resisting and marine parts.

PONTALLOR
Degussa AG
Precious metal.
Precious metal alloy for dentistry and dental engineering.

POPES ISLAND METAL
French manufacture
Cu 70, Zn 15, Ni 14, Sn 1.
For tableware, base for plated ware; generic name of a series
of French alloys.

POREX
Moraine Mfg., Inc.
Sn 5-11, bal Cu.
For filters, porous membranes, diffusers; powdered metals.

PORO BRONZE
Poro Metals Ltd.
Sb 13, Cu 7, bal Sn.
For bearings, bushings. Babbitt.

POROSINT BRONZE
Sintered Products Ltd.
Bronze, sintered. Seven grades (A-G) having various porosity
for filtering; maximum particle size from 0.0001-0.0024 in.

POROSINT CUPRO-NICKEL
Sintered Products Ltd.
Cu 67-70, Sn 6.5-7.5, bal Ni.
Sintered; for filtering; Grade CN 4 will pass 0.001 in.
maximum particle size.

POROSINT RIGID MESH
Sintered Products Ltd.
Similar to Porosint stainless steel but produced in sheet with
wide range of porosity. Meets AISI 316, but others (Monel,
AISI 310 and 95/5 Bronze) available.

POROSINT STAINLESS STEEL
Sintered Products Ltd.
AISI 316L powders, sintered. Six grades (P/21/2-P/40)
having various porosity for filtering; maximum particle size to
pass from 0.0001-0.0016 in.

POSITIVE-ROD NO. 105
Marquette Corp.
C 0.08-0.12, Mn 0.45-0.65, Si 0.15-0.25, S 0.03, P 0.03, bal
Fe.
Welded: 62,000-70,000 TS; 50,000-55,000 YS; 25-35 El; 40-45
RA. For all-position general purpose welding rod; AWS-
E6010.

POT METAL
Manufacturer not listed
Cu 67-80, Pb 20-33.
For bearings; heavy duty.

POTASSIUM
Atomergic Chemetals Corp.
K.
Purities: special distilled grade 99.99%, 99.9% (low sodium),
commercial grade. Packaging: bottles, ampules, cylinders
(under vacuum, inert gas, petroleum distillate).

POTASSIUM AMALGAM
Manufacturer not listed
Hg24K.
For organic reduction.

POTERIE D'ETAIN
Manufacturer not listed
Sn 90, Sb 9, Cu 1.
For tableware, bearings; anti-friction.

POTINGRIS
Manufacturer not listed
Potinjaune plus Pb, Sn.
For hardware.

POTINJAUNE
Manufacturer not listed
Cu 72, Zn 2, Pb 2, Sn 1.2.
For hardware; French yellow brass.

POTOMAC
AL Tech Specialty Steel Corp.
C 0.35, Si 0.85, W 1.25, Cr 5, V 0.2, Mo 1.5, bal Fe.
Air hardening hot work die steel for hot forging dies, hot nut
forming tools, and hot extrusion tooling. AISI H-12.

POTOMAC 2V
Allegheny Ludlum Steel
C 0.6, Mn 0.65, Si 1, Cr 6.5, V 2.15, Mo 1.4, bal Fe.
Air hardening hot work tool and die steel. *Obsolete*

POTOMAC 4V
Allegheny Ludlum Steel
C 1.15, Mn 0.35, Si 1, Cr 5.25, V 4, Mo 1.3, bal Fe.
Air hardening hot work tool and die steel. *Obsolete*

POTOMAC A
AL Tech Specialty Steel Corp.
C 0.4, Mn 0.3, Si 0.9, Cr 5, Mo 1.3, V 0.5, bal Fe.
Air hardenable to 290,000 psi TS; 55 Rock C. Maintains high
strength to 1000°F. For aircraft and missile cases, rocket
cases; also for general hot work tooling. AISI H-11.

POTOMAC M
AL Tech Specialty Steel Corp.
C 0.4, Si 1, Cr 5.25, V 1, Mo 1, bal Fe.
General purpose hot work die steel for aluminum and zinc
die casting dies, hot extrusion and forging; also cold heading
and plastic molding. Air hardening. AISI H-13.

POTOSI SILVER
Manufacturer not listed
Ni, Cu.
For ornamental and corrosion resistant parts; nickel silver.

POTTS GROUND FLAT STOCK
Horace T. Potts Co.
C 1.2, Mn 0.6-0.75, Cr 0.55, Mo 0-0.33, bal Fe.
For punches, dies, gauges, gears; ground flat stock, oil or
water hardened. *Obsolete*

POTTS SPECIAL DRILL ROD
Horace T. Potts Co.
C 0.95-1.05, bal Fe.
Rolled: 80,000-90,000 TS; 40,000-45,000 YS; 20 El; 160-190
Brin. For drills, arbors, roller bearings, broaches; uniform
quality. *Obsolete*

POTTS SUPERIOR
Horace T. Potts Co.
C, alloy, bal Fe.
For high speed cutting tools; high speed steel. *Obsolete*

POWDER BRAZE
Charles Hardy & Co.
Copper. Zn, bal Cu.
For brazing; powdered metals.

POWDIRON 56-L
GKN Powder Met Inc.
2.5% graphite, bal Fe.
Sintered: 5,000 TS; 5.6 density; 28% oil by volume. For
bearings for low cost, light duty motor applications. *Obsolete*

POWDIRON 59 I, ETC.
Now GKN 59-I, ETC.

POWDIRON 59-FM
GKN Powder Met Inc.
Fe 99.5, 0.5% other.
Sintered: 17,000 TS; 5.9 density; 25% oil by volume. For low
and medium load bearings requiring machining, staking or
spinning; good ductility. ASTM B310-67, Type 1, Class A;
SAE 850; PMPMA F-0000-N. *Obsolete*

POWDIRON 59-I
GKN Powder Met Inc.
Fe 100.
Sintered: 12,000 TS; 5.9 dDensity; 25% oil by volume. For low
cost bearings for wheels, toys, and general bearing
applications. ASTM B439-67 Gr. 1; SAE 850; PMPMA F-0000-
N. *Obsolete*

POWDIRON 59-PC
GKN Powder Met Inc.
Cu 4, 0.8% C (comb.), bal Fe.
Sintered: 34,000 TS; 5.9 density; 20% oil by volume. For structural parts requiring ductility and shock resistance. ASTM B426-65, Type 1, Gr. 2; SAE 865 A. *Obsolete*

POWDIRON 59-T
GKN Powder Met Inc.
0.8% C (comb.), bal Fe.
Sintered: 24,000 TS; 5.9 density; 20% oil by volume. For structural parts requiring good hardness and wear resistance. ASTM B310-67, Type 1, Class C; SAE 852. *Obsolete*

POWDIRON 61-P
GKN Powder Met Inc.
Cu 10, bal Fe.
Sintered: 30,000 TS; 6.1 density; 25% oil by volume. High strength, general purpose bearing for automotive, farm equipment, wheel bearings, and machine tool applications. ASTM B439-67, Gr. 3; SAE 862; PMPMA FC-1000-N. *Obsolete*

POWDIRON TCU
GKN Powder Met Inc.
Cu 20, 0.8% C (comb.), bal Fe.
Sintered: 85,000 TS; 7.3 density. For structural parts, particularly gears. ASTM B303-67, Class C; SAE 872; PMPMA FX-2010-T. *Obsolete*

POWER
Magnolia Metal Corp.
Cu, Sb, bal Sn.
Cast: 17,500 TS; 10,500 YS; 27 Brin. For bearings; heavy loads. *Obsolete*

POWER 101
Manufacturer not listed
Si 2.25-2.75, bal Fe.
For laminations for power units; high permeability.

POWER 117
Manufacturer not listed
Si 1, bal Fe.
For laminations for power units; high permeability.

POWER 130
Manufacturer not listed
Si 0.5-0.75, bal Fe.
For laminations for power units; high permeability.

POWER 145
Manufacturer not listed
Si 0.25-0.5, bal Fe.
For laminations for power units; high permeability.

POWER 165
Manufacturer not listed
Under 0.25 Si, bal Fe.
For laminations for power units; high permeability.

POWER 52
Manufacturer not listed
Si 4-5, bal Fe.
For laminations for power units; high permeability.

POWER 58
Manufacturer not listed
Si 4.25-4.75, bal Fe.
For laminations for power units; high permeability.

POWER 65
Manufacturer not listed
Si 4.24-4.75, bal Fe.
For laminations for power units; high permeability.

POWER 72
Manufacturer not listed
Si 3.75-4.25, bal Fe.
For laminations for power units; high permeability.

POWER 82
Manufacturer not listed
Si 3-3.5, bal Fe.
For laminations for power units; high permeability.

POWER NICKLE GENUINE BABBITT
Magnolia Metal Corp.
Sn 83-85, Cu, bal Sb.
Cast: 17,500 TS; 10,500 YS; 27 Brin. For bearings; for high pressures and temperatures.

POWERSTEEL
Crobalt Inc.
C 0.9, Co 4, bal Fe.
For tools and cutters. Cast.

POWHATAN
Atlas Specialty Steels
C 0.83, W 18, Cr 4, V 1.7, Co 10.5, Mo 1, bal Fe.
For cutting tools, lathe and planer tools; high speed steel. *Obsolete*

PRASEODYMIUM
Atomergic Chemetals Corp.
Pr.
Purities: 99.9%, 99.5+%. Forms: ingot, lump, sheet, rod, wire, foil, turnings.

PRECEDENT 71 GRADE A
Now USCO-PRECEDENT 71A (771.2).

PRECEDENT 71 GRADE B
Now USCO-PRECEDENT 71B (B771.2).

PRECEDENT NO. 356
U.S. Reduction Co.
Now USCO 356.1. A03561

PRECEDENT NO. 356A
U.S. Reduction Co.
Now USCO 356.2. A13562

PRECISION
Cold Metal Products Company, Inc.
C 0.3, bal Fe.
For structural parts. *Obsolete*

PRECISION A-12
Precision Casting Co.
Si 11-13, Cu 0-0.6, Cu 0-2, Fe 0-2, Ni 0-0.5, bal Al.
Cast: 33,000 TS; 1.5 El; 80 Brin. For light alloy die casting; corrosion resistant.

PRECISION A-218
Precision Casting Co.
Mg 8, Fe 0-1.8, Cu 0-0.2, Mn 0-0.3, Ni 0-0.1, bal Al.
For die casting; corrosion resistant.

PRECISION A-360
Precision Casting Co.
Si 8-10, Mg 0.5, Fe 0-2, bal Al.
Die cast: 42,000 TS; 23,000 YS; 1.8 El. For die castings; corrosion resistant.

PRECISION A-50
Precision Casting Co.
Si 4.5-6, Cu 0-0.6, Fe 0-2, bal Al.
Cast: 29,000 TS; 4 El; 70 Brin. For die casting; high impact strength.

PRECISION A-54
Precision Casting Co.
Si 4.5-5.5, Cu 3.4-4.5, Fe 0-2, bal Al.
Cast: 32,000 TS; 2 El; 75 Brin. For die casting; high strength.

PRECISION A-74
Gilby-Fodor S.A.
Si 7.5, Cu 4, bal Al.
For die castings. *Obsolete*

PRECISION A-94
Precision Casting Co.
Si 35-45, Cu 30-40, Fe 0-13, Zn 0-6, bal Al.
Die cast: 38,000 TS; 2 El; 77 Brin. For die casting; high strength.

PRECISION BR-1
Precision Casting Co.
Cu 63-67, Zn 32-36, Si 0.7-1.3.
Die cast: 70,000 TS; 35,000 YS; 25 El; 120 Brin. For die castings; high strength.

PRECISION BR-2
Precision Casting Co.
Cu 81.5, Zn 14.5, Si 4.
Die cast: 85,000 TS; 50,000 YS; 8 El; 170 Brin. For die casting; high strength.

PRECISION BR-3
Precision Casting Co.
Cu 83, Zn 10, Si 5, Mn 1, Al 1.
Die cast: 105,000 TS; 60,000 YS; 5 El; 100 Brin. For die castings; high strength.

PRECISION M-13
Precision Casting Co.
Al 8.3-9.7, Zn 0.4-1, bal Mg.
Die cast: 29,000-34,000 TS; 20,000 YS; 2-5 El; 66 Brin. For die castings.

PRECISION ZN-4
Gilby-Fodor S.A.
Al 1.5, Cu 1, Mg 0.4, bal Zn.
For die castings. *Obsolete*

PRECISION ZN-5
Precision Casting Co.
Al 3.5-4.3, Cu 0.75-1.25, bal Zn.
Cast: 42,000 TS; 4 El; 75 Brin. For die castings.

PRECISION ZN-6
Precision Casting Co.
Al 3.5-4.3, Cu 0-0.1, bal Zn.
Cast: 36,000 TS; 5 El; 65 Brin. For die castings.

PRECISION ZN-7
Precision Casting Co.
Al 3.5-4.3, Cu 2.5-3.5, bal Zn.
Cast: 45,000 TS; 5 El; 80 Brin. For die castings.

PREGA
Uddeholm Corp.
C 0.45, Cr 3, Mo 0.45, Mn 0.6, bal Fe.
For drop forging dies, hobbing dies; hot work steel, oil hardened. *Obsolete*

PREMABRAZE 130
Handy & Harmon
Precious metal. Au 82, Ni 18.
MP and FP: 1742°F. For brazing stainless steel, Inconel X, other nickel base alloys; good oxidation resistance to 1500°F. AWS BAu-4.

PREMABRAZE 615
Handy & Harmon
Precious metal. Ag 61.5, Cu 24, In 14.5.
MP: 1155-1305°F. For ferrous and non-ferrous and in vacuum tube brazing.

PREMABRAZE 616 (VTG)
Handy & Harmon
Ag 61.5, Cu 24, In 14.5.
MP: 1155°F; FP: 1305°F. For ferrous and nonferrous alloys used in moderate temperature vacuum tubes and systems.

PREMAG
Murex Ltd.
Ag 0.8, bal Cu.
For welding rod; copper welds.

PREMALOY 301
Handy & Harmon
Precious metal. Pt, Au, Ag, Cu, bal Pd.
For resistors; high electrical resistance.

PREMET
Allegheny Ludlum Steel
Co 30, W 19, C 0.12, V 2, Mo 5, Mn 0.2-0.5, bal Fe.
For cutting tools; cast alloy. *Obsolete*

PREMIER
H.K. Porter Co., Inc.
Cr 15, Ni 62, bal Fe.
Annealed: 95,000; 35,000 YS; 40 El; 60 RA. For resistances; heat resistant; now Alray C. *Obsolete*

PREMIER "C"
H.K. Porter Co., Inc.
Fe 24, Ni 60, Cr 16.
For rheostat and resistance units; used up to 1700 F. *Obsolete*

PREMIER 3 1/2 NI WELD
U.S. Steel Corp.
C 0.15-0.25, Mn 0.4-0.6, Ni 3.2-3.7, bal Fe.
For welding wire. *Obsolete*

PREMIER AW WELD
U.S. Steel Corp.
C 0.14-0.2, Mn 0.9-1.2, bal Fe.
For welding wire; acetylene. *Obsolete*

PREMIER EA WELD
U.S. Steel Corp.
C 0-0.06, Mn 0-0.17, bal Fe.
For welding wire; electric. *Obsolete*

PREMIER EB WELD
U.S. Steel Corp.
C 0.13-0.18, Mn 0.4-0.6, bal Fe.
For welding wire; electric. *Obsolete*

PREMIER EP WELD
U.S. Steel Corp.
C 0.13-0.18, Mn 0.4-0.6, bal Fe.
For welding wire; electric. *Obsolete*

PREMIER GA WELD
U.S. Steel Corp.
C 0-0.06, Mn 0-0.17, bal Fe.
For welding wire; acetylene. *Obsolete*

PREMIER GS WELD
U.S. Steel Corp.
C 0-0.1, Mn 0.4-0.6, bal Fe.
For welding wire; acetylene. *Obsolete*

PREMIER H S WELD
U.S. Steel Corp.
C 0.13-0.18, Mn 0.4-0.6, Ni 1.1-1.4, Cr 0.55-0.75, bal Fe.
For welding wire; acetylene welding. *Obsolete*

PREMIER HC WELD
U.S. Steel Corp.
C 0.95-1.1, Mn 0.4-0.6, bal Fe.
For welding wire; electric or gas. *Obsolete*

PREMIER MB SPRING
U.S. Steel Corp.
C 0.45-0.7, Mn 0.6-1.2, bal Fe.
For springs; oil hardening. *Obsolete*

PREMIER NICKEL CHROME
H.K. Porter Co., Inc.
Ni 60-62, Fe 25, Cr 14-16, C 0.1.
Wire: 100,000 TS; 60,000 YS; 25 El; 50 RA; 182 Brin. For rheostats, dipping baskets, heating units, wire cloth; corrosion resisting. *Obsolete*

PREMIER NO. 1
Manufacturer not listed
C 0.1-0.2, Cr 0.65-0.85, V 0-0.1, bal Fe.
For case hardened parts; carburizing steel.

PREMIER NO. 6
English manufacture
C 0.6-0.7, Cr 0.7-0.9, V 0-0.1, bal Fe.
For hammers, dies, blacksmith tools; water hardened.

PREMIER SPRING
U.S. Steel Corp.
C 0.45-0.7, Mn 0.6-1.2, bal Fe.
Drawn: 145,000-260,000 TS. For springs; oil hardening. *Obsolete*

PREMO
Uddeholm Corp.
C 0.04, Cr 4, Mo 0.5, Si 0.1, Mn 0.1, bal Fe.
For plastic mold dies, cold hubbed dies; case hardened, oil or water hardened.

PRESCOLOY
Pressed Steel Car Co.
C 0.2, Ni 0.9, bal Fe.
90,000 TS; 60,000 YS; 25 El; 40 RA. For truck side frames, bolsters, freight car castings. High strength.

PRESNEAL
Heppenstall Co.
Tool material. C 0.3, Ni, Mo, bal Fe.
Annealed. For dies, punches; hot work, press and upsetter.

PRESS E-Z
Jessop Steel Co.
C 0.05, Mn 0.2, bal Fe.
For plastic molds, die casting dies. *Obsolete*

PRESS-DIE
A. Finkl & Sons Co.
Tool material. C 0.18-0.23, Mn 0.6-0.8, Ni 3-3.5, Cr 0.2, Mo 3.25-3.45, V 0.05-0.08, bal Fe.
Precipitation hardened steel. Prehardened: 37-46 Rock C. For forging press dies; flat open die forging dies, upsetter dies, inserts and piercing and punching dies.

PRESS-X
A. Finkl & Sons Co.
Tool material.
Now PRESS-DIE.

PRESSANT
Otto Wolff Handelsgesellschaft
C 0.15, Si 0.2, Mn 0.4, bal Fe.
Annealed: 70,000 TS; 40,000 YS; 25 El; 60 RA; 145 Brin. For gears, pinions, cams, camshafts; case hardening steel.

PRESSANT ECR80
Otto Wolff Handelsgesellschaft
C 0.15, Si 0.25, Mn 0.5, Cr 0.65, bal Fe.
Annealed: 70,000 TS; 40,000 YS; 25 El; 60 RA; 145 Brin. For gears, pinions, cams, camshafts; case hardening steel.

PRESSANT EM212
Otto Wolff Handelsgesellschaft
C 0.2, Mn 1.25, Cr 1.15, bal Fe.
For gears, cams, camshafts; case hardening steel.

PRESSCO NO. 1
Pressco Casting & Mfg. Corp.
Cu 85-88, Fe 1-1.5, Al 10-11.
Die cast: 75,000-85,000 TS; 36,000 YS; 17-26 El; 140-180 Brin. For marine hardware, propellers; heat treatable; Al-bronze.

PRESSCO NO. 10
Pressco Casting & Mfg. Corp.
Cu 0-60, 1.0 Si min, bal Zn.
Cast: 74,000 TS; 38,000 YS; 19 El; 130 Brin. For hardware, die castings; high strength.

PRESSCO NO. 2
Pressco Casting & Mfg. Corp.
Cu 80-83, Zn 13-16, Si 3.5-4.5.
Die cast: 75,000-85,000 TS; 40,000 YS; 160-200 Brin. For gears, shafts, pump parts; silicon bronze.

PRESSCO NO. 3
Pressco Casting & Mfg. Corp.
Cu 55-59, Mn 1, Fe 0-0.2, Sn 0-1, Pb 0-0.5, bal Zn.
Cast: 76,000 TS; 33,000 YS; 15 El; 102 Brin. For fittings, spiders, brackets; SAE 43; manganese bronze.

PRESSCO NO. 4
Pressco Casting & Mfg. Corp.
Cu 58-62, Zn 36-41, Pb 0-1, Sn 0-1.
Die cast: 50,000-60,000 TS; 35,000 YS; 10-12 El. For die castings; corrosion resistant.

PRESSCO NO. 5
Pressco Casting & Mfg. Corp.
Ni 3.5-5.5, Fe 3-5, Al 10-11, Mn 0-3.5, bal Cu.
Cast: 90,000 TS; 40,000 YS; 10 El; 157 Brin. For pump parts, bushings, bearings; Al bronze, heat treatable.

PRESSCO NO. 6
Pressco Casting & Mfg. Corp.
Al 0-0.5, 59.0 Cu min, bal Zn.
Cast 40,000 TS; 11,000 YS; 25 El; 79 Brin. For hardware, die castings; yellow brass.

PRESSCO NO. 7
Pressco Casting & Mfg. Corp.
Cu 61.5-63.5, Zn 35-38, Pb 0.75-1.25, Al 0.9-1.2, Sn 0-0.5, Fe 0.5.
Cast: 63,000 TS; 35,000 YS; 30 El; 102 Brin. For hardware, die castings; free-cutting.

PRESSCO NO. 8
Pressco Casting & Mfg. Corp.
Fe 4.5-6.5, Ni 6.5-7.5, Al 11.5-12.5, bal Cu.
Cast: 115,000 TS; 82,000 YS; 3 El; 220 Brin. For hardware, die castings; Al-bronze.

PRESSCO NO. 9
Pressco Casting & Mfg. Corp.
Pb 0-1.3, 58.0 Cu min, bal Zn.
Cast: 49,000 TS; 12,000 YS; 5 El; 70 Brin. For hardware, die castings; SAE 41; leaded brass.

PRESSOFOND S5C
Montecatini Settore Alluminio
Cu 4, Si 5.5, bal Al.
Cast: 28,000-32,000 TS; 20,000-23,000 YS; 4-1.5 El; 55-75 Brin. For light alloy parts; heat treatable.

PRESSOFOND S9CF
Montecatini Settore Alluminio
Si 8.5, Cu 3.5, Fe 0.9, bal Al.
Cast: 41,000-47,000 TS; 24,000-29,000 YS; 2.5-5 El; 80-100 Brin. For light alloy parts; corrosion resistant.

PRESSURDIE "C"
CCS Braeburn Alloy Steel
C 0.38-0.43, Cr 4-4.5, W 4-4.5, Co 4-4.5, V 1.9-2.2, Mo 0.4-0.7, bal Fe.
For dies, dummy blocks, valve dies, extrusion mandrels. Hot work steel, air hardened. *Obsolete*

PRESSURDIE 1
CCS Braeburn Alloy Steel
C 0.3-0.4, Si 0.8-1, W 5, Co 6, Mo 0.25, Cr 4.75-5.25, V 2, bal Fe.
For extrusion dies, dummy blocks, punches. Hot work steel, oil hardened. *Obsolete*

PRESSURDIE 16
CCS Braeburn Alloy Steel
C 0.55, Si 0.9, W 1.25, Cr 5, Mo 1.25, bal Fe.
For trimming and forming dies, forging die inserts. Hot work steel, oil hardened. *Obsolete*

PRESSURDIE 2

CCS Braeburn Alloy Steel
C 0.35, Mo 1.5, W 1.25, Cr 5, V 0.15, bal Fe.
For extrusion and forging dies, punches. Type H12 hot work steel. *Obsolete*

PRESSURDIE 3

CCS Braeburn Alloy Steel
C 0.3-0.4, Si 0.85-1, Cr 4.9-5.25, V 0.1-0.15, Mo 0.45-0.6, bal Fe.
For tools, aluminum die casting dies. Oil hardening. *Obsolete*

PRESSURDIE 3-L

CCS Braeburn Alloy Steel
C 0.37, Mn 0.35, Si 1, Cr 5.5, V 0.45, Mo 1.15, bal Fe.
For dies. Oil hardening. *Obsolete*

PRESSURDIE 5

CCS Braeburn Alloy Steel
C 0.38, Si 1, Cr 3.5, V 1, W 1.25, Mo 1, Mn 0.6, bal Fe.
Heat treated: 268,500 TS; 243,000 YS; 8.5 El; 28 RA; 520 Brin. At 900 F: 191,400 TS; 172,000 YS; 14 El; 41 RA; 420 Brin. For aluminum extrusion dies and mandrels, upset punches; hot work steel, tough. *Obsolete*

PRESSURDIE 6

CCS Braeburn Alloy Steel
C 1, Cr 3.2, V 0.3, Mo 2.5, Co 0.4, Mn 0.5, bal Fe.
For aluminum extrusion dies and mandrels, hot shears. Hot work steel, tough. *Obsolete*

PRESSURDIE NO. 7

CCS Braeburn Alloy Steel
C 0.5, Si 1, Cr 5, V 2, Mo 1.3, bal Fe.
Air or oil hardening hot work tool and die steel. *Obsolete*

PRESSURE DIE CASTING

Cerro Metal Products Co.
Cu 62, Pb 0.75, Sn 0.5, bal Zn.
Die cast: 58,000 TS; 36,000 YS; 6 El; 7 RA; 124 Brin. For plumbing fixtures, hardware and pressure die casting. CA 858.

PRESSURE TITE IRON

Janney Cylinder Co.
TC 2.85, CC 0.65, Ni 1.25, Mn 0.8, Cr 0.4, Si 1.6, bal Fe.
Cast: 55,000 TS; 285 Brin. For liners, engine sleeves, oil pumps; centrifugal cast.

PRESTEM

Creusot-Loire
C 0.2, Ni 3.15, Mo 3.4, bal Fe.
Hot work tool steel for forging dies. AFNOR 20 DN 33-12.

PRESTEM

Heppenstall Co.
Tool material. C 0.3, Ni, Mo, bal Fe.
For dies punches; hot work, press and upsetter, prehardened.

PRESTO

Carpenter Technology Corp.
Cr 1.4, C 1.05, bal Fe.
For ball and roller bearings. *Obsolete*

PRESTO

Remystahl
C 1.35, Cr 4.3, Co, V, W, bal Fe.
For engraving tools, form cutters, forming dies; high speed steel. *Obsolete*

PRESTO 2 KRONEN

Remystahl
C 0.74, Cr 4.1, V 1.1, W 18.5, bal Fe.
For lathe and planer tools, reamers, broaches, taps; high speed steel. AISI T1. W.-Nr. 1.3355.

PRESTO B

Remystahl
C 0.82, Cr 4.1, Mo 0.85, W 8.7, V 1.6, bal Fe.
For lathe and planer tools, reamers, drills, taps; high speed steel. *Obsolete*

PRESTO C 3

Remystahl
C 1, Cr 4, Mo 2.5-2.8, V 2.2-2.5, W 2.7-3, bal Fe.
High speed steel cutting tool. W.-Nr. 1.3333.

PRESTO CHROME BEARING STEEL

Carpenter Technology Corp.
C 1, Mn 0.3, Si 0.25, Cr 1.4, bal Fe.
Oil or water hardenable to 60-66 Rock C. For ball and roller bearings and races. G15216

PRESTO ENORM

Remystahl
C 0.82, Cr 4.1, Mo 0.85, V 1.6, W 8.7, bal Fe.
For lathe and planer tools, drills, taps, hobs; high speed steel. *Obsolete*

PRESTO MO 10

Remystahl
C 0.9, Mo 8.6, Cr 4, V 2, W 1.7, Co 8.25, bal Fe.
High speed tool steel. AISI M34. W.-Nr. 1.3249.

PRESTO MO 166

Remystahl
C 1.2, Cr 4, Mo 5, W 6.3, V 3, bal Fe.
High speed tool steel. AISI M3 class 2. W.-Nr. 1.3344.

PRESTO MO 8

Remystahl
C 1.05-1.12, Mo 9-10, W 1.2-1.8, V 1-1.3, Co 7.5-8.5, Cr 3.6-4.4, bal Fe.
High speed tool steel. AISI M42. W.-Nr. 1.3247.

PRESTO MO PLUS

Remystahl
C 0.86-0.94, Cr 3.8-4.5, Mo 4.7-5.2, W 6-6.7, V 1.7-2, bal Fe.
High speed steel. For cutting tools. AISI M2. W.-Nr. 1.3343.

PRESTO MO PLUS CO

Remystahl
C 0.9, Cr 4.2, Mo 5, W 6.3, V 2, Co 5, bal Fe.
High speed tool steel. Similar to AISI M35. W.-Nr. 1.3243.

PRESTO MO PLUS K

Remystahl
C 1, Cr 4.2, Mo 5, W 6.5, V 2, bal Fe.
High speed tool steel. Similar to AISI M-41. W.-Nr. 1.3246.

PRESTO MO SUPERIOR

Remystahl
C 0.8, Cr 4, Mo 8.6, W 1.7, V 1.2, bal Fe.
Mo high speed steel. For cutting tools. AISI M1. W.-Nr. 1.3346.

PRESTO STEEL NO. 9-266

Carpenter Technology Corp.
C 1.05, Cr 1.4, bal Fe.
Hardened. For ball races; see "Presto Tool." *Obsolete*

PRESTO UNIKUM

Remystahl
C 0.79, Co 4.7, Cr 4.3, Mo 0.75, V 1.5, W 18, bal Fe.
For lathe and planer tools, reamers, broaches, taps; high speed steel. W.-Nr. 1.3255.

PRESTO UNIKUM 125

Remystahl
C 0.8, Cr 4.3, Co, W, V, Mo, bal Fe.
For lathe and planer tools, reamers, broaches, hobs; high speed steel. *Obsolete*

PRESTO UNIKUM 3

Remystahl
C 0.86, Co 2.8, Cr 4.3, Mo 0.85, V 2.1, W 12, bal Fe.
For lathe and planer tools, reamers, broaches, hobs; high speed steel. *Obsolete*

PRESTO UNIKUM 5

Remystahl
C 0.79, Co 4.7, Cr 4.3, Mo 0.75, V 1.5, W 18, bal Fe.
For lathe and planer tools, reamers, broaches, drills; high speed steel. *Obsolete*

PRESTO UNIKUM SPEZIAL

Remystahl
C 1.25, Cr 4.2, Mo 3.7, W 10, V 3.2, Co 10, bal Fe.
High speed tool steel. W.-Nr. 1.3207.

PRESTO UNIKUM SUPERB

Remystahl
C 0.76, Co 10, Cr 4.2, Mo 0.8, V 1.6, W 18, bal Fe.
For lathe and planer tools, form cutters; high speed steel. W.-Nr. 1.3265.

PRESTO V205

Remystahl
C 1.3, Cr 4.3, Mo 0.85, V 3.8, W 12, bal Fe.
For engraving tools, form cutters, forming dies; high speed steel. W.-Nr. 1.3302.

PRESTO V210

Remystahl
C 1.3-1.45, Cr 4.2, W 12, Mo 0.7-1, V 3.75, Co 4.8, bal Fe.
High speed tool steel. AISI T15. W.-Nr. 1.3202.

PRESTO V22

Remystahl
C 0.95, Cr 4.1, Mo 0.8, V 2.5, W 12, bal Fe.
For lathe and planer tools, reamers, broaches, hobs; high speed steel. W.-Nr. 1.3318.

PRESTO VA MO SUPERIOR

Remystahl
C 1.02, Cr 4.2, Mo 8.5, W 1.8, V 2, bal Fe.
High speed tool steel. AISI M7. W.-Nr. 1.3348.

PREUSS ALLOY

Manufacturer not listed
Co 30, Si 0.2-1.5, C 0.04-0.06, bal Fe.
For magnetic circuits; high permeability.

PREWESTA 1

Stahlwerk Stahlschmidt GmbH & Co.
C 0.55, Ni 1, V 0.18, W 1.8, bal Fe.
For upsetters, crimpers, cold punches; oil hardened, tough.

PREWESTA 2

Stahlwerk Stahlschmidt GmbH & Co.
C 0.45, Ni 1, V 0.18, W 1.85, bal Fe.
For upsetters, crimpers, cold punches; oil hardened, tough.

PREWESTA 3

Stahlwerk Stahlschmidt GmbH & Co.
C 0.35, Cr 1.05, W 1.85, V 0.1, bal Fe.
For upsetters, crimpers, cold punches; oil hardened, tough.

PREXI

Uddeholm Corp.
C 0.15, Si 0.25, Mn 1, Cr 1.2, Mo 0.25, bal Fe.
For die casting dies, plastic mold dies; case hardened, oil hardened.

PRG 35

Otto Wolff Handelsgesellschaft
C 0.5, Cr 1.05, Ni 3.25, bal Fe.
For gears, bolts, springs, crankshafts; oil hardened, shock resistant.

964 / WOLDMAN'S ENGINEERING ALLOYS

PRIMA 100
Bergische Stahl Industrie
C 0.9, Si 0.25-0.5, Mn 0.3-0.8, bal Fe.
Annealed: 95,000 TS; 50,000 YS; 25 El; 45 RA; 190 Brin. For springs, tools, cutters, punches, drills, taps; Type W1; water hardened. *Obsolete*

PRIMA 2H
Stahlwerke R. & H. Plate
C 1.3, Mn 0-0.25, Si 0-0.25, bal Fe.
For engravers tools, form cutters; water hardened; Type W1. *Obsolete*

PRIMA 3MH
Stahlwerke R. & H. Plate
C 1.15, Si 0-0.25, Mn 0-0.35, bal Fe.
For drills, taps, hobs, cutters, lathe tools; Type W1; water hardened. *Obsolete*

PRIMA 45
Otto Wolff Handelgesellschaft
C 0.45, Si 0.5, Mn 0.6, bal Fe.
Hot rolled: 98,000 TS; 59,000 YS; 24 El; 45 RA; 212 Brin. For gears, bolts, shafts; water hardened.

PRIMA 4ZH
Westa-Westdeutsche
C 1.1, Si 0-0.25, Mn 0-0.25, bal Fe.
For drills, taps, hobs, cutters, lathe tools; Type W1; water hardened.

PRIMA 55
Otto Wolff Handelgesellschaft
C 0.55, Si 0.2, Mn 0.6, bal Fe.
Heat treated: 150,000 TS; 110,000 YS; 14 El; 45 RA; 320 Brin. For gears, springs crimpers; axles, shafts; water hardened.

PRIMA 5ZAH
Westa-Westdeutsche
C 0.85, Si 0-0.25, Mn 0-0.25, bal Fe.
Heat treated: 190,000 TS; 145,000 YS; 10 El; 30 RA; 400 Brin. For springs, tools, punches, drills; Type W1; water hardened.

PRIMA 6 SEHR ZAH
Stahlwerke R. & H. Plate
C 0.7, Si 0-0.25, Mn 0-0.25, bal Fe.
Heat treated: 175,000 TS; 128,000 YS; 12 El; 37 RA; 355 Brin. For springs, axes, rails, tools, punches; Type W1; water hardened. *Obsolete*

PRIMA 60
Otto Wolff Handelgesellschaft
C 0.6, Si 0.5, Mn 0.6, bal Fe.
Heat treated: 160,000 TS; 113,000 YS; 12 El; 40 RA; 325 Brin. For springs, punches, crimpers, axes, hammers; water hardened.

PRIMA 65
Bergische Stahl Industrie
C 0.45, Si 0.5, Mn 0.6, bal Fe.
Hot rolled: 98,000 TS; 60,000 YS; 24 El; 45 RA; 215 Brin. For gears, bolts, machine tool parts, shafts; water hardened. *Obsolete*

PRIMA 75
Otto Wolff Handelgesellschaft
C 0.75, Si 0.4, Mn 0.6, bal Fe.
Heat treated: 180,000 TS; 140,000 YS; 14 El; 36 RA; 375 Brin. For crimpers, punches, springs, rails; water hardened.

PRIMA 85
Bergische Stahl Industrie
C 0.6, Si 0.4, Mn 0.6, bal Fe.
Heat treated: 160,000 TS; 115,000 YS; 12 El; 40 RA; 325 Brin. For gears, bolts, hammers, axles, shafts; water hardened. *Obsolete*

PRIMA 90
Bergische Stahl Industrie
C 0.67, Si 0.25-0.5, Mn 0.3-0.8, bal Fe.
Heat treated: 175,000 TS; 128,000 YS; 12 El; 37 RA; 355 Brin. For rails, springs, hammers, axes; water hardened. *Obsolete*

PRIMA 90
Otto Wolff Handelgesellschaft
C 0.9, Si 0.4, Mn 0.6, bal Fe.
Heat treated: 190,000 TS; 145,000 YS; 10 El; 30 RA; 40 Brin. For tools, cutters, springs, drills; water hardened; Type W1.

PRIMA 95
Bergische Stahl Industrie
C 0.75, Si 0.25-0.5, Mn 0.3-0.8, bal Fe.
Heat treated: 185,000 TS; 140,000 YS; 15 El; 40 RA; 395 Brin. For springs, rails, punches, tools, crimpers; water hardened. *Obsolete*

PRIMA H
J.C. Soding & Halbach
C 0.9, Si 0.25-0.5, Mn 0.3-0.8, bal Fe.
Heat treated: 190,000 TS; 145,000 YS; 10 El; 30 RA; 400 Brin. For springs, tools, cutters, punches, drills, taps; Type W1; water hardened. *Obsolete*

PRIMA H
Plettenberger Gusstahlfabrik
C 1.3, Si 0-0.25, Mn 0-0.25, bal Fe.
For engraver's tools, drills, reamers, form cutters; Type W1; water hardened.

PRIMA HARTE 3
Idealstahl Breidenbach KG
C 0.9, Si 0.4, Mn 0.6, bal Fe.
Heat treated: 190,000 TS; 145,000 YS; 10 El; 30 RA; 400 Brin. For springs, tools, cutters, dies; Type W1; water hardened.

PRIMA HARTE 3/4
Idealstahl Breidenbach KG
C 0.75, Si 0.4, Mn 0.6, bal Fe.
Heat treated: 175,000 TS; 126,000 YS; 10 El; 35 RA; 375 Brin. For springs, tools, hammers, rails, axes; Type W1; water hardened.

PRIMA HARTE 4
Idealstahl Breidenbach KG
C 0.6, Si 0.4, Mn 0.6, bal Fe.
Heat treated: 160,000 TS; 113,000 YS; 12 El; 40 RA; 325 Brin. For crankshafts, axes, axles, crimpers, punches, gears; water hardened.

PRIMA HARTE 5
Idealstahl Breidenbach KG
C 0.45, Si 0.4, Mn 0.6, bal Fe.
Hot rolled: 98,000 TS; 59,000 YS; 24 El; 45 RA; 215 Brin. For gears, pinions, bolts, studs, crankshafts; water hardened.

PRIMA HARTE 6
Idealstahl Breidenbach KG
C 0.35, Si 0.3-0.5, Mn 0.3-0.8, bal Fe.
Hot rolled: 85,000 TS; 54,000 YS; 30 El; 53 RA; 185 Brin. For gears, bolts, fasteners, crankshafts; water hardened.

PRIMA MH
J.C. Soding & Halbach
C 0.75, Si 0.25-0.5, Mn 0.3-0.8, bal Fe.
Heat treated: 175,000 TS; 130,000 YS; 10 El; 35 RA; 380 Brin. For springs, rails, hammers, tools, punches; Type W1; water hardened. *Obsolete*

PRIMA MH
Plettenberger Gusstahlfabrik
C 1.15, Si 0-0.25, Mn 0-0.25, bal Fe.
Annealed: 100,000 TS; 56,000 YS; 20 El; 40 RA; 210 Brin. For drills, taps, reamers, hobs, cutters; Type W1; water hardened.

PRIMA MH
Kind & Co. Edelstahlwerk
C 0.9, Si 0.25-0.5, Mn 0.3-0.8, bal Fe.
Heat treated: 190,000 TS; 145,000 YS; 10 El; 30 RA; 400 Brin. For springs, tools, drills, taps, reamers, hobs; Type W1; water hardened.

PRIMA MH
Hoffman & Co. KG
C 0.9, Si 0.25-0.5, Mn 0.3-0.8, bal Fe.
Heat treated: 190,000 TS; 145,000 YS; 10 El; 30 RA; 400 Brin. For springs, tools, drills, taps, reamers, hobs; Type W1; water hardened.

PRIMA MH 100
Vereinigte Edelstahlwerke
C 1, Si 0-0.25, Mn 0-0.25, bal Fe.
Annealed: 100,000 TS; 53,000 YS; 21 El; 42 RA; 200 Brin. For springs, tools, cutters, reamers; Type W1; water hardened. *Obsolete*

PRIMA MH 115
Vereinigte Edelstahlwerke
C 1.15, Si 0-0.25, Mn 0-0.25, bal Fe.
Annealed: 110,000 TS; 56,000 YS; 18 El; 40 RA; 210 Brin. For drills, taps, reamers, broaches; Type W1; water hardened. *Obsolete*

PRIMA ZAHWEICH
Vereinigte Edelstahlwerke
C 0.45, Si 0-0.25, Mn 0-0.25, bal Fe.
Hot rolled: 98,000 TS; 59,000 YS; 24 El; 54 RA; 212 Brin. For gears, bolts, fasteners, shafts; water hardened. *Obsolete*

PRIMA ZH I
Vereinigte Edelstahlwerke
C 1, Si 0-0.25, Mn 0-0.25, bal Fe.
Annealed: 100,000 TS; 53,000 YS; 21 El; 42 RA; 200 Brin. For cutters, drills, punches, springs, reamers, taps; Type W1; water hardened. *Obsolete*

PRIMA ZH II
Vereinigte Edelstahlwerke
C 0.9, Si 0-0.25, Mn 0-0.25, bal Fe.
Heat treated: 190,000 TS; 145,000 YS; 10 El; 30 RA; 400 Brin. For cutters, taps, springs, drills, hobs; Type W1; water hardened. *Obsolete*

PRIMA ZH III
Vereinigte Edelstahlwerke
C 0.85, Si 0-0.25, Mn 0-0.25, bal Fe.
Heat treated: 188,000 TS; 143,000 YS; 12 El; 35 RA; 320 Brin. For drills, punches, taps, springs, cutters; Type W1; water hardened. *Obsolete*

PRIMER GILDING
Manufacturer not listed
Cu 97, Zn 3, traces Pb, Fe.
For primers, base for fire-enameled parts.

PRINCE
A. Milne & Co.
C 1, Cr 1, bal Fe.
For ball bearings; water hardening.

PRINCES METAL NO. 1
Manufacturer not listed
Cu 61-83, Zn 17-39.
For hardware; brass.

PRINCES METAL NO. 2
Manufacturer not listed
Cu 85, Zn 15.
For flexible hose; good workability.

PRIZE RIBBON
Hewitt Metals Corp.
Cu, Sb, bal Sn.
Cast: 15,200 TS; 34.4 Brin. For bearings. Babbitt.

PRK 33 COBALTCROM
Ziv Steel & Wire Co.
C 1.4, Cr 13, Co 3.3, Ni 0.5, Mo 0.6, bal Fe.
For blanking and forming dies, trimmers, punches; air hardened, non-deforming, wear resistant.

PRK-33
Darwin & Milner Inc.
Co 3.7, Cr 13.6, Fe 79.5, Mo 0.84, C 1.5, Si 0.6, V 0.25, Mn 0.3, Ni 0.5.
Air hardened: 602-654 Brin. For milling cutters, twist drills, press tools; non-scaling steel, non-deforming.

PRK-33
Detroit Alloy Steel
Co 3.7, Cr 13.6, Fe 79.5, Mo 0.84, C 1.5, Si 0.6, V 0.25, Mn 0.3, Ni 0.5.
Air hardened: 602-654 Brin. For milling cutters, twist drills, press tools; non-scaling steel, non-deforming.

PRK-SH
Darwin & Milner Inc.
Alloy steel. C, Cr, W, bal Fe.
For self hardening welding electrodes; for hardened tool and die steels.

PRN2
Eagle & Globe Steel Ltd.
Tool material. C 1.05, Mn 1.3, Cr 1.3, Si 0.25, bal Fe.
Cold work tool and die steel. Oil hardened. For roller bearings, crusher rolls, blanking and forming tools. AS1239 L100A.

PROBEDIT
Russian manufacture
WC.
For cutting tools and dies; sintered.

PRODUCTION-ROD NO. 110
Marquette Corp.
C 0.06-0.1, P 0.01, Si 0.1-0.25, Mn 0.55, S 0.03, bal Fe.
Welded: 68,000-75,000 TS; 55,000-65,000 YS; 23-27 El; 50-55 RA. For all-purpose welding rod; AWS-E6013.

PROFERALL
Textron Inc.
Si 2.2-2.35, Mn 0.5-0.6, Cr 0.8-1, Mo 0.4-0.5, Ni 0.4-0.5, TC 3.15, CC 0.55-1, bal Fe.
60,000-75,000 TS; 250-300 Brin. For diesel engine parts, crankshafts, cylinders, pistons, refrigeration parts; high test cast iron; wear resistant.

PROFERALL A
Textron Inc.
C 3.3, Ni 1.5, Cr 0.9, Mo 0.2, Si 2, bal Fe.
Cast: 35,000 TS; 207-241 Brin. For oil pump gears; cast iron.

PROGEN
Seaboard Steel Co. of America
C 0.33-0.4, Mo 0.5-0.7, Cr 0.55-0.75, Cu 0.45-0.65, Si 0.5-0.7, Ti 0.1, bal Fe.
Annealed: 97,000 TS; 57,000 YS; 22 El; 42 RA; 197 Brin.
Hardened: 285,000 TS; 245,000 YS; 11 El; 40 RA; 555 Brin.
For tools, chisels, picks, dies for die casting; tough, shock resistant, self-tempering tool steel.

PROJECT-70
Carpenter Technology Corp.
C 0-0.15, Cr 12-14, S 0.15, bal Fe.
Annealed: 103,000 TS; 75,000 YS; 22 El; 216 Brin. For fittings, gears, housings, valve stems, valve trim. Corrosion resistant, free cutting. *Obsolete*

PROJECTILE STEEL
American manufacture
Cr 2.4, C 0.8, Mn 0.4, Si 0.2, bal Fe.
110,000 TS; 80,000 YS; 18 El. For projectiles; oil hardening.

PROMAL
F.M.C. Corp.
Si 1, Mn 0.3, 1.8 total C, 0.4 combined C, bal Fe.
Cast: 65,000-70,000 TS; 45,000-50,000 YS; 10-14 El; 170-190 Brin. For cast chains, sprockets, parts, valve parts, wrench gears, brake drums; malleable cast iron.

PROMAL
Capitol Castings Inc.
TC 1.8, CC 0.4, Si 1, bal Fe.
Cast: 65,000-70,000 TS; 45,000-50,000 YS; 10-14 El; 170-190 Brin. For sprockets, gears, chains, brake drums; malleable iron. *Obsolete*

PROMET 100
American Crucible Products Co.
Cu 55-60, Fe 0.5-2, Al 0.5-2, Mn 0.25-1.5, Sn 0-1, bal Zn.
Cast: 65,000-80,000 TS; 28,000-40,000 YS; 20-40 El; 115-150 Brin. For valves, engine framing, valve stems, lever arms, gears, liners, bearings, marine hardware. Alpha-beta manganese bronze, SAE 43. *Obsolete*

PROMET 104 AB
American Crucible Products Co.
Al 10-11, Fe 3.5, Ni 0-1, Mn 0-0.5, 83.0 Cu min.
Sand cast: 80,000 TS min; 30,000 YS min; 12 El min; 145 Brin min. Centrifugal: 90,000 TS min; 32,000 YS min; 12 El min; 150 Brin min. For gears, worm wheels, rolling mill bearings, and slippers, bearing races. Aluminum bronze, wear and fatigue and corrosion resistant.

PROMET 115-N
American Crucible Products Co.
Fe 3-5, Ni 3-5.5, Al 10-11.5, Mn 0-3.5, 78.0 Cu min.
Cast: 100,000 TS; 45,000 YS; 8 El; 195 Brin. Heat treated: 118,000 TS; 80,000 YS; 7 El; 240 Brin. For bearings, gears, cams, valve and pump parts, castings. Corrosion resistant, heat treatable. ASM 4880.

PROMET 150
American Crucible Products Co.
Cu 60-68, Sn 0-0.2, Pb 0-0.2, Fe 2-4, Al 3.5-5, Mn 2.5-5, bal Zn.
Cast: 90,000 TS min; 45,000 YS min; 18 El min. For gears, cams, hydraulic cylinder parts, bearings.

PROMET 18
American Crucible Products Co.
Pb 15, Sn 4.5, Ni 1.5, bal Cu.
Cast: 30,000 TS; 19,000 YS; 18 El; 60 Brin. For bearings, bushings, liners, sleeves; heavy duty. *Obsolete*

PROMET 200
American Crucible Products Co.
Cu 60-68, Al 3.5-7, Fe 2-4, Mn 2.5-5, Sn 0-0.2, Pb 0-0.2, bal Zn.
Sand cast: 110,000-119,000 TS; 60,000-80,000 YS; 12-14 El; 200-225 Brin. For valve stems, gears, cams, bridge parts, heavy load bearings, hydraulic cylinder parts. SAE 430B ASTM B147-8C.

PROMET 22
American Crucible Products Co.
Cu 86, Sn 10, Pb 2, Zn 2.
Sand cast: 43,000 TS; 19,000 YS; 18-40 El; 68 Brin. For heavy duty bearings, nuts, gears, worm wheels. *Obsolete*

PROMET 2S
American Crucible Products Co.
Cu 70-75, Sn 5, Pb 19-23, Ni 0.2-0.8.
Sand cast: 27,000 TS; 15,000 YS; 16 El; 45-55 Brin. For bearings, bushings, sleeves. AMS 4840; ASTM B144-3E.

PROMET 406
American Crucible Products Co.
Cu 62, Sn 1, Pb 1, 36.0 Zn (nominal).
Cast: 40,000 TS; 14,000 YS; 15 El. For bushings, hardware fittings, ornaments. QQ-B-621 Class A.

PROMET 6CR
American Crucible Products Co.
Cu 76-80, Sn 7-9, Pb 11-14, Ni 0.25-1.
Cast: 37,000 TS; 20,000 YS; 20 El; 65 Brin. For bearings, wear parts. Corrosion resistant, good embeddability. *Obsolete*

PROMET 6SK
American Crucible Products Co.
Cu 75-79, Sn 6.5-8, Pb 12.5-15.5, Ni 1.25-1.75.
Sand Cast: 36,000 TS; 19,000 YS; 18 El; 60 Brin. For bearings, bushings, liners, sleeves, pump bodies; free-cutting.

PROMET 712
American Crucible Products Co.
Cu 67-72, Sn 8.5-11, Pb 18-22, Ni 0.25-1.
Sand cast: 33,000 TS; 17,000 YS; 18 El; 53-63 Brin. For bearings, especially under boundary or mixed film lubricants.

PROMET 782
American Crucible Products Co.
Cu 65-72, Sn 7-9, Pb 20-24, Ni 0.25-1.
Sand cast: 30,000 TS; 17,000 YS; 15 El; 45-60 Brin. High lead tin bronze; for bearings, especially under boundary or mixed film lubricants. Can deform to correct bad fit.

PROMET 80A
American Crucible Products Co.
Cu 78-82, Sn 9-11, Pb 8-11, Zn 0-0.75, Ni 0-0.75.
Sand cast: 39,000 TS; 18,000 YS; 30 El; 67 Brin. High leaded tin bronze for bearing applications.

PROMET 83A
American Crucible Products Co.
Cu 81-85, Sn 6.25-7.5, Pb 6-8, Zn 2-4, Ni 0-0.5.
Sand cast: 34,000 TS; 18,000 YS; 15 El; 65 Brin. Leaded, general utility bearing bronze for bearings, bushings. SAE 660. ASTM B144-3B.

PROMET 85A
American Crucible Products Co.
Cu 84-86, Sn 4-6, Pb 4-6, Zn 4-6, Ni 0-0.8.
Sand Cast: 35,000 TS; 15,000 YS; 32 El; 54-67 Brin. For bearings, bushings. SAE 40; ASTM B145-4A.

PROMET 86P
American Crucible Products Co.
Cu 78-82, Sn 2.25-3.5, Pb 6-8, Zn 7-10, Ni 0-1.
Sand cast: 34,000 TS; 15,000 YS; 24 El; 55 Brin. Leaded semi-red brass. For general hardware fittings, pump bodies, low pressure valves and fittings. *Obsolete*

PROMET 89-S
American Crucible Products Co.
Cu 86-89, Sn 7-9, Ni 0.25-0.75, Pb 2-4.
Centrifugal chill cast: 40,000 TS; 20,000 YS; 30 El. For heavy duty bearings, gears, worm wheels.

PROMET 91-CK
American Crucible Products Co.
Sn 9-10.5, Pb 1.3-3, Ni 0.75-1.5, Zn 0-1.5, 83.0 Cu min.
Centrifugal chill cast: 50,000 TS; 22,000 YS; 16 El. For heavy duty bearings, gears, worm wheels.

PROMET 91-SK
Now PROMET 91-CK.

PROMET 91-SK
American Crucible Products Co.
Pb 1.5-2.5, Sn 9-10, Ni 1-1.5, bal Cu.
Cast: 50,000 TS; 25,000 YS; 26 El; 75 Brin. For bearings, bushings, liners, sleeves; heavy duty. *Obsolete*

PROMET 93 AB
American Crucible Products Co.
Al 8.5-9.5, Fe 2.5-4, 86.0 Cu min.
Sand cast: 79,000 TS; 28,000 YS; 30 El; 130 Brin. Alpha phase aluminum bronze. For worm wheels, bearings and bushings, slides, pump parts, utility machine parts.

PROMET BRONZE 53
American Crucible Products Co.
Sn 11.5-12.5, P 0-0.02, 0.5% others max, bal Cu.
Centrifugal cast: 50,000 TS; 25,000 YS; 12 El; 95 Brin. For worm gears and wheels. *Obsolete*

PROMET BRONZE 53L
American Crucible Products Co.
Sn 9.75-11.5, P 0-0.3, Pb 0.25, Ni 1.25-1.75, Cu 85-89, 0.5% others.
Centrifugal cast: 35,000 TS; 12 El; 85 Brin. For worm gears and wheels. *Obsolete*

PROMET BRONZE 53N
American Crucible Products Co.
Sn 11-12, P 0-0.25, Ni 1.25-1.75, 0.5% others max, bal Cu.
Centrifugal cast: 50,000 TS; 28,000 YS; 12 El; 100 Brin. For worm gears and wheels. *Obsolete*

PROMET M-640
American Crucible Products Co.
Cu 84-88, Sn 10-12, Pb 0.5-1.25, Ni 1.25-1.75, P 0.1-0.3.
Centrifugal chill cast: 55,000 TS; 27,000 YS; 12 El. For heavy duty bearings, gears, worm wheels.

PROMET NO. 1
American Crucible Products Co.
Cu 66-71, Sn 2-4, Pb 24.5-29.5, Ni 0.5-1.
Cast: 18,000 TS; 11,000 YS; 10 El; 30 Brin. For bearings, wear parts, castings; sand and centrifugally cast. *Obsolete*

PROMET NO. 101-AB
American Crucible Products Co.
Cu 86-90, Fe 0.75-1.5, Al 9-11, Mn 0-0.5, Ni 0-0.5.
Cast: 85,000 TS; 45,000 YS; 25 El. For gears, propellers, shafts; Al-bronze, corrosion resistant.

PROMET NO. 3
American Crucible Products Co.
Cu 85-89, Sn 7.5-9, Pb 0.5-1.5, Ni 0.5-1, Zn 2.5-5.
Bearing and gear bronze. Sand cast: 40,000 TS; 20,000 YS; 28 El; 70 Brin. For high duty bearings, gears, pump impellers, piston rings, steam fittings.

PROMET NO. 6
American Crucible Products Co.
Cu 76-79, Sn 4.5-6.5, Pb 12.5-15.5, Ni 1-1.7.
Cast: 30,000 TS; 20,000 YS; 16 El; 64 Brin. For bearings, bushings, castings; heavy duty.

PROMET P-48N
American Crucible Products Co.
Sn 9.25-10.75, Pb 2-3.5, Ni 3-4, 1.25 others, bal Cu.
Leaded nickel tin bronze. Sand cast: 50,000 TS; 28,000 YS; 18 El; 85 Brin. For heavy duty bearings, nuts, gears, worm wheels, ASTM B427; Alloy D.

PROMET X
American Crucible Products Co.
Sb 75, Cu 3.5, bal Sn.
Cast: 10,900 TS; 8 El. For bearings, bushings; Babbitt; C.S. 13625.

PROMET XX
American Crucible Products Co.
Pb 7.5, Sb 15, Sn 10.
Cast: 10,500 TS; 4 El. For bearings, bushings; Babbitt; C.S. 17150.

PROMET XXX
American Crucible Products Co.
Sb 17.5, Sn 1.4, As 1.5, Cu 0.5, bal Pb.
Cast: 10,350 TS; 2 El. For bearings, bushings; Babbitt; C.S. 14200.

PROMETAL
Ste de Produits Metallurgiques
C, Cr, Ni, bal Fe.
For stainless parts; heat and corrosion resistant.

PROMETAL CHROME INOXYDABLE R.F.
Ste de Produits Metallurgiques
C, Cr, Ni, bal Fe.
For stainless parts; heat and corrosion resistant.

PROMETAL INOXYDABLE R.F.-3
Ste de Produits Metallurgiques
C, Cr, Ni, bal Fe.
For stainless parts; heat and corrosion resistant.

PROMETHIUM
Manufacturer not listed
Cu 67, Zn 30, Al 3.
For condenser tubes; high ductility.

PROOF AG
Sheffield Smelting Co. Ltd.
Ag 99.9.
Melting range: 960-960°C. Maximum stress: 12.6 kgf/mm^2; 45 El. For vacuum brazing, fuses, electronic applications.

PROPELLER BUSHING
Manufacturer not listed
Zn 69, Sn 19, Sb 7, Cu 5.
For propeller bushings.

PROPLATINUM
Manufacturer not listed
Ni 72, Ag 23.6, Bi 3.7, Au 0.7.
For corrosion resistant parts; platinum substitute.

PROSPECTOR
J.F. Jelenko & Co.
Precious metal. Au 20, Pd 20, Ag 38.
80,000 psi TS; 36,500 psi YS; 8 El; 135 Brin. Type III hard dental alloy. For inlays, crowns, and fixed bridgework.

PROTAL /-BD
Siemens AG
Aluminum.
Aluminum alloy AlCuMg/-Pb.

PROTECTIVE DECK PLATE
Manufacturer not listed
Ni 3.5, Cr 1.5, C 0.2-0.3, bal Fe.
For armor plate; case-hardened.

PROTECTIVE NETTING
Manufacturer not listed
Ni 27.8, C 0.4, bal Fe.
For armor steel for torpedo defense, netting; corrosion resistant.

PROTOCHROME
Manufacturer not listed.
C 2-2.5, Cr 24-27, bal Fe.
Welded: 600 Brin. For hard surfacing electrode; heat and abrasion resistant.

PROTOLOY
Molecu-Wire Corp.
Ni 80, Cr 20.
Resistance wire. Electrical resistivity: 650 ohms/circular mil ft. Annealed: 100,000 psi TS. Hard drawn: 225,000 psi TS. For resistors and heating elements. Maximum operating temperature: 1100°C.

PROTOLOY B
Molecu-Wire Corp.
Ni 80, Cr 20.
Resistance wire. Electrical resistivity: 680 ohms/circular mil ft. Annealed: 100,000 psi TS. Hard drawn: 225,000 psi TS. For resistors. Maximum operating temperature: 1100°C *Obsolete*

PRP-13-8 MO
Precision Rolled Products, Inc.
Stainless steel. C 0-0.05, Cr 12.7, Ni 8.2, Al 1.1, Mo 2, bal Fe.
Precipitation hardenable stainless steel for fasteners, valve parts, shaft pins, washers and landing gear parts. H-950: 225,000 TS; 205,000 YS; 12 El; 50 RA; 443 Brin. Meets AMS 5629. S13800

PRP-15-5 PH
Precision Rolled Products, Inc.
Stainless steel. C 0.07, Cr 15, Ni 4.5, Cu 3.4, bal Fe.
Precipitation hardenable stainless steel for fasteners, valve parts, shafts, gears, and chemical processing equipment. H-900: 190,000 TS; 175,000 YS; 14 El; 54 RA; 390 Brin. Meets AMS 5659. S15500

PRP-15-7 MO
Precision Rolled Products, Inc.
Stainless steel. C 0.07, Cr 15, Ni 7, Al 1.1, bal Fe.
Precipitation hardenable stainless steel for fasteners, valve parts, and chemical processing equipment. H-950: 220,000 TS; 200,000 YS; 10 El; 30 RA; 440 Brin. Meets AMS 5520. S15700

PRP-17-4 PH
Precision Rolled Products, Inc.
Stainless steel. C 0.04, Cr 16, Ni 4.25, Cu 3.3, bal Fe.
Precipitation hardenable stainless steel for fasteners, valves and pump parts, chains and gears. H-900: 149,000-203,000 TS; 132,000-186,000 YS; 10-11 El; 30-50 RA; 410 Brin. Meets AMS 5643; AMS 5622. S17400

PRP-188
Precision Rolled Products, Inc.
C 0.1, Cr 22, Ni 22, W 14, bal Co.
Work strengthened columbium-nickel-chromium-tungsten alloy for valves, springs, jet engine rings and parts. Solution treated: 37,000-130,000 TS; 24,000-69,500 YS; 56-72 El. Meets AMS 5772. R30188

PRP-304
Precision Rolled Products, Inc.
Stainless steel. C 0-0.08, Ni 9.5, Cr 19, bal Fe.
Type 304 stainless steel for chemical and food processing equipment. Annealed: 85,000 TS; 35,000 YS; 60 El; 70 RA; 160 Brin. Cold drawn: 180,000 TS; 125,000 YS; 10 El; 330 Brin. Meets AMS 5639; QQ-S-763. S30400

PRP-304L
Precision Rolled Products, Inc.
Stainless steel. C 0-0.03, Ni 9.5, Cr 19, bal Fe.
Type 304 stainless steel for welded assemblies without subsequent heat treatment. Annealed: 85,000 TS; 35,000 YS; 60 El; 70 RA; 160 Brin. Meets AMS 5647; QQ-S-763. S30403

PRP-310
Precision Rolled Products, Inc.
Stainless steel. C 0-0.25, Ni 20.5, Cr 25, bal Fe.
Type 310 stainless steel for jet engine rings and parts, furnace components. Annealed: 100,000 TS; 45,000 YS; 50 El; 65 RA; 180 Brin. Meets AMS 5651; QQ-S-763. S31000

PRP-316
Precision Rolled Products, Inc.
Stainless steel. C 0-0.1, Ni 12, Cr 17, Mo 2.5, bal Fe.
Type 316 stainless steel for chemical, paper, and oil refining industry equipment. Annealed: 80,000 TS; 30,000 YS; 60 El; 80 RA; 140 Brin. Cold drawn: 150,000 TS; 135,000 YS; 6 El; 300 Brin. Meets AMS 5648; QQ-S-763. S31600

PRP-316L
Precision Rolled Products, Inc.
Stainless steel. C 0-0.03, Ni 12, Cr 17, Mo 2.5, bal Fe.
Type 316 stainless steel for welded construction where intergranular carbide precipitation must be avoided. Annealed: 80,000 TS; 30,000 YS; 60 El; 80 RA; 140 Brin. Cold drawn: 150,000 TS; 135,000 YS; 6 El; 300 Brin. Meets AMS 5648; QQ-S-763. S31603

PRP-321
Precision Rolled Products, Inc.
Stainless steel. C 0-0.08, Ni 9.5, Cr 18, Ti 0.5, bal Fe.
Type 321 stainless steel for airplane exhaust manifolds, expansion joints, and flexible couplings. Annealed: 85,000 TS; 35,000 YS; 55 El; 65 RA; 165 Brin. Cold drawn: 95,000 TS; 60,000 YS; 40 El; 60 RA; 185 Brin. Meets AMS 5645; QQ-S-763. S32100

PRP-330

Precision Rolled Products, Inc.

C 0.06, Ni 35.5, Cr 18.5, Si 1.1, Fe 44.
Iron-nickel-chromium alloy for furnace and heat exchanger parts. Annealed: 85,000 TS; 36,000 YS; 47 El; 65 RA; 150 Brin. Meets AMS 5716. N08330

PRP-333

Precision Rolled Products, Inc.

Nickel. C 0.04, Fe 17, Cr 25.5, Mo 3.2, W 3.2, Co 3, bal Ni.
Alloy for parts and fixtures for furnaces, heat exchangers and carburizing containers. Solution treated: 57,000-100,000 TS; 28,000-50,000 YS; 45-50 El; 41-57 RA; 172 Brin. Meets AMS 5717. N06333

PRP-347

Precision Rolled Products, Inc.

Stainless steel. C 0-0.08, Ni 10.5, Cr 18, Cb 1, bal Fe.
Type 347 stainless steel for jet engine rings and parts, furnace components, heat exchanger tubing. Annealed: 90,000 TS; 40,000 YS; 50 El; 70 RA; 160 Brin. Meets AMS 5646; AMS 5654; QQ-S-763. S34700

PRP-350

Precision Rolled Products, Inc.

Stainless steel. C 0.12, Cr 16.5, Ni 4.25, Mo 2.9, N 0.1, bal Fe.
Precipitation hardenable stainless steel for jet engine compressor blades, valves, springs, and knife blades. Heat treated: 161,000 TS; 45,000 YS; 22 El; 200 Brin. Meets AMS 5745. S35000

PRP-3500

Precision Rolled Products, Inc.

C 0.75, Co 40, Cr 3.8, W 4.8, bal Fe.
Permanent magnet alloy for magnetic and electrical equipment.

PRP-36NI-FE

Precision Rolled Products, Inc.

Fe 64, Ni 36.
Alloy with a low coefficient of expansion to 150°C for thermostats and temperature regulators. Annealed: 70,000 TS; 24,000 YS; 36 El; 68 RA; 143 Brin. K93600

PRP-400

Precision Rolled Products, Inc.

Nickel. Cu 31.5, Ni 66.5.
Alloy for corrosion resistant valves, shafts and marine fixtures. Annealed: 70,000 TS; 25,000 YS; 60 El; 110 Brin. Cold drawn: 84,000 TS; 55,000 YS; 40 El; 160 Brin. Meets AMS 4675; QQ-N-281. N04400

PRP-410

Precision Rolled Products, Inc.

Stainless steel. C 0-0.15, Cr 12, bal Fe.
Type 410 stainless steel for jet engine rings and parts, steam turbine buckets and blades, compressor blades, valves, and fasteners. Annealed: 75,000 TS; 40,000 YS; 35 El; 70 RA; 155 Brin. Cold drawn: 100,000 TS; 85,000 YS; 17 El; 60 RA; 205 Brin. Hardened: 157,000 TS; 145,000 YS; 13 El; 69 RA; 300 Brin. Meets AMS 5613; QQ-S-763. S41000

PRP-42NI-FE

Precision Rolled Products, Inc.

Fe 58, Ni 42.
Alloy with a low coefficient of expansion to 650°F for thermostats and temperature regulators. Annealed: 70,000 TS; 50,000 YS; 35 El. K94200

PRP-430

Precision Rolled Products, Inc.

Stainless steel. C 0-0.12, Cr 16, bal Fe.
Type 430 stainless steel for combustion chambers, chemical industry applications and automotive trim. Annealed: 70,000 TS; 40,000 YS; 30 El; 55 RA; 140 Brin. Cold drawn: 130,000 TS; 120,000 YS; 2 El. Meets AMS 5627; QQ-S-763. S43000

PRP-450

Precision Rolled Products, Inc.

Stainless steel. C 0-0.05, Cr 14.5, Ni 5.5, Cu 1.5, bal Fe.
Precipitation hardenable stainless steel for fasteners and valves; corrosion and oxidation resistant. Annealed: 144,000 TS; 117,000 YS; 14 El; 48 RA; 271 Brin. Heat treated: 196,000 TS; 184,000 YS; 14 El; 57 RA; 400 Brin. Meets AMS 5763; AMS 5773. S45000

PRP-455

Precision Rolled Products, Inc.

Stainless steel. C 0.03, Cr 11.75, Ni 8.5, Cu 2.25, bal Fe.
Precipitation hardenable stainless steel for fasteners and valves; corrosion and oxidation resistant. Annealed: 140,000 TS; 115,000 YS; 14 El; 60 RA; 294 Brin. Heat treated: 190,000 TS; 175,000 YS; 15 El; 55 RA; 371 Brin. Meets AMS 5617. S45500

PRP-5616

Precision Rolled Products, Inc.

C 0.15, Cr 13, Ni 2, W 3, bal Fe.
Martensitic chromium-nickel-tungsten alloy for fasteners, jet engine compressor parts, steam turbine buckets and blades. Hardened: 95,000-160,000 TS; 84,000-135,000 YS; 16-17 El; 45-64 RA; 327 Brin. Meets AMS 5616. S41800

PRP-600

Precision Rolled Products, Inc.

Nickel. C 0.04, Ni 74, Cr 15.5, Fe 8.
Alloy for jet engine rings and parts, furnace heat treat fixtures, nuclear reactor parts. Annealed: 80,000 TS; 30,000 YS; 50 El; 120 Brin. Cold drawn: 105,000 TS; 80,000 YS; 30 El; 180 Brin. Meets AMS 5665; MIL-N-6710. N06600

PRP-625

Precision Rolled Products, Inc.

Nickel. C 0.05, Ni 32, Cr 21.5, Mo 9, 3.65 Cb + Ta.
Alloy for jet engine rings and parts, furnace heat treat fixtures, nuclear reactor parts. Annealed: 80,000 TS; 30,000 YS; 50 El; 120 Brin. Cold drawn: 105,000 TS; 80,000 YS; 30 El; 180 Brin. Meets AMS 5665; MIL-N-6710. N06625

PRP-718

Precision Rolled Products, Inc.

Nickel. C 0.04, Ni 52.5, Cr 18.6, Fe 18.5, Mo 3, Ti 0.9, Al 0.5, 5.1 Cb + Ta.
Alloy for fasteners, hot shear knives, jet engine rings, shafts, buckets and compressor blades. Solution treated and aged: 140,000-180,000 TS; 125,000-150,000 YS; 6 El; 8 RA; 331 Brin. Meets AMS 5662; AMS 5663; AMS 5664. N07718

PRP-722

Precision Rolled Products, Inc.

Nickel. C 0.04, Ni 73, Cr 15.5, Fe 7, Ti 2.5, Al 0.7.
Alloy for jet engine rings and parts. Solution treated and aged: 59,000-158,000 TS; 46,000-82,000 YS; 24-26 El. Meets AMS 5714. N07722

PRP-73-3.5CR

Precision Rolled Products, Inc.

C 1, Cr 3.5, bal Fe.
Permanent magnet alloy for magnetic and electrical equipment.

PRP-750

Precision Rolled Products, Inc.

Nickel. C 0.04, Ni 72, Cr 15.5, Fe 7, Ti 2.5, Al 0.7, 0.95 Cb + Ta.
Alloy for nuclear reactor parts, fasteners, jet engine rings and parts, and steam turbine blades. Solution treated and aged: 120,000-162,000 TS; 82,000-92,000 YS; 9-24 El. Meets AMS 5667; AMS 5668; AMS 5620; AMS 5670; AMS 5671; AMS 5747. N07750

PRP-800

Precision Rolled Products, Inc.

C 0.05, Ni 32.5, Cr 21, Fe 45.
Iron-nickel-chromium alloy for furnace and heat exchanger parts. Annealed: 85,000 TS; 36,000 YS; 48 El; 138 Brin. Cold drawn: 112,000 TS; 100,000 YS; 17 El; 64 RA. Meets AMS 5871; AMS 5766. N08800

PRP-825

Precision Rolled Products, Inc.

Nickel. C 0.03, Ni 42, Cr 21.5, Mo 3, Cu 2, bal Fe.
Alloy for pickle tank heaters and hooks, fixtures and parts exposed to phosphoric acid. Cold drawn: 85,000 TS; 35,000 YS; 30 El. Annealed: 67,500 TS; 30,900 YS; 62 El; 68 RA. N08825

PRP-903

Precision Rolled Products, Inc.

Nickel. C 0.01, Ni 38, Co 15, Cb 3, Ti 1.4, Al 0.7, bal Fe.
Alloy for steam turbine blades, rocket thrust chambers, jet engine seals. Solution treated and aged: 180,000 TS; 150,000 YS; 10 El. Meets AMS 5806. N19903

PRP-A286

Precision Rolled Products, Inc.

C 0.05, Cr 15, Ni 26, Mo 1.2, Ti 2.1, Al 0.25, bal Fe.
Precipitation hardenable iron-nickel alloy for fasteners, jet engine rings, wheels, compressor blades and casings. Solution treated and aged: 135,000 TS; 85,000 YS; 30 El; 55 RA. Meets AMS 5525; AMS 5731; AMS 5732; AMS 5734; AMS 5735; AMS 5737. K66286

PRP-B1-18CO

Precision Rolled Products, Inc.

C 0.75, Co 18.5, W 4.8, Cr 3.8, bal Fe.
Permanent magnet alloy for magnetic and electrical equipment.

PRP-B3-3CO

Precision Rolled Products, Inc.

C 1.1, Co 3.2, Cr 4, bal Fe.
Permanent magnet alloy for magnetic and electrical equipment.

PRP-C276

Precision Rolled Products, Inc.

Nickel. C 0.02, Cr 15.5, Mo 16, Fe 5.5, W 3.8, bal Ni.
Alloy for jet engine rings and parts, diesel engine combustion parts, fixtures exposed to nitric acid and organic salts. Solution treated: 114,000 TS; 53,000 YS; 59 El; 172 Brin. Meets ASTM B-574. N10276

PRP-CP-TI-1

Precision Rolled Products, Inc.

Titanium. C 0-0.1, Fe 0-0.2, N 0-0.03, O 0-0.18, H 0-0.0125.
Commercially pure titanium for airframes, chemical, desalination and marine parts. Annealed: 22,000-48,000 TS; 14,000-35,000 YS; 30-32 El; 55-80 RA. Meets MIL-T-9047. R50250

PRP-CP-TI-2

Precision Rolled Products, Inc.

Titanium. C 0-0.1, Fe 0-0.3, N 0-0.03, O 0-0.25, H 0-0.0125.
Commercially pure titanium for jet engine rings and parts, airframes. Annealed: 28,000-63,000 TS; 17,000-50,000 YS; 28-35 El; 50-75 RA. Meets MIL-T-9047. R50400

PRP-CP-TI-4

Precision Rolled Products, Inc.

Titanium. C 0-0.1, Fe 0-0.5, N 0-0.05, O 0-0.4, H 0-0.0125.
Commercially pure titanium for jet engine rings and parts, high speed fans. Annealed: 45,000-96,000 TS; 25,000-85,000 YS; 20-25 El; 40-70 RA. Meets MIL-T-9047; AMS 4921. R50700

PRP-F75

Precision Rolled Products, Inc.

C 0.08, Cr 27, Mo 5.8, bal Co.
Alloy for surgical implants. Solution treated: 170,000 TS; 90,000 YS; 30 El; 280 Brin.

PRP-GSA

Precision Rolled Products, Inc.

Fe 53, Ni 29, Co 17.
Controlled expansion alloy for glass to metal seals; power and microwave tubes. Annealed: 85,000 TS; 55,000 YS; 25 El; 185 Brin. Meets AMS 7727; MIL-I-23011. K94610

PRP-HN

Precision Rolled Products, Inc.
Nickel. C 0.06, Ni 74, Cr 7, Mo 16.5.
Alloy for jet engine rings and parts, fixtures exposed to hot molten fluoride salts. Annealed: 86,400 TS; 37,300 YS; 17 El. Meets AMS 5771. N10003

PRP-HW

Precision Rolled Products, Inc.
Nickel. C 0.06, Ni 64, Cr 5, Mo 24.5, Fe 5.5.
Alloy for weld wire for high temperature components. Solution treated: 158,000 TS; 82,000 YS; 26 El. Meets AMS 5758. N10004

PRP-HX

Precision Rolled Products, Inc.
Nickel. C 0.1, Cr 22, Fe 18.5, Co 1.5, Mo 9, W 0.6, bal Ni.
Alloy for fasteners, combustion liners, jet engine afterburner casings, rings and seals. Solution treated: 83,000-113,000 TS; 40,700-55,800 YS; 37-44 El. Meets AMS 5754. N06002

PRP-K500

Precision Rolled Products, Inc.
Nickel. Cu 28, Ni 66.5, Al 3, Ti 0.62.
Alloy for corrosion resistant impellers, pump shafts, springs and valve trim. Annealed: 90,000 TS; 40,000 YS; 45 El; 140 Brin. Aged: 130,000 TS; 85,000 YS; 35 El; 250 Brin. Meets AMS 4676; QQ-N-286. N05500

PRP-L605

Precision Rolled Products, Inc.
C 0.1, Co 52, Cr 20, Ni 10, W 15.
Work strengthened columbium-nickel-chromium-tungsten alloy for valves, springs, jet engine rings and parts. Annealed: 155,000 TS; 70,000 YS; 50 El; 130 Brin. Meets AMS 5759. R30605

PRP-MP35N

Precision Rolled Products, Inc.
C 0-0.025, Cr 20, Ni 35, Co 35.
Alloy for fasteners, springs, marine and medical applications requiring stress corrosion resistance. Annealed: 140,000 TS; 61,000 YS; 70 El; 179 Brin. Work strengthened: 286,000 TS; 225,000 YS; 11 El; 49 RA; 481 Brin. Meets AMS 5758; AMS 5844; AMS 5845. R30035

PRP-N155

Precision Rolled Products, Inc.
C 0.12, Cr 21, Ni 20, Co 20, Mo 3, W 2.5, N 0.15, 1.0 Cb + Ta, bal Fe.
Iron-columbium-nickel-chromium alloy for fasteners, exhaust manifolds, tailpipes and combustion chambers. Solution treated: 80,000-119,000 TS; 43,000-58,000 YS; 32-40 El; 43-48 RA. Meets AMS 5768; AMS 5769. R30155

PRP-RENE 41

Precision Rolled Products, Inc.
Nickel. C 0.09, Co 11, Cr 19, Mo 9.8, Ti 3.2, Al 1.6, bal Ni.
Alloy for fasteners, jet engine wheels, rings, buckets, compressor blades and vanes. Solution treated and aged: 16,000 TS; 120,000 YS; 14 El. Meets AMS 5712; AMS 5713. N07041

PRP-TI-5AL-2.5SN

Precision Rolled Products, Inc.
Titanium. C 0-0.08, Al 5, O 0-0.2, Fe 0.5, Sn 2.5, H 0-0.02.
Weldable alloy for jet engine rings and parts, steam turbine blades. Annealed: 82,000-125,000 TS; 65,000-117,000 YS; 16-18 El; 40-45 RA. Meets MIL-T-9047; AMS 4926; AMS 4966. R54520

PRP-TI-6AL-2SN-4ZR-2MO

Precision Rolled Products, Inc.
Titanium. C 0-0.05, Al 6, Mo 2, Zr 4, Sn 2, H 0-0.0125.
Heat treatable alloy for jet engine rings and parts requiring high strength. Solution treated and aged: 112,000-142,000 TS; 85,000-130,000 YS; 15-16 El; 35-42 RA. Meets MIL-T-9047; AMS 4975; AMS 4976. R54620

PRP-TI-6AL-2SN-4ZR-6MO

Precision Rolled Products, Inc.
Titanium. C 0-0.05, Al 6, Mo 6, Zr 4, Sn 2, H 0-0.0125.
Heat treatable alloy for jet engine rings and parts requiring high strength. Solution treated and aged: 148,000-184,000 TS; 122,000-170,000 YS; 10-18 El; 23-55 RA. Meets MIL-T-9047; AMS 4981. R56260

PRP-TI-6AL-4V

Precision Rolled Products, Inc.
Titanium. C 0-0.08, Al 6, V 4, H 0-0.0125.
Heat treatable alloy for fasteners, jet engine rings, compressor blades and discs, structural forgings and steam turbine blades. Annealed: 100,000-138,000 TS; 90,000-128,000 YS; 14 El; 30-35 RA. Solution treated and aged: 130,000-170,000 TS; 110,000-155,000 YS; 10 El; 25-28 RA. Meets MIL-T-9047; AMS 4928M; AMS 4965; AMS 4967. R56400

PRP-TI-6AL-4V (ELI)

Precision Rolled Products, Inc.
Titanium. C 0-0.08, O 0-0.13, Al 6, V 4, H 0-0.0125.
Heat treatable alloy for high pressure cryogenic vessels operating down to -320°F. Annealed: 130,000-220,000 TS; 120,000-205,000 YS; 14-15 El; 35 RA. Meets MIL-T-9047; AMS 4930; AMS 4931. R56401

PRP-TI-6AL-6V-2SN

Precision Rolled Products, Inc.
Titanium. C 0-0.05, Al 6, V 6, Sn 2, H 0-0.0125.
Heat treatable alloy for structural aircraft parts, rocket motor cases, jet engine rings and parts. Annealed: 135,000-155,000 TS; 117,000-145,000 YS; 14-18 El; 30-42 RA. Solution treated and aged: 142,000-185,000 TS; 130,000-170,000 YS; 10-12 El; 20-28 RA. Meets MIL-T-9047; AMS 4971; AMS 4978; AMS 4979. R56620

PRP-TI-8AL-1MO-1V

Precision Rolled Products, Inc.
Titanium. C 0-0.08, Al 8, Mo 1, Fe 0-0.3, V 1, H 0-0.015.
For jet engine rings and parts requiring high tensile and creep rupture strength. Duplex annealed: 115,000-145,000 TS; 90,000-138,000 YS; 15-20 El; 28-38 RA. Meets MIL-T-9047; AMS 4972; AMS 4973. R54810

PRP-V57

Precision Rolled Products, Inc.
C 0.05, Cr 14.8, Ni 27, Mo 1.2, Ti 3, Al 0.25, bal Fe.
Precipitation hardenable iron-nickel alloy for extrusion tooling. Solution treated and aged: 120,000-175,000 TS; 125,000 YS; 21 El; 35 RA.

PRP-WASPALOY

Precision Rolled Products, Inc.
Nickel. C 0.07, Co 13.5, Cr 19.5, Mo 4.3, Ti 3, Al 1.4, bal Ni.
Alloy for fasteners, jet engine wheels, rings, buckets, compressor blades and vanes. Solution treated and aged: 117,000-188,000 TS; 99,000-115,000 YS; 28 El; 25-41 RA; 375 Brin. Meets AMS 5706; AMS 5707; AMS 5708; AMS 5709. N07001

PRV 13

Vallourec S.A.
C 0.11, Cr 13, Mn 0.7, Si 2, bal Fe.
Resistant to oxidation reactions in sulfurous media up to 850°C. For heat exchangers, pyrometric sheaths.

PRV 18

Vallourec S.A.
C 0.1, Cr 18.5, Mn 0.6, Si 1.85, bal Fe.
Resistant to sulfurous media to 1000°C. For heat exchangers, heating cells.

PRV 18-10

Vallourec S.A.
C 0.07, Cr 18, Ni 11, Ti 0.5, bal Fe.
Resistant to creep rupture and oxidation to 850°C. For heat exchangers, reactors, heating cells. Similar to AISI 321.

PRV 25

Vallourec S.A.
C 0.1, Cr 24.5, Mn 1.2, Si 2, bal Fe.
Resistant to sulfurous media to 1100°C. For heat exchangers, pyrometric sheaths.

PRV 25-12

Vallourec S.A.
C 0.15, Cr 23, Ni 14, bal Fe.
Resistant to creep rupture and oxidation up to 1050°C. For heat exchangers, reactors. Similar to AISI 309.

PRV 25-20

Vallourec S.A.
C 0.12, Cr 25, Ni 20, bal Fe.
For high temperature operation to 1150°C. Similar to AISI 310.

PRV 29

Vallourec S.A.
C 0.11, Cr 29, Si 1, bal Fe.
Resistant to sulfurous media to 1150°C. For heat exchangers, pyrometric sheaths.

PRV 33-20

Vallourec S.A.
C 0.1, Cr 21, Ni 32.5, bal Fe.
For high temperature operation to 1150°C.

PRV 75-15

Vallourec S.A.
C 0.06, Cr 15.5, Ni 72, bal Fe.
Oxidation resistant to 1200°C. For furnace equipment, pyrometer sheaths, heat exchangers.

PRYM 210

William Prym-Werke GmbH & Co. KG
Copper. bal Cu.
SE-Cu (oxygen free high conductivity copper). DIN 2.0070; ASTM C 10200; BS C 103.

PRYM 211

William Prym-Werke GmbH & Co. KG
Copper. bal Cu.
E-Cu (ETP copper). DIN 2.0060; DIN 2.0065; ASTM C 1100; BS C 101.

PRYM 212

William Prym-Werke GmbH & Co. KG
Copper. Fe 2, P, bal Cu.
Low alloyed copper. DIN 2.1310; ASTM C 19400.

PRYM 214

William Prym-Werke GmbH & Co. KG
Copper. bal Cu.
Low alloyed copper. DIN 2.1489.

PRYM 216

William Prym-Werke GmbH & Co. KG
Copper.
Copper-tin alloy.

PRYM 217

William Prym-Werke GmbH & Co. KG
Copper. Ag 0.1, bal Cu.
Low alloyed copper. DIN 2.1203.

PRYM 218

William Prym-Werke GmbH & Co. KG
Copper.
Copper-zinc alloy. DIN 2.0205.

PRYM 219

William Prym-Werke GmbH & Co. KG
Copper.
SF-Cu (phosphorous deoxidized) DIN 2.0090; ASTM C 12200; BS C 106.

PRYM 220
William Prym-Werke GmbH & Co. KG
Copper.
SW-Cu. DIN 2.0076.

PRYM 221
William Prym-Werke GmbH & Co. KG
Cu 95, Zn 5.
Lead free copper-zinc alloy. DIN 2.0220; ASTM C 21000; BS CZ 125.

PRYM 222
William Prym-Werke GmbH & Co. KG
Cu 90, Zn 10.
Lead free copper-zinc alloy. DIN 2.0230; ASTM C 22000; BS CZ 101.

PRYM 223
William Prym-Werke GmbH & Co. KG
Cu 85, Zn 15.
Lead free copper-zinc alloy. DIN 2.0240; ASTM C 23000; BS CZ 102.

PRYM 224
William Prym-Werke GmbH & Co. KG
Cu 80, Zn 20.
Lead free copper-zinc alloy. DIN 2.0250; ASTM C 24000; BS CZ 103.

PRYM 225
William Prym-Werke GmbH & Co. KG
Cu 72, Zn 28.
Lead free copper-zinc alloy. DIN 2.0261.

PRYM 226
William Prym-Werke GmbH & Co. KG
Cu 70, Zn 30.
Lead free copper-zinc alloy. DIN 2.0265; ASTM C 26000; BS CZ 106.

PRYM 227
William Prym-Werke GmbH & Co. KG
Cu 67, Zn 33.
Lead free copper-zinc alloy. DIN 2.0280; ASTM C 26800; BS CZ 107.

PRYM 228
William Prym-Werke GmbH & Co. KG
Cu 75, Zn 25.
Lead free copper-zinc alloy. ASTM C 25000.

PRYM 230
William Prym-Werke GmbH & Co. KG
Zn 31, Si, bal Cu.
Copper-zinc alloy. DIN 2.0490.

PRYM 231
William Prym-Werke GmbH & Co. KG
Copper. bal Cu.
Welding and brazing alloy. DIN 2.0366.

PRYM 232
William Prym-Werke GmbH & Co. KG
Copper. Zn 40, bal Cu.
Welding and brazing alloy. DIN 2.0367.

PRYM 233
William Prym-Werke GmbH & Co. KG
Zn 39, Mn, Si, Sn, bal Cu.
Copper-zinc alloy.

PRYM 233
William Prym-Werke GmbH & Co. KG
Copper. bal Cu.
Welding and brazing alloy. DIN 2.0531.

PRYM 234
William Prym-Werke GmbH & Co. KG
Copper. bal Cu.
Welding and brazing alloy.

PRYM 235
William Prym-Werke GmbH & Co. KG
Zn 37, Si, bal Cu.
Copper-zinc alloy.

PRYM 236
William Prym-Werke GmbH & Co. KG
Copper. Ni 10, Zn 42, bal Cu.
Welding and brazing alloy. DIN 2.0711.

PRYM 237
William Prym-Werke GmbH & Co. KG
Copper. Zn 39, bal Cu.
Welding and brazing alloy. DIN 2.0533.

PRYM 242
William Prym-Werke GmbH & Co. KG
Copper. Cu 63, Zn 37.
Copper-zinc alloy.

PRYM 243
William Prym-Werke GmbH & Co. KG
Cu 63, Zn 37.
Lead free copper-zinc alloy. DIN 2.0321; ASTM C 27200; BS CZ 108.

PRYM 244
William Prym-Werke GmbH & Co. KG
Copper. Cu 64, Zn 36.
Copper-zinc alloy.

PRYM 245
William Prym-Werke GmbH & Co. KG
Copper. Cu 63, Zn 37.
Copper-zinc alloy.

PRYM 246
William Prym-Werke GmbH & Co. KG
Copper. Cu 64, Zn 36.
Copper-zinc alloy.

PRYM 247
William Prym-Werke GmbH & Co. KG
Copper. Cu 65, Zn 35.
Copper-zinc alloy.

PRYM 248
William Prym-Werke GmbH & Co. KG
Cu 64, Zn 36.
Lead free copper-zinc alloy. DIN 2.0335; ASTM C 27000; BS CZ 108.

PRYM 250
William Prym-Werke GmbH & Co. KG
Copper. Zn 39, Pb 2, Cu 58.
Leaded copper-zinc alloy. DIN 2.0380; ASTM C 37700; BS CZ 122.

PRYM 251
William Prym-Werke GmbH & Co. KG
Copper. Zn 40, Pb 2, Cu 58.
Leaded copper-zinc alloy. DIN 2.0402.

PRYM 252
William Prym-Werke GmbH & Co. KG
Copper. Zn 39, Pb 3, Cu 58.
Leaded copper-zinc alloy. DIN 2.0401; ASTM C 38500; BS CZ 121.

PRYM 253
William Prym-Werke GmbH & Co. KG
Copper. Zn 39, Pb 3, Cu 58.
Leaded copper-zinc alloy. DIN Prym alloy.

PRYM 254
William Prym-Werke GmbH & Co. KG
Copper. Zn 36, Pb 3, Cu 60.
Leaded copper-zinc alloy. ASTM C 36000.

PRYM 255
William Prym-Werke GmbH & Co. KG
Copper. Zn 38, Pb 4, bal Cu.
Copper-zinc alloy.

PRYM 256
William Prym-Werke GmbH & Co. KG
Cu 58, Zn 40, Pb 2.
Copper-zinc alloy. DIN 2.0402; BS CZ 122.

PRYM 259
William Prym-Werke GmbH & Co. KG
Copper. Zn 38, Pb 1.5, Cu 60.
Leaded copper-zinc alloy. DIN 2.0371; ASTM C 37700; BS CZ 128.

PRYM 260
William Prym-Werke GmbH & Co. KG
Cu 60, Zn 40.
Lead free copper-zinc alloy. DIN 2.0360; ASTM C 28000; BS CZ 109.

PRYM 261
William Prym-Werke GmbH & Co. KG
Copper. Zn 36, Pb 3, bal Cu.
Leaded copper-zinc alloy. DIN 2.0375; ASTM C 36000; BS CZ 124.

PRYM 262
William Prym-Werke GmbH & Co. KG
Copper. Cu 63, Zn 36, Pb 1.5.
Leaded copper-zinc alloy. DIN 2.0331; ASTM C 35300; BS CZ 119.

PRYM 263
William Prym-Werke GmbH & Co. KG
Copper. Zn 36, Pb 1.7, bal Cu.
Copper-zinc alloy.

PRYM 265
William Prym-Werke GmbH & Co. KG
Cu 60.5, Zn 39, Pb 0.5.
Copper-zinc alloy. DIN 2.0372; ASTM C 36500; BS CZ 123.

PRYM 280
William Prym-Werke GmbH & Co. KG
Copper. Ni 12, Zn 24, bal Cu.
Copper-nickel-zinc alloy. DIN 2.0730; ASTM C 75700; BS (Ns 104).

PRYM 281
William Prym-Werke GmbH & Co. KG
Copper. Ni 18, Zn 20, bal Cu.
Copper-nickel-zinc alloy. DIN 2.0740; ASTM C 76400; BS (Ns 106).

PRYM 282
William Prym-Werke GmbH & Co. KG
Copper. Sn 6, bal Cu.
Copper-tin alloy. DIN 2.1020; ASTM C 51900; BS PB 103.

PRYM 283
William Prym-Werke GmbH & Co. KG
Copper. Sn 8, bal Cu.
Copper-tin alloy. DIN 2.1030; ASTM C 52100; BS PB 104.

PRYM 284
William Prym-Werke GmbH & Co. KG
Copper. Sn 4, bal Cu.
Copper-tin alloy. DIN 2.1016; ASTM C 51100; BS PB 101.

PRYM 286
William Prym-Werke GmbH & Co. KG
Copper. Ni 9, Sn 2, bal Cu.
Copper-nickel alloy. DIN 2.0875.

PSI NO. 1

Peterson Steels Inc.
C 0.92-1.02, Mn 0.95-1.2, Si 0.5-0.7, Cr 0.9-1.15, bal Fe.
For master tools, gauges, ceramic molds; non-deforming, wear resistant.

PSI NO. 2

Peterson Steels Inc.
C 0.87-0.97, Mn 1.4-1.7, Si 0.6-0.8, Cr 1.4-1.7, bal Fe.
For master tools and gauges, clutch liners, pump sleeves; non-deforming, wear resistant.

PSP

Thyssen Edelstahlwerke AG
C 0.5, Si 0.25, Mn 0.5, Cr 1, Ni 3.25, bal Fe.
For coining dies, gears, shafts; oil hardened, tough.
Obsolete

PSU

Thyssen Edelstahlwerke AG
C 0.08, Mn 0.4, Si 0.05, bal Fe.
Welded: 68,000 TS; 45,000 YS; 30 El; 60 RA. For welding rod.
Obsolete

PTM-45

J.F. Jelenko & Co.
Precious metal. Au 45, Pd 40, Ag 5.
115,000 psi TS; 83,000 psi YS; 20 El; 200 Brin. Dental alloy for fusing porcelain to metal.

PTM-88

J.F. Jelenko & Co.
Precious metal. Pd 88.
115,000 psi TS; 83,000 psi YS; 25 El; 210 Brin. Dental alloy for fusing porcelain to metal.

PULSUS

Saarstahl AG
C 0.35-0.42, Cr 4.7-5, Ni 1.5, Mo 0.5, V 0.2, bal Fe.
For tools, dies, and hot work steel. *Obsolete*

PUMA

Hufnagel GmbH
C 0.65, Cr 4.25, Mo 0.8, W 18, V 1.6, Co 15, bal Fe.
Co-W high speed steel; for rough machining, especially on austenitic steels.

PUMP COCKS

Manufacturer not listed
Zn 72, Cu 7, Sn 21.
For pump cocks; castings.

PURALLOY

Lobdell Co.
C 3, Si, Mn, bal Fe.
For rolls for paper and allied industries; cast iron.

PURCO CHISEL

A.R. Purdy Co. Inc.
C 0.7, Si 2, Mn 0.9, bal Fe.
For chisels, punches; tough.

PURE ORE A.D., 95

Manufacturer not listed
C, Cr, W, Mo, Si, bal Fe.
For jewelers dies, coining dies. *Obsolete*

PURE ORE AIRCHROME

Now AIR CHROME.

PURE ORE CLIPPER

Now CLIPPER.

PURE ORE COBALT

Manufacturer not listed
C 0.76, Cr 4.2, W 14, Mo 0.5, V 2.25, Co 5, bal Fe.
For special tools; high speed steel. *Obsolete*

PURE ORE COLD HEADING

Manufacturer not listed
C 0.7, V 2, bal Fe.
For header dies. *Obsolete*

PURE ORE D-52

Manufacturer not listed
bal Fe.
For extrusion dies. *Obsolete*

PURE ORE D-C-33

Manufacturer not listed
C 0.33, Cr 5, W 1.4, Mo 1.6, Si 0.9, Mn 0.3, bal Fe.
For hammer dies, press tools, punches, hot shears; hot work steel, oil hardened. *Obsolete*

PURE ORE DIE CASTING

Manufacturer not listed
C, Cr, W, Mo, Si, bal Fe.
For die casting molds. *Obsolete*

PURE ORE E-N-97

Manufacturer not listed
C 0.58, Cr 0.9, Mo 4, Ni 3.2, bal Fe.
For cold forging, dies. *Obsolete*

PURE ORE EXTRA

Manufacturer not listed.
C as specified, bal Fe.
For tools, dies; water hardening; AISI W1.

PURE ORE FINISHING

Manufacturer not listed
C 1.2, W 0.5, bal Fe.
For cutting tools; wear resistant. *Obsolete*

PURE ORE HI-RUN

Now HI-RUN.

PURE ORE HIGH SPEED

Manufacturer not listed
C 0.7, W 18, Cr 4, V 1, bal Fe.
For special tools, high speed cutters; high speed steel.
Obsolete

PURE ORE NO. 10

Manufacturer not listed
C, Cr, V, Mn, bal Fe.
For tools, dies. *Obsolete*

PURE ORE PRIOR

Manufacturer not listed
C 0.8, Cr 4.5, W 18-19, Mo 0.6, V 2, bal Fe.
For cutting tools, dies; high speed steel. *Obsolete*

PURE ORE PRIOR EXTRA

Manufacturer not listed
C 0.7, W 18, Cr 4, V 1, bal Fe.
For high speed cutters; high speed steel. *Obsolete*

PURE ORE QUARRY TOOL STEEL

Manufacturer not listed
C 0-1, bal Fe.
For quarry tools, planing tools, plug drills, hammers; tough.
Obsolete

PURE ORE ROLMO

Manufacturer not listed
C 1, Cr 1.2, Mo 0.4, bal Fe.
For tools, dies, rolls, hammer; wear resisting. *Obsolete*

PURE ORE SOLID DRILL

Manufacturer not listed
C 1.2, bal Fe.
For rock drills; water hardened. *Obsolete*

PURE ORE SPECIAL

Manufacturer not listed.
C as specified, bal Fe.
Water hardening tool steel; for dies and special tools; AISI W1.

PURE ORE STANDARD

Manufacturer not listed.
C as specified; bal Fe.
Water hardening tool steel; for tools, dies, AISI W2.

PURE ORE SUPERALLOY

Now SUPERALLOY.

PURE ORE V-995

Manufacturer not listed
bal Fe.
For tools, dies. *Obsolete*

PURE ORE Z-457 DIE CAST STEEL

Manufacturer not listed
bal Fe.
For tools, dies. *Obsolete*

PURE SILVER SOLDER

Manufacturer not listed
Ag 72, Cu 28.
For silver solder; corrosion resistant.

PURE-ORE KLS-44

Manufacturer not listed
C 0.57, Mn 0.5, Cr 1, Mo 0.9, Ni 1.4, bal Fe.
For hot work dies; hot work steel. *Obsolete*

PURE-ORE PRODUCTION HI-CR, HI-C

Manufacturer not listed
C 1.5, Cr 13, bal Fe.
For tools, dies. *Obsolete*

PURETUNG

GTE Products Corp.
W 99.9.
For welding electrodes.

PURO 17/7

Remystahl
Stainless steel. C 0-0.12, Si 0-1, Mn 0-2, Cr 16-18, Ni 6-9, bal Fe.
Austenitic, stainless steel; work hardens easily; for springs. W.-Nr. 1.4310; similar to AISI 301.

PURO 18/10 L SUPRA

Remystahl
Stainless steel. C 0-0.03, Si 0-1, Mn 0-2, Cr 16.5-18.5, Ni 11-14, Mo 2-2.5, bal Fe.
Austenitic, stainless; for equipment in chemical and cellulose industries. W.Nr. 1.4404; AISI 316L.

PURO 18/10 SUPRA

Remystahl
Stainless steel. C 0-0.07, Si 0-1, Mn 0-2, Cr 16.5-18.5, Ni 10.5-13.5, Mo 2-2.5, bal Fe.
Austenitic, stainless; for equipment in chemical and textile industries. W.-Nr. 1.4401; AISI 316.

PURO 18/12 LM SUPRA

Remystahl
Stainless steel. C 0-0.03, Si 0-1, Mn 0-2, Cr 16.5-18.5, Ni 12.5-15, Mo 2.5-3, bal Fe.
Austenitic, stainless steel; improved corrosion resistance for chemical equipment. W.-Nr. 1.4435; similar to low carbon. AISI 317.

PURO 18/14 M4

Remystahl
Stainless steel. C 0-0.07, Si 0-1, Mn 0-2, Cr 16-18, Ni 12.5-14.5, Mo 4-5, bal Fe.
Austenitic, stainless steel; extra resistance to corrosion and pitting for chemical equipment. W.-Nr. 1.4449; similar to AISI 317.

PURO 18/18 M NB

Remystahl
Stainless steel. C 0-0.07, Cr 17.5, Ni 20, Cu 2, Mo 2.2, Cb = 8 x C min, bal Fe.
Stabilized, austenitic, stainless steel. W.-Nr. 1.4505.

PURO 18/8

Remystahl
Stainless steel. C 0-0.15, Cr 18, Ni 8.5, bal Fe.
Annealed: 80,000 TS; 35,000 YS; 55 El; 75 RA; 150 Brin. For chemical plant equipment, tanks; Type 302; stainless, austenitic. *Obsolete*

PURO 18/8 E

Remystahl
Stainless steel. C 0-0.08, Si 0-1, Mn 0-2, Cr 17-19, Ni 9-12, Ti = 5 x C min (0.80 max), bal Fe.
Stabilized, austenitic, stainless steel; weldable. W.-Nr. 1.4541; similar to AISI 321.

PURO 18/8 L SUPRA

Remystahl
Stainless steel. C 0-0.03, Si 0-1, Mn 0-2, Cr 18-20, Ni 10-12.5, bal Fe.
Austenitic, stainless; for food industries. W.-Nr. 1.4306; AISI 304L.

PURO 18/8 NB

Remystahl
Stainless steel. C 0-0.08, Si 0-1, Mn 0-2, Cr 17-19, Nb 9-12, Nb = 10 x C min (1.0 max), bal Fe.
Stabilized, austenitic, stainless steel; weldable. W.-Nr. 1.4550; similar to AISI 347.

PURO 18/8 NB SG

Remystahl
Stainless steel. C 0-0.07, Cr 17-20, Ni 9-11.5, Mo 0-0.2, Nb = 10 x C min, bal Fe.
Stabilized, austenitic, stainless steel. W.-Nr. 1.4543; AISI 347.

PURO 18/8 S

Mannesmannrohren-Werke AG
C 0-0.15, Si 0-1, Mn 0-2, Cr 17-19, Ni 8-10, S 0.15-0.35, bal Fe.
Free machining type austenitic stainless. W.-Nr. 1.4305; AISI 303. *Obsolete*

PURO 18/8 SUPRA

Remystahl
Stainless steel. C 0-0.07, Si 0-1, Mn 0-2, Cr 17-19, Ni 8.5-10.5, bal Fe.
Austenitic, stainless steel. W.-Nr. 1.4301; AISI 304.

PURO 19/11

Remystahl
Stainless steel. C 0-0.07, Si 0-1, Mn 0-2, Cr 17-19, Ni 11-13, bal Fe.
Austenitic, stainless steel; for chemical industry. W.-Nr. 1.4303; similar to AISI 305.

PURO 25/25 M

Remystahl
Stainless steel. C 0-0.07, Cr 25, Ni 25, Mo 2.2, Ti = 10 x C min (0.60 max), bal Fe.
Austenitic, stainless steel; for chemical equipment. W.-Nr.1.4577.

PURO 275 N

Remystahl
Stainless steel. C 0.1, Cr 26-28, Ni 4-5, Mo 1.3-2, bal Fe.
Stainless steel; for high temperature operations. W.-Nr. 1.4460.

PURO ANTIOXYDUR

Remystahl
Stainless steel. C 0.15-0.23, Si 0-1, Mn 0-1, Cr 16-18, Ni 1.5-2.5, bal Fe.
Martensitic, stainless steel. W.-Nr. 1.4057; similar to AISI 431.

PURO CR

Remystahl
Stainless steel. C 0-0.08, Si 0-1, Mn 0-1, Cr 15.5-17.5, bal Fe.
Ferritic, stainless steel. W.-Nr. 1.4016; AISI 430.

PURO CRTI

Remystahl
Stainless steel. C 0-0.07, Cr 16-18, Ti = 7 x C min, bal Fe.
Stabilized, stainless steel. Not hardenable by heat treating; weldable. W.Nr. 1.4510.

PURO H

Remystahl
Stainless steel. C 0.42-0.5, Cr 12.5-14.5, bal Fe.
Annealed: 100,000 TS; 55,000 YS; 25 El; 55 RA; 200 Brin. For valve trim, gages, valves, turbine blades; Type 420; corrosion resistant. W.-Nr. 1.4034.

PURO H EXTRA

Remystahl
Stainless steel. C 0.9, Cr 17-19, Mo 1-1.3, V 0.07-0.12, bal Fe.
Martensitic, stainless steel. W.Nr. 1.4112; similar to AISI 440B.

PURO H MO EXTRA

Remystahl
Stainless steel. C 0.95-1.2, Cr 16-18, Mo 0.6, bal Fe.
Martensitic, stainless steel. For stainless ball bearings. W-Nr. 1.4125; AISI 440C.

PURO H MO2

Remystahl
Stainless steel. C 0.33-0.45, Cr 16.5, Mo 1.15, bal Fe.
Martensitic, stainless steel. W.-Nr. 1.4122.

PURO M

Remystahl
Stainless steel. C 0.17-0.25, Cr 12-14, bal Fe.
Annealed: 95,000 TS; 50,000 YS; 28 El; 58 RA; 195 Brin. For valve trim, gages, cutlery, turbine blades; Type 420; stainless, hardenable. W.-Nr. 1.4021.

PURO M 2

Remystahl
Stainless steel. C 0.12-0.17, Si 0-1, Mn 0-1, Cr 12-14, bal Fe.
Martensitic, stainless steel. W.-Nr. 1.4024.

PURO W

Remystahl
Stainless steel. C 0.1, Cr 13, bal Fe.
Annealed: 75,000 TS; 40,000 YS; 35 El; 70 RA; 155 Brin. Cold drawn: 100,000 TS; 85,000 YS; 17 El; 60 RA; 200 Brin. For valve trim, cutlery, surgical instruments; Type 410; stainless.

PURO W2

Remystahl
Stainless steel. C 0-0.08, Mn 0-1, Si 0-1, Cr 12-14, bal Fe.
Chromium, stainless steel. W.-Nr. 1.4000.

PURO W2 AL

Remystahl
Stainless steel. C 0-0.08, Si 0-1, Mn 0-1, Cr 12-14, Al 0.1-0.3, bal Fe.
Ferritic, stainless steel. AISI 405; W.-Nr. 1.4002.

PURO WA

Remystahl
Stainless steel. C 0-0.15, Si 0-1, Mn 0-1, Cr 12-13, S 0.15-0.25, bal Fe.
Martensitic, stainless steel. Free machining. W.-Nr. 1.4005; AISI 416.

PURO WS

Remystahl
Stainless steel. C 0.1-0.17, Cr 15.5-17.5, Mo 0.2-0.6, S 0.15-0.35, Mn 0-1.5, bal Fe.
Free machining; ferritic, stainless steel. W.-Nr. 1.4104; AISI 430F.

PURO WW

Remystahl
Stainless steel. C 0-0.08, Si 0-1, Mn 0-1, Cr 13-15, bal Fe.
Chromium, stainless steel. W.-Nr. 1.4001.

PURO-MMO

Remystahl
C 0.2, Cr 13, Mo 1, bal Fe.
Annealed: 95,000 TS; 50,000 YS; 25 El; 92 Rock B. Heat treated: 250,000 TS; 215,000 YS; 8 El; 52 Rock C. For cutlery, surgical instruments, gears, shafts, needle valves, gages. Corrosion resistant, hardenable.

PURO-MMO

Remystahl
C 0.2, Cr 13, Mo 1, bal Fe.
Annealed: 95,000 TS; 50,000 YS; 25 El; 92 Rock B. Heat treated: 250,000 TS; 215,000 YS; 8 El; 52 Rock C. For cutlery, surgical instruments, gears, shafts, needle valves, gages. Corrosion resistant, hardenable.

PURO-MMO 2

Remystahl
C 0.15, Cr 13, Mo 1, bal Fe.
Annealed: 70,000 TS; 35,000 YS; 30 El; 80 Rock B. Cold Drawn: 95,000 TS; 80,000 YS; 15 El; 92 Rock B. For chemical plant and oil refinery equipment. Corrosion resistant.

PURON

Westinghouse Electric Corp.
Fe 99.95.
For spectroscopic and magnetic standards; high purity iron. *Obsolete*

PURPLE CUT STEEL

Allegheny Ludlum Steel
C 1.2, Cr 1.2, Mo 1.2, V 0.25, Si 1.2, W 0.4, Mn 0.3, bal Fe.
For taps, threading dies, broaches, milling cutters, drills, and tools, punches; retains cutting edge at purple heat. *Obsolete*

PURPLE LABEL

Wallace Murray Corp.
C 0.5-1.1, Mn 0.2-0.9, bal Fe.
Water hardening. For tools.

PURPLE LABEL (AISI T4)

Jessop Steel Co.
C 0.74, Co 5, W 18.5, Cr 4.3, V 1.5, bal Fe.
For high speed cutting tools, lathe and planer tools; high speed steel.

PURPLE LABEL EXTRA

Jessop Steel Co.
C 0.78, Mo 0.8, W 18.5, Co 8, V 2, Cr 4, bal Fe.
Annealed: 228-241 Brin. For heavy roughing cutters, lathe and planer tools; Type T5; high speed steel.

PURPLE LABEL SPECIAL

Jessop Steel Co.
C 0.79, Cr 4, V 2, Mo 0.75, W 14, Co 5, bal Fe.
For lathe and planer tools, drills, broaches; high speed steel. *Obsolete*

PW 190R

Plattawerke GmbH
C 0.35, Cr 2.8, W 9, V 0.3, bal Fe.
For extrusion rams and liners; hot-work steel, oil hardened.

PW2

Thyssen Edelstahlwerke AG
C 0.55, Cr 1.25, W 2.75, V 0.2, bal Fe.
Heat treated: 231,000-315,000 TS; 209,000-275,000 YS; 9-11 El; 27-42 RA; C 48-56 Rock. For header and swaging dies, chipping chisels, rivet busters, track tools. Type S1 shock resisting tool steel, good fatigue resistance. *Obsolete*

PWA 1030
Cannon-Muskegon Corp.
C 0.06, Mn 0.08, S 0.01, P 0.004, Si 0.1, Cr 20.5, Co 14.2, Mo 4.2, Ti 3, Al 1.5, Zr 0.03, B 0.003, Cu 0.05, bal Ni.
Aged: 186,000 TS; 138,000 YS; 26 El; 39 Rock C. For jet engine turbine buckets and discs, high temperature bolts, missile systems. Precipitation hardened. Waspalloy. High temperature strength.

PWA 651A
Pratt & Whitney Cutting Tool & Gage
Nickel. Cr 23.5-26.5, Mo 2-4, Fe 0-5, Co 10-15, W 6-8, bal Ni.
For jet engine components; heat and corrosion resistant.

PWA 652
Cannon-Muskegon Corp.
C 0-0.1, Mn 0-0.5, Si 0-0.75, Cr 18-21, Mo 3.5-5, bal Ni.
Heat treated: 188,000 TS; 115,000 YS; 28 El; 25 RA. For jet engine and gas turbine buckets and discs, high temperature bolts. Creep and heat resistant. Similar to Waspalloy.

PWA 653
Cannon-Muskegon Corp.
C 0.4-0.5, W 10-12, Cr 20-22, Ni 0-1, Fe 0-0.6, 1.5-2.0 Cb + Ta, bal Co.
Cast: 120,000 TS; 90,000 YS; 38 Rock C. At 1800°F: 37,000 TS; 27,000 YS; 24 El. For gas turbine engine components, turbine nozzle vanes. High heat and oxidation resistant in the 1000-2000°F range.

PWA 656B
Cannon-Muskegon Corp.
C 0.05-0.09, Mn 0-0.15, Si 0-0.2, Cr 14-15.5, Mo 3.9-4.9, bal Ni.
Rolled: 205,000 TS; 140,000 YS; 17 El; 20 RA. For turbine and jet engine components, combustion chambers. Creep and oxidation resistant. Similar to U-700.

PWA 663
Cannon-Muskegon Corp.
C 0.1, Cr 8, Co 10, Mo 6, Ta 4.25, Al 6, Ti 1, Zr, B, bal Co.
At room temperature: 141,000 TS; 120,000 YS; 8 El. At 1400°F: 138,000 TS; 117,000 YS; 4 El. For high temperature applications, turbines, jet engines. Heat resistant. Similar to B1900.

PWA 664
Cannon-Muskegon Corp.
C 0.12-0.17, Cr 8-10, W 11.5-13.5, Co 9-11, Cb 0.75-1.25, Ti 2, Zr 0.05, B 0.01-0.02, Al 4.7-5.2, bal Ni.
As cast: 135,000 TS; 120,000 YS; 7 El. At 1400°F: 135,000 TS; 122,000 YS; 3.5 El. For cast turbine blades and vanes in aircraft gas turbines. Precipitation hardening, high temperature strength.

PWA 786A
Pratt & Whitney Cutting Tool & Gage
C 0-0.25, Cr 15.5-17.5, Ni 3.5-5, Cu 3.5-4.5, Ti 0.2, bal Fe.
For chemical plant equipment; heat and corrosion resistant.

PWA-649
Cannon-Muskegon Corp.
C 0.03-0.1, Mn 0-0.35, Mo 2.8-3, Ni 50-55, Cr 17-21, bal Fe.
At room temperature: 200,000 TS; 170,000 YS; 20 El; 25 RA; 42 Rock C. At 1200°F: 165,000 TS; 140,000 YS; 20 El; 25 RA. For aircraft and jet engine components, high temperature bolting. Similar to alloy 718C. Corrosion and oxidation resistance.

PYR-AIR-DIE
Pyramid Steel Company
C, alloy, bal Fe.
For blanking and forming dies, cutter blades; air hardened, non-deforming.

PYR-AIR-DIE MODIFIED
Pyramid Steel Company
C 0.55, Mn 0.9, Si 0.25, Cr 3.5, Ni 0.25, V 0.25, Mo 1.4, bal Fe.
Air or oil hardened tool steel.

PYR-OIL-DIE
Pyramid Steel Company
C, alloy, bal Fe.
For dies, punches, crimpers, knives; oil hardened, non-deforming.

PYRAD
Now AUBERT & DUVAL PYRAD.

PYRAD 38 D
Aubert & Duval
Nickel. C 0.05, Si 0-0.3, Mn 0-0.2, Cr 18, Mo 5.2, Al 0.8, Ti 2.3, Fe 35, Zr 0-0.2, Ni 38.
Wrought: 1080 MPa TS; 720 MPa YS. Equivalent to P.E. 11.

PYRAD 49 D
Aubert & Duval
Alloy steel. C 0-0.1, Cr 22, Fe 19, Mo 9, bal Ni.
Air cooled: 790 MPa TS; 54 El. For jet engine components, housings, blades, hot gas manifolds, combustion chamber parts.

PYRAD 53 N
Aubert & Duval
Alloy steel. C 0.1, Cr 18, Fe 18, Nb 5, Mo 3, Ti 1, Al 0.5, bal Ni.
Solution treated: 1350 MPa TS; 20 El. For marine and land machines or engines.

PYRAL
Creusot-Loire
C 0-0.12, Cr 6, Mo, Al, bal Fe.
Annealed: 70,000 TS; 30,000 YS; 28 El; 65 RA; 160 Brin. For oil refinery equipment; Type 501; corrosion and creep resistant. *Obsolete*

PYRALLOY
Dunford Hadfields Ltd.
C 0.3, Cr 12-16, bal Fe.
69,000 TS. For furnace parts, valves for internal combustion engines; heat and corrosion resistant to 800 C. *Obsolete*

PYRAMID BABBITT
Magnolia Metal Corp.
Sn, Sb, bal Pb.
Cast: 17,800 TS; 8,870 Brin. For bearings, Babbitts; pouring temperature 875-1000°F.

PYRASTEEL NO. 14
Chicago Steel Foundry
C 0.2, Cr 18, Mo 0.5, bal Fe.
At 70°F: 110,000 TS; 75,000 YS; 18 El; 32 RA; 215 Brin. At 1000°F: 70,000 TS; 49,000 YS; 24 El; 64 RA. For corrosion and heat resistant parts, oil refineries; corrosion and heat resistant.

PYRASTEEL NO. 18
Chicago Steel Foundry
Ni 25-28, Cr 16-18, Si 2.5, bal Fe.
For chain conveyors, lead pots; resistant to 1800°F.

PYRASTEEL NO. 20
Chicago Steel Foundry
C 0.2, Cr 17-18, Ni 35-38, Si 2.5, bal Fe.
For heat resistant parts, furnace parts; corrosion and heat resistant.

PYRASTEEL NO. 2000
Chicago Steel Foundry
Cr 25-27, Ni 12-14, Si 1.5, C, bal Fe.
For castings, kilns, clinker coolers; heat and corrosion resistant to 2000°F.

PYRASTEEL SPECIAL
Chicago Steel Foundry
Ni 14, Cr 25-28, Si 2.5, bal Fe.
For salt pots; resistant to 2000°F.

PYRASTEEL-A
Chicago Steel Foundry
C 0.3, Ni 4-8, Cr 8, bal Fe.
For carburizing boxes, furnace parts and grids, heat and resistant parts; heat resistant.

PYRASTEEL-B
Chicago Steel Foundry
C 0.3, Ni 15, Cr 14, bal Fe.
For carburizing boxes, furnace parts; heat resistant.

PYRASTEEL-C
Chicago Steel Foundry
C 0.3, Ni 25, Cr 20, bal Fe.
Cast: 68,000 TS; 38,000 YS; 17 El; 160 Brin. For furnace parts, torch nozzles, trays; heat resistant.

PYRASTEEL-D
Chicago Steel Foundry
C 0.3, Ni 35, Cr 25, bal Fe.
Cast: 70,000 TS; 40,000 YS; 10 El; 12 RA; 170 Brin. For carburizing boxes, pots; heat resistant.

PYRISTA
Firth-Vickers Stainless Steels Ltd.
C 0.09, Cr 29, Ni 1.8, bal Fe.
Ferritic stainless steel bar.

PYRO DIE
Crucible Materials Corp.
Tool material. C 0.4, Cr 1, V 0.25, bal Fe.

PYROCAST
Pacific Foundry Co.
C 1.75-2, Cr 24-28, bal Fe.
Cast: 70,000 TS; 70,000 YS; 0 El; 0 RA; 300 Brin. For heat treating and carburizing boxes, furnace parts; high heat resistance.

PYROCHROM 20
Pose-Marre Edelstahlwerk G.m.b.H.
C 0-0.05, Cr 25, Ni 20, bal Fe.
Drawn wire: 60-75 kp/mm² TS; 30 El min. For electric heating elements. *Obsolete*

PYROCHROM 30
Pose-Marre Edelstahlwerk G.m.b.H.
C 0-0.05, Cr 20, Ni 34, bal Fe.
Drawn wire: 60-75 kp/mm² TS; 30 El min. For electric heating elements. *Obsolete*

PYROCHROM 60
Pose-Marre Edelstahlwerk G.m.b.H.
C 0-0.05, Cr 15, Ni 64, bal Fe.
Drawn wire: 60-75 kp/mm² TS; 30 El min. For electric heating elements. *Obsolete*

PYROCHROM 80
Pose-Marre Edelstahlwerk G.m.b.H.
C 0-0.05, Cr 18, Ni 80, bal Fe.
Drawn wire: 60-75 kp/mm² TS; 30 YS min. For electric heating elements. *Obsolete*

PYRODIE
British Steel plc
C 0.4, Si 1, Cr 5, Mo 1.3, V 1, bal Fe.
Obsolete

PYRODIE
Heppenstall Co.
C, alloy, bal Fe.
For dies. *Obsolete*

PYRODUR 10
Bergische Stahl Industrie
C 0.6, Cr 22.5, Si 1.6, bal Fe.
As cast: 200-280 Brin. Elevated temperature applications; resistant to scaling to about 1050°C. W.-Nr. 1.4745; DIN G-X40CrSi23.

PYRODUR 12

Bergische Stahl Industrie
C 0.6, Cr 29, Si 1.6, bal Fe.
As cast: 220-300 Brin. Elevated temperature applications; resistant to scaling to about 1180°C. W.-Nr. 1.4776; DIN G-X40CrSi29.

PYRODUR 12R

Bergische Stahl Industrie
C 1.4, Cr 29, Si 1.6, bal Fe.
Cast, annealed: 250-350 Brin. Elevated temperature applications; resistant to scaling to about 1150°C. W.-Nr. 1.4777; DIN G-X130CrSi29.

PYRODUR 30 CN 38 NB

Gusstahl-Handels GmbH
C 0.1-0.2, Si 1-1.2, Mn 9-1.5, Cr 24-27, Ni 19-21, bal Fe.
Castings for high temperature operation. W.-Nr. 1.4849; DIN G-X 15 CrNiSi 25 20.

PYRODUR 30-CN 10

Bergische Stahl Industrie
C 0.3, Cr 22, Ni 10, Si 1.25, bal Fe.
As cast: 71,000-91,000 TS; 12 min El. Elevated temperature operation: resistant to scaling to about 1050°C. W.-Nr. 1.4826; ASTM A297 Gr.HF. DIN G-X40CrNiSi229.

PYRODUR 30-CN 13

Bergische Stahl Industrie
C 0.3, Cr 25, Ni 12.5, Si 1.25, bal Fe.
As cast: 71,000-92,000 TS; 10 min El. Elevated temperature operations; resistant to scaling to about 1150°C. W.-Nr. 1.4837; DIN G-X35CrNiSi2512; ASTM A297 Gr.HH.

PYRODUR 30-CN 38

Bergische Stahl Industrie
C 0.25, Ni 37, Cr 17.5, Si 1.25, bal Fe.
As cast: 64,000-85,000 TS; 8 El min. Elevated temperature operations; resistant to scaling to about 1100°C. W. Nr. 1.4865; DIN G-X40NiCrSi3616. ASTM A297 Gr.HU.

PYRODUR 40-CN 20

Bergische Stahl Industrie
C 0.3, Cr 25, Ni 20, Si 1.25, bal Fe.
As cast: 64,000-85,000 TS; 28,000 YS min; 8 El min. Elevated temperature operations; resistant to scaling to about 1150°F. W.-Nr. 1.4848; DIN G-X40CrNiSi2520; ASTM A297 Fr.HK.

PYRODUR 40-CN5

Bergische Stahl Industrie
C 0.35, Cr 27, Ni 4, Si 1.25, bal Fe.
As cast: 64,000-85,000 TS; 42,500 YS min; 4 El min. Elevated temperature applications; resistant to scaling to about 1150°F. W.-Nr. 1.4823; DIN G-X40CrNiSi27 4. ASTM A297 Gr.HC.

PYRODUR 7

Bergische Stahl Industrie
C 0.3, Cr 6, Si 2.25, bal Fe.
Cast, annealed: 71,000-100,000 TS; 6 El min. Elevated temperature applications; resistant to scaling to about 700°C. W.-Nr. 1.4710; DIN G-X30CrSi6.

PYRODUR 8

Bergische Stahl Industrie
C 0.6, Cr 13, Si 1.25, bal Fe.
Cast, annealed: 71,000-100,000 TS; 4 El min. Elevated temperature applications; resistant to scaling to about 800°C. W.-Nr. 1.4729; DIN G-X45CrSi13.

PYRODUR 9

Bergische Stahl Industrie
C 0.6, Cr 17, Si 1.25, bal Fe.
Cast, annealed: 71,000-100,000 TS; 2 El min. Elevated temperature applications; resistant to scaling to about 900°C. W.-Nr. 1.4740; DIN G-X40CrSi17.

PYRODUR 9R

Bergische Stahl Industrie
C 1.5, Cr 19, Si 1.25, bal Fe.
Cast, annealed: 250-350 Brin. Elevated temperature applications; resistant to scaling to about 900°C. W.-Nr. 1.4743; DIN G-X160CrSi18.

PYRODUR CO 50

Bergische Stahl Industrie
C 0-0.08, Si 0.4, Mn 0.4, Cr 27-29, Mo 2.2, Co 48-50, bal Fe.
Castings for high temperature operation. W.-Nr. 2.4778; DIN G-CoCr 28.

PYRODUR CO 51

Bergische Stahl Industrie
C 0-0.08, Si 0.4, Mn 0.4, Cr 27-29, Mo 2.2, Co 48-50, Cb, bal Fe.
Castings for high temperature operations. W.Nr. 1.4779; DIN G-CoCr 28 Nb.

PYRODUR P-CN 15

Bergische Stahl Industrie
C 0.3, Cr 27, Ni 15.5, Si 1.25, bal Fe.
As cast: 71,000-92,000 TS; 8 minimum El. Elevated temperature operations; resistant to scaling to about 1150 C. Werkstoff Nr. 1.4846; DIN G-X40CrNiSi2614; ASTM A297; Gr.HI. *Obsolete*

PYRODUR P-CN 15H

Bergische Stahl Industrie
C 0.7, Cr 27, Ni 15.5, Si 1.25, bal Fe.
As cast: 71,000-92,000 TS; 3 minimum El. Elevated temperature operations; resistant to scaling to about 1150 C. DIN G-X70CrNiSi2715. *Obsolete*

PYRODUR P-CN 15T

Bergische Stahl Industrie
C 0.3, Cr 19.5, Ni 15, Si 1.25, bal Fe.
As cast: 71,000-92,000 TS; 12 minimum El. Elevated temperature operations; resistant to scaling to about 1050 C. DIN G-X25CrNiSi2014; Werkstoff Nr. 1.4832. *Obsolete*

PYRODUR P-CN 15TW

Bergische Stahl Industrie
C 0.12, Cr 19.5, Ni 15, Si 1, bal Fe.
Solution annealed & aged: 64,000-85,000 TS; 28,000 minimum YS; 20 minimum El. Elevated temperature operations; resistant to scaling to about 1050 C. DIN G-X15CrNiSi2015. *Obsolete*

PYRODUR P-CN 15W

Bergische Stahl Industrie
C 0.12, Cr 27, Ni 15.5, Si 1.25, bal Fe.
Solution annealed & aged: 64,000-85,000 TS; 28,000 minimum YS; 20 minimum El. Elevated temperature operations; resistant to scaling to about 1150 C. DIN G-X12CrNiSi2715. *Obsolete*

PYRODUR P-CN 20H

Bergische Stahl Industrie
C 0.7, Cr 25, Ni 20, Si 1.75, bal Fe.
As cast: 71,000-92,000 TS; 3 minimum El. Elevated temperature operations; resistant to scaling to about 1150 C. DIN G-X70CrNiSi2520. *Obsolete*

PYRODUR P-CN 20W

Bergische Stahl Industrie
C 0.15, Cr 25, Ni 21, Si 0.75, bal Fe.
Solution annealed & aged: 64,000-85,000 TS; 28,000 minimum YS; 20 minimum El. Elevated temperature operations; resistant to scaling to about 1150 C. Werkstoff Nr. 1.4849; DIN G-X15CrNiSi2520. ASTM A351; Gr.CK20. *Obsolete*

PYRODUR P-CN 35

Bergische Stahl Industrie
C 0.3, Ni 35, Cr 20.5, Si 1.75, bal Fe.
As cast: 64,000-85,000 TS; 10 minimum El. Elevated temperature operations; resistant to scaling to about 1150 C. DIN G-X30NiCrSi3520. *Obsolete*

PYRODUR P-CN 35 W

Bergische Stahl Industrie
C 0.1, Ni 35, Cr 20.5, Si 1.75, bal Fe.
Solution annealed & aged: 64,000-85,000 TS; 28,000 minimum YS; 20 minimum El. Elevated temperature operations; resistant to scaling to about 1150 C. DIN G-X10NiCrSi3520. *Obsolete*

PYRODUR P-CN 60

Bergische Stahl Industrie
C 0.35, Ni 53, Cr 21, bal Fe.
As cast: 57,000-78,000 TS; 4 minimum El. Elevated temperature operations; resistant to scaling to about 1050 C. DIN G-X35NiCr5422. *Obsolete*

PYRODUR P-CN 80

Bergische Stahl Industrie
C 0.15, Ni 78, Cr 14, bal Fe.
Solution annealed & aged: 64,000-85,000 TS; 31,000 minimum YS; 12 minimum El. Elevated temperature operations; resistant to scaling to about 1020 C. DIN G-X15NiCr7814. *Obsolete*

PYROFERAL

Czechoslovakian manufacture
C, Cr, bal Fe.
For furnace parts and equipment; heat resistant.

PYROFIX-1

TradeARBED Inc.
Stainless steel. C 0-0.25, Mn 0-2, Si 0-1, Cr 24-26, Ni 19-22, bal Fe.
Austenitic stainless steel; for elevated temperature service. AISI 310. *Obsolete*

PYROFIX-2

TradeARBED Inc.
Stainless steel. C 0.08, Mn 0-2, Si 0.75-1.5, Cr 17-20, Ni 34-37, bal Fe.
Austenitic stainless steel; for high temperature equipment. AISI 330. *Obsolete*

PYROFIX-3

TradeARBED Inc.
Stainless steel. C 0-0.2, Mn 0-2, Si 0-1, Cr 22-24, Ni 12-15, bal Fe.
Austenitic stainless steel; for elevated temperature equipment. AISI 309. *Obsolete*

PYROFIX-5

TradeARBED Inc.
Stainless steel.
Austenitic stainless steel. *Obsolete*

PYROMET

Now CARPENTER PYROMET.

PYROMET 860

Cannon-Muskegon Corp.
C 0.05, Mn 0.25, Si 0.1, Cr 13, Ni 44, Co 4, Mo 6, Ti 3, Al 1, B 0.01, bal Fe.
For turbine engine parts.

PYROMET ALLOY CTX-3

Carpenter Technology Corp.
C 0-0.06, Mn 0-1, Si 0-0.5, P 0-0.015, S 0-0.015, Cr 0-0.2, Ni 38, Co 13, Ti 1.5, Al 0-0.35, Cu 0-0.5, B 0-0.012, 4.80 Cb + Ta, bal Fe.
RT: 830 MPa YS; 1180 MPa TS; 14 El; 23 RA. High strength precipitation hardenable alloy for compressor and exhaust casings, seal and other gas turbine engine components, ordnance hardware, gauge blocks, rocket engine thrust chambers, steam turbine blades, springs, die casting dies. Pyromet Alloy 907.

PYROMET ALLOY CTX-909
Carpenter Technology Corp.
C 0-0.06, Mn 0-0.5, Si 0.4, P 0-0.015, S 0-0.015, Cr 0-0.5, Ni 38, Co 14, Ti 1.6, Al 0-0.15, Cu 0-0.5, B 0-0.012, 4.90 Cb + Ta, bal Fe.
Solution treated and aged, RT: 883-1083 MPa YS; 1193-1338 MPa TS; 10-13 El; 14-30 RA. High strength, precipitation hardenable alloy with a low coefficient of thermal expansion, high hot hardness and good thermal fatigue resistance. For compressor and exhaust casings, seals and other gas turbine engine components, ordnance hardware, gauge blocks, rocket engine thrust chambers, steam turbine blades, springs, die casting dies.

PYROMET ALLOY V-57
Carpenter Technology Corp.
C 0-0.08, Mn 0-0.35, Si 0-0.5, P 0-0.015, S 0-0.015, Cr 13.5-16, Ni 25.5-28.5, Mo 1-1.5, Ti 2.7-3.2, V 0-0.5, Al 0.1-0.35, B 0.005-0.012, bal Fe.
Heat treated, RT: 814 MPa YS; 1186 MPa TS; 18 El; 24 RA; 321 Brin. Austenitic precipitation hardening alloy with high strength and good corrosion resistance for service up to 760°C in turbine wheels, spacers, buckets, high temperature bolting. AISI No. 663.

PYROMIC
Telcon Metals Ltd.
Ni 80, Cr 20.
For electrical resistance; formerly PYROMIC NO. 2.

PYROMIC NO. 1
Telcon Metals Ltd.
Ni, Cr, bal Fe.
For electrical resistance. *Obsolete*

PYRON 10
SWB Stahlformguss Gesellschaft mbH
C 0-0.12, Si 2, Cr 18, bal Fe.
Annealed: 85,000 TS; 45,000 YS; 25 El; 50 RA; 160 Brin. For furnace parts and equipment; heat and corrosion resistant. *Obsolete*

PYRON 100
PYRON Corp.
Mn 0.5, bal Fe.
Iron powder for production of high strength, low density P/M parts and bearings.

PYRON 10A1
SWB Stahlformguss Gesellschaft mbH
C 0-0.12, Si 1, Al 1, Cr 18, bal Fe.
Annealed: 85,000 TS; 45,000 YS; 25 El; 50 RA; 160 Brin. For oil refinery equipment; heat and corrosion resistant. *Obsolete*

PYRON 12A1
SWB Stahlformguss Gesellschaft mbH
C 0-0.12, Si 1.5, Al 1.5, Cr 24, bal Fe.
Annealed: 85,000 TS; 50,000 YS; 30 El; 55 RA; 180 Brin. For oil refinery and furnace equipment; heat resistant. *Obsolete*

PYRON 16.36
SWB Stahlformguss Gesellschaft mbH
C 0.12, Cr 16, Ni 25, bal Fe.
For furnace parts, chemical plant equipment; corrosion and heat resistant. *Obsolete*

PYRON 20.13
SWB Stahlformguss Gesellschaft mbH
C 0.12, Cr 20, Ni 13, bal Fe.
Annealed: 80,000 TS; 35,000 YS; 50 El; 65 RA; 160 Brin. For chemical plant equipment; stainless, austenitic. *Obsolete*

PYRON 2010
SWB Stahlformguss Gesellschaft mbH
C 0.15, Cr 19.5, Ni 9.5, bal Fe.
Annealed: 80,000 TS; 35,000 YS; 55 El; 75 RA; 150 Brin. For chemical plant equipment, tanks, mixers; Type 302; stainless, austenitic. *Obsolete*

PYRON 2419
SWB Stahlformguss Gesellschaft mbH
C 0.15, Si 2, Cr 24, Ni 19, bal Fe.
Annealed: 100,000 TS; 45,000 YS; 50 El; 65 RA; 185 Brin. For furnace parts, valves, pumps, turbine parts; Type 310; stainless, austenitic. *Obsolete*

PYRON 25.04 SPEZIAL
SWB Stahlformguss Gesellschaft mbH
C 0.12, Cr 25, Ni 4, bal Fe.
For furnace parts, valves, heat treating boxes; heat resistant. *Obsolete*

PYRON 2504
SWB Stahlformguss Gesellschaft mbH
C 0.2, Si 1.2, Cr 25, Ni 4, bal Fe.
For furnace parts, valves, heat treating boxes; heat resistant. *Obsolete*

PYRON 6
SWB Stahlformguss Gesellschaft mbH
Si 2-3, Cr 6, C, bal Fe.
For oil refinery equipment; creep and heat resistant. *Obsolete*

PYRON 7
SWB Stahlformguss Gesellschaft mbH
Si 2-3, Cr 6, C, bal Fe.
For oil refinery equipment; creep and heat resistant. *Obsolete*

PYRON 8
SWB Stahlformguss Gesellschaft mbH
C 0-0.12, Si 2-3, Cr 6, bal Fe.
For oil refinery equipment; heat and creep resistant. *Obsolete*

PYRON 8A1
SWB Stahlformguss Gesellschaft mbH
C 0-0.12, Si 0.8, Al 0.8, Cr 6.5, bal Fe.
For oil refinery equipment; heat and creep resistant. *Obsolete*

PYRON 9
SWB Stahlformguss Gesellschaft mbH
C 0-0.12, Si 2.2, Cr 13, bal Fe.
Annealed: 80,000 TS; 40,000 YS; 35 El; 65 RA; 160 Brin. For valves, pumps, oil refinery equipment; heat and corrosion resistant. *Obsolete*

PYRON 9A1
SWB Stahlformguss Gesellschaft mbH
C 0-0.12, Si 1.2, Al 1, Cr 13, bal Fe.
Annealed: 80,000 TS; 40,000 YS; 35 El; 65 RA; 160 Brin. For valves, pumps, oil refinery equipment; heat and corrosion resistant. *Obsolete*

PYRON AC-325
PYRON Corp.
Iron powder (95% -325 mesh); for chemical applications requiring high specific surface and reactivity.

PYRON D-63
PYRON Corp.
Mn 0.5, bal Fe.
Iron powder, more compressible than PYRON 100; particularly suited for manufacture of iron-copper-carbon P/M parts.

PYRON G105
SWB Stahlformguss Gesellschaft mbH
C 0.6, Si 1.5, Cr 22, bal Fe.
For furnace parts, heat treating boxes; heat resistant. *Obsolete*

PYRON G115
SWB Stahlformguss Gesellschaft mbH
C 0.6, Si 1.5, Cr 29, bal Fe.
For furnace parts, heat treating boxes; heat resistant. *Obsolete*

PYRON G2210
SWB Stahlformguss Gesellschaft mbH
C 0.4, Si 2, Cr 22, Ni 9.5, bal Fe.
Cast: 85,000 TS; 45,000 YS; 35 El; 165 Brin. For furnace parts, heat treating boxes, valves; corrosion and heat resistant. *Obsolete*

PYRON G2519
SWB Stahlformguss Gesellschaft mbH
C 0.4, Si 2, Cr 25, Ni 19, bal Fe.
Cast: 75,000 TS; 50,000 YS; 17 El; 170 Brin. For pumps, valves, furnace parts, heat treating boxes; corrosion and heat resistant. *Obsolete*

PYRON G27.04
SWB Stahlformguss Gesellschaft mbH
C 0.4, Si 1.3, Cr 27, Ni 4, bal Fe.
Cast: 90,000 TS; 65,000 YS; 2 El; 212 Brin. For furnace and turbine parts; heat resistant. *Obsolete*

PYRON G55.24NI
SWB Stahlformguss Gesellschaft mbH
C 0.55, Si 1.3, Cr 55, Ni 24, bal Fe.
For furnace parts; heat and oxidation resistant. *Obsolete*

PYRON G85
SWB Stahlformguss Gesellschaft mbH
C 0.3, Si 2.2, Cr 6, bal Fe.
For oil refinery equipment; heat and creep resistant. *Obsolete*

PYRON G95
SWB Stahlformguss Gesellschaft mbH
C 0.5, Si 1.5, Cr 17, bal Fe.
For furnace parts, chemical plant equipment; heat and corrosion resistant. *Obsolete*

PYRON LD 80
PYRON Corp.
Iron powder, used in friction applications where a finer material with higher apparent density than R-12 is required. Apparent density 1.75-2.10 g/cm³.

PYRON M-IRON
PYRON Corp.
Mo 1, bal Fe.
Iron powder used in low density P/M parts where high sintered hardness is required.

PYRON R-12
PYRON Corp.
Iron powder, low density, high green strength, high surface area used in friction applications. Apparent density 1.0-1.5 g/cm³.

PYRON R-80
PYRON Corp.
Iron powder; low density, high green strength iron powder used in friction material formulations.

PYRONEAL
Heppenstall Co.
Tool material. C 0.55, Cr 0.9, Mo 0.7, Si 0.6, Ni 2.2, bal Fe.
For hot work dies, punches; oil or air hardened; hot work, press and upsetter dies.

PYROPHORIC ALLOY
American manufacture
Mn 10, Sb 10, Cr 20, Ti 15, bal Fe.
For lighters for gas stoves, cigarette lighters; produces violent sparks when abrased.

PYROS
Creusot-Loire
Ni 82, Cr 7, Fe 3, W 5, Mn 3.
For dilatometric temperature indicating or control devices; heat resisting alloy. *Obsolete*

PYROTEM

Heppenstall Co.
Tool material. C 0.55, Ni 2.1, Cr 0.87, Mo 0.73, bal Fe.
Annealed: 250 Brin. Heat treated: 570 Brin. For hot work dies, cold shear knives; oil hardened; shock resistant.

PYROTHERM 14

Pose-Marre Edelstahlwerk G.m.b.H.
C 0-0.12, Cr 13, bal Fe.
Annealed: 55-70 kp/mm^2 TS; kp/mm^2 YS min; 15 El min. Heat resistant, rolled and wrought alloy. For furnace engineering in cement, glass, metallurgical, and petrochemical industries. *Obsolete*

PYROTHERM 15/16

Pose-Marre Edelstahlwerk G.m.b.H.
C 0-0.05, Cr 15, Ni 64, bal Fe.
As quenched: 50-70 kp/mm^2 TS; 18 kp/mm^2 YS min; 30 El min. Heat resistant, rolled and wrought alloy. For furnace engineering in cement, glass, metallurgical, and petrochemical industries. *Obsolete*

PYROTHERM 18

Pose-Marre Edelstahlwerk G.m.b.H.
C 0-0.12, Cr 18, bal Fe.
Annealed: 55-70 kp/mm^2 TS; 35 kp/mm^2 YS min; 15 min El. Heat resistant, rolled and wrought alloy. For furnace engineering in cement, glass, metallurgical, and petrochemical industries. *Obsolete*

PYROTHERM 18/36

Pose-Marre Edelstahlwerk G.m.b.H.
C 0-0.15, Cr 18, Ni 36, bal Fe.
As quenched: 50-70 kp/mm^2 TS; kp/mm^2 YS min; 30 El min. Heat resistant rolled and wrought alloy. For furnace engineering in cement, glass, metallurgical, and petrochemical industries. *Obsolete*

PYROTHERM 18/36 NB

Pose-Marre Edelstahlwerk G.m.b.H.
C 0.15, Cr 18, Ni 36, Nb 1.2, bal Fe.
As quenched: 50-70 kp/mm^2 TS; 20 kp/mm^2 YS min; 30 El min. Heat resistant, rolled and wrought alloy. For furnace engineering in cement, glass, metallurgical and petrochemical industries. *Obsolete*

PYROTHERM 18/8

Pose-Marre Edelstahlwerk G.m.b.H.
C 0-0.12, Cr 18, Ni 9, bal Fe.
As quenched: 55-70 kp/mm^2 TS; 22 kp/mm^2 YS min; 40 El min. Heat resistant, rolled and wrought alloy. For furnace engineering in cement, glass, metallurgical and petrochemical industries. *Obsolete*

PYROTHERM 18/8 TI

Pose-Marre Edelstahlwerk G.m.b.H.
C 0-0.15, Cr 18, Ni 9, Ti = 5 x C, bal Fe.
As quenched: 55-70 kp/mm^2 TS; 22 kp/mm^2 YS min; 40 El min. Heat resistant, rolled and wrought alloy. For furnace engineering in cement, glass, metallurgical, and petrochemical industries. *Obsolete*

PYROTHERM 20/15

Pose-Marre Edelstahlwerk G.m.b.H.
C 0-0.13, Cr 20, Ni 15, bal Fe.
As quenched: 55-70 kp/mm^2 TS; 25 kp/mm^2 YS min; 40 El min. Heat resistant, rolled and wrought alloy. For furnace engineering in cement, glass, metallurgical, and petrochemical industries. *Obsolete*

PYROTHERM 20/15 MO

Pose-Marre Edelstahlwerk G.m.b.H.
C 0-0.13, Cr 20, Ni 15, Mo 1.2, bal Fe.
As quenched: 55-70 kp/mm^2 TS; 25 kp/mm^2 YS min; 40 El min. Heat resistant, rolled and wrought alloy. For furnace engineering in cement, glass, metallurgical, and petrochemical industries. *Obsolete*

PYROTHERM 20/15 MONB

Pose-Marre Edelstahlwerk G.m.b.H.
C 0-0.13, Cr 20, Ni 15, Mo 1.2, Nb 1.2, bal Fe.
As quenched: 55-70 kp/mm^2 TS; 25 kp/mm^2 YS min; 40 El min. Heat resistant, rolled and wrought alloy. For furnace engineering in cement, glass, metallurgical, and petrochemical industries. *Obsolete*

PYROTHERM 20/33 ALTI

Pose-Marre Edelstahlwerk G.m.b.H.
C 0.05-0.1, Cr 20, Ni 33, Al 0.2, Ti 0.2, bal Fe.
As quenched: 50-70 kp/mm^2 TS; 20 kp/mm^2 YS min; 30 El min. Heat resistant, rolled and wrought alloy. For furnace engineering in cement, glass, metallurgical, and petrochemical industries. *Obsolete*

PYROTHERM 22/10

Pose-Marre Edelstahlwerk G.m.b.H.
C 0-0.2, Cr 21, Ni 11, bal Fe.
As quenched: 60-75 kp/mm^2 TS; 30 kp/mm^2 YS min; 40 El min. Heat resistant, rolled and wrought alloy. For furnace engineering in cement, glass, metallurgical and petrochemical industries. *Obsolete*

PYROTHERM 22/14

Pose-Marre Edelstahlwerk G.m.b.H.
C 0-0.1, Cr 22, Ni 14, bal Fe.
As quenched: 55-70 kp/mm^2 TS; 22 kp/mm^2 YS min; 40 El min. Heat resistant, rolled and wrought alloy. For furnace engineering in cement, glass, metallurgical, and petrochemical industries. *Obsolete*

PYROTHERM 24/24 NB

Pose-Marre Edelstahlwerk G.m.b.H.
C 0.35, Cr 24, Ni 24, Nb 1.5, bal Fe.
As quenched: 60-80 kp/mm^2 TS; 30 kp/mm^2 YS min; 20 El min. Heat resistant, rolled and wrought alloy. For furnace engineering in cement, glass, metallurgical and petrochemical industries. *Obsolete*

PYROTHERM 25

Pose-Marre Edelstahlwerk G.m.b.H.
C 0-0.12, Si 1.5, Cr 24, Al 1.5, bal Fe.
For oil refinery equipment; heat and corrosion resistant. *Obsolete*

PYROTHERM 25/12

Pose-Marre Edelstahlwerk G.m.b.H.
C 0-0.2, Cr 23, Ni 12, bal Fe.
As quenched: 60-75 kp/mm^2 TS; 30 kp/mm^2 YS min; 40 El min. Heat resistant, rolled and wrought alloy. For furnace engineering in cement, glass, metallurgical, and petrochemical industries. *Obsolete*

PYROTHERM 25/20

Pose-Marre Edelstahlwerk G.m.b.H.
C 0-0.15, Cr 25, Ni 20, Si 1.8, bal Fe.
As quenched: 55-75 kp/mm^2 TS; 30 kp/mm^2 YS min; 40 El min. Heat resistant, rolled and wrought alloy. For furnace engineering in cement, glass, metallurgical, and petrochemical industries. *Obsolete*

PYROTHERM 25/20 H

Pose-Marre Edelstahlwerk G.m.b.H.
C 0.4, Cr 25, Ni 20, bal Fe.
As quenched: 55-75 kp/mm^2 TS; 30 kp/mm^2 YS min; 20 El min. Heat resistant, rolled and wrought alloy. For furnace engineering in cement, glass, metallurgical and petrochemical industries. *Obsolete*

PYROTHERM 25/20 NB

Pose-Marre Edelstahlwerk G.m.b.H.
C 0-0.4, Cr 25, Ni 20, Nb 1.2, bal Fe.
As cast: 71,100-99,000 TS; 35,000 minimum YS; 10 minimum El. Austenitic stainless; for heat exchanger tubes, annealing furnace grates, and castings for petroleum industry. *Obsolete*

PYROTHERM 25/35 H

Pose-Marre Edelstahlwerk G.m.b.H.
C 0.45, Cr 25, Ni 35, bal Fe.
As quenched: 55-75 kp/mm^2 TS; 30 kp/mm^2 YS min; 12 El min. Heat resistant, rolled and wrought alloy. For furnace engineering in cement, glass, metallurgical, and petrochemical industries. *Obsolete*

PYROTHERM 25/4

Pose-Marre Edelstahlwerk G.m.b.H.
C 0.2, Cr 25, Ni 4, bal Fe.
As quenched: 60-75 kp/mm^2 TS; 40 kp/mm^2 YS; 26 El min. Heat resistant, rolled and wrought alloy. For furnace engineering in cement, glass, metallurgical and petrochemical industries. *Obsolete*

PYROTHERM 28

Pose-Marre Edelstahlwerk G.m.b.H.
C 0-0.12, Cr 28, bal Fe.
Annealed: 55-70 kp/mm^2 TS; 40 kp/mm^2 YS min; 12 min El. Heat resistant, rolled and wrought alloy. For furnace engineering in cement, glass, metallurgical and petrochemical industries. *Obsolete*

PYROTHERM 28/48/5

Pose-Marre Edelstahlwerk G.m.b.H.
C 0-0.15, Cr 28, Ni 48, W 5, bal Fe.
As quenched: 55-80 kp/mm^2 TS; 25 kp/mm^2 YS min; 20 El min. Heat resistant, rolled and wrought alloy. For furnace engineering in cement, glass, metallurgical and petrochemical industries. *Obsolete*

PYROTHERM 28/5

Pose-Marre Edelstahlwerk G.m.b.H.
C 0.2, Cr 28, Ni 5, bal Fe.
As quenched: 60-75 kp/mm^2 TS; 40 kp/mm^2 YS min; 26 El min. Heat resistant, rolled and wrought alloy. For furnace engineering in cement, glass, metallurgical, and petrochemical industries. *Obsolete*

PYROTHERM 5 MO

Pose-Marre Edelstahlwerk G.m.b.H.
C 0-0.15, Cr 5, Mo 0.5, bal Fe.
As tempered: 50-60 kp/mm^2 TS; 25 kp/mm^2 YS min; 23 El min. Heat resistant, rolled and wrought alloy. For furnace engineering in cement, glass, metallurgical and petrochemical industries. *Obsolete*

PYROTHERM 6

Pose-Marre Edelstahlwerk G.m.b.H.
C 0-0.12, Cr 6, bal Fe.
Annealed: 55-70 kp/mm^2 TS; 35 kp/mm^2 YS min; 15 min El. Heat resistant, rolled and wrought alloy. For furnace engineering in cement, glass, metallurgical and petrochemical industries. *Obsolete*

PYROTHERM G 13 MO

Pose-Marre Edelstahlwerk G.m.b.H.
C 0-0.1, Cr 13, Mo 0.4, bal Fe.
Annealed: 600-800 N/mm^2 TS; 15 El min. Heat and corrosion resistant, cast alloy. For elevated temperature equipment.

PYROTHERM G 16/13 NB

Pose-Marre Edelstahlwerk G.m.b.H.
C 0.08, Cr 16, Ni 13, Nb = 8 x C, bal Fe.
As cast: 390-590 N/mm^2 TS; 175 N/mm^2 YS min; 25 El min. Heat resistant, cast alloy. For elevated temperature equipment.

PYROTHERM G 17/53 NB

Pose-Marre Edelstahlwerk G.m.b.H.
C 0.45, Cr 17, Ni 53, Nb 1.2, bal Fe.
As cast: 420-620 N/mm^2 TS; 220 N/mm^2 YS min; 5 min El. Heat resistant, cast alloy. For elevated temperature equipment.

PYROTHERM G 18
Pose-Marre Edelstahlwerk G.m.b.H.
C 0.4, Cr 17, bal Fe.
Annealed: 300 Brin max. Heat resistant, cast alloy. For elevated temperature equipment.

PYROTHERM G 18 H
Pose-Marre Edelstahlwerk G.m.b.H.
C 1.5, Cr 18, bal.
Annealed, heat resistant, cast alloy. For elevated temperature equipment.

PYROTHERM G 18/36 H
Pose-Marre Edelstahlwerk G.m.b.H.
C 0.4, Cr 18, Ni 36, bal Fe.
As cast: 390-590 N/mm^2 TS; 205 N/mm^2 YS min; 8 min El. Heat resistant, cast alloy. For elevated temperature equipment.

PYROTHERM G 18/36 NB
Pose-Marre Edelstahlwerk G.m.b.H.
C 0.4, Cr 18, Ni 36, Nb 1.2, bal Fe.
As cast: 440-640 N/mm^2 TS; 230 N/mm^2 YS min; 8 min El. Heat resistant, cast alloy. For elevated temperature equipment.

PYROTHERM G 18/36H
Pose-Marre Edelstahlwerk G.m.b.H.
C 0-0.4, Cr 18, Ni 36, bal Fe.
As cast: 56,900-85,300 TS; 29,900 minimum YS; 8 minimum El. Austenitic stainless; for grates, carburizing boxes in annealing and carburizing furnaces. ACI Type HU. *Obsolete*

PYROTHERM G 18/8
Pose-Marre Edelstahlwerk G.m.b.H.
C 0.3, Cr 18, Ni 8, bal Fe.
As cast: 440-640 N/mm^2 TS; 205 N/mm^2 YS min; 12 El min. Heat resistant. cast alloy. For elevated temperature equipment.

PYROTHERM G 20/15
Pose-Marre Edelstahlwerk G.m.b.H.
C 0.2, Cr 20, Ni 15, bal Fe.
As cast: 440-640 N/mm^2 TS; 215 N/mm^2 YS min; 20 El min. Heat resistant, cast alloy. For elevated temperature equipment.

PYROTHERM G 20/32 NB
Pose-Marre Edelstahlwerk G.m.b.H.
C 0.1, Cr 20, Ni 32, Nb 1.1, bal Fe.
As cast: 440-640 N/mm^2 TS; 180 N/mm^2 YS min; 20 El min. Heat resistant, cast alloy. For elevated temperature equipment.

PYROTHERM G 20/33
Pose-Marre Edelstahlwerk G.m.b.H.
C 0-0.15, Cr 20, Ni 33, bal Fe.
As cast: 56,900-85,300 TS; 25,600 minimum YS; 12 minimum El. Austenitic stainless; for cracking tubes, heat exchanger tubes, similar high temperature equipment. *Obsolete*

PYROTHERM G 20/33H
Pose-Marre Edelstahlwerk G.m.b.H.
C 0-0.4, Cr 20, Ni 33, bal Fe.
As cast: 56,900-85,300 TS; 29,900 minimum YS; 8 minimum El. Austenitic stainless; for cracking tubes, heat exchanger tubes, similar high temperature equipment. *Obsolete*

PYROTHERM G 20/33NB
Pose-Marre Edelstahlwerk G.m.b.H.
C 0-0.1, Cr 20, Ni 33, Nb 1.5, bal Fe.
As cast: 56,900-85,300 TS; 28,500 minimum YS; 8 minimum El. Austenitic stainless; for heat exchanger tubes, cracking equipment in petroleum industry, and similar parts for elevated temperature operation. *Obsolete*

PYROTHERM G 22/10
Pose-Marre Edelstahlwerk G.m.b.H.
C 0.4, Cr 22, Ni 10, bal Fe.
As cast: 440-640 N/mm^2 TS; 205 N/mm^2 YS min; 12 El min. Heat resistant, cast alloy. For elevated temperature equipment.

PYROTHERM G 22/14
Pose-Marre Edelstahlwerk G.m.b.H.
C 0.15, Cr 22, Ni 14, bal Fe.
As cast: 390-590 N/mm^2 TS; 195 N/mm^2 YS min; 14 El min. Heat resistant, cast alloy. For elevated temperature equipment.

PYROTHERM G 22/14 H
Pose-Marre Edelstahlwerk G.m.b.H.
C 0.45, Cr 22, Ni 14, bal Fe.
At cast: 540-740 N/mm^2 TS; 220 N/mm^2 YS min; 4 min El. Heat resistant, cast alloy. For elevated temperature equipment.

PYROTHERM G 24/24 NB
Pose-Marre Edelstahlwerk G.m.b.H.
C 0.35, Cr 24, Ni 24, Nb 1.5, bal Fe.
As cast: 490-690 N/mm^2 TS; 230 N/mm^2 YS min; 10 El min. Heat resistant, cast alloy. For elevated temperature equipment.

PYROTHERM G 25
Pose-Marre Edelstahlwerk G.m.b.H.
C 0.4, Cr 25, bal Fe.
Annealed: 300 Brin max. Heat resistant, cast alloy. For elevated temperature equipment.

PYROTHERM G 25 H
Pose-Marre Edelstahlwerk G.m.b.H.
C 1.3, Cr 25, bal Fe.
Heat resistant, cast alloy (as cast). For elevated temperature equipment.

PYROTHERM G 25/12
Pose-Marre Edelstahlwerk G.m.b.H.
C 0.4, Cr 25, Ni 12, bal Fe.
As cast: 440-640 N/mm^2 TS; 220 N/mm^2 YS min; 10 El min. Heat resistant, cast alloy. For elevated temperature equipment.

PYROTHERM G 25/20
Pose-Marre Edelstahlwerk G.m.b.H.
C 0.15, Cr 25, Ni 20, bal Fe.
As cast: 440-640 N/mm^2 TS; 205 N/mm^2 YS min; 15 El min. Heat resistant, cast alloy. For elevated temperature equipment.

PYROTHERM G 25/20 H
Pose-Marre Edelstahlwerk G.m.b.H.
C 0.4, Cr 25, Ni 20, bal Fe.
As cast: 490-690 N/mm^2 TS; 230 N/mm^2 YS min; 10 El min. Heat resistant, cast alloy. For elevated temperature equipment.

PYROTHERM G 25/35 COWNB
Pose-Marre Edelstahlwerk G.m.b.H.
C 0.45, Cr 26, Ni 35, Co 16, W 5, Nb 1, bal Fe.
As cast: 440-640 N/mm^2 TS; 230 N/mm^2 YS min; 4 El min. Heat resistant, cast alloy. For elevated temperature equipment.

PYROTHERM G 25/35 H
Pose-Marre Edelstahlwerk G.m.b.H.
C 0.45, Cr 25, Ni 35, bal Fe.
As cast: 490-690 N/mm^2 TS; 230 N/mm^2 YS min; 8 min El. Heat resistant, cast alloy. For elevated temperature equipment.

PYROTHERM G 25/35 NB
Pose-Marre Edelstahlwerk G.m.b.H.
C 0.4, Cr 25, Ni 35, Nb 1.5, bal Fe.
As cast: 490-690 N/mm^2 TS; 230 N/mm^2 YS min; 8 min El. Heat resistant, cast alloy. For elevated temperature equipment.

PYROTHERM G 25/4
Pose-Marre Edelstahlwerk G.m.b.H.
C 0.4, Cr 25, Ni 4, bal Fe.
As cast: 550-750 N/mm^2 TS; 295 N/mm^2 YS min; 2 min El. Heat resistant, cast alloy. For elevated temperature equipment.

PYROTHERM G 25/4 H
Pose-Marre Edelstahlwerk G.m.b.H.
C 1.3, Cr 25, Ni 4, bal Fe.
Heat resistant, cast alloy (as cast). For elevated temperature equipment.

PYROTHERM G 26/32 MO
Pose-Marre Edelstahlwerk G.m.b.H.
C 0.4, Cr 26, Ni 32, Mo 1.5, bal Fe.
As cast: 45-65 kp/mm^2 TS; 22 kp/mm^2 YS min; 8 min El. Heat and corrosion resistant, cast alloy. For elevated temperature equipment. *Obsolete*

PYROTHERM G 26/36 MO
Pose-Marre Edelstahlwerk G.m.b.H.
C 0.45, Cr 25, Ni 35, Mo 1.5, bal Fe.
As cast: 440-640 N/mm^2 TS; 230 N/mm^2 YS; 8 El min. Heat resistant, cast alloy. For elevated temperature equipment.

PYROTHERM G 26/36 SO
Pose-Marre Edelstahlwerk G.m.b.H.
C 0.45, Cr 26, Ni 36, Mo, bal Fe.
As cast: 440-640 N/mm^2 TS; 230 N/mm^2 YS min; 8 min El. Heat resistant, cast alloy. For elevated temperature equipment.

PYROTHERM G 26/9
Pose-Marre Edelstahlwerk G.m.b.H.
C 0.35, Cr 26, Ni 9, bal Fe.
As cast: 45-65 kp/mm^2 TS; 21 kp/mm^2 YS min; 12 El min. Heat and corrosion resisting cast alloy. For elevated temperature equipment. *Obsolete*

PYROTHERM G 27/10 AL
Pose-Marre Edelstahlwerk G.m.b.H.
C 0.4, Cr 27, Ni 10, Al 0.7, bal Fe.
As cast: 440-640 N/mm^2 TS; 205 N/mm^2 YS min; 10 El min. Heat resistant, cast alloy. For elevated temperature equipment.

PYROTHERM G 28
Pose-Marre Edelstahlwerk G.m.b.H.
C 0.5, Cr 28, bal Fe.
Annealed: 300 Brin max. Heat resistant, cast alloy. For elevated temperature equipment.

PYROTHERM G 28 H
Pose-Marre Edelstahlwerk G.m.b.H.
C 1.3, Cr 28, bal Fe.
Heat resistant, cast alloy (as cast). For elevated temperature equipment.

PYROTHERM G 28/15 W
Pose-Marre Edelstahlwerk G.m.b.H.
C 0.6, Cr 28, Ni 15, W 2, bal Fe.
As cast: 440-640 N/mm^2 TS; 230 N/mm^2 YS min; 5 min El. Heat resistant, cast alloy. For elevated temperature equipment.

PYROTHERM G 28/20 AL
Pose-Marre Edelstahlwerk G.m.b.H.
C 0.4, Cr 28, Ni 20, Al 0.4, bal Fe.
As cast: 490-690 N/mm^2 TS; 230 N/mm^2 YS min; 10 El min. Heat resistant, cast alloy. For elevated temperature equipment.

PYROTHERM G 28/20 MO
Pose-Marre Edelstahlwerk G.m.b.H.
C 0.4, Cr 28, Ni 20, Mo 1.2, bal Fe.
As cast: 50-70 kp/mm^2 TS; 25 kp/mm^2 YS min; 10 El min. Heat and corrosion resistant, cast alloy. For elevated temperature equipment. *Obsolete*

PYROTHERM G 28/33 SI
Pose-Marre Edelstahlwerk G.m.b.H.
C 0.4, Cr 28, Ni 33, Si 2.5, bal Fe.
As cast: 490-690 N/mm² TS; 230 N/mm² YS min; 5 min El.
Heat resistant, cast alloy. For elevated temperature
equipment.

PYROTHERM G 28/48 5
Pose-Marre Edelstahlwerk G.m.b.H.
C 0.4, Cr 28, Ni 48, W 5, bal Fe.
As cast: 440-640 N/mm² TS; 270 N/mm² YS min; 3 min El.
Heat resistant, cast alloy. For elevated temperature
equipment.

PYROTHERM G 28/5
Pose-Marre Edelstahlwerk G.m.b.H.
C 0.4, Cr 28, Ni 5, bal Fe.
As cast: 550-750 N/mm² TS; 295 N/mm² YS min; 2 min El.
Heat resistant, cast alloy. For elevated temperature
equipment.

PYROTHERM G 30/30
Pose-Marre Edelstahlwerk G.m.b.H.
C 0.5, Cr 30, Ni 30, bal Fe.
As cast: 490-690 N/mm² TS; 230 N/mm² YS min; 6 min El.
Heat resistant, cast alloy. For elevated temperature
equipment.

PYROTHERM G 5 MO
Pose-Marre Edelstahlwerk G.m.b.H.
C 0.1, Cr 5, Mo 0.5, bal Fe.
Annealed: 640-840 N/mm² TS; 410 N/mm² YS; 18 El min.
Heat resistant, cast alloy. For elevated temperature
equipment.

PYROTHERM G 50/50 NB
Pose-Marre Edelstahlwerk G.m.b.H.
C 0-0.05, Cr 49, Ni 49, Nb 1.5, bal Fe.
As cast: 540-740 N/mm² TS; 270 N/mm² YS min; 10 El min.
Heat resistant, cast alloy. For elevated temperature
equipment.

PYROTHERM G 6
Pose-Marre Edelstahlwerk G.m.b.H.
C 0.3, Cr 6, bal Fe.
Annealed: 540-740 N/mm² TS; 295 /mm² YS min; 3 El min.
Heat resistant, cast alloy. For elevated temperature
equipment.

PYROTHERM G 60/40
Pose-Marre Edelstahlwerk G.m.b.H.
C 0-0.05, Cr 60, Ni 39, bal Fe.
As cast: 765-965 N/mm² TS. Heat resistant, cast alloy. For
elevated temperature equipment.

PYROTHERM G UMCO 50
Pose-Marre Edelstahlwerk G.m.b.H.
C 0.2, Cr 28, Co 50, bal Fe.
As cast: 490-740 N/mm² TS; 235 N/mm² YS min; 6 min El.
Heat equipment. For elevated temperature equipment.

PYROTHERM G UMCO 51
Pose-Marre Edelstahlwerk G.m.b.H.
C 0.3, Cr 28, Co 50, Nb 2, bal Fe.
As cast: 490-740 N/mm² TS; 235 N/mm² YS min; 3 min El.
Heat resistant, cast alloy. For elevated temperature
equipment.

PYROTHERM G14
Pose-Marre Edelstahlwerk G.m.b.H.
C 0.5, Si 2.2, Cr 14, bal Fe.
For valves, cutlery, turbine blades, knives; Type 420;
stainless, hardenable. *Obsolete*

PYROTHERM G17
Pose-Marre Edelstahlwerk G.m.b.H.
C 0.5, Si 1.5, Cr 17, bal Fe.
Annealed: 80,000 TS; 50,000 YS; 25 El; 50 RA; 150 Brin. Cold
drawn: 130,000 TS; 120,000 YS; 2 El; 185 Brin. For chemical
plant equipment, furnace parts; corrosion resistant; Type
CB50. *Obsolete*

PYROTHERM G25/35
Pose-Marre Edelstahlwerk G.m.b.H.
C 0.5, Si 1.8, Cr 25, Ni 30, bal Fe.
For furnace parts, fixtures, conveyors, heat treat boxes;
corrosion and heat resistant. *Obsolete*

PYROTHERM G25H
Pose-Marre Edelstahlwerk G.m.b.H.
C 0.6, Si 1.5, Cr 22, bal Fe.
Cast: 70,000 TS; 65,000 YS; 2 El; 190 Brin. For furnace parts,
heat treat boxes, skids, housings; heat resistant. *Obsolete*

PYROTHERM G26/14
Pose-Marre Edelstahlwerk G.m.b.H.
C 0.4, Si 2, Cr 26, Ni 14, bal Fe.
Cast: 75,000 TS; 47,000 YS; 17 El; 25 RA; 200 Brin. For
furnace parts, salt pots, stack dampers, grate bars; Type HH;
corrosion and heat resistant. *Obsolete*

PYROTHERM G28/14
Pose-Marre Edelstahlwerk G.m.b.H.
C 0.4, Cr 28, Ni 14, bal Fe.
As cast: 440-640 N/mm² TS; 220 N/mm² YS min; 8 min El.
Heat resistant, cast alloy. For elevated temperature
equipment.

PYROTHERM UMCO 50
Pose-Marre Edelstahlwerk G.m.b.H.
C 0-0.1, Cr 28, Co 50, bal Fe.
As quenched: 70-90 kp/mm² TS; 25 kp/mm² YS min; 10 El
min. Heat resistant, rolled and wrought alloy. For furnace
engineering in cement, glass, metallurgical and
petrochemical industries. *Obsolete*

PYROTHERM UMCO 51
Pose-Marre Edelstahlwerk G.m.b.H.
C 0-0.1, Cr 28, Co 50, Nb 2.2, bal Fe.
As quenched: 80-100 kp/mm² TS; 30 kp/mm² YS min; 6 El
min. Heat resistant, rolled and wrought alloy. For furnace
engineering in cement, glass, metallurgical and
petrochemical industries. *Obsolete*

PYROTOOL A
Now CARPENTER PYROTOOL A.

PYROTOOL ALLOY V (HOT-HARD)
Carpenter Technology Corp.
Tool material. C 0.04, Mn 0.25, Si 0.25, Cr 14.5, Ni 27, Mo
1.25, Ti 3, V 0.2, bal Fe.
Age hardened, RT: 682 MPa YS; 1207 MPa TS; 37-39 Rock C.
Austenitic, precipitation hardenable alloy with high strength
and ductility up to 704°C. For high temperature tooling,
extrusion and forging dies, liners, mandrels, rams, dummy
blocks, holders, rings.

PYROTOUGH
British Steel Corp.
C 0.35, Cr 2.75, V 0.25, W 9.5, bal Fe.
Air or oil hardening high tungsten tool steel for hot work; for
tube mandrels, hot punches, hot heading dies, brass
stamping dies, plastic mold cores. *Obsolete*

PYROTOUGH 78 HOT WORK DIE STEEL
Carpenter Technology Corp.
Tool material. C 0.4, Mn 0.45, S 0-0.005, Si 1, Cr 4.45, Mo
2.05, V 0.8, bal Fe.
Heat treated and air cooled, RT: 1585-1655 MPa YS;
1790-1860 MPa TS; 9-12 El; 43-46 RA; 50.5-51.5 Rock C. For
hot working tools including mandrels for extruding copper,
brass, stainless steel and nickel base superalloys and dies for
extruding copper and brass.

PYROVAN
Latrobe Steel Co.
C 0.75, Si 1, Cr 5.25, V 2.5, Mo 1.1, bal Fe.
Hardened: 55-63 Rock C. Hot work die steel with extra high
carbon for higher hardness and wear resistance. For forging
dies, hot press and forming dies. Air hardenable, good
temperature resistance. *Obsolete*

PYTHON
AL Tech Specialty Steel Corp.
C 0.8-1.2, V 0.2, Mn 0.25, Si 0.25, bal Fe.
Typical heat treated hardness: 56-60 Rock C. Water
hardening tool steel; for blanking and forming dies, solid cold
heading dies, machine parts. AISI W-2.

Q-ALLOY "C-1"

Alloy Engineering & Casting Co.
Cr 26-30, bal Fe.
Cast: 80,000 TS; 40,000 YS; 3 El; 4 RA; 180 Brin. Heat treated: 40,000-55,000 TS; 30,000-45,000 YS; 0-1 El; 0-2 RA; 160-550 Brin. For acid mine water, nitric acid and bad sulfuric acid furnace conditions, and heat treating boxes. Corrosion and heat resistant.

Q-ALLOY "C-2"

Alloy Engineering & Casting Co.
Cr 16-18, C 0.4, bal Fe.
Cast: 60,000 TS; 55,000 YS; 12 El; 17 RA; 205 Brin. Heat treated: 80,000 TS; 60,000 YS; 20 El; 35 RA. Suited for acid mine water and nitric acid corrosion resisting parts; heat treating boxes. Corrosion resistant; heat resistant. *Obsolete*

Q-ALLOY "C-3"

Alloy Engineering & Casting Co.
Cr 28-30, Ni 0-3, high C, bal Fe.
Cast: 30,000-45,000 TS; 0 El; 0 RA; 300-600 Brin. For rolling mill guides, wear plates for crusher parts, and chute plates. Heat and wear resistant. *Obsolete*

Q-ALLOY "C.N.-1"

Alloy Engineering & Casting Co.
Cr 24-26, Ni 11-13, bal Fe.
Cast: 65,000-85,000 TS; 40,000 YS; 10-27 El; 15-45 RA; 180-220 Brin. Heat treated: 50,000-65,000 TS; 35,000-50,000 YS; 0-11 El; 0-12 RA. For furnace parts where temperature is not excessive and heat treating boxes. Heat and corrosion resistant; 2100°F maximum operating temperature. *Obsolete*

Q-ALLOY "C.N.-2"

Alloy Engineering & Casting Co.
Cr 18-20, Ni 8-10, bal Fe.
Cast: 60,000-80,000 TS; 35,000-40,000 YS; 10-50 El; 10-40 RA; 160-210 Brin. Heat treated: 80,000-100,000 TS; 35,000-50,000 YS; 5-15 El; 10-20 RA. For heat treating boxes, and furnace parts. Corrosion and heat resistant; 2000°F maximum operating temperature. *Obsolete*

Q-ALLOY A

Alloy Engineering & Casting Co.
C 0.35-0.75, Cr 15-19, Ni 64-68, bal Fe.
Cast: 65,000 TS; 36,000 YS; 9 El; 176 Brin. For lead pots, carburizing boxes, furnace fixtures, and cyanide pots. Heat and corrosion resistant. *Obsolete*

Q-ALLOY B

Alloy Engineering & Casting Co.
C 0.35-0.75, Ni 58-62, Cr 10-14, bal Fe.
Cast: 68,000 TS; 36,000 YS; 4 El; 185 Brin. For furnace parts, carburizing and heat treating boxes, gas retorts, and muffles. Heat resistant; 2000°F maximum operating temperature. *Obsolete*

Q-ALLOY B-SI

Alloy Engineering & Casting Co.
C 0.36-0.5, Ni 58-66, Cr 16-23, Si, bal Fe.
For heat and corrosion resistant parts. Heat and corrosion resistant. *Obsolete*

Q-ALLOY C-1SC

Alloy Engineering & Casting Co.
C 0-0.25, Cr 28, Ni 0-3, bal Fe.
For salt pots and furnace equipment. High corrosion resistance; brittle.

Q-ALLOY CHROME C-1

Alloy Engineering & Casting Co.
Cr 25-30, Ni 0-3, C 0.2, bal Fe.
For corrosion resistant castings. Resists mine water corrosion.

Q-ALLOY CHROME C-1A

Alloy Engineering & Casting Co.
C 0.21-0.35, Cr 24-30, bal Fe.
For heat and corrosion resistant parts. Heat and corrosion resistant. *Obsolete*

Q-ALLOY CHROME C-2

Alloy Engineering & Casting Co.
C, Cr, bal Fe.
For chemical equipment. Resists nitric acid corrosion. *Obsolete*

Q-ALLOY CHROME C-2A

Alloy Engineering & Casting Co.
C 0.13-0.2, Cr 12-15, bal Fe.
For corrosion resistant parts. Corrosion resistant.

Q-ALLOY CHROME C-2B

Alloy Engineering & Casting Co.
C 0.21-0.35, Cr 12-15, bal Fe.
For corrosion resistant parts. Corrosion resistant. *Obsolete*

Q-ALLOY CHROME C-2C

Alloy Engineering & Casting Co.
C 0-0.12, Cr 16-23, bal Fe.
For corrosion resistant parts. Heat and corrosion resistant. *Obsolete*

Q-ALLOY CHROME C-2D

Alloy Engineering & Casting Co.
C 0.13-0.2, Cr 16-23, bal Fe.
For corrosion resistant parts. Heat and corrosion resistant. *Obsolete*

Q-ALLOY CHROME C-2E

Alloy Engineering & Casting Co.
C 0.21-0.35, Cr 16-23, bal Fe.
For corrosion resistant parts. Heat and corrosion resistant. *Obsolete*

Q-ALLOY CHROME C-3

Alloy Engineering & Casting Co.
C, alloy, bal Fe.
500 Brin. For mill guides. Heat resistant to 2000°F. *Obsolete*

Q-ALLOY CHROME C-3A

Alloy Engineering & Casting Co.
C 0.81-1.1, Cr 16-23, bal Fe.
For heat and corrosion resistant parts. Heat and corrosion resistant.

Q-ALLOY CHROME CN1-A

Alloy Engineering & Casting Co.
C 0.21-0.35, Cr 24-30, Ni 7-11, bal Fe.
For heat and corrosion resistant parts. Heat and corrosion resistant. *Obsolete*

Q-ALLOY CHROME CN1-B

Alloy Engineering & Casting Co.
C 0.13-0.2, Cr 24-30, Ni 12-15, bal Fe.
For heat and corrosion resistant parts. Heat and corrosion resistant. *Obsolete*

Q-ALLOY CHROME CN1-S

Alloy Engineering & Casting Co.
C 0.36-0.5, Cr 24-30, Ni 7-11, bal Fe.
For heat and corrosion resistant parts. Heat and corrosion resistant. *Obsolete*

Q-ALLOY CHROME CN2-A

Alloy Engineering & Casting Co.
C 0.21-0.35, Cr 16-23, Ni 7-11, W, bal Fe.
For heat and corrosion resistant parts. Heat and corrosion resistant. *Obsolete*

Q-ALLOY CHROME J

Alloy Engineering & Casting Co.
C 0-0.12, Cr 12-15, bal Fe.
For corrosion resistant parts. Corrosion resistant. *Obsolete*

Q-ALLOY CHROME K-1

Alloy Engineering & Casting Co.
Ni, Cr, bal Fe.
For furnace parts not subject to high sulfur fumes. High heat resistance up to 2200°F. *Obsolete*

Q-ALLOY CHROME KA-2

Alloy Engineering & Casting Co.
C 0.13-0.2, Cr 16-23, Ni 7-11, bal Fe.
For heat and corrosion resistant parts. Stainless. *Obsolete*

Q-ALLOY CHROME KA-2H

Alloy Engineering & Casting Co.
C 0.21-0.35, Cr 16-23, Ni 7-11, bal Fe.
For heat and corrosion resistant parts. Stainless. *Obsolete*

Q-ALLOY CHROME KA-2MO

Alloy Engineering & Casting Co.
C 0.13-0.2, Cr 16-23, Ni 7-11, Mo, bal Fe.
For heat and corrosion resistant parts. Heat and corrosion resistant. *Obsolete*

Q-ALLOY CHROME KA-2S

Alloy Engineering & Casting Co.
C 0-0.12, Cr 16-23, Ni 7-11, bal Fe.
For stainless parts. Stainless. *Obsolete*

Q-ALLOY CHROME KA-4

Alloy Engineering & Casting Co.
C 0.21-0.35, Cr 16-23, Ni 7-11, Mo, bal Fe.
For heat and corrosion resistant parts. Heat and corrosion resistant. *Obsolete*

Q-ALLOY CN-1 MO

Alloy Engineering & Casting Co.
Cr 24-26, Ni 11-13, Mo 1.5-3, bal Fe.
For pulp industry; exposure to hot sulfite liquors and bleaches. Corrosion and heat resistant. *Obsolete*

Q-ALLOY CN-1H-MO

Alloy Engineering & Casting Co.
C 0.13-0.2, Cr 24-30, Ni 7-11, Mo, bal Fe.
For heat and corrosion resistant parts. Heat and corrosion resistant. *Obsolete*

Q-ALLOY CN1-H

Alloy Engineering & Casting Co.
C 0.13-0.2, Cr 24-30, Ni 7-11, bal Fe.
For furnace chains, muffles, and retorts. Heat and corrosion resistant. *Obsolete*

Q-ALLOY CN2-MO

Alloy Engineering & Casting Co.
Cr 17-21, Ni 7-9, Mo 1-4, C 0.2, bal Fe.
For chemical equipment. Stainless; austenitic. *Obsolete*

Q-ALLOY KNC-3

Alloy Engineering & Casting Co.
C 0.13-0.2, Cr 24-30, Ni 16-23, Si, bal Fe.
For heat and corrosion resistant parts. Heat and corrosion resistant. *Obsolete*

Q-ALLOY X-2

Alloy Engineering & Casting Co.
Ni 30, Cr 30, C, bal Fe.
For high temperature applications. Corrosion and heat resistant. *Obsolete*

Q-ALLOY X-3

Alloy Engineering & Casting Co.
Ni 20, Cr 30, C, bal Fe.
For high temperature applications. Corrosion and heat resistant. *Obsolete*

Q-ALLOY X-4

Alloy Engineering & Casting Co.
Ni 20, Cr 25, C, bal Fe.
For high temperature applications. Corrosion and heat resistant. *Obsolete*

Q-ALLOY X-5

Alloy Engineering & Casting Co.
Ni 5, Cr 28, C, bal Fe.
For high temperature applications. Corrosion and heat resistant. *Obsolete*

Q-ALLOY X-6

Alloy Engineering & Casting Co.
Ni 16, Cr 28, C, bal Fe.
For high temperature applications. Corrosion and heat resistant. *Obsolete*

Q-ALLOY X-7

Alloy Engineering & Casting Co.
C 0.2, Cr 25, Ni 12, bal Fe.
For furnace rails and rollers, oil still tube supports. Austenitic; resists thermal shock and high stresses. *Obsolete*

Q-ALLOY X-8

Alloy Engineering & Casting Co.
Ni 10, Cr 20, C, bal Fe.
For high temperature applications. Corrosion and heat resistant.

Q-ALLOY, GRADE A+

Alloy Engineering & Casting Co.
Ni 66-68, Cr 19-21, bal Fe.
Cast: 80,000-90,000 TS; 45,000-50,000 YS. For carburizing boxes, retorts, solution pots, and furnace parts. Heat resistant to 2220°F. *Obsolete*

Q-TEMP 10B18Q

United States Steel Corp.
C 0.15-0.2, Mn 0.8-1.1, Si 0.2-0.35, 0.0005 B min, bal Fe.
Rolled: 70,400 TS; 38,700 YP; 25 El; 63 RA; 72 Rock B. For fasteners, bolts, screws, clips. Tough; water hardened.

Q-TEMP 10B21Q

United States Steel Corp.
C 0.18-0.23, Mn 0.8-1.1, Si 0.2-0.35, 0.0005 B min, bal Fe.
Rolled: 72,000 TS; 39,000 YS; 24 El; 62 RA; 73 Rock B. For fasteners, bolts, screws, clips. Tough; water hardened.

Q-TEMP 10B22Q

United States Steel Corp.
C 0.17-0.23, Mn 1-1.3, Si 0.2-0.35, 0.0005 B min, bal Fe.
Rolled: 75,000 TS; 40,000 YS; 22 El; 60 RA; 76 Rock B. For fasteners, screws, bolts, nuts. Tough; water hardened.

Q-TEMP 10B23Q

United States Steel Corp.
C 0.17-0.23, Mn 1.1-1.4, Si 0.2-0.35, 0.0005 B min, bal Fe.
Rolled: 80,000 TS; 44,000 YS; 20 El; 56 RA; 78 Rock B. For fasteners, screws, bolts, nuts. Tough; water hardened.

Q-TEMP 41B20 Q

United States Steel Corp.
C 0.18-0.23, Mn 0.75-1, Si 0.1-0.35, Cr 0.25-0.4, Mo 0.15-0.25, 0.0005 B min, bal Fe.
Oil or water hardened to 160 ksi min YS. For bolts, fasteners.

Q-TEMP 41BV20 Q

United States Steel Corp.
C 0.18-0.23, Mn 0.75-1, Si 0.1-0.35, Cr 0.25-0.4, Mo 0.15-0.25, V 0.03-0.08, 0.0005 B min, bal Fe.
Oil or water hardened to 160 ksi min YS. For bolts, fasteners.

Q-TEMP 50B23 Q

United States Steel Corp.
C 0.2-0.25, Mn 0.75-1, Si 0.1-0.35, Cr 0.25-0.4, 0.0005 B min, bal Fe.
Oil or water hardened to 165 ksi min YS. For bolts, fasteners.

Q-TEMP 60B20 Q

United States Steel Corp.
C 0.18-0.23, Mn 0.75-1, Si 0.1-0.35, Cr 0.25-0.4, V 0.03-0.08, 0.0005 B min, bal Fe.
Oil or water hardened to 160 ksi min YS. For bolts, fasteners.

QE 22 A

Fansteel/Wellman Dynamics
Magnesium. Ag 2-3, Zr 0.4-1, RE 1.75-2.25, bal Mg.
Cast: 35,000 psi TS; 25,000 psi YS; 2 El; 80 Brin. High strength magnesium castings.

QE 22A (STANDARD ALLOY)

Dow Chemical Co.
Magnesium. Zr 0.7, bal Mg.
T6-temper: 40,000 psi TS; 30,000 psi YS; 4 El. At 400°F: 28,000 psi TS; 25,000 psi YS; 22 El. For aircraft, jet engine and missile components; highest cast yield strength; age-hardenable. *Obsolete*

QH21A

Fansteel/Wellman Dynamics
Magnesium. RE 0.6-1.5, Ag 2-3, Zr 0.4-1, Th 0.6-1.6, bal Mg.
Magnesium casting alloy. Cast: 35,000 psi TS; 27,000 psi YS; 2 El; 80 Brin.

QRO 90 SUPREME

Uddeholm Corp.
C 0.4, Si 0.3, Mn 0.75, Cr 2.6, Mo 2.25, V 0.9, bal Fe.
245 ksi TS; 205 ksi YP; 9 El; 40 RA; 50 Rock C. High performance, chromium-molybdenum-vanadium alloyed hot work tool steel. For tools for die casting, extrusion and hot forging.

QT 100

Algoma Steel Corp. Ltd.
HSLA steel. C 0.2, Mn 1.5, P 0.035, S 0.035, Si 0.5, Cb 0.06, Mo 0-0.3, B 0.003, bal Fe.
115 ksi TS min; 100 ksi YS min; 18 El min; 240-300 Brin. Quenched and tempered plate. For use in construction equipment, bridges, tanks, off-highway equipment and offshore structures. Weldable and corrosion resistant.

QT 700

Algoma Steel Corp. Ltd.
HSLA steel. C 0.2, Mn 1.5, P 0.035, S 0.035, Si 0.5, Cb 0.06, Mo 0-0.3, B 0.003, bal Fe.
790 MPa TS min; 700 MPa YS min; 18 El min; 240-300 Brin. Quenched and tempered plate. For use in construction equipment, bridges, tanks, off-highway equipment and offshore structures. Weldable and corrosion resistant.

QT.131 GRADE A

British Steel plc
C 0-0.2, Si 0-0.5, Mn 0-1.6, Nb 0-0.5, bal Fe.
Quenched and tempered steel. *Obsolete*

QT.131 GRADE B

British Steel plc
C 0-0.2, Si 0.1-0.5, Mn 0-1.8, Nb 0-0.5, bal Fe.
Quenched and tempered steel. *Obsolete*

QT.445 (1)

British Steel plc
C 0-0.15, Si 0-0.25, Mn 0-1.2, Mo 0-0.35, Cr 0-0.4, Ni 0-1.6, V 0.08-0.12, bal Fe.
Used mainly for large pressure vessel or boiler construction.

QT.445 (2)

British Steel plc
C 0-0.15, Si 0-0.25, Mn 0-1.2, Mo 0-0.35, Cr 0-0.8, Ni 0-1.6, bal Fe.
Good impact strength at low temperatures; weldable. For crane parts, storage tanks, and mining equipment.

QT.445 GRADE A

British Steel plc
C 0.15-0.18, Si 0-0.8, Mn 0.8-1.1, Mo 0.18-0.4, Cr 0.5-0.8, Zr 0.05-0.15, B 0.0005-0.0025, O, bal Fe.
Quenched and tempered steel. For crane structural members, oil rigs, and bridges.

QT.445 GRADE B

British Steel plc
C 0.15-0.18, Si 0-0.9, Mn 0.8-1.1, Mo 0.25-0.6, Cr 0.5-0.8, Zr 0.05-0.15, B 0.0005-0.0025, bal Fe.
Quenched and tempered steel. For crane structural members, oil rigs, and bridges.

QUAKER

Gulf Steel Corp.
C 0.4, Mn 0.4, Cr 4.5, Mo 1, W 0.8, bal Fe.
For tools, dies; air or oil hardened.

QUAKER LC2

Quaker Alloy Casting Co.
C 0.18, Mn 0.65, Si 0.45, Ni 2.4, bal Fe.
Normalized, quenched and tempered: 70,000 psi TS; 40,000 psi YS; 24 El; 160 Brin. Steel castings for low temperature operations. ASTM A 352 Gr. LC2.

QUAKER LC3

Quaker Alloy Casting Co.
C 0.1, Mn 0.65, Si 0.45, Ni 3.5, bal Fe.
Normalized, quenched and tempered: 70,000 psi TS; 40,000 psi YS; 24 El; 170 Brin. Nickel steel castings for low temperature operations. ASTM A 352 Gr. LC3.

QUAKER LCA

Quaker Alloy Casting Co.
C 0.2, Mn 0.6, Si 0.45, bal Fe.
Normalized, quenched and tempered: 60,000 psi TS; 30,000 psi YS; 24 El; 150 Brin. Carbon steel castings for low temperature operations. ASTM A 352 Gr. LCA.

QUAKER LCB

Quaker Alloy Casting Co.
C 0.25, Mn 0.7, Si 0.45, bal Fe.
Normalized, quenched and tempered: 65,000 psi TS; 35,000 psi YS; 24 El; 160 Brin. Carbon steel castings for low temperature operations. ASTM A 352 Gr. LCB.

QUAKER LCC

Quaker Alloy Casting Co.
C 0.2, Mn 0.85, Si 0.45, bal Fe.
Normalized, quenched and tempered: 70,000 psi TS; 40,000 psi YS; 22 El; 160 Brin. Carbon steel castings for low temperature operations. ASTM A 352 Gr. LCC.

QUAKER Q05

Quaker Alloy Casting Co.
C 0.05, Mn 0.1, Si 0.6, bal Fe.
Cast, annealed: 50,000 psi TS; 28,000 psi YS; 30 El; 120 Brin. Similar to AISI 1005.

QUAKER Q105

Quaker Alloy Casting Co.
C 0.3, Mn 0.8, Si 0.45, Cr 0.6, Ni 0.6, Mo 0.2, bal Fe.
Normalized, quenched and tempered: 105,000 psi TS; 85,000 psi YS; 17 El; 230 Brin. Alloy steel castings for structural service. ASTM A148 Gr. 105-85. Similar to AISI 8630.

QUAKER Q120

Quaker Alloy Casting Co.
C 0.3, Mn 0.8, Si 0.45, Cr 0.6, Ni 0.6, Mo 0.2, bal Fe.
Normalized, quenched and tempered: 120,000 psi; 95,000 psi YS; 14 El; 260 Brin. Alloy steel castings for structural service. ASTM 148 Gr. 120-95. Similar to AISI 8630.

QUAKER Q150

Quaker Alloy Casting Co.
C 0.3, Mn 0.8, Si 0.45, Cr 0.6, Ni 0.6, Mo 0.2, bal Fe.
Normalized, quenched and tempered: 150,000 psi TS; 125,000 psi YS; 9 El; 330 Brin. Alloy steel castings for structural service. ASTM A148 Gr. 150-125. Similar to AISI 8630.

QUAKER Q1CM

Quaker Alloy Casting Co.
C 0.15, Mn 0.65, Si 0.45, Cr 1.25, Mo 0.5, bal Fe.
Normalized and tempered: 70,000 psi TS; 40,000 psi YS; 20 El; 150 Brin. Alloy steel castings for elevated temperature service. ASTM A 217 Gr. WC6; A 356 Gr. 6.

QUAKER Q2CM

Quaker Alloy Casting Co.
C 0.15, Mn 0.6, Si 0.45, Cr 2.4, Mo 1, bal Fe.
Normalized and tempered: 70,000 psi TS; 40,000 psi YS; 20 El; 170 Brin. Alloy steel castings for elevated temperature service. ASTM A 217 Gr. WC9; A 356 Gr.10.

QUAKER Q41

Quaker Alloy Casting Co.
C 0.4, Mn 0.7, Si 0.45, Cr 1, Mo 0.25, bal Fe.
Normalized and tempered: 100,000 psi TS; 70,000 psi YS; 15 El; 220 Brin. Alloy steel castings for structural service. AISI 4140.

QUAKER Q4340

Quaker Alloy Casting Co.
C 0.4, Mn 0.7, Si 0.45, Cr 0.8, Ni 2, Mo 0.4, bal Fe.
Normalized and tempered: 100,000 psi TS; 70,000 psi YS; 15 El; 220 Brin. Alloy steel castings for structural service. Similar to AISI 4340.

QUAKER Q5M

Quaker Alloy Casting Co.
C 0.15, Mn 0.6, Si 0.45, Cr 5.5, Mo 0.5, bal Fe.
Normalized and tempered: 90,000 psi TS; 60,000 psi YS; 18 El; 200 Brin. Alloy steel castings for elevated temperature service. ASTM A 217 Gr. C5; SAE 60502.

QUAKER Q60

Quaker Alloy Casting Co.
C 0.2, Mn 0.6, Si 0.45, bal Fe.
Normalized and tempered: 60,000 psi TS; 30,000 psi YS; 24 El; 140 Brin. Carbon steel castings. ASTM A27 Gr. 60-30.

QUAKER Q65

Quaker Alloy Casting Co.
C 0.2, Mn 0.65, Si 0.45, bal Fe.
Normalized and tempered: 65,000 psi TS; 35,000 psi YS; 24 El; 140 Brin. Carbon steel castings. ASTM A27 Gr. 65-35.

QUAKER Q70

Quaker Alloy Casting Co.
C 0.25, Mn 0.7, Si 0.45, bal Fe.
Normalized and tempered: 70,000 psi TS; 36,000 psi YS; 22 El; 150 Brin. Carbon steel castings. ASTM A27 Gr. 70-36.

QUAKER Q70-1

Quaker Alloy Casting Co.
C 0.2, Mn 0.85, Si 0.45, bal Fe.
Normalized and tempered: 70,000 psi TS; 40,000 psi YS; 22 El; 160 Brin. Carbon steel castings. ASTM A27 Gr. 70-40.

QUAKER Q80

Quaker Alloy Casting Co.
C 0.4, Mn 0.7, Si 0.45, bal Fe.
Normalized and tempered: 80,000 psi TS; 40,000 psi YS; 17 El; 180 Brin. Carbon steel castings. ASTM A148 Gr. 80-40.

QUAKER Q86

Quaker Alloy Casting Co.
C 0.15, Mn 0.8, Si 0.45, Cr 0.6, Ni 0.6, Mo 0.2, bal Fe.
Normalized and tempered: 80,000 psi TS; 50,000 psi YS; 22 El; 180 Brin. Alloy steel castings for structural service. ASTM A148 Gr. 80-50. Similar to AISI 8615.

QUAKER Q90

Quaker Alloy Casting Co.
C 0.2, Mn 0.8, Si 0.45, Cr 0.9, Ni 0.6, Mo 0.2, bal Fe.
Normalized and tempered: 90,000 psi TS; 60,000 psi YS; 20 El; 190 Brin. Alloy steel castings for structural service. ASTM A148 Gr. 90-60. Similar to AISI 8630.

QUAKER Q9M

Quaker Alloy Casting Co.
C 0.15, Mn 0.5, Si 0.45, Cr 9, Mo 1, bal Fe.
Normalized and tempered: 90,000 psi TS; 60,000 psi YS; 18 El; 200 Brin. Alloy steel castings for elevated temperature service. ASTM A 217 Gr. C12.

QUAKER QCCM

Quaker Alloy Casting Co.
C 0.15, Mn 0.65, Si 0.45, Cr 0.55, Mo 0.5, bal Fe.
Normalized and tempered: 70,000 psi TS; 40,000 psi YS; 22 El; 150 Brin. Alloy steel castings for elevated temperature service. ASTM A356 Gr. 5.

QUAKER QCN

Quaker Alloy Casting Co.
C 0.03, Mn 0.8, Si 0.4, Ni 0.3, Cu 0.69, Cb 1.
As cast: 60,000 psi TS; 32,000 psi YS; 20 El. Cu-Ni casting.

QUAKER QHB

Quaker Alloy Casting Co.
C 0.08, Mn 0.75, Si 0.4, Ni 66, Mo 28, V 0.35, bal Fe.
Solution annealed: 72,000 psi TS; 46,000 psi YS; 6 El. High alloy, corrosion resistant castings. Hastelloy B Type; ASTM A296 N-12M.

QUAKER QHC

Quaker Alloy Casting Co.
C 0.08, Mn 0.75, Si 0.4, Cr 16.5, Ni 57, Mo 17, W 4.25, V 0.3, bal Fe.
Solution annealed: 72,000 psi TS; 46,000 psi YS; 4 El. High alloy, corrosion resistant castings. Hastelloy C Type; ASTM A296 CW 12M.

QUAKER QINC

Quaker Alloy Casting Co.
C 0.05, Mn 0.75, Si 0.75, Cr 15.5, Ni 75, bal Fe.
As cast: 70,000 psi TS; 28,000 psi YS; 30 El. Ni-Cr-Fe alloy castings; corrosion resistant. ASTM A296 Cy-40.

QUAKER QMC

Quaker Alloy Casting Co.
C 0.2, Mn 0.65, Si 0.45, Mo 0.5, bal Fe.
Normalized and tempered: 65,000 psi TS; 35,000 psi YS; 24 El; 140 Brin. Alloy steel castings for elevated temperature service. ASTM A217 Gr. WC1; A356 Gr.2.

QUAKER QML

Quaker Alloy Casting Co.
C 0.15, Mn 1, Si 1.2, Ni 66, Cu 30.
As cast: 65,000 psi TS; 32,000 psi YS; 25 El. Monel castings; Monel 410. ASTM A296 M35.

QUAKER QML-CB

Quaker Alloy Casting Co.
C 0.15, Mn 1, Si 1.2, Ni 66, Cu 3, Cb 1.3.
As cast: 65,000 psi TS; 32,500 psi YS; 25 El. Monel type castings; similar to Monel 411.

QUAKER QNI

Quaker Alloy Casting Co.
C 0.15, Mn 1, Si 1, Ni 97, bal Fe.
As cast: 50,000 psi TS; 18,000 psi YS; 10 El. High nickel castings; corrosion resistant. ASTM A296 CZ100; Nickel 210.

QUALITAT-55

Aluminium-Zentral e.V.
Aluminum. Cu 3-5, Si 0.3-0.7, Mn 0.3-0.8, bal Al.
Heat treated: 50,000-60,000 TS; 18-23 El; 100-110 Brin. For rivets, hydraulic fittings, aircraft engine components. Heat treatable, high fatigue strength. *Obsolete*

QUALITAT-M

Aluminium-Zentral e.V.
Aluminum. Si 0.85, Mn 0.6-0.9, Mg 0.5-1, bal Al.
Heat treated: 50,000-60,000 TS; 3-8 El; 95-100 Brin. For general light structures, booms, scaffolds, transmission tires. Good strength and fabricability. Heat treatable. *Obsolete*

QUALITY CARBON DRILL ROD

Latrobe Steel Co.
C 1.1-1.25, Mn 0.15-0.4, Si 0.1-0.4, P 0.03, S 0.03, bal Fe.
Water hardening tool steel, AISI W1. Type W1 water hardening tool steel. *Obsolete*

QUALITY STEEL

Engelhard Corp.
Cr 1.2-1.5, Ni 4.5-5, Mn 0.3-0.5, C 0.3-0.35, bal Fe.
226,000 psi TS; 15 El; 45 RA. For gears; air hardening. *Obsolete*

QUARZAL

German manufacture
Cu 15, Si 0-10, Fe, Ni, Cr, Mo, bal Al.
For bearings; wear resisting.

QUARZAL Q12

German manufacture
Cu 5-15, Fe 0-1, 0.5 hardening agent, bal Al.
125 Brin. For bearings.

QUARZAL Q15

German manufacture
Cu 5-15, Fe 0-1, 0.5 hardening agent, bal Al.
125 Brin. For bearings.

QUARZAL Q2

German manufacture
Cu 2, Fe 0-1, 0.5 hardening agent, bal Al.
For bearings.

QUARZAL Q5

German manufacture
Cu 5-15, Fe 0-1, 0.5 hardening agent, bal Al.
125 Brin. For bearings.

QUARZAL Q8

German manufacture
Cu 5-15, Fe 0-1, 0.5 hardening agent, bal Al.
For bearings.

QUATRE ECLAIR SPECIAL

Creusot-Loire
C 0.85, W 19, Co 3, V 1.5, Mo 1, bal Fe.
For lathe and planer tools; oil hardened. *Obsolete*

QUATRE ECLAIRS

Creusot-Loire
C 0.85, W 18, V 1.6, Mo 0.6, bal Fe.
For lathe and planer tools, saws; oil hardened. *Obsolete*

QUEEN'S METAL

English manufacture
Sn 88.5, Sb 7, Cu 3.5, Zn 1.
For utensils; resists tarnishing.

QUEEN'S METAL "B"

English manufacture
Sn 50.5, Sb 16.5, Pb 16.5, Zn 16.5.
For type metal.

QUEEN'S METAL "C"

English manufacture
Sn 87, Sb 8.5, Cu 3.5, Zn 1.
For type metal.

QUEEN'S METAL "D"

English manufacture
Sn 73.36, Sb 8.88, Pb 8.88, Zn 8.88.
For type metal.

QUEEN'S METAL "E"

English manufacture
Sn 88.5, Sb 7, Cu 3.5, Bi 1.
For type metal.

QUICKSILVER SOLDER

Manufacturer not listed
Ag 57-63, Cu 21-25, Sn 3.8-6.2, Zn 10-12.
For silver solder; corrosion resistant.

QUINZE ECLAIRS

Creusot-Loire
C 0.98, W 20, Co 15, V 1.1, Mo 1, bal Fe.
For cutting and boring tools; oil hardened. *Obsolete*

R 2
Saarstahl AG
Now SAARSTAHL CK 22.

R 2800
Telcon Metals Ltd.
Ni-Fe.
Temperature compensating alloy.

R 3
Saarstahl AG
Now SAARSTAHL CK 35.

R 304 UD
Kawasaki Steel Corp.
Stainless steel. C 0.12, Mn 1, P 0-0.04, S 0-0.015, Si 0.5, Cu 2, Ni 7.2, Cr 14, Mo 0.4, bal Fe.
100,000 psi TS; 65 El. Austenitic stainless steel for deep drawing purposes.

R 3F
Saarstahl AG
Now SAARSTAHL CF 35.

R 4
Saarstahl AG
Now SAARSTAHL CK 45.

R 4
Japan Metal Industry Co. Ltd.
C 0.07, Cr 25.43, Ni 4.19, Mo 1.52, bal Fe.
Corrosion resistant alloy.

R 409 SR
Kawasaki Steel Corp.
Stainless steel. C 0.008, Mn 0.5, P 0-0.04, S 0-0.015, Si 1.4, Cr 11.5, Ti 0.25, bal Fe.
70,000 psi TS; 35 El. Heat resistant stainless steel for automotive catalytic exhaust gas converter systems.

R 410 L
Kawasaki Steel Corp.
Stainless steel. C 0.02, Mn 0.45, P 0-0.04, S 0-0.015, Si 0.5, Cr 12.5, bal Fe.
63,000 psi TS; 36 El. Heat resistant stainless steel for automotive exhaust gas systems.

R 410 UL
Kawasaki Steel Corp.
Stainless steel. C 0.006, Mn 0.45, P 0-0.04, S 0-0.015, Si 0.5, Cr 12.5, bal Fe.
60,000 psi TS; 40 El. Heat resistant stainless steel for automotive exhaust gas systems.

R 430 LT
Kawasaki Steel Corp.
Stainless steel. C 0.006, Mn 0.9, P 0-0.04, S 0-0.015, Si 0.45, Cr 16.3, Ti = 20 x C + N, bal Fe.
Corrosion resistant ferrite type stainless steel. 67,000 psi TS; 28 El.

R 434 LT (N)-1
Kawasaki Steel Corp.
Stainless steel. C 0.006, Mn 0.4, P 0-0.04, S 0-0.015, Si 0.045, Cr 16.3, Mo 0.95, Ti = 20 x C + N, Cb = 20 x C + N, bal Fe.
71,000 psi TS; 33 El. Corrosion resistant ferrite type stainless steel.

R 434 LT (N)-2
Kawasaki Steel Corp.
Stainless steel. C 0.006, Mn 0.45, P 0-0.04, S 0-0.01, Si 0.45, Cr 18.5, Mo 2, Ti = 20 x C + N, Cb = 20 x C + N, bal Fe.
77,000 psi TS; 32 El. Corrosion resistant ferrite type stainless steel.

R 4F
Saarstahl AG
C 0.45, Si 0.25, Mn 0.65, bal Fe.
Carbon steel for structural purposes. Fine grain steel. W.-Nr. 1193. *Obsolete*

R 5F
Saarstahl AG
Now SAARSTAHL CF 53.

R 7
Saarstahl AG
C 0.75, Si 0.25, Mn 0.7, bal Fe.
Water hardenable construction steel. *Obsolete*

R 7F
Saarstahl AG
C 0.7, Si 0.25, Mn 0.3, bal Fe.
Carbon steel for structural purposes. Fine grain steel. W.Nr. 1249. *Obsolete*

R BRAND COPPER
Climax Performance Materials Corp.
Fire refined copper ingot. For remelting copper for copper alloys.

R CUPRO-NICKEL (70/30)
Henry Wiggin & Co. Ltd.
Ni 30, Cu 70.
For heavy current resistances; low specific resistance. *Obsolete*

R CUPRO-NICKEL (80/20)
Henry Wiggin & Co. Ltd.
Ni 20, Cu 80.
For heavy current resistances; low specific resistance. *Obsolete*

R CUPRO-NICKEL (90/10)
Henry Wiggin & Co. Ltd.
Ni 10, Cu 90.
For heavy current resistances; low specific resistance. *Obsolete*

R Z 1
Saarstahl AG
C 1.05, Si 0.2, Mn 0.4, bal Fe.
For cold worked tension springs. W.-Nr. 1274. *Obsolete*

R Z 5
English manufacture
Zn 3.5-5, Zr 0.4-1, 0.75-1.75 Rare Earths, bal Mg.
Sand cast and precipitation treated: 200 MPa TS; 135 MPa YS; 4 El 65 Brin. BS 2970 MAG 5.

R-235
Now ALLVAC R-235.

R-303
Reynolds Metals Co.
Zn 6.5-7.5, Cu 1, Mg 2, Fe 0.3, Si 0.2, Cr 0.2, Ni 0.1, bal Al.
Annealed: 28,000 TS; 14,000 YS; 20 El. Heat treated: 82,000 TS; 75,000 YS; 12 El. For light alloy parts; high strength. *Obsolete*

R-41
Driver Harris Co.
Nickel. Cr 19, Co 11, Mo 10, Ti 3, Al 2, bal Ni.
Wire, rod, ribbon for welding wire (Rene 41), and fastener stock. *Obsolete*

R-63
Driver Harris Co.
Nickel. Mn 4, bal Ni.
Wire and ribbon for combustion engine electrodes. *Obsolete*

R-M-20
Ross-Meehan Foundries
C 0-0.07, Ni 29, Cr 20, Mo 2-3, Cu 4, bal Fe.
Cast: 65,000-75,000 TS; 28,000-38,000 YS; 35-50 El; 40-50 RA; 120-150 Brin. For chemical plant equipment, mixers; resists mixed acids, austenitic. *Obsolete*

R-MONEL
Now MONEL ALLOY R-405.

R-P
Darwin & Milner Inc.
Tool material. C 0.65, Mn 0.8, Si 1.95, bal Fe.
For tools; oil hardened.

R-S "A" METAL
Milwaukee Steel Foundry Co.
C 3.2, Mn, Si, bal Fe.
For machinery castings, gears, abrasion resistant, cast iron.

R-S "H" METAL
R-S Products Corp.
C, alloy, bal Fe.
For valves; abrasion and corrosion resistant.

R-THERM 20-12 MO
SWB Stahlformguss Gesellschaft mbH
C 0.2, Cr 12, Mo 1.2, bal Fe.
Annealed: 80,000 TS; 40,000 YS; 24 El; 94 Rock B. Hardened: 220,000 TS; 180,000 YS; 12 El; 48 Rock C. For gears, cutlery, surgical instruments, valves, bearings. Corrosion resistant, hardenable. *Obsolete*

R-THERM 2012 MOWV
SWB Stahlformguss Gesellschaft mbH
C 0.22, Cr 11.5, Mo 1, Ni 0.5, V 0.3, W 0.5, bal Fe.
Annealed: 90,000 TS; 40,000 YS; 25 El; 92 Rock B. Hardened: 235,000 TS; 200,000 YS; 10 El; 50 Rock C. For valves, cutlery, bearings, surgical instruments. Corrosion resistant, hardenable. *Obsolete*

R.B. CHISEL
CCS Braeburn Alloy Steel
C, Cr, bal Fe.
For tools, chisels. *Obsolete*

R.B. SPECIAL
CCS Braeburn Alloy Steel
C, alloy, bal Fe.
For cutters, tools, dies; oil hardened. *Obsolete*

R.C.C.B.
Societe des Acieries de Longwy
C 2.25, Cr 13, bal Fe.
For tools, dies, nondeforming.

R.C.F.
Johnson Matthey
Ni 1.5, C 0.3, Cr 0.9, bal Fe.
For wear resistant cast steel parts. *Obsolete*

R.C.W. 2
Societe des Acieries de Longwy
C 0.28, W 9.5, Cr 2.5, V 0.1, bal Fe.
For hot work dies and tools; hot work steels.

R.C.W. 2VA
Societe des Acieries de Longwy
C 0.6, W 9.5, Cr 2.5, V 0.1, bal Fe.
For hot work dies and tools; hot work steels.

R.C.W.M.
Societe des Acieries de Longwy
C 0.5, W 9.5, Cr 2.5, V 0.1, bal Fe.
For hot work dies and tools, hot work steels.

R.D.S.
Now CARPENTER R.D.S.

R.D.X.
Joseph T. Ryerson & Son Inc.
C 0.7-0.8, Mo 0.2-0.3, bal Fe.
For tools, chisels, header dies; water hardening. *Obsolete*

R.F.F.
Societe des Acieries de Longwy
C 0.5, Mn 0.7, Si 1.9, Mo 0.2, bal Fe.
For tools, dies, punches; tough.

R.H.1
Manufacturer not listed.
C 0.8-0.9, W 17-20, Cr 3.5-5, V 1, Mo 3, bal Fe.
620 Brin. For welding rod for high speed steel; heat and wear resistant.

R.N.D
Agil Chemie
C 0.08, Mn 1-1.25, bal Fe.
Welded: 55,000 TS; 45,000 YS; 33-40 El; 60-80 RA. For welding rod.

R.R. 250
Sterling International Technology Ltd.
Cu 4.5-5.5, Mn 0.2-0.3, Ni 0.8-1.2, Ti 0.15-0.25, Sb 0.1-0.4, Co 0.1-0.4, bal Al.
Sand cast: 31,400-35,800 TS; 29,000 YS; 0-2 El; 90-100 Brin; 0.6 Izod. Modified "Y" alloy for service up to 250°C.

R.R. 350
Sterling International Technology Ltd.
Cu 4.5-5.5, Mn 0.2-0.3, Ni 1.3-1.8, Ti 0.15-0.25, Sb 0.1-0.4, Co 0.1-0.4, Zr 0.1-0.3, bal Fe.
Sand cast: 31,400-35,800 TS; 29,000 YS; 0-2 El; 90-100 Brin; 0.6 Izod. Modified "Y" alloy for service up to 350°C.

R.R. 50
Sterling International Technology Ltd.
Cu 0.8-2, Mg 0.05-0.2, Si 1.5-2.8, Fe 0.8-1.4, Ni 0.8-1.7, Ti 0.05-0.3, bal Al.
Sand cast: 22,400-27,800 TS; 14,500-16,800 YS; 2-4 El; 60-70 Brin; 1.6 Izod. General purpose sand or chill cast aluminum. BS 1490 LM 23-P; BS 2L51.

R.R.ALLOY C.242
Rolls-Royce Mfg. Co.
Nickel. C 0.3, Cr 20, Co 10, Mo 10, Fe 0-1, bal Ni.
For high temperature applications; heat and corrosion resistant.

R.S.V.
Societe des Acieries de Longwy
C 0.7, W 18, Cr 4, V 1, bal Fe.
For tools, dies, cutters, reamers; high speed steel.

R.T. 10
Societe des Acieries de Longwy
C 0.9-1, bal Fe.
For tools, taps, drills, punches; water hardened.

R.T. 6
Societe des Acieries de Longwy
C 0.7-0.8, bal Fe.
For tools, punches, springs; water hardened.

R.T. 7
Societe des Acieries de Longwy
C 0.7-0.8, bal Fe.
For tools, punches, springs; water hardened.

R.T. 8
Societe des Acieries de Longwy
C 0.9-1, bal Fe.
For tools, drills, springs; water hardened.

R.T.W. 2
Societe des Acieries de Longwy
C 0.5, Cr 1.5, W 2.25, V 0.2, bal Fe.
For tools, hot work dies; hot work steel.

R.T.W. 3
Societe des Acieries de Longwy
C 1.3, Cr 4.5, bal Fe.
For tools, dies; air or oil hardened.

R.U.S.
Societe des Acieries de Longwy
C 0.95, Mn 1.2, Cr 0.5, W 0.5, bal Fe.
For tools, dies; nondeforming.

R.V.M.
British Steel Corp.
C 1.55, Cr 11.5, Mo 0.75, V 0.25, bal Fe.
Air or oil hardening non-distorting complex alloy steel; for hot shear blades, tube expanders, swaging dies, cold drawing dies, press tools. *Obsolete*

R/X AG
Reactor Experiments, Inc.
Precious metal. Ag 99.9.
Foil (0.5-25) microns for radiation filters.

R/X AL
Reactor Experiments, Inc.
Precious metal. Al 99.5.
Foil (0.8-25) microns for radiation filters.

R/X AU
Reactor Experiments, Inc.
Precious metal. Au 99.9.
Foil (0.25-20) microns for radiation filters.

R/X BI
Reactor Experiments, Inc.
Precious metal. Bi 99.97.
Foil (1.00-25) microns for radiation filters.

R/X CD
Reactor Experiments, Inc.
Precious metal. Cd 99.7.
Foil (1.00-25) microns for radiation filters.

R/X CO
Reactor Experiments, Inc.
Precious metal. Co 99.9.
Foil (1.00-25) microns for radiation filters.

R/X CR
Reactor Experiments, Inc.
Precious metal. Cr 99.99.
Foil (1.00-25) microns for radiation filters.

R/X CU
Reactor Experiments, Inc.
Precious metal. Cu 99.9.
Foil (0.5-25) microns for radiation filters.

R/X FE
Reactor Experiments, Inc.
Precious metal. Fe 99.85.
Foil (0.5-25) microns for radiation filters.

R/X HF
Reactor Experiments, Inc.
Precious metal. Hf 97.
Foil (4.0-25) microns for radiation filters.

R/X IN
Reactor Experiments, Inc.
Precious metal. In 99.8.
Foil (2.0-25) microns for radiation filters.

R/X MG
Reactor Experiments, Inc.
Precious metal. Mg 99.9.
Foil (3.0-25) microns for radiation filters.

R/X MN
Reactor Experiments, Inc.
Precious metal. Mn 98.7.
Foil (1.0-25) microns for radiation filters.

R/X MO
Reactor Experiments, Inc.
Precious metal. Mo 99.9.
Foil (4.0-25) microns for radiation filters.

R/X NB
Reactor Experiments, Inc.
Precious metal. Nb 99.9.
Foil (2.5-25) microns for radiation filters.

R/X NI
Reactor Experiments, Inc.
Precious metal. Ni 99.9.
Foil (0.25-25) microns for radiation filters.

R/X PB
Reactor Experiments, Inc.
Precious metal. Pb 99.99.
Foil (1.0-25) microns for radiation filters.

R/X PD
Reactor Experiments, Inc.
Precious metal. Pd 99.75.
Foil (0.5-25) microns for radiation filters.

R/X PT
Reactor Experiments, Inc.
Precious metal. Pt 99.85.
Foil (0.5-25) microns for radiation filters.

R/X RE
Reactor Experiments, Inc.
Precious metal. Re 99.99.
Foil (12.5) microns for radiation filters.

R/X RH
Reactor Experiments, Inc.
Precious metal. Rh 99.85.
Foil (0.5-2.0) microns for radiation filters.

R/X SB
Reactor Experiments, Inc.
Precious metal. Sb 95.
Foil (2.0-25) microns for radiation filters.

R/X SN
Reactor Experiments, Inc.
Precious metal. Sn 99.75.
Foil (1.0-25) microns for radiation filters.

R/X TA
Reactor Experiments, Inc.
Precious metal. Ta 99.9.
Foil (1.5-25) microns for radiation filters.

R/X TI
Reactor Experiments, Inc.
Precious metal. Ti 99.6.
Foil (2.0-25) microns for radiation filters.

R/X V
Reactor Experiments, Inc.
Precious metal. V 99.8.
Foil (4.0-25) microns for radiation filters.

R/X W
Reactor Experiments, Inc.
Precious metal. W 99.95.
Foil (4.0-25) microns for radiation filters.

R/X ZN
Reactor Experiments, Inc.
Precious metal. Zn 99.9.
Foil (2.5-25) microns for radiation filters.

R/X ZR
Reactor Experiments, Inc.
Precious metal. Zr 99.8.
Foil (2.0-25) microns for radiation filters.

R1008
Eagle & Globe Steel Ltd.
Carbon steel. C 0-0.1, Mn 0.2-0.5, P 0-0.04, S 0-0.05, bal Fe.
General purpose low carbon rimmed steel with good surface after cold rolling. BSS970 EN Series EN2A/1; BSS970 Part 1 1972 040A04.

R10F5K5
Russian manufacture
C 1.5, W 10, Cr 4, V 4.5, Co 6, bal Fe.
Hardened: 66-68 Rock C. For cutters, reamers, broaches, drills, textile needles, form tools; high-speed steel, high red-hardness.

R15K12
Russian manufacture
C 0.8, W 15, Cr 3, V 1.5, Co 12, bal Fe.
Hardened: 64-67 Rock C. For lathe and planer tools, drills, reamers, broaches, form tools, hobs, taps; high-speed steel, high red-hardness.

R15K5
Russian manufacture
C 0.8, W 15, Cr 3, V 1.5, Co 5, bal Fe.
Hardened: 64-66 Rock C. For lathe and planer tools, form cutters, drills, reamers, broaches; high-speed steel, high red-hardness.

R18
Russian manufacture
C 0.7, W 18, Cr 4, V 1, bal Fe.
Hardened: 64-66 Rock C. For cutters, reamers, hobs, lathe and planer tools, form and milling cutters; high red-hardness, Type T1 high speed steel.

R18F4K8M
Russian manufacture
C 1.25-1.4, Cr 4.4-5, W 15.5-17, V 3.2-3.8, Co 7.5-8.5, Mo 1.2-1.5, bal Fe.
Hardened: 64-67 Rock C. For cutters, reamers, broaches, taps, hobs, textile needles; high red-hardness, high speed steel, abrasion resistant.

R18K10
Russian manufacture
C 0.88, Cr 4, V 1.3, W 18.3, Co 9.6, bal Fe.
Hardened: 64-67 Rock C. For cutters, reamers, broaches, drills, taps, hobs; high-speed steel, high red-hardness.

R18K8F4M
Russian manufacture
C 1.3, W 17, Cr 4, V 4, Mo 1.4, Co 8, bal Fe.
Hardened: 64-67 Rock C. For reamers, broaches, drills, hobs, taps, form cutters; high-speed steel, high red-hardness.

R2799
Telcon Metals Ltd.
Fe 70, Ni 30.
For compensating shunts for electrical equipment; temperature sensitive, magnetic.

R43
Now HEPPENSTALL R43.

R448
English manufacture
Cr 10-12, C, bal Fe.
For gas turbine parts; corrosion and heat resistant.

R6F2K14M5
Russian manufacture
C 0.8, Cr 4, V 2, W 6, Mo 5, Co 14, bal Fe.
Hardened: 64-66 Rock C. For lathe and planer tools, form cutters, reamers, drills, broaches; high-speed steel, high red-hardness.

R6F2K8M5
Russian manufacture
C 0.8, Mo 5, Cr 4, V 2, W 6, Co 8, bal Fe.
Hardened: 64-67 Rock C. For drills, reamers, broaches, lathe and planer tools; high-speed steel, high red-hardness.

R6K14F2M5
C 0.8, W 6, Cr 4, V 2, Mo 5, Co 14, bal Fe.
Hardened: 64-67 Rock C. For form cutters, drills, reamers, broaches, lathe and planer tools; high-speed steel, high red-hardness.

R6K8F2M5
Russian manufacture
C 0.8, W 6, Cr 4, V 2, Mo 5, Co 8, bal Fe.
Hardened: 64-66 Rock C. For lathe and planer tools, form cutters, reamers, drills, broaches; high-speed steel, high red-hardness.

R9
Russian manufacture
C 0.9, W 8.5, Cr 4, V 2.1, bal Fe.
Hardened: 64-66 Rock C. For lathe and planer tools, reamers, broaches, form cutters; high-speed steel, high red-hardness.

R9K10
Russian manufacture
C 0.95, W 9.5, Cr 4, V 2.4, Co 10, bal Fe.
Hardened: 65-67 Rock C. For reamers, broaches, cut-off tools, form cutters, drills; high-speed steel, high red-hardness.

R9K30
Russian manufacture
C 0.9, Cr 4, W 10.2, V 1.6, Co 29.4, bal Fe.
Hardened: 64-68 Rock C. For heavy duty cutters, drills, reamers, broaches, lathe tools; high-speed steel, high red-hardness.

R9K5 (EI 705)
Russian manufacture
C 0.8-0.9, Cr 3.8-4.4, Co 5-6, W 9-10.5, V 1.6-2, bal Fe.
Hardened: 64-67 Rock C. For machining heat resistant steels, tools, cutters, planer and lathe tools, broaches, reamers, form tools; high-speed steel, high red-hardness.

R9K8F4M
Russian manufacture
C 1.3, W 8.8, Cr 4.2, V 3.6, Mo 1, Co 8.1, bal Fe.
Hardened: 65-67 Rock C. For precision tools and cutters, reamers, broaches, drills; high-speed steel, high red-hardness.

RA
Acciaierie Valbruna s.p.a.
Tool Material. C 0.82, Si 0-0.4, Mn 0-0.4, Cr 4.15, Mo 0.85, V 1.55, W 8.65, bal Fe.
High speed steel.

RA 200/201
Rolled Alloys
Nickel. Fe 0-0.4, C 0-0.02, Cu 0-0.25, Mn 0-0.35, S 0-0.01, Si 0-0.35, 99.0 Ni min.
Low carbon nickel used at 800-1200°F without graphite precipitation; corrosion resistant. For caustic manufacture and storage, chemical shipping containers, synthetic fiber production, fluorine electrolysis, food processing equipment, and magnetostrictive devices. 55,000 TS; 15,000 YS; 40 El.

RA 26-1
Rolled Alloys
C 0.02, Mn 0.3, Si 0.3, Cr 26, Ni 0.2, Mo 1, N 0.03, Ti 0.05, bal Fe.
Annealed: 75,000 TS; 25,000 YS; 30 El; 177 Brin. Ferritic, titanium stabilized, heat and corrosion resistant alloy; weldable and machinable. For recuperators, oil burner parts, furnace lining, combustion chambers, pressure vessels, petrochemical evaporators, and bleaching equipment for water purification facilities. *Obsolete*

RA 309
Rolled Alloys
C 0-0.15, Cr 22-24, Ni 12-15, Si 0-1, Mn 0-2, bal Fe.
Annealed: 85,000-95,000 TS; 40,000-50,000 YS; 45-55 El; 150-185 Brin. For furnace parts and equipment, refinery equipment; corrosion and heat resistant, austenitic.

RA 310
Rolled Alloys
C 0-0.15, Cr 24-26, Ni 19-22, Cu 0.5, Si 0-0.75, Mn 0-2, Mo 0-0.75, bal Fe.
Annealed: 80,000 TS; 35,000 YS; 52 El; 75-85 Rock B. For furnace parts and equipment, gas turbine and jet engine components; corrosion and heat resistant, austenitic.

RA 320LR
Rolled Alloys
C 0-0.025, Cr 19-21, Ni 32-36, Mo 2-3, Mn 1.5-2, Si 0-0.15, P 0-0.015, S 0-0.01, Cu 3-4, Cb + Ta = 8 x C min up to 0.40 max, bal Fe.
Weld filler.

RA 330-04 WELDING WIRE
Rolled Alloys
C 0.2, Mn 5.25, Si 0.8, Cr 19, Ni 35, bal Fe.
Welding wire for GMAW and GTAW RA 330.

RA 330-04-15
Rolled Alloys
C 0-0.2, Cr 19, Ni 35, Mn 5.25, Si 0.8, bal Fe.
Weld metal electrode. 85,500 TS; 59,000 YS; 44 El; 52 RA. Mechanical properties of weld metal deposited by electrode are similar to RA 330.

RA 333
Cannon-Muskegon Corp.
C 0.05, Mn 1.5, Si 1.25, Cr 25, Ni 45, Mo 3, Co 3, W 3, Fe 18.
Room temp: 100,000 TS; 50,000 YS; 50 El. 1400°F: 57,200 TS; 28,300 YS; 44.8 El. Austenitic, non-hardenable, heat and corrosion resistant alloy; weldable and machinable. For high temp. application; turbine parts, radiant tubes, heat treating fixtures, chemical plant equipment. ASM 5593; ASM-5517.

RA 333
Rolled Alloys
C 0.05, Mn 1.5, Si 1.25, Cr 25, Ni 45, Mo 3, Co 3, W 3, Fe 18.
Room temp: 100,000 TS; 50,000 YS; 50 El. 1400°F: 57,200 TS; 28,300 YS; 44.8 El. Austenitic, non-hardenable, heat and corrosion resistant alloy; weldable and machinable. For high temp. application; turbine parts, radiant tubes, heat treating fixtures, chemical plant equipment. ASM 5593; ASM-5517.

RA 333 WELDING WIRE
Rolled Alloys
C 0.05, Mn 1.5, Si 0.85, Cr 25, Ni 45, Mo 3, Co 3, W 3, bal Fe.
Welding wire for GMAW and GTAW RA 333.

RA 333-70-16
Rolled Alloys
C 0-0.05, Mn 1.5, Si 0.85, Cr 25, Ni 45, Mo 3, Co 3, W 3, bal Fe.
Electrode designed for welding RA 333. 98,600 TS; 47,500 YS; 28.5 El.

RA 430
Rolled Alloys
C 0-0.12, Cr 14-18, bal Fe.
Annealed: 70,000-90,000 TS; 40,000-55,000 YS; 20-30 El; 40-55 RA; 140-160 Brin. Cold rolled: 90,000-130,000 TS; 65,000-120,000 YS; 2-20 El; 105-220 Brin. For oil burners, septic tanks, furnace parts; corrosion and heat resistant, ferritic. *Obsolete*

RA 446
Rolled Alloys
C 0-0.15, Cr 23-27, N 0-0.25, Mn 0-1.5, Si 0-1, bal Fe.
Annealed: 77,000 TS; 50,000 YS; 30 El; 80-94 Rock B. For furnace parts, heat treat boxes, conveyor chains; heat resistant, ferritic.

RA 600
Rolled Alloys
Cr 15, Fe 8, Mn 0.7, C 0.05, bal Ni.
Hot rolled: 110,000 TS; 45,000 YS; 40 El; 175 Brin. Heat treated: 112,000 TS; 49,500 YS; 41 El; 175 Brin. For furnace parts, grids, gas turbine components, retorts. Resists oxidizing and carburizing atmospheres, corrosion resistant. *Obsolete*

RA 601
Rolled Alloys
Nickel. Ni 61.5, Cr 22.5, Fe 14, C 0.05, Al 1.4.
High strength and thermal fatigue, oxidation resistant to 2000°F.

RA 85H
Rolled Alloys
Ni 14.5, Cr 18.5, Si 3.5, Al 1, C 0.2, Mn 0.5, bal Fe.
Austenitic heat resistant alloy for heat treating fixtures and bar frame baskets, molten salt hangers for austempering, sleeves and saggers for baking carbon products, radiant tubes, waste incineration, fluidized beds, and combustion nozzles. 105,000 TS; 42,000 YS; 60 El; 65 RA; 92 Rock B.

RABW
Saarstahl AG
Now SAARSTAHL 1.2767.

RACO 13
Reid-Avery Co.
C 0.07, Mn 0.3, Si 0.1, S 0.03, P 0.03, bal Fe.
For mild steel welding electrodes; E-6013.

RACO 20
Reid-Avery Co.
C 0.08, Mn 0.35, Si 0.1, S 0.03, P 0.03, bal Fe.
For welding electrodes for mild steel; E-6020.

RACO 25
Reid-Avery Co.
C 0.15, Cr 1, Mn 0.5, bal Fe.
Welded: 200 Brin. For welding electrodes for mild steel; machinable.

RACO 45
Reid-Avery Co.
C 0.25, Cr 2.5, Mn 1, bal Fe.
Welded: 450 Brin. For hard surfacing electrodes; water resistant.

RACO 5
Reid-Avery Co.
C 0.1, Mn 0.35, bal Fe.
For welding electrodes; for mild steel; AWS-E6030.

RACO 55
Reid-Avery Co.
C 0.4, Cr 3, Mn 1.5, bal Fe.
Welded: 550 Brin. For hard surfacing electrodes; water resistant.

RACO 7
Reid-Avery Co.
C 0.07, Mn 0.35, Si 0.1, S 0.03, P 0.03, bal Fe.
For welding electrodes for mild steel; E-6010.

RACO HD 64
Reid-Avery Co.
C 0.08, Mo 0.5, bal Fe.
For welding electrodes; for high tensile steel.

RACO HD-82
Standard Brake Shoe & Foundry Co.
C 0.15, Cr 1, Mn 0.5, bal Fe.
For welding electrodes for wear resistant surfaces. *Obsolete*

RACO HD30
Reid-Avery Co.
C, bal Fe.
For welding electrodes; mild steel. *Obsolete*

RACO HD6
Reid-Avery Co.
C 0.08, Mn 0.3, bal Fe.
For welding electrodes. *Obsolete*

RACO NO. 8
Reid-Avery Co.
C 0.07, bal Fe.
For welding electrodes; for mild steel.

RACO RED LABEL
Reid-Avery Co.
C 0.1, Mn 0.4, bal Fe.
For welding rods; gas welding.

RACO TYPE M
Reid-Avery Co.
C 0.1, Mn 0.4, Si 0.01, S 0.03, P 0.03, bal Fe.
For welding electrodes; copper coated.

RACOLLOY MANGANESE
Reid-Avery Co.
Mn 14, Ni 3.5, C, bal Fe.
For hard facing electrodes; for Mn steel, coated, austenitic.

RADAX SPEZIAL
Otto Wolff Handelgesellschaft
C 1.45, Cr 1.4, Mn 0.6, bal Fe.
For bearings, bushings, cutters; oil or water hardened, wear resistant.

RADAX W10
Otto Wolff Handelgesellschaft
C 1.2, W 1, V 0.1, Cr 0.2, bal Fe.
For cutters, dies; water hardened, wear resistant.

RADECO
Westa-Westdeutsche
C 1.3, Cr 4.3, Mo 0.85, V 3.8, W 12, bal Fe.
For forming and blanking dies, punches; high speed steel.

RADIAMETAL
Allegheny Ludlum Steel
Ni 47, bal Fe.
For electrical equipment. *Obsolete*

RADIANITE
Latrobe Steel Co.
C 0.7-0.8, Cr 16.5-18, bal Fe.
For cutlery, stainless articles; keen cutting edge. *Obsolete*

RADIKAL EXTRA
J.C. Soding & Halbach
C 0.82, Cr 4.1, Mo 0.85, V 1.6, W 8.7, bal Fe.
For lathe and planer tools, drills, taps, reamers; high speed steel. *Obsolete*

RADIKAL EXTRA M
J.C. Soding & Halbach
C 0.95, Cr 4, Mo, W, V, bal Fe.
For lathe and planer tools, hobs, broaches; high speed steel. *Obsolete*

RADIKAL MO 5
J.C. Soding & Halbach
C 0.85, W 6.7, Cr 4.5, V 2, Mo 5.3, bal Fe.
Hardened: 64-65 Rock C. For lathe and planer tools, hobs, taps, drills, reamers, broaches. High-speed steel, high red-hardness.

RADIKAL MO 55
J.C. Soding & Halbach
C 0.85, W 6.7, Cr 4.5, Co 5.5, V 2, Mo 5.3, bal Fe.
Hardened: 64-66 Rock C. For lathe and planer tools, hobs, taps, reamers, drills, broaches. High-speed steel, high red-hardness.

RADIKAL MO 60
J.C. Soding & Halbach
C 1.25, W 6.7, Cr 4.5, V 3.4, Mo 5.3, bal Fe.
Hardened: 64-66 Rock C. For lathe and planer tools, reamers, broaches, hobs, taps, form cutters. High-speed steel, high red-hardness.

RADIKAL MO 9
J.C. Soding & Halbach
C 0.85, Cr 4, Mo 9.2, W 2, V 1.3, bal Fe.
Hardened: 64-65 Rock C. For milling cutters, lathe and planer tools, broaches, reamers, hobs, taps. High-speed steel, high red-hardness.

RADIKAL MO 92
J.C. Soding & Halbach
C 1.1, W 2, Cr 4.2, V 2.1, Mo 9.2, bal Fe.
Hardened: 64-65 Rock C. For reamers, broaches, taps, hobs, form cutters. High-speed steel, high red-hardness.

RADIKAL MO 95
J.C. Soding & Halbach
C 0.85, W 2.2, Cr 4.2, Co 5, V 1.3, Mo 8.8, bal Fe.
Hardened: C 64-66 Rock. For reamers, broaches, hobs, taps, lathe and planer tools. High-speed steel, high red-hardness. *Obsolete*

RADIKAL MO 98
J.C. Soding & Halbach
C 0.95, W 2.2, Cr 4.2, Co 8, V 2.2, Mo 8.8, bal Fe.
Hardened: C 64-67 Rock. For reamers, drills, hobs, taps, broaches, lathe and planer tools. High speed steel, high red-hardness. *Obsolete*

RADIKAL MO6
J.C. Soding & Halbach
C 0.85, Cr 4, Mo, W, V, bal Fe.
For milling cutters, broaches, reamers, taps; high speed steel. *Obsolete*

RADIO ALLOY NO. 180
Wilbur B. Driver Co.
Copper. Cu 78, Ni 22.
For electrical instruments, resistors; maximum working temperature 400°C.

RADIO ALLOY NO. 30
Wilbur B. Driver Co.
Copper. Cu 98, Ni 2.
For electrical instruments, rheostats, voltage control relays; maximum working temperature 350°C.

RADIO ALLOY NO. 60
Wilbur B. Driver Co.
Copper. Cu 95, Ni 3.5.
For electrical instruments, resistors; maximum working temperature 350°C.

RADIO ALLOY NO. 90
Wilbur B. Driver Co.
Copper. Cu 88, Ni 12.
For electrical instruments, resistors; maximum working temperature 400°C.

RADIO METAL "A"
Telcon Metals Ltd.
Fe 50, Ni 45, Cu 5.
For magnetic alloy for the cores of small transformers; high permeability. *Obsolete*

RADIOHM
Driver Harris Co.
Cr 10-17, Al 5, bal Fe.
Rolled: 85,000-128,000 TS; 5-33 El. For radio tube parts, resistors; heat resistant to 1000 C. *Obsolete*

RADIOMETAL 36
Telcon Metals Ltd.
Ni 36, bal Fe.
Soft magnetic alloy, high resistivity, low losses for transformers.

RADIOMETAL 50

Telcon Metals Ltd.
Ni 50, Fe 50.
Soft magnetic alloy, high initial permeability, high saturation for transformers and relays.

RADIONOX A 182

Krupp Stahl AG
stainless steel. C 0.04, Cr 18, Ni 12, Mo, B, bal Fe.
Austenitic stainless steel.

RADIUM A

Hewitt Metals Corp.
Sb 4.5, Cu 4.5, bal Sn.
Cast: 12,850 TS; 28 Brin. For heavy duty bearings and bushings. Babbitt; 437-699°F MP.

RAE 1

Saarstahl AG
C 0.15, Si 0.25, Mn 0.8, Cr 1.05, Ni 1.45, bal Fe.
Alloy carburizing steel. For gears, spline couplings, shafts, and bolts. W.-Nr. 5713. *Obsolete*

RAE 3

Saarstahl AG
Now SAARSTAHL 14 NICR 14.

RAE 40 C

English manufacture
Mg 0.55, Cu 2, Mn 2.5, Ni 4.4, Be 0.27, bal Al.
Wrought (at RT): 36,000 TS; 1.0 El. Wrought (at 752°F): 4120 TS: 59.5 El. For aircraft and jet engine parts. Good resistance to loads at moderately elevated temperatures.

RAE 55

English manufacture
Mg 0.71, Cu 1.84, Mn 2.02, Ni 2.34, Be 0.2, bal Al.
Wrought (at RT): 39,700 TS; 9.5 El. Wrought (at 752°F): 4520 TS; 79.0 El. For aircraft and jet engine parts. Good resistance to loads at moderately elevated temperatures.

RAEX 27E POLAR

Rautaruukki Oy
C 0-0.18, Si 0.1-0.3, Mn 0.6-1, P 0-0.04, S 0-0.015, Cu 0.1-0.3, Ni 0-0.4, Ce 0.005-0.1, Al 0.02-0.06, As 0-0.08, Cr 0-0.2, Mo 0-0.08, N 0-0.012.
Fine grain shipbuilding steel for marine construction. 265 N/mm² YP min; 400-510 N/mm² TS min; 22 El min.

RAEX 32E POLAR

Rautaruukki Oy
C 0-0.18, Si 0.1-0.3, Mn 0.9-1.15, P 0-0.04, S 0-0.015, Cu 0.1-0.35, Ni 0-0.4, Ce 0.005-0.1, Al 0.02-0.06, Nb 0.02-0.05, V 0.05-0.1, As 0-0.08, Cr 0-0.2, Mo 0-0.08, N 0-0.012.
Fine grain shipbuilding steel for marine construction. 315 N/mm² YP min; 470-590 N/mm² TS min; 22 El min.

RAEX 36E POLAR

Rautaruukki Oy
C 0-0.18, Si 0.1-0.35, Mn 0.9-1.15, P 0-0.04, S 0-0.015, Cu 0.1-0.5, Ni 0-0.4, Ce 0.01-0.1, Al 0.02-0.06, Nb 0.02-0.05, V 0.05-0.1, As 0-0.08, Cr 0-0.2, Mo 0-0.08, N 0-0.012.
Fine grain shipbuilding steel for marine construction. 355 N/mm² YP min; 490-620 N/mm² TS min; 22 El min.

RAFFINAL

Wieland-Werke AG Metallwerke
Al 99.9.
Annealed: 6,800 TS; 2,800 YS; 35 El. For electrical conductors. High electrical conductivity. *Obsolete*

RAFFINAL

Aluminium Industrie Aktiengesellschaft
99.99 Al min.
Soft: 7000 TS; 3000 YS; 60 El; 14 Brin. Hard: 16,000 TS; 15,000 YS; 6 El; 30 Brin. For window frames, door handles, chemical plant equipment; corrosion resistant.

RAFFINAL

Aluminium Walzwerke Singen GmbH
99.99 Al min.
Soft: 7000 TS; 3000 YS; 60 El; 14 Brin. Hard: 16,000 TS; 15,000 YS; 6 El; 30 Brin. For window frames, door handles, chemical plant equipment; corrosion resistant.

RAFFINAL

Alluminio SA
99.99 Al min.
Soft: 7000 TS; 3000 YS; 60 El; 14 Brin. Hard: 16,000 TS; 15,000 YS; 6 El; 30 Brin. For window frames, door handles, chemical plant equipment; corrosion resistant.

RAIL STEEL

English manufacture
C 0.5-0.89, Mn 0.6-0.9, P 0-0.04, 0.15 Si min, bal Fe.
For steel rails; wear resistant.

RAIL-ARC

Now WEAR-O-MATIC RAIL-ARC.

RAILENDER

Champion Rivet Co.
C 0.35, Cr 1.8, V 0.12, bal Fe.
320-336 Brin. For hard facing electrodes; for battered rail ends.

RAILITE

Chrylser Corp.
Fe 75, Cu 25.
For bearing, oil-cushion bearings; sintered, self lubricating. *Obsolete*

RAILROAD "A" BRONZE

English manufacture
Sn 0-20, Zn 0-22, Pb 0-20, P 0-0.5, bal Cu.
For axle bearings, gears, piston rods, side valves; heavy duty.

RAILROAD BRONZE "I"

Manufacturer not listed.
Cu 89, Sn 2.4, Zn 7.8.
For locomotive bearings; heavy duty.

RAILROAD BRONZE "L"

Manufacturer not listed.
Cu 74.1, Sn 3.7, Zn 22.2.
For locomotive pistons and piston rods; high strength.

RAILROAD BRONZE "M"

Manufacturer not listed.
Cu 84.5, Sn 10, Pb 5, P 0.5.
For locomotive side valves; high strength.

RAILROAD BRONZE "N"

Manufacturer not listed.
Cu 84, Sn 8.5, Zn 5, Pb 2.5.
For locomotive injectors; free-cutting.

RAILROAD BRONZE "O"

Manufacturer not listed.
Cu 88.5, Sn 10, P 0.5.
For bronze parts; high strength.

RAILROAD BRONZE "P"

Manufacturer not listed.
Cu 80, Sn 5, Pb 15.
For locomotive axle box bearings; heavy duty.

RAILROAD BRONZE "Q"

Manufacturer not listed.
Cu 75, Sn 5, Pb 20.
For bearings, Babbitts; anti-friction.

RAILROAD BRONZE DURABLE "K"

Manufacturer not listed.
Cu 73.5, Sn 9.5, Zn 9.5, Pb 7.5.
For locomotive bearings; heavy duty.

RAILROAD BRONZE DUTCH "F"

Manufacturer not listed.
Cu 85.25, Sn 12.75, Zn 2.
For eccentric straps for railroad; high strength.

RAILROAD BRONZE FRENCH "B"

French manufacture
Cu 82, Sn 10, Zn 8.
For axle bearings; high strength.

RAILROAD BRONZE FRENCH COMMON "C"

French manufacture
Cu 78, Sn 20, Zn 2.
For axle bearings; high strength.

RAILROAD BRONZE GEARING "H"

Manufacturer not listed.
Cu 88.8, Sn 8.5, Zn 2.7.
For railroad gearing; tough.

RAILROAD BRONZE GERMAN "J"

Manufacturer not listed.
Cu 81, Zn 15, Pb 4.
For locomotive bearings; heavy duty.

RAILROAD BRONZE HARD "E"

Manufacturer not listed.
Cu 87.05, Sn 7.88, Zn 5.07.
For hard axle bearings, high strength.

RAILROAD BRONZE LAFOND "D"

Manufacturer not listed.
Cu 80, Sn 18, Zn 2.
For axle bearings; high strength.

RAILROAD BRONZE LAFOND "G"

Manufacturer not listed.
Cu 84, Sn 14, Zn 2.
For eccentric straps for railroads; high strength.

RAILROAD ROD

Now AIRCO RAILROAD ROD.

RAILWAY AXLE BOX

A. Cohn Ltd.
Cu 83, Sn 7, Zn 6, Pb 4.
Cast: 30,000 TS; 12,000 YS; 18 El. For castings, bearings; tough.

RAILWEAR

Hollup Corp.
C 0.7, Mn 13, bal Fe.
Cast: 350-380 Brin. For hard surfacing electrodes; coated, for rail joints.

RAISED-LAY

Engelhard Industries
Precious metal tip on base metal blade for electrical bar contacts; laminated, composite.

RAJAH

J.F. Jelenko & Co.
Precious metal. Au 58, Pd 3.5, Ag 27.
Quenched: 62,000 psi TS; 42,000 psi YS; 28 El; 130 Brin. Hardened: 90,000 psi TS; 80,000 psi YS; 10 El; 190 Brin. Type III hard dental alloy. For inlays, crowns, and fixed bridgework.

RAJAH

Universal Cyclops
C, alloy, bal Fe.
For tools. *Obsolete*

RAKEL'S METAL

Manufacturer not listed.
Cu 88, Ni 10, Mn 1, Zn 1.
For corrosion resistant parts; corrosion resistant.

RAM BRAND SWEDISH IRON
Manufacturer not listed.
C 0-0.05, Si 0.02, P 0-0.07, S 0-0.005, bal Fe.
33 El; 65-70 RA. For chains, chain hooks, stay bolts, locomotive parts; resists shock. *Obsolete*

RAMAX S
Uddeholm Corp.
C 0.3, Si 0.35, Mn 1.35, Cr 16.7, S 0.12, bal Fe.
1050 N/mm^2 TS; 860 N/mm^2 YS; 9.5 El; 37 RA; 320 Brin. Chromium alloyed stainless holder steel. For holders and bolsters for plastic and rubber molds. AISI 420F.

RAMET 1
Fansteel Metals
Submicron dispersion-strengthened carbide. Hardness: 91.5 Rock A; transverse rupture strength: 400,000 psi; compressive strength: 525,000 psi; Young's modulus: 80 x 10^6 psi. For machining high strength alloys and most difficult-to-machine alloys.

RAMSOS 1
Robert-Leyer-Pritzkow & Co.
C 0.3, Cr 2.6, V 0.3, W 8.5, bal Fe.
For punches, shears, extrusion rams; hot work steel; 30WCrV3411.

RAMSOS 1C
Robert-Leyer-Pritzkow & Co.
C 0.35, W, Cr, V, bal Fe.
For punches, shears, crimpers, upsetters; hot work steel; 35WCrV4012.

RAMSOS 1MO
Robert-Leyer-Pritzkow & Co.
C 0.35, Mo 5, Cr 2.5, V, bal Fe.
For punches, shears, crimpers, upsetters; hot work steel; 35MoCrV179.

RAMSOS 1N
Robert-Leyer-Pritzkow & Co.
C 0.25, W, Cr, Ni, V, bal Fe.
For punches, shears, crimpers; hot work steel; 25WCrNiV3610.

RAMSOS 2A
Robert-Leyer-Pritzkow & Co.
C 0.6, Mo, Cr, W, V, bal Fe.
For punches, shears, crimpers; hot work steel; 60MoCrWV9515.

RAMSOS 2B
Robert-Leyer-Pritzkow & Co.
C 0.6, W, Mo, Cr, V, bal Fe.
For punches, shears, crimpers; hot work steel; 60WMoCrV2726.

RAMSOS 3
Robert-Leyer-Pritzkow & Co.
C 0.35, Cr, W, Mo, Si, V, bal Fe.
For extrusion rams, liners, punches; hot work steel; 35CrWMoSiV226.

RAMSOS 55
Robert-Leyer-Pritzkow & Co.
C 0.55, Si 0.1-0.4, Mn 0.5-0.7, bal Fe.
For gears, shafts, axles, tools; D1N-C55WS; water hardened.

RAMSOS 5A
Robert-Leyer-Pritzkow & Co.
C 0.45, Cr, Ni, W, Si, Mn, bal Fe.
For hot work tools, punches, crimpers; hot work steel; 45CrNiWSiMn525.

RAMSOS 6(S6)
Robert-Leyer-Pritzkow & Co.
C 0.3, Cr 1.1, V 0.18, W 3.75, bal Fe.
For hot work tools, punches, crimpers; hot work steel; 30WCrV15.

RAMSOS 6MO
Robert-Leyer-Pritzkow & Co.
C 0.3, Cr, Mo, V, bal Fe.
For hot work tools; hot work steel; 30CrMoV1328.

RAMSOS 7
Robert-Leyer-Pritzkow & Co.
C 0.45, Cr 1.3, Mo 0.45, V 0.8, W 0.45, bal Fe.
For hot work tools; hot work steel; 45CrVMoW58.

RAMSOS 7MO
Robert-Leyer-Pritzkow & Co.
C 0.3, Cr, Mo, Mn, V, bal Fe.
For hot work tools; 30 CrMoMnV710; hot work steel.

RAMSOS 8
Robert-Leyer-Pritzkow & Co.
C 0.4, Cr 2, Mn 1, Mo 0.25, bal Fe.
For header tools, container jackets; 40CrMnMo7; hot work steel.

RAMSOS 80
Robert-Leyer-Pritzkow & Co.
C 0.8, Si 0.1-0.4, Mn 0.5-0.7, bal Fe.
For tools, springs, cutters, drills; water hardened; Type W1.

RAMSOS 9
Robert-Leyer-Pritzkow & Co.
C 0.3, Cr 2.5, Mo 0.2, V 0.15, bal Fe.
For extrusion press parts, mandrels, liners; 30CrMoV9; hot work steel.

RAMSOS G1
Robert-Leyer-Pritzkow & Co.
C 0.55, Cr 0.7, Mo 0.18, Ni 1.65, V 0.2, bal Fe.
For hot work tools, punches, upsetters; 55NiCrMoV6; hot work steel.

RAMSOS G1 SPECIAL
Robert-Leyer-Pritzkow & Co.
C 0.5-0.6, Cr 1.2, Mo 0.35, Ni 1.6, V 0.2, bal Fe.
For hot work tools, punches, crimpers; 56NiCrMoV7; hot work steel.

RAMSOS G1M
Robert-Leyer-Pritzkow & Co.
C 0.5, Mn 0.8-1.1, Si 0.25-0.5, bal Fe.
Heat treated: 115,000-135,000 TS. For gears, shafts, punches, crimpers; 50Mn4; oil hardened.

RAMSOS G3
Robert-Leyer-Pritzkow & Co.
C 0.45, Si 0.25-0.5, Mn 0.3-0.8, bal Fe.
For gears, pinions, shafts, axles; water hardened; C45W3.

RAMSOS G4
Robert-Leyer-Pritzkow & Co.
C 0.6, Si 0.25-0.5, Mn 0.3-0.8, bal Fe.
For crimpers, punches, tools, dies; C60W3; water hardened.

RAMSOS G5
Robert-Leyer-Pritzkow & Co.
C 0.75, Si 0.25-0.5, Mn 0.3-0.8, bal Fe.
For crimpers, tools, springs, punches; C75W3; water hardened.

RAMSOS GE4
Robert-Leyer-Pritzkow & Co.
C 1, Si 0.25-0.5, Mn 0.3-0.8, bal Fe.
For drills, taps, punches, tools; Type W1; water hardened; C100W2.

RAMSOS KSZ
Robert-Leyer-Pritzkow & Co.
C 0.3, Mn 0.55, Si 0.25, Cr 2.5, V 0.15, bal Fe.
For extrusion press parts, mandrels, liners; 30CrMoV9; hot work steel.

RAMSOS MUK
Robert-Leyer-Pritzkow & Co.
C 0.45, Si 0.9, Mn 0.3, Cr 1, V 0.2, W 1.8, bal Fe.
For hot working tools, crimpers, upsetters; 45WCrV7; oil hardened.

RAMSOS O
Robert-Leyer-Pritzkow & Co.
C 0.82, Cr 4.1, Mo 0.85, V 1.6, W 8.7, bal Fe.
For lathe and planer tools, reamers, broaches, hobs; high speed steel; 82WV3419.

RAMSOS O/2
Robert-Leyer-Pritzkow & Co.
C 0.65, Cr 3.75, Mo 0.85, V 0.7, W 8.5, bal Fe.
For lathe and planer tools, drills, hobs, reamers; high speed steel; 65WMo348.

RAMSOS OK
Robert-Leyer-Pritzkow & Co.
C 0.65, Cr 3.75, Mo 0.85, W 8.5, V 0.7, bal Fe.
For lathe and planer tools, hobs, broaches; high speed steel; 65WMo348.

RAMSOS OO
Robert-Leyer-Pritzkow & Co.
C 0.95, Cr 4, V 1, Mo, W, bal Fe.
For cutters, drills, hobs, reamers; high speed steel; 95WMo1126.

RAMSOS S4
Robert-Leyer-Pritzkow & Co.
C 0.3, Mn 0.3, Cr 2.35, V 0.6, W 4.25, bal Fe.
Heat treated: 198,000-242,000 TS. For extrusion dies; rams and liners; hot work steel; 30WCrV179.

RAMSOS SGK
Robert-Leyer-Pritzkow & Co.
C 0.3, Cr, Mo, Si, V, bal Fe.
For punches, upsetters, crimpers; 30CrMoSiV2011; hot work steel.

RAMSOS SUPER A
Robert-Leyer-Pritzkow & Co.
C 0.4, W, Mo, Cr, V, bal Fe.
For tools, dies, upsetters, punches; 40WMoCrV1838; hot work steel.

RAMSOS SUPER K
Robert-Leyer-Pritzkow & Co.
C 0.3, W, Co, Cr, Mo, V, bal Fe.
For crimpers, dies, punches, shears; 30WCoCrMoV2424; hot work steel.

RAMSOS SUPER WN
Robert-Leyer-Pritzkow & Co.
C 0.55, Ni, W, Cr, Co, Si, V, bal Fe.
For hot work tools, dies; 55 NiWCrCoSiV484416; oil hardened.

RAMSOS VC
Robert-Leyer-Pritzkow & Co.
C 0.4, Ni, Cr, Mo, bal Fe.
For gears, axles, crankshafts; 40NiCrMo176; oil hardened, shock resistant.

RAN
Rochling Burbach GmbH
C 0.45, Cr 0.5, Ni 1.5, bal Fe.
For gears, shafts, crankshafts, bolts, studs; oil hardened, shock resistant. *Obsolete*

RAN 1 W/H
Rochling Burbach GmbH
C 0.28-0.35, Cr 0.5, Ni 1.5, bal Fe.
For gears, shafts, crankshafts, bolts, studs; oil hardened, shock resistant. *Obsolete*

RAN 2 W/H
Rochling Burbach GmbH
C 0.28-0.35, Cr 0.7, Ni 2.5, bal Fe.
For gears, shafts, crankshafts, bolts, studs; oil hardened, shock resistant. *Obsolete*

RAN 3 W/H
Rochling Burbach GmbH
C 0.22-0.3, Cr 0.7, Ni 3.5, bal Fe.
For gears, shafts, crankshafts, bolts, studs; oil hardened, shock resistant. *Obsolete*

RAN 5
Saarstahl AG
Now SAARSTAHL 35 NICR 18.

RAN 6
Rochling Burbach GmbH
C 0.35, Cr 1.3, Ni 4.5, bal Fe.
For gears, shafts, crankshafts, bolts, studs; oil hardened, shock resistant. *Obsolete*

RAN FH
Rochling-Burbach GmbH
C 0.23-0.3, Cr 0.7, Ni 3.5, bal Fe.
For gears, shafts, crankshafts, bolts, studs; oil hardened, shock resistant.

RANALLOY TYPE A
Bergstrom Alloys Corp.
C, Mo, Ni, Cr, bal Fe.
500 Brin. For hard surfacing electrodes; wear resistant.

RANALLOY TYPE A
United States Steel Corp.
C, alloy, bal Fe.
For hard surfacing electrodes; wear resistant. *Obsolete*

RANALLOY TYPE B
Bergstrom Alloys Corp.
C, Mo, Ni, Cr, bal Fe.
530 Brin. For hard surfacing electrodes; wear resistant.

RANALLOY TYPE B
United States Steel Corp.
C, alloy, bal Fe.
For hard surfacing electrodes; wear resistant. *Obsolete*

RANALLOY TYPE C
Bergstrom Alloys Corp.
C, Mo, Ni, Cr, bal Fe.
480 Brin. For hard surfacing electrodes; wear resistant.

RANALLOY TYPE C
United States Steel Corp.
C, alloy, bal Fe.
For hard surfacing electrodes; wear resistant. *Obsolete*

RANDALL
Randall Graphite Products Co.
Copper. Cu graphite.
For bearings; sintered.

RANDOLF METAL
Manufacturer not listed.
For dental alloy; corrosion resistant.

RANGER D.R.
Jessop Steel Co.
C 1.2, bal Fe.
For drills, taps, cutters; drill rod, water hardening. *Obsolete*

RANITE 1
Rankin Mfg. Co.
Cobalt base (Co-Cr-W) electrode. For non-magnetic, non-machinable hard-surfacing for resistance to extreme abrasion, corrosion, and high temperatures. 52 Rock C.

RANITE 12
Rankin Mfg. Co.
Cobalt base (Co-Cr-W) electrode. For non-magnetic, machinable build-up for good abrasion resistance and medium impact. 45 Rock C.

RANITE 5
Rankin Mfg. Co.
35% alloy, bal Fe.
Bare tube rod for gas application two-layer deposit for light to moderate impact and severe abrasion; 61 Rock C.

RANITE 6
Rankin Mfg. Co.
Cobalt base (Co-Cr-W) electrode. For non-magnetic, machinable build-up for shear blades, valves for good metal-to-metal wear, impact, and hot wear. 41 Rock C.

RANITE A
Rankin Mfg. Co.
TC 3, Mn, Cr, bal Fe.
All position electrode, AC or DC; for build-up of worn parts as tractor rails, idlers, shovel buckets; 42-44 Rock C.

RANITE BU
Rankin Mfg. Co.
TC 5, Mn, Cr, bal Fe.
All position electrode, AC or DC, straight or reverse polarity; for build-up of shafts, gears, idlers; machinable. 28-34 Rock C.

RANITE BX
Rankin Mfg. Co.
TC 9, Mn, Cr, bal Fe.
All position electrode, AC or DC; for multipass build-up of dredge pump shells, cone crusher mantles and liners. 44-50 Rock C.

RANITE C
Rankin Mfg. Co.
TC 14, Cr, Mo, bal Fe.
All position electrode, AC or DC; for 1-2 passes hardsurfacing for extreme abrasion and medium impact, such as plow shares, conveyor screws. 58-60 Rock C.

RANITE C-X
Rankin Mfg. Co.
C, Cr, Fe.
For hard surfacing electrodes; abrasion resistant. *Obsolete*

RANITE D
Rankin Mfg. Co.
TC 14, Mn, Cr, Mo, B, bal Fe.
All position electrode, AC or DC; for 1-2 passes for extreme abrasion resistance, such as brick dies, drill collars, dredge teeth. 63-68 Rock C.

RANITE F
Rankin Mfg. Co.
TC 18, Mn, Cr, Mo, bal Fe.
All position electrode, AC or DC; for hardfacing truck beds, earth moving equipment, guides and chutes. 52-57 Rock C. Wear and abrasion resistant.

RANITE G
Rankin Mfg. Co.
TC 10, Mn, Cr, Ni, Mo, bal Fe.
All purpose electrode, AC or DC; for general purpose build-up and hardfacing. 40-54 Rock C.

RANITE J
Rankin Mfg. Co.
TC 26, Mn, Cr, Mo, bal Fe.
All position electrode, AC or DC; for 2 pass hardsurfacing at fast deposit rate. Good wear resistance up to 800°F. 45-50 Rock C.

RANITE TYPE B
Rankin Mfg. Co.
C, Si, Ti, Cr, Mo, B, bal Fe.
Cast: 444 Brin. For hard facing electrodes; wear and impact resistant. *Obsolete*

RANITE TYPE NO. 4
Rankin Mfg. Co.
C, Si, Cr, Mo, Ti, B, bal Fe.
Welded: 520 Brin. For hard facing electrodes; wear and shock resistant.

RANKIN 308-16
Rankin Mfg. Co.
C 0.07, Cr 19.5, Ni 9.5, Mn 1.6, bal Fe.
AC-DC-16 coated electrode for welding 200-300 series stainless steels, up to and including 308. 84,000-96,000 psi TS; 36-46 El.

RANKIN 308L
Rankin Mfg. Co.
C 0.04, Cr 19, Ni 9.5, Mn 1.5, Si 0.3, bal Fe.
AC-DC-16 coated electrode for welding 304 and 304L stainless steels. 80,000-90,000 psi TS; 36-46 El.

RANKIN 309-16
Rankin Mfg. Co.
C 0.1, Cr 23, Ni 13, Mn 1.6, Si 0.5, bal Fe.
AC-DC-16 coated electrode for welding dissimilar metals and cladding 18 to mild steels. 85,000-95,000 TS; 35-45 El.

RANKIN 310-16
Rankin Mfg. Co.
C 0.2, Cr 26, Ni 21, Mn 1.8, Si 0.4, bal Fe.
AC-DC-16 coated electrode for welding 310 and dissimilar metals. 85,000-95,000 psi TS; 35-45 El.

RANKIN 312-16
Rankin Mfg. Co.
C 0.15, Cr 29, Ni 9.5, Mn 1.9, Si 0.5, bal Fe.
AC-DC-16 coated electrode for welding high tensile to high temperature heat resisting alloys; and for dissimilar metals. 100,000-110,000 psi TS; 22-25 El.

RANKIN 316-16
Rankin Mfg. Co.
C 0.07, Cr 18, Ni 13, Mo 2.25, Mn 1.7, Si 0.3, bal Fe.
AC-DC-16 coated electrode for welding type 316 stainless steel. 85,000-95,000 psi TS; 35-45 El.

RANKIN 316-16L
Rankin Mfg. Co.
C 0-0.04, Cr 18, Ni 13, Mo 2.25, Mn 1.7, Si 0.5, bal Fe.
AC-DC-16 coated electrode for welding type 316L stainless steel. 80,000-90,000 psi TS; 35-45 El.

RANMANG
Rankin Mfg. Co.
16% total Mn, Ni, bal Fe.
All position electrode, AC or DC; for welding and build-up of high manganese steel in rails, dredge shells, rolls. As deposited: 75-85 Rock B. Cold worked: 45-50 Rock C. 144,000 psi TS; 33 El.

RANMANG 2
Rankin Mfg. Co.
TC 36, Mn, Cr, Ni, Si, bal Fe.
All position electrode, AC or DC; for joining and build-up of high manganese and other steels, on hammers, shovels, dragline pins and links. As deposited: 17-22 Rock C. Work hardened: 43-48 Rock C.

RANOMATIC A
Rankin Mfg. Co.
C, 2.5% alloy, bal Fe.
Martensitic, non-work-hardening welding wire for overlay to give moderate abrasion and good impact resistance. 42-44 Rock C.

RANOMATIC BU
Rankin Mfg. Co.
TC4, Cr, Mn, Mo, bal Fe.
Open arc welding wire for machinable build-up on carbon and low alloy steel parts. 24-36 Rock C.

RANOMATIC BX-2
Rankin Mfg. Co.
TC 13.5, Mn, Cr, Mo, Cu, bal Fe.
Self-shielded, flux-cored wire for rebuilding and hardsurfacing carbon and manganese steel parts. As deposited, third pass; 43-45 Rock C. Cold worked: 60 Rock C.

RANOMATIC D
Rankin Mfg. Co.
TC 14, Cr, Mo, B, bal Fe.
Open-arc welding wire for depositing an extremely hard, abrasion resistant coating. 63-68 Rock C.

RANOMATIC F
Rankin Mfg. Co.
TC 18, Cr, Mo, bal Fe.
Open-arc welding wire for depositing a hard, abrasion resistant coating. 55-60 Rock C.

RANOMATIC H
Rankin Mfg. Co.
TC 23, Mn, Cr, bal Fe.
High chromium, open-arc wire for two-pass coating for abrasion and corrosion resistance. 55 Rock C.

RANOMATIC O
Rankin Mfg. Co.
TC 23.5, Mn, Cr, Mo, bal Fe.
Self-shielded, flux-cored wire for multipass build-up of wear resistant coatings. As deposited: 45-46 Rock C. Work hardened: 63 Rock C.

RANSCO-ALLOY
Rasmussen Mfg. Co.
C 2.5-3.25, Si 1-2, Ni 12-15, Cu 5-7, bal Fe.
Cast: 20,000-35,000 TS; 125-200 Brin. For diesel engine cylinder liners. Centrifugally cast.

RANTUNG
Rankin Mfg. Co.
Cast tungsten carbide with binder; electrode. Coated for arc; uncoated for acetylene. For extreme hard, wear resistant coating.

RAONEL
Rolled Alloys
C 0-0.15, Cr 14-17, Fe 6-10, Mn 0-1, 72% min Ni.
Annealed: 80,000-100,000 TS; 25,000-50,000 YS; 35-55 El; 60-70 RA. Cold drawn: 105,000-150,000 TS; 80,000-125,000 YS; 10-30 El; 30-60 RA. For high temperature applications, regenerators, coolers; heat resisting, good hot and cold workability. *Obsolete*

RAPID EXTRA
Styria-Stahl Steirische Gusstahlwerke AG
W 19, Cr 4, V 2.25, C, bal Fe.
For tools, dies, cutters; high speed steel. *Obsolete*

RAPID EXTRA MO
Vereinigte Edelstahlwerke
C 0.8, Cr 4, W 6.5, Mo 5, V 2, bal Fe.
For cutting tools; high speed steel. *Obsolete*

RAPID FINISH
Jessop Steel Co.
C 1.3, Cr 0.75, W 3.75, bal Fe.
For plug gauges, forming dies, punches; water hardening. *Obsolete*

RAPID FINISHING
Jessop-Saville Ltd.
C, alloy, bal Fe.
For fast finishing tools; water hardened.

RAPID PANTHER
Styria-Stahl Steirische Gusstahlwerke AG
C 0.7, W 19, Cr 4, V 2.25, bal Fe.
For broaches, taps, reamers, hobs, lathe cutters; high speed steel, oil hardened. *Obsolete*

RAPID SPECIAL BNX
Thyssen Edelstahlwerke AG
C 0.8, Cr 4, Mo 0.3, V 1.6, W 8, Co 10, bal Fe.
For tools, dies, cutters; high speed steel. *Obsolete*

RAPID SPEZIAL
Thyssen Edelstahlwerke AG
C 0.7, W 19, Cr 4, V 2.2, bal Fe.
For high speed cutters; high speed steel. *Obsolete*

RAPID SPEZIAL
Westfalische Stahlgesellschaft mbH
C 0.95, W, Mo, bal Fe.
For tools, dies; oil hardening.

RAPID SPEZIAL BN
Thyssen Edelstahlwerke AG
C 0.82, Cr 4.1, Mo 0.85, V 1.6, W 8.7, bal Fe.
For drills, taps, reamers, hobs; high speed steel. *Obsolete*

RAPIDE B
Aubert & Duval
C 0.65-0.8, Co 1, Cr 4-5, W 17-19, V 1-1.5, bal Fe.
For cutters, taps, reamers, hobs, broaches; high speed steel, oil hardened. *Obsolete*

RAPIDE C
Aubert & Duval
C 0.75-0.9, Cr 4-5, W 17-19, V 1-1.5, Co 0.5-1, bal Fe.
For lathe and planer tools, milling cutters, taps; high speed steel, oil hardened. *Obsolete*

RAPIDE J
Aubert & Duval
C 0.85, Cr 4.5, W 18.75, Co 9.5, V 1.5, bal Fe.
For lathe and planer tools, reamers, hobs, taps, broaches; high speed steel, oil hardened.

RAPIDE L
Aubert & Duval
C 0.7, W 19, Cr 4, V 2.2, bal Fe.
For cutters, taps, reamers, hobs, broaches, drills; high speed steel, oil hardened. *Obsolete*

RAPIDE M
Aubert & Duval
Alloy steel. C 0.85, W 5.75, Mo 5.75, Cr 4.6, V 1.7, bal Fe.
Annealed: 840 Vickers; 290 Brin. For reamers, broaches, milling cutters, punches, lathe tools, screw-taps and drills.

RAPIDEX-040
Swiss Aluminium Ltd.
Aluminum. Si 0.3-0.45, Fe 0.1-0.3, Cu 0.1, Mn 0.1, Mg 0.35-0.5, Cr 0.05, Zn 0.1, Ti 0.1, bal Al.
Extrusion age hardenable alloy in billet form. 190-235 N/mm² TS; 150-215 N/mm² YS; 10-12 El; 65 Brin.

RAPIDEX-041
Swiss Aluminium Ltd.
Aluminum. Si 0.3-0.6, Fe 0.1-0.3, Cu 0.02, Mg 0.3-0.5, Zn 0.15, bal Al.
Extrusion alloy in billet form for electrical conductors. 160-200 N/mm² TS; 120-180 N/mm² YS; 12-14 El; 55 Brin.

RAPIDEX-043
Swiss Aluminium Ltd.
Aluminum. Si 0.35-0.5, Fe 0.1-0.3, Cu 0.1, Mn 0.1, Mg 0.35-0.55, Cr 0.05, Zn 0.1, Ti 0.1, bal Al.
Extrusion age hardenable alloy in billet form. 210-255 N/mm² TS; 160-230 N/mm² YS; 10-12 El; 70 Brin.

RAPIDEX-050
Swiss Aluminium Ltd.
Aluminum. Si 0.3-0.6, Fe 0.1-0.3, Cu 0.1, Mn 0.1, Mg 0.35-0.6, Cr 0.05, Zn 0.1, Ti 0.1, bal Al.
Extrusion age hardenable alloy in billet form. 245-290 N/mm² TS; 195-270 N/mm² YS; 8-10 El; 80 Brin.

RAPIDEX-051
Swiss Aluminium Ltd.
Aluminum. Si 0.5-0.6, Fe 0.1-0.3, Cu 0.02, Mg 0.35-0.5, Zn 0.15, bal Al.
Extrusion alloy in billet form for electrical conductors. 215-280 N/mm² TS; 160-240 N/mm² YS; 10-12 El; 75 Brin.

RAPIDIT M
Thyssen Edelstahlwerke AG
C 1.1, Si 4.3, W 2.4, Mo 2.4, V 2.8, bal Fe.
For welding rod; to repair high speed tools. *Obsolete*

RAPIDIT MA
Thyssen Edelstahlwerke AG
C 1.45, Si 4.3, W 2.4, Mo 2.4, V 2.8, bal Fe.
For welding rod; to repair high speed tools. *Obsolete*

RAPIDIT W
Thyssen Edelstahlwerke AG
C 1.2, Mn 0.35, Cr 4.3, W 10.5, V 4.3, bal Fe.
For welding rod; to repair high speed steel. *Obsolete*

RAPIDIT WA
Thyssen Edelstahlwerke AG
C 1.6, Mn 0.3, Cr 4.3, W 10.5, V 4.3, bal Fe.
For welding electrodes; for repairing high speed tools. *Obsolete*

RARE EARTH SILICIDE
Cyprus Foote Mineral Co.
Ce 15-18, Si 28-33, 30-35 total rare earths, bal Fe.
Ferroalloy for deoxidation, desulfurization and sulfide shape control of steels; also a source of cerium and rare earths for gray and cast iron. *Obsolete*

RARE EARTH SILICIDE
Globe Metallurgical, Inc.
Miscellaneous nonferrous. Si 30-35, Fe 30-35, 30.0-35.0 total rare earth elements.
Master alloy used to neutralize sulfur, improve ductility and minimize inclusions in cast irons and steel. Also used to produce nodular and compacted graphite irons.

RAUCHBERG
German manufacture
Cu 66-75, Pb 15-19, Sn 5-10, Sb 1-5.
For bearings; corrosion resistant.

RAUENLOY
U.S. Spring & Bumper Co.
C, Cr, Mo, V, bal Fe.
For leaf springs; tough, oil hardened.

RAVEN MST
Raven Steel & Tool Co.
C 0.9, Mn 1.6, bal Fe.
For punches, dies, crimpers, chisels, cutters; Type O2; non-deforming, oil hardened.

RAVEN NBS
Raven Steel & Tool Co.
C 0.7, Cr 0.75, Ni 1.5, Mo 0.25, bal Fe.
For punches, dies, rollers, mandrels; Type L6; oil hardened.

RAXA 129
Heinrich Reining GmbH
Cr 3, C, bal Fe.
For oil refinery equipment; creep and heat resistant. *Obsolete*

RAXA ATV
Heinrich Reining GmbH
C 1.45, Cr 1.5, V, bal Fe.
For cutting tools, dies, mandrels, rolls, wear resistant; oil hardening.

RAXA BK 100
Heinrich Reining GmbH
C 1.06, Cr 13.5, Mo 0.5, Ni 0.3, bal Fe.
For cutting tools, shear blades, rolls, forming tools; wear resistant; air or oil hardening.

RAXA BK 110
Heinrich Reining GmbH
C 0.95, Cr 12, Mo 1, V 1, bal Fe.
For shear blades, cutting tools, wear resistant; air or oil hardening.

RAXA BK 12 G
Heinrich Reining GmbH
C 1.2, Cr 12, V, bal Fe.
For wear parts in the cement and coal mining industry; air or oil hardening. Special steel casting quality.

RAXA BK 121
Heinrich Reining GmbH
C 1.55, Cr 12, Mo 0.7, V 1, bal Fe.
For cutting tools, dies, forming rolls; high wear resistance; air or oil hardening. AISI D2.

RAXA BK 122
Heinrich Reining GmbH
C 2.2, Cr 12.5, Mo 0.9, V 2.2, bal Fe.
For cutting tools, drawing dies, forming rolls, pressing tools; high wear resistance; air or oil hardening. Similar to AISI D7.

RAXA BK 211 G
Heinrich Reining GmbH
C 1.9, Cr 21, Mo 0.7, W 1.5, bal Fe.
For rolling mill equipment and rolls; high wear resistance, special steel casting quality.

RAXA BK 5
Heinrich Reining GmbH
C 1, Cr 5, Mo 1, V 0.15, bal Fe.
For cutting tools, shear blades, threading rolls; wear resistant; air or oil hardening.

RAXA BK 60
Heinrich Reining GmbH
C 0.65, Cr 14, Mo 0.6, bal Fe.
For cutting tools, shear blades.

RAXA BKL
Heinrich Reining GmbH
C 2.1, Cr 12, W 0.8, bal Fe.
For cutting tools, shaping rolls, wood knives; high wear resistance, air or oil hardening.

RAXA BKLO
Heinrich Reining GmbH
C 2.1, Co 1, Cr 13, Mo 0.5, W 0.7, bal Fe.
For cutting tools for electric sheets; high wear resistance, air or oil hardening.

RAXA BKR
Heinrich Reining GmbH
C 1.65, Cr 12, Mo 0.6, W 0.5, bal Fe.
For cutting tools, stamping tools, metal saws; high wear resistance, air or oil hardening.

RAXA BKS
Heinrich Reining GmbH
C 2.1, Cr 12, bal Fe.
For cutting tools, shear blades, pressing tools; high wear resistance, air or oil hardening.

RAXA BKV
Heinrich Reining GmbH
C 1.65, Cr 12, V 0.2, bal Fe.
For stamping tools, wood cutters, metal saws, threading rolls; high wear resistance, oil hardening.

RAXA BKW
Heinrich Reining GmbH
C 1.2, Cr 12, Mo 1.4, V 1.7, W 2.5, bal Fe.
For high wear resistant cutting tools; air or oil hardening.

RAXA C 25
Heinrich Reining GmbH
C 2.5, Cr 23, Ni 0.4, V 0.3, bal Fe.
For high wear resistant special steel casting tools for cement and coal mining industries. W. Nr. 1.2288.

RAXA C 35
Heinrich Reining GmbH
C 3.2, Cr 25, V 0.3, bal Fe.
For extremely wear resistant special steel casting tools for the coal mining industry.

RAXA C10
Heinrich Reining GmbH
C 0.76, Co 10, Cr 4.2, Mo 0.8, V 1.8, W 18, bal Fe.
For lathe and planer tools, drills, reamers, taps; high speed steel. *Obsolete*

RAXA CM 4 G
Heinrich Reining GmbH
C 1.3, Co 8, Cr 21, Mo 6.5, bal Fe.
For rolling mill equipment, twist rolls, guide rolls. Special steel casting quality; hot wear resistant.

RAXA CM 8 G
Heinrich Reining GmbH
C 2.4, Si 1.5, Co 24, Cr 17, Mo 2.3, Ni 6, V 1.4, W 2.8, bal Fe.
For rolling mill equipment, twist rolls, guide rolls. Special steel casting quality; hot wear resistant.

RAXA CM 88
Heinrich Reining GmbH
C 0.2, Co 10, Cr 9, Mo 2.2, W 7, bal Fe.
For hot working tools in pressure casting molds; hot wear resistant, oil hardening.

RAXA CM 89
Heinrich Reining GmbH
C 0.45, Co 5, Cr 5, Mo 3, V 2, bal Fe.
For hot working tools in pressure casting molds; hot wear resistant, oil hardening.

RAXA CMD
Heinrich Reining GmbH
C 0.4, Si 1, Cr 5, Mo 1.2, V 0.4, bal Fe.
For forging dies, pressure casting molds, extrusion tools. Hot working tool steel, oil hardening. Similar to AISI H11.

RAXA CMD/W
Heinrich Reining GmbH
C 0.37, Si 1, Cr 5, Mo 1.5, V 0.25, W 1.3, bal Fe.
For pressure casting molds, hot pressing and forging tools. Hot working tool steel. Similar to AISI H12.

RAXA CMER
Heinrich Reining GmbH
C 0.2, Mn 1.25, Cr 1.15, bal Fe.
For gears, cams, camshafts, fasteners; case hardening steel. *Obsolete*

RAXA CMG
Heinrich Reining GmbH
C 2.6, Si 1.7, Co 10, Cr 23, Mo 8, bal Fe.
For rolling mill equipment, twist rolls. Special steel casting quality; hot wear resistant.

RAXA CMR
Heinrich Reining GmbH
C 0.35, Cr 5, Mo 3, V 0.7, bal Fe.
For forging dies, hot drawing dies, mandrels; hot working tool steel, oil hardening.

RAXA CMS
Heinrich Reining GmbH
C 0.4, Cr 5, Mo 1.4, V 1, S 0.12, bal Fe.
For pressure casting molds, recipient casings, punches; hot working tool steel, oil hardening with good machinability.

RAXA CMV
Heinrich Reining GmbH
C 0.4, Si 1, Cr 5, Mo 1.4, V 1, bal Fe.
For forging dies, pressure casting molds and extrusion tools; hot working tool steel, oil hardening.

RAXA CNV3
Heinrich Reining GmbH
C, alloy, bal Fe.
For gears, cams, machine tool parts; case hardening steel. *Obsolete*

RAXA CNV5
Heinrich Reining GmbH
C, alloy, bal Fe.
For gears, bolts, machine tool parts; case hardening steel. *Obsolete*

RAXA CNZ3
Heinrich Reining GmbH
C, alloy, bal Fe.
For gears, bolts, machine tool parts; case hardening steel. *Obsolete*

RAXA CNZ5
Heinrich Reining GmbH
C 0.19, Mo 0.2, Cr 1.25, Ni 3.75, bal Fe.
For gears, bolts, cams, camshafts; case hardening steel, shock resistant. *Obsolete*

RAXA CRO
Heinrich Reining GmbH
C 0.3, Si 1, Cr 1.1, V 0.18, W 3.75, bal Fe.
For cold work tools, upsetters, header dies; oil hardened, tough. *Obsolete*

RAXA CRW
Heinrich Reining GmbH
C 0.3, Cr 1.1, V 0.18, W 3.75, bal Fe.
For cold work tools, upsetters, header dies; oil hardened, tough. *Obsolete*

RAXA CWV5/W
Heinrich Reining GmbH
C, alloy, bal Fe.
For gears, shafts, crankshafts; oil hardened. *Obsolete*

RAXA DBV
Heinrich Reining GmbH
C 0.45, Si 1.5, Cr 1.5, V 0.15, bal Fe.
For drawing tools, cutting tools; cold working tool steel, oil hardening.

RAXA DBV/K
Heinrich Reining GmbH
C 0.6, Si 1, Mn 1, Cr 1.2, V 0.1, bal Fe.
For shear blades, stamping tools, forging dies, molds; cold working tool steel, oil hardening.

RAXA DBV/W
Heinrich Reining GmbH
C 0.4, Si 1.5, Cr 1.5, V 0.12, bal Fe.
For shear blades, drawing dies; cold working tool steel, water hardening.

RAXA EXTRA 1

Heinrich Reining GmbH
C 1.3, Si 0-0.25, Mn 0-0.25, bal Fe.
For engravers' tools, taps, reamers; Type W1; water
hardened. *Obsolete*

RAXA EXTRA 2

Heinrich Reining GmbH
C 1.15, Si 0-0.25, Mn 0-0.25, bal Fe.
For drills, taps, reamers, broaches; Type W1; water hardened.
Obsolete

RAXA EXTRA 3

Heinrich Reining GmbH
C 1, Si 0-0.25, Mn 0-0.25, bal Fe.
For drills, taps, reamers, cutters; Type W1; water hardened.
Obsolete

RAXA EXTRA 4

Heinrich Reining GmbH
C 0.85, Si 0-0.25, Mn 0-0.25, bal Fe.
For drills, taps, reamers, broaches; Type W1; water hardened.
Obsolete

RAXA EXTRA 5

Heinrich Reining GmbH
C 0.75, Si 0-0.25, Mn 0-0.25, bal Fe.
For springs, tools, punches, dies, crimpers; drills, rails,
hammers, tools, springs. *Obsolete*

RAXA EXTRA-EXTRA 1

Heinrich Reining GmbH
C 1.3, Si 0-0.25, Mn 0-0.25, bal Fe.
For engravers' tools, taps, reamers; Type W1; water
hardened. *Obsolete*

RAXA EXTRA-EXTRA 2

Heinrich Reining GmbH
C 1.1, Si 0-0.25, Mn 0-0.25, bal Fe.
Annealed: 110,000 TS; 58,000 YS; 18 El; 38 RA; 210 Brin. For
drills, taps, springs, tools, reamers; Type W1; water hardened.
Obsolete

RAXA EXTRA-EXTRA 3

Heinrich Reining GmbH
C 1, Mn 0-0.25, Si 0-0.25, bal Fe.
Annealed: 100,000 TS; 53,000 YS; 21 El; 42 RA; 200 Brin. For
drills, taps, cutters, broaches, hobs; Type W1; water
hardened. *Obsolete*

RAXA EXTRA-EXTRA 4

Heinrich Reining GmbH
C 0.85, Si 0-0.25, Mn 0-0.25, bal Fe.
Heat treated: 190,000 TS; 145,000 YS; 10 El; 30 RA; 400 Brin.
For drills, springs, punches, tools; Type W1; water hardened.
Obsolete

RAXA EXTRA-EXTRA 5

Heinrich Reining GmbH
C 0.7, Si 0-0.25, Mn 0-0.25, bal Fe.
Heat treated: 175,000 TS; 128,000 YS; 12 El; 37 RA; 355 Brin.
For springs, rails, tools, hammers, axes, cutters; Type W1;
water hardened. *Obsolete*

RAXA EXTRA-EXTRA 6

Heinrich Reining GmbH
C 0.61, Si 0.25, Mn 0.65, bal Fe.
Heat treated: 160,000 TS; 114,000 YS; 12 El; 40 RA; 325 Brin.
For rails, tools, springs, hammers, axes; water hardened.
Obsolete

RAXA G15K

Heinrich Reining GmbH
C 0.25, Cr 14.5, Si 0.7, bal Fe.
Annealed: 95,000 TS; 50,000 YS; 25 El; 55 RA; 200 Brin. For
turbine blades, valves, pumps, surgical instruments; Type
420; stainless. *Obsolete*

RAXA G17K

Heinrich Reining GmbH
C 0.25, Cr 17, Ni 0-1.8, bal Fe.
Annealed: 125,000 TS; 95,000 YS; 20 El; 55 RA; 260 Brin. For
pumps, valves, marine hardware; Type 431; corrosion and
heat resistant. *Obsolete*

RAXA G188K

Heinrich Reining GmbH
C 0.15, Si 1.5, Cr 18, Ni 8.5, bal Fe.
Annealed: 80,000 TS; 35,000 YS; 55 El; 75 RA; 150 Brin. Cold
drawn: 180,000 TS; 15,000 YS; 10 El; 250 Brin. For chemical
plant equipment, tanks, mixers; Type 302; stainless,
austenitic. *Obsolete*

RAXA G188SK

Heinrich Reining GmbH
C 0.15, Cr 18, Ni 8.5, S 0.2, bal Fe.
Annealed: 80,000 TS; 40,000 YS; 50 El; 70 RA; 150 Brin. For
screw machine products, shafts, fasteners; Type 303;
stainless, free-cutting. *Obsolete*

RAXA G199K

Heinrich Reining GmbH
C 0.15, Cr 18, Ni 9.5, Mo 2, bal Fe.
Annealed: 85,000 TS; 35,000 YS; 50 El; 65 RA; 160 Brin. For
acid resistant chemical plant equipment, tanks; Type 316;
stainless, austenitic. *Obsolete*

RAXA G199SK

Heinrich Reining GmbH
C 0.15, Cr 18, Ni 9.5, Mo 2, S 0.2, bal Fe.
Annealed: 85,000 TS; 35,000 YS; 50 El; 65 RA; 160 Brin. For
screw machine products, fasteners; stainless, austenitic,
free-cutting. *Obsolete*

RAXA G28NK

Heinrich Reining GmbH
C 0.4, Si 1.3, Cr 27, Ni 4, bal Fe.
Cast: 90,000 TS; 65,000 YS; 2 El; 212 Brin. Aged: 97,000 TS;
65,000 YS; 18 El; 210 Brin. For cylinder liners, bushings,
valve seats and bodies; heat resistant; Type CC50. *Obsolete*

RAXA G30K

Heinrich Reining GmbH
C 1.2, Si 1.3, Cr 29, bal Fe.
For rollers, crushers, rabble arms; heat and abrasion
resistant. *Obsolete*

RAXA G30MK

Heinrich Reining GmbH
C 1.2, Si 1.3, Cr 29, Mo 2, bal Fe.
For rollers, crushers, rabble arms, grate bars; heat and
abrasion resistant. *Obsolete*

RAXA GARANT 10

Heinrich Reining GmbH
C 0.9, Cr 4, Mo 1, V 2.5, W 12, bal Fe.
For profile cutters, broaches, taps; high speed steel.

RAXA GARANT 100

Heinrich Reining GmbH
C 1.3, Co 5, Cr 4, Mo 2, V 3.75, W 12, bal Fe.
For turning tools, milling cutters; high speed steel.

RAXA GARANT 18

Heinrich Reining GmbH
C 0.74, Cr 4.1, V 1.1, W 18.5, bal Fe.
For drills, hobs, reamers, taps, broaches; high speed steel.
AISI T2.

RAXA GARANT 200

Heinrich Reining GmbH
C 0.76, Co 10, Cr 4.2, Mo 0.8, V 1.8, W 18, bal Fe.
For lathe and planer tools, reamers, broaches; high speed
steel.

RAXA GARANT 300

Heinrich Reining GmbH
C 0.75, Co 15, Cr 4.3, Mo 0.7, V 1.2, W 18, bal Fe.
For turning tools for roughing, planing tools; high speed
steel.

RAXA GARANT 33

Heinrich Reining GmbH
C 0.86, Co 2.8, Cr 4.3, Mo 0.85, V 2.1, W 12, bal Fe.
For drills, hobs, reamers, taps, broaches; high speed steel.
Obsolete

RAXA GARANT 400

Heinrich Reining GmbH
C 1.25, Co 12, Cr 4.3, Mo 5, V 3.3, W 9, bal Fe.
For turning tools of all types; high speed steel.

RAXA GARANT 50

Heinrich Reining GmbH
C 0.79, Co 4.3, Cr 4.3, Mo 0.75, V 1.5, W 18, bal Fe.
For lathe and planer tools, hobs, reamers, taps; high speed
steel. *Obsolete*

RAXA GARANT 55

Heinrich Reining GmbH
C 0.79, W 18, Co 4.7, V 1.5, Cr 4.3, bal Fe.
For reamers, broaches, taps, form cutters; high speed steel.

RAXA GARANT 6 M

Heinrich Reining GmbH
C 0.88, Cr 4, Mo 5, V 1.8, W 6.5, bal Fe.
For twist drills, end mills, threading rolls; high speed steel.
AISI M2.

RAXA GARANT 6 MC

Heinrich Reining GmbH
C 0.9, Co 5, Cr 4, Mo 5, V 1.8, W 6.5, bal Fe.
For twist drills, end mills, thread milling cutters; high speed
steel.

RAXA GARANT 6 MCK

Heinrich Reining GmbH
C 1.1, Co 5, Cr 4, Mo 4, V 1.8, W 6.75, bal Fe.
For milling cutters of all types; high speed steel. AISI M41.

RAXA GARANT 6 MK

Heinrich Reining GmbH
C 1, Cr 4, Mo 5, V 1.8, W 6.5, bal Fe.
For milling cutters of all types; high speed steel. AISI M2.

RAXA GARANT 6 MV

Heinrich Reining GmbH
C 1.2, Cr 4, Mo 5, V 3, W 6.5, bal Fe.
For taps; high speed steel. AISI M3 Class 2.

RAXA GARANT 70

Heinrich Reining GmbH
C 1.3, Cr 4.3, Mo 0.85, V 3.8, W 12, bal Fe.
For engravers' tools, form cutters; high speed steel.

RAXA GARANT 8 M

Heinrich Reining GmbH
C 0.8, Cr 4, Mo 8, V 1.2, W 1.8, bal Fe.
For twist drills; high speed steel. AISI M1.

RAXA GARANT 8 M 34

Heinrich Reining GmbH
C 0.9, Co 8, Cr 4, Mo 9, V 2, W 2, bal Fe.
For milling cutters of all types; high speed steel. Similar to
AISI M 34.

RAXA GARANT 8 M 42

Heinrich Reining GmbH
C 1.1, Co 8, Cr 4, Mo 9.5, V 1.2, W 1.5, bal Fe.
For milling cutters of all types; high speed steel. AISI M42.

RAXA GARANT 8 MC
Heinrich Reining GmbH
C 0.95, Co 5, Cr 4, Mo 8.5, V 2, W 2, bal Fe.
For end mills, twist drills; high speed steel. Similar to AISI M 30.

RAXA GARANT II
Heinrich Reining GmbH
C 0.82, Cr 4.1, Mo 0.85, V 1.6, W 8.7, bal Fe.
For lathe and planer tools, reamers, broaches, hobs; high speed steel. *Obsolete*

RAXA GARANT III
Heinrich Reining GmbH
C 1, Cr 4, Mo 2.7, V 2.3, W 2.8, bal Fe.
For circular metal saws, threading tools, band saws; high speed steel.

RAXA GBS
Heinrich Reining GmbH
C 1.25, Si 1.15, Mn 0.7, Cr 1.2, bal Fe.
For bearings, sleeves, liners; water hardened, wear resistant. *Obsolete*

RAXA GC10
Heinrich Reining GmbH
C 0.6, Si 1.5, Cr 22, bal Fe.
For oil refinery equipment; heat resistant. *Obsolete*

RAXA GC11
Heinrich Reining GmbH
C 0.6, Si 1.5, Cr 29, bal Fe.
For oil refinery equipment; rabble arms; heat resistant. *Obsolete*

RAXA GC8
Heinrich Reining GmbH
C 0.3, Si 2.2, Cr 6, bal Fe.
For oil refinery equipment; heat and creep resistant. *Obsolete*

RAXA GC9
Heinrich Reining GmbH
C 0.5, Si 1.5, Cr 17, bal Fe.
For oil refinery equipment; heat resistant. *Obsolete*

RAXA GCN10
Heinrich Reining GmbH
C 0.4, Si 2, Cr 22, Ni 9.5, bal Fe.
Cast: 85,000 TS; 45,000 YS; 35 El; 165 Brin. For furnace parts, salt pots, retorts, heat treat boxes; Type HF; corrosion and heat resistant. *Obsolete*

RAXA GCN12
Heinrich Reining GmbH
C 0.4, Cr 27, Ni 4, Si 1.3, bal Fe.
Cast: 90,000 TS; 65,000 YS; 2 El; 212 Brin. Heat treated: 97,000 TS; 65,000 YS; 18 El; 210 Brin. For cylinder liners, bushings, valve seats; Type CC50; corrosion and heat resistant. *Obsolete*

RAXA GCN12X
Heinrich Reining GmbH
C 0.4, Si 2, Cr 25, Ni 19, bal Fe.
Cast: 75,000 TS; 50,000 YS; 17 El; 170 Brin. Aged: 85,000 TS; 50,000 YS; 10 El; 190 Brin. For furnace parts, retorts, rabble arms, pots; Type HK; corrosion and heat resistant. *Obsolete*

RAXA GCV11
Heinrich Reining GmbH
C 1.3, Si 1.5, Cr 29, bal Fe.
For rollers, crushers, grate bars; heat and wear resistant. *Obsolete*

RAXA GCV9
Heinrich Reining GmbH
C 0.5, Si 1.5, Cr 17, bal Fe.
For oil refinery equipment; heat resistant. *Obsolete*

RAXA GF CR 12
Heinrich Reining GmbH
C 2, Cr 12, Ni 1.2, V 0.5, bal Fe.
For rolling mill equipment, guides; special steel casting quality, high wear resistance.

RAXA GFG
Heinrich Reining GmbH
C 0.3, Cr 1.1, Ni 0.5, bal Fe.
For rolling mill equipment, guides; special steel casting quality, high wear resistance.

RAXA GL 2
Heinrich Reining GmbH
C 0.4, Mn 1.5, Cr 2, Mo 0.2, bal Fe.
For forging dies, upsetting tools, extrusion discs; hot working tool steel, air or oil hardening.

RAXA GL 3
Heinrich Reining GmbH
C 0.55, Cr 0.7, Mo 0.3, Ni 1.7, V 0.1, bal Fe.
For forging dies, plastic molds; hot working tool steel, oil hardening.

RAXA GL 4
Heinrich Reining GmbH
C 0.55, Cr 1.1, Mo 0.5, Ni 1.7, V 0.1, bal Fe.
For forging dies, extrusion punches; hot working tool steel, air or oil hardening.

RAXA GL 5
Heinrich Reining GmbH
C 0.6, Cr 1.4, Mo 0.8, Ni 1.7, V 0.1, bal Fe.
For forging dies, extrusion punches; hot working tool steel, air or oil hardening.

RAXA GL1
Heinrich Reining GmbH
C 0.53, Si 0.9, Mn 0.9, bal Fe.
For gears, bolts, machine tool parts; water hardened. *Obsolete*

RAXA HDS
Heinrich Reining GmbH
C 0.45, Si 1, Cr 1.2, V 0.2, W 1.95, bal Fe.
For shear blades, pneumatic tools, cutting tools; cold work tool steel, oil hardening.

RAXA HDS/H
Heinrich Reining GmbH
C 0.45, Cr 1, V 0.2, W 1.85, bal Fe.
For cold work tools, headers, upsetters; oil hardened, tough. *Obsolete*

RAXA HM
Heinrich Reining GmbH
C 0.45, Ni, Cr, W, bal Fe.
For cold work tools, upsetters, punches, shears; oil hardened, tough. *Obsolete*

RAXA HSS
Heinrich Reining GmbH
C 1.05, Mn 1, Cr 1, W 1.2, bal Fe.
For cutting tools, measuring tools; cold working tool steel, oil hardening.

RAXA HWA
Heinrich Reining GmbH
C 0.3, Cr 2.65, V 0.35, W 8.5, bal Fe.
For extrusion rams and liners, mandrels; oil hardened, hot work steel.

RAXA HWA 10
Heinrich Reining GmbH
C 0.6, Cr 4, Mo 1, V 0.7, W 9, bal Fe.
For high stress hot working tools; hot working tool steel, air or oil hardening.

RAXA HWA CO
Heinrich Reining GmbH
C 0.45, Co 4.5, Cr 4.5, Mo 0.5, V 2, W 4.5, bal Fe.
For high stress hot working tools; hot working tool steel, air or oil hardening. AISI H19.

RAXA HWA SPEZIAL
Now RAXE HWAS.

RAXA HWAS
Heinrich Reining GmbH
C 0.3, Co 2, Cr 2.5, V 0.25, W 8.5, bal Fe.
For tube extrusion tools, pressure casting molds; hot working tool steel, air or oil hardening.

RAXA KEV
Heinrich Reining GmbH
C, alloy, bal Fe.
For machine tool parts; oil hardened. *Obsolete*

RAXA KLW
Heinrich Reining GmbH
C, alloy, bal Fe.
For machine tool parts; oil hardened. *Obsolete*

RAXA KSM
Heinrich Reining GmbH
C 1, V 0.1, bal Fe.
For cold impact tools, stamping tools; cold working tool steel, oil hardening.

RAXA MKS
Heinrich Reining GmbH
C 1.05, Mn 1.1, Cr 1.1, bal Fe.
For cutting tools, shaping rolls; cold working tool steel, oil hardening.

RAXA MN12
Heinrich Reining GmbH
C 0.2, Mn 13, bal Fe.
Annealed: 95,000 TS; 50,000 YS; 25 El; 55 RA; 195 Brin. Cold drawn: 105,000 TS; 85,000 YS; 17 El; 50 RA; 215 Brin. For turbine blades, valves, cutlery, surgical instruments; Type 420; stainless. *Obsolete*

RAXA MSA
Heinrich Reining GmbH
C 0.9, Mn 2, V 0.15, bal Fe.
For cutting tools, measuring tools; cold working tool steel, oil hardening.

RAXA MWA
Heinrich Reining GmbH
C 0.3, Cr 2.35, V 0.6, W 4.25, bal Fe.
For extrusion dies and rams, upsetters, headers; oil hardened, tough.

RAXA MWH
Heinrich Reining GmbH
C, alloy, bal Fe.
For gears, shafts, axles, bolts; oil hardened, tough. *Obsolete*

RAXA MWV
Heinrich Reining GmbH
C 0.45, C, Ni, Cr, bal Fe.
For crankshafts, gears, bolts; oil hardened, shock resistant. *Obsolete*

RAXA NWP
Heinrich Reining GmbH
C 0.5, Cr 1, Ni 3.25, bal Fe.
For stamping tools, cutlery punching tools; cold working tool steel, air or oil hardening.

RAXA NWP/K
Heinrich Reining GmbH
C 0.45, Cr 1.4, Mo 0.3, Ni 4, bal Fe.
For high stressed stamping tools, billet shearing knives; cold working tool steel, air or oil hardening.

RAXA PCO

Heinrich Reining GmbH
C 0.4, Cr 13, bal Fe.
Annealed: 95,000 TS; 50,000 YS; 25 El; 55 RA; 196 Brin. For valves, cutlery, surgical and dental instruments; Type 420; stainless. *Obsolete*

RAXA PDH

Heinrich Reining GmbH
C 0.27, Cr 1.4, Mo 1.3, V 0.4, bal Fe.
For forging dies; hot working tool steel, oil hardening. W. Nr. 1.2353.

RAXA PDL

Heinrich Reining GmbH
C 0.3, Cr 0.8, Mo 0.6, Ni 2.5, V 0.3, bal Fe.
For pilger mandrels 70-140 mm in diameter, mandrel bars; hot working tool steel, air or oil hardening.

RAXA PDM

Heinrich Reining GmbH
C 0.3, Cr 0.4, Mo 1.7, Ni 4.5, V 0.25, bal Fe.
For pilger mandrels up to 70 mm in diameter, mandrel bars; hot working tool steel, air or oil hardening.

RAXA PDR

Heinrich Reining GmbH
C 0.3, Cr 3, Mo 2.7, V 0.5, bal Fe.
For forging dies, pressure casting mold tools; hot working tool steel, air or oil hardening.

RAXA PSV

Heinrich Reining GmbH
C 0.25, Cr 0.8, Mo 0.3, Ni 1.5, V 0.2, bal Fe.
For pilgrin mandrels 140-200 mm in diameter, mandrel bars; hot working tool steel, oil hardening.

RAXA PW 12

Heinrich Reining GmbH
C 1.2, Cr 2, W 1.3, bal Fe.
For pilger rolls, hot-work; special steel casting quality.

RAXA PW 8

Heinrich Reining GmbH
C 0.8, Cr 2, W 1.3, bal Fe.
For pilger rolls, hot work; special steel casting quality.

RAXA PW 9

Heinrich Reining GmbH
C 0.9, Cr 2, W 1, bal Fe.
For pilger rolls, hot work; special steel casting quality.

RAXA R10

Heinrich Reining GmbH
C, alloy, bal Fe.
For machine tool parts. *Obsolete*

RAXA R15

Heinrich Reining GmbH
C, alloy, bal Fe.
For machine tool parts. *Obsolete*

RAXA R60

Heinrich Reining GmbH
C, alloy, bal Fe.
For machine tool parts. *Obsolete*

RAXA R70

Heinrich Reining GmbH
C, alloy, bal Fe.
For machine tool parts. *Obsolete*

RAXA REC60

Heinrich Reining GmbH
C, Cr, bal Fe.
For machine tool parts. *Obsolete*

RAXA ROV

Heinrich Reining GmbH
C 0.45, Cr 4.5, Mo 1.1, Ni 12, V 1, W 9, bal Fe.
For extrusion dies; hot working tool steel.

RAXA RWS 5

Heinrich Reining GmbH
C 1.3, Cr 1, V 0.35, W 5, bal Fe.
For cold drawing dies; cold working tool steel, water hardening.

RAXA RWS 5/W

Heinrich Reining GmbH
C 1.3, W 5, bal Fe.
For cold drawing dies, turning knives; cold working tool steel, water hardening.

RAXA RZM

Heinrich Reining GmbH
C 1.4, V 0.1, Mn 0.3, Si 0.25, bal Fe.
For engravers tools, form cutters, taps, drills; water hardened; Type W2. *Obsolete*

RAXA SKL

Heinrich Reining GmbH
C 1, Cr 1.5, bal Fe.
For cold rolls, drawing dies, threading tools; cold working tool steel; oil or water hardening.

RAXA SL 35

Heinrich Reining GmbH
C 1.45, V 3.5, bal Fe.
For high wear resistant cold extrusion and cold impact tools; cold working tool steel, water hardening.

RAXA SPEZIAL 1

Heinrich Reining GmbH
C 0.9, Si 0.25-0.5, Mn 0.3-0.8, bal Fe.
Heat treated: 190,000 TS; 145,000 YS; 10 El; 30 RA; 400 Brin. For springs, tools, dies, drills, taps; Type W1; water hardened. *Obsolete*

RAXA SPEZIAL 2

Heinrich Reining GmbH
C 0.75, Si 0.25-0.5, Mn 0.3-0.8, bal Fe.
Heat treated: 185,000 TS; 140,000 YS; 15 El; 38 RA; 400 Brin. For springs, rails, hammers, dies, crimpers; Type W1; water hardened. *Obsolete*

RAXA SPEZIAL 3

Heinrich Reining GmbH
C 0.67, Si 0.25-0.5, Mn 0.3-0.8, bal Fe.
Heat treated: 172,000 TS; 125,000 YS; 14 El; 40 RA; 350 Brin. For springs, rails, hammers, axes, punches; water hardened. *Obsolete*

RAXA SPEZIAL 4

Heinrich Reining GmbH
C 0.6, Si 0.25-0.5, Mn 0.3-0.8, bal Fe.
Heat treated: 160,000 TS; 113,000 YS; 12 El; 40 RA; 325 Brin. For springs, bolts, hammers, axes, punches; water hardened. *Obsolete*

RAXA SPEZIAL 5

Heinrich Reining GmbH
C 0.45, Si 0.25-0.5, Mn 0.3-0.8, bal Fe.
Hot rolled: 98,000 TS; 59,000 YS; 24 El; 45 RA; 212 Brin. For gears, bolts, fasteners, crankshafts; water hardened. *Obsolete*

RAXA SPEZIAL 6

Heinrich Reining GmbH
C 0.35, Si 0.25-0.5, Mn 0.3-0.8, bal Fe.
Hot rolled: 85,000 TS; 54,000 YS; 30 El; 53 RA; 185 Brin. For gears, bolts, fasteners, machine tool parts; water hardened. *Obsolete*

RAXA SPS

Heinrich Reining GmbH
C 0.55, Si 1.7, Mo 0.3, Ni 0.4, bal Fe.
For clamping tools, beating bars; cold working tool steel, oil hardening.

RAXA SSWR

Heinrich Reining GmbH
C 0.9, Si 0.25, Cr 0.8, bal Fe.
Heat treated: 190,000 TS; 145,000 YS; 10 El; 30 RA; 400 Brin. For springs, tools, cutters, drills, taps; Type W1; water hardened. *Obsolete*

RAXA TWO 3

Heinrich Reining GmbH
C 0.45, Cr 1.5, Mo 0.5, V 0.8, W 0.5, bal Fe.
For forging tools; hot working tool steel, oil hardening.

RAXA UW10

Heinrich Reining GmbH
C, alloy, bal Fe.
For machine tool parts. *Obsolete*

RAXA UW13

Heinrich Reining GmbH
C, alloy, bal Fe.
For machine tool parts. *Obsolete*

RAXA VS6

Heinrich Reining GmbH
C, alloy, bal Fe.
For machine tool parts. *Obsolete*

RAXA WCH

Heinrich Reining GmbH
C 0.45, W, Cr, V, bal Fe.
For dies, tools; oil hardened. *Obsolete*

RAXA WCNU

Heinrich Reining GmbH
C, alloy, bal Fe.
For machine tool parts. *Obsolete*

RAXA WCNV

Heinrich Reining GmbH
C 0.5, Si 1.3, Cr 13, Ni 13, V 1, W 2, bal Fe.
For extrusion dies; hot working tool steel.

RAXA WCNV/M

Heinrich Reining GmbH
C 0.35, Si 1.3, Cr 13, Ni 6, V 1, W 2.8, bal Fe.
For extrusion dies, hot working tool steel.

RAXA WCR 8

Heinrich Reining GmbH
C 0.5, Si 1, Cr 8.5, Mo 1.2, W 1.2, bal Fe.
For wood chopping knives; cold working tool steel, oil or air hardening.

RAXA WF 18

Heinrich Reining GmbH
C 2, Co 0.7, Cr 22, Ni 1.9, W 1.8, bal Fe.
For rolling mill equipment; special steel casting quality.

RAXA WF 6

Heinrich Reining GmbH
C 1, Cr 1.2, W 0.6, bal Fe.
For rolling mill equipment; special steel casting quality.

RAXA WHW

Heinrich Reining GmbH
C 0.6, Cr 1, V 0.2, W 2, bal Fe.
For shear blades, cutting tools; cold working tool steel, oil hardening.

RAXA WP6

Heinrich Reining GmbH
C, alloy, bal Fe.
For machine tool parts; oil hardened. *Obsolete*

RAXA WS1
Heinrich Reining GmbH
C 1.2, Cr 0.2, V 0.1, W 1, bal Fe.
For cutters, dies, form tools; water hardened. *Obsolete*

RAXA WSM
Heinrich Reining GmbH
C, alloy, bal Fe.
For machine tool parts. *Obsolete*

RAXA WZR
Heinrich Reining GmbH
C, Cr, W, bal Fe.
For machine tool parts. *Obsolete*

RAXA ZHW
Heinrich Reining GmbH
C, alloy, bal Fe.
For machine tool parts. *Obsolete*

RAXA ZRWS
Heinrich Reining GmbH
C 2.7, Co 1.5, Cr 22, Ni 2.5, V 0.3, W 5, bal Fe.
For hot drawing dies; special steel casting quality.

RAXA ZVW
Heinrich Reining GmbH
C 1.42, W, V, bal Fe.
For bearings, engraving tools, cutters; oil hardened.
Obsolete

RAXIT 006 G
Heinrich Reining GmbH
C 0.1, Cr 13, bal Fe.
For steam turbines, fittings; stainless steel, casting quality. W. Nr. 1.4006.

RAXIT 008 G
Heinrich Reining GmbH
C 0.12, Cr 13, Ni 1, bal Fe.
For valve parts; stainless steel casting quality. W. Nr. 1.4008.

RAXIT 027 G
Heinrich Reining GmbH
C 0.2, Cr 14, bal Fe.
For steam turbines, fittings, valves, glass molds; stainless steel casting quality. W. Nr. 1.4027.

RAXIT 059 G
Heinrich Reining GmbH
C 0.22, Cr 17, Ni 1.5, bal Fe.
For fittings, pump parts, stainless steel casting quality. W. Nr. 1.4059.

RAXIT 085 G
Heinrich Reining GmbH
C 0.7, Cr 29, bal Fe.
For chemical industry, mining; stainless steel casting quality, wear resistant. W. Nr. 1.4085.

RAXIT 086 G
Heinrich Reining GmbH
C 1.2, Cr 29, bal Fe.
For chemical industry, mining; stainless steel casting quality, wear resistant. W. Nr. 4086.

RAXIT 136 G
Heinrich Reining GmbH
C 0.7, Cr 29, Mo 2, bal Fe.
For chemical industry, mining; stainless steel casting quality, wear resistant. W. Nr. 1.4136.

RAXIT 138 G
Heinrich Reining GmbH
C 1.2, Cr 29, Mo 2, bal Fe.
For chemical industry, mining; stainless steel casting quality, wear resistant.

RAXIT 1815 SI
Heinrich Reining GmbH
C 0.03, Cr 18, Ni 15, Si 4, bal Fe.
For chemical industry; stainless steel casting quality.

RAXIT 219 G
Heinrich Reining GmbH
C 0.25, Cr 1, Mo 0.25, bal Fe.
Special steel casting quality; tough at subzero temperatures. W. Nr. 1.7219.

RAXIT 306 G
Heinrich Reining GmbH
C 0.03, Cr 18, Ni 11, bal Fe.
For chemical industry; stainless steel casting quality. W. Nr. 1.4306.

RAXIT 308 G
Heinrich Reining GmbH
C 0.07, Cr 18, Ni 10, bal Fe.
For chemical industry; stainless steel casting quality. W. Nr. 1.4308.

RAXIT 312 G
Heinrich Reining GmbH
C 0.1, Cr 18, Ni 9, bal Fe.
For water turbines, pump wheels; stainless steel casting quality. W. Nr. 1.4312.

RAXIT 313 G
Heinrich Reining GmbH
C 0.05, Cr 13, Ni 5, bal Fe.
For water turbines, pump-wheels; stainless steel casting quality. W. Nr. 1.4313.

RAXIT 340 G
Heinrich Reining GmbH
C 0.4, Cr 27, Ni 4, bal Fe.
For chemical industry, mining, stainless steel casting quality, wear resistant. W. Nr. 1.4340.

RAXIT 347 G
Heinrich Reining GmbH
C 0.08, Cr 26, Ni 7, bal Fe.
For chemical industry, fittings; stainless steel casting quality. W. Nr. 1.4347.

RAXIT 404 G
Heinrich Reining GmbH
C 0.03, Cr 18, Mo 2.3, Ni 12, bal Fe.
For chemical industry; stainless steel casting quality. W. Nr. 1.4404.

RAXIT 408 G
Heinrich Reining GmbH
C 0.06, Cr 18, Mo 2.3, Ni 11, bal Fe.
For chemical and textile industries; stainless steel casting quality. W. Nr. 1.4408.

RAXIT 410 G
Heinrich Reining GmbH
C 0.1, Cr 18, Mo 2.3, Ni 10, bal Fe.
For chemical and textile industries; stainless steel casting quality. W. Nr. 1.4410.

RAXIT 437 G
Heinrich Reining GmbH
C 0.06, Cr 18, Mo 2.7, Ni 12, bal Fe.
For chemical industry; stainless steel casting quality. W. Nr. 1.4437.

RAXIT 448 G
Heinrich Reining GmbH
C 0.06, Cr 17, Mo 4.5, Ni 13, bal Fe.
For chemical and textile industries; stainless steel casting quality. W. Nr. 1.4448.

RAXIT 460 G
Heinrich Reining GmbH
C 0.08, Cr 27, Mo 1.8, Ni 5, bal Fe.
For cellulose industry; stainless steel casting quality

RAXIT 464 G
Heinrich Reining GmbH
C 0.4, Cr 27, Mo 2.3, Ni 5, bal Fe.
For chemical industry; stainless steel casting quality, wear resistant. W. Nr. 1.4464.

RAXIT 500 G
Heinrich Reining GmbH
C 0.07, Cr 20, Cu 2, Mo 3, Ni 25, Nb, bal Fe.
For chemical industry; stainless steel casting quality. W. Nr. 1.4500.

RAXIT 5419
Heinrich Reining GmbH
C 0.22, Mo 0.4, Mn 0.7, bal Fe.
For turbine industry; special steel casting quality with high strength at elevated temperatures.

RAXIT 542 G
Heinrich Reining GmbH
C 0.05, Cr 17, Cu 4, Ni 4, Nb, bal Fe.
For water power plants; stainless steel casting quality, wear resistant. W. Nr. 1.4542.

RAXIT 552 G
Heinrich Reining GmbH
C 0.07, Cr 18, Ni 10, Nb, bal Fe.
For chemical industry; stainless steel casting quality. W. Nr. 1.4552.

RAXIT 581 G
Heinrich Reining GmbH
C 0.07, Cr 18, Mo 2.3, Ni 10, Nb, bal Fe.
For textile, cellulose and dye industries; stainless steel casting quality. W. Nr. 1.4581.

RAXIT 585 G
Heinrich Reining GmbH
C 0.07, Cr 18, Mo 2.3, Ni 20, Nb, bal Fe.
For chemical industry; stainless steel casting quality. W. Nr. 1.4585.

RAXIT 588 G
Heinrich Reining GmbH
C 0.05, Cr 18, Cu 2, Mo 3, Ni 22, Nb, bal Fe.
For chemical industry; stainless steel casting quality. W. Nr. 1.4588.

RAXIT 593 G
Heinrich Reining GmbH
C 0.07, Cr 25, Cu 1.5, Mo 2.5, Ni 8, bal Fe.
For chemical industry, pump manufacturers; stainless steel casting quality.

RAXIT 602 G
Heinrich Reining GmbH
C 0.1, Cr 16, Mo 17, Ni 52, W 4, bal Fe.
For chemical industry; stainless steel casting quality.

RAXIT 619 G
Heinrich Reining GmbH
C 0.25, bal Fe.
Special steel casting quality with high strength at elevated temperatures. W. Nr. 1.0619.

RAXIT 7225
Heinrich Reining GmbH
C 0.42, Cr 1, Mo 0.25, Mn 0.7, bal Fe.
For dredging and excavating industry; special steel casting quality, oil hardening.

RAXIT 810 G
Heinrich Reining GmbH
C 0.05, Mo 28, bal Fe.
For chemical industry; stainless steel casting quality.

RAXIT 902 G
Heinrich Reining GmbH
C 0.06, Cr 18, Ni 10, bal Fe.
Special stainless steel casting quality, tough at subzero temperatures. W. Nr. 1.6902.

RAXIT 905 G

Heinrich Reining GmbH
C 0.1, Cr 18, Ni 10, Nb, bal Fe.
Special stainless steel casting quality, tough at subzero temperatures. W. Nr. 1.6905.

RAXIT 948 G

Heinrich Reining GmbH
C 0.06, Cr 18, Ni 11, bal Fe.
For pressure vessels in nuclear reactors; stainless steel casting quality with high strength at extremely high temperatures.

RAXOTHERM 150 G

Heinrich Reining GmbH
C 0.5, Si 1.5, Cr 15, Ni 0.5, bal Fe.
For rolling mill equipment, wear parts; heat resisting casting quality.

RAXOTHERM 250 G

Heinrich Reining GmbH
C 0.5, Si 1, Cr 25, bal Fe.
For pressure casting crucibles; heat resisting casting quality.

RAXOTHERM 710 G

Heinrich Reining GmbH
C 0.3, Si 1.5, Cr 7, bal Fe.
For furnace industry; heat resisting up to 1300°F (704°C). Heat resisting casting quality. W. Nr. 1.4710.

RAXOTHERM 729 G

Heinrich Reining GmbH
C 0.4, Si 1.5, Cr 13, bal Fe.
For furnace industry; heat resisting up to 1550°F (843°C). Heat resisting casting quality. W. Nr. 1.4729.

RAXOTHERM 740 G

Heinrich Reining GmbH
C 0.4, Si 1.5, Cr 17, bal Fe.
For furnace industry; heat resisting up to 1650°F (900°C). Heat resisting casting quality. W. Nr. 1.4740.

RAXOTHERM 743 G

Heinrich Reining GmbH
C 1.6, Si 1.5, Cr 18, bal Fe.
For furnace industry; wear resistant, heat resisting up to 1650°F (900°C). Heat resisting casting quality. W. Nr. 1.4743.

RAXOTHERM 745 G

Heinrich Reining GmbH
C 0.4, Si 1.5, Cr 23, bal Fe.
For furnace industry; heat resisting up to 1900°F (1038°C). Heat resisting casting quality. W. Nr. 1.4745.

RAXOTHERM 776 G

Heinrich Reining GmbH
C 0.4, Si 1.5, Cr 29, bal Fe.
For furnace industry; heat resisting up to 2100°F (1150°C). Heat resisting casting quality. W. Nr. 1.4776.

RAXOTHERM 777 G

Heinrich Reining GmbH
C 1.3, Si 1.5, Cr 29, bal Fe.
For furnace industry; heat resisting up to 2000°F (1093°C). Heat resisting and wear resisting casting quality. W. Nr. 1.4777.

RAXOTHERM 823 G

Heinrich Reining GmbH
C 0.4, Si 1.5, Cr 27, Ni 4, bal Fe.
For furnace industry; heat resisting up to 2000°F (1093°C). Heat resisting casting quality. W. Nr. 1.4823.

RAXOTHERM 825 G

Heinrich Reining GmbH
C 0.25, Si 1.5, Cr 18, Ni 9, bal Fe.
For furnace industry; heat resisting up to 1650°F (900°C). Heat resisting casting quality. W. Nr. 1.4825.

RAXOTHERM 826 G

Heinrich Reining GmbH
C 0.4, Si 1.5, Cr 22, Ni 9, bal Fe.
For furnace industry; heat resisting up to 1750°F (954°C). Heat resisting casting quality. W. Nr. 1.4826.

RAXOTHERM 832 G

Heinrich Reining GmbH
C 0.25, Si 1.5, Cr 20, Ni 14, bal Fe.
For furnace industry; heat resistant to 1750°F (954°C). Heat resisting casting quality. W. Nr. 1.4832.

RAXOTHERM 837 G

Heinrich Reining GmbH
C 0.4, Si 1.5, Cr 25, Ni 12, bal Fe.
For furnace industry; heat resisting up to 1920°F (1050°C). Heat resisting casting quality. W. Nr. 1.4837.

RAXOTHERM 848 G

Heinrich Reining GmbH
C 0.4, Si 2, Cr 25, Ni 21, bal Fe.
For furnace industry; heat resisting up to 2000°F (1093°C). Heat resisting casting quality. W. Nr. 1.4848.

RAXOTHERM 849 G

Heinrich Reining GmbH
C 0.4, Si 1.5, Cr 18, Ni 38, Nb 1.5, bal Fe.
For furnace industry; heat resisting up to 2000°F (1093°C). Heat resisting casting quality. W. Nr. 1.4849.

RAXOTHERM 857 G

Heinrich Reining GmbH
C 0.4, Si 2, Cr 25, Ni 35, bal Fe.
For oil and gas installation; heat resisting casting quality. W. Nr. 1.4857.

RAXOTHERM 865 G

Heinrich Reining GmbH
C 0.4, Si 2, Cr 18, Ni 36, bal Fe.
For furnace industry; heat resistant up to 2100°F (1150°C). Heat resisting casting quality. W. Nr. 1.4865.

RAXOTHERM 879 G

Heinrich Reining GmbH
C 0.5, Si 1, Cr 28, Ni 50, W 5, bal Fe.
For furnace industry, hearth rolls; heat resisting up to 2100°F (1150°C). Heat resisting casting quality. W. Nr. 1.4879.

RAXOTHERM CO 5 G

Heinrich Reining GmbH
C 0.1, Co 50, Cr 28, bal Fe.
For furnace industry; heat resisting up to 2150°F (1177°C). Heat resisting casting quality. W. Nr. 1.4778.

RAXOTHERM CO 6 G

Heinrich Reining GmbH
C 0.35, Co 50, Cr 28, Nb 2, bal Fe.
For furnace industry; heat resisting up to 2150°F (1177°C). Heat resisting casting quality.

RAXOTHERM CR NI 50/50

Heinrich Reining GmbH
C 0.07, Cr 50, Ni 50, Nb.
For oil refining plants, power plants, high heat resisting; stainless special steel casting quality.

RAXOTHERM WC MO/G

Heinrich Reining GmbH
C 0.48, Cr 1.5, Mo 0.8, V 0.3, bal Fe.
For pressure casting industry; high temperature casting quality.

RAYMUR

Manufacturer not listed.
For ornamental and corrosion resistant parts; corrosion resistant.

RAYO

Manufacturer not listed.
Ni 85, Cr 15.
For resistance alloy; heat resistant.

RB 10

Saarstahl AG
Now SAARSTAHL 1.1645.

RB 11

Saarstahl AG
Now SAARSTAHL 1.1654.

RB 6

Saarstahl AG
Now SAARSTAHL 1.1620.

RB 8

Saarstahl AG
Now SAARSTAHL 1.1625.

RBFF

Saarstahl AG
C 0.52, Si 0.25, Mn 1.75, V 0.1, bal Fe.
For leaf springs. W.-Nr. 0915. *Obsolete*

RC

Acciaierie Valbruna s.p.a.
C 0.75, Cr 4.05, W 18, V 1.1, Si 0-0.4, Mn 0-0.4, bal Fe.
Tungsten high speed steel. W. Nr. 1.3355; AISI T1.

RC2

Acciaierie Valbruna s.p.a.
Tool Material. C 0.85, Cr 4, W 18.25, V 2, Si 0-0.4, Mn 0-0.4, Mo 0-1, bal Fe.
Tungsten high speed steel. AISI T2.

RCA NO. 91

RCA Corporate Standards Engineering
Ni-Co.
For electronic equipment; heat resistant. *Obsolete*

RCA NO. 97

RCA Corporate Standards Engineering
Ni-Co.
For electronic equipment; heat resistant. *Obsolete*

RCA-N100

RCA Corporate Standards Engineering
C 0.1, W 2, Al 1, Mg 0.05, bal Ni.
For filaments. *Obsolete*

RCA-N9

RCA Corporate Standards Engineering
C 0.1, Mn 0.2, Si 0.16, Mg 0.05, bal Ni.
For filaments. *Obsolete*

RCA-N91

RCA Corporate Standards Engineering
Si 0.1, Mn 0.1, Co 20, 0.04% min C, bal Ni.
For rectifier filaments. *Obsolete*

RCA-N97

RCA Corporate Standards Engineering
Co 40, Si 0.16, Mn 0.1, 0.04% min C, bal Ni.
For rectifier filaments. *Obsolete*

RCC

Saarstahl AG
Now SAARSTAHL 1.2080.

RCC EXTRA

Saarstahl AG
Now SAARSTAHL 1.2436.

RCC SPEZIAL

Saarstahl AG
Now SAARSTAHL 1.2601.

RCC SUPRA

Saarstahl AG
Now SAARSTAHL 1.2379.

RCCM

Creusot-Loire
C 0.9, Mn 1.9, Cr 0.5, V 0.1, bal Fe.
Cold work steel; oil hardening. For punching and forming dies; reamers. AFNOR 90 MCV 8.

RCCV

Saarstahl AG
C 2.1, Cr 12, Mo 1.15, V 2.1, bal Fe.
Cold work tool steel. Stamping dies for dynamo and transformer sheets, and plastic molds. W.-Nr. 2378; similar to AISI D7. *Obsolete*

RCK10

Acciaierie Valbruna s.p.a.
Tool Material. C 0.78, Cr 4.15, W 18, V 1.6, Mo 0.65, Co 9.5, Si 0-0.4, Mn 0-0.4, bal Fe.
High speed tool steel. W. Nr. 1.3265; AISI T5.

RCK5

Acciaierie Valbruna s.p.a.
Tool Material. C 0.8, Cr 4, W 18, V 1.25, Co 5, Si 0-0.5, Mn 0-0.5, Mo 0-0.5, bal Fe.
High speed tool steel. W. Nr. 1.3255; AISI T4.

RCW 1

Saarstahl AG
C 0.58, Cr 3.75, Mo 0.7, V 0.7, W 9, bal Fe.
Hot work tool steel. For tube pressing mandrels up to 30 mm in diameter. W.-Nr. 2622; AFNOR Z 60 WC 09 04. *Obsolete*

RCW 2 H

Saarstahl AG
C 0.32, Cr 3.55, V 0.35, W 8.5, bal Fe.
Hot work steel. For hot extrusion dies, pressure casting molds, and forging dies. W.-Nr. 2581; AISI H21. *Obsolete*

RD ALLOY

American manufacture
Cu alloy.
For brazing of tools; age-hardenable.

RDC 1

Saarstahl AG
C 0.36, Si 1, Cr 5, Mo 1.4, V 0.25, W 1.35, bal Fe.
Hot work tool steel. For pressing and forging dies. W.-Nr. 2606; similar to AISI H 12. *Obsolete*

RDC 2

Saarstahl AG
Now SAARSTAHL 1.2343.

RDC 2V

Saarstahl AG
Now SAARSTAHL 1.2344.

RDC 6

Saarstahl AG
Now SAARSTAHL 1.2362X1.

RDC 8

Saarstahl AG
Now SAARSTAHL 1.2369.

RE-CRO

Midvale-Heppenstall Co.
C 1.2, Mn 0.25, Cr 0.5, Si 0.2, bal Fe.
For cold forming dies; water hardening. *Obsolete*

RE-MO 50/50

Chase Brass & Copper Co.
Rh 47.9, Mo 52.1.
Recrystallized: 150,000 TS; 123,000 YS; 350 VHN. Wrought: 240,000 TS; 210,000 YS; 600 VHN. For nuclear and space propulsion applications, thermocouple protection tubes, furnace resistance heaters and inductors. High heat resistance. *Obsolete*

REACTAL

J.J. Rieter & Co.
C 3.2, Si 2.2, Ni 1.5, Cr 0.8, bal Fe.
Cast: 42,000-55,000 TS; 187 Brin. For corrosion and heat resisting parts, corrosion and heat resistant castings.

READILY FUSIBLE ALLOY

Manufacturer not listed.
Zn 67, Cu 33.
For brazing; high strength.

READING 15 ALUMINUM-85 VANADIUM

Reading Alloys, Inc.
Reactive/refractive. Al 13-16, C 0.1, V 82-86.
Master alloy.

READING 35 ALUMINUM-65 VANADIUM

Reading Alloys, Inc.
Reactive/refractive. Al 34-39, C 0.18, V 60-65.
Master alloy.

READING 40 ALUMINUM-60 ZIRCONIUM

Reading Alloys, Inc.
Reactive/refractive. Al 39-45, C 0.1, Zr 55-61.
Master alloy.

READING 44 ALUMINUM-56 CHROMIUM

Reading Alloys, Inc.
Reactive/refractive. Al 43-45, Cr 55-57.
Master alloy.

READING 50 ALUMINUM-50 VANADIUM

Reading Alloys, Inc.
Reactive/refractive. Al 45-49, C 0.1, V 50-54.
Master alloy.

READING 50 MOLYBDENUM-50 NICKEL

Reading Alloys, Inc.
Reactive/refractive. Al 0.05, C 0.05, Mo 47-50, Ni 47-50, Si 2.2.
Master alloy.

READING 60 ALUMINUM-40 VANADIUM

Reading Alloys, Inc.
Reactive/refractive. Al 54-59, C 0.1, V 40-45.
Master alloy.

READING CHROMIUM-NICKEL

Reading Alloys, Inc.
Reactive/refractive. Al 0.75, Cr 78-82, Ni 18-22.
Master alloy.

READY-FLOW

Engelhard Minerals & Chemicals Corp.
Ag 56, Cu 44.
For brazing; M.P. 1165 F. *Obsolete*

READY-MARK

Brown & Sharpe Mfg. Co.
C 0.9, Mn 1.2, Cr 0.5, V 0.2, W 0.5, bal Fe.
AISI type O1 oil hardening tool steel.

READYWELD

Lincoln Electric Co.
C 0.2, bal Fe.
Cast: 60,000-70,000 TS; 16-22 El. For welding rods. *Obsolete*

RECIDAL

Italian manufacture
Al alloy.
For screw machine products, fasteners; free-cutting.

RECIDAL 55

Italian manufacture
Cu 5.3-5.7, Pb 0.2-0.3, Cd 0.2-0.3, Zn 0.4-0.6, Fe 0.3, Si 0.25, bal Al.
For screw machine products, fasteners; free-cutting, corrosion resistant.

RECKHAMMER SSW

German manufacture
C 0.9, Cr 0.8, Mn 0.3, bal Fe.
For bearings, liners, cutters, sleeves; water hardened, wear resistant.

RECN

Saarstahl AG
C 0.18, Si 0.25, Mn 0.5, Cr 1.95, Ni 1.95, bal Fe.
Alloy carburizing steel. For highly stressed automotive and tractor gears and spline couplings. W.-Nr. 5920. *Obsolete*

RECO 1

Usines Emile Henricott, SA
Ni 24, Al 13, Cu 4, bal Fe.
For permanent magnets, electrical and magnetic equipment. High permeability.

RECO 2A

Dutch manufacture
Al 7, Ni 20, Co 20, Cu 7, Ti 6.5, bal Fe.
For permanent magnets.

RECO 3A

Usines Emile Henricott, SA
Ni 19, Co 12, Al 10, Cu 6.
For permanent magnets, electrical and magnetic equipment. High permeability.

RECO II A

Usines Emile Henricott, SA
C, alloy, bal Fe.
For permanent magnets; high permeability.

RECORD

American Smelting & Refining Co.
Sb, Sn, bal Pb.
For bearings; Babbitt.

RECORD 66

Boyd-Wagner Co.
C 0.7, W 5, Mo 6.75, Cr 4.25, V 1.75, bal Fe.
For gear cutters, form cutters, cold or hot work dies; high speed steel.

RECORD EMINENT-28

Manufacturer not listed.
C 0.8, Cr 4.5, Mo 1, Co 10, W 18.5, V 1.6, bal Fe.
For milling cutters, lathe and planer tools, reamers; high speed steel.

RECORD EXTRA

Boyd-Wagner Co.
C 0.8, W 18, Cr 3.75, V 1.25, Co 2.25, bal Fe.
260 Brin. For dies, tools for high speed production for turning and heavy cutting, sealing material; high speed steel.

RECORD MO 50

Idealstahl Breidenbach KG
C 0.85, Cr, V, Mo, W, bal Fe.
For lathe and planer tools, reamers, broaches, hobs; high speed steel.

RECORD SELECT

Boyd-Wagner Co.
C 0.8, W 18, Cr 3.75, V 1.25, Co 5.5, bal Fe.
For tools, dies, cutters; high speed steel. *Obsolete*

RECORD STAR

Boyd-Wagner Co.
C 0.7, Cr 4, W 14, V 2, bal Fe.
For tools, cutters; high speed steel.

RECORD SUPERIOR

Boyd-Wagner Co.
C 0.8, W 18, Cr 4, V 1, bal Fe.
For tools, dies for boring, planing and slotting, taps, reamers, punches; high speed steel.

RECORD-EMINENT
Boyd-Wagner Co.
C 0.85, W 18.5, Cr 4.25, V 2.5, Co 10.5, bal Fe.
For tools, dies, cutters; high speed steel.

RECOVAC 100, BS
Vacuumschmelze GmbH
Ni 77.
Soft magnetic alloy with improved wear resistance for wear resistant magnet heads and relay components.

RED (EINHEITS) BRASS
German manufacture
Cu 83-89, Zn 5-12, Pb 3-10, Sn 2-5.
Hard: 80,000 TS; 71,000 YS; 135 El; 4 Brin. Soft: 41,000 TS; 18,000 YS; 47 El; 52 Brin. For tubing, hardware; Tombac, free-cutting.

RED ANCHOR
Teledyne Allvac
C 0.95-1.1, bal Fe.
For general tools, chasers; water hardened.

RED ANCHOR DRILL ROD
Teledyne Allvac
C 0.95-1.1, bal Fe.
80,000 TS; 60,000 YS; 25 El; 50 RA. For anvils, dental tools, spindles, precision shafts for motors; water hardened.

RED ARROW
Latrobe Steel Co.
Co-Cr-W-C.
For cutting tools, bits; cast-to-shape. *Obsolete*

RED ARROW 101
Latrobe Steel Co.
C, alloy, bal Fe.
For cutting tools; oil hardening. *Obsolete*

RED ARROW 202
Latrobe Steel Co.
Cr 25-35, W 10-25, Co 40-60.
For roughing and finishing tools, cutters; cast alloy. *Obsolete*

RED ARROW 222
Latrobe Steel Co.
Cr 25-35, W 10-25, Co 40-60.
For roughing and finishing cutting tools; cast alloy. *Obsolete*

RED ARROW 303
Latrobe Steel Co.
Cr 25-35, W 10-25, Co 40-60.
For roughing and finishing cutting tools; cast alloy. *Obsolete*

RED ARROW 333
Latrobe Steel Co.
Cr 25-35, W 10-25, Co 40-60.
For cutting tools; cast alloy. *Obsolete*

RED ARROW 48 S
Latrobe Steel Co.
Cr 25-35, W 10-25, Co 40-60.
For roughing and finishing cutting tools; heat resistant, cast alloy. *Obsolete*

RED ARROW WELDING ROD
Latrobe Steel Co.
W-Cr-Co-C.
For hard surfacing welding rod; cast alloy. *Obsolete*

RED BRASS
Anaconda Co.
Copper. Cu 85, Zn 15.
Tube and sheet: 42,000-50,000 TS; 20,000-40,000 YS; 4-43 El; 68 RA. For salt water pipes, plumbing, hardware. Corrosion resistant.

RED BRASS
Chase Brass & Copper Co.
Cu 80, Zn 20.
Plates and sheets: 43,000-85,000 TS; 18,000-60,000 YS; 4-50 El; 53-150 Brin. For pipes, hardware; corrosion resistant. *Obsolete*

RED BRASS
Chase Brass & Copper Co.
Cu 85, Zn 15.
Annealed: 42,000 TS; 14,000 YS; 45 El. Rolled: 70,000 TS; 58,000 YS; 5 El. For plumbing, pipe, tubes, hardware; corrosion resisting. *Obsolete*

RED BRASS
Anaconda Co.
Cu 85, Zn 15.
Tubes and sheets: 42,000-50,000 TS; 20,000-40,000 YS; 4-43 El; 68 RA. For salt water pipes, plumbing, hardware; corrosion resistant. *Obsolete*

RED BRASS 230
Anaconda Co.
Cu 85, Zn 15.
Ann: 39,000-47,000 TS; 8000-19,000 YS; 43-48 El. Cold rolled: 44,000-90,000 TS; 23,000-80,000 YS; 2-39 El. Golden color, stress-corrosion resistant brass. For jewelry, pen caps, pencil ferrules, steam iron tubes.

RED BRASS 230
Olin Corp.
Cu 85, Zn 15.
Ann: 39,000-47,000 TS; 8000-19,000 YS; 43-48 El. Cold rolled: 44,000-90,000 TS; 23,000-80,000 YS; 2-39 El. Golden color, stress-corrosion resistant brass. For jewelry, pen caps, pencil ferrules, steam iron tubes.

RED BRASS 24
Anaconda Co.
Copper. Cu 85, Zn 15.
Soft: 40,000 TS; 15,000 YS; 50 El. Hard: 69,000 TS; 55,000 YS; 10 El. For plumbing pipe, condenser tubes, radiator cores. Red brass.

RED BRASS, 85%
Chase Brass & Copper Co.
Cu 85, Zn 15.
Annealed: 42,000 TS; 14,000 YS; 45 El. Rolled: 70,000 TS; 58,000 YS; 5 El. For plumbing, pipe, tubes, hardware; corrosion resisting.

RED CASTING BRASS
General Motors Corp./Central Foundry
Cu 85, Sn 5, Pb 5, Zn 5.
Cast: 27,000 psi TS; 18 El. For castings, hardware; GM No. 4041 M.

RED CIRCLE
Hyde Park Foundry & Machine Co.
C 3.2, Si 1.5, bal Fe.
For cast iron rolls; chilled.

RED CIRCLE
Allied Steel & Tractor Products Inc.
C, alloy, bal Fe.
For maintenance and repair operations; oil hardening.

RED CUT COBALT
Teledyne Vasco
C 0.72, W 17.25, Cr 4, Co 4.5, V 1, Mo 0.5, bal Fe.
For cutting and forming tools, lathe tools, planer tools, reamers, taps; applicable for deep cutting, scaly material.

RED CUT COBALT B
Teledyne Vasco
C 0.78, W 18, Cr 4, Co 8.75, V 1.85, Mo 0.75, bal Fe.
For cutting and forming tools, reamers, taps; high speed steel.

RED CUT SUPERIOR
Teledyne Vasco
C 0.5-0.8, Si 0.25-0.4, W 17.5-18.5, Cr 3.75-4.25, V 0.95-1.15, bal Fe.
Annealed: 105,000 TS; 75,000 YS; 14 El; 17 RA; 255 Brin. For tools, blanking dies, chasers, forming tools and dies, lathe and planer tools, taps; high speed steel; water and abrasion resistant.

RED CUT SUPERIOR FM
Teledyne Vasco
C 0.5-0.8, W 17.5-18.5, Cr 3.75-4.25, V 0.95-1.15, bal Fe.
Annealed: 105,000 TS; 75,000 YS; 14 El; 17 RA; 235 Brin. For blanking and forming dies, lathe tools; high speed steel, free-cutting. *Obsolete*

RED CUT SUPERIOR TEMPER "J"
Teledyne Vasco
C 0.5-0.55, W 17.5-18.5, Cr 3.75-4.25, V 0.95-1.15, bal Fe.
For high speed cutting tools, blanking dies, forming tools, taps; wear and abrasion resistant. Type H26.

RED CUT SUPERIOR TEMPER "M"
Teledyne Vasco
C 0.61-0.65, W 17.5-18.5, Cr 3.75-4.25, V 0.95-1.15, bal Fe.
For high speed cutting tools, blanking dies, forming tools, taps; wear and abrasion resistant. *Obsolete*

RED CUT SUPERIOR TEMPER "O"
Teledyne Vasco
C 0.66-0.7, W 17.5-18.5, Cr 3.75-4.25, V 0.95-1.15, bal Fe.
For high speed cutting tools, blanking dies, forming tools, taps; wear and abrasion resistant. *Obsolete*

RED CUT SUPERIOR TEMPER "P"
Teledyne Vasco
C 0.71-0.75, W 17.5-18.5, Cr 3.75-4.25, V 0.95-1.15, bal Fe.
For high speed cutting tools, blanking dies, forming tools, taps; water and abrasion resistant.

RED DEVIL
Champion Rivet Co.
C 0.07, bal Fe.
Welded: 68,000 TS; 56,000 YS; 30 El; 50 RA. For welding rods; flux coated.

RED DEVIL 75
Champion Rivet Co.
C 0.1, Mo 0.6, bal Fe.
Welded: 74,000 TS; 62,000 YS; 32 El; 65 RA. For welding rods; flux coated.

RED DEVIL 85
Champion Rivet Co.
Mo 0.9, C, bal Fe.
Welded: 80,000 TS; 72,000 YS; 24 El; 47 RA. For welding rods; flux coated.

RED DIAMOND 10
Spartan Redheugh Ltd.
C 0.45, Mn 1.75, bal Fe.
925 N/mm^2 TS; 540 N/mm^2 YS; 15 El; 250-290 Brin. For applications involving moderate wear by gouging and high stress grinding abrasion such as liner plates and scraper blades.

RED DIAMOND 12
Spartan Redheugh Ltd.
Stainless steel. C 0-0.03, Mn 0-1.3, Cr 11.5, Ni 0-1, N 0-0.03, Ti = 5 x C + N max, bal Fe.
500 N/mm^2 TS; 350 N/mm^2 YS; 30 El; 150 Brin; 30 J IS (Charpy V-notch, RT). Tough, ductile, corrosion resistant. For handling materials with an abrasive action in a corrosive atmosphere.

RED DIAMOND 14H
Spartan Redheugh Ltd.
C 0.85, Mn 13, bal Fe.
800 N/mm^2 TS; 25 El; 250-320 Brin. IS: 2.8 kgf·m min
(Charpy V-notch, RT); 2.1 kgf·m min (Charpy V-notch, at
-50°C). Austenitic manganese steel that can work harden up
to 800 Brin yet retain core ductility. For liner plates in shot
blast cabinets and other applications involving heavy impact.

RED DIAMOND 14S
Spartan Redheugh Ltd.
C 0.85, Mn 13, bal Fe.
615 N/mm^2 TS; 50 El; 200-240 Brin. IS: 10.4 kgf·m min
(Charpy V-notch, RT); 6.9 kgf·m min (Charpy V-notch, at
-50°C). Softened austenitic manganese steel that can work
harden up to 800 Brin yet retain core ductility. For liner plates
in shot blast cabinets and other applications involving heavy
impact.

RED DIAMOND 20H
Spartan Redheugh Ltd.
C 0.25, Mn 0.75, Cr 1, Mo 0.2, bal Fe.
850 N/mm^2 TS; 680 N/mm^2 YS; 18 El; 250-270 Brin; 20 J IS
(Charpy V-notch, RT). Wear resistant alloy steel for liner
plates in skips, chutes and hoppers handling coal, coke,
sand, gravel, concrete, etc., for screed plates in road laying
machines and for mold boards in the agricultural industry.

RED DIAMOND 20S
Spartan Redheugh Ltd.
C 0.2, Mn 0.8, Cr 0.5, Mo 0.2, Ni 0.55, bal Fe.
695 N/mm^2 TS; 685 N/mm^2 YS; 25 El; 200-235 Brin. IS: 34 J
(Charpy V-notch, RT); 27 J (Charpy V-notch, at -20°C).
Ductile alloy steel for liner plates in skips, chutes and
hoppers handling coal, coke, sand, gravel, concrete, etc., for
screed plates in road laying machines and for mold boards in
the agricultural industry.

RED DIAMOND 21
Spartan Redheugh Ltd.
C 0.4, Mn 0.75, Cr 1.1, Mo 0.25, bal Fe.
1235 N/mm^2 TS; 1000 N/mm^2 YS; 10 El; 320-370 Brin. Wear
resistant alloy steel plate for use in applications involving
gouging and high stress grinding abrasion.

RED DIAMOND 320 N
Spartan Redheugh Ltd.
C 0.2, Mn 1.5, Cr 1.5, bal Fe.
1000 N/mm^2 TS; 800 N/mm^2 YS; 12 El; 300-340 Brin; 15 J
IS (Charpy V-notch, RT). General purpose abrasion resistant
plate.

RED DIAMOND 9H
Spartan Redheugh Ltd.
C 0.55, Mn 0.65, bal Fe.
695 N/mm^2 TS; 585 N/mm^2 YS; 18 El; 220-240 Brin. Wear
resistant carbon steel for liner plates, scraper blades and
other applications involving moderate wear by gouging and
high stress grinding abrasion.

RED DIAMOND 9S
Spartan Redheugh Ltd.
C 0.45, Mn 0.75, bal Fe.
615 N/mm^2 TS; 495 N/mm^2 YS; 20 El; 200-220 Brin. Ductile,
weldable carbon steel for liner plates, scraper blades and
other applications involving moderate wear by gouging and
high stress grinding abrasion.

RED DIAMOND ARCLAD
Spartan Redheugh Ltd.
Arclad layer: 40% chromium carbide max, bal Fe.
600 Brin max. Extra heavy duty wear plate consisting of a
weld deposited onto a mild steel backing plate manufactured
to BS 4360 Grade 43A.

RED FOX 30
British Steel Corp.
C 0.45, Cr 20, Ni 8, Si, W.
Normalized: 115,000 TS; 32 El; 42 RA. For exhaust valves;
maximum working temperature 1050 C. *Obsolete*

RED FOX 34
United Engineering Steels Ltd.
C 0-0.15, Cr 19, Ni 11, Si 2, bal Fe.
Good scaling resistance in air at temperatures up to 1000°C.
AISI 302 B.

RED FOX-309
United Engineering Steels Ltd.
C 0.2, Cr 23, Ni 14, bal Fe.
Annealed: 75,000 TS; 30,000 YS; 40 El; 220 Brin. For high
temperature service, kilns, combustion chambers, salt pots,
and brazing fixtures. AISI Type 309, austenitic. Heat and
corrosion resistant.

RED FOX-310
British Steel Corp.
C 0.2, Cr 25, Ni 21, bal Fe.
Annealed: 87,000 TS; 36,000 YS; 54 El; 85 Rock B. For high
temperature service, furnace equipment, carburizing boxes,
valves pumps. Heat and corrosion resistant. AISI Type 310,
austenitic.

RED GOLD
Manufacturer not listed.
Au 75, Cu 15.
For jewelry, ornaments; corrosion resistant.

RED INDIAN
Atlas Specialty Steels
C 0.35, W 4.5, Mo 0.3, V 0.3, Co 0.5, Si 1, bal Fe.
For die casting dies, extrusion dies and mandrels; hot work
steel. *Obsolete*

RED LABEL
Edgar T. Ward's Sons Co.
C 0.9-1.2, bal Fe.
For tools. *Obsolete*

RED LABEL (WARD'S)
Edgar T. Ward's Sons Co.
C, bal Fe.
For tools. *Obsolete*

RED LABEL EXTRA
Wallace Murray Corp.
Mn 0.25, Si 0.25, P 0-0.015, S 0-0.015, C as desired, V
optional, bal Fe.
AISI Type W2 water hardening tool steel. *Obsolete*

RED METAL
Manufacturer not listed.
Cu 70, Zn 20, Pb 6, Sn 4.
For hardware; free-cutting.

RED RAY
Manufacturer not listed.
Ni 85, Cr 15.
For heating elements; heat resistant.

RED SABRE
Bethlehem Steel Corp
C 1.5, W 12, V 5, Cr 4.75, Co 5, bal Fe.
For tool bits, cutters, dies; Type T15; high speed steel.
Obsolete

RED SEAL DIE STEEL
Teledyne Firth Sterling
C 1, Mn 0.25, bal Fe.
For tools, dies, drills; water hardened. *Obsolete*

RED SHADOW
Agawam Tool Co.
C 0.4, Cr 3.1, Mn 0.25, V 0.25, W 10, bal Fe.
For tools, dies; hot work steel.

RED SHADOW
Ziv Steel & Wire Co.
C 0.8, Cr 4, Mo 5, W 6.5, V 2, bal Fe.
For milling cutters, reamers, broaches, hobs, drills; high
speed steel, oil hardened.

RED STAR DIE
Teledyne Vasco
C 0.65-1.45, bal Fe.
For drills, dies. *Obsolete*

RED STAR TAP
Teledyne Vasco
C 1.1, bal Fe.
For taps, drills, cutters; highest quality carbon tool steel.

RED STAR TOOL
Teledyne Vasco
C 0.6-1.2, bal Fe.
For tools, drills, punches; standard carbon tool steel.

RED STAR TUNGSTEN
Teledyne Vasco
C 1.2, Si 0.3, Mn 0.25, W 1.6, Cr 0.7, V 0.2, Mo 0.25, bal Fe.
Oil or water hardening tool steel. For blanking dies, taps,
reamers, thread rolling dies, punches, finishing tools,
gauges. Modified AISI O7.

RED STAR TUNGSTEN 10 TEMPER
Teledyne Vasco
C 0.9-1, Cr 0.35-0.5, W 1.5-1.7, V 0.06-0.1, bal Fe.
Annealed: 90,000 TS; 65,000 YS; 22 El; 42 RA; 190 Brin. For
taps, reamers, punches, dies, finishing tools, gages; water
hardening.

RED STAR TUNGSTEN 12 TEMPER
Teledyne Vasco
C 1.25-1.15, Cr 0.6-0.8, W 1.5-1.7, V 0.1-0.25, Mo 0.25, bal
Fe.
Heat treated; 180,000-145,000 TS; 154,000-210,000 YS; 1-13
El; 5-36 420-302 Brin. For taps, reamers, punches, dies,
finishing tools, gages; water or oil hardening.

RED STAR TUNGSTEN DRILL ROD
Teledyne Allvac
C 1.2, W 1.6, Cr 0.7, Mo 0.25, V 0.2, bal Fe.
For drill rod; water hardened.

RED STREAK
Now SIMONDS RED STREAK.

RED STREAK ALNICO
Wallace Murray Corp.
C 0.1, Ni 17.5, Al 11, Co 13, bal Fe.
For permanent magnets; ALNICO NO. 2.

RED TIGER
Bethlehem Steel Corp.
C 1, W 18.5, Cr 4.5, V 2.6, Mo 0.65, bal Fe.
For tools, cutters, taps, reamers; high speed steel. *Obsolete*

RED TIP 201
Brown-Wales Co.
Zn, Pb, bal Cu.
For screw machine products; free cutting.

RED TIP 360
Mueller Brass Co.
Cu 61.5, Pb 3, bal Zn.
Half hard: 62,000 TS; 53,000 YS; 25 El. Annealed: 49,000 TS;
18,000 YS; 53 El. For screw machine products; free-cutting.

RED X-11
Apex International Alloys Inc.
Cu 1-2, Si 10.5-11.8, Mg 0.4-1, Zn 0-0.6, Mn 0.8, bal Al.
Cast-F: 34,000 TS; 23,000 YS; 2 El; 80 Brin. T6-Temper:
45,000 TS; 35,000 YS; 2 El; 100 Brin. T51-Temper: 39,000 TS;
31,000 YS; 1 El; 90 Brin. For light alloy castings; high fluidity.
Obsolete

RED, WHITE AND BLUE
Darwins Alloy Castings
C 0.9-1, bal Fe.
For tools, taps, drills; water hardened. *Obsolete*

RED-X-10
Apex International Alloys Inc.
Si 9.5-10.5, Cu 2, Mg 0.6, Mn 0.6, bal Al.
Permanent mold: 40,000 TS; 30,000 YS; 0.5 El; 105 Brin. For
light alloy parts, pistons; high fluidity. *Obsolete*

RED-X-13
Apex International Alloys Inc.
Cu 2, Mg 0.7, Mn 0.75, Si 12, bal Al.
Heat treated: 34,000-50,000 TS; 23,000-41,000 YS; 1.5-1.0 El;
85-110 Brin. For permanent mold castings, crank cases,
pistons; heat treatable. *Obsolete*

RED-X-20
Apex International Alloys Inc.
Cu 1.5, Si 20.5, Mg 0.6, Mn 0.4, Ni 0.7, bal Fe.
F temper: 25,000 TS; 1.0 El; 90 Brin. T7 temper: 38,000 TS;
1.0 El; 110 Brin. For permanent mold pistons; heat treatable.
Obsolete

RED-X-5
Apex International Alloys Inc.
Cu 1-2, Mg 0.2-0.5, Mn 0.2-0.6, Si 4.5-6.5, bal Al.
Permanent mold: 32,000 TS; 20,000 YS; 3.5 El; 65 Brin. Sand
cast: 26,000 TS; 16,000 YS; 2.5 El; 60 Brin. For forgings and
castings; age hardenable. *Obsolete*

RED-X-6
Apex International Alloys Inc.
Cu 2, Mg 0.5, Mn 0.4, Si 8, bal Al.
Cast: 33,000 TS; 20,000 YS; 3 El; 65 Brin. Heat treated:
42,000 TS; 26,000 YS; 1 El; 95 Brin. For light alloy parts,
permanent mold castings; heat treatable. *Obsolete*

RED-X-6
Sheepbridge Equipment Ltd.
Cu 1-2, Mg 0.2-0.5, Mn 0.4, Si 6, bal Fe.
For light alloy castings; age hardenable.

RED-X-8-5
Apex International Alloys Inc.
Cu 5, Mg 0.25, Si 8, Mn 0.3, bal Al.
For light alloy parts, pistons; age hardenable. *Obsolete*

REDALLOY
Chase Brass & Copper Co.
Cu 85, Zn 14, Sn 1.
Annealed: 42,000 TS; 48 El. For condenser tubes, heat
exchanger tubes; corrosion resistant. *Obsolete*

REDALOY
Cerro Metal Products Co.
Cu 55, Pb 0-1, Al 2.25, bal Zn.
Extruded: 85,000 TS; 25,000 YS; 20 El; 80 Rock B. Drawn:
90,000 TS; 35,000 YS; 15 El; 85 Rock B. For hardware,
plumbing, fixtures. Corrosion resistant. *Obsolete*

REDALOY NO. 1
Cerro Metal Products Co.
Cu 53.75-54.25, Pb 3.3-3.5, bal Zn.
77,000-88,000 TS; 24,000-64,000 YS; 7-17 El; 11-22 RA;
70-100 Brin. For screw machine parts; free cutting. *Obsolete*

REDALOY NO. 2
Cerro Metal Products Co.
Cu 57.5-58, Pb 3.3-3.5, bal Zn.
63,000-70,000 TS; 23,000-56,000 YS; 13-32 El; 16-29 RA;
65-100 Brin. For screw machine parts; free cutting. *Obsolete*

REDALOY NO. 3
Cerro Metal Products Co.
Cu 52-53.7, Pb 0.4-0.6, bal Zn.
For pressure die castings; high strength. *Obsolete*

REDALOY NO. 8
Cerro Metal Products Co.
Cu 54, Pb 3.4, bal Zn.
For forgings, hardware; free-cutting. *Obsolete*

REDHARD HS
Redhard Metals Inc.
C 0.7, W 18, Cr 4, V 2, bal Fe.
For cutting tools; high speed steel.

REDHARD NF
Redhard Metals Inc.
Co 45, Cr 32, W 18, C 2.5, 2.1 others.
Cast: 73,000 TS; 650 Brin. For cutting tools, form tools;
nonferrous, heat and abrasion resistant.

REDI H
Westa-Westdeutsche
C 0.45, Si, Cr, V, bal Fe.
For gears, bolts, springs, crankshafts; oil hardened, shock
resistant.

REDI SPEZIAL
Westa-Westdeutsche
C 0.61, Cr 1.18, V 0.1, Mn 0.75, bal Fe.
For springs, dies, tools; oil hardened, tough.

REDI-KOTE
United States Steel Corp.
Low C, Mn, bal Fe.
Steel sheet with remelted zinc coating.

REED BRASS
American manufacture
Cu 69, Zn 30, Sn 1.
For pipes, tubes, fittings, condenser tubes; corrosion
resistant.

REED BRASS
Waterbury Rolling Mills Inc.
Cu 67, Sn 1, Zn 32.
For pipes, tubes, fittings, condenser tubes; corrosion
resistant.

REEX
Italian manufacture
C 0.45-0.9, Cr 11, Ni 14, Si 1, W 2, bal Fe.
For valves; heat and corrosion resistant.

REFAX METAL
Watsontown Foundry
C 3.2, Ni 1, Cr 0.8, Si 2, bal Fe.
For valve bodies; cast iron.

REFLECTAL
Vereinigte Leichtmetallwerke, G.m.b.H.
Aluminum. Mg 1.2, bal Al.
Annealed: 21,000 TS; 8000 YS; 24 El. For boats, fan blades,
window frames; corrosion resistant.

REFLECTAL 20
Alluminio SA
Mg 1.5-2.5, bal Al.
1/2 H-temper: 31,500 TS; 27,300 YS; 5 El; 65 Brin. H-temper:
35,500 TS; 31,500 YS; 3 El; 70 Brin. For window frames, door
handles, costume jewelry, high corrosion resistance.

REFLECTAL 20
Aluminium Industrie Aktiengesellchaft
Mg 1.5-2.5, bal Al.
1/2 H-temper: 31,500 TS; 27,300 YS; 5 El; 65 Brin. H-temper:
35,500 TS; 31,500 YS; 3 El; 70 Brin. For window frames, door
handles, costume jewelry, high corrosion resistance.

REFLECTAL 20
Aluminium Walzwerke Singen GmbH
Mg 1.5-2.5, bal Al.
1/2 H-temper: 31,500 TS; 27,300 YS; 5 El; 65 Brin. H-temper:
35,500 TS; 31,500 YS; 3 El; 70 Brin. For window frames, door
handles, costume jewelry, high corrosion resistance.

REFLECTAL 5
Aluminium Industrie Aktiengesellschaft
Mg 0.3-1, bal Al.
Soft: 11,500 TS; 4500 YS; 35 El; 22 Brin. 1/4 hard temper:
13,500 TS; 13,000 YS; 20 El; 32 Brin.. For mirrors, jewelry,
reflectors, watch cases; corrosion resistant.

REFLECTAL 5
Aluminium Walzwerke Singen GmbH
Mg 0.3-1, bal Al.
Soft: 11,500 TS; 4500 YS; 35 El; 22 Brin. 1/4 hard temper:
13,500 TS; 13,000 YS; 20 El; 32 Brin.. For mirrors, jewelry,
reflectors, watch cases; corrosion resistant.

REFLECTAL 74
Alluminio SA
Mg 0.5-1, Si 0.2-1, bal Al.
TA-14 Temper: 20,000-28,400 TS; 11,400-21,500 YS; 20-35
El; 55-75 Brin. TA-16 Temper: 43,000 TS; 36,000 YS; 5 El; 100
Brin. For marine structures and hardware, architectural
applications. Heat treatable, corrosion resistant.

REFLECTAL 74
Lavorazione Leghe Leggere SpA
Mg 0.5-1, Si 0.2-1, bal Al.
TA-14 Temper: 20,000-28,400 TS; 11,400-21,500 YS; 20-35
El; 55-75 Brin. TA-16 Temper: 43,000 TS; 36,000 YS; 5 El; 100
Brin. For marine structures and hardware, architectural
applications. Heat treatable, corrosion resistant.

REFLECTAL-05
Alluminio SA
Aluminum. Mg 0.2-1, bal Al.
R-Temper: 10,000-15,000 TS; 5000 YS; 20-40 El; 25 Brin. 1/2
Hard: 16,000-20,000 TS; 15,000 YS; 12-15 El; 30 Brin. For
costume jewelry, mirrors, reflectors, watch cases, window
frames. Corrosion resistant.

REFLECTAL-050
Aluminium Walzwerke Singen GmbH
Aluminum. Mg 0.5, bal Al.
Annealed: 80 MPa TS; 25 El. Hard: 180 MPa TS; 2 El. Plate
for decorative purposes.

REFLECTAL-100
Swiss Aluminium Ltd.
Aluminum. Si 0.01, Cu 0.003, Mn 0.003, Mg 0.8-1.1, Cr 0.003,
Zn 0.01, 0.008 Ti + Fe, bal Al.
Extrusion alloy in billet form for electrical conductors.
100-130 N/mm^2 TS; 30-80 N/mm^2 YS; 13-15 El; 30 Brin.

REFLECTAL-107
Aluminium Industrie Aktiengesellschaft
Si 0.2-1, Mg 0.2-1, bal Al.
Heat treated: 28,500-42,500 TS; 21,500-35,500 YS; 15-5 El;
80-100 Brin. For ornamental trim, window frames, door
handles; age-hardenable, corrosion resistant.

REFLECTAL-107
Aluminium Walzwerke Singen GmbH
Si 0.2-1, Mg 0.2-1, bal Al.
Heat treated: 28,500-42,500 TS; 21,500-35,500 YS; 15-5 El;
80-100 Brin. For ornamental trim, window frames, door
handles; age-hardenable, corrosion resistant.

REFOR-14-14
Forjas Alavesas S.A.
C 0.44, Mn 0-1, Si 0-2, Cr 14, Ni 14, W 2.5, bal Fe.
Austenitic steel for valves. IHA F-321; BS EN 54; AFNOR Z 45
CNWS 15-14.

REFOR-25-20
Forjas Alavesas S.A.
C 0-0.2, Mn 0-2, Si 0-2.5, Cr 25, Ni 20, bal Fe.
Austenitic stainless steel for parts operating at high
temperature. IHA F-331; AISI 310; BS A-11.

REFOR-30

Forjas Alavesas S.A.
C 0-0.2, Mn 0-1, Si 0-2, Cr 28, bal Fe.
Stainless steel for furnace equipment that operates at high temperatures. AFNOR Z 15C27; similar to AISI 446.

REFOR-9-2

Forjas Alavesas S.A.
C 0.44, Mn 0.45, Si 3.1, Cr 9, bal Fe.
Martensitic steel for valves. IHA F-322; DIN X 45 CrSi9; BS EN 52.

REFORMEND

Arcos Alloys
C 0.2, bal Fe.
For welding electrodes; for cast iron. *Obsolete*

REFRACTALOY 26

Cannon-Muskegon Corp.
C 0-0.08, Cr 16-20, Ni 35-39, Co 16-22, Mo 3, Si 0.5-1.5, Mn 0.4-1, Ti 3, 0.3 max Al, bal Fe.
Hardened: 170,000 TS; 100,000 YS; 18 El; 20 RA; 241-311 Brin. For turbine blades and parts, bolts, springs; heat resistant to 1500°F; high creep strength.

REFRACTALOY 26

Westinghouse Electric Corp.
C 0-0.08, Cr 16-20, Ni 35-39, Co 16-22, Mo 3, Si 0.5-1.5, Mn 0.4-1, Ti 3, 0.3 max Al, bal Fe.
Hardened: 170,000 TS; 100,000 YS; 18 El; 20 RA; 241-311 Brin. For turbine blades and parts, bolts, springs; heat resistant to 1500°F; high creep strength.

REFRACTALOY 70

Westinghouse Electric Corp.
Ni 19-21, Co 28.5-31.5, Cr 19-21, Mo 8, W 4, Mn 2, Si 0.8, C 0-0.08, bal Fe.
Hardened: 132,000 psi TS; 87,000 psi YS; 3 El; 3 RA; 315 Brin. For high temperature applications; resists heat over 1200°F. *Obsolete*

REFRACTALOY 80

Westinghouse Electric Corp.
C 0-0.2, Co 28.5-31.5, Cr 19.5-21.5, Ni 19-21, Fe 0-16, Mo 9-11, W 4.5-5.5, Mn 0.8, Si 0.8.
Cast: 83,000 psi TS; 57,000 psi YS; 8 El; 11 RA; 180 Brin.
Hardened: 100,000 psi TS; 84,000 psi YS; 3 El; 4 RA; 220 Brin. For nozzle vanes, gas turbine parts; high ductility, creep strength and oxidation resistance. *Obsolete*

REFRACTALOY B

Westinghouse Electric Corp.
Cr 25, Ni 30, Mo 8, bal Fe.
For combustion chambers, exhaust cones; high heat and corrosion resistance. *Obsolete*

REFRACTORY

English manufacture
C 0.05, Co 30, Ni 21, Fe 14, Cr 20, Mo 8, W 4.
For jet engine components; high strength, heat resistant.

REFRACTORY SOLDER

English manufacture
Cu 50, Zn 50.
For solder; brazing.

REGAL

Remystahl
C 1.42, W, V, bal Fe.
For engravers' tools, forming and blanking dies; water or oil hardened. W.-Nr. 1.2562.

REGAL

Time Steel Service Inc.
C 1.55, Cr 11.5, V 1, Mo 0.8, bal Fe.
High carbon, high chromium tool and die steel; air or oil hardened. AISI D2.

REGAL SPEZIAL

Remystahl
C 1.3, W 5, bal Fe.
Cold work tool steel, for drawing dies. W.-Nr. 1.2453.

REGEL-METALL

English manufacture
Sn 83.3, Sb 11.1, Cu 5.6.
For bearings, bushings; anti-friction.

REGELMETALL WM 80

Manufacturer not listed.
Sn 80, Pb, Sb.
For bearings; antifriction metal.

REGENT

Dohlen-Stahl Gusstahl-Handels GmbH
C 2.1, Cr 11.5, Mn 0.3, bal Fe.
For forming and blanking dies, punches; oil hardened, non-deforming.

REGENT

Remystahl
C 0.9, Mn 2, V 0.1, Cr 0.35, bal Fe.
For blanking and forming dies; oil hardened, non-deforming. W.-Nr. 1.2842.

REGENT C.3.H.

Burys & Co. Ltd.
C 0.15, bal Fe.
For plastic dies, hobbed dies; case hardened.

REGENT G.14

Burys & Co. Ltd.
C 1.5, Cr 12, V 0.25, Mo 0.75, bal Fe.
For thread rolling dies, gauges, broaches; nondeforming, air hardening.

REGENT G.22

Burys & Co. Ltd.
C 1.5, Cr 1, V 0.5, Mo 0.2, Ni 3, Mn 0.5, bal Fe.
For shear blades, punches, dies; oil hardened, tough.

REGENT I.18

Burys & Co. Ltd.
C 2, Cr 12, bal Fe.
For punches, blanking and forming dies; oil hardened, nondeforming.

REGENT J.15

Burys & Co. Ltd.
C 1.15, W 1.5, V 0.15, bal Fe.
For taps, drills, reamers, cutters, threading dies; water hardened, wear resistant.

REGENT J.7

Burys & Co. Ltd.
C 0.5, Mn 1, Cr 1, Mo 0.35, bal Fe.
For hand chisels, sawes, cutters; tough, air hardened, nontempering.

REGENT L.5

Burys & Co. Ltd.
C 0.4, Mn 0.5, Ni 0.25, bal Fe.
For chisels, punches; tough, oil hardened.

REGENT N.1

Burys & Co. Ltd.
C 0.6, Si 2, V 0.25, Mo 0.25, Mn 0.74, bal Fe.
For punches, concrete breaker picks, chisels; oil hardened, tough.

REGENT N.1.H.

Burys & Co. Ltd.
C 0.35, Cr 1.25, Ni 4.25, bal Fe.
For plastic mold dies; air or oil hardened.

REGENT N.13

Burys & Co. Ltd.
C 0.4, Cr 1.25, Ni 1.75, Mo 0.2, bal Fe.
For plastic mold and die casting dies; oil hardened, tough.

REGENT N.3.C.

Burys & Co. Ltd.
C 0.12, Cr 1.25, Ni 4, bal Fe.
For plastic mold dies; case hardened, tough.

REGENT R.16

Burys & Co. Ltd.
C 0.1, Cr 13, bal Fe.
Annealed: 75,000 TS; 40,000 YS; 35 El; 70 RA; 155 Brin. For multiple dies, valves, cutlery; corrosion resistant; Type 410.

REGENT Z

Dohlen-Stahl Gusstahl-Handels GmbH
C 2.1, Cr 11.5, W 0.7, bal Fe.
For forming and blanking dies, punches: oil hardened, non-deforming.

REGIN 1

Uddeholm Corp.
C 0.42, Si 0.9, Cr 1, W 2.5, bal Fe.
For hot work tools, chisels; hot work steel. *Obsolete*

REGIN 3

Uddeholm Corp.
C 0.48, Si 0.9, Cr 1.15, Mo 0.25, W 2.25, V 0.15, bal Fe.
For chisels, dies; hot work steel. AISI S1.

REGIN 711

Manufacturer not listed.
C 0.5, Cr 1.15, W 2.5, V 0.2, bal Fe.
For shear blades, hot heading dies, chisels; hot work steel.

REGULAR

Latrobe Steel Co.
C 0.3-0.4, Cr 13-14, Mn 0-0.5, bal Fe.
Heat treated: 200,000-232,000 TS; 180,000-220,000 YS; 0-10 El; 0-20 RA; 400-505 Brin. Annealed: 90,000 TS; 48,000 YS; 28 El; 60 RA; 170 Brin. For cutlery, bayonets, dies, gages, knives, swords, needles, clippers, pump rods, pistons, cold sets, granite points; stainless when hardened and polished. *Obsolete*

REGULAR BOLT DIE

Midvale-Heppenstall Co.
C 1, Cr 4, bal Fe.
For hot work dies, mandrels; hot work dies, mandrels; hot work steel. *Obsolete*

REGULAR CARBON

Midvale-Heppenstall Co.
C 0.6-1.2, bal Fe.
For tools, taps, punches; water hardening. *Obsolete*

REGULAR POWER TRANSFORMER NO. 72

Follansbee Steel Co.
C 0.03, Si 3.75, bal Fe.
Annealed: 80,000 TS; 85,000 YS; 10 El. For radio and generators; good magnetic properties. *Obsolete*

REGULAR STAINLESS-420

Latrobe Steel Co.
C 0.3-0.4, Cr 13-14, bal Fe.
For chemical engineering equipment, cutlery, tools; stainless; Type 420. *Obsolete*

REGULAR STRAIGHT CARBON

Jessop-Saville Ltd.
C 0.7-1.3, bal Fe.
For tools, drills, taps; water hardening.

REGULAR TOOL

Midvale-Heppenstall Co.
C 0.6-1.4, bal Fe.
For tools, dies, machine parts, axes, chisels, rock drills, hammers; strong and tough. *Obsolete*

REGULAR-SS
Latrobe Steel Co.
C 0-0.35, Cr 13.5, Si 0.35, Mn 0.35, bal Fe.
Annealed: 90,000 TS; 48,000 YS; 28 El; 60 RA; 179 Brin. For valves, cutlery, surgical instruments; corrosion resistant.
Obsolete

REGULIT K
Now VEW K100.

REGULIT KNL
Now VEW K105.

REGULIT KR
Now VEW K107.

REGULUS
American manufacture
Pb 75, Sb 25.
For acid valves, cocks, flanges, chemical apparatus; hard lead.

REGULUS METAL "A"
Manufacturer not listed.
Sb 6-8, bal Pb.
For chemical plant equipment; hard lead.

REGULUS METAL "B"
Manufacturer not listed.
Sb 8-10, bal Pb.
For chemical equipment; hard lead.

REGULUS METAL "C"
Manufacturer not listed.
Sb 10-12, bal Pb.
For bearings; hard lead.

REGULUS METAL "D"
Manufacturer not listed.
Sb 12, bal Pb.
For bearings; hard lead.

REGULUS VENUS
English manufacture
Cu 50, Sb 50.

REICH'S BRONZE
German manufacture
Cu 85, Fe 7.5, Al 0.6, Mn 0.5.
For strong corrosion resistant parts.

REICHSBAHN "G"
German manufacture
Cu 86, Sn 10, Pb 4.
For railroad dynamo and cold rolled bearings; heavy duty.

REICHSBAHN "B"
German manufacture
Cu 85, Sn 11, Zn 4.
For railroad bearings; heavy duty.

REICHSBAHN "C"
German manufacture
Cu 85, Sn 9, Zn 6.
For railroad valves, slides; high strength.

REICHSBAHN "D"
German manufacture
Cu 91, Sn 5, Zn 4.
For railroad pipe flanges; tough.

REICHSBAHN "E"
German manufacture
Cu 82, Sn 10, Zn 7, Pb 1.
For machinery castings; free-cutting.

REICHSBAHN "F"
German manufacture
Cu 85, Sn 5, Zn 8, Pb 2.
For machinery castings; free-cutting.

REICHSBAHN "H"
German manufacture
Cu 77, Sn 8, Pb 15.
For railroad heavy duty bearings; heavy duty.

REICHSBAHN "I"
German manufacture
Cu 79, Sn 20, Pb 1.
For car journal bearings; heavy duty.

REICHSBAHN "J"
German manufacture
Cu 85, Sn 14, Pb 1.
For railroad heavy duty bearings.

REICHSBAHN "K"
German manufacture
Cu 89, Sn 10.
For worm wheels; tough.

REICHSBAHN, EINHEITS RED BRASS "A"
German manufacture
Cu 85, Sn 9, Zn 6.
For railroad hardware; high strength.

REIN-ALU 99.5
Kreidler Werke G.m.b.H.
Aluminum. 99.5 Al (pure aluminum).

REIN-KUPFER
Isabellenhuette
Copper. Cu 99.9.
Annealed: 200 N/mm^2 TS. For electrical equipment and instruments. Resistance alloy. Maximum working temperature to 150°C.

REIN-NICKEL
Isabellenhuette
Nickel. Ni 99.6.
Annealed: 450 N/mm^2 TS. For electrical equipment and instruments. Resistance alloy. Maximum working temperature to 700°C.

REIN-NICKEL GNI
Vereinigte Deutsche Nickel-Werke AG
Si 1.5, bal Ni.
For pump castings; corrosion and heat resistant. *Obsolete*

REINALUMINIUM
Vereinigte Leichtmetallwerke, G.m.b.H.
Aluminum. 99 Al or 99.5 Al.
Annealed: 9,000-16,000 TS; 30-40 El. Hard: 29,000 TS; 25,000 YS; 4 El. For chemical apparatus, tanks, dairy industry, electrical parts; commercially pure Al.

REINALUMINIUM 2S
Rorschach Aluminiumwerke
Aluminum. Al 99-99.4.
S-O temper: 14,000 TS; 7100 YS; 35 El. S-H temper: 25,600 TS; 24,200 YS; 6 El. For cooking utensils, food and chemical processing equipment; wrought, corrosion resistant.

REINALUMINIUM AL99
Swiss Aluminium Ltd.
Aluminum. Cu 0.05, Mn 0.05, Mg 0.05, Cr 0.05, Zn 0.1, Ti 0.05, 1.0 Si + Fe max, bal Al.
Extrusion alloy in billet form; not age hardenable. 75-110 N/mm^2 TS; 30-80 N/mm^2 YS; 18-20 El; 25 Brin.

REINALUMINIUM AL99
Swiss Aluminium Ltd.
Aluminum. Cu 0.05, Mn 0.05, Mg 0.05, Cr 0.05, Zn 0.1, Ti 0.05, 1.0 Si + Fe max, bal Al.
Wrought alloy in rolling slab form.

REINALUMINIUM AL99.5
Swiss Aluminium Ltd.
Aluminum. Si 0.25, Fe 0.4, Cu 0.05, Mn 0.05, Mg 0.05, Cr 0.03, Zn 0.07, Ti 0.05, bal Al.
Extrusion alloy in billet form; not age hardenable. 65-100 N/mm^2 TS; 20-60 N/mm^2 YS; 22-25 El; 20 Brin.

REINALUMINIUM AL99.5
Swiss Aluminium Ltd.
Aluminum. Si 0.25, Fe 0.4, Cu 0.05, Mn 0.05, Mg 0.05, Cr 0.03, Zn 0.07, Ti 0.05, bal Al.
Wrought alloy in rolling slab form.

REINALUMINIUM ALFESI
Swiss Aluminium Ltd.
Aluminum. Si 0.5-0.8, Fe 0.7-1, Cu 0.1, Mn 0.1, Mg 0.06, Cr 0.06, Zn 0.1, Ti 0.05, bal Al.
Wrought alloy in rolling slab form.

REINALUMINIUM E-AL99, 5E
Swiss Aluminium Ltd.
Aluminum. Si 0.25, Fe 0.4, Cu 0.02, Mg 0.03, Zn 0.05, bal Al.
Extrusion alloy in billet form for electrical conductors. 70-100 N/mm^2 TS; 20-60 N/mm^2 YS; 18-22 El; 20 Brin.

REINALUMINIUM E-AL99, 7E
Swiss Aluminium Ltd.
Aluminum. Si 0.1, Fe 0.25, Cu 0.02, Mg 0.02, Zn 0.05, bal Al.
Extrusion alloy in billet form for electrical conductors. 65-100 N/mm^2 TS; 20-60 N/mm^2 YS; 20-25 El; 20 Brin.

REINING BKV
Now RAXA BKV.

REINING MSM
Heinrich Reining GmbH
C 0.53, Si 0.9, Mn 0.9, bal Fe.
For punches, pneumatic tools, oil hardened, tough.
Obsolete

REINNICKEL
VDM Nickel-Technologie AG
Ni 99-99.7.
Annealed: 65,000 TS; 18,000 YS; 50 El; 90 Brin. Hard rolled: 113,000 TS; 108,000 YS; 2 El; 200 Brin. For chemical and food-handling equipment; pure Ni; corrosion resistant.
Obsolete

REISHOLZ A22
Mannesmannrohren-Werke AG
C 0.22, Si 0.25, Mn 0.45, bal Fe.
Annealed: 73,000 TS; 41,000 YS; 22 El; 58 RA; 140 Brin. For bolts, fasteners, gears, cams, camshafts; case hardening steel. *Obsolete*

REISHOLZ A22E
Mannesmannrohren-Werke AG
C 0.28, Mn, bal Fe.
Hot rolled: 80,000 TS; 50,000 YS; 30 El; 56 RA; 165 Brin. For bolts, fasteners, gears, machine tool parts; water hardened.
Obsolete

REISHOLZ A32E
Mannesmannrohren-Werke AG
C 0.36, Mn, bal Fe.
Hot rolled: 85,000 TS; 54,000 YS; 30 El; 54 RA; 185 Brin. For bolts, fasteners, machine tool parts, gears; water hardened.
Obsolete

REISHOLZ A35
Mannesmannrohren-Werke AG
C 0.35, Si 0.25, Mn 0.45, bal Fe.
Hot rolled: 85,000 TS; 54,000 YS; 30 El; 54 RA; 185 Brin. For gears, bolts, machine tool parts; water hardened. *Obsolete*

REISHOLZ B412
Mannesmannrohren-Werke AG
C 0.24, Si, Mn, Ni, bal Fe.
For gears, pinions, shafts, bolts; oil hardened, shock resistant. *Obsolete*

REISHOLZ B412 MO
Mannesmannrohren-Werke AG
C 0.24, Si, Mn, Ni, Mo, bal Fe.
For gears, shafts, bolts, studs, crankshafts; oil hardened, shock resistant. *Obsolete*

REISHOLZ B44
Mannesmannrohren-Werke AG
C 0.24, Ni 1.15, Cr 0-0.15, bal Fe.
For gears, pinions, fasteners, shafts; oil hardened, shock resistant. *Obsolete*

REISHOLZ B76
Mannesmannrohren-Werke AG
C 0.35, Ni 1.35, Cr 0-0.6, bal Fe.
For gears, pinions, fasteners, shafts, axles; oil hardened, shock resistant. *Obsolete*

REISHOLZ BCD53T
Mannesmannrohren-Werke AG
C 0.28, Cr 1.25, Mo 0.35, Ni 1.95, bal Fe.
For gears, shafts, crankshafts, machine tool parts; oil hardened, shock resistant. *Obsolete*

REISHOLZ BCD53W
Mannesmannrohren-Werke AG
C 0.34, Cr 1.45, Mo 0.25, Ni 1.55, bal Fe.
For gears, shafts, machine tool parts; oil hardened, shock resistant. *Obsolete*

REISHOLZ BCD54
Mannesmannrohren-Werke AG
C 0.25, Cr 1, Mo 0.18, Ni 1.5, bal Fe.
For gears, shafts, crankshafts, machine tool parts; oil hardened, shock resistant. *Obsolete*

REISHOLZ BCD66
Mannesmannrohren-Werke AG
C 0.28, Cr 1.15, Mo 0.25, Ni 1.15, bal Fe.
For gears, pinions, shafts, machine tool parts; oil hardened, shock resistant. *Obsolete*

REISHOLZ BCD66T
Mannesmannrohren-Werke AG
C 0.28, Cr 1.05, Mo 0.45, bal Fe.
For gears, pinions, shafts, machine tool parts; oil hardened, shock resistant. *Obsolete*

REISHOLZ CE36
Mannesmannrohren-Werke AG
C 0.36, Mn 1.25, Cr 1.15, bal Fe.
For forging dies, crankshafts, fasteners; oil hardened, tough. *Obsolete*

REISHOLZ D39K
Mannesmannrohren-Werke AG
C 0.17, Mo, V, bal Fe.
For gears, bolts, machine tool parts; case hardened. *Obsolete*

REISHOLZ DC54
Mannesmannrohren-Werke AG
C 0.24, Cr 1.15, Mo 0.25, Mn 0.55, bal Fe.
For gears, bolts, machine tool parts; oil hardened, tough. *Obsolete*

REISHOLZ DC84
Mannesmannrohren-Werke AG
C 0.42, Cr 1.1, Mo 0.2, Mn 0.65, bal Fe.
For gears, bolts, machine tool parts; oil hardened, tough. *Obsolete*

REISHOLZ DE20
Mannesmannrohren-Werke AG
C 0.2, Mn 1.1, Mo 0.25, Si 0.25, bal Fe.
For gears, cams, camshafts, machine tool parts; case hardened, tough. *Obsolete*

REISHOLZ DE46
Mannesmannrohren-Werke AG
C 0.2, Mo 0.25, Mn 1.1, Si 0.25, bal Fe.
For gears, cams, camshafts, machine tool parts; case hardened, tough. *Obsolete*

REISHOLZ E305
Mannesmannrohren-Werke AG
C 0.3, Si 0.37, Mn 0.45, bal Fe.
Hot rolled: 80,000 TS; 50,000 YS; 30 El; 56 RA; 165 Brin. For gears, bolts, machine tool parts; water hardened. *Obsolete*

REISHOLZ E36
Mannesmannrohren-Werke AG
C 0.36, Si 0.25, Mn 1.25, bal Fe.
For gears, cams, camshafts, fasteners; water hardened. *Obsolete*

REISHOLZ E46K
Mannesmannrohren-Werke AG
C 0.2, V 0.12, Mn 1.45, Si 0.25, bal Fe.
For gears, cams, camshafts, fasteners; case hardening steel, tough. *Obsolete*

REISHOLZ EF18
Mannesmannrohren-Werke AG
C 0.18, Si 0.5, Mn 1.15, bal Fe.
For gears, cams, machine tool parts; case hardened. *Obsolete*

REISHOLZ EF46
Mannesmannrohren-Werke AG
C 0.46, Si 0.9, Mn 0.9, bal Fe.
For gears, springs, punches, bolts, machine tool parts; water hardened. *Obsolete*

REISHOLZ EF55W
Mannesmannrohren-Werke AG
C 0.55, Cr, Mn, Si, bal Fe.
For gears, bolts, machine tool parts; oil hardened. *Obsolete*

REISHOLZ EF60
Mannesmannrohren-Werke AG
C 0.53, Si 0.8, Mn 1.05, bal Fe.
For gears, springs, bolts, fasteners; water hardened. *Obsolete*

REISHOLZ M1
Mannesmannrohren-Werke AG
C 0.12, Si 0.25, Mn 0.55, bal Fe.
Annealed: 65,000 TS; 48,000 YS; 28 El; 65 RA; 130 Brin. For nails, rivets, screws, bolts, bushings; case hardened. *Obsolete*

REISHOLZ M2
Mannesmannrohren-Werke AG
C 0-0.22, Si 0.2, Mn 0.3, bal Fe.
Annealed: 73,000 TS; 60,000 YS; 22 El; 58 RA; 150 Brin. For fan blades, bushings, gears, shafts, rivets; case hardened. *Obsolete*

REISHOLZ M3
Mannesmannrohren-Werke AG
C 0-0.25, Si 0.2, Mn 0.3, bal Fe.
Hot rolled: 70,000 TS; 45,000 YS; 31 El; 58 RA; 145 Brin. For gears, bolts, armature shafts, keys, brackets; water hardened. *Obsolete*

REISHOLZ NI
Mannesmannrohren-Werke AG
C 0.26, Cr, Mo, bal Fe.
For gears, bolts, machine tool parts; oil hardened, tough. *Obsolete*

REISHOLZ NIA
Mannesmannrohren-Werke AG
C 0.22, Si 0.25, Mn 0.6, Cr 2.25, bal Fe.
For oil refinery equipment; creep and heat resistant. *Obsolete*

REISHOLZ NIC
Mannesmannrohren-Werke AG
C 0.22, Si 0.2, Mn 0.6, Cr 1.5, bal Fe.
For oil refinery equipment, gears, shafts; heat and creep resistant. *Obsolete*

REISHOLZ NIN
Mannesmannrohren-Werke AG
C 0.24, Cr 2.5, Mo 0.25, Ni 0-0.8, bal Fe.
For oil refinery equipment; heat and creep resistant. *Obsolete*

REISHOLZ NIT
Mannesmannrohren-Werke AG
C 0.32, Cr, Ni, Mo, bal Fe.
For gears, pinions, shafts; oil hardened, shock resistant. *Obsolete*

REISHOLZ S22
Mannesmannrohren-Werke AG
C 0.22, Si 0.25, Mn 0.45, bal Fe.
Annealed: 75,000 TS; 45,000 YS; 22 El; 58 RA; 140 Brin. For screws, bolts, camshafts, bushings; water hardened. *Obsolete*

REISHOLZ S35
Mannesmannrohren-Werke AG
C 0.35, Si 0.25, Mn 0.55, bal Fe.
Hot rolled: 85,000 TS; 54,000 YS; 30 El; 53 RA; 185 Brin. For armature shafts, brackets, gears, bolts; water hardened. *Obsolete*

REISHOLZ S45
Mannesmannrohren-Werke AG
C 0.45, Si 0.25, Mn 0.65, bal Fe.
Hot rolled: 98,000 TS; 58,000 YS; 24 El; 45 RA; 212 Brin. For axles, gears, bolts, bushings; water hardened. *Obsolete*

REISHOLZ SK11H
Mannesmannrohren-Werke AG
C 0.2, Si 0.25, Mn 0.6, Mo 0.3, bal Fe.
Annealed: 75,000 TS; 42,000 YS; 20 El; 55 RA; 150 Brin. For gears, bolts, fasteners, rivets; case hardened. *Obsolete*

REISHOLZ SK11R
Mannesmannrohren-Werke AG
C 0.15, Si 0.25, Mn 0.65, Mo 0.3, bal Fe.
Annealed: 75,000 TS; 42,000 YS; 20 El; 55 RA; 150 Brin. For gears, bolts, fasteners, camshafts; case hardened. *Obsolete*

REISHOLZ SK12
Mannesmannrohren-Werke AG
C 0.16, Cr 1.05, Mo 0.45, Mn 0.65, bal Fe.
For camshafts, cams, bolts, gears; case hardened. *Obsolete*

REISHOLZ SK12E
Mannesmannrohren-Werke AG
C 0.24, Cr 1.25, Mo 0.45, Mn 0.55, bal Fe.
For gears, bolts, camshafts, crankshafts; oil hardened, tough. *Obsolete*

REISHOLZ SK12ES
Mannesmannrohren-Werke AG
C 0.24, Cr 1.35, Mo 0.55, V 0.2, bal Fe.
For die casting dies, gears, crankshafts; oil hardened, tough. *Obsolete*

REISHOLZ SK12H
Mannesmannrohren-Werke AG
C 0.2, Cr 1.05, Mo 0.45, Mn 0.65, bal Fe.
For gears, bolts, camshafts, cams; case hardened. *Obsolete*

REISHOLZ SK12R
Mannesmannrohren-Werke AG
C 0.13, Cr 0.85, Mo 0.45, Mn 0.55, bal Fe.
For gears, bolts, camshafts; case hardened. *Obsolete*

REISHOLZ SK14R
Mannesmannrohren-Werke AG
C 0.1, Si 1.1, Cr 1.8, Mo 0.3, V 0.3, bal Fe.
For camshafts, cams, plastic mold dies; case hardened. *Obsolete*

REISHOLZ SK16
Mannesmannrohren-Werke AG
C 0.1, Cr 2.25, Mo 1.1, Mn 0.5, bal Fe.
For plastic mold dies; case hardened. *Obsolete*

REISHOLZ SK20
Mannesmannrohren-Werke AG
C 0.22, Cr 11.5, Mo 1, Ni 0.5, V 0.3, bal Fe.
Annealed: 90,000 TS; 40,000 YS; 25 El; 92 Rock B.
Hardened: 240,000 TS, 205,000 YS; 10 El; 50 Rock C. For valves, bearings, cutlery, surgical instruments, pivots. Corrosion resistant, hardenable. *Obsolete*

REISHOLZ SK20W
Mannesmannrohren-Werke AG
C 0.2, Cr 12, Mo 1, Ni 0.4, V 0.3, W 0.5, bal Fe.
Annealed: 95,000 TS; 40,000 YS; 25 El; 92 Rock B.
Hardened: 235,000 TS, 100,000 YS; 11 El; 48 Rock C. For valves, bearings, cutlery, surgical instruments, gears. Corrosion resistant, hardenable. *Obsolete*

REISHOLZ SK30
Mannesmannrohren-Werke AG
C 1.4, Mn, bal Fe.
For bearings, cutters, engravers' tools; Type W1; water hardened. *Obsolete*

REISHOLZ SK31H
Mannesmannrohren-Werke AG
C 0.18, Cr 0.65, Mo 0.15, Mn 0.85, bal Fe.
For gears, bolts, camshafts, cams; case hardened. *Obsolete*

REISHOLZ SK31R
Mannesmannrohren-Werke AG
C 0.15, Cr 0.75, Mo 0.15, Mn 0.75, bal Fe.
For gears, bolts, camshafts, cams; case hardened. *Obsolete*

REISHOLZ SK32
Mannesmannrohren-Werke AG
C 0.16, Cr 1.05, Mo 0.25, V 0.2, bal Fe.
For gears, bolts, camshafts, cams; case hardened. *Obsolete*

REISHOLZ SK32R
Mannesmannrohren-Werke AG
C 0.13, Cr 1.05, Mo 0.25, Mn 0.55, bal Fe.
For gears, bolts, camshafts, cams; case hardened. *Obsolete*

REISHOLZ ST 35
Mannesmannrohren-Werke AG
C 0-0.17, Si 0-0.35, Mn 0-0.4, bal Fe.
Annealed: 70,000 TS; 40,000 YS; 25 El; 60 RA; 145 Brin. For rivets, nails, bolts, nuts, gears, shafts; case hardened. *Obsolete*

REISHOLZ ST 45
Mannesmannrohren-Werke AG
C 0-0.22, Si 0-0.35, Mn 0-0.45, bal Fe.
Annealed: 73,000 TS; 42,000 YS; 22 El; 58 RA; 150 Brin. For fan blades, gears, bolts, rivets, shafts; case hardened. *Obsolete*

REITHS ALLOY
German manufacture
Cu 74.5, Sn 11.6, Pb 0.9, Sb 4.9.
For bearings; heavy duty.

REKFORD EMINENT
Boyd-Wagner Co.
C, Mn, Cr, W, V, Mo, Co, bal Fe.
For tools, dies; high speed steel.

REKFORD EMINENT
Boyd-Wagner Co.
C, Mn, Cr, W, V, Mo, Co, bal Fe.
For tools, dies; high speed steel.

REKORD 1939
Manufacturer not listed.
C 0.8, Mo 0.2, W 12.5, Co 0.8, V 2.5, Cr 4, bal Fe.
For drills, reamers, broaches, taps, milling cutters; high speed steel.

REKORD BK03
Idealstahl Breidenbach KG
C 0.9, Cr 4, W, Co, V, bal Fe.
For lathe and planer tools, reamers, broaches; high speed steel.

REKORD BK05
Idealstahl Breidenbach KG
C 0.79, Co 4.75, Cr 4.3, Mo 0.75, V 1.5, W 18, bal Fe.
For lathe and planer tools, reamers, broaches; high speed steel.

REKORD BKO10
Idealstahl Breidenbach KG
C 0.76, Co 10, Cr 4.2, Mo 0.8, V 1.8, W 18, bal Fe.
For lathe and planer tools, reamers, broaches, hobs; high speed steel.

REKORD BRKO
Idealstahl Breidenbach KG
C 0.8, Co, Cr, Mo, V, W, bal Fe.
For lathe and planer tools, reamers, broaches, taps; high speed steel.

REKORD BVK4
Idealstahl Breidenbach KG
C 1.35, Co, Mo, V, W, bal Fe.
For engravers' tools, forming dies, broaches, taps; high speed steel.

REKORD BVN
Idealstahl Breidenbach KG
C 1.3, Cr 4.3, Mo 0.85, V 3.8, W 12, bal Fe.
For engravers' tools, forming dies, broaches, taps; high speed steel.

REKORD BW018
Idealstahl Breidenbach KG
C 0.74, Cr 4.1, V 1.1, W 18.5, bal Fe.
For lathe and planer tools, reamers, taps, hobs; high speed steel.

REKORD EXTRA-26
Manufacturer not listed.
C 0.8, Mo 1.2, Co 2.5, Cr 4.5, W 18.5, V 1.6, bal Fe.
For milling cutters, lathe and planer tools, reamers; high speed steel.

REKORD MO 90
Idealstahl Breidenbach KG
C 0.8, Cr, V, Mo, W, bal Fe.
For lathe and planer tools, reamers, taps, drills; high speed steel.

REKORD SELECT
Boyd-Wagner Co.
C, Mn, Cr, W, V, Mo, Cu, bal Fe.
For tools, dies.

REKORD SELECT-27
Manufacturer not listed.
C 0.8, Mo 1.2, Cr 4.5, W 18.5, Co 5.5, V 1.6, bal Fe.
For milling cutters, lathe and planer tools, drills; high speed steel.

REKORD SUPERIOR
Idealstahl Breidenbach KG
C 0.86, Cr 4.1, Mo 0.85, V 2.5, W 12, bal Fe.
For lathe and planer tools, milling cutters, drills; high speed steel.

REKORD-25
Manufacturer not listed.
C 0.7, Cr 4.5, W 18.5, Co 0-0.6, V 1.2, bal Fe.
For lathe and planer tools, drills, reamers, hobs; high speed steel.

RELAY 2
AL Tech Specialty Steel Corp.
Si 1-1.5, bal Fe.
For armatures, relays, solenoid switches; high permeability.

RELAY 2SS
AL Tech Specialty Steel Corp.
Si 1-1.5, bal Fe.
For armatures, relays, solenoid switches; high permeability.

RELAY 5
AL Tech Specialty Steel Corp.
Si 2.5, bal Fe.
For armatures, switches, solenoid switches; high permeability.

RELAY 5SS
AL Tech Specialty Steel Corp.
C 0.5, Cr 0.7, Mo 0.5, bal Fe.
For tools; oil hardening, shock resisting.

RELIABRAZE 2417
Alpha Metals Inc.
Paste of copper powder and vehicle for copper brazing.

RELIANCE
Emsco Derrick & Equipment Co.
C 0.07-0.13, Mn 0.2-0.5, Ni 3-3.5, Cu 1.4-1.6, bal Fe.
Rolled: 95,000 TS; 75,000 YS; 30 El; 60 RA; 183 Brin. For oil well sucker rods; for severe service.

RELIANCE
Teledyne Pittsburgh Tool Steel
C 0.95, Si 0.2, Mn 0.35, bal Fe.
Cold drawn flats and squares. AISI Type W1; water hardening tool steel.

RELIANCE
Reliance Steel Casting Co.
C 0.25-0.4, bal Fe.
For steel castings; water hardening.

RELIANITE
American Meter Co.
C 3.2, Si, Mn, bal Fe.
For housings, gears, shafts; nodular iron, ductile.

RELIT
Russian manufacture
WC + Co.
For cutting tools; hard sintered alloy.

RELLEUM
Mueller Brass Co.
Pb 2, Zn 31, bal Cu.
For hardware, bolts, screw machine products; free cutting. *Obsolete*

RELY "YZ"
Rely Metal Works
Sn 91.5, Sb 3.5, Cu 4.3, Ni 0.5.
For bearings.

RELY AWM1
Rely Metal Works
Sn 89, Sb 8, Cu 3.
For bearings; Babbitt.

RELY C.M. 1
Rely Metal Works
Cu 6, Sn 84, Sb 10.
For bearings; Babbitt.

RELY C.M.2
Rely Metal Works
Cu 6, Sn 72, Pb 12, Sb 10.
For bearings; Babbitt.

RELY C.M.3
Rely Metal Works
Cu 3, Sn 40, Pb 42, Sb 15.
For bearings; Babbitt.

RELY C.M.4
Rely Metal Works
Cu 3, Sn 31, Pb 50, Sb 16.
For bearings; Babbitt.

RELY C.M.5
Rely Metal Works
Cu 1, Sn 15, Pb 70, Sb 14.
For bearings; Babbitt.

RELY C.M.6
Rely Metal Works
Sn 6, Pb 80, Sb 14.
For bearings; Babbitt.

REM ARMS HD-1000 DENSITY 6.8
Remington Arms Co. Inc.
Fe.
Sintered. As sintered: 23,000 TS; 12 El; 40 Rock F. For low and medium stressed mechanical components; soft magnetic cores. MPIF F-0000-S; ASTM B310-70, Type III, Class A.

REM ARMS HD-1000 DENSITY 6.8
Remington Arms Co. Inc.
Iron, sintered. Theoretical density 86%; induction 11.4 kilogauss max; remanance 9.6 kilogauss; coercive force 1.65 oersted; permeability 2900 max; electrical resistance 12 micro-ohm/cm. For DC applications, generator rotors, etc. MPIF F-0000-R; ASTM B310, Type III, Class A.

REM ARMS HD-1000 DENSITY 7.0
Remington Arms Co. Inc.
Fe.
Sintered. As sintered: 26,500 TS; 15 El; 50 Rock F. For low and medium stressed mechanical components; soft magnetic cores. MPIF F-0000-S ASTM B310-70, Type IV, Class A.

REM ARMS HD-1000 DENSITY 7.2
Remington Arms Co. Inc.
Fe.
Sintered. As sintered: 29,500 TS; 20 El; 55 Rock F. For low and medium stressed mechanical components; soft magnetic cores. MPIF F-0000-T; ASTM B310-70, Type IV, Class A.

REM ARMS HD-1000 DENSITY 7.4
Remington Arms Co. Inc.
Fe.
Sintered. As sintered: 35,000 TS; 24 El; 20 Rock B. For low and medium stressed mechanical components; soft magnetic cores. MPIF F-0000-T ASTM B310-70, Type V, Class A.

REM ARMS HD-1000 DENSITY 7.6
Remington Arms Co. Inc.
Fe.
Sintered. As sintered: 40,000 TS; 29 El; 30 Rock B. For low and medium stressed mechanical components; soft magnet cores; this grade can be case hardened, plated and colored. MPIF F-0000-U; ASTM B310-70, Type V, Class A.

REM ARMS HD-1003 DENSITY 6.8
Remington Arms Co. Inc.
Medium carbon steel, sintered. As sintered: 32,000 TS; 9 El; 12 Rock B. For medium stressed structural parts. MPIF F-0000-S; ASTM B310-70; Type III, Class B.

REM ARMS HD-1003 DENSITY 7.2
Remington Arms Co. Inc.
Medium carbon steel, sintered. As sintered: 43,000 TS; 11 El; 37 Rock B. Can be heat treated to 50,000-80,000 psi. For medium stressed structural parts. MPIF F-0000-T; ASTM B310-70, Type IV, Class B.

REM ARMS HD-1007 DENSITY 6.8
Remington Arms Co. Inc.
High carbon steel, sintered. As sintered: 46,000 TS; 4 El; 44 Rock B. For highly stressed structural parts. Heat treatable to 65,000-100,000 psi. MPIF F-0008-S; ASTM B310-70, Type III, Class C.

REM ARMS HD-1007 DENSITY 7.2
Remington Arms Co. Inc.
High carbon steel, sintered. As sintered: 55,000 TS; 5 El; 58 Rock B. For highly stressed structural parts. Heat treatable to 65,000-100,000 psi. MPIF F-0008-T ASTM B310-70, Type IV, Class C.

REM ARMS HD-2020 DENSITY 6.8
Remington Arms Co. Inc.
Low Ni, Fe.
Sintered. As sintered: 38,000 TS; 8 El; 32 Rock B. Good strength and toughness; for small appliance parts. MPIF FN-0200-S; ASTM B484-70, Type II, Gr. 1, Class A.

REM ARMS HD-2020 DENSITY 7.0
Remington Arms Co. Inc.
Low Ni, Fe.
Sintered. As sintered: 41,000 TS; 11 El; 40 Rock B. Good strength and toughness; for small appliance parts. MPIF FN-0200-S; ASTM B484-70, Type II, Gr. 1, Class A.

REM ARMS HD-2020 DENSITY 7.2
Remington Arms Co. Inc.
Low Ni, Fe.
Sintered. As sintered: 47,000 TS; 14 El; 49 Rock B. Good strength and toughness; for gears, cams and small appliance parts. MPIF FN-0200-T; ASTM B484-70, Type III, Gr. 1, Class A.

REM ARMS HD-2020 DENSITY 7.4
Remington Arms Co. Inc.
Low Ni, Fe.
Sintered. As sintered: 53,000 TS; 18 El; 60 Rock B. Good strength and toughness; for gears, cams and small appliance parts. MPIF FN-0200-T; ASTM B484-70, Type III, Gr. 1, Class A.

REM ARMS HD-2027 DENSITY 6.8
Remington Arms Co. Inc.
Low nickel steel, sintered. As sintered: 72,000 TS; 3 El; 74 Rock B. Good strength and toughness; hardenable to 40-45 Rock C. MPIF FN-0208-S; ASTM B484-70, Type II, Gr. 1, Class C.

REM ARMS HD-2027 DENSITY 7.2
Remington Arms Co. Inc.
Low nickel steel, sintered. As sintered: 91,000 TS; 4 El; 86 Rock B. Good strength and toughness; hardenable to 40-45 Rock C. MPIF FN-0208-T; ASTM B484-70 Type III, Gr. 1, Class C.

REM ARMS HD-2040 DENSITY 6.8
Remington Arms Co. Inc.
Medium Ni, Fe.
Sintered. As sintered: 37,000 TS; 5 El; 35 Rock B. For case hardening. MPIF FN-0400-T; ASTM B484-70, Type II, Gr. 2, Class A.

REM ARMS HD-2040 DENSITY 7.2
Remington Arms Co. Inc.
Medium Ni, Fe.
Sintered. As sintered: 48,000 TS; 8 El; 58 Rock B. For case hardening. MPIF FN-0400-T; ASTM B484-70, Type III, Gr. 2, Class A.

REM ARMS HD-2047 DENSITY 6.8
Remington Arms Co. Inc.
Medium Ni, Fe.
Sintered. As sintered: 74,000 TS; 2 El; 80 Rock B. High strength, tough core; hardenable to 120,000-160,000 psi. MPIF FN-0408-S; ASTM B484-70, Type II, Gr. 2, Class C.

REM ARMS HD-2047 DENSITY 7.2
Remington Arms Co. Inc.
Medium nickel steel, sintered. As sintered: 95,000 TS; 3.5 El; 87 Rock B. High strength, tough core; hardenable to 120,000-160,000 psi. MPIF FN-0408-T; ASTM B484-70, Type III, Gr. 2, Class C.

REM ARMS HD-2070 DENSITY 6.8
Remington Arms Co. Inc.
High Ni, Fe.
Sintered. As sintered: 65,000 TS; 4.5 El; 64 Rock B. High strength, fair ductility; easily swaged or formed. MPIF FN-0700-S; ASTM B484-70, Type II, Gr. 3, Class A.

REM ARMS HD-2070 DENSITY 7.2
Remington Arms Co. Inc.
High Ni, Fe.
Sintered. As sintered: 76,000 TS; 8 El; 78 Rock B. High strength, fair ductility; easily swaged or formed. MPIF FN-0700-T; ASTM B484-70, Type III, Gr. 3, Class A.

REM ARMS HD-2076 DENSITY 6.8
Remington Arms Co. Inc.
High nickel steel, sintered. As sintered: 94,000 TS; 3.5 El; 86 Rock B. High strength; heat treatable. MPIF FN-0705-S; ASTM B484-70, Type II, Gr. 3, Class C.

REM ARMS HD-2076 DENSITY 7.2
Remington Arms Co. Inc.
High nickel steel, sintered. As sintered: 105,000 TS; 4 El; 20 Rock C. High strength; heat treatable. MPIF FN-0705-T; ASTM B484-70, Type III, Gr. 3, Class C.

REM ARMS HD-2108 DENSITY 7.4
Remington Arms Co. Inc.
Air hardening nickel steel, sintered. As sintered: 150,000 TS; 4 El; 40 Rock C. For jobs requiring strong, wear resisting parts.

REM ARMS HD-3050 DENSITY 7.0
Remington Arms Co. Inc.
Remington alloy, sintered. As sintered: 32,000 TS; 12 El; 30 Rock B. High ductility by single pressing.

REM ARMS HD-3050 DENSITY 7.2
Remington Arms Co. Inc.
Remington alloy, sintered. As sintered: 37,000 TS; 16 El; 35 Rock B. High density and high ductility by single pressing.

REM ARMS HD-3050 DENSITY 7.4
Remington Arms Co. Inc.
Remington alloy, sintered. As sintered: 47,000 TS; 25 El; 50 Rock B. Good impact strength; high density and high ductility; for case hardening, plating, coloring.

REM ARMS HD-3050 DENSITY 7.6
Remington Arms Co. Inc.
Remington alloy, sintered. As sintered: 53,000 TS; 30 El; 60 Rock B. Good impact strength; high density and high ductility; for case hardening, plating, coloring.

REM ARMS HD-4058 DENSITY 6.6
Remington Arms Co. Inc.
Copper steel, sintered. As sintered: 78,000 TS; 1 El; 82 Rock B. Medium strength components. ASTM B426-70, Gr. 2, Type III.

REM ARMS HD-4058 DENSITY 6.8
Remington Arms Co. Inc.
Copper steel, sintered. As sintered: 96,000 TS; 1 El; 90 Rock B. Medium strength components. ASTM B426-70, Gr. 2, Type IV.

REM ARMS HD-4058 DENSITY 7.0
Remington Arms Co. Inc.
Copper steel, sintered. As sintered: 105,000 TS; 2 El; 96 Rock B. Medium strength components. ASTM B426-70, Gr. 2, Type IV.

REM ARMS HD-4058 DENSITY 7.2
Remington Arms Co. Inc.
Copper steel, sintered. As sintered: 110,000 TS; 2 El; 25 Rock C. For medium strength components. ASTM B426-70, Gr. 2, Type V.

REM ARMS HDM-1000 DENSITY 6.4
Remington Arms Co. Inc.
Iron, sintered. Theoretical density 81%; induction 9.1 kilogauss max; remanance 7.4 kilogauss; coercive force 1.80 oersted; permeability 2350 max; electrical resistance 12 micro-ohm/cm. For pole pieces, cores, armatures, generator rotors, stator cores. MPIF F-0000-P; ASTM B310-70, Type II, Class A.

REM ARMS HDM-1000 DENSITY 7.2
Remington Arms Co. Inc.
Iron, sintered. Theoretical density 92%; induction 13.6 kilogauss max; remanance 11.8 kilogauss; coercive force 1.60 oersted; permeability 3700 max; electrical resistance 12 micro-ohm/cm. For DC applications, cores, rotors, stators. MPIF F-0000-S; ASTM B310-70, Type IV, Class A.

REM ARMS HDM-1000 DENSITY 7.4
Remington Arms Co. Inc.
Iron, sintered. Theoretical density 94%; induction 14.7 kilogauss max; remanance 12.9 kilogauss; coercive force 1.50 oersted; permeability 4200 max; electrical resistance 12 micro-ohm/cm. For DC applications, pole pieces, stators. MPIF F-0000-T; ASTM B310-70, Type V, Class A.

REM ARMS HDM-1000 DENSITY 7.6
Remington Arms Co. Inc.
Iron, sintered. Theoretical density 97%; induction 15.8 kilogauss max; remanance 14.0 kilogauss; coercive force 1.35 oersted; permeability 4750 max; electrical resistance 12 micro-ohm/cm. For DC applications, pole pieces, cores, stators. MPIF F-0000-U; ASTM B310-70, Type V, Class A.

REM ARMS HDM-2500 DENSITY 6.8
Remington Arms Co. Inc.
Fe-Ni.
Sintered. Theoretical density 83%; induction 9.3 kilogauss max; remanance 7.1 kilogauss; coercive force 0.26 oersted; electrical resistance 43 micro-ohm/cm. For torque motors, relays, missile components.

REM ARMS HDM-2500 DENSITY 7.1
Remington Arms Co. Inc.
Fe-Ni.
Sintered. Theoretical density 87%; induction 10.9 kilogauss max; remanance 8.0 kilogauss; coercive force 0.25 oersted; electrical resistance 43 micro-ohm/cm. For torque motors, relays, missile components.

REM ARMS HDM-2500 DENSITY 7.5
Remington Arms Co. Inc.
Fe-Ni.
Sintered. Theoretical density 91%; induction 12.7 kilogauss max; remanance 9.4 kilogauss; coercive force 0.24 oersted; permeability 21,000 max; electrical resistance 43 micro-ohm/cm. Very low coercive force and high permeability; for torque motors, relays, missile components.

REM ARMS HDM-2500 DENSITY 7.7
Remington Arms Co. Inc.
Fe-Ni.
Sintered. Theoretical density 94%; induction 13.3 kilogauss max; remanance 9.5 kilogauss; coercive force 0.22 oersted; permeability 22,000 max; electrical resistance 43 micro-ohm/cm. Very low coercive force and high permeability; for torque motors, relays, missile components.

REM ARMS HDM-3050 DENSITY 7.0
Remington Arms Co. Inc.
Soft magnetic alloy, sintered. Theoretical density 88%; induction 11.6 kilogauss max; remanance 6.7 kilogauss; coercivec force 1.60 oersted; electrical resistance 26 micro-ohm/cm. For DC applications.

REM ARMS HDM-3050 DENSITY 7.2
Remington Arms Co. Inc.
Soft magnetic alloy, sintered. Theoretical density: 90%; induction B max, kilogauss 13.5; remanance Br, kilogauss 8.1; coercive Hc force, oersted 1.50; electrical resistance, 26 micro-ohm/cm. For DC applications.

REM ARMS HDM-3050 DENSITY 7.4
Remington Arms Co. Inc.
Soft magnetic alloy, sintered. Theoretical density 93%; induction 13.8 kilogauss max; remanance 10.2 kilogauss; coercive force 1.10 oersted; electrical resistance 26 micro-ohm/cm. For DC applications.

REM ARMS HDM-3050 DENSITY 7.6
Remington Arms Co. Inc.
Soft magnetic alloy, sintered. Theoretical density 96%; induction 14.4 kilogauss max; remanance 10.5 kilogauss; coercive force 1.10 oersted; electrical resistance 26 micro-ohm/cm. For DC applications.

REM ARMS HDM-5010 DENSITY 7.2
Remington Arms Co. Inc.
Si-Fe.
Sintered. Theoretical density 92%; induction 13.6 kilogauss max; remanance 10.3 kilogauss; coercive force 1.5 oersted; permeability 3450 max; electrical resistance 22 micro-ohm/cm. For AC applications, solenoids, relays, armatures.

REM ARMS HDM-5020 DENSITY 7.1
Remington Arms Co. Inc.
Si-Fe.
Sintered. Theoretical density 92%; induction 13.2 kilogauss max; remanance 10.3 kilogauss; coercive force 1.20 oersted; electrical resistance 39 micro-ohm/cm. For AC applications, relays, armatures, cores.

REM ARMS HDM-5030 DENSITY 6.8
Remington Arms Co. Inc.
Si-Fe.
Sintered. Theoretical density 89%; induction 11.7 kilogauss max; remanance 9.4 kilogauss; coercive force 1.30 oersted; electrical resistance 46 micro-ohm/cm. For AC applications, solenoids, pole pieces, stators.

REM ARMS HDM-5030 DENSITY 7.0
Remington Arms Co. Inc.
Si-Fe.
Sintered. Theoretical density 91%; induction 13.1 kilogauss max; remanance 10.9 kilogauss; coercive force 1.25 oersted; electrical resistance 46 micro-ohm/cm. For AC applications, solenoids, armatures, cores.

REM ARMS HDM-5030 DENSITY 7.2
Remington Arms Co. Inc.
Si-Fe.
Sintered. Theoretical density 94%; induction 13.9 kilogauss max; remanance 11.8 kilogauss; coercive force 1.20 oersted; permeability 4860 max; electrical resistance 46 micro-ohm/cm. For AC applications, relays, actuator cores.

REM ARMS HDM-5030 DENSITY 7.4
Remington Arms Co. Inc.
Si-Fe.
Sintered. Theoretical density 96%; induction 14.4 kilogauss max; remanance 12.4 kilogauss; coercive force 1.00 oersted; electrical resistance 46 micro-ohm/cm. For AC applications, electromagnetic devices.

REM ARMS HDM-5040 DENSITY 7.3
Remington Arms Co. Inc.
Si-Fe.
Sintered. Theoretical density 96%; induction 14.2 kilogauss max; remanance 10.8 kilogauss; coercive force 1.00 oersted; permeability 5850 max; electrical resistance 59 micro-ohm/cm. For AC applications, solenoids, pole pieces.

REM ARMS HDM-5050 DENSITY 7.2
Remington Arms Co. Inc.
Si-Fe.
Sintered. Theoretical density 96%; induction 13.7 kilogauss max; remanance 8.7 kilogauss; coercive force 0.85 oersted; permeability 6400 max; electrical resistance 68 micro-ohm/cm. For AC applications, relays, armatures, solenoids.

REM ARMS HDM-9800 DENSITY 7.8
Remington Arms Co. Inc.
Ni-Fe-Mo.
Sintered. Theoretical density 90%; induction 7.2 kilogauss max; remanance 4.8 kilogauss; coercive force 0.07 oersted; permeability 37,000 max; electrical resistance 55 micro-ohm/cm. For high permeability uses.

REM ARMS HDS 430 DENSITY 6.6
Remington Arms Co. Inc.
Type 430 stainless (ferritic). As sintered: 35,000 TS; 2 El; 90 Rock B. Nonhardenable; fair corrosion resistance; ferromagnetic.

REM ARMS HDS-303L DENSITY 6.6
Remington Arms Co. Inc.
303L-stainless (austenitic). As sintered: 62,000 TS; 4 El; 63 Rock B, 2 ft·lb Charpy. Free machining, nonmagnetic, nonhardenable. MPIF SS-303-R.

REM ARMS HDS-303L DENSITY 6.8
Remington Arms Co. Inc.
303L-stainless (austenitic). As sintered: 69,000 TS; 12 El; 70 Rock B; 12 ft·lb Charpy. Free machining, nonmagnetic, nonhardenable. MPIF SS-303-R.

REM ARMS HDS-303L DENSITY 7.0
Remington Arms Co. Inc.
303L-stainless (austenitic). As sintered: 75,000 TS; 16 El; 75 Rock B; 40 ft·lb Charpy. Free machining, nonmagnetic, nonhardenable.

REM ARMS HDS-304L DENSITY 6.6
Remington Arms Co. Inc.
304L-stainless (austenitic). As sintered: 64,000 TS; 5 El; 70 Rock B; 3 ft·lb Charpy. Nonmagnetic, nonhardenable, stainless. ASTM B525-70, Type II, Gr. I.

REM ARMS HDS-304L DENSITY 6.8
Remington Arms Co. Inc.
304L-stainless (austenitic). As sintered: 71,000 TS; 12 El; 75 Rock B; 12 ft·lb Charpy. Nonmagnetic, nonhardenable, stainless. ASTM B525-70 Type II, Gr. I.

REM ARMS HDS-304L DENSITY 7.0
Remington Arms Co. Inc.
304L-stainless (austenitic). As sintered: 75,000 TS; 15 El; 78 Rock B; 42 ft·lb Charpy. Nonmagnetic, nonhardenable, stainless. ASTM B525-70 Type III, Gr. I.

REM ARMS HDS-316L DENSITY 6.6
Remington Arms Co. Inc.
316L-stainless (austenitic). As sintered: 60,000 TS; 4 El; 70 Rock B; 4 ft·lb Charpy. Nonmagnetic, nonhardenable, improved corrosion resistance. MPIF SS-316-R; ASTM B525-70, Type II, Gr. II.

REM ARMS HDS-316L DENSITY 6.8
Remington Arms Co. Inc.
316L-stainless (austenitic). As sintered: 68,000 TS; 12 El; 74 Rock B; 10 ft·lb Charpy. Nonmagnetic, nonhardenable, improved corrosion resistance. MPIF SS-316-R; ASTM B525-70, Type II, Gr. II.

REM ARMS HDS-316L DENSITY 7.0
Remington Arms Co. Inc.
316L-stainless (austenitic). As sintered: 73,000 TS; 16 El; 79 Rock B; 45 ft·lb Charpy. Nonmagnetic, nonhardenable, improved corrosion resistance. ASTM B525-70, Type III, Gr. II.

REM ARMS HDS-410 DENSITY 6.4
Remington Arms Co. Inc.
Type 410 stainless (martensitic). As sintered: 67,000 TS; 15 Rock C; 1 ft·lb Charpy. Hardenable to 50 Rock C equivalent; fair corrosion resistance; ferromagnetic. MPIF SS-410-P.

REM ARMS HDS-410 DENSITY 6.6
Remington Arms Co. Inc.
Type 410 stainless (martensitic). As sintered: 84,000 TS; 1 El; 20 Rock C; 1.5 ft·lb Charpy. Hardenable to 50 Rock C equivalent; fair corrosion resistance; ferromagnetic.

REM ARMS HDS-410 DENSITY 6.8
Remington Arms Co. Inc.
Type 410 stainless (martensitic). As sintered: 88,000 TS; 2 El; 20 Rock C; 1.5 ft·lb Charpy. Hardenable to 50 Rock C equivalent; fair corrosion resistance; ferromagnetic.

REM ARMS HDS-410 DENSITY 7.0
Remington Arms Co. Inc.
Type 410 stainless (martensitic). As sintered: 94,000 TS; 2 El; 25 Rock C; 2 ft·lb Charpy. Hardenable to 50 Rock C equivalent; fair corrosion resistance; ferromagnetic.

REM ARMS HDS-410 DENSITY 7.2
Remington Arms Co. Inc.
Type 410 stainless (martensitic). As sintered: 73,000 TS; 6 El; 75 Rock B; 3 ft·lb Charpy. Hardenable to 50 Rock C equivalent; fair corrosion resistance; ferromagnetic.

REM ARMS HDS-430 DENSITY 6.8
Remington Arms Co. Inc.
Type 430 stainless (ferritic). As sintered: 50,000 TS; 2 El; 85 Rock B. Nonhardenable; fair corrosion resistance; ferromagnetic.

REM ARMS HDS-430 DENSITY 7.0
Remington Arms Co. Inc.
Type 430 stainless (ferritic). As sintered: 60,000 TS; 4 El; 70 Rock B. Nonhardenable; fair corrosion resistance; ferromagnetic.

REM ARMS HDS-9400 DENSITY 6.8
Remington Arms Co. Inc.
High corrosion resistant stainless. As sintered: 50,000 TS; 4 El; 60 Rock B. Outstanding corrosion resistance to H_2SO_4, caustic solution brine, plating solutions.

REM ARMS HDS-9400 DENSITY 7.0
Remington Arms Co. Inc.
High corrosion resistant stainless. As sintered: 66,000 TS; 11 El; 70 Rock B. Outstanding corrosion resistance to H_2SO_4, caustic solutions, brine, plating solutions.

REMA
Hoyland Steel Co.
C 0.05, Mn 0.2, Si 0.1, bal Fe.
For hobbing molds; deep cavity molds for plastic industry.

REMA
Fagersta Bruks Aktiebolag
C 1, Mn 1.2, Cr 0.5, bal Fe.
For telephone switchboard equipment. *Obsolete*

REMA
Great Western Steel Co.
C 0.05, Mn 0.2, Si 0.1, bal Fe.
For hobbing molds; case hardening steel.

REMA HOBBING DIE
Brukskoncernen AB
C 1, Mn 1.2, Cr 0.5, bal Fe.
For hobbing dies for plastics; oil hardened. *Obsolete*

REMA HOBBING DIE
Fagersta Bruks Aktiebolag
C 1, Mn 1.2, Cr 0.5, bal Fe.
For hobbing dies for plastics; oil hardened. *Obsolete*

REMALLOY
Rennie Tool Co. Ltd.
C 0.7, Cr 4, W 18, V 2, Co 12, bal Fe.
For lathe and planer tools, reamers, broaches, taps, hobs; high speed steel, oil hardened.

REMALLOY
Arnold Engineering Co.
Mo 17, Co 12, bal Fe.
Bar: 126,000 TS; 50,000 transverse strength; 60 Rock C; residual flux density 10,000 gauss; coercive force 230 oersted. For permanent magnets; precipitation hardened. Fabricated parts only.

REMALLOY
Hoyland Steel Co.
C 0.1, Mn 0.5, Cr 0.6, Ni 1.5, bal Fe.
For plastic dies, hobbing molds; case-hardening.

REMALLOY
Great Western Steel Co.
C 0.1, Mn 0.5, Cr 0.6, Ni 1.5, bal Fe.
For plastic dies, hobbing molds; case-hardening.

REMALLOY 17
Wallace Murray Corp.
Mo 17, Co 12, bal Fe.
Maximum energy product 1.10; residual flux density 10,000; coercive force 250. Rolled or cast permanent magnet.

REMALLOY 17
Bell Telephone Laboratories
Mo 17, Co 12, Fe 71, Mn 3.
For permanent magnets; high coercive force.

REMALLOY 20
Wallace Murray Corp.
Mo 20, Co 12, bal Fe.
Maximum energy product 1.25; residual flux density 8550; coercive force 335. Hot rolled or cast permanent magnet.

REMALLOY 20
Western Electric Co.
Fe 67.7, Co 12, Mo 20, Mn 0.3.
For receiver magnets; permanent magnet. 355 coercive force.

REMALLOY 20
Wallace Murray Corp.
Fe 67.7, Co 12, Mo 20, Mn 0.3.
For receiver magnets; permanent magnet. 355 coercive force.

REMANIT
Thyssen Edelstahlwerke AG
C 0.2, Cr 13, Mo 1.2, bal Fe.
For surgical instruments, knives, cutlery; corrosion resistant, hardenable. *Obsolete*

REMANIT 11A
Thyssen Edelstahlwerke AG
C 0.2, Cr 25, Ni 21, bal Fe.
Annealed: 86,000 TS; 35,000 YS; 55 El; B 85 Rock. For furnace equipment, heat treating boxes, valves, pumps. Type 310 stainless steel, austenitic. Heat and corrosion resistant. *Obsolete*

REMANIT 1510
Remanit GmbH
C 0-0.15, Cr 11.5-13.5, bal Fe.
Heat treated: 120,000-135,000 TS; 110,000-117,000 YS; 15-16 El; 58-63 RA; 220-240 Brin. For cutlery, valves, and turbine blades. Type 410; corrosion resistant.

REMANIT 1510 AB
Thyssen Edelstahlwerke AG
C 0.1, Cr 15, Ni 0.5, bal Fe.
For corrosion resistant parts; corrosion resistant. *Obsolete*

REMANIT 1510 AL
Thyssen Edelstahlwerke AG
C 0.06, Cr 13, Al 0.2, bal Fe.
Annealed: 65,000 TS; 40,000 YS; 30 El; 160 Brin. For annealing boxes, quenching racks, oil burners, oil refining equipment. Type 405 stainless steel, ferritic. Good oxidation resistance. *Obsolete*

REMANIT 1510 H
Remanit GmbH
C 0-0.12, Cr 14-18, bal Fe.
Annealed: 70,000 TS; 40,000 YS; 30 El; 55 RA; 150 Brin. Cold drawn: 130,000 TS; 120,000 YS; 2 El; 185 Brin. For oil refinery equipment, bolts, and oil burners. Type 430; stainless; ferritic.

REMANIT 1520
Thyssen Edelstahlwerke AG
C 0.2, Cr 15, Ni 0.5, bal Fe.
For corrosion resistant parts; corrosion resistant. *Obsolete*

REMANIT 1540
Thyssen Edelstahlwerke AG
C 0.4, Cr 15, Mo 0.2, bal Fe.
For corrosion resistant parts; knives; corrosion resistant. *Obsolete*

REMANIT 1610
Thyssen Edelstahlwerke AG
C 0.1, Cr 16, Ni 0.2, bal Fe.
For corrosion resistant parts; corrosion resistant. *Obsolete*

REMANIT 1610 S
Thyssen Edelstahlwerke AG
C 0.1, Si 0.5, Cr 18, Ti, bal Fe.
For corrosion resistant parts; corrosion resistant. *Obsolete*

REMANIT 1620
Thyssen Edelstahlwerke AG
C 0.2, Cr 16, Ni 2, bal Fe.
For corrosion resistant parts; corrosion resistant. *Obsolete*

REMANIT 1710
Thyssen Edelstahlwerke AG
C 0.1, Cr 17, Ni 0.5, Mo 1.8, bal Fe.
For corrosion resistant parts; knives; corrosion resistant. *Obsolete*

REMANIT 1710A
Thyssen Edelstahlwerke AG
C 0.1, Cr 17, Ni 0.5, Mo 0.4, S 0.2, bal Fe.
For corrosion resistant parts; corrosion resistant. *Obsolete*

REMANIT 1790 V
Thyssen Edelstahlwerke AG
C 0.9, Cr 17.5, Mo 1.2, V 0.1, bal Fe.
Annealed: 110,000 TS; 65,000 YS; 18 El; 35 RA; 220 Brin. Heat treated: 280,000 TS; 270,000 YS; 3 El; 15 RA; 555 Brin. For bearings, valve parts, surgical instruments; stainless, hardenable. *Obsolete*

REMANIT 1810
Thyssen Edelstahlwerke AG
C 0.4, Cr 29, bal Fe.
Cast: 90,000 TS; 65,000 YS; 2 El; 212 Brin. For cylinder liners, grate bars, valve seats and bodies; Type CC50; heat resistant. *Obsolete*

REMANIT 1810 SST
Thyssen Edelstahlwerke AG
C 0.06, Cr 18, Ni 10, Mo 2, bal Fe.
Annealed: 90,000 TS; 40,000 YS; 40 El; B 85 Rock. For chemical plant equipment, agitators, digesters, tanks. Type 18-12 Mo stainless steel, austenitic, nonmagnetic. *Obsolete*

REMANIT 1810 SSW
Thyssen Edelstahlwerke AG
C 0.05, Cr 18, Ni 10, Mo 2.5, bal Fe.
Annealed: 85,000 TS; 35,000 YS; 45 El; B 85 Rock. For chemical plant equipment, agitators, digesters, tanks. Type 316 stainless steel, austenitic, acid resistant. *Obsolete*

REMANIT 1813 SSW
Thyssen Edelstahlwerke AG
C 0.07, Mn 1.3, Cr 17, Mo 4.5, Ni 13, bal Fe.
Annealed: 85,000 TS; 35,000 YS; 55 El; 70 RA; 160 Brin. For chemical plant equipment, mixers, agitators; Type 317, stainless, austenitic. *Obsolete*

REMANIT 1860 M
Thyssen Edelstahlwerke AG
C 0.12, Mn 9, Cr 18, Ni 5, bal Fe.
Annealed: 100,000 TS; 50,000 YS; 60 El; B 90 Rock. For dairy and chemical plant equipment, springs, tubular furniture, trim. Type 202 stainless steel, austenitic. *Obsolete*

REMANIT 1880
Thyssen Edelstahlwerke AG
C 0.1, Cr 18, Ni 8, bal Fe.
For corrosion resistant parts; austenitic, stainless. *Obsolete*

REMANIT 1880 A
Thyssen Edelstahlwerke AG
C 0.15, Mn 1.3, Cr 18, Ni 8.5, S 0.15, bal Fe.
Annealed: 80,000 TS; 40,000 YS; 40 El; 55 RA; 170 Brin. For screw machine products, hardware; stainless, austenitic, free-cutting. *Obsolete*

REMANIT 1880 FH
Thyssen Edelstahlwerke AG
C 0.15, Cr 18, Ni 8, bal Fe.
Annealed: 80,000 TS; 35,00 YS; 45 El; 65 RA; 160 Brin. For chemical plant equipment; Type 302, stainless, austenitic. *Obsolete*

REMANIT 1880 SEW
Thyssen Edelstahlwerke AG
C 0.02, Cr 18-20, Ni 8-12, bal Fe.
Annealed: 85,000 TS; 30,000 YS; 60 El; 140 Brin. For welded tanks, chemical plant equipment, kitchen apparatus. Type 304L stainless steel, austenitic. *Obsolete*

REMANIT 1880 SSEW
Thyssen Edelstahlwerke AG
C 0.03, Cr 16-18, Ni 10-14, Mo 2-3, bal Fe.
Annealed: 85,000 TS; 35,000 YS; 50 El; B 82 Rock. For chemical plant equipment, agitators, digesters, valve trim. Type 316L stainless steel, austenitic. *Obsolete*

REMANIT 1880 SST
Thyssen Edelstahlwerke AG
C 0.1, Mn 1.2, Cr 18, Mo 2.2, Ni 12, Ti 0.34, bal Fe.
Annealed: 80,000 TS; 30,000 YS; 60 El; 80 RA; 140 Brin. Cold drawn: 150,000 TS; 135,000 YS; 6 El; 300 Brin. For chemical plant equipment, digesters, evaporators; Type 316 Ti, stainless, austenitic. *Obsolete*

REMANIT 1880 SSW
Thyssen Edelstahlwerke AG
C 0.07, Mn 1.2, Cr 18, Ni 11, Mo 2.2, bal Fe.
Annealed: 80,000 TS; 30,000 YS; 60 El; 80 RA; 140 Brin. Cold drawn: 150,000 TS; 135,000 YS; 6 El; 300 Brin. For chemical plant equipment, digesters, evaporators; Type 316, stainless, austenitic. *Obsolete*

REMANIT 1880 SW
Thyssen Edelstahlwerke AG
C 0.07, Mn 1, Cr 18, Ni 10, bal Fe.
Annealed: 85,000 TS; 35,000 YS; 60 El; 70 RA; 150 Brin. Cold drawn: 180,000 TS; 125,000 YS; 10 El; 330 Brin. For welded structures, architectural trim; Type 304, stainless, austenitic. *Obsolete*

REMANIT 1880S
Thyssen Edelstahlwerke AG
C 0.09, Cr 18, Ni 8, Ta 1.5, bal Fe.
For corrosion resistant parts; wrought quality. *Obsolete*

REMANIT 1880SS
Remanit GmbH
C 0-0.1, Cr 16-18, Ni 10-14, Mo 2-3, bal Fe.
Annealed: 85,000-95,000 TS; 35,000-45,000 YS; 50-60 El; 60-75 RA; 150-190 Brin. For chemical plant equipment, mixers, agitators, and filters. Type 316; stainless; austenitic.

REMANIT 1880SS
Thyssen Edelstahlwerke AG
C 0.09, Cr 18, Ni 8, Mo 2, Ta 1, bal Fe.
For textile industry; corrosion resistant. *Obsolete*

REMANIT 1880ST
Remanit GmbH
C 0-0.08, Cr 17-19, Ni 8-11, Ti = 5 x C, bal Fe.
Normalized: 93,000 TS; 36,000 YS; 45 El; 60 RA; 165 Brin.
Annealed: 87,000 TS; 33,000 YS; 57 El; 73 RA; 155 Brin. For chemical plant equipment. Type 321; austenitic; stainless.

REMANIT 1990 S.-CR
Thyssen Edelstahlwerke AG
C 0.1, Si 1.7, Cr 18, Ni 8.5, bal Fe.
For stainless parts; stainless, austenitic. *Obsolete*

REMANIT 1990 SS.-CR
Thyssen Edelstahlwerke AG
C 0.1, Si 2.4, Cr 18, Ni 9.5, Mo 2, bal Fe.
For stainless parts; stainless, austenitic. *Obsolete*

REMANIT 2525 SST
Thyssen Edelstahlwerke AG
C 0.06, Mn 1.2, Cr 25, Mo 2.2, Ni 25, bal Fe.
For heat and corrosion resistant parts, furnace parts; heat and corrosion resistant. *Obsolete*

REMANIT 2804 G
Thyssen Edelstahlwerke AG
C 0.4, Si 1.2, Cr 27, Ni 4, bal Fe.
For furnace parts, heat treat boxes; heat resistant. *Obsolete*

REMANIT 2890 MO
Thyssen Edelstahlwerke AG
C 1.3, Si 1.2, Cr 28, Mo 2, bal Fe.
For valves, furnace parts; heat and corrosion resistant. *Obsolete*

REMANIT 4000
Thyssen Edelstahlwerke AG
Stainless steel. C 0-0.08, Cr 12-14, bal Fe.
Hot worked flat and long products. Werkstoff Nr. 1.4000 (DIN 17 440). 250-400 N/mm^2 YP; 400-700 N/mm^2 TS; 13-20 El; 180 Brin.

REMANIT 4002
Thyssen Edelstahlwerke AG
Stainless steel. C 0-0.08, Cr 12-14, Al 0.1-0.3, bal Fe.
Hot worked flat and long products. Werkstoff Nr. 1.4002 (DIN 17 440). 250-400 N/mm^2 YP; 400-700 N/mm^2 TS; 13-20 El; 180 Brin.

REMANIT 4006
Thyssen Edelstahlwerke AG
Stainless steel. C 0.08-0.12, Cr 12-14, bal Fe.
Hot worked flat and long products. Werkstoff Nr. 1.4006 (DIN 17 440). 250-420 N/mm^2 YP; 450-800 N/mm^2 TS; 12-20 El; 190 Brin.

REMANIT 4016
Thyssen Edelstahlwerke AG
Stainless steel. C 0-0.08, Cr 15.5-17.5, bal Fe.
Cold rolled flat products. Werkstoff Nr. 1.4016 (DIN 17 441). 340 N/mm^2 YP; 520 N/mm^2 TS; 24-35 El.

REMANIT 4021
Thyssen Edelstahlwerke AG
Stainless steel. C 0.17-0.25, Cr 12-14, bal Fe.
Cold rolled flat products. Werkstoff Nr. 1.4021 (DIN 17 441). 620 N/mm^2 TS; 22-25 El; 180 Brin.

REMANIT 4021
Thyssen Edelstahlwerke AG
Stainless steel. C 0.17-0.25, Cr 12-14, bal Fe.
Hot worked flat and long products. Werkstoff Nr. 1.4021 (DIN 17 440). 450-550 N/mm^2 YP; 650-950 N/mm^2 TS; 8-14 El; 220 Brin.

REMANIT 4024
Thyssen Edelstahlwerke AG
Stainless steel. C 0.12-0.17, Cr 12-14, bal Fe.
Hot worked flat and long products. Werkstoff Nr. 1.4024 (DIN 17 440). 450 N/mm^2 YP; 650-800 N/mm^2 TS; 14 El; 220 Brin.

REMANIT 4028
Thyssen Edelstahlwerke AG
Stainless steel. C 0.28-0.35, Cr 12-14, bal Fe.
Hot worked flat and long products. Werkstoff Nr. 1.4028 (DIN 17 440). 600 N/mm^2 YP; 780-1000 N/mm^2 TS; 11 El; 235 Brin.

REMANIT 4031
Thyssen Edelstahlwerke AG
Stainless steel. C 0.35-0.42, Cr 12.5-14.5, bal Fe.
Hot worked flat and long products. Werkstoff Nr. 1.4031 (DIN 17 440). 800 N/mm^2 TS max; 225 Brin; 52 Rock C.

REMANIT 4034
Thyssen Edelstahlwerke AG
Stainless steel. C 0.42-0.5, Cr 12.5-14.5, bal Fe.
Cold rolled flat products. Werkstoff Nr. 1.4034 (DIN 17 441). 580-780 N/mm^2 TS; 13-15 El min; 245 Brin min.

REMANIT 4034
Thyssen Edelstahlwerke AG
Stainless steel. C 0.42-0.5, Cr 12.5-14.5, bal Fe.
Hot worked flat and long products. Werkstoff Nr. 1.4034 (DIN 17 440). 800 N/mm^2 TS max; 225 Brin; 55 Rock C.

REMANIT 4057
Thyssen Edelstahlwerke AG
Stainless steel. C 0.14-0.23, Cr 15.5-17.5, Ni 1.5-2.5, bal Fe.
Hot worked flat and long products. Werkstoff Nr. 1.4057 (DIN 17 440). 550 N/mm^2 YP; 750-950 N/mm^2 TS; 5-14 El; 280 Brin.

REMANIT 4104
Thyssen Edelstahlwerke AG
Stainless steel. C 0.1-0.17, Cr 15.5-17.5, Mo 0.2-0.6, S 0.15-0.35, bal Fe.
Hot worked flat and long products. Werkstoff Nr. 1.4104 (DIN 17 440). 450 N/mm^2 YP; 540-840 N/mm^2 TS; 11-16 El; 230 Brin.

REMANIT 4105
Thyssen Edelstahlwerke AG
Stainless steel. C 0-0.06, Cr 16.5-18.5, Mo 0.2-0.6, S 0.15-0.35, bal Fe.
Hot worked flat and long products. Werkstoff Nr. 1.4105 (DIN 17 440). 270 N/mm^2 YP; 450-650 N/mm^2 TS; 20 El; 190 Brin.

REMANIT 4109
Thyssen Edelstahlwerke AG
Stainless steel. C 0.6-0.75, Cr 13-15, Mo 0.5-0.6, bal Fe.
Hot worked flat and long products. Werkstoff Nr. 1.4109. 950 N/mm^2 TS max; 260 Brin; 57 Rock C.

REMANIT 4112
Thyssen Edelstahlwerke AG
Stainless steel. C 0.85-0.95, Cr 17-19, Mo 0.9-1.3, V 0.07-0.12, bal Fe.
Hot worked flat and long products. Werkstoff Nr. 1.4112. 900 N/mm^2 TS; 255 Brin; 58 Rock C.

REMANIT 4113
Thyssen Edelstahlwerke AG
Stainless steel. C 0-0.08, Cr 16-18, Mo 0.9-1.3, bal Fe.
Cold rolled flat products. Werkstoff Nr. 1.4113 (DIN 17 441). 360 N/mm^2 YP; 580 N/mm^2 TS; 26-32 El.

REMANIT 4116

Thyssen Edelstahlwerke AG

Stainless steel. C 0.42-0.5, Cr 13.8-15, Mo 0.45-0.6, V 0.1-0.15, bal Fe.

Hot worked flat and long products. Werkstoff Nr. 1.4116 (DIN 17 440). 900 N/mm^2 TS; 270 Brin; 55 Rock C.

REMANIT 4120

Thyssen Edelstahlwerke AG

Stainless steel. C 0.17-0.22, Cr 12-14, Mo 0.9-1.3, Ni 0-1, bal Fe.

Hot worked flat and long products. Werkstoff Nr. 1.4120. 550 N/mm^2 YP; 750-900 N/mm^2 TS; 14 El; 225-245 Brin.

REMANIT 4122

Thyssen Edelstahlwerke AG

Stainless steel. C 0.33-0.43, Cr 15.5-17.5, Mo 0.8-1.3, Ni 0-1, bal Fe.

Hot worked flat and long products. Werkstoff Nr. 1.4122. 550 N/mm^2 YP; 750-900 N/mm^2 TS; 12 El; 260-275 Brin.

REMANIT 4125

Thyssen Edelstahlwerke AG

Stainless steel. C 0.95-1.2, Cr 16-18, Mo 0.4-0.8, bal Fe.

Hot worked flat and long products. Werkstoff Nr. 1.4125. 900 N/mm^2 TS; 270 Brin; 59 Rock C.

REMANIT 4301

Thyssen Edelstahlwerke AG

Stainless steel. C 0-0.07, Cr 17-19, Ni 8.5-10.5, bal Fe.

Cold rolled flat products. Werkstoff Nr. 1.4301 (DIN 17 441). 290-320 N/mm^2 YP; 680 N/mm^2 TS; 45-60 El.

REMANIT 4301

Thyssen Edelstahlwerke AG

Stainless steel. C 0-0.07, Cr 17-19, Ni 8.5-10.5, bal Fe.

Hot worked flat and long products. Werkstoff Nr. 1.4301 (DIN 17 440). 195-230 N/mm^2 YP; 500-700 N/mm^2 TS; 35-45 El.

REMANIT 4303

Thyssen Edelstahlwerke AG

Stainless steel. C 0-0.07, Cr 17-19, Ni 11-13, bal Fe.

Hot worked flat and long products. Werkstoff Nr. 1.4303 (DIN 17 440). 185-220 N/mm^2 YP; 490-690 N/mm^2 TS; 45 El.

REMANIT 4305

Thyssen Edelstahlwerke AG

Stainless steel. C 0-0.12, Cr 17-19, Ni 8-10, S 0.15-0.35, P 0-0.06, bal Fe.

Hot worked flat and long products. Werkstoff Nr. 1.4305 (DIN 17 441). 195-230 N/mm^2 YP; 500-700 N/mm^2 TS; 35 El.

REMANIT 4306

Thyssen Edelstahlwerke AG

Stainless steel. C 0-0.03, Cr 18-20, Ni 10-12.5, bal Fe.

Cold rolled flat products. Werkstoff Nr. 1.4306 (DIN 17 441). 285 N/mm^2 YP; 565 N/mm^2 TS; 55-70 El.

REMANIT 4306

Thyssen Edelstahlwerke AG

Stainless steel. C 0-0.03, Cr 18-20, Ni 10-12.5, bal Fe.

Hot worked flat and long products. Werkstoff Nr. 1.4306 (DIN 17 440). 180-350 N/mm^2 YP; 460-850 N/mm^2 TS; 20-45 El.

REMANIT 4311

Thyssen Edelstahlwerke AG

Stainless steel. C 0-0.03, Cr 17-19, Ni 8.5-11.5, N 0.12-0.22, bal Fe.

Hot worked flat and long products. Werkstoff Nr. 1.4311 (DIN 17 440). 270-305 N/mm^2 YP; 550-760 N/mm^2 TS; 30-40 El.

REMANIT 4313

Thyssen Edelstahlwerke AG

Stainless steel. C 0-0.05, Cr 12.5-14, Mo 0.4-0.7, Ni 3.5-4.5, Si 0-0.6, bal Fe.

Hot worked flat and long products. Werkstoff Nr. 1.4313 SEW 400. 550-850 N/mm^2 YP; 750-1200 N/mm^2 TS; 11-17 El; 270-310 Brin.

REMANIT 4401

Thyssen Edelstahlwerke AG

Stainless steel. C 0-0.07, Cr 16.5-18.5, Mo 2-2.5, Ni 10.5-13.5, bal Fe.

Cold rolled flat products. Werkstoff Nr. 1.4401 (DIN 17 441). 310 N/mm^2 YP; 610 N/mm^2 TS; 53-62 El.

REMANIT 4401

Thyssen Edelstahlwerke AG

Stainless steel. C 0-0.07, Cr 16.5-18.5, Mo 2-2.5, Ni 10.5-13.5, bal Fe.

Hot worked flat and long products. Werkstoff Nr. 1.4401 (DIN 17 440). 205-240 N/mm^2 YP; 500-710 N/mm^2 TS; 30-40 El.

REMANIT 4404

Thyssen Edelstahlwerke AG

Stainless steel. C 0-0.03, Cr 16.5-18.5, Mo 2-2.5, Ni 11-14, bal Fe.

Cold rolled flat products. Werkstoff Nr. 1.4404 (DIN 17 441). 310 N/mm^2 YP; 590 N/mm^2 TS; 51-60 El.

REMANIT 4404

Thyssen Edelstahlwerke AG

Stainless steel. C 0-0.03, Cr 16.5-18.5, Mo 2-2.5, Ni 11-14, bal Fe.

Hot worked flat and long products. Werkstoff Nr. 1.4404 (DIN 17 440). 190-350 N/mm^2 YP; 490-850 N/mm^2 TS; 20-40 El.

REMANIT 4406

Thyssen Edelstahlwerke AG

Stainless steel. C 0-0.03, Cr 16.5-18.5, Mo 2-2.5, Ni 10.5-13.5, N 0.12-0.22, bal Fe.

Hot worked flat and long products. Werkstoff Nr. 1.4406 (DIN 17 440). 280-315 N/mm^2 YP; 580-800 N/mm^2 TS; 35-40 El.

REMANIT 4418

Thyssen Edelstahlwerke AG

Stainless steel. C 0-0.05, Cr 15-16.5, Mo 0.9-1.5, Ni 5-6, bal Fe.

Hot worked flat and long products. Werkstoff Nr. 1.4418 SEW 400. 550-850 N/mm^2 YP; 830-1250 N/mm^2 TS; 11-16 El; 260-385 Brin.

REMANIT 4420

Thyssen Edelstahlwerke AG

Stainless steel. C 0-0.03, Cr 16.5-18.5, Mo 2.5-3, Ni 11.5-14.5, N 0.14-0.22, bal Fe.

Hot worked flat and long products. Werkstoff Nr. 1.4429 (DIN 17 440). 295-330 N/mm^2 YP; 580-800 N/mm^2 TS; 30-40 El.

REMANIT 4435

Thyssen Edelstahlwerke AG

Stainless steel. C 0-0.03, Cr 17-18.5, Mo 2.5-3, Ni 12.5-15, bal Fe.

Cold rolled flat products. Werkstoff Nr. 1.4435 (DIN 17 441). 320 N/mm^2 YP; 630 N/mm^2 TS; 48-58 El.

REMANIT 4435

Thyssen Edelstahlwerke AG

Stainless steel. C 0-0.03, Cr 17-18.5, Mo 2.5-3, Ni 12.5-15, bal Fe.

Hot worked flat and long products. Werkstoff Nr. 1.4435 (DIN 17 440). 190-350 N/mm^2 YP; 490-850 N/mm^2 TS; 20-40 El.

REMANIT 4436

Thyssen Edelstahlwerke AG

Stainless steel. C 0-0.07, Cr 16.5-18.5, Mo 2.5-3, Ni 11-14, S 0-0.025, bal Fe.

Cold rolled flat products. Werkstoff Nr. 1.4436 (DIN 17 441). 320 N/mm^2 YP; 630 N/mm^2 TS; 48-58 El.

REMANIT 4436

Thyssen Edelstahlwerke AG

Stainless steel. C 0-0.07, Cr 16.5-18.5, Mo 2.5-3, Ni 11-14, bal Fe.

Hot worked flat and long products. Werkstoff Nr. 1.4436 (DIN 17 440). 205-240 N/mm^2 YP; 510-710 N/mm^2 TS; 40 El.

REMANIT 4438

Thyssen Edelstahlwerke AG

Stainless steel. C 0-0.03, Cr 17.5-19.5, Mo 3-4, Ni 14-17, bal Fe.

Hot worked flat and long products. Werkstoff Nr. 1.4438 (DIN 17 440). 195-230 N/mm^2 YP; 490-690 N/mm^2 TS; 35 El.

REMANIT 4439

Thyssen Edelstahlwerke AG

Stainless steel. C 0-0.03, Cr 16.5-18.5, Mo 4-5, Ni 12.5-14.5, N 0.12-0.22, bal Fe.

Cold rolled flat products. Werkstoff Nr. 1.4439 (DIN 17 441). 315-345 N/mm^2 YP min; 600-800 N/mm^2 TS; 35-40 El.

REMANIT 4439

Thyssen Edelstahlwerke AG

Stainless steel. C 0-0.03, Cr 16.5-18.5, Mo 4-5, Ni 12.5-14.5, N 0.12-0.22, bal Fe.

Hot worked flat and long products. Werkstoff Nr. 1.4439 (DIN 17 440). 285-315 N/mm^2 YP; 580-800 N/mm^2 TS; 35 El.

REMANIT 4460

Thyssen Edelstahlwerke AG

Stainless steel. C 0-0.05, Cr 25-28, Mo 1.3-2, Ni 4.5-6, N 0.05-0.2, bal Fe.

Hot worked flat and long products. Werkstoff Nr. 1.4460 (DIN 17 440). 490 N/mm^2 YP; 600-800 N/mm^2 TS; 20 El; 220 Brin.

REMANIT 4462

Thyssen Edelstahlwerke AG

Stainless steel. C 0-0.03, Cr 21-23, Mo 2.5-3.5, Ni 4.5-6.5, N 0.08-0.2, bal Fe.

Cold rolled flat products. Werkstoff Nr. 1.4462 (SEW 400). 480-550 N/mm^2 YP min; 680-880 N/mm^2 TS; 25 El min.

REMANIT 4462

Thyssen Edelstahlwerke AG

Stainless steel. C 0-0.03, Cr 21-23, Mo 2.5-3.5, Ni 4.5-6.5, N 0.08-0.2, bal Fe.

Hot worked flat and long products. Werkstoff Nr. 1.4462 (DIN 17 440). 450-480 N/mm^2 YP; 680-880 N/mm^2 TS; 20-30 El.

REMANIT 4465

Thyssen Edelstahlwerke AG

Stainless steel. C 0-0.02, Cr 24-26, Mo 2-2.5, Ni 22-25, N 0.08-0.16, bal Fe.

Hot worked flat and long products. Werkstoff Nr. 1.4465 SEW 400. 255-295 N/mm^2 YP; 540-740 N/mm^2 TS; 30-40 El.

REMANIT 4505

Thyssen Edelstahlwerke AG

Stainless steel. C 0-0.05, Cr 16.5-18.5, Mo 2-2.5, Ni 19-21, Cu 1.8-2.2, Nb = 8 x C min, bal Fe.

Hot worked flat and long products. Werkstoff Nr. 1.4505 SEW 400. 225-265 N/mm^2 YP; 490-740 N/mm^2 TS; 30-40 El.

REMANIT 4509

Thyssen Edelstahlwerke AG

Stainless steel. C 0-0.03, Cr 17.5-19, Ti 0.1-0.5, Nb 0.6-0.9, bal Fe.

Cold rolled flat products. Werkstoff Nr. 1.4509. 320 N/mm^2 YP; 420-600 N/mm^2 TS; 20-25 El.

REMANIT 4510

Thyssen Edelstahlwerke AG

Stainless steel. C 0-0.05, Cr 16-18, Ti 0-1.2, Ti = 7 x C min, bal Fe.

Cold rolled flat products. Werkstoff Nr. 1.4510 (DIN 17 441). 310 N/mm^2 YP; 480 N/mm^2 TS; 30-45 El.

REMANIT 4510

Thyssen Edelstahlwerke AG

Stainless steel. C 0-0.08, Cr 16-18, Ti 0-1.2, Ti = 7 x C min, bal Fe.

Hot worked flat and long products. Werkstoff Nr. 1.4510 (DIN 17 440). 270 N/mm^2 YP; 450-600 N/mm^2 TS; 20 El; 170 Brin.

REMANIT 4511
Thyssen Edelstahlwerke AG
Stainless steel. C 0-0.05, Cr 16-18, Ti 0-1.2, Ti = 12 x C min, bal Fe.
Cold rolled flat products. Werkstoff Nr. 1.4511 (DIN 17 441). 350 N/mm² YP; 580 N/mm² TS; 22-34 El.

REMANIT 4512
Thyssen Edelstahlwerke AG
Stainless steel. C 0-0.05, Cr 10.5-12.5, Ti 0-1, Ti = 6 x C min, bal Fe.
Cold rolled flat products. Werkstoff Nr. 1.4512 (DIN 17 441). 280 N/mm² YP; 440 N/mm² TS; 29-42 El.

REMANIT 4512
Thyssen Edelstahlwerke AG
Stainless steel. C 0-0.05, Cr 10.5-12.5, Ti 0-1, Ti = 6 x C min, bal Fe.
Hot worked flat and long products. Werkstoff Nr. 1.4512 SEW 400. 220 N/mm² YP; 330-560 N/mm² TS; 20 El.

REMANIT 4522
Thyssen Edelstahlwerke AG
Stainless steel. C 0-0.02, N 0-0.2, Cr 17-19, Mo 1.8-2.3, Nb 0-1.2, Nb = 15 x (C + N) min, bal Fe.
Hot worked flat and long products. Werkstoff Nr. 1.4522 SEW 400. 320 N/mm² YP; 450-650 N/mm² TS; 20 El; 170 Brin.

REMANIT 4539
Thyssen Edelstahlwerke AG
Stainless steel. C 0-0.02, N 0-0.15, Cr 19-21, Mo 4-5, Ni 24-26, Cu 1-2, bal Fe.
Hot worked flat and long products. Werkstoff Nr. 1.4539 SEW 400. 220-250 N/mm² YP; 520-720 N/mm² TS; 35-40 El.

REMANIT 4541
Thyssen Edelstahlwerke AG
Stainless steel. C 0-0.08, Cr 17-19, Ni 9-12, Ti = 5 x C (0.8 min), bal Fe.
Cold rolled flat products. Werkstoff Nr. 1.4541 (DIN 17 441). 320-350 N/mm² YP; 620 N/mm² TS; 52-58 El.

REMANIT 4541
Thyssen Edelstahlwerke AG
Stainless steel. C 0-0.08, Cr 17-19, Ni 9-12, Ti 0-0.8, Ti = 5 x C min, bal Fe.
Hot worked flat and long products. Werkstoff Nr. 1.4541 (DIN 17 440). 200-240 N/mm² YP; 500-730 N/mm² TS; 30-40 El.

REMANIT 4542
Thyssen Edelstahlwerke AG
Stainless steel. C 0-0.07, Cr 15.5-17.5, Ni 3-5, Cu 3-5, Nb 0.15-0.45, bal Fe.
Hot worked flat and long products. Werkstoff Nr. 1.4542. 724-1172 N/mm² YP; 931-1310 N/mm² TS; 10-16 El; 266-380 Brin.

REMANIT 4550
Thyssen Edelstahlwerke AG
Stainless steel. C 0-0.08, Cr 17-19, Ni 9-12, Nb 0-1, Nb = 10 x C min, bal Fe.
Hot worked flat and long products. Werkstoff Nr. 1.4550 (DIN 17 440). 205-240 N/mm² YP; 510-740 N/mm² TS; 30-40 El.

REMANIT 4563
Thyssen Edelstahlwerke AG
Stainless steel. C 0-0.02, Cr 26-28, Mo 3-4, Ni 30-32, Cu 0.8-1.5, N 0-0.15, bal Fe.
Hot worked flat and long products. Werkstoff Nr. 1.4563 SEW 400. 210-240 N/mm² YP; 500-750 N/mm² TS; 35-40 El.

REMANIT 4571
Thyssen Edelstahlwerke AG
Stainless steel. C 0-0.08, Cr 16.5-18.5, Mo 2-2.5, Ni 10.5-13.5, Ti 0-0.8, Ti = 5 x C min, bal Fe.
Cold rolled flat products. Werkstoff Nr. 1.4571 (DIN 17 441). 320-350 N/mm² YP; 620 N/mm² TS; 50-56 El.

REMANIT 4571
Thyssen Edelstahlwerke AG
Stainless steel. C 0-0.08, Cr 16.5-18.5, Mo 2-2.5, Ni 10.5-13.5, Ti 0-0.8, Ti = 5 x C min, bal Fe.
Hot worked flat and long products. Werkstoff Nr. 1.4571 (DIN 17 440). 210-250 N/mm² YP; 500-730 N/mm² TS; 30-35 El.

REMANIT 4575
Thyssen Edelstahlwerke AG
Stainless steel. C 0-0.02, Cr 27-29, Mo 2-3, Ni 3-4.5, Nb + Zr = 5 x (C + N) min, bal Fe.
Cold rolled flat products. Werkstoff Nr. 1.4575. 500 N/mm² YP min; 600-750 N/mm² TS; 16 El min.

REMANIT 4575
Thyssen Edelstahlwerke AG
Stainless steel. C 0-0.02, Cr 27-29, Mo 2-3, Ni 3-4.5, Nb + Zr = 10 x (C + N) min, bal Fe.
Hot worked flat and long products. Werkstoff Nr. 1.4575. 500 N/mm² YP min; 600-750 N/mm² TS; 16 El.

REMANIT 4589
Thyssen Edelstahlwerke AG
Stainless steel. C 0-0.08, Cr 13.5-15.5, Mo 0.2-1.2, Ni 1-2.5, Ti 0.3-0.5, bal Fe.
Cold rolled flat products. Werkstoff Nr. 1.4589. 450 N/mm² YP; 650 N/mm² TS; 20-22 El.

REMANIT 8A
Thyssen Edelstahlwerke AG
C 0.06, Cr 18, Ni 11, Ti 0.3, bal Fe.
Annealed: 85,000 TS; 35,000 YS; 55 El; B 80 Rock. For welded structures, chemical plant equipment, tanks. Type 321 stainless steel, stabilized, austenitic. *Obsolete*

REMANIT ATS 1
Thyssen Edelstahlwerke AG
C 0.1, Mn 1-1.5, Si 0.3-0.6, Cr 15-17, Ni 12-14, Cb 1.2, bal Fe.
Annealed: 85,000 TS; 35,000 YS; 60 El; B 80 Rock. For welded structures and tanks, chemical plant equipment, digesters. Stainless steel, austenitic, stabilized for welding. *Obsolete*

REMANIT G FEDER HARD
Thyssen Edelstahlwerke AG
C 0.5, Cr 29, Mo 4.8, Co 62, bal Fe.
For heat and corrosion resistant castings; corrosion and heat resistant. *Obsolete*

REMANIT HB
Thyssen Edelstahlwerke AG
C 0.15, Si 1, Mn 1, Mo 32, Ni 64, bal Fe.
For heat and corrosion resistant castings; heat and corrosion resistant. *Obsolete*

REMANIT HC
Thyssen Edelstahlwerke AG
C 0.15, Cr 15, Mo 17, Ni 60, W 5, bal Fe.
For heat and corrosion resistant castings, pumps, valves. *Obsolete*

REMENDUR
Bell Telephone Laboratories
Co 49, V 2-5, bal Fe.
Remanence: 21,500 Gausses. Coercive force 20-60 oersteds. Residual induction 16,000-21,500 Gausses. For magnetic devices, telephone switches. Exhibits square hysteresis loop. Non-directional properties, high curie temperature. Permanent magnet. High residual induction.

REMENDUR 27
Wilbur B. Driver Co.
React and refract. V 2.7, Co 48.5, bal Fe.
For self latching reed switches. Saturation induction 20,000 Gausses. Coercive force 27 oersteds, magnetically semihard.

REMENDUR 38
Wilbur B. Driver Co.
React and refract. V 3.5, Co 48.5, bal Fe.
Bias magnets for latching relays. Saturation induction 20,000 Gausses. coercive force 38 oersteds, magnetically semihard.

REMENDUR 48
Wilbur B. Driver Co.
React and refract. V 4.2, Co 48, bal Fe.
Bias magnets. Saturation induction 20,000 Gausses. Coercive force 48 oersteds, magnetically semihard.

REMIRAL
German manufacture
Mg 0.5-2, bal Al.
H25 Temper: 30,000 TS; 25,000 YS. For reflectors, automotive trim; corrosion resistant. Bright aluminum sheets.

REMIRAL-100
Aluminium Walzwerke Singen GmbH
Aluminum. Mg 1, bal Al.
Annealed: 100 MPa TS; 25 El. Hard: 200 MPa TS; 2 El. For ornamental and decorative parts.

REMIRAL-100
Swiss Aluminium Ltd.
Aluminum. Si 0.06, Fe 0.04, Cu 0.01, Mn 0.03, Mg 0.8-1.1, Cr 0.01, Zn 0.04, Ti 0.01, bal Al.
Extrusion alloy in billet form for electrical conductors. 100-130 N/mm² TS; 30-80 N/mm² YS; 13-15 El; 30 Brin.

REMOUNT
George Cook & Co., Ltd.
C 1.3, Mn 0.4, Si 0.2, V 0.1, bal Fe.
For wire drawing dies, cold forming tools; water hardened; Type W2. *Obsolete*

REMY 1.3912
Remystahl
INVAR. For production, transportation and storage of liquid gases.

REMY 1.4539
Remystahl
C 0-0.03, Cr 20, Mo 4.5, Ni 25, Cu 1.5, bal Fe.
Corrosion resistant. AISI 904, ASTM B625.

REMY 1.4826
Remystahl
INCOLOY 800 H. Employed where maximum creep-rupture strength is required.

REMY 1.4864
Remystahl
INCOLOY DS. For furnace building industry.

REMY 1.4876
Remystahl
INCOLOY 800. For pigtails, headers, collectors and manifolds, transfer piping in steam and hydrocarbon reforming.

REMY 1.4980
Remystahl
C 0-0.08, Cr 15, Mo 1.25, Ni 25.5, V 0.3, Al 0-0.35, Ti 2.1, B 0.003-0.01, bal Fe.
For high temperatures, gas turbines. ASTM A-453. Grade 660.

REMY 1.4986
Remystahl
C 0.04-0.1, Cr 16.5, Mo 1.8, Ni 16.5, Nb + Ta = 10 x C min (1.20 max), bal Fe.
High temperature, steam boilers, turbines, especially nuts and bolts.

REMY 1.6562
Remystahl
C 0.4, Al 0.025, Cr 0.8, Mo 0.35, Ni 1.8, bal Fe.
Special bolts and nuts. AISI 4340.

REMY 1.6909
Remystahl
C 0-0.07, Mn 8.5, N 0.25, Cr 18, Ni 5.5, bal Fe.
Tough, austenitic with higher tensile strength, but slightly less corrosion resistant. For nuts, bolts, pressure vessels. Also for temperatures from +20 to -196°C.

REMY 2.4066
Remystahl
NICKEL 200. For chemical and food industry, loading plants, equipment for aircraft and rockets.

REMY 2.4068
Remystahl
NICKEL 201. For caustic evaporators, plating rods, combustion boats, chemical plants.

REMY 2.4360
Remystahl
MONEL 400. For valves and pumps, propeller shafts, electronic and electrical parts.

REMY 2.4375
Remystahl
MONEL K 500. For cardan and pump shafts, petroleum drilling, bolts, doctor blades, valve spindles.

REMY 2.4610
Remystahl
HASTELLOY C 4. For chemical process industry equipment and components for FGD systems.

REMY 2.4617
Remystahl
HASTELLOY B 2. For processes involving sulfuric, hydrochloric and phosphoric acids.

REMY 2.4619
Remystahl
HASTELLOY G 3. For flue gas desulfurization systems, air pollution control systems, evaporators, heat exchangers, tank liners, etc.

REMY 2.4634
Remystahl
NIMONIC 105. For chemical industry, aircraft and rockets.

REMY 2.4642
Remystahl
INCONEL 690. Good against corrosive attack by alkaline oxides and molten glasses and silicates.

REMY 2.4660
Remystahl
ALLOY 20 CB 3. For heat exchangers, pickling vats, pump shafts.

REMY 2.4663
Remystahl
INCONEL 617. For gas turbine components, air heaters, furnace parts and radiant tubes, high temperature sections of nuclear process heat prototype plants.

REMY 2.4665
Remystahl
HASTELLOY X. For casings and turbine blades for gas turbines, components for heat treatment furnaces.

REMY 2.4668
Remystahl
INCONEL 718. For high-strength structural parts in nuclear power plants, high-temperature components for gas turbines.

REMY 2.4669
Remystahl
INCONEL X 750. For structural parts and springs for nuclear power plants, high-temperature components for gas turbines.

REMY 2.4816
Remystahl
INCONEL 600. For muffels, heat-exchanger tubes, combustion chamber liners, nuclear reactors, plants for the food and chemical industry.

REMY 2.4819
Remystahl
HASTELLOY C 276. For components in organic chloride processes, pulp and paper industry, FGD plants.

REMY 2.4851
Remystahl
INCONEL 601. For heat treating baskets and fixtures, trays, muffels, retorts, flame shields, burner nozzles, etc.; space and processing industry.

REMY 2.4856
Remystahl
INCONEL 625. For flue gas scrubbers, superphosphoric acid production, nuclear waste evaporators, thrust reverser systems.

REMY 2.4858
Remystahl
INCOLOY 825. For oil and gas coolers, equipment and components in sour gas service.

REMY 2.4869
Remystahl
ALLOY 80/20. For electric furnaces and kilns.

REMY 2.4951
Remystahl
NIMONIC 75. For gas turbine casing and components in heat treatment furnaces.

REMY 2.4952
Remystahl
NIMONIC 80 A. For exhaust valves and other components for combustion engines, bolts for land-based gas turbines, components in jet engines.

REMY 2.4969
Remystahl
NIMONIC 90. For gas turbines blades, springs for application at high temperatures.

REMY AMN
Remystahl
C 0.9-1.05, Mn 1-1.2, Cr 0.5-0.7, W 0.5-0.7, V 0.1, bal Fe. Oil hardened, cold work tool steel. W.-Nr. 1.2510; AISI O1.

REMY CE6
Remystahl
C 0.15, Cr 0.65, Mn 0.5, bal Fe.
For gears, bolts, machine tool parts; case hardened. *Obsolete*

REMY CE8
Remystahl
C 0.16, Cr 0.95, Mn 1.15, bal Fe.
For gears, cams, camshafts; case hardened, tough. *Obsolete*

REMY CEMO10
Remystahl
C 0.24, Cr 1.15, Mo 0.25, Mn 0.55, bal Fe.
For gears, bolts, machine tool parts; oil hardened, shock resistant. *Obsolete*

REMY CEMO8
Remystahl
C 0.22, Cr 1.2, Mo 0.25, Mn 0.65, bal Fe.
For gears, bolts, machine tool parts; case hardened, tough. *Obsolete*

REMY CEN4
Remystahl
C 0.19, Cr 1.25, Mo 0.2, Ni 3.75, bal Fe.
For gears, bolts, camshafts, cams; case hardened, shock resistant. *Obsolete*

REMY CNLK
Remystahl
C 0.45, Cr 1.4, Mo 0.7, V 0.3, Mn 0.75, bal Fe.
For gears, bolts, machine tool parts; oil hardened, shock resistant. W.-Nr. 1.2323.

REMY DS
Remystahl
C 0.45, Cr, V, Si, bal Fe.
For springs, gears, crankshafts, bolts, studs; oil hardened, shock resistant. *Obsolete*

REMY EXTRA 100
Remystahl
C 1, Si 0-0.25, Mn 0-0.25, bal Fe.
Annealed: 100,000 TS; 53,000 YS; 21 El; 42 RA; 200 Brin. For drills, taps, cutters, springs; Type W1; water hardened. *Obsolete*

REMY EXTRA 120
Remystahl
C 1.15, Si 0-0.25, Mn 0-0.25, bal Fe.
Annealed: 110,000 TS; 56,000 YS; 18 El; 40 RA; 210 Brin. For springs, taps, reamers, cutters, hobs; Type W1; water hardened. *Obsolete*

REMY EXTRA 130
Remystahl
C 1.3, Si 0-0.25, Mn 0-0.25, bal Fe.
For engravers' tools, taps, drills, cutters; Type W1; water hardened. *Obsolete*

REMY EXTRA 50
Remystahl
C 0.45, Si 0.25-0.5, Mn 0.3-0.8, bal Fe.
Hot rolled: 98,000 TS; 59,000 YS; 24 El; 45 RA; 212 Brin. For gears, bolts, machine tool parts; water hardened. *Obsolete*

REMY EXTRA 60
Remystahl
C 0.6, Si 0.25-0.5, Mn 0.3-0.8, bal Fe.
Heat treated: 160,000 TS; 113,000 YS; 12 El; 40 RA; 325 Brin. For gears, rails, springs, machine tool parts; water hardened. *Obsolete*

REMY EXTRA 70
Remystahl
C 0.7, Si 0-0.25, Mn 0-0.25, bal Fe.
Heat treated: 175,000 TS; 128,000 YS; 12 El; 37 RA; 352 Brin. For rails, punches, crimpers, axes; Type W1; water hardened. *Obsolete*

REMY EXTRA 80
Remystahl
C 0.85, Si 0-0.25, Mn 0-0.25, bal Fe.
Heat treated: 190,000 TS; 145,000 YS; 12 El; 35 RA; 390 Brin. For tools, springs, cutters, drills, taps; Type W1; water hardened. *Obsolete*

REMY EXTRA EDEL 10
Remystahl
C 1, Si 0-0.25, Mn 0-0.25, bal Fe.
Annealed: 100,000 TS; 53,000 YS; 20 El; 40 RA; 200 Brin. For springs, tools, drills, taps, broaches; Type W1; water hardened. *Obsolete*

REMY EXTRA EDEL 10 SPEZIAL
Remystahl
C 1.25, Si 1.15, Cr 1.2, Mn 0.7, bal Fe.
For bearings, dies, cutters; water or oil hardened. *Obsolete*

REMY EXTRA EDEL 12
Remystahl
C 1.1, Si 0-0.25, Mn 0-0.25, bal Fe.
Annealed: 110,000 TS; 55,000 YS; 18 El; 40 RA; 210 Brin. For springs, tools, cutters, taps, reamers, drills; Type W1; water hardened. *Obsolete*

REMY EXTRA EDEL 7
Remystahl
C 0.7, Si 0-0.25, Mn 0-0.25, bal Fe.
Heat treated: 175,000 TS; 128,000 YS; 12 El; 37 RA; 355 Brin.
For rails, tools, punches, axes, crimpers; Type W1; water hardened. *Obsolete*

REMY EXTRA EDEL 8
Remystahl
C 0.85, Si 0-0.25, Mn 0-0.25, bal Fe.
Heat treated: 190,000 TS; 145,000 YS; 10 El; 30 RA; 400 Brin.
For cutters, drills, taps, springs, reamers; Type W1; water hardened. *Obsolete*

REMY KP
Remystahl
C 0.5, Cr 1.05, Ni 3.25, bal Fe.
For gears, bolts, crankshafts; oil hardened, shock resistant. W.-Nr. 1.2721.

REMY LDZ SPEZIAL
Remystahl
C, alloy, bal Fe.
For machine tool parts; oil hardened. *Obsolete*

REMY MS70
Remystahl
C 1.15, Cr 0.65, V 0.1, Mn 0.3, bal Fe.
For dies, tools, punches, bearings; water hardened, wear resistant. *Obsolete*

REMY MS75
Remystahl
C 1.15, Cr 0.65, V 0.1, bal Fe.
For cutters, forming and piercing dies; oil or water hardened, wear resistant. *Obsolete*

REMY NV50
Remystahl
C 0.42, Mn 1.75, V 0.1, bal Fe.
For punches, shears, gears, crankshafts; oil hardened, tough. *Obsolete*

REMY ONLK
Remystahl
C 0.45, Cr 1.4, Mo 0.7, Mn 0.7, bal Fe.
For dies, gears, bolts, crankshafts; oil hardened, tough. W.-Nr. 1.2323.

REMY P6R
Remystahl
C 0.62, Si 0.9, Mn 0.9, Cr 0.2-0.4, bal Fe.
For gears, bolts, upsetters, shears, crimpers; water or oil hardened, tough. W.-Nr. 1.2826.

REMY S30
Remystahl
C 0.7, Si 1.7, Mn 0.7, bal Fe.
Heat treated: 340,000 TS; 283,000 YS; 5 El; 20 RA; 600 Brin.
For punches, chisels, springs, shear blades; oil hardened. *Obsolete*

REMY SPEZIAL 1 GESTO
Remystahl
C 1.05, Mn 0.07, Si 0.25, Cr 1.1, bal Fe.
For bearings, liners, sleeves, cutters; water or oil hardened, wear resistant.

REMY SPEZIAL 12
Remystahl
C 1.25, Si 1.15, Mn 0.7, Cr 1.2, bal Fe.
For bearings, cold work tools, liners, sleeves; oil hardened. *Obsolete*

REMY SPEZIAL 13
Remystahl
C 1.05, Cr 1, Mn 0.3, Si 0.25, bal Fe.
For bearings, cold work tools, liners, sleeves; water hardened, wear resistant. *Obsolete*

REMY SPEZIAL 14
Remystahl
C 0.5, Mn 0.95, Cr 1.05, V 0.1, bal Fe.
For springs, gears, bolts, punches; oil hardened, shock resistant. *Obsolete*

REMY SPEZIAL 14H
Remystahl
C 0.58, Cr 1, V 0.09, Mn 0.95, bal Fe.
For springs, gears, bolts, crankshafts; oil hardened, shock resistant. *Obsolete*

REMY SPEZIAL 2
Remystahl
C 1.4-1.6, Mn 0.6, Cr 1.4, bal Fe.
For bearings, cutters, blanking and forming dies; oil hardened, wear and abrasion resistant. W.-Nr. 1.2063.

REMY SPEZIAL 3
Remystahl
C 1.05, Cr 1, W 1.15, Mn 0.8-1.1, bal Fe.
For heading and blanking dies; oil hardened. W.-Nr. 1.2419.

REMY SPEZIAL 4
Remystahl
C 1.4, Cr 0.3, V 0.1, Mn 0.3, bal Fe.
For blanking and forming dies, engravers' tools; water hardened, wear resistant. *Obsolete*

REMY SPEZIAL 6
Remystahl
C 0.35, Cr 1.05, V 0.18, W 1.95, Si 0.9, bal Fe.
For forging and heading dies, die casting dies; oil hardened, tough. W.-Nr. 1.2541.

REMY SPEZIAL 7
Remystahl
C 0.45, Si 0.9, Cr 1.05, V 0.18, W 1.95, bal Fe.
For forging and heading dies; oil hardened, tough. W.-Nr. 1.2542.

REMY SPEZIAL 8
Remystahl
C 1.05, Cr 1.2, bal Fe.
For cutters, bearings, liners, sleeves; oil or water hardened. W.-Nr. 1.2067.

REMY WM SUPRA
Remystahl
C 0.36-0.42, Si 0.9-1.2, Cr 4.8-5.5, Mo 1.1-1.4, V 0.25-0.5, bal Fe.
Hot work tool steel; forging dies. W.-Nr. 1.2343; similar to AISI H11.

REMY WM SUPRA V
Remystahl
C 0.4, Si 1, Cr 5.2, Mo 1.3, V 1, bal Fe.
Hot work tool steel; extrusion dies, piercing mandrels, forging dies. W.-Nr. 1.2344; similar to AISI H13.

REMY WM SUPRA WO
Remystahl
C 0.36, Si 1, Cr 5.3, Mo 1.5, W 1.3, V 0.3, bal Fe.
Air hardened, hot work tool steel. W.-Nr. 1.2606; AISI H12.

REMY WP 3
Remystahl
C 0.33, Cr 3, Mo 2.8, V 0.5, bal Fe.
Hot work tool steel; pressing dies. W.-Nr. 1.2365; similar to AISI H10.

REMY WP12
Remystahl
C 0.3, Cr 2.4, Co 2, V 0.25, W 8.5, bal Fe.
For extrusion rams, liners, punches; hot work steel, oil hardened. W.-Nr. 1.2662.

REMY WP4
Remystahl
C 0.3, Cr 1.1, V 0.18, W 3.75, Si 1, bal Fe.
For punches, forging dies, upsetters; oil hardened. W.-Nr. 1.2564.

REMY WPK45
Remystahl
C 0.3, Cr 2.35, V 0.6, W 4.2, bal Fe.
For extrusion rams and liners; oil hardened, tough. W.-Nr. 1.2567.

REMY WPN
Remystahl
C 0.3, Si 1, Cr 1.1, V 0.18, W 3.75, bal Fe.
For heading and forming dies, punches; oil hardened. *Obsolete*

REMY WS1
Remystahl
C 1.2, W, bal Fe.
For bearings, heading dies, shear blades; oil or water hardened. *Obsolete*

REMY WZK SPEZIAL
Remystahl
C 0.3, V 0.35, Cr 2.65, W 8.5, bal Fe.
For extrusion rams and liners, punches, shears; hot work steel, oil hardened. W.-Nr. 1.2581.

RENAL
J. & A. Erbsloh Aluminium
Aluminum. Al 99.99.
Soft: 13,000 TS; 5000 YS; 45 El; 23 Brin. For electrical conductors; high electrical conductivity. *Obsolete*

RENAULT ALLOY
French manufacture
Al 88, Cu 2, Zn 10.
For light alloy parts; non-hardenable.

RENAULT AT.12
Regie Nationale des Usines Renault
C 1.9, Mn 0.3, Ni 0.7, Cr 13, Mo 0.3, bal Fe.
For tools, dies; nondeforming.

RENAULT AT.13
Regie Nationale des Usines Renault
C 1.9, Mn 0.4, Ni 0.5, Cr 12.5, Mo 0.6, W 0.8, Co 3.2, bal Fe.
For tools, dies; nondeforming.

RENAULT AT.14
Regie Nationale des Usines Renault
C 0.58, Mn 0.67, Ni 1.45, Cr 0.67, Mo 0.3, V 0.12, bal Fe.
For forging dies, tools; oil hardening.

RENAULT AT.2
Regie Nationale des Usines Renault
C 0.8, Ni 0.3, Cr 5, Mo 0.7, W 18.5, V 1.3, bal Fe.
For broaches, cutters, drills; high speed steel.

RENAULT AT.25
Regie Nationale des Usines Renault
C 1.03, Mn 1.82, Ni 0.1, V 0.13, Cr 0.1, bal Fe.
For tools, dies; nondeforming.

RENAULT AT.26
Regie Nationale des Usines Renault
C 0.8, Mn 0.3, Ni 0.1, bal Fe.
For tools, drills, hobs, springs, punches; water hardened.

RENAULT AT.3
Regie Nationale des Usines Renault
C 0.8, Si 0.3, Ni 0.3, Cr 4.5, Mo 6, W 6.7, V 1.6, bal Fe.
For tools, cutters, reamers, broaches; high speed steel.

RENAULT AT.35
Regie Nationale des Usines Renault
C 0.3, Mn 0.6, Ni 0.6, Cr 3.3, Mo 0.5, W 12, V 0.1, Co 2.5, bal Fe.
For dies, tools; hot work steel.

RENAULT AT.4
Regie Nationale des Usines Renault
C 0.8, Ni 0.2, Cr 5, Mo 1.4, W 19.5, V 2, Co 5.2, bal Fe.
For tools, cutters, hobs, reamers; high speed steel.

RENAULT AT.7
Regie Nationale des Usines Renault
C 1.2, Mn 0.35, Ni 0.2, Cr 1.35, bal Fe.
For tools; water hardening.

RENAULT AT.9
Regie Nationale des Usines Renault
C 0.4, Mn 0.4, Ni 4.9, Cr 0.4, Mo 1.2, bal Fe.
For gears, shafts; tough.

RENAULT G
Regie Nationale des Usines Renault
C 0.45, Si 0-1.15, Mn 0.7, Ni 14, Cr 14, Mo 0.3, W 2.25, bal Fe.
For exhaust valves; heat resistant.

RENAULT H
Regie Nationale des Usines Renault
C 0.34, Si 0.4, Mn 0.75, Ni 1.4, Cr 0.92, bal Fe.
For gears, shafts, pinions; tough.

RENAULT P
Regie Nationale des Usines Renault
C 1.25, Mn 0.55, Ni 0.45, Cr 14, Mo 1.2, Co 1.2, bal Fe.
For valves; heat resistant.

RENAULT R
Regie Nationale des Usines Renault
C 0.55, Si 0 0.35, Mn 0.45, Ni 0.4, Cr 14, bal Fe.
For pump parts; corrosion resistant.

RENAULT S
Regie Nationale des Usines Renault
C 0.14, Mn 0.75, Ni 1.4, Cr 1, bal Fe.
For gears, shafts; case-hardening.

RENAULT WA
Regie Nationale des Usines Renault
C 0.4, Mn 0.7, Ni 0-0.5, Cr 1, Mo 0.3, bal Fe.
For gears, shafts; oil hardening.

RENAULT WB
Regie Nationale des Usines Renault
C 0.35, Mn 0.7, Ni 0-0.4, Cr 1, Mo 0.3, bal Fe.
For gears, shafts; oil hardening.

RENAULT WE
Regie Nationale des Usines Renault
C 0.36, Mn 0.7, Ni 0-0.4, Cr 1, Mo 0.3, bal Fe.
For gears, shafts; oil hardening.

RENAULT WZ
Regie Nationale des Usines Renault
C 0.42, Mn 0.7, Ni 0.5, Cr 1, Mo 0.3, bal Fe.
For gears, shafts; oil hardening.

RENE
Enpar Sonderwerkstoffe GmbH
Nickel.
Alternate manufacturer.

RENE 100
Cannon-Muskegon Corp.
Ti 4-4.4, Al 5.3-5.7, Cr 9-10, Co 14-16, Mo 2.7-3.3, Fe 0-1, B 0.02, C 0.2, V 1, Zr 0.03-0.09, bal Ni.
Cast: 147,000 TS; 123,000 YS; 9 El; 30-44 Rock C. At 1500°F: 144,000 TS; 118,000 YS; 6 El; 7.2 RA. For turbine blades operating to 1900°F. Excellent high temperature rupture strength and long time stability.

RENE 125
Cannon-Muskegon Corp.
C 0.1, Cr 9, Co 10, Mo 2, W 7, Ti 2.6, Al 4.8, B 0.015, Zr 0.05, Ta 3.8, Hf 1.6, bal Ni.
Turbine blade alloy.

RENE 41
Now ALLVAC RENE 41, CRUCIBLE RENE 41.

RENE 41
Cannon-Muskegon Corp.
C 0.12, Cr 19, Co 11, Mo 10, Ti 3, Al 1.5, bal Ni.
Heat treated: 160,000-206,000 TS; 120,000-154,000 YS; 14-18 El. For jet engine components, after-burners, turbine wheels, bolts. Age-hardenable, high heat and corrosion resistance.

RENE 41 WIRE
National Standard Co.. C 0.06-0.12, Cr 18-20, Mo 9-10.5, Fe 5, Co 10-12, Al 0.5-1.8, Ti 3-3.3, B 0-0.01, Ni 53.
For high temperature springs in 1100-1500°F range.

RENE 62
Manufacturer not listed.
Al 1.1-1.4, Cb 2.1-2.5, Ti 2.35-2.63, Mo 8.5-9.5, Cr 13.5-16.5, Fe 21-24, C 0.02-0.08, Mn 0-0.25, B 0.01, bal Ni.
At RT: 205,000 TS; 160,000 YS; 6 El. At 1400°F: 145,000 TS; 117,000 YS; 20 El. For turbine components, jet engines, turbine wheels and frames. Precipitation hardening, good weldability. High corrosion, heat and oxidation resistance.

RENE 63
Manufacturer not listed.
C 0.1, Cr 14, Co 15, Mo 6, W 3, Ti 2.5, Al 3.8, B 0.015, bal Ni.
Heat treated: 210,000 TS; 148,000 YS; 18 El. At 1400°F: 155,000 TS; 138,000 YS; 4-9 El. For jet engine and gas turbine parts, high temperature bolting. High rupture strength. Corrosion, heat and oxidation resistant.

RENE 77
Cannon-Muskegon Corp.
Cr 15, Mo 5, Ti 3.5, Al 4.4, B 0.02, C 0.06, Co 15, Zr 0.04, bal Ni.
Cast: 125,000 TS; 110,000 YS; 8 RA. At 1600°F: 85,000 TS; 75,000 YS; 12 RA. For jet engine turbine blades. High creep-rupture strength to 1800°F. Resists hot sulfidation corrosion.

RENE 80
Cannon-Muskegon Corp.
Cr 14, Co 9.5, Mo 4, W 4, Ti 5, Al 3, C 0.17, Zr 0.03, B 0.015, bal Ni.
Cast alloy; for turbine blades. Very good hot corrosion resistance.

RENE 85
Cannon-Muskegon Corp.
C 0.27, Cr 9.3, Co 15, Mo 3.25, W 5.35, Ti 3.25, Al 5.25, B 0.015, Zr 0.03, bal Ni.
Compressor disc alloy; cast.

RENE 95
Cannon-Muskegon Corp.
C 0.015, Cr 14, Co 8, Mo 3.5, W 3.5, Cb 3.5, Ti 2.5, Al 3.5, B 0.01, Zr 0.05, bal Ni.
Turbine or compressor disc alloy, cast.

RENNITE
Rennie Tool Co. Ltd.
C 0.7, Cr 4, W 18, V 2, Co 12, bal Fe.
For lathe and planer tools, hobs, taps, reamers, broaches; high speed steel, oil hardened.

RENNITE
B.A. Field Co.
C 0.9, W 3.5, bal Fe.
For cutting tools. Oil hardened.

RENNY D150 MEISSELSTAHL
Remystahl
C 0.38, Si, Cr, V, bal Fe.
For gears, bolts, machine tool parts; oil hardened, tough.
Obsolete

RENOWN
Latrobe Steel Co.
C 1, Cr 0.1, V 0.2, bal Fe.
For stamping dies, fine edged tools; water hardened.
Obsolete

RENYX, AL BASE
Allied Die Casting Corp.
Aluminum. Al 91.5, Ni 4, Cu 4, Si 0.5.
For die castings; corrosion resisting.

RENYX, ZN BASE
Allied Die Casting Corp.
Zinc. Al 4, Cu 3, Zn 92, Mn 0.1.
For die castings; corrosion resisting.

REO (W2, GRADE 3)
Time Steel Service Inc.
C 1, Mn 0.35, Si 0.2, bal Fe.
Water hardened tool steel; AISI W2.

REO EXTRA
Time Steel Service Inc.
C 1, Mn 0.35, Si 0.2, V 0.2, bal Fe.
Water hardened tool steel. AISI W2.

REO SPECIAL (W2, GRADE 1)
Time Steel Service Inc.
C 1, Mn 0.35, Si 0.2, bal Fe.
Water hardened tool steel, AISI W2.

REOSTENE
English manufacture
Ni, Fe.
For resistances; heat resistant.

REPAROX XFC
Eutectic Corp.
Electrode for torch joining steel, cast iron, copper base alloys. Bonding temperature: 1400-1600°F; 65,000 psi TS.
Obsolete

REPUBLIC 07
Republic Steel Corp.
C 0.8, W 2.5, bal Fe.
Formerly Republic Velvet. For finishing tools for non-ferrous alloys, blanking dies; semi-high speed steel. *Obsolete*

REPUBLIC 100-AR
Gulf States Steel, Inc.
HSLA steel. C 0.16, Mn 1.4, P 0.015, S 0.015, Si 0.25, Cu 0.7, Ni 1.4, Cr 1.4, Mo 0.45, Al 0.04, bal Fe.
Abrasion resistant high strength steel plate for wear and liner plates in mining equipment, agricultural equipment, truck and heavy equipment industries. 100,000 YS min; 140,000 TS min; 16 El min.

REPUBLIC 6-H-W
Republic Steel Corp.
C 0.6, Cr 4, Mo 0.5, V 0.8, Ni 0.3, bal Fe.
For gripper and forming dies, flying shears; hot work steel.
Obsolete

REPUBLIC 65
Republic Steel Corp.
C 0.12, Mn 0.45, Si 0.2, Cu 1, Ni 1.3, Mo 0.2, bal Fe.
Rolled: 89,000 TS; 67,000 YS; 18 El. For welded structures, truck trailers, mine cars; tough, good weldability. *Obsolete*

REPUBLIC 70
Gulf States Steel, Inc.
HSLA steel. C 0.2, Mn 1, P 0.04, S 0.04, Si 0.15, Cu 1-1.5, Mo 0.2-0.3, Ni 1.2-1.75, bal Fe.
Tough, atmospheric corrosion resistant precipitation hardening steel. 70,000 YS min; 90,000 TS min; 18 El.

REPUBLIC A 2
Republic Steel Corp.
C 1, Cr 5.25, V 0.25, Mo 1.15, bal Fe.
For cold work dies, forming rolls, shear blades. Type A2 air hardening tool steel. *Obsolete*

REPUBLIC A 4
Republic Steel Corp.
C 1, Mn 2, Cr 0.9, Mo 0.9, bal Fe.
Type A4 air hardening tool steel. *Obsolete*

REPUBLIC ACME
Republic Steel Corp.
C 0.3-0.5, Si 0.3, Mn 0.5, bal Fe.
For armature shafts, spindles, gears, pinions; water hardened. *Obsolete*

REPUBLIC D2
Republic Steel Corp.
C 1.55, Cr 12, Mo 0.75, V 0.25, bal Fe.
For drawing and blanking dies, punches, shear blades; cold work steel. *Obsolete*

REPUBLIC D3
Republic Steel Corp.
C 2.2, Cr 12, V 0.25, bal Fe.
For drawing and blanking dies, shear blades; nondeforming. *Obsolete*

REPUBLIC DOUBLE STRENGTH GRADE 1-A
Republic Steel Corp.
C 0-0.3, Mn 0.5-1, Cu 0.5-1.5, Ni 0.5-1.5, Mo 0-0.2, bal Fe.
Normalized: 85,000 TS; 70,000 YS; 40 El. For sheets, strips, plates for transportation equipment; resists atmospheric corrosion. *Obsolete*

REPUBLIC DOUBLE STRENGTH GRADE 1
Republic Steel Corp.
C 0-0.12, Mn 0.5-1, Cu 0.5-1, Ni 0.5-1, 0.10% Mo min, bal Fe.
Hot rolled: 70,000 TS; 50,000 YS; 22 El. For sheets, strips, plates for transportation equipment; resists atmospheric corrosion. *Obsolete*

REPUBLIC EXTRA SPECIAL
Republic Steel Corp.
C 0.65-1.4, bal Fe.
For machine shop tools, milling cutters, punches, dies, taps, lathe and planer tools, reamers; water hardened. *Obsolete*

REPUBLIC FAST FINISHING
Republic Steel Corp.
C 1.35, W 3.5, bal Fe.
For tools, cutters; for fast finishing cuts and smooth finish. *Obsolete*

REPUBLIC H12
Republic Steel Corp.
C 0.35, Si 1, Cr 5, V 0.25, W 1.25, Mo 1.5, bal Fe.
For punching, piercing, forming and gripper dies. Type H12, hot work tool steel. *Obsolete*

REPUBLIC H13
Republic Steel Corp.
C 0.4, Si 1, Cr 5, V 1, Mo 1.2, bal Fe.
For die casting dies, forging punches, mandrels; resists heat checking. Type H13 hot work tool steel. *Obsolete*

REPUBLIC H14
Republic Steel Corp.
C 0.4, Mn 0.3, Si 1.15, Cr 5, W 5, Mo 0.25, Co 0.5, bal Fe.
Ultra high strength metal for severe hot working. Type M14 hot work tool steel. *Obsolete*

REPUBLIC H21
Republic Steel Corp.
C 0.3, Cr 3.5, W 9, V 0.5, bal Fe.
For hot punches, dies; air hardening. Type H21 hot work tool steel. *Obsolete*

REPUBLIC H21M
Republic Steel Corp.
C 0.3, Cr 2.75, Ni 1.75, W 9.5, Mo 0.25, bal Fe.
For die casting dies, hot work dies, hot piercers. Type H21 hot work tool steel. *Obsolete*

REPUBLIC H24
Republic Steel Corp.
C 0.5, Cr 3, W 14, V 0.5, bal Fe.
For blanking and forming dies, hot punches, extrusion mandrels; hot work steel for severe service. Type H24 hot work tool steel. *Obsolete*

REPUBLIC H26
Republic Steel Corp.
C 0.58, Mn 0.3, Si 0.3, Cr 4, V 1.1, W 18.25, bal Fe.
Ultra high strength metal for severe hot working. Type H26 hot work tool steel. *Obsolete*

REPUBLIC H42
Republic Steel Corp.
C 0.6, Cr 4, V 2, W 6, Mo 5, bal Fe.
For hot extrusion dies, hot forming dies, hot upsetting and piercing tools, hot shear blades. Type H42 hot work tool steel. *Obsolete*

REPUBLIC HP9-4-25
Republic Steel Corp.
C 0.25, Ni 8.5, Co 3.75, Cr 0.5, Mo 0.5, V 0.08, Mn 0.2, bal Fe.
Heat treated: 200,000-250,000 TS; 190,000-220,000 YS; 12-15 El; 55-65 RA. For solid fuel rocket motor cases, submarine hulls, pressure vessels, aircraft structural components. Weldable after heat treatment. *Obsolete*

REPUBLIC HP9-4-45
Republic Steel Corp.
C 0.45, Ni 8, Co 3.75, Cr 0.3, Mo 0.3, V 0.1, bal Fe.
Heat treated: 250,000-290,000 TS; 235,000-250,000 YS; 7-9 El; 30-40 RA; 469-543 Brin. For highly stressed rocket motor cases, aircraft structural components, landing gears, thin walled forged sections. High yield strength and toughness. *Obsolete*

REPUBLIC L2
Republic Steel Corp.
C 0.45-0.55, Mn 0.5-0.8, Cr 0.8-1, V 0.15-0.25, bal Fe.
Heat treated: 185,000-260,000 TS; 165,000-240,000 YS; 6-15 El; 17-45 RA. For drive shafts, spindles, rear axles, knuckles; shock resistant. *Obsolete*

REPUBLIC M1
Republic Steel Corp.
C 0.8, Cr 3.75, V 1.15, W 1.5, Mo 8.75, bal Fe.
For drills, reamers, lathe tools, milling cutters. Type M1 high speed tool steel. *Obsolete*

REPUBLIC M10
Republic Steel Corp.
C 0.85, Cr 4, V 2, Mo 8, bal Fe.
For drills, reamers, taps, counter bores, broaches, shear blades, lathe tools. Type M10 high speed tool steel. *Obsolete*

REPUBLIC M2
Republic Steel Corp.
C 0.79-0.86, W 6-6.7, Cr 4, V 1.9, Mo 5, bal Fe.
For tools, reamers, cutters, lathe and planer tools, milling cutters; high speed steel. *Obsolete*

REPUBLIC M3
Republic Steel Corp.
C 1.1, Cr 4, V 3, W 6, Mo 5, bal Fe.
For drills, counter bores, reamers, taps, lathe tools, hobs, milling cutters. Type M3 high speed tool steel. *Obsolete*

REPUBLIC M4
Republic Steel Corp.
C 1.28, Cr 4.5, V 4, W 5.5, Mo 4.5, bal Fe.
For drills, reamers, taps, threading dies, lathe tools. Type M4 high speed tool steel. *Obsolete*

REPUBLIC M43
Republic Steel Corp.
C 1.1, Mn 0.3, Si 0.3, Cr 4.25, V 2, W 1.75, Mo 9.25, Co 8.25, bal Fe.
For lathe tools and other tools operating at higher speed and higher temperature; holds edge at higher temperature. Type M43 high speed tool steel. *Obsolete*

REPUBLIC M7
Republic Steel Corp.
C 1, Mo 8.75, W 1.75, Cr 4, V 2, bal Fe.
For drills, reamers, broaches, lathe tools. Type M7 high speed tool steel. *Obsolete*

REPUBLIC O1
Republic Steel Corp.
C 0.9, Mn 1.2, Cr 0.5, W 0.5, bal Fe.
For dies, taps, gauges, broaches, reamers, hobs; non-deforming, cold work steel. *Obsolete*

REPUBLIC PLASTIC DIE
Republic Steel Corp.
C 0-0.08, Mn 0-0.1, bal Fe.
For plastic mold dies, hubbed cavity tools; water hardened, easy to hub. *Obsolete*

REPUBLIC RS-100
Republic Steel Corp.
C 0-0.2, Mn 3, Al 1.5, bal Ti.
Annealed: 110,000 TS; 100,000 YS; 12 El. For aircraft and jet engine components; heat resistant to 1000 F. *Obsolete*

REPUBLIC RS-110
Republic Steel Corp.
Cr 3-4, Fe 1-2, bal Ti.
Rolled: 120,000-160,000 TS; 110,000-130,000 YS; 8-12 El; 270-340 Brin. For jet engine components; corrosion and heat resistant. *Obsolete*

REPUBLIC RS-110A
Republic Steel Corp.
C 0-0.2, Mn 8, bal Ti.
Annealed: 120,000 min TS; 110,000 min YS; 10 min El. For aircraft and jet engine components; heat resistant, low density. *Obsolete*

REPUBLIC RS-110B
Republic Steel Corp.
C 0-0.1, Mn 3, Al 2.5, bal Ti.
Annealed: 120,000 min TS; 110,000 min YS; 10 min El. For aircraft and jet engine components; heat resistant, low density. *Obsolete*

REPUBLIC RS-110C
Republic Steel Corp.
C 0.05, N 0.017, Fe 0.2, Al 5, Sn 2.5, bal Ti.
Annealed: 115,000 TS; 110,000 YS; 10 El. For jet engine components; weldable. *Obsolete*

REPUBLIC RS-110C
Republic Steel Corp.
Al 5, Sn 2.5, bal Ti.
Rolled: 125,000 TS; 120,000 YS; 18 El; 40 RA; 320 Brin. For compressor blades, welded rings, aircraft components; good weldability. *Obsolete*

REPUBLIC RS-120
Republic Steel Corp.
Mn 6-8, C 0-0.2, 0.7% max others, bal Ti.
Heat treated: 130,000-170,000 TS; 120,000-160,000 YS; 7-18 El; 280-380 Brin. For aircraft skins, ducts, shrouds, jet engine parts; age-hardenable. *Obsolete*

REPUBLIC RS-120A
Republic Steel Corp.
C 0-0.1, Fe 0.15, Al 6, V 4, bal Ti.
Annealed: 130,000 TS; 120,000 YS; 10 El; 30 RA. For jet engine components, fasteners; good weldability, high notch toughness. *Obsolete*

REPUBLIC RS-130

Republic Steel Corp.
Al 4, Mn 4, bal Ti.
Annealed: 140,000 min TS; 130,000 min YS; 10 min El; 30 min RA. For aircraft and jet engine components; heat resistant, low density. *Obsolete*

REPUBLIC RS-135

Republic Steel Corp.
C 0-0.1, Al 6.3-7.3, Mo 3.5-4.5, Fe 0.1-0.2, bal Ti.
Heat treated: 185,000 TS; 165,000 YS; 9 El; 28 RA. Annealed: 156,000 TS; 147,000 YS; 14 El. 44 RA. For jet engine and missile components; heat treatable, used up to 1100 F. *Obsolete*

REPUBLIC RS-140

Republic Steel Corp.
Cr 2.75, Fe 1.25, Al 5, C 0.08, bal Ti.
Annealed: 161,000 TS; 157,000 YS; 16 El; 40 RA; 370 Brin. For jet engine components; good hot strength to 1000 F. *Obsolete*

REPUBLIC RS-40

Republic Steel Corp.
C 0-0.2, bal Ti.
Annealed: 50,000 TS; 40,000 YS; 22 El. For non-structural aircraft parts; for moderate to severe forming. *Obsolete*

REPUBLIC RS-40

Republic Steel Corp.
Ti.
Rolled: 65,000 TS; 50,000 YS; 28 El; 50 RA. For low stressed aircraft parts; unalloyed titanium. *Obsolete*

REPUBLIC RS-55

Republic Steel Corp.
C 0-0.2, Fe 0.2, N 0.012, bal Ti.
Annealed: 65,000 TS; 55,000 YS; 20 El; 30 RA. For non-structural aircraft parts; for moderate to severe forming. *Obsolete*

REPUBLIC RS-55

Republic Steel Corp.
C 0-0.2, H 0-0.015, N 0-0.08, bal Ti.
Rolled: 75,000 TS; 65,000 YS; 25 El; 50 RA. For non-structural aircraft parts; unalloyed titanium, good formability. *Obsolete*

REPUBLIC RS-6AL-4V

Republic Steel Corp.
Al 6, V 4, bal Ti.
Annealed: 130,000 min TS; 120,000 min YS; 10 min El; 30 min RA. For aircraft and jet engine components; heat resistant, low density. *Obsolete*

REPUBLIC RS-70

Republic Steel Corp.
C 0-0.2, N 0.015, Fe 0.2, bal Ti.
Annealed: 80,000 TS; 70,000 YS; 15 El; 30 RA. For moderate stressed aircraft parts; resists corrosion fatigue in sea water. *Obsolete*

REPUBLIC RS-70

Republic Steel Corp.
C 0-0.15, H 0-0.015, N 0-0.07, bal Ti.
Rolled: 90,000 TS; 80,000 YS; 20 El; 40 RA. For moderately stressed aircraft parts; unalloyed titanium. *Obsolete*

REPUBLIC S1

Republic Steel Corp.
C 0.5, W 2.2, Cr 1.2, V 0.25, bal Fe.
For hot punches, and dies, shear blades, pneumatic tools; hot work steel. *Obsolete*

REPUBLIC S3

Republic Steel Corp.
C 0.45-0.55, Cr 0.9-1.1, W 1-1.25, bal Fe.
For hand chisels, rivet sets, beading tools, cold cutters, shear blades; shock resistant. *Obsolete*

REPUBLIC S4

Republic Steel Corp.
C 0.55, Mn 0.9, Si 1.8-2.2, Cr 0.3, V 0.3, bal Fe.
For cold punches, shears, pneumatic tools; tough, shock resistant. *Obsolete*

REPUBLIC T1

Republic Steel Corp.
C 0.72, Mn 0.3, W 18, Cr 4, V 1, bal Fe.
For lathe tools, milling cutters, taps. Type T1 high speed tool steel. *Obsolete*

REPUBLIC T2

Republic Steel Corp.
C 0.8, W 19, Cr 4, V 2, Mo 0.75, bal Fe.
For tools, lathe cutters, roll turning tools, reamers, broaches; high speed steel. *Obsolete*

REPUBLIC T4

Republic Steel Corp.
C 0.72, W 18, Cr 4, V 1, Co 5.5, Mo 0.5, bal Fe.
For cutters for cast iron and hard, gritty surfaces, high speed steel. *Obsolete*

REPUBLIC TOOL STEEL

Republic Steel Corp.
C 0.65-1, bal Fe.
For ordinary tools, drills. *Obsolete*

REPUBLIC UA-6

Republic Steel Corp.
C 0.6-0.7, Mn 0.6, Cr 1, W 0.3, bal Fe.
For punches, dies, collets, shear blades, broaches; tough, shock resistant. *Obsolete*

REPUBLIC UA-8

Republic Steel Corp.
C 1, Mn 0.35, Cr 1.35, Mo 0.35, bal Fe.
For forming and header dies, cold punches; Type L7; water hardened. *Obsolete*

REPUBLIC UA-8

Republic Steel Corp.
C 1, Mn 0.35, Cr 1.35, Mo 0.35, bal Fe.
For swaging rolls and dies, hot press work; high resistance to sinking and caving. *Obsolete*

REPUBLIC UA-8

Republic Steel Corp.
C 1, Mn 0.35, Cr 1.35, Mo 0.35, bal Fe.
For dies, swager rolls and dies, cold coining dies; high resistance to sinking. *Obsolete*

REPUBLIC UNIQUE ALLOY

Republic Steel Corp.
C 0.5, Cr 0.9, V 0.2, bal Fe.
For shafts, bolts, screws, gears, axles, piston rods; resists torsional strains. *Obsolete*

REPUBLIC VELVET

Republic Steel Corp.
C 0.8, W 2.5, bal Fe.
For finishing tools for non-ferrous alloys, blanking dies, taps; semi-high speed steel. *Obsolete*

REPUBLIC W1

Republic Steel Corp.
C 0.65-1.2, Mn 0.3, Si 0.25, bal Fe.
For cold heading and forming tools, drills, punches; Type W1; water hardened. *Obsolete*

REPUBLIC W2

Republic Steel Corp.
C 0.85-1.6, V 0.3, bal Fe.
Formerly Republic Dumost. For cutting tools, dies, chisels, drills, shear blades; resists shock, fatigue and impact. *Obsolete*

REPUBLIC W5

Republic Steel Corp.
C 1.1, Cr 0.6, bal Fe.
For dies, punches, cold work tools; water hardening. *Obsolete*

REPUBLIC YA-4

Republic Steel Corp.
C 0.5, Mn 0.6, Cr 1, W 0.2-0.3, bal Fe.
For tools, blacksmith tools, machine tool parts; shock resistant, tough. *Obsolete*

RESILIA

Bethlehem Steel Corp.
C 0.7, Si 2, Mn 1.1, bal Fe.
For springs. *Obsolete*

RESISCO

IMI Knoch Ltd.
Cu 91, Al 7, Ni 2.
Drawn: 100,000 TS; 85,000 YS; 11 El; 205 Brin. For condensers, coolers; high corrosion resistance.

RESISCO (ALUMINUM-BRONZE)

IMI Yorkshire Alloys Ltd.
Al 6.5-7.5, bal Cu.
Annealed: 68,100 psi TS; 27,500 psi YS; 60 El; 110 DPN (106 Brin). For condenser tube.

RESISTAC

American Manganese Bronze Co.
Copper. Cu 88, Al 10, Fe 2.
For chemical apparatus; Al-Bronze.

RESISTAC

American Manganese Bronze Co.
Copper. Cu 90, Al 9, Fe 1.
Cast: 75,000 TS; 37,000 YS; 15 El; 20 RA; 135 Brin. Acid resisting castings and forgings; abrasion resisting.

RESISTAC

American Manganese Bronze Co.
Copper. Cu 89, Al 10, Fe 1.
Hot rolled: 100,000 TS; 60,000 YS; 6 El; 6 RA. For equipment, chemical apparatus; Al-Bronze.

RESISTAC NO. 1

American Manganese Bronze Co.
Copper. Al 8-11, Fe 0-5, Ni 0-5, bal Cu.
Cast or rolled: 65,000 TS; 28,000 YS; 20 El; 20 RA; 120 Brin. For castings, gears; corrosion resistant.

RESISTAC NO. 2

American Manganese Bronze Co.
Copper. Al 8-11, Fe 0-5, Ni 0-5, bal Cu.
Cast or rolled: 75,000 TS; 32,000 YS; 25 El; 25 RA; 150 Brin. For castings, gears; corrosion resistant.

RESISTAC NO. 3

American Manganese Bronze Co.
Copper. Al 8-11, Fe 0-5, Ni 0-5, bal Cu.
Cast: 90,000 TS; 42,000 YS; 15 El; 15 RA; 185 Brin. Forged: 95,000 TS; 45,000 YS; 17 El; 17 RA; 195 Brin. For castings, gears, pumps, impellers; corrosion resistant.

RESISTAL

English manufacture
Cu 90, Al 9, Fe 1.
75,000 TS; 37,000 YS; 20 El. For gears, slides; corrosion resistant.

RESISTAL P

Degussa AG
Precious metal.
Nonprecious metal alloy for dentistry and dental engineering.

RESISTALOY

Cerro Metal Products Co.
Cu 59, Al 2, Ni 1, bal Zn.
Cold drawn: 112,750 TS; 85,000 YS; 8.5 El; 11.0 RA; 205 Brin.
For shafts, bearings, bolts, nuts, studs, and high tensile
forgings. Resists sea water corrosion. *Obsolete*

RESISTALOY NO. 12

Cerro Metal Products Co.
Cu 59, Ni 1, Al 2, bal Zn.
Rolled: 95,000 TS; 58,000 YS; 190 Brin. For propeller shafts,
boat hardware, and marine applications. Nickel brass.
Obsolete

RESISTANCE NO. 1

Manufacturer not listed.
Ag 67, Pt 33.
For resistances; heat resistant.

RESISTANCE NO. 2

Manufacturer not listed.
Cu 85, Mn 12, Fe 3.
For resistances; heat resistant.

RESISTANCE NO. 3

Manufacturer not listed.
Cu 57, Zn 26, Ni 18.
For resistances; corrosion resistant.

RESISTANCE NO. 4

Manufacturer not listed.
Cu 56, Ni 26, Zn 18, Fe 1.
For resistances; corrosion resistant.

RESISTANCE, HIGH MAGNETIC

Manufacturer not listed.
Cr 70, Ni 30.
For magnet; high permeability.

RESISTANCE, LUNGE

English manufacture
Cu 84-87, Mn 12-14, Fe 1.8-1.9.
For electrical resistors.

RESISTHERM

Isabellenhuette
Nickel. Ni 70, Cr 1, Fe 29.
Annealed: 600 N/mm^2 TS. For electrical equipment and
instruments. Resistance alloy. Maximum working temperature
to 800°C.

RESISTIN-1

English manufacture
Fe 1.8, Mn 11.7, Cu 86.5.
For resistance alloy.

RESISTIN-2

English manufacture
Fe 3, Mn 12, Cu 85.
For resistance alloy.

RESISTO

Uddeholm Corp.
C 0.6, Si 1.85, Mn 0.7, Mo 0.45, V 0.2, bal Fe.
For shear blades, pneumatic chisels; shock resistant, oil
hardened. *Obsolete*

RESISTO-CAST

Grand Northern Products Ltd.
C 3.5, Si 3.2, Mn 0.6, S 0.067, P 0.41, Cr 0.29, Ni 1.33, Mo
0.2, Cu 0.15, bal Fe.
For oxyacetylene welding repair in cast iron castings.

RESISTO-LOY

Grand Northern Products Ltd.
C 3.5, Si 0.46, Mn 1, Cr 29.3, Mo 4, Cu 0.3, bal Fe.
Coated, hardfacing electrode. As arc welded: 100,000 psi TS
approx; 58-60 Rock C. For overlay or buildup on shovel teeth,
dredging tools, plow shares, cement mill rings; valves for
handling corrosive chemicals; good resistance to abrasion
and corrosion.

RESISTVAR 1

Hamilton Technology Inc.
Ni 45, Cu 55.
Electrical resistance alloy with negative temperature
coefficient of resistance. Electrical resistivity (nominal) 53
microcentimeters. Temperature coefficient of resistance (-18
to 100 C) -24 to -75 PPM/C. Thermal e.m.f. against copper:
42 microvolts/C. *Obsolete*

RESOLUT 1

Georgsmarienwerke Selesiastahl GmbH
C 2.1, Cr 11.5, bal Fe.
For blanking and forming dies, punches; oil or air hardened,
nondeforming.

RESOLUT 1E

Georgsmarienwerke Selesiastahl GmbH
C 2.1, Cr 11.5, W 0.7, bal Fe.
For blanking, piercing, and forming dies, punches; oil or air
hardened, nondeforming.

RESOLUT 1W

Georgsmarienwerke Selesiastahl GmbH
C 1.65, Cr 11.5, V 0.1, bal Fe.
For blanking, piercing, and forming dies, punches; air
hardened, nondeforming.

RESOLUT 2

Georgsmarienwerke Selesiastahl GmbH
C 0.9, Mn 1.9, Si 0.25, V 0.1, bal Fe.
For punches, cutters, crimpers, forming dies; oil hardened,
nondeforming.

RESOLUT 3

Georgsmarienwerke Selesiastahl GmbH
C 1.25, Si 1.15, Mn 0.7, Cr 1.2, bal Fe.
For bearings, liners, cutters; water hardened, wear resistant.

RESOLUT 4

Georgsmarienwerke Selesiastahl GmbH
C 1.45, Cr 1.4, Mn 0.6, bal Fe.
For bearings, bushings, liners, cutters; water hardened, wear
resistant.

RESOURCE

Osborn Steels Ltd.
C 0.3, Cr 13, bal Fe.
For plastic mold dies; corrosion resistant. *Obsolete*

RESULPHURIZED STOCK

U.S. Steel Corp.
Mn 0.6-0.9, C, 0.075% S min, bal Fe.
For threading steel; bars and rods for threading purposes.
Obsolete

RETORT

Edgar Allen Balfour Ltd.
C 0.3-0.35, Si 0.4-0.8, Mn 0.6-0.8, bal Fe.
For machinery parts, gears; castings.

REVALON

Revere Copper Products, Inc.
Cu 76, Zn 22, Al 2, As 0.05.
Hard: 85,000 TS; 60,000 YS; 10 El. Soft: 60,000 TS; 27,000
YS; 55 El. For marine parts; hardware; tough. *Obsolete*

REVERE

Revere Copper Products, Inc.
Cu alloy.
For processing equipment. *Obsolete*

REVERE 0-1

Revere Copper Products, Inc.
Al 8.5, Mn 0.2, Zn 0.5, bal Mg.
Forged: 46,000 TS; 32,000 YS; 10 El; 65 Brin. Heat treated:
50,000 TS; 34,000 YS; 6 El; 72 Brin. For light alloy forgings;
responds to heat treatment. *Obsolete*

REVERE 1100

Revere Copper Products, Inc.
99 Al min.
O-temper: 13,000 TS; 5000 YS; 45 El; 23 Brin. H 18-temper:
24,000 TS; 22,000 YS; 15 El; 44 Brin. For trim, housings,
containers, chemical equipment; commercially pure
aluminum. *Obsolete*

REVERE 1145

Revere Copper Products, Inc.
99.45 Al min.
For wrappings, electrical condensers; foil. *Obsolete*

REVERE 2014

Revere Copper Products, Inc.
Mn 0.8, Mg 0.4, Si 0.8, Cu 4.4, bal Al.
O-temper: 27,000 TS; 14,000 YS; 18 El; 45 Brin. T 4-temper:
62,000 TS; 44,000 YS; 20 El; 105 Brin. T 6-temper: 70,000 TS;
60,000 YS; 13 El; 135 Brin. For aircraft forgings, hardware,
structural fittings; heat treatable, high strength. *Obsolete*

REVERE 2017

Revere Copper Products, Inc.
Mn 0.5, Mg 0.5, Cu 4, bal Al.
O-temper: 26,000 TS; 10,000 YS; 22 El; 45 Brin. T 4-temper:
62,000 TS; 40,000 YS; 22 El; 105 Brin. For aircraft structures,
fittings, screw machine products; heat treatable, good
formability. *Obsolete*

REVERE 2024

Revere Copper Products, Inc.
Mn 0.6, Mg 1.5, Cu 4.5, bal Al.
O-temper: 27,000 TS; 11,000 YS; 22 El; 47 Brin. T 4-temper:
68,000 TS; 48,000 YS; 19 El; 120 Brin. For aircraft structures;
heat treatable. *Obsolete*

REVERE 3003

Revere Copper Products, Inc.
Mn 1.2, bal Al.
O-temper: 16,000 TS; 6000 YS; 40 El; 28 Brin. H 14-temper:
31,500 TS; 19,000 YS; 16 El; 40 Brin. H 18-temper: 29,000 TS;
26,000 YS; 10 El; 55 Brin. For structural work, tank cars,
cooking utensils, heat exchangers; resists atmosphere
corrosion. *Obsolete*

REVERE 3004

Revere Copper Products, Inc.
Mn 1.2, Mg 1, bal Al.
O-temper: 26,000 TS; 10,000 YS; 25 El; 45 Brin. H-34 temper:
34,000 TS; 27,000 YS; 12 El; 63 Brin. H 38-temper: 40,000 TS;
34,000 YS; 6 El; 77 Brin. For fuel lines, fan blades, roofing,
hydraulic tubing; high resistance to weather corrosion.
Obsolete

REVERE 3005

Revere Copper Products, Inc.
Si 0.6, Fe 0.7, Cu 0.3, Mn 1-1.5, Mg 0.2-0.6, Zn 0.25, Ti 0.1,
bal Al.
Wrought aluminum alloy. AA 3005. *Obsolete*

REVERE 3150

Revere Copper Products, Inc.
99.45 Al min.
O-temper: 12,000 TS; 4000 YS. H-16 temper: 18,000 TS;
16,000 YS. For electrical conductors. Was Revere E.C.
Obsolete

REVERE 5005

Revere Copper Products, Inc.
Mg 0.8, bal Al.
O-temper: 18,000 TS; 6000 YS; 30 El. H 16-temper: 26,000
TS; 25,000 YS; 5 El. H 32-temper: 29,000 TS; 27,000 YS; 5 El.
For light alloy parts; strip and tube. *Obsolete*

REVERE 5050

Revere Copper Products, Inc.
Mg 1.4, bal Al.
O-temper: 21,000 TS; 8000 YS; 24 El. H 32-temper: 25,000
TS; 21,000 YS; 9 El. H 38-temper: 32,000 TS; 29,000 YS; 6 El.
For light alloy parts; strip and tube. *Obsolete*

REVERE 5051
Revere Copper Products, Inc.
Mg 0.7, bal Al.
O-temper: 18,000 TS; 6,000 YS; 30 El. H-16 temper: 26,000 TS; 25,000 YS; 5 El. H-32 temper: 29,000 TS; 27,000 YS; 5 El. For anodized parts. *Obsolete*

REVERE 5052
Revere Copper Products, Inc.
Mg 2.5, Cr 0.25, bal Al.
O-temper: 27,000 TS; 12,000 YS; 30 El; 45 Brin. H 34-temper: 37,000 TS; 31,000 YS; 14 El; 67 Brin. H 38-temper: 41,000 TS; 36,000 YS; 8 El; 85 Brin. For camera cases, deck housings, gasoline tanks, aircraft components; resists sea water corrosion. *Obsolete*

REVERE 53 S
Revere Copper Products, Inc.
Mg 1.3, Cr 0.25, Si 0.7, bal Al.
O-temper: 16,000 TS; 7,000 YS; 35 El; 26 Brin. T 4-temper: 33,000 TS; 20,000 YS; 30 El; 65 Brin. T 6-temper: 39,000 TS; 33,000 YS; 20 El; 80 Brin. For aircraft forgings, crankcase nose pieces; heat treatable. *Obsolete*

REVERE 6061
Revere Copper Products, Inc.
Mg 1, Cr 0.25, Si 0.6, Cu 0.25, bal Al.
O-temper: 18,000 TS; 8000 YS; 30 El; 30 Brin. T 4-temper: 35,000 TS; 21,000 YS; 25 El; 65 Brin. T 6-temper: 45,000 TS; 40,000 YS; 17 El; 95 Brin. For structures, marine equipment, engine baffles; heat treatable, good weldability. *Obsolete*

REVERE 6062
Revere Copper Products, Inc.
Cu 0.25, Si 0.6, Mg 1, Cr 0.06, bal Al.
O-temper: 18,000 TS; 8000 YS; 30 El. T 4-temper: 35,000 TS; 21,000 YS; 25 El. T 6-temper: 45,000 TS; 40,000 YS; 7 El. For extrusions; age hardened. *Obsolete*

REVERE 6063
Revere Copper Products, Inc.
Mg 0.7, Si 0.4, bal Al.
F-temper: 22,000 TS; 15,000 YS; 20 El; 42 Brin. T 5-temper: 30,000 TS; 25,000 YS; 12 El; 65 Brin. T 6-temper: 35,000 TS; 30,000 YS; 12 El; 73 Brin. For light alloy parts, moldings, trim, architecture; heat treatable, good corrosion resistance. *Obsolete*

REVERE 6101
Revere Copper Products, Inc.
Fe 0-0.4, Si 0.3-0.6, Mg 0.4-0.8, bal Al.
Rolled: 29,000 TS; 25,000 YS. For light alloy parts; corrosion resistant. *Obsolete*

REVERE 7% ALUMINUM BRONZE
Revere Copper Products, Inc.
Al 6-8, Fe 1.5-3.5, Mn 0-1, bal Cu.
Soft: 76,000 TS; 33,000 YS; 40 El. Rolled: 78,000 TS; 45,000 YS; 35 El. For heat exchangers, desalination equipment, valves, fittings, marine applications, corrosion resistant tanks and piping. C61400

REVERE 7% ALUMINUM BRONZE - MODIFIED
Revere Copper Products, Inc.
Fe 2-3, Sn 0.2-0.5, Al 6-7.5, bal Cu.
Soft: 76,000 TS; 33,000 YS; 40 El. Rolled: 78,000 TS; 45,000 YS; 35 El. For heat exchangers, desalination equipment, valves, fittings, marine applications, corrosion resistant tanks and piping. C61300

REVERE ALLOY 464
Revere Copper Products, Inc.
Cu 59-62, Sn 0.5-1, bal Zn.
Hard: 75,000 TS; 53,000 YS; 20 El; B 82 Rock. Soft: 57,000 TS; 25,000 YS; 64 El; B 55 Rock. For propeller and pump shafts, piston rods, marine hardware, bolts, valve stems. Naval brass. Corrosion resistant. *Obsolete*

REVERE ALLOY NO. 102
Revere Copper Products, Inc.
Now REVERE OXYGEN FREE COPPER. C10200

REVERE ALLOY NO. 115
Revere Copper Products, Inc.
Cu 99.9, P 0.1.
For welding and brazing. *Obsolete*

REVERE ALLOY NO. 116
Revere Copper Products, Inc.
Now REVERE TOUGH PITCH COPPER. C11600

REVERE ALLOY NO. 116
Revere Copper Products, Inc.
Si 0.35, bal Cu.
For welding and brazing. *Obsolete*

REVERE ALLOY NO. 122
Revere Copper Products, Inc.
Cu 90, Te 0.15, bal Zn.
For engraver's plates; special engraver's commercial bronze. *Obsolete*

REVERE ALLOY NO. 150
Revere Copper Products, Inc.
Cu 75, bal Zn.
Hard: 90,000 TS; 64,000 YS; 6 El. Soft: 47,000 TS; 16,000 YS; 62 El. For springs; Spring Brass. *Obsolete*

REVERE ALLOY NO. 173
Revere Copper Products, Inc.
Cu 63, bal Zn.
Hard: 88,000 TS; 57,000 YS; 15 El. Soft: 47,000 TS; 16,000 YS; 60 El. For jewelry; common brass. *Obsolete*

REVERE ALLOY NO. 220A
Revere Copper Products, Inc.
Now REVERE COMMERCIAL BRONZE. C22000

REVERE ALLOY NO. 226
Revere Copper Products, Inc.
Now REVERE JEWELRY BRONZE. C22600

REVERE ALLOY NO. 230
Revere Copper Products, Inc.
Now REVERE RED BRASS. C23000

REVERE ALLOY NO. 232
Revere Copper Products, Inc.
Cu 65, Pb 2, bal Zn.
For scales; free-cutting. *Obsolete*

REVERE ALLOY NO. 233
Revere Copper Products, Inc.
Cu 64, Pb 2.5, bal Zn.
Hard: 87,000 TS; 65,000 YS; 4 El. Soft: 45,000 TS; 12,000 YS; 61 El. For stamped and machined parts; free-cutting. *Obsolete*

REVERE ALLOY NO. 254
Revere Copper Products, Inc.
Cu 60.5, Pb 1.2, bal Zn.
Hard: 80,000 TS; 45,000 YS; 40 El. Soft: 54,000 TS; 20,000 YS; 6 El. For screw machine products; free-cutting tube. *Obsolete*

REVERE ALLOY NO. 260
Revere Copper Products, Inc.
Now REVERE CARTRIDGE BRASS. C26000

REVERE ALLOY NO. 268
Revere Copper Products, Inc.
Cu 66, Zn 34.
Hard: 74,000 TS; 60,000 YS; 8 El; 80 Rock B. Soft: 47,000 TS; 15,000 YS; 62 El; 64 Rock F. For rivets, eyelets, electrical sockets; yellow brass. Was Revere 170. *Obsolete*

REVERE ALLOY NO. 270
Revere Copper Products, Inc.
Cu 65, bal Zn.
Hard: 90,000 TS; 55,000 YS; 15 El. Soft: 42,000 TS; 13,000 YS; 60 El. For rivets, fasteners; high strength. Was Revere 170A. *Obsolete*

REVERE ALLOY NO. 280
Revere Copper Products, Inc.
Now MUNTZ METAL. C28000

REVERE ALLOY NO. 280C
Revere Copper Products, Inc.
Cu 61.3, Pb 0.25, bal Zn.
For stamped parts; yellow brass. Was Revere 238. *Obsolete*

REVERE ALLOY NO. 309
Revere Copper Products, Inc.
Cu 95, Sn 1, bal Zn.
For jewelry; corrosion resistant. *Obsolete*

REVERE ALLOY NO. 310
Revere Copper Products, Inc.
Sn 5, P 0.2, bal Cu.
Spring temper: 91,000-105,000 TS; 4 El; 95 Rock B. Soft temper: 40,000-55,000 TS; 57-64 El. For diaphragms, bellows, springs, fuse clips, lock washers, clutch discs. High fatigue and corrosion resistance. Grade A Phosphor Bronze. Was Revere 308. *Obsolete*

REVERE ALLOY NO. 317
Revere Copper Products, Inc.
Zn 38, Pb 2, bal Cu.
Soft: 52,000 TS; 20,000 YS; 45 El; 72 Brin. For hardware, plumbing goods, forgings; extruded, forging brass. Was Revere 280. *Obsolete*

REVERE ALLOY NO. 325
Revere Copper Products, Inc.
Cu 73, Zn 24.5, Pb 2.5.
Hard: 80,000 TS; 75,000 YS; 5 El. Soft: 50,000 YS; 15,000 YS; 55 El. For hardware, machinery parts; free-cutting. Was Revere 215. *Obsolete*

REVERE ALLOY NO. 332
Revere Copper Products, Inc.
Cu 66.5, Pb 1.6, bal Zn.
For bushings, screw machine products; tubing, free-machining. Was Revere 222. *Obsolete*

REVERE ALLOY NO. 340
Revere Copper Products, Inc.
Cu 64.5, Pb 1, bal Zn.
Hard: 90,000 TS; 66,000 YS; 5 El. Soft: 47,000 TS; 12,000 YS; 60 El. For hardware, bolts, gears; free-cutting. Was Revere 227. *Obsolete*

REVERE ALLOY NO. 342
Revere Copper Products, Inc.
Cu 62, Pb 1.75, bal Zn.
Hard: 80,000 TS; 53,000 YS; 15 El. Soft: 50,000 TS; 17,000 YS; 48 El. For rivets, gears, clock parts; free-cutting. Was Revere 247. *Obsolete*

REVERE ALLOY NO. 342
Revere Copper Products, Inc.
Cu 62.5-66.5, Pb 1.5-2.5, Fe 0-0.1, bal Zn.
Hard: 80,000 TS; 60,000 YS; 5 El; B 85 Rock. Soft: 50,000 TS; 18,000 YS; 55 El; B 25 Rock. For clock and watch backs, engraver's plates, gears, channel plates. High leaded brass, high machinability. *Obsolete*

REVERE ALLOY NO. 344
Revere Copper Products, Inc.
Cu 63, Pb 0.75, bal Zn.
Hard: 67,000 TS; 42,000 YS; 19 El. Soft: 46,000 TS; 14,000 YS; 55 El. For machined cold headed parts; free-cutting. Was Revere 250. *Obsolete*

REVERE ALLOY NO. 350

Revere Copper Products, Inc.
Cu 83.5, Sn 1.5, bal Zn.
Hard: 91,000 TS; 70,000 YS; 4 El. Soft: 50,000 TS; 12,000 YS; 61 El. For pens, instruments; pen metal. *Obsolete*

REVERE ALLOY NO. 362

Revere Copper Products, Inc.
Cu 61.5, Pb 3.8, bal Zn.
Hard: 63,000 TS; 44,000 YS; 13 El. Soft: 43,000 TS; 14,000 YS; 52 El. For valve bonnets, screw machine products; free-cutting. Was Revere 252. *Obsolete*

REVERE ALLOY NO. 365

Revere Copper Products, Inc.
Now LEADED MUNTZ METAL. C36500

REVERE ALLOY NO. 380

Revere Copper Products, Inc.
Now REVERE ARCHITECTURAL BRONZE. C38000

REVERE ALLOY NO. 385

Revere Copper Products, Inc.
Cu 58, Pb 2.75, bal Zn.
Soft: 60,000 TS; 20,000 YS; 30 El. For architectural trim, hardware; free-cutting. Was Revere 277. *Obsolete*

REVERE ALLOY NO. 385

Revere Copper Products, Inc.
Cu 60, Pb 0.3, Sn 0.75, bal Zn.
Medium hard: 86,000 TS; 59,000 YS; 10 El. Soft: 63,000 TS; 26,000 YS; 35 El. For marine applications; Low Leaded Naval Brass. *Obsolete*

REVERE ALLOY NO. 411

Revere Copper Products, Inc.
Cu 90, Zn 9.5, Sn 0.5.
Hard: 65,000 TS; 57,000 YS; 5 El; 140 Brin. Soft: 45,000 TS; 10,000 YS; 42 El; 100 Brin. For bearings, lamp connections, weatherstrip; heavy duty. Was Revere 325. *Obsolete*

REVERE ALLOY NO. 411

Revere Copper Products, Inc.
Zn 9.5, Sn 0.5, bal Cu.
Hard: 65,000 TS; 62,000 YS; 5 El; B 76 Rock. Soft: 45,000 TS; 14,000 YS; 42 El; F 55 Rock. For bushings, weatherstrips, fuse clips. Low coefficient friction. Good formability and good corrosion resistance. *Obsolete*

REVERE ALLOY NO. 415

Revere Copper Products, Inc.
Cu 90, Sn 2, bal Zn.
For weather strip; corrosion resistant. Was Revere 327. *Obsolete*

REVERE ALLOY NO. 425

Revere Copper Products, Inc.
Cu 88, Sn 2, bal Zn.
For jewelry; corrosion resistant. Was Revere 337. *Obsolete*

REVERE ALLOY NO. 428

Revere Copper Products, Inc.
Cu 95.5, Ni 0.6, bal Zn.
Hard: 87,000 TS; 6 El. Soft: 49,000 YS; 34 El. For flatware, paper mill screens. *Obsolete*

REVERE ALLOY NO. 432

Revere Copper Products, Inc.
Cu 86, Sn 0.5, bal Zn.
For jewelry; corrosion resistant. Was Revere 336. *Obsolete*

REVERE ALLOY NO. 432

Revere Copper Products, Inc.
Cu 78, Si 1, bal Zn.
Hard: 109,000 TS; 73,000 YS; 4 El. Soft: 56,000 TS; 14,000 YS; 61 El. For oil coolers; Evaporator Brass No. 1. *Obsolete*

REVERE ALLOY NO. 440

Revere Copper Products, Inc.
Cu 61.2, Pb 0.75, Sn 0.3, bal Zn.
For architectural panels and trim; Muntz Metal, sheathing. *Obsolete*

REVERE ALLOY NO. 444

Revere Copper Products, Inc.
Zn 4, Pb 4, Sn 4, bal Cu.
For bearings, bushings; Four Forty Four Bronze. *Obsolete*

REVERE ALLOY NO. 445

Revere Copper Products, Inc.
Cu 71, P 0.05, Sn 1, bal Zn.
Hard: 10,000 TS; 80,000 YS; 3 El. Soft: 53,000 TS; 22,000 YS; 65 El. For condenser tubes. Phosphorized Admiralty Metal. Was Revere 362. *Obsolete*

REVERE ALLOY NO. 447

Revere Copper Products, Inc.
Cu 70, Si 0.5, bal Zn.
Hard: 100,000 TS; 63,000 YS; 4 El. Soft: 49,000 TS; 11,000 YS; 63 El. For refrigerators, evaporators; Evaporator Brass No. 2. *Obsolete*

REVERE ALLOY NO. 456

Revere Copper Products, Inc.
Cu 58, Sn 1, Fe 1, Si 0.1, bal Zn.
For bronze welding rod; working temperature 1600 F; low fuming. *Obsolete*

REVERE ALLOY NO. 464A

Revere Copper Products, Inc.
Cu 60, Zn 39.25, Sn 0.75.
Hard: 82,000 TS; 55,000 YS; 20 El. Soft: 60,000 TS; 22,000 YS; 45 El. For propeller and pump shafts, piston rods. Roman Bronze. Was Revere 380. *Obsolete*

REVERE ALLOY NO. 464B

Revere Copper Products, Inc.
Cu 62, Sn 0.75, bal Zn.
Hard: 10,000 TS; 70,000 YS; 4 El. Soft: 58,000 TS; 21,000 YS; 36 El. For marine applications. Hard Naval Brass. Was Revere 370. *Obsolete*

REVERE ALLOY NO. 465A

Revere Copper Products, Inc.
Cu 60.5, Pb 0.25, Mn 0.07, Sn 0.07, bal Zn.
For coal screens; high strength. Was Revere 443. *Obsolete*

REVERE ALLOY NO. 467

Revere Copper Products, Inc.
Cu 58.5, Fe 0.65, Mn 0.25, Sn 1, bal Zn.
For high strength applications. *Obsolete*

REVERE ALLOY NO. 482

Revere Copper Products, Inc.
Cu 60, Pb 0.75, Sn 0.75, bal Zn.
Hard: 86,000 TS; 59,000 YS; 10 El. Soft: 63,000 TS; 26,000 YS; 35 El. For tube headers. Leaded Naval Brass. Was Revere 387. *Obsolete*

REVERE ALLOY NO. 485

Revere Copper Products, Inc.
Cu 60, Zn 37.25, Pb 2, Sn 0.75.
Hard: 90,000 TS; 70,000 YS; 5 El. Soft: 58,000 TS; 20,000 YS; 35 El. For forgings, screw machine products. Leaded Naval Brass, free-cutting. Was Revere 389. *Obsolete*

REVERE ALLOY NO. 505

Revere Copper Products, Inc.
Sn 0-4, P 0.2, Cu 98.75.
Spring temper: 68,000-83,000 TS; 4 El; 74-85 Rock B. Soft: 34,000-50,000 TS; 47-50 El; 50-67 Rock F. For diaphragms, bellows, springs, fuse clips, lock washers, clutch discs. High fatigue and corrosion resistance. Grade E Phosphor Bronze. Was Revere 305. *Obsolete*

REVERE ALLOY NO. 521

Revere Copper Products, Inc.
Sn 8, P 0.2, bal Cu.
Spring temper: 105,000-118,500 TS; 3 El; 100 Rock B. Soft: 53,000-67,000 TS; 65-70 El; 20-70 Rock B. For diaphragms, bellows, springs, fuse clips, lock washers, clutch discs. High fatigue and corrosion resistance. Grade D Phosphor Bronze. Was Revere 315. *Obsolete*

REVERE ALLOY NO. 524

Revere Copper Products, Inc.
Sn 10, P 0.2, bal Cu.
Hard temper: 94,000-109,000 TS; 7 El; 100 Rock B. Soft: 58,000-73,000 TS; 62-68 El; 25-75 Rock B. For diaphragms, bellows, springs, fuse clips, lock washers, clutch discs. High fatigue and corrosion resistance. Grade D Phosphor Bronze. Was Revere 317. *Obsolete*

REVERE ALLOY NO. 530

Revere Copper Products, Inc.
Cu 66, Mn 0.25, Ni 20, bal Zn.
Hard: 86,000 TS; 78,000 YS; 3 El. Soft: 52,000 TS; 17,000 YS; 38 El. For springs, jewelry; 20% Nickel Silver. *Obsolete*

REVERE ALLOY NO. 562

Revere Copper Products, Inc.
Cu 57.5, Mn 0.25, Ni 15, bal Fe.
For ornaments, springs; 15% Nickel Silver. *Obsolete*

REVERE ALLOY NO. 574

Revere Copper Products, Inc.
Cu 61.5, Pb 2, Ni 10, Mn 0.25, bal Fe.
For ornaments, panels, jewelry; Leaded B Nickel Silver. *Obsolete*

REVERE ALLOY NO. 576

Revere Copper Products, Inc.
Cu 66, Pb 1, Ni 8.5, Mn 0.25, bal Zn.
For washers, jewelry, panels; Leaded Nickel Silver. *Obsolete*

REVERE ALLOY NO. 577

Revere Copper Products, Inc.
Cu 66, Ni 8, Mn 0.25, bal Zn.
For washers, jewelry, panels; Panel Stock. *Obsolete*

REVERE ALLOY NO. 578

Revere Copper Products, Inc.
Cu 61, Pb 2, Ni 9, Mn 0.25, bal Zn.
For washers, jewelry, panels; Leaded 9% Nickel Silver. *Obsolete*

REVERE ALLOY NO. 639

Revere Copper Products, Inc.
Cu 91, Al 7, Si 2.
Hard: 95,000 TS; 53,000 YS; 25 El. Soft: 85,000 TS; 43,000 YS; 35 El. For hardware, fittings, valve stems. Al bronze. Was Revere 436. *Obsolete*

REVERE ALLOY NO. 675

Revere Copper Products, Inc.
Cu 58, Fe 0.75, Mn 0.5, Sn 1, bal Zn.
Hard: 90,000 TS; 55,000 YS; 10 El; 185 Brin. Soft: 65,000 TS; 27,000 YS; 35 El; 120 Brin. For valve stems, clutch disks, pump parts. Mn bronze. Was Revere 454. *Obsolete*

REVERE ALLOY NO. 677

Revere Copper Products, Inc.
Cu 56.7, Pb 0.75, Ni 1.75, Mn 0.15, Fe 1.2, As 0.6, bal Zn.
Hard: 85,000 TS; 62,000 YS; 10 El. Soft: 65,000 TS; 26,000 YS; 40 El. For valve stems. Arsenical Bronze. Was Revere 469. *Obsolete*

REVERE ALLOY NO. 700

Revere Copper Products, Inc.
Cu 55, Ni 18, Zn 27.
Hard: 100,000 TS; 85,000 YS; 3 El; 91 Rock B. Soft: 60,000 TS; 27,000 YS; 40 El; 90 Rock F. For springs, trim, plumbing fixtures. Nickel silver 18% B. Was Revere 555. *Obsolete*

REVERE ALLOY NO. 706

American Injector Co.
Copper. Fe 1.4, Mn 0.3, Ni 10, bal Cu.
For condenser and heat exchanger tubes; 90-10 Cupro-Nickel. Was Revere 508.

REVERE ALLOY NO. 735

Revere Copper Products, Inc.
Cu 72, Mn 0.15, Ni 18, bal Zn.
For hollow ware, cutlery, trim. Nickel silver. Deep drawing. Was Revere 536. *Obsolete*

REVERE ALLOY NO. 752

Revere Copper Products, Inc.
Cu 65.5, Ni 18, Zn 16.5.
Hard: 85,000 TS; 74,000 YS; 3 El; 87 Rock B. Soft: 58,000 TS; 25,000 YS; 40 El; 85 Rock F. For marine and auto trim, hardware. Nickel silver 18% A. Was Revere 533. *Obsolete*

REVERE ALLOY NO. 757

Revere Copper Products, Inc.
Cu 66, Ni 12, bal Zn.
Hard: 95,000 TS; 80,000 YS; 3 El. Soft: 53,000 TS; 17,000 YS; 46 El. For jewelry, springs. 12% nickel silver. Was Revere 540. *Obsolete*

REVERE ALLOY NO. 770

Revere Copper Products, Inc.
Cu 53.5-56.5, Ni 16.5-19.5, Fe 0-0.1, bal Zn.
Hard: 105,000 TS; 90,000 YS; 5 El; B 95 Rock. Soft: 65,000 TS; 28,000 YS; 35 El; B 55 Rock. For springs, marine hardware, jewelry, tableware, screw machine products. Corrosion resistant. *Obsolete*

REVERE ALLOY NO. 776

Revere Copper Products, Inc.
Cu 43, Ni 13, bal Zn.
For elevator trim, springs. 13% nickel silver. Was Revere 560. *Obsolete*

REVERE ALLOY NO. 788

Revere Copper Products, Inc.
Cu 65, Pb 2, Ni 10, Mn 0.25, bal Zn.
For washers, jewelry, panels. Leaded 10% nickel silver. Was Revere 580. *Obsolete*

REVERE ALLOY NO. 790

Revere Copper Products, Inc.
Cu 66, Pb 2, Ni 12, bal Zn.
Hard: 78,000 TS; 75,000 YS; 5 El. Soft: 55,000 TS; 18,000 YS; 40 El. For key blanks, hardware, stampings. Nickel silver. Was Revere 575. *Obsolete*

REVERE ALLOY NO. 99

Revere Copper Products, Inc.
Cu 99.4, As 0.5, bal P.
Hard: 52,000 TS; 35,000 YS; 5 El. Soft: 35,000 YS; 8,000 YS; 40 El. For commercial products; tough pitch. *Obsolete*

REVERE ARCHITECTURAL BRONZE

Revere Copper Products, Inc.
Cu 55-60, Pb 1.5-2.5, bal Zn.
Hard: 70,000 TS; 55,000 YS; 10 El. As extruded: 60,000 TS; 20,000 YS; 30 El. Architectural extrusions such as trim, thresholds, window frames. C38000

REVERE CARTRIDGE BRASS

Revere Copper Products, Inc.
Cu 68.5-71.5, bal Zn.
Half hard: 62,000 TS; 52,000 YS; 25 El. Soft: 30,000 TS; 17,000 YS; 57 El. For cartridge cases, radiator and heater cores and tanks, lamp fixtures, reflectors and bases, finish hardware articles. C26000

REVERE COMMERCIAL BRONZE

Revere Copper Products, Inc.
Cu 89-91, bal Zn.
Hard: 60,000 TS; 55,000 YS; 20 El. Soft: 40,000 TS; 10,000 YS; 50 El. For costume jewelry, ornamental trim, lipstick cases, fasteners, screws, escutcheons, kick plates. C22000

REVERE COPPER NICKEL, 10%

Revere Copper Products, Inc.
Fe 1-1.8, Ni 9-11, Mn 0-1, Zn 0-1, bal Cu.
As hot rolled: 45,000 TS; 20,000 YS; 40 El. For heat exchangers, desalination equipment, flanges, pumps, valves, piping systems, fittings, marine applications, corrosion resistant tanks and vessels. C70600

REVERE COPPER-NICKEL, 30%

Revere Copper Products, Inc.
Fe 0.4-0.7, Mn 0.4-1, Ni 29-33, Zn 0-0.25, bal Cu.
As hot rolled: 55,000 TS; 20,000 YS; 45 El. For heat exchangers, marine applications including protective sheathing, corrosion resistant vessels, valves, flanges, fittings, piping systems, pumps. C71500

REVERE ELECTROLYTIC TOUGH PITCH COPPER

Revere Copper Products, Inc.
99.90+ Cu.
Half hard: 42,000 TS; 36,000 YS; 14 El. Soft: 34,000 TS; 10,000 YS; 40 El. For roofing, flashing, gutters, downspouts, gaskets, anodes, bus bars, vats, rivets, nails, chemical process equipment, heat sinks. C11000

REVERE FREE CUTTING BRASS

Revere Copper Products, Inc.
Cu 60-63, Pb 2.5-3.7, bal Zn.
As extruded: 49,000 TS; 18,000 YS; 45 El. Quarter hard: 56,000 TS; 42,000 YS; 20 El. For extruded shapes that may requires some machining such as drilling, threading, milling, etc. C36000

REVERE FS

Revere Copper Products, Inc.
Zn 0.7-1.3, Al 2.5-3.5, Si 0-0.3, Cu 0-0.05, Ni 0-0.03, 0.20% min Mn, bal Mg.
Hard: 38,000 TS; 26,000 YS; 4 El. Annealed: 32,000 TS; 10 El. For light alloy parts, aircraft parts; wrought. *Obsolete*

REVERE FS-1

Revere Copper Products, Inc.
Al 9, Zn 1, Mn 0.3, bal Mg.
For light alloy parts, sheets. *Obsolete*

REVERE GILDING METAL

Revere Copper Products, Inc.
Cu 94-96, bal Zn.
Half hard: 56,000 TS; 49,000 YS; 5 El. Soft: 35,000 TS; 11,000 YS; 45 El. For coins, metals, tokens, bullet jackets, primers, jewelry. C21000

REVERE J-1

Revere Copper Products, Inc.
Al 6.5, Mn 0.2, Zn 1, bal Mg.
Rolled: 14 El; 60 Brin. For light alloy parts; sheets. *Obsolete*

REVERE JEWELRY BRONZE

Revere Copper Products, Inc.
Cu 86-88, bal Zn.
Hard: 66,000 TS; 56,000 YS; 5 El. Soft: 40,000 TS; 12,000 YS; 45 El. For eyelets, fasteners, compacts, emblems, costume jewelry. C22600

REVERE JS

Revere Copper Products, Inc.
Zn 0.4-1.3, Al 4.2-5.3, Si 0-3, Cu 0-0.05, Ni 0-0.03, 0.18% min Mn, bal Mg.
Hard: 40,000 TS; 27,000 YS; 3 El. Annealed: 37,000 TS; 8 El. For light alloy parts, aircraft parts; wrought. *Obsolete*

REVERE JS-1

Revere Copper Products, Inc.
Al 5, Zn 1, 0.2% min Mn, bal Mg.
40,000 TS; 23,000 YS; 18 El. For light alloy parts; sheets. *Obsolete*

REVERE LOW BRASS

Revere Copper Products, Inc.
Cu 79-81, bal Zn.
Half hard: 61,000 TS; 40,000 YS; 7 El. Soft: 47,000 TS; 15,000 YS; 47 El. For ornamental metal work, medallions, tokens, spandrels, clock dials. C24000

REVERE LOW OXYGEN ELECTROLYTIC COPPER

Revere Copper Products, Inc.
O 0-0.02, 99.90 Cu min.
Half hard: 42,000 TS; 36,000 YS; 14 El. Soft: 34,000 TS; 10,000 YS; 40 El. For applications that require good electrical and thermal conductivity but cannot tolerate the oxygen content of tough pitch copper and do not require the properties of oxygen free material. C10920

REVERE M

Revere Copper Products, Inc.
Mn 1.5, bal Mg.
Rolled: 37,000 TS; 28,000 YS; 7 El; 56 Brin. For light alloy parts; corrosion resistant. *Obsolete*

REVERE MZC COPPER

Revere Copper Products, Inc.
Cr 0.04-1, Mg 0.03-0.06, Zr 0.08-0.2, 98.7 Cu min.
For mold plates for continuous casting, collider energy targets, applications requiring high strength and good conductivity. C18100

REVERE NAVAL BRASS

Revere Copper Products, Inc.
Cu 50-62, Sn 0.5-1, bal Zn.
As hot rolled: 55,000 TS; 25,000 YS; 45 El. Quarter hard: 70,000 TS; 55,000 YS; 15 El. For petrochemical equipment, condenser tube plates and baffles, marine equipment, flanges, tool plates, architectural metal work. C46400

REVERE NICKEL ALUMINUM BRONZE, 10%

Revere Copper Products, Inc.
Fe 2-4, Al 9-11, Ni 4-5.5, Mn 0-1.5, bal Cu.
As hot rolled: 90,000 TS; 50,000 YS; 15 El. Cold rolled: 100,000 TS; 80,000 YS; 10 El. For heat exchangers, marine applications, corrosion resistant tanks and vessels, flanges, valves, piping systems, pumps, desalination equipment. C63000

REVERE NICKEL ALUMINUM BRONZE, 9%

Revere Copper Products, Inc.
Fe 3.5-4.3, Al 8.7-9.5, Mn 1.2-2, Ni 4-4.8, bal Cu.
As hot rolled: 90,000 TS; 50,000 YS; 20 El. Cold rolled: 100,000 TS; 80,000 YS; 12 El. For marine applications, corrosion resistant tanks and vessels, heat exchangers, valves, piping systems, flanges, pumps, desalination equipment. C63200

REVERE NICKEL SILVER, 45-10

Revere Copper Products, Inc.
Cu 43-47, Ni 9-11, bal Zn.
As extruded: 85,000 TS; 20 El. For extruded architectural shapes such as thresholds, trim on escalators and elevators, window frames. C77400

REVERE NO. 110

Revere Copper Products, Inc.
Now REVERE ELECTROLYTIC TOUGH PITCH COPPER. C11000

REVERE NO. 113

Revere Copper Products, Inc.
Cu 99.9, bal Ag.
Hard: 51,000 TS; 48,000 YS; 5-15 El; 58 Rock B. Soft 32,000 TS; 45 El. For conductors. Lake copper. Was Revere 101. *Obsolete*

REVERE NO. 120

Revere Copper Products, Inc.
P 0.007, bal Cu.
For high conductivity tubing. Was Revere 112. *Obsolete*

REVERE NO. 122
Revere Copper Products, Inc.
Now REVERE PHOSPHORUS DEOXIDIZED COPPER. C12200

REVERE NO. 142
Revere Copper Products, Inc.
As 0.3, 99.70 Cu + P.
Hard: 60,000 TS; 55,000 YS; 5 El; 228 Brin. Soft: 36,000 TS; 10,000 YS; 40 El; 63 Brin. For tubes. Arsenical copper. Was Revere 113. *Obsolete*

REVERE NO. 145
Revere Copper Products, Inc.
Te 0.5, bal Cu.
1/2-H temper: 38,000 TS; 30,000 YS; 15 El. H-temper: 44,000 TS; 38,000 YS; 10 El. F.C. copper. Was Revere 114. *Obsolete*

REVERE NO. 158
Revere Copper Products, Inc.
Cu 72, Zn 28.
Hard: 76,000 TS; 55,000 YS; 4 El; 188 Brin. Soft: 47,000 TS; 15,000 YS; 55 El; 62 Brin. For condenser tubes; best quality brass. *Obsolete*

REVERE NO. 210A
Revere Copper Products, Inc.
Now REVERE GILDING METAL. C21000

REVERE NO. 234
Revere Copper Products, Inc.
Cu 63.5, Zn 35, Pb 1.5.
Hard: 85,000 TS; 50,000 YS; 5 El; 150 Brin. Soft: 45,000 TS; 10,000 YS; 50 El; 60 Brin. For screw machine products, hardware, engravings; engravers brass, free cutting. *Obsolete*

REVERE NO. 240
Revere Copper Products, Inc.
Now REVERE LOW BRASS. C24000

REVERE NO. 260
Revere Copper Products, Inc.
Cu 60, Zn 38, Pb 2.
Hard: 70,000 TS; 31,000 YS; 10 El. Soft: 50,000 TS; 15,000 YS; 45 El. For hardware; forging rod, free cutting. *Obsolete*

REVERE NO. 262
Revere Copper Products, Inc.
Cu 68, Zn 32.
Hard: 90,000 TS; 64,000 YS; 3 El; 187 Brin. Soft: 47,000 TS; 11,000 YS; 62 El; 62 Brin. For drawn and spun parts; deep drawing and spinning brass. Was Revere 165. *Obsolete*

REVERE NO. 280B
Revere Copper Products, Inc.
Cu 60, Zn 40.
Hard: 88,000 TS; 80,000 YS; 12 El; 191 Brin. Soft: 59,000 TS; 15,000 YS; 52 El; 64 Brin. For architectural structures, panel sheets, condenser tubes. Muntz Metal. Was Revere 180. *Obsolete*

REVERE NO. 330
Revere Copper Products, Inc.
Cu 67, Zn 32.25, Pb 0.75.
Hard: 50,000 TS; 5 El. Soft: 44,000 TS; 45 El. For hardware, screw machine products. Leaded brass, free-cutting. Was Revere 224. *Obsolete*

REVERE NO. 342A
Revere Copper Products, Inc.
Cu 63, Zn 35, Pb 2.
Hard: 85,000 TS; 50,000 YS; 5 El; 150 Brin. Soft: 45,000 TS; 10,000 YS; 50 El; 60 Brin. For watch parts, locks. Heavy leaded brass. Was Revere 235. *Obsolete*

REVERE NO. 360
Revere Copper Products, Inc.
Now REVERE FREE CUTTING BRASS. C36000

REVERE NO. 380B
Revere Copper Products, Inc.
Cu 56, Zn 41.25, Pb 2.75.
Hard: 70,000 TS; 55,000 YS; 10 El; 144 Brin. Soft: 50,000 TS; 15,000 YS; 20 El; 64 Brin. For builders hardware, architectural work. Architectural bronze. Was Revere 285. *Obsolete*

REVERE NO. 381
Revere Copper Products, Inc.
Cu 60, Zn 39.25, Sn 0.75.
Hard: 70,000 TS; 55,000 YS; 25 El; 140 Brin. Soft: 54,000 TS; 15,000 YS; 50 El. For hardware, condenser tubes. Roman bronze. *Obsolete*

REVERE NO. 422
Revere Copper Products, Inc.
Cu 87, Zn 12, Sn 1.
Hard: 80,000 TS; 4 El; 150 Brin. Soft: 45,000 TS; 40 El; 100 Brin. For chains. Chain bronze. Was Revere 340. *Obsolete*

REVERE NO. 443
Revere Copper Products, Inc.
Cu 71, Zn 28, Sn 1, As 0.03.
Hard: 100,000 TS; 98,000 YS; 3 El; 215 Brin. Soft: 53,000 TS; 18,000 YS; 60 El; 53 Brin. For condenser tubes. Admiralty metal. Corrosion resistant. Was Revere 358. *Obsolete*

REVERE NO. 464
Revere Copper Products, Inc.
Now REVERE NAVAL BRASS. C46400

REVERE NO. 535
Revere Copper Products, Inc.
Cu 66, Zn 19, Ni 15.
Hard: 93,000 TS; 75,000 YS; 5 El; 192 Brin. Soft: 58,000 TS; 15,000 YS; 45 El; 73 Brin. For lighting fixtures, hollow ware, dishes, trays; Nickel Silver, 15%. *Obsolete*

REVERE NO. 548R
Revere Copper Products, Inc.
Cu 62, Zn 33, Ni 5.
Hard: 85,000 TS; 60,000 YS; 2 El; 150 Brin. Soft: 50,000 TS; 15,000 YS; 45 El; 70 Brin. For jewelry, decorative trim. Nickel silver, 5%. *Obsolete*

REVERE NO. 608
Revere Copper Products, Inc.
Cu 95, Al 5, As 0.035.
Hard: 70,000 TS; 50,000 YS; 25 El; 150 Brin. Soft: 57,000 TS; 10,000 YS; 55 El. For hardware, bushings. Al bronze. Was Revere 429. *Obsolete*

REVERE NO. 614
Revere Copper Products, Inc.
Now REVERE 7% ALUMINUM BRONZE. C61400

REVERE NO. 710
Revere Copper Products, Inc.
Cu 80, Ni 20, Fe 0.5, Mn 0-0.7.
Hard: 80,000 TS; 76,000 YS; 3 El; 152 Brin. Soft: 49,000 TS; 14,000 YS; 35 El; 64 Brin. For condenser tubes, corrosion resistant tanks. 20% cupro-nickel. Was Revere 505. *Obsolete*

REVERE NO. 715
Revere Copper Products, Inc.
Now REVERE COPPER-NICKEL, 30%. C71500

REVERE NO. 745
Revere Copper Products, Inc.
Cu 65, Zn 25, Ni 10.
Hard: 90,000 TS; 65,000 YS; 3 El; 152 Brin. Soft: 50,000 TS; 15,000 YS; 45 El; 70 Brin. For stampings, jewelry. Nickel silver, 10%. Was Revere 545. *Obsolete*

REVERE OXYGEN FREE COPPER (1)
Revere Copper Products, Inc.
99.95 Cu min.
Hard: 50,000 TS; 48,000 YS; 5 El. Soft: 32,000 YS; 8000 YS; 40 El. For bus conductors, electronic applications, glass-to-metal seals, vacuum applications, sputter plates, heat sinks. C10200

REVERE OXYGEN FREE COPPER (2)
Revere Copper Products, Inc.
99.95 Cu min, 0.027 Ag min.
Hard: 50,000 TS; 48,000 YS; 5 El. Soft: 32,000 TS; 9000 YS; 40 El. For bus bars, contacts, commutator segments and bars, chemical process equipment, applications requiring exposure to solder-type temperatures without softening. Silver bearing (8 oz min) copper. 10400

REVERE OXYGEN FREE COPPER (3)
Revere Copper Products, Inc.
99.95 Cu min, 0.085 Ag min.
Hard: 50,000 TS; 48,000 YS; 5 El. Soft: 32,000 TS; 9000 YS; 40 El. For bus bars, commutator segments and bars, chemical process equipment, fusion energy magnets; mold plates, applications requiring resistance to softening at temperatures in the 650-750°F range. Silver bearing (25 oz min) copper. 10700

REVERE OXYGEN FREE ELECTRONIC COPPER
Revere Copper Products, Inc.
99.99 Cu min.
Hard: 50,000 TS; 48,000 YS; 5 El. Soft: 32,000 TS; 9000 YS; 40 El. For bus conductors vacuum seals, anodes for vacuum tubes, transistor components, rectifiers. C10100

REVERE OXYGEN FREE EXTRA LOW PHOSPHORUS
Revere Copper Products, Inc.
P 0.001-0.005, 99.95 Cu min.
Hard: 50,000 TS; 48,000 YS; 5 El. Soft: 32,000 TS; 9000 YS; 40 El. For bus bars, electrical conductors, chemical process equipment, applications requiring good conductivity and welding or brazing properties. C10300

REVERE PHOSPHOR BRONZE A
Revere Copper Products, Inc.
Sn 5, P 0.05, bal Cu.
Hard: 70,000 TS; 58,000 YS; 25 El; 78 B Rock. For springs, fuse clips, bellows, diaphragms, corrosion resistant; Revere Alloy No. 308. *Obsolete*

REVERE PHOSPHOR BRONZE B
Revere Copper Products, Inc.
Sn 7, P 0.5, bal Cu.
For springs. *Obsolete*

REVERE PHOSPHOR BRONZE C
Revere Copper Products, Inc.
Sn 8, P 0.05, bal Cu.
Hard: 93,000 TS; 72,000 YS; 10 El; 93 Rb. Soft: 55,000 TS; 70 El; 75 Rf Brin. For springs, fuse clips, bellows, diaphragms; corrosion resistant, Revere Alloy No. 315. *Obsolete*

REVERE PHOSPHOR BRONZE D
Revere Copper Products, Inc.
Sn 10, P 0.05, bal Cu.
Hard: 100,000 TS; 13 El; B 97 Rock. Soft: 66,000 TS; 68 El; B 55 Rock. For springs, fuse clips, bellows, diaphragms, welding rod; corrosion resistant. *Obsolete*

REVERE PHOSPHOR BRONZE E
Revere Copper Products, Inc.
Sn 0-4, P 0.06, bal Cu.
Hard: 65,000 TS; 50,000 YS; 8 El; B 75 Rock. Soft: 40,000 TS; 14,000 YS; 48 El; F 60 Rock. For springs, fuse clips, bellows, diaphragms, electrical contacts, metal hose; corrosion resistant. *Obsolete*

REVERE PHOSPHORUS DEOXIDIZED COPPER
Revere Copper Products, Inc.
P 0.015-0.04, 99.90 Cu min.
Half hard: 42,000 TS; 36,000 YS; 14 El. Soft: 33,000 TS; 10,000 YS; 45 El. For non-electrical applications that may be brazed, welded or otherwise exposed to heat and hydrogen. C12200

REVERE PLATERS BRONZE
Revere Copper Products, Inc.
Cu 89-93, Sn 0.7-1.3, bal Zn.
Quarter hard: 49,000 TS; 40,000 YS; 25 El. Hard: 71,000 TS; 62,000 YS; 4 El. Plater bar. C41300

REVERE RED BRASS
Revere Copper Products, Inc.
Cu 84-86, bal Zn.
Half hard: 57,000 TS; 49,000 YS; 12 El. Soft: 41,000 TS; 14,000 YS; 46 El. For name plates, badges, tags, rouge boxes, fasteners, eyelets, dials. C23000

REVERE TIN BEARING TELLURIUM COPPER
Revere Copper Products, Inc.
Sn 0.04-0.15, Te 0.005-0.05, 99.90 Cu min.
Hard: 51,000 TS; 45,000 YS; 12 El. Soft: 32,000 TS; 9000 YS; 45 El. For radiators, heat exchangers, commutator bars, heat sinks, lead frames, applications requiring good resistance to softening at temperatures up to 850°F. C14420

REVERE TOUGH PITCH COPPER (1)
Revere Copper Products, Inc.
0.085 Ag min, 99.90 Cu min.
Hard: 52,000 TS; 48,000 YS; 5 El. Soft: 33,000 TS; 7000 YS; 40 El. For radiators, gaskets, commutator bars, electrical contacts, switches, bus bars. Silver bearing (25 oz min) copper. C11600

REVERE TOUGH PITCH COPPER (2)
Revere Copper Products, Inc.
99.90 Cu min, 0.034 Ag min.
Hard: 52,000 TS; 48,000 YS; 5 El. Soft: 33,000 TS; 9000 YS; 40 El. For radiators, gaskets, commutator bars, electrical contacts, switches, bus bars, applications requiring resistance to softening at solder temperatures. Silver bearing (10 oz min) copper. C11400

REVERE TOUGH PITCH COPPER (3)
Revere Copper Products, Inc.
99.90 Cu min, 0.054 Ag min.
Hard: 52,000 TS; 48,000 YS; 5 El. Soft: 33,000 TS; 9000 YS; 40 El. For radiators, gaskets, commutator bars, electrical contacts, switches, bus bars, applications requiring resistance to softening at elevated temperatures. Silver bearing (16 oz min) copper. C11500

REVERE ZIRCONIUM COPPER
Revere Copper Products, Inc.
Zr 0.1-0.2, 99.80 Cu min.
As hot rolled: 38,000 TS; 10,000 YS; 36 El. Cold rolled: 53,000 TS; 48,000 YS; 20 El. For high strength applications requiring good conductivity. C15000

REX
Westfalische Stahlgesellschaft mbH
C 0.56, Cr 0.3, Mo 0.18, Ni 1.65, V 0.1, bal Fe.
For forging and heading dies, punches; oil hardened, tough.

REX "A"
Colt Industries
C 0.4, W 18, Cr 4, bal Fe.
For punches, dies; hot work steel. *Obsolete*

REX 1059
Colt Industries
C 1.2, Cr 4, V 4, W 14, Mo 0.75, bal Fe.
For cutting tools, planer and shaper tools; high speed steel. *Obsolete*

REX 1092
Colt Industries
Cr 4, V 1.5, W 5.5, Mo 3.75, bal Fe.
For boring tools, broaches, chasers. *Obsolete*

REX 18-8
Colt Industries
C 0.1, Cr 18, Ni 8, bal Fe.
For stainless parts; stainless steel. *Obsolete*

REX 3-V
Colt Industries
C 1, Cr 4, W 14, V 3, Mo 0.75, bal Fe.
For turning tools; high speed steel. *Obsolete*

REX 326
Firth Brown Ltd.
C 0.25, Cr 17, Ni 17, Mo 2.5, Co 7, Cb 1.8, Mn 3, bal Fe.
For jet engine components; high heat resistance.

REX 326D
Firth Brown Ltd.
Co 0.45, C 14.3, Ni 14.6, W 2.2, Mo 2.2, Co 9.5, Cb 2.8, bal Fe.
For valves, pump parts; corrosion and heat resistant.

REX 326D
Universal Cyclops
Co 0.45, C 14.3, Ni 14.6, W 2.2, Mo 2.2, Co 9.5, Cb 2.8, bal Fe.
For valves, pump parts; corrosion and heat resistant.

REX 327
Firth Brown Ltd.
C 0.2, Cr 17, Ni 17, Mo 3, Co 7, Ti 0.8, Cu 3, bal Fe.
For valves, pump parts; heat and corrosion resistant.

REX 337 A
English manufacture
C, Ni, Cr, bal Fe.
For high temperature applications; heat and corrosion resistant.

REX 4-V
Colt Industries
C 1.25, Cr 4, W 18.5, V 4, Mo 0.75, bal Fe.
For broaches, reamers, rifling tools; high speed steel. *Obsolete*

REX 400
English manufacture
Ni 76, Cr 19, Al 0.6, Ti 2.1, bal Fe.
For high temp. applications; heat resistant.

REX 440
Colt Industries
C 0.8, Cr 4, W 19.5, V 2, Mo 0.6, Co 12, bal Fe.
For cutting tools, lathe and cut-off tools; high speed steel, heavy duty. *Obsolete*

REX 448
Youngstown Steel
C 0.15, Si 0.75, Mn 0.75, Cr 11.5, Mo 0.75, V 0.15, Cb 0.45, bal Fe.
Annealed: 70,000 TS; 35,000 YS; 30 El; 80 Rock B. For cutlery, surgical instruments, chemical plant equipment. Corrosion and heat resistant.

REX 448
Firth Brown Ltd.
C 0.1, Cr 11, Mo 0.8, Ni 0.8, Cb 0.5, V 0.2, Mn 1, bal Fe.
For aircraft and jet engine components, corrosion and heat resistant.

REX 467
Firth Brown Ltd.
C 0.2, Cr 14, Ni 10, Mo 2, Ti 0.8, Cu 2.5, bal Fe.
For valves, pump parts; corrosion and heat resistant.

REX 467
Firth-Vickers Stainless Steels Ltd.
C 0.2, Si 0.7, Mn 1, Ni 9.5, Cr 14.5, Cu 2.5, Mo 2, Ti 0.8, bal Fe.
For turbine blades in jet engines; high creep resistance, stainless.

REX 49
Crucible Materials Corp.
Tool material. C 1.1, W 6.75, Mo 3.75, Cr 4.25, V 2, Co 5, bal Fe.
Hardened: 64-69 Rock C. For special applications of twist drills, end mills, form cutters, lathe tools, reamers. AISI Type M41. High speed tool steel.

REX 539
Firth Brown Ltd. .
C 0.37, Si 1.6, Mn 1.6, Ni 1.8, Cr 0.11, Mo 0.5, V 0.2, bal Fe.
Heat treated: 270,000 TS; 230,000 YS. For aircraft parts, high stress applications; ultra high tensile steel. *Obsolete*

REX 648C
Firth Brown Ltd.
C 0-0.18, Si 0-0.1, Mn 0-0.1, Ni 17-19, Cr 0-0.2, Mo 4.6-5.2, Al 0.05-0.15, Co 7-8.5, Ti 0.3-0.6, bal Fe.
Maraging steel.

REX 78
Firth Brown Ltd.
C 0.07, Cr 14, Ni 18, Mo 3.75, Ti 0.65, Cu 3.6, bal Fe.
Heat treated: 94,000 TS; 49,000 YS; 39 El; 57 RA. For gas turbine disks and blades; corrosion and heat resistant.

REX 78
Firth-Vickers Stainless Steels Ltd.
C 0.07, Cr 14, Ni 18, Mo 3.75, Ti 0.65, Cu 3.6, bal Fe.
Heat treated: 94,000 TS; 49,000 YS; 39 El; 57 RA. For gas turbine disks and blades; corrosion and heat resistant.

REX 939
Colt Industries
C 1, W 18, Cr 4, V 3, bal Fe.
For reamers, hobs, taps, milling cutters; high speed steel. *Obsolete*

REX 95
Crucible Materials Corp.
Tool material. C 0.8, W 14, Cr 4, V 2, Mo 0.75, Co 5.25, bal Fe.
Hardened: 63-66 Rock C. For single point lathe tools, severe cutting operations, tools for stainless steels. AISI Type T8. High speed tool steel.

REX AA
Crucible Materials Corp.
Tool material. C 0.75, Mn 0.3, Cr 4, V 1.15, W 18, bal Fe.
Hardened: 62-65 Rock C. For the lathe cutting tools, drills, end mills, reamers, slotting saws, gear cutters. AISI Type T1. High speed tool steel.

REX AA-OX
Colt Industries
C, alloy, bal Fe.
For cutting tools; high speed steel. *Obsolete*

REX AA-PX
Colt Industries
C, alloy, bal Fe.
For cutting tools; high speed steel. *Obsolete*

REX AAA
Crucible Materials Corp.
Tool material. C 0.7, W 17-19, Mo 0.5, Cr 4, V 1, Co 5, bal Fe.
For cutters, reamers, drills, hobs, broaches, taps; Type T4; high speed steel.

REX BRONZE
Whipple & Choate Co.
Cu 80, Zn 10, Pb 10.
For castings.

REX CAST ALUMINUM
Hollup Corp.
Cu, bal Al.
For welding rod; for Al alloys.

REX CAST IRON

Hollup Corp.
C 3-3.5, Si 3-3.5, bal Fe.
For welding rod; for cast iron.

REX CHAMPION

Colt Industries
C 0.7, W 14, V 2, Cr 4, bal Fe.
Heat treated: 160,000 TS: 600 Brin. For cutting tools, drills, broaches, dies; high speed steel. *Obsolete*

REX COMPOSITION

Whipple & Choate Co.
Cu 78-86, Sn 3-6, Pb 2-6, Zn 5-14.
For castings; free-cutting.

REX LA

Colt Industries
C 0.9, Mn 0.3, Si 0.3, Cr 3, V 1.2, W 1.5, Mo 3.25, bal Fe.
Air or oil hardening tool steel; AISI M52. *Obsolete*

REX M-2

Crucible Materials Corp.
Tool material. C 0.83, W 6.4, Cr 4.1, V 1.9, Mo 5, bal Fe.
For tools, cutters, broaches, chasers, hobs; high speed steel.
AISI M2.

REX M-3

Crucible Materials Corp.
Tool material. C 1, Cr 4, V 2.5, W 5.7, Mo 5, bal Fe.
For broaches, form tools, reamers, cut-off tools; high speed steel, oil hardened.

REX M.M.

Colt Industries
C 0.8, W 5.5, Mo 4, Cr 4, V 1.5, bal Fe.
For hobs, drills, reamers, taps, lathe and planer tools; high speed steel. *Obsolete*

REX M2

Crucible Materials Corp.
C 0.85, Mn 0.3, Si 0.3, Cr 4.15, V 1.95, W 6.4, Mo 5.
High speed steel. For boring tools, milling cutters, twist drills.
AISI M2.

REX M2 HIGH CARBON

Crucible Materials Corp.
Tool material. C 1, W 6.4, Mo 5, Cr 4.15, V 1.95, bal Fe.
Hardens to 62.5-66.5 Rock C. For boring tools, twist drills, milling cutters, threading tools, thread chasers; improved wear resistance. AISI Type M2. High speed tool steel.

REX M2S

Crucible Materials Corp.
Tool material. C 0.85, W 6.4, Mo 5, Cr 4.15, V 1.95, bal Fe.
Hardens to 61-67 Rock C. Free machining; for form tools, drills, end mills, milling cutters, hobs, broaches. AISI Type M2. High speed tool steel.

REX M2S HIGH CARBON

Crucible Materials Corp.
Tool material. C 1, W 6.4, Mo 5, Cr 41.15, V 1.95, S 0.15, bal Fe.
Hardens to 62-66 Rock C. Free machining type high speed steel. For boring tools, drills, end mills, reamers, broaches, milling cutters, hobs. AISI Type M2. High speed tool steel.

REX M3-1

Crucible Materials Corp.
Tool material. C 1.05, Cr 4, V 2.4, W 6.25, Mo 6.25, bal Fe.
For lathe and planer tools, milling cutters, reamers; Type M3.
High speed steel.

REX M33

Crucible Materials Corp.
Tool material. C 0.9, Mo 9.5, W 1.75, Cr 3.75, V 1.25, Co 8, bal Fe.
Hardened: 63-68 Rock C. For tough machining and higher speed operations of drills, end mills, form cutters, lathe tools, taps, reamers. AISI Type M33. High speed tool steel.

REX M3S-2

Crucible Materials Corp.
Tool material. C 1.2, Mn 0.3, Si 0.3, Cr 4, W 6.25, Mo 6.25, S 0.15, bal Fe.
Hardened: 64-67 Rock C. For form tools, milling cutters, hobs, cut-off tools, reamers, lathe tools. High-speed steel, free-cutting. Resistance to softening at high temperatures.

REX M42

Crucible Materials Corp.
Tool material. C 1.1, W 1.5, Mo 9.5, Cr 3.75, V 1.15, Co 8, bal Fe.
Hardened: 64-69 Rock C. For special purpose drills, end mills, form cutters, reamers, lathe tools. AISI Type M42. High speed tool steel.

REX MMM

Colt Industries
C 0.85, Cr 4, V 1.6, W 5-6, Mo 4-5, Co 5-5.5, bal Fe.
For cutters, tools, reamers, drills; high speed steel. *Obsolete*

REX SUPER VAN

Colt Industries
C 0.7, W 18, Cr 4, V 2, Mo 0.65, bal Fe.
For cutting tools for hard materials, milling cutters; high speed steel. *Obsolete*

REX SUPERCUT

Colt Industries
C 0.77, Cr 4, W 18.5, Mo 0.65, V 2-0.5, Co 8, bal Fe.
For cutters, reamers, hobs, lathe and form tools; high speed steel, heavy duty. *Obsolete*

REX T15S

Crucible Materials Corp.
Tool material. C 1.55, Cr 4, V 5, W 12.25, Co 5, S 0.1, bal Fe.
High speed steel, free machining grade. For cutting tools as boring tools, end mills, broaches, gear cutters.

REX TMO-5

Colt Industries
C 0.7, Cr 3.25-4.25, W 1.25-2, V 0.75-1.5, Mo 7.5-9.5, Co, bal Fe.
For reamers, lathe tools, drills, broaches, dies, hot work dies, cutting tools; high speed steel; wear resistant. *Obsolete*

REX TMO-5

Crucible Specialty Metals
C 0.7, Cr 3.25-4.25, W 1.25-2, V 0.75-1.5, Mo 7.5-9.5, Co, bal Fe.
For reamers, lathe tools, drills, broaches, dies, hot work dies, cutting tools; high speed steel; wear resistant. *Obsolete*

REX TMO-8

Colt Industries
C 0.9, Cr 4, V 2, W 2, Mo 9, Co 8, bal Fe.
For cutting tools, reamers, milling cutters; high speed steel.
Obsolete

REX VM

Crucible Materials Corp.
Tool material. C 0.98, Mo 8, Cr 4, W 0.7, V 1.95, bal Fe.
Hardened to 58-66 Rock C. For broaches, center drills, lathe tools, end mills, thread chasers, hobs. AISI Type M10. High speed tool steel.

REX VM "PX" TEMPER

Crucible Materials Corp.
Tool material. C 0.55, Cr 4, Mo 8, V 2, bal Fe.
High speed steel, low carbon grade. For extrusion dies, piercer points, valve extrusion die rings and inserts.

REX VM DREADNAUGHT

Colt Industries
C 0.86, Cr 4, V 2, Mo 8, bal Fe.
For taps, drills, planer and boring tools; high speed steel.
Obsolete

REX VM DREADNAUGHT

Crucible Specialty Metals
C 0.86, Cr 4, V 2, Mo 8, bal Fe.
For taps, drills, planer and boring tools; high speed steel.
Obsolete

REX VM-S

Colt Industries
C 0.85, Cr 4, V 1.9, Mo 8, S 0.15, bal Fe.
For lathe and planer tools, reamers, hobs, broaches; high speed steel. *Obsolete*

REX Z METAL

Chain Belt Co.
C 2.4, Mn 0.75, Si 1, Cu 0.18, bal Fe.
Heat treated: 75,000 TS; 50,000 YS; 8 El; 200 Brin. For chains, buckets, sprockets. Pearlitic malleable iron.

REX-A-LITE

Sight Feed Generator Co.
C, Cr, Mn, Si, bal Fe.
Welded: 570 Brin. For hard facing electrodes; abrasion resistant.

REX-A-LITE 1

Sight Feed Generator Co.
Co, Cr, W.
Welded: 550 Brin. For hard facing electrodes; abrasion and corrosion resistant.

REX-A-LITE 6

Sight Feed Generator Co.
Co, Cr, W.
Welded: 440 Brin. For hard facing electrodes; shock and impact resistant.

REX-M7

Crucible Materials Corp.
Tool material. C 1, Mn 0.3, Si 0.3, Cr 3.75, W 1.75, Mo 8.75, V 2, bal Fe.
Hardened: 64-66 Rock C. For twist drills, end mills, cutting tools, lathe tools, threading tools, hobs, reamers, taps, chasers, form cutters. High speed steel. Improved wear resistance. Type M7 tool steel.

REX-TMO

Crucible Materials Corp.
Tool material. C 0.85, Cr 3.75, V 1.15, W 1.7, Mo 8.5, bal Fe.
For lathe and planer tools, milling cutters, hobs; high speed steel; Type M1.

REX-TMO-S

Colt Industries
C 0.85, Cr 3.75, V 1.15, W 1.55, Mo 8.7, S 0.15, bal Fe.
For lathe and planer tools, reamers, hobs, broaches; high speed steel, free-cutting. *Obsolete*

REXALLOY

Alloy Cast Products, Inc.
Cr 33, Co 44, W 17, C 2.2, B 0.2, Fe 0-1.
Cast nonferrous alloy for cutting tools and wear resistant applications. 54-62 Rock C.

REXALLOY 33

Alloy Cast Products, Inc.
Cr 33, Co 44, W 17, C 2.25, B 0.22, Mn 1, Si 0.5, Fe 0-2.
Cast nonferrous alloy for cutting tools and wear resistant applications. 55-63 Rock C; 50,000-60,000 TS.

REXALLOY 43

Alloy Cast Products, Inc.
Cr 35.5, Co 50, W 8.75, C 2.05, Mn 1, Si 0.5, Mo 0.5, Fe 0-2.
Cast nonferrous alloy for cutting tools and wear resistant applications. Graphite cast only: 54-58 Rock C.

REXALLOY A

Alloy Cast Products, Inc.
Cr 28, Co 64, W 4.5, C 1.2, Mn 1, Si 0.5, Fe 0-2.
Cast nonferrous alloy for cutting tools and wear resistant applications. 40-44 Rock C.

REXALLOY C
Alloy Cast Products, Inc.
Cr 28, Co 58, W 9, C 2.25, Mn 1, Si 0.5, Fe 0-1.
Cast nonferrous alloy for cutting tools and wear resistant applications. 50 Rock C.

REXALOY
Sight Feed Generator Co.
C, Cr, Ni, Cu, V, Mn, bal Fe.
For hard facing electrodes. *Obsolete*

REXARC
Sight Feed Generator Co.
C, Mn, bal Fe.
Welded: 380 Brin. For hard facing electrodes; wear resistant.

REXARC 1
Sight Feed Generator Co.
Ni.
For welding electrodes for cast iron; machinable welds.

REXARC 30
Sight Feed Generator Co.
C, Cr, Mn.
Welded: 30 Rock C. For hard facing electrodes; metallic coated.

REXARC 48
Sight Feed Generator Co.
C, Cr, bal Fe.
Welded: 48 Rock C. For hard facing electrodes.

REXARC 50
Sight Feed Generator Co.
C, Cr, bal Fe.
Welded: 50 Rock C. For hard facing electrodes; impact and abrasion resistant.

REXARC 520
Sight Feed Generator Co.
C, Cr, Mo, bal Fe.
Welded: 54 Rock C. For hard facing electrodes; abrasion resistant.

REXARC 55
Sight Feed Generator Co.
C, Cr, bal Fe.
Welded: 55 Rock C. For hard facing electrodes; shock resistant.

REXARC 65
Sight Feed Generator Co.
C, Cr, TiO, Mo, bal Fe.
Welded: 65 Rock C. For hard facing electrodes; abrasion resistant.

REXARC MANGANESE M1
Sight Feed Generator Co.
C, Mn, Ni, bal Fe.
Welded: 35 Rock C. For hard facing electrodes; build-up on Mn castings, wear resistant.

REXCOBALT
French manufacture
C, alloy, bal Fe.
For tools.

REXITE GRADE CR-7
Colt Industries
WC + Co.
For cutting tools; tough, wear resistant, sintered. *Obsolete*

REXITE GRADE CR-8
Colt Industries
WC + Co.
For cutting tools; high speed finishing, sintered. *Obsolete*

REXITE GRADE UC-14
Colt Industries
WC + Co.
For cutting tools; hard, wear resistant, sintered. *Obsolete*

REXITE GRADE UC-16
Colt Industries
WC + Co.
For cutting tools; shock resistant, sintered. *Obsolete*

REXITE UC-25
Colt Industries
WC-Co.
For cutting tools; for cast iron. *Obsolete*

REXOR
Allied Steel & Tractor Products Inc.
C 0.35, Mn 0.7, Cr 0.6, Mo 0.35, Si 0.15, Ni 0.2, bal Fe.
Annealed: 107,000 TS; 80,000 YS; 22 El; 248 Brin. For chisels, pneumatic tools; tough, wear resistant.

REXTOX
English manufacture
Fe-Mn-Cu.
For resistance alloy.

REXWELD 27
Colt Industries
C 1.25, Cr 28, bal Fe.
Welded: 310 Brin. For hard facing electrode; resists heat checking. *Obsolete*

REXWELD 33W
Colt Industries
C 2.2, Cr 33, W 18, Co 45.
Welded: 600 Brin. For hard facing electrode; high abrasion resistance. *Obsolete*

REXWELD 54
Colt Industries
Fe-Ni-Cr.
For hard surfacing; cast welding rod. *Obsolete*

REXWELD 57
Colt Industries
C 4, Cr 28, Mo 8, bal Fe.
Welded: 580 Brin. For hard facing electrode, pulverizers; abrasion and impact resistant. *Obsolete*

REXWELD 64
Colt Industries
C 0.75, Si 3.5, Cr 14.5, Ni 73, Fe 4, B 3.5.
For hard facing electrodes; corrosion and wear resistant. *Obsolete*

REXWELD 66
Colt Industries
Ni 55, Cr 16.5, Mo 17, W 4.5, Fe 6.
Welded: 200-400 Brin. For hard facing alloy for shear blades and forging dies; resists thermal shock. *Obsolete*

REXWELD A
Colt Industries
C 1, Cr 27.5, Ni 2, W 4.5, Fe 1.5, Co 65.
For hard facing electrode; corrosion and wear resistant. *Obsolete*

REXWELD B
Colt Industries
C 1.65, Cr 27.5, W 7.5, Mo 0.5, Fe 1.5, Co 60.
For hard facing electrode; corrosion and wear resistant. *Obsolete*

REXWELD C
Colt Industries
C 2.4, Cr 29.5, W 12, Mo 0.5, Fe 1.5, Co 55.
For hard facing electrode; corrosion and wear resistant. *Obsolete*

REXWELD C
Colt Industries
C 2, Cr 30, W 12, Co 51.
For hard facing rod; wear resistant. *Obsolete*

REXWELD VT
Colt Industries
C 0.3, Cr 27.5, Ni 2.5, Mo 5.5, Fe 1.5, Co 62.
For hard facing electrode; corrosion and wear resistant. *Obsolete*

REYNOLDS
Reynolds Aluminum
Aluminum.
All Reynolds Aluminum Alloys now identified by AA numbers.

REYNOLDS 108
Reynolds Metals Co.
Cu 4, Si 3, bal Al.
Cast: 21,000 TS; 14,000 YS; 2.5 El; 55 Brin. For sand castings; good machinability. *Obsolete*

REYNOLDS 112
Reynolds Metals Co.
Cu 7, Zn 1.7, bal Al.
Cast: 24,000 TS; 15,000 YS; 1.5 El; 70 Brin. For sand castings; non-hardenable. *Obsolete*

REYNOLDS 113
Reynolds Metals Co.
Cu 7, Si 2, Zn 1.7, bal Al.
Sand cast: 24,000 TS; 15,000 YS; 1.5 El; 70 Brin. For sand castings and permanent molds; non-hardenable. *Obsolete*

REYNOLDS 122
Reynolds Metals Co.
Cu 10, Mg 0.2, bal Al.
T2-temper: 27,000 TS; 20,000 YS; 1 El; 80 Brin. T61-temper: 40,000 TS; 30,000 YS; 4 El; 115 Brin. For sand castings; heat treatable. *Obsolete*

REYNOLDS 13
Reynolds Metals Co.
Si 12, bal Al.
Die cast: 37,000 TS; 18,000 YS; 1.8 El. For sand and die castings; corrosion resistant. *Obsolete*

REYNOLDS 138
Reynolds Metals Co.
Cu 10, Si 4, Mg 0.3, bal Al.
Permanent mold cast: 32,000 TS; 24,000 YS; 1.5 El; 100 Brin. For permanent mold castings; non-hardenable. *Obsolete*

REYNOLDS 142
Reynolds Metals Co.
Cu 4, Mg 1.5, Ni 2, bal Al.
T21-temper: 27,000 TS; 18,000 YS; 1 El; 70 Brin. T571-temper: 32,000 TS; 28,000 YS; 0.5 El; 85 Brin. T77-temper: 28,000 TS; 25,000 YS; 3 El; 75 Brin. For sand castings; heat treatable. *Obsolete*

REYNOLDS 14S
Reynolds Metals Co.
Si 0.8, Cu 4.4, Mn 0.8, Mg 0.4, bal Al.
O-temper: 27,000 TS; 14,000 YS; 18 El; 45 Brin. T4-temper: 62,000 TS; 42,000 YS; 22 El; 105 Brin. T6-temper: 70,000 TS; 60,000 YS; 13 El; 135 Brin. For aircraft forgings; wrought. *Obsolete*

REYNOLDS 152
Reynolds Metals Co.
Cu 7, Si 5.5, Mg 0.3, bal Al.
T524-temper: 29,000 TS; 18,000 YS; 1.0 El; 95 Brin. T74-temper: 35,000 TS; 26,000 YS; 0.5 El; 100 Brin. For permanent mold castings; heat treatable. *Obsolete*

REYNOLDS 17S
Reynolds Metals Co.
Cu 4, Mn 0.5, Mg 0.5, bal Al.
O-temper: 26,000 TS; 10,000 YS; 22 El; 45 Brin. T4-temper: 62,000 TS; 40,000 YS; 22 El; 105 Brin. For aircraft structures; heat treatable. *Obsolete*

REYNOLDS 18S
Reynolds Metals Co.
Cu 4, Mg 0.5, Ni 2, bal Al.
T61-temper: 62,000 TS; 48,000 YS; 12 El; 120 Brin. For pistons; heat treatable. *Obsolete*

REYNOLDS 195
Reynolds Metals Co.
Cu 4.5, Si 0.8, bal Al.
T4-temper: 32,000 TS; 16,000 YS; 8.5 El; 60 Brin. T6-temper: 36,000 TS; 24,000 YS; 5 El; 75 Brin. T62-temper: 40,000 TS; 30,000 YS; 2 El; 95 Brin. For sand castings; heat treatable. *Obsolete*

REYNOLDS 212
Reynolds Metals Co.
Cu 8, Si 1.2, bal Al.
Cast: 23,000 TS; 14,000 YS; 2 El; 65 Brin. For sand castings; non-hardenable. *Obsolete*

REYNOLDS 214
Reynolds Metals Co.
Mg 3.8, bal Al.
Cast: 25,000 TS; 12,000 YS; 9 El; 50 Brin. For sand castings; corrosion resistant. *Obsolete*

REYNOLDS 218
Reynolds Metals Co.
Mg 8, bal AL.
Die cast: 42,000 TS; 23,000 YS; 7 El. For die castings; corrosion resistant. *Obsolete*

REYNOLDS 220
Reynolds Metals Co.
Mg 10, bal Al.
T4-temper: 46,000 TS; 25,000 YS; 14 El; 75 Brin. For sand castings; heat treatable. *Obsolete*

REYNOLDS 24S
Reynolds Metals Co.
Cu 4.5, Mn 0.6, Mg 1.5, bal Al.
O-temper: 27,000 TS; 11,000 YS; 22 El; 47 Brin. T3-temper: 69,000 TS; 48,000 YS; 19 El; 120 Brin. T36-temper: 73,000 TS; 57,000 YS; 130 Brin. For aircraft structures; heat treatable. *Obsolete*

REYNOLDS 25S
Reynolds Metals Co.
Si 0.8, Cu 4.5, Mn 0.8, bal Al.
T6-temper: 58,000 TS; 37,000 YS; 19 El; 110 Brin. For propellers, engine parts, hardware; heat treatable. *Obsolete*

REYNOLDS 2S
Reynolds Metals Co.
99% Al min.
O-temper: 13,000 TS; 5,000 YS; 45 El; 23 Brin. H14-temper: 18,000 TS; 16,000 YS; 20 El; 32 Brin. H18-temper: 24,000 TS; 22,000 YS; 15 El; 44 Brin. For containers, tanks, auto body parts; wrought. *Obsolete*

REYNOLDS 319
Reynolds Metals Co.
Cu 3.5, Si 6, bal Al.
Cast: 27,000 TS; 18,000 YS; 2 El; 70 Brin. T6-temper: 36,000 TS; 24,000 YS; 2 El; 80 Brin. For sand castings; heat treatable. *Obsolete*

REYNOLDS 32S
Reynolds Metals Co.
Si 12.5, Cu 0.9, Mg 1, Ni 0.9, bal Al.
T6-temper: 55,000 TS; 46,000 YS; 9 El; 120 Brin. For aircraft pistons; heat treatable. *Obsolete*

REYNOLDS 333
Reynolds Metals Co.
Cu 3.8, Si 9, Mg 0.4, bal Al.
T533-temper: 32,000 TS; 25,000 YS; 1 El; 100 Brin. For permanent mold castings; heat treatable. *Obsolete*

REYNOLDS 356
Reynolds Metals Co.
Si 7, Mg 0.3, bal Al.
T51-temper: 25,000 TS; 20,000 YS; 2 El; 60 Brin. T6-temper: 33,000 TS; 24,000 YS; 4 El; 70 Brin. T7-temper: 34,000 TS; 30,000 YS; 2 El; 75 Brin. For sand castings; heat treatable. *Obsolete*

REYNOLDS 360
Reynolds Metals Co.
Si 9.5, Mg 0.5, bal Al.
Die cast: 42,000 TS; 23,000 YS; 1.8 El. For die castings; high fluidity. *Obsolete*

REYNOLDS 380
Reynolds Metals Co.
Cu 3.5, Si 8.5, bal Al.
Die cast: 45,000 TS; 25,000 YS; 2 El. For die castings; good machinability. *Obsolete*

REYNOLDS 3S
Reynolds Metals Co.
Mn 1.2, bal Al.
O-temper: 16,000 TS; 6,000 YS; 40 El; 28 Brin. H14-temper: 22,000 TS; 19,000 YS; 16 El; 40 Brin. H18-temper: 26,000 TS; 26,000 YS; 10 El; 55 Brin. For structures, tank cars, pipes; wrought, resists atmosphere corrosion. *Obsolete*

REYNOLDS 406
Reynolds Metals Co.
Mn 2, bal AL.
Cast: 19,000 TS; 9,000 YS; 12 El; 35 Brin. For sand castings; corrosion resistant. *Obsolete*

REYNOLDS 43
Reynolds Metals Co.
Si 5, bal Al.
Sand cast: 19,000 TS; 9,000 YS; 6 El; 40 Brin. Die cast: 30,000 TS; 14,000 YS; 7 El. For sand and die castings; corrosion resistant. *Obsolete*

REYNOLDS 52S
Reynolds Metals Co.
Mg 2.5, Cr 0.25, bal Al.
O-temper: 28,000 TS; 13,000 YS; 30 El; 45 Brin. H32-temper: 34,000 TS; 27,000 YS; 18 El; 62 Brin. H34-temper: 37,000 TS; 31,000 YS; 14 El; 67 Brin. For camera cases, deck housings, gasoline tanks; wrought, corrosion resistant. *Obsolete*

REYNOLDS 56S
Reynolds Metals Co.
Mn 0.1, Mg 5.2, Cr 0.1, bal Al.
O-temper: 42,000 TS; 22,000 YS; 35 El. H18-temper: 62,000 TS; 58,000 YS; 6 El. For rivets, fine wire; wrought. *Obsolete*

REYNOLDS 63S
Reynolds Metals Co.
Si 0.4, Mg 0.7, bal Al.
F-temper: 22,000 TS; 13,000 YS; 42 Brin. T5-temper: 31,000 TS; 24,000 YS; 65 Brin. T6-temper: 35,000 TS; 30,000 YS; 73 Brin. For light alloy parts; heat treatable. *Obsolete*

REYNOLDS 645
Reynolds Metals Co.
Cu 2.7, Zn 10.5, Fe 1.2, bal Al.
Cast: 29,000 TS; 17,000 YS; 4 El; 70 Brin. For sand castings; non-hardenable. *Obsolete*

REYNOLDS 750
Reynolds Metals Co.
Cu 1, Ni 1, Sn 6.5, bal Al.
T533-temper: 20,000 TS; 8,500 YS; 10 El; 45 Brln. For bearings, permanent mold castings; heat treatable. *Obsolete*

REYNOLDS 85
Reynolds Metals Co.
Cu 4, Si 5, bal Al.
Die cast: 40,000 TS; 22,000 YS; 3.5 El. For die castings; good machinability. *Obsolete*

REYNOLDS A-108
Reynolds Metals Co.
Cu 4.5, Si 5.5, bal Al.
Permanent mold cast: 28,000 TS; 16,000 YS; 2 El; 70 Brin. For permanent mold castings; good machinability. *Obsolete*

REYNOLDS A132
Reynolds Metals Co.
Cu 0.8, Si 12, Mg 1.2, Ni 2.5, bal Al.
T551-temper: 36,000 TS; 28,000 YS; 0.5 El; 105 Brin. T65-temper: 47,000 TS; 43,000 YS; 0.5 El; 125 Brin. For permanent mold castings; heat treatable. *Obsolete*

REYNOLDS A17S
Reynolds Metals Co.
Cu 2.5, Mg 0.3, bal Al.
T4-temper: 43,000 TS; 24,000 YS; 27 El; 70 Brin. For rivets, aircraft structures; heat treatable. *Obsolete*

REYNOLDS A214
Reynolds Metals Co.
Mg 3.8, Zn 1.8, bal Al.
Permanent mold cast: 27,000 TS; 16,000 YS; 7 El; 60 Brin. For permanent mold castings; corrosion resistant. *Obsolete*

REYNOLDS A51S
Reynolds Metals Co.
Si 1, Mg 0.6, Cr 0.25, bal Al.
T6-temper: 48,000 TS; 43,000 YS; 17 El; 100 Brin. For crankcases, connecting rods; heat treatable. *Obsolete*

REYNOLDS AS NO. 1
Reynolds Metals Co.
Al alloy.
For cold worked parts. *Obsolete*

REYNOLDS AS NO. 2
Reynolds Metals Co.
Al alloy.
For deep drawn parts. *Obsolete*

REYNOLDS B195
Reynolds Metals Co.
Cu 4.5, Si 2.5, bal Al.
T4-temper: 37,000 TS; 19,000 YS; 9 El; 75 Brin. T6-temper: 40,000 TS; 26,000 YS; 5 El; 90 Brin. T7-temper: 39,000 TS; 20,000 YS; 4.5 El; 80 Brin. For permanent mold castings; heat treatable. *Obsolete*

REYNOLDS B214
Reynolds Metals Co.
Si 1.8, Mg 3.8, bal Al.
Cast: 20,000 TS; 13,000 YS; 2 El; 50 Brin. For sand castings; corrosion resistant. *Obsolete*

REYNOLDS C113
Reynolds Metals Co.
Cu 7, Si 3.5, bal Al.
Permanent mold cast: 30,000 TS; 28,000 YS; 1 El; 80 Brin. For permanent mold castings; non-hardenable. *Obsolete*

REYNOLDS F214
Reynolds Metals Co.
Si 0.5, Mg 3.8, bal Al.
Cast: 20,000 TS; 12,000 YS; 3 El; 50 Brin. For sand castings; corrosion resistant. *Obsolete*

REYNOLDS J51S
Reynolds Metals Co.
Si 0.2-0.5, Cu 0.15-0.4, Mg 0.4-0.8, Zn 0-0.2, bal Al.
For brazed shelves; age-hardened. *Obsolete*

REYNOLDS METAL
Reynolds Metals Co.
Al.
For foil. *Obsolete*

REYNOLDS NO. 2364
Reynolds Metals Co.
Si 6, Cu 4, Fe, bal Al.
For alloying aluminum; master alloy. *Obsolete*

REYNOLDS NO. 2393
Reynolds Metals Co.
Si 9, Cu 3.5, Fe 0.75, bal Al.
For alloying aluminum; master alloy. *Obsolete*

REYNOLDS R317
Reynolds Metals Co.
Cu 3.5-4.5, Mn 0.4-1, Mg 0.2-0.8, Cr 0.25, Pb 0.3-0.7, Bi 0.3-0.7, bal Al.
T4-Temper: 62,000 TS; 40,000 YS; 22 El; 105 Brin. O-Temper: 26,000 TS; 10,000 YS; 22 El; 45 Brin. For light alloy parts, forging; free machining, wrought. *Obsolete*

REZISTAL "K.A. 2H"
Colt Industries
C 0-0.25, Ni 8-9.5, Cr 17.5-19, bal Fe.
Annealed: 85,000-95,000 TS; 30,000-40,000 YS; 55-60 El; 65-75 RA; 130-170 Brin. For welded construction which cannot receive Strauss heat treatment; corrosion resisting, austenitic. *Obsolete*

REZISTAL "K.A. 2M"
Colt Industries
C 0-0.15, Ni 8-10, Cr 18-20, Mo 2.5-3.5, bal Fe.
Heat treated: 90,000-100,000 TS; 45,000-50,000 YS; 55-60 El; 70-75 RA; 170-185 Brin. For use in pulp industry exposed to hot sulfite liquors and bleaches; corrosion resisting. *Obsolete*

REZISTAL "K.A.2S"
Colt Industries
C 0-0.07, Ni 8-9.5, Cr 17.5-19, bal Fe.
Annealed: 85,000-95,000 TS; 30,000-40,000 YS; 60-65 El; 65-75 RA; 170 Brin. For tanks, shafts, spokes, lamps, trim; corrosion resistant. *Obsolete*

REZISTAL 2 C
Colt Industries
C 0-0.2, Si 2-2.5, Ni 8-9.5, Cr 17.5-19, bal Fe.
Annealed: 90,000-100,000 TS; 45,000-55,000 YS; 55-65 El; 50-70 RA; 150-180 Brin. For stainless parts, piston rods, shafts, valves, acid pumps, annealing boxes; corrosion and scale resisting. *Obsolete*

REZISTAL 2600
Colt Industries
Cr 9, Ni 22.5, C 0.3, Cu 1.25, bal Fe.
Annealed: 85,000-95,000 TS; 45,000-55,000 YS; 30-40 El; 50-55 RA; 165-200 Brin. For furnace parts, oil pump parts, acid tanks; corrosion resisting. *Obsolete*

REZISTAL 2600-S
Colt Industries
C 0-0.15, Cr 7.5-9, Ni 21.5-23, Cu 1-1.5, bal Fe.
Annealed: 85,000-95,000 TS; 45,000-55,000 YS; 30-40 El; 50-55 RA; 165-200 Brin. For heat and corrosion resistant parts; heat and corrosion resistant. *Obsolete*

REZISTAL 3
Colt Industries
Cr 25, Ni 12, Si 2.25, C 0-0.2, bal Fe.
Annealed: 95,000-105,000 TS; 40,000-65,000 YS; 30-40 El; 45-50 RA; 160-200 Brin. For furnace parts; corrosion resisting. *Obsolete*

REZISTAL 301
Colt Industries
C 0.08-0.2, Cr 16-18, Ni 6-8, bal Fe.
Rolled: 150,000 TS; 110,000 YS; 15 El; 320 Brin. For stainless parts; stainless, austenitic. *Obsolete*

REZISTAL 302
Colt Industries
C 0.08-0.2, Cr 17-19, Ni 8-10, Mo 0.15-0.25, bal Fe.
Annealed: 95,000 TS; 40,000 YS; 60 El; 65 RA; 155 Brin. For architectural trim, chemical equipment; stainless, austenitic. *Obsolete*

REZISTAL 302 B
Colt Industries
C 0.08-0.2, Cr 17-19, Ni 8-10, Si 2-3, bal Fe.
Annealed: 95,000 TS; 40,000 YS; 55 El; 65 RA; 170 Brin. For furnace parts, air preheated; stainless, austenitic. *Obsolete*

REZISTAL 303
Colt Industries
C 0-0.15, Cr 17-19, Ni 8-10, Mo 0-0.6, S 0.07, bal Fe.
Annealed: 90,000 TS; 40,000 YS; 50 El; 55 RA; 175 Brin. For screw machine products; stainless, austenitic, free-cutting. *Obsolete*

REZISTAL 304
Colt Industries
C 0-0.08, Cr 17-19, Ni 8-11, bal Fe.
Annealed: 90,000 TS; 35,000 YS; 60 El; 65 RA; 170 Brin. For chemical and textile equipment; stainless, austenitic. *Obsolete*

REZISTAL 305
Colt Industries
C 0-0.12, Cr 17-19, Ni 10-13, bal Fe.
Annealed: 85,000 TS; 38,000 YS; 50 El; 55 RA; 160 Brin. For spun parts; stainless, austenitic. *Obsolete*

REZISTAL 308
Colt Industries
C 0-0.08, Cr 17-21, Ni 10-12, bal Fe.
Annealed: 85,000 TS; 30,000 YS; 55 El; 65 RA; 150 Brin. For stainless parts; stainless, austenitic. *Obsolete*

REZISTAL 309
Colt Industries
C 0-0.2, Cr 22-24, Ni 12-15, bal Fe.
Annealed: 95,000 TS; 40,000 YS; 55 El; 65 RA; 170 Brin. For air preheater baffles, aircraft heaters; stainless, austenitic. *Obsolete*

REZISTAL 309 B
Colt Industries
C 0-0.2, Cr 22-24, Ni 12-15, Si 2-3, bal Fe.
Annealed: 105,000 TS; 55,000 YS; 35 El; 45 RA; 185 Brin. For air preheater baffles, sulfite liquor equipment; stainless, austenitic. *Obsolete*

REZISTAL 310
Colt Industries
C 0-0.25, Cr 24-26, Ni 19-22, bal Fe.
Annealed: 100,000 TS; 45,000 YS; 50 El; 65 RA; 170 Brin. For heat exchangers, furnace doors, retorts; stainless, austenitic. *Obsolete*

REZISTAL 311
Colt Industries
C 0-0.25, Cr 18-20, Si 2-3, Ni 24-26, bal Fe.
Annealed: 90,000 TS; 45,000 YS; 45 El; 55 RA; 180 Brin. For annealing boxes, heat treating equipment; stainless, heat resistant. *Obsolete*

REZISTAL 314
Colt Industries
C 0-0.25, Cr 23-26, Ni 19-22, Si 1.5-3, bal Fe.
Annealed: 105,000 TS; 50,000 YS; 45 El; 60 RA; 180 Brin. For heat exchangers, furnace doors, retorts; heat resistant. *Obsolete*

REZISTAL 316
Colt Industries
C 0-0.1, Cr 16-18, Ni 10-14, Mo 2-3, bal Fe.
Annealed: 90,000 TS; 45,000 YS; 45 El; 55 RA; 180 Brin. For high temperature chemical handling equipment; corrosion and heat resistant. *Obsolete*

REZISTAL 317
Colt Industries
C 0-0.1, Cr 18-20, Ni 11-14, Mo 3-4, bal Fe.
Annealed: 85,000 TS; 40,000 YS; 50 El; 160 Brin. For high temperature chemical handling equipment; acid resistant. *Obsolete*

REZISTAL 318
Colt Industries
C 0-0.08, Cr 17-19, Ni 13-15, Mo 1.75-2.75, 8 x C = Cb, bal Fe.
Annealed: 90,000 TS; 45,000 YS; 45 El; 55 RA; 180 Brin. For welded applications; stabilized, stainless. *Obsolete*

REZISTAL 321
Colt Industries
C 0-0.08, Cr 17-19, Ni 8-11, 5 x C = Ti, bal Fe.
Annealed: 85,000 TS; 35,000 YS; 55 El; 65 RA; 160 Brin. For exhaust manifolds, boiler shells, expansion joints; stainless, austenitic. *Obsolete*

REZISTAL 325
Colt Industries
C 0.3-0.5, Cr 7-10, Ni 19-23, Cu 1-1.5, bal Fe.
Annealed: 100,000 TS; 50,000 YS; 35 El; 45 RA; 180 Brin. For oil pump shafts, mine water equipment; corrosion resistant. *Obsolete*

REZISTAL 329
Colt Industries
C 0.2, Cr 18, Ni 8, bal Fe.
For chemical plant equipment; stainless, austenitic. *Obsolete*

REZISTAL 347
Colt Industries
C 0-0.08, Cr 17-19, Ni 9-12, 10 x C = Cb, bal Fe.
Annealed: 90,000 TS; 40,000 YS; 50 El; 65 RA; 160 Brin. For exhaust manifolds, boiler shells; stainless, austenitic. *Obsolete*

REZISTAL 3C
Colt Industries
C 0.13-0.2, Cr 21-26, Ni 10-13, Si 2-3, bal Fe.
Annealed: 95,000-105,000 TS; 40,000-50,000 YS; 40 El; 50 RA; 170-200 Brin. For stainless parts; stainless and heat resistant. *Obsolete*

REZISTAL 4
Colt Industries
C 0-0.2, Si 2.25-3, Ni 25-26, Cr 17-20, bal Fe.
Annealed: 90,000-110,000 TS; 45,000-50,000 YS; 30-35 El; 35-45 RA; 160-185 Brin. For furnace parts, carburizing boxes, Diesel engine valves; corrosion and heat resisting. *Obsolete*

REZISTAL 403
Colt Industries
C 0-0.15, Cr 11.5-13, bal Fe.
Annealed: 75,000 TS; 40,000 YS; 30 El; 75 RA; 150 Brin. For steam turbine blades and parts; corrosion resistant. *Obsolete*

REZISTAL 405
Colt Industries
C 0-0.08, Cr 11.5-13.5, Al 1-3, bal Fe.
Annealed: 70,000 TS; 40,000 YS; 30 El; 60 RA; 150 Brin. For turbine blades, heat and corrosion resisting. *Obsolete*

REZISTAL 406
Colt Industries
C 0-0.15, Cr 12-14, Al 3.5-4.5, bal Fe.
Annealed: 75,000 TS; 45,000 YS; 25 El; 160 Brin. For electrical resistor elements; heat resistant. *Obsolete*

REZISTAL 410
Colt Industries
C 0-0.15, Cr 11.5-13.5, bal Fe.
Annealed: 75,000 TS; 45,000 YS; 30 El; 75 RA; 150 Brin. For coal screens and chutes, cutlery; corrosion resistant. *Obsolete*

REZISTAL 414

Colt Industries

C 0-0.15, Cr 11.5-13.5, Ni 1.25-2.5, bal Fe.
Annealed: 105,000 TS; 85,000 YS; 23 El; 65 RA; 225 Brin. For rules, mild springs, scraper knives; corrosion resistant
Obsolete

REZISTAL 416

Colt Industries

C 0-0.15, Cr 12-14, Mo 0-0.6, 0.07% min Se, bal Fe.
Annealed: 85,000 TS; 45,000 YS; 28 El; 65 RA; 180 Brin. For bolts, nuts, screws, golf clubs, valve trim; corrosion resistant, free-cutting. *Obsolete*

REZISTAL 420

Colt Industries

Cr 12-14, 0.15% min C, bal Fe.
Annealed: 95,000 TS; 60,000 YS; 30 El; 60 RA; 180 Brin. For bushings, cutlery, plastic molds, valve trim; corrosion resistant. *Obsolete*

REZISTAL 430

Colt Industries

C 0-0.12, Cr 14-18, bal Fe.
Annealed: 75,000 TS; 40,000 YS; 35 El; 60 RA; 160 Brin. For annealing boxes for brass, decorative trim; corrosion resistant. *Obsolete*

REZISTAL 431

Colt Industries

C 0-0.2, Cr 15-17, Ni 1.25-2.5, bal Fe.
Annealed: 120,000 TS; 95,000 YS; 23 El; 60 RA; 250 Brin. For aircraft fittings, bolts, paper machinery; corrosion resistant. *Obsolete*

REZISTAL 440A

Colt Industries

C 0.6-0.75, Cr 16-18, Mo 0-0.75, bal Fe.
Annealed: 100,000 TS; 60,000 YS; 25 El; 50 RA; 200 Brin. For bearings, cutlery, valve parts; corrosion resistant. *Obsolete*

REZISTAL 440B

Colt Industries

C 0.75-0.95, Cr 16-18, Mo 0-0.75, bal Fe.
Annealed: 105,000 TS; 65,000 YS; 20 El; 45 RA; 210 Brin. For bearings, cutlery, seam rollers; corrosion resistant. *Obsolete*

REZISTAL 440BM

Colt Industries

C 0.85-1.1, Cr 17-19, Mo 0.4-0.7, V 0.1-0.2, bal Fe.
Annealed: 110,000 TS; 65,000 YS; 15 El; 30 RA; 220 Brin. For bearings, cutlery; corrosion resistant. *Obsolete*

REZISTAL 440C

Colt Industries

C 0.95-1.2, Cr 16-18, Mo 0-0.75, bal Fe.
Annealed: 110,000 TS; 65,000 YS; 15 El; 30 RA; 220 Brin. For ball bearings, valve parts, bushings, cutlery; corrosion resistant. *Obsolete*

REZISTAL 442

Colt Industries

C 0-0.25, Cr 18-23, bal Fe.
Annealed: 85,000 TS; 45,000 YS; 20 El; 50 RA; 175 Brin. For nitric acid storage tanks and equipment; corrosion resistant. *Obsolete*

REZISTAL 446

Colt Industries

C 0-0.35, Cr 23-27, Ni 0-0.25, bal Fe.
Annealed: 85,000 TS; 55,000 YS; 25 El; 45 RA; 180 Brin. For salt bath electrodes, radio tube parts; corrosion resistant. *Obsolete*

REZISTAL 7

Colt Industries

C 0-0.25, Si 0.7-1.5, Ni 19-21, Cr 24-26, bal Fe.
Annealed: 100,000-110,000 TS; 45,000-55,000 YS; 45-55 El; 50-60 RA; 155-175 Brin. For furnace doors, skids, heat exchangers, pump parts, piston rods, retorts, valve parts; high heat and corrosion resistance. *Obsolete*

REZISTAL FM-18-8

Colt Industries

C 0.13-0.2, Cr 18, Ni 8, Mo, Se, bal Fe.
Annealed: 85,000-95,000 TS; 30,000-40,000 YS; 60-65 El; 65-75 RA; 130-170 Brin. For stainless parts, bolts, screws; stainless, heat and corrosion resistant. *Obsolete*

REZISTAL K A 2 MO

Colt Industries

Cr 18, Ni 8, C 0-0.15, Mo 3, bal Fe.
Annealed: 95,000 TS; 45,000 YS; 55 El; 70 RA; 170-200 Brin. For stainless parts; corrosion resisting. *Obsolete*

REZISTAL KA-2

Colt Industries

C 0.08-0.2, Cr 18, Ni 8, bal Fe.
Annealed: 85,000-95,000 TS; 30,000-40,000 YS; 55-60 El; 65-75 RA; 130-170 Brin. For stainless steel parts, tanks, shafts, spokes, lamps, trim; corrosion resistant; non-magnetic. *Obsolete*

REZISTAL KA-2-MO-T

Colt Industries

C 0.13-0.2, Cr 16-23, Ni 7-11, Mo, Ti, bal Fe.
For heat and corrosion resistant parts; heat and corrosion resistant. *Obsolete*

REZISTAL KA-2S-19-9

Colt Industries

C 0-0.07, Cr 19, Ni 9, bal Fe.
For stainless parts; stainless. *Obsolete*

REZISTAL KA-2S-20-10

Colt Industries

C 0-0.07, Cr 20, Ni 10, bal Fe.
Annealed: 95,000 TS; 35,000 YS; 57 El; 70 RA; 150 Brin. For stainless and heat resistant parts; stainless. *Obsolete*

REZISTAL KA-2S-MO

Colt Industries

C 0-0.07, Cr 18-20, Ni 8-12, Mo 2.5-3.5, bal Fe.
Annealed: 90,000-100,000 TS; 40,000-50,000 YS; 50-60 El; 60-75 RA; 170-200 Brin. For corrosion and heat resistant parts, tanks, tubes; corrosion and heat resistant. *Obsolete*

REZISTAL KA-2S-MO-T

Colt Industries

C 0-0.1, Cr 16-23, Ni 7-11, Mo, Ti, bal Fe.
For stainless parts; stainless. *Obsolete*

REZISTAL KA-2ST

Colt Industries

C 0-0.1, Cr 16-23, Ni 7-11, Ti, bal Fe.
Annealed: 80,000 TS; 35,000 YS; 57 El; 70 RA; 150 Brin. For stainless parts, tanks, tubes; stainless. *Obsolete*

REZISTAL KA-2ST SPECIAL 19-9

Colt Industries

C 0-0.07, Cr 19, Ni 9, Ti 0.35, bal Fe.
For stainless parts; stainless. *Obsolete*

REZISTAL KA-2T

Colt Industries

C 0.13-0.2, Cr 16-23, Ni 7-11, Ti, bal Fe.
For stainless parts; stainless. *Obsolete*

REZISTAL KA2

Krupp Stahl AG

stainless steel. C 0-0.15, Mn 0.4-0.65, Ni 8-9.5, Cr 17.5-19, bal Fe.
Heat treated: 85,000-95,000 TS; 30,000-40,000 YS; 55-60 El; 70-75 RA; 135-145 Brin. For all welded construction which cannot receive the Strauss heat treatment after welding; radiators, lamps, food and cooking utensils; non-magnetic, austenitic. *Obsolete*

REZISTAL KA2-19-9

Colt Industries

C 0.13-0.2, Cr 19, Ni 9, bal Fe.
For stainless parts, tanks, shafts, lamps, trim; stainless. *Obsolete*

REZISTAL KA2-20-10

Colt Industries

C 0.13-0.2, Cr 20, Ni 10, bal Fe.
For stainless parts; stainless. *Obsolete*

REZISTAL KA2SCB

Colt Industries

C 0-0.1, Cr 18, Ni 10.5, Cb = 8 x C, bal Fe.
Annealed: 85,000 TS; 35,000 YS; 57 El; 70 RA; 150 Brin. For heat and corrosion resisting parts; heat and corrosion resisting. *Obsolete*

REZISTAL NCR 124

Colt Industries

C 0.55, Mn 4.5, Cr 4, Ni 12, bal Fe.
For high temperature applications; heat resistant. *Obsolete*

REZISTAL NCR-238

Colt Industries

C 0.15, Cr 8, Ni 23, Cu 1, bal Fe.
For water fittings, beater blades, evaporators, shafting; heat and corrosion resistant. *Obsolete*

REZISTAL NCR-238

Crucible Specialty Metals

C 0.15, Cr 8, Ni 23, Cu 1, bal Fe.
For water fittings, beater blades, evaporators, shafting; heat and corrosion resistant. *Obsolete*

REZISTAL NO.2 CW

Colt Industries

C 0.13-0.2, Cr 16-23, Ni 7-11, W, Si, bal Fe.
For heat and corrosion resistant parts; heat and corrosion resistant. *Obsolete*

REZISTAL SAFWELD

Hawkridge Bros. Co.

C 0.1, Cr 18, Ni 8, bal Fe.
For welding rod; stainless, austenitic. *Obsolete*

REZISTAL STAINLESS 122

Colt Industries

C 0-0.15, Cr 12.5, Ni 1.75, bal Fe.
For corrosion resisting parts; corrosion resisting. *Obsolete*

REZISTAL STAINLESS 162

Colt Industries

C 0-0.2, Cr 16, Ni 2, bal Fe.
For propeller shafts, fittings, bolts; corrosion resisting. *Obsolete*

REZISTAL STAINLESS B-100

Colt Industries

C 0.81-1.1, Cr 16-23, bal Fe.
For cutlery, instruments, ball bearings; heat and corrosion resistant. *Obsolete*

REZISTAL STAINLESS B-80

Colt Industries

C 0.85, Cr 17, bal Fe.
For cutlery; corrosion resisting. *Obsolete*

REZISTAL STAINLESS BM

Colt Industries

C 0-1, Cr 16-18, Mo 0.5, bal Fe.
For cutlery; corrosion and wear resistant. *Obsolete*

REZISTAL STAINLESS IRON 12

Colt Industries

C 0-0.15, Mn 0.35-0.6, Si 0-0.5, Cr 15-18, bal Fe.
Oil treated: 106,000-182,000 TS; 86,000-179,000 YS; 18-24 El; 65-72 RA; 228-363 Brin. For turbine blading, golf clubs, pump rods, gate valves, cutlery, machine parts; resists atmosphere, salt water. *Obsolete*

REZISTAL STAINLESS IRON 12
Crucible Specialty Metals
C 0-0.15, Mn 0.35-0.6, Si 0-0.5, Cr 15-18, bal Fe.
Oil treated: 106,000-182,000 TS; 86,000-179,000 YS; 18-24 El; 65-72 RA; 228-363 Brin. For turbine blading, golf clubs, pump rods, gate valves, cutlery, machine parts; resists atmosphere, salt water. *Obsolete*

REZISTAL STAINLESS IRON 17
Colt Industries
C 0-0.12, Cr 17, bal Fe.
Annealed: 75,000-85,000 TS; 45,000-55,000 YS; 30-40 El; 50-60 RA; 150-190 Brin. For corrosion resistant parts, storage tanks; corrosion resistant. *Obsolete*

REZISTAL STAINLESS IRON 20
Colt Industries
C 0-0.35, Cr 20, bal Fe.
Annealed: 70,000-80,000 TS; 45,000-55,000 YS; 30-40 El; 50-60 RA; 165-185 Brin. For corrosion resistant parts, storage tanks; corrosion resistant. *Obsolete*

REZISTAL STAINLESS IRON 27
Colt Industries
C 0.21-0.35, Cr 27, bal Fe.
Rolled: 80,000-90,000 TS; 55,000-65,000 YS; 20-30 El; 40-50 RA; 170-200 Brin. For heat and corrosion resistant parts; heat and corrosion resistant. *Obsolete*

REZISTAL STAINLESS IRON F.M. 2
Colt Industries
C 0-0.15, Mn 0.3, Si 0.3-0.5, S 0.3-0.35, Mo 0.55, Cr 14-15.5, bal Fe.
Annealed: 90,000 TS; 55,000 YS; 28 El; 62 RA; 160 Brin. For stainless iron parts; stainless iron. *Obsolete*

REZISTAL STAINLESS IRON F.M. 2
Crucible Specialty Metals
C 0-0.15, Mn 0.3, Si 0.3-0.5, S 0.3-0.35, Mo 0.55, Cr 14-15.5, bal Fe.
Annealed: 90,000 TS; 55,000 YS; 28 El; 62 RA; 160 Brin. For stainless iron parts; stainless iron. *Obsolete*

REZISTAL STAINLESS STEEL GRADE "A"
Colt Industries
C 0.45, Cr 12-15, bal Fe.
Heat treated: 250,000 TS; 220,000 YS; 4 El; 6 RA. For stainless parts, cutlery; corrosion resistant. *Obsolete*

REZISTAL STAINLESS STEEL GRADE "A"
Crucible Specialty Metals
C 0.45, Cr 12-15, bal Fe.
Heat treated: 250,000 TS; 220,000 YS; 4 El; 6 RA. For stainless parts, cutlery; corrosion resistant. *Obsolete*

REZISTAL STAINLESS STEEL GRADE "B"
Colt Industries
C 0.6, Cr 15-18, bal Fe.
Heat treated: 430-600 Brin. For stainless parts, bearings, valves; corrosion resistant. *Obsolete*

REZISTAL STAINLESS STEEL GRADE "B"
Crucible Specialty Metals
C 0.6, Cr 15-18, bal Fe.
Heat treated: 430-600 Brin. For stainless parts, bearings, valves; corrosion resistant. *Obsolete*

REZISTAL TURBINE
Colt Industries
C 0-0.15, Cr 12, bal Fe.
For turbine parts; corrosion resisting. *Obsolete*

REZISTAL VT
Colt Industries
C 0.25, Mn 0.5, Ni 2.5, Cr 27, Mo 5.5, bal Co.
Cast: 96,000 TS; 70,000 YS; 8 El; 15 RA; 290 Brin. For jet engine components; heat resistant. *Obsolete*

REZISTAL WH
Colt Industries
C 0.55, Mn 4, Cr 18, Ni 9, bal Fe.
For railroad frogs, crane tracks and wheels; work hardened. *Obsolete*

REZISTAL X-40
Colt Industries
C 0.5, Ni 10.5, Cr 25.5, W 7.5, bal Co.
At 70 F: 110,000 TS; 64,000 YS; 15 El; 23.4 RA; 270 Brin. At 1500 F: 59,000 TS; 17.4 El. For valves, high temperature applications; heat resistant. *Obsolete*

REZISTAN
Uniworld Corp. of America
C 3.2, Cr, Ni, Cu, Mo, Al, bal Fe.
For pumps, valves, combustion chambers; alloy cast iron, corrosion resistant.

RF
Friedr. Lohmann GmbH
Stainless steel. C 0.45, Cr 13, Mn 0.3, bal Fe.
Annealed: 229 Brin. For valves, cutlery, surgical and dental instruments and for molds for processing of corrosive plastics. Type 420 stainless. S42000

RF-1
Forjas Alavesas S.A.
C 0.18, Mn 0.7, Si 0.25, Mo 0.5, bal Fe.
For tubing, vessels for elevated temperature. DIN 15Mo3; ASTM A182-F1; AFNOR 13 CD2.

RF-2
Forjas Alavesas S.A.
C 0.13, Mn 0.5, Cr 1.1, Mo 0.5, bal Fe.
For equipment for elevated temperature. ASTM A182-F12; AFNOR 10CD5.

RF-3
Forjas Alavesas S.A.
C 0.13, Mn 0.5, Si 0.4, Cr 2.3, Mo 1, bal Fe.
Fittings for oil refineries and other high temperature operations. ASTM A182-F22; AFNOR 10CD9.

RF-5
Forjas Alavesas S.A.
C 0.11, Mn 0.5, Si 0.25, Cr 5, Mo 0.5, bal Fe.
Equipment for oil refineries and other high temperature operations. Similar to SAE 51501; ASTM A182-F5/F5A; DIN 12CrMo195; BS625.

RFF
Saarstahl AG
Now SAARSTAHL 60 SICR 7.

RFFH
Saarstahl AG
Now SAARSTAHL 65 SI 7.

RFFL
Saarstahl AG
C 0.63, Si 1.65, Mn 0.8, Cr 0.25, bal Fe.
Steel for flat or coil springs. W.-Nr. 0960. *Obsolete*

RFFN
Saarstahl AG
Now SAARSTAHL 51 SI 7.

RFFO
Saarstahl AG
Now SAARSTAHL 55 SI 7.

RFFW
Saarstahl AG
Now SAARSTAHL 46 SI 7.

RFFWW
Saarstahl AG
C 0.38, Si 1.8, Mn 0.7, bal Fe.
Water hardening spring steel. W.-Nr. 0970. *Obsolete*

RG5
German manufacture
Cu 85, Sn 5, Zn 7, Pb 3.
For castings; pressure tight.

RG7H
Kawasaki Steel Corp.
High magnetic induction, grain oriented silicon sheet: 0.012-in. thick. Maximum permeability: 0.082 (H/M); resistivity: 45 micro-ohm·cm; core loss: (W15/50) 0.36 W/lb. Grain oriented silicon steel; for motors and/or transformers.

RGS 1
Saarstahl AG
Now SAARSTAHL 1.2713.

RGS 3
Saarstahl AG
Now SAARSTAHL 1.2743.

RGS 4
Saarstahl AG
Now SAARSTAHL 1.2714.

RGS 6
Saarstahl AG
Now SAARSTAHL 1.2777.

RGT 0
Saarstahl AG
C 0.13, Cr 19.5, Fe 0-5, Ti 0.4, bal Ni.
850 N/mm^2 TS (RT); 350 N/mm^2 YS (RT). Corrosion resistant and heat resistant alloy. For combustion chamber parts. W.-Nr. 2.4951. *Obsolete*

RGT 1
Saarstahl AG
Now SAARSTAHL 1.4980 and SAARSTAHL 1.4944.

RGT 101
Saarstahl AG
Now SAARSTAHL 2.4650.

RGT 12
Saarstahl AG
Now SAARSTAHL 2.4969 and SAARSTAHL 2.4622.

RGT 13
Saarstahl AG
C 0.06, Cr 20, Mo 4.5, Co 18, Fe 0-5, Al 1.5, Ti 2.5, bal Ni.
1250 N/mm^2 TS (RT); 800 N/mm^2 YS (RT); 30 El (RT). Corrosion resisting and heat resisting alloy. For parts for gas turbines and transmissions operating up to 900°C. W.-Nr. 2.4982. *Obsolete*

RGT 14
Saarstahl AG
Now SAARSTAHL 2.4983.

RGT 15
Saarstahl AG
Now SAARSTAHL 2.4973.

RGT 16
Saarstahl AG
Now SAARSTAHL 2.4634.

RGT 18
Saarstahl AG
C 0.15, Cr 15, Mo 4, Co 30, Fe 0-2, Al 3, Ti 2.5, bal Ni.
1150 N/mm^2 TS (RT); 750 N/mm^2 YS (RT); 15 El (RT). Corrosion resisting and heat resisting alloy. For parts for gas turbines and transmissions for operation up to 950°C. *Obsolete*

RGT 3
Saarstahl AG
Now SAARSTAHL 2.4952 and SAARSTAHL 2.4631.

RGT 32
Saarstahl AG
C 0.12, Cr 21, Ni 20, Mo 3, Co 20, W 2.5, N 0.15, 1.0 Nb + Ta, bal Fe.
800 N/mm² TS (RT); 350 N/mm² YS (RT); 35 El (RT).
Corrosion resisting and heat resisting alloy. For parts of gas turbines and transmissions for operation up to 850°C. W.-Nr. 1.4971. *Obsolete*

RGT 33
Saarstahl AG
C 0.45, Cr 20, Ni 20, Mo 4, Co 20, W 4, 4.0 Nb + Ta, bal Fe.
1000 N/mm² TS (RT); 550 N/mm² YS (RT); 10 El (RT).
Corrosion resisting and heat resisting alloy. For parts of gas turbines and transmissions for operation up to 850°C. W.-Nr. 1.4977 and 1.4978. *Obsolete*

RGT 35
Saarstahl AG
Now SAARSTAHL 2.4989.

RGT 36
Saarstahl AG
Now SAARSTAHL 2.4964.

RGT 4
Saarstahl AG
C 0.06, Cr 20, Mo 4.5, Fe 0-5, Al 1.4, Ti 2.4, bal Ni.
1200 N/mm² TS (RT); 750 N/mm² YS (RT); 20 El (RT).
Corrosion resisting and heat resisting alloy. For parts for gas turbines and transmissions. For use up to 850°C. W.-Nr. 2.4976. *Obsolete*

RGT 5
Saarstahl AG
C 0.1, Cr 22, Mo 9, Co 1.5, W 0.6, Fe 18.5, bal Ni.
800 N/mm² TS (RT); 350 N/mm² YS (RT); 40 El (RT).
Corrosion resisting and heat resisting alloy. Parts for gas turbines and transmissions, for use up to 850°C. W.-Nr. 2.4972. *Obsolete*

RGT 501
Saarstahl AG
Now SAARSTAHL 2.4654.

RGT 6
Saarstahl AG
Now SAARSTAHL 2.4668X1.

RGT 601
Saarstahl AG
Now SAARSTAHL 2.4668.

RGT 8
Saarstahl AG
Now SAARSTAHL 2.4662.

RGT 9
Saarstahl AG
C 0.06, Cr 15, Ni 45, Mo 4, W 4, Al 1, Ti 3, bal Fe.
1300 N/mm² TS (RT); 900 N/mm² YS (RT); 12 El (RT).
Corrosion resisting and heat resisting alloy. For parts operating up to 850°C. *Obsolete*

RH SPECIAL
Cytemp Specialty Steel Div.
C 0.55, Mn 0.35, Si 0.9, Cr 5, V 0.25, W 1.25, Mo 1.25, bal Fe.
AISI Type A8 air hardening tool steel. *Obsolete*

RHEINDUR
German manufacture
Aluminum. Cu 3.5-5, Si 0.5-1, Mn 0.5-1, Mg 0.5-1.2, bal Al.
Heat treated: 57,000-71,000 TS; 50,000-54,000 YS; 10-12 El; 130-140 Brin. For fasteners, aircraft and jet engine components. Age hardenable. High strength.

RHEINROHR 18/12 MS
Rheinische Rohrenwerke Atkiengesellschaf
C 0.1, Cr 16-18, Ni 10.5-13, Mo 2-2.5, Ti = 4 x C, bal Fe.
Rolled: 78,000-107,000 TS; 35,600 YS; 40-45 RA. For superheater tubes, boiler ends, chemical tanks; heat resistant to 850°C.

RHEINROHR 18/12S
Rheinische Rohrenwerke Atkiengesellschaf
C 0.08, Cr 16, Ni 13, Cb, bal Fe.
Rolled: 78,000-107,000 TS; 31,300 YS; 40-50 El. For superheater tubes, boiler ends, chemical tanks; heat resistant to 850°C.

RHEINROHR 230V
Rheinische Rohrenwerke Atkiengesellschaf
C 0.12, Ni 5, bal Fe.
Rolled: 85,000-102,000 TS; 58,000 YS; 20 El. For refrigeration components, bolts, fasteners; ductile at low temperature.

RHEINROHR 3CA2
Rheinische Rohrenwerke Atkiengesellschaf
C 0.15, Cr 0.75, Mo 0.15, bal Fe.
For gears, shafts, cams, camshafts; case hardened, tough.

RHEINROHR 3DB22
Rheinische Rohrenwerke Atkiengesellschaf
C 0.13, Cr 1, Mo 0.25, V 0.2, bal Fe.
For gears, shafts, cams, camshafts; case hardened, tough.

RHEINROHR 3HK5
Rheinische Rohrenwerke Atkiengesellschaf
C 0.2, Si 0.5, Cr 2-2.5, Mo 1, bal Fe.
Rolled: 71,000-92,500 TS; 42,700 YS; 12-18 El. For superheater tubes, boiler ends, chemical tanks; heat resistant to 580°F.

RHEINROHR 9F2
Rheinische Rohrenwerke Atkiengesellschaf
C 0.44, Cr, Si, Mn, bal Fe.
For gears, shafts, crankshafts, bolts, studs; water or oil hardened.

RHEINROHR C1203
Rheinische Rohrenwerke Atkiengesellschaf
C alloy, bal Fe.
For machine tool parts; oil hardened, tough.

RHEINROHR C73
Rheinische Rohrenwerke Atkiengesellschaf
C 0.35, Si 0.27, Mn 0.5, bal Fe.
Hot rolled: 85,000 TS; 54,000 YS; 30 El; 53 RA; 183 Brin. For gears, bolts, fasteners; water hardened.

RHEINROHR CS65S
Rheinische Rohrenwerke Atkiengesellschaf
C 0.12, Si 1.15, Cr 2.5, Mo 0.45, bal Fe.
Annealed: 65,000-85,000 TS; 36,000 YS; 21 El. For superheater tubes, boiler ends, drums; heat resistant to 650°F.

RHEINROHR CUNI 47
Rheinische Rohrenwerke Atkiengesellschaf
C 0.2, Cu, Ni, bal Fe.
For machine tool parts; case hardened.

RHEINROHR CUNI 52
Rheinische Rohrenwerke Atkiengesellschaf
C 0.2, Cu, Ni, Mn, bal Fe.
For machine tool parts.

RHEINROHR CUNI 52 MO
Rheinische Rohrenwerke Atkiengesellschaf
C 0.2, Cu, Ni, Mn, Mo, bal Fe.
For machine tool parts.

RHEINROHR CV18W
Rheinische Rohrenwerke Atkiengesellschaf
C 0.2, Si 1, Cr 1.5-2, V, Mo, bal Fe.
Rolled: 71,000-92,500 TS; 42,700 YS; 20-25 El. For superheater tubes, boiler ends, chemical tanks; heat resistant to 620°C.

RHEINROHR D33
Rheinische Rohrenwerke Atkiengesellschaf
C 0.17, Mn 1.05, Si 0.3, bal Fe.
For gears, bolts, machine tool parts; case hardened.

RHEINROHR D45
Rheinische Rohrenwerke Atkiengesellschaf
C alloy, bal Fe.
For machine tool parts; oil hardened, tough.

RHEINROHR D45V
Rheinische Rohrenwerke Atkiengesellschaf
C 0.1, Cr 1.5, V, bal Fe.
Rolled: 64,000-85,000 TS; 42,700 YS; 18-25 El. For petroleum cracking and ammonia synthesis equipment; heat resistant to 550°C.

RHEINROHR D73
Rheinische Rohrenwerke Atkiengesellschaf
C 0.4, Si 0.37, Mn 0.95, bal Fe.
For gears, bolts, machine tool parts; water hardened.

RHEINROHR D83
Rheinische Rohrenwerke Atkiengesellschaf
C 0.46, Si 0.35, Mn 0.8, bal Fe.
For gears, bolts, machine tool parts; water hardened.

RHEINROHR E33
Rheinische Rohrenwerke Atkiengesellschaf
C 0.14, Mn 1.05, Si 0.3, bal Fe.
For gears, shafts, machine tool parts; case hardening steel.

RHEINROHR E34SP
Rheinische Rohrenwerke Atkiengesellschaf
C 0.17, Mn 1.05, Si 0.3, bal Fe.
For gears, shafts, machine tool parts; case hardening steel.

RHEINROHR E35SP
Rheinische Rohrenwerke Atkiengesellschaf
C 0.19, Si 0.5, Mn 1.15, bal Fe.
For gears, machine tool parts; case hardening steel.

RHEINROHR E45
Rheinische Rohrenwerke Atkiengesellschaf
C 0.19, Mn 1.15, Si 0.5, bal Fe.
For gears, shafts, machine tool parts; case hardening steel.

RHEINROHR E45SP
Rheinische Rohrenwerke Atkiengesellschaf
C 0.22, Si 0.5, Mn 1.15, bal Fe.
For gears, shafts, machine tool parts; case hardening steel.

RHEINROHR E93
Rheinische Rohrenwerke Atkiengesellschaf
C 0.46, Mn 0.96, Si 0.35, bal Fe.
Hot rolled: 100,000 TS; 60,000 YS; 24 El; 45 RA; 215 Brin. For gears, bolts, shafts, fasteners; water hardened.

RHEINROHR F1203
Rheinische Rohrenwerke Atkiengesellschaf
C alloy, bal Fe.
For machine tool parts; oil hardened.

RHEINROHR H-205
Rheinische Rohrenwerke Atkiengesellschaf
C, bal Fe.
For boiler plates.

RHEINROHR H-215
Rheinische Rohrenwerke Atkiengesellschaf
C, bal Fe.
For boiler plates.

RHEINROHR H-215A
Rheinische Rohrenwerke Atkiengesellschaf
C, bal Fe.
For boiler plates.

RHEINROHR H-316
Rheinische Rohrenwerke Atkiengesellschaf
C, bal Fe.
For boiler plates.

RHEINROHR H-326
Rheinische Rohrenwerke Atkiengesellschaf
C, bal Fe.
For boiler plates.

RHEINROHR H-437
Rheinische Rohrenwerke Atkiengesellschaf
C, bal Fe.
For boiler plates.

RHEINROHR HSB40
Rheinische Rohrenwerke Atkiengesellschaf
C, bal Fe.
For machine tool parts; water hardened.

RHEINROHR HSB45
Rheinische Rohrenwerke Atkiengesellschaf
C, bal Fe.
For machine tool parts; water hardened.

RHEINROHR HSB50
Rheinische Rohrenwerke Atkiengesellschaf
C, bal Fe.
For machine tool parts; water hardened.

RHEINROHR HSB50R
Rheinische Rohrenwerke Atkiengesellschaf
C, bal Fe.
For machine tool parts; water hardened.

RHEINROHR HSB55
Rheinische Rohrenwerke Atkiengesellschaf
C, bal Fe.
For machine tool parts; water hardened.

RHEINROHR IK!
Rheinische Rohrenwerke Atkiengesellschaf
C 0.1, Ti 0.5, bal Fe.
Rolled: 48,000-60,000 TS; 21,400 YS; 22 El. For chromium
impregnation; for stainless surfaces.

RHEINROHR MN47MO
Rheinische Rohrenwerke Atkiengesellschaf
C 0.19, Mn, Mo, bal Fe.
For gears, bolts, camshafts, cams; case hardened.

RHEINROHR N10
Rheinische Rohrenwerke Atkiengesellschaf
C 0.21, Cr 3, Mo 0.4, V 0.8, W 0.37, bal Fe.
For die casting dies; oil hardened.

RHEINROHR N1K
Rheinische Rohrenwerke Atkiengesellschaf
C 0.16, Cr 2.25, bal Fe.
For oil refinery equipment; heat and creep resistant.

RHEINROHR N8N
Rheinische Rohrenwerke Atkiengesellschaf
C 0.17, Cr 2.7, Mo 0.25, V 0.15, bal Fe.
For camshafts, cams, oil refinery equipment; case hardened.

RHEINROHR N9
Rheinische Rohrenwerke Atkiengesellschaf
C 0.2, Cr 3.25, Mo 0.5, V 0.5, bal Fe.
For die casting dies; oil hardened.

RHEINROHR TH31
Rheinische Rohrenwerke Atkiengesellschaf
C 0.15, Mo 0.3, Mn 0.6, Si 0.25, bal Fe.
For gears, bolts, machine tool parts; case hardening steel.

RHEINROHR TH32
Rheinische Rohrenwerke Atkiengesellschaf
C 0.13, Cr 0.85, Mo 0.45, bal Fe.
For machine tool parts, gears, pinions; case hardening steel.

RHEINROHR TS53
Rheinische Rohrenwerke Atkiengesellschaf
C 0.2, Ti 0.5, bal Fe.
Rolled: 50,000-64,000 TS; 25,000 YS; 20-30 El. For steam
boilers; good creep strength.

RHEINROHR TS57
Rheinische Rohrenwerke Atkiengesellschaf
C alloy, bal Fe.
For machine tool parts; oil hardened.

RHEINROHR TSL
Rheinische Rohrenwerke Atkiengesellschaf
C alloy, bal Fe.
For machine tool parts; oil hardened.

RHEINROHR TT
Rheinische Rohrenwerke Atkiengesellschaf
C 0.1, Mn, Al, bal Fe.
Rolled: 58,000-72,000 TS; 36,000 YS min; 22 El min. For
refrigeration equipment; ductile to -80°C.

RHEINROHR TTA23
Rheinische Rohrenwerke Atkiengesellschaf
C 0.15, Mn 17, Cr 3, bal Fe.
Rolled: 100,000-128,000 TS; 38,500 YS; 30 El. For low
temperature applications; ductile to -185°C.

RHEINROHR-MV12
Mannesmannrohren-Werke AG
C 0.2, Cr 12, Mo 1, V 0.3, bal Fe.
Annealed: 95,000 TS; 40,000 YS; 25 El; B 92 Rock.
Hardened: 240,000 TS; 205,000 YS; 10 El; C 50 Rock. For
valves, cutlery, bearings, surgical instruments. Corrosion
resistant, hardenable. *Obsolete*

RHEINROHR-MVW12
Mannesmannrohren-Werke AG
C 0.2, Cr 12, Mo 1, V 0.3, W 0.5, bal Fe.
Annealed: 95,000 TS; 40,000 YS; 25 El; B 92 Rock.
Hardened: 235,000 TS; 195,000 YS; 10 El; C 48 Rock. For
valves, bearings, cutlery, surgical instruments. Corrosion
resistant, hardenable. *Obsolete*

RHENIUM
Vereinigte Chemische Fabriken
Re 99.5.
70,000 TS; 24 El. For constituent of alloys and chemicals;
electroplating.

RHENIUM
Atomergic Chemetals Corp.
Re.
Purities: 99.999%, 99.99%, 99.9%, 99+%. Forms: powder,
wire, sheet, foil, sintered bar, tubing, arc melted buttons,
single crystals.

RHEOTAN NO. 1
English manufacture
Cu 84, Mn 12, Zn 4.
For resistances.

RHEOTAN NO. 2
English manufacture
Cu 53, Zn 18, Ni 25, Fe 5.
For resistances.

RHEOTAN NO. 3
English manufacture
Cu 84, Mn 2, Zn 4, Fe 12.
For resistances.

RHF 15
Saarstahl AG
Now SAARSTAHL 1.7734.

RHF 32
Saarstahl AG
Now SAARSTAHL 1.6358 and SAARSTAHL 1.6354.

RHINO 3.3.3.
Commentryenne
C 0.95-1.03, Mn 0.2-0.4, Si 0.2-0.4, P 0-0.03, S 0-0.03, Cr
3.8-4.5, Mo 2.5-2.8, W 2.7-3, V 2.2-2.5, bal Fe.
For hand hacksaw blades and power blades.

RHINO CO10
Commentryenne
C 1.2-1.35, Si 0.2-0.4, Mn 0.2-0.4, S 0-0.03, P 0-0.03, W
9.5-11, Cr 3.8-4.5, Mo 3.5-4, V 3-3.5, Co 10-11, bal Fe.
Bar, circles, sheet, discs. High speed steel for tool bits.

RHINO CO2
Commentryenne
C 0.98-1.02, Mn 0.25-0.35, Si 0.2-0.4, P 0-0.03, S 0-0.03, Cr
4.1-4.4, Mo 4.75-5.25, W 6-6.75, V 1.75-2.05, Co 1.75-2.25,
bal Fe.
For cutters, broaches, tool bits, taps, special drills, saws, dies,
punches.

RHINO CO3
Commentryenne
C 0.95-1.03, Mn 0.2-0.4, Si 0.2-0.4, P 0-0.03, S 0-0.03, Cr
3.8-4.5, Mo 2.5-2.8, W 2.7-3, V 2.2-2.5, Co 2.25-2.75, bal Fe.
High speed steel.

RHINO M1
Commentryenne
C 0.82-0.85, Si 0.25-0.4, Mn 0.2-0.3, S 0-0.025, P 0-0.025, W
1.4-1.8, Cr 3.7-4, Mo 8.5-9, V 1.15-1.3, bal Fe.
Bar, blanks, sheet, discs. High speed steel for tools such as
drills, dies, cutters, saws, punches. AISI M1.

RHINO M2
Commentryenne
C 0.85-0.9, Si 0-0.35, Mn 0-0.35, S 0-0.025, P 0-0.025, W
6-6.75, Cr 3.9-4.3, Mo 4.75-5.25, V 1.75-2.05, bal Fe.
Bar, blanks, sheet, discs; replaces 18-4-1. High speed steel
for cutters, drills, taps, dies, chasers, saws, punches. AISI M2.

RHINO M3 TYPE 2
Commentryenne
C 1.17-1.2, Mn 0.2-0.4, Si 0.2-0.4, P 0-0.03, S 0-0.03, Cr 3.8-4,
Mo 4.7-5.1, W 6-6.5, V 2.8-3, bal Fe.
High speed steel for taps, broaches, cutters. AISI M3 Type 2.

RHINO M35
Commentryenne
C 0.89-0.94, Si 0.2-0.4, Mn 0.2-0.3, S 0-0.03, P 0-0.03, W
6-6.5, Cr 4.25-4.5, Mo 4.75-5.2, V 1.75-1.8, Co 4.5-5, bal Fe.
Bar, circles, sheet, discs. High speed steel for cutters,
broaches, tool bits, special drills, saws, dies, punches. AISI
M35.

RHINO M4
Commentryenne
C 1.25-1.3, Si 0.25-0.35, Mn 0.2-0.3, S 0-0.03, P 0-0.03, W
5.25-5.75, Cr 4.25-4.75, Mo 4.25-4.75, V 3.75-4.25, bal Fe.
Bar, blanks, sheet, discs. High speed steel for lathe and
planer tools, hacksaws, punches. AISI M4.

RHINO M42
Commentryenne
C 1.05-1.1, Si 0.15-0.3, Mn 0.15-0.3, S 0-0.025, P 0-0.025, W
1.3-1.7, Cr 3.5-4, Mo 9.25-9.75, V 1.05-1.25, Co 7.75-8.25, bal
Fe.
Bar, forgings, sheet, discs. Super high speed steel for milling
cutters, tool bits, drills, reamers, broaches. AISI M42.

RHINO M7
Commentryenne
C 0.98-1.03, Si 0.2-0.3, Mn 0.2-0.3, S 0-0.03, P 0-0.03, W
1.5-2, Cr 3.5-4, Mo 8.4-9.1, V 1.9-2.2, bal Fe.
Bar, blanks, sheet, discs. High speed steel for drills, taps,
cutters. AISI M7.

RHINO MCO
Commentryenne
C 0.88-0.92, Mn 0.2-0.3, Si 0.4-0.55, P 0-0.25, S 0-0.25, Cr 3.6-3.85, Mo 4.75-5, V 1.1-1.4, Co 2.05-2.55, bal Fe.

RHINO MV2
Commentryenne
C 0.86-0.9, Mn 0.2-0.3, Si 0.2-0.4, P 0-0.025, S 0-0.025, Cr 4-4.25, Mo 4.25-4.75, W 1-1.3, V 1.75-2.05, bal Fe.
Steel used mainly for drills. AISI M52.

RHINO SUPERAL
Commentryenne
C 1.1-1.15, Mn 0.25-0.3, Si 0.3-0.4, P 0-0.025, S 0-0.025, Cr 3.7-4, Ni 0-0.2, Mo 4.9-5.1, W 6-6.2, V 1.75-1.8, Cu 0-0.2, Al 1-1.3, bal Fe.
For milling cutters.

RHODIUM
Atomergic Chemetals Corp.
Rh.
Purities: 99.999%, 99.99%, 99.9%. Forms: sponge, powder, wire, foil, sheet, rod, single crystals.

RHOTANIUM
English manufacture
Au 60-90, Pd 10-40.
For jewelry; corrosion resistant.

RI PREMIUM
Sandvik Steel Co.
WC.
For cutting tools, for superalloys. Sintered carbide, abrasion resistant. *Obsolete*

RICH GOLD METAL
Manufacturer not listed.
Cu 90, Zn 10.
For window screen wire; commercial bronze.

RICH LOW BRASS
Anaconda Co.
Copper. Zn 15, Pb, Fe, bal Cu.
For hardware, fittings. Red brass.

RICHARD PLASTIC BABBITT
English manufacture
Sn 82.43, Sb 9.77, Cu 8.1.
For Babbitt, bearings; antifriction.

RICHARDS
English manufacture
Sn 71.5, Zn 25, Al 3.5.
For aluminum solder.

RICHARDS ALLOY
English manufacture
Zn 96, Al 4.
For die castings.

RICHARDS BRONZE
English manufacture
Zn 42, Fe 1, Al 1-2, bal Cu.
For marine hardware; corrosion resistant, high strength.

RICHARDSON'S SPECULUM
English manufacture
Cu 65, Sn 30, As 2, Si 2, Zn 0.7.
For reflectors; high polish.

RICHLOY HI TENSILE SOLDER NO. 7
National Cable & Metal Co.
Sn 65, Pb 35.
Cast: 9577 TS; 32 El. For soft solder; melting point 360°F.

RICHLOY HI-TENSILE SOLDER NO. 3
National Cable & Metal Co.
Sn 20, Pb 80.
Cast: 7183 TS; 47 El. For soft solder; melting point 416°F.

RICHLOY HI-TENSILE SOLDER NO. 4
National Cable & Metal Co.
Sn 30, Pb 70.
Cast: 7883 TS; 42 El. For soft solder; melting point 390°F.

RICHLOY HI-TENSILE SOLDER NO. 5
National Cable & Metal Co.
Sn 40, Pb 60.
Cast: 9000 TS; 36 El. For soft solder; melting point 370°F.

RICHLOY HI-TENSILE SOLDER NO. 6
National Cable & Metal Co.
Sn 50, Pb 50.
Cast: 9412 TS; 34 El. For soft solder; melting point 358°F.

RIGOR
Uddeholm Corp.
C 1, Si 0.2, Mn 0.6, Cr 5.3, Mo 1.1, V 2, bal Fe.
For dies; for cold work. AISI A2.

RIKEN 201
Riken Metal Mfg. Co.
Magnesium. Al 5-7, Mn 0.1-0.5, Zn 0-1.5, bal Mg.
Rolled: 31,000-37,000 TS; 3-10 El. For light parts; wrought alloy.

RIKEN 202
Riken Metal Mfg. Co.
Magnesium. Al 8-11, Mn 0.1-0.5, Zn 0-1, bal Mg.
Rolled: 37,00-42,500 TS; 5-7 El. For light parts; wrought alloy.

RIKEN 203
Riken Metal Mfg. Co.
Magnesium. Al 0-0.5, Mn 0.5-2.5, Zn 0-0.5, bal Mg.
Rolled: 25,500-30,000 TS; 2 El. For light parts; wrought alloy.

RIKEN 501
Riken Metal Mfg. Co.
Magnesium. Al 3.5-6.5, Mn 0.1-0.5, Zn 2.5-3.5, bal Mg.
Cast: 25,000 TS; 5 El. For light parts; sand casting.

RIKEN 502A
Riken Metal Mfg. Co.
Magnesium. Al 8-11, Mn 0.1-0.5, bal Mg.
Cast: 21,300 TS. For light parts; sand casting.

RIKEN 502B
Riken Metal Mfg. Co.
Magnesium. Al 8-11, Mn 0.1-0.5, bal Mg.
Cast: 28,500-30,000 TS. For light parts; sand casting.

RIKEN 601
Riken Metal Mfg. Co.
Magnesium. Al 8-11, Mn 0.1-1, Zn 0.2-1, bal Mg.
Die cast. For light die castings; high strength.

RIM-ROLL NO. 9
Manufacturer not listed.
C 1.05, Cr 0.5, bal Fe.
For rolls; water hardened.

RIOK 5F5
Russian manufacture
C 1.5, W 10, Cr 4, V 4.5, Co 6, bal Fe.
Hardened: 65-68 Rock C. For broaches, reamers, knitting and textile needles, form cutters; high-speed steel, high red-hardness.

RITA ALLOY
Cannon-Stein Steel Co.
C 0.5, Mn 0.9, Cr 0.8, Ni 1.5, bal Fe.
Oil quenched: 220,000 TS; 12 El; 40 RA; 444 Brin. For structural purposes, gears; oil hardening. *Obsolete*

RITA ALLOY NO. 2
Cannon-Stein Steel Co.
C 0.2, Mn 1.15, Ni 0.5, Cr 0.5, bal Fe.
Rolled: 85,000 TS; 174 Brin. For gears, shafts; case hardening. *Obsolete*

RITA ALLOY NO. 4
Cannon-Stein Steel Co.
C 0.4, Mn 0.9, Cr 0.5, Ni 0.5, bal Fe.
Heat treated: 223-461 Brin. For gears, shafts; oil hardening. *Obsolete*

RITA ALLOY NO. 5
Cannon-Stein Steel Co.
C 0.5, Mn 1.2, Ni 0.5, Cr 0.6, bal Fe.
Heat treated: 269-629 Brin. For gears, bolts, studs, axles; tough. *Obsolete*

RITA ALLOY NO. 7
Cannon-Stein Steel Co.
C, Mn, Cr, Ni, bal Fe.
Oil treated: 230,000 TS; 200,000 YS; 10 El; 35 RA; 650 Brin.
For tools, dies; oil hardened. *Obsolete*

RITA CHROME VANADIUM
Cannon-Stein Steel Co.
C 0.6, Cr 1.5, V 0.2, bal Fe.
For tools, taps, drills; oil hardened. *Obsolete*

RITA COBALT HIGH SPEED
Cannon-Stein Steel Co.
C 0.7, W 18, Cr 4, V 1, Co 5, bal Fe.
For tools, high speed cutters; high speed steel. *Obsolete*

RITA ELECTRIC TOOL
Cannon-Stein Steel Co.
C 0.7, Si 1.8, Cr 0.5, bal Fe.
For tools, rock drills, chisels; oil hardened. *Obsolete*

RITA ENDUR
Cannon-Stein Steel Co.
C, Si, V, Ni, Mo, bal Fe.
For hobs, cutters; oil hardening. *Obsolete*

RITA EXTRA SPECIAL
Cannon-Stein Steel Co.
C 0.7, W 14, Cr 4, V 1-0, bal Fe.
For tools, taps, drills; high speed steel. *Obsolete*

RITA EXTRA TOOL
Cannon-Stein Steel Co.
C 1, bal Fe.
For tools, taps, dies, punches; water hardened. *Obsolete*

RITA HIGH SPEED
Cannon-Stein Steel Co.
C 0.7, W 18, Cr 4, V 1, bal Fe.
For tools, high speed cutters; high speed steel. *Obsolete*

RITA HOT FORGING DIE NO. 2
Cannon-Stein Steel Co.
C 0.5, W 3, V 0.2, bal Fe.
For hot forging dies; hot work steel *Obsolete*

RITA HOT-FORGING DIE NO. 1
Cannon-Stein Steel Co.
C 0.5, Cr 3-5, W 1.5, bal Fe.
For hot forging dies; hot work steel. *Obsolete*

RITA NOCHANGE OIL HARDENING
Cannon-Stein Steel Co.
C 0.9, Mn 1-2, Cr 0.5, bal Fe.
For tools, dies, broaches, punches, taps; non-deforming. *Obsolete*

RITA SPECIAL TOOL
Cannon-Stein Steel Co.
C 1, V 0.2, bal Fe.
For tools, dies, punches, taps; water hardened. *Obsolete*

RITA STANDARD DRILL ROD
Cannon-Stein Steel Co.
C 1-2, bal Fe.
For tools, drills, taps, reamers, punches, pins; water hardened. *Obsolete*

RITA STANDARD TOOL
Cannon-Stein Steel Co.
C 1, bal Fe.
For tools, rock drills, die blocks; water or oil hardened.
Obsolete

RITA THREE POINT
Cannon-Stein Steel Co.
C, W, Cr, Ni, bal Fe.
For tools, dies; non-deforming. *Obsolete*

RITA VANADIUM DRILL ROD
Cannon-Stein Steel Co.
C 1.2, Cr 0.5, V 0.2, bal Fe.
Oil treated: 230,000 TS; 205,000 YS; 10 El; 40 RA; 444 Brin.
For tools, drills, punches; oil or water hardened. *Obsolete*

RIVALOY NO. 24
Riverside Metals Corp.
Ni 20, Ag 5, bal Cu.
For jewelry novelties; corrosion resistant. *Obsolete*

RIVER ACE 100
Kawasaki Steel Corp.
C 0.1, Mn 0.54, P 0.014, S 0.007, Si 0.23, Cu 0.47, Ni 1.54, Cr 0.51, Mo 0.66, V 0.046, B 0.0018, bal Fe.
143,400 psi TS; 135,100 psi YS; 23 El. High-strength low-alloy quenched and tempered steel for weldable structures and construction equipment.

RIVER ACE 60
Kawasaki Steel Corp.
C 0.12, Mn 1.23, P 0.015, S 0.016, Si 0.35, Ni 0.31, Mo 0.05, V 0.041, bal Fe.
94,000 psi TS; 80,000 psi YS; 35 El. High-strength low-alloy quenched and tempered steel for weldable structures, penstock and pressure vessels.

RIVER ACE 60L
Kawasaki Steel Corp.
C 0.12, Mn 1.37, P 0.012, S 0.007, Si 0.36, Ni 0.86, Mo 0.128, V 0.017, bal Fe.
94,700 psi TS; 82,600 psi YS; 29 El. High-strength low-alloy quenched and tempered steel for low temperature pressure vessels and structures.

RIVER ACE 70
Kawasaki Steel Corp.
C 0.12, Mn 1.04, P 0.016, S 0.011, Si 0.25, Cu 0.24, Ni 0.58, Cr 0.43, Mo 0.28, V 0.031, B 0.0031, bal Fe.
105,000 psi TS; 96,000 psi YS; 25 El. High-strength low-alloy quenched and tempered steel for weldable structures and pressure vessels.

RIVER ACE 70L
Kawasaki Steel Corp.
C 0.15, Mn 1.12, P 0.016, S 0.006, Si 0.29, Cu 0.24, Ni 0.08, Cr 0.4, Mo 0.29, V 0.027, B 0.0028, bal Fe.
105,400 psi TS; 92,500 psi YS; 27 El. High-strength low-alloy quenched and tempered steel for low temperature pressure vessels and structures.

RIVER ACE K-O
Kawasaki Steel Corp.
C 0.15, N 1.14, P 0.012, S 0.015, Si 0.27, Cu 0.21, N 0.28, Cr 0.45, Mo 0.31, V 0.015, B 0.0022, bal Fe.
123,000 psi TS; 114,000 psi YS; 24 El. High-strength low-alloy quenched and tempered steel for weldable structures and pressure vessels.

RIVER ACE K-O M
Kawasaki Steel Corp.
C 0.12, N 0.97, P 0.006, S 0.008, Si 0.29, Cu 0.21, Ni 1.08, Cr 0.44, Mo 0.36, V 0.03, B 0.003, bal Fe.
122,300 psi TS; 112,400 psi YS; 24 El. High-strength low-alloy quenched and tempered steel for weldable structures; penstock and pressure vessel purposes.

RIVER ACE K-OL
Kawasaki Steel Corp.
C 0.16, N 0.88, P 0.009, S 0.01, Si 0.28, Cu 0.24, Ni 1, Cr 0.58, Mo 0.53, V 0.05, B 0.0035, bal Fe.
126,700 psi TS; 116,500 psi YS; 24 El. High-strength low-alloy quenched and tempered steel for low temperature pressure vessels and structures.

RIVER TEN 41
Kawasaki Steel Corp.
C 0.12, N 0.52, P 0.065, S 0.02, Cu 0.3, bal Fe.
62,300 TS; 42,800 psi YS; 28 El. Low-alloy steel with good atmospheric corrosion resistance.

RIVER TEN 45K
Kawasaki Steel Corp.
C 0.12, Mn 0.54, P 0.01, S 0.015, Si 0.27, C 0.25, Ni 0.13, Cr 0.28, bal Fe.
84,800 psi TS; 67,800 psi YS; 32 El. Low-alloy steel, resistant to sulfuric acid corrosion.

RIVER TEN 50
Kawasaki Steel Corp.
C 0.1, Mn 0.85, P 0.018, S 0.012, Si 0.38, Cu 0.3, Ni 0.27, Cr 0.42, Cb 0.013, bal Fe.
77,100 psi TS; 57,300 psi YS; 25 El. High-strength low-alloy steel, good atmospheric corrosion resistance, for weldable structures.

RIVER TEN 50M
Kawasaki Steel Corp.
C 0.08, Mn 0.5, P 0.086, S 0.017, Si 0.3, Cu 0.4, Ni 0.39, Cr 0.72, Cb 0.028, bal Fe.
81,600 psi TS; 65,600 psi YS; 30 El. High-strength low-alloy steel with good atmospheric corrosion resistance.

RIVER TEN 53
Kawasaki Steel Corp.
C 0.13, Mn 0.67, P 0.019, S 0.008, Si 0.38, Cu 0.27, Ni 0.34, Cr 0.49, Cb 0.035, bal Fe.
81,100 psi TS; 64,400 psi YS; 36 El. High-strength low-alloy steel, good corrosion resistance, for weldable structures.

RIVER TEN 62
Kawasaki Steel Corp.
C 0.13, Mn 0.56, P 0.022, S 0.01, Si 0.5, Cu 0.33, Ni 0.33, Cr 0.49, Mo 0.065, bal Fe.
93,200 psi TS; 79,100 psi YS; 29 El. Quenched and tempered high-strength low-alloy steel, good atmospheric corrosion resistance, for weldable structures.

RIVER TEN R
Kawasaki Steel Corp.
C 0.09, Mn 0.43, P 0.086, S 0.009, Si 0.35, Cu 0.32, Ni 0.32, Cr 0.46, bal Fe.
73,800 psi TS; 57,700 psi YS; 37 El. High-strength low-alloy steel with good atmospheric corrosion resistance.

RIVER-H360
Kawasaki Steel Corp.
C 0.16, Mn 1, P 0.016, S 0.01, Si 0.24, Cr 0.66, Mo 0.3, V 0.032, B 0.002, bal Fe.
380 Brin min. Low-alloy steel with minimum hardness for abrasion resistant applications.

RIVERSIDE 109
Riverside Metals Corp.
Cu 92, Zn 5.85, Sn 2.
For switches, relays, contacts. Good corrosion resistance. *Obsolete*

RIVERSIDE 110
Riverside Metals Corp.
Cu 87.5, Zn 10.25, Sn 2.25.
For ring traveler wire on textile machines. High strength wire, corrosion resistant. *Obsolete*

RIVERSIDE 111
Riverside Metals Corp.
Cu 90, Zn 8.25, Sn 1.75.
For electrical contacts, clips, fuses. High strength and corrosion resistance. *Obsolete*

RIVERSIDE 61
Riverside Metals Corp.
Cu 70, Zn 30.
Strip: 76,000 TS; 63,000 YS; 8 El; 156 Brin. Wire: 124,000 TS; 4 El. For cartridge cases, lamp bodies, rivets, eyelets. High strength and ductility. Good corrosion resistance. *Obsolete*

RIVERSIDE 62
Riverside Metals Corp.
Cu 85, Zn 15.
Strip: 70,000 TS; 57,000 YS; 5 El; 140 Brin. Wire: 88,000 TS; 6 El. For jewelry, badges, tags, dials, hardware. High ductility, corrosion resistant. *Obsolete*

RIVERSIDE 63
Riverside Metals Corp.
Cu 90, Zn 10.
Strip: 61,000 TS; 54,000 YS; 5 El; 125 Brin. Wire: 74,000 TS; 4 El. For jewelry, compacts, weather stripping, fasteners, screens. Highly ductile, excellent cold workability, corrosion resistant. *Obsolete*

RIVERSIDE 64
Riverside Metals Corp.
Zn 5, Cu 95.
Hard strip: 56,000 TS; 50,000 YS; 4 El; 115 Brin. Wire: 65,000 TS. For jewelry, emblems, medals, bullet jackets. Excellent workability. Corrosion resistant. *Obsolete*

RIVERSIDE 66
Riverside Metals Corp.
Cu 80, Zn 20.
Strip: 74,000 TS; 59,000 YS; 7 El; 156 Brin. Wire: 107,000 TS; 5 El. For medallions, battery caps, clock dials. Good ductility, corrosion resistant. *Obsolete*

RIVERSIDE 70
Riverside Metals Corp.
Cu 87.5, Zn 12.5.
Strip: 66,000 TS; 56,000 YS; 5 El; 132 Brin. Wire: 83,000 TS; 5 El. For jewelry, compacts, zippers, channels. Highly ductile, good cold workability; corrosion resistant. *Obsolete*

RIVERSIDE 72-28
Riverside Foundry & Galvanizing Co.
Cu 70, Sn 2, Pb 28.
For bearings; heavy duty. *Obsolete*

RIVERSIDE 74-26
Riverside Foundry & Galvanizing Co.
Cu 70, Sn 4, Pb 26.
For bearings; heavy duty. *Obsolete*

RIVERSIDE 78
Riverside Metals Corp.
Cu 92, Zn 7, Sn 1.
For base metal for gold plated electronic components. Plater metal. *Obsolete*

RIVERSIDE 78-22
Riverside Foundry & Galvanizing Co.
Cu 70, Sn 8, Pb 22.
For bearings; heavy duty. *Obsolete*

RIVERSIDE 80-20
Riverside Foundry & Galvanizing Co.
Cu 70, Sn 10, Pb 20.
For bearings; heavy duty. *Obsolete*

RIVERSIDE 80-20 HARD
Riverside Foundry & Galvanizing Co.
Cu 70, Sn 10, 20% Pb + Ni.
For bearings; heavy duty. *Obsolete*

RIVERSIDE GRADE A PHOSPHOR BRONZE
Riverside Metals Corp.
Sn 3.5-5.8, P 0.03-0.35, bal Cu.
For springs; tough. *Obsolete*

RIVERSIDE GRADE E PHOSPHOR BRONZE
Riverside Metals Corp.
Sn 2.7-3.3, P 0.03-0.35, bal Cu.
For springs; tough. *Obsolete*

RIVERSIDE NICKEL SILVER NO. 18
Riverside Metals Corp.
Zn 15, Ni 18, bal Cu.
For springs for telephone, radio and electrical equipment,
silverware, cutlery, hardware; corrosion resistant. *Obsolete*

RIVERSIDE NO. 1
Riverside Metals Corp.
Ni 30, Zn 14, bal Cu.
Soft: 70,000 TS; 42,000 YS; 102 Brin. Hard: 100,000 TS;
85,000 YS; 195 Brin. For watch cases, stampings; corrosion
resisting. *Obsolete*

RIVERSIDE NO. 10
Riverside Metals Corp.
Ni 10, Zn 25, bal Cu.
Rolled: 50,000-105,000 TS; 33,000-71,000 YS. For drawn and
spun parts; nickel silver. *Obsolete*

RIVERSIDE NO. 101
Riverside Metals Corp.
Cu 87.5, Sn 2.25, Zn 10.25.
Soft: 49,000 TS; 60 Brin. For jewelry; corrosion resistant.
Obsolete

RIVERSIDE NO. 102
Riverside Metals Corp.
Cu 87, Zn 12, Sn 1.
Hard: 75,000 TS; 153 Brin. For valves, carburetors, novelties,
buckles; corrosion resistant. *Obsolete*

RIVERSIDE NO. 105
Riverside Metals Corp.
Cu 95, Zn 4, Sn 1.
Rolled: 38,000-80,000 TS. For fuse clips, electrical contacts;
38% electrical conductivity. *Obsolete*

RIVERSIDE NO. 108
Riverside Metals Corp.
Cu 88, Zn 10.25, Sn 1.75.
Soft: 49,000 TS; 62 Brin. 1\2 Hard: 62,000 TS; 127 Brin.
Spring: 93,000 TS; 190 Brin. For conveyor chains, jewelry,
novelties; wear and corrosion resistant. *Obsolete*

RIVERSIDE NO. 11
Riverside Metals Corp.
Ni 16, Zn 16, bal Cu.
Annealed. For spun and drawn parts; nickel silver. *Obsolete*

RIVERSIDE NO. 12
Riverside Metals Corp.
Ni 15, Zn 23, bal Cu.
Rolled. For stampings, fishing reels; nickel silver. *Obsolete*

RIVERSIDE NO. 120
Riverside Metals Corp.
Cu 97.7, Be 1.95, Co 0.35.
Soft: 60,000 TS; 80 Brin. Hard: 120,000 TS; 250 Brin. For
bearings, bolts, springs, contact brushes; age-hardenable,
corrosion resistant. *Obsolete*

RIVERSIDE NO. 120
Riverside Metals Corp.
Be 1.95, Co 0.35, bal Cu.
Annealed: 70,000 TS; 53 El; 108 Brin. Hard: 120,000 TS; 9 El;
250 Brin. For springs, clips, bearings; corrosion resistant,
age-hardenable. *Obsolete*

RIVERSIDE NO. 13
Riverside Metals Corp.
Ni 15, Pb 1, Zn 19, bal Cu.
Hard: 86,000 TS; 172 Brin. For stampings; nickel silver, free
cutting. *Obsolete*

RIVERSIDE NO. 14
Riverside Metals Corp.
Ni 12, Zn 26, bal Cu.
For bolts, shafts; nickel silver. *Obsolete*

RIVERSIDE NO. 15
Riverside Metals Corp.
Ni 10, bal Cu.
For coins, jewelry; corrosion resistant. *Obsolete*

RIVERSIDE NO. 16
Riverside Metals Corp.
Ni 10, Pb 2, Zn 23, bal Cu.
Hard: 87,000 TS; 176 Brin. For hardware, marine parts; nickel
silver, free cutting. *Obsolete*

RIVERSIDE NO. 17
Riverside Metals Corp.
Ni 5, bal Cu.
For coins, medals. *Obsolete*

RIVERSIDE NO. 18
Riverside Metals Corp.
Ni 5, Pb 2, bal Cu.
For hardware. *Obsolete*

RIVERSIDE NO. 2
Riverside Metals Corp.
Ni 25, Zn 13, bal Cu.
Rolled: 58,000-115,000 TS; 41,000-84,000 YS; 107-205 Brin.
For stampings, drawing; nickel silver. *Obsolete*

RIVERSIDE NO. 20
Riverside Metals Corp.
Ni 12, Pb 2, Zn 21, bal Cu.
Hard: 86,000 TS; 222 Brin. For keys, hardware; nickel silver,
free cutting. *Obsolete*

RIVERSIDE NO. 205
Riverside Metals Corp.
Cu 91.6, Sn 8, P 0.4.
For welding rod; P-bronze. *Obsolete*

RIVERSIDE NO. 209
Riverside Metals Corp.
Sn 10, P 0.05, bal Cu.
Soft: 65,000 TS; 83 Brin. Hard: 91,000 TS; 210 Brin. For
springs, bearings, diaphragms, welding rod; P-bronze,
corrosion resistant. *Obsolete*

RIVERSIDE NO. 209D
Riverside Metals Corp.
Sn 9-11, P 0.03-0.25, bal Cu.
soft: 58,000 TS; 71 Brin. Hard: 109,000 TS; 250 Brin. For
springs, diaphragms, valves, pump rods; P-bronze, corrosion
resistant. *Obsolete*

RIVERSIDE NO. 21
Riverside Metals Corp.
Ni 21, bal Cu.
For tubes, condensers; corrosion resistant. *Obsolete*

RIVERSIDE NO. 23
Riverside Metals Corp.
Ni 16, Pb 0.5, Zn 15.5, bal Cu.
For screw machine parts; free machining. *Obsolete*

RIVERSIDE NO. 27
Riverside Metals Corp.
Cu 70, Ni 30.
Soft: 57,000-70,000 TS; B 25-55 Rock. 1/2 Hard:
66,000-80,000 TS; B 76-85 Rock. Spring: 84,000-94,000 TS; B
87-91 Rock. For process machinery parts, heat exchangers,
marine hardware. Corrosion resistant, tough. *Obsolete*

RIVERSIDE NO. 3
Riverside Metals Corp.
Ni 18, Zn 17, bal Cu.
Rolled: 50,000-104,000 TS; 35,000-88,000 YS; 71-190 Brin.
For watch cases, jewelry, emblems; nickel silver. *Obsolete*

RIVERSIDE NO. 30
Riverside Metals Corp.
Sn 3.8-5.2, P 0.03-0.05, bal Cu.
Soft: 40,000 TS; 56 Brin. Hard: 87,000 TS; 185 Brin. For
springs, gears, bolts, bushings; P-bronze, corrosion resistant.
Obsolete

RIVERSIDE NO. 31
Riverside Metals Corp.
P 0.2, Sn 4.3, Pb 1-2, bal Cu.
Rolled: 50,000-100,000 TS. For gears, pinions, bolts; free
machining. *Obsolete*

RIVERSIDE NO. 32
Riverside Metals Corp.
Cu 97.65, Be 2, Ni 0.35.
Heat treated: 65,000-200,000 TS; 0.5-50 El; 350 Brin. For
springs, gears; heat treatable. *Obsolete*

RIVERSIDE NO. 33
Riverside Metals Corp.
Sn, P, bal Cu.
For bearings, bushings; heavy duty. *Obsolete*

RIVERSIDE NO. 330
Riverside Metals Corp.
Cu 95.65, Sn 4, P 0.35.
For welding rod; P-bronze. *Obsolete*

RIVERSIDE NO. 39
Riverside Metals Corp.
Cu 97, Sn 3.
Soft: 46,000 TS; 58 Brin. Hard: 71,000 TS; 150 Brin. For
springs, pump rods, marine hardware; P-bronze, corrosion
resistant. *Obsolete*

RIVERSIDE NO. 4
Riverside Metals Corp.
Ni 18, Zn 22, bal Cu.
Hard: 112,000 TS. For springs, keys, fasteners, hardware;
corrosion resistant, nickel silver. *Obsolete*

RIVERSIDE NO. 40
Riverside Metals Corp.
Cu 97, Sn 3.
Soft: 48,000 TS; 64 Brin. Hard: 75,000 TS; 159 Brin. Spring:
94,000 TS; 200 Brin. For springs, clips; corrosion resistant.
Obsolete

RIVERSIDE NO. 400
Riverside Metals Corp.
Ni 12, Zn 21.5, Pb 1.5, bal Cu.
For screw machines parts; free cutting. *Obsolete*

RIVERSIDE NO. 401
Riverside Metals Corp.
Ni 10, Zn 23.5, Pb 1.5, bal Cu.
For screw machine parts; free cutting. *Obsolete*

RIVERSIDE NO. 402
Riverside Metals Corp.
Ni 12, Zn 23, bal Cu.
Soft: 60,000 TS; 32,000 YS; 73 Brin. For marine parts,
hardware; nickel silver, corrosion resistant. *Obsolete*

RIVERSIDE NO. 403
Riverside Metals Corp.
Ni 15, Zn 20, bal Cu.
Soft: 60,000 TS; 31,000 YS; 73 Brin. For marine parts,
hardware; nickel silver, corrosion resistant. *Obsolete*

RIVERSIDE NO. 404
Riverside Metals Corp.
Ni 30, Zn 22, bal Cu.
For resistance wire; nickel silver. *Obsolete*

RIVERSIDE NO. 405
Riverside Metals Corp.
Ni 10, Zn 20, bal Cu.
Soft: 54,000 TS; 32,000 YS; 70 Brin. For fasteners, bolts, marine hardware; nickel silver, corrosion resistant. *Obsolete*

RIVERSIDE NO. 409
Riverside Metals Corp.
Cu 56.5, Ni 12.1, Zn 31.5.
Hard: 90,000 TS; 185 Brin. Spring: 122,000 TS; 240 Brin. For marine hardware, pump parts, cutlery; nickel silver, corrosion resistant. *Obsolete*

RIVERSIDE NO. 44
Riverside Metals Corp.
Sn 4, Pb 4, bal Cu.
For bearings, castings; free-cutting, heavy duty. *Obsolete*

RIVERSIDE NO. 5
Riverside Metals Corp.
Ni 18, Zn 15.9, Pb 1.1, bal Cu.
Rolled: 50,000-90,000 TS. For screw machine products; nickel silver. *Obsolete*

RIVERSIDE NO. 6
Riverside Metals Corp.
Ni 18, Zn 10, bal Cu.
Rolled: 50,000-90,000 TS; 32,000-76,000 YS. For deep drawn or spun parts; nickel silver. *Obsolete*

RIVERSIDE NO. 7
Riverside Metals Corp.
Ni 18, Zn 27, bal Cu.
Rolled: 70,000-123,000 TS; 37,000-83,000 YS. For springs, bolts, studs; nickel silver. *Obsolete*

RIVERSIDE NO. 8
Riverside Metals Corp.
Ni 2, Pb 2, bal Cu.
Rolled: 250 Brin. For hardware, marine parts; free cutting, corrosion resistant. *Obsolete*

RIVERSIDE NO. 80
Riverside Metals Corp.
Cu 88, Sn 4, Zn 4, Pb 4.
For screw machine parts; free cutting. *Obsolete*

RIVERSIDE NO. 9
Riverside Metals Corp.
Ni 16, bal Cu.
For hardware; corrosion resistant. *Obsolete*

RIVERSIDE PHOSPHOR BRONZE NO. 209
Riverside Metals Corp.
Cu 90, Sn 10, Trace P.
135,000 TS; 115,000 YS; 1-70 El. For screw propellers, fittings exposed to sea water, bearings; high resistance to corrosion and fatigue. *Obsolete*

RIVERSIDE PHOSPHOR BRONZE NO. 30
Riverside Metals Corp.
Sn 3.8-4.8, P 0.05-0.35, Zn 0.3, Fe 0.1, 94.4% Cu min.
Soft: 40,000-55,000 TS; 40 El. Hard: 71,000-88,000 TS; 100 Brin. For springs and contacts in radio and electrical equipment, ignition systems, worm wheels; resists salt water corrosion; see Omega Brand Phosphor Bronze No. 30. *Obsolete*

RIVERSIDE PHOSPHOR BRONZE NO. 47
Riverside Metals Corp.
Sn 7.5-8.5, P 0.05-0.25, Zn 0.2, Fe 0.1, 91% Cu min.
Soft: 53,000-67,000 TS; 70 El. Hard: 84,000-101,000 TS; For spring contacts, gun fittings, fuse clips, screw propellers, worm wheels; used where forming and endurance requirements are severe; see Omega Brand Phosphor Bronze No. 47. *Obsolete*

RK10
Russian manufacture
C 0.8, W 18, Cr 4, V 1.5, Mo 0.5, Co 10, bal Fe.
Hardened: 64-68 Rock C. For cutters, lathe and planer tools, reamers, hobs, form cutters; high-speed steel, high red-hardness.

RK15
Russian manufacture
C 0.8, W 18, Cr 4, V 1, Mo 0.6, Co 14, bal Fe.
Hardened: 64-66 Rock C. For reamers, broaches, form cutters, lathe and planer tools, hobs, milling cutters; high-speed steel, high red-hardness.

RK5
Russian manufacture
C 0.7, W 18, Cr 4, V 1.2, Mo 0.4, Co 5, bal Fe.
Hardened: 64-66 Rock C. For lathe and planer tools, reamers, cut-off tools and drills, broaches, hobs; high-speed steel, high red-hardness.

RKCM
Saarstahl AG
Now SAARSTAHL 1.2363.

RM 210
IMI Rolled Metals Ltd.
Cu 90, Zn 10.
Known as 90/10 Gilding metal and as "commercial bronze." For ornaments, imitation jewelry. BS 2870 CZ.101; ASTM B36-220.

RM 215
IMI Rolled Metals Ltd.
Cu 85, Zn 15.
Known as 85/15 Gilding metal or as "red brass." For ornamental work, vanity cases, piping for carrying liquids, diaphragms and bellows. BS 2870 CZ.102; ASTM B36-230.

RM 220
IMI Rolled Metals Ltd.
Cu 80, Zn 20.
Known as 80/20 Gilding metal and as "low brass." For architectural work, costume jewelry, strip for slide fasteners, diaphragms, bellows, wire for Fourdrinier wire cloth. BS 2870 CZ.103; ASTM B36-240.

RM 230 (CARTRIDGE BRASS)
IMI Rolled Metals Ltd.
Cu 70, Zn 30.
High ductility for deep drawing into cartridge cases, rivets, screws, lamp fixtures, pen and pencil parts, and many similar drawn or formed parts. BS 2870 CZ.106; ASTM B36-260.

RM 421
IMI Rolled Metals Ltd.
Cu 63, Pb 1, bal Zn.
Called matrix and key brass. For key blanks, watch, clock and instrumental parts; free machining. *Obsolete*

RM 433
IMI Rolled Metals Ltd.
Cu 62, Pb 2, bal Zn.
Strip is called "clock brass," wire is heading and drilling wire. For mild cold work; free machining; clock and instrument work. BS 2800 CZ.119; ASTM B121-353. *Obsolete*

RM 444
IMI Rolled Metals Ltd.
Cu 61, Pb 1, bal Zn.
Free machining brass sheet called "engraving brass." For ornamental work, plaques. *Obsolete*

RM 445
IMI Rolled Metals Ltd.
Cu 61, Pb 1.5, bal Zn.
Sheet and strip, called "rule brass." For engraving, key blanks, watch and clock parts. *Obsolete*

RM 656
IMI Rod & Wire
Cu 95, Sn 5, P.
Phosphor bronze sheet, various tempers; for springs and stampings in textile, paper and chemical plant operations. ASTM B 103 Al.

RM 656
IMI Rolled Metals Ltd.
Cu 95, Sn 5, P.
Phosphor bronze sheet, various tempers; for springs and stampings in textile, paper and chemical plant operations. ASTM B 103 Al.

RM 848 (CUPRO-NICKEL)
IMI Rolled Metals Ltd.
Cu 75, Ni 25.
Coinage alloy. BS 2870 CN.104.

RM SPECIAL SEHR HART
Vereinigte Edelstahlwerke
C 1.5, Cr 0.65, W 7.5, V 0.4, bal Fe.
For tools, dies, cutters; oil hardening. *Obsolete*

RM1
Acciaierie Valbruna s.p.a.
Tool Material. C 0.83, Si 0-0.4, Mn 0-0.4, Cr 3.75, Mo 8.7, V 1.15, W 1.75, bal Fe.
High speed steel. AISI M1; W. Nr. 1.3346.

RM2
Acciaierie Valbruna s.p.a.
Tool Material. C 0.83, Si 0-0.3, Mn 0-0.3, Cr 4.15, Mo 5, V 1.95, W 6.1, bal Fe.
High speed steel. AISI M2; W. Nr. 1.3343.

RM3
Acciaierie Valbruna s.p.a.
Tool Material. C 1.05, Si 0-0.45, Mn 0-0.4, Cr 4.15, Mo 6, V 2.5, W 6.2, bal Fe.
High speed steel. AISI M3.

RM42
Acciaierie Valbruna s.p.a.
Tool Material. C 1.1, Si 0-0.4, Mn 0-0.4, Co 8.25, Cr 3.85, Mo 9.5, V 1.15, W 1.5, bal Fe.
High speed steel. AISI M42; W. Nr. 1.3247.

RM7
Acciaierie Valbruna s.p.a.
Tool Material. C 1.05, Si 0-0.45, Mn 0-0.4, Cr 3.75, Mo 8.7, V 2, W 1.75, bal Fe.
High speed steel. AISI M7; W. Nr. 1.3348.

RMI 0.2% PD
RMI Co.
Titanium. C 0-0.08, Fe 0-0.3, N 0-0.03, Pd 0-0.2, O 0-0.02, bal Ti. Sheet: 0.015 H. Bar: 0.0125 H.
RT, wrought, min: 50,000 TS; 40,000 YS; 20 El. At 800°F, typical: 26,000 TS; 15,000 YS; 25 El. At 1000°F, typical: 18,000 TS; 11,000 YS; 32 El. For corrosion resistance in the chemical industry where media is mildly reducing or varies between oxidizing and reducing.

RMI 13V-11CR-3AL
RMI Co.
Titanium. C 0-0.06, N 0-0.05, H 0-0.0175, Fe 0-0.35, V 12.5-14.5, Cr 10-12, Al 2.5-4, bal Ti.
RT, min: 130,000 TS; 125,000 YS; 10 El. At 1000°F: 100,000 TS; 75,000 YS; 35 El. Beta alloy. For high strength aircraft structures.

RMI 1AL-8V-5FE
RMI Co.
Titanium. C 0-0.05, N 0-0.08, O 0.25-0.5, Al 0.8-1.8, Fe 4-6, V 7.5-8.5, H 0-0.0125, bal Ti.
RT, as heat treated, min: 210,000 TS; 200,000 YS; 6 El; For high strength fasteners. Alpha-beta alloy. *Obsolete*

RMI 25
RMI Co.
Titanium. C 0-0.08, Fe 0-0.2, N 0-0.03, O 0-0.18, bal Ti. Sheet: 0.015 H. Bar: 0.0125 H. Billet: 0.010 H.
RT, min: 35,000 TS; 25,000 YS; 24 El at 800°F, typical: 20,000 TS; 13,000 YS; 26 El; 70-80 RA. For corrosion resistance in the chemical and marine industries and for airframe construction.

RMI 30
RMI Co.
Titanium. C 0-0.08, Fe 0-0.25, N 0-0.05, bal Ti. Billet: 0.010 H max.
RT, wrought, min: 40,000 TS; 30,000 YS; 25 El. At 800°F, typical: 20,000 TS; 13,000 YS; 26 El. Commercially pure titanium, wrought. For corrosion resistance in the chemical and marine industries; in airframe construction. *Obsolete*

RMI 3AL-2.5V
RMI Co.
Titanium. C 0-0.05, N 0-0.02, Fe 0-0.3, O 0-0.12, V 2-3, Al 2.5-3.5, bal Ti. Sheet: 0.015 H. Bar: 0.0125 H.
Cold worked, stress relieved tubing at RT: 125,000 TS; 105,000 YS; 10 El. At 800°F: 55,000 TS; 40,000 YS; 22 El. For aircraft hydraulic tubing and fittings. Alpha-beta alloy.

RMI 3AL-8V-6CR-4ZR-4MO
RMI Co.
Titanium. C 0-0.05, N 0-0.03, Al 3-4, V 7.5-8.5, Cr 5.5-6.5, Fe 0-0.03, O 0-0.14, Mo 3.5-4.5, Zr 3.5-4.5, bal Ti.
Beta alloy. For heavy section, high strength forgings for airframe, cold headed rivets, high shear fasteners.

RMI 40
RMI Co.
Titanium. C 0-0.08, Fe 0-0.25, N 0-0.03, O 0-0.2, bal Ti. Billet: 0.010 H. Sheet: 0.015 H. Bar: 0.0125 H.
RT, wrought, min: 50,000 TS; 40,000 YS; 20 El. At 800°F, typical: 26,000 TS; 15,000 YS; 25 El. For corrosion resistance in the chemical and marine industries; in aircraft construction.

RMI 4AL-3MO-1V
RMI Co.
Titanium. C 0-0.08, Fe 0-0.25, N 0-0.05, Al 3.75-4.75, Mo 2.3-3.5, V 0.5-1.5, H 0-0.015, bal Ti.
RT, min: 125,000 TS; 115,000 YS; 10 El. At 1000°F: 65,000 TS; 55,000 YS; 35 El. Plate, sheet, strip; alpha-beta alloy. For aircraft uses requiring high strength and elevated temperature stability. *Obsolete*

RMI 55
RMI Co.
Titanium. C 0-0.08, Fe 0-0.5, N 0-0.05, O 0-0.3, bal Ti. Sheet: 0.015 H. Bar: 0.0125 H. Billet: 0.010 H.
RT, wrought, min: 65,000 TS; 55,000 YS; 18 El. At 800°F, typical: 29,000 TS; 17,000 YS; 18 El. At 1000°F, typical: 22,000 TS; 13,000 YS; 33 El. For corrosion resistance in the chemical and marine industries and for aircraft construction.

RMI 5AL-2.5SN
RMI Co.
Titanium. C 0-0.08, Fe 0-0.5, N 0-0.05, O 0-0.2, Al 4-6, Sn 2-3, bal Ti. Billet: 0.175 H. Bar, sheet: 0.020 H.
RT, wrought, min: 120,000 TS; 115,000 YS; 10 El. At 800°F, typical: 78,000 TS; 59,000 YS; 18 El. At 1000°F, typical: 67,000 TS; 55,000 YS; 19 El. For parts requiring weldability, oxidation resistance, stability and strength at elevated temperatures. Alpha alloy.

RMI 5AL-2.5SN ELI
RMI Co.
Titanium. C 0-0.08, Fe 0-0.25, N 0-0.03, O 0-0.13, Al 4.7-5.75, Sn 2-3, bal Ti. Bar, billet: 0.125 H. Sheet: 0.015 H.
RT, wrought, min: 105,000 TS; 100,000 YS; 10 El. At -320°F, typical: 180,000 TS; 168,000 YS; 16 El. At -423°F, typical: 229,000 TS; 206,000 YS; 15 El. For use in high pressure cryogenic vessels at temperatures below -320°F. Alpha alloy.

RMI 5AL-5SN-2ZR-2MO-SI
RMI Co.
Titanium. C 0-0.05, N 0-0.03, Fe 0-0.15, O 0-0.13, Al 4.5-5.5, Sn 4.5-5.5, Zr 1.75-2.25, Mo 1.75-2.25, Si 0.2-0.3, H 0-0.0125, bal Ti.
RT, typical: 140,000 TS; 126,000 YS; 12 El; 35 RA. At 600°F, typical: 118,000 TS; 85,000 YS; 18 El; 41 RA. At 800°F, typical: 108,000 TS; 81,000 YS; 20 El; 60 RA. At 1000°F, typical: 106,000 TS; 80,000 YS; 20 El; 50 RA. For use where high creep strength and elevated temperature stability are required, such as in jet engine components.

RMI 5AL-5SN-2ZR-2MO-SI
RMI Co.
Titanium. C 0-0.05, N 0-0.03, Fe 0-0.15, O 0-0.13, H 0-0.0125, Sn 4.5-5.5, Zr 1.75-2.25, Mo 1.75-2.25, Si 0.2-0.3, Al 4.5-5.5, bal Ti.
At 800°F: 108,000 TS; 81,000 YS; 20 El. For use where high creep strength and elevated temperature strength are required, as in jet engine components. Alpha alloy.

RMI 5AL-6SN-2ZR, 1MO-SI
RMI Co.
Titanium.
Now RMI 5AL-5SN-2ZR-2MO-SI.

RMI 6AL-2CB-1TA-1MO
RMI Co.
Titanium. C 0-0.05, N 0-0.03, Fe 0-0.25, O 0-0.1, H 0-0.0125, Cb 1.5-2.5, Ta 0.5-1.5, Mo 0.5-1, Al 5.5-6.5, bal Ti.
RT, wrought: 115,000 TS; 95,000-100,000 YS; 10 El. At 1000°F, typical: 70,000 TS; 55,000 YS; 15 El. For parts requiring toughness, strength, weldability and resistance to seawater. Alpha alloy.

RMI 6AL-2SN-2ZR-2MO-2CR-0.25SI
RMI Co.
Titanium. C 0-0.05, Fe 0-0.25, N 0-0.03, Al 5.25-6.25, Sn 1.75-2.25, Zr 1.75-2.25, Mo 1.75-2.25, Cr 1.75-2.25, Si 0.2-0.3, O 0-0.14, H 0-0.0125, bal Ti.
Solution treated and aged, RT: 170,000 TS; 160,000 YS; 12 El; 20 RA. At 800°F: 130,000 TS; 102,000 YS; 21 El; 40 RA. For air frame and jet engine parts requiring high strength and fracture toughness. Alpha-beta alloy.

RMI 6AL-2SN-4ZR-2MO-0.10SI
RMI Co.
Titanium. C 0-0.08, N 0-0.05, O 0-0.12, Fe 0-0.25, Al 5.5-6.5, Sn 1.75-2.25, Zr 3.5-4.5, Mo 1.75-2.25, Si 0-0.1, bal Ti. Billet: 0.010 H. Bar: 0.125 H.
RT, min: 130,000 TS; 120,000 YS; 10 El. At 1000°F: 75,000 TS; 60,000 YS; 25 El. Alpha-beta alloy. Good high temperature creep strength; for jet engine components.

RMI 6AL-2SN-4ZR-2MO-SI
RMI Co.
Titanium.
Now RMI 6AL-2SN-4ZR-2MO-0.10SI.

RMI 6AL-2SN-4ZR-6MO
RMI Co.
Titanium. C 0-0.04, N 0-0.04, Fe 0-0.15, Al 5.5-6.5, Sn 1.75-2.25, Zr 3.5-4.5, Mo 5.5-6.5, O 0-0.15, H 0-0.0125, bal Ti.
Solution treated and aged, RT: 170,000 TS; 160,000 YS; 8-10 El; 15-20 RA. At 1000°F: 125,000 TS; 95,000 YS; 15 El, 50 RA. For jet engine components requiring high tensile strength with intermediate creep strength. Alpha-beta alloy.

RMI 6AL-4V
RMI Co.
Titanium. C 0-0.08, Fe 0-0.25, N 0-0.05, Al 5.5-6.75, V 3.5-4.5, O 0-0.2, bal Ti. Billet: 0.010 H. Sheet: 0.015 H. Bar: 0.0125 H.
RT, min: 130,000 TS; 120,000 YS; 10 El. At 800°F: 90,000 TS; 75,000 YS; 18 El. For compressor blades, discs, rings for jet engines, aircraft components, pressure vessels. Alpha-beta alloy.

RMI 6AL-4V ELI
RMI Co.
Titanium. C 0-0.08, O 0-0.13, Al 5.5-6.5, V 3.5-4.5, Fe 0-0.25, bal Ti. Billet: 0.010 H. Sheet: 0.015 H. Bar: 0.0125 H.
RT, min: 130,000 TS; 120,000 YS; 10 El. At -320°F: 220,000 TS; 205,000 YS; 14 El. At -423°F: 265,000 TS; 250,000 YS; 6 El. Extra low interstitials permit use for high pressure cryogenic vessels down to -320°F. Alpha-beta alloy.

RMI 6AL-6V-2SN
RMI Co.
Titanium. C 0-0.08, N 0-0.04, Al 5-6, V 5-6, Sn 1.5-2.5, Fe 0.35-1, Cu 0.35-1, O 0-0.2, bal Ti. Billet: 0.0125 H. Bar, sheet, plate: 0.015 H.
RT, min: 150,000 TS; 140,000 YS; 8-20 El. At 600°F: 135,000 TS; 117,000 YS. For rocket engine cases, ordnance components, aircraft parts. Alpha-beta alloy.

RMI 70
RMI Co.
Titanium. C 0-0.08, Fe 0-0.5, N 0-0.05, O 0-0.4, bal Ti. Sheet: 0.015 H. Bar: 0.0125 H. Billet: 0.010 H.
RT, wrought, min: 80,000 TS; 70,000 YS; 15 El. At 800°F, typical: 34,000 TS; 21,000 YS; 22 El. At 1000°F, typical: 27,000 TS; 16,000 YS; 29 El. For corrosion resistance in the chemical and marine industries and for aircraft construction.

RMI 7AL-4MO
RMI Co.
Titanium. C 0-0.08, Fe 0-0.25, N 0-0.05, Al 6.5-7.5, Mo 3.5-4.5, bal Ti. Billet: 0.010 H.
RT, min: 145,000 TS; 135,000 YS; 10 El. At 1000°F: 105,000 TS; 88,000 YS; 20 El. For aircraft and jet engines. Alpha-beta alloy.

RMI 8AL-1MO-1V
RMI Co.
Titanium. C 0-0.08, N 0-0.05, Al 7.5-8.5, Mo 0.75-1.25, V 0.75-1.25, O 0-0.12, bal Ti. Billet: 0.0125 H. Sheet: 0.015 H. Bar: 0.0125 H.
RT, min: 135,000 TS; 120,000 YS; 10 El. At 1000°F: 80,000 TS; 60,000 YS; 20 El. Alpha-beta alloy. Good creep resistance. For aircraft and jet engine parts.

RMI 8MN
RMI Co.
Titanium. C 0-0.08, N 0-0.05, Mn 6.5-9, O 0-0.2, bal Ti. Sheet: 0.015 H.
RT, min: 120,000 TS; 110,000 YS; 10 El. At 800°F: 87,000 TS; 66,000 YS; 20 El. Sheet and plate for aircraft skin and structural components. Alpha-beta alloy.

RMI TI 3AL-8V-4ZR-4MO
RMI Co.
Titanium.
Now RMI 3AL-8V-6CR-4ZR-4MO.

RMI TI-5AL-2SN-2ZR-4MO-4CR
RMI Co.
Titanium. C 0.05, N 0.03, Fe 0.15, Al 4.5-5.5, Sn 1.75-2.25, Zr 1.75-2.25, Mo 3.5-4.5, Cr 3.5-4.5, O 0-0.13, bal Ti. Billet: 0.010 H.
Solution treated and aged, RT: 180,000 TS; 170,000 YS; 7 El; 15 RA. For aircraft frames and jet engine components. Alpha-beta alloy.

RMI TI-6AL-2SN-2ZR-2MO-2CR-0.25 SI
RMI Co.
Titanium.
Now RMI 6AL-2SN-2ZR-2MO-2CR-0.25SI.

RMI TI-6AL-2SN-4ZR-6MO
RMI Co.
Titanium.
Now RMI 6AL-2SN-4ZR-6MO.

RMK5
Acciaierie Valbruna s.p.a.
Tool Material. C 0.8, Cr 4.15, Mo 5, W 6.35, V 2, Co 5, Si 0-0.4, Mn 0-0.4.
Mo-W-Co high speed tool steel. W. Nr. 1.3243; AISI M35.

RNC CARIMPHY
Creusot-Loire
Ni, Cr, bal Fe.
For heating elements, electrical resistances; good to 1100 C.
Obsolete

RNC SUPERIMPHY
Creusot-Loire
Ni 80, Cr 20.
For heating elements, radiators, electrical resistances; good to 1100 C. *Obsolete*

RNO 100
Saarstahl AG
C 1.05, Si 0.4, Mn 0.3, Cr 17, Mo 0.5, bal Fe.
For corrosion resistant roller bearings. DIN X 105 CrMo 17. Similar to AISI 440C. *Obsolete*

RNO MOV
Saarstahl AG
Now SAARSTAHL 1.4923.

RNOD CO
Saarstahl AG
Now SAARSTAHL 1.4911.

RNOD NI
Saarstahl AG
Now SAARSTAHL 1.4939.

RO NI 9
Remystahl
C 0-0.1, Ni 9, bal Fe.
Steel; tough at subzero temperatures. W.-Nr. 1.5662.

RO SI
Saarstahl AG
C 0.1, Si 0.25, Mn 0.4, bal Fe.
Carburizing steel for piston pins, camshafts, levers. W.-Nr. 1121. *Obsolete*

ROBERTS-AUSTEN (PURPLE GOLD)
English manufacture
Au 79, Al 21.
For jewelry; corrosion resistant.

ROBINS C.I. 25
Robins Engineers & Constructors Inc.
TC 3.4, CC 0.5, Si 2.5, bal Fe.
Cast: 28,000 TS; 170-190 Brin. For gray iron castings. Transverse strength 2000; transverse deflection 0.25 in.

ROBINS C.I. 35G
Robins Engineers & Constructors Inc.
TC 3.3, CC 0.55, Si 2.3, Cr 0.3, bal Fe.
Cast: 35,000 TS; 190-230 Brin. For cast iron gears. Transverse strength 2200; transverse deflection 0.3 in.

ROBINS C.I. 35SG
Robins Engineers & Constructors Inc.
TC 3.3, CC 0.65, Si 2.3, Ni 1.1, bal Fe.
Cast: 35,000 TS; 210-240 Brin. For cast iron gears. Transverse strength 2300; transverse deflection 0.3 in.

ROBINS C.I. 50
Robins Engineers & Constructors Inc.
TC 3.1, CC 0.65, Si 2.4, Ni 0.75, Cr 0.2, Mo 0.5, bal Fe.
Cast: 50,000 TS; 230 Brin. Heat treated: 70,000 TS; 270 Brin. For cast iron castings, gears. Transverse strength 3000; transverse deflection 0.4 in.

ROBINS COPPER-FREE NI-RESIST
Robins Engineers & Constructors Inc.
TC 2.5-3, Si 2-2.5, Ni 13.5-14.5, bal Fe.
Cast: 25,000 TS; 120-140 Brin. Transverse strength 2100; transverse deflection 0.6 in.

ROBINS E-5
Robins Engineers & Constructors Inc.
C 3.3-3.5, Si 2-2.2, Mn 0.55-0.7, Ni 0.9-1.1, Mo 0.2-0.3, Cr 0.1-0.3, bal Fe.
Cast: 35,000 TS; 190-220 Brin. For Al melting pots; cast iron. For extrusions, forgings and pressings; corrosion resistant.

ROBINS HIGH CHROME NI-RESIST
Robins Engineers & Constructors Inc.
TC 2.5-3, Si 2-2.5, Ni 13.5-14.5, Cu 5.5-6.5, Cr 4-6, bal Fe.
Cast: 30,000 TS; 160-200 Brin. For stoker gates, flue dampers. Transverse strength 2400; transverse deflection 0.45 in.

ROBINS NI-HARD
Robins Engineers & Constructors Inc.
C 3.3-3.6, Si 0.7-1, Mn 0.7-1, Ni 4.25-4.75, Cr 1.25-1.7, bal Fe.
Cast: 575-650 Brin. For rolling mill guides, crusher segments, liners, rolls; cast iron.

ROBINS STANDARD NI-RESIST
Robins Engineers & Constructors Inc.
TC 2.5-3, Si 2-2.5, Ni 13.5-14.5, Cu 5.5-6.5, Cr 2-2.5, Mn 0.9-1.1, bal Fe.
Cast: 25,000 TS; 130-160 Brin. For castings; nonmagnetic; corrosion resistant.

ROBUST
Saarstahl AG
C 0.5, W 2.25, Cr 1.5, V 0.2, bal Fe.
For tools, dies, hot work dies; hot work steel.

ROBUST
Edelstahlwerk Rochling A.G.
C 0.7, W 18, Cr 4, V 1, bal Fe.
For tools, cutters; high speed steel.

ROBUST 35
Saarstahl AG
C 0.32-4, Si 1.2-1.5, Cr 1.2-1.5, V 0.05-1.1, bal Fe.
For tools, dies; hot work steel. *Obsolete*

ROBUST 40
Saarstahl AG
C 0.4-0.5, Mn 0.4-0.6, Si 1.2-1.5, Cr 1.2-1.5, V 0.05-1.1, bal Fe.
For tools, dies, punches; shock resistant. *Obsolete*

ROBUST 50
Saarstahl AG
C 0.6, Cr 1.18, V 0.1, Si 0.8, Mn 0.75, bal Fe.
For gears, springs, crankshafts, fasteners; oil hardened, shock resistant.

ROBUST M
Saarstahl AG
C 0.45, Si 1.1, Cr 1.2, W 2.5, bal Fe.
For tools, dies; tough. *Obsolete*

ROC
Forjas Alavesas S.A.
C 1.9, Mn 0.4, Si 0-0.5, Cr 12, W 0.7, bal Fe.
High carbon-high chromium, air hardening tool steel; for cold work stamping dies. IHA F-521; DIN X210CrW12.

ROC EXTRA
Forjas Alavesas S.A.
C 1.55, Mn 0.4, Si 0-0.5, Cr 12, Mo 0.8, V 0.25, bal Fe.
High carbon-high chromium, air hardening tool steel; for cold work stamping dies. DIN X165 CrMoV12; AFNOR Z160CD12.

ROCAN
Revere Copper Products, Inc.
Cu.
For conduits, sheeting. *Obsolete*

ROCH 2
Rochling Burbach GmbH
C 0.6, Cr 1.18, V 0.1, bal Fe.
For bolts, springs, crankshafts; oil hardened, shock resistant.
Obsolete

ROCH 3
Rochling Burbach GmbH
C 0.55, Cr 1.05, V 0.18, W 1.85, bal Fe.
For punches, crimpers, upsetters, dies; oil hardened.
Obsolete

ROCH 4
Rochling Burbach GmbH
C 0.5, Cr 1.05, Ni 3.25, bal Fe.
For gears, shafts, bolts, studs, crankshafts; oil hardened, shock resistant. *Obsolete*

ROCH 5
Rochling Burbach GmbH
C 2.1, Cr 11.5, bal Fe.
For blanking and forming dies, punches, shears; oil or air hardened, non-deforming. *Obsolete*

ROCH 6
Rochling Burbach GmbH
C 0.9, Mn 1.9, V 0.1, bal Fe.
For punches, dies, crimpers, cutters; oil hardened, non-deforming. *Obsolete*

ROCH 7
Rochling Burbach GmbH
C 0.3, W, Cr, V, bal Fe.
For extrusion press rams, liners, upsetters; hot work steel, oil hardened. *Obsolete*

ROCH 8
Rochling Burbach GmbH
C 0.3, Si 1, Cr 1.1, V 0.18, W 3.75, bal Fe.
For extrusion rams, upsetters, punches; hot work steel, oil hardened. *Obsolete*

ROCHETTE-UMCO50
Laminoire de la Rochette
C 0.05-0.12, Ti 0.18, Cb 0.6, Cr 26-30, Co 47-52, bal Fe.
Wrought: 132,000 TS; 61,000 YS; 7 El; 6 RA; 350 Brin. For furnace baffles, burner tips, sintering grates, quench baskets. Corrosion, heat, thermal shock resistant.

ROCHLING D SPEZIAL
Rochling Burbach GmbH
C 1.25, Si 1.15, Mn 0.7, Cr 1.2, bal Fe.
For bearings, sleeves, liners, bushings; water hardened, wear resistant. *Obsolete*

ROCHLING ECMO100
Rochling Burbach GmbH
C 0.2, Cr 1.15, Mn 1.25, bal Fe.
For gears, bolts, machine tool parts; case hardened, tough.
Obsolete

ROCHLING ECN180
Rochling Burbach GmbH
C 0.18, Cr 2, Ni 2, Mn 0.5, bal Fe.
For gears, bolts, crankshafts; case hardened, shock resistant.
Obsolete

ROCHLING ECV150
Rochling Burbach GmbH
C, alloy, bal Fe.
For machine tool parts; oil hardened. *Obsolete*

ROCHLING ES114
Rochling Burbach GmbH
C 1.4, Si 0.25, Mn 0.3, Cr 0.3, V 0.1, bal Fe.
For bearings, bushings, cutters; oil hardened, wear resistant.
Obsolete

ROCHLING F1620
Rochling Burbach GmbH
C 0.3, Cr, V, bal Fe.
For gears, shafts, bolts, fasteners; oil hardened, tough.
Obsolete

ROCHLING HOWG
Rochling Burbach GmbH
C 0.55, Mo 0.18, Ni 1.65, Cr 0.7, bal Fe.
For forging dies, headers, shears, punches; oil hardened, shock resistant. *Obsolete*

ROCHLING K
Rochling Burbach GmbH
C 1.05, Cr 1, W 1.15, Mn 0.9, bal Fe.
For dies, shears, cutters; water hardened, tough. *Obsolete*

ROCHLING M10C
Rochling Burbach GmbH
C 0.15, Cr 0.65, Mn 0.5, bal Fe.
For gears, bolts, machine tool parts; case hardened. *Obsolete*

ROCHLING MC15
Rochling Burbach GmbH
C 0.16, Cr 0.95, Mn 1.15, Si 0.25, bal Fe.
For gears, bolts, camshafts, fasteners; case hardening steel. *Obsolete*

ROCHLING MC20
Rochling Burbach GmbH
C 0.2, Cr 1.15, Mn 1.25, Si 0.25, bal Fe.
For gears, cams, camshafts, bolts; case hardening steel. *Obsolete*

ROCHLING MFR
Rochling Burbach GmbH
C 0.4, Cr, Mn, Mo, bal Fe.
For gears, bolts, machine tool parts; oil hardened, tough. *Obsolete*

ROCHLING MM
Rochling Burbach GmbH
C 0.5, Cr 1.05, Ni 3.25, Mn 0.5, bal Fe.
For gears, bolts, crankshafts, fasteners, studs; oil hardened, shock resistant. *Obsolete*

ROCHLING MO 15
Rochling Burbach GmbH
C 0.15, Cr, Mo, bal Fe.
For gears, cams, machine tool parts; case hardening steel, tough. *Obsolete*

ROCHLING MO 20
Rochling Burbach GmbH
C 0.2, Cr, Mo, bal Fe.
For gears, cams, machine tool parts; case hardening steel, tough. *Obsolete*

ROCHLING MO 230
Rochling Burbach GmbH
C 0.3, Cr 2.5, Mo 0.2, V 0.15, bal Fe.
For gears, bolts, crimpers, crankshafts; oil hardened, tough. *Obsolete*

ROCHLING MO 240
Rochling Burbach GmbH
C 0.42, Cr, Mo, bal Fe.
For gears, bolts, crankshafts; oil hardened, tough. *Obsolete*

ROCHLING MO 25
Rochling Burbach GmbH
C 0.25, Cr 1, Mn 0.65, Mo 0.2, bal Fe.
For gears, bolts, machine tool parts, shafts; water or oil hardened, tough. *Obsolete*

ROCHLING MO 35
Rochling Burbach GmbH
C 0.33, Mo 0.2, Mn 0.65, Cr 1, bal Fe.
For gears, bolts, machine tool parts, shafts; oil hardened, tough. *Obsolete*

ROCHLING MO 40
Rochling Burbach GmbH
C 0.42, Cr 1, Mo 0.2, Mn 0.65, bal Fe.
For gears, bolts, machine tool parts, shafts; oil hardened, tough. *Obsolete*

ROCHLING MO 50
Rochling Burbach GmbH
C 0.5, Mn 0.65, Cr 1, Mo 0.25, bal Fe.
For gears, bolts, crankshafts, fasteners; oil hardened, tough. *Obsolete*

ROCHLING NG5A
Rochling Burbach GmbH
C 0.45, Cr 1.35, Mo 0.45, V 0.8, W 0.45, bal Fe.
For forging and header dies, punches, shears; oil hardened, shock resistant. *Obsolete*

ROCHLING NGS
Rochling Burbach GmbH
C 0.3, Cr, Mo, Mn, V, bal Fe.
For machine tool parts; oil hardened, tough. *Obsolete*

ROCHLING NH11G
Rochling Burbach GmbH
C 0.4, Cr 22, Ni 9.5, bal Fe.
Cast: 85,000 TS; 45,000 YS; 35 El; 165 Brin. For heat treat boxes, baskets, burner tips, conveyors; corrosion and heat resistant. *Obsolete*

ROCHLING NH22G
Rochling Burbach GmbH
C 0.4, Si 2, Cr 25, Ni 19, bal Fe.
Cast: 75,000 TS; 50,000 YS; 17 El; 170 Brin. For furnace parts, retorts, rabble arms, dampers; Type HK; corrosion and heat resistant. *Obsolete*

ROCHLING NH25G
Rochling Burbach GmbH
C 0.4, Cr 26, Ni 14, bal Fe.
Cast: 75,000 TS; 47,000 YS; 17 El; 25 RA; 200 Brin. For heat treat boxes, furnace parts, retorts, salt pots; Type HH; corrosion and heat resistant. *Obsolete*

ROCHLING NH4
Rochling Burbach GmbH
C 0.2, Si 1.2, Cr 25, Ni 4, bal Fe.
Cast: 90,000 TS; 65,000 YS; 2 El; 212 Brin. For cylinder liners, bushings, valve seats and bodies; corrosion and heat resistant. *Obsolete*

ROCHLING NH40G
Rochling Burbach GmbH
C 0.2, Ni 38, Cr 18, bal Fe.
For furnace parts, heat treat boxes, retorts, salt pots; corrosion and heat resistant. *Obsolete*

ROCHLING NH4G
Rochling Burbach GmbH
C 0.4, Si 1.3, Cr 27, bal Fe.
Cast: 70,000 TS; 65,000 YS; 2 El; 190 Brin. For furnace parts, grate bars, salt pots, skids; heat resistant. *Obsolete*

ROCHLING NH60
Rochling Burbach GmbH
C 0.15, Cr, Ni, bal Fe.
For chemical plant equipment; corrosion resistant. *Obsolete*

ROCHLING NH8
Rochling Burbach GmbH
C 0.27, Cr 27, Ni 11, Si 1.3, bal Fe.
Cast: 75,000 TS; 47,000 YS; 10 El; 25 RA; 205 Brin. For salt pots, retorts, heat treat boxes, furnace parts; corrosion and heat resistant. *Obsolete*

ROCHLING NH80
Rochling Burbach GmbH
C 0.15, Ni, Cr, bal Fe.
For chemical plant equipment; corrosion resistant. *Obsolete*

ROCHLING OCE12
Rochling Burbach GmbH
C 0.76, Ni, bal Fe.
For tools, dies. *Obsolete*

ROCHLING OCE34
Rochling Burbach GmbH
C 0.79, Ni, bal Fe.
For tools, dies. *Obsolete*

ROCHLING OCE57
Rochling Burbach GmbH
C 0.83, Ni, bal Fe.
For tools, dies. *Obsolete*

ROCHLING PG1
Rochling Burbach GmbH
C 0.35, Cr, Mo, V, bal Fe.
For forging and heading dies, punches, crimpers; oil hardened, tough. *Obsolete*

ROCHLING PG3
Rochling Burbach GmbH
C 0.3, Cr, Mo, V, bal Fe.
For forging and heading dies, punches, crimpers; oil hardened, tough. *Obsolete*

ROCHLING PRIMA-H
Rochling Burbach GmbH
C 1.3, Si 0-0.25, Mn 0-0.25, bal Fe.
For engravers' tools, textile needles, taps; Type W1; water hardened. *Obsolete*

ROCHLING PRIMA-MH
Rochling Burbach GmbH
C 1.15, Si 0-0.25, Mn 0-0.25, bal Fe.
Annealed: 110,000 TS; 58,000 YS; 18 El; 38 RA; 210 Brin. For springs, taps, tools, cutters, reamers; Type W1; water hardened. *Obsolete*

ROCHLING PRIMA-ZAH
Rochling Burbach GmbH
C 0.85, Si 0-0.25, Mn 0-0.25, bal Fe.
Heat treated: 200,000 TS; 150,000 YS; 10 El; 30 RA; 400 Brin. For springs, taps, reamers, drills, hobs; Type W1; water hardened. *Obsolete*

ROCHLING PRIMA-ZH
Rochling Burbach GmbH
C 1, Si 0-0.25, Mn 0-0.25, bal Fe.
Heat treated: 200,000 TS; 150,000 YS; 8 El; 28 RA; 410 Brin. For springs, taps, reamers, broaches; Type W1; water hardened. *Obsolete*

ROCHLING R 2/3
Rochling Burbach GmbH
C 0.28, Si 0.25, Mn 0.45, bal Fe.
Hot rolled: 80,000 TS; 50,000 YS; 30 El; 56 RA; 163 Brin. For gears, bolts, machine tool parts; water hardened. *Obsolete*

ROCHLING R 6/7
Rochling Burbach GmbH
C 0.67, Si 0.25, Mn 0.65, bal Fe.
Heat treated: 170,000 TS; 125,000 YS; 10 El; 35 RA; 360 Brin. For gears, springs, rails, punches, axes; water hardened. *Obsolete*

ROCHLING R1
Rochling Burbach GmbH
C 0.15, Si 0.25, Mn 0.37, bal Fe.
Annealed: 70,000 TS; 40,000 YS; 25 El; 60 RA; 145 Brin. For gears, bolts, machine tool parts; case hardened. *Obsolete*

ROCHLING R5
Rochling Burbach GmbH
C 0.53, Si 0.25, Mn 0.65, bal Fe.
Normalized: 100,000 TS; 65,000 YS; 19 El; 26 RA; 220 Brin. For gears, bodies, machine tool parts; water hardened. *Obsolete*

ROCHLING R6
Rochling Burbach GmbH
C 0.6, Si 0.25, Mn 0.65, bal Fe.
Heat treated: 160,000 TS; 113,000 YS; 12 El; 40 RA; 325 Brin. For gears, springs, hammers, rails, punches; water hardened. *Obsolete*

ROCHLING R8
Rochling Burbach GmbH
C 0.8, Si 0.25, Mn 0.65, bal Fe.
Heat treated: 188,000 TS; 143,000 YS; 12 El; 35 RA; 390 Brin.
For springs, tools, drills, taps; Type W1; water hardened.
Obsolete

ROCHLING RA2
Rochling Burbach GmbH
C 0.15, Si 0.25, Mn 0.37, bal Fe.
Annealed: 70,000 TS; 40,000 YS; 25 El; 60 RA; 145 Brin. For
gears, bolts, machine tool parts; case hardened. *Obsolete*

ROCHLING RA2W
Rochling Burbach GmbH
C 0.1, Si 0.25, Mn 0.37, bal Fe.
Annealed: 100,000 TS; 53,000 YS; 21 El; 42 RA; 200 Brin. For
gears, fasteners, machine tool parts; case hardened.
Obsolete

ROCHLING RAB1
Rochling Burbach GmbH
C 0.4, Ni, Cr, Mo, bal Fe.
For gears, bolts, crankshafts; oil hardened, shock resistant.
Obsolete

ROCHLING RAE2
Rochling Burbach GmbH
C 0.13, Cr 0.7, Ni 2.5, bal Fe.
For gears, bolts, camshafts; case hardened, shock resistant.
Obsolete

ROCHLING RAE3B
Rochling Burbach GmbH
C 0.13, Cr 0.7, Ni 3.5, bal Fe.
For gears, bolts, camshafts, cams; case hardened, shock
resistant. *Obsolete*

ROCHLING RAE5
Rochling Burbach GmbH
C 0.13, Si 0-0.35, Mn 0-0.5, Cr 1.1, Ni 4.5, bal Fe.
For gears, bolts, camshafts, cams; case hardened, shock
resistant. *Obsolete*

ROCHLING RB9
Rochling Burbach GmbH
C 0.55, Si 0.1-0.4, Mn 0.5-0.7, bal Fe.
Heat treated: 115,000-160,000 TS; 77,000-113,000 YS; 12-23
El; 40-54 RA; 229-321 Brin. For crankshafts, gears, girders,
bushings, bolts; water hardened. *Obsolete*

ROCHLING RBF
Rochling Burbach GmbH
C 0.40, bal Fe.
Hot rolled: 98,000 TS; 59,000 YS; 24 El; 45 RA; 212 Brin. For
axles, gears, bolts, crankpins; water hardened. *Obsolete*

ROCHLING RBH
Rochling Burbach GmbH
C 0.7, Si 0-0.25, Mn 0-0.25, bal Fe.
Heat treated: 122,000-174,000 TS; 82,000-128,000 YS; 12-22
El; 37-52 RA; 241-352 Brin. For springs, clutch disks, rails,
wheels; water hardened. *Obsolete*

ROCHLING RCCK
Rochling Burbach GmbH
C 2.1, Cr 11.5, W 0.7, bal Fe.
For forming and blanking dies, punches; oil hardened, non-
deforming. *Obsolete*

ROCHLING RCCW
Rochling Burbach GmbH
C 1.65, Cr 11.5, V 0.1, bal Fe.
For forming and blanking dies, punches; air hardened,
nondeforming. *Obsolete*

ROCHLING RCNH
Rochling Burbach GmbH
C 0.5, Cr 1.05, Ni 3.25, bal Fe.
For gears, bolts, crankshafts; oil hardened, shock resistant.
Obsolete

ROCHLING RCW2
Rochling Burbach GmbH
C 0.3, Cr 2.65, V 0.35, W 8.5, bal Fe.
For extrusion press rams, dies and liners; oil hardened,
tough. *Obsolete*

ROCHLING RCW3
Rochling Burbach GmbH
C 0.28, W, Cr, Ni, V, bal Fe.
For forging dies, upsetters, crimpers; oil hardened. *Obsolete*

ROCHLING RE2
Rochling Burbach GmbH
C 0.15, Si 0.25, Mn 0.37, bal Fe.
Annealed: 70,000 TS; 55,000 YS; 25 El; 60 RA; 143 Brin. For
screws, bolts, nuts, rivets, gears, shafts; case hardened.
Obsolete

ROCHLING RE2C
Rochling Burbach GmbH
C 0.15, Cr 0.65, Mn 0.37, bal Fe.
For gears, bolts, camshafts, cams, bushings; case hardened.
Obsolete

ROCHLING RE2CW
Rochling Burbach GmbH
C 0.13, Mn 0.37, Si 0.25, Cr 0.5, bal Fe.
For gears, bolts, camshafts, cams, bushings; case hardened.
Obsolete

ROCHLING RE2W
Rochling Burbach GmbH
C 0.1, Si 0.25, Mn 0.37, bal Fe.
Annealed: 64,000 TS; 48,000 YS; 28 El; 65 RA; 131 Brin. For
nails, rivets, case hardened parts, gears; case hardened.
Obsolete

ROCHLING RECN1
Rochling Burbach GmbH
C 0.15, Cr 1.55, Ni 1.55, bal Fe.
For camshafts, cams, bolts, bushings; case hardened, tough.
Obsolete

ROCHLING RECNW
Rochling Burbach GmbH
C 0.15, Cr 1.55, Ni 1.55, bal Fe.
For camshafts, cams, bearings; case hardened, shock
resistant. *Obsolete*

ROCHLING REN2
Rochling Burbach GmbH
C 0.13, Cr, Ni, bal Fe.
For machine tool parts; oil hardened. *Obsolete*

ROCHLING RF6
Rochling Burbach GmbH
C 0.56, Mn 0.55, Si 0.3, bal Fe.
Heat treated: 115,000-160,000 TS; 77,000-113,000 YS; 12-23
El; 40-54 RA; 229-321 Brin. For wheels, die blocks, rails,
girders; water hardened. *Obsolete*

ROCHLING RFA
Rochling Burbach GmbH
C 0.55, Si 1.7, Mn 0.7, bal Fe.
For springs, punches; oil hardened, tough. *Obsolete*

ROCHLING RFAW
Rochling Burbach GmbH
C 0.46, Si 1.7, Mn 0.65, bal Fe.
For springs, punches; oil hardened, tough. *Obsolete*

ROCHLING RG10
Rochling Burbach GmbH
C 1, Si 0-0.25, Mn 0-0.25, bal Fe.
Heat treated: 130,000-185,000 TS; 80,000-120,000 YS; 10-21
El; 30-47 RA; 270-375 Brin. For cutters, taps, springs, drills,
reamers; Type W1; water hardened. *Obsolete*

ROCHLING RG11
Rochling Burbach GmbH
C 1.15, Si 0-0.25, Mn 0-0.25, bal Fe.
Annealed: 110,000 TS; 55,000 YS; 18 El; 38 RA; 210 Brin. For
springs, taps, reamers, drills, broaches; Type W1; water
hardened. *Obsolete*

ROCHLING RG12
Rochling Burbach GmbH
C 1.3, Si 0-0.25, Mn 0-0.25, bal Fe.
For forming and blanking dies, engravers tools; Type W1;
water hardened. *Obsolete*

ROCHLING RG6
Rochling Burbach GmbH
C 0.7, Si 0-0.25, Mn 0-0.25, bal Fe.
Heat treated: 122,000-174,000 TS; 82,000-128,000 YS; 12-22
El; 37-52 RA; 241-352 Brin. For springs, clutch disks, die
blocks, rails; water hardened; Type W1. *Obsolete*

ROCHLING RG7
Rochling Burbach GmbH
C 0.85, Si 0-0.25, Mn 0-0.25, bal Fe.
Heat treated: 129,000-188,000 TS; 87,000-143,000 YS; 12-21
El; 35-50 RA; 255-388 Brin. For drills, springs, reamers, taps,
broaches; water hardened. *Obsolete*

ROCHLING RG8
Rochling Burbach GmbH
C 0.88, Si 0-0.25, Mn 0-0.25, bal Fe.
Heat treated: 130,000-190,000 TS; 88,000-145,000 YS; 12-20
El; 35-50 RA; 260-390 Brin. For drills, springs, reamers, taps,
broaches, cutters; Type W1; water hardened. *Obsolete*

ROCHLING RGS2
Rochling Burbach GmbH
C 0.55, Cr 0.7, Mo 0.18, Ni 1.65, bal Fe.
For forging dies, gears, bolts, punches; oil hardened, shock
resistant. *Obsolete*

ROCHLING RH15
Rochling Burbach GmbH
C 1.2, Mn 12.5, Si 0.4, bal Fe.
For wear plates, rails, frogs, dipper teeth; wear and abrasion
resistant. *Obsolete*

ROCHLING RHB9
Rochling Burbach GmbH
C 0.55, Si 0.1-0.4, Mn 0.5-0.7, bal Fe.
Annealed: 100,000 TS; 55,000 YS; 15 El; 22 RA; 180 Brin. For
crankshafts, bolts, gears, crankpins; water hardened.
Obsolete

ROCHLING RKMV
Rochling Burbach GmbH
C 1, Si 0.2, Mn 0.25, V 0.1, bal Fe.
Heat treated: 130,000-184,000 TS; 80,000-118,000 YS; 10-20
El; 30-45 RA; 270-375 Brin. For drills, springs, taps, broaches,
reamers; Type W1; water hardened. *Obsolete*

ROCHLING RLB
Rochling Burbach GmbH
C 0.15, Si 0.35, Mn 0.35, bal Fe.
Annealed: 70,000 TS; 55,000 YS; 25 El; 60 RA; 145 Brin. For
screws, bolts, gears, rivets, bushings; case hardened.
Obsolete

ROCHLING RM2
Rochling Burbach GmbH
C 0.22, Si 0.25, Mn 0.45, bal Fe.
Annealed: 75,000 TS; 63,000 YS; 22 El; 58 RA; 150 Brin. For
fan blades, bushings, bolts, gears; case hardened. *Obsolete*

ROCHLING RM2/3
Rochling Burbach GmbH
C 0.28, Si 0.25, Mn 0.45, bal Fe.
Annealed: 80,000 TS; 50,000 YS; 30 El; 56 RA; 165 Brin. For
camshafts, gears, bolts, keys, brackets; water hardened.
Obsolete

ROCHLING RM3
Rochling Burbach GmbH
C 0.35, Si 0.25, Mn 0.55, bal Fe.
Hot rolled: 85,000 TS; 54,000 YS; 30 El; 53 RA; 185 Brin. For gears, shafts, axles, bolts, fishplates; water hardened.
Obsolete

ROCHLING RM4
Rochling Burbach GmbH
C 0.45, Si 0.25, Mn 0.65, bal Fe.
Hot rolled: 98,000 TS; 60,000 YS; 24 El; 45 RA; 215 Brin. For axles, gears, bolts, bushings, tie rods; water hardened.
Obsolete

ROCHLING RM5
Rochling Burbach GmbH
C 0.53, Si 0.25, Mn 0.65, bal Fe.
Annealed: 96,000 TS; 52,000 YS; 16 El; 23 RA; 170 Brin. For axles, bolts, gears, shafts; water hardened. *Obsolete*

ROCHLING RM6
Rochling Burbach GmbH
C 0.61, Si 0.25, Mn 0.65, bal Fe.
Heat treated: 115,000-160,000 TS; 77,000-113,000 YS; 12-23 El; 40-54 RA; 230-320 Brin. For wheels, die blocks, girders, rails, springs; water hardened. *Obsolete*

ROCHLING RM6/7
Rochling Burbach GmbH
C 0.8, Si 0.25, Mn 0.65, bal Fe.
Heat treated: 130,000-188,000 TS; 87,000-143,000 YS; 12-21 El; 35-50 RA; 255-388 Brin. For springs, taps, drills, reamers, cutters; Type W1; water hardened. *Obsolete*

ROCHLING RNO
Rochling Burbach GmbH
C 0.2, Si 0.4, Cr 12-14, bal Fe.
Annealed: 95,000 TS; 50,000 YS; 25 El; 55 RA; 196 Brin. For turbine blades, valves, cutlery; Type 420; stainless. *Obsolete*

ROCHLING RNO SPEZIAL G
Rochling Burbach GmbH
C 0.4, Si 1.3, Cr 27, Ni 4, bal Fe.
Cast: 90,000 TS; 65,000 YS; 2 El; 212 Brin. For cylinder liners, bushings, valve seats and bodies; corrosion and heat resistant. *Obsolete*

ROCHLING RNO18
Rochling Burbach GmbH
C 0-0.1, Cr 17.5, Ti = 7 x C, bal Fe.
Annealed: 125,000 TS; 95,000 YS; 20 El; 55 RA; 260 Brin. For welded chemical plant equipment; Type 430 Ti; corrosion resistant. *Obsolete*

ROCHLING RNO25
Rochling Burbach GmbH
C 0-0.15, Mo 2.5, Ti 1.8, Cr, bal Fe.
For chemical plant equipment; corrosion and heat resistant.
Obsolete

ROCHLING RNO30EG
Rochling Burbach GmbH
C 1.2, Si 1.3, Cr 29, Mo 2, bal Fe.
For crushers, rollers, baffles; heat and abrasion resistant.
Obsolete

ROCHLING RNO30G
Rochling Burbach GmbH
C 1.2, Si 1.3, Cr 29, bal Fe.
For crushers, rollers, baffles; heat and abrasion resistant.
Obsolete

ROCHLING RNOA17
Rochling Burbach GmbH
C 0.12, Cr 16.5, Mo 0.25, S 0.2, bal Fe.
Annealed: 80,000 TS; 50,000 YS; 25 El; 50 RA; 150 Brin. For screw machine products, bolts, screws; Type 430 F; corrosion resistant. *Obsolete*

ROCHLING RNOB
Rochling Burbach GmbH
C 0.8, Cr 14, bal Fe.
For bearings, pivots, cutlery; corrosion and wear resistant.
Obsolete

ROCHLING RNOC
Rochling Burbach GmbH
C 0.9, Cr 18, V 0.1, Mo 1.2, bal Fe.
For bearings, cutlery, valves; corrosion and wear resistant.
Obsolete

ROCHLING RNOF
Rochling Burbach GmbH
C 0.3, Cr 12-14, bal Fe.
Annealed: 100,000 TS; 55,000 YS; 20 El; 55 RA; 200 Brin. For valves, cutlery, surgical and dental instruments; Type 420; corrosion resistant. *Obsolete*

ROCHLING RNOG
Rochling Burbach GmbH
C 0.25, Si 0.7, Cr 14.5, Ni 0-1, bal Fe.
Annealed: 100,000 TS; 55,000 YS; 20 El; 55 RA; 200 Brin. For valves, cutlery, surgical and dental instruments; corrosion resistant. *Obsolete*

ROCHLING RNOH
Rochling Burbach GmbH
C 0.4, Si 0.4, Cr 13, bal Fe.
Annealed: 100,000 TS; 55,000 YS; 20 El; 55 RA; 200 Brin. For valves, cutlery, surgical and dental instruments; Type 420; corrosion resistant. *Obsolete*

ROCHLING RNOM
Rochling Burbach GmbH
C 0.17, Cr, Mo, bal Fe.
For gears, pinions, camshafts, cams; case hardened.
Obsolete

ROCHLING RNOS
Rochling Burbach GmbH
C 0.22, Cr 17, Ni 1.5, Si 0.4, bal Fe.
Annealed: 125,000 TS; 95,000 YS; 20 El; 55 RA; 260 Brin. For pumps, marine hardware, valves; Type 431; corrosion and heat resistant. *Obsolete*

ROCHLING RNOSG
Rochling Burbach GmbH
C 0.25, Si 0.7, Cr 17, Ni 0-1.8, bal Fe.
Cast: 85,000 TS; 50,000 YS; 5 El; 10 RA; 200 Brin. For furnace parts, heat treating boxes; corrosion resistant.
Obsolete

ROCHLING RNOT
Rochling Burbach GmbH
C 0.17, Cr, Ni, bal Fe.
For chemical plant equipment; stainless. *Obsolete*

ROCHLING RNOW
Rochling Burbach GmbH
C 0-0.12, Si 0.4, Cr 13, bal Fe.
Annealed: 75,000 TS; 40,000 YS; 35 El; 70 RA; 155 Brin. For turbine blades, valves, cutlery; Type 410; corrosion resistant.
Obsolete

ROCHLING RNOW-W
Rochling Burbach GmbH
C 0-0.12, Cr 14-18, bal Fe.
Annealed: 80,000 TS; 50,000 YS; 25 El; 50 RA; 150 Brin. For oil refinery equipment, sinks, oil burners; Type 430; corrosion resistant. *Obsolete*

ROCHLING RNOX
Rochling Burbach GmbH
C 0.35, Cr 16.5, Mo 1.15, bal Fe.
For furnace parts, oil refinery equipment; heat and corrosion resistant. *Obsolete*

ROCHLING ROT 40
Rochling Burbach GmbH
C 0.28, Cr 1, Ni 1, Mo 0.2, bal Fe.
For oil refinery equipment; corrosion and heat resistant.
Obsolete

ROCHLING ROT 60
Rochling Burbach GmbH
C 0.34, Cr 1.55, Ni 1.55, Mo 0.2, bal Fe.
For oil refinery equipment, valves; corrosion and creep resistant. *Obsolete*

ROCHLING RPDW
Rochling Burbach GmbH
C 0.33, Cr 1, Mn 0.65, Mo 0.2, bal Fe.
For gears, bolts, oil refinery equipment; heat and creep resistant. *Obsolete*

ROCHLING RPDZ
Rochling Burbach GmbH
C 0.41, Cr 1, Mn 0.65, bal Fe.
For gears, bolts, shafts; oil hardened, tough. *Obsolete*

ROCHLING RPDZW
Rochling Burbach GmbH
C 0.41, Mn 0.65, Cr 1, bal Fe.
For gears, bolts, shafts; oil hardened, tough. *Obsolete*

ROCHLING RR3
Rochling Burbach GmbH
C 0.35, Si 0.25-0.5, Mn 0.3-0.8, bal Fe.
Hot rolled: 85,000 TS; 54,000 YS; 30 El; 53 RA; 180 Brin. For gears, shafts, axles, bolts, screws; water hardened. *Obsolete*

ROCHLING RR4
Rochling Burbach GmbH
C 0.45, Si 0.25-0.5, Mn 0.3-0.8, bal Fe.
Hot rolled: 98,000 TS; 58,000 YS; 24 El; 45 RA; 212 Brin. For axles, gears, bolts, crankshafts; water hardened. *Obsolete*

ROCHLING RR6
Rochling Burbach GmbH
C 0.6, Si 0.25-0.5, Mn 0.3-0.8, bal Fe.
Heat treated: 170,000 TS; 120,000 YS; 10 El; 35 RA; 350 Brin. For axles, die blocks, rails, crankshafts; water hardened.
Obsolete

ROCHLING RR7
Rochling Burbach GmbH
C 0.75, Si 0.25-0.5, Mn 0.3-0.8, bal Fe.
Heat treated: 180,000 TS; 140,000 YS; 15 El; 40 RA; 375 Brin. For springs, clutch discs, rails, girders; water hardened.
Obsolete

ROCHLING RR8
Rochling Burbach GmbH
C 0.9, Si 0.25-0.5, Mn 0.3-0.8, bal Fe.
Heat treated: 185,000 TS; 120,000 YS; 10 El; 30 RA; 380 Brin. For springs, taps, reamers, drills, cutters; Type W1; water hardened. *Obsolete*

ROCHLING RSH
Rochling Burbach GmbH
C 1.05, Cr 1, Mn 0.3, bal Fe.
For bearings, bushings, liners; water hardened, wear resistant. *Obsolete*

ROCHLING RT EXTRA
Rochling Burbach GmbH
C 1, Si 0-0.25, Mn 0-0.25, bal Fe.
Annealed: 100,000 TS; 53,000 YS; 21 El; 42 RA; 200 Brin. For springs, taps, reamers, hobs, drills, cutters; Type W1; water hardened. *Obsolete*

ROCHLING RT10EE
Rochling Burbach GmbH
C 1, Si 0.25, Mn 0.3, V 0.1, bal Fe.
Heat treated: 190,000 TS; 120,000 YS; 10 El; 30 RA; 400 Brin. For springs, drills, taps, reamers, cutters; water hardened; Type W1. *Obsolete*

ROCHLING RT11 CR
Rochling Burbach GmbH
C 1.1, Cr 0.4, Si 0.2, Mn 0.3, bal Fe.
Annealed: 100,000 TS; 53,000 YS; 21 El; 42 RA; 200 Brin. For springs, drills, taps, cutters; water hardened; Type W1.
Obsolete

ROCHLING RT11 EXTRA
Rochling Burbach GmbH
C 1.1, Si 0-0.25, Mn 0-0.25, bal Fe.
Annealed: 100,000 TS; 53,000 YS; 21 El; 42 RA; 200 Brin. For springs, drills, reamers, cutters, taps; water hardened; Type W1. *Obsolete*

ROCHLING RT12 EXTRA
Rochling Burbach GmbH
C 1.3, Si 0-0.25, Mn 0-0.25, bal Fe.
For blanking and forming dies, reamers, taps; water hardened; Type W1. *Obsolete*

ROCHLING RT14 CR
Rochling Burbach GmbH
C 1.4, Cr 0.3, V 0.1, bal Fe.
For blanking and forming dies, reamers, taps; Type W1; water hardened. *Obsolete*

ROCHLING RT14 EXTRA
Rochling Burbach GmbH
C 1.3, Si 0-0.25, Mn 0-0.25, bal Fe.
For blanking and forming dies; Type W1; water hardened. *Obsolete*

ROCHLING RT7 EXTRA
Rochling Burbach GmbH
C 0.75, Si 0-0.25, Mn 0-0.25, bal Fe.
Heat treated: 175,000 TS; 128,000 YS; 12 El; 37 RA; 350 Brin. For springs, die blocks, rails, girders; water hardened. *Obsolete*

ROCHLING RT9
Rochling Burbach GmbH
C 0.85, Si 0-0.25, Mn 0-0.25, bal Fe.
Heat treated: 190,000 TS; 145,000 YS; 12 El; 35 RA; 390 Brin. For reamers, drills, springs, taps, broaches; water hardened. *Obsolete*

ROCHLING RTC20
Rochling Burbach GmbH
C 0.85, Cr 1.75, Mn 0.35, bal Fe.
Annealed: 110,000 TS; 60,000 YS; 15 El; 30 RA; 200 Brin. For forming dies, cutters, bearings; oil hardened, abrasion resistant. *Obsolete*

ROCHLING RTK10
Rochling Burbach GmbH
C 0.9, Cr 0.8, Mn 0.3, bal Fe.
Heat treated: 190,000 TS; 145,000 YS; 10 El; 30 RA; 400 Brin. For bearings, liners, sleeves, cutters; oil or water hardened, wear resistant. *Obsolete*

ROCHLING RTW2
Rochling Burbach GmbH
C 0.35, Cr 1.05, V 0.18, W 1.85, Si 0.9, bal Fe.
For heading and blanking dies, cold work tools; oil hardened, tough. *Obsolete*

ROCHLING RTW3
Rochling Burbach GmbH
C 1.3, W 4.75, Cr 0-0.2, bal Fe.
For forming and drawing dies, reamers; water hardened, tough. *Obsolete*

ROCHLING RVE
Rochling Burbach GmbH
C 0.45, Cr, Si, bal Fe.
For bolts, gears, machine tool parts; oil hardened, tough. *Obsolete*

ROCHLING RVE EXTRA
Rochling Burbach GmbH
C 0.55, Cr, Ni, W, bal Fe.
For forging and heading dies, shears; oil hardened, tough. *Obsolete*

ROCHLING RVE42
Rochling Burbach GmbH
C 0.45, Cr, Ni, W, bal Fe.
For forging and heading dies, shears; oil hardened, tough. *Obsolete*

ROCHLING RW2
Rochling Burbach GmbH
C 1.05, Cr 1, Mn 0.3, bal Fe.
For bearings, cutters, liners, sleeves; water hardened, wear resistant. *Obsolete*

ROCHLING RW3A
Rochling Burbach GmbH
C 1, Cr 1.1, Mn 0.07, bal Fe.
For bearings, cutters, liners, sleeves; water hardened, wear resistant. *Obsolete*

ROCHLING RWS2
Rochling Burbach GmbH
C 0.3, V 0.18, Cr 1.1, W 3.75, bal Fe.
For header dies, shears, punches; oil hardened, tough. *Obsolete*

ROCHLING SGM
Rochling Burbach GmbH
C 0.2, Cr 13, Mn 0.4, bal Fe.
Annealed: 95,000 TS; 50,000 YS; 25 El; 55 RA; 195 Brin. For turbine blades, valves, cutlery; Type 420; stainless. *Obsolete*

ROCHLING SGM2
Rochling Burbach GmbH
C 0.3, Co 2, Cr 2.4, V 0.25, W 8.5, bal Fe.
For hot work tools, punches, upsetters; hot work steel, oil hardened. *Obsolete*

ROCHLING VC140
Rochling Burbach GmbH
C 0.41, Cr 1.1, Mn 0.7, Si 0.25, bal Fe.
For gears, bolts, shafts, axles; oil hardened, tough. *Obsolete*

ROCHLING VC170
Rochling Burbach GmbH
C 0.36, Cr 1.6, Mn 0.45, Si 0.25, bal Fe.
For gears, bolts, shafts, axles; oil hardened, tough. *Obsolete*

ROCHLING VM100
Rochling Burbach GmbH
C 0.4, Si 0.38, Mn 0.95, bal Fe.
Hot rolled: 90,000 TS; 58,000 YS; 27 El; 50 RA; 200 Brin. For gears, bolts, shafts, axles; water hardened. *Obsolete*

ROCHLING VM125
Rochling Burbach GmbH
C 0.3, Mn 1.25, Cr 0-0.3, bal Fe.
For gears, bolts, camshafts, crankshafts; water hardened, tough. *Obsolete*

ROCHLING VM80
Rochling Burbach GmbH
C 0.2, Mn 0.95, Si 0.38, bal Fe.
For gears, bolts, camshafts, cams; case hardened. *Obsolete*

ROCHLING VM90
Rochling Burbach GmbH
C 0.3, Mn 0.95, Si 0.38, bal Fe.
Hot rolled: 85,000 TS; 55,000 YS; 30 El; 55 RA; 180 Brin. For gears, bolts, shafts, axles; water hardened. *Obsolete*

ROCHLING VME
Rochling Burbach GmbH
C 0.45, Si, Mn, bal Fe.
Hot rolled: 98,000 TS; 59,000 YS; 24 El; 45 RA; 212 Brin. For axles, gears, bolts, shafts; water hardened. *Obsolete*

ROCHLING VMS140
Rochling Burbach GmbH
C 0.46, Si 0.9, Mn 0.9, bal Fe.
For gears, dies, fasteners, crankshafts; oil hardened, tough.
Obsolete

ROCHLING VMS160
Rochling Burbach GmbH
C 0.53, Si 0.9, Mn 0.9, bal Fe.
For dies, upsetters, crankshafts; oil hardened, tough.
Obsolete

ROCHLING VMV125
Rochling Burbach GmbH
C 0.27, Mn, Cr, V, bal Fe.
For gears, bolts, crankshafts; oil hardened, tough. *Obsolete*

ROCHLING VMV140
Rochling Burbach GmbH
C 0.42, Mn 1.75, V 0.1, bal Fe.
Hot rolled: 97,000 TS; 59,000 YS; 40 El; 200 Brin. For punches, forming dies, upsetters; oil hardened, tough. *Obsolete*

ROCHLING WP3
Rochling Burbach GmbH
C 0.13, Si 0-0.35, Mn 0-0.5, bal Fe.
Annealed: 70,000 TS; 55,000 YS; 25 El; 60 RA; 145 Brin. For nails, rivets, screws, camshafts; case hardened. *Obsolete*

ROCHLING ZK
Saarstahl AG
C 1.45, Si 0.25, Mn 0.6, Cr 1.4, bal Fe.

ROCHLINGSTAHL 30
Rochling Burbach GmbH
C 0.34, Cr 1.4, Al 1.1, Mn 0.6, bal Fe.
For oil refinery parts; creep and heat resistant. *Obsolete*

ROCHLINGSTAHL 32
Rochling Burbach GmbH
C 0.5, Ni 3.25, Cr 1.05, Mn 0.5, bal Fe.
For gears, bolts, crankshafts, machine tool parts; oil hardened, shock resistant. *Obsolete*

ROCHLINGSTAHL 34
Rochling Burbach GmbH
C 0.9, Mn 1.9, V 0.1, bal Fe.
For punches, shears, form tools, upsetters; oil hardening, nondeforming. *Obsolete*

ROCHLINGSTAHL B.S.2
Rochling Burbach GmbH
C 0.75-0.83, Mn 0.5-0.7, Si 0.25-0.4, bal Fe.
For punches, springs, tools; water hardening. *Obsolete*

ROCHLINGSTAHL BP21
Rochling Burbach GmbH
C 0.15, Si 0.15-0.35, Mn 0.25-0.5, bal Fe.
For gears, fasteners, shafts, machine tool parts; case hardening steel. *Obsolete*

ROCHLINGSTAHL BP22
Rochling Burbach GmbH
C 0.15, Si 0.25, Mn 0.5, Cr 0.65, bal Fe.
For gears, fasteners, machine tool parts; case hardening steel. *Obsolete*

ROCHLINGSTAHL BP25
Rochling Burbach GmbH
C 0.2, Cr 1.15, Mn 1.25, bal Fe.
For gears, cams, camshafts, fasteners; case hardening steel, tough. *Obsolete*

ROCHLINGSTAHL BP29
Rochling Burbach GmbH
C 0.55, Mo 0.18, Cr 0.7, V 0.1, Ni 1.65, bal Fe.
For gears, bolts, crankshafts; oil hardened, shock resistant. *Obsolete*

ROCHLINGSTAHL BP29 EXTRA
Rochling Burbach GmbH
C 0.56, Ni 1.8, Cr 0.8, Mo 0.2, bal Fe.
For gears, bolts, crankshafts, machine tool parts; oil
hardened, shock resistant. *Obsolete*

ROCHLINGSTAHL D SPECIAL
Rochling Burbach GmbH
C 0.96-1, Cr 1.9-2.1, bal Fe.
For tools, dies, bearings; oil hardening. *Obsolete*

ROCHLINGSTAHL ES
Rochling Burbach GmbH
C 1.3-1.5, Mn 0.3, Cr 0.5-0.8, bal Fe.
For tools, dies, drills; water hardening. *Obsolete*

ROCHLINGSTAHL FVC
Rochling Burbach GmbH
C, Cr, V, Mn, bal Fe.
For gears, bolts, crankshafts; oil hardened. *Obsolete*

ROCHLINGSTAHL FVM45
Rochling Burbach GmbH
C 0.42, Si 0.25, Mn 1.75, V 0.1, bal Fe.
For springs, punches, crimpers, upsetters; oil hardened,
tough. *Obsolete*

ROCHLINGSTAHL MFR
Rochling Burbach GmbH
C 0.42-0.5, Mn 0.8-1.1, Cr 0.8-1.1, Mo 0.25-0.4, bal Fe.
For tools, dies; hot work steel. *Obsolete*

ROCHLINGSTAHL NGS
Rochling Burbach GmbH
C 0.25-0.33, Mn 0.9, Cr 1.8, Mo 1, V 0.6, bal Fe.
For tools, dies; hot work steel. *Obsolete*

ROCHLINGSTAHL PG3
Rochling Burbach GmbH
C 0.27, Cr 2.9, Mo 2.5, V 0.4, bal Fe.
For tools, dies; hot work steel. *Obsolete*

ROCHLINGSTAHL PWD
Rochling Burbach GmbH
C 0.23-0.28, Cr 0.35-0.5, Ni 4-4.5, Mo 1.15-1.25, bal Fe.
For tools, dies; hot work steel. *Obsolete*

ROCHLINGSTAHL PWD13
Rochling Burbach GmbH
C 0.22-0.3, Cr 0.6-0.9, Ni 1.2-1.5, Mo 0.2-0.4, V 0.2, bal Fe.
For tools, dies; hot work steel. *Obsolete*

ROCHLINGSTAHL PWD6
Rochling Burbach GmbH
C 0.24-0.32, Cr 0.6-0.9, Ni 2.2-2.5, Mo 0.6, V 0.3, bal Fe.
For tools, dies; hot work steel. *Obsolete*

ROCHLINGSTAHL PWDN
Rochling Burbach GmbH
C 0.25-0.32, Cr 0.6-0.8, Ni 4.5, W 3.8-4.3, Mo 0.65-0.85, bal
Fe.
For tools, dies; hot work steel. *Obsolete*

ROCHLINGSTAHL R.R.7
Rochling Burbach GmbH
C 0.7-0.8, Mn 0.6-0.8, Si 0.3, bal Fe.
For punches, springs, tools; water hardening. *Obsolete*

ROCHLINGSTAHL RAB1
Rochling Burbach GmbH
C 0.35-0.38, Cr 1.4, Ni 4.5, bal Fe.
For tools, dies, gears, shafts; shock resistant. *Obsolete*

ROCHLINGSTAHL RAN
Rochling Burbach GmbH
C 0.4-0.5, Mn 0.5-0.8, Cr 1.2-1.5, Ni 1.1-1.4, bal Fe.
For tools, dies; hot work steel. *Obsolete*

ROCHLINGSTAHL RB7
Rochling Burbach GmbH
C 0.68-0.75, Mn 0.2-0.35, Si 0.1-0.2, bal Fe.
For springs, tools; water hardening. *Obsolete*

ROCHLINGSTAHL RCNW
Rochling Burbach GmbH
C 0.33-0.38, Cr 1.4, Ni 4-4.5, W 0.9-1.1, bal Fe.
For tools, dies; hot work steel. *Obsolete*

ROCHLINGSTAHL RCW2
Rochling Burbach GmbH
C 0.24-0.3, Cr 2.2-2.8, W 9-10, bal Fe.
For tools, dies; hot work steel. *Obsolete*

ROCHLINGSTAHL RKMV
Rochling Burbach GmbH
C 0.95-1, bal Fe.
For tools, dies, taps, drills; water hardening. *Obsolete*

ROCHLINGSTAHL RLB
Rochling Burbach GmbH
C 0.11-0.18, Mn 0.25-0.4, Si 0.2-0.3, bal Fe.
For gears, pinions, shafts; case hardening. *Obsolete*

ROCHLINGSTAHL RSH
Rochling Burbach GmbH
C 1-1.1, Cr 0.8-1, bal Fe.
For tools, dies, bearings; water or oil hardening. *Obsolete*

ROCHLINGSTAHL RSV
Rochling Burbach GmbH
C 0.78-0.85, Cr 4.5, W 12.5, Mo 8, V 1.9, bal Fe.
For tools, dies, cutters; high speed steel. *Obsolete*

ROCHLINGSTAHL RSZ
Rochling Burbach GmbH
C 0.72-0.8, Cr 4.5, W 11.5, Mo 0.5, V 0.8-1, bal Fe.
For tools, dies, cutter; high speed steel. *Obsolete*

ROCHLINGSTAHL RSZ SPECIAL
Rochling Burbach GmbH
C 0.75-0.83, Cr 4.5, W 10.5-11.5, Mo 0.7, V 1.4, bal Fe.
For tools, dies, cutters; high speed steel. *Obsolete*

ROCHLINGSTAHL RT11
Rochling Burbach GmbH
C 1.05-1.15, Mn 0.15-0.25, Si 0.2, bal Fe.
For tools, drills, taps; water hardening. *Obsolete*

ROCHLINGSTAHL RT11 CR
Rochling Burbach GmbH
C 1-1.1, Cr 0.2-0.4, bal Fe.
For tools, dies; water hardening. *Obsolete*

ROCHLINGSTAHL RT14 CR
Rochling Burbach GmbH
C 1.35-1.45, Cr 0.4-0.6, bal Fe.
For tools, dies; oil hardening. *Obsolete*

ROCHLINGSTAHL RT7
Rochling Burbach GmbH
C 0.7-0.8, Mn 0.22-0.32, Si 0.08-0.15, bal Fe.
For tools, springs, punches; water hardening. *Obsolete*

ROCHLINGSTAHL RT7 CR
Rochling Burbach GmbH
C 0.7-0.79, Cr 0.45-0.55, bal Fe.
For tools, dies, punches; water hardening. *Obsolete*

ROCHLINGSTAHL RTC14 MO
Rochling Burbach GmbH
C 1.1-1.2, Cr 1.15-1.3, Mo 0.2-0.3, bal Fe.
For tools, dies, bearings; oil or water hardening. *Obsolete*

ROCHLINGSTAHL RTC20
Rochling Burbach GmbH
C 0.88-0.95, Cr 2, bal Fe.
For tools, dies; oil hardening. *Obsolete*

ROCHLINGSTAHL RTK10
Rochling Burbach GmbH
C 1-1.1, Cr 1-1.25, bal Fe.
For tools, dies; oil hardening. *Obsolete*

ROCHLINGSTAHL RTW2
Rochling Burbach GmbH
C 0.33-0.4, Si 1, Cr 1.1, W 2.6-3, bal Fe.
For tools, dies; tough. *Obsolete*

ROCHLINGSTAHL RTW3 SPECIAL
Rochling Burbach GmbH
C 1.3-1.4, Cr 0.7, W 4.6-5, bal Fe.
For tools, dies, cutters; oil hardening. *Obsolete*

ROCHLINGSTAHL RUS2
Rochling Burbach GmbH
C 1-1.1, Cr 0.9-1.1, Si 1.2-1.4, bal Fe.
For tools, dies, chisels, hammers; tough, shock resistant.
Obsolete

ROCHLINGSTAHL RWM
Rochling Burbach GmbH
C 0.5-0.58, Si 0.5, Mn 0.5, bal Fe.
For tools, dies; for hot work steel. *Obsolete*

ROCHLINGSTAHL RWS
Rochling Burbach GmbH
C 0.33-0.36, Cr 3.4, W 2.8, V 0.3, bal Fe.
For tools, dies; hot work steel. *Obsolete*

ROCHLINGSTAHL RWS2
Rochling Burbach GmbH
C 0.24-0.28, Cr 1.2, W 4.5, bal Fe.
For tools, dies; hot work steel. *Obsolete*

ROCHLINGSTAHL RWS4
Rochling Burbach GmbH
C 0.32-0.38, Si 0.9, Cr 1.5, W 4.5, Mo 0.5, bal Fe.
For tools, dies; hot work steel. *Obsolete*

ROCHLINGSTAHL S
Rochling Burbach GmbH
C 0.25-1.35, Mn 0.3, Si 0.15, bal Fe.
For tools, jigs, fixtures; water hardening. *Obsolete*

ROCHLINGSTAHL SGM4
Rochling Burbach GmbH
C 0.24-0.3, Cr 1.3-1.6, W 5, Mo 0.7, Co 5, Cu 1.5, bal Fe.
For tools, dies; hot work steel. *Obsolete*

ROCHLINGSTAHL SKSV
Rochling Burbach GmbH
C 0.65-0.73, W 0.6, V 0.15, bal Fe.
For tools, dies, punches, crimpers; tough. *Obsolete*

ROCHLINGSTAHL T76
Rochling Burbach GmbH
C 1.45-1.55, Cr 0.6-0.8, W 7-9, V 0.4, bal Fe.
For tools, dies, cutters; oil hardening. *Obsolete*

ROCHLINGSTAHL WNC
Rochling Burbach GmbH
C 0.35-0.4, Cr 1.3-1.5, Ni 3.2-3.6, W 5.5, bal Fe.
For tools, dies; hot work steel. *Obsolete*

ROCHLINGSTAHL ZK
Rochling Burbach GmbH
C 1.6, Cr 2, bal Fe.
For tools, dies; oil hardening. *Obsolete*

ROCKALOY GRADE A
Industries Trading Co.
WC + Co.
For cutting tools; carbide.

ROCKALOY NO. 110
Industries Trading Co.
WC + Co.
For cutting tools; carbide.

ROCKALOY NO. 111

Industries Trading Co.
WC + Co.
For cutting tools; carbides.

ROCKALOY NO. 112

Industries Trading Co.
WC + Co.
For cutting tools; carbides.

ROCKET

Lehigh Steel Corp.
C, alloy, bal Fe.
For tools, dies; wear and abrasion resistant, air or oil hardening.

ROCKRITE

Tube Reducing Corp.
C 1, Cr 1, bal Fe.
For bearings, liners; SAE-52100 tubing.

ROCKWELL

Time Steel Service Inc.
C 0.6, Mn 0.7, Si 1.85, V 0.2, Mo 0.45, bal Fe.
Shock resistant tool steel; oil or water hardened. AISI S5.

ROCOLOY

American manufacture
C 0.39, Mn 0.7, Cr 1.1, Si 1, Co 1, Mo 0.25, V 0.15, bal Fe.
Modified AISI 4137 with Co added.

ROCOLOY 270

American manufacture
C 0.4, Si 1.3, Cr 1.35, Ni 0.8, Mo 0.5, W 0.3, V 0.15, Co 1.35, bal Fe.
Heat treated: 325,000 TS; 270,000 YS; 7 El; 55 Rock C. For high strength, corrosion resistant parts, fasteners, rocket motor cases. Heat treatable. High strength.

RODAR

Wilbur B. Driver Co.
Superalloy. Ni 29, Co 17, Mn 0.3, bal Fe.
Annealed: 80,000 TS; 30,000 YS; 82 Rock B. Cold drawn: 90,000 TS; 50,500 YS; 250 Brin. For metal to glass seals; resists thermal shock.

RODDRAW 305

Teledyne Rodney Metals
C 0-0.12, Cr 18, Ni 11.5, bal Fe.
Annealed: 90,000 psi (620.5 MPa) TS; 38,000 psi (262 MPa) YS; 50 El; 80 Rock B. For deep drawn parts, aerosol tops, gaskets. AMS 5514.

RODFLEX 270

Teledyne Rodney Metals
C 0.15, Cr 16-18, Ni 6-8, Mn 0-2, Si 0-1, Cu 0-0.5, Mo 0-0.5, bal Fe.
Cold rolled: 270,000 psi TS min. "High Yield" Cr-Ni stainless steel strip. For springs, retainer rings, fasteners.

RODFLEX 290

Teledyne Rodney Metals
C 0-0.15, Cr 16-18, Ni 6-8, Mn 0-2, Si 0-1, Cu 0-0.5, Mo 0-0.5, bal Fe.
Cold rolled: 290,000 psi TS min. "High Yield" Cr-Ni stainless strip. For springs, retainer rings, fasteners.

RODFOR

Forjas Alavesas S.A.
C 1, Mn 0.3, Si 0-0.4, Cr 1.5, bal Fe.
Oil hardening drill rod; for drills, hardened shafts, drift pins.
IHA F523-F131; AFNOR 100Cr6; AISI 52100.

RODINOX 301

Teledyne Rodney Metals
C 0-0.15, Cr 17, Ni 7, bal Fe.
Annealed: 100,00 psi (689.5 MPa) TS; 40,000 psi (275.8 MPa) YS; 60 El; 85 Rock B. For name plates, helicopter blades, stamped parts. MIL-3-5059.

RODINOX 302

Teledyne Rodney Metals
C 0-0.15, Cr 18, Ni 9, bal Fe.
Annealed: 90,000 psi (620.5 MPa) TS; 40,000 psi (275.8 MPa) YS; 60 El; 85 Rock B. For laminated couplings, camera parts, diaphragms. MIL-S-5059.

RODINOX 304

Teledyne Rodney Metals
C 0-0.08, Cr 19, Ni 10, bal Fe.
Annealed: 90,000 psi (620.5 MPa) TS; 40,000 psi (275.8 MPa) YS; 50 El; 80 Rock B. For shim stock, deep drawing parts, watch bands. MIL-S-5059.

RODINOX 304L

Teledyne Rodney Metals
C 0-0.03, Cr 19, Ni 10, bal Fe.
Annealed: 90,000 psi (620.5 MPa) TS; 40,000 psi (275.8 MPa) YS; 50 El; 80 Rock B. For bellows, flexible metal hose, diaphragms. AMS 5511.

RODINOX 410

Teledyne Rodney Metals
C 0-0.15, Cr 12, bal Fe.
Annealed: 70,000 psi (482.6 MPa) TS; 45,000 psi (310.3 MPa) YS; 25 El; 80 Rock B. Heat treated: 180,000 psi (1241 MPa) TS; 140,000 psi (965 MPa) YS; 40 Rock C. Annealed: fasteners, valves. Heat treated: valves, shims, cutlery, springs. AMS 5504.

RODINOX 430

Teledyne Rodney Metals
C 0-0.12, Cr 17, bal Fe.
Annealed: 75,000 psi (517 MPa) TS; 45,000 psi (310 MPa) YS; 25 El; 80 Rock B. For gaskets, trim, chemical handling; non-hardenable by heat treatment.

RODMAG 80

Teledyne Rodney Metals
C 0.06, Ni 79, Mo 4.5, bal Fe.
For magnetic shielding of electrical components. MIL-N-47037.

RODNEY TYPE R215

Teledyne Rodney Metals
C 0.1, Mn 15, Si 0.65, Ni 0.75, Cr 18, N 0.4, bal Fe.
Annealed: 125,000 TS; 70,000 YS; 45 El; B 98 Rock. For nonmagnetic cables, computer components, vacuum chambers. Austenitic, high strength, nonmagnetic. Corrosion and heat resistant. *Obsolete*

RODRESIST 316

Teledyne Rodney Metals
C 0.08, Cr 18, Ni 13.5, Mo 2.5, bal Fe.
Annealed: 90,000 psi (620.5 MPa) TS; 40,000 psi (275.8 MPa) YS; 50 El; 80 Rock B. Bellows, surgical implants, diaphragms, marine applications. AMS 5524.

RODRESIST 316L

Teledyne Rodney Metals
C 0-0.03, Cr 17, Ni 13, Mo 2.5, bal Fe.
Annealed: 90,000 psi (620.5 MPa) TS; 40,000 psi (275.8 MPa) YS; 50 El; 80 Rock B. For bellows, turbine heat exchangers, welded parts. AMS 5507.

RODSEAL 29-17

Teledyne Rodney Metals
C 0-0.06, Ni 29, Co 17, bal Fe.
Coefficient thermal expansion: 4.38×10^{-6} in./in.·°F, 70-1000°F. For glass seals and hermetic seals. ASTM F-15.

RODSEAL 36

Teledyne Rodney Metals
C 0.04, Ni 36, bal Fe.
Coefficient thermal expansion: 1.6×10^{-6} in./in.·°F, 70-1000°F. For precision instruments, bi-metals, optical equipment.

RODSEAL 42

Teledyne Rodney Metals
C 0.05, Ni 41.5, bal Fe.
Coefficient thermal expansion: 4.8×10^{-6} in./in.·°F, 70-1000°F. For glass seals. ASTM F-30.

RODTEMP 309

Teledyne Rodney Metals
C 0-0.08, Cr 23, Ni 13.5, bal Fe.
Annealed: 90,000 psi (620.5 MPa) TS; 45,000 psi (310.3 MPa) YS; 45 El; 85 Rock B. Heat exchangers, furnace parts, and for high temperature operation. AMS 5523.

RODTEMP 310

Teledyne Rodney Metals
C 0-0.08, Cr 25, Ni 20.5, bal Fe.
Annealed: 90,000 psi (620.5 MPa) TS; 45,000 psi (310.3 MPa) YS; 45 El; 85 Rock B. Jet engine parts, oil refinery equipment, furnaces. AMS 5521.

RODWELD 321

Teledyne Rodney Metals
C 0-0.08, Cr 18, Ni 10.5, Ti = 5 x C min, bal Fe.
Annealed: 95,000 psi (655 MPa) TS; 35,000 psi (241.3 MPa) YS; 45 El; 80 Rock B. For jet engine insulation blankets, welded parts. AMS 5510.

RODWELD 347

Teledyne Rodney Metals
C 0-0.08, Cr 18, Ni 10.5, Cb + Ta = 10 x C min, bal Fe.
Annealed: 95,000 psi (655 MPa) TS; 40,000 psi (275.8 MPa) YS; 45 El; 85 Rock B. For jet engine seals, bellows, welded parts. AMS 5512.

ROESCH

English manufacture
Zn 50, Sn 49, Sb 0.7, Cu 0.2.
For aluminum solder.

ROGER METAL

Ste de Produits Metallurgiques
C, Cr, Ni, bal Fe.
For stainless parts; corrosion resistant.

ROHMIUM

Manufacturer not listed.
Al alloy.
For light alloy parts.

ROHN

Manufacturer not listed.
Ni 50, Cr 30, Fe 17.
For resistances, heat and corrosion resistant parts; non-scaling, heat resistant.

ROHRENWERK HMS 42

Rheinische Rohrenwerke Atkiengesellschaf
C 0.18, bal Fe.
Rolled: 57,000-72,000 TS; 34,000-40,000 YS. For gears, cams, camshafts; case hardening steel.

ROKBORE

British Steel Corp.
C 1.2, bal Fe.
For hollow mine drills. *Obsolete*

ROL-MAN

Manganese Steel Forge Co.
Mn 11-14, C 1-1.3, bal Fe.
130,000-160,000 TS; 35-50 El; 180 Brin. For woven wire screens, wear plates, pins, bushings, stokers; austenitic steel, non-magnetic.

ROLEY 275

Robert-Leyer-Pritzkow & Co.
C 0.95, Cr 4, V 1, W, Mo, bal Fe.
For lathe and planer tools, reamers, broaches, drills, hobs; high speed steel.

ROLEY 300
Robert-Leyer-Pritzkow & Co.
C 0.82, Cr 4.1, Mo 0.85, V 1.6, W 8.7, bal Fe.
For lathe and planer tools, drills, hobs; high speed steel.

ROLEY B SPEZIAL
Robert-Leyer-Pritzkow & Co.
C 0-0.1, Si 2, Cr 18, Ni 8.5, bal Fe.
Annealed: 80,000 TS; 35,000 YS; 55 El; 75 RA; 150 Brin. Cold drawn: 180,000 TS; 150,000 YS; 10 El; 250 Brin. For chemical plant equipment, tanks, filters, mixers; Type 302; stainless austenitic.

ROLEY B SUPER
Robert-Leyer-Pritzkow & Co.
C 0-0.1, Si 2, Mo 2, Cr 18, Ni 8.5, bal Fe.
Annealed: 85,000 TS; 35,000 YS; 50 El; 65 RA; 160 Brin. Cold drawn: 150,000 TS; 135,000 YS; 6 El; 300 Brin. For acid resistant chemical plant equipment; type 316; stainless, austenitic.

ROLEY BA
Robert-Leyer-Pritzkow & Co.
C 0-0.15, Cr 18, Ni 8.5, bal Fe.
Annealed: 80,000 TS; 35,000 YS; 55 El; 75 RA; 150 Brin. Cold drawn: 180,000 TS; 150,000 YS; 10 El; 250 Brin. For chemical plant equipment, tanks, mixers, filters; Type 302; stainless austenitic.

ROLEY BA88
Robert-Leyer-Pritzkow & Co.
C 0-0.15, Cr 18, Ni 8.5, S 0.2, bal Fe.
For screw machine products, fasteners; Type 303; stainless, free-cutting.

ROLEY CM2
Robert-Leyer-Pritzkow & Co.
C 0.35, C 16.5, Mo 1.15, bal Fe.
For oil refinery and chemical plant equipment; corrosion resistant.

ROLEY CME2
Robert-Leyer-Pritzkow & Co.
C 0.2, Mn 1.25, Cr 1.15, bal Fe.
For gears, cams, camshafts; case hardened, tough.

ROLEY CME3
Robert-Leyer-Pritzkow & Co.
C 0.15, Cr 0.65, Mn 0.5, bal Fe.
For gears, bolts, machine tool parts; case hardened.

ROLEY CN12
Robert-Leyer-Pritzkow & Co.
C 0-0.1, Cr 12.5, Ni 12, bal Fe.
For valves; corrosion and heat resistant.

ROLEY CNM
Robert-Leyer-Pritzkow & Co.
C 0-0.07, Cr 17, Mo 4.75, Ni 13, bal Fe.
Annealed: 90,000 TS; 40,000 YS; 45 El; 60 RA; 170 Brin. For acid resistant chemical plant equipment, tanks; Type 317; stainless, austenitic.

ROLEY CRP2
Robert-Leyer-Pritzkow & Co.
C 0.4, Cr, Mn, Mo, bal Fe.
For machine tool parts, bolts, gears; oil hardened, tough.

ROLEY CT
Robert-Leyer-Pritzkow & Co.
C 0-0.12, Cr 18, Ni 9.5, Ti = 4 x C, bal Fe.
Annealed: 80,000 TS; 35,000 YS; 55 El; 75 RA; 150 Brin. Cold drawn: 180,000 TS; 150,000 YS; 10 El; 250 Brin. For chemical plant equipment, tanks; Type 321; stainless, austenitic.

ROLEY CT2N
Robert-Leyer-Pritzkow & Co.
C 0-0.12, Cr 18, Ni 9.5, Cb = 8 x C, bal Fe.
Annealed: 90,000 TS; 45,000 YS; 56 El; 65 RA; 160 Brin. For welded chemical plant equipment; Type 347; stainless, austenitic.

ROLEY CU
Robert-Leyer-Pritzkow & Co.
C 0-0.07, Cr 17.5, Mo 2, Ni 17.5, Ti = 7 x C, bal Fe.
For valves, chemical plant equipment, welded parts; austenitic, stainless.

ROLEY DSA
Robert-Leyer-Pritzkow & Co.
C 1.42, W, V, bal Fe.
For cutters, engravers' tools; water hardened, wear resistant.

ROLEY DSA SPEZIAL
Robert-Leyer-Pritzkow & Co.
C 1.3, W 4.75, Cr 0-0.2, Mn 0.3, bal Fe.
For engravers' tools, forming and blanking dies; water hardened, wear resistant.

ROLEY ECR15
Now VEW E410.

ROLEY ECR20
Now VEW E400.

ROLEY EFW
Robert-Leyer-Pritzkow & Co.
C 0.15, Si 0.25, Mn 0.37, bal Fe.
Annealed: 70,000 TS; 40,000 YS; 25 El; 60 RA; 145 Brin. For gears, bolts, fasteners, machine tool parts; case hardened.

ROLEY ERJ 15
Robert-Leyer-Pritzkow & Co.
C 0.15, Cr 1.55, Mn 1.55, bal Fe.
For gears, cams, camshafts; case hardened, tough.

ROLEY ERJ 20
Robert-Leyer-Pritzkow & Co.
C 0.18, Cr 2, Ni 2, Mn 0.5, bal Fe.
For gears, cams, camshafts; case hardened, tough.

ROLEY ERJ 25
Robert-Leyer-Pritzkow & Co.
C 0.13, Cr 0.7, Ni 2.5, Mn 0-0.5, bal Fe.
For gears, cams, camshafts; case hardened, tough.

ROLEY ERJ 35
Robert-Leyer-Pritzkow & Co.
C 0.14, Ni 3.5, Cr 0.7, bal Fe.
For gears, cams, camshafts; case hardened, tough. *Obsolete*

ROLEY ERJ 45
Robert-Leyer-Pritzkow & Co.
C 0.13, Cr 1.1, Ni 4.5, bal Fe.
For gears, bolts, camshafts, fasteners; case hardened, tough. *Obsolete*

ROLEY ERM 15
Robert-Leyer-Pritzkow & Co.
C 0.15, Cr, Mo, bal Fe.
For gears, cams, camshafts; case hardened, tough. *Obsolete*

ROLEY ERM 20
Now VEW E300.

ROLEY ESPH
Robert-Leyer-Pritzkow & Co.
C 1.3, Si 0-0.25, Mn 0-0.25, bal Fe.
For engraving tools, cutters, broaches; Type W1; water hardened.

ROLEY ESPZH
Robert-Leyer-Pritzkow & Co.
C 1, Si 0-0.25, Mn 0-0.25, bal Fe.
Annealed: 100,000 TS; 53,000 YS; 21 El; 42 RA; 200 Brin. For drills, taps, reamers, hobs, cutters; Type W1; water hardened.

ROLEY EXTRA HART
Robert-Leyer-Pritzkow & Co.
C 1, Si 0-0.25, Mn 0-0.25, bal Fe.
Annealed: 100,000 TS; 53,000 YS; 21 El; 42 RA; 200 Brin. For springs, drills, taps, reamers; Type W1; water hardened.

ROLEY EXTRA MH
Robert-Leyer-Pritzkow & Co.
C 1.15, Si 0-0.25, Mn 0-0.25, bal Fe.
Annealed: 110,000 TS; 58,000 YS; 18 El; 38 RA; 210 Brin. For springs, drills, reamers, taps; Type W1; water hardened.

ROLEY EXTRA SEHR ZAH
Robert-Leyer-Pritzkow & Co.
C 0.7, Si 0-0.25, Mn 0-0.25, bal Fe.
Heat treated: 175,000 TS; 128,000 YS; 12 El; 37 RA; 355 Brin. For springs, rails, cutters, tools, axes, hammers; Type W1; water hardened.

ROLEY EXTRA SPEZIAL MH
Robert-Leyer-Pritzkow & Co.
C 1.1, Mn 0-0.25, 0.25 Si max, bal Fe.
Annealed: 110,000 TS; 58,000 YS; 18 El; 40 RA; 210 Brin. For springs, taps, reamers, hobs, broaches; Type W1; water hardened.

ROLEY EXTRA SPEZIAL SEHR ZAH
Robert-Leyer-Pritzkow & Co.
C 0.7, Si 0-0.25, Mn 0-0.25, bal Fe.
Heat treated: 175,000 TS; 128,000 YS; 12 El; 37 RA; 355 Brin. For springs, rails, hammers, tools, axes; Type W1; water hardened.

ROLEY EXTRA SPEZIAL ZAH
Robert-Leyer-Pritzkow & Co.
C 0.85, Si 0-0.25, Mn 0-0.25, bal Fe.
Heat treated: 190,000 TS; 145,000 YS; 10 El; 30 RA; 400 Brin. For springs, drills, cutters, taps, punches; Type W1; water hardened.

ROLEY EXTRA SPEZIAL ZH
Now VEW K988.

ROLEY EXTRA ZAH
Now VEW K930.

ROLEY EXTRA ZH
Now VEW K990.

ROLEY EZH
Now VEW K990.

ROLEY FAM
Now VEW F180.

ROLEY GW
Robert-Leyer-Pritzkow & Co.
C 1.2, W, V, bal Fe.
For cutters, forming dies; water hardened.

ROLEY GWA
Robert-Leyer-Pritzkow & Co.
C 1.05, Cr 1.2, bal Fe.
For bearings, bushings, sleeves, liners; water hardened, wear resistant.

ROLEY GWK
Robert-Leyer-Pritzkow & Co.
C 1.25, Si 1.15, Mn 0.7, Cr 1.2, bal Fe.
For bearings, cutters, bushings, liners; water or oil hardened.

ROLEY GWV
Robert-Leyer-Pritzkow & Co.
C 1.15, Cr 0.65, V 0.1, bal Fe.
For bearings, bushings, liners; water hardened.

ROLEY HIR15 W/R
Robert-Leyer-Pritzkow & Co.
C 0.28-0.35, Cr 0.5, Ni 1.5, bal Fe.
For gears, bolts, fasteners; oil hardened, shock resistant. *Obsolete*

ROLEY HIR25
Robert-Leyer-Pritzkow & Co.
C 0.36, Cr 0.7, Ni 2.5, Mn 0.6, bal Fe.
For gears, bolts, machine tool parts; oil hardened, shock resistant. *Obsolete*

ROLEY HIR35
Robert-Leyer-Pritzkow & Co.
C 0.22-0.34, Cr 0.7, Ni 3.5, bal Fe.
For gears, bolts, machine tool parts; oil hardened, shock resistant.

ROLEY HIR45
Robert-Leyer-Pritzkow & Co.
C 0.35, Mn 0.6, Cr 1.3, Ni 4.5, bal Fe.
For gears, bolts, crankshafts; oil hardened, shock resistant.

ROLEY HMN15
Robert-Leyer-Pritzkow & Co.
C 0.3, Mn 1.35, Si 0.25, bal Fe.
For gears, bolts, machine tool parts; water hardened, tough.

ROLEY HMV35
Now VEW V762.

ROLEY HMV42
Now VEW V742.

ROLEY HMV50
Now VEW F550.

ROLEY HRIM20
Now VEW V145.

ROLEY HRM15
Now VEW V340.

ROLEY HRM35
Now VEW V330.

ROLEY HRM40
Now VEW V320.

ROLEY HRM45
Robert-Leyer-Pritzkow & Co.
C 0.42, Cr, Mo, bal Fe.
For gears, bolts, machine tool parts; oil hardened, tough.

ROLEY HRMV25
Robert-Leyer-Pritzkow & Co.
C 0.3, Cr 2.5, Mo 0.2, V 0.15, bal Fe.
For forging dies; upsetters, header dies; oil hardened, tough.

ROLEY KGS
Robert-Leyer-Pritzkow & Co.
C 1, Cr 1.55, Mn 0.35, bal Fe.
For bearings, cutters, bushings; oil or water hardened, wear resistant.

ROLEY KL3
Now VEW R100.

ROLEY KLR
Now VEW K505.

ROLEY KLS
Now VEW R100.

ROLEY KSA
Robert-Leyer-Pritzkow & Co.
C 0.9, Mn 1.9, V 0.1, Si 0.25, bal Fe.
For punches, dies, cutters, shears; oil hardened, non-deforming.

ROLEY L1
Robert-Leyer-Pritzkow & Co.
C 0 0.12, Si 0.4, Cr 13, bal Fe.
Annealed: 75,000 TS; 40,000 YS; 35 El; 70 RA; 155 Brin. For turbine blades, cutlery, valves; Type 410; corrosion resistant.

ROLEY L2
Robert-Leyer-Pritzkow & Co.
C 0-0.12, Cr 13, Si 0.4, bal Fe.
Annealed: 75,000 TS; 40,000 YS; 35 El; 70 RA; 155 Brin. For turbine blades, cutlery, valves; Type 410; corrosion resistant.

ROLEY L3
Robert-Leyer-Pritzkow & Co.
C 0-0.12, Si 0.4, Cr 13, bal Fe.
Annealed: 75,000 TS; 40,000 YS; 35 El; 70 RA; 155 Brin. For turbine blades, cutlery, valves, surgical instruments; Type 410; corrosion resistant.

ROLEY M1
Robert-Leyer-Pritzkow & Co.
C 0.2, Si 0.4, Cr 13, bal Fe.
Annealed: 95,000 TS; 50,000 YS; 25 El; 55 RA; 195 Brin. For turbine blades, valves, cutlery, surgical instruments; Type 420; corrosion resistant.

ROLEY M2
Robert-Leyer-Pritzkow & Co.
C 0.22, Cr 17, Ni 1.5, bal Fe.
Annealed: 125,000 TS; 95,000 YS; 20 El; 55 RA; 260 Brin. For pumps, marine hardware, valves; Type 431.

ROLEY M3
Robert-Leyer-Pritzkow & Co.
C 0.4, Si 0.4, Cr 13, bal Fe.
Annealed: 100,000 TS; 55,000 YS; 20 El; 50 RA; 200 Brin. For valves, cutlery, surgical instruments; Type 420; corrosion resistant.

ROLEY M3E
Robert-Leyer-Pritzkow & Co.
C 0.9, Cr 18, Mo 1.15, V 1, bal Fe.
Annealed: 107,000 TS; 62,000 YS; 18 El; 35 RA; 220 Brin. Heat treated: 280,000 TS; 270,000 YS; 3 El; 15 RA; 555 Brin. For cutlery, valves, ball bearings, surgical instruments; Type 440 B; corrosion resistant.

ROLEY M4
Robert-Leyer-Pritzkow & Co.
C 0.12, Cr 16.5, Mo 0.25, S 0.2, bal Fe.
For screw machine products, bolts, shafts, fasteners; Type 416; corrosion resistant, free-cutting.

ROLEY MA3
Now VEW V960.

ROLEY MA4
Now VEW V945.

ROLEY MA5
Now VEW V935.

ROLEY MA6
Now VEW V920.

ROLEY MO-1
Robert-Leyer-Pritzkow & Co.
C 0.7, Mo, Cr, W, V, bal Fe.
For lathe and planer tools, reamers, drills, taps; high speed steel.

ROLEY MO-2
Robert-Leyer-Pritzkow & Co.
C 0.85, W, Mo, bal Fe.
For lathe and planer tools, reamers, drills, taps; high speed steel.

ROLEY MO-3
Robert-Leyer-Pritzkow & Co.
C 0.88, Co, W, Mo, Cr, V, bal Fe.
For lathe and planer tools, drills, high speed steel.

ROLEY MO-4
Robert-Leyer-Pritzkow & Co.
C 1.15, Co, W, Mo, Cr, V, bal Fe.
For engravers' tools, form cutters, broaches, reamers; high speed steel.

ROLEY NALH
Robert-Leyer-Pritzkow & Co.
C 0.34, Cr 1.4, Al 1.1, bal Fe.
For oil refinery equipment; heat and creep resistant. *Obsolete*

ROLEY NALJ
Robert-Leyer-Pritzkow & Co.
C 0.33, Al 1.1, Cr 1.7, Mn 0.5, Ni 1.1, bal Fe.
For oil refinery equipment; heat and creep resistant. *Obsolete*

ROLEY NALM
Robert-Leyer-Pritzkow & Co.
C 0.32, Al 1.1, Cr 1.1, Mo 0.18, bal Fe.
For oil refinery equipment; heat and creep resistant. *Obsolete*

ROLEY NALW
Robert-Leyer-Pritzkow & Co.
C 0.27, Al 1.1, Cr 1.4, Mn 0.6, bal Fe.
For oil refinery equipment; heat and creep resistant. *Obsolete*

ROLEY NMV
Now VEW V304.

ROLEY P1D
Robert-Leyer-Pritzkow & Co.
C 0.46, Ni, Cr, W, bal Fe.
For forging and heading dies; upsetters; oil hardened, tough.

ROLEY P2
Robert-Leyer-Pritzkow & Co.
C 0.5, Ni 3.25, Cr 1.05, Mn 0.5, bal Fe.
For gears, bolts, crankshafts, fasteners; oil hardened, shock resistant.

ROLEY P3
Robert-Leyer-Pritzkow & Co.
C 0.45, Mn 0.7, V 0.3, Cr 1.4, Mo 0.7, bal Fe.
For gears, bolts, machine tool parts; oil hardened, tough.

ROLEY P6
Robert-Leyer-Pritzkow & Co.
C 1.1, Cr 0.4, Mn 0.3, Si 0.2, bal Fe.
For bearings, cutters, bushings; water hardened, wear resistant.

ROLEY P7
Robert-Leyer-Pritzkow & Co.
C 1.05, Cr 1, Mn 0.3, Si 0.25, bal Fe.
For bearings, bushings, liners; water hardened, wear resistant.

ROLEY P8
Robert-Leyer-Pritzkow & Co.
C 0.9, Cr 0.8, Mn 0.3, Si 0.3, bal Fe.
For bearings, bushings, liners; water hardened, wear resistant.

ROLEY PRIMA SEHR ZAH
Robert-Leyer-Pritzkow & Co.
C 0.45, Mn 0.3-0.8, Si 0.25-0.5, bal Fe.
Hot rolled: 98,000 TS; 60,000 YS; 24 El; 55 RA; 212 Brin. For axles, gears, bolts, bushings, crankshafts; water hardened.

ROLEY PRIMA WEICH
Robert-Leyer-Pritzkow & Co.
C 0.35, Si 0.25-0.5, Mn 0.25-0.8, bal Fe.
Hot rolled: 84,000 TS; 54,000 YS; 30 El; 53 RA; 183 Brin. For gears, shafts, axles, bolts, screws; water hardened.

ROLEY PRIMA ZAH
Robert-Leyer-Pritzkow & Co.
C 0.6, Si 0.25-0.5, Mn 0.25-0.8, bal Fe.
Heat treated: 115,000-160,000 TS; 77,000-113,000 YS; 12-23 El; 40-54 RA; 230-320 Brin. For wheels, die blocks, rails, girders; water hardened.

ROLEY SAO
Robert-Leyer-Pritzkow & Co.
C 0.9, V 0.1, Mn 1.9, bal Fe.
For punches, forming and blanking dies, shear blades; oil hardened, non-deforming.

ROLEY SF
Robert-Leyer-Pritzkow & Co.
C 0.65, Si 1.7, Mn 0.7, bal Fe.
For springs, punches, upsetters; oil hardened, tough.

ROLEY SK
Robert-Leyer-Pritzkow & Co.
C 0.55, Si 0.9, Mn 0.3, Cr 1.05, V 0.18, W 1.85, bal Fe.
For cold heading and forming dies; oil hardened, tough.

ROLEY SP2
Robert-Leyer-Pritzkow & Co.
C 0.7, Si 1.7, Mn 0.7, bal Fe.
For springs, chisels, punches, pneumatic tools; oil hardened, shock resistant.

ROLEY SP2W
Robert-Leyer-Pritzkow & Co.
C 0.53, Si 0.8, Mn 1.05, bal Fe.
For crankshafts, bolts, gears, axles; water hardened, tough.

ROLEY SPEZIAL
Robert-Leyer-Pritzkow & Co.
C 0.86, Cr 4.1, Mo 0.85, V 2.5, W 12, bal Fe.
For lathe and planer tools, drills; high speed steel.

ROLEY SUPER
Robert-Leyer-Pritzkow & Co.
C 1.3, Cr 4.3, Mo 0.85, V 3.8, W 12, bal Fe.
For forming and blanking dies, cutters; high speed steel.

ROLEY SUPER 3
Robert-Leyer-Pritzkow & Co.
C 0.74, Cr 4.1, V 1.1, W 18.5, bal Fe.
For lathe and planer tools, reamers, broaches; high speed steel.

ROLEY SUPER SPEZIAL
Robert-Leyer-Pritzkow & Co.
C 0.65, Cr 4, W, Co, V, Mo, bal Fe.
For lathe and planer tools, reamers, punches, taps; high speed steel.

ROLEY SUPERIOR
Robert-Leyer-Pritzkow & Co.
C 1.05, Cr 1, W 1.15, Mn 0.9, bal Fe.
For heading and blanking dies; form tools; oil hardened, wear resistant.

ROLEY SVA
Robert-Leyer-Pritzkow & Co.
C 1, V 0.1, Mn 0.25, Si 0.2, bal Fe.
For blanking and forming dies; Type W2; water hardened.

ROLEY T70
Robert-Leyer-Pritzkow & Co.
C 0-0.1, Cr 17.5, Ti = 7 x C, bal Fe.
Annealed: 80,000 TS; 50,000 YS; 25 El; 50 RA; 150 Brin. For oil refinery equipment, dairy and food equipment; Type 430 Ti, corrosion resistant.

ROLEY TO SUPER
Robert-Leyer-Pritzkow & Co.
C 0.76, Co 10, Cr 4.2, Mo 0.8, V 1.8, W 18, bal Fe.
For lathe and planer tools, form cutters, drills; high speed steel.

ROLEY TO1
Robert-Leyer-Pritzkow & Co.
C 1.35, W, Co, bal Fe.
For engravers' tools, forming and blanking dies; oil hardened, wear resistant.

ROLEY TO2
Robert-Leyer-Pritzkow & Co.
C 0.79, Co 4.7, Cr 4.3, Mo 0.75, V 1.5, W 18, bal Fe.
For lathe and planer tools, cutters, reamers, drills; high speed steel.

ROLEY TO3
Robert-Leyer-Pritzkow & Co.
C 0.86, Co 2.8, Cr 4.3, Mo 0.85, V 2.1, W 12, bal Fe.
For lathe and planer tools, cutters, reamers, taps; high speed steel.

ROLEY TO4
Robert-Leyer-Pritzkow & Co.
C 0.8, Cr 4, Co, Mo, W, V, bal Fe.
For lathe and planer tools, broaches; high speed steel.

ROLEY UDL
Robert-Leyer-Pritzkow & Co.
C 0.38, Cr 1.05, V 0.1, Mn 0.95, bal Fe.
For gears, bolts, springs, crankshafts; oil hardened, shock resistant.

ROLEY UDW
Robert-Leyer-Pritzkow & Co.
C 0.38, Si, Cr, V, bal Fe.
For gears, bolts, springs; oil hardened, shock resistant.

ROLEY UDWM
Robert-Leyer-Pritzkow & Co.
C 0.61, Cr 1.18, V 0.1, Mn 0.75, Si 0.85, bal Fe.
For springs, punches, crankshafts; oil hardened, shock resistant.

ROLEY UK
Robert-Leyer-Pritzkow & Co.
C 0.55, Si 0.9, Cr 1.05, V 0.18, W 1.85, bal Fe.
For heading and blanking dies; cold work tools; oil hardened, tough.

ROLEY UKL
Robert-Leyer-Pritzkow & Co.
C 0.67, Si 1.3, Mn 0.5, Cr 0.5, bal Fe.
For punches, springs, pneumatic tools; oil hardened, tough.

ROLEY UKS
Robert-Leyer-Pritzkow & Co.
C 0.45, Si, Cr, V, bal Fe.
For gears, bolts, crankshafts, springs; oil hardened, shock resistant.

ROLEY UKW(D)
Robert-Leyer-Pritzkow & Co.
C 0.35, Si 0.9, Cr 1.05, V 0.18, W 1.85, bal Fe.
For heading and forming dies, cold work tools; oil hardened, tough.

ROLEY WB10
Robert-Leyer-Pritzkow & Co.
C 0.25, Cr 1.15, Mo 0.25, Mn 0.55, bal Fe.
For gears, bolts, crankshafts, axles; oil hardened, tough.

ROLEY WB20
Robert-Leyer-Pritzkow & Co.
C 0.24, Cr 1.35, Mo 0.55, V 0.2, bal Fe.
For die casting dies, punches; oil hardened, tough.

ROLEY ZC10
Robert-Leyer-Pritzkow & Co.
C 0.41, Si 0.25, Mn 0.7, Cr 1.1, bal Fe.
For gears, bolts, machine tool parts; oil hardened, tough.

ROLEY ZCMO10
Robert-Leyer-Pritzkow & Co.
C 0.5, Cr 1, Mo 0.2, Mn 0.65, Si 0.25, bal Fe.
For gears, bolts, machine tool parts; oil hardened, tough.

ROLEY ZIM EXTRA
Robert-Leyer-Pritzkow & Co.
C 1.4, Cr 0.3, V 0.1, Mn 0.3, bal Fe.
For engravers' tools, cutters, bearings; water or oil hardened, wear resistant.

ROLEY ZIM SPEZIAL
Robert-Leyer-Pritzkow & Co.
C 1.4, Cr 0.3, V 0.1, Mn 0.3, bal Fe.
For engravers' tools, cutters, form tools; oil hardened, wear resistant.

ROLLE MA-356
Rolle Mfg. Co.
Si 6.5-7.5, Mg 0.2-0.75, Be 0.03-0.25, Ti 0.2, Cu 0.2, Zn 0.1, bal Al.
Cast: 40,000 TS; 30,000 YS; 5 El; 60 Brin. Heat treated: 50,000 TS; 40,000 YS; 2 El; 70 Brin. For gear cases, housings, crankcases, oil pans, airframe fittings; age hardenable, high strength, corrosion resistant.

ROLLED AUGER
Ziv Steel & Wire Co.
C, bal Fe.
For tools; water hardened.

ROLLO
British Steel Corp.
C 0.9, V 0.2, bal Fe.
For turning tools. *Obsolete*

ROLLODUR 1
Rollschen Eisenwerke, A.G.
Cr 13, W 10, C 1.5, bal Fe.
For milling cutters, tools; cast to shape.

ROLLODUR 108-32
Rollschen Eisenwerke, A.G.
C 1.8, W 10, Cr, Mo, Co, V, bal Fe.
For cutting tools; abrasion resistant.

ROLLODUR 2
Rollschen Eisenwerke, A.G.
Cr 13, W 4, C 1.5, bal Fe.
For milling cutters, tools; cast to shape.

ROLLODUR 44-22
Rollschen Eisenwerke, A.G.
C 1.4, W 4, Cr, Mo, Co, V, bal Fe.
For cutting tools, milling cutters; cast to shape, abrasion resistant.

ROLLOY
Dresser Industries
TC 2.5-3.5, Ni 4.5, Cr 1.5, bal Fe.
Cast: 570-650 Brin. For wear resistant castings; wear resistant.

ROLLSCHEN 1 CR 10 NI
Rollschen Eisenwerke, A.G.
C 0.2, Cr 0.1, Ni 1, bal Fe.
Heat treated: 107,000-148,000 TS; 85,000 YS; 12 El; 45 RA; 210-250 Brin. For gears, shafts, cams, pinions, camshafts; case hardening.

ROLLSCHEN 1 CR 15 NI
Rollschen Eisenwerke, A.G.
C 0.2, Cr 0.1, Ni 1.5, bal Fe.
Heat treated: 135,000-176,000 TS; 99,600 YS; 10 El; 40 RA; 220-260 Brin. For gears, shafts, pinions, camshafts; case hardening.

ROLLSCHEN 1 CR 20 NI
Rollschen Eisenwerke, A.G.
C 0.2, Cr 0.1, Ni 2, bal Fe.
Heat treated: 162,000-205,000 TS; 114,000 YS; 7 El; 35 RA; 235-300 Brin. For gears, shafts, camshafts; tough, case hardening.

ROLLSCHEN 1 CR 5 NI
Rollschen Eisenwerke, A.G.
C 0.2, Cr 1, Ni 0.5, bal Fe.
Heat treated: 78,000-125,000 TS; 50,000 YS; 16 El; 55 RA; 170-250 Brin. For gears, shafts, pinions, cams, camshafts; tough, case hardening steel.

ROLLSCHEN 2 MO 4
Rollschen Eisenwerke, A.G.
C 0.2, Mo 0.4, bal Fe.
Heat treated: 78,300-98,500 TS; 50,000 YS; 16 El; 60 RA. For gears, shafts, pinions; case hardening.

ROLLSCHEN 3 CR 10 MO 10
Rollschen Eisenwerke, A.G.
C 0.3, Cr 1, Mo 1, bal Fe.
Heat treated: 121,000-142,000 TS; 107,000 YS; 15 El; 50 RA. For gears, shafts, axles; oil hardening, tough.

ROLLSCHEN 3 CR 10 NI
Rollschen Eisenwerke, A.G.
C 0.4, Cr 0.3, Ni 1, bal Fe.
Heat treated: 99,000-121,000 TS; 71,000 YS; 15 El; 55 RA; 230-250 Brin. For gears, shafts, crankshafts; oil hardening, tough.

ROLLSCHEN 3 CR 15 NI
Rollschen Eisenwerke, A.G.
C 0.4, Cr 0.3, Ni 1.5, bal Fe.
Heat treated: 107,000-128,000 TS; 85,000 YS; 14 El; 55 RA; 240-260 Brin. For gears, shafts, crankshafts; oil hardening, tough.

ROLLSCHEN 3 CR 15 NI 12 MO AL
Rollschen Eisenwerke, A.G.
C 0.3, Cr 1.5, Ni 0.8, Mo 0.3, Al 1, bal Fe.
Heat treated: 114,000-142,000 TS; 85,000 YS; 14 El; 40 RA; 250 Brin. For gears, pinions, cams, shafts; nitriding steel.

ROLLSCHEN 3 CR 20 NI
Rollschen Eisenwerke, A.G.
C 0.4, Cr 0.3, Ni 2, bal Fe.
Heat treated: 121,000-142,000 TS; 100,000 YS; 14 El; 50 RA; 260-280 Brin. For gears, shafts, crankshafts; oil hardening, tough.

ROLLSCHEN 3 CR 5 NI
Rollschen Eisenwerke, A.G.
C 0.4, Cr 0.3, Ni 0.5, bal Fe.
Heat treated: 85,500-99,500 TS; 57,000 YS; 19 El; 55 RA; 210-230 Brin. For gears, shafts, crankshafts; oil hardening, tough.

ROLLSCHEN 3 CR NI 10 MO
Rollschen Eisenwerke, A.G.
Cr 0.3, Mo 1, Ni, bal Fe.
Heat treated: 128,000-148,000 TS; 107,000 YS; 12 El; 50 RA; 230-260 Brin. For gears, shafts, crankshafts; oil hardening, tough.

ROLLSCHEN 3 CR NI 15 MO
Rollschen Eisenwerke, A.G.
Cr 0.3, Mo 1, C, Ni, bal Fe.
Heat treated: 148,000-176,000 TS; 121,000 YS; 9 El; 45 RA; 245 Brin. For gears, shafts, crankshafts; oil hardening, tough.

ROLLSCHEN 3 CR NI 20 MO
Rollschen Eisenwerke, A.G.
Cr 0.3, Mo 1, Ni, bal Fe.
Heat treated: 169,000-197,000 TS; 141,000 YS; 9 El; 40 RA; 260 Brin. For gears, shafts, axles; oil hardening, tough.

ROLLSCHEN 4 CR 10 MO 4
Rollschen Eisenwerke, A.G.
C 0.4, Cr 1, Mo 0.4, bal Fe.
Heat treated: 128,000-148,000 TS; 107,000 YS; 12 El; 45 RA. For gears, shafts, axles; oil hardening, tough.

ROLLSCHEN 5 CR SI
Rollschen Eisenwerke, A.G.
C 0.5, Mn 0.7, Si 1.2, bal Fe.
Heat treated: 220,000-250,000 TS; 185,000 YS; 4 El; 18 RA. For springs; oil hardening.

ROLLSCHEN 5 MN SI
Rollschen Eisenwerke, A.G.
C 0.5, Mn 0.7, Si 1.2, bal Fe.
Heat treated: 184,000-214,000 TS; 148,000 YS; 7 El; 20 RA. For springs; oil hardening.

ROLLSCHEN 5 NI 12 CR 8 MO 3
Rollschen Eisenwerke, A.G.
C 0.5, Cr 1, Ni 1.7, Mo 0.3, bal Fe.
Heat treated: 190,000-200,000 TS. For gears, shafts, crankshafts, axles; oil hardening, shock resistant.

ROLLSCHEN 5 NI 12 CR 8 MO 7
Rollschen Eisenwerke, A.G.
C 0.5, Cr 1, Ni 1.7, Mo 0.7, bal Fe.
Heat treated: 220,000-240,000 TS. For gears, shafts, crankshafts, axles; oil hardening, shock resistant.

ROLLSCHEN WA EXTRA
Rollschen Eisenwerke, A.G.
C 0.5, Si 0.7, Cr 1.5, Mo 0.5, V 0.3, bal Fe.
Heat treated: 156,000-184,000 TS. For gears, shafts, cranshafts, axles; oil hardening, shock resistant.

ROLLSCHEN WAG46
Rollschen Eisenwerke, A.G.
C 0.4, Si 1, Cr 5.5, Mo 0.8, V 0.5, bal Fe.
Heat treated: 185,000-205,000 TS. For tools, dies, punches; oil hardening, shock resistant.

ROMA BRONZE
English manufacture
Cu 59, Zn 41, Pb 0.4, Al 0.2, traces Fe.
For hardware, nuts, bolts; free-cutting.

ROMAN BRONZE
Revere Copper Products, Inc.
Zn 39, Sn 0.75, bal Cu.
For hardware, welding rod; high strength. *Obsolete*

ROMANIUM
English manufacture
Cu 0.25, Ni 1.75, Sn 0.15, W 0.17, Al 97.43.
Hard rolled: 47,500 TS; 5 El. Annealed: 34,000 TS; 19.4 El.
For light alloy parts.

ROMILLY BRASS
English manufacture
Zn 28, Sn 0.2, Pb 0.8, bal Cu.
For fittings, hardware; free-cutting, corrosion resistant.

RONCAN
Revere Copper Products, Inc.
Cu.
For roofing, building; refined copper. *Obsolete*

RONENSIL
Saarstahl AG
Mn 16, Cr 7-9, Mo 1, C, bal Fe.
For stainless steel parts; stainless.

RONENSIL
Rochling Burbach GmbH
Cr 8-9, Mn 12, bal Fe.
For tableware, household utensils; non-magnetic. *Obsolete*

RONIA METAL
Manufacturer not listed.
Cu, Zn, Co, Mn, P.
For hardware; corrosion resistant.

ROOFLOY
American Smelting & Refining Co.
Ca 0.02, Mg 0.02, Sn 0.02, bal Pb.
For roofing; rolled sheet.

RORSCHACH 23S
Rorschach Aluminiumwerke
Aluminum. Cu 3.5-4.5, Si 0.2-0.9, Mn 0.8-1.5, Mg 0.8-1.5, bal Al.
For light alloy parts; age hardened.

RORSCHACH 26S
Rorschach Aluminiumwerke
Aluminum. Cu 3.5-4.5, Si 0-1, Mg 0.2-1, Mn 0.2-1, bal Al.
For light alloy parts; age hardened.

RORSCHACH 65S
Rorschach Aluminiumwerke
Aluminum. Si 0.5-1, Mn 0.2-1, Mg 0.5-1, 0.15-0.40 Cr or Cu, bal Al.
For light alloy parts.

RORSCHACH A4S
Rorschach Aluminiumwerke
Aluminum. Mg 0-1, Mn 0.5-1, bal Al.
For light alloy parts.

RORSCHACH A56S
Rorschach Aluminiumwerke
Aluminum. Mg 4-6, 0.4 Mn or Cr max, bal Al.
For light alloy parts; corrosion resistant.

RORSCHACH A57S
Rorschach Aluminiumwerke
Aluminum. Mg 0-2, 0.5 Mn or Cr max, bal Al.
For light alloy parts; corrosion resistant.

RORSCHACH AR-A4S
Rorschach Aluminiumwerke
Aluminum. Mn 1, Mg 0.7, bal Al.
O-temper: 17,000 TS; 5700 YS; 25 El; 40 Brin. 1/2 H-temper: 28,500 TS; 21,000 YS; 7 El; 60 Brin. H-temper: 31,400 TS; 30,000 YS; 5 El; 70 Brin. For roofing; good formability.

RORSCHACH AR-A57S
Rorschach Aluminiumwerke
Aluminum. Mg 1.5, bal Al.
O-temper: 18,500 TS; 7100 YS; 35 El; 30 Brin. 1/2 H-temper: 30,000 TS; 28,500 YS; 6 El; 60 Brin. H-temper: 35,000 TS; 34,000 YS; 4 El; 70 Brin. For chemical drums, architectural trim; good formability.

RORSHACH 17S
Rorschach Aluminiumwerke
Aluminum. Cu 4, Mg 0.5, Mn 0.5, bal Al.
O-temper: 32,700 TS; 32,700 YS; 18 El; 45 Brin. T6-temper: 64,000 TS; 42,700 YS; 17 El; 110 Brin. For aircraft and transport equipment and parts, structures; age hardenable.

RORSHACH 17S CLAD
Rorschach Aluminiumwerke
Aluminum. Cu 4, Mg 0.5, Mn 0.5, bal Al.
O-temper: 31,000 TS; 14,200 YS; 18 El; 45 Brin. T6-temper: 59,800 TS; 39,800 YS; 17 El; 110 Brin. RT-temper: 64,000 TS; 45,500 YS; 11 El; 115 Brin. For aircraft structures, fuselage, wings; clad, age hardenable.

RORSHACH 24S
Rorschach Aluminiumwerke
Aluminum. Cu 4.2, Mg 1.5, Mn 0.5, bal Al.
O-temper: 37,000 TS; 18,500 YS; 14 El; 45 Brin. T-temper: 69,700 TS; 49,800 YS; 15 El; 110 Brin. RT-temper: 71,100 TS; 59,800 YS; 9 El; 115 Brin. For aircraft structures, fuselage, wings; age hardenable.

RORSHACH 24S CLAD

Rorschach Aluminiumwerke
Aluminum. Cu 4.2, Mg 1.5, Mn 0.5, bal Al.
O-temper: 34,200 TS; 14,200 YS; 14 El. T-temper: 66,800 TS;
45,500 YS; 15 El. RT-temper: 68,300 TS; 47,000 YS; 9 El. For
aircraft structures, fuselage, wings; age hardenable, clad.

RORSHACH 3S

Rorschach Aluminiumwerke
Aluminum. Mn 1-2, bal Al.
Soft: 18,500 TS; 10,000 YS; 30 El; 25 Brin. Hard: 33,500 TS;
31,500 YS; 4 El; 60 Brin. For food, textile and marine
equipment; corrosion resistant.

RORSHACH 4S

Rorschach Aluminiumwerke
Aluminum. Mg 0.5-1.5, Mn 0.5-1.5, bal Al.
Soft: 28,500 TS; 10,000 YS; 17 El; 40 Brin. Hard: 45,500 TS;
34,200 YS; 4 El; 90 Brin. For transport and marine
equipment; stronger than RORSHACH 3S.

RORSHACH 51S

Rorschach Aluminiumwerke
Aluminum. Si 1, Mg 0.5-1, Mn 0.2-1, bal Al.
O-temper: 18,500 TS; 11,400 YS; 22 El; 25 Brin. W-temper:
38,400 TS; 24,400 YS; 10 El; 90 Brin. T-temper: 47,000 TS;
38,400 YS; 11 El; 100 Brin. For ship building and
transportation industries; corrosion resistant.

RORSHACH 57S

Rorschach Aluminiumwerke
Aluminum. Mg 2.5, Cr 0.25, bal Al.
O-temper: 31,300 TS; 19,900 YS; 25 El; 45 Brin. H-temper:
45,500 TS; 39,800 YS; 4 El; 95 Brin. For architectural trim,
refrigerator trays; wrought.

RORSHACH 61S

Rorschach Aluminiumwerke
Aluminum. Mg 0.6, Si 1, Cr 0.25, bal Al.
O-temper: 19,900 TS; 11,400 YS; 20 El; 45 Brin. W-temper:
42,700 TS; 35,500 YS; 17 El; 95 Brin. RT-temper: 54,100 TS;
45,500 YS; 7 El; 105 Brin. For structures; age hardenable.

ROSE 3X

Schmidt & Clemens Edelstahlwerke
C 0.85, Si 0.4, Mn 0.6, bal Fe.
Heat treated: 130,000-190,000 TS; 90,000-145,000 YS; 12-20
El; 35-50 RA; 255-390 Brin. For tools, springs, punches,
gauges; Type W1; water hardened. Obsolete

ROSE FUSIBLE ALLOY-1

English manufacture
Bi 50, Sn 22, Pb 28.
For safety plugs, fuses; melting point 79°C.

ROSE FUSIBLE ALLOY-2

English manufacture
Bi 35, Sn 30, Pb 35.
For safety plugs, fire extinguishers; melting point 98°C

ROSE LABEL

Allan & Son
Tool Steel. Mn 0.27, C, bal Fe.
For tools, cutters; water hardened.

ROSEIN

English manufacture
Ni 40, Al 30, Sn 20, Ag 10.
For jewelry, ornaments; corrosion resistant.

ROSENHAIN-ARCH-BUTT ALLOY

Manufacturer not listed.
Al 72, Zn 25, Cu 3.
For strong light alloy parts; non-hardenable.

ROSIN

English manufacture
Ni 40, Al 30, Sn 20, Ag 10.
For jewelry and ornamental articles.

ROSS ALLOY

English manufacture
Cu 68, Sn 32.
For reflectors; "Rotguss."

ROSSLYN METAL

American Clad Metals Inc.
Copper. Stainless clad copper.
Rolled: 80,000-86,000 TS; 37,000-50,000 YS; 31-60 El. For
cooking utensils, kettles, and heat exchangers. Formerly
"Amclad."

ROSTHORN-WILN BRASS

English manufacture
Zn 31.9, bal Cu.
For condenser tubes, stampings, hardware; high ductility.

ROSTODUR R3

Hoffman Elektrogusstahlwerk, Alb.
C 0.2, Cr 13, Mo 1, bal Fe.
Annealed: 95,000 TS; 40,000 YS; 25 El; 82 Rock B.
Hardened: 235,000 TS; 200,000 YS; 10 El; 50 Rock C. For
valves, cutlery, bearings, surgical instruments. Corrosion
resistant, hardenable.

ROSTODUR R7

Hoffman Elektrogusstahlwerk, Alb.
C 0.15, Cr 13, Mo 1, bal Fe.
Annealed: 80,000 TS; 40,000 YS; 15 El; 93 Rock B. For
springs, chemical plant and oil refinery equipment, table
flatware. Corrosion resistant.

ROTAS

Worthington Steel & Annealing Co.
C 0.15-0.25, bal Fe.
For case-hardened parts. Case-hardening steel.

ROTELLOY 2

Telcon Metals Ltd.
Fe-Co-V.
Soft magnetic alloy, high strength, for high-speed generators.

ROTELLOY 3

Telcon Metals Ltd.
Fe-Co-V.
Soft magnetic alloy, high strength, for high speed generators.

ROTGUSS

German manufacture
Cu 82-93, Sn 4-10, Zn 3-10, Pb 1-2.
For cast fittings, valves, hardware; free-cutting.

ROTGUSS 10

Oederlin & Co. Ltd.
Copper. Zn, bal Cu.
Cast: 29,000-36,000 TS; 12 El; 70 Brin. For hardware; brass.

ROTGUSS 4

Oederlin & Co. Ltd.
Copper. Zn, bal Cu.
Cast: 26,000-29,000 TS; 15 El; 50 Brin. For hardware; brass.

ROTGUSS 5

Oederlin & Co. Ltd.
Copper. Zn, bal Cu.
Cast: 21,000-29,000 TS; 10 El; 60 Brin. For hardware; brass.

ROTGUSS 8

Oederlin & Co. Ltd.
Copper. Zn, bal Cu.
Cast: 21,000-29,000 TS; 8 El; 70 Brin. For hardware; brass.

ROTGUSS RED BRONZE

German manufacture
Sn 4-10, Zn 3-10, Pb, bal Cu.
For hardware, fittings; free-cutting, castings.

ROTOR LDP

Haeckerstahl GmbH
C 1.05, W 1.15, Mn 0.9, Cr 1, bal Fe.
For bearings, cutters, bushings; oil or water hardened, wear
resistant.

ROTOR N

Haeckerstahl GmbH
C 1.05, Cr 0.65, V 0.1, Mn 0.3, bal Fe.
For bearings, bushings, cutters; water hardened, tough.

ROTOR SPEZIAL N

Haeckerstahl GmbH
C 1.45, Cr 1.4, Mn 0.6, bal Fe.
For bearings, cutters, liners, forming dies; oil or water
hardened.

ROTOR-11

Dohlen-Stahl Gusstahl-Handels GmbH
C 1.2, Cr 0.2, V 0.1, W 1, bal Fe.
For blanking and forming dies, headers; oil hardened,
abrasion resistant.

ROTORENALUMINIUM

Swiss Aluminium Ltd.
Aluminum. Si 0.3, Fe 0.4, Cu 0.02, Mg 0.01, Zn 0.07, bal Al.
Primary foundry alloy in ingot form. 70-120 N/mm^2 TS; 20-40
N/mm^2 YS; 10-50 El; 15-25 Brin.

ROTOXIT

English manufacture
Si 4, bal Cu.
For chemical apparatus; corrosion resistant.

ROTPUNKT

SWB Stahlformguss Gesellschaft mbH
C 0.9, Cr 4, V 1, W, Mo, bal Fe.
For lathe and planer tools, hobs, taps, reamers; high speed
steel. Obsolete

ROTPUNKT DMO5

SWB Stahlformguss Gesellschaft mbH
C 0.7, Cr 4, V 1, W, Mo, bal Fe.
For lathe and planer tools, hobs, broaches; high speed steel.
Obsolete

ROTPUNKT DMO5CO

SWB Stahlformguss Gesellschaft mbH
C 0.7, Cr 4, W, Mo, V, bal Fe.
For lathe and planer tools, drills, reamers, taps; high speed
steel. Obsolete

ROTPUNKT K5

SWB Stahlformguss Gesellschaft mbH
C 0.7, Cr 4, W, Mo, V, bal Fe.
For lathe and planer tools, hobs, broaches; high speed steel.
Obsolete

ROTUNG D

Plansee, Metallwerk Gesellschaft
W, Cu.
For electrical contacts. Tungsten impregnated with copper.

ROTUNG P

Plansee, Metallwerk Gesellschaft
W, Cu.
For electrical contacts. Tungsten impregnated with copper.

ROULZ

German manufacture
Cu 35-50, Ni 25-30, Ag 20-40.
For ornamental and corrosion resistant pens.

ROXO

Marsh Bros. & Co. Ltd.
C 0.4, Si 2.4, bal Fe.
For castings; cast iron.

RPG 3
Saarstahl AG
Now SAARSTAHL 1.2365X1.

RQ AR-321
Bethlehem Steel Corp.
C 0.25-0.32, Mn 0.4-0.65, Si 0.2-0.35, Cr 0.8-1.15, Mo 0.15-0.25, bal Fe.
Quenched and tempered to 321 Brin min. Abrasion resistant steel plate to 3 in. thick.

RQ AR-340
Bethlehem Steel Corp.
C 0.25-0.32, Mn 0.4-0.65, S 0.2-0.35, Cr 0.8-1.15, Mo 0.15-0.25, bal Fe.
Quenched and tempered to 340 Brin min. Abrasion resistant steel plate to 2 in. thick.

RQ AR-360
Bethlehem Steel Corp.
C 0.25-0.32, Mn 0.4-0.65, Si 0.2-0.35, Cr 0.8-1.15, Mo 0.15-0.25, bal Fe.
Quenched and tempered to 360 Brin min. Abrasion resistant steel plate to 1-1/2 in. thick.

RQ AR-400
Bethlehem Steel Corp.
C 0.25-0.32, Mn 0.4-0.65, Si 0.2-0.35, Cr 0.8-1.15, Mo 0.15-0.25, bal Fe.
Quenched and tempered to 400 Brin min. Abrasion resistant steel plate, 3/8 to 1 in. thick.

RQ-100
Bethlehem Steel Corp.
C 0.12-0.21, N 0.45-0.7, S 0.2-0.35, B 0.001-0.005, Mo 0.45-0.6, Ni 1.2-1.5, Cr 0.85-1.2, bal Fe.
Roller quenched and tempered: 115,000-135,000 psi TS; 100,000 psi YS min; 18 El; 235-293 Brin. Structural steel plate, to 2-1/2 in. thick.

RQ-100A
Bethlehem Steel Corp.
C 0.12-0.21, N 0.45-0.7, S 0.2-0.35, B 0.001-0.005, Mo 0.5-0.65, bal Fe.
Roller quenched and tempered: 115,000-135,000 psi TS; 100,000 psi YS; 18 El; 235-293 Brin. Structural steel plate, to 1-1/4 in. thick.

RQ-100B
Bethlehem Steel Corp.
C 0.12-0.21, Mn 0.45-0.7, Si 0.2-0.35, B 0.001-0.005, Mo 0.45-0.6, Ni 1.2-1.5, bal Fe.
Roller quenched and tempered: 115,000-135,000 psi TS; 100,000 psi YS min; 18 El; 235-293 Brin. Structural steel plate, to 2 in. thick.

RQ-321
Bethlehem Steel Corp.
C 0.12-0.21, Mn 0.45-0.7, Si 0.2-0.35, Ni 1.2-1.5, Cr 0.85-1.2, Mo 0.45-0.6, B 0.001-0.005, bal Fe.
Quenched and tempered to 321 Brin min. Abrasion resistant steel plate to 4 in. thick.

RQ-321A
Bethlehem Steel Corp.
C 0.12-0.21, Mn 0.45-0.7, Si 0.2-0.3, Mo 0.5-0.65, B 0.001-0.005, bal Fe.
Quenched and tempered to 321 Brin min. Abrasion resistant steel plate to 1-1/4 in. thick.

RQ-321B
Bethlehem Steel Corp.
C 0.12-0.21, Mn 0.45-0.7, Si 0.2-0.35, Ni 1.2-1.5, Mo 0.45-0.6, B 0.001-0.005, bal Fe.
Quenched and tempered to 321 Brin min. Abrasion resistant steel plate to 2 in. thick.

RQ-340
Bethlehem Steel Corp.
C 0.12-0.21, N 0.45-0.7, Si 0.2-0.35, Ni 1.2-1.5, Cr 0.85-1.2, Mo 0.45-0.6, B 0.001-0.005, bal Fe.
Quenched and tempered to 340 Brin min. Abrasion resistant steel plate to 2 in. thick.

RQ-340A
Bethlehem Steel Corp.
C 0.12-0.21, Mn 0.45-0.7, Si 0.2-0.35, Mo 0.5-0.65, B 0.001-0.005, bal Fe.
Quenched and tempered to 340 Brin min. Abrasion resistant steel plate to 1 in. thick.

RQ-340B
Bethlehem Steel Corp.
C 0.12-0.21, Mn 0.45-0.7, Si 0.2-0.35, Ni 1.2-1.5, Mo 0.45-0.6, B 0.001-0.005, bal Fe.
Quenched and tempered to 340 Brin min. Abrasion resistant steel plate to 1-1/2 in. thick.

RQ-360
Bethlehem Steel Corp.
C 0.12-0.21, Mn 0.45-0.7, Si 0.2-0.35, Ni 1.2-1.5, Cr 0.85-1.2, Mo 0.45-0.6, B 0.001-0.005, bal Fe.
Quenched and tempered to 360 Brin min. Abrasion resistant steel plate to 1-1/2 in. thick.

RQ-360A
Bethlehem Steel Corp.
C 0.12-0.21, Mn 0.45-0.7, Si 0.2-0.35, Mo 0.5-0.65, B 0.001-0.005, bal Fe.
Quenched and tempered to 360 Brin min. Abrasion resistant steel plate to 1/2 in. thick.

RQ-360B
Bethlehem Steel Corp.
C 0.12-0.21, Mn 0.45-0.7, Si 0.2-0.35, Ni 1.2-1.5, Mo 0.45-0.6, B 0.001-0.005, bal Fe.
Quenched and tempered to 360 Brin min. Abrasion resistant steel plate to 1 in. thick.

RQC-100
Bethlehem Steel Corp.
C 0.12-0.21, Mn 0-1.5, Si 0.15-0.3, 0.0005 B min, bal Fe.
Quenched and tempered: 110,000 psi TS min; 100,000 psi YS min; 18 El min. Steel plate for structural purposes; 3/16 to 1-1/4 in. thick.

RQC-321
Bethlehem Steel Corp.
C 0-0.28, Mn 0-1.5, Si 0.2-0.6, 0.0050 B min, bal Fe.
Quenched and tempered to 321 Brin min. Abrasion resistant steel plate to 1-1/4 in. thick.

RQC-340
Bethlehem Steel Corp.
C 0-0.28, Mn 0-1.5, Si 0.2-0.6, 0.005 B min, bal Fe.
Quenched and tempered to 340 Brin min. Abrasion resistant steel plate, 1/4 to 1 in. thick.

RQC-60
Bethlehem Steel Corp.
C 0-0.2, Mn 0.7-1.35, Si 0.15-0.5, bal Fe.
As normalized: 70,000-90,000 psi TS; 50,000 YS min; 22 El. Quenched and tempered: 80,000-100,000 psi TS; 60,000 YS min; 22 El. For pressure vessels, weldable.

RQC-80
Bethlehem Steel Corp.
C 0-0.2, Mn 0-1.35, S 0.15-0.3, B, bal Fe.
Quenched and tempered: 95,000 psi TS min; 80,000 psi YS min; 18 El min. Steel plate for structural applications; 3/16 to 1-1/2 in. thick.

RQC-90
Bethlehem Steel Corp.
C 0-0.2, Mn 0-1.35, S 0.15-0.3, 0.0005 B min, bal Fe.
Quenched and tempered: 100,000 psi TS min; 90,000 psi YS min; 18 El min. Steel plate for structural purposes; 3/16 to 1-1/2 in. thick.

RQT 501
British Steel plc
C 0-0.16, Si 0-0.5, P 0-0.035, S 0-0.015, Mn 0-1.5, V 0-0.08, Mo 0-0.2, bal Fe.
Quenched and tempered: 470 N/mm^2 YS.

RQT 601
British Steel plc
C 0-0.2, Si 0-0.5, P 0-0.035, S 0-0.015, Mn 0-1.5, Nb 0-0.06, B 0-0.003, Ti 0-0.04, bal Fe.
Quenched and tempered: 620 N/mm^2 YS.

RQT 701
British Steel plc
C 0-0.2, Si 0-0.5, S 0-0.035, P 0-0.015, Mn 0-1.5, Nb 0-0.06, B 0-0.003, Ti 0-0.04, bal Fe.
Quenched and tempered: 690 N/mm^2 YS.

RQTUF 501
British Steel plc
C 0-0.16, Si 0-0.5, P 0-0.02, S 0-0.005, Mn 0-1.5, V 0-0.08, Mo 0-0.25, Ni 0-0.7, bal Fe.
Similar to RQT 501. For offshore situations.

RR 5
Saarstahl AG
C 0.55, Si 0-0.15, Mn 0.4, bal Fe.
Carbon tool steel; stone working tools. W.-Nr. 1820. *Obsolete*

RR.90-227
Sintered Products Ltd.
C, Cu-Fe, sintered.
Double press, double sintered. 70 g/cm^3 density; 71 kg/mm^2 TS; 2-5 El. For gears, sprockets, levers.

RRAC7
High Duty Alloys Ltd.
Sn 4.6, Ni 1.6-2, Mn 0.7-0.9, Sb 0.4-0.8, Si 0.4-0.6, Mg 0.35-0.5, bal Al.
For bearings; for engine bearings. *Obsolete*

RRAC9
High Duty Alloys Ltd.
Sn 55.5-70, Ni 1.5-1.8, Cu 0.6-0.9, Mg 0.7-1, Si 0.15-0.35, Fe 0.2-0.45, bal Al.
For bearings; connecting rods. *Obsolete*

RS 1 SPEZIAL
Thyssen Edelstahlwerke AG
C, bal Fe.
For steel trimming and weld steel. *Obsolete*

RS 115
Renewal Services Inc.
Cu 85, Sn 5, Pb 5, Zn 5.
Bar: 32,000-43,000 TS; 15,000-24,000 YS; 20-35 El; 55-65 Brin. For bearings, bushings, sleeves, liners. Ounce metal, free machining.

RS 123
Renewal Services Inc.
Cu 81, Sn 3, Pb 7, Zn 9.
Bar: 29,000-39,000 TS; 13,000-17,000 YS; 18-30 El; 50-60 Brin. For bearings, bushings, liners, sleeves. Good embeddability.

RS 194
Renewal Services Inc.
Cu 81, Sn 19.
Bar: 40,000-48,000 TS; 150-170 Brin. For bearings, bushings, linings. Bridge bronze, corrosion resistant.

RS 197
Renewal Services Inc.
Cu 85, Sn 14, Zn 1.
Bar: 30,000-39,000 TS; 1-3 El. For bearings, bushings, sleeves. Corrosion resistant.

RS 198

Renewal Services Inc.
Cu 83, Sn 14, Zn 3.
Bar: 32,000 TS; 23,000 YS; 12 El; 75 Brin. For bearings, bushings, sleeves. Navy H, corrosion resistant.

RS 2 SPEZIAL

Thyssen Edelstahlwerke AG
C, bal Fe.
For steel trimming and weld steel. *Obsolete*

RS 205

Renewal Services Inc.
Cu 89, Sn 10, Pb 2.
Bar: 35,000 TS; 21,000 YS; 12 El; 85 Brin. For worm wheels, gears, shafts. Gear bronze, corrosion resistant, free-machining.

RS 206

Renewal Services Inc.
Cu 88, Sn 10, Pb 2.
Bar: 36,000-46,000 YS; 18,000-26,000 YS; 15-25 El; 65-80 Brin. For bearings, bushings, liners. Leaded gun metal, good embeddability.

RS 210

Renewal Services Inc.
Cu 88, Sn 10, Zn 2.
Bar: 40,000-50,000 TS; 20,000-25,000 YS; 20-35 El; 60-75 Brin. For bearings, bushings, liners, gears. Gun metal, corrosion resistant.

RS 215

Renewal Services Inc.
Cu 87, Sn 10, Pb 1, Zn 2.
Bar: 37,000-47,000 TS; 17,000-22,000 YS; 18-25 El. For bearings, bushings, liners. Commerical gun metal, corrosion resistant.

RS 225

Renewal Services Inc.
Cu 88, Sn 8, Zn 4.
Bar: 47,000-51,000 TS; 18,000-23,000 YS; 65-80 Brin; 25-35 El. For bearings, bushings, gears. Navy G, bronze, corrosion resistant.

RS 230

Renewal Services Inc.
Cu 87, Sn 8, Pb 1, Zn 4.
Bar: 36,000-45,000 TS; 16,000-24,000 YS; 18-30 El; 60-75 Brin. For gears, bushings, bearings. Leaded Navy G, bronze.

RS 245

Renewal Services Inc.
Cu 88, Sn 6, Pb 1, Zn 4.
Bar: 36,000-48,000 TS; 16,000-21,000 YS; 25-40 El; 60-72 Brin. For bearings, bushings, liners. Navy M, bronze, corrosion resistant.

RS 3

Thyssen Edelstahlwerke AG
C, bal Fe.
For steel trimming and weld steel. *Obsolete*

RS 305

Renewal Services Inc.
Cu 80, Sn 10, Pb 10.
Bar: 27,000-37,000 TS; 15,000-22,000 YS; 10-20 El; 55-70 Brin. For bearings, bushings, liners. Corrosion resistant, good embeddability.

RS 311

Renewal Services Inc.
Cu 84, Sn 8, Pb 8.
Bar: 25,000-32,000 TS; 7-11 El. For bearings, bushings, liners. Corrosion resistant, good embeddability.

RS 315

Renewal Services Inc.
Cu 83, Sn 7, Pb 7, Zn 3.
Bar: 30,000-38,000 TS; 17,000-21,000 YS; 12-20 El; 55-65 Brin. For bearings, bushings, liners. Bearing bronze, corrosion resistant.

RS 319

Renewal Services Inc.
Cu 78, Sn 6, Pb 16.
Bar: 25,000-33,000 TS; 14,000-20,000 YS; 10-18 El; 50-60 Brin. For bearings, bushings, liners. Semi-plastic bronze, good embeddability.

RS 321

Renewal Services Inc.
Cu 71, Sn 7, Pb 21.
Bar: 24,000-31,000 TS; 15,000 YS; 15 El; 55 Brin. For bearings, bushings, liners. Hi-leaded tin bronze.

RS 323

Renewal Services Inc.
Cu 70, Sn 5, Pb 25.
Bar: 23,000-30,000 TS; 11,000-16,000 YS; 7-16 El; 42-45 Brin. For bearings, bushings, liners. Hi-leaded tin bronze.

RS 326

Renewal Services Inc.
Cu 85, Sn 5, Pb 9, Zn 1.
Bar: 25,000-32,000 TS; 12,000-18,000 YS; 8-12 El; 60 Brin. For bearings, bushings, liners. Corrosion resistant.

RS 327

Renewal Services Inc.
Cu 75, Sn 5, Pb 20.
Bar: 22,000-30,000 TS; 7-10 El. For bearings, bushings, liners. Corrosion resistant. Good embeddability.

RS 4

Thyssen Edelstahlwerke AG
C, bal Fe.
For steel trimming and weld steel. *Obsolete*

RS 410

Renewal Services Inc.
Cu 56, Sn 2, Pb 9, Zn 20, Ni 12.
Bar: 34,000-40,000 TS; 17,000-20,000 YS; 25 El; 50-60 Brin. For bushings, bearings, liners. Nickel silver, corrosion resistant.

RS 415 C

Renewal Services Inc.
Cu 83, Ni 2, Fe 4, Al 11.
Bar: 85,000-95,000 TS; 30,000-41,000 YS; 12-20 El; 160-195 Brin. For bearings, liners, bushings, shafting. Aluminum bronze, corrosion resistant.

RS 415 D

Renewal Services Inc.
Cu 81, Ni 4, Fe 4, Al 11.
Bar: 95,000 TS; 45,000 YS; 7 El; 195 Brin. For bushings, bearings, liners, shafting. Corrosion resistant. Aluminum bronze.

RS 415A

Renewal Services Inc.
Cu 87, Fe 3, Al 9.
Bar: 60,000-75,000 TS; 23,000-28,000 YS; 15-25 El; 90-125 Brin. For gears, bushings, bearings, shafts. Aluminum bronze, corrosion resistant, tough.

RS 415B

Renewal Services Inc.
Cu 88, Fe 1, Al 10.
Bar: 70,000-85,000 TS; 25,000-30,000 YS; 20-35 El; 100-140 Brin. For gears, bushings, bearings, shafts. Aluminum bronze, corrosion resistant, tough.

RS 421

Renewal Services Inc.
Cu 58, Zn 40, Fe 1, Al 1, Mn 0.25.
Bar: 70,000-88,000 TS; 28,000-40,000 YS; 20-35 El; 90-120 Brin. For bearings, bushings, sleeves, shafting, impellers. Manganese bronze, corrosion resistant.

RS 423

Renewal Services Inc.
Cu 64, Zn 24, Fe 3, Al 5, Mn 4.
Bar: 90,000-100,000 TS; 45,000-55,000 YS; 18-30 El; 160-200 Brin. For bushings and bearings, shafting. Manganese bronze, corrosion resistant.

RS 424

Renewal Services Inc.
Cu 64, Zn 24, Fe 3, Al 5, Mn 4.
Bar: 11,000-120,000 TS; 65,000-90,000 YS; 12-20 El; 170-225 Brin. For bushings and bearings, shafts. Corrosion resistant. Manganese bronze.

RS 500

Renewal Services Inc.
Zn 5, Ni 1.5, Fe 2.5, Al 1.5, Si 3-5, Mn 0.25, bal Cu.
Bar: 45,000-65,000 TS; 20,000-35,000 YS; 15-40 El; 55-90 Brin. For bushings and bearings, shafts. Silicon bronze, corrosion resistant.

RS LEGIERUNG

Aluminium-Zentral e.V.
Aluminum. Si 0.7-1.2, Mn 0.7-1.2, Mg 0.8-1.2, bal Al.
Heat treated: 40,000-51,000 TS; 28,000-43,000 YS; 5-10 El; 75-95 Brin. For general structures, scaffolds, rails, booms, transmission towers. Age hardenable, good strength and fabricability. *Obsolete*

RSD13

Eagle & Globe Steel Ltd.
Tool material. C 1, Mn 0.5, Cr 5, Mo 1, V 0.25, Si 0.25, bal Fe. Cold work tool and die steel. Air hardened. For dies, punches, rolls, gauges, shear blades, master hobs. AS1239 A2A; AISI A2; Werkstoff 1.2363.

RT 10

Saarstahl AG
Now SAARSTAHL 1.1645.

RT 6

Saarstahl AG
Now SAARSTAHL 1.1620.

RT 8

Saarstahl AG
Now SAARSTAHL 1.1625.

RTC 14

Saarstahl AG
Now SAARSTAHL 1.2067.

RTS

Saarstahl AG
Now SAARSTAHL 1.2210.

RTS

Now JESSOP RTS.

RTW 1

Saarstahl AG
Now SAARSTAHL 1.2516.

RTW 2 H

Saarstahl AG
C 0.45, Si 1, Cr 1.1, V 0.18, W 2, bal Fe.
Cold work tool steel; for punches, stamping tools, chisels, pneumatic tools. W.-Nr. 2542; AFNOR 45 Wc 20. *Obsolete*

RTWK

Saarstahl AG
Now SAARSTAHL 1.2550.

RUBBON
Now ALL-STATE NO. 55 RUBBON.

RUBEL BRONZE "D.W."
Allgemeines Deutsches Metallwerk GmbH
Cu 40-55, Ni 3-15, Mn 1-3, Fe 1-2, Al 0.5-3, bal Zn.
Cast: 60,000 TS; 30 El; 90 Brin. Hot pressed: 75,000 TS; 30
El; 100 Brin. For printing rolls. Corrosion resistant.

RUBEL BRONZE "L"
Allgemeines Deutsches Metallwerk GmbH
Cu 40-55, Ni 3-15, Mn 1-3, Fe 1-2, Al 0.5-3, bal Zn.
Cast: 60,000 TS; 30 El; 90 Brin. Hot pressed: 80,000 TS; 35
El; 115 Brin. For bearings, bushings. Corrosion resistant.

RUBEL BRONZE "L.W."
Allgemeines Deutsches Metallwerk GmbH
Cu 40-55, Ni 3-15, Mn 1-3, Fe 1-2, Al 0.5-3, bal Zn.
Cast: 60,000 TS; 30 El; 120 Brin. For bearings, bushings.
Corrosion resistant.

RUBEL BRONZE H-1
Allgemeines Deutsches Metallwerk GmbH
Cu 41, Zn 48, 11 Mn + Ni + Fe + Al.
Cast: 99,000 TS; 8 El; 160 Brin. Hot pressed: 105,000 TS; 12
El; 190 Brin. For propellers, piston rings, valves, rotor covers,
shrinking rings. High strength, noncorrosive.

RUBEL BRONZE H-2
Allgemeines Deutsches Metallwerk GmbH
Copper. Cu 50, Zn 42, 8 Mn + Ni + Fe + Al.
Cast: 82,000 TS; 8 El; 150 Brin. Hot pressed: 105,000 TS; 15
El; 170 Brin. For bearing bushings, piston rods, gears, valve
stems. High strength, noncorrosive.

RUBEL BRONZE H-2 "W"
Allgemeines Deutsches Metallwerk GmbH
Cu 40-55, Ni 3-15, Mn 1-3, Fe 1-2, Al 0.5-3, bal Zn.
Cast: 70,000 TS; 10 El; 140 Brin. Hot pressed: 90,000 TS; 25
El; 160 Brin. For bearings, bushings. Corrosion resistant.

RUBEL BRONZE W-1
Allgemeines Deutsches Metallwerk GmbH
Copper. Cu 50, Zn 41, 9 Mn + Ni + Fe + Al.
Cast: 70,000 TS; 20 El; 100 Brin. Hot pressed: 85,000 TS; 25
El; 110 Brin. For construction material for turbines, marine
engines and locomotives. High strength, noncorrosive.

RUBEL BRONZE W-2
Allgemeines Deutsches Metallwerk GmbH
Copper. Cu 51, Zn 43, 6 Mn + Ni + Fe + Al.
Cast: 70,000 TS; 18 El; 110 Brin. Hot pressed: 85,000 TS; 22
El; 120 Brin. For propellers, centrifugal drums, bearing
bushings, steam and safety jackets. High strength,
noncorrosive.

RUBEL BRONZE W-2
Allgemeines Deutsches Metallwerk GmbH
Cu 40-55, Ni 3-15, Mn 1-3, Fe 1-2, Al 0.5-3, bal Zn.
Cast: 65,000 TS; 15 El; 105 Brin. Hot pressed: 75,000 TS; 20
El; 115 Brin. For die forgings, rod, tubes, propellers.
Corrosion resistant.

RUBEL METAL
English manufacture
Cu 50-55, Zn 40, 4-5 Fe + Mn.
60,000 TS. For fittings, hardware; tough with high ductility.

RUBELIT
Allgemeines Deutsches Metallwerk GmbH
Cu 40-55, Mn 1-3, Ni 3-15, Fe 1-2, Al 0.5-3, bal Zn.
Hot pressed: 69,000-76,000 TS; 5-20 El; 130-190 Brin. For
bushings for bearings, piston bolts, valve guides. Corrosion
resistant.

RUBIDIUM
Atomergic Chemetals Corp.
Rb.
Purities: 99.9+%, 99.5%. Packaging: glass ampules, steel
containers (under vacuum inert gas, oil).

RUBIS BA1
Georgsmarienwerke Selesiastahl GmbH
C 0.5, Cr 1.05, Ni 3.25, bal Fe.
For gears, bolts, forging dies, crankshafts; oil hardened,
shock resistant, nondeforming.

RUBIS HKP
Georgsmarienwerke Selesiastahl GmbH
C 1, V 0.1, Mn 0.25, Si 0.2, bal Fe.
For blanking dies, bearings, cutters; Type W2; water
hardened.

RUHRLIT KOBALT 135
Friedr. Lohmann GmbH
C 1.3, Cr 4.3, Mo 0.85, V 3.8, W 12, bal Fe.
For engraving tools, lathe and planer cutters, hobs. High
speed steel. *Obsolete*

RUHRLIT NH18
Friedr. Lohmann GmbH
Now NH18V.

RUHRLIT SS20
Friedr. Lohmann GmbH
C 0.82, Cr 4.1, Mo 0.85, V 1.6, W 8.7, bal Fe.
For lathe and planer tools, drills, reamers, hobs. High speed
steel. *Obsolete*

RUHRLIT SS32
Friedr. Lohmann GmbH
C 0.8, Co 2.6, Cr 4, Mo 0.8, V 1.75, W 11.6, bal Fe.
For lathe and planer tools, reamers, broaches, drills. High
speed steel. *Obsolete*

RUHRSTAHL MI
Ruhrstahl A.G.
C 0-0.17, Si 0.3, Mn 0.3, bal Fe.
Annealed: 70,000 TS; 55,000 YS; 25 El; 60 RA; 140 Brin. For
gears, pinions, machine tool parts, rivets; case hardened.

RUHRSTAHL MII
Ruhrstahl A.G.
C 0-0.23, Si 0.3, Mn 0.3, bal Fe.
Annealed: 75,000 TS; 60,000 YS; 22 El; 58 RA; 150 Brin. For
fan blades, bushings, bolts, nuts, rivets; water hardened.

RUINELLA
Continental Copper & Steel Ind.Inc.
Cu alloy.
For castings. *Obsolete*

RULE BRASS 238
Anaconda Co.
Copper. Zn 35, Pb 2.5, bal Cu.
Soft: 45,000 TS; 50 El. Hard: 73,000 TS; 7 El. For screw
machine products, bolts, nuts, screws. Free cutting.

RUNE
Uddeholm Corp.
C 1.4, Cr 0.45, bal Fe.
For razors; water hardened. *Obsolete*

RUPEE (SILVER)
Manufacturer not listed.
Ag 91.6, Cu 8.3.
For coinage; corrosion resistant.

RUS
Saarstahl AG
Now SAARSTAHL 1.2842.

RUS 3
Saarstahl AG
Now SAARSTAHL 1.2510.

RUS 4
Saarstahl AG
C 1.05, Mn 0.9, Cr 1, W 1.15, bal Fe.
Cold work tool steel; for cutting and stamping sheet steel,
hand taps, thread chasers. W.-Nr. 2410. *Obsolete*

RUSCAR
Charles Carr Ltd.
Pb, Sn, Sb.
For antifriction bearings; Babbitt.

RUSSIAN PACKING
Russian manufacture
Zn 98.5, Sn 0.98, Pb 0.32, Fe 0.16.
For bearings; will not resist heat or live steam.

RUSSIAN-1
Russian manufacture
Cu 64, Ni 18, Zn 18, Fe 0.3, Pb 0.3.
For ornaments, corrosion resisting white metal parts.

RUSSIAN-2
Russian manufacture
Cu 1.3, Zn 26.5, Sn 72.2.
For bearings; anti-friction.

RUSSIAN-3
Russian manufacture
Cu 66-78, Zn 21-34, Sn 0-1.
For condenser tubes; high ductility.

RUSTLESS 12
Armco Steel Corp.
C 0-0.15, Cr 11.5-13.5, Mn 0-1, Si 0-1, bal Fe.
Annealed: 78,000 TS; 40,000 YS; 35 El; 73 RA; 148 Brin. For
valves, bolts; corrosion resistant. *Obsolete*

RUSTLESS 12 AL
Armco Steel Corp.
C 0-0.08, Cr 11.5-13.5, Al 0.1-0.3, bal Fe.
Annealed: 75,000 TS; 45,000 YS; 30 El; 70 RA; 165 Brin. For
turbine blades; corrosion resistant. *Obsolete*

RUSTLESS 12 FM
Armco Steel Corp.
C 0-0.15, Cr 12-14, Mo 0-0.6, S 0.18-0.35, bal Fe.
Annealed: 78,000 TS; 40,000 YS; 31 El; 62 RA; 148 Brin. For
screw machine parts; free cutting. *Obsolete*

RUSTLESS 12 T
Armco Steel Corp.
C 0-0.15, Cr 11.5-13, Mn 0-1, Si 0-1, bal Fe.
Annealed: 78,000 TS; 40,000 YS; 35 El; 73 RA; 148 Brin. For
turbine blades; corrosion resistant. *Obsolete*

RUSTLESS 12-2
Armco Steel Corp.
C 0-0.15, Cr 11.5-13.5, Ni 1.25-2.5, bal Fe.
Annealed: 115,000 TS; 93,000 YS; 20 El; 65 RA; 245 Brin. For
valves, bolts; corrosion resistant. *Obsolete*

RUSTLESS 13 C 35
Armco Steel Corp.
Cr 12-14, Si 0-1, Mn 0-1, C, bal Fe.
Annealed: 98,000 TS; 60,000 YS; 28 El; 65 RA; 170 Brin.
Hardened: 250,000 TS; 225,000 YS; 8 El; 25 RA; 500 Brin. For
surgical and dental instruments; corrosion resistant.
Obsolete

RUSTLESS 13HC
Armco Steel Corp.
C 0-0.4, Cr 13-15, bal Fe.
For cutlery, ball bearings, valve trim; corrosion resistant.
Obsolete

RUSTLESS 13HC35
Armco Steel Corp.
C 0-0.4, Cr 12-14, Ni 0-5, bal Fe.
Annealed: 85,000 TS; 40,000 YS; 34 El; 68 RA; 160 Brin.
Hardened: 250,000 TS; 190,000 YS; 3 El; 5 RA; 500 Brin. For
cutlery, scissors, surgical instruments; corrosion resistant.
Obsolete

RUSTLESS 15
Armco Steel Corp.
C 0-0.12, Cr 14-16, bal Fe.
Annealed: 75,000 TS; 45,000 YS; 35 El; 70 RA; 152 Brin. Cold drawn: 110,000 TS; 95,000 YS; 12 El; 65 RA; 220 Brin. For cold upset screw products, auto and refrigerator trim; non-hardenable, weldable. *Obsolete*

RUSTLESS 16-2
Armco Steel Corp.
C 0-0.2, Cr 15-17, Ni 1.25-1.75, Mn 0-1, Si 0-1, bal Fe.
Annealed: 120,000 TS; 95,000 YS; 21 El; 60 RA; 245 Brin. For bolts, shafting; corrosion resistant. *Obsolete*

RUSTLESS 16-6
Armco Steel Corp.
C 0.12-0.2, Cr 15-17, Ni 5-7, bal Fe.
Stainless. *Obsolete*

RUSTLESS 17
Armco Steel Corp.
C 0-0.12, Cr 14-18, Ni 0-0.5, bal Fe.
Annealed: 75,000 TS; 40,000 YS; 35 El; 75 RA; 140 Brin.
Worked: 100,000 TS; 10 El; 60 RA; 250 Brin. For trim, ornaments, screws; stainless, non-hardening. *Obsolete*

RUSTLESS 17-7
Armco Steel Corp.
C 0.1-0.2, Cr 16-18, Ni 6-8.5, Mn 0-2, Si 0-1, bal Fe.
Annealed: 105,000 TS; 45,000 YS; 55 El; 68 RA; 165 Brin.
Worked: 325,000 TS; 300,000 YS. For stainless parts, springs, wire cloth; stainless, non-magnetic. *Obsolete*

RUSTLESS 17-C-100
Armco Steel Corp.
C 0.95, Cr 16-18, Mn 0-1, Si 0-1, bal Fe.
Annealed: 110,000 TS; 60,000 YS; 15 El; 30 RA; 230 Brin.
Hardened: 288,000 TS; 275,000 YS; 2 El; 10 RA; 580 Brin. For cutlery; wear resistant. *Obsolete*

RUSTLESS 17-C-60
Armco Steel Corp.
C 0.6-0.7, Cr 16-18, Mn 0-1, Si 0-1, bal Fe.
Annealed: 105,000 TS; 65,000 YS; 23 El; 50 RA; 205 Brin. For cutlery; wear resistant. *Obsolete*

RUSTLESS 17-C-80
Armco Steel Corp.
C 0.7-0.9, Cr 16-18, Mn 0-1, Si 0-1, bal Fe.
Annealed: 107,000 TS; 62,000 YS; 19 El; 40 RA; 220 Brin. For cutlery; wear resistant. *Obsolete*

RUSTLESS 17FM
Armco Steel Corp.
C 0-0.12, Cr 14-18, Ni 0-0.5, 0.15% S min, bal Fe.
Annealed: 75,000 TS; 34 El; 62 RA; 140 Brin. Worked: 100,000 TS; 8 El; 58 RA; 250 Brin. For screw, bolts; free cutting, stainless. *Obsolete*

RUSTLESS 17HC60
Armco Steel Corp.
C 0.65, Cr 14-18, Ni 0-0.5, bal Fe.
Heat treated: 100,000-270,000 TS; 2-27 El; 3-55 RA; 190-575 Brin. For cutlery, scissors, valve trim; stainless, hardenable. *Obsolete*

RUSTLESS 17HC90
Armco Steel Corp.
C 1, Cr 14-18, Ni 0-0.5, bal Fe.
Heat treated: 105,000-285,000 TS; 1.5-21 El; 2-42 RA; 200-625 Brin. For cutlery, scissors, valve trim; stainless, hardenable. *Obsolete*

RUSTLESS 18-12-3 MO
Armco Steel Corp.
C 0-0.1, Cr 16-18, Ni 10-14, Mo 1.75-2.5, bal Fe.
Annealed: 85,000 TS; 30,000 YS; 60 El; 70 RA; 150 Brin. For process industries, stainless parts; heat and corrosion resistant. *Obsolete*

RUSTLESS 18-12-4 MO
Armco Steel Corp.
C 0-0.1, Cr 17.5-20, Ni 10-14, Mo 3-4, bal Fe.
Annealed: 85,000 TS; 30,000 YS; 60 El; 70 RA; 150 Brin. For stainless parts; heat and corrosion resistant. *Obsolete*

RUSTLESS 18-12MD
Armco Steel Corp.
C 0-0.09, Cr 16-20, Ni 0-14, Mo 2-4, bal Fe.
85,000-225,000 TS; 50,000-200,000 YS; 5-55 El; 55-70 RA; 140-400 Brin. For heat resistant parts; heat and corrosion resistant. *Obsolete*

RUSTLESS 18-8
Armco Steel Corp.
C 0.08-0.2, Mn 0-2, Si 0-1, Cr 17-19, Ni 8-10, bal Fe.
Annealed: 85,000 TS; 30,000 YS; 60 El; 70 RA; 150 Brin.
Worked: 250,000 TS; 225,000 YS. For screws, bolts, stainless parts; stainless. *Obsolete*

RUSTLESS 18-8 CB
Armco Steel Corp.
C 0-0.1, Cr 17-19, Ni 9-12, Cb = 8 x C, bal Fe.
Annealed: 87,000 TS; 32,000 YS; 50 El; 70 RA; 165 Brin. For welded parts; stainless. *Obsolete*

RUSTLESS 18-8 TI
Armco Steel Corp.
C 0-0.1, Cr 17-19, Ni 8-11, Ti = 4 x C, bal Fe.
Annealed: 85,000 TS; 30,000 YS; 55 El; 70 RA; 150 Brin. For welded parts; stainless. *Obsolete*

RUSTLESS 18-8FM (SELENIUM)
Armco Steel Corp.
C 0-0.2, Mn 0-1, Se 0.1-0.35, Cr 17-19, Ni 8-10, bal Fe.
Annealed: 88,000 TS; 33,000 YS; 54 El; 60 RA; 165 Brin. For shafting, valve parts; stainless, free cutting. *Obsolete*

RUSTLESS 18-8FM (SULPHUR)
Armco Steel Corp.
C 0-0.2, Mn 0-2, Si 0-1, S 0.18-0.35, Cr 17-19, Ni 8-10, bal Fe.
Annealed: 88,000 TS; 33,000 YS; 54 El; 60 RA; 165 Brin. For shafting, valve parts; stainless, free cutting. *Obsolete*

RUSTLESS 20-10
Armco Steel Corp.
C 0.08, Cr 20-22, Ni 10-12, Mn 0-2, Si 0-1, bal Fe.
Annealed: 95,000 TS; 60 El; 70 RA; 140 Brin. Worked: 225,000 TS; 5 El; 5.5 RA; 400 Brin. For welding wire, milk and food handling equipment; austenitic; Stainless Type 307. *Obsolete*

RUSTLESS 25-12
Armco Steel Corp.
C 0-0.2, Cr 22-24, Ni 12-15, Mn 0-2, bal Fe.
Annealed: 95,000 TS; 50,000 YS; 50 El; 60 RA; 170 Brin. For high temperature service and severe chemical corrosion; heat and corrosion resistant. *Obsolete*

RUSTLESS 25-20
Armco Steel Corp.
C 0-0.25, Mn 0-2, Si 0-1.5, Cr 24-26, Ni 19-22, bal Fe.
Annealed: 90,000 TS; 40,000 YS; 50 El; 66 RA; 170 Brin. For welding electrodes, furnace parts; resists heat to 2000 F. *Obsolete*

RUSTLESS 27
Armco Steel Corp.
C 0-0.35, Cr 23-27, Mn 0-1, Si 0-1, bal Fe.
Annealed: 80,000 TS; 40,000 YS; 27 El; 60 RA; 175 Brin. For exhaust manifolds, furnace parts; heat resistant. *Obsolete*

RUSTLESS 29-9
Armco Steel Corp.
C 0-0.25, Cr 27-31, Ni 8-10, bal Fe.
Stainless Type 312. *Obsolete*

RUSTLESS 5
Armco Steel Corp.
C 0-0.1, Cr 4-6, Mn 0-1, Si 0-1, bal Fe.
Annealed: 70,000 TS; 26,000 YS; 36 El; 69 RA; 135 Brin.
Worked: 90,000 TS; 80,000 YS; 20 El; 70 RA; 190 Brin. For bolts, nuts; corrosion resistant. *Obsolete*

RUSTLESS 8-20
Armco Steel Corp.
C 0-0.25, Cr 7-10, Ni 19-23, bal Fe.
Annealed: 75,000 TS; 30,000 YS; 50 El; 70 RA; 125 Brin. For valves; heat and corrosion resistant. *Obsolete*

RUSTLESS RR-11
Armco Steel Corp.
C 0-0.07, Cr 10-12, bal Fe.
Annealed: 75,000 TS; 55,000 YS; 28 El; 65 RA; 170 Brin.
Tempered: 100,000 TS; 80,000 YS; 28 El; 65 RA; 207 Brin. For railroad and coal industry equipment; corrosion resistant, readily weldable. *Obsolete*

RUTHENIUM
Atomergic Chemetals Corp.
Ru.
Purities: 99.999%, 99.99%, 99.9%. Forms: sponge, powder, rod, disc, single crystals.

RV4
Acciaierie Valbruna s.p.a.
Tool Material. C 1.25, Cr 4.15, Mo 0.85, W 12, V 3.75, Si 0-0.4, Mn 0-0.4, bal Fe.
High speed tool steel. W. Nr. 1.3302.

RV5K
Acciaierie Valbruna s.p.a.
Tool Material. C 1.55, Cr 4.5, W 12.5, V 5, Co 5, Si 0-0.4, Mn 0-0.4, bal Fe.
W-Co high speed tool steel. W. Nr. 1.3202; AISI T15.

RW 1
Saarstahl AG
C 1.05, Si 0.25, N 0.35, Cr 0.5, bal Fe.
For ball and roller bearings. W.-Nr. 3501; DIN 105 Cr 2. *Obsolete*

RW 3
Saarstahl AG
Now SAARSTAHL 100 CR 6.

RW 4
Saarstahl AG
Now SAARSTAHL 100 CRMN 6.

RW 5
Saarstahl AG
C 1, Si 0.3, Mn 0.7, Cr 1.8, Mo 0.3, bal Fe.
For ball and roller bearings. W.-Nr. 3536. *Obsolete*

RW 50
Saarstahl AG
Now SAARSTAHL 1.3552.

RW 60
Saarstahl AG
Now SAARSTAHL 1.3343.

RW 7
Saarstahl AG
C 0.97, Si 0.5, Mn 1, Cr 1.95, Mo 0.55, bal Fe.
For ball and roller bearings. W.-Nr. 3539; DIN 100 CrMnMo 8. *Obsolete*

RWA
Saarstahl AG
C 0.3, Cr 2.6, V 0.6, W 4.25, bal Fe.
Hot work tool steel; for pressure casting molds, forging and pressing dies. W.-Nr. 2567; AFNOR 30 WCV 40 12. *Obsolete*

RWVM

Acciaierie Valbruna s.p.a.
Tool Material. C 1, Si 0-0.45, Mn 0-0.4, Cr 4.15, Mo 2.65, V 2.35, W 2.85, bal Fe.
High speed steel. W. Nr. 1.3333.

RXM15

Kawasaki Steel Corp.
Stainless steel. C 0.06, Mn 1.2, P 0-0.04, S 0-0.015, Cr 20, Si 3.2, Ni 14, 0.08 rare earth metals, bal Fe.
96,000 psi TS; 64 El. Heat resistant stainless steel for automotive thermal reactor systems.

RY-ALLOY

Joseph T. Ryerson & Son Inc.
C 0.85-1, Mn 1-1.2, V 0.2, W 0.5, Cr 0.5, bal Fe.
Oil-hardened tool steel. AISI Type O1.

RY-ALLOY DRILL

Joseph T. Ryerson & Son Inc.
C 0.95, Mn 1.6, Si 0.25, Mo 0.25, bal Fe.
AISI Type O2 oil hardening tool steel. *Obsolete*

RY-ALLOY FLAT

Joseph T. Ryerson & Son Inc.
C 0.9, Mn 1.2, Si 0.3, Cr 0.5, W 0.5, bal Fe.
AISI type O1 oil hardening tool steel. *Obsolete*

RY-ARM

Joseph T. Ryerson & Son Inc.
C, alloy, bal Fe.
Rolled: 100,000 TS; 65,000 YS; 20 El; 50 RA; 200 Brin. For armature shafts; preheat treated. *Obsolete*

RY-AX

Joseph T. Ryerson & Son Inc.
C 0.41, Mn 1.55, S 0.127, bal Fe.
Heat treated: 100,000-125,000 TS; 60,000-75,000 YS; 16-20 El; 45-50 RA; 225-228 Brin. For axles, shafts, spindles; heat treated alloy steel. *Obsolete*

RY-DIE

Joseph T. Ryerson & Son Inc.
C 1, Mn 0.6, Cr 5.2, Mo 1.1, V 0.25, bal Fe.
Heat treated: 154,000-251,000 TS; 5-13 El; 3-36 RA; 360-500 Brin. For blanking and forming dies, shear blades; Type A2; air hardened, non-deforming. *Obsolete*

RYANITE

Allyne-Ryan Foundry Co.
C 3.2, Si 2.4, Ni 1.5, Cr 4, bal Fe.
Cast: 38,000-40,000 TS; 225-250 Brin. Heat treated: 60,000-70,000 TS; 425-450 Brin. For dies, brake drums, dies for sheet metal fabrication. Cast iron.

RYANITE NO. 2

Allyne-Ryan Foundry Co.
C 3.4, Ni 1.5, Cr 0.5, bal Fe.
For crankcases, cylinder blocks, cylinder barrels. Cast iron.

RYCASE

Joseph T. Ryerson & Son Inc.
C 0.14-0.2, Mn 1-1.3, P 0-0.04, S 0.08-0.13, Si 0-0.1, bal Fe.
Heat treated: 82-119 ksi TS; 64-99 ksi YS; 19-22 El; 38-62 RA; 154-198 Brin. For screw stock, carburized parts; free-cutting carburizing steel. AISI 1117.

RYCO 44

Joseph T. Ryerson & Son Inc.
C 0.4-0.48, Mn 1.35-1.65, bal Fe.
Hot rolled: 80,000 TS; 70,000 YS; 20 El; 50 RA; 225 Brin. Cold drawn: 125,000 TS; 100,000 YS; 10 El; 30 RA; 269 Brin. For spindles, axles, gears, studs, tie rods; tough, shock resistant. *Obsolete*

RYCROME

Joseph T. Ryerson & Son Inc.
C 0.4, Mn 0.9, Cr 0.9, Mo 0.2, Si 0.2, bal Fe.
Heat treated: 105-135 ksi TS; 105 ksi YS; 16-20 El; 54-61 RA; 269-302 Brin. For crankshafts, pinions, bolts, studs, machinery parts requiring moderate strength. AISI 4140-4142.

RYCUT 20

Joseph T. Ryerson & Son Inc.
C 0.2, Pb 0.15-0.35, Cr 0.5, Mo 0.2, Ni 0.6, Mn 0.8, bal Fe.
Rolled: 92,000 TS; 65,000 YS; 24 El; 61 RA; 192 Brin. For gears, spindles, cams, shafts, bushings; free-cutting, case hardened steel. AISI 86L20.

RYCUT 40 ANNEALED

Joseph T. Ryerson & Son Inc.
C 0.4, Cr 0.9, Mo 0.2, Mn 0.9, Pb 0.15-0.35, bal Fe.
Annealed: 91 ksi TS; 63 ksi YS; 27 El; 58 RA; 187 Brin. Free-machining steel; for gears, shafts, cams, studs, wrenches, arbors. AISI 41L40.

RYCUT 47

Joseph T. Ryerson & Son Inc.
C 0.47, Pb 0.25, Mn 0.9, Ni 0.6, Cr 0.5, Mo 0.2, bal Fe.
Annealed: 98,000 TS; 57,000 YS; 22 El; 49 RA; 197 Brin. For arbors, axles, gears, shafts; free-cutting, oil hardened. *Obsolete*

RYCUT 50 ANN

Joseph T. Ryerson & Son Inc.
C 0.5, Mn 0.8, Cr 0.9, Mo 0.2, Pb 0.15-0.35, bal Fe.
Annealed: 103 ksi TS; 69 ksi YS; 23 El; 51 RA; 212 Brin. Free cutting leaded steel; for shafts, collets, brake dies, gears, boring bars. AISI 41L47-41L50.

RYCUT HT

Joseph T. Ryerson & Son Inc.
C 0.47, Mn 0.9, Cr 1, Mo 0.2, Pb 0.15-0.35, bal Fe.
Heat treated: 130 ksi TS; 110 ksi YS; 18 El; 52 RA. For machinery parts, shafts, gears. Free-cutting. AISI 41L42, 41L47, 41L50.

RYCUT RS ANN

Joseph T. Ryerson & Son Inc.
C 0.5, Mn 0.8, Cr 0.9, Mo 0.2, S 0.06-0.1, bal Fe.
Annealed: 103 ksi TS; 69 ksi YS; 23 El; 51 RA; 212 Brin. For shafts, brake dies, gears, machinery parts; free-cutting, resulfurized 4150.

RYCUT RS HT

Joseph T. Ryerson & Son Inc.
C 0.51, Mn 1, Cr 0.8, O 0.2, S 0.06-0.1, bal Fe.
Heat treated: 130 ksi TS; 105 ksi YS; 16 El; 48 RA. For shafts, brake dies, gears, machinery parts; free-cutting, resulfurized 4150.

RYERSON ABRASION RESISTING STEEL

Now AR-360.

RYERSON CARBON

Joseph T. Ryerson & Son Inc.
C 1, Mn, Si, bal Fe.
AISI Type W1 water hardening tool steel. *Obsolete*

RYERSON H.T.M.

Joseph T. Ryerson & Son Inc.
C 0.4, Cr 1.12, Ni 2.3, Mo 0.4, bal Fe.
Heat treated: 140,000-298,000 TS; 105,000-237,000 YS; 12-20 El; 38-56 RA; 321-578 Brin. For gears, worms, shafts, forming rolls; tough, shock resistant. *Obsolete*

RYERSON SHOCK TOOL

Joseph T. Ryerson & Son Inc.
C, alloy, bal Fe.
For punches, shears, concrete busters, track tools; shock resistant. *Obsolete*

RYERSON V.D.

Now V.D. TOOL.

RYNALLOY

Teledyne Ryan Aeronautical
C 0-2.5, Ni 20, Si 0-6, Mn 0-1, 1.8 Cr min, bal Fe.
For ball and socket joints in aircraft exhaust systems; antigalling; for use up to 1800°F.

RYOLITE "B.F.D."

Joseph T. Ryerson & Son Inc.
C, Cr, W, V, bal Fe.
For forming dies, blanking dies, plug gauges, drill bushings; oil hardening. *Obsolete*

RYOLITE CHROMIUM HOT WORK

Joseph T. Ryerson & Son Inc.
C, Cr, V, bal Fe.
For hot heading dies, hot shearing tools. *Obsolete*

RYOLITE TUNGSTEN HOT WORK

Joseph T. Ryerson & Son Inc.
C, Cr, V, bal Fe.
For hot heading dies, hot shearing tools. *Obsolete*

RYOLITE X

Joseph T. Ryerson & Son Inc.
C, V, bal Fe.
For hand tools, punches, bushings, hammers, arbors, drills. *Obsolete*

RYOLITE XX

Joseph T. Ryerson & Son Inc.
C, V, bal Fe.
For punches, dies, hand chisels, shear blades, turning tools, drills. *Obsolete*

RYOLITE XXX

Joseph T. Ryerson & Son Inc.
C, V, bal Fe.
For dies, tools, punches, drills, reamers, taps. *Obsolete*

RYOLITE, DIAMOND B

Joseph T. Ryerson & Son Inc.
C, bal Fe.
For hammers, chisels, crowbars, wedges, sledges, drills, tools, wrenches; tough and strong. *Obsolete*

RYOLITE-4-POINT

Joseph T. Ryerson & Son Inc.
C, Si, Mn, bal Fe.
For hand tools, chisels, caulking tools, pneumatic tools. *Obsolete*

RYTENSE

Joseph T. Ryerson & Son Inc.
C 0.4, N 1.55, P 0-0.04, S 0.08-0.13, Si 0-0.1, bal Fe.
Rolled: 99 ksi TS; 61 ksi YS; 25 El; 51 RA. For gears, cams, arbors, rollers, bushings, machinery parts; free-cutting. AISI 1141.

RYTENSE 44

Joseph T. Ryerson & Son Inc.
C 0.47, Mn 1.45, P 0-0.04, S 0.24-0.33, bal Fe.
Rolled: 102 ksi TS; 57 ksi YS; 24 El; 48 RA. For machinery parts, shafts, gears. Free-cutting. AISI 1144.

RYTENSE AA

Now RYTENSE.

RZF

Saarstahl AG
C 0.67, Si 1.3, Mn 0.5, Cr 0.5, bal Fe.
For valve springs, coil springs, torsion bar springs. W.-Nr. 7103. *Obsolete*

S & J SPECIAL
Colt Industries
C 1.25, bal Fe.
For drill rod, tools, taps, reamers; water hardening. *Obsolete*

S 1117
VILLARES
C 0.17, Mn 1.15, P 0-0.04, S 0.08-0.13, bal Fe.
Free cutting steel. Similar to SAE 1117.

S 1118
VILLARES
C 0.17, Mn 1.45, P 0-0.04, S 0.08-0.13, bal Fe.
Free cutting steel. Similar to SAE 1118.

S 1120
VILLARES
C 0.21, Mn 0.85, P 0-0.04, S 0.08-0.13, bal Fe.
Free cutting steel. *Obsolete*

S 1126
VILLARES
C 0.26, Mn 0.85, P 0-0.04, S 0.08-0.13, bal Fe.
Free cutting steel. *Obsolete*

S 1137
VILLARES
C 0.36, Mn 1.5, P 0-0.04, S 0.08-0.13, bal Fe.
Free cutting steel. Similar to SAE 1137.

S 1141
VILLARES
C 0.41, Mn 1.5, P 0-0.04, S 0.08-0.13, bal Fe.
Free cutting steel. Similar to SAE 1141.

S 1144
VILLARES
C 0.44, Mn 1.5, P 0-0.04, S 0.24-0.33, bal Fe.
Free cutting steel. Similar to SAE 1144.

S 1146
VILLARES
C 0.46, Mn 0.85, P 0-0.04, S 0.08-0.13, bal Fe.
Free cutting steel. Similar to SAE 1146.

S 18/8
Sintered Products Ltd.
Cr 18, Ni 8, bal Fe.
Sintered. 6.7-7.1 g/cm^3 density; 44 kg/mm^2 TS; 10 El min; 30 Rock B. BSS A520; AISI 316L.

S 18/8 3-020
Sintered Products Ltd.
Austenitic 18-8 stainless steel, sintered. 6.7-7.1 g/cm^3 density; 10 El min; 36.2-44 kg/mm^2 TS; 130 Vickers.

S 333
Bergische Stahl Industrie
C 0.95-1.03, Cr 3.8-4.5, W 2.7-3, V 2.2-2.5, Mo 2.5-2.8, bal Fe.
High speed steel. W.-Nr. 1.3333.

S 410
Sintered Products Ltd.
C 0.1, Cr 12, bal Fe.
Sintered. 6.7-7.1 g/cm^3 density; 50 kg/mm^2 TS; 7 El min; 50 Rock B. BSS A530; AISI 410.

S METAL
Koppers Co. Inc.
High C, alloy, bal Fe.
For piston rings. *Obsolete*

S NO. 2
McInnes Steel Co.
C, Cr, W, bal Fe.
For milling cutters, dies, taps, master tools; nondeforming. *Obsolete*

S NO. 4
McInnes Steel Co.
C, Cr, W, Mo, bal Fe.
For rivet sets, shear blades, pneumatic tools. *Obsolete*

S-1112
VILLARES
C 0-0.13, Mn 0.85, P 0.07-0.12, S 0.16-0.23, bal Fe.
Free cutting steel. Similar to SAE 1212.

S-495 ALLOY
Allegheny Ludlum Steel
C 0.43, Ni 19-24, Cr 14-18, W 4, Mo 3.5, Cb 3.6, bal Fe.
Rolled: 135,000 TS; 82,000 YS; 13 El; 15 RA. For engine parts; heat resistant. *Obsolete*

S-497 ALLOY
Allegheny Ludlum Steel
C 0.4-0.5, Co 19-21, Cr 15, Ni 20, Mo 4, W 4, Cb 4, bal Fe.
For gas turbine parts; heat resistant. *Obsolete*

S-588 ALLOY
Allegheny Ludlum Steel
C 0.45, Mn 0.5, Cr 20, Ni 20, Mo 4, W 4, Cb 4, bal Fe.
For jet engines, turbines; heat resistant to 1500 F. *Obsolete*

S-590
Now UNITEMP S-590.

S-590 ALLOY
Allegheny Ludlum Steel
C 0.43, Mn 1.6, Cr 20, Ni 20, Co 20, Mo 4.5, W 4, Cb 4, bal Fe.
For jet engines, turbines, wheels and buckets; heat resistant to 1500 F. *Obsolete*

S-816
Now UDIMET S-816, UNITEMP S-816.

S-816 ALLOY
Allegheny Ludlum Steel
C 0.3-0.45, Ni 18-22, Cr 18-22, Mo 3.5-5, W 3.5-8, Cb 2.5-4, Fe 4, 38% Co min.
For jet engine parts, buckets for gas turbines; heat resistant, forging alloy. *Obsolete*

S-BABBITT
American Smelting & Refining Co.
Sb 15, As 1, Sn 1, Cu 0.5, bal Pb.
At 70°F: 10,350 psi TS; 2 El; 20 Brin. At 392°F: 126 95 El. For bearings, liners; Babbitt.

S-G 13
Symington-Gould Corp.
C 0.25, Mn 0.5, bal Fe.
Normalized: 60,000 TS; 30,000 YS; 24 El; 35 RA. For general usage; case hardened.

S-G 17
Symington-Gould Corp.
C 0.28, Mn 0.6, bal Fe.
Normalized: 65,000 TS; 35,000 YS; 20 El; 30 RA. For general usage; case hardened.

S-G 30
Sonken-Galamba Corp.
Cu 6-8, Si 1.5-2.5, Zn 1-1.5, Ti 0.2, bal Al.
Cast. For general light alloy castings; modified Alcoa 212 alloy.

S-G 80-10-10
Sonken-Galamba Corp.
Sn 9.25-10.5, Pb 9-11, Zn 0-0.75, bal Cu
Cast: 25,000-35,000 TS; 15,000-22,000 YS; 6-12 El; 50-60 Brin. For bearings, acid resistant castings; corrosion resistant.

S-G 81-3-6-10
Sonken-Galamba Corp.
Sn 2.5-3.5, Pb 5-7, Zn 9-11, bal Cu.
Cast: 28,000-36,000 TS; 13,000-16,000 YS; 15-28 El; 50-60 Brin. For plumbing goods, pressure valves, fittings; free-cutting.

S-G 83-4-6-7
Sonken-Galamba Corp.
Sn 3.5-4.2, Pb 5.7-6.7, Zn 6-8, bal Cu.
Cast: 30,000-35,000 TS; 12,000-16,000 YS; 15-25 El; 50-60 Brin. For plumbing goods, valves, gas and water fittings; red brass, free-cutting.

S-G 83-7-7-3
Sonken-Galamba Corp.
Sn 6.5-7.5, Pb 6.5-8, Zn 2.5-4, bal Cu.
Cast: 28,000-35,000 TS; 15,000-20,000 YS; 10-15 El; 55-65 Brin. For bearings, bushings; free-cutting.

S-G 85-5-5-5
Sonken-Galamba Corp.
Sn 4-6, Pb 4-6, Zn 4-6, bal Cu.
Cast: 30,000-40,000 TS; 15,000-25,000 YS; 15-33 El; 55-65 Brin. For gears, valve bodies, plumbing; free-cutting bronze.

S-G 9
Symington-Gould Corp.
C 0.26, Mn 0.75, bal Fe.
Normalized: 70,000 TS; 36,000 YS; 22 El; 30 RA. For general usage; case hardened.

S-G 90
Sonken-Galamba Corp.
Cu 3.7-4.7, Si 2.7-3.7, Zn 0.5, Fe 1, Ni 0.4, bal Al.
Cast. For general light alloy castings; modified Alcoa 108 alloy.

S-G TYPE 108
Sonken-Galamba Corp.
Cu 2.5-4.7, Si 2.5-4.5, Mn 0.4, Fe 1.1, Mg 0.3-1, Ti 0.2, Ni, bal Al.
Sand cast: 21,000 TS; 14,000 YS; 2.5 El; 60 Brin. For utility sand castings; good castability and machinability.

S-G TYPE 12
Sonken-Galamba Corp.
Cu 6-8.5, Si 1-3, Mn 0.4, Fe 0.8-1.5, Mg 0.05, Zn 0.2-2.2, Ni 0.4, Ti 0.2, bal Al.
Sand cast: 19,000 TS. For low strength light weight castings.

S-G TYPE 122
Sonken-Galamba Corp.
Cu 9.2-10.8, Si 1-1.7, Mn 0.2, Fe 1.5, Mg 0.2, Zn 0.5, Ni 0.3, Ti 0.2, bal Al.
Heat treated: 41,000-48,000 TS; 36,000-40,000 YS; 0.5 El; 115-140 Brin. For pistons, valve guides, bearings; age hardenable, wear resistant.

S-G TYPE 13
Sonken-Galamba Corp.
Cu 0.6, Si 9.5-13, Fe 2.2, Mg 0.1, Mn 0.3, Ni 0.5, bal Al.
Die cast: 37,000 TS; 18,000 YS; 1.8 El. For meter cases, switch boxes, low stressed castings; thin wall castings.

S-G TYPE 132
Sonken-Galamba Corp.
Cu 0.5-2, Si 11-13, Mn 0.6, Fe 1.2, Mg 1.2, Zn 0.2, Ni 0.5-3, Ti 0.2, bal Al.
Heat treated: 36,000-47,000 TS; 28,000-43,000 YS; 0.5 El; 105-125 Brin. For automotive engine pistons and sleeves; low expansion alloy.

S-G TYPE 138
Sonken-Galamba Corp.
Cu 9-11, Si 3.5-4.5, Fe 1.2, Mn 0.3, Mg 0.2, Zn 0.5, bal Al.
Permanent mold: 30,000 TS; 24,000 YS; 15 El; 90 Brin. For pistons, cylinder heads; similar to Alcoa 138.

S-G TYPE 142
Sonken-Galamba Corp.
Cu 3.5-4.5, Si 0.7, Mn 0.2, Fe 0.9, Mg 1.6, Zn 0.5, Ni 2, Ti 0.2, bal Al.
Heat treated: 27,000-47,000 TS; 18,000-42,000 YS; 0.5-1.0 El; 70-110 Brin. For pistons, cylinder heads; age-hardenable, strength at elevated temperatures.

S-G TYPE 195
Sonken-Galamba Corp.
Cu 3-5, Si 1.2-2.5, Fe 1, Mn 0.5, Zn 0.4, Ti 0.2, bal Al.
Heat treated: 32,000-40,000 TS; 16,000-34,000 YS; 1.5-8.5 El; 60-95 Brin. For crank cases, aircraft components; age hardenable.

S-G TYPE 212
Sonken-Galamba Corp.
Cu 6-8.5, Si 1-3, Mn 0.4, Fe 0.8-1.5, Zn 0.2-2.2, bal Al.
Sand cast: 19,000-24,000 TS; 11,000-14,000 YS; 1.0-2.5 El; 50-65 Brin For general utility castings; similar to Alcoa 212.

S-G TYPE 214
Sonken-Galamba Corp.
Cu 0.1, Si 0.4, Fe 0.5, Mg 3.25-4.5, Zn 0.25, Ti 0.2, bal Al.
Sand cast: 25,000 TS; 12,000 YS; 9 El; 50 Brin. For light weight castings; similar to Alcoa 214.

S-G TYPE 218
Sonken-Galamba Corp.
Cu 0.2, Si 0.3, Mn 0.3, Fe 1.8, Mg 7.5-8.5, Zn 0.1, bal Al.
Die cast: 42,000 TS; 23,000 YS; 7.0 El; 75 Brin. For airplane wheel flanges and brake shoes; similar to Alcoa 218.

S-G TYPE 220
Sonken-Galamba Corp.
Cu 0.2, Si 0.2, Mg 9.5-10.6, Zn 0.1, bal Al.
Heat treated: 46,000 TS; 25,000 YS; 14 El; 75 Brin. For railway and bus equipment; similar to Alcoa 220; age hardenable.

S-G TYPE 355
Sonken-Galamba Corp.
Cu 1-1.8, Si 4.5-5.5, Mg 0.5, Ti 0.2, Mn 0.2, bal Al.
Heat treated: 38,000-43,000 TS; 25,000-36,000 YS; 0.5-4 El; 80-90 Brin. For aircraft engine castings, cylinder heads; similar to Alcoa 355; age hardenable.

S-G TYPE 356
Sonken-Galamba Corp.
Cu 0.3, Si 0.5-7.5, Mn 0.4, Fe 0.5, Mg 0.3, bal Al.
Heat treated: 32,000-40,000 TS; 24,000-30,000 YS; 2-5 El; 70-90 Brin. For leakproof castings; similar to Alcoa 356; age hardenable.

S-G TYPE 43
Sonken-Galamba Corp.
Cu 0.1-0.6, Si 4.5-6, Mn 0.2, Fe 0.8-2, Zn 0.3, Ni 0.3, Ti 0.2, bal Al.
Die cast: 30,000 TS; 14,000 YS; 5.0 El; 40 Brin. For marine castings, manifolds, cooking utensils; corrosion resistant.

S-G TYPE 81
Sonken-Galamba Corp.
Cu 6-8.5, Si 1-3.5, Mn 0.4, Fe 2-3, Zn 1.8-2, Ti 0.2, Ni 0.4, bal Al.
Die cast: 33,000 TS; 25,000 YS; 1 El; 80 Brin. For instrument housings, cases; high fluidity.

S-G TYPE 85
Sonken-Galamba Corp.
Cu 3-4.5, Si 4.5-9.5, Mn 0.4, Fe 1.3-2.3, Zn 0.6-1, Ni 0.4, Cr 0.2, bal Al.
Die cast: 38,000 TS; 22,000 YS; 2.5 El; 85 Brin. For general die castings; high strength.

S-G TYPE A108
Sonken-Galamba Corp.
Cu 4-5, Si 5-6, Fe 1, Mn 0.3, Ti 0.2, Zn 0.5, bal Al.
Permanent mold: 28,000 TS; 16,000 YS; 2 El; 65 Brin. For utility castings; intricate shapes.

S-G TYPE AXS-679
Sonken-Galamba Corp.
Cu 3-4, Si 4.5-9.5, Fe 1.3, Zn 0.6, Ni 0.5, Cr 0.2, bal Al.
Die cast. For general die castings; corrosion resistant.

S-G TYPE B195
Sonken-Galamba Corp.
Cu 4.5, Si 2-3, Mn 0.05-4, Fe 1.2, Zn 0.5, Ni 0.2, Ti 0.2, bal Al.
Heat treated: 37,000-42,000 TS; 19,000-28,000 YS; 6-9 El; 75-90 Brin. For aircraft castings; age hardenable.

S-G TYPE C113
Sonken-Galamba Corp.
Cu 6-8.5, Si 1-6, Mn 0.4, Fe 1.5, Zn 1-2.5, Ti 0.2, Ni 0.2, bal Al.
Permanent mold: 30,000 TS; 24,000 YS; 1 El; 70 Brin. For utility castings; similar to Alcoa C113.

S-G TYPE SG 300
Sonken-Galamba Corp.
Al alloy.
Cast: 32,000 TS min; 5 El min. For light alloy castings.

S-G VALVE BRONZE
Sonken-Galamba Corp.
Sn 5.5-6.5, Pb 1-2, Zn 3-5, bal Cu.
Cast: 34,000 TS max; 22 El. For valve bodies; composition M alloy, free-cutting.

S-G YELLOW BRASS
Sonken-Galamba Corp.
Sn 0.5-1.2, Pb 3-5, Fe 0-0.5, Cu 78-81, bal Zn.
Cast: 28,000-35,000 TS; 10,000-15,000 YS; 25-35 El; 25-35 Brin. For plumbing fixtures, ferrules, battery terminals; free-cutting brass.

S-LESS STEEL (BREARLY)
English manufacture
Cr 13, C 0.3, bal Fe.
For corrosion resisting parts.

S-M
Carpenter Technology Corp.
C 0.6, Mn 0.8, Si 1.9, bal Fe.
For punches, chisels; shock resistant. *Obsolete*

S-M ALLOY
S-M Metal Works
Aluminum. Si 5, bal Al.
Al welding rod for welding and soldering of Al; corrosion resistant.

S-TREATED
Heppenstall Co.
Tool material. C 0.6, Cr 0.3, bal Fe.
Prehardened: 201-235 Brin. For die blocks; hot work, hammer press, short runs.

S-TREATED 14A60
Heppenstall Co.
Tool material. C 0.6, Mn 0.8, Si 0.25, Cr 0.25, bal Fe.
Water hardened tool steel; AISI W4.

S. A. BRILLIANT
Swedish American Steel Corp.
C 0.7, W 18, Cr 4, V 1, bal Fe.
For drills, lathe and planer tools; high speed steel.

S. A. DURABIL
Swedish American Steel Corp.
C 0.45, Cr 1.25, W 2.25, V 0.2, bal Fe.
For hot work dies; impact resistant; hot work steel.

S. A. M. B.
Manufacturer not listed.
Al 0.2-1, Cu 56-59, Fe 0.8-1.2, Mn 0.2-2, Sn 0.8-1.5, bal Zn.
75,000 TS; 35,000 YS; 20 El; 140 Brin. For extrusions; corrosion resistant.

S. SIRIUS
French manufacture
C 0.25, Ni 16, Cr 17, W 3, Co 12, Ti 2, bal Fe.
For valves; heat and corrosion resistant.

S.A.P. 865
Aluminium Industrie Aktiengesellschaft
13-14 Al_2O_2, bal Al.
Sintered: 49,000 TS; 35,000 YS; 8 El; 105 Brin. At 930°F: 16,000 TS; 13,500 YS; 2 El. For compressor blades, engine components, heat exchangers, pistons, cylinder heads. High creep resistant. Sintered powder. Dispersion hardened.

S.A.P. 865
Swiss Aluminium Ltd.
13-14 Al_2O_2, bal Al.
Sintered: 49,000 TS; 35,000 YS; 8 El; 105 Brin. At 930°F: 16,000 TS; 13,500 YS; 2 El. For compressor blades, engine components, heat exchangers, pistons, cylinder heads. High creep resistant. Sintered powder. Dispersion hardened.

S.A.P. 865
Nurnberg Aluminiumwerke GmbH
13-14 Al_2O_2, bal Al.
Sintered: 49,000 TS; 35,000 YS; 8 El; 105 Brin. At 930°F: 16,000 TS; 13,500 YS; 2 El. For compressor blades, engine components, heat exchangers, pistons, cylinder heads. High creep resistant. Sintered powder. Dispersion hardened.

S.A.P. 895
Aluminium Industrie Aktiengesellschaft
10-11 Al_2O_2, bal Al.
At 70°F: 43,000 TS; 29,000 YS; 16 El; 100 Brin. For compressor blades, engine components, heat exchangers, pistons, cylinder heads. Sintered powder. Dispersion hardened.

S.A.P. 895
Nurnberg Aluminiumwerke GmbH
10-11 Al_2O_2, bal Al.
At 70°F: 43,000 TS; 29,000 YS; 16 El; 100 Brin. For compressor blades, engine components, heat exchangers, pistons, cylinder heads. Sintered powder. Dispersion hardened.

S.A.P. 895
Swiss Aluminium Ltd.
10-11 Al_2O_2, bal Al.
At 70°F: 43,000 TS; 29,000 YS; 16 El; 100 Brin. For compressor blades, engine components, heat exchangers, pistons, cylinder heads. Sintered powder. Dispersion hardened.

S.A.P. 930
Aluminium Industrie Aktiengesellschaft
7 Al_2O_2, bal Al.
At 70°F: 36,000 TS; 23,000 YS; 20 El; 90 Brin. At 750°F: 12,100 TS; 11,200 YS; 5.7 El. For compressor blades, engine components, heat exchangers, cylinder heads, pistons. Strong high temperature material. Sintered powder. Dispersion hardened.

S.A.P. 930
Nurnberg Aluminiumwerke GmbH
7 Al_2O_2, bal Al.
At 70°F: 36,000 TS; 23,000 YS; 20 El; 90 Brin. At 750°F: 12,100 TS; 11,200 YS; 5.7 El. For compressor blades, engine components, heat exchangers, cylinder heads, pistons. Strong high temperature material. Sintered powder. Dispersion hardened.

S.A.P. 930
Swiss Aluminium Ltd.
7 Al_2O_2, bal Al.
At 70°F: 36,000 TS; 23,000 YS; 20 El; 90 Brin. At 750°F: 12,100 TS; 11,200 YS; 5.7 El. For compressor blades, engine components, heat exchangers, cylinder heads, pistons. Strong high temperature material. Sintered powder. Dispersion hardened.

S.A.T.T. SPECIAL
Swedish American Steel Corp.
C 1.2, Cr 0.75, W 1.5, Mo 0.2, bal Fe.
For tools, cutters; oil hardening.

S.B. STEEL
Colt Industries
C, Cr, bal Fe.
For tools. *Obsolete*

S.C.C.-210
Sigmund Cohn Corp.
Al 0.1, C 0.34, Co 0.62, Fe 0.09, Mn 0.06, Mg 0.21, bal Ni.
For filaments for battery tubes. *Obsolete*

S.C.C.-213
Sigmund Cohn Corp.
Al 1.9, U 2.1, W 1, bal Ni.
For filaments and grids. *Obsolete*

S.C.C.-213M
Sigmund Cohn Corp.
Al 1, C 0.25, Mo 2, bal Ni.
For grid wire. *Obsolete*

S.C.C.-531
Sigmund Cohn Corp.
Co 19, Fe 2.5, Si 1.5, bal Ni.
For rectifier filaments. *Obsolete*

S.C.C.-925C
Sigmund Cohn Corp.
Al 1.9, Trace C, bal Ni.
For filaments for battery tubes. *Obsolete*

S.C.C.-C43
Sigmund Cohn Corp.
Mn 0.1, Si 3.5, bal Ni.
For rectifier filaments. *Obsolete*

S.C.C.-W4
Sigmund Cohn Corp.
C 0.04, Co 0.9, Fe 0.03, Cu 0.02, bal Ni.
For filaments and grids; heat resistant. *Obsolete*

S.C.S.S.
British Steel Corp.
C 0.7, Mn 0.9, bal Fe.
Heat treated: 180,000-190,000 TS; 8-12 El; 20-30 RA; 402-430 Brin. For springs; tough. *Obsolete*

S.D.S.
British Steel Corp.
C 0.2, bal Fe.
Annealed: 45,000-56,000 TS; 27,000-33,000 YS; 30-36 El; 61-69 RA. For cold riveting, to replace wrought iron; dead soft welding quality. *Obsolete*

S.D.T.-T.C.O.
Societe Nouvelle du Saut-du-Tarn
C 0.7, W 18, Co 5, Cr 4, V 1.3, bal Fe.
For tools, dies, cutters; high speed steel. *Obsolete*

S.G. IRON
Fromson Co. Inc.
C 3.3, Si 2.5, Ni 1.5, Mg 0.05, bal Fe.
Cast: 160-300 Brin. For high strength and tough castings; nodular cast iron.

S.G. IRON
Sheepbridge Engineering Ltd.
C 3.3, Si 2.5, Ni 1.5, Mg 0.05, bal Fe.
Cast: 160-300 Brin. For high strength and tough castings; nodular cast iron.

S.G.N1-RESIST D2C (MODIFIED)
International Nickel Inc.
Cast iron. C 2.2-2.6, Si 1.9-2.6, Mn 3.7-4.4, Ni 21-24, Cr 0-0.2, P 0-0.1, bal Fe.
For pumps, valves, compressors, turbo-expanders. Austenitic iron for subzero service. High notch toughness. *Obsolete*

S.G.N1-RESIST D2M
International Nickel Inc.
Cast iron. C 2.2-2.6, Si 1.9-2.6, Mn 3.7-4.4, Ni 21-24, Mg 0.06-0.12, Cr 0.2, P 0-0.1, bal Ni.
For pumps, valves, compressors, turbo-expanders, low temperature equipment. Cryogenic toughness and ductility. *Obsolete*

S.I.F. NO. 23
Suffolk Iron Foundry Ltd.
Al 6, Zn 1, bal Mg.
Welded: 36,000-40,000 TS; 16,000-20,000 YS; 10-12 El; 65-70 Brin. For Mg welding rod. *Obsolete*

S.I.F. NO. 24
Suffolk Iron Foundry Ltd.
Mn 1.5, bal Mg.
Welded: 20,000-28,000 TS; 14,000-18,000 YS; 2-3 El; 40-50 Brin. For Mg welding rod. *Obsolete*

S.I.P. NO. 25
Suffolk Iron Foundry Ltd.
Ni 67, Cu 28.
Welded: 66,000 TS. For Monel welding rod; corrosion resistant. *Obsolete*

S.K. SILVER STEEL
Now SILVER STEEL.

S.K.F. 100
S.K.F. Industries Inc.
C 0.7-1, bal Fe.
For stamping, cold heading and trimming dies, general tools; water hardened.

S.K.F. DRILL
S.K.F. Industries Inc.
C, bal Fe.
For drills.

S.K.F. NO. 100
S.K.F. Industries Inc.
C, Cr, W, V, bal Fe.
For hot punching and piercing dies. *Obsolete*

S.K.F. NO. 114
S.K.F. Industries Inc.
C, V, bal Fe.
For tools. *Obsolete*

S.K.F. NO. 13
S.K.F. Industries Inc.
C 1.1, Cr 1, bal Fe.
For balls, rollers, spindles; water hardened.

S.K.F. NO. 3 TUBING
S.K.F. Industries Inc.
C 1, Cr 1, bal Fe.
For bearing races, bushings; water hardened.

S.K.F. NO. 48 STEEL
S.K.F. Industries Inc.
high C, high Cr, Mo, bal Fe.
For cold trimming, drawing, punching and blanking dies, gauges; air hardening steel.

S.K.F. NO. 50
S.K.F. Industries Inc.
C, Cr, W, bal Fe.
For blanking and moulding dies, gages, cutters, broaches; nondeforming. *Obsolete*

S.K.F. NO. 685
S.K.F. Industries Inc.
C, alloy, bal Fe.
For tools. *Obsolete*

S.K.F. NO. 7
S.K.F. Industries Inc.
C 1.1, Cr 0.6, bal Fe.
For balls, rollers, spindles; water hardened.

S.K.F. NO. 711
S.K.F. Industries Inc.
C 0.5, W 2.5, Cr 1.2, Si 0.7, V 2, bal Fe.
For tools, shear blades, dies, cutters, chisels; good wear resistance.

S.K.F. NO. 9
S.K.F. Industries Inc.
C 1.1, Cr 0.9, bal Fe.
For balls, rollers, spindles; water hardened.

S.K.F. NO. 99
S.K.F. Industries Inc.
C, Cr, bal Fe.
For hot heading tools and dies. *Obsolete*

S.K.F. SPECIAL
S.K.F. Industries Inc.
C, bal Fe.
For tools. *Obsolete*

S.K.F. STEEL NO. 1
S.K.F. Industries Inc.
C 1, Mn 1.05, Si 0.45, Cr 1.05, bal Fe.
For ball bearings, races and balls; wear resistant.

S.K.F. STEEL NO. 22
S.K.F. Industries Inc.
C 1, Mo 0.35, Cr 1.1, bal Fe.
For trimming and stamping dies, drills; wear resistant.

S.K.F. STEEL NO. 3
S.K.F. Industries Inc.
C 1, Cr 1.5, bal Fe.
For ball bearings, races and balls; wear resistant.

S.K.F. STEEL NO. 36
S.K.F. Industries Inc.
C 2.25, Cr 11.5-12.5, V 0.15, Mn 1.2, Si 0.2, bal Fe.
For cold trimming, drawing, punching, blanking dies, gages; formerly known as "VH-63 Steel".

S.L.B.
Logan Iron & Steel Co.
C 0.45, bal Fe.
For castings.

S.L.V. STEEL
Firth Brown Ltd.
C 0.45, Cr 8, Si 3.5, bal Fe.
Oil treated: 145,000-156,000 TS; 100,000-112,000 YS; 20-15 El; 43-53 RA; 255-285 Brin. For valves for automobile and aircraft engines; heat and corrosion resistant; Firth-Brown J-181

S.L.V. STEEL
Firth-Vickers Stainless Steels Ltd.
C 0.45, Cr 8, Si 3.5, bal Fe.
Oil treated: 145,600-156,800 TS; 100,000-112,000 YS; 20-15 El; 43-53 RA; 255-285 Brin. For valves for automobile and aircraft engines; heat and corrosion resistant; Firth-Brown J-181.

S.M. 40, ETC.
Now JESSOP STEELS.

S.M. STEEL

Carpenter Technology Corp.
Si 2, Mn 0.8, C 0.5-0.6, bal Fe.
For machinery parts, gears, shafts. Water hardening.
Obsolete

S.M.L. ALLOY (MONEL)

English manufacture
Ni 68, Cu 28, Fe 2.5, Mn 2.5.
For parts requiring high corrosion resistance and strength at elevated temperatures.

S.M.S. ALLOY 1

SMS Corp.
Cu-W.
Bar: 135,000 TS; 130 Brin. For resistance welding electrode; 35% conductivity, for stainless steel and brass.

S.M.S. ALLOY 10

SMS Corp.
Cu-W.
Bar: 160,000 TS; 205 Brin. For resistance welding electrode; 28% conductivity, for inserts and facings.

S.M.S. ALLOY 100W

SMS Corp.
Cu-W.
Bar: 200,000 TS; 76 Rock A. For resistance welding electrode; 30% conductivity, for red brass.

S.M.S. ALLOY 101

SMS Corp.
Cu alloy.
Bar: 50,000-60,000 TS; 15,000-20,000 YS; 25-20 El; 100-116 Brin. For resistance welding electrode; 80% conductivity, for coated metals.

S.M.S. ALLOY 103

SMS Corp.
Cu alloy.
Bar: 55,000-65,000 TS; 25,000-35,000 YS; 15 El; 116-137 Brin. For resistance welding electrode; 75% conductivity, for carbon steel and brass.

S.M.S. ALLOY 116

SMS Corp.
Cu alloy.
Cast: 65,000-75,000 TS; 12,000-16,000 YS; 2-10 El; 116-165 Brin. For resistance welding electrode.

S.M.S. ALLOY 20

SMS Corp.
Cu-W.
Bar: 170,000 TS; 228 Brin. For resistance welding electrode; 27% conductivity, for heavy projection welds.

S.M.S. ALLOY W-2

SMS Corp.
Cu alloy.
Cast: 90,000 TS; 68,000 YS; 1 El; 330 Brin. Rolled: 140,000 TS; 85,000 YS; 2 El; 330 Brin. For resistance welding electrode.

S.M.S. ALLOY W5

SMS Corp.
Be 0.4, Co 2.6, bal Cu.
Cast: 90,000 TS; 55,000 YS; 10-15 El. Rolled: 125,000 TS; 65,000 YS; 8-15 El. For springs, bellows, diaphragms; age-hardenable, corrosion resistant.

S.M.Z. ALLOY

Union Carbide Corp.
Si 60-65, Mn 5-7, Zr 5-7, bal Fe.
For graphitizing alloy for cast iron; ladle addition.

S.N.C.

Latrobe Steel Co.
C 0.56, Si 2, Mn 0.9, Cr 0.25, V 0.3, bal Fe.
For tools, punches, dies. *Obsolete*

S.N.C.G.

British Steel Corp.
C 0.4, Ni 3, bal Fe.
Heat treated: 232,000 TS; 14 El; 39 RA; 512 Brin. For gears; shock resistant. *Obsolete*

S.O.B.V. ALLOY

Osborn Steels Ltd.
Cr 4.5, Co 10, W 19, V 1.2, C 0.75, Mo 0.75, Si 0.25, Mn 0.3, bal Fe.
Hardened: 350,000 TS. For cutting tools for machining hard metals; super high speed steel. *Obsolete*

S.O.D. DIE STEEL

CCS Braeburn Alloy Steel
C 0.9, Mn 1.65, bal Fe.
Heat treated: 570-650 Brin. For hobs, cutters, blanking and forming dies; oil hardened, non-deforming. *Obsolete*

S.S. TOOL

Colt Industries
C 1, Cr 1.2, Mo 0.3, bal Fe.
For rolls, tools, blanking dies; oil hardening. *Obsolete*

S/100

Eagle & Globe Steel Ltd.
Steel. C 0.9-1, S 0-0.025, P 0-0.025, Mn 0.31-0.51, bal Fe.
Cold rolled strip steel: for high duty springs, spring clips, hobby knives. Hardened and tempered strip steel: for high duty springs, doctor blades, handsaws, trowels. UK BS1449 CS.100; SAE 1095; DIN 1.7222 MK.101.

S/100C

Eagle & Globe Steel Ltd.
Steel. C 0.9-1, S 0-0.025, P 0-0.025, Mn 0.31-0.51, Cr 0.4-0.6, bal Fe.
Hardened and tempered strip steel. For tobacco knives, textile knives.

S/125C

Eagle & Globe Steel Ltd.
Steel. C 1.15-1.25, S 0-0.025, P 0-0.025, Mn 0.4-0.6, Cr 0.5-0.7, Mo 1.2-1.5, Si 1.2-1.5, bal Fe.
Cold rolled strip steel. For hacksaw blades.

S/39

Eagle & Globe Steel Ltd.
Steel. C 0.34-0.39, S 0-0.025, P 0-0.025, Mn 0.7-0.9, bal Fe.
Cold rolled strip steel. For seat belt fittings, transmission chain. UK BS1449 CS.40; SAE 1037; DIN 1.7222 C.35.

S/70

Eagle & Globe Steel Ltd.
Steel. C 0.65-0.7, S 0-0.025, P 0-0.025, Mn 0.65-0.85, bal Fe.
Cold rolled strip steel: for formed springs, steel toe caps, high duty transmission chain sideplates. Hardened and tempered strip steel: for high duty springs. UK BS1449 CS.70; SAE 1070; DIN 1.7222 CK.67.

S/83

Eagle & Globe Steel Ltd.
Crinoline steel. C 0.75-0.83, S 0-0.025, P 0-0.025, Mn 0.65-0.85, bal Fe.
Hardened and tempered strip steel. For springs, trowels, handsaws, narrow bandsaws. UK BS1449 CS.80; SAE 1080; DIN 1.7222 MK.75.

S/NCM

Eagle & Globe Steel Ltd.
Steel. C 0.75-0.82, S 0-0.025, P 0-0.025, Mn 0.25-0.45, Ni 0.55-0.75, Cr 0.2-0.3, Mo 0.1-0.2, bal Fe.
Hardened and tempered strip steel. For high duty handsaws, circular saws, bandknives.

S/S FR. VIII

Paul Bergsoe & Son
Sn 80, Sb 11.5, Cu 5.5, Pb 3.
Cast: 17,800 TS; 35 Brin. MP: 355-675°F. For steam turbine and generator bearings. Wear resistant.

S1010

Eagle & Globe Steel Ltd.
Carbon steel. C 0.08-0.13, Mn 0.3-0.6, P 0-0.05, S 0-0.05, bal Fe.
370 MPa TS; 13 El. General purpose low carbon steel for small case hardened parts requiring a hard work surface; readily welded and cold worked. BSS970 EN Series EN32A; BSS970 Part 1 1972 045M10.

S1040

Eagle & Globe Steel Ltd.
Carbon steel. C 0.37-0.44, Mn 0.6-0.9, Si 0-0.35, P 0-0.05, S 0-0.05, bal Fe.
For agricultural implement manufacture. AS1442 S1040; BS970 080M40 (EN8); AISI 1040.

S1045

Eagle & Globe Steel Ltd.
Carbon steel. C 0.43-0.5, Mn 0.6-0.9, Si 0-0.35, P 0-0.05, S 0-0.05, bal Fe.
For agricultural implement manufacture. AS1442 S1045; BS970 080A47 (EN43B); AISI 1045.

S1214

Eagle & Globe Steel Ltd.
Alloy steel. C 0-0.15, Mn 0.8-1.2, P 0.04-0.09, S 0.25-0.35, bal Fe.
450 MPa TS; 9 El. Machinability and free cutting properties. or automatic and semiautomatic machines for repetition parts. Rephosphorized, resulfurized, leaded steel can be case hardened.

S12L14

Eagle & Globe Steel Ltd.
Alloy steel. C 0-0.15, Mn 0.8-1.2, P 0.04-0.09, S 0.25-0.35, Pb 0.15-0.35, bal Fe.
450 MPa TS; 9 El. Machinability and free cutting properties. For automatic and semiautomatic machines for repetition parts. Rephosphorized, resulfurized, leaded steel can be case hardened.

S2-5-1

Westig (U.K.) Ltd.
C 0.95, Si 0-0.45, Mn 0-0.4, P 0-0.03, S 0-0.03, Cr 3.7, Mo 5, V 1.3, W 1.8, bal Fe.
High speed steel.

S6-3-2

Westig (U.K.) Ltd.
C 0.9, Si 0-0.45, Mn 0-0.4, P 0-0.03, S 0-0.03, Cr 4.1, Mo 3, V 2, W 6.3, bal Fe.
High speed steel.

S6-5-2

Westig (U.K.) Ltd.
C 0.9, Si 0-0.45, Mn 0-0.4, P 0-0.03, S 0-0.03, Cr 4.1, Mo 5, V 1.8, W 6.3, bal Fe.
High speed steel equivalent to JIS SKH 9; BS BM2; AISI M2.

S6-5-2-5

Westig (U.K.) Ltd.
C 0.92, Si 0-0.45, Mn 0-0.4, P 0-0.03, S 0-0.03, Cr 4.1, Mo 5, V 1.8, W 6.3, Co 4.8, bal Fe.
High speed steel equivalent to JIS SKH 55.

SA

Thyssen Edelstahlwerke AG
C 0.6, Si 2, Mn 9, bal Fe.
For springs. *Obsolete*

SA 22

Now CRUCIBLE SA 22.

SA 300

Thyssen Edelstahlwerke AG
C 0.7, Cr 4, V 1, Mo 9, W 1.5, bal Fe.
For heavy duty drills, saw segments; high speed steel.
Obsolete

SA GREY LABEL
Swedish American Steel Corp.
C 0.5, bal Fe.
For drop forge dies.

SA HY-PRO
Swedish American Steel Corp.
C 0.7, W 18, Cr 4, V 1, Co 5, bal Fe.
For cutting tools; high speed steel.

SA HY-PRO DIE STEEL
Swedish American Steel Corp.
C 1.5, Cr 13, bal Fe.
For dies; non-deforming.

SA LILY BRAND
Swedish American Steel Corp.
C, alloy, bal Fe.
For brass working dies.

SA N-N
Swedish American Steel Corp.
C 0.5, W 2, bal Fe.
For gripper dies, piercers; hot work steel.

SA NON PA-REIL
Swedish American Steel Corp.
C 0.9, Mn 1.2, bal Fe.
For dies, punches, upsetters, crimpers; Type O1; non-deforming.

SA-32
Sterling Alloy Casting Corp.
C 0.2, Cr 20, Ni 12, bal Fe.
For furnace parts; resists heat to 1400°F.

SA-50
GTE Sylvania
W alloy.
For glass sealing-in wire. *Obsolete*

SA-BEST EXTRA STANDARD
Disston Inc.
C 0.7-1.2, bal Fe.
For tools and dies; water hardening. *Obsolete*

SA-CN
Swedish American Steel Corp.
C 0.5, W 2.5, bal Fe.
For bolt heading dies, hot work punches; hot work steel.

SA-N-65
Swedish American Steel Corp.
C 0.5, Cr 0.9, V 0.2, bal Fe.
For rivet sets; punches; shock resistant.

SA-N-88
Swedish American Steel Corp.
C, alloy, bal Fe.
For hobs, punches; tough.

SA-WHITE LABEL
Swedish American Steel Corp.
C 0.9, bal Fe.
For taps, drills, forming and blanking dies; water hardening.

SA-YELLOW LABEL
Swedish American Steel Corp.
C 0.8, Cr 0.9, V 0.2, bal Fe.
For punches, dies, shear blades; oil hardening.

SA/O
Thyssen Edelstahlwerke AG
C, Si, Mn, bal Fe.
For springs; oil hardening. *Obsolete*

SA/W
Thyssen Edelstahlwerke AG
C, Si, Mn, bal Fe.
For springs; oil hardening. *Obsolete*

SA/W
Thyssen Edelstahlwerke AG
C, Si, Mn, bal Fe.

SA200
Thyssen Edelstahlwerke AG
C 0.8, W 12, Cr 4, V 2.5, Mo 0.85, bal Fe.
For milling cutters, lathe and planer tools, reamers; high speed steel, oil hardened. *Obsolete*

SA500
Thyssen Edelstahlwerke AG
C 1.3, Cr 4, Mo 0.8, V 4, W 12, bal Fe.
For milling cutters, lathe and planer tools, drills, hobs; high speed steel, oil hardened. *Obsolete*

SA900
Thyssen Edelstahlwerke AG
C 1.5, Cr 4.25, Mo 0.65, V 4.25, W 12.5, Co 5, bal Fe.
For tools, dies; cutters; high speed steel. *Obsolete*

SAARSTAHL 1.1620
Saarstahl AG
C 0.7, Si 0.2, Mn 0.25, bal Fe.
Carbon tool steel. For stamping and forming dies, and woodworking tools. DIN 1.1620.

SAARSTAHL 1.1625
Saarstahl AG
C 0.8, Si 0.2, Mn 0.25, bal Fe.
Carbon tool steel. For cutting and stamping tools; burring and trimming tools. DIN 1.1625.

SAARSTAHL 1.1645
Saarstahl AG
C 1.05, Si 0.2, Mn 0.25, bal Fe.
Carbon tool steel. For hand tools as shears; numbering stamps. DIN 1.1645.

SAARSTAHL 1.1654
Saarstahl AG
C 1.15, Si 0.2, Mn 0.25, bal Fe.
Carbon tool steel. For milling cutters, taps, broaches, reamers, and shears. DIN 1.1654.

SAARSTAHL 1.1730
Saarstahl AG
C 0.47, Si 0.3, Mn 0.7, bal Fe.
Carbon tool steel; for hand tools such as axes, hatchets, knives, hand saws. DIN 1.1730.

SAARSTAHL 1.1740
Saarstahl AG
C 0.58, Si 0.25, Mn 0.7, bal Fe.
Carbon tool steel; for hand tools, for saws, hammers, screw drivers. DIN 1.1740.

SAARSTAHL 1.2067
Saarstahl AG
C 1, Cr 1.5, bal Fe.
Cold work tool steel; for drills, threading dies, lathe centers, reamers, drift pins. AFNOR 100 C 6. DIN 1.2067.

SAARSTAHL 1.2080
Saarstahl AG
C 2.05, Cr 11.5, bal Fe.
Cold work tool steel. For heavy duty stamping and punching; shear blades. Similar to AISI D3. DIN 1.2080.

SAARSTAHL 1.2103
Saarstahl AG
C 0.58, Si 1.8, Mn 0.75, Cr 0.4, bal Fe.
Cold work tool steel. For chuck jaws, rivet sets, springs. AFNOR 60 SC 7. DIN 1.2103.

SAARSTAHL 1.2162
Saarstahl AG
C 0.2, Mn 1.25, Cr 1.15, bal Fe.
Case hardenable steel for plastic molds. DIN 1.2162.

SAARSTAHL 1.2210
Saarstahl AG
C 1.25, Cr 0.65, V 0.1, bal Fe.
Cold work tool steel; for taps, counterbores, broaches, punches. AFNOR 120 C 3. DIN 1.2210.

SAARSTAHL 1.2242
Saarstahl AG
C 0.56, Mn 1.05, Cr 1.15, Ni 0.15, V 0.12, bal Fe.
For tools, dies, crimpers, punches; oil hardening. Tough. DIN 1.2242.

SAARSTAHL 1.2312
Saarstahl AG
C 0.4, Mn 1.5, Cr 1.95, Mo 0.2, V 0.1, S 0.06, bal Fe.
Hardenable steel for plastic molds. DIN 1.2312.

SAARSTAHL 1.2316
Saarstahl AG
C 0.42, Mn 0.55, Cr 15.5, Mo 1.05, bal Fe.
Corrosion resistant steel for molding corrosive plastics. DIN 1.2316.

SAARSTAHL 1.2323X2
Saarstahl AG
C 0.45, Cr 1.4, Mo 0.7, V 0.3, bal Fe.
Hot work tool steel for extrusion press tools. AFNOR 45 CDV 7. DIN 1.2323.

SAARSTAHL 1.2343
Saarstahl AG
C 0.38, Si 1, Cr 5, Mo 1.25, V 0.45, bal Fe.
Hot work tool steel. For pressure casting molds for light metal; extrusion press tools. AISI H 11. DIN 1.2343.

SAARSTAHL 1.2344
Saarstahl AG
C 0.38, Si 1, Cr 5, Mo 1.25, V 0.95, bal Fe.
Hot work tool steel. For extruding and piercing mandrels and forging dies. Similar to AISI H 13. DIN 1.2344.

SAARSTAHL 1.2362X1
Saarstahl AG
C 0.63, Si 1.1, Cr 5.25, Mo 1.4, V 0.5, bal Fe.
Hot or cold work tool steel. For trimming dies, punching dies, shears, and plastic molds. AFNOR Z60 CDV5.

SAARSTAHL 1.2363
Saarstahl AG
C 1, Cr 5.25, Mo 1.1, V 0.2, bal Fe.
Cold work tool steel; for punching, cutting and trimming tools. Similar to AISI A2. DIN 1.2363.

SAARSTAHL 1.2365X1
Saarstahl AG
C 0.3, Cr 3, Mo 2.85, V 0.5, bal Fe.
Hot work tool steel; pressing dies, die inserts, punches, die casting dies. Similar to AISI H10. DIN 1.2365X1.

SAARSTAHL 1.2369
Saarstahl AG
C 0.81, Cr 4, Mo 4.25, V 1, bal Fe.
Cold work tool steel. For coining and stamping dies, reamers, and broaches. AFNOR 80 DCV 42. DIN 1.2369.

SAARSTAHL 1.2379
Saarstahl AG
C 1.55, Cr 11.5, Mo 0.75, V 0.95, bal Fe.
Cold work tool steel. For punching and stamping dies, thread rolling dies, and reamers. Similar to AISI D2. DIN 1.2379.

SAARSTAHL 1.2436
Saarstahl AG
C 2.05, Cr 11.5, W 0.7, bal Fe.
Cold work tool steel. For heavy duty stamping and cutting tools; punches, reamers, and broaches. AFNOR Z 200 CW 12. DIN 1.2436.

SAARSTAHL 1.2510
Saarstahl AG
C 0.93, Mn 1.1, Cr 0.6, V 0.1, W 0.6, bal Fe.
Oil hardening cold work tool steel; for taps, milling cutters, punches, shears. Similar to AISI O1. DIN 1.2510.

SAARSTAHL 1.2516
Saarstahl AG
C 1.2, Cr 0.25, V 0.1, W 1, bal Fe.
Cold work tool steel; for drills, counter sinks, rotary files, milling cutters. AFNOR 120 W 12. DIN 1.2516.

SAARSTAHL 1.2550
Saarstahl AG
C 0.6, Si 0.6, Cr 1.1, V 0.18, W 2, bal Fe.
Cold work tool steel; for punches, cold shears, chisels. AFNOR 60 WC 20. DIN 1.2550.

SAARSTAHL 1.2601
Saarstahl AG
C 1.65, Cr 11.5, Mo 0.6, V 0.1, W 0.5, bal Fe.
Cold work tool steel. For blanking and trimming dies, thread rolling dies, reamers, and plastic molds. AFNOR Z 165 CDV 12. DIN 1.2601.

SAARSTAHL 1.2713
Saarstahl AG
C 0.55, Cr 0.7, Mo 0.3, Ni 1.5, V 0.1, bal Fe.
Hot work tool steel. For forging dies and metal extrusion tools. DIN 1.2713.

SAARSTAHL 1.2714
Saarstahl AG
C 0.56, Cr 1.1, Mo 0.5, Ni 1.75, V 0.1, bal Fe.
Hot or cold work tool steel. For drop hammer dies, punches, and extrusion dies. AFNOR 55 NCD 07 05. DIN 1.2714.

SAARSTAHL 1.2738X1
Saarstahl AG
C 0.37, Mn 1.1, Cr 2.1, Mo 0.45, Ni 0.5, V 0.08, bal Fe.
Oil hardenable steel for plastic molds.

SAARSTAHL 1.2743
Saarstahl AG
C 0.58, Cr 1.15, Mo 0.35, Ni 2.75, V 0.1, bal Fe.
Hot or cold work tool steel. For forging and embossing dies, shear blades, and molds. AFNOR 60 NCD 12. DIN 1.2743.

SAARSTAHL 1.2764
Saarstahl AG
C 0.19, Mn 0.4, Cr 1.3, Mo 0.2, Ni 4, bal Fe.
Air hardenable, case hardening steel with low distortion. For plastic molding. May be nitrided and polished. DIN 1.2764.

SAARSTAHL 1.2767
Saarstahl AG
C 0.45, Cr 1.35, Mo 0.25, Ni 4, V 0.05, bal Fe.
Cold work tool steel. For cover dies, embossing dies, and bending tools, shear blades. AFNOR 45 NCD 17. DIN 1.2767.

SAARSTAHL 1.2777
Saarstahl AG
C 0.2, Mo 3.35, Ni 3.1, bal Fe.
Hot work tool steel. For hot pressing dies. AFNOR 20 DN 32-12. DIN 1.2777.

SAARSTAHL 1.2782
Saarstahl AG
C 0.08, Si 2.2, Mn 1.35, Cr 25, Ni 20.5, bal Fe.
Heat resisting steel. For annealing pots, enameling grates, and glass-molds. Similar to AISI 310. DIN 1.2782.

SAARSTAHL 1.2786X2
Saarstahl AG
C 0-0.15, Si 1.9, Cr 19, Ni 35, bal Fe.
Heat resisting steel. For furnace parts and glass-molds.

SAARSTAHL 1.2842
Saarstahl AG
C 0.9, Mn 2, Cr 0.35, V 0.12, bal Fe.
Cold work tool steel; for cutting and punching tools and dies, shear blades, reamers, molds. Similar to AISI O2. DIN 1.2842.

SAARSTAHL 1.3243
Saarstahl AG
C 0.92, Cr 4.25, Mo 4.8, V 1.8, W 6.35, Co 4.6, bal Fe.
High speed steel. For special twist drills, milling cutters, and sectional tools. Similar to AISI M35. DIN 1.3243.

SAARSTAHL 1.3246
Saarstahl AG
C 1.1, Cr 4.25, Mo 3.8, V 1.85, W 6.85, Co 5, bal Fe.
High speed steel. For drills, milling cutters, reamers, taps, counter sinks, lathe tools. DIN 1.3246.

SAARSTAHL 1.3247
Saarstahl AG
C 1.05, Cr 4, Mo 9.2, V 1.1, W 1.4, bal Fe.
High speed steel with very high cutting performance. For drills, milling cutters, and lathe tools. AISI M2, high carbon type. DIN 1.3247.

SAARSTAHL 1.3343
Saarstahl AG
C 0.9, Si 0.3, Mn 0.3, cr 4.25, Mo 5, V 1.85, W 6.4, bal Fe.
For special roller bearings; high speed steel. Similar to AISI M2. DIN 1.3343.

SAARSTAHL 1.3343
Saarstahl AG
C 0.88, Cr 4.25, Mo 4.8, V 1.8, W 6.35, bal Fe.
High speed steel. For drills, lathe tools, milling cutters, and rough machining. DIN 1.3343.

SAARSTAHL 1.3344
Saarstahl AG
C 1.25, Cr 4.25, Mo 4.8, V 2.8, W 6.35, bal Fe.
High speed steel for heavy duty broaches, reamers, and milling cutters. AISI M3 Class 2. DIN 1.3344.

SAARSTAHL 1.3346
Saarstahl AG
C 0.82, Cr 4, Mo 8.4, V 1.1, W 1.7, bal Fe.
High speed steel. For twist drills, lathe tools, milling cutters, and rough machining. AISI M1. DIN 1.3346.

SAARSTAHL 1.3348
Saarstahl AG
C 1, Cr 4, Mo 8.2, V 1.9, W 1.7, bal Fe.
High speed steel. For milling cutters, lathe tools, and cutting hard materials. AISI M7. DIN 1.3348.

SAARSTAHL 1.3552
Saarstahl AG
C 0.83, Si 0.2, Mn 0.25, Cr 4.2, Mo 4.25, V 1, bal Fe.
For roller bearings. Similar to AISI M 50. DIN 1.3552.

SAARSTAHL 1.4828
Saarstahl AG
C 0.15, Si 2.2, Mn 1.4, Cr 20, Ni 11.5, bal Fe.
Heat resisting steel. For annealing trays, hardening boxes. DIN 1.4828.

SAARSTAHL 1.4911
Saarstahl AG
C 0.08, Cr 10.5, Ni 0.7, Co 6.5, V 0.25, Mo 0.8, Nb 0.25, bal Fe.
1100 N/mm^2 TS (RT); 850 N/mm^2 YS (RT). Corrosion resistant and heat resistant alloy for operation up to 600°C. DIN 1.4911.

SAARSTAHL 1.4923
Saarstahl AG
C 0.21, Si 0.2, Mn 0.7, Cr 12, Mo 1, Ni 0.7, V 0.3, bal Fe.
Heat treated: 785-930 N/mm^2 TS. For parts requiring corrosion resistance and heat resistance up to 580°C. DIN 1.4923.

SAARSTAHL 1.4939
Saarstahl AG
C 0.1, Cr 11.5, Ni 2.6, Mo 1.65, V 0.3, bal Fe.
1100 N/mm^2 TS (RT); 800 N/mm^2 YS (RT). Corrosion resistant and heat resistant alloy for operation up to 600°C. DIN 1.4939.

SAARSTAHL 1.4980, SAARSTAHL 1.4944
Saarstahl AG
C 0.06, Cr 15, Ni 25.5, Mo 1.3, V 0.25, Ti 2.1, bal Fe.
1100 N/mm^2 TS (RT); 750 N/mm^2 YS (RT). Corrosion resisting and heat resisting alloy. Parts for gas and steam turbines up to 750°C. DIN 1.4980; DIN 1.4944.

SAARSTAHL 1.5752
Saarstahl AG
C 0.14, Mn 0.55, Cr 0.75, Ni 3.45, bal Fe.
Case hardening steel. For plastic molds. DIN 1.5752.

SAARSTAHL 1.6356
Saarstahl AG
C 0.01, Co 12, Mo 4, Ni 18, Ti 1.6, bal Fe.
Heat treated: 2350 N/mm^2 TS; 2260 N/mm^2 YS; 4 El min. Vacuum melted, high strength maraging steel. DIN 1.6356.

SAARSTAHL 1.6358
Saarstahl AG
C 0.01, Co 9, Mo 5, Ni 18.5, Ti 0.7, bal Fe.
Heat treated: 1960 N/mm^2 TS; 1910 N/mm^2 YS; 5 El min. Vacuum melted, high strength maraging steel. DIN 1.6358.

SAARSTAHL 1.6358, 1.6354
Saarstahl AG
C 0.01, Co 9, Mo 5, Ni 18.5, Ti 0.75, bal Fe.
High strength maraging steel. 2000 N/mm^2 TS. AFNOR Z 2 NKD 18-09. DIN 1.6358; 1.6354.

SAARSTAHL 1.6359
Saarstahl AG
C 0.02, Co 7.5, Mo 4.8, Ni 18, Ti 0.4, bal Fe.
Heat treated: 1720 N/mm^2 TS min; 1620 N/mm^2 YS min. Vacuum melted, high strength maraging steel. AFNOR Z 2 Ni Co Mo 18 85. DIN 1.6359.

SAARSTAHL 1.7734
Saarstahl AG
C 0.15, Si 0.15, Mn 0.95, Cr 1.35, Mo 0.9, V 0.25, bal Fe.
Quenched and tempered: 1080 N/mm^2 TS; 930 N/mm^2 YS; 10 El. High strength low carbon steel; case hardenable. DIN 1.7734.

SAARSTAHL 10 CRMO G10
Saarstahl AG
Cr 0.12, Si 0.2, Mn 0.5, Cr 2.2, Mo 1, bal Fe.
Heat treated: 450-600 N/mm^2 TS. Resistant to temperatures up to 530°C. For pressure vessels and boiler parts. DIN 1.7380.

SAARSTAHL 100 CR 6
Saarstahl AG
C 1, Si 0.25, Mn 0.35, Cr 1.5, bal Fe.
For ball and roller bearings. DIN 1.3505.

SAARSTAHL 100 CRMN 6
Saarstahl AG
C 0.95, Si 0.6, Mn 1.1, Cr 1.5, bal Fe.
For ball and roller bearings. DIN 1.3520.

SAARSTAHL 13 CRMO 44
Saarstahl AG
C 0.15, Si 0.25, Mn 0.6, Cr 0.9, Mo 0.5, bal Fe.
Heat treated: 440-590 N/mm^2 TS. For temperatures up to 530°C. For boiler tubes and parts; pressure vessels. DIN 1.7335.

SAARSTAHL 14 MOV 63
Saarstahl AG
C 0.15, Si 0.25, Mn 0.4, Cr 0.5, Mo 0.6, V 0.28, bal Fe.
Heat treated: 490-690 N/mm^2 TS. Resistant to temperatures up to 560°C. For boiler parts. Weldable. DIN 1.7715.

SAARSTAHL 14 NICR 14
Saarstahl AG
C 0.14, Si 0.25, Mn 0.55, Cr 0.75, Ni 3.45, bal Fe.
Alloy carburizing steel. For automotive gears, spline couplings, and universal couplings. W.-Nr. 5752. DIN 1.5752.

SAARSTAHL 15 MO 3
Saarstahl AG
C 0.17, Si 0.25, Mn 0.6, Mo 0.3, bal Fe.
Heat treated: 440-580 N/mm^2 TS. Resistant to temperatures up to 530°C. For pressure vessels and boiler parts. DIN 1.5415.

SAARSTAHL 15 NICUMONB 5
Saarstahl AG
C 0.15, Si 0.35, Mn 1.1, Mo 0.35, Ni 1.2, Cu 0.65, Nb 0.02, bal Fe.
Heat treated: 580-740 N/mm^2 TS. For pressure vessels and boiler parts. Weldable. DIN 1.6368.

SAARSTAHL 16 CRMO 4
Saarstahl AG
C 0.18, Si 0.25, Mn 0.85, Cr 1, Mo 0.25, bal Fe.
Alloy carburizing steel. For gears, pinions, shafts. DIN 1.7242.

SAARSTAHL 16 MNCR 5
Saarstahl AG
C 0.16, Si 0.25, Mn 1.25, Cr 1.05, bal Fe.
Carburizing steel for camshafts, piston pins, and cog wheels. DIN 1.7131.

SAARSTAHL 16 MNCRB 5
Saarstahl AG
C 0.16, Si 0.25, Mn 1.25, Cr 1.05, B, bal Fe.
Carburizing steel for piston pins, camshafts, and cog wheels. DIN 1.7160.

SAARSTAHL 17 CR 3
Saarstahl AG
C 0.18, Si 0.17, Mn 0.6, Cr 0.8, bal Fe.
Carburizing steel for roller bearing, camshafts, piston pins, cog wheels. DIN 1.7016.

SAARSTAHL 17 CRNIMO 6
Saarstahl AG
C 0.17, Si 0.25, Mn 0.5, Cr 1.65, Mo 0.3, Ni 1.55, bal Fe.
Alloy carburizing steel for highly stressed case hardened parts. DIN 1.6587.

SAARSTAHL 2.4537
Saarstahl AG
C 0.08, Cr 16.5, Mo 17, W 4, 52.0 Ni min, bal Fe.
Corrosion resistant alloy. DIN 2.4537.

SAARSTAHL 2.4634
Saarstahl AG
C 0.15, Cr 15, Mo 5, Co 20, Al 4.5, Ti 1.2, bal Ni.
1200 N/mm^2 TS (RT); 800 N/mm^2 YS (RT); 16 El (RT).
Corrosion resisting and heat resisting alloy. For parts for gas turbines and transmissions for operation up to 950°C. DIN 2.4634.

SAARSTAHL 2.4650
Saarstahl AG
C 0.06, Cr 20, Co 20, Mo 3.8, Ti 2.2, bal Fe.
1000 N/mm^2 TS (RT); 650 N/mm^2 YS (RT). Corrosion resistant and heat resistant alloy for operation up to 750°C. DIN 2.4650.

SAARSTAHL 2.4654
Saarstahl AG
C 0.06, Cr 20, Mo 4.3, Co 14, Al 1.4, Ti 3, bal Ni.
Corrosion resisting and heat resisting alloy, for compressor and turbine parts. DIN 2.4654.

SAARSTAHL 2.4662
Saarstahl AG
C 0.04, Cr 12.5, Ni 43.5, Mo 6, Ti 3, bal Fe.
1200 N/mm^2 TS (RT); 850 N/mm^2 YS (RT); 15 El (RT).
Corrosion resisting and heat resisting alloy. For parts for aircraft engines and gas turbines. DIN 2.4662.

SAARSTAHL 2.4668
Saarstahl AG
C 0.05, Cr 18.5, Ni 53, Mo 3, Al 0.6, Ti 0.8, Nb 5.3, bal Fe.
1400 N/mm^2 TS (RT); 1150 N/mm^2 YS (RT); 15 El (RT).
Corrosion resisting and heat resisting alloy, for parts operating up to 750°C: Hot work tools. DIN 2.4668.

SAARSTAHL 2.4668X1
Saarstahl AG
C 0.06, Cr 15, Fe 7, Al 0.7, Ti 2.5, Nb 0.05, bal Ni:
1100 N/mm^2 TS (RT); 650 N/mm^2 YS (RT); 25 El (RT).
Corrosion resisting and heat resisting alloy, for parts operating up to 850°C.

SAARSTAHL 2.4952, SAARSTAHL 2.4631
Saarstahl AG
C 0.06, Cr 20, Al 1.4, Ti 2.4, bal Ni.
1200 N/mm^2 TS (RT); 750 N/mm^2 YS (RT); 20 El (RT).
Corrosion resisting and heat resisting alloy. Parts for gas turbines and transmissions, for use up to 850°C. DIN 2.4952; DIN 2.4631.

SAARSTAHL 2.4964
Saarstahl AG
C 0.1, Cr 20, Ni 10, Co 50, W 15.
850 N/mm^2 TS (RT); 350 N/mm^2 YS (RT); 35 El (RT).
Corrosion resisting and heat resisting alloy. For parts of gas turbines and transmissions for operation up to 950°C. DIN 2.4964.

SAARSTAHL 2.4969, SAARSTAHL 2.4622
Saarstahl AG
C 0.04, Cr 20, Co 16, Al 1.5, Ti 2.5, bal Ni.
1250 N/mm^2 TS (RT); 800 N/mm^2 YS (RT); 25 El (RT).
Corrosion resisting and heat resisting alloy. For parts for gas turbines and transmissions for use up to 900°C. DIN 2.4969; DIN 2.4622.

SAARSTAHL 2.4973
Saarstahl AG
C 0.1, Cr 19, Mo 10, Co 11, Al 1.5, Ti 3, bal Ni.
1350 N/mm^2 TS (RT); 1000 N/mm^2 YS (RT); 12 El (RT).
Corrosion resisting and heat resisting alloy. For parts for gas turbines and transmissions operating up to 950°C and hot working tools. DIN 2.4973.

SAARSTAHL 2.4983
Saarstahl AG
C 0.07, Cr 18.5, Mo 4, Co 18.5, Fe 0-4, Al 3.2, Ti 3.1, bal Ni.
1300 N/mm^2 TS (RT); 800 N/mm^2 YS (RT); 15 El (RT).
Corrosion resisting and heat resisting alloy, for parts for gas turbines and transmissions operating up to 950°C. DIN 2.4983.

SAARSTAHL 2.4989
Saarstahl AG
C 0.38, Cr 20, Ni 20, Mo 4, Co 42, W 4, Nb 4, bal Ni.
1000 N/mm^2 TS (RT); 500 N/mm^2 YS (RT); 25 El (RT).
Corrosion resisting and heat resisting alloy. For parts for gas turbines and transmissions for operation up to 900°C. DIN 2.4989.

SAARSTAHL 20 CRMONIV 47
Saarstahl AG
C 0.2, Si 0.2, Mn 0.7, Cr 1.3, Mo 1, Ni 0.7, V 0.3, bal Fe.
Heat treated: 700-850 N/mm^2 TS. For steam turbine parts. DIN 1.6979.

SAARSTAHL 20 CRMOV 135
Saarstahl AG
C 0.2, Si 0.25, Mn 0.4, Cr 3.15, Mo 0.55, V 0.5, bal Fe.
Heat treated: 640-785 N/mm^2 TS. For high pressure hydrogenation vessels. DIN 1.7779.

SAARSTAHL 20 MN 5
Saarstahl AG
C 0.2, Si 0.5, Mn 1.2, bal Fe.
Heat treated: 490-640 N/mm^2 TS. For pressure vessels and boiler parts. Weldable. DIN 1.1133.

SAARSTAHL 20 MNCR 5
Saarstahl AG
C 0.2, Si 0.25, Mn 1.25, Cr 1.25, bal Fe.
Carburizing steel for spindles, camshafts, shafts, piston pins, and gears. DIN 1.7147.

SAARSTAHL 20 MNMONI 55
Saarstahl AG
C 0.2, Si 0.25, Mn 1.4, Mo 0.5, Ni 0.75, bal Fe.
Heat treated: 550-700 N/mm^2 TS. For pressure vessels and boiler parts. Weldable. DIN 1.6310.

SAARSTAHL 20 NICRMO 145
Saarstahl AG
C 0.22, Si 0.25, Mn 0.4, Cr 1.4, Mo 0.35, Ni 3.8, bal Fe.
Heat treated: 900-1150 N/mm^2 TS. For bolts, nuts, and high pressure vessels. DIN 1.6772.

SAARSTAHL 21 CRMO 10
Saarstahl AG
C 0.2, Si 0.25, Mn 0.5, Cr 2.3, Mo 0.5, bal Fe.
Heat treated: 650-850 N/mm^2 TS. For centrifugally casting molds. DIN 1.2313.

SAARSTAHL 21 CRMOV 511
Saarstahl AG
C 0.21, Si 0.5, Mn 0.4, Cr 1.35, Mo 1.1, V 0.3, bal Fe.
Heat treated: 690-835 N/mm^2 TS. Resistant to temperatures up to 530°C. For bolts and nuts. DIN 1.8070.

SAARSTAHL 21 CRMOV 57
Saarstahl AG
C 0.21, Si 0.2, Mn 0.5, Cr 1.35, Mo 0.75, V 0.3, bal Fe.
Heat treated: 700-850 N/mm^2 TS. Resistant to temperatures up to 550°C. For nuts and bolts. DIN 1.7709.

SAARSTAHL 21 MNSI 5
Saarstahl AG
C 0.21, Si 0.4, Mn 1.4, bal Fe.
For welded chains. DIN 1.0471.

SAARSTAHL 21 NICRMO 2
Saarstahl AG
C 0.21, Si 0.25, Mn 0.8, Cr 0.5, Mo 0.2, Ni 0.5, bal Fe.
Alloy carburizing steel. For stressed case hardened parts; automotive parts. AISI 8620. DIN 1.6523.

SAARSTAHL 22 NIMOCR 37
Saarstahl AG
C 0.18, Si 0.25, Mn 0.8, Cr 0.4, Mo 0.65, Ni 0.7, bal Fe.
Heat treated: 580-710 N/mm^2 TS. For parts for boilers. Weldable. DIN 1.6751.

SAARSTAHL 23 CRNIMO 747
Saarstahl AG
C 0.22, Si 0.15, Mn 0.75, Cr 1.9, Mo 0.75, Ni 1.05, bal Fe.
Heat treated: 750-900 N/mm^2 TS. For steam turbine parts. DIN 1.6749.

SAARSTAHL 23 MNNIMOCR 64
Saarstahl AG
C 0.24, Si 0.2, Mn 1.5, Cr 0.27, Mo 0.52, Ni 1.05, bal Fe.
Heat treated: 1080 min N/mm^2 TS. For strong, heavy, welded chains. DIN 1.6753.

SAARSTAHL 24 CRMO 10
Saarstahl AG
C 0.23, Si 0.25, Mn 0.65, Cr 2.5, Mo 0.3, bal Fe.
Heat treated: 540-690 N/mm^2 TS. For high pressure hydrogenation vessels. DIN 1.7273.

SAARSTAHL 24 CRMO 5
Saarstahl AG
C 0.26, Si 0.25, Mn 0.7, Cr 1.1, Mo 0.25, bal Fe.
Heat treated: 590-735 N/mm^2 TS. Resistant to temperatures up to 530°C. For bolts and nuts. DIN 1.7285.

SAARSTAHL 26 NICRMOV 115
Saarstahl AG
C 0.28, Si 0.1, Mn 0.25, Cr 1.65, Mo 0.45, Ni 3, V 0.1, bal Fe.
Heat treated: 850-1000 N/mm^2 TS. For generator and turbine parts. DIN 1.6948.

SAARSTAHL 26 NICRMOV 145
Saarstahl AG
C 0.28, Si 0.1, Mn 0.25, Cr 1.65, Mo 0.45, Ni 3.6, V 0.1, bal Fe.
Heat treated: 900-1050 N/mm^2 TS. For generator and turbine parts. DIN 1.6957.

SAARSTAHL 26 NICRMOV 85
Saarstahl AG
C 0.27, Si 0.1, Mn 0.25, Cr 1.35, Mo 0.45, Ni 2.1, V 0.1, bal Fe.
Heat treated: 750-900 N/mm^2 TS. For generator and turbine parts. DIN 1.6931.

SAARSTAHL 27 MNSI 5
Saarstahl AG
C 0.27, Si 0.4, Mn 1.4, bal Fe.
For welded chains. DIN 1.0412.

SAARSTAHL 28 CRMONIV 49
Saarstahl AG
C 0.29, Si 0.1, Mn 0.7, Cr 1.3, Mo 0.95, Ni 0.7, V 0.3, bal Fe.
Heat treated: 700-850 N/mm^2 TS. For steam turbine parts. DIN 1.6985.

SAARSTAHL 30 CRMONIV 511
Saarstahl AG
C 0.3, Si 0.1, Mn 0.75, Cr 1.3, Mo 1.1, Ni 0.7, V 0.3, bal Fe.
Heat treated: 700-850 N/mm^2 TS. For steam turbine parts. DIN 1.6346.

SAARSTAHL 30 CRMOV 9
Saarstahl AG
C 0.3, Si 0.25, Mn 0.55, Cr 2.4, Mo 0.2, V 0.15, bal Fe.
Deep hardening alloy steel for structural part on heavy equipment; crankshafts, bolts. DIN 1.7707.

SAARSTAHL 30 CRNIMO 8
Saarstahl AG
C 0.3, Si 0.2, Mn 0.5, Cr 1.95, Mo 0.45, Ni 1.95, bal Fe.
Deep hardening steel for highly stressed parts in automotive and heavy equipment. DIN 1.6580.

SAARSTAHL 30 NICRMO 22
Saarstahl AG
C 0.3, Si 0.3, Mn 0.8, Cr 0.5, Mo 0.2, Ni 0.55, bal Fe.
Alloy steel hardenable to 850-1000 N/mm^2 TS. For automotive equipment. AISI 8630. DIN 1.6545.

SAARSTAHL 32 CRMOV 1210
Saarstahl AG
C 0.32, Si 0.25, Mn 0.4, Cr 3.1, Mo 1, V 0.3, bal Fe.
Alloy steel hardenable to 1520-1620 N/mm^2 TS; 1300 N/mm^2 YS. For highly stressed parts; deep hardening. DIN 1.7765.

SAARSTAHL 34 CR 4
Saarstahl AG
C 0.34, Si 0.25, Mn 0.85, Cr 1, bal Fe.
Heat treatable steel for axles, crankshafts. Similar to AISI 5135. DIN 1.7033.

SAARSTAHL 34 CRNIMO 6
Saarstahl AG
C 0.35, Si 0.25, Mn 0.5, Cr 1.5, Mo 0.25, Ni 1.55, bal Fe.
Alloy steel hardenable to 1200 N/mm^2 TS. For automotive equipment and engine parts. DIN 1.6582.

SAARSTAHL 35 NICR 18
Saarstahl AG
C 0.35, Si 0.25, Mn 0.5, Cr 1.4, Ni 4.35, bal Fe.
Deep hardening alloy steel for structural automotive parts, and heavy equipment. DIN 1.5864.

SAARSTAHL 37 MNSI 5
Saarstahl AG
C 0.37, Si 1.25, Mn 1.25, bal Fe.
Heat treatable steel; for shafts and arbors. DIN 1.5122.

SAARSTAHL 40 NICRMO 186
Saarstahl AG
C 0.42, Si 0.25, Mn 0.4, Cr 1.55, Mo 0.45, Ni 4.4, bal Fe.
Deep hardening alloy steel. Hardenable to 1500 N/mm^2 TS. DIN 1.6738.

SAARSTAHL 42 CRMO 4
Saarstahl AG
C 0.42, Si 0.25, Mn 0.7, Cr 1, Mo 0.2, bal Fe.
Heat treated: 700-900 N/mm^2 TS. For engineering parts. DIN 1.7725.

SAARSTAHL 46 SI 7
Saarstahl AG
C 0.46, Si 1.8, Mn 0.7, bal Fe.
For springs, laminated springs. Water hardened. DIN 1.5024.

SAARSTAHL 50 CRV 4
Saarstahl AG
C 0.5, Si 0.25, Mn 0.9, Cr 1.1, V 0.1, bal Fe.
For highly stressed flat or coiled springs. Similar to AISI 6150. DIN 1.8159.

SAARSTAHL 51 CRMOV 4
Saarstahl AG
C 0.5, Si 0.25, Mn 0.9, Cr 1.1, Mo 0.25, V 0.1, bal Fe.
For highly stressed flat or coil springs. DIN 1.7701.

SAARSTAHL 51 SI 7
Saarstahl AG
C 0.51, Si 1.75, Mn 0.7, bal Fe.
Water hardening spring steel. DIN 1.5025.

SAARSTAHL 55 CR 3
Saarstahl AG
C 0.57, Si 0.3, Mn 0.85, Cr 0.85, bal Fe.
For flat and coil springs. W.-Nr. 7176. DIN 1.7176.

SAARSTAHL 55 SI 7
Saarstahl AG
C 0.58, Si 1.75, Mn 0.8, bal Fe.
Water hardening steel for flat or spiral springs. DIN 1.5026.

SAARSTAHL 58 CRV 4
Saarstahl AG
C 0.58, Si 0.25, Mn 0.95, Cr 1.1, V 0.1, bal Fe.
For highly stressed flat or coil springs. DIN 1.8161.

SAARSTAHL 60 CRMOV 135
Saarstahl AG
C 0.57, Si 0.25, Mn 0.4, Cr 3.1, Mo 0.7, V 0.15, bal Fe.
Heat treated: 60-75 ShC. For hot rolls for aluminum and back up rolls. DIN 1.2356.

SAARSTAHL 60 SICR 7
Saarstahl AG
C 0.62, Si 1.75, Mn 0.85, Cr 0.4, bal Fe.
For flat or coil springs. DIN 1.7108.

SAARSTAHL 65 MNCRMO 4
Saarstahl AG
C 0.63, Si 0.4, Mn 0.9, Cr 0.9, Mo 0.25, bal Fe.
Heat treated: 700-900 N/mm^2 TS. For grooved hot rolls for steel rolling mills. DIN 1.2309.

SAARSTAHL 65 SI 7
Saarstahl AG
C 0.65, Si 1.75, Mn 0.8, bal Fe.
Steel for flat or spiral springs. DIN 1.5028.

SAARSTAHL 86 CRMOV 7
Saarstahl AG
C 0.25, Si 0.25, Mn 0.7, Cr 1.9, Mo 0.25, V 0.07, bal Fe.
Heat treated: 850-1000 N/mm^2 TS. For wear resistant hot rolls for steel rolling mills. DIN 1.2327.

SAARSTAHL CF 35
Saarstahl AG
C 0.35, Si 0.25, Mn 0.55, bal Fe.
Carbon steel for structural purposes. Fine grain steel. DIN 1.1183.

SAARSTAHL CF 53
Saarstahl AG
C 0.53, Si 0.25, Mn 0.55, bal Fe.
Carbon steel for structural purposes. Fine grain steel. DIN 1.1213.

SAARSTAHL CK 22
Saarstahl AG
C 0.2, Si 0.25, Mn 0.5, bal Fe.
Heat treated: 410-540 N/mm^2 TS. Weldable. For forgings and tubing DIN 1.1151.

SAARSTAHL CK 35
Saarstahl AG
C 0.35, Si 0.25, Mn 0.7, bal Fe.
Heat treated: 500-650 N/mm^2 TS. For bolts, nuts, and forgings. DIN 1.1181.

SAARSTAHL CK 45
Saarstahl AG
C 0.45, Si 0.25, Mn 0.65, bal Fe.
Hardened: 590-740 N/mm^2 TS. For bolts, nuts, and engineering parts. DIN 1.1191.

SAARSTAHL SAE 1137
Saarstahl AG
C 0.36, Mn 1.5, P 0-0.04, S 0.105, bal Fe.
Free-cutting steel. DIN 1.9737.

SAARSTAHL SAE 1144
Saarstahl AG
C 0.44, Mn 1.5, P 0-0.04, S 0.29, bal Fe.
Free-cutting steel.

SAARSTAHL SAE 1146
Saarstahl AG
C 0.46, Si 0.85, P 0-0.04, S 0.105, bal Fe.
Free-cutting steel.

SAARSTAHL SAE 11L37
Saarstahl AG
C 0.36, Mn 1.5, P 0-0.04, S 0.105, Pb 0.25, bal Fe.
Free-cutting steel.

SAARSTAHL SAE 1215
Saarstahl AG
C 0-0.08, Mn 0.9, P 0.065, S 0.31, bal Fe.
Free-cutting steel. DIN 1.9704.

SAARSTAHL SAE 12L14
Saarstahl AG
C 0-0.15, Mn 1, P 0.065, S 0.31, bal Fe.
Free-cutting steel. DIN 1.9718.

SAARSTAHL ST E 355
Saarstahl AG
C 0.2, Si 0.25, Mn 1.45, bal Fe.
Weldable fine grain steel for construction purposes. DIN 1.0562.

SABECO METAL
Manco Products, Inc.
Cu 70, Sn 9, Pb 21.
For machine bearings. *Obsolete*

SABECO NO. 11
Saginaw Bearing Co.
Cu 69-71, Sn 10.5-11.5, Pb 18-20.
26,500 TS; 6 El; 6.5 RA; 66 Brin. For extra heavy pressure bearings; bearing bronze.

SABECO NO. 11 H G
Saginaw Bearing Co.
Cu 69-71, Sn 10.5-11.5, Pb 18-20.
27,300 TS; 4.3 El; 3.5 RA; 78 Brin. For worm wheels, clutch shifter shoes, heavy bearings; bearing bronze.

SABECO NO. 14
Saginaw Bearing Co.
Cu 70, Sn 14, Pb 16.
Cast: 27,300 TS; 4.2 El; 3.5 RA; 78 Brin. For bearings, bushings.

SABECO NO. 16
Manco Products, Inc.
Sn 15-16.5, Pb 13.5-14.5, bal Cu.
Cast: 32,250 TS; 1 El; 1.5 RA; 119 Brin. For friction rings, heavy duty bearings; Babbitt.

SABECO NO. 16
Saginaw Bearing Co.
Sn 15-16.5, Pb 13.5-14.5, bal Cu.
Cast: 32,250 TS; 1 El; 1.5 RA; 119 Brin. For friction rings, heavy duty bearings; Babbitt.

SABECO NO. 5
Saginaw Bearing Co.
Cu 69-71, Sn 4.5-5.5, Pb 24-26.
22,600 TS; 10 El; 9.5 RA; 52 Brin. For light or medium load bearings, or water lubricated bearings; bearing bronze. *Obsolete*

SABECO NO. 9
Saginaw Bearing Co.
Cu 69-71, Sn 8.5-9.5, Pb 20-22.
25,100 TS; 9.5 El; 8.3 RA; 60 Brin. For heavy load bearings; bearing bronze.

SABEN
Sanderson Kayser Ltd.
C 0.7, W 14, Cr 4, V 1, bal Fe.
For punches, nail dies, drills, taps; high speed steel, oil hardened.

SABEN 6-5-2
Sanderson Kayser Ltd.
C 0.83, Cr 4.1, Mo 5, W 6.4, V 1.9, bal Fe.
High speed tool steel. B.S. 4659 Type BM 2; AISI M2.

SABEN EXTRA
Sanderson Kayser Ltd.
C 0.75, Cr 4.1, W 18, V 1.1, bal Fe.
High speed tool steel. B.S. 4659 Type BT1; AISI T1.

SABEN HC
Sanderson Kayser Ltd.
C 1.25, Cr 4.5, W 13.5, Mo 0.3, V 3.7, bal Fe.
W-V type high speed steel.

SABEN TENCO
Sanderson Kayser Ltd.
C 0.75, Cr 5, Mo 0.6, W 18.5, V 1.75, Co 9.5, bal Fe.
Cobalt high speed steel. B.S. 4659 Type BT 5; AISI T5.

SABEN WUNDA
Sanderson Kayser Ltd.
C 0.75, Cr 4.25, Mo 0.6, W 18.5, V 1.1, Co 5.5, bal Fe.
Cobalt high speed steel. B.S. 4659 Type BT4; AISI T4.

SABEX
Sanderson Kayser Ltd.
C 0.28, Cr 0.3, Mn 0.3, Ni 2.25, Cr 2.5, W 9.5, V 0.15, bal Fe.
Hot work tool steel. B.S. 4659 Type BH21A.

SABRE
Atlas Specialty Steels
C 1.25, Si 0.3, Mn 0.3, W 10, Cr 4.25, Mo 2.5, V 4.3, Co 5.5, bal Fe.
For tool bits, form tools, form and blank dies; high red hardness, high speed steel. *Obsolete*

SABUS
Sanderson Kayser Ltd.
C 0.7, W 22, bal Fe.
For glass cutting knives, drills, milling cutters, saws; super high speed steel.

SAE 300 (7Q5)
Sterling Metals Ltd.
Cu 6.25-7.75, Mg 0.15-0.4, Si 5-6, Fe 0-0.7, Mn 0-0.6, Ni 0-0.5, bal Al.
Chill cast: 42,500-49,000 TS; 40,000-47,000 YS; 0-1 El; 100-160 Brin; 0.5 IZ. General purpose, chill cast only, good strength but low shock resistance; heat treatable. *Obsolete*

SAE EX1
International Nickel Inc.
Alloy steel. C 0.15-0.21, Mn 0.35-0.6, Si 0.3, Ni 4.8-5.3, Mo 0.2-0.3, bal Fe.
Hardened: 200,000 TS; 155,000 YS; 15 El; 60 RA; 43 Rock C. For heavy duty parts in earth moving equipment, and trucks, rotary and percussion rock bits, ring gears, transmissions. High strength and toughness. Shock resistant. Heavy duty carburizing steel. *Obsolete*

SAE EX2
International Nickel Inc.
Alloy Steel. C 0.65-0.75, Mn 0.25-0.45, Si 0.2-0.35, Ni 0.7-1, Cr 0.15-0.3, Mo 0.08-0.15, bal Fe.
Heat treated: 60 Rock C min. For stress-bearing applications, spindles, rolls, bearings. Wear resistant, shock resistant. *Obsolete*

SAFE 100C6
SAFE (Societe des Aciers Fins de L'est)
C 1, Mn 0.3, Si 0.2, Cr 1.5, bal Fe.
For bearing rollers and balls.

SAFE 10NC6
SAFE (Societe des Aciers Fins de L'est)
C 0.1, Mn 0.75, Si 0.25, Ni 1.5, Cr 1, bal Fe.
For bolts, shafts; case hardened for gears, pinions.

SAFE 14NC11
SAFE (Societe des Aciers Fins de L'est)
C 0.14, Mn 0.5, Si 0.25, Ni 2.75, Cr 0.75, bal Fe.
Case hardened for large, highly stressed gears.

SAFE 16MC5
SAFE (Societe des Aciers Fins de L'est)
C 0.16, Mn 1.2, Si 0.3, Cr 1, bal Fe.
Low alloy/low carbon steel; usually case-hardened for gears, shafts, universal joints.

SAFE 16NC6
SAFE (Societe des Aciers Fins de L'est)
C 0.16, Mn 0.75, Si 0.25, Ni 1.5, Cr 1, bal Fe.
For bolts, shafts; case hardened for gears, pinions.

SAFE 18CD4
SAFE (Societe des Aciers Fins de L'est)
C 0.18, Mn 0.75, Si 0.25, Cr 1, Mo 0.25, bal Fe.
For shafts, bolts; carburized for gears.

SAFE 18NCD3
SAFE (Societe des Aciers Fins de L'est)
C 0.18, Mn 0.8, Si 0.25, Ni 1.5, Cr 1, Mo 0.25, bal Fe.
For cogwheels, pinions; usually case hardened. Similar to AISI 4320.

SAFE 20MC5
SAFE (Societe des Aciers Fins de L'est)
C 0.2, Mn 1.2, Si 0.3, Cr 1.2, bal Fe.
Low alloy/low carbon steel; usually case-hardened for sprockets, cardan joints.

SAFE 20NC6
SAFE (Societe des Aciers Fins de L'est)
C 0.2, Mn 0.75, Si 0.25, Ni 1.5, Cr 1, bal Fe.
For bolts, shafts, case hardened for gears, cams.

SAFE 20NCD2
SAFE (Societe des Aciers Fins de L'est)
C 0.2, Mn 0.8, Si 0.25, Ni 0.55, Cr 0.55, Mo 0.2, bal Fe.
For bolts, flanges; case hardened for gears. AISI 8720.

SAFE 25CD4
SAFE (Societe des Aciers Fins de L'est)
C 0.25, Mn 0.75, Si 0.25, Cr 1, Mo 0.25, bal Fe.
For shafts, axles, pinions.

SAFE 30CAD6-12
SAFE (Societe des Aciers Fins de L'est)
C 0.3, Mn 0.7, Si 0.3, Cr 1.65, Mo 0.35, Al 1.15, bal Fe.
For nitrided highly stressed sliding parts.

SAFE 30CD12
SAFE (Societe des Aciers Fins de L'est)
C 0.3, Mn 0.5, Si 0.25, Cr 3, Mo 0.4, bal Fe.
For highly stressed mechanical parts.

SAFE 30CD4
SAFE (Societe des Aciers Fins de L'est)
C 0.3, Mn 0.75, Si 0.25, Cr 1, Mo 0.25, bal Fe.
For shafts, bolts, pinions.

SAFE 30NC11
SAFE (Societe des Aciers Fins de L'est)
C 0.3, Mn 0.5, Si 0.25, Ni 2.75, Cr 0.75, bal Fe.
For cogwheels, crankshafts, gears.

SAFE 32C4
SAFE (Societe des Aciers Fins de L'est)
C 0.32, Mn 0.75, Si 0.3, Cr 1, bal Fe.
For arbors, axles, automotive parts.

SAFE 35CD4
SAFE (Societe des Aciers Fins de L'est)
C 0.35, Mn 0.75, Si 0.25, Cr 1, Mo 0.25, bal Fe.
For crankshafts, shafts, pinions.

SAFE 35NC6
SAFE (Societe des Aciers Fins de L'est)
C 0.35, Mn 0.75, Si 0.25, Ni 1.5, Cr 1, bal Fe.
For shafts, bolts, spindles, axles.

SAFE 35NCD16
SAFE (Societe des Aciers Fins de L'est)
C 0.35, Mn 0.45, Si 0.25, Ni 4, Cr 1.8, Mo 0.4, bal Fe.
Deep hardening steel for crankshafts, large, highly stressed components.

SAFE 35NCD6
SAFE (Societe des Aciers Fins de L'est)
C 0.35, Mn 0.8, Si 0.25, Ni 1.5, Cr 1.5, Mo 0.25, bal Fe.
For crankshafts, arbors, axles.

SAFE 38C2
SAFE (Societe des Aciers Fins de L'est)
C 0.38, Mn 0.75, Si 0.3, Cr 0.45, bal Fe.
For automotive parts.

SAFE 38C4
SAFE (Societe des Aciers Fins de L'est)
C 0.38, Mn 0.75, Si 0.3, Cr 1, bal Fe.
For shafts, axles, steering columns.

SAFE 40CAD6-12
SAFE (Societe des Aciers Fins de L'est)
C 0.4, Mn 0.7, Si 0.3, Cr 1.65, Mo 0.35, Al 1.15, bal Fe.
For nitrided highly stressed sliding parts.

SAFE 40NCD3
SAFE (Societe des Aciers Fins de L'est)
C 0.4, Mn 0.7, Si 0.25, Ni 0.85, Cr 0.75, Mo 0.25, bal Fe.
For gears, shafts, axles, bolts, flanges.

SAFE 42C2
SAFE (Societe des Aciers Fins de L'est)
C 0.42, Mn 0.75, Si 0.3, Cr 0.45, bal Fe.
For shafts, arbors.

SAFE 42C4

SAFE (Societe des Aciers Fins de L'est)
C 0.42, Mn 0.75, Si 0.3, Cr 1, bal Fe.
For axles, shafts, hand tools.

SAFE 42CD4

SAFE (Societe des Aciers Fins de L'est)
C 0.42, Mn 0.75, Si 0.25, Cr 1, Mo 0.25, bal Fe.
For cogwheels, connecting rods, pinions.

SAFE 50CV4

SAFE (Societe des Aciers Fins de L'est)
C 0.5, Mn 0.9, Si 0.3, Cr 1, V 0.15, bal Fe.
Chrome-vanadium steel for shafts, gears, axles, hand tools.

SAFE CND8

SAFE (Societe des Aciers Fins de L'est)
C 0.3, Mn 0.45, Si 0.25, Ni 2, Cr 2, Mo 0.4, bal Fe.
For crankshafts, highly stressed shafting.

SAFETY-SILV 25

J.W. Harris Co., Inc.
Ag 25, Cu 43, Zn 30, Sn 2.
Melting point: 1265°F solidus; 1430°F liquidus. AWS A5.8.

SAFETY-SILV 30

J.W. Harris Co., Inc.. Ag 30, Cu 38, Zn 32.
Melting point: 1250°F solidus; 1410°F liquidus. AWS A5.8,
BAg20.

SAFETY-SILV 35

J.W. Harris Co., Inc.
Ag 35, Cu 32, Zn 33.
Melting point: 1260°F solidus; 1350°F liquidus. AWS A5.8.

SAFETY-SILV 40

J.W. Harris Co., Inc.
Ag 40, Cu 30.5, Zn 29.5.
Melting point: 1250°F solidus; 1350°F liquidus. AWS A5.8.

SAFETY-SILV 40T

J.W. Harris Co., Inc.
Ag 40, Cu 30, Zn 28, Sn 2.
Melting point: 1200°F solidus; 1310°F liquidus. AWS A5.8,
Class BAg 28.

SAFETY-SILV 45

J.W. Harris Co., Inc.
Ag 45, Cu 30, Zn 25.
Melting point: 1250°F solidus; 1370°F liquidus. AWS A5.8,
Class BAg 5.

SAFETY-SILV 50

J.W. Harris Co., Inc.
Ag 50, Cu 34, Zn 16.
Melting point: 1270°F solidus; 1425°F liquidus. AWS A5.8,
Class BAg6.

SAFETY-SILV 50N

J.W. Harris Co., Inc.
Ag 50, Cu 20, Zn 28, Ni 2.
Melting point: 1220°F solidus; 1305°F liquidus. AWS A5.8,
Class BAg 24.

SAFETY-SILV 54

J.W. Harris Co., Inc.
Ag 54, Cu 40, Zn 4, Ni 1.
Melting point: 1325°F solidus; 1575°F liquidus. AWS A5.8,
Class BAg 13.

SAFETY-SILV 56 BARE & FLUX COATED

J.W. Harris Co., Inc.
Ag 56, Cu 22, Zn 17, Sn 5.
Melting range: 1145-1205°F solidus; 1145-1205°F liquidus.
For ferrous and non-ferrous alloys. Often used to braze
stainless steel for food service applications. AWS A5.8 BAg 7.

SAFETY-SILV 63T

J.W. Harris Co., Inc.
Ag 63, Cu 28.5, Ni 2.5, Sn 6.
Melting point: 1275°F solidus; 1475°F liquidus. AWS A5.8,
Class BAg 21.

SAFETY-SILV 72

J. W. Harris Co., Inc
Ag 72, Cu 28.
Melting point: 1435°F solidus; 1435°F liquidus. AWS A5.8,
Class BAg 8.

SAFETY-SILV HARRIS 40 NI 2

J.W. Harris Co., Inc.
High strength silver brazing rod, cadmium free. For use on
light gage materials. AWS AS.8 BAg 4.

SAFFALLOY

Cerro Metal Products Co.
Now CERROSAFE.

SAFLEX

Engelhard Corp.
Bimetal. For thermostatic bimetals; active from 500-800°F.
Obsolete

SAGAMORE

AL Tech Specialty Steel Corp.
C 1, Cr 5, Mo 1, V 0.3, bal Fe.
Usually at 58-62 Rock C. General purpose air hardening cold
work die steel; good wear resistance and toughness. For
blanking and forming dies, punches, rolls and trim dies. AISI
A-2.

SAGAMORE EZ

AL Tech Specialty Steel Corp.
C 1, Cr 5, Mo 1, V 0.3, S 0.12, bal Fe.
Same as SAGAMORE except free-machining grade.

SAGAMORE V

Allegheny Ludlum Steel
C 2.4, Cr 5.25, V 4.25, Mo 1.1, bal Fe.
Air hardening tool steel, good wear resistance; AISI A7.
Obsolete

SAHIB

Hardenite Steel Co. Ltd.
C, bal Fe.
For tools.

SAINT JUERY EXTRA

Societe Nouvelle du Saut-du-Tarn
C 0.7, W 18, Cr 4, V 1, bal Fe.
For tools, dies, cutters; high speed steel. *Obsolete*

SALAMANDRE D. 1

Compagnie Ateliers et Forges de la Loire
C 1, Cr 1.65, Ni 0.5, V 0.15, bal Fe.
Heat treated: 200,000 TS; 185,000 YS; 3 El; 444 Brin. For
rollers, bearings, cams, bushings, sleeves, cutting tools,
punches. Cold work steel, oil hardening.

SALGE ANTIFRICTION METAL

English manufacture
Zn 86, Sn 9.9, Cu 4, Pb 1.1.
For bearings; anti-friction.

SALLIT'S SPECULUM

Manufacturer not listed.
Cu 65, Sn 31, Ni 4.
For metallic mirrors; corrosion resistant.

SALVO

Midvale-Heppenstall Co.
C 0.6, Cr 1, Ni 2, V 0.2, bal Fe.
For rivet sets, tools; tough, great depth hardness. *Obsolete*

SAM ALLOY

Manufacturer not listed.
Mi 11, Cu 1.5, Ni 1.25, Mn 1, Cr 0.3, Ti 0.02, bal Al.
Cast: 15,000 TS; 0.5 El; 52 Brin. Forged: 10.0 El; 62 Brin. For
aircraft engine components; high temperature resistance to
600°F.

SAMARIUM

Atomergic Chemetals Corp.
Sa.
Purities: special distilled grade 99.9%, 99.5%. Forms: ingot,
lump, rod, filings, wire, sheet, foil, powder.

SAMLEGIERUNG

Manufacturer not listed.
Mn 13, Si 10.2, Al 5.8, C 2.5, Cu 0.3, bal Fe.

SAMPSON

English manufacture
C 0.43, Mn 0.43, Ni 1.22, Cr 0.43, bal Fe.
For tools, gears; oil hardened.

SAMPSON METAL

Manufacturer not listed.
Zn 88, Cu 4, Al 8.
For bearings; anti-friction.

SAMSON

Carpenter Technology Corp.
C 0.4, Cr 0.6, Ni 1.25, bal Fe.
Heat treated: 110,000-240,000 TS; 90,000-210,000 YS; 8-22
El; 25-60 RA; 230-440 Brin. For tempered set screws, shanks
for high speed tools. *Obsolete*

SAMSON EXTRA

Carpenter Technology Corp.
C 0.1, Mn 0.3, Si 0.2, Cr 2.3, bal Fe.
Case hardened: 108,000-120,000 TS; 69,000-91,000 YS;
200-250 Brin. For plastic molds; hobbing steel. *Obsolete*

SAMSON NO. 158

Carpenter Technology Corp.
C 0.1, Ni 3.5, Cr 1.5, bal Fe.
For gears, shafts, clutch parts; case hardening steel.
Obsolete

SAMSON NO. 2-547

Carpenter Technology Corp.
C 0.2, Ni 3.5, bal Fe.
For gears, shafts; carburizing steel. *Obsolete*

SAMSON NO. 3-427

Carpenter Technology Corp.
C 0.3, Cr 1, Mo 0.2, bal Fe.
For aircraft and automotive parts; water hardening. *Obsolete*

SAMSON NO. 3-547

Carpenter Technology Corp.
C 0.3, Ni 3.5, bal Fe.
For shafts, gears, bolts; oil hardening. *Obsolete*

SAMSON NO. 4-408

Carpenter Technology Corp.
C 0.4, Ni 3, Cr 0.75, bal Fe.
For clutches, shafts; oil hardening. *Obsolete*

SAMSON NO. 436

Carpenter Technology Corp.
C 0.15, Ni 1.75, Mo 0.25, bal Fe.
For gears, shafts. *Obsolete*

SAMSON NO. 5-317

Carpenter Technology Corp.
C 0.5, Ni 1.75, Cr 1, bal Fe.
For clutches, gears, shafts; oil hardening. *Obsolete*

SAMSON NO. 5-720

Carpenter Technology Corp.

C 0.5, Cr 0.9, V 0.2, bal Fe.

For leaf and coil springs, gears, shafts; oil hardening. *Obsolete*

SAMSON NO. 500

Carpenter Technology Corp.

C 0.1, Ni 5, bal Fe.

For turbine blades, gears; carburizing steel. *Obsolete*

SANBOLD CVS

Sanderson Kayser Ltd.

C 0.5, Cr 1, V 0.15, bal Fe.

Chrome-vanadium steel.

SANBOLD NA 35

Sanderson Kayser Ltd.

Ni-Fe high permeability alloy. Maximum permeability: 12,000-20,000; maximum flux density: 13,000 gauss. For magnetic screens, relays in automatic telephones.

SANBOLD NA 47

Sanderson Kayser Ltd.

Ni 40-50, bal Fe.

Maximum permeability: 15,000-30,000; maximum flux density: 16,000 gauss. For transformers and chokes.

SANBOLD NA 76

Sanderson Kayser Ltd.

Ni 76-77, plus others, bal Fe.

Maximum permeability: 50,000-150,000; maximum flux density: 8000 gauss. Low hysteresis loss. For relay devices, chokes, transformers.

SANBRON AA

Sanderson Kayser Ltd.

C 0.1, Cr 18, Ni 8, bal Fe.

Annealed: 85,000 TS; 35,000 YS; 40 El; 55 RA; 150 Brin. For trim, cabinets, hospital equipment; Type 302; stainless, austenitic. *Obsolete*

SANBRON BB

Sanderson Kayser Ltd.

C 0.15-0.35, Ni 1, 12% min Cr, bal Fe.

Heat treated: 100,000-116,000 TS; 20 minimum El; 45 minimum RA; 207-235 Brin. For propeller blades, pump rods, valve parts; corrosion resistant, hardenable. *Obsolete*

SANBRON CC

Sanderson Kayser Ltd.

C 0.15, Ni 1, 12% min Cr, bal Fe.

Heat treated: 103,000-116,000 TS; 20 minimum El; 45 minimum RA; 207-235 Brin. For chemical plant equipment; corrosion resistant, hardenable. *Obsolete*

SANBRON DD

Sanderson Kayser Ltd.

C 0-0.25, Cr 16-20, 1% min Ni, bal Fe.

Heat treated: 123,000 TS; 20 minimum El; 25 minimum RA; 241 Brin. For chemical plant equipment; corrosion resistant, hardenable. *Obsolete*

SANBRON EE

Sanderson Kayser Ltd.

C 0-0.15, Ni 0-1, Cr 0-12, bal Fe.

Heat treated: 67,000-90,000 TS; 15 minimum El; 150 Brin. For corrosion resistant parts; corrosion resistant, hardenable. *Obsolete*

SANBRON FF

Sanderson Kayser Ltd.

C 0.25-0.35, Cr 12-12.5, bal Fe.

Heat treated: 177,000 minimum TS; 8 minimum El; 25 minimum RA; 341-415 Brin. For surgical instruments, cutlery, knives; corrosion resistant, hardenable. *Obsolete*

SANBRON H.S. NO. 1

Sanderson Kayser Ltd.

C 0.65, W 10, Cr 3-3.5, V 0.25, bal Fe.

For cold heading dies; high speed steel, oil hardened. *Obsolete*

SANBRON H.S. NO. 2

Sanderson Kayser Ltd.

C 0.45, W 10, Cr 3.5, V 0.25, bal Fe.

For hot heading dies; hot work steel, oil hardened. *Obsolete*

SANCY

Aubert & Duval

C 1.7, Cr 13, Mo 0.5, W 0.5, bal Fe.

For forming and blanking dies, punches; oil or air hardened, non-deforming.

SANDERSON

Colt Industries

C 0.2, Ni 3.5, bal Fe.

For case-hardened parts, gears, pinions; forging steel, *Obsolete*

SANDERSON 3890

Sanderson Kayser Ltd.

C 0.4-0.5, Mn 0.5-0.7, bal Fe.

Annealed: 92,000 TS; 50,000 YS; 22 El; 40 RA; 200 Brin. Heat treated: 105,000 TS; 67,000 YS; 25 El; 48 RA; 204 Brin. For crankshafts, axles, steering rods, bolts; water hardened. *Obsolete*

SANDERSON 4379 NON-STAIN

Sanderson Kayser Ltd.

C 0.28-0.35, Cr 12.75-13.5, Ni 0.45-0.55, bal Fe.

Heat treated: 168,000-228,000 TS; 153,000-219,000 YS; 17 El; 53 RA; 321 Brin. For surgical instruments, cutlery, ball and roller bearings; corrosion resistant, hardenable. *Obsolete*

SANDERSON 476

Sanderson Kayser Ltd.

C 1.5, Cr 11.5, Mo 0.75, bal Fe.

For blanking and forming dies, punches; air hardened, non-deforming.

SANDERSON 476 HEAVY DUTY

Sanderson Kayser Ltd.

C 1.85, Cr 11-12, V 0.33, Mn 0.7, bal Fe.

For blanking and forming dies, threading dies; oil hardened, non-deforming.

SANDERSON 476 SPECIAL

Sanderson Kayser Ltd.

C 2.25, Cr 13, bal Fe.

For blanking and forming dies, punches; oil hardened, non-deforming.

SANDERSON B.883

Sanderson Kayser Ltd.

C 0.35-0.4, Ni 2.75-3.1, Cr 0.5, Mo 0.25, bal Fe.

Annealed: 112,000 TS; 90,000 YS; 25 El; 63 RA; 200 Brin. Heat treated: 266,000 TS; 221,000 YS; 14 El; 39 RA; 500 Brin. For gears, shafts, crankshafts, bolts; tough, shock resistant. *Obsolete*

SANDERSON CHROME-VANADIUM

Sanderson Kayser Ltd.

C 0.28-0.35, Cr 0.2-0.3, V 0.25, bal Fe.

Annealed: 90,000 TS; 50,000 YS; 25 El; 45 RA; 180 Brin. For valve plates for high pressure cylinders; water hardened, tough. *Obsolete*

SANDERSON D70 MAGNET STEEL

Sanderson Kayser Ltd.

C 0.7, W 6, bal Fe.

For permanent magnet. *Obsolete*

SANDERSON DOUBLE SPECIAL

Colt Industries

C 1.3, W 3.6, Cr 0.25, bal Fe.

For finishing tools, drawing dies; also Crucible Double Special. *Obsolete*

SANDERSON EXTRA

Crucible Materials Corp.

Tool material. C 1-1.1, Mn 0.25, bal Fe.

For punches, dies, cutters, general tools; water hardening.

SANDERSON EXTRA 3-1/2

Colt Industries

C 1, bal Fe.

For dies; cold heading dies. *Obsolete*

SANDERSON EXTRA 4-1/2

Colt Industries

C 1, bal Fe.

For dies for wood screws; water hardening. *Obsolete*

SANDERSON N.2832

Sanderson Kayser Ltd.

C 0.15-0.2, Ni 2.75-3.5, Mn 0.5, bal Fe.

Annealed: 80,000 TS; 67,000 YS; 38 El; 52 RA; 163 Brin. Heat treated: 100,000 TS; 71,000 YS; 32 El; 68 RA; 217 Brin. For gears, shafts, machine tool parts, fixtures; case hardening steel, shock resistant. *Obsolete*

SANDERSON N.2834

Sanderson Kayser Ltd.

C 0.3-0.35, Mn 0.6, Ni 4.75-5, bal Fe.

Heat treated: 145,000-237,000 TS; 103,000-224,000 YS; 14-22 El; 47-61 RA; 311-444 Brin. For gears, axles, crankshafts, propeller shafts; oil hardened, shock resistant. *Obsolete*

SANDERSON N.3603

Sanderson Kayser Ltd.

C 0.15, Si 0.15, Mn 0.6, Ni 5, bal Fe.

Hot rolled: 96,000 TS; 71,000 YS; 34 El; 48 RA; 217 Brin. Heat treated: 179,000 TS; 156,000 YS; 14 El; 42 RA; 350 Brin. For shafts, camshafts, cams, fixtures; case hardening steel, shock resistant. *Obsolete*

SANDERSON N.5366

Sanderson Kayser Ltd.

C 0.3-0.35, Mn 0.6, Ni 2.75-3.25, bal Fe.

Heat treated: 127,000-134,000 TS; 112,000-123,000 YS; 20-23 El; 55-60 RA; 228-255 Brin. For valves, gears, crankshafts, axles; oil hardened, tough. *Obsolete*

SANDERSON NK2237

Sanderson Kayser Ltd.

C 0.35, Si 0.4, Mn 0.5, Ni 1.25, Cr 1, bal Fe.

Normalized: 107,000 TS; 78,000 YS; 25 El; 42 RA; 215 Brin. Heat treated: 256,000 TS; 417 Brin. For gears, shafts, crankshafts; oil hardened, tough and wear resistant. *Obsolete*

SANDERSON NK2833

Sanderson Kayser Ltd.

C 0.28-0.35, Mn 0.7, Ni 1.9-2.2, Cr 0.95-1.05, bal Fe.

Heat treated: 134,000-266,000 TS; 89,000-224,000 YS; 5-25 El; 6-55 RA; 247-277 Brin. For gears, shafts, crankshafts; oil hardened, tough and wear resistant. *Obsolete*

SANDERSON NK8053

Sanderson Kayser Ltd.

C 0.13-0.17, Mn 0.6, Ni 3-3.5, Cr 0.6-0.75, bal Fe.

Heat treated: 112,000-134,000 TS; 89,000-112,000 YS; 13-15 El; 30-35 RA; 217-285 Brin. For gears, chain wheels, cams, shafts; case hardening steel, shock and wear resistant. *Obsolete*

SANDERSON NK8055

Sanderson Kayser Ltd.

C 0.3-0.35, Mn 0.6, Ni 3-3.5, Cr 0.7-0.9, bal Fe.

Annealed: 96,000 TS; 60,000 YS; 24 El; 48 RA; 196 Brin. Heat treated: 134,000 TS; 116,000 YS; 21 El; 59 RA; 275 Brin. For turnbuckles, crankshafts, bolts, gears; shock resistant, oil hardened. *Obsolete*

SANDERSON NK8073
Sanderson Kayser Ltd.
C 0.3-0.35, Mn 0.5, Ni 4.25-4.75, Cr 1.25-1.5, bal Fe.
Oil hardened: 256,000 TS; 224,000 YS; 10 El; 25 RA; 500 Brin. Air hardened: 246,000 TS; 212,000 YS; 15 El; 30 RA; 447 Brin. For inlet valves, gears, bolts, crankshafts; shock resistant, oil or air hardened. *Obsolete*

SANDERSON NO. 1
Sanderson Kayser Ltd.
C 0.1-0.14, Si 0.08-0.12, Mn 0.7-0.9, bal Fe.
Heat treated: 61,000-78,500 TS; 45,000-55,000 YS; 44-49 El; 33-67 RA; 163-180 Brin. For gears, pinions, shafts, machinery parts; case hardening steel. *Obsolete*

SANDERSON NO. 2
Sanderson Kayser Ltd.
C 0.15-0.18, Si 0.1-0.15, Mn 0.7-0.9, bal Fe.
Hot rolled: 83,000 TS; 67,000 YS; 37 El; 55 RA; 163 Brin. Heat treated: 95,000 TS; 67,000 YS; 28 El; 60 RA; 180 Brin. For gears, cams, shafts, machinery parts; case hardening steel. *Obsolete*

SANDERSON SPECIAL
Colt Industries
C 1.1, Cr 0.2, bal Fe.
For cutting tools, taps, reamers; water hardening. *Obsolete*

SANDERSON SPECIAL
Sanderson Kayser Ltd.
C 0.9, bal Fe.
For tools, drills; water hardened. *Obsolete*

SANDERSON SPECIAL DRILL ROD
Crucible Materials Corp.
Tool material. C 1.25, Mn 0.25, bal Fe.
For tools, punches; drill rod.

SANDERSON SS8085
Sanderson Kayser Ltd.
C 0.5, Si 1.2-1.75, Mn 0.6-0.8, bal Fe.
Heat treated: 212,000-235,000 TS; 190,000-212,000 YS; 8-12 El; 25-30 RA. For auto springs; tough, shock resistant. *Obsolete*

SANDERSON TOOL
Colt Industries
C 0.07-1.2, bal Fe.
For punches, dies, general tools; water hardening. *Obsolete*

SANDERSON TUNGSTEN TAP
Colt Industries
C 1.1, W 1.4, bal Fe.
For cutting tools; fast finishing. *Obsolete*

SANDERSON XX
Crucible Materials Corp.
C 0.9, Cr 1, bal Fe.
For burnishing tools; water hardening.

SANDOW
LTV Steel
C 0.4, Cr 0.9, Mo 0.2, bal Fe.
Heat treated: 125,000 psi TS; 110,000 psi YS; 19 El; 47 RA; 293 Brin. For axles, gears, crane hooks, and mandrels; tough, oil hardened.

SANDUSKY 1N
Sandusky Foundry & Machine Co.
Cu 84-87, Sn 4-6, Zn 3-5, Pb 4-6, Ni 0-1.
30,000 TS; 14,000 YS; 20 El; 60 Brin. For rolls, cylinders, liners, shaft sleeves, bushings, centrifugal castings.

SANDUSKY 55
Sandusky Foundry & Machine Co.
Fe 1.3, Ni 20.1, Mn 1, bal Cu.
Cast: 75,000 TS; 55,000 YS; 10 El; 140 Brin. Tubes for marine propulsion shaft sleeves and for paper machine roll applications.

SANDUSKY ALLOY 63
Sandusky Foundry & Machine Co.
C 0-0.08, Mn 0-2, Si 0-2, Cr 19.5-23.5, Ni 7-11, Mo 1-3, bal Fe.
Cast: 88,800 psi TS; 50,000 psi YS; 35 El; 182 Brin. Rolls for paper machinery.

SANDUSKY ALLOY 70
Sandusky Foundry & Machine Co.
C 0-0.06, Mn 0-1, Si 0-1, Cr 11-14, Ni 3.5-4.5, Mo 1-1.75, bal Fe.
Cast: 100,000 psi TS; 90,000 psi YS; 15 El; 260 Brin. Rolls for paper machinery.

SANDUSKY ALLOY 75
Sandusky Foundry & Machine Co.
C 0-0.1, Mn 0-1.5, Si 0-1.5, Cr 24.5-27, Ni 6-7.5, bal Fe.
Cast: 100,000 psi TS; 51,000 psi YS; 10 El; 235 Brin. Rolls for paper machinery.

SANDUSKY ALLOY 86
Sandusky Foundry & Machine Co.
C 0-0.1, Mn 0-2, Si 0-1.5, Cr 25-27, Ni 5-7.5, Cu 1.5-3.5, bal Fe.
100,000 TS; 51,000 YS; 10 El.

SANDUSKY NO. 5
Sandusky Foundry & Machine Co.
Ni 1.25-1.5, C, bal Fe.
Cast: 25,000-40,000 TS; 180-250 Brin. For cast iron rolls; centrifugal casting. *Obsolete*

SANDUSKY SA-351
Sandusky Foundry & Machine Co.
C 0-0.08, Cr 18-21, Ni 8-11, bal Fe.
Cast: 75,000 TS; 40,000 YS; 55 El; 156 Brin. Annealed: 77,000 TS; 37,000 YS; 55 El; 140 Brin. For pumps, valves, impellers, food processing equipment, digesters, mixers; corrosion resistant, austenitic, non-magnetic. *Obsolete*

SANDVIK 10L
Sandvik Steel Co.
C 0.5, Si 0.25, Mn 0.5, bal Fe.
Normalized: 100,000 TS; 63,000 YS; 28 El; 41 RA; 205 Brin. For tubes, gears, shafts, water hardened.

SANDVIK 10RA50
Sandvik Steel Co.
C 0.08, Si 0.65, Mn 1.6, P 0-0.03, S 0.3, Cr 17.5, Ni 9.5, Mo 0.5, bal Fe.
Austenitic free-machining stainless steel. For shafts, pins, bolts, studs, screws, etc., that require much machining but must be stainless.

SANDVIK 10RE20
Sandvik Steel Co.
C 0.08, Cr 26.3, Ni 5, bal Fe.
Heat resisting steel.

SANDVIK 10RE51
Sandvik Steel Co.
Stainless steel. C 0.08, Si 0.45, Mn 0.45, Cr 26.3, Ni 5, Mo 1.45, bal Fe.
Annealed: 105,000 TS; 80,000 YS; 25 El; 230 Brin. For valves, valve fittings, pump parts. Type 329; heat resistant steel. S32900

SANDVIK 11
Sandvik Steel Co.
C 0.6, Si 0.2, Mn 0.3, bal Fe.
Heat treated: 115,000-160,000 TS; 77,000-113,000 YS; 12-23 El; 40-54 RA; 229-320 Brin. For wheels, die blocks, girders, rails, discs; water hardened.

SANDVIK 11L
Sandvik Steel Co.
C 0.6, Si 0.25, Mn 0.5, bal Fe.
Normalized: 107,000 TS; 63,000 YS; 26 El; 37 RA; 225 Brin. For springs, typewriter parts; water hardened, for cold drawing.

SANDVIK 11R51
Sandvik Steel Co.
C 0.09, Si 8, Cr 17, Si 1.1, Mn 1.25, Mo 0.7, bal Fe.
Rolled: 299,000 TS; 277,000 YS; 2-4 El. Tempered: 341,000 TS; 327,000 YS; 1-4 El. For springs for instruments; high fatigue life and tensile strength.

SANDVIK 12
Sandvik Steel Co.
C 0.63, Si 0.25, Mn 0.35, bal Fe.
Normalized: 114,000 TS; 64,000 YS; 24 El; 33 RA; 240 Brin. For tools, mechanical parts; water hardened.

SANDVIK 12 W12-C 1
Sandvik Steel Co.
C 0.65, W 5.5-6, bal Fe.
For magnetic parts; magnetic steel.

SANDVIK 12C27
Sandvik Steel Co.
C 0.58, Si 0.35, Mn 0.35, Cr 14, bal Fe.
Chromium stainless strip steel; hardenable. For cutlery, surgical knives, food processing knives; in thin strips for razor blades.

SANDVIK 12R10
Sandvik Steel Co.
C 0.1, Si 0.45, Mn 0.45, Cr 18, Ni 9, bal Fe.
Hard drawn stainless steel wire for springs; 0.013-0.125 in. diameter: 256,000-313,000 psi TS. W. Nr. 4300; AISI 302.

SANDVIK 12R10HV
Sandvik Steel Co.
C 0.1, Si 0.45, Mn 0.45, Cr 18, Ni 9, bal Fe.
Vacuum melted austenitic hard drawn spring wire: 0.013-0.125 in. diameter has 256,000-313,000 psi TS; good fatigue properties. W. Nr. 4300; AISI 302.

SANDVIK 12R11
Sandvik Steel Co.
C 0.1, Si 0.45, Mn 0.45, Cr 18, Ni 7.7, bal Fe.
Annealed: 110,000 TS; 40,000 YS; 60 El; 85 Rock B. Available in eight other tempers. For aircraft structural members, trailer bodies, wheel covers, diaphragms, springs. Type 301; stainless, austenitic. S30103

SANDVIK 12R72
Sandvik Steel Co.
Stainless steel. C 0.1, Si 0.5, Mn 1.8, Cr 15, Ni 15, Mo 1.2, Ti 0.45, bal Fe.
1% proof stress at 750°F: 27,700 psi min. 1% proof stress at 1200°F: 25,600 psi min. Austenitic Mo-Ti stabilized Ni-Cr stainless for steam or superheater tubes or other parts at temperatures up to about 1380°F.

SANDVIK 13
Sandvik Steel Co.
C 0.7, Si 0.3, Mn 0.45, bal Fe.
Normalized: 117,000 TS; 64,000 YS; 23 El; 30 RA; 250 Brin. For piston rings, umbrella ribs, clicker dies; water hardened.

SANDVIK 13LM
Sandvik Steel Co.
C 0.75, Mn 0.65, bal Fe.
Annealed: 175 Brin. For springs; oil or water hardened.

SANDVIK 13M
Sandvik Steel Co.
C 0.75, Si 0.2, Mn 0.65, bal Fe.
Bainite hardened strip steel: 192,000-228,000 TS; 171,000-199,000 YS; 6-8.5 El; 41-47 Rock C. For stamped and lightly formed parts. W. Nr. 0605/1.1248; AISI 1074.

SANDVIK 14BSA
Sandvik Steel Co.
C 0.68, Si 0.25, Mn 0.85, P 0-0.015, S 0.15, bal Fe.
Free machining, water hardening, carbon steel; for shafts, pins, screws, studs, miscellaneous hardware.

SANDVIK 14N3
Sandvik Steel Co.
C 0.75, Ni 2.6, bal Fe.
Heat treated: 199,000 TS. For wood band saws; oil hardened.

SANDVIK 14NC
Sandvik Steel Co.
C 0.75, Cr 0.15, Ni 0.5, bal Fe.
Heat treated: 192,000 TS. For wood band saws, chain saws; oil hardened.

SANDVIK 14P
Sandvik Steel Co.
C 0.75, Pb 0.2, bal Fe.
Rolled: 120,000 TS. For clock and watch parts; free-cutting.

SANDVIK 14S1C1
Sandvik Steel Co.
C 0.75, Cr 0.35, Si 1.4, bal Fe.
Heat treated: 127,000 TS. For springs, friction saws, flapper valves; oil hardened.

SANDVIK 15RE10
Sandvik Steel Co.
Stainless steel. C 0.12, Si 0.5, Mn 1.7, Cr 24.5, Ni 20.5, bal Fe.
0.2% offset yield at 1290°F: 17,000 psi min. Austenitic stainless, weldable, good resistance to oxidation at elevated temperatures; for heat exchanger tubes, furnace parts. W. Nr. 1.4845; AISI 310. S31000

SANDVIK 15RE12
Sandvik Steel Co.
C 0.12, Si 0.45, Mn 1.8, P 0-0.015, S 0-0.015, Cr 26, Ni 21, bal Fe.
Wire for MIG, TIG, and submerged arc welding of Type 310, stainless steel. W. Nr. 4842; AISI 310.

SANDVIK 15VDT
Sandvik Steel Co.
C 0.8, V 0.1, bal Fe.
Drawn: 100,000 TS. For wire for latch needles; water hardened.

SANDVIK 16CIV
Sandvik Steel Co.
C 0.9, Si 0.25, Mn 0.55, Cr 0.6, V 0.1, bal Fe.
Oil hardening cold work tool steel; for small drills, punches.

SANDVIK 17
Sandvik Steel Co.
C 1, Si 0.25, Mn 0.45, bal Fe.
For tools, springs, valves, doctor blades; water hardening.

SANDVIK 17AP
Sandvik Steel Co.
C 1, Pb 0.2, S 0.05, P 0.025, bal Fe.
Annealed: 82,000 TS. For watch parts, machinery parts; free-cutting, oil hardened.

SANDVIK 17C
Sandvik Steel Co.
C 1, Cr 0.15, Si 0.3, Mn 0.3, bal Fe.
Annealed: 85,000 TS; 185 Brin. For tools, saws, springs, piston skirt expanders; water hardening.

SANDVIK 17VDT
Sandvik Steel Co.
C 1, V 0.1, bal Fe.
For needle wire; oil hardened.

SANDVIK 18C283
Sandvik Steel Co.
C 1.05, Cr 14, Mo 0.5, Co 0.5, Cu 0.25, bal Fe.
Annealed: 125,000 TS; 250 Brin. For knives, ball bearings, ball point pens; stainless, hardenable.

SANDVIK 18C29
Sandvik Steel Co.
C 0.95, Cr 13.5, bal Fe.
Air or oil hardening tool steel. Corrosion resistant.

SANDVIK 18RM11
Sandvik Steel Co.
C 0.15, Si 0.4, Mn 6, Cr 18, Ni 8, bal Fe.
Austenitic,stainless steel wire for metallizing and for welding. W. Nr. 4370.

SANDVIK 1B
Sandvik Steel Co.
C 0-0.06, bal Fe.
Normalized: 50,000 TS; 38,000 YS; 44 El; 67 RA; 100 Brin. For flexible tubing, deep drawn parts, tin plate containers; ductile.

SANDVIK 1C27
Sandvik Steel Co.
Stainless steel. C 0-0.08, Si 0.3, Mn 0.3, Cr 13, bal Fe.
Ferritic-martensitic chromium corrosion resistant steel; weldable. For heat exchanger tubes. AISI 410.

SANDVIK 1C34
Sandvik/Coromant
C 0.05, Si 0.3, Mn 0.3, Cr 17, bal Fe.
Chromium stainless steel wire for cold heading; Werkstoff Nr. 4016; AISI 430. Also used as MIG, TIG welding wire.

SANDVIK 1C345
Sandvik/Coromant
C 0.08, Si 0.5, Mn 0.5, Cr 16.7, Mo 1.8, Ti 0.6, bal Fe.
Ferritic stabilized stainless tubing. For heat exchangers, gas preheaters, evaporators. DIN X8 Cr Mo Ti 17; Werkstoff Nr. 1.4523.

SANDVIK 1DT
Sandvik Steel Co.
C 0-0.07, Si 0.03, Mn 0.15, bal Fe.
Normalized: 46,000 TS; 31,000 YS; 52 El; 79 RA; 90 Brin. For lamp globes, housings, deep drawn parts; ductile.

SANDVIK 1DTR
Sandvik Steel Co.
C 0-0.07, Si 0.03, Mn 0.15, bal Fe.
For magnetic purposes; low coercive force.

SANDVIK 1HS49
Sandvik Steel Co.
C 0.05, Si 0.5, Mn 1.9, Ni 2, Mo 0.55, bal Fe.
Wire as filler metal for welding low carbon, low alloy steels.

SANDVIK 1S15
Sandvik Steel Co.
C 0-0.05, Si 1.5, Cu 0.1, bal Fe.
Annealed: 64,000 TS; 36 El. For relays, small transformers and electric motors; dynamo iron, high permeability.

SANDVIK 1S35
Sandvik Steel Co.
C 0.06, Si 3.4, bal Fe.
Annealed: 85,000 TS; 36 El. For relay cores, transformers; magnetic iron, high permeability.

SANDVIK 1XR17
Sandvik Steel Co.
C 0-0.01, Si 0.4, Mn 1.8, P 0-0.015, S 0-0.015, Cr 20, Ni 9.5, bal Fe.
Drawn wire for MIG, TIG, and submerged arc welding of 18-8 stainless steel.

SANDVIK 2 C-34
Sandvik Steel Co.
C 0.13, Cr 17, bal Fe.
Annealed: 78,000 TS; 47,000 YS; 36 El; 66 RA; 165 Brin. At 800°C: 6720 TS; 90 El; 99 RA. For use where material is to be subjected to hot bending or welding, superheater tubes; cutlery; resists corrosion of HNO_3 but not of H or H_2SO_4; stainless.

SANDVIK 2 N-3 CL
Sandvik Steel Co.
C 0.08-0.13, Mn 0.35-0.6, Cr 0.5-0.8, Ni 2.75-3.5, bal Fe.
Annealed: 80,000-95,000 TS; 38,000-55,000 YS; 32-38 El; 60-70 RA; 160-195 Brin. Hardened: 130,000-156,000 TS; 100,000-130,000 YS; 18-24 El; 58-65 RA; 285-350 Brin. For gears, wheels, piston rods, cam shafts, steering arms, levers; case-hardening steel.

SANDVIK 20
Sandvik Steel Co.
C 1.15, Si 0.25, Mn 0.35, bal Fe.
For needles, cutters, tools; water hardening.

SANDVIK 20 WIW
Sandvik Steel Co.
C 1.15, W 0.5, V 1, bal Fe.
For tools, dies; water hardening.

SANDVIK 20.25.5.LCU
Sandvik Steel Co.
Weld metal. C 0.014, Si 0.37, Mn 1.5, P 0.005, S 0.01, Cr 19.5, Ni 25, Mo 4.5, Cu 1.5, N 0.05, bal Fe.
RT: 78,000 psi TS; 46,000 psi YS; 37 El; 160 Brin. Used for MIG, TIG, plasma-arc and submerged-arc welding. W08904

SANDVIK 20.25.5.LCU-15
Sandvik Steel Co.
Weld metal. C 0.03, Cr 20, Ni 25.5, Mo 4.7, Mn 1.8, Si 0.5, P 0-0.025, S 0-0.02, Cu 1.5, bal Fe.
RT: 79,750 psi TS; 50,750 psi YS; 35 El. Lime type electrode with a globular transfer and fast freezing slag designed for out of position welding. Suited for pipe welding.

SANDVIK 20.25.5.LCU-16
Sandvik Steel Co.
Weld metal. C 0.03, Cr 20, Ni 25.5, Mo 4.7, Mn 1.6, Si 0.4, P 0.02, S 0.02, Cu 1.5, bal Fe.
RT: 79,750 psi TS; 50,750 psi YS; 35 El. Rutile type electrode with a globular transfer.

SANDVIK 20P
Sandvik Steel Co.
C 1.15, Pb 0.2, bal Fe.
For reeds in musical boxes; water hardened.

SANDVIK 20W
Sandvik Steel Co.
C 1.15, W 0.15, bal Fe.
For watch mainsprings; oil hardened.

SANDVIK 21C
Sandvik Steel Co.
C 1.25, Cr 0.15, Si 0.2, Mn 0.35, bal Fe.
For razor blades, band saws; water hardening.

SANDVIK 21T10
Sandvik Steel Co.
C 1.25, Cr 0.25, W 1.75, V 0.1, bal Fe.
210 Brin. For tools, cutters, saw blades; fast finishing steel.

SANDVIK 21T10P
Sandvik Steel Co.
C 1.25, Cr 0.25, W 1.75, V 0.1, Pb 0.2, bal Fe.
For dental burs, saw blades; Ledloy.

SANDVIK 22.12.HT
Sandvik Steel Co.
Weld metal. C 0.08, Si 1.6, Mn 0.5, P 0.018, S 0.01, Cr 21, Ni 10, N 0.15, rare earth metals, bal Fe.
RT: 84,000 psi TS; 52,000 psi YS; 40 El; 180 Brin. Used for MIG, TIG and plasma-arc welding. S30815

SANDVIK 22.12.HTR
Sandvik Steel Co.
Weld metal. C 0.05, Cr 21.5, Ni 10.5, Mn 0.8, Si 1.3, P 0.02, S 0.02, N 0.18, bal Fe.
RT: 94,250 psi TS; 65,250 psi YS; 35 El. Rutile-basic type electrode with normal recovery for welding stainless heat resistant steel.

SANDVIK 22.8.3.L
Sandvik Steel Co.
Weld metal. C 0.015, Si 0.5, Mn 1.5, P 0.015, S 0.002, Cr 22.5, Ni 8, Mo 3, N 0.12, bal Fe.
RT: 109,000 psi TS; 87,000 psi YS; 25 El; 240 Brin. Used for MIG, TIG, plasma-arc and submerged-arc welding. W30853

SANDVIK 22.9.3.L-15
Sandvik Steel Co.
Weld metal. C 0.03, Cr 22, Ni 9.5, Mo 3, Mn 1.6, Si 0.3, P 0.03, S 0.025, N 0.15, bal Fe.
RT: 110,000 psi TS; 80,000 psi YS; 25 El. Lime type electrode with a globular transfer and fast freezing slag designed for out of position welding. Suited for pipe welding.

SANDVIK 22.9.3.L-16
Sandvik Steel Co.
Weld metal. C 0.03, Cr 22, Ni 9.5, Mo 3, Mn 1.6, Si 0.4, P 0.03, S 0.025, N 0.15, bal Fe.
RT: 110,000 psi TS; 80,000 psi YS; 27 El. Rutile type electrode with a globular type transfer.

SANDVIK 22C
Sandvik Steel Co.
C 1.29, Si 0.2, Mn 0.3, Cr 0.15, bal Fe.
Strip steel, designed for metal band saws, hardenable.

SANDVIK 25.20.L
Sandvik Steel Co.
Weld metal. C 0.015, Si 0.2, Mn 1.6, P 0.01, S 0.01, Cr 24, Ni 20, N 0.05, bal Fe.
RT: 86,000 psi TS; 55,000 psi YS; 37 El; 170 Brin. Used for MIG, TIG, plasma-arc and submerged-arc welding.

SANDVIK 25.22.2.LMNB
Sandvik Steel Co.
Weld metal. C 0.03, Cr 25, Ni 22, Mo 2.3, Mn 3.5, Si 0.4, P 0-0.02, S 0-0.02, N 0.15, bal Fe.
RT: 87,000 psi TS; 60,900 psi YS; 30 El. Lime type electrode with a globular transfer and fast freezing slag designed for out of position welding. Suited for pipe welding.

SANDVIK 25.22.2LMN
Sandvik Steel Co.
Weld metal. C 0.015, Si 0.1, Mn 4, P 0.011, S 0.009, Cr 24.5, Ni 21.5, Mo 2.1, N 0.11, bal Fe.
RT: 83,000 psi TS; 49,000 psi YS; 42 El; 170 Brin. Used for MIG, TIG, plasma-arc and submerged-arc welding.

SANDVIK 253MA
Sandvik Steel Co.
C 0.08, Si 1.7, Mn 0.3, P 0-0.04, S 0-0.03, Cr 21, Ni 11, N 0.17, 0.04 Ce, bal Fe.
Austenitic, stainless steel. 600-850 N/mm^2 TS; 310 N/mm^2 YS min (0.2% offset); 40 El min; 190 Vickers approx.

SANDVIK 27.31.4.LCU
Sandvik Steel Co.
Weld metal. C 0.015, Si 0.15, Mn 1.7, P 0.01, S 0.005, Cr 27, Ni 31, Mo 3.5, Cu 1, N 0.05, bal Fe.
RT: 78,000 psi TS; 52,000 psi YS; 35 El; 160 Brin. Used for MIG, TIG, plasma-arc and submerged-arc welding. W08028

SANDVIK 27.31.4.LCUR
Sandvik Steel Co.
Weld metal. C 0.025, Cr 27, Ni 31, Mo 3.5, Mn 1, Si 0.8, P 0.02, S 0.02, Cu 1, N 0.09, bal Fe.
RT: 79,750 psi TS; 50,750 psi YS; 35 El. Rutile-basic type electrode with normal recovery for welding austenitic steel requiring corrosion resistance.

SANDVIK 2C344
Sandvik Steel Co.
C 0.1, Si 0.9, Mn 0.3, Cr 17.5, Al 1, bal Fe.
Ferritic, aluminum-alloyed stainless tubing. For gas preheaters, furnace conveyor rollers, soot blower tubes; good scaling resistance. W. Nr. 1.4742.

SANDVIK 2LS
Sandvik Steel Co.
C 0.1, Si 0.15, Mn 0.4, Cr 0.2, Cu 0-0.3, bal Fe.
Normalized: 57,000 TS; 40,000 YS; 45 El; 70 RA; 110 Brin. For machine parts; case hardening steel.

SANDVIK 2N9
Sandvik Steel Co.
C 0.08, Si 0.2, Mn 0.5, P 0-0.02, S 0-0.02, Ni 9.3, bal Fe.
RT: 108,000 TS; 99,000 YS; 25 El. -320°F: 149,000 TS; 135,000 YS; 25 El. Weldable, for containers for low temperature operation, liquefied gases.

SANDVIK 2R12
Sandvik Steel Co.
Stainless steel. C 0-0.02, Si 0-0.1, Mn 0.8, Cr 18.5, Ni 11, bal Fe.
For processing nuclear fuel; heat exchangers, piping nitric acid. S30403

SANDVIK 2R16
Sandvik Steel Co.
C 0-0.02, Si 0.3, Mn 1, P 0-0.015, S 0-0.015, Cr 19.5, Ni 10.5, bal Fe.
Drawn wire for MIG, TIG, and submerged arc welding of 18-8 stainless steel. W. Nr. 4302/4316; AISI 304L.

SANDVIK 2R17
Sandvik Steel Co.
C 0-0.02, Si 0.4, Mn 1.8, P 0-0.015, S 0-0.015, Cr 20.5, Ni 10, bal Fe.
Drawn wire for MIG, TIG and submerged arc welding of austenitic, stainless steels.

SANDVIK 2R61
Sandvik Steel Co.
C 0-0.02, Si 0.4, Mn 1.4, P 0.015, S 0.015, Cr 18.5, Ni 10.5, Mo 2.7, bal Fe.
Drawn wire for MIG, TIG and submerged arc welding of stainless steel. AISI 316L.

SANDVIK 2R62
Sandvik Steel Co.
C 0-0.02, Si 0.35, Mn 1.8, P 0-0.015, S 0-0.015, Cr 19.5, Ni 13, Mo 2.2, bal Fe.
Drawn wire for MIG, TIG and submerged arc welding of stainless steel. Similar to AISI 316/317.

SANDVIK 2R63
Sandvik Steel Co.
C 0-0.02, Si 0.35, Mn 1.8, P 0-0.015, Cr 1, Ni 12, Mo 2.8, bal Fe.
Drawn wire for MIG, TIG and submerged arc welding of 316 type stainless steel. W. Nr. 4403/4430; AISI 316L.

SANDVIK 2RE10
Sandvik Steel Co.
C 0-0.02, Si 0-0.3, Mn 1.7, Cr 24.5, Ni 20.5, bal Fe.
Austenitic, stainless steel with special resistance to oxidizing conditions as in nitric acid; used in chemical plants.

SANDVIK 2RE69
Sandvik Steel Co.
Stainless steel. C 0-0.02, Si 0-0.4, Mn 1.7, Cr 25, Ni 22, Mo 2.1, N 0.12, bal Fe.
At 68°F: 84-113 ksi TS; 43 ksi YS (1% offset) 30 El min; 170 Vickers. Tube, for fertilizer industry; nitric acid production, urea strippers. High resistance to corrosion.

SANDVIK 2RK65
Sandvik Steel Co.
Stainless steel. C 0-0.02, Mn 1.8, Cr 20, Ni 25, Si 0.4, Mo 4.5, Cu 1.5, bal Fe.
Tube, annealed: 71,000-107,000 TS; 31,000 YS min; 40 El min. Austenitic, stainless with high resistance to sulfuric acid, phosphoric acid; for heating coils, heat exchanger tubes, in chemical tanks. N08904

SANDVIK 2RN65
Sandvik Steel Co.
C 0-0.02, Si 0.45, Mn 1.8, Cr 17.5, Ni 24, Mo 4.7, bal Fe.
Tube, annealed: 71,000-107,000 TS; 31,000 YS min; 40 El min; 165 Vickers approx. Austenitic, stainless steel tube; high resistance to general corrosion, especially sulfuric acid and chlorides; for textile, pulp and paper industries.

SANDVIK 2S
Sandvik Steel Co.
C 0.1, Si 0.15, Mn 0.4, bal Fe.
Normalized: 57,000 TS; 39,000 YS; 45 El; 70 RA; 110 Brin. For boiler and locomotive tubes; also case hardening steel.

SANDVIK 3 MO
Sandvik Steel Co.
C 0.15, Si 0.2, Mn 0.7, P 0-0.02, S 0-0.02, Mo 0.33, bal Fe.
For tube exposed to steam temperature. Yield point at 570°F: 31,500 psi min. Yield point at 930°F: 21,500 psi min. DIN 15Mo3.

SANDVIK 307-16
Sandvik Steel Co.
Weld metal. C 0.1, Cr 20, Ni 9.5, Mo 1, Mn 4.2, Si 0.4, P 0.02, S 0.005, bal Fe.
RT: 90,000 psi TS; 64,000 psi YS; 35 El. Rutile type electrode designed to operate on direct current, reversed polarity and alternating current. Also exhibit good arc stability and slag removal. W30710

SANDVIK 307H
Sandvik Steel Co.
Weld metal. C 0.12, Si 0.4, Mn 6, P 0.011, S 0.01, Cr 18, Ni 8, N 0.05, bal Fe.
RT: 94,000 psi TS; 67,000 psi YS; 42 El; 200 Brin. Used for MIG, TIG, plasma-arc and submerged-arc welding. W30740

SANDVIK 307L/307-15
Sandvik Steel Co.
Weld metal. C 0.1, Cr 20, Ni 9.5, Mo 1, Mn 4.2, Si 0.4, P 0.02, S 0.005, bal Fe.
RT: 90,000 psi TS; 64,000 psi YS; 35 El. Lime covering containing less than 9% TiO$_2$ as specified in MIL-E-13080. For use on direct current, reversed polarity only. W30710

SANDVIK 308/308H
Sandvik Steel Co.
Weld metal. C 0.05, Si 0.3, Mn 1.7, P 0.009, S 0.01, Cr 20, Ni 9.5, N 0.05, bal Fe.
RT: 90,000 psi TS; 61,000 psi YS; 41 El; 180 Brin. Used for MIG, TIG, plasma-arc and submerged-arc welding. W30840

SANDVIK 308/308H-15
Sandvik Steel Co.
Weld metal. C 0.06, Cr 20.5, Ni 9.5, Mn 1.7, Si 0.5, P 0.02, S 0.005, bal Fe.
RT: 94,000 psi TS; 67,000 psi YS; 42 El. Lime type electrode with a globular transfer and fast freezing slag designed for out of position welding. Suited for pipe welding. W30810

SANDVIK 308/308H-16
Sandvik Steel Co.
Weld metal. C 0.06, Cr 20.5, Ni 9.5, Mn 1.5, Si 0.4, P 0.02, S 0.005, bal Fe.
RT: 88,000 psi TS; 62,000 psi YS; 45 El. Rutile type electrode with a globular type transfer. W30810

SANDVIK 308/308H-16SA
Sandvik Steel Co.
Weld metal. C 0.06, Cr 20, Ni 9.5, Mo 0.3, Mn 0.6, Si 0.8, P 0.02, S 0.005, bal Fe.
RT: 88,000 psi TS; 62,000 psi YS; 45 El. Rutile type electrode with a spray type transfer and a high silicon, low manganese weld deposit. W30810

SANDVIK 308L
Sandvik Steel Co.
Weld metal. C 0.02, Si 0.3, Mn 1.7, P 0.012, S 0.01, Cr 20, Ni 10, N 0.05, bal Fe.
RT: 87,000 psi TS; 57,000 psi YS; 34 El; 160 Brin. Used for MIG, TIG, plasma-arc and submerged-arc welding. W30843

SANDVIK 308L-15
Sandvik Steel Co.
Weld metal. C 0.03, Cr 20, Ni 10, Mn 1.6, Si 0.6, P 0.02, S 0.005, bal Fe.
RT: 93,000 psi TS; 70,000 psi YS; 38 El. Lime type electrode with a globular transfer and fast freezing slag designed for out of position welding. Suited for pipe welding. W30813

SANDVIK 308L-16
Sandvik Steel Co.
Weld metal. C 0.03, Cr 20, Ni 10, Mn 1.6, Si 0.4, P 0.02, S 0.005, bal Fe.
RT: 85,000 psi TS; 60,000 psi YS; 46 El. Rutile type electrode with a globular type transfer. W30813

SANDVIK 308L-16SA
Sandvik Steel Co.
Weld metal. C 0.02, Cr 19, Ni 10, Mo 0.2, Mn 0.6, Si 0.7, P 0.02, S 0.005, bal Fe.
RT: 85,000 psi TS; 60,000 psi YS; 46 El. Rutile type electrode with a spray type transfer and a high silicon, low manganese weld deposit. W30813

SANDVIK 308LSI
Sandvik Steel Co.
Weld metal. C 0.02, Si 0.8, Mn 1.6, P 0.01, S 0.011, Cr 20, Ni 10.5, N 0.04, bal Fe.
RT: 87,000 psi TS; 57,000 psi YS; 42 El; 160 Brin. Filler metal suited for MIG welding. Also used for TIG and plasma-arc welding. W30848

SANDVIK 308MOL-15
Sandvik Steel Co.
Weld metal. C 0.03, Cr 20, Ni 9.5, Mo 2.5, Mn 1.5, Si 0.6, P 0.03, S 0.005, bal Fe.
RT: 90,000 psi TS; 62,000 psi YS; 38 El. Lime type electrode with a globular transfer and fast freezing slag designed for out of position welding. Suited for pipe welding. W30823

SANDVIK 308MOL 16
Sandvik Steel Co.
Weld metal. C 0.03, Cr 20, Ni 9.5, Mo 2.5, Mn 1.5, Si 0.4, P 0.02, S 0.005, bal Fe.
RT: 83,000 psi TS; 62,000 psi YS; 45 El. Rutile type electrode with a globular type transfer. W30823

SANDVIK 309
Sandvik Steel Co.
Weld metal. C 0.05, Si 0.4, Mn 1.7, P 0.01, S 0.011, Cr 23, Ni 13, N 0.05, bal Fe.
RT: 86,000 psi TS; 57,000 psi YS; 47 El; 160 Brin. Used for MIG, TIG, plasma-arc and submerged-arc welding. W30940

SANDVIK 309-15
Sandvik Steel Co.
Weld metal. C 0.07, Cr 23, Ni 13.5, Mn 1.6, Si 0.6, P 0.02, S 0.005, bal Fe.
RT: 85,500 psi TS; 70,000 psi YS; 38 El. Lime type electrode with a globular transfer and fast freezing slag designed for out of position welding. Suited for pipe welding. W30910

SANDVIK 309-16
Sandvik Steel Co.
Weld metal. C 0.07, Cr 23, Ni 13, Mn 1.8, Si 0.4, P 0.02, S 0.005, bal Fe.
RT: 88,000 psi TS; 60,000 psi YS; 40 El. Rutile type electrode with a globular type transfer. W30910

SANDVIK 309-16SA
Sandvik Steel Co.
Weld metal. C 0.07, Cr 24, Ni 13, Mn 0.6, Si 0.8, P 0.02, S 0.005, bal Fe.
RT: 88,000 psi TS; 60,000 psi YS; 40 El. Rutile type electrode with a spray type transfer and a high silicon, low manganese weld deposit. W30910

SANDVIK 309CB-15
Sandvik Steel Co.
Weld metal. C 0.08, Cr 23, Ni 12.8, Cb 0.8, Mn 1.8, Si 0.6, P 0.02, S 0.005, bal Fe.
RT: 98,000 psi TS; 80,000 psi YS; 32 El. Lime type electrode with a globular transfer and fast freezing slag designed for out of position welding. Suited for pipe welding. W30917

SANDVIK 309CB-16
Sandvik Steel Co.
Weld metal. C 0.07, Cr 23, Ni 12.8, Cb 0.8, Mn 1.8, Si 0.5, P 0.02, S 0.005, bal Fe.
RT: 93,000 psi TS; 73,000 psi YS; 35 El. Rutile type electrode with a globular type transfer. W30917

SANDVIK 309L
Sandvik Steel Co.
Weld metal. C 0-0.015, Si 0.4, Mn 2, P 0.012, S 0.01, Cr 23.5, Ni 13.5, N 0.06, bal Fe.
RT: 87,000 psi TS; 58,000 psi YS; 40 El; 160 Brin. Used for MIG, TIG, plasma-arc and submerged-arc welding. W30943

SANDVIK 309L-15
Sandvik Steel Co.
Weld metal. C 0.03, Cr 23, Ni 13, Mn 1.9, Si 0.6, P 0.02, S 0.005, bal Fe.
RT: 87,000 psi TS; 63,800 psi YS; 39 El. Lime type electrode with a globular transfer and fast freezing slag designed for out of position welding. Suited for pipe welding. W30913

SANDVIK 309L-16
Sandvik Steel Co.
Weld metal. C 0.03, Cr 23, Ni 13.5, Mn 1.8, Si 0.4, P 0.02, S 0.005, bal Fe.
65-600°F: 67,000-80,000 psi TS; 44,000-57,000 psi YS; 22-40 El. Rutile type electrode with a globular type transfer. W30913

SANDVIK 309L-16SA
Sandvik Steel Co.
Weld metal. C 0.02, Cr 22.5, Ni 13, Mo 0.3, Mn 0.6, Si 0.8, P 0.02, S 0.005, bal Fe.
RT: 80,000 psi TS; 57,000 psi YS; 40 El. Rutile type electrode with a spray type transfer and a high silicon, low manganese weld deposit. W30913

SANDVIK 309LHF
Sandvik Steel Co.
Weld metal. C 0.02, Si 0.4, Mn 1.6, P 0.012, S 0.009, Cr 24, Ni 12, N 0.04, bal Fe.
RT: 87,000 psi TS; 59,000 psi YS; 41 El; 160 Brin. Used for MIG, TIG, plasma-arc and submerged-arc welding. W30943

SANDVIK 309MO-15
Sandvik Steel Co.
Weld metal. C 0.08, Cr 20, Ni 12.5, Mo 2.5, Mn 1.8, Si 0.6, P 0.02, S 0.005, bal Fe.
RT: 96,000 psi TS; 68,000 psi YS; 35 El. Lime type electrode with a globular transfer and fast freezing slag designed for out of position welding. Suited for pipe welding. W30920

SANDVIK 309MO-16
Sandvik Steel Co.
Weld metal. C 0.07, Cr 23, Ni 12.5, Mo 2.5, Mn 1.9, Si 0.4, P 0.02, S 0.005, bal Fe.
RT: 96,000 psi TS; 68,000 psi YS; 35 El. Rutile type electrode with a globular type transfer. W30920

SANDVIK 309SI
Sandvik Steel Co.
Weld metal. C 0.07, Si 0.8, Mn 1.6, P 0.008, S 0.011, Cr 23, Ni 13, N 0.05, bal Fe.
RT: 93,000 psi TS; 58,000 psi YS; 35 El; 180 Brin. Filler metal suited for MIG welding. Also used for TIG and plasma-arc welding. W30941

SANDVIK 310
Sandvik Steel Co.
Weld metal. C 0.1, Si 0.5, Mn 1.7, P 0.012, S 0.01, Cr 27, Ni 21, N 0.05, bal Fe.
RT: 86,000 psi TS; 57,000 psi YS; 43 El; 160 Brin. Used for MIG, TIG, plasma-arc and submerged-arc welding. W31040

SANDVIK 310-15
Sandvik Steel Co.
Weld metal. C 0.11, Cr 26, Ni 20.5, Mn 1.7, Si 0.6, P 0.02, S 0.005, bal Fe.
RT: 88,000 psi TS; 59,000 psi YS; 40 El. Lime type electrode with a globular transfer and fast freezing slag designed for out of position welding. Suited for pipe welding. W31010

SANDVIK 310-16
Sandvik Steel Co.
Weld metal. C 0.11, Cr 26, Ni 21, Mn 1.8, Si 0.4, P 0.02, S 0.005, bal Fe.
RT: 89,000 psi TS; 59,000 psi YS; 41 El. Rutile type electrode with a globular type transfer. W31010

SANDVIK 310H
Sandvik Steel Co.
Weld metal. C 0.42, Si 1, Mn 1.2, P 0.009, S 0.01, Cr 24.5, Ni 20.5, N 0.05, bal Fe.
RT: 101,000 psi TS; 65,000 psi YS; 27 El; 220 Brin. Used for MIG, TIG, and plasma-arc welding.

SANDVIK 310H-15
Sandvik Steel Co.
Weld metal. C 0.4, Cr 26, Ni 20.5, Mn 1.7, Si 0.6, P 0.02, S 0.005, bal Fe.
RT: 110,000 psi TS; 75,000 psi YS; 18 El. Lime type electrode with a globular transfer and fast freezing slag designed for out of position welding. Suited for pipe welding. W31015

SANDVIK 310H-16
Sandvik Steel Co.
Weld metal. C 0.4, Cr 26, Ni 21, Mn 1.8, Si 0.4, P 0.02, S 0.005, bal Fe.
RT: 110,000 psi TS; 75,000 psi YS; 18 El. Rutile type electrode with a globular type transfer. W31015

SANDVIK 312
Sandvik Steel Co.
Weld metal. C 0.1, Si 0.4, Mn 1.6, P 0.013, S 0.013, Cr 29, Ni 9, N 0.05, bal Fe.
RT: 106,000 psi TS; 75,000 psi YS; 25 El; 200-250 Brin. Used for MIG, TIG, plasma-arc and submerged-arc welding.

SANDVIK 312-16
Sandvik Steel Co.
Weld metal. C 0.11, Cr 29, Ni 9, Mn 1.7, Si 0.5, P 0.02, S 0.005, bal Fe.
RT: 120,000 psi TS; 104,000 psi YS; 24 El. Rutile type electrode with a globular type transfer. W31310

SANDVIK 312-16SA
Sandvik Steel Co.
Weld metal. C 0.11, Cr 29, Ni 9, Mn 0.6, Si 0.8, P 0.02, S 0.005, bal Fe.
RT: 120,000 psi TS; 104,000 psi YS; 24 El. Rutile type electrode with a spray type transfer and a high silicon, low manganese weld deposit. W31310

SANDVIK 316/316H
Sandvik Steel Co.
Weld metal. C 0.05, Si 0.4, Mn 1.7, P 0.012, S 0.01, Cr 19.5, Ni 11.5, Mo 2.2, N 0.05, bal Fe.
RT: 88,000 psi TS; 59,000 psi YS; 40 El; 180 Brin. Used for MIG, TIG, plasma-arc and submerged-arc welding. W31640

SANDVIK 316/316H-15
Sandvik Steel Co.
Weld metal. C 0.07, Cr 19.5, Ni 11.5, Mo 2.0, Mn 1.8, Si 0.6, P 0.02, S 0.005, bal Fe.
RT: 92,000 psi TS; 65,000 psi YS; 39 El. Lime type electrode with a globular transfer and fast freezing slag designed for out of position welding. Suited for pipe welding. W31610

SANDVIK 316/316H-16
Sandvik Steel Co.
Weld metal. C 0.06, Cr 19.5, Ni 11.5, Mo 2.3, Mn 1.8, Si 0.4, P 0.02, S 0.005, bal Fe.
RT: 86,000 psi TS; 61,000 psi YS; 43 El. Rutile type electrode with a globular type transfer. W31610

SANDVIK 316/316H-16SA
Sandvik Steel Co.
Weld metal. C 0.06, Cr 19, Ni 11.5, Mo 2.3, Mn 0.6, Si 0.8, P 0.02, S 0.005, bal Fe.
RT: 86,000 psi TS; 61,000 psi YS; 43 El. Rutile type electrode with a spray type transfer and a high silicon, low manganese weld deposit. W31610

SANDVIK 316L
Sandvik Steel Co.
Weld metal. C 0.015, Si 0.3, Mn 1.8, P 0.01, S 0.01, Cr 18.5, Ni 12, Mo 2.6, N 0.05, bal Fe.
RT: 88,000 psi TS; 59,000 psi YS; 35 El; 160 Brin. Used for MIG, TIG, plasma-arc and submerged-arc welding. W31643

SANDVIK 316L-15
Sandvik Steel Co.
Weld metal. C 0.03, Cr 19, Ni 11.5, Mo 2.3, Mn 1.6, Si 0.6, P 0.02, S 0.005, bal Fe.
RT: 90,000 psi TS; 62,000 psi YS; 38 El. Lime type electrode with a globular transfer and fast freezing slag designed for out of position welding. Suited for pipe welding. W31613

SANDVIK 316L-16
Sandvik Steel Co.
Weld metal. C 0.03, Cr 19, Ni 11.5, Mo 2.3, Mn 1.8, Si 0.4, P 0.02, S 0.005, bal Fe.
RT: 83,000 psi TS; 62,000 psi YS; 45 El. Rutile type electrode with a globular type transfer. W31613

SANDVIK 316L-16SA
Sandvik Steel Co.
Weld metal. C 0.02, Cr 18.5, Ni 11.3, Mo 2.3, Mn 0.6, Si 0.8, P 0.02, S 0.005, bal Fe.
RT: 83,000 psi TS; 62,000 psi YS; 45 El. Rutile type electrode with a spray type transfer and a high silicon, low manganese weld deposit. W31613

SANDVIK 316LSI
Sandvik Steel Co.
Weld metal. C 0.02, Si 0.9, Mn 1.7, P 0.01, S 0.013, Cr 18.5, Ni 11.5, Mo 2.6, N 0.06, bal Fe.
RT: 88,000 psi TS; 58,000 psi YS; 37 El; 160 Brin. Filler metal suited for MIG welding. Also used for TIG and plasma-arc welding. W31648

SANDVIK 317L
Sandvik Steel Co.
Weld metal. C 0.02, Si 0.4, Mn 1.7, P 0.01, S 0.012, Cr 18.5, Ni 13, Mo 3.5, N 0.05, bal Fe.
RT: 87,000 psi TS; 55,000 psi YS; 47 El; 160 Brin. Used for MIG, TIG, plasma-arc and submerged-arc welding. W31743

SANDVIK 317L-15
Sandvik Steel Co.
Weld metal. C 0.03, Cr 19, Ni 12.5, Mo 3.5, Mn 1.7, Si 0.6, P 0.02, S 0.005, bal Fe.
RT: 88,000 psi TS; 69,000 psi YS; 40 El. Lime type electrode with a globular transfer and fast freezing slag designed for out of position welding. Suited for pipe welding. W31713

SANDVIK 317L-16
Sandvik Steel Co.
Weld metal. C 0.03, Cr 19.5, Ni 12.5, Mo 3.5, Mn 1.7, Si 0.4, P 0.02, S 0.005, bal Fe.
RT: 89,000 psi TS; 73,000 psi YS; 41 El. Rutile type electrode with a globular type transfer. W31713

SANDVIK 318
Sandvik Steel Co.
Weld metal. C 0.03, Si 0.4, Mn 1.3, P 0.012, S 0.011, Cr 18, Ni 11, Mo 2.6, Cb 0.55, N 0.05, bal Fe.
RT: 88,000 psi TS; 58,000 psi YS; 36 El; 220 Brin. Used for MIG, TIG, plasma-arc and submerged-arc welding. W31940

SANDVIK 318-15
Sandvik Steel Co.
Weld metal. C 0.06, Cr 19.5, Ni 11.5, Mo 2.3, Cb 0.7, Mn 1.9, Si 0.6, P 0.02, S 0.005, bal Fe.
RT: 90,000 psi TS; 62,000 psi YS; 38 El. Lime type electrode with a globular transfer and fast freezing slag designed for out of position welding. Suited for pipe welding. W31910

SANDVIK 318-16
Sandvik Steel Co.
Weld metal. C 0.06, Cr 19.5, Ni 11.5, Mo 2.3, Cb 0.7, Mn 1.9, Si 0.4, P 0.02, S 0.005, bal Fe.
RT: 90,000 psi TS; 62,000 psi YS; 38 El. Rutile type electrode with a globular type transfer. W31910

SANDVIK 318SI
Sandvik Steel Co.
Weld metal. C 0.4, Si 0.8, Mn 1.7, P 0.009, S 0.011, Cr 18, Ni 12, Mo 2.6, Cb 0.55, N 0.05, bal Fe.
RT: 88,000 psi TS; 58,000 psi YS; 35 El; 160 Brin. Filler metal suited for MIG welding. Also used for TIG and plasma-arc welding.

SANDVIK 320-15
Sandvik Steel Co.
Weld metal. C 0.06, Cr 19.5, Ni 33, Mo 2.5, Cb 0.6, Mn 1.5, Si 0.5, P 0.02, S 0.01, Cu 3.5, bal Fe.
RT: 84,000 psi TS; 51,000 psi YS; 38 El. Lime type electrode with a globular transfer and fast freezing slag designed for out of position welding. Suited for pipe welding. W88021

SANDVIK 320-16
Sandvik Steel Co.
Weld metal. C 0.06, Cr 19.5, Ni 33, Mo 2.5, Cb 0.5, Mn 1.5, Si 0.5, P 0.02, S 0.01, Cu 3.5, bal Fe.
RT: 84,000 psi TS; 51,000 psi YS; 38 El. Rutile type electrode with a globular type transfer. W88021

SANDVIK 330-15
Sandvik Steel Co.
Weld metal. C 0.21, Cr 16.5, Ni 35.5, Mn 1.8, Si 0.6, P 0.02, S 0.01, bal Fe.
RT: 89,000 psi TS; 62,000 psi YS; 35 El. Lime type electrode with a globular transfer and fast freezing slag designed for out of position welding. Suited for pipe welding. W88331

SANDVIK 330-16
Sandvik Steel Co.
Weld metal. C 0.21, Cr 16.5, Ni 35.5, Mn 1.8, Si 0.6, P 0.02, S 0.01, bal Fe.
RT: 89,000 psi TS; 62,000 psi YS; 35 El. Rutile type electrode with a globular type transfer. W88331

SANDVIK 347
Sandvik Steel Co.
Weld metal. C 0.03, Si 0.4, Mn 1.2, P 0.013, S 0.013, Cr 19.5, Ni 9.5, Cb 0.55, N 0.05, bal Fe.
RT: 88,000 psi TS; 58,000 psi YS; 42 El; 160 Brin. Used for MIG, TIG, plasma-arc and submerged-arc welding. W34740

SANDVIK 347-15
Sandvik Steel Co.
Weld metal. C 0.05, Cr 20, Ni 9.5, Cb 0.7, Mn 1.7, Si 0.6, P 0.02, S 0.005, bal Fe.
RT: 93,000 psi TS; 73,000 psi YS; 38 El. Lime type electrode with a globular transfer and fast freezing slag designed for out of position welding. Suited for pipe welding. W34710

SANDVIK 347-16
Sandvik Steel Co.
Weld metal. C 0.06, Cr 20, Ni 9.5, Cb 0.7, Mn 1.7, Si 0.4, P 0.02, S 0.005, bal Fe.
RT: 94,000 psi TS; 74,000 psi YS; 40 El. Rutile type electrode with a globular type transfer. W34710

SANDVIK 347SI
Sandvik Steel Co.
Weld metal. C 0.04, Si 0.8, Mn 1.6, P 0.013, S 0.01, Cr 19, Ni 10, Cb 0.5, N 0.05, bal Fe.
RT: 88,000 psi TS; 58,000 psi YS; 36 El; 220 Brin. Filler metal suited for MIG welding. Also used for TIG and plasma-arc welding. W34748

SANDVIK 3HS32
Sandvik Steel Co.
C 0.15, Si 0.15, Mn 0.95, Cr 1.35, Mo 0.9, V 0.25, bal Fe.
Tube: 135,000-164,000 TS; 114,000 YS; 10 El; 290-340 Vickers. Bainitic high strength steel tube. For mechanically loaded tubular construction. W. Nr. 1.7734.

SANDVIK 3L7
Sandvik Steel Co.
C 0.12, Si 0.15, Mn 0.7, bal Fe.
Normalized: 64,000 TS; 44,000 YS; 42 El; 72 RA; 125 Brin. For bus bodies; good welding properties.

SANDVIK 3LS
Sandvik Steel Co.
C 0.15, Si 0.15, Mn 0.5, bal Fe.
Normalized: 60,000 TS; 40,000 YS; 41 El; 65 RA; 115 Brin. For machine parts, gears, shafts; case hardened steel.

SANDVIK 3M01
Sandvik Steel Co.
C 0.15, Mo 0.5, bal Fe.
Annealed: 68,000 TS; 47,000 YS; 38 El; 135 Brin. For boiler and superheater tubes for steam; for steam temperatures up to 900°F.

SANDVIK 3R12
Sandvik Steel Co.
Stainless steel. C 0-0.03, Si 0.4, Mn 1.3, Cr 18.5, Ni 10.5, bal Fe.
1% offset yield, 210°F: 24,200 psi min. 1% offset yield, 1110°F: 15,600 psi min. Low carbon austenitic, stainless, weldable, for tubing for food and dairy plants, breweries, chemical and nuclear power plants. W. Nr. 1.4306, 1.4316; AISI 304L. S30403

SANDVIK 3R13
Sandvik Steel Co.
C 0.03, Si 0.45, Mn 0.85, Cr 18, Ni 12.5, bal Fe.
Austenitic, stainless steel wire for cold-heading; W. Nr. 4306; AISI 305.

SANDVIK 3R19
Sandvik Steel Co.
C 0.03, Si 0.6, Mn 1.3, Cr 18.5, Ni 9.5, N 0.18, bal Fe.
0.2% offset yield, 68°F: 40 ksi min. 0.2% offset yield, 660°F: 19.2 ksi min. Used largely for tubing carrying gas or fluid under pressure. W. Nr. 1.4311. ASTM 304L, N.

SANDVIK 3R60
Sandvik Steel Co.
Stainless steel. C 0-0.03, Si 0.6, Mn 1.7, Cr 17, Ni 13.6, Mo 2.6, bal Fe.
1% offset yield at 390°F: 24 ksi min. Austenitic, stainless tube steel, improved corrosion resistance, weldable. Miscellaneous high purity tubing applications where corrosion resistance is major requirement. W. Nr. 1.4435; AISI 316L. S31603

SANDVIK 3R64
Sandvik Steel Co.
Stainless steel. C 0.03, Si 0.6, Mn 1.7, Cr 18.5, Ni 14.5, Mo 3.1, bal Fe.
Annealed: 85,000 TS; 45,000 YS; 60 El; 80 Rock B. For paper and pulp mill equipment, chemical plant equipment, acid tanks, digesters, evaporators. Austenitic, stainless, Type 317L. S31703

SANDVIK 3R67
Sandvik Steel Co.
C 0-0.025, Si 0.4, Mn 1.8, P 0-0.015, Cr 19, Ni 13.5, Mo 3.3, bal Fe.
Drawn wire for MIG, TIG and submerged arc welding of Type 317 stainless steel. AISI 317.

SANDVIK 3R69
Sandvik Steel Co.
C 0-0.03, Si 0.6, Mn 1.7, P 0-0.03, S 0-0.03, Cr 17.5, Ni 13.5, Mo 2.8, N 0.18, bal Fe.
Austenitic, stainless ELC-steel. 600-800 N/mm^2 TS; 300 N/mm^2 YS min (0.2% offset); 30 El min; 40 El min; 165 Vickers (nominal). ASTM 316L(+N).

SANDVIK 3RE13
Sandvik Steel Co.
C 0-0.025, Si 0.4, Mn 1.8, P 0-0.015, S 0-0.015, Cr 24, Ni 13.5, bal Fe.
Wire for MIG, TIG and submerged arc welding of stainless steel. AISI 309.

SANDVIK 3RE60
Sandvik Steel Co.
Stainless steel. C 0-0.03, Si 1.7, Mn 1.5, Cr 18.5, Ni 4.9, Mo 2.7, N 0.07, bal Fe.
Tube: 92,000-128,000 TS; 64,000 YS min; 30 El; 260 Vickers. Ferritic-austenitic stainless with high resistance to stress corrosion and intergranular corrosion; largely for heat exchanger tubes. S31500

SANDVIK 3RN14
Sandvik Steel Co.
C 0.03, Si 0.45, Mn 0.85, Cr 16.5, Ni 18, bal Fe.
Austenitic, stainless steel wire for cold-heading.

SANDVIK 3RS17
Sandvik Steel Co.
C 0-0.025, Si 0.85, Mn 1.8, Cr 20.6, Ni 9.7, bal Fe.
Austenitic, stainless steel wire for metallizing; also for welding using MIG methods.

SANDVIK 3RS61
Sandvik Steel Co.
C 0-0.025, Si 0.85, Mn 1.7, P 0-0.015, Cr 18.5, Ni 10.5, Mo 2.7, bal Fe.
Drawn wire for MIG welding of Type 316 stainless steel. W. Nr. 4403/4430.

SANDVIK 3RS62
Sandvik Steel Co.
C 0-0.025, Si 0.85, Mn 1.8, Cr 19.5, Ni 13, Mo 2.3, bal Fe.
Austenitic, stainless steel wire for metallizing; similar to AISI 316L. Also used for MIG welding.

SANDVIK 410-15
Sandvik Steel Co.
Weld metal. C 0.08, Cr 12.5, Mn 0.6, Si 0.5, P 0.02, S 0.01, bal Fe.
RT: 86,000 psi TS; 64,000 psi YS; 27 El. Lime type electrode with a globular transfer and fast freezing slag designed for out of position welding. Suited for pipe welding. W41010

SANDVIK 410-16
Sandvik Steel Co.
Weld metal. C 0.08, Cr 12.5, Mn 0.6, Si 0.4, P 0.02, S 0.01, bal Fe.
RT: 86,000 psi TS; 64,000 psi YS; 27 El. Rutile type electrode with a globular type transfer. W41010

SANDVIK 410NIMO-15
Sandvik Steel Co.
Weld metal. C 0.05, Cr 11.5, Ni 4.5, Mo 0.5, Mn 0.6, Si 0.5, P 0.02, S 0.01, bal Fe.
RT: 110,000 psi TS; 85,000 psi YS; 15 El. Lime type electrode with a globular transfer and fast freezing slag designed for out of position welding. Suited for pipe welding. W41016

SANDVIK 410NIMO-16
Sandvik Steel Co.
Weld metal. C 0.05, Cr 11.5, Ni 4.5, Mo 0.5, Mn 0.6, Si 0.4, P 0.02, S 0.01, bal Fe.
RT: 110,000 psi TS; 85,000 psi YS; 15 El. Rutile type electrode with a globular type transfer. W41016

SANDVIK 430
Sandvik Steel Co.
Weld metal. C 0.04, Si 0.7, Mn 0.4, P 0.015, S 0.005, Cr 17, Ti 0.3, bal Fe.
160-200 Brin. Used for MIG, TIG, plasma-arc and submerged-arc welding. W43040

SANDVIK 430-15
Sandvik Steel Co.
Weld metal. C 0.08, Cr 15.5, Mn 0.6, Si 0.5, P 0.02, S 0.01, bal Fe.
RT: 65,000 psi TS; 45,000 psi YS; 21 El. Lime type electrode with a globular transfer and fast freezing slag designed for out of position welding. Suited for pipe welding. W43010

SANDVIK 430-16
Sandvik Steel Co.
Weld metal. C 0.08, Cr 15.5, Mn 0.6, Si 0.5, P 0.02, S 0.01, bal Fe.
RT: 65,000 psi TS; 45,000 psi YS; 21 El. Rutile type electrode with a globular type transfer. W43010

SANDVIK 4C27A
Sandvik Steel Co.
C 0.2, Ni 0.9, Mn 1, S 0.18, Cr 12-14, Mo 1.3, bal Fe.
Annealed: 95,000 TS; 50,000 YS; 25 El; 55 RA; 195 Brin. Cold drawn: 105,000 TS; 85,000 YS; 17 El; 50 RA; 215 Brin. For cutlery, valves, turbine blades, gages, gears; stainless, hardenable.

SANDVIK 4C54
Sandvik Steel Co.
Stainless steel. C 0.18, Si 0.5, Mn 0.8, Cr 26.5, N 0.15, bal Fe.
Ferritic chromium stainless steel. Yield strength at 1110°F: 15,500 psi min. Good resistance to scaling, sulfurous gases, oil-ash corrosion at elevated temperatures; for pyrometer protection tubes, soot-blowing tubes, heat exchangers. W. Nr. 1.4083; AISI 446. S44600

SANDVIK 4HSL31
Sandvik Steel Co.
C 0.18, Si 0.25, Mn 0.75, Cr 1.5, Mo 0.4, Al 0.03, B 0.004, bal Fe.
Tube: 121,000-142,000 TS; 100,000 YS; 15 El. Bainitic high strength steel tube. For mechanically loaded tubular construction.

SANDVIK 4L7
Sandvik Steel Co.
C 0.2, Mn 0.7, bal Fe.
Normalized: 71,000 TS; 51,000 YS; 33 El; 67 RA; 140 Brin. For bake oven tubes, boiler and superheater tubes for steam; steam temperatures up to 850°F.

SANDVIK 4LM
Sandvik Steel Co.
C 0.2, Mn 1.45, bal Fe.
Normalized: 80,000 TS; 48,000 YS; 28 El; 160 Brin. For weldable tubes, engineering purposes; tough.

SANDVIK 4LS
Sandvik Steel Co.
C 0.2, Si 0.2, Mn 0.1, bal Fe.
Normalized: 65,000 TS; 46,000 YS; 38 El; 60 RA; 130 Brin. For gears, pinions, shafts; case hardening steel.

SANDVIK 4N2C34
Sandvik Steel Co.
C 0.2, Si 0.3, Mn 0.5, Cr 16.6, Ni 2.2, bal Fe.
Martensitic, stainless steel; hardenable to about 205,000 psi TS. For stainless structural parts requiring high strength; good to about 900°F. W. Nr. 1.4057; AISI 431.

SANDVIK 5LMV
Sandvik Steel Co.
C 0.25, Mn 1.4, V 0.1, bal Fe.
Normalized: 60,000-75,000 TS; 45,000 YS min; 22 El; 180-220 Brin. Hardened: 78,000-95,000 TS; 70,000 YS min; 13 El; 240-280 Brin. Mechanical tubing, weldable, for tubular load carrying structures. W. Nr. 1.5213.

SANDVIK 5R10
Sandvik Steel Co.
Stainless steel. C 0-0.05, Si 0.6, Mn 1.3, Cr 18.5, Ni 9, bal Fe.
Annealed: 102,000 TS; 50,000 YS; 55 El. Half hard: 156,000 TS; 128,000 YS; 17 El. General purpose stainless steel for food, soft drink, brewery, dairy plant and for chemical plants. W. Nr. 4301; AISI 304. S30409

SANDVIK 5R60
Sandvik Steel Co.
Stainless steel. C 0-0.05, Si 0.6, Mn 1.7, Cr 17, Ni 12.5, Mo 2.6, bal Fe.
Austenitic, stainless with excellent corrosion resistance; for use in food and beverage plants, textile, chemical plants. W. Nr. 4436; AISI 316. S31609

SANDVIK 5R60
Sandvik Steel Co.
C 0.55, Si 0.6, Mn 1.7, P 0-0.02, S 0-0.02, Cr 16.5, Ni 13.5, Mo 2.3, bal Fe.
1% proof stress at 750°F: 22,800 psi min. 1% proof stress at 1290°F: 18,500 psi min. Austenitic, stainless with molybdenum. Good hot strength and resistance to scaling; for steam pipes and superheater tubes. ASTM A213 Gr. TP316H, etc.

SANDVIK 5RA50
Sandvik Steel Co.
Stainless steel. C 0-0.05, Si 0.4, Mn 1.7, S 0.2, Cr 18, Ni 9.5, Mo 0.5, bal Fe.
Austenitic; for food and chemical processing equipment; free machining type. W. Nr. 1.4305; AISI 303. S30300

SANDVIK 630-15
Sandvik Steel Co.
Weld metal. C 0.03, Cr 16.5, Ni 4.8, Mn 0.5, Si 0.5, P 0.02, S 0.01, Cu 3.6, bal Fe.
RT: 190,000 psi TS; 164,000 psi YS; 11 El. Lime type electrode with a globular transfer and fast freezing slag designed for out of position welding. Suited for pipe welding. W37410

SANDVIK 630-16
Sandvik Steel Co.
Weld metal. C 0.03, Cr 16.5, Ni 4.8, Mn 0.5, Si 0.5, P 0.02, S 0.01, Cu 3.6, bal Fe.
RT: 190,000 psi TS; 164,000 psi YS; 11 El. Rutile type electrode with a globular type transfer. W37410

SANDVIK 6C27
Sandvik Steel Co.
C 0.3, Cr 14, bal Fe.
Annealed: 104,000 TS; 74,000 YS; 28 El; 55 RA; 255 Brin. For electric razors; stainless, martensitic, hardenable.

SANDVIK 6HS63
Sandvik Steel Co.
C 0.3, Si 0.25, Mn 0.55, Cr 1, Ni 3.3, Mo 0.25, bal Fe.
Annealed (approx): 107,000 TS; 230 Vickers. Hardened: 156,000-199,000 TS; 128,000-149,000 YS; 10-12 El. Steel tubing, hardenable. For internal gears, female splined couplings. W. Nr. 1.5755.

SANDVIK 6LM
Sandvik Steel Co.
C 0.3, Si 0.2, Mn 1.05, bal Fe.
Normalized: 92,000 TS; 62,000 YS; 32 El; 61 RA; 180 Brin. For bicycle tubes, hollow blooms; water hardening, for cold drawing.

SANDVIK 6R42
Sandvik Steel Co.
C 0-0.06, Si 0.4, Mn 1.3, P 0-0.025, S 0-0.02, Cr 19.5, Ni 9.5, Nb 0.9, bal Fe.
Niobium stabilized wire for MIG, TIG, and submerged arc welding of austenitic stainless steel. W. Nr. 4551. AISI 347.

SANDVIK 6R60
Sandvik Steel Co.
C 0-0.04, Si 0.6, Mn 1.7, Cr 17.5, Ni 12.6, Mo 2.8, bal Fe.
Austenitic, stainless hard drawn wire for springs; 0.013-0.124 in. diameter wire has 220,000-277,000 psi TS; very good corrosion resistance. W. Nr. 4436; AISI 316.

SANDVIK 6R81
Sandvik Steel Co.
C 0-0.06, Si 0.4, Mn 1.3, P 0-0.025, S 0-0.02, Cr 19, Ni 11, Mo 2.7, bal Fe.
Niobium stabilized wire for MIG, TIG, and submerged arc welding of austenitic, stainless steel. W. Nr. 4576.

SANDVIK 6X10RN90
Sandvik Steel Co.
Cr 20, Ni 30, Al 1.6, Si 0-1, bal Fe.
Good creep strength at elevated temperatures.

SANDVIK 700
Sandvik Steel Co.
Now SANDVIK 2RK65.

SANDVIK 7C27MO2
Sandvik Steel Co.
C 0.35, Cr 13.5, Ni 0.4, Mo 1.2, bal Fe.
Annealed: 95,000 TS; 50,000 YS; 25 El; 55 RA; 195 Brin. For valves, cutlery, oil refinery equipment; corrosion resistant.

SANDVIK 7L
Sandvik Steel Co.
C 0.35, Si 0.2, Mn 0.5, bal Fe.
Normalized: 82,000 TS; 54,000 YS; 31 El; 50 RA; 170 Brin. For mechanical parts, axles, hydraulic cylinders; water hardened.

SANDVIK 7RE10
Sandvik Steel Co.
Stainless steel. C 0.055, Mn 1.7, Si 1.2, Cr 24.5, Ni 20.5, bal Fe.
Austenitic stainless steel. AISI 310S. S31008

SANDVIK 7RE12
Sandvik Steel Co.
C 0.055, Si 0.3, Mn 1.5, P 0-0.015, S 0-0.015, Cr 25.5, Ni 21, bal Fe.
Wire for MIG, TIG, and submerged arc welding of Type 310, stainless steel. W. Nr. 4842; AISI 310.

SANDVIK 8R30
Sandvik Steel Co.
Stainless steel. C 0-0.08, Si 0.6, Mn 1.5, Cr 17.5, Ni 10.5, Ti = 5 x C min, bal Fe.
1% proof stress at 750°F: 23,500 psi min. 1% proof stress at 1290°F: 17,800 psi min. Austenitic stabilized stainless for steam pipes and superheater tubes for temperatures up to about 800°C (1470°F). W. Nr. 4541. S32109

SANDVIK 8R40
Sandvik Steel Co.
Stainless steel. C 0.06, Si 0.4, Mn 1.7, Cr 17.5, Ni 11, Nb + Ta = 10 x C min, bal Fe.
1% proof stress at 750°F: 27,000 psi min. 1% proof stress at 1290°F; 20,500 psi min. Austenitic stabilized stainless for superheater tubes for temperatures up to about 1290°F. W.Nr. 1.4550. S34700

SANDVIK 8R41
Sandvik Steel Co.
Stainless steel. C 0.06, Si 0.4, Mn 1.3, Cr 15.5, Ni 13, Nb = 10 x C, bal Fe.
1% proof stress at 750°F: 27,000 psi min. 1% proof stress at 1290°F: 20,500 psi min. Austenitic stabilized stainless; mainly used for steam pipes and superheater tubes to temperatures of about 1290°F. W. Nr. 4961.

SANDVIK 8R42
Sandvik Steel Co.
C 0-0.06, Si 0.4, Mn 1.3, Cr 19.5, Ni 9.5, Nb 0.9, bal Fe.
Stabilized austenitic stainless steel wire for metallizing.

SANDVIK 8R70
Sandvik Steel Co.
Stainless steel. C 0-0.08, Si 0.5, Mn 1.7, Cr 17, Ni 13, Mo 2.1, Ti = 5 x C min, bal Fe.
Stabilized austenitic stainless steel; very good corrosion resistance even at elevated temperatures; for food and chemical industries. W. Nr. 1.4571. T-316+T.

SANDVIK 8R80
Sandvik Steel Co.
C 0-0.08, Cr 17.5, Ni 13.4, Mo 2.45, Cb = 10 x C, bal Fe.
Austenitic stainless steel. Similar to AISI 317.

SANDVIK 8R81
Sandvik Steel Co.
C 0-0.06, Si 0.45, Mn 1.3, Cr 19, Ni 11, Mo 2.8, Nb 0.9, bal Fe.
Austenitic stainless steel wire for metallizing.

SANDVIK CORONA 62
Sandvik Steel Co.
C 0.8, Si 0.3, Mn 0.3, Cr 4, Mo 5, W 6.3, V 2, bal Fe.
High speed steel metal band saw steel. AISI M2.

SANDVIK CORONA 67
Sandvik Steel Co.
C 0.92, Si 0.3, Mn 0.3, Cr 4, Mo 2, W 1, V 2.25, bal Fe.
Air or oil hardening metal band strip steel.

SANDVIK CT 525
Sandvik Steel Co.
16 WC, 37 TiC, 6 Ta(Nb)C, 9 Mo_2C, 12 TiN, 12 Co, 8 Ni.
Cermet grade.

SANDVIK CT515
Sandvik Steel Co.
16 WC, 38 TiC, 6 Ta(Nb)C, 10 Mo_2C, 12 TiN, 8 Co, 6 Ni,, 4 VC.
Titanium base carbide; resistance to oxidation; for surface finish in steel and stainless steel applications. 1650 Vickers.

SANDVIK CT520
Sandvik Steel Co.
16 WC, 37 TiC, 6 Ta(Nb)C, 9 Mo_2C, 12 TiN, 12 Co, 8 Ni.
Titanium base cermet for high speed machining of steel and stainless steel. 1450 Vickers.

SANDVIK GC 1025
Sandvik Steel Co.
Titanium coated carbide.

SANDVIK GC 215
Sandvik Steel Co.
86 WC, 2 TiC, 6 Ta(Nb)C, 6 Co.
Coated carbide: TiC + TiN.

SANDVIK GC-A
Sandvik Steel Co.
84 WC, 2 TiC, 6 Ta(Nb)C, 10 Co.
Coated carbide: TiN + TiC + TiN.

SANDVIK H10
Sandvik Steel Co.
93.5 WC, 0.5 Ta(Nb)C, 6 Co.
Carbide. For machining of aluminum. Good edge sharpness. 1750 Vickers.

SANDVIK H10F
Sandvik Steel Co.
89.5 WC, 0.5 Cr_3C_2, 10 Co.
Carbide. For machining of heat resistant and titanium alloys. Good edge strength. 1650 Vickers.

SANDVIK H13A
Sandvik Steel Co.
94 WC, 6 Co.
Carbide. For medium to rough turning of heat resistant and titanium alloys. 1600 Vickers.

SANDVIK H1P
Sandvik Steel Co.
85.5 WC, 7.5 TiC, 1 Ta(Nb)C, 6 Co.
Carbide. Finishing and light roughing of cast iron. For alloyed cast iron, bronze and brass. High cutting speeds and moderate feeds. 1850 Vickers.

SANDVIK HARDFLEX 11L
Sandvik Steel Co.
C 0.55, Si 0.25, Mn 0.5, bal Fe.
Bainite hardened strip steel: 128,000-164,000 TS; 107,000-142,000 YS; 10-14 El; 27-36 Rock C. For stamped and lightly formed parts. W. Nr. 1.1210; AISI 1055.

SANDVIK HARDFLEX 13LM
Sandvik Steel Co.
C 0.65, Si 0.2, Mn 0.75, bal Fe.
Bainite hardened strip steel: 192,000-228,000 TS; 171,000-199,000 YS; 6-8.5 El; 41-47 Rock C. For stamped and lightly formed parts. W. Nr. 0603/1.1231; AISI 1065.

SANDVIK HM
Sandvik Steel Co.
85 WC, 7.5 TiC, 7.5 Co.
Carbide. Uncoated grade for general purpose cast iron milling. Recommended for finishing. 1700 Vickers.

SANDVIK HML
Sandvik Steel Co.
94 WC, 6 Co.
Carbide. 1500 Vickers.

SANDVIK HT3
Sandvik Steel Co.
C 0-0.1, Cr 5, bal Fe.
Annealed: 70,000-72,000 TS; 28,000-30,000 YS; 28-36 El; 65-70 RA; 135-160 Brin. For oil refinery equipment; Type 502; heat and creep resistant.

SANDVIK HT5
Sandvik Steel Co.
C 0.12, Cr 0.9, Mo 0.55, bal Fe.
Annealed: 74,000 TS; 46,000 YS; 30 El; 65 RA; 170 Brin. For boiler and superheater tubes; for steam temperatures up to 950°F.

SANDVIK HT7
Sandvik Steel Co.
C 0.1, Mn 0.5, Cr 9, Mo 1, Si 0.7, bal Fe.
Annealed: 92,000 TS; 67,000 YS; 30 El; 73 RA; 195 Brin. For tubes in oil refineries and superheaters.

SANDVIK HT8 NB
Sandvik Steel Co.
C 0.08, Cr 2.25, Mo 1, Nb 1.001, bal Fe.
Annealed: 75,000 TS; 50,000 YS; 33 El; 76 RA; 170 Brin. For boiler and superheater tubes; for steam temperatures up to 1005°F.

SANDVIK HT9, HT9SB
Sandvik Steel Co.
C 0.2, Si 0.4, Mn 0.5, Cr 11.5, Ni 0.5, Mo 1, V 0.3, W 0.5, bal Fe.
At 570°F: 57,000 YS min. At 930°F: 43,000 YS min. Ferritic chromium steel for superheater tubes for temperatures up to about 1200°F. W. Nr. 1.4935.

SANDVIK HT91
Sandvik Steel Co.
C 0.2, Si 0.4, Mn 0.5, Cr 11.5, Ni 0.5, Mo 1, V 0.3, bal Fe.
0.2% proof stress at 570°F: 57,000 psi. 0.2% proof stress at 930°F: 43,000 psi. Ferritic chromium steel for superheater tubes to operate up to about 1200°F. W. Nr. 1.4922.

SANDVIK NI-CL
Sandvik Steel Co.
Weld metal. C 0.5, Ni 93, Mn 0.3, Si 1, Fe 3.
Extrusion coated electrode for welding of gray iron grades to themselves or to dissimilar metals where phosphorus levels are not too high. W82001

SANDVIK NICR-3
Sandvik Steel Co.
Weld metal. C 0.02, Si 0.2, Mn 3, P 0.008, S 0.01, Cr 20, Ni 72, Ti 0.2, Fe 1, Cb 2.5, bal Fe.
RT: 96,000 psi TS; 57,000 psi YS; 45 El; 160 Brin. Used for MIG, TIG, plasma-arc and submerged-arc welding. N06082

SANDVIK NICR-3HP
Sandvik Steel Co.
Weld metal. C 0.02, Si 0.2, Mn 3, P 0.005, S 0.004, Cr 20.5, Ni 72, Ti 0.2, Fe 0.5, Cb 2.5, bal Fe.
RT: 96,000 psi TS; 57,000 psi YS; 45 El; 180 Brin. Used for MIG, TIG, plasma-arc and submerged-arc welding. N06082

SANDVIK NICRFE-2
Sandvik Steel Co.
Weld metal. C 0.02, Cr 16.5, Ni 70, Mo 1.2, Cb 1.7, Mn 2.3, Si 0.4, P 0.015, S 0.005, Fe 7.
RT: 94,000 psi TS; 56,000 psi YS; 45 El. Lime-titania based electrode with a globular type transfer. W86133

SANDVIK NICRFE-3
Sandvik Steel Co.
Weld metal. C 0.03, Cr 16.5, Ni 70, Cb 1.4, Mn 6.5, Si 0.5, P 0.01, S 0.004, Fe 6.
RT: 93,000 psi TS; 57,000 psi YS; 50 El. Lime-titania based electrode with a globular type transfer. W86182

SANDVIK NICRFE-3H
Sandvik Steel Co.
Weld metal. C 0.05, Cr 15.5, Ni 69, Cb 1.8, Mn 6.4, Si 0.4, P 0.01, S 0.004, Fe 6.
RT: 103,000 psi TS; 72,000 psi YS; 40 El. Lime-titania based electrode with a globular type transfer.

SANDVIK NICRMO-3
Sandvik Steel Co.
Weld metal. C 0.02, Si 0.2, Mn 0.2, P 0.009, S 0.007, Cr 22, Ni 60, Mo 9, Fe 2.5, Cb 3.4, bal Fe.
RT: 97,000 psi TS; 62,000 psi YS; 42 El; 220 Brin. Used for MIG, TIG, plasma-arc and submerged-arc welding. N06625

SANDVIK NICRMO-3
Sandvik Steel Co.
Weld metal. C 0.03, Cr 21, Ni 61, Mo 9, Cb 3.3, Mn 0.5, Si 0.6, P 0.02, S 0.005, Fe 5.
RT: 112,000 psi TS; 72,000 psi YS; 35 El. Lime-titania based electrode with a globular type transfer. W86112

SANDVIK NIFE-CL
Sandvik Steel Co.
Weld metal. C 1, Ni 54, Mn 0.7, Si 0.5, Fe 44, Cu 1.
Extrusion coated electrode for welding of ductile iron and castings to themselves or to dissimilar metals, such as mild steel, stainless steel, or nickel base alloys. W82002

SANDVIK ON50HV
Sandvik Steel Co.
C 0-0.02, Si 0-0.08, Mn 0.45, Ni 51, bal Fe.
Soft magnetic iron-nickel alloy. Low coercive force; high permeability; coefficient of expansion same as lead glass; for reed relays.

SANDVIK R4
Sandvik Steel Co.
86.5 WC, 1 Ta(Nb)C, 12.5 Co.
Carbide. Uncoated grade for steel and stainless milling. 1250 Vickers.

SANDVIK S10T
Sandvik Steel Co.
55.5 WC, 19 TiC, 16 Ta(Nb)C, 9.5 Co.
Carbide. Threading grade for steel, cast steel and cast iron. 1650 Vickers.

SANDVIK S1P
Sandvik Steel Co.
55.5 WC, 19 TiC, 16 Ta(Nb)C, 9.5 Co.
Carbide. Finish and light roughing of steel and steel castings. High cutting speeds, moderate feeds. 1550 Vickers.

SANDVIK S6
Sandvik Steel Co.
77 WC, 4 TiC, 8 Ta(Nb)C, 11 Co.
Carbide. Roughing of steel, stainless steel, steel castings and milling. 1400 Vickers.

SANDVIK SANICRO 28
Sandvik Steel Co.
C 0-0.02, Si 0-1, Mn 0-2, P 0-0.02, S 0-0.015, Cr 27, Ni 31, Mo 3.5, Cu 1, bal Fe.
Austenitic, stainless ELC alloy. Seamless tube and pipe; materials for oil and gas production. N08028

SANDVIK SANICRO 30
Sandvik Steel Co.
Nickel. C 0-0.03, Si 0.5, Mn 0.5, Cr 22, Ni 34, Ti 0.5, bal Fe.
Nickel-iron-chromium alloy tube, wire. Good resistance to oxidation, stress corrosion, gases at elevated temperatures. W. Nr. 1.4876; ASTM B163; Alloy 800. N08800

SANDVIK SANICRO 31H
Sandvik Steel Co.
Nickel. C 0.07, Si 0.5, Mn 0.5, Cr 21, Ni 31, Ti 0.35, Al 0.35, bal Fe.
Nickel-iron-chromium alloy bar, tube, strip, wire. High resistance to scaling in air, to carburization, to nitrogen absorption, to stress corrosion; used in petroleum industry. W. Nr. 1.4876; ASTM B163, B407, B408, B409. N08810

SANDVIK SANICRO 70
Sandvik Steel Co.
Nickel. C 0-0.03, Si 0.4, Mn 0.8, Cr 16.5, Ni 72.5, Fe 9, Al 0.2, Ti 0.3.
Austenitic nickel base alloy; good resistance to scaling, carburizing and steam at elevated temperatures. W. Nr. 2.4816; ASTM B163, B166, B167. N06600

SANDVIK SANICRO 71
Sandvik Steel Co.
Nickel. C 0.05, Si 0.4, Mn 0.8, Cr 16, Ni 72.5, Ti 0.35, Al 0.25.
Austenitic nickel base alloy; good resistance to attack by most gases at elevated temperatures. W. Nr. 2.4816; ASTM B163, B166, B167. N06600

SANDVIK SANICRO 72
Sandvik Steel Co.
C 0-0.03, Si 0.2, Mn 3, P 0-0.01, S 0-0.01, Cr 20, Ni 72.5, Ti 0.32, Fe 0-1, Nb 2.5.
Wire for MIG, TIG, and submerged arc welding of nickel base alloys, stainless steel and carbon steel.

SANDVIK SM
Sandvik Steel Co.
70 WC, 9.5 TiC, 9.5 Ta(Nb)C, 11 Co.
Carbide. 1600 Vickers.

SANDVIK SM30
Sandvik Steel Co.
70 WC, 4 TiC, 16 Ta(Nb)C, 10 Co.
Carbide. Uncoated grade for steel milling. Better resistance to thermal cracking than SMA. 1550 Vickers.

SANDVIK SMA
Sandvik Steel Co.
64.5 WC, 12.5 TiC, 13 Ta(Nb)C, 10 Co.
Carbide. Uncoated grade for all purpose milling of steel. High wear resistance and used for mixed steel and cast iron milling. 1600 Vickers.

SANDVIK TITANIUM 20
Sandvik Steel Co.
Fe 0-0.3, C 0-0.05, N 0-0.03, O 0-0.2, H 0-0.015, bal Ti.
High purity titanium; for chemical industry, pulp and paper industry, electroplating racks, food industry, heat exchangers.

SANDVIK TITANIUM 9
Sandvik Steel Co.
Fe 0-0.15, C 0-0.035, N 0-0.01, O 0-0.09, H 0-0.005, bal Ti.
Very high purity titanium; for parts and tubing for nuclear reactors, sea water, acids, plating equipment; also used in wire form for welding titanium.

SANDVIK ZIRCALOY 2
Sandvik Steel Co.
Sn 1.2-1.7, Fe + Cr + Ni = 0.18-0.38, bal Zr (high purity).
For fuel element canning tubes in nuclear reactors. ASTM B353.

SANDVIK ZIRCALOY 4
Sandvik Steel Co.
Sn 1.2-1.7, Fe + Cr = 0.28-0.37, bal Zr (high purity).
For fuel element canning tubes in nuclear reactors. ASTM B353.

SANICRO
Now SANDVIK SANICRO.

SANS RIVAL
Societe Nouvelle du Saut-du-Tarn
C 0.7, W 18, Cr 4, V 1, bal Fe.
For tools, dies, cutters; high speed steel. *Obsolete*

SAPHIR
Ugine Aciers
C 1-1.2, bal Fe.
For drills, taps, reamers, shears, punches; Type W1, water hardened.

SARANAC
Colt Industries
C 0.9, Cr 0.9, bal Fe.
For tools. *Obsolete*

SARATOGA
AL Tech Specialty Steel Corp.
C 0.9, Si 0.38, W 0.5, Mn 1.2, Cr 0.5, bal Fe.
Usually at 56-60 Rock C. Oil hardening, non-deforming tool steel, suitable for blanking and forming dies, punches, plug gages and trim dies. AISI 0-1.

SARDO
Hidalgo Steel Co. Inc.
C 1.5, Cr 13, bal Fe.
For tools, dies; non-deforming, air hardening.

SASHCHAIN
Manufacturer not listed.
Cu 92-95, Sn 5-8.
For sash chain; bronze.

SATAN
A. Milne & Co.
C 0.23, Mn 0.6, Si 1.25, Cr 10, Ni 0.75, V 1, W 0.45, Mo 0.45, bal Fe.
Hot work tool steel; chromium type.

SATCO
NL Industries
Pb 97.5, Sn, bal Ca.
27 Brin. For bearings, crosshead gibs, trailer and driving boxes; white metal, for high speeds and heavy loads. *Obsolete*

SATMUMETAL
Telcon Metals Ltd.
Ni 53-58, bal Fe.
Maximum permeability 240,000, flux density 15,000, coercive force 0.025. Initial permeability 65,000. For small distribution transformers, instrument transformers, ground leakage protective devices, torroids. Soft magnetic alloy.

SATURN
Cytemp Specialty Steel Div.
C 1.25, W 3.5, bal Fe.
For wire drawing plates, dies, cutting tools, chasers, drills; fast finishing steel. *Obsolete*

SAUBRON HOT DIE
Sanderson Kayser Ltd.
W 9.5, Cr 2.5, V 0.1, C, bal Fe.
For hot dies; hot die steel. *Obsolete*

SAUREFEST ALLOYS
Now JUNKER.

SAUT-DU-TARN 18.CD.4F
Societe Nouvelle du Saut-du-Tarn
C 0.14-0.25, Cr 0.8-1.2, Mo 0.15-0.3, bal Fe.
Heat treated: 165,000-222,000 TS; 92,000 YS; 8 RA. For gears, bolts, cams, camshafts; oil hardened, tough.

SAUT-DU-TARN 25.CD.4F
Societe Nouvelle du Saut-du-Tarn
C 0.22-0.3, Cr 0.8-1.2, Mo 0.15-0.3, bal Fe.
Heat treated: 186,000-242,000 TS; 107,000 YS; 3 RA. For gears, bolts, crankshafts; oil hardened, tough.

SAUT-DU-TARN 30.NC.11F
Societe Nouvelle du Saut-du-Tarn
C 0.25-0.35, Cr 0.6-0.9, Ni 2.5-3, bal Fe.
Heat treated: 128,000-157,000 TS; 107,000 YS; 8 RA. For gears, bolts, crankshafts; oil hardened, shock resistant.

SAUT-DU-TARN 35.CD.4F
Societe Nouvelle du Saut-du-Tarn
C 0.3-0.39, Cr 0.8-1.2, Mo 0.15-0.3, bal Fe.
Heat treated: 221,000-245,000 TS; 2 RA. For gears, bolts, crankshafts; oil hardened, tough.

SAUT-DU-TARN 35.NC.15H
Societe Nouvelle du Saut-du-Tarn
C 0.3-0.38, Cr 1.5-1.9, Ni 3.5-4, bal Fe.
Heat treated: 242,000 TS; 5 RA. For gears, bolts, crankshafts; oil hardened, shock resistant.

SAUT-DU-TARN 35.NCD.12H
Societe Nouvelle du Saut-du-Tarn
C 0.25-0.35, Cr 0.7-1.2, Ni 2.7-3.3, Mo 0.25-0.45, bal Fe.
Heat treated: 150,000 TS; 127,000 YS. For gears, bolts, crankshafts; oil hardened, shock resistant.

SAUT-DU-TARN 35.NCD.6F
Societe Nouvelle du Saut-du-Tarn
C 0.3-0.38, Cr 0.8-1.2, Ni 1.2-1.6, Mo 0.15-0.3, bal Fe.
Heat treated: 165,000-200,000 TS; 144,000 YS. For gears, bolts, crankshafts; oil hardened, shock resistant.

SAUT-DU-TARN 45.S7
Societe Nouvelle du Saut-du-Tarn
C 0.4-0.5, Si 1.6-2.1, Mn 0.4-0.8, bal Fe.
Heat treated: 192,000-242,000 TS; 150,000 YS; 3 El; 241 Brin. For springs, torsion bars; oil hardened, tough.

SAUT-DU-TARN 55.S6
Societe Nouvelle du Saut-du-Tarn
C 0.55, Si 1.6-2.1, Mn 0.6-1, bal Fe.
Heat treated: 186,000-213,000 TS; 170,000 YS; 2 El; 270 Brin. For springs, torsion bars; oil hardened, tough.

SAUT-DU-TARN NO. 1
Societe Nouvelle du Saut-du-Tarn
C 0.9-1, bal Fe.
For tools, drills, punches; water hardened.

SAUT-DU-TARN NO. 2
Societe Nouvelle du Saut-du-Tarn
C 1.1, bal Fe.
For tools, drills; water hardened.

SAVA-Y
Alluminio SA
Aluminum. Cu 3.8-4.2, Mg 1.3-1.7, Ni 1.8-2.3, bal Al.
Cast: 36,000-47,000 TS; 32,700-42,700 YS; 0.5 El; 100-115 Brin. For pistons, cylinder heads; age hardenable, heat resistant.

SAVBIT
Multicore
Pb 48.5, Sn 50, Cu 1.5.
Solder to save copper erosion. MP 361°F. FP 419°F.

SAVILLE B 1
J.J. Saville & Co. Ltd.
C 0.55, Mn 0.6, bal Fe.
For rifle barrels, gear wheels, housings; water hardening.

SAVILLE BEST CAST STEEL
J.J. Saville & Co. Ltd.
C 0.7-1.2, Mn 0.3, bal Fe.
For blanking and forming dies, shear blades; water hardening.

SAVILLE CAST
J.J. Saville & Co. Ltd.
C 0.7-1.2, Mn 0.3, bal Fe.
For punches, shear blades, chisels, files; water hardening.

SAVILLE CAST STEEL WARRANTED
J.J. Saville & Co. Ltd.
C 0.7-1.2, Mn 0.3, bal Fe.
For wood working tools, punches, drills, taps; water hardening.

SAVILLE CROWN
J.J. Saville & Co. Ltd.
C 0.44, Mn 0.35, Cr 1.3, V 0.15, W 2.3, bal Fe.
For punches, chipping chisels, pneumatic tools; tough, shock resistant.

SAVILLE F 2
J.J. Saville & Co. Ltd.
C 0.57, Cr 0.7, V 0.2, bal Fe.
For die casting dies; hot die steel.

SAVILLE F.7
J.J. Saville & Co. Ltd.
C 0.48, Cr 1.5, W 1.2, V 0.12, Ni 0.7, bal Fe.
For chisels, rivet snaps, beading tools; shock resistant.

SAVILLE G-0
J.J. Saville & Co. Ltd.
C 0.33, Mn 0.5, Ni 3.5, Cr 0.75, bal Fe.
For bearing caps, valve rockers, propeller shafts; shock resistant.

SAVILLE G-1 SPECIAL
J.J. Saville & Co. Ltd.
C 0.32, Mn 0.5, Ni 4.1, Cr 1.2, Mo 0.25, bal Fe.
For gears, pinions, shafts; tough.

SAVILLE G-14
J.J. Saville & Co. Ltd.
C 0.39, Ni 4.1, Cr 1.3, Mo 0.3, bal Fe.
For chisels, cold punches, press tools; shock resistant.

SAVILLE G-18 B
J.J. Saville & Co. Ltd.
C 0.4, Si 1, Ni 13, Cr 13, Co 0.1, Nb 3, W 2.5, Mo 2, bal Fe.
At 70°F: 92,000 TS; 40 El. At 1300°F: 56,000 TS; 12 El. At 1500°F: 44,000 TS; 9 El. For aircraft engine exhaust valves; austenitic, high hot strength.

SAVILLE G-2
J.J. Saville & Co. Ltd.
C 0.42, Si 1.3, Ni 13, Cr 13, W 2.5, bal Fe.
At 70°F: 106,000 TS; 31 El. At 1100°F: 66,000 TS; 35 El. At 1500°F: 32,000 TS; 55 El. For aircraft engine exhaust valves; austenitic, resists shock at high temperature.

SAVILLE H-27
J.J. Saville & Co. Ltd.
C 0.37, Mn 0.6, Cr 3, Mo 0.9, V 0.2, bal Fe.
190,000 TS; 12 El; 435 Brin. For crankshafts, air screen shafts; shock resistant, nitriding steel.

SAVILLE H-30
J.J. Saville & Co. Ltd.
C 0.6, Mn 0.6, Cr 0.6, bal Fe.
Heat treated: 110,000-130,000 TS; 80,000-90,000 YS; 12-15 El; 300-341 Brin. For cylinder liners, gears, axles; water hardening.

SAVILLE H-42
J.J. Saville & Co. Ltd.
C 1.65, Mn 0.45, Cr 13, Mo 0.7, V 0.3, bal Fe.
For burnishing rolls, shear blades, master hobs; air hardening, nondeforming.

SAVILLE H-44
J.J. Saville & Co. Ltd.
C 1, Mn 0.7, Cr 1.45, bal Fe.
For press tools, shear blades, cold rolls; cold work steel.

SAVILLE J 23
J.J. Saville & Co. Ltd.
C 0.36, Si 0.4, Cr 2.6, V 1, W 1.8, Mo 4.3, bal Fe.
For die casting dies, swaging dies; hot die steel.

SAVILLE J-32
J.J. Saville & Co. Ltd.
C 0.32, Ni 4, Cr 1.5, W 5.8, V 0.35, bal Fe.
For hot punches, stamping dies; hot work steel.

SAVILLE K 18
J.J. Saville & Co. Ltd.
C 0.32, Si 0.95, Cr 5, W 1.4, Mo 1.7, bal Fe.
For die casting dies; hot die steel.

SAVILLE KSA
J.J. Saville & Co. Ltd.
C 0.9, Mn 1.85, Cr 0.45, W 0.45, bal Fe.
For gages, taps, dies, chasers, cams, pawls; nondeforming, oil hardening.

SAVILLE O.K.
J.J. Saville & Co. Ltd.
C 0.44, Cr 1.3, W 2.3, Si 0.7, V 0.15, bal Fe.
For dies and punches for hot work; resists heat checking.

SAVILLE PERMABRITE
J.J. Saville & Co. Ltd.
C 0.1-0.14, Cr, Ni, bal Fe.
For stainless steel parts; austenitic, super stainless.

SAVILLE RUSTLESS IRON
J.J. Saville & Co. Ltd.
C 0.15, Cr, Ni, bal Fe.
For stainless parts; stainless, dead soft type.

SAVILLE S.V.L.
J.J. Saville & Co. Ltd.
C 0.32, Ni 4.1, Cr 1.3, Mo 0.2, bal Fe.
Hardened: 215,000 TS; 190,000 YS; 12 El; 477 Brin. For plastic molding dies; shock resistant, air hardening.

SAVILLE STAINLESS STEEL
J.J. Saville & Co. Ltd.
C 0.28-0.32, Cr, Ni, bal Fe.
For stainless parts; stainless, hardening quality.

SAVILLE WPS
J.J. Saville & Co. Ltd.
C 2.3, Mn 0.35, Cr 13, bal Fe.
For gages, shear blades, thread rollers; resists wear and corrosion, nondeforming.

SAVILLE XLD
J.J. Saville & Co. Ltd.
C 0.82, Cr 5, Mo 0.6, V 1.7, W 22, Co 17, bal Fe.
For turning, planing and boring tools; high speed steel.

SAXONIA METAL
English manufacture
Zn 84.8, Sn 5.3, Cu 6, Pb 3, Al 0.2.
For bearings; will not resist heat or live steam.

SC
Thyssen Edelstahlwerke AG
C, Cr, Si, Mn, bal Fe.

SC SPECIAL
Teledyne Vasco
C 0.5, Mn 0.3, Cr 3, W 14, V 0.5, Si 0.3, bal Fe.
For hot work and extrusion dies; oil hardening, hot work steel.
Type H24.

SC-6
American Hoist & Derrick Co.
C 0.5, Mn 0.7, Ni 1.5, Cr 0.75, Mo 0.5, bal Fe.
Normalized: 180,000 TS; 140,000 YS; 8 El; 12 RA; 240 Brin.
For gears, pinions, shafts, axles; oil hardened, shock resistant.

SC1
Acciaierie Valbruna s.p.a.
C 0.53, Si 1.35, Mn 0.75, P 0-0.035, S 0-0.035, Cr 1, bal Fe.
Spring steel.

SC2
Acciaierie Valbruna s.p.a.
C 0.67, Si 1.3, Mn 0.5, P 0-0.035, S 0-0.035, Cr 0.5, bal Fe.
Spring steel. W. Nr. 1.7103.

SC23
Eagle & Globe Steel Ltd.
Tool material. C 2, Si 0.3, Mn 0.2, Cr 12, bal Fe.
Annealed: 248 Brin max. Oil hardening, high carbon, high chromium steel resistant to abrasive wear. For drawing dies, gauges, brick mold liners, thread rolling dies, cold forming rolls, burnishing tools, hobbing dies, lamination dies and dies for molding abrasive powders and corrosive plastics.
AS1239 D3A; AISI D3; BS4659 BD3; Werkstoff 1.2080.

SC25
Eagle & Globe Steel Ltd.
Tool material. C 1.5, Si 0.3, Mn 0.2, Cr 12, V 0.25, Mo 0.75, bal Fe.
Annealed: 241 Brin max. Air hardening cold work die steel. For blanking, molding, forming, thread rolling and drawing dies, shear blades, gauges, burnishing rolls. AS1239 D2A; AISI D2; BS4659 BD2; Werkstoff 1.2601.

SC46
Eagle & Globe Steel Ltd.
Tool material. C 2.25, Mn 0.7, Cr 5.25, Mo 1, W 1.35, V 4.25, bal Fe.
Cold work tool and die steel. Air hardened, for abrasion resisting applications. For brick and ceramic molds, punches and dies. AISI A7.

SC65
Delta Enfield Metals Ltd.
Ag 6-7, bal Cu.
Forged: 61,000-67,000 TS; 5-12 El; 115-140 Brin. 82-86% electrical conductivity. For heavy duty seam welding wheels. BS 4577 and ISO 5182 alloy A/4/3. *Obsolete*

SCANDIUM
Atomergic Chemetals Corp.
Sc.
Purities: special distilled 99.9%, 99.5%. Forms: ingot, lump, sheet, wire, foil.

SCASON
Pechiney/Eurotungstene
WC plus alloy.
Anti-skid and wear resistant alloy for anti-skid studs for tires; studded straps for snow and ice chains.

SCB OR SCB ALLOYS
Stauffer Chemical Corp.
All alloys inactive; Stauffer no longer supplies.

SCB-291
Now FANSTEEL SCB-291.

SCB-885
Now COLUMBIUM D-36.

SCB-990
Now FANSTEEL 80.

SCEPTRE BRASS
Manufacturer not listed.
Cu 62, Zn 36, Fe 1.4, Al 1.1, Pb 0.07.
For ornamental parts, hardware; high strength.

SCF-19 ALLOY
Carpenter Technology Corp.
Stainless steel. C 0.03, Mn 5, Si 0.4, P 0.025, S 0.003, Cr 20, Ni 18, Mo 5, N 0.35, bal Fe.
Annealed, RT: 434 MPa YS; 827 MPa TS; 52 El; 60 RA; 93 Rock B. Austenitic, nitrogen strengthened stainless steel with good stress/corrosion/cracking resistance, high strength and low magnetic permeability. For oil and gas well drilling components such as drill collars, stabilizers, MWD housings.

SCH 3
Creusot-Loire
C 0.45, Cr 9, Si 3, bal Fe.
Heat treated (quenched and tempered): 880 N/mm² psi TS.
Valve steel. AFNOR Z45CS9; W.-Nr. 1.4718.

SCH 4
Creusot-Loire
C 0.4, Cr 9.5, Si 2.3, Mo 2.3, bal Fe.
Heat treated (quenched and tempered): N/mm² psi TS min.
Valve steel. AFNOR Z40 CSD 10.

SCHENK DGZNA14 (Z400)
Schenk Leichtgusswerke
Al 3.5-4.3, Cu 0-0.6, Mg 0.02-0.05, bal Zn.
Die cast: 36,000 TS; 1.5 El; 70 Brin. For ornaments, instrument housings; high strength.

SCHENK DGZNA14CUL (Z400)
Schenk Leichtgusswerke
Al 3.5-4.3, Cu 0.6-1, Mg 0.02-0.05, bal Zn.
Die cast: 38,000 TS; 2 El; 80 Brin. For ornaments, instrument housings; high strength.

SCHENK DMGA1911
Schenk Leichtgusswerke
Al 7.6-10, Zn 0.1-1, Mn 0.1-0.6, bal Mg.
Die cast: 22,000-31,000 TS; 0.2-1 RA; 55 Brin. For instrument housings.

SCHENK GA1CUSI NO. 223
Schenk Leichtgusswerke
Cu 4-7, Si 2-4, bal Al.
Sand cast: 23,000-30,000 TS; 17,000-23,000 YS; 0.5-2 El; 75-100 Brin. For aircraft castings; age-hardenable.

SCHENK GA1MG3(CU) NO. 241
Schenk Leichtgusswerke
Mg 1.5-4, bal Al.
Sand cast: 20,000-27,000 TS; 11,000-14,500 YS; 2-6 El; 50-65 Brin. For marine castings; corrosion resistant.

SCHENK GA1MG3HY31
Schenk Leichtgusswerke
Mg 1.5-3.5, bal Al.
Sand cast: 19,900-27,000 TS; 11,400-14,300 YS; 3-8 El; 50-60 Brin. For marine castings; corrosion resistant.

SCHENK GA1MG5HYD51
Schenk Leichtgusswerke
Mg 4.5-5.5, Mn 0.1-0.5, bal Al.
Sand cast: 22,800-28,000 TS; 12,000-14,000 YS; 2-5 El; 55-70 Brin. For marine castings; corrosion resistant.

SCHENK GA1MGMN
Schenk Leichtgusswerke
Mg 1.5-3, Mn 0.6-1.3, Fe 0.5, bal Al.
Sand cast: 20,000-27,000 TS; 11,400-14,300 YS; 3-8 El; 50-60 Brin. For marine and chemical industry castings; corrosion resistant.

SCHENK GA1SI
Schenk Leichtgusswerke
Si 11-13.5, Mn 0.3-0.5, Ti 0.15, Fe 0.6, bal Al.
Sand cast: 24,200-31,000 TS; 11,400-12,800 YS; 4-8 El; 50-60 Brin. For automotive castings; corrosion resistant.

SCHENK GA1SI(CU) NO. 231
Schenk Leichtgusswerke
Si 11-13, bal Al.
Sand cast: 15,000-31,000 TS; 11,000-14,500 YS; 1-4 El; 50-60 Brin. For pressure proof castings; corrosion resistant.

SCHENK GA1SI5MG
Schenk Leichtgusswerke
Si 0.4-0.6, Mg 0.5-0.8, Mn 0.2-0.6, bal Al.
Sand cast: 21,000-27,000 TS; 14,000-19,000 YS; 1-3 El; 60-70 Brin. For light alloy parts.

SCHENK GA1SI6CU3 NO. 225
Schenk Leichtgusswerke
Si 5.5-7, Cu 2-4, Mn 0.4-0.6, bal Al.
Sand cast: 23,000-30,000 TS; 11,000-23,000 YS; 1-3 El; 65-80 Brin. For pressure tight castings; corrosion resistant.

SCHENK GA1SI9(CU) NO. 232
Schenk Leichtgusswerke
Si 7-11, Mn 0.2-0.5, bal Al.
Sand cast: 21,000-29,000 TS; 14,000-20,000 YS; 1-3 El; 65-85 Brin. For pressure tight castings; corrosion resistant.

SCHENK GA1SICU NO. 234
Schenk Leichtgusswerke
Si 5-6, Cu 1.2-1.6, Mg 0.4-0.6, bal Al.
Sand cast: 23,100-31,000 TS; 14,000-20,000 YS; 1-3 El; 65-85 Brin. For instrument housings; corrosion resistant.

SCHENK GA1SIMG
Schenk Leichtgusswerke
Si 9-10, Mg 0.25-0.4, Mn 0.3-0.5, Ti 0.15, Fe 0.6, bal Al.
Sand cast: 20,000-39,000 TS; 12,700-15,700 YS; 2-5 El; 55-65 Brin. For automotive castings; corrosion resistant.

SCHENK GA1SIMG(CU) NO. 233
Schenk Leichtgusswerke
Si 8.5-10.5, Mg 0.2-0.4, bal Al.
Sand cast: 27,000-35,000 TS; 13,000-16,000 YS; 2-5 El; 55-65 Brin. For pressure tight castings; corrosion resistant.

SCHENK GA1ZNCU NO. 212
Schenk Leichtgusswerke
Cu 3-5, Zn 4-6, Fe 1.2, Mn 0.6, Si 0-1.5, bal Al.
Sand cast: 21,000-30,000 TS; 13,000-19,000 YS; 1-3 El; 65-80 Brin. For light alloy parts.

SCHENK GDA1MG9
Schenk Leichtgusswerke
Mg 6-10, Mn 0.2-0.7, bal Al.
Die cast: 27,000-38,000 TS; 1-3 El; 65-85 Brin. For marine castings; corrosion resistant.

SCHENK GDA1MGSI
Schenk Leichtgusswerke
Mg 0.3-2.5, Si 1.5-5, Mn 0-1.5, bal Al.
Die cast: 22,800-27,000 TS; 1-3 El; 55-70 Brin. For marine castings, housings; corrosion resistant.

SCHENK GDA1SI(CU) NO. 231
Schenk Leichtgusswerke
Si 11-13, Mn 0.5, bal Al.
Die cast: 26,000-36,000 TS; 1-3 El; 55-75 Brin. For pressure tight castings; corrosion resistant.

SCHENK GDA1SI13
Schenk Leichtgusswerke
Si 11-13, Mn 0.2-0.7, Mg 0-0.5, bal Al.
Die cast: 28,000-36,000 TS; 1-3 El; 60-80 Brin. For thin wall castings; corrosion resistant.

SCHENK GDA1SI7
Schenk Leichtgusswerke
Si 6-10, Mn 0.2-0.7, Mg 0-0.5, bal Al.
Die cast: 24,000-35,000 TS; 1-3 El; 55-75 Brin. For thin wall castings; corrosion resistant.

SCHENK GDA1SICU NO. 311
Schenk Leichtgusswerke
Si 5-6.5, Cu 2-3, Mn 0.2-0.6, bal Al.
Die cast: 27,000-33,000 TS; 1-2.5 El; 55-75 Brin. For instrument housings; corrosion resistant.

SCHENK GMGA13ZN
Schenk Leichtgusswerke
Al 2.6-3.4, Zn 0.6-1.4, Mn 0.1-0.5, bal Mg.
Sand cast: 20,000-24,200 TS; 7000-8500 YS; 5-10 El; 40-50 Brin. For light alloy parts; Type AZ31.

SCHENK GMGA14ZN
Schenk Leichtgusswerke
Al 3.3-4.1, Zn 2.3-3.1, Mn 0.1-0.5, bal Mg.
Sand cast: 24,000-31,000 TS; 10,000-12,800 YS; 3 El; 45 Brin. For light alloy parts; Type AZF.

SCHENK GMGA16ZN
Schenk Leichtgusswerke
Al 5.3-6.1, Zn 2.3-3.1, Mn 0.1-0.5, bal Mg.
Sand cast: 18,600 TS; 11,500 YS; 1.5 El; 60 Brin. For light alloy parts; Type AZG.

SCHENK GMGA19
Schenk Leichtgusswerke
Al 7.8-8.8, Zn 0.1-0.8, Mn 0.1-0.5, bal Mg.
Sand cast: 24,000 TS; 13,000 YS; 4 El; 55 Brin. For light alloy parts; Type A9V.

SCHIRM ZAH
Dorrenberg Edelstahl GmbH
C 0.85, Si 0-0.25, Mn 0-0.25, bal Fe.
Heat treated: 190,000 TS; 140,000 YS; 10 El; 30 RA; 400 Brin. For springs, taps, reamers, broaches, drills; Type W1; water hardened.

SCHIRM ZH
Dorrenberg Edelstahl GmbH
C 1, Si 0-0.25, Mn 0-0.25, bal Fe.
Annealed: 100,000 TS; 53,000 YS; 21 El; 42 RA; 200 Brin. For springs, taps, reamers, drills, lathe tools; Type W1; water hardened.

SCHMIDT LOCOMOTIVE BEARING
English manufacture
Cu 86, Sn 14.
For bearings; bronze.

SCHMIDT TSS
Karl Schmidt Co.
Aluminum. Mg 2.4-4, Mn 0-0.4, Cr 0-0.3, bal Al.
Soft: 28,000 TS; 13,000 YS; 30 El; 47 Brin. Hard: 40,000 TS; 35,000 YS; 10 El; 73 Brin. For aircraft tanks and fittings, fuel lines, marine parts; resists sea water corrosion.

SCHMIDTCLEMENS PHM, ETC.
Now MARKER or MAERKER.

SCHNEIDER 0.5 FO8
Creusot-Loire
C 0.15, Si 0.25, Mn 0.5, Cr 0.55, Mo 0.55, bal Fe.
Normalized: 65,000 TS; 40,000 YS; 22 El. For turbine vanes, oil refinery equipment; tough, shock resistant. *Obsolete*

SCHNEIDER 1
Creusot-Loire
C 0-0.15, Cr 11.5-13.5, bal Fe.
Heat treated: 120,000-135,000 TS; 110,000-117,000 YS; 15-16 El; 58-63 RA; 220-240 Brin. For cutlery, valves, turbine blades; Type 410; corrosion resistant. *Obsolete*

SCHNEIDER 1 (NI)
Creusot-Loire
C 0-0.15, Cr 12-14, Ni 1.25-2.5, 0.07% P min, S, or Se, bal Fe.
Annealed: 75,000 TS; 40,000 YS; 30 El; 60 RA; 155 Brin. Cold drawn: 100,000 TS; 85,000 YS; 13 El; 50 RA; 205 Brin. For screw machine products, valve trim, pump shafts; Type 416; free-cutting, stainless. *Obsolete*

SCHNEIDER 1.1 FO8
Creusot-Loire
C 0.13, Si 0.25, Mo 0.5, Cr 1, Mo 0.55, bal Fe.
Normalized: 65,000 TS; 40,000 YS; 22 El. For oil refinery equipment, heat exchanger tubes; tough, creep resistant. *Obsolete*

SCHNEIDER 1.3 FOV
Creusot-Loire
C 0.24, Si 0.3, Mn 0.5, Cr 1.25, Mo 0.6, V 0.25, bal Fe.
Heat treated: 121,000 TS; 101,000 YS; 12 El. For bolts, fasteners, turbine rotors and discs; operating temperature 530 C maximum.- *Obsolete*

SCHNEIDER 1.4DF3
Creusot-Loire
Mn 0.6-0.9, Cr 0.8-1.2, Ni 1.2-1.6, 0.35% C min, bal Fe.
Hardened: 170,000-235,000 TS; 148,000-215,000 YS; 6-10 El. For gears, shafts; tough. *Obsolete*

SCHNEIDER 1.4DF5
Creusot-Loire
Mn 0.6-0.9, Cr 0.8-1.2, Ni 1.2-1.6, 0.25% C min, bal Fe.
Hardened: 158,000-225,000 TS; 135,000-191,000 YS; 6-11 El. For tools, gears, shafts; tough. *Obsolete*

SCHNEIDER 1.4DF6
Creusot-Loire
Mn 0.6-0.9, Cr 0.8-1.2, Ni 1.2-1.6, 0.2% C min, bal Fe.
Hardened: 142,000-200,000 TS; 122,000-170,000 YS; 7-12 El. For case hardened parts, gears, shafts; tough. *Obsolete*

SCHNEIDER 1.4DF8C
Creusot-Loire
Mn 0.6-0.9, Cr 0.8-1.2, Ni 1.2-1.6, 0.10% C min, bal Fe.
Hardened: 121,000-162,000 TS; 92,000-140,000 YS; 9-14 El. For case hardened parts; tough. *Obsolete*

SCHNEIDER 11
Creusot-Loire
C 0.08-0.2, Mn 0-2, Cr 17-19, Ni 8-10, bal Fe.
Annealed: 85,000 TS; 35,000 YS; 60 El; 70 RA; 150 Brin. Cold drawn: 125,000 TS; 95,000 YS; 22 El; 55 RA; 277 Brin. For oil refinery chemical plant equipment; Type 302; stainless, austenitic. *Obsolete*

SCHNEIDER 11DS
Creusot-Loire
C 0-0.08, Cr 18-20, Ni 8-11, Mn 0-2, bal Fe.
Annealed: 90,000 TS; 45,000 YS; 60 El; 135 Brin. Cold drawn: 180,000 TS; 150,000 YS; 10 El; 330 Brin. For chemical plant equipment, welded structures; Type 304; stainless, austenitic. *Obsolete*

SCHNEIDER 13
Creusot-Loire
C 0.08-0.2, Cr 17-19, Ni 8-10, Si 2-3, Mn 0-2, bal Fe.
Annealed: 90,000 TS; 40,000 YS; 50 El; 65 RA; 160 Brin. For chemical plant equipment; austenitic, stainless; Type 302B. *Obsolete*

SCHNEIDER 13 FF
Creusot-Loire
C 2.25, Cr 13, bal Fe.
For tools, dies; non-deforming. *Obsolete*

SCHNEIDER 14
Creusot-Loire
C 0-0.1, Cr 16-18, Ni 10-14, Mo 2-3, bal Fe.
Annealed: 85,000-95,000 TS; 35,000-45,000 YS; 50-60 El; 60-75 RA; 150-190 Brin. For acid-resistant chemical plant equipment; Type 316; stainless, austenitic. *Obsolete*

SCHNEIDER 14DF7C
Creusot-Loire
Mn 0.6-0.9, Cr 0.8-1.2, Ni 1.2-1.6, 0.16% C min, bal Fe.
Hardened: 128,000-184,000 TS; 110,000-155,000 YS; 8-13 El; 60 RA. For case hardened parts, gears, shafts; tough. *Obsolete*

SCHNEIDER 14S
Creusot-Loire
C 0-0.1, Cr 18-20, Ni 11-14, Mo 3-4, bal Fe.
Annealed: 85,000-95,000 TS; 35,000-45,000 YS; 50-60 El; 60-75 RA; 150-190 Brin. For acid resistant chemical plant equipment; Type 317; stainless, austenitic. *Obsolete*

SCHNEIDER 14S-TI
Creusot-Loire
C 0-0.1, Cr 16-18, Ni 10-14, Mo 1.7-2.7, Ti = 10 x C, bal Fe.
Annealed: 85,000-95,000 TS; 35,000-45,000 YS; 50-60 El; 60-75 RA; 150-190 Brin. For acid resistant chemical plant equipment; Type 316 Ti; stainless, austenitic. *Obsolete*

SCHNEIDER 15
Creusot-Loire
C 0-0.08, Cr 17-19, Ni 8-11, Ti = 5 x C, bal Fe.
Normalized: 93,000 TS; 36,000 YS; 45 El; 60 RA; 165 Brin. Annealed: 87,000 TS; 36,000 YS; 57 El; 73 RA; 155 Brin. For chemical plant equipment, welded structures; Type 321; austenitic, stainless. *Obsolete*

SCHNEIDER 1A
Creusot-Loire
Cr 12-14, 0.15% C min, bal Fe.
Annealed: 88,000 TS; 40,000 YS; 32 El; 68 RA; 170 Brin. Heat treated: 256,000 TS; 190,000 YS; 6 El; 10 RA; 540 Brin. For cutlery, valve trim, springs, turbine blades; Type 420; stainless, hardenable. *Obsolete*

SCHNEIDER 1F 3/4
Creusot-Loire
Si 0-0.4, Mn 0.6-0.9, Cr 0.7-1.1, 0.28% C min, bal Fe.
Hardened: 182,000-270,000 TS; 142,000-210,000 YS; 4-9 El. For gears, shafts; water hardening. *Obsolete*

SCHNEIDER 1F 4/5
Creusot-Loire
Si 0-0.4, Mn 0.6-0.9, Cr 0.7-1.1, 0.33% C min, bal Fe.
Hardened: 148,000-220,000 TS; 128,000-189,000 YS; 5-10 El; 50 Brin. For tools, gears, shafts; water hardening. *Obsolete*

SCHNEIDER 1S
Creusot-Loire
C 0-0.08, Cr 11.5-13, Al 0.1-0.3, bal Fe.
Normalized: 71,000 TS; 42,600 YS; 22 El; 70 RA; 150 Brin.
Heat treated: 175,000 TS; 145,000 YS; 21 El; 64 RA; 352 Brin.
For oil refinery and chemical plant equipment; Type 405; corrosion resistant. *Obsolete*

SCHNEIDER 2
Creusot-Loire
Cr 12-14, 0.15% C min, bal Fe.
Annealed: 88,000 TS; 40,000 YS; 32 El; 68 RA; 170 Brin. Heat treated: 256,000 TS; 190,000 YS; 6 El; 10 RA; 540 Brin. For cutlery, valve trim, springs, turbine blades; Type 420; stainless, hardenable. *Obsolete*

SCHNEIDER 2.2 FO
Creusot-Loire
C 0.1, Si 0.3, Mn 0.5, Cr 2.2, Mo 1, bal Fe.
Heat treated: 60,000 TS; 30,000 YS; 20 El. For heat exchanger tubes, oil refinery equipment; creep and shock resistant. *Obsolete*

SCHNEIDER 2DFO3
Creusot-Loire
C 0.25, Ni 2.25, Cr 0.55, Mo 0.4, bal Fe.
Heat treated: 120,000 TS; 100,000 YS; 14 El. For bolts, fasteners, gears, shafts; oil hardened, shock resistant. *Obsolete*

SCHNEIDER 3
Creusot-Loire
C 0-0.15, Cr 12-14, bal Fe.
Annealed: 88,000 TS; 40,000 YS; 32 El; 68 RA; 170 Brin. Heat treated: 256,000 TS; 190,000 YS; 6 El; 10 RA; 540 Brin. For cutlery, valve trim, turbine blades; Type 420; stainless, hardenable. *Obsolete*

SCHNEIDER 31
Creusot-Loire
C 0.25, Cr 23-30, Ni 0-5, bal Fe.
Normalized: 93,000 TS; 71,000 YS; 20 El; 48 RA; 210 Brin.
For valves, pumps, furnace equipment; Type 327; heat resistant. *Obsolete*

SCHNEIDER 3FDO1
Creusot-Loire
Mn 0.6-1, Cr 2.5-3, Ni 1.2-1.6, Mo 0.2-0.4, 0.33% C min, bal Fe.
Hardened: 185,000-255,000 TS; 162,000-225,000 YS; 5-8 El.
For gears, pinions, shafts; tough. *Obsolete*

SCHNEIDER 3FO
Creusot-Loire
C 0.25, Cr 3, Mo 0.5, bal Fe.
Normalized: 100,000 TS; 72,000 YS; 16 El. For turbine rotors, oil refinery equipment; good creep properties. *Obsolete*

SCHNEIDER 4
Dunford Hadfields Ltd.
  Obsolete

SCHNEIDER 5.1 B
Creusot-Loire
C 0-0.12, Cr 14-18, bal Fe.
Annealed: 70,000 TS; 40,000 YS; 30 El; 55 RA; 150 Brin. Cold drawn: 130,000 TS; 120,000 YS; 2 El; 185 Brin. For oil refinery equipment; bolts, kitchen sinks; Type 430; stainless, ferritic. *Obsolete*

SCHNEIDER 60
Creusot-Loire
C 0-0.2, Cr 22-24, Ni 12-15, Mn 0-2, Si 0-1, bal Fe.
Annealed: 85,000-95,000 TS; 40,000-50,000 YS; 45-55 El; 150-185 Brin. For heat treating boxes, oil refinery and chemical plant equipment; Type 309; austenitic, heat resistant. *Obsolete*

SCHNEIDER 61
Creusot-Loire
C 0-0.25, Cr 24-26, Ni 19-22, bal Fe.
Annealed: 95,000 TS; 45,000 YS; 50 El; 65 RA; 180 Brin. At 1200 F: 57,000 TS; 22,000 YS; 32 El; 45 RA. For furnace parts and equipment, valves, pumps; Type 310; austenitic, heat resistant. *Obsolete*

SCHNEIDER 63
Creusot-Loire
C 0-0.25, Cr 23-26, Ni 19-22, Si 1.5-3, bal Fe.
Annealed: 100,000 TS; 50,000 YS; 45 El; 60 RA; 185 Brin. For furnace parts, radiant tubes, heat treating boxes; Type 314; oxidation resistant. *Obsolete*

SCHNEIDER A10/11B
Creusot-Loire
C 0.16, Si 0.2, Mn 0.55, bal Fe.
Annealed: 53,000 TS; 31,000 YS; 26 El. For gears, cams, machine tool parts; case hardening steel. *Obsolete*

SCHNEIDER A11B
Creusot-Loire
C 0.14, Si 0.2, Mn 0.45, bal Fe.
Annealed: 47,000 TS; 29,000 YS; 30 El. For gears, cams, machine tool parts; case hardening steel. *Obsolete*

SCHNEIDER A8B
Creusot-Loire
C 0.22, Si 0.35, Mn 0.65, bal Fe.
Annealed: 70,000 TS; 38,000 YS; 21 El. For gears, cams, machine tool parts; case hardening steel. *Obsolete*

SCHNEIDER A9/10B
Creusot-Loire
C 0.18, Si 0.25, Mn 0.6, bal Fe.
Annealed: 61,000 TS; 35,000 YS; 24 El. For gears, cams, machine tool parts; case hardening steel. *Obsolete*

SCHNEIDER C.10
Creusot-Loire
C 1.05-1.15, bal Fe.
For tools, cutters; water hardening. *Obsolete*

SCHNEIDER C.12
Creusot-Loire
C 1.2, bal Fe.
For tools, cutters; water hardening. *Obsolete*

SCHNEIDER C.9
Creusot-Loire
C 0.9-1, bal Fe.
For tools, cutters, drills; water hardening. *Obsolete*

SCHNEIDER C1
Creusot-Loire
C 0.7-1.2, bal Fe.
For tools, dies, drills; water hardened. *Obsolete*

SCHNEIDER C2-C2B
Creusot-Loire
C 1.1, bal Fe.
For tools, drills, taps; water hardened. *Obsolete*

SCHNEIDER C4
Creusot-Loire
C 0.95, bal Fe.
For tools, punches, springs; water hardened. *Obsolete*

SCHNEIDER C5
Creusot-Loire
C 0.75, bal Fe.
For tools; springs; water hardened. *Obsolete*

SCHNEIDER C6-C6S
Creusot-Loire
C 0.75, bal Fe.
For tools; springs; water hardened. *Obsolete*

SCHNEIDER CMP
Creusot-Loire
C 0.95, Mn 1.2, Cr 0.5, W 0.5, bal Fe.
For tools, dies, cutters; non-deforming. *Obsolete*

SCHNEIDER DFO
Creusot-Loire
C 0.3, Ni 2.5, Cr 0.7, Mo 0.3, bal Fe.
Annealed: 100,000 TS; 72,000 YS; 14 El. For rotors and rotor discs; ferritic, shock resistant. *Obsolete*

SCHNEIDER F.P.
Creusot-Loire
C 0.5, Cr 1.5, W 2.2, V 0.2, bal Fe.
For hot work dies and tools; hot work steel. *Obsolete*

SCHNEIDER F.P.B.
Creusot-Loire
C 0.5, Cr 1.5, W 2.2, V 0.2, bal Fe.
For hot work dies and tools; hot work steel. *Obsolete*

SCHNEIDER FLO6
Creusot-Loire
C 0.15, Si 1.25, Mn 0.5, Cr 1, Mo 0.55, bal Fe.
Normalized: 83,000 TS; 58,000 YS; 20 El. For oil refinery equipment, heat exchangers; creep and shock resistant. *Obsolete*

SCHNEIDER FLOV
Creusot-Loire
C 0.18, Si 1, Mn 0.5, Cr 1.25, Mo 1.1, V 0.45, bal Fe.
Heat treated: 121,000 TS; 101,000 YS; 12 El. For bolts, turbine discs and rotors; used up to 575 C. *Obsolete*

SCHNEIDER FO 5/6
Creusot-Loire
Mn 0.6-0.9, Cr 0.8-1.2, Mo 0.15-0.35, 0.25% C min, bal Fe.
Hardened: 142,000-225,000 TS; 107,000-180,000 YS; 6-12 El; 50 RA. For gears, shafts; tough. *Obsolete*

SCHNEIDER FO 6/7 S
Creusot-Loire
C 0.18-0.25, Mn 0.4-0.7, Cr 0.8-1.2, Mo 0.15-0.35, bal Fe.
Hardened: 100,000-128,000 TS; 89,000-106,000 YS; 13-15 El; 53 RA. For case hardened parts, gears, shafts; tough. *Obsolete*

SCHNEIDER FO3
Creusot-Loire
Mn 0.6-0.9, Cr 0.8-1.2, Mo 0.15-0.35, 0.42% C min, bal Fe.
Hardened: 185,000 TS; 155,000 YS; 8 El; 40 RA. For gears, pinions, shafts; tough. *Obsolete*

SCHNEIDER FO4
Creusot-Loire
Mn 0.6-0.9, Cr 0.8-1.2, Mo 0.15-0.35, 0.34% C min, bal Fe.
Hardened: 162,000-245,000 TS; 141,000-201,000 YS; 5-10 El; 45 RA. For gears, pinions, shafts; tough. *Obsolete*

SCHNEIDER FO7C
Creusot-Loire
Mn 0.6-0.9, Cr 0.8-1.2, Mo 0.15-0.35, 0.18% C min, bal Fe.
Hardened: 130,000-198,000 TS; 114,000-161,000 YS; 7-13 El.
For case hardened parts, gears, shafts; tough. *Obsolete*

SCHNEIDER FO8C
Creusot-Loire
Mn 0.6-0.9, Cr 0.8-1.2, Mo 0.15-0.35, 0.12% C min, bal Fe.
Hardened: 160,000 TS; 100,000-130,000 YS; 9-14 El. For gears, shafts; case hardening steel. *Obsolete*

SCHNEIDER FOV
Creusot-Loire
C 0.25, Si 0.3, Mn 0.6, Cr 1, Mo 1, V 0.25, bal Fe.
Normalized: 101,000 TS; 78,000 YS; 16 El. For turbine rotors and discs; used up to 575 C. *Obsolete*

SCHNEIDER FR
Creusot-Loire
C 0.95-1.1, Si 0-0.4, Mn 0-0.4, Cr 1.3-1.6, bal Fe.
Hardened: 620 Brin. For tools, dies, cutters; oil or water
hardening. *Obsolete*

SCHNEIDER M.L.
Creusot-Loire
C 0.5, Mn 0.7, Si 1.9, Mo 0.2, bal Fe.
Hardened: 190,000 TS; 160,000 YS; 7 El; 40 RA. For tools,
dies, punches; tough. *Obsolete*

SCHNEIDER M.L.S.
Creusot-Loire
C 0.5, Mn 0.7, Si 1.9, Mo 0.2, bal Fe.
For tools, dies, punches; tough. *Obsolete*

SCHNEIDER MO7
Creusot-Loire
C 0.15, Mn 1, Si 0.25, Mo 0.55, bal Fe.
Normalized: 70,000 TS; 36,000 YS; 20 El. For camshafts,
gears, oil refinery equipment; case-hardening, tough.
Obsolete

SCHNEIDER P2
Creusot-Loire
C 1.3, W 4.5, bal Fe.
For tools, dies; oil hardened. *Obsolete*

SCHNEIDER S.3
Creusot-Loire
C 0.8, W 18, Cr 4, V 1, bal Fe.
For drills, reamers, broaches, form tools, cutters, lathe and
planer tools. Type T1 high-speed steel. *Obsolete*

SCHNEIDER S.3V
Creusot-Loire
C 0.8, W 19, Cr 4, V 2.25, bal Fe.
For tools, dies, cutters, hobs, drills, reamers, broaches. Type
T2 high speed steel. *Obsolete*

SCHNEIDER S.I.S.
Creusot-Loire
W 9.5, Cr 2.5, V 0.1, C, bal Fe.
For hot work dies and tools; hot work steel. *Obsolete*

SCHNEIDER S.I.S.B.
Creusot-Loire
W 9.5, Cr 2.5, V 0.1, C, bal Fe.
For hot work dies and tools; hot work steel. *Obsolete*

SCHNEIDER S.I.S.B.-S.I.
Creusot-Loire
W 9.5, Cr 2.5, V 0.1, C, bal Fe.
For hot work dies and tools; hot work steel. *Obsolete*

SCHNEIDER S.S. 8
Creusot-Loire
C 0.7, W 18, Co 9, Mo 1, Cr 4, V 2, bal Fe.
For tools, dies, cutters; high speed steel. *Obsolete*

SCHNEIDER SO3-M2
Creusot-Loire
C 0.7, Cr 3.9-4.3, W 6.6-6.75, Mo 4.75-5.25, V 2, bal Fe.
For tools, cutters; high speed steel. *Obsolete*

SCHNEIDER SS4
Creusot-Loire
C 0.7, W 18, Co 5, Cr 4, V 1.3, bal Fe.
For lathe and planer tools, drills, taps, reamers, hobs; high
speed steel, oil hardened. *Obsolete*

SCHOELLER ARWA
Now VEW N208.

SCHOELLER ARWKN
Now VEW N206.

SCHOELLER MAO
Now VEW A500.

SCHOELLER MAO SUPERIOR
Now VEW A604.

SCHOELLER MAOY
Now VEW A500.

SCHOELLER MASD
Now VEW A122.

SCHOELLER MASN
Now VEW A350.

SCHOELLER MASO
Now VEW A120.

SCHOELLER MASO SUPERIOR
Now VEW A200.

SCHOELLER MASOY
Now VEW A100.

SCHOELLER MASOY SUPERIOR
Now VEW A205.

SCHOELLER MASWY
Now VEW A305.

SCHOELLER R1
Now VEW H525.

SCHOELLER R9
Now VEW H550.

SCHOLLER 1116S
Vereinigte Edelstahlwerke
C 0.75, Si 0.4, Mn 0.6, bal Fe.
Heat treated: 175,000 TS; 128,000 YS; 12 El; 37 RA; 370 Brin.
For punches, crimpers, rails, springs; water hardened.
Obsolete

SCHOLLER AMC
Vereinigte Edelstahlwerke
C 0.5, Cr 1, Mo 0.2, bal Fe.
For gears, pinions, crankshafts, bolts, studs; oil hardened,
shock resistant. *Obsolete*

SCHOLLER ARG
Now VEW W501.

SCHOLLER ARH HART
Now VEW N540.

SCHOLLER ARH ZAH
Now VEW N324.

SCHOLLER ARKS
Now VEW N320.

SCHOLLER ARL
Now VEW N350.

SCHOLLER ARW
Now VEW N100.

SCHOLLER ARWDT
Now VEW N238.

SCHOLLER ARWK
Now VEW N200.

SCHOLLER ARWKT
Now VEW N205.

SCHOLLER ARZ
Now VEW N320.

SCHOLLER AS2
Vereinigte Edelstahlwerke
C 0.25, Cr 18, Ni 8, bal Fe.
Annealed: 85,000 TS; 40,000 YS; 50 El; 70 RA; 170 Brin. For
chemical plant equipment; Type 302; stainless, austenitic.
Obsolete

SCHOLLER AS2W
Vereinigte Edelstahlwerke
C 0.07, Cr 18, Ni 8, bal Fe.
Annealed: 80,000 TS; 35,000 YS; 55 El; 75 RA; 150 Brin. For
chemical plant equipment, tanks, mixers; Type 304; stainless,
austenitic. *Obsolete*

SCHOLLER AS4W
Vereinigte Edelstahlwerke
C 0.05, Cr 17.5, Ni 11, Mo 2.7, bal Fe.
Annealed: 85,000 TS; 35,000 YS; 50 El; 65 RA; 160 Brin. For
acid resistant chemical plant equipment, mixers; Type 316;
stainless, austenitic. *Obsolete*

SCHOLLER ASA
Vereinigte Edelstahlwerke
C 0.08, Cr 12, Ni 12, bal Fe.
For valves; stainless. *Obsolete*

SCHOLLER BM25
Now VEW V920.

SCHOLLER BM35
Now VEW V935.

SCHOLLER BM45
Now VEW V945.

SCHOLLER BM60
Now VEW V960.

SCHOLLER BRU1
Vereinigte Edelstahlwerke
C 0.45, Cr, Ni, bal Fe.
For gears, bolts, crankshafts, fasteners; oil hardened, shock
resistant. *Obsolete*

SCHOLLER CVS
Now VEW K506.

SCHOLLER DHS
Now VEW S600.

SCHOLLER DS
Now VEW V320.

SCHOLLER DSB
Vereinigte Edelstahlwerke
C 0.42, Cr 1, Mo 0.2, bal Fe.
For gears, pinions, bolts, studs, shafts; oil hardened, tough.
Obsolete

SCHOLLER DW
Vereinigte Edelstahlwerke
C 0.3, Cr 1.1, V 0.18, W 3.75, bal Fe.
For cold work tools, header dies, punches; oil hardened,
tough. *Obsolete*

SCHOLLER EC1
Vereinigte Edelstahlwerke
C 0.15, Cr 0.65, Mn 0.5, bal Fe.
For bolts, gears, machine tool parts; case hardening steel.
Obsolete

SCHOLLER ECM
Now VEW E300.

SCHOLLER ED
Now VEW E304.

SCHOLLER EN1.5
Vereinigte Edelstahlwerke
C 0.13, Cr 0.2, Ni 1.5, bal Fe.
For gears, bolts, fasteners, shafts; case hardened, shock resistant. *Obsolete*

SCHOLLER ENC1
Now VEW E230.

SCHOLLER ENC2
Now VEW E220.

SCHOLLER ENC3
Vereinigte Edelstahlwerke
C 0.13, Cr 0.7, Ni 3.5, bal Fe.
For gears, cams, camshafts, machine tool parts; case hardening steel, shock resistant. *Obsolete*

SCHOLLER ENC4
Now VEW E204.

SCHOLLER EXTRA MH
Vereinigte Edelstahlwerke
C 1.1, Si 0-0.25, Mn 0-0.25, bal Fe.
Annealed: 110,000 TS; 58,000 YS; 18 El; 40 RA; 210 Brin. For springs, tools, cutters, taps, drills; Type W1; water hardened. *Obsolete*

SCHOLLER EXTRA WEICH
Vereinigte Edelstahlwerke
C 0.7, Si 0-0.25, Mn 0-0.25, bal Fe.
Heat treated: 175,000 TS; 128,000 YS; 12 El; 37 RA; 355 Brin. For springs, rails, tools, punches; Type W1; water hardened. *Obsolete*

SCHOLLER EXTRA ZAH
Now VEW K980.

SCHOLLER EXTRA ZH
Now VEW K990.

SCHOLLER FCR
Now VEW K245.

SCHOLLER FLA
Vereinigte Edelstahlwerke
C 0.13, Si 0-0.35, Mn 0-0.5, Cr 0.2, Ni 1.5, bal Fe.
For gears, shafts, machine tool parts; case hardening steel, tough. *Obsolete*

SCHOLLER FSH
Now VEW E920.

SCHOLLER HCN2
Vereinigte Edelstahlwerke
C 0.27, Mn, Cr, bal Fe.
For gears, bolts, machine tool parts; oil or water hardened. *Obsolete*

SCHOLLER HCN3 W/H
Vereinigte Edelstahlwerke
C 0.2-0.35, Cr 0.7, Ni 3.5, Mn 0.6, bal Fe.
For gears, bolts, machine tool parts; oil hardened, shock resistant. *Obsolete*

SCHOLLER HCN4 W/H
Vereinigte Edelstahlwerke
C 0.22, Mn 0.6, Cr 0.7, Ni 3.5, bal Fe.
For gears, bolts, crankshafts, fasteners; oil hardened. *Obsolete*

SCHOLLER HLDH
Vereinigte Edelstahlwerke
C 0.38, Si, Cr, V, bal Fe.
For gears, bolts, crankshafts; oil hardened, shock resistant. *Obsolete*

SCHOLLER HLDW
Vereinigte Edelstahlwerke
C 0.45, Si, Cr, V, bal Fe.
For gears, bolts, crankshafts, springs; oil hardened, shock resistant. *Obsolete*

SCHOLLER HMC
Vereinigte Edelstahlwerke
C 0.33, Cr 1, Mo 0.2, Mn 0.65, bal Fe.
For gears, bolts, fasteners, shafts; oil hardened, shock resistant. *Obsolete*

SCHOLLER HNC1.5
Now VEW V224.

SCHOLLER KHP
Now VEW W326.

SCHOLLER KLI
Vereinigte Edelstahlwerke
C 1, Cr 1.55, Mn 0.35, Si 0.3, bal Fe.
For bearings, cutters, liners; water or oil hardened, wear resistant. *Obsolete*

SCHOLLER KW10
Vereinigte Edelstahlwerke
C 0.09, Cr 14, bal Fe.
Annealed: 75,000 TS; 50,000 YS; 35 El; 70 RA; 155 Brin. For turbine blades, valves, cutlery; Type 403; corrosion resistant. *Obsolete*

SCHOLLER KW10 SPEC
Vereinigte Edelstahlwerke
C 0.2, Cr 14, Mo 1, bal Fe.
Annealed: 95,000 TS; 50,000 YS; 25 El; 55 RA; 195 Brin. For turbine blades, valves, cutlery; Type 420Mo; corrosion resistant, hardenable. *Obsolete*

SCHOLLER KW100
Vereinigte Edelstahlwerke
C 0.95-1.2, Cr 16-18, bal Fe.
Annealed: 110,000 TS; 65,000 YS; 15 El; 30 RA; 240 Brin. For cutlery, valves, ball bearings, surgical instruments; Type 440C; corrosion resistant, hardenable. *Obsolete*

SCHOLLER KW20
Vereinigte Edelstahlwerke
C 0.07, Cr 17.5, bal Fe.
Annealed: 80,000 TS; 50,000 YS; 25 El; 50 RA; 150 Brin. For oil refinery equipment, sinks, soot blowers; Type 430; corrosion resistant. *Obsolete*

SCHOLLER KW40
Vereinigte Edelstahlwerke
C 0.4, Cr 13.5, bal Fe.
Annealed: 100,000 TS; 55,000 YS; 23 El; 52 RA; 200 Brin. For valves, cutlery, surgical and dental instruments; Type 420; corrosion resistant. *Obsolete*

SCHOLLER KW80
Vereinigte Edelstahlwerke
C 0.75-0.95, Cr 16-18, bal Fe.
Annealed: 107,000 TS; 62,000 YS; 18 El; 35 RA; 220 Brin. Heat treated: 280,000 TS; 270,000 YS; 3 El; 15 RA; 555 Brin. For cutlery, valves, ball bearings, surgical instruments; Type 440B; corrosion resistant, hardenable. *Obsolete*

SCHOLLER KWA
Vereinigte Edelstahlwerke
C 0.08, Cr 17, bal Fe.
Annealed: 80,000 TS; 50,000 YS; 25 El; 50 RA; 150 Brin. For oil refinery equipment, sinks, soot blowers; Type 430; corrosion resistant. *Obsolete*

SCHOLLER KWB
Vereinigte Edelstahlwerke
C 0.16, Cr 17, Ni 2, Mo 1.2, bal Fe.
Annealed: 125,000 TS; 95,000 YS; 20 El; 55 RA; 260 Brin. For pumps, marine hardware, valves; Type 431; corrosion resistant. *Obsolete*

SCHOLLER KWZ
Vereinigte Edelstahlwerke
C 0.12, Cr 14-18, S, Mo, bal Fe.
Annealed: 80,000 TS; 50,000 YS; 25 El; 50 RA; 150 Brin. For screw machine products; corrosion resistant, free-cutting. *Obsolete*

SCHOLLER LES
Now VEW E920.

SCHOLLER LM1
Now VEW E200.

SCHOLLER LM2
Now VEW E216.

SCHOLLER LM3 W/H
Vereinigte Edelstahlwerke
C 0.22-0.3, Cr 0.7, Ni 3.5, bal Fe.
For gears, bolts, shafts, fasteners; oil hardened, tough. *Obsolete*

SCHOLLER MA0
Now VEW A500.

SCHOLLER MA1
Now VEW A505.

SCHOLLER MA3
Now VEW A522.

SCHOLLER MACDW
Vereinigte Edelstahlwerke
C 0-0.07, Cr 17.5, Mo 2, Ni 17.5, Cu 2, Ti = 7 x C, bal Fe.
For valves, pump parts; corrosion and heat resistant. *Obsolete*

SCHOLLER MAG
Vereinigte Edelstahlwerke
C 0.15, Si 1.5, Cr 18, Ni 8.5, bal Fe.
Annealed: 80,000 TS; 35,000 YS; 55 El; 75 RA; 150 Brin. Cold drawn: 180,000 TS; 150,000 YS; 10 El; 250 Brin. For chemical plant equipment, tanks, mixers, filters; Type 302, stainless, austenitic. *Obsolete*

SCHOLLER MAN
Now VEW A750.

SCHOLLER MAS
Vereinigte Edelstahlwerke
C 0-0.07, Cr 18, Mo 2, Ni 10.5, bal Fe.
Annealed: 85,000 TS; 35,000 YS; 50 El; 60 RA; 165 Brin. For acid resistant chemical plant equipment, tanks; Type 316; stainless, austenitic. *Obsolete*

SCHOLLER MASBW
Now VEW A102.

SCHOLLER MASW
Now VEW A300.

SCHOLLER MASWS
Vereinigte Edelstahlwerke
C 0-0.12, Cr 18, Mo 2, Ni 10.5, Ti = 4 x C, bal Fe.
Annealed: 85,000 TS; 35,000 YS; 50 El; 65 RA; 160 Brin. For welded acid resistant chemical plant equipment; Type 316 Ti; stainless, austenitic. *Obsolete*

SCHOLLER MAT
Now VEW A700.

SCHOLLER MNF
Now VEW F180.

SCHOLLER MS
Now VEW K465.

SCHOLLER NSW
Now VEW W107.

SCHOLLER OF56
Vereinigte Edelstahlwerke
C 0.56, Si 0.3, Mn 0.55, bal Fe.
Heat treated: 115,000-160,000 TS; 77,000-113,000 YS; 12-23 El; 40-54 RA; 230-320 Brin. For die blocks, girders, clutch discs; water hardened. *Obsolete*

SCHOLLER PM
Vereinigte Edelstahlwerke
C 1.4, Si 0.25, Mn 0.3, V 0.1, bal Fe.
For engravers' tools, cutters, forming and blanking dies; Type W2; water hardened. *Obsolete*

SCHOLLER PRIMA 100
Vereinigte Edelstahlwerke
C 1.3, Si 0.25, Mn 0.25, bal Fe.
For engravers' tools, cutters, forming and blanking dies; Type W1; water hardened. *Obsolete*

SCHOLLER PRIMA WEICH
Vereinigte Edelstahlwerke
C 0.7, Si 0.2, Mn 0.2, bal Fe.
Heat treated: 122,000-175,000 TS; 82,000-128,000 YS; 12-22 El; 37-52 RA; 240-350 Brin. For wheels, die blocks, girders, springs, rails; Type W1; water hardened. *Obsolete*

SCHOLLER PRIMA ZAH
Now VEW K980.

SCHOLLER PRIMA ZH
Now VEW K990.

SCHOLLER S200
Vereinigte Edelstahlwerke
C 0.25, Cr 1, Mo 0.2, Mn 0.65, bal Fe.
For gears, bolts, studs, machine tool parts; oil hardened, tough. *Obsolete*

SCHOLLER S200H
Now VEW V330.

SCHOLLER S200W
Now VEW V340.

SCHOLLER SAS2
Vereinigte Edelstahlwerke
C 0.05, Cr 18, Ni 9, Cb, bal Fe.
Annealed: 80,000 TS; 35,000 YS; 55 El; 75 RA; 150 Brin. For chemical plant equipment, mixers, agitators; Type 347; stainless, austenitic. *Obsolete*

SCHOLLER SAS4
Vereinigte Edelstahlwerke
C 0.07, Cr 17.5, Ni 9.5, Mo 1.5, Cb, bal Fe.
Annealed: 85,000 TS; 35,000 YS; 50 El; 65 RA; 160 Brin. For welded acid resistant chemical plant equipment; stainless, austenitic. *Obsolete*

SCHOLLER SAS4MN
Vereinigte Edelstahlwerke
C 0.07, Cr 17.5, Ni 11.5, Mo 2.7, Cb, bal Fe.
Annealed: 85,000 TS; 35,000 YS; 50 El; 65 RA; 160 Brin. For welded acid resistant chemical plant equipment; Type 316 Cb; stainless, austenitic. *Obsolete*

SCHOLLER SAS5
Vereinigte Edelstahlwerke
C 0.05, Cr 17.5, Ni 14.5, Mo 4.5, bal Fe.
Annealed: 90,000 TS; 40,000 YS; 45 El; 60 RA; 180 Brin. For acid resistant chemical plant equipment; Type 317; stainless, austenitic. *Obsolete*

SCHOLLER SAST2
Vereinigte Edelstahlwerke
C 0.07, Cr 18, Ni 8.5, Ti, bal Fe.
Annealed: 85,000 TS; 35,000 YS; 55 El; 65 RA; 150 Brin. For chemical plant equipment; Type 321; stainless, austenitic. *Obsolete*

SCHOLLER SAST4
Vereinigte Edelstahlwerke
C 0.07, Cr 17.5, Ni 11, Mo 2.7, Ti, bal Fe.
Annealed: 85,000 TS; 35,000 YS; 50 El; 65 RA; 160 Brin. For welded acid resistant chemical plant equipment; Type 316 Ti; stainless, austenitic. *Obsolete*

SCHOLLER SAWM
Now VEW B404.

SCHOLLER SC1
Now VEW K240.

SCHOLLER SC2
Vereinigte Edelstahlwerke
C 1.25, Si 1.15, Mn 0.7, Cr 1.2, bal Fe.
For cutters, bearings, upsetters; oil hardened, tough. *Obsolete*

SCHOLLER SG
Now VEW W100.

SCHOLLER SGHV
Now VEW W100.

SCHOLLER SHM
Now VEW F124.

SCHOLLER SKVL
Vereinigte Edelstahlwerke
C 0.08, Cr 17.5, Mo 1.2, bal Fe.
Annealed: 120,000 TS; 90,000 YS; 22 El; 57 RA; 250 Brin. For pumps, marine hardware, valves; corrosion resistant. *Obsolete*

SCHOLLER SNK3
Vereinigte Edelstahlwerke
C 1, V 0.1, Mn 0.25, Si 0.2, bal Fe.
For cutters, dies, tools; Type W2; water hardened. *Obsolete*

SCHOLLER SPZ
Now VEW K245.

SCHOLLER TAMB
Vereinigte Edelstahlwerke
C 1.2, Cr 0.2, V 0.1, W 1, bal Fe.
For cold work tools, heading dies; oil hardened, tough. *Obsolete*

SCHOLLER U3
Now VEW K455.

SCHOLLER V132S
Vereinigte Edelstahlwerke
C 0.42, Si 0.25, Mn 1.75, V 0.1, bal Fe.
For punches, dies, crimpers, gauges; oil hardened, tough. *Obsolete*

SCHOLLER V230
Now VEW K505.

SCHOLLER VC135
Now VEW V510.

SCHOLLER VCMV
Now VEW V350.

SCHOLLER VCV230
Vereinigte Edelstahlwerke
C 0.58, Cr 1.05, V 0.1, Mn 0.95, bal Fe.
For springs, punches, forging and heading dies; oil hardened, shock resistant. *Obsolete*

SCHOLLER VNC1
Now VEW V155.

SCHOLLER VNC2
Now VEW V145.

SCHOLLER VPN
Now VEW V500.

SCHOLLER WKL
Now VEW K200.

SCHOLLER WU
Vereinigte Edelstahlwerke
C 0.3, Co 2, Cr 2.4, V 0.25, W 8.5, bal Fe.
For extrusion dies, rams and liners, punches; hot work steel, oil hardened. *Obsolete*

SCHOMBERG ALLOY
Manufacturer not listed.
Zn 87, Sn 10, Cu 3.
For bearings; anti-friction.

SCHOMBERG BEARING
Manufacturer not listed.
Zn 59, Sn 40, Cu 0.4, Pb 0.2, Fe 0.2.
For bearings; anti-friction.

SCHULZ
German manufacture
Co 31, Cr 15, W 35, Fe 2, Mo 10, C 0.7.
For tools; corrosion and heat resistant.

SCHULZ ALLOY
Manufacturer not listed.
Zn 91, Cu 6, Al 3.
For die castings, ornamental parts; free-cutting.

SCHWABENSTAHL
Schwaebische Huettenwerke GmbH
Cold drawn steel. Pb, C, alloy, bal Fe.
For machine tool parts. SAE 1113. G12144

SCHWARZPUNKT
SWB Stahlformguss Gesellschaft mbH
C 1, Cr, W, Mo, V, bal Fe.
For cutters, dies, reamers; oil hardened. *Obsolete*

SCHWEISTAHL 5H
Vereinigte Edelstahlwerke
C 0.55, Si 0.1-0.4, Mn 0.5-0.7, bal Fe.
Heat treated: 155,000 TS; 110,000 YS; 14 El; 45 RA; 320 Brin. For gears, bolts, crankshafts, axles; water hardened. *Obsolete*

SCHWERMETALL
Plansee, Metallwerk Gesellschaft
W alloy.
Sintered: 72,000-85,000 TS; 220-250 Brin. For gyros, counterweights, and radioactive shields. Sintered; heavy alloy.

SCIMITARS
Barker & Allen Ltd.
Ni 18, Zn 30, bal Cu.
For domestic utensils, ornaments; nickel silver.

SCLERON
Vereinigte Leichtmetallwerke, G.m.b.H.
Aluminum. Cu 1.5, Si 4, bal Al.
Heat treated: 58,000-72,000 TS; 42,000-50,000 YS; 10-15 El; 100-135 Brin. For small structural parts; age-hardened.

SCLERON NO. 1
Metallbank AG
Li 0.1, Cu 4, bal Al.
42,500 TS; 15-20 El. For connecting rods, machine construction; similar to "Aeron."

SCLERON NO. 1
Vereinigte Leichtmetallwerke, G.m.b.H.
Li 0.1, Cu 4, bal Al.
42,500 TS; 15-20 El. For connecting rods, machine construction; similar to "Aeron."

SCLERON NO. 2
Metallbank AG
Zn 12, bal Al.
57,000-71,000 TS; 10-15 El. For machinery parts; age-hardened. *Obsolete*

SCLERON NO. 2
Vereinigte Leichtmetallwerke, G.m.b.H.
Zn 12, bal Al.
57,000-71,000 TS; 10-15 El. For machinery parts; age-hardened. *Obsolete*

SCLERON NO. 3
Metallbank AG
Li 0.1, Cu 3, Zn 12, Si 0.5, Mn 0.6, Fe 0.4, bal Al.
Aged: 57,000-64,000 TS; 38,000 YS; 15 El. For machinery parts; age-hardened. *Obsolete*

SCLERON NO. 3
Vereinigte Leichtmetallwerke, G.m.b.H.
Li 0.1, Cu 3, Zn 12, Si 0.5, Mn 0.6, Fe 0.4, bal Al.
Aged: 57,000-64,000 TS; 38,000 YS; 15 El. For machinery parts; age-hardened. *Obsolete*

SCM
Acciaierie Valbruna s.p.a.
C 0.46, Si 1.5, Mn 0.65, P 0-0.035, S 0-0.035, Cr 0.65, Mo 0.2, bal Fe.
Spring steel.

SCN 1
Acciaierie Valbruna s.p.a.
C 0.55, Si 1.4, Mn 0.85, Cr 0.87, Ni 0.6, P 0.035, S 0.035, bal Fe.
Silicon alloy steel.

SCOOTER
British Rolling Mills Ltd.
C 0.15, S 0.3, bal Fe.
Drawn: 89,600 TS; 15 El; 45 RA. General purpose, free-cutting steel.

SCOOTER CASE
British Rolling Mills Ltd.
C 0.1, Mn 1-1.5, S 0.4, bal Fe.
For cams, gears, fasteners; case hardened.

SCOTT
Time Steel Service Inc.
C 2.25, Mn 0.35, Si 0.5, Cr 11.5, V 0.2, Mo 0.8, bal Fe.
High carbon, high chromium tool and die steel; air or oil hardened. AISI D4.

SCOTTS ACME
Manufacturer not listed.
C 0.4, bal Fe.
For shafting; water hardening.

SCOVILL 100
Century Brass Products Inc.
Cu 64, Ni 18, Zn 18.
Soft: 55,000 TS; 21,000 YS; 48 El. Hard: 98,000 TS; 93,000 YS; 6 El. For resistance wire springs, cutlery, hardware; Nickel-Silver. *Obsolete*

SCOVILL 105
Century Brass Products Inc.
Cu 64, Ni 10, Zn 26.
Soft: 55,000 TS; 210,000 YS; 54 El. Hard: 102,000 TS; 100,000 YS; 7 El. For resistance wire springs, cutlery, hardware; Nickel-Silver. *Obsolete*

SCOVILL 118
Century Brass Products Inc.
Cu 64, Pb 1, Ni 12, Zn 20.
Soft: 60,000 TS; 28,000 YS; 43 El. Hard: 104,000 TS; 101,000 YS; 4 El. For watch blanks, screw machine parts; Nickel-Silver. *Obsolete*

SCOVILL 322 HIGH LEADED BRASS 67%
Century Brass Products Inc.
Cu 67, Pb 1.8, Zn 31.2.
Tube. Free-cutting. *Obsolete*

SCOVILL 325 LANCASHIRE BRASS
Century Brass Products Inc.
Cu 72-75, Pb 2.5-3, bal Zn.
Annealed: 43,000 TS; 50 El. Hard: 73,000 TS; 8 El. For watch and clock parts; free cutting. *Obsolete*

SCOVILL 342 HIGH LEADED BRASS-65%
Century Brass Products Inc.
Cu 65, Pb 2.2, Zn 32.8.
For screw machine products, hardware; free-cutting. *Obsolete*

SCOVILL 350
Century Brass Products Inc.
Pb 1, Cu 61, bal Zn.
Soft: 54,000 TS; 40 El. Hard: 80,000 TS; 6 El. For screw machine products; free-cutting. *Obsolete*

SCOVILL 403 GILDING BRONZE
Century Brass Products Inc.
Cu 95, Sn 2, Zn 3. *Obsolete*

SCOVILL 405 PENNY BRONZE
Century Brass Products Inc.
Cu 95, Sn 1, Zn 4.
For coins; corrosion resistant. *Obsolete*

SCOVILL 434 SPRING BRONZE 0.8%
Century Brass Products Inc.
Cu 85, Sn 0.8, Zn 14.2.
For springs; corrosion resistant. *Obsolete*

SCOVILL 651 LOW SILICON BRONZE TYPE B
Century Brass Products Inc.
Cu 98, Si 2. *Obsolete*

SCOVILL 7% NICKEL-SILVER
Century Brass Products Inc.
Cu 65, Ni 7, Zn 28.
Soft: 54,000 TS; 24,00 YS; 65 El. Hard: 98,000 TS; 4 El. For wire products; corrosion resistant. *Obsolete*

SCOVILL 706 CUPRO NICKEL 10%
Century Brass Products Inc.
Cu 88.6, Fe 1.4, Ni 10. *Obsolete*

SCOVILL 735 NICKEL SILVER 72-18
Century Brass Products Inc.
Cu 72, Ni 18, Zn 10.
Corrosion resistant. *Obsolete*

SCOVILL 745
Century Brass Products Inc.
Cu 64, Ni 10, Zn 26.
Soft: 55,000 TS; 210,000 YS; 54 El. Hard: 102,000 TS; 100,000 YS; 7 El. For resistance wire springs, cutlery, hardware; Nickel-Silver. *Obsolete*

SCOVILL 760
Century Brass Products Inc.
Cu 64, Pb 1, Ni 12, Zn 23.
Soft: 60,000 TS; 28,000 YS; 43 El. Hard: 104,000 TS; 101,000 YS; 4 El. For watch blanks, screw machine parts; Nickel-Silver. *Obsolete*

SCOVILL ADMIRALTY 445
Century Brass Products Inc.
Fe 0 0.06, Pb 0-0.075, 0.90% min Sn, 70% min Cu, bal Zn.
Soft: 55,000 TS; 65 El; 50 Brin. Hard: 105,000 TS; 2 El; 175 Brin. For condenser tubes for polluted fresh or sea water; corrosion resistant. *Obsolete*

SCOVILL ALLOY 169 YELLOW BRASS, 65% 270
Century Brass Products Inc.
Cu 65, bal Zn.
Annealed: 49,000 TS; 57 El. Hard: 74,000 TS; 8 El. For socket shells, pins, rivets; excellent cold working. *Obsolete*

SCOVILL ALLOY 280 MUNTZ METAL
Century Brass Products Inc.
Cu 61.5, Pb 0-0.3, bal Zn.
Annealed: 54,000 TS; 45 El. Hard: 70,000 TS; 10 El. For heat exchanger tubes; excellent hot working. *Obsolete*

SCOVILL ALLOY 314
Century Brass Products Inc.
Cu 89, Pb 2, Ni 1, Zn 8.
Soft: 37,000 TS. Hard: 62,000 TS. For hardware; leaded commercial bronze, free-cutting. *Obsolete*

SCOVILL ALLOY 320
Century Brass Products Inc.
Cu 85, Pb 1.8, Zn 13.2.
Soft: 40,000 TS. Hard: 65,000 TS. For hardware; leaded red brass, free-cutting. *Obsolete*

SCOVILL ALLOY 323 TIN BRASS (1.75%)
Century Brass Products Inc.
Cu 90.5, Sn 1.75, bal Zn.
Annealed: 42,000 TS; 45 El. Hard: 59,000 TS; 4 El. For costume jewelry; for decorative purposes. *Obsolete*

SCOVILL ALLOY 360
Century Brass Products Inc.
Pb 2.5-3.75, Cu 60-63, bal Zn.
Annealed: 49,000 TS; 53 El; 58 RA. 1/2 Hard: 58,000 TS; 25 El. For screw machine parts; free cutting. *Obsolete*

SCOVILL ALLOY 413 TIN BRASS (1%)
Century Brass Products Inc.
Cu 90.5, Sn 1, bal Zn.
Annealed: 41,000 TS; 42 El. Hard: 67,000 TS; 4 El. For costume jewelry, springs; for decorative purposes. *Obsolete*

SCOVILL ALLOY 415 TIN BRASS (2%)
Century Brass Products Inc.
Cu 90.5, Sn 2, bal Zn.
Annealed: 45,000 TS; 55 El. Hard: 70,000 TS; 3 El. For costume jewelry; spring properties. *Obsolete*

SCOVILL ALLOY 419
Century Brass Products Inc.
Cu 89, Zn 6.5, Sn 4.5.
Annealed: 40,000 TS; 20 El. For optical instruments; spring properties. *Obsolete*

SCOVILL ALLOY 430 SPRING BRONZE
Century Brass Products Inc.
Cu 86.5, Sn 1-2, bal Zn.
Annealed: 50,000 TS; 55 El. Hard: 94,000 TS; 1 El. For springs, diaphragms; tough. *Obsolete*

SCOVILL ALLOY 445
Century Brass Products Inc.
Cu 71, Sn 1, Zn 28, P 0.3.
Soft: 50,000 TS; Rockwell B54. Hard: 84,000 TS; Rockwell B84. For springs, diaphragms; Phosphorized Admiralty. *Obsolete*

SCOVILL ALLOY 464 NAVAL BRASS (A)
Century Brass Products Inc.
Cu 60, Sn 0.75, Pb 0.2, bal Zn.
Annealed: 58,000 TS; 45 El. Hard: 80,000 TS; 20 El. For marine hardware, valve stems, rivets; corrosion resistant; "Ocobronze." *Obsolete*

SCOVILL ALLOY 482 NAVAL BRASS (B)
Century Brass Products Inc.
Cu 60.5, Sn 0.75, Pb 0.75, bal Zn.
Annealed: 63,000 TS; 35 El. Hard: 88,000 TS; 10 El. For marine hardware; corrosion resistant; low leaded. *Obsolete*

SCOVILL ALLOY 485 NAVAL BRASS (C)
Century Brass Products Inc.
Cu 60, Sn 0.75, Pb 1.75, bal Zn.
High leaded. Annealed: 58,000 TS; 29 El. Hard: 88,000 TS; 9 El. For screw machine products; corrosion resistant, free-cutting. *Obsolete*

SCOVILL ALLOY 516 NICKEL SILVER (8%)
Century Brass Products Inc.
Cu 64.8, Ni 8, bal Zn.
Annealed: 47,000 TS; 53 El. Hard: 93,000 TS; 3 El. For strip products; excellent cold working. *Obsolete*

SCOVILL ALLOY 521
Century Brass Products Inc.
Cu 92, Sn 8, P 0-0.25.
Soft: 58,000 TS; Rockwell F80. Hard: 95,000 TS; Rockwell B93. For springs; tough; P-Bronze. *Obsolete*

SCOVILL ALLOY 675 MAGANESE BRONZE (A)
Century Brass Products Inc.
Cu 58.5, Sn 0.1, Fe 0.1, Mn 0.3, bal Zn.
Annealed: 65,000 TS; 33 El. Hard: 84,000 TS; 19 El. For valve stems, welding rod, condenser plates; excellent hot working. *Obsolete*

SCOVILL ALLOY 710
Century Brass Products Inc.
Cu 80, Ni 20.
Soft: 49,000 TS; Rockwell F73. Hard: 72,000 TS; Rockwell B81. For condenser tubing; corrosion resistant. *Obsolete*

SCOVILL ALLOY 715
Century Brass Products Inc.
Cu 70, Ni 30.
Soft: 58,000 TS; Rockwell F80. Hard: 80,000 TS; Rockwell B84. For condenser tubing; corrosion resistant. *Obsolete*

SCOVILL ALLOY 743 NICKEL SILVER (8%)
Century Brass Products Inc.
Cu 64.8, Ni 8, bal Zn.
Annealed: 47,000 TS; 53 El. Hard: 93,000 TS; 3 El. For strip products; excellent cold working. *Obsolete*

SCOVILL ALLOY 752
Century Brass Products Inc.
Cu 65, Ni 18, Zn 17.
Soft: 58,000 TS; Rockwell F85. Hard: 85,000 TS; Rockwell B87. For hardware; nickel silver. *Obsolete*

SCOVILL ALLOY 754 NICKEL SILVER (15%)
Century Brass Products Inc.
Cu 64.8, Ni 15, bal Zn.
Annealed: 54,000 TS; 42 El. Hard: 90,000 TS; 2 El. For costume jewelry; corrosion resistant. *Obsolete*

SCOVILL ALLOY 757 NICKEL SILVER (12%)
Century Brass Products Inc.
Cu 64.8, Ni 12, bal Zn.
Annealed: 53,000 TS; 46 El. Hard: 88,000 TS; 3 El. For hollow are; corrosion resistant. *Obsolete*

SCOVILL ALLOY 770 (B)
Century Brass Products Inc.
Cu 55, Ni 18, bal Zn.
Annealed: 60,000 TS; 40 El. Hard: 100,000 TS; 3 El. For resistance wire, springs, optical goods; corrosion resistant. *Obsolete*

SCOVILL ALLOY 774 ALUMINUM BRASS TYPE B
Century Brass Products Inc.
Cu 78, Al 2.1, As 0.035, bal Zn.
Annealed: 60,000 TS; 55 El. For condensers, distiller tubes; corrosion resistant. *Obsolete*

SCOVILL ALLOY 787 ALUMINUM BRASS TYPE B
Century Brass Products Inc.
Cu 78, Al 2.1, As 0.035, bal Zn.
Annealed: 60,000 TS; 55 El. For condensers, distiller tubes; corrosion resistant. *Obsolete*

SCOVILL ALUMINUM BRASS TYPE A
Century Brass Products Inc.
Cu 78, Al 2.1, As 0.07, bal Zn.
Annealed: 60,000 TS; 55 El. For heat exchanger, evaporator tubes; corrosion resistant. *Obsolete*

SCOVILL CARTRIDGE BRASS, 260
Century Brass Products Inc.
Cu 70, Zn 30.
Annealed: 47,000 TS; 62 El. Hard: 76,000 TS; 8 El. For cartridge cases; high conductivity. *Obsolete*

SCOVILL COMMERCIAL BRONZE, ALLOY 220
Century Brass Products Inc.
Cu 89-91, bal Zn.
Annealed: 37,000 TS; 45 El; Hard: 60,000 TS; 5 El. For hardware, bullet jackets; deep drawing. *Obsolete*

SCOVILL EXTRA HIGH LEADED BRASS 356
Century Brass Products Inc.
Cu 63, Pb 2.5, bal Zn.
Annealed: 49,000 TS; 52 El. Hard: 74,000 TS; 7 El. For clock gears, plates; free-cutting. *Obsolete*

SCOVILL FORGING BRASS, ALLOY 377
Century Brass Products Inc.
Cu 58-61, Pb 1.5-2.5, bal Zn.
Forged: 55,000 TS; 40 El. For hardware, forgings. *Obsolete*

SCOVILL HIGH BRASS, 270
Century Brass Products Inc.
Cu 64.5-66.5, Pb 0-0.3, bal Zn.
Annealed: 48,000 TS; 60 El. Hard: 74,000 TS; 8 El; 150 Brin. For deep drawing, spinning, stamping; high ductility. *Obsolete*

SCOVILL LEADED MUNTZ METAL 380
Century Brass Products Inc.
Cu 61, Pb 1, bal Zn.
Annealed: 54,000 TS; 40 El. Hard: 80,000 TS; 6 El. For screw machine products; good machinability. *Obsolete*

SCOVILL LEADED ROD 360
Century Brass Products Inc.
Cu 61.5, Pb 3, bal Zn.
Soft: 48,000 TS; 18,000 YS; 55 El. Hard: 90,000 TS; 88,000 YS; 6 El. For screw machine parts. *Obsolete*

SCOVILL LOW BRASS, ALLOY 240
Century Brass Products Inc.
Cu 79-81, Pb 0-0.05, bal Zn.
Annealed: 43,000 TS; 50 El; 159 Brin. Hard: 73,000 TS; 8 El; 156 Brin. For drawn and formed parts; corrosion resistant. *Obsolete*

SCOVILL LOW LEADED BRASS (64%) ALLOY 330
Century Brass Products Inc.
Cu 67, Pb 0.6, bal Zn.
Annealed: 47,000 TS; 57 El. Hard: 75,000 TS; 8 El. For pipe, pump lines, trap tubes, tubes; good machinability. *Obsolete*

SCOVILL LOW LEADED BRASS (67%) ALLOY 335
Century Brass Products Inc.
Cu 65, Pb 0.5, bal Zn.
Annealed: 49,000 TS; 57 El. Hard: 74,000 TS; 8 El. For butts, hinges; good cold working. *Obsolete*

SCOVILL MATRIX BRASS 340
Century Brass Products Inc.
Cu 64.5, Pb 1, bal Zn.
Annealed: 49,000 TS; 54 El. Hard: 74,000 TS; 7 El. For gears, rivets, screws; good cold working. *Obsolete*

SCOVILL MUNTZ 280
Century Brass Products Inc.
Pb 0-0.35, Fe 0-0.1, 60% min Cu, bal Zn.
Soft: 45,000 TS; 20,000 YS; 60 El. Hard: 100,000 TS; 98,000 YS; 2 El. For condenser tubes for pure fresh water; corrosion resistant. *Obsolete*

SCOVILL NAVAL BRASS 464
Century Brass Products Inc.
Cu 59-61, Sn 0.5-1, Fe 0-0.06, Pb 0-0.2, bal Zn.
Soft: 54,000-65,000 TS; 25,000-27,000 YS; 25-40 El. Hard: 105,000 TS; 99,000 YS; 7 El. For boat shafting, turn buckles, welding rod; "Scovill No. 16 Alloy"; resists salt water corrosion. *Obsolete*

SCOVILL NICKEL SILVER (5%)
Century Brass Products Inc.
Cu 64.8, Ni 5, bal Zn.
Annealed: 52,000 TS; 72 El. Hard: 98,000 TS; 4 El. For costume jewelry, dials, plates; excellent cold working. *Obsolete*

SCOVILL NO. 1032
Century Brass Products Inc.
S 0.25, bal Cu.
Cold drawn: 53,000 TS; 50,000 YS; 10 El; 107 Brin. For cable clamps, radio parts, prongs; free-cutting, high electrical conductivity. *Obsolete*

SCOVILL NO. 3105
Century Brass Products Inc.
Mn 0.4, Mg 0.5, bal Al.
O-Temper: 18,000 TS; 7,000 YS; 25 El; 32 Brin. H14-Temper: 25,000 TS; 23,000 YS; 6 El; 45 Brin. For siding, general purpose sheet. Non-heat treatable. *Obsolete*

SCOVILL NO. 99 ALLOY
Century Brass Products Inc.
Cu 95, Sn 5.
Soft: 50,000 TS; 22,000 YS; 58 El. Hard: 95,000 TS; 94,000 YS; 5 El. For springs, switches; Rockwell B 32-93. Government bronze 510. *Obsolete*

SCOVILL RED BRASS, 85%, ALLOY 230
Century Brass Products Inc.
Cu 85, bal Zn.
Annealed: 41,000 TS; 48 El. Hard: 79,000 TS; 6 El. For screw shells, compacts, costume jewelry; resists season cracking and dezincification. *Obsolete*

SCOVILL TIN BRASS (5%) 419
Century Brass Products Inc.
Cu 90.5, Sn 5, bal Zn.
Annealed: 49,000 TS; 74 El. For costume jewelry; spring properties; "Optical Metal." *Obsolete*

SCOVILLE 7% NICKEL-SILVER 743
Century Brass Products Inc.
Cu 65, Ni 7, Zn 28.
Soft: 54,000 TS; 24,000 YS; 65 El. Hard: 98,000 TS; 4 El. For wire products; corrosion resistant. *Obsolete*

SCOVILLS HARDWARE BRONZE
Century Brass Products Inc.
Cu 84-86, Pb 1.2-2.5, bal Zn.
Annealed: 35,000 TS; 25 El; 71 RA. Drawn: 45,000 TS, 19,000 YS; 10 El; 60 RA. For hardware parts, screw machine parts; "Scovill No. 197 Alloy;" free from season cracking. *Obsolete*

SCR3
Sumitomo Metal America Inc.
C 0-0.03, Si 1.8, Mn 1.5, Ni 25, Cr 25, V 1.5, Ti 0.3, bal Fe.
82,500 psi TS; 38,400 psi YS; 60 El. Austenitic stainless steel resistant to stress corrosion cracking in high temperature and high pressure water containing chlorides. Good formability and weldability.

SCREEN PLATES
American manufacture
Cu 58, Zn 41, Sn 0.75, Pb 0.25.
For screen plates; high strength.

SCREW BRONZE
American manufacture
Zn 5, Sn 1, Pb 0.5, bal Cu.
For screws, bolts, nuts, hardware; free-cutting.

SCREW METAL
American manufacture
Cu 60, Pb 1.5, bal Zn.
For screws, bolts, studs; free-cutting.

SCREW NUT BRONZE
American manufacture
Cu 86, Sn 11, bal Zn.
For screws, nuts, bolts; high strength.

SCRIBE-IT
Marshall Steel Co.
C 0.95, Mn 1.2, Si 0.25, Cr 0.5, V 0.2, W 0.5, bal Fe.
Oil hardening tool steel; AISI O1.

SCV.5
Spencer Clark Metal Industries Ltd.
C 1, Mn 0.5, Si 0.3, Mo 0.95, Cr 5, V 0.25, bal Fe.
Chromium cold work tool steel.

SCX SERIES
Sumitomo Metal America Inc.
C 0.12, Si 0-1.45, Mn 0-1.6, Cb, bal Fe.
0.6-2.3 mm thick cold rolled sheet or coil. Available in two strength grades. SCX 50: 50 ksi YP min. SCX 60: 60 ksi YP min. High strength low alloy steel for cold formed parts, automotive and appliances.

SCX15
Eagle & Globe Steel Ltd.
Tool material. C 0.52, Si 0.8, Mn 0.45, Cr 1.3, W 2.15, V 0.2, bal Fe.
Annealed: 235 Brin max. Oil hardening, shock resisting steel. For hand and pneumatic chisels, heavy duty punches, screwdriver bits, shear blades, boilermakers' tools and swaging dies. AS1239 S1A-5; AISI S1; BS4659 BS1; Werkstoff 1.2547.

SCXD SERIES
Sumitomo Metal America Inc.
HSLA steel. C 0.07, Si 0.4, Mn 2.9, Cb, bal Fe.
0.6-2.3 mm thick cold rolled sheet or coil. Available in three strength grades. SCXD 80: 80 ksi TS min. SCXD 90: 90 ksi TS min. SCXD 100: 100 ksi TS min. High strength low alloy steel for cold formed parts, automotive and appliances. Good formability and spot weldability.

SD-NICKEL
Metal & Thermit Corp.
Ni 99.9.
For anodes. Improves plating.

SD111
Acciaierie Valbruna s.p.a.
Tool Material. C 1.05, Mn 1, Cr 1, Si 0.5, bal Fe.
Tool steel, for special purposes. W. Nr. 1.2127.

SD3
Teledyne Firth Sterling
Tool material. Ni 13.
Combination of cobalt binder and alloy carbide for cutting metals yielding continuous chips producing craters on top rake surfaces. 93.2 Rock A.

SD5
Teledyne Firth Sterling
Tool material. Ni 15.
Combination of cobalt binder and alloy carbide for cutting metals yielding continuous chips producing craters on top rake surfaces. 91.8 Rock A.

SD8
Acieries et Forges d'Anor
C 0.6-0.65, Si 0.3-1.1, Mn 0.2-0.4, Cr 7.5-8.5, Ni 1.2-1.4, Mo 1-1.2, W 0.4-0.5, V 0.4-0.5, bal Fe.
High alloy tool steel. AFNOR Z60 CND 8.

SDM
Acciaierie Valbruna s.p.a.
Tool Material. C 0-0.88, Mn 2, V 0.15, Si 0-0.4, Cr 0-0.35, bal Fe.
Oil hardening cold work tool steel. W. Nr. 1.2842; AISI O2.

SDW
Acciaierie Valbruna s.p.a.
Tool Material. C 1, Mn 1, Cr 1, W 1.1, Si 0-0.35, bal Fe.
Cold work tool steel. W. Nr. 1.2419.

SEA STEEL
Alloy Engineering & Casting Co.
C 0.2, Cr 18, Ni 8, bal Fe.
For marine fittings, hardware, and pulleys. Stainless; austenitic. *Obsolete*

SEA WATER ALLOY-1
American manufacture
Ni 17, Mn 5, C 0.7, bal Fe.
For marine construction; stainless and corrosion resistant.

SEA WATER ALLOY-2
American manufacture
Cr 1.2, Ni 24, Mn 0.6, C 0.5, bal Fe.
For marine construction; stainless and corrosion resistant.

SEA WATER BRONZE
English manufacture
Ni 33, Sn 16, Zn 5.5, Bi 1, bal Cu.
For marine parts, hardware; resists sea water corrosion.

SEACO
Stulz Sickles Steel Co.
C 0.7-1.2, bal Fe.
For tools, drills, taps. *Obsolete*

SEACO "70-20"
Stulz Sickles Steel Co.
C 0.7, V 0.2, bal Fe.
For drills, tools. *Obsolete*

SEACO BLUE DIAMOND
Stulz Sickles Steel Co.
C, alloy, bal Fe.
For tools, drills, taps. *Obsolete*

SEACO CHROME VANADIUM
Stulz Sickles Steel Co.
C 0.5, Ni 1.5, Cr 0.8, bal Fe.
For tools, dies, gears. *Obsolete*

SEACO EXTRA
Stulz Sickles Steel Co.
C 0.6, V 0.2, bal Fe.
For chisels, rock drills. *Obsolete*

SEACO HIGH SPEED
Stulz Sickles Steel Co.
C 0.7, W 18, Cr 4, V 1, bal Fe.
For cutting tools; high speed steel.

SEACO HOT DIE
Stulz Sickles Steel Co.
C 0.5, Cr 3.5, bal Fe.
For hot dies, tools. *Obsolete*

SEACO NO-LABEL
Stulz Sickles Steel Co.
C 0.9-1.1, bal Fe.
For drills, taps, cutters. *Obsolete*

SEACO NON-SHRINKABLE
Stulz Sickles Steel Co.
C 0.9, Mn 1.2, Cr 0.5, bal Fe.
For tools, punches; non-shrinking. *Obsolete*

SEACO O.H. NO. 4 NON-SHRINK
Stulz Sickles Steel Co.
C 0.9, Mn 1.2, Cr 0.5, bal Fe.
For dies, cutters, punches; non-deforming. *Obsolete*

SEACO RED DIAMOND
Stulz Sickles Steel Co.
C, alloy, bal Fe.
For drills, tools, taps. *Obsolete*

SEACO STANDARD
Stulz Sickles Steel Co.
C 0.7, bal Fe.
For tools, chisels, rock drills. *Obsolete*

SEACO YELLOW DIAMOND
Stulz Sickles Steel Co.
C, alloy, bal Fe.
For dies. *Obsolete*

SEALALLOY
Manufacturer not listed.
Bi alloy.
For sealing metal to glass.

SEALCOR
Now ALL-STATE SEALCOR.

SEALMET 1
Allegheny Ludlum Steel
Cr 28, bal Fe plus Ni.
Cold drawn: 88,000 TS; 69,000 YS; 23 El; 162 Brin. For glass-to-metal seals; controlled expansion. *Obsolete*

SEALMET 4
Allegheny Ludlum Steel
Ni 42, Cr 6, bal Fe.
Cold drawn: 80,000 TS; 40,000 YS; 31 El; 127 Brin. For glass-to-metal seals; controlled expansion. *Obsolete*

SEALMET HC-1
Allegheny Ludlum Steel
Cr 28, bal Fe.
For glass to metal seals; coefficient expansion of glass. *Obsolete*

SEALMET HC4
Allegheny Ludlum Steel
Ni 42, bal Fe.
Annealed: 80,000 TS; 30,000 YS; 30 El; 125 Brin. Hard: 140,000 TS; 228 Brin. For instruments, glass-to-metal seals; controlled expansion. *Obsolete*

SEALVAC A
Colt Industries
Co 17, Ni 29, bal Fe.
For glass to metal seals, sealing leads in electric light bulbs and radio tubes. Controlled expansion. *Obsolete*

SEALVAC A
Vacuum Metals Corp.
Co 17, Ni 29, bal Fe.
For glass to metal seals, sealing leads in electric light bulbs and radio tubes. Controlled expansion. *Obsolete*

SEALVAR

Pfizer Inc.
Ni 29, Co 17, bal Fe.
For glass to metal sealing; hermetic seals with the harder glasses and ceramics in electronic industries. Controlled expansion.

SEAMLESS TUBING ALLOY

American manufacture
Cu 60-61.5, Zn 38.5-40.
For tubing; high strength.

SEARCHER

Osborn Steels Ltd.
C 0.45, W 2, Cr 1, Mn 0.4, Si 1, bal Fe.
For punches, crimpers, upsetters; oil hardened, shock resistant. *Obsolete*

SEARCHER A.1

Osborn Steels Ltd.
C 0.63, W 1.6, Cr 0.4, Si 1.1, bal Fe.
For punches, crimpers; oil hardened. *Obsolete*

SEARING BEARING

English manufacture
Cu 86, Sn 14.
For bearings; bronze.

SEARS HARD FACING

Stulz Sickles Steel Co.
Mn 11-13.5, C, bal Fe.
Welded: 500-600 Brin. For hard facing electrodes for build-up work; austenitic, work hardened.

SECAERO

Seaboard Steel Co. of America
C 1, Mn 0.65, Cr 5, V 0.3, Mo 1.1, bal Fe.
Air hardening tool steel, cold work type; AISI A2.

SECO II

Swedish manufacture
TiC, WC.
For tools, dies; sintered carbides.

SECO XXX

Seaboard Steel Co. of America
C 0.5, Mn 0.9, bal Fe.
For chisels, hammers; tough.

SECOBALT

Seaboard Steel Co. of America
C 0.75, Cr 5, Co 11, V 1.5, W 18, Mo 0.8, bal Fe.
For cutters, tools; high speed steel.

SECODIE

Seaboard Steel Co. of America
C 1.35, Cr 0.6, V 0.35, Mo 1.5, Si 1.2, bal Fe.
For dies; non-deforming.

SECOLEO

Seaboard Steel Co. of America
C 0.75, Mn 0.45, Si 0.2, Cr 1.1, Ni 1.65, bal Fe.
Oil hardening tool and die steel.

SECOLOY

Seaboard Steel Co. of America
C 2.5, Cr 33, Co 44, W 17, Fe 2.
For tool bits, cutters, cast.

SECON 1% SILICON ALUMINUM UBG

Secon Metals Corp.
Si 0.85-1.15, very low Fe, Cu, Mg, bal Al.
For ultrasonic bonding.

SECON 422

Secon Metals Corp.
Platinum-rhodium-ruthenium alloy. Used in potentiometers.

SECON 430

Secon Metals Corp.
Mo, bal Pd.
Wire: 215,000 TS. For resistors, potentiometers; high electrical resistance.

SECON 436

Secon Metals Corp.
Tungsten-platinum alloy. For use in potentiometers and strain gages.

SECON 436D

Secon Metals Corp.
W, Mo, bal Pt.
Wire: 300,000 TS. For resistors, potentiometers; high electrical resistance.

SECON 445

Secon Metals Corp.
Mo, bal Pd.
Wire: 250,000 TS. For resistors, potentiometers; high electrical resistance.

SECON 449

Secon Metals Corp.
Rh, bal Pt.
Wire: 240,000 TS. For resistors, potentiometers; high electrical resistance.

SECON 486

Secon Metals Corp.
Mo, bal Pd.
Wire: 250,000 TS. For resistors, potentiometers; high electrical resistance.

SECOVAN

Seaboard Steel Co. of America
C 0.7, Cr 4.5, V 2.25, W 18, Mo 0.65, bal Fe.
For tools, cutters; high speed steel.

SECRETAN

English manufacture
Cu 91-95, Al 5-9, Mg 1.5, P 0.5.
For strong corrosion resistant parts.

SECURIT LDS

Haeckerstahl GmbH
C 0.9, Mn 1.9, V 0.1, bal Fe.
For punches, dies, crimpers, form tools; oil hardened, nondeforming.

SEFCO NO. 2

Sessions Foundry Co.
C 3.2, Si 2.5, Ni 1.5, bal Fe.
Cast: 66,000 TS; 230 Brin. For gears, machinery parts; cast iron.

SEGMENT STEEL

Disston Inc.
C 0.85, Cr 0.75, bal Fe.
For tools, dies; oil hardening. *Obsolete*

SEIBEL WSM

W. Seibel AG
Al alloy.
For light alloy parts.

SEIFERT

English manufacture
Sn 73, Zn 21, Pb 5, P 0.5, Sb 0.5.
For ornaments, bearings; anti-friction.

SEL 1

Manufacturer not listed.
C 0.08, Cr 15, Co 26, Mo 4.5, Al 4.4, Ti 2.3, B 0.015, Fe 0-1, bal Ni.
For gas turbines, high temperature parts, jet engine components. Cast alloy, high heat and corrosion resistance.

SEL 15

Cannon-Muskegon Corp.
C 0.07, Cr 11, Co 14.5, Mo 6.5, W 1.5, Al 5.4, Ti 2.5, Fe 0-0.5, B 0.015, 0.4 Cb + Ta, bal Ni.
For gas turbine components, high temperature parts. Cast alloy, high heat and corrosion resistant.

SELECT

Latrobe Steel Co.
C 0.9, Cr 3.5, Mo 6, V 0.2, bal Fe.
For hot work tools, rivet dies; hot work steel. *Obsolete*

SELECT B

Latrobe Steel Co.
C 1, Cr 5, V 0.25, Mo 1, Si 0.3, Mn 0.75, bal Fe.
For blanking and forming dies, punches, plastic dies; wear resistant, air hardening.

SELECT B-FM

Latrobe Steel Co.
C 1, Cr 5.25, V 0.3, Mo 1.1, S 0.15, bal Fe.
For blanking and forming dies; air hardening, nondeforming, self lubricating. AISI A2. *Obsolete*

SELECT M

Latrobe Steel Co.
C 0.9, Cr 3.5, Mo 0.6, bal Fe.
For hot work tools and dies; hot work steel. *Obsolete*

SELENIUM COPPER-948

Anaconda Co.
Copper. Cu 99.5, Se 0.5.
Soft: 32,000 TS; 10,000 YS; 45 El. Hard: 45,000 TS; 40,000 YS; 15 El. For electrical conductors. Hot work qualities.

SELF HARDENING

H. Boker & Co.
C 2.15, Cr 1.15, Mn 1.65, W 10.5, bal Fe.
For tack dies; fatigue resistant. *Obsolete*

SELF LUBE BRONZE

Now KEYSTONE C-64, etc.

SELF-HARDENING

Crucible Materials Corp.
Tool material. C 1.9, Cr 4, Mo 4.1, bal Fe.

SELF-HARDENING

Edgar T. Ward's Sons Co.
C, alloy, bal Fe.
For tools. *Obsolete*

SELF-HARDENING

Columbia Tool Steel Co.
C 1.6, Cr 3.75, W 9.5, bal Fe.

SELF-HARDENING

H. Boker & Co.
C 1.6, Cr 3.75, W 9.5, bal Fe.

SELF-HARDENING

Edgar Allen Balfour Ltd.
C, W, bal Fe.
For tools. *Obsolete*

SELF-HARDENING

Colt Industries
C, Mo, bal Fe.
For cutting tools. *Obsolete*

SELFLUBE IRON

Keystone Carbon Co.
Fe 90, Cu 10. *Obsolete*

SELFLUBE IRON

Welding Alloy Mfg. Co.
Fe 95, Cu 5.

SELVA METAL
English manufacture
Zn, bal Cu.
For hardware; a high tension brass.

SEMALLOY NO. 100
Semi-Alloys
Bi 43, Pb 22, Sn 8, Cd 5, In 18, Hg 4.
Liquidus 110°F; solidus 100°F (38-43°C). Note: Semalloys No. 100 through 2400 are low melting alloys used for fusible elements in automatic sprinklers, solders, encapsulating electronic equipment, anchoring small assemblies, patterns, models, low melting castings, fusible cores, dental work.

SEMALLOY NO. 1001
Semi-Alloys
In 52, Sn 48.
MP 243°F; 117°C.

SEMALLOY NO. 1002
Semi-Alloys
Ga 62.5, In 21.5, Sn 16.
MP 51°F; 11°C, eutectic alloy.

SEMALLOY NO. 1003
Semi-Alloys
Ga 75.5, In 24.5.
MP 61°F; 16°C, eutectic alloy.

SEMALLOY NO. 1004
Semi-Alloys
Ga 82, Sn 12, Zn 6.
MP 63°F; 17°C, eutectic alloy.

SEMALLOY NO. 1005
Semi-Alloys
Ga 92, Sn 8.
MP 68°F; 20°C, eutectic alloy.

SEMALLOY NO. 1006
Semi-Alloys
Cs 100.
MP 86°F; 30°C.

SEMALLOY NO. 1007
Semi-Alloys
In 55, Pb 45.
MP 419°F; 215°C; eutectic alloy.

SEMALLOY NO. 1008
Semi-Alloys
Ga 100.
MP 86°F; 30°C.

SEMALLOY NO. 1009
Semi-Alloys
Rb 100.
MP 102°F; 30°C.

SEMALLOY NO. 1010
Semi-Alloys
Pb 93, In 5, Ag 2.
MP 541°F; 283°C; eutectic alloy.

SEMALLOY NO. 1011
Semi-Alloys
P 100.
MP 111°F; 44°C.

SEMALLOY NO. 1012
Semi-Alloys
Ga 95, In 5.
MP 77°F; 25°C.

SEMALLOY NO. 1013
Semi-Alloys
In 90, Ag 10.
Liquidus 459°F; solidus 288°F (141-237°C).

SEMALLOY NO. 1015
Semi-Alloys
Bi 45, Pb 23, In 19, Sn 8, Cd 5.
MP 117°F; 47°C.

SEMALLOY NO. 102
Semi-Alloys
Bi 45, Pb 23, Sn 11, Cd 5, In 16.
Liquidus 126°F; solidus 117°F (47-52°C).

SEMALLOY NO. 103
Semi-Alloys
Bi 49, Pb 18, Sn 11, Cd 1, In 21.
Liquidus 133°F; solidus 129°F (54-56°C).

SEMALLOY NO. 104
Semi-Alloys
Bi 49.3, Pb 26.3, Sn 13.2, Cd 9.8, Ga 1.4.
Liquidus 151°F; solidus 149°F (65-66°C).

SEMALLOY NO. 1040
Semi-Alloys
Bi 49, In 21, Pb 18, Sn 12.
MP 136°F; 58°C; eutectic alloy.

SEMALLOY NO. 105
Semi-Alloys
Bi 48, Pb 25, Sn 13, Cd 9, In 5.
Liquidus 149°F; solidus 134°F (57-65°C).

SEMALLOY NO. 1050
Semi-Alloys
Bi 49, In 21, Pb 18, Cd 1.
MP 136°F; 58°C; eutectic alloy.

SEMALLOY NO. 106
Semi-Alloys
Bi 48, Pb 26, Sn 13, Cd 9, In 4.
Liquidus 149°F; solidus 142°F (61-65°C).

SEMALLOY NO. 1060
Semi-Alloys
In 51, Bi 33, Sn 16.
MP 142°F; 61°C; eutectic alloy.

SEMALLOY NO. 107
Semi-Alloys
Bi 43, Pb 23, Sn 12, Cd 9, Hg 13.
Liquidus 151°F; solidus 131°F (55-66°C).

SEMALLOY NO. 1075
Semi-Alloys
In 62, Bi 30, Cd 8.
MP 144°F; 62°C; eutectic alloy.

SEMALLOY NO. 108
Semi-Alloys
Bi 49, Pb 18, Sn 15, In 18.
Liquidus 156°F; solidus 136°F (58-69°C).

SEMALLOY NO. 1090
Semi-Alloys
Bi 50, Pb 27, Sn 13, Cd 10.
MP 158°F; 70°C; eutectic alloy.

SEMALLOY NO. 110
Semi-Alloys
Bi 51, Pb 28, Sn 12, Cd 9.
Liquidus 163°F; solidus 158°F (70-73°C).

SEMALLOY NO. 1100
Semi-Alloys
Bi 52, Pb 26, In 22.
MP 158°F; 70°C; eutectic alloy.

SEMALLOY NO. 111
Semi-Alloys
Bi 50, Pb 25, Sn 12.5, Cd 12.5.
Liquidus 165°F; solidus 158°F (70-74°C)

SEMALLOY NO. 112
Semi-Alloys
Bi 50, Pb 24.95, Sn 12.5, Cd 12.5, Ag 0.05.
Liquidus 165°F; solidus 158°F (70-74°C).

SEMALLOY NO. 1125
Semi-Alloys
In 66, Bi 34.
MP 162°F; 72°C; eutectic alloy.

SEMALLOY NO. 1127
Semi-Alloys
Bi 48.5, In 41.5, Cd 10.
MP 172°F; 78°C, eutectic alloy.

SEMALLOY NO. 113
Semi-Alloys
Bi 48, Sn 20, Pb 19, Cd 13.
Liquidus 168°F; solidus 158°F.

SEMALLOY NO. 114
Semi-Alloys
Bi 50, Pb 35, Sn 9, Cd 6.
Liquidus 173°F; solidus 158°F (70-78°C).

SEMALLOY NO. 115
Semi-Alloys
Bi 42, Pb 35, Sn 13, Cd 10.
Liquidus 176°F; solidus 158°F (70-80°C).

SEMALLOY NO. 1150
Semi-Alloys
Bi 57, Sn 17, In 26.
MP 174°F; 79°C; eutectic alloy.

SEMALLOY NO. 116
Semi-Alloys
Bi 50, Pb 39, Sn 3, Cd 8.
Liquidus 180°F; solidus 170°F (77-82°C).

SEMALLOY NO. 117
Semi-Alloys
Bi 50, Pb 39, Sn 1, Cd 8, In 2.
Liquidus 185°F; solidus 178°F (81-85°C).

SEMALLOY NO. 1175
Semi-Alloys
Bi 54.02, In 29.68, Sn 16.38.
MP 178°F; 81°C.

SEMALLOY NO. 118
Semi-Alloys
Bi 43, Pb 38, Sn 10, Cd 9.
Liquidus 190°F; solidus 160°F (71-78°C).

SEMALLOY NO. 119
Semi-Alloys
Bi 50, Pb 39, Sn 2, Cd 8, In 1.
Liquidus 192°F; solidus 176°F (80-89°C).

SEMALLOY NO. 120
Semi-Alloys
Bi 51, Pb 31, Sn 15, Cd 1, In 2.
Liquidus 192°F; solidus 176°F (80-89°C).

SEMALLOY NO. 121
Semi-Alloys
Bi 51, Pb 40, Cd 8, In 1.
Liquidus 192°F; solidus 188°F (87-89°C).

SEMALLOY NO. 122
Semi-Alloys
Bi 40, Pb 37, Sn 13, Cd 10.
Liquidus 185°F; solidus 158°F (70-85°C).

SEMALLOY NO. 1220
Semi-Alloys
Bi 52, Pb 40, Cd 8.
MP 198°F; 92°C; eutectic alloy.

SEMALLOY NO. 123
Semi-Alloys
Bi 52, Pb 32, Sn 15, Cd 1.
Liquidus 198°F; solidus 181°F (89-92°C).

SEMALLOY NO. 124
Semi-Alloys
Bi 51.5, Pb 31.5, Sn 15, In 2.
Liquidus 200°F; solidus 190°F (88-93°C).

SEMALLOY NO. 1245
Semi-Alloys
In 44, Sn 42, Cd 14.
MP 200°F; 93°C; eutectic alloy.

SEMALLOY NO. 125
Semi-Alloys
Bi 50, Pb 39, Sn 4, Cd 7.
Liquidus 200°F; solidus 165°F (74-93°C).

SEMALLOY NO. 126
Semi-Alloys
Bi 45.3, Sn 24.5, Pb 17.9, Cd 12.3.
Liquidus 190°F; solidus 158°F (70-88°C).

SEMALLOY NO. 127
Semi-Alloys
Bi 38.4, Pb 30.8, Sn 15.4, Cd 15.4.
Liquidus 207°F; solidus 158°F (70-97°C).

SEMALLOY NO. 1270
Semi-Alloys
Bi 52.5, Pb 32, Sn 15.5.
MP 203°F; 95°C; eutectic alloy.

SEMALLOY NO. 1275
Semi-Alloys
Bi 52, Pb 30, Sn 18.
MP 205°F; 96°C; eutectic alloy.

SEMALLOY NO. 1280
Semi-Alloys
Bi 52, Pb 32, Sn 16.
MP 205°F; 96°C; eutectic alloy.

SEMALLOY NO. 1285
Semi-Alloys
Bi 46, Sn 34, Pb 20.
MP 212°F; 100°C; eutectic alloy.

SEMALLOY NO. 129
Semi-Alloys
Bi 50, Pb 31, Sn 19.
Liquidus 210°F; solidus 200°F (93-99°C).

SEMALLOY NO. 1290
Semi-Alloys
Bi 50, Pb 28, Sn 22.
Melting range: 205-229°F; 96-109°C.

SEMALLOY NO. 130
Semi-Alloys
Bi 41, Pb 28, Sn 22, Cd 9.
Liquidus 215°F; solidus 158°F (70-102°C).

SEMALLOY NO. 131
Semi-Alloys
Bi 50, Pb 25, Sn 25.
Liquidus 239°F; solidus 203°F (95-115°C).

SEMALLOY NO. 132
Semi-Alloys
Bi 52, Pb 31.7, Sn 15.3, In 1.
Liquidus 201°F; solidus 195°F (91-94°C).

SEMALLOY NO. 133
Semi-Alloys
Pb 38, Bi 37, Sn 25.
Liquidus 261°F; solidus 199°F (93-127°C).

SEMALLOY NO. 1330
Semi-Alloys
Bi 54, Sn 26, Cd 20.
MP 217°F; 103°C; eutectic alloy.

SEMALLOY NO. 134
Semi-Alloys
Bi 56, Pb 22, Sn 22.
Liquidus 219°F; solidus 203°F (95-104°C).

SEMALLOY NO. 135
Semi-Alloys
Bi 51.6, Sn 37.4, In 6, Pb 5.
Liquidus 264°F; solidus 203°F (95-129°C).

SEMALLOY NO. 136
Semi-Alloys
Bi 52, Pb 38, Sn 10.
Liquidus 221°F; solidus 208°F (98-105°C).

SEMALLOY NO. 137
Semi-Alloys
Bi 35, Pb 35, Sn 20, Cd 10.
Liquidus 221°F; solidus 158°F (70-105°C).

SEMALLOY NO. 138
Semi-Alloys
Bi 45, Pb 35, Sn 20.
Liquidus 225°F; solidus 205°F (96-107°C).

SEMALLOY NO. 138A
Semi-Alloys
Bi 55, Pb 39, Sn 6.
Liquidus 226°F; solidus 215°F (102-108°C).

SEMALLOY NO. 139
Semi-Alloys
Bi 46, Pb 34, Sn 20.
Liquidus 227°F; solidus 203°F (95-108°C).

SEMALLOY NO. 140
Semi-Alloys
Bi 52, Pb 41, Sn 7.
Liquidus 234°F; solidus 208°F (98-112°C).

SEMALLOY NO. 1400
Semi-Alloys
Bi 67, In 33.
MP 229°F; 109°C; eutectic alloy.

SEMALLOY NO. 1405
Semi-Alloys
S 100.
MP 235°F; 113°C.

SEMALLOY NO. 141
Semi-Alloys
Bi 54, Pb 44, Sn 1, Cd 1.
Liquidus 235°F; solidus 219°F (104-113°C).

SEMALLOY NO. 142
Semi-Alloys
Bi 40, Pb 33, Sn 14, Cd 13.
Liquidus 235°F; solidus 162°F (92-113°C).

SEMALLOY NO. 143
Semi-Alloys
Bi 50, Pb 30, Sn 20.
Liquidus 239°F; solidus 203°F (95-104°C).

SEMALLOY NO. 144
Semi-Alloys
Pb 70, Bi 25, Sn 5.
Liquidus 240°F; solidus 212°F (100-116°C).

SEMALLOY NO. 145
Semi-Alloys
Bi 50, Cd 25, Sn 25.
Liquidus 235°F; solidus 217°F (103-113°C).

SEMALLOY NO. 146
Semi-Alloys
Bi 53, Pb 42, Sn 5.
Liquidus 243°F; solidus 217°F (103-117°C).

SEMALLOY NO. 147
Semi-Alloys
Bi 38, Pb 26, Sn 32, Cd 3, Sb 1.
Liquidus 244°F; solidus 167°F (75-118°C).

SEMALLOY NO. 148
Semi-Alloys
Bi 54, Pb 43, Sn 3.
Liquidus 246°F; solidus 226°F (108-119°C).

SEMALLOY NO. 149
Semi-Alloys
Bi 55, Pb 44, Sn 1.
Liquidus 248°F; solidus 242°F (117-120°C).

SEMALLOY NO. 150
Semi-Alloys
Bi 55, Pb 44, In 1.
Liquidus 250°F; solidus 248°F (120-121°C).

SEMALLOY NO. 151
Semi-Alloys
Bi 57, Pb 41, Cd 2.
Liquidus 250°F; solidus 197°F (92-121°C).

SEMALLOY NO. 1510A
Semi-Alloys
In 74, Cd 26.
MP 253°F; 123°C; eutectic alloy.

SEMALLOY NO. 152
Semi-Alloys
Bi 31, Pb 46, Sn 18, Cd 5.
Liquidus 253°F; solidus 158°F (70-123°C).

SEMALLOY NO. 153
Semi-Alloys
Bi 52.98, Pb 42.49, Sn 4.53.
Liquidus 243°F; solidus 217°F (103-117°C).

SEMALLOY NO. 1530
Semi-Alloys
Bi 55, Pb 45.
MP 255°F; 124°C; eutectic alloy.

SEMALLOY NO. 154
Semi-Alloys
Bi 58, Pb 42.
Liquidus 259°F; solidus 255°F.

SEMALLOY NO. 155
Semi-Alloys
Sn 50, In 50.
Liquidus 257°F; solidus 243°F (117-125°C).

SEMALLOY NO. 1555
Semi-Alloys
Sn 46, Te 37, Cd 17.
MP 262°F; 128°C; eutectic alloy.

SEMALLOY NO. 1580
Semi-Alloys
Bi 56, Sn 40, Zn 4.
MP 266°F; 130°C; eutectic alloy.

SEMALLOY NO. 159
Semi-Alloys
In 95, Ga 5.
Liquidus 266°F; solidus 256°F.

SEMALLOY NO. 1590
Semi-Alloys
Sn 52, In 48.
Liquidus 268°F; solidus 243°F (117-131°C).

SEMALLOY NO. 1592
Semi-Alloys
Bi 40, Sn 40, Pb 20.
Liquidus 266°F; solidus 250°F (121-130°C).

SEMALLOY NO. 1594
Semi-Alloys
In 70, Sn 15, Pb 9.6, Cd 5.4.
MP 257°F; 125°C; eutectic alloy.

SEMALLOY NO. 1596
Semi-Alloys
Bi 58, Pb 42.
Liquidus 261°F; solidus 257°F (125-127°C).

SEMALLOY NO. 1598
Semi-Alloys
In 95, Bi 5.
Liquidus 302°F; solidus 257°F (125-150°C).

SEMALLOY NO. 160
Semi-Alloys
Bi 57, Pb 2, Sn 41.
Liquidus 271°F; solidus 262°F (128-133°C).

SEMALLOY NO. 161
Semi-Alloys
Bi 32, Pb 34, Sn 34.
Liquidus 271°F; solidus 205°F (96-133°C).

SEMALLOY NO. 1610
Semi-Alloys
Sn 37.5, Pb 37.5, In 25.
Liquidus 358°F; solidus 273°F (134-181°C).

SEMALLOY NO. 162
Semi-Alloys
Bi 38, Pb 31, Sn 31.
Liquidus 275°F; solidus 205°F (96-135°C).

SEMALLOY NO. 163
Semi-Alloys
Bi 38.41, Pb 30.77, Sn 30.77, Ag 0.5.
Melting range: 205-275°F; 96-135°C.

SEMALLOY NO. 1630
Semi-Alloys
Bi 57, Pb 1, Sn 42.
MP 275°F; 135°C; eutectic alloy.

SEMALLOY NO. 164
Semi-Alloys
Bi 55, Pb 5, Sn 40.
Liquidus 277°F; solidus 250°F.

SEMALLOY NO. 164
Semi-Alloys
Bi 36.45, Pb 31.75, Sn 31.5, Cd 0.25, Ag 0.05.
Liquidus 277°F; solidus 203°F (95-136°C).

SEMALLOY NO. 1640
Semi-Alloys
Bi 55, Sn 40, Pb 5.
Melting range: 250-277°F; 121-136°C.

SEMALLOY NO. 165
Semi-Alloys
Bi 36, Pb 32, Sn 31, Ag 1.
Liquidus 277°F; solidus 203°F (95-136°C).

SEMALLOY NO. 166
Semi-Alloys
Bi 36, Pb 32, Sn 32.
Liquidus 279°F; solidus 203°F (95-137°C).

SEMALLOY NO. 167
Semi-Alloys
Bi 28.5, Pb 43, Sn 28.5.
Liquidus 279°F; solidus 205°F (96-137°C).

SEMALLOY NO. 1680
Semi-Alloys
Bi 58, Sn 42.
MP 281°F; 138°C; eutectic alloy.

SEMALLOY NO. 169
Semi-Alloys
Bi 5, Pb 32, Sn 45, Cd 18.
Liquidus 282°F; solidus 270°F (132-139°C).

SEMALLOY NO. 1690
Semi-Alloys
Sn 51, Pb 31, Cd 18.
MP 288°F; 142°C; eutectic alloy.

SEMALLOY NO. 170
Semi-Alloys
Bi 31, Pb 38, Sn 31.
Liquidus 282°F; solidus 205°F (96-139°C).

SEMALLOY NO. 171
Semi-Alloys
Bi 33.33, Pb 33.34, Sn 33.33.
Melting range: 205-289°F; 96-143°C.

SEMALLOY NO. 1710
Semi-Alloys
Bi 60, Cd 40.
MP 291°F; 144°C; eutectic alloy.

SEMALLOY NO. 1715
Semi-Alloys
In 97, Ag 3.
MP 290°F; 143°C.

SEMALLOY NO. 172
Semi-Alloys
Bi 34, Pb 33, Sn 33.
Melting range: 205-289°F; 96-143°C.

SEMALLOY NO. 1725
Semi-Alloys
Sn 51.2, Pb 30.6, Cd 18.2.
MP 291°F; 144°C.

SEMALLOY NO. 173
Semi-Alloys
Sn 58, In 42.
Liquidus 292°F; solidus 242°F.

SEMALLOY NO. 1730
Semi-Alloys
Sn 58, In 42.
Liquidus 292°F; solidus 243°F (117-144°C).

SEMALLOY NO. 1735
Semi-Alloys
Sn 70, Pb 18, Cu 12.
MP 324°F; 162°C.

SEMALLOY NO. 174
Semi-Alloys
Bi 21, Pb 42, Sn 37.
Liquidus 306°F; solidus 248°F.

SEMALLOY NO. 1740
Semi-Alloys
Pb 42, Sn 37, Bi 21.
Liquidus 306°F; solidus 248°F (120-152°C).

SEMALLOY NO. 1741
Semi-Alloys
In 99.4, Ga 0.6.
MP 306°F; 152°C; eutectic alloy.

SEMALLOY NO. 1742
Semi-Alloys
In 99, Cu 1.
MP 307°F; 153°C; eutectic alloy.

SEMALLOY NO. 1743
Semi-Alloys
In 99.6, Ga 0.4.
MP 307°F; 153°C; eutectic alloy.

SEMALLOY NO. 1744
Semi-Alloys
In 99.5, Ga 0.5.
MP 307°F; 153°C; eutectic alloy.

SEMALLOY NO. 1746
Semi-Alloys
Sn 50, Pb 32, Cd 18.
MP 295°F; 146°C; eutectic alloy.

SEMALLOY NO. 175
Semi-Alloys
In 80, Pb 15, Ag 5.
Liquidus 315°F; solidus 315°F.

SEMALLOY NO. 1760
Semi-Alloys
100 In.
MP 313°F; 156°C.

SEMALLOY NO. 1765
Semi-Alloys
In 80, Pb 15, Ag 5.
MP 301°F; 149°C; eutectic alloy.

SEMALLOY NO. 177
Semi-Alloys
Bi 45, Pb 33, Cd 22.
Liquidus 316°F; solidus 197°F.

SEMALLOY NO. 1770
Semi-Alloys
Bi 45, Pb 33, Cd 22.
Liquidus 316°F; solidus 199°F (92-150°C).

SEMALLOY NO. 1775
Semi-Alloys
Sn 50, Bi 25, Pb 25.
Liquidus 320°F; solidus 293°F (145-160°C).

SEMALLOY NO. 178
Semi-Alloys
Bi 45, Pb 55.
Liquidus 320°F; solidus 253°F.

SEMALLOY NO. 1780
Semi-Alloys
Pb 55, Bi 45.
Liquidus 320°F; solidus 253°F (123-160°C).

SEMALLOY NO. 179
Semi-Alloys
Sn 50, Pb 25, Cd 25.
Liquidus 320°F; solidus 293°F.

SEMALLOY NO. 1790
Semi-Alloys
Sn 50, Pb 25, Cd 25.
Liquidus 320°F; solidus 290°F (145-160°C).

SEMALLOY NO. 180
Semi-Alloys
Sn 70, Pb 18, In 12.
Liquidus 324°F; solidus 324°F.

SEMALLOY NO. 181
Semi-Alloys
Bi 16, Pb 36, Sn 48.
Liquidus 324°F; solidus 284°F.

SEMALLOY NO. 1810
Semi-Alloys
Sn 48, Pb 36, Bi 16.
Liquidus 324°F; solidus 284°F (140-162°C).

SEMALLOY NO. 1815
Semi-Alloys
In 95, Ag 5.
Liquidus 324°F; solidus 319°F (159-162°C).

SEMALLOY NO. 182
Semi-Alloys
Bi 14, Pb 43, Sn 43.
Liquidus 325°F; solidus 291°F.

SEMALLOY NO. 1820
Semi-Alloys
Pb 43, Sn 43, Bi 14.
Liquidus 325°F; solidus 291°F (144-163°C).

SEMALLOY NO. 183
Semi-Alloys
Sn 50, Pb 40, Bi 10.
Liquidus 330°F; solidus 248°F.

SEMALLOY NO. 1830
Semi-Alloys
Sn 50, Pb 40, Bi 10.
Liquidus 332°F; solidus 248°F (120-167°C).

SEMALLOY NO. 1836
Semi-Alloys
Sn 48.8, Pb 41, Bi 10.2.
Liquidus 331°F; solidus 288°F (142-166°C).

SEMALLOY NO. 184
Semi-Alloys
Bi 22, Pb 51, Sn 27.
Liquidus 338°F; solidus 268°F.

SEMALLOY NO. 1840
Semi-Alloys
Pb 51, Sn 27, Bi 22.
Liquidus 338°F; solidus 268°F (131-170°C).

SEMALLOY NO. 185
Semi-Alloys
Bi 40, Sn 60.
Liquidus 338°F; solidus 281°F.

SEMALLOY NO. 1850
Semi-Alloys
Sn 60, Bi 40.
Liquidus 338°F; solidus 281°F (138-170°C).

SEMALLOY NO. 186
Semi-Alloys
Bi 20, Pb 50, Sn 30.
Liquidus 343°F; solidus 266°F.

SEMALLOY NO. 1860
Semi-Alloys
Pb 50, Sn 30, Bi 20.
Liquidus 343°F; solidus 266°F (130-173°C).

SEMALLOY NO. 187
Semi-Alloys
Bi 13, Pb 47, Sn 40.
Liquidus 349°F; solidus 294°F.

SEMALLOY NO. 1870
Semi-Alloys
Pb 47, Sn 40, Bi 13.
Liquidus 349°F; solidus 294°F (146-176°C).

SEMALLOY NO. 1873
Semi-Alloys
Sn 68, Cd 32.
MP 351°F; 177°C; eutectic alloy.

SEMALLOY NO. 188
Semi-Alloys
Sn 61.5, Pb 35.5, Ag 3.
Liquidus 355°F; solidus 355°F.

SEMALLOY NO. 1880
Semi-Alloys
Sn 62.5, Pb 36, Ag 1.5.
MP 355°F; 179°C; eutectic alloy.

SEMALLOY NO. 189
Semi-Alloys
Bi 25, Pb 15, Sn 60.
Liquidus 356°F; solidus 205°F.

SEMALLOY NO. 1890
Semi-Alloys
Sn 60, Bi 25, Pb 15.
Liquidus 356°F; solidus 205°F (96-180°C).

SEMALLOY NO. 1910
Semi-Alloys
Sn 62, Pb 38.
MP 361°F; 183°C; eutectic alloy.

SEMALLOY NO. 1914
Semi-Alloys
Sn 63, Pb 37.
MP 361°F; 183°C; eutectic alloy.

SEMALLOY NO. 1916
Semi-Alloys
Sn 65, Pb 35.
Liquidus 364°F; solidus 361°F (183-184°C).

SEMALLOY NO. 1918
Semi-Alloys
Sn 65, Pb 34, Sb 1.
Liquidus 370°F; solidus 363°F (184-188°C).

SEMALLOY NO. 1920
Semi-Alloys
Sn 60, Pb 40.
Liquidus 374°F; solidus 361°F (183-190°F).

SEMALLOY NO. 1923
Semi-Alloys
Sn 70, Pb 30.
Liquidus 378°F; solidus 361°F (183-192°C).

SEMALLOY NO. 1927
Semi-Alloys
Sn 75, Pb 25.
Liquidus 383°F; solidus 361°F (183-195°C).

SEMALLOY NO. 1930
Semi-Alloys
Pb 55.5, Sn 40.5, Bi 4.
Liquidus 388°F; solidus 338°F (170-198°C).

SEMALLOY NO. 1933
Semi-Alloys
Sn 91, Zn 9.
MP 390°F; 199°C; eutectic alloy.

SEMALLOY NO. 1934
Semi-Alloys
Sn 92, Zn 8.
Liquidus 392°F; solidus 388°F (198-200°C).

SEMALLOY NO. 1935
Semi-Alloys
Sn 55, Pb 45.
Liquidus 392°F; solidus 361°F (183-200°C).

SEMALLOY NO. 1936
Semi-Alloys
Sn 80, Pb 20.
Liquidus 390°F; solidus 361°F (183-199°C).

SEMALLOY NO. 1937
Semi-Alloys
Sn 50, Pb 47, Sb 3.
Liquidus 399°F; solidus 365°F (185-204°C).

SEMALLOY NO. 1938
Semi-Alloys
Sn 85, Pb 15.
Liquidus 403°F; solidus 361°F (183-206°C).

SEMALLOY NO. 1940
Semi-Alloys
Pb 55, Sn 44, Ag 1.
Liquidus 410°F; solidus 350°F (177-210°C).

SEMALLOY NO. 1952
Semi-Alloys
Sn 90, Pb 10.
Liquidus 415°F; solidus 361°F (183-213°C).

SEMALLOY NO. 1956
Semi-Alloys
In 50, Pb 50.
Liquidus 408°F; solidus 356°F (180-209°C).

SEMALLOY NO. 1960
Semi-Alloys
Pb 50, Sn 50.
Liquidus 421°F; solidus 361°F (183-216°C).

SEMALLOY NO. 1968
Semi-Alloys
Pb 52, Sn 48.
Liquidus 424°F; solidus 361°F (183-218°C).

SEMALLOY NO. 1970
Semi-Alloys
Pb 55, In 45.
Liquidus 428°F; solidus 343°F (173-220°C).

SEMALLOY NO. 1971
Semi-Alloys
100 Se.
MP 428°F; 220°C.

SEMALLOY NO. 1978
Semi-Alloys
Sn 96.5, Ag 3.5.
MP 430°F; 221°C.

SEMALLOY NO. 1980
Semi-Alloys
Sn 95.5, Ag 3.5, Cd 1.
Liquidus 430°F; solidus 425°F (218-221°C).

SEMALLOY NO. 1982
Semi-Alloys
Sn 95, Pb 5.
Liquidus 432°F; solidus 361°F (183-222°C).

SEMALLOY NO. 1990
Semi-Alloys
Sn 97.5, Ag 2.5.
Liquidus 438°F; solidus 430°F (221-226°C).

SEMALLOY NO. 1994
Semi-Alloys
Pb 55, Sn 45.
Liquidus 441°F; solidus 361°F (183-227°C).

SEMALLOY NO. 2000
Semi-Alloys
Bi 48, Pb 28.5, Sn 14.5, Sb 9.
Liquidus 440°F; solidus 217°F (103-227°C).

SEMALLOY NO. 2003
Semi-Alloys
Sn 61, Pb 36, Ag 3.
Liquidus 478°F; solidus 354°F (179-248°C).

SEMALLOY NO. 2006
Semi-Alloys
Sn 99.25, Cu 0.75.
MP 440°F; 227°C; eutectic alloy.

SEMALLOY NO. 2010
Semi-Alloys
Sn 99, Ga 1.
MP 442°F; 228°C; eutectic alloy.

SEMALLOY NO. 2012
Semi-Alloys
Sn 99, Te 1.
Liquidus 842°F; solidus 450°F (232-450°C).

SEMALLOY NO. 2025
Semi-Alloys
Pb 58, Sn 40, Sb 2.
Liquidus 448°F; solidus 365°F (185-231°C).

SEMALLOY NO. 2030
Semi-Alloys
100 Sn.
MP 450°F; 232°C.

SEMALLOY NO. 2040
Semi-Alloys
Pb 60, Sn 37, Ag 0.
Liquidus 450°F; solidus 355°F (179-232°C).

SEMALLOY NO. 2050
Semi-Alloys
Pb 68, Sn 23, Cd 9.
Liquidus 455°F; solidus 293°F (145-235°C).

SEMALLOY NO. 2053
Semi-Alloys
Sn 99, Sb 1.
MP 456°F; 236°C; eutectic alloy.

SEMALLOY NO. 2055
Semi-Alloys
Pb 63, Sn 35, Sb 2.
Liquidus 468°F; solidus 365°F (185-242°C).

SEMALLOY NO. 2060
Semi-Alloys
Pb 60, Sn 40.
Liquidus 455°F; solidus 361°F (183-235°C).

SEMALLOY NO. 2070
Semi-Alloys
Pb 62, Sn 37, As 1.
Liquidus 460°F; solidus 361°F (183-238°C).

SEMALLOY NO. 2080
Semi-Alloys
Sn 97, Sb 3.
Liquidus 460°F; solidus 450°F (232-238°C).

SEMALLOY NO. 2090
Semi-Alloys
Sn 95, Sb 5.
Liquidus 464°F; solidus 452°F (233-240°C).

SEMALLOY NO. 2092
Semi-Alloys
Pb 62, Sn 38.
Liquidus 468°F; solidus 361°F (183-242°C).

SEMALLOY NO. 2093
Semi-Alloys
Sn 72, Cd 28.
Liquidus 358°F; solidus 351°F (177-181°C).

SEMALLOY NO. 2094
Semi-Alloys
Pb 63.2, Sn 35, Sb 1.8.
Liquidus 470°F; solidus 365°F (185-243°C).

SEMALLOY NO. 2095
Semi-Alloys
Sn 60, Pb 36, Ag 4.
Liquidus 473°F; solidus 354°F (179-245°C).

SEMALLOY NO. 2096
Semi-Alloys
Pb 65, Sn 35.
Liquidus 477°F; solidus 361°F (183-247°C).

SEMALLOY NO. 2097
Semi-Alloys
Sn 61.5, Pb 35.5, Ag 3.
Liquidus 478°F; solidus 354°F (179-248°C).

SEMALLOY NO. 2098
Semi-Alloys
Pb 83, Cd 17.
MP 478°F; 248°C; eutectic alloy.

SEMALLOY NO. 2100
Semi-Alloys
Bi 95, Sn 5.
Liquidus 483°F; solidus 273°F (134-251°C).

SEMALLOY NO. 2120
Semi-Alloys
Pb 85, Sb 10, Sn 5.
Liquidus 493°F; solidus 464°F (240-256°C).

SEMALLOY NO. 2125
Semi-Alloys
Pb 63, Sn 34, Zn 3.
Liquidus 493°F; solidus 338°F (170-256°C).

SEMALLOY NO. 2126
Semi-Alloys
Pb 82, Bi 14.5, Cd 2.5, Sb 1.
Liquidus 498°F; solidus 261°F (127-259°C).

SEMALLOY NO. 2127
Semi-Alloys
Sn 75, Zn 25.
Liquidus 572°F; solidus 388°F (198-300°C).

SEMALLOY NO. 2128
Semi-Alloys
Sn 50, Pb 47, Ag 3.
Liquidus 500°F; solidus 354°F (179-260°C).

SEMALLOY NO. 2129
Semi-Alloys
Pb 73.7, Sn 25, Sb 1.3.
Liquidus 504°F; solidus 363°F (184-262°C).

SEMALLOY NO. 2130
Semi-Alloys
Pb 90, Sb 10.
Liquidus 500°F; solidus 486°F (252-260°C).

SEMALLOY NO. 2131
Semi-Alloys
Cd 82.5, Zn 17.5.
MP 509°F; 265°C; eutectic alloy.

SEMALLOY NO. 2133
Semi-Alloys
Pb 91, Sb 9.
Liquidus 509°F; solidus 486°F (252-265°C).

SEMALLOY NO. 2135
Semi-Alloys
Pb 75, Sn 25.
Liquidus 514°F; solidus 361°F (183-268°C).

SEMALLOY NO. 2136
Semi-Alloys
Pb 79, In 20, Sb 1.
Liquidus 517°F; solidus 363°F (184-269°C).

SEMALLOY NO. 2137
Semi-Alloys
Pb 75, Sb 15, Sn 10.
Liquidus 514°F; solidus 404°F (240-268°C).

SEMALLOY NO. 2138
Semi-Alloys
Pb 75, In 25.
Liquidus 508°F; solidus 440°F (227-264°C).

SEMALLOY NO. 2139
Semi-Alloys
Sn 80, Zn 20.
Liquidus 518°F; solidus 390°F (199-270°C).

SEMALLOY NO. 2140
Semi-Alloys
Bi 100.
MP 520°F; 271°C.

SEMALLOY NO. 2153
Semi-Alloys
Sn 90, Ag 10.
Liquidus 563°F; solidus 430°F (221-295°C).

SEMALLOY NO. 2160
Semi-Alloys
Pb 92.5, Sn 5, Ag 2.5.
MP 536°F; 280°C; eutectic alloy.

SEMALLOY NO. 2165
Semi-Alloys
Pb 81, In 19.
Liquidus 536°F; solidus 518°F (270-280°C).

SEMALLOY NO. 2170
Semi-Alloys
Pb 92, Sn 5, Sb 3.
Liquidus 545°F; solidus 463°F (239-285°C).

SEMALLOY NO. 2172
Semi-Alloys
Pb 85, Sb 11.5, Sn 3.5.
MP 464°F; 240°C; eutectic alloy.

SEMALLOY NO. 2173
Semi-Alloys
Pb 94, Sb 6.
Liquidus 545°F; solidus 486°F (252-285°C).

SEMALLOY NO. 2175
Semi-Alloys
Pb 92, Sb 8.
Liquidus 520°F; solidus 486°F (252-271°C).

SEMALLOY NO. 2177
Semi-Alloys
Pb 93, Sn 5, Ag 2.
Liquidus 568°F; solidus 530°F (277-298°C).

SEMALLOY NO. 2178
Semi-Alloys
Cd 60, Zn 30, Sn 10.
Liquidus 550°F; solidus 315°F (157-288°C).

SEMALLOY NO. 2180
Semi-Alloys
Pb 85, Sn 15.
Liquidus 553°F; solidus 407°F (225-289°C).

SEMALLOY NO. 2187
Semi-Alloys
Pb 57, Sn 40, Ag 3.
Liquidus 543°F; solidus 354°F (179-284°C).

SEMALLOY NO. 2188
Semi-Alloys
Pb 90, In 5, Ag 5.
Liquidus 590°F; solidus 554°F (290-310°C).

SEMALLOY NO. 2190
Semi-Alloys
Pb 90, Sn 5, Ag 5.
MP 558°F; 292°C; eutectic alloy.

SEMALLOY NO. 2191
Semi-Alloys
Pb 90, Ag 10.
Liquidus 842°F; solidus 579°F (304-450°C).

SEMALLOY NO. 2200
Semi-Alloys
Sn 95, Ag 5.
Liquidus 473°F; solidus 430°F (221-245°C).

SEMALLOY NO. 2210
Semi-Alloys
Pb 95, Sb 5.
Liquidus 563°F; solidus 486°F (252-295°C).

SEMALLOY NO. 2213
Semi-Alloys
Cd 90, Zn 10.
Liquidus 570°F; solidus 509°F (265-299°C).

SEMALLOY NO. 2215
Semi-Alloys
Pb 96, Sb 4.
Liquidus 570°F; solidus 486°F (252-299°C).

SEMALLOY NO. 2216
Semi-Alloys
Sn 97, Cu 3.
Liquidus 572°F; solidus 441°F (227-300°C).

SEMALLOY NO. 2217
Semi-Alloys
Pb 90, In 10.
Liquidus 572°F; solidus 561°F (294-300°C).

SEMALLOY NO. 2218
Semi-Alloys
Pb 90, Sn 10.
Liquidus 573°F; solidus 514°F (268-301°C).

SEMALLOY NO. 2220
Semi-Alloys
Sn 60, Pb 34, Ag 6.
Liquidus 579°F; solidus 351°F (177-304°C).

SEMALLOY NO. 2230
Semi-Alloys
Pb 95.5, Ag 2.5, Sn 2.
Liquidus 580°F; solidus 570°F (299-304°C).

SEMALLOY NO. 2235
Semi-Alloys
Pb 92.5, In 5, Ag 2.5.
MP 585°F; 307°C; eutectic alloy.

SEMALLOY NO. 2240
Semi-Alloys
Pb 97.5, Ag 1.5, Se 1.
MP 588°F; 309°C; eutectic alloy.

SEMALLOY NO. 2245
Semi-Alloys
Sn 70, Zn 30.
Liquidus 592°F; solidus 390°F (199-311°C).

SEMALLOY NO. 2253
Semi-Alloys
Pb 97.5, Ag 1.75, Sn 0.75.
MP 590°F; 310°C; eutectic alloy.

SEMALLOY NO. 2260
Semi-Alloys
Pb 70, Sn 27, Ag 3.
Liquidus 594°F; solidus 354°F (179-312°C).

SEMALLOY NO. 2261
Semi-Alloys
Pb 95, In 5.
Liquidus 597°F; solidus 594°F (312-314°C).

SEMALLOY NO. 2262
Semi-Alloys
Pb 92, In 8.
Liquidus 581°F; solidus 572°F (300-305°C).

SEMALLOY NO. 2264
Semi-Alloys
Pb 98, Sb 1.2, Ga 0.8.
MP 599°F; 315°C; eutectic alloy.

SEMALLOY NO. 2265
Semi-Alloys
Cd 78.4, Zn 16.6, Ag 5.
Liquidus 600°F; solidus 480°F (249-316°C).

SEMALLOY NO. 2269
Semi-Alloys
100 Pb.
MP 621°F; 327°C.

SEMALLOY NO. 2270
Semi-Alloys
Pb 96.5, Ag 3.5.
Liquidus 621°F; solidus 579°F (304-327°C).

SEMALLOY NO. 2275
Semi-Alloys
Pb 98, Sb 2.
Liquidus 608°F; solidus 572°F (300-320°C).

SEMALLOY NO. 2278
Semi-Alloys
Pb 99, Sb 1.
Liquidus 615°F; solidus 594°F (312-324°C).

SEMALLOY NO. 2284
Semi-Alloys
100 Cd.
MP 610°F; 321°C.

SEMALLOY NO. 2287
Semi-Alloys
Pb 98.5, Sb 1.5.
Liquidus 612°F; solidus 583°F (306-322°C).

SEMALLOY NO. 2290
Semi-Alloys
Sn 98, As 2.
Liquidus 626°F; solidus 448°F (231-330°C).

SEMALLOY NO. 2291
Semi-Alloys
Zn 60, Cd 40.
Liquidus 635°F; solidus 509°F (265-335°C).

SEMALLOY NO. 2292
Semi-Alloys
Sn 60, Zn 40.
Liquidus 645°F; solidus 390°F (198-341°C).

SEMALLOY NO. 2294
Semi-Alloys
Sn 99, Ge 1.
Liquidus 653°F; solidus 450°F (232-345°C).

SEMALLOY NO. 2295
Semi-Alloys
Pb 83, Sb 15, Sn 1, As 1.
Liquidus 667°F; solidus 594°F (312-353°C).

SEMALLOY NO. 2300
Semi-Alloys
Pb 95, Ag 5.
Liquidus 687°F; solidus 579°F (304-364°C).

SEMALLOY NO. 2306
Semi-Alloys
Zn 70, Sn 30.
Liquidus 707°F; solidus 388°F (198-375°C).

SEMALLOY NO. 2310
Semi-Alloys
Sn 99.5, P 0.5.
Liquidus 707°F; solidus 450°F (232-375°C).

SEMALLOY NO. 2313
Semi-Alloys
Zn 95, Al 5.
MP 720°F; 382°C; eutectic alloy.

SEMALLOY NO. 2315
Semi-Alloys
Cd 95, Ag 5.
Liquidus 740°F; solidus 649°F (343-393°C).

SEMALLOY NO. 2317
Semi-Alloys
Zn 90, Cd 10.
Liquidus 750°F; solidus 511°F (266-399°C).

SEMALLOY NO. 2320
Semi-Alloys
Pb 99.
Liquidus 752°F; solidus 621°F (327-400°C).

SEMALLOY NO. 2330
Semi-Alloys
Cd 94.5, Ag 5.5.
Liquidus 752°F; solidus 649°F (343-400°C).

SEMALLOY NO. 2340
Semi-Alloys
Pb 98, Zn 2.
Liquidus 784°F; solidus 604°F (318-418°C).

SEMALLOY NO. 2345
Semi-Alloys
Zn 100.
MP 787°F; 419°C.

SEMALLOY NO. 2347
Semi-Alloys
Ge 55, Al 45.
MP 795°F; 424°C.

SEMALLOY NO. 2350
Semi-Alloys
Sn 95, As 5.
Liquidus 797°F; solidus 448°F (231-425°C).

SEMALLOY NO. 2380
Semi-Alloys
Sn 97, P 3.
Liquidus 932°F; solidus 450°F (232-500°C).

SEMALLOY NO. 2384
Semi-Alloys
Al 75, Si 10, Zn 10, Cu 4, Fe 1.
Liquidus 1040°F; solidus 960°F (516-560°C).

SEMALLOY NO. 2388
Semi-Alloys
Mg 83, Zn 5, Al 12.
Melting range: 1125-1145°F; 607-618°C. Note: SEMALLO alloys, available in various forms, are used essentially in brazing operations, but also for fusable links for safety devices, special castings and other special applications.

SEMALLOY NO. 2389
Semi-Alloys
100 Te.
MP 842°F; 450°C.

SEMALLOY NO. 2391
Semi-Alloys
Al 95, Si 5.
Liquidus 1165°F; solidus 1071°F (577-629°C).

SEMALLOY NO. 2395
Semi-Alloys
Al 86, Si 10, Cu 4.
Liquidus 1085°F; solidus 970°F (521-585°C).

SEMALLOY NO. 2400
Semi-Alloys
100 Bi.
MP 520°F; 271°C.

SEMALLOY NO. 2401
Semi-Alloys
Al 92.5, Si 7.5.
Liquidus 1135°F; solidus 1071°F (577-613°C).

SEMALLOY NO. 2403
Semi-Alloys
Al 88.3, Si 11.7.
MP 1071°F; 577°C; eutectic alloy.

SEMALLOY NO. 2405
Semi-Alloys
Al 90, Si 10.
Liquidus 1094°F; solidus 1071°F (577-590°C).

SEMALLOY NO. 2406
Semi-Alloys
Al 88, Si 12.
Liquidus 1080°F; solidus 1071°F (577-582°C).

SEMALLOY NO. 2411
Semi-Alloys
Mg 89, Al 9, Zn 2.
Liquidus 1110°F; solidus 770°F (410-599°C).

SEMALLOY NO. 2414
Semi-Alloys
Sb 100.
MP 1168°F; 631°C.

SEMALLOY NO. 2420
Semi-Alloys
Ag 45, Cu 15, Zn 16, Cd 24.
Melting range: 1125-1145°F; 607-618°C.

SEMALLOY NO. 2425
Semi-Alloys
Ag 44, Cu 27, Cd 15, Zn 13, P 1.
Liquidus 1205°F; solidus 110°F (599-652°C).

SEMALLOY NO. 2430
Semi-Alloys
Ag 50, Cu 15.5, Zn 16.5, Cd 18.
Melting range: 1160-1175°F; 627-635°C.

SEMALLOY NO. 2435
Semi-Alloys
Ag 51, Cu 22, Zn 19, Cd 7, Sn 1.
Liquidus 1180°F; solidus 1161°F (627-638°C).

SEMALLOY NO. 2437
Semi-Alloys
Ag 49, Zn 23, Cu 16, Mn 7.5, Ni 4.5.
Liquidus 1300°F; solidus 1161°F (627-704°C).

SEMALLOY NO. 2440
Semi-Alloys
Ag 50, Cu 15, Zn 25, Cd 10.
Melting range: 1160-1185°F; 627-640°C.

SEMALLOY NO. 2445
Semi-Alloys
Ag 41, Cd 24, Zn 18, Cu 17.
Liquidus 1161°F; solidus 1125°F (607-627°C).

SEMALLOY NO. 2450
Semi-Alloys
Ag 45, Cu 18, Zn 21, Cd 16.
Melting range: 1141-1190°F; 616-643°C.

SEMALLOY NO. 2454
Semi-Alloys
Mg 100.
MP 1202°F; 650°C.

SEMALLOY NO. 2460
Semi-Alloys
Ag 56, Cu 22, Zn 17, Sn 5.
Melting range: 1145-1205°F; 618-627°C.

SEMALLOY NO. 2465
Semi-Alloys
Ag 81, Ge 19.
MP 1206°F; 651°C.

SEMALLOY NO. 2467
Semi-Alloys
Al 100.
MP 1220°F; 660°C.

SEMALLOY NO. 2469
Semi-Alloys
Mg 99, Mn 1.
Liquidus 1200°F; 649°C.

SEMALLOY NO. 2470
Semi-Alloys
Ag 61, Cu 24, In 15.
Melting range: 1166-1265°F; 630-685°C.

SEMALLOY NO. 2475
Semi-Alloys
Ag 61.5, Cu 24, In 14.5.
Liquidus 1305°F; solidus 1155°F (624-707°C).

SEMALLOY NO. 2477
Semi-Alloys
Cu 75, Ag 17.75, P 7.25.
MP 1189°F; 643°C; eutectic alloy.

SEMALLOY NO. 2480
Semi-Alloys
Cu 82.7, Ag 10, Pd 7.3.
Melting range: 1190-1265°F; 643-685°C.

SEMALLOY NO. 2485
Semi-Alloys
Ag 50, Cu 15.5, Cd 16, Ni 3, Zn 15.5.
Melting range: 1170-1270°F; 632-688°C.

SEMALLOY NO. 2490
Semi-Alloys
Ag 70, In 30.
Melting range: 1130-1279°F; 610-693°C.

SEMALLOY NO. 2500
Semi-Alloys
Ag 35, Cu 26, Zn 21, Cd 18.
Melting range: 1125-1295°F; 607-702°C.

SEMALLOY NO. 2503
Semi Alloys
Ag 45, Cu 30, Zn 12, Mn 13.
MP 1298°F; 703°C; eutectic alloy.

SEMALLOY NO. 2507
Semi-Alloys
Cu 80, Ag 15, P 5.
Melting range: 1190-1300 F. *Obsolete*

SEMALLOY NO. 2510
Semi-Alloys
Cu 80, Ag 15, Pd 5.
Melting range: 1185-1300°F; 640-704°C.

SEMALLOY NO. 2515
Semi-Alloys
Ag 60, Cu 20, Zn 7, Cd 10, Sn 3.
Melting range: 1270-1300°F; 688-704°C.

SEMALLOY NO. 2520
Semi-Alloys
Ag 30, Cu 29, Zn 21, Cd 20.
Melting range: 1125-1300°F; 607-704°C.

SEMALLOY NO. 2521
Semi-Alloys
Ag 30, Cu 27, Zn 23, Cd 20.
Liquidus 1310°F; solidus 1125°F (607-710°C).

SEMALLOY NO. 2523
Semi-Alloys
Ag 54.5, Cu 15.5, Zn 14, Cd 15, Sn 1.
Liquidus 1190°F; solidus 1130°F (610-643°C).

SEMALLOY NO. 2525
Semi-Alloys
Ag 45, Zn 19, Cu 18, Cd 18.
Liquidus 1190°F; solidus 1130°F (610-643°C).

SEMALLOY NO. 2527
Semi-Alloys
Ag 40, Cd 27, Cu 18, Zn 15.
Liquidus 1195°F; solidus 1130°F (610-646°C).

SEMALLOY NO. 2530
Semi-Alloys
Ag 60, Cu 27, In 13.
Melting range: 1175-1300°F; 635-704°C.

SEMALLOY NO. 2534
Semi-Alloys
Ba 100.
MP 1310°F; 710°C.

SEMALLOY NO. 2535
Semi-Alloys
Ag 66.7, Cu 22.3, Zn 11.
Melting range: 1280-1320°F; 693-716°C.

SEMALLOY NO. 2539
Semi-Alloys
Ag 40, Cu 27, Zn 18, Cd 15.
Liquidus 1325°F; solidus 1161°F (627-718°C).

SEMALLOY NO. 2540
Semi-Alloys
Sn 96, Se 4.
Liquidus 1310°F; solidus 450°F (232-710°C).

SEMALLOY NO. 2543
Semi-Alloys
Ag 45, Cd 20.5, Cu 17, Zn 16.5, Sn 0.5, Pb 0.5.
Liquidus 1150°F; solidus 1135°F (613-621°C).

SEMALLOY NO. 2545
Semi-Alloys
Ag 50, Cd 19, Zn 16, Cu 15.
Liquidus 1160°F; solidus 1145°F (618-627°C).

SEMALLOY NO. 2550
Semi-Alloys
Ag 60, Cu 25, Zn 15.
Melting range: 1245-1325°F; 674-718°C.

SEMALLOY NO. 2560
Semi-Alloys
Ag 60, Cu 30, Sn 10.
Melting range: 1115-1325°F; 602-718°C.

SEMALLOY NO. 2565
Semi-Alloys
Ag 65, Cu 20, Zn 15.
Melting range: 1240-1325°F; 671-718°C.

SEMALLOY NO. 2570
Semi-Alloys
Cu 87, P 7, Ag 6.
Melting range: 1190-1320°F; 643-718°C.

SEMALLOY NO. 2573
Semi-Alloys
Ag 60.5, Cu 22.5, Zn 7, Cd 10.
Melting range: 1285-1335°F; 696-724°C.

SEMALLOY NO. 2577
Semi-Alloys
Ag 60, Cu 20, Cd 10, Zn 10.
Melting range: 1285-1335°F; 696-724°C.

SEMALLOY NO. 2580
Semi-Alloys
Ag 50, Cu 28, Zn 22.
Melting range: 1250-1340°F; 677-727°C.

SEMALLOY NO. 2585
Semi-Alloys
Ag 50, Zn 28, Cu 20, Ni 2.
Melting range: 1220-1301°F; 660-705°C.

SEMALLOY NO. 2590
Semi-Alloys
Ag 40, Cu 30.5, Zn 29.5.
Melting range: 1245-1340°F; 674-727°C.

SEMALLOY NO. 2600
Semi-Alloys
Ag 75, Zn 25.
Melting range: 1300-1345°F; 704-729°C.

SEMALLOY NO. 2610
Semi-Alloys
Ag 57.3, Cu 32.7, Sn 7, Mn 3.
Melting range: 1120-1345°F; 604-729°C.

SEMALLOY NO. 2617
Semi-Alloys
Ag 55, Cu 31.5, Zn 11.7, Ni 1.8.
Melting range: 1300-1355°F; 704-735°C.

SEMALLOY NO. 2620
Semi-Alloys
Ag 60, Cu 30, Zn 10.
Melting range: 1245-1382°F; 674-750°C.

SEMALLOY NO. 2630
Semi-Alloys
Ag 45, Cu 30, Zn 25.
Melting range: 1250-1370°F; 677-743°C.

SEMALLOY NO. 2633
Semi-Alloys
Ag 61, Cu 24, Zn 15.
Melting range: 1260-1375°F; 682-745°C.

SEMALLOY NO. 2637
Semi-Alloys
Ag 98, Li 2.
Melting range: 1274-1382°F; 690-750°C.

SEMALLOY NO. 2640
Semi-Alloys
Cu 34, Ag 31.5, Zn 15.5, Cd 19.
Melting range: 1165-1390°F; 629-754°C.

SEMALLOY NO. 2650
Semi-Alloys
Ag 70, Cu 20, Zn 10.
Melting range: 1275-1360°F; 691-738°C.

SEMALLOY NO. 2653
Semi-Alloys
Ag 66.7, Cu 27.7, Zn 5.6.
Melting range: 1350-1390°F; 732-754°C.

SEMALLOY NO. 2655
Semi-Alloys
Ag 66.7, Cu 28.25, Zn 5.05.
Melting range: 1360-1395°F; 738-757°C.

SEMALLOY NO. 2657
Semi-Alloys
Ag 72.15, Cu 22.8, Zn 5.05.
Melting range: 1345-1400°F; 729-760°C.

SEMALLOY NO. 2660
Semi-Alloys
Ag 68, Cu 27, Zn 5.
Melting range: 1370-1400°F; 743-760°C.

SEMALLOY NO. 2663
Semi-Alloys
Ag 68, Cu 26.6, Zn 0.4, Sn 5.
Melting range: 1370-1400°F; 743-760°C.

SEMALLOY NO. 2665
Semi-Alloys
Ag 40, Cu 36, Zn 24.
Melting range: 1229-1405°F; 665-763°C.

SEMALLOY NO. 2667
Semi-Alloys
Cu 48, Zn 32, Ag 20.
Melting range: 1370-1410°F; 743-766°C.

SEMALLOY NO. 2668
Semi-Alloys
Zn 34.2, Cu 32.3, Ag 28.5, Mn 5.
Melting range: 1305-1385°F; 707-752°C.

SEMALLOY NO. 2670
Semi-Alloys
Cu 38, Zn 32, Ag 30.
Melting range: 1250-1410°F; 677-766°C.

SEMALLOY NO. 2675
Semi-Alloys
Ag 72, Cu 28, Li 0.2.
MP 1410°F; 760°C; eutectic alloy plus lithium.

SEMALLOY NO. 2680
Semi-Alloys
Cu 93, P 7.
Melting range: 1305-1485°F; 707-807°C.

SEMALLOY NO. 2684
Semi-Alloys
100 Sr.
MP 1418°F; 770°C.

SEMALLOY NO. 2685
Semi-Alloys
Ag 75, Cu 20, Zn 5.
Melting range: 1350-1425°F; 732-744°C.

SEMALLOY NO. 2690
Semi-Alloys
Ag 50, Cu 34, Zn 16.
Melting range: 1270-1425°F; 688-724°C.

SEMALLOY NO. 2695
Semi-Alloys
Cu 40.15, Zn 32.85, Ag 27.
Melting range: 1350-1430°F; 732-777°C.

SEMALLOY NO. 2700
Semi-Alloys
Ag 40, Cu 30, Zn 28, Ni 2.
Melting range: 1290-1435°F; 671-778°C.

SEMALLOY NO. 2710
Semi-Alloys
Ag 72, Cu 28.
Melting range: 1435-1435°F; eutectic alloy.

SEMALLOY NO. 2720
Semi-Alloys
Ag 63, Cu 27, In 10.
Melting range: 1265-1346°F; 685-730°C.

SEMALLOY NO. 2727
Semi-Alloys
Ag 68, Cu 27, Sn 5.
Melting range: 1369-1400°F; 743-760°C.

SEMALLOY NO. 2735
Semi-Alloys
Zn 64.75, Cu 27.25, Sn 7.5, Pb 0.5.
Melting range: 1385-1400°F; 752-782°C.

SEMALLOY NO. 2740
Semi-Alloys
Cu 91, P 7, Ag 2.
Melting range: 1185-1455°F; 643-791°C.

SEMALLOY NO. 2750
Semi-Alloys
Ag 71.15, Cu 28.1, Ni 0.75.
Melting range: 1436-1463°F; 780-795°C.

SEMALLOY NO. 2753
Semi-Alloys
Ag 75, Cu 22, Zn 3.
Melting range: 1365-1450°F; 741-788°C.

SEMALLOY NO. 2755
Semi-Alloys
Ag 68, Cu 27, Pd 5.
Melting range: 1480-1490°F; 804-810°C.

SEMALLOY NO. 2757
Semi-Alloys
Cu 92.75, P 7.25.
Melting range: 1310-1456°F; 710-790°C.

SEMALLOY NO. 2758
Semi-Alloys
Ag 71.15, Cu 28.1, Ni 0.75.
Melting range: 1436-1463°F.

SEMALLOY NO. 2760
Semi-Alloys
Cu 50, Ag 40, Mn 10.
MP 1472°F; 800°C; eutectic alloy.

SEMALLOY NO. 2762
Semi-Alloys
Ag 71.5, Cu 28, Ni 0.5.
Melting range: 1435-1450°F; 779-788°C.

SEMALLOY NO. 2765
Semi-Alloys
Cu 80, Ag 15, P 5.
Melting range: 1190-1475°F.

SEMALLOY NO. 2770
Semi-Alloys
Ag 63, Cu 28.5, Sn 6, Ni 2.5.
Melting range: 1275-1475°F; 690-802°C.

SEMALLOY NO. 2772
Semi-Alloys
Ag 75, Cu 24.5, Ni 0.5.
Melting range: 1435-1475°F; 779-802°C.

SEMALLOY NO. 2774
Semi-Alloys
Cu 88.75, P 6.25, Ag 5.
Melting range: 1190-1480°F.

SEMALLOY NO. 2775
Semi-Alloys
Cu 89, Ag 5, P 6.
Melting range: 1189-1480°F; 643-804°C.

SEMALLOY NO. 2776
Semi-Alloys
Ag 52.2, Cu 38.4, Zn 9.4.
Melting range: 1345-1435°F; 729-779°C.

SEMALLOY NO. 2778
Semi-Alloys
Ag 80, Cu 16, Zn 4.
Melting range: 1345-1490°F; 729-810°C.

SEMALLOY NO. 2780
Semi-Alloys
Ag 68.4, Cu 26.6, Pd 5.
Melting range: 1485-1490°F; 807-810°C.

SEMALLOY NO. 2783
Semi-Alloys
Cu 88, Ag 6, P 6.
Melting range: 1189-1480°F; 643-804°C.

SEMALLOY NO. 2800
Semi-Alloys
Cu 45, Zn 30, Ag 20, Cd 5.
Liquidus 1500°F; solidus 1140°F (615-815°C).

SEMALLOY NO. 2805
Semi-Alloys
Cu 90, Ag 15, P 5.
Melting range: 1189-1450°F; 643-788°C.

SEMALLOY NO. 2810
Semi-Alloys
Cu 45, Zn 35, Ag 20.
Melting range: 1430-1500°F; 777-816°C.

SEMALLOY NO. 2813
Semi-Alloys
Cu 47.75, Zn 32.8, Ag 19.45.
Melting range: 1440-1500°F; 782-816°C.

SEMALLOY NO. 2817
Semi-Alloys
Cu 47, Zn 35, Ag 18.
Melting range: 1440-1500°F; 782-816°C.

SEMALLOY NO. 2819
Semi-Alloys
As 100.
MP 1503°F; 817°C.

SEMALLOY NO. 2845
Semi-Alloys
Cu 80, Sn 20.
Melting range: 1470-1635°F; 798-890°C.

SEMALLOY NO. 2852
Semi-Alloys
Ag 95, Al 5.
Melting range: 1440-1510°F; 782-821°C.

SEMALLOY NO. 2857
Semi-Alloys
Ag 60, Cu 40.
Melting range: 1434-1535°F; 779-835°C.

SEMALLOY NO. 2858
Semi-Alloys
Ag 60, Zn 20, Cu 18, Ni 2.
Melting range: 1434-1580°F; 779-860°C.

SEMALLOY NO. 2860
Semi-Alloys
Ag 77, Cu 21, Ni 2.
Melting range: 1435-1525°F; 779-829°C.

SEMALLOY NO. 2865
Semi-Alloys
Ag 60, Zn 20, Cu 15, Ni 5.
Melting range: 1230-1545°F; 666-841°C.

SEMALLOY NO. 2874
Semi-Alloys
Ca 100.
MP 1562°F; 850°C.

SEMALLOY NO. 2890
Semi-Alloys
Ag 65, Cu 28, Mn 5, Ni 2.
Melting range: 1385-1560°F; 752-849°C.

SEMALLOY NO. 2893
Semi-Alloys
Cu 87.75, Ge 12, Ni 0.25.
Melting range: 1508-1769°F; 820-965°C.

SEMALLOY NO. 2895
Semi-Alloys
Ag 58.5, Cu 31.5, Pd 10.
Melting range: 1515-1566°F; 824-852°C.

SEMALLOY NO. 2900
Semi-Alloys
Cu 53, Zn 38, Ag 9.
Melting range: 1450-1565°F; 788-852°C.

SEMALLOY NO. 2910
Semi-Alloys
Ag 58, Cu 32, Pd 10.
Melting range: 1521-1565°F; 827-852°C.

SEMALLOY NO. 2915
Semi-Alloys
Ag 68.8, Cu 26.7, Ti 4.5.
Melting range: 1526-1562°F; 830-850°C.

SEMALLOY NO. 2920
Semi-Alloys
Ag 54, Cu 40, Zn 5, Ni 1.
Melting range: 1325-1575°F; 718-857°C.

SEMALLOY NO. 2922
Semi-Alloys
Cu 52.5, Ag 25, Zn 22.5.
Melting range: 1250-1575°F; 677-857°C.

SEMALLOY NO. 2924
Semi-Alloys
Ag 40, Cu 30, Zn 25, Ni 5.
Melting range: 1220-1580°F; 660-860°C.

SEMALLOY NO. 2926
Semi-Alloys
Cu 50, Zn 40, Ag 10.
Melting range: 1495-1590°F; 813-866°C.

SEMALLOY NO. 2928
Semi-Alloys
Cu 52.5, Zn 47.5.
Melting range: 1570-1595°F; 854-868°C.

SEMALLOY NO. 2930
Semi-Alloys
Sn 96, S 4.
Liquidus 1576°F; solidus 450°F (232-858°C).

SEMALLOY NO. 2940
Semi-Alloys
Ag 90, Cu 10.
Melting range: 1434-1598°F; 779-870°C.

SEMALLOY NO. 2945
Semi-Alloys
Cu 52, Zn 38, Ag 10.
Melting range: 1470-1575°F; 799-857°C.

SEMALLOY NO. 2950
Semi-Alloys
Cu 58, Zn 37, Ag 5.
Melting range: 1575-1615°F; 841-879°C.

SEMALLOY NO. 2952
Semi-Alloys
Cu 56, Zn 37, Ag 7.
Melting range: 1575-1600°F; 857-871°C.

SEMALLOY NO. 2953
Semi-Alloys
Cu 90, Sn 10.
Melting range: 1550-1830°F; 843-999°C.

SEMALLOY NO. 2954
Semi-Alloys
Cu 51.5, Zn 45, Sn 3.5.
Melting range: 1585-1610°F; 863-877°C.

SEMALLOY NO. 2955
Semi-Alloys
Ag 82, Ga 9, Pd 9.
Melting range: 1553-1616°F; 845-880°C.

SEMALLOY NO. 2956
Semi-Alloys
Ni 89, P 11.
MP 1610°F; 880°C.

SEMALLOY NO. 2958
Semi-Alloys
Cu 57.5, Ag 25, Zn 17.5.
Melting range: 1255-1625°F; 880-885°C.

SEMALLOY NO. 2960
Semi-Alloys
Ag 62.5, Cu 32.5, Ni 5.
Melting range: 1435-1590°F; 779-866°C.

SEMALLOY NO. 2965
Semi-Alloys
Cu 56, Zn 40, Sn 4.
Melting range: 1590-1630°F; 866-888°C.

SEMALLOY NO. 2968
Semi-Alloys
Ni 77, Cr 13, P 10.
MP 1630°F; 888°C.

SEMALLOY NO. 2970
Semi-Alloys
Cu 56, Zn 40, Sn 1, Fe 1, Mn 1, Ni 1.
Melting range: 1590-1630°F; 866-888°C.

SEMALLOY NO. 2973
Semi-Alloys
Cu 55, Zn 44.75, Mn 0.25.
Melting range: 1610-1635°F; 877-891°C.

SEMALLOY NO. 2975
Semi-Alloys
Cu 60, Zn 39.2, Sn 0.8.
Melting range: 1625-1652°F; 885-900°C.

SEMALLOY NO. 2980
Semi-Alloys
Ag 92.5, Cu 7.3, Li 0.2.
Melting range: 1400-1635°F; 760-891°C.

SEMALLOY NO. 2985
Semi-Alloys
Cu 51.5, Zn 44, Ag 4.5.
Melting range: 1410-1635°F; 766-891°C.

SEMALLOY NO. 2987
Semi-Alloys
Ag 92.5, Cu 7.3, Li 0.2.
Melting range: 1435-1634°F.

SEMALLOY NO. 2990
Semi-Alloys
Ag 92.5, Cu 7.5.
Melting range: 1490-1634°F; 810-890°C.

SEMALLOY NO. 2993
Semi-Alloys
Au 79.5, Cu 20.5.
Melting range: 1635-1635°F; eutectic alloy.

SEMALLOY NO. 2996
Semi-Alloys
Ag 56, Cu 42, Ni 2.
Melting range: 1420-1640°F; 771-893°C.

SEMALLOY NO. 2997
Semi-Alloys
Ni 76.85, Cr 13, P 10, Mn 0.15.
Melting range: 1615-1640°F; 879-893°C.

SEMALLOY NO. 3000
Semi-Alloys
Au 75, Cu 20, Ag 5.
Melting range: 1625-1645°F; 885-895°C.

SEMALLOY NO. 3010
Semi-Alloys
Ag 52, Cu 28, Pd 20.
Melting range: 1614-1648°F; 879-898°C.

SEMALLOY NO. 3020
Semi-Alloys
Cu 95, P 5.
Melting range: 1305-1650°F; 707-899°C.

SEMALLOY NO. 3025
Semi-Alloys
Cu 57, Zn 42, Sn 1.
Melting range: 1630-1650°F; 888-999°C.

SEMALLOY NO. 3027
Semi-Alloys
Cu 59, Zn 40, Sn 1.
Melting range: 1630-1650°F; 888-899°C.

SEMALLOY NO. 3030
Semi-Alloys
Ag 65, Cu 20, Pd 15.
Melting range: 1562-1652°F; 850-900°C.

SEMALLOY NO. 3043
Semi-Alloys
Cu 60, Zn 40.
Melting range: 1645-1648°F; 895-898°C.

SEMALLOY NO. 3050
Semi-Alloys
Cu 60, Ag 40.
Melting range: 1434-1670°F; 779-910°C.

SEMALLOY NO. 3063
Semi-Alloys
Ni 42, Fe 34.5, Co 7.5, Cu 5.
Melting range: 1600-1675°F; 871-913°C.

SEMALLOY NO. 3071
Semi-Alloys
Ag 99.5, Li 0.5.
Melting range: 1616-1697°F; 880-925°C.

SEMALLOY NO. 3073
Semi-Alloys
Cu 47, Zn 42, Ni 10.5, Ag 0.5.
Melting range: 1685-1710°F; 818-932°C.

SEMALLOY NO. 3074
Semi-Alloys
Cu 48, Zn 42, Ni 10.
Melting range: 1690-1715°F; 921-935°C.

SEMALLOY NO. 3075
Semi-Alloys
100 Ge.
MP 1719°F; 937°C.

SEMALLOY NO. 3077
Semi-Alloys
Ti 71.5, Ni 28.5.
MP 1728°F; 942°C.

SEMALLOY NO. 3080
Semi-Alloys
Ag 99, As 1.
Melting range: 1670-1742°F; 910-950°C.

SEMALLOY NO. 3090
Semi-Alloys
Ag 54, Pd 25, Cu 21.
Melting range: 1654-1742°F; 900-950°C.

SEMALLOY NO. 3110
Semi-Alloys
Cu 70, Ag 30.
Melting range: 1434-1742°F; 779-950°C.

SEMALLOY NO. 3120
Semi-Alloys
Au 82, Ni 18.
MP 1742°F; 950°C.

SEMALLOY NO. 3123
Semi-Alloys
Cu 47, Zn 43, Ni 10.
Melting range: 1660-1750°F; 905-955°C.

SEMALLOY NO. 3127
Semi-Alloys
Ag 85, Sn 15.
Melting range: 1335-1526°F; 724-830°C.

SEMALLOY NO. 3130
Semi-Alloys
100 Ag.
MP 1761°F; 961°C.

SEMALLOY NO. 3140
Semi-Alloys
Ag 96, Mn 4.
MP 1762°F; 961°C.

SEMALLOY NO. 3150
Semi-Alloys
W 62, Ag 35, Ni 3.
MP 1764°F; 962°C.

SEMALLOY NO. 3170
Semi-Alloys
Ag 85, Mn 15.
Melting range: 1760-1780°F; 960-971°C.

SEMALLOY NO. 3190
Semi-Alloys
Cu 85, Sn 8, Ag 7.
Melting range: 1225-1805°F; 663-985°C.

SEMALLOY NO. 3201
Semi-Alloys
Fe 91.75, Si 4.5, B 3, Ti 0.75.
Melting range: 1795-1820°F; 979-993°C.

SEMALLOY NO. 3202
Semi-Alloys
Ni 91.25, Si 4.5, B 2.9, Fe 1.35.
Melting range: 1795-1820°F; 979-993°C.

SEMALLOY NO. 3203
Semi-Alloys
Ni 83.35, Cr 6, B 3, Fe 2.5, Mn 0.15, Si 5.
Melting range: 1750-1825°F; 954-996°C.

SEMALLOY NO. 3204
Semi-Alloys
Ni 82, Cr 7, Si 4.5, B 2.9, Fe 3.6.
Melting range: 1780-1830°F; 971-999°C.

SEMALLOY NO. 3205
Semi-Alloys
Cu 80, Ag 20.
Melting range: 1434-1796°F; 779-980°C.

SEMALLOY NO. 3206
Semi-Alloys
Ni 82.5, Cu 7, B 3, Si 4.5, Fe 3.
Melting range: 1780-1830°F; 971-999°C.

SEMALLOY NO. 3209
Semi-Alloys
Ni 72.5, Cr 16, B 3.5, Fe 3, Si 5.
Melting range: 1825-1840°F; 996-1005°C.

SEMALLOY NO. 3210
Semi-Alloys
Ag 95, Pd 5.
Melting range: 1778-1850°F; 970-1010°C.

SEMALLOY NO. 3220
Semi-Alloys
Cu 60, Au 40.
Melting range: 1805-1850°F.

SEMALLOY NO. 3225
Semi-Alloys
Au 40, Cu 60.
Melting range: 1814-1850°F.

SEMALLOY NO. 3240
Semi-Alloys
Cu 62.5, Au 37.5.
Melting range: 1814-1859°F.

SEMALLOY NO. 3241
Semi-Alloys
Ni 76.3, Cr 7, W 6, Si 4.5, Fe 3, B 3.2.
Melting range: 1790-1900°F; 977-1038°C.

SEMALLOY NO. 3243
Semi-Alloys
Cu 63, Au 37.
Melting range: 1815-1860°F; 990-1016°C.

SEMALLOY NO. 3247
Semi-Alloys
Mn 60, Ni 40.
MP 1864°F; 1018°C; eutectic alloy.

SEMALLOY NO. 3250
Semi-Alloys
Cu 95.8, Si 3.1, Mn 1.1.
Melting range: 1742-1866°F; 950-1019°C.

SEMALLOY NO. 3252
Semi-Alloys
Ag 48.6, Pd 22.5, Cu 18.9, Ni 10.
Melting range: 1796-1850°F; 980-1010°C.

SEMALLOY NO. 3255
Semi-Alloys
Cu 91.2, Al 7, Si 1.8.
Melting range: 1800-1840°F; 982-1004°C.

SEMALLOY NO. 3260
Semi-Alloys
Cu 65, Au 35.
Melting range: 1832-1868°F.

SEMALLOY NO. 3263
Semi-Alloys
Mn 67, Co 16, Ni 16, B 1.
Melting range: 1840-1870°F; 1005-1021°C.

SEMALLOY NO. 3265
Semi-Alloys
Ni 73.2, Cr 13.5, Si 4.5, Fe 4.5, B 3.5.
Melting range: 1760-1875 F. *Obsolete*

SEMALLOY NO. 3267
Semi-Alloys
Ni 91.8, Si 4.5, B 3.5, Mn 0.2.
Melting range: 1800-1875°F; 982-1024°C.

SEMALLOY NO. 3269
Semi-Alloys
Ni 65.5, Mn 23, Si 7, Cu 4.5.
Melting range: 1800-1850°F; 982-1010°C.

SEMALLOY NO. 3270
Cu 62, Au 35, Ni 3.
Melting range: 1787-1886°F; 975-1030°C.

SEMALLOY NO. 3272
Semi-Alloys
Ni 92.5, Si 4.5, B 3.
Melting range: 1800-1900°F; 982-1038°C.

SEMALLOY NO. 3280
Semi-Alloys
Pb 96, S 4.
Liquidus 1868°F; solidus 621°F (327-1020°C).

SEMALLOY NO. 3290
Semi-Alloys
Cu 97, Si 3.
Melting range: 1778-1877°F; 970-1025°C.

SEMALLOY NO. 3293
Semi-Alloys
Ni 89.35, P 6.15, Si 4.5.
Melting range: 1616-1880°F; 879-1027°C.

SEMALLOY NO. 3295
Semi-Alloys
Cu 92, Sn 8.
Melting range: 1616-1868°F; 880-1020°C.

SEMALLOY NO. 3296
Semi-Alloys
Cu 93, Si 2.8-4, Sn 1.5, Mn 1.5.
Melting range: 1780-1880°F; 971-1027°C.

SEMALLOY NO. 3311
Semi-Alloys
Mn 68, Ni 32.
Melting range: 1859-1886°F; 1015-1030°C.

SEMALLOY NO. 3315
Semi-Alloys
Ni 74.85, Mn 17, Si 8, C 0.15.
Melting range: 1850-1890°F; 1010-1032°C.

SEMALLOY NO. 3320
Semi-Alloys
Cu 70, Au 30.
Melting range: 1859-1895°F.

SEMALLOY NO. 3322
Semi-Alloys
Ni 70.55, Cr 16.5, Si 4.5, Fe 4, B 3.85, C 0.6.
Melting range: 1780-1900°F; 1010-1066°C.

SEMALLOY NO. 3324
Semi-Alloys
Ni 7.4 ?, Cr 12.5, Fe 4.5, Si 4.5, B 3.3, C 0.8.
Melting range: 1790-1900°F; 977-1038°C.

SEMALLOY NO. 3327
Semi-Alloys
Ni 91, Si 4.5, Fe 1.5, B 3.
Melting range: 1800-1900°F; 982-1038°C.

SEMALLOY NO. 3328
Semi-Alloys
Cu 86, Al 11, Fe 3.
Melting range: 1880-1900°F; 1027-1038°C.

SEMALLOY NO. 3331
Semi-Alloys
Cu 92, Al 8.
MP 1904°F; 1040°C.

SEMALLOY NO. 3332
Semi-Alloys
Cu 89, Al 10, Fe 1.
Melting range: 1904-1913°F; 1040-1045°C.

SEMALLOY NO. 3333
Semi-Alloys
Cu 95, Sn 5.
Melting range: 1750-1920°F; 954-1049°C.

SEMALLOY NO. 3335
Semi-Alloys
Cu 97, As 3.
Melting range: 1724-1922°F; 940-1055°C.

SEMALLOY NO. 3340
Semi-Alloys
Mn 70, Ni 30.
Melting range: 1886-1922°F; 1030-1050°C.

SEMALLOY NO. 3341
Semi-Alloys
Cu 98.5, Si 1.5.
Melting range: 1890-1940°F; 1032-1060°C.

SEMALLOY NO. 3343
Semi-Alloys
Cu 97, Ge 3.
Melting range: 1894-1922°F; 1035-1050°C.

SEMALLOY NO. 3345
Semi-Alloys
Cu 95, Al 5.
Melting range: 1920-1945°F; 1049-1063°C.

SEMALLOY NO. 3360
Semi-Alloys
Ni 94.6, Si 3.5, B 1.9.
Melting range: 1814-1931°F; 990-1055°C.

SEMALLOY NO. 3370
Semi-Alloys
Ni 81.35, Cr 15, B 3.5, C 0.15.
MP 1931°F; 1055°C.

SEMALLOY NO. 3380
Semi-Alloys
Cu 95, Ag 5.
Melting range: 1832-1940°F; 1000-1060°C.

SEMALLOY NO. 3400
Semi-Alloys
Cu 97, Ga 3.
Melting range: 1922-1949°F; 1050-1065°C.

SEMALLOY NO. 3410
Semi-Alloys
Au 72, Ni 22, Cr 6.
Melting range: 1785-1950°F.

SEMALLOY NO. 3415
Semi-Alloys
Ni 93.8, Si 3.5, Fe 1.5, B 1.2.
Melting range: 1800-1950°F; 932-1066°C.

SEMALLOY NO. 3420
Semi-Alloys
Ag 90, Pd 10.
Melting range: 1835-1950°F; 1002-1066°C.

SEMALLOY NO. 3430
Semi-Alloys
Pt 71, Sn 29.
MP 1958°F; 1070°C.

SEMALLOY NO. 3445
Semi-Alloys
Ni 74, Cr 14, Fe 4.5, Si 4.5, B 3.
Melting range: 1780-1900°F; 971-1038°C.

SEMALLOY NO. 3447
Semi-Alloys
Ni 74.5, Cr 15, Si 4.5, Fe 3, B 3.
Melting range: 1780-1970°F; 971-1077°C.

SEMALLOY NO. 3448
Semi-Alloys
Cu 98.7, Sn 1.3.
Melting range: 1900-1970°F; 1035-1075°C.

SEMALLOY NO. 3449
Semi-Alloys
Ni 68.5, W 12, Cr 10, Fe 3.5, Si 3.5, B 2.5.
Melting range: 1780-2000°F; 971-1093°C.

SEMALLOY NO. 3450
Semi-Alloys
Cu 99.7, Si 0.3.
MP 1981°F; 1083°C.

SEMALLOY NO. 3460
Semi Alloys
Cu 100.
MP 1083°C; 1981°F.

SEMALLOY NO. 3470
Semi-Alloys
Cu 82, Pd 18.
Melting range: 1985-2012°F; 1085-1100°C.

SEMALLOY NO. 3475
Semi-Alloys
Ni 71, Cr 18, B 11.
Melting range: 1975-2075°F; 1079-1135°C.

SEMALLOY NO. 3479
Semi-Alloys
Ni 61, Cr 19, Si 10, Mn 10.
Melting range: 1975-2075°F; 1079-1135°C.

SEMALLOY NO. 3480
Semi-Alloys
Ni 81.35, Cr 15, B 3.5, Mn 0.15.
Melting range: 1900-1990°F; 1038-1088°C.

SEMALLOY NO. 3490
Semi-Alloys
Cu 35, Pd 30, Ni 20, Mn 15.
MP 1994°F; 1090°C.

SEMALLOY NO. 3500
Semi-Alloys
Ni 63, W 16, Cr 11.5, Fe 3.75, Si 3.25, B 2.5.
Melting range: 1790-2020°F; 971-1104°C.

SEMALLOY NO. 3505
Semi-Alloys
Cu 55, Pd 20, Ni 15, Mn 10.
MP 2020°F; 1104°C.

SEMALLOY NO. 3510
Semi-Alloys
Ag 75, Pd 20, Mn 5.
Melting range: 1960-2050°F; 1071-1121°C.

SEMALLOY NO. 3520
Semi-Alloys
Ni 70.85, Cr 19, Si 10, C 0.15.
Melting range: 1975-2075°F; 1080-1135°C.

SEMALLOY NO. 3521
Semi-Alloys
Ni 79.75, Cr 11, Si 3.5, Fe 3.5, B 2.25.
Melting range: 1780-2120°F; 971-1160°C.

SEMALLOY NO. 3530
Semi-Alloys
Ni 71, Cu 19, Si 10.
Melting range: 1975-2075°F; 1079-1135°C.

SEMALLOY NO. 3540
Semi-Alloys
Cu 88.7, Ni 10, Fe 1.3.
Melting range: 2012-2102°F; 1100-1150°C.

SEMALLOY NO. 3550
Semi-Alloys
Ni 82.55, Cr 10, Si 2.5, Fe 2.5, B 2, C 0.45.
Melting range: 1780-2120°F; 971-1160°C.

SEMALLOY NO. 3560
Semi-Alloys
Ag 73, Pt 27.
Melting range: 1823-2120°F; 995-1160°C.

SEMALLOY NO. 3565
Semi-Alloys
Co 50.8, Cr 19, Ni 17, Si 8, W 4, B 0.8, C 0.4.
Melting range: 2025-2100°F; 1107-1149°C.

SEMALLOY NO. 3570
Semi-Alloys
Ag 80, Pd 20.
Melting range: 1958-2102°F; 1070-1150°C.

SEMALLOY NO. 3580
Semi-Alloys
Cu 75, Ni 25.
Melting range: 2102-2210°F; 1150-1210°C.

SEMALLOY NO. 3590
Semi-Alloys
Pd 65, Co 35.
Melting range: 2246-2255°F; 1230-1235°C.

SEMALLOY NO. 3592
Semi-Alloys
Cu 79, Ni 21.
Melting range: 2100-2190°F; 1149-1199°C.

SEMALLOY NO. 3600
Semi-Alloys
Ag 64, Pd 33, Mn 3.
Melting range: 2100-2250°F; 1149-1232°C.

SEMALLOY NO. 3610
Semi-Alloys
Pd 60, Ni 40.
MP 2260°F; 1237°C.

SEMALLOY NO. 3615
Semi-Alloys
Cu 70, Ni 30.
Melting range: 2140-2260°F; 1171-1238°C.

SEMALLOY NO. 3620
Semi-Alloys
Cu 69.4, Ni 30, Fe 0.6.
Melting range: 2138-2264°F; 1170-1240°C.

SEMALLOY NO. 3624
Semi-Alloys
Mn 100.
MP 2271°F; 1244°C.

SEMALLOY NO. 3625
Semi-Alloys
Cu 60, Pt 40.
Melting range: 2165-2221°F; 1185-1216°C.

SEMALLOY NO. 3630
Semi-Alloys
Cu 60, Pd 40.
Melting range: 2057-2102°F; 1125-1150°C.

SEMALLOY NO. 3640
Semi-Alloys
Pd 54, Ni 36, Cr 10.
Melting range: 2250-2300°F; 1232-1260°C.

SEMALLOY NO. 3644
Semi-Alloys
Be 100.
MP 2343°F; 1284°C.

SEMALLOY NO. 3650
Semi-Alloys
Cu 55, Ni 45.
Melting range: 2246-2363°F; 1230-1295°C.

SEMALLOY NO. 3660
Semi-Alloys
Ni 70, Pd 30.
Melting range: 2354-2408°F; 1290-1320°C.

SEMALLOY NO. 3670
Semi-Alloys
Ni 53.5, Mo 46.5.
MP 2399°F; 1315°C.

SEMALLOY NO. 3680
Semi-Alloys
Co 63, Mo 37.
MP 2440°F; 1340°C.

SEMALLOY NO. 3690
Semi-Alloys
Ni 67, Cu 28.5, Fe 2.5, Mg 2.
Melting range: 2370-2460°F; 1299-1349°C.

SEMALLOY NO. 3700
Semi-Alloys
Ni 67, Cu 29.25, Al 3, Sn 0.75.
Melting range: 2400-2460°F; 1316-1349°C.

SEMALLOY NO. 3710
Semi-Alloys
Ni 72.85, Cr 17, Fe 10, C 0.15.
Melting range: 2540-2600°F; 1393-1427°C.

SEMALLOY NO. 3714
Semi-Alloys
Si 100.
MP 2574°F; 1412°C.

SEMALLOY NO. 3720
Semi-Alloys
Ni 94, Al 4.75, Sn 1, C 0.25.
Melting range: 2615-2635°F; 1435-1446°C.

SEMALLOY NO. 3730
Semi-Alloys
Fe 64, Ni 36.
Melting range: 2633-2642°F; 1445-1450°C.

SEMALLOY NO. 3740
Semi-Alloys
100 Ni.
MP 1453°C; 2647°F.

SEMALLOY NO. 3744
Semi-Alloys
Co 100.
MP 2718°F; 1492°C.

SEMALLOY NO. 3750
Semi-Alloys
Fe 99.1, Mn 0.45, Si 0.25, C 0.2.
MP 2760°F; 1516°C.

SEMALLOY NO. 3754
Semi-Alloys
Fe 100.
MP 2798°F; 1537°C.

SEMALLOY NO. 3760
Semi-Alloys
100 Pd.
MP 1552°C; 2826°F.

SEMALLOY NO. 3765
Semi-Alloys
100 Ti.
MP 1660°C; 3020°F.

SEMALLOY NO. 3774
Semi-Alloys
100 V.
MP 3380°F; 1860°C.

SEMALLOY NO. 3775
Semi-Alloys
Pt 95, Ir 5.
Melting range: 3216-3220°F; 1769-1771°C.

SEMALLOY NO. 3780
Semi-Alloys
100 Pt.
MP 1773°C; 3216°F.

SEMALLOY NO. 3782
Semi-Alloys
Pt 90, Ir 10.
Melting range: 3236-3272°F; 1780-1800°C.

SEMALLOY NO. 3784
Semi-Alloys
Cr 100.
MP 3362°F; 1850°C.

SEMALLOY NO. 3785
Semi-Alloys
Pt 85, Ir 15.
Melting range: 3272-3308°F; 1800-1820°C.

SEMALLOY NO. 3786
Semi-Alloys
Pt 95, Rh 5.
Melting range: 3290-3299°F; 1810-1815°C.

SEMALLOY NO. 3790
Semi-Alloys
Pt 90, Rh 10.
Melting range: 3326-3362°F; 1830-1850°C.

SEMALLOY NO. 3794
Semi-Alloys
Pt 80, Ir 20.
Melting range: 3326-3362°F; 1830-1850°C.

SEMALLOY NO. 3800
Semi-Alloys
100 Zr.
MP 1852°C; 3366°F.

SEMALLOY NO. 3805
Semi-Alloys
Pt 80, Rh 20.
Melting range: 3425-3452°F; 1885-1900°C.

SEMALLOY NO. 3810
Semi-Alloys
Mo 80, Ru 20.
MP 3452°F; 1900°C.

SEMALLOY NO. 3815
Semi-Alloys
Pt 70, Rh 30.
Melting range: 3497-3506°F; 1925-1930°C.

SEMALLOY NO. 3820
Semi-Alloys
Pt 60, Rh 40.
Melting range: 3515-3533°F; 1835-1945°C.

SEMALLOY NO. 3830
Semi-Alloys
100 Rh.
MP 1966°C; 3571°F.

SEMALLOY NO. 3840
Semi-Alloys
Pt 60, Ir 40.
Melting range: 3632-3659°F; 2000-2015°C.

SEMALLOY NO. 3850
Semi-Alloys
Mo 90, Si 10.
Melting range: 3848-3902°F; 2120-2150°C.

SEMALLOY NO. 3854
Semi-Alloys
100 Hf.
MP 4028°F; 2220°C.

SEMALLOY NO. 3856
Semi-Alloys
100 B.
MP 3812°F; 2100°C.

SEMALLOY NO. 3860
Semi-Alloys
100 Ir.
MP 2450°C; 4442°F.

SEMALLOY NO. 3870
Semi-Alloys
100 Nb.
MP 2468°C; 4474°F.

SEMALLOY NO. 3880
Semi-Alloys
Tungsten carbide.
MP 2480°C; 4496°F.

SEMALLOY NO. 3890
Semi-Alloys
100 Ru.
MP 2500°C; 4532°F.

SEMALLOY NO. 3894
Semi-Alloys
100 Mo.
MP 4748°F; 2620°C.

SEMALLOY NO. 3900
Semi-Alloys
100 Ta.
MP 2996°C; 5425°F.

SEMALLOY NO. 3904
Semi-Alloys
100 Os.
MP 4892°F; 2700°C.

SEMALLOY NO. 3910
Semi-Alloys
100 Re.
MP 3180°C; 5756°F.

SEMALLOY NO. 3920
Semi-Alloys
100 W
MP 6116°F; 3380°C.

SEMALLOY NO. 3930
Semi-Alloys
100 C.
MP 9032°F; 5000°C.

SEMALLOY NO. A2297
Semi-Alloys
Au 60, Ag 20, Sn 20.
MP 680°F; 360°C; eutectic alloy.

SEMALLOY NO. A2360
Semi-Alloys
Au 75, In 25.
Liquidus 856°F; solidus 844°F (451-485°C).

SEMALLOY NO. A2370
Semi-Alloys
Au 80, In 20.
Liquidus 905°F; solidus 844°F (451-485°C).

SEMALLOY NO. A2495
Semi-Alloys
Au 38, Ag 26, Cu 19, Cd 16, Zn 1.
Melting range: 1175-1300°F; 635-704°C.

SEMALLOY NO. A2531
Semi-Alloys
Ag 32, Cu 25, Au 23, Cd 19, Zn 1.
Melting range: 1200-1285°F; 649-696°C.

SEMALLOY NO. A2567
Semi-Alloys
Ag 31, Au 29, Cu 20, Cd 19, Zn 1.
Melting range: 1280-1400°F; 693-760°C.

SEMALLOY NO. A2820
Semi-Alloys
Au 50, Ag 29.5, Cu 20.5.
Melting range: 1460-1510°F; 793-821°C.

SEMALLOY NO. A2840
Semi-Alloys
Au 49, Ag 27.5, Cu 23.5.
Melting range: 1460-1515°F; 793-824°C.

SEMALLOY NO. A2850
Semi-Alloys
Au 53, Ag 23.5, Cu 23.5.
Melting range: 1480-1520°F; 804-877°C.

SEMALLOY NO. A2870
Semi-Alloys
Au 58.3, Ag 24, Cu 17.7.
Melting range: 1500-1550°F; 816-843°C.

SEMALLOY NO. A2880
Semi-Alloys
Au 60, Ag 20, Cu 20.
Melting range: 1535-1553°F; 835-845°C.

SEMALLOY NO. A2925
Semi-Alloys
Au 42, Ag 32, Cu 16, Zn 10.
Melting range: 1335-1380°F; 724-749°C.

SEMALLOY NO. A2927
Semi-Alloys
Au 48, Ag 29, Cu 17, Zn 6.
Melting range: 1350-1450°F; 732-788°C.

SEMALLOY NO. A3040
Semi-Alloys
Au 60, Cu 37, In 3.
Melting range: 1580-1652°F; 860-900°C.

SEMALLOY NO. A3060
Semi-Alloys
Au 81.5, Cu 15.5, Ni 3.
Melting range: 1652-1670°F.

SEMALLOY NO. A3067
Semi-Alloys
Au 66.7, Cu 33.3.
Melting range: 1655-1685°F; 903-920°C.

SEMALLOY NO. A3070
Semi-Alloys
Au 58.3, Cu 39.6, Ag 2.1.
Melting range: 1660-1690°F; 905-921°C.

SEMALLOY NO. A3085
Semi-Alloys
Au 81.5, Cu 16.5, Ni 2.
Melting range: 1670-1697°F; 910-925°C.

SEMALLOY NO. A3160
Semi-Alloys
Cu 55.5, Au 41.7, Ag 2.8.
Melting range: 1695-1765°F; 924-963°C.

SEMALLOY NO. A3180
Semi-Alloys
Au 50, Ag 50.
Melting range: 1742-1787°F; 950-975°C.

SEMALLOY NO. A3200
Semi-Alloys
Au 94, Cu 6.
Melting range: 1769-1814°F; 965-990°C.

SEMALLOY NO. A3207
Semi-Alloys
Au 35, Ag 65.
Melting range: 1796-1832°F; 980-1000°C.

SEMALLOY NO. A3208
Semi-Alloys
Au 43, Cu 57.
Melting range: 1796-1832°F; 980-1000°C.

SEMALLOY NO. A3212
Semi-Alloys
Au 50, Cu 40, Ni 10.
Melting range: 1780-1839°F; 971-1004°C.

SEMALLOY NO. A3214
Semi-Alloys
Au 50, Ag 50.
Melting range: 1832-1868°F; 1000-1020°C.

SEMALLOY NO. A3230
Semi-Alloys
Cu 77, Au 20, In 3.
Melting range: 1778-1859°F; 970-1015°C.

SEMALLOY NO. A3253
Semi-Alloys
Cu 55, Au 45.
Melting range: 1749-1780°F; 954-971°C.

SEMALLOY NO. A3260
Semi-Alloys
Au 73.8, Ni 26.2.
Melting range: 1796-1850°F; 980-1010°C.

SEMALLOY NO. A3265
Semi-Alloys
Au 70, Ni 22, Pd 8.
Melting range: 1841-1899°F; 1005-1037°C.

SEMALLOY NO. A3275
Semi-Alloys
Ag 60, Au 40.
Melting range: 1814-1841°F; 990-1005°C.

SEMALLOY NO. A3297
Semi-Alloys
Cu 80, Au 20.
Melting range: 1940-1976°F; 1060-1080°C.

SEMALLOY NO. A3310
Semi-Alloys
Au 99, Ga 1.
Melting range: 1877-1886°F; 1025-1030°C.

SEMALLOY NO. A3330
Semi-Alloys
Au 99.4, Sb 0.6.
Liquidus 1886°F; solidus 680°F (360-1030°C).

SEMALLOY NO. A3350
Semi-Alloys
Au 75, Pd 20, Mn 5.
Melting range: 1846-1962°F; 1008-1072°C.

SEMALLOY NO. A3440
Semi-Alloys
Au 65, Ni 35.
Melting range: 1769-1967°F; 965-1075°C.

SEMALLOY NO. A3455
Semi-Alloys
Cu 90, Au 10.
Melting range: 1945-1972°F; 1063-1078°C.

SEMALLOY NO. A3465
Semi-Alloys
Au 98.5, Pd 1.5.
Melting range: 1967-2012°F; 1075-1100°C.

SEMALLOY NO. A3485
Semi-Alloys
Au 50, Ni 25, Pd 25.
Melting range: 2016-2050°F; 1102-1121°C.

SEMALLOY NO. A3575
Semi-Alloys
Au 96.5, Pd 3.5.
Melting range: 2057-2102°F; 1125-1150°C.

SEMALLOY NO. A3585
Semi-Alloys
Ni 36, Pd 34, Au 30.
Melting range: 2075-2136°F; 1135-1169°C.

SEMALLOY NO. A3595
Semi-Alloys
Au 94, Pd 6.
Melting range: 2102-2192°F; 1150-1200°C.

SEMALLOY NO. A3635
Semi-Alloys
Au 91, Pd 9.
Melting range: 2192-2282°F; 1200-1250°C.

SEMALLOY NO. A3655
Semi-Alloys
Au 82, Pd 18.
Melting range: 2417-2462°F; 1325-1350°C.

SEMALLOY NO. A3665
Semi-Alloys
Au 74, Pd 26.
Melting range: 2534-2561°F; 1390-1405°C.

SEMALLOY NO. A3675
Semi-Alloys
Au 62, Pd 38.
Melting range: 2606-2642°F; 1430-1450°C.

SEMALLOY NO. A3735
Semi-Alloys
Au 60, Pd 40.
Melting range: 2642-2665°F; 1450-1463°C.

SEMALLOY NO. A3745
Semi-Alloys
Pd 60, Au 40.
Melting range: 2723-2732°F; 1495-1500°C.

SEMALLOY NO. A3770
Semi-Alloys
Pt 75, Pd 20, Au 5.
Melting range: 2993-3083°F; 1645-1695°C.

SEMALLOY NO. A902
Semi-Alloys
Pb 85, Au 15.
MP 419°F; 215°C; eutectic alloy.

SEMALLOY NO. A903
Semi-Alloys
Sn 90, Au 10.
MP 423°F; 217°C; eutectic alloy.

SEMALLOY NO. A904
Semi-Alloys
Bi 82, Au 18.
MP 466°F; 241°C; eutectic alloy.

SEMALLOY NO. A905
Semi-Alloys
Au 80, Sn 20.
MP 536°F; 280°C; eutectic alloy.

SEMALLOY NO. A906
Semi-Alloys
Sn 71, Au 29.
Liquidus 541°F; solidus 486°F (252-283°C).

SEMALLOY NO. A907
Semi-Alloys
Pb 68, Au 32.
Liquidus 568°F; solidus 489°F (254-298°C).

SEMALLOY NO. A908
Semi-Alloys
Cd 87, Au 13.
MP 588°F; 309°C; eutectic alloy.

SEMALLOY NO. A909
Semi-Alloys
Sn 55, Au 43.
Liquidus 644°F; solidus 588°F (309-340°C).

SEMALLOY NO. A910
Semi-Alloys
Au 84, Ga 16.
MP 646°F; 341°C; eutectic alloy.

SEMALLOY NO. A911
Semi-Alloys
Au 88, Ge 12.
MP 673°F; 356°C; eutectic alloy.

SEMALLOY NO. A912
Semi-Alloys
Au 75, Sb 25.
MP 680°F; 360°C; eutectic alloy.

SEMALLOY NO. A913
Semi-Alloys
Au 98, Si 2.
Liquidus 1724°F; solidus 698°F (370-940°C).

SEMALLOY NO. A914
Semi-Alloys
Au 96.9, Si 3.1.
MP 698°F; 370°C; eutectic alloy.

SEMALLOY NO. A915
Semi-Alloys
Te 83, Au 17.
MP 781°F; 416°C; eutectic.

SEMALLOY NO. A916
Semi-Alloys
Au 58, Te 42.
MP 837°F; 447°C.

SEMALLOY NO. A917
Semi-Alloys
Au 73, In 27.
MP 844°F; 451°C.

SEMALLOY NO. A918
Semi-Alloys
Au 68, Ga 32.
MP 844°F; 451°C; eutectic alloy.

SEMALLOY NO. A919
Semi-Alloys
Sb 55, Au 45.
MP 860°F; 460°C; eutectic alloy.

SEMALLOY NO. A920
Semi-Alloys
Au 81, In 19.
Liquidus 910°F; solidus 896°F (480-488°C).

SEMALLOY NO. A921
Semi-Alloys
Au 59, In 41.
MP 921°F; 494°C; eutectic alloy.

SEMALLOY NO. A922
Semi-Alloys
Cd 58, Au 42.
MP 932°F; 500°C; eutectic alloy.

SEMALLOY NO. A923
Semi-Alloys
Au 96, Al 4.
MP 977°F; 525°C; eutectic alloy.

SEMALLOY NO. A9231
Semi-Alloys
Ag 44, Au 38, Ge 18.
Liquidus 977°F; solidus 959°F (515-525°C).

SEMALLOY NO. A924
Semi-Alloys
Au 53, Cd 47.
MP 1004°F; 540°C; eutectic alloy.

SEMALLOY NO. A925
Semi-Alloys
Au 50, In 50.
Liquidus 1004°F; solidus 921°F (494-540°C).

SEMALLOY NO. A9251
Semi-Alloys
Ag 45, Au 38, Ge 17.
MP 977°F; 525°C; eutectic alloy.

SEMALLOY NO. A926
Semi-Alloys
Au 91, Al 9.
MP 1056°F; 569°C; eutectic alloy.

SEMALLOY NO. A927
Semi-Alloys
Mg 61, Au 39.
MP 1069°F; 576°C; eutectic alloy.

SEMALLOY NO. A928
Semi-Alloys
Au 72, Cd 28.
MP 1134°F; 612°C; eutectic alloy.

SEMALLOY NO. A929
Semi-Alloys
Au 69, Zn 31.
MP 1159°F; 626°C; eutectic alloy.

SEMALLOY NO. A930
Semi-Alloys
Al 95, Au 5.
MP 1188°F; 642°C; eutectic alloy.

SEMALLOY NO. A931
Semi-Alloys
Au 85, Zn 15.
MP 1188°F; 642°C.

SEMALLOY NO. A933
Semi-Alloys
Au 78.5, Mg 21.5.
MP 1429°F; 776°C; eutectic alloy.

SEMALLOY NO. A934
Semi-Alloys
Au 80.5, Mg 19.5.
MP 1436°F; 780°C; eutectic alloy.

SEMALLOY NO. A935
Semi-Alloys
Au 89, Th 11.
MP 1490°F; 810°C; eutectic alloy.

SEMALLOY NO. A936
Semi-Alloys
Au 94, Mg 6.
MP 1521°F; 827°C; eutectic alloy.

SEMALLOY NO. A9524
Semi-Alloys
Au 67, Cu 33.
Melting range: 1650-1684°F; 899-918°C.

SEMALLOY NO. A9540
Semi-Alloys
Au 80, Cu 20.
MP 1632°F; 889°C; eutectic alloy.

SEMALLOY NO. A9570
Semi-Alloys
Au 89, Mn 11.
MP 1760°F; 960°C.

SEMALLOY NO. A9580
Semi-Alloys
Au 90, Mn 10.
Melting range: 1760-1787°F; 960-975°C.

SEMALLOY NO. A9585
Semi-Alloys
Cu 62, Au 38.
Melting range: 1814-1859°F; 990-1015°C.

SEMALLOY NO. A9590
Semi-Alloys
Cu 60, Au 40.
Melting range: 1796-1841°F; 980-1005°C.

SEMALLOY NO. A961
Semi-Alloys
Au 60, Cu 40.
Melting range: 1679-1724°F; 915-940°C.

SEMALLOY NO. A9610
Semi-Alloys
Au 72, Ni 22, Cr 6.
Melting range: 1785-1950°F; 974-1066°C.

SEMALLOY NO. A9614
Semi-Alloys
Cu 65, Au 35.
Melting range: 1832-1868°F; 1000-1020°C.

SEMALLOY NO. A9615
Semi-Alloys
Cu 78, Au 20, In 2.
Melting range: 1787-1877°F; 975-1025°C.

SEMALLOY NO. A963
Semi-Alloys
Au 75, Ag 25.
Melting range: 1886-1904°F.

SEMALLOY NO. A9640
Semi-Alloys
Cu 70, Au 30.
Melting range: 1859-1895°F; 1015-1035°C.

SEMALLOY NO. A965
Semi-Alloys
Au 80, Ag 20.
Melting range: 1900-1918°F; 1038-1048°C.

SEMALLOY NO. A9650
Semi-Alloys
Au 83, Ag 17.
Melting range: 1904-1922°F; 1040-1050°C.

SEMALLOY NO. A966
Semi-Alloys
Au 100.
MP 1945°F; 1063°C.

SEMALLOY NO. A9670
Semi-Alloys
Au 56, Mn 44.
MP 1963°F; 1073°C; eutectic alloy.

SEMALLOY NO. A9690
Semi-Alloys
Au 54, Mn 46.
MP 1985°F; 1085°C; eutectic alloy.

SEMALLOY NO. A9700
Semi-Alloys
Au 59, Mn 41.
MP 2003°F; 1095°C; eutectic alloy.

SEMALLOY NO. A971
Semi-Alloys
Au 55, Pt 45.
Melting range: 2236-2786°F; 1280-1530°C.

SEMALLOY NO. A972
Semi-Alloys
Au 92, Pd 8.
Melting range: 2174-2264°F; 1190-1240°C.

SEMALLOY NO. A973
Semi-Alloys
Au 87, Pd 13.
Melting range: 2300-2381°F; 1260-1305°C.

SEMALLOY NO. A974
Semi-Alloys
Au 75, Pt 25.
Melting range: 2210-2570°F; 1210-1410°C.

SEMALLOY NO. A975
Semi-Alloys
Au 75, Pd 25.
Melting range: 2507-2552°F; 1375-1400°C.

SEMALLOY NO. A976
Semi-Alloys
Au 65, Pd 35.
Melting range: 2590-2624°F; 1421-1440°C.

SEMALLOY NO. A977
Semi-Alloys
Au 20, Pt 80.
Melting range: 2570-3002°F; 1410-1650°C.

SEMALLOY NO. A978
Semi-Alloys
Pt 91, Au 9.
Melting range: 2912-3110°F; 1600-1710°C.

SEMALLOY NO. AG1.5
Semi-Alloys
Pb 97.5, Ag 1.5, Sn 1.
MP 588°F; 309°C; eutectic alloy.

SEMALLOY NO. AG2.5
Semi-Alloys
Pb 97.5, Ag 2.5.
MP 580°F; 304°C; eutectic alloy.

SEMALLOY NO. AG5.5
Semi-Alloys
Pb 94.5, Ag 5.5.
Liquidus 689°F; solidus 579°F (304-365°C).

SEMALLOY NO. PB65
Semi-Alloys
Pb 64.5, Sn 34.5.
Liquidus 475°F; solidus 360°F (182-246°C).

SEMALLOY NO. PB70
Semi-Alloys
Pb 70, Sn 30.
Liquidus 491°F; solidus 361°F (183-255°C).

SEMALLOY NO. PB80
Semi-Alloys
Pb 80, Sn 20.
Liquidus 535°F; solidus 361°F (183-279°C).

SEMALLOY NO. SB5
Semi-Alloys
Sn 94.8, Sb 5, Pb 0.2.
Liquidus 464°F; solidus 450°F (232-240°C).

SEMALLOY NO. SN10
Semi-Alloys
Pb 87.9, Zn 10, Ag 2.1.
Liquidus 570°F; solidus 514°F (268-299°C).

SEMALLOY NO. SN20
Semi-Alloys
Pb 79, Sn 20, Sb 1.
Liquidus 530°F; solidus 360°F (182-277°C).

SEMALLOY NO. SN30
Semi-Alloys
Pb 68.4, Sn 30, Sb 1.6.
Liquidus 490°F; solidus 360°F (182-254°C).

SEMALLOY NO. SN35
Semi-Alloys
Pb 63, Sn 35, Sb 1.8, Bi 0.2.
Liquidus 475°F; solidus 360°F (182-246°C).

SEMALLOY NO. SN40
Semi-Alloys
Pb 59.4, Sn 40.
Liquidus 460°F; solidus 360°F (182-238°C).

SEMALLOY NO. SN5
Semi-Alloys
Pb 95, Sn 5.
Liquidus 594°F; solidus 518°F (270-312°C).

SEMALLOY NO. SN50
Semi-Alloys
Sn 50, Pb 49.4.
Liquidus 420°F; solidus 360°F (182-216°C).

SEMALLOY NO. SN60
Semi-Alloys
Sn 60, Pb 39.4.
Liquidus 375°F; solidus 360°F (182-191°C). QQ S-571D.

SEMALLOY NO. SN62
Semi-Alloys
Sn 62, Pb 35.4, Ag 2.
Liquidus 372°F; solidus 350°F (177-189°C). QQ S-571D.

SEMALLOY NO. SN63
Semi-Alloys
Sn 63, Pb 36.4.
MP 360°F; 182°C. QQS-571D.

SEMALLOY NO. SN70
Semi-Alloys
Sn 70, Pb 29.4.
Liquidus 380°F; solidus 360°F (182-193°C). QQ S-591D.

SEMALLOY SN96
Semi-Alloys
Sn 96, Ag 4.
Liquidus 432°F; solidus 430°F (221-222°C).

SEMDEX
Heppenstall Co.
Tool material. C 0.6, Si 0.25, Mn 0.8, Cr 0.25, bal Fe.
Water hardened tool steel for trimmer dies, saw blocks.

SEMI HIGH
Disston Inc.
C 1.1, Cr 0.3, W 1.1, V 0.2, bal Fe.
For tools; semi-high speed. *Obsolete*

SEMI-ALLOYS NO. 157
Semi-Alloys
Pb 38, Bi 37, Sn 25.
Liquidus 260 F; Solidus 200 F. *Obsolete*

SEMI-ALLOYS NO. 158
Semi-Alloys
Bi 52, Pb 5, Sn 37, In 6.
Liquidus 264 F; Solidus 203 F. *Obsolete*

SEMI-PLASTIC BRONZE
English manufacture
Cu 75-79, Pb 13.5-16.5, Sn 7-9.
20,000 TS; 10 El. For machine bearings; leaded bronze.

SEMI-STEEL
English manufacture
Si 1.4-1.8, TC 3-3.5, CC 0.5-0.81, bal Fe.
For frames, housings, cast gears; cast iron.

SEMINOLE HARD
AL Tech Specialty Steel Corp.
C 0.52, Cr 1.3, W 2, V 0.25, Si 0.9, bal Fe.
Oil hardening, shock resisting tool steel for moderate hot work; for shear blades, dies, punches, chisels, pneumatic tools. AISI S-1.

SEMINOLE MEDIUM
AL Tech Specialty Steel Corp.
C 0.42, Cr 1.3, W 2, Mo 0.25, Si 0.25, bal Fe.
Oil hardening, shock resisting tool steel for moderate hot work; for shear blades, dies, punches, chisels, pneumatic tools. Lower carbon version of AISI S-1. *Obsolete*

SEMINOLE STEEL
Allegheny Ludlum Steel
W 2, Cr 1.3, V 0.2, C 0.4, bal Fe.
Annealed: 79,000 TS; 33 El, 56 RA; 153 Brin. Heat treated: 300,000 TS; 250,000 YS; 6 El; 12 RA; 527 Brin. For axle, tools, chisels, punches, swages, dies; shock resisting tool steel. *Obsolete*

SEMPERAL
Metallwerke, A.G.
Cu 4, Zn 1, Mg 0.3, Mn 0.6, Fe 0.7, bal Al.
Soft: 36,000 TS; 22 El; 60 Brin. Rolled: 71,000 TS; 2 El; 170 Brin. For light alloy parts; heat treatable.

SENDUST
Siemans & Halske AG.
Refractory. Si 9.6, Al 5.4, Co 85.
For cores in electromagnetic apparatus. Brittle, cast alloy.

SENDUST
Japanese manufacture
Si 9.6, Al 5.4, Co 85.
For cores of electromagnetic apparatus; brittle, cast alloy.

SENECA
Atlas Specialty Steels
C 0.35, Cr 3.2, W 9.5, V 0.4, bal Fe.
Hot work tool steel. Similar to AISI H21.

SENECA
Seneca Wire & Mfg. Co.
C 0.7, bal Fe.
For wire.

SENECA HOT DIE
Atlas Specialty Steels
C 0.28, W 9.5, Cr 3.2, V 0.4, bal Fe.
Heat treated: 243,000 TS; 215,000 YS; 12 El; 37 RA; 50 Rock C. For punches, hot dies, swaging dies; hot work. *Obsolete*

SERAING PISTON
English manufacture
Cu 89, Zn 9, Sn 2.
Cast: 35,000 TS; 16 El. For piston rings; corrosion resistant.

SERMALLOY A2
SERMAG
Co 24, Al 7.5, Ni 12, Cu 2, Si 0.2, bal Fe.
Remanance 14,300 gauss; coercive force 500 oersted; energy product 5,000,000 max. For magnetometry. High residual induction, Alnico type alloy. Permanent magnet.

SERMALLOY A2GD
SERMAG
Co 24, Al 7.5, Ni 12, Cu 2, Si 0.2, bal Fe.
Remanance 14,400 gauss; coercive force 610 oersted; energy product 6,600,000 max. For magnetometry. Columnar Alnico type permanent magnet.

SERMALLOY A$_1$
SERMAG
Fe, Co, Ni, Al.
Permanent magnet material. Coercive force 2000 +/-100 oersted. Energy product 5.5 +/-0.5 MG.Oe max. Remanance 7400 +/-200 gauss. Saturation field 6000 oersted (approximate).

SERMALLOY A$_2$
SERMAG
Al-Ni-Co.
Permanent magnet material. Coercive force 500 oersted. Energy product 5 MG.Oe max. Remanance 14,200 +/-200 gauss. Saturation field 1800 oersted (approximate).

SERMALLOY A$_2$ GD
SERMAG
Al-Ni-Co.
Permanent magnet material. Coercive force 600 oersted. Energy product 5,500,000 max. Remanance 14,300 gauss. Saturation field 2000 oersted.

SERMALLOY A$_2$GDH
SERMAG
Co 24, Al 7.5, Ni 12, Cu 2, Si 0.2, bal Fe.
Remanance 15,000 gauss; coercive force 280 oersted. For hysteresis drives. Alnico type permanent magnet.

SERMALLOY AL
SERMAG
Co 38, Al 7.5, Ni 14, Cu 3, Ti 8, bal Fe.
Cast: remanant induction; 7400 gauss; coercive force 2100 oersted; 9482 magnetic saturation; 874°C; Curie temperature. For magnetic and electrical equipment. Permanent magnet. Magnetically anisotropic.

SEVA (70/30 BRASS)
IMI Yorkshire Alloys Ltd.
Cu 69-71, As 0.02-0.035, bal Zn.
Annealed: 40,300-47,000 TS; 18,000-24,000 YS; 60-75 El; 54-73 Brin. For condensers, coolers, evaporators, juice heaters, vacuum pan tubes for the sugar industry. Malleable and ductile brass. Corrosion resistant.

SEVENTY-THIRTY
English manufacture
Cu 70, Ni 30.
For turbine blades, condenser tubes; corrosion resistant.

SEYCAST
Seymour Products Co.
Ni 99.
For anodes; controlled grain.

SEYMOUR 1008
Seymour Products Co.
Ni 8, Cu 61.5, bal Zn.
Soft: 60,000 TS; 40 El. Hard: 110,000 TS; 7 El. Spring: 130,000 TS; 2 El. For cutlery, hardware, trim, appliances, hospital equipment; nickel silver, corrosion resistant.

SEYMOUR 1009
Seymour Products Co.
Cu 61.5, Ni 8, Pb 2, bal Zn.
For keys, hardware, cutlery, trim; leaded nickel silver.

SEYMOUR 1011
Seymour Products Co.
Cu 65, Ni 10, bal Zn.
Soft: 60,000 TS; 40 El. Hard: 110,000 TS; 7 El. Spring: 130,000 TS; 2 El. For cutlery, hardware, trim, appliances, hospital equipment. Nickel Silver Gr. A; corrosion resistant.

SEYMOUR 1017
Seymour Products Co.
Cu 71.5, Ni 10, bal Zn.
Soft: 55,000 TS. Hard: 102,000 TS. Spring: 120,000 TS. For cutlery, hardware, trim, appliances, hospital equipment; nickel silver, corrosion resistant.

SEYMOUR 1091
Seymour Products Co.
Cu 61.5, Ni 10, Pb 1, bal Zn.
Rolled: 90,000 TS. For hardware, cutlery, trim, hospital equipment; leaded nickel silver.

SEYMOUR 1092
Seymour Products Co.
Cu 61.5, Ni 10, Pb 2, bal Zn.
For keys, hardware, cutlery; leaded nickel silver.

SEYMOUR 1095
Seymour Products Co.
Cu 61.5, Ni 10, Pb 0.5, bal Zn.
For hardware, trim, cutlery, hospital equipment; leaded nickel silver.

SEYMOUR 10A7
Seymour Products Co.
Ni 10, Zn 18.5, Cu 71.5.
Soft: 54,000 TS; 67 Brin. 1/2 H-temper: 70,000 TS; 144 Brin. H-temper: 80,000 TS; 165 Brin. For hollow ware, flatware, zipper parts; nickel silver, good formability.

SEYMOUR 1211
Seymour Products Co.
Cu 65, Ni 12, bal Zn.
Soft: 60,000 TS; 45 El. Hard: 115,000 TS; 7 El. Spring: 136,000 TS; 2 El. For cutlery, hardware, trim, appliances, hospital equipment; nickel silver, corrosion resistant.

SEYMOUR 1214
Seymour Products Co.
Cu 56.5, Ni 12, bal Zn.
Soft: 60,000 TS; 47 El. Hard: 103,000 TS; 7 El. Spring: 117,000 TS; 2 El. For cutlery, hardware, trim, appliances, hospital equipment; nickel silver, corrosion resistant.

SEYMOUR 1291
Seymour Products Co.
Cu 61.5, Ni 12, Pb 1, bal Zn.
For hardware, trim, cutlery; leaded nickel silver.

SEYMOUR 1292
Seymour Products Co.
Cu 61.5, Ni 12, Pb 2, bal Zn.
For keys, hardware, cutlery, trim; leaded nickel silver.

SEYMOUR 1295
Seymour Products Co.
Cu 61.5, Ni 12, Pb 0.5, bal Zn.
For hardware, cutlery, hospital equipment; leaded nickel silver.

SEYMOUR 12A4
Seymour Products Co.
Cu 56.5, Ni 12, Zn 31.5.
1/4 H: 65,000 TS; 108 Brin. Hard: 90,000 TS; 185 Brin. Spring: 109,000 TS; 222 Brin. For springs, clips; nickel silver, corrosion resistant.

SEYMOUR 1511
Seymour Products Co.
Cu 65, Ni 15, bal Zn.
Soft: 60,000 TS; 47 El. Hard: 103,000 TS; 7 El. Spring: 117,000 TS; 2 El. For cutlery, hardware, trim, appliances, hospital equipment; nickel silver, corrosion resistant.

SEYMOUR 1591
Seymour Products Co.
Cu 61.5, Ni 15, Pb 1, bal Zn.
For hardware, cutlery, hospital equipment; leaded nickel silver.

SEYMOUR 18 AL
Seymour Products Co.
Ni 18, Zn 17, Cu 65.
Hard: 80,000-120,000 TS; 3-12 El; 185-210 Brin. Soft: 50,000-55,000 TS; 40-45 El; 75-77 Brin. For springs, bellows, hardware, welding rod; nickel silver, good formability.

SEYMOUR 1811
Seymour Products Co.
Cu 65, Ni 18, bal Zn.
Soft: 60,000 TS; 47 El. Hard: 103,000 TS; 7 El. Spring: 117,000 TS; 2 El. For cutlery, hardware, trim, appliances, hospital equipment; nickel silver, corrosion resistant.

SEYMOUR 1814
Seymour Products Co.
Cu 55, Ni 18, bal Zn.
Soft: 70,000 TS; 45 El. Hard: 123,000 TS; 6 El. Spring: 145,000 TS; 2 El. For cutlery, hardware, trim, appliances, hospital equipment; nickel silver, corrosion resistant.

SEYMOUR 1815
Seymour Products Co.
Cu 65, Ni 18, Ag 0.15, bal Zn.
Soft: 60,000 TS; 47 El. Hard: 103,000 TS; 7 El. Spring: 117,000 TS; 2 El. For cutlery, hardware, trim, appliances, hospital equipment; nickel silver, corrosion resistant.

SEYMOUR 1817
Seymour Products Co.
Cu 72, Ni 18, bal Zn.
For cutlery, hardware, trim, appliances, hospital equipment; nickel silver.

SEYMOUR 1891
Seymour Products Co.
Cu 61.5, Ni 18, Pb 1, bal Zn.
For hardware, cutlery, hospital equipment; leaded nickel silver.

SEYMOUR 1895
Seymour Products Co.
Cu 61.5, Ni 18, Pb 0.5, bal Zn.
For hardware, cutlery, hospital equipment; leaded nickel silver.

SEYMOUR 1897
Seymour Products Co.
Cu 65, Ni 18, Ag 0.15, Pb 1, bal Zn.
For hardware, cutlery, hospital equipment; leaded nickel silver.

SEYMOUR 2265
Seymour Products Co.
Cu 66, bal Zn.
1/2 H-temper: 60,000 TS. Hard temper: 73,000 TS. Spring: 90,500 TS. For electrical components, hardware, watch parts, jewelry; high brass.

SEYMOUR 2270
Seymour Products Co.
Cu 70, bal Zn.
1/2 H-temper: 62,000 TS; 121 Brin. Hard temper: 76,000 TS; 159 Brin. Spring: 95,400 TS; 190 Brin. For electrical components, hardware, jewelry, watches, lamps; cartridge brass.

SEYMOUR 2280
Seymour Products Co.
Cu 80, bal Zn.
1/2 H-temper: 60,000 TS; 117 Brin. Hard temper: 72,500 TS; 150 Brin. Spring: 89,000 TS; 185 Brin. For electrical components, hardware, jewelry, watches, lamps; low brass.

SEYMOUR 2285
Seymour Products Co.
Cu 85, bal Zn.
1/2 H-temper: 56,000 TS; 114 Brin. Hard temper: 67,500 TS; 139 Brin. Spring: 82,000 TS; 165 Brin. For electrical components, hardware, jewelry, watches, lamps; rich low brass.

SEYMOUR 2287
Seymour Products Co.
Cu 87, bal Zn.
1/2 H-temper: 54,000 TS; 116 Brin. Hard temper: 65,000 TS; 137 Brin. Spring: 75,000 TS; 156 Brin. For electrical components, hardware, jewelry, watches, lamps; Nugild.

SEYMOUR 2290
Seymour Products Co.
Cu 90, bal Zn.
1/2 H-temper: 52,000 TS; 104 Brin. Hard temper: 61,500 TS; 125 Brin. Spring: 73,000 TS; 153 Brin. For electrical components, hardware, jewelry, watches, lamps; commercial bronze.

SEYMOUR 2295
Seymour Products Co.
Cu 95, bal Zn.
1/2 H-temper: 47,000 TS; 83 Brin. Hard temper: 54,500 TS; 110 Brin. Spring: 64,000 TS; 130 Brin. For electrical parts, hardware, jewelry, watches, lamps; gilding bronze.

SEYMOUR 2401
Seymour Products Co.
Cu 90, Sn 0.5, Zn 9.5.
Soft: 38,000 TS, 12,000 YS; 45 El. For weather stripping; brass, corrosion resistant.

SEYMOUR 2402
Seymour Products Co.
Cu 90, Sn 2, Zn 8.
Soft: 40,000 TS; 42 El. For hardware, costume jewelry; corrosion resistant.

SEYMOUR 2405
Seymour Products Co.
Cu 90.5, Sn 5, Zn 4.5.
For hardware, costume jewelry; corrosion resistant.

SEYMOUR 2451
Seymour Products Co.
Cu 95, Sn 1, Zn 4.
For hardware, weather stripping; corrosion resistant.

SEYMOUR 265
Seymour Products Co.
Cu 65, Zn 35.
For hardware, plumbing fixtures; yellow brass.

SEYMOUR 270
Seymour Products Co.
Cu 70, Zn 30.
Hard: 76,000 TS; 62,000 YS; 10 El; 159 Brin. Soft: 47,000 TS; 17,000 YS; 65 El; 62 Brin. For tanks, cartridge cases, vessels; cartridge brass.

SEYMOUR 280
Seymour Products Co.
Cu 80, Zn 20.
Hard: 91,000 TS; 55,000 YS; 10 El; 190 Brin. Soft: 46,000 TS; 15,000 YS; 50 El; 57 Brin. For hardware, fasteners; low brass.

SEYMOUR 285
Seymour Products Co.
Cu 85, Zn 15.
Hard: 84,000 TS; 55,000 YS; 2 El; 170 Brin. Soft: 43,000 TS; 15,000 YS; 42 El; 56 Brin. For diaphragms, fasteners; red brass.

SEYMOUR 2875
Seymour Products Co.
Cu 87.5, Sn 2.25, Zn 10.25.
Soft: 45,000 TS; 57 El. Hard: 102,000 TS; 7 El. For hardware, costume jewelry; corrosion resistant.

SEYMOUR 2876
Seymour Products Co.
Cu 87, bal Zn.
1/2 H-temper: 54,000 TS; 116 Brin. Hard temper: 65,000 TS; 138 Brin. Spring: 75,000 TS; 160 Brin. For electrical components, hardware, watches, jewelry; high brass.

SEYMOUR 2877
Seymour Products Co.
Cu 86.5, Sn 0.5, Zn 13.
For hardware, chains, fasteners, costume jewelry; jewelers' bronze, corrosion resistant.

SEYMOUR 290
Seymour Products Co.
Cu 90, Zn 10.
Hard: 72,000 TS; 47,000 YS; 6 El; 145 Brin. Soft: 39,000 TS; 12,000 YS; 40 El; 54 Brin. For hardware, fasteners; commercial brass.

SEYMOUR 2963
Seymour Products Co.
Cu 63.5, Pb 0.75, bal Zn.
1/4 H-temper: 60,000 TS; 50,000 YS; 23 El; 117 Brin. Hard: 73,000 TS; 60,000 YS; 8 El; 150 Brin. Spring: 90,500 TS; 185 Brin. For hardware, hinges, watch backs, fasteners; low leaded brass.

SEYMOUR 2964
Seymour Products Co.
Cu 64, Pb 2, bal Zn.
1/4 H-temper: 60,000 TS; 117 Brin. Hard: 73,000 TS; 150 Brin. Spring: 90,500 TS; 185 Brin. For clock parts, gears, wheels, channel plates; high leaded brass, corrosion resistant.

SEYMOUR 2965
Seymour Products Co.
Cu 62, Pb 2.5, bal Zn.
1/4 H-temper: 60,000 TS; 117 Brin. Hard: 73,000 TS; 150 Brin. Spring: 90,500 TS; 185 Brin. For clock parts, gears, wheels, channel plates; extra high leaded brass, free-cutting.

SEYMOUR 430
Seymour Products Co.
Sn 5, Pb 5, P 0.25, bal Cu.
For bushings, gears, keys, bearings; leaded P-bronze.

SEYMOUR 5770
Seymour Products Co.
Cu 70, Ni 30.
For process equipment; cupro-nickel, corrosion resistant.

SEYMOUR 5778
Seymour Products Co.
Ni 22, bal Cu.
For process equipment; cupro-nickel, corrosion resistant.

SEYMOUR 5780
Seymour Products Co.
Cu 80, Ni 20.
For process equipment; cupro-nickel, corrosion resistant.

SEYMOUR 5788
Seymour Products Co.
Ni 10, bal Cu.
For process equipment; cupro-nickel, corrosion resistant.

SEYMOUR 5794
Seymour Products Co.
Ni 3.5, bal Cu.
For process equipment; cupro-nickel, corrosion resistant.

SEYMOUR 5798
Seymour Products Co.
Ni 2, bal Cu.
For process equipment; cupro-nickel, corrosion resistant.

SEYMOUR 8120
Seymour Products Co.
Ag 0.05, bal Cu.
For engraving plates; engravers' copper.

SEYMOUR 8121
Seymour Products Co.
O 0.04, 99.9 Cu min.
1/4 H-temper: 38,000 TS; 30,000 YS; 25 El; 64 Brin. 1/2 H-temper: 42,000 TS; 36,000 YS; 14 El; 75 Brin. Hard: 50,000 TS; 45,000 YS; 6 El; 83 Brin. For gutters, radio parts, terminals, ball floats; tough pitch copper.

SEYMOUR 8122
Seymour Products Co.
99.92 Cu min.
1/4 H-temper: 38,000 TS; 30,000 YS; 35 El; 64 Brin. Hard: 50,000 TS; 45,000 YS; 12 El; 83 Brin. For bus bars and bus conductors, wave guides; OFHC copper.

SEYMOUR 9230
Seymour Products Co.
Sn 8, P 0.3, bal Cu.
For springs, fuse clips, diaphragms; P-bronze.

SEYMOUR 9231
Seymour Products Co.
Sn 6, P 0.3, bal Cu.
For instruments; special Fourdrinier wire, P-bronze.

SEYMOUR 9370
Seymour Products Co.
Sn 3, P 0.23, bal Cu.
For springs, diaphragms, electrical contacts; Fourdrinier wire, P-bronze.

SEYMOUR 9430
Seymour Products Co.
Sn 5, Pb 5, P 0.15, bal Cu.
For bearings, bushings, liners; bearing bronze.

SEYMOUR 9444
Seymour Products Co.
Sn 4, Pb 4, Zn 4, P 0.07, bal Cu.
For bearings, bushings; leaded bronze.

SEYMOUR 9494
Seymour Products Co.
Sn 5, Pb 1, P 0.07, bal Cu.
For bearings, liners, sleeves; leaded P-bronze.

SEYMOUR 971
Seymour Products Co.
Sn 3, P 0.25, bal Cu.
1/2 H-temper: 60,000 TS; 125 Brin. H-temper: 72,000 TS; 150 Brin. Spring temper: 87,000 TS; 176 Brin. For springs, clips, diaphragms, contacts. Phosphor Bronze Grade E.

SEYMOUR 9910
Seymour Products Co.
Sn 10, P 0.1, bal Cu.
For springs, discs, diaphragms, bellows; P-bronze.

SEYMOUR 9928
Seymour Products Co.
Sn 8, P 0.1, bal Cu.
For springs, thermostats, diaphragms, microphones; P-bronze.

SEYMOUR 9950
Seymour Products Co.
Sn 5, P 0.15, bal Cu.
1/2 H-temper: 55,000-70,000 TS. Hard temper: 72,000-87,000 TS. For springs, switches, diaphragms, clutch discs, chemical hardware; P-bronze, electrical conductivity: 16%.

SEYMOUR 9971
Seymour Products Co.
Sn 3, P 0.1, bal Cu.
1/2 H-temper: 55,000-65,000 TS. Hard: 67,000-77,000 TS. For low-duty springs, diaphragms, electrical contacts; P-bronze, electrical conductivity: 22%.

SEYMOUR 9985
Seymour Products Co.
Sn 1.35, P 0.07, bal Cu.
For low-duty springs, diaphragms, electrical contacts; P-bronze.

SEYMOUR NICKEL SILVER GRADE A
Seymour Products Co.
Zn 23, Ni 19, bal Cu.
100,000 TS. For costume jewelry; corrosion resistant.

SEYMOUR NO. 10A1
Seymour Products Co.
Ni 10, Zn 23, Cu 67.
Hard: 109,000 TS; 1.5 El; 188 Brin. Soft: 52,000 TS; 40 El; 70 Brin. For hardware; electrical conductivity: 9% (sheet form); 10% Nickel Silver.

SEYMOUR NO. 10X1
Seymour Products Co.
Ni 10, Zn 28, Pb 1, bal Cu.
Hard: 75,000 TS; 5 El; 176 Brin. Soft: 54,000 TS; 30 El; 75 Brin. For hardware, screw machine products; nickel silver, free-cutting, corrosion resistant.

SEYMOUR NO. 10X2
Seymour Products Co.
Ni 12, Zn 25, Pb 2, bal Cu.
Hard: 85,000 TS; 5 El; 172 Brin. For marine hardware, screw machine products; nickel silver, free-cutting, corrosion resistant.

SEYMOUR NO. 12A1
Seymour Products Co.
Ni 12, Zn 23, bal Cu.
Hard: 106,000 TS; 2 El; 210 Brin. For marine hardware, fasteners; nickel silver, corrosion resistant.

SEYMOUR NO. 12X1
Seymour Products Co.
Ni 12, Pb 1, Zn 25, bal Cu.
Hard: 90,000 TS; 5 El; 172 Brin. Soft: 56,000 TS; 35 El; 75 Brin. For screw machine products, marine hardware; corrosion resistant, free-cutting.

SEYMOUR NO. 12X2
Seymour Products Co.
Ni 12, Zn 26, Pb 2, bal Cu.
Hard: 95,000 TS; 3 El; 195 Brin. Soft: 57,000 TS; 45 El; 76 Brin. For marine hardware, keys, fasteners; nickel silver, free-cutting, corrosion resistant.

SEYMOUR NO. 15 AL
Seymour Products Co.
Ni 15, Cu 65, bal Zn.
Hard: 97,000 TS; 195 Brin. For hardware, marine parts; nickel silver, corrosion resistant.

SEYMOUR NO. 15A
Seymour Products Co.
Ni 15, Zn 21, Cu 64.
Hard: 93,000 TS; 5 El; 190 Brin. Soft: 58,000 TS; 40 El; 75 Brin. For hardware; corrosion resistant.

SEYMOUR NO. 15B
Seymour Products Co.
Ni 15, Zn 27, Cu 58.
Hard: 100,000 TS; 2 El; 204 Brin. Soft: 55,000 TS; 36 El; 77 Brin. For hardware; corrosion resistant.

SEYMOUR NO. 15X1
Seymour Products Co.
Ni 15, Pb 1, Zn 23, bal Cu.
Hard: 85,000 TS; 10 El; 150 Brin. Soft: 58,000 TS; 33 El; 80 Brin. For marine hardware, screw machine products; corrosion resistant, free-cutting.

SEYMOUR NO. 15X2
Seymour Products Co.
Ni 15, Zn 23, Pb 2, bal Cu.
Hard: 92,000 TS; 4 El; 185 Brin. Soft: 58,000 TS; 45 El; 75 Brin. For hardware, screw machine products; nickel silver, free-cutting, corrosion resistant.

SEYMOUR NO. 18A3
Seymour Products Co.
Ni 18, Zn 17, Cu 65.
Hard: 100,000 TS; 3 El; 190 Brin. Soft: 54,000 TS; 40 El; 78 Brin. For hardware; corrosion resistant.

SEYMOUR NO. 18A4
Seymour Products Co.
Ni 18, Zn 27, bal Cu.
Hard: 115,000 TS; 2 El; 210 Brin. Soft: 65,000 TS; 45 El; 83 Brin. For springs, marine hardware; nickel silver, corrosion resistant.

SEYMOUR NO. 18A7
Seymour Products Co.
Ni 18, Zn 7, Cu 75.
Hard: 85,000 TS; 4 El; 180 Brin. Soft: 50,000 TS; 32 El; 65 Brin. For hardware; corrosion resistant.

SEYMOUR NO. 18X1
Seymour Products Co.
Ni 18, Pb 1, Zn 19, Cu 62.
Hard: 85,000 TS; 13 El; 150 Brin. For screw machine products; 18% leaded Nickel Silver; free cutting.

SEYMOUR NO. 25
Seymour Products Co.
Ni 25, Zn 19, Cu 56.
Hard: 110,000 TS; 4 El; 204 Brin. Soft: 72,000 TS; 30 El; 110 Brin. For hardware; corrosion resistant.

SEYMOUR NO. 287
Seymour Products Co.
Zn 13, bal Cu.
Hard: 65,000 TS; 137 Brin. For jewelry, trim; Nugild.

SEYMOUR NO. 444
Seymour Products Co.
Cu 88, Sn 4, Pb 4, Zn 4.
Hard: 60,000 TS; 20 El. Soft: 48,000 TS. For bearings, studs, shafts, gears, bolts; free-cutting. Phosphor Bronze Grade B2.

SEYMOUR NO. 494
Seymour Products Co.
Cu 94, Sn 5, Pb 1.
Hard: 61,500 TS; 20 El. Soft: 50,000 TS; 40 El. For bolts, screws, bearings, gears, shafts. Phosphor Bronze Grade B1.

SEYMOUR NO. 5
Seymour Products Co.
Ni 5, Zn 34, Cu 61.
Hard: 135,000 TS; 2 El. Soft: 50,000 TS. For screw machine parts; electrical conductivity: 12% (wire form).

SEYMOUR NO. 5A1
Seymour Products Co.
Ni 5, Zn 27, bal Cu.
Hard: 100,000 TS; 3 El; 185 Brin. For hardware, springs; nickel silver, corrosion resistant.

SEYMOUR NO. 780
Seymour Products Co.
Ni 20, Cu 80.
Hard: 85,000 TS; 2 El; 165 Brin. Soft: 50,000 TS; 30 El; 72 Brin. For condenser tubes; corrosion resistant.

SEYMOUR NO. 785
Seymour Products Co.
Ni 15, Cu 85.
Hard: 70,000 TS; 3 El; 160 Brin. Soft: 45,000 TS; 30 El; 70 Brin. For condenser tubes; corrosion resistant.

SEYMOUR NO. 895
Seymour Products Co.
Cu 89.5, Sn 10.5, 0.10 P min.
For welding rod to build up bearings; Grade C arc welding.

SEYMOUR NO. 910
Seymour Products Co.
Sn 10, P 0.2, bal Cu.
Soft: 65,000 TS; 65 El; 75 Brin. Spring: 122,000 TS; 5 El; 240 Brin. For springs, diaphragms, fasteners, cams, bellows, shafts; P-bronze, Gr. D.

SEYMOUR NO. 9225
Seymour Products Co.
Cu 91.7, Sn 8.25, P 0.25.
For welding rod for bronze, brass and steel; Grade C arc welding.

SEYMOUR NO. 928
Seymour Products Co.
Sn 8, P 0.25, bal Cu.
Hard: 112,000 TS; 38 El; 240 Brin. Soft: 60,000 TS; 55 El; 75 Brin. For springs, bellows, diaphragms, hardware, bolts; P-bronze, Gr. C; tough.

SEYMOUR NO. 950
Seymour Products Co.
Sn 5, P 0.25, bal Cu.
Hard: 100,000 TS; 1.5 El; 210 Brin. Soft: 50,000 TS; 50 El; 75 Brin. For springs, diaphragms, bolts, fasteners, hardware; P-bronze Gr. A; tough.

SEYMOUR NO. 9522
Seymour Products Co.
Cu 95, Sn 5, 0.22 P min.
For welding rod for general purpose; Grade A arc welding.

SEYMOURCO
Seymour Products Co.
Zn 34, Ni 5, bal Cu.
For costume jewelry; nickel silver.

SEYMOURITE
Seymour Products Co.
Cu 64, Ni 18, Zn 18.
For ornaments, electrical equipment; high corrosion resistant.

SF 67
United Engineering Steels Ltd.
C 0.7, Cr 13, bal Fe.
Razor blade strip.

SFC 2/61/0
Sintered Products Ltd.
Fe (some Cu-C-S, sintered).
Free machining grade of FC2/61/0.

SFS
Styria-Stahl Steirische Gusstahlwerke AG
C 0.9, V 0.1-0.2, bal Fe.
For tools, punches, drills; water hardened. *Obsolete*

SG-100 NICKEL
Manufacturer not listed.
Co 0.18, Cu 0.004, Fe 0.02, S 0.0025, C 0.007, 99.93 Ni + Co.
Annealed Strip: 50,000 TS; 12,000 YS; 45 El. For cathode sleeves and shields, heat exchanger fins, getter flags, passive cathodes, magneto-strictive transducers. High purity nickel from powder metallurgy.

SG-101 NICKEL
Manufacturer not listed.
Cu 0.005, Fe 0.01, C 0.01, S 0.0004, 99.9 Ni + Co.
High purity nickel strip.

SG124
Contact Technologies, Inc.
Ag 80.
Silver graphite. Density: 5.5 g/cm^3; specific resistivity: 0.10 micro-ohm-m; hardness: 15W44; strength: 5,000 psi. For sliding contacts for slip ring assemblies; brushes for electric motors.

SG161
Contact Technologies, Inc.
Ag 50.
Silver graphite. Density: 3.2 g/cm^3; specific resistivity: 1cx10^{-4} ohm/in.; hardness: 15W34; strength: 3500 psi. Brushes for electric motors.

SGM 4
Saarstahl AG
C 0.43, Cr 4.25, Mo 0.5, V 2, W 4.25, Co 4.25, bal Fe.
Hot work tool steel; for hot extruding dies, forging dies, brass casting molds. W.-Nr. 2678; AISI H19. *Obsolete*

SGS-TANTALUM
National Research Corp.
100 Ta + Yttrium.
300% stronger than ordinary Ta at 1650°C. For electronic and electrical components, vacuum furnaces, aerospace and nuclear systems. Fine controlled grain. No oxygen embrittlement caused by grain coarsening to 2300°C.

SH 50
Heppenstall Co.
C 0.5, Cr 0.75, V 0.15, bal Fe.
For die blocks. *Obsolete*

SH-IS-54
Mitsubishi Metal America Corp.
C 0.12-0.17, Mn 0.8-1.1, Si 0.35-0.6, Cu 0-0.4, bal Fe.
Rolled: 77,000-88,000 TS; 50,000 YP min; 20 El min. For railroad and mine cars, bridges, agricultural equipment. High-tensile, low alloy constructional steel.

SH-IS-60
Mitsubishi Metal America Corp.
C 0.15-0.21, Mn 0.95-1.25, Si 0.5-0.8, bal Fe.
Rolled: 85,000-100,000 TS; 54,000 YP min; 17 El min. For railroad and mine cars, bridges, agricultural equipment. Constructional steel, tough.

SH-KH 15
Russian manufacture
C 0.95-1.1, Cr 1.3-1.6, Mn 0.2, Ni 0-0.2, bal Fe.
For ball bearings, races, pivots; water hardened.

SHAKUDO (SHAKDO)
Japanese manufacture
Cu 94-96, Au 3.7-4.2, Ag 0.1-1.6, traces of Pb, Fe, As.
For ornamental parts; corrosion resistant.

SHAPERITE
LaSalle Steel Co.
C 0.4, Mn, bal Fe.
Cold drawn: 100,000 TS; 90,000 YS; 13 El; 45 RA; 217 Brin.
For machinery parts; Modified X-1340. *Obsolete*

SHARALLOY
Sharon Steel Corp.
C 0.1, Mn 0.4, Si, Cb, bal Fe.
HSLA hot rolled sheet and strip. 60,000 psi TS min; 45,000 psi YS min; 22 El min. SAE 945A, C.

SHARALLOY 45
Sharon Steel Corp.
C 0.1, Mn 0.4, 0.01 Cb min, Si, bal Fe.
HSLA hot rolled sheet and strip. 60,000 psi TS min; 45,000 psi YS min; 25 El min. For automotive and structural applications. SAE 945X.

SHARALLOY 50
Sharon Steel Corp.
C 0.16, Mn 0.4, 0.01 Cb min, Si, bal Fe.
HSLA hot rolled sheet and strip. 65,000 psi TS min; 50,000 psi YS min; 22 El min. For automotive and structural applications. SAE 950X.

SHARALLOY 55
Sharon Steel Corp.
C 0.16, Mn 0.5, 0.01 Cb min, Si, bal Fe.
HSLA hot rolled sheet and strip. 70,000 psi TS min; 55,000 psi YS min; 20 El min. For automotive and structural applications. SAE 955X.

SHARALLOY 60
Sharon Steel Corp.
C 0.16, Mn 0.5, 0.01 Cb min, Si, bal Fe.
HSLA hot rolled sheet and strip. 75,000 psi TS min; 60,000 psi YS min; 18 El min. For automotive and structural applications. SAE 960X.

SHARON 0126-8
Sharon Steel Corp.
C 1.21-1.35, Mn 0.1-0.25, Si 0.1-0.25, Cr 0.1-0.25, bal Fe.
Sheet or strip; usually for band saws.

SHARON 0127-5
Sharon Steel Corp.
C 1.1-1.25, Mn 0.3-0.5, Si 0.18-0.33, Cr 0.25-0.5, bal Fe.
Sheet and strip; for razor blades.

SHARON 0130-5
Sharon Steel Corp.
C 1.2-1.5, Mn 12-15, Si 0-0.55, bal Fe.
Sheet, plate or strip; for steel helmets. Hadfield Helmet Steel.

SHARON 0136-5
Sharon Steel Corp.
C 0.9-1.1, Mn 0.15-0.4, Si 0.2-0.35, Mo 0.55-0.7, bal Fe.
Sheet and strip; for hack saw blades.

SHARON 0137-5
Sharon Steel Corp.
C 0.5-0.6, Mn 1.3-1.65, Si 0.15-0.3, Mo 0.1-0.2, bal Fe.
Sheet, plate or strip; for hand shears.

SHARON 0141-8
Sharon Steel Corp.
C 0.7-0.81, Mn 0.3-0.45, Si 0.15-0.3, Ni 2.25-2.65, bal Fe.
Sheet, plate or strip; for large saw blades.

SHARON 0143-5
Sharon Steel Corp.
C 0-0.08, Mn 0.3-0.45, Si 0.1-0.25, Ti = 5 x C min, bal Fe.
Sheet, plate or strip; for nuclear applications.

SHARON 0166-5
Sharon Steel Corp.
C 0.79-0.9, Mn 0.2-0.35, Si 0.2-0.35, Cr 0.35-0.55, Ni 0.6-0.9, bal Fe.
Typical use: saw bars.

SHARON 0168-5
Sharon Steel Corp.
C 0.1-0.15, Mn 0.5-0.7, Si 0.1-0.2, Cr 0.2-0.4, Ni 0.2-0.4, bal Fe.
Typical use: needle bearings.

SHARON 0170-6
Sharon Steel Corp.
C 0.95-1.1, Mn 0.3-0.5, Si 0.1-0.25, Cr 0.4-0.6, Mo 0.15-0.25, bal Fe.
Typical use: cutlery.

SHARON 0175-6
Sharon Steel Corp.
C 0.7-0.81, Mn 0.3-0.45, Si 0.15-0.3, Ni 2.25-2.65, V 0.05-0.1, bal Fe.
Typical use: saw chain links.

SHARON 0181-5
Sharon Steel Corp.
C 0.6-0.7, Mn 0.3-0.5, Si 0.2-0.35, Cr 0.25-0.4, Ni 1.2-1.5, Mo 0.08-0.15, bal Fe.
Typical use: circular saws.

SHARON 0182-5
Sharon Steel Corp.
C 0.61-0.72, Mn 0.35-0.55, Si 0.2-0.35, Cr 0.4-0.6, Ni 0.6-0.9, Mo 0.1-0.2, bal Fe.
Typical use: saw chain links.

SHARON 0182-5EF
Sharon Steel Corp.
C 0.62-0.71, Mn 0.35-0.5, Si 0.2-0.3, Ni 0.6-0.9, Cr 0.4-0.6, Mo 0.1-0.2, bal Fe.
Hardenable to 58-63 Rock C. Typical use: chain saw cutters and drive links.

SHARON 0183-5
Sharon Steel Corp.
C 0.69-0.8, Mn 0.25-0.4, Si 0.15-0.3, Cr 0.1-0.25, Ni 0.45-0.75, Mo 0.1-0.2, bal Fe.
Typical use: saw bars.

SHARON 0184-5
Sharon Steel Corp.
C 0.7-0.81, Mn 0.35-0.55, Si 0.2-0.35, Cr 0.4-0.6, Ni 0.6-0.9, Mo 0.1-0.2, bal Fe.
Typical use: circular saws.

SHARON 0186-5
Sharon Steel Corp.
C 0.64-0.75, Mn 0.4-0.6, Si 0.2-0.35, Cr 0.3-0.5, Ni 0.7-1, Mo 0.08-0.15, bal Fe.
Typical use: circular saws.

SHARON 0188-5
Sharon Steel Corp.
C 0.79-0.9, Mn 0.25-0.4, Si 0.15-0.3, Cr 0.1-0.25, Ni 0.6-0.9, Mo 0.13-0.2, bal Fe.
Typical use: circular saws.

SHARON 0190-6
Sharon Steel Corp.
C 0.28-0.33, Mn 0.45-0.65, Si 0.45-0.75, Cr 1-1.5, Mo 0.4-0.6, V 0.2-0.3, bal Fe.
Typical use: AMS 6385 aircraft brakes.

SHARON 0191-6
Sharon Steel Corp.
C 0.95-1.1, Mn 0.2-0.4, Si 0.1-0.25, Cr 0.4-0.6, Mo 0.1-0.2, V 0.1-0.2, bal Fe.
Typical use: cutlery.

SHARON 18-8
Castalloy Co., Inc.
C 0.13-0.2, Cr 18, Ni 8, bal Fe.
Annealed: 90,000 TS; 30,000 YS; 50 El; 65 RA. Cold rolled: 175,000 TS; 150,000 YS; 10 El; 45 RA. For stainless parts; stainless; austenitic, resists scaling to 1550 F. *Obsolete*

SHARON 18-8
Sharon Steel Corp.
C 0.13-0.2, Cr 18, Ni 8, bal Fe.
Annealed: 90,000 TS; 30,000 YS; 50 El; 65 RA. Cold rolled: 175,000 TS; 150,000 YS; 10 El; 45 RA. For stainless parts; stainless; austenitic, resists scaling to 1550 F. *Obsolete*

SHARON 3100 SERIES
Sharon Steel Corp.
C 0.13-0.53, Mn 0.4-0.9, Si 0.2-0.35, Ni 1.1-1.4, Cr 0.55-0.9, bal Fe.
Composition as desired. Oil hardenable. Typical uses: axles, drive shafts, nuts and bolts, tubing, valve tips, washers.

SHARON 4000 SERIES
Sharon Steel Corp.
C 0.09-0.5, Mn 0.7-1, Si 0.2-0.35, Mo 0.15-0.3, bal Fe.
Composition as desired. Oil hardenable. Typical uses: springs; flat and coiled, mower blades, pruning shear blades, clutch plates.

SHARON 4100 SERIES A
Sharon Steel Corp.
C 0.18-0.4, Mn 0.49-0.9, Si 0.2-0.35, Cr 0.4-1.1, Mo 0.08-0.25, bal Fe.
Composition as desired. Oil hardenable. Typical uses: seat belt hardware, fittings.

SHARON 4100 SERIES B
Sharon Steel Corp.
C 0.38-0.65, Mn 0.75-1, Si 0.2-0.35, Cr 0.7-1.1, Mo 0.15-0.35, bal Fe.
Composition as desired. Oil hardenable. Typical uses: aircraft couplings, landing gear, trencher knives.

SHARON 4112
Sharon Steel Corp.
C 0.1-0.15, Mn 0.3-0.6, Si 0.15-0.3, Cr 0.3-0.5, Mo 0.15-0.25, bal Fe.
Typical use: wheel bearings.

SHARON 5100 SERIES A
Sharon Steel Corp.
C 0.13-0.33, Mn 0.7-0.9, Si 0.2-0.35, Cr 0.7-1.1, bal Fe.
Composition as desired. Oil hardenable. Typical uses: fasteners, case hardened parts.

SHARON 5100 SERIES B
Sharon Steel Corp.
C 0.3-0.51, Mn 0.6-0.95, Si 0.2-0.35, Cr 0.7-1.15, bal Fe.
Composition as desired. Oil hardenable. Typical uses: transmission gears, roller chain, spline shafts.

SHARON 5100 SERIES C
Sharon Steel Corp.
C 0.48-0.64, Mn 0.7-1, Si 0.2-0.35, Cr 0.7-0.9, bal Fe.
Composition as desired. Oil hardenable. Typical uses: coil and flat springs, cutlery.

SHARON 6100 SERIES A
Sharon Steel Corp.
C 0.16-0.21, Mn 0.5-0.7, Si 0.2-0.35, Cr 0.5-0.7, V 0.1-0.15, bal Fe.
For case hardened parts.

SHARON 6100 SERIES B
Sharon Steel Corp.
C 0.48-0.53, Mn 0.7-0.9, Si 0.2-0.35, Cr 0.8-1.1, 0.15 V min, bal Fe.
Typical uses: springs, circular saw blades, business machine parts.

SHARON 8600 SERIES A
Sharon Steel Corp.
C 0.13-0.33, Mn 0.7-0.9, Si 0.2-0.35, Ni 0.4-0.7, Cr 0.4-0.6, Mo 0.15-0.25, bal Fe.
Composition as desired. Oil hardenable. Typical uses: carburized gears, spline shafts, bearings and races.

SHARON 8600 SERIES B
Sharon Steel Corp.
C 0.35-0.64, Mn 0.75-1, Si 0.2-0.35, Ni 0.4-0.7, Cr 0.4-0.6, Mo 0.15-0.25, bal Fe.
Composition as desired. Oil hardenable. Typical uses: spline shafts, helical springs.

SHARON STAINLESS 18-8
Sharon Steel Corp.
C 0.07, Cr 18, Ni 8, bal Fe.
Annealed: 85,000 TS; 30,000 YS; 45 El; 65 RA; 150 Brin. For chemical plant equipment; austenitic, stainless; Type 304. *Obsolete*

SHARON UT-1
Sharon Steel Corp.
C 0-0.08, Mn 0-1, Si 0-1, Cr 10.5-11.75, Ti = 6 x 0.75 Al max, bal Fe.
Typical use: utility stainless.

SHARON UT-3
Sharon Steel Corp.
C 0-0.08, Mn 0-1, Si 0-1, Cr 11.5-14.5, Al 0.1-0.3, bal Fe.
Typical use: utility stainless.

SHARON UT-9
Sharon Steel Corp.
C 0-0.03, Mn 0-1, Si 0-1, Cr 12-14.5, bal Fe.
Typical use: utility stainless.

SHARON-SHARP KN-44
Sharon Steel Corp.
C 1, Cr 12, bal Fe.
For knives, cutlery; air hardening, stainless. *Obsolete*

SHARPEALOY
Brown & Sharpe Mfg. Co.
C 1, W 1.75, Cr 3.85, V 1.1, Mo 8.35, Co 3.85, bal Fe.
High speed tool steel; cutting tools.

SHAWINIGAN "20"
Shawinigan Chemicals Ltd.
Ni 29, Cr 19, Mo 3, Cu 4, Si 1, C 0-0.07, bal Fe.
Annealed: 65,000-75,000 TS; 26,000-32,000 YS; 35-50 El; 40-60 RA; 120-150 Brin. For sulfuric acid equipment; corrosion resistant to H_2SO_4.

SHAWINIGAN CA
Shawinigan Chemicals Ltd.
C 0-0.2, Cr 12-14, Ni 0-1, bal Fe.
Annealed: 100,000 TS; 85,000 YS; 20 El; 35 RA; 200 Brin. For valve seats; wear resisting, corrosion resisting.

SHAWINIGAN CF-SE
Shawinigan Chemicals Ltd.
C 0.1, Cr 18-20, Ni 8-10, Se 0.25, bal Fe.
Annealed: 80,000 TS; 40,000 YS; 45 El; 60 RA; 180 Brin. For stainless parts; free-cutting, austenitic, stainless.

SHAWINIGAN HC
Shawinigan Chemicals Ltd.
C 0.2-0.3, Cr 26-30, Ni 0-3, bal Fe.
Cast: 112,000 TS; 75,000 YS; 25 El; 25 RA: 225 Brin. For furnace parts; heat resisting to sulfur atmosphere.

SHAWINIGAN HH
Shawinigan Chemicals Ltd.
C 0.25-0.35, Cr 24-26, Ni 11-13, bal Fe.
Cast: 85,000 TS; 40,000 YS; 25 El; 45 RA; 180 Brin. For furnace parts; heat resisting.

SHAWINIGAN HL
Shawinigan Chemicals Ltd.
C 0.25-0.35, Ni 20-22, Cr 26-30, bal Fe.
Cast: 75,000 TS; 45,000 YS; 10 El; 180 Brin. For furnace parts; heat resisting.

SHAWINIGAN HR-1
Shawinigan Chemicals Ltd.
C 0.13-0.2, Cr 24-30, bal Fe.
For heat and corrosion resistant parts; heat and corrosion resistant. *Obsolete*

SHAWINIGAN HR-2
Shawinigan Chemicals Ltd.
C 0.13-0.2, Cr 24-30, Ni 12-15, bal Fe.
For heat and corrosion resistant parts; heat and corrosion resistant. *Obsolete*

SHAWINIGAN HR-3

Shawinigan Chemicals Ltd.
C 0.3-0.5, Cr 16-23, Ni 31-39, bal Fe.
For heat and corrosion resistant parts; heat and corrosion resistant. *Obsolete*

SHAWINIGAN HT

Shawinigan Chemicals Ltd.
C 0.2-0.35, Ni 35-40, Cr 15-18, bal Fe.
Cast: 70,000 TS; 35,000 YS; 12 El; 15 RA. For furnace parts; heat resisting to 2100°F.

SHAWINIGAN KA-2-MO

Shawinigan Chemicals Ltd.
C 0-0.1, Cr 18-20, Ni 8-10, Mo 2.5-3.5, Si 0.5, Mn 0.4, bal Fe.
Annealed: 85,000 TS; 45,000 YS; 46 El; 60 RA; 180 Brin. For apparatus in the sulfite industry and chemical plants with strong acid corrosion resistance.

SHAWINIGAN KA2

Shawinigan Chemicals Ltd.
Cr 18, Ni 8, C, bal Fe.
For corrosion resistant and stainless parts; stainless. *Obsolete*

SHAWINIGAN NIROSTA

Shawinigan Chemicals Ltd.
C 0.21-0.35, Cr 31-39, Ni 12-15, bal Fe.
For heat and corrosion resistant parts; heat and corrosion resistant. *Obsolete*

SHAWINIGAN SSS

Shawinigan Chemicals Ltd.
Ni 35, Cr 15, C, bal Fe.
For heat and corrosion resistant castings and equipment; heat and corrosion resistant.

SHAWINIGAN X

Shawinigan Chemicals Ltd.
C 0.2, Cr 22, Ni 14, W, bal Fe.
For jet engine parts; heat resistant

SHAWNEE

AL Tech Specialty Steel Corp.
C 0.4, Cr 4.25, W 4.25, Co 4.25, V 2, bal Fe.
Heavy duty hot die steel with superior hot hardness and resistance to heat checking. Applications include press forging dies, brass extrusion dies and valve extrusion inserts. AISI H-19.

SHEARCUT

Gulf Steel Corp.
C 0.55, Mn 0.8, Cr 0.25, Si 2, Mo 0.4, V 0.3, bal Fe.
For shear blades, punches, jaws, chisels; Type S5; oil hardened.

SHEARCUT

Pennsylvania Steel Corp.
C 0.55, Mn 0.8, Cr 0.25, Si 2, Mo 0.4, V 0.3, bal Fe.
For shear blades, punches, jaws, chisels; Type S5; oil hardened.

SHEARTOUGH

British Steel Corp.
C 0.45, Si 0.25, Mn 0.3, Ni 2.25, Cr 1.15, V 0.2, bal Fe.
Oil hardening alloy tool steel; for shear blades, guillotine blades, shafts. *Obsolete*

SHEATHING BRONZE

English manufacture
Ni 32.5, Sn 16, Zn 5.5, Bi 1.5, bal Cu.
For sheathing, marine construction; resists water corrosion.

SHEEPBRIDGE STOKES MARK III

Sheepbridge Alloy Castings Ltd
C 0.3, Ni 1.5, Cr 0.9, bal Fe.
For centrifugal castings; tough.

SHEEPBRIDGE STOKES MARK III

Sheepbridge Engineering Ltd.
C 0.3, Ni 1.5, Cr 0.9, bal Fe.
For centrifugal castings; tough.

SHEETWELD

Westinghouse Electric Corp.
C 0.2, bal Fe.
For welding electrodes for light sheet metal. *Obsolete*

SHEF-LO-TEMP

Armco Steel Co.
C 0.2, Mn 0.7-1.35, Si 0.15-0.3, bal Fe.
Heat treated: 70,000 TS; 50,000 YS; 21 El. For structural members, pressure vessels, derricks, booms, bridges. Tough, shock resistant.

SHEF-LO-TEMP

Armco Steel Corp.
C 0.2, Mn 0.7-1.35, Si 0.15-0.3, bal Fe.
Heat treated: 70,000 TS; 50,000 YS; 21 El. For structural members, pressure vessels, derricks, booms, bridges. Tough, shock resistant.

SHEF-SUPER-LO-TEMP

Armco Steel Co.
C 0.2, Mn 1.2-1.45, Si 0.2-0.5, bal Fe.
Heat treated: 80,000 TS; 60,000 YS; 21 El. For mine and railroad cars, pressure vessels, derricks, bridges. Wear resistant, shock resistant.

SHEF-SUPER-LO-TEMP

Armco Steel Corp.
C 0.2, Mn 1.2-1.45, Si 0.2-0.5, bal Fe.
Heat treated: 80,000 TS; 60,000 YS; 21 El. For mine and railroad cars, pressure vessels, derricks, bridges. Wear resistant, shock resistant.

SHEF-TEN STEEL

Armco
C 0-0.25, Mn 1.4, Ni 0-0.4, 0.2 Cu min, bal Fe.
Rolled: 70,000 TS; 40,000 YS; 20 El. For mine and railroad cars, shovels, structures; good formability. *Obsolete*

SHEFFALLOY

New England Collapsible Tube Co.
Pb, Sn.
For collapsible tubes.

SHEFFIELD

Jessop-Saville Ltd.
C 0.82-0.85, Mn 0.3, Cr 0.5-0.7, Mo 0.18-0.22, bal Fe.
For coining and cutlery dies; water hardening.

SHEFFIELD HARD ALLOY

English manufacture
Cu 46, Sn 31, Zn 20.
For ornamental parts; corrosion resistant.

SHEFFIELD HI-STRENGTH-B

Armco Steel Co.
C 0-0.22, Mn 0-1.25, 0.20% min Cu, 0.02% min V, bal Fe.
Rolled: 67,000-70,000 TS; 46,000-50,000 YS; 18-24 El. For structures, bus bodies, mine cars, derricks, booms. High-strength, low-alloy steel. Good weldability. *Obsolete*

SHEFFIELD HI-STRENGTH-B

Armco Steel Corp.
C 0-0.22, Mn 0-1.25, 0.20% min Cu, 0.02% min V, bal Fe.
Rolled: 67,000-70,000 TS; 46,000-50,000 YS; 18-24 El. For structures, bus bodies, mine cars, derricks, booms. High-strength, low alloy steel. Good weldability. *Obsolete*

SHEFFIELD HIGH STRENGTH-A

Armco Steel Co.
C 0.09, Cu 0.4, Ni 0.7, Cr 0.6, Mo 0.12, bal Fe.
Rolled: 66,000 TS; 46,000 YS; 21 El. For structural members, auto and railroad cars; good formability. *Obsolete*

SHEFFIELD NICKEL SILVER

English manufacture
Cu 55-63, Zn 17-37, Ni 11-19, Pb 0.3.
Used as base for plated tableware; nickel silver.

SHEFFIELD NO. 10

North American Steel Corp.
C 0.3, Ni, Mo, Cr, bal Fe.
Heat treated: 154,000 TS; 139,000 YS; 16 El; 55 RA; 300 Brin. For gears, cams, arbors, shafts, axles, mandrels; preheat treated, tough.

SHEFFIELD NO. 11

North American Steel Corp.
C 0.4, bal Fe.
Annealed: 90,000 TS; 50,000 YS; 175 Brin. For mandrels, springs, tongs, rivets, shackles; forging quality.

SHEFFIELD NO. 12

North American Steel Corp.
C, alloy, bal Fe.
For gears, picks, lathe centers, shovel teeth; shock and fatigue resistant.

SHEFFIELD NO. 20

North American Steel Corp.
C, alloy, bal Fe.
Heat treated: 154,000 TS; 139,000 YS; 16 El; 55 RA; 300 Brin. For shafting, pump rods; oil hardened.

SHEFFIELD P.B.

North American Steel Corp.
C, alloy, bal Fe.
For press brake dies; oil hardened.

SHEFFIELD PLATE

North American Steel Corp.
C, alloy, bal Fe.
Heat treated: 150,000 TS; 140,000 YS; 14 El; 55 RA; 365 Brin. For chute liners, conveyors, baffle plates; work hardened, fatigue resistant.

SHEFFIELD PRESS BRAKE

North American Steel Corp.
C 0.53, Mn 0.95, Si 0.31, Cr 1, Ni 0.5, V 0.24, Mo 0.25, bal Fe.
Oil hardened steel for miscellaneous tools.

SHEFFIELD SSS-100

Armco Steel Co.
C 0.12-0.2, Mn 0.4-0.7, B 0.0015-0.005, Cr 1.4-2, Cu 0.2-0.4, Mo 0.4-0.6, Ti 0.04-0.1, bal Fe.
Plate: 115,000-135,000 TS; 100,000 YS; 18 minimum El; 50 minimum RA. For pressure vessels, bridges, mine cars, power shovels, cranes, trucks, trailers. Shock and wear resistant constructional steel. Quenched and tempered. *Obsolete*

SHEFFIELD SSS-100A

Armco Steel Co.
C 0.13-0.2, Mn 0.4-0.7, Si 0.2-0.35, B 0.0015-0.005, Cr 0.85-1.2, Cu 0.2-0.4, Mo 0.15-0.25, Ti 0.04-0.1, bal Fe.
Plate: 115,00-135,000 TS; 100,000 YS; 18 El minimum; 50 RA minimum. For pressure vessels, bridges, mine cars, power shovels, cranes, trucks, trailers. Shock and wear resistant constructional steel. *Obsolete*

SHEFFIELD SSS-100A-AR

Armco Steel Co.
C 0.13-0.2, Mn 0.6, Si 0.25, B 0.005, Cr 0.85-1.2, Cu 0.2-0.4, Mo 0.15-0.25, Ti 0.04-0.1, bal Fe.
Heat treated: 162,000 TS; 148,000 YS; 15 El; 43 RA. For bridges, booms, dipper sticks, penstocks, chutes, wear plates, truck bodies, floor plates. Quenched and tempered at mill. Tough. Wear and abrasion resistant. *Obsolete*

SHEFFIELD SSS-100AR
Armco Steel Co.
C 0.2, Mn 0.4-0.7, B 0.005, Cr 1.4-2, Cu 0.2-0.4, Mo 0.4-0.6, Ti 0.04-0.1, bal Fe.
Heat treated: 161,000 TS; 146,000 YS; 16 El; 45 RA. For bridges, booms, dipper sticks, penstocks, chutes, wear plates, truck bodies, floor plates. Quenched and tempered at mill. Tough. Wear and abrasion resistant. *Obsolete*

SHEFFIELD SSS-100B
Armco Steel Co.
C 0.13-0.2, Mn 0.4-0.7, Si 0.2-0.35, Cr 1.15-1.65, Mo 0.25-0.4, Ti 0.04-0.1, Cu 0.2-0.4, B 0.0015-0.005, bal Fe.
Plate: 105,000-135,000 TS; 90,000-100,000 YS; 16-18 El. For pressure vessels, bridges, mine cars, power shovels, cranes, trailers. Shock and wear resistant constructional steel. *Obsolete*

SHEFFIELD SX25
Sheffield Smelting Co. Ltd.
Ag, Cu, Pb.
Cast: 5,600 TS; 45 El. For soft solder; M.P. 302-305 C; general purpose solder. *Obsolete*

SHEFFIELD SX4
Sheffield Smelting Co. Ltd.
Ag-Sn.
Cast: 7,800 TS; 48 El. For soft silver solder for sanitary and domestic purposes; M P 221-224 C; corrosion resistant. *Obsolete*

SHEFFIELD SX5
Sheffield Smelting Co. Ltd.
Ag, Cd, Zn.
Cast: 29,200 TS; 8 El. For silver solder where strength at high temperature is important; M.P. 270-285 C; corrosion resistant. *Obsolete*

SHEFFIELD-AR
Armco Steel Co.
C 0.43, Mn 1.5-2, Si 0.35, bal Fe.
Hot rolled: 225 Brin. For structural members, gears, shafts, countershafts. Abrasion resistant, water harden. *Obsolete*

SHELLDIE
A. Finkl & Sons Co.
Tool material. C 0.33-0.39, Mn 0.5-0.7, Si 0.75-1.1, Cr 4.75-5.25, Mo 1.7-2, V 0.26-0.28, bal Fe.
Heat treated (as required): 37-46 Rock C. For forging die inserts, punches, mandrels, heading dies. Resists heat checking. AISI H-11.

SHELLEX
A. Finkl & Sons Co.
C 0.4, Si 1, Mn 0.5, Cr 3.3, Mo 2.4, V 0.4, bal Fe.
For forging dies, inserts, punches, extrusion tools; oil hardened, tough, resists heat check. *Obsolete*

SHEMTSCHUSHNY
English manufacture
Fe 1.3-4.45, Mn 11.1-13.8, Cu 84.2-86.8, Si 0.5-0.6.
For electrical resistances.

SHIBU-ICHI
Japanese manufacture
Cu 67.3, Ag 32.1, Fe 0.5, trace Au.
For ornaments; corrosion resistant.

SHIBU-ICHI
Japanese manufacture
Cu 51.1, Ag 48.9, Au 0.12.
For ornaments, jewelry; corrosion resistant.

SHIELD-ARC 100
Lincoln Electric Co.
C 0.12, bal Fe.
Welded: 95,000-105,000 TS; 14-20 El. For welding rod; E-10010. *Obsolete*

SHIELD-ARC 65T
Lincoln Electric Co.
C, bal Fe.
Steel arc welding electrode. As welded: 75,000-87,000 psi TS. AWS E7010-G.

SHIELD-ARC 85
Lincoln Electric Co.
C 0.12, Mn 0.6, P 0.03, S 0.04, Si 0.4, Mo 0.5, bal Fe.
Steel arc welding electrode. AWS Class E7010-Al.

SHIELD-ARC 85P
Lincoln Electric Co.
C, bal Fe.
Steel arc welding electrode. AWS Class E7010-G.

SHIELD-ARC HYP
Lincoln Electric Co.
C, bal Fe.
Steel arc welding electrode.

SHIELD-ARC X70
Lincoln Electric Co.
C, bal Fe.
Steel arc welding electrode. As welded: 92,000-93,000 psi TS. AWS E8010-G.

SHIELD-O-MATIC
Hollup Corp.
C 0.15, bal Fe.
For electrodes for welding; coated-arc.

SHIGA WHITE GOLD
Japanese manufacture
Au 60-90, Ni 5-20, Cr 5-20.
For ornaments, jewelry; corrosion resistant.

SHIP BRAND BLUE LABEL
Webb Wire Works
C 0.1, Cr 13.5, bal Fe.
Heat treated: 200,000 TS. Rolled: 100,000 TS. For springs; corrosion resistant. *Obsolete*

SHIP BRAND ORANGE LABEL
Webb Wire Works
C 0.07, Cr 18-21, Ni 8-11, bal Fe.
Rolled: 120,000 TS. Heat treated: 190,000-290,000 TS. For springs. Corrosion resistant.

SHIP BRAND RED LABEL
Webb Wire Works
C 0.35, Cr 13.5, bal Fe.
Rolled: 100,000 TS. For springs. Corrosion resistant.

SHIP NAIL ALLOY
English manufacture
Sn 50, Pb 33, Sb 17.
For bearings; anti-friction.

SHIP NAIL BRASS
English manufacture
Cu 64, Zn 25, Pb 8.5, Sn 2.5.
For ship nails, hardware; free-cutting.

SHOCK PROOF
Wheeling-Pittsburgh Steel Corp.
C 3, Si 1.5, bal Fe.
Cast: 53,000 TS. For castings, gears; malleable iron. *Obsolete*

SHOCK RESISTING
Edgar T. Ward's Sons Co.
C 0.6, Mn 0.9, bal Fe.
For punches, shears; tough.

SHOCK RESISTING
Jessop-Saville Ltd.
C, alloy, bal Fe.
For tools, dies; shock resistant.

SHOCK STEEL
Jessop Steel Co.
C 0.54, Mn 0.95, Si 2, Mo 0.5, bal Fe.
For punches, rivet sets, headers; shock and impact resistant. *Obsolete*

SHOCK-DIE
Columbia Tool Steel Co.
Tool material. C 0.5, Mn 0.75, Mo 1.4, Cr 3.2, bal Fe.
Air hardened; shock resistant tool steel. For punches, blanking dies, chisels, gripper dies. AISI S7.

SHOCK-RITE
St. Lawrence Steel Co.
C 0.65, Mn 0.85, Si 2, Cr 0.25, Mo 0.25, V 0.2, bal Fe.
Oil or water hardened, shock resistant tool steel; for shear blades, pneumatic tools. AISI S5.

SHOE NAIL BRASS
American manufacture
Cu 63, Zn 37.
For hardware, shoe nails; yellow brass.

SHOE TIP METAL
English manufacture
Cu 88, Zn 12.
For shoe tips; red brass.

SHOW CASE METAL
English manufacture
Cu 58-59.5, Zn 22.5-24, Ni 8-18.
For show cases, architectural parts; corrosion resistant.

SHUNT STEEL
Japan Special Steel Co.
Fe 70, Ni 30.
For compensating shunts for electrical equipment; temperature-sensitive, magnetic.

SHURBOND 0
Anchor Alloys Inc.
Copper. Cu 92.8, P 7.2.
Melting range: 1300-1450°F. Economical, highly fluid, self-brazing on copper.

SHURBOND 02
Anchor Alloys Inc.
Copper. Ag 2, Cu 91, P 7.
Melting range: 1185-1270°F. For brazing copper, brass and steel.

SHURBOND 06
Anchor Alloys Inc.
Copper. Ag 6, Cu 86.5, P 7.5.
Melting range: 1185-1350°F. For brazing ferrous and non-ferrous alloys. AWS BCuP-4.

SHURBOND 120
Anchor Alloys Inc.
Copper. Ag 20, Cu 45, Zn 35.
For silver solder; MP 1430-1500°F.

SHURBOND 13
Anchor Alloys Inc.
Copper. Ag 30, Cu 38, Zn 32.
For silver solder; MP 1370-1410°F.

SHURBOND 14
Anchor Alloys Inc.
Ag 40, Cu 36, Zn 24.
For silver solder; MP 1330-1445°F.

SHURBOND 145
Anchor Alloys Inc.
Ag 45, Cu 30, Zn 25.
For silver solder; MP 1250-1370°F. Type BAg5.

SHURBOND 15
Anchor Alloys Inc.
Copper. Cu 80, P 5, Ag 15.
Melting range: 1185-1460°F. For brazing copper base alloys with wide gaps. AWS-BCuP-5.

SHURBOND 2
Anchor Alloys Inc.
Copper. Cu 91, P 7, Ag 2.
Melting range: 1190-1425°F. For brazing copper and copper alloys.

SHURBOND 20
Anchor Alloys Inc.
Copper. Ag 20, Cu 45, Zn 30, Cd 5.
For silver solder; MP 1140-1500°F.

SHURBOND 240
Anchor Alloys Inc.
Ag 40, Cu 30, Zn 28, Ni 2.
For silver solder; MP 1240-1435°F; Type BAg4.

SHURBOND 250
Anchor Alloys Inc.
Ag 50, Cu 34, Zn 16.
For silver solder; MP 1270-1425°F; Type BAg6.

SHURBOND 254
Anchor Alloys Inc.
Ag 40, Cu 30, Zn 25, Ni 5.
For silver solder; MP 1240-1560°F.

SHURBOND 28
Anchor Alloys Inc.
Ag 50, Cu 28, Zn 22.
For silver solder; MP 1250-1340°F.

SHURBOND 31
Anchor Alloys Inc.
Copper. Cu 34, Ag 31.5, Cd 19, Zn 15.5.
Melting range: 1185-1390°F. For torch and induction brazing ferrous and non-ferrous alloys.

SHURBOND 33
Anchor Alloys Inc.
Ag 60, Cu 25, Zn 15.
For silver solder; MP 1260-1325°F.

SHURBOND 35
Anchor Alloys Inc.
Ag 35, Cu 26, Zn 21, Cd 18.
For silver solder; MP 1125-1295°F. Type BAg2.

SHURBOND 350
Anchor Alloys Inc.
Ag 50, Cu 15.5, Zn 15.5, Ni 3.
For silver solder; MP 1170-1270°F.

SHURBOND 4
Anchor Alloys Inc.
Copper. Ag 10, Cu 52, Zn 38.
For silver solder; MP 1450-1565°F.

SHURBOND 45
Anchor Alloys Inc.
Ag 45, Cu 15, Zn 16, Cd 24.
For silver solder; MP 1125-1145°F; Type BAg1.

SHURBOND 5 BRAZE
Anchor Alloys Inc.
Copper. Cu 89, P 6, Ag 5.
Melting range: 1185-1485°F. For brazing copper base alloys where poor fit-ups exist.

SHURBOND 5 SOLDER
Anchor Alloys Inc.
Ag 5, Zn 16.5, Cd 78.5.
Melting range: 500-725°F. Solder for copper, brass and steel.

SHURBOND 50
Anchor Alloys Inc.
Ag 50, Cu 15.5, Zn 16.5, Cd 18.
For silver solder; MP 1160-1175°F; Type BAg1a.

SHURBOND 54
Anchor Alloys Inc.
Ag 54, Cu 40, Zn 5, Ni 1.
For silver solder; MP 1325-1575°F.

SHURBOND 56
Anchor Alloys Inc.
Ag 56, Cu 22, Zn 17, Sn 5.
For silver solder; MP 1145-1205°F.

SHURBOND 5X
Anchor Alloys Inc.
Ag 5, Cd 95.
Melting range: 620-725°F. Solder for copper, brass and steel.

SHURBOND 6
Anchor Alloys Inc.
Copper. Cu 88, P 6, Ag 6.
Melting range: 1185-1480°F. For brazing ferrous and non-ferrous alloys; ductile joints.

SHURBOND 60
Anchor Alloys Inc.
Ag 60, Cu 30, Sn 10.
For silver solder; MP 1115-1325°F.

SHURBOND 65
Anchor Alloys Inc.
Ag 65, Cu 20, Zn 15.
For silver solder; MP 1235-1310°F.

SHURBOND 6F
Anchor Alloys Inc.
Copper. Cu 86.75, P 7.25, Ag 6.
Melting range: 1190-1300°F. Very fluid, for brazing copper base alloys with tight fits.

SHURBOND 70
Anchor Alloys Inc.
Ag 70, Cu 20, Zn 10.
For silver solder; MP 1275-1360°F.

SHURBOND 72
Anchor Alloys Inc.
Ag 72, Cu 28.
For silver solder; MP 1435-1435°F.

SHURBOND 75
Anchor Alloys Inc.
Ag 75, Cu 22, Zn 3.
For silver solder; MP 1365-1450°F.

SHURBOND 79
Anchor Alloys Inc.
Copper. Ag 25, Cu 52.5, Zn 22.5.
For silver solder; MP 1500-1575°F.

SHURBOND 85
Anchor Alloys Inc.
Ag 85, Mn 15.
Melting range: 1760-1778°F. AMS 4766; ASTM B-260-52T BAg-Mn.

SHX SERIES
Sumitomo Metal America Inc.
HSLA steel. C 0-0.18, Si 0-0.55, Mn 0-1.5, Cr, Mo, Cb, V, Ti added if necessary, bal Fe.
1.57-12.7 mm thick hot rolled sheet or coil. Five strength grades. SHX 50: 50 ksi YP min. SHX 55: 55 ksi YP min. SHX 60: 60 ksi YP min. SHX 70: 70 ksi YP min. SHX 80: 80 ksi YP min. High strength low alloy steel for cold worked parts, automotive and appliances.

SHXF SERIES
Sumitomo Metal America Inc.
HSLA steel. C 0-0.12, Si 0-0.55, Mn 0-1.65, Mo, Cb, V, and Ti added if necessary, bal Fe.
1.57-4.76 mm thick hot rolled sheet or coil. Six strength grades. SHXF 50: 50 ksi YP min. SHXF 60: 60 ksi YP min. SHXF 70: 70 ksi YP min. SHXF 80: 80 ksi YP min. SHXF 100: 100 ksi YP min. High strength low alloy steel for cold worked parts, automotive and appliances. Good formability and weldability.

SIBLEY ALLOY
English manufacture
Al 67, Zn 33.
For strong light alloy parts; non-hardenable.

SIBLEY CASTING ALLOY
English manufacture
Al 80, Zn 20.
For strong light alloy castings; non-hardenable.

SICRIMPHY 1
Creusot-Loire
C, alloy, bal Fe.
For valves. *Obsolete*

SICRIMPHY 2
Creusot-Loire
C, alloy, bal Fe.
For valves. *Obsolete*

SICRIMPHY 3
Creusot-Loire
Si 2.5, Cr 10, Mo 0.7, C, bal Fe.
For corrosion resistant parts; corrosion resistant. *Obsolete*

SICROMAL 10
Rheinische Rohrenwerke Atkiengesellschaf
Cr 17.5-18.5, Si 0.8-1.1, Al 0.8-1.1, C 0-0.12, Mn 0-1, bal Fe.
Air annealed: 71,000-92,500 TS; 43,000 YS; 12 El; 140-180 Brin. For superheaters, annealing ovens, pots, bolts; resists heat to 1000°C.

SICROMAL 12
Rheinische Rohrenwerke Atkiengesellschaf
Cr 23-25, Al 1.3-1.6, Mn 0-1, C 0-0.12, Si 1.3-1.6, bal Fe.
Air annealed: 71,000-92,500 TS; 43,000 YS; 10 El; 140-180 Brin. For superheaters, pots, bolts, annealing ovens; resists heat to 1200°C.

SICROMAL 12H
Rheinische Rohrenwerke Atkiengesellschaf
C 0.15, Cr 23, Si 1, Al 2.5, bal Fe.
Annealed: 92,000 TS; 57,000 YS; 10 El. For resistances, heating elements, heat resistant to 1200°C.

SICROMAL 130
Rheinische Rohrenwerke Atkiengesellschaf
Cr, Si, Al, bal Fe.
Annealed: 107,000 TS; 57,000 YS; 10 El. For resistance wire; resists heat to 1250°C.

SICROMAL 17/17 KMS
Rheinische Rohrenwerke Atkiengesellschaf
C 0.1, Si 0.3-0.5, Mn 0-2, Cr 17-18, Cu 1.8-2.2, Ni 17-18, Mo 1.8-2.2, Ti = 7 x C, bal Fe.
Water quenched: 78,200-106,700 TS; 35,500 YS; 40 El; 155-210 Brin. For sulfuric acid equipment; corrosion and acid resistant.

SICROMAL 17/17 KNS
Rheinische Rohrenwerke Atkiengesellschaf
C 0-0.7, Cr 17.5, Mo 2, Ni 17.5, Cu 2, Ti = 7 x C, bal Fe.
For valves and pump parts, furnace equipment; corrosion and heat resistant.

SICROMAL 18-8 MS
Rheinische Rohrenwerke Atkiengesellschaf
C 0-0.12, Cr 17.5-18.5, Ni 10-11, Mo 1.8-2.2, Mn 0-2, Ti = 4 x C, bal Fe.
Water quenched: 78,000-107,000 TS; 38,000 YS; 40 El; 155-210 Brin. For chemical and textile industries; austenitic, stainless.

SICROMAL 18-8 S
Rheinische Rohrenwerke Atkiengesellschaf
C 0-0.12, Cr 17.5-18.5, Ni 9-10, Mn 0-2, Ti = 4 x C, bal Fe.
Water quenched: 78,000-107,000 TS; 38,000 YS; 40 El; 155-210 Brin. For chemical industries; austenitic, stainless.

SICROMAL 18/12 MS
Rheinische Rohrenwerke Atkiengesellschaf
C 0-0.1, Mn 0-2, Cr 16, Ni 13, Mo 2-2.5, 1.0 Ta/Nb, bal Fe.
Water quenched: 78,200-106,700 TS; 38,400 YS; 40 El; 150-210 Brin. For stainless parts, oil refinery equipment; stainless, austenitic.

SICROMAL 18/12 S
Rheinische Rohrenwerke Atkiengesellschaf
C 0-0.1, Mn 0-2, Cr 18, Ni 10, 1.0 Ta/Nb, bal Fe.
Water quenched: 78,200-106,700 TS; 38,400 YS; 40 El; 150-210 Brin. For stainless parts, oil refinery equipment; stainless, austenitic.

SICROMAL 18/8
Rheinische Rohrenwerke Atkiengesellschaf
C 0-0.15, Cr 18, Ni 8.5, bal Fe.
Annealed: 80,000 TS; 35,000 YS; 55 El; 75 RA. 150 Brin. Cold drawn: 180,000 TS; 150,000 YS; 10 El; 250 Brin. For chemical plant equipment, tanks, mixers, filters; Type 302; stainless, austenitic.

SICROMAL 19 MS
Rheinische Rohrenwerke Atkiengesellschaf
C 0-0.1, Cr 16.5-17.5, Mo 1.6-1.9, Mn 0-1, Ti = 7 x C, bal Fe.
Air annealed: 71,000-92,500 TS; 43,000 YS; 20 El; 140-180 Brin. For chemical and textile industries; stainless, resists HNO_3.

SICROMAL 20-10
Rheinische Rohrenwerke Atkiengesellschaf
C 0.1-0.2, Cr 19-20, Ni 9-10, Si 1.8-2.3, Mn 0-2, bal Fe.
Water quenched: 85,000-107,000 TS; 43,000 YS; 40 El; 165-210 Brin. For furnace parts, crucibles, autoclaves; austenitic, resists heat to 1050°C.

SICROMAL 23-20
Rheinische Rohrenwerke Atkiengesellschaf
C 0.1-0.2, Cr 23-25, Ni 20-22, Si 1.8-2.3, Mn 0-2, bal Fe.
Water quenched: 85,000-107,000 TS; 43,000 YS; 40 El; 165-210 Brin. For recuperators, crucibles, autoclaves; austenitic, resists heat to 1200°C.

SICROMAL 8
Rheinische Rohrenwerke Atkiengesellschaf
Cr 6.3-6.8, Al 0.6-0.9, C 0-0.12, Si 0.6-0.9, bal Fe.
Air annealed: 64,000-85,500 TS; 35,000 YS; 20 El; 125-165 Brin. For superheaters, recuperators, annealing ovens, pots, bolts; resists heat to 800°C.

SICROMAL 85
Rheinische Rohrenwerke Atkiengesellschaf
C 0.15, Cr 6.5, Si 2, bal Fe.
Rolled: 78,000-107,000 TS; 56,900 YS; 18 El. For steam boilers, superheater tubes, recuperators; heat resistant to 850°C.

SICROMAL 9
Rheinische Rohrenwerke Atkiengesellschaf
Cr 12.5-13.5, Si 1-1.3, Al 0.8-1.1, C 0-0.12, Mn 0-1, bal Fe.
Air annealed: 71,900-92,500 TS; 43,000 YS; 15 El; 140-180 Brin. For annealing ovens, pyrometer tubes, superheaters; resists heat to 900°C.

SICROMAL C 7A
Rheinische Rohrenwerke Atkiengesellschaf
Cr, Si, Al, bal Fe.
Annealed: 78,000 TS; 57,000 YS; 10 El. For resistance wire; resists heat to 1100°C.

SICROMAL CS 65 S
Rheinische Rohrenwerke Atkiengesellschaf
C 0-0.15, Si 1.3-1.7, Mn 0-1, Cr 2-2.5, Mo 0.4-0.5, bal Fe.
Normalized: 64,000-85,500 TS; 35,500 YS; 18 El; 125 Brin. For oil refining equipment, steam boilers.

SICROMAL CV 18 W
Rheinische Rohrenwerke Atkiengesellschaf
C 0.1, Si 1.2-1.4, Mn 0-1, Cr 1.5-2, Mo 0.2-0.3, V 0.2-0.3, bal Fe.
Normalized: 64,000-85,500 TS; 42,700 YS; 20 El; 125-165 Brin. For steam boilers.

SICROMAL D
Rheinische Rohrenwerke Atkiengesellschaf
C 0-0.15, Si 0.3-0.5, Mn 0-1, Cr 4.5-5.5, Mo 0.45-0.55, bal Fe.
Annealed: 64,000-85,300 TS; 42,600 YS; 20 El; 125-165 Brin. For ammonia synthesis equipment; corrosion resistant.

SICROMAL D 16 S
Rheinische Rohrenwerke Atkiengesellschaf
C 0-0.1, Cr 17-18, Mn 0-1, Ti = 7 x C, bal Fe.
Air annealed: 64,000-85,500 TS; 43,000 YS; 20 El; 125-165 Brin. For chemical industries; stainless, resists HNO_3.

SICROMAL D 45 V
Rheinische Rohrenwerke Atkiengesellschaf
C 0-0.12, Si 0.4-0.8, Mn 0-1, Cr 1.4-1.8, V 0.25-0.35, bal Fe.
Normalized: 64,000-85,300 TS; 42,670 YS; 18 El; 125-165 Brin. For ammonia synthesis apparatus.

SICROMAL D12
Rheinische Rohrenwerke Atkiengesellschaf
C 0-0.12, Cr 12-13, Si 0.4, bal Fe.
Rolled: 71,000-92,500 TS; 42,700 YS; 20 El. For chemical plant equipment; corrosion resistant.

SICROMAL D16
Rheinische Rohrenwerke Atkiengesellschaf
C 0.12, Cr 18, Si 0.4, bal Fe.
Rolled: 64,000-85,000 TS; 42,700 YS; 20 El. For nitric acid and soap making equipment; corrosion resistant.

SICROMAL D7
Rheinische Rohrenwerke Atkiengesellschaf
C 0.2, Cr 7, Mo, bal Fe.
Rolled: 64,000 TS; 35,600 YS; 20 El. For petroleum cracking equipment; heat resistant to 700°C.

SICROMAL D9
Rheinische Rohrenwerke Atkiengesellschaf
C 0.15, Cr 9, Mo 0.5, bal Fe.
Rolled: 64,000 TS; 35,600 YS; 20 El. For hydrogenating and ammonia synthesis equipment; heat resistant to 750°C.

SICROMAL NO. 10
Vereinigte Stahlwerke
C 0-0.12, Si 1, Al 1, Cr 18, S 0-0.005, bal Fe.
At 20 C: 85,200 TS; 50,400 YS; 10 El; 63 RA. At 800 C: 14,500 TS; 7,200 YS; 53 El; 98 RA. For pyrometer tubes, anchor bolts, furnace parts; resists oxidizing, reducing and sulfur gases. *Obsolete*

SICROMAL NO. 12
Vereinigte Stahlwerke
C 0-0.12, Si 1.5, Al 1.5, Cr 24, bal Fe.
Annealed: 71,000-92,500 TS; 43,000 YS; 12 El; 185 Brin. For pyrometer tubes, tuyere pipes, annealing boxes; resists oxidizing, reducing and sulfur gases. *Obsolete*

SICROMAL NO. 8
Vereinigte Stahlwerke
C 0-0.12, Si 0.8, Cr 4-5, Al 0.5, bal Fe.
At 20 C: 67,000 TS; 43,000 YS; 20 El; 75 RA; 140 Brin. At 800 C: 10,000 TS; 6,000 YS; 38 El; 93 RA. For superheater and recuperator tubes, annealing furnaces; resists oxidation and sulfur fumes. *Obsolete*

SICROMAL NO. 9
Vereinigte Stahlwerke
C 0-0.12, Si 1.2, Al 1, Cr 13, S 0-0.005, bal Fe.
At 20 C: 78,400 TS; 40,000 YS; 15 El; 68 RA; 160 Brin. At 800 C: 11,400 TS; 7,200 YS; 50 El; 90 RA. For superheater and recuperator tubes, pipes, fittings; resists oxidation and sulfur gases. *Obsolete*

SICROMAL TS 57
Rheinische Rohrenwerke Atkiengesellschaf
C 0-0.1, Si 0.7-1.1, Mn 0-1, Ti = 5 x C, bal Fe.
Normalized: 48,300-68,300 TS; 28,500 YS; 20 El; 95-135 Brin. For construction steel.

SICROMAL TSL
Rheinische Rohrenwerke Atkiengesellschaf
C 0.1, Ti 0.7, bal Fe.
Rolled: 48,000-60,000 TS; 21,400 YS; 22 El. For steam boilers and equipment in soap industries; resists caustic embrittlement.

SICROMO 1
Timken Co.
C 0-0.15, Si 1-1.4, Cr 0.75-1.25, Mo 0.45-0.65, bal Fe.
Annealed: 60,000 TS; 25,000 YS; 30 El; 50 RA; 163 Brin. For high temperature tubing; resists heat to 1150 F. *Obsolete*

SICROMO 2
Timken Co.
C 0-0.15, Mo 0.5, Si 1-1.5, Cr 1.75-2.25, bal Fe.
Annealed: 60,000 TS; 25,000 YS; 30 El; 163 Brin. For refinery tubing; resists heat to 1200 F. *Obsolete*

SICROMO 2-1/2
Timken Co.
C 0-0.15, Mn 0-0.5, Si 0.5-1, Cr 2.25-2.75, Mo 0.4-0.6, bal Fe.
At 85 F: 74,400 TS; 40,650 YS; 38.5 El; 72.6 RA. At 1400 F: 12,900 TS; 7,500 YS; 75.0 El; 96.0 RA. For refinery tubes, still tubes, condenser tubes; resists oxidation and corrosion to oils. *Obsolete*

SICROMO 3
Timken Co.
C 0-0.15, Mo 0.5, Si 1-1.5, Cr 2.75-3.25, bal Fe.
Annealed: 60,000 TS; 25,000 YS; 30 El; 50 RA; 160 Brin. For high temperature tubing, cracking stills; resists heat to 1200 F. *Obsolete*

SICROMO 5
Timken Co.
C 0-0.15, Mo 0.5, Si 1-2, Cr 4-6, bal Fe.
Annealed: 60,000 TS; 25,000 YS; 30 El; 163 Brin. For tubing; resists heat to 1300 F. *Obsolete*

SICROMO 5M
Timken Co.
C 0-0.15, Mn 0-0.5, Si 0.5-1, Cr 4-6, Mo 0.9-1.1, bal Fe.
For gasoline and petroleum refineries; corrosion resistant. *Obsolete*

SICROMO 5MS
Timken Co.
C 0-0.15, Mn 0-0.5, Si 1-1.5, Cr 4-6, Mo 0.9-1.1, bal Fe.
Rolled: 60,000 TS; 25,000 YS; 30 El; 163 Brin. For high temperature tubing; high surface stability. *Obsolete*

SICROMO 5S
Timken Co.
C 0-0.15, Si 1-2, Cr 4-6, Mo 0.44-0.65, bal Fe.
Annealed: 60,000 TS; 25,000 YS; 30 El; 50 RA; 163 Brin. For high temperature tubing in oil refineries; heat resistant to 1500°F.

SICROMO 7
Timken Co.
C 0-0.15, Si 0.5-1, Cr 6-8, Mo 0.44-0.65, bal Fe.
Annealed: 60,000 TS; 25,000 YS; 30 El; 50 RA; 179 Brin. For high temperature tubing in oil refineries; corrosion resistant.

SICROMO 7M
Timken Co.
C 0-0.15, Mn 0-0.5, Si 0.5-1, Cr 6-8, Mo 0.9-1.1, bal Fe.
For heat and corrosion resistant parts; heat and corrosion resistant. *Obsolete*

SICROMO 9
Timken Co.
C 0-0.15, Mn 0-0.5, Si 0.5-1, Cr 8-10, Mo 0.4-0.6, bal Fe.
For heat and corrosion resistant parts; heat and corrosion resistant. *Obsolete*

SICROMO 9CR 1.00 MOM
Timken Co.
C 0-0.15, Si 0.25-1, Cr 8-10, Mo 0.9-1.1, V 0.25-0.35, Cb 0.05-0.15, 0.007 B, 0.01 N, 0.01 Zr added, bal Fe.

SICROMO 9CR 1MO
Timken Co.
C 0-0.15, Si 0.25-1, Cr 8-10, Mo 0.9-1.1, bal Fe.

SICROMO 9CR 1MO CA2
Timken Co.
C 0-0.15, Si 0.25-1, Cr 8-10, Mo 0.9-1.1, S 0-0.015, bal Fe.

SICROMO 9M
Timken Co.
C 0-0.15, Si 0.5-1, Cr 7-9, Mo 0.9-1.1, bal Fe.
Annealed: 60,000 TS; 25,000 YS; 30 El; 50 RA; 179 Brin. For high temperature tubing in oil refineries; corrosion resistant. *Obsolete*

SIDERAPHITE
English manufacture
Fe 62, Ni 23, Cu 5, Al 5, W 4.
For acid resisting vessels and apparatus; stainless and corrosion resistant.

SIEMENS-HALSKE (1)
Siemens & Halske AG.
Nickel. Ni 90-95, Ta 5-10.
For heat and corrosion resistant parts; stainless and corrosion resisting.

SIEMENS-HALSKE (2)
Siemens & Halske AG.
Nickel. Ni 70, Ta 30.
For spark plug electrodes; stainless and corrosion resisting.

SIEMENS-HALSKE (3)
Siemens & Halske AG.
Zn 48, Sb 5, Cd 47.
For solder; stainless and corrosion resisting.

SIF MAGNESIUM NO. 23
Sifbronze
Magnesium. Al 7.5-9, Zn 0.7, Mn 0.0, bal Mg.
For oxy-acetylene or TIG welding magnesium alloy extrusions, forgings and castings. BS 2901 MAG1.

SIF SILVER SOLDER NO. 38
Sifbronze
Silver. Ag 34, Cu 25, Zn 20, Cd 21.
Silver solder for brazing ferrous and non-ferrous alloys except aluminum alloys. 440 N/mm^2 TS; 125 Brin.

SIF SILVER SOLDER NO. 39
Sifbronze
Silver. Ag 38, Cu 31, Zn 29, Sn 2.
Cadium free silver solder for brazing ferrous and non-ferrous alloys except aluminum alloys. 460 N/mm^2 TS; 120 Brin.

SIFALBRONZE NO. 32
Sifbronze
Copper. Al 10, Fe 1, bal Cu.
Aluminum bronze rod suitable for welding similar materials; corrosion resistant. 550 N/mm^2 TS; 150 Brin. BS 2901 C13.

SIFALUMIN NO. 14
Sifbronze
Aluminum. Al 99.5.
Welded: 90 N/mm^2 TS; 15 Brin. Aluminum rod.

SIFALUMIN NO. 15
Sifbronze
Aluminum. Si 5, bal Al.
Welded: 120 N/mm^2 TS; 40 Brin. Rod for welding Duralumin.

SIFALUMIN NO. 16
Sifbronze
Aluminum. Si 10-12, bal Al.
Welded: 150 N/mm^2 TS; 50 Brin. Rod for welding and brazing.

SIFALUMIN NO. 27
Sifbronze
Aluminum. Mg 5, Mn 1, bal Al.
Welded: 250 N/mm^2 TS; 60 Brin. Aluminum wire for welding magnesium bearing aluminum alloys.

SIFALUMIN NO. 29
Suffolk Iron Foundry Ltd.
Cu 5, bal Al.
Welded: 22,900 TS. For welding rod for Al. *Obsolete*

SIFALUMIN NO. 36
Sifbronze
Aluminum. Si 12, bal Al.
Aluminum rod for low temperature brazing. 150 N/mm^2 TS; 50 Brin.

SIFALUMIN NO. 37
Sifbronze
Aluminum. Mg 5.3, Mn 0.8, Cr 0.1, Ti 0.1, bal Al.
Aluminum alloy rod for welding 5083 (N8) and for military applications. 300 N/mm^2 TS; 70 Brin. BS 2901 5556.

SIFBRASS NO. 6
Suffolk Iron Foundry Ltd.
Zn 45-50, bal Cu.
For brazing alloy for cast iron and copper; M.P. 850 C. *Obsolete*

SIFBRONZE
Suffolk Iron Foundry Ltd.
Cu 60-70, Zn 20-30, traces of Sn, Fe, Mn.
64,000 TS; 22 El; 156 Brin. For welding rods; for welding cast iron or malleable steel. *Obsolete*

SIFBRONZE NO. 1
Sifbronze
Copper. Cu 60, Si 0.3, bal Zn.
430 N/mm^2 TS; 120 Brin. Rod for brazing and bronze welding.

SIFBRONZE NO. 10
Sifbronze
Copper. Cu 60, Mn 0.2, Si 0.3, bal Zn.
460 N/mm^2 TS; 130 Brin. Manganese-bronze rod for brazing; resists sea water corrosion.

SIFBRONZE NO. 102 (FLUXCOATED)
Sifbronze
Copper.
Now SIFREDICOTE NO. 2.

SIFBRONZE NO. 104 (FLUXCOATED)
Sifbronze
Copper.
Now SIFREDICOTE NO. 1.

SIFBRONZE NO. 2
Sifbronze
Copper. Cu 48, Ni 10, Si 0.3, bal Zn.
540 N/mm^2 TS; 200 Brin. Rod for building up worn components like gear teeth, bearings and valve seats; wear resistant.

SIFBRONZE NO. 3
Sifbronze
Copper. Ni 15, bal Cu.
Welded: 66,000 TS; 36,000 YS; 24 El; 30 RA; 124 Brin. For welding rod; corrosion resistant. *Obsolete*

SIFBRONZE NO. 4
Sifbronze
Copper. Sn, bal Cu.
Welded: 60,000 TS; 32,000 YS; 28 El; 35 RA; 117 Brin. For welding rod for Aluminum bronze; flux coated. *Obsolete*

SIFBRONZE NO. 5
Suffolk Iron Foundry Ltd.
Sn, bal Cu.
For welding rod; bronze. *Obsolete*

SIFCOLOY
Spuck Iron & Foundry Co.
TC 3.1, Si 1.14, Cu 0.62, Mo 0.76, bal Fe.
Cast: 58,000 TS; 250 Brin. For nozzle castings, lock gate valves; erosion resistant.

SIFCUPRON NO. 17
Sifbronze
Copper. P 7, bal Cu.
Low melting copper phosphorous alloy rod for joining non-ferrous metals. Used in the refrigeration and electrical industries for copper tubing, switchgear, dynamos, motors and cable joints. BS 1845 CP3.

SIFCUPRON NO. 17-2AG
Sifbronze
Copper. Ag 2, P 6, bal Cu.
Low melting brazing alloy used in the production of hot water cylinders and electric motors; resistant to corrosion and dezincification. BS 1845 CP2.

SIFMIG G1B
Sifbronze
Aluminum. 99.5 Al min.
Rod produces a ductile weld equal in strength to that of the base metal. 90 N/mm^2 TS; 15 Brin. BS 2901 1050.

SIFMIG NG 21
Sifbronze
Aluminum. Si 5, bal Al.
Aluminum alloy for welding Duralumin, cast and wrought alloys. 120 N/mm^2 TS; 40 Brin. BS 2901 4043.

SIFMIG NG 6
Sifbronze
Aluminum. Mg 5, bal Al.
Aluminum alloy for welding magnesium bearing aluminum alloys. 250 N/mm^2 TS; 60 Brin. BS 2901 5056.

SIFMIG NG 61
Sifbronze
Aluminum. Mg 5.3, Mn 0.8, Cr 0.1, Ti 0.1, bal Al.
Aluminum alloy spooled wire for welding 5083 (N8) and for military applications. 300 N/mm^2 TS; 70 Brin. BS 2901 5556.

SIFONITE NO. 18
Suffolk Iron Foundry Ltd.
C, alloy, bal Fe.
Welded: 610 Brin. For hard surfacing electrode; wear and heat resistant. *Obsolete*

SIFPHOSPHOR BRONZE NO. 8
Sifbronze
Copper. Sn 7, bal Cu.
Rod for fusion welding of phosphor bronze castings; good color match. BS 2901 C11.

SIFREDICOTE NO. 1

Sifbronze
Copper. Cu 60, Si 0.3, bal Zn.
For brazing all classes of ferrous metals. Flux coated type of SIFBRONZE NO. 1.

SIFREDICOTE NO. 2

Sifbronze
Copper. Cu 48, Ni 10, Si 0.3, bal Zn.
Rod for high strength production and maintenance applications. Fluxcoated type of SIFBRONZE NO. 2.

SIFSERRATE

Sifbronze
Copper. Cu 60, Si 0.3, bal Zn.
Flux impregnated brazing rod for brazing clean mild steel components. 430 N/mm^2 TS; 120 Brin. BS 1845 CZ6-1453 C2.

SIFSERRATE NO. 10

Sifbronze
Copper. Cu 60, Mn 0.2, Si 0.3, bal Zn.
Flux impregnated brazing rod. 460 N/mm^2 TS; 130 Brin. BS 1845 CZ7-1453 C4.

SIFSILCOPPER NO. 7

Sifbronze
Copper. Ag 1, bal Cu.
Rod for full fusion welding of deoxidized copper; easy flowing. BS 1453 C1.

SIFSILCOPPER NO. 968

Sifbronze
Copper. Mn 1, Si 3, bal Cu.
For fusion welding similar alloys. BS 2901 C9.

SIFSILCOPPER NO. 985

Sifbronze
Copper. Mn 0.25, Si 0.25, bal Cu.
Copper rod suitable for TIG welding of copper. 340 N/mm^2 TS; 70 Brin. BS 2901 C7.

SIFSTEEL A15

Sifbronze
C 0.1, Si 0.6, Mn 1.3, Al 0.2, bal Fe.
Copper coated, triple deoxidized mild steel rod. Applications include pipe welding and root runs on heavy vessels. 440 N/mm^2 TS; 120 Brin. BS 2901 A15.

SIFSTEEL A31

Sifbronze
Alloy steel. C 0.1, Si 0.7, Mn 1.8, Mo 0.5, bal Fe.
Copper coated, alloy steel rod. Suitable on low temperature pressure vessel and pipe work applications. 460 N/mm^2 TS; 180 Brin. BS 1453 and 2901 A31.

SIFSTEEL A32

Sifbronze
Alloy steel. C 0.1, Si 0.5, Mn 1, Cr 1.3, Mo 0.5, bal Fe.
Copper coated, alloy steel rod. Suitable for creep resistant steels of similar composition. 500 N/mm^2 TS; 180 Brin. BS 1453 and 2901 A32.

SIFSTEEL A33

Sifbronze
Alloy steel. C 0.1, Si 0.5, Mn 1, Cr 2.4, Mo 1, bal Fe.
Copper coated, alloy steel rod. Suitable for high temperature and pressure applications on materials of similar composition. 525 N/mm^2 TS; 200 Brin. BS 1453 and 2901 A33.

SIFSTEEL AMS NO. 13

Suffolk Iron Foundry Ltd.
C 0.1, bal Fe.
Welded: 46,000 TS; 32,000 YS; 13 El; 20 RA. For welding rod for steel; M P 1450 C. *Obsolete*

SIFSTEEL NO. 11

Sifbronze
Stainless steel. C 0.06, Mn 0.4, bal Fe.
Low carbon, mild steel rod for welding; copper coated. 350 N/mm^2 TS; 120 Brin. BS 1453 A1.

SIFSTEEL NO. 12 (WEAR RESISTING)

Sifbronze
C, Si, Mn, Cr, Fe.
Weld filler metal for building up worn surfaces of cams, rail points, steel pins. BS 1453 A5. *Obsolete*

SIFSTEEL NO. 19

Sifbronze
Stainless steel. C 0.27, Si 0.4, Mn 1.4, bal Fe.
Copper coated, medium carbon alloy steel rod for high strength, ductile weld. 525 N/mm^2 TS; 150 Brin. BS 1453 A3 2901 A16.

SIFSTEEL NO. 22

Sifbronze
Stainless steel. C 0.15, Si 0.2, Mn 1.3, bal Fe.
Rod for oxy-acetylene welding of pipelines and pressure vessels; high strength. 450 N/mm^2 TS; 140 Brin. BS 1453 A2.

SIFSTEEL NO. 26

Suffolk Iron Foundry Ltd.
Ni 3.5, C, bal Fe.
For welding rod; tough. *Obsolete*

SIFSTEEL NO. 34

Suffolk Iron Foundry Ltd.
Ni 80, Cr 20.
For resurfacing valves, joining heating elements; heat resistant. *Obsolete*

SIFSTEEL STAINLESS NO. 20

Sifbronze
Stainless steel. C 0.04, Si 0.4, Mn 1.5, Ni 10, Cr 20, Nb 0.6, bal Fe.
Niobium stabilized stainless steel filler rod for use on 18/8 stainless steel. Applications include aircraft fabrication, stainless steel pipelines, tanks, fittings, and hospital equipment. 650 N/mm^2 TS; 180 Brin. BS 1453 and 2901 347 S96.

SIFSTEEL STAINLESS NO. 21

Sifbronze
Stainless steel. C 0.02, Si 0.4, Mn 1.5, Ni 12, Cr 19, Mo 2, bal Fe.
Corrosion resistant stainless steel filler rod for welding molybdenum bearing stainless steel. 650 N/mm^2 TS; 180 Brin. BS 1453 and 2901 316 S92.

SIFSTEEL STAINLESS NO. 22 (NB)

Sifbronze
Stainless steel.
Now SIFSTEEL STAINLESS NO. 20.

SIFSTEEL STAINLESS NO. 26

Suffolk Iron Foundry Ltd.
Cr 18, Ni 8, C, bal Fe.
Welded: 10,000 TS; 34,000 YS; 56 El; 60 RA. For welding rod for stainless steels; stainless. *Obsolete*

SIFSTEEL STAINLESS NO. 33

Sifbronze
Stainless steel. C 0.02, Si 0.4, Mn 1.5, Ni 10, Cr 21, bal Fe.
Stainless steel filler rod suitable for welding austenitic stainless steels; corrosion and wear resistant. 650 N/mm^2 TS; 180 Brin. BS 1453 and 2901 308 S92.

SIFSTEEL STAINLESS NO. 34

Sifbronze
Stainless steel. C 0.1, Si 0.4, Mn 1.5, Ni 13, Cr 24, bal Fe.
Stainless steel rod suitable for joining similar materials and dissimilar stainless steels. 650 N/mm^2 TS; 180 Brin. BS 1453 and 2901 309 S94.

SIFSTEEL STAINLESS NO. 35

Sifbronze
Stainless steel. C 0.1, Si 0.4, Mn 1.5, Ni 21, Cr 26, bal Fe.
25/20 type stainless steel rod suitable for heat resistant, austenitic stainless steels. 650 N/mm^2 TS; 180 Brin. BS 1453 and 2901 310 S94.

SIFSTEEL W.R. NO. 12

Suffolk Iron Foundry Ltd.
C, Si, Cr, Mn, bal Fe.
Welded: 250 Brin. For hard facing electrodes; wear resistant. *Obsolete*

SIGAL

German manufacture
Al 87, Si 13.

SIGERON

Prescott Co.
C 3.2, Si 2.2, Ni 1.5, bal Fe.
For castings. Nickel cast iron.

SIGMA ST 1325-1470

Krupp Stahl AG
tool material. C 0.48, Mn 0.7, Si 1.8, Cr 0.4, bal Fe.
W. Nr. 1.8863. *Obsolete*

SIGMA ST 835-1030

Krupp Stahl AG
tool material. C 0.7, Mn 1.3, Si 0.75, bal Fe.
For pliers. W.Nr. 1.8861. *Obsolete*

SIGMALUMIN

Aluminium Belge, S.A.
Si 0.8, Cu 3.8, Mn 0.7, bal Al.
Soft: 25,000-29,000 TS; 11,400-15,900 YS; 15-20 El. For light alloy parts; age hardenable.

SIL 1

American manufacture
C 0.45, Mn 0.4, Si 3.3, Cr 8.5, bal Fe.
1400°F: 20,000 psi TS. Exhaust valve steel. SAE HNV 3.

SIL 10

Now SILCROME X10.

SIL 10

American manufacture
C 0.38, Mn 1, Si 3, Cr 19, Ni 8, bal Fe.
1400°F: 41,200 psi TS; 28,400 psi YS; 18 El. Exhaust valve steel. SAE EV 5.

SIL 10N

American manufacture
C 0.38, Mn 1, Si 3, Cr 19, Ni 8, N 0.15-0.25, bal Fe.
1400°F: 41,200 psi TS; 28,400 psi YS; 18 El. Exhaust valve steel. SAE EV 6.

SIL 20

United Wire & Supply Co.
Ag 20, Cu 45, Zn 35.
For silver brazing; MP 1430-1500°F.

SIL 20C

United Wire & Supply Co.
Ag 20, Cu 45, Zn 30, Cd 5.
For silver brazing; MP 1140-1500°F.

SIL 30

United Wire & Supply Co.
Ag 30, Cu 38, Zn 32.
For silver brazing; MP 1370-1410°F.

SIL 45

United Wire & Supply Co.
Ag 45, bal Cu.
For silver solder. MP 1250-1370°F; AWS-BAg 5.

SIL 746
American manufacture
C 0.7, Mn 6.3, Si 0.55, Cr 21, Ni 0-1.9, N 0.23, bal Fe.
1400°F: 50,000 psi TS; 30,000 psi YS; 8 El. Exhaust valve steel. SAE EV 11.

SIL 9
United Wire & Supply Co.
Ag 9, Cu 53, Zn 38.
For silver brazing; MP 1510-1575°F.

SIL NO. 1
Now CARPENTER SIL NO.1 and EMS-1.

SIL-25
United Wire & Supply Co.
Ag 25, Cu 52.5, Zn 22.5.
For silver solder; MP 1500-1575°F.

SIL-35
United Wire & Supply Co.
Ag 35, Cu 32, Zn 33.
Melting range: 1260-1370°F. Brazing: 1370-1550°F. 72,000 psi TS. For various brazing operations.

SIL-40
United Wire & Supply Co.
Ag 40, Cr 30.5, Zn 29.5.
For silver solder. MP 1080-1340°F.

SIL-40N
United Wire & Supply Co.
Ag 40, bal Cu.
For silver solder. MP 1220-1435°F; AWS-BAg 4.

SIL-40T
United Wire & Supply Co.
Ag 40, Cu 30, Zn 28, Sn 2.
Melting range: 1220-1320°F. For various brazing operations.

SIL-5
United Wire & Supply Co.
Ag 5, Zn 37, Cu 58.
For silver solder; MP 1575-1600°F.

SIL-50
United Wire & Supply Co.
Ag 50, bal Cu.
For silver solder. MP 1275-1425°F; AWS-BAg 6.

SIL-54N
United Wire & Supply Co.
Ag 54, bal Cu.
For silver solder; MP 1225-1275°F.

SIL-56T
United Wire & Supply Co.
Ag 56, bal Cu.
For silver solder. MP 1145-1205°F; AWS-BAg 7.

SIL-5C
United Wire & Supply Co.
Ag 5, Cd 95.
For silver solder; MP 625-740°F.

SIL-60
United Wire & Supply Co.
Ag 60, Cu 25, Zn 15.
For silver solder. MP 1260-1325°F.

SIL-60T
United Wire & Supply Co.
Ag 60, Cu 30, Sn 10.
Melting range: 1095-1325°F. Brazing: 1325-1500°F. 70,000 psi TS; 36 El; 80 Rock 15-T. For various brazing operations. Formerly SIL-6T.

SIL-65
United Wire & Supply Co.
Ag 65, bal Cu.
For silver solder. MP 1280-1325°F; AWS-BAg 9.

SIL-70
United Wire & Supply Co.
Ag 70, bal Cu.
For silver solder. MP 1335-1390°F; AWS-BAg 10.

SIL-72
United Wire & Supply Co.
Ag 72, bal Cu.
For silver solder. MP 1435°F; AWS-BAg 8.

SIL-7T
United Wire & Supply Co.
Ag 7, Cu 85, Sn 8.
For silver solder; MP 1225-1805°F.

SIL-80
United Wire & Supply Co.
Ag 80, Cu 16, Zn 4.
For silver solder. MP 1360-1490°F.

SIL-85M
United Wire & Supply Co.
Ag 85, bal Cu.
For silver solder. MP 1745-1760°F.

SIL-AID
United Wire & Supply Co.
Ag 35, Cu 26, Zn 21, Cd 18.
Cast: 60,000 TS. For silver brazing all metals; MP 1125-1300°F.

SIL-BOND 30
United Wire & Supply Co.
Ag 30, Cu 27, Zn 33, Cd 20.
Melting range: 1125-1310°F. Brazing: 1310-1550°F. 66,000 psi TS; 35 El; 82 Rock 15-T. For various brazing operations.

SIL-BOND 31
United Wire & Supply Co.
Ag 31.5, Cu 34, Zn 15.5, Cd 19.
For silver brazing. MP 1165-1390°F; fillet type.

SIL-BOND 35
United Wire & Supply Co.
Ag 35, Cu 26, Zn 21, Cd 18.
Melting range: 1125-1295°F. Brazing: 1295-1550°F. 65,000 psi TS; 28 El; 80 Rock 15-T. For various silver brazing operations.

SIL-BOND 45
United Wire & Supply Co.
Ag 45, Cu 15, Zn 16, Cd 24.
Cast: 70,000 TS. For silver brazing. MP 1125-1145°F; general purpose.

SIL-BOND 50
United Wire & Supply Co.
Ag 50, Cu 15.5, Zn 16.5, Cd 18.
Cast: 65,000 TS. For silver brazing. MP 1160-1175°F; general purpose.

SIL-BOND 50N
United Wire & Supply Co.
Ag 50, Cu 15.5, Zn 15.5, Cd 16, Ni 3.
For silver brazing for carbide tipped tools. MP 1195-1270°F, resists chloride corrosion.

SIL-CO
United Wire & Supply Co.
Ag 70, Cu 20, Zn 10.
For silver solder. MP 1235-1360°F.

SIL-CON
United Wire & Supply Co.
Ag 50, Cu 34, Zn 16.
For silver solder. MP 1260-1410°F.

SIL-EEN
United Wire & Supply Co.
Ag 40, Cu 36, Zn 24.
For silver solder. MP 1330-1445°F.

SIL-EX
United Wire & Supply Co.
Ag 50, Cu 15.5, Zn 16.5, Cd 18.
For silver brazing. MP 1160-1175°F.

SIL-FIL
United Wire & Supply Co.
Ag 50, Cu 15.5, Zn 15.5, Cd 16, Ni 3.
For silver solder. MP 1170-1270°F.

SIL-FOS
Handy & Harmon
Copper. Ag 15, Cu 80, P 5.
MP: 1185°F; FP: 1300°F. For brazing copper and copper based alloys where joints are wide. AWS BCuP-5.

SIL-FOS
Johnson Matthey plc
Ag 14.5, Cu 80.85, P 4.65.
Phosphorus bearing silver brazing alloy; 644-800°C MP; DIN 8513 L-Ag15P.

SIL-FOS 18
Handy & Harmon
Ag 18, Cu 74.75, P 7.25.
MP: 1190°F; FP: 1190°F. Ternary eutectic filler metal for joints where good fit-up can be maintained and low melting point is important. Clearance: 0.001-0.003 in. Very fast flow.

SIL-FOS 2
Handy & Harmon
Ag 2, Cu 91, P 7.
MP: 1190°F; FP: 1450°F. Filler metal comparable to Fos-Flo 7. Recommended joint clearance: 0.001-0.005 in. Medium flow.

SIL-FOS 2M
Handy & Harmon
Ag 2, Cu 91.4, P 6.6.
MP: 1190°F; FP: 1495°F. Fills moderate gaps in poorly fitted joints. More ductile than Fos-Flo 7 or Sil-Fos 2. Used on copper tube headers and similar applications where sleeve fit is not practical. Recommended joint clearance: 0.002-0.005 in. Slow flow.

SIL-FOS 5
Johnson Matthey plc
Ag 5, Cu 89, P 6.
Phosphorus bearing silver brazing alloy; 644-815°C MP; DIN 8513 L-Ag5P.

SIL-FOS 6
Handy & Harmon
Ag 6, Cu 86.75, P 7.25.
MP: 1190°F; FP: 1325°F. Fluid filler metal for close fit-up work. Low melting range for use where temperature is a factor. Recommended joint clearance: 0.001-0.003 in. Fast flow.

SIL-FOS 6
Johnson Matthey plc
Ag 6, Cu 86.75, P 7.25.
Phosphorus bearing silver brazing alloy; 644-718°C MP; DIN 8513.

SIL-FOS 6M
Handy & Harmon
Ag 6, Cu 88, P 6.
MP: 1190°F; FP: 1460°F. Recommended for use where close fit-up cannot be maintained. Has the ability to fill gaps and form fillets without affecting joint strength. Recommended joint clearance: 0.002-0.005 in. Slow flow.

SIL-FOS-5
Handy & Harmon
Copper. Ag 5, Cu 89, P 6.
MP: 1185°F; FP: 1300°F. For brazing copper and copper alloys. AWS BCuP-3.

SIL-GON
United Wire & Supply Co.
Ag-Cu.
For silver solder. MP 1275-1425°F.

SIL-LO
United Wire & Supply Co.
Ag 15, Cu 80, P 5.
For silver brazing nonferrous metals. MP 1185-1300°F.

SIL-LON
United Wire & Supply Co.
Cu 30.5, Zn 29.5, Ag 40.
For silver brazing all metals. Cd free; MP 1250-1350°F.

SIL-LOY
United Wire & Supply Co.
Ag 65, Cu 20, Zn 15.
For silver brazing; Cd free; MP 1265-1325°F.

SIL-MAG NO. 1
SiMETCO, Inc.
Si 42-47, Mg 7.5-10, bal Fe.
For cast iron inoculant; for ductile iron. *Obsolete*

SIL-MAG NO. 2
SiMETCO, Inc.
Si 60-65, Mg 16-20, bal Fe.
For cast iron inoculant; for ductile iron. *Obsolete*

SIL-MAG-M NO. 1
SiMETCO, Inc.
Si 42-47, Mg 7.5-10, 1 Mischmetal, bal Fe.
For cast iron inoculant; for ductile iron. *Obsolete*

SIL-MAG-M NO. 2
SiMETCO, Inc.
Si 60-65, Mg 16-20, 1 Mischmetal, bal Fe.
For cast iron inoculant; for ductile iron. *Obsolete*

SIL-MAN
Teledyne Vasco
C 0.55, Si 2.1, Mn 0.85, Cr 0.25, V 0.3, bal Fe.
205,000 TS; 175,000 YS; 8 El; 16 RA; 400 Brin. For shock tools, punches, chisels, shear blades; shock resisting tool steel; silico-manganese steel.

SIL-NIK
United Wire & Supply Co.
Ag 40, Cu 30, Zn 28, Sn 2.
For silver solder. MP 1240-1435°F.

SIL-O-ACID NO. 7
Manufacturer not listed.
Copper. Co, Si, Mg, bal Cu.
Cast: 52,000 TS; 21 El; 20 RA; 78 Brin. For acid proof parts; resists acetic, carbolic, sulfuric and tartaric acid. *Obsolete*

SIL-OID
United Wire & Supply Co.
Ag 31.5, Cu 34, Zn 15.5, Cd 19.
For silver brazing all metals. MP 1165-1390°F.

SIL-TEN S
Now USS SIL-TEN.

SIL-TEX
United Wire & Supply Co.
Ag 60, Cu 25, Zn 15.
For silver brazing all metals. Cd free; MP 1230-1330°F.

SIL-TIN
United Wire & Supply Co.
Ag 56, Cu 22, Zn 17, Sn 5.
For silver solder. MP 1145-1205°F.

SIL-TITE
United Wire & Supply Co.
Ag 45, Cu 30, Zn 25.
For silver brazing all metals. Cd free; MP 1240-1370°F.

SIL-TRODE
Ampco Pittsburgh Corp.
Si, bal Cu.
For welding electrodes; for repairing silicon bronze parts.
Obsolete

SIL-TRON
United Wire & Supply Co.
Ag 80, Cu 16, Zn 4.
For silver brazing. Cd free; MP 1375-1480°F.

SIL-UTEC
United Wire & Supply Co.
Ag 72, Cu 28.
For silver solder. MP 1435°F.

SIL-X
Chromium Mining & Smelting Corp. Ltd.
ferrosilicon.
For steel deoxidizer; compounded with an oxidizing agent.

SILAFOND BETA
Soc. Alluminio Veneto per Azioni
Aluminum. Si 11.5-13, Mg 0.3, Mn 0.5, bal Al.
Cast: 34,000-46,000 TS; 28,000-40,000 YS; 1-4 El; 80-110 Brin. For light alloy parts; corrosion resistant.

SILAFONT 1
Aluminium GmbH
Si 11-13.5, Mn 0.3-0.5, bal Al.
Sand cast: 25,000-32,000 TS; 11,400-12,000 YS; 4-8 El; 50-60 Brin. Permanent mold: 28,000-37,000 TS; 12,000-16,000 YS; 3-7 El; 60-70 Brin. For chemical industry, ship and auto construction; high corrosion resistance.

SILAFONT 1
Oederlin & Co. Ltd.
Si 11-13.5, Mn 0.3-0.5, bal Al.
Sand cast: 25,000-32,000 TS; 11,400-12,000 YS; 4-8 El; 50-60 Brin. Permanent mold: 28,000-37,000 TS; 12,000-16,000 YS; 3-7 El; 60-70 Brin. For chemical industry, ship and auto construction; high corrosion resistance.

SILAFONT 3
Aluminium GmbH
Si 8.5-10.5, Mg 0.2-0.4, Mn 0-0.5, bal Al.
Sand cast: 45,000-54,000 TS; 31,000-37,000 YS; 3-7 El; 90-100 Brin. Permanent mold: 48,000-58,000 TS; 35,000-40,000 YS; 4-8 El; 95-110 Brin. For chemical industry, ship and auto construction; high corrosion resistance.

SILAFONT 3
Oederlin & Co. Ltd.
Si 8.5-10.5, Mg 0.2-0.4, Mn 0-0.5, bal Al.
Sand cast: 45,000-54,000 TS; 31,000-37,000 YS; 3-7 El; 90-100 Brin. Permanent mold: 48,000-58,000 TS; 35,000-40,000 YS; 4-8 El; 95-110 Brin. For chemical industry, ship and auto construction; high corrosion resistance.

SILAFONT CUPRO
Soc. Alluminio Veneto per Azioni
Aluminum. Cu 0.8, Si 12-13.3, Mn 0.3, bal Al.
Cast: 31,000-37,000 TS; 17,000-20,000 YS; 1.5-3.0 El; 60-80 Brin. For light alloy parts; corrosion resistant.

SILAFONT MN
Soc. Alluminio Veneto per Azioni
Aluminum. Si 9, Mg 0.35, Mn 0.5, bal Al.
Cast: 35,000-43,000 TS; 28,000-37,000 YS; 3.5-5.5 El; 80-95 Brin. For light alloy parts; corrosion resistant.

SILAFONT NORMAL
Soc. Alluminio Veneto per Azioni
Aluminum. Si 12.75-13.25, bal Al.
Cast: 31,000-34,000 TS; 14,000-17,000 YS; 1.5-3.0 El; 60-80 Brin. For light alloy parts; corrosion resistant.

SILAFONT-09
Swiss Aluminium Ltd.
Aluminum. Si 9.5-10.6, Fe 0.4, Cu 0.01, Mn 0.4, Mg 0.05, Zn 0.1, Ti 0.05, bal Al.
Primary foundry alloy in ingot form. 240-280 N/mm^2 TS; 140-180 N/mm^2 YS; 5-10 El; 60-80 Brin.

SILAFONT-1
Aluminium Industries, AG
Si 12-13, Mn 0.3-0.6, bal Al.
Sand cast: 27,000 TS; 10,000 YS; 6 El; 55 Brin. Chill cast: 33,000 TS; 14,000 YS; 5 El; 60 Brin. For chemical industry, ship and auto construction; heat treatable.

SILAFONT-1
Swiss Aluminium Ltd.
Si 12-13, Mn 0.3-0.6, bal Al.
Sand cast: 27,000 TS; 10,000 YS; 6 El; 55 Brin. Chill cast: 33,000 TS; 14,000 YS; 5 El; 60 Brin. For chemical industry, ship and auto construction; heat treatable.

SILAFONT-1
Aluminium Industrie Aktiengesellschaft
Si 12-13, Mn 0.3-0.6, bal Al.
Sand cast: 27,000 TS; 10,000 YS; 6 El; 55 Brin. Chill cast: 33,000 TS; 14,000 YS; 5 El; 60 Brin. For chemical industry, ship and auto construction; heat treatable.

SILAFONT-12 DV
Swiss Aluminium Ltd.
Aluminum. Si 10.5-13.5, Fe 0.15, Cu 0.01, Mn 0.02, Mg 0.02, Zn 0.07, Ti 0.05, bal Al.
Primary foundry alloy in ingot form. 150-210 N/mm^2 TS; 90-120 N/mm^2 YS; 9-15 El; 50-60 Brin.

SILAFONT-13
Swiss Aluminium Ltd.
Aluminum. Si 10.5-13.5, Fe 0.15, Cu 0.01, Mn 0.02, Mg 0.02, Zn 0.07, Ti 0.05, bal Al.
Primary foundry alloy in ingot form. 150-280 N/mm^2 TS; 90-180 N/mm^2 YS; 5-13 El; 50-80 Brin.

SILAFONT-14
Swiss Aluminium Ltd.
Aluminum. Si 10.5-13.5, Fe 0.5, Cu 0.05, Mn 0.4, Mg 0.1, Zn 0.1, Ti 0.05, bal Al.
Primary foundry alloy in ingot form. 140-230 N/mm^2 TS; 80-150 N/mm^2 YS; 2-10 El; 50-60 Brin.

SILAFONT-2
Aluminium Industries, AG
Si 12-13, Mg 0.2-0.4, Mn 0.3-0.6, bal Al.
Sand cast: 40,000 TS; 34,000 YS; 2.5 El; 95 Brin. Chill cast: 43,000 TS; 38,000 YS; 2 El; 100 Brin. For chemical industry, ship and auto construction; heat treatable.

SILAFONT-2
Swiss Aluminium Ltd.
Si 12-13, Mg 0.2-0.4, Mn 0.3-0.6, bal Al.
Sand cast: 40,000 TS; 34,000 YS; 2.5 El; 95 Brin. Chill cast: 43,000 TS; 38,000 YS; 2 El; 100 Brin. For chemical industry, ship and auto construction; heat treatable.

SILAFONT-2
Aluminium Industrie Aktiengesellschaft
Si 12-13, Mg 0.2-0.4, Mn 0.3-0.6, bal Al.
Sand cast: 40,000 TS; 34,000 YS; 2.5 El; 95 Brin. Chill cast: 43,000 TS; 38,000 YS; 2 El; 100 Brin. For chemical industry, ship and auto construction; heat treatable.

SILAFONT-20
Swiss Aluminium Ltd.
Aluminum. Si 10.5-13.5, Fe 0.15, Cu 0.01, Mn 0.02, Mg 0.1-0.45, Zn 0.07, Ti 0.05, bal Al.
Primary foundry alloy in ingot form. 170-300 N/mm^2 TS; 100-220 N/mm^2 YS; 1-11 El; 50-90 Brin.

SILAFONT-3
Aluminium Industries, AG
Si 9-10, Mg 0.2-0.4, Mn 0.3-0.6, bal Al.
Sand cast: 38,000 TS; 30,000 YS; 7 El; 80 Brin. Chill cast: 39,000 TS; 33,000 YS; 6 El; 85 Brin. For chemical industry, ship and auto construction; heat treatable.

SILAFONT-3
Swiss Aluminium Ltd.
Si 9-10, Mg 0.2-0.4, Mn 0.3-0.6, bal Al.
Sand cast: 38,000 TS; 30,000 YS; 7 El; 80 Brin. Chill cast: 39,000 TS; 33,000 YS; 6 El; 85 Brin. For chemical industry, ship and auto construction; heat treatable.

SILAFONT-3
Aluminium Industrie Aktiengesellschaft
Si 9-10, Mg 0.2-0.4, Mn 0.3-0.6, bal Al.
Sand cast: 38,000 TS; 30,000 YS; 7 El; 80 Brin. Chill cast: 39,000 TS; 33,000 YS; 6 El; 85 Brin. For chemical industry, ship and auto construction; heat treatable.

SILAFONT-30
Swiss Aluminium Ltd.
Aluminum. Si 9-10, Fe 0.15, Cu 0.01, Mn 0.05, Mg 0.3-0.45, Zn 0.07, Ti 0.05, bal Al.
Primary foundry alloy in ingot form. 220-340 N/mm^2 TS; 140-280 N/mm^2 YS; 2-7 El; 70-115 Brin.

SILAFONT-31 DV
Swiss Aluminium Ltd.
Aluminum. Si 9-10, Fe 0.15, Cu 0.01, Mn 0.05, Mg 0.3-0.45, Zn 0.07, Ti 0.05, bal Al.
Primary foundry alloy in ingot form. 160-300 N/mm^2 TS; 80-270 N/mm^2 YS; 2-6 El; 50-110 Brin.

SILAFONT-35
Swiss Aluminium Ltd.
Aluminum. Si 9-10.5, Fe 0.4, Cu 0.03, Mn 0.4, Mg 0.25-0.4, Zn 0.1, Ti 0.1, bal Al.
Primary foundry alloy in ingot form. 210-340 N/mm^2 TS; 180-280 N/mm^2 YS; 1-4 El; 80-120 Brin.

SILAFONT-4
Aluminium Industrie Aktiengesellschaft
Si 12-13, Cu 0.5-1, Mn 0.3, bal Al.
Sand cast: 31,000 TS; 13,000 YS; 2-5 El; 55 Brin. Permanent mold: 35,000 TS; 14,000 YS; 2-4 El; 65 Brin. For instrument housings, cases, intricate castings; corrosion resistant, good castability.

SILAFONT-4
Swiss Aluminium Ltd.
Si 12-13, Cu 0.5-1, Mn 0.3, bal Al.
Sand cast: 31,000 TS; 13,000 YS; 2-5 El; 55 Brin. Permanent mold: 35,000 TS; 14,000 YS; 2-4 El; 65 Brin. For instrument housings, cases, intricate castings; corrosion resistant, good castability.

SILAFONT-5
Swiss Aluminium Ltd.
Aluminum. Si 9-9.8, Mg 0.2-0.35, Co 0.45-0.55, bal Al.
As cast: 26,000-33,000 TS; 14,000 YS; 3-6 El; 50-60 Brin. Heat treated: 42,000 TS; 37,000 YS; 2-5 El; 70-95 Brin. For general sand and die castings, housings, instrument cases. Impact and shock resistant. Heat treatable. *Obsolete*

SILAFONT-6
Swiss Aluminium Ltd.
Aluminum. Si 8.5-11, Mg 0.2-0.35, bal Al.
Die cast: 40,000 TS; 26,000 YS; 1 El; 80 Brin. For general die castings, instrument cases, housings. Good strength and corrosion resistant. *Obsolete*

SILAFONT-7
Swiss Aluminium Ltd.
Aluminum. Si 12.8-13.4, Mg 0.9-1.2, Cu 1-1.4, Ni 0.9-1.2, bal Al.
Heat treated: 44,000 TS; 37,000 YS; 0.5 El; 135 Brin. For engine pistons. Sand and die casting alloy, good hot crack resistance. Heat treatable. *Obsolete*

SILAFONT-70
Swiss Aluminium Ltd.
Aluminum. Si 11-13.5, Fe 0.15, Cu 0.8-1.3, Mn 0.05, Mg 0.8-1.3, Zn 0.1, Ti 0.1, Ni 0.8-1.3, bal Al.
Primary foundry alloy in ingot form. 220-400 N/mm^2 TS; 200-390 N/mm^2 YS; 0.3-2 El; 110-160 Brin.

SILAFONT-90
Swiss Aluminium Ltd.
Aluminum. Si 16-18, Fe 0.3, Cu 4-5, Mn 0.15, Mg 0.5-0.6, Zn 0.1, Ti 0.2, bal Al.
Primary foundry alloy in ingot form. 165-275 N/mm^2 TS; 160-245 N/mm^2 YS; 0.4-1 El; 105-120 Brin.

SILAFONT-92
Swiss Aluminium Ltd.
Aluminum. Si 17-19, Fe 0.15, Cu 0.8-1.3, Mn 0.05, Mg 0.8-1.3, Zn 0.1, Ti 0.1, Ni 0.8-1.3, bal Al.
Primary foundry alloy in ingot form. 180-220 N/mm^2 TS; 170-200 N/mm^2 YS; 0.2-0.7 El; 90-125 Brin.

SILAL
Sheepbridge Alloy Castings Ltd.
C 2.5, Al 0-1, Mn 0.96, Si 5, P 0.16, Cr 0.05, bal Fe.
Cast: 20,000 TS; 160 Brin. For stove and furnace parts, fire bars; cast iron, scale resistant.

SILAL
Sheepbridge Engineering Ltd.
C 2.5, Al 0-1, Mn 0.96, Si 5, P 0.16, Cr 0.05, bal Fe.
Cast: 20,000 TS; 160 Brin. For stove and furnace parts, fire bars; cast iron, scale resistant.

SILAL 53
Eduard Hueck
Mg 2-4, Mn 0-0.4, Cr 0-0.3, bal Al.
Soft: 28,000 TS; 13,000 YS; 30 El; 47 Brin. Hard: 40,000 TS; 35,000 YS; 10 El; 73 Brin. For aircraft tanks and fittings, fuel lines, marine parts. Resists seawater corrosion. *Obsolete*

SILAL V
Eduard Hueck
Al 97.2, Mg 1.2, Mn 0.8, Si 0.5, Ti 0.3.
For light alloy parts, rolling and extrusion. *Obsolete*

SILANCA
English manufacture
Ag 92.5-91.5, Zn 0-2.5, Sb 4-4.5, Cd 1-3.
For silverware; corrosion resistant.

SILBEREISEN NO. 1
Friedrich Wilhelms-Hutte
Si 1-1.5, Mn 0.9-1.5, P 0.07-0.09, S 0.02-0.04, C 2.8, bal Fe. 36,000-50,000 TS. For steamship and locomotive cylinders, turbine housings, pump bodies, high pressure valves; high strength machine and centrifugal castings.

SILBEREISEN NO. 2
Friedrich Wilhelms-Hutte
Si 1.7-2.5, Mn 0.4-0.7, P 0.07-0.09, S 0.02-0.04, C 2.8, bal Fe. 36,000-50,000 TS. For steamship and locomotive cylinders, turbine housings, pump bodies, high pressure valves; malleable iron and chilled iron castings.

SILBEREISEN NO. 3
Friedrich Wilhelms-Hutte
Si 1-1.3, Mn 0.4-0.7, P 0.3-0.8, S 0.02-0.04, C 2.8, bal Fe. 36,000-50,000 TS. For steamship and locomotive cylinders, turbine housings, pump bodies, high pressure valves; heavy castings, hard iron and chilled castings.

SILBEREISEN NO. 4
Friedrich Wilhelms-Hutte
Si 1.3-1.7, Mn 0.4-0.7, P 0.3-0.8, S 0.02-0.04, C 2.8, bal Fe. 36,000-50,000 TS. For steamship and locomotive cylinders, turbine housings, pump bodies, high pressure valves; ordinary machinery castings.

SILBEREISEN NO. 5
Friedrich Wilhelms-Hutte
Si 1.7-2.5, Mn 0.4-0.7, P 0.3-0.8, S 0.02-0.04, C 2.8, bal Fe. 36,000-50,000 TS. For steamship and locomotive cylinders, turbine housings, pump bodies, high pressure valves; thin sections, common machinery castings.

SILBEREISEN NO. 6
Friedrich Wilhelms-Hutte
Si 1.7-2.5, Mn 0.4-0.7, C 0-2.8, Ni 1.2, Cr 0.5-1, bal Fe.
For cast iron cylinders, pump bodies, acid resisting castings; cast iron.

SILBEREISEN NO. 7
Friedrich Wilhelms-Hutte
Si 1.7-2.5, Mn 0.4-0.7, C 0-2.8, Ni 1-2, Cr 0.5-1, bal Fe.
For cast iron cylinders, pump bodies, acid resisting castings; cast iron.

SILBERKREUZ 1
Plettenberger Gusstahlfabrik
C 0.86, Cr 4.1, Mo 0.85, V 2.5, W 12, bal Fe.
For lathe and planer tools, reamers, drills, hobs; high speed steel.

SILBERKREUZ 2
Plettenberger Gusstahlfabrik
C 0.82, Cr 4.1, Mo 0.85, V 1.6, W 8.7, bal Fe.
For lathe and planer tools, reamers, broaches, hobs; high speed steel.

SILBERKREUZ SPEZIAL
Plettenberger Gusstahlfabrik
C 0.74, Cr 4.1, V 1.1, W 18.5, bal Fe.
For lathe and planer tools, drills, taps, reamers; high speed steel.

SILBERLOT
German manufacture
Brazing alloys.
Series of brazing alloys used in Germany. German standard Nos.

SILBERPUNKT
SWB Stahlformguss Gesellschaft mbH
C 0.74, Cr 4.1, V 1.1, W 18, bal Fe.
For lathe and planer tools, reamers, broaches, taps; high speed steel. *Obsolete*

SILBLOCK 25
Foote Mineral Co.
Si 25, bal Fe.
For blocking open hearth and electric furnace steel; controls carbon content. *Obsolete*

SILBLOCK 50
Foote Mineral Co.
Si 50, bal Fe.
For blocking open hearth and electric furnace steel; controls carbon content. *Obsolete*

SILBRALLOY
Johnson Matthey plc
Ag 2, Cu 91.5, P 6.5.
Phosphorus bearing silver brazing alloy; 644-825°C MP; DIN 8513 L-Ag2P.

SILBRASS
Manufacturer not listed.
Si 4-5, Zn 10-12, bal Cu.
For corrosion resistant parts; corrosion resistant.

SILBRAX

Arcos Alloys
Cu.
For welding rod for deoxidized copper; self-fluxing. *Obsolete*

SILCALFA

Alfa Romeo
Si 11-14, bal Al.
For automotive castings; corrosion resistant. *Obsolete*

SILCAZ ALLOY 3

Union Carbide Corp.
Si 35-40, Ca 9-11, Al 6-8, Zr 3-5, Ti 9-11, B 0.55-0.75, bal Fe.
For addition agent in steel making; alloying agent.

SILCHROM 20

Thyssen Edelstahlwerke AG
C 0.8, Si 2, Cr 20, Ni 1.5, bal Fe.
For valves; heat and corrosion resistant.

SILCHROM 25

Thyssen Edelstahlwerke AG
C 0.45, Si 1.2, Mn 1.1, Cr 23, Mo 2.8, Ni 5, bal Fe.
For valves; heat and corrosion resistant.

SILCHROM I

Thyssen Edelstahlwerke AG
C 0.4, Si 3.5, Cr 9.5, bal Fe.
For valves, valve seats; heat resistant. *Obsolete*

SILCHROM II

Thyssen Edelstahlwerke AG
C 0.4, Si 4, Cr 3, bal Fe.
For valves; heat resistant. *Obsolete*

SILCHROM IX

Thyssen Edelstahlwerke AG
C 0.45, Si 3.2, Cr 8.5, bal Fe.
For valves; heat and corrosion resistant. *Obsolete*

SILCHROM V

Thyssen Edelstahlwerke AG
C 0.45, Si, Cr, bal Fe.
For bolts, gears, fasteners; oil hardened. *Obsolete*

SILCHROME

Allegheny Ludlum Steel
Cr 9.5, Si 4, C 0.5, bal Fe.
For heat and corrosion resisting parts; heat resistant. *Obsolete*

SILCHROME

Allegheny Ludlum Steel
Fe 64, Ni 21, Cr 6.2, Si 6.1, Mn 0.8, C 0.14.
For heat and corrosion resisting parts; heat and corrosion resistant. *Obsolete*

SILCHROME 10

AL Tech Specialty Steel Corp.
Stainless Steel. Cu 0.37, Cr 19, Ni 8, Si 3, Mn 1, N 0.1, Mo 0.3, Cu 0.3, bal Fe.
For exhaust valves. Austenitic, stainless, heat resistant.

SILCHROME 25-12

AL Tech Specialty Steel Corp.
C 0.2, Cr 25, Ni 12, bal Fe.
Annealed: 95,000 TS; 48,000 YS; 42 El; 50 RA; 167 Brin. For corrosion and heat resistant parts; heat and corrosion resistant.

SILCHROME 25-20

AL Tech Specialty Steel Corp.
Cr 25, Ni 20, C, bal Fe.
For heat and corrosion resistant parts; heat and corrosion resistant.

SILCHROME KA2M

Allegheny Ludlum Steel
Cr 18, Ni 8, Mo 3, C, bal Fe.
For corrosion resistant parts; corrosion resistant. *Obsolete*

SILCHROME TPA

AL Tech Specialty Steel Corp.
C 0.4-0.5, Si 0.3-0.8, Ni 13-15, Cr 13-15, W 2-3, bal Fe.
170-210 Brin. For airplane valves; heat resistant.

SILCHROME WIRE

Allegheny Ludlum Steel
Cr 18, Si 3, W 3, C 0.3, bal Fe.
For heat and corrosion resisting wire; heat and corrosion resistant. *Obsolete*

SILCHROME XB

AL Tech Specialty Steel Corp.
C 0.75-0.85, Si 1.5-2.5, Cr 19-23, Ni 1-2, bal Fe.
270-530 Brin. For valves and valve seats; heat resistant.

SILCORO 60

GTE Products Corp./Wesgo Div.
Au 60, Cu 20, Ag 20.
Brazing alloy available in foil, flexibraze, wire, powder, extrudable paste and preform. Liquidus 1553°F. Solidus 1535°F.

SILCORO 60

Western Gold & Platinum Co.
Au 60, Cu 20, Ag 20.
For brazing. Melting point: 835-845°C.

SILCORO 75

GTE Products Corp./Wesgo Div.
Au 75, Cu 20, Ag 5.
Brazing alloy available in foil, flexibraze, wire, powder, extrudable paste and preform. Liquidus 1643°F. Solidus 1625°F.

SILCORO 75

Western Gold & Platinum Co.
Au 75, Cu 20, Ag 5.
For brazing brasses. Melting point: 1625-1643°F; corrosion resistant.

SILCOZ ALLOY

Union Carbide Corp.
Si 35-40, Ca 9-11, Al 6-8, Zr 3-5, Ti 9-11, B 0.55-0.75.
For boron additions to steel; increased hardenability.

SILCROME 12

Allegheny Ludlum Steel
C 0-0.12, Cr 11.5-13, bal Fe.
Annealed: 65,000 TS; 35,000 YS; 30 El; 60 RA; 156 Brin. Heat treated: 130,000 TS; 95,000 YS; 15 El; 35 RA; 280 Brin. For applications requiring moderate corrosion resistance and high physical properties; corrosion resistant; scaling temperature 1100 F. *Obsolete*

SILCROME 12-EZ

Allegheny Ludlum Steel
C 0-0.12, Cr 12-15, bal Fe.
For stainless parts; easy machining, stainless. *Obsolete*

SILCROME 46 M

Allegheny Ludlum Steel
C 0-0.25, Cr 4-6, Mo 0.5, bal Fe.
Annealed: 68,000 TS; 29,000 YS; 32 El; 70 RA; 142 Brin. Heat treated: 117,000 TS; 101,000 YS; 18 El; 64 RA; 262 Brin. For machinery parts; scaling temperature 1000 F; IZ 67-72. *Obsolete*

SILCROME CC

Allegheny Ludlum Steel
C 0-0.12, Cr 13, Si 1, Cu 1, bal Fe.
For stainless parts; stainless. *Obsolete*

SILCROME H-17

Allegheny Ludlum Steel
C 1, Cr 17, bal Fe.
Annealed: 120,000 TS; 60,000 YS; 10 El; 15 RA; 240 Brin. Heat treated: 175,000 TS; 140,000 YS; 7 El; 10 RA; 350 Brin. For heat and corrosion resistant parts; hard, heat and corrosion resistant. *Obsolete*

SILCROME H-17-EZ

Allegheny Ludlum Steel
Cr 17, C, bal Fe.
For heat and corrosion resistant parts; corrosion and heat resistant. *Obsolete*

SILCROME KA-2

Allegheny Ludlum Steel
C 0.09-0.2, Cr 18, Ni 8, bal Fe.
Annealed: 90,000 TS; 30,000 YS; 55 El; 65 RA; 150 Brin. Hard drawn: 200,000 TS; 150,000 YS; 25 El; 45 RA; 315 Brin. For heat and corrosion resistant parts; heat and corrosion resistant; scaling temperature 1550 F. *Obsolete*

SILCROME KA-2

Chemalloy Electronics Corp.
C 0.09-0.2, Cr 18, Ni 8, bal Fe.
Annealed: 90,000 TS; 30,000 YS; 55 El; 65 RA; 150 Brin. Hard drawn: 200,000 TS; 150,000 YS; 25 El; 45 RA; 315 Brin. For heat and corrosion resistant parts; heat and corrosion resistant; scaling temperature 1550 F. *Obsolete*

SILCROME KA-2B

Allegheny Ludlum Steel
C 0.13-0.2, Cr 16-23, Ni 7-11, bal Fe.
For heat and corrosion resistant parts; stainless, heat and corrosion resistant. *Obsolete*

SILCROME KA2-EZ

Allegheny Ludlum Steel
C 0.08-0.2, Cr 18, Ni 8, bal Fe.
For heat and corrosion resistant parts; free machining; heat and corrosion resistant. *Obsolete*

SILCROME KA2S

Allegheny Ludlum Steel
C 0-0.08, Cr 18, Ni 8, bal Fe.
For corrosion resistant parts; corrosion resistant. *Obsolete*

SILCROME KA2T

Allegheny Ludlum Steel
Cr 18, Ni 8, C, Ti, bal Fe.
For corrosion resistant parts; corrosion resistant. *Obsolete*

SILCROME L-12

Allegheny Ludlum Steel
C 0.3, Cr 12, bal Fe.
For corrosion resistant parts; corrosion resistant. *Obsolete*

SILCROME M-17

Allegheny Ludlum Steel
C 0.6, Cr 17, bal Fe.
For heat and corrosion resistant parts; heat and corrosion resistant. *Obsolete*

SILCROME NO. 9

Acme Steel Co.
C 0.5, Ni 14, Cr 14, Si 3, W 2.5, Mo 0.5, bal Fe.
For valve seats; heat and corrosion resistant. *Obsolete*

SILCROME RA

Allegheny Ludlum Steel
C 0.12, Cr 16, Si 1, Cu 1, bal Fe.

SILCROME RA-EZ

Allegheny Ludlum Steel
C 0.12, Cr 16, Si 1, Cu 1, bal Fe.
For corrosion resisting screw machine parts; easy machining. *Obsolete*

SILCROME X

Allegheny Ludlum Steel
C 0.8, Si 2, Cr 20, bal Fe.
For truck, airplane and auto engine valves; heat resistant. *Obsolete*

SILCROME X10

Allegheny Ludlum Steel
C 0.3-0.45, Cr 18, Ni 8, Mn 0.5-1.5, Si 2.5-3.5, bal Fe.
For exhaust valves; heavy duty. *Obsolete*

SILCROME X142
Allegheny Ludlum Steel
C 0.4-0.5, Si 0.3-0.8, Ni 13-15, Cr 13-15, W 2-3, bal Fe.
170-210 Brin. For airplane valves; heat resistant. *Obsolete*

SILCROME X9
Allegheny Ludlum Steel
C 0.7, Si 3, W 10, bal Fe.
For valves, valve seats; heat resistant. *Obsolete*

SILCROME XCR
Allegheny Ludlum Steel
C 0.4-0.5, Si 1, Mn 1, Cr 23-24, Ni 4.5-5, Mo 2.5-3, bal Fe.
300-370 Brin. For valves and valve seats; heat resistant.
Obsolete

SILCROME-1
Allegheny Ludlum Steel
C 0.4-0.5, Si 3-3.5, Cr 8-9, bal Fe.
240-550 Brin. For supercharger wheels for jet engines,
engine valves; heat and corrosion resistant. *Obsolete*

SILCRY
Creusot-Loire
C, alloy, bal Fe.
For valves. *Obsolete*

SILECTRON
Allegheny Ludlum Steel
Si 3, bal Fe.
Rolled: 50,000 TS; 44,000 YS; 17 El. For distribution power
transformers, cores. Grain oriented, high permeability, low
core loss as high induction. *Obsolete*

SILECTRON
George W. Prentiss & Co.
Mn 4, Si 1, bal Ni.
For spark plugs. *Obsolete*

SILESIA D610PW
Georgsmarienwerke Selesiastahl GmbH
C 0.3, Cr 2.5, Mo 0.2, Mn 0.55, bal Fe.
For gears, dies, forging dies; oil hardened, tough.

SILESIA E60
Georgsmarienwerke Selesiastahl GmbH
C 0.15, Cr 0.65, Mn 0.5, Si 0.25, bal Fe.
For gears, bolts, machine tool parts; case hardened.

SILESIA EK20
Georgsmarienwerke Selesiastahl GmbH
C 0.2, Si 0.25, Cr 1.25, Mn 1.15, bal Fe.
For gears, bolts, camshafts, cams; case hardening steel.

SILESIA F6
Georgsmarienwerke Selesiastahl GmbH
C 0.7, Si 1.7, Mn 0.7, bal Fe.
For springs; oil hardened, tough.

SILESIA R2
Georgsmarienwerke Selesiastahl GmbH
C 1.3, Si 0-0.25, Mn 0-0.25, bal Fe.
For engravers' tools, textile needles, reamers, drills; Type W1;
water hardened.

SILESIA R3
Georgsmarienwerke Selesiastahl GmbH
C 1.15, Si 0-0.25, Mn 0-0.25, bal Fe.
Annealed: 110,000 TS; 58,000 YS; 18 El; 38 RA; 210 Brin. For
springs, drills, reamers, broaches, Type W1; water hardened.

SILESIA R4
Georgsmarienwerke Selesiastahl GmbH
C 1, Si 0-0.25, Mn 0-0.25, bal Fe.
Annealed: 100,000 TS; 53,000 YS; 21 El; 42 RA; 200 Brin. For
springs, taps, drills, reamers, broaches; Type W1; water
hardened.

SILESIA R5
Georgsmarienwerke Selesiastahl GmbH
C 0.85, Si 0-0.25, Mn 0-0.25, bal Fe.
Heat treated: 190,000 TS; 145,000 YS; 10 El; 30 RA; 400 Brin.
For springs, taps, drills, cutters, reamers; Type W1; water
hardened.

SILESIA R6
Georgsmarienwerke Selesiastahl GmbH
C 0.7, Si 0-0.25, Mn 0-0.25, bal Fe.
Heat treated: 175,000 TS; 128,000 YS; 12 El; 37 RA; 355 Brin.
For rails, punches, springs, girders; Type W1; water
hardened.

SILESIA RWA
Georgsmarienwerke Selesiastahl GmbH
C 0.45, Cr 1.4, Mo 0.7, V 0.3, bal Fe.
For forging and die casting dies; oil hardened.

SILESIA RWA EXTRA
Georgsmarienwerke Selesiastahl GmbH
C 0.4, Si 0.25, Cr, Mn, Mo, bal Fe.
For gears, bolts, crankshafts; oil hardened, tough.

SILESIA RWA SPEZIAL
Georgsmarienwerke Selesiastahl GmbH
C 0.3, Cr 1.1, V 0.18, W 3.75, Si 1, bal Fe.
For header and forging dies, punches; oil hardened, tough.

SILESIA SC
Georgsmarienwerke Selesiastahl GmbH
C 0.67, Si 1.3, Mn 0.5, Cr 0.5, bal Fe.
For springs, chisels, punches; oil hardened, tough.

SILESIA SCN
Georgsmarienwerke Selesiastahl GmbH
C 0.45, Ni, Cr, bal Fe.
For gears, bolts, shafts, axles; oil hardened, shock resistant.

SILESIA TCN SPEZIAL
Georgsmarienwerke Selesiastahl GmbH
C 0.5, Cr 1.05, Ni 3.25, bal Fe.
For gears, bolts, crankshafts, axles; oil hardened, shock
resistant.

SILESIA TCS
Georgsmarienwerke Selesiastahl GmbH
C 0.61, Cr 1.18, V 0.1, Mn 0.75, Si 0.85, bal Fe.
For forging dies, punches, upsetters; oil hardened, tough.

SILESIA TCV
Georgsmarienwerke Selesiastahl GmbH
C 1.15, Cr 0.65, V 10, bal Fe.
For bearings, cold work tools; oil hardened, wear resistant.

SILESIA TFE
Georgsmarienwerke Selesiastahl GmbH
C 1.05, Cr 1, W 1.15, Mn 0.9, bal Fe.
For bearings, sleeves, liners; water hardened, wear resistant.

SILESIA TFE24
Georgsmarienwerke Selesiastahl GmbH
C 1.42, W, V, bal Fe.
For engravers' tools, form cutters; water hardened.

SILESIA TKL1
Georgsmarienwerke Selesiastahl GmbH
C 1, Cr 1.55, Mn 0.35, bal Fe.
For bearings, liners, sleeves; oil or water hardened, wear
resistant.

SILESIA TKL2
Georgsmarienwerke Selesiastahl GmbH
C 1.05, Cr 1, Mn 0.3, bal Fe.
For bearings, liners, sleeves; water hardened, wear resistant.

SILESIA TKL3
Georgsmarienwerke Selesiastahl GmbH
C 1.1, Cr 0.4, Mn 0.3, Si 0.2, bal Fe.
For bearings, liners, cutters; water hardened, wear resistant.

SILESIA TSPE
Georgsmarienwerke Selesiastahl GmbH
C 1.2, V 0.1, W 1, Mn 0.28, bal Fe.
For blanking and heading dies; oil hardened, wear resistant.

SILESIA TV3
Georgsmarienwerke Selesiastahl GmbH
C 1, V 0.1, Mn 0.25, Si 0.2, bal Fe.
For bearings, heading dies; Type W2; water hardened.

SILESIA TWA
Georgsmarienwerke Selesiastahl GmbH
C 0.3, Cr 2.65, V 0.35, W 8.5, bal Fe.
For extrusion rams and liners, punches; oil hardened, hot
work steel.

SILESIA TWA SPEZIAL
Georgsmarienwerke Selesiastahl GmbH
C 0.3, Co 2, Cr 2.4, V 0.25, W 8.5, bal Fe.
For extrusion rams and punches, shears; hot work steel, oil
hardened.

SILESIA TWA24
Georgsmarienwerke Selesiastahl GmbH
C 0.3, Cr 2.35, V 0.6, W 4.25, Mn 0.3, bal Fe.
For extrusion rams, forging dies, punches; oil hardened, hot
work steel.

SILESIA TZ
Georgsmarienwerke Selesiastahl GmbH
C 1.4, Cr 0.3, V 0.1, Mn 0.3, bal Fe.
For engravers' tools, blanking and forming dies; oil
hardened, wear resistant.

SILESIA V201
Georgsmarienwerke Selesiastahl GmbH
C 0.55, Cr 0.7, Mo 0.18, Ni 1.65, V 0.1, bal Fe.
For forging and heading dies, punches; oil hardened, tough.

SILESIA V201 SPEZIAL
Georgsmarienwerke Selesiastahl GmbH
C 0.56, Cr 0.9, Mo 0.2, V 0.1, Ni 2, bal Fe.
For forging and die casting dies; oil hardened, tough.

SILESIA W35
Georgsmarienwerke Selesiastahl GmbH
C 0.35, Si 0.25-0.5, Mn 0.3-0.8, bal Fe.
Hot rolled: 85,000 TS; 54,000 YS; 30 El; 54 RA; 185 Brin. For
gears, shafts, axles, bolts; water hardened.

SILESIA W45
Georgsmarienwerke Selesiastahl GmbH
C 0.45, Si 0.25-0.5, Mn 0.3-0.8, bal Fe.
Hot rolled: 98,000 TS; 58,000 YS; 24 El; 45 RA; 212 Brin. For
axles, gears, bolts, shafts, tie rods; water hardened.

SILESIA W60
Georgsmarienwerke Selesiastahl GmbH
C 0.6, Si 0.25-0.5, Mn 0.3-0.8, bal Fe.
Heat treated: 160,000 TS; 113,000 YS; 12 El; 40 RA; 320 Brin.
For wheels, die blocks, girders, rails, bushings; water
hardened.

SILESIA W75
Georgsmarienwerke Selesiastahl GmbH
C 0.75, Si 0.25-0.5, Mn 0.3-0.8, bal Fe.
Heat treated: 180,000 TS; 135,000 YS; 12 El; 36 RA; 370 Brin.
For springs, rails, girders, clutch discs; Type W1; water
hardened.

SILESIA W90
Georgsmarienwerke Selesiastahl GmbH
C 0.9, Si 0.25-0.5, Mn 0.3-0.8, bal Fe.
Heat treated: 180,000 TS; 120,000 YS; 10 El; 30 RA; 380 Brin.
For springs, taps, reamers, drills, broaches; Type W1; water
hardened.

SILESIA WG
Georgsmarienwerke Selesiastahl GmbH
C 0.53, Si 0.9, Mn 0.9, bal Fe.
For punches, upsetters, bolts; water hardened.

SILESIA Z3
Georgsmarienwerke Selesiastahl GmbH
C 0.35, Si 0.9, Cr 1.05, V 0.18, W 1.85, bal Fe.
For header and forging dies, punches; oil hardened, tough.

SILESIA Z4 SPEZIAL
Georgsmarienwerke Selesiastahl GmbH
C 0.45, W, Cr, V, bal Fe.
For header and forging dies, die casting dies; oil hardened, tough.

SILESIA Z5
Georgsmarienwerke Selesiastahl GmbH
C 0.55, Si 0.9, Cr 1.05, V 0.18, W 1.85, bal Fe.
For header and forging dies, punches; oil hardened, tough.

SILESIA Z6
Georgsmarienwerke Selesiastahl GmbH
C 0.38, Si, Cr, V, bal Fe.
For gears, bolts, crankshafts; oil hardened, tough.

SILESIA Z7
Georgsmarienwerke Selesiastahl GmbH
C 0.45, Si, Cr, V, bal Fe.
For springs, gears, bolts, crankshafts; oil hardened, shock resistant.

SILEX-65
Forjas Alavesas S.A.
C 0.65, Mn 0.8, Si 1.8, bal Fe.
Shock resistant tool steel; picks and shovels.

SILFERAL
Aluminiumwerke Maulbronn
Si, bal Al.
Chill cast: 25,000-29,000 TS; 2-3 El; 65 Brin. For cylinder heads, motorcycle cylinders; non-hardenable.

SILFLO "5" ETC
Now ALL-STATE SILFLO "5" ETC.

SILFRAM
Stoody Company
C 2, Si 2, W 2, Ni 10, Cr 31, bal Fe.
Weld: 360 Brin. For welding of rod, hard facing for agricultural equipment; corrosion, abrasion and wear resisting. *Obsolete*

SILICAL
Swedish manufacture
Si 11-14, bal Al.
For exhaust valves; austenitic.

SILICARB
Bethlehem Steel Corp.
C 0.55, Mn 0.75, Mo 0.2, Si 2, bal Fe.
For impact tools; shock resistant, water or oil hardened. *Obsolete*

SILICO ALLOY (S5)
Columbia Tool Steel Co.
Tool material. C 0.58, Mn 0.85, Si 1.95, Mo 0.45, Cr 0.3, bal Fe.
Annealed: 108,000 TS; 64,000 YS; 27 El; 212 Brin. For shear blades, pneumatic tools, mandrels, bending dies, pipe cutters, chisels, punches. Oil or water hardened. Type S5; tool steel. Shock resistant. Tough.

SILICO I
Krupp Stahl AG
alloy steel. C 0.45, Cr, Si, bal Fe.
For gears, shafts, crankshafts, fasteners; water hardened. *Obsolete*

SILICO II
Krupp Stahl AG
alloy steel. C 0.45, Cr, Si, bal Fe.
For gears, bolts, shafts, fasteners; water hardened. *Obsolete*

SILICO MANGANESE SPRING
U.S. Steel Corp.
C 0.55-0.65, Mn 0.7-1, Si 1.8-2, bal Fe.
For springs; oil hardening. *Obsolete*

SILICO MANGANESE SPRING WIRE
Crucible Materials Corp.
Alloy steel. C 0.5, Mn 0.7, Si 2, bal Fe.
For coil and flat springs, automatic springs; high fatigue resistance.

SILICO NICKEL
Gilby-Fodor S.A.
Cu 96, Si 2.5, Mn 1.5.
For cathodes and filaments in electronic tubes. Heat resistant.

SILICO-CHROMIUM STEEL
American manufacture
Cr 9-12, Si 5, C 1.2, bal Fe.
For valves, heat resistant parts.

SILICO-MANGANESE
Foote Mineral Co.
C 3, Mn 65-70, bal Si. Modifications: 2.5% C, 65-70% Mn, bal Si; 2.0% C,, 65-70% Mn, bal Si; 1.0% max C, 65-70% Mn, bal Si.
For metallurgical applications in steel. *Obsolete*

SILICO-MANGANESE BRASS
American manufacture
Zn 40, Mn 1-2, Fe 0.2-0.3, Si 0.05-1, bal Cu.
For marine parts, architectural applications; corrosion resistant.

SILICO-MANGANESE STEEL
American manufacture
Mn 0.45-1.5, Si 0.4-1.9, bal Fe.
For springs.

SILICO-SPIEGEL
Union Carbide Corp.
Mn 15-20, Si 8-15, P 0.15, C, bal Fe.
For steel making; Si addition.

SILICON
Atomergic Chemetals Corp.
Si.
Purities: semiconductor (N and P types), hyper-pure (99.9999%), optical (99.999%), 99.99%, 99.9%. Forms: polycrystalline, powder, float zoned and czochraiski single crystal granules, epitaxial, boules, deposited rod, chunk, etc.

SILICON ALUMINUM VANADIUM
Foote Mineral Co.
Si 80, Al 10, V 10.
For use in manufacture of non-ferrous alloys. *Obsolete*

SILICON BRONZE ELECTRODE
J.W. Harris Co., Inc.
Silicon bronze electrode; AC-DC. AWS A5-6 ECuSi. For surfacing steel and for joining cast iron to steel and malleable iron.

SILICON COPPER 943
Anaconda Co.
Si 0.25, bal Cu.
For welding rod; for copper to steel. *Obsolete*

SILICON COPPER E-1232
Accurate Brass Co.
Si 2.75-3.25, bal Cu.
Forged: 50,000-55,000 TS; 30,000-35,000 YS; 25-30 El; 95-120 Brin. For electrical applications; forgings, high conductivity.

SILICON CORE IRON
Now CARPENTER SILICON CORE IRON.

SILICON CORE IRON B-FM
Carpenter Technology Corp.
C 0-0.05, Mn 0-0.4, Si 2.25-2.75, P 0.08-0.15, bal Fe.
Coercive force from 10,000 Gausses: 0.65 Oersted. Residual induction from 10,000 Gausses: 5600-7600 Gausses. Saturation 20,000 Gausses at H = 200. For solenoid switches, armatures, pole pieces, relays. Free machining, high permeability.

SILICON MONEL.METAL 2.5%
International Nickel Inc.
Ni 64-71, Cu 28-30, Fe 1-2, Si 2.5-3, Mn 0.5-0.75, C 0.15-0.25.
Cast: 85,000 TS; 50,000 YS; 16 El; 250 Brin. For valves, power plant equipment; combines some hardness with moderate ductility. *Obsolete*

SILICON NICKEL A
Wilbur B. Driver Co.
Ni 95, Si 2.5, Mn 2.5.
Annealed. For electrical resistances. *Obsolete*

SILICON NICKEL B
Wilbur B. Driver Co.
Ni 97, Si 3.
For filaments; manganese free. *Obsolete*

SILICON RED BRASS
Chase Brass & Copper Co., Inc.
Copper. Cu 81.5, Zn 14.5, Si 4.
Annealed: 85,000 psi TS; 43,000 psi YS; 25 El. Drawn (7%): 100,000 psi TS; 57,000 psi YS; 21 El. For high strength valve stems.

SILICON RED BRASS-6942
Anaconda Co.
Copper. Cu 82, Zn 17, Si 1.
Soft: 55,000 TS; 20,000 YS; 60 El. Hard: 90,000 TS; 50,000 YS; 8 El. For resistance welding.

SILICON STEEL-M15
Allegheny Ludlum Steel
Si 4, bal Fe.
Rolled: 61,000-79,000 TS; 60,000-66,000 YS; 4-8 El; Rock B 75. For motors, generators, transformers, lamination. High magnetic permeability. Transformer M58 Grade. *Obsolete*

SILICON STEEL-M15
U.S. Steel Corp.
Si 4, bal Fe.
Rolled: 61,000-79,000 TS; 60,000-66,000 YS; 4-8 El; Rock B 75. For motors, generators, transformers, lamination. High magnetic permeability. Transformer M58 Grade. *Obsolete*

SILICON STEEL-M15
Armco Steel Corp.
Si 4, bal Fe.
Rolled: 61,000-79,000 TS; 60,000-66,000 YS; 4-8 El; Rock B 75. For motors, generators, transformers, lamination. High magnetic permeability. Transformer M58 Grade. *Obsolete*

SILICON STRUCTURAL STEEL
Bethlehem Steel Corp.
C 0.25, Si 0.5, bal Fe.
Rolled: 45,000 TS. For bridges, towers, structures; not weldable. *Obsolete*

SILICON TIN BRONZE 5072
Anaconda Co.
Copper. Cu 97.5, Sn 1.75, Si 0.75.
Forged: 45,000 TS; 14,000 YS; 40 El. For electrical apparatus. Corrosion resistant, forgeable.

SILICON VANADIUM
Foote Mineral Co.
V 40, Si 60.
For use in manufacture of non-ferrous alloys. *Obsolete*

SILICON-FERROCHROME
American manufacture
Cr 45-55, Fe 30-50, Si 1-17, C 2.5-7.
For steel making; Cr-addition agent.

SILICON-MANGANESE
Union Carbide Corp.
Mn 68, Si 20, Fe 11, C 0.7.
For steel making; Mn addition.

SILICON-NICKEL
Union Carbide Corp.
Ni 80-40, Si 16-18, Fe 2.5-30.
For steel making; Ni addition.

SILIMAZ
Krupp Stahl AG
alloy steel. C 0.6, Si 15.5, Mn 2.6, bal Fe.
Cast: 15,000 TS; 50 Rock C. For pumps, valves, drains,
agitators, fittings, castings. Acid resistant. Brittle as cast.
Sensitive to thermal shock. *Obsolete*

SILIMO
Bethlehem Steel Corp.
C 0.5, Mn 0.4, Mo 0.5, V 0.2, Si 1.1, bal Fe.
For impact tools; shock resistant, water or oil hardened.
Obsolete

SILKORRIT-2
Georgsmarienwerke Selesiastahl GmbH
C 0-0.07, Cr 18, Ni 9.5, bal Fe.
Annealed: 85,000 TS; 35,000 YS; 60 El; 75 RA; 150 Brin. Cold
drawn: 180,000 TS; 125,000 YS; 10 El; 330 Brin. For chemical
plant equipment, welded structures; Type 304; stainless,
austenitic.

SILKORRIT-2T
Georgsmarienwerke Selesiastahl GmbH
C 0-0.1, Cr 18, Ni 8.5, bal Fe.
Annealed: 80,000 TS; 35,000 YS; 55 El; 75 RA; 150 Brin. Cold
drawn: 180,000 TS; 130,000 YS; 10 El; 250 Brin. For chemical
plant equipment, tanks, mixers, agitators; Type 302; stainless,
austenitic.

SILKORRIT-2U
Georgsmarienwerke Selesiastahl GmbH
C 0-0.12, Cr 18, Ni 9.5, Ti = 4 x C, bal Fe.
Annealed: 85,000 TS; 35,000 YS; 55 El; 65 RA; 150 Brin. Cold
drawn: 95,000 TS; 60,000 YS; 40 El; 60 RA; 185 Brin. For
welded chemical plant equipment, tanks, mixers; Type 321;
stainless, austenitic.

SILKORRIT-4
Georgsmarienwerke Selesiastahl GmbH
C 0-0.07, Cr 18, Mo 2, Ni 10.5, bal Fe.
Annealed: 85,000 TS; 35,000 YS; 50 El; 65 RA; 160 Brin. Cold
drawn: 150,000 TS; 135,000 YS; 6 El; 300 Brin. For acid
resistant chemical plant equipment; Type 316; stainless,
austenitic.

SILKORRIT-4T
Georgsmarienwerke Selesiastahl GmbH
C 0-0.1, Cr 18, Mo 2.2, Ni 9.5, bal Fe.
Annealed: 85,000 TS; 35,000 YS; 50 El; 65 RA; 160 Brin. Cold
drawn: 150,000 TS; 135,000 YS; 6 El; 300 Brin. For acid
resistant chemical plant equipment; Type 316; stainless,
austenitic.

SILKORRIT-4U
Georgsmarienwerke Selesiastahl GmbH
C 0-0.12, Cr 18, Ni 10.5, Mo 2, Ti = 4 x C, bal Fe.
Annealed: 85,000 TS; 35,000 YS; 45 El; 60 RA; 160 Brin. For
welded acid resistant chemical plant equipment; Type 316 Ti;
stainless, austenitic.

SILLMAN BRONZE
English manufacture
Al 9.7, Fe 3.9, bal Cu.
For jewelry, bullet shells; corrosion resistant, Al-bronze.

SILMA
Deutsche Gold und Silber Scheide
Ag 85, Mn 15, 0.0-0.5 others.
For silver solder brazing; melting point 1790°F; shearing
strength, 27000; for high temperature service.

SILMA
Gold Und Silber Scheide Anstalt
Ag 85, Mn 15.
For solder; high temperature service; melting point 1790°F.

SILMAL
French manufacture
Mg 1, Si 1.75, bal Al.
Rolled: 38,000-43,000 TS; 25-30 El. For light alloy parts;
same as "Silumin."

SILMALAR
Aluminium Wire & Cable Co., Ltd.
Mg 0.5, Si 0.5, bal Al.
Rolled: 43,000 TS. For conductors; conductivity 53.5 I ACS.
Obsolete

SILMALEC
Alcan-Booth Industries, Ltd.
Mg 0.5, Si 0.5, bal Al.
For light alloy parts; age-hardening alloy. *Obsolete*

SILMANAL
Colt Industries
Ag 86.75, Mn 8.8, Al 4.45.
Bar: 110,000 TS; 230 Brin. 590 Br, 6300 Hc, 292 Bo. For
electrical and magnetic equipment. High coercive force.
Permanent magnet. *Obsolete*

SILMANAL
Indiana General
Ag 86.75, Mn 8.8, Al 4.45.
Bar: 110,000 TS; 230 Brin. 590 Br, 6300 Hc, 292 Bo. For
electrical and magnetic equipment. High coercive force.
Permanent magnet. *Obsolete*

SILMANAL
General Electric Co.
Ag 86.75, Mn 8.8, Al 4.45.
For permanent magnets; high coercive force. *Obsolete*

SILMELEC
French manufacture
Al 97.8, Mg 0.6, Mn 0.6, Si 1.
For electrical conductor wires; high electrical conductivity.

SILMET
Barker & Allen Ltd.
Ni-Cu-Zn alloys.
For deep stamping in electroplate trade; deep stamping and
spinning.

SILMO
Timken Co.
C 0-0.15, Si 1.15-1.65, Mo 0.45-0.65, bal Fe.
Annealed: 60,000 TS; 30,000 YS; 30 El; 50 RA; 103 Brin. For
high temperature tubing; heat resistant to 1000 F. *Obsolete*

SILNIC BRONZE (COPPER ALLOY NO. 647)
Chase Brass & Copper Co., Inc.
Copper. Ni 1.6-2.2, Si 0.45-0.75, bal Cu.
Aged: 90,000 psi TS; 80,000 psi YS; 12 El. For fasteners,
bolts, structural components; high strength, corrosion
resistant.

SILUMIN
Vereinigte Leichtmetallwerke, G.m.b.H.
Aluminum. Si 12-13.5, bal Al.
Hard: 36,000 TS; 2 El. For cylinders; wrought.

SILUMIN
Vereinigte Deutsche Metallwerke AG
Si 11-14, bal Al.
Annealed: 21,000 TS; 8500 YS; 20 El; 50 Brin. For automotive
castings, instruments, housings; corrosion resistant, easy to
cast.

SILUMIN
Metallgesellschaft A.G.
Si 11-14, bal Al.
Annealed: 21,000 TS; 8500 YS; 20 El; 50 Brin. For automotive
castings, instruments, housings; corrosion resistant, easy to
cast.

SILUMIN
Maywood Chemical Works
Si 11-14, bal Al.
Sand cast: 30,000 TS; 13,000 YS; 8 El; 60 Brin. For
automotive castings; good castability, corrosion resistant.
Obsolete

SILUMIN-BETA
Metallgesellschaft A.G.
Si 12, Mg 0.3, Mn 0.5, bal Al.
Sand cast: 25,000 TS; 14,000 YS; 2-4 El; 60 Brin. For
automotive castings; high fluidity, corrosion resistant.

SILUMIN-BETA
Vereinigte Deutsche Metallwerke AG
Si 12, Mg 0.3, Mn 0.5, bal Al.
Sand cast: 25,000 TS; 14,000 YS; 2-4 El; 60 Brin. For
automotive castings; high fluidity, corrosion resistant.

SILUMIN-GAMMA
Metallgesellschaft A.G.
Si 12, Mn 0.5, Mg 0.3, bal Al.
Cast: 28,000 TS; 16,000 YS; 3 El; 60 Brin. For automotive
castings; good castability, corrosion resistant.

SILUMIN-GAMMA
Vereinigte Deutsche Metallwerke AG
Si 12, Mn 0.5, Mg 0.3, bal Al.
Cast: 28,000 TS; 16,000 YS; 3 El; 60 Brin. For automotive
castings; good castability, corrosion resistant.

SILVALOY
Gilby-Fodor S.A.
Si 3, Ni 97.
Annealed: 75,000 TS; 55,000 YS; 25 El. For cathodes and
filaments in electronic tubes. Heat resistant.

SILVALOY
Pyramid Steel Company
C, alloy, bal Fe.
Heat treated: 164,000 TS; 141,000 YS; 17 El; 57 RA; 280 Brin.
For shafting, pump rods; preheat treated steel.

SILVALOY 105
Engelhard Corp.
Ag 45, Cu 30, Zn 12, Mn 13.
For brazing; good strength to 700°F; 1298°F MP.

SILVALOY 15
Engelhard Corp.
Ag 15, P 5, Cu 80.
For silver brazing for Cu alloys; 1185-1280°F MP.

SILVALOY 20
Engelhard Corp.
Ag 20, Cu 45, Zn 30, Cd 5.
For silver brazing; 1130-1500°F MP.

SILVALOY 250
Engelhard Corp.
Ag 40, Cu 30, Zn 28, Ni 2.
For silver brazing for stainless steel; 1240-1435°F MP.

SILVALOY 254
Engelhard Corp.
Ag 40, Cu 30, Zn 25, Ni 5.
For brazing steels; hardening and brazing in one operation.
1220-1580°F MP.

SILVALOY 301
Engelhard Corp.
Ag 72, Cu 28.
For silver brazing; 1435°F MP.

SILVALOY 35
Engelhard Corp.
Ag 35, Cu 26, Zn 21, Cd 18.
For silver brazing; 1125-1295°F MP.

SILVALOY 355
Engelhard Corp.
Ag 56, Cu 22, Zn 17, Sn 5.
For brazing. 1145-1205°F MP.

SILVALOY 377
Engelhard Corp.
Ag 5, Cd 95.
Uses: Brazing alloy. 640-740°F MP.

SILVALOY 40
Engelhard Minerals & Chemicals Corp.
Ag 40, Cu 18, Zn 15, Cd 27.
For silver brazing; MP 1135-1205 F. *Obsolete*

SILVALOY 45
Engelhard Corp.
Ag 45, Cu 15, Zn 16, Cd 24.
For silver brazing; 1145-1190°F MP.

SILVALOY 45M
Engelhard Minerals & Chemicals Corp.
Ag 45, Cu 15, Zn 16, Cd 24.
For silver solder; M.P. 1125-1145 F. *Obsolete*

SILVALOY 5
Engelhard Corp.
Ag 5, Cu 88.5, P 6.5.
For brazing or joining copper; 1185-1280°F MP.

SILVALOY 50
Engelhard Corp.
Ag 50, Cu 15.5, Zn 16.5, Cd 18.
For silver brazing; 1160-1175°F MP.

SILVALOY 503
Engelhard Corp.
Ag 50, Cu 15.5, Zn 15.5, Cd 16, Ni 3.
For silver brazing of carbide tipped tools; 1195-1270°F MP.

SILVALOY 54
Engelhard Corp.
Ag 54, Cu 40, Zn 5, Ni 1.
For brazing. 1325-1575°F MP.

SILVALOY 6
Engelhard Corp.
Ag 6, Cu 88, P 6.
For brazing alloy. 1190-1300°F MP.

SILVALOY 60
Engelhard Corp.
Ag 60, Cu 30, Sn 10.
For atmosphere furnace brazing. 600-725°C MP.

SILVALOY 850
Engelhard Corp.
Ag 85, Mn 15.
For silver brazing; 1760-1778°F MP.

SILVALOY A-11
Engelhard Corp.
Ag 20, Cu 45, Zn 35.
For silver brazing; 1315-1500°F MP.

SILVALOY A-13
Engelhard Corp.
Ag 30, Cu 38, Zn 32.
For silver brazing; 1370-1410°F MP.

SILVALOY A-14
Engelhard Corp.
Ag 40, Cu 36, Zn 24.
For silver brazing; 1235-1415°F MP.

SILVALOY A-18
Engelhard Corp.
Ag 45, Cu 30, Zn 25.
For silver brazing; 1250-1370°F MP.

SILVALOY A-25
Engelhard Corp.
Ag 50, Cu 34, Zn 16.
For silver brazing; 1272-1425°F MP.

SILVALOY A-28
Engelhard Corp.
Ag 50, Cu 28, Zn 22.
For silver brazing; 1250-1340°F MP.

SILVALOY A-33
Engelhard Corp.
Ag 60, Cu 25, Zn 15.
For silver brazing; 1260-1325°F MP.

SILVALOY A-4
Engelhard Corp.
Ag 9, Cu 53, Zn 38.
For silver brazing; 1410-1565°F MP.

SILVALOY A-49
Engelhard Corp.
Ag 80, Cu 16, Zn 4.
For silver brazing; 1360-1490°F MP.

SILVALOY A-79
Engelhard Corp.
Ag 25, Cu 52.5, Zn 22.5.
For brazing alloy; 1250-1575°F MP.

SILVALOY AE-100
Engelhard Corp.
Ag 92.5, Cu 7.3, Li 0.2.
For atmosphere furnace brazing. 1435-1635°C MP.

SILVALOY AE-102
Engelhard Corp.
Ag 72, Cu 27.8, Li 0.2.
For brazing. 1410°F MP. Eutectic alloy.

SILVALOY AT-600
Engelhard Corp.
Ag 98, Li 2.
For brazing. 1290-1400°F MP.

SILVALOY AT-601
Engelhard Corp.
Ag 97, Li 3.
For brazing. 1115-1235°F MP.

SILVALOY B-70
Engelhard Corp.
Ag 7, Cu 85, Sn 8.
For brazing. 1225-1805°F MP.

SILVALOY EASY
Engelhard Corp.
Ag 65, Cu 20, Zn 15.
For silver solder; 1280-1325°F MP.

SILVALOY HARD
Engelhard Corp.
Ag 75, Cu 22, Zn 3.
For silver solder; 1365-1450°F MP.

SILVALOY I-401
Engelhard Corp.
Ag 5, Zn 16.6, Cd 78.4.
For brazing. 480-600°F MP.

SILVALOY K-427
Engelhard Corp.
Ag 75, Zn 25.
For brazing. 1300-1345°F MP.

SILVALOY MEDIUM
Engelhard Corp.
Ag 70, Cu 20, Zn 10.
For silver solder; 1275-1360°F MP.

SILVALOY S-475
Engelhard Corp.
Ag 95, Al 5.
For brazing. 1550-1600°F MP.

SILVALOY T-51
Engelhard Corp.
Ag 75, Cu 24.5, Ni 0.5.
For brazing alloy. 1435-1475°F MP.

SILVALOY T-53
Engelhard Corp.
Ag 71.5, Cu 28, Ni 5.
For brazing. 1435-1450°F MP.

SILVALOY T50
Engelhard Corp.
Ag 62.5, Cu 32.5, Ni 5.
For silver brazing for precipitation hardening steels;
1435-1590°F MP; good wetting.

SILVALOY T52
Engelhard Corp.
Ag 77, Cu 21, Ni 2.
For silver brazing for precipitation hardening steels;
1435-1525°F MP; good wetting.

SILVALOY VTG 301
Engelhard Corp.
Ag 72, Cu 28.
For brazing filler metal for electronic applications. 780°C MP.

SILVALOY VTG 303
Engelhard Corp.
Ag 50, Cu 50.
Annealed wire: 60,8000 psi TS; 13 El. For brazing filler metals
for electronic applications. 780-855°C MP.

SILVALOY VTG 60
Engelhard Corp.
Ag 60, Cu 30, Sn 10.
For brazing filler metals for electronic applications.
600-725°C MP.

SILVALOY VTG-AQ602
Engelhard Corp.
Ag 61.5, Cu 24, In 14.5.
Annealed wire: 58,700 psi TS; 37.5 El. For brazing filler
metals for electronic applications. 630-705°C MP.

SILVANITE
Columbia Tool Steel Co.
Tool material. C 1.3, W 8, Cr 4, V 0.5, bal Fe.
Special purpose high speed tool steel; hardened: 60-64 Rock
C; ready to use. For woodworking tools, scrapers, paper
knives. Type F8.

SILVAZ
Union Carbide Corp.
B 0.6, Si 38, Al 6, Ti 10, Zr 6, V 10, bal Fe.
For deoxidizer; for steel making.

SILVAZ 3
Union Carbide Corp.
B 0.5, Si 35-40, Al 6, Ti 10, Zr 6, V 10, bal Fe.
For hardening constituent in steel; adding B to steel.

SILVEL
Manufacturer not listed.
Cu 67.5, Zn 16, Ni 6.5, Fe 2.2.
Hard drawn: 130,000 TS; 2 El. For tableware, plumbing
fixtures, electric contact springs; German Silver.

SILVEL

Manufacturer not listed.
Cu 73, Mn 12, Sn 12, Fe 1.8, Pb 0.5, Al 0.3.
Annealed: 73,000 TS; 32 El. For tableware, plumbing fixtures, electric contact springs; German Silver.

SILVER

Atomergic Chemetals Corp.
Ag.
Purities, zone refined: 99.9999%, 99.999%, 99.99%, 99.9%.
Forms: rod, ingot, shot, sheet, powder, wire, foil, single crystals.

SILVER ALLOYED COPPER 110

Anaconda Co.
Copper. Ag 1, Mn 0.1, bal Cu.
For welding rod for oxyacetylene welding of copper.

SILVER BEARING COPPER 111

Anaconda Co.
Copper. Cu 99.9, bal Ag.
Hard: 48,000 TS; 40,000 YS; 6 El; 83 Brin. Soft: 33,000 TS; 10,000 YS; 45 El; 42 Brin. For electrical apparatus. High conductivity.

SILVER BEARING COPPER, COPPER NO. 113

Anaconda Co.
99.7 + Cu, 8 oz/ton Ag.
Sheet-annealed or cold rolled. For core and fin stock for radiators, assemblies for tin dipping. Higher softening temperature than pure copper. ASTM B152.

SILVER BEARING COPPER, COPPER NO. 113

Chase Brass & Copper Co.
99.7 + Cu, 8 oz/ton Ag.
Sheet-annealed or cold rolled. For core and fin stock for radiators, assemblies for tin dipping. Higher softening temperature than pure copper. ASTM B152.

SILVER BEARING COPPER, COPPER NO. 116

Anaconda Co.
99.7 + Cu, 25 oz/ton Ag.
Sheet-annealed or cold rolled. For core and fin stock for radiators, commutator segments, tin dipped assemblies. Higher softening temperature than pure copper. ASTM B152.

SILVER BEARING COPPER, COPPER NO. 116

Chase Brass & Copper Co.
99.7 + Cu, 25 oz/ton Ag.
Sheet-annealed or cold rolled. For core and fin stock for radiators, commutator segments, tin dipped assemblies. Higher softening temperature than pure copper. ASTM B152.

SILVER BEARING COPPER-112

Anaconda Co.
Copper. Cu 99.9, 10-15 oz Ag/ton.
Soft: 33,000 TS; 10,000 YS; 35 El. Hard: 46,000 TS; 40,000 YS; 5 El. For electrical apparatus. For high softening temperature.

SILVER BEARING COPPER-114

Anaconda Co.
Copper. Cu 99.9, Ag 0.034-0.051.
Hard: 48,000 TS; 40,000 YS; 6 El; 50 Rock B. Soft: 33,000 TS; 10,000 YS; 45 El; 45 Rock F. For electrical and electronic components. Electrical conductivity: 100%.

SILVER BRONZE

English manufacture
Zn 13-23, Ni 16-18, Al 0-2, Pb 0-2, Si 0.3, bal Cu.
For bearings, bushings, hardware; corrosion resistant.

SILVER CHISEL

Nagle/Sybron Corp.
C, alloy, bal Fe.
For welding electrode for chisels; tough, shock resistant.

SILVER COPPER

Criterion Metals, Inc.
Copper. Cu 99.8, P 0.05, Ag 0-0.034, Mg 0-0.11.
Thin gauge sheet, various tempers: 15-72 ksi YS min; 34-76 ksi TS min. C15500

SILVER FOIL-1

American manufacture
Sn 90, Zn 10.
For foil for wrapping purposes.

SILVER FOIL-2

American manufacture
Sn 98, Cu 2.
For foil for wrapping purposes.

SILVER FOIL-3

American manufacture
Sn 91, Zn 8.3, Pb 0.4.
For foil for wrapping purposes.

SILVER FOX 10

British Steel Corp.
C 0.1, Cr 12-13, bal Fe.
Air hardened: 85,000-166,000 TS; 66,000-142,000 YS; 13-32 El; 37-67 RA. For turnbuckles, rivets, split pins, shrouding for turbine blading; stainless iron. *Obsolete*

SILVER FOX 11

British Steel Corp.
C 0.15, Cr 12-13, bal Fe.
Heat treated: 107,000 TS; 30 El; 50 RA. For taper pins, hinge pins, surgical instruments, scissors; stainless, martensitic. *Obsolete*

SILVER FOX 12

British Steel Corp.
C 0.2, Cr 12-13, bal Fe.
Heat treated: 107,000 TS; 30 El; 50 RA. For valve parts, precision instruments; stainless martensitic. *Obsolete*

SILVER FOX 13

British Steel Corp.
C 0.3, Cr 12-14, bal Fe.
Heat treated: 112,000 TS; 28 El; 50 RA. For table cutlery, special tools, exhaust valves; heat resistant; high strength at elevated temperature. *Obsolete*

SILVER FOX 14

British Steel Corp.
C 0.4, Cr 14, bal Fe.
Air hardened: 89,500-235,000 TS; 10-35 El; 20-70 RA; 183-477 Brin. For dies and molds for plastic materials; corrosion resistant; IZ 20-100. *Obsolete*

SILVER FOX 15

British Steel Corp.
C 0.7, Cr 14, bal Fe.
For stainless parts, cutlery; stainless, hardenable. *Obsolete*

SILVER FOX 17

British Steel Corp.
C 0-0.12, Cr 16-18, bal Fe.
Annealed: 78,000-101,000 TS; 28 El; 60 RA. For nitric acid manufacture apparatus; non-hardenable; resists corrosion. *Obsolete*

SILVER FOX 18

British Steel Corp.
C 0-0.2, Cr 18, Ni 2, bal Fe.
Tempered: 123,000-145,000 TS; 22 El; 40 RA. For precision instruments, valves pump rods; stainless, martensitic. *Obsolete*

SILVER FOX 24

British Steel Corp.
Cr 12, Ni 12, C, alloys, bal Fe.
Annealed: 90,000-135,000 TS; 26,000-45,000 YS; 60-80 El; 60-80 RA. For centrifugal separators, dryers; resists sulphuric acid corrosion; stainless. *Obsolete*

SILVER FOX 321

British Steel Corp.
C 0-0.08, Cr 17-19, Ni 9-11, Ti, bal Fe.
Annealed: 85,000 TS, 30,000 YS; 55 El, B 80 Rock. For welded structures, tanks, chemical plant equipment, digesters, evaporators. Stabilized, non-magnetic. Type 321 stainless, austenitic. *Obsolete*

SILVER FOX 6

British Steel Corp.
C 0.1, Cr 14, bal Fe.
Annealed: 78,400 TS; 31 El; 70 RA. For stainless parts; not hardenable. *Obsolete*

SILVER FOX 610

British Steel Corp.
C 0.1, Cr 13, bal Fe.
Tempered: 89,600 TS; 30 El; 50 RA. For stainless parts; stainless, free-cutting. *Obsolete*

SILVER FOX 611

British Steel Corp.
C 0.15, Cr 13, bal Fe.
Tempered: 107,500 TS; 30 El; 50 RA. For stainless parts; free-cutting, stainless. *Obsolete*

SILVER FOX 612

British Steel Corp.
C 0.2, Cr 13, bal Fe.
Tempered: 107,500 TS; 30 El; 50 RA. For stainless parts; free-cutting, stainless. *Obsolete*

SILVER FOX 613

British Steel Corp.
C 0.3, Cr 13, bal Fe.
Tempered: 112,000 TS; 28 El; 50 RA. For stainless parts; free-cutting, stainless, hardenable. *Obsolete*

SILVER HARDTEM

Heppenstall Co.
C 0.35, Mn 0.3, Cr 5, Mo 2, Si 0.1, Ag 0.2, bal Fe.
For zinc die casting dies, hot work dies; prehardened. *Obsolete*

SILVER LABEL

Peninsular Steel Co.
C 0.6, Mn 0.8, Mo 0.45, V 0.2, Si 1.85, bal Fe.
For punching and shearing dies; oil or water hardening. AISI S5 tool steel.

SILVER LABEL

Wallace Murray Corp.
C 0.9, Cr 0.5, Mn 1.1, W 0.5, bal Fe.
For hobs, taps, reamers, dies; nondeforming.

SILVER LEAF

American manufacture
Zn 8.25, Sn 91, Pb 0.35.
For foil for wrapping purposes.

SILVER MAG NICKEL

Engelhard Corp.
Mg 0-0.28, Ni 0-0.18, 99.4 Ag min.
Annealed: 32,000-38,000 psi TS; 12-27 El; 75 Knoop. Hardened: 68,000-74,000 psi TS; 2-6 El; 169 Knoop. For springs, switches. Vacuum melted, corrosion resistant. Hardened by oxidation. *Obsolete*

SILVER METAL-1

English manufacture
Zn 67, Ag 33.
For ornaments.

SILVER METAL-2

English manufacture
Cu 53.3, Zn 8.3, Ni 31.6, Fe 0.8, Sn 1.2, Pb 3.6, Mn 0.3.
For corrosion resistant parts.

SILVER SOLDER 154
Metals & Controls, Inc.
Ag-Cu alloy.
For brazing; M.P. 1130-1150 F. *Obsolete*

SILVER SOLDER AK NO. 5
Texas Instruments Inc./Materials Control
Ag 35, Cu 42, Zn 11, Cd 12.
For brazing; MP 1190-1180°F; easy flow. *Obsolete*

SILVER SOLDER BH-1
Metals & Controls, Inc.
Ag-Cu alloy.
For brazing; M.P. 1495-1590 F. *Obsolete*

SILVER SOLDER CH-1
Metals & Controls, Inc.
Ag-Cu alloy.
For brazing; M.P. 1340-1405 F. *Obsolete*

SILVER SOLDER CLC
Metals & Controls, Inc.
Ag-Cu alloy.
For brazing; M.P. 1165-1245 F. *Obsolete*

SILVER SOLDER CM-3
Metals & Controls, Inc.
Ag-Cu alloy.
For brazing; M.P. 1295-1345 F. *Obsolete*

SILVER SOLDER KA
Texas Instruments Inc./Materials Control
Ag 53.5, Cu 31.5, Zn 15.
For brazing; MP 1280-1320°F; medium flow. *Obsolete*

SILVER SOLDER KC NO. 4
Texas Instruments Inc./Materials Control
Ag 54, Cu 41.1, Zn 3.9, Ni 1.
For brazing; MP 1290-1400°F; hard flow. *Obsolete*

SILVER SOLDER KH-105
Metals & Controls, Inc.
Ag-Cu alloy.
For brazing; M.P. 1170-1185 F. *Obsolete*

SILVER SOLDER KH-7
Metals & Controls, Inc.
Ag-Cu alloy.
For brazing; M.P. 1160-1175 F. *Obsolete*

SILVER SOLDER KL NO. 1
Texas Instruments Inc./Materials Control
Ag 52, Cu 16, Zn 15.5, Cd 16.5.
For brazing; MP 1150-1160°F; easy flow. *Obsolete*

SILVER SOLDER LH-3
Metals & Controls, Inc.
Ag-Cu alloy.
For brazing; M.P. 1440-1500 F. *Obsolete*

SILVER SOLDER LM-1
Metals & Controls, Inc.
Ag-Cu alloy.
For brazing; M.P. 1350-1430 F. *Obsolete*

SILVER SOLDER MK NO. 3
Texas Instruments Inc./Materials Control
Ag 75, Zn 24, Cd 1.
For brazing; MP 1380-1430°F; medium flow. *Obsolete*

SILVER SOLDER SB-2
Metals & Controls, Inc.
Ag-Cu alloy.
For brazing; M.P. 1285-1335 F. *Obsolete*

SILVER SOLDER SC NO. 1
Texas Instruments Inc./Materials Control
Ag 63.3, Cu 25.6, Zn 11.1.
For brazing; MP 1280-1340°F; easy flow. *Obsolete*

SILVER SOLDER SC NO. 4
Texas Instruments Inc./Materials Control
Ag 63.8, Cu 29.7, Zn 5.71, Ni 0.75.
For brazing; MP 1260-1430°F; medium flow. *Obsolete*

SILVER SOLDER SC-2
Metals & Controls, Inc.
Ag-Cu alloy.
For brazing; M.P. 1265-1315 F. *Obsolete*

SILVER SOLDER SH-2
Metals & Controls, Inc.
Ag-Cu alloy.
For brazing; M.P. 1190-1240 F. *Obsolete*

SILVER SOLDER SH-7
Metals & Controls, Inc.
Ag-Cu alloy.
For brazing; M.P. 1270-1310 F. *Obsolete*

SILVER SOLDER SM-1
Metals & Controls, Inc.
Ag-Cu alloy.
For brazing; M.P. 1360-1395 F. *Obsolete*

SILVER SOLDERS
American manufacture
Ag 32, Cu 23, Zn 17, Cd 18.
For solders, silver solders; corrosion resistant.

SILVER STEEL
Eagle & Globe Steel Ltd.
Tool material. C 1, Mn 0.35, Cr 0.3, bal Fe.
Annealed: 210 Brin approx. For dowel and location pins, small punches, gauges, spindles for small motors, meters and instruments. ASGIO-1966 W5A-10; BS1407.

SILVER STEEL
E.C. Atkins & Co.
C 1, bal Fe.
For saws; water hardened.

SILVER STEEL
Sanderson Kayser Ltd.
C 1.2, Mn 0.4, Cr 0.4, bal Fe.
Water hardening carbon tool steel.

SILVER STRIPE HIGH SPEED
Great Western Steel Co.
C 0.7, Cr 4, W 18, V 1, bal Fe.
For tools, cutters, boring and chasing tools; high speed steel.

SILVER TIP
LaSalle Steel Co.
C 0.08-0.13, Mn 0.6-0.9, Si 0.16-0.23, bal Fe.
Cold drawn: 83,000 TS; 76,000 YS; 15 El; 50 RA; 195 Brin.
For screw machine parts; modified SAE-1112. *Obsolete*

SILVER X NICKEL SILVER 18 ALLOY 163
Olin Corp.
Ni 18, Zn 10, bal Cu.
Hard: 79,000 TS; 4 El. Soft: 51,000 TS; 45 El. For deep drawn shells and vessels, vases, candlestick holders; deep drawing, corrosion resistant. *Obsolete*

SILVER, COINAGE U.S.A
American manufacture
Ag 90, Cu 10.
For coins; corrosion resistant.

SILVER-BEARING COPPER
Copper Range Co.
25 oz/ton Ag, bal Cu.
Hard: 50,000 TS; 40,000 YS; 6 El; 50 Rock B. Soft: 34,000 TS; 10,000 YS; 45 El; 45 Rock F. For electrical and electronic components. Electrical conductivity: 100.5. Superior elevated temperature properties.

SILVER-FLO 12
Johnson Matthey plc
Ag 12, Cu 48, Zn 40.
Cadmium-free bearing silver brazing alloy; 800-830°C MP; DIN 8513 L-Ag12.

SILVER-FLO 20
Johnson Matthey plc
Ag 20, Cu 44, Zn 35.9, Sn 0.1.
Cadmium-free bearing silver brazing alloy; 776-815°C MP; DIN 8513 L-Ag20.

SILVER-FLO 25
Johnson Matthey plc
Ag 25, Cu 41, Zn 34.
Cadmium-free bearing silver brazing alloy; 700-800°C MP; DIN 8513 L-Ag25.

SILVER-FLO 302
Johnson Matthey plc
Ag 30, Cu 36, Zn 32, Sn 2.
Cadmium-free bearing silver brazing alloy; 665-755°C MP; DIN 8513.

SILVER-FLO 33
Johnson Matthey plc
Ag 33, Cu 33.5, Zn 33.5.
Cadmium-free bearing silver brazing alloy; 700-740°C MP; DIN 8513.

SILVER-FLO 34
Johnson Matthey plc
Ag 34, Cu 36.75, Zn 27, Sn 2.25.
Cadmium-free bearing silver brazing alloy; 630-730°C MP; DIN 8513 L-Ag34 Sn.

SILVER-FLO 38
Johnson Matthey plc
Ag 38, Cu 31, Zn 29, Sn 2.
Cadmium-free bearing silver brazing alloy; 660-720°C MP; DIN 8513.

SILVER-FLO 40
Johnson Matthey plc
Ag 40, Cu 30, Zn 28, Sn 2.
Cadmium-free bearing silver brazing alloy; 650-710°C MP; DIN 8513.

SILVER-FLO 44
Johnson Matthey plc
Ag 44, Cu 30, Zn 26.
Cadmium-free bearing silver brazing alloy; 675-735°C MP; DIN 8513 L-Ag44.

SILVER-FLO 452
Johnson Matthey plc
Ag 45, Cu 27.75, Zn 25, Sn 2.25.
Cadmium-free bearing silver brazing alloy; 640-680°C MP; DIN 8513 L-Ag45Sn.

SILVER-FLO 5
Johnson Matthey plc
Ag 5, Cu 55, Zn 39.9, Sn 0.1.
Cadmium-free bearing silver brazing alloy; 830-870°C MP; DIN 8513 L-Ag5.

SILVER-FLO 55
Johnson Matthey plc
Ag 55, Cu 21, Zn 22, Sn 2.
Cadmium-free bearing silver brazing alloy; 630-660°C MP; DIN 8513 L-Ag55Sn.

SILVER-FLO 56
Johnson Matthey plc
Ag 56, Cu 22, Zn 17, Sn 5.
Cadmium-free bearing silver brazing alloy; 618-652°C MP; DIN 8513.

SILVER-MAGNESIUM-NICKEL
Handy & Harmon
Ag 99.5, Mg 0.3, Ni 0.2, bal Fe.
Heat treated: 70,000 TS; 57,000 YS; 5-11 El; 70-130 Brin. For electrical contacts, cables, connectors; high electrical and thermal conductivity, air hardened. *Obsolete*

SILVERBOND
Jessop Steel Co.

Stainless clad steel. For structural applications; stainless. *Obsolete*

SILVERCOTE BERYLLIUM COPPER
Little Falls Alloy Inc.
Be 2, Co 0-0.5, bal Cu.
Heat treated: 200,000 TS; 450 Brin. For springs, lead wires, pins; age hardenable, low creep and corrosion resistance.

SILVERCOTE BERYLLIUM COPPER NO. 10
Little Falls Alloy Inc.
Be 0.5, Co 2.5, bal Cu.
Cold drawn: 120,00-130,000 TS. For lead wires, connectors; age hardenable, high fatigue resistance.

SILVERIN 405
Vereinigte Deutsche Nickel-Werke AG
Nickel. Fe 1.5, C 0.05, Mn 1.9, S 0.04, Si 0.1, 64.5 Ni + Co, bal Cu.
Strip, wire, bar, forged and turned parts for the optical and fastener industries.

SILVERINE
Stone Manganese - J. Stone & Co. Ltd.
Cu 71-80, Ni 16-17, Zn 1-8, Sn 1-2.8, Co 1-2, Fe 1-1.5.
For ornamental and sanitary fittings; corrosion resistant.

SILVERITE
English manufacture
Cu, Ni,
For corrosion resistant parts; nickel silver.

SILVEROID
Henry Wiggin & Co. Ltd.
Cu 54, Ni 45, Mn 1.
Annealed: 60,000 TS; 48 El; 90 Brin. For cutlery; resists corrosion. *Obsolete*

SILVERTEM
Heppenstall Co.
Tool material. C 0.35, Cr 4.75, W 1.25, Mo 1.5, Si 1, Ag 0.2, bal Fe.
For hot work dies; prehardened, die steel.

SILVERY PIG IRON 16%
Cyprus Foote Mineral Co.
Si 16, C 0.75, Mn 1, bal Fe.
Ferrosilicon for alloying in cast iron and some steels. *Obsolete*

SILVERY PIG IRON 20%
Cyprus Foote Mineral Co.
Si 20, C 0.25, Mn 1, bal Fe.
Ferrosilicon for alloying in cast iron and some steels. *Obsolete*

SILVERY PIG IRON PULVERIZED
Cyprus Foote Mineral Co.
Si 15.01-15.5, C 0.8, Mn 1, bal Fe.
For benefication of certain ores and minerals. *Obsolete*

SILVIUM
FSB Inc.
Lead. Sb, bal Pb.
For grids for batteries; corrosion resistant.

SILVORE
English manufacture
Cu 61.9, Zn 19.5, Ni 18.4, Fe 0.3.
For show cases and architectural trim; corrosion resistant.

SILVUNG 240
Plansee, Metallwerk Gesellschaft
W, Ag.
For contacts for interrupters. Tungsten impregnated with Ag.

SILVUNG D
Plansee, Metallwerk Gesellschaft
W, Ag.
For contacts for voltage regulators. Tungsten impregnated with Ag.

SILVUNG P
Plansee, Metallwerk Gesellschaft
W, Ag.
For facing contacts. Tungsten impregnated with Ag.

SIMAF
Pechiney Electrometallurgie
Miscellaneous nonferrous. Mg 5-6.5, Ca 0.4-0.6, La 0-0.4, Si 45-50, Al 0.8-1.3, bal Fe.
Cast iron nodularizer.

SIMAGAL
Kreidler Werke G.m.b.H.
Aluminum.
Aluminum alloy AlMgSi1. *Obsolete*

SIMAGAL 200
Kreidler Werke G.m.b.H.
Mg 0.6-1.4, Si 0.6-1.6, Mn 0.6-1, Cr 0-0.3, bal Al.
Annealed: 21,000 TS; 8,000 YS; 24 El. For window frames, fan blades, gutters, boats; good weldability and formability. *Obsolete*

SIMALLOY 322
Wallace Murray Corp.
C 0-0.08, Si 0-1, Cr 17, Ni 7, Mo 0.7, bal Fe.
Modified AISI 301; austenitic, stainless steel.

SIMALLOY 600
Wallace Murray Corp.
C 0.15, Mn 1, Si 0.5, Cr 15.5, Fe 8, 72.0 Ni min.
For heat exchangers, jet engine parts. Good high temperature properties. ASTM B-166, B-168; Inconel 600.

SIMALLOY 718
Now INCONEL ALLOY 718.

SIMALLOY 722
Now INCONEL ALLOY 722.

SIMALLOY 800
Now INCOLOY 800.

SIMALLOY A-286
Now A-286.

SIMALLOY HX
Wallace Murray Corp.
Fe 18.5, Cr 22, Mo 9, Co 1.5, W 0.6, C 0.1, bal Ni.
High temperature alloy. AMS 5536, 5754; AISI 680.

SIMALLOY L-605
Now ALLOY L-605.

SIMALLOY N-155
Now N-155.

SIMALLOY X 750
Wallace Murray Corp.
C 0.8, Mn 1, Si 0.5, Cr 15, Cb 1, Ti 2.4, Al 0.7, Fe 6.5, 70.0 Ni min.
Age hardenable high temperature alloy. ASTM A 637; Inconel X 750,

SIMANAL
Ohio Ferro Alloys Corp.
Si 20, Mn 20, Al 20, bal Fe.
For steel making; deoxidizer. *Obsolete*

SIMBRAX
Manufacturer not listed.
Cu 85-97, Fe 1.5, Si 1-3.5, Zn 2-7.
Soft: 45,000 TS; 20,000 YS; 50 El; 40 Brin. Hard: 120,000 TS; 2 El; 250 Brin. For pressure gauges; corrosion resistant.

SIMETEORA
English manufacture
C 0.76, Mn 0.17, Cr 2.87, W 16.86, bal Fe.
For tools, cutters, drills; high speed steel.

SIMGAL
Now ALCAN GB-50S.

SIMILARGENT
English manufacture
Cu, Ni.
For white metal parts; nickel silver.

SIMILOR
English manufacture
Cu 80-89, Zn 9-20, Sn 0-7.
For hardware, pipe fittings; corrosion resistant.

SIMO
Aubert & Duval
C 0.4, Si 2.5, Cr 10, Mo 0.9, bal Fe.
Rolled: 141,000 TS; 114,000 YS; 15 El. For high temperature valves; corrosion and heat resistant.

SIMO
Acciaierie Valbruna s.p.a.
Tool Material. C 0.6, Si 1.9, Mn 0.8, Mo 0.35, Cr 0-0.3, bal Fe.
Shock resisting type tool steel. AISI S5,

SIMOCH
Teledyne Vasco
C 0.46-0.5, Si 0.8-1, Cr 3-3.5, Mo 1.4, V 0.3, bal Fe.
For shear knives and blades, punches, rivet sets; air hardened, shock resistant. Type S7.

SIMONDS
Now SIMONDS TEENAX 46.

SIMONDS
Wallace Murray Corp.
Ni 71, Cr 2.8, Si 3.67, bal Fe.
For radiant tubes, furnace muffles, conveyor belts, high temperature equipment. Corrosion and oxidation resistant. Heat resistant.

SIMONDS "T.A.S."
Wallace Murray Corp.
C, W, Cr, bal Fe.
For shear blades, heavy duty cutters, dies; oil hardening. *Obsolete*

SIMONDS 1050 ROTARY RULE
Wallace Murray Corp.
C 0.5, Mn 0.8, bal Fe.
Cold rolled strip; as rolled or heat treated.

SIMONDS 1064 RULE DIE
Wallace Murray Corp.
C 0.64, Mn 0.6, bal Fe.
Cold rolled strip; as rolled or heat treated.

SIMONDS 1070 CIRCULAR SAW
Wallace Murray Corp.
C 0.7, Mn 0.8, bal Fe.
Cold rolled strip; as rolled or heat treated.

SIMONDS 12225
Wallace Murray Corp.
C 2.15, Cr 12, bal Fe.
For gages, knives, blanking and drawing dies; Type D3; oil hardened. *Obsolete*

SIMONDS 139
Wallace Murray Corp.
Ni 39, bal Fe.
For hypodermic syringes, precision machinery components, bimetallics. Controlled expansion. *Obsolete*

SIMONDS 142
Wallace Murray Corp.
Ni 42, bal Fe.
For bimetals. Controlled expansion. *Obsolete*

SIMONDS 149
Wallace Murray Corp.
Ni 49, bal Fe.
For bimetals. Controlled expansion. *Obsolete*

SIMONDS 168
Wallace Murray Corp.
C 0.85, Cr 12, V 0.8, Mo 0.7, bal Fe.
Air or oil hardening tool and die steel, chromium type. AISI D2. *Obsolete*

SIMONDS 17% COBALT MAGNET STEEL
Wallace Murray Corp.
Co 17, bal Fe.
For permanent magnets.

SIMONDS 260 MBS
Wallace Murray Corp.
C 1.25, Cr 0.3, bal Fe.
Cold rolled strip; as rolled or heat treated.

SIMONDS 30% NICKEL STEEL
Wallace Murray Corp.
C 0.1, Ni 30, bal Fe.
For temperature compensators; controlled expansion.

SIMONDS 35% COBALT MAGNET STEEL
Wallace Murray Corp.
Co 35, bal Fe.
For permanent magnets.

SIMONDS 3500
Wallace Murray Corp.
Co 38, Cr 3.8, W 5, C 0.75, bal Fe.
Maximum energy product, 0.982; residual flux density 10,400; coercive force 230. Hot rolled or cast: 38-45 Rock C. Hardened: 60-63 Rock C.

SIMONDS 38-7FM
Wallace Murray Corp.
Ni 38, bal Fe.
For bourdon tubes, control devices, tuning forks. Constant modulus, controlled expansion.

SIMONDS 40% COBALT MAGNET STEEL
Wallace Murray Corp.
Co 40, bal Fe.
For permanent magnets.

SIMONDS 42-43
Wallace Murray Corp.
Ni 42, bal Fe.
For glass to metal seals, sealing leads in electric light bulbs and radio tubes. Controlled expansion.

SIMONDS 47
Wallace Murray Corp.
C 0.5, Cr 1.4, V 0.25, W 2, Mo 0.25, Si 0.8, bal Fe.
For chisels; shock resistant.

SIMONDS 48-50
Wallace Murray Corp.
Ni 48, bal Fe.
For sealing leads in electric light bulbs and radio tubes, glass to metal seals. Controlled expansion.

SIMONDS 6150 EB BACKER
Wallace Murray Corp.
C 0.5, Mn 0.8, Cr 1, bal Fe.
Cold rolled strip; as rolled or heat treated. Similar to AISI 6150.

SIMONDS 73-3.5 CO
Wallace Murray Corp.
Cr 3.5, C 1, bal Fe.
Maximum energy product 0.302; residual flux density 9700; coercive force 81. Rolled or cast permanent magnet.

SIMONDS 74
Wallace Murray Corp.
Cr 6, bal Fe.
Hardened: 300,000-400,000 TS; 62-66 Rock C. For permanent magnets. *Obsolete*

SIMONDS 81-18.5 CO
Wallace Murray Corp.
Co 8.5, Cr 3.75, W 5, C 0.75, bal Fe.
Maximum energy product 0.690; residual flux density 10,700; coercive force 160. Rolled or cast permanent magnet.

SIMONDS 81C
Wallace Murray Corp.
C 0.7, Cu 3, Cr 4, Mn 0.7, Co 20, bal Fe.
Annealed: 25-35 Rock C. Hardened: 300,000 TS; 60-65 Rock C. For permanent magnets in hysteresis motors, speedometers, meters, gages. Quench hardened alloy. Non-directional magnetic properties.

SIMONDS 83-3 CO
Wallace Murray Corp.
Co 3.25, Cr 4, C 1, bal Fe.
Maximum energy product 0.382; residual flux density 9700; coercive force 81. Rolled or cast permanent magnet.

SIMONDS AIRTRUE 51
Wallace Murray Corp.
C 1, Cr 5, V 0.25, Mo 1, bal Fe.
For gages, knives, blanking and forming dies; Type A2; air hardening; nondeforming.

SIMONDS ALNICO NO. 2, 2C, 5, ETC.
Now ALNICO 2, 2C, 5, ETC.

SIMONDS BLUE LABEL
Wallace Murray Corp.
C 0.8-1.25, bal Fe.
For edge holding tools; water hardened. *Obsolete*

SIMONDS BLUE STREAK
Wallace Murray Corp.
C 0.1, Ni 22, Al 12, Co 3, bal Fe.
For magnets; permanent. *Obsolete*

SIMONDS CCM
Wallace Murray Corp.
C 1.5, Cr 12, Mo 0.9, V 0.9, bal Fe.
For tools, dies, cold work dies; air hardening; nondeforming. AISI D2.

SIMONDS CHROMIUM MAGNET STEEL
Wallace Murray Corp.
Cr 3.5, C 1, bal Fe.
For magnets; permanent. *Obsolete*

SIMONDS CHROMIUM-MOLYBDENUM MAGNET STEEL
Wallace Murray Corp.
Cr 4, Mo, bal Fe.
For magnets; permanent.

SIMONDS D.N.V.
Wallace Murray Corp.
C 0.35, Cr 4, V 0.35, Mo 0.4, W 9.5, bal Fe.
For bolt and rivet dies, forging mandrels, piercing punches; Type H21; hot work steel. *Obsolete*

SIMONDS DIAMOND "S"
Wallace Murray Corp.
C 0.1-1.4, bal Fe.
For cutting tools, drills, taps; water hardening. *Obsolete*

SIMONDS ELINVAR
Now ELINVAR.

SIMONDS FRONTIER
Wallace Murray Corp.
C 0.7, W 18, Cr 4, V 1, bal Fe.
For cutting tools, reamers; high speed steel. *Obsolete*

SIMONDS FRONTIER SPECIAL
Wallace Murray Corp.
C 0.7, W 18, Cr 0.4, V 1, Co 5, bal Fe.
For heavy duty high speed tools; high speed steel. *Obsolete*

SIMONDS GREEN STREAK
Wallace Murray Corp.
Al 12, Ni 25, Si 0.4, bal Fe.
For permanent magnets. *Obsolete*

SIMONDS H-12
Wallace Murray Corp.
C 0.35, Cr 5, V 0.4, W 1.5, Mo 1.5, bal Fe.
For punches, extrusion dies, upsetters; Type H12; hot work steel. *Obsolete*

SIMONDS H-13
Wallace Murray Corp.
C 0.35, Cr 5, V 1, Mo 1.5, bal Fe.
For hot work tools; hot work steel; Type H13. *Obsolete*

SIMONDS HIGH SPEED
Now SIMONDS RED STREAK.

SIMONDS INVAR 136
Wallace Murray Corp.
Ni 36, bal Fe.
Low expansion alloy; often used for bimetals.

SIMONDS MOLVA
Wallace Murray Corp.
C 0.8, Cr 4, V 1.75, Mo 8.5, bal Fe.
For cutting tools, drills, reamers; high speed steel. *Obsolete*

SIMONDS MOLVA-T
Now MOLVA-T.

SIMONDS NO. 116 PAPER KNIFE
Wallace Murray Corp.
C 1.18, Mn 0.5, Cr 0.5, Mo 0.9, bal Fe.
Oil hardening tool steel for paper cutters, shears.

SIMONDS NO. 12150
Wallace Murray Corp.
C 1.5, Cr 12, V 0.15, Mo 0.8, bal Fe.
For punches, dies; abrasion and corrosion resistant.

SIMONDS NO. 174 PLANER KNIFE
Wallace Murray Corp.
C 1.15, Cr 8.5, Si 1, Mo 1.5, W 1.5, V 2, bal Fe.
Air hardening; cold work tool steel for planer knives and blades.

SIMONDS NO. 176 CHIPPER KNIFE
Wallace Murray Corp.
C 0.48, Si 1.1, Cr 8.5, Mo 1.5, W 1.2, bal Fe.
Air hardening; tool steel; for chipper knives.

SIMONDS NO. 3400
Wallace Murray Corp.
Co 33, bal Fe.
For permanent magnets; high permeability. *Obsolete*

SIMONDS NO. 8 WOOD SAW
Wallace Murray Corp.
C 0.75, Ni 2.5, bal Fe.
Oil hardening steel for wood-cutting saws.

SIMONDS NO. 864
Wallace Murray Corp.
C 0.9, Mn 1.6, bal Fe.
For punches, trimmers, crimpers, blanking dies; Type O1; oil hardened, non-deforming. *Obsolete*

SIMONDS NON-SHRINKING
Wallace Murray Corp.
C 0.9, Mn 1.2, Cr 0.5, bal Fe.
For dies, punches, shears; non-deforming. *Obsolete*

SIMONDS O.H.D. NO. 106
Wallace Murray Corp.
C 1.2, Cr 0.6, V 0.25, W 1.65, bal Fe.
For punches, dies, crimpers; oil hardened, tough. *Obsolete*

SIMONDS PM DIE
Wallace Murray Corp.
C 0.35, Cr 13, bal Fe.
Air hardening; tool steel for plastic mold dies. AISI 420.

SIMONDS RED LABEL
Wallace Murray Corp.
C 0.8-1.2, Mn 0.3, Si 0.25, bal Fe.
For taps, drills, reamers, hobs, punches; Type W1; water hardening.

SIMONDS RED LABEL SPECIAL
Wallace Murray Corp.
C 0.6-1.4, Mn 0.25, Si 0.25, bal Fe.
For hot or cold work dies, shear blades, punches; water hardened. *Obsolete*

SIMONDS RED STREAK
Wallace Murray Corp.
C 0.75, Cr 4, V 1, W 18, bal Fe.
For drills, reamers, cutting tools; high speed steel.

SIMONDS RED STREAK AIR HARDENING
Wallace Murray Corp.
C 1, Cr 5, V 0.25, Mo 1, bal Fe.
For dies, gauges, knives, forming and drawing dies; non-deforming, air hardened. *Obsolete*

SIMONDS RED STREAK OIL HARDENING
Wallace Murray Corp.
C 0.9, Mo 0.3, Cr 1.5, bal Fe.
For dies, gages, knives, tools; oil hardened, shock resistant. *Obsolete*

SIMONDS S.T.M.
Wallace Murray Corp.
C 0.8, Cr 3.75, V 1.1, W 1.0, Mo 0.6, bal Fe.
For cutting tools, drills, reamers; high speed steel.

SIMONDS SLICER KNIFE
Wallace Murray Corp.
C 1.3, Cr 1, bal Fe.
Oil hardening steel; for circular slicing knives.

SIMONDS SUPER COBALT
Wallace Murray Corp.
C 0.8, Cr 4, Co 8, V 2, W 18, Mo 0.7, bal Fe.
For lathe and planer tools, reamers, hobs, milling cutters; Type T5; high speed steel. *Obsolete*

SIMONDS SUPER INVAR TYPE A
Wallace Murray Corp.
Ni 31, Co 4.5, bal Fe.
For bimetals; controlled expansion.

SIMONDS TEENAX 46
Wallace Murray Corp.
C 0.9, Cr 0.5, Mn 1.25, V 0.15, W 0.5, bal Fe.
For taps, punches, shear blades, chisels, upsetters; Type O1; oil hardening; nondeforming.

SIMONDS TUNCO
Wallace Murray Corp.
C 0.7, Cr 4, Co 5, V 1, W 18, bal Fe.
For drills, reamers, cutting tools; high speed steel. *Obsolete*

SIMONDS TUNGSTEN MAGNET STEEL
Wallace Murray Corp.
C 1, W 3, bal Fe.
For permanent magnets.

SIMPLEX
Union Carbide Corp.
C 0-0.01, Si 5-7, Cr 63-66, bal Fe.
For making stainless steel; alloy additive.

SIMPLEX "CH"
Colt Industries
C 0.2, Ni 1.25, Cr 0.6, bal Fe.
Heat treated: 90,000 TS; 80,000 YS; 16 El; 45 RA; 260 Brin.
For gears, pinions, shafts; carburizing steel. *Obsolete*

SIMPLEX FORGING
Colt Industries
Ni 1.25, Cr 0.6, C 0.4, bal Fe.
Heat treated: 130,000 TS; 110,000 YS; 17 El; 45 RA; 290 Brin.
For gears, high strength machine parts; forging steel. *Obsolete*

SIN-CHU
Japanese manufacture
Cu 66.5, Zn 33.4, Fe 0.1.
For hardware; brass.

SINIMAX
Allegheny Ludlum Steel
Ni 42, Si 3, Fe 55.
For electrical equipment, auto transformers, tape recording heads; magnetically soft. *Obsolete*

SINTAG 852
Handy & Harmon
Precious metal. Ag 85, 15.0 CdO.
For electric contacts. Sintered, corrosion resistant.

SINTAG 903
Handy & Harmon
Precious metal. Ag 90, 10.0 CdO.
For electric contacts. Sintered, corrosion resistant.

SINTALOY B-901
Dixon Sintaloy Inc.
Copper. Cu 90, Sn 10.
Sintered bronze. As sintered: 30,000 TS; 2 El. General purpose structural parts.

SINTALOY BB-901
Dixon Sintaloy Inc.
Copper. Cu 88, Sn 10, C 1.75.
Sintered bronze. As sintered: 15,000 TS; 1 El. General purpose structural parts. SAE 901.

SINTALOY CI-100 DENSITY 5.1-6.1
Dixon Sintaloy Inc.
Fe 99, C 1.
Sintered steel, low density. Sintered: 16,000-24,000 TS; 0.2 El; 10-50 Rock B. Low cost, quality structural parts.

SINTALOY CI-100 DENSITY 6.1-6.5
Dixon Sintaloy Inc.
Fe 99, C 1.
Sintered steel, medium density. As sintered: 19,000-30,000 TS; 0.3 El; 20-50 Rock B. Low cost, quality structural parts. ASTM B310 Type 1 Class A.

SINTALOY CI-100-HT
Dixon Sintaloy Inc.
Fe 99, C 1.
Sintered steel, heat treated. HT: 30,000-40,000 TS; 10-35 Rock C. Improved strength and hardness.

SINTALOY CI-100-ST
Dixon Sintaloy Inc.
Fe 99, C 1.
Sintered steel, steam treated. As treated: 19,000-30,000 TS; 50-90 Rock B. Improved hardness, wear and corrosion resistance.

SINTALOY CX-140
Dixon Sintaloy Inc.
Ni 1.75, Mo 0.5, Cu 1.5, C 0.6, bal Fe.
Sintered: 60,000 TS; 2-4 El; 60-80 Rock B. High strength, low alloy steel characteristics.

SINTALOY CX-140-HT
Dixon Sintaloy Inc.
Ni 1.75, Mo 0.5, Cu 1.5, C 0.6, bal Fe.
Sintered and heat treated: 100,000 TS; 30-40 Rock C. Good strength, hardness, wear resistance.

SINTALOY EC-100
Dixon Sintaloy Inc.
Copper. Cu 100.
Sintered copper. As sintered: 17,000-34,000 TS; 6-20 El. Excellent electrical conductivity.

SINTALOY F-1003
Dixon Sintaloy Inc.
Fe 96, Ni 3, C 1.
Sintered: 60,000 TS min; 2-6 El; 60-90 Rock B. Good strength alloy.

SINTALOY F-1003-HT
Dixon Sintaloy Inc.
Fe 96, Ni 3, C 1.
Sintered and heat treated: 100,00-150,000 TS; 1-4 El; 20-45 Rock C. High strength alloy, good wear resistance.

SINTALOY HDM
Dixon Sintaloy Inc.
Fe 100.
Sintered iron. As sintered: 25,000-40,000 TS; 2-15 El. For magnetic applications.

SINTALOY HDS
Dixon Sintaloy Inc.
Fe 99, C 1.
Sintered steel, high density. As sintered: 45,000-70,000 TS; 2-12 El; 30-70 Rock B. Similar characteristics to annealed steel.

SINTALOY HDS-HT
Dixon Sintaloy Inc.
Fe 99, C 1.
Sintered steel, heat treated, high density. HT: 60,000-120,000 TS; 0.2 El; 20-50 Rock C. Hardened steel for wear parts.

SINTALOY L-1004
Dixon Sintaloy Inc.
Fe 95, Cu 4, C 1.
Sintered: 25,000-45,000 TS; 0.5 El; 20-60 Rock B. Moderate strength structural parts.

SINTALOY L-1007
Dixon Sintaloy Inc.
Fe 92, Cu 7, C 1.
Sintered: 30,000-50,000 TS; 0.5 El; 25-70 Rock B. Moderate strength structural parts. ASTM B222.

SINTALOY LI-1010
Dixon Sintaloy Inc.
Fe 74, Cu 25, C 1.
Sintered: 55,000 TS; 0.5 El; 85-95 Rock B. Infiltrated alloy; good mechanical properties. ASTM B303.

SINTALOY LI-1010 HT
Dixon Sintaloy Inc.
Fe 74, Cu 25, C 1.
Sintered and heat treated: 90,000 TS; 30-40 Rock C. Good strength, hardness and wear.

SINTALOY N-6418
Dixon Sintaloy Inc.
Cu 64, Zn 18, Ni 18.
Sintered nickel silver. As sintered: 19,000-36,000 TS; 5-18 El. White alloy with excellent corrosion resistance and good wear. *Obsolete*

SINTALOY SS-316L
Dixon Sintaloy Inc.
Stainless steel. Cr 16, Ni 12, Mo 2, bal Fe.
Sintered: 35,000-45,000 TS; 4-6 El; 55-70 Rock B. Authentic; good corrosion resistant properties.

SINTALOY SS-410L
Dixon Sintaloy Inc.
Cr 12.5, bal Fe.
Sintered: 34,000-49,000 TS; 5-8 El; 50-75 Rock B. Corrosion resistant; heat treatable.

SINTALOY SS-410L-HT
Dixon Sintaloy Inc.
Stainless steel. Cr 12.5, C, bal Fe.
Sintered and heat treated: 100,000-140,000 TS; 1 El; 25-37 Rock C. High strength.

SINTALOY Z-703
Dixon Sintaloy Inc.
Cu 70, Zn 30.
Sintered: 23,000-30,000 TS; 5-10 El. For hardware, machine tool parts; powder metallurgy. *Obsolete*

SINTALOY Z-703P
Dixon Sintaloy Inc.
Cu 70, Zn 29.7, P 0.3.
Sintered: 28,000-36,000 TS; 20-30 El. For hardware, structural and machine parts; powder metallurgy. *Obsolete*

SINTALOY Z-802
Dixon Sintaloy Inc.
Copper. Cu 80, Zn 20.
Sintered: 29,000-36,0000 TS; 10-18 El. For hardware, machine tool parts; yellow brass, powder metallurgy.

SINTALOY Z-802 PB
Dixon Sintaloy Inc.
Copper. Cu 78.5, Zn 20, Pb 1.5.
Sintered: 27,000-34,000 TS; 10-18 El. For hardware, machine tool parts; leaded yellow brass.

SINTALOY Z-802 PB-P
Dixon Sintaloy Inc.
Copper. Cu 78.5, Zn 19.7, Pb 1.5, P 0.3.
Sintered: 29,000-36,000 TS; 18-32 El. For hardware, machine tool parts; powder metallurgy.

SINTALOY Z-851
Dixon Sintaloy Inc.
Cu 85, Zn 15.
Sintered: 29,000-35,000 TS; 8-15 El. For hardware, machine tool parts; red brass, powder metallurgy. *Obsolete*

SINTALOY Z-901
Dixon Sintaloy Inc.
Copper. Cu 90, Zn 10.
Sintered: 20,000-28,000 TS; 8-15 El. For hardware, structural parts; high electrical conductivity.

SINTALOY Z-901 PB
Dixon Sintaloy Inc.
Copper. Cu 88.5, Zn 10, Pb 1.5.
Sintered: 18,000-27,000 TS; 8-15 El. For hardware, structural parts; free-cutting, powder metallurgy.

SINTALOY Z-901P
Dixon Sintaloy Inc.
Copper. Cu 90, Zn 9.5, P 0.5.
Sintered: 25,000-30,000 TS; 18-25 El. For hardware, structural parts; powder metallurgy.

SINTEEL M
Electro Metal Corp.
Steel impregnated with Cu.
Sintered: 100,000 TS. Heat treated: 170,000 YS. For machinery parts; sintered.

SINTEEL R
Electro Metal Corp.
Steel impregnated with Cu.
For machinery parts; sintered.

SINTEEL S
Electro Metal Corp.
Steel impregnated with Cu.
For machinery parts; sintered.

SINTERBRONZE
Plansee, Metallwerk Gesellschaft
Copper. Sn, bal Cu.
Sintered. For bearings.

SINTERLOY A
Charles Hardy & Co.
C 0.15, bal Fe.
For gears, cams, washers; sintered alloy.

SINTERLOY B
Charles Hardy & Co.
C 0.4, bal Fe.
For gears, cams, washers; sintered alloy.

SINTERLOY C
Charles Hardy & Co.
C 0.8, bal Fe.
For gears, cams; sintered alloy.

SINTHERM
Plansee, Metallwerk Gesellschaft
TiC, Zr, B.
For evaporation boats.

SINTOX
Sintox Corp. of America
Al_2O_3.
For cutting tools; sintered.

SINTRAMANT
Metallwerk Plansee Gesellschaft
WC.
For hard facing electrode; sintered. *Obsolete*

SINTREX D
Easton Metal Powder Inc.
99.0 Fe min, 0.12 other nonvolatile max.
Electrolytic iron powder. 100 mesh standard, annealed powder for sintering to soft magnetic iron parts.

SINTREX E
Easton Metal Powder Inc.
99.0 Fe min, 0.12 other nonvolatile max.
Electrolytic iron powder. 100 mesh sponge, annealed powder, for sintering to soft magnetic iron parts.

SINTREX F
Easton Metal Powder Inc.
99.0 Fe min, 0.12 other nonvolatile max.
Electrolytic iron powder. 200 mesh annealed powder for sintering into soft magnetic iron parts.

SIOUX H-HARDENED RINGS
Sioux Tools Inc.
C 0.76, Mn 0.54, Si 1.38, Cr 3.8, Mo 5.53, bal Fe.
Hardened: 200,000-225,000 TS; 9-11 El; 28-35 RA; 47 Rock C. For valve rings, valve seat inserts.

SIOUX SUPER CAST RINGS
Sioux Tools Inc.
Mo 0.45-0.55, Si 1.1-1.2, C 1.6, Mn 0.6-0.7, Ni 1-2, Cr 0.55-0.65, bal Fe.
At 75°F: 75,000 TS; 46,000 YS; 35 El; 1.8 RA. At 1000°F: 43,000 TS; 33,000 YS; 14 El; 13.3 RA. For valve seat inserts, valve seat rings. Heat and scale resistant.

SIRIUS 30
Compagnie Ateliers et Forges de la Loire
Stainless steel. C 0.19, Cr 19, Ni 7, W 4, Si 1.5, Mn 1.5, bal Fe.
Heat treated: 107,000 TS; 43,000 YS; 30 El. For turbine blades and discs, catalyzer equipment, jet engine parts. Austenitic stainless steel. Heat resistant to 1000°C.

SIRIUS 345
Now NICRAL H.

SIRIUS 35
Now NICRAL C.

SIRIUS 36
Creusot-Loire
C 0.52, Cr 21, Ni 4, Mn 9, Nb 0.15, N 0.35, bal Fe.
Solution treated: 1030 N/mm^2 psi TS min. Valve steel.
AFNOR Z52CMN 21-9 plus N; SAE EV 8.

SIRIUS 40
Compagnie Ateliers et Forges de la Loire
Stainless steel. C 0.32, Cr 14, Ni 14, W 2.5, Si 1.5, bal Fe.
Annealed: 114,000 TS; 64,000 YS; 40 El. For valves, exhaust valves for automobile and aircraft engines. Austenitic, stainless, heat resistant.

SIRIUS HT
Creusot-Loire
C 0.25, Ni 16, Cr 17, W 3, Ti 2, bal Fe.
For gas turbine components; heat and creep resistant. *Obsolete*

SIRIUS SUPER-HT
Compagnie Ateliers et Forges de la Loire
Stainless steel. C 0.2, Cr 16, Ni 14, Co 10, W 2.5, Ti 2, Si 1, bal Fe.
Heat treated: 100,000 TS; 72,000 YS; 20 El. For gas turbine component parts. Austenitic, stainless. Resists oxidation up to 950°C.

SIRIUS-10 CAST
Compagnie Ateliers et Forges de la Loire
Stainless steel. C 0.25, Cr 18, Ni 9.5, Si 1.3, bal Fe.
For furnace parts, superheaters, heat treating boxes. Austenitic, stainless. Resists sulfur atmosphere.

SIRIUS-5 CAST
Compagnie Ateliers et Forges de la Loire
Stainless steel. C 0.3, Cr 16, Ni 12, Si 3.6, Mn 1.2, bal Fe.
For furnace parts, heat treating boxes, superheaters, carburizing boxes, salt pots. Austenitic, stainless; useful to 1050-1100°C in slightly sulfur atmosphere.

SIRONZE
Wilbur B. Driver Co.
Cu 96, Si 3, Fe 0.1.
Annealed: 70,000 TS; 17 El; 26 Brin. Drawn: 110,000 TS; 83,000 YS; 15 El; 22 RA; 189 Brin. For bolts, nuts, screws; corrosion resistant. *Obsolete*

SIS
Manufacturer not listed.
Prefix used on some Swedish alloy standards.

SIS 22-40
Swedish manufacture
C 0.32, Cr 2.4-2.6, Mo 0.3, V 0.3, Ni 0.5, bal Fe.
This steel is usually nitrided; for gears, cams, pinions, camshafts.

SISCO
Swedish Iron & Steel Corp.
C 0.9, bal Fe.
For tools, drills; water hardened. *Obsolete*

SISCO OIL HARDENING CHISEL
Swedish Iron & Steel Corp.
C 0.6, Si 2, Mn 1, bal Fe.
For chisels; tough. *Obsolete*

SISERSKITE
Russian manufacture
Os 57, Ru 8, 34 Rh + Ir, Pt, Pd, Au, Cu.
For fountain pen points; neutral osmiridium mined in U.S.S.R.

SISTAL
Teledyne Vasco
C 0.47, Si 0.9, Mn 0.3, Cr 8.25, W 1.2, Mo 1.35, V 0.3, bal Fe.
Air hardenable to about 58 Rock C. Hot work tool steel. For extrusion presses, upsetting rams, pressure dies.

SITKIN NO. 100
Sitkin Smelting & Refining Co.
Sn 4.5-5.7, Pb 2.75-4, Zn 1-9.5, bal Cu.
For bearings and pump impellers; free cutting.

SITKIN NO. 101
Sitkin Smelting & Refining Co.
Sn 4.5-5.75, Pb 2-3, Zn 9.5-16, bal Cu.
For bushings, bearings, and pump impellers; free-cutting.

SITKIN NO. 110
Sitkin Smelting & Refining Co.
Sn 4.5-5.5, Pb 4.25-7, Zn 2-7, bal Cu.
For pump impellers, bushings, and bearings. Free-cutting; red brass.

SITKIN NO. 115
Sitkin Smelting & Refining Co.
Sn 4-5.5, Pb 4.25-7, Zn 4-10, bal Cu.
Cast: 33,000-46,000 TS; 17,000-24,000 YS; 15-35 El; 12-32 RA; 55-65 Brin. For hardware and plumbing. Red brass; free-cutting.

SITKIN NO. 120
Sitkin Smelting & Refining Co.
Sn 3-4.5, Pb 6-10, Zn 5-10, bal Cu.
Cast: 30,000-38,000 TS; 12,000-17,000 YS; 15-27 El; 12-25 RA; 50-60 Brin. For hardware and plumbing. Red brass; free-cutting.

SITKIN NO. 123
Sitkin Smelting & Refining Co.
Sn 2-4, Pb 5-8, Zn 6-18, bal Cu.
Cast: 29,000-39,000 TS; 13,000-17,000 YS; 18-30 El; 15-27 RA; 50-60 Brin. For hardware and plumbing. Plumbers brass; free-cutting.

SITKIN NO. 125
Sitkin Smelting & Refining Co.
Sn 2-4, Pb 5-8, Zn 6-18, bal Cu.
Cast: 30,000-40,000 TS; 12,000-16,000 YS; 20-35 El; 15-30 RA; 50-60 Brin. For hardware and plumbing. Plumbers brass; free-cutting.

SITKIN NO. 130
Sitkin Smelting & Refining Co.
Sn 1-3, Pb 4-10, Zn 8-20, bal Cu.
For hardware and plumbing. Free-cutting.

SITKIN NO. 131
Sitkin Smelting & Refining Co.
Sn 1-2, Pb 1-2, Zn 2-8, bal Cu.
For hardware and plumbing. Free cutting.

SITKIN NO. 132
Sitkin Smelting & Refining Co.
Sn 1-3, Pb 1-3, Cu 86-88, bal Zn.
For hardware and plumbing. Free-cutting.

SITKIN NO. 193
Sitkin Smelting & Refining Co.
Sn 21-23, 0.0-0.5 others, bal Cu.
For bearings.

SITKIN NO. 194
Sitkin Smelting & Refining Co.
Sn 18-20, bal Cu.
For bearings.

SITKIN NO. 195
Sitkin Smelting & Refining Co.
Sn 16-18, bal Cu.
For bearings.

SITKIN NO. 196
Sitkin Smelting & Refining Co.
Sn 14-16, bal Cu.
For bearings.

SITKIN NO. 197
Sitkin Smelting & Refining Co.
Sn 13-15, Zn 0-1.5, Pb 0.2, bal Cu.
For bearings.

SITKIN NO. 198
Sitkin Smelting & Refining Co.
Sn 12.5-14.5, Zn 2.5-4.5, Pb 0-1, bal Cu.
For bearings.

SITKIN NO. 199
Sitkin Smelting & Refining Co.
Sn 12-14, bal Cu.
For bearings.

SITKIN NO. 200
Sitkin Smelting & Refining Co.
Sn 10.75-12.25, Pb 0.75-1.25, Zn 0-0.1, bal Cu.
For bearings; free-cutting.

SITKIN NO. 201
Sitkin Smelting & Refining Co.
Sn 10.75-12, Pb 0.8-1.2, Zn 1-4, bal Cu.
For bearings.

SITKIN NO. 205
Sitkin Smelting & Refining Co.
Sn 9.75-12, bal Cu.
For bearings.

SITKIN NO. 206
Sitkin Smelting & Refining Co.
Sn 9-11, Pb 1-2.5, bal Cu.
For bearings.

SITKIN NO. 210
Sitkin Smelting & Refining Co.
Sn 9-10.7, Zn 1-7, Pb 0-0.2, bal Cu.
For bearings.

SITKIN NO. 215
Sitkin Smelting & Refining Co.
Sn 9-10.7, Zn 1-7, Pb 0.8-1.2, bal Cu.
For bearings.

SITKIN NO. 220
Sitkin Smelting & Refining Co.
Sn 9-10.7, Pb 1.2-2.7, Zn 1-5, bal Cu.
For bearings.

SITKIN NO. 221
Sitkin Smelting & Refining Co.
Sn 9-10.7, Pb 2.7-4.2, Zn 1-5, bal Cu.
For bearings.

SITKIN NO. 225
Sitkin Smelting & Refining Co.
Sn 7.5-9.7, Zn 1-7, Pb 0.2-0.4, bal Cu.
For bearings.

SITKIN NO. 230
Sitkin Smelting & Refining Co.
Sn 7.5-9, Pb 0.8-1.2, Zn 1-7, bal Cu.
Cast: 33,000-43,000 TS; 16,000-24,000 YS; 18-30 El; 15-30 RA; 60-75 Brin. For bearings and gun metal.

SITKIN NO. 235
Sitkin Smelting & Refining Co.
Sn 7.5-9, Pb 1.2-2.7, Zn 1-6, bal Cu.
For bearings.

SITKIN NO. 240
Sitkin Smelting & Refining Co.
Sn 7.5-9, Pb 2.7-4.2, Zn 1-6, bal Cu.
For bearings.

SITKIN NO. 241
Sitkin Smelting & Refining Co.
Sn 7.5-9, Pb 4.2-6.2, Zn 1-6, bal Cu.
For bearings.

SITKIN NO. 242
Sitkin Smelting & Refining Co.
Sn 6-8, Zn 0-0.5, bal Cu.
For bearings.

SITKIN NO. 245
Sitkin Smelting & Refining Co.
Sn 5.7-7.5, Pb 1-2, Zn 1-7, bal Cu.
Cast: 34,000-42,000 TS; 16,000-21,000 YS; 20-35 El; 16-30 RA; 60-70 Brin. For valves and pump parts. Free-cutting.

SITKIN NO. 250
Sitkin Smelting & Refining Co.
Sn 5.7-7.5, Pb 2-3.2, Zn 2-6, bal Cu.
For bearings.

SITKIN NO. 251
Sitkin Smelting & Refining Co.
Sn 5.7-7.5, Pb 3.2-5.2, Zn 1-6, bal Cu.
For bearings.

SITKIN NO. 253
Sitkin Smelting & Refining Co.
Sn 4.5-5.7, Pb 1-2, Zn 1-9, bal Cu.
For bearings.

SITKIN NO. 255
Sitkin Smelting & Refining Co.
Sn 4-5.7, Pb 2-2.7, Zn 1-4, bal Cu.
For bearings.

SITKIN NO. 256
Sitkin Smelting & Refining Co.
Sn 3-4.5, Pb 1.7-3, Zn 1-9, bal Cu.
For bearings.

SITKIN NO. 257
Sitkin Smelting & Refining Co.
Sn 3-4, Pb 0.5-1, Cu 85-90, bal Zn.
For bearings.

SITKIN NO. 295
Sitkin Smelting & Refining Co.
Sn 15-17, Pb 4-6, Zn 0.25-1, bal Cu.
For bearings.

SITKIN NO. 296
Sitkin Smelting & Refining Co.
Sn 12-14, Pb 14-16, Zn 0-0.5, Ni 0.5-1, bal Cu.
For bearings.

SITKIN NO. 296.5
Sitkin Smelting & Refining Co.
Sn 10.5-12.5, Pb 7-10, bal Cu.
For bearings.

SITKIN NO. 297
Sitkin Smelting & Refining Co.
Sn 9.5-11.5, Pb 4.2-6, 0.75-1.50 others, bal Cu.
For bearings.

SITKIN NO. 298
Sitkin Smelting & Refining Co.
Sn 9-11, Pb 9-11.5, Zn 0.15-0.75, bal Cu.
For bearings.

SITKIN NO. 299
Sitkin Smelting & Refining Co.
Sn 9-11, Pb 9-11, Zn 0.15-0.75, bal Cu.
For bearings.

SITKIN NO. 300
Sitkin Smelting & Refining Co.
Sn 9-10.7, Pb 8.5-11.7, Sb 0-0.25, bal Cu.
For bearings.

SITKIN NO. 305
Sitkin Smelting & Refining Co.
Sn 9-10.7, Pb 8.5-21.7, 0.75-1.50 others, bal Cu.
Cast: 27,000-37,000 TS; 15,000-22,000 YS; 6-12 El; 5-11 RA;
55-70 Brin. For bearings; bearing bronze.

SITKIN NO. 310
Sitkin Smelting & Refining Co.
Sn 7.5-9, Pb 7.2-8.7, 0.75-1.50 others, bal Cu.
For bearings.

SITKIN NO. 311
Sitkin Smelting & Refining Co.
Sn 7.5-9, Pb 7.2-8.7, bal Cu.
For bearings.

SITKIN NO. 312
Sitkin Smelting & Refining Co.
Sn 7.5-9, Pb 8.7-11, Zn 0-3, 0.75-1.50 others, bal Cu.
For bearings.

SITKIN NO. 313
Sitkin Smelting & Refining Co.
Sn 7.5-9, Pb 12-16, Ni 0.25-0.5, bal Cu.
For bearings.

SITKIN NO. 314
Sitkin Smelting & Refining Co.
Sn 7.5-9, Pb 12-16, 0.25-0.75 others, bal Cu.
For bearings.

SITKIN NO. 315
Sitkin Smelting & Refining Co.
Sn 5.5-7.5, Pb 6-10.5, Zn 0-4, 0.75-1.5 others, bal Cu.
Cast: 30,000-38,000 TS; 17,000-21,000 YS; 12-20 El; 10-22
RA; 55-65 Brin. For bearings and bushings. Bearing bronze.

SITKIN NO. 319
Sitkin Smelting & Refining Co.
Sn 5.7-7.5, Pb 12-16, 0.75-2.00 others, bal Cu.
Cast: 25,000-30,000 TS; 14,000-20,000 YS; 10-18 El; 8-15 RA;
50-60 Brin. For bearings and bushings. Bearing bronze.

SITKIN NO. 320
Sitkin Smelting & Refining Co.
Sn 5.7-7.5, Pb 16.7-29.7, Zn 0-3, bal Cu.
For bearings.

SITKIN NO. 321
Sitkin Smelting & Refining Co.
Sn 5.7-7.5, Pb 16-29, 0.75-1.50 others, bal Cu.
For bearings.

SITKIN NO. 322
Sitkin Smelting & Refining Co.
Sn 4-5.7, Pb 21-26, 0.00-1.50 others, bal Cu.
Cast: 23,000-30,000 TS; 11,000-15,000 YS; 7-16 El; 5-12 RA;
42-55 Brin. For bearings and bushings. Bearing bronze.

SITKIN NO. 323
Sitkin Smelting & Refining Co.
Sn 4-6, Pb 21.7-26, Zn 0-0.5, bal Cu.
For bearings.

SITKIN NO. 324
Sitkin Smelting & Refining Co.
Sn 4-5.7, Pb 26-32, 0.75-1.50 impurities, bal Cu.
For bearings.

SITKIN NO. 325
Sitkin Smelting & Refining Co.
Sn 4.5-5.7, Pb 11.7-21.7, Zn 0-3, bal Cu.
For bearings.

SITKIN NO. 326
Sitkin Smelting & Refining Co.
Sn 3.5-5.5, Pb 8-10, Zn 0-4, bal Cu.
For bearings.

SITKIN NO. 400
Sitkin Smelting & Refining Co.
Sn 0.5-1.7, Pb 2-4, Cu 70-74, bal Zn.
Cast: 35,000-40,000 TS; 12,000-14,000 YS; 25-40 El; 20-40
RA; 40-55 Brin. For plumbing and hardware. Leaded yellow
brass; free-cutting.

SITKIN NO. 403
Sitkin Smelting & Refining Co.
Cu 66-70, Sn 0.5-1.7, Pb 2-4, bal Zn.
Cast: 30,000-38,000 TS; 11,000-15,000 YS; 20-35 El; 15-30
RA; 40-60 Brin. For plumbing and hardware. Yellow brass;
free-cutting.

SITKIN NO. 405
Sitkin Smelting & Refining Co.
Cu 0-66, Sn 0.5-1.7, Pb 2-4, bal Zn.
For hardware and screw machine products. Free-cutting;
yellow brass.

SITKIN NO. 405.1
Sitkin Smelting & Refining Co.
Cu 0-65, Sn 0.5-1.5, Pb 0-0.5, bal Zn.
For hardware and screw machine products. Yellow brass.

SITKIN NO. 405.2
Sitkin Smelting & Refining Co.
Cu 0-65, Sn 0.5-1.5, Pb 0.5-1, bal Zn.
For hardware and screw machine products. Free-cutting;
yellow brass.

SITKIN NO. 406
Sitkin Smelting & Refining Co.
Cu 0-66, Sn 0.5-1, Pb 1-1.5, bal Zn.
For hardware and screw machine products. Free-cutting;
yellow brass.

SITKIN NO. 407
Sitkin Smelting & Refining Co.
Cu 66-72, Sn 0-0.5, Pb 0-0.5, bal Zn.
For hardware and screw machine products. Yellow brass.

SITKIN NO. 407.5
Sitkin Smelting & Refining Co.
Cu 90-95, Sn 0-0.2, Pb 0-0.5, bal Zn.
For clips, fasteners, and springs. Good corrosion resistance.

SITKIN NO. 408
Sitkin Smelting & Refining Co.
Cu 83-86, Sn 0-0.2, Pb 0-0.5, bal Zn.
For clips, fuse parts, and hardware. Good corrosion
resistance.

SITKIN NO. 409
Sitkin Smelting & Refining Co.
Cu 0-61, Pb 1.5-2, bal Zn.
For screw machine products and hardware. Free-cutting;
yellow brass.

SITKIN NO. 410
Sitkin Smelting & Refining Co.
Sn 0-5, Pb 0-10, Ni 9-14, bal Cu + Zn.
Cast: 30,000-40,000 TS; 15,000-20,000 YS; 10-25 El; 7-20 RA;
50-60 Brin. Nickel silver; leaded.

SITKIN NO. 411
Sitkin Smelting & Refining Co.
Sn 0-5, Pb 0-10, Ni 14-18, bal Cu + Zn.
Cast: 35,000-45,000 TS; 17,000-24,000 YS; 15-30 El; 15-30
RA; 65-80 Brin. For bearings. Nickel silver; leaded.

SITKIN NO. 412
Sitkin Smelting & Refining Co.
Sn 0-5, Pb 0-10, Ni 18-22, bal Cu + Zn.
Cast: 40,000-60,000 TS; 17,000-30,000 YS; 15-25 El; 11-22
RA; 76-120 Brin. For bearings. Nickel silver; leaded.

SITKIN NO. 413
Sitkin Smelting & Refining Co.
Sn 0-5, Pb 0-10, Ni 22-27, bal Cu + Zn.
For bearings. Nickel silver; leaded.

SITKIN NO. 414
Sitkin Smelting & Refining Co.
Sn 0-5, Pb 0-10, Ni 27-33, bal Cu + Zn.
For bearings. Nickel silver; leaded.

SITKIN NO. 415
Sitkin Smelting & Refining Co.
Sn 0-1, Pb 0-0.1, Al 8-13, bal Cu + Zn.
Cast: 70,000-87,000 TS; 25,000-30,000 YS; 2-38 El; 20-36 RA;
110-140 Brin. For hardware, propellers, and gears. Al bronze.

SITKIN NO. 420
Sitkin Smelting & Refining Co.
Zn, Mn, bal Cu.
Cast: 60,000-65,000 TS. For hardware, propellers, and gears.
Mn bronze.

SITKIN NO. 421
Sitkin Smelting & Refining Co.
Cu 56-69, Fe 0.75-2, Pb 0-0.3, Ni 0-0.5, bal Zn.
Cast: 65,000-68,000 TS; 28,000-40,000 YS; 20-35 El; 20-40
RA; 115 Brin. For propellers, gears, and hardware. Bronze.

SITKIN NO. 422
Sitkin Smelting & Refining Co.
Zn, Mn, bal Cu.
Cast: 80,000-90,000 TS. For hardware, gears, and propellers.
Mn bronze.

SITKIN NO. 423
Sitkin Smelting & Refining Co.
Zn, Mn, bal Cu.
Cast: 90,000-100,000 TS. For hardware, gears, and
propellers. Mn bronze.

SITKIN NO. 424
Sitkin Smelting & Refining Co.
Cu 60-68, Fe 2-4, Ni 0-1, Pb 0-0.1, bal Zn.
Cast: 100,000-120,000 TS; 65,000-90,000 YS; 12-18 El; 5-18
RA; 190-235 Brin. For hardware, gears, and propellers. Mn
bronze.

SITKIN NO. 500
Sitkin Smelting & Refining Co.
Sn 0-2, Si 2-5.5, Fe 0-2.5, Mn 0-1.5, 90.0 Cu min.
For bolts, gears, and fasteners. Si bronze.

SIVAN
Colt Industries
C 0.6, Si 1.2, V 0.2, bal Fe.
For tools; water hardening. *Obsolete*

SIVAR 48
Thyssen Edelstahlwerke AG
C 0.03, Ni 29, Co 18, bal Fe.
For sealing glass to metal; controlled expansion. *Obsolete*

SIVYER "DYNAMO"

Sivyer Steel Corp.

C 0.12, Mn 0.1, Si 0.2, bal Fe.

Normalized: 55,000 TS; 35,000 YS; 30 El; 60 RA; 120 Brin. For electrical parts; high magnetic permeability. *Obsolete*

SIVYER "POLE"

Sivyer Steel Corp.

C 0.18, Mn 0.55, Si 0.25, bal Fe.

For cast hardened parts; carburizing steel. *Obsolete*

SIVYER 120 CM

Sivyer Steel Corp.

C 0.28, Mn 1.2, Cr 0.45, Mo 0.19, bal Fe.

Low alloy steel. 90-120 TS; 60-95 YS; 187-302 Brin.

SIVYER 20

Sivyer Steel Corp.

C 0-0.07, Ni 29, Cr 20, Mo 2-3, Cu 4, bal Fe.

Cast: 65,000-75,000 TS; 28,000-38,000 YS; 35-50 El; 40-50 RA; 120-150 Brin. For chemical plant equipment, valves, mixers, tanks; resists mixed acids, austenitic. *Obsolete*

SIVYER 30 V

Sivyer Steel Corp.

C 0.28, Mn 0.8, Cr 0.6, Ni 0.5, Mo 0.2, bal Fe.

Low alloy steel for castings requiring moderate to high strength. 90-150 TS; 60-135 YS; 187-363 Brin. Meets SAE 8630.

SIVYER 5% CR

Sivyer Steel Corp.

C 0-0.25, Cr 4.5-6.5, Mo 0.45-0.65, bal Fe.

Normalized: 100,000-110,000 TS; 75,000-90,000 YS; 18-22 El; 229-241 Brin. For oil refinery and chemical plant equipment; Type 502; corrosion resistant. *Obsolete*

SIVYER 75 CM

Sivyer Steel Corp.

C 0.28, Mn 0.75, Cr 0.47, Mo 0.18, bal Fe.

Low alloy steel. 90-120 TS; 60-95 YS; 187-302 Brin. Meets SAE 4128.

SIVYER ARMOR

Sivyer Steel Corp.

C 0.27, Mn 1.25, Cr 0.55, Ni 0.55, Mo 0.43, bal Fe.

Low alloy steel for castings requiring high strength, impact resistance, and hardenability. 90-160 TS; 60-145 YS; 187-415 Brin. Meets SAE 8627 Mod; Mil-A-11356E, ASTM A-148.

SIVYER B-30

Sivyer Steel Corp.

C 0.31, Mn 0.85, Cr 0.5, Ni 0.65, Mo 0.35, B 0.004, bal Fe.

Low alloy steel. 60-135 TS; 90-150 YS; 187-363 Brin. Meets SAE 86B30 Mod.

SIVYER CHROME ELECTRIC STEEL

Sivyer Steel Corp.

C 0.22, Cr 5, Mn 0.75, Si 0.4, bal Fe.

Annealed: 95,000 TS; 55,000 YS; 18 El; 25 RA; 200 Brin. Heat treated: 100,000-200,000 TS; 70,000-165,000 YS; 3-19 El; 4-35 RA; 200-550 Brin. For miscellaneous castings; corrosion resistant. *Obsolete*

SIVYER CHROME NICKEL ELECTRIC STEEL

Sivyer Steel Corp.

C 0.4, Cr 0.9, Ni 3, bal Fe.

Annealed: 100,000 TS; 65,000 YS; 20 El; 35 RA; 200 Brin. Heat treated: 115,000-210,000 TS; 85,000-173,000 YS; 6-18 El; 10-43 RA; 200-600 Brin. For gears, shafts; oil hardening. *Obsolete*

SIVYER CNM-20

Sivyer Steel Corp.

C 0.19, Mn 0.75, Cr 0.6, Ni 0.48, Mo 0.4, bal Fe.

Low alloy steel for castings requiring moderate to high strength and good weldability. 90-120 TS; 60-95 YS; 187-302 Brin. Meets SAE 8620 Mod.

SIVYER CNM-30

Sivyer Steel Corp.

C 0.28, Mn 0.8, Cr 0.6, Ni 0.48, Mo 0.4, bal Fe.

Low alloy steel for castings requiring moderate to high strength and depth hardness. 90-150 TS; 60-135 YS; 187-363 Brin. Meets SAE 8628 Mod.

SIVYER HIGH CARBON STEEL

Sivyer Steel Corp.

C 0.7, bal Fe.

Annealed: 80,000 TS; 40,000 YS; 20 El; 30 RA; 175 Brin. Heat treated: 85,000-150,000 TS; 55,000-90,000 YS; 3-20 El; 4-35 RA; 175-400 Brin. For tools, springs; water hardening. *Obsolete*

SIVYER HIGH SILICON

Sivyer Steel Corp.

C 0.09-0.12, Mn 0.15, Si 1.75, bal Fe.

For electrical equipment; high permeability. *Obsolete*

SIVYER M-25

Sivyer Steel Corp.

C 0.21, Mn 1.3, bal Fe.

Carbon-manganese steel for general purpose castings requiring low to moderate strength and good weldability. 70-90 TS; 36-60 YS; 140-240 Brin. Meets ASTM A-27, A-148, 1E627, QQS-681, and SAE 1521.

SIVYER M-40

Sivyer Steel Corp.

C 0.35, Mn 1.35, bal Fe.

Carbon-manganese steel for general purpose castings requiring moderate strength or surface hardening to 50 Rock C. 80-120 TS; 40-95 YS; 163-302 Brin. Meets SAE 1535.

SIVYER MANGANESE-CARBON ELECTRIC STEEL

Sivyer Steel Corp.

C 0.32, Mn 1.45, Si 0.4, bal Fe.

Annealed: 95,000 TS; 55,000 YS; 22 El; 40 RA; 200 Brin. Heat treated: 95,000 TS; 65,000 YS; 22-2.5 El; 3.5-50 RA; 200-450 Brin. For gears, shafts; oil hardening. *Obsolete*

SIVYER MM-20

Sivyer Steel Corp.

C 0.19, Mn 1.3, Mo 0.23, bal Fe.

Low alloy steel for general purpose castings requiring weldability, moderate strength and machinability. 80-120 TS; 50-95 YS; 160-302 Brin. Meets SAE 4020 Mod.

SIVYER MM-30

Sivyer Steel Corp.

C 0.3, Mn 1.3, Mo 0.3, bal Fe.

Low alloy steel for castings requiring moderate to high strength and hardenability. 90-135 TS; 60-115 YS; 187-320 Brin. Meets SAE 4030 Mod.

SIVYER MN-1

Sivyer Steel Corp.

C 1.18, Mn 13.5, bal Fe.

Hatfields manganese for castings requiring wear resistance in high impact applications such as power shovel buckets, lips, corners, railroad switches and frogs. Meets ASTM-A-128.

SIVYER MN-2

Sivyer Steel Corp.

C 1.07, Mn 13.5, bal Fe.

Hatfields manganese for castings requiring wear resistance in high impact applications such as power shovel buckets, lips, corners, railroad switches and frogs. Meets ASTM-A-128.

SIVYER MN-3

Sivyer Steel Corp.

C 1.18, Mn 14, Cr 1.8, bal Fe.

Hatfields manganese for castings requiring wear resistance in high impact applications such as power shovel buckets, lips, corners, railroad switches and frogs. Meets ASTM-A-128.

SIVYER MN-4

Sivyer Steel Corp.

C 1.05, Mn 13.5, Mo 1, bal Fe.

Hatfields manganese for castings requiring wear resistance in high impact applications such as power shovel buckets, lips, corners, railroad switches and frogs. Meets ASTM-A-128.

SIVYER NITRO

Sivyer Steel Corp.

C 0.3, Mn 0.75, Si 0.4, Cr 1.5, V 0.6, bal Fe.

For gears, pinions, shafts, cams; nitriding steel. *Obsolete*

SIVYER NO. 60

Sivyer Steel Corp.

Cr 18, Ni 8, Mo 0.5, C 0.1, bal Fe.

Cast: 70,000-80,000 TS; 30,000-40,000 YS; 50-60 El; 50-70 RA; 135-150 Brin. For stainless parts; corrosion resistant. *Obsolete*

SIVYER NO. 62

Sivyer Steel Corp.

Cr 23-25, Ni 11-13, Si 1, Mn 0.5, C 0.18, bal Fe.

Cast: 75,000-85,000 TS; 30,000-40,000 YS; 35-45 El; 35-45 RA. For furnace parts, heat and corrosion resistant parts; heat and corrosion resistant. *Obsolete*

SIVYER NO. 64

Sivyer Steel Corp.

C 0.25, Mn 0.5, Si 1, Cr 30, Ni 9.5, bal Fe.

For corrosion and heat resistant parts; corrosion and heat resistant. *Obsolete*

SIVYER NO. 66

Sivyer Steel Corp.

Cr 12-14, Mn 0.5, Si 1, C 0.1, bal Fe.

Cast: 75,000-170,000 TS; 45,000-130,00 YS; 8-25 El; 15-50 RA; 175-350 Brin. For furnace parts, corrosion resistant parts; corrosion resistant. *Obsolete*

SIVYER NO. 67

Sivyer Steel Corp.

Cr 16-18, C 0.25, Mn 0.5, Si 1, bal Fe.

Cast: 85,000-105,000 TS; 55,000-70,000 YS; 10-20 El; 15-35 RA; 160-190 Brin. For furnace parts, corrosion resistant parts; corrosion resistant. *Obsolete*

SIVYER NO. 70

Sivyer Steel Corp.

Cr 15-17, Ni 35-37, Si 1, C 0.45, Mn 0.5, bal Fe.

Cast: 60,000-70,000 TS; 1-7 El; 1-8 RA; 200-225 Brin. For furnace parts, heat and corrosion resistant parts; corrosion and heat resistant. *Obsolete*

SIVYER OBLALLOY 28

Sivyer Steel Corp.

C 0.25, Mn 0.9, Cr 0.75, Ni 1.75, Mo 0.47, B 0.003, bal Fe.

Low alloy steel for castings requiring high strength and impact resistance at low temperatures such as drive wheels and tumblers, and track shoes. 85-135 TS; 105-160 YS; 217-450 Brin. Meets SAE 43B25 Mod.

SIVYER S-20

Sivyer Steel Corp.

C 0.15, Mn 0.8, bal Fe.

Plain carbon steel for general purpose castings requiring low strength and good weldability. Ferritic and martensitic valve castings for low temperature service. 60-70 TS; 30-40 YS; 131-187 Brin. Meets ASTM A-27, A-216, A-352, AAR-201, QQS-681, and SAE 1016.

SIVYER S-30

Sivyer Steel Corp.

C 0.24, Mn 0.65, bal Fe.

Plain carbon steel for general purpose castings requiring low strength and good weldability. Ferritic and martensitic valve castings for low temperature service. 60-70 TS; 30-40 YS; 131-187 Brin. Meets ASTM A-27, A-216, A-352, AAR-201, 1E-3, 1E527, and SAE 1024.

SIVYER S-50
Sivyer Steel Corp.
C 0.45, Mn 0.7, bal Fe.
Plain carbon steel for castings requiring response to surface hardening to 57 Rock C. 80 TS; 40 YS; 160-235 Brin. Meets SAE 1045.

SIVYER TRI 50 SB
Sivyer Steel Corp.
C 0.33, Mn 0.85, Cr 0.7, Ni 0.48, Mo 0.5, bal Fe.
Low alloy steel for castings requiring high hardness for shearblades and knives. 85-135 TS; 105-160 YS; 217-580 Brin. Meets SAE 8633 Mod.

SIVYER TRI 50-X
Sivyer Steel Corp.
C 0.32, Mn 0.85, Cr 0.65, Ni 0.48, Mo 0.55, bal Fe.
Low alloy steel for castings requiring high strength, toughness and hardenability. 85-135 TS; 105-160 YS; 217-580 Brin. Meets SAE 8632 Mod.

SIVYER TRI 50-XB
Sivyer Steel Corp.
C 0.32, Mn 0.85, Cr 0.65, Ni 0.48, Mo 0.55, bal Fe.
Low alloy steel for castings requiring high strength and hardenability such as automotive shredder mill components. 85-135 TS; 105-150 YS; 217-580 Brin. Meets SAE 86B32.

SIVYER VANADIUM ELECTRIC STEEL
Sivyer Steel Corp.
C 0.45, V 0.2, Mn 0.75, Si 0.4, bal Fe.
Annealed: 85,000 TS; 50,000 YS; 22 El; 40 RA; 180 Brin. For shafts, gears, pinions; water hardening. *Obsolete*

SIWI
Darwin & Milner Inc.
Tool material. C, alloy, bal Fe.
For dies; hot die steel.

SIX EIGHTY ALLOY
Johnson Matthey plc
65.0 Ag.
For dental amalgams.

SIX-MAX
Atlas Steels Ltd.
C 0.8, W 6, Cr 4, V 1.5, Mo 6, bal Fe.
For cutting tools; high speed steel. *Obsolete*

SIXIX
Atlas Specialty Steels
C 0.8, Cr 4, W 6.5, Mo 5, V 2, bal Fe.
For cutters, tools; high speed steel. *Obsolete*

SIXTY N
United States Steel Corp.
C 0.2, Mn 1.35, Si 0.35, N 0.02, V 0.04-0.11, bal Fe.
Normalized: 80 ksi TS; 60 ksi YS; 23 El. Readily formed and welded. For stressed structures at low temperatures; arctic and marine structures. ASTM A633 Grade E.

SJ CALKING LEAD
Manufacturer not listed.
Ag 0.0005, Cd 0.0003, Pb 99.99-0.
For plumbing applications. Exceeds ASTM B29-55.

SJMFA-1000
Manufacturer not listed.
Ca 0.1, Ag 0-0.001, bal Pb (Bi, Sn, Sb, Cu, As, Zn, Cd, Ni, Fe 0.0005, max each).
Cast and aged: 5400-5800 TS; 30-45 El; 623-625°F melting range; 65-80 Rock R. Cast grids for maintenance-free batteries.

SJMFA-300
Manufacturer not listed.
Ca 0.03, Ag 0-0.001, bal Pb (Bi, Sn, Sb, Cu, As, Zn, Cd, Ni, Fe 0.0005, max each).
Cast and aged: 3300-3700 TS; 40-50 El; 422°F (approx) melting range; 40-45 Rock R. Cast grids for industrial stand-by batteries.

SJMFA-700
Manufacturer not listed.
Ca 0.07, Ag 0-0.001, bal Pb (Bi, Sn, Sb, Cu, As, Zn, Cd, Ni, Fe 0.0005, max each).
Cast and aged: 5200-5600 TS; 30-45 El; 623°F (approx) melting range; 70-80 Rock R. Cast lead products of moderate strength.

SJMFA-707
Manufacturer not listed.
Sn 0.7, Cd 0.07, Ag 0-0.001, bal Pb (Bi, Sb, Cu, As, Zn, Cd, Ni, Fe 0.0005 max, each).
Cast and aged: 7000-7500 TS; 30-40 El; 619-633°F melting range; 85-95 Rock R. All purpose casting; battery grids.

SJMFA-903
Manufacturer not listed.
Sn 0.3, Ca 0.1, Ag 0-0.001, bal Pb (Bi, Sb, Cu, As, Zn, Cd, Ni, Fe 0.0005, max each).
Cast and aged: 6000-6500 TS; 25-35 El; 622-640°F melting range; 80-85 Rock R. Cast parts and grids for maintenance-free batteries.

SJMFA-905
Manufacturer not listed.
Sn 0.5, Ca 0.1, Ag 0-0.001, bal Pb (Bi, Sb, Cu, As, Zn, Cd, Ni, Fe 0.0005, max each).
Cast and aged: 6500-7000 psi TS; 25-35 El; 621-637°F melting range; 85-90 Rock R. All purpose casting; grids for storage batteries.

SJMFA-910
Manufacturer not listed.
Sn 1, Ca 0.1, Ag 0-0.001, bal Pb (Bi, Sb, Cu, As, Zn, Cd, Ni, Fe 0.0005 max, each).
Cast and aged: 7500-8000 TS; 20-35 El; 617-630°F melting range; 90-95 Rock R. Cast grids for storage batteries.

SK 42
Wheeling-Pittsburgh Steel Corp.
C 0.27, Mn 1.2, bal Fe.
Plate: 60,000 TS; 42,000 YS; 19 El. For mine and railroad cars, booms, derricks, bridges, pressure vessels. Good fabricability and weldability. *Obsolete*

SK A45
Wheeling-Pittsburgh Steel Corp.
C 0.22, Mn 1.25, Si 0.1, 0.02 V min, bal Fe.
Rolled: 65,000 TS; 45,000 YS; 19 El. For railroad and mine cars, booms, bridges, derricks, pressure vessels. Good fabricability and weldability. *Obsolete*

SK A50
Wheeling-Pittsburgh Steel Corp.
C 0.26, Mn 1.3, Si 0.1, 0.02 V min, bal Fe.
Rolled: 70,000 TS; 50,000 YS; 18 El. For railroad and mine cars, automobile and bus bodies, bridges, derricks, pressure vessels. Good fabricability and weldability. *Obsolete*

SK H8
Japanese manufacture
C 0.65, W 16, Cr 4.3, V 0.76, Co 2.64, bal Fe.
Hardened: 64-66 Rock C. For tools and cutters, lathe and planer tools, drills; high-speed steel, high red-hardness.

SK-3
Thyssen Edelstahlwerke AG
C 0.05, Cr 1, bal Fe.
For bearing balls and ball races; case hardened. *Obsolete*

SKEFKO NO. 1
Hofors Steel Works
C 0.95-1.05, Cr 1.4-1.65, Mn 0.25-0.35, bal Fe.
Heat treated: 90,000-160,000 TS; 60,000-140,000 YS; 6-20 El; 12-45 RA; 190-450 Brin. For ball bearings; water hardening.

SKEFKO NO. 2
Hofors Steel Works
C 1.1, Cr 0.4-0.6, Mo 0.3-0.4, bal Fe.
For ball bearings; water hardening.

SKEFKO NO. 3
Hofors Steel Works
C 0.95-1.05, Cr 1-1.25, Mo 0.3-0.4, bal Fe.
For ball bearings; water hardening.

SKELTER PLATE
Associated Steel Corp.
C 0.39, Mn 2.3, Cr 0.9, Mo 0.34, bal Fe.
Heat treated to 139,000 psi YS.

SKF NO. 2
S.K.F. Industries Inc.
C 0.87-0.97, Mn 1.4-1.7, Cr 1.4-1.7, bal Fe.
For trimming and stamping dies; oil hardened.

SKF NO. 6
S.K.F. Industries Inc.
C 0.9, Cr 1.1, Mn 0.6, bal Fe.
For ball bearings, rollers, spindles; water hardened.

SKF NO. 85
S.K.F. Industries Inc.
C, alloy, bal Fe.
For chisels; water hardened.

SKF STEEL NO. 46
S.K.F. Industries Inc.
C, Cr, W, V, bal Fe.
For tools, cutters, dies; non-deforming.

SKHLF STEEL
Russian manufacture
C 0.2, Si 0.3, Mn 0.5, Cr 0.55, Ni 0.45, Cu 0.68, bal Fe.
For welded structures; good weldability.

SL250
Sheffield Smelting Co. Ltd.
Ag 2.5, Cu, Pb.
Melting range: 302-305°C. Maximum stress: 3.9 kgf/mm^2; 45 El. For soft soldering; high temperature grade.

SLATER AISI TYPE 304
Slater Steels
Stainless steel. C 0-0.08, Mn 0-2, Si 0-1, Cr 18-20, Ni 8-11, bal Fe.
Annealed: 85,000 TS; 35,000 YS; 60 El; 149 Brin. Austenitic, non-magnetic, corrosion resistant, hardenable by cold work only. For shafts, dairy fittings, meat hangers and hooks, food processing equipment. AISI 304; AMS 5639; ASTM A193.

SLATER ALLOY 21CR-6NI-9MN
Slater Steels
Stainless steel. Cr 21, Ni 6, Mn 9, bal Fe.
Austenitic stainless steel with high yield strength and good corrosion resistance. Good elevated temperature properties and retains high strength and toughness at subzero temperatures. Readily forgeable. Can be welded by conventional welding methods. Typical applications: aircraft parts such as ducting and bellows systems, tailpipes, exhaust systems, clamps, flanges, and hydraulic tubing. Also cryogenic applications.

SLATER ALLOY 400
Slater Steels
Stainless steel. C 0.12, Mn 0.9, Si 0.15, S 0.005, Cu 31.5, Fe 1.35, 66.0 Ni + Co.
High strength, good ductility and corrosion resistance. Can be both hot and cold worked followed by a low temperature treatment. Typical applications: various shafting applications, rods, valves, standard fasteners, turbine blading, heater tubes, tanks, springs, carburetor parts, heat exchangers, condensers, coolers, spray nozzles, sterilizers, food service equipment.

SLATER ALLOY 600
Slater Steels

Stainless steel. C 0.1, Mn 1, P 0.03, S 0.007, Si 0.5, Cr 14-17, Cu 0.5, Fe 6-10, Co 0.1, Al 0.35, Ti 0.1-0.5, B 0.007, Mg 0.06, 72.0 Ni min.
Nonmagnetic, high corrosion and heat-resistant nickel base alloy. High strength, hot and cold workability, resistance to ordinary forms of corrosion, good heat resistance, and freedom from aging or stress corrosion throughout the range from annealed to heavily cold worked material. Good mechanical properties at cryogenic as well as elevated temperatures. Material is highly resistant to stress corrosion cracking in contact with boiling water, pressurized water and steam. Produced in electroslag remelting furnace. Typical applications: pumps and control rod drive mechanisms both inside and outside the core of nuclear reactors, primary heat exchanger tubing, springs, furnace muffles, chemical and food processing equipment, carburizing baskets, fixtures and rotors.

SLATER ALLOY 800
Slater Steels

Stainless steel. C 0-0.1, Mn 0-1.5, S 0-0.015, Si 0-1, Cr 19-23, Ni 30-35, Cu 0-0.75, Al 0.15-0.6, Ti 0.15-0.6, 39.5 Fe min.
Nickel, iron, chromium alloy that has good oxidation and corrosion resistance at elevated temperatures. Good resistance to thermal cycling. Does not exhibit embrittlement after exposure to intermediate temperatures. Good workability and welding properties and resists corrosion in different atmospheres. Resists chloride stress corrrosion cracking. Typical applications: Furnace parts and other heat treating equipment, heat exchangers, manifolds, heating element tubes, gaskets, electric ranges and other domestic appliances, petro chemical equipment, process piping, power plant steam tubing and related components.

SLATER ALLOY 800H
Slater Steels

Stainless steel. C 0.05-0.1, Mn 0-1.5, S 0-0.015, Si 0-1, Cr 19-23, Ni 30-35, Cu 0-0.75, Al 0.15-0.6, Ti 0.15-0.6, 39.5 Fe min.
Similar to Slater Alloy 800 with better high temperature strength, creep and stress rupture properties above 1100°F.

SLATER ALLOY A-286
Slater Steels

Stainless steel. C 0.08, Mn 0.35, P 0.02, S 0.01, Si 0.3, Cr 13.5-16, Ni 24-27, Mo 1-1.5, B 0.003-0.01, Ti 2.15-2.3, V 0.1-0.5, Al 0.35, bal Fe.
Age-hardenable, austenitic, iron-nickel-chromium alloy designed for service up to 1300°F. High strength and good corrosion resistance at the operating temperature. High ductility in notched sections. Can be precipitation (age) hardened and strengthened by heat treatment. Typical applications: afterburner parts, bolts, fasteners, casings, frames, and turbine wheels and blades in jet engines and superchargers.

SLATER ALLOY HX
Slater Steels

Stainless steel. C 0.05-0.1, Mn 1, P 0.04, S 0.03, Si 1, Cr 20.5-23, Mo 8-10, Co 0.5-2.5, W 0.2-1, Fe 17-20, B 0.01, bal Ni.
Nickel-chromium-iron-molybdenum alloy with good oxidation resistance and strength at temperatures up to 2200°F. Resistant to thermal shock. Better fabricability than most other high temperature alloys. Can be welded by conventional processes by procedures used with the austenitic stainless steels. Typical applications: aircraft engine parts such as combustion chambers, afterburners, tailpipes. Also used for fans, roller hearths, and support members in industrial furnaces.

SLATER AM350
Slater Steels

Stainless steel. C 0.1, Mn 0.8, P 0.03, S 0.01, Si 0.25, Cr 16.5, Ni 4.3, Mo 2.75, N 0.1, bal Fe.
Annealed form is relatively soft and ductile. High strength-to-weight ratios. Can be heat treated to very high strength levels. Water quenched and annealed: (1900°F) 145,000 TS; 60,000 YS (0.2% offset); 40 El; 20 RA. Typical applications: structural applications, blades, valves, and springs.

SLATER AM355
Slater Steels

Stainless steels. C 0.13, Mn 0.95, P 0.03, S 0.01, Si 0.25, Cr 15.5, Ni 4.3, Mo 2.75, N 0.1, bal Fe.
Annealed form is relatively soft and ductile. High strength-to-weight ratios. Can be heat treated to very high strength levels. Reduces likelihood of ferrite formation in the structure. Water quenched and annealed: (1900°F) 160,000 TS; 60,000 YS (0.2% offset); 30 El; 15 RA. Typical applications: structural applications, blades, valves, and springs.

SLATER FW 20
Slater Steels

Stainless steel. C 0-0.07, Mn 0-2, P 0-0.035, S 0-0.035, Si 0-1, Cr 19-21, Ni 30-38, Mo 2-3, Cu 3-4, 1.0 Cb + Ta max, bal Fe.
Austenitic stainless with good corrosion resistance to sulfuric acid. Widely used in chemical industry. Increased formability can be obtained by using a higher annealing temperature. Parts should not be welded or heated over 1000°F. Can be readily hot or cold worked. Typical applications: acid handling systems and related parts, heat exchangers, valves, fittings, pump shafts, fasteners, pharmaceutical equipment, chemical handling equipment, mixing tanks and equipment, oil and refining equipment.

SLATER TOOL STEEL TYPE A-2
Slater Steels

Stainless steel. C 0.95-1.05, Mn 0-1, Si 0-0.5, Cr 4.75-5.5, Mo 0.9-1.4, V 0.15-0.5, bal Fe.
High carbon, medium alloy, air hardening tool steel. Minimum distortion and good resistance to cracking in hardening. Moderate resistance to softening at elevated temperature. Used for blanking dies, thread roll dies, rolls, trim dies, form dies, punches, and coining dies. Dimensional stability makes it suitable for precision tools and gauges.

SLATER TOOL STEEL TYPE D-2
Slater Steels

Stainless steel. C 1.4-1.6, Mn 0-0.6, Si 0-0.6, Cr 11-13, Mo 0.7-1.2, V 0-1.15, bal Fe.
High carbon, high chromium, air hardening, cold working tool steel. Deep hardening, offers good wear resistance, as well as resistance to tempering at elevated temperatures. High chromium properties gives it mild corrosion resistance in hardened condition. Typical applications: blanking dies, thread roll dies, rolls, trim dies, and form roll tooling, shear and slitter knives.

SLATER TOOL STEEL TYPE H-13
Slater Steels

Stainless steel. C 0.32-0.45, Mn 0.2-0.5, Si 0.8-1.2, Cr 4.75-5.5, Mo 1.1-1.75, V 0.8-1.2, bal Fe.
Medium carbon, chromium tool steel designed primarily for applications requiring toughness and good red hardness. Used mainly in hot work applications, but can be used as a cold work tool steel as well. Typical applications: forging dies, aluminum extrusion tooling, aluminum die cast dies, hot punches, and cold header casings.

SLATER TYPE 15-5
Slater Steels

Stainless steel. C 0-0.07, Mn 0-1, Si 0-1, Cr 14-15.5, Ni 3.5-5.2, Cu 2.5-4.5, 0.15-0.45 Cb + Ta, bal Fe.
Precipitation hardening stainless steel. Condition: H 900: 200,000 psi TS; 185,000 psi YS; 14 El; 450 Brin. For valves, shafts, gears, bolts, studs.

SLATER TYPE 17-4
Slater Steels

Stainless steel. C 0-0.07, Mn 0-1, Si 0-1, Cr 15.5-17.5, Ni 3-5, Cu 3-5, 0.15-0.45 Cb + Ta, bal Fe.
Precipitation hardening stainless steel. Condition: H 900: 200,000 psi TS; 185,000 psi YS; 14 El; 420 Brin. For valves, shafts, gears, fasteners, studs.

SLATER TYPE 2205
Slater Steels

Stainless steel. C 0-0.03, Mn 0-2, P 0-0.03, S 0-0.02, Si 0-1, Cr 21-23, Ni 4.5-6.5, Mo 2.5-3.5, N 0.08-0.2, bal Fe.
Very low carbon duplex stainless steel. High strength and corrosion resistance. Superior resistance to chloride stress-corrosion cracking than Slater Type 316 or Type 317, but with twice the yield strength. Typical applications: parts for heat exchangers, pressure vessels, desalination plants, rotors, fans, and shafting where corrosion is important. Also used in the chemical, refining, power plant, and oil and gas industries. Annealed: 105,000 psi TS; 90,000 psi YS (0.2% offset); 35 El (in 2 in.); 65 RA; 255 Brin.

SLATER TYPE 303 FORGING QUALITY
Slater Steels

Stainless steel. C 0-0.15, Mn 0-2, Si 0-1, Cr 17-19, Ni 8-10, Mo 0-0.6, 0.15 S min, bal Fe.
Annealed: 90,000 TS; 35,000 YS; 50 El; 160 Brin. Forging grade, austenitic, free machining, hardenable only by cold work, non-magnetic. For hot headed bolts, aircraft fittings, forged and machined fittings, and fasteners. AISI 303; AMS 5640 Type 1; SAE 30303.

SLATER TYPE 303 MAX MACHINABILITY
Slater Steels

Stainless steel. C 0-0.15, Mn 0-2, Si 0-1, Cr 17-19, Ni 8-10, Mo 0-0.6, 0.15 S min, bal Fe.
Annealed: 90,000 TS; 35,000 YS; 50 El; 160 Brin. Austenitic, free machining, non-magnetic, hardenable only by cold work, corrosion resistant. For studs, bolts, nuts, gears, hardware fittings, screw machine parts. AISI 303; AMS 5640 Type 1; SAE 30303.

SLATER TYPE 304 MAX MACHINABILITY
Slater Steels

Stainless steel. C 0-0.08, Mn 0-2, S 0-0.03, Si 0-1, Cr 18-20, Ni 8-12, bal Fe.
Annealed: 85,000 TS; 35,000 YS; 60 El; 149 Brin. Austenitic, non-magnetic, corrosion resistant, hardenable by cold work only. For shafts, dairy fittings, valve parts, food and laundry equipment. AISI 304; AMS 5639; SAE 30304.

SLATER TYPE 304 N
Slater Steels

Stainless steel. C 0-0.08, Mn 0-2, Si 0-1, Cr 18-20, Ni 8-10, N 0.18-0.3, bal Fe.
Annealed: 115,000 TS; 70,000 YS; 45 El; 65 RA; 229 Brin. Higher strength than AISI 304.

SLATER TYPE 304L
Slater Steels

Stainless steel. C 0-0.03, Mn 0-2, Si 0-1, Cr 18-20, Ni 8-11, bal Fe.
Annealed: 80,000 TS; 30,000 YS; 60 El; 140 Brin. Austenitic, weldable, non-magnetic, hardenable only by cold work, corrosion resistant. For weld rod, welded assemblies, food processing equipment. AISI 304L; AMS 5647; SAE 30304L.

SLATER TYPE 309
Slater Steels

Stainless steel. C 0-0.2, Mn 0-2, Si 0-1, S 0-0.03, Cr 22-24, Ni 12-15, bal Fe.
Annealed: 95,000 TS; 40,000 YS; 45 El; 160 Brin. Austenitic, corrosion resistant, good resistance to attack at high temperatures to 2000°F. For furnace parts, heat treating baskets and equipment, weld rod, pump parts. AISI 309; ASTM A276; SAE 30309.

SLATER TYPE 309S
Slater Steels

Stainless steel. C 0-0.08, Mn 0-2, Si 0-1, S 0-0.03, Cr 22-24, Ni 12-15, bal Fe.
Annealed: 95,000 TS; 40,000 YS; 45 El; 160 Brin. Austenitic, weldable, good high temperature resistance to corrosion and oxidation. For furnace parts, heat treating equipment, cement kiln chain, weld rod.

SLATER TYPE 310
Slater Steels

Stainless steel. S 0-0.25, Mn 0-2, Si 0-1.5, S 0-0.03, Cr 24-26, Ni 19-22, bal Fe.
Annealed: 95,000 TS; 45,000 YS; 50 El; 185 Brin. Austenitic, corrosion resistant, weldable, non-magnetic, good heat resistance. For furnace parts, gas turbine parts, pump parts, welded assemblies, weld rod. AISI 310; AMS 5694; SAE 30310.

SLATER TYPE 310S
Slater Steels

Stainless steel. C 0-0.08, Mn 0-2, Si 0-1.5, S 0-0.03, Cr 24-26, Ni 19-22, bal Fe.
Annealed: 95,000 TS; 45,000 YS; 50 El; 185 Brin. Austenitic, corrosion resistant, weldable, non-magnetic, good heat resistance. For furnace parts, oil refinery equipment, weld rod, jet engine rings. AISI 310 S; AMS 5651; SAE 30310 S.

SLATER TYPE 316 MAX MACHINABILITY
Slater Steels

Stainless steel. C 0-0.08, Mn 0-2, Si 0-1, S 0-0.03, Cr 16-18, Ni 10-14, Mo 2-3, bal Fe.
Annealed: 80,000 TS; 30,000 YS; 60 El; 149 Brin. Austenitic, non-magnetic, improved machinability, excellent corrosion resistance, weldable. For automatic screw machine parts for food, chemical, paper and textile equipment. AMS 5648; ASTM A276; SAE 30316.

SLATER TYPE 316F
Slater Steels

Stainless steel. C 0-0.08, Mn 1-2, P 0.11-0.17, S 0.1-0.2, Si 0-1, Cr 17-19, Ni 12-14, Mo 1.75-2.5, bal Fe.
Annealed: 80,000 TS; 30,000 YS; 55 El; 149 Brin. Austenitic, non-magnetic, free machining, excellent corrosion resistance. For automatic screw machine parts requiring good corrosion resistance.

SLATER TYPE 316L
Slater Steels

Stainless steel. C 0-0.03, Mn 0-2, Si 0-1, S 0-0.03, Cr 16-18, Ni 10-14, Mo 2-3, bal Fe.
Annealed: 80,000 TS; 30,000 YS; 60 El; 140 Brin. Austenitic, non-magnetic, excellent corrosion resistance, weldable. Parts for special application in food, paper, photographic and chemical processing equipment. AISI 316 L; AMS 5653; SAE 30316 L.

SLATER TYPE 317
Slater Steels

Stainless steel. C 0-0.08, Mn 0-2, Si 0-1, S 0-0.03, Cr 18-20, Ni 11-15, Mo 3-4, bal Fe.
Annealed: 85,000 TS; 40,000 YS; 50 El; 160 Brin. Austenitic, excellent corrosion resistance, useful to 1700°F, weldable. For parts for paper, food, chemical, refining, photographic and textile industries. AISI 317; ASTM A479; SAE 30317.

SLATER TYPE 317L
Slater Steels

Stainless steel. C 0-0.03, Mn 0-2, Si 0-1, S 0-0.03, Cr 18-21, Ni 11-15, Mo 3-4, bal Fe.
Annealed: 85,000 TS; 40,000 YS; 50 El; 160 Brin. Austenitic, excellent corrosion resistance, useful to 1700°F; weldable. For parts for food, paper, chemical, refining, photographic and textile industries.

SLATER TYPE 321
Slater Steels

Stainless steel. C 0-0.08, Mn 0-2, Si 0-1, S 0-0.03, Cr 17-19, Ni 9-12, Ti = 5 x C min, bal Fe.
Annealed: 85,000 TS; 35,000 YS; 55 El; 149 Brin. Austenitic, non-magnetic, weldable. For welded assemblies, for parts operated at temperatures up to 1600°F AISI 321; AMS 5645; SAE 30321.

SLATER TYPE 330
Slater Steels

Stainless steel. C 0-0.08, Mn 0-2, Si 0.75-1.5, Cr 17-20, Ni 34-37, bal Fe.
Annealed: 85,000 TS; 45,000 YS; 35 El; 179 Brin. Austenitic, non-magnetic, weldable, good at elevated temperatures. For furnace parts, heat treating boxes and baskets, jet engine parts, petroleum plants. AMS 5716; SAE 30330.

SLATER TYPE 347
Slater Steels

Stainless steel. C 0-0.08, Mn 0-2, Si 0-1, Cr 17-19, Ni 9-13, Ta + Cb = 10 x C min, bal Fe.
Annealed: 90,000 TS; 35,000 YS; 50 El; 160 Brin. Austenitic, non-magnetic, weldable. For parts subject to temperatures to 1600°F, welded assemblies, weld rods. AISI 347; AMS 5646, 5680; SAE 30347.

SLATER TYPE 348
Slater Steels

Stainless steel. C 0-0.08, Mn 0-2, Si 0-1, Cr 17-19, Ni 9-13, Ta 0-0.1, Co 0-0.2, Cb + Ta = 10 x C min, bal Fe.
Annealed: 90,000 TS; 35,000 YS; 50 El; 160 Brin. Austenitic, non-magnetic, heat resistant, weldable. For welded construction for radioactive systems, and operating temperatures 800-1500°F. AISI 348; SAE 30348; MIL-S-23195.

SLATER TYPE 403
Slater Steels

Stainless steel. C 0-0.15, Mn 0-1, Si 0-0.5, Cr 11.5-13, bal Fe.
Annealed: 75,000 TS; 40,000 YS; 35 El; 155 Brin. Quenched and tempered: 1200°F; 100,000 TS 85,000 YS; 23 El; 225 Brin. Magnetic, heat treatable, weldable. For jet engine parts, corrosion resistant hardware, steam and gas turbine blading. AISI 403; AMS 5614; SAE 51403.

SLATER TYPE 405
Slater Steels

Stainless steel. C 0-0.08, Mn 0-1, Si 0-1, Cr 11.5-14.5, Al 0.1-0.3, bal Fe.
Annealed: 70,000 TS; 40,000 YS; 30 El; 149 Brin. Magnetic, not normally hardenable, oxidation resistant up to 1500°F. For annealing boxes, forged fittings, quench racks. AISI 405; SAE 51405.

SLATER TYPE 410
Slater Steels

Stainless steel. C 0-0.15, Mn 0-1, Si 0-1, Cr 11.5-13.5, bal Fe.
Annealed: 75,000 TS; 40,000 YS; 35 El; 155 Brin. Quenched and tempered: 1200°F; 100,000 TS; 85,000 YS; 35 El; 225 Brin. Magnetic, heat treatable, weldable, corrosion resistant. For cutlery, corrosion resistant hardware, steam and gas turbine parts, pump parts. AISI 410; AMS 5613; SAE 51410.

SLATER TYPE 410 CB
Slater Steels

Stainless steel. C 0-0.15, Cr 11.5-13.5, Cb 0-0.2, Si 0-0.5, bal Fe.
Annealed: 75,000 TS; 40,000 YS; 35 El; 155 Brin. Quenched and tempered: 1200°F; 124,000 TS; 100,000 YS; 20 El. Magnetic, hardenable, weldable, corrosion resistant. For steam turbine blades, valve parts, aircraft and missile components. AMS 5609.

SLATER TYPE 410 LOW CARBON
Slater Steels

Stainless steel. C 0-0.08, Mn 0-1, Si 0-1, Cr 12.5-14, bal Fe.
Annealed: 75,000 TS; 40,000 YS; 35 El; 155 Brin. Magnetic weldable corrosion resistant, only partially hardenable. For welded assemblies.

SLATER TYPE 416 FORGING QUALITY
Slater Steels

Stainless steel. C 0-0.15, Mn 0-1.25, P 0-0.06, Si 0-1, Cr 12-14, Ni 0-0.5, Mo 0-0.6, 0.15 S min, bal Fe.
Processed to produce good quality forgings. Hardenable to 35 Rock C min. Magnetic corrosion resistant, free machining. For hot headed bolts, forged and machined parts. AISI 416; AMS 5610; SAE 51416.

SLATER TYPE 416 HIGH HARDENABILITY
Slater Steels

Stainless steel. C 0-0.15, Mn 0-1.25, Si 0-1, Cr 12-14, Ni 0-0.5, Mo 0-0.6, 0.15 S min, bal Fe.
Annealed, cold finished: 90,000 TS; 80,000 YS; 15 El; 197 Brin. Guaranteed hardenable to 35 Rock C min. (average: 41 Rock C). Magnetic, corrosion resistant, free machining. For fasteners, shafts, impellers gears. AISI 416; AMS 5610; SAE 51416.

SLATER TYPE 416 MAX MACHINABILITY
Slater Steels

Stainless steel. C 0-0.15, Mn 0-1.25, Si 0-1, Cr 12-14, Ni 0-0.5, Mo 0-0.6, 0.15 S min, bal Fe.
Annealed, cold finished: 90,000 TS; 80,000 YS; 15 El; 197 Brin. Quenched and tempered: 1200°F: 110,000 TS; 85,000 YS; 18 El; 225 Brin. Magnetic, hardenable, corrosion resistant, free machining. For pump shafts, motor shafts, valve trim, valve stems, fasteners, fittings. AISI 416; AMS 5610; SAE 51416.

SLATER TYPE 416 XS
Slater Steels

Stainless steel. C 0-0.1, Mn 0-1, P 0-0.06, Si 0-1, Cr 13-15, Ni 0-0.5, Mo 0.3-0.6, 0.35 S min, bal Fe.
Annealed, cold finished: 90,000 TS; 80,000 YS; 15 El; 197 Brin. Magnetic, corrosion resistant, free machining, not recommended for welding or quench hardening. For oil burner nozzles, motor shafts, screw machine parts.

SLATER TYPE 420
Slater Steels

Stainless steel. Mn 0-1, Si 0-1, Cr 12-14, over 0.15 C, bal Fe.
Annealed: 95,000 TS; 50,000 YS; 25 El; 195 Brin. Quenched and tempered: 600°F: 230,000 TS; 195,000 YS; 8 El; 500 Brin. Magnetic, corrosion resistant. For cutlery, surgical and dental instruments, valves, shafts, gears, knife blades, springs. AISI 420; AMS 5621; SAE 51420.

SLATER TYPE 420 F
Slater Steels

Stainless steel. Mn 0-1.25, P 0-0.06, Si 0-1, Cr 12-14, Mo 0-0.6, 0.15 S min, over 0.15 C, bal Fe.
Annealed: 95,000 TS; 50,000 YS; 25 El; 195 Brin. Quenched and tempered: 600°F: 230,000 TS; 195,000 YS; 8 El; 500 Brin. Magnetic, corrosion resistant, free machining. For gears, shafts, surgical and dental instruments, knife blades. AISI 420; AMS 5620; SAE 51420.

SLATER TYPE 422
Slater Steels

Stainless steel. C 0.2-0.25, Mn 0-1, Si 0-1, Cr 12-14, Ni 0.5-1, Mo 0.75-1.25, W 0.75-1.25, V 0.2-0.5, bal Fe.
Martensitic type corrosion resisting steel. Heat treated at 1000°F; 220,000 psi TS; 158,000 psi YS; 460 Brin. For steam turbine blades and buckets, valves, high temperature bolting. AISI 422.

SLATER TYPE 430
Slater Steels

Stainless steel. C 0-0.12, Mn 0-1, Si 0-1, Cr 14-18, bal Fe.
Annealed: 75,000 TS; 45,000 YS; 30 El; 156 Brin. Ferritic type magnetic, corrosion resistant, not hardenable. For automotive fasteners, kitchen utensils, cold headed parts, food handling equipment. AISI 430; AMS 5627; SAE 51430.

SLATER TYPE 430 F
Slater Steels

Stainless steel. C 0-0.12, Mn 0-1.25, P 0-0.06, Si 0-1, Cr 14-18, Mo 0-0.6, 0.15 S min, bal Fe.
Annealed: 80,000 TS; 45,000 YS; 25 El; 163 Brin. Ferritic, magnetic, free machining, corrosion resistant, not hardenable, poor weldability. For machined parts, bolts and nuts, fasteners, shafts, burner parts. AISI 430 F; SAE 51430 F.

SLATER TYPE 430 F SOLENOID QUALITY
Slater Steels

Stainless steel. C 0-0.06, C 0-0.5, Mn 0-0.5, S 0.25-0.35, Si 0-0.5, Cr 17.25-18.25, Ni 0-0.3, Mo 0-0.5, Cu 0-0.25, bal Fe.
Annealed: 80,000 TS; 45,000 YS; 25 El; 163 Brin. Magnetic, ferritic, free machining. For solenoid valve plungers.

SLATER TYPE 440-C

Slater Steels

Stainless steel. C 0.95-1.2, Mn 0-1, P 0-0.04, S 0-0.03, Si 0-1, Cr 16-18, Mo 0-0.75, bal Fe.

High carbon stainless steel providing stainless properties and maximum hardness. Attains highest hardness of any stainless steel when heat treated. Primarily used as a bearing steel. Also considered for valve seats, needle valves, pump parts, ball studs, bushings, ball check valves, cutlery. 285,000 psi TS; 275,000 psi YS (0.2% offset); 2 El (in 2 in.); 10 RA; 580 Brin.

SLATER TYPE 615 (GREEK ASCOLOY)

Slater Steels

Stainless steel. C 0.15-0.2, Mn 0-0.5, Si 0-0.5, Cr 12-14, Ni 1.8-2.2, Mo 0-0.5, W 2.5-3.5, bal Fe.

Annealed: 140,000 TS; 130,000 YS; 15 El; 277 Brin.
Quenched and tempered: 1200°F: 135,000 TS; 108,000 YS; 20 El; 277 Brin. Martensitic, magnetic, hardenable to 45 Rock C min. For steam turbine parts, gas turbine parts, jet engine parts, compressor blades and wheels.

SLICKER SOLDER

American manufacture

Sn 66, Pb 34.

Commercial lead solder; melting point 356°F.

SLIDE VALVE BRONZE

English manufacture

Sn 2.5, Zn 9, bal Cu.

For slide valves; high strength.

SLIDEX

Jones & Rooke Ltd.

Ni 10-18, C, Zn, bal Cu.

Nickel silver alloy; for slide fasteners.

SLOCRODE

Sheepbridge Alloy Castings Ltd.

C 3.2, Ni 0.75-1.25, Cr 0.5-0.75, bal Fe.

Oil treated: 36,000-40,000 TS; 200-500 Brin. For castings for high duty purposes; cast iron.

SLOCRODE

Sheepbridge Engineering Ltd.

C 3.2, Ni 0.75-1.25, Cr 0.5-0.75, bal Fe.

Oil treated: 36,000-40,000 TS; 200-500 Brin. For castings for high duty purposes; cast iron.

SLT2N

Sumitomo Metal America Inc.

C 0-0.14, Si 0-0.3, Mn 0-0.7, Ni 2.1-2.5, bal Fe.

6-32 mm thick plate: 65,000-77,000 TS; 37,000 YP min. Nickel steel for low temperature service up to -70°C for pressure vessels, cryogenic equipment. High notch toughness. Conforms to ASTM A203 Gr. A.

SLT3NA

Sumitomo Metal America Inc.

C 0-0.14, Si 0-0.3, Mn 0-0.7, Ni 3.25-3.75, bal Fe.

6-50 mm thick plate: 65,000-81,000 TS; 37,000 or 40,000 YP min. Nickel normalized steel for low temperature service to -101°C. High notch toughness. Available in strength grades: SLT3N-26A, SLT3N-28A. Conforms to ASTM A203 Gr. D or E.

SLT3NB

Sumitomo Metal America Inc.

C 0-0.14, Si 0-0.3, Mn 0-0.9, Ni 3.25-3.75, Mo 0-0.25, Cr 0-0.5, bal Fe.

6-50 mm thick plate: 79,000-100,000 TS; 64,000 YP min. Nickel steel, quenched and tempered for low temperature service to -101°C. High notch toughness.

SLT9N-53

Sumitomo Metal America Inc.

C 0-0.12, Si 0.15-0.3, Mn 0-0.9, Ni 8.5-9.5, bal Fe.

6-26 mm thick plate: 100,000-120,000 TS; 75,000 YP min. Double normalized and tempered nickel steel for low temperature service to -196°C. For LNG storage tanks. Meets ASTM-A353.

SLT9N-60

Sumitomo Metal America Inc.

C 0-0.12, Si 0.15-0.3, Mn 0-0.9, Ni 8.5-9.5, bal Fe.

6-50 mm thick plate: 100,000-120,000 TS; 85,000 YP min. Quenched and tempered nickel steel for low temperature service to -196°C. For LNG storage tanks. Meets ASTM-A553 Type I.

SM 100

Colt Industries

C 1, bal Fe.

For drills, punches; drill rod, water hardening. *Obsolete*

SM 200

Martin-Marietta Corp.

C 0.15, Cr 9, W 12.5, Co 10, Cb 1, Al 5, Ti 2, Zr 0.05, B 0.015, Fe 0.25, bal Ni.

RT: 135,000 TS; 120,000 YS; 7 El. At 1900°F: 80,000 TS; 50,000 YS; 5 El. For cast gas turbine blades, jet engine components. Precipitation hardening. Good oxidation resistance, useful strength to 1900°F.

SM 211

Martin-Marietta Corp.

C 0.15, Cr 9, Co 10, Mo 2.5, W 5.5, Cb 2.7, Ti 2, Al 5, B 0.015, Zr 0.05, bal Ni.

For high temperature applications, fasteners, jet engine and gas turbine components. High rupture strength, heat and corrosion resistant.

SM 237

Sterling International Technology Ltd.

Mg 0.2-0.4, Si 6.5-7.5, Ti 0-0.2, bal Al.

Chill cast: 200 MPa TS; 95 MPa YS; 7 El.

SM 302

Martin-Marietta Corp.

C 0.78-0.93, Cr 20-23, W 9-11, Ta 8-10, Zr 0.1-0.3, Fe 0-1.5, B 0-0.01, bal Co.

Cast: 140,000 TS; 100,000 YS; 2 El. At 1800°F: 40,000 TS; 32,000 YS; 15 El. For aircraft gas turbine guide vanes and buckets. High oxidation resistance to 2000°F.

SM 322

Martin-Marietta Corp.

C 0.9-1.1, Cr 20-23, W 8-10, Ta 4-5, Zr 2-2.5, Ti 0.65-0.85, Fe 0-1.5, bal Co.

Cast: 121,000 TS; 91,000 YS; 3.2 El; 4 RA. At 1400°F: 86,000 TS; 51,00 YS; 6.3 El; 9.3 RA. For turbine blades and vanes. High heat and oxidation resistant. Vacuum melted and vacuum cast superalloy.

SM 3630

Svenska Metallverken A.B.

Copper. Cu 96, Si 3, Mn 1.

Annealed: 57,000 TS; 17,000 YS; 50 El; 50 RA; 85 Brin. Hard: 71,000 TS; 43,000 YS; 20 El; 30 RA; 150 Brin. For sheet, rod; silicon bronze.

SM 40

Sterling International Technology Ltd.

Mg 0.5-0.75, Zn 4.8-5.7, Cr 0.4-0.6, bal Al.

Sand cast: 215 MPa TS; 170 MPa YS; 4 El; 60 Brin.

SM FERROCHROME (HIGH CARBON)

Union Carbide Corp.

Cr 60-65, Si 4-6, Mn 4-6, C 4-6, bal Fe.

For Cr additions to steel or iron; Si and Mn protects the Cr.

SM-0005

Svenska Metallverken A.B.

Copper. Cu 99.9.

Annealed: 30,000 TS; 6000 YS; 45 El; 70 RA; 50 Brin. For electrical equipment; high conductivity.

SM-0010

Svenska Metallverken A.B.

Copper. Cu 99.9.

Cold drawn: 61,000 TS; 57,000 YS; 3 El; 50 RA; 70 Brin. For electrical equipment; high conductivity.

SM-0020

Svenska Metallverken A.B.

Copper. Cu 99.9.

Wire: 61,000 TS; 57,000 YS; 3 El; 50 RA; 70 Brin. Hard drawn: 37,000 TS; 29,000 YS; 10 El; 40 RA. For electrical equipment; fire refined.

SM-0021

Svenska Metallverken A.B.

Copper. Cu 99.8.

Annealed: 30,000 TS; 6000 YS; 45 El; 70 RA; 50 Brin. Rolled: 39,000 TS; 33,000 YS; 20 El; 60 RA; 75 Brin. For sheet, tubes; deoxidized copper.

SM-0710

Svenska Metallverken A.B.

Copper. Sn 1, bal Cu.

Annealed: 37,000 TS; 9000 YS; 45 El; 60 RA; 55 Brin. Hard: 46,000 TS; 38,000 YS; 15 El; 40 RA; 95 Brin. For sheet, wire, rod; phosphor bronze.

SM-0720

Svenska Metallverken A.B.

Copper. Sn 2, bal Cu.

Annealed: 40,000 TS; 12,000 YS; 50 El; 60 RA; 60 Brin. Hard: 51,000 TS; 44,000 YS; 15 El; 40 RA; 110 Brin. For sheet; phosphor bronze.

SM-1063

Svenska Metallverken A.B.

Copper. Cu 63, bal Zn.

Annealed: 47,000 TS; 14,000 YS; 60 El; 60 RA; 65 Brin. Hardened: 60,000 TS; 45,000 TS; 25 El; 40 RA; 115 Brin. For sheet, wire, tubes; yellow brass.

SM-1067

Svenska Metallverken A.B.

Copper. Cu 67, Zn 23.

Drawn: 60,000 TS; 45,000 YS; 20 El; 45 RA. For tubing; yellow brass

SM-1072

Svenska Metallverken A.B.

Copper. Cu 72, bal Zn.

Annealed: 45,000 TS; 14,000 YS; 65 El; 70 RA; 60 Brin. For sheet, cartridge cases; cartridge brass.

SM-1076

Svenska Metallverken A.B.

Copper. Cu 76, bal Zn.

Annealed: 43,000 TS; 11,000 YS; 50 El; 65 RA; 60 Brin. For wire; ductile brass.

SM-1080

Svenska Metallverken A.B.

Copper. Cu 80, bal Zn.

Annealed: 40,000 TS; 11,000 YS; 55 El; 75 RA; 55 Brin. Rolled: 50,000 TS; 35,000 YS; 25 El; 65 RA; 110 Brin. For sheet; low brass.

SM-1085

Svenska Metallverken A.B.

Copper. Cu 85, bal Zn.

Annealed: 37,000 TS; 11,000 YS; 55 El; 75 RA; 55 Brin. Rolled: 51,000 TS; 40,000 YS; 20 El; 65 RA; 110 Brin. For wire, tube, rod; red brass.

SM-1092

Svenska Metallverken A.B.

Copper. Cu 92, bal Zn.

Annealed: 34,000 TS; 11,000 YS; 50 El; 75 RA; 55 Brin. Rolled: 45,000 TS; 33,000 YS; 20 El; 65 RA; 100 Brin. For sheet; commercial bronze.

SM-1258

Svenska Metallverken A.B.

Copper. Pb 1, Cu 58, bal Zn.

Forged: 64,000 TS; 17,000 YS; 20 El; 20 RA; 110 Brin. For architectural applications, trim, hardware; leaded brass, free cutting.

SM-1361

Svenska Metallverken A.B.
Copper. Pb 1.5, Cu 61, bal Zn.
Annealed: 50,000 TS; 17,000 YS; 40 El; 40 RA; 80 Brin. Cold drawn: 55,000 TS; 35,000 YS; 25 El; 30 RA; 100 Brin. For hardware, bolts, nuts, screws, fasteners; leaded brass, free cutting.

SM-1459

Svenska Metallverken A.B.
Copper. Pb 2, Cu 59, bal Zn.
Annealed: 53,000 TS; 17,000 YS; 30 El; 25 RA; 90 Brin. Cold drawn: 68,000 TS; 57,000 YS; 8 El; 15 RA; 140 Brin. For hardware, bolts, nuts, screws, fasteners; leaded brass, free cutting.

SM-1658

Svenska Metallverken A.B.
Copper. Pb 3, Cu 59, bal Zn.
Cold drawn: 61,000 TS; 31,000 YS; 20 El; 15 RA; 110 Brin. For screw machine products, fasteners; free cutting, leaded brass.

SM-2060

Svenska Metallverken A.B.
Copper. Cu 60, Al 0.5, Pb 0.5, Sn 1, Mn 2, Fe 0.8, bal Zn.
Cold drawn: 68,000 TS; 43,000 YS; 20 El; 25 RA; 130 Brin. For vessels, tanks, containers; brass.

SM-2265

Champion Rivet Co.
Al 2.2, Pb 2, Cu 65, bal Zn.
Cast: 64,000 TS; 43,000 YS; 25 El; 30 RA; 120 Brin. For welding rod; Al-brass.

SM-2266

Svenska Metallverken A.B.
Copper. Cu 66, Al 4.5, Fe 2, Mn 3.5, bal Zn.
Forged: 100,000 TS; 57,000 YS; 15 El; 20 RA; 180 Brin. For rods, forgings; aluminum brass.

SM-2276

Svenska Metallverken A.B.
Copper. Cu 76, Al 2, bal Zn.
Hard: 61,000 TS; 40 El. For tubes; admiralty brass.

SM-2359

Svenska Metallverken A.B.
Copper. Cu 59, Mn 2, bal Zn.
Drawn: 64,000 TS; 43,000 YS; 30 El; 35 RA; 120 Brin. For wire, rod; manganese brass.

SM-2770

Svenska Metallverken A.B.
Copper. Cu 70, Sn 1, bal Zn.
Drawn: 57,000 TS; 30 El. For tubes, condenser tubes; admiralty brass.

SM-3010

Svenska Metallverken A.B.
Copper. Cu 82, Al 10, Ni 4, Fe 4.
Forged: 92,000 TS; 50,000 YS; 15 El; 160 Brin. For rod, shapes, propellers; aluminum bronze.

SM-3080

Svenska Metallverken A.B.
Copper. Cu 85, Al 8, Ni 1, Fe 2, Mn 3, Sn 1.
Forged: 71,000 TS; 28,000 YS; 35 El; 40 RA; 120 Brin. For wire, rod, shapes; aluminum bronze.

SM-3090

Svenska Metallverken A.B.
Copper. Cu 91, Al 9.
Rolled: 71,000 TS; 28,000 YS; 30 El; 35 RA; 100 Brin. For sheet, rod; aluminum bronze.

SM-3099

Svenska Metallverken A.B.
Copper. Cu 81.5, Al 10, Ni 1.5, Fe 3, Mn 3, Sn 1.
Forged: 92,000 TS; 57,000 YS; 15 El; 20 RA; 160 Brin. For rods, shapes; aluminum bronze.

SM-3617

Svenska Metallverken A.B.
Copper. Cu 98, Si 1.7, Mn 0.3.
Annealed: 38,000 TS; 11,000 YS; 35 El. Drawn: 50,000 TS; 28,000 YS; 15 El. For wire, rod; silicon bronze.

SM-3730

Svenska Metallverken A.B.
Copper. Sn 3, bal Cu.
Annealed: 43,000 TS; 14,000 YS; 55 El; 60 RA; 70 Brin. Rolled: 54,000 TS; 46,000 YS; 30 El; 40 RA; 115 Brin. For sheet; phosphor bronze.

SM-3750

Svenska Metallverken A.B.
Copper. Sn 5, bal Cu.
Annealed: 47,000 TS; 17,000 YS; 65 El; 65 RA; 70 Brin. Rolled: 63,000 TS; 54,000 YS; 25 El; 50 RA; 130 Brin. For sheet; phosphor bronze.

SM-3770

Svenska Metallverken A.B.
Copper. Sn 7, bal Cu.
Rolled: 100,000 TS; 90,000 YS; 8 El; 35 RA; 200 Brin. Drawn: 135,000 TS. For sheet, wire; phosphor bronze.

SM-3790

Svenska Metallverken A.B.
Copper. Sn 9, bal Cu.
Hard: 75,000 TS; 57,000 YS; 35 El; 50 RA; 170 Brin. For sheet, tubes; phosphor bronze.

SM-3840

Svenska Metallverken A.B.
Copper. Sn 4, Pb 4, Zn 4, bal Cu.
Rolled: 57,000 TS; 48,000 YS; 20 El; 35 RA; 125 Brin. For sheet; free cutting.

SM-4220

Svenska Metallverken A.B.
Copper. Cu 80, Ni 20.
Drawn: 60,000 TS; 43,000 YS; 5 El; 30 RA; 120 Brin. For tubes, condenser tubes; cupro-nickel.

SM-4230

Svenska Metallverken A.B.
Copper. Cu 70, Ni 30.
Drawn: 80,000 TS; 64,000 YS; 5 El; 30 RA; 155 Brin. For tubes, condenser tubes; cupro-nickel.

SM-4312

Svenska Metallverken A.B.
Copper. Cu 64, Ni 12, bal Zn.
Annealed: 57,000 TS; 21,000 YS; 35 El; 60 RA; 80 Brin. Hard: 80,000 TS; 64,000 YS; 8 El; 40 RA; 150 Brin. For sheet, tubes, wire, shapes; nickel silver.

SM-4318

Svenska Metallverken A.B.
Copper. Cu 60, Ni 18, bal Zn.
Annealed: 60,000 TS; 26,000 YS; 30 El; 50 RA; 90 Brin. Hard: 83,000 TS; 67,000 YS; 8 El; 35 RA; 150 Brin. For sheet, tubes, wires, shapes; nickel silver.

SM-5023

Svenska Metallverken A.B.
Aluminum. Al 99.7.
Drawn: 27,000 TS. For wire.

SM-5025

Svenska Metallverken A.B.
Aluminum. Al 99.5.
Drawn: 25,000 TS. For wire.

SM-5027

Svenska Metallverken A.B.
Aluminum. Al 99.3.
Rolled: 18,000 TS; 16,000 YS; 4 El; 45 Brin. For sheet.

SM-5078

Svenska Metallverken A.B.
Aluminum. Mn 1.2, bal Al.
Annealed: 14,000 TS; 6000 YS; 30 El; 35 Brin. Hard: 26,000 TS; 23,000 YS; 5 El; 60 Brin. For sheet, tube, shapes; 3 S alloy.

SM-5083

Svenska Metallverken A.B.
Aluminum. Mg 0.6, bal Al.
Annealed: 14,000 TS; 6000 YS; 30 El; 35 Brin. Hard: 26,000 TS; 23,000 YS; 5 El; 60 Brin. For sheet, tubes, shapes; 63 S alloy.

SM-6307

Svenska Metallverken A.B.
Aluminum. Mg 0.8, Mn 0.6, Cu 4.2, Si 0.5, bal Al.
Age hardened: 57,000 TS; 35,000 YS; 14 El; 120 Brin. For sheet, tube, rod, shapes; Al 17 S alloy; age hardenable.

SM-6315

Svenska Metallverken A.B.
Aluminum. Mg 1.5, Mn 0.6, Cu 4.5, bal Al.
Age hardened: 63,000 TS; 40,000 YS; 12 El; 130 Brin. For sheet, tube, rod, shapes; Al 24 S alloy; age hardenable.

SM-6508

Svenska Metallverken A.B.
Aluminum. Mg 0.8, Mn 0.7, Ni 1, bal Al.
Age hardened: 43,000 TS; 35,000 YS; 10 El; 100 Brin. For sheet, tube, rod, shapes; age hardenable.

SM-6525

Svenska Metallverken A.B.
Aluminum. Mg 2.5, Cr 0.25, bal Al.
Annealed: 26,000 TS; 11,000 YS; 18 El; 50 Brin. Hard: 35,000 TS; 33,000 YS; 4 El; 90 Brin. For sheet, tube, shapes; 52 S alloy.

SM-6536

Svenska Metallverken A.B.
Aluminum. Mg 3.5, bal Al.
Annealed: 32,000 TS; 11,000 YS; 18 El; 60 Brin. Hard: 43,000 TS; 35,000 YS; 6 El; 100 Brin. For sheet, wire, shapes; A 214 alloy.

SM-6821

Svenska Metallverken A.B.
Mg 1.5, Mn 0.6, Cu 4.5, bal Al.
Heat treated: 57,000 TS; 37,000 YS; 12 El; 130 Brin. For sheet, aircraft structures; age hardenable.

SM-6958

Svenska Metallverken A.B.
Aluminum. Mg 2.5, Mn 0.1, Cu 1.6, Zn 5.75, Cr 0.2, Ti 0.05, bal Al.
Age hardened: 78,000 TS; 68,000 YS; 6 El; 155 Brin. 75 S alloy; age hardenable.

SM116

Birmingham Aluminum Casting Co.
Aluminum. Mg 5.5-6.5, Si 11-13, Ni 1-2, bal Al.
44,000 TS; 36,000 YS; 10 El; 125 Brin. For light alloy parts, die castings; hardenable, high temperature applications.

SM340

Contact Technologies, Inc.
Ag 75, 20.0 graphite, 5.0 MoS$_2$.
Silver-graphite-molybdenum disulfide. Density: 5.2 g/cm^3; specific resistivity: 0.015 micro-ohm-m; hardness: 15W30; strength: 5,000 psi. Electrical contacts for circuit breaker applications.

SM476

Contact Technologies, Inc.
Ag 85, 3.0 graphite, 12.0 MoS$_2$.
Silver-graphite-molybdenum disulfide. Density: 7.9 g/cm^3; specific resistivity: 0.064 micro-ohm-m; hardness: 30T45; strength: 12,000 psi. For sliding contacts for slip ring assemblies and brushes for electrical motors used in high-altitude conditions.

SM4F

Friedr. Lohmann GmbH
C 0.45, Si 0.25-0.5, Mn 0.3-0.8, bal Fe.
Annealed: 229 Brin. For gears, bolts, machine tool parts; build-up material for tools and agricultural tools of all kinds, bolster steel. SAE Type 1045; water hardened.

SM7CR

Friedr. Lohmann GmbH
C 0.75, Si 0.25-0.5, Mn 0.3-0.8, Cr 0.3-0.4, bal Fe.
Annealed: 235 Brin. For springs, rails, punches, hammers. Type W1; water hardened.

SM9CR

Friedr. Lohmann GmbH
C 0.9, Si 0.25-0.5, Mn 0.3-0.8, Cr 0.5-0.6, bal Fe.
Annealed: 235 Brin. For springs, dies, tools, cutters, punches. Oil hardened.

SMC-155

Colt Industries
C 0.5, Si 1.1, Mo 0.5, bal Fe.
For tools; shock resistant. *Obsolete*

SMENA NO. 1

Central Institute of Metals
W 20.6, Cr 31.4, Ni 33.5, C 4.2, Fe 8.7, 1.6 Mn + Si.
For cutting tools, drills, reamers, hobs. High speed steel.

SMENA NO. 2

Central Institute of Metals
Co-Cr-W.
To replace Stellite for resurfacing. Hard alloy, wear and corrosion resistant.

SMG-2

J.M. Ney Co.
Au 87, Pt 7, Pd 5.
Dental alloy, ceramic gold. Melting range: 2010-2220°F; casting temperature 2430°F. Cast: 69,000 psi TS (nominal); 49,000 YS (0.1% offset, nominal), 185 HV; 195 DWT/in.³. For all porcelain to gold restorations; light gold color, hard.

SMG-2 NEY CERAMIC SOLDER

J.M. Ney Co.
Now SMG-2 SOLDER.

SMG-2 SOLDER

J.M. Ney Co.
Au 85.5, Ag 5, Pt 4, Pd 3.
Dental solder, gold color. For use with SMG-2, IMAGE, OPTION, NEYDIUM GOLD CERAMIC casting golds. Flows at 1925°F.

SMG-3

J.M. Ney Co.
Au 81, Pt 6, Pd 11.
Dental alloy, ceramic gold. Melting range: 2106-2282°F; casting temperature 2430°F. Cast: 102,000 psi TS (nominal); 79,000 psi YS (0.1% offset, nominal); 240 HV; 186 DWT/in.³ density. For all porcelain to gold restorations; white gold color; extra hard.

SMG-3 NEY CERAMIC SOLDER

J.M. Ney Co.
Now SMG-3 SOLDER.

SMG-3 SOLDER

J.M. Ney Co.
Dental solder, white color. For use with SMG-3, ECLIPSE, ENCORE, ORION casting golds. Flows at 2100°F.

SMG-W

J.M. Ney Co.
Dental alloy; white gold color. Melting range: 2090-2230°F; 205 Brin. *Obsolete*

SMG-Y

J.M. Ney Co.
Dental alloy, ceramic gold. Melting range: 2080-2285°F; 180 Brin. Gold color; bond strength. *Obsolete*

SMG-YW NEY CERAMIC SOLDER

J.M. Ney Co.
Now SMG-YW SOLDER.

SMG-YW SOLDER

J.M. Ney Co.
Au 82, Ag 8, Pd 8.
Dental solder, white color. For use with NEYDIUM GOLD CERAMIC, IMAGE, TEMPO, APPLAUSE casting golds. Flows at 2000°F.

SMITH NO. 10

Solar Basic Industries
Cr 37.5, Al 7.5, bal Fe.
200-235 Brin. For electrical resistors for heating furnaces, heating elements; heat resistant; electrical resitivity: 1000 ohms; maximum operating temperature 2400°F.

SMITH SW-15

A.O. Smith Co.
bal Fe.
For welding electrodes; mild steel. *Obsolete*

SMITH SW-17

A.O. Smith Co.
C, bal Fe.
For welding electrodes; mild steel. *Obsolete*

SMITHWAY

Solar Basic Industries
C alloy, bal Fe.
For welding electrodes.

SMITTER LENIAN

English manufacture
Cu 72, Ni 13, Zn 9.8, Sn 2.3, Fe 2, Bi 1.
For corrosion resistant parts.

SMO 10 CO 8

Bergische Stahl Industrie
C 1.05-1.12, Cr 3.6-4.4, W 1.2-1.8, V 1-1.3, Mo 9-10, Co 7.5-8.5, bal Fe.
High speed steel. W.-Nr. 1.3247.

SMO 5

Bergische Stahl Industrie
C 0.84-0.92, Cr 3.8-4.5, W 6-6.7, V 1.7-2, Mo 4.7-5.2, bal Fe.
High speed steel. W.-Nr. 1.3343.

SMO 5 CO 5

Bergische Stahl Industrie
C 0.88-0.96, Cr 3.8-4.5, W 6-6.7, V 1.7-2, Mo 4.7-5.2, Co 4.5-5, bal Fe.
High speed steel. W.-Nr. 1.3243.

SMO 5 H

Bergische Stahl Industrie
C 0.95-1.05, Cr 3.8-4.5, W 6-6.7, V 1.2-2, Mo 4.7-5.2, bal Fe.
High speed steel. W.-Nr. 1.3342.

SMO 5 V 3

Bergische Stahl Industrie
C 1.17-1.27, Cr 3.8-4.5, W 6-6.7, V 2.7-3.2, Mo 4.7-5.2, bal Fe.
High speed steel. W.-Nr. 1.3344.

SMO 6 CO 5

Bergische Stahl Industrie
C 1.13-1.2, Cr 3.8-4.5, W 6-6.75, V 3-3.5, Mo 4.8-5.3, Co 4.5-5, bal Fe.
High speed steel. W.-Nr. 1.3244.

SMO 9

Bergische Stahl Industrie
C 0.78-0.86, Cr 3.5-4.2, W 1.5-2, V 1-1.3, Mo 8-9.2, bal Fe.
High speed steel. W.-Nr. 1.3346.

SMO 9 CO 5

Bergische Stahl Industrie
C 0.9-0.98, Cr 3.4-4.2, W 1.5-2, V 1.8-2.2, Mo 8-9.2, Co 4.5-5, bal Fe.
High speed steel. W.-Nr. 1.3248.

SMO 9 V

Bergische Stahl Industrie
C 0.97-1.07, Cr 3.5-4.2, W 1.5-2, V 1.8-2.2, Mo 8-9.2, bal Fe.
High speed steel. W.-Nr. 1.3348.

SMOOTH-CUT

Columbia Tool Steel Co.
C 0.7, Cr 1.1, Mn 2, S, bal Fe.
Annealed: 180 Brin. Heat treated: 293,000 TS; 265,000 YS; 1 El; 55 Rock C. For blanking and forming dies, cutters, shear blades, punches, coining dies. Type A6 oil or air hardening, non-deforming. *Obsolete*

SMOOTHCOTE NO. 34

Now ALL-STATE NO. 34.

SMOOTHCUT

Now ATMODIE SMOOTHCUT.

SMOOTHOLE

Midvale-Heppenstall Co.
C 0.85, bal Fe.
For mining hollow drills; mandrel rolled. *Obsolete*

SMOOTHOLE GRANITE DRILL

Midvale-Heppenstall Co.
C 0.87, V 0.15, Mo 0.1, bal Fe.
For hollow drills; water hardened. *Obsolete*

SN 5 1/2

Sintered Products Ltd.
Nickel alloy steel, sintered. 6.6 g/cm³ density; 47 kg/mm² TS; 40 Rock C approx. BSS A400; MPIE FN-0505-T.

SO 7.1002 (304 + SI)

Sterling International Technology Ltd.
Cu 3.8-4.5, Si 1-1.5, bal Al.
Sand cast: 40,000-45,000 TS; 27,000-29,000 YS; 4-6 El; 90-100 Brin; 3.5 Izod. Sand or chill cast aluminum alloy. Hardenable; good strength and less hot short.

SO-LUMINUM

Manufacturer not listed.
Sn 55, Zn 33, Al 11, Cu 1.
For aluminum solder.

SOBU

Osborn Steels Ltd.
C 0.7, W 18, Co 9, Mo 1, Cr 4, V 2, bal Fe.
For lathe and planer tools, reamers, hobs, drills, broaches; high speed steel, oil hardened. *Obsolete*

SODERFORS 12/1.0

Manufacturer not listed.
C 1, Si 0.2, Mn 0.3, V 0.1, bal Fe.
Annealed: 100,000 TS; 55,000 YS; 20 El; 40 RA; 220 Brin. For heading and forming dies, punches; Type W2; water hardened.

SODERFORS 63

Manufacturer not listed.
C 1.5, Cr 12, Mo 0.8, V 0.2, bal Fe.
For shears, punches, blanking and forming dies; air-hardened, non-deforming.

SODERFORS ANVIL

Manufacturer not listed.
C 0.8, Mn 0.9, bal Fe.
For tools; water hardened.

SODERFORS ARROW
Manufacturer not listed.
Mn 1.2, W 0.5, Cr 0.5, V 0.1, High C, bal Fe.
For tools, dies, thread and block gages, master tools, taps; non-shrinking, non-warping.

SODERFORS CROWN "W"
Manufacturer not listed.
C 0.9, Mn 0.3, Cr 1.8, W 4, V 0.2, bal Fe.
For tools, for turning chilled iron rolls and all hard metals; non-warping.

SODERFORS DOUBLE CRANE
Manufacturer not listed.
C 0.9-1.2, bal Fe.
For tools, drills, stone tools; water hardened.

SODERFORS DRILL
Manufacturer not listed.
C 1-1.2, bal Fe.
For drills, tools, hollow and solid drills; water hardened.

SODERFORS NO. 05
Manufacturer not listed.
C 0.06, Si 0.1, Mn 0.2, bal Fe.
For case hardened gears, cams, camshafts, plastic molds. Case hardening steel, wear resistant.

SODERFORS NO. 1
Manufacturer not listed.
C 0.6, Mn 0.3, bal Fe.
For tools, punches; water hardening.

SODERFORS NO. 10
Manufacturer not listed.
C 0.6-1.45, Si 0.2, Mn 0.3, Cr 1, bal Fe.
For tools, drills, springs; water hardening.

SODERFORS NO. 11
Manufacturer not listed.
C 0.6-1.45, Si 0.2, Mn 0.3, Cr 1.5, bal Fe.
For tools, drills, springs; water hardening.

SODERFORS NO. 11A
Manufacturer not listed.
C 1, Cr 2, bal Fe.
For tools, dies; water hardening.

SODERFORS NO. 12
Manufacturer not listed.
C 0.8, Si 0.2, Mn 0.3, Cr 0.45, V 0.1, bal Fe.
For tools, drills, taps; water hardening.

SODERFORS NO. 12/0.8
Manufacturer not listed.
C 0.75-0.85, V 0.05-0.15, bal Fe.
Annealed: 88,000 TS; 52,000 YS; 20 El; 190 Brin. For cutters, shears, slitters, gripper and forming dies, coining and embossing dies. Type W2 water hardening steel, wear resistant.

SODERFORS NO. 13
Manufacturer not listed.
C 0.95, Si 0.2, Mn 0.3, Cr 0.45, V 0.1, bal Fe.
For tools, drills, cutters; water hardening.

SODERFORS NO. 14
Manufacturer not listed.
C 1.05, Si 0.2, Mn 0.3, Cr 0.45, V 0.1, bal Fe.
For tools, drills, cutters; water hardening.

SODERFORS NO. 15
Manufacturer not listed.
C 0.9-1.05, Si 1.35-1.65, Mn 0.6-0.9, Cr 0.9-1.15, bal Fe.
For bearings, liners, cutters, bushings, cold heading tools. Oil hardening, wear and shock resistant.

SODERFORS NO. 16
Manufacturer not listed.
C 0.95, Si 0.2, Mn 1.25, Cr 0.5, W 0.5, V 0.1, bal Fe.
For tools, dies, cutters; non-shrinking. AISI O1.

SODERFORS NO. 17
Manufacturer not listed.
C 1.15, Si 0.2, Mn 0.3, V 0.1, W 0.5, bal Fe.
For drills, taps, hobs, broaches; water hardened.

SODERFORS NO. 17A
Manufacturer not listed.
C 1.1, Cr 0.25, V 0.13, W 1.1, bal Fe.
Annealed: 100,000 TS; 60,000 YS; 15 El; 210 Brin. For punches, slitters, bearings, textile needles, broaches, reamers. Type F1 tool steel. Water or oil hardening. Abrasion resistant.

SODERFORS NO. 18
Manufacturer not listed.
C 0.45, Si 0.25, Cr 1.4, W 2.2, V 0.2, Mo 0.2-0.3, bal Fe.
For chisels, punches, shock resisting.

SODERFORS NO. 19
Manufacturer not listed.
C 1.15, Si 0.1, Cr 0.5, W 1.5, bal Fe.
For tools, cutters; keen cutting edge.

SODERFORS NO. 2
Manufacturer not listed.
C 0.7, Si 0.2, Mn 0.3, bal Fe.
For tools, punches, dies; water hardening.

SODERFORS NO. 20
Manufacturer not listed.
C 1.3, Cr 1.8, W 4, V 0.2, bal Fe.
For tools, dies; oil hardening.

SODERFORS NO. 21
Manufacturer not listed.
C 1.3, Cr 0.5, W 5, bal Fe.
For tools, dies, cutters; oil hardening.

SODERFORS NO. 214
Manufacturer not listed.
C 1.05, V 0.2, Si 0.2, Mn 0.3, bal Fe.
Annealed: 90,000 TS; 55,000 YS; 18 El; 200 Brin. For shears, slitters, lathe and planer tools, drills, hobs. Water hardening, wear resistant.

SODERFORS NO. 215
Manufacturer not listed.
C 1, Si 1.3, Mn 0.8, Cr 1, bal Fe.
For bearings, liners, bushings, cutters, punches, cold heading and drawing dies. Oil hardening, abrasion resistant.

SODERFORS NO. 22
Manufacturer not listed.
C 0.5, Cr 3, W 9, V 0.3, bal Fe.
For die casting dies, extrusion mandrels and liners; hot work steel, oil hardened.

SODERFORS NO. 23
Manufacturer not listed.
C 0.45, Cr 3, W 14, bal Fe.
For tools, dies; hot work steel.

SODERFORS NO. 24
Manufacturer not listed.
C 0.45, Cr 3, V 0.3, W 9, bal Fe.
For hot punches, extrusion dies; hot piercers and grippers. Hot work tool steel, oil hardening, tough and shock resistant.

SODERFORS NO. 25
Manufacturer not listed.
C 0.7, Cr 4, W 18, V 1, Co 0.4, bal Fe.
For drills, reamers, broaches, hobs, lathe cutters; high speed steel, oil hardened.

SODERFORS NO. 26
Manufacturer not listed.
C 0.8, Cr 5.5, W 18, V 1.5, Mo 1.5, Co 2, bal Fe.
For tools, cutters; high speed steel. AISI T2.

SODERFORS NO. 27
Manufacturer not listed.
C 0.8, Cr 5.5, W 18, V 1.5, Mo 1.5, Co 5.5, bal Fe.
For tools, cutters; high speed steel. AISI T4.

SODERFORS NO. 28
Manufacturer not listed.
C 0.75, Cr 5.5, W 18, V 1.5, Mo 1, Co 10, bal Fe.
For tools, cutters; high speed steel. AISI T5.

SODERFORS NO. 29
Manufacturer not listed.
C 0.77-0.87, Cr 3.5-4.5, Mo 4.5-5.5, V 1.7-2.1, W 6-7, bal Fe.
Hardened: 64-67 Rock C. For lathe and planer tools, form cutters, reamers, broaches, drills, hobs. Type M2 high-speed steel, high red-hardness.

SODERFORS NO. 29A
Manufacturer not listed.
C 0.82-0.9, Cr 3.5-4.5, Mo 2.8-3.6, V 1.9-2.3, W 6-7, bal Fe.
Hardened: 64-66 Rock C. For lathe and planer tools, hobs, reamers, drills, taps. High-speed steel, high red-hardness.

SODERFORS NO. 29B
Manufacturer not listed.
C 1.45, Cr 4, Mo 5, V 5, W 6.5, bal Fe.
For textile needles, reamers, broaches, chasers, hobs. High speed steel, high red-hardness.

SODERFORS NO. 3
Manufacturer not listed.
C 0.8, Si 0.2, Mn 0.3, bal Fe.
For tools, springs; water hardening.

SODERFORS NO. 3 1/2
Manufacturer not listed.
C 0.9, Si 0.2, Mn 0.3, bal Fe.
For tools, springs, drills; water hardening.

SODERFORS NO. 30
Manufacturer not listed.
C 1.1, Co 9.5, Cr 4.3, Mo 5, V 2.6, W 6.5, bal Fe.
Hardened: 64-68 Rock C. For broaches, reamers, twist drills, lathe and milling cutters, boring tools. Type M36 high-speed steel, high red hardness.

SODERFORS NO. 31
Manufacturer not listed.
C 1, Cr 3, bal Fe.
For dies, tools; oil hardening.

SODERFORS NO. 32
Manufacturer not listed.
C 0.65, W 6, bal Fe.
For hot work tools; hot work steel.

SODERFORS NO. 323
Manufacturer not listed.
C 0.3, Si 0.3, Co 4.8, Cr 1.5, Mo 0.5, V 0.1, W 5.3, bal Fe.
For hot work tools, punches, hot shears. Oil hardening, shock resistant.

SODERFORS NO. 364
Manufacturer not listed.
C 1.45-1.65, Cr 11-13, Mo 0.7-0.9, V 0.7-1, bal Fe.
Heat treated: 278,000 TS; 214,000 YS; 1 El; 57 Rock C. For blanking and drawing dies, broaches, hobs, shear blades, punches, gauges. Type D2 air-hardening steel. Non-deforming, shock resistant.

SODERFORS NO. 4
Manufacturer not listed.
C 1, Si 0.2, Mn 0.3, bal Fe.
For tools, drills, taps; water hardening.

SODERFORS NO. 41
Manufacturer not listed.
C 0.5-1.3, Si 0.2, Mn 0.55, bal Fe.
For tools, dies, punches; water hardening.

SODERFORS NO. 42
Manufacturer not listed.
C 0.45-0.7, Si 0.7, Mn 0.9, bal Fe.
For tools, dies, punches; water hardening.

SODERFORS NO. 423
Manufacturer not listed.
C 1.15, Cr 4, Mo 5, V 3, W 6, bal Fe.
Hardened: 64-68 Rock C. For drills, reamers, broaches, form cutters, lathe and planer tools, hobs. Type M3 high-speed steel. High red hardness.

SODERFORS NO. 424
Manufacturer not listed.
C 1.1, Co 5.3, Cr 4.3, Mo 5, V 2.6, W 6.5, bal Fe.
Hardened: 64-68 Rock C. For broaches, reamers, taps, chasers, drills, lathe and planer tools. Type M35 high-speed steel, high red-hardness.

SODERFORS NO. 43
Manufacturer not listed.
C 0.6, Si 1.9, Mn 0.9, bal Fe.
For tools, springs; oil hardening.

SODERFORS NO. 431
Manufacturer not listed.
C 0.8, Cr 4, Mo 9, V 1.2, W 1.5, bal Fe.
Hardened: 64-66 Rock C. For lathe and planer tools, form cutters, drills, hobs, reamers, gauges. Type M1 high-speed steel, oil hardening. High red-hardness.

SODERFORS NO. 433
Manufacturer not listed.
C 1, Cr 4, Mo 9, V 2, W 1.8, bal Fe.
Hardened: 64-67 Rock C. For drills, end mills, thread rolling dies, chasers, punches, taps. Type M7 high-speed steel. High red-hardness.

SODERFORS NO. 434
Manufacturer not listed.
C 1.2, Co 5, Cr 4, Mo 9, V 3, W 1.8, bal Fe.
For reamers, broaches, drills, form cutters, chasers. High-speed steel, good red-hardness.

SODERFORS NO. 46
Manufacturer not listed.
C 0.5, Cr 1.1, V 0.15, bal Fe.
For gears, shafts; oil hardened, tough.

SODERFORS NO. 481
Manufacturer not listed.
C 0.8, Cr 4.5, Mo 8, V 2, W 9, bal Fe.
For chasers, taps, drills, reamers, broaches, lathe and planer tools. High-speed steel, good red-hardness.

SODERFORS NO. 5
Manufacturer not listed.
C 1.1, Si 0.2, Mn 0.3, bal Fe.
For tools, drills, taps; water hardening.

SODERFORS NO. 502
Manufacturer not listed.
C 0.1, Cr 14, bal Fe.
Annealed: 90,000 TS; 50,000 YS; 40 El; 50 RA; 160 Brin. For cutlery, surgical instruments, chemical plant equipment; Type 410; corrosion resistant.

SODERFORS NO. 504
Manufacturer not listed.
C 0-0.08, Cr 17.5, Ni 12, Mo 2.75, bal Fe.
Annealed: 80,000 TS; 30,000 YS; 60 El; 80 Rock B. For chemical plant equipment, evaporators, acid tanks, digesters. Corrosion resistant, austenitic, non-hardenable.

SODERFORS NO. 506
Manufacturer not listed.
C 0-0.1, Cr 27, Ni 5.3, Mo 1.6, bal Fe.
For heat treating boxes, furnace parts, retorts, gas burners, combustion chambers. Corrosion and heat resistant.

SODERFORS NO. 508
Manufacturer not listed.
C 0-0.1, Cr 27, Ni 5.3, Mo 1.6, bal Fe.
For heat treating boxes, furnace parts, retorts, gas burners, combustion chambers. Corrosion and heat resistant.

SODERFORS NO. 509
Manufacturer not listed.
C 0.12, Cr 12, Ni 0.6, Mo 0.5, bal Fe.
Annealed: 80,000 TS; 42,000 YS; 30 El; 160 Brin. For chemical plant and oil refinery equipment, tableware, knives. Heat and corrosion resistant.

SODERFORS NO. 510
Manufacturer not listed.
C 0.15, Cr 14, Ni 0.3, bal Fe.
Annealed: 75,000 TS; 40,000 YS; 35 El; 70 RA; 155 Brin. Cold drawn: 100,000 TS; 85,000 YS; 17 El; 60 RA; 205 Brin. For cutlery, pump parts, valves; Type 410; corrosion resistant.

SODERFORS NO. 511
Manufacturer not listed.
C 0.2, Cr 14, Ni 0.3, bal Fe.
For pump-shafts, cutlery, valves; corrosion resistant.

SODERFORS NO. 512
Manufacturer not listed.
C 0.35, Cr 14, Ni 0.3, bal Fe.
Annealed: 95,000 TS; 50,000 YS; 25 El; 55 RA; 196 Brin. Heat treated: 250,000 TS; 215,000 YS; 8 El; 25 RA; 512 Brin. For cutlery, knives, surgical instruments, gears, shafts; Type 420; corrosion resistant.

SODERFORS NO. 514
Manufacturer not listed.
C 0.53-0.58, Cr 13.5-14, Ni 0.3, bal Fe.
Annealed: 100,000 TS; 50,000 YS; 90 Rock B. For surgical instruments, cutlery, knives, gears, shafts. Corrosion and heat resistant.

SODERFORS NO. 516
Manufacturer not listed.
C 0.2, Cr 14, Ni 0.3, Mo 0.3, bal Fe.
Annealed: 95,000 TS; 50,000 YS; 25 El; 92 Rock B. For cutlery, surgical instruments, gauges, needle valves, bearings. Corrosion and heat resistant.

SODERFORS NO. 517
Manufacturer not listed.
C 1, Cr 14, Mo 0.2, Co 0.6, V 0.1, bal Fe.
For bearings: corrosion resistant.

SODERFORS NO. 522
Manufacturer not listed.
C 0-0.08, Cr 17.5, Mn 0.5, bal Fe.
Annealed: 80,000 TS; 50,000 YS; 25 El; 50 RA; 150 Brin. For household utensils, sinks, oil refinery equipment; stainless, ferritic; Type 430.

SODERFORS NO. 523
Manufacturer not listed.
C 0-0.1, Cr 17, Ni 0.35, Mo 1.75, bal Fe.
Annealed: 90,000 TS; 50,000 YS; 20 El; 165 Brin. For oil burners, hardware, marine fittings, food handling equipment. Corrosion and heat resistant.

SODERFORS NO. 525
Manufacturer not listed.
C 0.15-0.25, Cr 16-18, Ni 1.25-2.5, bal Fe.
Annealed: 125,000 TS; 90,000 YS; 20 El; 260 Brin. For pump shafts, marine hardware, valve trim, aircraft components. Type 431; corrosion and heat resistant.

SODERFORS NO. 526
Manufacturer not listed.
C 0.15, Cr 17.5, Mo 1.8, bal Fe.
Annealed: 85,000 TS; 48,000 YS; 25 El; 160 Brin. For oil burner parts, hardware, marine fixtures, food handling equipment. Corrosion and heat resistant.

SODERFORS NO. 527
Manufacturer not listed.
C 0.15, Cr 17, Ni 1.5, Mo 0.5, Si 0.2, bal Fe.
For cutlery, tableware; corrosion resistant, free-cutting.

SODERFORS NO. 532
Manufacturer not listed.
C 0.2, Cr 25, bal Fe.
For stainless parts; corrosion resistant.

SODERFORS NO. 541
Manufacturer not listed.
C 0.41, Si 3, Cr 9, Mo 0.3, bal Fe.
For valves; heat and corrosion resistant.

SODERFORS NO. 552
Manufacturer not listed.
C 0-0.03, Cr 17-20, Ni 9-12, bal Fe.
Annealed: 80,000 TS; 30,000 YS; 60 El; 74 Rock B. For architectural molding and trim, processing equipment, welded tanks, evaporators. Type 304L stainless steel, austenitic.

SODERFORS NO. 553
Manufacturer not listed.
C 0.05, Cr 18, Ni 9, bal Fe.
For chemical and textile plant equipment; Type 304; austenitic, stainless.

SODERFORS NO. 554
Manufacturer not listed.
C 0-0.08, Cr 18, Ni 9, bal Fe.
For stainless parts; austenitic, stainless.

SODERFORS NO. 555
Manufacturer not listed.
C 0.13, Cr 18, Ni 9, bal Fe.
Annealed: 90,000 TS; 35,000 YS; 50 El; 16 RA; 140 Brin. For chemical and plastic plant equipment. Type 302; stainless, austenitic.

SODERFORS NO. 556
Manufacturer not listed.
C 0-0.1, Cr 18, Ni 9, S 0.2, bal Fe.
Rolled: 80,000-100,000 TS; 40,000-60,000 YS; 150-200 Brin. For screw machine products, fasteners, machinery parts. Type 303; austenitic, free-cutting, stainless.

SODERFORS NO. 558
Manufacturer not listed.
C 0-0.08, Cr 17-19, Ni 9-13, Ti = 5 x C, bal Fe.
Annealed: 85,000 TS; 30,000 YS; 58 El; 82 Rock B. For welded structures, chemical plant equipment, tanks, evaporators. Type 321 stainless steel, austenitic, stabilized, welding grade.

SODERFORS NO. 558 CB
Manufacturer not listed.
C 0-0.08, Cr 17-19, Ni 9-13, Cb 0-1.2, bal Fe.
Annealed: 85,000 TS; 35,000 YS; 60 El; 80 Rock B. For welded structures, chemical plant equipment, digesters. Type 347 stainless steel, austenitic, stabilized, welding grade.

SODERFORS NO. 558 NB
Manufacturer not listed.
C 0-0.08, Cr 17-19, Ni 9-13, Cb 0-1.2, bal Fe.
Annealed: 90,000 TS; 40,000 YS; 55 El; 82 Rock B. For welded structures, chemical plant equipment, digesters. Type 347 stainless steel, austenitic, stabilized, welding grade.

SODERFORS NO. 558 TI
Manufacturer not listed.
C 0.06, Cr 18, Ni 11, Ti 0.4, bal Fe.
Annealed: 85,000 TS; 30,000 YS; 55 El; 80 Rock B. For welded structures and tanks, chemical plant equipment, jet aircraft components, exhaust systems. Type 321 stainless steel, austenitic, stabilized.

SODERFORS NO. 559
Manufacturer not listed.
C 0-0.12, Cr 14, Ni 13, bal Fe.
For valves; corrosion and heat resistant.

SODERFORS NO. 562
Manufacturer not listed.
C 0-0.05, Cr 18, Ni 10, Mo 1.5, bal Fe.
For chemical plant equipment; acid resistant, austenitic.

SODERFORS NO. 562 EL
Manufacturer not listed.
C 0-0.03, Cr 17-18, Ni 12-14, Mo 2.3-3.8, bal Fe.
Annealed: 85,000 TS; 35,000 YS; 50 El; 82 Rock B. For chemical plant equipment, acid tanks, agitators, valve trim. Type 316L stainless steel, austenitic, acid resistant.

SODERFORS NO. 562 N
Manufacturer not listed.
C 0-0.06, Cr 17.5, Ni 10.5-11.5, Mo 2.25, bal Fe.
Annealed: 80,000 TS; 30,000 YS; 60 El; 80 Rock B. For chemical plant equipment, agitators, digesters, evaporators. Type 316, austenitic, stainless.

SODERFORS NO. 563
Manufacturer not listed.
C 0-0.08, Cr 18, Ni 10, Mo 1.5, bal Fe.
For chemical plant equipment; acid resistant, austenitic.

SODERFORS NO. 564
Manufacturer not listed.
C 0-0.05, Cr 18, Ni 12, Mo 2.7, Mn 1.2, bal Fe.
For chemical plant equipment; stainless, austenitic.

SODERFORS NO. 565
Manufacturer not listed.
C 0-0.08, Mn 1.2, Cr 18, Ni 12, Mo 2.7, bal Fe.
For stainless parts; austenitic, stainless.

SODERFORS NO. 566
Manufacturer not listed.
C 0-0.1, Cr 27, Ni 5, Mo 1.6, bal Fe.
For chemical plant equipment, sulphuric acid tanks. Type 329; stainless steel.

SODERFORS NO. 571
Manufacturer not listed.
C 0.45, Cr 14, Ni 14, Mo 0.3, W 2.2, bal Fe.
For engine valves; austenitic, stainless.

SODERFORS NO. 576
Manufacturer not listed.
C 0-0.15, Si 1.2, Cr 23, Ni 21, bal Fe.
For furnace parts, heat treating boxes; heat resistant to 1100°C.

SODERFORS NO. 584
Manufacturer not listed.
C 0-0.12, Mn 7.5-10, Cr 17-19, Ni 4-6, N 0-0.25, bal Fe.
Annealed: 100,000 TS; 50,000 YS; 60 El; 90 Rock B. For dairy and chemical plant equipment springs, refrigerator trays, cabinets, architectural trim. Type 202 stainless steel, austenitic.

SODERFORS NO. 6
Manufacturer not listed.
C 1.2, Si 0.2, Mn 0.3, bal Fe.
For tools, cutters, drills; water hardening.

SODERFORS NO. 61
Manufacturer not listed.
C 2.2, Cr 13, 1 W-Mo-Ni-Co, bal Fe.
For tools, dies; non-deforming.

SODERFORS NO. 62
Manufacturer not listed.
C 2.2, Cr 12.5, bal Fe.
For dies; non-deforming. AISI D3.

SODERFORS NO. 65
Manufacturer not listed.
C 0.95-1.05, Cr 5-5.5, Mo 1-1.2, V 0.15-0.25, bal Fe.
Annealed: 103,000 TS; 51,000 YS; 26 El; 18 Rock C. Heat treated: 255,000 TS; 200,000 YS; 3 El; 57 Rock C: For blanking and trimming dies, cutters, gauges, punches, burnishing tools, shear blades, broaches. Type A2 air hardening steel, non-deforming.

SODERFORS NO. 67
Manufacturer not listed.
C 0.35-0.42, Si 0.8-1.2, Cr 5-5.5, Mo 1.2-1.6, V 0.85-1.15, bal Fe.
Annealed: 98,000 TS; 74,000 YS; 28 El; 216 Brin. Heat treated: 290,000 TS; 228,000 YS; 3 El; 55 Rock C. For forging and heading dies, compression tools, die casting and extrusion dies. Type H13 hot work steel. High hot hardness. Shock resistant.

SODERFORS NO. 7
Manufacturer not listed.
C 1.3, Si 0.2, Mn 0.3, bal Fe.
For tools, cutters; water hardening.

SODERFORS NO. 8
Manufacturer not listed.
C 1.45, Si 0.2, Mn 0.3, bal Fe.
For tools, cutters; water hardening.

SODERFORS NO. 81
Manufacturer not listed.
C 0.13, Cr 0.7, Ni 3, bal Fe.
For gears, shafts; case hardening.

SODERFORS NO. 82
Manufacturer not listed.
C 0.3, Cr 0.6, Ni 2, bal Fe.
For shafts, gears, pinions; tough.

SODERFORS NO. 83
Manufacturer not listed.
C 0.3, Cr 0.8, Ni 3, bal Fe.
For gears, pinions, shafts; tough.

SODERFORS NO. 84
Manufacturer not listed.
C 0.6, Ni 1.8, Cr 0.7, Mo 0.7, bal Fe.
For drop forging dies; oil hardened, tough.

SODERFORS NO. 85
Manufacturer not listed.
C 0.5, Cr 0.8, Ni 3.25, Mo 0.2, bal Fe.
For gears, dies, tools; tough.

SODERFORS NO. 9
Manufacturer not listed.
C 0.6-1.45, Si 0.2, Mn 0.3, Cr 0.45, bal Fe.
For tools, punches; water hardening.

SODERFORS NO. 9A
Manufacturer not listed.
C 0.6-1.45, Si 0.2, Mn 0.3, Cr 0.25, bal Fe.
For tools, punches; water hardening.

SODERFORS NO. 9B
Manufacturer not listed.
C 0.6-1.45, Si 0.2, Mn 0.3, Cr 0.15, bal Fe.
For tools, springs, dies; water hardening.

SODING ADK
J.C. Soding & Halbach
C 2.1, Cr 12, bal Fe.
For blanking and forming dies, punches, gages; oil hardening, non-deforming. DIN X210Cr12.

SODING ADK SPEZIAL
J.C. Soding & Halbach
C 2.1, Cr 12, bal Fe.
For blanking and forming dies, gages, punches; air hardened, non-deforming.

SODING ADP
J.C. Soding & Halbach
C 1.65, Cr 11.5, V 0.1, bal Fe.
For blanking and forming dies, gauges, punches; air hardened, non-deforming. *Obsolete*

SODING ADP SPEZIAL
J.C. Soding & Halbach
C 1.7, Cr 12, Mo 0.6, W 0.5, bal Fe.
For punches, crimping and forming dies, blanking and trimming dies, punches. Air hardening, non-deforming. Cold-work tool steel. DIN X165CrMoV12; Werkstoff Nr. 1.2601.

SODING ADT
J.C. Soding & Halbach
C 2.1, Cr 11.5, bal Fe.
For blanking and forming dies, gauges, punches; oil or air hardened, non-deforming. *Obsolete*

SODING CNF
J.C. Soding & Halbach
C 0.15, Ni, Cr, bal Fe.
For gears, bolts, machine tool parts; case hardened, tough. *Obsolete*

SODING CNL
J.C. Soding & Halbach
C 0-0.15, Cr 18, Ni 8.5, bal Fe.
Annealed: 80,000 TS; 35,000 YS; 55 El; 75 RA; 150 Brin. For chemical plant equipment, tanks, mixers, filters; Type 302; stainless, austenitic. *Obsolete*

SODING CNL EXTRA
J.C. Soding & Halbach
C 0-0.12, Cr 18, Ni 9.5, Ti = 4 x C, bal Fe.
Annealed: 85,000 TS; 35,000 YS; 55 El; 65 RA; 150 Brin. For welded chemical plant equipment; Type 321; stainless, austenitic. *Obsolete*

SODING CNL SPEZIAL
J.C. Soding & Halbach
C 0-0.07, Cr 18, Ni 9.5, bal Fe.
Annealed: 85,000 TS; 35,000 YS; 60 El; 70 RA; 150 Brin. For chemical plant equipment, tanks, vessels; Type 304; stainless, austenitic. *Obsolete*

SODING CNM
J.C. Soding & Halbach
C 0.15, Ni, Cr, bal Fe.
For chemical plant equipment; stainless. *Obsolete*

SODING CNP
J.C. Soding & Halbach
C 0-0.07, Cr 17.5, Ni 17.5, Mo 2, Cu 2, Ti = 4 x C, bal Fe.
For chemical plant equipment, welded structures; corrosion resistant. *Obsolete*

SODING CNU EXTRA
J.C. Soding & Halbach
C 0-0.12, Cr 18, Mo 2, Ni 10.5, Ti = 4 x C, bal Fe.
Annealed: 90,000 TS; 40,000 YS; 50 El; 60 RA; 160 Brin. For welded acid resistant chemical plant equipment; Type 316 Ti; stainless, austenitic. *Obsolete*

SODING CNU SPEZIAL
J.C. Soding & Halbach
C 0-0.07, Cr 18, Ni 10.5, Mo 2, bal Fe.
Annealed: 85,000 TS; 35,000 YS; 55 El; 65 RA; 160 Brin. For acid resistant chemical plant equipment; Type 316; stainless, austenitic. *Obsolete*

SODING CORROCHROM 13 T
J.C. Soding & Halbach
C 0.2, Si 1, Mn 1, Cr 13, Mo 1.15, bal Fe.
Air hardening corrosion resistant steel for plastic molds, heat and abrasion resistant. DIN X20CrMo13; Werkstoff Nr. 1.4120. *Obsolete*

SODING CRWC
J.C. Soding & Halbach
C 0.3, W 10, Cr 2.5, Co 2.5, V 0.3, bal Fe.
For extrusion dies, mandrels, die casting dies, hot shears, gripper dies. Hot work tool steel, oil hardening. *Obsolete*

SODING CS1
J.C. Soding & Halbach
C 0.62, Si 1, Mn 1, Cr 0.6, bal Fe.
Oil hardening cold work tool steel; for shear knives, cold punches, machine knives. DIN 62SiMnCr4; Werkstoff Nr. 1.2101. *Obsolete*

SODING CSJ
J.C. Soding & Halbach
C 0.67, Si 1.3, Mn 0.5, Cr 0.5, bal Fe.
For punches, crimpers, upsetters; oil hardened, shock resistant. *Obsolete*

SODING EXTRA HART
J.C. Soding & Halbach
C 1, Si 0-0.25, Mn 0-0.25, bal Fe.
Annealed: 100,000 TS; 53,000 YS; 21 El; 42 RA; 200 Brin. For springs, taps, drills, crimpers; Type W1; water hardened. *Obsolete*

SODING EXTRA MH
J.C. Soding & Halbach
C 1.15, Si 0-0.25, Mn 0-0.25, bal Fe.
Annealed: 110,000 TS; 57,000 YS; 18 El; 40 RA; 210 Brin. For springs, drills, cutters, taps, reamers, hobs; Type W1; water hardened. *Obsolete*

SODING EXTRA SPEZIAL 35
J.C. Soding & Halbach
C 1.15, V 0.1, bal Fe.
For cold working tools, drills, lathe and planer tools, hobs, headers, forming and drawing dies. Water hardening, wear resistant. DIN 100V1; Werkstoff Nr. 1.2833. *Obsolete*

SODING EXTRA WEICH
J.C. Soding & Halbach
C 0.7, Si 0-0.25, Mn 0-0.25, bal Fe.
Heat treated: 175,000 TS; 128,000 YS; 12 El; 37 RA; 355 Brin. For springs, rails, punches, hammers, axes, tools; Type W1; water hardened. *Obsolete*

SODING EXTRA ZAH
J.C. Soding & Halbach
C 0.85, Si 0-0.25, Mn 0-0.25, bal Fe.
Heat treated: 190,000 TS; 145,000 YS; 10 El; 30 RA; 400 Brin. For drills, taps, springs, reamers; Type W1; water hardened. *Obsolete*

SODING EXTRA ZH
J.C. Soding & Halbach
C 1, Si 0-0.25, Mn 0-0.25, bal Fe.
Annealed: 100,000 TS; 53,000 YS; 21 El; 42 RA; 200 Brin. For springs, drills, taps, reamers, broaches; Type W1; water hardened. *Obsolete*

SODING GCNS
J.C. Soding & Halbach
C 0.15, Si 1.5, Cr 18, Ni 8.5, bal Fe.
Annealed: 80,000 TS; 35,000 YS; 55 El; 75 RA; 150 Brin. For chemical plant equipment, tanks, mixers; Type 302; stainless, austenitic. *Obsolete*

SODING GCNV
J.C. Soding & Halbach
C 0.15, Cr 18, Mo 2, Ni 9.5, bal Fe.
Annealed: 85,000 TS; 35,000 YS; 50 El; 65 RA; 160 Brin. For acid resistant chemical plant equipment; Type 316; stainless, austenitic. *Obsolete*

SODING GCR17SI
J.C. Soding & Halbach
C 0.5, Si 1.5, Cr 17, bal Fe.
For oil refinery equipment; creep and heat resistant. *Obsolete*

SODING GCR23
J.C. Soding & Halbach
C 0.6, Si 1.5, Cr 22, bal Fe.
For oil refinery equipment; creep and heat resistant. *Obsolete*

SODING GCR27NI
J.C. Soding & Halbach
C 0.4, Cr 26, Ni 14.
ast: 75,000 TS; 47,000 YS; 17 El; 25 RA; 200 Brin. For hea OMPOSITION al Fe. *Obsolete*

SODING GCR28
J.C. Soding & Halbach
C 0.4, Cr 27, Ni 4, bal Fe.
Cast: 90,000 TS; 65,000 YS; 2 El; 212 Brin. Aged: 97,000 TS; 65,000 YS; 18 El; 210 Brin. For cylinder liners, bushings, valve seats and bodies; heat resistant. *Obsolete*

SODING GCR28NI
J.C. Soding & Halbach
C 0.4, Cr 27, Ni 4, bal Fe.
Cast: 90,000 TS; 65,000 YS; 2 El; 212 Brin. Aged: 97,000 TS; 65,000 YS; 18 El; 210 Brin. For cylinder liners, bushings, valves; heat resistant. *Obsolete*

SODING GCR30
J.C. Soding & Halbach
C 1.2, Cr 29, bal Fe.
For wear parts, rollers, crushers, grates; heat and abrasion resistant. *Obsolete*

SODING GCR30SI
J.C. Soding & Halbach
C 1.3, Cr 29, Si 1.5, bal Fe.
For wear parts, rollers, grates, crushers; heat and abrasion resistant. *Obsolete*

SODING GCR30W
J.C. Soding & Halbach
C 0.6, Si 1.5, Cr 29, bal Fe.
For grates, valve seats, cylinder liners; heat and abrasion resistant. *Obsolete*

SODING GCR6
J.C. Soding & Halbach
C 0.3, Si 2.2, Cr 6, bal Fe.
For oil refinery equipment; creep and heat resistant. *Obsolete*

SODING GMN
J.C. Soding & Halbach
C 0.55, Mo 0.18, Ni 1.65, V 0.1, Cr 0.7, bal Fe.
For gears, bolts, crankshafts; oil hardened, shock resistant. *Obsolete*

SODING GMN EXTRA
J.C. Soding & Halbach
C 0.56, Cr 0.75, Mo 0.2, Ni 1.8, V 0.1, bal Fe.
For bolts, gears, crankshafts; oil hardened, shock resistant. *Obsolete*

SODING GNRSN
J.C. Soding & Halbach
C 0.25, Cr 17, Ni 0-1.8, bal Fe.
Annealed: 125,000 TS; 90,000 YS; 20 El; 55 RA; 260 Brin. For pumps, marine hardware, valves; Type 431; corrosion and heat resistant. *Obsolete*

SODING GONC
J.C. Soding & Halbach
C 0.4, Cr 25, Ni 19, bal Fe.
Cast: 75,000 TS; 50,000 YS; 17 El; 170 Brin. For furnace parts, retorts, stack dampers; Type HK. *Obsolete*

SODING GONF
J.C. Soding & Halbach
C 0.5, Cr 25, Ni 30, bal Fe.
For retorts, furnace parts, rabble arms; heat resistant. *Obsolete*

SODING H
J.C. Soding & Halbach
C 0.45, Si 0.25-0.5, Mn 0.3-0.8, bal Fe.
Hot rolled: 98,000 TS; 59,000 YS; 24 El; 45 RA; 212 Brin. For gears, bolts, machine tool parts; water hardened. *Obsolete*

SODING HBC/NI
J.C. Soding & Halbach
C 0.12, Si 1.5, Mn 1, Cr 29.5, bal Fe.
For parts such as furnace parts, steam boiler equipment, and heat treat boxes that operate at temperatures up to 1150 C. DIN X10CrSi29; Werkstoff Nr. 1.4772. *Obsolete*

SODING HID1
J.C. Soding & Halbach
C 0.55, Cr 0.9, V 0.18, W 1.85, bal Fe.
For chisels, air hammers, punches, shears; cold work steel, oil hardened, shock resistant. *Obsolete*

SODING HID2
J.C. Soding & Halbach
C 0.45, Si 0.9, Cr 1.7, V 0.2, W 1.85, bal Fe.
For chisels, punches, pneumatic tools, crimpers; shock resistant, oil hardened, cold work steel. *Obsolete*

SODING HID3
J.C. Soding & Halbach
C 0.35, Si 0.9, Cr 1, V 0.18, W 1.85, bal Fe.
For chisels, punches, pneumatic tools, crimpers; shock resistant, oil hardened. *Obsolete*

SODING HID4
J.C. Soding & Halbach
C 0.45, Si 1, Cr 1.1, V 0.2, W 2, bal Fe.
Cold work tool steel, for chisels, pneumatic tools, rivet hammers. DIN 45WCrV7; Werkstoff Nr. 1.2542. *Obsolete*

SODING HPM
J.C. Soding & Halbach
C 0.45, Cr, V, Si, bal Fe.
For springs, gears, crankshafts; oil hardened, shock resistant. *Obsolete*

SODING KLR
J.C. Soding & Halbach
C 1, Cr 1.1, Mn 0.07, Si 0.25, bal Fe.
For bearings, liners, bushings, cutters; water hardened, wear resistant. *Obsolete*

SODING LKW
J.C. Soding & Halbach
C 0.9, Cr 0.8, bal Fe.
For cold working tools, embossing tools, cams, hand tools, woodworking tools. DIN 90 Cr3; Werkstoff Nr. 1.2056. *Obsolete*

SODING LMD
J.C. Soding & Halbach
C 1.2, V 0.1, Cr 0.2, W 1, bal Fe.
For cutters, bearings; water hardened, wear resistant. *Obsolete*

SODING LME
J.C. Soding & Halbach
C 1.05, Cr 1, W 1.15, bal Fe.
For cutters, bearings; water hardened, wear resistant. *Obsolete*

SODING MSW
J.C. Soding & Halbach
C 0.35, Cr 1.05, V 0.18, W 1.85, Si 0.9, bal Fe.
For forging and header dies, punches, upsetters; oil hardening, hot work tool steel; tough. *Obsolete*

SODING MW

J.C. Soding & Halbach

C 0.45, Cr 13, Ni 13, W 1.2, V 1.2, bal Fe.
Austenitic hot-work tool steel for extrusion dies and tube presses handling heavy metals. Hot-work steel DIN X50NiCrWV1313.

SODING MW 575

J.C. Soding & Halbach

C 0.35, Cr 13, Ni 5.75, W 2.75, V 0.7, bal Fe.
Alloy hot-work steel; for extrusion dies.

SODING NR17T

J.C. Soding & Halbach

C 0-0.1, Cr 17.5, Ti = 7 x C, bal Fe.
Annealed: 80,000 TS; 50,000 YS; 25 El; 50 RA; 150 Brin. For welded oil refinery equipment; corrosion resistant. *Obsolete*

SODING NR18T

J.C. Soding & Halbach

C 0-0.1, Cr 17, Mo 1.8, Ti = 7 x C, bal Fe.
Annealed: 85,000 TS; 52,000 YS; 23 El; 53 RA; 160 Brin. For welded acid resistant equipment; corrosion resistant. *Obsolete*

SODING NRSA

J.C. Soding & Halbach

C 0.12, Cr 16.5, Mo 0.25, S 0.2, bal Fe.
Annealed: 80,000 TS; 50,000 YS; 25 El; 50 RA; 150 Brin. For screw machine products; free-cutting; Type 430 F; stainless. *Obsolete*

SODING NRSB 1320

J.C. Soding & Halbach

C 0.2, Si 0.4, Mn 0.3, Cr 13, bal Fe.
Air hardening steel for pressure chambers for light metal die casting; heat, abrasion and corrosion resistant. DIN X20Cr13; Werkstoff Nr. 1.2082. *Obsolete*

SODING NRSB 1340

J.C. Soding & Halbach

C 0.4, Si 0.4, Mn 0.3, Cr 13, bal Fe.
Air or oil hardenable; for heavy duty plastic molds; good resistance to abrasion and corrosion. DIN X40Cr13; Werkstoff Nr. 1.2083. *Obsolete*

SODING NRSE

J.C. Soding & Halbach

C 0.9, Cr, Mo, bal Fe.
For valves, cutlery; oil hardened. *Obsolete*

SODING NRSF

J.C. Soding & Halbach

C 0.08, Cr 17, Si 0.4, bal Fe.
Annealed: 80,000 TS; 50,000 YS; 25 El; 50 RA; 150 Brin. For oil refinery equipment, oil burners, sinks; Type 430; stainless. *Obsolete*

SODING NRSG

J.C. Soding & Halbach

C 0.2, Si 0.4, Cr 13, bal Fe.
Annealed: 95,000 TS; 50,000 YS; 25 El; 55 RA; 195 Brin. For cutlery, valves, surgical instruments; Type 420; stainless. *Obsolete*

SODING NRSM

J.C. Soding & Halbach

C 0.4, Cr 13, Si 0.4, bal Fe.
Annealed: 100,000 TS; 55,000 YS; 22 El; 52 RA; 200 Brin. For cutlery, valves, surgical and dental instruments; Type 420; stainless, hardenable. *Obsolete*

SODING NRSN

J.C. Soding & Halbach

C 0.22, Si 0.4, Cr 17, Ni 1.5, bal Fe.
Annealed: 125,000 TS; 95,000 YS; 20 El; 55 RA; 260 Brin. For pumps, marine hardware, valves; Type 431; stainless. *Obsolete*

SODING NRST

J.C. Soding & Halbach

C 0.2, Cr 13, Mo 1.15, bal Fe.
Annealed: 100,000 TS; 55,000 YS; 22 El; 52 RA; 200 Brin. For chemical plant equipment, valves, cutlery; Type 420 Mo; stainless. *Obsolete*

SODING NRSW

J.C. Soding & Halbach

C 0-0.12, Si 0.4, Cr 13, bal Fe.
Annealed: 75,000 TS; 40,000 YS; 35 El; 70 RA; 155 Brin. For turbine blades, valves, cutlery; Type 410; stainless. *Obsolete*

SODING OA1310

J.C. Soding & Halbach

C 0.12, Si 1.2, Mn 0.3, Al 1, Cr 13, bal Fe.
For parts in glass, porcelain and enameling industry that operate at temperatures up to 950 C. DIN X10CrAl13; Werkstoff Nr. 1.4724. *Obsolete*

SODING OA1810

J.C. Soding & Halbach

C 0.12, Si 1, Mn 0.3, Al 1, Cr 18, bal Fe.
For parts for furnaces, petroleum industry, heat treating, power plants operating at temperatures up to 1050 C. DIN X10CrAl18; Werkstoff Nr. 1.4742. *Obsolete*

SODING OA2415

J.C. Soding & Halbach

C 0.12, Si 1.4, Mn 0.3, Al 1.5, Cr 24, bal Fe.
For parts for furnaces, heat treat equipment, petroleum cracking, glass making that operate at temperatures up to 1200 C. DIN X10CrAl24; Werkstoff Nr. 1.4762. *Obsolete*

SODING OA708

J.C. Soding & Halbach

C 0.12, Si 0.8, Mn 0.3, Al 0.8, Cr 6.5, bal Fe.
For parts of furnaces, steam plants, heat treating equipment that operate at temperatures up to 800 C. DIN X10CrAl7; Werkstoff Nr. 1.4713. *Obsolete*

SODING ON 189

J.C. Soding & Halbach

C 0.15, Si 0.3, Mn 1.2, Cr 18, Ni 10, Ti 0.7, bal Fe.
For parts operating at temperatures up to 800 C as hot plates, brackets for heat exchangers, furnace parts; corrosion resistant. DIN X12 CrNiTi189; Werkstoff Nr. 1.4878. *Obsolete*

SODING ON16 36

J.C. Soding & Halbach

C 0.15, Si 1.8, Mn 1.2, Cr 16, Ni 35.5, bal Fe.
For carburizing boxes, annealing boxes, petroleum processing equipment, heat exchanger tubes that operate at temperatures up to 1100 C; resistant to carburizing gases. DIN 12NiCrSi3616; Werkstoff Nr. 1.4864. *Obsolete*

SODING ON20 12

J.C. Soding & Halbach

C 0.2, Si 2, Mn 1.2, Cr 20, Ni 12, bal Fe.
For parts operating at temperatures up to 1050 C in petroleum plants, glass plants, enameling plants, heat treating equipment, corrosion resistant. DIN X15CrNiSi2012; Werkstoff Nr. 1.4828. *Obsolete*

SODING ON25 20

J.C. Soding & Halbach

C 0.2, Si 2, Mn 1.2, Cr 25, Ni 20, bal Fe.
For parts for furnaces, heat treating equipment, jet engines, petroleum plants, glass factories that operate at temperatures up to 1200 C; corrosion resistant. DIN X15CrNiSi 25 20; Werkstoff Nr. 1.4841. *Obsolete*

SODING ON25 4

J.C. Soding & Halbach

C 0.2, Si 1.2, Mn 1.2, Cr 25, Ni 4, bal Fe.
For muffles, retorts, hot plates and other parts that operate at temperatures up to 1100 C. DIN X20CrNiSi25 4; Werkstoff Nr. 1.4821. *Obsolete*

SODING ONC

J.C. Soding & Halbach

C 0.15, Si 2, Cr 24, Ni 19, bal Fe.
Annealed: 100,000 TS; 45,000 YS; 50 El; 65 RA; 185 Brin. For furnace parts, pumps, valves, jet parts; Type 310; stainless, austenitic. *Obsolete*

SODING ONIA

J.C. Soding & Halbach

C 0.15, Si 2, Cr 19.5, Ni 9.5, bal Fe.
Annealed: 80,000 TS; 35,000 YS; 55 El; 75 RA; 150 Brin. Cold drawn: 180,000 TS; 150,000 YS; 10 El; 250 Brin. For chemical plant and oil refinery equipment; Type 302; stainless, austenitic. *Obsolete*

SODING PERMANENT

J.C. Soding & Halbach

C 0.75, Cr 4.5, W 18.5, V 1.2, bal Fe.
Hardenable to 62-65 Rock C. For roughing and finishing cuts of all kinds, lathe tools, milling cutters, taps, drills, reamers. High-speed steel B18.

SODING PERMANENT 120

J.C. Soding & Halbach

C 1.35, Cr 4.5, W 9, Mo 3.7, V 3.75, Co 10.5, bal Fe.
Hardenable to 63-68 Rock C. For lathe tools, form tools for automatics, relieving tools and form cutters for planing and slotting operations. High-speed steel EW9Co10.

SODING PERMANENT 35

J.C. Soding & Halbach

C 0.8, Cr 4.5, W 18.5, Mo 0.8, V 1.7, Co 5, bal Fe.
Hardenable to 63-66 Rock C. For heavy and roughing cuts on lathes, planers, and slotters on cast iron, steel and cast steel including stainless grades. High-speed steel E18Co5.

SODING PERMANENT 70

J.C. Soding & Halbach

C 1.4, Cr 4.5, W 12.5, Mo 1, V 4, Co 5, bal Fe.
Hardenable to 63-66 Rock C. For form tools for automatics, lathe tools, relieving tools; medium and finishing cuts at high speed. High-speed steel.

SODING PK35

J.C. Soding & Halbach

C 0.5, Cr 1.05, Ni 3.25, Mn 0.5, bal Fe.
For gears, bolts, crankshafts, axles; oil hardened, shock resistant. *Obsolete*

SODING PRIMA WEICH

J.C. Soding & Halbach

C 0.35, Si 0.1-0.4, Mn 0.5-0.7, bal Fe.
Hot rolled: 85,000 TS; 54,000 YS; 30 El; 53 RA; 180 Brin. For crankshafts, gears, shafts, fasteners; water hardened. *Obsolete*

SODING PRIMA ZAH

J.C. Soding & Halbach

C 0.45, Si 0.1-0.4, Mn 0.5-0.7, bal Fe.
Hot rolled: 98,000 TS; 59,000 YS; 24 El; 45 RA; 212 Brin. For axles, gears, bolts, tie-rods, bushings; water hardened. *Obsolete*

SODING PRIMA ZH

J.C. Soding & Halbach

C 0.6, Si 0.1-0.4, Mn 0.5-0.7, bal Fe.
Heat treated: 115,000-160,000 TS; 77,000-113,000 YS; 12-23 El; 40-54 RA; 230-320 Brin. For wheels, die blocks, girders; water hardened. *Obsolete*

SODING RADIKAL MO5

J.C. Soding & Halbach

C 0.85, Cr 4.5, W 6.7, Mo 5.3, V 2, bal Fe.
Hardenable to 62-66 Rock C. For milling cutters, twist drills, reamers, form cutters and lathe tools; for heavy and rough cuts; good toughness. High-speed steel DMo5; AISI M2.

SODING RADIKAL MO55

J.C. Soding & Halbach
C 0.85, Cr 4.5, W 6.7, Mo 5.3, V 2, Co 5.5, bal Fe.
Hardenable to 62-66 Rock C. For lathe, planer and slotter tools, milling cutters, twist drills; for difficult machining and drilling operations. High-speed steel E Mo5Co5.

SODING RADIKAL MO60

J.C. Soding & Halbach
C 1.25, Cr 4.5, W 6.7, Mo 5.3, V 3.4, bal Fe.
Hardenable to 62-66 Rock C. For form tools on medium-sized lathes and automatics, milling cutters, twist drills, chasers, broaches, reamers. High-speed steel EMo5V3.

SODING RADIKAL MO9

J.C. Soding & Halbach
C 0.85, Cr 4.2, W 2, Mo 9.2, V 1.3, bal Fe.
Hardenable to 62-66 Rock C. For twist drills, taps, reamers, milling cutters and tools for woodworking machines. High-speed steel BMo9; AISI M1.

SODING RADIKAL MO92

J.C. Soding & Halbach
C 1.1, Cr 4.2, W 2, Mo 9.2, V 2.1, bal Fe.
Hardenable to 62-66 Rock C. For twist drills, reamers, lathe tools, milling cutters; and tools for woodworking machines. High-speed steel BMo9V.

SODING SCHM

J.C. Soding & Halbach
C 0.4, Cr, Mn, Mo, bal Fe.
For gears, shafts, bolts, axles; oil hardened, tough. *Obsolete*

SODING SHCM EXTRA

J.C. Soding & Halbach
C 0.45, Cr 1.4, Mo 0.7, V 0.3, Mn 0.7, bal Fe.
For forging and upsetting dies; oil hardened, tough. *Obsolete*

SODING SPEZIAL DIAMANT

J.C. Soding & Halbach
C 1.42, W, V, bal Fe.
For blanking and forming dies, cutters; water or oil hardened. *Obsolete*

SODING SPEZIAL MH

J.C. Soding & Halbach
C 1.1, Si 0-0.25, Mn 0-0.25, bal Fe.
Annealed: 100,000 TS; 53,000 YS; 21 El; 42 RA; 200 Brin. For springs, taps, drills, reamers; Type W1; water hardened. *Obsolete*

SODING SPEZIAL WEICH

J.C. Soding & Halbach
C 0.7, Si 0-0.25, Mn 0-0.25, bal Fe.
For springs, punches, girders, rails, die blocks; Type W1; water hardened. *Obsolete*

SODING SPEZIAL ZAH

J.C. Soding & Halbach
C 0.85, Si 0-0.25, Mn 0-0.25, bal Fe.
Heat treated: 190,000 TS; 145,000 YS; 10 El; 30 RA; 400 Brin. For springs, taps, drills, reamers, tools; Type W1; water hardened. *Obsolete*

SODING SPEZIAL ZH

J.C. Soding & Halbach
C 1, Si 0-0.25, Mn 0-0.25, bal Fe.
Annealed: 100,000 TS; 53,000 YS; 21 El; 42 RA; 200 Brin. For springs, taps, drills, reamers, broaches; Type W1; water hardened. *Obsolete*

SODING SPG

J.C. Soding & Halbach
C 1.25, Si 1.15, Cr 1.2, Mn 0.7, bal Fe.
For bearings, cutters, liners, sleeves; water or oil hardened, wear resistant. *Obsolete*

SODING SPH 12

J.C. Soding & Halbach
C 0.5, Cr 4, Ni 12, W 10, Co 1.5, bal Fe.
Alloy hot work tool steel; for extrusion dies.

SODING SPH 7

J.C. Soding & Halbach
C 0.32, Cr 2.9, Mo 2.8, V 0.5, bal Fe.
Oil or salt bath hardenable to 40-51 Rock C. For extrusion ram heads for light and heavy metals, hot upsetting tools, pressure die casting dies for brass, light metals and zinc. Hot-work tool steel X32CrMoV33.

SODING SPH 8

J.C. Soding & Halbach
C 0.3, Cr 2.8, Mo 2.8, Co 2.6, V 0.5, bal Fe.
Oil hardenable to 40-50 Rock C. For tools for metal extrusion and tube presses, extrusion dies for heavy metals, pressure die casting dies. Hot-work steel.

SODING SPH 9

J.C. Soding & Halbach
C 0.4, Cr 4, W 4, V 2, Co 1.5, bal Fe.
Hot-work tool steel; for extrusion press dies, hot upsetting tools, sliding cores in dies.

SODING SPH2

J.C. Soding & Halbach
C 0.3, Si 1, Cr 1.1, V 0.18, W 3.75, bal Fe.
For extrusion dies, rams, liners, punches; hot work steel, oil hardened. *Obsolete*

SODING SPH3

J.C. Soding & Halbach
C 0.3, V 0.6, Cr 2.35, W 4.25, bal Fe.
For extrusion dies, rams, liners, punches; hot work steel, oil hardened. *Obsolete*

SODING SPHX

J.C. Soding & Halbach
C 0.3, V 0.35, Cr 2.65, W 8.5, bal Fe.
For extrusion dies, rams, liners, punches; hot work steel, oil hardened. *Obsolete*

SODING STA

J.C. Soding & Halbach
C 1.45, Cr 1.4, Mn 0.6, bal Fe.
For bearings, cutters, engravers' tools; oil or water hardened. *Obsolete*

SODING UNB

J.C. Soding & Halbach
C 1.1, Cr 0.9, Mn 1.1, bal Fe.
For cold working tools, threading tools for hand operation, light punching dies. DIN 105 MnCr4; Werkstoff Nr. 1.2127. *Obsolete*

SODING UNIVERSAL

J.C. Soding & Halbach
C 0.9, Mn 2, V 0.1, bal Fe.
For cold working tools, punches. DIN 90 MnV8; Werkstoff Nr. 1.2842. *Obsolete*

SODING W15

J.C. Soding & Halbach
C 0.38, Cr 1.5, V 0.1, Si 1.5, Mn 0.4, bal Fe.
Cold work tool steel; for compressed air tools, blacksmith's chisels, trimming tools. DIN 38 SiCrV6; Werkstoff Nr. 1.2248. *Obsolete*

SODING W35

J.C. Soding & Halbach
C 0.4, Cr 5.5, Mo 1.4, V 0.3, bal Fe.
Air or oil hardenable to 40-54 Rock C. For impact and pressure forging dies, pressure die casting dies, extrusion dies, piercing and punching dies. Hot-work tool steel X38CrMoV51; AISI H11.

SODING W36

J.C. Soding & Halbach
C 0.37, Cr 4.8, W 1.4, Si 0.9, Mn 0.6, V 0.2, bal Fe.
Oil or air hardenable to 40-54 Rock C. For pressure die casting dies for light metal, tin, zinc and lead alloys when long life is important. For forging die inserts. Hot-work steel X37CrMoW51 AISI H12.

SODING W37

J.C. Soding & Halbach
C 0.4, Cr 5, Mo 1.5, V 1, bal Fe.
Air or oil hardenable to 40-54 Rock C. For pressure die casting dies for light metals, tin, zinc and lead alloys; also for forging die inserts, piercing mandrels and extrusion mandrels. Hot-work tool steel X40CrMoV51; AISI H13.

SODING WM

J.C. Soding & Halbach
C 0.45, Mn 0.8, bal Fe.
Water hardening steel for small, lightly loaded dies. DIN C45W3; Werkstoff Nr. 1.1730. *Obsolete*

SODING ZWAT

J.C. Soding & Halbach
C 2.1, Cr 11.5, bal Fe.
For blanking and forming dies, punches; oil or air hardened, non-deforming. *Obsolete*

SODIUM

Atomergic Chemetals Corp.
Na.
Purities: distilled 99.99%; 99.95%, (nuclear grade). Packaging: bottles, ampules, cylinder (under vacuum, inert gas, petroleum distillate).

SODIUM-LEAD ALLOY

Grasselli Chemical Co.
Pb 98, Na 2.
For nonferrous alloy deoxidizer.

SODIUM-ZINC ALLOY

Grasselli Chemical Co.
Zn 98, Na 2.
For nonferrous alloy deoxidizer. Brittle.

SOFT

Electro-Steel Co.
C 0.5, bal Fe.
For tools, shafts, gears; water hardened.

SOFTITE

Wheeling-Pittsburgh Steel Corp.

Galvanized steel for structural components.

SOFTWELD

Lincoln Electric Co.
C, bal Fe.
Arc welding electrode. AWS Class EN: C1.

SOHO SELF-HARDENING

Hobson, Houghton & Co.
C 0.9, Cr 4, V 0.2, bal Fe.
For tools, dies; oil hardened.

SOLABRAZE C-6601

International Harvester Co.
Zr 43, Ni 12, Be 2, bal Ti.
For brazing titanium to itself or to ceramic parts in vacuum or oxygen-free inert gas. Melting range: 1470-1500°F. Braze at 1650°F.

SOLABRAZE C-6602

International Harvester Co.
Zr 45, Ni 8, Be 2, bal Ti.
For brazing titanium to itself or to ceramic parts in high vacuum or oxygen-free inert gas. MP: 1650°F. Braze at 1700°F.

SOLABRAZE C-6603
International Harvester Co.
Zr 47, Be 5, bal Ti.
For brazing titanium to itself or to ceramic parts in high vacuum or oxygen-free inert gas. Melting range: 1640-1660°F. Braze at 1725°F.

SOLABRAZE C-6604
International Harvester Co.
Zr 47, Be 5, Al 5, bal Ti.
For brazing titanium to itself or to ceramic parts in high vacuum or oxygen-free inert gas. MP: 1700°F. Braze at 1750°F.

SOLAR
Now CARPENTER SOLAR.

SOLAR
Westa-Westdeutsche
C 1.05, Cr 1, W 1.15, Mn 0.9, bal Fe.
For cutters, reamers, forming dies; water or oil hardened.

SOLAR ECLIPSE
C. Joselin & Co.
C, alloy, bal Fe.
For cold hobbing dies; oil hardened.

SOLBISKYS ALLOY
German manufacture
Zn 1.4, Ni 0.5-1, Cd 0-3, Sn 0.3, bal Al.
For light alloy parts; non-hardenable.

SOLDALOY 2
CMW Inc.
Cu alloy.
For soldering iron tip; high thermal conductivity 40%.
Obsolete

SOLDALOY NO. 1
CMW Inc.
Cu alloy.
65,000 TS; 20 El; 93-137 Brin. For soldering iron tips; 85% electrical conductivity. *Obsolete*

SOLDALOY NO. 8
CMW Inc.
Cu alloy.
90,000 TS; 10 El; 165, 210, 137 Brin. For soldering iron tips; 50% electrical conductivity. *Obsolete*

SOLDER HC
J.L. Snowbar
Pb, Sn.
20,000 TS. For solder for Al.

SOLDER NO. 1
English manufacture
Sn 30-40, Sb 2-3, bal Pb.
6890 TS; 117 El; 12 Brin. For commercial lead solders; melting point 401°F.

SOLDER NO. 2
English manufacture
Sn 50, Pb 50.
6400 TS; 96 El; 11 Brin. For commercial lead solders; melting point 446°F.

SOLDERAL
Non-Corrodal Alloys Ltd.
Si 5, bal Al.
For Al solder, castings. Non-corroding.

SOLDERZIT
L & R Mfg. Co.
Pb, Sn.
For soft soldering; powdered soft solder and flux combination.

SOLDIER A26
Burys & Co. Ltd.
C 0.3, W 9, Cr 3.25, V 0.2, bal Fe.
For rivet dies, extrusion and press dies, mandrels; hot work steel, oil hardened.

SOLDIER A26N
Burys & Co. Ltd.
C 0.25, W 8.5, Cr 3, V 0.25, Ni 2.5, bal Fe.
For rivet dies, extrusion and press dies, mandrels; hot work steel, oil hardened.

SOLDIER CHROME H.33
Burys & Co. Ltd.
C 0.95, Cr 4, bal Fe.
For hot shears, upsetters, punches; hot work steel for non-ferrous metals.

SOLDIER M.314
Burys & Co. Ltd.
C 0.35, Cr 1, Ni 3, Mo 0.3, Mn 0.5, bal Fe.
For die casting dies, stamping die blocks; hot work steel, oil hardened.

SOLDIER P.23
Burys & Co. Ltd.
C 0.4, Si 1, Cr 5, W 4, Mo 0.4, bal Fe.
For hot dies, punches, mandrels, gripper dies; hot work steel, air or oil hardened.

SOLDIER SCT C.16.B.
Burys & Co. Ltd.
C 0.5, Cr 1.5, W 2, V 0.2, Si 1, bal Fe.
For shear blades, swaging dies; hot work steel, shock resistant.

SOLEIL 1.2
Societe des AFY
Cr 12-14, 0.15 C min, bal Fe.
Annealed: 88,000 TS; 40,000 YS; 32 El; 68 RA; 170 Brin. Heat treated: 256,000 TS; 190,000 YS; 6 El; 10 RA; 540 Brin. For cutlery, valve trim, turbine blades; Type 420; stainless, hardenable.

SOLEIL 16
Compagnie Ateliers et Forges de la Loire
Stainless steel. C 0.06, Cr 16.5, bal Fe.
Annealed: 70,000 TS; 40,000 YS; 30 El; 140 Brin. For food and chemical industries, nitric acid equipment, cooking utensils. Ferritic stainless steel. Resists nitric acid, non-hardenable.

SOLEIL 17
Compagnie Ateliers et Forges de la Loire
Stainless steel. C 0.2, Cr 17, bal Fe.
Annealed: 80,000 TS; 45,000 YS; 30 El; 130 Brin. For chemical plant and oil refinery equipment, fixtures, pump parts. Ferritic stainless steel, non-hardenable.

SOLEIL 18
Societe des AFY
C 0-0.12, Cr 14-18, bal Fe.
Annealed: 70,000 TS; 40,000 YS; 30 El; 55 RA; 150 Brin. Cold drawn: 130,000 TS; 120,000 YS; 2 El; 185 Brin. For oil refinery equipment, bolts, kitchen sinks; Type 430; stainless, ferritic.

SOLEIL 2B
Compagnie Ateliers et Forges de la Loire
Stainless steel. C 0.2, Cr 13, bal Fe.
Annealed: 90,000 TS; 45,000 YS; 30 El; 90 Rock B. Heat treated: 225,000 TS; 160,000 YS; 8-16 El. For cutlery, turbine blades, valves, pumps, furnace parts, pistons. Type 420 stainless steel. Hardenable. Corrosion and heat resistant.

SOLEIL 3
Societe des AFY
C 0-0.15, Cr 11.5-13.5, bal Fe.
Heat treated: 120,000-135,000 TS; 110,000-117,000 YS; 15-16 El; 58-63 RA; 220-240 Brin. For cutlery, valves, turbine blades; Type 410; corrosion resistant.

SOLEIL 4S
Compagnie Ateliers et Forges de la Loire
Stainless steel. C 0.05, Cr 13, Al 0.2, bal Fe.
Annealed: 68,000 TS; 38,000 YS; 30 El. For oil refinery equipment, cracking stills, tubes for the petroleum industry. Ferritic stainless steel. Non-hardenable.

SOLEIL A1M
Creusot-Loire
C 0-0.07, Mn 0-1, Si 0-1, Ni 1.5-2.5, Cr 11-13.5, Mo 0-0.75, bal Fe.
Stainless casting, ferritic type; 200-250 Brin. Parts for steam turbines. AFNOR Z4CN 13.2 M.

SOLEIL A2
Creusot-Loire
C 0.12, Cr 12.5, bal Fe.
Martensitic stainless steel; weldable. For kitchen equipment, knife handles, pump bodies, impellers. AFNOR Z12C13; AISI 410.

SOLEIL A21M
Creusot-Loire
C 0.08-0.15, Mn 0-1, Si 0-1, Ni 0.5-1.5, Cr 12-14, bal Fe.
Stainless casting; annealed: 170-290 Brin. For pumps and faucets; food industries; cold dilute organic acids. AFNOR Z12CN13M; W.Nr. 4008.

SOLEIL A2M
Creusot-Loire
C 0-0.15, Mn 0-1, Si 0-1, Cr 11.5-14, bal Fe.
Stainless casting; annealed: 180-220 Brin. For petroleum and food industries. AFNOR Z12C13M; ACI CA 15.

SOLEIL A2U
Creusot-Loire
C 0-0.15, Cr 13, Mo 0-0.6, 0.15 S min, bal Fe.
Quenched and tempered: 640-830 N/mm^2 psi TS. Free machining stainless steel. AFNOR Z12CF13; similar to AISI 416.

SOLEIL A3
Creusot-Loire
C 0-0.12, Mn 0.5, Si 0-0.6, Cr 13, bal Fe.
Martensitic type stainless steel; weldable. AFNOR Z15C13; AWS ER 410.

SOLEIL A4
Creusot-Loire
C 0.2, Cr 13, bal Fe.
Martensitic stainless steel; air or oil hardenable to 740-880 N/mm^2 psi TS. For pump bodies, valves, control rods. AFNOR Z20C13; AISI 420.

SOLEIL A4M
Creusot-Loire
C 0.18-0.25, Mn 0-1, Si 0-1, Cr 12.5-14.5, bal Fe.
Stainless casting; annealed: 200-250 Brin. Pumps and faucets for food industries. AFNOR Z20C13M; similar to ACI CA-40.

SOLEIL A5
Creusot-Loire
C 0.3, Cr 13, bal Fe.
Martensitic stainless steel; air or oil hardenable to 830-980 N/mm^2 psi TS. For axles, shafts, valves, pump parts. AFNOR Z30C13; AISI 420.

SOLEIL A5M
Creusot-Loire
C 0.25-0.35, Mn 0-1, Si 0-1, Cr 13-15, bal Fe.
Stainless casting; annealed: 250-300 Brin. Faucets and pumps for paper industries. AFNOR Z30C13M; similar to ACI CA-40.

SOLEIL A5U
Creusot-Loire
C 0.3, Cr 13, Ni 0-1, Mo 0-0.6, 0.15 S min, bal Fe.
Quenched and tempered: 830-980 N/mm^2 psi TS. Free machining martensitic stainless steel. AFNOR Z30CF13; AISI 420 F.

SOLEIL A6M
Creusot-Loire
C 0.35-0.45, Mn 0-1, Si 0-1, Cr 13-15, bal Fe.
Stainless casting; annealed: 260-300 Brin. For pump parts in paper industries. AFNOR Z38C13M; W.Nr. 1.4034.

SOLEIL A8
Creusot-Loire
C 0.7, Mn 0.5, Si 0-0.6, Cr 15, bal Fe.
Martensitic stainless. AFNOR Z70C15; AISI 440A.

SOLEIL B2
Creusot-Loire
C 0-0.08, Cr 12.5, bal Fe.
Ferritic type stainless steel; not usually hardenable. Weldable. 440-640 N/mm^2 psi TS. For parts subject to water and steam. AFNOR Z6C13.

SOLEIL B3
Creusot-Loire
C 0-0.08, Cr 12.5, Al 0.2, bal Fe.
Ferritic type stainless steel; weldable; not hardenable by heat treating. 440-640 N/mm^2 psi TS. For use with water or steam; petroleum industries. AFNOR Z6 CA13; AISI 405.

SOLEIL B4
Creusot-Loire
C 0-0.1, Cr 17, bal Fe.
Ferritic type stainless; not hardenable by heat treating. 440-640 N/mm^2 psi TS. Tableware, forks and spoons, kitchenware, soap and dyeing industries. AFNOR Z8C17; AISI 430.

SOLEIL B4U
Creusot-Loire
C 0.12, Cr 17, Mo 0-0.6, 0.15 S min, bal Fe.
Wrought, annealed: 440-640 N/mm^2 psi TS. Free machining ferritic type stainless steel. AFNOR Z10CF17; AISI 430 F.

SOLEIL C1
Creusot-Loire
C 0-0.12, Cr 16, Ni 3, Cu 1.5, bal Fe.
Martensitic stainless steel; air or oil hardenable to 740-880 N/mm^2 psi TS. For pump parts in organic acid or neutral briny solutions. AFNOR Z10CNU 17-04.

SOLEIL C2M
Creusot-Loire
C 0.15-0.25, Mn 0-1, Si 0-1, Ni 1.5-3, Cr 15-18, bal Fe.
Stainless casting; annealed: 275-310 Brin. For parts subject to sea water or dilute organic acids. AFNOR Z20Cn17.2M; similar to AISI 431.

SOLEIL C3M
Creusot-Loire
C 0-0.06, Mn 0-1, Si 0-1, Ni 3.5-5, Cr 15-17.5, Cu 3-4.5, 0.15 Nb min, bal Fe.
Stainless casting; heat treated: 365-410 Brin. Good corrosion resistance at elevated temperature. AFNOR Z4CNUNB 16.4M.

SOLEIL C4
Creusot-Loire
Ni 5, Mo 0.9, C 0-0.06, Cr 16, bal Fe.
Martensitic stainless; air or oil hardenable to 880-1080 N/mm^2 psi TS. Bars; good strength and corrosion resistance to sea water, organic acids. AFNOR Z4CND 16-05; see also VIRGO 3.

SOLEIL C4M
Creusot-Loire
C 0-0.06, Mn 0-1, Si 0-1, Ni 4-5.5, Cr 15.5-17.5, Mo 0.5-1.5, bal Fe.
Stainless casting. For water turbines, pumps and other parts in sea water. AFNOR Z4 CND 17.4M.

SOLEIL C5M
Creusot-Loire
C 0-0.06, Mn 0-1, Si 0-1, Ni 3.5-5, Cr 12-13.5, Mo 0.4-0.7, bal Fe.
Stainless casting. Parts for pumps, turbines, nuclear energy. AFNOR Z4CND13 4M; ACI CA6NM.

SOLEIL D9
Creusot-Loire
C 1.1, Mn 0.5, Si 0-0.6, Cr 16.5, Mo 0.5, bal Fe.
Martensitic stainless. AFNOR Z110CD17; AISI 440C.

SOLEIL NO. 2
Creusot-Loire
C 0.32, Cr 13, bal Fe.
Annealed: 95,000 TS; 50,000 YS; 25 El; 55 RA; 195 Brin. For valves, cutlery, surgical instruments; corrosion resistant. *Obsolete*

SOLEIL O
Compagnie Ateliers et Forges de la Loire
Stainless steel. C 0.44, Cr 14, bal Fe.
Annealed: 95,000 TS; 50,000 YS; 25 El; '92 Rock B. For cutlery, turbine blades, diesel engine pistons, surgical instruments. Martensitic stainless steel. Hardenable. Corrosion and heat resistant.

SOLEIL T1
Compagnie Ateliers et Forges de la Loire
C 0.2, Cb 0.4, Cr 11, Mo 0.75, V 0.5, Ni 0.3, bal Fe.
Annealed: 95,000 TS; 40,000 YS; 25 El; 92 Rock B. Heat treated: 240,000 TS; 205,000 YS; 9 El; 50 Rock C. For surgical instruments, hardware, knives, gears, shafts. Corrosion resistant, hardenable.

SOLEIL T2
Compagnie Ateliers et Forges de la Loire
C 0.2, Cr 11, Mo 1, Ni 0.8, V 0.35, W 0.6, bal Fe.
Annealed: 95,000 TS; 40,000 YS; 25 El; 92 Rock B. Hardened: 240,000 TS; 205,000 YS; 9 El; 50 Rock C. For valves, bearings, cutlery, surgical instruments. Corrosion resistant, hardenable.

SOLEIL-1
Creusot-Loire
C 0.38-0.43, Cr 13, bal Fe.
For valves, cutlery, surgical and dental instruments; Type 420; corrosion resistant. *Obsolete*

SOLEIL-2
Creusot-Loire
C 0.19-0.25, Cr 13, bal Fe.
Annealed: 95,000 TS; 50,000 YS; 25 El; 55 RA; 196 Brin. For turbine blades, valves, cutlery, surgical instruments; Type 420; corrosion resistant. *Obsolete*

SOLEIL-30
Creusot-Loire
C 0.1, Cr 21, Ni 1.5, bal Fe.
For furnace parts and equipment; heat resistant. *Obsolete*

SOLEIL-4
Creusot-Loire
C 0.14, Cr 13, Mo 1, bal Fe.
For turbine blades, valves, cutlery, surgical instruments; corrosion resistant. *Obsolete*

SOLID DRILL
Agawam Tool Co.
C 1.05, Mn 0.35, bal Fe.
For granite drills; water hardened.

SOLID DRILL
Horace T. Potts Co.
C 1.1-1.2, bal Fe.
For tools, drills. *Obsolete*

SOLOCOAT SC820
Wall Colmonoy Corp.
Hard, stainless alloy coating with good wear and corrosion resistance. Ideal for coating non-copper bearing surfaces or journals, cylinder liners, capstans, pump seals, or to build up parts for press fits. Can be finished by grinding. 30 Rock C; 0.030 in. thickness max.

SOLOCOAT SC840
Wall Colmonoy Corp.
General-purpose, stainless-type coating with good wear and corrosion resistance at high temperatures. Good for metal-to-metal sliding friction and erosive applications and for build-up on low carbon, low alloy steels and machinable stainless steels. For pump parts and water turbine blades. Can be machined with carbide tools or ground. 80 Rock B.

SOLOCOAT SC850
Wall Colmonoy Corp.
Aluminum-bronze alloy with good wear resistance on soft bearing applications. Can be machined smooth using carbide tools. Good for heavy build-up on mismachined or worn down copper alloy and low carbon steel components. For pump seals, valve parts, and marine parts. Rock 50 B.

SOLOCOAT SC870
Wall Colmonoy Corp.
Nickel-aluminum-molybdenum alloy for general purpose and bond coat use. Yields good wear resistance on hard bearing applications where particle erosion and fretting wear is encountered. Recommended for rebuilding machine bedways, wear rings, press-fit parts. Can be carbide-tool machined or ground or feather-edged. 80 Rock B.

SOLUMANG 75B
Cyprus Foote Mineral Co.
Al 25, bal pure electrolytic Mn.
For use in steel melting. *Obsolete*

SOLUMANG 90PF
Cyprus Foote Mineral Co.
10.0 alkali halides flux, bal pure electrolytic Mn.
For rapid addition of manganese in alloying and deoxidation of steel. *Obsolete*

SOLUSAL
Electro-Steel Co.
Cu, Si, bal Al.
34,000 TS; 16 El. For airplane, automobile and marine parts to resist corrosion; light alloy.

SOLUSAL
Non-Corrodal Alloys Ltd.
Cu, Si, bal Al.
34,000 TS; 16 El. For airplane, automobile and marine parts to resist corrosion; light alloy.

SONA-METAL
Manos Ltd.
Si 0.2-0.7, Cu 3-4, Sn 0.7, Mg 0.5, Fe 1, Ni 0.2, bal Al.
36,000-50,000 TS; 4-7 El; 60-90 Brin. For light alloy parts; age-hardenable.

SONDERLEGIERUNG
Vereinigte Leichtmetallwerke, G.m.b.H.
Aluminum. Mn 1, Si 0.2, Fe 0.5, bal Al.
Annealed: 19,000 TS; 14,000 YS; 30 El. For deep drawing parts; corrosion resistant.

SONDERMESSING
German manufacture
Cu, alloy, bal Zn.
For nuts, bolts, hardware; special alloy high tensile brass.

SONDERSTAHL DAUER D-X
Stahlwerke Kabel, C.
C 2.1, Cr 11.5, Si 0.35, Mn 0.3, bal Fe.
For blanking and forming dies, punches; oil hardened, nondeforming.

SONDERSTAHL MIME EXTRA ZH
Stahlwerke Kabel, C.
C 1.05, Si 0.25, Cr 1, W 1.15, Mn 0.8, bal Fe.
For blanking and forming dies; oil hardened, tough.

SONDERSTAHL TCK
Stahlwerke Kabel, C.
C 1, Cr 1.1, Mn 0.07, Si 0.25, bal Fe.
For bearings, liners, sleeves, cutters; water hardened, wear resistant.

SONDERSTAHLE E 5
Kind & Co. Edelstahlwerk
C 0.55, Si 0-0.15, Mn 0.4, bal Fe.
Unalloyed tool steel. W. Nr. 1820.

SONDERSTAHLE E 60 R
Kind & Co. Edelstahlwerk
C 0.6, Si 0.35, Mn 0.7, bal Fe.
Unalloyed tool steel. W. Nr. ~1740.

SONDERSTAHLE EMM 100
Kind & Co. Edelstahlwerk
C 0.75, Si 0.25, Mn 0.7, bal Fe.
Unalloyed tool steel.

SONDERSTAHLE EMM 60
Kind & Co. Edelstahlwerk
C 0.46, Si 0.25, Mn 0.65, bal Fe.
Unalloyed tool steel.

SONDERSTAHLE EMM 70
Kind & Co. Edelstahlwerk
C 0.61, Si 0.25, Mn 0.65, bal Fe.
Unalloyed tool steel.

SONDERSTAHLE F 100
Kind & Co. Edelstahlwerk
C 0.75, Si 0.25, Mn 0.7, bal Fe.
Unalloyed tool steel.

SONDERSTAHLE F 80
Kind & Co. Edelstahlwerk
C 0.61, Si 0.25, Mn 0.65, bal Fe.
Unalloyed tool steel.

SONDERSTAHLE K 85 CR
Kind & Co. Edelstahlwerk
C 0.85, Si 0.25, Mn 0.25, Cr 0.2, bal Fe.
Unalloyed tool steel.

SORCERY B1912, ETC.
Now B1912, ETC.

SOREL'S ALLOY-1
English manufacture
Zn 80, Cu 10, Fe 10.
For bearings.

SOREL'S ALLOY-2
English manufacture
Zn 98, Cu 1, Fe 1.
For bearings.

SORELMETAL D-1
Qit-Fer et Titane Inc.
Cast iron. C 2.85, Si 0.18, Mn 0.009, P 0.027, S 0.02, bal Fe.
Pig iron for ductile iron castings.

SORELMETAL F-1
Qit-Fer et Titane Inc.
Cast iron. C 4.3, Si 0.18, Mn 0.009, P 0.027, S 0.006, bal Fe.
Pig iron for ductile iron castings.

SORELMETAL S-100
Qit-Fer et Titane Inc.
Cast iron. C 4.05, Si 1.1, Mn 0.009, P 0.027, S 0.008, bal Fe.
Pig iron for ductile iron castings.

SORMITE
Central Institute of Metals
Cr 30, Si 4.5, Mn 1.1, C 3.2, Ni 5.3, bal Fe.
To replace certain Stellite alloys, hard facing. Hard alloy, wear and corrosion resistant.

SORMITE
Russian manufacture
C 1.75, Cr 15.5, Ni 2, Si 2, bal Fe.
For blanking and forming dies, gages, punches; air hardened, non-deforming.

SOS6-6
Russian manufacture
C, alloy, bal Fe.
Bearings for high speeds and pressures; anti-friction.

SOUDINOX 1330
Creusot-Loire
C 0-0.03, Mn 1.5, Si 0-1, Cr 20.5, Ni 10.2, Mo 2.9, bal Fe.
Austenitic stainless steel; weldable grade. AFNOR Z2CND20-10; W. Nr. 1.4428.

SOUDINOX 308L
Creusot-Loire
C 0-0.03, Mn 1, Si 0-1, Cr 19.5, Ni 10.75, bal Fe.
Austenitic stainless steel; weldable grade. AFNOR Z2CN 20-10; low carbon modification of AISI 308.

SOUDINOX 308L TM
Creusot-Loire
C 0-0.025, Mn 1.75, Si 0-1, Cr 19.5, Ni 10.75, bal Fe.
Austenitic stainless steel; weldable grade. AFNOR Z1CN 20-10; AWS ER 308L.

SOUDINOX 308W
Creusot-Loire
C 0-0.02, Mn 1, Si 0-1, Cr 19.5, Ni 11, bal Fe.
Austenitic stainless steel; weldable grade. AFNOR Z1CN20-10; AWS ER308L.

SOUDINOX 309
Creusot-Loire
C 0-0.08, Mn 2, Si 0-1, Cr 24, Ni 13.5, bal Fe.
Austenitic stainless steel; weldable grade. AFNOR Z10CN24-13; AWS ER309.

SOUDINOX 309L
Creusot-Loire
C 0-0.03, Mn 1.5, Si 0-1, Cr 24.5, Ni 13, bal Fe.
Austenitic stainless steel; weldable grade. AFNOR Z2CN24-13; AWS ER309L.

SOUDINOX 310
Creusot-Loire
C 0-0.1, Mn 2.2, Si 0-1, Cr 27, Ni 21, bal Fe.
Austenitic stainless steel; weldable grade. AFNOR Z12CN25-20; AWS ER310.

SOUDINOX 312
Creusot-Loire
C 0-0.15, Mn 2, Si 0-1, Cr 30, Ni 9.5, bal Fe.
Austenitic stainless steel, weldable grade. AFNOR Z12CN30-10; AWS ER312.

SOUDINOX 316L
Creusot-Loire
C 0-0.03, Mn 1.2, Si 0-1, Cr 17.5, Ni 11, Mo 2.6, bal Fe.
Austenitic stainless steel; weldable grade. AFNOR Z2CND18.11; AISI 316L.

SOUDINOX 316L TM
Creusot-Loire
C 0-0.02, Mn 1.75, Si 0-1, Cr 19, Ni 13.25, Mo 2.6, bal Fe.
Austenitic stainless steel, weldable grade. AFNOR Z2CND19-13; AWS ER316L.

SOUDINOX 410
Creusot-Loire
C 0-0.05, Mn 0.5, Si 0-0.6, Cr 13, bal Fe.
Chrome corrosion resisting steel, low carbon type; weldable.
For corrosion resistance and heat resistance. AFNOR Z3C14.

SOUDINOX 65
Creusot-Loire
C 0-0.02, Mn 7, Si 0-1, Cr 25.5, Ni 20, bal Fe.
Austenitic stainless steel; weldable grade. AFNOR Z2CNM 25-20.

SOUDINOX A
Creusot-Loire
C 0-0.03, Mn 1, Si 0-1, Cr 20.5, Ni 8, Mo 2.5, Cu 1.5, bal Fe.
Austenitic stainless steel; weldable grade. AFNOR Z5CNDU21-08.

SOUDINOX B6
Creusot-Loire
C 0-0.02, Mn 1.2, Si 0-1, Cr 20, Ni 25.5, Mo 4.3, Cu 1.6, bal Fe.
Austenitic stainless steel; weldable grade. AFNOR Z1NCDU25-20.

SOUDINOX S1
Creusot-Loire
C 0-0.02, Mn 6.5, Si 3.7, Cr 19.5, Ni 14, bal Fe.
Austenitic stainless steel; weldable grade. AFNOR Z2CNMS20-14.

SP
W. Ossenberg & Cie Edelstahlwerke
C 0.7, Mn 0.7, Si 1.7, bal Fe.
Water hardenable to about 54 Rock C. For collets, chuck jaws, clamps, screw drivers. W.-Nr. 1.2823.

SP 35X
Hewitt Metals Corp.
Sn 35, bal Pb.
Block solder. Melt range: 358-460°F.

SP 40X
Hewitt Metals Corp.
Sn 40, bal Pb.
Block solder. Melt range: 358-448°F.

SP-3
TRW Inc.
C 2.5, Si 0-1, Cr 20, Ni 40, W 6, Co 12, bal Fe.
For diesel engine valves.

SP10
Le Bronze Industriel
Sn 10, Pb 9, bal Cu.
Leaded bronze; cast or wrought: 75 Brin. AFNOR UE10 Pb8; SAE 64.

SP10
Sintered Products Ltd.
Alloy steel, sintered. 5.9-6.2 g/cm^3 density; 44 kg/mm^2 TS; 1.52-2 El; 40 Rock B. Meets SAE 866A; ASTM B426 Type 1 Grade 3.

SP15
Le Bronze Industriel
Sn 7.5, Pb 15, bal Cu.
Copper-tin-lead alloy; cast or wrought: 45 Brin. AFNOR U Pb15 E8.

SP20
Sintered Products Ltd.
Cu-Ni-C steel, sintered. 6.6-6.8 g/cm^3 density; 39-47 kg/mm^2 TS; 1-3 El; 200 Vickers. Hardenable to 35 Rock C approx. For clutch transmission parts, hubs, levers. BSS A401; MPIE FN-0310-T.

SP22
Le Bronze Industriel
Sn 5, Pb 18, bal Cu.
Copper-lead-tin alloy; cast or wrought: 40 Brin. AFNOR UPb20 E5.

SP22

Sintered Products Ltd.
Cu-Ni-C steel, sintered. 6.8-6.9 g/cm³ density; 63 kg/mm² TS; 1 El; 300 Vickers. High duty structural components. BSS A402; MPIE FN-0508-T.

SP24

Sintered Products Ltd.
Nickel alloy steel, sintered. 6.8 g/cm³ density min; 39 kg/mm² TS; 4 El min; 40 Rock B. Can be hardened or case hardened. BSS A400; MPIE FN-0550-T.

SP25

Sintered Products Ltd.
Nickel alloy steel, sintered. 6.8-7.2 g/cm³ density; 41 kg/mm² TS; hardenable to 35 Rock C approx. BSS A401; MPIE FN-0310-T.

SP26

Sintered Products Ltd.
Cu-Ni-C steel, sintered. 6.8 g/cm³ density; 47 kg/mm² TS; 1-3 El; 200 Vickers. Gears for oil pumps, diesel engines. BSS A400; MPIE FN-0500-T.

SP5

Le Bronze Industriel
Sn 10, Pb 5, 0.3 P min, bal Cu.
Leaded bronze; cast or wrought: 75 Brin. AFNOR UE10 PB5.

SPA ALLOY

Abex Corp.
Stainless steel. C 1.6, Cr 28, Ni 2, Mo 2, Cu 1, Si 0-2, M 0-1.5, bal Fe.
Cast alloy, hardenable to 600 Brin. Very good abrasion resistance; excellent corrosion-erosion resistance in alumina slurry at pH 2.5 to pH 11.

SPANAL 320

Kreidler Werke G.m.b.H.
Cu 2.5-5, Mg 0.2-1.8, Mn 0.3-1.5, Pb 0.5-2.5, Sn, Cd, Bi, bal Al.
Annealed: 27,000 TS; 11,000 YS; 22 El; 47 Brin. Heat treated: 72,000 TS; 57,000 YS; 130 Brin. For screw machine products, aircraft parts; free-cutting, age-hardenable.

SPANASIL

Kreidler Werke G.m.b.H.
Aluminum. Mg 0.8, Si 1, Pb 1, bal Al.
20-28 kp/mm² TS; 10-18 kp/mm² YS; 8-10 El. Al Mg Si Pb.

SPANG CHALFANT 1

National Supply Co.
C 0.15, Cr 2, Mn 0.5, Mo 0.5, bal Fe.
Rolled: 66,500 TS; 40,500 YS; 40 El; 131 Brin. For gears, shafts, rolls; case hardened.

SPANG CHALFANT 2

National Supply Co.
C 0.1-0.2, Mn 0.5, bal Fe.
Rolled: 62,400 TS; 36 El; 126 Brin. For gears, shafts, rolls; case hardened.

SPANG CHALFANT 3

National Supply Co.
C 0.1-0.2, Mo 0.45-0.65, bal Fe.
Rolled: 64,000 TS; 32,500 YS; 37 El; 126 Brin. For gears, shafts, rolls; case hardened.

SPANVAL 320

English manufacture
C, alloy, bal Fe.
For engineering structures; strong.

SPARE'S MANGANESE BRONZE

American Manganese Bronze Co.
Mn, bal Cu.
70,000 TS; 35,000 YS; 28 El; 30 RA. For propeller blades, gun mounts, gears, worm wheels; CYP 28,000. *Obsolete*

SPARK PLUG METAL

International Nickel Inc.
Ni 94, Mn 2-6, traces Fe + Cu.
74,000 TS; 24,000 YS; 51 El; 60 RA. For spark plug wire; P.L. 20,000; "D" Nickel. *Obsolete*

SPARKALOY

Wilbur B. Driver Co.
Si 1, Mn 4, bal Fe.
For spark plugs; wear resistant. *Obsolete*

SPARKONITE 1

CMW Inc.
Cu 95, 5% BaCo₃.
Sintered material for EDM electrodes. (Electrical discharge machining). *Obsolete*

SPARKONITE 10

CMW Inc.
W 68, Cu 32.
Sintered and infiltrated material for electrodes for EDM (electrical discharge machining). *Obsolete*

SPARKONITE 2A

CMW Inc.
Cu 97.2, 2.8% ZrO₂.
Sintered material for EDM electrodes. *Obsolete*

SPARKONITE 3

CMW Inc.
Special graphite base composition. For EDM electrodes. *Obsolete*

SPARKONITE 7

CMW Inc.
W 70, Cu 30.
Sintered material for EDM electrodes. *Obsolete*

SPARTA (A2)

Cytemp Specialty Steel Div.
Tool material. C 1.05, Cr 5, Mo 0.9-1.25, V 0.25, bal Fe.
For cutting tools, dies, punches; air hardened.

SPARTA 80

Cytemp Specialty Steel Div.
C 0.8, Mn 0.65, Cr 5.25, Mo 1.1, V 0.25, bal Fe.
For blanking and forming dies, punches; air hardening. *Obsolete*

SPARTA CV

Cytemp Specialty Steel Div.
C 2.35, Cr 5.25, V 4, Mo 1, W 1, bal Fe.
Hardened: 65-67 Rock C. For blanking and forming dies, lamination and coining dies; cold shear blades, plug gauges. Type A7 cold work die steel. Abrasion resistant. *Obsolete*

SPARTAN REDHEUGH 30 OAK TYPE 1

Spartan Redheugh Ltd.
C 0-0.22, Si 0.1-0.6, Mn 0-1.5, Cr 0-0.5, S 0-0.04, P 0-0.04, Nb 0-0.08, bal Fe.
61.4 kgf/mm² TS min; 44.1 kgf/mm² YS min; 17 El; 2.8 kgf·m IS (Charpy V-notch, at 0°C). General structural steel suitable for fusion welding. For crane jibs, earthmoving equipment, pontoon bridges.

SPARTAN REDHEUGH 30 OAK TYPE 2

Spartan Redheugh Ltd.
C 0-0.22, Si 0.1-0.6, Mn 0-1.5, Cr 0-0.5, S 0-0.04, P 0-0.04, Nb 0-0.08, bal Fe.
56.7 kgf/mm² TS min; 39.4 kgf/mm² YS min; 20 El; 8.3 kgf·m IS (Charpy V-notch, at 0°C). General structural steel suitable for fusion welding. For crane jibs, earthmoving equipment, pontoon bridges.

SPARTAN REDHEUGH 301S21

Spartan Redheugh Ltd.
Stainless steel. C 0-0.15, Si 0-1, Mn 0-2, P 0-0.045, S 0-0.03, Cr 16-18, Ni 6-8, bal Fe.
Austenitic stainless steel.

SPARTAN REDHEUGH 304S11

Spartan Redheugh Ltd.
Stainless steel. C 0-0.03, Si 0-1, Mn 0-2, P 0-0.045, S 0-0.03, Cr 17-19, Ni 9-12, bal Fe.
Austenitic stainless steel.

SPARTAN REDHEUGH 304S15

Spartan Redheugh Ltd.
Stainless steel. C 0-0.06, Si 0-1, Mn 0-2, P 0-0.045, S 0-0.03, Cr 17.5-19, Ni 8-11, bal Fe.
Austenitic stainless steel.

SPARTAN REDHEUGH 304S16

Spartan Redheugh Ltd.
Stainless steel. C 0-0.06, Si 0-1, Mn 0-2, P 0-0.045, S 0-0.03, Cr 17.5-19, Ni 9-11, bal Fe.
Austenitic stainless steel.

SPARTAN REDHEUGH 304S31

Spartan Redheugh Ltd.
Stainless steel. C 0-0.07, Si 0-1, Mn 0-2, P 0-0.045, S 0-0.03, Cr 17-19, Ni 8-11, bal Fe.
Austenitic stainless steel.

SPARTAN REDHEUGH 305S19

Spartan Redheugh Ltd.
Stainless steel. C 0-0.1, Si 0-1, Mn 0-2, P 0-0.045, S 0-0.03, Cr 17-19, Ni 11-13, bal Fe.
Austenitic stainless steel.

SPARTAN REDHEUGH 309S24

Spartan Redheugh Ltd.
Stainless steel. C 0-0.15, Si 0-1, Mn 0-2, P 0-0.045, S 0-0.03, Cr 22-25, Ni 13-16, bal Fe.
Austenitic stainless steel.

SPARTAN REDHEUGH 310S24

Spartan Redheugh Ltd.
Stainless steel. C 0-0.15, Si 0-1, Mn 0-2, P 0-0.045, S 0-0.03, Cr 23-26, Ni 19-22, bal Fe.
Austenitic stainless steel.

SPARTAN REDHEUGH 315S16

Spartan Redheugh Ltd.
Stainless steel. C 0-0.07, Si 0-1, Mn 0-2, P 0-0.045, S 0-0.03, Cr 16.5-18.5, Mo 1.25-1.75, Ni 9-11, bal Fe.
Austenitic stainless steel.

SPARTAN REDHEUGH 316S11

Spartan Redheugh Ltd.
Stainless steel. C 0-0.03, Si 0-1, Mn 0-2, P 0-0.045, S 0-0.03, Cr 16.5-18.5, Mo 2-2.5, Ni 11-14, bal Fe.
Austenitic stainless steel.

SPARTAN REDHEUGH 316S13

Spartan Redheugh Ltd.
Stainless steel. C 0-0.03, Si 0-1, Mn 0-2, P 0-0.045, S 0-0.03, Cr 16.5-18.5, Mo 2.5-3, Ni 11.5-14.5, bal Fe.
Austenitic stainless steel.

SPARTAN REDHEUGH 316S31

Spartan Redheugh Ltd.
Stainless steel. C 0-0.07, Si 0-1, Mn 0-2, P 0-0.045, S 0-0.03, Cr 16.5-18.5, Mo 2-2.5, Ni 10.5-13.5, bal Fe.
Austenitic stainless steel.

SPARTAN REDHEUGH 316S33

Spartan Redheugh Ltd.
Stainless steel. C 0-0.07, Si 0-1, Mn 0-2, P 0-0.045, S 0-0.03, Cr 16.5-18.5, Mo 2.5-3, Ni 11-14, bal Fe.
Austenitic stainless steel.

SPARTAN REDHEUGH 317S12

Spartan Redheugh Ltd.
Stainless steel. C 0-0.03, Si 0-1, Mn 0-2, P 0-0.045, S 0-0.03, Cr 17.5-19.5, Mo 3-4, Ni 14-17, bal Fe.
Austenitic stainless steel.

SPARTAN REDHEUGH 317S16
Spartan Redheugh Ltd.
Stainless steel. C 0-0.06, Si 0-1, Mn 0-2, P 0-0.045, S 0-0.03, Cr 17.5-19.5, Mo 3-4, Ni 12-15, bal Fe.
Austenitic stainless steel.

SPARTAN REDHEUGH 320S31
Spartan Redheugh Ltd.
Stainless steel. C 0-0.08, Si 0-1, Mn 0-2, P 0-0.045, S 0-0.03, Cr 16.5-18.5, Mo 2-2.5, Ni 11-14, Ti = 5 x C (0.8 max), bal Fe.
Austenitic stainless steel.

SPARTAN REDHEUGH 320S33
Spartan Redheugh Ltd.
Stainless steel. C 0-0.08, Si 0-1, Mn 0-2, P 0-0.045, S 0-0.03, Cr 16.5-18.5, Mo 2.5-3, Ni 11.5-14.5, Ti = 5 x C (0.80 max), bal Fe.
Austenitic stainless steel.

SPARTAN REDHEUGH 321S31
Spartan Redheugh Ltd.
Stainless steel. C 0-0.08, Si 0-1, Mn 0-2, P 0-0.045, S 0-0.03, Cr 17-19, Ni 9-12, Ti = 5 x C (0.80 max), bal Fe.
Austenitic stainless steel.

SPARTAN REDHEUGH 347S31
Spartan Redheugh Ltd.
Stainless steel. C 0-0.08, Si 0-1, Mn 0-2, P 0-0.045, S 0-0.03, Cr 17-19, Ni 9-12, Nb = 10 x C (1.0 max), bal Fe.
Austenitic stainless steel.

SPARTAN REDHEUGH 403S17
Spartan Redheugh Ltd.
Stainless steel. C 0-0.08, Si 0-1, Mn 0-1, P 0-0.4, S 0-0.03, Cr 12-14, Ni 0-1, bal Fe.
Ferritic stainless steel.

SPARTAN REDHEUGH 405S17
Spartan Redheugh Ltd.
Stainless steel. C 0-0.08, Si 0-1, Mn 0-1, P 0-0.4, S 0-0.03, Cr 12-14, Ni 0-1, Al 0.1-0.3, bal Fe.
Ferritic stainless steel.

SPARTAN REDHEUGH 409S19
Spartan Redheugh Ltd.
Stainless steel. C 0-0.08, Si 0-1, Mn 0-1, P 0-0.4, S 0-0.03, Cr 10.5-12.5, Ni 0-1, Ti = 6 x C (1.0 max), bal Fe.
Ferritic stainless steel.

SPARTAN REDHEUGH 410S21
Spartan Redheugh Ltd.
Stainless steel. C 0.09-0.15, Si 0-1, Mn 0-1, P 0-0.4, S 0-0.03, Cr 11.5-13.5, Ni 0-1, bal Fe.
Martensitic stainless steel.

SPARTAN REDHEUGH 420S45
Spartan Redheugh Ltd.
Stainless steel. C 0.28-0.36, Si 0-1, Mn 0-1, P 0-0.4, S 0-0.03, Cr 12-14, Ni 0-1, bal Fe.
Martensitic stainless steel.

SPARTAN REDHEUGH 430S17
Spartan Redheugh Ltd.
Stainless steel. C 0-0.08, Si 0-1, Mn 0-1, P 0-0.4, S 0-0.03, Cr 16-18, Ni 0-1, bal Fe.
Ferritic stainless steel.

SPARTAN REDHEUGH 434S17
Spartan Redheugh Ltd.
Stainless steel. C 0-0.08, Si 0-1, Mn 0-1, P 0-0.4, S 0-0.03, Cr 16-18, Mo 0.9-1.3, Ni 0-1, bal Fe.
Ferritic stainless steel.

SPARTAN-5
Atlas Specialty Steels
C 0.5, W 18, Cr 4, V 1, bal Fe.
For cutting tools, shear blades; high speed steel, hot work tools. *Obsolete*

SPARTAN-7
Atlas Specialty Steels
C 0.7, Mn 0.25, W 18, Cr 4, V 1.2, bal Fe.
For hot shear blades, drills, reamers, tips, broaches; high speed steel. *Obsolete*

SPEAR "5-6-2"
Spear & Jackson (Industrial) Ltd.
C 0.83, Cr 4.25, W 6, Mo 5, V 2, bal Fe.
Molybdenum-tungsten high speed steel for cutting tools. AISI M-2.

SPEAR 3S
Spear & Jackson (Industrial) Ltd.
C 0.8, S 0.25, Mn 0.3, Ni 2.4, Cr 0.2, bal Fe.
Plate or sheet for saw blades.

SPEAR 50
Spear & Jackson (Industrial) Ltd.
C 0.55, Mn 0.6, bal Fe.
Oil or water hardening steel; for clipper tools, bolsters, pressure plates.

SPEAR 75
Spear & Jackson (Industrial) Ltd.
C 0.75, Mn 0.7, bal Fe.
Oil hardening steel; for press brakes, shear blades, drifts, machine parts.

SPEAR A.H.C.
Spear & Jackson (Industrial) Ltd.
C 0.5, Si 1, Cr 1.5, W 2, V 0.12, bal Fe.
Oil hardening steel for pneumatic tools, chisels and shock resisting tools and parts. AISI S1.

SPEAR B.1
Spear & Jackson (Industrial) Ltd.
C 0.32, Ni 4.25, Cr 1.25, Mo 0.3, bal Fe.
Oil or air hardenable tool steel for plastic molds.

SPEAR B.4
Spear & Jackson (Industrial) Ltd.
C 0.32, Si 0.3, Mn 0.6, Cr 2.1, Mo 0.25, V 0.15, bal Fe.
Alloy steel for nitriding.

SPEAR CRH
Spear & Jackson (Industrial) Ltd.
C 0.75, Si 0.25, Mn 0.6, Cr 0.4, bal Fe.
Plate or sheet for saw blades.

SPEAR CV
Spear & Jackson (Industrial) Ltd.
C 0.78, Si 0.25, Mn 0.6, Cr 0.45, V 0.15, bal Fe.
Plate or sheet for saw blades.

SPEAR D.1
Spear & Jackson (Industrial) Ltd.
C 0.5, Mn 0.7, Cr 1, V 0.25, bal Fe.
Oil hardening steel for die casting dies, stamping and forging dies, hot shear blades. AISI L2.

SPEAR D.12
Spear & Jackson (Industrial) Ltd.
C 2, Cr 13, bal Fe.
High carbon, high chrome type tool steel. Oil hardenable; for abrasion and corrosion resisting tools and dies. AISI D3.

SPEAR D.13
Spear & Jackson (Industrial) Ltd.
C 1.9, Cr 13.5, Mo 1.3, bal Fe.
High carbon, high chrome type tool steel. Air or oil hardening for heavy duty use, abrasion and corrosion resisting tools and dies. Similar to AISI D3.

SPEAR D.14
Spear & Jackson (Industrial) Ltd.
C 1.3, Cr 13, Mo 1.5, Co 2.75, bal Fe.
High carbon, high chrome type tool steel. Air or oil hardening for heavy duty use, abrasion and corrosion resisting tools and dies. AISI D5.

SPEAR D.15
Spear & Jackson (Industrial) Ltd.
C 1, Mn 0.7, Cr 5, Mo 1, V 0.25, bal Fe.
Air hardening tool steel for blanking and forming dies, shear blades, gauges. AISI A2.

SPEAR D.17
Spear & Jackson (Industrial) Ltd.
C 1.5, Cr 12, V 0.9, Mo 0.9, bal Fe.
High carbon, high chromium type tool steel. Air or oil hardening for abrasion and corrosion resisting tools and dies. AISI D2.

SPEAR D.5
Spear & Jackson (Industrial) Ltd.
C 0.35, Si 1, Cr 5.75, V 1, Mo 1.5, bal Fe.
Oil hardening hot work tool steel; for die casting dies, gripping and forming tools, hot shear blades. Chromium type hot work tool steel. AISI H13.

SPEAR D.7
Spear & Jackson (Industrial) Ltd.
C 0.32, Si 1, W 1.5, Cr 5, V 0.4, Mo 1.5, bal Fe.
Air hardening tool steel for gripper dies, mandrels, header and trimmer dies. Chromium type hot work tool steel. AISI H12.

SPEAR D.9
Spear & Jackson (Industrial) Ltd.
C 0.3, W 9.5, Cr 3.2, V 0.3, bal Fe.
Tungsten type hot work tool steel for forging dies, header tools, extrusion dies. AISI H21.

SPEAR D.X
Spear & Jackson (Industrial) Ltd.
C 0.28, Cr 0.85, W 2.2, V 0.25, Mo 0.5, Ni 2.25, bal Fe.
Oil hardening hot work tool steel for tools requiring maximum toughness.

SPEAR EXM
Spear & Jackson (Industrial) Ltd.
C 0.73, Si 0.25, Mn 0.4, Ni 1.4, Cr 0.2, bal Fe.
Plate or sheet for saw blades.

SPEAR M.1
Spear & Jackson (Industrial) Ltd.
C 0.8, Cr 4.25, W 1.25, Mo 8.5, V 1.15, bal Fe.
Molybdenum high speed steel for cutting tools. AISI M1.

SPEAR M.42
Spear & Jackson (Industrial) Ltd.
C 1.05, Cr 4.25, W 1, Mo 9, V 1.1, Co 8, bal Fe.
Cobalt-molybdenum high speed steel for cutting tools. AISI M42.

SPEAR N.S.
Spear & Jackson (Industrial) Ltd.
C 0.95, Mn 1.2, Cr 0.5, W 0.5, V 0.15, bal Fe.
Oil hardening tool steel for jigs and gauges, cold trimming dies, punches, shear blades. AISI O1.

SPEAR NO. 2 GRADE 1
Spear & Jackson (Industrial) Ltd.
C 1.05, Mn 0.2, bal Fe.
Water hardening tool steel; for taps, twist drills, lathe centers, shear blades. AISI W1.

SPEAR NO. 2 GRADE 2
Spear & Jackson (Industrial) Ltd.
Cu 0.95, Mn 0.2, bal Fe.
Water hardening tool steel; for cold heading dies, punches, woodworking tools. AISI W1.

SPEAR NO. 2 GRADE 3
Spear & Jackson (Industrial) Ltd.
Cu 0.85, Mn 0.2, bal Fe.
Water hardening tool steel; for pneumatic hammers and chisels, press tools, mason's tools. AISI W1.

SPEAR NO. 2 VANADIUM
Spear & Jackson (Industrial) Ltd.
C 1, Mn 0.2, V 0.25, bal Fe.
Water hardening tool steel; for blanking dies, punches, hand stamps, chuck jaws.

SPEAR P.S.
Spear & Jackson (Industrial) Ltd.
C 1.05, Mn 0.6, Cr 1.5, Mo 0.25, bal Fe.
Oil hardening tool steel; for forming and blanking tools, gauges, taps, bushings. AISI L4.

SPEAR TC
Spear & Jackson (Industrial) Ltd.
C 0.45, Si 0.25, Mn 0.5, Ni 1.5, Cr 0.15, Mo 0.3, bal Fe.
Plate or sheet for saw blades.

SPEAR VANCHIP
Spear & Jackson (Industrial) Ltd.
C 0.68, Si 1, Cr 8, Ni 1.5, V 1.4, Mo 1.4, bal Fe.
Air hardenable tool steel for chipper knives, veneer knives, shear blades.

SPECIAL
Edgar T. Ward's Sons Co.
C 0.7, W 18, Cr 4, V 2, bal Fe.
For tools, cutters; high speed steel.

SPECIAL
Manufacturer not listed.
C 0.7, Cr 4, V 1, W 1.5, Mo 9.5, bal Fe.
For cutting tools, drills; high speed steel.

SPECIAL
Electro-Steel Co.
C 0.8, V 0.2, Cr 0.8, bal Fe.
For tools, cutters; oil hardened.

SPECIAL
Hidalgo Steel Co. Inc.
C, W, bal Fe.
For dies; high duty shafts; hot dies; oil hardened.

SPECIAL
Jessop Steel Co.
C 0.7, W 18, Cr 4, V 1, bal Fe.
For drilling tools, reamers, cutters, broaches, lathe tools; high speed steel. *Obsolete*

SPECIAL "H.M." OIL HARDENING STEEL
Jessop Steel Co.
C 0.9, Mn 1.7, V 0.2, bal Fe.
For cutting tools, drawing and blanking dies, punches, form cutters; maximum hardness and superior cutting qualities.
Obsolete

SPECIAL 10 T
Teledyne Vasco
C 1, V 0.12, bal Fe.
Annealed: 70,000 TS; 45,000 YS; 25 El; 57 RA; 170 Brin. Heat treated: 258,000 TS; 240,000 YS; 5 RA; 465 Brin. For tools, drills, taps; water hardening. *Obsolete*

SPECIAL 8 T
Teledyne Vasco
C 0.8, V 0.12, bal Fe.
Annealed: 70,000 TS; 45,000 YS; 25 El; 57 RA; 170 Brin. Heat treated: 258,000 TS; 240,000 YS; 5 RA; 465 Brin. For tools for general purposes; water hardening. *Obsolete*

SPECIAL ALLOY
Edgar T. Ward's Sons Co.
C 0.7, Mn 1, Cr 0.4, bal Fe.
For tools, dies; water hardened.

SPECIAL ALLOY (DISSTON)
Disston Inc.
C 0.45, Cr 0.8, bal Fe.
For tools, punches, dies; water hardened. *Obsolete*

SPECIAL ALLOYS
Wallace Murray Corp.
C, alloy, bal Fe.
For tools.

SPECIAL ALUMINUM
American manufacture
Al 10.5-10.7, Fe 3.1-4.25, Ni 3.6, Mn 0-1.6, bal Cu.
200-300 Brin. For worm wheels, gears, trolley shoes; heat treatable.

SPECIAL ARDHO N.H.O.
Spencer Clark Metal Industries Ltd.
C 0.35, Si 0.25, Mn 0.5, Ni 3.2, Cr 0.78, bal Fe.
Nickel-chromium tool steel, for smiths' tools, punches, dies.

SPECIAL BB
Jessop-Saville Ltd.
C 0.28, Cr 3.2, V 0.3, W 9.5, bal Fe.
Hardened: 243,000 TS; 215,000 YP; 12 El; 37 RA; 50 Rock C.
For die casting dies for brass and aluminum bronze, extrusion dies and liners, hot forging dies. Oil or air hardening; AISI Type H21.

SPECIAL BOLT DIE
Midvale-Heppenstall Co.
C 0.35, Cr 1.8, W 11, bal Fe.
For hot work dies, bolt headers, gripper and forging dies; hot work steel. *Obsolete*

SPECIAL CARBON
Latrobe Steel Co.
C 1, Mn 0.25, Cr 0-0.1, bal Fe.
For drills, tools, taps; water hardened. *Obsolete*

SPECIAL CARBON
Midvale-Heppenstall Co.
C 0.6-1.2, bal Fe.
For tools, drills, punches; water hardened. *Obsolete*

SPECIAL CARBON SSC
Darwin & Milner Inc.
Tool material. C 0.95, Ni 0.35, V, bal Fe.
For tools, punches; water hardened.

SPECIAL CHROME VANADIUM
Edgar Allen Balfour Ltd.
C 0.6, Cr 1, V 0.2, Mn 0.5, bal Fe.
Oil hardened. For rivet sets and dies, hot punch and diework, thread rolling; shock resistant. *Obsolete*

SPECIAL CHROME VANADIUM
A. Milne & Co.
C 0.6, Cr 0.9-1, V 0.2, bal Fe.
For tools, punches; oil hardening.

SPECIAL COLD PRESSING VANADIUM STEEL
Duraloy Blaw-Knox/Union Steel Casting
C 1.1, Si 0.2, Mn 0.3, V 0.2, bal Fe.
For cold heading dies; swaging dies; water hardened, shock resistant.

SPECIAL CONQUEROR
Joseph Beardshaw & Son Ltd.
C 0.85-1.1, bal Fe.
Heat treated: 190,000 TS; 145,000 YS; 10 El; 30 RA; 400 Brin.
For rock drills, punches, taps, shear blades; water hardened; Type W1.

SPECIAL D
Hidalgo Steel Co. Inc.
C, W, bal Fe.
For tools; oil hardened.

SPECIAL DOUBLE CROWN
British Steel Corp.
C 0-0.15, Mn 0.6, Si 0.2, 0.03% max S + P, bal Fe.
Rolled: 50,000-63,000 TS; 30,000-36,000 YS; 25 El. For locomotive fire box parts; good weldability. *Obsolete*

SPECIAL DRACO
Cytemp Specialty Steel Div.
C 0.7, Cr 1.4, V 0.2, bal Fe.
For tools, dies, drills; water hardened. *Obsolete*

SPECIAL DYNAMO
Follansbee Steel Co.
C 0.04, Si 3.2, bal Fe.
Annealed: 75,000 TS; 55,000 YS; 12 El. For induction and power and radio transformers; low loss. *Obsolete*

SPECIAL FA71
Creusot-Loire
C 0.4, Cr 3.15, Mo 1, V 0.2, bal Fe.
Hot work tool steel for forging and hot forming dies; good thermal shock resistance. AFNOR 40 CDV 12; similar to AISI S7.

SPECIAL FA82
Creusot-Loire
C 0.38, Si 1, Cr 5.25, Mo 1.35, V 0.5, bal Fe.
Hot work tool steel for forging and upsetting dies, and forming dies. AFNOR Z38 CDV5; AISI H 11.

SPECIAL G
D.G. Gautier & Co.
C 0.7-1.2, bal Fe.
For tools; water hardened.

SPECIAL HAN-10
Teledyne Firth Sterling
Tool material. WC 90, Ni 10.
For seal rings. 89.5 Rock A.

SPECIAL HAN-6
Teledyne Firth Sterling
Tool material. WC 94, Ni 6.
For corrosion wear applications. 90.5 Rock A.

SPECIAL HAR
Teledyne Firth Sterling
Tool material.
For platelet wear applications. 91.5 Rock A.

SPECIAL HAR-13
Teledyne Firth Sterling
Tool material.
For platelet wear applications. 88.5 Rock A.

SPECIAL HAR-7
Teledyne Firth Sterling
Tool material.
For platelet wear applications. 92.5 Rock A.

SPECIAL HDS
J.J. Saville & Co. Ltd.
C 0.28, Cr 3.2, V 0.3, W 9.5, Mo 0.3, bal Fe.
For casting dies, hot swaging dies; hot die steel.

SPECIAL HIGH CARBON 51-V
Jessop Steel Co.
C 1.1, Cr 3, W 1.25, V 0.2, bal Fe.
For twist drills, taps, dies, milling cutters; oil hardening.
Obsolete

SPECIAL HOT DIE
Darwins Alloy Castings
W 9.5, Cr 2.5, V 0.1-0.15, bal Fe.
For cutting tools, dies; hot-die steels. *Obsolete*

SPECIAL HS-55
Bethlehem Steel Corp.
C 0.5, Cr 4, V 1, W 18, bal Fe.
Air hardened: 60-62 Rock C. Oil hardened: 60-64 Rock C. For extrusion and forging dies, swaging and pressing dies, punches, hammer tools. Type H26; hot-work steel, high red-hardness and toughness. *Obsolete*

SPECIAL HS55
Now BETHLEHEM SPECIAL HS.

SPECIAL INLAY

J.F. Jelenko & Co.
Precious metal. Au 83, Pd 1, Ag 10.
Quenched: 50,000 psi TS; 15,000 psi YS; 35 El; 73 Brin.
Dental alloy, Type I; for soft inlays.

SPECIAL K

Houghton & Richards Inc.
C 2.2, Cr 12, Si 0.6, Mn 0.3, bal Fe.
For punch and die work tools, cutting dies, plug gages;
nondeforming.

SPECIAL K

George Cook & Co., Ltd.
C 1.9, Cr 13.5, Si 0.65, Mn 0.3, bal Fe.
For punches, dies, swaging and forming tools; air or oil
hardened; non-deforming.

SPECIAL K

Vereinigte Edelstahlwerke
C 1.9, Cr 13.5, Si 0.65, Mn 0.3, bal Fe.
For punches, dies, swaging and forming tools; air or oil
hardened; non-deforming.

SPECIAL K

U.N. Alloy Steel Corp.
C 1.9, Cr 13.5, Si 0.65, Mn 0.3, bal Fe.
For punches, dies, swaging and forming tools; air or oil
hardened; non-deforming.

SPECIAL K5

Westinghouse Electric Corp.
C 2, Cr 5, bal Fe.
For tools, dies, forming and blanking dies; air hardened,
nondeforming. *Obsolete*

SPECIAL K5

U.N. Alloy Steel Corp.
C 1, Mn 0.7, Si 0.3, Cr 5.3, V 0.15, Mo 1, bal Fe.
Annealed: 103,000 TS; 51,000 YS; 26 El; 18 Rock C. Heat
treated: 253,000 TS; 200,000 YS; 3 El; 53 Rock C. For
blanking and trimming dies, shear blades, broaches, cutters,
punches. Type A2; air hardened tool steel, non-deforming.

SPECIAL KCO

U.N. Alloy Steel Corp.
C 1.27, Cr 12, V 0.35, Co 3, bal Fe.
Heat treated: 180,000-200,000 TS; 1 El. For broaches,
burnishing tools, slitter cutters, shears, chasers, reamers,
hobs, milling cutters. Type D5 tool steel, air hardening, non-
deforming. *Obsolete*

SPECIAL KCO

Vereinigte Edelstahlwerke
C 1.27, Cr 12, V 0.35, Co 3, bal Fe.
Heat treated: 180,000-200,000 TS; 1 El. For broaches,
burnishing tools, slitter cutters, shears, chasers, reamers,
hobs, milling cutters. Type D5 tool steel, air hardening, non-
deforming. *Obsolete*

SPECIAL KMV

Now VEW K110.

SPECIAL KN

Raven Steel & Tool Co.
C 1.5, Cr 12, Mo 1, bal Fe.
For blanking and forming dies, punches; Type D2; air
hardened, non-deforming.

SPECIAL KR

Raven Steel & Tool Co.
C 2.25, Cr 12, bal Fe.
For lamination and blanking dies, punches; Type D3; oil
hardened, non-deforming.

SPECIAL KRM

U.N. Alloy Steel Corp.
C 2.15, Cr 12, V 0.2, Mo 0.8, bal Fe.
Heat treated: C 63-67 Rock. For thread rolling dies, extrusion
dies, gauges, knurlers, shear blades. Type D4 tool steel. Heat
and abrasion resistant. Oil hardening. *Obsolete*

SPECIAL KRM

Vereinigte Edelstahlwerke
C 2.15, Cr 12, V 0.2, Mo 0.8, bal Fe.
Heat treated: C 63-67 Rock. For thread rolling dies, extrusion
dies, gauges, knurlers, shear blades. Type D4 tool steel. Heat
and abrasion resistant. Oil hardening. *Obsolete*

SPECIAL LOHYS

Sankey & Sons Ltd.
Si 0.75, Fe 99.25.
Initial permeability 400. Maximum permeability 5350.
Coercive force 0.85. For electrical equipment, magnetic
instruments. Soft magnetic material.

SPECIAL M

Allegheny Ludlum Steel
C 0.7, Alloy, bal Fe.
For tools, taps, drills. *Obsolete*

SPECIAL M-O HIGH SPEED

Republic Steel Corp.
C 0.79-0.86, W 6-6.75, Cr 4, Mo 5, V 2, bal Fe.
For lathe and planer tools, reamers, broaches, hobs; high
speed steel. *Obsolete*

SPECIAL MANGANESE NICKEL

American Chain & Cable
C 0.4, 12-14 Mn + Ni, bal Fe.
For welding electrode and hard facing. Shielded arc.

SPECIAL MD

Creusot-Loire
C 0.55, Si 1, Cr 1.1, W 2, bal Fe.
Cold work tool steel; shock resisting type. For chisels, rivet
sets, pneumatic tools. AFNOR 55 WCS 20; similar to AISI S1.

SPECIAL MOTOR

Follansbee Steel Co.
C 0.04, Si 2.5, bal Fe.
Annealed: 67,000 TS; 50,000 YS; 15 El. For rotating
equipment, motors. *Obsolete*

SPECIAL MS

Marathon Specialty Steels Inc.
C 0.28, Si 0.5, Mn 0.3, Cr 1.2, Mo 1.4, V 0.25, W 2.3, Co 2.3,
bal Fe.
For die casting dies; extrusion press components. Thermal
shock resistance.

SPECIAL NO. 0

Edgar T. Ward's Sons Co.
C, alloy, bal Fe.
For tools. *Obsolete*

SPECIAL NO. 13

Allegheny Ludlum Steel
C 0.8, bal Fe.
For tools. *Obsolete*

SPECIAL NO. 18

Boyd-Wagner Co.
C 0.45, Cr 1.4, W 2.2, V 0.2, Mo 0.3, bal Fe.
180 Brin. For pneumatic tools, shear blades, chisels, rivet
sets; shock tool steel.

SPECIAL NO. 18

Boyd-Wagner Co.
C 0.45, Cr 1.4, W 2.2, V 0.2, Mo 0.3, bal Fe.
180 Brin. For pneumatic tools, shear blades, chisels, rivet
sets; shock tool steel.

SPECIAL NO. 18S

Boyd-Wagner Co.
C 0.45, Cr 1.4, W 2.2, Mo 0.3, Si 1, V 0.2, bal Fe.
For shear blades, rivet busters, chisels; shock resistant.

SPECIAL NO. 18S

Boyd-Wagner Co.
C 0.45, Cr 1.4, W 2.2, Mo 0.3, Si 1, V 0.2, bal Fe.
For shear blades, rivet busters, chisels; shock resistant.

SPECIAL NO. 19

Boyd-Wagner Co.
C, W, Mn, Cr, bal Fe.
200 Brin. For taps, thread dies; oil hardened.

SPECIAL NO. 20

Boyd-Wagner Co.
C, Cr, W, Mn, V, bal Fe.
240 Brin. For finishing tools; oil hardened.

SPECIAL OIL HARDENING

Edgar T. Ward's Sons Co.
C 0.9, Mn 1.2, bal Fe.
For tools, dies; non-deforming.

SPECIAL OIL HARDENING

Jessop Steel Co.
C 0.9, Mn 1.7, Cr 0.2, bal Fe.
Annealed: 190 Brin. For spindles, hobs, press tools, punches,
crimpers; non-deforming, oil hardened. *Obsolete*

SPECIAL OIL HARDENING

Teledyne Firth Sterling
C 0.9, Mn 1.8, bal Fe.
For tools, dies, taps, hobs, gauges; oil hardening,
nondeforming. *Obsolete*

SPECIAL OILWAY

H. Boker & Co.
C 0.9, Mn 1.6, bal Fe.
For tools, dies, punches, cutters; Type O2; oil hardened.
Obsolete

SPECIAL PISTON STEEL

Teledyne Firth Sterling
Steel. C 1.05, Mn 0.25, bal Fe.
For pistons; water hardened. *Obsolete*

SPECIAL POWER TRANSFORMER NO. 65

Follansbee Steel Co.
C 0.03, Si 4, bal Fe.
Annealed: 73,000 TS; 62,000 YS; 8 El. For power and radio
transformers; high permeability. *Obsolete*

SPECIAL PROCESSED

Jessop Steel Co.
C 1, bal Fe.
For cold heading dies and punches; water hardened.
Obsolete

SPECIAL PUNCH

Latrobe Steel Co.
C 1, W 2, bal Fe.
For punches; tough. *Obsolete*

SPECIAL PUNCH

Midvale-Heppenstall Co.
C 0.6, Si 1.2, bal Fe.
For punches, cold punches. *Obsolete*

SPECIAL PUNCH & CHISEL STEEL

McInnes Steel Co.
C 0.5, Mn 0.25, W 2.15, Cr 1.5, V 0.25, bal Fe.
For punches, chisels, pneumatic tools; tough, shock
resistant. *Obsolete*

SPECIAL PURPOSE ALLOY

Sharon Steel Corp.
C 1.18-1.31, Mn 0.15-0.35, Si 0.25, Cr 0.1-0.25, bal Fe.
Hot rolled: 30-35 Rock C. Cold rolled, annealed: 20 Rock C
max. Hardenable to 65 Rock C max. For hack saw blades,
razor blades, cutlery.

SPECIAL R-2

Jessop-Saville Ltd.
C 0.2, Cr 13, bal Fe.
Hardened: 210,000-140,000 TS; 195,000-120,000 YS; 10-15
El; 461-300 Brin. For plastic molding dies; corrosion resistant.

SPECIAL R-2
Saville & Co. Ltd., J.J.
C 0.2, Cr 13, bal Fe.
Hardened: 210,000-140,000 TS; 195,000-120,000 YS; 10-15
El; 461-300 Brin. For plastic molding dies; corrosion resistant.

SPECIAL SEHR HART
Vereinigte Edelstahlwerke
C 1.4, Cr 0.65, W 4.75, V 0.25, bal Fe.
For tools, dies, cutters; oil hardening. *Obsolete*

SPECIAL SHELL TURNING
English manufacture
C 0.7, W 18, Cr 4, V 1, bal Fe.
For high speed cutting; high speed steel.

SPECIAL SIFBRONZE 22% NICKEL
Suffolk Iron Foundry Ltd.
Ni 22, bal Cu.
Welded: 70,000 TS; 54,000 YS; 4 El; 216 Brin. For welding
rod for stainless steel and cast iron; wear resistant. *Obsolete*

SPECIAL TOOL
Colt Industries
C 0.7-1.2, bal Fe.
For cutters, drills, broaches, dies; water hardened. *Obsolete*

SPECIAL U1
Creusot-Loire
C 0.9, Mn 1.9, Cr 0.5, V 0.1, bal Fe.
Cold work tool steel; oil hardening. For punching and
forming dies, reamers. AFNOR 90 MCV 8.

SPECIAL V
CCS Braeburn Alloy Steel
C 0.6-1.2, V 0.2, Mn 0.25, Si 0.25, bal Fe.
For drills, reamers, punches, taps, knives, razors; Type W1;
water hardened. *Obsolete*

SPECIAL VANADIUM
Colt Industries
C 0.67, Cr 0.45, V 0.2, bal Fe.
For dies, cold punches, impact tools; tough. *Obsolete*

SPECIAL VANADIUM
Columbia Tool Steel Co.
C, alloy, bal Fe.
For tools, dies; oil hardening. *Obsolete*

SPECIAL VANADIUM
A. Milne & Co.
C 0.8-1.1, V 0.1-0.2, bal Fe.
For tools; water hardening.

SPECIAL VANADIUM
Kidd Drawn Steel Co.
C, bal Fe.
For dies, cold punches, impact tools, shear blades; tough
and shock resistant.

SPECIAL VERY HARD
Houghton & Richards Inc.
C 1.1, W 2.8, bal Fe.
For fast finishing tools; oil or water hardening.

SPECIAL W 2
Thyssen Edelstahlwerke AG
C 0.3, Cr 2.2-2.5, V 0.3-0.5, W 2, bal Fe.
For tools, dies; for extrusions. *Obsolete*

SPECIAL W10D1
Columbia Tool Steel Co.
C 0.7-1.2, Cr 0.15, Mn 0.25, bal Fe.
For forming dies, reamers, roll thread dies; water hardened.
Obsolete

SPECIAL W5
Thyssen Edelstahlwerke AG
C 0.3, Cr 2.4, V 0.55, W 4.5, bal Fe.
For tools, dies, extrusion press mandrels, hot punching dies.
Hot work steel, tough. *Obsolete*

SPECIAL WATER-HARDENING TOOL STEEL
Carpenter Technology Corp.
C 0.7-1.25, Mn 0.25, Si 0.25, bal Fe.
Water hardening. For punches, drills, stamps, jigs, bushings.
AISI W1. G10900

SPECIAL WIRE DRAWING DIE
Columbia Tool Steel Co.
C 2, Mn 1.8, Cr 1.85, W 10.5, Mo 0.5, bal Fe.
For drawing dies; for wire. *Obsolete*

SPECIAL WORTLE DIE
Allegheny Ludlum Steel
C 1.2, W 2.5-3.5, Mn 0.7, bal Fe.
For tools, dies. *Obsolete*

SPECIAL-BLOC
Creusot-Loire
C 0.55, Mn 0.75, Ni 0.55, Cr 1, Mo 0.45, V 0.05, bal Fe.
Oil hardening hot work tool steel; for hot forming dies.
AFNOR 55 CNDV 4.

SPECIAL-PRESSE
Creusot-Loire
C 0.2, Ni 3.15, Mo 3.4, bal Fe.
Hot work tool steel for forging dies. AFNOR 20 DN 33-12.

SPECIAL-PURPOSE 0125-8
Sharon Steel Corp.
C 1.18-1.31, Mn 0.15-0.35, Si 0.25, Cr 0.1-0.25, bal Fe.
For razor blades, hack saw blades.

SPECIALLOY
Specialty Steel Co. of America
C 0.51, Mn 0.93, Cr 0.98, Ni 0.52, V 0.22, Mo 0.25, bal Fe.
Low alloy medium carbon oil hardened tool steel.

SPECIALLOY 1026
Specialloy Inc.
Copper. Fe 10, Cu 90.
Low temperature introduction of iron to copper alloys.

SPECIALLOY 1029
Specialloy Inc.
99.9 Cu min.
For iron and steel alloying; addition agent.

SPECIALLOY 1103
Specialloy Inc.
Copper. Li 2, Cu 98.
Deoxidizer for copper alloys.

SPECIALLOY 1105
Specialloy Inc.
Copper. B 2, Cu 98.
Deoxidizer for copper alloys.

SPECIALLOY 1124
Specialloy Inc.
Copper. Cr 1, Cu 99.
Conductivity copper casting alloy.

SPECIALLOY 1124CC
Specialloy Inc.
Copper. Cr 1, Cu 99.
Continous cast extrusion or forging stock of CDA 182, RWMA
Class 2 Alloy.

SPECIALLOY 1213CC
Specialloy Inc.
Aluminum bronze alloy, extrusion and forging stock,
continuous cast billet.

SPECIALLOY 1214
Specialloy Inc.
Si 4, Mn 1.1, bal Cu.
For brass alloying; addition agent; Everdur. *Obsolete*

SPECIALLOY 1224
Specialloy Inc.
Copper. Cr 5, Cu 95.
Master alloy for conductive copper castings.

SPECIALLOY 1314
Specialloy Inc.
Copper. Si 10, Cu 90.
Silicon source for copper alloy addition.

SPECIALLOY 1327
Specialloy Inc.
Copper. Co 10, Cu 90.
Cobalt addition for copper alloys.

SPECIALLOY 1328
Specialloy Inc.
Copper. Nb 1, Ni 10, Cu 89.
Cupro-nickel casting alloy.

SPECIALLOY 1328CC
Specialloy Inc.
Copper. Ni 10, Cu 90.
Continuous cast cupro-nickel extrusion and forging stock;
CDA 706.

SPECIALLOY 1340
Specialloy Inc.
Copper. Zr 13, Cu 87.
Zirconium addition to copper alloys.

SPECIALLOY 1414
Specialloy Inc.
Copper. Si 15, Cu 85.
Silicon addition to copper alloys.

SPECIALLOY 1512
Specialloy Inc.
Copper. Mg 20, Cu 80.
Magnesium additions to copper alloys.

SPECIALLOY 1514
Specialloy Inc.
Copper. Si 20, Cu 80.
Silicon additions to copper alloys.

SPECIALLOY 1714
Specialloy Inc.
Copper. Si 30, Cu 70.
Silicon additions to copper alloys.

SPECIALLOY 1722
Specialloy Inc.
Copper. Ti 30, Cu 70.
Titanium additions to copper alloys.

SPECIALLOY 1725
Specialloy Inc.
Copper. Mn 30, Cu 70.
For brass alloying; addition agent.

SPECIALLOY 1726
Specialloy Inc.
Copper. Fe 30, Cu 70.
Iron additions for copper alloys.

SPECIALLOY 1727
Specialloy Inc.
Copper. Co 30, Cu 70.
Cobalt additions for copper alloys.

SPECIALLOY 1728
Specialloy Inc.
Copper. Nb 1, Ni 30, Cu 69.
High strength weldable grade cupro-nickel casting alloy.

SPECIALLOY 1728CC
Specialloy Inc.
Copper. Ni 30, Cu 70.
Continuous cast cupro-nickel extrusion and forging stock.
CDA 715.

SPECIALLOY 1733
Specialloy Inc.
Copper. As 30, Cu 70.
Arsenic additions to copper alloys.

SPECIALLOY 1826
Specialloy Inc.
Fe 50, Cu 50.
Iron additions for copper alloys.

SPECIALLOY 1827
Specialloy Inc.
Co 50, Cu 50.
Cobalt additions for copper alloys.

SPECIALLOY 1848
Specialloy Inc.
Cd 50, Cu 50.
Cadmium additions for copper alloys.

SPECIALLOY 1852
Specialloy Inc.
Te 50, Cu 50.
Tellurium additions for copper alloys.

SPECIALLOY 1928CC
Specialloy Inc.
Cr 2.5, Ni 30, Cu 67.5.
Continuous cast, chromium modified, cupro-nickel extrusion
and forging stock. CDA 719.

SPECIALLOY 2029
Specialloy Inc.
Cu 30, Ni 70.
Copper plus nickel additions for iron alloys.

SPECIALLOY 2829
Specialloy Inc.
Cu 50, Ni 50.
Nickel addition to copper alloys; also copper-nickel additions
to iron alloys.

SPECIALLOY 2841
Specialloy Inc.
Nb 68, Ni 40.
Niobium (columbium) additions to cupro-nickel alloys.

SPECIALLOY 2842
Specialloy Inc.
Mo 50, Ni 50.
Molybdenum additions to iron alloys.

SPECIALLOY 2929
Specialloy Inc.
Si 1, Mn 1, Fe 2, Cu 30, Ni 66.
Corrosion resistant high strength nickel copper casting alloy.

SPECIALLOY 2929A
Specialloy Inc.
60% min Ni, 30% min Cu.
For iron and steel alloying; addition agent. *Obsolete*

SPECIALLOY 2929B
Specialloy Inc.
Si 2, Mn 1, Fe 2, Cu 30, Ni 65.
Medium hardness nickel copper casting alloy.

SPECIALLOY 2929C
Specialloy Inc.
Si 3, Mn 1, Fe 2, Cu 30, Ni 64.
High hardness nickel copper casting alloy.

SPECIALLOY 2929D
Specialloy Inc.
Si 4, Mn 1, Fe 2, Cu 30, Ni 63.
High hardness nickel copper casting alloy.

SPECIALLOY 2929W
Specialloy Inc.
Nb 2, Mn 1, Fe 2, Si 1, Cu 30, Ni 64.
Weldable grade, corrosion resistant, high strength nickel
copper casting alloy.

SPECIALLOY 2942
Specialloy Inc.
Mo 25, Fe 25, Ni 50.
Molybdenum addition to iron alloys.

SPECIALLOY 3028
Specialloy Inc.
Ni 45, Cr 15, bal Fe.
For iron and steel alloying; addition agent. *Obsolete*

SPECIALLOY 3028A
Specialloy Inc.
Ni 70, Cr 15, bal Fe.
For iron and steel alloying; addition agent. *Obsolete*

SPECIALLOY 3841
Specialloy Inc.
Nb 60, Fe 40.
Niobium additions to ferrous and nonferrous alloys.

SPECIALLOY 3928
Specialloy Inc.
Cr 8, Cu 28, Ni 56, bal Fe.
Nickel-copper-chromium additions to iron alloys.

SPECIALLOY 5025 MANGANINGOT TM
Specialloy Inc.
99.5 Mn min.
Fused alloy elemental manganese addition for ferrous and
nonferrous alloys.

SPECIALLOY 5813 ALUMANINGOT TM
Specialloy Inc.
Al 40, Mn 60.
High purity aluminum plus manganese addition to iron
alloys.

SPECIFICATION 55-ALLOY STEEL
Colt Industries
C 0.68, Mn 0.6, Si 0.3, Ni 0.65, Cr 0.7, Mo 0.2, bal Fe.
Oil hardenable to 62 Rc. For chain saw links, cutters, drive
links, power saw blades. *Obsolete*

SPECULUM
Manufacturer not listed.
Cu 68.25, Sn 31.75.
For reflectors, mirrors; Cu_4, Sn.

SPECULUM METAL
American manufacture
Cu 66-68, Sn 32-34.
Rolled: 4,000-6,000 TS; 0 El. For telescope reflectors, mirrors;
resists tarnish; approximately Cu_3Sn.

SPEDEX
Barker & Allen Ltd.
Ni 8-20, Pb 2, Zn 25, bal Cu.
For keys, electroplaters, sundries, gas and electric meters,
carburetors, screws; nickel silver; free machining.

SPEED ALLOY
LTV Steel
C 0.35, Cr 1.04, B 0.002, Mn 1.15, bal Fe.
Rolled: 105,000 psi TS; 65,000 psi YS; 20 El; 207 Brin. For
cams, gears, racks, yokes, platens, and die casting dies. Oil
hardening.

SPEED CASE
Escaut & Meuse
C 0.2, Mn 1.25, Si 0.25, P 0.02, bal Fe.
Rolled: 70,000 TS; 45,000 YS; 30 El; 54 RA. For carburized
gears, cams, shafts, camshafts, jigs, figures. Free-machining,
case-hardening.

SPEED CASE
Jones & Laughlin Steel Co.
C 0.2, Mn 1.25, Si 0.25, P 0.02, bal Fe.
Rolled: 70,000 TS; 45,000 YS; 30 El; 54 RA. For carburized
gears, cams, shafts, camshafts, jigs, figures. Free-machining,
case-hardening.

SPEED CASE X-1525
Monarch Steel Corp.
C 0.2, bal Fe.
Hot rolled: 70,000 TS; 52,000 YS; 37.5 El; 60 RA; 137 Brin.
Cold rolled: 85,000 TS; 80,000 YS; 26 El; 48 RA; 156 Brin. For
case hardened parts, gears, shafts, pinions. Carburizing
steel.

SPEED STAR
Now CARPENTER SPEED STAR.

SPEED TREAT
LTV Steel
C 0.42-0.53, Mn 1-1.25, S 0.2-0.3, P 0-0.045, bal Fe.
Hot rolled plate: 45,000 psi YS. Oil or water hardenable to
197,000 psi YS max. Free machining plate for structural
parts.

SPEED TREAT
Ste d'Escaut & Meuse
C 0.45, Mn 1.25, bal Fe.
Hot rolled: 90,000 TS; 60,000 YS; 23 El; 40 RA. Cold finished:
100,000-105,000 TS; 80,000-85,000 YS; 18-19 El; 30-37 RA.
For gears, shafts, pinions, lead screws; X1545 analysis.

SPEED TREAT X-1535
Monarch Steel Corp.
C 0.3-0.4, S 0.25, bal Fe.
For screw machine parts. Free machining.

SPEED TREAT X-1535
Fitzsimmons Steel Co.
C 0.3-0.4, bal Fe.
For screw machine parts. *Obsolete*

SPEED TREAT X-1545
Monarch Steel Corp.
C 0.4-0.5, S 0.25, bal Fe.
For screw machine parts. Free machining.

SPEED-ALLOY 105
Teledyne McKay
C 0.08, Mn 1.35, Si 0.55, Mo 0.3, Ni 2, bal Fe.
Low alloy flux cored gas shielded welding wire. Welded:
108,000 psi TS; 98,000 psi YS; 22 El; 58 RA. AWS E100T5-K3.
For welding low-alloy high-strength steel in the 100,000 psi
TS range; approved for armor plate.

SPEED-ALLOY 105-D2
Teledyne McKay
C 0.09, Mn 1.9, Si 0.55, Mo 0.4, bal Fe.
Low alloy flux cored gas shielded welding wire. Welded:
110,000 psi TS; 95,000 psi YS; 25 El; 55 RA. AWS E100T5-D2.
For 100,000 psi TS min steel and 1.75% Mn, 0.35% Mo (D2)
steel.

SPEED-ALLOY 111
Teledyne McKay
C 0.06, Mn 1.45, Si 0.6, Mo 0.35, Ni 2.1, bal Fe.
Low alloy flux cored gas shielded welding wire. Welded:
115,000 psi TS; 100,000 psi YS; 19 El; 42 RA. AWS E110T1-
K3. For welding low-alloy high-strength steel in the 110,000
psi TS range.

SPEED-ALLOY 111-V

Teledyne McKay
C 0.05, Mn 1.7, Si 0.4, Mo 0.4, Ni 2, bal Fe.
Low alloy flux cored gas shielded welding wire. Welded: 115,000 psi TS; 103,000 psi YS; 18 El; 53 RA. AWS E110T1-K3.

SPEED-ALLOY 115

Teledyne McKay
C 0.06, Mn 1.5, Si 0.55, Mo 0.3, Ni 2.2, bal Fe.
Low alloy flux cored gas shielded welding wire. Welded: 115,000 psi TS; 105,000 psi YS; 20 El; 55 RA. AWS E110T5-K3. For welding low-alloy high-strength steel in the 110,000 psi TS range.

SPEED-ALLOY 12-H

Teledyne McKay
C 0.18, Mn 1.4, Si 0.55, Mo 0.3, Cr 0.45, Ni 1.25, bal Fe.
Low alloy flux cored gas shielded welding wire. Welded: 135,000 psi TS; 125,000 psi YS; 9 El; 14 RA. AWS E120T1-K5. For repair and reclamation of AISA 4130 and similar castings.

SPEED-ALLOY 125

Teledyne McKay
C 0.06, Mn 1.4, Si 0.55, Mo 0.35, Ni 2.2, Cr 0.4, bal Fe.
Low alloy flux cored gas shielded welding wire. Welded: 128,000 psi TS; 118,000 psi YS; 19 El; 55 RA. AWS E120T5-K4. Similar to Speed-Alloy 115.

SPEED-ALLOY 4130-V

Teledyne McKay
C 0.19, Mn 1.2, Si 0.8, Mo 0.35, Cr 1.1, bal Fe.
Low alloy flux cored gas shielded welding wire. AWS E121T1-G.

SPEED-ALLOY 5025

Teledyne McKay
C 0.05, Mn 0.9, Si 0.55, Mo 0.55, Cr 5, bal Fe.
Low alloy flux cored gas shielded welding wire. Welded: 142,000 psi TS; 122,000 psi YS; 15 El; 35 RA. AWS A5.22 E502T-1. For welding 5Cr-0.5Mo plate and pipe.

SPEED-ALLOY 66

Teledyne McKay
C 0.08, Mn 1.5, Si 0.35, bal Fe.
Mild steel flux cored gas shielded welding wire. Welded: 77,000 psi TS; 65,000 psi YS; 21 El. For deep groove semiautomatic welding. AWS A5.26.

SPEED-ALLOY 71

Teledyne McKay
C 0.08, Mn 1.35, Si 0.5, bal Fe.
Mild steel flux cored gas shielded welding wire. Welded: 87,000 psi TS; 75,000 psi YS; 25 El. General purpose for single and multiple pass welding. AWS E70T-1.

SPEED-ALLOY 71-A1

Teledyne McKay
C 0.8, Mn 0.9, Si 0.55, Mo 0.55, bal Fe.
Low alloy flux cored gas shielded welding wire. Welded: 85,000 psi TS; 74,000 psi YS; 22 El. AWS E80T1-A1. For welding carbon and 0.5% Mo steel.

SPEED-ALLOY 71-V

Teledyne McKay
C 0.07, Mn 1.35, Si 0.45, bal Fe.
Mild steel flux cored gas shielded welding wire. Welded: 88,000 psi TS; 75,000 psi YS; 26 El. AWS E71T-1.

SPEED-ALLOY 71A1-V

Teledyne McKay
C 0.06, Mn 1.1, Si 0.65, Mo 0.5, bal Fe.
Low alloy flux cored gas shielded welding wire. Welded: 96,000 psi TS; 86,000 psi YS; 22 El. AWS E81T1-A1.

SPEED-ALLOY 73

Teledyne McKay

Typical deposit analysis: 0.10%C, 1.85% Mn, 0.95% Si. Flux cored weld wire. Welded: 120,000 TS; 112,000 YS; 18 El. For welding low alloy, high strength steel. AWS E70T-3. *Obsolete*

SPEED-ALLOY 75

Teledyne McKay
C 0.08, Mn 1.25, Si 0.5, bal Fe.
Mild steel flux cored gas shielded welding wire. Welded: 77,000 psi TS; 65,000 psi YS; 32 El. AWS E-70T-5.

SPEED-ALLOY 75-A1

Teledyne McKay
C 0.08, Mn 0.85, Si 0.5, Mo 0.55, bal Fe.
Low alloy flux cored gas shielded welding wire. Welded: 78,000 psi TS; 66,000 psi YS; 21 El. For welding carbon and 0.5% Mo steel. AWS E70T5-A1.

SPEED-ALLOY 77

Teledyne McKay
C 0.07, Mn 1.4, Si 0.4, bal Fe.
Mild steel flux cored gas shielded welding wire. Welded: 84,000 psi TS; 72,000 psi YS; 26 El. AWS E70T-G.

SPEED-ALLOY 81-W

Teledyne McKay
C 0.07, Mn 1.05, Si 0.55, Cr 0.55, Ni 0.6, Cu 0.5, bal Fe.
Low alloy flux cored gas shielded welding wire. Welded: 90,000 psi TS; 80,000 psi YS; 25 El; 60 RA. AWS E80T-W.

SPEED-ALLOY 81-W-V

Teledyne McKay
C 0.05, Mn 1.2, Si 0.5, Cr 0.6, Ni 0.7, Cu 0.6, bal Fe.
Low alloy flux cored gas shielded welding wire. Welded: 97,000 psi TS; 87,000 psi YS; 21 El; 59 RA. AWS E80T1-W.

SPEED-ALLOY 81B2L-V

Teledyne McKay
C 0.03, Mn 0.85, Si 0.55, Mo 0.5, Cr 1.15, bal Fe.
Low alloy flux cored gas shielded welding wire. Welded: 111,000 psi TS; 97,000 psi YS; 22 El; 59 RA. AWS E81T1-B2.

SPEED-ALLOY 81NI1-V

Teledyne McKay
C 0.05, Mn 1.2, Si 0.3, Ni 0.95, bal Fe.
Low alloy flux cored gas shielded welding wire. Welded: 97,000 psi TS; 89,000 psi YS; 24 El; 60 RA. AWS E81T1-Ni1.

SPEED-ALLOY 81NI2-V

Teledyne McKay
C 0.06, Mn 1, Si 0.35, Ni 2, bal Fe.
Low alloy flux cored gas shielded welding wire. Welded: 93,000 psi TS; 83,000 psi YS; 25 El; 59 RA. AWS E81T1-Ni2.

SPEED-ALLOY 85

Teledyne McKay
C 0.08, Mn 1, Si 0.55, Mo 0.15, Ni 1.3, bal Fe.
Low alloy flux cored gas shielded welding wire. Welded: 86,000 psi TS; 75,000 psi YS; 26 El; 66 RA. AWS E80T5-K2.

SPEED-ALLOY 85-C1

Teledyne McKay
C 0.08, Mn 0.95, Si 0.55, Ni 2.5, bal Fe.
Low alloy flux cored gas shielded welding wire. Welded: 90,000 psi TS; 80,000 psi YS; 25 El; 60 RA. AWS E80T5-Ni2.

SPEED-ALLOY 85-C2

Teledyne McKay
C 0.08, Mn 1, Si 0.55, Ni 3.3, bal Fe.
Low alloy flux cored gas shielded welding wire. Welded: 100,000 psi TS; 90,000 psi YS; 24 El; 55 RA. AWS E90T5-Ni3. For welding 3% nickel steel.

SPEED-ALLOY 85-C3

Teledyne McKay
C 0.05, Mn 0.95, Si 0.55, Mo 0.55, Ni 0.95, bal Fe.
Low alloy flux cored gas shielded welding wire. Welded: 88,000 psi TS; 80,000 psi YS; 25 El; 62 RA. AWS E80T5-K1. For welding low-alloy high-strength steel in the 80,000 psi TS range; for welding 1% nickel steel.

SPEED-ALLOY 85B2L

Teledyne McKay
C 0.04, Mn 0.8, Si 0.55, Mo 0.55, Cr 1.4, bal Fe.
Low alloy flux cored gas shielded welding wire. Welded: 95,000 psi TS; 85,000 psi YS; 24 El; 60 RA. AWS E80T5-B2L.

SPEED-ALLOY 91

Teledyne McKay
C 0.06, Mn 1, Si 0.55, Ni 1.5, bal Fe.
Low alloy flux cored gas shielded welding wire. Welded: 98,000 psi TS; 90,000 psi YS; 21 El; 57 RA. AWS E90T1-K2. For welding low-alloy high-strength steel in the 90,000 psi TS range.

SPEED-ALLOY 91-B3

Teledyne McKay
C 0.1, Mn 0.9, Si 0.55, Mo 1.05, Cr 2.4, bal Fe.
Low alloy flux cored gas shielded welding wire. Welded: 125,000 psi TS; 115,000 psi YS; 17 El; 42 RA. AWS E100T1-B3. For 2.5 Cr, 1.0 Mo steel.

SPEED-ALLOY 91-B3L

Teledyne McKay
C 0.03, Mn 0.8, Si 0.55, Mo 1.05, Cr 2.4, bal Fe.
Low alloy flux cored gas shielded welding wire. Welded: 105,000 psi TS; 92,000 psi YS; 16 El; 39 RA. AWS E90T1-B3L. Low carbon version of Speed-Alloy 91-B3.

SPEED-ALLOY 91B3-V

Teledyne McKay
C 0.09, Mn 1.1, Si 0.6, Mo 1.05, Cr 2.2, bal Fe.
Low alloy flux cored gas shielded welding wire. Welded: 151,000 psi TS; 105,000 psi YS; 17 El; 46 RA. AWS E91T1-B3.

SPEED-ALLOY 91B3L-V

Teledyne McKay
C 0.04, Mn 0.9, Si 0.5, Mo 1.05, Cr 2.15, bal Fe.
Low alloy flux cored gas shielded welding wire. Welded: 127,000 psi TS; 105,000 psi YS; 20 El; 56 RA. AWS E91T1-B3.

SPEED-ALLOY 95

Teledyne McKay
C 0.08, Mn 1.25, Si 0.5, Mo 0.2, Ni 1.6, bal Fe.
Low alloy flux cored gas shielded welding wire. Welded: 97,000 psi TS; 87,000 psi YS; 23 El; 65 RA. AWS E90T5-K2. For welding low-alloy high-strength steel in the 90,000 psi TS range.

SPEED-ALLOY 95-D1

Teledyne McKay
C 0.07, Mn 1.65, Si 0.4, Mo 0.45.
Welded: 105,000 TS; 97,000 YS; 22 El. Flux cored alloy weld wire for welding Mn-Mo steels in the 90,000-100,000 psi TS range. AWS E90T. *Obsolete*

SPEEDALOY

Great Western Steel Co.
C 2, Cr 32, W 18, Co 48.
For dies, cutters, tool bits; cast alloy.

SPEEDALOY

Hoyland Steel Co.
C 2, Cr 32, W 18, Co 48.
For dies, cutters, tool bits; cast alloy.

SPEEDALOY

Tungsten Alloy Mfg. Co., Inc.
C 2, Cr 32, W 18, Co 48.
For dies, cutters, tool bits; cast alloy.

SPEEDICUT 14

Firth Brown Ltd.
C 0.6, W 14, Cr 3.5, V 0.65, bal Fe.
Hardened: 600-700 Brin. For high speed tools, cutters; "Firth Brown K-200", "Browns Atlas"; high speed steel. *Obsolete*

SPEEDICUT LEDA

Firth Brown Ltd.
C 0.75, Cr 4, Mo 7, V 1.4, W 18, Co 5.2, bal Fe.
Hardened: 600-700 Brin. For high speed tools, cutters; super high speed steel for severe service; "Firth Brown K-203," "Browns Atlas AEI." *Obsolete*

SPEEDICUT MAXIMUM
Firth Brown Ltd.
C 0.73, W 22, Cr 4.8, V 1.5, bal Fe.
Hardened: 600-700 Brin. For high speed tools, cutters; for severe service; "Firth Brown K-202". *Obsolete*

SPEEDICUT MAXIMUM-18
Firth Brown Ltd.
C 0.75, Cr 4.25, V 1.4, W 18, bal Fe.
For high speed tools, cutters; high speed steel "Firth Brown K-201"; "Browns Atlas Extra." *Obsolete*

SPEEDICUT SIXLEDA
Firth Brown Ltd.
C 0.7, Co 15, W 20, Cr 5.5, Mo 0.7, bal Fe.
For lathe and planer tools, form cutters, turning tools; high speed steel, oil hardened. *Obsolete*

SPEEDICUT SUPERLEDA
Firth Brown Ltd.
C 0.78, Cr 4.8, Mo 1, V 2, W 18.8, Co 9.3, bal Fe.
For cutting tools, reamers; high speed steel. *Obsolete*

SPELTER WIRE
English manufacture
Cu 64, Zn 36, Pb 0.4, Fe 0.2.
For brazing.

SPENARD
Spencer Clark Metal Industries Ltd.
C 0.32, Si 0.3, Mn 0.5, Ni 4.1, Cr 1.3, Mo 0.3, bal Fe.
Nickel-chromium tool steel, for punches, trimming dies, plastic molds.

SPERRY ALLOY 37
Sperry Gyroscope Co.
Si 4.3, Mg 1.8, Zn 3.5, Cu 0.1, Cr 0.2, Fe 0.3, Mn 0.4, bal Al.
For die castings, gimbal rings; high yield strength.

SPEZIAL BRONZE
Neubrandenburg Eisenwerke GmbH
Copper. Sn, bal Cu.
For hardware; corrosion resistant.

SPEZIAL EINSATZSTAHL ES2
Kind & Co. Edelstahlwerk
C 0.15, Si 0.1-0.4, Mn 0.25-0.8, bal Fe.
Annealed: 70,000 TS; 55,000 YS; 25 El; 60 RA; 145 Brin. For gears, bolts, camshafts, cams, rivets. Case hardened. *Obsolete*

SPEZIAL NO. 5
Thyssen Edelstahlwerke AG
C, W, Cr, bal Fe.
For steel trimming and weld steel. *Obsolete*

SPEZIAL W
Thyssen Edelstahlwerke AG
C 0.3, Cr 2.65, V 0.35, W 8.5, bal Fe.
For extrusion rams and dies, punches; oil hardened, tough. *Obsolete*

SPEZIAL WSF
Thyssen Edelstahlwerke AG
C 0.3, Cr 2.35, V 0.6, W 4.25, bal Fe.
For extrusion rams and liners, punches; oil hardened, tough. *Obsolete*

SPHERIX
Pechiney Electrometallurgie
Miscellaneous nonferrous. Al 0.9, Bi 1, Ca 1.5, Si 72, 0.5 RE, bal Fe.
Cast iron inoculant.

SPHERULITE
Vulcan Foundry Co.
C 3, Si 2.5, bal Fe.
For housings, gears, shafts; cast iron.

SPHINX
Ugine Aciers
C 0.8, Mn 2, V 0.2, bal Fe.
For punches, dies, low temperature appliances; oil hardened.

SPHYNX
Ugine Aciers
C 0.8, Mn 2, V 0.2, W 0.5, bal Fe.
For punches, crimpers, headers, cutters; oil hardened, nondeforming.

SPIAUTER
English manufacture
Zn 90, Sb 8, Cu 2.
For bearings; hard zinc.

SPIEGELEISEN
Now ELECTROMET SPIEGELEISEN.

SPIN 320
Centrifugal Products Inc.
C 0-0.1, Cr 20-23, Mo 8-10, Fe 0-5, Co 0-1, 3.15-4.15 Cb + Ta, bal Ni.
Cast: 75,000 psi TS min; 40,000 psi YS min; 18 El min; 30 Rock C max. For strength and oxidation resistance at elevated temperature (INCONEL 625).

SPIN 518-A
Centrifugal Products Inc.
C 3.5, Si 2.2, bal Fe.
Annealed: 65,000 psi TS; 45,000 psi YS; 12 El; 149-187 Brin.
Ductile iron with ferrite matrix. ASTM A536 65-45-12.

SPIN 518-N
Centrifugal Products Inc.
C 3.5, Si 2.2, bal Fe.
Normalized: 80,000 psi TS; 55,000 psi YS; 6 El; 202-269 Brin.
Ductile iron with pearlite-ferrite matrix. ASTM A536 80-55-06.

SPIN 522
Centrifugal Products Inc.
C 3.5, Si 2.2, Ni 0.9, Mo 0.5, bal Fe.
Normalized: 100,000 psi TS; 70,000 psi YS; 3 El; 241-285 Brin. Ductile iron with pearlite matrix. ASTM A536 100-70-03.

SPIN 775
Centrifugal Products Inc.
C 0-0.15, Mn 0-1, Si 0-1.5, Cr 11.5-14, Mo 0-0.5, Ni 0-1, bal Fe.
Centrifugally cast; heat treatable to 363-428 Brin and 160,000 psi TS (before tempering). ACI CA-15.

SPIN 776
Centrifugal Products Inc.
C 0.32, Cr 11.5-14, Mn 0-1, Si 0-1.5, Ni 0-1, Mo 0-0.5, bal Fe.
Centrifugally cast; heat treatable to about 500 Brin (before tempering). ACI CA-40.

SPIN 803PH
Certified Alloy Products Inc.
C 0.05, Cr 16, Ni 4, Cu 3, Cb 0.2, bal Fe.
Aged: 190,000 TS; 170,000 YS; 12 El; 35 RA; 415 Brin. High strength stainless; good galling resistance. For landing gear bushings, sleeves, pressure vessels. *Obsolete*

SPIN 902
Centrifugal Products Inc.
C 0.2-0.6, Mn 0-2, Si 0-2, Cr 24-28, Ni 18-22, Mo 0-0.5, bal Fe.
Centrifugally cast stainless steel. ACI CK-20.

SPIN 902 HK
Certified Alloy Products Inc.
C 0.4, Cr 25, Ni 20, bal Fe.
As cast: 87,000 TS; 45,000 YS; 20 El; 16 RA. Creep strength and oxidation resistance to 2000°F; for furnace tubes, retorts. *Obsolete*

SPIN 914
Centrifugal Products Inc.
C 0-0.15, Mn 0-1, Si 0-1.5, Cr 11.5-14, Ni 0-1, Mo 0-1, S, bal Fe.
Centrifugally cast; free machining grade of SPIN 775 (and ACI CA-15).

SPIN 935
Centrifugal Products Inc.
C 0.25-0.35, Mn 0-1, Si 0-1, Cr 23-25, Ni 23-25, Cb 1.4-1.8, bal Fe.
Centrifugally cast alloy. (Inconel 519).

SPINNING SILVER
English manufacture
Cu 67, Zn 17, Ni 16.
For spoons, forks, knives; spun and drawn.

SPINZWEL
Henry Wiggin & Co. Ltd.
Ni 10, Cu 64, Zn 26.
For hardware; Ni silver. *Obsolete*

SPOOLALLOY
Pacific Welding Alloys Mfg. Co.
C, alloy, bal Fe.
For heliarc welding wire and steel welding.

SPOOLARC ALUMINUM NO. 4043
Chemetron Corp.
Continuous bare electrode wire, AA 4043 composition, for general welding of many aluminum alloys (5% Si). AWS Class ER4043.

SPOOLARC ALUMINUM NO. 4047
Chemetron Corp.
Continuous bare electrode wire, AA 4047 composition, for welding many aluminum alloys (12% Si). (Was No. 718). AWS Class ER 4047; BA1Si-4.

SPOOLARC ALUMINUM NO. 5183
Chemetron Corp.
Continuous bare electrode wire, AA 5183 composition, for welding many aluminum alloys containing magnesium. AWS Class ER 5183.

SPOOLARC ALUMINUM NO. 5356
Chemetron Corp.
Continuous bare electrode wire, AA 5356 composition, for welding many aluminum alloys containing magnesium. AWS Class ER 5356.

SPOOLARC ALUMINUM NO. 5554
Chemetron Corp.
Continuous bare electrode wire, AA 5554 composition, for welding many aluminum alloys containg magnesium. Recommended for elevated temperature service. AWS ER 5554, AA 5554.

SPOOLARC ALUMINUM NO. 5556
Chemetron Corp.
Continuous bare electrode, AA 5556 composition for welding many aluminum alloys containing magnesium. AWS Class ER 5556.

SPOOLARC ALUMINUM NO.1100
Chemetron Corp.
Continuous bare electrode wire, AA 1100 composition, for welding 1100 and 3003 aluminum. AWS Class ER1100.

SPOOLARC DEOXIDIZED COPPER WELDING ELECT
Chemetron Corp.
As welded: up to 35,000 psi TS; 40-50 El; 25-40 Rock F. For welding copper to itself and to mild steel or to overlay steel; with MIG equipment. AWS A5.6-69 (E Cu).

SPOOLARC PHOSPHOR "BRONZEC" WELDING EL
Chemetron Corp.
As welded: up to 65,000 psi TS; up to 37,000 psi YS; 42-50 El;
80-100 Brin. For welding or build-up on copper base alloys
and cast iron; with MIG equipment. AWS A5.6-69 (ECuSnC).

SPOOLARC SILICON BRONZE WELDING ELECTROD
Chemetron Corp.
As welded: up to 58,000 psi TS; up to 25,000 psi YS; 53-55 El;
80-100 Brin. For welding silicon bronze, copper alloys and
some iron base metals; with MIG equipment. AWS A5.6-69
(ECuSi).

SPOON METAL
Manufacturer not listed.
Cu, Ni.
For spoons; nickel silver.

SPRABABBITT A
METCO Inc.
Sb 7.5, Cu 3.5, bal Sn.
For wire for metal spraying; Babbitt.

SPRABRASS Y
METCO Inc.
Zn 34, bal Cu.
For wire for metal spraying; brass.

SPRABRONZE C
METCO Inc.
Sn 10, bal Cu.
For wire for metal spraying; bronze.

SPRABRONZE M
METCO Inc.
Zn 40, Cu 58, Sn 2, Fe, Mn.
For wire for metal spraying; bronze.

SPRADRONZE P
METCO Inc.
Sn 5, P 0.25, bal Cu.
For wire for metal spraying; P-bronze.

SPRABRONZE T
METCO Inc.
Zn 39, Sn 1, bal Cu.
For wire for metal spraying; Tobin bronze.

SPRAIRON A
METCO Inc.
Fe.
For metal spraying; iron wire.

SPRAM 14112
Eutectic Corp.
Alloy powder for flame spray to develop surface for severe
abrasive resistance. Matrix 59-62 Rock C, Diamix C particles.
Obsolete

SPRAM 14493
Eutectic Corp.
Alloy powder for flame spray for ductile, unlimited buildup;
resistant to cracking. 25-30 Rock C. *Obsolete*

SPRAM 14494
Eutectic Corp.
Alloy powder for flame spray; gives machinable, corrosion
and impact resistant coating. All purpose. 35-40 Rock C.
Obsolete

SPRAM 14495
Eutectic Corp.
Alloy powder for flame spray; machinable with carbide tools.
45-50 Rock C. *Obsolete*

SPRAM 14496
Eutectic Corp.
Alloy powder for flame spray; gives hard, smooth deposits;
finish by grinding; use against seals, packing, sliding
surfaces. 55-60 Rock C. *Obsolete*

SPRAM 14496 DIAMAX
Eutectic Corp.
Alloy powder for flame spray; best resistance to fine particle
abrasion. Requires grind finishing. Matrix: 60-63 Rock C,
DIAMIX A particles. *Obsolete*

SPRASTEEL 10
METCO Inc.
C 0.1, bal Fe.
For metal spraying; steel wire.

SPRASTEEL 120
METCO Inc.
C 1.2, bal Fe.
For metal spraying; steel wire.

SPRASTEEL 25
METCO Inc.
C 0.25, bal Fe.
For metal spraying; steel wire.

SPRASTEEL 40
METCO Inc.
C 0.4, bal Fe.
For metal spraying; steel wire.

SPRASTEEL 80
METCO Inc.
C 0.8, bal Fe.
For metal spraying; steel wire.

SPREEMETALL
Allgemeine Elektrizitats-Gesellschaft
Zn 43, Mn 1.5, Pb 0.5, bal Cu.
Cold rolled: 86,000 TS; 50,000 YS; 5 El; 140 Brin. For sheet,
bar, forgings; resists sea water corrosion.

SPRING BRASS
Chase Brass & Copper Co.
Cu 66-72, bal Zn.
Rolled: 95,000-120,000 TS; 70,000-80,000 YS; 2-4 El. For
springs; withstands drastic treatment. *Obsolete*

SPRING BRONZE 0.8%
Century Brass Products Inc.
Cu 85, Sn 0.8, Zn 14.2.
Sheet or strip, rolled. CDA 434; ASTM 591.

SPRING BRONZE, 1%
Century Brass Products Inc.
Cu 87.5, Sn 1, Zn 11.5.
Sheet or strip, rolled. CDA 422; ASTM B591.

SPRING GOLD
Manufacturer not listed.
Cu 50, Au 25, Ag 25.
For jewelry, dental applications; corrosion resistant.

SPRING OREIDE
Century Brass Products Inc.
Cu 92, Sn 8.
Soft: 55,000 TS; 25,000 YS; 75 El. Hard: 120,000 TS; 117,000
YS; 1 El. For spring contacts, switches, diaphragms; "Scovill
No. 151 Alloy." *Obsolete*

SPRING SILVER
Manufacturer not listed.
Cu 54.6, Zn 27.3, Ni 18.
For corrosion resistant springs; corrosion resistant.

SPRING SILVER
Engelhard Minerals & Chemicals Corp.
Mg 0-0.28, Ni 0-0.18, 99.4 Ag min.
Air Hardened: 58,000 min TS; 5-15 El. For spring contacts,
switches, relays. Air hardens by internal oxidation; 70% elect.
cond. Good spring properties.

SPRING SILVER
Engelhard Industries
Mg 0-0.28, Ni 0-0.18, 99.4 Ag min.
Air Hardened: 58,000 min TS; 5-15 El. For spring contacts,
switches, relays. Air hardens by internal oxidation; 70% elect.
cond. Good spring properties.

SPRING STEEL
Manufacturer not listed.
C 0.9-1.1, Mn 0.4-0.8, bal Fe.
For springs; oil hardening.

SPRING STEEL SHEET
Eagle & Globe Steel Ltd.
Carbon steel. C 0.65-0.7, S 0 0.05, P 0-0.05, Mn 0.65-0.85,
bal Fe.
B70 warranted spring steel. For flat springs, formed springs,
motor springs, high strength structural components,
fasteners and cutting tools.

SPRING WIRE BRASS
American manufacture
Cu 70-74, Zn 26-30.
100,000 TS. For springs.

SPS 245
Atlas Specialty Steels
C 0.4, Mn 0.75, Cr 0.6, Mo 0.15, bal Fe.
Annealed: 100,000 TS; 68,000 YS; 30 El; 58 RA; 187 Brin. For
shafts, gears, bolts, studs; 45 Izod.

SPZ
Le Bronze Industriel
Sn 7, Zn 4, Pb 6, bal Cu.
Leaded bronze, wrought: 75 Brin. AFNOR UE7 Z5 Pbb.

SQUARE 50
Magnetic Metals Co.
Ni 47-50, Fe 50.
Grain oriented alloy processed for maximum B-H loop
squareness.

SQUARE 80
Magnetic Metals Co.
80 Ni, Fe, Mo.
High maximum flux density, high gain, high squareness
ratio; soft magnetic alloy.

SR
Bergische Stahl Industrie
C 1.65-1.75, Si 0.25-0.4, Mn 0.2-0.4, Cr 11-12, V 0.07-0.12,
bal Fe.
Cold work tool steel. W.-Nr. 1.2201.

SR SPEZIAL
Bergische Stahl Industrie
C 1.55-1.75, Si 0.25-0.4, Mn 0.2-0.4, Cr 11-12, Mo 0.5-0.7, W
0.4-0.6, V 0.07-0.12, bal Fe.
Hot work tool steel. W.-Nr. 1.2601.

SR STEEL IIA-1121
Uniworld Corp. of America
C, Cr, Ni, bal Fe.
Cast: 112,000 TS; 44,000 YS; 23 El; 16 RA; 200 Brin. Rolled:
95,000 TS; 75,000 YS; 50 El; 65 RA. For pumps, valves,
propellers, heat exchangers; stainless, austenitic.

SR STEEL IIA-1123
Uniworld Corp. of America
C, Cr, Ni, bal Fe.
For chemical plant equipment; stainless, austenitic.

SR STEEL IIA-1132
Uniworld Corp. of America
C 0-0.12, Ni 13-15, Cr 11-13, Mo 6-7, Cu 5.5-6.5, bal Fe.
Annealed: 95,000 TS; 75,000 YS; 50 El; 65 RA; 180 Brin.
32,500 TS (at 1500°F). For chemical plant and food
processing equipment; stainless, austenitic.

SR STEEL UST
Uniworld Corp. of America
C 0.1, Ni 20, Cr 20, 10.0 Cu + Mo max, bal Fe.
33,800 TS (at 1600°F); 10 El; 17 RA. For chemical plant
equipment; austenitic, stainless.

SR-NICKEL
Vereinigte Deutsche Nickel-Werke AG
Ni 99.8.
For electroplating anodes; chemical equipment. *Obsolete*

SRA 100 (1144)
Joseph T. Ryerson & Son Inc.
C 0.45, Mn 1.5, P 0-0.04, S 0.24-0.33, Si 0.25, bal Fe.
Cold drawn, strain hardened: 125 ksi TS (typical); 100 ksi YS
min; 12 El; 34 RA. For moderately stressed parts; free
machining with minimum distortion.

SRMV
Bergische Stahl Industrie
C 1.5-1.6, Si 0.3-0.5, Mn 0.3-0.5, Cr 11.5-12.5, Mo 0.6-0.8, V
0.9-1.1, bal Fe.
Cold work tool steel. W.-Nr. 1.2379.

SRS
Eagle & Globe Steel Ltd.
Tool material. C 0.6, Si 1.6, Mn 0.8, Cr 0.35, Mo 0.4, V 0.15,
bal Fe.
Annealed: 235 Brin max. Oil hardening shock resisting tool
steel. For punches and dies, shear blades and cropping
tools, pneumatic and hand chisels, collets, feed fingers and
vise jaws, jigs, fixtures. AS1239 S5A; AISI S5; BS4659 BS5.

SS 1 EXTRA EXTRA
Thyssen Edelstahlwerke AG
C 1.05, bal Fe.
For cutting tools; high hardness. *Obsolete*

SS 1-6
Thyssen Edelstahlwerke AG
C 0.7-1.4, bal Fe.
For jaws, tools, springs; water hardening. *Obsolete*

SS 2 EXTRA EXTRA
Thyssen Edelstahlwerke AG
C 1, bal Fe.
For scrapers, hard tools, drawing dies, water hardening.
Obsolete

SS 3 EXTRA EXTRA
Thyssen Edelstahlwerke AG
C 1, bal Fe.
For medium hard tools, milling machine tools; water
hardening. *Obsolete*

SS 4 EXTRA EXTRA
Thyssen Edelstahlwerke AG
C 1, bal Fe.
For punches, edge engraving tools; tough. *Obsolete*

SS 410 3.020
Sintered Products Ltd.
Ferritic 410 stainless steel, sintered. 6.7-7.1 g/cm³ density; 7
El; 42.5-49 kg/mm² TS; 170 Vickers.

SS 5 EXTRA EXTRA
Thyssen Edelstahlwerke AG
C 1, bal Fe.
For hand hammer, center punches; tough. *Obsolete*

SS 511
W. Ossenberg & Cie Edelstahlwerke
C 1.3, Mn 0.3, Si 0.2, Cr 0.15, W 5, bal Fe.
Hardenable to 63-65 Rock C. For drawing dies. W.-Nr.
1.2453.

SS 6 EXTRA EXTRA
Thyssen Edelstahlwerke AG
C 1, Si 1.5, bal Fe.
For hot chisels, soft tools, riveting set; shock resistant.
Obsolete

SS 931
Sheepbridge Alloy Castings Ltd.
C 3, Si 2.2, Cr 1.3, bal Fe.
Cast: 260-310 Brin. For valve seats, heat resistant castings;
cast iron.

SS 931
Sheepbridge Engineering Ltd.
C 3, Si 2.2, Cr 1.3, bal Fe.
Cast: 260-310 Brin. For valve seats, heat resistant castings;
cast iron.

SS CUT A
Wallace Murray Corp.
C 0.7, Mn 0.75, Si 0.75, Mo 0.5, Cr 17.5, bal Fe.
Martensitic chromium stainless steel for knives, cutlery. AISI
440A.

SS CUT B
Wallace Murray Corp.
C 0.9, Mn 0.75, Si 0.75, Mo 0.5, Cr 17.5, bal Fe.
Martensitic chromium stainless steel for knives, cutlery. AISI
440B.

SS CUT C
Wallace Murray Corp.
C 1.05, Mn 0.75, Si 0.75, Mo 0.5, Cr 17.5, bal Fe.
Martensitic chromium stainless steel for knives, cutlery. AISI
440C.

SS EXTRA COLD HEADING
Hoyland Steel Co.
C 1-1.1, bal Fe.
For cold heading dies.

SS-6
Chemetron Corp.
Mn 0.59, Si 0.21, weld metal; 0.08 C, bal Fe.
As welded: 76,000 psi TS; 69,000 psi YS; 26 El. AC-DC,
straight polarity electrode for high speed welding of
horizontal fillets, especially in shipyards. ABS Filler Grade
No.H2.

SS-DEMI-DUR
Compagnie Ateliers et Forges de la Loire
C 0.45, Si 1.9, bal Fe.
For chisels, punches.

SS15 QUALITY
Delta Metal (BW) Ltd.
Ni 15, Cu, bal Fe.
Extruded: 84,000 TS; 54,000 YS; 18 El; 10 RA; 150 Brin.
Drawn: 90,000 TS; 58,000 YS; 15 El; 10 RA; 145 Brin. For
stampings, hardware; silver bronze. *Obsolete*

SS150
Friedr. Lohmann GmbH
C 1.35, Co 4.8, Cr 4.1, Mo 0.8, V 3.8, W 12, bal Fe.
For reamers, broaches, milling cutters. Type T15; high speed
steel.

SS2
Thyssen Edelstahlwerke AG
C 0.7-1.3, bal Fe.
For cutting tools, drills, taps; Type W1, water hardened.
Obsolete

SS3
Thyssen Edelstahlwerke AG
C 1.1, Si 0-0.25, Mn 0-0.25, bal Fe.
Annealed: 110,000 TS; 58,000 YS; 19 El; 40 RA; 210 Brin. For
springs, drills, reamers, broaches; Type W1; water hardened.
Obsolete

SS315
Friedr. Lohmann GmbH
C 0.95, Cr 4.1, Mo 2.6, V 2.3, W 2.8, bal Fe.
For reamers, broaches, milling cutters. High speed steel.

SS4
Thyssen Edelstahlwerke AG
C 1, Si 0-0.25, Mn 0-0.25, bal Fe.
Heat treated: 185,000 TS; 120,000 YS; 10 El; 30 RA; 390 Brin.
For springs, taps, reamers, hobs, cutters; Type W1; water
hardened. *Obsolete*

SS460
Friedr. Lohmann GmbH
C 1.3, Cr 4.1, Mo 0.85, V 3.8, W 12, bal Fe.
For engraving tools, milling cutters, broaches. High speed
steel.

SS5
Thyssen Edelstahlwerke AG
C 0.85, Si 0-0.25, Mn 0-0.25, bal Fe.
Heat treated: 190,000 TS; 140,000 YS; 10 El; 30 RA; 395 Brin.
For springs, taps, reamers, cutters, drills; Type W1; water
hardened. *Obsolete*

SS6
Thyssen Edelstahlwerke AG
C 0.7-1.3, bal Fe.
For cutting tools, drills, taps; Type W1, water hardening.
Obsolete

SSB
Manufacturer not listed.
Cu 4.5-5.5, bal Al.
For light alloy parts; super duralumin type.

SSB
Thyssen Edelstahlwerke AG
C, bal Fe.
For spindles, axles; water hardened. *Obsolete*

SSF
Thyssen Edelstahlwerke AG
C, Mn, bal Fe.
For axles, mandrels; oil hardening. *Obsolete*

SSM
Thyssen Edelstahlwerke AG
C, Mn, bal Fe.
For axles, mandrels; water hardened. *Obsolete*

SSS-100
Now ARMCO SSS 100.

ST 35.8
Krupp Stahl AG
C 0-0.17, Si 0-0.35, Mn 0.4-0.8, P 0-0.04, S 0-0.04, bal Fe.
DIN 17175/79; W. Nr. 1.0305.

ST 400
Sheffield Smelting Co. Ltd.
Ag 4, Sn.
Melting range: 221-224°C. Maximum stress: 5.5 kgf/mm²; 48
El. For soft soldering; corrosion resistant.

ST-10 SOLDER
American Smelting & Refining Co.
Pb 87.5, Sn 10, Ag 1.5, Bi 0.5, Sb 0.5.
Cast: 5335 psi TS; 11.5 El. For solder; 554°F MP.

ST-15
Alpha Metals Inc.
Pb 82.75, Sn 15, Ag 1.25, Bi 0.5, Sb 0.55.
For soft solder; M.P. 278-532°F.

ST-15 SOLDER
American Smelting & Refining Co.
Pb 82.75, Sn 15, Ag 1.25, Bi 0.5, Sb 0.5.
Cast: 6180 psi TS; 11.5 El. For solder; 532°F MP.

ST-20 SOLDER
American Smelting & Refining Co.
Pb 77.75, Sn 20, Ag 1.25, Bi 0.5, Sb 0.5.
Cast: 6840 psi TS; 14 El. For solder; 518°F MP.

ST-20A SOLDER
American Smelting & Refining Co.
Pb 76.75, Sn 20, Ag 1.25, Bi 1.5, Sb 0.5.
Cast: 6720 psi TS; 13 El. For solder; 514°F MP.

ST-20B
Alpha Metals Inc.
Pb 73.25, Sn 20, Ag 1.25, Bi 5, Sb 0.5.
For soft solder; M.P. 258-486°F.

ST-20B SOLDER
American Smelting & Refining Co.
Pb 73.25, Sn 20, Ag 1.25, Bi 5, Sb 0.5.
Cast: 6810 psi TS; 27 El. For solder; 486°F MP.

ST-30
Alpha Metals Inc.
Pb 67.75, Sn 30, Sb 1.25, Bi 0.5, Sb 0.5.
For soft solder; M.P. 248-478°F.

ST-30 SOLDER
American Smelting & Refining Co.
Pb 67.75, Sn 30, Ag 1.25, Bi 0.5, Sb 0.5.
Cast: 7625 psi TS; 15 El. For solder; 478°F MP.

ST-48
Deutsche Industrie
C 0.4, bal Fe.
For gears, shafts; water hardened.

ST-51
Bliss & Laughlin Steel Co.
C 0.48-0.55, Mn 1.35-1.65, S 0.08-0.13, bal Fe.
Cold drawn, stress relieved: 140,000 TS min; 125,000 YS min; 285 Brin. For machine parts, shafts. Free machining, eliminates heat treatment.

ST-52
German manufacture
Si 0.5-1.1, Mn 0.75-1.6, Cu 0.25-1, Cr 0-0.6, Mo 0-0.25, C 0.12-0.15, bal Fe
89,600 TS; 65,000 YS; 24 El. For structural parts; excellent for welding.

ST. 37.161
Swiss manufacture
C 0.15, S 0.06, P 0.06, bal Fe.
Annealed: 70,000 TS; 40,000 YS; 25 El, 60 RA; 135 Brin. For bridge construction; good weldability.

ST. 40.21
German manufacture
C 0.18, bal Fe.
For welded sluice gates; excellent weldability.

ST. 52
German manufacture
C 0.2, Si 0.15, Mn 0.5, Cu 0.5, bal Fe.
For railway cars, bus and truck bodies; high strength structural steel.

ST. 60
Swiss manufacture
C 0.1, bal Fe.
Annealed: 64,000 TS; 41,000 YS; 28 El; 65 RA; 130 Brin. For welded structures; good weldability.

ST. JOE CHEMICAL LEAD
Manufacturer not listed.
Cu 0.05, Ag 0.005, Cd 0.0003, Ni 0.0003, bal Pb.
Cast: 2400-2600 TS; 40-50 El; 5 Brin. 618°F melting point; 0.41 lbs/in^3 density. Chemical equipment; acid containers. Exceeds ASTM B29-55.

ST. Z1
Swiss manufacture
C 0.13, S 0.05, P 0.05, bal Fe.
Rolled: 52,000-64,000 TS; 34,000 YS; 25 El. For bridge construction; good weldability.

ST52 CARBON STEEL HOLLOW BARS
Eagle & Globe Steel Ltd.
Carbon steel. C 0-0.2, Si 0-0.5, Mn 0-1.5, P 0-0.045, S 0-0.045, bal Fe.
Hollow bars: 490 MPa TS; 325 MPa YS; 19 El.
BS120M19/150M19; SS 2172; DIN St52; AFNOR 20M5; SAE 1518; Werkstoff 1.0841.

STA 8
Sheepbridge Engineering Ltd.
C 3.2, Si 1.6-2.8, P 0.2, bal Fe.
Cast: 210-260 Brin. For thin section castings. Cast iron.

STA-GLOSS "A"
Jessop Steel Co.
C 0.3-0.4, Cr 12-14, bal Fe.
Annealed: 100,000 TS; 75,000 YS; 25-30 El; 55-65 RA; 175 Brin. Hardened: 270,000 TS; 525 Brin. For valves, airplane parts, arch supports, ball bearings, knives, dies, gages; stainless and corrosion resisting. *Obsolete*

STA-GLOSS B
Jessop Steel Co.
C 0.6, Mn 0.35, Cr 17, bal Fe.
Heat treated: 280,000 TS; 240,000 YS; 3 El; 3 RA; 525 Brin.
For knives, surgical instruments, rolls, piston rods, valves; stainless, corrosion resistant. *Obsolete*

STA-GLOSS C
Jessop Steel Co.
C 0.2, Cr 13, bal Fe.
For cutlery, valves, surgical instruments; Type 420. *Obsolete*

STABIL
Peter A. Frasse & Co.
C, alloy, bal Fe.
For tools. *Obsolete*

STABIL
Westa-Westdeutsche
C 0.9, Mn 1.9, V 0.1, bal Fe.
For punches, cutters, forming dies; oil hardened, nondeforming.

STABIL SPEZIAL
Westa-Westdeutsche
C 1.05, Cr 1, W 1.15, Mn 0.9, bal Fe.
For cutters, bearings, forming dies; oil or water hardened.

STABILOR
Degussa AG
Precious metal.
Precious metal alloy for dentistry and dental engineering.

STABLE-ARC
Lincoln Electric Co.
C, bal Fe.
Steel arc welding electrode. AWS Class E4510.

STABON E120
Stackpole Magnet Division
Permanent magnet. Residual flux density: 2250 Gauss.
Coercive force: 2000 Oersted.

STABON E140
Stackpole Magnet Division
Permanent magnet. Residual flux density: 2450 Gauss.
Coercive force: 2200 Oersted.

STABON IM140
Stackpole Magnet Division
Permanent magnet. Residual flux density: 2450 Gauss.
Coercive force: 2200 Oersted.

STABON IM160
Stackpole Magnet Division
Permanent magnet. Residual flux density: 2550 Gauss.
Coercive force: 2300 Oersted.

STABON IM180
Stackpole Magnet Division
Permanent magnet. Residual flux density: 2700 Gauss.
Coercive force: 2280 Oersted.

STABRITE
Garrett Brass & Aluminum Foundry Co.
Ni alloy.
For castings. Corrosion resistant.

STACKPOLE 2331
Stackpole Magnet Division
Permanent magnet. Residual flux density: 1980-2360 Gauss.
Coercive force: 1650-1930 Oersted.

STACKPOLE 2532 SLEEVES
Stackpole Magnet Division
Permanent magnet. Residual flux density: 2575 Gauss.
Coercive force: 2130 Oersted.

STACKPOLE 2732 SEGMENTS
Stackpole Magnet Division
Permanent magnet. Residual flux density: 2850 Gauss.
Coercive force: 2400 Oersted.

STACKPOLE 3540
Stackpole Magnet Division
Permanent magnet. Residual flux density: 3620 Gauss.
Coercive force: 3400 Oersted.

STACKPOLE 3547
Stackpole Magnet Division
Permanent magnet. Residual flux density: 3620 Gauss.
Coercive force: 3400 Oersted.

STACKPOLE 3831
Stackpole Magnet Division
Permanent magnet. Residual flux density: 3920 Gauss.
Coercive force: 3175 Oersted.

STACKPOLE 3838
Stackpole Magnet Division
Permanent magnet. Residual flux density: 3920 Gauss.
Coercive force: 3520 Oersted.

STACKPOLE 4130
Stackpole Magnet Division
Permanent magnet. Residual flux density: 4150 Gauss.
Coercive force: 3000 Oersted.

STACKPOLE 7
Stackpole Magnet Division
Permanent magnet. Residual flux density: 3400 Gauss.
Coercive force: 3250 Oersted.

STACKPOLE 8
Stackpole Magnet Division
Permanent magnet. Residual flux density: 3850 Gauss.
Coercive force: 3050 Oersted.

STACKPOLE FW-41
Stackpole Carbon Co.
Ag-W.
For electric contacts; sintered. *Obsolete*

STAFLO 761
Handy & Harmon
Precious metal. Cu 7.3, Li 0.2, 20.0 Ni + 80.0 Lithobraze 925
(Lithobraze 925 is 92.5 Ag, 7.3 Cu, 0.2 Li).
Duplex brazing alloy for brazing honeycomb panels of
stainless, helps control fillets.

STAG
Edgar Allen Balfour Ltd.
C 1.1, bal Fe.
For tools, drills, taps; water hardened.

STAG 14 AIR HARDENING
Edgar Allen Balfour Ltd.
C 0.67, W 15, Cr 3.5, V 0.37, bal Fe.
For woodworking tools, end mills, drills; high-speed steel.

STAG AIR-HARDENING
Edgar Allen Balfour Ltd.
C, W, Cr, V, bal Fe.
For twist drills, milling cutters, hot punches, reamers, chisels;
high speed steel. *Obsolete*

STAG ALLENITE
Edgar Allen Balfour Ltd.
WC + Co.
For tipped cutting tools; fast cutting on cast iron; sintered
carbide.

STAG ATHYWELD 60/1/13
Edgar Allen Balfour Ltd.
WC.
52 Rock C. For hard facing brick making machines, mixer
knives. Wear and abrasion resistant.

STAG EXTRA SPECIAL
Edgar Allen Balfour Ltd.
C 0.8, W 19, Co 6, Cr 5, V 1.5, Mo 0.5, bal Fe.
For lathe and planer cutters, hobs, reamers, taps; high-
speed steel, oil hardened.

STAG MAJOR
Edgar Allen Balfour Ltd.
C 0.8, W 21, Co 11, Cr 5, V 1.5, Mo 0.5, bal Fe.
For drills, reamers, hobs, lathe and planer cutters; high-
speed steel, oil hardened.

STAG MO
Edgar Allen Balfour Ltd.
C 0.8, W 6.5, Mo 4, Cr 4, V 1.5, bal Fe.
For tools, cutters, reamers, broaches, taps; high speed- steel.

STAG MO562
Edgar Allen Balfour Ltd.
C 0.85, Cr 4.75, Mo 5, W 6, V 2, bal Fe.
For drills, taps, reamers, form tools; high-speed steel.

STAG SPECIAL
Edgar Allen Balfour Ltd.
C 0.7, W 19, Cr 4, V 1, bal Fe.
For planing and shaping tools, drills, reamers, hobs; high-
speed steel, oil hardened.

STAHLSCHMIDT C1V
Stahlwerk Stahlschmidt GmbH & Co.
C 0.5, Cr 1, V 0.09, bal Fe.
For springs, gears, machine tool parts, bolts; oil hardened,
shock resistant.

STAHLSCHMIDT C2V
Stahlwerk Stahlschmidt GmbH & Co.
C 0.58, Cr 1, V 0.09, bal Fe.
For springs, gears, crankshafts; oil hardened, shock resistant.

STAHLSCHMIDT CE10
Stahlwerk Stahlschmidt GmbH & Co.
C 0.2, Mn 1.25, Cr 1.15, bal Fe.
For gears, cams, camshafts; case hardened, tough.

STAHLSCHMIDT CE6
Stahlwerk Stahlschmidt GmbH & Co.
C 0.15, Cr 0.65, Mn 0.5, bal Fe.
For gears, fasteners, machine tool parts; case hardened,
shock resistant.

STAHLSCHMIDT CE8
Stahlwerk Stahlschmidt GmbH & Co.
C 0.16, Cr 0.95, Mn 1.15, bal Fe.
For gears, cams, camshafts; case hardened, tough.

STAHLSCHMIDT CNL
Stahlwerk Stahlschmidt GmbH & Co.
C 0.5, Cr 1.05, Ni 3.25, bal Fe.
For gears, bolts, crankshafts; oil hardened, shock resistant.

STAHLSCHMIDT E16
Stahlwerk Stahlschmidt GmbH & Co.
C 0.15, Si 0.25, Mn 0.37, bal Fe.
Annealed: 70,000 TS; 40,000 YS; 25 El; 60 RA; 145 Brin. For
gears, bolts, machine tool parts; case hardening steel.

STAHLSCHMIDT E25
Stahlwerk Stahlschmidt GmbH & Co.
C 0.22, Si 0.25, Mn 0.45, bal Fe.
Annealed: 75,000 TS; 42,000 YS; 22 El; 58 RA; 140 Brin. For
gears, shafts, fasteners, bolts, screws; water hardened.

STAHLSCHMIDT EXTRA
Stahlwerk Stahlschmidt GmbH & Co.
C 0.82, Cr 4.1, Mo 0.85, V 1.6, W 8.7, bal Fe.
For lathe and planer tools, reamers, broaches; high speed
steel.

STAHLSCHMIDT G500
Stahlwerk Stahlschmidt GmbH & Co.
C 1, Si 0.2, Mn 0.25, V 0.1, bal Fe.
Annealed: 100,000 TS; 53,000 YS; 21 El; 42 RA; 200 Brin. For
drills, taps, reamers, springs; Type W2; water hardened.

STAHLSCHMIDT HS23
Stahlwerk Stahlschmidt GmbH & Co.
C 0.95, W, Mo, bal Fe.
For cutters, dies; oil hardened, tough.

STAHLSCHMIDT HS55
Stahlwerk Stahlschmidt GmbH & Co.
C 0.85, W, Mo, Cr, V, bal Fe.
For cutters, dies; oil hardened, tough.

STAHLSCHMIDT HWG
Stahlwerk Stahlschmidt GmbH & Co.
C 0.56, Ni, Cr, Mo, V, bal Fe.
For upsetters, extrusion dies; oil hardened, tough.

STAHLSCHMIDT HWGW
Stahlwerk Stahlschmidt GmbH & Co.
C 0.55, Cr 0.7, Mo 0.18, Ni 1.65, V 0.1, bal Fe.
For forging dies, upsetters, shears; oil hardened, tough.

STAHLSCHMIDT MS5
Stahlwerk Stahlschmidt GmbH & Co.
C 0.37, Si 1.25, Mn 1.25, bal Fe.
For gears, punches, machine tool parts; oil hardened, tough.

STAHLSCHMIDT MV7
Stahlwerk Stahlschmidt GmbH & Co.
C 0.42, Mn 1.75, Si 0.25, V 0.1, bal Fe.
For punches, gears, bolts, shafts; oil hardened, tough.

STAHLSCHMIDT NE15
Stahlwerk Stahlschmidt GmbH & Co.
C 0.15, Cr 1.55, Ni 1.55, bal Fe.
For die casting dies, gears, cams; case hardened, tough.

STAHLSCHMIDT NE25
Stahlwerk Stahlschmidt GmbH & Co.
C 0.13, Cr 0.7, Ni 2.5, bal Fe.
For gears, bolts, machine tool parts; case hardened, shock
resistant.

STAHLSCHMIDT NE35
Stahlwerk Stahlschmidt GmbH & Co.
C 0.13, Cr 0.7, Ni 3.5, bal Fe.
For gears, bolts, camshafts, cams; case hardened, shock
resistant.

STAHLSCHMIDT NE45
Stahlwerk Stahlschmidt GmbH & Co.
C 0.13, Cr 1.1, Ni 4.5, Mn 0.5, bal Fe.
For gears, bolts, camshafts, cams; case hardened, shock
resistant.

STAHLSCHMIDT NV15
Stahlwerk Stahlschmidt GmbH & Co.
C 0.28-0.35, Cr 0.5, Ni 1.5, Mn 0.6, bal Fe.
For bolts, gears, machine tool parts; oil hardened, shock
resistant.

STAHLSCHMIDT NV25
Stahlwerk Stahlschmidt GmbH & Co.
C 0.28-0.35, Mn 0.6, Cr 0.7, Ni 2.5, bal Fe.
For gears, bolts, crankshafts, axles; oil hardened, shock
resistant.

STAHLSCHMIDT NV35
Stahlwerk Stahlschmidt GmbH & Co.
C 0.22-0.3, Mn 0.6, Cr 0.7, Ni 3.5, bal Fe.
For gears, bolts, crankshafts, axles; oil hardened, shock
resistant.

STAHLSCHMIDT NV45
Stahlwerk Stahlschmidt GmbH & Co.
C 0.35, Ni 4.5, Cr 1.3, Mn 0.6, bal Fe.
For gears, dies, crankshafts, bolts; oil hardened, shock
resistant.

STAHLSCHMIDT PD EXTRA
Stahlwerk Stahlschmidt GmbH & Co.
C 0.38, Si, Cr, V, bal Fe.
For gears, bolts, springs, crankshafts; oil hardened, shock
resistant.

STAHLSCHMIDT RCM
Stahlwerk Stahlschmidt GmbH & Co.
C 0.45, Cr 1.4, Mo 0.7, V 0.3, Mn 0.7, bal Fe.
For forging dies, punches, gears, crankshafts; oil hardened,
tough.

STAHLSCHMIDT SC15
Stahlwerk Stahlschmidt GmbH & Co.
C 0.67, Si 1.3, Cr 0.5, Mn 0.5, bal Fe.
For shears, rivet sets, upsetters; oil hardened, tough.

STAHLSCHMIDT SCE
Stahlwerk Stahlschmidt GmbH & Co.
C 0.15, Cr 0.65, Mn 0.5, bal Fe.
For gears, bolts, oil refinery equipment; case hardened,
creep resistant.

STAHLSCHMIDT SCS
Stahlwerk Stahlschmidt GmbH & Co.
C 1.05, Cr 1.15, bal Fe.
For liners, sleeves, cutters, bearings; water or oil hardened,
wear resistant.

STAHLSCHMIDT SE1
Stahlwerk Stahlschmidt GmbH & Co.
C 0.2, Mn 1.25, Cr 1.25, bal Fe.
For gears, bolts, crankshafts, cams; case hardened, tough.

STAHLSCHMIDT SGK
Stahlwerk Stahlschmidt GmbH & Co.
C 1.45, Cr 1.4, Mn 0.6, bal Fe.
For blanking and forming dies; water or oil hardened.

STAHLSCHMIDT SGS2
Stahlwerk Stahlschmidt GmbH & Co.
C 1.3, Si 0-0.25, Mn 0-0.25, bal Fe.
For blanking and forming dies, cutters, reamers; Type W1; water hardened.

STAHLSCHMIDT SGS3
Stahlwerk Stahlschmidt GmbH & Co.
C 1.15, Si 0-0.25, Mn 0-0.25, bal Fe.
Annealed: 110,000 TS; 58,000 YS; 18 El; 40 RA; 210 Brin. For springs, drills, reamers, taps, broaches; Type W1; water hardened.

STAHLSCHMIDT SGS4
Stahlwerk Stahlschmidt GmbH & Co.
C 1, Si 0-0.25, Mn 0-0.25, bal Fe.
Annealed: 100,000 TS; 53,000 YS; 21 El; 42 RA; 200 Brin. For springs, tools, drills, taps, reamers; Type W1; water hardened.

STAHLSCHMIDT SGS5
Stahlwerk Stahlschmidt GmbH & Co.
C 0.85, Si 0-0.25, Mn 0-0.25, bal Fe.
Heat treated: 125,000-185,000 TS; 85,000-140,000 YS; 14-24 El; 37-52 RA; 230-400 Brin. For springs, drills, reamers, taps, cutters; Type W1; water hardened.

STAHLSCHMIDT SGS6
Stahlwerk Stahlschmidt GmbH & Co.
C 0.7, Si 0-0.25, Mn 0-0.25, bal Fe.
Heat treated: 122,000-174,000 TS; 82,000-128,000 YS; 12-22 El; 37-52 RA; 240-352 Brin. For wheels, die blocks, rails, springs; Type W1; water hardened.

STAHLSCHMIDT SNR
Stahlwerk Stahlschmidt GmbH & Co.
C 1.65, Si 11.5, V 0.1, bal Fe.
For blanking and forming dies, punches; air hardened, nondeforming.

STAHLSCHMIDT SNS EXTRA
Stahlwerk Stahlschmidt GmbH & Co.
C 2.1, Cr 11.5, Mn 0.3, bal Fe.
For blanking and forming dies, punches; oil hardened, nondeforming.

STAHLSCHMIDT SNS SPEZIAL
Stahlwerk Stahlschmidt GmbH & Co.
C 2.1, Cr 11.5, W 0.7, bal Fe.
For blanking and forming dies; oil hardened, nondeforming.

STAHLSCHMIDT SP EXTRA
Stahlwerk Stahlschmidt GmbH & Co.
C 0.7, Si 1.7, Mn 0.7, bal Fe.
For punches, shears, upsetters; oil hardened, tough.

STAHLSCHMIDT SPEZIAL ZAH
Stahlwerk Stahlschmidt GmbH & Co.
C 0.74, Cr 4.1, V 1.1, W 18.5, bal Fe.
For lathe and planer tools, reamers, broaches, taps; high speed steel.

STAHLSCHMIDT SS II
Stahlwerk Stahlschmidt GmbH & Co.
C 0.9, Mn 1.9, V 0.1, bal Fe.
For blanking and forming dies, cutters, punches; oil hardened, nondeforming

STAHLSCHMIDT SSF
Stahlwerk Stahlschmidt GmbH & Co.
C 0.6, Si 0.25-0.5, Mn 0.3-0.8, bal Fe.
Heat treated: 115,000-160,000 TS; 77,000-113,000 YS; 12-23 El; 40-54 RA; 230-320 Brin. For wheels, die blocks, rails, girders, springs; water hardened.

STAHLSCHMIDT STANDARD
Stahlwerk Stahlschmidt GmbH & Co.
C 0.79, Co 4.75, Cr 4.3, Mo 0.75, V 1.5, W 18, bal Fe.
For lathe and planer tools, reamers, drills, hobs; high speed steel.

STAHLSCHMIDT STANDARD SPEZIAL
Stahlwerk Stahlschmidt GmbH & Co.
C 0.76, Co 10, Cr 4.2, Mo 0.8, V 1.8, W 18, bal Fe.
For lathe and planer tools, drills, reamers; high speed steel.

STAHLSCHMIDT STS3
Stahlwerk Stahlschmidt GmbH & Co.
C 1.1, Si 0-0.25, Mn 0-0.25, bal Fe.
Heat treated: 140,000-200,000 TS; 90,000-125,000 YS; 10-20 El; 30-45 RA; 275-400 Brin. For springs, reamers, drills, taps; Type W1; water hardened.

STAHLSCHMIDT STS4
Stahlwerk Stahlschmidt GmbH & Co.
C 1, Si 0-0.25, Mn 0-25, bal Fe.
Heat treated: 130,000-185,000 TS; 80,000-120,000 YS; 10-20 El; 30-45 RA; 270-390 Brin. For springs, hobs, cutters, reamers, taps; Type W1; water hardened.

STAHLSCHMIDT STS5
Stahlwerk Stahlschmidt GmbH & Co.
C 0.85, Si 0-0.25, Mn 0-0.25, bal Fe.
Heat treated: 130,000-180,000 TS; 80,000-118,000 YS; 12-24 El; 32-47 RA; 260-370 Brin. For springs, taps, reamers, drills; Type W1; water hardened.

STAHLSCHMIDT STS6
Stahlwerk Stahlschmidt GmbH & Co.
C 0.7, Si 0-0.25, Mn 0-0.25, bal Fe.
Heat treated: 122,000-174,000 TS; 82,000-128,000 YS; 12-22 El; 37-52 RA; 240-352 Brin. For wheels, die blocks, rails, springs; Type W1; water hardened.

STAHLSCHMIDT SUPERIOR
Stahlwerk Stahlschmidt GmbH & Co.
C 0.86, Cr 4.1, Mo 0.85, V 2.5, W 12, bal Fe.
For lathe and planer tools, reamers, broaches; high speed steel.

STAHLSCHMIDT SZP 120
Stahlwerk Stahlschmidt GmbH & Co.
C 0.3, Co 2, V 0.25, W 8.5, Cr 2.4, bal Fe.
For extrusion dies, rams and liners, punches, shear blades; oil hardened, hot work steel.

STAHLSCHMIDT SZP 40
Stahlwerk Stahlschmidt GmbH & Co.
C 0.3, Cr 2.35, V 0.6, W 4.25, bal Fe.
For shear blades, pneumatic tools, punches; oil hardened, hot work steel.

STAHLSCHMIDT SZP 80
Stahlwerk Stahlschmidt GmbH & Co.
C 0.3, Cr 2.65, V 0.35, W 8.5, bal Fe.
For extrusion dies, rams and liners; oil hardened, hot work steel.

STAHLSCHMIDT UE10
Stahlwerk Stahlschmidt GmbH & Co.
C 0.2, Cr, Mo, bal Fe.
For gears, bolts, camshafts, cams; case hardened, tough.

STAHLSCHMIDT UE8
Stahlwerk Stahlschmidt GmbH & Co.
C 0.15, Cr, Mo, bal Fe.
For gears, bolts, camshafts, cams; case hardened, tough.

STAHLSCHMIDT UV12
Stahlwerk Stahlschmidt GmbH & Co.
C 0.25, Mn 0.65, Cr 1, Mo 0.2, bal Fe.
For gears, bolts, camshafts; oil hardened, tough.

STAHLSCHMIDT UV13
Stahlwerk Stahlschmidt GmbH & Co.
C 0.33, Cr 1, Mo 0.2, Mn 0.65, bal Fe.
For gears, bolts, machine tool parts; oil hardened, tough.

STAHLSCHMIDT UV14
Stahlwerk Stahlschmidt GmbH & Co.
C 0.42, Cr 1, Mo 0.2, bal Fe.
For gears, bolts, crankshafts, dies; oil hardened, tough.

STAHLSCHMIDT UV24
Stahlwerk Stahlschmidt GmbH & Co.
C 0.5, Cr 1, Mo 0.2, Mn 0.65, bal Fe.
For gears, bolts, crankshafts, fasteners; oil hardened, shock resistant.

STAHLSCHMIDT V35
Stahlwerk Stahlschmidt GmbH & Co.
C 0.35, Mn 0.55, Si 0.25, bal Fe.
Hot rolled: 85,000 TS; 54,000 YS; 30 El; 53 RA; 183 Brin. For gears, bolts, crankshafts, axles, screws; water hardened.

STAHLSCHMIDT V45
Stahlwerk Stahlschmidt GmbH & Co.
C 0.45, Mn 0.5, Si 0.25, bal Fe.
Hot rolled: 98,000 TS; 59,000 YS; 24 El; 45 RA; 212 Brin. For axles, gears, bolts, crankshafts; water hardened.

STAHLSCHMIDT V60
Stahlwerk Stahlschmidt GmbH & Co.
C 0.61, Si 0.25, Mn 0.55, bal Fe.
Heat treated: 115,000-100,000 TS; 77,000-113,000 YS; 12-23 El; 40-54 RA; 230-320 Brin. For girders, rails, springs, die blocks; water hardened.

STAIN-ROD NO. 308
Marquette Corp.
C 0-0.07, Ni 9, Cr 19, bal Fe.
Welded: 85,000-95,000 TS; 40-50 El. For stainless steel welding rod; austenitic, stainless.

STAINALOY
Acme Electric Welder Co.
Copper. Cu alloy.
For resistance and spot welding electrodes.

STAINLEND 14/75
Arcos Alloys
Now ARCOS 4NiA.

STAINLEND 15/35
Now STAINLEND 330.

STAINLEND 15/35
Arcos Alloys
Cr 15, Ni 35, C, bal Fe.
For welding electrodes; stainless, austenitic. *Obsolete*

STAINLEND 19/9
Now STAINLEND 347.

STAINLEND 25/12
Now STAINLEND 309 CB.

STAINLEND 308
Arcos Alloys
Now ARCOS 308.

STAINLEND 308HC
Arcos Alloys
Now ARCOS 308HC.

STAINLEND 308L
Arcos Alloys
Now ARCOS 308L.

STAINLEND 309
Arcos Alloys
Now ARCOS 309.

STAINLEND 309 CB
Arcos Alloys
Stainless steel. C 0.08, Mn 1.9, Si 0.5, Cr 23, Ni 13, 0.8 Cb + Ta, bal Fe.
For welding electrodes; Cb-stabilized stainless steel for root pass welding in Type 347 clad steel. *Obsolete*

STAINLEND 309 MO
Arcos Alloys
Now ARCOS 309 MO.

STAINLEND 310
Arcos Alloys
Now ARCOS 310.

STAINLEND 310 CB
Arcos Alloys
Now ARCOS 310 CB.

STAINLEND 310 HC
Arcos Alloys
Now ARCOS 310 HC.

STAINLEND 310 MO
Arcos Alloys
Now ARCOS 310 MO.

STAINLEND 312
Arcos Alloys
Now ARCOS 312.

STAINLEND 316
Arcos Alloys
Now ARCOS 316.

STAINLEND 316L
Arcos Alloys
Now ARCOS 316L.

STAINLEND 317
Arcos Alloys
Now ARCOS 317L.

STAINLEND 318
Arcos Alloys
Now ARCOS 318.

STAINLEND 330
Arcos Alloys
Now ARCOS 330.

STAINLEND 347
Arcos Alloys
Now ARCOS 347.

STAINLEND 349
Arcos Alloys
Stainless steel. C 0.09, Mn 1.8, Si 0.5, Cr 19.5, Ni 9, Mo 0.5, Cb 0.9, W 1.4, bal Fe.
For welding electrodes; superalloy electrode, stainless steel type 349. *Obsolete*

STAINLEND 410
Arcos Alloys
Now ARCOS 410.

STAINLEND 430
Arcos Alloys
Stainless steel. C 0.08, Mn 0.8, Si 0.4, Cr 16.5, Ni 0.4, bal Fe.
For welding electrodes; ferritic stainless steel, type 430. *Obsolete*

STAINLEND 8N12
Arcos Alloys
Now ARCOS 8N12.

STAINLEND KMOCB
Arcos Alloys
Cr 18, Ni 12, C, Mo, Cb, bal Fe.
For welding electrodes; stainless, austenitic. *Obsolete*

STAINLEND HCN
Now STAINLEND 310.

STAINLEND KMO
Now STAINLEND 316.

STAINLEND KMOCB
Now STAINLEND 318.

STAINLESS "X"
Teledyne Vasco
C 0.33, Cr 14, bal Fe.
108,000-231,000 TS; 5-23 El; 12-54 RA; 228-477 Brin. For cutlery, valves. *Obsolete*

STAINLESS INVAR
American manufacture
Fe 36.5, Cr 9.5, Co 54.
For instruments; low expansion.

STAINLESS INVAR
Japanese manufacture
Co 5, Ni 31.5, Fe 63.5.
For stainless and corrosion resistant parts and equipment.

STAINLESS IRON
H.K. Porter Co., Inc.
C 0-0.12, Cr 12-23, bal Fe.
For corrosion resistant parts; corrosion resistant. *Obsolete*

STAINLESS IRON "N"
Latrobe Steel Co.
C 0.1, Cr 16.5, bal Fe.
66,000-149,000 TS; 44,000-121,000 YS; 11-25 El; 32-65 RA; 165-315 Brin. For stainless iron parts; stainless iron; IZ 20-80. *Obsolete*

STAINLESS IRON (AMERICAN STAINLESS)
American Stainless Steel Co.
Stainless steel. C 0-0.12, Cr 8-60, bal Fe.
For hardware, tools, chemical equipment; stainless.

STAINLESS IRON 2 FM
Colt Industries
Cr 14-15, Mo 0.55, C 0-0.11, bal Fe.
90,000-140,000 TS; 55,000-100,000 YS; 19-28 El; 46-62 RA; 160-300 Brin. For stainless parts; corrosion resisting. *Obsolete*

STAINLESS IRON C-1
American Stainless Steel Co.
Stainless steel. C 0-0.12, Cr 0-15, bal Fe.
70,000-190,000 TS; 16-35 El; 140-387 Brin. For stampings, automobile trim moldings, machine parts, cutlery; corrosion resistant.

STAINLESS IRON C-2
American Stainless Steel Co.
Stainless steel. C 0-0.12, Cr 15-18, bal Fe.
75,000-90,000 TS; 28 El; 170 Brin. For exterior trim for buildings and automobiles, cold rivets, nitric acid equipment, stampings; corrosion resistant.

STAINLESS IRON C-3
American Stainless Steel Co.
Stainless steel. C 0-0.35, Cr 18-23, bal Fe.
100,000 TS; 22 El; 196 Brin. For screw machine products, wire paper mill products, chemical apparatus; corrosion resistant.

STAINLESS IRON C-4
American Stainless Steel Co.
Stainless steel. C 0-0.35, Cr 23-30, bal Fe.
95,000 TS; 18 El; 187 Brin. For furnace parts, valve parts, conveyor chain bolts and nuts, heat resisting parts; corrosion and heat resistant.

STAINLESS IRON NO. 16
Colt Industries
C 0-0.1, Mn 0.35-0.6, Cr 15-18, bal Fe.
Annealed: 75,000-85,000 TS; 45,000-55,000 YS; 35-45 El; 60-70 RA; 155-170 Brin. To resist nitric acid and gunpowder; austenitic, corrosion resistant. *Obsolete*

STAINLESS IRON NO. 18
Colt Industries
C 0-0.1, Mn 0.35-0.6, Cr 18-23, bal Fe.
Annealed: 70,000-80,000 TS; 45,000-55,000 YS; 30-40 El; 50-60 RA; 165-185 Brin. To resist mixed acid conditions as nitric acid-sulfuric acid and high temperatures; austenitic, corrosion resistant. *Obsolete*

STAINLESS IRON NO. 24
Colt Industries
C 0.25, Mn 0.35-0.6, Cr 24-26, bal Fe.
Rolled: 80,000-90,000 TS; 55,000-65,000 YS; 20-30 El; 40-50 RA; 170-200 Brin. To resist mixed acids and high temperatures; stainless; austenitic. *Obsolete*

STAINLESS IRON TYPE F.I.
Firth-Vickers Stainless Steels Ltd.
C 0.1, Cr 11-13, Si 0-1, Mn 0-1, bal Fe.
Heat treated: 68,000-83,000 TS; 40,000-56,000 YS; 30-40 El; 50-60 RA; 150-180 Brin. For turbine blades, shrouding strip, aircraft fittings; Type 403 and 410; corrosion resistant.

STAINLESS IRON TYPE FI 17
Firth-Vickers Stainless Steels Ltd.
C 0.1, Si 0.2, Mn 0.2, Cr 17, bal Fe.
Annealed: 75,000 TS; 48,000 YS; 32 El; 50 RA; 160 Brin. For corrosion resistant parts; brittle welds, corrosion resistant.

STAINLESS M.S. NO. 2
Carpenter Technology Corp.
C 0.3, Cr 13, bal Fe.
For plastic mold dies; oil or air hardened. *Obsolete*

STAINLESS NO. 20
Duriron Co. Inc.
C 0-0.07, Mn 0.75, Si 1, Cr 20, Ni 29, Mo 2, Cu 3, bal Fe.
Rolled: 85,000 TS; 35,000 YS; 35-50 El; 50-70 RA; 150-180 Brin. For wrought stainless parts; corrosion resistant to H$_2$SO$_4$. *Obsolete*

STAINLESS S-20
Stainless Foundry & Engineering Inc.
C 0-0.07, Ni 29, Cr 20, Mo 3-4, Cu 4, bal Fe.
Cast: 65,000-75,000 TS; 28,000-38,000 YS; 35-50 El; 45-50 RA; 120-160 Brin. For chemical plant equipment, valves, agitators, tanks; resists mixed acid austenitic.

STAINLESS STEEL
Wilbur B. Driver Co.
C 0-0.07, Cr 16-23, Ni 7-11, bal Fe.
For stainless parts; stainless. *Obsolete*

STAINLESS STEEL (THOMAS STEEL CO)
Thomas Steel Strip Co.
Cr 18, Ni 8, C, bal Fe.
For stainless parts; stainless. *Obsolete*

STAINLESS STEEL A
American Stainless Steel Co.
Stainless steel. Cr 8-16, 0.12 C min, bal Fe.
Hardened: 260,000 TS; 10 El; 512 Brin. For cutlery, springs, balls, bearing races, surgical instruments, hypodermic needles; stainless and corrosion resistant.

STAINLESS STEEL A.S.S
American Stainless Steel Co.
Stainless steel. Cr 8-60, Mn 0.4, 0.12 C min, bal Fe.
Rolled: 80,000-250,000 TS; 24,000-150,000 YS; 10-15 El; 30-35 RA; 1 Brin. For cutlery, surgical instruments, valve seats, pump shafts; corrosion and abrasion resisting.

STAINLESS STEEL B
American Stainless Steel Co.
Stainless steel. Cr 16-18, 0.12 C min, bal Fe.
Hardened: 601 Brin. For cutlery, bearing races, surgical instruments; stainless and corrosion resistant.

STAINLESS STEEL CLAD ALUMINUM

Texas Instruments Inc./Materials Control
Type 434 or 301 stainless steel clad to 1100, 3003, or 5052 aluminum in various ratios of stainless steel to aluminum. For automotive trim and truck bumpers. Available in H or O tempers.

STAINLESS STEEL GRADE A

Hawkridge Bros. Co.
C 0.1, Cr 18, Ni 8, bal Fe.
For corrosion resistant parts; stainless, austenitic. *Obsolete*

STAINLESS STEEL GRADE H.C.

Hawkridge Bros. Co.
C 0.2, Cr 18, Ni 8, bal Fe.
For stainless parts; stainless, austenitic. *Obsolete*

STAINLESS STEEL TYPE F.I.

Firth-Vickers Stainless Steels Ltd.
C 0-0.12, Si 0-0.8, Mn 0-1, Cr 11.5-13.5, bal Fe.
Martensitic type stainless steel. BS 970 (PT4) 410S21; similar to AISI 403, 410.

STAINLESS STEEL TYPE F.T.V.

Firth-Vickers Stainless Steels Ltd.
C 0.04, Si 0.4, Mn 0.8, Cr 20, bal Fe.
Annealed: 72,000 TS; 49,300 YS; 30 El; 160 Brin. For sealing to soft glass; corrosion resistant. *Obsolete*

STAINLESS STEEL TYPE FAL

Firth-Vickers Stainless Steels Ltd.
C 0.1, Si 0.4, Mn 0.6, Cr 12, Al 4.5, bal Fe.
Heat treated: 88,000 TS; 62,000 YS; 25 El. For resistances, rheostats; maximum service temperature 700°C.

STAINLESS STEEL TYPE FG

Firth Brown Ltd.
C 0.25, Cr 12-14, Si 0.4, Mn 0.3, bal Fe.
Heat treated: 90,000-112,000 TS; 56,000-78,500 YS; 30-20 El; 60-50 RA; 220-240 Brin. For turbine blades, high pressur steam valves; Type 420; corrosion resistant.

STAINLESS STEEL TYPE FG

Firth-Vickers Stainless Steels Ltd.
C 0.25, Cr 12-14, Si 0.4, Mn 0.3, bal Fe.
Heat treated: 90,000-112,000 TS; 56,000-78,500 YS; 30-20 El; 60-50 RA; 220-240 Brin. For turbine blades, high pressur steam valves; Type 420; corrosion resistant.

STAINLESS STEEL TYPE FH

Firth-Vickers Stainless Steels Ltd.
C 0.3, Cr 12-14, bal Fe.
Heat treated: 230,000 TS; 450-550 Brin. For cutlery, surgical instruments, knives, edge tools, rolls, springs, ball bearings and races; magnetic, stainless; Firth Brown I-162, Firth FH Stainless.

STAINLESS STEEL TYPE FHM

Firth-Vickers Stainless Steels Ltd.
C 0.8, Si 0.2, Mn 0.2, Cr 17, Mo 0.5, bal Fe.
Annealed: 120,000 TS; 98,000 YS; 260 Brin. For ball bearings; corrosion resistant.

STAINLESS W

United States Steel Corp.
C 0.06, Mn 0.7, Si 1.25, Ni 7, Cr 17, Ti 1.1, Al 0.5.
Heat treated: 200 ksi TS; 170 ksi YS; 10 El. Ultra high strength corrosion resistant steel. For pump and valve parts, wind tunnels.

STAINTEC 10670

Eutectic Corp.
Nickel base alloy powder for overlays on steel and nickel alloys. *Obsolete*

STAINTIN 157PA

Eutectic Corp.
For soldering stainless steel. 425°F MP.

STAINTRODE A

Eutectic Corp.
Electrode for AC-DC welding of 300 series stainless. 90,000 psi TS.

STAINTRODE A-MO

Eutectic Corp.
Electrode for AC-DC welding of 303, 315, 316 and 329 stainless steels. 80,000 psi TS. *Obsolete*

STAINTRODE A-MO-L

Eutectic Corp.
Electrode for AC-DC welding of 316, 316L and 318 stainless steels. 90,000 psi TS.

STAINTRODE B

Eutectic Corp.
Now STAINTRODE B-L. *Obsolete*

STAINTRODE B-L

Eutectic Corp.
For AC-DC metallic arc welding of 301, 302, 304, 305, 306 and 308 stainless steels; low carbon content; 85,000 psi TS.

STAINTRODE B-MO

Eutectic Corp.
Now STAINTRODE BMO-L. *Obsolete*

STAINTRODE BMO-L

Eutectic Corp.
Electrode for AC-DC welding of several 300 series stainless steels; low carbon content; 90,000 psi TS.

STAINTRODE D

Eutectic Corp.
Electrode for AC-DC welding of type 310 stainless. 95,000 psi TS.

STAINWELD 308-15

Lincoln Electric Co.
C 0.08, Cr 18, Ni 9, Mn 2.5, Si 0.9, P 0.04, S 0.03, bal Fe.
Stainless steel, arc welding electrode. AWS Class E308-15.

STAINWELD 308-16

Lincoln Electric Co.
C 0.08, Cr 18, Ni 9, Mn 2.5, Si 0.9, P 0.04, S 0.03, bal Fe.
Stainless steel, arc welding electrode. AWS Class E308-16.

STAINWELD 308L-16

Lincoln Electric Co.
C 0.08, Cr 18, Ni 9, Mn 2.5, Si 0.9, P 0.04, S 0.03, bal Fe.
Stainless steel, welding electrode. AWS Class E308L-16.

STAINWELD 309-16

Lincoln Electric Co.
C 0.08, Cr 23.9, Ni 12.4, bal Fe.
Stainless steel, arc welding electrode. AWS E309-16.

STAINWELD 310-15

Lincoln Electric Co.
C 0.2, Cr 25, Ni 20, Mn 2.5, Si 0.75, bal Fe.
Stainless steel, arc welding electrode. AWS Class E310-15.

STAINWELD 310-16

Lincoln Electric Co.
C 0.2, Cr 25, Ni 20, Mn 2.5, Si 0.75, bal Fe.
Stainless steel, arc welding electrode. AWS Class E310-16.

STAINWELD 316L-16

Lincoln Electric Co.
C 0.04, Cr 17, Ni 11, Mo 2, Mn 2.5, Si 0.9, bal Fe.
Stainless steel, arc welding electrode. AWS Class E316L-16.

STAINWELD 347-15

Lincoln Electric Co.
C 0.08, Cr 18, Ni 9, Mn 2.5, Si 0.9, bal Fe.
Stainless steel, arc welding electrode. AWS Class E347-15.

STAINWELD 347-16

Lincoln Electric Co.
C 0.08, Cr 18, Ni 9, Mn 2.5, Si 0.9, bal Fe.
Stainless steel, arc welding rod. AWS Class E347-16.

STAINWELD B

Lincoln Electric Co.
Cr 25, Ni 12, C 0.2, Cb, bal Fe.
For stainless steel welding rods; coated, shielded arc. *Obsolete*

STAINWELD-A

Lincoln Electric Co.
Cr 18, Ni 8, C, Cb, bal Fe.
For hard facing welding rod for corrosion resistance; corrosion resistant. *Obsolete*

STALCREST

British Steel plc
C 0-0.17, Si 0-0.4, Mn 0-1, P 0.07-0.1, V 0-0.1, Cr 0.7-1, Cu 0.25-0.55, 0.025 Al min, bal Fe.
Weldable steel with good weathering properties. Properties meet BS 4360 WR 50. *Obsolete*

STALINIT NO. 2

Russian manufacture
Ti-Cr-Fe.
For cutting tools; hard sintered alloy.

STALINITE

Russian manufacture
Cr 9.5, Mn 11.5, C, bal Fe.
For machine parts, dredging, oil drilling, agricultural and kindred equipment; sintered alloy.

STALLOY

English manufacture
Si 3-4, C, bal Fe.
For telephone and loud speaker diaphragms, electrical machinery, instruments; max permeability 8,000.

STAMINAL

Midvale-Heppenstall Co.
C 0.5, Si 1, Ni 2.7, Mo 0.3, Cr 0.4, bal Fe.
For hollow drills; oil hardening. *Obsolete*

STAMINAL

Latrobe Steel Co.
C 0.55, Cr 0.4, Ni 2.7, Mn 0.9, V 0.13, Mo 0.45, Si 1, bal Fe.
For chisels, cold cutters, die blocks, punches; oil or air hardened; tough.

STAN-FOS

Johnson Matthey plc
Cu 86.25, P 6.75, Sn 7.
Phosphorus bearing silver brazing alloy; 640-680°C MP; DIN 8513.

STANDALLOY

Degussa AG
Silver.
Silver alloy for mixing amalgams for dentistry and dental engineering.

STANDALLOY F

Degussa AG
Silver.
Silver alloy for mixing amalgams for dentistry and dental engineering.

STANDARD

Boyd-Wagner Co.
C 0.9, Mn 0.3, bal Fe.
For general tools, drills, stone tools, dies; water hardened.

STANDARD

Boyd-Wagner Co.
C 0.9, Mn 0.3, bal Fe.
For general tools, drills, stone tools, dies; water hardened.

STANDARD (DISSTON)

Disston Inc.
C 0.7-1.2, bal Fe.
For tools, drills, taps; water hardened. *Obsolete*

STANDARD (WARDS)

Edgar T. Ward's Sons Co.
C 0.7-1.4, bal Fe.
For tools. *Obsolete*

STANDARD ADMIRALTY BRONZE

Belmont Metals Inc.
Cu 88, Sn 10, Zn 2.
For bronze castings, gears, worm wheels, trolley wheels; tough, wear resistant.

STANDARD ALLOY "H.R.-1"

Michigan Standard Alloy Inc.
Ni 20, Cr 20, C 0.4, Si 1.5, Mn 0.7, bal Fe.
For furnace parts; heat and corrosion resistant.

STANDARD ALLOY "H.R.-2"

Michigan Standard Alloy Inc.
Ni 25, Cr 20, Si 1.5, C 0.4, Mn 0.7, bal Fe.
For furnace parts; heat and corrosion resistant.

STANDARD ALLOY "H.R.-3"

Michigan Standard Alloy Inc.
Ni 36, Cr 15, C 0.4, Mn 0.7, Si 1.5, bal Fe.
Cast: 62,000 TS; 35,000 YS; 10 El; 11 RA; 190 Brin. For lead pots, carburizing boxes, trays; heat resistant.

STANDARD ALLOY "H.R.-4"

Michigan Standard Alloy Inc.
Ni 62, Cr 14, C 0.4, Mn 0.7, Si 1.5, bal Fe.
Cast: 65,000 TS; 35,000 YS; 5 El; 5 RA; 180 Brin. For carburizing boxes, retorts; heat resistant.

STANDARD ALLOY "H.R.-5"

Michigan Standard Alloy Inc.
Ni 20, Cr 25, C 0.4, Mn 0.7, Si 1.7, bal Fe.
73,000 TS; 46,000 YS; 17 El. For furnace parts; heat resistant.

STANDARD ALLOY "H.R.-6"

Michigan Standard Alloy Inc.
Ni 13, Cr 25, C 0.4, Mn 0.7, Si 1.7, bal Fe.
Cast: 85,000 TS; 50,000 YS; 15 El; 15 RA; 170 Brin. For furnace rails, stack dampers; heat resistant.

STANDARD ALLOY "H.R.-7"

Michigan Standard Alloy Inc.
Stainless steel. Ni 8, Cr 18, C 0.3, Mn 1, Si 1, bal Fe.
Cast: 97,000 TS; 45,000 YS; 25 El; 20 RA. For stainless castings; stainless.

STANDARD ALLOY "H.R.-8"

Michigan Standard Alloy Inc.
Cr 28-30, Ni 3, Cr 0.4, Si 1.8, Mn, bal Fe.
Cast: 50,000 TS; 40,000 YS; 1 El; 0 RA; 196 Brin. For acid resistant equipment, grate bars; acid and corrosion resistant.

STANDARD ALLOY CO. H R-5 M

Michigan Standard Alloy Inc.
Fe 50, Cr 25, Ni 20, Mo 2.5-4, Mn 0.4, C 0.3.
Cast: 62,000 TS; 51,000 YS; 18 El; 13 RA; 190-210 Brin. For furnace parts; heat and corrosion resistant.

STANDARD ALLOY CR-1

Michigan Standard Alloy Inc.
C 0-0.08, Mn 1.5, Si 2, Cr 18-21, Ni 8-11, bal Fe.
Cast: 70,000 TS; 28,000 YS; 55 El; 150 Brin. For chemical and food processing equipment, furnace parts; ACI-CF8; corrosion and heat resistant.

STANDARD ALLOY CR-2

Michigan Standard Alloy Inc.
C 0-0.08, Mn 1.5, Si 0-1.5, Cr 18-21, Ni 9-12, Mo 2-3, bal Fe.
Cast: 75,000 TS; 30,000 YS; 50 El; 160 Brin. For chemical plant equipment; ACI-CF8M; corrosion resistant.

STANDARD ALLOY CR-3

Michigan Standard Alloy Inc.
C 0-0.2, Mn 0-1.5, Si 0-2, Cr 18-21, Ni 8-11, bal Fe.
Cast: 75,000 TS; 30,000 YS; 50 El; 160 Brin. For chemical and food processing equipment; ACI-CF20; corrosion resistant.

STANDARD ALLOY HR-10

Michigan Standard Alloy Inc.
C 0.21-0.35, Cr 24-30, Ni 7-11, Si, bal Fe.
For heat and corrosion resistant parts; heat and corrosion resistant.

STANDARD ALLOY HR-11

Michigan Standard Alloy Inc.
C 0.21-0.35, Cr 16-23, Ni 12-15, bal Fe.
For heat and corrosion resistant parts; heat and corrosion resistant.

STANDARD ALLOY HR-9

Michigan Standard Alloy Inc.
Cr 18, C, bal Fe.
For corrosion resistant parts; corrosion resistant.

STANDARD ALLOY K.A. 4

Michigan Standard Alloy Inc.
Cr 17-20, Ni 8-10, C 0.25, Mo 2-4, bal Fe.
Annealed: 105,000 TS; 49,000 YS; 55 El; 62 RA. For pipe fittings, pump casings, impellers, shafts; austenitic; will not respond to heat treatment.

STANDARD ALLOY, SPECIAL

Michigan Standard Alloy Inc.
Cr 20, Ni 25, C 0.4, bal Fe.
62,100 TS; 3 El; 177 Brin. For furnace parts; heat and corrosion resistant.

STANDARD ALNICO

Harsco Corp. (Taylor-Wharton Div.)
Al 11-13, Ni 19-23, Co 4.5-5.5, bal Fe.
For magnets; Alnico I. *Obsolete*

STANDARD AR-1

Michigan Standard Alloy Inc.
C 0.2, Ni 9, Cr 19, Si 3, bal Fe.
Cast. For castings, valves and fittings for handling H_2SO_4; acid resistant.

STANDARD BEARING

Manufacturer not listed.
Pb 58, Sn 26, Sb 15, Cu 1.
For bearings; anti-friction.

STANDARD BEARING

Bethlehem Steel Corp.
C 1, Mn 0.35, Cr 1.5, bal Fe.
For bearings, rollers, master gauges; water hardening. *Obsolete*

STANDARD CADMIUM

American manufacture
Ag 92.5, Cu 5.75, Cd 1.75.
For solder, brazing; silver solder.

STANDARD CARBON

Republic Steel Corp.
C 0.65-1, Mn 0.3, Si 0.15-0.3, bal Fe.
For cold work dies; water hardening. *Obsolete*

STANDARD CARBON

Latrobe Steel Co.
Now LATROBE CARBON.

STANDARD DIEHARD

Firth Brown Ltd.
C 1.5, Si 0.3, Mn 0.3, Cr 12, Mo 0.9, V 1.15, bal Fe.
For dies; nondeforming. *Obsolete*

STANDARD DRACO

Cytemp Specialty Steel Div.
C 0.7-1.2, V 0.2, bal Fe.
For tools, dies, punches; water hardened. *Obsolete*

STANDARD DRACO DV

Cytemp Specialty Steel Div.
C 1, V 0.5, bal Fe.
For drills, taps, reamers, broaches; Type W3; oil or water hardened. *Obsolete*

STANDARD G

D.G. Gautier & Co.
C 0.7-1.4, bal Fe.
For chisels, tools; water hardened.

STANDARD GLYCO

Joseph T. Ryerson & Son Inc.
Sn, Sb, bal Pb.
For bearings; Babbitt.

STANDARD GOLD

Manufacturer not listed.
Au 90, Cu 10.
For money, jewelry; corrosion resistant.

STANDARD GOLD (GREAT BRITAIN)

English manufacture
Au 92, Cu 8.
For coinage; corrosion resistant.

STANDARD H.R. ALLOYS

Michigan Standard Alloy Inc.
Cr 16-25, Ni 20-60, C, bal Fe.
For furnace parts; heat resistant.

STANDARD MISCO

Michigan Steel Casting Co.
Fe 46, Ni 35, Cr 15, Mn 0.5, C 0.6.
Cast: 75,000 TS; 70,000 YS; 3 El; 6 RA; 180 Brin. Rolled: 100,000 TS; 81,000 YS; 42 El; 52 RA; 190 Brin. For heat treating boxes, furnace parts, valves, stills; heat resisting; wear and corrosion resistant.

STANDARD NICKEL BRONZE

Belmont Metals Inc.
Cu 88, Sn 5, Zn 2, Ni 5.
For valves, corrosion resistant parts; corrosion resistant.

STANDARD NO. 4 BABBITT

Hoyt Metal Co.
Pb, Sb, Sn.
At 70°F: 14.3 Brin. At 212°F: 6.4 Brin. For bearings for line shafts; Babbitt, heavy loads.

STANDARD NO. 4 BABBITT

NL Industries
Pb, Sb, Sn.
At 70°F: 14.3 Brin. At 212°F: 6.4 Brin. For bearings for line shafts; Babbitt, heavy loads.

STANDARD PHOSPHOR BRONZE

American manufacture
Cu 80, Sn 10, Pb 10, up to 0.25 P.
Cast: 28,000 TS; 6-9 El; 55-65 Brin. For high speed bearings; heavy duty.

STANDARD SILVER

Manufacturer not listed.
Ag 92.5, Cu 7.5.
For coinage; corrosion resistant.

STANDARD SILVER STERLING

Manufacturer not listed.
Ag 90, Cu 10.
For cutlery, ornaments, utensils.

STANDARD TOOL
Colt Industries
C 0.07-1.2, bal Fe.
For general tools; see Black Diamond. *Obsolete*

STANDFEST SPEZIAL
Stahlwerke Sudwestfalen
C 0.9, Si 0.25, Mn 1.9, V 0.1, bal Fe.
For forming and blanking dies, punches; oil hardened, non-deforming. *Obsolete*

STANELEC
Empire Sheet & Tin Plate Co.
Si 0.25, bal Fe.
For motors, armatures; high permeability.

STANFIRE
Standard Brake Shoe & Foundry Co.
C 3.2, Ni 1.5, Cr 0.8, bal Fe.
For fire box grates; heat resisting cast iron.

STANLEY GRADE A
Stanley Works
C 0-0.12, Cr 12-15, bal Fe.
For corrosion resistant parts; corrosion resistant. *Obsolete*

STANLEY GRADE B
Stanley Works
C 0-0.12, Cr 15-18, bal Fe.
For corrosion resistant parts; corrosion resistant. *Obsolete*

STANLEY GRADE C
Stanley Works
C 0.13-0.35, Cr 18-23, bal Fe.
For corrosion resistant parts; corrosion resistant. *Obsolete*

STANLEY GRADE D
Stanley Works
C 0.13-0.2, Cr 17-19, Ni 7-9, bal Fe.
For stainless, heat and corrosion resistant parts; stainless, heat and corrosion resistant. *Obsolete*

STANLEY, GRADE E
Stanley Works
C 0.13-0.2, Cr 12-15, Ni 5-7, Mn, bal Fe.
For corrosion resistant parts; corrosion resistant. *Obsolete*

STANNIOL
Manufacturer not listed.
Sn 96, Pb 2.4, Cu 1, Ni 0.3, Fe 0.1.
For bearings, wrapping foil.

STANNIOL
Manufacturer not listed.
Sn 98.9, Pb 0.7, Cu 0.33.
For bearings, wrapping foil.

STANNUM METAL
Lumen Bearing Co.
Sn 90, Sb 6, Cu 4.
Cast. For machine bearings; high tin Babbitt.

STANTUFONT
Soc. Alluminio Veneto per Azioni
Aluminum. Cu 0.5-1, Si 12.4-13, Mg 0.7-1, Ni 2-2.4, bal Al
Heat treated: 40,000-50,000 TS; 38,000-43,000 YS; 0.3-0.5 El; 120-130 Brin. For light alloy parts; age hardened.

STAR
A. Oohn Ltd.
Cu 78, Sn 6, Zn 10, Pb 6.
For castings; free cutting.

STAR BLUE CHIP
Teledyne Firth Sterling
Steel. C 0.7, W 14, Cr 4, V 2, bal Fe.
For cutters, drills, reamers, hobs, lathe and planer tools; high speed steel. *Obsolete*

STAR COLUMBIUM M-8
Carpenter Technology Corp.
C 0.8, Cr 4.25, W 5.5, Mo 4.5, V 1.5, Cb 1.25, bal Fe.
For cold work dies, blanking dies; high speed steel. *Obsolete*

STAR ETD
Vereinigte Edelstahlwerke
C 0.5, Cr 1.5, W 2.25, V 0.25, bal Fe.
For hot work tools and dies; hot work steel. *Obsolete*

STAR J METAL
Now HAYNES STELLITE STAR J.

STAR-BORON
Carpenter Technology Corp.
C 0.5-0.6, Cr 4-4.5, W 1.5-2, V 1.5-2, Mo 8-9, Co 7.75-8.5, B 0.4-0.6, bal Fe.
For tools, cutters; high speed steel. *Obsolete*

STAR-MAX
Now CARPENTER STAR MAX.

STAR-MO
Teledyne Firth Sterling
Steel. C 0.8, W 6.5, Mo 5, Cr 4, V 2, bal Fe.
For cutting tools; high speed steel. *Obsolete*

STAR-MO F.M.
Teledyne Firth Sterling
Steel. C 0.83, W 6.4, Mo 5, Cr 4, V 2, S 0.12, bal Fe.
For reamers, broaches, chasers, lathe and form tools, milling cutters; high speed steel, free machining. *Obsolete*

STAR-MO MOD
Teledyne Firth Sterling
Steel. C 0.65, Cr 4.2, Mo 6, W 6, V 2, bal Fe.
For reamers, drills, lathe and planer tools; high speed steel. *Obsolete*

STAR-MO-M2
Teledyne Firth Sterling
C 0.82, W 6.5, Mo 5, Cr 4.2, V 1.9, bal Fe.
For cutting tools, taps, reamers, drills, broaches, thread chasers; high speed steel. *Obsolete*

STAR-MOLY
Carpenter Technology Corp.
Cr 3.25-4.25, W 1.25-2, V 0.75-1.5, Mo 7.5-9.5, High C, (Co), bal Fe.
For reamers, lathe tools, drills, broaches, dies, hot work dies, cutting tools; high speed steel, wear resistant. *Obsolete*

STAR-ZENITH
Now CARPENTER STAR-ZENITH.

STARK
Latrobe Steel Co.
C 1.32, W 5.5, Cr 4.5, V 4, Mo 4.5, bal Fe.
Hardened: 64-66 Rock C. For drills, reamers, lathe and planer tools, chisels, plastic core pins, chasers. Type M-4 tool steel. Abrasion resistant and tough. High speed steel.

STARKAD
Uddeholm Corp.
C 1.3, Cr 0.25, W 5.5, bal Fe.
For drawing dies; oil hardened. *Obsolete*

STARLI
Paul Bergsoe & Son
Sn 85, Sb 10, Cu 5.
Cast: 19,900 TS; 28 Brin. MP: 440-630°F. For engine bearings. Shock resistant.

STARO
Stahlwerk Stahlschmidt GmbH & Co.
C 0.4, Si 0.4, Mn 0.3, Cr 13, bal Fe.
Annealed: 100,000 TS; 55,000 YS; 20 El; 50 RA; 210 Brin. For valves, cutlery, surgical and dental instruments; Type 420; stainless.

STARRETT NO. 496
Starrett Co.
C 0.9, Mn 12.2, Cr 0.5, W 0.5, V 0.2, bal Fe.
For punches, crimpers, upsetters, headers; oil hardened, non-deforming.

STARRETT NO. 496
York Machine & Supply Co.
C 0.9, Mn 12.2, Cr 0.5, W 0.5, V 0.2, bal Fe.
For punches, crimpers, upsetters, headers; oil hardened, non-deforming.

STARRETT NO. 497
York Machine & Supply Co.
C 0.95-1.05, Mn 0.6, Si 0.4, Cr 5-5.5, Mo 1, V 0.2, bal Fe.
For thread roller dies, master hubs, and forming dies. Air hardening and non-deforming.

STARRETT NO. 497
L.S. Starrett Co.
C 0.95-1.05, Mn 0.6, Si 0.4, Cr 5-5.5, Mo 1, V 0.2, bal Fe.
For thread roller dies, master hubs, forming dies; air hardening, nondeforming.

STARRETT NO. 498
L.S. Starrett Co.
C 0.18.
Free machining low carbon steel for precision ground flat stock.

STATO
Uddeholm Corp.
C 0.35, Cr 1.1, Mo 0.2, bal Fe.
For gears, shafts, mandrels; oil hardened. *Obsolete*

STATO 21
Uddeholm Corp.
C 0.15, Si 0.25, Mn 0.6, Mo 0.32, bal Fe.
For gears, cams, camshafts; case hardened, tough. *Obsolete*

STATO 23
Uddeholm Corp.
C 0.15, Mn 0.8, Cr 1, Mo 0.35, bal Fe.
For gears, shafts; case hardened. *Obsolete*

STATO 28
Uddeholm Corp.
C 0.1, Cr 2.5, Mo 1, bal Fe.
For gears, cams, camshafts; case hardened, tough. *Obsolete*

STATO 3
Uddeholm Corp.
C 0.15, Cr 1.2, Mo 0.25, bal Fe.
For gears, shafts, machinery parts; case hardened. *Obsolete*

STATO 5
Uddeholm Corp.
C 0.25, Mn 0.65, Cr 1.05, Mo 0.2, bal Fe.
For gears, shafts, bolts; tough. *Obsolete*

STATO 8
Uddeholm Corp.
C 0.4, Mn 0.65, Cr 1.1, Mo 0.25, bal Fe.
For gears, shafts, bolts; tough. *Obsolete*

STATOS EXTRA
Now VEW K720.

STATOS SPEZIAL
Now VEW K100.

STATOS SPEZIAL GW
Now VEW K116.

STATOS SPEZIAL W
Now VEW K107.

STATOS SUPERIOR
Bohler Gesellschaft M.B.H.
C 1.05, Cr 1, W 1.15, bal Fe.
For cold working tools, headers, punches, bearings; water hardened, wear resistant.

STATOS V
Bohler Gesellschaft M.B.H.
C 1.45, Cr 1.4, bal Fe.
For punches, drawing dies, bearings, liners; water hardened.

STATUARY BRONZE
Manufacturer not listed.
Cu 88-95, Sn 1.4-10, Zn 9.5, Pb 0-6, trace P.
For statuary, castings.

STATUARY BRONZE MELANCHTON, WITTENBERG
Manufacturer not listed.
Cu 89.55, Sn 2.99, Zn 7.45.
For statuary; good castability.

STATUARY BRONZE, AUGSBURG "A"
Manufacturer not listed.
Cu 89.43, Sn 8.17, Pb 1.07, P 0.34, Ni 0.19.
For statuary; corrosion resistant.

STATUARY BRONZE, AUGSBURG "B"
Manufacturer not listed.
Cu 94.74, Sn 1.64, Zn 0.54, Pb 6.24, Ni 0.71.
For statuary; corrosion resistant.

STATUARY BRONZE, BACCHUS POTSDAM
Manufacturer not listed.
Cu 89.34, Sn 7.5, Zn 1.63, Pb 1.21, P 0.18.
For statuary; corrosion resistant.

STATUARY BRONZE, BAVARIA MUNICH
Manufacturer not listed.
Cu 91.55, Sn 1.77, Zn 5.5, Pb 1.3.
For statuary; good castability.

STATUARY BRONZE, COLUMN VENDOME
Manufacturer not listed.
Cu 89.2, Sn 10.2, Zn 0.5, Pb 0.1.
For statuary; good castability.

STATUARY BRONZE, MUNICH
Manufacturer not listed.
Cu 92.88, Sn 4.18, Zn 0.44, Pb 2.31, P 0.15.
For statuary; good castability.

STAUFFER 90TA-10W
Stauffer Chemical Corp.
Ta 90, W 10.
Annealed: 80,000 TS; 67,000 YS; 25 El; 50 RA. Cold drawn: 160,000 TS; 1 El. For rocket engine parts, nozzles, flame barriers; useful in temperature range of -200 F to +5200 F. *Obsolete*

STAUFFER SCB-291
Stauffer Chemical Corp.
Ta 10, W 10, bal Cb.
For high temperature applications; heat resistant. *Obsolete*

STAUFFER SM 291
Stauffer Chemical Corp.
Ta 10, W 10, bal Cb.
At 2200 F: 37,100 TS; 34,200 YS. At 3000 F: 11,800 TS; 10,500 YS. For space vehicles, nuclear reactors. High temperature strength. *Obsolete*

STAUFFER STA 880
Stauffer Chemical Corp.
W 12.5, bal Ta.
At room temperature: 102,000 TS; 85,000 YS; 23 El (annealed). At 5300 F: 930 TS; 880 YS. For nozzles, space equipment, rocket components, chemical equipment. Develops optimum strength above 4000 F. Melted by electron beam. *Obsolete*

STAUFFER STA-900
Stauffer Chemical Corp.
W 10, Ta 90.
For high temperature applications; corrosion and heat resistant. *Obsolete*

STAVANGER N250
Stavanger Co.
Stainless steel. C 0.25, Cr 23-30, bal Fe.
Annealed: 85,000 TS; 55,000 YS; 25 El; 83 Rock B. For annealing boxes, oil burners, glass molds, furnace parts and fittings, exhaust manifolds. Type 446 stainless steel, heat and corrosion resistant.

STAVANGER N330
Stavanger Co.
C 0-0.35, Cr 23-27, bal Fe.
Annealed: 90,000 TS; 60,000 YS; 20 El; 45 RA; 180 Brin. For furnace parts, heat treating boxes; Type 446; heat resistant.

STAVANGER S128M SPECIAL
Stavanger Co.
Stainless steel. C 0-0.06, Cr 18, Ni 9, Mo 1.5, bal Fe.
Annealed: 85,000 TS; 35,000 YS; 50 El; 65 RA; 160 Brin. For acid resistant chemical plant equipment; Type 316; stainless.

STAVANGER S178
Stavanger Co.
Stainless steel. C 0.08-0.2, Mn 0-2, Cr 17-19, Ni 8-10, bal Fe.
Annealed: 85,000 TS; 35,000 YS; 60 El; 70 RA; 150 Brin. Cold drawn: 125,000 TS; 95,000 YS; 22 El; 55 RA; 277 Brin. For oil refinery and chemical plant equipment; Type 302; stainless, austenitic.

STAVANGER S178 IISP
Stavanger Co.
Stainless steel. C 0.08, Cr 19, Ni 11, bal Fe.
Annealed: 80,000 TS; 30,000 YS; 60 El; 80 Rock B. For chemical plant equipment, digesters, evaporators, tanks. Type 304; stainless steel, austenitic.

STAVANGER S178 SUPRA
Stavanger Co.
Stainless steel. C 0.03, Cr 18-20, Ni 9-12, bal Fe.
Annealed: 77,000 TS; 30,000 YS; 60 El; 75 Rock B. For chemical plant equipment, evaporators, digesters, tanks. Type 304L; stainless steel, austenitic.

STAVANGER S1C SPEC
Stavanger Co.
Stainless steel. C 0.08, Cr 14, bal Fe.
Annealed: 75,000 TS; 36,000 YS; 30 El; 80 Rock B. For flat springs, knives, tableware, chemical plant equipment. Type 410; stainless steel, heat and corrosion resistant.

STAVANGER S1CH-NI
Stavanger Co.
C 0-0.15, Cr 13, Ni 2, bal Fe.
For oil refinery equipment, valves; corrosion resistant.

STAVANGER S1CHMO
Otto Wolff Handelgesellschaft
C 0.12, Cr 13, Mo 1.1, bal Fe.
Annealed: 80,000 TS; 35,000 YS; 25 El; 92 Rock B. For springs, shafts, table flatware, oil refinery and chemical plant equipment. Corrosion resistant.

STAVANGER S23/20
Stavanger Co.
Stainless steel. C 0.2, Si 2, Cr 25, Ni 20, bal Fe.
Annealed: 100,000 TS; 50,000 YS; 45 El; 90 Rock B. For furnace parts, annealing boxes, heat treating fixtures. Type 314; stainless steel, austenitic, heat and corrosion resistant.

STAVANGER S268M
Stavanger Co.
C 0.2, Cr 23-28, Ni 2.5-5, Mo 1-2, bal Fe.
Normalized: 103,000 TS; 78,000 YS; 18 El; 45 RA; 235 Brin.
Annealed: 95,000 TS; 41,000 YS; 29 El; 60 RA; 225 Brin. For valves, pumps, furnace parts; Type 327; heat resistant.

STAVANGER S278
Stavanger Co.
C 0.12, Cr 17-19, Ni 8-10, bal Fe.
Annealed: 90,000 TS; 40,000 YS; 50 El; 58 Rock B. For chemical plant equipment, vessels, tanks, mixers, agitators. Type 302 stainless steel, austenitic.

STAVANGER SC-140
Stavanger Co.
Stainless steel. Cr 12-14, 0.15 C min, bal Fe.
Annealed: 88,000 TS; 40,000 YS; 32 El; 68 RA; 170 Brin. For cutlery, valve trim, turbine blades; Type 420; stainless, hardenable.

STAVANGER SC140L
Stavanger Co.
Stainless steel. C 0.2, Cr 14, bal Fe.
eat treated: 225,000 TS; 200,000 YS; 12 El; 48 Rock C. For cutlery, surgical instruments, scissors, knives, gears, shafts. Type 420; stainless steel, hardenable. Heat and corrosion resistant.

STAVANGER SC140MO
Stavanger Co.
C 0.14, Cr 14.5, Mo 1, bal Fe.
Annealed: 80,000 TS; 35,000 YS; 25 El; 92 Rock B. For springs, shafts, table flatware, oil refinery and chemical plant equipment. Corrosion resistant, hardenable.

STAVANGER SCN192
Stavanger Co.
C 0-0.25, Cr 17, Ni 2, bal Fe.
For oil refinery equipment; Type 341; corrosion and heat resistant.

STAVANGER SIC
Stavanger Co.
C 0-0.08, Cr 11.5-13, Al 0.1-0.3, bal Fe.
Annealed: 71,000 TS; 42,600 YS; 22 El; 70 RA; 150 Brin. Heat treated: 175,000 TS; 145,000 YS; 21 El; 64 RA; 352 Brin. For oil refinery and chemical plant equipment; Type 405; corrosion resistant.

STAVANGER SIC16
Stavanger Co.
Stainless steel. C 0-0.12, Cr 14-18, bal Fe.
Annealed: 70,000 TS; 40,000 YS; 30 El; 55 RA; 150 Brin. Cold drawn: 130,000 TS; 120,000 YS; 2 El; 185 Brin. For oil refinery equipment, bolts, kitchen sinks; Type 430; stainless, ferritic.

STAVANGER SIC18
Stavanger Co.
Stainless steel. C 0-0.12, Cr 14-18, bal Fe.
Annealed: 70,000 TS; 40,000 YS; 30 El; 55 RA; 150 Brin. Cold drawn: 130,000 TS; 120,000 YS; 2 El; 185 Brin. For oil refinery equipment, bolts, kitchen sinks; Type 430; stainless, ferritic.

STAVANGER SICH
Stavanger Co.
C 0-0.15, Cr 11.5-13.5, Mn 0-1, bal Fe.
Annealed: 80,000 TS; 40,000 YS; 35 El; 70 RA; 155 Brin. Cold drawn: 100,000 TS; 85,000 YS; 17 El; 60 RA; 205 Brin. For valve parts, turbine blades, cutlery, knives; Type 410; corrosion resistant.

STAVANGER SN12-12
Stavanger Co.
C 0.08-0.16, Cr 12, Ni 12, bal Fe.
For valves; corrosion and heat resistant.

STAVANGER SN128, SPEC
Stavanger Co.
Stainless steel. C 0-0.1, Cr 16-18, Ni 10-14, Mo 1.7-2.7, Cb = 10 x C, bal Fe.
Annealed: 85,000-95,000 TS; 35,000-45,000 YS; 50-60 El; 60-75 RA; 150-190 Brin. For acid resistant chemical plant equipment; Type 316Cb; stainless, austenitic.

STAVANGER SN128IIISP

Stavanger Co.
Stainless steel. C 0.06, Cr 16-18, Ni 10-14, Mo 2-3, bal Fe.
Annealed: 80,000 TS; 30,000 YS; 60 El; 78 Rock B. For chemical plant equipment, vessels, tanks, digesters, agitators. Type 316; stainless steel, austenitic.

STAVANGER SN128M

Stavanger Co.
Stainless steel. C 0-0.1, Cr 16-18, Ni 10-14, Mo 2-3, bal Fe.
Annealed: 85,000-95,000 TS; 35,000-45,000 YS; 50-60 El; 60-75 RA; 150-190 Brin. For acid resistant chemical plant equipment; Type 316; stainless, austenitic.

STAVANGER SN128MIIISP

Stavanger Co.
Stainless steel. C 0.08, Cr 17, Ni 12, Mo 2, bal Fe.
Annealed: 80,000 TS; 30,000 YS; 60 El; 78 Rock B. For chemical plant equipment, tanks, digesters, agitators, mixers. Type 316; stainless steel, austenitic.

STAVANGER SN129

Stavanger Co.
Stainless steel. C 0-0.08, Cr 18-20, Ni 8-11, Mn 0-2, bal Fe.
Annealed: 90,000 TS; 45,000 YS; 60 El; 135 Brin. Cold drawn: 180,000 TS; 150,000 YS; 10 El; 330 Brin. For chemical plant equipment, welded structures; Type 304; stainless, austenitic.

STAVANGER SNW

Stavanger Co.
C 0-0.25, Cr 24-26, Ni 19-22, bal Fe.
Annealed: 95,000 TS; 45,000 YS; 50 El; 65 RA; 185 Brin. At 1200°F: 57,000 TS; 22,000 YS; 32 El; 45 RA. For furnace parts and equipment, heat treating boxes; Type 310; austenitic, heat resistant.

STAVANGER SS286M

Stavanger Co.
Stainless steel. C 0.12, Cr 25-30, Ni 3.5-6, Mo 1-2, bal Fe.
Annealed: 105,000 TS; 80,000 YS; 25 El; 230 Brin. For valves, valve fittings, pump parts, furnace equipment. Precipitation hardening. Type 329; stainless steel, austenitic, corrosion and heat resistant.

STAVAX ESR

Uddeholm Corp.
C 0.35, Mn 0.45, Si 0.45, Cr 13.6, bal Fe.
Annealed: 210-220 Brin. High strength, deep hardened plastic mold steel with ultra high polish potential; oil hardened, non-deforming. AISI Type 420; corrosion resistant steel.

STAY-BRITE 8 SOLDER

J.W. Harris Co., Inc.
Ag 0-6, bal Sn.
Melting range: 430-535°F liquidus. Similar to Stay-Brite, Plastic range useful in bridging wider gaps. NSF accepted.

STAY-BRITE SOLDER

J.W. Harris Co., Inc.
Ag 0-4, bal Sn.
Melting range: 430°F liquidus. Low temperature solder for all metals except aluminum. Used in refrigeration joints.

STAY-SILV 15

J.W. Harris Co., Inc.
Cu 80, P 5, Ag 15.
Melting range: 1190-1480°F solidus; 1190-1480°F liquidus. For brazing ferrous and non-ferrous alloys. AWS A5.8 BCuP-5.

STAY-SILV 2

J.W. Harris Co., Inc.
Cu 91, P 7, Ag 2.
Melting range: 1190-1450°F solidus; 1190-1450°F liquidus. Filler metal for brazing copper in plumbing, air conditioning, electrical connections. AWS A5.8 BCuP6; BCuP6.

STAY-SILV 5

J.W. Harris Co., Inc.
Cu 89, P 6, Ag 5.
Melting range: 1190-1500°F solidus; 1190-1500°F liquidus. For brazing copper base alloys where close fit-ups cannot be maintained. AWS A5.8 BCuP-3.

STAY-SILV 6

J.W. Harris Co., Inc.
Cu 87.5, P 6.5, Ag 6.
Melting range: 1190-1425°F solidus; 1190-1425°F liquidus. For brazing ferrous and non-ferrous alloys. AWS A5.8 BCuP-3; ASTM B260-62T BCuP-3.

STAY-SILV 6HP

J.W. Harris Co., Inc.
Cu 86.8, P 7.2, Ag 6.
Melting range: 1190-1335°F solidus; 1190-1335°F liquidus. For brazing ferrous and non-ferrous alloys. AWS A5.8 BCuP-4.

STAY-SILV O

J.W. Harris Co., Inc.
Cu 92.9, P 7.1.
Melting range: 1310-1475°F solidus; 1310-1475°F liquidus. Filler metal for brazing copper in plumbing, air conditioning, electrical connections. AWS A5.8 BCuP-2; BCuP-2.

STAYBLADE MAX

Firth-Vickers Stainless Steels Ltd.
C 0.12, Cr 18.5, Ni 8.5, Ti 0.8, Al 1.4, bal Fe.
For turbine blades, boiler drums; austenitic; resists heat to 900°C.

STAYBLADE STEEL

Firth-Vickers Stainless Steels Ltd.
C 0.22, Si 1, Mn 0.6, Cr 20, Ni 8.5, Ti 1.2, bal Fe.
Heat treated: 95,000 TS; 58,000 YS; 30 El; 50 RA; 200 Brin. For heat and corrosion resistant parts; heat and corrosion resistant.

STAYBOLT

F.A.C.A.
C 0.05, Si 0.15, 2 SiO$_2$, bal Fe.
For staybolts; wrought iron.

STAYBRITE 17/7

Firth-Vickers Stainless Steels Ltd.
C 0.1, Si 0.6, Mn 1.3, Cr 17.5, Ni 7.5, bal Fe.
Soft sheet: 108,000 TS; 38,000 YP; 55 El. Hard sheet: 180,000 TS; 146,000 YP; 10 El. For springs, aircraft structural parts, diaphragms, household utensils, trim. Type 301; stainless steel, nonmagnetic, austenitic.

STAYBRITE 254

Firth-Vickers Stainless Steels Ltd.
C 0.07, Si 0.4, Mn 0.8, Cr 18, Ni 18, Ti 0.6, Mo 3.7, Cu 2.4, bal Fe.
Annealed: 84,000 TS; 38,000 YS; 47 El; 66 RA; 160 Brin. For stainless parts; austenitic, stainless, stabilized. *Obsolete*

STAYBRITE 317 (L)

Firth-Vickers Stainless Steels Ltd.
C 0.03, Si 0.2-1, Mn 0.5-2, Cr 18-19.5, Ni 14-16, Mo 3-4, bal Fe.
Annealed: 72,000 TS min; 28,000 YS min; 40 El min. Weldable without subsequent heat treatment. Excellent corrosion resistance and resistance to pitting corrosion; for parts in textile, paper pulp and food industries. AISI 317L.

STAYBRITE D.D.Q.

Firth-Vickers Stainless Steels Ltd.
C 0.1, Cr 12, Ni 12, Si 0.3, Mn 0.8, bal Fe.
Annealed: 78,000-90,000 TS; 32,000-37,000 YS; 40-60 El; 40-60 RA; 130-150 Brin. For domestic table and holloware, decorative trim; austenitic, stainless.

STAYBRITE EMS

Firth-Vickers Stainless Steels Ltd.
C 0.12, Si 0.8, Mn 1.6, S 0.25, Cr 18, Ni 10, Mo 0.28, Ti 0.6, bal Fe.
Annealed: 82,000 TS; 36,000 YS; 53 El; 54 RA; 175 Brin. For stainless parts; free-cutting, austenitic, stainless.

STAYBRITE F.C.B. (H) STEEL

Firth-Vickers Stainless Steels Ltd.
C 0.04-0.09, Cr 17-19, Ni 9-12, Nb = 10 x C, bal Fe.
Stabilized, austenitic stainless steel. Similar to AISI 347.

STAYBRITE F.D.P.

Firth-Vickers Stainless Steels Ltd.
C 0-0.1, Si 0.2-1, Mn 0.5-2, Cr 17-19, Ni 8.5-10.5, Ti = 5 x C min, bal Fe.
Ann: 78,000 TS min; 30,000 YS min; 40 El min; 180 Vickers. Weldable stainless; for welded assemblies; for operation at elevated temperatures. BS EN.58B, S129, S520, S521; AISI 321.

STAYBRITE F.D.P. (H) STEEL

Firth-Vickers Stainless Steels Ltd.
C 0.04-0.09, Cr 17-19, Ni 9-12, Mo 0.7, Ti = 5 x C min, bal Fe.
Stabilized, austenitic stainless steel. Similar to AISI 321.

STAYBRITE F.D.P. (L)

Firth-Vickers Stainless Steels Ltd.
C 0.05, Si 0.8, Mn 0.8, Cr 18, Ni 11, Ti = 5 x C min, bal Fe.
Annealed: 90,000 TS; 40,000 YP; 50 El; 50 RA; 160 Brin. For welded equipment in chemical, plastic and textile plants, tanks. Type 321; stainless, austenitic, welding grade. Stabilized.

STAYBRITE F.M.B.

Firth-Vickers Stainless Steels Ltd.
C 0.07, Cr 18, Ni 8, Mo 3, bal Fe.
Annealed: 90,000-112,000 TS; 36,000-42,500 YS; 40-60 El; 40-60 RA; 160-190 Brin. For chemical apparatus to resist acetic and other acids at elevated temperatures; rust and acid resisting; non-magnetic; Firth Brown I-166.

STAYBRITE F.M.B. (H) STEEL

Firth-Vickers Stainless Steels Ltd.
C 0.04-0.09, Cr 16-18, Ni 10-13, Mo 2-2.75, bal Fe.
Austenitic stainless steel. Similar to AISI 316.

STAYBRITE F.M.B. (L)

Firth-Vickers Stainless Steels Ltd.
C 0.03, Si 0.4, Mn 1.5, Cr 17.5, Ni 12, Mo 2.8, bal Fe.
Annealed: 88,000 TS; 35,000 YP; 50 El; 50 RA; 160 Brin. For chemical, textile, plastic and pharmaceutical plant equipment. Type 316L; stainless steel, austenitic, nonmagnetic.

STAYBRITE F.M.B.3 T

Firth-Vickers Stainless Steels Ltd.
C 0.06, Cr 18-20, Ni 10-14, Mo 3-4, bal Fe.
Annealed: 85,000 TS; 45,000 YS; 60 El; B 80 Rock. For chemical plant equipment, paper and pulp mill components, digesters, evaporators, mixers. Type 317 stainless steel, austenitic, acid resistant. *Obsolete*

STAYBRITE F.M.S.

Firth-Vickers Stainless Steels Ltd.
C 0.06, Si 0.9, Mn 0.55, Cr 19.5, Ni 8.7, Mo 2.8, Nb 0.8, bal Fe.
Austenitic stainless steel spun castings, weldable. Cast: 90,000 TS; 45,000 YS (0.2%); 35 El. For cast parts (requiring welding) for use in food and chemical industries. *Obsolete*

STAYBRITE F.M.X.

Firth-Vickers Stainless Steels Ltd.
C 0.05, Si 0.6, Mn 0.95, Cr 23, Ni 9.4, Mo 3.4, bal Fe.
Austenitic stainless steel spun castings. Cast: 98,000 TS; 56,000 YS (0.2%); 30 El. For cast parts used in food and chemical industries. *Obsolete*

STAYBRITE F.S.L. (L)

Firth-Vickers Stainless Steels Ltd.
C 0.03, Si 0.6, Mn 1, Cr 18.5, Ni 11, bal Fe.
Annealed: 85,000 TS; 36,000 YS; 60 El; 65 RA; 160 Brin. For chemical, textile and pharmaceutical processing equipment, tanks, agitators, vessels, filters. Type 304L; stainless, austenitic, nonmagnetic.

STAYBRITE F.S.L. CO. TA

Firth-Vickers Stainless Steels Ltd.
C 0-0.03, Mn 0.5-2, Si 0.2-1, Cr 18-19.5, Ni 10-11.5, Co, Ta, bal Fe.
Austenitic stainless steel; for elevated temperature and corrosion resistant operations. Weldable.

STAYBRITE F.S.T.

Firth-Vickers Stainless Steels Ltd.
C 0-0.1, Cr 18, Ni 8, Mn 0.8, Si 0.6, bal Fe.
Annealed: 83,000-101,000 TS; 34,000-41,000 YS; 40-60 El; 40-60 RA; 160-180 Brin. For vats, tanks, ship fittings, trim, reflectors; Type 302; austenitic, stainless.

STAYBRITE F.S.T. (FC)

Firth-Vickers Stainless Steels Ltd.
C 0-0.1, Si 0.2-1, Mn 1-2, S 0.15-0.3, Cr 17-19, Mo 0.2-0.5, Ni 8.5-10.5, bal Fe.
Annealed: 78,000 TS min; 30,000 YS min; 40 El min; 175 Vickers. Free machining austenitic stainless for machined parts as studs, cap screw shafts. EN 58 AM; AISI 303.

STAYBRITE F.S.T. (L)

Firth-Vickers Stainless Steels Ltd.
C 0.06, Si 0.6, Mn 1.5, Cr 18.5, Ni 10, bal Fe.
Annealed: 90,000 TS; 36,000 YP; 60 El; 65 RA; 160 Brin. For architectural trim, kitchen equipment, chemical and textile processing equipment. Type 304; stainless steel, nonmagnetic, austenitic.

STAYBRITE F.S.T. (L) (H) STEEL

Firth-Vickers Stainless Steels Ltd.
C 0.04-0.09, Cr 17.5-19, Ni 8-11, bal Fe.
Austenitic stainless steel. Weldable; for elevated temperature operation.

STAYBRITE FCB

Firth-Vickers Stainless Steels Ltd.
C 0.12, Si 0.6, Mn 0.4, Cr 18, Ni 10, Cb 1.2, bal Fe.
Annealed: 84,000 TS; 37,000 YS; 58 El; 65 RA; 175 Brin. For stainless parts; austenitic, stainless. Similar to AISI 347.

STAYBRITE FCB (T)

Firth-Vickers Stainless Steels Ltd.
C 0.13, Si 0.6, Mn 1.2, Cr 18, Ni 12, Cb 1.3, bal Fe.
Heat treated: 85,000 TS; 38,000 YS; 58 El; 65 RA; 175 Brin.
For stainless parts; stainless, austenitic.

STAYBRITE FG

Firth-Vickers Stainless Steels Ltd.
C 0.2, Cr 18, Ni 8, bal Fe.
Annealed: 80,000 TS; 35,000 YS; 55 El; 75 RA; 150 Brin. For chemical plant equipment, tanks, filters; stainless, austenitic; Type 302. *Obsolete*

STAYBRITE FH

Firth-Vickers Stainless Steels Ltd.
C 0.22, Cr 12-14, bal Fe.
Annealed: 95,000 TS; 50,000 YS; 25 El; B 92 Rock. Heat treated: 250,000 TS; 215,000 YS; 8 El; C 50 Rock. For cutlery, surgical instruments, scissors, knives, gears, shafts. Type 420 stainless steel, hardenable. Heat and corrosion resistant. *Obsolete*

STAYBRITE FJ

Firth-Vickers Stainless Steels Ltd.
C 0.1, Cr 11.5-13.5, bal Fe.
Annealed: 70,000 TS; 35,000 YS; 30 El; B 80 Rock. For flat springs, knives, tableware, chemical plant equipment. Type 410 stainless steel, heat and corrosion resistant. *Obsolete*

STAYBRITE FMB (FC)

Firth-Vickers Stainless Steels Ltd.
C 0-0.08, Mn 0-2, Si 0-1, Cr 16-18, Mo 1.75-2.5, Ni 10-14, P 0.2, 0.10 S min, bal Fe.
Free machining austenitic stainless steel.

STAYBRITE FMB (L) CO

Firth-Vickers Stainless Steels Ltd.
C 0-0.03, Mn 0.5-2, Si 0.2-1, Cr 16.5-18, Ni 12-14, Mo 2.25-3, Co, bal Fe.
Austenitic stainless steel.

STAYBRITE FMB (TI)

Firth-Vickers Stainless Steels Ltd.
C 0.07, Si 0.3, Mn 0.3, Cr 18, Ni 12-14, Mo 3.25, Ti = 4 x C min, bal Fe.
Annealed: 85,000 TS; 38,000 YS; 50 El; 50 RA; 170 Brin. For chemical plant equipment, evaporators; austenitic, stainless.

STAYBRITE FMB(V4A)

Oederlin & Co. Ltd.
C, Cr, Ni, bal Fe.
Annealed: 58,000-78,000 TS; 29,000-35,000 YS; 12-18 El; 170-200 Brin. For chemical plant equipment, dye plant equipment; stainless, austenitic. *Obsolete*

STAYBRITE FML

Firth-Vickers Stainless Steels Ltd.
C 0.07, Si 0.6, Mn 0.8, Cr 18, Ni 9.5, Mo 1.25, bal Fe.
Annealed: 85,000 TS; 38,000 YS; 50 El; 50 RA; 170 Brin. For stainless parts; austenitic, stainless.

STAYBRITE FSL

Firth-Vickers Stainless Steels Ltd.
C 0.05, Si 0.5, Mn 0.4, Cr 19, Ni 10, bal Fe.
Annealed: 84,000 TS; 35,000 YS; 60 El; 65 RA; 160 Brin. For stainless parts; austenitic, stainless. *Obsolete*

STAYBRITE FSM (2)

Firth-Vickers Stainless Steels Ltd.
C 0.1, Cr 18, Ni 5, Mn 8, bal Fe.
Rolled: 112,000 TS; 56,000 YS; 50 El; 200 Brin. For domestic utensils and equipment; stainless. *Obsolete*

STAYBRITE STEEL

Firth Brown Ltd.
Cr 18, Ni 8-9, C, bal Fe.
100,000 TS; 35,000 YS; 50 El; 40 RA; 153 Brin. For stainless steel parts; corrosion resisting. *Obsolete*

STEAM BRONZE-1

American manufacture
Cu 85, Zn 5, Pb 5, Sn 5.
26,000 TS; 16 El; 20 RA. For valves and fittings, automotive parts; castings.

STEAM BRONZE-2

American manufacture
Cu 88, Pb 2, Sn 8.
32,000 TS; 22 El; 25 RA. For valves, fittings, automobile parts; free-cutting.

STEAM METAL

Belmont Metals Inc.
Cu 88, Sn 6, Pb 3, Zn 3.
For injectors, valves, steam specialties; free-cutting.

STEAM METAL, AJAX

Ajax Metal
Copper. Cu 87, Sn 6, Zn 5, Pb 2.

STEAM METAL, AJAX

H. Kramer & Co.
Copper. Cu 87, Sn 6, Zn 5, Pb 2.

STEAM VALVE BRONZE

Manufacturer not listed.
Sn, Zn, Pb, bal Cu.
For steam valves; pressure tight castings.

STEB METAL

Samuel Taylor Ltd.
Ag, Cu.
For enameling and electrical contacts. Silver rolled into copper.

STEDMETAL

Stedman Foundry & Machine Co.
C 3.2, Si 2.5, Ni 1.5, Mn 0.7, bal Fe.
For crushers, grinders, pulverizers. Ni-cast iron.

STEEL 15 CR 3

German manufacture
C 0.12-0.18, Mn 0.5, Si 0.2, Cr 0.5-0.8, bal Fe.
Heat treated: 85,000-120,000 TS. For gears, shafts, machine tool parts; case hardening steel, water hardened.

STEEL 15 CR NI 6

German manufacture
C 0.12-0.17, Mn 0.5, Si 0.2, Ni 1.4-1.7, Cr 1.4-1.7, bal Fe.
Heat treated: 140,000-185,000 TS. For gears, cams, shafts, camshafts; case-hardening steel, oil hardened.

STEEL 16 MN CR 5

German manufacture
C 0.14-0.19, Mn 1-1.3, Si 0.25, Cr 0.8-1.1, bal Fe.
Heat treated: 115,000-155,000 TS. For gears, shafts, machine tool parts, case hardening steel, oil hardened.

STEEL 18 CR NI 8

German manufacture
C 0.15-0.2, Mn 0.5, Si 0.2, Ni 1.8-2.1, Cr 1.8-2.1, bal Fe.
Heat treated: 170,000-205,000 TS. For gears, shafts, cams, pinions, camshafts; case hardening steel, oil hardened.

STEEL 1KH14ND

Russian manufacture
C 0.05, Cr 13.5, Ni 1.5, Cu 1.37, bal Fe.
For stainless parts; stainless, austenitic.

STEEL 1KH18N9T

Russian manufacture
C, Ni, Cr, Ti, bal Fe.
For stainless parts; stainless.

STEEL 20 MN CR 5

German manufacture
C 0.17-0.22, Mn 1.1-1.4, Si 0.2, Cr 1-1.3, bal Fe.
Heat treated: 127,000-170,000 TS. For gears, pinions, cams, shafts, camshafts; case hardening steel, oil hardened.

STEEL 20-XGP

Russian manufacture
C, alloy, bal Fe.
Rolled: 144,000 TS; 114,000 YS. For machine tool parts; case hardened.

STEEL 20-XHM

Russian manufacture
C, alloy, bal Fe.
For machine tool parts; case hardened.

STEEL 25 CR MO 4

German manufacture
C 0.22-0.29, Mn 0.7, Si 0.25, Cr 0.9-1.2, Mo 0.15-0.25, bal Fe.
Heat treated: 115,000-135,000 TS. For construction parts and equipment; oil hardened, shock resistant.

STEEL 30 CR MO V 9

German manufacture
C 0.26-0.34, Mn 0.6, Si 0.25, Cr 2.3-2.7, Mo 0.15-0.25, V 0.1-0.2, bal Fe.
Heat treated: 180,000-207,000 TS. For construction equipment, gears, crankshafts, axles; oil hardened, shock resistant.

STEEL 30 CR NI MO 8

German manufacture
C 0.26-0.34, Ni 1.8-2.1, Cr 1.8-2.1, Mo 0.25-0.35, bal Fe.
Heat treated: 177,000-207,000 TS. For gears, shafts, crankshafts, axles; oil hardened, shock resistant.

STEEL 30 MN 5
German manufacture
C 0.27-0.34, Mn 1.2-1.5, Si 0.15-0.35, bal Fe.
Heat treated: 115,000-135,000 TS. For gears, shafts, machine tools, axles; oil hardened.

STEEL 34 CR 4
German manufacture
C 0.3-0.37, Mn 0.5-0.8, Si 0.15-0.35, Cr 0.9-1.2, bal Fe.
Heat treated: 127,000-150,000 TS. For construction parts and equipment; oil hardened.

STEEL 34 CR MO 4
German manufacture
C 0.3-0.37, Mn 0.7, Si 0.25, Cr 0.9-1.2, Mo 0.15-0.25, bal Fe.
Heat treated: 127,000-135,000 TS. For construction parts and equipment; oil hardened, shock resistant.

STEEL 34 CR NI MO 6
German manufacture
C 0.3-0.38, Ni 1.4-1.7, Cr 1.4-1.7, Mo 0.15-0.25, bal Fe.
Heat treated: 157,000-185,000 TS. For gears, shafts, crankshafts, axles; oil hardening, shock resistant.

STEEL 36 CR NI MO 4
German manufacture
C 0.32-0.4, Ni 0.9-1.2, Cr 0.9-1.2, Mo 0.15-0.25, bal Fe.
Heat treated: 142,000-170,000 TS. For construction parts, gears, crankshafts; oil hardened, shock resistant.

STEEL 37 MN SI 5
German manufacture
C 0.33-0.61, Mn 1.1-1.4, Si 1.1-1.4, bal Fe.
Heat treated: 127,000-150,000 TS. For construction parts, machine tools; oil hardened.

STEEL 40 MN 4
German manufacture
C 0.36-0.44, Mn 0.8-1.1, Si 0.25-0.5, bal Fe.
Heat treated: 115,000-135,000 TS. For gears, shafts, machine tools, axles; oil hardening.

STEEL 41 CR 4
German manufacture
C 0.38-0.44, Mn 0.5-0.8, Si 0.15-0.35, Cr 0.9-1.2, bal Fe.
Heat treated: 127,000-150,000 TS. For construction parts and equipment; oil hardened.

STEEL 42 CR MO 4
German manufacture
C 0.38-0.45, Mn 0.7, Si 0.25, Cr 0.9-1.2, Mo 0.15-0.25, bal Fe.
Heat treated: 142,000-170,000 TS. For construction parts and equipment; oil hardened, shock resistant.

STEEL 42 MN V 7
German manufacture
C 0.38-0.45, Mn 1.6-1.9, Si 0.15-0.35, V 0.07-0.12, bal Fe.
Heat treated: 142,000-170,000 TS. For construction parts and equipment; oil hardened.

STEEL 50 CR MO 4
German manufacture
C 0.46-0.54, Mn 0.7, Si 0.25, Cr 0.9-1.2, Mo 0.15-0.25, bal Fe.
Heat treated: 157,000-185,000 TS. For construction parts and equipment; oil hardened, shock resistant.

STEEL 50 KHFA
Russian manufacture
C, bal Fe.
For springs; water hardened.

STEEL 55C2
Russian manufacture
C, alloy, bal Fe.
For machine tool parts.

STEEL CLAD CU & STEEL CLAD CU CLAD STEEL
Texas Instruments Inc./Materials Control
1008 AK steel clad to copper in various ratios for support arms in hermetic thermostats for motors. Tempers 1-5 available.

STEEL EC30
German manufacture
C 0.1-0.16, Cr 0.3-0.5, Mn 0.4-0.6, Si 0-0.35, bal Fe.
Heat treated: 78,000-100,000 TS. For gears, pinions, shafts, cams; case hardening, water hardened.

STEEL EC60
German manufacture
C 0.12-0.18, Cr 0.6-0.9, Mn 0.5, Si 0-0.35, bal Fe.
Heat treated: 100,000-127,000 TS. For gears, pinions, shafts, cams; case hardening, water hardened.

STEEL ECMO 100
German manufacture
C 0.18-0.23, Mn 0.9-1.2, Cr 1.1-1.4, Mo 0.3, bal Fe.
Heat treated: 155,000-205,000 TS. For gears, pinions, shafts, cams; case hardening, oil hardened.

STEEL ECMO 80
German manufacture
C 0.13-0.17, Cr 1-1.3, Mo 0.2-0.3, Mn 0.8-1.1, bal Fe.
Heat treated: 120,000-155,000 TS. For gears, shafts, pinions, cams; case hardening, oil hardened.

STEEL ECN25
German manufacture
C 0.1-0.17, Ni 2.25-2.75, Cr 0.55-0.95, bal Fe.
Heat treated: 115,000-155,000 TS. For gears, pinions, shafts, cams; case hardening, shock resistant.

STEEL ECN35
German manufacture
C 0.1-0.17, Ni 3.25-3.75, Cr 0.55-0.95, bal Fe.
Heat treated: 127,000-170,000 TS. For gears, pinions, shafts, cams, crankshafts; case hardening, shock resistant.

STEEL ECN45
German manufacture
C 0.1-0.17, Ni 4.25-4.75, Cr 0.9-1.35, bal Fe.
Heat treated: 170,000-200,000 TS. For gears, pinions, shafts, cams, camshafts; case hardening, shock resistant.

STEEL EI-257
Russian manufacture
C 0.11, Cr 13, Ni 12, W 2.5, Mo 0.68, bal Fe.
For steam tubes, boilers.

STEEL EN15
German manufacture
C 0.1-0.17, Ni 1.25-1.75, Cr 0-0.2, bal Fe.
Heat treated: 85,000-115,000 TS. For gears, pinions, shafts, cams; case hardening, shock resistant, water hardened.

STEEL OILITE A
Chrylser Corp.
C 0-0.25, 97.75% Fe min.
Sintered: 30,000 TS; 7.0 min density; 27,000 shear; 24,000 fatigue. ASTM B310-67 Type IV C1.A. PMPMA F-0000-S. *Obsolete*

STEEL OILITE AH
Chrylser Corp.
Cu 1-0, C 0.6-1, 94.0% Fe min.
Sintered, heat treated: 120,000 TS; 7.0 min density; 100,000 shear; 28,000 fatigue. *Obsolete*

STEEL-COTE NITRA-ALLOY
Fusion Inc.
For solder; for soft spot protection during heat treatment. *Obsolete*

STEEL-COTE NO. G
Fusion Inc.
For solder; general duty. *Obsolete*

STEEL-COTE NO. L
Fusion Inc.
For solder; for lead or Terne Plate. *Obsolete*

STEEL-COTE NO. T
Fusion Inc.
For solder; for tin plate. *Obsolete*

STEEL-COTE NO. W
Fusion Inc.
For solder; for weld coating. *Obsolete*

STEEL-COTE NO. Z
Fusion Inc.
For solder; for zinc or galvanized parts. *Obsolete*

STEEL-OILITE
Chrysler Corp.
Fe.
Sintered: 65,000 TS. For gears, cams, brackets, lever arms; sintered. *Obsolete*

STEELARC
Now ALL-STATE STEELARC.

STEELMET 100
CMW Inc.
Mn 1, Ni 1, bal Fe.
Sintered: 50,000-55,000 TS; 30,000-35,000 YS; 20-30 El; 100-110 Brin. For machine tool parts; prealloyed metal powders. *Obsolete*

STEELMET 101
CMW Inc.
Ni 9, Mn 1, bal Fe.
Sintered: 78,000-82,000 TS; 68,000-70,000 YS; 8-12 El; 176-185 Brin. For machine tool parts; prealloyed metal powders. *Obsolete*

STEELMET 302
CMW Inc.
C 0.2, Cr 18, Ni 8, bal Fe.
Sintered: 75,000-90,000 TS; 35,000-37,000 YS; 15-30 El; 150-185 Brin. For stainless components; stainless, austenitic. *Obsolete*

STEELMET 600
CMW Inc.
Cu 2, Ni 25, bal Fe.
Sintered: 70,000 TS; 55,000 YS; 16 El; 155 Brin. Heat treated: 80,000 TS; 88,000 YS; 3 El; 216 Brin. For machine tool parts; prealloyed alloy powders. *Obsolete*

STEELTECTIC
Eutectic Corp.
Now STEELTECTIC N. *Obsolete*

STEELTECTIC N
Eutectic Corp.
Electrode for AC-DC metallic arc pass over pass welding on mild steel without slag chipping; thin sections; 80,000 psi TS.

STEELTON L.P. IRON
Bethlehem Steel Corp.
P 0-0.035, S 0-0.035, Pig iron.
For steel making; for acid O.H. and electric furnaces. *Obsolete*

STEFANITE A2
Usines Emile Henricott, SA
Carbide.
For cutting tools; for high speed finishing.

STEFANITE A3
Usines Emile Henricott, SA
Carbide.
For cutting tools; for medium speed cutting.

STEFANITE A4
Usines Emile Henricott, SA
Carbide.
For cutting tools; medium speed machining.

STEFANITE A6
Usines Emile Henricott, SA
Carbide.
For cutting tools; for roughing and finishing.

STEFANITE A9
Usines Emile Henricott, SA
Carbide.
For cutting tools; for low speed rough turning.

STEFANITE F6
Usines Emile Henricott, SA
Carbide.
For cutting tools; for cutting cast iron.

STEIERISCHE NO. 779
Styria-Stahl Steirische Gusstahlwerke AG
C 0.7, W 18, Co 9, Mo 1, Cr 4, V 2, bal Fe.
For lathe and planer tools, milling cutters, hobs, taps; Type T4; high speed steel. *Obsolete*

STELCO CB 45-60
Stelco Steel
C 0.2, Mn 1.2, 0.005 Cb min, bal Fe.
Rolled, CB45: 60,000 TS; 45,000 YS; 25 El. CB50: 65,000 TS; 50,000 YS; 22 El. CB55: 70,000 TS; 55,000 YS; 20 El. CB60: 75,000 TS; 60,000 YS; 18 El. For mine and railroad cars, pressure vessels, derricks, booms, bridges. Good fabricability and weldability.

STELCO COLUMBIUM
Stelco Steel
Now STELCO CB 45-60.

STELCO VANADIUM 441
Stelco Steel
HSLA steel. C 0.2, Mn 1.2, 0.02 V min, bal Fe.
Rolled: 65,000 TS; 50,000 YS; 22 El. For bridges, booms, pressure vessels, mine and railroad cars. Tough, good weldability and formability. *Obsolete*

STELCOLOY 50
Stelco Steel
HSLA steel. C 0-0.15, Mn 0-1.35, Si 0.15-0.3, Cu 0.2-0.5, Ni 0.2-0.5, bal Fe.
Steel resistant to brittle fracture and atmosphere corrosion for welded, riveted or bolted construction, especially welded bridges and buildings. Rolled: 70,000 TS; 50,000 YS; 22 El.

STELCOLOY 60
Stelco Steel
HSLA steel. C 0.2, Mn 1.5, Si 0.15-0.3, Cu 0.2-0.5, Cr 0.3-0.5, Ni 0.25-0.5, V 0.01-0.12, bal Fe.
Wrought: 80,000 TS; 60,000 YS; 18 El. Atmosphere corrosion resistant.

STELCOLOY 70
Stelco Steel
HSLA steel. C 0-0.22, Mn 0-1.5, Si 0.15-0.3, Cu 0.2-0.5, Ni 0.25-0.5, Cb 0-0.05, V 0.02-0.1, bal Fe.
Steel for the transportation industry. Rolled: 90,000 TS; 70,000 YS; 14 El (8 in.).

STELLAR
Specialty Steel Co. of America
C 0.34, Mn 0.8, Si 0.5, Cr 0.9, Mo 0.35, Cu 0.33.
Low alloy, oil hardened tool steel.

STELLIT TTS NO. 1
Russian manufacture
Co-W-Cr.
For cutting tools; hard sintered alloy.

STELLIT TTS NO. 2
Russian manufacture
Co-W-Cr.
For cutting tools; hard sintered alloy.

STELLITE
Enpar Sonderwerkstoffe GmbH
Cobalt.
Alternate manufacturer.

STELLITE 1
Deloro Stellite Ltd.
C 2.4, Cr 33, W 13, bal Co.
Cast: 95,000 TS; 56 Rock C. For hardfacing applications. Heat, corrosion and wear resistant.

STELLITE 100
Deloro Stellite Ltd.
Cr 34, W 19, C 2, bal Co.
Cast: 56,000 TS; 61-66 Rock C. For tool bits, milling cutters, blades.

STELLITE 12
Deloro Stellite Ltd.
C 0.2, Cr 27, Mo 2, Co 63, Ni 2.
Cast: 120,000 TS; 30-35 Rock C. For gas turbine blades, brass casting dies, extrusion dies. Cast alloy. High temperature strength and excellent corrosion resistance.

STELLITE 12
Deloro Stellite Ltd.
C 1.8, W 9, Cr 29, bal Co.
Cast: 10,000 TS; 510 Brin. For hardfacing electrodes. Good high temperature strength. Heat and wear resistant. DTD 4734.

STELLITE 12P
Deloro Stellite Ltd.
Cr 31, W 9, C 1.4, bal Co.
Hardfacing alloy; as deposited 43-48 Rock C.

STELLITE 190
Deloro Stellite Ltd.
Cobalt. Cr 26, C 3.3, W 14, Ni 1, Si 1, Fe 8, Mn 0.5, bal Co.
Standard cobalt base alloy.

STELLITE 20
Deloro Stellite Ltd.
Cr 33, W 18, C 2.5, bal Co.
Cast: 80,000 TS; 55-59 Rock C. High abrasion and corrosion resistance. DTD 4737.

STELLITE 208
Deloro Stellite Ltd.
Co 49, Cr 26, Mo 3, Fe 20.
556 MN/m^2; 10 El; 24-29 Rock C; 210-245 DPH. Casting alloy for gas turbine blades, brass casting dies and extrusion dies.

STELLITE 238
Deloro Stellite Ltd.
Cobalt. Cr 26, C 0.1, Mo 3, Si 1, Fe 20, Mn 1, bal Co.
Standard cobalt base alloy.

STELLITE 250
Deloro Stellite Ltd.
Cr 28, C 0.1, Fe 20, bal Co.
As cast: 541 MN/m^2 TS; 309 MN/m^2 YS; 8 El; 19-29 Rock C. Casting alloy for turbine blades, brass casting dies, extrusion dies.

STELLITE 251
Deloro Stellite Ltd.
Cr 28, C 0.3, Fe 18, Nb 2, bal Co.
As cast: 618 MN/m^2 TS; 463 MN/m^2 YS; 4 El; 23-35 Rock C. Casting alloys for turbine blades, brass casting dies, extrusion dies.

STELLITE 3
Deloro Stellite Ltd.
C 2.4, Cr 30, W 13, bal Co.
Cast: 90,000 TS; 51-58 Rock C. For pump sleeves, rotary seal rings, wear pads, bearing sleeves. High abrasion and corrosion resistant. DTD 4732.

STELLITE 306
Deloro Stellite Ltd.
Cobalt. Cr 25, C 0.4, W 2, Ni 6, Si 1, Fe 4, Mn 1, Nb 5, bal Co.
Standard cobalt base alloy.

STELLITE 306
Deloro Stellite Ltd.
Co 60, Cr 25, W 2, C 0.4, Ni 5, Nb 6.
As cast: 34-39 Rock C; 345-380 DPH. Hard facing alloy having good thermal and mechanical shock.

STELLITE 31
Deloro Stellite Ltd.
Cr 26, W 7, C 0.5, Ni 10, bal Co.
Cast: 96,000 TS; 30-35 Rock C. Corrosion resistant, high temperature strength and resistance to thermal shock. DTD 4824.

STELLITE 314
Deloro Stellite Ltd.
Cobalt. Cr 32, C 1.9, W 6, Ni 6, Si 1, Fe 5, Mn 1, Nb 6, Cu 2, bal Co.
Standard cobalt base alloy.

STELLITE 4
Deloro Stellite Ltd.
Cr 33, W 14, C 1, bal Co.
Cast: 130,000 TS; 45-49 Rock C. High temperature strength and wear resistant castings.

STELLITE 50
Cabot Corporation
Co 56, Cr 27, W 14, C 2, bal Fe.
For hard facing rod, welding; abrasion resistant. *Obsolete*

STELLITE 506
Deloro Stellite Ltd.
Co 55, Cr 35, W 7.5, C 1.6.
As cast: 39-43 Rock C; 380-425 DPH. For hard surfacing.

STELLITE 6
Deloro Stellite Ltd.
Cr 26, W 5, C 1, bal Co.
Cast: 116,000 TS; 1 El; 400 Brin. For weld deposits on steam and chemical valves, shear blades, tong bits. Corrosion and heat resistant. DTD 4733.

STELLITE 694
Deloro Stellite Ltd.
Cobalt. Cr 28, C 1, W 19, Ni 5, Si 1, Fe 2.5, Mn 1, V 1, bal Co.
Standard cobalt base alloy.

STELLITE 7
Deloro Stellite Ltd.
Cr 26, W 6, C 0.4, bal Co.
Cast: 120,000 TS; 30-35 Rock C. For gas turbine blades, brass casting dies, extrusion dies. Resistant to thermal shock. Excellent corrosion resistance and high temperature strength. DTD 4736.

STELLITE 8
Deloro Stellite Ltd.
Now STELLITE 21.

STELLITE 98M2 ALLOY

Deloro Stellite, Inc.

C 1.7-2.2, Si 0-1, Ni 2-5, Fe 0-2.5, Cr 28-32, Mo 0-0.08, Mn 0-1, B 0.7-1.5, V 3.7-4.7, W 17-20, bal Co.
Sintered: 8.45 density; 150,000 psi transverse strength; 115,000 psi TS; hardness 58 Rock C. For high temperature operations. Note: this alloy is no longer a cast alloy.

STELLITE ALLOY 1

Deloro Stellite, Inc.

Reactive & refractory. Cr 33, W 13, C 2.5, bal Co.
Cast: 80,000 psi TS; 51-58 Rock C. For hardfacing electrodes; heat and wear resistant.

STELLITE ALLOY 100

Deloro Stellite, Inc.

Reactive & refractory. Cr 34, W 19, C 2, bal Co.
Cast: 56,000 psi TS; 61-66 Rock. For tool bits, milling cutter blades, etc.

STELLITE ALLOY 12

Deloro Stellite, Inc.

Reactive & refractory. Cr 29, W 9, C 1.8, bal Co.
Cast: 108,000 psi TS; 47-51 Rock C. For hardfacing electrodes; heat and wear resistant.

STELLITE ALLOY 20

Deloro Stellite, Inc.

Reactive & refractory. Cr 33, W 18, C 2.5, bal Co.
Cast: 80,000 psi TS; 55-59 Rock C. High abrasion and corrosion resistance.

STELLITE ALLOY 250

Deloro Stellite, Inc.

Reactive & refractory. C 0.1, Cr 28, Fe 20, Nb, bal Co.
As cast: 541 MN/m^2 TS; 309 MN/m^2 YS; 8 El; 19-29 Rock C. Corrosion resistant; high temperature strength, ductility, and good resistance to thermal shock; machinable. For turbine blades, brass casting dies, and extrusion dies.

STELLITE ALLOY 3

Deloro Stellite, Inc.

Reactive & refractory. Cr 30, W 13, C 2.4, bal Co.
Cast: 80,000 psi TS; 51-58 Rock C. For castings; high abrasion and corrosion resistant.

STELLITE ALLOY 4

Deloro Stellite, Inc.

Reactive & refractory. Cr 31, W 14, C 1, bal Co.
Cast: 130,000 psi TS; 45-49 Rock C. High temperature strength and wear resistant castings.

STELLITE ALLOY 6

Deloro Stellite, Inc.

Reactive & refractory. Cr 26, W 5, C 1, bal Co.
Cast: 116,000 psi TS; 39-43 Rock C. For hardfacing electrodes; heat, shock, and wear resistant.

STELLITE ALLOY 7

Deloro Stellite, Inc.

Reactive & refractory. Cr 26, W 6, C 0.4, bal Co.
Cast: 120,000 psi TS; 30-35 Rock C. Corrosion resistant; high temperature strength with good ductility and resistance to thermal shock.

STELLITE ALLOY 8

Deloro Stellite, Inc.

Reactive & refractory. Cr 27, Ni 2, C 0.2, Mo 6, bal Co.
Cast: 120,000 psi TS; 30-35 Rock C. Thermal shock resistant; high temperature strength; corrosion resistant.

STELLITE ALLOY NO. 1 (POWDER)

Deloro Stellite, Inc.

C 2.5, Cr 30, Si 1, Fe 0-3, Ni 0-3, W 12, bal Co.
For plasma arc weld surfacing; 48 Rock C. Excellent metal-to-metal wear.

STELLITE ALLOY NO. 1 (SOLID)

Deloro Stellite, Inc.

C 2.2-2.5, Cr 30, W 12, bal Co.
Bare or covered electrode for hard facing. As welded: 46-54 Rock C. Excellent metal-to-metal wear.

STELLITE ALLOY NO. 1016 (POWDER)

Deloro Stellite, Inc.

C 2.5, Cr 32, Fe 0-3, Ni 0-2.5, W 17, bal Co.
For plasma arc surfacing; 61 Rock C. Excellent erosion and corrosion resistance and metal-to-metal wear.

STELLITE ALLOY NO. 12 (POWDER)

Deloro Stellite, Inc.

C 1.4, Cr 29, Si 1.4, Fe 0-3, Ni 0-3, W 8, bal Co.
For plasma arc surfacing; 41 Rock C. Excellent erosion, corrosion resistance and good metal-to-metal wear.

STELLITE ALLOY NO. 12 (SOLID)

Deloro Stellite, Inc.

C 1.25-1.35, Cr 29, W 8, bal Co.
Bare or covered electrode, or tube wire, for hard facing. As welded: 40-47 Rock C. Excellent metal-to-metal wear.

STELLITE ALLOY NO. 156

Deloro Stellite, Inc.

C 1.6, Cr 28, Si 1.1, Ni 0-3, W 4, bal Co.
Powder for plasma arc surfacing; 43 Rock C.

STELLITE ALLOY NO. 157

Deloro Stellite, Inc.

C 0.07, Cr 21, Si 1.6, B 2.4, W 4.5, bal Co.
Powder for flame spray, plasma arc surfacing, or manual torch; 52-56 Rock C.

STELLITE ALLOY NO. 158

Deloro Stellite, Inc.

C 0.75, Cr 26, Si 1.2, Fe 0.75, Ni 0-3, B 0.7, W 5.5, bal Co.
Powder for plasma arc surfacing; 43 Rock C.

STELLITE ALLOY NO. 19

Deloro Stellite, Inc.

C 1.5-2.1, Si 0-1, Ni 0-3, Fe 0-3, Cr 29.5-32.5, W 9.5-11.5, Mn 0-1, B 0-1, bal Co.
Sintered: 8.35 density; 275,000 psi transverse strength; 150,000 psi TS; 51 Rock C. For high temperature operation. Note: this alloy is no longer a cast alloy.

STELLITE ALLOY NO. 190 PM

Deloro Stellite, Inc.

C 3-3.5, Si 0-1, Ni 0-3, Fe 0-5, Cr 24-28, W 13-15, Mn 0-1, B 0-1, bal Co.
Sintered: 8.50 density; 135,000 psi transverse strength; 90,000 psi TS; 58 Rock C. For high temperature operation.

STELLITE ALLOY NO. 21 (SOLID)

Deloro Stellite, Inc.

C 0.25, Cr 27, Mo 5, Ni 2.8, bal Co.
Cast rod, tube wire and covered electrode for hard-facing purposes. Soft as cast. Work hardens to 38-45 Rock C.

STELLITE ALLOY NO. 3 PM

Deloro Stellite, Inc.

C 2-2.7, Si 0-1, Ni 0-3, Fe 0-3, Cr 29-33, W 11-14, Mn 0-1, B 0-1, bal Co.
Sintered: 8.40 density; 140,000 psi transverse strength; 125,000 psi TS; hardness 54 Rock C. Good strength to above 1400°F.

STELLITE ALLOY NO. 4

Cabot Corporation

C, Cr, W, bal Co.
Cast: 39,100 TS; 39,000 YS; 1 El; 1 RA; 450 Brin. For tools, cutters; wear and corrosion resistant. *Obsolete*

STELLITE ALLOY NO. 40

Now HAYNES ALLOY NO. 40.

STELLITE ALLOY NO. 6 (POWDER)

Deloro Stellite, Inc.

C 1.1, Cr 28, Si 1, Fe 0-3, Ni 0-3, W 4, bal Co.
For plasma arc weld surfacing; 37 Rock C. Excellent erosion, corrosion resistance and good metal-to-metal wear.

STELLITE ALLOY NO. 6 (SOLID)

Deloro Stellite, Inc.

C 1-1.1, Cr 28, W 4, bal Co.
Bare or covered electrode, or tube wire for hard facing. As welded: 37-41 Rock C; work hardens to 49 Rock C. Excellent metal-to-metal wear.

STELLITE ALLOY NO. 6 PM

Deloro Stellite, Inc.

C 0.9-1.4, Si 0-1.5, Ni 0-3, Fe 0-3, Cr 27-31, W 3.5-5.5, Mo 0-1.5, Mn 0-1, B 0-1, bal Co.
Sintered: 8.20 density; 250,000 psi transverse strength; 130,000 psi TS; 48 Rock C. Good strength to above 1400°F.

STELLITE F

Deloro Stellite Ltd.

C 2, Mn 0.3, Si 1, Cr 25, Ni 22, Fe 1, Mo 0.6, W 12, bal Co.
For hardfacing of combustion engine valves to give enhanced resistance to corrosion and erosion; 40-45 Rock C equivalent.

STELLITE HS-30-422

Cabot Corporation

C 0.4, Cr 26, Ni 15, Co 49.5, Mo 6, Fe 1.
For blades for gas engines, jet engines and locomotive turbines; high temperature applications, heat resistant. *Obsolete*

STELLITE MALLEABLE

Cabot Corporation
Co-Cr-W.
Cast. For welding rod for hard facing wearing parts, valves, bushings. *Obsolete*

STELLITE NO. 100

Cabot Corporation
Ni-Co-Cr-W.
For cutting tools. *Obsolete*

STELLITE NO. 2

Cabot Corporation
Cr, WC, bal Co.
Cast: 67,000 TS. For cutting tools. *Obsolete*

STELLITE NO. 21 (POWDER)

Deloro Stellite, Inc.

C 0.25, Cr 27, Mo 5.5, Fe 0-2, Ni 2.8, bal Co.
For plasma arc surfacing; 28 Rock C. Good erosion and impact resistance and metal-to-metal wear.

STELLITE NO. 2400

Cabot Corporation

C 2.3, Mn 0.7, Si 0.5, Cr 31.3, W 17.5, Co 40, B 0.2, Mo 0.3, V 2.4, Fe 4.2.
For cutting tools; tough. *Obsolete*

STELLITE SF1

Deloro Stellite Ltd.

Co 45, Cr 19, W 13, C 1.3, Ni 13, Si 3, B 2.5.
As cast: 54-58 Rock C. For hardfacing; wear and abrasion resistant. DTD 4655 A.

STELLITE SF12

Deloro Stellite Ltd.

Co 52, Cr 19, W 9, C 1, Ni 13, Si 2.5, B 1.5.
Cast: 48-50 Rock C. For hardfacing; wear and abrasion resistant; machinable with difficulty. DTD 4757.

STELLITE SF20

Deloro Stellite Ltd.

Co 42, Cr 19, W 15, C 1.5, Ni 13, Si 3, B 3.
Cast: 60-62 Rock C. For hardfacing; wear and abrasion resistant; not machinable. DTD 4784.

STELLITE SF6
Deloro Stellite Ltd.
Co 50, Cr 19, W 8, C 0.7, Ni 13, Si 3, B 1.7.
Cast: 43-46 Rock C. For hardfacing; wear and abrasion resistant; machinable. DTD 4756.

STELLITE STAR J-METAL PM
Deloro Stellite, Inc.
C 2.2-2.7, Ni 0-2.5, Si 0-1, Fe 0-3, W 16-19, Mn 0-1, B 0-1, Cr 31-34, bal Co.
Sintered: 8.58 density; 125,000 psi transverse strength; 75,800 psi TS; 60 Rock C. For high temperature operations. Note: this alloy is no longer a cast alloy.

STELLITE X40
Deloro Stellite Ltd.
Now STELLITE 31.

STELLITE, ETC.
Now HAYNES STELLITE and DELORO STELLITE.

STELLRAM
Stellram Ltd.
Tool material. WC.
For cutting tools; sintered carbide.

STELMAX 120
Stelco Steel
HSLA steel. C 0.15, Mn 1.5, N 0.02, 0.005 Cb min, 0.01 V min, bal Fe.
Wrought: 120,000 TS; 120,000 YS; 8 El. For automotive bumpers and reinforcements.

STELMAX 45
Stelco Steel
HSLA steel. C 0.15, Mn 1.5, 0.005 Cb min, 0.01 V min, bal Fe.
Steel for high strength sheet with good formability and weldability. Wrought: 55,000 TS; 45,000 YS; 25 El. Meets ASTM A715, SAE J1392.

STELMAX 50
Stelco Steel
HSLA steel. C 0.15, Mn 1.5, 0.01 V min, 0.005 Cb min, bal Fe.
Wrought: 60,000 TS; 50,000 YS; 24 El. Steel for high strength sheet with good formability and weldability. Meets ASTM A715, SAE J1392.

STELMAX 60
Stelco Steel
HSLA steel. C 0.15, Mn 1.5, 0.005 Cb min, 0.01 V min, bal Fe.
Wrought: 70,000 TS; 60,000 YS; 22 El. Steel for high strength sheet with good formability and weldability. Meets ASTM A715, SAE J1392.

STELMAX 70
Stelco Steel
HSLA steel. C 0.15, Mn 1.5, N 0.02, 0.01 V min, 0.005 Cb min, bal Fe.
Wrought: 80,000 TS; 70,000 YS; 20 El. Steel for high strength sheet with good formability and weldability. Meets ASTM A715, SAE J1392.

STELMAX 80
Stelco Steel
HSLA steel. C 0.15, Mn 1.5, N 0.02, 0.005 Cb min, 0.01 V min, bal Fe.
Wrought: 90,000 TS; 80,000 YS; 18 El. Steel for high strength sheet with good formability and weldability. Meets ASTM A715, SAE J1392.

STELVETITE
British Steel plc
Mild steel; plastic coated.

STEMALLOY
Lunkenheimer Co.
Sn, alloy, bal Cu.
For valve stems and seats, globe valves; valve bronze, corrosion and wear resistant.

STENTOR
Now CARPENTER STENTOR.

STENTOR
Rochling Burbach GmbH
C 0.55, Cr 0.7, Mo 0.18, Ni 1.65, V 0.1, bal Fe.
For heading and forging dies, die casting dies; oil hardened, tough. *Obsolete*

STENTOR
Carpenter Technology Corp.
C 0.85-0.95, Mn 1.5-1.8, bal Fe.
For tools, hobs, broaches, gages, dies, cutters; oil hardening tool steel. *Obsolete*

STEPHENSON LOCOMOTIVE BEARING
English manufacture
Cu 79.5, Sn 7.5, Zn 5, Pb 8.
For bearings for locomotives; heavy duty.

STEPHENSON PISTON RINGS
English manufacture
Cu 84, Zn 8.3, Sn 2.9, Pb 4.3, Fe 0.4.
For piston rings; tough.

STEPHENSON-1
English manufacture
Cu 84, Zn 8.3, Pb 2.9, Fe 0.4.
30,000 TS; 17 El; 19 RA. For piston rings; Proportional limit 16,000.

STEPHENSON-2
English manufacture
Sn 31, Fe 31, Cu 19, Zn 19.
For castings.

STERCON-1000
Teledyne Firth Sterling
Steel. C 0.07, Cr 19, Ni 14, Mo 4.3, Ti 3, Al 1.3, Fe 1, bal Ni.
For high temperature applications; corrosion and heat resistant. *Obsolete*

STEREOTYPE METAL
Manufacturer not listed.
Sn 60, Pb 35, Sb 5.
For type metal.

STEREOTYPE METAL
Manufacturer not listed.
Pb 70, Sn 7, Sb 23.
For type.

STEREOTYPE METAL
Manufacturer not listed.
Pb 82, Sb 14.8, Sn 3.2.
12,000 TS; 4 El; 22 Brin. For type metal; M.P. 468°F.

STEREOTYPE METAL
Manufacturer not listed.
Pb 82, Sn 6, Sb 12.
For type.

STEREOTYPE METAL
Manufacturer not listed.
Pb 67, Sn 17, Sb 16.
For type.

STEREOTYPE METAL
Manufacturer not listed.
Pb 76, Sn 4, Sb 20.
For type.

STERLIN
Manufacturer not listed.
Cu 69, Ni 18, Zn 13, Pb 0.8.
For base for plated tableware; corrosion resistant.

STERLINE
Manufacturer not listed.
Cu 68-68.5, Zn 12.8-13.3, Ni 18, Fe 0.75-0.8, Pb 0-0.8.
For tableware, architectural trim; corrosion resistant.

STERLING
Castings Corp.
C 0.1, Cr 18, Ni 8, bal Fe.
Cast: 75,000 TS; 30,000 YS; 25 El; 40 RA; 150 Brin. For chemical plant equipment; Type 302; stainless, austenitic.

STERLING
English manufacture
Sn 62, Zn 15, Al 11, Pb 8.3, Cu 2.5, Sb 1.2.
For solder.

STERLING "M" DRILL STEEL
Teledyne Firth Sterling
Steel. C 1.25, Mn 0.3, Cr 0.25, bal Fe.
For drills, taps; water hardened. *Obsolete*

STERLING "V"
Teledyne Firth Sterling
C 1, Mn 0.35, Cr 0.5, V 0.2, bal Fe.
For shear knives; water or oil hardened. *Obsolete*

STERLING "V"
Teledyne Firth Sterling
C 0.9, Mn 0.35, Cr 0.15, V 0.15, bal Fe.
For punches, springs, shear blades, taps; water hardened. *Obsolete*

STERLING "V" (TAP)
Teledyne Firth Sterling
C 1.1, Mn 0.3, Cr 0.15, V 0.25, bal Fe.
For taps, drills; water hardened. *Obsolete*

STERLING 17-12 MO
Teledyne Firth Sterling
C 0-0.1, Cr 16-18, Ni 10-14, Mo 1.75-2.5, bal Fe.
For stainless and heat resistant parts; heat and corrosion resistant. *Obsolete*

STERLING 18-8FC
Teledyne Firth Sterling
C 0-0.2, Cr 17-19, Ni 8-10, Se 0.25, bal Fe.
For bolts, screws, pump shafts; stainless, free-cutting. *Obsolete*

STERLING 18-8S
Teledyne Firth Sterling
Stainless steel. C 0-0.08, Cr 18-20, Ni 8-10, bal Fe.
For stainless and heat resistant parts; stainless, malleable. *Obsolete*

STERLING 18-9 TI
Teledyne Firth Sterling
C 0-0.1, Cr 17-19, Ni 8-11, 4 x C = Ti, bal Fe.
For welded stainless parts; resists welding intergranular corrosion. *Obsolete*

STERLING 2 L 91
Sterling International Technology Ltd.
Cu 4-5, Si 0-0.25, Fe 0-0.25, Ti 0-0.25, bal Al.
Sand cast: 220 MPa TS; 165 MPa YS; 7 El; 60 Brin. BS 1490 LM 11.

STERLING 2 L 99
Sterling International Technology Ltd.
Mg 0.2-0.45, Si 6.5-7.5, Ti 0-0.2, bal Al.
Sand cast: 230 MPa TS; 185 MPa YS; 2 El; 80 Brin. BS 1490 LM 27.

STERLING 3 L 51
Sterling International Technology Ltd.
Cu 0.8-2, Mg 0.05-0.2, Si 1.5-2.8, Ni 0.8-1.7, Fe 0.8-0.14, Al.
Sand cast: 160 MPa TS; 125 MPa YS; 2 El; 60 Brin. BS 1490 LM 23.

STERLING 300
Sterling Metals Ltd.
Mg 9.5-11, bal Al.
Sand cast: 40,000-47,000 TS; 21,300-23,500 YS; 8-12 El;
70-80 Brin; 10 IZ. Sand or chill cast aluminum; good proof
stress, resistance to shock and corrosion. Gen.Eng.BS
1490:LM10-W, 11-W, 11-WP. *Obsolete*

STERLING 304 + SI
Now SO 7.1002 (304 + SI).

STERLING 305/3051
Sterling Metals Ltd.
Cu 2-4, Si 4-6, Mn 0.3-0.7, bal Fe.
Sand cast: 20,000-21,400 TS; 9,000-13,000 YS; 1.5-2 El;
70-80 Brin; 0.8 IZ. General purpose sand or chill cast
aluminum. Gen.Eng.BS 1490: LM 22-W. *Obsolete*

STERLING 356
Sterling Metals Ltd.
Mg 0.2-0.45, Si 6.5-7.5, bal Al.
Sand cast: 17,900-20,200 TS; 10,000-11,200 YS; 2-3 El; 50-60
Brin, 1.1 IZ. Heat treated: 33,600-36,000 TS; 29,200-31,400
YS, 0-1 El; 80-90 Brin; 1.4 IZ. Good general purpose heat
treatable sand or chill cast aluminum alloy; corrosion
resistant and pressure tight. Gen.Eng.B.S. 1490 LM 25-M,
25-P, 25-W, 25-WP. *Obsolete*

STERLING 4 L 35
Sterling International Technology Ltd.
Cu 3.5-4.5, Mg 1.2-1.7, Ni 1.8-2.3, bal Al.
Chill cast: 280 MPa TS; 230 MPa YS; 100 Brin. BS 1490 LM
14.

STERLING A.356
Sterling Metals Ltd.
Mg 0.2-0.45, Si 6.5-7.5, bal Al. (lower residuals than Sterling
356).
Chill cast bar: 40,000 TS; 29,000 YS; 5 El; 80-90 Brin; 2.0 IZ.
Premium quality aluminum casting, sand or gravity die.
(AIRCRAFT) D.T.D.Draft M184 (England). *Obsolete*

STERLING D.R.
Teledyne Firth Sterling
C 1.2, bal Fe.
For drills, taps, cutters; drill rod, water hardening. *Obsolete*

STERLING METAL
Manufacturer not listed.
Cu 66, Zn 27-33, Fe 0.7, Pb 0-2, Sn.
For hardware, fittings; free-cuttings.

STERLING NITRARD NO. 1
Teledyne Firth Sterling
C 1.5, Mn 0.2, Cr 12, Mo 1, V 1, bal Fe.
For drawing dies, hot working dies; nitriding steel. *Obsolete*

STERLING SILVER
Handy & Harmon
Precious metal. Ag 92.5, Cu 7.5.
Sand cast: 30,000 TS; 18,000 YS; 41 El; 55 RA. Chill cast:
24,000 TS; 19,000 YS; 7 El; 16 RA. For silverware, coins,
jewelry; corrosion resistant.

STERLING SILVER SOLDER
Manufacturer not listed.
Ag 80, Zn 18, Cu 2.5.
For silver solder; corrosion resistant.

STERLING STAINLESS "NI-T"
Teledyne Firth Sterling
Stainless steel. C 0-0.15, Cr 11-13, Ni 1.2-2.5, bal Fe.
For gun barrels, pistons, bolts, pump rods; corrosion
resistant, hardenable. *Obsolete*

STERLING STAINLESS 302
Teledyne Firth Sterling
C 0.08-0.2, Cr 7-10, Ni 8-10, bal Fe.
Annealed: 85,000-95,000 TS; 30,000-35,000 YS; 55-60 El;
70-75 RA; 145-160 Brin. For stainless parts, pumps and valve
parts, pistons, nozzles; stainless, austenitic. *Obsolete*

STERLING STAINLESS 303
Teledyne Firth Sterling
C 0-0.15, Mn 0-2, Cr 17-19, Ni 8-10, 0.07% P min, S or Se,
0.6% Mo or Zr max, bal Fe.
Water quenched: 90,000 TS; 35,000 YS; 50 El; 55 RA; 160
Brin. For screw machine products; free-cutting, stainless,
austenitic. *Obsolete*

STERLING STAINLESS 304
Teledyne Firth Sterling
C 0-0.08, Mn 0-2, Cr 18-20, Ni 8-11, bal Fe.
Water quenched: 85,000 TS; 35,000 YS; 60 El; 70 RA; 150
Brin. For springs, oil refinery equipment; austenitic, stainless.
Obsolete

STERLING STAINLESS 316
Teledyne Firth Sterling
C 0-0.1, Mn 0-2, Cr 16-18, Ni 10-14, Mo 2-3, bal Fe.
Water quenched: 80,000 TS; 30,000 YS; 60 El; 70 RA; 150
Brin. For marine applications; austenitic, stainless. *Obsolete*

STERLING STAINLESS 321
Teledyne Firth Sterling
C 0-0.08, Mn 0-2, Cr 17-19, Ni 8-11, Ti = 5 x C, bal Fe.
Water quenched: 85,000 TS; 35,000 YS; 55 El; 65 RA; 150
Brin. For welded construction and equipment; austenitic,
stainless. *Obsolete*

STERLING STAINLESS 347
Teledyne Firth Sterling
C 0-0.08, Mn 0-2, Cr 17-19, Ni 9-12, Cb = 10 x C, bal Fe.
Water quenched: 90,000 TS; 35,000 YS; 50 El; 65 RA; 160
Brin. For welded structures and equipment; resists
intergranular corrosion, austenitic. *Obsolete*

STERLING STAINLESS 403
Teledyne Firth Sterling
C 0-0.15, Mn 0-1, Cr 11.5-13, bal Fe.
Hardened: 110,000 TS; 85,000 YS; 23 El; 67 RA; 225 Brin. For
steam turbine blades; corrosion resistant. *Obsolete*

STERLING STAINLESS 410
Teledyne Firth Sterling
Stainless steel. C 0-0.15, Mn 0-1, Cr 11.5-13.5, bal Fe.
Hardened: 190,000 TS; 145,000 YS; 15 El; 55 RA; 390 Brin.
For valve parts, pump rods, pistons, shafts, bolts; corrosion
resistant, hardenable; Type 410. *Obsolete*

STERLING STAINLESS 414
Teledyne Firth Sterling
C 0-0.15, Mn 0-1, Cr 11.5-13, Ni 1.25-2.5, bal Fe.
Hardened: 200,000 TS; 150,000 YS; 15 El; 55 RA; 410 Brin.
For corrosion resistant parts; corrosion resistant. *Obsolete*

STERLING STAINLESS 416
Teledyne Firth Sterling
Stainless steel. C 0-0.15, Mn 0-1.25, Cr 12-14, P or Se, 0.6 Mo
or Zr max, 0.07 S min, bal Fe.
Hardened: 110,000 TS; 85,000 YS; 18 El; 55 RA; 230 Brin. For
screw machine products; corrosion resistant, hardenable,
free cutting; Type 416. *Obsolete*

STERLING STAINLESS 420
Teledyne Firth Sterling
Stainless steel. Mn 0-1, Cr 12-14, 0.15 C min, bal Fe.
Hardened: 230,000 TS; 195,000 YS; 8 El; 25 RA; 500 Brin. For
cutlery, needles, optical mirrors; corrosion resistant,
hardenable; Type 420. *Obsolete*

STERLING STAINLESS 430
Teledyne Firth Sterling
Stainless steel. C 0-0.12, Mn 0-1, Cr 14-18, bal Fe.
Cold drawn: 85,000 TS; 70,000 YS; 20 El; 65 RA; 185 Brin.
For corrosion and heat resistant parts; nonhardenable,
corrosion resistant; Type 430. *Obsolete*

STERLING STAINLESS 431
Teledyne Firth Sterling
C 0-0.2, Mn 0-1, Cr 15-17, Ni 1.25-2.5, bal Fe.
Hardened: 205,000 TS; 155,000 YS; 15 El; 55 RA; 415 Brin.
For corrosion and heat resistant parts; hardenable, corrosion
resistant. *Obsolete*

STERLING STAINLESS 440-A
Teledyne Firth Sterling
Stainless steel. C 0.6-0.75, Cr 16-18, Mo 0-0.75, bal Fe.
Hardened: 260,000 TS; 240,000 YS; 5 El; 20 RA; 510 Brin. For
valve parts, knives, ball bearings; corrosion resistant,
hardenable; Type 440-A. *Obsolete*

STERLING STAINLESS 440-B
Teledyne Firth Sterling
Stainless steel. C 0.75-0.95, Cr 16-18, Mo 0-0.75, bal Fe.
Hardened: 280,000 TS; 270,000 YS; 3 El; 15 RA; 555 Brin. For
valve parts, knives, ball bearings; corrosion resistant,
hardenable; Type 440-B. *Obsolete*

STERLING STAINLESS 440-BMO
Teledyne Firth Sterling
Stainless steel. C 0.85, Mn 0.4, Cr 17, V 0.2, Mo 2.5, bal Fe.
Hardened: 280,000 TS; 270,000 YS; 3 El; 15 RA; 590 Brin. For
oil refineries, valve parts; corrosion resistant, hardenable.
Obsolete

STERLING STAINLESS 440-BV
Teledyne Firth Sterling
Stainless steel. C 0.75-0.95, Mn 0.3-0.5, Cr 16-18, Mo 0.4-0.6,
V 0.15-0.25, bal Fe.
Hardened: 280,000 TS; 270,000 YS; 3 El; 15 RA; 590 Brin. For
cutlery, knives; corrosion resistant, hardenable. *Obsolete*

STERLING STAINLESS 440-CM
Teledyne Firth Sterling
Stainless steel. C 0.95-1.2, Mn 0-1, Cr 16-18, Mo 0-0.75, bal
Fe.
Hardened: 285,000 TS; 275,000 YS; 2 El; 10 RA; 580 Brin. For
bushings, ball races, valve parts, cutlery; corrosion resistant,
hardenable; Type 440-C. *Obsolete*

STERLING STAINLESS 440-CX
Teledyne Firth Sterling
Stainless steel. C 0.95-1.2, Mn 0-1, Cr 16-18, Mo 0-0.75, bal
Fe.
Hardened: 285,000 TS; 275,000 YS; 2 El; 10 RA; 610 Brin. For
ball bearings, cutlery; corrosion resistant, hardenable; Type
440-C. *Obsolete*

STERLING STAINLESS A
Teledyne Firth Sterling
Stainless steel. C 0.35, Mn 0.35, Cr 12-15, bal Fe.
Oil treated: 240,000-150,000 TS; 200,000-125,000 YS; 4-12
El; 8-40 RA; 500-300 Brin. Annealed: 100,000 TS; 65,000 YS;
27 El; 60 RA; 185 Brin. For cutlery, sharp edged or pointed
parts, scissors, surgical and dental instruments; stain
resistant, tough and hard; maximum operating temperature
1650°F. *Obsolete*

STERLING STAINLESS B
Teledyne Firth Sterling
Stainless steel. Cr 16, Mn 0.35, C 0.65, bal Fe.
Oil treated: 130,000-270,000 TS; 100,000-245,000 YS; 2-12
El; 3-30 RA; 285-545 Brin. Annealed: 95,000 TS; 54,000 YS;
27 El; 45 RA; 185 Brin. For cutlery, surgical and dental
instruments, ball bearings, valve seats; maximum operating
temperature 1650°F; requires heat treatment for stainless
properties. *Obsolete*

STERLING STAINLESS BHH
Teledyne Firth Sterling
Stainless steel. C 1.05, Cr 17.5, bal Fe.
For ball races, valve parts, bearing surfaces, machine parts;
heat and corrosion resistant; hard grade. *Obsolete*

STERLING STAINLESS BHHX
Teledyne Firth Sterling
C 0.95-1.2, Mn 0-1, Si 0-1, Cr 16-18, Mo 0.7, bal Fe.
For bearings; stainless, hardenable. *Obsolete*

STERLING STAINLESS H

Teledyne Firth Sterling
C 0.8, Cr 13.5, bal Fe.
For stainless articles; stainless when heat treated. *Obsolete*

STERLING STAINLESS M

Teledyne Firth Sterling
Stainless steel. C 0-0.12, Cr 15-18, bal Fe.
Annealed: 100,000 TS; 70,000 YS; 25 El; 65 RA; 200 Brin. For heat resisting parts, corrosion resisting parts; stain resisting without heat treatment; 40 Izod. *Obsolete*

STERLING STAINLESS MG

Teledyne Firth Sterling
Stainless steel. C 0-0.35, Mn 0.5, Cr 18-23, bal Fe.
For shafts, pump rods, furnace parts, glass molds; high temperature resistance; corrosion resistant. *Obsolete*

STERLING STAINLESS T

Teledyne Firth Sterling
Stainless steel. Cr 12.5, Ni 0.35, Mn 0.35, C 0.12, bal Fe.
Oil treated: 120,000-185,000 TS; 100,000-160,000 YS; 17-22 El; 60-70 RA; 240-395 Brin. Normalized: 145,000 TS; 125,000 YS; 20 El; 63 RA; 241 Brin. For pump rods, machinery parts, turbine blades, shafts, pistons, gun barrels; resists stain without heat treatment; maximum operating temperature 1650°F. *Obsolete*

STERLING TX

Teledyne Firth Sterling
C 0.15, Cr 11.5-13, bal Fe.
For turbine blades. *Obsolete*

STERLING XX

Teledyne Firth Sterling
Stainless steel. C 0.95, Mn 0.3, Cr 0.05, bal Fe.
For tools, dies; water hardened. *Obsolete*

STERLITE

Sterlite Foundry & Mfg. Co.
Copper. Cu 53, Ni 25, Zn 20, Mn.
53,650 TS; 27,250 YS; 33 El; 35 RA; 99 Brin. For ship fittings, laundry and dairy machinery, surgical instruments; tough and corrosion resisting.

STERMET GR. 1

Sterling Alloy Casting Corp.
Ni 65-70, Cr 17-20, bal Fe.
For heat resistant castings, heat treating and carburizing pots; for heat shock.

STERMET GR. 10

Sterling Alloy Casting Corp.
Ni 8-10, Cr 18-20, bal Fe.
Cast: 80,000-95,000 TS; 30,000-45,000 YS; 40-60 El; 50-60 RA; 180 Brin. For food and dairy equipment; heat and corrosion resistant.

STERMET GR. 2

Sterling Alloy Casting Corp.
Ni 58-62, Cr 10-14, bal Fe.
Cast: 60,000-75,000 TS; 30,000-40,000 YS; 2-5 El; 2-6 RA; 185 Brin. For castings, rolls, plates, enameling fixtures; for heat shock; heat resistant.

STERMET GR. 3

Sterling Alloy Casting Corp.
Ni 35-40, Cr 16-20, bal Fe.
For castings, rolls, rails, furnace parts; heat resistant.

STERMET GR. 4

Sterling Alloy Casting Corp.
Ni 33-37, Cr 13-17, bal Fe.
Cast: 60,000-75,000 TS; 35,000-50,000 YS; 4-6 El; 4-7 RA; 185 Brin. For furnace parts, conveyors, belts, burner parts; heat resistant, castings.

STERMET GR. 5

Sterling Alloy Casting Corp.
Ni 8-10, Cr 28-30, bal Fe.
For stoker parts, dampers, skid rails, castings; heat resistant.

STERMET GR. 6

Sterling Alloy Casting Corp.
Ni 10-14, Cr 24-26, bal Fe.
Cast: 80,000-100,000 TS; 40,000-55,000 YS; 10-18 El; 10-20 RA, 175 Brin. For heat resistant castings; for severe sulfur atmospheres and abrasive conditions.

STERMET GR. 7

Sterling Alloy Casting Corp.
Ni 18-22, Cr 23-27, bal Fe.
For castings, furnace parts; heat and corrosion resistant.

STERMET GR. 8

Sterling Alloy Casting Corp.
Ni 8-12, Cr 26-30, bal Fe.
For furnace parts, skid rails, castings; heat resistant.

STERMET GR. 9

Sterling Alloy Casting Corp.
Ni 0-3, Cr 27-30, bal Fe.
Cast: 40,000-55,000 TS; 35,000-45,000 YS; 1.0 El; 1.5 RA; 180 Brin. For castings, salt pots; heat resistant.

STEROPES

Universal Cyclops
C, alloy, bal Fe.
For tools. *Obsolete*

STERRO METAL

English manufacture
Cu 55-60, Zn 38-42, Fe 1.8-4.7.
Forged: 78,200 TS; 27,800 YS; 39 El. Cast: 60,500 TS. For hydraulic cylinders subjected to high pressures; pressure-tight castings.

STERVAC 1000

Teledyne Firth Sterling
Stainless steel. C 0.08, Cr 19.5, Mo 4.3, Co 13.5, Al 1.3, Ti 3.1, Fe 1.25, bal Ni.
At 70°F: 188,000 TS; 115,000 YS; 28 El; 25 RA. At 1400°F: 117,000 TS; 99,000 YS; 28 El; 41 RA. For turbo blades and rotor discs, high temperature bolts; heat resistant to 1500°F. *Obsolete*

STERVAC 2000

Teledyne Firth Sterling
Stainless steel. C 0.15, Cr 19, Mo 9.75, Co 10, Al 1.1, Ti 2.5, Fe 1.25, Zr 0.06, Be 0.005, bal Ni.
At 70°F: 170,000 TS; 98,000 YS; 20 El; 22 RA. At 1400°F: 120,000 TS; 92,000 YS; 20 El. For jet engine and gas turbine buckets; heat resistant to 1500°F. *Obsolete*

STERVAC 3000

Teledyne Firth Sterling
C 0.08, Cr 14, Ni 33, Mo 4, Al 0.25, Zr 0.25, Ti 2, W 6.5, bal Fe.
At 1200°F: 134,000 TS; 121,000 YS; 14 El; 20 RA. High temperature steel for jet engine components; high heat resistance to 1350°F. *Obsolete*

STERVAC 4000

Teledyne Firth Sterling
C 0.09, Cr 19, Ni 52.5, Mo 9.75, Co 11, Al 1.65, Ti 3.15, B 0.005, bal Fe.
At 1400°F: 120,000 TS; 110,000 YS; 8 El; 12 RA. High temperature steel for jet engine components; useful up to 1650°F. *Obsolete*

STERVAC 5000

Teledyne Firth Sterling
C 0-0.15, Al 3, Ti 3, Mo 4, Cr 18, Co 17, Fe 3, bal Ni.
At 70°F: 197,000 TS; 18 El; 22 RA. At 1200°F: 175,000 TS; 8 El; 16 RA. At 1800°F: 4,600 TS; 22 El; 11 RA. High temperature steel for gas turbine components, bolts, valves; useful up to 1600°F, age hardenable. *Obsolete*

STEWART ALLOY NO. 10

Stewart-Warner (Die Casting Div.)
Cu 3.5-4.5, Si 4.5-5.5, bal Al.
Cast: 37,000 TS; 18,000 YS; 2.5 El; 78 Brin. For castings; die cast. *Obsolete*

STEWART ALLOY NO. 10R

Stewart-Warner (Die Casting Div.)
Cu 3-4, Si 8.5-9.5, bal Al.
Cast: 44,000 TS; 21,000 YS; 2 El; 80 Brin. For castings; die cast. *Obsolete*

STEWART ALLOY NO. 12

Stewart-Warner (Die Casting Div.)
Cu 8, Si 2.5, bal Al.
Cast: 43,000 TS; 19,000 YS; 2 El; 80 Brin. For light castings; casting alloy. *Obsolete*

STEWART ALLOY NO. 13X

Stewart-Warner (Die Casting Div.)
Cu 0-0.15, Si 10-12, bal Al.
Die cast: 38,000 TS; 20,000 YS; 4 El; 2.5 RA; 79 Brin. For die castings; high fluidity. *Obsolete*

STEWART ALLOY NO. 16

Stewart-Warner (Die Casting Div.)
Cu 0-0.25, Si 0-1, bal Al.
Die cast: 32,000 TS; 16,000 YS; 3.5 El; 3.0 RA; 70 Brin. For die castings; high fluidity. *Obsolete*

STEWART ALLOY NO. 18

Stewart-Warner (Die Casting Div.)
Ni 1.3, Mg 7.5-8.5, bal Al.
Die cast: 37,000 TS; 21,000 YS; 4.5 El; 9 RA; 80 Brin. For die castings; high corrosion resistant. *Obsolete*

STEWART ALLOY NO. 3

Stewart-Warner (Die Casting Div.)
Si 12.5, bal Al.
Cast: 36,000 TS; 16,000 YS; 2.0 El; 80 Brin. For intricate castings; casting alloy. *Obsolete*

STEWART LUMITE NO. 10

Stewart-Warner (Die Casting Div.)
Cu 4, Si 5, bal Al.
Cast: 32,000 TS; 18,000 YS; 2 El; 60 Brin. For cylinder heads; casting alloy. *Obsolete*

STEWART LUMITE NO. 5

Stewart-Warner (Die Casting Div.)
Cu 4, Si 1.75, Ni 4, bal Al.
Cast: 30,000 TS; 19,000 YS; 1.5 El; 80 Brin. For pistons, cylinder heads; casting alloy. *Obsolete*

STEWART MAGNESIUM R

Stewart-Warner (Die Casting Div.)
Al 8.3-9.7, Si 0-0.5, Zn 0.4-1, 0.13% Mn min, bal Mg.
Cast: 33,000 TS; 3 El; 60 Brin. For castings; die cast. *Obsolete*

STEWART WHITE BRASS NO. 19

Stewart-Warner (Die Casting Div.)
Al, Cu, bal Zn.
Die cast: 45,000 TS; 3 El; 85 Brin. For die castings, hardware. *Obsolete*

STEWART WHITE BRASS NO. 3

Stewart-Warner (Die Casting Div.)
Al 3.5-4.5, Cu 0-0.1, bal Zn.
Die cast: 35,000 TS; 6 El; 70 Brin. For die castings, hardware, housings, covers, radiator caps. *Obsolete*

STEWART ZN NO. 2

Stewart-Warner (Die Casting Div.)
Cu 2.5-3.5, Al 3.7-4.2, Mg 0.04, bal Zn.
Cast: 47,900 TS; 5 El; 83 Brin. For castings; die cast. *Obsolete*

STEWART ZN NO. 3

Stewart-Warner (Die Casting Div.)
Cu 0-0.1, Al 3.75-4.25, Mg 0.04, bal Zn.
Cast: 40,300 TS; 4.7 El; 74 Brin. For castings; die cast. *Obsolete*

STEWART ZN NO. 5
Stewart-Warner (Die Casting Div.)
Cu 0.75-1.2, Al 3.7-4.2, Mg 0.04, bal Zn.
Cast: 45,400 TS; 3 El; 79 Brin. For castings; die cast.
Obsolete

STIRLING CAST
Manufacturer not listed.
Cu 66.2, Zn 33.11, Pb 0.02, Fe 0.66.
For hardware; high strength.

STITCHING WIRE
Manufacturer not listed.
Ni 74.5, Cr 18, Mn 7.5.
For electrical resistances; resistance alloy.

STM
Now SIMMONDS STM.

STM
Japan Special Steel Co.
C 0-0.15, Cr 11.5-14, Mo 0.2-0.6, Ni 0-0.6, bal Fe.
Annealed: 80,000 TS; 38,000 YS; 25 El; 92 Rock B. Cold
drawn: 100,000 TS; 80,000 YS; 15 El; 96 Rock B. For springs,
shafts, table flatware, oil refinery equipment. Corrosion
resistant.

STMCO M-33
Wallace Murray Corp.
Co 0.88, Cr 3.75, W 1.75, Mo 9.5, V 1.2, Co 8.25, bal Fe.
Co-Mo high speed steel for cutting tools; good red hardness.
AISI M33.

STOL 76
KM-kabelmetal AG
Copper. Ni 1.5, Si 0.3, P 0.03, bal Cu.
Electronics, lead frames, connectors and other electronical
purposes. Product forms: strip. Cold rolled: 85 ksi TS; 81 ksi
YS; 4 El.

STOL 77
KM-kabelmetal AG
Copper. Sn 0.15, Mg 0.1, Ag 0.05, P 0.05, bal Cu.
Electronics, lead frames and other electrical purposes.
Product forms: strip. Cold rolled: 70 ksi TS; 63 ksi YS; 4 El.

STOL 78
KM-kabelmetal AG
Copper. Mg 0.6, bal Cu.
Electronics, connectors, and other electrical purposes.
Product forms: strip. Cold rolled: 90 ksi TS; 84 ksi YS; 4 El.

STOL 79
KM-kabelmetal AG
Copper. Zn 0.1, Fe 2.3, Mg 0.1, P 0.03, bal Cu.
Electronics, lead frames, connectors and other electrical
purposes. Product forms: strip. Cold rolled: 75 ksi TS; 65 ksi
YS; 5 El.

STOLLBERG BRASS
English manufacture
Zn 32.8, Sn 0.4, Pb 2, bal Cu.
For forgings, hardware, fittings; good workability, free-cutting.

STONE BRONZE
Stone Manganese - J. Stone & Co. Ltd.
Cu 58, Zn 39, Fe 1.5, Al 0.8, Mn 0.5, bal Cu.
For marine propellers, nuts, bolts; corrosion resistant.

STONES "608" SUPERHEAT BRONZE
Stone Manganese - J. Stone & Co. Ltd.
Ni 7, bal Cu.
For parts exposed to superheated steam, bucket wheels,
pistons, locomotive slide valves, cocks, resists corrosion and
erosion.

STONES ENGLISH GEAR BRONZE
Stone Manganese - J. Stone & Co. Ltd.
Cu 89, Sn 11, trace P.
Cast: 39,000-42,000 TS; 21,000-24,000 YS; 6-10 El; 7-9 RA;
80 Brin. For gears, bearings; used for severe service; CEL
16,000.

STONES GEAR BRONZE
Belmont Metals Inc.
Sn 11, Pb 0.25, P 0.25, bal Cu.
For gears, worm wheels; tough, wear resistant.

STONES Z-METAL
Stone Manganese - J. Stone & Co. Ltd.
Sn, Ni, bal Cu.
For fire box stays; heat resistant.

STONEWELL BABBITT
United American Metals Corp.
Pb, Sb, Sn.
Cast: 10,650 psi TS; 7550 psi YS. For bearings for heavy
loads and pressures, machinery bearings, Babbitt metal.

STOODEX
Stoody Company
C 2, Cr 24, W 8, Mo 4, Co 0.6, bal Fe.
Cast: 78,000 TS; 500 Brin. For hard facing electrodes; impact
resistant. *Obsolete*

STOODITE
Stoody Company
C 4, Si 1.5, Cr 31, Mn 4, bal Fe.
Welded: 600 Brin. For hard facing rod for farm tools,
crushers; abrasion resistant.

STOODITE 45
Stoody Company
C 2.3, Cr 16, W 10, Mo 1, Co 38, V 2, bal Fe.
Cast: 101,000 TS; 450 Brin. For hard facing electrodes;
tough. *Obsolete*

STOODITE 54
Stoody Company
C 3, Cr 11, W 10, Mo 4, Co 30, V 2, bal Fe.
Cast: 75,000 TS; 540 Brin. For hard facing electrodes; for tool
making. *Obsolete*

STOODITE 63
Stoody Company
C 2.7, Cr 10, W 20, Mo 9, Co 20, bal Fe.
Cast: 81,000 TS; 630 Brin. For hard facing electrodes; for tool
making. *Obsolete*

STOODITE K
Stoody Company
C 1.5, Mn 1, W 3, Mo 6.5, Si 1, bal Fe.
580 Brin. For hard facing electrodes. *Obsolete*

STOODY 1
Stoody Company
C 2.5, Cr 26, W 10.5, Co 60, Fe 1.
Welded: 50,000 TS. For hard facing electrodes; abrasion,
wear and heat resistant.

STOODY 100
Stoody Company
36 (Cr, Mo, Mn, C, Si), bal Fe.
Cast: 520-580 Brin. For crushers, tool joints; hard facing
electrode.

STOODY 100 HC
Stoody Company
C 5, Cr 28, Mo 1, bal Fe.
Open arc wire electrode. Hard facing for earth abrasion.

STOODY 101
Stoody Company
C 2, Cr 28, Ni 8, W 1.5, Mn 0.6, Si 1, bal Fe.
Cast: 350-400 Brin. For electrodes for submerged arc
welding; hard facing.

STOODY 102
Stoody Company
C 0.3, Cr 5, W 1.5, Mn 1.5, Mn 0.75, Si 1, bal Fe.
Cast: 520-550 Brin. For electrodes for submerged arc
welding; hard facing.

STOODY 1027
Stoody Company
C 1.5, Cr 4.5, Mn 1.2, Si 1, bal Fe.
For hard facing electrodes; impact resistant.

STOODY 103
Stoody Company
C 4, Cr 30.5, Mn 3.5, Si 1.2, bal Fe.
560 Brin. For welding electrodes; hard facing.

STOODY 104
Stoody Company
3.2 (Mn, Mo, Si, C), bal Fe.
Cast: 270-320 Brin. For tractor rollers, idlers, trunnions; hard
facing electrode.

STOODY 105
Stoody Company
8 (Cr, Mo, V, Mn, Si, C), bal Fe.
Forged: 200,000 TS; 14 El; 14 RA; 470 Brin. For hard facing
electrodes, rollers and idlers for tractors and shovels;
abrasion and impact resistant.

STOODY 105B
Stoody Company
C 0.25, Cr 3, Ni 1.5, Mo 0.5, V 0.3, Mn 1.25, Si 0.5, bal Fe.
Cast: 500-520 Brin. For electrode for submerged arc; hard
facing.

STOODY 106
Stoody Company
13 (Ni, Mo, Si, Mn, C), bal Fe.
Cast: 140,000 TS; 400 Brin. For hard facing electrode for steel
mill roll necks, cable drums; abrasion and impact resistant.

STOODY 107
Stoody Company
6.5 (Cr, Mn, Mo, Si, C), bal Fe.
Cast: 360-410 Brin. For hard facing electrode.

STOODY 110
Stoody Company
Mn 16, Cr 16, bal Fe.
Open arc wire electrode. For joining and rebuilding dissimilar
metals, carbon and austenitic manganese steel.

STOODY 120
Stoody Company
C 2.5, Cr 15, Mn 2, Si 1, Zr 0.25, bal Fe.
Cast: 540 Brin. For electrode for open or submerged arc;
hard facing. *Obsolete*

STOODY 121
Stoody Company
C 2.5, Cr 15, Mn 2, Si 1, Zr 0.25, bal Fe.
Cast: 560 Brin. For electrodes for submerged arc welding;
hard facing.

STOODY 122
Stoody Company
C 0.75, Cr 8, Mn 2.5, Si 1, Zr 0.05, bal Fe.
Cast: 530 Brin. For electrode for submerged arc; hard facing.
Obsolete

STOODY 130
Stoody Company
60 WC in 40 steel matrix.
For electrodes for open arc welding; hard facing.

STOODY 131
Stoody Company
C 2, Cr 28, Mo 4, bal Fe.
Open arc wire electrode. Hard facing for earth abrasion.

STOODY 133
Stoody Company
C 2, Cr 28, Ni 4, bal Fe.
Open arc wire electrode. For crack free deposits, earth abrasion hard facing.

STOODY 2110
Stoody Company
Mn 16, Cr 16, bal Fe.
Coated electrode. For joining and rebuilding dissimilar metals, carbon and austenitic manganese steel.

STOODY 31
Stoody Company
C 2, Cr 28, Mo 4, bal Fe.
Coated electrode for out of position welding. Hard facing for earth abrasion.

STOODY 33
Stoody Company
C 2, Cr 28, Ni 4, bal Fe.
Coated electrode for hard facing. For crack free deposits, earth abrasion hard facing.

STOODY 6
Stoody Company
C 1.2, Si 1, Cr 28, W 5, Co 70, Fe 1.
Welded: 125,000 TS; 2 El; 400 Brin. For hard facing electrodes; impact and corrosion resistant.

STOODY 60
Stoody Company
Nickel-base hard-facing powder. As deposited: 50 Rock C.

STOODY 63
Stoody Company
Nickel-base hard-facing powder. As deposited: 19 Rock C.

STOODY 64
Stoody Company
Nickel-base hard-facing powder. As deposited: 40 Rock C.

STOODY 65
Stoody Company
Nickel-base hard-facing powder. As deposited: 50 Rock C.

STOODY 85
Stoody Company
WC in STOODY 60 powder. For severe earth abrasion and hard facing; matrix: 63 Rock C.

STOODY AC
Stoody Company
C 1, Cr 3, bal Fe.
540 Brin. For hard facing electrodes. *Obsolete*

STOODY DYNAMANG
Stoody Company
Mn 14, Ni 4, Cr 4, bal Fe.
Coated rod and open arc welding wire. For joining and rebuilding austenitic manganese steel.

STOODY MANGANESE
Stoody Company
C 0.7, Mn 14, Ni 5, bal Fe.
Cast: 82,000 TS; 60,000 YS; 15 El; 12 RA. For welding electrodes for Mn steel; austenitic, wear resistant.

STOODY MANGANESE NICKEL
Stoody Company
20 (Mn, Ni, Si, C), bal Fe.
For hard facing electrode; for manganese steel parts. *Obsolete*

STOODY SELF-HARDENING
Stoody Company
C 1, Mo 0.1, Mn 2, Cr 6, Si 1.2, bal Fe.
Welded: 550 Brin. For welding rod, hard surfacing; hard facing, tough; non-machinable.

STOODY SELF-HARDENING 21
Stoody Company
C 3, Cr 14, Mn 2.5, Si 1.2, bal Fe.
For welding electrodes, hard facing rod; severe abrasion resistance.

STOPYT
GTE Products Corp./Wesgo Div.
Rare earth composition applied to a metal surface to produce a barrier to the flow of molten brazing alloys.

STOPYT 62A
GTE Products Corp./Wesgo Div.
Rare earth composition that is least reactive with titanium and titanium-containing compositions; applied to superalloys produces a barrier to the flow of molten brazing alloys.

STOVALL
Manufacturer not listed.
Au 72, Cu 16.5, Pt 2, Ni 13.5.
For dental alloy; corrosion resistant.

STRAIN TEMPERED 1045
Bliss & Laughlin Steel Co.
C 0.43-0.5, Mn 0.6-0.9, bal Fe.
Cold drawn, stress relieved. 125,000 TS min; 100,000 YS min. For machine parts, shafts. Eliminates heat treatment.

STRAIN TEMPERED 1050
Bliss & Laughlin Steel Co.
C 0.48-0.55, Mn 0.6-0.9, bal Fe.
125,000 TS min, 100,000 YS min. Cold drawn, stress relieved. For machine parts, shafts. Eliminates heat treatment.

STRAIN TEMPERED 1141
Bliss & Laughlin Steel Co.
C 0.37-0.45, Mn 1.35-1.65, S 0.08-0.13, bal Fe.
Cold drawn, stress relieved. 125,000 TS min; 100,000 YS min. For machine parts, shafts. Free machining, eliminates heat treating.

STRAIN TEMPERED 1144
Bliss & Laughlin Steel Co.
C 0.4-0.48, Mn 1.35-1.65, S 0.24-0.33, bal Fe.
Cold drawn, stress relieved. 125,000 TS min; 100,000 YS min. For machine parts, shafts. Free machining, eliminates heat treating.

STRAIN TEMPERED 1151
Bliss & Laughlin Steel Co.
C 0.48-0.55, Mn 0.7-1, S 0.08-0.13, bal Fe.
Cold drawn, stress relieved. 125,000 TS min; 100,000 YS min. For machine parts, shafts. Free machining, eliminates heat treating.

STRAIN TEMPERED 4140
Bliss & Laughlin Steel Co.
C 0.38-0.43, Mn 0.75-1, Si 0.2-0.35, Cr 0.8-1.1, Mo 0.15-0.25, bal Fe.
Cold drawn, stress relieved. 125,000 TS min; 105,000 YS min. For machine parts, shafts. Eliminates heat treatment.

STRAIN TEMPERED 41L45
Bliss & Laughlin Steel Co.
C 0.43-0.48, Mn 0.75-1, Si 0.2-0.35, Cr 0.8-1.1, Mo 0.15-0.25, Pb 0.15-0.35, bal Fe.
Cold drawn, stress relieved. 150,000 TS min; 130,000 YS min; 10 El; 35 RA; 302 Brin. For machine parts, shafts. Eliminates heat treatment.

STRAIN-TEMPERED STEEL
Bliss & Laughlin Steel Co.
C 0.15, Mn 0.84, S 0.13, bal Fe.
For machinery parts; free machining. *Obsolete*

STRAINFREE ELASTUF PENN
Horace T. Potts Co.
C 0.35-0.45, Mn 1.25, bal Fe.
Cold drawn: 120,000-135,000 TS; 110,000-120,000 YS; 10-20 El; 30-45 RA; 269-300 Brin. For gears, worms, shafts, spindles; oil hardening. *Obsolete*

STRAUSS METAL
Chemalloy Electronics Corp.
tungsten carbide W-C-Co.
For high speed cutting tools; cemented alloy. *Obsolete*

STRAUSS METAL
Ziv Steel & Wire Co.
tungsten carbide W-C-Co.
For high speed cutting tools; cemented alloy. *Obsolete*

STRENES A
Advance Foundry Co.
C 2.8, Si 1.5, bal Fe.
Cast: 500 Brin. For dies, tools; white cast iron. *Obsolete*

STRENES B
Advance Foundry Co.
C, Cr, Ni, Mo, bal Fe.
For dies, bushings.

STRENES C
Advance Foundry Co.
C 3, Si 2.2, Mn 0.9, Mo 0.4, Cr 0.4, bal Fe.
Cast: 50,000 TS; 190 Brin. Heat treated: 75,000 TS; 550 Brin. For pump impellers, fixtures, bushings; alloy cast iron.

STRENES D
Advance Foundry Co.
C, Cr, Ni, Mo, bal Fe.
For dies, bushings; heat resistant.

STRENES E
Advance Foundry Co.
C 3, Si 1.5, Ni 14, Cr 3, Cu 6, bal Fe.

STRENICOR
Revere Copper Products, Inc.
Ni 3.5-5, Si 0.7-2, Fe 0.3-1, bal Cu.
Cast: 90,000 TS; 70,000 YS; 1 El; 8 RA. Forged: 107,000 TS; 83,000 YS; 5 El; 7 RA. For transformer terminal eye-bolts. Corrosion resistant, no season cracking. *Obsolete*

STRENLITE
Stelco Steel
C 0-0.28, Mn 1.1-1.6, Si 0-0.31, 0.2 Cu min, bal Fe.
High strength steel for construction of riveted or bolted bridges and buildings and other structural purposes. Rolled: 70,000 TS; 50,000 YS; 18 El.

STRESS-PROOF NO. 1
LaSalle Steel Co.
C 0.35-0.45, Mn 1.35-1.65, S 0.2-0.3, bal Fe.
Rolled: 125,000 TS. For studs, bolts, shafts; free machining. *Obsolete*

STRESS-PROOF NO. 2
LaSalle Steel Co.
C 0.35-0.45, Mn 1.35-1.65, S 0.2-0.3, bal Fe.
Rolled: 115,000 TS; 110,000 YS; 16 El; 50 RA; 268 Brin. For studs, bolts, shafts; free machining. *Obsolete*

STRESS-PROOF NO. 3
LaSalle Steel Co.
C 0.35-0.45, Mn 1.35-1.65, S 0.2-0.3, bal Fe.
Rolled: 80,000 TS. For studs, bolts, shafts; free machining. *Obsolete*

STRESSITE COPPER
Chrylser Corp.
Cu.
For electric equipment; high conductivity. *Obsolete*

STRESSITE IRON
Chrylser Corp.
Fe.
For gears, retainers, ball bearings; sintered. *Obsolete*

STRESSPROOF
LaSalle Steel Co.
C 0.4-0.48, Mn 1.35-1.65, P 0-0.4, S 0.24-0.33, Si 0.15-0.3, bal Fe.
Steel bar. 130,000 psi TS (mean); 100,000 psi YS min; 12 El (mean); 34 RA (mean); 83% of 1212 machining characteristics.

STRESSPROOF
Steel Co. of Canada Ltd.
C 0.4-0.48, Mn 1.35-1.65, P 0-0.04, S 0.24-0.33, Si 0.15-0.3, Cu 0-0.25, bal Fe.
Cold finished: 132,000 min TS; 100,000 min YS; 12 El; 83% machinability. For bolts, shafts, studs; free machining.

STRESSPROOF
LaSalle Steel Co.
C 0.4-0.48, Mn 1.35-1.65, P 0-0.04, S 0.24-0.33, Si 0.15-0.3, Cu 0-0.25, bal Fe.
Cold finished: 132,000 min TS; 100,000 min YS; 12 El; 83% machinability. For bolts, shafts, studs; free machining.

STRIKING DIE
Colt Industries
C 0.95, Cr 0.4, V 0.25, bal Fe.
For tools, cold forging dies; water or oil hardening. *Obsolete*

STROH
SPS Industries Inc.
C 0.4, bal Fe.
For steel castings, gears; wear and shock resistant.

STROLOY 5C
Now B & W STROLDY 5C.

STRONGER-THAN-STEEL
Aurora Industries
Cu 89, Al 10, Fe 1.
Cast: 80,000 TS; 140 Brin. Heat treated: 100,000 TS; 160 Brin. For die castings; resists corrosion. *Obsolete*

STRONGSET
Now ALL-STATE NO. 509.

STRONTIUM
Atomergic Chemetals Corp.
Sr.
Purities: 99.95+%, 99.5-99.7%, 98%. Forms: rod, lump, wire.

STRUCTURAL 1311/1315
VDM Aluminium GmbH
Aluminum.
Aluminum alloy AlSi5Mg.

STRUCTURAL SILICON (SIL-TEN)
U.S. Steel Corp.
C 0-0.4, P 0-0.04, S 0-0.05, o.60% Mn min, 0.20% Si min, bal Fe.
Rolled: 80,000-95,000 TS; 45,000 YS; 30 El. For structural parts; add 0.20% Cu for corrosion resistance. *Obsolete*

STRUCTURAL SILICON STEEL
U.S. Steel Corp.
C 0-0.4, P 0-0.04, S 0-0.05, 0.2% Si min, bal Fe.
Hot rolled: 80,000-95,000 TS; 45,000 YS; 30 El. For derricks and bridge work; high elastic plates and shapes. *Obsolete*

STUDAL
French manufacture
Al 97.7, Mg 1, Mn 1.3.
For light alloy parts.

STUDENT ALLOY
Johnson Matthey plc
Ag 12.
Dental casting alloy for teaching purposes.

STUFFING BOX ALLOY
English manufacture
Cu 61.5, Ni 15.5, Zn 11, Pb 10, Sn 2.
For stuffing boxes; corrosion resistant.

STURDICAST
J.F. Jelenko & Co.
Precious metal. Au 60, Pd 4, Ag 22.
Quenched: 70,000 psi TS; 43,500 psi YS; 34 El; 135 Brin.
Hardened: 128,000 psi TS; 97,500 psi YS; 3 El; 225 Brin.
Type IV extra-hard dental alloy. For hard inlays, thin crowns, fixed bridgework, and partial dentures.

STURDY FORTY
LaSalle Steel Co.
C 0.43-0.5, Mn 0.6-0.9, bal Fe.
Cold rolled: 102,500 TS; 93,500 YS; 15 El; 52 RA; 202 Brin.
For machinery parts, gears, shafts; water hardening. *Obsolete*

STYLUS METAL
Manufacturer not listed.
Sn 5, Cu 20, Sb 12, Pb 63.

STYLUS METAL
Manufacturer not listed.
Pb 85, Sn 5, Sb 10.
For bearings; Babbitt.

STYRIA 1818
Vereinigte Edelstahlwerke
C, bal Fe.
For machine tool parts. *Obsolete*

STYRIA 2
Vereinigte Edelstahlwerke
C, bal Fe.
For gears, bolts, machine tool parts; water hardened. *Obsolete*

STYRIA 2028
Vereinigte Edelstahlwerke
C, bal Fe.
For machine tool parts. *Obsolete*

STYRIA 3
Vereinigte Edelstahlwerke
C, bal Fe.
For gears, bolts, machine tool parts; water hardened. *Obsolete*

STYRIA 4
Vereinigte Edelstahlwerke
C, bal Fe.
For gears, bolts, machine tool parts; water hardened. *Obsolete*

STYRIA 5
Vereinigte Edelstahlwerke
C 0.9, Si 0.4, Mn 0.8, bal Fe.
Heat treated: 130,000-185,000 TS; 80,000-120,000 YS; 10-20 El; 30-47 RA; 269-375 Brin. For springs, tools, cutters, taps, drills; Type W1; water hardened. *Obsolete*

STYRIA 6
Vereinigte Edelstahlwerke
C 0.75, Si 0.5, Mn 0.8, bal Fe.
Heat treated: 122,000-175,000 TS; 82,000-130,000 YS; 12-20 El; 35-50 RA; 240-355 Brin. For springs, punches, crimpers; Type W1; water hardened. *Obsolete*

STYRIA 779
Now VEW S302.

STYRIA A-1
Vereinigte Edelstahlwerke
C 1.4, Cr 0.3, V 0.1, bal Fe.
For engraving tools, cutters; water hardened, keen edge.
Obsolete

STYRIA A-2
Vereinigte Edelstahlwerke
C 1.3, Mn 0.35, Si 0.25, bal Fe.
For engraving tools, reamers, taps; Type W1; water hardened.
Obsolete

STYRIA A-3
Vereinigte Edelstahlwerke
C 1.15, Si 0.25, Mn 0.35, bal Fe.
Annealed: 105,000 TS; 55,000 YS; 20 El; 40 RA; 205 Brin. For drills, taps, reamers, broaches, hobs; Type W1; water hardened. *Obsolete*

STYRIA ACM
Now VEW V800.

STYRIA ACM-N
Now VEW V810.

STYRIA B100
Now VEW M112.

STYRIA BCM
Now VEW M-152.

STYRIA BHM
Vereinigte Edelstahlwerke
C 0.28, Ni, Mo, bal Fe.
For gears, bolts, shafts, axles; oil hardened, shock resistant.
Obsolete

STYRIA BN3
Now VEW M120.

STYRIA BN4 SPECIAL
Now VEW M130.

STYRIA BSP
Now VEW P906.

STYRIA CE
Vereinigte Edelstahlwerke
C 0.15, Cr 0.65, Mn 0.5, bal Fe.
For gears, shafts, fasteners, bolts; case hardened, water hardened. *Obsolete*

STYRIA CM EXTRA
Vereinigte Edelstahlwerke
C, bal Fe.
For gears, shafts, fasteners, machine tool parts; case hardened, water hardened. *Obsolete*

STYRIA CM EXTRA K
Vereinigte Edelstahlwerke
C, alloy, bal Fe.
For machine tool parts; oil hardened. *Obsolete*

STYRIA CM SPEZIAL
Now VEW V304.

STYRIA CMG
Now VEW W331.

STYRIA CMH
Now VEW V320.

STYRIA CMHH
Vereinigte Edelstahlwerke
C 0.42, Cr, Mo, bal Fe.
For gears, bolts, machine tool parts; oil hardened, tough.
Obsolete

STYRIA CMM
Now VEW V330.

STYRIA CMS
Now VEW V350.

STYRIA CMW
Now VEW V304.

STYRIA CMZ
Now VEW V320.

STYRIA CMZH
Now VEW V310.

STYRIA CR-V-SILBERSTAHL
Now VEW K510.

STYRIA CR35
Now VEW V510.

STYRIA CR35H
Now VEW V500.

STYRIA CRD
Now VEW K116.

STYRIA CRD SPEZIAL
Now VEW K105.

STYRIA CRLS
Now VEW K305.

STYRIA CRS 815W
Now VEW K301.

STYRIA CSF
Now VEW F200.

STYRIA CV3
Vereinigte Edelstahlwerke
C, alloy, bal Fe.
For machine tool parts; tough, oil hardened. *Obsolete*

STYRIA CV3M
Vereinigte Edelstahlwerke
C 0.4, Cr 0.6, 0.7% Si + W, bal Fe.
For dies and molds. Water hardening, tough. *Obsolete*

STYRIA CV5
Now VEW F550.

STYRIA CVF
Now VEW F550.

STYRIA CVZ
Vereinigte Edelstahlwerke
C 0.5, Cr 1.05, V 0.1, Mn 0.95, bal Fe.
For springs, gears, bolts, crankshafts; oil hardened, shock resistant. *Obsolete*

STYRIA DGM
Now VEW W322.

STYRIA DPK
Now VEW K247.

STYRIA DPKH
Vereinigte Edelstahlwerke
C 0.61, Si 0.85, Mn 0.75, Cr 1.18, V 0.1, bal Fe.
For springs; oil hardened, shock resistant. *Obsolete*

STYRIA E100
Now VEW E300.

STYRIA E30
Vereinigte Edelstahlwerke
C 0.13, Cr, bal Fe.
For gears, machine tool parts; case hardening steel. *Obsolete*

STYRIA E60
Now VEW E525.

STYRIA E80
Now VEW E304.

STYRIA ECR
Now VEW K204.

STYRIA ECR EXTRA
Now VEW K205.

STYRIA ECR2
Now VEW K201.

STYRIA EE
Now VEW E204.

STYRIA EHH
Vereinigte Edelstahlwerke
C, alloy, bal Fe.
For machine tool parts. *Obsolete*

STYRIA EM12
Now VEW K701.

STYRIA EMH
Vereinigte Edelstahlwerke
C 1, Si 0-0.25, Mn 0-0.2, bal Fe.
For drills, taps, cutters, springs; Type W1; water hardened. *Obsolete*

STYRIA EN1W
Vereinigte Edelstahlwerke
C 0.13, Ni 1.5, Cr 0.2, bal Fe.
For gears, cams, camshafts, fasteners; case hardening steel, tough. *Obsolete*

STYRIA EN36
Styria-Stahl Steirische Gusstahlwerke AG
Ni 36, bal Fe.
Annealed: 70,000 TS; 24,000 YP; 36 El; 143 Brin. Cold drawn: 90,000 TS; 70,000 YP; 20 El; 185 Brin. For variable condensers, thermostatic and temperature control devices, bimetals. Similar to Invar. Low coefficient of expansion. *Obsolete*

STYRIA EN36
Vereinigte Edelstahlwerke
Ni 36, bal Fe.
Annealed: 70,000 TS; 24,000 YP; 36 El; 143 Brin. Cold drawn: 90,000 TS; 70,000 YP; 20 El; 185 Brin. For variable condensers, thermostatic and temperature control devices, bimetals. Similar to Invar. Low coefficient of expansion. *Obsolete*

STYRIA EN3M
Vereinigte Edelstahlwerke
C, alloy, bal Fe.
For machine tool parts; oil hardened. *Obsolete*

STYRIA EN3W
Vereinigte Edelstahlwerke
C, alloy, bal Fe.
For machine tool parts; oil hardened. *Obsolete*

STYRIA EN5M
Vereinigte Edelstahlwerke
C, alloy, bal Fe.
For machine tool parts; oil hardened. *Obsolete*

STYRIA EN5W
Vereinigte Edelstahlwerke
C, alloy, bal Fe.
For machine tool parts; oil hardened. *Obsolete*

STYRIA ENC10
Vereinigte Edelstahlwerke
C, alloy, bal Fe.
For machine tool parts; oil hardened. *Obsolete*

STYRIA ENC15
Now VEW E230.

STYRIA ENC2
Now VEW E220.

STYRIA ESPA
Now VEW K721.

STYRIA ESPAS
Now VEW K722.

STYRIA ESS
Now VEW K978.

STYRIA EWP
Vereinigte Edelstahlwerke
C, alloy, bal Fe.
For machine tool parts; oil hardened. *Obsolete*

STYRIA EWP398
Now VEW W100.

STYRIA EWP970
Vereinigte Edelstahlwerke
C 0.3, Co 2, Cr 2.4, V 0.25, W 8.5, bal Fe.
For extrusion press rams and liners; hot work steel, oil hardened. *Obsolete*

STYRIA EWPX
Now VEW W100.

STYRIA EWS
Vereinigte Edelstahlwerke
C, alloy, bal Fe.
For machine tool parts; oil hardened. *Obsolete*

STYRIA EWS SPEZIAL
Vereinigte Edelstahlwerke
C 1.35, W, Cr, V, bal Fe.
For cutters, forming and blanking dies; water hardened, wear resistant. *Obsolete*

STYRIA EWS1
Now VEW K405.

STYRIA EXTRA EXTRA
Vereinigte Edelstahlwerke
C 1.4, Cr 0.3, V 0.1, bal Fe.
For bearings, bushings, liners; water hardened, wear resistant. *Obsolete*

STYRIA EXTRA HH
Now VEW K993.

STYRIA EXTRA MH
Now VEW K992.

STYRIA EXTRA ZH
Now VEW K988.

STYRIA EXTRA ZH SPEZIAL
Now VEW K760.

STYRIA EXTRA ZW
Now VEW K976.

STYRIA EZH
Styria-Stahl Steirische Gusstahlwerke AG
C 1, Si 0-0.25, Mn 0-0.25, bal Fe.
Annealed: 100,000 TS; 53,000 YS; 21 El; 42 RA; 200 Brin. For springs, tools, drills, taps, reamers; Type W1; water hardened. *Obsolete*

STYRIA EZH
Vereinigte Edelstahlwerke
C 1, Si 0-0.25, Mn 0-0.25, bal Fe.
Annealed: 100,000 TS; 53,000 YS; 21 El; 42 RA; 200 Brin. For springs, tools, drills, taps, reamers; Type W1; water hardened. *Obsolete*

STYRIA EZH SPEZIAL
Styria-Stahl Steirische Gusstahlwerke AG
C 1, V 0.1, Mn 0.25, bal Fe.
For cutters, drills, reamers, taps, hobs; Type W2; water hardened. *Obsolete*

STYRIA EZH SPEZIAL
Vereinigte Edelstahlwerke
C 1, V 0.1, Mn 0.25, bal Fe.
For cutters, drills, reamers, taps, hobs; Type W2; water hardened. *Obsolete*

STYRIA EZW
Vereinigte Edelstahlwerke
C 0.7, Si 0-0.25, Mn 0-0.25, bal Fe.
Heat treated: 174,000 TS; 128,000 YS; 12 El; 37 RA; 355 Brin. For rails, axes, hammers, punches, crimpers; Type W1; water hardened. *Obsolete*

STYRIA FLP
Now VEW K607.

STYRIA G10
Now VEW V900.

STYRIA G2
Vereinigte Edelstahlwerke
C, alloy, bal Fe.
For machine tool parts; oil hardened. *Obsolete*

STYRIA G3
Vereinigte Edelstahlwerke
C, alloy, bal Fe.
For machine tool parts; oil hardened. *Obsolete*

STYRIA G3F
Now VEW F912.

STYRIA G4
Bohler Gesellschaft M.B.H.
C 0.75, Si 0.25, Mn 0.55, bal Fe.
For rails, punches, hammers, axes, springs; water hardened.

STYRIA G4F
Styria-Stahl Steirische Gusstahlwerke AG
C 0.7, Mn 0.6, Si 0.2, bal Fe.
Hardened: 170,000-215,000 TS; 150,000 YS min; 5 El; 35 RA.
For springs, hammers, punches, cutters, drills, saws. Similar to SAE 1070, water harden. *Obsolete*

STYRIA G4F
Vereinigte Edelstahlwerke
C 0.7, Mn 0.6, Si 0.2, bal Fe.
Hardened: 170,000-215,000 TS; 150,000 YS min; 5 El; 35 RA.
For springs, hammers, punches, cutters, drills, saws. Similar to SAE 1070, water harden. *Obsolete*

STYRIA G5
Now VEW V960.

STYRIA G5F
Vereinigte Edelstahlwerke
C 1.3, bal Fe.
For files, textile needles. Salt or water hardening, wear resistant. *Obsolete*

STYRIA G6
Now VEW V955.

STYRIA G6S
Vereinigte Edelstahlwerke
C 0.5, bal Fe.
For agricultural equipment, gears, shafts. Water harden. *Obsolete*

STYRIA G7
Now VEW V945.

STYRIA G7S
Now VEW V935.

STYRIA G7W
Styria-Stahl Steirische Gusstahlwerke AG
C 0.22, Si 0.15, Mn 0.35, 0.06% S and P, bal Fe.
Annealed: 75,000 TS; 42,000 YS; 22 El; 58 RA; 145 Brin. For gears, shafts, machine tool parts; case hardening steel. *Obsolete*

STYRIA G7W
Vereinigte Edelstahlwerke
C 0.22, Si 0.15, Mn 0.35, 0.06% S and P, bal Fe.
Annealed: 75,000 TS; 42,000 YS; 22 El; 58 RA; 145 Brin. For gears, shafts, machine tool parts; case hardening steel. *Obsolete*

STYRIA G8
Now VEW V920.

STYRIA GBK
Now VEW K200.

STYRIA GWR
Vereinigte Edelstahlwerke
C 1.5, Cr, Si, bal Fe.
For bearings, bushings, sleeves, liners; water hardened, wear resistant. *Obsolete*

STYRIA HW18
Now VEW H120.

STYRIA HW7
Now VEW H160.

STYRIA HWD
Vereinigte Edelstahlwerke
C 0.2, Si 0.4, Cr 21, Mo 1.5, bal Fe.
Annealed: 112,000 TS; 78,000 YS; 15 El; 140 Brin. For nitric acid storage tanks, combustion chambers, heat treating equipment. Heat resistant to 1050 C, Type 442, non-hardenable. *Obsolete*

STYRIA HWDA
Now VEW H100.

STYRIA HWDN
Now VEW H300.

STYRIA KCN
Now VEW V204.

STYRIA KCR
Vereinigte Edelstahlwerke
C 1.05, Si 0.2, Mn 0.3, Cr 0.4, bal Fe.
For bearings, cutters, bushings, liners; water hardened, wear resistant. *Obsolete*

STYRIA KLR
Now VEW K505.

STYRIA KLS
Now VEW R100.

STYRIA KS45
Vereinigte Edelstahlwerke
C 0.87, Ni 0.7, W 0.7, bal Fe.
For cold heading dies. Water hardening. *Obsolete*

STYRIA KWS
Styria-Stahl Steirische Gusstahlwerke AG
C 1, Cr 1.1, Mn 0.07, bal Fe.
For bearings, cutters, bushings; water hardened, wear resistant. *Obsolete*

STYRIA KWS
Vereinigte Edelstahlwerke
C 1, Cr 1.1, Mn 0.07, bal Fe.
For bearings, cutters, bushings; water hardened, wear resistant. *Obsolete*

STYRIA KWSD
Now VEW K331.

STYRIA LBS
Styria-Stahl Steirische Gusstahlwerke AG
C 0.4, Ni, Cr, Mo, bal Fe.
For gears, bolts, machine tool parts; oil hardened, shock resistant. *Obsolete*

STYRIA LBS
Vereinigte Edelstahlwerke
C 0.4, Ni, Cr, Mo, bal Fe.
For gears, bolts, machine tool parts; oil hardened, shock resistant. *Obsolete*

STYRIA LH
Now VEW M252.

STYRIA LHP
Now VEW W502.

STYRIA M-80
Now VEW E410.

STYRIA M-90
Now VEW E400.

STYRIA M1H
Styria-Stahl Steirische Gusstahlwerke AG
C 0.53, Si 0.9, Mn 0.9, bal Fe.
For machine tool parts, shafts; water hardened. *Obsolete*

STYRIA M1H
Vereinigte Edelstahlwerke
C 0.53, Si 0.9, Mn 0.9, bal Fe.
For machine tool parts, shafts; water hardened. *Obsolete*

STYRIA M3MO
Vereinigte Edelstahlwerke
C, bal Fe.
For machine tool parts. *Obsolete*

STYRIA MAN
Now VEW E930.

STYRIA MCV
Now VEW V564.

STYRIA MN2
Vereinigte Edelstahlwerke
C, alloy, bal Fe.
For machine tool parts; oil hardened. *Obsolete*

STYRIA MNC
Vereinigte Edelstahlwerke
C, alloy, bal Fe.
For machine tool parts; oil hardened. *Obsolete*

STYRIA MNS
Now VEW V762.

STYRIA MNV
Now VEW V742.

STYRIA MP
Bohler Gesellschaft M.B.H.
C 0.5, Cr 1, Ni 3.25, Mo, bal Fe.
For cold working tools, upsetters, forging dies, gears, bolts; oil hardened, shock resistant.

STYRIA MSH
Now VEW F105.

STYRIA MSS
Vereinigte Edelstahlwerke
C 0.53, Si 0.9, Mn 0.9, bal Fe.
For gears, bolts, punches, fasteners; oil or water hardened. *Obsolete*

STYRIA MSW
Now VEW F100.

STYRIA NCZ
Now VEW V157.

STYRIA NH2H
Vereinigte Edelstahlwerke
C 0.35, Si 0-0.35, Mn 0-0.5, Cr 0.5, Ni 1.5, bal Fe.
For gears, bolts, shafts, machine tool parts; oil hardened, tough. *Obsolete*

STYRIA NH2W
Vereinigte Edelstahlwerke
C 0.28, Cr 0.5, Ni 1.5, bal Fe.
For gears, bolts, shafts, machine tool parts; oil hardened, tough. *Obsolete*

STYRIA NH3
Vereinigte Edelstahlwerke
C 0.22-0.3, Cr 0.7, Ni 3.5, bal Fe.
For gears, bolts, machine tool parts; oil hardened, shock resistant. *Obsolete*

STYRIA NHM
Vereinigte Edelstahlwerke
C 0.18, Cr 0.5, Ni 1.5, bal Fe.
For gears, cams, camshafts, fasteners; case hardened, shock resistant. *Obsolete*

STYRIA NHP
Now VEW W501.

STYRIA NHP SPECIAL
Now VEW W500.

STYRIA NW1
Vereinigte Edelstahlwerke
C, Ni, Cr, bal Fe.
For machine tool parts; oil hardened, tough. *Obsolete*

STYRIA NW2
Vereinigte Edelstahlwerke
C 0.13, Cr 0.7, Ni 2.5, bal Fe.
For gears, cams, camshafts, fasteners; case hardened, tough. *Obsolete*

STYRIA NW3
Now VEW E200.

STYRIA O5K
Now VEW V953.

STYRIA O6K
Vereinigte Edelstahlwerke
C 0.71, Si 0.22, Mn 0.27, bal Fe.
Heat treated: 122,000-174,000 TS; 82,000-128,000 YS; 12-22 El; 37-52 RA; 240-350 Brin. For springs, rails, clutch discs, die blocks; water hardened; Type W1. *Obsolete*

STYRIA O7K
Now VEW V946.

STYRIA OM EXTRA
Vereinigte Edelstahlwerke
C 0.42, Cr 1.6, V 1, bal Fe.
For springs, bolts, gears, crankshafts; oil hardened, shock resistant. *Obsolete*

STYRIA OMS
Vereinigte Edelstahlwerke
C 0.37, Si 1.25, Mn 1.25, bal Fe.
For punches, rivet sets, shears, upsetters; oil hardened, tough. *Obsolete*

STYRIA P2
Styria-Stahl Steirische Gusstahlwerke AG
C 1.3, Si 0.2, Mn 0.2, bal Fe.
For blanking and forming dies, punches; Type W1; water hardened. *Obsolete*

STYRIA P2
Vereinigte Edelstahlwerke
C 1.3, Si 0.2, Mn 0.2, bal Fe.
For blanking and forming dies, punches; Type W1; water hardened. *Obsolete*

STYRIA P3
Styria-Stahl Steirische Gusstahlwerke AG
C 1.15, Si 0-0.25, Mn 0-0.25, bal Fe.
For springs, taps, reamers, broaches, drills; Type W1; water hardened. *Obsolete*

STYRIA P3
Vereinigte Edelstahlwerke
C 1.15, Si 0-0.25, Mn 0-0.25, bal Fe.
For springs, taps, reamers, broaches, drills; Type W1; water hardened. *Obsolete*

STYRIA P4
Styria-Stahl Steirische Gusstahlwerke AG
C 1, Si 0.25, Mn 0.2, bal Fe.
Heat treated: 130,000-185,000 TS; 80,000-120,000 YS; 10-20 El; 30-46 RA; 270-375 Brin. For springs, taps, cutters, drills, reamers; Type W1; water hardened. *Obsolete*

STYRIA P4
Vereinigte Edelstahlwerke
C 1, Si 0.25, Mn 0.2, bal Fe.
Heat treated: 130,000-185,000 YS; 80,000-120,000 YS; 10-20 El; 30-46 RA; 270-375 Brin. For springs, taps, cutters, drills, reamers; Type W1; water hardened. *Obsolete*

STYRIA P5
Styria-Stahl Steirische Gusstahlwerke AG
C 0.85, Si 0.2, Mn 0.2, bal Fe.
Heat treated: 130,000-190,000 TS; 88,000-144,000 YS; 12-21 El; 35-50 RA; 255-390 Brin. For springs, taps, cutters, drills, tools, hammers; Type W1; water hardened. *Obsolete*

STYRIA P5
Vereinigte Edelstahlwerke
C 0.85, Si 0.2, Mn 0.2, bal Fe.
Heat treated: 130,000-190,000 TS; 88,000-144,000 YS; 12-21 El; 35-50 RA; 255-390 Brin. For springs, taps, cutters, drills, tools, hammers; Type W1; water hardened. *Obsolete*

STYRIA P5M
Styria-Stahl Steirische Gusstahlwerke AG
C 0.86, Si 0.2, Mn 0.2, bal Fe.
Heat treated: 130,000-190,000 TS; 88,000-144,000 YS; 12-21 El; 35-50 RA; 255-390 Brin. For springs, taps, cutters, drills, reamers: Type W1; water hardened. *Obsolete*

STYRIA P5M
Vereinigte Edelstahlwerke
C 0.86, Si 0.2, Mn 0.2, bal Fe.
Heat treated: 130,000-190,000 TS; 88,000-144,000 YS; 12-21 El; 35-50 RA; 255-390 Brin. For springs, taps, cutters, drills, reamers; Type W1; water hardened. *Obsolete*

STYRIA P6
Styria-Stahl Steirische Gusstahlwerke AG
C 0.7, Si 0.2, Mn 0.2, bal Fe.
Heat treated: 122,000-174,000 TS; 82,000-128,000 YS; 12-22 El; 37-52 RA; 240-355 Brin. For springs, rails, clutch discs, die blocks; Type W1; water hardened. *Obsolete*

STYRIA P6
Vereinigte Edelstahlwerke
C 0.7, Si 0.2, Mn 0.2, bal Fe.
Heat treated: 122,000-174,000 TS; 82,000-128,000 YS; 12-22 El; 37-52 RA; 240-355 Brin. For springs, rails, clutch discs, die blocks; Type W1; water hardened. *Obsolete*

STYRIA P973
Bohler Gesellschaft M.B.H.
C 0.85, Ni 0.6, bal Fe.
For machine tool parts, cold heading dies, punches. Water hardening, wear resistant.

STYRIA PECO
Now VEW K244.

STYRIA PFM
Now VEW K612.

STYRIA PK
Styria-Stahl Steirische Gusstahlwerke AG
C, alloy, bal Fe.
For machine tool parts; oil hardened. *Obsolete*

STYRIA PK
Vereinigte Edelstahlwerke
C, alloy, bal Fe.
For machine tool parts; oil hardened. *Obsolete*

STYRIA PS2
Styria-Stahl Steirische Gusstahlwerke AG
C 1.3, Si 0-0.2, Mn 0-0.2, bal Fe.
For blanking and forming dies, reamers; Type W1; water hardened. *Obsolete*

STYRIA PS2
Vereinigte Edelstahlwerke
C 1.3, Si 0-0.2, Mn 0-0.2, bal Fe.
For blanking and forming dies, reamers; Type W1; water hardened. *Obsolete*

STYRIA PS3
Styria-Stahl Steirische Gusstahlwerke AG
C 1.15, Cr 0.2, Mn 0.2, bal Fe.
For reamers, cutters, bearings; water hardened, wear resistant. *Obsolete*

STYRIA PS3
Vereinigte Edelstahlwerke
C 1.15, Cr 0.2, Mn 0.2, bal Fe.
For reamers, cutters, bearings; water hardened, wear resistant. *Obsolete*

STYRIA PS4
Styria-Stahl Steirische Gusstahlwerke AG
C 1, Si 0-0.25, Mn 0-0.25, bal Fe.
Heat treated: 130,000-185,000 TS; 80,000-118,000 YS; 10-21 El; 30-47 RA; 270-375 Brin. For springs, drills, taps, reamers, cutters; Type W1; water hardened. *Obsolete*

STYRIA PS4
Vereinigte Edelstahlwerke
C 1, Si 0-0.25, Mn 0-0.25, bal Fe.
Heat treated: 130,000-185,000 TS; 80,000-118,000 YS; 10-21 El; 30-47 RA; 270-375 Brin. For springs, drills, taps, reamers, cutters; Type W1; water hardened. *Obsolete*

STYRIA RA
Now VEW A505.

STYRIA RA1
Now VEW A120.

STYRIA RA1 EXTRA
Styria-Stahl Steirische Gusstahlwerke AG
C 0-0.07, Cr 17.5, Ni 17.5, Mo 2, Cu 2, Ti = 7 x C, bal Fe.
For valves, pumps, acid resistant equipment; corrosion resistant. *Obsolete*

STYRIA RA1 EXTRA
Vereinigte Edelstahlwerke
C 0-0.07, Cr 17.5, Ni 17.5, Mo 2, Cu 2, Ti = 7 x C, bal Fe.
For valves, pumps, acid resistant equipment; corrosion resistant. *Obsolete*

STYRIA RA1 SPEZIAL
Now VEW A350.

STYRIA RA1T
Now VEW A300.

STYRIA RA3
Now VEW A100.

STYRIA RA3 EXTRA
Now VEW A354.

STYRIA RAF
Now VEW H525.

STYRIA RAFK
Now VEW H550.

STYRIA RAS (NB)
Now VEW A750.

STYRIA RAT
Now VEW A700.

STYRIA RAW
Now VEW A500.

STYRIA RAZ
Now VEW A506.

STYRIA RCP
Styria-Stahl Steirische Gusstahlwerke AG
C 0.45, W, Cr, V, bal Fe.
For forging and heading dies; oil hardened, tough. *Obsolete*

STYRIA RCP
Vereinigte Edelstahlwerke
C 0.45, W, Cr, V, bal Fe.
For forging and heading dies; oil hardened, tough. *Obsolete*

STYRIA RK15
Styria-Stahl Steirische Gusstahlwerke AG
C 0.9, W 18, V 0.1, Mo 1.1, bal Fe.
For hot work tools and dies; oil hardened, wear resistant.
Obsolete

STYRIA RK15
Vereinigte Edelstahlwerke
C 0.9, W 18, V 0.1, Mo 1.1, bal Fe.
For hot work tools and dies; oil hardened, wear resistant.
Obsolete

STYRIA RKH
Now VEW N688.

STYRIA RP1K
Now VEW W326.

STYRIA RS1
Now VEW N540.

STYRIA RS2
Now VEW N530.

STYRIA RS2M
Now VEW N330.

STYRIA RS2W
Now VEW N220.

STYRIA RS3
Now VEW N315.

STYRIA RS3N
Styria-Stahl Steirische Gusstahlwerke AG
C 0.22, Cr 17, Ni 1.5, bal Fe.
Annealed: 125,000 TS; 95,000 YS; 20 El; 55 RA; 260 Brin. For
pumps, marine hardware, valves; Type 431; stainless.
Obsolete

STYRIA RS3N
Vereinigte Edelstahlwerke
C 0.22, Cr 17, Ni 1.5, bal Fe.
Annealed: 125,000 TS; 95,000 YS; 20 El; 55 RA; 260 Brin. For
pumps, marine hardware, valves; Type 431; stainless.
Obsolete

STYRIA RS3S
Styria-Stahl Steirische Gusstahlwerke AG
C 0-0.1, Cr 17, Mo 1.8, Ti = 7 x C, bal Fe.
Annealed: 125,000 TS; 95,000 YS; 20 El; 55 RA; 260 Brin. For
welded oil refinery equipment; heat and corrosion resistant.
Obsolete

STYRIA RS3S
Vereinigte Edelstahlwerke
C 0-0.1, Cr 17, Mo 1.8, Ti = 7 x C, bal Fe.
Annealed: 125,000 TS; 95,000 YS; 20 El; 55 RA; 260 Brin. For
welded oil refinery equipment; heat and corrosion resistant.
Obsolete

STYRIA RS3T
Styria-Stahl Steirische Gusstahlwerke AG
C 0-0.1, Cr 17.5, Ti = 7 x C, bal Fe.
Annealed: 80,000 TS; 50,000 YS; 25 El; 50 RA; 150 Brin. For
welded oil refinery equipment; corrosion and heat resistant.
Obsolete

STYRIA RS3T
Vereinigte Edelstahlwerke
C 0-0.1, Cr 17.5, Ti = 7 x C, bal Fe.
Annealed: 80,000 TS; 50,000 YS; 25 El; 50 RA; 150 Brin. For
welded oil refinery equipment; corrosion and heat resistant.
Obsolete

STYRIA RSK
Now VEW N200.

STYRIA RSMO
Now VEW N242.

STYRIA RSN
Now VEW N351.

STYRIA RSV
Styria-Stahl Steirische Gusstahlwerke AG
C, alloy, bal Fe.
For machine tool parts; oil hardened. *Obsolete*

STYRIA RSV
Vereinigte Edelstahlwerke
C, alloy, bal Fe.
For machine tool parts; oil hardened. *Obsolete*

STYRIA RSW
Now VEW N100.

STYRIA RSZ
Now VEW N310.

STYRIA S-SPECIAL
Vereinigte Edelstahlwerke
C 1.2, W 1.3, 0.2% V + Cr, bal Fe.
For cold working tools, bearings. Water hardening, abrasion
resistant. *Obsolete*

STYRIA S434
Now VEW K760.

STYRIA SCV
Now VEW K510.

STYRIA SIMPLEX
Now VEW K608.

STYRIA SKL
Styria-Stahl Steirische Gusstahlwerke AG
C 1, Cr 1.55, Mn 0.35, bal Fe.
For bearings, liners, sleeves, forming and blanking dies; oil
hardened, abrasion resistant. *Obsolete*

STYRIA SKL
Vereinigte Edelstahlwerke
C 1, Cr 1.55, Mn 0.35, bal Fe.
For bearings, liners, sleeves, forming and blanking dies; oil
hardened, abrasion resistant. *Obsolete*

STYRIA SP5H
Now VEW K970.

STYRIA SP5W
Now VEW K960.

STYRIA SP6H
Now VEW K950.

STYRIA SP6W
Now VEW K945.

STYRIA SPAH
Now VEW E920.

STYRIA SPAW
Now VEW E900.

STYRIA SPEZIAL CN
Styria-Stahl Steirische Gusstahlwerke AG
C 0.18, Cr 1.25, Mo 0.2, Ni 3.75, bal Fe.
For camshafts, cams, gears, bolts; case hardened, shock
resistant. *Obsolete*

STYRIA SPEZIAL CN
Vereinigte Edelstahlwerke
C 0.18, Cr 1.25, Mo 0.2, Ni 3.75, bal Fe.
For camshafts, cams, gears, bolts; case hardened, shock
resistant. *Obsolete*

STYRIA SPG EXTRA
Now VEW W300.

STYRIA SPG EXTRA V
Now VEW W302.

STYRIA SPG SPECIAL W
Now VEW W304.

STYRIA SPMK
Styria-Stahl Steirische Gusstahlwerke AG
C, alloy, bal Fe.
For machine tool parts; oil hardened, tough. *Obsolete*

STYRIA SPMK
Vereinigte Edelstahlwerke
C, alloy, bal Fe.
For machine tool parts; oil hardened, tough. *Obsolete*

STYRIA SPMW
Now VEW V904.

STYRIA SS 652
Now VEW S600.

STYRIA SS 653 SPECIAL
Bohler Gesellschaft M.B.H.
C 1.2, W 6, Mo 5, V 3, Cr 4, bal Fe.

STYRIA SS1200 ULTRA
Vereinigte Edelstahlwerke
C 1.2, W 7.5, Cr 4, Mo 5.8, V 3, Co 12.5, bal Fe.
For lathe and planer tools, drills, reamers, broaches, form
cutters. High speed steel, high red-hardness. *Obsolete*

STYRIA SS25K
Now VEW S610.

STYRIA SS39KN
Now VEW S203.

STYRIA SS50KN
Styria-Stahl Steirische Gusstahlwerke AG
C 1.35, Cr 4, W, Co, V, bal Fe.
For blanking and forming dies, engravers' tools; high speed
steel. *Obsolete*

STYRIA SS50KN
Vereinigte Edelstahlwerke
C 1.35, Cr 4, W, Co, V, bal Fe.
For blanking and forming dies, engravers' tools; high speed steel. *Obsolete*

STYRIA SS921
Now VEW S401.

STYRIA SSF
Now VEW K240.

STYRIA SW1
Now VEW K465.

STYRIA SWS
Now VEW K460.

STYRIA SX3 TANNEN
Styria-Stahl Steirische Gusstahlwerke AG
C 0.55, Si 0.1-0.4, Mn 0.3-0.5, bal Fe.
Annealed: 100,000 TS; 55,000 YS; 15 El; 22 RA; 180 Brin. For machine tool parts, gears, bolts; water hardened. *Obsolete*

STYRIA SX3 TANNEN
Vereinigte Edelstahlwerke
C 0.55, Si 0.1-0.4, Mn 0.3-0.5, bal Fe.
Annealed: 100,000 TS; 55,000 YS; 15 El; 22 RA; 180 Brin. For machine tool parts, gears, bolts; water hardened. *Obsolete*

STYRIA TENIT-KL
Vereinigte Edelstahlwerke
C 0.6, W 2.2, Cr 1, bal Fe.
For cold work tools, mandrels, punches. Cold work steel, tough. *Obsolete*

STYRIA US-SILBERSTAHL
Now VEW K992.

STYRIA V101
Now VEW H800.

STYRIA V66
Now VEW H802.

STYRIA VCN
Now VEW V214.

STYRIA VCN15
Styria-Stahl Steirische Gusstahlwerke AG
C 0.34, Cr 1.55, Ni 1.55, Mn 0.55, bal Fe.
For forging and upsetting dies; oil hardened, tough. *Obsolete*

STYRIA VCN15
Vereinigte Edelstahlwerke
C 0.34, Cr 1.55, Ni 1.55, Mn 0.55, bal Fe.
For forging and upsetting dies; oil hardened, tough. *Obsolete*

STYRIA VCV2
Styria-Stahl Steirische Gusstahlwerke AG
C 0.3, Cr, Ni, Mo, bal Fe.
For gears, bolts, machine tool parts; oil hardened, shock resistant. *Obsolete*

STYRIA VCV2
Vereinigte Edelstahlwerke
C 0.3, Cr, Ni, Mo, bal Fe.
For gears, bolts, machine tool parts; oil hardened, shock resistant. *Obsolete*

STYRIA VK20N
Now VEW H730.

STYRIA VKS
Now VEW H700.

STYRIA VKSN
Styria-Stahl Steirische Gusstahlwerke AG
C 0.8, Si 2, Cr 20, Ni 1.4, bal Fe.
Heat treated: 280,000 TS; 270,000 YS; 3 El; 15 RA; 555 Brin.
For cutlery, valves, ball bearings, furnace wear plates; corrosion resistant, hardenable. *Obsolete*

STYRIA VKSN
Vereinigte Edelstahlwerke
C 0.8, Si 2, Cr 20, Ni 1.4, bal Fe.
Heat treated: 280,00 TS; 270,000 YS; 3 El; 15 RA; 555 Brin.
For cutlery, valves, ball bearings, furnace wear plates; corrosion resistant, hardenable. *Obsolete*

STYRIA VMS
Styria-Stahl Steirische Gusstahlwerke AG
C 0.36, Si 1.25, Mn 1.25, bal Fe.
For punches, upsetters, riveters; oil hardened. *Obsolete*

STYRIA VMS
Vereinigte Edelstahlwerke
C 0.36, Si 1.25, Mn 1.25, bal Fe.
For punches, upsetters, riveters; oil hardened. *Obsolete*

STYRIA VNC10
Now VEW V165.

STYRIA VNC15
Now VEW V155.

STYRIA VNC2
Now VEW V145.

STYRIA W-SILBERSTAHL
Now VEW K405.

STYRIA WB
Now VEW M312.

STYRIA WCM
Now VEW D330.

STYRIA WCMD
Styria-Stahl Steirische Gusstahlwerke AG
C 0.24, Cr 1.25, Mo 0.45, Mn 0.55, bal Fe.
For gears, bolts, crankshafts, axles; oil hardened, tough. *Obsolete*

STYRIA WCMD
Vereinigte Edelstahlwerke
C 0.24, Cr 1.25, Mo 0.45, Mn 0.55, bal Fe.
For gears, bolts, crankshafts, axles; oil hardened, tough. *Obsolete*

STYRIA WCMDW
Styria-Stahl Steirische Gusstahlwerke AG
C 0.18, Cr 1, Mo 0.45, bal Fe.
For gears, bolts, fasteners, cams, camshafts. Carburizing, water or oil hardening. *Obsolete*

STYRIA WCMDW
Vereinigte Edelstahlwerke
C 0.18, Cr 1, Mo 0.45, bal Fe.
For gears, bolts, fasteners, cams, camshafts. Carburizing, water or oil hardening. *Obsolete*

STYRIA WCMV
Now VEW D240.

STYRIA WCMW
Now VEW V340.

STYRIA WD30
Now VEW D500.

STYRIA WK5K
Now VEW W105.

STYRIA WKM
Styria-Stahl Steirische Gusstahlwerke AG
C, alloy, bal Fe.
For machine tool parts; oil hardened, tough. *Obsolete*

STYRIA WKM
Vereinigte Edelstahlwerke
C, alloy, bal Fe.
For machine tool parts; oil hardened, tough. *Obsolete*

STYRIA WKM 33
Now VEW W320.

STYRIA WPN EXTRA
Vereinigte Edelstahlwerke
C 0.45, Cr 13.5, Ni 13, W 1.3, V 1.2, bal Fe.
For valves, furnace parts. Heat and corrosion resistant. *Obsolete*

STYRIA WPS
Now VEW W706.

STYRIAN BLUE LABEL
Skoda Works National Corp.
C 0.9, V 0.2, bal Fe.
For cutting tools; water hardened.

SU-16
NGK Metals Corp.
W 11, Mo 3, Hf 2, C 0.08, bal Cb.
At 78°F: 112,000 TS; 90,000 YS; 25 El; 43 RA. At 1112°F: 90,500 TS; 64,500 YS; 15 El; 45 RA. For jet engine and missile components. High heat resistant. Good rupture strength.

SU-31
Now COLUMBIUM SU-31.

SU-PYR-LOY
Pyramid Steel Company
C, alloy, bal Fe.
Rolled: 153,000 TS; 120,000 YS; 241 Brin. Heat treated: 264,000 TS; 210,000 YS; 530 Brin. For shear blades, chisels, punches; water hardened, non-tempering.

SU-VENEER
Superior Steel Corp.
Clad steel.
For corrosion resistant stampings and drawn parts; corrosion resistant.

SUB 66
Sanderson Kayser Ltd.
W 5.5, Mo 4, Cr 4, V 1.5, C, bal Fe.
For tools, cutters; high speed steel. *Obsolete*

SUBSTITUTE
Manufacturer not listed.
Cu 90, Sn 6.5, Zn 3, 0.5 Pb.

SUDAL
Griesogen Griesheimer Autogen
Pb, Sn.
For Al solders.

SUDALON "A"
Griesogen Griesheimer Autogen
Pb, Sn.
For Al solders.

SUDALON "B"
Griesogen Griesheimer Autogen
Pb, Sn.
For Al solders.

SUDALON "C"
Griesogen Griesheimer Autogen
Pb, Sn.
For Al solders.

SUEDOIS-S1

Compagnie Ateliers et Forges de la Loire
C 1, Mn 0.4, Si 0.2, bal Fe.
Hardened: 216,000 TS; 152,000 YP; 11 El; 600 Brin. For precision tools, drills, cutters, chisels, punches, taps, springs, bumper bars, bearings. Water hardening, wear resistant.

SUEDOIS-S2

Compagnie Ateliers et Forges de la Loire
C 0.8, Mn 0.7, Si 0.2, bal Fe.
Hardened: 183,000 TS; 135,000 YP; 13 El; 40 Rock C. For precision tools, drills, cutters, chisels, taps, punches. Water hardening, wear resistant.

SUHLER-WHITE COPPER

Manufacturer not listed.
Cu 40, Ni 32, Zn 25, Pb 2.6.
For ornamental parts; corrosion resistant.

SUJ2-N

Fujikoshi Steel Industry Co. Ltd.
C 1.2, Cr 1, Mn 0.6, Si 0.4, bal Fe.
Heat treated: 225,000 TS; 210,000 YS; 444 Brin. For ball bearings, balls and races, liners, bushings. Water hardening. Similar to 52100 Steel.

SULFOR

Forjas Alavesas S.A.
C 0.1, Mn 1.1, Si 0-0.06, P 0-0.05, S 0.3, Pb 0.2, bal Fe.
Free machining, low carbon steel. IHA F-212; AISI 12L14.

SULGERS ANTIFRICTION

English manufacture
Zn 83.6, Sn 9.9, Cu 4, Pb 1.2.
For bearings; will not resist heat or live steam.

SULPH-COPPER

Olin Brass, Indianapolis
S 0.3, Cu 99.7.
1/2 H-temper: 42,000 psi TS; 39,000 psi YS; 20-35 El; 77 Brin. H-temper: 48,000 psi TS; 46,000 psi YS; 18-20 El; 81 Brin. For nozzles, welding tips, and motor parts; free-cutting; high electrical and thermal conductivity. *Obsolete*

SULZER-2 13 CR CO MO

Sulzer Brothers, Inc.
C 0.14, Si 0.6, Mn 0.7, Cr 13, Mo 0.8, Co 10, bal Fe.
Cast, heat treated: 950-1150 N/mm^2 TS; 750 N/mm^2 YS min; 12 El min (or stronger if desired). For cast pump impellers, wheels. Corrosion resistant, high strength steel. DIN GX 14CrCoMo1310.

SUMET BRONZE SM-2

Sumet Corp.
Pb 28-32, Sn 0-1, Ni 0-0.05, bal Cu.
10,500 TS; 9 El; 32 Brin. For light duty and high speed bearings, packing rings. *Obsolete*

SUMET BRONZE SM-4

Sumet Corp.
Pb 26-29, Sn 1-3.5, Ni 0-0.05, bal Cu.
Soft: 18,600 TS; 9,560 YS; 16 El; 33 Brin. For light and medium duty and high speed bearings. *Obsolete*

SUMET BRONZE SM22

Sumet Corp.
Pb 10, Sn, bal Cu.
For bearings; heavy duty.

SUMIALLOY E202

Sumitomo Metal America Inc.
C 0.2, Si 0.25, Mn 1.3, P 0.025, S 0.02, Cr 0.6, B 0.002, bal Fe.
25-100 mm diameter bar. Case hardening steel with good hardenability; for automotive components such as gears.

SUMISTRONG 60 Q-B

Sumitomo Metal America Inc.
HSLA steel. C 0-0.18, Si 0-0.5, Mn 0-1.5, Cu 0-0.3, Mo 0-0.3, V 0-0.15, Ni 0-0.6, Cr 0-0.2, bal Fe.
85,000 TS min; 71,000 YS min; 20 El min. High strength low alloy steel pipe for marine structural services. Good weldability.

SUMISTRONG 80 Q-C

Sumitomo Metal America Inc.
HSLA steel. C 0-0.16, Si 0-0.35, Mn 0-1.2, Cu 0-0.3, Ni 0.4-1.2, Cr 0.4-0.8, Mo 0.3-0.6, V 0-0.1, bal Fe.
114,000-135,000 TS; 99,600 YS min; 18 El min. High strength low alloy steel pipe for marine structural services. Good weldability.

SUMITEN 100 CLASS

Sumitomo Metal America Inc.
C 0-0.18, Si 0-0.55, Mn 0-1.5, Cu 0.15-0.5, Ni 0.3-2.75, Cr 0.3-1.2, Mo 0.1-0.9, V 0-0.1, B 0-0.006, bal Fe.
6-50 mm thick plate: 134,000-164,000 TS; 128,000 YP min. For heavy welded structures, bridges, pressure vessels. Two types: SUMITEN 100, 100 W.

SUMITEN 60 CLASS

Sumitomo Metal America Inc.
HSLA steel. C 0-0.2, Si 0-0.55, Mn 0.9-1.6, Cu 0-0.5, Ni 0-0.6, Cr 0-0.5, Mo 0-0.35, V 0-0.1, bal Fe.
6-100 mm thick plate: 86,000-103,000 TS; 66,000 YP min. High strength low alloy steel. Five types: SUMITEN 60, 60 K, 60 W, 60 F, 60 LT.

SUMITEN 70R

Sumitomo Metal America Inc.
C 0.15, Si 0.32, Mn 1.36, P 0.025, S 0.018, Cu 0.05, Cr 0.02, V 0.09, bal Fe.
20 mm thick plate: 105,000 TS; 87,800 YS; 32 El. High strength low alloy steel for structural purposes. *Obsolete*

SUMITEN 80 CLASS

Sumitomo Metal America Inc.
HSLA steel. C 0-0.18, Si 0-0.55, Mn 0-1.2, Cu 0.15-0.5, Ni 0-1.5, Cr 0.3-1.3, Mo 0.1-0.7, V 0-0.1, B 0-0.006, bal Fe.
6-100 mm thick plate: 114,000-136,000 TS; 100,000 YP min. High strength low alloy steel for bridges, marine structures, penstocks, pressure vesssels. Four types: SUMITEN 80, 80 W, 80.S, 80.SW.

SUMITOMO ESD

Sumitomo Metal America Inc.
Cu 2, Mn 0.5, Mg 1.5, Zn 7.5, bal Al.
For aircraft parts; age-hardenable. *Obsolete*

SUMMIT

British Steel plc
Mild steel.

SUN BRONZE

English manufacture
Co 40-60, Al 10, Cu 30-50.
For high temperature fittings; heat resistant.

SUN BRONZE-1

English manufacture
Cu 95, Al 5.
For pump rods, bushings, propeller blade bolts; corrosion resistant.

SUN BRONZE-2

American manufacture
Co 40-60, Cu 30-50, Al 10.
For high temperature fittings; heat resistant.

SUNCAST

J.F. Jelenko & Co.
Precious metal. Au 50, Pd 4, Ag 35.
Quenched: 72,000 psi TS; 44,200 psi YS; 35 El; 120 Brin. Hardened: 100,000 psi TS; 84,000 psi YS; 10 El; 210 Brin. Type III hard dental alloy. For inlays, crowns, and fixed bridgework.

SUNCAST-D

J.F. Jelenko & Co.
Precious metal. Au 55, Pd 5, Ag 28.5.
Quenched: 75,000 psi TS; 47,000 psi YS; 34 El; 150 Brin. Hardened: 102,000 psi TS; 85,000 psi YS; 13 El; 205 Brin. Type III hard dental alloy. For inlays, crowns, and fixed bridgework.

SUNDEEL

Manufacturer not listed.
Au 80, W 10, Ni 7.
Used as dental alloy; corrosion resistant.

SUNRAY GOLD

Jones, Rd., Ltd.
Ni, bal Cu.
For imitation gold, jewelry. Corrosion resistant.

SUNSHINE METAL

Manufacturer not listed.
C 2.9, Si 2, bal Fe.
For rubber tire molds; semi-steel cast iron.

SUPARD CHROME

Janney Cylinder Co.
C 0.4, Cr 11.5-14, Ni 0-1, Mo 0-0.5, bal Fe.
Heat treated: 240,000 TS; 175,000 YS; 5 El; 7 RA; 450 Brin. For pump liners, plunger sleeves, bushings; resists H_2SO_4.

SUPER

Agawam Tool Co.
C 1.1, Cr 3.75, Mn 0.3, V 3.25, W 17.5, Mo 0.8, bal Fe.
For lathe and planer tools; high speed steel.

SUPER 3

Now HOWMET SUPER 3.

SUPER 6

Now HOWMET SUPER 6.

SUPER A

Fusion Inc.
For solder; tins without special cleaning. *Obsolete*

SUPER A MEEHANITE

Meehanite Metal Corp.
C 3, Cu 0.5, bal Fe.
50,000 TS. For gears, piston rings, engines; cast iron. *Obsolete*

SUPER ALLOY

Ingersoll Steel Co.
C 0.55-0.65, Ni 1-1.5, Cr 0.45-0.75, bal Fe.
Heat treated: 225,000 TS; 220,000 YS; 6 El; 20 RA; 470 Brin. For plow and harrow discs; tough. *Obsolete*

SUPER ANTINIT B

Vereinigte Edelstahlwerke
Ni 64, Mo 28, bal Fe.
Annealed: 64,000 TS; 51,200 YS; 210 Brin. For chemical plant equipment; resists HCl. *Obsolete*

SUPER ANTINIT C

Now VEW L318.

SUPER ASCOLOY

Allegheny Ludlum Steel
Cr 17-20, Ni 7-10, <0.2% C, bal Fe.
Annealed: 90,000 TS; 45,000 YS; 61 El; 75 RA; 170 Brin. Hot rolled: 110,900 TS; 81,970 YS; 43 El; 66 RA; 223 Brin. At 900 C: 13,590 TS; 36 El; 57 RA. Used for all applications where exceptional resistance to most forms of corrosion must be coupled with high tensile properties; austenitic alloy; max operating temperature 1700 F. *Obsolete*

SUPER AUSTENITE

T. Inman & Co. Ltd.
C 0.75, Cr 4.5, W 18, Mo 0.5, V 1, bal Fe.
"18-4-1" type high speed tool steel. For lathe and planing tools, drills, reamers, broaches, cores for pressure die casting.

SUPER AVIONAL

Aluminium Industrie Aktiengesellschaft
Cu 4.5, Si 0.8, Mg 1, Mn 1.2, bal Al.
Aged: 68,000 TS; 48,000 YS; 15 El; 130 Brin. For aircraft parts, structural members; age-hardenable.

SUPER AVIONAL

Lavorazione Leghe Leggere SpA
Cu 4.5, Si 0.8, Mg 1, Mn 1.2, bal Al.
Aged: 68,000 TS; 48,000 YS; 15 El; 130 Brin. For aircraft parts, structural members; age-hardenable.

SUPER BRONZOCHROM 10186

Eutectic Corp.
Bronze-nickel alloy powder for steel, nickel alloys, copper alloys. *Obsolete*

SUPER BUFFALOY

Buffalo Wire Works Co. Inc.
High C, bal Fe.
Wire for screening abrasive material.

SUPER CAPITAL

Eagle & Globe Steel Ltd.
Tool material. C 1.3, Cr 4.25, Mo 3.1, W 9, V 3.5, Co 9.5, bal Fe.
High speed, high performance steel in abrasive and heavy cutting conditions. AS1239 M100A; Werkstoff 1.3207.

SUPER CAST 4

Superior Die Casting Corp.
Al 4, Mg 0.02, bal Zn.
For die castings.

SUPER CAST ALLOY NO. 6

Now SUPER CAST 4.

SUPER CAST S102

Superior Die Casting Corp.
Si 10.5, Cu 2.5, bal Al.
For aluminum die castings.

SUPER CAST S49

Superior Die Casting Corp.
Si 8.5, Cu 3.5, bal Al.
For aluminum die castings.

SUPER CESCO

Crucible Electric Steel Co.
C 0.55, Mn 1, Mo 1.3, V 0.35, Si 2, bal Fe.
For tools; shock resistant.

SUPER CESCO DIAMOND

Crucible Electric Steel Co.
C 0.5-0.65, Si 2, Mn 1, Mo 1.25, V 0.3, bal Fe.
Heat treated: 236,400-338,000 TS; 231,000-279,200 YS; 4-10.5 El; 7.7-31 RA; 612 Brin. For chisels, concrete busters, drills, caulking and beading tools.

SUPER CHLOR

Duriron Co. Inc.
Alloy steel. Cr 4-5, Si 14.2-14.7, C 0.95-1.1, bal Fe.
Cast alloy for handling H_2SO_4, HNO_3 and HCl.

SUPER CHNC

British Steel Corp.
C 0.15, Mn 0.4, Ni 4.2, Cr 1.2, Mo 0.25, bal Fe.
Oil harden: 193,000-204,000 TS; 14-16 El; 50-55 RA; 40-55 IZ.
For aircraft engine gears, supercharger gears, heavy duty worms, cams, camshafts. Tough, shock resistant. Case hardening. *Obsolete*

SUPER CHNC

British Steel Corp.
C 0.15, Mn 0.4, Ni 4.2, Cr 1.2, Mo 0.25, bal Fe.
Oil harden: 193,000-204,000 TS; 14-16 El; 50-55 RA; 40-55 IZ.
For aircraft engine gears, supercharger gears, heavy duty worms, cams, camshafts. Tough, shock resistant. Case hardening. *Obsolete*

SUPER CHROME

H. Boker & Co.
C 2.25, Cr 13-14, Ni 0.5-0.75, bal Fe.
For drawing and blanking dies, punches, form dies; oil hardened, non-deforming. *Obsolete*

SUPER CHROME

Time Steel Service Inc.
C 1.4, Mn 0.3, Si 0.6, Cr 13, Ni 0.5, Mo 0.6, Co 3.3, bal Fe.
High carbon, high chromium tool and die steel; air or oil hardened. AISI D5.

SUPER CHROME VANADIUM

Sanderson Kayser Ltd.
C 0.95-1.05, Cr 0.9-1, V 0.15-0.25, bal Fe.
For cutlery dies, drawing and stamping dies; deep hardening, tough. *Obsolete*

SUPER COAT

Yawata Iron & Steel Co., Ltd.
Sheet metal coated with 0.03-0.06 microns chromium and 0.014-0.2 microns CrO_3. For beverage cans, crown caps, pails, dry cells, motor oil cans. Resists staining and corrosion.

SUPER COBALT

Latrobe Steel Co.
C 0.85, V 1.9, Co 8.25, W 18.75, Cr 4.1, bal Fe.
For tools, high speed cutting tools; high speed steel. AISI T5. *Obsolete*

SUPER DBL

Allegheny Ludlum Steel
C 0.8, W 5.5, Mo 4.2, Co 7.7, Cr 4, V 1.7, bal Fe.
For lathe and boring tools, form tools, cutters for heavy duty cutting; wear resistant high speed steel; AISI M36. *Obsolete*

SUPER DREADNOUGHT

Colt Industries
C 0.7, Cr 4, V 2, W 14, bal Fe.
For high speed cutting tools, roughing cuts on hard material; high speed steel. *Obsolete*

SUPER DREADNOUGHT

Crucible Specialty Metals
C 0.7, Cr 4, V 2, W 14, bal Fe.
For high speed cutting tools, roughing cuts on hard material; high speed steel. *Obsolete*

SUPER DURALUMIN

Alcan-Booth Industries, Ltd.
Cu 4.3, Mg 0.5, Si 0.78, Mn 0.49, bal Al.
59,000-75,500 TS; 13-25 El. For airplane and dirigible parts; age-hardenable. *Obsolete*

SUPER ECLAIR

Creusot-Loire
C 0.62, W 14, bal Fe.
For hot work tools; oil hardened. *Obsolete*

SUPER HALLAMITE

Hallamshire Steel Co.
C 0.7, W 18, Cr 4, V 1, Co 5, bal Fe.
For hacksaw blades, saws, cutting tools; high speed steel, oil hardened.

SUPER HARDTEM

Heppenstall Co.
Tool material. C 0.43, Ni 2.6, Cr 1.35, Mo 0.55, V 0.18, bal Fe.
Annealed: 250 Brin. Heat treated: 470 Brin. For die blocks; oil hardened; shock resistant.

SUPER HI-PHY KIRKSITE

NL Industries
Al 4, Cu 3, Mg 0.01, Be 0.007, Pb 0-0.005, Sn 0-0.002, Cd 0-0.002, Fe 0-0.05, bal Zn.
Sand cast: 40,000 TS; 3.0 El; 34 impact strength; 109 Brin. *Obsolete*

SUPER HY-TUF

Colt Industries
C 0.32-0.43, Mn 1.3, Si 2.3, Cr 1.4, V 0.2, Mo 0.35, bal Fe.
Heat treated: 294,000 TS; 241,000 YS; 10 El; 35 RA; 540 Brin.
For aircraft structures, landing gears, cylinders; oil hardened, high toughness. *Obsolete*

SUPER HY-TUF HIGH CARBON

Colt Industries
C 0.47, Mn 1.3, Si 2.4, Cr 1.1, V 0.25, Mo 0.4, bal Fe.
Heat treated: 325,000 TS; 24 El. For gears, shafts, machine tool parts; tough, shock resistant. *Obsolete*

SUPER HYDRA

Osborn Steels Ltd.
C 0.75, W 18, Cr 4.5, V 1.2, bal Fe.
For lathe and planer tools, drills, taps, hobs; high speed steel. *Obsolete*

SUPER IMPACTO

Rio Algom Corp.
C 0.12, Mn 0.5, Cr 1.5, Ni 3.25, bal Fe.
Heat treated: 175,000 TS; 135,000 YS; 16 El; 59 RA; 341 Brin; 48 IS (Izod). For heavy duty gears, bearings, clutch dogs, broach holders, plastic molds, transmission shafts. Tough and fatigue resistant.

SUPER IMPACTO

Atlas Steels Ltd.
C 0.12, Mn 0.5, Cr 1.5, Ni 3.25, bal Fe.
Heat treated: 175,000 TS; 135,000 YS; 16 El; 59 RA; 341 Brin; 48 IS (Izod). For heavy duty gears, bearings, clutch dogs, broach holders, plastic molds, transmission shafts. Tough and fatigue resistant.

SUPER INCOMPARABLE

Flockton, Tompkin & Co., Ltd.
C 0.8, W 19-20, Cr 4.6, V 1.6, Co 5.5, Mo, bal Fe.
For continuous cutting tools for cast iron; high speed steel.

SUPER INMANITE

T. Inman & Co. Ltd.
C 0.8, Cr 4.5, W 22, Mo 1, V 1.5, Co 10, bal Fe.
Cobalt-tungsten high speed tool steel. For heavy duty machining operations.

SUPER INVINCIBLE 5% COBALT

Joseph Beardshaw & Son Ltd.
C 0.75, W 18, Cr 4, V 1.5, Co 5, bal Fe.
For tools, cutters, reamers, hobs, chasers; high speed steel.

SUPER INVINCIBLE ADVANCE

Joseph Beardshaw & Son Ltd.
C 0.75, W 19, Cr 4.25, V 1.75, Co 10, bal Fe.
For lathe and turning tools, drills, broaches; high speed steel.

SUPER INVINCIBLE TB

Joseph Beardshaw & Son Ltd.
C 0.75, W 14, Cr 4, V 2, Co 5, bal Fe.
For shear blades, tools, cutters, punches; high speed steel.

SUPER JACKSBERG

Joseph Jackman & Co. Ltd.
C, alloy, bal Fe.
For high speed tools and cutters. High speed steel.

SUPER KINITE

H. Boker & Co.
C 1.5, Cr 12.7, Mo 0.85, V 0.65, Ni 0.16, Co 3.5, bal Fe.
For tools, dies, cutting tools; air hardening; abrasion resistant. *Obsolete*

SUPER LA-LED
LaSalle Steel Co.
C 0-0.13, Mn 0.85-1.35, Pb 0.15-0.35, S 0.5, bal Fe.
Cold drawn: 70,000 TS; 60,000 YS; 12 El; 150 Brin. For screw
machine products, bolts, fasteners, shafts. Free-machining.

SUPER LEDA
Firth Brown Ltd.
C 0.7, Cr, W, V, bal Fe.
For lathe and planer tools, reamers, broaches, hobs; high
speed steel. *Obsolete*

SUPER LION
Burys & Co. Ltd.
C 0.75, W 19, Cr 4.5, V 1.25, Co 10, Mo 0.75, bal Fe.
For lathe and planer tools, drills, reamers; high speed steel.

SUPER LION EXTRA
Burys & Co. Ltd.
C 0.75, W 20, Cr 4, V 1, Co 10, Mo 0.75, bal Fe.
For lathe and planer tools, hobs, drills, reamers; high speed
steel.

SUPER LION SPECIAL
Burys & Co. Ltd.
C 0.8, W 19, Cr 4.5, V 1.25, Co 15, Mo 0.75, bal Fe.
For lathe and planer tools, drills, reamers; high speed steel.

SUPER LMW
Allegheny Ludlum Steel
C 0.7, Cr 4, V 2, Mo 8, Co 8, bal Fe.
For tools, cutters, milling, lathe and planer tools; high speed
steel. *Obsolete*

SUPER LMW EXTRA
AL Tech Specialty Steel Corp.
C 0.9, Cr 3.75, V 1.15, W 1.65, Mo 9.3, Co 8.25, bal Fe.
Heavy duty cobalt bearing high speed steel used mainly for
cutting tools requiring exceptional red hardness. AISI M-33.
Obsolete

SUPER LMW SPECIAL
Allegheny Ludlum Steel
C 0.9, Cr 3.75, V 2, W 1.65, Mo 8.7, Co 8.25, bal Fe.
High speed steel for cutting tools, molybdenum-cobalt type;
AISI M34. *Obsolete*

SUPER MANGANESE BRONZE
American Manganese Bronze Co.
Copper. Cu 69, Zn 20, Al 6.5, Fe 2.5, Mn 2.
Cast: 110,000 TS; 200 Brin. For propeller blades, marine
parts; corrosion resistant.

SUPER MENDUR
Arnold Engineering Co.
Fe 49, Co 49, V 2.
Soft magnetic material, high permeability. Fabricated parts
only.

SUPER MO CHIP HIGH SPEED
Teledyne Firth Sterling
Tool steel. C 0.7, Cr 4, Co 8, V 1.5, Mo 8, bal Fe.
For tools, cutters, reamers; high speed steel. *Obsolete*

SUPER MOTUNG
Cytemp Specialty Steel Div.
C 0.8, Cr 3.75, Co 5, V 1.25, W 1.5, Mo 8.5, bal Fe.
For cutting tools, form tools, taps, broaches; high speed
steel. *Obsolete*

SUPER MOTUNG 33
Cytemp Specialty Steel Div.
C 0.88, Mo 9.5, W 1.75, Cr 4, V 1.25, Co 8.25, bal Fe.
Hardened: 64-67 Rock C. For cutters, drills, taps, end mills,
lathe tools, milling cutters, chasers. Good red-hardness and
wear resistance. AISI M33. High speed steel. *Obsolete*

SUPER MOTUNG 34
Cytemp Specialty Steel Div.
C 0.9, Mo 8.5, W 1.75, Cr 4, V 2, Co 8.25, bal Fe.
Hardened: 64-67 Rock C. For cutters, drills, taps, chasers,
lathe tools, milling cutters, form tools. Good cutting
properties and red-hardness. AISI M34. High speed steel.
Obsolete

SUPER NERVA
Cie Francais des Metaux
C 0.7, W 18, Cr 4, V 1, bal Fe.
For tools, cutters; high speed steel.

SUPER NICKEL-701
Anaconda Co.
Copper. Cu 70, Ni 30.
Hard: 70,000 TS; 60,000 YS; 10 El. For condenser tubes.
Corrosion resistant.

SUPER NILVAR
Driver Harris Co.
Ni 31, Co 7, bal Fe.
For instruments; zero expansion. *Obsolete*

SUPER OILITE
Chrysler Corp.
Cu 18-22, bal Fe.
Sintered: 22,000 TS; 5.8-6.2 density. Oilite bearing material.
ASTM B439-67 Grade 4; SAE 863.

SUPER OILITE 16
Chrysler Corp.
Cu 18-22, C 0.6-1, bal Fe.
Sintered: 40,000 TS; 5.8-6.2 density; Oilite bearing material,
self lubricating bearings.

SUPER OILITE 2-1
Chrysler Corp.
Cu 1.75-3, C 0.5-1, bal Fe.
Sintered: 50,000 TS; 6.1-6.4 density; 15,000 fatigue strength.
ASTM B426-65, Type 11 Gr 1 Class C; SAE 864B.

SUPER OILITE 2-1 HARDENED
Chrysler Corp.
C 0.8, Cu 2, Fe 97.
Sintered and hardened: 3 ranges from 55,000-80,000 TS;
50,000-75,000 YS; 25-40 Rc; Density 6.1-7.2. MPIF FC-0208-P
or R or S; ASTM B-426 Grade 1 Type II or IV.

SUPER OILITE 5-1
Chrysler Corp.
Cu 4-6, C 0.5-1, bal Fe.
Sintered: 50,000 TS; 6.1-6.4 density; 15,000 fatigue strength.
ASTM B426-65 Type 11 Gr 2, Class C; SAE 865B.

SUPER OILITE 9
Chrysler Corp.
Cu 7-11, Fe 86.5-0, C 0-0.3.
Sintered: 29,500-34,000 TS; 5.8-6.2 density; 17,000 shear;
13,000 fatigue. ASTM B439-67 Gr 3; SAE 862.

SUPER OILITE 9-1
Chrysler Corp.
Cu 6.5-7.5, Fe 89-0, C 0.6-1.
Sintered: 40,000-50,000 TS; 5.8-6.2 density. ASTM B426-65
Gr 3 Cl.C; SAE 866 A; PMPMA FC-0710-N.

SUPER OILITE-M
Chrysler Corp.
C 0.5, S 0.5, Cu 2, Fe 97.
Sintered: 40,000-50,000 TS; 36,000-45,000 YS; 1.0 El; Density
6.1-6. Machinable grade Oilite bearing material.

SUPER PANTHER
Allegheny Ludlum Steel
C 0.8, W 19, Cr 4, V 2, Co 8, Mo 0.7, bal Fe.
For lathe and planer tools, reamers, broaches, form cutters;
high speed steel; Type T5.

SUPER PERMALLOY "C"
ITT Components Group Europe
Nickel. Ni 78, Fe 13, Mo 4, Cu 5.
For telecommunications, magnetic amplifiers, wide band
transformers, current transformers, computers. High initial
permeability with low losses.

SUPER PURITY
British Aluminium Co., Ltd.
99.99 Al min.
O-temper: 8500 TS. 1/2 H-temper: 11,000 TS. H-temper:
14,500 TS. Extruded: 8000 TS. For roofing flashings,
reflectors; high corrosion resistance.

SUPER PYRONEAL
Heppenstall Co.
Tool material. C 0.42-0.47, Mn 0.3-0.6, Si 0.5-0.7, Ni 4.1-4.5,
Cr 1.25-1.65, Mo 0.7-0.8, V 0.12-0.16, bal Fe.
For press and hammer dies, upsetter and insert dies. Oil
hardened. High compressive strength. Resists softening at
high working temperatures.

SUPER PYROTEM
Heppenstall Co.
Tool material. C 0.42-0.47, Mn 0.3-0.6, Si 0.5-0.7, Ni 4.1-4.5,
V 0.12-0.16, Cr 1.25-1.65, Mo 0.7-0.8, bal Fe.
For press and hammer dies, upsetter and insert dies. Oil
hardened; high compressive strength. Resists softening at
high working temperatures.

SUPER RADIOMETAL
Telcon Metals Ltd.
Ni 50, Fe 50.
Annealed: 58,000 TS; 19,000 YS; 27 El (magnetic annealed).
For electrical and magnetic equipment, core material for
audio and instrument transformers, magnetic shields.
Vacuum melted; soft magnet. High magnetic permeability
and low losses.

SUPER RAPID
George Cook & Co., Ltd.
C 0.8, Cr 4, W 18, V 1.2, bal Fe.
For lathe and planer tools, broaches, reamers, drills; high
speed steel.

SUPER RAPID EXTRA
Raven Steel & Tool Co.
C 0.7, W 18, Cr 4, V 1, bal Fe.
For cutters, taps, broaches, drills; high speed steel.

SUPER RAPID EXTRA
Now VEW S200.

SUPER RAPID EXTRA 214A
Vereinigte Edelstahlwerke
C 0.7, W 9.5, Cr 3.7, V 2.8, Co 2.8, bal Fe.
For tools, dies, cutters; high speed steel. *Obsolete*

SUPER RAPID EXTRA 214N
Vereinigte Edelstahlwerke
C 0.86, Co 2.8, Cr 4.3, Mo 0.85, V 2.1, W 12, bal Fe.
For lathe and planer tools, reamers, broaches, drills; high
speed steel. *Obsolete*

SUPER RAPID EXTRA 500
Now VEW S305.

SUPER RAPID EXTRA 500N
Vereinigte Edelstahlwerke
C 0.85, Cr 4, W 13, Mo 1, V 2.5, Co 5, bal Fe.
For cutting tools; high speed steel. *Obsolete*

SUPER RAPID EXTRA A
Vereinigte Edelstahlwerke
C 0.65, W 8.5, Cr 3.8, V 1.8, bal Fe.
For tools, dies, cutters; high speed steel. *Obsolete*

SUPER RAPID EXTRA HVA
Vereinigte Edelstahlwerke
C 0.9, W 9.5, Cr 3.8, V 2.5, bal Fe.
For tools, dies, cutters; high speed steel. *Obsolete*

SUPER RAPID EXTRA HVN
Now VEW S203.

SUPER RAPID EXTRA MO
Now VEW S600.

SUPER RAPID EXTRA N
Now VEW S205.

SUPER RAPID EXTRA V
Now VEW S201.

SUPER RUSTFREE SR STEEL
Uniworld Corp. of America
Cr, Ni, Mo, bal Fe.
Cold worked: 90,000-105,000 TS; 75,000-100,000 YS; 4-50 El;
28-65 RA; 156-410 Brin. Annealed: 95,300 TS; 75,000 YS. For
pickling equipment, turbine compressors, gasoline cracking
plants; corrosion resistant, "Super Rustfree Steels;" austenitic.

SUPER SAMSON
Now CARPENTER SUPER SAMSON.

SUPER SILICON CAST IRON NO. 9
Sifbronze
C 3.3, Si 2.6, Mn 0.7, S 0.1, P 0.5, bal Fe.
Rod for full fusion welding cast iron; cast iron. 200 N/mm^2
TS; 180 Brin.

SUPER SIRIUS
Creusot-Loire
C, Ni, Cr, W, Co, Ti, bal Fe.
For high temperature applications; high heat, creep and
oxidation resistance. *Obsolete*

SUPER SIRIUS HT
Creusot-Loire
C, Ni, Cr, W, Ti, bal Fe.
For gas turbine components; heat and creep resistant.
Obsolete

SUPER SPECIAL EXPRESS
Leadbeater & Scott Ltd.
C 0.7, W 18, Cr 4, V 1, bal Fe.
For lathe and planer tools, hobs, reamers, taps; high speed
steel.

SUPER SQUAREMU 79
Magnetic Metals Co.
Ni-Fe.
For magnetic and electrical equipment; magnetic alloy.
Obsolete

SUPER STAR ZENITH
Carpenter Technology Corp.
C 0.8, Cr 4, V 2, W 18, bal Fe.
For cutters, hobs, drills, taps, reamers. High speed steel.
Type T2.

SUPER STAR-MO M2-5
Teledyne Firth Sterling
Tool steel. C 0.85, W 6, Cr 4.1, V 2, Mo 5, Co 5, bal Fe.
For boring tools, broaches, hobs, milling cutters; high speed
steel. *Obsolete*

SUPER STRENGTH 100
Armco
C 0.15, Mn 0.5, Si 0.35, Cr 1.65, Cu 0.25, Mo 0.5, B 0.002, Ti
0.07, bal Fe.
For mine cars, chutes, bus and trailer bodies, structures,
derricks. Low-alloy high-strength structural steel. Shock
resistant. *Obsolete*

SUPER STRENGTH MALLEABLE
Kencroft Malleable Co.
Cast Iron. C 3, Si 1.2, Mn 0.7, bal Fe.
Cast: 57,000 TS; 38,000 YS; 18 El; 120 Brin. For machinery
parts; malleable cast iron.

SUPER T-H24
Southwire Co.
0.50 Fe min, 0.50 Co min, 0.06 Si min, bal Al.
19.0 ksi TS; 17.0 ksi YS; 15 El; 61% IACS min. For aircraft
wire, high temperature wire.

SUPER TERRIFIC
Wardlows Ltd.
W 18, Co 5, Cr 4, V 1.3, C, bal Fe.
For tools, dies, cutters; high speed steel.

SUPER TIGER
Bethlehem Steel Corp.
C 0.8, W 18.5, Cr 4.5, Mo 0.8, V 1.7, Co 7.5, bal Fe.
For tools, drills, cutters; high speed steel. *Obsolete*

SUPER TRICENT
Bethlehem Steel Corp.
C 0.55, Ni 3.6, Cr 0.9, Mo 0.5, Mn 0.8, Si 2.1, bal Fe.
Heat treated: 342,000 TS. For aircraft structures, landing
gears; tough, shock resistant. *Obsolete*

SUPER TUFF
Time Steel Service Inc.
C 0.5, Mn 0.7, Cr 3.25, Mo 1.4, bal Fe.
Shock resistant tool steel; air or oil hardened. AISI S7.

SUPER UGIMAX
Ugine Aciers
Ni 14, Al 8, Co 24-35, Cu 3, bal Fe.
Residual induction 13,400 gauss; coercive force 800 oersted;
energy product 8 max. For magnets in loud speakers,
microphones, measuring instruments. Permanent magnet,
high permeability.

SUPER VAN DREADNOUGHT
Colt Industries
C 0.8, W 18, V 2, Cr 4, Mo 0.6, bal Fe.
For lathe and planer tools, reamers, hobs, drills, taps; Type
T2; high speed steel. *Obsolete*

SUPER VAN DREADNOUGHT
Hawkridge Bros. Co.
C 0.8, W 18, V 2, Cr 4, Mo 0.6, bal Fe.
For lathe and planer tools, reamers, hobs, drills, taps; Type
T2; high speed steel. *Obsolete*

SUPER VNCA
British Steel Corp.
C 0.3, Mn 0.6, Ni 3, Cr 0.75, Mo 0.25, bal Fe.
Oil hardened: 124,000-140,000 TS; 103,000-132,000 YS;
20-24 El; 59-66 RA; 62-100 IZ. For shafts, spindles,
crankshafts, axles, bolts, connecting rods. Oil hardening,
tough, shock resistant. *Obsolete*

SUPER WH MEEHANITE
Meehanite Metal Corp.
C 3, Si 2, bal Fe.
For ball mill liners, muller tires; wear resisting cast iron.
Obsolete

SUPER X LEADED NICKEL SILVER 8%
Olin Corp.
Ni 8, Zn 28, Pb 1.5, bal Cu.
Hard: 85,000 TS; 2 El. For key and cylinder blocks, lock parts;
nickel silver, corrosion resistant. *Obsolete*

SUPER X NICKEL SILVER 12 ALLOY NO. 176
Olin Corp.
Ni 12, Zn 18, bal Cu.
Hard: 79,000 TS; 4 El. Soft: 52,000 TS; 50 El. For coins,
emblems, hardware, jewelry; corrosion resistant, good cold
workability. *Obsolete*

SUPER X NICKEL SILVER 18 ALLOY 164
Olin Corp.
Ni 18, Zn 17, bal Cu.
Hard: 85,000 TS; 75,000 YS; 3 El. Soft: 58,000 TS; 28,000 YS;
40 El. For coins, jewelry, hardware; corrosion resistant, good
cold workability. *Obsolete*

SUPER X NICKEL SILVER 18 ALLOY 168
Olin Corp.
Ni 18, Zn 27, bal Cu.
Hard: 100,000 TS; 85,000 YS; 3 El. Soft: 60,000 TS; 27,000
YS; 40 El. For springs, switch blades, washers, radio tubes,;
corrosion resistant, good spring qualities. *Obsolete*

SUPER X PHOSPHOR BRONZE GR. C ALLOY 113
Olin Corp.
Sn 8, P 0.05-0.35, bal Cu.
Hard: 93,000 TS; 72,000 YS; 40 El. Soft: 55,000 TS; 21,000
YS; 70 El. For springs, contacts, diaphragms, fuse clips;
fatigue and wear resistant. *Obsolete*

SUPER X PHOSPHOR BRONZE GR. E ALLOY 101
Olin Corp.
Sn 3.5, P 0.05-0.35, bal Cu.
Hard: 65,000 TS; 50,000 YS; 8 El. Soft: 40,000 TS; 14,000 YS;
48 El. For fuse clips, springs, clutch plates; corrosion and
wear resistant. *Obsolete*

SUPER ZORITE
Michiana Products Corp.
Ni 37-40, Cr 17-21, bal Fe.
For beams, rails, trays, furnace parts; heat resisting to
1800°F.

SUPER-ALLOY
Manufacturer not listed.
C 0.5, Si 0.75, Cr 1.15, W 2.5, V 0.2, bal Fe.
Oil hardening; for tools, dies. AISI S1. Shock resisting tool
steel.

SUPER-COR
Airco Vacuum Metals
Mn 1.5, Si 0.49, Ni 0.05, Cr 0.05, Mo 0.01, V 0.02, bal Fe.
Welded: 83,600 psi TS; 72,000 psi YS; 27 El. Good impact
strength. Weld deposit: 0.052 C. AWS A5.20 Class E 70T-1.

SUPER-DE LAVAUD
United States Pipe & Foundry Co.
C 3, Si 2.5, bal Fe.
For gears, shafts, housings; cast iron. *Obsolete*

SUPER-DE LAVAUD METAL
United States Pipe & Foundry Co.
TC 3.5-3.75, Mn 0.3-5, Si 1.5-1.85, bal Fe.
Cast: 32,000 TS; 0.5 El. For pipes; centrifugal cast iron.
Obsolete

SUPER-DUCT F-3
Industrial Steels Inc.
C 0.4, Ni 1.5, Cr 0.8, Mo 0.2, bal Fe.
For machinery parts; high strength and tough.

SUPER-GENUINE BABBITT
Now ADAMANT SUPER-GENUINE BABBITT.

SUPER-HIGH SPEED
American manufacture
C 0.75-0.9, W 17-22, Cr 4.5-6, V 1-2.5, Mo 0.75-1.5, Co 8-15,
bal Fe.
For high speed tools; high speed steel; has great red
hardness.

SUPER-HOLFOS W.W
Holcroft Castings & Forgings Ltd.
Copper. Cu 88.3, Sn 11.5, P 0.2.
Centrifuge cast: 45,000 TS; 27,000 YS; 8 El; 100 Brin. Good
wear and abrasion resistance. For worm and helical gears,
drive shafts, splined couplings. BS 1400 PB2-C; BS 421; SAE
65.

SUPER-INVAR
Westinghouse Electric Corp.
Ni 31, Co 5, bal Fe.
For instruments; zero expansion. *Obsolete*

SUPER-KUT

Great Western Steel Co.

C 0.7, Cr 4, W 18, Mo 0.5, Cr 5, V 1, bal Fe.

For tools, cutters, broaches, chasers; high speed steel.

SUPER-LOY

General Steel Industries

C 0.2, bal Fe.

For screws, filter cloth.

SUPER-MUMETAL 50

Telcon Metals Ltd.

Ni 77, bal Fe.

Rolled: 132,000 TS; 290 Brin. Annealed: 78,000 TS; 61,000 YS; 110 Brin. For electronic and electrical equipment, motors; soft magnet, high permeability. *Obsolete*

SUPER-RAPID EXTRA

Houghton & Richards Inc.

C 0.7, W 18, V 1, Cr, bal Fe.

For punches, dies, cutting tools; high speed steel.

SUPER-SAFETY

M.E. Cunningham Co.

C, alloy, bal Fe.

For steel stamps.

SUPER-SHOCK

Allied Steel & Tractor Products Inc.

C 0.5, Si 0.75, Cr 1.15, W 2.5, V 0.2, bal Fe.

For cold battering and chipping tools, caulking and beading tools; shock resistant, wear resistant.

SUPER-TM-2

Timken Co.

C 0.38-0.43, Ni 1.85-2.25, Cr 1-1.3, Mo 0.35-0.55, bal Fe.

Heat treated: 271,000-280,000 TS; 221,000-235,000 YS; 7.0-10.5 El; 24.5-38.5 RA; 510-530 Brin. For aircraft structural parts, landing gears, impeller shafts; oil hardened, shock resistant. *Obsolete*

SUPER-X-PHOSPHOR BRONZE GRADE D

Olin Corp.

Cu 90, Sn 10, P.

Hard: 103,000 TS; 13 El. Soft: 65,000 TS; 60 El. For springs, diaphragms; tough. *Obsolete*

SUPER-Y

Chicago Malleable Casting Co.

C 3.3, Ni 1.5, Cr 0.5, bal Fe.

Annealed: 60,000 TS; 44,000 YS; 18 El; 143 Brin. For castings; malleable iron.

SUPERALLOYS ALLOY V57

Eagle & Globe Steel Ltd.

Superalloy. O 0-0.08, Cr 14.8, Ni 27, Mo 1.25, Ti 3, Al 0.25, V 0-0.5, B 0.01, bal Fe.

Austenitic, precipitation hardenable iron base alloy. For turbine wheels, spacers, buckets, and high temperature bolting. High temperature tooling, liners, extrusion dies, mandrels, rams, forging dies, dummy blocks, holders, rings. AISI 663.

SUPERALUMAG T.35

Trefileries & Laminoirs du Havre

Zn 3, Mg 1.5, Cr 0.2, Mn 0.4, bal Al.

Annealed: 26,000 TS; 10,000 YS; 20 El; 38 Brin. Hardened: 50,000 TS; 35,000 YS; 16 El; 70 Brin. For engine parts; corrosion resistant.

SUPERALUMAG T.45

Trefileries & Laminoirs du Havre

Zn 6, Mg 2, Cr 0.3, Mn 0.5, bal Al.

Annealed: 29,000 TS; 10,000 YS; 20 El; 50 Brin. Hardened: 69,000 TS; 50,000 YS; 12 El; 150 Brin. For aircraft parts; hardenable.

SUPERALUMAG T.60

Trefileries & Laminoirs du Havre

Zn 8, Mg 2, Cu 1.5, Cr 0.3, Mn 0.5, bal Fe.

Annealed: 45,000 TS; 10,000 YS; 16 El; 55 Brin. Hardened: 86,000 TS; 72,000 YS; 8 El; 180 Brin. For aircraft structures; hardenable.

SUPERB

Latrobe Steel Co.

C 0.7, Cr 0.95, V 0.18, Si 0.25, Mn 0.25, bal Fe.

For tools, dies, chasers; water or oil hardened.

SUPERBASIQUE

Ste de Produits Metallurgiques

Stainless steel. C, Cr, Ni, bal Fe.

For corrosion resistant parts; stainless, austenitic.

SUPERBOLT

Timken Co.

C 0.35-0.45, Mn 0.5-0.8, Ni 1.5-2, Mo 0.2-0.3, 0.25% min Cr, bal Fe.

For crosshead bolts; tough. *Obsolete*

SUPERBRONZE

English manufacture

Cu 57-69, Zn 21-38, Mn 3-3.2, Fe 1.3-2, Al 1.2-5.

For marine parts, propeller blades; corrosion resisting.

SUPERCASE

Supersteels Inc.

C 0.08-0.15, bal Fe.

Hot rolled: 60,000 TS; 33 El; 60 RA; 110 Brin. Heat treated: 81,000 TS; 24 El; 69 RA; 217 Brin. For camshafts, gears, bolts, transmission shafts, worms, sprockets. Case hardening steel, free machining.

SUPERCHITONAL

Lavorazione Leghe Leggere SpA

For aircraft structures. Aluminum clad Superavional.

SUPERDIE

Columbia Tool Steel Co.

Tool material. C 2.2, Cr 12, W 0.8, Si 1, bal Fe.

Annealed: 269 Brin. For blanking, trimming dies, burnishing tools, extrusion and drawing dies; resistant to abrasive wear. AISI D3; tool steel.

SUPERDIE NO. 1

Belmont Metals Inc.

Cu 3, Al 4, bal Zn.

37,000 psi TS. For cast dies and molds, for decorative and industrial castings; melting point 717°F.

SUPERDIE NO. 2

Belmont Metals Inc.

Cu 3, Al 4, Ti, bal Zn.

39,500 psi TS. For decorative and industrial casting; melting point 720°F.

SUPERDIE NO. 3

Belmont Metals Inc.

Cu 3, Al 4, Ti, Be, bal Zn.

40,000 psi TS. For high strength cast dies and molds, for high strength decorative and industrial castings; melting point 722°F.

SUPERDRAW

Edgar Allen Balfour Ltd.

C 1.3, Mn 0.3, W 2.25, bal Fe.

For cold-drawing dies; water hardened.

SUPERELSO

Creusot-Loire

C 0-0.18, Mn 0.9, Si 0.3, Cr 1.3, Cu 0.2, Mo 0.25, Ni 1.5, V, bal Fe.

Plate: 114,000-135,000 psi TS; 97,000 psi YS; 15 El min. For mine cars, chutes, bus and trailer bodies, derricks, structures, booms. Low-alloy high strength structural steel, shock resistant.

SUPERELSO 52

Creusot-Loire

C 0-0.2, Ni 0.4, Mo 0.3, Cr 0.4, Mn 1.2, bal Fe.

Heat treated: 86,000-103,000 TS; 68,000 YS min. For mine cars, derricks, truck and bus bodies, chutes. Low-alloy, high strength structural steel; suitable for case-hardening. *Obsolete*

SUPERELSO 60

Creusot-Loire

C 0-0.2, Mn 1.2, Mo 0.4, bal Fe.

Heat treated: 100,000 TS min; 85,000 YS min. For mine cars, chutes, derricks, truck and bus bodies, booms. Low-alloy, high strength structural steel, shock resistant. *Obsolete*

SUPERELSO 600

Creusot-Loire

C 0-0.12, Mn 1.15, Si 0-0.5, Ni 0.8, Cr 1.75, Mo 0.25, bal Fe.

Quenched and tempered: 700 N/mm² psi TS min; 600 N/mm² psi YS min. Weldable; for general civil construction.

SUPERELSO 70

Creusot-Loire

C 0-0.18, Mn 1.3, Si 0-0.5, Ni 1, Cr 1, Mo 0.35, bal Fe.

Weldable; for civil construction.

SUPERFINE IX

Kelly Foundry & Machine Co.

C, alloy, bal Fe.

Cast: 32,000 TS; 190 Brin. For cast iron castings; Tr. S. 2500. *Obsolete*

SUPERFINE MM

Kelly Foundry & Machine Co.

C, alloy, bal Fe.

Cast: 44,000 TS; 180 Brin. For cast iron castings; Tr. S. 7000. *Obsolete*

SUPERFINE MOLD

Kelly Foundry & Machine Co.

C, alloy, bal Fe.

Cast: 26,000 TS; 160 Brin. For glass mold castings; cast iron. *Obsolete*

SUPERFINE MX

Kelly Foundry & Machine Co.

C, alloy, bal Fe.

Cast: 62,500 TS; 220 Brin. For cast iron castings; Tr. S. 8620. *Obsolete*

SUPERFINE XXX

Kelly Foundry & Machine Co.

C, alloy, bal Fe.

Cast: 43,000 TS; 200 Brin. For cast iron castings; Tr. S. 8380. *Obsolete*

SUPERFORM 30

LTV Steel

C 0.03-0.08, Mn 0.25-0.5, Si 0.01-0.06, Zr 0.05-0.12, Al 0.02-0.06, bal Fe.

30,000 psi YS min. Good cold-forming and welding. For railroad car components, disc brake parts.

SUPERFORM 40

LTV Steel

C 0.14-0.2, Mn 0.4-0.9, Si 0.01-0.02, Zr 0.05-0.12, Al 0.02-0.06, bal Fe.

40,000 psi YS min. For wheel spiders and bumpers.

SUPERIMPHY

Creusot-Loire

Ni 80, Cr 20.

For electrical resistances, heating elements; heat and corrosion resistant. *Obsolete*

SUPERIOR

CCS Braeburn Alloy Steel

C, alloy, bal Fe.

For tools. *Obsolete*

SUPERIOR
H.K. Porter Co., Inc.
Ni 76-84, Cr 16-23, C.
For heat and corrosion resistant parts; heat and corrosion resistant. *Obsolete*

SUPERIOR
Bethlehem Steel Corp.
C 0.7-1.3, V 0.15-0.25, bal Fe.
For taps, reamers, cutters, punches, dies, broaches, chisels; tool steel. *Obsolete*

SUPERIOR
American manufacture
Cr 19.5, Fe 0.5, Mn 2, bal Ni.
For resistance; resistance alloy.

SUPERIOR
Superior Steel Corp.
C 0.2, Cr 18, Ni 8, bal Fe.
For stainless steel parts; stainless.

SUPERIOR 1
CCS Braeburn Alloy Steel
C 2.05-2.2, Cr 11.5-13.5, V 0.55-0.65, bal Fe.
For tools, cutters, blanking dies, shear knives, ceramic tools. Non-deforming. *Obsolete*

SUPERIOR 2
CCS Braeburn Alloy Steel
C 1.45-1.6, Cr 11-12, V 0.15-0.25, Mo 0.7-0.8, bal Fe.
For tools, cutters, coining and forming and blanking dies. Oil hardening. *Obsolete*

SUPERIOR 3
CCS Braeburn Alloy Steel
C 1.5, Cr 12, V 0.87, Mo 1, bal Fe.
For shear blades, blanking and forming dies. Cold worked steel. *Obsolete*

SUPERIOR 5
CCS Braeburn Alloy Steel
C 1.35, Mn 0.3, Si 1, Cr 6.25, V 6, Mo 1, bal Fe.
For cold steel extrusion dies, cold heading dies, form punches. Tough and wear resistant. Oil or salt bath hardening. *Obsolete*

SUPERIOR A NICKEL CHROME
H.K. Porter Co., Inc.
Ni 80, Cr 20.
For electrical resistances; non-magnetic. *Obsolete*

SUPERIOR ARK
Jessop-Saville Ltd.
C 0.7, W 19, Cr 4, V 2, bal Fe.
For high speed cutting tools; high speed steel.

SUPERIOR ARK HIGH SPEED
Jessop Steel Co.
C 0.7-0.75, W 18, Cr 4, V 1, bal Fe.
For boring tools, threading dies, gear cutters; high speed steel. *Obsolete*

SUPERIOR ARK HIGH SPEED
Jessop-Saville Ltd.
C 0.7-0.75, W 18, Cr 4, V 1, bal Fe.
For boring tools, threading dies, gear cutters; high speed steel. *Obsolete*

SUPERIOR COPPER STEEL
Superior Metal Co.
C, alloy, Cu, bal Fe.
For replacing solid copper; Cu plated corrosion resistant steel.

SUPERIOR NO. 1
Superior Foundry Co.
TC 3-3.25, Si 1.9-2.1, Ni 0.75-1.25, Cr 0.3-0.6, Mo 0.6-0.8, bal Fe.
Cast: 50,000-60,000 TS; 235-269 Brin. For diesel pistons; cast iron.

SUPERIOR NO. 2
Superior Foundry Co.
TC 3-3.25, Si 2-2.25, Ni 0.75, Mo 0.3-0.5, M 0.6-0.8, bal Fe.
Cast: 40,000 TS; 196-235 Brin. Transverse strength 3000 lb; Transverse deflection 0.30 in.; cast iron.

SUPERIOR NO. 3
Superior Foundry Co.
TC 3-3.25, Si 2-2.25, Mn 0.7-0.9, Cu 0.75-1, bal Fe.
Cast: 38,000 TS; 187-217 Brin. For machine tools; Transverse strength 2900 lb; Transverse deflection 0.30 in.

SUPERIOR NO. 4
Superior Foundry Co.
TC 3-3.3, Si 2-2.2, Mn 0.6-0.8, Cr 0.3, bal Fe.
Cast: 35,000 TS; 187-217 Brin. For elevator sheaves; Transverse strength 2800 lb; Transverse deflection 0.28 in.

SUPERIOR NO. 5
Superior Foundry Co.
TC 3-3.25, Si 2.1-2.4, Mn 0.6-0.8, Mo 0.3-0.4, bal Fe.
Cast: 38,000 TS; 211-248 Brin. For automotive manifold; cast iron.

SUPERIOR NO. 6
Superior Foundry Co.
TC 3-3.3, Si 2-2.2, Mn 0.65-0.8, Cr 0.3, Mo 0.3, bal Fe.
Cast: 40,000 TS; 207-241 Brin. For pressure castings; Transverse strength 3000 lb; Transverse deflection 0.30 in.

SUPERIOR NO. 7
Superior Foundry Co.
TC 3-3.25, Si 2-2.2, Mn 0.6-0.9, Ni 1.25-1.75, Cr 0.4-0.6, Mo 0.4-0.6, bal Fe.
Cast: 50,000-60,000 TS; 207-255 Brin. For lathe beds, dies; suitable for flame hardening.

SUPERIOR NO. 8
Superior Foundry Co.
TC 3-3.25, Si 1.8-2, Mn 0.7-0.9, Ni 1, Cr 0.25-0.35, Mo 0.3-0.4, bal Fe.
Cast: 40,000-55,000 TS; 207-217 Brin. For diesel heads; resists heat shock and wear.

SUPERIOR OIL
Ackerlind Steel Co., Inc.
C 0.92, Mn 1.15, Cr 0.5, W 0.5, V 0.2, bal Fe.
For tools, dies, cutters; nondeforming, oil hardening. *Obsolete*

SUPERIOR OIL HARDENING
Jessop-Saville Ltd.
C 0.9, Mn 1.2, Cr 0.5, W 0.5, bal Fe.
For press tools, dies, gauges, punches; oil hardened, non-deforming.

SUPERIOR SHAFTING
Atlas Specialty Steels
C 0.35, Mn 1.5, Si 0.25, Cr 2, S 0-0.1, bal Fe.
For shafting; free cutting.

SUPERIOR TOOL
Colt Industries
Cr 0.48, C, bal Fe.
For die casting dies; water hardened. *Obsolete*

SUPERIOR TOOL STEEL
McInnes Steel Co.
C 0.9, Mn 0.4, Si 3, bal Fe.
For chisels, trimming dies, punches, general tools; water hardening. *Obsolete*

SUPERIOR X-3012
Superior Tube Company
Zr 0.1, W 2, bal Ni.
For vacuum tube cathodes; excellent electron emission characteristics. *Obsolete*

SUPERKORE BB
U.S. Steel Corp.
C 0-0.25, Mn 0.55, Ni 1.8, Mo 0.25, 0.005% B min, bal Fe.
Heat treated: 190,000 TS; 155,000 YS; 13 El; 50 RA; 370 Brin.
For gears, pinions, shafts; case hardening steel. *Obsolete*

SUPERLA
Boyd-Wagner Co.
C 1.3, V, bal Fe.
For lathe and planer tools, reamers, taps, drills; Type W2; water hardened.

SUPERLOY
Acme Electric Welder Co.
Copper. Cu alloy.
For resistance and spot welding electrodes.

SUPERLOY
Resisto-Loy Company, Inc.
Cr 30, Mo 8, Co 8, B 0.05, W 5, C 0.2, bal Fe.
Welded: 200,000 TS; 1,000 Brin. For hard facing electrodes; wear resistant, bond composition. *Obsolete*

SUPERLOY
Birdsboro Corp.
C 3.2, Si 1.4, Mn 0.8, bal Fe.
For rolls; cast iron. *Obsolete*

SUPERLOY 30
Washington Iron Works, Inc.
C 0.25-0.35, Mn 1.2-1.3, Si 0-0.75, bal Fe.
Annealed: 60,000 TS; 30,000 YS; 24 El; 35 RA; 150 Brin. For castings, gears, shafts; water hardened.

SUPERLOY 42
Washington Iron Works, Inc.
C 0.25-0.35, Mn 1.1-1.5, Si 0-0.75, bal Fe.
Annealed: 80,000 TS; 40,000 YS; 17 El; 25 RA; 175 Brin. For castings, gears, shafts; water hardened.

SUPERLOY 53
Washington Iron Works, Inc.
C 0.3-0.4, Mn 0.7-0.9, Si 0.3-0.5, Cr 0.6-0.7, Ni 1.25-1.75, Mo 0.25-0.35, bal Fe.
Annealed: 85,000 TS; 50,000 YS; 20 El; 30 RA; 190 Brin. For castings, gears, shafts, crankshafts; tough, shock resistant.

SUPERLOY 530
Washington Iron Works, Inc.
C, alloy, bal Fe.
For machine tool parts; high strength and shock resistant.

SUPERLOY 64
Washington Iron Works, Inc.
C 0.35-0.45, Mn 1.35-1.65, Si 0.3-5, Cr 0.7-0.9, Ni 1.25-1.75, Mo 0.3-0.4, bal Fe.
Annealed: 120,000 TS; 90,000 YS; 10 El; 15 RA; 260 Brin. For castings, gears, shafts, crankshafts; tough, shock resistant.

SUPERLOY K2MO
Washington Iron Works, Inc.
C 0.15, Cr 18, Ni 8, Mo 3.5.
Annealed: 85,000 TS; 40,000 YS; 40 El; 60 RA; 160 Brin. For chemical plant equipment; stainless; Type 317. *Obsolete*

SUPERLOY MANGANESE STEEL
Washington Iron Works, Inc.
C 0.7, Mn 13, bal Fe.
For railroad frogs, tracks; abrasion resistant. *Obsolete*

SUPERLOY NO. 10
Washington Iron Works, Inc.
C 0.3, Ni 1.5, Cr 0.8, Mo 0.2, bal Fe.
For gears, housings, shafts; shock resistant. *Obsolete*

SUPERLOY NO. 4
Washington Iron Works, Inc.
C 0.4, Ni 1.5, Cr 0.9, Mo 0.2, bal Fe.
For gears and castings; shock resistant. *Obsolete*

SUPERLOY NO. 7
Washington Iron Works, Inc.
C 0.7, Mn 12, bal Fe.
For liner plates, crusher jaws, dipper teeth; abrasion resistant.
Obsolete

SUPERMAGALUMA
Aluminium Belge, S.A.
Mg 5.4, Mn 0.15, bal Al.
Soft: 36,000-42,500 TS; 14,500-21,000 YS; 20-25 El. Hard:
50,000-65,000 TS; 42,500-58,000 YS; 4-6 El. For light alloy
parts; heat treatable, corrosion resistant.

SUPERMAL
Dresser Industries
TC 1.7, Mn 0.3, Si 1.2, bal Fe.
Heat treated: 75,000 TS; 55,000 YS; 8 El; 187 Brin. For
chains, buckets, conveyor equipment; pearlitic, wear
resistant; malleable iron.

SUPERMALLOY
Magnetic Metals Co.
80 Ni, Fe, Mo.
Soft magnetic alloy processed for high initial permeability.

SUPERMALLOY
Arnold Engineering Co.
Fe 15.7, Mo 5, Mn 0.3, bal Ni.
For transformers, communication and radar equipment;
magnetically soft. Fabricated parts only.

SUPERMALOY
Western Electric Co.
Ni 79, Mo 5, Fe 15.7, Mn 0.3.
For communication transformers, magnets; high
permeability.

SUPERMENDUR
Telcon Metals Ltd.
Soft magnetic alloy, high saturation, square loop, for special
transformers.

SUPERMENDUR
Spang Specialty Metals
Fe 49, Co 49, V 2.
For amplifiers, pulse and power transformers; high
permeability. Soft magnet.

SUPERMENDUR
Magnetic Metals Co.
Fe 49, Co 49, V 2.
For amplifiers, pulse and power transformers; high
permeability. Soft magnet.

SUPERMENDUR
Westinghouse Electric Corp.
Fe 49, Co 49, V 2.
For amplifiers, pulse and power transformers; high
permeability. Soft magnet.

SUPERMENDUR
Bell Telephone Laboratories
Fe 49, Co 49, V 2.
For amplifiers, pulse and power transformers; high
permeability. Soft magnet.

SUPERMETAL
Parkersburg Steel Co.
Coated low carbon steel.
For oven linings; corrosion and heat resistant. *Obsolete*

SUPERMETONIC
Pechiney/Trefimetaux
Ni 30, Fe 2, Mn 2, bal Cu.
Cu-Ni tubing for heat exchanger applications; Cu Ni 30 Fe 2
Mn 2.

SUPERMOLD
Latrobe Steel Co.
C 0.3, Cr 1.7, Mo 0.4, bal Fe.
Molds for plastics, die casting dies. AISI P20. *Obsolete*

SUPERMU 10
Now SUPERPERM 49.

SUPERMU 40
Now SUPERPERM 80.

SUPERMUMETAL
Telcon Metals Ltd.
Fe 14, Cu 5, Mo 4, bal Ni.
For electrical and magnetic equipment. Vacuum melted.
High magnetic permeability and low electrical losses.

SUPERNICKIMPHY
Creusot-Loire
C 0.02, Mn 0.1, Cu 0.02, bal Ni.
For radio tube cathodes; heat and corrosion resistant.
Obsolete

SUPERPERM 49
Magnetic Metals Co.
50 Ni, Fe.
Soft magnetic alloy. High initial permeability and high
maximum flux.

SUPERPERM 80
Magnetic Metals Co.
80 Ni, Fe, Mo.
Soft magnetic alloy processed for high permeability and low
magnetizing force.

SUPERSQUARE 80
Magnetic Metals Co.
80 Ni, Fe, Mo.
Higher maximum flux density, high gain, high squareness
ratio; soft magnetic alloy.

SUPERSTAR
Now CARPENTER SUPERSTAR.

SUPERSTON 10
Ampco Pittsburgh Corp.
Al 8, Fe 3, Mn 12, Ni 2, bal Cu.
Sand cast: 98,000 TS; 48,000 YS; 26 El; 28 RA; 185 Brin.
Centrifugal cast: 105,000 TS; 50,000 YS; 30 El; 30 RA; 190
Brin. Heat treated: 125,000 TS; 75,000 YS; 12 El; 255 Brin.
For valves, pumps, propellers, gears, slides; corrosion and
wear resistant.

SUPERSTON 40
Stone Manganese - J. Stone & Co. Ltd.
Al 8, Fe 3, Mn 12, Ni 2, bal Cu.
Sand cast: 98,000 TS; 48,000 YS; 26 El; 28 RA; 185 Brin.
Centrifugal cast: 105,000 TS; 50,000 YS; 30 El; 30 RA; 190
Brin. Heat treated: 125,000 TS; 75,000 YS; 12 El; 255 Brin.
For valves, pumps, propellers, gears, slides; corrosion and
wear resistant.

SUPERSTON 60
Stone Manganese - J. Stone & Co. Ltd.
Mn 11-14, Al 8.5-9, Fe 2-4, Ni 1.5-4.5, Sn 0-1, Pb 0-0.02, bal
Cu.
Sand cast: 107,500 TS; 56,000 YS; 10 El. Chill cast: 127,000
TS; 83,000 YS; 5.5 El. For marine propellers, shafts,
hardware. High strength, tough, corrosion resistant.

SUPERSTON 70
Stone Manganese - J. Stone & Co. Ltd.
Cu 72, Mn 15, Fe 3, Ni 2.5, Al 7.5.
Aluminum bronze casting for marine applications.

SUPERSTON L-189
Stone Manganese - J. Stone & Co. Ltd.
Al 8-12, Ni 4-6, Fe 4-6, bal Cu.
Cast: 90,000 TS; 45,000 YS; 15-20 El; 180 Brin. For engine
parts; Aluminum bronze, corrosion resistant.

SUPERTEMP
Bethlehem Steel Corp.
C 0.35, Mn 0.8, Cr 0.45, W 1, Mo 0.6, bal Fe.
Heat treated: 140,000 TS; 120,000 YS; 18 El; 50 RA; 321 Brin.
At 1000 F: 110,000 TS; 90,000 YS; 20 El; 70 RA; 241 Brin. For
bolts, studs, forgings; creep resistant, oil hardened. *Obsolete*

SUPERTHERM
Abex Corp.
Stainless steel. Ni 35, Cr 26, Co 15, W 5, Si 1.6, C 0.5, bal Fe.
Cast: 75,000 TS; 48,000 YS; 5 El; 215 Brin. For furnace parts,
cement kilns, smelting and calcining equipment. High heat
resistance to 2300°F. Good hot ductility.

SUPERTOUGH B.20
British Steel Corp.
C 0.2, Mn 1.5, bal Fe.
Oil hardened: 89,000-112,000 TS; 60,000-92,000 YS; 21-26
El; 59 RA; 94-118 IZ. For welded structures, crankshafts,
steering levers, shafts, spindles. Oil or water hardening.
Obsolete

SUPERTOUGH B.25
British Steel Corp.
C 0.25, Mn 1.5, bal Fe.
Oil hardened: 89,000-112,000 TS; 60,000-92,000 YS; 21-26
El; 59 RA; 94-118 IZ. For welded structures, crankshafts,
steering levers, shafting, spindles. Oil or water hardening.
Obsolete

SUPERTOUGH B.35
British Steel Corp.
C 0.35, Mn 1.5, bal Fe.
Oil hardened: 96,000-120,000 TS; 70,000-88,000 YS; 22-29
El; 54-68 RA; 60-92 IZ. For axles, shafts, crankshafts,
connecting rods. Oil or water hardening. *Obsolete*

SUPERTOUGH C.20
British Steel Corp.
C 0.18, Mn 1.6, Ni 0.6, Mo 0.25, bal Fe.
Oil hardened: 97,000-103,000 TS; 78,000-88,000 YS; 22-30
El; 61-67 RA; 70-109 IZ. For welded structures, axles, railway
equipment, dredging and earth moving equipment. Oil or
water hardening. *Obsolete*

SUPERTOUGH C.40
British Steel Corp.
C 0.35, Mn 1.6, Mo 0.28, bal Fe.
Oil hardened: 106,000-146,000 TS; 79,000-125,000 YS; 18-28
El; 52-66 RA; 45-100 IZ. For crankshafts, bolts, axles, shafts,
studs, levers, connecting rods. Tough shock resistant. Oil
hardening. *Obsolete*

SUPERTOUGH C.40M
British Steel Corp.
C 0.35, Mn 1.6, Mo 0.45, bal Fe.
Oil hardened: 116,000-153,000 TS; 90,000-135,000 YS; 19-23
El; 50-61 RA; 49-74 IZ. For crankshafts, connecting rods,
axles, levers, bolts, fasteners. Tough, impact resistant, oil
harden. *Obsolete*

SUPERTOUGH D
British Steel Corp.
C 0.4, Mn 1.35, Ni 0.75, Cr 0.45, Mo 0.2, bal Fe.
Oil hardened: 112,000-153,000 TS; 88,000-140,000 YS; 20-27
El; 52-64 RA; 42-85 IZ. For bolts and studs, crankshafts,
gears, boring bars, collets. Oil hardening, shock resistant.
Obsolete

SUPERTOUGH G
Sheffield Forgemasters Ltd.
C 0.15-0.23, Si 0-0.35, Mn 0.75-1.25, Cr 0.75-1.25, Mo
0.4-0.6, Ni 0.5-1, bal Fe.
High strength low alloy steel suitable for large forgings and
fabricated pressure vessels. Weldable.

SUPERWELD NO. 55
Edgcomb Metals Co.
C 0.2, bal Fe.
Deposited: 65,000 TS, 45,000 YS; 30 El. For welding
electrode for heavy plate welding. *Obsolete*

SUPERWELD NO. 66
Edgcomb Metals Co.
C, bal Fe.
Deposited: 60,000-70,000 TS; 45,000-55,000 YS; 25-30 El.
For welding electrode for light and heavy gauge material.

SUPERWELD NO. 77
Edgcomb Metals Co.
C 0.08, Mn 0.45, Si 0.3, bal Fe.
Deposited: 75,000 TS; 65,000 YS; 20 El. AC-DC welding
electrode for light and heavy gauge carbon steel; production
and maintenance welding. Conforms to AWS Class E-6012.

SUPERWELD NO. 77A
Edgcomb Metals Co.
C 0.1, Mn 0.35, Si 0.4, bal Fe.
Deposited: 70,000 TS; 60,000 YS; 20 El. AC-DC welding
electrode for carbon sheet steel; improved weld appearance;
easy clean-up. Conforms to AWS Class E-6013.

SUPRA
English manufacture
Si 20, Cu 5, Mn 2, Fe 0.7, bal Al.
Cast: 21,000 TS; 0.5 El; 100-130 Brin. For pistons; light
weight.

SUPRA RECORD III
Stahlwerke Sudwestfalen
C 0.76, Co 10, Cr 4.2, Mo 0.8, V 1.8, W 18, bal Fe.
For lathe and planer tools, reamers, broaches, hobs; high
speed steel. *Obsolete*

SUPRA REKORD I
Stahlwerke Sudwestfalen
C 0.86, Co 2.8, Cr 4.3, Mo 0.85, V 2.1, W 12, bal Fe.
For lathe and planer tools, drills, reamers, taps; high speed
steel. *Obsolete*

SUPRA REKORD II
Stahlwerke Sudwestfalen
C 0.79, Co 4.7, Cr 4.3, Mo 0.75, V 1.5, W 18, bal Fe.
For lathe and planer tools, reamers, broaches, taps; high
speed steel. *Obsolete*

SUPRA VENIVICI
Stahlwerke Sudwestfalen
C 0.86, Cr 4.1, Mo 0.85, V 2.5, W 12, bal Fe.
For lathe and planer tools, milling cutters, hobs; high speed
steel. *Obsolete*

SUPRA VENIVICI EXTRA
Stahlwerke Sudwestfalen
C 0.74, Cr 4.1, V 1.1, W 18.5, bal Fe.
For lathe and planer tools, drills, taps, hobs; high speed
steel. *Obsolete*

SUPRAKOLBENMETALL
Manufacturer not listed.
Si 20, Cu 5, Mn 2, Fe 0.7, bal Al.
For light alloy parts, pistons.

SUPRAL 100
Superform Metals Ltd.
Aluminum. Cu 6, Zr 0.4, bal Al.
Alloy for superplastic forming. As formed: 18,000 YS; 26,000
TS; 7 El. Heat treated: 44,000 YS; 60,000 TS; 5 El. AMS 4208.

SUPRAL 150
Superform Metals Ltd.
Aluminum. Cu 6, Zr 0.4, bal Al.
SUPRAL 100 clad with 8% thickness of 99.7% aluminum for
corrosion resistant superplastic forming. As formed: 16,000
YS; 23,000 TS; 8 El. Heat treated: 39,000 YS; 57,000 TS; 6 El.
AMS 4209.

SUPRAL 5000
Superform Metals Ltd.
Aluminum. bal Al.
Aluminum alloy for corrosion resistant superplastic forming.
As formed: 19,000 YS; 29,000 TS; 15 El.

SUPRALOY
Bergstrom Alloys Corp.
C, B, bal Fe.
640 Brin. For hard surfacing electrodes; for metal to metal
wear.

SUPRALOY
United States Steel Corp.
C, alloy, bal Fe.
For hard surfacing electrodes; wear resistant. *Obsolete*

SUPREME
Latrobe Steel Co.
C, Cr, V, bal Fe.
For tools, dies. *Obsolete*

SUPREMUS (AISI T1)
Jessop Steel Co.
C 0.73, W 18, Cr 4, V 1.1, bal Fe.
For high speed tools, cutters, planer tools, broaches, drills;
high speed steel.

SUPREMUS EXTRA (AISI T2)
Jessop Steel Co.
C 0.85, Mo 0.7, W 18.5, Cr 4, V 2, bal Fe.
For cutters, drills, lathe centers, broaches, reamers; high
speed steel; Type T1.

SUPREMUS T-1
Jessop Steel Co.
C 0.73, Cr 4, W 18, V 1.1, Mo 0-0.6, bal Fe.
For reamers, drills, lathe and planer tools; high speed steel.
Obsolete

SUPRIMPACTO
Atlas Specialty Steels
C 0.12, Cr 1.5, Ni 3.75, bal Fe.
For large gears; deep hardening carburizing steel.

SUPRIMPACTO
Atlas Specialty Steels
C 0.12, Mn 5, Cr 1.5, Ni 3.75, bal Fe.
For plastic mold dies, carburized parts, machine cut cavities;
case hardened.

SURA 13-CKN-30
Surahammars Bruk
C 0.12, Cr 0.75, Ni 3, bal Fe.
For gears, shafts, camshafts; case hardened, shock resistant.
Obsolete

SURA 18-CKN-30
Surahammars Bruk
C 0.18, Cr 0.75, Ni 3, bal Fe.
For gears, shafts; case hardened, shock resistant. *Obsolete*

SURA 25-CKM-10
Surahammars Bruk
C 0.25, Cr 1.05, Mo 0.2, bal Fe.
For machinery parts; water hardening. *Obsolete*

SURA 30-CKMN-32
Surahammars Bruk
C 0.3, Cr 1.05, Ni 3.25, Mo 0.25, bal Fe.
For gears, shafts; oil hardened, shock resistant. *Obsolete*

SURA 30-CMK-27
Surahammars Bruk
C 0.3, Cr 2.7, Mo 0.5, bal Fe.
For machinery parts; oil hardened. *Obsolete*

SURA 35-CKN-25
Surahammars Bruk
C 0.35, Cr 1.15, Ni 2.6, bal Fe.
For shafts, gears, pinions; oil hardened, tough. *Obsolete*

SURA 40-CKN-12
Surahammars Bruk
C 0.4, Mn 0.8, Cr 0.8, Ni 1.25, bal Fe.
For shafts, gears, pinions; oil hardened, tough. *Obsolete*

SURA 40-CMK-10
Surahammars Bruk
C 0.4, Cr 1.1, Mo 0.25, bal Fe.
For machinery parts; oil hardened. *Obsolete*

SURA 50-CG-12
Surahammars Bruk
C 0.5, Mn 1.2, bal Fe.
For gears, shafts, machinery parts; oil hardened, tough.
Obsolete

SURA 50-CK-5
Surahammars Bruk
C 0.5, Cr 0.5, bal Fe.
For shafts; water hardening. *Obsolete*

SURA CK-26
Surahammars Bruk
Si 2.5-3.5, bal Fe.
For large motors and generators; nonoriented silicon steel
lamination. Meets AISI M-15.

SURA CK-27
Surahammars Bruk
Si 2.5-3.5, bal Fe.
For large motors and generators; nonoriented silicon steel.
Meets AISI M-19.

SURA CK-30
Surahammars Bruk
Si 2.5-3.5, bal Fe.
For large motors and generators; nonoriented silicon steel.
Meets AISI M-22.

SURA CK-33
Surahammars Bruk
Si 2.5-3.5, bal Fe.
For large electrical motors; nonoriented silicon steel.

SURA CK-37
Surahammars Bruk
Si 2.5-3, bal Fe.
For electrical motors; nonoriented silicon steel. Meets AISI
M-36.

SURA CK-40
Surahammars Bruk
Si 2-2.5, bal Fe.
For electrical motors; nonoriented silicon steel. Meets AISI
M-43.

SURA DK-59
Surahammars Bruk
Si 1.3-2, bal Fe.
For electrical motors; nonoriented silicon steel laminations.
Meets AISI M-45.

SURA DK-70
Surahammars Bruk
Si 1-1.5, bal Fe.
For small electrical motors; nonoriented silicon steel
laminations. Meets AISI M-47.

SURA M-3
Surahammars Bruk
Si 3-3.5, bal Fe.
For transformers; grain oriented silicon steel.

SURA M-4
Surahammars Bruk
Si 3-3.5, bal Fe.
For transformers; grain oriented silicon steel.

SURA M-5
Surahammars Bruk
Si 3-3.5, bal Fe.
For transformers; grain oriented silicon steel.

SURA M-6
Surahammars Bruk
Si 3-3.5, bal Fe.
For transformers; grain oriented silicon steel.

SURA M-OH
Surahammars Bruk
Si 3-3.5, bal Fe.
For transformers; high permeability, grain oriented silicon steel.

SURA-A30
Surahammars Bruk
Now SURA M-6.

SURA-AT26
Surahammars Bruk
Now SURA M-5.

SURA-C33
Surahammars Bruk
Now SURA M-OH.

SURA-C37
Surahammars Bruk
Now SURA DK-70.

SURA-C43
Surahammars Bruk
Now SURA DK-59.

SURA-C44
Surahammars Bruk
Now SURA CK-40.

SURA-CT28
Surahammars Bruk
Now SURA CK-37.

SURA-CT30
Surahammars Bruk
Now SURA CK-33.

SURA-D57
Surahammars Bruk
Now SURA CK-30.

SURA-D59
Surahammars Bruk
Now SURA CK-27.

SURA-D60
Surahammars Bruk
Now SURA CK-26.

SURA-D66
Surahammars Bruk
Si 1.3, bal Fe.
For small electrical motors; silicon steel laminations. *Obsolete*

SURA-D68
Surahammars Bruk
Si 1, bal Fe.
For small electrical motors; silicon steel laminations. *Obsolete*

SURA-D70
Surahammars Bruk
Si 1, bal Fe.
For electrical machines; silicon steel. *Obsolete*

SURA-D80
Surahammars Bruk
Si 0.5, bal Fe.
For armatures, motors, generators; silicon steel laminations. *Obsolete*

SURAHAMMAR 15-CMK-6
Surahammars Bruk
C 0.15, Cr 0.8, Mo 0.6, bal Fe.
For machinery parts, gears; case hardened. *Obsolete*

SUREWELD A
Hollup Corp.
C 0.07, bal Fe.
Welded: 67,000 TS; 53,000 YS; 30 El. For welding electrodes; coated.

SUREWELD B
Hollup Corp.
C 0.06, bal Fe.
Welded: 70,000 TS; 55,000 YS; 30 El. For welding electrodes; coated.

SUREWELD C
Hollup Corp.
C, bal Fe.
Welded: 67,000 TS; 25 El. For welding electrodes; coated.

SUREWELD F
Hollup Corp.
C 0.09, bal Fe.
Welded: 67,000 TS; 55,000 YS; 27 El; 35 RA. For welding electrodes; coated.

SUREWELD H
Hollup Corp.
C, 12-15 alloy, bal Fe.
For hard facing electrodes; wear resistant.

SUREWELD H1
Hollup Corp.
C, 15 alloy, bal Fe.
Welded: 550-650 Brin. For hard facing electrodes; impact resistance.

SUREWELD H2
Hollup Corp.
C, 18 alloy, bal Fe.
Welded: 500-575 Brin. For hard facing electrodes; wear resistant.

SUREWELD MLY
Hollup Corp.
C 0.06, bal Fe.
Welded: 80,000 TS; 62,000 YS; 25 El. For welding electrodes; coated.

SUREWELD N
Hollup Corp.
C 0.9, bal Fe.
Welded: 75,000 TS; 55,000 YS; 25 El; 20 RA. For welding electrodes; coated.

SUREWELD NO. 450
Hollup Corp.
C, alloy, bal Fe.
For welding electrodes; coated, wear and shock resistant.

SUREWELD NO. 520
Hollup Corp.
Mn 12-14, C, bal Fe.
600 Brin. For Mn steel welding electrodes; hard surfacing.

SUREWELD NO. 550
Hollup Corp.
C, alloy, bal Fe.
600-675 Brin. For hard surfacing electrodes; abrasion and wear resistant.

SUREWELD NO. 650
Hollup Corp.
C, alloy, bal Fe.
700-800 Brin. For hard-surfacing electrodes; severe abrasion resistance.

SUREWELD TD
Hollup Corp.
C, 2 alloy, bal Fe.
Welded: 600-780 Brin. For hard facing electrodes for cutting tools and dies; wear resistant.

SURF-HARD
Rio Algom Corp.
C 0.4, Mn 0.6, Cr 1.6, Mo 0.35, Al 1.05, bal Fe.
Annealed: 87,000 TS; 54,000 YS; 20 El; 179 Brin. For ejector pins for plastic molds, bushings, sleeves, shafts, gears. Nitriding steel, wear and abrasion resistant, nitrided case.

SURGALOY
Davis & Geck Inc.
C 0-0.2, Cr 18, Ni 8, bal Fe.
For sutures for surgery; stainless.

SUSINI
Japanese manufacture
Cu 1.5-4.5, Mn 1-8, Zn 0.5-1.5, bal Al.
For light alloy parts; non-hardenable.

SUTTONITE H.S.S.
Welding Equipment & Supply Co.
C 0.7, W 18, Cr 4, V 1, bal Fe.
For high speed steel welding electrodes. *Obsolete*

SUWEFA 12M
Stahlwerke Sudwestfalen
C 1.2, Si 0.4, Mn 12.5, bal Fe.
For wear plates, rails, crushers, rollers; wear and abrasion resistant. Work hardened. *Obsolete*

SUWEFA A12
Stahlwerke Sudwestfalen
C 0-0.1, Cr 12.5, Ni 12, bal Fe.
For chemical plant equipment, valves; heat and corrosion resistant. *Obsolete*

SUWEFA A18
Stahlwerke Sudwestfalen
C 0-0.15, Cr 18, Ni 9, bal Fe.
Annealed: 80,000 TS; 35,000 YS; 55 El; 75 RA; 150 Brin. For chemical plant equipment, tanks, mixers; Type 302; stainless, austenitic. *Obsolete*

SUWEFA A182Z
Stahlwerke Sudwestfalen
C 0-0.12, Cr 18, Ni 10.5, Mo 2, Ti = 4 x C, bal Fe.
Annealed: 85,000 TS; 35,000 YS; 55 El; 65 RA; 150 Brin. For welded chemical plant equipment, tanks, vessels; Type 321; stainless, austenitic. *Obsolete*

SUWEFA A182ZN
Stahlwerke Sudwestfalen
C 0-0.12, Cr 18, Ni 10.5, Mo 2, Cb = 8 x C, bal Fe.
Annealed: 90,000 TS; 45,000 YS; 56 El; 65 RA; 160 Brin. For welded chemical plant equipment, tanks, vessels; Type 347; stainless, austenitic. *Obsolete*

SUWEFA A18Z
Stahlwerke Sudwestfalen
C 0-0.12, Cr 18, Ni 9.5, Ti = 4 x C, bal Fe.
Annealed: 85,000 TS; 35,000 YS; 55 El; 65 RA; 150 Brin. For welded chemical plant equipment, tanks; Type 321; stainless, austenitic. *Obsolete*

SUWEFA A18ZN
Stahlwerke Sudwestfalen
C 0-0.12, Cr 18, Ni 9.5, Cb = 8 x C, bal Fe.
Annealed: 90,000 TS; 45,000 YS; 56 El; 65 RA; 160 Brin. For welded chemical plant equipment, tanks, vessels; Type 347; stainless, austenitic. *Obsolete*

SUWEFA A2182Z
Stahlwerke Sudwestfalen
C 0.07, Cr 17.5, Mo 2, Ni 17.5, Cu 2, Ti = 7 x C, bal Fe.
For chemical plant equipment; heat and corrosion resistant. *Obsolete*

SUWEFA AS175
Stahlwerke Sudwestfalen
C 0-0.07, Cr 17, Mo 4.75, Ni 13, bal Fe.
Annealed: 90,000 TS; 40,000 YS; 55 El; 65 RA; 160 Brin. For acid resistant chemical plant equipment, tanks; Type 317; stainless, austenitic. *Obsolete*

SUWEFA AS18
Stahlwerke Sudwestfalen
C 0-0.07, Cr 18, Ni 9.5, bal Fe.
Annealed: 85,000 TS; 35,000 YS; 60 El; 70 RA; 150 Brin. Cold drawn: 180,000 TS; 125,000 YS; 10 El; 330 Brin. For welded chemical plant equipment, tanks, mixers; Type 304; stainless, austenitic. *Obsolete*

SUWEFA AS182
Stahlwerke Sudwestfalen
C 0-0.07, Cr 18, Ni 10.5, Cb = 8 x C, bal Fe.
Annealed: 90,000 TS; 45,000 YS; 56 El; 65 RA; 160 Brin. Cold drawn: 100,000 TS; 65,000 YS; 40 El; 60 RA; 205 Brin. For welded chemical plant equipment, tanks, mixers; Type 347; stainless, austenitic. *Obsolete*

SUWEFA BEM
Stahlwerke Sudwestfalen
C 0.4, Si 0.38, Mn 0.95, bal Fe.
Hot rolled: 90,000 TS; 58,000 YS; 27 El; 50 RA; 200 Brin. For gears, shafts, bolts, fasteners; water hardened. *Obsolete*

SUWEFA BKM
Stahlwerke Sudwestfalen
C 0.3, Si 0.25, Mn 1.35, bal Fe.
For cams, bolts, camshafts, machine tool parts; oil hardened, tough. *Obsolete*

SUWEFA C12
Stahlwerke Sudwestfalen
C 2.1, Mn 0.3, Cr 11.5, bal Fe.
For blanking and forming dies, punches; oil hardened, non-deforming. *Obsolete*

SUWEFA C12 SPEZIAL
Stahlwerke Sudwestfalen
C 2.1, Mn 0.3, Cr 11.5, W 0.7, bal Fe.
For blanking and forming dies, punches; oil hardened, non-deforming. *Obsolete*

SUWEFA C12CO
Stahlwerke Sudwestfalen
C 1.65, Mn 0.3, Cr 11.5, Co, bal Fe.
For blanking and forming dies, punches; air hardened, non-deforming. *Obsolete*

SUWEFA C1VV
Stahlwerke Sudwestfalen
C 0.5, Cr 1, V 0.09, Mn 0.85, bal Fe.
For springs, gears, crankshafts, bolts; oil hardened, shock resistant. *Obsolete*

SUWEFA C45
Stahlwerke Sudwestfalen
C 0.41, Cr 1, Mn 0.65, bal Fe.
For springs, gears, crankshafts, bolts, studs; oil hardened, shock resistant. *Obsolete*

SUWEFA CG
Stahlwerke Sudwestfalen
C 0.4, Cr, Mn, Mo, bal Fe.
For gears, bolts, machine tool parts; oil hardened, tough. *Obsolete*

SUWEFA CIVV
Stahlwerke Sudwestfalen
C 0.5, Cr 1, V 0.09, bal Fe.
For springs, gears, bolts, studs; oil hardened, tough. *Obsolete*

SUWEFA CME
Stahlwerke Sudwestfalen
C 0.15, Cr, Mo, bal Fe.
For gears, bolts, machine tool parts; case hardened, tough. *Obsolete*

SUWEFA CMM
Stahlwerke Sudwestfalen
C 0.2, Cr, Mo, bal Fe.
For gears, cams, machine tool parts; case hardened, tough. *Obsolete*

SUWEFA CMV SPEZIAL
Stahlwerke Sudwestfalen
C 0.5, Cr, Mo, bal Fe.
For gears, bolts, machine tool parts; oil hardened, tough. *Obsolete*

SUWEFA CMV2
Stahlwerke Sudwestfalen
C 0.25, Cr 1, Mo 0.2, Mn 0.65, bal Fe.
For gears, bolts, machine tool parts; oil hardened, tough. *Obsolete*

SUWEFA CMV3
Stahlwerke Sudwestfalen
C 0.33, Cr 1, Mo 0.2, Mn 0.65, bal Fe.
For gears, fasteners, machine tool parts; oil hardened, tough. *Obsolete*

SUWEFA CMV4
Stahlwerke Sudwestfalen
C 0.42, Cr 1, Mo 0.2, Mn 0.65, bal Fe.
For gears, bolts, machine tool parts; oil hardened, tough. *Obsolete*

SUWEFA CN1V
Stahlwerke Sudwestfalen
C 0.28-0.35, Cr 0.5, Ni 1.5, bal Fe.
For gears, bolts, machine tool parts; oil hardened, tough. *Obsolete*

SUWEFA CN2E
Stahlwerke Sudwestfalen
C 0.13, Cr 0.7, Ni 2.5, bal Fe.
For gears, cams, camshafts; case hardened, tough. *Obsolete*

SUWEFA CN2V
Stahlwerke Sudwestfalen
C 0.28-0.35, Cr 0.7, Ni 2.5, bal Fe.
For gears, bolts, crankshafts; oil hardened, tough. *Obsolete*

SUWEFA CN3E
Stahlwerke Sudwestfalen
C 0.13, Cr 0.7, Ni 3.5, bal Fe.
For gears, cams, camshafts, machine tool parts; case hardened, shock resistant. *Obsolete*

SUWEFA CN3V
Stahlwerke Sudwestfalen
C 0.22-0.3, Cr 0.7, Ni 3.5, bal Fe.
For gears, bolts, shafts, studs, machine tool parts; oil hardened, shock resistant. *Obsolete*

SUWEFA CN4E
Stahlwerke Sudwestfalen
C 0.13, Cr 1.1, Ni 4.5, bal Fe.
For gears, bolts, cams, shafts, camshafts; case hardening steel. *Obsolete*

SUWEFA CN4V
Stahlwerke Sudwestfalen
C 0.35, Cr 1.3, Ni 3.5, bal Fe.
For gears, bolts, fasteners, crankshafts; oil hardened, shock resistant. *Obsolete*

SUWEFA CO 1000
Stahlwerke Sudwestfalen
C 0.76, Co 10, Cr 4.2, Mo 0.8, V 1.8, W 18, bal Fe.
For lathe and planer tools, milling cutters, hobs; high speed steel. *Obsolete*

SUWEFA CO 300
Stahlwerke Sudwestfalen
C 0.86, Co 2.8, Cr 4.3, Mo 0.85, V 2.1, W 12, bal Fe.
For lathe and planer tools, hobs, reamers, drills, taps; high speed steel. *Obsolete*

SUWEFA CO 500
Stahlwerke Sudwestfalen
C 0.79, Co 4.75, Cr 4.3, Mo 0.75, V 1.5, W 18, bal Fe.
For lathe and planer tools, reamers, broaches, taps; high speed steel. *Obsolete*

SUWEFA CSV4
Stahlwerke Sudwestfalen
C 0.38, Si 0.85, Cr 1, V 0.1, bal Fe.
For springs, gears, bolts, studs; oil hardened, shock resistant. *Obsolete*

SUWEFA CSV5
Stahlwerke Sudwestfalen
C 0.45, Cr 1, V 0.1, bal Fe.
For springs, gears, bolts, studs; oil hardened, shock resistant. *Obsolete*

SUWEFA CSV6
Stahlwerke Sudwestfalen
C 0.61, Cr 1.18, V 0.1, Mn 0.75, bal Fe.
For springs, gears, bolts, studs; oil hardened, shock resistant. *Obsolete*

SUWEFA CV
Stahlwerke Sudwestfalen
C 0.41, Cr 1, Mn 0.65, bal Fe.
For gears, bolts, crankshafts; oil hardened, tough. *Obsolete*

SUWEFA CV50
Stahlwerke Sudwestfalen
C 0.41, Sn 0.25, Mn 0.65, Cr 1, bal Fe.
For gears, bolts, studs, shafts, fasteners; oil hardened, shock resistant. *Obsolete*

SUWEFA DMO 10
Stahlwerke Sudwestfalen
C 0.24, Cr 1.15, Mo 0.25, bal Fe.
For gears, bolts, fasteners, machine tool parts; oil hardened, tough. *Obsolete*

SUWEFA DMO 11
Stahlwerke Sudwestfalen
C 0.15, Mn 0.65, Mo 0.3, bal Fe.
For gears, fasteners, machine tool parts; case hardening steel. *Obsolete*

SUWEFA DMO 14
Stahlwerke Sudwestfalen
C 0.13, Cr 0.85, Mo 0.45, bal Fe.
For gears, fasteners, machine tool parts; case hardening steel. *Obsolete*

SUWEFA DMO 15
Stahlwerke Sudwestfalen
C 0.24, Cr 1.25, Mo 0.45, bal Fe.
For gears, machine tool parts; oil hardened, tough. *Obsolete*

SUWEFA DMO 20
Stahlwerke Sudwestfalen
C 0.24, Cr 1.35, Mo 0.55, V 0.2, bal Fe.
For gears, shafts, machine tool parts; oil hardened, tough. *Obsolete*

SUWEFA E50
Stahlwerke Sudwestfalen
C 0.35, Si 0.25, Mn 0.55, bal Fe.
For gears, bolts, machine tool parts; water hardened. *Obsolete*

SUWEFA EC3
Stahlwerke Sudwestfalen
C 0.15, Si 0.25, Mn 0.5, Cr 0.65, bal Fe.
For gears, bolts, machine tool parts; case hardened. *Obsolete*

SUWEFA ECM
Stahlwerke Sudwestfalen
C 0.2, Si 0.25, Mn 1.25, Cr 1.15, bal Fe.
For gears, cams, camshafts, fasteners; case hardened, shock resistant. *Obsolete*

SUWEFA EH50
Stahlwerke Sudwestfalen
C 0.15, Si 0.25, Mn 0.37, bal Fe.
Annealed: 70,000 TS; 40,000 YS; 25 El; 60 RA; 145 Brin. For gears, fasteners, machine tool parts; case hardening steel. *Obsolete*

SUWEFA EHC40
Stahlwerke Sudwestfalen
C 0.13, Cr 0.5, bal Fe.
For gears, bolts, oil refinery equipment; case hardening steel. *Obsolete*

SUWEFA EHC70
Stahlwerke Sudwestfalen
C 0.15, Cr 0.65, Mn 0.5, bal Fe.
For gears, shafts, oil refinery equipment; case hardening steel. *Obsolete*

SUWEFA EHC90
Stahlwerke Sudwestfalen
C 0.16, Cr 0.95, Mn 1.15, bal Fe.
For gears, bolts, camshafts, cams; case hardening steel. *Obsolete*

SUWEFA ELCS
Stahlwerke Sudwestfalen
C 1.45, Mn 0.6, Cr 1.4, bal Fe.
For bearings, liners, sleeves, bushings; water hardened, wear resistant. *Obsolete*

SUWEFA EMC5
Stahlwerke Sudwestfalen
C 0.16, Cr 0.95, Mn 1.15, bal Fe.
For gears, cams, camshafts, fasteners; case hardening steel, tough. *Obsolete*

SUWEFA EMC5H
Stahlwerke Sudwestfalen
C 0.2, Mn 1.25, Cr 1.15, bal Fe.
For gears, cams, camshafts, fasteners; case hardening steel, tough. *Obsolete*

SUWEFA ENC14
Stahlwerke Sudwestfalen
C 0.13, Cr 0.7, Ni 3.5, bal Fe.
For gears, cams, camshafts; case hardening steel, tough. *Obsolete*

SUWEFA ENC18
Stahlwerke Sudwestfalen
C 13, Cr 1.1, Ni 4.5, bal Fe.
For gears, bolts, machine tool parts; case hardening steel, tough. *Obsolete*

SUWEFA EW3
Stahlwerke Sudwestfalen
C 0.15, Mn 0.5, Cr 0.65, bal Fe.
For gears, cams, camshafts, machine tool parts; case hardening steel. *Obsolete*

SUWEFA EW5H
Stahlwerke Sudwestfalen
C 0.2, Cr 1.15, Mn 1.25, bal Fe.
For gears, cams, camshafts; case hardening steel, tough. *Obsolete*

SUWEFA EX15
Stahlwerke Sudwestfalen
C 0.15, Cr 1.55, Ni 1.55, bal Fe.
For gears, cams, camshafts; case hardened, tough. *Obsolete*

SUWEFA EX20
Stahlwerke Sudwestfalen
C 0.18, Cr 2, Ni 2, bal Fe.
For gears, cams, camshafts; case hardened, tough. *Obsolete*

SUWEFA EX25
Stahlwerke Sudwestfalen
C 0.13, Cr 0.7, Ni 2.5, bal Fe.
For gears, bolts, machine tool parts; case hardened, shock resistant. *Obsolete*

SUWEFA EXTRA
Stahlwerke Sudwestfalen
C 1, V 0.1, Mn 0.25, bal Fe.
For cutters, taps, springs, drills, reamers; Type W2; water hardened. *Obsolete*

SUWEFA EXTRA MH
Stahlwerke Sudwestfalen
C 1.1, Si 0-0.25, Mn 0-0.25, bal Fe.
Annealed: 110,000 TS; 56,000 YS; 20 El; 40 RA; 210 Brin. For springs, tools, cutters, taps, drills; Type W1; water hardened. *Obsolete*

SUWEFA EXTRA WEICH
Stahlwerke Sudwestfalen
C 0.7, Si 0-0.2, Mn 0-0.25, bal Fe.
Heat treated: 174,000 TS; 128,000 YS; 12 El; 37 RA; 355 Brin. For springs, rails, punches, axes, hammers; Type W1; water hardened. *Obsolete*

SUWEFA EXTRA ZAH
Stahlwerke Sudwestfalen
C 0.7, Si 0-0.25, Mn 0-0.25, bal Fe.
Heat treated: 175,000 TS; 128,000, YS; 12 El; 37 RA; 355 Brin. For springs, rails, axes, punches; Type W1; water hardened. *Obsolete*

SUWEFA EXTRA ZH
Stahlwerke Sudwestfalen
C 1, Si 0-0.25, Mn 0.25, bal Fe.
Annealed: 100,000 TS; 53,000 YS; 21 El; 42 RA; 200 Brin. For cutters, taps, springs, drills, broaches; Type W1; water hardened. *Obsolete*

SUWEFA F11Z
Stahlwerke Sudwestfalen
C 0-0.1, Cr 17.5, Ti = 7 x C, bal Fe.
Annealed: 80,000 TS; 50,000 YS; 25 El; 50 RA; 150 Brin. For oil refinery equipment, sinks; Type 430 Ti; stainless. *Obsolete*

SUWEFA F13
Stahlwerke Sudwestfalen
C 0-0.12, Cr 13, bal Fe.
Annealed: 75,000 TS; 40,000 YS; 35 El; 70 RA; 155 Brin. For turbine blades, valves, cutlery; Type 410; stainless. *Obsolete*

SUWEFA F17
Stahlwerke Sudwestfalen
C 0.8, Cr 17, bal Fe.
Annealed: 107,000 TS; 62,000 YS; 18 El; 35 RA; 220 Brin. For cutlery, valves, ball bearings, surgical instruments; Type 440B; corrosion resistant. *Obsolete*

SUWEFA F171Z
Stahlwerke Sudwestfalen
C 0-0.1, Cr 17, Mo 1.8, Ti = 7 x C, bal Fe.
Annealed: 125,000 TS; 95,000 YS; 20 El; 55 RA; 260 Brin. For welded chemical plant equipment, pump parts; corrosion resistant; Type 431 Ti. *Obsolete*

SUWEFA F17A
Stahlwerke Sudwestfalen
C 0.12, Cr 16.5, Mo 0.25, S 0.2, bal Fe.
Annealed: 80,000 TS; 50,000 YS; 25 El; 50 RA; 150 Brin. For screw machine products, oil refinery equipment; Type 430Γ; corrosion resistant. *Obsolete*

SUWEFA FBV
Stahlwerke Sudwestfalen
C 1.15, Cr 0.65, V 0.1, bal Fe.
For bearings, bushings, liners; water or oil hardened, wear resistant. *Obsolete*

SUWEFA FPC4
Stahlwerke Sudwestfalen
C 0.4, Cr, Mn, V, bal Fe.
For gears, bolts, crankshafts, fasteners; oil hardened, shock resistant. *Obsolete*

SUWEFA GB1
Stahlwerke Sudwestfalen
C 0.15, Si 0.25, Mn 0.37, bal Fe.
Annealed: 70,000 TS; 40,000 YS; 25 El; 60 RA; 145 Brin. For gears, bolts, machine tool parts; case hardened. *Obsolete*

SUWEFA GB11
Stahlwerke Sudwestfalen
C 0.15, Si 0.25, Mn 0.37, bal Fe.
Annealed: 70,000 TS; 40,000 YS; 25 El; 60 RA; 145 Brin. For gears, bolts, machine tool parts; case hardened. *Obsolete*

SUWEFA GB12
Stahlwerke Sudwestfalen
C 0.22, Si 0.25, Mn 0.45, bal Fe.
Cold drawn: 78,000 TS; 68,000 YS; 20 El; 55 RA; 160 Brin. For machine tool parts; case hardened. *Obsolete*

SUWEFA GB13
Stahlwerke Sudwestfalen
C 0.35, Si 0-0.25, Mn 0-0.55, bal Fe.
Hot rolled: 85,000 TS; 54,000 YS; 30 El; 53 RA; 185 Brin. For gears, machine tool parts; water hardened. *Obsolete*

SUWEFA GB14
Stahlwerke Sudwestfalen
C 0.45, Si 0.25, Mn 0.65, bal Fe.
Hot rolled: 98,000 TS; 59,000 YS; 24 El; 45 RA; 212 Brin. For gears, bolts, machine tool parts; water hardened. *Obsolete*

SUWEFA GB16
Stahlwerke Sudwestfalen
C 0.6, Si 0.38, Mn 0.65, bal Fe.
Heat treated: 160,000 TS; 113,000 YS; 12 El; 40 RA; 321 Brin. For machine tool parts, axes, shafts; water hardened. *Obsolete*

SUWEFA GB18
Stahlwerke Sudwestfalen
C 0.75, Si 0.38, Mn 0.7, bal Fe.
Heat treated: 185,000 TS; 142,000 YS; 15 El; 40 RA; 390 Brin. For rails, springs, axes, hammmers, punches; Type W1; water hardened. *Obsolete*

SUWEFA GB2
Stahlwerke Sudwestfalen
C 0.22, Si 0.25, Mn 0.45, bal Fe.
Annealed: 73,000 TS; 42,000 YS; 21 El; 57 RA; 145 Brin. For machine tool parts, gears, fasteners; case hardened. *Obsolete*

SUWEFA GB3
Stahlwerke Sudwestfalen
C 0.35, Si 0.25, Mn 0.55, bal Fe.
Hot rolled: 85,000 TS; 54,000 YS; 30 El; 53 Ra; 185 Brin. For gears, bolts, machine tool parts; water hardened. *Obsolete*

SUWEFA GB4
Stahlwerke Sudwestfalen
C 0.4, Si 0.25, Mn 0.65, bal Fe.
Hot rolled: 98,000 TS; 60,000 YS; 24 El; 45 RA; 212 Brin. For gears, bolts, machine tool parts; water hardened. *Obsolete*

SUWEFA GB5
Stahlwerke Sudwestfalen
C 0.53, Si 0.25, Mn 0.65, bal Fe.
Normalized: 100,000 TS; 55,000 YS; 18 El; 28 RA; 200 Brin. For gears, bolts, machine tool parts; water hardened. *Obsolete*

SUWEFA GB5H
Stahlwerke Sudwestfalen
C 0.56, Si 0.3, Mn 0.55, bal Fe.
Normalized: 102,000 TS; 56,000 YS; 18 El; 25 RA; 205 Brin.
For gears, bolts, machine tool parts; water hardened.
Obsolete

SUWEFA GB6
Stahlwerke Sudwestfalen
C 0.61, Si 0.25, Mn 0.65, bal Fe.
Heat treated: 160,000 TS; 113,000 YS; 12 El; 40 RA; 320 Brin.
For rails, axes, hammers, springs; water hardened. *Obsolete*

SUWEFA GB9
Stahlwerke Sudwestfalen
C 0.9, Si 0.37, Mn 1, bal Fe.
For springs, cutters; Type W1; water hardened. *Obsolete*

SUWEFA GBS40
Stahlwerke Sudwestfalen
C 0.45, Ni, Cr, bal Fe.
For gears, bolts, machine tool parts; oil hardened, tough.
Obsolete

SUWEFA GBS45
Stahlwerke Sudwestfalen
C 0.45, Ni, Cr, W, bal Fe.
For gears, bolts, machine tool parts; oil hardened, tough.
Obsolete

SUWEFA GBS50
Stahlwerke Sudwestfalen
C 0.5, Cr 1.05, Ni 3.25, Mn 0.5, bal Fe.
For gears, bolts, machine tool parts; oil hardened, shock
resistant. *Obsolete*

SUWEFA GC2
Stahlwerke Sudwestfalen
C 1.45, Cr 1.4, Mn 0.6, bal Fe.
For bearings, bushings, liners; water hardened, wear
resistant. *Obsolete*

SUWEFA GCS12
Stahlwerke Sudwestfalen
C 1.25, Cr 1.2, Si 1.15, Mn 0.7, bal Fe.
For bearings, bushings, liners; water hardened, wear
resistant. *Obsolete*

SUWEFA GCS9
Stahlwerke Sudwestfalen
C 0.9, Cr 1.2, Si 1.15, Mn 0.7, bal Fe.
For bearings, bushings, liners; water hardened, wear
resistant. *Obsolete*

SUWEFA GCV
Stahlwerke Sudwestfalen
C 0.38, Si 1.2, Cr, V, bal Fe.
For bolts, gears, machine tool parts; oil hardened, tough.
Obsolete

SUWEFA GDW
Stahlwerke Sudwestfalen
C 0.4, Cr, Mn, Mo, bal Fe.
For gears, bolts, machine tool parts; oil hardened, shock
resistant. *Obsolete*

SUWEFA GF2
Stahlwerke Sudwestfalen
C 1.3, Si 0-0.25, P 0-0.25, bal Fe.
For gears, shafts, machine tool parts, bolts; water hardened.
Obsolete

SUWEFA GF43
Stahlwerke Sudwestfalen
C 0.45, Si 0.25-0.5, Mn 0.3-0.8, bal Fe.
For gears, shafts, fasteners, machine tool parts; water
hardened. *Obsolete*

SUWEFA GFD3
Stahlwerke Sudwestfalen
C 0.46, Mn 1.7, bal Fe.
For punches, crimpers, shear blades; oil hardened. *Obsolete*

SUWEFA GFD4
Stahlwerke Sudwestfalen
C 0.46, Si 1.7, Mn 0.65, bal Fe.
For upsetters, riveters, punches; shock resistant, oil
hardened. *Obsolete*

SUWEFA GFD5
Stahlwerke Sudwestfalen
C 0.65, Si 1.75, Mn 0.7, bal Fe.
For springs; tough, shock resistant. *Obsolete*

SUWEFA GFD5W
Stahlwerke Sudwestfalen
C 0.55, Si 1.7, Mn 0.7, bal Fe.
For upsetters, springs, punches; tough, shock resistant.
Obsolete

SUWEFA GFD7
Stahlwerke Sudwestfalen
C 0.65, Si 1.15, Mn 1, bal Fe.
For springs; shock resistant. *Obsolete*

SUWEFA GFD7R
Stahlwerke Sudwestfalen
C 0.7, Mn 0.7, Si 1.7, bal Fe.
For springs; shock resistant. *Obsolete*

SUWEFA GFD8
Stahlwerke Sudwestfalen
C 0.67, Si 1.3, Mn 0.5, Cr 0.5, bal Fe.
For springs; shock resistant. *Obsolete*

SUWEFA GHE SPEZIAL
Stahlwerke Sudwestfalen
C 0.56, Ni, Cr, Mo, V, bal Fe.
For gears, bolts, machine tool parts; oil hardened, shock
resistant. *Obsolete*

SUWEFA GKLO
Stahlwerke Sudwestfalen
C 1, Cr 1.55, Mn 0.35, Si 0.3, bal Fe.
For bearings, liners, sleeves; water hardened. *Obsolete*

SUWEFA GR
Stahlwerke Sudwestfalen
C 1.05, Cr 1, W 1.15, Mn 0.9, bal Fe.
For cutters, fast finishing tools; water hardened. *Obsolete*

SUWEFA GR3
Stahlwerke Sudwestfalen
C 1.42, W, V, bal Fe.
For fast finishing cutters, form tools; water hardened.
Obsolete

SUWEFA GS
Stahlwerke Sudwestfalen
C 0.55, Cr 0.7, Mo 0.18, Ni 1.65, V 0.1, bal Fe.
For gears, bolts, crankshafts; oil hardened, shock resistant.
Obsolete

SUWEFA GS7
Stahlwerke Sudwestfalen
C 0.8, Si 0.1-0.4, Mn 0.5-0.7, bal Fe.
Heat treated: 188,000 TS; 143,000 YS; 12 El; 35 RA; 390 Brin.
For springs, taps, drills, hobs, reamers; Type W1; water
hardened. *Obsolete*

SUWEFA GS8
Stahlwerke Sudwestfalen
C 0.8, Si 0.1-0.4, Mn 0.5-0.7, bal Fe.
Heat treated: 188,000 TS; 143,000 YS; 12 El; 35 RA; 390 Brin.
For springs, cutters, drills, taps; Type W1; water hardened.
Obsolete

SUWEFA GSI
Stahlwerke Sudwestfalen
C 1.15, Si 0.2, Mn 0.3, Cr 0.65, V 0.1, bal Fe.
For bearings, bushings, liners; oil hardened, wear resistant.
Obsolete

SUWEFA GSP5
Stahlwerke Sudwestfalen
C 0.53, Si 0.9, Mn 0.9, bal Fe.
For gears, bolts, studs, fasteners; water hardened. *Obsolete*

SUWEFA GSP6
Stahlwerke Sudwestfalen
C 0.7, Si 1.7, Mn 0.7, bal Fe.
For springs; oil hardened, tough. *Obsolete*

SUWEFA GSS
Stahlwerke Sudwestfalen
C 0.9, Mn 1.9, V 0.1, bal Fe.
For punches, dies, upsetters, cutters; oil hardened, non-
deforming. *Obsolete*

SUWEFA GSS111
Stahlwerke Sudwestfalen
C 1.05, Cr 1, W 1.15, Mn 0.9, bal Fe.
For fast finishing tools, bearings, cutters; water hardened.
Obsolete

SUWEFA GSS2
Stahlwerke Sudwestfalen
C 0.9, Mn 1.9, V 0.1, bal Fe.
For punches, dies, upsetters, cutters; oil hardened, non-
deforming. *Obsolete*

SUWEFA GVS6
Stahlwerke Sudwestfalen
C 0.55, Si 0.1-0.4, Mn 0.5-0.7, bal Fe.
Normalized: 100,000 TS; 55,000 YS; 18 El; 26 RA; 200 Brin.
For gears, bolts, shafts, studs; water hardened. *Obsolete*

SUWEFA GW10
Stahlwerke Sudwestfalen
C 1, Si 0-0.25, Mn 0-0.25, bal Fe.
Annealed: 100,000 TS; 53,000 YS; 21 El; 42 RA; 200 Brin. For
springs, cutters, drills, taps, reamers; Type W1; water
hardened. *Obsolete*

SUWEFA GW10E
Stahlwerke Sudwestfalen
C 1, Si 0-0.25, Mn 0-0.25, bal Fe.
Annealed: 100,000 TS; 53,000 YS; 21 El; 42 RA; 200 Brin. For
springs, cutters, drills, taps,reamers; Type W1; water
hardened. *Obsolete*

SUWEFA GW10V
Stahlwerke Sudwestfalen
C 1, Si 0.2, Mn 0.25, V 0.1, bal Fe.
For springs, cutters, drills, taps, reamers; Type W2; water
hardened. *Obsolete*

SUWEFA GW11
Sudwestfalen Stahlwerke AG
C 1.15, Si 0-0.25, Mn 0-0.25, bal Fe.
Annealed: 110,000 TS; 56,000 YS; 18 El; 40 RA; 210 Brin. For
springs, taps, cutters, broaches; Type W1; water hardened.

SUWEFA GW11E
Stahlwerke Sudwestfalen
C 1.1, Si 0-0.25, Mn 0-0.25, bal Fe.
Annealed: 110,000 TS; 56,000 YS; 20 El; 40 RA; 205 Brin. For
springs, taps, reamers, broaches; Type W1; water hardened.
Obsolete

SUWEFA GW13
Stahlwerke Sudwestfalen
C 1.3, Si 0-0.25, Mn 0-0.25, bal Fe.
For engravers' tools, reamers, broaches, taps; Type W1; water
hardened. *Obsolete*

SUWEFA GW23
Stahlwerke Sudwestfalen
C 0.15, Si 0.25-0.5, Mn 0.3-0.7, bal Fe.
Annealed: 70,000 TS; 40,000 YS; 25 El; 60 RA; 145 Brin. For gears, bolts, machine tool parts; case hardened. *Obsolete*

SUWEFA GW33
Stahlwerke Sudwestfalen
C 0.35, Si 0.25-0.5, Mn 0.3-0.7, bal Fe.
Hot rolled: 85,000 TS; 54,000 YS; 30 El; 53 RA; 185 Brin. For gears, bolts, machine tool parts; water hardened. *Obsolete*

SUWEFA GW43
Stahlwerke Sudwestfalen
C 0.45, Si 0.25-0.5, Mn 0.3-0.7, bal Fe.
Hot rolled: 98,000 TS; 60,000 YS; 24 El; 45 RA; 215 Brin. For gears, bolts, machine tool parts; water hardened. *Obsolete*

SUWEFA GW63
Stahlwerke Sudwestfalen
C 0.6, Si 0.25-0.5, Mn 0.3-0.7, bal Fe.
Heat treated: 160,000 TS; 113,000 YS; 12 El; 40 RA; 32 Brin. For gears, rails, springs, hammers; water hardened. *Obsolete*

SUWEFA GW7
Stahlwerke Sudwestfalen
C 0.7, Si 0-0.25, Mn 0-0.25, bal Fe.
Heat treated: 175,000 TS; 128,000 YS; 12 El; 37 RA; 355 Brin. For rails, springs, axes, hammers, punches; water hardened. *Obsolete*

SUWEFA GW73
Stahlwerke Sudwestfalen
C 0.67, Si 0.25-0.5, Mn 0.3-0.7, bal Fe.
Heat treated: 170,000 TS; 125,000 YS; 14 El; 38 RA; 340 Brin. For rails, springs, axes, crimpers; water hardened. *Obsolete*

SUWEFA GW7E
Stahlwerke Sudwestfalen
C 0.7, Si 0-0.25, Mn 0-0.25, bal Fe.
Heat treated: 175,000 TS; 128,000 YS; 12 El; 37 RA; 355 Brin. For rails, springs, axes, hammers, punches; water hardened. *Obsolete*

SUWEFA GW8
Stahlwerke Sudwestfalen
C 0.85, Si 0-0.25, Mn 0-0.25, bal Fe.
Heat treated: 190,000 TS; 145,000 YS; 10 El; 30 RA; 400 Brin. For springs, taps, cutters, drills; Type W1; water hardened. *Obsolete*

SUWEFA GW8E
Stahlwerke Sudwestfalen
C 0.85, Si 0-0.25, Mn 0-0.25, bal Fe.
Heat treated: 190,000 TS; 145,000 YS; 10 El; 30 RA; 400 Brin. For springs, taps, cutters, drills; Type W1; water hardened. *Obsolete*

SUWEFA GW93
Stahlwerke Sudwestfalen
C 0.9, Si 0.25-0.5, Mn 0.3-0.7, bal Fe.
Heat treated: 190,000 TS; 145,000 YS; 10 El; 30 RA; 400 Brin. For springs, tools, cutters, drills; Type W1; water hardened. *Obsolete*

SUWEFA GZB10A
Stahlwerke Sudwestfalen
C 0.15, Si 2, Cr 19.5, Ni 9.5, bal Fe.
Annealed: 80,000 TS; 35,000 YS; 55 El; 75 RA; 150 Brin. For chemical plant equipment, tanks, mixers; Type 302; stainless, austenitic. *Obsolete*

SUWEFA GZB10F
Stahlwerke Sudwestfalen
C 0-0.12, Al 0.95, bal Fe.
For oil refinery equipment; heat and creep resistant. *Obsolete*

SUWEFA GZB11FA
Stahlwerke Sudwestfalen
C 0.2, Cr 25, Si 1.2, Ni 4, bal Fe.
Cast: 90,000 TS; 65,000 YS; 18 El; 210 Brin. For cylinder liners, valve seats and bodies; corrosion and heat resistant. *Obsolete*

SUWEFA GZB12A
Stahlwerke Sudwestfalen
C 0.15, Cr 24, Ni 19, bal Fe.
Annealed: 100,000 TS; 45,000 YS; 50 El; 65 RA; 185 Brin. For furnace parts, valves, pumps, turbines; Type 310; corrosion and heat resistant. *Obsolete*

SUWEFA GZB12F
Stahlwerke Sudwestfalen
C 0-0.12, Al 1.5, Cr 24, bal Fe.
For oil refinery equipment; heat and creep resistant. *Obsolete*

SUWEFA GZB8F
Stahlwerke Sudwestfalen
C 0-0.12, Si 0.7, Mn 0.3, Cr 6.5, Al 0.75, bal Fe.
For oil refinery equipment; creep and heat resistant. *Obsolete*

SUWEFA GZW
Stahlwerke Sudwestfalen
C 1.4, V 0.1, Cr 0.3, Mn 0.3, bal Fe.
For bearings, liners, cutters; water hardened, wear resistant. *Obsolete*

SUWEFA H10
Stahlwerke Sudwestfalen
C 0.21, Cr 3, Mo 0.4, V 0.8, W 0.37, bal Fe.
For upsetters, crimpers; oil hardened. *Obsolete*

SUWEFA HE15
Stahlwerke Sudwestfalen
C 0.15, Si 0.25, Mn 0.37, bal Fe.
Annealed: 70,000 TS; 40,000 YS; 25 El; 60 RA; 145 Brin. For gears, bolts, machine tool parts; case hardened. *Obsolete*

SUWEFA HEC10
Stahlwerke Sudwestfalen
C 0.15, Si 0.25, Mn 0.5, Cr 0.65, bal Fe.
For gears, machine tool parts; case hardened. *Obsolete*

SUWEFA HEC5
Stahlwerke Sudwestfalen
C 0.13, Cr, bal Fe.
For machine tool parts. *Obsolete*

SUWEFA HEN15
Stahlwerke Sudwestfalen
C 0.13, Cr 0.2, Ni 1.5, bal Fe.
For gears, bolts, camshafts; case hardened, shock resistant. *Obsolete*

SUWEFA HH
Stahlwerke Sudwestfalen
C 1.2, Mn, bal Fe.
For cutters, wear parts; oil hardened. *Obsolete*

SUWEFA HKC10
Stahlwerke Sudwestfalen
C 0.33, Cr 1, Mo 0.2, Mn 0.65, bal Fe.
For gears, bolts, crankshafts, fasteners; oil hardened, tough. *Obsolete*

SUWEFA HSB2
Stahlwerke Sudwestfalen
C 0.75, Si 0-0.25, Mn 0-0.25, bal Fe.
Heat treated: 180,000-130,000 TS; 135,000-90,000 YS; 10-20 El; 35-50 RA; 370-245 Brin. For springs, tools, cutters, rails, die blocks; Type W1; water hardened. *Obsolete*

SUWEFA HSB3
Stahlwerke Sudwestfalen
C 0.85, Cr 0.5, bal Fe.
For bearings, bushings, cutters; water hardened. *Obsolete*

SUWEFA HZCV
Stahlwerke Sudwestfalen
C 0.5, Mn 0.95, Cr 1.05, V 0.1, bal Fe.
For springs, gears, bolts, shafts; oil hardened, shock resistant. *Obsolete*

SUWEFA KL3
Stahlwerke Sudwestfalen
C 1, Cr 1.55, Mn 0.35, bal Fe.
For bearings, bushings, liners; water hardened, wear resistant. *Obsolete*

SUWEFA KMC
Stahlwerke Sudwestfalen
C 0.2, Cr 1.15, Mn 1.25, bal Fe.
For bearings, liners, races, bushings; case hardened, tough. *Obsolete*

SUWEFA KZM
Stahlwerke Sudwestfalen
C 1.4, Mn 0.3, V 0.1, bal Fe.
For engravers tools, punches, header dies; water hardened; Type W2. *Obsolete*

SUWEFA LCN
Stahlwerke Sudwestfalen
C 0.5, Cr 1.05, Ni 3.25, bal Fe.
For gears, shafts, countershafts, bolts, studs; oil hardened, shock resistant. *Obsolete*

SUWEFA LCV
Stahlwerke Sudwestfalen
C 0.5, Cr 1.05, V 0.1, Mn 0.95, bal Fe.
For gears, springs, bolts, studs, shafts; oil hardened, shock resistant. *Obsolete*

SUWEFA LCW1
Stahlwerke Sudwestfalen
C 0.35, Si 0.9, Cr 1.05, V 0.18, W 1.85, bal Fe.
For header dies, upsetters, crimpers; oil hardened, tough. *Obsolete*

SUWEFA LCW2
Stahlwerke Sudwestfalen
C 0.35, Cr 1.05, V 0.18, W 1.85, bal Fe.
For header dies, punches, crimpers, upsetters; oil hardened, tough. *Obsolete*

SUWEFA LCW3
Stahlwerke Sudwestfalen
C 0.55, Si 0.9, Cr 1.05, V 0.18, W 1.85, bal Fe.
For header dies, punches, shears, crimpers; oil hardened, tough. *Obsolete*

SUWEFA LST
Stahlwerke Sudwestfalen
C 0.45, Cr, V, Si, bal Fe.
For gears, springs, bolts, studs, shafts; oil hardened, shock resistant. *Obsolete*

SUWEFA M-M1
Stahlwerke Sudwestfalen
C 0.9, Cr 18, Mo 1.15, V 1, bal Fe.
Heat treated: 280,000 TS; 270,000 YS; 3 El; 15 RA; 555 Brin. For cutlery, valves, ball bearings; hardenable, corrosion resistant. *Obsolete*

SUWEFA M13
Stahlwerke Sudwestfalen
C 0.4, Si 0.4, Cr 13, bal Fe.
Annealed: 95,000 TS; 50,000 YS; 25 El; 55 RA; 195 Brin. For valves, cutlery, surgical and dental instruments; Type 420; stainless. *Obsolete*

SUWEFA MO-333
Stahlwerke Sudwestfalen
C 0.95, Cr, W, Mo, bal Fe.
For lathe and planer tools, reamers, broaches, taps; high speed steel. *Obsolete*

SUWEFA MO-560
Stahlwerke Sudwestfalen
C 0.85, Cr, W, Mo, bal Fe.
For lathe and planer tools, drills, reamers, hobs; high speed steel. *Obsolete*

SUWEFA MO-900
Stahlwerke Sudwestfalen
C 0.8, Mo, Cr, W, V, bal Fe.
For lathe and planer tools, reamers, broaches, taps; high speed steel. *Obsolete*

SUWEFA MO1
Stahlwerke Sudwestfalen
C 0.95, W, Mo, bal Fe.
For dies, tools, cutters; oil hardened. *Obsolete*

SUWEFA MOG
Stahlwerke Sudwestfalen
C 0.45, Mn 0.7, Cr 1.4, Mo 0.7, V 0.3, bal Fe.
For forging dies, punches, shears; oil hardened, tough. *Obsolete*

SUWEFA MV EXTRA
Stahlwerke Sudwestfalen
C 0.42, Mn 1.75, Si 0.25, V 0.1, bal Fe.
For punches, upsetters, shears; oil hardened, tough. *Obsolete*

SUWEFA MVC20
Stahlwerke Sudwestfalen
C 0.3, Cr, V, bal Fe.
For gears, bolts, crankshafts; oil hardened, tough. *Obsolete*

SUWEFA NC6
Stahlwerke Sudwestfalen
C 0.27, Mn 0.6, Al 1.1, Cr 1.4, bal Fe.
For oil refinery equipment; heat and creep resistant. *Obsolete*

SUWEFA NC6H
Stahlwerke Sudwestfalen
C 0.34, Al 1.1, Cr 1.4, Mn 0.6, bal Fe.
For oil refinery equipment; heat and creep resistant. *Obsolete*

SUWEFA NCMO4
Stahlwerke Sudwestfalen
C 0.32, Cr 1.1, Al 1.1, Mo 0.18, bal Fe.
For oil refinery equipment; heat and creep resistant. *Obsolete*

SUWEFA NCV9 SPEZIAL
Stahlwerke Sudwestfalen
C 0.31, Mn 0.6, Cr 2.35, Mo 0.18, V 0.13, bal Fe.
For oil refinery equipment; heat and creep resistant. *Obsolete*

SUWEFA NFKC
Stahlwerke Sudwestfalen
C 1.42, W, V, bal Fe.
For forming and blanking dies, cutters; oil hardened, wear resistant. *Obsolete*

SUWEFA NG
Stahlwerke Sudwestfalen
C 0.55, Cr 0.7, Mo 0.18, Ni 1.65, V 0.1, bal Fe.
For forging and heading dies, upsetters; oil hardened, tough. *Obsolete*

SUWEFA NG2 SUPRA
Stahlwerke Sudwestfalen
C 0.56, Mo 0.2, Ni 1.8, V 0.1, Cr 0.8, bal Fe.
For forging and heading dies, upsetters; oil hardened, tough. *Obsolete*

SUWEFA NI 2
Stahlwerke Sudwestfalen
C 0.13, Cr 0.2, Ni 1.5, bal Fe.
For gears, bolts, camshafts; case hardened. *Obsolete*

SUWEFA NIC 3
Stahlwerke Sudwestfalen
C 0.13, Ni 2.5, Cr 0.7, bal Fe.
For gears, bolts, crankshafts, cams; case hardened, shock resistant. *Obsolete*

SUWEFA NIC 4
Stahlwerke Sudwestfalen
C 0.13, Cr 0.7, Ni 3.5, bal Fe.
For gears, bolts, camshafts, cams; case hardened, shock resistant. *Obsolete*

SUWEFA NIC 4H
Stahlwerke Sudwestfalen
C 0.22-0.3, Cr 0.7, Ni 3.5, bal Fe.
For gears, bolts, crankshafts, fasteners; oil hardened, shock resistant. *Obsolete*

SUWEFA NIC 5
Stahlwerke Sudwestfalen
C 0.13, Cr 1.1, Ni 4.5, bal Fe.
For gears, bolts, camshafts, cams; case hardened, shock resistant. *Obsolete*

SUWEFA NIC 5H
Stahlwerke Sudwestfalen
C 0.35, Cr 1.3, Ni 4.5, bal Fe.
For gears, bolts, machine tool parts; oil hardened, shock resistant. *Obsolete*

SUWEFA NIC2
Stahlwerke Sudwestfalen
C 0.25-0.35, Cr 0.7, Ni 2.5, bal Fe.
For gears, bolts, fasteners, crankshafts; oil hardened, shock resistant. *Obsolete*

SUWEFA NSI
Stahlwerke Sudwestfalen
C 0.7, Si 1.7, Mn 0.7, bal Fe.
For springs, pneumatic tools, chisels; oil hardened, tough. *Obsolete*

SUWEFA PIL80
Stahlwerke Sudwestfalen
C 0.28, Ni, Cr, Mo, V, bal Fe.
For forging and heading dies; oil hardened, tough. *Obsolete*

SUWEFA PRIMA WEICH
Stahlwerke Sudwestfalen
C 0.7, Si 0-0.25, Mn 0-0.25, bal Fe.
Heat treated: 175,000-122,000 TS; 128,000-82,000 YS; 12-22 El; 37-52 RA; 350-240 Brin. For springs, rails, clutch discs, girders, rails; Type W1; water hardened. *Obsolete*

SUWEFA PRIMA ZAH
Stahlwerke Sudwestfalen
C 0.85, Si 0-0.25, Mn 0-0.25, bal Fe.
Heat treated: 190,000 TS; 145,000 YS; 10 El; 32 RA; 400 Brin. For drills, taps, reamers, springs, hobs; Type W1; water hardened. *Obsolete*

SUWEFA PRIMA ZH
Stahlwerke Sudwestfalen
C 1.05, Si 0-0.25, Mn 0-0.25, bal Fe.
Heat treated: 200,000 TS; 125,000 YS; 8 El; 28 RA; 400 Brin. For springs, drills, reamers, taps, broaches; Type W1; water hardened. *Obsolete*

SUWEFA REA
Stahlwerke Sudwestfalen
C 0-0.15, Cr 18, Ni 8.5, bal Fe.
Annealed: 80,000 TS; 35,000 YS; 55 El; 75 RA; 150 Brin. For chemical plant equipment, tanks; Type 302; stainless, austenitic. *Obsolete*

SUWEFA REA2
Stahlwerke Sudwestfalen
C 0-0.15, Cr 18, Ni 8.5, bal Fe.
Annealed: 80,000 TS; 35,000 YS; 55 El; 75 RA; 150 Brin. For chemical plant equipment, tanks, mixers; Type 302; stainless, austenitic. *Obsolete*

SUWEFA REA2-K
Stahlwerke Sudwestfalen
C 0-0.12, Cr 18, Ni 9.5, Ti = 4 x C, bal Fe.
Annealed: 85,000 TS; 35,000 YS; 55 El; 65 RA; 150 Brin. For welded chemical plant equipment, mixers, tanks; Type 321; stainless, austenitic. *Obsolete*

SUWEFA REA2-KN
Stahlwerke Sudwestfalen
C 0-0.12, Cr 18, Ni 9.5, Cb = 8 x C, bal Fe.
Annealed: 90,000 TS; 45,000 YS; 56 El; 65 RA; 160 Brin. For welded chemical plant equipment, tanks, vessels; Type 347; stainless, austenitic. *Obsolete*

SUWEFA REA2-KS
Stahlwerke Sudwestfalen
C 0.07, Cr 18, Ni 9.5, bal Fe.
Annealed: 85,000 TS; 35,000 YS; 60 El; 70 RA; 150 Brin. For chemical plant equipment, tanks, vessels; Type 304; stainless, austenitic. *Obsolete*

SUWEFA REA4-K
Stahlwerke Sudwestfalen
C 0-0.12, Cr 18, Mo 2, Ni 10.5, Ti = 4 x C, bal Fe.
Annealed: 90,000 TS; 40,000 YS; 45 El; 60 RA; 170 Brin. For welded acid resistant equipment, tanks, mixers; Type 316 Ti; stainless, austenitic. *Obsolete*

SUWEFA REA4-KN
Stahlwerke Sudwestfalen
C 0-0.12, Cr 18, Ni 10.5, Mo 2, Cb = 8 x C, bal Fe.
Annealed: 90,000 TS; 40,000 YS; 45 El; 60 RA; 70 Brin. For welded acid resistant chemical plant equipment; Type 316 Cb; stainless, austenitic. *Obsolete*

SUWEFA REA4-KS
Stahlwerke Sudwestfalen
C 0-0.07, Cr 18, Ni 10.5, Mo 2, bal Fe.
Annealed: 85,000 TS; 35,000 YS; 50 El; 65 RA; 160 Brin. Cold drawn: 150,000 TS; 135,000 YS; 6 El; 300 Brin. For acid resistant chemical plant equipment; Type 316; stainless, austenitic. *Obsolete*

SUWEFA REH
Stahlwerke Sudwestfalen
C 0.4, Si 0.4, Cr 13, bal Fe.
Annealed: 100,000 TS; 55,000 YS; 20 El; 50 RA; 205 Brin. For valves, cutlery, surgical and dental instruments; Type 420; stainless. *Obsolete*

SUWEFA REH EXTRA
Stahlwerke Sudwestfalen
C 0.9, Cr, V, bal Fe.
For bearings, blanking and forming dies; oil hardened, wear resistant. *Obsolete*

SUWEFA REHD
Stahlwerke Sudwestfalen
C 0.2, Cr 13, Mo 1.15, bal Fe.
Annealed: 100,000 TS; 55,000 YS; 22 El; 52 RA; 200 Brin. For valves, cutlery, chemical plant equipment; Type 420 Mo; stainless. *Obsolete*

SUWEFA REM
Stahlwerke Sudwestfalen
C 0.2, Cr 13, Si 0.4, Mn 0.3, bal Fe.
Annealed: 95,000 TS; 50,000 YS; 25 El; 55 RA; 196 Brin. For surgical and dental instruments, valves, cutlery; Type 420; stainless. *Obsolete*

SUWEFA RES
Stahlwerke Sudwestfalen
C 0.65, Si 1.3, Mn 0.5, Cr 0.5, bal Fe.
For springs; oil hardened, tough. *Obsolete*

SUWEFA RESO
Stahlwerke Sudwestfalen
C 0.65, Si 1.15, Mn 1, bal Fe.
For springs; oil hardened, tough. *Obsolete*

SUWEFA REW

Stahlwerke Sudwestfalen
C 0-0.12, Si 0.4, Cr 13, bal Fe.
Annealed: 75,000 TS; 40,000 YS; 35 El; 70 RA; 155 Brin. For turbine blades, valves, cutlery, surgical instruments; Type 410; stainless. *Obsolete*

SUWEFA REXH

Stahlwerke Sudwestfalen
C 0.22, Cr 17, Ni 1.5, bal Fe.
Annealed: 125,000 TS; 95,000 YS; 20 El; 55 RA; 250 Brin. For pumps, marine hardware, valves; Type 431; heat resistant. *Obsolete*

SUWEFA REXH EXTRA

Stahlwerke Sudwestfalen
C 0.22, Si 0.4, Cr 17, Ni 1.5, bal Fe.
Annealed: 125,000 TS; 95,000 YS; 20 El; 55 RA; 260 Brin. Cold drawn: 130,000 TS; 110,000 YS; 15 El; 35 RA; 270 Brin. For pumps, marine hardware, valves; Type 431; heat resistant. *Obsolete*

SUWEFA REXN

Stahlwerke Sudwestfalen
C 0-0.1, Cr 17.5, Ti = 7 x C, bal Fe.
Annealed: 80,000 TS; 50,000 YS; 25 El; 50 RA; 150 Brin. For oil refinery equipment, oil burner heaters; Type 430 Ti; stainless. *Obsolete*

SUWEFA REXW

Stahlwerke Sudwestfalen
C 0.8, Cr 18, bal Fe.
Annealed: 110,000 TS; 65,000 YS; 18 El; 35 RA; 220 Brin. Heat treated: 280,000 TS; 270,000 YS; 3 El; 15 RA; 555 Brin. For bearings, cutlery, valves; stainless, hardenable. *Obsolete*

SUWEFA SBH

Stahlwerke Sudwestfalen
C 1.25, Si 1.15, Mn 0.7, Cr 1.2, bal Fe.
For blanking and forming dies, bearings; oil hardened, wear resistant. *Obsolete*

SUWEFA SBN

Stahlwerke Sudwestfalen
C 1.1, Cr 0.4, Mn 0.3, bal Fe.
For bearings, cutters, liners; water or oil hardened. *Obsolete*

SUWEFA SCHE

Stahlwerke Sudwestfalen
C 0.61, Si 0.85, Mn 0.75, Cr 1.18, V 0.1, bal Fe.
For springs, gears, bolts, crankshafts; oil hardened, shock resistant. *Obsolete*

SUWEFA SER

Stahlwerke Sudwestfalen
C 0.67, Si 1.3, Mn 0.5, Cr 0.5, bal Fe.
For springs, punches, pneumatic tools; oil hardened, tough. *Obsolete*

SUWEFA SIW

Stahlwerke Sudwestfalen
C 1.2, Cr 0.2, V 0.1, W 1, bal Fe.
For bearings, sleeves, liners; water hardened, wear resistant. *Obsolete*

SUWEFA SJCV

Stahlwerke Sudwestfalen
C 1.15, Si 0.2, Mn 0.3, Cr 0.65, V 0.1, bal Fe.
For bearings, sleeves, liners; water hardened. *Obsolete*

SUWEFA SJI

Stahlwerke Sudwestfalen
C 0.40, Si 1.7, Mn 0.65, bal Fe.
For springs, punches, chisels; oil hardened, tough. *Obsolete*

SUWEFA SJII

Stahlwerke Sudwestfalen
C 0.55, Si 1.7, Mn 0.7, bal Fe.
For springs, punches, chisels, pneumatic tools; oil hardened, tough. *Obsolete*

SUWEFA SJIII

Stahlwerke Sudwestfalen
C 0.65, Si 1.7, Mn 0.7, bal Fe.
For springs, punches, chisels, pneumatic tools; oil hardened, tough. *Obsolete*

SUWEFA T13

Stahlwerke Sudwestfalen
C 0.2, Si 0.4, Cr 13, bal Fe.
Annealed: 95,000 TS; 50,000 YS; 25 El; 50 RA; 195 Brin. For valves, cutlery, surgical and dental instruments; Type 420; stainless. *Obsolete*

SUWEFA T131

Stahlwerke Sudwestfalen
C 0.2, Si 0.4, Cr 13, Mo 1.15, bal Fe.
Annealed: 100,000 TS; 55,000 YS; 20 El; 45 RA; 200 Brin. For oil refinery equipment, valves, cutlery; Type 430 Mo; stainless. *Obsolete*

SUWEFA T17

Stahlwerke Sudwestfalen
C 0.22, Si 0.4, Cr 17, Ni 1.5, bal Fe.
Annealed: 125,000 TS; 95,000 YS; 20 El; 50 RA; 260 Brin. For pumps, marine hardware, valves; Type 431; stainless. *Obsolete*

SUWEFA T171

Stahlwerke Sudwestfalen
C 0.35, Cr 1.65, Mo 1.15, bal Fe.
For oil refinery equipment, heat treating boxes; corrosion resistant. *Obsolete*

SUWEFA TC4

Stahlwerke Sudwestfalen
C 0.33, Si 0.25, Mn 0.65, Cr 1, bal Fe.
For gears, bolts, crankshafts, axles; oil hardened, tough. *Obsolete*

SUWEFA TC4 SPEZIAL

Stahlwerke Sudwestfalen
C 0.41, Cr 1, Mn 0.65, bal Fe.
For gears, bolts, crankshafts, axles; oil hardened, tough. *Obsolete*

SUWEFA TCMO4

Stahlwerke Sudwestfalen
C 0.34, Cr 1.05, V 0.2, Mn 0.65, bal Fe.
For gears, bolts, springs, crankshafts; oil hardened, shock resistant. *Obsolete*

SUWEFA TCMO4H

Stahlwerke Sudwestfalen
C 0.42, Cr 1.05, V 0.2, Mn 0.65, bal Fe.
For gears, bolts, springs, crankshafts; oil hardened, shock resistant. *Obsolete*

SUWEFA TCMO4W

Stahlwerke Sudwestfalen
C 0.25, Cr 1, Mo 0.2, Mn 0.65, bal Fe.
For gears, bolts, crankshafts; oil hardened, tough. *Obsolete*

SUWEFA TCMO5

Stahlwerke Sudwestfalen
C 0.5, Cr 1, Mo 0.2, Mn 0.65, bal Fe.
For gears, bolts, crankshafts; oil hardened, tough. *Obsolete*

SUWEFA TCV4

Stahlwerke Sudwestfalen
C 0.5, Cr 1, V 0.09, bal Fe.
For springs, bolts, crankshafts; oil hardened, shock resistant. *Obsolete*

SUWEFA TCV5

Stahlwerke Sudwestfalen
C 0.58, Cr 1, V 0.09, bal Fe.
For springs, shafts, gears, crankshafts; oil hardened, shock resistant. *Obsolete*

SUWEFA TCV9 SPEZIAL

Stahlwerke Sudwestfalen
C 0.3, Cr 2.5, Mo 0.2, V 0.15, bal Fe.
For gears, shafts, crankshafts, pinions, bolts; oil hardened, shock resistant. *Obsolete*

SUWEFA TFH

Stahlwerke Sudwestfalen
C 0.58, Cr, V, bal Fe.
For springs, bolts, crankshafts; oil hardened, shock resistant. *Obsolete*

SUWEFA TM4

Stahlwerke Sudwestfalen
C 0.4, Si 0.37, Mn 0.95, bal Fe.
Hot rolled: 92,000 TS; 58,000 YS; 27 El; 50 RA; 200 Brin. For gears, bolts, shafts, axles, screws; water hardened. *Obsolete*

SUWEFA TM5

Stahlwerke Sudwestfalen
C 0.3, Si 0.25, Mn 1.35, bal Fe.
For gears, bolts, shafts, axles, crankshafts; water hardened. *Obsolete*

SUWEFA TMCV4

Stahlwerke Sudwestfalen
C 0.27, Mn, Cr, V, bal Fe.
For gears, bolts, machine tool parts; oil hardened, tough. *Obsolete*

SUWEFA TMS4

Stahlwerke Sudwestfalen
C 0.53, Si 0.8, Mn 1.05, bal Fe.
For gears, bolts, machine tool parts; water hardened. *Obsolete*

SUWEFA TMS5

Stahlwerke Sudwestfalen
C 0.37, Si 1.25, Mn 1.25, bal Fe.
Hot rolled: 97,000 TS; 59,000 YS; 25 El; 52 RA; 200 Brin. For punches, chisels, pneumatic tools; oil hardened, shock resistant. *Obsolete*

SUWEFA TMV7

Stahlwerke Sudwestfalen
C 0.42, Mn 1.75, V 0.1, bal Fe.
For punches, chisels, pneumatic tools, gears; oil hardened, tough. *Obsolete*

SUWEFA TNC14H

Stahlwerke Sudwestfalen
C 0.22-0.3, Cr 0.7, Ni 3.5, bal Fe.
For gears, bolts, crankshafts, fasteners; oil hardened, shock resistant. *Obsolete*

SUWEFA TNC18

Stahlwerke Sudwestfalen
C 0.35, Cr 1.3, Ni 4.5, bal Fe.
For gears, bolts, crankshafts, axles; oil hardened, shock resistant. *Obsolete*

SUWEFA TNC6

Stahlwerke Sudwestfalen
C 0.28-0.35, Cr 0.5, Ni 1.5, bal Fe.
For gears, bolts, crankshafts, fasteners; oil hardened, shock resistant. *Obsolete*

SUWEFA TX10

Stahlwerke Sudwestfalen
C 0.36, Ni 1, Cr 1, Mo 0.2, bal Fe.
For die casting and plastic mold dies; oil hardened, tough. *Obsolete*

SUWEFA TX15

Stahlwerke Sudwestfalen
C 0.35, Ni 1.55, Cr 1.55, Mo 0.2, bal Fe.
For die casting and plastic mold dies; oil hardened, tough. *Obsolete*

SUWEFA TX20
Stahlwerke Sudwestfalen
C 0.3, Cr 2, Mo 0.3, Ni 2, bal Fe.
For die casting and plastic mold dies; oil hardened, tough.
Obsolete

SUWEFA UCN
Stahlwerke Sudwestfalen
C 0.19, Cr 1.25, Mo 0.2, Ni 3.75, bal Fe.
For gears, bolts, camshafts, cams; oil hardened, tough.
Obsolete

SUWEFA UER
Stahlwerke Sudwestfalen
C 0.4, Cr 13, Mn 0.3, bal Fe.
Annealed: 100,000 TS; 55,000 YS; 20 El; 50 RA; 200 Brin. For valves, cutlery, surgical and dental instruments; Type 420; corrosion resistant. *Obsolete*

SUWEFA UH50
Stahlwerke Sudwestfalen
C 0.15, Si 0.25-0.5, Mn 0.3-0.8, bal Fe.
Annealed: 70,000 TS; 40,000 YS; 25 El; 60 RA; 143 Brin. For gears, bolts, cams, screws, fan blades; case hardened.
Obsolete

SUWEFA UH60
Stahlwerke Sudwestfalen
C 0.15, Cr 0.65, Si 0.25, Mn 0.5, bal Fe.
For gears, bolts, fan blades; case hardened. *Obsolete*

SUWEFA UMC
Stahlwerke Sudwestfalen
C 0.2, Cr 1.15, Mn 1.25, Si 0.25, bal Fe.
For gears, bolts, camshafts, cams; case hardened. *Obsolete*

SUWEFA UNIVERSAL
Stahlwerke Sudwestfalen
C 0.53, Si 0.9, Mn 0.9, bal Fe.
For springs, bolts, gears, machine tool parts; oil hardened, tough. *Obsolete*

SUWEFA URV
Stahlwerke Sudwestfalen
C 1.42, W, V, bal Fe.
For engravers' tools, cutters, forming dies; oil hardened, wear resistant. *Obsolete*

SUWEFA USS
Stahlwerke Sudwestfalen
C 0.9, Mn 1.9, V 0.1, bal Fe.
For punches, blanking and forming dies; oil hardening, non-deforming. *Obsolete*

SUWEFA V200
Stahlwerke Sudwestfalen
C 0.82, Cr 4.1, Mo 0.85, V 1.6, W 8.7, bal Fe.
For lathe and planer tools, reamers, broaches; high speed steel. *Obsolete*

SUWEFA V300
Stahlwerke Sudwestfalen
C 0.86, Cr 4.1, Mo 0.85, V 2.5, W 12, bal Fe.
For lathe and planer tools, broaches, taps, drills; high speed steel. *Obsolete*

SUWEFA V400
Stahlwerke Sudwestfalen
C 1.3, Cr 4.3, Mo 0.85, V 3.8, W 12, bal Fe.
For engravers' tools, form cutters, reamers, taps; high speed steel. *Obsolete*

SUWEFA VC11
Stahlwerke Sudwestfalen
C 1.1, Cr 0.4, Mn 0.3, bal Fe.
For bearings, liners, sleeves, cutters; water or oil hardened.
Obsolete

SUWEFA VC12
Stahlwerke Sudwestfalen
C 2.1, Cr 11.5, Mn 0.3, bal Fe.
For blanking and forming dies; oil hardened, non-deforming.
Obsolete

SUWEFA VCN188
Stahlwerke Sudwestfalen
C 0.45, Cr, Ni, W, bal Fe.
For forging and heading dies; oil hardened, tough. *Obsolete*

SUWEFA VCS2
Stahlwerke Sudwestfalen
C 0.45, Si, Cr, bal Fe.
For machine tool parts; oil hardened, tough. *Obsolete*

SUWEFA VCS9
Stahlwerke Sudwestfalen
C 0.45, Cr, Si, bal Fe.
For machine tool parts; oil hardened, tough. *Obsolete*

SUWEFA VMS
Stahlwerke Sudwestfalen
C 0.37, Si 1.25, Mn 1.25, bal Fe.
For machine tool parts; oil hardened, tough. *Obsolete*

SUWEFA VRV
Stahlwerke Sudwestfalen
C 1.3, Si 0.25, Mn 0.3, Cr 0-0.2, W 4.75, bal Fe.
For cutters, engravers' tools, reamers; fast-finishing tool steel.
Obsolete

SUWEFA W SPEZIAL 4
Stahlwerke Sudwestfalen
C 0.45, Si 0.25-0.5, Mn 0.3, bal Fe.
Hot rolled: 98,000 TS; 58,000 YS; 24 El; 45 RA; 212 Brin. For machine tool parts, gears; water hardened. *Obsolete*

SUWEFA W SPEZIAL H
Stahlwerke Sudwestfalen
C 0.9, Si 0.25-0.5, Mn 0.3-0.8, bal Fe.
Heat treated: 190,000 TS; 145,000 YS; 10 El; 30 RA; 400 Brin. For springs, taps, cutters, reamers, drills; Type W1; water hardened. *Obsolete*

SUWEFA W SPEZIAL MH
Stahlwerke Sudwestfalen
C 0.75, Si 0.25-0.5, Mn 0.3-0.8, bal Fe.
Heat treated: 175,000 TS; 128,000 YS; 12 El; 37 RA; 352 Brin. For springs, rails, clutch discs, girders; Type W1; water hardened. *Obsolete*

SUWEFA W SPEZIAL WEICH
Stahlwerke Sudwestfalen
C 0.35, Si 0.25-0.5, Mn 0.3-0.8, bal Fe.
Hot rolled: 85,000 TS; 54,000 YS; 30 El; 53 RA; 183 Brin. For gears, bolts, shafts, axles; water hardened. *Obsolete*

SUWEFA W SPEZIAL ZAH
Stahlwerke Sudwestfalen
C 0.45, Si 0.25-0.5, Mn 0.3-0.8, bal Fe.
Hot rolled: 98,000 TS; 59,000 YS; 24 El; 45 RA; 212 Brin. For axles, gears, bolts, tie rods; water hardened. *Obsolete*

SUWEFA W SPEZIAL ZH
Stahlwerke Sudwestfalen
C 0.6, Si 0.25-0.5, Mn 0.3-0.8, bal Fe.
Heat treated: 160,000 TS; 113,000 YS; 12 El; 40 RA; 320 Brin. For wheels, die blocks, girders, springs; water hardened.
Obsolete

SUWEFA W18
Stahlwerke Sudwestfalen
C 0.74, Cr 4.1, V 1.1, W 18.5, bal Fe.
For lathe and planer tools, reamers, broaches, drills; high speed steel. *Obsolete*

SUWEFA W6
Stahlwerke Sudwestfalen
C 0.6, Si 0.38, Mn 0.65, bal Fe.
Heat treated: 160,000 TS; 115,000 YS; 12 El; 40 RA; 325 Brin. For wheels, die blocks, girders, rails, springs; water hardened.
Obsolete

SUWEFA WCV4
Stahlwerke Sudwestfalen
C 0.5, Cr 1, Mn 0.85, bal Fe.
For gears, bolts, crankshafts; oil hardened, tough. *Obsolete*

SUWEFA WCV5
Stahlwerke Sudwestfalen
C 0.58, Cr 1, V 0.09, Mn 0.95, bal Fe.
For springs, gears, bolts, machine tool parts; oil hardened, shock resistant. *Obsolete*

SUWEFA WCV9 SPEZIAL
Stahlwerke Sudwestfalen
C 0.3, Cr, Mo, V, bal Fe.
For forging and heading dies, punches; oil hardened, tough.
Obsolete

SUWEFA WEL
Stahlwerke Sudwestfalen
C 1.05, Cr, bal Fe.
For bearings, cutters, header dies; water hardened, wear resistant. *Obsolete*

SUWEFA WES
Stahlwerke Sudwestfalen
C 1.2, V 0.1, Cr 0.2, Si 0.28, W 1, bal Fe.
For header and blanking dies; oil hardened, tough. *Obsolete*

SUWEFA WF8
Stahlwerke Sudwestfalen
C 0.67, Si 1.3, Mn 0.5, Cr 0.5, bal Fe.
For springs, punches, chisels, upsetters; oil hardened, tough.
Obsolete

SUWEFA WGKL
Stahlwerke Sudwestfalen
C 1, Si 0.3, Mn 0.35, Cr 1.55, bal Fe.
For bearings, cutters, blanking dies; oil hardened, abrasion resistant. *Obsolete*

SUWEFA WKL
Stahlwerke Sudwestfalen
C 1, Cr 1.1, Mn 0.07, Si 0.25, bal Fe.
For bearings, header dies; water hardened, wear resistant.
Obsolete

SUWEFA WM13
Stahlwerke Sudwestfalen
C 0.4, Si 0.4, Mn 0.3, Cr 13, bal Fe.
Annealed: 100,000 TS; 55,000 YS; 25 El; 55 RA; 200 Brin. For valves, cutlery, surgical and dental instruments; Type 420; stainless. *Obsolete*

SUWEFA WMS
Stahlwerke Sudwestfalen
C 0.3, Cr 2.65, V 0.33, W 8.5, bal Fe.
For extrusion rams and liners, punches; hot work steel, oil hardened. *Obsolete*

SUWEFA WMS SPEZIAL
Stahlwerke Sudwestfalen
C 0.3, Co 2, Cr 2.4, V 0.25, W 8.5, bal Fe.
For hot work tools, dies, punches; hot work steel, oil hardened. *Obsolete*

SUWEFA WMS5
Stahlwerke Sudwestfalen
C 0.3, Cr 2.35, V 0.6, W 4.25, Mn 0.3, bal Fe.
For extrusion rams, upsetters, punches, shears; hot work steel, oil hardened. *Obsolete*

SUWEFA WMV
Stahlwerke Sudwestfalen
C 0.45, Cr 1.35, Mo 0.45, V 0.8, W 0.45, bal Fe.
For forging and heading dies; die casting dies; oil hardened, tough. *Obsolete*

SUWEFA WRL
Stahlwerke Sudwestfalen
C 0.9, Cr 0.8, Mn 0.3, Si 0.25, bal Fe.
For bearings, cutters, sleeves, liners; water hardened, wear resistant. *Obsolete*

SUWEFA WRL HART
Stahlwerke Sudwestfalen
C 1.05, Cr 1, Mn 0.3, Si 0.25, bal Fe.
For bearings, cutters, sleeves, liners; water hardened, wear resistant. *Obsolete*

SUWEFA WS1
Stahlwerke Sudwestfalen
C 1.2, W, bal Fe.
For cutters, dies; oil hardened, wear resistant. *Obsolete*

SUWEFA WT13
Stahlwerke Sudwestfalen
C 0.2, Si 0.4, Mn 0.3, Cr 13, bal Fe.
Annealed: 73,000 TS; 40,000 YS; 22 El; 58 RA; 140 Brin. For turbine blades, valves, cutlery; Type 420; stainless. *Obsolete*

SUWEFA WZK SPEZIAL
Stahlwerke Sudwestfalen
C 0.3, Cr 2.65, V 0.35, W 8.5, bal Fe.
For extrusion rams and liners, punches, upsetters; hot work steel, oil hardened. *Obsolete*

SUWEFA ZGW
Stahlwerke Sudwestfalen
C 1.65, Cr 11.5, V 0.1, bal Fe.
For forming and blanking dies; air hardened, non-deforming. *Obsolete*

SUWEFA ZH12
Stahlwerke Sudwestfalen
C 2.1, Cr 11.5, Mn 0.3, Si 0.25, bal Fe.
For forming and blanking dies, punches; oil hardened, non-deforming. *Obsolete*

SUWEFA ZH120
Stahlwerke Sudwestfalen
C 2.1, Cr 11.5, Mn 0.3, Si 0.35, bal Fe.
For blanking and forming dies, punches; oil hardened, non-deforming. *Obsolete*

SV
Friedr. Lohmann GmbH
C 0.8, Cr 0.6, V 0.2, bal Fe.
For bearings, cutters, form dies, punches. Oil or water hardened, wear resistant.

SV 2
Bergische Stahl Industrie
C 0.78-0.86, Cr 3.8-4.5, W 8.3-9, V 1.4-1.7, Mo 0.7-1, bal Fe.
High speed steel. W.-Nr. 1.3316.

SV 4
Bergische Stahl Industrie
C 1.2-1.35, Cr 3.8-4.5, W 11.5-12.5, V 3.5-4, Mo 0.7-1, bal Fe.
High speed steel. W.-Nr. 1.0002.

SV 4 CO 5
Bergische Stahl Industrie
C 1.3-1.45, Cr 3.8-4.5, W 11.5-12.5, V 3.5-4, Co 4.5-5, Mo 0.7-1, bal Fe.
High speed steel. W.-Nr. 1.3202.

SV-RHF 10
Saarstahl AG
C 0.38, Si 0.25, Mn 0.65, Cr 0.8, Ni 1.8, Mo 0.35, V 0.1, bal Fe.
Quenched and tempered: 1770 N/mm^2 TS; 1470 N/mm^2 YS; 6 El. For highly stressed parts; vacuum melted. W.-Nr. 1.6926. *Obsolete*

SV-RHF 20
Saarstahl AG
C 0.4, Si 0.9, Mn 0.3, Cr 5, Mo 1.3, V 0.5, bal Fe.
Quenched and tempered: 1960 N/mm^2 TS; 1570 N/mm^2 YS; 6 El. For highly stressed parts; vacuum melted. W.-Nr. 1.7783. *Obsolete*

SV-RHF 30
Saarstahl AG
Now SAARSTAHL 1.6359.

SV-RHF 32
Saarstahl AG
Now SAARSTAHL 1.6358.

SV-RHF 33
Saarstahl AG
Now SAARSTAHL 1.6356.

SVEA METAL
George W. Prentiss & Co.
pure Fe.
44,000 TS; 23,000 YS; 40 El; 65 RA; 86 Brin. For radio tubes, X-ray equipment, neon lights, mercury tubes; for magnetic and electronic application. *Obsolete*

SVENSKA SM-0010
Svenska Metallverken A.B.
Copper. Cu 99.99, 0.03 O$_2$.
Annealed: 33,000 TS; 70,000 YS; 45 El. 1/2 H-temper: 40,000 TS; 36,000 YS; 15 El; 80 Brin. For electrical terminals; tough pitch copper.

SVENSKA SM-1092
Svenska Metallverken A.B.
Copper. Zn 8, Cu 92.
Annealed: 36,000 TS; 10,000 YS; 45 El; 50 Brin. H-temper: 64,000 TS; 56,000 YS; 5 El; 125 Brin. For hardware; commercial bronze.

SVENSKA SM-1160
Svenska Metallverken A.B.
Copper. Cu 60, Pb 0.4, bal Zn.
Annealed: 52,000 TS; 18,000 YS; 40 El; 100 Brin. 1/4 H-temper: 57,000 TS; 43,000 YS; 20 El; 137 Brin. For architectural trim, condenser tubes, hardware; muntz metal.

SVENSKA SM-1163
Svenska Metallverken A.B.
Copper. Cu 63, Pb 0.3, bal Zn.
Annealed: 51,000 TS; 18,000 YS; 50 El; 64 Brin. H-temper: 86,000 TS; 65,000 YS; 5 El; 176 Brin. For screw machine products; low leaded brass.

SVENSKA SM-1261
Svenska Metallverken A.B.
Copper. Cu 61, Pb 1, bal Zn.
Annealed: 52,000 TS; 18,000 YS; 40 El; 75 Brin. H-temper: 70,000 TS; 45,000 YS; 20 El; 137 Brin. For screw machine products; leaded brass.

SVENSKA SM-1263
Svenska Metallverken A.B.
Copper. Cu 63, Pb 1, bal Zn.
Annealed: 51,000 TS; 18,000 YS; 40 El; 69 Brin. 1/2 H-temper: 65,000 TS; 50,000 YS; 20 El; 132 Brin. For screw machine products; leaded brass.

SVENSKA SM-1661
Svenska Metallverken A.B.
Copper. Cu 61.5, Pb 3, bal Zn.
1/2 H-temper: 49,000 TS; 18,000 YS; 40 El; 61 Brin. H-temper: 68,000 TS; 52,000 YS; 20 El; 150 Brin. For screw machine products; leaded brass.

SVENSKA SM-1956
Svenska Metallverken A.B.
Copper. Cu 56, Pb 0.5, Al 0.4, bal Zn.
For architectural trim; brass.

SVENSKA SM-2276
Svenska Metallverken A.B.
Copper. Cu 76, Al 2, As 0.05, bal Zn.
Annealed: 65,000 TS; 42,000 YS; 40 El; 107 Brin. For condenser tubes; aluminum brass.

SVENSKA SM-2771
Svenska Metallverken A.B.
Copper. Cu 71, Sn 1, As 0.4, bal Zn.
Annealed: 65,000 TS; 42,000 YS; 35 El; 107 Brin. For condenser tubes, evaporator and heat exchanger tubes; inhibited admiralty brass.

SVENSKA SM-4210
Svenska Metallverken A.B.
Copper. Cu 88.4, Ni 10, Fe 1.2, Mn 0.4, bal Zn.
Cold drawn: 44,000-60,000 TS; 22,000-57,000 YS; 15-45 El; 64-120 Brin. For condenser tubes; corrosion resistant.

SVERKER 1
Uddeholm Corp.
C 2, Mn 0.7, Cr 13, V 0.2, bal Fe.
For dies; for cold work, non-deforming.

SVERKER 2
Uddeholm Corp.
C 1.5, Cr 12, Mo 0.08, V 0.2, bal Fe.
For dies; for cold work, non-deforming.

SVERKER 21
Uddeholm Corp.
C 1.55, Cr 12, Mo 0.8, V 0.8, Si 0.3, Mn 0.3, bal Fe.
Hardened: 278,000 TS; 214,000 YS; 1 El; 56 Rock C. For hobs, plastic molds, punches. Type D2; air hardened steel, wear resistant.

SVERKER 3
Uddeholm Corp.
C 2.05, Mn 0.8, Cr 12.5, W 1.3, bal Fe.
For cold work dies; air or oil hardened; non-deforming. AISI D6.

SVM
U.N. Alloy Steel Corp.
C 0.55, Mn 0.75, Si 1.85, V 0.25, Mo 0.5, bal Fe.
Heat treated: 275,000 TS; 247,000 YS; 9 El; 514 Brin. For pneumatic tools, shear blades, punches, chisels, caulking tools. Type S5 tool steel; tough, shock resistant.

SW 12 CO 3
Bergische Stahl Industrie
C 0.77-0.85, Cr 3.8-4.5, W 11.5-12.5, V 1.7-2, Mo 0.7-1, Co 2.5-3, bal Fe.
High speed steel. W.-Nr. 1.3211.

SW 12 V 2
Bergische Stahl Industrie
C 0.9-1, Cr 3.8-4.5, W 11.5-12.5, V 2.3-2.6, Mo 0.7-1, bal Fe.
High speed steel. W.-Nr. 1.3318.

SW 18
Bergische Stahl Industrie
C 0.7-0.78, Cr 3.8-4.5, W 17.5-18.5, V 1-1.2, bal Fe.
High speed steel. W.-Nr. 1.3355.

SW 18 CO 10

Bergische Stahl Industrie
C 0.72-0.8, Cr 3.8-4.5, W 17.5-18.5, V 1.4-1.7, Mo 0.5-0.8, bal Fe.
High speed steel. W.-Nr. 1.3265.

SW 18 CO 5

Bergische Stahl Industrie
C 0.75-0.83, Cr 3.8-4.5, W 17.5-18.5, V 1.4-1.7, Mo 0.5-0.8, Co 4.5-5, bal Fe.
High speed steel. W.-Nr. 1.3255.

SW 9 CO 10

Bergische Stahl Industrie
C 1.2-1.35, Cr 3.8-4.5, W 9.5-11, V 3-3.5, Mo 3.5-4, Co 10-11, bal Fe.
High speed steel. W.-Nr. 1.3207.

SW-14

Chemetron Corp.
Mn 0.5, Si 0.2, P 0.012, S 0.021, weld metal: 0.09 C, bal Fe.
As welded: 73,500 psi TS; 60,000 psi YS; 24 El. All-position, covered electrode, AC-DC, reverse polarity; for welding mild steel. AWS Class E6011.

SW-15

Chemetron Corp.
Mn 0.47, Si 0.45, weld metal: 0.10 C, bal Fe.
As welded: 74,000 psi TS; 63,000 psi YS; 22 El. Covered electrode, AC-DC, for general purpose welding and repair. AWS E6013.

SW-15-IP

Chemetron Corp.
Mn 0.5, Si 0.26, weld metal: 0.09 C, bal Fe.
As welded: 73,000 psi TS; 61,000 psi YS; 25 El. AC-DC electrode with an iron powder coating to improve deposition rate for general welding and repair. AWS E7014.

SW-44

Chemetron Corp.
Mn 0.95, Si 0.83, weld metal: 0.09 C, bal Fe.
As welded: 81,500 psi TS; 72,000 psi YS; 22 El. High speed, heavy coated, iron-powder, AC-DC, electrode for high deposition. AWS Class E7024.

SW-47

Chemetron Corp.
Mn 0.88, Si 0.63, weld metal: 0.07 C, bal Fe.
As welded: 81,500 psi TS; 67,000 psi YS; 33 El. All-position, iron-powder, low-hydrogen, AC-DC, reverse polarity electrode for welding difficult to weld steels. AWS Class E7018.

SW-610

Chemetron Corp.
Mn 0.3, P 0.01, S 0.022, Si 0.2, weld metal: 0.10 C, bal Fe.
As welded: 71,000 psi TS; 60,000 psi YS; 24 El. Covered electrode for welding mild steel; all-position, DC reverse polarity. AWS Class E6010.

SW-612

Chemetron Corp.
Mn 0.42, Si 0.23, P 0.014, S 0.028, weld metal: 0.07 C, bal Fe.
As welded: 71,500 psi TS; 61,500 psi YS; 24 El. Covered electrode, AC-DC, straight polarity for welding mild steel with poor fit-up. AWS Class E6012.

SWB 14N6

SWB Stahlformguss Gesellschaft mbH
C 0.13, Cr 0.2, Ni 1.5, bal Fe.
For gears, pinions, shafts, cams, camshafts; case hardened, shock resistant. *Obsolete*

SWB 15C3

SWB Stahlformguss Gesellschaft mbH
C 0.15, Cr 0.65, bal Fe.
For gears, pinions, shafts, cams, camshafts; case hardened, shock resistant. *Obsolete*

SWB 15CN6

SWB Stahlformguss Gesellschaft mbH
C 0.15, Cr 1.5, Ni 1.5, bal Fe.
For gears, pinions, camshafts, cams, shafts; case hardened, shock resistant. *Obsolete*

SWB 15MNC32

SWB Stahlformguss Gesellschaft mbH
C 0.15, bal Fe.
Cold drawn: 72,000 TS; 60,000 YS; 22 El; 58 RA; 145 Brin.
For gears, cams, camshafts, fasteners; case hardening steel. *Obsolete*

SWB 16M5

SWB Stahlformguss Gesellschaft mbH
C 0.16, Mn 1.1, Si, bal Fe.
For gears, cams, camshafts; case hardening. *Obsolete*

SWB 16M8

SWB Stahlformguss Gesellschaft mbH
C 0.16, Mn 1.2, Si, bal Fe.
For gears, cams, camshafts; case hardening. *Obsolete*

SWB 16MC5

SWB Stahlformguss Gesellschaft mbH
C 0.16, Si 0.25, Mn 1.15, Cr 0.95, bal Fe.
For gears, cams, camshafts, machine tool parts; case hardening. *Obsolete*

SWB 7SC5

SWB Stahlformguss Gesellschaft mbH
C 0.67, Si 1.3, Mn 0.5, Cr 0.5, bal Fe.
For springs; oil hardened, good resiliency. *Obsolete*

SWB-100C4

SWB Stahlformguss Gesellschaft mbH
C 1, Cr 1.1, Mn 0.07, bal Fe.
For bearings, liners, sleeves; water hardened, wear resistant. *Obsolete*

SWB-100C6

SWB Stahlformguss Gesellschaft mbH
C 1, Mn 0.35, Cr 1.55, bal Fe.
For bearings, liners, races, sleeves; water hardened, wear resistant. *Obsolete*

SWB-105C4

SWB Stahlformguss Gesellschaft mbH
C 1.05, Si 0.25, Mn 0.3, Cr 1, bal Fe.
For bearings, liners, sleeves, cutters; water hardened, wear resistant. *Obsolete*

SWB-110C2

SWB Stahlformguss Gesellschaft mbH
C 1.1, Cr 0.4, Mn 0.3, bal Fe.
For bearings, liners, races, sleeves; water hardened, wear resistant. *Obsolete*

SWB-120C5

SWB Stahlformguss Gesellschaft mbH
C 1.2, Mn 0.3, Si 0.2, bal Fe.
For cutters, drills, reamers, hobs, taps; Type W1; water hardened. *Obsolete*

SWB-130C1

SWB Stahlformguss Gesellschaft mbH
C 1.25, Si 0.2, Mn 0.3, Cr 1, bal Fe.
For bearings, races, taps, reamers; water hardened, wear resistant. *Obsolete*

SWB-13C2

SWB Stahlformguss Gesellschaft mbH
C 0.13, Cr, bal Fe.
For gears, cams, machine tool parts; case hardening. *Obsolete*

SWB-140C3

SWB Stahlformguss Gesellschaft mbH
C 1.4, Si 0.2, Mn 0.3, Cr 1, bal Fe.
For bearings, sleeves, liners, races; water hardened, wear resistant. *Obsolete*

SWB-150C6

SWB Stahlformguss Gesellschaft mbH
C 1.45, Si 0.25, Mn 0.6, Cr 1.4, bal Fe.
For bearings, sleeves, liners, races; water hardened, wear resistant. *Obsolete*

SWB-18CN8

SWB Stahlformguss Gesellschaft mbH
C 0.18, Cr 2, Ni 2, bal Fe.
For gears, shafts, machine tool parts; case hardening. *Obsolete*

SWB-200C8

SWB Stahlformguss Gesellschaft mbH
C 2, Cr 13, bal Fe.
For blanking and forming dies; oil or air hardened, non-deforming. *Obsolete*

SWB-20MC5

SWB Stahlformguss Gesellschaft mbH
C 0.2, Mn 1.25, Cr 1.15, bal Fe.
For gears, shafts, machine tool parts; case hardening. *Obsolete*

SWB-22CV4

SWB Stahlformguss Gesellschaft mbH
C 0.22, Cr 1.1, V 0.2, bal Fe.
For gears, cams, camshafts; case hardening. *Obsolete*

SWB-22M3

SWB Stahlformguss Gesellschaft mbH
C 0.22, Mn 1, bal Fe.
For gears, cams, camshafts; case hardening. *Obsolete*

SWB-24CMO5

SWB Stahlformguss Gesellschaft mbH
C 0.24, Cr 1.15, Mo 0.25, bal Fe.
For gears, bolts, fasteners, shafts; water or oil hardened, tough. *Obsolete*

SWB-24CMO54

SWB Stahlformguss Gesellschaft mbH
C 0.24, Cr 1.25, Mo 0.45, bal Fe.
For gears, shafts, bolts, studs, fasteners; water or oil hardened, tough. *Obsolete*

SWB-24CMOV55

SWB Stahlformguss Gesellschaft mbH
C 0.25, Cr 1.35, Mo 0.55, V 0.2, bal Fe.
For gears, machine tool parts, bolts, shafts; oil hardened, shock resistant. *Obsolete*

SWB-25CMO4

SWB Stahlformguss Gesellschaft mbH
C 0.25, Cr 1, bal Fe.
For gears, shafts, bolts, studs; water hardened. *Obsolete*

SWB-27MCV5

SWB Stahlformguss Gesellschaft mbH
C 0.27, Mn, Cr, V, bal Fe.
For gears, shafts, machine tool parts; water hardened, tough. *Obsolete*

SWB-30M5

SWB Stahlformguss Gesellschaft mbH
C 0.3, Mn 1.35, bal Fe.
For gears, shafts, machine tool parts; water or oil hardened. *Obsolete*

SWB-30MOV9

SWB Stahlformguss Gesellschaft mbH
C 0.3, Cr 2.5, Mo 0.2, V 0.15, bal Fe.
For gears, axles, machine tool parts, bolts; oil hardened, shock resistant. *Obsolete*

SWB-31CMOV9

SWB Stahlformguss Gesellschaft mbH
C 0.31, Cr 2.35, Mo 0.18, V 0.13, bal Fe.
For gears, shafts, machine tool parts, bolts; oil hardened, shock resistant. *Obsolete*

SWB-32CAMO4
SWB Stahlformguss Gesellschaft mbH
C 0.32, Al 1.1, Cr 1.1, Mo 0.18, bal Fe.
For oil refinery equipment, fasteners; creep resistant.
Obsolete

SWB-33CAN7
SWB Stahlformguss Gesellschaft mbH
C 0.33, Al 1.1, Cr 1.7, Ni 1, bal Fe.
For oil refinery equipment, fasteners; creep resistant.
Obsolete

SWB-34C4
SWB Stahlformguss Gesellschaft mbH
C 0.33, Cr 1, Mo 0.65, bal Fe.
For gears, bolts, machine tool parts; water hardened.
Obsolete

SWB-34CA6
SWB Stahlformguss Gesellschaft mbH
C 0.34, Al 1.1, Cr 1, Mn 0.05, bal Fe.
For oil refinery equipment, fasteners; creep resistant.
Obsolete

SWB-34CNMO6
SWB Stahlformguss Gesellschaft mbH
C 0.34, Mn 0.55, Cr 1.5, Mo 0.2, Ni 1.5, bal Fe.
For gears, shafts, machine tool parts, bolts; oil hardened,
shock resistant. *Obsolete*

SWB-35CMO4
SWB Stahlformguss Gesellschaft mbH
C 0.33, Mn 0.65, Cr 1, Mo 0.2, bal Fe.
For gears, shafts, machine tool parts, bolts; oil hardened,
shock resistant. *Obsolete*

SWB-35MC72
SWB Stahlformguss Gesellschaft mbH
C 0.4, Mn 1.5, Cr, bal Fe.
For gears, shafts, machine tool parts, crimpers; oil hardened,
tough. *Obsolete*

SWB-36C6
SWB Stahlformguss Gesellschaft mbH
C 0.36, Mn 0.45, Cr 1.6, bal Fe.
For gears, shafts, machine tool parts; water hardened.
Obsolete

SWB-36CNMO4
SWB Stahlformguss Gesellschaft mbH
C 0.36, Mn 0.65, Cr 1, Mo 0.2, Ni 1, bal Fe.
For gears, shafts, machine tool parts, bolts; oil hardened,
shock resistant. *Obsolete*

SWB-37MS5
SWB Stahlformguss Gesellschaft mbH
C 0.37, Si 1.25, Mn 1.25, bal Fe.
For machine tool parts, upsetters; oil hardened, shock
resistant. *Obsolete*

SWB-37MSG
SWB Stahlformguss Gesellschaft mbH
C 0.35, Si 1.25, Mn 1.25, bal Fe.
For machine tool parts, upsetters; oil hardened, shock
resistant. *Obsolete*

SWB-40M4
SWB Stahlformguss Gesellschaft mbH
C 0.4, Mn 0.95, bal Fe.
Hot rolled: 95,000 TS; 60,000 YS; 25 El; 50 RA; 205 Brin. For
gears, shafts, machine tool parts, bolts; water hardened.
Obsolete

SWB-41C4
SWB Stahlformguss Gesellschaft mbH
C 0.41, Cr 1, Mn 0.65, bal Fe.
For gears, shafts, machine tool parts; water hardened.
Obsolete

SWB-42CMO4
SWB Stahlformguss Gesellschaft mbH
C 0.42, Mn 0.65, Cr 1, Mo 0.2, bal Fe.
For gears, shafts, crimpers, bolts, punches; oil hardened,
tough. *Obsolete*

SWB-42CV6
SWB Stahlformguss Gesellschaft mbH
C 0.42, Cr, V, bal Fe.
For gears, shafts, bolts, axes, fasteners; oil hardened, shock
resistant. *Obsolete*

SWB-42MV7
SWB Stahlformguss Gesellschaft mbH
C 0.42, Mn 1.75, V 0.1, bal Fe.
For punches, crimpers, upsetters; shock resistant. *Obsolete*

SWB-46M7
SWB Stahlformguss Gesellschaft mbH
C 0.46, Mn 1.75, V 0.1, bal Fe.
For punches, crimpers, upsetters; shock resistant. *Obsolete*

SWB-46MSG
SWB Stahlformguss Gesellschaft mbH
C 0.46, Si 1.2, Mn 1.25, bal Fe.
For punches, upsetters, riveters; oil hardened, shock
resistant. *Obsolete*

SWB-50CMO4
SWB Stahlformguss Gesellschaft mbH
C 0.5, Mn 0.65, Cr 1, Mo 0.2, bal Fe.
For bolts, studs, machine tool parts; oil hardened, shock
resistant. *Obsolete*

SWB-50CV4
SWB Stahlformguss Gesellschaft mbH
C 0.5, Mn 0.85, Cr 1, V 0.09, bal Fe.
For bolts, springs, shafts, crankshafts; oil hardened, shock
resistant. *Obsolete*

SWB-50CVG
SWB Stahlformguss Gesellschaft mbH
C 0.5, Mn 0.85, Cr 1, V 0.09, bal Fe.
For bolts, fasteners, crankshafts, axles, springs; oil hardened,
shock resistant. *Obsolete*

SWB-53MS4
SWB Stahlformguss Gesellschaft mbH
C 0.53, Si 0.8, Mn 1.05, bal Fe.
For springs, bolts, shafts; tough, oil hardened. *Obsolete*

SWB-58CV4
SWB Stahlformguss Gesellschaft mbH
C 0.58, Mn 0.95, Cr 1, V 0.09, bal Fe.
For springs, crankshafts, axles; tough, oil hardened.
Obsolete

SWB-70WS7
SWB Stahlformguss Gesellschaft mbH
C 0.7, Si 1.7, Mn 0.7, bal Fe.
For dies, punches, upsetters, springs; oil hardened, shock
resistant. *Obsolete*

SWB-74NC2
SWB Stahlformguss Gesellschaft mbH
C 0.75, Ni, Cr, bal Fe.
For punches, upsetters, crimpers; oil hardened, shock
resistant. *Obsolete*

SWB-85C7
SWB Stahlformguss Gesellschaft mbH
C 0.85, Cr 1.75, Mn 0.35, bal Fe.
For bearings, liners, sleeves; water or oil hardened, wear
resistant. *Obsolete*

SWB-A10
SWB Stahlformguss Gesellschaft mbH
C 1, Si 0.25, Mn 0.25, bal Fe.
Annealed: 100,000 TS; 53,000 YS; 21 El; 42 RA; 200 Brin. For
drills, taps, reamers, cutters, springs; Type W1; water
hardened. *Obsolete*

SWB-A10 SPEZIAL
SWB Stahlformguss Gesellschaft mbH
C 1, Si 0.2, Mn 0.25, V 0.1, bal Fe.
Annealed: 100,000 TS; 53,000 YS; 21 El; 42 RA; 200 Brin. For
drills, taps, springs, reamers; Type W2; water hardened.
Obsolete

SWB-A11
SWB Stahlformguss Gesellschaft mbH
C 1.1, Si 0.25, Mn 0.25, bal Fe.
Annealed: 105,000 TS; 55,000 YS; 20 El; 40 RA; 200 Brin. For
drills, hobs, reamers, broaches; Type W1; water hardened.
Obsolete

SWB-A13
SWB Stahlformguss Gesellschaft mbH
C 1.3, Si 0.25, Mn 0.25, bal Fe.
For engravers' tools, taps, reamers; Type W1; water
hardened. *Obsolete*

SWB-A7
SWB Stahlformguss Gesellschaft mbH
C 0.4, Si 0.25, Mn 0.25, bal Fe.
Hot rolled: 92,000 TS; 58,000 YS; 27 El; 50 RA; 205 Brin. For
gears, bolts, shafts, studs, fasteners; water hardened.
Obsolete

SWB-A8
SWB Stahlformguss Gesellschaft mbH
C 0.85, Si 0.25, Mn 0.25, bal Fe.
Heat treated: 190,000 TS; 145,000 YS; 12 El; 35 RA; 390 Brin.
For drills, hobs, springs, cutters; Type W1; water hardened.
Obsolete

SWB-CMO4
SWB Stahlformguss Gesellschaft mbH
C 0.33, Mn 0.65, Cr 1, Mo 0.2, bal Fe.
For gears, shafts, crankshafts, machine tool parts, bolts; oil
hardened, tough. *Obsolete*

SWB-CNMO8
SWB Stahlformguss Gesellschaft mbH
C 0.3, Cr 2, Mo 0.3, Ni 2, bal Fe.
For gears, bolts, machine tool parts; water or oil hardened,
shock resistant. *Obsolete*

SWEAT-ON PASTE
Now COLMONOY SWEAT-ON PASTE.

SWED-OIL
Manufacturer not listed.
C 0.9, Mn 1.2, Cr 0.5, W 0.5, V 0.2, bal Fe.
Oil hardening tool and die steel.

SWEDESTEEL "A"
Paragon Steel Co.
C 0.8, bal Fe.
For rock drills, concrete busters; water hardening.

SWEDESTEEL "B"
Paragon Steel Co.
C 0.8, Mo 0.2, bal Fe.
For scarifier teeth, chisel blanks; water hardening.

SWEDESTEEL "C"
Paragon Steel Co.
C 0.5, Cr 1, Ni 1.5, bal Fe.
For shovel teeth; oil hardening.

SWEDESTEEL "D"
Paragon Steel Co.
C 0.5, Cr 1.5, V 0.2, bal Fe.
For diggings bars.

SWEDESTEEL "E"
Paragon Steel Co.
C 0.7, W 18, Cr 4, V 1, bal Fe.
For tool bits, drills, taps; high speed steel.

SWEDISH STEEL NO. 4
Boyd-Wagner Co.
C 0.9-1, bal Fe.
For embossing dies, wood and machine screw coldheader dies; water hardened.

SWEDISH STEEL NO. 41
Boyd-Wagner Co.
C, high Si, high Mn, bal Fe.
For chipping chisels; retains keen cutting edge.

SWEDISH STEEL NO. 9
Boyd-Wagner Co.
C, alloy, bal Fe.
For large cold header dies; water hardened.

SWEDISH VANADIUM
P.F. McDonald & Co.
C 0.7, V 0.2, bal Fe.
For tools and dies, punches; water hardened.

SWEFA GZB9F
Stahlwerke Sudwestfalen
C 0-0.12, Si 1.15, Mn 0.3, Cr 13, Al 0.95, bal Fe.
For oil refinery equipment; creep and heat resistant.
Obsolete

SWESCO E
Swedish Crucible Steel Co.
C 0.1, Cr 17-20, bal Fe.
Heat treated: 75,000 TS; 40,000 YS; 50 El. For heat treatment castings; resists oxidation to 1550°F.

SWESCO E-2
Swedish Crucible Steel Co.
C 0.2-0.3, Mn 0.6-0.85, Si 0.35-0.5, bal Fe.
Annealed: 72,000 TS; 42,000 YS; 29 El; 49 RA; 147 Brin.
Normalized: 78,500 TS; 44,500 YS; 30 El; 48 RA; 163 Brin.
For cast gears, shafts, machinery parts; water hardened.

SWESCO E-3
Swedish Crucible Steel Co.
C 0.3-0.35, Mn 1, Mo 0.3, Cu 0.8, bal Fe.
Annealed: 97,000 TS; 58,500 YS; 24 El; 39 RA; 192 Brin. Heat treated: 113,000 TS; 100,000 YS; 20 El; 55 RA; 245 Brin. For cast gears, shafts, machinery parts; water hardened, tough.

SWESCO E-4
Swedish Crucible Steel Co.
C 0.45-0.5, Mn 1, Mo 0.3, Cr 1.25, Cu 0.9, bal Fe.
Cast: 107,000 TS; 78,250 YS; 19 El; 41 RA; 217 Brin. For pumps, valves.

SWESCO E-5
Swedish Crucible Steel Co.
C 0.4-0.5, Mn 0.7-0.9, B 0.003, Si 0.3-0.5, bal Fe.
Normalized: 98,500 TS; 49,500 YS; 22 El; 32 RA; 179 Brin.
For dies, gears, sprockets, runways; cast to shape.

SWESCO F
Swedish Crucible Steel Co.
C 0.4, Cr 23-28, Ni 9-13, bal Fe.
Cast: 70,000 TS; 45,000 YS; 15 El. For heat and corrosion resistant castings; resists oxidation to 2100°F.

SWESCO G
Swedish Crucible Steel Co.
C 0.1, Ni 35-37, Cr 13-17, bal Fe.
Cast: 60,000 TS; 35,000 YS; 5 El. For heat resistant castings; resists oxidation to 2100°F.

SWIFTWELD
Lincoln Electric Co.
C 0.2, bal Fe.
Cast: 45,000-55,000 TS; 5-10 El. For welding electrodes; for mild steel.

SWITCH COPPER
Revere Copper Products, Inc.
Cu.
Rolled: 36,000 TS; 15 El. For switch blades, electrical parts; high electrical conductivity. *Obsolete*

SXZ 500
Sheffield Smelting Co. Ltd.
Ag 5, Zn, Cd.
Melting range: 270-285°C. Maximum stress: 20.5 kgf/mm^2; 8 El. For soft soldering; high strength.

SYCOB
Aubert & Duval
Tool material. C 2, Cr 12.5, Co 0.5, Mo 0.4, V 0.2, bal Fe.
For cutting, pressing and stamping tools, drawing dies; oil or air hardened, non-deforming.

SYL-CARB
GTE Products Corp.
Tungsten carbide powders used in the manufacture of cemented carbide products such as metal cutting tools, dies, drilling tools.

SYLCUM
English manufacture
Cu 7.3, Mn 0.5, Ni 1.4, Fe 0.5, Si 9, bal Al.
For pistons; cast, corrosion resistant.

SYLVALOY
Wilbur B. Driver Co.
Ni 97, Si 3.
Annealed: 75,000 TS; 35 RA. For vacuum tube filaments; magnetic. *Obsolete*

SYLVAN STAR
Teledyne Firth Sterling
Tool steel. C 1.05, Cr 0.1, V 0.2, bal Fe.
Annealed: 100,000 TS; 53,000 YP; 21 El; 197 Brin. Heat treated: 104,000-216,000 TS; 74,000-152,000 YP; 11-23 El; 32-50 RA; 200-600 Brin. For cold work dies, tools, cutters, axes, chisels, punches. Type W2 water hardening tool steel.
Obsolete

SYLVANIA A
GTE Products Corp.
Hf 0.5, C 0.02, bal W.
Stress relieved: 64,200 TS at 2000°F. For gyros, high density materials. Very little weldability. Highest creep strength at 3200°F, brittle. High ductile to brittle transition temperature.

SYLVANIA MT-104
GTE Products Corp.
Ti 0.5, Zr 0.08, C 0.025, bal Mo.
Rolled: 145,000 TS; 117,000 YS; 21 El; 52 RA. Recrystallized: 85,000 TS; 56,000 YS; 40 El; 34 RA. Stress relieved: 133,000 TS; 100,000 YS; 31 El; 59 RA. For heat engines, heat exchangers, nuclear reactors, radiation shields, rocket nozzles, vector controls. Good high temperature properties. Powder metallurgy.

SYLVANIA NC73
Now SYLVANIA WN-103.

SYLVANIA NF 22
Now SYLVANIA WN-102.

SYLVANIA NO. 4
GTE Products Corp.
Ni 42, Cr 5.6, Mn 0.2, Si 0.25, Al 0-0.15, bal Fe.
Annealed: 73,000 TS; 30,000 YS; 40 El; 107 Brin. Cold drawn: 140,000 TS; 135,000 YS; 2 El; 210 Brin. For glass to metal seals. Controlled expansion.

SYLVANIA NR-106
GTE Products Corp.
8-10 Eu_2O_3 + Gd_2O_3 + Sm_2O_3, bal Ni.
RT: 56,000 TS; 28,000 YS; 15 El; 230 DPH. At 1800°F: 7000 TS; 5000 YS; 14 El. For nuclear reactor control structures. Neutron flux control. Used up to 2000°F. Sintered powders. Dispersion alloy.

SYLVANIA NS55
GTE Products Corp.
Na 0.0018, K 0.0049, Ca 0.0003, Si 0.0008, Cr 0.0008, Fe 0.0025, Al 0.0004, C 0.02, Mg 0-0.0004, 0.00002 cu, bal Fe.
For high temperature filaments and instruments. High purity iron.

SYLVANIA WN-102
GTE Products Corp.
W 95, Ni 2.5, Fe 2.5.
Sintered: 125,000 TS; 100,000 YS; 11 El; 26-31 Rock C. For counterweights, radioactive shielding, electrical contacts, flywheels, governors, gyroscopes. Sintered heavy alloy.

SYLVANIA WN-103
GTE Products Corp.
W 90, Ni 7, Cu 3.
Sintered: 110,000 TS; 95,000 YS; 8 El; 20-25 Rock C. For counterweights, radioactive shielding, electrical contacts, gyros, flywheels, governors. Sintered heavy metal.

SYLVANITE F8
National Supply Co.
C 1.25, Cr 4, W 8, V 0.5, bal Fe.
For woodworking knives, scrapers, woodworking tools. High speed steel, good red-hardness.

SYMPHONY
J.F. Jelenko & Co.
Precious metal. Au 2, Pd 27, Ag 61.
Quenched: 85,000 psi TS; 35,000 psi YS; 10 El; 150 Brin.
Hardened: 90,000 psi TS; 65,000 psi YS; 10 El; 155 Brin.
Type III hard dental alloy. For inlays, crowns, and fixed bridgework.

SYRACUSE GENUINE BABBITT
United American Metals Corp.
Cu, Sb, Pb, bal Sn.
70,000 psi TS. For bearings to resist shock and high temperature; Babbitt metal; shock resistant.

T 18
Bergische Stahl Industrie
C 1-1.1, Si 0.15-0.3, Mn 1-1.2, Cr 0.7-1, bal Fe.
Cold work tool steel. W.-Nr. 1.2127.

T 4
Saarstahl AG
Now SAARSTAHL 1.1730.

T 5
Saarstahl AG
Now SAARSTAHL 1.1740.

T 6
Saarstahl AG
C 0.67, Si 0.25, Mn 0.7, bal Fe.
Carbon tool steel; for knives, hand saws, wood working tools. W.-Nr. 1744. *Obsolete*

T 7
Saarstahl AG
C 0.75, Si 0.25, Mn 0.7, bal Fe.
Carbon tool steel; for hand tools, planer blades, wood chisels, wood working tools. W.-Nr. 1750. *Obsolete*

T ALLOY "A"
CCS Braeburn Alloy Steel
C 0.33, Cr 3.5, W 9.6, V 0.5, bal Fe.
For tools, forming dies. Hot work steel. *Obsolete*

T E M
Now TRIPLE EXTRA M; T.E.R. now TRIPLE EXTRA; T.E.D. now TRIPLE EXTRA D.

T G S
Now CARPENTER T G S.

T H 60B
Rheinische Rohrenwerke Atkiengesellschaf
Cu, Ni.
Rolled: 79,650 TS; 52,625 YS; 20 El. For thick walled, seamless high pressure drums; good strength at high temperatures.

T H 61B
Rheinische Rohrenwerke Atkiengesellschaf
Cu, Ni.
Rolled: 85,340 TS; 55,470 YS; 23 El. For thick walled, seamless high pressure drums; good strength at high temperatures.

T H 62B
Rheinische Rohrenwerke Atkiengesellschaf
Cu, Ni.
Rolled: 99,560 TS; 61,160 YS; 20 El. For thick walled, seamless high pressure drums; good strength at high temperatures.

T H-30
Rheinische Rohrenwerke Atkiengesellschaf
C, bal Fe.
At 20°C: 61,000 TS; 42,500 YS; 20 El. At 600°C: 34,000 TS; 15,500 YS. For boiler and superheater tubes; good strength at high temperature; corrosion resistant.

T H-31
Rheinische Rohrenwerke Atkiengesellschaf
C, bal Fe.
At 20°C: 71,700 TS; 44,000 YS; 18 El. At 600°C: 35,000 TS; 23,000 YS. For boiler tubes, superheater tubes; maintains strength at high temperature.

T QUALITY
George Cook & Co., Ltd.
C 0.7-1.1, Mn 0.35, Si 0.35, bal Fe.
Water hardening tool steel. For masonry and wood working tools, files, blacksmith tools. AISI W1.

T-1 STEEL
United States Steel Corp.
C 0.17, Mn 0.8, Si 0.25, Ni 0.8, Cr 0.55, Mo 0.5, Cu 0.3, B, V, bal Fe.
Water quenched and tempered: 125 ksi TS; 118 ksi YS; 21 El. May be machined, torch cut, formed, welded. For pressure vessels, earth moving equipment, heavy vehicles, cranes, low temperature parts subject only to atmosphere. ASTM A514, A517.

T-1 STEEL 321
United States Steel Corp.
C 0.17, Mn 0.8, Si 0.25, Ni 0.8, Cr 0.55, Mo 0.5, B, V, Cu, bal Fe.
Water quenched and tempered: 321 Brin. Machine or torch cut, welded, mild forming. For heavy impact and abrasion; power shovel buckets, dump trucks, chutes. ASTM A678.

T-1 STEEL 340
United States Steel Corp.
C 0.17, Mn 0.8, Si 0.25, Ni 0.8, Cr 0.55, Mo 0.5, B, V, Cu, bal Fe.
Water quenched and tempered: 340 Brin. Machine or torch cut, welded, mild forming. For heavy impact and abrasion; power shovel buckets, dump trucks, chutes. ASTM A678.

T-1 STEEL 360
United States Steel Corp.
C 0.17, Mn 0.8, Si 0.25, Ni 0.8, Cr 0.55, Mo 0.5, B, V, Cu, bal Fe.
Water quenched and tempered: 360 Brin. Machine or torch cut, welded, mild forming. For heavy impact and abrasion; power shovel buckets, dump trucks, chutes. ASTM A678.

T-1 TYPE A STEEL
United States Steel Corp.
C 0.17, Mn 0.8, Si 0.25, Cr 0.55, Mo 0.2, B, V, bal Fe.
Water quenched and tempered: 125 ksi TS; 118 ksi YS; 21 El. May be machined, torch cut, formed, welded. For pressure vessels, earth moving equipment, heavy vehicles, cranes, low temperature parts subject only to atmosphere. ASTM A514, A517.

T-1 TYPE A STEEL 321
United States Steel Corp.
C 0.17, Mn 0.8, Si 0.25, Cr 0.55, Mo 0.2, B, V, bal Fe.
Water quenched and tempered: 321 Brin. Machine or torch cut, welded, mild forming. For heavy impact and abrasion; power shovel buckets, dump trucks, chutes. ASTM 678.

T-1 TYPE A STEEL 340
United States Steel Corp.
C 0.17, Mn 0.8, Si 0.25, Cr 0.55, Mo 0.2, B, V, bal Fe.
Water quenched and tempered: 340 Brin. Machine or torch cut, welded, mild forming. For heavy impact and abrasion; power shovel buckets, dump trucks, chutes. ASTM A678.

T-1 TYPE A STEEL 360
United States Steel Corp.
C 0.17, Mn 0.8, Si 0.25, Cr 0.55, Mo 0.2, B, V, bal Fe.
Water quenched and tempered: 360 Brin. Machine or torch cut, welded, mild forming. For heavy impact and abrasion; power shovel buckets, dump trucks, chutes. ASTM A678.

T-1 TYPE B STEEL
United States Steel Corp.
C 0.17, Mn 1.15, Si 0.25, Ni 0.5, Cr 0.55, Mo 0.25, B, V, bal Fe.
Water quenched and tempered: 125 ksi TS; 118 ksi YS. May be machined, torch cut, formed, welded. For pressure vessels, earth moving equipment, heavy vehicles, cranes, low temperature parts subject only to atmosphere. ASTM A514, A517.

T-1 TYPE B STEEL 321
United States Steel Corp.
C 0.17, Mn 1.15, Si 0.25, Ni 0.5, Cr 0.55, Mo 0.25, B, V, bal Fe.
Water quenched and tempered: 321 Brin. Machined, torch cut, welded, mild forming. For heavy impact and abrasion; power shovel buckets, dump trucks, chutes. ASTM A678.

T-1 TYPE B STEEL 340
United States Steel Corp.
C 0.17, Mn 1.15, Si 0.25, Ni 0.5, Cr 0.55, Mo 0.25, B, V, bal Fe.
Water quenched and tempered: 340 Brin. Machined, torch cut, welded, mild forming. For heavy impact and abrasion; power shovel buckets, dump trucks, chutes. ASTM A678.

T-1 TYPE B STEEL 360
United States Steel Corp.
C 0.17, Mn 1.15, Si 0.25, Ni 0.5, Cr 0.55, Mo 0.25, B, V, bal Fe.
Water quenched and tempered: 360 Brin. Machined, torch cut, welded, mild forming. For heavy impact and abrasion; power shovel buckets, dump trucks, chutes. ASTM A678.

T-1-ALLOY
National Bronze & Aluminum Foundry
Aluminum. Cu 4, Ti 0.2, bal Al.
Heat treated: 30,000-34,000 TS; 7-9 El. For light alloy parts; age-hardenable.

T-10-0-2
Techni-Cast Corp.
Copper. Cu 88, Sn 10, Zn 2.
Tin bronze. 45,000 psi TS; 22,000 psi YS; 25 El.

T-10-10-0
Techni-Cast Corp.
Copper. Cu 80, Sn 10, Pb 10.
High leaded tin bronze. 35,000 psi TS; 20,000 psi YS; 18 El.

T-10-2-0
Techni-Cast Corp.
Copper. Cu 88, Sn 10, Pb 2.
42,000 psi TS; 23,000 psi YS; 18 El.

T-11-0-0
Techni-Cast Corp.
Copper. Cu 88.8, Sn 11, P 0.2.
Tin bronze. 50,000 psi TS; 26,000 psi YS; 16 El.

T-111
Now WESTINGHOUSE T-111.

T-12
Teledyne Firth Sterling
Tool material. WC 72, Co 8, 20.0 TiC + TaC + NbC.
Combination of cobalt binder and alloy carbide for cutting metals yielding continuous chips producing craters on top rake surfaces. 91.0 Rock A.

T-12-0-0 + 1.5 NI
Techni-Cast Corp.
Copper. Cu 86.3, Sn 12, Ni 1.5, P 0.2.
Tin bronze. 52,000 psi TS; 30,000 psi YS; 16 El.

T-14
Teledyne Firth Sterling
Tool material. WC 68.5, Co 10.5, 21.0 TiC + TaC + NbC.
Combination of cobalt binder and alloy carbide for cutting metals yielding continuous chips producing craters on top rake surfaces. 91.1 Rock A.

T-14-0-1
Techni-Cast Corp.
Copper. Cu 85, Sn 14, Zn 1.
Tin bronze. 35,000 psi TS; 28,000 psi YS; 2 El.

T-16-0-0
Techni-Cast Corp.
Copper. Cu 83.8, Sn 16, P 0.2.
Tin bronze. 36,000 psi TS; 28,000 psi YS; 2 El.

T-17-4PH
Techni-Cast Corp.
Stainless steel. Cr 16, Si 0.75, Ni 4, Cu 3, Nb 0.25, bal Fe.
190,000 psi YS; 168,000 psi YS; 11 El; 35 RA; 393 Brin.

T-19-0-0
Techni-Cast Corp.
Copper. Cu 80.8, Sn 19, P 0.2.
Tin bronze. 36,000 psi TS; 30,000 psi YS; 1 El.

T-222
Now WESTINGHOUSE T-222.

T-25
Teledyne Firth Sterling
Tool material. WC 69.6, Co 5.5, 24.9 TiC + TaC + NbC.
Combination of cobalt binder and alloy carbide for cutting
metals yielding continuous chips producing craters on top
rake surfaces. 92.4 Rock A.

T-3
Atrax Cemented Carbide
Sintered carbide. 230,000 transverse strength; 13.2-13.5
g/cm^3 density; 90.0-91.0 RA. Industry code: C-5; ISO P-40.

T-35
Atrax Cemented Carbide
Sintered carbide. 220,000 transverse strength; 12.7-13.0
g/cm^3 density; 90.0-91.0 RA.

T-5-20-0
Techni-Cast Corp.
Copper. Cu 75, Sn 5, Pb 20.
High leaded tin bronze. 28,000 psi TS; 15,000 psi YS; 15 El.

T-5-25-0
Techni-Cast Corp.
Copper. Cu 70, Sn 5, Pb 25.
High leaded tin bronze. 27,000 psi TS; 13,000 psi YS; 12 El.

T-5-5-5
Techni-Cast Corp.
Copper. Cu 85, Sn 5, Pb 5, Zn 5.
35,000 psi TS; 20,000 psi YS; 25 El.

T-50
Atrax Cemented Carbide
Sintered carbide. 230,000 transverse strength; 12.6-12.8
g/cm^3 density; 91.5-92.2 RA. Industry code: C-5-6; ISO
P25,30.

T-51
Atrax Cemented Carbide
Sintered carbide. 230,000 transverse strength; 12.25-12.45
g/cm^3 density; 90.0-91.0 RA. Industry code: C-5.

T-56
Atrax Cemented Carbide
Sintered carbide. 280,000 transverse strength; 13.15-13.30
g/cm^3 density; 90.8-91.8 RA. ISO P-35.

T-6
Atrax Cemented Carbide
Sintered carbide. 175,000 transverse strength; 11.0-11.2
g/cm^3 density; 91.8-92.4 RA.

T-6-2-4
Techni-Cast Corp.
Copper. Cu 88.5, Sn 6, Pb 1.5, Zn 4.
40,000 psi TS; 18,000 psi YS; 30 El.

T-63
Atrax Cemented Carbide
Sintered carbide. 200,000 transverse strength; 11.25-11.45
g/cm^3 density; 91.7-92.3 RA. Industry code: C-6.

T-64
Atrax Cemented Carbide
Sintered carbide. 220,000 transverse strength; 12.25-12.45
g/cm^3 density; 91.2-92.2 RA. Industry code: C-6.

T-65
Atrax Cemented Carbide
Sintered carbide. 210,000 transverse strength; 12.75-13.0
g/cm^3 density; 91.7-92.3 RA. Industry code: C-7.

T-7-15-0
Techni-Cast Corp.
Copper. Cu 78, Sn 7, Pb 15.
High leaded tin bronze. 32,000 psi TS; 18,000 psi YS; 18 El.

T-7-7-3
Techni-Cast Corp.
Copper. Cu 83, Sn 7, Pb 7, Zn 3.
High leaded tin bronze. 35,000 psi TS; 18,000 psi YS; 20 El.

T-70-30
Techni-Cast Corp.
Copper. Cu 67, Fe 0.75, Si 0.35, Ni 30, Mn 0.75, Nb 1.15.
Nickel. As cast: 72,000 psi TS; 40,000 psi YS; 35 El; 135 Brin.

T-76
Atrax Cemented Carbide
Sintered carbide. 200,000 transverse strength; 12.27-12.4
g/cm^3 density; 92.4-93.2 RA. Industry code: C-7; ISO P-10,
20.

T-78
Atrax Cemented Carbide
Sintered carbide. 200,000 transverse strength; 12.9-13.1
g/cm^3 density; 93.0-94.0 RA. ISO P-05.

T-8
Atrax Cemented Carbide
Sintered carbide. 140,000 transverse strength; 5.8-6.0 g/cm^3
density; 92.5-93.5 RA. Industry code: C-8; ISO P-01.

T-8-0-4
Techni-Cast Corp.
Copper. Cu 88, Sn 8, Zn 4.
Tin bronze. 45,000 psi TS; 21,000 psi YS; 30 El.

T-8-1-4
Techni-Cast Corp.
Copper. Cu 87.25, Sn 8, Pb 0.75, Zn 4.
40,000 psi TS; 20,000 psi YS; 25 El.

T-8-8-0
Techni-Cast Corp.
Copper. Cu 84, Sn 8, Pb 8.
High leaded tin bronze. 33,000 psi TS; 17,000 psi YS; 20 El.

T-90-10
Techni-Cast Corp.
Copper. Cu 87, Fe 1.25, Si 0.2, Ni 10, Mn 0.75, Nb 0.8.
Nickel. As cast: 50,000 psi TS; 30,000 psi YS; 28 El.

T-ALLOY
CCS Braeburn Alloy Steel
C 0.38, Cr 3.5, W 10.5, V 0.4, bal Fe.
Hardened: 48-52 Rock C. For brass forging dies, hot forming
dies, hot punches, spreading punches, dummy blocks. Type
H22 hot work tool steel. *Obsolete*

T-ALLOY "N"
CCS Braeburn Alloy Steel
C 0.28, Mo 0.25, W 10.5, Cr 3, V 0.25, Ni 1.6, bal Fe.
For tools, dies; shock resistant. *Obsolete*

T-ALLOY B
CCS Braeburn Alloy Steel
C 0.5, Cr 3, W 15, V 0.5, bal Fe.
For dies, punching and forming dies. Hot work steel.
Obsolete

T-ALLOY C
CCS Braeburn Alloy Steel
C 0.25, Cr 4, W 15, V 0.5, bal Fe.
For punching and forming dies. Hot work steel. *Obsolete*

T-ALLOY TOOL
CCS Braeburn Alloy Steel
C 0.35, Cr 3.5, V 0.4, W 11, bal Fe.
For hot shear knives, hot forming dies; hot work steel.
Obsolete

T-C 20
Acme Aluminum Alloys Inc.
W 20, Co 80.
For dies, cutlery, and bearings. Corrosion resistant.

T-C 25
Acme Aluminum Alloys Inc.
W 25, Co 75.
For dies, cutlery, and bearings. Corrosion resistant.

T-C 30
Acme Aluminum Alloys Inc.
W 30, Co 70.
For dies, cutlery, and bearings. Corrosion resistant.

T-C 35
Acme Aluminum Alloys Inc.
W 35, Co 65.
For dies, cutlery, and bearings. Corrosion resistant.

T-CA15
Techni-Cast Corp.
Stainless steel. Cr 12, C 0.12, bal Fe.
115,000 psi TS; 100,000 psi YS; 22 El; 55 RA; 235 Brin.

T-CA40
Techni-Cast Corp.
Stainless steel. Cr 12, C 0.3, bal Fe.
130,000 psi TS; 105,000 psi YS; 15 El; 55 RA; 269 Brin.

T-CA6MN
Techni-Cast Corp.
Stainless steel. Cr 13, Ni 4, Mo 0.75, bal Fe.
120,000 psi TS; 95,000 psi YS; 24 El; 60 RA; 255 Brin.

T-CF3
Techni-Cast Corp.
Stainless steel. Cr 19, Ni 9, C 0-0.03, bal Fe.
82,000 psi TS; 40,000 psi YS; 55 El; 169 Brin.

T-CF3M
Techni-Cast Corp.
Stainless steel. Cr 19, Ni 10, Mo 2.5, C 0-0.03, bal Fe.
85,000 psi TS; 42,000 psi YS; 50 El; 176 Brin.

T-CF8
Techni-Cast Corp.
Stainless steel. Cr 19, Ni 9, bal Fe.
80,000 psi TS; 40,000 psi YS; 50 El; 165 Brin.

T-CF8C
Techni-Cast Corp.
Stainless steel. Cr 19, Ni 10, Nb 0.75, bal Fe.
78,000 psi TS; 38,000 psi YS; 39 El; 159 Brin.

T-CF8M
Techni-Cast Corp.
Stainless steel. Cr 19, Ni 10, Mo 2.5, bal Fe.
80,000 psi TS; 42,000 psi YS; 50 El; 165 Brin.

T-CG8M
Techni-Cast Corp.
Stainless steel. Cr 19, Ni 11, Mo 3.5, bal Fe.
83,000 psi TS; 44,000 psi YS; 45 El; 172 Brin.

T-CN7M
Techni-Cast Corp.
Stainless steel. Cr 20, Ni 29, Cu 3.5, Mo 2.5, bal Fe.
69,000 psi TS; 32,000 psi YS; 48 El; 144 Brin.

T-CR-CU
Techni-Cast Corp.
Copper. Cu 99, Cr 1.
Heat treated: 51,000 psi TS; 40,000 psi YS; 17 El; 105 Brin.

T-K ALLOY
Now CARPENTER T-K.

T-LOY 34

National Castings, Inc.

C 0.3-0.38, Mn 0.8, Si 0.4, Ni 0.6-0.9, Cr 0.4-0.6, Mo 0.2, bal Fe.

Heat treated: 90,000 TS; 60,000 YS; 20 El; 40 RA; 187 Brin. For shovel bottoms, gears, shafts, racks; oil hardened, shock resistant.

T-LOY 42

National Castings, Inc.

C 0.38-0.46, Mn 0.8, Ni 0.9-1.1, Cr 0.6-0.8, Mo 0.25-0.35, bal Fe.

Normalized: 100,000 TS; 75,000 YS; 12 El; 25 RA; 250 Brin. Oil hardened: 175,000 TS; 145,000 YS; 10 El; 25 RA; 250 Brin. For tractor treads, wear parts, dies; tough, oil hardened.

T.2

Acieries de Champagnole

C 0.35, Cr 2.3, W 3, V 0.2, Mo 0.3, Mn 0.3, Si 0.3, bal Fe. Oil hardening tool and die steel. For press and forging dies, shears, and rivet hammers. AFNOR: 05 WOD 00.10.

T.35

Acieries de Champagnole

C 1.03, Cr 0.7, W 3.7, V 0.3, Mo 0.4, Mn 0.55, Si 0.3, bal Fe. Oil hardening tool steel. For drills, boring tools, taps, and threading tools. AFNOR: 100 WC 35.03; AISI F2.

T.6

Acieries de Champagnole

C 0.33, Cr 3, W 10.5, V 0.4, Mo 0.4, Mn 0.3, Si 0.3, bal Fe. Air or oil hardening tool and die steel; for hot forging, thread rolling dies, and extrusion dies for brass and bronze. AFNOR: Z 30 WC 10.03; AISI H21/22.

T.A.W

Now FANSTEEL 61 METAL.

T.C. FERRY

Henry Wiggin & Co. Ltd.

Ni 45, Cu 55.

Annealed: 73,000 TS; 34,000 YS; 47 El; 77 RA; 155 Brin. For thermocouple wires and compensating leads. *Obsolete*

T.C.12

Acieries de Champagnole

C 0.33, Cr 12, W 12, V 1, Mn 0.3, Si 0.5, bal Fe. Air or oil hardening tool and die steel. For extrusion dies, hot forging and threading dies and molds for brass and bronze castings. AFNOR: Z 35 WCV 12.12; AISI H23.

T.C.610

Acieries de Champagnole

C 0.2, Cr 10, W 6.5, Mo 2, Mn 0.3, Si 0.4, Co 10, bal Fe. Air or oil hardening tool and die steel. For hot extrusion dies; withstands hot contact for longer time. AFNOR: Z 20 CKWD 10.10.07.02.

T.E.D

Acieries de Champagnole

C 1.3, Cr 4.3, W 12, Mo 0.9, V 4, bal Fe. Air or oil hardens to 63-66 Rock C. For slotting tools, gear cutters, form cutters, and punches. AFNOR: Z 130 WVD 12.04.01. Germany: S 12.1.4 (EV 4) W.Nr. 1.3302.

T.E.D.C

Acieries de Champagnole

C 0.8, Cr 4.2, W 12.5, Mo 1.8, V 1.8, Co 1.2, bal Fe. Air or oil hardens to 63-66 Rock C. For lathe tools, twist drills, reamers, milling cutters, and broaches. AFNOR: Z 80 WDVK 12.02.02.

T.E.M

Acieries de Champagnole

C 1, Cr 4.2, W 6.5, Mo 5, V 1.9, bal Fe. Air or oil hardens to 63-66 Rock C. For twist drills, reamers, lathe cutters, and broaches. AFNOR: Z 100 WDV 06.05.02. AISI M2 (high carbon) high speed steel.

T.E.M.02.C

Acieries de Champagnole

C 1, Cr 4.2, W 6.5, Mo 5, V 1.9, Co 2, bal Fe. Air or oil hardens to 63-67 Rock C. For drills, reamers, lathe cutting tools, taps, and gear cutters. AFNOR: Z 100 WDVK 06.05.02.02.

T.E.M.02.V

Acieries de Champagnole

C 1.08, Cr 4.2, W 6.3, Mo 5.3, V 2.4, bal Fe. Air or oil hardens to 62-67 Rock C. For drills, taps, threading form tools, lathe tools, hobs, and broaches. AFNOR: Z 110 WDV 06.05.02. AISI M3(1).

T.E.M.03.V

Acieries de Champagnole

C 1.2, Cr 4, W 6.3, Mo 5, V 3, bal Fe. Air or oil hardens to 62-66.5 Rock C. For drills, reamers, gear cutters, broaches, lathe tools, and thread chasers. AFNOR: Z 120 WDV 06.05.03. AISI M 3(2). Germany: S 6.5.3(EMo5V3); W.-Nr. 1.0044.

T.E.M.05.C

Acieries de Champagnole

C 1, Cr 4.3, W 7.7, Mo 5, V 2.3, Co 5, bal Fe. Air or oil hardens to 63-67 Rock C. For drills, taps, reamers, and turning tools for machining hard metals. AFNOR: Z 100 WDKV 07.05.05.02. AISI M41.

T.E.M.08.C

Acieries de Champagnole

C 1.1, Cr 4, W 5, Mo 6.7, V 2, Co 8, bal Fe. Air or oil hardens to 63-69 Rock C. For drills and lathe tools machining abrasive materials, and finishing cuts. AFNOR: Z 110 KDWV 08.07.05.02.

T.E.M.60

Acieries de Champagnole

C 0.6, Cr 4, W 6.5, Mo 5, V 1.9, Mn 0.3, Si 0.3, bal Fe. Air or oil hardening tool and die steel. For cold extrusion dies and hot forming dies. AFNOR: Z 60 WDCV.06.05.04.02; AISI H42.

T.E.M.92

Acieries de Champagnole

C 0.92, Cr 4.2, W 6.5, Mo 5, V 1.9, bal Fe. Air or oil hardens to 64-66 Rock C. For drills, broaches, reamers, lathe, and threading tools. Good wear resistance. AFNOR Z 92 WDV 06.05.02. AISI: M2. Germany: S 6.5.2 (DMo5); W. Nr. 1.3343.

T.E.M.C

Acieries de Champagnole

C 0.85, Cr 4.3, W 6.5, Mo 5, V 1.9, Co 5, bal Fe. Air or oil hardens to 62-67 Rock C. For drills and lathe tools for special work requiring high temperature, high speed special alloys. AFNOR: Z 85 WDKV 06.05.05.02. AISI M 35. Germany: S 6.5.2.5 (EMo 5 Co 5); W.-Nr. 1.3243.

T.E.M.C V

Acieries de Champagnole

C 1.57, Cr 4.3, W 6.7, Mo 5, V 4.8, Co 5, bal Fe. Air or oil hardens to 62-68 Rock C. For high speed finish cutting on hard and tough materials such as stainless steels. AFNOR: Z 150 WDKV 06.05.05.05. AISI-M15.

T.E.M.C V.11

Acieries de Champagnole

C 1.65, Cr 4.3, W 6.8, Mo 5.3, V 4.8, Co 11, bal Fe. Air or oil hardens to 62-69 Rock C. For tool bits for finishing and high speed machining of stainless steels and non-metallics. AFNOR: Z 160 KWDV 10.06.05.05.

T.E.M.P

Acieries de Champagnole

C 1.1, Cr 4.3, W 7, Mo 4, V 2, Co 5, bal Fe. Air or oil hardens to 63-67 Rock C. For reamers, broaches, lathe tools, and thread chasers. AFNOR: Z 110 WDKV 07.04.05.02. AISI: M 41. Germany: S 7.4.2.5; W.Nr. 1.3246.

T.E.M.V

Acieries de Champagnole

C 1.28, Cr 4.3, W 5.6, Mo 4.6, V 4, bal Fe. Air or oil hardens to 63-66.5 Rock C. For drills, reamers, and finish cutting lathe tools; by special temper for punches, files, and chisels. AFNOR: Z 130 WDV 06.05.04. AISI M4.

T.E.M.V.11

Acieries de Champagnole

C 1.25, Cr 4.3, W 7.5, Mo 6.3, V 3, Co 11, bal Fe. Air or oil hardens to 62-69 Rock C. For form tools, lathe tools, and climb milling cutters for machining hard and tough steels. AFNOR: Z 125 KWDV 11.08.06.03. Germany: S 10.4.3.10 (EW9Co10); W.Nr. 1.3207.

T.P.A.

TRW Inc.

C 0.45, Ni 13-15, Cr 13-15, Mo 0.5, W 1.75-3, bal Fe. At 70 F: 120,600 TS; 84,000 YS; 30 El; 44 RA; 215 Brin. At 1600 F: 23,000 TS; 20 El; 27 RA. For valves, valve seats, exhaust valves, heat resistant, austenitic. *Obsolete*

T.R.I.

Tin Research Institute

Sn 21, Cu 1.8, Si 0.15, Fe 0.13, bal Al. Cast: 15,500 TS; 4.8 El; 43 Brin. For railroad bearings.

T.R.S.

Allied Steel & Tractor Products Inc.

C, Cr, Mo, W, bal Fe. Rolled: 134,000 TS; 16 El; 37 RA; 269-550 Brin. For tools, fixtures; general purpose tools, shock resistant.

T.S.B.

Russian manufacture

C 0.7, W 18, Cr 4, V 1, bal Fe. For cutting tools, reamers; high speed steel.

T.S.R

Acieries de Champagnole

C 1.05, Cr 0.85, W 0.9, Mo 0.15, Mn 0.95, Si 0.25, bal Fe. Oil (or salt bath) hardening tool steel. For drills, cutting tools, forming tools, and gages. AFNOR: 100 MCW 04; AISI O1.

T.S.S. 3 ALLOY

Karl Schmidt Co.

Aluminum. Mg 2.5, Ti 0.3, Si 0.7, bal Al. Heat treated: 33,000 TS; 20,600 YS; 4 El; 5 RA; 75-90 Brin. For furniture, interior light fixtures, wire, castings; resists sea water corrosion.

T.S.S. 5 ALLOY

Karl Schmidt Co.

Aluminum. Mg 5, Ti 0.3, bal Al. Sand cast: 27,000-33,000 TS; 14,000 YS; 5.5-12 El; 3-15 RA; 65 Brin. For furniture, interior light fixtures, wire, castings; resists sea water corrosion.

T.S.S. 8 ALLOY

Karl Schmidt Co.

Aluminum. Mg 7.5, Ti 0.3, bal Al. Heat treated: 30,000-37,000 TS; 20,000 YS; 4.5-8 El; 2-20 RA; 80 Brin. For furniture, interior light fixtures, wire, castings; resists sea water corrosion.

T.T.Q. HIGH SPEED

Jessop Steel Co.

C 0.73, W 22, Cr 5, Co 11, V 1.6, bal Fe. For lathe and planer tools, reamers, hobs, broaches, taps; high speed steel, oil hardened. *Obsolete*

T.V.K. 45

Acieries de Champagnole

C 0.4, Cr 4.5, W 4.5, V 2.2, Mo 0.4, Mn 0.3, Si 0.3, bal Fe. Air or oil hardening tool and die steel. For hot forging and thread rolling copper, brass and bronze, hot punches, and molds for brass. AFNOR: Z 35 WKCV 05.05.04.02; AISI H 19.

T05, 6.4 MIN
Keystone Carbon Co.
6.4 g/cm^3 density min; 55,000 psi TS. For corrosion resistant structural parts. AISI Type 304 stainless steel. ASTM B525-83A, Grade 1.

T08, 6.4 MIN
Keystone Carbon Co.
6.4 g/cm^3 density min; 65,000 psi TS; 90 Rock B. For corrosion resistant structural parts that require high hardness. AISI Type 410 stainless steel.

T100
Acciaierie Valbruna s.p.a.
C 0-0.2, Ni 3.5, Si 0-0.35, Mn 0-0.9, P 0.035, S 0.04, bal Fe.
Steel, usually for forgings requiring notch toughness; ASTM A350-LF3; W. Nr. 1.5637.

T150
Acciaierie Valbruna s.p.a.
C 0.4, Ni 3.5, bal Fe.
Alloy steel for low temperature service. Similar to ASTM A320 L9. *Obsolete*

T15K6
Russian manufacture
WC, TiC, Co.
For cutting tools; sintered carbide, abrasion resistant.

T200
Acciaierie Valbruna s.p.a.
C 0-0.1, Ni 9, Si 0-0.35, Mn 0.55, P 0.035, S 0.035, bal Fe.
Nickel steel for low temperature service. ASTM A353; W. Nr. 1.5662.

T3
Acieries de Champagnole
C 1.3, Cr 4.5, bal Fe.
For tools and dies. Air or oil hardened.

T45
Acciaierie Valbruna s.p.a.
C 0.19, Mn 1.05, Si 0-0.3, P 0.035, S 0.035, bal Fe.
Steel, usually for forgings requiring notch toughness. ASTM A350 LF2; W. Nr. 1.0482.

T50
Acciaierie Valbruna s.p.a.
C 0-0.18, Mn 0-0.7, Ni 2.3, bal Fe.
Nickel low carbon steel. 18 Ni9. *Obsolete*

T51
Now HEPPENSTALL T51.

T5K12V
Russian manufacture
Co 12, 5.0 TiC, 83.0 WC.
For hard cutting tools, wear resistant dies; sintered carbide, wear resistant.

T60
Acciaierie Valbruna s.p.a.
C 0-0.3, Si 0.3, Mn 0-1.35, P 0-0.035, S 0-0.04, Ni 1.5, bal Fe.
Tough at subzero. ASTM A350-LF5; W. Nr. 1.5622.

TA QUALITY
George Cook & Co., Ltd.
C 1.15, Mn 0.35, Si 0.2, bal Fe.
For punches, dies, drill jigs, cutting tools; water hardened.
Obsolete

TA-1
Techni-Cast Corp.
Aluminum. Al 99.98, bal others.

TA-10 W
Now FANSTEEL 60 METAL.

TA-12.5 W
Fansteel Metals
W 12.5, bal Ta.
At RT: 102,000 TS; 85,000 YS. At 4900 F: 1,900 TS; 1,400 YS. For high temperature applications. Good high temperature strength and cryogenic ductility. *Obsolete*

TA-135
National Castings, Inc.
C, alloy, bal Fe.
For gear sectors, castings.

TA-195
Techni-Cast Corp.
Aluminum. Al 94.5, Si 1, Cu 4.5.
T4 temper: 33,000 psi TS; 17,000 psi YS; 9 El; 70 Brin. T6 temper: 38,000 psi TS; 25,000 psi YS; 6 El; 85 Brin.

TA-214
Techni-Cast Corp.
Aluminum. Al 96, Si 4.
F temper: 26,000 psi TS; 14,000 psi YS; 10 El; 50 Brin.

TA-319
Techni-Cast Corp.
Aluminum. Al 90.5, Si 6, Cu 3.5.
F temper: 34,000 psi TS; 19,000 psi YS; 2.5 El; 85 Brin. T6 temper: 40,000 psi TS; 27,000 psi YS; 3.5 El; 95 Brin.

TA-354
Techni-Cast Corp.
Aluminum. Al 88.7, Si 9, Cu 1.8, Mg 0.5.
T61 temper: 55,000 psi TS; 41,000 psi YS; 6 El; 105 Brin.

TA-355
Techni-Cast Corp.
Aluminum. Al 92.25, Si 6, Cu 1.25, Mg 0.5.
T6 temper: 42,000 psi TS; 27,000 psi YS; 4 El; 100 Brin.

TA-356
Techni-Cast Corp.
Aluminum. Al 92.65, Si 7, Mg 0.35.
T6 temper: 38,000 psi TS; 30,000 psi YS; 5 El; 95 Brin.

TA-43
Techni-Cast Corp.
Aluminum. Al 95, Si 5.
F temper: 23,000 psi TS; 9,000 psi YS; 10 El; 50 Brin.

TA-850
Techni-Cast Corp.
Aluminum. Al 89.5, Ni 1.25, Cu 2, Mg 0.75, Sn 6.5.
T5 temper: 32,000 psi TS; 23,000 psi YS; 5 El; 70 Brin.

TA-8W-2 HF
Fansteel Metals
W 8, Hf 2, bal Ta.
At RT: 112,000 TS; 96,000 YS. At 24000 F: 40,000 TS; 33,000 YS. At 3500 F: 11,000 TS; 11,000 YS. For high temperature applications. Good high temperature strength and cryogenic ductility. *Obsolete*

TA-NI (TANTALUM-NICKEL)
Fansteel Metals
Reactive/Refractory. Ni plus Ta.
115,000 TS. For electronic tube parts, chemical applications.
Obsolete

TA-TENZ
Techni-Cast Corp.
Aluminum. Al 91.4, Cu 0.7, Mg 0.4, Zn 7.5.
T5 temper: 40,000 psi TS; 28,000 psi YS; 6 El; 75 Brin.

TA-W (TANTALUM-TUNGSTEN)
Fansteel Metals
Ta plus W.
For electronic tube parts, rectifier electrodes, chemical applications; harder and stiffer than pure tantalum. *Obsolete*

TAB-1
Techni-Cast Corp.
Aluminum. Cu 87.5, Fe 3.5, Al 9.
Bronze. As cast: 75,000 psi TS; 30,000 psi YS; 35 El; 147 Brin.

TAB-2
Techni-Cast Corp.
Aluminum. Cu 88.25, Fe 1.25, Al 10.5.
Bronze. As cast: 77,000 psi TS; 30,000 psi YS; 25 El; 156 Brin. Heat treated: 85,000 psi TS; 42,000 psi YS; 15 El; 179 Brin.

TAB-3
Techni-Cast Corp.
Aluminum. Cu 86, Fe 3.5, Al 10.5.
Bronze. As cast: 85,000 psi TS; 36,000 psi YS; 18 El; 179 Brin.

TAB-4
Techni-Cast Corp.
Aluminum. Cu 85.75, Fe 3.5, Al 10.75.
Bronze. Heat treated: 100,000 psi TS; 50,000 psi YS; 10 El; 210 Brin.

TAB-5
Techni-Cast Corp.
Aluminum. Cu 81, Fe 3.25, Al 10.75, Ni 5.
Bronze. As cast: 104,000 psi TS; 55,000 psi YS; 15 El; 217 Brin. Heat treated: 112,000 psi TS; 72,000 psi YS; 10 El; 235 Brin.

TAB-5 N1
Techni-Cast Corp.
Aluminum. Cu 81, Fe 4, Al 9, Ni 5, Mn 1.
Bronze. As cast: 95,000 psi TS; 50,000 psi YS; 18 El; 201 Brin.

TAB-6
Techni-Cast Corp.
Aluminum. Cu 91, Al 7, Si 2.
Bronze. As cast: 75,000 psi TS; 35,000 psi YS; 18 El; 156 Brin.

TAB-7
Techni-Cast Corp.
Aluminum. Cu 80.25, Fe 3.25, Al 11.25, Ni 5, Mn 0.25.
Bronze. Heat treated: 145,000 psi TS; 120,000 psi YS; 3 El; 286 Brin.

TAB-8
Techni-Cast Corp.
Aluminum. Cu 78.25, Fe 4, Al 10.75, Ni 5, Mn 2.
Bronze. Heat treated: 135,000 psi TS; 100,000 psi YS; 7 El; 277 Brin.

TAB-9
Techni-Cast Corp.
Aluminum. Cu 75, Fe 3, Al 8, Ni 2, Mn 12.
As cast: 100,000 psi TS; 46,000 psi YS; 25 El; 217 Brin.

TACK DIE
Colt Industries
C 2, Cr 2, W 12, bal Fe.
For tools, dies; oil hardened. *Obsolete*

TACK DIE
Disston Inc.
C 2.05, Mn 1.05, Si 0.85, Cr 1.25, W 9.5, bal Fe.
For tack dies; oil hardening. *Obsolete*

TACONY ALLOY DIE
Tacony Steel Co.
C, alloy, bal Fe.
For dies. *Obsolete*

TACONY ALLOY STEEL
Tacony Steel Co.
C, alloy, bal Fe.
For tools. *Obsolete*

TACONY ALLOY TRIMMER

Tacony Steel Co.
C, alloy, bal Fe.
For tools, dies. *Obsolete*

TACONY BIT

Tacony Steel Co.
C 0.6-0.8, bal Fe.
For bits, tools. *Obsolete*

TACONY CARBON

Tacony Steel Co.
C 0.7-1.2, bal Fe.
For tools. *Obsolete*

TACONY CHISEL AND DRILL

Tacony Steel Co.
C 0.7-0.9, bal Fe.
For chisels, drills. *Obsolete*

TACONY GOL CHISEL

Tacony Steel Co.
C, alloy, bal Fe.
For chisels. *Obsolete*

TACONY JAR

Tacony Steel Co.
C 0.8-1.2, bal Fe.
For tools. *Obsolete*

TACONY REGULAR DIE

Tacony Steel Co.
C, bal Fe.
For dies. *Obsolete*

TACONY REGULAR TRIMMER

Tacony Steel Co.
C 0.7-1, bal Fe.
For tools, dies. *Obsolete*

TADANAC ALLOYS

Cominco Metals
Lead.
See COMINCO ALLOYS.

TAFA BONDARC 75B

TAFA Inc.
Ni 95, Al 5.
Nickel-aluminum alloy for use in arc spray systems. MP 3000°F. 20-25 Rock C. Bond tensile strength: 9110 psi clean surface; 9746 psi blasted surface. Coating tensile strength: 18,000 psi.

TAFALOY 200

TAFA Inc.
Zinc wire for mold applications. MP 770°F. 13-24 Rock H.

TAFALOY 204M

TAFA Inc.
Kirksite type wire for arc spraying. MP 792°F. 52 Rock H.

TAFALOY 205

TAFA Inc.
Tin-zinc metal sprayed tooling for molds used in the manufacture of polyurethane, PVC, ABS, fiberglass, and polycarbonate products. MP 450°F. 15-20 Rock H. Bond strength: 2895 psi blasted steel; coating tensile strength: 13,000 psi; shrink: 0.001 in./in.

TAFALOY 45CT

TAFA Inc.
Ti 0-1, Cr 45, bal Ni.
Coatings resistant to corrosive vanadium and sulfur gases in boiler atmospheres. MP 2700°F. Macrohardness: 313 Knoop (100). Bond strength: 6900 psi; 24,000 psi TS.

TAIFUN

Dorrenberg Edelstahl GmbH
C 1.05, Mn 0.9, Cr 1, W 1.15, bal Fe.
For cold work tools, heading dies; oil hardened; tough.

TAIL SHAFT BRONZE

Janney Cylinder Co.
Cu 87, Sn 6, Pb 3, Zn 4.
For bearings, pump liners; centrifugal casting.

TALABOT 10 NC 6

Societe Nouvelle du Saut-du-Tarn
C 0.07-0.12, Mn 0.6-0.9, Ni 1.2-1.6, Cr 0.85-1.15, bal Fe.
Low alloy low carbon structural steel. AFNOR 10 NC 6; Werkstoff Nr. 1.5713.

TALABOT 12 CD 4

Societe Nouvelle du Saut-du-Tarn
C 0.08-0.14, Mn 0.5-0.8, Cr 0.85-1.15, Mo 0.15-0.3, bal Fe.
Low alloy low carbon structural steel. AFNOR 12 CD 4;
Werkstoff Nr. 1.7262.

TALABOT 14 NC 11

Societe Nouvelle du Saut-du-Tarn
C 0.11-0.17, Mn 0.35-0.6, Ni 2.5-3, Cr 0.6-0.9, bal Fe.
Alloy steel for carburizing. AFNOR 14 NC 11; Werkstoff Nr. 1.5732.

TALABOT 15 ND 8

Societe Nouvelle du Saut-du-Tarn
C 0.13-0.18, Mn 0.2-0.5, Ni 1.8-2.3, Mo 0.15-0.3, bal Fe.
Alloy structural steel; also for carburizing. AFNOR 15 ND 8.

TALABOT 16 MC 5

Societe Nouvelle du Saut-du-Tarn
C 0.14-0.19, Mn 1-1.3, Cr 0.8-1.1, bal Fe.
Low alloy low carbon structural steel. AFNOR 16 MC 5;
Werkstoff Nr. 1.7131.

TALABOT 18 CD 4

Societe Nouvelle du Saut-du-Tarn
C 0.18-0.22, Mn 0.6-0.9, Cr 0.85-1.15, Mo 0.15-0.3, bal Fe.
AFNOR 18 CD 4; Werkstoff Nr. 1.7264.

TALABOT 20 MC 5

Societe Nouvelle du Saut-du-Tarn
C 0.17-0.22, Mn 1.1-1.4, Cr 1-1.3, bal Fe.
AFNOR 20 MC 5; Werkstoff Nr. 1.7147.

TALABOT 25 CD 4

Societe Nouvelle du Saut-du-Tarn
C 0.22-0.29, Mn 0.6-0.9, Cr 0.85-1.15, Mo 0.15-0.3, bal Fe.
AFNOR 25 CD 4; Werkstoff Nr. 1.7218.

TALABOT 30 NC 11

Societe Nouvelle du Saut-du-Tarn
C 0.27-0.34, Mn 0.35-0.6, Ni 2.5-3, Cr 0.6-0.9, bal Fe.
AFNOR 30 NC 11; Werkstoff Nr. 1.5736.

TALABOT 32 C 4

Societe Nouvelle du Saut-du-Tarn
C 0.3-0.35, Mn 0.6-0.9, Cr 0.85-1.15, bal Fe.
AFNOR 32 C 4; Werkstoff Nr. 1.7033.

TALABOT 35 CD 4

Societe Nouvelle du Saut-du-Tarn
C 0.3-0.37, Mn 0.6-0.9, Cr 0.85-1.15, Mo 0.15-0.3, bal Fe.
Structural steel. AFNOR 35 CD 4; Werkstoff Nr. 1.7220.

TALABOT 35 NC 15

Societe Nouvelle du Saut-du-Tarn
C 0.3-0.38, Mn 0.2-0.5, Ni 3.5-4, Cr 1.5-1.9, bal Fe.
Alloy structural steel. AFNOR 35 NC 15.

TALABOT 35 NCD 14

Societe Nouvelle du Saut-du-Tarn
C 0.3-0.4, Mn 0.2-0.5, Ni 3.2-3.7, Cr 1.2-1.6, Mo 0.2-0.4, bal Fe.
Alloy structural steel. AFNOR 35 NCD 14. Werkstoff Nr. 1.6746.

TALABOT 40 NC 17

Societe Nouvelle du Saut-du-Tarn
C 0.37-0.45, Mn 0.15-0.55, Ni 4-4.5, Cr 1.5-2, bal Fe.
Deep hardening alloy constructional steel. AFNOR 40 NC 17;
Werkstoff Nr. 1.5864.

TALABOT 45 C 4

Societe Nouvelle du Saut-du-Tarn
C 0.41-0.48, C 0.6-0.9, Cr 0.85-1.15, bal Fe.
Constructional steel. AFNOR 45 C 4; similar to AISI 5145.

TALABOT 50 CV 4

Societe Nouvelle du Saut-du-Tarn
C 0.47-0.55, Mn 0.7-1.1, Cr 0.85-1.15, V 0.1-0.2, bal Fe.
Chromium-vanadium structural steel. AFNOR 50 CV 4; AISI 6150

TALABOT A.C.T. NO. 1

Societe Nouvelle du Saut-du-Tarn
C 0.5, Cr 1.5, W 2.2, V 0.2, bal Fe.
For hot work dies and tools; hot work steel.

TALABOT A.C.T. NO. 2

Societe Nouvelle du Saut-du-Tarn
C 0.5, Cr 1.5, W 2.2, V 0.2, bal Fe.
For hot work dies and tools; hot work steel.

TALABOT F.C. 2

Societe Nouvelle du Saut-du-Tarn
C 2.25, Cr 13, bal Fe.
For tools, dies; non-deforming. *Obsolete*

TALABOT MNS

Societe Nouvelle du Saut-du-Tarn
C 0.5, Mn 0.7, Si 1.9, Mo 0.2, bal Fe.
For tools, dies, punches; tough.

TALABOT SPECIAL OM

Societe Nouvelle du Saut-du-Tarn
C 0.95, Mn 1.2, Cr 0.5, W 0.5, bal Fe.
For tools, dies, cutters; nondeforming.

TALIDE 101

Metal Carbides Corp.
W, C, Co.
Sintered: 840 Brin. For drawing dies, blast nozzles; tranverse strength 190,000; crushing strength 810,000; sintered carbides.

TALIDE C-75

Metal Carbides Corp.
WC.
Sintered: 85 Rock A. For cutting tools; sintered carbide.

TALIDE C-7525

Metal Carbides Corp.
W 66, Ta 5, Co 25, C 6.
Sintered: 210,000 TS; 650 Brin. For cold working tools, blanking dies; sintered carbide.

TALIDE C-80

Metal Carbides Corp.
WC.
Sintered: 87 Rock A. For cutting tools; sintered carbide.

TALIDE C-8020

Metal Carbides Corp.
C 5, W 75, Co 20.
Sintered: 190,000 TS; 680 Brin. For punching and shearing tools; sintered carbides.

TALIDE C-85

Metal Carbides Corp.
WC.
Sintered: 88 Rock A. For cutting tools; sintered carbide.

TALIDE C-8515

Metal Carbides Corp.
W 79.8, C 5.2, Co 15.
Sintered: 175,000 TS; 700 Brin. For impact tools; sintered carbides.

TALIDE C-88

Metal Carbides Corp.
Co 9, WC 91.
Sintered: 91 Rock A; 275,000 tranverse strength; 275,000 crushing strength. For extrusion and swaging dies, draw punches. Wear and abrasion resistant.

TALIDE C-89

Metal Carbides Corp.
W 85.4, C 5.5, Co 10.
Sintered: 150,000 TS; 90 Rock A. For draw dies, cutting tools; sintered carbide.

TALIDE C-907

Metal Carbides Corp.
For cutting tools on finish cutting operations, cold work dies for processing aluminum; sintered carbide tool material.

TALIDE C-91

Metal Carbides Corp.
W 86.4, C 5.6, Co 8.
Sintered: 140,000 TS; 91.5 Rock A. For draw dies, cutting tools; sintered carbide.

TALIDE C-95

Metal Carbides Corp.
W 91, C 6, Co 3.
Sintered: 810 Brin. 200,000 tranverse strength; 775,000 crushing strength. For precision boring and finishing cutters; sintered carbides.

TALIDE C-99

Metal Carbides Corp.
WC.
Sintered: 91 Rock A. For cutting tools; sintered carbide.

TALIDE C93

Metal Carbides Corp.
W 86, Ta 2.3, C 5.8, Co 6.
Sintered: 130,000 TS; 790 Brin. For wear plates, guides, dies; heat resistant to 2000°F.

TALIDE CT-20

Metal Carbides Corp.
For dies for cold extrusion of non-ferrous metals; sintered carbide tool material.

TALIDE CT-28

Metal Carbides Corp.
For cutting tools for flash trimming operations; sintered carbide tool material.

TALIDE S-88

Metal Carbides Corp.
W 75, Ti 14, Ta 5.6, C 6.3, Co 9.
93 Rock A. For cutting tools, cold forming dies; sintered carbides.

TALIDE S-880

Metal Carbides Corp.
For cutting tools for rough cutting operations on steel; sintered carbide tool material.

TALIDE S-90

Metal Carbides Corp.
W 73, Ti 8, Ta 2, C 7, Co 10.
91 Rock A. For cutting tools; sintered carbides.

TALIDE S-901

Metal Carbides Corp.
For cutting tools for finish cutting operations on steel; sintered carbide tool material.

TALIDE S-92

Metal Carbides Corp.
W 69.5, Ti 9.6, Ta 5.6, C 7.3, Co 8.
92 Rock A. For cutting tools; sintered carbides.

TALIDE S-94

Metal Carbides Corp.
W 65.7, Ti 12, Ta 7.5, C 7.8, Co 7.
Sintered: 810 Brin. For precision boring cutters; 180,000 tranverse strength; 750,000 crushing strength; sintered carbides.

TALLY-HO 100

British Steel Corp.
Low C, bal Fe.
Hardened: 89,000 TS; 61,000 YS; 31 El; 59 RA. For case-hardened parts, hot working dies; carburizing steel. *Obsolete*

TALLY-HO 200

British Steel Corp.
Low C, alloy, bal Fe.
Hardened: 191,000-198,000 TS; 165,000-179,000 YS; 15.5-16.5 El; 47-57 RA. For dies, inserts, hot working dies; carburizing steel. *Obsolete*

TALLY-HO 300

British Steel Corp.
C, high alloy, bal Fe.
Heat treated: 250,000-265,000 TS; 206,000-213,000 YS; 12.5-13.5 El; 34 RA; 400 Brin. For plastic industry, dies, hot working dies; oil and air-hardened. *Obsolete*

TALMI GOLD-1

Japanese manufacture
Cu 90, Zn 8.9, Au 0.9.
For cheap jewelry; Au welded on by rolling.

TALMI GOLD-2

Japanese manufacture
Cu 86, Zn 12, Sn 1.1, Fe 0.3.
For hardware, cheap jewelry; corrosion resistant.

TALON

Edgar Allen Balfour Ltd.
C 0.9, bal Fe.
For tools.

TAM

NL Industries
Ti 12-15, Al 18-22, Si 2-4, Cr 4.5.
For refining steel; grain refiner. *Obsolete*

TAM CUPRO-TITANIUM

NL Industries
Ti 25-30, Fe 0-0.2, Si, C, bal Cu.
Used in adding Ti to Cu alloys; hardener. *Obsolete*

TAM FOUNDRY FERRO TITANIUM

NL Industries
C 0-0.1, Si 20-24, Ti 28-32, Al 0.5-1.5, bal Fe.
For addition agent for alloy cast iron; inoculant. *Obsolete*

TAM LOW CARBON FERRO TITANIUM 25% GRADE

NL Industries
C 0-0.1, Si 3-4.5, Ti 26-29, Al 4.5-6.5, bal Fe.
For addition agent for alloy steel; inoculant. *Obsolete*

TAM LOW CARBON FERRO-TITANIUM 40% GRADE

NL Industries
Ti 38-42, Al 6-9, Si 3-5, C 0.1, bal Fe.
For source of Ti for alloy steels; Ti-additions. *Obsolete*

TAM MAGNESIUM-ZIRCONIUM K-50

NL Industries
Zr 50, Mg 50.
Addition agent to Mg alloys; for alloying. *Obsolete*

TAM MANGANESE-TITANIUM REGULAR GRADE

NL Industries
C 0-0.1, Ti 25-28, Al 8-10, Si 1.5-3.5, Fe 20, bal Mn.
For steel deoxidizer; grain refiner. *Obsolete*

TAM MANGANESE-TITANIUM SPECIAL GRADE

NL Industries
C 0-0.1, Ti 27-31, Al 8-10, Si 2-4, Fe 2-5, bal Mn.
For steel deoxidizer; grain refiner. *Obsolete*

TAM MED. CARBON FERRO-CARBON TITANIUM

NL Industries
Ti 17-21, C 3-5, 2.0-3.0 Si approx, 1.3 Al approx, bal Fe.
For steel metallurgical applications; for extra low C rimming steel. *Obsolete*

TAM METALLIC TITANIUM

NL Industries
Ti 98, Fe 0.2, Si 0.2, C 0.2, Al 0.2.
For making of special titanium alloys; Ti-addition. *Obsolete*

TAM MOLYBDENUM TITANIUM

NL Industries
Mo 48, Ti 16, Al 2, Si 7, Fe 27.
For ladle additions to cast iron; Mo-addition. *Obsolete*

TAM NICKEL TITANIUM

NL Industries
Ni 62, Ti 25, Al 8, Si 1, Fe 3.5.
For ladle additions to steel; Ni-addition. *Obsolete*

TAM ORIGINAL FERROCARBON TITANIUM

NL Industries
Ti 15-18, C 6-8, Si 2.65, Al 1.9, bal Fe.
For deoxidizer in steel making. *Obsolete*

TAM SILICO TITANIUM

NL Industries
Ti 44-47, Si 42-45, C 0.03-0.1, Al 1.5-3, bal Fe.
For cast iron inoculant, steel deoxidizer; low MP. *Obsolete*

TAMCO

NL Industries
C 0.2, Ti 0.2, bal Fe.
For machinery parts; corrosion resistant. *Obsolete*

TANDEM

Manufacturer not listed.
Pb 78, Sb 17, Sn 5.
For bearings; anti-friction.

TANDEM B.M

English manufacture
Zn 66, Sn 21.5, Cu 7, Pb 4.8, Al 0.4.
For bearings; will not resist heat or live steam.

TANK BRAND

Manufacturer not listed.
C 0.3-0.5, bal Fe.
For machinery parts; water hardening.

TANNE 3M

Vereinigte Edelstahlwerke
C 0.55, Si 0.1-0.4, Mn 0.5-0.7, bal Fe.
Heat treated: 155,000 TS; 112,000 YS; 15 El; 45 RA; 310 Brin. For crankshafts, tie rods, axles, bushings; water hardened. *Obsolete*

TANNENBAUM

Sudwestfalen Stahlwerke AG.
C 0.55, Si 0.1-0.4, Mn 0.3-0.7, bal Fe.
Annealed: 100,000 TS; 55,000 YS; 15 El; 22 RA; 180 Brin. For machine tool parts, bolts, gears; water hardened. *Obsolete*

TANNENBAUM

Stahlwerke Sudwestfalen
C 0.55, Si 0.1-0.4, Mn 0.3-0.7, bal Fe.
Annealed: 100,000 TS; 55,000 YS; 15 El; 22 RA; 180 Brin. For machine tool parts, bolts, gears; water hardened. *Obsolete*

TANNENBAUM

Vereinigte Edelstahlwerke
C 0.55, Si 0.3, Mn 0.6, bal Fe.
Heat treated: 160,000 TS; 115,000 YS; 13 El; 42 RA; 340 Brin. For gears, shafts, tools, mandrels; water hardened. *Obsolete*

TANTALOY

Now FANSTEEL 61 METAL.

TANTALOY

Fansteel Metals
TaC + Co.
For cutting tools; for corrosion resistant steels. *Obsolete*

TANTALUM

Atomergic Chemetals Corp.
Ta.
Purities, zone refined: 99.995%, premium grade 99.95%, 99.8% melting grade, 99.7% sintering grade, capacitor grades. Forms: powder, bar, sheet, foil, wire, pellet, rod, tubing, ingot, single crystals.

TANTIRON

Bethlehem Foundry & Machine Co.
Fe 84-87, Si 13.5, C 1, Mn 0.4.
For chemical apparatus and corrosion resisting machine parts, vessels, kettles, pipes; corrosion and acid resisting. *Obsolete*

TANTUNG

Fansteel Metals
Co 45-50, Cr 27-32, Mn 1-3, W 14-19, Ta 2-7, C 2-4, Fe 2-5, Cb.
For cutting tools, mixer blades, hot swaging dies; wear and corrosion resistant. *Obsolete*

TANTUNG 144

Fansteel Metals
Tool material. Co 43-48, Cr 25-30, W 16-21, C 2-4, Cb 3-8, Mn 1-3, Fe 0-1.
Cast cobalt base alloy. Chill cast: 64 Rock C; 275,000 transverse strength. For cutting tools, wear and corrosion resistant applications; good toughness and excellent abrasion resistance.

TANTUNG 148

Fansteel Metals
Co 45-50, Cr 27-32, W 14-19, C 2-4, Mn 1-3, Fe 2-5, 2-7% Ta or Cb.
Cast: 350 Brin. Heat treated: 460 Brin. For cutting tools, burner plates; shock, wear and corrosion resistant. *Obsolete*

TANTUNG 162A

Fansteel Metals
Co 45-50, Cr 27-32, W 14-19, C 2-4, Mn 1-3, Fe 2-5, 2-7% Ta or Cb.
Cast: 490 Brin. For cutting tools, knives; high red hardness and transverse rupture strength. *Obsolete*

TANTUNG 166A

Fansteel Metals
Co 45-50, Cr 27-32, W 14-19, C 2-4, Mn 1-3, Fe 2-5, 2-7% Ta or Cb.
Cast: 65,000 TS; 550 Brin. For cutting tools, mold plungers; high red hardness and transverse rupture strength. *Obsolete*

TANTUNG 171

Fansteel Metals
Co 45-50, Cr 27-32, W 14-19, C 2-4, Mn 1-3, Fe 2-5, 2-7% Ta or Cb.
Cast: 300 Brin. For cutting tools, cast parts; corrosion and shock resistant. *Obsolete*

TANTUNG G

Fansteel Metals
Tool material. Co 45-50, Cr 27-32, W 14-19, C 2-4, Cb 2-7, Mo 1-3, Fe 0-1.
Cast cobalt base alloy. Chill cast: 62 Rock C; 300,000 traverse strength. For cutting tools, wear and corrosion resistant applications; high toughness and good abrasion resistance.

TANTUNG G2

Fansteel Metals
Cr-W-Co, TaC-CbC.
For cutting tools; cast. *Obsolete*

TAP STEEL SPECIAL

Midvale-Heppenstall Co.
C, Cr, V, W, bal Fe.
For tools, taps, dies; oil hardened. *Obsolete*

TAPDIE

Columbia Tool Steel Co.
Tool material. C 1.25, W 1.4, Cr 0.45, V 0.2, bal Fe.
Oil or water hardened; cold work tool steel. For taps, threading tools, broaches, reamers. AISI O7.

TARNAC

Manufacturer not listed.
Cu, Mn, Ni.
For electrical instruments; v. "Manganin."

TARPON

Allegheny Ludlum Steel
C, alloy, bal Fe.
For tools, fixtures; water hardening. *Obsolete*

TATA BB

Tata Iron & Steel Co.
C 1, Cr 1.5, Mo 0.2, bal Fe.
For ball bearings, races, cold rolls; wear resistant. *Obsolete*

TATA CRDY

Tata Iron & Steel Co.
C 0.5, Cr 1, bal Fe.
For die blocks, rifle barrels; wear resistant. *Obsolete*

TATA CRW

Tata Iron & Steel Co.
C 0.4, Si 1, Cr 1, W 2, bal Fe.
For pneumatic tools, boiler snaps; tough, shock resistant. *Obsolete*

TATA DYH

Tata Iron & Steel Co.
C 1.35, W 4, Cr 1, bal Fe.
For cold working dies, plug gauges, brass cutting tools; oil hardening. *Obsolete*

TATA G-30-N

Tata Iron & Steel Co.
C 0.14, Ni 3.5, bal Fe.
Normalized: 100,000 TS; 18 El. For gears, pinions, shafts; case hardening. *Obsolete*

TATA G-35-CRN

Tata Iron & Steel Co.
C 0.12, Ni 3.5, Cr 1, bal Fe.
Rolled: 125,000 TS; 14 El. For gears, pinions, shafts; tough. *Obsolete*

TATA G-45-N

Tata Iron & Steel Co.
C 0.12, Ni 4.5, bal Fe.
Rolled: 90,000 TS; 50,000 YS; 20 El. For case hardened parts, gears, shafts; tough. *Obsolete*

TATA HC-12-CR

Tata Iron & Steel Co.
C 0.35, Cr 14, bal Fe.
For cutlery, corrosion resistant. *Obsolete*

TATA HCRH

Tata Iron & Steel Co.
C 2, Cr 12, bal Fe.
For wortle plates, blanking dies; wear resistant. *Obsolete*

TATA HCRM

Tata Iron & Steel Co.
C 1, Cr 12, bal Fe.
For shear blades, thread rollers, blanking dies; wear resistant. *Obsolete*

TATA HD

Tata Iron & Steel Co.
C 0.35, W 10, Cr 4, bal Fe.
For dies, mandrels, piercing and hot heading dies; hot work steel. *Obsolete*

TATA HMV

Tata Iron & Steel Co.
C 0.95, Mn 1.7, Mo 0.2, V 2, bal Fe.
For press tools, gages, taps, dies; non-deforming. *Obsolete*

TATA LA 55

Tata Iron & Steel Co.
C 0.14-0.18, Mn 1.3-1.6, Nb 0-0.1, V 0-0.18, Si 0-0.5, bal Fe.
Rolled: 42 kgf/mm^2 YS min; 55 kgf/mm^2 TS min; 20 El min. For structural members in bridges and buildings, transportation, and earth moving equipment.

TATA LA 60

Tata Iron & Steel Co.
C 0.17-0.21, Mn 1.4-1.6, Nb 0-0.1, V 0-0.18, Si 0-0.5, bal Fe.
Normalized: 45 kgf/mm^2 YS min; 60 kgf/mm^2 TS min; 20 El min. For structural members in bridges and buildings, transportation, and earth moving equipment.

TATA LC-12-CR

Tata Iron & Steel Co.
C 0.15, Cr 12, bal Fe.
Rolled: 78,000 TS; 25 El; 150-255 Brin. For surgical instruments, forceps; corrosion resistant. *Obsolete*

TATA LSB

Tata Iron & Steel Co.
C 0.6, Cr 1, Mo, V, bal Fe.
For shear blades; tough, wear resistant. *Obsolete*

TATA NCMA

Tata Iron & Steel Co.
C 0.3, Ni 4, Cr 1, Mo 0.2, bal Fe.
Hardened: 220,000 TS; 180,000 YS; 8 El; 444 Brin. For gears, gudgeon pins; air hardening. *Obsolete*

TATA NCMB

Tata Iron & Steel Co.
C 0.4, Ni 1.5, Cr 1, Mo 0.2, bal Fe.
For piston rods, gears, crankshafts; air hardening. *Obsolete*

TATA NCRDY

Tata Iron & Steel Co.
C 0.5, Ni 1.2, Cr 0.8, bal Fe.
For die blocks; shock resistant. *Obsolete*

TATA NDY

Tata Iron & Steel Co.
C 0.6, Ni 1.5, bal Fe.
Heat treated: 212-269 Brin. For die blocks, forging dies; oil hardening. *Obsolete*

TATA NIMN

Tata Iron & Steel Co.
C 1, Mn 12, Ni 4, bal Fe.
Normalized: 110,000 TS; 60,000 YS; 40 El; 60 RA; 150-277 Brin. For wearing plates, axle box liners; abrasion resistant austenitic. *Obsolete*

TATA P-15-N

Tata Iron & Steel Co.
C 0.12, Ni 2, bal Fe.
Normalized: 58,000 TS; 35,000 YS; 25 El; 60 RA. Boiler stays, case hardened parts, tough. *Obsolete*

TATA P-35-N
Tata Iron & Steel Co.
C 0.4, Ni 3.5, bal Fe.
Normalized: 110,000 TS; 85,000 YS; 20 El; 223-302 Brin. For crankshafts, gears, pistons; tough. *Obsolete*

TATA SPL. "A" GRADE
Tata Iron & Steel Co.
C 0-0.05, Mn 0-0.2, S 0-0.015, P 0-0.025, Si 0-0.05, Al killed, bal Fe.
Rolled and annealed. Permeability at 1.5-2 oersteds: 2000 min. Total induction at 200-250 oersteds: 20,000 Gauss min. Soft magnet yokes. Rolled plates, Galvanizer's pot.

TATA SS
Tata Iron & Steel Co.
C 0.15, Ni 8, Cr 18, bal Fe.
Rolled: 80,000 TS; 34,000 YS; 20 El. For chemical plant equipment; stainless. *Obsolete*

TATA T.C.S.10
Tata Iron & Steel Co.
C 0.95-1.05, bal Fe.
For tapes, reamers, chisels, mint dies; water hardening. *Obsolete*

TATA T.C.S.11
Tata Iron & Steel Co.
C 1.05-1.15, bal Fe.
For twist drills, reamers, punches, cutters; water hardening. *Obsolete*

TATA T.C.S.12
Tata Iron & Steel Co.
C 1.15-1.25, bal Fe.
For cutting tools, hack saw blades, lathe centers; water hardening. *Obsolete*

TATA T.C.S.7
Tata Iron & Steel Co.
C 0.65-0.75, bal Fe.
For hammers, spindles, pressing and stamping dies; water hardening. *Obsolete*

TATA T.C.S.8
Tata Iron & Steel Co.
C 0.75-0.85, bal Fe.
For hand chisels, shear blades, punches; water hardening. *Obsolete*

TATA T.C.S.9
Tata Iron & Steel Co.
C 0.85-0.95, bal Fe.
For cold heading dies, mandrels, chisels, cutters; water hardening. *Obsolete*

TATA T.H.S.1
Tata Iron & Steel Co.
C 0.65-0.75, W 14, Cr 4, V 1, bal Fe.
For taps, reamers, piercing and blanking dies, shears; high speed steel. *Obsolete*

TATA T.H.S.3.
Tata Iron & Steel Co.
C 0.7-0.8, W 18, Cr 4, V 0-2, bal Fe.
For form tools, broaches, drills, reamers; high speed steel. *Obsolete*

TATA T.H.S.5
Tata Iron & Steel Co.
C 0.7-0.8, W 20, Cr 4, V 2, Co 10, bal Fe.
For cutting tools, planing and shaping tools; high speed steel. *Obsolete*

TATA THS-4
Tata Iron & Steel Co.
C 0.7, W 18, C 4.5, V 1, Co 5, bal Fe.
For turning tools, boring and milling cutters; high speed steel. *Obsolete*

TATA THS2
Tata Iron & Steel Co.
C 0.7, W 18, Cr 4, V 1, bal Fe.
For drills, hobs, reamers, broaches, form cutters; high speed steel. *Obsolete*

TATA THS6
Tata Iron & Steel Co.
C 0.7, W 22, Cr 4, V 1, bal Fe.
For cutters; tools; high speed steel. *Obsolete*

TATA TS
Tata Iron & Steel Co.
C 1.2, W 1, bal Fe.
For taps, drills, reamers, threading dies; water hardening, wear resistant. *Obsolete*

TATA W6F
Tata Iron & Steel Co.
C 1.25, W 6, bal Fe.
For finishing tools, cutters; fast finishing steel. *Obsolete*

TATMO
Now ELECTRITE TATMO.

TATMO
Latrobe Steel Co.
C 0.83, Cr 3.75, W 1.5, V 1.05, Mo 8.5, bal Fe.
For reamers, lathe tools, drills, broaches, dies, hot work dies, cutting tools; high speed steel; wear resistant.

TATMO COBALT
Latrobe Steel Co.
C 0.9, W 1.5, Cr 3.75, V 2, Mo 8, Co 8.25, bal Fe.
Desegatized super high-speed steel for heavy duty machining of hard or heat treated material. Similar to AISI M-34.

TATMO V
Now ELECTRITE TATMO V.

TATMO V
Latrobe Steel Co.
C 1, W 1.75, Cr 3.75, V 2, Mo 8.75, bal Fe.
Hardened: 65-68 Rock C. For cutters, drills, reamers, lathe and planer tools, form tools, broaches. Type M-7 high speed steel. Good red-hardness and edge toughness.

TATMO V-N
Latrobe Steel Co.
C 1.02, W 1.75, Cr 3.75, V 1.85, Mo 8.5, Ni.
M7 grade high speed steel. For twist drills, reamers and taps.

TAURUS BRONZE MARK I
David Brown Foundries Co.
Cu 87.7, Sn 12, P 0.3.
Sand cast: 24,700-40,300 TS; 15,700-24,700 YS; 5-25 El. For high quality gear bronze; IZ 25-40.

TAURUS BRONZE MARK II
David Brown Foundries Co.
Cu 89.7, Sn 10, P 0.3.
Sand cast: 27,000-47,000 TS; 15,700-22,300 YS; 6-28 El. For bearings requiring some plasticity; IZ 27-44.

TAURUS BRONZE MARK III
David Brown Foundries Co.
Cu 86, Sn 14, trace P.
Sand cast: 29,000-33,200 TS; 24,700-29,000 YS; 3-1 El. For bearings for turn tables, movable bridges; IZ 13-17.

TAURUS BRONZE MARK IV
David Brown Foundries Co.
Cu 88.25, Sn 10.5, P 0.25, Ni 1.
Sand cast: 31,300-49,000 TS; 17,800-31,300 YS; 8-30 El. For sand castings requiring high strength, toughness and wearing qualities; IZ 40-60.

TAURUS BRONZE MARK IX
David Brown Foundries Co.
Cu 76.7, Sn 10, P 0.3, Ni 1, Pb 12.
Sand cast: 31,300-37,800 TS; 20,700-27,000 YS; 8-13 El. For heavy duty plastic bronze bearings; IZ 60-80.

TAURUS BRONZE MARK V
David Brown Foundries Co.
Cu 88, Sn 11.2, P 0.3, Ni 0.5.
Centrifugal: 40,300-49,000 TS; 27,000-36,000 YS; 5-15 El. For gear bronze; IZ 24-32.

TAURUS BRONZE MARK VI
David Brown Foundries Co.
Cu 88, Sn 10, Zn 2.
Sand cast: 31,300-44,700 TS; 17,800-22,300 YS; 10-25 El. For general bronze castings; similar to Admiralty Metal.

TAURUS BRONZE MARK VII
David Brown Foundries Co.
Cu 81.7, Sn 12, P 0.3, Ni 1, Pb 5.
Sand cast: 29,000-40,000 TS; 22,300-31,000 YS; 4-10 El. For cogging and rolling mill bearings, heavy duty bearings; IZ 40-50.

TAURUS BRONZE MARK VIII
David Brown Foundries Co.
Cu 80.7, Sn 10, P 0.3, Ni 1, Pb 8.
Sand cast: 31,300-37,800 TS; 22,300-27,000 YS; 6-12 El. For bearings, running against hard shafts; IZ 80-90.

TAURUS BRONZE MARK X
David Brown Foundries Co.
Cu 84, Sn 8, Pb 4, Zn 4.
Sand cast: 29,000-42,600 TS; 13,500-15,700 YS; 18-26 El. For dense general purpose castings, pump bodies; IZ 90-120.

TAURUS BRONZE MARK XI
David Brown Foundries Co.
Cu 85, Sn 5, Pb 5, Zn 5.
Sand cast: 29,000-40,300 TS; 13,500-15,700 YS; 20-30 El. For castings not requiring hardness, bushes and bearings; IZ 70-120.

TAURUS BRONZE MARK XII
David Brown Foundries Co.
Cu 88, Sn 6, Zn 6.
Sand cast: 36,000-42,600 TS; 15,700-17,800 YS; 30-60 El. For light, thin section, tough castings; IZ 120.

TAURUS MARK XIV
David Brown Foundries Co.
Cu 86, Sn 7, Zn 5, Pb 2.
Cast: 31,300-40,300 TS; 13,500-17,900 YS; 8-12 El. For general purpose castings; free-cutting.

TAURUS MARK XIX
David Brown Foundries Co.
Cu 88.5, Sn 10.5, Ni 0.75, P 0.25.
Centrifugal cast: 31,400-49,400 TS; 17,900-31,400 YS; 8-30 El. For centrifugal castings; wear resistant.

TAURUS MARK XVI
David Brown Foundries Co.
Cu 90.75, Sn 9, Ni 0.5, P 0.25.
Cast: 44,800-53,800 TS; 22,400-31,400 YS; 21-30 El. For general purpose castings; corrosion resistant.

TAURUS MARK XX
David Brown Foundries Co.
Cu 80.5, Ni 4, Mn 2, Fe 4, Al 9.5.
Centrifugal cast: 89,600-105,000 TS; 35,200-56,000 YS; 12-16 El. For highly stressed bearings and structures; fatigue resistant.

TAURUS MARK XXI
David Brown Foundries Co.
Cu 55, Mn 0.75, Fe 1, Al 1, bal Zn.
Sand cast: 67,000-78,500 TS; 25,900-31,400 YS; 20-28 El. For castings; corrosion resistant.

TAURUS MARK XXII
David Brown Foundries Co.
Cu 65, Mn 2, Fe 2, Al 5, bal Zn.
Sand cast: 85,000-94,200 TS; 40,300-49,300 YS; 15-18 El. For castings; corrosion resistant.

TAURUS MARK XXIII
David Brown Foundries Co.
Cu 60, Mn 2, Fe 2, Al 5.5, bal Zn.
Sand cast: 107,500-112,000 TS; 58,200-67,200 YS; 12-15 El. For castings; corrosion resistant.

TAURUS MARK XXIV
David Brown Foundries Co.
Cu 73, Sn 5, Ni 17, Pb 5.
Centrifugal cast: 48,800-56,500; 35,840-44,800 YS; 1-2 El. For bushings, castings; good machinability.

TAURUS MARK XXV
David Brown Foundries Co.
Cu 94.5, Sn 1.5, Ni 4.
Centrifugal cast: 38,200-49,300 TS; 20,300-26,800 YS; 25-42 El. For slip rings for electrical machines; high conductivity.

TAURUS MARK XXVI
David Brown Foundries Co.
Cu 44, Sn 11, Ni 38, Pb 7.
Centrifugal cast: 49,300-53,800 TS; 22,400-31,400 YS; 2-4 El. For valve seats; for severe service.

TAURUS MARK XXX
David Brown Foundries Co.
Cu 88, Sn 2, Zn 10.
Sand cast: 26,900-31,300 TS; 30-35 El. For brazing brass; high strength.

TAURUS MARK XXXI
David Brown Foundries Co.
Cu 59, Zn 40, Pb 1.
Sand cast: 40,300-44,800 TS; 15 El. For valves, roll and ball bearing castings; free-cutting.

TAWILCO
Taylor-Wilson Mfg. Co.
Cast Iron. C 3, alloy, bal Fe.
Cast: 50,000 TS; 45,000 YS; 0.1 El; 174-500 Brin. For machinery parts, gears, housings; heat and abrasion resistant.

TAYLOR 751
Robert Taylor & Co.
C 3.3, Si 2, Mn, alloy, bal Fe.
For brick mold die liners; cast iron, hard, wear resistant.

TAYLOR CRA
Robert Taylor & Co.
C 3.3, Si 1.8, Mn, alloy, bal Fe.
For castings for clay working industries; cast iron, hard, wear resistant.

TAYLOR NI-MO STEEL
S.G. Taylor Chain Co.
C 0.2, Ni 1.5, Mo 0.2, bal Fe.
For hoist and drag chains.

TAYLOR TM ALLOY
S.G. Taylor Chain Co.
C 0.18-0.23, Ni 0.4-0.7, Mo 0.15-0.25, Cr 0.4-0.6, bal Fe.
For chains; shock resistant.

TAYLOR-WHITE
English manufacture
W 8.5, Cr 3, C 0.75-1, bal Fe.
For tools, dies, cutters; semi high speed steel.

TAZ-8
Cannon-Muskegon Corp.
Ta 8, Cr 6, Al 6, W 4, Mo 4, V 2.5, B 0.004, Zr 1, C 0.125, bal Ni.
At 1900°F: 56,000 TS. For high speed aircraft nose cones, turbine buckets. High stress rupture strength. Cast alloy. Highest strength at 1900°F. Hardenable.

TAZ-8A
Cannon-Muskegon Corp.
Ta 8, Cr 6, Al 6, Mo 4, W 4, Cb 2.5, Zr 1, C 0.125, B 0.004, bal Ni.
As cast: 130,000 TS. Rolled: 240,000 TS. For low-stress components, stator vanes, turbine buckets, after-burner liners, transition ducts. High oxidation resistance and high temperature strength.

TAZ-8B
Cannon-Muskegon Corp.
Ta 8, Cr 6, Al 6, Mo 4, W 4, Cb 1.5, Zr 1, Co 5, C 0.125, B 0.004, bal Ni.
At room temperature: 150,000 TS; 3 El. For turbine and jet engine buckets. High creep and stress rupture strength. High temperature strength. Heat and oxidation resistant.

TB-20
Teledyne Firth Sterling
Tool material. WC 77.8, Co 8.5, 13.7 TiC + TaC + NbC.
Combination of cobalt binder and alloy carbide for cutting metals yielding continuous chips producing craters on top rake surfaces. 91.4 Rock A.

TBE-10
Techni-Cast Corp.
Copper. Cu 96.8, Co 2.6, Be 0.6.
Heat treated: 100,000 psi TS; 70,000 psi YS; 10 El; 195 Brin.

TBE-20
Techni-Cast Corp.
Copper. Cu 97.25, Si 0.25, Co 0.5, Be 2.
Heat treated: 185,000 psi TS; 170,000 psi YS; 3 El; 372 Brin.

TBS 600, TBS 1000
Now TIMKEN TBS 600, TBS 1000.

TC QUALITY
George Cook & Co., Ltd.
C 0.88, Mn 0.35, Si 0.2, bal Fe.
For cutting tools, punches, dies; water hardened. *Obsolete*

TCS (TERNE COATED STAINLESS STEEL)
Follansbee Steel Co.
Coating: 80 Pb, 20 Sn. (Terne) Base: Type 304 (18 Cr, 8 Ni) stainless steel. Soft: 80,000 TS; 30,000 YS. For roofing, gutters, flashings, architectural brackets. Easy to form and to solder.

TD NICKEL
Cannon-Muskegon Corp.
Ni 98, ThO$_2$.
At R.T.: 65,000 TS; 50,000 YS; 23 El; 90 RA. At 2400°F: 11,000 TS; 10,000 YS; 0.4 El; 5 RA. At 1200°F: 32,000 TS; 30,000 YS; 10 El; 45 RA. For aircraft gas turbines, afterburners, high temperature applications, fasteners. Dispersion hardening alloy. Good characteristics at 900-1200°F.

TD NICKEL
DuPont de Nemours & Co., E.I.
Ni 98, ThO$_2$.
At R.T.: 65,000 TS; 50,000 YS; 23 El; 90 RA. At 2400°F: 11,000 TS; 10,000 YS; 0.4 El; 5 RA. At 1200°F: 32,000 TS; 30,000 YS; 10 El; 45 RA. For aircraft gas turbines, afterburners, high temperature applications, fasteners. Dispersion hardening alloy. Good characteristics at 900-1200°F.

TD NICKEL
Fansteel
Ni 98, ThO$_2$.
At R.T.: 65,000 TS; 50,000 YS; 23 El; 90 RA. At 2400°F: 11,000 TS; 10,000 YS; 0.4 El; 5 RA. At 1200°F: 32,000 TS; 30,000 YS; 10 El; 45 RA. For aircraft gas turbines, afterburners, high temperature applications, fasteners. Dispersion hardening alloy. Good characteristics at 900-1200°F.

TD NICR
Cannon-Muskegon Corp.
Cr 20, 2 ThO$_2$, bal Ni.
At R.T.: 125,000 TS; 80,000 YS; 18 El. (in 1"). At 1500°F: 35,000 TS; 30,000 YS; 7 El (in 1"). For high temperature applications, jet engines, aerospace structural components, furnace hardware. Dispersion hardening alloy. High oxidation and sulfidation resistance.

TD NICR
DuPont de Nemours & Co., E.I.
Cr 20, 2 ThO$_2$, bal Ni.
At R.T.: 125,000 TS; 80,000 YS; 18 El. (in 1"). At 1500°F: 35,000 TS; 30,000 YS; 7 El (in 1"). For high temperature applications, jet engines, aerospace structural components, furnace hardware. Dispersion hardening alloy. High oxidation and sulfidation resistance.

TD NICR
Fansteel
Cr 20, 2 ThO$_2$, bal Ni.
At R.T.: 125,000 TS; 80,000 YS; 18 El. (in 1"). At 1500°F: 35,000 TS; 30,000 YS; 7 El (in 1"). For high temperature applications, jet engines, aerospace structural components, furnace hardware. Dispersion hardening alloy. High oxidation and sulfidation resistance.

TD12ZRE
Ugine Aciers
Mo 11.5, Zr 6, Sn 4.5, bal Ti.
Quenched and tempered: 1450 N/mm^2 TS; 1360 N/mm^2 YS; 4 El. Metastable beta type alloy, good properties in as-quenched condition; very strong after tempering.

TE
American manufacture
C 0.1, Mn 0.7, Cr 20, Ni 30, Mo 4, W 4, Ta 1.9, 0.15 N$_2$, bal Fe.
For jet engine parts; high heat and oxidation resistant.

TEA LEAD
English manufacture
Pb 98, Sn 2.
For lead foil for packing tea.

TEC
Now BRAZE 053 (TEC).

TEC
Handy & Harmon
Low melting.
Now BRAZE 053.

TEC 4
Handy & Harmon
Ag 4, Cu 1, Zn 16.6, bal Cd.
For brazing joints requiring higer shear strengths than lead-tin solders. MP 500-625 F. *Obsolete*

TEC Z
Now BRAZE 056 (TEC Z).

TECH-TRONIC 32
Techalloy Co. Inc.
C 0.15, Mn 11-14, Si 1, Cr 16.5-19, N 0.2-0.45, Ni 0.5-2.5, bal Fe.
Wire for springs, screens, cages, racks.

TECH-TRONIC 50
Techalloy Co. Inc.
C 0.06, Mn 4-6, Si 1, Cr 20.5-23.5, Ni 11.5-13.5, Mo 1.5-3, Cb 0.1-0.3, V 0.1-0.3, N 0.2-0.4, bal Fe.
Wire for AC power equipment, marine hardware, springs, fasteners.

TECHALLOY 294 (45-55)
Techalloy Co. Inc.
Ni 43-45, C 0.25, Mn 2, Fe 2.5, Si 1, bal Cu.
For resistance wiring operating up to 110°F (595°C).

TECHALLOY 302
Techalloy Co. Inc.
C 0-0.2, Cr 18, Ni 8, bal Fe.
Annealed: 90,000 TS; 40,000 YS; 50 El; 160 Brin. Cold drawn: 180,000 TS; 150,000 YS; 10 El; 215 Brin. For chemical plant equipment; stainless, austenitic.

TECHALLOY 304
Techalloy Co. Inc.
C 0.08, Cr 18, Ni 8, bal Fe.
Annealed: 85,000 TS; 35,000 YS; 60 El; 70 RA; 150 Brin. Cold drawn: 180,000 TS; 125,000 YS; 10 El; 330 Brin. For chemical and textile plant equipment, tanks, vessels; for welded structures, stainless, austenitic.

TECHALLOY 308
Techalloy Co. Inc.
C 0.08, Mn 2, Cr 19-21, Ni 10-12, bal Fe.
Welding wire with enriched Cr-Ni content.

TECHALLOY 308L
Techalloy Co. Inc.
C 0.03, Mn 2, Si 1, Cr 19-21, Ni 10-12, bal Fe.
Low carbon austenitic stainless steel welding wire.

TECHALLOY 309
Techalloy Co. Inc.
C 0.1, Cr 25, Ni 12, bal Fe.
Annealed: 90,000 TS; 40,000 YS; 50 El; 83 Rock B. For refinery and chemical plant equipment, furnace parts; austenitic, corrosion and heat resistant.

TECHALLOY 309L
Techalloy Co. Inc.
C 0.03, Mn 2, Si 1, Cr 19-21, Ni 10-12, bal Fe.
Low carbon austenitic stainless steel welding wire for welding 309 stainless.

TECHALLOY 310
Techalloy Co. Inc.
C 0-0.25, Cr 25, Ni 20, bal Fe.
Annealed: 100,000 TS; 45,000 YS; 50 El; 65 RA; 185 Brin. For furnace parts, heat treat boxes, furnace linings; austenitic, corrosion and heat resistant.

TECHALLOY 314
Techalloy Co. Inc.
C 0.2, Cr 25, Ni 20, Si 2-3, bal Fe.
Annealed: 100,000 TS; 50,000 YS; 45 El; 60 RA; 180 Brin. For furnace parts, heat treat boxes, fixtures; heat resistant, austenitic.

TECHALLOY 316
Techalloy Co. Inc.
C 0.1, Cr 18, Ni 12, Mo 2-3, bal Fe.
Annealed: 80,000 TS; 30,000 YS; 60 El; 80 RA; 135 Brin. Cold drawn: 150,000 TS; 135,000 YS; 6 El; 300 Brin. For chemical plant equipment, mixers, tanks, agitators; austenitic, acid resistant.

TECHALLOY 317
Techalloy Co. Inc.
C 0.1, Cr 18, Ni 12, Mo 3-4, bal Fe.
Annealed: 85,000-90,000 TS; 40,000 YS; 50-45 El; 160 Brin. For chemical plant equipment, mixers, tanks, agitators; austenitic, acid resistant.

TECHALLOY 330
Techalloy Co. Inc.
C 0.2, Cr 15, Ni 35, bal Fe.
Annealed: 80,000 TS; 50,000 YS; 30 El; 30 RA; 185 Brin. For furnace parts, heat treat boxes; austenitic, heat resistant.

TECHALLOY 347
Techalloy Co. Inc.
C 0.08, Cr 18, Ni 10, Cb = 10 x C, bal Fe.
Annealed: 90,000 TS; 35,000 YS; 50 El; 65 RA; 110 Brin. Cold drawn: 100,000 TS; 65,000 YS; 40 El; 60 RA; 210 Brin. For chemical plant equipment, tanks, vessels; welded structures, austenitic, stainless.

TECHALLOY 36 (INVAR)
Techalloy Co. Inc.
C 0.04, Mn 0.35, Si 0.12, Ni 36, bal Fe.
Very low thermal expansion rate.

TECHALLOY 430
Techalloy Co. Inc.
C 0.1, Cr 14-18, bal Fe.
Annealed: 75,000 TS; 45,000 YS; 30 El; 65 RA; 155 Brin. Cold drawn: 130,000 TS; 120,000 YS; 2 El; 185 Brin. For screw machine products, pump and valve parts; corrosion resistant.

TECHALLOY A (80-20)
Techalloy Co. Inc.
Ni 77-79, C 0.15, Mn 2.5, Fe 1, Si 1, Cr 19-21.
For electrical resistance elements operating up to 2150°F (1175°C).

TECHALLOY C (62-16)
Techalloy Co. Inc.
Ni 57, C 0.15, Mn 1, Si 1, Cr 14-18, bal Fe.
For electrical resistance elements operating up to 1700°F (925°C).

TECHALLOY C-20
Techalloy Co. Inc.
C 0-0.07, Cr 29, Ni 20, Si 1, Mo 2, 2.0 Cu min, bal Fe.
Cast: 65,000 TS; 30,000 YS; 45 El; 50 RA; 130 Brin.. Rolled: 85,000 TS; 35,000 YS; 35 El; 50 RA; 180 Brin. For chemical plant equipment, mixers, agitators; resists mixed acids, austenitic.

TECHALLOY D (35-19)
Techalloy Co. Inc.
Ni 34-37, C 0.15, Mn 1, Si 1-3, Cr 18-21, bal Fe.
For resistance wiring operating up to 1400-1600°F (760-870°C).

TECHALLOY GLASSEAL 29-17
Techalloy Co. Inc.
C 0.06, Ni 29, Mn 0.5, Co 17, Mg 0.5, Si 0.2, bal Fe.
For sealing metal to hard glass.

TECHALLOY GLASSEAL 42
Techalloy Co. Inc.
C 0.03, Mn 0.8, Si 0.3, Ni 42, bal Fe.
For sealing metal to 1075 glass, also 0120 and 0010 glasses, as-sealed beam headlights, electronic tubes, industrial lamps.

TECHALLOY GLASSEAL 52
Techalloy Co. Inc.
C 0.05, Mn 0.5, Si 0.3, Ni 50.5, bal Fe.
For sealing metal to soft glass.

TECHNALLOY A
Gulf & Western Mfg. Co.
C 1.2-1.3, bal Fe.
For rolls; for steel mills.

TECHNALLOY AX
Gulf & Western Mfg. Co.
C, bal Fe.
For heavy duty rolls; for blooming mills.

TECHNALLOY B
Gulf & Western Mfg. Co.
C 1.8-2, bal Fe.
For rolls; for steel mills.

TECHNALLOY C
Gulf & Western Mfg. Co.
C 1.5-1.8, bal Fe.
For heavy duty rolls; for structural mills.

TECHNALLOY D
Gulf & Western Mfg. Co.
C 2.2-2.3, bal Fe.
For rolls; for steel mills.

TECHNALLOY E
Gulf & Western Mfg. Co.
C 2.4-2.6, bal Fe.
For rolls; for steel mills.

TECHNALLOY STANDARD
Gulf & Western Mfg. Co.
C 1, bal Fe.
For rolls; for steel mills.

TECHNICAL IRON
Russian manufacture
C 0-0.025, Mn 0-0.03, Si 0-0.03, bal Fe.
For magnetic circuits, deflectors; high permeability.

TECHNIGALVA
Pasminco Europe (Mazak) Ltd.
Ni, bal Zn.
For galvanizing.

TECO
Eisler Electric Co.
WC, VC, CrC, ThC, MoC, Co, Ni.
For tools, wearing parts, cutters; a series of refractory alloys.

TECO 38
Tungsten Electric Corp.
WC.
For hard facing electrodes; mild steel tube filled with WC.

TECO 469
Tungsten Electric Corp.
WC + Co.
For cutting tools; cemented.

TECO 51
Tungsten Electric Corp.
WC, Co.
For cutting tools; sintered.

TECO 610 G
Tungsten Electric Corp.
WC, Co.
For cutting tools; sintered.

TECO 946 X
Tungsten Electric Corp.
WC, Co.
For cutting tools; sintered.

TECO A
Tungsten Electric Corp.
WC, Co.
For cutting tools; sintered.

TECO A-1
Tungsten Electric Corp.
WC, Co.
For cutting tools; sintered.

TECO B
Tungsten Electric Corp.
WC, Co.
For cutting tools; sintered.

TECO C
Tungsten Electric Corp.
WC, Co.
For cutting tools; sintered.

TECO CF
Tungsten Electric Corp.
WC, Co.
For cutting tools; sintered.

TECO CHD
Tungsten Electric Corp.
WC, Co.
For cutting tools; sintered.

TECO CMD
Tungsten Electric Corp.
WC, Co.
For cutting tools; sintered.

TECO GRADE 23
Super Tool Co.
WC.
For cutting tools; 90.8 RA. *Obsolete*

TECO GRADE 607
Super Tool Co.
WC.
For cutting tools; 91.8 RA. *Obsolete*

TECO GRADE 610
Super Tool Co.
WC.
For cutting tools; 92.0 RA. *Obsolete*

TECO GRADE 625
Super Tool Co.
WC.
For cutting tools; 91.0 RA. *Obsolete*

TECO GRADE 68
Super Tool Co.
WC.
For cutting tools; 91.3 RA. *Obsolete*

TECO GRADE 946
Super Tool Co.
WC.
For cutting tools; 90.2 RA. *Obsolete*

TECO GRADE 946X
Super Tool Co.
WC.
For cutting tools; 91.5 RA. *Obsolete*

TECO GRADE 964
Super Tool Co.
WC.
For cutting tools; 91.0 RA. *Obsolete*

TECO GRADE K-65
Super Tool Co.
WC.
For cutting tools; 91.5 RA. *Obsolete*

TECO GRADE KT-13
Super Tool Co.
WC.
For cutting tools; 89.0 RA. *Obsolete*

TECO GRADE T-13
Super Tool Co.
WC.
For cutting tools; 88.0 RA. *Obsolete*

TECO KT 13
Tungsten Electric Corp.
WC, Co.
For cutting tools; sintered.

TECO TH
Tungsten Electric Corp.
Carbide.
For cold work dies; sintered.

TECO-ATS1
Tungsten Electric Corp.
Mo 10, bal TiC.
For tools, bearings, seals, wear parts. Resists heat and wear, sintered carbide.

TEELO
Sanderson Kayser Ltd.
High C, bal Fe.
For gauges, taps, threading dies; non-deforming, oil hardened. *Obsolete*

TEENAX 46
Now SIMONDS TEENAX 46.

TEGO
Goldschmidt A.G.
Pb 78.83, Sb 15-18, Sn 1.5-3, Cu 1-2, As 0.3-0.8.
For bearings. Antifriction.

TELASTIC MOLY
Now MOLY TELASTIC.

TELCALLOY 1
Telcon Metals Ltd.
Cu-Ni.
Electrical resistant alloy.

TELCALLOY 1.5
Telcon Metals Ltd.
Cu-Ni.
Electrical resistant alloy.

TELCALLOY 2
Telcon Metals Ltd.
Cu-Ni.
Electrical resistant alloy.

TELCALLOY 3
Telcon Metals Ltd.
Cu-Ni.
Electrical resistant alloy.

TELCALLOY 4
Telcon Metals Ltd.
Cu-Ni.
Electrical resistant alloy.

TELCON 140
Telcon Metals Ltd.
Ni 36, bal Fe/Ni, Mn, Fe.
Thermostatic bimetal. ASTM flexivity/$^{\circ}$F: 15.5×10^{-6}.

TELCON 15
Telcon Metals Ltd.
Ni 36, bal Fe/Ni.
Thermostatic bimetal. ASTM flexivity/$^{\circ}$F: 10.3×10^{-6}.

TELCON 188
Telcon Metals Ltd.
Ni, Fe, Cr/Ni, Fe, Cr.
Thermostatic bimetal. ASTM flexivity/$^{\circ}$F: 9.8×10^{-6}.

TELCON 200
Telcon Metals Ltd.
Ni 36, bal Fe/Mn, Cu, Ni.
Thermostatic bimetal. ASTM flexivity/$^{\circ}$F: 21.4×10^{-6}.

TELCON 36/64
Telcon Metals Ltd.
Fe 36, Ni 64.
For pulse and radar transformers, relay cores; soft magnet, high permeability.

TELCON 400
Telcon Metals Ltd.
Ni 42, bal Fe/Ni, Mn, Fe.
Thermostatic bimetal. ASTM flexivity/$^{\circ}$F: 13.1×10^{-6}.

TELCON 41
Telcon Metals Ltd.
Ni/Ni, Cr, Fe.
Thermostatic bimetal. ASTM flexivity/$^{\circ}$F: 4.6×10^{-6}.

TELCON 75
Telcon Metals Ltd.
Ni 58, bal Fe/Ni, Mn, Fe.
Thermostatic bimetal. ASTM flexivity/$^{\circ}$F: 7.2×10^{-6}.

TELCON 79
Telcon Metals Ltd.
Ni 77, Fe 14, Cu 5, Mo 4.
Soft magnetic alloy, high permeability at low field strengths for transformers, chokes.

TELCON BRONZE
Telcon Metals Ltd.
Be 2, Co 0.25, bal Cu.
Annealed: 80,000 TS; 30,000 YS; 50 El; 100 Brin. Heat treated: 180,000 TS; 165,000 YS; 3 El; 352 Brin. For springs, bushings, diaphragms, clips, electrical contacts; hardenable, tough, corrosion resistant. *Obsolete*

TELCON CU BE 250
Now CU BE 250.

TELCON CU-BE MOLD
Telcon Metals Ltd.
Be 2.7, bal Cu.
Heat treated: 168,000-179,000 TS; 2 El; 400-480 Brin. For injection molds, plastic molds; age-hardenable, wear and corrosion resistant. *Obsolete*

TELCON E140
Telcon Metals Ltd.
Ni 36, bal Fe/Ni, Cr, Fe.
Thermostatic bimetal. ASTM flexivity/$^{\circ}$F: 14.4×10^{-6}.

TELCON E400
Telcon Metals Ltd.
Ni 42, bal Fe/Ni, Cr, Fe.
Thermostatic bimetal. ASTM flexivity/$^{\circ}$F: 11.7×10^{-6}.

TELCON MARAGING STEEL
Telcon Metals Ltd.
For high strength applications and MIG and TIG welding.

TELCON R.2799
Telcon Metals Ltd.
Fe 70, Ni 30.
Rolled: 90,100 TS; 35,100 YS. For magnetic shunts, speedometers, tachometers; soft magnet, Curie temperature at 60°C.

TELCON STAN
Telcon Metals Ltd.
Cu-Ni.
Electrical resistance alloy.

TELCOSEAL
Telcon Metals Ltd.
Fe 54, Ni 29, Co 17.
For sealing to borosilicate glasses in thermionic valve and cathode ray tube construction. Coefficient of expansion same as glass.

TELCOSEAL 1
Telcon Metals Ltd.
Ni 29, Co 17, bal Fe.
For glass/ceramic to metal sealing. Coefficient of expansion, 20-500°C: 6.2×10^{-6}. For use with borosilicate glass.

TELCOSEAL 2

Telcon Metals Ltd.
Ni 42, bal Fe.
For glass/ceramic to metal sealing. Coefficient of expansion, 20-500°C: 8.0 x 10⁻⁶.

TELCOSEAL 3

Telcon Metals Ltd.
Ni 42, Cr 6, bal Fe.
For glass/ceramic to metal sealing. Coefficient of expansion, 20-500°C: 11.7 x 10⁻⁶.

TELCOSEAL 3/2

Telcon Metals Ltd.
Ni 47, Cr 5, bal Fe.
For glass/ceramic to metal sealing. Coefficient of expansion, 20-500°C: 10.7 x 10⁻⁶.

TELCOSEAL 4

Telcon Metals Ltd.
Ni 46, bal Fe.
For glass/ceramic to metal sealing. Coefficient of expansion, 20-500°C: 8.4 x 10⁻⁶.

TELCOSEAL 6

Telcon Metals Ltd.
Ni 49, bal Fe.
For glass/ceramic to metal seals. Coefficient of expansion, 20-500°C: 9.5 x 10⁻⁶.

TELCOSEAL 6/2

Telcon Metals Ltd.
Ni 48, bal Fe.
For glass/ceramic to metal sealing. Coefficient of expansion, 20-500°C: 9.2 x 10⁻⁶.

TELCOSEAL 6/3

Telcon Metals Ltd.
Ni 50, bal Fe.
For glass/ceramic to metal sealing. Coefficient of expansion, 20-500°C: 9.7 x 10⁻⁶.

TELCOSEAL 6/4

Telcon Metals Ltd.
Ni 51, bal Fe.
For glass/ceramic to metal sealing. Coefficient of expansion, 20-500°C: 10.0 x 10⁻⁶.

TELCUMAN

Telcon Metals Ltd.
Cu alloy.

TELECTAL

Metallgesellschaft Reuterweg
Si 1.5, Li 0.1, bal Al.
43,000-47,000 TS; 6 El; 80-100 Brin. For light-alloy parts, electrical conductors; non-hardenable.

TELECUT

Teledyne Vasco
C 1.07, Si 0.5, Mn 0.25, S 0-0.03, P 0-0.03, W 1.5, V 1.15, Mo 9.5, Cr 3.75.
For cutting tools including end mills, milling cutters, broaches, and gear cutters. High speed steel.

TELEDIUM

Goodlass Wall & Lead Industries Ltd.
Te 0.02-0.1, bal Pb.
Cold rolled: 3800-5000 TS; 8 Brin. For chemical plant equipment, cable sheathing, batteries, water supply pipes. Corrosion resistant.

TELEGRAPH BRONZE

English manufacture
Cu 80, Pb 7.5, Zn 7.5, Sn 5.
For machine parts, switches; free-cutting.

TELEMET

Allegheny Ludlum Steel
C 0-0.25, Cr 16-23, Mn 0-2, Si 0-1, bal Fe.
Rolled: 60,000-75,000 TS; 40,000-45,000 YS; 30-35 El; 178-185 Brin. For television picture tubes, metal to glass seals; heat and corrosion resistant.

TELLOY 4140

Inland Steel Co.
Alloy steel. C 0.38-0.43, Mn 0.75-1, Cr 1, Mo 0.2, P 0-0.035, S 0.04-0.06, 0.01 Te min, bal Fe.
Free machining alloy steel.

TELLURIUM COPPER

Chase Brass & Copper Co., Inc.
Copper. Te 0.5, bal Cu.
Annealed: 33,000 psi TS; 8,000 psi YS; 3 El. Drawn: 48,000 psi TS; 46,000 YS; 18 El. For general use; high conductivity.

TELLURIUM COPPER 112

Olin Brass, Indianapolis
Copper. Te 0.5, Cu 99.5.
For welding torch and soldering iron tips; high conductivity.

TELLURIUM COPPER 145

Anaconda Co.
Copper. Cu 99.49, Te 0.5, P 0.01.
Hard rod: 48,000 TS; 40,000 YS; 12 El; 50 Rock B. Soft rod: 32,000 TS; 10,000 YS; 45 El; 45 Rock B. For transformer and circuit breaker terminals, studs, bolts, current carrying parts. Electrical conductivity: 95%. Free cutting, not subject to hydrogen embrittlement.

TELLURIUM COPPER 1452

Anaconda Co.
Copper. Te 0.5, bal Cu.
Hard: 48,000 TS; 40,000 YS; 12 El; 83 Brin. Soft: 32,000 TS; 10,000 YS; 45 El; 42 Brin. For electrical apparatus, screw machine parts. High conductivity, free cutting.

TELLURIUM COPPER TYPE B

Chase Brass & Copper Co.
Te 0.5, Ni 1, P 0.2, bal Cu.
Forged: 60,000 TS; 40,000 YS; 28 El. Heat treated: 75,000 TS; 50,000 YS; 15 El. For machinery parts; heat treatable. *Obsolete*

TELLURIUM LEAD

NL Industries
Te 0.06, Cu 0.06, bal Pb.
Rolled. For pipes, chemical industry; greater corrosion resistance to H_2SO_4. *Obsolete*

TELPHY

Creusot-Loire
Nickel alloy.
For motors, electrical equipment; non-magnetic. *Obsolete*

TEMCROSS

Ingersoll Steel Co.
C 0.4, Cr 1, V 0.2, bal Fe.
For gears, shafts; oil hardening. *Obsolete*

TEMP AIR 8

Gulf Steel Corp.
C 0.6, Cr 3.6, V 1.75, Mo 8.5, bal Fe.
For drills, cutters, extrusion dies; hot work steel, oil hardened.

TEMP ALLOY

Continental Copper & Steel Ind.Inc.
C 0.5, Ni, Cr, bal Fe.
For steel castings. *Obsolete*

TEMPALLOY

Atlas Foundry Co.
C, alloy, bal Fe.
For heat resisting parts. *Obsolete*

TEMPALOY "A"

Anaconda Co.
Cu 95, Ni 4, Si 1.
Cold rolled: 120,000 TS; 105,000 YS; 12 El; 110 Brin. For corrosion resistant parts; corrosion resistant. *Obsolete*

TEMPALOY 630

Anaconda Co.
Copper. Cu 82, Al 9.5, Mn 1, Ni 5, Fe 2.5.
Hard: 105,000 TS; 60,000 YS; 12 El. For propeller shafting. Heat treatable.

TEMPALOY 841

Anaconda Co.
Cu 91.2, Ni 4, Al 4, Si 0.8.
Cast: 70,000-80,000 TS; 45,000-60,000 YS; 8-15 El. For castings; high strength. *Obsolete*

TEMPALTO

Bull's Metal & Marine Ltd.
Cu 56, Ni 14, Pb 30, P.
For metallic packing; for high temperatures.

TEMPER TOUGH

Darwin & Milner Inc.
Tool material. C 0.75, Mn 0.6, Si 1.15, Cr 0.8, V 0.15, Mo 0.3, bal Fe.
Oil hardened tool steel; AISI L6.

TEMPERA 1

Robert-Leyer-Pritzkow & Co.
C 0-0.12, Si 0.8, Al 0.8, Cr 6.5, bal Fe.
Annealed: 70,000 TS; 30,000 YS; 28 El; 65 RA; 150 Brin. For oil refinery equipment; corrosion and creep resistant.

TEMPERA 2

Robert-Leyer-Pritzkow & Co.
C 0-0.12, Si 1.2, Al 1, Cr 13, bal Fe.
Annealed: 75,000 TS; 40,000 YS; 35 El; 70 RA; 155 Brin. For oil refinery equipment; corrosion and creep resistant.

TEMPERA 3

Robert-Leyer-Pritzkow & Co.
C 0-0.12, Si 1.5, Al 1.5, Cr 24, bal Fe.
Annealed: 75,000-95,000 TS; 46,000-60,000 YS; 25-35 El; 46-65 RA; 160-200 Brin. For furnace equipment, heat treat boxes; heat resistant.

TEMPERA 4

Robert-Leyer-Pritzkow & Co.
C 0.15, Si 2, Cr 20, Ni 9.5, bal Fe.
Annealed: 75,000 TS; 35,000 YS; 50 El; 65 RA; 160 Brin. For chemical plant equipment; Type 302; stainless, austenitic.

TEMPERA 5

Robert-Leyer-Pritzkow & Co.
C 0.15, Si 2, Cr 24, Ni 19, bal Fe.
Annealed: 100,000 TS; 45,000 YS; 50 El; 65 RA; 185 Brin. For furnace parts, heat treat boxes; Type 310; corrosion and heat resistant.

TEMPERA A

Robert-Leyer-Pritzkow & Co.
C 0-0.12, Si 1, Al 1, Cr 18, bal Fe.
Annealed: 90,000 TS; 55,000 YS; 20 El; 40 RA; 160 Brin. For chemical plant equipment, furnace parts; corrosion and creep resistant.

TEMPERA B

Robert-Leyer-Pritzkow & Co.
C 0.2, Si 1.2, Cr 25, Ni 4, bal Fe.
Annealed: 95,000 TS; 60,000 YS; 25 El; 45 RA; 180 Brin. For furnace parts, heat treat boxes; heat and creep resistant.

TEMPERATURE COMPENSATOR 32

Now CARPENTER TEMPERATURE COMPENSATOR 32.

TEMPERATURE COMPENSATOR ALLOY NO. 30
English manufacture
Fe-Ni.
For compensating shunts for electrical equipment; temperature-sensitive, magnetic.

TEMPERDIE
Pennsylvania Steel Corp.
C 0.4, Mn 1, Cr 1.5, V 0.25, Mo 1, bal Fe.
For hot header dies, punches, hot press dies; hot work steel oil hardened.

TEMPERIM
Chain Belt Co.
C 3.2, Si, Mn, other alloy elements, bal Fe.
Cast: 400 Brin. For wear resistant parts. Cast iron, abrasion resistant.

TEMPERITE
General Cable Corp.
Pb, Sn, Cd.
For temperature indicating alloys; low melting series of alloys.

TEMPEST
Osborn Steels Ltd.
C 0.16, Cr 17, Ni 1.8, bal Fe.
For furnace parts, cutlery, valves, pumps; corrosion resistant. *Obsolete*

TEMPLATE
Bull's Metal & Marine Ltd.
Zn alloy.
For general castings; white metal.

TEMPLESS
Grayborn Steel Co.
O 0.9, alloy, bal Fe.
For punches, dies, forming and blanking dies, upsetters; no temper required.

TEMPLESS NO. 11
Grayborn Steel Co.
C 0.3-0.37, Mn 0.4, Cr 0.7-0.9, Mo 0.4-0.6, bal Fe.
Rolled: 130,000 TS; 16 El; 50 RA; 220 Brin. For pneumatic tools, axes, chisels, caulking tools; shock resistant.

TEMPLEX
Crane Co.
C 0.35-0.5, Mn 0.4-0.7, Cr 0.8-1.1, Mo 0.3-0.4, V 0.2-0.3, bal Fe.
Heat treated: 125,000-145,000 TS; 105,000-120,000 YS; 14-16 El; 45-50 RA. For high temperature bolting; ASTM-A193 G.B. *Obsolete*

TEMPO
J.M. Ney Co.
Pd 55, Ag 35.
White gold color alloy for porcelain fused-to-metal restorations. Melting range: 2140-2280°F; casting temperature: 2500°F; 200 Brin; 240 HV.

TEMPO
Pennsylvania Steel Corp.
C 0.9, Mn 1.15, Cr 0.5, W 0.5, bal Fe.
For gauges, hobs, punches, master tools; oil hardened, non-deforming.

TEN STAR
Now CARPENTER TEN STAR.

TENAILLE NO 2 1/2
Chiers-Chatillon
C 0.8-1, bal Fe.
For tools, drills; water hardened. *Obsolete*

TENAX METAL
Westa-Westdeutsche
Al 4.2-4.6, Cu 2.2-3, Pb 1.2, Fe 0.35, bal Zn.
For solder for aluminum ware; strong.

TENAZ-1
Forjas Alavesas S.A.
C 0.38, Mn 0.4, Si 1, Cr 1.15, W 2, V 0.15, bal Fe.
For chisels, punches, pneumatic tools. IHA F-525; Din 35 WCrV7.

TENAZ-2
Forjas Alavesas S.A.
C 0.5, Mn 0.4, Si 1, Cr 1.15, W 2, V 0.15, bal Fe.
Shock resisting tool steel; for punches, pneumatic tools, chisels. IHA F-524; similar to AISI S1.

TENBOR
British Steel plc
C 0.25-0.3, Si 0.25, Mn 1.25, S 0.005, P 0.025, Ti 0.03, B 0.002, bal Fe.
Heat-treatable. For wear-resistant components.

TENCO
Sanderson Kayser Ltd.
W 18, Co 9, Mo 1, Cr 4, V 2, C, bal Fe.
For tools, dies, cutters; high speed steel.

TENELON
United States Steel Corp.
C 0.08-0.12, Cr 17-18.5, Mn 14.5-16, Si 0.3-1, Ni 0-0.75, 0.35 N min, bal Fe.
Manganese austenitic steel. Work hardened. For parts subject to extreme wear and abrasion.

TENEM
NGK Metals Corp.
Sn, Be, bal Cu.
For corrosion and wear resistant parts; high strength.

TENFORM PK
British Steel plc
C 0-0.08, Mn 0-0.6, P 0-0.1, S 0-0.03, bal Fe.
225-270 N/mm^2 YS min.

TENFORM XF
British Steel plc
C 0-0.08, Mn 0-1.2, Ti 0-0.3, bal Fe.
300-550 N/mm^2 YS min.

TENFORM XK
British Steel plc
C 0-0.08, Mn 0-1.2, Nb 0-0.3, bal Fe.
300-450 N/mm^2 YS min. Cold rolled strip: 300-350 N/mm^2 YS min.

TENIT 1720
Styria-Stahl Steirische Gusstahlwerke AG
C 0.3, Si 1, Mn 0.4, Cr 1.1, W 3.75, bal Fe.
For tools, dies, punches, headers; hot work steel, oil hardened. *Obsolete*

TENIT 1720 SPEZIAL
Now VEW W106.

TENIT KL
Styria-Stahl Steirische Gusstahlwerke AG
C 0.5, V 0.2, Cr 1.05, W 1.85, Si 0.9, bal Fe.
For hot work dies, punches, chisels; hot work steel, oil or water hardened. *Obsolete*

TENIT KL SPEZIAL
Steirische Gusstahlwerke, A.G.
C 0.55, Cr 1, V 0.18, W 1.8, bal Fe.
For hot work tools and dies; hot work steel, oil hardened.

TENIT W
Styria-Stahl Steirische Gusstahlwerke AG
C 0.35, V 0.2, W 1.85, Cr 1.05, Si 0.9, bal Fe.
For hot work dies, punches, chisels; hot work steel, oil hardened. *Obsolete*

TENNAX
Edgar T. Ward's Sons Co.
C, alloy, bal Fe.
For tools; non-deforming. *Obsolete*

TENNESEAL
United States Steel Corp.
Low C, Mn, bal Fe.
Pre-formed, zinc coated steel sheet for roofing and siding.

TENNESSEE SPECIAL
U.S. Steel Corp.
C 0.95, bal Fe.
For general tools, broaches; water hardened. *Obsolete*

TENS-50
Navan Inc.
Si, Mg, Be, N, bal Al.
Sand cast: 46,000 TS; 36,000 YS; 5 El. Permanent mold: 50,000 TS; 44,000 YS; 6 El. For aircraft and missile components, gear cases; age-hardenable.

TENSIL-COR
Airco Vacuum Metals
C 0.04, Mn 1.63, Si 0.5, Mo 0.5, bal Fe.
Welded: 112,000 psi TS; 104,700 psi YS; 24 El. CO$_2$ shielded, cored wire for single or multipass welding of high tensile low alloy steel. Good toughness down to -75°F. AWS E 70T-G.

TENSIL-COR
Airco Vacuum Metals
Mn 1.63, Si 0.5, Mo 0.46, V 0.03, Cr 0.04, bal Fe.
Welded: 108,500 psi TS; 103,000 psi YS; 23 El. Good impact strength. Weld deposit: 0.072 C. AWS A5.20 class E70T-G.

TENSILE-FLEX
International Wire Products
Oxygen free, copper-based, precipitation-type alloy wire. High strength, resistant to hydrogen embrittlement, long flex life, high electrical conductivity.

TENSILEND 100
Arcos Alloys
C 0.06, Mn 0.55, Si 0.4, Ni 1.8, Mo 0.6, V 0.1, bal Fe.
Welded: 100,000-110,000 TS; 91,000 YS; 22 El; 58 RA; 230 Brin. For welding electrodes for low alloy steel; low H$_2$. *Obsolete*

TENSILEND 120
Arcos Alloys
C 0.07, Mn 0.9, Si 0.25, Ni 1.8, Mo 0.8, V 0.5, bal Fe.
Welded: 120,000-130,000 TS; 107,000 YS; 19 El; 52 RA; 260 Brin. For welding rods for marine applications; low H$_2$. *Obsolete*

TENSILEND 70
Arcos Alloys
C 0.06, Mn 0.55, Si 0.35, bal Fe.
Welded: 70,000-80,000 TS; 70,000 YS; 26 El; 60 RA; 180 Brin. For welding electrodes for low alloy steel; low H$_2$. *Obsolete*

TENSILEND 80
Arcos Alloys
C 0.06, Mn 0.8, Si 0.4, Ni 1, bal Fe.
For welding electrodes; class E8016-C3, 1% nickel steel. *Obsolete*

TENSILIA
English manufacture
C 0.48, Mn 0.51, V 0.18, bal Fe.
For tools, gears, pinions, shafts; water hardening.

TENSILITE
American Manganese Bronze Co.
Copper. Cu 64-67, Zn 24-29, Mn 2.5-3.8, Al 3.1-4.4, Fe 0-1.2. 103,100 TS; 32,500 YS; 19 El; 23 RA. For gears, worm wheels; tough.

TENSILITE
American Manganese Bronze Co.
Copper. Cu 68.26, Zn 21.04, Mn 3.2, Si 0.04, Pb 0.14, Al 4.8, Fe 2.03.
For gears, worm wheels; wear resistant.

TENSILOY
British Steel plc
Electrical steel; high permeability. For high magnetic flux density applications.

TENSILOY BORON
Allied Steel & Chemical Co.
C, alloy, bal Fe.
For machinery parts, gears, pinions, shafts; pre-heat-treated, abrasion and shock resistant.

TENSILOY EXTRA
Allied Steel & Chemical Co.
C 0.5, Si 1.3, Mo 0.3, W 0.5, bal Fe.
For gears, crimpers, punches, dies, upsetters; oil hardened, shock resistant.

TENSILOY H.C.C.
Allied Steel & Chemical Co.
C 1.5, Cr 12, V 1, Mo 1, bal Fe.
For blanking and forming dies, punches; Type D2; air hardened, nondeforming.

TENSILOY H.S.S.
Allied Steel & Chemical Co.
C 0.8, Cr 4, V 2, W 18, Mo 0.65, bal Fe.
For reamers, taps, broaches, drills, hobs, lathe tools; Type T2; high speed steel.

TENSILOY NITRIDING DIE
Allied Steel & Chemical Co.
C 0.8, Cr 4, V 1, W 1.5, Mo 8, bal Fe.
For reamers, broaches, dies, hobs; Type M1; high speed steel.

TENSILOY NON-TEMPERING
Allied Steel & Chemical Co.
C 0.33, Cr 0.75, Cu 0.75, Mo 0.75, bal Fe.
For tools, dies; water hardened, nontempering.

TENSILOY O.H. DIE
Allied Steel & Chemical Co.
C 0.9, Mn 1.15, Cr 0.5, W 0.5, bal Fe.
For punches, dies, crimpers, cutting tools; oil hardened, nondeforming.

TENSITE
Gilby-Fodor S.A.
Ni 98, Al 2.
Annealed: 57,000 TS; 18,000 YS; 45 El. For cathodes and filaments in electronic tubes. Heat resistant.

TENSITE
Wilbur B. Driver Co.
Al 2, Ni 98.
Annealed: 56,500 TS. For vacuum tube filaments. *Obsolete*

TENUAL 1
National Bronze & Aluminum Foundry
Aluminum. Si 5, bal Al.
Cast: 19,000 TS; 8000 YS; 8 El; 40 Brin. For instrument cases, thin wall pressure tight castings; corrosion resistant.

TENUAL 50
National Bronze & Aluminum Foundry
Aluminum. Cu 6-8, Zn 0-2.5, Fe 0-1.5, Si 0-2, bal Al.
For crankcases, oil pans, cylinder heads.

TENUAL NO. 226
National Bronze & Aluminum Foundry
Aluminum. Cu 9.25-10.75, Fe 0.9-1.5, Mg 0.15-0.35, 87 Al min.
Cast: 21,000-24,000 TS; 115-127 Brin. For pistons, valve tappets, camshaft bearings, guides; same as SAE-34.

TENUAL T
National Bronze & Aluminum Foundry
Aluminum. Ti 0.2, Mn 1, bal Al.
For aircraft structures, cowling; corrosion resistant.

TENUAL T TITANITE
National Bronze & Aluminum Foundry
Aluminum. Ti 0.2, Mn 1.5, bal Al.
For light alloy parts; corrosion resistant.

TENZALOY
American Smelting & Refining Co.
Cu 0.8, Mg 0.4, Zn 8, bal Al.
Aged: 32,000 psi TS; 22,000 psi YS; 3 El; 65 Brin. For light castings, housings, machinery parts; high strength, requires no heat treatment.

TEPAZ
Haeckerstahl GmbH
C 0-0.12, Si 1.15, Cr 13, Al 0.95, bal Fe.
For oil refinery equipment; corrosion and creep resistant.

TEPAZ 1020
Haeckerstahl GmbH
C 0.15, Si 2, Cr 19.5, Ni 9.5, bal Fe.
Annealed: 80,000 TS; 35,000 YS; 55 El; 65 RA; 170 Brin. For chemical plant equipment; Type 302; stainless, austenitic.

TEPAZ 2025
Haeckerstahl GmbH
C 0.15, Cr 24, Ni 19, bal Fe.
Annealed: 100,000 TS; 45,000 YS; 50 El; 65 RA; 180 Brin. For furnace parts, heat treat boxes; heat resistant, austenitic; Type 310.

TERBIUM
Atomergic Chemetals Corp.
Tb.
Purities: 99.9% special distilled grade, 99.5+%. Forms: ingot, lump, sheet, foil, rod, wire, filings, sponge, single crystals.

TERMACID
Myrens Verkstec A/S
C 0.1-0.25, Cr 27, Ni 4.5, Mo, Mn, bal Fe.
For furnace parts, heat treating boxes; heat resistant to 2000°F.

TERMAFOND 245C
Montecatini Settore Alluminio
Cu 3.75-4.25, Si 14-15, Mg 0.6-1, Ni 3.7-4.3, Co 0.5-0.9, bal Al.
Heat treated: 24,000-33,000 TS; 22,000-21,000 YS; 0.2-0.5 El; 115-140 Brin. For light alloy parts; age-hardened, corrosion resistant.

TERMAFOND 245NT
Montecatini Settore Alluminio
Cu 4.3-4.7, Si 14.5-15.5, Mg 0.7-1.1, Mn 0.7, Ti 0.2, Ni 2.3-2.7, bal Al.
Heat treated: 21,000-28,500 TS; 20,000-27,000 YS; 0.5-2 El; 110-130 Brin. For light alloy parts; age-hardened.

TERMAFOND C10
Montecatini Settore Alluminio
Cu 9.3-10.7, Fe 0.5-1.3, Mg 0.2-0.4, bal Al.
Heat treated: 40,000-51,200 TS; 34,000-37,000 YS; 0.2-0.5 El; 125 150 Brin. For pistons; age-hardenable.

TERMAFOND C12N
Montecatini Settore Alluminio
Cu 11-12.5, Fe 0.6-0.8, Ti 0.1-0.2, Ni 0.4-0.6, bal Al.
Heat treated: 40,000-65,000 TS; 24,000-27,000 YS; 1-1.5 El; 110-130 Brin. For pistons; age-hardenable.

TERMAFOND C12T
Montecatini Settore Alluminio
Cu 11-12.5, Ti 0.2, Fe 1, bal Al.
Heat treated: 38,000-42,700 TS; 22,800-25,600 YS; 1-1.5 El; 110-125 Brin. For pistons; age-hardenable.

TERMAFOND C3
Montecatini Settore Alluminio
Cu 2.9-3.2, Fe 1.6, Si 0.7, Mg 0.6, Ti 0.2, Ni 0.6, bal Al.
Heat treated: 47,000-60,000 TS; 43,000-51,000 YS; 1-5 El; 115-150 Brin. For light alloy parts; age-hardenable.

TERMAFOND C46
Montecatini Settore Alluminio
Cu 9.5-10.5, Si 0.8-1.2, Mg 0.3, Ti 0.2, Ni 1.3-1.7, bal Fe.
Heat treated: 45,000-54,000 TS; 37,000-48,000 YS; 0.5-1 El; 115-140 Brin. For engine cylinder heads, pistons; age-hardenable.

TERMAFOND S10
Montecatini Settore Alluminio
Cu 2-2.5, Si 9.5-10.5, Mg 1.4, Ni 0.8-1.2, bal Al.
Heat treated: 36,000-50,000 TS; 34,000-42,000 YS; 0.3-0.5 El; 95-125 Brin. For light alloy parts; for high temperature service.

TERMAFOND S121
Montecatini Settore Alluminio
Cu 2.25, Fe 0.8-1.2, Si 11-13, Mg 1.5, Mn 0.5, bal Al.
Heat treated: 41,000-50,000 TS; 38,000-43,000 YS; 0.3-0.5 El; 120-135 Brin. For light alloy parts; corrosion resistant.

TERMAFOND S122
Montecatini Settore Alluminio
Cu 0.5-1.1, Si 13, Mg 1.2, Ti 0.15, Ni 2.2, bal Al.
Heat treated: 40,000-50,000 TS; 38,000-43,000 YS; 0.3-0.5 El; 120-130 Brin. For light alloy parts; for high temperature service.

TERMAFOND Y
Montecatini Settore Alluminio
Cu 3.8-4.2, Mg 1.3-1.7, Ni 1.8-2.3, bal Al.
Heat treated: 36,000-47,000 TS; 32,000-43,000 YS; 0.3-0.5 El; 95-115 Brin. For engine cylinder heads, pistons; age-hardenable.

TERMAFOND YT
Montecatini Settore Alluminio
Cu 4, Mg 1.5, Ni 2, bal Al.
For engine cylinders, pistons; age-hardenable.

TERNAL
Aluminium Laufen AG
Aluminum.
Aluminum alloy AlSi5.

TERNALLOY 5
Apex International Alloys, Inc.
Mg 1.6, Zn 3, Cr 0.3, Mn 0.5, bal Al.
SC-F temper: 29,000 TS; 13,000 YS; 12 El; 50 Brin. SC-T 51 temper: 30,000 TS; 18,000 YS; 7 El; 60 Brin. PM-Age temper: 43,000 TS; 22,000 YS; 16 El; 70 Brin. For castings; natural aging. *Obsolete*

TERNALLOY 6
Apex International Alloys, Inc.
Mn 0.2-0.4, Zn 3.4-3.9, Cr 0.2-0.4, Mg 0.6-2, bal Al.
Cast: 30,000 TS; 20,000 YS; 2 El; 70 Brin. Aged: 35,000 TS; 23,000 YS; 4 El. For light alloy castings; heat treatable.

TERNALLOY 6
Sheepbridge Equipment Ltd.
Mn 0.2-0.4, Zn 3.4-3.9, Cr 0.2-0.4, Mg 0.6-2, bal Al.
Cast: 30,000 TS; 20,000 YS; 2 El; 70 Brin. Aged: 35,000 TS; 23,000 YS; 4 El. For light alloy castings; heat treatable.

TERNALLOY 7

Apex International Alloys, Inc.
Cr 0.2-0.4, Cu 0-0.2, Mg 1.8-2.4, Mn 0.2-0.4, Zn 4-4.5, bal Al.
F-Temper: 30,000 TS; 19,000 YS; 5 El; 65 Brin. T6-Temper:
44,000 TS; 40,000 YS; 1.5 El; 80 Brin. For light alloy castings;
responds to heat treatment. *Obsolete*

TERNALLOY 8

Apex International Alloys, Inc.
Cr 0.2-0.4, Cu 0-0.2, Mg 1.8-2.4, Mn 0.2-0.4, Zn 4.9, bal Al.
SC-F-Temper: 33,000 TS; 22,000 YS; 3 El; 70 Brin. SC-T6-
Temper: 57,000 TS; 55,000 YS; 100 Brin. PM-6 Temper:
58,000 TS; 50,000 YS; 3 El; 110 Brin. For light alloy castings;
responds to heat treatment. *Obsolete*

TERNE METAL

NL Industries
Sn 15-30, Pb 70-85, Sb 0-2.
For terne plate, roofing, gasoline and oil tanks; corrosion
resistant bearings. *Obsolete*

TERNE ROOFING

Follansbee Steel Co.
Base: copper bearing steel. Coating: 80 Pb, 20 Sn (Terne).
Soft: 45,000 TS. For roofing, gutters, flashings, down drains,
termite shields. Easy to form and solder.

TERNEX

British Steel plc
Mild steel.

TERRA

Ludlow Steel Corp.
C, alloy, bal Fe.
Hardenable to 60-61 Rock C. For ball races, blanking and
forming dies, cutlery dies, collets, drills, gauges, pistons,
rings, taps, vise jaws; oil hardening, non-deforming.

TERRIFIC

Wardlows Ltd.
W 18, Cr 4, V 1, C, bal Fe.
For tools, dies, cutters; high speed steel.

TETMAJER ALUMINUM BRONZE

English manufacture
Al 4-10, Si 1-3, Fe 0.7-1, bal Cu.
For bushings, fittings; tough, high strength.

TETON

AL Tech Specialty Steel Corp.
C 1, Cr 1.3, bal Fe.
For bearing races and balls, compression dies, machinery
parts. Furnished vacuum melted for aircraft bearing
components. Oil or water hardening.

TEW 15 CR 3, ETC.

Thyssen Edelstahlwerke AG
See DIN 15 Cr 3, etc.
Structural and case hardening grades. 82 grades.

TEX NON-3110CK

Sanderson Kayser Ltd.
C, Cr, W, bal Fe.
For pneumatic tools, rivet sets, punches; oil hardened, shock
resistant. *Obsolete*

TEXALLOY HEAT RESISTANT

Texas Electric Steel Casting Co.
Cr 18, Ni 8, C, bal Fe.
For stainless and heat resistant parts; stainless and heat
resistant. *Obsolete*

THALASSAL

Manufacturer not listed.
Mn 2.5, Mg 2.3, Sb 0.2, Si, bal Al.
Sand cast: 24,000-31,000 TS. For dirigible and airplane parts,
marine parts; resists sea water corrosion.

THALLIUM

Atomergic Chemetals Corp.
Tl.
Purities, zone refined: 99.9999%, 99.999%, 99.99%, 99+%.
Forms: rod, stick, ingot, shot, wire, single crystals.

THERLO

Harrison Alloys Inc.
Superalloy.
Now HAI-373 ALLOY.

THERMAL B

Allied Steel & Tractor Products Inc.
C 0.4, Cr 3-4, W 14, V 0.6, bal Fe.
For tools, cutters, hot shear blades; hot work steel.

THERMAL BB

Allied Steel & Tractor Products Inc.
C 0.58, Cr 3-4, W 10-14, V 0.4, bal Fe.
For tools, cutters; high speed steel.

THERMALLOY

Now AMSCO THERMALLOY.

THERMALLOY "A"

Abex Corp.
C 0.5, Ni 64-66, Cr 17-20, Si 1.5, Mn 1.5, bal Fe.
Cast: 65,000 TS; 50,000 YS; 1-2 El; 1-2 RA; 241 Brin. For
carburizing boxes, retorts, cyanide pots; max operating
temperature 2100 F; heat and corrosion resisting. *Obsolete*

THERMALLOY "B"

Abex Corp.
Ni 38-40, Cr 15-18, Si 1.5, C 0.4-0.6, bal Fe.
Cast: 65,000 TS; 50,000 YS; 1-2 El; 1-2 RA; 241 Brin. At 850
C: 34,000 TS; 4 El; 10 RA. For furnace parts, rollers, furnace
rails, grids, lead and cyanide pots; max operating
temperature 2050 F; heat and corrosion resisting. *Obsolete*

THERMALLOY "C"

Abex Corp.
C 0.6, Ni 0-2, Cr 25-30, Mn 0.8, Si 1, bal Fe.
Cast: 50,000 TS; 40,000 YS; 1 El; 1-2 RA; 196 Brin. For sulfur
roasting furnace parts, rabble arms and blades; heat and
corrosion resistant. *Obsolete*

THERMALLOY "D"

Abex Corp.
C 0.3-0.6, Ni 2-5, Cr 25-30, Mn 0.8, Si 1, bal Fe.
Casting: 50,000 TS; 40,000 YS; 1 El; 1-2 RA; 196 Brin. For
annealing boxes, lead pots, rabble arms, pyrometer tubes;
heat resistant. *Obsolete*

THERMALLOY "E"

Abex Corp.
C 0.3-0.6, Ni 8-12, Cr 24-28, Mn 0.8, Si 1.5, bal Fe.
For cement mill and oil still parts, mushroom valves, skid
rails, annealing boxes; heat and corrosion resistant.
Obsolete

THERMALLOY 18-8

Abex Corp.
C 0.13-0.2, Cr 18, Ni 8, bal Fe.
For stainless parts; stainless. *Obsolete*

THERMALLOY 20

Abex Corp.
Stainless steel. C 0-1, Cr 20, Ni 0-3, bal Fe.
Aged: 115,000 TS; 80,000 YS; 18 El. Cast: 75,000 TS; 60,000
YS; 2 El; 190 Brin. For furnace parts, grate bars, salt pots; ACI
Type HC; heat resistant.

THERMALLOY 28

Abex Corp.
Stainless steel. C 0.2-0.4, Cr 28, Ni 3, bal Fe.
Cast: 90,000 TS; 50,000 YS; 1 El; 1 RA; 240 Brin. Annealed:
67,000 TS; 48,000 YS; 4 El; 4 RA; 190 Brin. For salt pots,
sintering bars, grate bars, dampers; Type HC; corrosion
resistant to sulfur atmosphere.

THERMALLOY 30

Abex Corp.
Stainless steel. C 0.2-0.4, Cr 18-21, Ni 9-11, bal Fe.
Cast: 93,000 TS; 50,000 YS; 32 El; 34 RA; 176 Brin.
Annealed: 97,000 TS; 50,000 YS; 35 El; 38 RA; 173 Brin. For
conveyor belts, oil stills, furnace parts; Type HF; heat
resistant to 1600°F.

THERMALLOY 32

Abex Corp.
Stainless steel. C 0.4, Ni 3, Cr 29, Mn 0.3, Si 2, bal Fe.
Cast heat resistant alloy.

THERMALLOY 34

Abex Corp.
Stainless steel. C 0.5, Ni 5, Cr 28, Mn 0.5, Si 1.5, bal Fe.
Cast heat resistant alloy.

THERMALLOY 38

Abex Corp.
Stainless steel. C 0.4, Ni 12, Cr 28, Mn 0.5, Si 1.5, bal Fe.
Cast heat resistant alloy.

THERMALLOY 38E

Abex Corp.
Cr 28, Ni 11, C, bal Fe.
For heat resistant castings; corrosion and heat resistant.
Obsolete

THERMALLOY 4

Abex Corp.
C, Cr, Ni, bal Fe.
For hard facing electrode; resists high temperature and
thermal shock.

THERMALLOY 40A2

Abex Corp.
Cr 25, Ni 15, Cb 1, C, bal Fe.
For heat resistant castings; corrosion and heat resistant.
Obsolete

THERMALLOY 40B (HH)

Abex Corp.
Stainless steel. C 0.3, Ni 12, Cr 25, Mn 0.5, Si 1.5, bal Fe.
Cast heat resistant alloy. ACI Grade HH Type II.

THERMALLOY 40E

Abex Corp.
Stainless steel. C 1, Ni 12, Cr 26, Mn 0.5, Si 1.5, bal Fe.
Cast heat resistant alloy.

THERMALLOY 43

Abex Corp.
C 0.2-0.5, Cr 26-30, Ni 14-18, bal Fe.
Cast: 80,000 TS; 45,000 YS; 12 El; 160 Brin. Aged: 90,000 TS;
65,000 YS; 6 El; 200 Brin. For hearth plates, lead pots, retorts;
Type HI; corrosion and heat resistant. *Obsolete*

THERMALLOY 45

Abex Corp.
Stainless steel. C 0.4, Ni 25, Cr 20, Mn 0.5, Si 1.5, bal Fe.
Cast heat resistant alloy. ACI Grade HN.

THERMALLOY 47 (HK)

Abex Corp.
Stainless steel. C 0.4, Ni 20, Cr 26, Mn 0.5, Si 1.5, bal Fe.
Cast heat resistant alloy. ACI Grade HK.

THERMALLOY 47D

Abex Corp.
C 0.5, Cr 27, Ni 20, bal Fe.
For heat resistant castings; corrosion and heat resistant.
Obsolete

THERMALLOY 48

Abex Corp.
Stainless steel. C 0.4, Ni 20, Cr 30, Mn 0.5, Si 1.5, bal Fe.
Cast heat resistant alloy.

THERMALLOY 50 (HT)
Abex Corp.
Stainless steel. C 0.5, Ni 35, Cr 18, Mn 0.5, Si 2, bal Fe.
Cast heat resistant alloy. ACI Grade HT.

THERMALLOY 50 CQ
Abex Corp.
Stainless steel. C 0.5, Ni 35, Cr 16, Mn 0.5, Si 2, Cb 1, bal Fe.
Cast heat resistant alloy.

THERMALLOY 58
Abex Corp.
Stainless steel. C 0.35-0.75, Cr 17-19, Ni 37-40, bal Fe.
Cast: 72,000 TS; 40,000 YS; 4 El; 3 RA; 200 Brin. Annealed:
73,000 TS; 40,000 YS; 9 El; 9 RA; 200 Brin. For retorts,
muffles, cyanide and lead pots; Type HU; heat resistant.

THERMALLOY 58 CQ
Abex Corp.
Stainless steel. C 0.5, Ni 38, Cr 18, Mn 0.5, Si 2, Cb 1, bal Fe.
Cast heat resistant alloy.

THERMALLOY 58B
Abex Corp.
Ni 38, Cr 18, bal Fe.
Cast. For carburizing equipment, retorts, cyanide pots; heat
resistant. *Obsolete*

THERMALLOY 63 (HP)
Abex Corp.
Stainless steel. C 0.35-0.75, Ni 34-38, Cr 25-28, Si 2, bal Fe.
Cast: 68,000 TS; 32,000 YS; 19 El; 18 RA; 170 Brin.
Annealed: 78,000 TS; 46,000 YS; 6 El; 5 RA; 175 Brin. For
calcining tubes, furnace shafts, heat treating boxes; heat
resistant to 2100°F. ACI HP.

THERMALLOY 72
Abex Corp.
Superalloy. C 0.4-0.6, Ni 58-62, Cr 12, Mn 0.5, Si 2, bal Fe.
Cast: 59,000-73,000 TS; 30,000-37,000 YS; 3-9 El; 4-9 RA;
166-198 Brin. For retorts, muffles, heat treating boxes; heat
and corrosion resistant.

THERMALLOY 72M
Abex Corp.
Ni 60-62, Cr 12-15, Mo 1-2.5, bal Fe.
For carburizing boxes, rolls, grids, furnace castings; heat
resistant. *Obsolete*

THERMALLOY 85
Abex Corp.
Superalloy. C 0.35-0.75, Cr 17-19, Ni 65-85, bal Fe.
Cast: 78,000-88,000 TS; 35,000-44,000 YS; 7-10 El; 8-15 RA;
190-220 Brin. For retorts, muffles, burner parts, heat treating
boxes; Type HX; heat resistant to 2100°F in the absence of
sulfur.

THERMALLOY 85 EX
Abex Corp.
Nickel. C 0.5, Ni 68, Cr 17, Mn 0.5, Si 2, Mo 2, bal Fe.
Cast heat resistant alloy.

THERMALLOY 85E
Abex Corp.
Cr 15, Ni 65, Mo 2, C, bal Fe.
For heat resistant castings; corrosion and heat resistant.
Obsolete

THERMALLOY H
Abex Corp.
Cr 18-21, Ni 9-12, C, bal Fe.
For pump parts, paper mill and chemical mixer castings; heat
and corrosion resistant castings. *Obsolete*

THERMALLOY HC-250
Abex Corp.
Cast iron. Cr 25, high C, bal Fe.
Cast and heat treated: 70,000-100,000 TS; 50,000-80,000 YS;
0.2-0.5 El; 550-750 Brin. For castings, pumps, stokes, crusher
rolls, grinding rolls; abrasion resistant, oxidation resistant to
1850°F.

THERMALLOY J
Abex Corp.
Cr 27-30, Ni 7-10, C, bal Fe.
For dampers, valves, chemical machinery parts, skid rails;
heat and corrosion resistant castings. *Obsolete*

THERMALLOY K
Abex Corp.
Cr 25-28, Ni 34-37, C, bal Fe.
For retorts, muffles, pots, carburizing boxes, furnace parts;
heat and corrosion resistant up to 2000 F; castings. *Obsolete*

THERMALLOY L
Abex Corp.
Cr 24-27, Ni 19-22, C, bal Fe.
For castings for high temperature and carburizing conditions
with S present; heat and corrosion resistant, castings.
Obsolete

THERMANIT 1 M
Thyssen Edelstahlwerke AG
C 0.07, Cr 1.3, Mo 1, V 0.3, bal Fe.
Coated electrode for overlaying tool steels. Werkstoff Nr.
1.7339.

THERMANIT 13 CO
Thyssen Edelstahlwerke AG
C 0.06-0.15, Cr 28-29, Co 50, bal Fe.
Coated electrodes for joining and overlaying. Werkstoff Nr.
2.4778. 400 N/mm² YS; 600 N/mm² TS; 5 El.

THERMANIT 13/04
Thyssen Edelstahlwerke AG
C 0.03-0.05, Cr 12.5-13, Ni 4-4.5, Mo 0.5, bal Fe.
Coated electrode for overlaying on water turbine
construction. Werkstoff Nr. 1.4351. 600 N/mm² YS; 800
N/mm² TS; 15 El.

THERMANIT 13/65 TT
Thyssen Edelstahlwerke AG
C 0.05, Mn 3.5, Cr 15, Mo 5.5, Nb 1.1, W 1.2, Fe 8.5, bal Ni.
Coated electrodes for joining and overlaying. 410 N/mm²
YS; 660 N/mm² TS; 35 El.

THERMANIT 14 H
Thyssen Edelstahlwerke AG
C 0.06, Cr 14-14.5, Si 1.5, bal Fe.
Coated electrodes for joining and overlaying. Werkstoff Nr.
1.4733. 380 N/mm² YS; 570 N/mm² TS; 25 El.

THERMANIT 14K
Thyssen Edelstahlwerke AG
C 0.06-0.1, Cr 12.5-14, Ni 1.3, bal Fe.
Coated electrodes for joining, welding and overlaying.
Werkstoff Nr. 1.4008. 450 N/mm² YS; 650 N/mm² TS; 15 El.

THERMANIT 1520
Thyssen Edelstahlwerke AG
C 0.2-0.22, Cr 13-13.5, Mo 1.1, bal Fe.
Coated electrode for overlaying on water turbine
construction. Werkstoff Nr. 1.4119. 440 N/mm² YS; 590
N/mm² TS; 15 El.

THERMANIT 16/36
Thyssen Edelstahlwerke AG
C 0.22-0.25, Cr 18.5, Ni 35, bal Fe.
Coated electrodes for joining and overlaying. Werkstoff Nr.
1.4863. 320 N/mm² YS; 550 N/mm² TS; 15 El.

THERMANIT 1610
Thyssen Edelstahlwerke AG
C 0.06-0.065, Cr 17-17.5, Ti = 8 x C min, bal Fe.
Coated electrodes for joining and overlaying. Werkstoff Nr.
1.4502. 295 N/mm² YS; 490 N/mm² TS; 20 El.

THERMANIT 17
Thyssen Edelstahlwerke AG
C 0.05-0.07, Cr 17-18, bal Fe.
Coated electrode for overlaying on water turbine
construction. Werkstoff Nr. 1.4015. 340 N/mm² YS; 540
N/mm² TS; 20 El.

THERMANIT 17/06
Thyssen Edelstahlwerke AG
C 0.03-0.04, Cr 17, Mo 1, Ni 5.5, bal Fe.
Coated electrodes for joining and overlaying. Werkstoff Nr.
1.4405. 570 N/mm² YS; 800 N/mm² TS; 15 El.

THERMANIT 17/15 TT
Thyssen Edelstahlwerke AG
C 0.2, Mn 10.5, Cr 17-17.5, Ni 13-14, W 3.5, bal Fe.
Coated electrodes for joining and overlaying. 470 N/mm²
YS; 600 N/mm² TS; 30 El.

THERMANIT 1720
Thyssen Edelstahlwerke AG
C 0.18-0.2, Cr 16-16.5, Mo 1.1, bal Fe.
Coated electrodes for joining and overlaying. Werkstoff Nr.
1.4115. 500 N/mm² YS; 700 N/mm² TS; 15 El.

THERMANIT 1720
Thyssen Edelstahlwerke AG
C 0.18-0.2, Cr 16-16.5, Mo 1.1, bal Fe.
Coated electrode for overlaying on water turbine
construction. Werkstoff Nr. 1.4115. 500 N/mm² YS; 700
N/mm² TS; 15 El.

THERMANIT 1740
Thyssen Edelstahlwerke AG
C 0.3-0.38, Cr 16-16.5, Mo 1.1, bal Fe.
Coated electrode for overlaying on water turbine
construction. Werkstoff Nr. 1.4122. 600 N/mm² YS; 800
N/mm² TS; 12 El.

THERMANIT 18/15 E
Thyssen Edelstahlwerke AG
C 0.025-0.035, Cr 18, Ni 16.5, Mo 3.7, bal Fe.
Coated electrodes for joining and overlaying. Werkstoff Nr.
1.4438. 320 N/mm² YS; 520 N/mm² TS; 25 El.

THERMANIT 18/17 E
Thyssen Edelstahlwerke AG
C 0.025-0.035, Cr 18, Ni 17.5, Mo 4.5, bal Fe.
Coated electrodes for joining and overlaying. Werkstoff Nr.
1.4440. 320 N/mm² YS; 570 N/mm² TS; 34 El.

THERMANIT 19/15
Thyssen Edelstahlwerke AG
C 0.025-0.04, Cr 20-20.5, Ni 15.5-16.5, Mo 3, Mn 6-7.5, N
0.18, bal Fe.
Coated electrodes for joining and overlaying. Werkstoff Nr.
1.4455. 430 N/mm² YS; 650 N/mm² TS; 30 El.

THERMANIT 19/15 H, 21/17 E
Thyssen Edelstahlwerke AG
C 0.025-0.04, Cr 20-21, Ni 15.5-17.5, Mo 2.8-3, Mn 6-7, N
0.12-0.18, bal Fe.
Coated electrodes for welded consumables in urea synthesis
plants. Werkstoff Nr. 1.4455. 340-430 N/mm² YS; 520-650
N/mm² TS; 30-35 El.

THERMANIT 20/10
Thyssen Edelstahlwerke AG
C 0.05, Si 0.5-1, Cr 20-20.5, Ni 10.5, Mo 3.3, bal Fe.
Coated electrodes for joining and overlaying. Werkstoff Nr.
1.4431. 450 N/mm² YS; 640 N/mm² TS; 25 El.

THERMANIT 20/16
Thyssen Edelstahlwerke AG
C 0-0.04, Cr 20.5-21, Ni 16-17, Mo 3, Mn 5-7, N 0.18, bal Fe.
Coated electrodes for joining and overlaying. Werkstoff Nr.
1.3954. 430 N/mm² YS; 640 N/mm² TS; 30 El.

THERMANIT 20/16/510
Thyssen Edelstahlwerke AG
C 0-0.04, Cr 25, Ni 21, Mo 3.6, Mn 5.5, N 0.38, bal Fe.
Coated electrodes for joining and overlaying. Werkstoff Nr.
1.3984. 500 N/mm² YS; 700 N/mm² TS; 30 El.

THERMANIT 20/25 CU
Thyssen Edelstahlwerke AG
C 0.02-0.03, Cr 20, Mo 4.3, Ni 25, Cu 1.5, bal Fe.
Coated electrodes for joining and overlaying. Werkstoff Nr.
1.4539. 350 N/mm^2 YS; 550 N/mm^2 TS; 35 El.

THERMANIT 21/10 E
Thyssen Edelstahlwerke AG
C 0.025-0.04, Cr 21.5, Ni 10.5, bal Fe.
Coated electrodes for joining and overlaying. Werkstoff Nr.
1.4331. 350 N/mm^2 YS; 550 N/mm^2 TS; 30 El.

THERMANIT 21/10 E/RR
Thyssen Edelstahlwerke AG
C 0.025, Si 0.2, Mn 1.4, Cr 21.5, Ni 10.5, bal Fe.
Coated electrodes for joints on clad products. Werkstoff Nr.
1.4331.

THERMANIT 21/10 ENB/RR
Thyssen Edelstahlwerke AG
C 0.025, Si 0.2, Mn 1.4, Cr 21.5, Ni 11, Nb 0.7, bal Fe.
Coated electrodes for joints on clad products. Werkstoff Nr.
1.4555.

THERMANIT 21/33
Thyssen Edelstahlwerke AG
C 0.15, Cr 22, Ni 33, bal Fe.
Coated electrodes for joining and overlaying. Werkstoff Nr.
1.4850. 380 N/mm^2 YS; 600 N/mm^2 TS; 25 El.

THERMANIT 22/09
Thyssen Edelstahlwerke AG
C 0.025-0.04, Cr 22.5-23, Ni 9, Mo 3, bal Fe.
Coated electrodes for joining and overlaying. Werkstoff Nr.
1.4462. 480 N/mm^2 YS; 680 N/mm^2 TS; 25 El.

THERMANIT 23/11 MOZL
Thyssen Edelstahlwerke AG
C 0-0.035, Cr 22.5-23, Ni 13.5-14, Mo 2.3-2.6, bal Fe.
Coated electrodes for joining and overlaying. Werkstoff Nr.
1.4459. 450 N/mm^2 YS; 620 N/mm^2 TS; 30 El.

THERMANIT 23/11 ZL
Thyssen Edelstahlwerke AG
C 0-0.04, Cr 23.5, Ni 12.5, Mo 1, bal Fe.
Coated electrodes for joining and overlaying. Werkstoff Nr.
1.4556. 350 N/mm^2 YS; 600 N/mm^2 TS; 25 El.

THERMANIT 24/12 ENB
Thyssen Edelstahlwerke AG
C 0-0.04, Cr 24, Ni 12.5, Nb = 8-10 x C min, bal Fe.
Coated electrodes for joining and overlaying. Werkstoff Nr.
1.4556. 400 N/mm^2 YS; 600 N/mm^2 TS; 28 El.

THERMANIT 24/12 ENB/RR
Thyssen Edelstahlwerke AG
C 0-0.04, Si 0.2-0.8, Mn 1-1.7, Cr 24, Ni 12.5, Nb 0-0.65, Nb =
8-10 x C min, bal Fe.
Coated electrodes for joints on clad products. Werkstoff Nr.
1.4556. 400 N/mm^2 YS; 600 N/mm^2 TS; 28 El.

THERMANIT 25/14 E
Thyssen Edelstahlwerke AG
C 0-0.04, Cr 23.5-24.5, Ni 12.5-13, bal Fe.
Coated electrodes for joining and overlaying. Werkstoff Nr.
1.4332. 400 N/mm^2 YS; 550 N/mm^2 TS; 30 El.

THERMANIT 25/14 E/RR
Thyssen Edelstahlwerke AG
C 0-0.04, Si 0.2-0.8, Mn 0.8-1.2, Cr 23.5-24.5, Ni 12.5-13, bal
Fe.
Coated electrodes for joints on clad products. Werkstoff Nr.
1.4332. 400 N/mm^2 YS; 550 N/mm^2 TS; 30 El.

THERMANIT 25/22 H
Thyssen Edelstahlwerke AG
C 0.025-0.035, Cr 24.5-25, Ni 22, Mo 2.2, Mn 5-6, N
0.12-0.15, bal Fe.
Coated electrodes for joining and overlaying. Werkstoff Nr.
1.4465. 320-400 N/mm^2 YS; 550-600 N/mm^2 TS; 30 El.

THERMANIT 25/24 R
Thyssen Edelstahlwerke AG
C 0.3, Cr 25, Ni 24, Nb 1.2, Si 1, bal Fe.
Coated electrodes for joining and overlaying. Werkstoff Nr.
1.4830. 400 N/mm^2 YS; 600 N/mm^2 TS; 10 El.

THERMANIT 25/35
Thyssen Edelstahlwerke AG
C 0.4, Cr 26, Ni 35, Si 1, bal Fe.
Coated electrodes for joining and overlaying. Werkstoff Nr.
1.4857. 400 N/mm^2 YS; 600 N/mm^2 TS; 10 El.

THERMANIT 25/35 R
Thyssen Edelstahlwerke AG
C 0.4, Cr 26, Ni 35, Nb 1.3, Si 1, bal Fe.
Coated electrodes for joining and overlaying. Werkstoff Nr.
1.4853. 400 N/mm^2 YS; 600 N/mm^2 TS; 8 El.

THERMANIT 30
Thyssen Edelstahlwerke AG
C 0.05, Cr 30, Ni 1.5, Si 1.5, bal Fe.
Coated electrodes for joining and overlaying. Werkstoff Nr.
1.4773.

THERMANIT 30/10
Thyssen Edelstahlwerke AG
C 0.1-0.11, Cr 29-30, Ni 9, bal Fe.
Coated electrodes for joining and overlaying. Werkstoff Nr.
1.4337. 500 N/mm^2 YS; 750 N/mm^2 TS; 20 El.

THERMANIT 30/40 E
Thyssen Edelstahlwerke AG
C 0.02-0.03, Cr 28-29, Mo 3.8, Ni 36, Cu 1.8, bal Fe.
Coated electrodes for joining and overlaying. Werkstoff Nr.
2.4653. 350 N/mm^2 YS; 550 N/mm^2 TS; 25 El.

THERMANIT 30/50
Thyssen Edelstahlwerke AG
C 0.1-0.5, Cr 28.5, Ni 49, W 4.5, Si 1.3, bal Fe.
Coated electrodes for joining and overlaying. Werkstoff Nr.
2.4870. 450 N/mm^2 YS; 650 N/mm^2 TS; 5 El.

THERMANIT 5 M
Thyssen Edelstahlwerke AG
C 0.08, Cr 6, Mo 0.6, bal Fe.
Coated electrode for overlaying mangle rolls. Werkstoff Nr.
1.7373.

THERMANIT 50/50 NB
Thyssen Edelstahlwerke AG
C 0-0.1, Cr 50, Nb 1.8, Fe 0-1, bal Ni.
Coated electrodes for joining and overlaying. Werkstoff Nr.
2.4680. 650 N/mm^2 YS; 800 N/mm^2 TS; 3 El.

THERMANIT 625
Thyssen Edelstahlwerke AG
C 0.03-0.04, Cr 21.5-22, Mo 9-9.5, Nb 3-3.5, Fe 0-4, bal Ni.
Coated electrodes for joining and overlaying. Werkstoff Nr.
2.4621. 420 N/mm^2 YS; 760 N/mm^2 TS; 30 El.

THERMANIT 625/RR
Thyssen Edelstahlwerke AG
C 0-0.04, Si 0.2-0.4, Mn 0.3-1, Cr 21.5-22, Mo 9-9.5, Fe 0-3,
Nb 3-3.5, bal Ni.
Coated electrodes for joining and overlaying. Werkstoff Nr.
2.4831. 420 N/mm^2 YS; 760 N/mm^2 TS; 30 El.

THERMANIT 9
Thyssen Edelstahlwerke AG
C 0.07, Cr 9, bal Fe.
Coated electrodes for joining and overlaying. Werkstoff Nr.
1.4716. 550 N/mm^2 YS; 700 N/mm^2 TS; 18 El.

THERMANIT A
Thyssen Edelstahlwerke AG
C 0.06-0.07, Si 0.4-0.9, Cr 19-19.5, Ni 12, Mo 2.8, Nb = 8-12
x C, bal Fe.
Coated electrodes for joining and overlaying. Werkstoff Nr.
1.4576. 380 N/mm^2 YS; 550 N/mm^2 TS; 30 El.

THERMANIT ATS 115 W 160
Thyssen Edelstahlwerke AG
C 0.1, Cr 20, Ni 10, Mo 15, bal Co.
Coated electrodes for joining and overlaying. Werkstoff Nr.
2.4964. 350 N/mm^2 YS; 800 N/mm^2 TS; 15 El.

THERMANIT ATS 15
Thyssen Edelstahlwerke AG
C 0.1, Cr 16.5, Ni 17, Mo 1.8, Nb = 10 x C, bal Fe.
Coated electrodes for joining and overlaying. Werkstoff Nr.
1.4981. 340 N/mm^2 YS; 540 N/mm^2 TS; 20 El.

THERMANIT ATS 4
Thyssen Edelstahlwerke AG
C 0.05, Cr 18.5, Ni 9.5, bal Fe.
Coated electrodes for joining and overlaying. Werkstoff Nr.
1.4948. 320 N/mm^2 YS; 500 N/mm^2 TS; 35 El.

THERMANIT C
Thyssen Edelstahlwerke AG
C 0.13, Cr 25, Ni 20, Si 1, bal Fe.
Coated electrodes for joining and overlaying. Werkstoff Nr.
1.4829. 320 N/mm^2 YS; 550 N/mm^2 TS; 25 El.

THERMANIT CM
Thyssen Edelstahlwerke AG
C 0.12, Cr 25, Ni 20, Mn 4, Si 1, bal Fe.
Coated electrodes for joining and overlaying. Werkstoff Nr.
1.4842. 380 N/mm^2 YS; 550 N/mm^2 TS; 30 El.

THERMANIT CR
Thyssen Edelstahlwerke AG
C 0.4, Cr 25.5, Ni 21.5, Si 1, bal Fe.
Coated electrodes for joining and overlaying. Werkstoff Nr.
1.4846. 400 N/mm^2 YS; 600 N/mm^2 TS; 10 El.

THERMANIT D
Thyssen Edelstahlwerke AG
C 0.11, Cr 22, Ni 11, Si 1.2, bal Fe.
Coated electrodes for joining and overlaying. Werkstoff Nr.
1.4829. 320 N/mm^2 YS; 550 N/mm^2 TS; 30 El.

THERMANIT G
Thyssen Edelstahlwerke AG
C 0.05-0.06, Si 0.4-0.9, Cr 18.5-19, Ni 11.5, Mo 2.8, bal Fe.
Coated electrodes for joining and overlaying. Werkstoff Nr.
1.4403. 320 N/mm^2 YS; 550 N/mm^2 TS; 35 El.

THERMANIT GE
Thyssen Edelstahlwerke AG
C 0.025-0.04, Si 0.4-0.9, Cr 18.5-19, Ni 11.5, Mo 2.8, bal Fe.
Coated electrodes for joining and overlaying. Werkstoff Nr.
1.4430. 320 N/mm^2 YS; 550 N/mm^2 TS; 35 El.

THERMANIT GE/RR
Thyssen Edelstahlwerke AG
C 0-0.04, Si 0.2-0.8, Mn 1-1.8, Cr 18.5-19, Ni 11.5-12.5, Mo
2.8, bal Fe.
Coated electrodes for joining and overlaying. Werkstoff Nr.
1.4430. 320 N/mm^2 YS; 550 N/mm^2 TS; 35 El.

THERMANIT H
Thyssen Edelstahlwerke AG
C 0.06-0.07, Si 0.4-0.9, Cr 19.5, Ni 9.5, Nb = 8-12 x C min,
bal Fe.
Coated electrodes for joining and overlaying. Werkstoff Nr.
1.4551. 380 N/mm^2 YS; 550 N/mm^2 TS; 30 El.

THERMANIT HE
Thyssen Edelstahlwerke AG
C 0.025-0.035, Si 0.4-0.9, Cr 19-20, Ni 10-10.5, Nb = 8-13 x C
min, bal Fe.
Coated electrodes for joining and overlaying. Werkstoff Nr.
1.4551. 400 N/mm^2 YS; 600 N/mm^2 T3; 30 El.

THERMANIT HE/RR
Thyssen Edelstahlwerke AG
C 0-0.035, Si 0.2-0.8, Mn 1-1.5, Cr 19-20, Ni 10-10.5, Nb
0-0.65, Nb = 13 x C min, bal Fe.
Coated electrodes for joining and overlaying. Werkstoff Nr.
1.4551. 400 N/mm^2 YS; 600 N/mm^2 TS; 30 El.

THERMANIT J

Thyssen Edelstahlwerke AG
C 0.05-0.06, Si 0.4-0.9, Cr 19.5-20, Ni 9.5-10, bal Fe.
Coated electrodes for joining and overlaying. Werkstoff Nr.
1.4302. 320 N/mm^2 YS; 550 N/mm^2 TS; 35 El.

THERMANIT JE

Thyssen Edelstahlwerke AG
C 0.025-0.04, Si 0.4-0.9, Cr 19.5-20, Ni 9.5-10, bal Fe.
Coated electrodes for joining and overlaying. Werkstoff Nr.
1.4316. 320 N/mm^2 YS; 550 N/mm^2 TS; 35 El.

THERMANIT JE/RR

Thyssen Edelstahlwerke AG
C 0-0.04, Si 0.2-0.9, Mn 0.8-1.7, Cr 19.5-20, Ni 9.5-10, bal Fe.
Coated electrodes for joining and overlaying. Werkstoff Nr.
1.4316. 320 N/mm^2 YS; 550 N/mm^2 TS; 35 El.

THERMANIT L

Thyssen Edelstahlwerke AG
C 0.06, Cr 26, Ni 4.5, bal Fe.
Coated electrodes for joining and overlaying. Werkstoff Nr.
1.4820. 500 N/mm^2 YS; 700 N/mm^2 TS; 20 El.

THERMANIT LMS

Thyssen Edelstahlwerke AG
C 0.07, Cr 24.5, Ni 8.5, Mo 1.5, bal Fe.
Coated electrodes for joining and overlaying. Werkstoff Nr.
1.4582. 500 N/mm^2 YS; 700 N/mm^2 TS; 15 El.

THERMANIT MTS 4

Thyssen Edelstahlwerke AG
C 0.18-0.26, Cr 11.5, Mo 1, Ni 0-0.7, W 0.5, V 0.3, bal Fe.
Coated electrodes for joining and overlaying. Werkstoff Nr.
1.4937. 590 N/mm^2 YS; 700 N/mm^2 TS; 15 El.

THERMANIT NI 60

Thyssen Edelstahlwerke AG
C 0.4, Si 3, Cr 15.5, Ni 61, bal Fe.
Welding consumables for welding valve stem ends. Meets
DIN 8555.

THERMANIT NI 60 A

Thyssen Edelstahlwerke AG
C 0.7, Si 3, Cr 15.5, Ni 61, bal Fe.
Welding consumables for welding valve stem ends. Meets
DIN 8555.

THERMANIT NICRO 182

Thyssen Edelstahlwerke AG
C 0-0.05, Cr 16, Mn 6.5, Nb 2, Fe 6, 67.0 Ni min.
Coated electrodes for joining and overlaying. Werkstoff Nr.
2.4620. 370 N/mm^2 YS; 620 N/mm^2 TS; 35 El.

THERMANIT NICRO 182/RR

Thyssen Edelstahlwerke AG
C 0-0.05, Si 0.4, Mn 6.5, Cr 16, Fe 6, Nb 2, 67.0 Ni min.
Coated electrodes for joining and overlaying. Werkstoff Nr.
2.4620. 370 N/mm^2 YS; 620 N/mm^2 TS; 35 El.

THERMANIT NICRO 82

Thyssen Edelstahlwerke AG
C 0.03-0.05, Cr 19.5, Mn 2.8-4, Nb 2, Fe 0.03-4, 67.0 Ni min.
Coated electrodes for joining and overlaying. Werkstoff Nr.
2.4648. 380 N/mm^2 YS; 620 N/mm^2 TS; 35 El.

THERMANIT NICRO 82/RR

Thyssen Edelstahlwerke AG
C 0-0.05, Si 0.2-0.4, Mn 2.8-4, Cr 19.5, Fe 0-3, Nb 2.1, Ti 0.3, bal Ni.
Coated electrodes for joining and overlaying. Werkstoff Nr.
2.4648. 380 N/mm^2 YS; 620 N/mm^2 TS; 35 El.

THERMANIT NIMO C

Thyssen Edelstahlwerke AG
C 0.015-0.02, Si 0.1-0.3, Cr 19-20, Mo 15, Fe 0-1.5, bal Ni.
Coated electrodes for joining and overlaying. Werkstoff Nr.
2.4657. 450 N/mm^2 YS; 700 N/mm^2 TS; 40 El.

THERMANIT P 35 A

Thyssen Edelstahlwerke AG
C 0.4, Si 2.5, Cr 21, Co 35, Ni 35, W 2.5, Fe 0-2.
Welding consumables for welding valve stem ends. Meets
DIN 8555.

THERMANIT P.33

Thyssen Edelstahlwerke AG
C 0.12, Mn 19.5, Si 1.5, Cr 12.5, Ni 1.5, bal Fe.
Welded: 100,000 TS; 57,000 YS; 40 El; 45 RA. For welding
rod; austenitic. *Obsolete*

THERMANIT X

Thyssen Edelstahlwerke AG
C 0.1, Cr 18.5, Ni 8, Mn 7, bal Fe.
Coated electrodes for joining and overlaying. Werkstoff Nr.
1.4370. 320 N/mm^2 YS; 600 N/mm^2 TS; 40 El.

THERMANIT XS

Thyssen Edelstahlwerke AG
C 0.2, Cr 19, Ni 8, Mn 5, bal Fe.
Coated electrode for overlaying on rails. Werkstoff Nr. 1.4370.
350 N/mm^2 YS; 600 N/mm^2 TS; 35 El.

THERMAX 10 A

Thyssen Edelstahlwerke AG
C 0.1, Si 2.35, Mn 1.25, Cr 19, Ni 10, bal Fe.
For corrosion and heat resistant parts; corrosion and heat
resistant, austenitic. *Obsolete*

THERMAX 102

Sheepbridge Alloy Castings Ltd.
C 0-0.25, Si 0.5-1.5, Mn 0-1, Ni 10-14, Cr 20-25, W 2.5-3.5,
bal Fe.
Cast, annealed: 35 tsi TS; 10 El; 240 Brin max. For furnace
parts and heat treating trays. B.S. 1648 Grade E.

THERMAX 104

Sheepbridge Alloy Castings Ltd.
C 0-0.25, Si 0.5-1.5, Mn 0.5-1.5, Ni 10-14, Cr 20-25, W 0-1,
bal Fe.
Cast, annealed: 35 tsi TS; 10 El; 200-240 Brin. For furnace
parts and heat treating trays. B.S. 1648 Grade E.

THERMAX 10A

Remanit GmbH
C 0-0.2, Cr 22-24, Ni 12-15, Mn 0-2, Si 0-1, bal Fe.
Annealed: 85,000-95,000 TS; 40,000-50,000 YS; 45-55 El;
150-185 Brin. For heat treating boxes, oil refinery and
chemical plant equipment. Type 309; austenitic; heat
resistant.

THERMAX 10FAL

Thyssen Edelstahlwerke AG
C 0.1, Si 1, Cr 18, Al 8, bal Fe.
For chemical and plastic plant equipment; corrosion
resistant. *Obsolete*

THERMAX 11 A

Thyssen Edelstahlwerke AG
C 0.15, Si 2.3, Mn 1.25, Cr 23.5, Ni 19.5, bal Fe.
For corrosion and heat resistant parts; corrosion and heat
resistant, austenitic. *Obsolete*

THERMAX 11 FN

Thyssen Edelstahlwerke AG
C 0.2, Si 0.5, Mn 0.6, Cr 25, Ni 2.5-3, bal Fe.
For corrosion and heat resistant parts; corrosion and heat
resistant. *Obsolete*

THERMAX 115

Sheepbridge Alloy Castings Ltd.
C 0.08-0.15, Mn 0.5-1.25, P 0-0.04, S 0-0.04, Cr 10-12.5, Ni
0-1.5, Mo 0.5-1, V 0.1-0.5, Nb 0.2-0.6, Si 0-1, bal Fe.
Cast, heat treated: 60 tsi TS min; 50 tsi YS min; 10 El;
286-331 Brin. Good creep resistance; for use to 650°C for
gas turbine applications. Modified AISI 410.

THERMAX 11A

Remanit GmbH
C 0-0.25, Cr 24-26, Ni 19-22, bal Fe.
Annealed: 95,000 TS; 45,000 YS; 50 El; 65 RA; 180 Brin. At
1200°F: 57,000 TS; 22,000 YS; 32 El; 45 RA. For furnace
parts, heat treating boxes, valves, and pumps. Type 310;
austenitic; heat resistant.

THERMAX 11AS G

Thyssen Edelstahlwerke AG
C 0.4, Si 1.8, Cr 26, Ni 14, bal Fe.
For heat treat boxes, pots, furnace parts; heat resistant.
Obsolete

THERMAX 11F

Hochfrequenz-Tiegelstahl GmbH
C 0.6, Si 1.5, Cr 22, bal Fe.
For valve and pump parts, furnace equipment; corrosion and
heat resistant.

THERMAX 12 F.-CR

Thyssen Edelstahlwerke AG
C 0.15, Si 1.75, Mn 1, Cr 25, Ni 0.5, bal Fe.
For corrosion and heat resistant parts; corrosion and heat
resistant. *Obsolete*

THERMAX 12FAL

Thyssen Edelstahlwerke AG
C 0.1, Si 1.5, Cr 24, Al 1.5, bal Fe.
For furnace parts, chemical plant equipment; corrosion and
heat resistant. *Obsolete*

THERMAX 12FH G

Thyssen Edelstahlwerke AG
C 1.3, Si 1.8, Cr 29, bal Fe.
For chemical plant equipment, furnace parts; corrosion and
heat resistant. *Obsolete*

THERMAX 12FN G

Thyssen Edelstahlwerke AG
C 0.4, Si 1.2, Cr 27, Ni 4, bal Fe.
For furnace parts, heat treat boxes; corrosion and heat
resistant. *Obsolete*

THERMAX 13 CO-G

Thyssen Edelstahlwerke AG
C 0.1, Ti 0.18, Cb 0.6, Cr 26-30, Co 47-52, bal Fe.
Cast: 135,000 TS; 48,000 YS; 10 El; 10 RA; 250 Brin. For
furnace baffles, burner tips, sintering grates, quench baskets.
Corrosion, heat and thermal shock resistant. *Obsolete*

THERMAX 25

Sheepbridge Alloy Castings Ltd.
C 0.2-0.5, Si 0-2, Mn 0-2, Cr 23-26, Ni 18-22, Mo 0-0.5, bal
Fe.
At 800°C: 11.3 tsi. 0.2% proof stress. Austenitic corrosion and
heat resisting alloy. Cracking tubes in the gas and petro-
chemical industry and for furnace parts; cast alloy. Similar to
ASTM A297 HK.

THERMAX 30

Sheepbridge Alloy Castings Ltd.
C 0.2-0.35, Cr 28-32, bal Fe.
Annealed: 62,800 TS; 2 El; 200-240 Brin. For furnace parts,
heat treating boxes; Britain 1648 B; scale resistant to 1100°C.

THERMAX 37

Sheepbridge Alloy Castings Ltd.
C 0.2-0.5, Si 0-2, Mn 0-2, Cr 15-19, Ni 33-37, bal Fe.
At 800°C: 15.4 tsi. TS; 16 El. Cast alloy. Austenitic alloy for
high temperature equipment such as furnace parts,
carburizing containers.

THERMAX 45

Sheepbridge Alloy Castings Ltd.
C 0-0.3, Si 0-2.5, Mn 0-1.5, Cr 20, Ni 45, bal Fe.
Cast: 70,000 TS; 6 El; 170-200 Brin. For furnace parts such as
heat resisting trays and rollers; good resistance to scaling
and thermal shock.

THERMAX 4713, ETC.
Thyssen Edelstahlwerke AG
See Werkstoff Nr. 1.4713.
Heat resisting steels; 11 grades.

THERMAX 519
Sheepbridge Alloy Castings Ltd.
C 0.25-0.35, Mn 0-1, Si 0-0.3, P 0-0.03, Ni 23-25, Nb 1.4-1.8, S 0-0.04, Cr 23-25, bal Fe.
36-38 tsi TS; 13-15 tsi YS (0.1% proof stress); 18-25 El. Heat resisting alloy for centrifugal cast tubes in the reformer and petrochemical industry; good stress rupture properties.

THERMAX 531
Sheepbridge Alloy Castings Ltd.
C 0.35-0.5, Si 2, Mn 2, Cr 23-26, Ni 32-35, W 1-1.5, bal Fe.
Stress to rupture at 1038°C: 2,060 psi (1000 h). Austenitic heat resisting cast alloy with good oxidation resistance, creep strength, and stress to rupture at elevated temperatures.

THERMAX 532
Sheepbridge Alloy Castings Ltd.
C 0.08-0.12, Mn 0.5-0.15, Si 0.3-1, Ni 29-33, Cr 18-22, Mo 0-0.5, Nb 0.5-0.6, bal Fe.
At 800°C: 12.2 tsi TS; 6.2 tsi YS; 56 El. Austenitic heat resisting cast alloy.

THERMAX 533
Sheepbridge Alloy Castings Ltd.
C 0.3-0.4, Mn 0-1.5, Si 0-2, Cr 23-26, Ni 33-37, Nb 0.2-1.5, bal Fe.
At 800°C: 17.3 tsi TS; 9.8 tsi YS; 18 El. Austenitic cast alloy for high temperature operation.

THERMAX 60
Sheepbridge Alloy Castings Ltd.
C 0-0.5, Cr 15-20, Ni 60-65, bal Fe.
Annealed: 67,200 TS; 3 El; 180-210 Brin. For furnace parts, muffles, heat treating equipment; corrosion and heat resistant.

THERMAX 638
Sheepbridge Alloy Castings Ltd.
C 0.4-0.55, Si 0-1, Mn 0-1, Ni 34-36, Cr 25-27, Co 14.16, W 4-5.5, Nb 0.7-1.3, bal Fe.
At 800°C: 24.8 tsi TS; 13.4 tsi YS; 15 El. Austenitic heat resisting cast alloy.

THERMAX 657
Sheepbridge Alloy Castings Ltd.
C 0-0.1, Fe 0-1, Si 0-0.5, Cr 48-52, Nb 1.4-1.7, 0.16 N_2 max, 0.2 C max + N_2, bal Ni.
39-52 tsi TS; 16-27 tsi YS (0.2% proof stress); 10-30 El. High strength cast Ni-Cr-Nb alloy; resistant to ash products of low grade fuel; good creep-rupture properties.

THERMAX 70
Sheepbridge Alloy Castings Ltd.
C 0.4-0.5, Si 1.2, Mn 1.5, Ni 45-50, Cr 26-29, W 4-6, bal Fe.
Austenitic heat resisting cast alloy with good stress rupture properties and thermal fatigue resistance.

THERMAX 75
Sheepbridge Alloy Castings Ltd.
C 0.1, Si 0-1, Mn 0-1, Fe 0-5, Ti 0.4, Cr 20, bal Ni.
For furnace castings and parts for gas turbines; resists scaling and thermal shock.

THERMAX 8A
Thyssen Edelstahlwerke AG
C 0.1, Mn 1, Cr 18, Ni 10, bal Fe.
Annealed: 85,000 TS; 40,000 YS; 50 El; 65 RA; 160 Brin. For chemical plant equipment; stainless, austenitic. *Obsolete*

THERMAX 8FAL
Thyssen Edelstahlwerke AG
C 0.1, Cr 6.5, Al 0.75, bal Fe.
For oil refinery equipment; heat resistant. *Obsolete*

THERMAX 90
Sheepbridge Alloy Castings Ltd.
C 0.1, Si 0-1, Mn 0-1, Al 0.9, Ti 1.6, Fe 0-5, Cr 20, Co 18, bal Ni.
For gas turbines and jet engine components; high creep properties in the range 1100-1650°F.

THERMCO 50
Otto Junker GmbH
C 0.1, Ti 0.18, Cb 0.6, Cr 26-30, Co 47-52, bal Fe.
Cast: 135,000 TS; 48,000 YS; 10 El; 10 RA; 250 Brin. For furnace baffles, burner tips, sintering grates, quench baskets. Corrosion, heat and thermal shock resistant. *Obsolete*

THERMELAST 4002
Vacuumschmelze GmbH
Fe, Ni, Cr, Ti, Al.
Constant modulus alloy, hardenable, for balance springs, leaf springs, diaphragms. 400 Vickers min; 1300 N/mm² YS min; 1350 N/mm² TS min.

THERMELAST 4290
Vacuumschmelze GmbH

Modulus alloy for springs in scales, tuning forks, and oscillators. 400 Vickers min; 1300 N/mm² YS min; 1450 N/mm² TS min.

THERMELAST 5409
Vacuumschmelze GmbH
Modulus alloy for springs in scales, tuning forks, and oscillators. 450 Vickers min; 1350 N/mm² YS min; 1500 N/mm² TS min.

THERMENOL
Colt Industries
Al 16, Mo 4-5, bal Fe.
Cold drawn: 62,000-135,000 TS; 3 El. For aircraft applications; high resistance to oxidation. *Obsolete*

THERMIMPHY
Creusot-Loire
C 0.12, Cr 7, Mo 0.5, V 0.3, bal Fe.
Quenched and tempered: 590-730 N/mm² psi TS; 440 N/mm² psi YS min; 15 El. For high temperature bolts and parts. AFNOR Z 12 CDV 7; ASTM A 182 Gr. F7.

THERMISILID
Krupp Stahl AG
cast iron. Si 14-16, special cast iron.
350-290 Brin. For chemical equipment construction, acid plants; acid proof and heat resisting to H_2SO_4. *Obsolete*

THERMISILID EXTRA
English manufacture
Si 14-18, bal Fe.
Cast: 250 Brin. For acid plants, explosives manufacturing; corrosion and acid resistant.

THERMIT
Goldschmidt A.G.
Sb 14-16, Sn 5-7, Cu 0.8-1.2, Ni 0.7-1.5, Pb 72-78.5, As 0.3-0.8, Cd 0.7-1.5.
29 Brin. For bearings for pumps, railroads, engines. Standard German white metal; cast alloy.

THERMKON 62
CMW Inc.
W 90, Cu 10.
Electrical conductivity: 32% IACS. Hardness: 27 Rock C. Density: 17.17 g/cm³. 116,000 psi YS; 125,000 psi TS; 3.2 x 10^{-6} in./in.·°F thermal expansion; 91 Btu/h/ft·°F thermal conductivity; 180,000 psi flexual strength; 37 x 10^6 psi modulus of elasticity. Custom material engineered for specific applications.

THERMKON 68
CMW Inc.
W 85, Cu 15.
Electrical conductivity: 35% IACS. Hardness: 25 Rock C. Density: 16.60 g/cm³. 105,000 psi YS; 120,000 psi TS; 3.6 x 10^{-6} in./in.·°F thermal expansion; 97 Btu/h/ft·°F thermal conductivity; 175,000 psi flexural strength; 36 x 10^6 psi modulus of elasticity. Custom material engineered for specific applications.

THERMKON 76
CMW Inc.
W 80, Cu 20.
Electrical conductivity: 41% IACS. Hardness: 103 Rock B. Density: 15.56 g/cm³. 89,000 psi YS; 115,000 psi TS; 4.2 x 10^{-6} in./in.·°F thermal expansion; 104 Btu/h/ft·°F thermal conductivity; 170,000 psi flexural strength; 35 x 10^6 psi modulus of elasticity. Custom material engineered for specific applications.

THERMKON 83
CMW Inc.
W 75, Cu 25.
Electrical conductivity: 45% IACS. Hardness: 98 Rock B. Density: 14.84 g/cm³. 76,000 psi YS; 100,000 psi TS; 4.6 x 10^{-6} in./in.·°F thermal expansion; 110 Btu/h/ft·°F thermal conductivity; 150,000 psi flexural strength; 34 x 10^6 psi modulus of elasticity. Custom material engineered for specific applications.

THERMO
Crucible Specialty Metals
C, alloy, bal Fe.
For tools, dies; oil hardening.

THERMO 50
Now JUNKER THERMO 50.

THERMO DIE
Colt Industries
C 0.4, Cr 2, V 0.2, bal Fe.
Oil hardened. For hot work, dies, die casting dies for aluminum work; no heat checking. *Obsolete*

THERMO DIE
Hawkridge Bros. Co.
C 0.4, Ni 2.4, Mn 0.65, V 0.2, bal Fe.
For Zn die casting dies; resists heat checking. *Obsolete*

THERMO DIE
Crucible Specialty Metals
C 0.4, Cr 2, V 0.2, bal Fe.
Oil hardened. For hot work, dies, die casting dies for aluminum work; no heat checking. *Obsolete*

THERMO-ELECTRIC NICKEL BABBITT
Duraloy Blaw-Knox
Cu 2.25, Sn 90, Ni 0.25, Sb 7.5.
10,660 TS; 13 El; 24 Brin. For Babbitt service, bearings; C.E.L. 6300. *Obsolete*

THERMO-LECTRIC COPPER-HARDENED BABBITT
Duraloy Blaw-Knox
Sn 20, Pb 68, Sb 11, Cu 1.
9,050 TS; 1.7 El; 18 Brin. For Babbitt bearings; C.E.L. 4000. *Obsolete*

THERMOCROM
Willworthy Piston Ring Ltd.
C 3.2, Si 2.2, Mo 0.2, Cr 0.8, bal Fe.
For piston rings; cast iron.

THERMODIC HW
Bethlehem Steel Corp.
C 0.6, Mn 0.5, Cr 1.05, Mo 0.9, Ni 1.5.
For hot work dies for punching, shearing and forming; hot work steel, oil hardened. *Obsolete*

THERMODIE
Bethlehem Steel Corp.
C 0.6, Cr 1, Mo 0.9, Ni 1.4, bal Fe.
For tools, dies, punches; tough. *Obsolete*

THERMODIT 101
Creusot-Loire
C 0.38, Si 1, Cr 5.25, Mo 1.35, V 0.5, bal Fe.
Hot work tool steel for forging, upsetting and hot forming dies. AFNOR Z 38 CDV 5; AISI H 11.

THERMODUR 10 CR 20 32 NB
Bergische Stahl Industrie
C 0.1, Ni 32, Cr 20, Nb 1, bal Fe.
Austenitic cast steel for elevated temperature service. SEW 1.4859; G-X 10 NiCrNb 32 20.

THERMODUR 10CN5050
Bergische Stahl Industrie
C 0-0.1, Si 0.1-0.3, Mn 0.2-0.5, Cr 50-52, bal Ni.
Cast: 64,000 min TS; 42,000 min YS; 5 min El. Rupture strength: 2,400 psi, 10,000 h, 800 C. Resistant to scaling in oxidizing and ashfree and sulfur free atmosphere to 1110 C. DIN G-X10CrNi5050; ASTM A560-50/50. *Obsolete*

THERMODUR 10CN6040
Bergische Stahl Industrie
C 0-0.1, Si 0.1-0.3, Mn 0.2-0.5, Cr 58-61, Fe 0-2, bal Ni.
Cast: 100,000 min TS; 71,000 min YS. Rupture strength: 1,850 psi, 10,000 h, 800 C. Resistance to scaling in oxidizing and ashfree and sulfur free atmosphere to 1110 C. DIN G-X10CrNi6040; ASTM A560-60/40. *Obsolete*

THERMODUR 15 CN 13
Bergische Stahl Industrie
C 0.1-0.15, Si 0.5-1, Mn 1-1.5, Cr 22.5-24, Ni 12-13, bal Fe.
Cast: 70,000 min TS; 30,000 YS min; 30 El min. Rupture strength: 185 psi, 10,000 h, 800°C. Resistance to scaling in oxidizing and ashfree and sulfur free atmosphere to 1150°C. DIN-X15CrNiSi2512; ASTM A351-CH20.

THERMODUR 15 CN 15 T
Bergische Stahl Industrie
C 0.1-0.15, Si 0.5-1, Mn 1-1.5, Cr 19-20, Ni 14.5-15.5.
Cast: 64,000-85,000 TS; 28,000 YS min; 20 El min. Rupture strength: 1850 psi, 10,000 h, 800°C. Resistance to scaling in oxidizing and ashfree and sulfur free atmosphere to 1050°C. DIN G-X15CrNiSi2014.

THERMODUR 15 CN 20
Bergische Stahl Industrie
C 0.1-0.15, Si 0.5-1, Mn 1-1.5, Cr 23.5-25, Ni 20-21, bal Fe.
Cast: 64,000-92,000 TS; 30,000 YS min; 30 El min. Rupture strength: 2300 psi, 10,000 h, 800°C. Resistance to scaling in oxidizing and ashfree and sulfur free atmosphere to 1150°C. SEW-Nr.1.4849; DIN G-X15CrNiSi2520-ASTM A351-CK20.

THERMODUR 15CN1775
Bergische Stahl Industrie
C 0.1-0.15, Si 0.5-1, Mn 1-1.5, Cr 15-16, Ni 72-74, Fe 8-10.
Cast: 68,000-82,000 TS; 31,000 min YS; 12 min El. Rupture strength: 4,500 psi, 10,000 h, 800 C. Resistance to scaling in oxidizing and ashfree and sulfur free atmosphere to 950 C. DIN G-X15NiCrSi7517. *Obsolete*

THERMODUR 15CN2035
Bergische Stahl Industrie
C 0.1-0.15, Si 0.5-1, Mn 1-1.5, Cr 20-22, Ni 34-35, bal Fe.
Cast: 64,000-78,000 TS; 28,000 min YS; 20 min El. Rupture strength: 2,700 psi, 10,000 h, 800 C. Resistance to scaling in oxidizing and ashfree and sulfur free atmosphere to 1100 C. DIN G-X15NiCrSi3520. *Obsolete*

THERMODUR 15CN2535
Bergische Stahl Industrie
C 0.1-0.15, Si 0.5-1, Mn 1-1.5, Cr 25-26.5, Ni 34-35, bal Fe.
Cast: 64,000-92,000 TS; 30,000 min YS; 10 min El. Rupture strength: 2,700 psi, 10,000 h, 800 C. Resistance to scaling in oxidizing and ashfree and sulfur free atmosphere to 1150 C. DIN G-X15CrNiSi3525. *Obsolete*

THERMODUR 20 CMW 5
Bergische Stahl Industrie
C 0.2, Mn 0.6, Cr 2.3, Mo 1, bal Fe.
Ferritic cast steel for elevated temperature service. SEW 1.7379; GS-18 CrMo 9 10.

THERMODUR 30 CN 13
Bergische Stahl Industrie
C 0.25-0.3, Si 0.5-1, Mn 1-1.5, Cr 24-25, Ni 12-13, bal Fe.
Cast: 64,000-92,000 TS; 10 El min. Rupture strength: 2400 psi, 10,000 h, 800°C. Resistance to scaling in oxidizing and ashfree and sulfur free atmosphere to 1150°C. SEW-Nr. 1.4837; DIN G-X35CrNiSi2512; ASTM A447-II.

THERMODUR 30 CN 24 24 NB
Bergische Stahl Industrie
C 0.3, Cr 24, Ni 24, Si 2, Nb, bal Fe.
Austenitic cast steel for elevated temperature service. SEW 1.4855; G-X 30 CrNiSiNb 24 24.

THERMODUR 30 CN 2535
Bergische Stahl Industrie
C 0.25-0.3, Si 0.5-1, Mn 1-1.5, Cr 25-26.5, Ni 34-35, bal Fe.
Cast: 64,000-92,000 TS; 35,000 YS min; 10 El min. Rupture strength: 5100 psi, 10,000 h, 800°C. Resistance to scaling in oxidizing and ashfree and sulfur free atmosphere to 1150°C. SEW-Nr. 1.4857; DIN G-X30NiCrSi3525.

THERMODUR 30CN15T
Bergische Stahl Industrie
C 0.25-0.3, Si 0.5-1, Mn 0.5-1, Cr 19-20, Ni 14.5-15.5, bal Fe.
Cast: 64,000-92,000 TS; 28,000 min YS; 12 min El. Rupture strength: 3,000 psi, 10,000 h, 800 C. Resistance to scaling in oxidizing and ashfree and sulfur free atmosphere to 1050 C. SEW-Nr. 1.4832; DIN G-X25CrNiSi2014. *Obsolete*

THERMODUR 30CN20
Bergische Stahl Industrie
C 0.25-0.3, Si 0.5-1, Mn 1-1.5, Cr 23.5-25, Ni 20-21, bal Fe.
Cast: 64,000-92,000 TS; 35,000 min YS; 10 min El. Rupture strength: 4,700 psi, 10,000 h, 800 C. Resistance to scaling in oxidizing and ashfree and sulfur free atmosphere to 1150 C. SEW-Nr. 1.4848; DIN G-X35CrNiSi2520; ASTM A351-HK30. *Obsolete*

THERMODUR 30CN2035
Bergische Stahl Industrie
C 0.25-0.3, Si 0.5-1, Mn 1-1.5, Cr 20-22, Ni 34-35, bal Fe.
Cast: 64,000-92,000 TS; 10 min El. Rupture strength: 5,100 psi, 10,000 h, 800 C. Resistance to scaling in oxidizing and ashfree and sulfur free atmosphere to 1100 C. DIN G-X30NiCrSi3520. *Obsolete*

THERMODUR 30CN2055
Bergische Stahl Industrie
C 0.25-0.35, Si 0.5-1, Mn 1-1.5, Cr 20-22, Ni 52-54, bal Fe.
Cast: 57,000-78,000 TS; 28,000 min YS; 4 min El. Rupture strength: 4,800 psi, 10,000 h, 800 C. Resistance to scaling in oxidizing and ashfree and sulfur free atmosphere to 1000 C. DIN G-X35NiCr5422. *Obsolete*

THERMODUR 40 CN 20
Bergische Stahl Industrie
C 0.2-0.5, Si 1-2.5, Mn 0-1.5, Cr 24-27, Ni 19-21, bal Fe.
Austenitic cast steel for elevated temperature service. SEW 1.4848; G-X 40 CrNiSi 25 20.

THERMODUR 40 CN CW 25
Bergische Stahl Industrie
C 0.35-0.4, Si 0.5-1, Mn 0.5-1, Cr 25-26.5, Ni 34-35, Co 14-15, W 5-5.5, bal Fe.
Cast: 71,000-100,000 TS; 38,000 YS min; 4 El min. Rupture strength: 2800 psi, 10,000 h, 1000°C. Resistance to scaling in oxidizing and ashfree and sulphur free atmosphere to 1200°C. DIN G-X40NiCrCoW352515.

THERMODUR 50 CN 30 30
Bergische Stahl Industrie
C 0.5, Ni 30, Cr 30, bal Fe.
Austenitic cast steel for elevated temperature service. SEW 1.4868; G-X 50 CrNi 30 30.

THERMODUR 50 CNW 30 50
Bergische Stahl Industrie
C 0.45-0.55, Si 0.5-1, Mn 0.5-1, Cr 27-28.5, Ni 46-48, W 4.5-5, bal Fe.
Cast: 57,000-85,000 TS; 35,000 YS min; 4 El min. Rupture strength: 2400 psi, 10,000 h, 1000°C. Resistance to scaling in oxidizing and ashfree and sulfur free atmosphere to 1150°C. SEW-Nr. 2.4879; DIN G-X50NiCrW4828.

THERMODUR C18 8ESS
Bergische Stahl Industrie
C 0-0.08, Si 0.5-1, Mn 0.5-1, Cr 17-18, Ni 10.5-11.5, Mo 2-2.2, Nb = 8 x C, bal Fe.
Cast: 64,000-92,000 TS; 27,000 YS min; 20 El min. Rupture strength: 48,000 psi, 10,000 h, 550°C. Resistance to scaling in oxidizing and ashfree and sulfur free atmosphere to 750°C. SEW-Nr. 1.4581; DIN G-X7CrNiMoNb1810.

THERMODUR C18.8ES
Bergische Stahl Industrie
C 0-0.07, Si 0.5-1, Mn 0.5-1, Cr 18-19, Ni 10-11, Mo 2-2.2, bal Fe.
Cast: 64,000-92,000 TS; 35,000 YS min; 35 El min. Rupture strength: 47,000 psi, 10,000 h, at 550°C. Resistance to scaling in oxidizing and ashfree atmosphere without sulfur compounds to 750°C. SEW-Nr. 1.4408; DIN G-X6CrNiMo1810; ASTM A351-CF8M.

THERMODUR C18.8S
Bergische Stahl Industrie
C 0-0.07, Si 0.5-1, Mn 0.5-1, Cr 18-19, Ni 9-10, Mo 0-0.75, bal Fe.
Cast: 64,000-92,000 TS; 30,000 YS min; 35 El min. Rupture strength: 30,000 psi, 10,000 h, at 500°C. Resistance to scaling in oxidizing and ashfree atmosphere without sulfur compounds to 750°C. SEW-Nr. 1.4308; DIN G-X6CrNi189; ASTM A351-CF8.

THERMODUR C18.8SS
Bergische Stahl Industrie
C 0-0.08, Si 0.5-1, Mn 0.5-1, Cr 18-19, Ni 9.5-10.5, Mo 0-0.75, Nb = 8 x C, bal Fe.
Cast: 64,000-92,000 TS; 30,000 YS min; 30 El min. Rupture strength: 42,000 psi, 10,000 h, at 550°C. Resistance to scaling in oxidizing and ashfree atmosphere without sulfur compounds to 750°C. SEW-Nr. 1.4552; DIN G-X7CrNiNb189; ASTM A351-CF8C.

THERMODUR CMVW11
Bergische Stahl Industrie
C 0.22, Cr 12, Ni 0.85, Mo 1.1, V 0.3, W 0.5, bal Fe.
Cast, heat treated: 100,000-128,000 TS; 85,000 min YS; 16 min El. For elevated temperature operation. Werkstoff Nr. 1.4932; DIN G-X22CrMoWV121. *Obsolete*

THERMODUR CMVW5
Bergische Stahl Industrie
C 0.15-0.2, Si 0.3-0.5, Mn 0.5-0.8, Cr 1.2-1.5, Mo 0.9-1.1, V 0.2-0.3, bal Fe.
Cast: 85,000-112,000 TS; 64,000 YS min; 15 El min. Resistance to scaling in oxidizing and ashfree atmosphere without sulfur compounds. SEW-Nr. 1.7706; DIN GS-17CrMoV511.

THERMODUR CMW11
Bergische Stahl Industrie
C 0.2-0.26, Si 0.2-0.4, Mn 0.5-0.7, Cr 11.3-12.2, Ni 0.7-1, Mo 1-1.2, V 0.25-0.35, bal Fe.
Cast: 100,000-127,000 TS; 85,000 YS min; 15 El min. Resistance to scaling in oxidizing and ashfree atmosphere without sulfur compounds. SEW-Nr. 1.4931; DIN G-X22CrMoV121.

THERMODUR CMW3
Bergische Stahl Industrie
C 0.15-0.2, Si 0.3-0.5, Mn 0.5-0.8, Cr 1-1.5, Mo 0.45-0.55, bal Fe.
Cast: 71,000-92,000 TS; 45,000 YS min; 20 El min. Resistance to scaling in oxidizing and ashfree atmosphere without sulfur compounds to 570°C. SEW-Nr. 1.7357; DIN GS-17CrMo55; ASTM A217-WC6.

THERMODUR CMW3K

Bergische Stahl Industrie
C 0.17, Mn 0.75, Cr 0.75, Mo 0.35, bal Fe.
Cast, heat treated: 78,000 min TS; 42,500 min YS; 20 min El.
Resistant to scaling at elevated temperatures. DIN
GS-17CrMo33. *Obsolete*

THERMODUR CMW5

Bergische Stahl Industrie
C 0.1-0.15, Si 0.3-0.5, Mn 0.5-0.7, Cr 2-2.5, Mo 0.9-1.1, bal Fe.
Cast: 71,000-100,000 TS; 45,000 YS min; 20 El min.
Resistance to scaling in ashfree atmosphere without sulfur compounds to 580°C. SEW-Nr. 1.7380; DIN GS-12CrMo910; ASTM A217-WC9.

THERMODUR CW11

Bergische Stahl Industrie
C 0.22, Cr 12, Ni 0.85, V 0.12, bal Fe.
Cast, heat treated: 92,000-113,000 TS; 71,000 min YS; 16 min El. For elevated temperature operation. DIN G-X22CrV12. *Obsolete*

THERMODUR MVW2

Bergische Stahl Industrie
C 0.17, Mo 0.5, V 0.3, bal Fe.
Cast, heat treated: 71,000-92,000 TS; 42,500 min YS; 20 min El. Resistant to scaling at elevated temperature. Werkstoff Nr. 1.5402; DIN GS-17MoV53. *Obsolete*

THERMODUR MW 4

Bergische Stahl Industrie
C 0.2, Mn 0.7, Mo 0.6, bal Fe.
Low alloy cast steel. GS-20 Mo 5.

THERMODUR MW2

Bergische Stahl Industrie
C 0.18-0.23, Si 0.3-0.5, Mn 0.5-0.8, Cr 0-0.3, Mo 0.35-0.4, bal Fe.
Cast: 64,000-85,000 TS; 35,000 YS min; 22 El min.
Resistance to scaling in oxidizing and ashfree atmosphere without sulfur compounds to 560°C. SEW-Nr. 1.5419; DIN GS-22Mo4; ASTM A217-WC1.

THERMODUR P6M

Bergische Stahl Industrie
C 0.08-0.15, Si 0.3-0.5, Mn 0.4-0.7, Cr 4.5-5.5, Mo 0.45-0.55, bal Fe.
Cast: 92,000-120,000 TS; 60,000 YS min; 18 El min.
Resistance to scaling in oxidizing and ashfree atmosphere without sulfur compounds to 600°C. SEW-Nr. 1.7363; DIN GS-12CrMo195; ASTM A217-C5.

THERMODUR P7M

Bergische Stahl Industrie
C 0.08-0.15, Si 0.3-0.5, Mn 0.5-0.8, Cr 9-10, Mo 1.1-1.2, bal Fe.
Cast: 92,000-120,000 TS; 60,000 YS min; 18 El min.
Resistance to scaling in oxidizing and ashfree atmosphere without sulfur compounds to 630°C. SEW-Nr. 1.7389; DIN G-X12CrMo101; ASTM A217-C12.

THERMODUR T-C 16.13 S

Bergische Stahl Industrie
C 0.06, Cr 16.5, Ni 12.5, bal Fe.
Cast, annealed: 64,000-85,000 TS; 28,000 min YS; 20 min El.
Austenitic stainless casting; for elevated temperature operation. DIN G-X6CrNi1613. *Obsolete*

THERMODUR T-C 16.13 SS

Bergische Stahl Industrie
C 0.07, Cr 16.5, Ni 12.5, Nb, bal Fe.
Cast, annealed: 64,000-85,000 TS; 28,000 min YS; 20 min El.
Austenitic stainless casting; for elevated temperature operation. Werkstoff Nr. 1.4968; DIN G-X7CrNiNb1613. *Obsolete*

THERMODUR T-C 16.16 ES

Bergische Stahl Industrie
C 0.07, Cr 16.5, Ni 16, Mo 2.25, bal Fe.
Cast, annealed: 64,000-85,000 TS; 28,000 min YS; 20 min El.
Austenitic stainless casting; for elevated temperature operation. DIN G-X8CrNiMo1616. *Obsolete*

THERMODUR T-C 16.16 ESS

Bergische Stahl Industrie
C 0.07, Cr 16.5, Ni 16, Mo 2.25, Nb, bal Fe.
Cast, annealed: 64,000-85,000 TS; 28,000 min YS; 20 min El.
Austenitic stainless casting; for elevated temperature operation. DIN G-X7CrNiMoNb1616. *Obsolete*

THERMODUR T-C 18.8

Bergische Stahl Industrie
C 0.1, Cr 18, Ni 8.5, bal Fe.
Cast, annealed: 70,000 min TS; 30,000 min YS; 25 min El.
Austenitic stainless casting; for elevated temperature operation. Werkstoff Nr. 1.4312; DIN G-X10CrNi188. *Obsolete*

THERMODUR T-C 18.8 ES

Bergische Stahl Industrie
C 0.07, Cr 17.5, Ni 10.5, Mo 2.25, bal Fe.
Cast, annealed: 70,000 min TS; 30,000 min YS; 25 min El. Austenitic stainless casting; for elevated temperature operation. Werkstoff Nr. 1.4408; DIN G-X6CrNiMo1810. ASTM A351 Gr. CF8M. *Obsolete*

THERMODUR T-C 18.8 ESS

Bergische Stahl Industrie
C 0.08, Cr 17.5, Ni 11, Mo 2.25, Nb, bal Fe.
Cast, annealed: 70,000 min TS; 30,000 min YS; 20 min El.
Austenitic stainless casting; for elevated temperature operation. Werkstoff Nr. 1.4581; DIN G-X7CrNiMoNb1810. *Obsolete*

THERMODUR T-C 18.8 S

Bergische Stahl Industrie
C 0.07, Cr 18, Ni 9.5, bal Fe.
Cast, annealed: 70,000 min TS; 30,000 min YS; 25 min El.
Austenitic stainless casting; for elevated temperature operation. Werkstoff Nr. 1.4308; DIN G-X6CrNi189. ASTM A351 Gr. CF8. *Obsolete*

THERMODUR T-C 18.8 SS

Bergische Stahl Industrie
C 0.08, Cr 18, Ni 10, Nb, bal Fe.
Cast, annealed: 70,000 min TS; 30,000 min YS; 20 min El.
Austenitic stainless casting; for elevated temperature operation. Werkstoff Nr. 1.4552; DIN G-X7CrNiNb189; ASTM A351 Gr. CF8C. *Obsolete*

THERMODUR T-CN 13W

Bergische Stahl Industrie
C 0.12, Cr 23, Ni 12.5, bal Fe.
Cast, annealed: 64,000-85,000 TS; 28,000 min YS; 20 min El.
Austenitic stainless casting; for elevated temperature operation. DIN G-X12CrNiSi2312; ASTM A351 Gr. CH20. *Obsolete*

THERMODUR T-CN 15 TW

Bergische Stahl Industrie
C 0.15, Cr 19.5, Ni 15, bal Fe.
Cast, annealed: 64,000-85,000 TS; 28,000 min YS; 20 min El.
Austenitic stainless casting; for elevated temperature operation. DIN G-X15CrNiSi2015. *Obsolete*

THERMODUR T-CN 20

Bergische Stahl Industrie
C 0.3, Cr 25, Ni 20, bal Fe.
As cast: 64,000-85,000 TS; 28,000 min YS; 8 min El.
Austenitic stainless casting; for high temperature operation. Werkstoff Nr. 1.4848; DIN G-X40CrNiSi2520; ASTM A297 Gr. HK. *Obsolete*

THERMODUR T-CN 20 W

Bergische Stahl Industrie
C 0.15, Cr 25, Ni 21, bal Fe.
Cast, annealed: 64,000-85,000 TS; 28,000 min YS; 20 min El.
Austenitic stainless casting; for elevated temperature operation. Werkstoff Nr. 1.4849; DIN G-X15CrNiSi2520. ASTM A351 Gr. CK-20. *Obsolete*

THERMODUR T-CN 35 W

Bergische Stahl Industrie
C 0.1, Ni 35, Cr 20.5, bal Fe.
Cast, annealed: 64,000-85,000 TS; 28,000 min YS; 20 min El.
Austenitic stainless casting; for high temperature operation. DIN G-X10NiCrSi3520. *Obsolete*

THERMODUR T-CN 60

Bergische Stahl Industrie
C 0.35, Ni 53, Cr 21, bal Fe.
Cast, annealed: 64,000-85,000 TS; 28,000 min YS; 4 min El.
Austenitic stainless casting; for high temperature operation. DIN G-X35NiCr5422. *Obsolete*

THERMODUR T-CN 80

Bergische Stahl Industrie
C 0.15, Ni 78, Cr 14, bal Fe.
Cast, annealed: 64,000-85,000 TS; 28,000 min YS; 12 min El.
Austenitic stainless casting; for high temperature operation. DIN G-X15NiCr7814. *Obsolete*

THERMODUR W1

Bergische Stahl Industrie
C 0.18-0.23, Si 0.3-0.5, Mn 0.5-0.8, Cr 0-0.3, bal Fe.
Cast: 64,000-85,000 TS; 35,000 YS min; 22 El min.
Resistance to scaling in oxidizing and ashfree atmosphere without sulfur compounds to 560°C. SEW-Nr. 1.0619; DIN GS-C25; ASTM A216-WCA.

THERMODUR W2

Bergische Stahl Industrie
C 0.27-0.3, Si 0.3-0.5, Mn 0.6-0.8, Cr 0-0.3, bal Fe.
Cast: 70,000 TS min; 40,000 YS min; 20 El min. Resistance to scaling in oxidizing and ashfree atmosphere without sulfur compounds to 560°C. DIN GS-C30; ASTM A216-WCB.

THERMODYNE

American Smelting & Refining Co.
Sb, Pb, bal Sn.
For bearings; Babbitt.

THERMOFLEX

Wm. Wilkinsen
For thermostatic and contact alloy; bi-metal.

THERMOFLUX

Vacuumschmelze GmbH
Soft magnetic 30% nickel-iron alloy for temperature compensation in permanent magnet systems.

THERMOLD 75

Cytemp Specialty Steel Div.
C 0.35, Si 1, Mn 0.3, Cr 3.5, V 0.75, Mo 2.5, Co 2, bal Fe.
Heat treated: 296,000 TS; 247,000 YS; 9 El; 16 RA; 9 Charpy; 59 Rock C. For aluminum die casting dies and inserts, cores, plungers, sleeves, slides, extrusion and forging dies. Hot work steel. Air hardening. Good thermal fatigue resistance.

THERMOLD A

Now THERMOLD H11.

THERMOLD AV

Now THERMOLD H13.

THERMOLD B

Now THERMOLD H12.

THERMOLD H10
Cytemp Specialty Steel Div.
C 0.4, Mn 0.55, Si 1, Cr 3.25, V 0.4, Mo 2.5, bal Fe.
Heat treated: 233,000-301,000 TS; 201,000-260,000 YS; 8-10 El; 26-32 RA; 46-55 Rock C. For mandrels, dies, bolsters, dummy blocks, punches, gripper and header dies, hot shears, aluminum die casting dies. Resists softening at high temperatures. Type H10 hot work steel. *Obsolete*

THERMOLD H11
Cytemp Specialty Steel Div.
C 0.35, Mn 0.4, Si 1, Cr 5, V 0.45, Mo 1.5, bal Fe.
Heat treated: 225,000-305,000 TS; 195,000-255,000 YS; 8-15 El; 29-50 RA; 45-56 Rock C. For die casting dies, punches, piercing tools, mandrels, extrusion tooling, forging dies. Type H11 hot work steel. High hardenability. Good resistance to softening.

THERMOLD H12
Cytemp Specialty Steel Div.
C 0.35, Mn 0.4, Si 1, Cr 5, W 1.5, V 0.4, Mo 1.5, bal Fe.
Heat treated: 192,000-263,000 TS; 175,000-241,000 YS; 8-15 El; 25-43 RA; 41-52 Rock C. For extrusion dies, dummy blocks, gripper and header dies, forging die inserts, punches, mandrels. High hardenability, tough. Type H12 hot work steel. *Obsolete*

THERMOLD H13
Cytemp Specialty Steel Div.
C 0.35, Mn 0.4, Si 1, Cr 5, V 1, Mo 1.5, bal Fe.
Heat treated: 189,000-258,000 TS; 170,000-238,000 YS; 8-18 El; 28-48 RA; 41-55 Rock C. For die casting dies and inserts, cores, ejector pins, plungers, sleeves, extrusion and forging dies. High hardenability and toughness. Type H13 hot work steel. Air hardened.

THERMOLD H19
Cytemp Specialty Steel Div.
C 0.4, Cr 4.25, W 4.25, V 2, Co 4.25, bal Fe.
Heat treated: 233,000 TS; 217,000 YS; 5.6 El; 17 RA; 58 Rock C. For extrusion dies and inserts, dummy blocks, punches, mandrels, forging dies and inserts. Good red- hardness and shock resistance. Type H19 hot work steel. *Obsolete*

THERMOLD H21
Cytemp Specialty Steel Div.
C 0.35, Mn 0.25, Si 0.35, Cr 3.5, W 9.5, V 0.5, bal Fe.
Heat treated: 242,500 TS; 220,500 YS; 7.4 El; 27 RA; 54 Rock C. For mandrels, hot punches, and blanking dies, fly shear blades, extrusion and gripper dies, piercing points, dummy blocks. Resists softening. Type H21 hot work steel. *Obsolete*

THERMOLD H22
Cytemp Specialty Steel Div.
C 0.38, Mn 0.25, Si 0.35, Cr 3, W 11, V 0.4, bal Fe.
Heat treated: 225,000 TS; 200,500 YS; 8.8 El; 30 RA; 55 Rock C. For mandrels, hot blanking dies, hot punches, flying shear blades, extrusion dies, dummy blocks, gripper dies. Shock resistant. Resists softening. Type H22 hot work steel. *Obsolete*

THERMOLD H23
Cytemp Specialty Steel Div.
C 0.32, Mn 0.35, Si 0.5, Cr 12, W 12, V 1, bal Fe.
Heat treated: (at 1000°F) 167,500 TS; 135,000 YS; 7.4 El; 13.5 RA; 34 Rock C. For extrusion dies, die casting dies, mandrels, hot punches, gripper dies. High resistance to softening. Retains hot hardness. Type H23 hot work steel. *Obsolete*

THERMOLD H26
Cytemp Specialty Steel Div.
C 0.5, Mn 0.25, Si 0.35, Cr 4, W 18, V 1, bal Fe.
Heat treated: 258,000 TS; 235,500 YS; 3.6 El; 6 RA; 57 Rock C. For mandrels, hot blanking dies and punches, flying shear blades, extrusion and trim dies. Maximum hot strength and resistance to softening. Type H26 hot work steel. *Obsolete*

THERMOLD J COLD WORK
Cytemp Specialty Steel Div.
C 0.5, Mn 0.4, Si 1.1, Cr 5, V 1, Mo 1.4, Ni 1.5, bal Fe.
For cold heading dies, die inserts, coining dies, forming rolls. AISI Type A9 air hardening cold work tool steel.

THERMOLD J HOT WORK
Cytemp Specialty Steel Div.
C 0.5, Mn 0.4, Si 1, Cr 5, V 1, Mo 1.4, Ni 1.5, bal Fe.
For punches, piercing tools, forging dies, gripper dies, extrusion tooling, coining dies. AISI Type A9 air hardening tool steel.

THERMOLD KL
Cytemp Specialty Steel Div.
C 0.35, Mn 0.6, Si 1.5, Cr 7.25, W 7.25, bal Fe.
Heat treated: 232,000 TS; 203,000 YS; 10.5 El; 34 RA. For extrusion and gripper dies, forming and blanking dies, hot punches. Good toughness and red-hardness. Hot work steel. Oil hardening. *Obsolete*

THERMOLD Z
Cytemp Specialty Steel Div.
C 0.37, Mn 0.8, Si 0.3, Cr 1.25, V 0.15, Mo 0.35, B 0.001, bal Fe.
Hardenable to 35-54 Rock C. For plastic molds, zinc die casting molds. AISI P20 Mold Steel. *Obsolete*

THERMON 12
SWB Stahlformguss Gesellschaft mbH
C 0.32, Mn 0.3, Si 0.3, Cr 3, Mo 2.8, V 0.5, bal Fe.
For hot dies. Hot work tool steel, tough, shock resistant. *Obsolete*

THERMON 15
SWB Stahlformguss Gesellschaft mbH
C 0.3, Cr 1.1, V 0.18, W 3.75, bal Fe.
For upsetters, punches, extrusion press parts; oil hardened, tough. *Obsolete*

THERMON 17
SWB Stahlformguss Gesellschaft mbH
C 0.3, Cr 2.35, V 0.6, W 4.25, bal Fe.
For extrusion press parts, rams, punches; hot work steel, oil hardened. *Obsolete*

THERMON 17 EXTRA
SWB Stahlformguss Gesellschaft mbH
C 0.3, Cr, Co, V, W, bal Fe.
For extrusion press parts, rams, punches; hot work steel, oil hardened. *Obsolete*

THERMON 21
SWB Stahlformguss Gesellschaft mbH
C 0.4, Cr, Mo, V, W, bal Fe.
For hot work tools, extrusion press parts; oil hardened, hot work steel. *Obsolete*

THERMON 34
SWB Stahlformguss Gesellschaft mbH
C 0.3, Cr 2.65, V 0.35, W 8.5, bal Fe.
For extrusion rams and liners, punches, upsetters; oil hardened, hot work steel. *Obsolete*

THERMON 34 EXTRA
SWB Stahlformguss Gesellschaft mbH
C 0.3, Co 2, Cr 2.4, V 0.25, W 8.5, bal Fe.
For extrusion press parts, rams, punches; hot work steel, oil hardened. *Obsolete*

THERMON 4630, ETC.
Thyssen Edelstahlwerke AG
See Werkstoff Nr. 1.4630.
High temperature alloys; 11 grades.

THERMON 7
SWB Stahlformguss Gesellschaft mbH
C 0.45, Cr 1.05, V 0.2, W 1.85, bal Fe.
For upsetters, heading dies, punches; oil hardened, tough. *Obsolete*

THERMON 7H
SWB Stahlformguss Gesellschaft mbH
C 0.55, Cr 1.05, V 0.2, W 1.85, bal Fe.
For upsetters, riveters, heading dies; oil hardened, tough. *Obsolete*

THERMON 7W
SWB Stahlformguss Gesellschaft mbH
C 0.35, Cr 1.05, V 0.2, W 1.85, bal Fe.
For upsetters, punches, heading dies; oil hardened, tough. *Obsolete*

THERMON A
SWB Stahlformguss Gesellschaft mbH
C 0.45, Cr, Ni, W, bal Fe.
For hot work tools, dies, punches; hot work steel, oil hardened. *Obsolete*

THERMON RR
SWB Stahlformguss Gesellschaft mbH
C 0.2, Cr 13, bal Fe.
Annealed: 95,000 TS; 50,000 YS; 25 El; 55 RA; 195 Brin. Cold drawn: 105,000 TS; 85,000 YS; 17 El; 50 RA; 215 Brin. For turbine blades, cutlery, valves, surgical instruments; Type 420; stainless. *Obsolete*

THERMON SS
SWB Stahlformguss Gesellschaft mbH
C 0.65, Cr 3.75, Mo 0.85, V 0.7, W 8.5, bal Fe.
For cutting tools, drills, taps, hot work tools; high speed steel. *Obsolete*

THERMON U
SWB Stahlformguss Gesellschaft mbH
C 0.45, Cr 1.35, Mo 0.45, V 0.8, W 0.45, bal Fe.
For upsetters, riveters, punches; oil hardened. *Obsolete*

THERMONEAL 10
Heppenstall Co.
Tool material. C 0.3-0.36, Mn 0.2-0.4, Si 0.8-1.2, Cr 4.25-5.25, Mo 1.75-2, bal Fe.
At 80°F: 217,000 TS; 184,000 YS; 13 El; 40 RA. At 1000°F: 145,000 TS; 105,000 YS; 20 El; 65 RA. For upsetting and forging dies, aluminum die casting dies, extrusion dies. Type H11 hot work tool steel. Air hardened.

THERMONEAL 11
Heppenstall Co.
Tool material. C 0.37-0.42, Cr 5-5.5, Mn 0.23-0.38, Si 0.85-1.1, Mo 1-1.2, V 0.45-0.55, bal Fe.
At 80°F: 217,000 TS; 184,000 YS; 13 El; 40 RA. At 1000°F: 145,000 TS; 105,000 YS; 20 El; 65 RA. For die casting dies, upsetting and press forging dies, extrusion dies. Type H11 hot work tool steel.

THERMONEAL 13
Heppenstall Co.
Tool material. C 0.37-0.42, Cr 5-5.5, Mn 0.23-0.38, Mo 1-1.3, Si 0.85-1.1, V 0.9-1, bal Fe.
At 80°F: 217,000 TS; 184,000 YS; 13 El; 40 RA. At 1000°F: 145,000 TS; 105,000 YS; 20 El; 65 RA. For upsetting and press forging dies, extrusion dies. Type H13 hot work steel.

THERMOPERM
Krupp Stahl AG
iron. Fe 70, Ni 30.
For compensating shunts for electrical equipment; temperature-sensitive, magnetic. *Obsolete*

THERMOSIL
Cyprus Foote Mineral Co.
61 Si approx, bal Fe.
Ferrosilicon used as graphitizing inoculant for gray and ductile iron when added to ladle; also as an additive. *Obsolete*

THERMOTEM
Heppenstall Co.
Tool material. C 0.33, Cr 4.75, Mo 1.87, bal Fe.
Annealed: 250 Brin. Heat treated: 510 Brin. For extrusion parts, dies, shear knives; oil hardened.

THERMOTEM 10

Heppenstall Co.
Tool material. C 0.3-0.36, Cr 4.25-5.25, Si 0.8-1.2, Mn 0.2-0.4, Mo 1.75-2, bal Fe.
At 80°F: 217,000 TS; 184,000 YS; 13 El; 40 RA. At 1000°F: 145,000 TS; 105,000 YS; 20 El; 65 RA. For upsetting and forging dies, aluminum die casting dies, extrusion dies. Type H11 hot work steel. Air hardened.

THERMOTEM-11

Heppenstall Co.
Tool material. C 0.4, Si 1, Cr 5.25, Mo 1.1, V 0.5, bal Fe.
For hot work dies and punches; hot work steel, air hardened.

THERMOTEM-12

Heppenstall Co.
Tool material. C 0.37, Si 14, Cr 5, Mo 1.45, W 1.25, V 0.3, bal Fe.
For hot work dies and punches; hot work steel, air hardened.

THERMOTEM-13

Heppenstall Co.
Tool material. C 0.4, Si 1, Cr 5.25, Mo 1.1, V 0.95, bal Fe.
For hot work dies and punches; hot work steel, air hardened.

THERMOWEAR HOT WORK DIE STEEL (RED-HARD)

Carpenter Technology Corp.
Tool material. C 0.6, Mn 0.5, Si 1, Cr 4, Mo 2.5, V 1, Cb 1.5, Ti 0.1, bal Fe.
As hardened: 55 Rock C. Wear resistant alloy with high red-hardness, good resistance to abrasion and fair toughness. For hot work compression tools such as dies, inserts, blades, dummy blocks, punches.

THESSCAL A

Sheffield Smelting Co. Ltd.
Ag 5, bal Zn.
Melt range: 420-450 C. Yield point: 8.4 tsi; 5 El. For torch brazing wrought aluminum. *Obsolete*

THESSCAL P12

Sheffield Smelting Co. Ltd.
Zn, Al.
Melt range: 440-510 C. Yield point; 18.6 tsi; 6 El. For torch brazing aluminum castings. *Obsolete*

THESSCO 625

Sheffield Smelting Co. Ltd.
Au, Ag, Cu.
Wrought: 13.7 density; 160 HV hardness; 12% electrical conductivity. For tarnish resistant slip ring or brush spring alloy. *Obsolete*

THESSCO E5

Sheffield Smelting Co. Ltd.
Ag 20, Cu, Zn.
Melt range: 770-810 C. Yield point: 30.0 tsi; 4 El. For silver brazing, high melt point, high strength, low cost. *Obsolete*

THESSCO GD 25

Sheffield Smelting Co. Ltd.
Ag, CdO (internally oxidized).
Wrought: 10.0 density; 60 HV hardness; 8.2% electrical conductivity. For anti-welding contacts. *Obsolete*

THESSCO GD.35

Sheffield Smelting Co. Ltd.
Ag, CdO (internally oxidized).
Wrought: 9.9 density; 65 HV hardness; 75% electrical conductivity. For anti-welding contacts. *Obsolete*

THESSCO H12

Sheffield Smelting Co. Ltd.
Ag 72, Cu.
Melting range: 780-780°C. 19.0 tsi YP; 10 El. Silver brazing, high quality joints in protective atmosphere.

THESSCO HT5

Sheffield Smelting Co. Ltd.
Ag 5, Cd, Zn.
Melt range: 270-285 C. Yield point: 10 tsi; 8 El. Silver bearing soft solder, high melting point, good strength. *Obsolete*

THESSCO HX0

Sheffield Smelting Co. Ltd.
Ag 60, Cu, Sn.
Melt range: 600-720 C. Yield point: 18.5 tsi; 5 El. Silver brazing, for use in protective atmospheres. *Obsolete*

THESSCO L3

Sheffield Smelting Co. Ltd.
Ag 10, Cu, Zn.
Melting range: 840-855°C; 31.5 tsi YP; 7 El. For silver brazing, high melt point, high strength.

THESSCO L7

Sheffield Smelting Co. Ltd.
Ag 14, Cu, Zn.
Melting range: 810-835°C; 33.5 tsi YP; 3 El. For silver brazing, high melt point, high strength.

THESSCO LX13

Sheffield Smelting Co. Ltd.
Ag 20, Cu, Zn, Cd.
Melting range: 720-775°C; 20.5 tsi YP; 9 El. For silver brazing, high melt point.

THESSCO LX16

Sheffield Smelting Co. Ltd.
Ag 23, Cu, Zn, Cd.
Melting range: 610-720°C; 17.0 tsi YP; 18 El. For silver brazing, general purpose, wide melting range.

THESSCO LX18

Sheffield Smelting Co. Ltd.
Ag 25, Cu, Zn, Cd.
Melting range: 605-710°C; 20.0 tsi YP; 14 El. For silver brazing, general purpose, wide melting range.

THESSCO M1

Sheffield Smelting Co. Ltd.
Ag 31, Cu, Zn.
Melting range: 715-755°C; 30.5 tsi YP; 23 El. For silver brazing, high melting point, high strength.

THESSCO MX0

Sheffield Smelting Co. Ltd.
Ag 30, Cu, Zn, Cd.
Melting range: 605-680°C; 15 tsi YP; 10 El. For silver brazing, general purpose.

THESSCO MX12

Sheffield Smelting Co. Ltd.
Ag 42, Cu, Zn, Cd.
Melting range: 610-620°C; 17.0 tsi YP; 27 El. For silver brazing, low melting point.

THESSCO MX18

Sheffield Smelting Co. Ltd.
Ag 48, Cu, Zn, Cd.
Melting range: 630-640°C; 19.5 tsi YP; 3 El. For silver brazing, extremely fluid.

THESSCO MX18 PLUS

Sheffield Smelting Co. Ltd.
Ag 48, Ni 3, Cu, Zn, Cd.
Melt range: 640-660 C. Yield point; 20.0 tsi; 35 El. For silver brazing tungsten carbide and hard metals. *Obsolete*

THESSCO MX4

Sheffield Smelting Co. Ltd.
Ag 34, Cu, Zn, Cd.
Melting range: 610-665°C; 15.0 tsi YP; 15 El. For silver brazing, general purpose.

THESSCO MX8

Sheffield Smelting Co. Ltd.
Ag 38, Cu, Zn, Cd.
Melting range: 605-650°C; 16.0 tsi YP; 24 El. For silver brazing, general purpose, low melting point.

THESSCO PHOSPHALLOY NO. 1

Sheffield Smelting Co. Ltd.
Ag 15, Cu, P.
Melt range: 645-700+ C. Yield point: 30.5 tsi; 25 El. For silver brazing, self fluxing on copper. *Obsolete*

THESSCO PHOSPHALLOY NO. 2

Sheffield Smelting Co. Ltd.
Ag 2, Cu, P.
Melt range: 645-740+ C. Yield point: 16.5 tsi; 5 El. For silver brazing, self fluxing on copper. *Obsolete*

THESSCO PHOSPHALLOY NO. 5

Sheffield Smelting Co. Ltd.
Ag 5, Cu, P.
Melt range: 645-700+ C. Yield point: 25.0 tsi; 6 El. For silver brazing, self fluxing on copper. *Obsolete*

THESSCO PROOF SILVER

Sheffield Smelting Co. Ltd.
Ag 99.99.
MP: 960°C. 2.9 tsi YP; 45 El. For brazing.

THESSCO REDIBRAZE

Sheffield Smelting Co. Ltd.
Bimetal of either GD. 25 or GD. 35 on silver.

THESSCO SILBRAZE

Sheffield Smelting Co. Ltd.
Ag 1, Cu, Zn, Si.
Melting range: 885-895°C. 14.2 tsi YP; 43 El. For brazing; silver bearing brass.

THESSCO SX25

Sheffield Smelting Co. Ltd.
Ag 2.5, Pb, Cu.
Melt point: 302-305 C. Yield point: 1.4 tsi; 45 El. Soft solder, high melting point. *Obsolete*

THESSCO SX4

Sheffield Smelting Co. Ltd.
Ag 4, Sn.
Melt range: 221-224 C. Yield point: 3 tsi; 48 El. Soft solder, excellent corrosion resistance. *Obsolete*

THESSCO TRIMETAL

Sheffield Smelting Co. Ltd.
Ag 48, Ni 3, Cu, Zn, Cd, on copper.
Melt range: 640-660 C. Silver brazing sandwich for carbide tool tipping. *Obsolete*

THESSCO TYPE 7

Sheffield Smelting Co. Ltd.
Cu, P.
Melt range: 705-800 C. Yield point: 20.0 tsi; 2 El. Brazing alloy, self fluxing on copper. *Obsolete*

THESSCONITE G.4

Sheffield Smelting Co. Ltd.
Ag, W.
Sintered: 15.0 density; 225 Vickers; electrical conductivity: 43%. For heavy duty arc resistant contacts.

THESSCONITE G.7

Sheffield Smelting Co. Ltd.
Ag, W.
Sintered: 13.9 density; 165 Vickers; electrical conductivity: 50%. For heavy duty arc resistant contacts.

THESSCONITE G.9

Sheffield Smelting Co. Ltd.
Ag, W.
Sintered: 13.5 density; 140 Vickers; electrical conductivity: 55%. For heavy duty arc resistant contacts.

THESSCONITE GC.15
Sheffield Smelting Co. Ltd.
Ag, C.
Sintered: 10.2 density; 64 Vickers; electrical conductivity: 85%. For sliding electrical contacts.

THESSCONITE GD.10
Sheffield Smelting Co. Ltd.
Ag, CdO.
Sintered: 10.0 density; 60 Vickers; electrical conductivity: 82%. For anti-welding contacts.

THESSCONITE GN.1
Sheffield Smelting Co. Ltd.
Ag, Ni.
Sintered: 9.9 density; 90 Vickers; electrical conductivity: 66%. For medium duty electrical contacts.

THESSCONITE HW.10
Sheffield Smelting Co. Ltd.
Ag, WC.
Sintered: 12.5 density; 130 HV hardness; 60% electrical conductivity. For arc resistant and anti-welding contacts. *Obsolete*

THESSCONITE U.3
Sheffield Smelting Co. Ltd.
Cu, W.
Sintered: 14.9 density; 240 HV hardness; 28% electrical conductivity. For oil immersed circuit breaker contacts. *Obsolete*

THESSCONITE U.4
Sheffield Smelting Co. Ltd.
Cu, W.
Sintered: 14.6 density; 220 HV hardness; 35% electrical conductivity. For oil immersed circuit breaker contacts. *Obsolete*

THESSCONITE U.8
Sheffield Smelting Co. Ltd.
Cu, W.
Sintered: 13.2 density; 180 HV hardness; 40% electrical conductivity. For oil immersed circuit breaker contacts. *Obsolete*

THETALOY
Pratt & Whitney Cutting Tool & Gage
C 0.38, Cr 25, W 7, Mo 3, Co 12.5, Fe 6, Mn 2.5, bal Ni. For high temperature applications; corrosion and heat resistant.

THIN FIN
Chase Brass & Copper Co.
Cu 99.9, Cd 0.1.
Sheet: 65,000 TS; 62,000 YS; 1.5 El. 98%/ACS Electrical conductivity. 54,000 psi TS after 10 min, anneal at 650 F. Automotive radiators. *Obsolete*

THOMASTRIP
Thomas Steel Strip Co.
C 0.2, bal Fe.
For gears, retainers, ball bearings, structural usage; sintered, cold rolled strip steel. *Obsolete*

THOR
Now THERMOLD H23.

THOR DOPPELKREUZ EXTRA ZH
Stahlwerke Kabel, C.
C 0.6, Si 0.25-0.5, Mn 0.2-0.8, bal Fe.
For gears, rails, shafts, axles, crankshafts; water hardened.

THOR DOPPELKREUZ H
Stahlwerke Kabel, C.
C 0.9, Si 0.25-0.5, Mn 0.2-0.8, bal Fe.
Heat treated: 190,000 TS; 145,000 YS; 10 El; 30 RA; 400 Brin. For drills, tools, springs, cutters; Type W1; water hardened.

THOR DOPPELKREUZ MH
Stahlwerke Kabel, C.
C 0.75, Si 0.25-0.5, Mn 0.3-0.8, bal Fe.
Heat treated: 185,000 TS; 140,000 YS; 14 El; 38 RA; 375 Brin. For rails, springs, tools, hammers, axes; Type W1; water hardened.

THOR DOPPELKREUZ SEHR ZAH
Stahlwerke Kabel, C.
C 0.15, Si 0.15-0.35, Mn 0.25-0.5, bal Fe.
Cold drawn: 72,000 TS; 60,000 YS; 22 El; 58 RA; 145 Brin. For gears, pinions, cams, bolts, camshafts; case hardening steel.

THOR DOPPELKREUZ ZAH
Stahlwerke Kabel, C.
C 0.35, Si 0.25-0.5, Mn 0.3-0.8, bal Fe.
Hot rolled: 85,000 TS; 54,000 YS; 30 El; 53 RA; 185 Brin. For gears, bolts, fasteners, shafts; water hardened.

THOR DOPPELKREUZ ŻH
Stahlwerke Kabel, C.
C 0.45, Si 0.25-0.5, Mn 0.3-0.8, bal Fe.
Hot rolled: 98,000 TS; 58,000 YS; 24 El; 45 RA; 212 Brin. For gears, bolts, fasteners, shafts; water hardened.

THOR EXTRA EXTRA ZH
Stahlwerke Kabel, C.
C 1.1, Si 0-0.25, Mn 0-0.25, bal Fe.
Annealed: 100,000 TS; 53,000 YS; 21 El; 42 RA; 200 Brin. For drills, taps, reamers, cutters, hobs; Type W1; water hardened.

THOR EXTRA SEHR ZAH
Stahlwerke Kabel, C.
C 0.75, Si 0-0.25, Mn 0-0.25, bal Fe.
Heat treated: 185,000 TS; 140,000 YS; 14 El; 38 RA; 375 Brin. For springs, rails, axes, shafts; Type W1; water hardened.

THOR EXTRA ZAH
Stahlwerke Kabel, C.
C 0.85, Si 0-0.25, Mn 0-0.25, bal Fe.
Heat treated: 190,000 TS; 145,000 YS; 12 El; 35 RA; 390 Brin. For springs, tools, drills, cutters; Type W1; water hardened.

THOR EXTRA ZH
Stahlwerke Kabel, C.
C 1, Si 0-0.25, Mn 0-0.25, bal Fe.
Annealed: 100,000 TS; 53,000 YS; 21 El; 42 RA; 200 Brin. For drills, taps, cutters, springs, tools; Type W1; water hardened.

THOR PRIMA EXTRA ZH
Stahlwerke Kabel, C.
C 1.15, Si 0-0.25, Mn 0-0.25, bal Fe.
Annealed: 110,000 TS; 58,000 YS; 18 El; 38 RA; 215 Brin. For drills, taps, cutters, springs; Type W1; water hardened.

THOR PRIMA MH
Stahlwerke Kabel, C.
C 1.3, Mn 0-0.25, Si 0-0.25, bal Fe.
For cutting and engraving tools, taps; Type W1; water hardened.

THOR PRIMA SEHR ZAH
Stahlwerke Kabel, C.
C 0.7, Si 0-0.25, Mn 0-0.25, bal Fe.
Heat treated: 175,000 TS; 130,000 YS; 12 El; 36 RA; 355 Brin. For rails, springs, tools, axes, hammers; Type W1; water hardened.

THOR PRIMA ZAH
Stahlwerke Kabel, C.
C 0.85, Si 0-0.25, Mn 0-0.25, bal Fe.
Heat treated: 190,000 TS; 145,000 YS; 10 El; 30 RA; 400 Brin. For drills, taps, springs, tools, reamers; Type W1; water hardened.

THOR PRIMA ZH
Stahlwerke Kabel, C.
C 1, Si 0-0.25, Mn 0-0.25, bal Fe.
Annealed: 100,000 TS; 53,000 YS; 21 El; 42 RA; 200 Brin. For drills, taps, hobs, saws, cutters, springs; Type W1; water hardened.

THORAN
German manufacture
W 95.85, C 3.94, W_2C + WC plus molybdenum carbide, Mo. For hard cutting tools and dies, tips for high speed tools; cast alloy.

THORIUM
Atomergic Chemetals Corp.
Th.
Purities: 99.99+%, nuclear grade 99.8+%, commercial grade 98/99%. Forms: powder, crystal, bar, sintered pellets, sheet, foil, ingot, rod.

THREAD GAUGE STEEL
Carpenter Technology Corp.
C 0.2, Mn 1.3, Si 0.2, bal Fe.
For thread gauges. Case hardening. *Obsolete*

THREADING BRASS-223
Anaconda Co.
Copper. Cu 65, Zn 34.75, Pb 0.25.
Soft: 45,000 TS; 17,000 YS; 55 El. Hard: 73,000 TS; 60,000 YS; 10 El. For drawing, forming, switch plates. High ductility.

THREDWELL
Horace T. Potts Co.
C 0.08-0.25, S 0.075-0.15, bal Fe.
Hot rolled: 50,000-60,000 TS; 25,000-35,000 YS; 20-35 El; 35-60 RA. For bolts, studs, machine parts; rapid machinability. *Obsolete*

THREE STAR
Midvale-Heppenstall Co.
C 0.72, Cr 4, W 18.5, V 1.2, Co 4, Mo 0.75, bal Fe. For cutting tools; high speed steel. *Obsolete*

THREE STAR
Midvale-Heppenstall Co.
C 0.72, Cr 4, W 18.5, V 1.2, Co 4, Mo 0.75, bal Fe. For cutting tools; high speed steel.

THREE-TWENTY
English manufacture
Al 77, Zn 20, Cu 3.
For light alloy parts; non-hardenable.

THULIUM
Atomergic Chemetals Corp.
Tm.
Purities: 99.9% special distilled grade, 99.5+%. Forms: ingot, lump, turnings, sheet, rod, foil, wire, sponge.

THURSTON BRASS
American manufacture
Zn 44.5, Sn 0.5, bal Cu.
For architectural trim; corrosion resistant.

THURSTON BUTTON
English manufacture
Zn 33-37, Sn 1.5-6, Pb 12-15, bal Cu.
For ornamental and architectural parts; free-cutting.

THYRAPID 3202, ETC.
Thyssen Edelstahlwerke AG
See Werkstoff Nr. 1.3202, etc.
High speed tool steels; 10 grades.

THYRODUR 1525, ETC.
Thyssen Edelstahlwerke AG
See Werkstoff Nr. 1.1525, etc.
Cold work tool steels; 25 grades.

THYROPLAST 2162, ETC.

Thyssen Edelstahlwerke AG
See Werkstoff Nr. 1.2162, etc.
Plastic mold steels; 8 grades.

THYROTHERM 2323, ETC.

Thyssen Edelstahlwerke AG
See Werkstoff Nr. 1.2323, etc.
Hot work tool steels; 12 grades.

TI 227

British Aluminium Co., Ltd.
Cu 0.15, Mg 6.5-7.5, Si 0.6, Fe 0.7, Mn 1, Ti 0.2, bal Al.
O-temper: 44,800 TS; 20,200 YS: 18 El. 1/2 H-temper: 56,000
TS; 36,000 YS; 5 El. For structural members; high strength,
corrosion resistant. *Obsolete*

TI 3-2.5

Ulbrich Stainless Steels/Special Metals
Titanium. C 0-0.08, N 0-0.05, H 0-0.0125, Fe 0-0.25, Ti 92, O
0-0.12, Al 2.5-3.5, V 2-3, 0.40 others max.
Alpha-beta alloy; weldable. Good strength. May be
strengthened by cold-working.

TI 441

British Aluminium Co., Ltd.
Cu 0.15, Mg 0.4-1.5, Si 0.75-1.3, Fe 0.6, Mn 1, Ti 0.2, bal Al.
W-temper: 27,000 TS; 15,700 YS; 18 El. WP-temper: 40,000
TS; 33,600 YS; 10 El. For structural members; age-hardened,
corrosion resistant. *Obsolete*

TI 551

British Aluminium Co., Ltd.
Cu 3-5, Mg 0.4-1.2, Si 0.7, Fe 0.7, Mn 0.4-1.2, Ti 0.3, bal Al.
Heat treated: 53,800-56,000 TS; 31,400-33,600 YS; 10-15 El.
For structural members; age-hardened, corrosion resistant.
Obsolete

TI 663

British Aluminium Co., Ltd.
Cu 1-2, Mg 0.5-1.2, Si 0.75-1.25, Fe 0.75, Mn 1, Ti 0.3, bal Al.
W-temper: 38,100 TS; 22,400 YS; 15 El. WP-temper: 56,000
TS; 44,800 YS; 8 El. For structural members; age-hardened,
corrosion resistant. *Obsolete*

TI 775

British Aluminium Co., Ltd.
Cu 1.5, Mg 2-3.5, Si 0.5, Fe 0.5, Mn 0.8, Zn 4.5-6.5, Ti 0.3, bal
Al.
WP-temper: 72,000-78,400 TS; 60,500-67,200 YS; 5-7 El. For
structural members; age-hardened, high strength. *Obsolete*

TI 8

TRW Inc.
Ti base, Mo, Ni sintered carbide. 175,000 transverse strength;
93 Rock A; density: 5.83. For fast and general to light
machining of steel, particularly where shock and scale are
absent.

TI 886

British Aluminium Co., Ltd.
Cu 1.8-2.5, Mg 0.65-1.2, Si 0.55-1.25, Ni 0.6-1.4, Fe 0.6-1.2,
Ti 0.2, bal Al.
WP-temper: 51,500-56,000 TS; 38,000-42,600 YS; 6-8 El. For
structural members; age-hardened, corrosion resistant.
Obsolete

TI PD

TIMET
Titanium. Pd 0.15-0.2, bal Ti.
Plate: 62,000 TS; 46,000 YS; 27 El; 38 RA; 200 Brin. For
chemical industry equipment, for special corrosion
applications. Alpha type. Resists seawater corrosion.
Obsolete

TI-0.15 PD

TIMET
Pd 0.15, bal Ti.
At room temperature: 55,000 TS; 45,000 YS; 22 El; 35 RA
(min). At 800 F: 19,800 TS; 11,700 YS; 25 El; 76 RA. For
chemical industry applications where environments are
moderately reducing. Resists sea water corrosion. All alpha-
type alloy. *Obsolete*

TI-100A

TIMET
Titanium. C 0-0.1, N 0-0.05, H 0-0.01, Fe 0-0.3, O 0-0.4, bal
Ti.
Annealed: 85,000 TS; 70,000 YS; 15 El; 30 RA; 295 Brin. For
high temperature applications; commercially pure titanium.

TI-10V-2FE-3AL

Astro
Titanium. C 0-0.05, N 0-0.05, O 0-0.13, Fe 1.6-2.2, Al 2.6-3.4,
V 9-11, 0.015 H_2.
170,000 psi TS (ultimate); 160,000 psi YS (0.2% offset); 6 El
(in 2 in., sheet > 0.025 in. thick); density 0.174 lb/in.[3]; AMS
4983, 4984, 4986, 4987. For high-strength air frame
components.

TI-10V-2FE-3AL

TIMET
Titanium.
Now TI-10V-2FE-3AL STA.

TI-10V-2FE-3AL (STOA)

TIMET
Titanium. O 0.13, Fe 1.6-2.2, H 0.015, C 0.05, N 0.05, Al
2.6-3.4, V 9-11, bal Ti.
Near-beta alloy. STOA, RT min: 135 ksi TS; 125 ksi YS; 20 El;
15 RA. Ingot, billet, bar. For airframe forging applications.

TI-10V-2FE-3AL STA

TIMET
Titanium. Al 2.6-3.4, V 9-11, O 0-0.13, Fe 1.6-2.2, H 0-0.015,
C 0-0.05, N 0-0.05, bal Ti.
Heat treated: 140,000-180,000 psi TS; 130,000-170,000 psi
YS; 8-10 El; 15-20 RA. Good strength to 600°F.

TI-111

British Aluminium Co., Ltd.
Cu 0.15, Si 0.6, Fe 0.7, Mn 1-1.5, bal Al.
O-temper: 15,000 TS; 30 El. 1/2 H-temper: 26,000 TS; 7 El.
H-temper: 26,000 TS; 3 El. For structural members; corrosion
resistant. *Obsolete*

TI-13V-11CR-3AL

Astro
Titanium. C 0-0.05, N 0-0.05, Fe 0-0.35, V 12.5-14.5, Cr 10-12,
Al 2.5-3.5, 0.17 O_2, bal Ti. Bar, billet: 0.0250 H_2.
170,000 psi TS (ultimate); 160,000 psi YS (0.2% offset); 4 El
(in 2 in., sheet > 0.025 in. thick); density 0.175 lb/in.[3]; AMS
4959. For high-strength air frame components, including wire
springs.

TI-13V-11CR-3AL (STA)

TIMET
Titanium. O 0.17, Fe 0.35, H 0.025, C 0.05, N 0.05, Al 2.5-3.5,
V 12.5-14.5, Cr 10-12, bal Ti.
Beta alloy. STA, RT min: 170 ksi TS; 160 ksi YS; 4 El. Ingot,
plate, sheet, strip. For some airframe sheet metal applications
and for springs.

TI-140A

TIMET
C 0-0.1, Fe 1.5-2.5, Cr 1.5-2.5, Mo 1.5-2.5, bal Ti.
Annealed: 130,000-150,000 TS; 120,000-140,000 YS; 15 min
El; 286-336 Brin. For jet engine components; high strength,
low weight. *Obsolete*

TI-150B

TIMET
C 0.02, Cr 5, Fe 5, Mo 0.5, W 0-0.2, bal Ti.
Annealed: 160,000 TS; 130,000 YS; 10 El; 322 Brin. For jet
engine and missile components; high strength, low weight.
Obsolete

TI-155A

TIMET
Fe 1.1-1.7, Cr 1.1-1.7, Mo 0.9-1.5, Al 5-6, bal Ti.
Annealed: 155,000 TS; 140,000 YS; 15 El; 35 RA; 360 Brin.
Heat treated: 180,000 TS; 173,000 YS; 15 El; 25 RA; 400 Brin.
For aircraft propeller blades, cylinders and pistons; heat
treatable. *Obsolete*

TI-15V-3CR-3AL-3SN

TIMET
Titanium. O 0.13, Fe 0.25, H 0.015, C 0.05, N 0.05, Al 2.5-3.5,
V 14-16, Sn 2.5-3.5, Cr 2.5-3.5, bal Ti.
Beta-alloy. STA, RT min: 145 ksi TS; 140 ksi YS; 7 El. Ingot,
billet, plate, sheet, strip, welded tube. For sheet metal
applications.

TI-15V-3CR-3SN-3AL

Astro
Titanium. C 0-0.05, N 0-0.03, Fe 0-0.3, O 0-0.13, V 14-16, Cr
2.5-3.5, Sn 2.5-3.5, Al 2.5-3.5, 0.0150 H_2, bal Ti.
145,000 psi TS (ultimate); 140,000-170,000 psi YS (0.2%
offset); 7 El (in 2 in., sheet > 0.025 in. thick); density 0.172
lb/in.[3]; AMS 4914. For high-strength aircraft and aerospace
components.

TI-222

British Aluminium Co., Ltd.
Cu 0.15, Mg 1.7-2.7, Si 0.6, Fe 0.7, Mn 0.5, Ti 0.2, Cr 0.5, bal
Al.
O-temper: 31,000 TS; 20,000 YS; 18 El. 1/2 H-temper: 33,600
TS; 27,000 YS; 5 El. For structural members; corrosion
resistant. *Obsolete*

TI-223

British Aluminium Co., Ltd.
Cu 0.15, Mg 3-4, Si 0.6, Fe 0.7, Mn 1, Cr 0.5, Ti 0.2, bal Al.
O-temper: 34,500 TS; 13,500 YS; 18 El. 1/4 H-temper: 38,100
TS; 24,500 YS; 8 El. For structural members; corrosion
resistant. *Obsolete*

TI-225

British Aluminium Co., Ltd.
Cu 0.15, Mg 4.5-5.5, Si 0.6, Fe 0.7, Mn 1, Cr 0.5, Ti 0.2, bal
Al.
O-temper: 38,000 TS; 18,000 YS; 18 El. 1/4 H-temper: 42,600
TS; 31,400 YS; 8 El. For structural members; corrosion
resistant. *Obsolete*

TI-35A

TIMET
Titanium. C 0-0.1, N 0-0.03, H 0-0.15, Fe 0-0.02, O 0-0.18, bal
Ti.
Annealed: 35,000 TS; 25,000 YS; 22 El; 35 RA; 170 Brin. For
high temperature applications; commercially pure titanium.

TI-35A WITH PD

TIMET
Titanium. O 0.18, Fe 0.2, H 0.015, C 0.1, N 0.03, Pd
0.12-0.25, bal Ti.
Commercially pure. Annealed, RT min: 35 ksi TS; 25 ksi YS;
25 El; 35 RA. Ingot, billet, bar, plate, sheet, strip, welded tube;
corrosion resistant.

TI-3AL-2.5V

Teledyne Wah Chang Albany
Titanium. Al 3, V 2.5, bal Ti.
Cold worked and stress relieved. RT: 895 MPa TS; 760 MPa
YS; 19 El (in 50 mm). High corrosion resistance and strength.

TI-3AL-2.5V

Bishop Tube Co.
Al 3, V 2.5, bal Ti.
Cold drawn tube: 105,000-135,000 TS; 75,000-120,000 YS;
3-30 El. For aerospace equipment, airplane components.
High strength, ductility and fabricability. Used at
temperatures up to 600°F.

TI-3AL-2.5V
TIMET

Titanium. O 0.12, Fe 0.3, H 0.015, C 0.05, N 0.02, Al 2.5-3.5, V 2-3, bal Ti.

Alpha-beta alloy. Annealed, RT min: 90 ksi TS; 70 ksi YS; 15 El; 15-27 Rock C. Ingot, billet, plate, sheet, strip. For tubing in aircraft and engine hydraulic systems and as foil for honeycomb applications.

TI-3AL-2.5V
Astro

Titanium. C 0-0.05, N 0-0.02, Fe 0-0.3, O 0-0.12, Al 2.5-3.5, V 2-3, bal Ti. Sheet: 0.0150 H; bar: 0.0125 H.

Mill annealed: 90,000 psi TS (ultimate); 75,000 psi YS (0.2% offset); 15 El (in 2 in., sheet >0.025 in. thick); density 0.162 lb/in.^{3} Cold-worked stress relieved tubing: 125,000 psi TS (ultimate); 105,000 psi YS (0.2% offset); 10 El (in 2 in., sheet >0.025 in. thick); density 0.162 lb/in.^{3}; AMS 4943, 4944; ASTM B337 Gr 9, ASTM B381 Gr 9, ASTM B338 Gr 9. Used for air frames and jet engines; hydraulic and fuel lines.

TI-3AL-8MO-8V-2FE
TIMET

Al 3, Mo 8, V 8, Fe 2, bal Ti.

Heat treated: 184,000 TS; 167,000 YS; 8 El. 13 RA. For high temperature applications. Good notch-toughness, age-hardenable. Good weldability. *Obsolete*

TI-3AL-8V-6CR-4ZR-4MO
Astro

Titanium. C 0-0.05, N 0-0.03, O 0-0.14, Al 3-4, V 7.5-8.5, Cr 5.5-6.5, Zr 3.5-4.5, Mo 3.5-4.5, bal Ti.

180,000 psi TS (ultimate); 170,000 psi YS (0.2% offset); 6 El (in 2 in., sheet >0.025 in. thick); density 0.174 lb/in.^{3}; AMS 4957, 4958. For air frame high strength fasteners, rivets, torsion bars, springs; pipe for oil industry and geothermal applications.

TI-3AL-8V-6CR-4ZR-4MO
TIMET

Titanium. O 0.14, Fe 0.3, H 0.02, C 0.05, N 0.03, Al 3-4, V 7.5-8.5, Zr 3.5-4.5, Cr 5.5-6.5, bal Ti.

Beta alloy. STA, RT min: 170 ksi TS; 160 ksi YS; 6 El; 19 RA. Ingot, billet. For springs, fasteners, torsion bars, sheet, and foil applications.

TI-444
British Aluminium Co., Ltd.

Cu 0.15, Mg 0.4-1.5, Si 0.7-1.3, Fe 0.6, Mn 1, Cr 0.5, Ti 0.2, bal Al.

Heat treated: 29,200-42,600 TS; 15,700-33,600 YS; 8-15 El. For structural members; age-hardened. *Obsolete*

TI-445
British Aluminium Co., Ltd.

Cu 0.15, Mg 0.4-1.5, Si 0.7-1.3, Fe 0.6, Mn 1, Ti 0.2, bal Al. W-temper: 29,000 TS; 15,700 YS; 13 El; WP-temper: 40,400 TS; 31,400 YS; 18 El. For structural members; age-hardened, high strength. *Obsolete*

TI-45A
TIMET

C 0.1, N 0.05, bal Ti.

Annealed: 68,000 TS; 48,000 YS; 23 El. For non-structural members, fasteners. Corrosion resistant. *Obsolete*

TI-4AL-3MO-1V
TIMET

Al 4, Mo 3, V 1, bal Ti.

Bar: 125,000 TS; 90,000 YS; 16 El. For high temperature applications. *Obsolete*

TI-4AL-4MO-2SN (TI-550)
TIMET

Titanium. O 0.16-0.2, Fe 0.12, H 0.015, C 0.05, N 0.03, Al 3.5-4.5, Sn 1.5-2.5, Mo 3.5-4.5, Si 0.4-0.6, bal Ti.

Alpha-beta alloy. STA, RT min: 160 ksi TS; 139 ksi YS; 9 El; 20 RA. Ingot, billet. For engine compressor blades and discs.

TI-50A
TIMET

Titanium. O 0.25, Fe 0.2, H 0.015, C 0.1, N 0.03, bal Ti.

Commercially pure. Annealed, RT min: 50 ksi TS; 40 ksi YS; 20 El; 35 RA; 160 Brin. For ingot, billet, bar, plate, sheet, strip, welded tube; corrosion resistant.

TI-50A
TIMET

C 0-0.08, Fe 0-0.2, H 0-0.0125, N 0-0.05, bal Ti.

Annealed: 50,000 TS; 40,000 YS; 22 El; 35 RA; 200 Brin. For non-structural parts, fasteners. Corrosion resistance. *Obsolete*

TI-50A WITH PD
TIMET

Titanium. O 0.25, Fe 0.2, H 0.015, C 0.1, N 0.03, Pb 0.12-0.25, bal Ti.

Commercially pure. Annealed, RT min: 50 ksi TS; 40 ksi YS; 20 El; 35 RA. Ingot, billet, bar, plate, sheet, strip, welded tube; corrosion resistant.

TI-554
British Aluminium Co., Ltd.

Cu 3.5-4.5, Mn 0.4-1.2, Si 0.7, Fe 0.7, Mn 0.4-1.2, bal Al. Heat treated: 53,800 TS; 30,400 YS; 15 El. For structural members; age hardened. *Obsolete*

TI-55A
TIMET

Titanium. C 0-0.2, H 0-0.015, bal Ti.

Bar: 75,000 TS; 65,000 YS; 25 El. For non-structural parts, fasteners. Corrosion resistant. *Obsolete*

TI-5AL, 2.5SN, 2.5V, 1.3CB,, 1.3TA
TIMET

Al 5, Sn 2.5, V 2.5, Cb 1.3, Ta 1.3, bal Ti.

Sheet: At 70 F: 134,000 TS; 121,000 YS; At -320 F: 210,000 TS; 199,000 YS; 14 El; For structural applications from ambient to liquid hydrogen temperatures. Good formability and weldability. Good low temperature properties. *Obsolete*

TI-5AL-2.5SN
Astro

Titanium. C 0-0.08, Fe 0-0.5, N 0-0.05, Al 4-6, Sn 2-3, 0.20 O$_2$, bal Ti. Forging, bar, sheet: 0.02 H$_2$.

120,000 psi TS (ultimate); 115,000 psi YS (0.2% offset); 10 El (in 2 in., sheet >0.025 in. thick); 25 RA (bar); 10-15 ft·lb IS (Charpy V-notch, at room temperature); density 0.162 lb/in.^{3}; AMS 4910, 4926, 4966. Used for air frame and jet engine applications requiring good weldability, stability and strength at elevated temperatures.

TI-5AL-2.5SN
TIMET

Titanium. Al 4-6, Sn 2-3, O 0-0.2, Fe 0-0.5, H 0-0.02, C 0-0.1, N 0-0.05, bal Ti.

Annealed: 120,000 TS; 113,000 YS; 10 El; 25 RA. For airframes, high temperature applications; resists oxidation to 1200°F.

TI-5AL-2.5SN ELI
TIMET

Titanium. Al 4.5-5.75, Sn 2-3, Fe 0-0.25, O 0-0.12, H 0-0.0125, C 0-0.05, N 0-0.035, bal Ti.

At -423°F: 229,000 TS; 206,000 YS; 10 El; 1.03 notch ratio. Annealed: 105,000 TS; 95,000 YS; 10 El; 36 Rock C. For high temperature applications, fasteners, airframes. Extra low interstitials; improved notch strength at -423°F.

TI-5AL-2SN-4MO-2ZR-4CR (TI-17)
TIMET

Titanium. O 0.08-0.13, Fe 0.25, H 0.0125, C 0.05, N 0.05, Al 4.5-5.5, Sn 1.6-2.4, Mo 3.5-4.5, Zr 1.6-2.4, Cr 3.5-4.5, bal Ti. Near-beta alloy. STA, RT min: 163 ksi TS; 153 ksi YS; 5 El; 10 RA. Ingot, billet. For engine compressor blades and discs.

TI-5AL-4FE-CR
TIMET

Al 4.75-5.75, Fe 0.75-1.75, Cr 2.25-3.25, bal Ti.

At room temperature: 150,000 TS; 140,000 YS; 10 El; 25 RA; Rock C 32-38. At 800 F: 109,500 TS; 79,800 YS; 15 El; 22 RA. For airframe components. Alpha-beta type alloy. Age hardenable. *Obsolete*

TI-5AL-5SN-5ZR
TIMET

Al 5, Sn 5, Zr 5, bal Ti.

Alpha anneal: 128,300 TS; 118,000 YS; 18 El; 37.3 RA. Alpha-beta anneal: 132,300 TS; 122,700 YS; 15 El; 32 RA. At 1000 F: 84,500 TS; 67,400 YS; 20 El; 37 RA. For high temperature applications, turbine engines, airframes. Alpha type titanium alloy for use up to 1100 F. Minimum creep. *Obsolete*

TI-5AL-6SN-2ZR-MO-0.25SI
TIMET

Al 5, Sn 6, Zr 2, Si 0.25, Mo, bal Ti.

Heat treated: 158,000 TS; 144,000 YS; 12 El; 37 RA. For forged engine discs. Corrosion resistant, high strength, good heat resistance. *Obsolete*

TI-6.5AL-2CB-1TA-1.2MO
TIMET

Al 6.5, Cb 2, Ta 1, Mo 1.1, Fe 0.05, bal Ti.

Rolled: 131,000 TS; 109,000 YS; 12 El; 25 RA. Stress-relieved: 132,000 TS; 113,000 YS; 13 El; 23 RA. For pressure hulls. Tough, shock resistant. *Obsolete*

TI-6.5AL-3MO-1V
TIMET

Al 6.5, Mo 3, V 1, bal Ti.

Bar: 155,000 TS; 150,000 YS; 17 El. For high temperature applications. *Obsolete*

TI-65A
TIMET

Titanium. C 0-0.1, Fe 0-0.2, H 0-0.015, N 0-0.05, O 0-0.35, bal Ti.

Annealed: 65,000 TS; 55,000 YS; 18 El; 35 RA; 225 Brin. For aircraft structural members. Corrosion resistant.

TI-666
British Aluminium Co., Ltd.

Cu 3.5-4.8, Mg 0.8, Si 0.9, Fe 1, Mn 1.2, Ti 0.3, bal Al. W-temper: 54,800 TS; 31,400 YS; 15 El. For structural members; age-hardened, high strength. *Obsolete*

TI-667
British Aluminium Co., Ltd.

Cu 3.5-4.8, Mg 0.8, Si 0.9, Fe 1, Mn 1.2, Ti 0.3, bal Al. W-temper: 54,800 TS; 31,400 YS; 15 El. For structural members; age-hardened, high strength. *Obsolete*

TI-679
IMI Knoch Ltd.

Al 2-2.25, Sn 10.5-11, Zr 4-5, Mo 0.8-1, Si 0.15-0.21, bal Ti. Heat treated: 140,000 TS; 130,000 YS; 10 El; 20 RA. For jet engine compressor wheels and blades. High stress stability at 900°F, good creep and short time strength.

TI-6AL-2CB-1TA-0.8MO
TIMET

Al 6, Cb 2, Ta 1, Mo 0.8, bal Ti.

Rolled: 122,000 TS; 103,000 YS; 13 El; 28 RA. Stress-relieved: 124,000 TS; 106,000 YS; 13 El; 27 RA. For pressure hulls. Improved resistance to salt water cracking. Tough, shock resistant. *Obsolete*

TI-6AL-2SN-4ZR-2MO
Astro

Titanium. C 0-0.1, Fe 0-0.25, N 0-0.5, Al 5.5-6.5, Sn 1.8-2.2, Zr 3.6-4.4, Mo 1.8-2.2, 0.15 O$_2$, bal Ti. Bar: 0.125 H$_2$.

130,000 psi TS (ultimate); 120,000 psi YS (0.2% offset); 10 El (in 2 in., sheet >0.025 in. thick); 25 RA (bar); density 0.164 lb/in.^{3}; AMS 4976. For use where optimum creep strength and high strength at elevated temperature are required.

TI-6AL-2SN-4ZR-2MO

TIMET. Al 5.5-6.5, Sn 1.8-2.2, Zr 3.6-4.4, Mo 1.8-2.2, C 0-0.05, O 0-0.15, Fe 0-0.25, H 0-0.0125, N 0-0.05, bal Ti. Annealed: 135,000 TS; 125,000 YS; 8 El; 25 RA; 36 Rock C. For high temperature components to 1050°F; compressor wheels, missiles, space craft. High temperature alloy. Alpha type. Good creep and tensile strength to 1050°F.

TI-6AL-2SN-4ZR-2MO-0.08SI

TIMET
Titanium. Al 5.5-6.5, Sn 1.8-2.2, Zr 3.6-4.4, Mo 1.8-2.2, Si 0-0.1, N 0-0.05, O 0-0.15, Fe 0-0.25, H 0-0.0125, C 0-0.05, bal Ti.
Duplex annealed: 135,000 TS; 125,000 YS; 8 El. For engine discs. For 800-1000°F service, good heat resistance.

TI-6AL-2SN-4ZR-6MO

Astro
Titanium. C 0-0.1, Fe 0-0.15, N 0-0.04, O 0-0.15, Al 5.5-6.5, Sn 1.8-2.2, Zr 3.6-4.4, Mo 5.5-6.5, bal Ti. Bar: 0.0125 H. 170,000 psi TS (ultimate); 160,000 psi YS (0.2% offset); 10 El (in 2 in., sheet >0.025 in. thick); 20 RA (bar); density 0.169 lb/in.^{3}; AMS 4981. Used for jet engine components requiring high tensile strength and intermediate creep strength.

TI-6AL-2SN-4ZR-6MO

TIMET
Titanium. Al 5.5-6.5, Sn 1.75-2.25, Zr 3.5-4.5, Mo 5.5-6.5, C 0-0.04, O 0-0.15, Fe 0-0.15, H 0-0.0125, N 0-0.04, bal Ti. Alpha-beta alloy. Duplex annealed: 160,000-170,000 psi TS; 150,000-160,000 psi YS; 6-10 El; 12-20 RA. Triplex annealed: 170,000 psi TS; 155,000 psi YS; 10 El; 20 RA. Used for gas turbine applications.

TI-6AL-4V

Astro
Titanium. C 0-0.08, Fe 0-0.25, N 0-0.05, O 0-0.2, Al 5.5-6.75, V 3.5-4.5, bal Ti. Sheet: 0.0150 H; bar: 0.0125 H; billet: 0.0100 H.
130,000 psi TS (ultimate); 120,000 psi YS (0.2% offset); 10 El (in 2 in., sheet >0.025 in. thick); 25 RA (bar); 10-14 ft·lb IS (Charpy V-notch, at room temperature); density 0.160 lb/in.^{3}; AMS 4911, 4928, 4935, 4965, 4967; ASTM B265 Gr 5, ASTM B348 Gr 5, ASTM B381 Gr 5, ASTM F467 Gr 5, ASTM F468 Gr 5. Used for jet engine components; air frame forgings; sheet; plate; air frame components.

TI-6AL-4V

TIMET
Titanium. Al 5.5-6.75, V 3.5-4.5, O 0-0.2, Fe 0-0.4, H 0-0.015, C 0-0.1, N 0-0.05, bal Ti.
Annealed: 130,000 TS; 120,000 YS; 10 El; 36 Rock C. For high temperature applications; stable under stress to 950°F; heat treatable.

TI-6AL-4V (STA)

TIMET
Titanium. O 0.2, Fe 0.4, H 0.015, C 0.1, N 0.05, Al 5.5-6.75, V 3.5-4.5, bal Ti.
Alpha-beta alloy. STA, RT min: 150 ksi TS; 140 ksi YS; 10 El. Ingot, billet, bar, plate, sheet, strip. For turbine and airframe uses.

TI-6AL-4V ELI

TIMET
Titanium. Al 5.5-6.5, V 3.5-4.5, Fe 0-0.25, O 0-0.13, H 0-0.0125, C 0-0.08, N 0-0.05, bal Ti.
Annealed: 130,000 TS; 120,000 YS; 10 El; 32 Rock C. At -423°F: 263,000 TS. For high temperature applications, jet engine components, fasteners, aircraft forgings. Extra low interstitials. Improved notch strength at -423°F.

TI-6AL-6V-2SN

Astro
Titanium. C 0-0.05, N 0-0.04, O 0-0.2, Al 5-6, V 5-6, Sn 1.5-2.5, Fe 0.35-1, Cu 0.35-1, bal Ti. Bar, sheet, plate: 0.0150 H$_2$.
150,000 psi TS (ultimate); 140,000 psi YS (0.2% offset); 10 El (in 2 in., sheet >0.025 in. thick); 20 RA (bar); 10-14 ft·lb IS (Charpy V-notch, at room temperature); density 0.164 lb/in.^{3}; AMS 4918, 4936, 4971, 4978, 4979. Used for rocket engine case air frame applications.

TI-6AL-6V-2SN

TIMET
Titanium. Al 5-6, V 5-6, Sn 1.5-2.5, Fe 0.35-1, O 0-0.2, H 0-0.05, C 0-0.05, N 0-0.04, bal Ti.
Annealed: 150,000 TS; 140,000 YS; 8 El; 15 RA. For high strength pressure vessels operating at ambient temperatures. Heat treatable, alpha-beta type alloy.

TI-6AL-6V-2SN (STA)

TIMET
Titanium. O 0.2, Fe 0.35-1, H 0.015, C 0.05, N 0.04, Al 5-6, V 5-6, Sn 1.5-2.5, Cu 0.35-1, bal Ti.
Alpha-beta alloy. STA, RT min: 175 ksi TS; 160 ksi YS; 6 El; 15 RA. Billet, bar, plate, sheet, strip. Plate and forging applications for airframes.

TI-75A

TIMET
Titanium. C 0-0.1, Fe 0-0.3, H 0-0.015, N 0-0.05, O 0-0.4, bal Ti.
Annealed: 80,000 TS; 70,000 YS; 15 El; 30 RA; 250 Brin. For moderately stressed aircraft parts and structures. Corrosion resistant.

TI-7AL-12ZR

TIMET
Al 7, Zr 12, bal Ti.
Annealed: 135,300 TS; 123,800 YS; 15.5 El; 24.8 RA. At 1000 F: 87,100 TS; 68,900 YS; 18 El; 38.8 RA. At RT: 137,500 TS; 127,400 YS; 15.2 El. For high temperature applications, turbine engines, airframe components, fasteners. Alpha type titanium alloy use up to 1100 F. High creep strength. *Obsolete*

TI-7AL-2.5MO

TIMET
Al 7, Mo 2.5, bal Ti.
Plate: 129,000 TS; 109,000 YS; 10 El; 25 RA. Heat treated: 148,000 TS; 124,000 YS; 6 El; 10 RA. For hulls of deep-diving vehicles. Heat treatable. Resists sea water corrosion. *Obsolete*

TI-7AL-2CB-1TA

TIMET
Al 7, Cb 2, Ta 1, bal Ti.
Plate: CYS 115,000. For submarine hulls. Tough, high strength, corrosion resistant. *Obsolete*

TI-7AL-4MO

TIMET
Titanium. Al 6.5-7.3, Mo 3.5-4.5, O 0-0.2, Fe 0-0.3, H 0-0.13, C 0-0.1, N 0-0.05, bal Ti.
Annealed: 145,000 TS; 135,000 YS; 10 El; 20 RA. For aircraft gas turbine engines, compressor wheels, hub shafts; creep resistant, high temperature strength.

TI-7AL-4MO (STA)

TIMET
Titanium. O 0.2, Fe 0.3, H 0.013, C 0.1, N 0.05, Al 6.5-7.3, Mo 3.5-4.5, bal Ti.
Alpha-beta alloy. STA, RT min: 170 ksi TS; 160 ksi YS; 8 El; 20 RA. Ingot, billet, bar. For horns on ultrasonic welding equipment.

TI-8AL-10V

Republic Steel Corp.
Al 8, V 10, bal Ti.
Sheet heat treated: 208,000-223,000 TS; 199,000-214,000 YS; 1.0-5.0 El; Rock C 47. For rocket motor cases. Heat treatable. Corrosion resistant. *Obsolete*

TI-8AL-1MO-1V

Astro
Titanium. C 0-0.08, Fe 0-0.3, N 0-0.05, O 0-0.12, Al 7.35-8.35, Mo 0.75-1.25, V 0.75-1.25, bal Ti. Bar: 0.0125 H$_2$; forging: 0.0150 H$_2$.
130,000 psi TS (ultimate); 120,000 psi YS (0.2% offset); 15-25 ft·lb IS (Charpy V-notch, at room temperature); density 0.158 lb/in.^{3}; AMS 4916, 4925, 4972, 4973. Used for jet engine components requiring good creep strength, high strength at elevated temperatures.

TI-8AL-1MO-1V

TIMET
Titanium. Al 7.35-8.35, Mo 0.75-1.25, V 0.75-1.25, O 0-0.12, Fe 0-0.3, H 0-0.015, C 0-0.08, N 0-0.05, bal Ti.
Annealed, RT min: 130,000 TS; 120,000 psi YS; 10 El; 20 RA; 36 Rock C. For jet engine components to 900°F; excellent creep and rupture life.

TI-8AL-2CB-1TA

TIMET
Al 8, Cb 2, Ta 1-0, bal Ti.
Annealed: 126,000 TS; 120,000 YS; 17 El. At 2000 F: 1,051 TS; 552 YS; 40 El. At 600 F: 100,000 TS; 81,000 YS; 25 El. For parts requiring high impact strength, deep diving vehicles. Resists sea water corrosion. *Obsolete*

TI-AL

Chemalloy Co., Inc.
Titanium.
Titanium aluminum alloy.

TI-BRUSH 120-AM

Gould Inc.
Al 3.5-4.5, Mn 3.5-4.5, bal Ti.
At 70 F: 135,000 TS; 120,000 YS; 15 El; 30 RA; 300 Brin. At 600 F: 96,000 TS; 74,000 YS; 11 El; 26 RA. For jet engine components, fasteners; heat treatable, low density, heat resistant. *Obsolete*

TI-BRUSH 40

Gould Inc.
C 0-0.1, O 0-0.2, Mn 0-0.1, Fe 0-0.2, bal Ti.
Rolled: 60,000 TS; 40,000 YS; 30 El; 50 RA; 141 Brin. For aircraft and missile components; high fatigue-strength ratio. *Obsolete*

TI-BRUSH 50

Gould Inc.
C 0-0.1, O 0-0.2, Al 0.8-1, N 0-0.02, bal Ti.
At 70 F: 70,000 TS; 55,000 YS; 27 El; 41 RA; 160 Brin. At 800 F: 26,000 TS; 18,000 YS; 21 El; 49 RA. For aircraft and missile components; good weldability, low density. *Obsolete*

TI-BRUSH 65A

Gould Inc.
Al 1.9-2, C 0-0.1, O 0-0.2, N 0-0.02, bal Ti.
At 70 F: 85,000 TS; 65,000 YS; 20 El; 40 RA; 200 Brin. At 1000 F: 32,250 TS; 25,100 YS; 33 RA. For aircraft and missile components; heat resistant, low density. *Obsolete*

TI CAD K4

Texas Instruments Inc./Materials Control
Ag 89-91, Co 0.001-0.005, 9.2-11.0 CdO.
Electrical contact material providing resistance to erosion and welding.

TI-CAD K5

Texas Instruments Inc./Materials Control
Ag 84-86, Co 0.001-0.005, 14.5-15.8 CdO.
Electrical contact material providing resistance to erosion and welding.

TI-LOY 30

Chemalloy Co., Inc.
Titanium.
30% ferro titanium.

TI-LOY 40
Chemalloy Co., Inc.
Titanium.
40% ferro titanium.

TI-LOY 70
Chemalloy Co., Inc.
Titanium.
70% ferro titanium.

TI-LOY 85 BRIQUETTES
Chemalloy Co., Inc.
Titanium.
Titanium briquettes.

TI-LOY 99 POWDER
Chemalloy Co., Inc.
Titanium.
Titanium powders.

TI-NAMEL
Inland Steel Co.
HSLA steel. C 0.06, Mn 0.3, Cu 0-0.12, Al 0.05, Ti 0.3, bal Fe.
Hot rolled: 55,000 TS; 33,000 YS; 27 El; 121 Brin. For sheet
for enameling; specially prepared.

TI-SIL
Chemalloy Co., Inc.
Titanium.
Titanium silicon alloy.

TI-TAX
Thomas Prosser & Sons
WC, Co.
For cutting tools. *Obsolete*

TI.01
British Aluminium Co., Ltd.
Cu 4, Mg 0.6, Mn 0.5, bal Al.
Heat treated: 58,300 TS; 41,000 YS; 8-12 El. For aircraft
structures; age hardenable. *Obsolete*

TI.03
British Aluminium Co., Ltd.
Cu 1, Mg 1, Si 1, bal Al.
W(T4): 38,100 TS; 22,400 YS; 15 El. WP(T6): 56,000 TS;
44,800 YS; 6 El. For structures, marine trim; age hardenable.
Obsolete

TI.04
British Aluminium Co., Ltd.
Cu 4, Mg 0.6, Mn 0.5, bal Al.
Heat treated: 53,800 TS; 31,400 YS; 15 El. For aircraft
structures; age hardenable. *Obsolete*

TI.05
British Aluminium Co., Ltd.
Mg 5, bal Al.
O-temper: 38,100 TS; 18,000 YS; 18 El. 1/2 H-temper: 40,300
TS; 31,400 YS; 5 El. For marine parts; resists sea water
corrosion. *Obsolete*

TI.07
British Aluminium Co., Ltd.
Mg 7, bal Al.
O-temper: 44,800 TS; 20,200 YS; 18 El. 1/2 H-temper: 56,000
TS; 35,900 YS; 5 El. H-temper: 58,300 TS; 40,300 YS. For
marine parts; resists sea water corrosion. *Obsolete*

TI.11
British Aluminium Co., Ltd.
Mn 1.25, bal Al.
O-temper: 13,400-16,500 TS; 25-30 El. H-temper:
26,000-28,000 TS; 3-5 El. For tank cars, storage bins; resists
reducing and oxidizing atmospheres. *Obsolete*

TI.22
British Aluminium Co., Ltd.
Mg 2, bal Al.
O-temper: 24,600-31,400 TS; 18 El; 45 Brin. 1/2 H-temper:
33,600 TS; 26,900 YS; 5 El; 70 Brin. For deckhousings,
gasoline tanks, marine parts; corrosion resistant. *Obsolete*

TI.33
British Aluminium Co., Ltd.
Mg 3, bal Al.
O-temper: 31,400 TS; 13,500 YS; 18 El. 1/4 H-temper: 38,100
TS; 24,700 YS; 8 El. For light weight structural parts;
corrosion resistant, not heat treatable. *Obsolete*

TI.40
British Aluminium Co., Ltd.
Mg 0.75, Si 0.5, bal Al.
M-temper: 15,700 TS; 15 El. W(T4): 20,200 TS; 11,200 YS; 18
El. WP(T6): 26,900 TS; 22,400 YS; 12 El. For architectural and
building applications; corrosion resistant. *Obsolete*

TI.44
British Aluminium Co., Ltd.
Mg 0.75, Si 1, bal Al.
T4: 29,100 TS; 15,700 YS; 15 El. T6: 42,600 TS; 33,600 YS; 8
El. O-temper: 24,600 TS; 20 El. For light weight parts; good
workability. *Obsolete*

TI.53
British Aluminium Co., Ltd.
Cu 3.4, Mg 0.7, Sb 0.5, Sn 0.2, bal Al.
M-temper: 35,900 TS; 15,700 YS; 10 El. For screw machine
products; free-cutting. *Obsolete*

TI.55
British Aluminium Co., Ltd.
Cu 2.6, Si 1, Mn 0.3, Fe 1, Ni 0.7, bal Al.
WP(T6): 56,000 TS; 42,600 YS; 8 El. For general purpose
forgings; age hardenable. *Obsolete*

TI.56
British Aluminium Co., Ltd.
Cu 2, Mg 0.9, Si 0.8, Fe 0.9, Ni 1, Ti 0.1, bal Al.
WP(T6): 58,200-60,500 TS; 44,800-47,000 YS; 10 El. For
aircraft structures; age hardenable. *Obsolete*

TI.66
British Aluminium Co., Ltd.
Cu 4.4, Mg 0.6, Si 0.7, Mn 0.6, bal Al.
W(T4): 53,800-58,300 TS 31,400-35,900 YS; 12-15 El. WP(T6):
62,800 TS; 51,500 YS; 8 El. For aircraft structures; age
hardenable. *Obsolete*

TI.77
British Aluminium Co., Ltd.
Cu 0.4, Mg 2.7, Mn 0.5, Zn 5.3, bal Al.
WP-temper: 67,000-72,000 TS; 58,000-60,500 YS; 8 El. For
aircraft structures; Al coated, age hardenable. *Obsolete*

TI.88
British Aluminium Co., Ltd.
Cu 1, Mg 2.7, Mn 0.5, Zn 5.3, bal Al.
WP(T6): 78,400-85,100 TS; 67,200-74,000 YS; 5 El. For high
strength forgings; age hardenable. *Obsolete*

TIAL6SN2ZR4MO2
Titan-Aluminium-Feinguss GmbH
Fe 0-0.25, C 0-0.1, Al 5.5-6.5, Sn 1.8-2.2, Zr 3.6-4.4, Mn
1.8-2.2, bal Ti.
Investment casting alloy. 830 N/mm^2 YS; 900 N/mm^2 TS; 8
El. ASTM B367-38; 3.7144.4.

TIBELOY 637
Alloy Metals Inc.
Al 4, Pd 3.5, bal Ag.
For precious metal brazing. 1770-1800°F brazing
temperature.

TIBELOY 692
Alloy Metals Inc.
Al 8, Pd 4, bal Ag.
For precious metal brazing. 1500-1550°F brazing
temperature.

TIBON
Titusville Forge Co.
C, bal Fe.
For die blocks.

TICO
English manufacture
Ni 27.5-30.4, Mn 1.12, Cu 1.1, bal Fe.
For electrical resistances.

TICODE-12
TIMET
Titanium. Ni 0.6-0.9, Mo 0.2-0.4, H 0-0.015, O 0-0.25, Fe
0-0.3, C 0-0.08, N 0-0.03, bal Ti.
70,000 psi TS; 50,000 psi YS; 18 El. Excellent corrosion
resistance and good thermal conductivity; good strength at
600°F. For heat exchangers.

TICONAL
Philips Electronic & Associated Ind.
Al 6-9, Ni 12-18, Ti 0-2.5, Cu 2-5, Co 20-30, bal Fe.
For permanent magnet, sand and precision cast; similar to
Alnico 5.

TICONAL 1500
Philips Electronic & Associated Ind.
Co 35, Ni 13.5, Al 7.5, Ti 5, Cu 3, bal Fe.
Cast: 10,000 TS. 8500 Br, 1450 Hc, 5 BH max. For magnets
in electric clocks, electrical and magnetic equipment,
magnetic separators, generators. Permanent magnet. Lowest
losses after stabilization. Same as Alnico VIII.

TICONAL 2000
Philips Electronic & Associated Ind.
Co 40, Ti 7.5, bal Fe.
Coercive force 2000. For electric motors and magnetic
equipment. Permanent magnet, high magnetic permeability.

TICONAL 3A
Philips Electronic & Associated Ind.
Ni 19, Co 12, Al 10, Cu 6, bal Fe.
For permanent magnets in electrical and magnetic
equipment. High magnetic permeability.

TICONAL 600
Philips Electronic & Associated Ind.
Al 8, Ni 14, Co 24, Cu 3, bal Fe.
Cast: 5450 Transverse strength; 50 Rock C. 1200 Br; 560 Hc,
5,600,000 BH max. For relays, motors, voltage regulators,
magnetos, generators. Similar to Alnico V permanent
magnet.

TICONAL 700
Philips Electronic & Associated Ind.
Ni 14, Al 8, Co 24-35, Cu 3, bal Fe.
11,600 Br, 700 Hc, 5 BH max, 48 Rock C. Magnets for
measuring instruments, electrical and magnetic equipment.
Permanent magnet, high magnetic permeability.

TICONAL 750
Philips Electronic & Associated Ind.
Ni 14, Al 8, Co 24-35, Cu 3, bal Fe.
11,000 Br, 750 Hc, 4.4 BH max, 52 Rock C. For magnets in
electrical apparatus and magnetic equipment. Permanent
magnet, high magnetic permeability.

TICONAL 800
Philips Electronic & Associated Ind.
Ni 14, Al 8, Co 24-35, Cu 3, bal Fe.
10,600 Br, 800 Hc, 4.3 BH max, 52 Rock C. For magnets in
electronic, electrical and magnetic equipment. Same as
Alcomax IV. Permanent magnet. Thermal stability and
resistant to demagnetizing fields.

TICONAL C
Philips Electronic & Associated Ind.
Co 24, Ni 13.5, Al 8, Cu 3, Cb 0.5, bal Fe.
For permanent magnets in electrical and magnetic equipment. High magnetic permeability.

TICONAL D
Philips Electronic & Associated Ind.
Co 24, Ni 14, Al 8, Cu 3, Ti 1, bal Fe.
For permanent magnets in electrical and magnetic equipment. High magnetic permeability.

TICONAL E
Philips Electronic & Associated Ind.
Al 8, Ni 14, Co 24, Cu 3, bal Fe.
For permanent magnets; high permeability.

TICONAL E
Usines Emile Henricott, SA
Ni 15, Al 8, Co 25, Cu 3, bal Fe.
For permanent magnets; high permeability.

TICONAL F
Philips Electronic & Associated Ind.
Co 24, Ni 14, Al 8, Cu 3, Ti 0.5, bal Fe.
For permanent magnets in electrical and magnetic equipment. High magnetic permeability.

TICONAL GG
Philips Electronic & Associated Ind.
Al 8, Ni 14, Co 24, Cu 3, bal Fe.
For permanent magnets; high permeability.

TICONAL GX
Philips Electronic & Associated Ind.
Co 24, Ni 14, Al 8, Cu 3, bal Fe.
For permanent magnets in metering devices, electrical and magnetic equipment. High magnetic permeability. Strong permanent magnet.

TICONAL H
Philips Electronic & Associated Ind.
Co 24, Ni 13.5, Al 8, Cu 3, Cb 2, bal Fe.
For permanent magnets in electrical and magnetic equipment. High magnetic permeability.

TICONAL K
Philips Electronic & Associated Ind.
Al 6, Ni 18, Co 34, Ti 8, bal Fe.
For permanent magnets; high permeability.

TICONAL L
Philips Electronic & Associated Ind.
Co 19.5, Ni 14.5, Al 7, Cu 1.5, Si 0.8, bal Fe.
For permanent magnets in electrical and magnetic equipment. High magnetic permeability.

TICONAL S
Philips Electronic & Associated Ind.
Co 24, Ni 14, Al 8, Cu 3, bal Fe.
For permanent magnets in electrical and magnetic equipment. High magnetic permeability.

TICONAL X
Philips Electronic & Associated Ind.
Co 34, Ni 14.5, Al 7, Cu 4.5, Ti 5, bal Fe.
Max energy product 13,400,000. 8800 Br, 1500 Hc, 5,300,000 BH max. For permanent magnets, in electrical equipment. High magnetic permeability.

TICONAL XX
Philips Electronic & Associated Ind.
Al 7, Ni 15, Cu 4, Ti 5, Co 34, bal Fe.
For permanent magnets; highly directional crystal oriented.

TICONAL-G
Philips Electronic & Associated Ind.
Co 28, Ni 13, Al 8, Cu 3, bal Fe.
For permanent magnets. 12000 Br, 700 Hc.

TICONIUM
Consolidated Car Heating Co.
Ni 35, Co 31, Cr 23, Mo 6, C 0.01, bal Fe.
Cast: 110,000 TS; 65,000 YS; 6 El; 350 Brin. For bone surgery. Corrosion resistant.

TICUNI
GTE Products Corp./Wesgo Div.
Cu 15, Ni 15, Ti 70.
Brazing alloy available in foil, flexibraze, extrudable paste and preform. Liquidus 1760°F. Solidus 1670°F.

TICUNI
Western Gold & Platinum Co.
Ti 70, Cu 15, Ni 15.
For brazing titanium and stainless steel components.

TICUNI
GTE Sylvania
Ti 70, Cu 15, Ni 15.
For brazing titanium and stainless steel components.

TICUSIL
GTE Products Corp./Wesgo Div.
Ag 68.8, Cu 26.7, Ti 4.5.
Brazing alloy available in foil, flexibraze, wire, powder, extrudable paste and preform. Liquidus 1562°F. Solidus 1526°F.

TIERS ARGENT
English manufacture
Al 66, Ag 33.
For ornamental parts; corrosion resistant.

TIFICO
Titusville Forge Co.
C, alloy, bal Fe.
For engine crankshafts, die blocks, marine service parts; oil hardened.

TIG-TECTIC 182
Eutectic Corp.
Inert arc for joining copper base alloys and steel.

TIG-TECTIC 183
Eutectic Corp.
For inert arc welding of copper. *Obsolete*

TIG-TECTIC 1851
Eutectic Corp.
Aluminum bronze filler rod for inert arc joining and overlaying on bronze and other copper alloys.

TIG-TECTIC 2-24
Eutectic Corp.
High nickel filler metal for inert arc welds for cast iron. *Obsolete*

TIG-TECTIC 21
Eutectic Corp.
Inert arc overlays for aluminum.

TIG-TECTIC 21FC
Eutectic Corp.
Inert arc overlays for aluminum. *Obsolete*

TIG-TECTIC 222
Eutectic Corp.
Inert arc electrode for nickel alloys. *Obsolete*

TIG-TECTIC 224
Eutectic Corp.
Electrode for TIG welding gray cast iron. Machinable; 50,000 psi TS.

TIG-TECTIC 23
Eutectic Corp.
Inert arc overlays for aluminum.

TIG-TECTIC 5-AH
Eutectic Corp.
Inert arc or torch overlays on gauges, blanking dies, and forming dies.

TIG-TECTIC 5-HSS
Eutectic Corp.
Inert arc or torch high speed steel overlays.

TIG-TECTIC 5-HW
Eutectic Corp.
Inert arc or tooth overlays for hot work tools and dies.
Obsolete

TIG-TECTIC 5-OH
Eutectic Corp.
Inert arc or torch overlays for oil hardening steels. *Obsolete*

TIG-TECTIC 5-WH
Eutectic Corp.
Inert arc or torch overlays for water hardening steels.
Obsolete

TIG-TECTIC 66
Eutectic Corp.
Inert arc electrode for medium carbon steel welding and repair.

TIG-TECTIC 670
Eutectic Corp.
Inert arc welding of alloy and stainless steel.

TIG-TECTIC 680
Eutectic Corp.
Inert arc electrode for welding alloy steel; 120,000 TS.

TIG-TECTIC 6800
Eutectic Corp.
Inert arc electrode for corrosion resistant welds on nickel alloys.

TIG-TECTIC A
Eutectic Corp.
Columbium-stabilized stainless electrode for inert arc welding stainless steel.

TIG-TECTIC A-MO
Eutectic Corp.
Now TIG-TECTIC AMO-L. *Obsolete*

TIG-TECTIC AMO-L
Eutectic Corp.
Mo bearing stainless electrode for inert arc welding stainless steel.

TIG-TECTIC B
Eutectic Corp.
For inert arc welding 18-8 stainless. *Obsolete*

TIG-TECTIC D
Eutectic Corp.
25/20 stainless rod (Type 310) for inert arc welding 309-310 type stainless or dissimilar stainless. *Obsolete*

TIGER
Bisset Steel Co.
W 18, Cr 4, V 1, C, bal Fe.
For cutting tools, reamers, gages; high speed steel. *Obsolete*

TIGER BRAND
Bethlehem Steel Corp.
C 0.7-0.9, W 18, Cr 4, V 1, bal Fe.
For tools, cutters, reamers; high speed steel. *Obsolete*

TIGER BRONZE
Abex Corp.
Cu 75, Sn 6, Pb 16, Zn 3.
Sand cast: 20,000 psi TS; 14,000 psi YS; 7 El; 50 Brin. Similar to CDA 939.

TIGER SPECIAL
Bethlehem Steel Corp.
C 0.75, Cr 4, W 18, Mo 0.5, V 1, Co 5, bal Fe.
For punches, tools, cutters; tough, shock resistant. *Obsolete*

TIGER SPECIAL
Bisset Steel Co.
C 0.75, Cr 4.2, W 18, Mo 0.5, V 1.1, Co 5, bal Fe.
For lathe and planer tools, hobs, drills; high speed steel.
Obsolete

TIGER VAN
Bisset Steel Co.
C 0.8, Cr 4.2, W 18.5, Mo 0.5, V 2.1, bal Fe.
For drills, hobs, cutters, reamers, broaches, taps; high speed
steel. *Obsolete*

TIGERLOY
Massillon Steel Casting Co.
C 0.3, Ni 1.5, Mo 0.2, bal Fe.
Heat treated: 100,000 TS; 60,000 YS; 25 El; 45 RA; 207 Brin.
For castings. Impact resistant.

TIGERVAN
Bethlehem Steel Corp.
C 0.8, Cr 4.25, W 18.5, Mo 0.5, V 2.1, bal Fe.
For cutting tools; high speed steel. *Obsolete*

TILOY
Titusville Forge Co.
Cr, Mo, C, bal Fe.
For die blocks; oil hardened.

TIMANG (TISCO NO. 4)
Harsco Corp. (Taylor-Wharton Div.)
C 0.6-0.8, Mn 12-15, Ni 3, bal Fe.
Rolled: 135,000-155,000 TS; 45,000-55,000 YS; 50-80 El;
35-50 RA; 170-210 Brin. Annealed: 100,000-115,000 TS. For
welding rods for welding high Mn steels, woven screens,
commercial castings, elevator buckets; austenitic, wear
resisting steel. *Obsolete*

TIMANG 11-7
Harsco Corp. (Taylor-Wharton Div.)
Mn 11, Ni 7, C, bal Fe.
For earth moving equipment; wear resistant, nonmagnetic,
work hardens. *Obsolete*

TIMANG 13-3
Harsco Corp. (Taylor-Wharton Div.)
Mn 13, Ni 3, C, bal Fe.
For earth moving equipment; wear resistant, nonmagnetic,
work hardens. *Obsolete*

TIMAXX 5
Harsco Corp.
C 0.6-0.8, Mn 13-15, Si 1.25-1.75, Ni 2.75-3.25, Cr 4-4.5, bal
Fe.
Cast: 100,000 TS; 40,000 YS; 40 El; 30 RA; 180 Brin. For
castings; wear and abrasion resistant to 800°F.

TIME-GRAPH
Time Steel Service Inc.
C 1.45, Mn 0.8, Si 1.2, Cr 0.2, Mo 0.25, bal Fe.
Oil hardened tool steel. AISI O6.

TIMET TI-1100
TIMET
Titanium.
Developmental, near-alpha alloy. Beta forged plus 1100°F/8
h. At 1000°F: 99 ksi TS; 84 ksi YS; 11 El; 28 RA. Billet, bar,
sheet. For elevated temperature applications.

TIMKEN 0.50 MO
Timken Co.
C 0.1-0.2, Mn 0.3-0.8, Mo 0.44-0.65, bal Fe.
Annealed: 55,000 TS; 30,000 YS; 30 El; 150 Brin. For
cracking furnace tubes; to 1050°F.

TIMKEN 16-15-6
Timken Co.
C 0-0.08, Mn 6.5-8.5, Cr 15-17.5, Ni 14-17, Mo 5-7, bal Fe.
Rolled: 140,000 TS; 90,000 YS; 35 El. For gas turbine wheels
and shafting, bolts, high temperature components. High heat
and oxidation resistant. *Obsolete*

TIMKEN 16-25-6
Timken Co.
C 0-0.1, Mn 2, Si 1, Cr 15-17.5, Ni 24-27, Mo 5-7, N 0.1-0.2,
bal Fe.
Forged: 115,000 TS; 60,000 YS; 35 El; 201 Brin. Cold worked:
150,000 TS; 120,000 YS; 19 El; 40 RA; 300 Brin. For high
temperature applications, gas turbines; heat resistant,
austenitic. *Obsolete*

TIMKEN 17-22 "A" V
Timken Co.
C 0.25-0.3, Mn 0.6-0.9, Si 0.55-0.75, Cr 1-1.5, Mo 0.4-0.6, V
0.75-0.95, bal Fe.
For high temperature applications; good creep strength.

TIMKEN 17-22 (A)
Timken Co.
C 0.41-0.48, Mn 0.45-0.65, Si 0.55-0.75, Cr 1-1.5, V 0.2-0.3,
Mo 0.4-0.6, bal Fe.
Normalized: 125,000 TS; 105,000 YS; 16 El; 50 RA; 250 Brin.
For high temperature bolting; air hardening.

TIMKEN 17-22 (A) S
Timken Co.
C 0.28-0.33, Cr 1-1.5, Mo 0.4-0.6, V 0.2-0.3, Si 0.55-0.75, Mn
0.45-0.65, bal Fe.
Normalized: 125,000 TS; 105,000 YS; 16 El; 50 RA; 250 Brin.
For bolting, flanges, valves; resists thermal cracking.

TIMKEN 18-8 CB (347 TP)
Timken Co.
C 0-0.08, Mn 0-2, Cr 17-20, Ni 9-13, Cb 0-1, bal Fe.
Quenched: 75,000 TS; 30,000 YS; 35 El; 200 Brin. For high
temperature tubing; corrosion and heat resistant. *Obsolete*

TIMKEN 18-8 TI (321 TP)
Timken Co.
C 0-0.08, Mn 0-2, Cr 17-20, Ni 9-13, Ti 0-0.6, bal Fe.
Quenched: 75,000 TS; 30,000 YS; 35 El; 200 Brin. For tubing;
heat and corrosion resistant. *Obsolete*

TIMKEN 2 CR-1/2 MO
Timken Co.
C 0-0.15, Cr 1.75-2.25, Mo 0.45-0.65, bal Fe.
Annealed: 60,000 TS; 25,000 YS; 30 El; 163 Brin. For heat
exchangers; resists heat to 1150 F. *Obsolete*

TIMKEN 2-1/4 CR-1-MO
Timken Co.
C 0-0.15, Cr 2-2.5, Mo 0.9-1.1, bal Fe.
Annealed: 60,000 TS; 25,000 YS; 30 El; 50 RA; 163 Brin. For
high temperature tubing in oil refineries; resist heat to
1150°F.

TIMKEN 25-12 (309 TP)
Timken Co.
C 0-0.2, Mn 0-2, Si 0-1, Cr 22-24, Ni 12-15, bal Fe.
Quenched: 75,000 TS; 30,000 YS; 35 El; 200 Brin. For tubing;
heat and corrosion resistant. *Obsolete*

TIMKEN 25-20 (310 TP)
Timken Co.
C 0-0.15, Mn 0-2, Si 0-0.75, Cr 24-26, Ni 19-22, bal Fe.
Quenched: 75,000 TS; 30,000 YS; 35 El; 200 Brin. For tubing;
heat and corrosion resistant. *Obsolete*

TIMKEN 301
Timken Co.
C 0.08-0.15, Cr 16-18, Ni 6-8, bal Fe.
Annealed: 110,000 TS; 40,000 YS; 60 El; 165 Brin. Cold
drawn: 185,000 TS; 140,000 YS; 8 El; 410 Brin. For stainless
structures, household utensils, trailer bodies; Type 301;
stainless, austenitic. *Obsolete*

TIMKEN 302
Timken Co.
C 0.08-0.15, Cr 17-19, Ni 8-10, bal Fe.
Annealed: 85,000 TS; 35,000 YS; 45 El; 67 RA; 150 Brin. For
chemical plant equipment; Type 302; stainless, austenitic.
Obsolete

TIMKEN 303
Timken Co.
C 0-0.15, Cr 17-19, Ni 8-10, 0.07% min P, S, Se, bal Fe.
Annealed: 85,000 TS; 40,000 YS; 40 El; 60 RA; 160 Brin. For
screw machine products; Type 303; free cutting, stainless.
Obsolete

TIMKEN 304
Timken Co.
C 0-0.08, Cr 18-20, Ni 8-11, bal Fe.
Annealed: 85,000 TS; 35,000 YS; 60 El; 70 RA; 150 Brin. Cold
drawn: 180,000 TS; 125,000 YS; 10 El; 330 Brin. For
architectural molding and trim, chemical plant equipment;
Type 304; stainless, austenitic. *Obsolete*

TIMKEN 309
Timken Co.
C 0-0.2, Cr 22-24, Ni 12-15, bal Fe.
Annealed: 90,000 TS; 40,000 YS; 50 El; 65 RA; 160 Brin. For
furnace and oil burner parts, heat treat equipment; Type 309;
austenitic, heat resistant. *Obsolete*

TIMKEN 310
Timken Co.
C 0-0.25, Cr 24-26, Ni 19-22, bal Fe.
Annealed: 100,000 TS; 45,000 YS; 50 El; 65 RA; 185 Brin. For
furnace equipment, heat treat boxes, valves, pumps; Type
310; austenitic, heat resistant. *Obsolete*

TIMKEN 316
Timken Co.
C 0-0.1, Cr 16-18, Ni 11-14, Mo 2-3, bal Fe.
Annealed: 80,000 TS; 30,000 YS; 60 El; 80 RA; 135 Brin. Cold
drawn: 150,000 TS; 135,000 YS; 6 El; 300 Brin. For chemical
plant equipment, tanks, vessels, agitators, pumps; Type 316;
stainless, austenitic. *Obsolete*

TIMKEN 4-6% CR
Timken Co.
C 0-0.15, Mn 0-0.5, Si 0-0.5, Cr 4-6, bal Fe.
At 85 F: 74,400 TS; 40,000 YS; 30 El; 73 RA; 149 Brin. For
machinery parts, still tubes, oil refinery tubing; corro
Obsolete

TIMKEN 4-6% CR W
Timken Co.
C 0-0.15, Mn 0-0.5, Si 0-0.5, Cr 4-6, W 0.75-1.25, bal Fe.
At 85 F: 66,500 TS; 25,000 YS; 36 El; 71.4 RA; 130 Brin. For
machinery parts, still tubes, oil refinery tubing; corrosion
resistant to oil. *Obsolete*

TIMKEN 4140 BOLTING
Timken Co.
C 0.38-0.45, Mn 0.75-1, Cr 0.8-1.1, Mo 0.15-0.25, bal Fe.
Heat treated: 125,000 TS; 105,000 YS; 16 El; 50 RA; 250 Brin.
For high temperature bolting; to 850 F. *Obsolete*

TIMKEN 4422
Timken Co.
C 0.2-0.25, Mn 0.7-0.9, Mo 0.35-0.45, bal Fe.
For bearings, case hardening.

TIMKEN 4427
Timken Co.
C 0.24-0.29, Mo 0.35-0.45, Mn 0.7-0.9, bal Fe.
For bearings, liners. Case hardening.

TIMKEN 4620
Timken Co.
C 0.17-0.22, Mn 0.45-0.65, Ni 1.6-2, Mo 0.2-0.3, bal Fe.
At 850°F: 72,400 TS; 52,500 YS; 37.5 El; 70.0 RA. At 1200°F:
24,800 TS; 12,000 YS; 59.3 El; 79.9 RA. For gears, pinions,
shafts, high temperature steam service; case hardening steel;
corrosion resistant.

TIMKEN 4718
Timken Co.
C 0.16-0.21, Cr 0.35-0.55, Ni 0.9-1.2, Mo 0.3-0.4, Mn 0.7-0.9, bal Fe.
Heat treated: 210,000 TS; 154,000 YS; 13.5 El; 50 RA; 470 Brin. For gears, bolts, camshafts, cams; case hardened, shock resistant.

TIMKEN 5% CR-MO PLUS TI
Timken Co.
C 0-0.12, Cr 4-6, Mo 0.45-0.65, Ti = 4 x C min, bal Fe.
Annealed: 60,000 TS; 25,000 YS; 30 El; 163 Brin. For high temperature tubing; to 1700 F; corrosion resistant. *Obsolete*

TIMKEN 5.00 CR .50 MO
Timken Co.
C 0-0.15, Mn 0.3-0.6, Si 0-0.5, Cr 4-6, Mo 0.45-0.65, bal Fe.
At 850°F: 66,600 TS; 26,000 YS; 39 El. At 1400°F: 13,300 TS; 7250 YS; 65 El. For still tubes, oil refinery tubing; age-hardenable, resists oil corrosion and heat to 1200°F.

TIMKEN 52100
Timken Co.
C 0.98-1.1, Mn 0.25-0.45, Cr 1.3-1.6, bal Fe.
Annealed: 94,500 TS; 62,000 YS; 27 El; 62 RA; 179 Brin. Cold drawn: 107,000 TS; 87,000 YS; 17 El; 55 RA; 229 Brin. For bearings, races, balls, bushings; water or oil hardenable.

TIMKEN CARBON
Timken Co.
C 0.1-0.2, Mn 0.3-0.6, bal Fe.
Annealed: 48,000-61,500 TS; 30,000-45,000 YS; 28-31 El; 70 RA; 125-132 Brin. For oil refinery tubing, cracking furnace tubes, condenser tubes, heat exchanger tubes; for services to 900°F. *Obsolete*

TIMKEN CBS-600
Timken Co.
C 0.16-0.22, Si 0.9-1.25, Cr 1.25-1.65, Mo 0.9-1.1, bal Fe.
Heat treated: 98,000-177,500 TS; 25,000-118,250 YS; 18-29 El; 45-72 RA; 197-341 Brin. For jet engine and rocket components, anti-friction bearings; case hardening.

TIMKEN DM
Timken Co.
C 0-0.15, Mn 0.3-0.6, Si 0.5-1, Cr 1-1.5, Mo 0.44-0.65, bal Fe.
Annealed: 60,000 TS; 25,000 YS; 30 El; 50 RA; 163 Brin. For high temperature tubing; heat resistant to 1150°F.

TIMKEN DM-15
Timken Co.
C 0.15, Mn 0.45, Si 1, Cr 1.25, Mo 0.5, bal Fe.
At 850 F: 66,500 TS; 35,200 YS; 36.5 El; 72.7 RA. At 1400 F: 13,800 TS; 7,000 YS; 72 El; 98.5 RA. For tubes, case hardened parts, oil refinery tubes; case hardened steel, high creep strength. *Obsolete*

TIMKEN DM2
Timken Co.
C 0-0.15, Mn 0.3-0.6, Si 0-0.5, Cr 0.8-1.25, Mo 0.44-0.65, bal Fe.
Annealed: 60,000 TS; 25,000 YS; 30 El; 163 Brin. For superheaters, steam pipes, resists heat to 1150°F.

TIMKEN FINE GRAINED CARBON-MOLYBDENUM
Timken Co.
C 0.15, Mn 0.45, Mo 0.5, bal Fe.
At 850 F: 64,000 TS; 32,500 YS; 37.0 El; 62.7 RA. At 1400 F: 11,700 TS; 5,900 YS; 82.5 El; 89.0 RA. For oil refinery tubing, still tubes; case hardening steel; Ch. 40-64. *Obsolete*

TIMKEN HS 250 (4741)
Timken Co.
C 0.37-0.44, Mn 0.65-0.95, Si 0.4-0.6, Cr 0.8-1.1, Ni 0.7-1.1, Mo 0.3-0.4, bal Fe.
Heat treated: 187,000 TS; 182,000 YS; 14.5 El; 51 RA; 388 Brin. For gears, shafts, bolts, crankshafts; oil hardened, shock resistant.

TIMKEN IRON-CHROMIUM (TYPE 430)
Timken Co.
C 0-0.12, Mn 0-1, Si 0-1, Cr 14-18, bal Fe.
Non-hardenable stainless; for parts requiring stainless properties or for high temperature service. AISI 430. *Obsolete*

TIMKEN MM-9
Timken Co.
C 0-0.15, Mn 1.1-1.4, Si 0.15-0.3, Mo 0.2-0.3, bal Fe.
At 850 F: 72,100 TS; 51,200 YS; 36 El; 71.8 RA. At 1400 F: 13,400 TS; 6,100 YS; 80.5 El; 86.5 RA. For gears, pinions, shafts, high temperature steam service; high strength; case hardening steel. *Obsolete*

TIMKEN NITRIDING NO. 2 M
Timken Co.
C 0.3-0.4, Mn 0.6-1.1, Cr 0.9-1.4, Mo 0.15-0.25, Al 0.85-1.2, S 0.08-0.13, bal Fe.
Nitriding steel; for gears, cams, shafts; free-machining. *Obsolete*

TIMKEN NITRIDING STEEL NO. 2
Timken Co.
C 0.3-0.4, Mn 0.4-0.7, Cr 0.9-1.4, Mo 0.15-0.25, Al 0.85-1.2, bal Fe.
Nitriding steel; for gears, shafts. *Obsolete*

TIMKEN NITRIDING STEEL NO. 3
Timken Co.
C 0.38-0.43, Mn 0.5-0.7, Cr 1.4-1.8, Mo 0.3-0.4, Al 0.95-1.3, bal Fe.
Nitriding steel; for gears, shafts, cylinders.

TIMKEN QV
Timken Co.
C 0-0.15, Mn 1.85-2.15, Si 0.15-0.3, Mo 0.25-0.35, V 0.18-0.25, bal Fe.
For machinery parts; corrosion resistant. *Obsolete*

TIMKEN SPECIAL ROLL STEEL
Timken Co.
C 0.68-0.78, Mn 0.85-1, Cr 0.4-0.6, Ni 0.4-0.6, Mo 0.4, bal Fe.
For cold work rolls, machine tool parts; oil hardening. *Obsolete*

TIMKEN T-1 (TM-NO. 1)
Timken Co.
C 0.95-1.1, Cr 1.2-1.5, Mo 0.2-0.3, bal Fe.
For anti-friction bearings; deep hardening. *Obsolete*

TIMKEN T-2 (TM-NO. 2)
Timken Co.
C 0.95-1.1, Mn 1.05-1.35, Cr 1.2-1.5, Mo 0.45-0.55, bal Fe.
Heat treated. For anti-friction bearings; deep hardening. *Obsolete*

TIMKEN TBA-2
Timken Co.
C 0.7-0.8, Mn 1.05-1.35, Cr 0.9-1.2, Ni 1.3-1.65, Mo 1.2-1.4, bal Fe.
Air hardening steel for large bearings and machinery parts. Air hardenable to 56-60 Rock C in a 6 in. round.

TIMKEN TBS 1000
Timken Co.
C 0.7-0.9, Mn 0.4-0.6, Si 0.4-0.6, Cr 0.9-1.2, Mo 4.75-5.25, V 0.9-1.2, bal Fe.
For dies, cutters, punches, headers; oil or air hardening.

TIMKEN TBS-600
Timken Co.
C 0.95-1.1, Cr 1.25-1.65, Mo 0.25-0.35, Si 0.85-1.2, bal Fe.
For jet engine and rocket components; anti-friction bearings; hard and abrasion resistant; for service up to 600°F.

TIMKEN TBS-9
Timken Co.
C 0.89-1.01, Mn 0.5-0.8, Cr 0.4-0.6, Mo 0.08-0.15, bal Fe.
Through hardening bearing steel.

TIMKEN TDS-30
Timken Co.
C 0.17-0.22, Mn 0.45-0.65, Cr 0.4-0.6, Ni 1.65-2, Mo 0.2-0.3, bal Fe.
Tubular drill steel, must be carburized to harden.

TIMKEN TDS-70
Timken Co.
C 0.25-0.31, Mn 0.8-1.2, Si 0.5-0.8, Cr 1.9-2.4, Mo 0.25-0.35, bal Fe.
Tubular drill steel, air hardenable.

TIMKEN TDS-90
Timken Co.
C 0.23-0.28, Mn 0.4-0.6, Cr 3-3.5, Mo 0.45-0.6, bal Fe.
Tubular drill steel, air hardenable.

TIMKEN TP-304L
Timken Co.
C 0-0.035, Cr 18-20, Ni 8-13, bal Fe.
Annealed: 70,000 TS; 25,000 YS; 35 El. For corrosion resistant and welded tubes; heat treatment not required after welding. *Obsolete*

TIMKEN TP-316
Timken Co.
C 0.04-0.1, Cr 16-18, Ni 11-14, Mo 2-3, bal Fe.
Annealed: 75,000 TS; 30,000 YS; 35 El. For superheater tubing, main steam lines; Type 316L; suited at 1100-1200 F, stainless. *Obsolete*

TIMKEN TP-321H
Timken Co.
C 0.04-0.1, Cr 17-20, Ni 9-13, Ti = 4 x C, bal Fe.
Annealed: 75,000 TS; 30,000 YS; 35 El. For superheater tubing, heat exchanger tubes; heat resistant, for use above 1000 F. *Obsolete*

TIMKEN TW
Timken Co.
C 1-1.1, V 0.05, bal Fe.
For header dies; general purpose. *Obsolete*

TIMKEN TYPE 302B STAINLESS
Timken Co.
C 0.08-0.15, Mn 0-2, Si 2-3, Cr 17-19, Ni 8-10, bal Fe.
For heat resistant and stainless parts; oxidation resistant, stainless, austenitic. *Obsolete*

TIMKEN TYPE 308 STAINLESS
Timken Co.
C 0-0.08, Mn 0-2, Cr 19-21, Ni 10-12, bal Fe.
For corrosion and heat resistant parts; austenitic. *Obsolete*

TIMKEN TYPE 321 STAINLESS
Timken Co.
C 0-0.1, Mn 0-2, Cr 17-19, Ni 8-11, Ti = 5 x C min, bal Fe.
For welded structures; stainless, austenitic. *Obsolete*

TIMKEN TYPE 347 STAINLESS
Timken Co.
C 0-0.1, Mn 0-2, Cr 17-19, Ni 9-12, Cb = 10 x C min, bal Fe.
For welded structures; stainless, austenitic. *Obsolete*

TIMKEN TYPE 403 STAINLESS
Timken Co.
C 0-0.15, Mn 0-1, Si 0-0.5, Cr 11.5-13, bal Fe.
For steam turbine blades; corrosion resistant; hardenable. *Obsolete*

TIMKEN TYPE 405 STAINLESS
Timken Co.
C 0-0.08, Mn 0-1, Cr 11.5-13.5, Al 0.1-0.3, bal Fe.
For welded structures; corrosion resistant. *Obsolete*

TIMKEN TYPE 410 STAINLESS
Timken Co.
C 0-0.15, Mn 0-1, Cr 11.5-13.5, bal Fe.
For corrosion resistant parts; corrosion resistant, hardenable. *Obsolete*

TIMKEN TYPE 414 STAINLESS
Timken Co.
C 0-0.15, Mn 0-1, Cr 11.5-13.5, Ni 1.25-2.5, bal Fe.
For stainless tubes; corrosion resistant, hardenable.
Obsolete

TIMKEN TYPE 416 STAINLESS
Timken Co.
C 0-0.15, P 0-0.6, Cr 12-14, 0.15% min S, Se or 0.6% max Zr, Mo, bal Fe.
For corrosion resistant parts; free-cutting, corrosion resistant.
Obsolete

TIMKEN TYPE 420 STAINLESS
Timken Co.
Cr 12-14, 0.15% min C, bal Fe.
For valve parts; corrosion resistant, hardenable. *Obsolete*

TIMKEN TYPE 430 STAINLESS
Timken Co.
C 0-0.12, Cr 14-18, bal Fe.
For corrosion resistant parts; corrosion resistant, not hardenable. *Obsolete*

TIMKEN TYPE 430F STAINLESS
Timken Co.
C 0-0.12, P 0-0.6, Cr 14-18, 0.15% min S, Se or 0.6% max Zr, Mo, bal Fe.
For corrosion resistant parts; free-cutting. *Obsolete*

TIMKEN TYPE 431 STAINLESS
Timken Co.
C 0-0.2, Cr 15-17, Ni 1.25-2.5, bal Fe.
For corrosion resistant parts; corrosion and heat resistant, hardenable. *Obsolete*

TIMKEN TYPE 440 A STAINLESS
Timken Co.
C 0.6-0.75, Cr 16-18, Mo 0-0.75, bal Fe.
For stainless parts; hardenable, corrosion resistant. *Obsolete*

TIMKEN TYPE 440 B STAINLESS
Timken Co.
C 0.75-0.95, Cr 16-18, Mo 0-0.75, bal Fe.
For bearings, cutlery; hardenable, corrosion resistant.
Obsolete

TIMKEN TYPE 440 C STAINLESS
Timken Co.
C 0.95-1.2, Cr 16-18, Mo 0-0.75, bal Fe.
For bearings, hardenable; corrosion resistant. *Obsolete*

TIMKEN TYPE 501 STAINLESS
Timken Co.
Mn 0-1, Cr 4-6, Mo 0.45-0.65, 0.10% min C, bal Fe.
Annealed: 60,000 TS; 25,000 YS; 30 El; 163 Brin. For corrosion resistant parts; corrosion and heat resistant.
Obsolete

TIMKEN TYPE TP 405 STAINLESS
Timken Co.
C 0-0.08, Mn 0-1, Cr 11.5-13.5, Ni 0-0.5, Al 0.1-0.3, bal Fe.
For welded structures; corrosion resistant. *Obsolete*

TIMKEN TYPE TP 410 STAINLESS
Timken Co.
C 0-0.15, Mn 0-1, Cr 11.5-13.5, Ni 0-0.5, bal Fe.
For stainless tubes; corrosion resistant. *Obsolete*

TIMKEN TYPE TP 420 STAINLESS
Timken Co.
Cr 12-14, Ni 0-0.5, 0.15% min C, bal Fe.
For tubes; corrosion and heat resistant, hardenable.
Obsolete

TIMKEN TYPE TP 443 STAINLESS
Timken Co.
C 0-0.2, Mn 0-1, Cr 18-23, Cu 0.9-1.25, bal Fe.
For corrosion resistant tubing; not hardenable, corrosion resistant. *Obsolete*

TIMKEN WHS-100
Timken Co.
C 0.13-0.21, Mn 1-1.3, Cr 0.65-0.9, Ni 0.4-0.7, Mo 0.15-0.25, V 0.03-0.08, B 0.003, bal Fe.
100,000 psi YS; weldable.

TIMKEN WHS-130
Timken Co.
C 0.2-0.27, Mn 0.6-0.8, Cr 0.7-0.9, Ni 1.55-2, Mo 0.2-0.3, bal Fe.
130,000 psi YS min; weldable.

TIN
Atomergic Chemetals Corp.
Sn.
Purities, zone refined: 99.9999%, 99.999%, 99.99%, 99.9%.
Forms: rod, bar, shot, powder, wire, sheet, foil, single crystals.

TIN BRASS 1% TIN
Chase Brass & Copper Co., Inc.
Cu 87.5, Zn 11.5, Sn 1.
Annealed: 45,000 TS; 19,000 YS; 45 El. Rolled hard: 73,000 TS; 68,000 YS; 9 El. For weather strips, springs, clips, switches, terminals.

TIN BRASS 2% TIN
Chase Brass & Copper Co., Inc.
Cu 88.5, Zn 9.5, Sn 2.
Annealed: 44,000 TS; 18,000 YS; 48 El. Rolled hard: 76,000 TS; 73,000 YS; 4 El. For weather strip, springs, clips, switches, terminals.

TIN FOIL
Manufacturer not listed.
Sn 88, Pb 8, Cu 4, Sb 0.5.
For bearings, packing foil.

TIN WELD 1
Eutectic Corp.
Paint on solder paste combined with flux. All purpose; dissimilar metals. Bonding temperature 375°F; 7000 psi TS.

TIN WELD 3
Eutectic Corp.
Paint on solder paste containing neutral type flux for joining and tinning copper. Bonding temperature: 360°F; 8000 psi TS. *Obsolete*

TIN-BEARING COMMERCIAL BRONZE
Chase Brass & Copper Co.
Cu 90, Zn 9.5, Sn 0.5.
Annealed: 40,000 TS; 10,000 YS; 40 El. Hard rolled: 60,000 TS; 57,000 YS; 5 El. For bushings, bearings; heavy duty.
Obsolete

TINEA CLASSIC
Compagnie Francaise de l'Etain
Sn 45, Pb 42, Sb 10, Cu 3.
For bearings; anti-friction metal.

TINEA DIESEL
Compagnie Francaise de l'Etain
Sn 83, Sb 11, Cu 6.
Cast: 26 Brin. For bearings; MP 380°C; antifriction.

TINEA DIESEL R.S.
Compagnie Francaise de l'Etain
Pb 76.5, Sn 10, Sb 13.5, Cu 1.
Cast: 26.7 Brin. For bearings; MP 400°C; antifriction.

TINEA MARINE EK
Compagnie Francaise de l'Etain
Sn 72.5, Pb 16, Sb 7, Cu 4.5.
For bearings; MP 380°C; antifriction.

TINEA SUPER MARINE
Compagnie Francaise de l'Etain
Sn 90, Sb 6.5, Cu 3.5.
Cast: 23.7 Brin. For automobile bearings; MP 330°C; antifriction.

TINICOSIL NO. 10
Cerro Metal Products Co.
Zn, Ni, bal Cu.
For hardware. Corrosion resistant. *Obsolete*

TINICOSIL NO. 14
Cerro Metal Products Co.
Cu 42.5, Ni 15, bal Zn.
Rolled: 105,000 TS; 70,000 YS; 20 El; 215 Brin. For valve parts and instrument parts. Nickel silver; corrosion resistant.
Obsolete

TINICOSIL NO. 15
Cerro Metal Products Co.
Cu 41.75-42.25, Ni 14.75-15.25, bal Zn.
Hard: 215,000 TS; 192,000 YS; 8 El; 10 RA; 140 Brin. Soft: 118,000 TS; 84,000 YS; 11 El; 10 RA; 122 Brin. For hot forgings and hardware. Corrosion resistant. *Obsolete*

TINICOSIL NO. 20
Cerro Metal Products Co.
Cu 46-47, Pb 1.4-1.6, Ni 15-16, bal Zn.
For pressure die castings. Corrosion resistant. *Obsolete*

TINICOSIL NO. 21
Cerro Metal Products Co.
Cu 50, Ni 10, Pb 1, Fe 1, Mn 3, bal Zn.
Extrusion: 140 Brin. For valve parts, instrument parts, extrusions, and forgings. Nickel silver; corrosion resistant.
Obsolete

TINICOSIL NO. 53
Cerro Metal Products Co.
Cu 46, Ni 10, bal Zn.
Hard: 105,000 TS; 80,000 YS; 20 El; 15 RA; 158 Brin. Forged: 92,000 TS; 42,000 YS; 26 El; 25 RA; 140 Brin. For forgings, hardware parts, and plumbing fixtures. Corrosion resistant.
Obsolete

TINICOSIL NO. 54
Cerro Metal Products Co.
Cu 46.5, Pb 2.2, Ni 10, Mn 2, bal Zn.
Cold drawn: 88,000 TS; 68,000 YS; 30 El; 20 RA; 150 Brin. For screw machine products and fishing tackle. Leaded nickel silver; corrosion resistant; CA 798.

TINIDUR
Krupp Stahl AG
superalloy. C 0-0.15, Ni 30, Cr 15, Ti 1.8, bal Fe.
Rolled: 135,000-160,000 TS; 85,000-115,000 YS; 30-20 El; 55-30 RA. For gas turbine blades, jet engine components; heat and corrosion resistant. *Obsolete*

TINIDUR 1650
Krupp Stahl AG
stainless steel.
See CRONIDUR 4980

TINIDUR 1875
Krupp Stahl AG
stainless steel. C 0.04, Cr 14.7, Ni 26, Ti 2-3, Mn 1, Al 0.15, bal Fe.
For jet engine components; heat and corrosion resistant.
Obsolete

TINITE
H. Kramer & Co. (Ajax Metal Div.)
Pb, Sb, bal Sn.
For bearings; hard and tough without being brittle. *Obsolete*

TINMAN'S SOLDER
Manufacturer not listed.
Sn 66.5, Pb 33.5.
For solder; soft.

TINSEL
Manufacturer not listed.
Sn 60, Pb 40.
For tinsel for decorative purposes.

TIOGA
AL Tech Specialty Steel Corp.
C 0.67, Cr 0.65, Ni 1.4, Mo 0.2, bal Fe.
For lathe centers, clutch parts, cams, arbors; oil hardened, tough. AISI L6.

TIONA
Cytemp Specialty Steel Div.
C 1, Cr 0.85, Mn 2, Mo 0.85, bal Fe.
For blanking dies; air hardened. *Obsolete*

TIPALOY 100
Tipaloy Inc.

Cd-Cu alloy electrode. Wrought: 65,000 psi TS; 85% electrical conductivity. For spot welding light metal alloys, terne plate, galvanized and similar materials.

TIPALOY 130
Tipaloy Inc.
Cu-Cr alloy.
Cast: 55,000 TS; 31,000 YS; 22 El; 125 Brin. Wrought: 80,000 TS; 62,000 YS; 17 El; 165 Brin. For resistance-welder electrodes; wear resistant.

TIPALOY 200
Tipaloy Inc.
Co-Ag-Be, bal Cu.
Cast: 90,000 TS; 70,000 YS; 8-10 El; 190 Brin. Wrought: 100,000 TS; 80,000 YS; 8-10 El; 210 Brin. For resistance-welder electrodes; wear resistant.

TIPALOY T-100W
Tipaloy Inc.

100% W welding electrode. Density: 19.0 g/cm^3; 50,000 psi TS; 32% electrical conductivity. For spot welding copper and copper base alloys.

TIPALOY T-10W
Tipaloy Inc.
Cu-W welding electrode. Density: 14.3 g/cm^3; 90,000 psi TS; 28-32% electrical conductivity. For projection welding dies; for flash and butt welding dies.

TIPALOY T-10W53
Tipaloy Inc.
Cu alloy-W welding electrode. Heat treated: 160,000 psi TS; Density: 14.3 g/cm^3; 26-30% electrical conductivity. For high strength electrodes.

TIPALOY T-1W
Tipaloy Inc.
Cu-W welding electrode. Density: 12.6 g/cm^3; 63,000 psi TS; 35-42% electrical conductivity. Facings or electrodes for flash and butt welding electrodes and projection welding electrodes.

TIPALOY T-20W
Tipaloy Inc.
Cu-W welding electrode. Density: 14.7 g/cm^3; 95,000 psi TS; 27-31% electrical conductivity. For heavy duty projection welding.

TIPALOY T-253
Tipaloy Inc.
Ni, Be, bal Cu.
Cast: 80,000 psi TS; 40% electrical conductivity. Wrought: 100,000 psi TS; 40% electrical conductivity. Backing material for welding set-ups.

TIPALOY T-30W
Tipaloy Inc.
Cu-W welding electrode. Density: 14.6 g/cm^3; 100,000 psi TS; 26-30% electrical conductivity. For projection welding dies and cross wire welding.

TIPALOY T-35S
Tipaloy Inc.
Ag-W welding electrode. 50,000 psi TS; 51% electrical conductivity.

TIPALOY T-3W
Tipaloy Inc.
Cu-W welding electrode. Density: 13.5 g/cm^3; 75,000 psi TS; 31-38% electrical conductivity. For projection welding electrodes.

TIPALOY T-3W53
Tipaloy Inc.
Cu alloy-W welding electrode. Heat treated: 120,000 psi TS. Density: 13.5 g/cm^3; 28-32% electrical conductivity. For high strength electrodes.

TIPALOY T-4
Tipaloy Inc.
Cu base electrode. Cast: 110,000 psi TS; 20% electrical conductivity. Wrought: 170,000 psi TS; 23% electrical conductivity. Strong, wear resistant electrode.

TIPALOY T-5
Tipaloy Inc.
Al-Cu alloy electrode. Cast: 70,000 psi TS; 15% electrical conductivity. For certain flash and butt welding operations.

TIPALOY T-5W
Tipaloy Inc.
Cu-W welding electrode. Density: 13.8 g/cm^3; 85,000 psi TS; 28-33% electrical conductivity. For spot and projection welding electrodes.

TIPALOY T-G12
Tipaloy Inc.
Ag-WC welding electrode. Density: 11.8 g/cm^3; 35,000 psi TS; 50-60% electrical conductivity. Good conductivity; for light loads.

TIPALOY T-G13
Tipaloy Inc.
Ag-WC welding electrode. Density: 12.2 g/cm^3; 40,000 psi TS; 45-55% electrical conductivity. Good conductivity; for light loads.

TIPALOY T-G14
Tipaloy Inc.
Ag-WC welding electrode. Density: 13.1 g/cm^3; 55,000 psi TS; 30-40% electrical conductivity.

TIPALOY T-G17
Tipaloy Inc.
Ag-Mo welding electrode. Density: 10.1 g/cm^3; 60,000 psi TS; 45-50% electrical conductivity.

TIPALOY T-G18
Tipaloy Inc.
Ag-Mo welding electrode. Density: 10.2 g/cm^3; 45,000 psi TS; 50-55% electrical conductivity.

TIPALOY T-TC10
Tipaloy Inc.
Cu-W-WC welding electrode. Density: 11.5 g/cm^3; 75,000 psi TS; 40-45% electrical conductivity. For heavy duty projection welding and electro-forging.

TIPALOY T-TC20
Tipaloy Inc.
Cu-W-Wc welding electrode. Density: 12.2 g/cm^3; 85,000 psi TS; 30-35% electrical conductivity. For electro-forging and upsetting.

TIPALOY T-TC5
Tipaloy Inc.
Cu-W-WC welding electrode. Density: 11.0 g/cm^3; 70,000 psi TS; 45-50% electrical conductivity. For projection welding dies; light loads but needs abrasion resistance.

TIPALOY T-TC53
Tipaloy Inc.
Cu alloy-W-WC welding electrode. Heat treated: 150,000 psi TS. Density: 12.3 g/cm^3; 18-23 electrical conductivity. For electro-forging and upsetting where hardness and abrasion resistance are required.

TIPALOY T20S
Tipaloy Inc.
Ag-W welding electrode. 70,000 psi TS; 49% electrical conductivity.

TIPALOY-T-100M
Tipaloy Inc.
100% Mo welding electrode. Density: 9.9 g/cm^3; 80,000 psi TS; 32% electrical conductivity. For welding or electro-brazing non-ferrous metals having relatively high electrical conductivity.

TIRANO EXTRA
Vereinigte Edelstahlwerke
C 0.5, Cr 1.5, W 2.25, V 0.25, bal Fe.
For tools, dies, hot dies; hot work steel. *Obsolete*

TIRANO PRIMA
Vereinigte Edelstahlwerke
C 0-0.5, Mn 0.7, Si 1.8, Mn 0.2, bal Fe.
For tools, chisels; tough. *Obsolete*

TIRFING 1
Uddeholm Corp.
C 0.55, Si 1.75, Mn 0.75, Cr 0.2, bal Fe.
For springs, chisels; oil hardened.

TIRFING 2
Uddeholm Corp.
C 0.6, Si 1.5, Mn 0.5, Cr 0.5, bal Fe.
For springs, punches; oil hardened.

TIRFING 3
Uddeholm Corp.
C 0.5, Mn 0.85, Cr 1.05, V 0.15, bal Fe.
For springs, dies; oil hardened.

TISCO "TAPPI" CORROSION RESISTING STEEL
Harsco Corp. (Taylor-Wharton Div.)
C 0.13-0.35, Cr 16-23, Ni 7-11, Mo, bal Fe.
Annealed: 70,000-75,000 TS; 50-60 El; 45-60 RA; 130-160 Brin. For machinery parts in paper and pulp machinery; now Tisco No. 110. *Obsolete*

TISCO 101
Harsco Corp. (Taylor-Wharton Div.)
C 0.13-0.26, Cr 16-20, Si 0-2, Mn 0-1, Ni 7-11, bal Fe.
70,000-75,000 TS; 60-50 El; 60-45 RA; 130-150 Brin. For heat and corrosion resistant parts; heat and corrosion resistant, stainless. *Obsolete*

TISCO 101A
Harsco Corp. (Taylor-Wharton Div.)
C 0.13-0.16, Cr 16-20, Si 0-2, Mn 0-1, Ni 7-11, Se 0.2-0.3, bal Fe.
70,000-75,000 TS; 60-50 El; 45-60 RA; 130-150 Brin. For heat and corrosion resistant parts; heat and corrosion resistant, stainless. *Obsolete*

TISCO 102
Harsco Corp. (Taylor-Wharton Div.)
C 0-0.07, Cr 16-20, Si 0-2, Mn 0-1, Ni 7-11, bal Fe.
65,000-70,000 TS; 50-60 El; 45-60 RA; 130-150 Brin. For heat and corrosion resistant parts; heat and corrosion resistant, stainless. *Obsolete*

TISCO 102A
Harsco Corp. (Taylor-Wharton Div.)
C 0-0.07, Cr 16-20, Si 0-2, Mn 0-1, Ni 7-11, Se, bal Fe.
65,000-70,000 TS; 50-60 El; 45-60 RA; 130-150 Brin. For heat and corrosion resistant parts; heat and corrosion resistant, stainless. *Obsolete*

TISCO 103
Harsco Corp. (Taylor-Wharton Div.)
C 0.13-0.2, Cr 16-20, Si 0-2, Mn 0-1, Ni 7-11, bal Fe.
70,000-75,000 TS; 50-60 El; 45-60 RA; 130-150 Brin. For heat and corrosion resistant parts; heat and corrosion resistant, stainless. *Obsolete*

TISCO 103A

Harsco Corp. (Taylor-Wharton Div.)
C 0.13-0.2, Cr 16-20, Si 0-2, Mn 0-1, Ni 7-11, Se, bal Fe.
70,000-75,000 TS; 50-6- El; 45-60 RA; 130-150 Brin. For heat and corrosion resistant parts; heat and corrosion resistant, stainless. *Obsolete*

TISCO 104

Harsco Corp. (Taylor-Wharton Div.)
C 0.13-0.16, Cr 18-22, Ni 7-11, Mo 2-4, bal Fe.
70,000-75,000 TS; 50-60 El; 45-60 RA; 130-150 Brin. For heat and corrosion resistant parts; heat and corrosion resistant, stainless. *Obsolete*

TISCO 104A

Harsco Corp. (Taylor-Wharton Div.)
C 0.13-0.16, Cr 18-22, Ni 7-11, Mo 2-4, Se 0.2-0.3, bal Fe.
70,000-75,000 TS; 50-60 El; 45-60 RA; 130-150 Brin. For stainless parts; stainless, heat and corrosion resistant. *Obsolete*

TISCO 105

Harsco Corp. (Taylor-Wharton Div.)
C 0-0.07, Cr 18-22, Ni 7-11, Mo 2-4, bal Fe.
65,000-70,000 TS; 50-60 El; 45-60 RA; 130-150 Brin. For stainless parts; stainless, heat and corrosion resistant. *Obsolete*

TISCO 105

Harsco Corp. (Taylor-Wharton Div.)
C 0.05, Cr 19.5, Ni 10.5, Mo 2.5, bal Fe.
Cast: 80,000 TS; 42,000 YS; 50 El; 160 Brin. For chemical plant equipment, tanks, mixers; Type CF8M; stainless, austenitic. *Obsolete*

TISCO 105A

Harsco Corp. (Taylor-Wharton Div.)
C 0-0.07, Cr 18-22, Ni 7-11, Mo 2-4, Se 0.2-0.3, bal Fe.
65,000-70,000 TS; 50-60 El; 45-60 RA; 130-150 Brin. For stainless parts; stainless, heat and corrosion resistant. *Obsolete*

TISCO 106

Harsco Corp. (Taylor-Wharton Div.)
C 0.13-0.2, Cr 18-22, Ni 7-11, Mo 2-4, bal Fe.
70,000-75,000 TS; 50-60 El; 45-60 RA; 130-150 Brin. For stainless parts; stainless, heat and corrosion resistant. *Obsolete*

TISCO 106A

Harsco Corp. (Taylor-Wharton Div.)
C 0.13-0.2, Cr 18-22, Ni 7-11, Mo 2-4, Se 0.2-0.3, bal Fe.
70,000-75,000 TS; 50-60 El; 45-60 RA; 130-150 Brin. For stainless parts; stainless, heat and corrosion resistant. *Obsolete*

TISCO 107

Harsco Corp. (Taylor-Wharton Div.)
C 0.21-0.3, Cr 26-30, Ni 8-12, bal Fe.
75,000-85,000 TS; 12-18 El; 8-12 RA; 150-160 Brin. For boat and corrosion resistant parts; heat and corrosion resistant. *Obsolete*

TISCO 108

Harsco Corp. (Taylor-Wharton Div.)
C 0.13-0.2, Cr 23-27, Ni 17-21, bal Fe.
80,000-90,000 TS; 15-25 El; 15-25 RA; 160-180 Brin. For heat and corrosion resistant parts; heat and corrosion resistant. *Obsolete*

TISCO 109

Harsco Corp. (Taylor-Wharton Div.)
C 0-0.2, Ni 22-26, Cr 18-22, Si 0-1.5, bal Fe.
60,000-70,000 TS; 10-15 El; 20-30 RA. For heat and corrosion resistant parts; heat and corrosion resistant. *Obsolete*

TISCO 110

Harsco Corp. (Taylor-Wharton Div.)
C 0.13-0.35, Cr 16-23, Ni 20-30, Mo 2-4, bal Fe.
70,000-75,000 TS; 50-65 El; 45-65 RA; 130-150 Brin. For stainless parts; stainless, heat and corrosion resistant. *Obsolete*

TISCO 120

Harsco Corp. (Taylor-Wharton Div.)
C 0.5, Mn 1, Si 2, Cr 15-18, Ni 35-40, bal Fe.
150 Brin. For heat and corrosion resistant parts; maximum operating temperature 2000 F. *Obsolete*

TISCO 130

Harsco Corp. (Taylor-Wharton Div.)
C 0-0.1, Cr 26-30, Mn 0.4-0.6, Si 0-2, bal Fe.
40,000-60,000 TS; 0-3 El; 0-5 RA. For heat and corrosion resistant parts; heat and corrosion resistant. *Obsolete*

TISCO 131

Harsco Corp. (Taylor-Wharton Div.)
C 0-0.1, Cr 16-18, Mn 0.4-0.6, Si 0-2, bal Fe.
75,000-100,000 TS; 5-12 El; 6-15 RA; 190-210 Brin. For corrosion resistant parts; corrosion resistant. *Obsolete*

TISCO 132

Harsco Corp. (Taylor-Wharton Div.)
C 0-0.1, Cr 12-16, Mn 0.4-0.6, Si 0-2, bal Fe.
70,000-90,000 TS; 18-24 El; 25-45 RA; 160-180 Brin. For corrosion resistant parts; corrosion resistant. *Obsolete*

TISCO 132

Harsco Corp. (Taylor-Wharton Div.)
C 0.1, Cr 12.5, Mn 0.5, bal Fe.
Heat treated: 200,000 TS; 150,000 YS; 7 El; 390 Brin. For oil refinery and chemical plant equipment; Type CA5; stainless. *Obsolete*

TISCO 133

Harsco Corp. (Taylor-Wharton Div.)
C 0.3, Cr 12.5, Mn 0.5, bal Fe.
Heat treated: 220,000 TS; 165,000 YS; 2 El; 470 Brin. For chemical and textile plant equipment; Type CA40; stainless. *Obsolete*

TISCO 150

Harsco Corp. (Taylor-Wharton Div.)
C 2.5-3.2, Mn 0-0.5, Si 0-2, Cr 28-32, Ni 1-1.5, bal Fe.
Heat treated: 400-600 Brin. For centrifugal pumps handling sand and other materials in acid mine waters; unmachinable. *Obsolete*

TISCO 150 X

Harsco Corp. (Taylor-Wharton Div.)
C 3-3.75, Mn 0.4, Si 1-1.25, Ni 1-1.6, Cr 12-13, bal Fe.
Oil hardened: 600-650 Brin. For wheel abrator blades; hard, brittle, unmachinable. *Obsolete*

TISCO 150-Y

Harsco Corp. (Taylor-Wharton Div.)
C 3.4, Mn 0.4, Si 1.1, Cr 12.5, Mo 0.2, bal Fe.
Heat treated: 700 minimum Brin. For brick making equipment, mixing paddles; abrasion and wear resistant. *Obsolete*

TISCO 160

Harsco Corp. (Taylor-Wharton Div.)
C 2.5-3.5, Mn 1-1.5, Si 1.25-2, Cr 1.5-4, Ni 12-15, Cu 5-7, bal Fe.
20,000-35,000 TS; 120-170 Brin. For heat and corrosion resistant parts; heat and corrosion resistant; Ni-Resist. *Obsolete*

TISCO 17

Harsco Corp. (Taylor-Wharton Div.)
C 0.15, Mn 0.6, Si 0.45, bal Fe.
Cast: 65,000 TS; 35,000 YS; 30 El; 50 RA; 120 Brin. For machine tool castings, gears, housing; ASTM-A-27; water hardened. *Obsolete*

TISCO 23 W

Harsco Corp. (Taylor-Wharton Div.)
C 0.3-0.4, Mn 1.35, Cr 1.35, Mo 0.2, Si 0.4, bal Fe.
Cast and hardened: 162,500 TS; 156,500 YS; 4 El; 11.1 RA; 364 Brin. For high strength castings; shock and fatigue resistant. *Obsolete*

TISCO 5

Harsco Corp. (Taylor-Wharton Div.)
C 0.7, Si 1.5, Ni 3, Cr 4.25, 14% Min, bal Fe.
Cast and normalized: 100,000 TS; 40,000 YS; 25 El; 20 RA; 180 Brin. For shovels, dippers, dredging pumps; resists heat to 750 F; abrasion resistant. *Obsolete*

TISCO 52

Harsco Corp. (Taylor-Wharton Div.)
C 0.85, Cr 2.25, Mo 0.35, Mn 0.8, Si 0.4, bal Fe.
325 Brin. For high temperature applications; ASTM-A-148; good to 1100 F. *Obsolete*

TISCO 54

Harsco Corp. (Taylor-Wharton Div.)
C 0.45-0.55, Mn 0.6-0.8, Si 0.3-0.5, Cr 2.75-3.25, Mo 0.65-0.75, bal Fe.
Cast: 275-325 Brin. For castings; high abrasion resistance. *Obsolete*

TISCO 60

Harsco Corp. (Taylor-Wharton Div.)
C 0.3, Cr 5, Mo 0.5, V 0.25, Mn 0.8, bal Fe.
325 Brin. For high temperature applications; ASTM-A-148. *Obsolete*

TISCO 90 HIGH STRENGTH

Harsco Corp. (Taylor-Wharton Div.)
C 2.7, Ni 4.4, Cr 1.7, bal Fe.
Cast: 50,000 TS; 525 Brin. For liners, pulverizers, pump plungers; martensitic, white cast iron. *Obsolete*

TISCO 90 REGULAR

Harsco Corp. (Taylor-Wharton Div.)
C 3.3, Mn 0.55, Si 0.5, Ni 4.4, Cr 1.7, bal Fe.
Cast: 45,000 TS; 550 Brin. For liners, pulverizers, pump plungers, grinding balls; martensitic, white cast iron. *Obsolete*

TISCO HIGH CHROME IRON

Harsco Corp. (Taylor-Wharton Div.)
C 0-0.1, Mn 0.4-0.6, Si 0-2, Cr 26-30, bal Fe.
Cast: 40,000-60,000 TS; 30,000-40,000 YS; 0-3 El; 0-5 RA; 200-220 Brin. For apparatus to resist sulfur rich atmospheres at high temperature; now Tisco No. 130. *Obsolete*

TISCO HIGH CHROME IRON

Harsco Corp. (Taylor-Wharton Div.)
C 0-0.1, Mn 0.4-0.6, Si 0-2, Cr 26-30, bal Fe.
Cast: 40,000-60,000 TS; 30,000-40,000 YS; 0-3 El; 0-5 RA; 200-220 Brin. For apparatus to resist sulfur rich atmospheres at high temperature; now Tisco No. 130. *Obsolete*

TISCO LOW CHROME IRON

Harsco Corp. (Taylor-Wharton Div.)
C 0-0.1, Mn 0.4-6, Si 0-2, Cr 11.5-16, bal Fe.
Cast: 70,000-90,000 TS; 40,000-55,000 YS; 18-24 El; 25-45 RA; 160-180 Brin. For apparatus and vessels to resist alkaline liquors, food stuffs, steam and some acids; now Tisco No. 132. *Obsolete*

TISCO MEDIUM CHROME IRON

Harsco Corp. (Taylor-Wharton Div.)
C 0-0.1, Mn 0.4-0.6, Si 0-2, Cr 16-19, bal Fe.
Cast: 75,000-100,000 TS; 45,000-60,000 YS; 10-20 El; 15-35 RA; 190-210 Brin. For apparatus and vessels to resist nitric acid, food stuffs, sea water, steam; now Tisco No. 131. *Obsolete*

TISCO NIROSTA 28-11

Harsco Corp. (Taylor-Wharton Div.)
C 0-0.3, Mn 0-1, Si 0.5-1.5, Cr 26-30, Ni 8-12, bal Fe.
Normalized: 75,000-85,000 TS; 45,000-50,000 YS; 12-18 El;
8-12 RA; 150-160 Brin. For heat and corrosion resistant parts;
now Tisco No. 107. *Obsolete*

TISCO NO. 17

Harsco Corp. (Taylor-Wharton Div.)
C 0.1-0.19, Mn 0.6-0.8, Si 0.25-0.5, bal Fe.
Annealed: 60,000-70,000 TS; 30,000-40,000 YS; 25-35 El;
40-60 RA; 116-136 Brin. For motor frames, case-hardened
castings, drag links; free machining. *Obsolete*

TISCO NO. 19

Harsco Corp. (Taylor-Wharton Div.)
C 0.25-0.35, Mn 0.6-0.8, Si 0.3-0.5, bal Fe.
Annealed: 65,000-80,000 TS; 35,000-45,000 YS; 25-35 El;
35-55 RA; 126-156 Brin. For commercial castings. *Obsolete*

TISCO NO. 2

Harsco Corp. (Taylor-Wharton Div.)
C 1-1.4, Si 0-1, Cr 1.9-2.1, 11% min Mn, bal Fe.
Hardened: 125,900 TS; 45,000 YS; 25 El; 20 RA; 220 Brin. For
wear resisting parts, shovels, dippers; wear resisting.
Obsolete

TISCO NO. 22

Harsco Corp. (Taylor-Wharton Div.)
C 0.2-0.3, Mn 1.15-1.75, Si 0.25-0.5, bal Fe.
Heat treated: 80,000-90,000 TS; 50,000-60,000 YS; 25-35 El;
50-60 RA; 172-192 Brin. For tractor and moving machinery
parts; shock and fatigue resistant. *Obsolete*

TISCO NO. 23X

Harsco Corp. (Taylor-Wharton Div.)
C 0.2-0.25, Mn 1.3-1.6, Cr 0.6-0.9, Mo 0.2-0.3, bal Fe.
Heat treated: 90,000-110,000 TS; 68,000-90,000 YS; 20-30 El;
42-65 RA; 210-260 Brin. For caps, levers, castings; high
strength. *Obsolete*

TISCO NO. 40

Harsco Corp. (Taylor-Wharton Div.)
C 0.27-0.37, Mn 0.6-0.8, Si 0.3-0.5, Cr 0.65-0.85, bal Fe.
Annealed: 78,000-92,000 TS; 42,000-55,000 YS; 20-30 El;
30-45 RA; 180-350 Brin. For tractor shoes. *Obsolete*

TISCO NO. 41

Harsco Corp. (Taylor-Wharton Div.)
C 0.45-0.55, Mn 0.6-0.8, Si 0.3-0.5, Cr 1-1.25, bal Fe.
Heat treated: 115,000-130,000 TS; 60,000-80,000 YS; 6-12
El; 8-20 RA; 220-235 Brin. For liners for ball mills; hard, wear
resistant with moderate toughness. *Obsolete*

TISCO NO. 50

Harsco Corp. (Taylor-Wharton Div.)
C 0.45-0.55, Mn 0.8-1, Si 0.3-0.5, Cr 0.9-1, Mo 0.25-0.35, bal
Fe.
Noramlized: 130,000-150,000 TS; 95,000-115,000 YS; 5-12 El;
8-25 RA; 250-280 Brin. For roll shells; machinable. *Obsolete*

TISCO NO. 51

Harsco Corp. (Taylor-Wharton Div.)
C 0.75-0.85, Mn 0.7-0.9, Si 0.3-0.5, Cr 1.5-2, Mo 0.35-0.45,
bal Fe.
Normalized: 140,000-170,000 TS; 90,000-115,000 YS; 2-5 El;
4-8 RA; 300-400 Brin. For ball mill liners; resists abrasive wear
with little shock. *Obsolete*

TISCO NO. 53

Harsco Corp. (Taylor-Wharton Div.)
C 0.15-0.35, Mn 0.45-0.85, Si 0.2-0.6, Cr 4-6.5, Mo 0.4-0.65,
bal Fe.
Heat treated: 110,000-125,000 TS; 75,000-90,000 YS; 16-20
El; 30-55 RA; 215-255 Brin. For oil still valves and fittings;
resists H 2 S gas; low creep at high temperature. *Obsolete*

TISCO NO. 6

Harsco Corp. (Taylor-Wharton Div.)
C 0.2-0.3, Mn 10-12, Si 0-1, Ni 6-8, Cr 0-0.15, bal Fe.
Cast: 67,500 TS; 27,400 YS; 27.5 El; 37.5 RA; 112-150 Brin.
For castings; non-magnetic. *Obsolete*

TISCO NO. 65

Harsco Corp. (Taylor-Wharton Div.)
C 0.2-0.3, Mn 0.8-1, Si 0.3-0.5, Ni 0.75-1, bal Fe.
Annealed: 65,000-75,000 TS; 40,000-50,000 YS; 20-35 El;
40-65 RA; 130-150 Brin. For tractor and moving machinery
parts; shock and fatigue resistant. *Obsolete*

TISCO NO. 70

Harsco Corp. (Taylor-Wharton Div.)
C 0.3-0.4, Mn 0.6-0.8, Si 0.3-0.5, Cr 0.8-1, Ni 2.5-3, bal Fe.
Cast: 95,000-120,000 TS; 65,000-85,000 YS; 10-15 El; 10-15
RA; 210-230 Brin. For pinions and gears; resists wear and
repeated high stresses. *Obsolete*

TISCO NO. 71

Harsco Corp. (Taylor-Wharton Div.)
C 0.3-0.4, Mn 0.6-0.8, Si 0.3-0.5, Cr 0.6-0.8, Ni 1.25, bal Fe.
Cast: 95,000-110,000 TS; 55,000-65,000 YS; 15-25 El; 20-40
RA; 197-217 Brin. For pinions and racks for steam shovel;
tough. *Obsolete*

TISCO NO. 72

Harsco Corp. (Taylor-Wharton Div.)
C 0.45-0.55, Mn 0.6-0.8, Si 0.3-0.5, Cr 0.8-1, Ni 2.75-3.25, bal
Fe.
Cast: 95,000-115,000 TS; 55,000-75,000 YS; 12-22 El; 15-40
RA; 207-227 Brin. For track work, frogs, crossings, and railway
track switches; machinable. *Obsolete*

TISCO NO. 80

Harsco Corp. (Taylor-Wharton Div.)
C 0.45, Mn 0.7, Ni 1.75-2.25, Cr 1.5, Mo 0.3, bal Fe.
Heat treated: 115,000-215,000 TS; 80,000-200,000 YS; 4-12
El; 4-25 RA; 300-450 Brin. For sand pumps, dredges;
abrasion resistant, oil hardened. *Obsolete*

TISCO NO. 81

Harsco Corp. (Taylor-Wharton Div.)
C 0.5-0.6, Mn 0.6-0.8, Si 0.4-0.6, Cr 1.4-1.6, Ni 1.75-2.25, Mo
0.3-0.4, bal Fe.
Normalized: 232,000 TS; 230,000 YS; 1.6 El; 2.2 RA; 300-650
Brin. For centrifugal pumps handling sand; hard and brittle;
air hardened. *Obsolete*

TISCO NO. 90

Harsco Corp. (Taylor-Wharton Div.)
C 3-3.5, Mn 0.4-0.7, Si 0.4-0.7, Cr 1.25-1.75, Ni 4-4.5, bal Fe.
Heat treated: 250,000 TS; 550-600 Brin. For wearing parts,
sand pumps; hard, abrasive resistant cast iron. *Obsolete*

TISCO TAPPI CORROSION RESISTING STEEL

Harsco Corp.
C 0.13-0.35, Cr 16-23, Ni 7-11, Mo, bal Fe.
Annealed: 70,000-75,000 TS; 50-60 El; 45-60 RA; 130-160
Brin. For machinery parts in paper and pulp machinery; now
Tisco No. 110.

TISCON

Tata Iron & Steel Co.
C 0.19-0.23, Mn 0.75-0.85, bal Fe.
Twisted rolled bar: 49.5 kgf/mm^2 TS; 42.5 kgf/mm^2 YS; 14.5
El. Reinforcing bars.

TISCOR

Tata Iron & Steel Co.
C 0.1, Cr 0.7, Cu 0.35, bal Fe.
Rolled: 78,000 TS; 45,000 YS; 25 El. For structural work;
weldable. *Obsolete*

TISCRAL

Tata Iron & Steel Co.
C 0-0.22, Mn 1.2-1.5, P 0-0.08, Cr 0.5-0.8, V 0.02-0.1, Ti
0.005-0.015, Al 0.2-0.4, bal Fe.
Rolled: 45 kgf/mm^2 YS; 62 kgf/mm^2 TS; 18 El; 217 Brin.
Wear and abrasion resistant steel.

TISCROM

Tata Iron & Steel Co.
C 0.22, Mn 1, Cr 0.55, Cu 0.3, Si 0.1, bal Fe.
Rolled: 90,000 TS; 50,000 YS; 18 El. High tensile structural
steel; riveting grade, not suitable for welding. *Obsolete*

TISKA NIROSTA KA-2

Harsco Corp. (Taylor-Wharton Div.)
C 0-0.16, Cr 18, Ni 8, Mn 1, Si 2, bal Fe.
Annealed: 65,000-75,000 TS; 25,000-35,000 YS; 50-60 El;
45-60 RA; 130-145 Brin. For stainless parts, valves, fittings,
marine parts. *Obsolete*

TISKA NIROSTA KA-2-H

Harsco Corp. (Taylor-Wharton Div.)
Cr 18, Ni 8, Mn 1, Si 2, 0.16% min C, bal Fe.
Annealed: 65,000-75,000 TS; 25,000-35,000 YS; 50-60 El;
45-60 RA; 130-145 Brin. For stainless parts, valves, fittings,
marine parts. *Obsolete*

TISKA NIROSTA KA-2-H-MO

Harsco Corp. (Taylor-Wharton Div.)
C 0.16, Cr 18-22, Ni 7-10, Mo 2-4, Mn 1, bal Fe.
For stainless parts, valves, fittings. *Obsolete*

TISKA NIROSTA KA-2-MO

Harsco Corp. (Taylor-Wharton Div.)
C 0.16, Cr 18-22, Ni 7-10, Mo 2-4, Mn 1, bal Fe.
For stainless parts, valves, fittings. *Obsolete*

TISKA NIROSTA KA-2-S

Harsco Corp. (Taylor-Wharton Div.)
C 0-0.07, Cr 18, Ni 8, Mn 1, Si 2, bal Fe.
Annealed: 65,000-75,000 TS; 25,000-35,000 YS; 50-60 El;
45-60 RA; 130-145 Brin. For stainless parts, valves, fittings,
marine parts. *Obsolete*

TISKA NIROSTA KA-2-S-MO

Harsco Corp. (Taylor-Wharton Div.)
C 0.07, Cr 18-22, Ni 7-10, Mo 2-4, Mn 1, bal Fe.
For stainless parts, valves, fittings. *Obsolete*

TISKA NIROSTA KNC-3

Harsco Corp. (Taylor-Wharton Div.)
C 0-0.2, Cr 25, Ni 19, Mn 1, Si 2, bal Fe.
At 20 C: 85,300 TS; 38,000 YS; 15 El; 15 RA; 160 Brin. At 900
C: 17,500 TS; 17,000 YS; 25 El; 25 RA. For heat and
corrosion resistant parts. *Obsolete*

TISSIERS METAL-1

English manufacture
Cu 97, Zn 2, As 1.
For hardware; arsenical bronze; good conductivity.

TISSIERS METAL-2

English manufacture
Cu 97, Zn 2, Sn 0.5.
For hardware; arsenical bronze; good conductivity.

TITAN

Carpenter Technology Corp.
C 1, bal Fe.
For tools. Water hardening. *Obsolete*

TITAN 100

Adamas Carbide Corp.
TiC base sintered carbide. Density: 5.50; Hardness: 93.3
Rock A; transverse rupture strength 170,000 psi (1200
N/mm^2). For high speed semi-finish and finish machining of
alloy steels, tool steels, stainless steels and cast iron; for
speeds exceeding 800 SFPM. ISO P01 - P 05.

TITAN 50

Adamas Carbide Corp.

TiC base sintered carbide. Density: 5.82; Hardness: 90.5 Rock A; transverse rupture strength 275,000 psi (1940 N/mm^2). High toughness cemented titanium carbide grade for general purpose and semi-rough machining of steel. ISO P20 - P30.

TITAN 60

Adamas Carbide Corp.

TiC, Co.

For cutting tools in turning, boring and milling operations. Sintered carbide. Wear and crater resistant.

TITAN 80

Adamas Carbide Corp.

TiC, MoC, Ni.

Sintered: 92.7 Rock A; 200,000 transverse strength min. For cutting tools in high speed machining. Wear and crater resistant. Sintered carbide.

TITAN BEARING BRONZE CA 673

Cerro Metal Products Co.

Cu 59.25, Pb 1.1, Si 1.15, Mn 2.25, bal Zn.
For bushings, bearings, and gears. Free-cutting.

TITAN BRONZE

Cerro Wire & Cable Company

Cu 60-60.5, Sn 0.65-0.85, bal Zn.
Annealed: 74,700 TS; 40,300 YS; 34 El; 37 RA; 101 Brin. Drawn: 83,500 TS; 52,500 YS; 31 El; 44 RA. For marine shafts, fittings; corrosion resisting. *Obsolete*

TITAN BRONZE "A"

Cerro Metal Products Co.

Cu 53.5, Zn 46, Fe 0.3, Al 0.3.
Hot rolled: 75,000-85,000 TS; 40,000-48,000 YS; 25-30 El; 45-50 RA. Cast: 50,000-55,000 TS; 30,000-35,000 YS; 15-20 El. For gears, pinions, roller bearings, and pump rods. Non-corrosive. *Obsolete*

TITAN BRONZE "B"

Cerro Metal Products Co.

Cu 60.5, Sn 1.02, Zn 38.5.
Annealed: 74,740 TS; 40,277 YS; 34 El; 37 RA; 109 Brin. Hard drawn: 73,560 TS; 52,443 YS; 31 El; 44 RA. For marine shafts, underwater fittings. Non-corrosive. *Obsolete*

TITAN BRONZE WELDING ROD

Cerro Metal Products Co.

Cu 58.75, Sn 0.75, bal Zn.
Cold drawn: 94,000 TS; 75,000 YS; 14 El; 80 RA; 174 Brin. Welded: 52,000 TS; 40,000 YS; 15 El; 104 Brin. For welding rod; 1615°F MP; non-forming; CA 470.

TITAN MANGANESE BRONZE

Cerro Metal Products Co.

Cu 59, Pb 0.8, Sn 0.9, Mn 0.1, Fe 0.9, bal Zn.
Rolled: 75,000 psi TS; 35 El. For machine parts; free cutting; CA 767. *Obsolete*

TITAN MANGANESE BRONZE

Cerro Metal Products Co.

Cu 59.2, Fe 0.9, Sn 0.9, Mn 0.1, Si 0.1, bal Zn.
Cold drawn: 89,000 TS; 58,700 YS; 21.5 El; 40 RA; 159 Brin. For bolts, rods, and forgings. Corrosion resistant; high tensile; CA 675.

TITAN MANGANESE BRONZE WELDING ROD

Cerro Metal Products Co.

Cu 58.72, Mn 0.3, Fe 0.5, Ni 0.5, Sn 0.9, Si 0.12, bal Zn.
General purpose manganese bronze rod with nickel. For brass, copper, bronze, nickel alloys, steel and cast iron. CA 680.

TITAN MANGANESE BRONZE WELDING ROD

Cerro Metal Products Co.

Cu 59, Sn 0.8, Fe 0.8, Si 0.1, bal Zn.
Cold drawn: 96,000 TS; 79,000 YS; 16 El; 81 RA; 180 Brin. Welded: 53,000 TS; 38,000 YS; 12 El; 107 Brin. For welding rod for cast iron. 1620°F MP; general purpose rod; CA 681.

TITAN MANGANESE STEEL

Osborn Steels Ltd.

C 1.2, Si 0.5, Mn 12.5, S 0.03, P 0.05, bal Fe.
For wear parts; wear resistant. *Obsolete*

TITAN MUNTZ METAL

Cerro Metal Products Co.

Cu 61, bal Zn.
Cold drawn: 72,500 TS; 52,500 YS; 30 El; 66 RA; 135 Brin. For forgings, rod. Brass; CA 280.

TITAN NAVAL BRONZE

Cerro Metal Products Co.

Cu 60.25, Sn 0.75, Zn 39.
Annealed: 62,000 TS; 31,150 YS; 41 El; 95 Brin. Cold drawn: 69,000 TS; 46,000 YS; 32 El; 46 RA; 137 Brin. For ship construction and high strength forgings. Corrosion resistant; CA 464.

TITAN NAVAL BRONZE WELDING ROD

Cerro Metal Products Co.

Cu 59, Sn 0.75, bal Zn.
Cold drawn: 90,000 TS; 70,000 YS; 16 El; 78 RA; 170 Brin. Welded: 50,000 TS; 15 El; 104 Brin. For welding rod. 1625°F MP; CA 470.

TITAN NICKEL SILVER WELDING ROD

Cerro Metal Products Co.

Cu 48, Ni 10, Si 0.1, bal Zn.
Low fuming rod with nickel silver color and high mechanical properties. 1680°F MP; CA 773.

TITAN NO. 25

Cerro Metal Products Co.

Cu 59-59.5, Pb 0.9-1.1, Sn 0.5-0.7, bal Zn.
Hard: 68,400 TS; 47,600 YS; 30 El; 20 RA; 99 Brin. Soft: 66,000 TS; 35,050 YS; 38 El; 25 RA; 93 Brin. For hardware, screw machine products. Free-cutting. *Obsolete*

TITAN NO. 35

Cerro Metal Products Co.

Cu 64, Pb 1, bal Zn.
Hard: 45,500 TS; 24,000 YS; 54 El; 73 RA; 66 Brin. Soft: 42,000 TS; 20,000 YS; 62 El; 74 RA; 60 Brin. For hardware and screw machine products. Free-cutting swaging brass; CA 340.

TITAN NO. 36

Cerro Metal Products Co.

Cu 59, Pb 2, bal Zn.
Extruded: 55,000 TS; 23,000 YS; 40 El; 80 Brin. For hardware and screw machine products. Forging brass; free-cutting; CA 377.

TITAN NO. 4

Cerro Metal Products Co.

Cu 65, Zn 35.
For cold headed products, rivets, and bolts. Cold heading brass; good ductility; CA 270.

TITAN NO. 4 FORGE

Cerro Metal Products Co.

Cu 59-59.5, Pb 0.9-1.1, Sn 0.5-0.7, bal Zn.
63,000-68,000 TS; 35,000-47,000 YS; 30-44 El; 20-32 RA; 86-107 Brin. For screw machine parts. Free cutting. *Obsolete*

TITAN NO. 51

Cerro Metal Products Co.

Cu 59, Pb 0.7, Sn 0.9, Fe 0.9, Mn 0.2, bal Zn.
Soft: 65,000 TS; 30,000 YS; 30 El; 116 Brin. 1/2 H-temper: 78,000 TS; 52,000 YS; 22 El; 150 Brin. H-temper: 85,000 TS; 60,000 YS; 12 El; 165 Brin. For screw machine products and marine hardware. Leaded manganese bronze; CA 676.

TITAN NO. 52

Cerro Metal Products Co.

Cu 57.25, Pb 2.5, bal Zn.
For screw machine products. Free-cutting; CA 385.

TITAN NO. 65

Cerro Metal Products Co.

Cu 59, Sn 0.9, Fe 0.9, Mn 0.2, bal Zn.
Soft: 65,000 TS; 35,000 YS; 30 El; 116 Brin. 1/2 H-temper: 78,000 TS; 52,000 YS; 22 El; 150 Brin. H-temper: 85,000 TS; 60,000 YS; 12 El; 165 Brin. For propellers, marine hardware, and bolts. Corrosion resistant; manganese bronze; CA 675.

TITAN NO. 68

Cerro Metal Products Co.

Cu 56-60, Fe 0.8-1.3, Al 0.7-1.2, Mn 0-0.5, bal Zn.
Cast: 80,000 TS; 40,000 YS; 30 El; 137 Brin. For propellers, shafts, and marine hardware. Manganese bronze. *Obsolete*

TITAN NO. 6A

Cerro Metal Products Co.

Cu 60-60.5, Sn 0.65-0.85, bal Zn.
Hard: 63,000 TS; 36,000 YS; 42 El; 46 RA; 70 Brin. Soft: 62,000 TS; 29,000 YS; 48 El; 53 RA; 72 Brin. For hardware, marine parts. Corrosion resistant; CA 464.

TITAN NO. 73

Cerro Metal Products Co.

Si 1.9, Al 6.9, bal Cu.
Cast: 85,000 TS; 43,000 YS; 30 El; 135 Brin. For impellers and marine hardware. Aluminium-silicon bronze; CA 642.

TITAN NO. 74

Cerro Metal Products Co.

Cu 81.5, Si 4, bal Zn.
Cold drawn: 90,000 TS; 45,000 YS; 25 El; 170 Brin. For valve stems, line pole hardware, die castings and forgings. CA 694.

TITAN NO. 75

Cerro Metal Products Co.

Cu 70, Zn 30.
For cold headed products, rivets, and bolts. Cartridge brass; high ductility; CA 260.

TITAN NO. 83

Cerro Metal Products Co.

Cu 56, Pb 0.5, bal Zn.
For marine hardware. High strength. *Obsolete*

TITAN NO. 92

Cerro Metal Products Co.

Cu 58, Si 0.8, Al 1.6, Mn 2.5, Pb 0.3, bal Zn.
Cast: 85,000 TS; 45,000 YS; 18 El; 156 Brin. For propellers, impellers, and marine hardware. High manganese bronze; corrosion resistant; CA 674.

TITAN NO. 95

Cerro Metal Products Co.

Cu 63.5, Sn 0.6, bal Zn.
Soft: 56,000 TS; 24,000 YS; 37 El; 104 Brin. 1/2 H: 64,000 TS; 32,000 YS; 32 El; 121 Brin. Hard: 70,000 TS; 46,000 YS; 24 El; 150 Brin. For marine hardware and propeller shafts. Corrosion resistant; CA 462.

TITAN NO. 96

Cerro Metal Products Co.

Cu 90, Pb 2, Ni 1, bal Zn.
Soft: 38,000 TS; 12,000 YS; 45 El; 57 Brin. 1/2 H: 52,000 TS; 48,000 YS; 20 El; 107 Brin. For marine hardware, screw machine products. Free cutting; corrosion and wear resistant. *Obsolete*

TITAN P-1

Cerro Wire & Cable Company

Cu 60.5-62.5, Pb 0.75-1.25, Sn 0.5-0.7, bal Zn.
For pressure die castings; yellow brass. *Obsolete*

TITAN PENN BRONZE WELDING ROD

Cerro Metal Products Co.
Cu 58.75, bal Zn.
Welded: 57,000 TS; 35,000 YS; 16 El; 107 Brin. Cold drawn: 103,000 TS; 89,000 YS; 13 El; 83 RA; 207 Brin. For welding rod. 1628°F MP; high strength; ductile; CA 000-16.

TITAN PRESSURE CASTING ALLOY

Cerro Metal Products Co.
Cu 65, Pb 0.1, Si 0.9, bal Zn.
For pressure die castings. High strength; CA 879.

TITAN SILICON BRONZE

Cerro Metal Products Co.
Cu 95.9, Mn 0.8, Si 3.3.
High copper rod or wire for welding or cold heading. CA 655.

TITAN SWAGING BRASS

Cerro Metal Products Co.
Cu 64-64.5, Pb 0.7-1, bal Zn.
42,000-51,000 TS; 20,000-34,000 YS; 41-62 El; 71-74 RA; 58-84 Brin. For swagings, spinnings, and stampings. High strength. *Obsolete*

TITAN-SEEWASSER

German manufacture
Mg 2-4, Mn 1.2, 0-0.2 Sb or Ti, bal Al.
Sand cast: 20,000-27,000 TS; 11,400-14,300 YS; 3-8 El; 50-60 Brin. For marine hardware, chemical plant equipment; resists salt water corrosion.

TITANAL

English manufacture
Cu 12, Mg 0.8, Si 0.5, Fe 0.5, bal Al, for pistons; cast.

TITANALOY

Mathieson-Heglar Co.
Ti 0.12, Cu 1, bal Zn.
Sheet: 24,000 TS; 10 El; 55 Brin. For roofing, gutters, trim, housings, fuses, curtain walls; high creep resistance.

TITANIA

Bridgeport Rolling Mills Co.
Cu 0.8, Ti 0.15, bal Zn.
Sheet: 32,000-42,000 TS; 21-38 El. For architectural and industrial products, giftware, electrical and electronic components. Good workability. *Obsolete*

TITANIC CARBON STEEL

Osborn Steels Ltd.
C 0.55-1.6, Si 0.15-0.3, Mn 0.2-0.4, bal Fe.
For cutting tools; water hardening. *Obsolete*

TITANIT

American Cutting Alloys Inc.
Mo-Ti-W-carbides.
For cutting tools, high speed cutters; "Cutanit," high resistance to wear and abrasion. *Obsolete*

TITANIT

Thyssen Edelstahlwerke AG
Mo-Ti-W-carbides.
For cutting tools, high speed cutters, "Cutanit," high resistance to wear and abrasion. *Obsolete*

TITANIT F TI 1

Thyssen Edelstahlwerke AG
Carbides. For cutting tools; sintered carbide. *Obsolete*

TITANIT FM

Plansee, Metallwerk Gesellschaft
WC.
Sintered. For cutting tools; sintered carbides.

TITANIT G 1

Plansee, Metallwerk Gesellschaft
Carbide.
For cutting tools; sintered carbide.

TITANIT G 2

Plansee, Metallwerk Gesellschaft
For cutting tools; sintered carbide.

TITANIT G 3

Plansee, Metallwerk Gesellschaft
Carbide.
For cutting tools; sintered carbide.

TITANIT G 4

Plansee, Metallwerk Gesellschaft
Carbide.
For cutting tools; sintered carbide.

TITANIT G 5

Plansee, Metallwerk Gesellschaft
Carbide.
For cutting tools; sintered carbide.

TITANIT G 6

Plansee, Metallwerk Gesellschaft
Carbide.
For cutting tools; sintered carbide.

TITANIT H 1

Plansee, Metallwerk Gesellschaft
Carbide.
For cutting tools; sintered carbide.

TITANIT S TI 1

Thyssen Edelstahlwerke AG
Carbides.
For cutting tools; sintered carbide. *Obsolete*

TITANIT S TI 2

Thyssen Edelstahlwerke AG
Carbides.
For cutting tools; sintered carbide. *Obsolete*

TITANIT S TI 3

Thyssen Edelstahlwerke AG
Carbides.
For cutting tools; sintered carbide. *Obsolete*

TITANIT S TI 4

Thyssen Edelstahlwerke AG
Carbides.
For cutting tools; sintered carbide. *Obsolete*

TITANIT S2T

Plansee, Metallwerk Gesellschaft
WC.
Sintered. For cutting tools; sintered carbides.

TITANIT S3T

Plansee, Metallwerk Gesellschaft
WC.
Sintered. For cutting tools; sintered carbides.

TITANIT S4T

Plansee, Metallwerk Gesellschaft
WC.
Sintered. For cutting tools; sintered carbides.

TITANIT SIT

Plansee, Metallwerk Gesellschaft
WC.
Sintered. For cutting tools; sintered carbides.

TITANIT U

Plansee, Metallwerk Gesellschaft
Carbide.
For cutting tools; sintered carbide.

TITANIT WZ1

Plansee, Metallwerk Gesellschaft
WC.
Sintered. For cutting tools, sintered carbides.

TITANIT WZ2

Plansee, Metallwerk Gesellschaft
WC.
Sintered. For cutting tools; sintered carbides.

TITANITE

Aluminum Smelting & Refining Co., Inc.
Cu 1, Zn 4.5, Ti 0.2, Mn 1.2, Mg 0.5, bal Al.
Cast: 30,000 TS; 4-5 El; 55 Brin. Aged: 35,000 TS; 4-5 El; 65 Brin. For aircraft castings; self-aging, corrosion resistant. *Obsolete*

TITANITE

Titanite Alloys Corp.
Cu 4, Ti 0.2, bal Al.
Heat treated: 30,000-37,000 TS; 15,000-24,000 YS; 6-5 El; 60-75 Brin. For flywheel and axle housings, aircraft wheels, fittings; age-hardenable.

TITANITE

Wellman Dynamics Corp.
Cu 4, Ti 0.2, bal Al.
Heat treated: 30,000-37,000 TS; 15,000-24,000 YS; 6-5 El; 60-75 Brin. For flywheel and axle housings, aircraft wheels, fittings; age-hardenable.

TITANITE

National Bronze & Aluminum Foundry
Cu 4, Ti 0.2, bal Al.
Heat treated: 30,000-37,000 TS; 15,000-24,000 YS; 6-5 El; 60-75 Brin. For flywheel and axle housings, aircraft wheels, fittings; age-hardenable.

TITANIUM

Chase Brass & Copper Co., Inc.
Ti (CP). Annealed: 55,000 TS; 45,000 YS; 25 El. Tube for chemical processing. Corrosion resistant.

TITANIUM

Atomergic Chemetals Corp.
Ti.
Purities, zone refined: 99.995%, special crystal grade 99.95%, crystal bar and electrorefined 99.9+%, commercial 99.5%.
Forms: rods, sponge, granule, powder, wire, sheet, foil, ingot, tubing castings, single crystals.

TITANIUM 120

IMI Knoch Ltd.
Ti alloy.
Annealed: 67,200 TS; 20 El. For aircraft frames; low density.

TITANIUM 130

IMI Knoch Ltd.
Ti alloy.
Annealed: 58,000-89,600 TS; 40,400 YS; 15 El. For aircraft frames; low density.

TITANIUM 160

IMI Knoch Ltd.
Ti alloy.
Annealed: 89,600 TS; 67,200 YS; 12 El. For aircraft frames; low density.

TITANIUM 3 AL-2.5 V

Hamilton Technology Inc.
Al 2.5-3.5, V 2-3, O 0-0.12, N 0-0.045, C 0-0.05, H 0-0.015, Fe 0-0.3, bal Ti.
Annealed: 98,000 psi TS; 88,000 psi YS; 22 El. 75% cold worked: 160,000 psi TS; 148,000 psi YS; 5 El. Good formability; used for honeycomb structures.

TITANIUM 314A

IMI Knoch Ltd.
Al 4, Mn 4, bal Ti.
Annealed: 139,000 TS; 127,000 YS; 10 El. Hot rolled: 154,000 TS; 132,700 YS; 21 El; 33 RA. For jet engine and guided missile components; high temperature applications.

TITANIUM 314C
IMI Knoch Ltd.
Al 2, Mn 2, bal Ti.
Annealed: 89,600 TS; 67,200 YS; 15 El. Heat treated: 94,900 TS; 72,400 YS; 23 El; 47 RA. For aircraft and missile components; high temperature use.

TITANIUM 317
IMI Knoch Ltd.
Al 5, Sn 2.5, bal Ti.
Annealed: 112,000 TS; 101,000 YS; 10 El. For welded rings, compressor blades; weldable, good creep and fatigue strength to 800°F.

TITANIUM 318A
IMI Knoch Ltd.
Al 6, V 4, bal Ti.
Annealed: 139,000 TS; 130,000 YS; 10 El. For jet engine components, airframe forgings; weldable, good hot strength.

TITANIUM 371
IMI Knoch Ltd.
Sn, Al, bal Ti.
Heat treated: 148,000 TS; 126,600 YS; 17 El; 19 RA. For high temperature applications to 1000°F; heat treatable.

TITANIUM A-40
Ulbrich Stainless Steels/Special Metals
Titanium. C 0-0.1, N 0-0.03, H 0.005, Fe 0-0.3, 0.2500 O_2 max, bal Ti.
Alpha grade titanium; properties similar to austenitic stainless steel; per ASTM B-265 Grade 2.

TITANIUM A-55
Ulbrich Stainless Steels/Special Metals
Titanium. C 0-0.1, N 0-0.05, H 0.006, Fe 0-0.3, O 0-0.35, bal Ti.
Alpha phase titanium; moderate strength with good formability. Used in airframes, chemical and similar applications; per ASTM B-265 Grade 3.

TITANIUM A-70
Ulbrich Stainless Steels/Special Metals
Titanium. C 0-0.1, N 0-0.05, H 0.006, Fe 0-0.5, O 0-0.4, bal Ti.
Alpha phase titanium; better strength than A-55. For airframes and similar applications; per ASTM B-265 Grade 4.

TITANIUM ALUMINUM BRONZE NO. 1
Frontier Bronze Corp.
Copper. Cu 90, Al 10.
65,000-80,000 TS; 22,000-26,000 YS; 15-25 El; 16-24 RA; 90-100 Brin. For gears: 19,000 CYP.

TITANIUM ALUMINUM BRONZE NO. 5
Frontier Bronze Corp.
Copper. Cu 89, Al 10, Fe 1.
65,000-80,000 TS; 23,000-28,000 YS; 15-30 El; 21-29 RA; 92-100 Brin. For gears: 19,000 CYP.

TITANIUM BEARING STEEL
Inland Steel Co.
HSLA steel. C 0.2, Ti 0.3, bal Fe.
For jet engine parts; heat resistant; coated with porcelain enamel.

TITANIUM DISILICIDE
NL Industries
Ti 44.7, Si 51.8, Al 1.5, Fe 2.1.
For special alloy applications; cermet. *Obsolete*

TITANIUM EP 20-2
Manufacturer not listed.
Al 20, V 2, bal Ti.
At 70°F: 190,000 TS; 171,000 YS; 6 El; 380 Brin. At 800°F: 150,000 TS; 130,000 YS; 12 El. At 1100°F: 120,000 TS; 105,000 YS; 15 El. For high temperature applications, aircraft and missile components; high temperature strength, weldable.

TITANIUM EP 90-10
Manufacturer not listed.
Cr 9.8-10.2, Fe 0.005-0.05, bal Ti.
At 70°F: 201,000 TS; 175,000 YS; 6 El. At 800°F: 150,000 TS; 121,000 YS; 12 El. For pressure vessels, fasteners, clamps, airplane skins; heat treatable, high strength.

TITANIUM GR. NDA
Du Pont Co.
Fe 0-0.3, C 0.15, 97.0 Ti min, 0.2 N_2.
For ladle additions to stainless steels; unsintered pellets.

TITANIUM STEEL
American manufacture
Ti 0.3-9, C 0.1-0.8, bal Fe.
For machinery parts, tools; water hardening.

TITANIUM TI-175A
TIMET
C 0.02, Cr 3, Fe 1.5, N 0.04, W 0-0.08, bal Ti.
Annealed: 175,000 TS; 160,000 YS; 10 El; 30 RA. For aircraft members, engine components; heat treatable. *Obsolete*

TITANIUM TYPE TI-100 A
TIMET
C 0-0.1, W 0-0.02, bal Ti.
Annealed: 100,000 TS; 75,000 YS; 20-25 El; 45 Brin. For aircraft parts; not heat treatable. *Obsolete*

TITANIUM TYPE TI-125 A
TIMET
Cr 1.8, Fe 0.9, C 0.02, W 0.08, N 0.04, bal Ti.
For aircraft parts; work hardens. *Obsolete*

TITANIUM TYPE TI-150A
TIMET
Cr 2.7, Fe 1.3, N 0.02, W 0-0.04, C 0.02, bal Ti.
Annealed: 150,000 TS; 135,000 YS; 12-20 El; 40 RA. For bars, forgings; heat treatable. *Obsolete*

TITANIUM TYPE TI-75 A
TIMET
Fe 0.1, N 0.02, C 0-0.04, W 0-0.08, bal Ti.
Annealed: 70,000-80,000 TS; 45,000-55,000 YS; 20-30 El; 45-70 RA; 175 Brin. For aircraft parts; not heat treatable. *Obsolete*

TITANIUM-3 ALUMINUM- 2.5 VANADIUM
Chase Brass & Copper Co., Inc.
Al 3, V 2.5, bal Ti.
Extruded and annealed: 95,000 psi TS; 80,000 psi YS; 10 El. Aircraft hydraulic tube. High strength to weight ratio.

TITEM
Titusville Forge Co.
C, Cr, Mo, bal Fe.
For die blocks; oil hardened. *Obsolete*

TITUS
Cytemp Specialty Steel Div.
C 0.7, Cr 0.75, V 0.2, bal Fe.
Heat treated: 220,000-125,000 TS; 200,000-100,000 YS: 5-20 El; 5-53 RA; 487-320 Brin. For punches and chisels, axes, ball races, dies, shear blades; "Cyclops Titus." *Obsolete*

TIVAN
Titusville Forge Co.
C, Cr, Mo, bal Fe.
For fish tail bits; water hardened. *Obsolete*

TIWZ-12
Plansee, Metallwerk Gesellschaft
TiC.
For gas turbines, jet engines, and nuclear power plants. Sintered; oxidation resistant.

TIWZ-12
American Electro Metals Corp.
TiC.
For gas turbines, jet engines, nuclear power plants; sintered, oxidation resistant.

TIZIRBE
GTE Sylvania
Ti 48, Zr 48, Be 4.
For brazing titanium stainless steel components. Corrosion resistant.

TIZIRBE
Western Gold & Platinum Co.
Ti 48, Zr 48, Be 4.
For brazing titanium stainless steel components. Corrosion resistant.

TIZIT
Plansee, Metallwerk Gesellschaft
W 40-80, Fe 3-40, Ti 4-15, C 2-4, Ce 1-5, Cr 4.
Cast. For hard cutting tools and dies.

TIZIT A
Plansee, Metallwerk Gesellschaft
Carbide.
For cutting tools, dies; sintered carbides.

TIZIT CR1
Plansee, Metallwerk Gesellschaft
CrC, bal Ni.
For valves and pump parts. Corrosion resistant.

TIZIT CR2
Plansee, Metallwerk Gesellschaft
WC, bal Co.
For valves and pump parts. Corrosion resistant.

TIZIT FM
Plansee, Metallwerk Gesellschaft
WC, TiC, TaC, Co.
For cutting tools and dies; carbides. Wear resistant.

TIZIT G1
Plansee, Metallwerk Gesellschaft
WC, Co.
For cutting tools and dies; carbides. Wear resistant.

TIZIT G2
Plansee, Metallwerk Gesellschaft
WC, Co.
For cutting tools and dies; carbides. Wear resistant.

TIZIT G3
Plansee, Metallwerk Gesellschaft
WC, Co.
For cutting tools and dies; carbides. Wear resistant.

TIZIT G4
Plansee, Metallwerk Gesellschaft
WC, Co.
For cutting tools and dies; carbides. Wear resistant.

TIZIT G5
Plansee, Metallwerk Gesellschaft
WC, Co.
For cutting tools and dies; carbides. Wear resistant.

TIZIT G6
Plansee, Metallwerk Gesellschaft
WC, Co.
For cutting tools and dies; carbides. Wear resistant.

TIZIT H1
Plansee, Metallwerk Gesellschaft
WC, Co.
For cutting tools and dies; carbides. Wear resistant.

TIZIT H2
Plansee, Metallwerk Gesellschaft
WC, Co.
For cutting tools and dies; carbides. Wear resistant.

TIZIT H3
Plansee, Metallwerk Gesellschaft
WC, Co.
For cutting tools and dies; carbides. Wear resistant.

TIZIT S2T
Plansee, Metallwerk Gesellschaft
WC, TiC, TaC, Co.
For cutting tools and dies; carbides. Wear resistant.

TIZIT S3T
Plansee, Metallwerk Gesellschaft
Carbide.
For cutting tools. Extreme wear resistance; sintered.

TIZIT S4T
Plansee, Metallwerk Gesellschaft
WC, TiC, TaC, Co.
For cutting tools and dies; carbides. Wear resistant.

TIZIT SIT
Plansee, Metallwerk Gesellschaft
WC, TiC, TaC, Co.
For cutting tools and dies; carbides. Wear resistant.

TIZIT-U
Plansee, Metallwerk Gesellschaft
Carbide.
For cutting tools. Wear resistant; sintered.

TL-ALLOY
Anaconda Co.
Copper. Mn 3.2-3.8, bal Cu.
For tachometer drag cups. Corrosion resistant, high electrical
conductivity.

TM
Colt Industries
C 0.35, Cr 1.5, W 4, Co 5, bal Fe.
For extrusion dies; hot work steel. *Obsolete*

TM QUALITY
George Cook & Co., Ltd.
C 0.58, Mn 0.4, Si 0.2, bal Fe.
For spring collets, feed chucks; water hardened. *Obsolete*

TMB-1
Techni-Cast Corp.
Copper. Cu 58, Sn 0.75, Zn 38.75, Fe 1, Al 1, Mn 0.5.
75,000 psi TS; 32,000 psi YS; 30 El; 156 Brin.

TMB-2
Techni-Cast Corp.
Copper. Cu 64.5, Zn 24.75, Fe 3, Al 4.5, Mn 3.25.
100,000 psi TS; 55,000 psi YS; 20 El; 210 Brin.

TMB-3
Techni-Cast Corp.
Copper. Cu 64, Zn 24, Fe 3, Al 5.75, Mn 3.25.
115,000 psi TS; 80,000 psi YS; 16 El; 228 Brin.

TMB-4
Techni-Cast Corp.
Copper. Cu 59, Sn 0.75, Pb 1, Zn 36.75, Fe 1, Al 1, Mn 0.5.
66,000 psi TS; 26,000 psi YS; 25 El; 132 Brin.

TMC
Jessop Steel Co.
C 0.7, Mo 9.5, W 1.5, Cr 4, V 1, bal Fe.
Hardened: 64-67 Rock C. For cutting tools, lathe and planer
cutters, drills, taps, reamers. Type M1 high-speed steel.
Obsolete

TMO DREADNOUGHT
Colt Industries
C, alloy, bal Fe.
For tools; oil hardening. *Obsolete*

TMO DREADNOUGHT
Crucible Specialty Metals
C, alloy, bal Fe.
For tools; oil hardening. *Obsolete*

TMS-20
Thomas Foundries, Inc.
11.0 Mn min, C, Cr, Si, bal Fe.
Austenitic manganese steel casting. Similar to ASTM A 128.

TMS-30
Thomas Foundries, Inc.
11.0 Mn min, C, Cr, Si, bal Fe.
Austenitic manganese steel casting; more Cr than TMS-20.
Similar to ASTM A 128.

TMS-40
Thomas Foundries, Inc.
11.0 Mn min, C, Cr, Si, bal Fe.
Austenitic manganese steel casting; more Cr than TMS-30.
Similar to ASTM A 128.

TMS-ASH
Thomas Foundries, Inc.
11.0 Mn min, C, Mo, Si, bal Fe.
Austenitic manganese steel casting. Similar to ASTM A 128.

TMS-MO
Thomas Foundries, Inc.
11.0 Mn min, C, Mo, Si, bal Fe.
Austenitic manganese steel casting. Similar to ASTM A 128.

TMS-NI
Thomas Foundries, Inc.
11.0 Mn min, C, Ni, Si, bal Fe.
Austenitic manganese steel casting. Similar to ASTM A 128
Grade D.

TMS-O
Thomas Foundries, Inc.
11.0 Mn min, C, Si, bal Fe.
Austenitic manganese steel casting. Similar to ASTM A 128.

TN-12
Techni-Cast Corp.
Copper. Cu 65, Sn 2, Pb 6, Zn 15, Ni 12.
Nickel silver. 40,000 psi TS; 19,000 psi YS; 30 El.

TN-20
Techni-Cast Corp.
Copper. Cu 64, Sn 4, Pb 4, Zn 8, Ni 20.
Nickel silver. 43,000 psi TS; 24,000 psi YS; 25 El.

TN-25
Techni-Cast Corp.
Copper. Cu 66, Sn 5, Pb 2, Zn 2, Ni 25.
Nickel silver. 55,000 psi TS; 30,000 psi YS; 16 El.

TNC-A
Techni-Cast Corp.
Nickel. Cu 31, Fe 0.75, Si 1.75, Ni 65.75, Mn 0.75.
Copper-silicon (Monel). As cast: 75,000 psi TS; 40,000 psi YS;
35 El; 141 Brin.

TNC-B
Techni-Cast Corp.
Nickel. Cu 31, Fe 1, Si 3.25, Ni 64, Mn 0.75.
Copper-silicon (Monel). As cast: 115,000 psi TS; 70,000 psi
YS; 15 El; 250 Brin.

TNC-C
Techni-Cast Corp.
Nickel. Cu 30, Fe 1.25, Si 3.75, Ni 64, Mn 1.
Copper-silicon (Monel). As cast: 130,000 psi TS; 90,000 psi
YS; 12 El; 276 Brin.

TNC-D
Techni-Cast Corp.
Nickel. Cu 30, Fe 1.25, Si 4.25, Ni 63.5, Mn 1.
Copper-silicon (Monel). As cast: 332 Brin.

TNC-E
Techni-Cast Corp.
Nickel. Cu 31, Fe 1, Si 1.75, Ni 64, Mn 0.75, Nb 1.5.
Copper-silicon (Monel). As cast: 75,000 psi TS; 40,000 psi YS;
35 El; 141 Brin.

TNW
Latrobe Steel Co.
C 0.87, Cr 4, V 1.9, Mo 8, bal Fe.
For form cutters, taps, drills; high speed steel.

TO1, FILTER GRADE
Keystone Carbon Co.
Sintered stainless steel manufactured to AISI 316
composition. 5.5 nominal density; 10,000 psi TS min. Pore
size: 30 microns (0.0012 in.). Permeability: 5.0 CFM air/in.^{2}
in a 10 psi press. diff. for 0.100 in. filter thickness.

TO2, FILTER GRADE
Keystone Carbon Co.
Sintered stainless steel manufactured to AISI 316
composition. 5.7 nominal density; 15,000 psi TS min. Pore
size: 10 microns (0.0004 in.). Permeability: 1.8 CFM air/in.^{2}
in a 10 psi press. diff. for 0.100 in. filter thickness.

TO3, FILTER GRADE
Keystone Carbon Co.
Sintered stainless steel manufactured to AISI 316
composition. 6.0 nominal density; 15,000 psi TS min. Pore
size: 5 microns (0.0002 in.). Permeability: 0.7 CFM air/in.^{2} in
a 10 psi press. diff. for 0.100 in. filter thickness.

TO4
Keystone Carbon Co.
Sintered stainless steel manufactured to AISI 316
composition. 6.40 min density; 55,000 psi TS; 7.0 El; 55 Rock
B. For corrosion resistant structural parts. ASTM B525-83A,
Type 2.

TO6
Keystone Carbon Co.
Sintered stainless steel manufactured to AISI 304
composition. 6.40 density min; 55,000 psi TS; 7.0 El; 55 Rock
B. For corrosion resistant structural parts.

TO9, FILTER GRADE
Keystone Carbon Co.
Sintered stainless steel manufactured to AISI 316
composition. 4.5 nominal density; 5,000 psi TS min. Pore
size 60 microns (0.0024 in). Permeability: 14.0 CFM air/in.^{2}
in a 10 psi press. diff. for 0.100 in. filter thickness.

TOBIN BRONZE 452
Now TOBIN BRONZE 4641.

TOBIN BRONZE 4641
Anaconda Co.
Copper. Cu 60, Sn 0.75, bal Zn.
Hard: 63,000 TS; 35,000 YS; 35 El; 100 Brin. Soft: 56,000 TS;
22,000 YS; 45 El; 59 Brin. For piston rods, propeller shafts.
Corrosion resistant

TOBIN BRONZE 470
Anaconda Co.
Copper. Cu 59, Sn 0.6, bal Zn.
For welding rod for oxyacetylene welding of steel, cast iron
and copper alloys.

TOBIN BRONZE NO. 1875
Manufacturer not listed.
Cu 58.2, Zn 39.5, Sn 2.3.
For piston rods, propeller shafts, bolts, nuts; corrosion
resistant to sea water.

TOH

Eagle & Globe Steel Ltd.
Tool material. C 0.93, Si 0.2, Mn 1.2, Cr 0.5, W 0.5, V 0.1, bal Fe.
Annealed: 223 Brin max. Oil hardening tool steel. For gauges, taps, cold punches, cold blanking and forming dies, reamers, drill bushes, straight edges, deep drawing and needle dies, engraving dies and templates. AS1239 O1A; AISI O1; BS4659 B01; Werkstoff 1.2510.

TOKUSHU-RHM10

Tokoshu Seiko Co. Ltd.
C 0.08-0.18, Cr 12-13.5, Mo 0.3-0.6, Ni 0-0.6, bal Fe.
Annealed: 70,000 TS; 35,000 YS; 30 El; 80 Rock B. Heat treated: 180,000 TS; 140,000 YS; 15 El; 375 Brin. For chemical plant and oil refinery equipment. Corrosion resistant.

TOKUSHU-RHM11

Tokoshu Seiko Co. Ltd.
C 0.1-0.2, Cr 11.5-13, Mo 0.6-1.5, Ni 0-0.5, bal Fe.
Annealed: 75,000 TS; 38,000 YS; 28 El; 82 Rock B. Heat treated: 135,000 TS; 105,000 YS; 10 El; 29 Rock C. For oil refinery and chemical plant equipment. Corrosion resistant.

TOKUSHU-RHM37

Tokoshu Seiko Co. Ltd.
C 0.08-0.18, Cr 11.5-14, Mo 0.3-0.6, Ni 0-0.6, bal Fe.
Annealed: 70,000 TS; 35,000 YS; 30 El; 80 Rock B. For oil refinery equipment, chemical and rubber processing apparatus. Corrosion resistant.

TOKUSHU-RHM5

Tokoshu Seiko Co. Ltd.
C 0.25-0.35, Cr 12-14, Mo 0.3-0.5, bal Fe.
Annealed: 75,000 TS; 40,000 YS; 25 El; 78 Rock B. Heat treated: 200,000 TS; 150,000 YS; 10 El; 45 Rock C. For oil refinery and chemical plant equipment, cutlery, tableware, knives. Corrosion resistant, hardenable.

TOKUSHU-RHM7

Tokoshu Seiko Co. Ltd.
C 0.15-0.2, Cr 11-12.5, Mo 0.9-1.2, Ni 0-0.7, bal Fe.
Annealed: 70,000 TS; 35,000 YS; 30 El; 80 Rock B. Heat treated: 190,000 YS; 145,000 YS; 15 El; 40 Rock C. For chemical plant and oil refinery equipment. Corrosion resistant, hardenable.

TOKUSHU-RHM8

Tokoshu Seiko Co. Ltd.
C 0.12-0.17, Cr 11-12.5, Mo 0.9-1.2, Ni 0-0.6, bal Fe.
Annealed: 70,000 TS; 35,000 YS; 30 El; 80 Rock B. Heat treated: 135,000 TS; 105,000 YS; 10 El; 29 Rock C. For chemical plant and oil refinery equipment. Corrosion resistant.

TOKUSHU-RHM9

Tokoshu Seiko Co. Ltd.
C 0-0.2, Cr 11.5-14, Mo 0.8-1.3, Ni 0-1, bal Fe.
Annealed: 70,000 TS; 35,000 YS; 30 El; 80 Rock B. Heat treated: 145,000 TS; 115,000 YS; 20 El; 300 Brin. For oil refinery and chemical plant equipment. Corrosion resistant, hardenable.

TOKUSHU-RHMV4

Tokoshu Seiko Co. Ltd.
C 0.2-0.25, Cr 11-12.5, Mo 0.9-1.2, Ni 0.4-0.8, V 0.25-0.35, bal Fe.
Annealed: 95,000 TS; 40,000 YS; 25 El; 92 Rock B.
Hardened: 240,000 TS; 205,000 YS; 9 El; 50 Rock C. For valves, cutlery, bearings, surgical instruments. Corrosion resistant, hardenable.

TOKUSHU-RHMW1

Tokoshu Seiko Co. Ltd.
C 0.2-0.25, Cr 11-12.5, Mo 0.9-1.25, Ni 0.5-0.9, V 0.25, W 1.1, Al 0.05, Co 0.25, Sn 0.04, Ti 0.05, bal Fe.
Annealed: 90,000 TS; 40,000 YS; 25 El; 94 Rock B.
Hardened: 250,000 TS; 210,000 YS; 8 El; 52 Rock C. For valves, cutlery, surgical instruments, bearings. Corrosion resistant, hardenable.

TOKUSHU-RHMW2

Tokoshu Seiko Co. Ltd.
C 0.18-0.23, Cr 12-14, Mo 0.75-1.25, Ni 0.5-1, V 0.2-0.5, W 0.8-1.2, bal Fe.
Annealed: 85,000 TS; 40,000 YS; 25 El; 94 Rock B. Heat treated: 240,000 TS; 205,000 YS; 9 El; 50 Rock C. For surgical instruments, valves, bearings, cutlery, shafts. Corrosion resistant, hardenable.

TOLEDO

Specialty Steel Co. of America
C 0.9, Mn 1.2, Cr 0.5, V 0.2, W 0.5, bal Fe.
Oil hardened cold work tool steel. AISI O1.

TOLEDO

Andrews Toledo Ltd.
C 1.47, Mn 0.48, W 4.16, bal Fe.
For tools, dies; oil hardened.

TOLEDO 1% CHROME

Darwin Tools Ltd.
C 0.35-0.45, Cr 0.85-1.15, Mn 0.6-0.95, bal Fe.
Heat treated: 102,000-133,000 TS; 72,000-92,000 YS; 18-22 El; 200-302 Brin. For gears, bolts, crankshafts, machine tool parts; Brit. En18, oil hardened.

TOLEDO 1% CHROME-MOLYBDENUM

Darwin Tools Ltd.
C 0.35-0.45, Cr 0.9-1.2, Mo 0.2-0.35, Mn 0.5-0.8, bal Fe.
For gears, bolts, crankshafts; Brit. En19A-B-C, oil hardened, tough.

TOLEDO 1% NI-CR

Darwin Tools Ltd.
C 0.3-0.4, Ni 1-1.5, Cr 0.45-0.75, bal Fe.
Heat treated: 100,000-135,000 TS; 72,000-103,000 YS; 17-22 El; 200-232 Brin. For gears, bolts, machine tool parts; Brit. En 111, oil hardened, shock resistant.

TOLEDO 1% NI-CR-MO

Darwin Tools Ltd.
C 0.35-0.45, Ni 1.2-1.6, Cr 0.9-1.4, Mo 0.1-0.2, bal Fe.
Heat treated: 112,000-157,000 TS; 81,000-123,000 YS; 15-20 El; 223-237 Brin. For gears, bolts, crankshafts, machine tool parts; Brit. En110, oil hardened, shock resistant.

TOLEDO 13

Darwin Tools Ltd.
C 0.15-0.25, Mo 0.15-0.35, Ni 0.4-0.7, Mn 1.4-1.8, bal Fe.
Heat treated: 90,000 TS; 22 El; 180-230 Brin. For gears, cams, camshafts; Brin. En13, case hardened.

TOLEDO 15

Darwin Tools Ltd.
C 0.15-0.25, Mn 0.4-0.6, bal Fe.
Annealed: 73,000 TS; 41,000 YS; 22 El; 58 RA; 140 Brin. For nails, rivets, gears, cams, bushings; Brit. EN2C, case hardened.

TOLEDO 160

Darwin Tools Ltd.
C 0.35-0.45, Ni 1.5-2, Mo 0.2-0.35, bal Fe.
Heat treated: 102,000-134,000 TS; 72,000-104,000 YS; 17-22 El; 200-300 Brin. For gears, bolts, crankshafts, machine tool parts; Brit. EN160, oil hardened, shock resistant.

TOLEDO 20

Darwin Tools Ltd.
C 0.2-0.25, Si 0.05-0.35, Mn 0.6-1, bal Fe.
Cold drawn: 60,000 TS; 25 El; 140 Brin. For gears, bolts, fasteners, brackets; Brit. EN3C, water hardened.

TOLEDO 206

Darwin Tools Ltd.
C 0.12-0.17, Cr 0.3-0.5, Mn 0.3-0.5, bal Fe.
For gears, fasteners, machine tool parts; Brit. EN206, case hardened.

TOLEDO 207

Darwin Tools Ltd.
C 0.16-0.21, Cr 0.6-0.8, Mn 0.6-0.8, bal Fe.
For gears, fasteners, machine tool parts; Brit. EN207, case hardened.

TOLEDO 25

Darwin Tools Ltd.
C 0.25-0.3, Si 0.05-0.35, Mn 0-1, bal Fe.
Normalized: 80,000 TS; 25 El; 179 Brin. For gears, bolts, shafts, keys, brackets; Brit. EN4, water hardened.

TOLEDO 28

Darwin Tools Ltd.
C 0.25-0.4, Ni 3-4.5, Cr 0.75-1.25, Mo 0.2-0.65, bal Fe.
Heat treated: 135,000-180,000 TS; 103,000-144,000 YS; 25-35 El; 270-415 Brin. For gears, bolts, machine tool parts; Brit. EN 28, oil hardened, shock resistant.

TOLEDO 3% CHROME MOLYBDENUM

Darwin Tools Ltd.
C 0.15-0.35, Cr 2.5-3.5, Mo 0.3-0.7, bal Fe.
Heat treated: 101,000-224,000 TS; 72,000-180,000 YS; 10-22 El; 200-205 Brin. For oil refinery equipment; Brit. En29, creep resistant.

TOLEDO 3% NI-CR-MO

Darwin Tools Ltd.
Ni 3, C, Cr, Mo, bal Fe.
For gears, bolts, camshafts, cams; Brit. S107, case hardened.

TOLEDO 3/4% NICKEL

Darwin Tools Ltd.
C 0.5-0.6, Ni 0.5-0.8, Mn 0.5-0.8, bal Fe.
For gears, bolts, cranskshafts, machine tool parts; oil or water hardened, tough.

TOLEDO 30

Darwin Tools Ltd.
C 0.25-0.35, Si 0.05-0.35, Mn 0.6-1, bal Fe.
Hot rolled: 80,000 TS; 50,000 YS; 30 El; 56 RA; 163 Brin. For armature shafts, gears, bolts, axles, screws; Brit. EN5, water hardened.

TOLEDO 325

Darwin Tools Ltd.
C 0.22, Ni 1.5-2, Cr 0.4-0.6, Mo 0.2-0.3, bal Fe.
Heat treated: 123,000 TS; 15 El. For gears, bolts, cams, machine tool parts; Brit EN325, case hardened.

TOLEDO 351

Darwin Tools Ltd.
C 0-0.2, Ni 0.6-1, Cr 0.4-0.8, Mo 0-0.1, bal Fe.
Heat treated: 101,000 TS; 18 El. For gears, bolts, cams, machine tool parts; Brit. EN351, case hardened, shock resistant.

TOLEDO 352

Darwin Tools Ltd.
C 0-0.2, Ni 0.85-1.25, Cr 0.6-1, Mo 0-0.1, bal Fe.
Heat treated: 123,000 TS; 15 El. For gears, bolts, cams, machine tool parts; Brit. EN352, case hardened, shock resistant.

TOLEDO 353

Darwin Tools Ltd.
C 0-0.2, Ni 1-1.5, Cr 0.75-1.25, Mo 0.08-0.15, bal Fe.
Heat treated: 146,000 TS; 12 El. For gears, bolts, cams, machine tool parts; Brit. EN353, case hardened, shock resistant.

TOLEDO 354

Darwin Tools Ltd.
C 0-0.2, Ni 1.5-2, Cr 0.75-1.25, Mo 0.1-0.2, bal Fe.
Heat treated: 168,000 TS; 12 El. For gears, bolts, cams, machine tool parts; Brit. EN354, case hardened, shock resistant.

TOLEDO 355
Darwin Tools Ltd.
C 0-0.2, Ni 1.8-2.2, Cr 1.4-1.7, Mo 0.15-0.25, bal Fe.
Heat treated: 190,000 TS; 12 El. For gears, bolts, machine tool parts; Brit. EN355, case hardened, shock resistant.

TOLEDO 361
Darwin Tools Ltd.
C 0.13-0.17, Ni 0.4-0.7, Cr 0.55-0.85, Mo 0.08-0.15, bal Fe.
Heat treated: 100,000 TS; 18 El. For gears, bolts, machine tool parts; Brit. En361, case hardened, shock resistant.

TOLEDO 362
Darwin Tools Ltd.
C 0.18-0.23, Ni 0.4-0.7, Cr 0.55-0.8, Mo 0.08-0.15, bal Fe.
Heat treated: 123,000 TS; 15 El. Heat gears, bolts, camshafts, cams; Brit. EN362, case hardened, shock resistant.

TOLEDO 363
Darwin Tools Ltd.
C 0.22-0.26, Ni 0.4-0.7, Cr 0.55-0.8, Mo 0.08-0.15, bal Fe.
Heat treated: 146,000 TS. For gears, bolts, crankshafts, axles; Brit. EN363, case hardened, shock resistant.

TOLEDO 40
Darwin Tools Ltd.
C 0.35-0.45, Si 0.05-0.35, Mn 0.6-1, bal Fe.
Hot rolled: 91,000 TS; 58,000 YS; 27 El; 50 RA; 200 Brin. For gears, shafts, bolts, axles, fasteners; Brit. EN8, water hardened.

TOLEDO 55
Darwin Tools Ltd.
C 0.5-0.65, Si 0.05-0.35, Mn 0.6-0.8, bal Fe.
Cold drawn: 110,000-140,000 TS; 12 El; 223-302 Brin. For wheels, die blocks, rails, girders, springs; Brit. EN9, water hardened.

TOLEDO 60 CARBON CHROME
Darwin Tools Ltd.
C 0.5-0.7, Cr 0.5-0.8, Mn 0.5-0.8, bal Fe.
Heat treated: 123,000-155,000 TS; 15-21 El. For axes, hammers, punches, die blocks; Brit. EN11, water or oil hardened.

TOLEDO ALLOY 131
National Castings, Inc.
C 0.22-0.28, Cu 1.1, V 0.06, Si 0.5, Mn 0.8, bal Fe.
Cast: 90,000-100,000 TS. For machinery parts, gears, bolts; water hardened.

TOLEDO ALLOY 135
National Castings, Inc.
C 0.4-0.5, Mn 0.8, Cr 1, Cu 1, Mo 0.3, bal Fe.
Cast: 120,000 TS; 217 Brin. For guides, bolts, gears, housings; water or oil hardened.

TOLEDO B.C.H.
Darwin Tools Ltd.
C 0-0.18, Ni 3-3.7, Cr 0.6-1.1, Mo 0-0.25, bal Fe.
Heat treated: 123,000-156,000 TS; 13 El. For gears, camshafts, cams; Brit. EN36, case hardened, shock resistant.

TOLEDO B.C.M.F.
Darwin Tools Ltd.
C 0.27-0.44, Ni 2.3-2.8, Cr 0.5-0.8, Mo 0.4-0.7, bal Fe.
Heat treated: 123,000-224,000 TS; 92,000-180,000 YS; 10-18 El; 248-500 Brin. For gears, bolts, shafts, axles, machine tool parts; Brit. EN25 and 26, oil hardened, shock resistant.

TOLEDO B.C.M.O.
Darwin Tools Ltd.
C 0.25-0.35, Ni 3-3.7, Cr 0.5-1.3, Mo 0.2-0.65, bal Fe.
Heat treated: 123,000-157,000 TS; 92,000-123,000 YS; 15-18 El; 248-375 Brin. For gears, bolts, machine tool parts, axles; Brit. EN27, oil hardened, shock resistant.

TOLEDO B.C.N.
Darwin Tools Ltd.
C 0.25-0.35, Cr 0.5-1, Ni 2.75-3.5, Mo 0-0.65, bal Fe.
Heat treated: 112,000-146,000 TS; 81,000-112,000 YS; 16-20 El; 223-340 Brin. For gears, bolts, crankshafts, machine tool parts; Brit. EN23, oil hardened, shock resistant.

TOLEDO B.R.
Darwin Tools Ltd.
C 0.9-1.2, Mn 0.3-0.75, Cr 1-1.6, bal Fe.
For bearings, races, gauges, cutters; Brit. EN31, wear resistant, water hardened.

TOLEDO B.S.N.1
Darwin Tools Ltd.
C 0.3-0.45, Ni 0.6-1, Mn 1.5, Si 0.1-0.35, bal Fe.
Normalized: 90,000 TS; 22 El; 179-229 Brin. For gears, bolts, crankshafts, machine tool parts; Brit. EN12, oil hardened, shock resistant.

TOLEDO B.S.N.3
Darwin Tools Ltd.
C 0.25-0.35, Ni 2.75-3.25, Cr 0-0.3, bal Fe.
Rolled: 100,000 TS; 70,000 YS; 22 El; 200-235 Brin. For gears, bolts, crankshafts, axles; Brit. EN21, oil hardened, shock resistant.

TOLEDO B.S.N.35
Darwin Tools Ltd.
C 0.35-0.45, Ni 3.25-3.75, Mn 0.5-0.8, Cr 0-0.3, bal Fe.
Rolled: 110,000-125,000 TS; 75,000-88,000 YS; 18-20 El; 220-302 Brin. For gears, bolts, crankshafts, axles; Brit. EN22, oil hardened, shock resistant.

TOLEDO CARBON FILE
Darwin Tools Ltd.
C 0.8, bal Fe.
For files; water hardened.

TOLEDO CARBON SPRING
Darwin Tools Ltd.
C 0.45-0.9, Si 0.1-0.4, Mn 0.55-0.9, bal Fe.
For laminated and coil springs; Brit. EN42 and 43, water hardened.

TOLEDO CARBON-MANGANESE
Darwin Tools Ltd.
C 0.15-0.25, Mn 1.3-1.7, Cr 0.25, Ni 0-0.4, bal Fe.
Normalized: 80,000 TS; 20 El; 152-207 Brin. For gears, cams, camshafts, fasteners; Brit. EN14A, water hardened.

TOLEDO CHROME FILE
Darwin Tools Ltd.
C 0.8, Cr, bal Fe.
For files, water hardened.

TOLEDO CHROME SPRING
Darwin Tools Ltd.
C 0.45-0.6, Si 1-1.6, Mn 0.5-0.9, Cr 0.55-1.4, bal Fe.
For springs; Brit. EN48, oil or water hardened.

TOLEDO CHROME VANADIUM VALVE SPRING
Darwin Tools Ltd.
C 0.4-0.5, Cr 1-1.5, Mn 0.5-0.7, 0.15 V min, bal Fe.
For valve springs; Brit. EN50, oil hardened, tough.

TOLEDO EXTRA
Andrews Toledo Ltd.
C, bal Fe.
For tools, fixtures, jigs; water hardened.

TOLEDO G 105
Darwin Tools Ltd.
C 0.35-0.45, Ni 1.3-1.6, Cr 0.9-1.4, Mo 0.2-0.35, bal Fe.
Heat treated: 112,000-224,000 TS; 81,000-179,000 YS; 8-20 El; 223-500 Brin. For gears, bolts, crankshafts, machine tool parts; Brit. EN24, oil hardened, shock resistant.

TOLEDO G 110
Darwin Tools Ltd.
C 0.26-0.34, Ni 3.9-4.3, Cr 1.1-1.4, Mo 0-0.4, bal Fe.
Heat treated: 224,000 TS; 180,000 YS; 10 El; 440-500 Brin. For gears, bolts, crankshafts; Brit. EN30A, oil hardened, shock resistant.

TOLEDO H-O SILICOMANGANESE
Darwin Tools Ltd.
C 0.5-0.6, Si 1.5-2, Mn 0.7-1, bal Fe.
For laminated springs, torsion bars; Brit. EN45 and 46, water or oil hardened, tough.

TOLEDO HARD DRAWING CARBON SPRING
Darwin Tools Ltd.
C 0.45-0.85, Mn 0-1, Si 0-0.3, bal Fe.
For springs; Brit. EN49, water hardened.

TOLEDO HIGH CARBON SPRING
Darwin Tools Ltd.
C 0.9-1.2, Si 0-0.3, Mn 0.45-0.7, bal Fe.
For laminated and coil springs; Brit. EN44, water hardened.

TOLEDO HIGH SPEED
Andrews Toledo Ltd.
C 0.7, W 18, Cr 4, V 1, bal Fe.
For cutting tools, broaches; high speed steel.

TOLEDO LANTERN
Darwin Tools Ltd.
C 0-0.15, Si 0.05-0.35, Mn 0.4-0.7, bal Fe.
Rolled: 70,000 TS; 20 El; 140 Brin. For gears, bolts, rivets, nails, fan blades; Brit. EN32A, case hardened.

TOLEDO LOW ALLOY STEEL
Darwin Tools Ltd.
C 0.25-0.45, Mn 1.2-1.5, Ni 0.5-1, Cr 0.3-0.6, Mo 0.15-0.25, bal Fe.
For gears, bolts, machine tool parts; Brit. EN100, oil hardened, shock resistant.

TOLEDO MANGANESE MOLYBDENUM
Darwin Tools Ltd.
C 0.25-0.4, Mn 1.3-1.8, Mo 0.2-0.55, bal Fe.
Heat treated: 101,000-146,000 TS; 72,000-112,000 YS; 16-22 El; 200-234 Brin. For gears, bolts, crankshafts, machine tool parts; Brit. EN 16 and 17, oil hardened, shock resistant.

TOLEDO N.T.R.1
Darwin Tools Ltd.
C 0.1-0.3, Ni 0-0.4, Cr 2.9-3.5, Mo 0.4-0.7, bal Fe.
Heat treated: 100,000-135,000 TS; 72,000-103,000 YS; 17-22 El; 200-277 Brin. For gears, bolts, shafts, machine tool parts; Brit. EN40A, nitriding steel.

TOLEDO N.T.R.2
Darwin Tools Ltd.
C 0.3-0.5, Ni 0-0.4, Cr 2.5-3.5, Mo 0.7-1.2, bal Fe.
Heat treated: 190,000 TS; 153,000 YS; 10 El; 375-444 Brin. For gears, bolts, shafts, machine tool parts; Brit. EN40C nitriding steel.

TOLEDO N.T.R.3
Darwin Tools Ltd.
C 0.25-0.45, Ni 0-0.4, Cr 1.4-1.8, Mo 0.1-0.25, bal Fe.
Heat treated: 100,000-123,000 TS; 72,000-92,000 YS; 17-20 El; 200-302 Brin. For gears, bolts, shafts, machine tool parts; Brit. EN41, nitriding steel.

TOLEDO NO. 3
National Castings, Inc.
C 0.15, Ni 1.5, Mo 0.2, bal Fe.
Hardened: 110,000-115,000 TS; 70,000-80,000 YS; 24-26 El; 40-50 RA. For mining equipment, carburized parts; case-hardened steel.

TOLEDO NO. 4
National Castings, Inc.
C 0.7, Si 2, Mo 0.2, bal Fe.
90,000-115,000 TS; 60,000-75,000 YS; 207-228 Brin. For coal mining equipment, pug mill parts, paddles, liners; abrasion resistant.

TOLEDO NO. 6
National Castings, Inc.
C 0.9, Mn 1.2, bal Fe.
For dies, trimming, stamping, forming and embossing; nondeforming, cast to shape.

TOLEDO NO. 7
National Castings, Inc.
C 0.5, Cr 1.2, V 0.2, bal Fe.
90,000-95,000 TS; 60,000-67,000 YS; 22-25 El; 40-50 RA; 179-200 Brin. For railway locomotive parts; tough, shock resistant.

TOLEDO NO. 8
National Castings, Inc.
C 0.4, Ni 1.5, Cr 0.8, Mo 0.2, bal Fe.
110,000-120,000 TS; 75,000-85,000 YS; 20-26 El; 40-50 RA; 228-241 Brin. For automotive and aircraft parts, gears, short-run dies; oil hardening.

TOLEDO NON-SHRINKING
Andrews Toledo Ltd.
C 0.9, Mn 1.2, V 0.2, bal Fe.
For tools and dies; non-deforming.

TOLEDO R.S.2
Darwin Tools Ltd.
C 0.14-0.28, Ni 1.5-2, Mo 0.2-0.3, Mn 0.3-0.6, bal Fe.
Heat treated: 123,000 TS; 15 El. For machine tool parts, gears, bolts, cams; Brit. EN34 and 35, case hardened.

TOLEDO R.S.3
Darwin Tools Ltd.
C 0.1-0.15, Ni 2.75-3.5, C 0-0.3, bal Fe.
Rolled: 100,000 TS; 18 El. For gears, bolts, camshafts, cams; Brit. EN33, case hardened.

TOLEDO R.S.5
Darwin Tools Ltd.
C 0-0.16, Ni 4.5-5.5, Cr 0-0.3, bal Fe.
Rolled: 90,000 TS; 20 El. For gears, cams, camshafts; Brit. EN37, case hardened.

TOLEDO S.82
Darwin Tools Ltd.
C 0.12-0.18, Ni 3.8-4.5, Cr 1-1.4, Mo 0-0.35, bal Fe.
Heat treated: 190,000 TS; 12 El. For gears, bolts, camshafts, cams; Brit. EN39, case hardened, shock resistant.

TOLEDO S.90
Darwin Tools Ltd.
C 0-0.16, Ni 4.5-5.5, Mo 0.15-0.35, Cr 0-0.3, bal Fe.
Heat treated: 146,000 TS; 13 El. For gears, cams, camshafts; Brit. EN38, case hardened.

TOLEDO SCIMITAR
Andrews Toledo Ltd.
C, W, Cr, bal Fe.
For tools, dies; oil hardened.

TOLEDO SILICO CHROME SPRING
Darwin Tools Ltd.
C, Si, Cr, bal Fe.
For spiral and coil springs; Brit. STA.2D, oil hardened, tough.

TOLEDO SPECIAL
Andrews Toledo Ltd.
C 0.7-1.2, bal Fe.
For general tools; water hardened.

TOLEDO SUPERIOR
Andrews Toledo Ltd.
C 0.7-1.2, bal Fe.
For general tools; water hardened.

TOLEDO T.P.C.
Darwin Tools Ltd.
C 0.45-0.55, Cr 0.8-1.2, 0.15 V min, bal Fe.
For springs; Brit. EN47, water hardened.

TOLEDO X.B.P.
Darwins Alloy Castings
C 0.65-1.25, bal Fe.
For drills, taps, reamers, hobs; water hardened. *Obsolete*

TOLEDO-MOLYBDENUM BOLT STEEL
Darwin Tools Ltd.
C 0.2-0.45, Cr 0.5-1.5, Mo 0.5-0.9, Mn 0.4-0.7, bal Fe.
Heat treated: 123,000-146,000 TS; 92,000-112,000 YS; 16-18 El; 236-248 Brin. For gears, bolts, machine tool parts; Brit. EN20A-B, oil hardened, tough.

TOM ARC 10018MM
Chemetron Corp.
Mn 1.77, Si 0.68, Mo 0.44, weld metal: 0.06 C, bal Fe.
As welded: 106,000 psi TS; 101,000 psi YS; 22 El. Iron powder, low hydrogen electrode for welding low-alloy, high-tensile steels requiring weld strengths of 100,000 psi TS. AWS Class E 10018-D2.

TOMBAC LUDENSCHEIDT
Manufacturer not listed.
Cu 82.3, Zn 17.5.
For pipes; red brass.

TOMBAC, ARCET
Manufacturer not listed.
Cu 82.3, Zn 17.7.
For water pipes; red brass.

TOMBAC, CAST
Manufacturer not listed.
Cu 87, Zn 13.
For hardware; red brass.

TOMBAC, GALDEN
Manufacturer not listed.
Cu 87, Zn 17.5.
For pipes, hardware; red brass.

TOMBAC, OKER
Manufacturer not listed.
Cu 85.3, Zn 14.7.
For pipes; red brass.

TOMBAC, RED BRASS
Manufacturer not listed.
Zn 6, Cu 85, Sn 9.
Hard: 75,000 TS; 18,000 YS; 4 El; 135 Brin. Soft: 42,000 TS; 7,000 YS; 43 El; 52 Brin. For hardware, ornaments, tubing, marine parts; corrosion resistant.

TOMBAC, RED VIENNA
Australian manufacture
Zn 2, Cu 98.
For jewelry, bearings; corrosion resistant.

TOMBAC, SHEET, PARIS
Manufacturer not listed.
Cu 84-92, Zn 8-16.
For hardware, tubes; red brass.

TOMBAC-A
French manufacture
Zn 20, Cu 80.
38,000 TS; 30 El. For jewelry, bearings; corrosion resistant.

TOMBAC-B
French manufacture
Zn 17-20, Sn 0.3, Cu 80.
For jewelry, bearings; corrosion resistant.

TOMBASIL
Illingworth Steel Co.
Si 4, Zn 13, bal Cu.
Sand cast: 65,000 TS; 30,000 YS; 20 El; 130 Brin. Die cast: 90,000 TS; 50,000 YS; 25 El; 150 Brin. Corrosion resistant bronze; for valve stems, brush holders, pump impellers, structural castings.

TOMBASIL
Cerro Metal Products Co.
Cu 81, Si 4, bal Zn.
Rolled: 98,000 TS; 45,000 YS; 20 El; 22 RA; 185 Brin. For gears, shafts, and propellers. Corrosion and wear resistant. *Obsolete*

TONCAN IRON
Republic Steel Corp.
C 0.03, Mn 0.12, S 0.035, P 0.01, Mo 0.07, Si 0.005, Cu 0.45, bal Fe.
Normalized: 50,000 TS; 40,000 YS; 35 El; 110 Brin. For highway culverts, roofing, siding. Atmosphere corrosion resistant. Was Toncan Copper Molybdenum. *Obsolete*

TOOL
Manufacturer not listed.
C 0.9, V 0.2, bal Fe.
For drills, chisels, crow-bars; water or oil hardened.

TOOL HOLDER
Midvale-Heppenstall Co.
C 1.2, Cr 3, W 13, Mo 1.2, bal Fe.
For cutting tools. *Obsolete*

TOOL STEEL TYPE S-7
Slater Steels
Stainless steel. C 0.45-0.55, Mn 0.2-0.8, Si 0.2-1, Cr 3-3.5, Mo 1.3-1.8, V 0.2-0.3, bal Fe.
Medium carbon, air hardening, tool steel having high strength, good impact resistance, and high toughness. Good resistance to softening at moderately high temperatures. Useful in a variety of both cold work and hot work tool steel applications, such as chisels, rivet sets, cold punches, hot punches and shear knives.

TOOL-AGE
Now MCKAY TOOL-AGE.

TOOL-ALLOY
Now MCKAY TOOL-ALLOY.

TOOL-ARC AIR HARDENING
Chemetron Corp.
C 1, Cr 5, Mo 1, V 0.2, bal Fe.
For tool steel welding electrodes; for air hardening die steels.

TOOL-ARC HIGH SPEED
Chemetron Corp.
C 0.7, W 1.5, Mo 9.5, V 1, bal Fe.
For tool steel welding electrodes; for cutting tools.

TOOL-ARC HOT-WORK
Chemetron Corp.
C 0.3, Cr 5, Mo 1.3, W 1, bal Fe.
For tool steel welding electrodes; for hot die steels.

TOOL-ARC OIL HARDENING
Chemetron Corp.
C 0.9, Mn 1.2, Cr 0.5, W 0.5, bal Fe.
For tool steel welding electrodes; for non-deforming steel tools and dies.

TOOL-ARC WATER HARDENING
Chemetron Corp.
C 0.9, V 2, bal Fe.
For tool steel welding electrodes; for water hardening steel parts.

TOOL-FORGE
Now MCKAY TOOL-FORGE.

TOOL-N-DIE NO. 10-WH

Hobart Bros. Co.
High C, bal Fe.
Heat treated: 600 Brin. For welding electrodes for C tool steels; coated. *Obsolete*

TOOL-N-DIE NO. 15-AH

Hobart Bros. Co.
Cr 5, C, bal Fe.
Heat treated: 610 Brin. For welding electrodes for air hardening steels; coated, air hardening. *Obsolete*

TOOL-N-DIE NO. 16-HS

Hobart Bros. Co.
W 6, Mo 5, C, V, bal Fe.
Heat treated: 650 Brin. For welding electrodes for Mo high speed steel; coated. *Obsolete*

TOOL-N-DIE NO. 24-HA

Hobart Bros. Co.
C, alloy, bal Fe.
Welded: 250 Brin. For welding electrodes for alloy steels; coated, austenitic. *Obsolete*

TOOL-N-DIE NO. 34-DA

Hobart Bros. Co.
C, alloy, bal Fe.
Welded: 350 Brin. For welding electrodes for alloy steels; coated, austenitic. *Obsolete*

TOOL-N-DIE NO. 70-OH

Hobart Bros. Co.
High C, bal Fe.
Heat treated: 620 Brin. For welding electrodes for tool steel; coated. *Obsolete*

TOOL-N-DIE NO. 71-OH

Hobart Bros. Co.
Mn 1.2, High C, bal Fe.
Heat treated: 600 Brin. For welding electrodes for Mn tool steel; coated. *Obsolete*

TOOL-N-DIE NO. 72-HW

Hobart Bros. Co.
C, Cr, Mo, W, bal Fe.
Heat treated: 540 Brin. For welding electrodes for hot work steel; coated, hot work steel. *Obsolete*

TOOL-N-DIE NO. 73-HW

Hobart Bros. Co.
C, Mo, Cr, bal Fe.
Heat treated: 580 Brin. For welding electrodes for hot work steels; coated, hot work steel. *Obsolete*

TOOL-ROD NO. 650

Marquette Corp.
C 0.5-0.7, W 3.4, Cr 6-8, V 0.8, Mn 0.4, bal Fe.
Welded: 500 Brin. For hard facing electrode; for tool steels, for wear resistance.

TOOLCRAFT

E.M.F. Electric Co. Ltd.
C 1, Cr 1, bal Fe.
Welded: 600 Brin. For hard facing electrodes, cutting tools; wear resistant.

TOOLTEC 10675

Eutectic Corp.
Nickel base alloy powder for steel, cast iron, stainless, nickel alloys. *Obsolete*

TOOLTECTIC 6 HSS

Eutectic Corp.
For high speed steel overlays on tools by AC-DC reverse arc. 62 Rock C.

TOOLTECTIC 60H

Eutectic Corp.
For wear resistant overlays on steel machine parts. 59-62 Rock C.

TOOLTECTIC 623

Eutectic Corp.
For arc build-up on cast iron. 58,000 TS. *Obsolete*

TOOLTECTIC 6HW

Eutectic Corp.
For hard overlays on hot working punches. 52-55 Rock C.

TOOLTECTIC 6SH

Eutectic Corp.
For hard overlays on steels. 53-61 Rock C.

TOOLTECTIC 6WH

Eutectic Corp.
For hard overlays on water hardening tools. 56-59 Rock C.
Obsolete

TOOLWELD A & O

Lincoln Electric Co.
Cr 5, C, bal Fe.
Welded: 650 Brin. For hard surfacing electrodes; wear resistant; shielded arc.

TOP NOTCH (AISI S1)

Jessop Steel Co.
C 0.5, W 2.5, Cr 1.25, V 0.25, bal Fe.
For tools, chisels, track tools; shock resistant.

TOPAL "W"

Allgemeines Deutsches Metallwerk GmbH
Copper. Mn 0.5, Fe 1, Al 9, bal Cu.
Cast: 50,000-60,000 TS; 15-30 El; 130-150 Brin. Hot pressed: 60,000-80,000 TS; 15-30 El; 150-170 Brin. For shafts, bearings, dies, gears. Resists steam at 550°F.

TOPAL H

Allgemeines Deutsches Metallwerk GmbH
Copper. Cu 80, 20 Mn + Fe + Al.
Chill cast: 90,000 TS; 5 El; 170 Brin. Hot pressed: 114,000 TS; 5 El; 200 Brin. For shafts, bearings, dies, pump rods, steering nuts, gears. Noncorrosive.

TOPHAL X

Gilby-Fodor S.A.
Cr 35, Al 3, Fe 62.
Annealed: 85,000 TS; 57,000 YS; 20 El. For heating elements and induction furnaces.

TOPHAL Y

Gilby-Fodor S.A.
Cr 12, Al 5, bal Fe.
Annealed: 80,000 TS; 65,000 YS; 15 El. For heating elements. Heat resistant.

TOPHAL Z

Gilby-Fodor S.A.
Cr 22, Al 5, bal Fe.
Annealed: 85,000 TS; 65,000 YS; 20 El. For heating elements and electric furnaces.

TOPHEL

Wilbur B. Driver Co.
Nickel. Cr 10, bal Ni.
Wire: 88,000 TS; 30,000 YS; 35 El. For positive thermoelement for the standard Type K thermocouple. Oxidation resistant and stable.

TOPHEL ALLOY

Carpenter Technology Corp.
Ni 90, Cr 10.
At 20°C: 30,000 psi YS; 88,000 psi TS; 45 El. For thermocouples, extension wires.

TOPHEL II

Wilbur B. Driver Co.
Ni 90, Cr 10.
For positive thermoelement for the standard Type K thermocouple. Resists preferential oxidation of chromium. Oxidation resistant and stable. *Obsolete*

TOPHET "A"

Wilbur B. Driver Co.
Nickel. Ni 77-79, Cr 19-20.5, C 0-0.15, Mn 2.
Hard: 160,000 TS; 90,000 YS; 3 El; 45 RA. Soft: 120,000 TS; 70,000 YS; 30 El; 45 RA; 190 Brin. For resistance wire, heating elements, percolators, toasters, rheostats; tough to machine; maximum operating temperature 2100°F.

TOPHET "C"

Wilbur B. Driver Co.
Nickel. Ni 57-60, Cr 14-18, C 0-0.15, bal Fe.
Hard: 100,000-140,000 TS; 85,000 YS; 3-100 El; 16 RA. Soft: 100,000 TS; 60,000 YS; 25 El; 45 RA; 165 Brin. For resistance wire, heating elements, percolators, toasters, rheostats; resists some common acids, maximum operating temperature 1850°F.

TOPHET 30

Wilbur B. Driver Co.
Nickel. Ni 70, Cr 30.
For heating elements and furnace belts; maximum operating temperature 2300°F. Resists green rot.

TOPHET A

Gilby-Fodor S.A.
Ni 80, Cr 20.
Annealed; 105,000 TS; 70,000 YS; 40 El. For heating elements. For high temperatures.

TOPHET A-CB

Wilbur B. Driver Co.
Nickel. Cr 19, Cb 1.3, Si 1.5, bal Ni.
For high temperature furnace belts.

TOPHET ALLOY 30

Carpenter Technology Corp.
Ni 70, Cr 30.
896 MPa TS; 483 MPa YS; 35 El. Resists oxidation and green rot, maintains high strength at temperatures up to 1260°C. For use in industrial furnaces with exothermic, hydrogen, air and dissociated ammonia atmospheres.

TOPHET ALLOY A

Carpenter Technology Corp.
Ni 80, Cr 20.
827 MPa TS; 414 MPa YS; 35 El. Resists acid and alkaline solutions and carburizing atmospheres. For use in resistance heating applications and in baskets, belts and other hardware for the chemical industry. ASTM B-344, B-267.

TOPHET ALLOY C

Carpenter Technology Corp.
Ni 60, Cr 16, bal Fe.
758 MPa TS; 379 MPa YS; 35 El. Resists heat up to 1010°C. For domestic appliances, heavy duty rheostats and controls, edge wound power resistors. ASTM B-344, B-267.

TOPHET ALLOY D

Carpenter Technology Corp.
Ni 35, Cr 20, bal Fe.
621 MPa TS; 345 MPa YS; 35 El. Used in furnace elements between 816 and 982°C and in appliances and heating applications such as baseboard heaters operating near black heat. Resists green rot. ASTM B-344.

TOPHET B

Wilbur B. Driver Co.
C 0.1-0.15, Ni 68-71, Cr 19-20.5, bal Fe.
Annealed: 100,000 TS; 60,000 YS; 30 El; 45 RA. Drawn: 160,000 TS; 85,000 YS; 3 El. For heat and corrosion resistant parts; conveyor belts; heat and corrosion resistant. *Obsolete*

TOPHET C

Gilby-Fodor S.A.
Ni 60, Cr 16, bal Fe.
Annealed: 105,000 TS; 70,000 YS; 30 El. For heating elements and domestic use.

TOPHET D

Wilbur B. Driver Co.
Superalloy. Ni 34, Cr 20, C 0.15, bal Fe.
Annealed: 75,000 TS; 20 El. For moderate temperature service; used up to 1600°F.

TOPHET D-CB

Wilbur B. Driver Co.
Superalloy. Ni 35, Cr 20, Si 1.5, Cb 1.25, bal Fe.
For high temperature furnace belts.

TOPHET E

Gilby-Fodor S.A.
Ni 75.5, Cr 20, Ti 2.5, Si 0.5, Mn 0.5, Al 0.5.
Hardened: 280,000 TS; 210,000 YS; 2.5 El. For springs and gas turbine accessories. Age-hardenable.

TOPHET I-304

Gilby-Fodor S.A.
C 0-0.08, Cr 18-20, Ni 8-10, bal Fe.
Annealed: 85,000 TS; 30,000 YS; 60 El; 70 RA; 140-160 Brin. For tubing for chemical and food industries. Stainless; austenitic.

TOPHET I-316

Gilby-Fodor S.A.
C 0-0.1, Cr 16-18, Ni 10-14, Mo 2.5, bal Fe.
Annealed: 80,000 TS; 30,000 YS; 50 El; 70 RA; 140-160 Brin. For tubing for chemical and food industries. Stainless; austenitic.

TOPHET M

Gilby-Fodor S.A.
Ni 8, Cr 18, bal Fe.
290,000 TS; 1 El. For magnetic recording wire. Corrosion resistant.

TOPHET M-5

Wilbur B. Driver Co.
Ni 65, Cr 30, Mo 5.
Wire: 130,000-190,000 TS; 66,00-155,000 YS; 25-39 El. For pickling hooks, wire cloth, containers, baskets, reaction vessels. Corrosion resistant in strong oxidizing acids. *Obsolete*

TOPHET X

Wilbur B. Driver Co.
C 0.13-0.2, Ni 58-66, Cr 7-11, bal Fe.
For heat and corrosion resistant parts; heat and corrosion resistant. *Obsolete*

TOPHET-F

Gilby-Fodor S.A.
Ni 36-38, Cr 17-19, Si 2, bal Fe.
For heating elements. Useful operating temperature up to 1000°C.

TOPHET-F

Wilbur B. Driver Co.
Nickel. Ni 77, Cr 20, Al 3.
Wire: 100,000-200,000 TS; 60,000-175,000 YS; 2-30 El. For furnace belts in strongly reducing atmospheres at 1800-2100°F. High heat and corrosion resistance.

TOPHET-H

Gilby-Fodor S.A.
Ni 75-77, Cr 19-21, Al 3.5.
For heating elements. Useful operating temperature up to 1250°C.

TORDAL

VDM Nickel-Technologie AG
Cu 4, Pb 0.5, Bi 0.5, bal Al.
T3-temper: 55,000 TS; 43,000 YS; 15 El; 95 Brin. T6-temper: 57,000 TS; 39,000 YS; 17 El; 97 Brin. For screw machine products, machinery parts, fasteners; free-cutting. *Obsolete*

TOREADOR

Thomas Andrews & Co. Ltd.
C, alloy, bal Fe.
For high speed tools. High speed steel.

TOREADOR SUPRA

Thomas Andrews & Co. Ltd.
C, alloy, bal Fe.
For high speed tools. High speed steel.

TORMANC

British Steel Corp.
C 0.35, Mn 1.5, bal Fe.
Oil hardened: 84,000-101,000 TS; 54,000-67,000 YS; 27-30 El; 50-60 RA. For nuts, bolts; oil hardening; IZ 30-70. *Obsolete*

TORMANC MAJOR

British Steel Corp.
C 0.25-0.4, Mn 1.5, Mo 0.3, bal Fe.
Oil hardened: 123,000-156,000 TS; 105,000-139,000 YS; 18-26 El; 50-65 RA. For automobile crankshafts, axles, drive shafts, studs, bolts; oil hardening, IZ 50-90. *Obsolete*

TORMANC SPECIAL

British Steel Corp.
C 0.15-0.25, Mn 1.4-1.8, Ni 0.4-0.7, Mo 0.15-0.35, bal Fe.
Hardened: 90,000-107,000 TS; 67,000-80,000 YS; 22-30 El; 50-70 Brin. For gears, shafts, axles; oil hardened. *Obsolete*

TORMANC SPECIAL

British Steel Corp.
C 0.2, Mn 1.5, Ni, Mo, bal Fe.
Heat treated: 105,000 TS; 92,000 YS; 22 El; 64 RA. For pistons, couplings, connecting rods; for highly stressed parts. *Obsolete*

TORMOL

British Steel Corp.
C 0.3, Ni 2.5, Cr 0.7, Mo 0.5, bal Fe.
Oil hardened: 134,500-202,000 TS; 119,000-190,000 YS; 13-23 El; 45-63 RA. For crankshafts, gears, axle shafts, connecting rods; IZ 15-70. *Obsolete*

TORPEDO

Lehigh Steel Corp.
C 0.9, Cr 0.5, Mn 1.3, W 0.5, bal Fe.
For dies; punches, rivet sets; nondeforming, oil hardened.

TORPEDO BRONZE

English manufacture
Sn 0.5-1.5, Cu 59-62, bal Zn.
For torpedo parts; corrosion resistant.

TORRADAL

Vereinigte Metall. Ranshofen-Berndorf
Aluminum.
Aluminum alloy AlMgSiPb.

TORRADUR

Vereinigte Metall. Ranshofen-Berndorf
Aluminum.
Aluminum alloy AlCuMgPb.

TORRADUR B

Vereinigte Metall. Ranshofen-Berndorf
Aluminum.
Aluminum alloy AlCuBiPb.

TOTO HS-A1

Toto Steel Co. Ltd.
C 0.15-0.18, Mn 1.1-1.4, Si 0.35-0.6, 0.20 Cu min, bal Fe.
Rolled: 70,000-81,000 TS; 46,000 YP min; 20 El min. For agricultural equipment, structural members, buildings, bridges. Structural steel.

TOTO HS-B

Toto Steel Co. Ltd.
C 0.19-0.26, Mn 1.1-1.4, Si 0-0.4, 0.20 Cu min, bal Fe.
Rolled: 85,000-100,000 TS; 54,000 YP min; 17 El min. For structural members, buildings, bridges, agricultural equipment. Structural steel.

TOUCAS

Manufacturer not listed.
Cu 36, Ni 29, Fe 7.1, Pb 7.1, Sn 7.1, Sb 7.1, Zn 7.1.
For ornamental white metal parts; corrosion resistant.

TOUGH

Electro-Steel Co.
C 0.6, bal Fe.
For tools, dies; water or oil hardened.

TOUGH DEVIL NO. 1

Champion Rivet Co.
Mn 13, Ni 3, C, bal Fe.
Cast: 135,000 TS; 45,000 YS; 80 El; 50 RA; 180-200 Brin. For welding rods for high Mn steels; wear resistant; austenitic.

TOUGH DEVIL NO. 2

Champion Rivet Co.
C 0.8, Mn 14, Mo 0.2, bal Fe.
520 Brin. For hard surfacing electrodes; wear and abrasion resistant.

TOUGH HARD

Electro-Steel Co.
C 0.9, Mn 1, bal Fe.
For tools, dies; water or oil hardened.

TOUGH M

Bethlehem Steel Corp.
C 0.45, Mn 0.55, Si 0.2, Cr 0.95, V 0.2, bal Fe.
Low-alloy oil-hardened tool steel, AISI L2. *Obsolete*

TOUGH ONE

Wardlows Ltd.
C 1-1.15, bal Fe.
For tools, drills, taps, chasers; water hardened.

TOUGH THREE

Wardlows Ltd.
C 0.7-0.8, bal Fe.
For tools, springs, punches; water hardened.

TOUGH TWO

Wardlows Ltd.
C 0.9-1, bal Fe.
For tools, drills, chasers; water hardened.

TOUGH-ITE

Toughite Process Co.
C 0.7, W 18, Cr 4, V 1, bal Fe.
For cutting tool bits; high speed steel.

TOUGHDIE

Ziv Steel & Wire Co.
Tool material. C 0.45-0.55, Mn 0.6-0.7, Si 0.25, Cr 3-3.25, Mo 1.3-1.4, bal Fe.
AISI Type S7 shock resisting tool steel.

TOUGHITE NO. 5

Vulcan Iron Works
C 0.7, Cr 0.9, V 0.2, bal Fe.
For tires and rollers in kilns, coolers and dryers; oil hardened.

TOUGHTEM

Uddeholm Corp.
C 0.45, Si 0.3, Mn 0.9, Cr 1.5, Ni 0.4, Mo 0.2, V 0.1, micro alloy, bal Fe.
1225-1260 MPa TS; 1105-1150 MPa YS; 10.5-11.7 El; 40-44 RA. Chromium-nickel-molybdenum low alloy, hot work steel. For forging and general tooling applications.

TOURNAY METAL

French manufacture
Cu 82.5, Zn 17.5.
For cheap jewelry; brass.

TOURUN LEONARD'S METAL

Manufacturer not listed.
Sn 90, Cu 10.
For bearings; bronze, tough.

TP

Thyssen Edelstahlwerke AG
C 1, Cr 0.3, V 0.9, W 1, bal Fe.
For steel trimming and weld steel; oil hardening. *Obsolete*

TPM

American manufacture
C 0.05, Mn 2.3, Si 0.05, Cr 16, Co 0.5, Ti 3.1, Al 0.05, bal Ni.
For exhaust valves; SAE HEV 2.

TRABUK

English manufacture
Ni 5.5, Sb 5, Bi 2, bal Sn.
For food utensils; resists vegetable acids.

TRACKWEAR

Now AIRCO TRACKWEAR.

TRAFOPERM N2

Vacuumschmelze GmbH
Soft magnetic 3% silicon-iron alloy for pole pieces, relay
components and measuring systems; high saturation flux
density.

TRAKALOY

Bergstrom Alloys Corp.
Ti, Mn, Ni, Cr.
300 Brin. For hard facing electrodes; for rails and wheels.

TRAN-COR A-5

Armco
Si, bal Fe.
For audio transformer cores, low induction electrical
equipment; for radio use. *Obsolete*

TRAN-COR A-6

Armco
Si, bal Fe.
For audio transformers, electrical equipment; low induction.
Obsolete

TRAN-COR M-14

Armco
Si 4.5, bal Fe.
For transformers; core loss 52; high permeability. *Obsolete*

TRAN-COR M-15

Armco
Si 4-4.4, bal Fe.
For distribution transformers, rotating equipment; core loss
58; high permeability. *Obsolete*

TRAN-COR M-17

Armco
Si 4-4.4, bal Fe.
For transformers, motors, magnetic cores; core loss 65; high
permeability. *Obsolete*

TRAN-COR M-19

Armco
Si 3-3.4, bal Fe.
For generators, transformers, magnetic cores; core loss 72;
high permeability. *Obsolete*

TRAN-COR M-22

Armco
Si 3.2, bal Fe.
For transformers, motors, magnetic cores; high permeability;
core loss 82. *Obsolete*

TRAN-COR M-27

Armco
Si 2.8, bal Fe.
For transformers, motors, magnetic cores; high permeability;
core loss 101. *Obsolete*

TRAN-COR M-36

Armco
Si 1, bal Fe.
For magnetic cores, rotating machines, electrical equipment;
high permeability; core loss 117. *Obsolete*

TRAN-COR M-43

Armco
Si 0.5, bal Fe.
For armatures, rotating machines, motors; high permeability;
core loss 130. *Obsolete*

TRAN-COR T

Armco
Si, bal Fe.
For thin sheet for electric generators; high permeability.
Obsolete

TRAN-COR T-O

Armco
Si, bal Fe.
For electric generators, armatures; high permeability.
Obsolete

TRAN-COR T-O-S

Armco
Si, bal Fe.
For wound type transformers and reactors; high inductance,
low core loss. *Obsolete*

TRANCE PHOSPHOR BRONZE

Seymour Products Co.
Cu 90, Sn 5, Pb 5, trace P.
For hardware; free cutting.

TRANCOR 3 W

Armco
Si, bal Fe.
For transformer cores; oriented grains. *Obsolete*

TRANCOR 4 W

Armco
Si, bal Fe.
For transformer cores; oriented grains. *Obsolete*

TRANCOR H-0

Armco
High permeability iron-silicon electrical steel; grain oriented;
thickness 9 mils.

TRANCOR H-1

Armco
High permeability iron-silicon electrical steel; grain oriented;
thickness 11 mils.

TRANELEC A

Empire Sheet & Tin Plate Co.
Si, bal Fe.
For motors; high permeability. *Obsolete*

TRANELEC B

Empire Sheet & Tin Plate Co.
Si, bal Fe.
For motors; high permeability. *Obsolete*

TRANELEC C

Empire Sheet & Tin Plate Co.
Si, bal Fe.
For motors; high permeability. *Obsolete*

TRANSIL

British Steel plc
Electrical steel.

TRANSMISSION GLYCO

Joseph T. Ryerson & Son Inc.
Sn, Sb, bal Pb.
For bearings; Babbitt.

TRANSPARENT

Haeckerstahl GmbH
C 0.4, Cr 13, Mn 0.3, bal Fe.
Annealed: 95,000 TS; 50,000 YS; 25 El; 55 RA; 196 Brin. For
cutlery, valves, surgical and dental instruments; corrosion
resistant; Type 420.

TRANSPARENT 3003

Haeckerstahl GmbH
C 0.2, Cr 13, Mn 0.3, Si 0.4, bal Fe.
Cold drawn: 105,000 TS; 85,000 YS; 17 El; 50 RA; 215 Brin.
For cutlery, turbine blades, knives; Type 420; corrosion
resistant.

TRANSWELD

Lincoln Electric Co.
C 0.2, bal Fe.
Cast: 85,000 TS; 20 El. For welding rods; for mild steel.
Obsolete

TRANTINYL

Youngstown Alloy Castings Co.
C 0.7, Cr 1, bal Fe.
For guide shoes, tools; abrasion resisting.

TREM BRONZE

Greenleaf Corp.
Sn 2, bal Cu.
Annealed: 90,000 TS; 35,000 YS; 50 El; 65 RA; 160 Brin. For
springs, clips, hardware, fasteners. Tough, corrosion
resistant.

TRENITE

Trenite Foundry Corp.
C 0.1, Cr 18, Ni 8, bal Fe.
Annealed: 85,000 TS; 35,000 YS; 25 El; 45 RA; 150 Brin. For
chemical plant equipment, evaporators, tanks, Type 302;
austenitic, stainless.

TRENITE P

Trenite Foundry Corp.
C 0.25, Cr 20, Ni 10.
Cast: 45,000 TS; 300 Brin. For lead pots, heat treating
equipment; resists heat to 1650°F.

TRENITE WX

Trenite Foundry Corp.
C 0.3, Cr 14, bal Fe.
500 Brin. For coal grinders, hot coke car sides and mill liners;
abrasion resistant.

TRENITE-H

Trenite Foundry Corp.
C 0.2, Cr 18, Ni 8, bal Fe.
45,000 TS; 0.4 El; 250 Brin. For heat, wear, corrosion resistant
parts; heat, wear, corrosion resistant.

TRI-ACK

Teledyne Vasco
C 1.5, Cr 12, Mo 0.8, Co 0.4, bal Fe.
For forming dies, thread gauges, reamers, taps, master tools;
nondeforming. *Obsolete*

TRI-ALLOY

Ford Motor Co.
Pb 35-40, Ag 4.5, bal Cu.
For engine bearing for trucks and crankshafts; for heavy load
and high speed. *Obsolete*

TRI-CLOVER

Tri-Clover
C 0.3-0.5, bal Fe.
For machinery parts; water hardened.

TRI-CORE

Alpha Metals Inc.
Pb 40-60, bal Sn.
For solder; self-fluxing soft solder.

TRI-LOK
Tri-Lok Co.
C 0.4, bal Fe.
For gears, shafts; water hardened.

TRI-MO
Uddeholm Corp.
C 1.5, Cr 12, V 2, Mo 0.8, bal Fe.
Annealed: 102,000 TS; 14 El; 25 Brin. For tools, dies, reamers, gages, forming and blanking dies, thread rolling dies. Type D2; air hardened, non-deforming tool steel. *Obsolete*

TRI-MO AIR-HARDENING
Adams & Osgood Steel Co.
C, Cr, Mo, bal Fe.
For heavy dies for severe requirements; air hardened.

TRI-MO NO. 19
Peninsular Steel Co.
C 1.5, Cr 12, V 0.2, Mo 0.9, bal Fe.
For forming and blanking dies; air hardened, non-deforming.

TRI-PLY
Cytemp Specialty Steel Div.
Three ply composite steel with steel center, and Uniloy backing.
For heat and corrosion resistant parts; stainless clad steels. *Obsolete*

TRI-STEEL
Inland Steel Co.
HSLA steel. C 0.22, Mn 1.25, Si 0.3, V 0.02, bal Fe.
Rolled: 70,000 TS; 50,000 YS; 22 El. For railroad and agriculture equipment, mine cars; high-strength, low alloy construction steel.

TRI-TEN
United States Steel Corp.
C 0-0.22, Mn 0-1.25, Cu 0.2-0.6, 0.02 V min, bal Fe.
Rolled: 70,000 TS; 50,000 YS; 25 El; 140 Brin. For cranes, shovels, derricks, mine cars, truck bodies. Shock resistant. Good resistance to atmospheric corrosion. ASTM A441.

TRI-TEN E
U.S. Steel Corp.
C 0.25, Mn 1.3, Si 0.2, Ni 0.5-1, 0.2% min Cu, bal Fe.
Rolled: 70,000 TS; 50,000 YS; 22 El; 140 Brin. For structural parts, truck bodies, rail cars; high strength, good weldability. *Obsolete*

TRI-TUNG
Now UHB TRITUNG.

TRI-VAN
Uddeholm Corp.
C 2.25, Cr 12, bal Fe.
Hardened: 64-66 Rock C. For blanking and drawing dies, punches, shears, slitters, lamination and coining dies, gauges. Oil hardened, abrasion resistant, Type D4; die steel.

TRIALLOY
Tri-Clover
Ni alloy.
For sanitary fittings and valves; corrosion resistant.

TRIANGLE "X"
Uddeholm Corp.
C 1-1.2, bal Fe.
For rock drill pistons, punches, dies, broaches; water hardened. *Obsolete*

TRIANGLE 10
Triangle Conduit & Cable Co.
Cu 65, Zn 25, Ni 10.
Hard: 115,000 TS; 5 El. Soft: 55,000 TS; 45 El; 20,000 YS. For hardware, costume jewelry, optical parts, holloware. Nickel silver, electrical conductivity 9%; corrosion resistant.

TRIANGLE 12
Triangle Conduit & Cable Co.
Cu 65, Zn 23, Ni 12.
Hard: 115,000 TS; 3 El. Soft: 58,000 TS; 45 El; 20,000 YS. For slide fasteners, hardware, optical parts. Electrical conductivity 8%; nickel silver, corrosion resistant.

TRIANGLE 125
Triangle Conduit & Cable Co.
Cu 90, Zn 9, Sn 1.
CDA C41300.

TRIANGLE 15
Triangle Conduit & Cable Co.
Cu 65, Zn 20, Ni 15.
Hard: 103,000 TS; 5 El. Soft: 58,000 TS; 20,000 YS; 45 El. For camera goods, optical equipment, jewelry, etching stock. Nickel silver, electrical conductivity 7%; corrosion resistant.

TRIANGLE 170
Triangle Conduit & Cable Co.
Cu 70, Zn 30, P 0.02-0.05.
Hard: 110,000 TS; 7 El. Soft: 50,000 TS; 50 El. For condensers, heat exchangers, ferrules. Electrical conductivity 28%, inhibited.

TRIANGLE 171
Triangle Conduit & Cable Co.
Cu 70, Zn 30, As 0.02-0.05.
Hard: 110,000 TS; 7 El. Soft: 50,000 TS; 50 El. For marine hardware, condensers, heat exchangers. Electrical conductivity 28%, inhibited. Free from dezincification.

TRIANGLE 18
Triangle Conduit & Cable Co.
Cu 65, Zn 17, Ni 18.
Hard: 103,000 TS; 3 El; 90,000 YS. Soft: 58,000 TS; 45 El; 25,000 YS. For rivets, screws, zippers, hollow ware, truss wire, costume jewelry. Electrical conductivity 6%; nickel silver, corrosion resistant.

TRIANGLE 180
Triangle Conduit & Cable Co.
Ni 23, Cu 77.
CDA C71100.

TRIANGLE 30
Triangle Conduit & Cable Co.
Cu 98, Ni 2.
Hard: 67,000 TS; 4 El. Soft: 37,000 TS; 46 El. For resistances. Electrical conductivity 34%.

TRIANGLE 60
Triangle Conduit & Cable Co.
Cu 94, Ni 6.
Hard: 73,000 TS; 4 El. Soft: 37,000 TS; 46 El. For resistances. Electrical conductivity 17%.

TRIANGLE 70/30
Triangle Conduit & Cable Co.
Cu 70, Zn 30.
Hard: 110,000 TS; 7 El. Soft: 50,000 TS; 50 El; 18,000 YS. For fasteners, terminal plugs, plumbing goods, cartridge cases. Electrical conductivity 28%; high strength and ductility.

TRIANGLE 80
Triangle Conduit & Cable Co.
Cd 1, Cu 99.
Hard: 70,000 psi TS; 15 El. Soft: 42,000 psi TS; 45 El. Good electrical conductivity. CDA C16200.

TRIANGLE 85
Triangle Conduit & Cable Co.
Cd 0.8, Cu 99.2.
Hard: 69,000 psi TS; 15 El. Soft: 42,000 psi TS; 45 El. Good electrical conductivity. CDA C16200.

TRIANGLE 90
Triangle Conduit & Cable Co.
Cu 89, Ni 11.
Hard: 83,000 TS; 5 El. Soft: 47,000 TS; 46 El; 16,000 YS; 15 Rock B. For condensers, condenser plates, distiller tubes, ferrules, piping. Electrical conductivity 9%; corrosion resistant.

TRIANGLE 91/09
Triangle Conduit & Cable Co.
Cu 91.5, Zn 8, Sn 0.5.
Hard: 77,000 TS; 6 El. Soft: 44,000 TS; 40 El. For metal hose braids. Electrical conductivity 39%.

TRIANGLE 91X
Triangle Conduit & Cable Co.
Zn 9.5, Sn 0.5, bal Cu.
CDA C41100.

TRIANGLE X-EXTRA
Uddeholm Corp.
C 0.7, bal Fe.
For tools, drills, taps; water hardened. *Obsolete*

TRIANGLE X-SPECIAL
Uddeholm Corp.
C 0.9-1.2, bal Fe.
For tools, punches; water hardened. *Obsolete*

TRIBALOY ALLOY T-100
Deloro Stellite, Inc.
Co 55, Mo 35, Si 10.
Powder for powder metals or plasma spraying. Two-phase composition for wear resistance. Corrosion resistant; low coefficient of friction. Hardness of intermetallic: 1100 VHN (Vickers). For valves, pistons, vanes, and arbors.

TRIBALOY ALLOY T-400
Deloro Stellite, Inc.
Co 62, Mo 28, Cr 8, Si 2.
Rod or powder for plasma spray, powder metal, hard facing or remelt casting stock. Wear resistant; corrosion resistant over wide temperature range. For valves, tappet inserts, and pump parts.

TRIBALOY ALLOY T-700
Deloro Stellite, Inc.
Ni 50, Mo 32, Cr 15, Si 3.
Rod or powder for plasma spray, powder metal, hard facing or remelt casting stock. Resists mechanical wear and severe corrosion over wide temperature range. For thrust washers, ball and roller bearings, sleeve bearings, and valves.

TRIBALOY ALLOY T-800
Deloro Stellite, Inc.
Co 52, Mo 28, Cr 17, Si 3.
Rod or powder for plasma spray, powder metal, hard facing, or remelt casting stock. Resists mechanical wear and severe corrosion over wide temperature range. For valves, seals, piston rings, and tappet inserts.

TRIBALOY T400
Deloro Stellite Ltd.
Cobalt. Cr 8, C 0.1, Mo 28, Ni 1, Si 2.4, Fe 1, bal Co.
Standard cobalt base alloy.

TRIBALOY T700
Deloro Stellite Ltd.
Nickel. Co 1, Cr 15, C 0.1, Mo 32, Si 3.3, Fe 1, bal Ni.
Standard nickel base alloy.

TRIBALOY T800
Deloro Stellite Ltd.
Cobalt. Cr 17, C 0.1, Mo 28, Ni 1, Si 3.2, Fe 1, bal Co.
Standard cobalt base alloy.

TRICENT

International Nickel Inc.
Alloy steel. C 0.43, Si 1.6, Mn 0.8, Ni 1.8, Cr 0.85, Mo 0.38, V 0.08, bal Fe.
Heat treated: 290,000 TS; 250,000 YS; 10 El; 36 RA; 54 Brin.
For aircraft structures, landing gears; tough, shock resistant. *Obsolete*

TRICRANK

Titusville Forge Co.
C 0.4, Ni 1.5, Cr 0.8, bal Fe.
Normalized: 94,800 TS; 22 El; 48 RA. Heat treated: 89,000 TS; 27 El; 58 RA. For crankshafts, gears, pinions, axles; oil hardened. *Obsolete*

TRIDENT

Latrobe Steel Co.
C 0.57, Si 0.65, Mn 0.3, Cr 0.6, Mo 0.35, V 0.2, bal Fe.
For cold forming dies, punches, mandrels, chisels; cold work die steel. *Obsolete*

TRIMCO

Trimout Mfg. Co.
C 0.6, Ni 1.5, Cr 0.8, bal Fe.
For pipe wrenches; oil hardening.

TRIMET

GTE Sylvania
W alloy.
For wire for sealing into glass; heat resistant. *Obsolete*

TRIMET 104

Handy & Harmon
Copper.
MP: 1170°F; FP: 1270°F. Easy-Flo 3 on both sides of copper in 1-4-1 ratio.

TRIMET 177

Handy & Harmon
Copper.
MP: 1125°F; FP: 1295°F. Easy-Flo 35 on both sides of copper in 1-2-1 ratio. For brazing carbide tips.

TRIMET 201

Handy & Harmon
Copper.
MP: 1220°F; FP: 1435°F. BRAZE 403 on both sides of copper in 1-2-1 ratio. For brazing carbide tips.

TRIMET 202

Handy & Harmon
Copper.
MP: 1220°F; FP: 1580°F. BRAZE 404 on both sides of copper in 1-2-1 ratio. For brazing carbide tips.

TRIMET 258

Handy & Harmon
Copper.
MP: 1170°F; FP: 1270°F. Easy-Flo 3 on both sides of copper in 1-2-1 ratio. For brazing carbide tips.

TRIMMAX

Heppenstall Co.
Tool material. C 0.95, Cr 3.75, Mo 0.23, V 0.16, bal Fe.
Annealed: 220 Brin. Heat treated: 510 Brin. For hot and cold trimmers; oil hardened.

TRIMMAX, H41

Heppenstall Co.
Tool material. C 0.95, Mn 0.3, Si 0.25, Cr 3.75, V 0.17, Mo 0.2, bal Fe.
Air hardened tool and die steel for cold work operations.

TRIMMER DIE

P.F. McDonald & Co.
C, alloy, bal Fe.
For tools; tough.

TRIMO

Uddeholm Corp.
C 1.5, Mn 0.3, Si 0.4, Cr 12, Mo 0.8, V 0.9, bal Fe.
Annealed: 200-210 Brin. For dies, punches, shear blades, cold forming dies, exceptional wear resistance; air and oil hardened. Type D2; cold work tool steel. *Obsolete*

TRIPLE A ALMAG 55

Acme Aluminum Alloys Inc.
Mg 10-11.5, Be 0.2-1, B 0.0001-0.05, bal Al.
Cast: 50,000-60,000 TS; 32,000-35,000 YS; 20-30 El; 80-100 Brin. For light alloy castings and meters. Sand or permanent mold.

TRIPLE ALLOY PUNCH

Colt Industries
C 0.6, V 0.15-0.25, Cr 1.3, W 0.5, bal Fe.
For punches, dies; oil hardened. *Obsolete*

TRIPLE CONQUEROR

Joseph Beardshaw & Son Ltd.
C 1.3, W 4, Cr 1, Mo 0.3, bal Fe.
For drawing and extrusion dies, broaches; finishing steel, wear resistant.

TRIPLE CRESCENT

Spencer Clark Metal Industries Ltd.
C 1.25, Cr 1.25, W 4.3, V 0.3, bal Fe.
Tungsten-chromium tool steel, for reamers, broaches, finishing tools, gauges.

TRIPLE E 8176-H24

Southwire Co.
Fe 0.4-1, Si 0.03-0.15, bal Al.
17.5 ksi TS; 13.5 YS; 15 El; 61% IACS min. For building wire, magnet wire, automotive wire, communication cable.

TRIPLE ECLAIR

Compagnie Ateliers et Forges de la Loire
C 0.78, W 18.5, V 1, Mo 0.4, bal Fe.
For lathe and planer tools, hobs, milling cutters, taps, thread chasers, broaches, drills. High-speed steel. Oil hardening. Good red hardness.

TRIPLE ECLAIR

Creusot-Loire
C 0.78, W 18, Mo 0.6, V 0.6, bal Fe.
For lathe and planer tools; oil hardened. *Obsolete*

TRIPLE EXPRESS "A"

Hidalgo Steel Co. Inc.
C 0.7, W 6, Mo 6, Co 5, Cr 4, bal Fe.
For high speed lathe tools, twist drills, reamers; high speed steel.

TRIPLE EXPRESS "B"

Hidalgo Steel Co. Inc.
C 0.7, Mo 8, W 1, Cr 4, Co 5, bal Fe.
For high speed lathe tools, twist drills, reamers; high speed steel.

TRIPLE EXPRESS 3X

Compagnie Ateliers et Forges de la Loire
C 1.3, W 10.5, V 3.1, Mo 0.5, bal Fe.
Hardened: 64-68 Rock C. For finishing cutters, tools, deburrers, drills, broaches, milling cutters. High speed steel, oil hardening, good red hardness.

TRIPLE EXPRESS B

Compagnie Ateliers et Forges de la Loire
C 0.8, W 18, Co 5, V 1.1, Mo 0.9, bal Fe.
Hardened: 64-66 Rock C. For turning tools, lathe and planer finishing cutters, hobs, reamers, broaches, taps, drills. High-speed steel, oil hardening; high red hardness.

TRIPLE EXPRESS SUPER-4X

Compagnie Ateliers et Forges de la Loire
C 1.55, W 13, Co 5, V 5, Cr 5, Mo 0.5, bal Fe.
Hardened: 64-68 Rock C. For lathe and planer finishing tools, engraving tools, milling cutters, textile needles. High-speed steel, oil hardening. High red hardness.

TRIPLE EXPRESS T.M.

Compagnie Ateliers et Forges de la Loire
C 0.82, W 6, Mo 4.5, V 1.4, bal Fe.
Hardened: 64-66 Rock C. For lathe and planer tools, drills, saws, form cutters, broaches, milling cutters. High-speed steel; oil hardening, high red hardness.

TRIPLE EXPRESS V

Compagnie Ateliers et Forges de la Loire
C 0.85, W 18.5, V 1.9, Mo 0.4, bal Fe.
Hardened: 64-66 Rock C. For turning tools, lathe and planer cutters, drills, thread cutters, broaches, reamers, hobs. High-speed steel; oil hardening, high red hardness.

TRIPLE EXTRA

Midvale-Heppenstall Co.
C, alloy, bal Fe.
For high speed cutters. *Obsolete*

TRIPLE EXTRA (T.E.R.)

Acieries de Champagnole
C 0.78, Cr 4.5, W 18, Mo 0.6, V 1.25, bal Fe.
Air or oil hardens to 63-66 Rock C. For cutting tools, taps, thread chasers, and turning tools. AFNOR: Z 80 W 18. AISI T1 high speed steel. Germany: S 18.01 (B-18); W.Nr. 1.3355.

TRIPLE EXTRA 2 (T.E.R.2)

Acieries de Champagnole
C 0.82, Cr 4.5, W 18, Mo 0.6, V 2, bal Fe.
Air or oil hardens to 63-66 Rock C. For cutting tools, lathe, gear cutters, and broaches. AFNOR: Z 85 WV 18.02. AISI T2 high speed steel. Germany: W.Nr. 1.3357.

TRIPLE EXTRA 60 (T.E.R.60)

Acieries de Champagnole
C 0.65, Cr 4, W 18, Mo 0.4, V 1, bal Fe.
Air or oil hardens to 60-63 Rock C. For cutting tools or cold work tools. AFNOR: Z 65 W 18. AISI T1 high speed steel.

TRIPLE EXTRA D (T.E.D.)

Acieries de Champagnole
C 0.8, Cr 4.3, W 12.5, Mo 1.8, V 1.8, bal Fe.
Air or oil hardens to 62-66 Rock C. For lathe tools, drills, taps, reamers, and cut-off tools. AFNOR: Z 80 WDV 12.02.02. AISI T7 High speed steel. Germany: S 12.1.2. (0); W. Nr. 1.3318.

TRIPLE EXTRA DIE

Midvale-Heppenstall Co.
C, Cr, V, bal Fe.
For tools. *Obsolete*

TRIPLE EXTRA LION

Burys & Co. Ltd.
C 0.75, W 19, Cr 4.25, V 1.25, Co 5, Mo 0.75, bal Fe.
For lathe and planer tools, drills, hobs, broaches; high speed steel.

TRIPLE EXTRA M (T.E.M.)

Acieries de Champagnole
C 0.83, Cr 4.2, W 6.5, Mo 5, V 1.9, bal Fe.
Air or oil hardens to 63-66 Rock C. For twist drills, center drills, reamers, taps, cutters, and broaches. AFNOR: Z 85 WDV 06.05.02. AISI M2 high speed steel. Germany: S 6.5.2 (DMo5); W.nr. 1.3343.

TRIPLE GRIFFIN

Darwins Alloy Castings
C 1.35, Mn 0.4, Cr 0.35, W 2.75, Si 0.3, bal Fe.
For cold drawing dies, cutters; water hardened, hard case and tough core. *Obsolete*

TRIPLE LIFE D-C

Darwin & Milner Inc.
Tool material. C 1.2, Cr 4, bal Fe.
For dies, reamers; air hardened.

TRIPLE MERMAID

Spear & Jackson (Industrial) Ltd.
C 0.78, Cr 4.25, W 18, V 1.25, Co 5, bal Fe.
Cobalt high speed steel for cutting tools. AISI T4.

TRIPLE MUSHET
Osborn Steels Ltd.
W 19, Cr 4, V 2.25, C, bal Fe.
For tools, dies, cutters; high speed steel. *Obsolete*

TRIPLE SATAN
Creusot-Loire
C 0.32, W 9, Cr 3.3, bal Fe.
For extrusion rams, dies and liners, punches; hot work steel, oil hardened. *Obsolete*

TRIPLE VELOS
Spencer Clark Metal Industries Ltd.
C 0.75, W 18, Cr 4.25, V 1.1, Co 5, bal Fe.
Medium cobalt high speed steel.

TRIPLEX
Crane Co.
C 0.35-0.45, Mn 0.6-0.9, Cr 0.8-1.1, Mo 0.15-0.25, bal Fe.
Heat treated: 105,000-125,000 TS; 85,000-100,000 YS; 15-17 El; 50-55 RA. For high temperature bolting; ASTM-A193, Gr. B7. *Obsolete*

TRIPLI-CAST
Precious Metals Research Works Inc.
C alloy, bal Fe.
Rolled: 69,000-109,600 TS; 1.5-5 El; 132-205 Brin. For shafts; MP 1710°F.

TRISTELLE TS1
Deloro Stellite Ltd.
Iron. Co 12, Cr 30, C 1, Ni 10, Si 5, bal Fe.
Standard iron base alloy.

TRISTELLE TS2
Deloro Stellite Ltd.
Iron. Co 12, Cr 35, C 2, Ni 10, Si 5, bal Fe.
Standard iron base alloy.

TRISTELLE TS3
Deloro Stellite Ltd.
Iron. Co 12, Cr 35, C 3, Ni 10, Si 5, bal Fe.
Standard iron base alloy.

TRITEX
LaSalle Steel Co.
C 0.25-0.55, Mn, bal Fe.
Cold drawn: 70,000-120,000 TS; 13-20 El; 45-65 RA; 165-255 Brin. For splines, shafts, gears; water hardened. *Obsolete*

TRITEX BROACHRITE
LaSalle Steel Co.
C 0.35-0.45, Mn, bal Fe.
Cold drawn: 100,000 TS; 90,000 YS; 13 El; 45 RA; 217 Brin. For machinery parts, gears, shafts; water hardened. *Obsolete*

TRITEX NO. 1
LaSalle Steel Co.
C 0.32-0.39, Mn 1.35-1.65, S 0.08-0.13, bal Fe.
Cold rolled: 103,500 TS; 93,000 YS; 15 El; 56 RA; 217 Brin. For gears, shafts, machinery parts; water hardening. *Obsolete*

TRITEX NO. 2
LaSalle Steel Co.
C 0.4-0.5, Mn 1.3-1.6, Si 0.24-0.33, bal Fe.
Rolled: 90,000 TS; 60,000 YS; 26 El; 58 RA; 193 Brin. For gears, shafts, pinions; wear resistant. *Obsolete*

TRITEX PLUS
LaSalle Steel Co.
C 0.35-0.45, Mn, bal Fe.
For motor shafts, gears; water hardened. *Obsolete*

TRITON
CCS Braeburn Alloy Steel
C 0.5-0.55, Si 0.9-1.1, Mo 0.6, bal Fe.
For tools, cutters, punches, pneumatic chisels, rivet sets; shock resistant. *Obsolete*

TRITON A
CCS Braeburn Alloy Steel
C 0.5, Si 1, Mn 0.35, bal Fe.
For tools, chisels, punches; tough. *Obsolete*

TRITUNG
Uddeholm Corp.
C 2, Mn 0.75, Si 0.3, Cr 13, W 1.25, bal Fe.
Annealed: 230-260 Brin. For cold and hot trimming dies; exceptional wear resistance; air and oil hardened; non-deforming. Type D6; cold work tool steel. *Obsolete*

TRIUMPH
J.J. Saville & Co. Ltd.
C 0.68, Cr 3.75, Mo 0.6, V 0.6, W 14, bal Fe.
For slitting and hacksaw blades, punches; high speed steel.

TRIUMPH
Jessop-Saville Ltd.
C 0.68, Cr 3.75, V 0.6, W 14, bal Fe.
Hardened: 64-66 Rock C. For slitting saws, hacksaws, lathe centers, cold punches, reamers. High speed steel for cutting operations under conditions of heavy vibration.

TRIUMPH 200
Jessop-Saville Ltd.
C 0.7, Cr, W, V, bal Fe.
For lathe and planer tools, reamers, hobs, drills, taps; high speed steel.

TRIUMPH SUPERB
Jessop-Saville Ltd.
C 0.76, Cr 4.25, Mo 0.6, V 1.4, W 18, bal Fe.
For drills, broaches, reamers, hobs, gear cutters; high speed steel.

TRIUMPH SUPERB
Saville & Co. Ltd., J.J.
C 0.76, Cr 4.25, Mo 0.6, V 1.4, W 18, bal Fe.
For drills, broaches, reamers, hobs, gear cutters; high speed steel.

TRIUMPH SUPERB 1000
J.J. Saville & Co. Ltd.
C 0.8, Cr 4.75, Mo 0.6, V 1.6, W 18.5, Co 5.7, bal Fe.
For turning and boring tools, milling cutters, hobs; high speed steel.

TRIUMPH SUPERB DOUBLE 1000
Jessop-Saville Ltd.
C 0.82, Cr 4.75, Mo 0.6, V 1.6, W 20, Co 10.5, bal Fe.
For tools, cutters, hobs; high speed steel.

TRIUMPH SUPERB DOUBLE 1000
Saville & Co. Ltd., J.J.
C 0.82, Cr 4.75, Mo 0.6, V 1.6, W 20, Co 10.5, bal Fe.
For tools, cutters, hobs; high speed steel.

TRIUMPH SUPERB EXTRA
Jessop-Saville Ltd.
C 0.78, Cr 4.25, Mo 0.6, V 1.4, W 22, bal Fe.
For form cutters, tools for chilled roll turning; high speed steel.

TRIUMPH SUPERB EXTRA
Saville & Co. Ltd., J.J.
C 0.78, Cr 4.25, Mo 0.6, V 1.4, W 22, bal Fe.
For form cutters, tools for chilled roll turning; high speed steel.

TRIUMPHATOR
Now VEW K100.

TRIUMPHATOR M
Now VEW K116.

TRIUMPHATOR W
Now VEW K.

TRIUNFADOR
Vereinigte Edelstahlwerke
C 2.25, Cr 13, bal Fe.
For tools, dies; non-deforming. *Obsolete*

TRIUNFADOR Z
Vereinigte Edelstahlwerke
C 1.5, Cr 11.5, Mo 0.75, bal Fe.
For tools, dies; non-deforming. *Obsolete*

TRIVAN
Adams & Osgood Steel Co.
C, Cr, V, bal Fe.
For blanking and coining dies; oil hardened.

TRODALOY NO. 1
Manufacturer not listed.
Co 2.6, Be 0.4, bal Cu.
Heat treated: 90,000 TS; 45,000 YS; 10 El, 20 RA; 220 Brin. For resistance welding electrodes, springs; age-hardened. *Obsolete*

TRODALOY NO. 7
Manufacturer not listed.
Cr 0.4, Be 0.1, bal Cu.
Heat treated: 30,000-35,000 TS; 10-15 El; 80-85 Brin. For soldering tips, resistance welding electrodes; hardenable. *Obsolete*

TROJAN
Atlas Specialty Steels
C 0.8, W 18, Cr 4, V 2, Mo 0.5, bal Fe.
For tools, dies, hobs, cutters; high speed steel. *Obsolete*

TROJAN
American Crucible Products Co.
C 0.05, bal Fe.
For pipe; wrought iron. *Obsolete*

TROJAN BABBITT
Hoyt Metal Co.
Sb, Sn, Pb.
At 70°F: 27.4 Brin. At 212°F: 11.2 Brin. For bearing for steam and internal combustion engines; Babbitt, antifriction.

TROJAN BABBITT
NL Industries
Sb, Sn, Pb.
At 70°F: 27.4 Brin. At 212°F: 11.2 Brin. For bearing for steam and internal combustion engines; Babbitt, antifriction.

TROJAN SHANK
Atlas Steels Ltd.
C 0.35, Mn 0.7, Cr 0.5, Mo 0.3, Ni 3.25, bal Fe.
For mining bit shanks. *Obsolete*

TROJAN STEEL
Carpenter Technology Corp.
Cr 1, Ni 2, C, bal Fe.
Tempered: 225,000 TS; 220,000 YS; 10 El; 35 RA; 444 Brin. Annealed: 85,000 TS; 60,000 YS; 26 El; 62 RA; 163 Brin. For transmission gears. *Obsolete*

TROJAN STEEL NO. 8
Horace T. Potts Co.
C 0.75-0.85, bal Fe.
Rolled: 80,000-90,000 TS; 40,000-45,000 YS; 15-25 El; 170-190 Brin. For tools, chisels, track tools, forming dies; welds and forges. *Obsolete*

TROJAN STEEL NO. 9 1/2
Horace T. Potts Co.
C 0.9-1, bal Fe.
Annealed: 80,000-100,000 TS; 40,000-50,000 YS; 15-20 El; 25-40 RA; 180-210 Brin. For tools, dies, brakes, forming dies; hardens well. *Obsolete*

TROMALIT
Ugine Aciers
Co 11, Ni 24, Al 11, Cu 3.5, bal Fe.
For electrical and magnetic equipment. Permanent magnet; high permeability.

TROMALIT-ALNI
Ugine Aciers
Ni 27, Al 13, bal Fe.
For permanent magnets in electrical and magnetic equipment. High magnetic permeability.

TROMALIT-ALNI090
Ugine Aciers
Ni 22, Al 12, bal Fe.
For permanent magnets in electrical and magnetic equipment. High magnetic permeability.

TROMALIT-ALNICO160
Ugine Aciers
Ni 24, Al 11, Co 9.5, Cu 3.5, bal Fe.
For permanent magnets in electrical and magnetic equipment. High magnetic permeability.

TROMALIT-III
Ugine Aciers
Ni 12, Al 10, Co 4-20, Cu 2, bal Fe.
Bakelite bonded. Residual induction 3500 gauss; coercive force 6800 oersted; energy product 0.8 max. For magnets in speedometers and small battery motors. Permanent magnet, high permeability.

TROMALIT-IV
Ugine Aciers
Ni 12, Al 10, Co 4-20, Cu 2, bal Fe.
Bakelite bonded. Residual induction 3500 gauss; coercive force 1000 oersted; energy product 1 max. For magnets in speedometers and in small battery motors. Permanent magnet, high permeability.

TRONAMANG
Kerr-McGee Chemical Corp.
Mn 100.
For an addition agent in the melting of steel and non-ferrous alloys. Electrolytic manganese.

TRONAMANG 75
Kerr-McGee Chemical Corp.
Mn 75, Al 25.
Briquettes used in alloying aluminum.

TROPIC HOT DIE STEEL
Osborn Steels Ltd.
C 0.35, Si 0.3, Mn 0.4, W 9.5, Cr 3, V 0.5, bal Fe.
For hot dies; hot work steel. *Obsolete*

TROXEIT GRADE 85
Fansteel Metals
Cu, W.
150 Brin. For large welding dies, spot welding electrodes. *Obsolete*

TROXEIT GRADE 95
Fansteel Metals
Cu-W.
200 Brin. For projection welding tips and faces, spot welding, butt welding dies. *Obsolete*

TROXEIT, GRADE 100
Fansteel Metals
Cu-W.
For welding where extreme hardness is required. *Obsolete*

TROY
Time Steel Service Inc.
C 1.25, Mn 0.3, Si 0.35, Cr 0.4, V 0.2, W 1.4, bal Fe.
Oil hardened tool steel. AISI O7.

TRS SPECIALITY
Allied Steel & Tractor Products Inc.
Low C, Cr, Mo, W, bal Fe.
Heat treated: 140,000 TS; 260-300 Brin. For tools, dies, machine parts, gears, cams, pump shafts, clutches; tough, wear and fatigue resistant.

TRU-CAST
Manco Products Co.
Be, bal Cu.
Cast: 125,000 TS; 110,000 YS; 200 Brin. Hardened: 180,000 TS; 140,000 YS; 460 Brin. For stamping and casting dies, molds; hardenable.

TRU-COR
Kidd Drawn Steel
C 0.9-1.05, Mn 0.3-0.5, Si 0.15-0.3, bal Fe.
For machine tool parts, drills, boring tools; type W1; water hardened.

TRU-COR DRILL ROD
Precision-Kidd Steel Co.
C 0.95-1.05, Mn 0.3-0.5, bal Fe.
Water hardening drill rod; AISI W1.

TRU-GROUND
Jessop Steel Co.
C, alloy, bal Fe.

TRU-HEADERDIE
Jessop Steel Co.
C 1.4, V 3.5, bal Fe.

TRU-WEAR FM (AISI D7)
Jessop Steel Co.
C 2.2, Cr 12.5, V 4, Mo 1.1, bal Fe.
For header and forming dies; cold work steel.

TRUALLOY BEARING BRONZE
True Alloys Inc.
Sn, Pb, bal Cu.
Cast: 25,000 TS; 12,000 YS; 8 El; 60-80 Brin. For bearings; high compressive strength.

TRUALOY ALUMINUM
True Alloys Inc.
Si 5, bal Al.
Cast: 22,000 TS; 2 El; 60 Brin. For light alloy castings; die or sand castings.

TRUALOY ALUMINUM BRONZE
True Alloys Inc.
Cu 90, bal Al.
Cast: 65,000 TS; 25,000 YS; 20 El; 115 Brin. For sand castings.

TRUALOY COPPER
True Alloys Inc.
Cu.
For current conductors; high conductivity 90%.

TRUCOTE NO. 101 FC, ETC.
Now ALL STATE NO. 101 FC, etc.

TRUDIE
Champion Steel Co.
C 1.5, Cr 12, Mo 1, V 1, bal Fe.
Air hardening cold work tool steel; high carbon, high chrome type.

TRUDIE SPECIAL
Champion Steel Co.
C 1.5, Cr 12, Mo 1, Co 3, bal Fe.
Air hardening cold work tool steel.

TRUE-DIE
Allied Steel & Tractor Products Inc.
C, alloy, bal Fe.
For dies; non-shrinking.

TRUEWEAR
Jessop Steel Co.
C 2.45, Cr 12.25, V 4.25, Mo 1.1, bal Fe.
For dies, punches, pins, tools and forming rolls. Extreme wear resistance. AISI D7; cold work tool steel.

TRUFLEX A-1
Texas Instruments Inc./Materials Control
For thermostatic bimetal; maximum temperature 350°F.

TRUFLEX A-2
Texas Instruments Inc./Materials Control
Ni-Cr-Fe.
For thermostatic bimetal; maximum temperature 300°F.
Obsolete

TRUFLEX B-1
Texas Instruments Inc./Materials Control
For thermostatic bimetal. Differential expansion. Maximum operating temperature 1000°F. ASTM TM1.

TRUFLEX D-15
Texas Instruments Inc./Materials Control
Ni-Cr-Fe.
For thermostatic bimetal; maximum temperature 1100°F.
Obsolete

TRUFLEX B-2
Texas Instruments Inc./Materials Control
For thermostatic bimetal; maximum temperature 1000°F. ASTM TM6.

TRUFLEX B-20
Texas Instruments Inc./Materials Control
Ni-Cr-Fe.
For thermostatic bimetal; maximum temperature 1100°F.
Obsolete

TRUFLEX B-30
Texas Instruments Inc./Materials Control
Ni-Cr-Fe.
For thermostatic bimetal; maximum temperature 1100°F.
Obsolete

TRUFLEX B-35
Texas Instruments Inc./Materials Control
Ni-Cr-Fe.
For thermostatic bimetal; maximum temperature 1100°F.
Obsolete

TRUFLEX B-40
Texas Instruments Inc./Materials Control
Ni-Cr-Fe.
For thermostatic bimetal; maximum temperature 1100°F.
Obsolete

TRUFLEX B-47
Texas Instruments Inc./Materials Control
Ni-Cr-Fe.
For thermostatic bimetal; maximum temperature 1100°F.
Obsolete

TRUFLEX B12 1/2
Texas Instruments Inc./Materials Control
Ni-Cr-Fe.
For thermostatic bimetal; maximum temperature 1100°F.
Obsolete

TRUFLEX B3
Texas Instruments Inc./Materials Control
For thermostats (0-650°F); bimetal maximum 1000°F.

TRUFLEX D
Texas Instruments Inc./Materials Control
Ni-Cr-Fe.
For thermostatic bimetal; maximum temperature 1200°F.
Obsolete

TRUFLEX E-1
Texas Instruments Inc./Materials Control
For thermostatic bimetal; maximum temperature 1000°F.

TRUFLEX E-2
Texas Instruments Inc./Materials Control
Ni-Cr-Fe.
For thermostatic bimetal; maximum temperature 1200°F.
Obsolete

TRUFLEX E3
Texas Instruments Inc./Materials Control
For thermostats (0-650°F); bimetal maximum 1000°F. ASTM TM3.

TRUFLEX E4
Texas Instruments Inc./Materials Control
For thermostats (0-750°F); bimetal maximum 1000°F. ASTM TM4.

TRUFLEX E5
Texas Instruments Inc./Materials Control
For thermostats (0-750°F); bimetal maximum 1000°F. ASTM TM5.

TRUFLEX F
Texas Instruments Inc./Materials Control
Ni-Cr-Fe.
For thermostatic bimetal; maximum temperature 1200°F.
Obsolete

TRUFLEX G-1
Texas Instruments Inc./Materials Control
For thermostatic bimetal; maximum temperature 1000°F.
ASTM TM20.

TRUFLEX G-2
Texas Instruments Inc./Materials Control
Ni-Cr-Fe.
For thermostatic bimetal; maximum temperature 1200°F.
Obsolete

TRUFLEX H
Texas Instruments Inc./Materials Control
Ni-Cr-Fe.
For thermostatic bimetal; maximum temperature 1200°F.
Obsolete

TRUFLEX J
Texas Instruments Inc./Materials Control
Ni-Cr-Fe.
For thermostatic bimetal; maximum temperature 500°F.
Obsolete

TRUFLEX J1
Texas Instruments Inc./Materials Control
For thermostats (-50 to 400°F); bimetal maximum 625°F.

TRUFLEX J7
Texas Instruments Inc./Materials Control
For thermostats; corrosion resistant, bimetal maximum
temperature 625°F.

TRUFLEX K
Texas Instruments Inc./Materials Control
Ni-Cr-Fe.
For thermostatic bimetal; maximum temperature 1200°F.
Obsolete

TRUFLEX L
Texas Instruments Inc./Materials Control
Ni-Cr-Fe.
For thermostatic bimetal; maximum temperature 800°F.
Obsolete

TRUFLEX L1
Texas Instruments Inc./Materials Control
Ni-Cr-Fe.
For thermostats (-50 to 400°F); bimetal maximum 1000°F.
Obsolete

TRUFLEX M
Texas Instruments Inc./Materials Control
Ni-Cr-Fe.
For thermostatic bimetal; maximum temperature 1200°F.
Obsolete

TRUFLEX N
Texas Instruments Inc./Materials Control
Ni-Cr-Fe.
For thermostatic bimetal; maximum temperature 1000°F.
Obsolete

TRUFLEX NI
Texas Instruments Inc./Materials Control
Ni 36, bal Fe.
Nickel bonded to Invar. For thermostats. Differential
expansion bimetal. Low electrical resistance. *Obsolete*

TRUFLEX O
Texas Instruments Inc./Materials Control
Ni-Cr-Fe.
For thermostatic bimetal; maximum temperature 400°F.
Obsolete

TRUFLEX P675R
Texas Instruments Inc./Materials Control
Ni-Cr-Fe.
For thermostats, thermal cutouts and circuit breakers;
bimetal, high electrical resistance. ASTM TM2.

TRUFLEX P850R
Texas Instruments Inc./Materials Control
For thermal cutouts and circuit breakers; bimetal, high
electrical resistance. Maximum temperature 800°F.

TRUFLEX Q
Texas Instruments Inc./Materials Control
Ni-Cr-Fe.
For thermostatic bimetal; maximum temperature 1000°F.
Obsolete

TRUFLEX R
Texas Instruments Inc./Materials Control
Ni-Cr-Fe.
For thermostatic bimetal; maximum temperature 700°F.
Obsolete

TRUFLEX T
Texas Instruments Inc./Materials Control
Ni-Cr-Fe.
For thermostatic bimetal; maximum temperature 800°F.
Obsolete

TRUFLEX-25
Texas Instruments Inc./Materials Control
Ni-Cr-Fe.
For thermostatic bimetal; maximum temperature 1100°F.
Obsolete

TRUFORM
Jessop Steel Co.
C 0.9, Mn 1.2, W 0.5, Cr 0.5, bal Fe.
For dies, gages, tools, taps, reamers, screw dies;
nondeforming tool steel. AISI O1.

TRUFORM SPECIAL
Edgar T. Ward's Sons Co.
C 0.9, Mn 1.2, bal Fe.
For tools, dies; non-deforming.

TRUFORM SPECIAL
Jessop-Saville Ltd.
C, alloy, bal Fe.
For tools, dies; non-deforming.

TRUMPET BRASS
Waterbury Rolling Mills Inc.
Cu 83, Sn 1.5, bal Zn.

TRUMPET BRASS-435
Anaconda Co.
Copper. Cu 81, Zn 18, Sn 1.
Soft: 48,000 TS; 60 El. Hard: 80,000 TS; 10 El. For Bourdon
gauge tubes.

TRW 1800
Cannon-Muskegon Corp.
C 0.09, Cr 13, W 9, Cb 1.5, Ti 0.6, Al 6, B 0.07, Zr 0.07, bal Ni.
High temperature alloy; for gas turbine blades, space
vehicles.

TRW 1800
TRW Inc.
C 0.09, Cr 13, W 9, Cb 1.5, Ti 0.6, Al 6, B 0.07, Zr 0.07, bal Ni.
High temperature alloy; for gas turbine blades, space
vehicles.

TRW 1900
Cannon-Muskegon Corp.
Co 10, C 0.11, W 9, Cb 1.5, B 0.03, Zr 0.1, Cr 10.3, bal Ni.
For jet engine and gas turbine components, space vehicles.
High heat and oxidation resistance.

TRW 1900
TRW Inc.
Co 10, C 0.11, W 9, Cb 1.5, B 0.03, Zr 0.1, Cr 10.3, bal Ni.
For jet engine and gas turbine components, space vehicles.
High heat and oxidation resistance.

TRW 361
TRW Inc.
WC 90, Co 10.
400,000 transverse strength; 89.0 Rock A; density 14.52.
Sintered carbide tool material.

TRW 362
TRW Inc.
WC 90, Co 10.
400,000 transverse strength; 88.0 Rock A; density 14.53.
Sintered carbide tool material.

TRW 363
TRW Inc.
WC 88.5, Co 11, 0.5 TaC.
435,000 transverse strength; 89.8 Rock A; density 14.48.
Sintered carbide tool material.

TRW 364
TRW Inc.
WC 87, Co 13.
400,000 transverse strength; 88.5 Rock A; density 14.23.
Sintered carbide tool material.

TRW 366
TRW Inc.
WC 84, Co 16.
400,000 transverse strength; 87.0 Rock A; density 13.90.
Sintered carbide tool material.

TRW MOD 1900
TRW Inc.
C 0.13, Cr 10.3, Co 10, W 9, Ti 1, Al 6.3, Cb 1.5, B 0.03, Zr
0.1, Hf 0.5, Ta 0.5, V 0.5, bal Ni.
High temperature alloy.

TRW VI A
Cannon-Muskegon Corp.
C 0.13, Cr 6.1, Ti 1, Al 5.4, Mo 2, Co 7.5, W 5.8, Re 0.5, Hf
0.43, Zr 0.13, B 0.02, Ta 9, Cb 0.5, bal Ni.
Cast, at 1600°F: 126,000 TS; 112,000 YS; 2.5 El. For cast
turbine blades.

TRW VI A
TRW Inc.
C 0.13, Cr 6.1, Ti 1, Al 5.4, Mo 2, Co 7.5, W 5.8, Re 0.5, Hf
0.43, Zr 0.13, B 0.02, Ta 9, Cb 0.5, bal Ni.
Cast, at 1600°F: 126,000 TS; 112,000 YS; 2.5 El. For cast
turbine blades.

TS 12A

British Steel plc
C 0.17-0.23, Si 0.1-0.35, Mn 0.7-1.5, P 0-0.05, S 0-0.05, Ni 0-0.3, Cu 0-0.25, Nb 0.06-0.1, Sn 0-0.035, 0.02 Al (solution) min, bal Fe.
490-600 N/mm^2 TS; 325 N/mm^2 YS min. For boiler feed pipes.

TSB-1

Techni-Cast Corp.
Copper. Cu 92, Zn 4, Si 4.
60,000 psi TS; 25,000 psi YS; 35 El.

TSB-2

Techni-Cast Corp.
Copper. Cu 82, Zn 14, Si 4.
67,000 psi TS; 30,000 psi YS; 28 El.

TSR BRAKE DIE

St. Lawrence Steel Co.
C 0.5, Mn 1, Cr 1.2, V 0.25, Mo 0.3, bal Fe.
Oil hardened, low alloy tool steel.

TT6

Creusot-Loire
C 0.85, Cr 4.25, Mo 5, V 1.9, W 6, bal Fe.
High speed steel for lathe tools, milling cutters, drills. AFNOR Type 6-5-2; AISI M2.

TT6 C5

Creusot-Loire
C 0.82, Cr 4.25, Mo 5, V 1.85, W 6.35, Co 4.75, bal Fe.
High speed steel, good red hardness, for lathe tools, milling cutters, drills. AFNOR Type 6-5-2-5. Was TRIPLE EXTRA TM-K5.

TT7K12

Russian manufacture
Co 12, 81.0 WC, 4.0 TiC, 3.0 TaC.
For hard cutting tools, wear plates; sintered carbide, abrasion resistant.

TT7K15

Russian manufacture
Co 15, 78.0 WC, 4.0 TiC, 3.0 TaC.
For hard cutting tools, wear plates; sintered carbide, abrasion resistant.

TTV 1

Vallourec S.A.
O 0-0.18, N 0-0.03, H 0-0.015, Fe 0-0.2, C 0-0.1, bal Ti.
Resistant to pitting and corrosion.

TTV 2

Vallourec S.A.
O 0-0.25, N 0-0.03, H 0-0.015, Fe 0-0.3, C 0-0.1, bal Ti.
Resistant to pitting and corrosion.

TTV 3

Vallourec S.A.
O 0-0.35, N 0-0.03, H 0-0.015, Fe 0-0.3, C 0-0.1, bal Ti.
Resistant to corrosion and pitting.

TTV 7

Vallourec S.A.
O 0-0.25, N 0-0.03, H 0-0.015, Fe 0-0.3, C 0-0.1, P 0-0.18, bal Ti.
Corrosion resistant.

TUBE BORIUM

Stoody Company
WC 60, 40.0 steel.
For welding electrodes, hard facing rod; abrasion resistant.

TUBE BRASS

Anaconda Co.
Cu 60-70, Zn 30-40.
Annealed: 44,000-74,400 TS; 11,000-61,000 YS; 19-76 El; 65-77 RA. For condenser tubes. *Obsolete*

TUBE MANDREL

A. Milne & Co.
C 1.15-1.35, Cr 0.35, bal Fe.
For tools, mandrels, cutters; tough.

TUBE-ALLOY

Now MCKAY TUBE-ALLOY.

TUBE-ALLOY 219-O

Teledyne McKay
C 1, Mn 20, Si 0.6, Cr 4.5, bal Fe.
Flux cored open arc surfacing wire. 137,000 psi TS; 91,000 psi YS; 34 El; 16-23 Rock C (as deposited). For railroad track maintenance applications.

TUBE-ALLOY 236-S

Teledyne McKay
C 0.15, Mn 1.6, Si 0.6, Ni 5.3, Mo 5.5, bal Fe.
Metal cored submerged arc surfacing wire. 36-42 Rock C (as deposited). Hardfacing overlay.

TUBE-ALLOY 240-O

Teledyne McKay
C 3.2, Mn 1.8, Si 1.9, Cr 15.5, bal Fe.
Flux cored open arc surfacing wire. 35-52 Rock C (as deposited). For final overlays on roll crusher shells, hammermill hammers, cone crushers, shovel teeth, and augers.

TUBE-ALLOY 242-O

Teledyne McKay
C 0.14, Mn 1.6, Si 0.8, Cr 2.5, bal Fe.
Flux cored open arc surfacing wire. 29-45 Rock C (as deposited). For build-up or surfacing of mild and low alloy steel components.

TUBE-ALLOY 242-S

Teledyne McKay
C 0.16, Mn 1.9, Si 0.8, Cr 1.6, Mo 0.6, V 0.22, bal Fe.
Metal cored submerged arc surfacing wire. 29-45 Rock C (as deposited). Low alloy; medium hardness; martensitic.

TUBE-ALLOY 244-O

Teledyne McKay
C 2.5, Mn 1.6, Si 2, Cr 9, Mo 1.5, Cu 0.5, bal Fe.
Flux cored open arc surfacing wire. 24-40 Rock C (as deposited). For automatic rebuilding of dredge pump shells.

TUBE-ALLOY 250-S

Teledyne McKay
C 0.2, Mn 1.3, Si 0.8, Cr 11, bal Fe.
Metal cored submerged arc surfacing wire. 44-50 Rock C (as deposited). Modified 420 stainless steel composition.

TUBE-ALLOY 252-S

Teledyne McKay
C 0.18, Mn 2.1, Si 0.9, Cr 3.5, bal Fe.
Metal cored submerged arc surfacing wire. 30-48 Rock C (as deposited). For hardsurfacing overlays.

TUBE-ALLOY 255-O

Teledyne McKay
C 4.5, Mn 0.9, Si 0.5, Cr 26.5, bal Fe.
Flux cored open arc surfacing wire. 48-58 Rock C (as deposited). For coke pusher shoes, muller tires, dredging equipment, tamping tools, and fan blades.

TUBE-ALLOY 255-S

Teledyne McKay
C 4.5, Mn 3, Si 0.8, Cr 29, bal Fe.
Metal cored submerged arc surfacing wire. 50-58 Rock C (as deposited). High carbon, high chromium white iron.

TUBE-ALLOY 258-O

Teledyne McKay
C 0.45, Mn 1.4, Si 0.8, Cr 6, Mo 1.5, W 1.5, bal Fe.
Flux cored open arc surfacing wire. 54-57 Rock C (as deposited). For machine components, tools, and sliding metal parts.

TUBE-ALLOY 258-S

Teledyne McKay
C 0.25, Mn 1.6, Si 0.5, Cr 6, Mo 1.5, W 1.4, bal Fe.
Metal cored submerged arc surfacing wire. 46-54 Rock C (as deposited). Martensitic; for overlays on steel mill rolls.

TUBE-ALLOY 258-TIC-O

Teledyne McKay
C 2.1, Mn 1.3, Si 1.8, Cr 7, Mo 1.6, Ti 6, bal Fe.
Flux cored open arc surfacing wire. 48-54 Rock C (as deposited). Martensitic.

TUBE-ALLOY 263-O

Teledyne McKay
C 6, Mn 1, Si 0.5, Cr 23, bal Fe.
Flux cored open arc surfacing wire. 57-63 Rock C (as deposited). High chromium carbide surfacing alloy.

TUBE-ALLOY 821-S

Teledyne McKay
C 0.16, Mn 1.2, Si 0.6, Cr 6, Mo 1.4, W 1.1, bal Fe.
Metal cored submerged arc surfacing wire. 38-45 Rock C (as deposited). Martensitic; for overlays on steel mill rolls.

TUBE-ALLOY 829-O

Teledyne McKay
C 2, Mn 13, Si 0.6, Cr 3.2, Ti 3.5, bal Fe.
Flux cored open arc surfacing wire. 25-31 Rock C (as deposited). Austenitic manganese alloy steel containing titanium carbides. For build-up and surfacing.

TUBE-ALLOY 861-S

Teledyne McKay
C 0.15, Mn 0.9, Si 0.55, Cr 1.5, Mo 0.55, bal Fe.
Metal cored submerged arc surfacing wire. 21-30 Rock C (as deposited). For build-up or overlay for steel mill roll applications.

TUBE-ALLOY 865-S

Teledyne McKay
C 0.18, Mn 1.1, Si 0.4, Cr 12, Mo 1, Ni 2.7, V 0.2, Cb 0.2, bal Fe.
Metal cored submerged arc surfacing wire. 45-48 Rock C (as deposited). For steel mill rolls.

TUBE-ALLOY A250-S

Teledyne McKay
C 0.19, Mn 1, Si 0.6, Cr 11.4, bal Fe.
Metal cored submerged arc surfacing wire. 44-50 Rock C (as deposited). For steel mill rolls.

TUBE-ALLOY AP-O

Teledyne McKay
C 0.42, Mn 16.5, Si 0.3, Cr 13, bal Fe.
Flux cored open arc surfacing wire. 124,000 psi TS; 83,000 psi YS; 40 El; 18-24 Rock C (as deposited). Work hardening austenitic manganese steel alloy.

TUBE-ALLOY BU-A2

Teledyne McKay
C 0.5, Mn 18.4, Si 1.5, Cr 12.8, bal Fe.
Flux cored open arc surfacing wire. 124,000 psi TS; 84,000 psi YS; 37 El; 18-24 Rock C (as deposited). Work hardening austenitic manganese steel alloy. For build-up or overlay

TUBE-ALLOY BU-C1

Teledyne McKay
C 0.07, Mn 1, Si 0.6, Ti 1, bal Fe.
Flux cored open arc surfacing wire. 88,000 psi TS; 70,000 psi YS; 22 El; 20-30 Rock C (as deposited). Low alloy steel for build-up on low alloy pump shells.

TUBE-ALLOY BU-O

Teledyne McKay
C 0.06, Mn 0.9, Si 0.4, Cr 1, Ti 0.5, bal Fe.
Flux cored open arc surfacing wire. 16-22 Rock C (as deposited). Low alloy. For steel mill wobblers, coupling boxes, gear teeth, tractor rails, tractor rollers, and mine car wheels.

TUBE-ALLOY BU-S

Teledyne McKay
C 0.12, Mn 1.8, Si 0.8, Cr 0.7, bal Fe.
Metal cored submerged arc surfacing wire. 20-35 Rock C (as deposited). Low alloy; for build-up on mild and low alloy steel components.

TUBE-ALLOY M932-O

Teledyne McKay
C 0.11, Mn 1.33, Si 0.63, Cr 1.27, Mo 0.48, Ti 0.8, bal Fe.
Flux cored open arc surfacing wire. 37-39 Rock C (as deposited). For rebuilding mild and low alloy steel components including carbon steel railroad track components.

TUBELOY

American Smelting & Refining Co.
Ca 0.02, Mg 0.02, Sn 0.02, bal Pb.
Extruded: 40,000 TS; 15 El; 8 Brin. For water service pipe; corrosion resistant.

TUC-TUR-1

English manufacture
Cu 63, Ni 15, Zn 22.
For machine parts, hardware; close grained.

TUC-TUR-2

English manufacture
Cu 59-61, Sn 13-18, Zn 21-29, Fe 0.3.
For fittings, machine parts.

TUCUNSIL

GTE Sylvania
Ti 10, Ag, bal Cu.
For brazing titanium and stainless steel components.

TUCUNSIL

Western Gold & Platinum Co.
Ti 10, Ag, bal Cu.
For brazing titanium and stainless steel components.

TUF-COR

Airco Vacuum Metals
Mn 1.4, Ni 0.035, Cr 0.05, Mo 0.01, Si 0.58, bal Fe.
Welded: 89,000 psi TS; 78,000 psi YS; 25 El. Good impact strength. Weld deposit: 0.073 C. AWS A5.20 Class E70T-1.

TUF-CUT

Plew Tool Co.
C, Mo, bal Fe.
Water hardened. For tools.

TUF-STUF 224C

Now TUF-STUF 6181(224 C).

TUF-STUF 224E-30

Now MUELLER 6181.

TUF-STUF 224E-75

Now MUELLER 6180.

TUF-STUF 224H

Now TUF-STUF 6240(224 H).

TUF-STUF 224K

Now TUF-STUF 6300 (224 K).

TUF-STUF 6240 (224 H)

Mueller Brass Co.
Cu 86-87, Al 10.5-11.5, Fe 2-3.
Forged: 80,000 TS; 45,000 YS; 10 El; 170 Brin. Heat treated: 90,000 TS; 70,000 YS; 3 El; 200 Brin. For cams, rollers, gears, bearings, trolley shoes, shifter forks. Aluminum bronze. Heat treatable. Wear and corrosion resistant. *Obsolete*

TUF-STUF 6300 (224 K)

Mueller Brass Co.
Cu 78-82, Fe 2-3.5, Ni 4.5-5.5, Mn 0-1.5, Al 9.7-10.9.
Forged: 90,000 TS; 50,000 YS; 20 El; 35 RA; 180-200 Brin.
Heat treated: 105,000 TS; 65,000 YS; 15 El; 20 RA; 200-240 Brin. For valves, gears, worms, shafts, molds. High strength, non-galling, wear resistant; aluminum bronze. Heat treatable. *Obsolete*

TUF-STUFF 224-C, 6230 (224 C)

Mueller Brass Co.
Cu 87.5, Al 9.75, Fe 2.75.
Forged: 60,000-107,000 TS; 35,000-82,000 YS; 15-35 El; 140-225 Brin. Annealed: 75,000-90,000 TS; 40,000-52,000 YS; 15-25 El; 30 RA; 150 Brin. For pickling equipment, piston rods, valve stems, gears, pumps, nuts, impellers, screws, valve seats in oil refineries; corrosion and wear resistant. *Obsolete*

TUF-TIMBER

Carpenter Technology Corp.
C 0.6-1.1, bal Fe.
For general forgings. Water hardening. *Obsolete*

TUF-WEAR 20

Oklahoma Steel Castings Co.
C 0-0.07, Mn 0-1.5, Si 0-1.5, Cr 19-22, Ni 27.5-30.5, Mo 2-3, Cu 3-4, bal Fe.
Corrosion resistant steel casting. ACI CN-7 M.

TUF-WEAR 30

Oklahoma Steel Castings Co.
C 0-0.04, Mn 0-1, Si 0-1, Cr 25-27, Ni 4.75-6, Mo 1.75-2.25, Cu 2.75-3.25, bal Fe.
Corrosion resistant steel casting. ACI CD-4 M Cu.

TUFALOY

Fort Pitt Steel Castings
Cast Iron. C 0.25-0.35, Mn 1.25-1.35, bal Fe.
Normalized: 95,000 TS; 60,000 YS; 25 El; 55 RA; 180 Brin.
Heat treated: 160,000 TS; 110,000 YS; 12 El; 25 RA; 350 Brin.
For gears, shafts, high pressure castings; shock and wear resistant.

TUFALOY

Simpson Bros. Machine Works
C 3.3, Si 2.1, bal Fe.
For rolls; cast iron.

TUFANHARD 150

Hobart Welding Products
C 0.7, Mn 14, Si 0.5, Cr 4, Ni 4, bal Fe.
Weld metal. As welded: 14-18 Rock C; work hardens to 45-50 Rock C. For build-up on manganese steel crusher rolls, dipper buckets, power shovel teeth. DCEP or AC. *Obsolete*

TUFANHARD 160

Hobart Welding Products
C 0.75, Mn 4, Si 0.5, Cr 20, Ni 9, bal Fe.
Weld metal. As welded: 15-20 Rock C; work hardens to 50 Rock C. For overlay of railroad switches, frogs, railends. DCEP or AC. *Obsolete*

TUFANHARD 250

Hobart Welding Products
C 0.2, Mn 0.12, Cr 1.05, bal Fe.
Weld metal. As welded: 23-26 Rock C. For build-up on mild steel where good impact resistance is required. DCEP or DCEN. *Obsolete*

TUFANHARD 320

Hobart Welding Products
C 0.2, Mn 1, Si 0.4, Cr 0.9, Ni 0.3, bal Fe.
Weld metal. As welded: 26-34 Rock C. For miscellaneous build-up. DCEP or AC. *Obsolete*

TUFANHARD 375

Hobart Welding Products
C 0.23, Mn 0.69, Si 0.23, Cr 2.32, Mo 0.18, bal Fe.
Weld metal. As welded: 29-40 Rock C (depending on number of passes). For wear resistance under medium impact conditions. DCEN or DCEP. *Obsolete*

TUFANHARD 400

Hobart Bros. Co.
C, Cr, Mo, bal Fe.
400 Brin. For hard facing electrodes, surface hardening; shielded arc, coated, for moderate abrasion and high impact. *Obsolete*

TUFANHARD 450

Hobart Welding Products
C 2, Mn 1.9, Si 0.5, Cr 2, bal Fe.
Weld metal. As welded: 46-52 Rock C. For bulldozer blades, crusher jaws, bucket lips and teeth. DCEP or AC. *Obsolete*

TUFANHARD 550

Hobart Welding Products
C 3.5, Mn 2, Si 1.5, Cr 12, Mo 0.4, V 0.6, bal Fe.
Weld metal. As welded: 47-54 Rock C. For overlay on hammer mill hammers, coke machinery, grader blades. DCEP or AC. *Obsolete*

TUFANHARD 580

Hobart Welding Products
C 6, Mn 2, Si 2, Cr 26, V 2, bal Fe.
Weld metal. As welded: 48-65 Rock C. For overlay on parts requiring abrasion and corrosion resistance. DCEP or AC. *Obsolete*

TUFANHARD 600

Hobart Welding Products
C 0.29, Mn 0.84, Cr 4.31, Mo 0.28, bal Fe.
Weld metal. As welded: 52-58 Rock C. For overlay on parts subject to severe abrasion. DCEN, DCEP, or AC. *Obsolete*

TUFCUT

Allied Steel & Tractor Products Inc.
C, alloy, bal Fe.
For tools; water hardening.

TUFF-CAST

Tuff-Hard Corp.
C 0.7, W 18, Cr 4, V 1, bal Fe.
For cutting tools, hobs, drills; high speed steel.

TUFFALOY

Rennie Tool Co. Ltd.
C 0.8, V 0.2, bal Fe.
For tools, dies, fixtures; water hardened.

TUFFALOY 100W

Welding Sales & Engineering Co.
Cu-W.
Bar: 200,000 psi TS; 76 Rock A. For resistance welding electrode for nonferrous metals; 30% conductivity, for red brass and copper.

TUFFALOY 10W3

Welding Sales & Engineering Co.
23Cu-77W.
Bar: 160,000 psi TS; 205 Brin. For resistance welding electrode; 25% conductivity. For inserts and facings.

TUFFALOY 1W3

Welding Sales & Engineering Co.
Cu alloy.
Rolled: 135,000 psi TS; 130 Brin. For resistance welding electrode; 35% conductivity. For stainless steel and brass.

TUFFALOY 20W

Airco Welding Products
Cu-W.
For heavy duty projection welding electrodes; Class 12 alloy
Obsolete

TUFFALOY 20W3
Welding Sales & Engineering Co.
Cu-W.
Bar: 170,000 psi TS; 228 Brin. For resistance welding electrode; for heavy projection welding.

TUFFALOY 44
Welding Sales & Engineering Co.
Cu alloy.
Cast: 90,000 psi TS; 60,000 psi YS; 1 El; 330 Brin. Rolled: 140,000 TS; 85,000 psi YS; 2 El; 330 Brin. For resistance welding electrode; for flash, butt and projection welds.

TUFFALOY 53
Airco Welding Products
Cu-W.
Cast: 100,000 TS; 20 El. For resistance welding electrodes; RWMA 3, heat treatable. *Obsolete*

TUFFALOY 55
Airco Welding Products
Cu alloy.
Rolled: 100,000 TS; 50,000 YS; 10 El; 185 Brin. For resistance welding electrodes, current carrying shafts; heat treatable. *Obsolete*

TUFFALOY 55
Welding Sales & Engineering Co.
Be 0.4, Co 2.6, bal Cu.
Cast: 90,000 psi TS; 55,000 psi YS; 10-15 El. Rolled: 125,000 psi TS; 65,000 psi YS; 8-15 El. For springs, bellows, diaphragms; age-hardenable, corrosion resistant.

TUFFALOY 55S
Airco Welding Products
Cu alloy.
For resistance welding electrodes, current carrying shafts; heat treatable. *Obsolete*

TUFFALOY 66
Welding Sales & Engineering Co.
Cu alloy.
Cast: 65,000-75,000 psi TS; 12,000-16,000 psi YS; 10-2 El; 116-165 Brin. For resistance welding electrode; for high strength backing material.

TUFFALOY 77
Welding Sales & Engineering Co.
Cu alloy.
Rolled: 55,000-65,000 psi TS; 25,000-35,000 psi YS; 15 El. Cast: 45,000 psi TS; 20,000 psi YS; 12 El. For resistance welding electrode; 75% conductivity; for steel and brass.

TUFFALOY 88
Welding Sales & Engineering Co.
Cu alloy.
Bar: 50,000-60,000 psi TS; 15,000-20,000 psi YS; 25-20 El; 110-116 Brin. For resistance welding electrode; for coated metals, terneplate, galvanized stock.

TUFKUT
Great Western Steel Co.
C 0.5, Mn 0.7, Si 1.5, Mo 0.35, bal Fe.
For tools, chisels; tough, shock resistant.

TUFTEST
Medart Engineering & Equipment Co.
C 3, Si, Mn, bal Fe.
For grinding discs; cast iron, abrasion resistant. *Obsolete*

TUFWEAR NO. 2
Oklahoma Steel Castings Co.
C, Ni, Mn, bal Fe.
For sleeves, steel castings; oil well drilling machines.

TUFWEAR NO. 3
Oklahoma Steel Castings Co.
C, Ni, Cr, Mo, bal Fe.
For jaw clutches; oil well drilling machines.

TUFWEAR NO. 4
Oklahoma Steel Castings Co.
C, Ni, Mo, bal Fe.
For oil well equipment; tough.

TULA
Manufacturer not listed.
Ag with a small amount of Cu and Pb.
For jewelry, silverware; corrosion resistant.

TULIPE 10
Societe Nouvelle du Saut-du-Tarn
C, Mn, Si, bal Fe.
Heat treated: 50,000-92,000 TS; 30,000 YS; 20 El. For punchings, bolts, rivets; water hardened, AFNOR XC109.

TULIPE 12
Societe Nouvelle du Saut-du-Tarn
C, Mn, Si, bal Fe.
Heat treated: 99,000-171,000 TS; 53,000 YS; 6 El. For bolts, fasteners, forgings; water hardened, AFNOR XC12G.

TULIPE 18
Societe Nouvelle du Saut-du-Tarn
C, Mn, Si, bal Fe.
Heat treated: 78,000-107,000 TS; 54,000 YS; 16 El. For bolts, fasteners, forgings; water hardened, AFNOR XC18G.

TULIPE 18S
Societe Nouvelle du Saut-du-Tarn
C, Mn, Si, bal Fe.
Heat treated: 74,000-103,000 TS; 50,000 YS; 16 El. For bolts, fasteners, forgings; water hardened, AFNOR XC18S

TULIPE 32
Societe Nouvelle du Saut-du-Tarn
C, Mn, Si, bal Fe.
Heat treated: 107,000-128,000 TS; 82,000 YS; 12 El. For shunts, connecting rods, spindles; water hardened, AFNOR XC32G.

TULIPE 35
Societe Nouvelle du Saut-du-Tarn
C, Mn, Si, bal Fe.
Heat treated: 107,000-135,000 TS; 85,000 YS; 11 El. For connecting rods, spindles; water hardened, AFNOR XC35G.

TULIPE 38
Societe Nouvelle du Saut-du-Tarn
C, Mn, Si, bal Fe.
Heat treated: 114,000-144,000 TS; 89,000 YS; 112 El. For machinery parts, gears, bolts, shafts; water hardened, AFNOR XC38G.

TULIPE 42
Societe Nouvelle du Saut-du-Tarn
C, Mn, bal Fe.
Heat treated: 128,000-157,000 TS; 101,000 YS; 9 El. For bolts, fasteners; water hardened, AFNOR XC42G.

TULIPE 48
Societe Nouvelle du Saut-du-Tarn
C, Mn, Si, bal Fe.
Heat treated: 150,000 TS; 96,000 YS; 9 El. For agriculture and machine tool parts; water hardened, AFNOR XC48G.

TULIPE 55
Societe Nouvelle du Saut-du-Tarn
C, Mn, Si, bal Fe.
Heat treated: 135,000-164,000 TS; 98,00 YS; 6 El. For agriculture and machine tool parts; water hardened, AFNOR XC55G.

TULIPE 65
Societe Nouvelle du Saut-du-Tarn
C 0.6-0.8, Mn 0.65-0.9, Si 0-0.4, bal Fe.
Heat treated: 125,000 TS; 65,000 YS. For agricultural equipment; water hardened, AFNOR XC65G.

TULIPE 70
Societe Nouvelle du Saut-du-Tarn
C 0.67-0.75, Mn 0.6-0.85, Si 0-0.4, bal Fe.
Heat treated: 128,000 TS; 70,000 YS. For agricultural equipment; water hardened, AFNOR XC70G.

TULIPE 80
Societe Nouvelle du Saut-du-Tarn
C 0.75-0.85, Mn 0.5-0.75, Si 0-0.4, bal Fe.
Heat treated: 142,000 TS; 72,000 YS. For tool equipment; water hardened, AFNOR XC80G.

TULIPE EXTRA NO. 10
Societe Nouvelle du Saut-du-Tarn
C 0-0.17, Si 0.15-0.4, Mn 0.3-0.6, bal Fe.
Low carbon tool steel. AFNOR Type Y_3 12; Werkstoff Nr. 1.1705.

TULIPE EXTRA NO. 3
Societe Nouvelle du Saut-du-Tarn
C 0.9, Mn 0.4-0.6, bal Fe.
Water hardening tool steel. AFNOR Type Y_3 90; Werkstoff Nr. 1.1760; AISI W1.

TULIPE EXTRA NO. 4
Societe Nouvelle du Saut-du-Tarn
C 0.75, Mn 0.6-0.9, bal Fe.
Water hardening tool steel. AFNOR Type Y_3 75; Werkstoff Nr. 1.1750.

TULIPE EXTRA NO. 5
Societe Nouvelle du Saut-du-Tarn
C 0.65, Mn 0.6-0.9, bal Fe.
Water hardening tool steel. AFNOR Type Y_3 65; Werkstoff Nr. 1.1744.

TULIPE EXTRA NO. 6
Societe Nouvelle du Saut-du-Tarn
C 0.55, Mn 0.6-0.9, bal Fe.
Water hardening tool steel. AFNOR Type Y_3 55; Werkstoff Nr. 1.1735.

TULIPE EXTRA NO. 7
Societe Nouvelle du Saut-du-Tarn
C 0.45, Mn 0.6-0.9, bal Fe.
Water hardening tool steel. AFNOR Type Y_3 45; Werkstoff Nr. 1.1730.

TULIPE EXTRA NO. 8
Societe Nouvelle du Saut-du-Tarn
C 0.35, Mn 0.5-0.7, bal Fe.
Water hardening tool steel. AFNOR Type Y_3 35; Werkstoff Nr. 1.1720.

TUNCRO
Atlantic Steel Corp.
C 0.5, Cr 1.4, V 0.25, W 2, bal Fe.
For chisels, punches, pneumatic tools, hot work dies; hot work steel.

TUNG-ALLOY
Resisto-Loy Company, Inc.
C 0.1, W 35, Mo 16, Co 8, B 0.05, bal Fe.
Welded: 225,000 TS; 675 Brin. For hard surfacing electrodes; high wear resistant; resists heat and abrasion. *Obsolete*

TUNGALOY
Boyd-Wagner Co.
C 1.2, Cr 0.6, Mo 0.3, bal Fe.
For blanking and forming dies; oil or water hardened.

TUNGAY
English manufacture
Zn, Si, Al, Ni, bal Cu.
For corrosion resistant parts; alloyed by special process.

TUNGO
Teledyne Vasco
C 0.5, W 2, Cr 1.65, V 0.25, bal Fe.
Heat treated: 180,000 TS; 160,000 YS; 12 El; 45 RA; 400 Brin.
For punches, chisels, shear knives. *Obsolete*

TUNGROD
Abex Corp.
WC.
For hard facing electrodes; WC encased in a steel tube.

TUNGSIL
Engelhard Corp.
Ag 27-75, bal W.
For electrical contacts for circuit breakers; sintered. *Obsolete*

TUNGSIT
Tungsit Electro-Metals Works Ltd.
W_2C + WC.
For hard facing and rebuilding bits; wear resist.

TUNGSITE
Pyramid Steel Company
C 0.33, Cr 0.94, Mo 0.48, W 0.5, bal Fe.
Heat treated: 164,000 TS; 136,500 YS; 14 El; 43 RA; 275 Brin.
For punches, chisels, gages, shear blades, gears; oil hardened, tough.

TUNGSTEEL
St. Lawrence Steel Co.
C 0.35, Mn 0.65, Si 0.35, Cr 0.75, W 0.45, Mo 0.4, bal Fe.
Oil or water hardened steel for shock resistant applications, as riveting hammers.

TUNGSTEN
Atomergic Chemetals Corp.
W.
Purities: 99.999+%, 99.995% degassed, 99.99%, 99.95%, commercial purity 99+%. Forms: powder, wire, sheet, foil, tubing, crucibles.

TUNGSTEN (COARSE GRAIN)
Fansteel Metals
Tungsten. W 99.9.
Swaged rod: 29 Rock C; electrical conductivity 31% IACS. For electrical contacts. *Obsolete*

TUNGSTEN (FINE GRAIN)
Fansteel Metals
Tungsten. W 99.9.
Swaged rod: 41 Rock C; electrical conductivity 30% IACS. For electrical contacts. *Obsolete*

TUNGSTEN 09991
Fansteel Metals
Tungsten. Ni 1, W 99.
Sintered: electrical conductivity 19% IACS. For electrical contracts. *Obsolete*

TUNGSTEN 21
Ludlow Steel Corp.
C, W, alloy, bal Fe.
Hardenable in oil or water to 54-56 Rock C. Used for arbors, axles, boring bars, mandrels.

TUNGSTEN 26 RHENIUM
Hoskins Mfg.Co.
Tungsten. W 74, Re 26.
For thermocouples to 5200°F.

TUNGSTEN 3 RHENIUM
Hoskins Mfg.Co.
Tungsten. W 97, Re 3.
For thermocouples to 5200°F.

TUNGSTEN 5 RHENIUM
Hoskins Mfg.Co.
Tungsten. W 95, Rc 5.
For thermocouples to 5200°F.

TUNGSTEN 6%
Indiana General
W 6, C 0.75, bal Fe.
For permanent magnets; Br-10300; Hc-65. *Obsolete*

TUNGSTEN ALLOY 10F
Tungsten Alloy Mfg. Co., Inc.
Tungsten.
Submicron grade alloy for use on multiple point solid carbide tools such as reamers, end mills, burrs, drills, slitting saws, and stamping dies. 91.8 Rock A.

TUNGSTEN ALLOY 10T
Tungsten Alloy Mfg. Co., Inc.
Tungsten.
Sintered carbide tool material for cutting tools for general purpose machining of carbon and alloy steels.

TUNGSTEN ALLOY 11C
Tungsten Alloy Mfg. Co., Inc.
Tungsten.
Sintered carbide tool material for light to medium impact applications in mining and rock drilling. *Obsolete*

TUNGSTEN ALLOY 11H
Tungsten Alloy Mfg. Co., Inc.
Tungsten.
Sintered carbide tool material for cutting tools for planing steel when cratering is a factor. *Obsolete*

TUNGSTEN ALLOY 11MH
Tungsten Alloy Mfg. Co., Inc.
Tungsten.
For deep drawing of aluminum cans with thin walls such as beer and soft drink cans. 89.7 Rock A.

TUNGSTEN ALLOY 11T
Tungsten Alloy Mfg. Co., Inc.
Tungsten.
Sintered carbide tool material for cutting tools for roughing operations on carbon and alloy steels. *Obsolete*

TUNGSTEN ALLOY 12C
Tungsten Alloy Mfg. Co., Inc.
Tungsten.
Sintered carbide tool material for tools for mining and rock drilling applications.

TUNGSTEN ALLOY 12H
Tungsten Alloy Mfg. Co., Inc.
Tungsten.
Sintered carbide tool material for cutting tools where cratering and shock are factors as in shaping and planing.

TUNGSTEN ALLOY 12Y
Tungsten Alloy Mfg. Co., Inc.
Tungsten.
For applications requiring medium impact and severe wear resistance such as silicon steel lamination blanking dies, coining and sizing operations, cold extrusion, deep draw dies. 88.8 Rock A.

TUNGSTEN ALLOY 13H
Tungsten Alloy Mfg. Co., Inc.
Tungsten.
Sintered carbide tool material for cutting tools for special applications, such as machining uranium, removal of hot flashwelds, extremely heavy roughing where cratering is a factor.

TUNGSTEN ALLOY 15C
Tungsten Alloy Mfg. Co., Inc.
Tungsten.
Sintered carbide tool material for percussion applications subject to medium to heavy impact.

TUNGSTEN ALLOY 15F
Tungsten Alloy Mfg. Co., Inc.
Tungsten.
Submicron grade alloy for use on multiple point solid carbide tools where heavy shock is present. 89.8 Rock A.

TUNGSTEN ALLOY 16G
Tungsten Alloy Mfg. Co., Inc.
Tungsten.
Sintered carbide tool material for cutting tools for special applications such as machining hot flash from steel welds.

TUNGSTEN ALLOY 20H
Tungsten Alloy Mfg. Co., Inc.
Tungsten.
Sintered carbide tool material for parts and tools subject to heavy impact, galling conditions.

TUNGSTEN ALLOY 25H
Tungsten Alloy Mfg. Co., Inc.
Tungsten.
Sintered carbide tool material for parts and tools subject to very heavy impact, nongalling conditions.

TUNGSTEN ALLOY 30H
Tungsten Alloy Mfg. Co., Inc.
Tungsten.
Sintered carbide tool material for parts and tools subject to extra heavy impact, nongalling conditions.

TUNGSTEN ALLOY 4B
Tungsten Alloy Mfg. Co., Inc.
Tungsten.
Sintered carbide tool material for cutting tools for light finishing cast iron, nonferrous metals and nonmetallics.

TUNGSTEN ALLOY 5H
Tungsten Alloy Mfg. Co., Inc.
Tungsten.
Sintered carbide tool material for cutting cast iron when cratering is a factor. *Obsolete*

TUNGSTEN ALLOY 5S
Tungsten Alloy Mfg. Co., Inc.
Tungsten.
Sintered carbide tool material for fine finishing and precision boring of carbon and alloy steel.

TUNGSTEN ALLOY 6B
Tungsten Alloy Mfg. Co., Inc.
Tungsten.
Sintered carbide tool material for cutting tools for general purpose machining of cast iron, nonferrous metals and nonmetallics.

TUNGSTEN ALLOY 6BH
Tungsten Alloy Mfg. Co., Inc.
Tungsten.
C2 category grade for applications where galling or pickup is encountered. 92.5 Rock A.

TUNGSTEN ALLOY 8H
Tungsten Alloy Mfg. Co., Inc.
Tungsten.
Sintered carbide tool material for high speed planing cast iron where cratering is a factor; also for heat and wear resistant parts as gages.

TUNGSTEN ALLOY 8T
Tungsten Alloy Mfg. Co., Inc.
Tungsten.
Sintered carbide tool material for finishing cuts on steel.

TUNGSTEN ALLOY 9
Tungsten Alloy Mfg. Co., Inc.
Tungsten.
Now TUNGSTEN ALLOY A6.

TUNGSTEN ALLOY 9A
Tungsten Alloy Mfg. Co., Inc.
Tungsten.
Now TUNGSTEN ALLOY A12.

TUNGSTEN ALLOY 9A30

Tungsten Alloy Mfg. Co., Inc.
Tungsten.
Sintered carbide tool material to withstand heavy impact.
Obsolete

TUNGSTEN ALLOY 9B

Tungsten Alloy Mfg. Co., Inc.
Tungsten.
Sintered carbide tool material for cutting tools for the fine finishing and precision boring of cast iron, nonferrous metals and nonmetallics. *Obsolete*

TUNGSTEN ALLOY 9C

Tungsten Alloy Mfg. Co., Inc.
Tungsten.
Now TUNGSTEN ALLOY 4B.

TUNGSTEN ALLOY 9H

Tungsten Alloy Mfg. Co., Inc.
Tungsten.
Now TUNGSTEN ALLOY 6B.

TUNGSTEN ALLOY 9K

Tungsten Alloy Mfg. Co., Inc.
Tungsten.
Sintered carbide tool material for cutting tools for general purpose machining of steel on nonrigid machines. *Obsolete*

TUNGSTEN ALLOY 9M

Tungsten Alloy Mfg. Co., Inc.
Tungsten.
Now TUNGSTEN ALLOY A9.

TUNGSTEN ALLOY 9S

Tungsten Alloy Mfg. Co., Inc.
Tungsten.
Sintered carbide tool material for general purpose and heavy duty machining of carbon and alloy steels.

TUNGSTEN ALLOY 9T

Tungsten Alloy Mfg. Co., Inc.
Tungsten.
Sintered carbide tool material for cutting tools for semifinish cuts on carbon and alloy steels. *Obsolete*

TUNGSTEN ALLOY A12

Tungsten Alloy Mfg. Co., Inc.
Tungsten.
Sintered carbide tool material for wear resistant applications including heavy shock.

TUNGSTEN ALLOY A15

Tungsten Alloy Mfg. Co., Inc.
Tungsten.
Sintered carbide tool material for wear resistant applications, including heavy shock.

TUNGSTEN ALLOY A20

Tungsten Alloy Mfg. Co., Inc.
Tungsten.
Sintered carbide tool material to withstand heavy impact.

TUNGSTEN ALLOY A25

Tungsten Alloy Mfg. Co., Inc.
Tungsten.
Sintered carbide tool material to withstand heavy impact.

TUNGSTEN ALLOY A6

Tungsten Alloy Mfg. Co., Inc.
Tungsten.
Sintered carbide tool material for rough cutting cast iron, nonferrous metals and nonmetallics; wear resistant applications, light shock.

TUNGSTEN ALLOY A9

Tungsten Alloy Mfg. Co., Inc.
Tungsten.
Sintered carbide tool material for wear resistant applications subject to only light shock

TUNGSTEN BRASS

English manufacture
Cu 60, W 2-4, Ni 1-14, Al 0-3, Sn 0-0.2, bal Zn.
For hardware, fittings; corrosion resistant.

TUNGSTEN BRONZE

English manufacture
W 10, bal Cu.
For electrical contacts; wear resistant.

TUNGSTEN DIAMOND

George Cook & Co., Ltd.
C 1.38, W 4.5, Si 0.25, Cr 0.75, Mn 0.35, bal Fe.
For form tools, chasers, reamers, hobs; oil hardened, wear resistant.

TUNGSTEN HACK SAW

Colt Industries
C 1.3, W 1.4, bal Fe.
For hack saw blades; oil hardening. *Obsolete*

TUNGSTEN HACK SAW

Disston Inc.
C 1.2, V 0.1, W 1, bal Fe.
For hack saws; water hardened. *Obsolete*

TUNGSTEN HACKSAW

Jessop Steel Co.
C 1.15, W 1.1, Mn 0.3, bal Fe.
For hacksaws; oil or water hardened, tough. *Obsolete*

TUNGSTEN HOT WORK

A. Milne & Co.
C 0.43, Cr 2.5, V 0.1, W 9, bal Fe.
For hot shearing, hot punches, hot drawing; hot work steel.

TUNGSTEN MAGNET

Allegheny Ludlum Steel
C 0.7, Cr 0.5, W 5, bal Fe.
Permanent magnet. *Obsolete*

TUNGSTEN MAGNET

Latrobe Steel Co.
C 0.7, W 5.3, Cr 0.8, bal Fe.
For permanent magnets. *Obsolete*

TUNGSTEN MAGNET M-5

Colt Industries
C 0.7, W 5.2, bal Fe.
For permanent magnets. *Obsolete*

TUNGSTEN MAGNET M-6

Colt Industries
C 0.7, W 6, bal Fe.
For permanent magnets. *Obsolete*

TUNGSTEN METAL

Foote Mineral Co.
100% and 97% W.
For special metallurgical applications. *Obsolete*

TUNGSTEN NICKEL

Harrison Alloys Inc.
Nickel.
Now HAI-NI 634.

TUNGSTEN SPECIAL

Associated Steel Corp.
C 0.34, Mo 0.5, W 0.5, Cr 1, Mn 0.8, bal Fe.
Heat treated: 167,000 TS; 137,000 YS; 13 El; 43 RA; 280 Brin.
For shafts, chuck jaws, chisels, punches, gears; shock resistant, oil or water hardened.

TUNGSTEN TAP

Colt Industries
C 1.2, W 1.25, bal Fe.
For taps; oil hardening. *Obsolete*

TUNGSTEN TAP

Manufacturer not listed.
C 0.9, W 3.5, bal Fe.
For tools, dies; oil hardened.

TUNGSTEN TAP STEEL

Teledyne Firth Sterling
C 1.15, W 1.4, Cr 0.25, V 0.15, bal Fe.
For taps; water hardened. *Obsolete*

TUNGSTEN-COPPER

Remington Arms Co. Inc.
W 75, 25 C (or as ordered).
Sintered. For use as EDM and ECM electrodes.

TUNGSTEN-RHC

Chase Brass & Copper Co., Inc.
Re 4, Hf 0.35, C 0.25, bal W.
At 3500°F: 75,000 psi TS min. For components for electrical equipment and die casting machines, furnace heating elements, rocket engines. High heat and oxidation resistance.

TUNGTEC 10112

Eutectic Corp.
Tungsten carbide and nickel base alloy powder for hard surfacing.

TUNGTIC TC1

Toshiba Tungaloy Co., Ltd.
Mo 5, bal TiC.
Sintered: 120,000 transverse strength; 93.5 Rock A. For tools, bearings, seals. Resists heat and wear.

TUNGTIC TC2

Toshiba Tungaloy Co., Ltd.
Mo 10, bal TiC.
Sintered: 170,000 transverse strength; 92.5 Rock A. For tools, bearings, seals. Resists heat and wear

TUNGTUBE

Now AIRCO TUNGTUBE.

TUNGUM

Tungum Hydraulics Ltd.
Aluminium-nickel-silicon-brass. Al 0.7-1.2, Cu 81-86, Fe 0-0.25, Ni 0.8-1.4, Si 0.8-1.3, 0.75 total impurities max, bal Zn.
Solution treated and precipitation hardened: Tube: 61,900 TS; 31,400 YS; 40 El; 120 HV. Bar: 67,000 TS; 33,200 YS; 20 El; 120 HV. Forgings: 67,000 TS; 42,300 YS; 25 El; 120 HV. For ambient and cryogenic applications requiring strength together with good fatigue and corrosion resistance.

TUNGUM C

Tungum Hydraulics Ltd.
Cu 75, Al 3, Ni 2, Zn 20.
Cast: 78,500 TS. For castings in marine applications, hardware, shafts. High strength and corrosion resistant. *Obsolete*

TUNGWELD C

Lincoln Electric Co.
WC alloy.
For hard facing electrodes; abrasion resistant.

TUNGWELD F

Lincoln Electric Co.
WC alloy.
For hard facing electrodes; abrasion resistant.

TUNGWIN

Baldwin Steel Co.
C 0.34, Mn 0.78, Si 0.28, Cr 0.82, Mo 0.4, W 0.51, bal Fe.
Heat treated: 194,000-267,000 TS; 185,000-216,000 YS; 37-51 El; 9-15 RA; 395-498 Brin. For arbors, dies, gears, shafts, spindles; oil or water hardened, non-tempering.

TURBADIUM BRONZE
Baldwin-Lima-Hamilton Corp.
Copper. Cu 50.96, Sn 0.47, Pb 0.3, Mn 2.21, Ni 1.75, Zn 43,
Fe 1.36.
For propellers.

TURBADIUM BRONZE
Delta Metal (BW) Ltd.
Cu 48, Sn 0.5, Pb 0.1, Al 0.2, Mn 1.75, Ni 2, Zn 46.5.
Cast: 84,000 TS; 50,000 YS; 18 El; 150 Brin. For propellers,
turbine runners, impellers; non-corrosive, high strength
casting alloy; resists erosion. *Obsolete*

TURBADIUM BRONZE
Baldwin-Lima-Hamilton Corp.
Copper. Cu 50.96, Sn 0.47, Pb 0.3, Mn 2.21, Ni 1.75, Zn 43,
Fe 1.36.
For propellers.

TURBIDE R34
English manufacture
TiC, Cr_3C_2, bal Ni.
For gas turbine blades; sintered, oxidation resistant.

TURBIDE R45
Manufacturer not listed.
TiC, Cr_3C_2, bal Ni.
Sintered: 670 Brin. For gas turbine blades; sintered, oxidation
resistant.

TURBINE ALLOY
Teledyne Vasco
C 0.18-0.23, W 0.75-1.25, Cr 12-14, V 0.4, Mo 0.75-1.25, Ni
0.5-1, bal Fe.
Rolled: 144,000 TS; 122,000 YS; 16 El; 42 RA; 302 Brin. For
turbine buckets and blades, discs, shafts, bolts; corrosion
and heat resistant. *Obsolete*

TURBINE BLADING-1
Manufacturer not listed.
Cu 82.1, Ni 14.7, Al 2.5, Zn 0.7, Si 0.04.
For turbine blades; corrosion resistant.

TURBINE BLADING-2
Manufacturer not listed.
Cu 78-81, Ni 19-21, Fe 0-0.75.
For turbine blades; corrosion resistant.

TURBINE BRASS
English manufacture
Zn 22-32, Pb 0.25, Fe 0.4, bal Cu.
For turbine parts.

TURBINE BUSHING
English manufacture
Zn 11, Ni 15, Sn 2, Pb 10, bal Cu.
For turbine bearings and bushings; heavy duty.

TURBINE METAL
American Manganese Bronze Co.
Copper. Cu 55, Zn 35, Ni 3, Al, Mn, bal Fe.
Cast: 80,000 TS; 40,000 YS; 20 El; 20 RA; 125-160 Brin. For
turbine runners; corrosion and erosion resisting.

TURBISTON BRONZE
American manufacture
Cu 55, Zn 41, Ni 2, Al 1, Fe 0.84, Mn 0.16.
For pistons; highly resistant to sea water; also called a brass.

TURBISTONS BRASS
English manufacture
Zn 41, Fe 0.5, Ni 2, Al 1, Mn 0.2, bal Cu.
For hardware, fittings; corrosion resistant, high strength.

TURBO
Allgemeines Deutsches Metallwerk GmbH
Copper. Cu 77, Al 10, Ni 1.
Cast: 78,000-98,000 TS; 8-15 El; 140-170 Brin. Hot pressed:
98,000-121,000 TS; 10-20 El; 170-190 Brin. For worm wheels,
pumps, spindles, worm shafts. White alloy; resists acid
corrosion.

TURBO "S"
Allgemeines Deutsches Metallwerk GmbH
Copper. Ni 8, Al 12, bal Cu.
Cast: 69,000-86,000 TS; 8-15 El; 150-130 Brin. Hot pressed:
100,800-127,700 TS; 5-15 El; 180-220 Brin. For worm wheels,
pumps, spindles, worm shafts. White alloy; resists acid
corrosion; resists steam at 550°F.

TURBO GLYCO
Joseph T. Ryerson & Son Inc.
Sn, Sb, bal Pb.
For turbine bearings; Babbitt.

TURBOTHERM 12
Vereinigte Edelstahlwerke
C 0-0.1, Si 1, Cr 16, Ni 15, Mo 2, Ta, Nb, bal Fe.
Water quenched: 90,000 TS; 50,000 YS; 39 El; 50 RA. For jet
engine turbochargers; austenitic, heat resistant. *Obsolete*

TURBOTHERM 13CO-10
Vereinigte Edelstahlwerke
C 0.4, Cr 13, Ni 13, Mo 2, Co 10, W 2.5, Cb 3, bal Fe.
Heat treated: 92,500 TS; 49,800 YS. For gas turbine blades,
jet engine parts; heat resistant to 1560 F. *Obsolete*

TURBOTHERM 15M
Vereinigte Edelstahlwerke
C 0.15, Cr 13, Mo 1, bal Fe.
Heat treated: 106,700 TS; 78,300 YS. For gas turbine blades,
jet engine parts; heat resistant to 1076 F. *Obsolete*

TURBOTHERM 1613MV
Now VEW T250.

TURBOTHERM 1613NB
Now VEW T275.

TURBOTHERM 1616M
Now VEW T255.

TURBOTHERM 1810NB
Vereinigte Edelstahlwerke
C 0.1, Cr 18, Ni 10, Nb, bal Fe.
Annealed: 90,000 TS; 45,000 YS; 56 El; 65 RA; 160 Brin. For
corrosion and heat resistant parts; stainless, austenitic.
Obsolete

TURBOTHERM 20 MV
Now VEW T550.

TURBOTHERM 20 MVNB
Now VEW T560.

TURBOTHERM 20 MVW
Now VEW T502.

TURBOTHERM 20CO-20S
Vereinigte Edelstahlwerke
C 0.4, Cr 20, Ni 20, Mo 4, Co 20, W 4, Cb 4, bal Fe.
Heat treated: 113,800 TS; 71,200 YS. For gas turbine blades,
jet engine parts; heat resistant to 1560 F. *Obsolete*

TURBOTHERM 20CO-45
Now VEW T550.

TURBOTHERM 20M
Vereinigte Edelstahlwerke
C 0.2, Cr 13, Mo 1.2, bal Fe.
Heat treated: 106,700 TS; 78,300 YS. For water or steam
units; heat and corrosion resistant. *Obsolete*

TURBOTHERM 35CO-20
Vereinigte Edelstahlwerke
C 0.14, Cr 17, Ni 35, W 4, Mo 4, Co 20, Cb 2.
For gas turbine blades, jet engine parts; heat resistant.
Obsolete

TUREX 4
Societe Nouvelle des Acieries de Pompey
C 0.3, Si 1, Cr 1.3, V 0.2, W 4, bal Fe.
For forging dies, hot shears, rivet sets; hot work steel, oil
hardened.

TURNING-1
Manufacturer not listed.
Cu 58.5, Zn 29, Ni 12, Pb 0-5.
For white metal parts; corrosion resistant.

TURNING-2
Manufacturer not listed.
Cu 65, Zn 22, Ni 12, Pb 1.
For white metal parts; corrosion resistant.

TUSCO
Republic Steel Corp.
C 0.15-0.45, Mn 0.45-0.8, Si 0.1-0.2, Cr 0.4-1.75, Ni 1.25-3.75,
bal Fe.
For machinery parts, gears, shafts. *Obsolete*

TUT-A-BRASE
Allied Steel & Tractor Products Inc.
C, alloy, bal Fe.
For wear resistant plate; preheat treated; abrasion resistant.

TUTANIA, CAST
English manufacture
Sn 92, Sb 4.7, Cu 2.5, Pb 0.3.
For bearings, table and kitchenware; anti-friction.

TUTANIA, ENGLISH
English manufacture
Sn 80, Sb 16, Cu 2.7, Zn 1.3.
For dishes, ornamental pieces; corrosion resistant.

TUTANIA, PLATE
English manufacture
Sn 90-91, Pb 6-8, Cu 0.7-2.7, Zn 0.3-1.3.
For household ware, platters, dishes; free-cutting.

TUTENAG
Manufacturer not listed.
Co 4.54428e+030-1.09002e+027.
For roofing, alloying.

TV2 ALLOY
Russian manufacture
Ti alloy.
For high temperature applications; corrosion resistant.

TWCA-HF-MS-1
Teledyne Wah Chang Albany
Refractory. Zr 0-4.5, bal Hf.
65 MPa TS; 25 MPa YS (0.2% offset); 18 El min (in 2 in.).
Hafnium rod and wire for nuclear industry.

TWIN MO-CO
H. Boker & Co.
C 0.8, Cr 4, V 2, W 6, Mo 5, Co 8, bal Fe.
For lathe and planer tools, hobs, broaches; Type M36; high
speed steel. *Obsolete*

TWIN MO-VA 3
H. Boker & Co.
C, alloy, bal Fe.
For cutting tools; oil hardened. *Obsolete*

TWIN SIX
Allegheny Ludlum Steel
C 0.7, W 6, Mo 6, bal Fe.
For tools, drills, hobs. *Obsolete*

TWIN VAN 96
CCS Braeburn Alloy Steel
C 0.96, Cr 4.25, V 2.1, W 18.5, Mo 0.65, bal Fe.
High speed steel for cutting tools, tungsten type. AISI T2.
Obsolete

TWIN-MO

H. Boker & Co.
C 0.85, Cr 4, W 5.5, Mo 4.5, V 1.5, bal Fe.
For reamers, hobs, drills, taps, broaches; high speed steel;
Type M2. *Obsolete*

TWO STAR SPECIAL

Midvale-Heppenstall Co.
C 0.8, Cr 4, V 2, W 18, Mo 0.55, bal Fe.
For tools, cutters, reamers; high speed steel. *Obsolete*

TWO-TO-ONE

Manufacturer not listed.
Cu 66.7, Zn 33.3.
For drawn or spun brass parts; high ductility.

TWO-TONE H.C.

Resisto-Loy Company, Inc.
C, Cr, Ni, Mo, V, B, bal Fe.
For hard facing electrodes; wear resistant. *Obsolete*

TWO-TONE T.H.

Resisto-Loy Company, Inc.
C 2, Cr 12, Mo 2, Ni 8, B 1, Ti 4, bal Fe.
Welded: 60,000 TS; 540 Brin. For hard facing electrodes;
wear resistant. *Obsolete*

TWOSCORE

Dunford Hadfields Ltd.
C 0-0.25, Cr 16-20, Ni 1-3, bal Fe.
Heat treated: 100,000-168,000 TS; 15-20 El; 250-400 Brin. For
propeller shafts, spindles, rams, seaplane hardware;
corrosion and heat resistant. *Obsolete*

TY-LOY

W.S. Tyler Inc.
C 0.7, Cr 1.2, bal Fe.
For wire screens; abrasion resistant.

TYB-1

Techni-Cast Corp.
Copper. Cu 61, Sn 0.75, Zn 37.5, Al 0.75.
62,000 psi TS; 25,000 psi YS; 40 El.

TYB-2

Techni-Cast Corp.
Copper. Cu 63, Sn 1, Pb 1, Zn 34.75, Al 0.25.
50,000 psi TS; 18,000 psi YS; 40 El.

TYB-3

Techni-Cast Corp.
Copper. Cu 67, Sn 1, Pb 3, Zn 29.
37,000 psi TS; 14,000 psi YS; 38 El.

TYB-4

Techni-Cast Corp.
Copper. Cu 72, Sn 1, Pb 3, Zn 24.
38,000 psi TS; 15,000 psi YS; 38 El.

TYCON

Carpenter Technology Corp.
C 0.4, Mn 0.65, Cr 0.65, Ni 1.75, Mo 0.35, bal Fe.
For gears, shafts, axles; oil hardening. *Obsolete*

TYCROME

Now RYCROME.

TYLER

Time Steel Service Inc.
C 1.1, Mn 0.3, Si 0.25, Cr 0.6, bal Fe.
Water hardened tool steel; AISI W5.

TYLERITE

W.S. Tyler Inc.
C 3.2, Si 2.2, Ni 1.5, bal Fe.
For molding machine parts, match plates; cast iron.
Obsolete

TYPE 5 NICKEL-COBALT

Hanson-Van Winkle-Munning Co.
Ni-Co.
For anodes; electroplating.

TYPE 5 STEEL

Emsco Derrick & Equipment Co.
C 0.18-0.23, Mn 0.7-0.9, Ni 1.6-2, Mo 0.2-0.3, bal Fe.
Rolled: 88,000 TS; 70,000 YS; 35 El; 65 RA; 174 Brin. For oil
well sucker rods; case hardened.

TYPE F

A. Finkl & Sons Co.
Tool material. C 0.55, Cr, Ni, Mo steel.
Prehardened to ordered hardness. Economy hot work steel
for short runs. For hammer dies, saw blocks, and other hot
work tools.

TYPE METAL

American manufacture
Pb 50, Sb 28, Bi 22.
For type casting.

TYPE METAL, COMMON

American manufacture
Pb 56-60, Sb 4.5-30, Sn 10-40.
For type metal.

TYPE METAL, ENGLISH

English manufacture
Pb 55-78, Sb 5-30, Sn 2-35, Cu 0-1.
For type metal.

TYPE METAL, FRENCH

French manufacture
Pb 55-78, Sb 5-30, Sn 2-35, Cu 0-1.
For type metal.

TYPE METAL, GERMAN

German manufacture
Pb 55-78, Sb 5-30, Sn 2-35, Cu 0-1.
For type metal.

TYPE METAL, MONOTYPE

American manufacture
Pb 76, Sb 16, Sn 8.
For type casting.

TYPE METAL, STANDARD

American manufacture
Pb 79, Sb 16, Sn 5.
For type casting.

TYPE METAL, STANDARD

American manufacture
Pb 58, Sb 15, Sn 26, Cu 1.
For type casting.

TYPE METAL, STEREOTYPE

American manufacture
Pb 83.75, Sb 11.75, Sn 4, Cu 0.4.
For type casting.

TYPE SM-1

GTE Sylvania
Ag-Mo.
For electrical contacts; high conductivity facing material.
Obsolete

TYPE SM-2

GTE Sylvania
Ag-Mo.
For electrical contacts; high conductivity facing material
Obsolete

TYPE SM-3

GTE Sylvania
Ag-Mo.
For electrical contacts; high conductivity facing material.
Obsolete

TYPEWRITER METAL

Manufacturer not listed.
Cu 57, Ni 20, Zn 20, Al 3.
For typewriter parts; corrosion resistant.

TYPLEX

Ziv Steel & Wire Co.
C 0.4, Ni 4, Cr 1.5, Mo 0.8, Mn 0.4, bal Fe.
For hot forging dies, gauges; air or oil hardened, shock
resistant.

TYRANN EXTRA

Now VEW K466.

TYSELEY ALLOY

English manufacture
Cu 3.5, Al 8.7, Zn 87, Si 0.3.
For castings, ornaments.

TZ 6

English manufacture
Zn 5-6, Zr 0.4-1, Th 1.5-2.3, bal Mg.
Sand cast and precipitation treated: 255 MPa TS; 155 MPa
YS; 5 El 65 Brin. BS 2970 MAG 9.

TZC

AMAX Corp.
Ti 1.25, Zr 0.3, C 0.15, bal Mo.
Stress Relieved: 144,000 TS; 105,000 YS; 22 El; 36 RA. At
2000°F: 93,000 TS. At 2400°F: 60,000 TS. For aerospace
equipment and components. High hot strength, heat
treatable.

TZC

Universal Cyclops
Ti 1.25, Zr 0.3, C 0.15, bal Mo.
Stress Relieved: 144,000 TS; 105,000 YS; 22 El; 36 RA. At
2000°F: 93,000 TS. At 2400°F: 60,000 TS. For aerospace
equipment and components. High hot strength, heat
treatable.

TZM

AMAX Corp.
Ti 0.5, Zr 0.08, C 0.015, bal Mo.
Recrystallized: 80,000 TS; 55,000 YS; 20 El. At 2000°F:
73,000 TS; 150 DPH hard. At 2400°F: 53,500 TS. For heat
engines, heat exchangers, nuclear reactors, radiation shields,
extrusion dies, boring bars. High strength and hardness at
high temperatures, heat and corrosion resistant.

TZM

Universal Cyclops
Ti 0.5, Zr 0.08, C 0.015, bal Mo.
Recrystallized: 80,000 TS; 55,000 YS; 20 El. At 2000°F:
73,000 TS; 150 DPH hard. At 2400°F: 53,500 TS. For heat
engines, heat exchangers, nuclear reactors, radiation shields,
extrusion dies, boring bars. High strength and hardness at
high temperatures, heat and corrosion resistant.

U-10 ETC.
Now UNIMET U-10 ETC.

U-20
Aluminium Industrie Aktiengesellschaft
Cu 13, Mg 0.25, Mn 1.5, Ni 1, bal Al.
Heat treated: 40,000-43,000 TS; 0.5 El; 120-130 Brin. For pistons; heat resistant.

U-E3-S
French manufacture
Cu alloy.
100 Brin. For aircraft construction.

U-LOY
Republic Steel Corp.
C 0.06-0.1, Mn 0.25-0.4, Cu 0.25, bal Fe.
Rolled: 42,000-45,000 TS; 27,000-30,000 YS; 34-36 El; 42-47 RA. For culverts, roofing; sheets. *Obsolete*

U.B. DOUBLE BOLT
Uddeholm Corp.
C 0.4-0.5, bal Fe.
Annealed: 2 El. For tools, chisels; water hardened. *Obsolete*

U.C.A. 2-2
Uddeholm Corp.
C 0.9, Mn 1.6, Cr 1.5, bal Fe.
640 Brin. For blanking and forming dies, pressing dies; oil hardened, non-deforming. *Obsolete*

U.M.A. (AGATHON) NO. 2
Republic Steel Corp.
C 0.28-0.33, Mn 0.6-0.8, Cr 0.8-1, bal Fe.
Heat treated: 110,000-215,000 TS; 90,000-195,000 YS; 12-42 El; 26-64 RA; 225-430 Brin. For axles, steering knuckles and arms, crankshafts, connecting rods; for general forgings. *Obsolete*

U.M.A. (AGATHON) NO. 3
Republic Steel Corp.
C 0.33-0.38, Mn 0.7-0.9, Cr 0.8-1.1, bal Fe.
Heat treated: 23,000-125,000 TS; 95,000-210,000 YS; 12-27 El; 47-63 RA; 230-460 Brin. For axles, steering arms, steering knuckles; oil hardening. *Obsolete*

U.M.A. (AGATHON) NO. 4
Republic Steel Corp.
C 0.42-0.47, Mn 0.7-0.9, Cr 0.85-1.1, bal Fe.
Heat treated: 115,000-245,000 TS; 90,000-215,000 YS; 10-29 El; 30-63 RA; 255-500 Brin. For gears, machine parts, agricultural implements; oil hardening. *Obsolete*

U.M.A. (AGATHON) NO. 5
Republic Steel Corp.
C 0.47-0.52, Mn 0.7-0.9, Cr 0.85-1.1, bal Fe.
Heat treated: 150,000-265,000 TS; 130,000-235,000 YS; 12-25 El; 31-62 RA; 290-515 Brin. For tools, dies, shear blades, rolls; deep hardening. *Obsolete*

U.M.A. (AGATHON) NO.1
Republic Steel Corp.
C 0.1-0.2, Mn 0.35-0.65, Cr 0.55-0.75, bal Fe.
Heat treated: 85,000-150,000 TS; 60,000-120,000 YS; 10-30 El; 47-72 RA; 275 Brin. Rolled: 65,000-75,000 TS; 149-179 Brin. For case-hardened parts; for carburized work. *Obsolete*

U.M.A. (AGATHON) SPRING
Republic Steel Corp.
C 0.47-0.52, Mn 0.8-1, Cr 1-1.2, bal Fe.
165,000-255,000 TS; 145,000-222,000 YS; 13-20 El; 32-58 RA; 355-490 Brin. For leaf springs, lock nuts; flexible and shock resistant. *Obsolete*

U.S.S. 12
United States Steel Corp.
C 0-0.15, Mn 0-1, Si 0-1, Cr 11.5-13, bal Fe.
Annealed: 75,000 TS; 40,000 YS; 35 El; 70 RA; 155 Brin. For turbine blades, pump rods, mine pumps, valve stems, bolts, coal screens, golf clubs; moderate corrosion and heat resistance.

U.S.S. 12 AL
United States Steel Corp.
C 0-0.08, Cr 11.5-14.5, Al 0.1-0.3, bal Fe.
Annealed: 60,000 TS; 30,000 YS; 25 El; 163 Brin. Drawn: 85,000 TS; 70,000 YS; 20 El; 60 RA; 185 Brin. For oil and chemical industries, petroleum equipment; corrosion and oxidation resistant.

U.S.S. 12 AL
U.S. Steel Corp.
C 0-0.08, Cr 11.5-13.5, Al 0.1-0.3, bal Fe.
Annealed: 70,000 TS; 40,000 YS; 30 El; 60 RA; 150 Brin. Drawn: 85,000 TS; 70,000 YS; 20 El; 60 RA; 185 Brin. For petroleum industry, welded construction; non-air hardening in welding. Type 405. *Obsolete*

U.S.S. 12 F.M.
United States Steel Corp.
C 0-0.15, Mn 1.25, 0.6 max Mo or Zr, 0.15 Si min, bal Fe.
Annealed: 75,000 TS; 40,000 YS; 30 El; 60 RA; 155 Brin.
Heat treated: 110,000 TS; 85,000 YS; 18 El; 55 RA; 212 Brin.
For golf clubs, bolts, shafts, nuts, screws; good machinability, moderate corrosion and heat resistance. Type 416.

U.S.S. 12 TURBINE
United States Steel Corp.
C 0-0.15, Cr 11.5-13, bal Fe.
Annealed: 75,000 TS; 40,000 YS; 35 El; 70 RA; 155 Brin. For turbine blading; corrosion and heat resistant. Type 403.

U.S.S. 17
United States Steel Corp.
C 0-0.12, Mn 0-1, Si 0-1, Cr 16-18, bal Fe.
Annealed: 75,000 TS; 40,000 YS; 30 El; 50 RA; 212 Brin. For tanks for manufacture of HNO_3, auto tire covers, interior decoration; heat and corrosion resistant. Type 430.

U.S.S. 17 HIGH-CARBON
U.S. Steel Corp.
C 0.6-0.75, Mn 0-1, Si 0-1, Cr 16-18, Mo 0-0.75, bal Fe.
Annealed: 105,000 TS; 60,000 YS; 20 El; 45 RA; 215 Brin.
Heat treated: 260,000 TS; 240,000 YS; 5 El; 20 RA; 510 Brin.
For cutlery, valves; heat treatable, corrosion and wear resistant. *Obsolete*

U.S.S. 17-FM
United States Steel Corp.
C 0-0.12, Mn 0-1.25, Si 0-1, Cr 16, 0.6 max Mo or Zr, 0.15 S min, bal Fe.
Annealed: 80,000 TS; 55,000 YS; 25 El; 60 RA; 170 Brin.
Cold drawn: 90,000 TS; 80,000 YS; 15 El; 55 RA; 190 Brin.
For stainless parts; free cutting, stainless.

U.S.S. 18 10 CB
U.S. Steel Corp.
C 0-0.15, Mn 0-0.5, Si 0-0.75, Cr 17-19, Ni 8-12, Cb = 6 to 10% x C, bal Fe.
Annealed: 85,000-100,000 TS; 25,000-40,000 YS; 50-65 El; 60-75 RA. For welded structures in which intergranular corrosion must be avoided; stainless. *Obsolete*

U.S.S. 18-12
U.S. Steel Corp.
Cr 17-19, Ni 11-12.5, C 0.05-0.15, 0.50 Si + Mn, bal Fe.
Annealed: 80,000 TS; 40,000 YS; 60 El; 65 RA; 135 Brin.
Cold worked: 275,000 TS; 30 RA; 000 Brin. For stainless and heat resisting parts for cold working; austenitic. *Obsolete*

U.S.S. 18-8 CB-TA
United States Steel Corp.
C 0-0.08, Cr 17-19, Ni 9-12, Mn 0-2, Si 0-1, Cb = 10 x C, bal Fe.
Annealed: 95,000 TS; 40,000 YS; 45 El. For exhaust stacks, airplane gasoline tanks, welded structures; resists intergranular corrosion. Type 347.

U.S.S. 18-8 COLD ROLLED
U.S. Steel Corp.
C 0-0.12, Mn 0-0.7, Si 0-0.75, 17% min Cr, 7% min Ni, bal Fe.
1/4 hard: 125,000 TS; 75,000 YS; 25 El; 1/2 hard: 150,000 TS; 110,000 YS; 10 El; 3/4 hard: 175,000 TS; 135,000 YS; 5 El; hard: 185,000 TS; 140,000 YS; 4 El. For use in ship, aviation, railroad, bus line, tank truck construction; corrosion resistant. *Obsolete*

U.S.S. 18-8 COLUMBIUM BEARING TYPE 345
U.S. Steel Corp.
C 0-0.15, 17% Cr min, 7% Ni min, Cb = 6 to 10 x C, bal Fe.
Rolled: 80,000 TS; 30,000 YS; 50 El. For exhaust stacks, airplane gasoline tanks, welded structures; resists intergranular corrosion. *Obsolete*

U.S.S. 18-8 COLUMBIUM BEARING TYPE 346
U.S. Steel Corp.
C 0-0.15, 18% min Cr, 8% min Ni, Cb = 6 to 10 x C, bal Fe.
Rolled: 80,000 TS; 30,000 YS; 50 El. For exhaust stacks, airplane gasoline tanks, welded structures; resists intergranular corrosion. *Obsolete*

U.S.S. 18-8 F.M.
United States Steel Corp.
C 0-0.15, Mn 0-2, Si 0-1, Ni 8-10, Cr 17-19, Mo 0-0.6, 0.15 S min, bal Fe.
Annealed: 85,000 TS; 35,000 YS; 35 El; 50 RA; 217 Brin.
Drawn: 110,000 TS; 75,000 YS; 30 El; 50 RA; 240 Brin. For bolts, nuts, shafts, couplings, piston rods, propellers; heat and corrosion resistant and good machinability. Type 303.

U.S.S. 18-8 FS
United States Steel Corp.
C 0-0.12, Mn 0-0.2, Si 0-0.95, Cr 17-19, Ni 10-13, bal Fe.
Annealed: 85,000 TS; 38,000 YS; 50 El. For food and dairy equipment; cold headed parts; corrosion resistant. Type 305.

U.S.S. 18-8 MO
United States Steel Corp.
C 0-0.08, Mn 0-2, Si 0-1, Ni 10-14, Cr 16-18, Mo 2-3, bal Fe.
Annealed: 85,000 TS; 35,000 YS; 50 El; 55 RA; 150 Brin. For textile industry, paper and pulp mill equipment, tubes; corrosion resistant. Type 316.

U.S.S. 18-8 MOLYBDENUM BEARING
U.S. Steel Corp.
C 0-0.08, Mn 0-0.7, Si 0-0.75, Mo 2-4, 17% min Cr, 7% min Ni, bal Fe.
Minimum. 80,000 TS; 30,000 YS; 50 El. For wool dyeing in the textile industry, paper and pulp mill equipment; acid resistant. *Obsolete*

U.S.S. 18-8 S
United States Steel Corp.
C 0-0.08, Si 2, Mn 2, Ni 8-12, Cr 18-20, bal Fe.
Annealed: 85,000 TS; 35,000 YS; 50 El; 135 Brin. For stainless parts, acid resisting apparatus, food and dairy, chemical and oil industries; corrosion resistant; minimized intergranular corrosion. Type 304.

U.S.S. 18-8 S MO
U.S. Steel Corp.
C 0-0.11, Mn 0.5, Si 0-0.75, Cr 17-19, Ni 7-9, Mo 2-4, bal Fe.
Annealed: 95,000-110,000 TS; 35,000-50,000 YS; 40-55 El; 50-65 RA. For heat and corrosion resistant parts; increased resistance to acid corrosion. *Obsolete*

U.S.S. 18-8 SI
United States Steel Corp.
C 0-0.15, Mn 0-2, Si 2-3, Cr 17-19, Ni 8-10, bal Fe.
Annealed: 90,000 TS; 40,000 YS; 50 El; 65 RA; 85 Rock B.
For furnace parts, air preheaters, still liners; oxidation and
heat resistant.

U.S.S. 18-8 STABILIZED
U.S. Steel Corp.
C 0-0.1, Mn 0-2.5, Si 0-1.75, Ni 8-12, Cr 17-20, Ti = 4 x C,
bal Fe.
Annealed: 80,000-95,000 TS; 35,000-45,000 YS; 40-60 El;
135-185 Brin. For food-dairy-oil chemical industries;
resistant to corrosion and oxidation, adaptable to welding.
Obsolete

U.S.S. 18-8 STABILIZED TYPE 320
U.S. Steel Corp.
C 0-0.12, Mn 0.7, Si 0.75, 17% min Cr, 7% min Ni, 5-1/2
(C% - 0.02) Ti, bal Fe.
Minimum: 80,000 TS; 30,000 YS; 50 El. For exhaust stacks,
airplane gasoline tanks, welded structures; resists
intergranular corrosion. *Obsolete*

U.S.S. 18-8 STABILIZED TYPE 321
U.S. Steel Corp.
C 0-0.12, Mn 0.7, Si 0.75, 18% min Cr, 8% min Ni, 5-1/2
(C% - 0.02) Ti, bal Fe.
Minimum: 80,000 TS; 30,000 YS; 50 El. For exhaust stacks,
airplane gasoline tanks, welded structures; resists
intergranular corrosion. *Obsolete*

U.S.S. 18-8 TI
United States Steel Corp.
C 0-0.08, Mn 0-2, Si 0-1, Ni 9-12, Cr 17-19, Ti = 5 x C, bal
Fe.
Annealed: 90,000 TS; 35,000 YS; 50 El; 60 RA; 202 Brin. For
tubes for stills, smoke stacks, gasoline tanks, oil tanks,
aircraft exhaust; heat and corrosion resistance
(1000-1600°F) and when not practical to anneal after
welding. Type 321.

U.S.S. 18-8 TYPE 302
U.S. Steel Corp.
C 0-0.12, Mn 0-0.7, Si 0-0.75, 17% min Cr, 7% min Ni, bal
Fe.
Annealed: 80,000 TS; 30,000 YS; 50 El. For food and dairy
equipment; corrosion resistant. *Obsolete*

U.S.S. 18-8 TYPE 304
U.S. Steel Corp.
C 0-0.08, Mn 0-0.7, Si 0-0.75, 17% min Cr, 7% min Ni, bal
Fe.
Annealed: 80,000 TS; 30,000 YS; 50 El. For food and dairy
equipment; corrosion resistant. *Obsolete*

U.S.S. 18-8 TYPE 306
U.S. Steel Corp.
C 0-0.08, Mn 0-0.7, Si 0-0.75, 18% min Cr, 8% min Ni, bal
Fe.
Annealed: 80,000 TS; 30,000 YS; 50 El. For food and dairy
equipment; corrosion resistant. *Obsolete*

U.S.S. 18-8 TYPE 348
U.S. Steel Corp.
C 0-0.15, 18% min Cr, 8% min Ni, Cb = over 10 x C, bal Fe.
Rolled: 80,000 TS; 30,000 YS; 50 El. For exhaust stacks,
airplane gasoline tanks, welded structures; resists
intergranular corrosion. *Obsolete*

U.S.S. 18-85
United States Steel Corp.
C 0-0.15, Mn 0-2, Si 0-0.75, Ni 8-10, Cr 17-19, bal Fe.
Annealed: 90,000 TS; 40,000 YS; 50 El; 50 RA; 207 Brin. For
food containers, sinks, pipes, shafts, bolts, railroad
passenger cars, inside trim, architectural; heat and
corrosion resistant. Type 302.

U.S.S. 19-9
U.S. Steel Corp.
C 0-0.2, Mn 0-0.75, Si 0-0.75, Ni 8-10, Cr 18-20, bal Fe.
Annealed: 80,000 TS; 35,000 YS; 35 El; 50 RA; 217 Brin. For
trim, lockers, decking on ocean vessels; heat and corrosion
resistance. *Obsolete*

U.S.S. 19-9
U.S. Steel Corp.
C 0-0.2, Mn 0.5, Si 0.75, 8% min Ni, 18% min Cr, bal Fe.
90,000 TS; 50,000 YS; 60 El; 140 Brin. For stainless parts,
chemical apparatus; corrosion and acid resistant. *Obsolete*

U.S.S. 19-9 MO
United States Steel Corp.
C 0-0.08, Mn 0-2, Si 0-1, Cr 18-20, Ni 11-14, Mo 3-4, bal Fe.
Annealed: 90,000 TS; 40,000 YS; 45 El; 85 Rock B. For pulp
and paper mill equipment, chemical industry; corrosion,
heat and creep resistant.

U.S.S. 19-9 TI
U.S. Steel Corp.
C 0-0.2, Mn 0-0.75, Si 0-0.75, Ni 8-12, Cr 18-20, Ti = 4 x
C%, bal Fe.
Stabilized: 80,000 TS; 35,000 YS; 35 El; 50 RA; 217 Brin. For
gasoline and oil tanks for ocean vessels, seamless tubes;
heat and corrosion resistant between 1000-1600 F.
Obsolete

U.S.S. 20-10
U.S. Steel Corp.
C 0-0.2, Mn 0.5, Si 0.75, 9% Ni min, 19% Cr min, bal Fe.
For stainless parts, chemical apparatus; corrosion and acid
resistant. *Obsolete*

U.S.S. 20-10 S
United States Steel Corp.
C 0-0.08, Mn 0-2, Si 0-1.5, Ni 10-12, Cr 19-21, bal Fe.
Annealed: 80,000 TS; 35,000 YS; 35 El; 50 RA; 217 Brin. For
trim and structural purposes, furnace parts, weld rods; heat
and corrosion resistance. Type 308.

U.S.S. 21
United States Steel Corp.
C 0-0.2, Cr 18-23, Si 1, Mn 2, bal Fe.
Annealed: 75,000 TS; 40,000 YS; 22 El; 50 RA; 212 Brin. For
chemical apparatus, oil burner parts; corrosion and heat
resistant.

U.S.S. 25-12
United States Steel Corp.
C 0-0.2, Mn 0-2, Si 0-1, Ni 12-15, Cr 22-24, bal Fe.
Annealed: 90,000 TS; 45,000 YS; 45 El; 45 RA; 217 Brin. For
preheaters, furnace linings, furnace doors, annealing
boxes; heat and corrosion resistant.

U.S.S. 25-12 TI
U.S. Steel Corp.
C 0-0.2, Mn 0-2, Si 0-1, Ni 10-16, Cr 20-26, 4 + C% = Ti, bal
Fe.
Stabilized: 90,000 TS; 40,000 YS; 35 El; 45 RA; 217 Brin. For
rider sheets, preheaters, walking beams, furnace lining
furnace parts, annealing boxes; heat and corrosion
resistant between 1000-2100 F. *Obsolete*

U.S.S. 25-20
United States Steel Corp.
C 0-0.25, Mn 0-2, Cr 24-26, Ni 19-22, Si 0-1.5, bal Fe.
Annealed: 90,000-110,000 TS; 40,000-60,000 YS; 35-50 El;
45-60 RA. For corrosion and heat resisting parts; high heat
resisting.

U.S.S. 25-20 SI
United States Steel Corp.
C 0-0.25, Mn 0-2, Si 1.5-3, Cr 23-26, Ni 19-22, bal Fe.
Annealed: 100,000 TS; 50,000 YS; 40 El; 85 Rock B. For
heat exchangers, furnace doors, carburizing boxes;
oxidation and heat resistant.

U.S.S. 27
United States Steel Corp.
C 0-0.2, Mn 0-1, Si 0-1, Cr 23-27, bal Fe.
Annealed: 80,000 TS; 50,000 YS; 20 El; 50 RA; 217 Brin. For
furnace linings, furnace parts, gas preheaters, annealing
boxes; heat resistant 1500-2000°F. Type 446.

U.S.S. 4/6 CR-MO
U.S. Steel Corp.
C 0-0.2, Cr 4-6, Mn 0.3-0.6, Si 0.15-0.5, Mo 0.45-0.65, bal
Fe.
Annealed: 65,000 TS; 25,000 YS; 20 El; 50 RA; 187 Brin.
Heat treated: 175,000 TS; 110,000 YS; 30 El; 50 RA; 187
Brin. For condensers, hot oil lines, oil still tubes; corrosion
resistant. *Obsolete*

U.S.S. 4/6 CR-MO STEEL
U.S. Steel Corp.
C 0-0.15, Mn 0-0.5, Si 0-0.5, Cr 4-6, Mo 0.45-0.65, bal Fe.
Annealed (minimum): 60,000 TS; 25,000 YS; 30 El; 163 Brin
max. For oil cracking stills, hot oil piping, high temperature
steam piping; higher creep strength than 4/6 Cr Steel.
Obsolete

U.S.S. 4/6 CR-MO STEEL
U.S. Steel Corp.
C 0-0.2, Mn 0-0.5, Si 0-0.5, Cr 4-6, Mo 0.45-0.65, bal Fe.
Annealed (minimum): 60,000 TS; 25,000 YS; 30 El; 170 Brin
max. For oil cracking stills, hot oil piping, high temperature
steam piping; higher creep strength than 4/6 Cr Steel.
Obsolete

U.S.S. 4/6 CR-MO-TI
U.S. Steel Corp.
C 0-0.2, Cr 4-6, Mo 0.45-0.65, Ti = approx 5 + C%, bal Fe.
Hot rolled: 65,000 TS; 25,000 YS; 30 El; 50 RA; 202 Brin. For
oil still tubes and apparatus; for heat and corrosion
resistance. *Obsolete*

U.S.S. 4/6 CR-MO-TI STEEL
U.S. Steel Corp.
C 0-0.15, Mn 0-0.5, Si 0-0.5, Cr 4-6, Mo 0.45-0.65, Ti = 4 x
C, bal Fe.
Annealed (minimum): 60,000 TS; 25,000 YS; 30 El; 163 Brin
max. For oil cracking still tubes, superheater tubes, high
temperature steam piping; does not air harden; higher
corrosion resistance than 4/6 Cr or 4/6 Cr-Mo Steel.
Obsolete

U.S.S. 4/6 CR-TI
U.S. Steel Corp.
C 0-0.15, Cr 4-6, Mn 0.5, Si 0.5, Ti 0.6-0.9, bal Fe.
Annealed: 60,000 TS; 30,000 YS; 40 El. For corrosion
resisting parts; corrosion resistant. *Obsolete*

U.S.S. 5
U.S. Steel Corp.
Mn 0.3-1, Si 0.1-1, Cr 4-6, 0.1% C min, bal Fe.
Annealed: 60,000-75,000 TS; 25,000-35,000 YS; 30-40 El;
60-75 RA. For improved corrosion resistance qualities and
higher strength than medium carbon steels; corrosion
resistant. *Obsolete*

U.S.S. 5 MO
United States Steel Corp.
Cr 4-6, Mo 0.4-0.65, 0.1 C min, bal Fe.
Annealed: 70,000 TS; 30,000 YS; 28 El; 85 RA; 160 Brin.
Heat treated: 115,000 TS; 90,000 YS; 20 El; 60 RA; 240 Brin.
For petroleum industry; high temperature service. Type 501,
scaling temperature 1150°F.

U.S.S. 5S MO
United States Steel Corp.
C 0-0.1, Cr 4-6, Mo 0.4-0.65, bal Fe.
Annealed: 65,000 TS; 25,000 YS; 150 Brin. For petroleum
industry; high temperature service. Type 502, scaling
temperature 1150°F.

U.S.S. 9260
U.S. Steel Corp.
C 0.55-0.65, Mn 0.6-0.9, P 0-0.04, S 0-0.05, Si 1.8-2.2, bal Fe.
Annealed: 200 Brin max. For springs; tough. *Obsolete*

U.S.S. AIR-TEN
U.S. Steel Corp.
C 0.2-1, Mn 0-0.6, P 0-0.04, S 0-0.05, bal Fe.
Rolled: 40,000-100,000 TS; 25,000-100,000 YS. For welded structures; thin gauge sheets. *Obsolete*

U.S.S. C-1210 CU
U.S. Steel Corp.
C 0.07-0.15, Mn 0.3-0.6, P 0-0.11, S 0-0.06, 0.20% min Cu, bal Fe.
Drawn: 60,000-150,000 TS. For wire for galvanized fences. *Obsolete*

U.S.S. COPPER STEEL
U.S. Steel Corp.
C, 0.2% Cu min, bal Fe.
For culverts, building and general fabrication; hot rolled annealed and galvanized sheets. *Obsolete*

U.S.S. COR-TEN
U.S. Steel Corp.
C 0-0.12, Mn 0.2-0.5, P 0.07-0.15, Ni 0.5, Si 0.25-0.75, Cu 0.2-0.5, bal Fe.
Rolled: 70,000-63,000 TS; 50,000-43,000 YS; 22 El. For light weight construction, buses, trucks; corrosion resisting properties.

U.S.S. COR-TEN-B
United States Steel Corp.
C 0-0.1, Mn 0.1-0.3, Si 0.5-1, Cr 0.5-1.5, Cu 0.3-0.5, bal Fe.
Normalized: 70,000 TS; 50,000 YS; 22 El. For pipes where strength superior to low C steel as well as better resistance to corrosion is desired.

U.S.S. DYNAMO
U.S. Steel Corp.
Approx 3.2% Si, bal Fe.
For silicon steel sheets for electrical purposes, dynamo parts; resistivity 50. *Obsolete*

U.S.S. ELECTRICAL
U.S. Steel Corp.
Approx 1.4% Si, bal Fe.
For silicon steel sheets for electrical purposes, motors; resistivity 24. *Obsolete*

U.S.S. FIELD
U.S. Steel Corp.
Approximately 0.25% Si, bal Fe.
For silicon steel sheets for small motors; resistivity 15. *Obsolete*

U.S.S. LYNORE
U.S. Steel Corp.
Cu 0.2, Cr, Ni, bal Fe.
For structural parts; sheets. *Obsolete*

U.S.S. MAN-TEN
U.S. Steel Corp.
C 0-0.25, Mn 1.1-1.6, P 0-0.045, S 0-0.05, 0.15 Si min, 0.20 Cu min, bal Fe.
Rolled: 75,000-65,000 TS; 50,000-40,000 YS; 20 El. For light weight construction, truck and trailer frames, crane masts; high strength, resists atmospheric corrosion.

U.S.S. MANGANESE-NICKEL-COPPER
U.S. Steel Corp.
C 0-0.25, Mn 0-1.5, Ni 0.5-1, Cu 0.25-0.5, bal Fe.
Rolled: 70,000 TS; 50,000 YS; 22 El. For road and agricultural equipment; now Tri-Ten. *Obsolete*

U.S.S. MOTOR
United States Steel Corp.
2.7 Si approx, bal Fe.
For silicon steel sheet for rotating electrical machinery; resistivity 40.

U.S.S. NO. 18-8 STABILIZED
Now U.S.S. 18-8 Ti.

U.S.S. OPEN HEARTH IRON
U.S. Steel Corp.
C 0.02, Mn 0.03, P 0.01, S 0.02, bal Fe.
For culverts, drainage products. *Obsolete*

U.S.S. POLE
U.S. Steel Corp.
C 0.11-0.15, Approximately 0.60% Si, bal Fe.
For poles in rotating electrical machinery; resistivity 18, formerly "Pole Steel." *Obsolete*

U.S.S. RADIO TRANSFORMER 52
U.S. Steel Corp.
Si 4.7, bal Fe.
For audio transformers; sheets, high permeability. *Obsolete*

U.S.S. RADIO TRANSFORMER 72
U.S. Steel Corp.
Approximately 4% Si, bal Fe.
For silicon steel sheets for radio apparatus. *Obsolete*

U.S.S. SIL-TEN
U.S. Steel Corp.
C 0-0.4, 0.6 Mn min, 0.2 Si min, 0.2 Cu min, bal Fe.
Hot rolled: 80,000-95,000 TS; 45,000 YS; 18-30 El. For structural parts, mine cars, bus bodies; 0.20 min Cu added as ordered to increase corrosion resistance.

U.S.S. STRUX
U.S. Steel Corp.
C 0.43, Ni 0.75, V 0.05, Cr 0.9, Mo 0.55, Mn 0.9, B, bal Fe.
Heat treated: 280,000 TS. For aircraft structural parts, bolts, landing gears; tough, shock resistant. *Obsolete*

U.S.S. SUPER-KORE A
U.S. Steel Corp.
C 0.1, Ni 1.8, Cr 0.5, Mo 0.25, 0.3% V min, 0.0005% B min, bal Fe.
Oil hardened: 168,000-180,000 TS; 120,000-150,000 YS; 15-18 El; 52-58 RA; 352 Brin. For gears, shafts, pinions; case-hardening steel. *Obsolete*

U.S.S. SUPER-KORE AA
U.S. Steel Corp.
C 0.15, Ni 1.8, Cr 0.5, Mo 0.25, Mn 0.85, 0.0005% B min, bal Fe.
Oil hardened: 180,000 TS; 160,000 YS; 13 El; 50 RA; 363 Brin. For gears, pinions, shafts; case-hardening, heavily loaded applications. *Obsolete*

U.S.S. SUPER-KORE B
U.S. Steel Corp.
C 0.15, Mn 0.6, Ni 1.8, Mo 0.25, 0.0005% B min, bal Fe.
Oil hardened: 160,000-185,000 TS; 145,000-155,000 YS; 14 El; 51-53 RA; 352 Brin. For gears, shafts, pinions; case-hardening, heavily loaded applications. *Obsolete*

U.S.S. SUPER-KORE C
U.S. Steel Corp.
C 0.15, Ni 0.55, Cr 0.5, Mo 0.2, 0.0005% B min, bal Fe.
Oil hardened: 170,000 TS; 135,000 YS; 11 El; 50 RA; 341 Brin. For gears, pinions, shafts; case-hardening. *Obsolete*

U.S.S. TRANSFORMER 52
United States Steel Corp.
Si 4.5, bal Fe.
For transformers; sheet, high permeability.

U.S.S. TRANSFORMER 58
United States Steel Corp.
4.3 Si approx, bal Fe.
For silicon steel sheet for transformers in rotating electrical equipment radio parts; resistivity 60.

U.S.S. TRANSFORMER 65
United States Steel Corp.
3.8 Si approx, bal Fe.
For silicon steel sheet for transformers in rotating electrical equipment radio parts; resistivity 59.

U.S.S. TRANSFORMER 72
United States Steel Corp.
3.5 Si approx, bal Fe.
For silicon steel sheet for transformers, rotating machinery, electrical purposes; resistivity 57.

U.S.S. VITRENAMEL
United States Steel Corp.
C 0.2, bal Fe.
For enameling purposes, sheet.

U.S.S.12W
U.S. Steel Corp.
C 0-0.12, Cr 12, W, bal Fe.
For corrosion resistant parts; corrosion resistant. *Obsolete*

U.S.S.16-6
U.S. Steel Corp.
C 0.13-0.2, Cr 16, Ni 6, bal Fe.
For corrosion resistant parts; corrosion resistant. *Obsolete*

U.S.S.17W
U.S. Steel Corp.
C 0-0.12, Cr 17, W, bal Fe.
For corrosion and heat resistant parts; corrosion and heat resistant. *Obsolete*

U.S.S.18-8 S MO
U.S. Steel Corp.
C 0-0.11, Cr 18, Ni 8, Mo 2.4, bal Fe.
Annealed: 80,000 TS; 35,000 YS; 50 El; 150 Brin. For stainless and heat resistant parts in pulp mills; stainless, heat resistant. *Obsolete*

U.S.S.19-9S
U.S. Steel Corp.
C 0-0.11, Cr 19, Ni 9, bal Fe.
Annealed: 90,000 TS; 40,000 YS; 55 El; 150 Brin. For heat and corrosion resistant parts; stainless, heat and corrosion resistant. *Obsolete*

U.S.S.20-10S
U.S. Steel Corp.
C 0-0.08, Cr 20, Ni 10, bal Fe.
Annealed: 70,000 TS; 30,000 YS; 55 El; 175 Brin. For chemical and petroleum processing; heat and corrosion resistant, stainless. *Obsolete*

U.S.S.25-20
U.S. Steel Corp.
C 0.13-0.35, Cr 25, Ni 20, bal Fe.
For heat and corrosion resistant parts; heat and corrosion resistant, stainless. *Obsolete*

U.S.S.5W
U.S. Steel Corp.
Cr 5, C, W, bal Fe.
For corrosion resistant parts; corrosion resistant. *Obsolete*

U222
TRW Inc.
Titanium carbide coated carbide. Density: 14.95; 92.0 Rock A. Very high resistance to abrasion; good shock resistance. For semi-roughing and finishing, particularly of austenitic stainless steel.

U225

TRW Inc.
Titanium carbide coated carbide. Density: 12.6; 91.3 Rock A. For semi-roughing, including milling to general purpose machining for improvement in tool life.

U227

TRW Inc.
Titanium carbide coated carbide. Density 12.60; 92.6 Rock A. High resistance to crater and abrasion; for light, high production machining.

UC12

Acciaierie Valbruna s.p.a.
Tool Material. C 0.4, Si 0.4, Mn 1.5, Cr 1.9, Mo 0.2, bal Fe. Hot work tool steel. W. Nr. 1.2312.

UC6

TRW Inc.
WC 93.5, Ni 6.5.
91.2 Rock A; density 14.95. Sintered carbide tool material.

UCAR-75

British Steel Corp.
Mn 75, Al 25.
Hardener for aluminum alloys. Hardener dissolves in aluminum at 1290 F. *Obsolete*

UCHATINS BRONZE

English manufacture
Sn 8, Cu 92.
For bearings, gears, worm wheels; wear resistant.

UCV-60

Lukens Steel
C 0.25, Mn 1.5, Si 0-0.3, Cu 0.25-0.4, Cr 0.4-0.65, V 0.02-0.1, bal Fe.
70,000 psi TS; 63,000 psi YS; 19 El. High-strength, low-alloy steel.

UCV-65

Lukens Steel
C 0.23, Mn 1.55, Si 0.3, V 0.015, bal Fe.
80,000 psi TS; 60,000 psi YS; 15 El. High-strength, low-alloy steel. ASTM 572.

UDDCO 1

Uddeholm Corp.
C 0.14, Cr 5, bal Fe.
Annealed: 70,000 TS; 30,000 YS; 28 El; 65 RA; 160 Brin. For petroleum and oil refinery equipment. Type 501; creep resistant. *Obsolete*

UDDCO 2

Uddeholm Corp.
C 0.14, Cr 5, Mo 0.5, bal Fe.
Annealed: 75,000 TS; 35,000 YS; 25 El; 60 RA; 170 Brin. For petroleum and oil refinery equipment. Type 502; creep resistant. *Obsolete*

UDDCO 3

Uddeholm Corp.
C 0.06, Cr 3, Mo 0.5, bal Fe.
Annealed: 65,000 TS; 30,000 YS; 35 El; 70 RA; 150 Brin. For petroleum and oil refinery equipment. Type 502; creep resistant. *Obsolete*

UDDCO 6

Uddeholm Corp.
C 0.18, Cr 6, Si, bal Fe.
Annealed: 70,000 TS; 35,000 YS; 25 El; 60 RA; 180 Brin. For petroleum and oil refinery equipment. Type 501; creep resistant. *Obsolete*

UDDCO-12

Uddeholm Corp.
C 0.2, Cr 12, Mo 1, V 0.3, W 0.5, bal Fe.
Annealed: 90,000 TS; 40,000 YS; 25 El; 92 Rock B. Hardened: 240,000 TS; 205,000 YS; 9 El; 50 Rock C. For cutlery, valves, bearings, surgical instruments, hardware. Corrosion resistant, hardenable. *Obsolete*

UDDEHOLM 121

Uddeholm Corp.
C 0-0.08, Cr 11.5-13, Al 0.1-0.3, bal Fe.
Annealed: 71,000 TS; 42,600 YS; 22 El; 70 RA; 150 Brin. Heat treated: 175,000 TS; 145,000 YS; 21 El; 64 RA; 352 Brin. For oil refinery and chemical plant equipment. Type 405; corrosion resistant. *Obsolete*

UDDEHOLM 5.75

Uddeholm Corp.
C 0.35, Cr 23-27, bal Fe.
Annealed: 90,000 TS; 60,000 YS; 20 El; 45 RA; 180 Brin. For furnace parts, heat treating boxes; Type 446; heat resistant. *Obsolete*

UDDEHOLM 6.31.26

Uddeholm Corp.
Cr 12-14, 0.15 C min, bal Fe.
Annealed: 88,000 TS; 40,000 YS; 32 El; 68 RA; 170 Brin. For cutlery, valve trim, turbine blades; Type 420; stainless, hardenable. *Obsolete*

UDDEHOLM 63

Uddeholm Corp.
C 0-0.08, Cr 17-19, Ni 9-12, Cb = 10 x C, bal Fe.
Annealed: 85,000-95,000 TS; 35,000-45,000 YS; 50-55 El; 175 Brin. For welded structures, chemical plant equipment. Type 347; corrosion and heat resistant. *Obsolete*

UDDEHOLM STAINLESS 16

Uddeholm Corp.
C 0.27, Cr 13.5, Mo 1.5, bal Fe.
Annealed: 80,000 TS; 40,000 YS; 24 El; 94 Rock B. Hardened: 240,000 TS; 210,000 YS; 9 El; 52 Rock C. For cutlery, surgical instruments, hardware, bearings, valves, gears. Corrosion resistant, hardenable. *Obsolete*

UDDEHOLM STAINLESS 51

Uddeholm Corp.
C 0.1, Cr 14, Mo 1, bal Fe.
Annealed: 80,000 TS; 38,000 YS; 25 El; 93 Rock B. For springs, shafts, table flatware, oil refinery equipment. Corrosion resistant. *Obsolete*

UDDEHOLM STAINLESS 851

Uddeholm Corp.
C 0.2, Cr 11.5, Mo 1, V 0.3, bal Fe.
Annealed: 95,000 TS; 40,000 YS; 25 El; 92 Rock B. Hardened: 240,000 TS; 105,000 YS; 10 El; 50 Rock C. For valves, bearings, cutlery, gears, surgical instruments, shafts. Corrosion resistant, hardenable. *Obsolete*

UDIMAR B-250

Special Metals Corp.
Ferrous. C 0.02, Ni 18, Co 8, Fe 68.5, Mo 5, Al 0.1, Ti 0.4.
At 70°F: 262 ksi TS; 252 ksi YS; 8 El.

UDIMAR B-300

Special Metals Corp.
Ferrous. C 0.02, Ni 18.5, Co 9, Fe 66.7, Mo 5, Al 0.1, Ti 0.07.
At 70°F: 297 ksi TS; 290 ksi YS; 7 El.

UDIMET

Enpar Sonderwerkstoffe GmbH
Nickel.
Alternate manufacturer.

UDIMET

Special Metals Corp.
C 0.05, Cr 18, Fe 18, Mo 3, Cb 5.2, Ti 1, Al 0.06, B 0.004, bal Ni.
Heat treated: At 70°F: 190,000 TS; 165,000 YS; 20 El. At 1200°F: 170,000 TS; 140,000 YS; 18 El. For components of gas turbines, missile and booster assemblies; heat resistant; weldable; -423°F to +1300°F applications.

UDIMET 105

Special Metals Corp.
C 0.2, Mn 1, Si 1, Cr 13.5-15.75, Co 18-22, Mo 4.5-5.5, B 0.01, Al 4.5-4.9, Ti 0.9-1.5, Fe 2, bal Ni.
At 70°F: 166 ksi TS; 112 ksi YS; 12 El. At 1600°F: 9 ksi TS; 67 ksi YS; 24 El. Vacuum melted, precipitation hardened, high temperature alloy.

UDIMET 115

Special Metals Corp.
C 0.2, Mn 1, Si 1, Cr 14-16, Co 13.5-16.5, Mo 3-5, Al 4.5-5.5, Ti 3.5-4.5, Fe 1, bal Ni.
At 70°F: 178 ksi TS; 120 ksi YS; 25 El. At 1600°F: 120 ksi TS; 84 ksi YS; 18 El. At 1800°F; 75 ksi YS; 38 ksi YS; 23 El. Vacuum melted, precipitation hardened, high temperature alloy.

UDIMET 200

Special Metals Corp.
C 0-0.1, Al 0.35, Ti 2.5-3, Mo 5-7, Cr 11-14, Co 0-1, Ni 40-45, Mn 0-2, B 0.01-0.02, bal Fe.
At 70°F: 180 ksi TS; 130 ksi YS; 25 El. At 1200°F: 124 ksi TS; 93 ksi YS; 10 El. Vacuum melted, precipitation hardened, high temperature alloy. Similar to Incoloy 901.

UDIMET 41

Special Metals Corp.
C 0.09, Cr 19, Co 11, Mo 10, Ti 3, Al 1.5, Fe 0-5, bal Ni.
Heat treated: 206,000 TS; 154,000 YS; 14 El; 400 Brin. At 1200°F: 194,000 TS; 145,000 YS; 14 El. At 1700°F: 58,000 TS; 50,000 YS; 26 El. For jet engine components, afterburner parts, bolting; for severely stressed high temperature applications.

UDIMET 500

Special Metals Corp.
Nickel. C 0.08, Cr 18, Ni 54, Co 18.5, Mo 4, Ti 3, Al 3, B 0.006, Zr 0.05.
At 70°F: 190 ksi TS; 122 ksi YS; 32 El. Wrought.

UDIMET 500

Special Metals Corp.
C 0-0.15, Al 2.5-3.2, Ti 2.5-3.2, Mo 3-5, Cr 15-20, Co 13-20, Fe 0-4, bal Ni.
At 70°F: 188,000 psi TS; 15 El; 22 RA. At 1000°F: 185,000 psi TS; 11 El; 16 RA. At 1800°F: 46,000 psi TS; 22 El; 41 RA. For jet engine components; high heat resistance.

UDIMET 500

Cannon-Muskegon Corp.
C 0-0.15, Al 2.5-3.2, Ti 2.5-3.2, Mo 3-5, Cr 15-20, Co 13-20, Fe 0-4, bal Ni.
At 70°F: 188,000 psi TS; 15 El; 22 RA. At 1000°F: 185,000 psi TS; 11 El; 16 RA. At 1800°F: 46,000 psi TS; 22 El; 41 RA. For jet engine components; high heat resistance.

UDIMET 520

Special Metals Corp.
Nickel. C 0.05, Cr 19, Ni 57, Co 12, Mo 6, W 1, Ti 3, Al 2, B 0.005.
At 70°F: 190 ksi TS; 125 ksi YS; 21 El. Wrought.

UDIMET 520

Special Metals Corp.
C 0.05, Cr 19, Mo 6, Ti 3, Al 2, Co 12, W 1, B 0.005, bal Ni.
Heat treated at 70°F: 190,000 TS; 135,000 YS; 21 El. Heat treated at 1200°F: 170,000 TS; 115,000 YS; 17 El. Heat treated at 1800°F: 45,000 TS; 40,000 YS; 22 El. For high temperature applications, jet engine parts; high heat resistance.

UDIMET 520

Cannon-Muskegon Corp.
C 0.05, Cr 19, Mo 6, Ti 3, Al 2, Co 12, W 1, B 0.005, bal Ni.
Heat treated at 70°F: 190,000 TS; 135,000 YS; 21 El. Heat treated at 1200°F: 170,000 TS; 115,000 YS; 17 El. Heat treated at 1800°F: 45,000 TS; 40,000 YS; 22 El. For high temperature applications, jet engine parts; high heat resistance.

UDIMET 625

Special Metals Corp.
Nickel. C 0.05, Mn 0.2, Si 0.2, Ni 61, Cr 21.5, Fe 2.5, Mo 9, Cb 3.6, Ti 0.2, Al 0.2.
At 70°F: 140 ksi TS; 71 ksi YS; 50 El. Good oxidation and corrosion resistance; high strength and toughness from cryogenic temperatures to 2000°F.

UDIMET 630

Cannon-Muskegon Corp.
C 0-0.04, Mn 0-0.2, Si 0-0.2, Cr 17, Co 0-1, Mo 3.1, W 3, Cb 6, Ti 1.1, Al 0.6, B 0.005, Fe 17.5, bal Ni.
For parts operating up to 1000°F.

UDIMET 700

Cannon-Muskegon Corp.
C 0-0.15, Al 3.7-4.7, Ti 3-4, Mo 4.5-5.7, Cr 13-17, Co 17-20, B, bal Ni.
Rolled: 204,000 TS; 140,000 YS; 17 El; 20 RA. Heat treated at 1200°F: 180,000 TS; 140,000 YS; 16 El. Heat Treated at 1400°F: 150,000 TS; 125,000 YS; 33 El. Heat Treated at 1800°F: 50,000 TS; 45,000 YS; 28 El. For turbine blades, discs, combustion chambers; high creep and oxidation resistance.

UDIMET 700

Special Metals Corp.
C 0-0.15, Al 3.7-4.7, Ti 3-4, Mo 4.5-5.7, Cr 13-17, Co 17-20, B, bal Ni.
Rolled: 204,000 TS; 140,000 YS; 17 El; 20 RA. Heat treated at 1200°F: 180,000 TS; 140,000 YS; 16 El. Heat Treated at 1400°F: 150,000 TS; 125,000 YS; 33 El. Heat Treated at 1800°F: 50,000 TS; 45,000 YS; 28 El. For turbine blades, discs, combustion chambers; high creep and oxidation resistance.

UDIMET 700 (POWDER)

Special Metals Corp.
Nickel. C 0.03, Cr 15, Ni 55.4, Co 17, Mo 5, Ti 3.5, Al 4, B 0.02, Zr 0.04.
At 70°F: 136 ksi YS; 28 El. Powder.

UDIMET 700 (WROUGHT)

Special Metals Corp.
Nickel. C 0.08, Cr 15, Ni 53, Co 18.5, Mo 5.2, Ti 3.5, Al 4.3, B 0.03.
At 70°F: 204 ksi TS; 140 ksi YS; 17 El. Wrought.

UDIMET 700 B

Special Metals Corp.
Nickel. C 0.07, Cr 14.5, Ni 57.7, Co 15, Mo 4.2, Ti 3.5, Al 4.3, B 0.015.
Cast.

UDIMET 706

Special Metals Corp.
Nickel. C 0.03, Mn 0.2, Si 0.2, Cr 16, Ni 41.5, Fe 40, Cb 2.9, Ti 1.8, Al 0.2.
At 70°F: 190 ksi TS; 146 ksi YS; 20 El. Wrought.

UDIMET 710

Cannon-Muskegon Corp.
C 0.13, Cr 18, Co 15, Mo 3, W 1.5, Ti 5, Al 2.5, B 0.02, Zr 0.08, bal Ni.
Heat treated (for optimum stress rupture properties): At 1200°F: 190,000 TS; 130,000 YS; 8 El. At 1600°F: 105,000 TS; 95,000 YS; 32 El. At 1800°F: 60,000 TS; 40,000 YS; 30 El. Land based gas turbine blades, high temperature application to 1800°F.

UDIMET 710

Special Metals Corp.
C 0.13, Cr 18, Co 15, Mo 3, W 1.5, Ti 5, Al 2.5, B 0.02, Zr 0.08, bal Ni.
Heat treated (for optimum stress rupture properties): At 1200°F: 190,000 TS; 130,000 YS; 8 El. At 1600°F: 105,000 TS; 95,000 YS; 32 El. At 1800°F: 60,000 TS; 40,000 YS; 30 El. Land based gas turbine blades, high temperature application to 1800°F.

UDIMET 713

Special Metals Corp.
Nickel. C 0.12, Cr 12.5, Ni 74, Mo 4.2, Cb 2, Ti 0.8, Al 6.1, B 0.012, Zr 0.1.
At 70°F: 123 ksi TS; 107 ksi YS; 8 El. Cast.

UDIMET 713 L.C.

Special Metals Corp.
Nickel. C 0.05, Cr 12, Ni 75, Mo 4.5, Cb 2, Ti 0.6, Al 5.9, B 0.01, Zr 0.1.
At 70°F: 130 ksi TS; 109 ksi YS; 15 El. Cast.

UDIMET 718 (CAST)

Special Metals Corp.
Nickel. C 0.05, Mn 0.2, Si 0.2, Cr 19, Ni 52.5, Fe 18.5, Mo 3, Cb 5.2, Ti 0.8, Al 0.6, B 0.006.
At 70°F: 158 ksi TS; 133 ksi YS; 11 El. Cast.

UDIMET 718 (WROUGHT)

Special Metals Corp.
Nickel. C 0.04, Mn 0.2, Si 0.2, Cr 19, Ni 52.5, Fe 18.5, Mo 3, Cb 5.1, Ti 0.9, Al 0.5.
At 70°F: 108 ksi TS; 163 ksi YS; 21 El. Wrought.

UDIMET 720 (BLADE)

Special Metals Corp.
Nickel. C 0.035, Cr 18, Ni 55, Co 15, Mo 3, W 1.25, Ti 5, Al 2.5, B 0.035, Zr 0.035.
At 70°F: 160 ksi TS; 123 ksi YS; 7 El. Wrought.

UDIMET 720 (DISC)

Special Metals Corp.
Nickel. C 0.035, Cr 18, Ni 55, Co 15, Mo 3, W 1.25, Ti 5, Al 2.5, B 0.035, Zr 0.035.
At 70°F: 225 ksi TS; 177 ksi YS; 17 El. Wrought.

UDIMET 738

Special Metals Corp.
Nickel. C 0.17, Cr 16, Ni 61, Co 8.5, Mo 1.7, W 2.6, Cb 0.9, Ti 3.4, Al 3.4, B 0.1, Zr 0.1, Ta 1.7.
At 70°F: 159 ksi TS; 138 ksi YS. Cast.

UDIMET 75

Special Metals Corp.
C 0.08-0.15, Mn 1, Si 1, Cr 18-21, Ti 0.2-0.6, Fe 5, bal Ni.
At 70°F: 108 ksi TS; 40 ksi YS; 42 El. At 1200°F: 76 ksi TS; 30 ksi YS; 44 El. Vacuum melted high temperature alloy.

UDIMET 80A

Special Metals Corp.
Nickel. C 0.06, Mn 0.3, Si 0.3, Cr 19.5, Ni 76, B 0.003, Al 1.4, Ti 2.4, Zr 0.06.
At 70°F: 145 ksi TS; 90 ksi YS; 39 El. Vacuum melted, precipitation hardened, high temperature alloy.

UDIMET 90

Special Metals Corp.
Nickel. C 0.07, Mn 0.3, Si 0.3, Cr 19.5, Ni 59, Co 16.5, B 0.003, Al 1.5, Ti 2.5, Zr 0.06.
At 70°F: 179 ksi TS; 117 ksi YS; 33 El. Vacuum melted, precipitation hardened, high temperature alloy.

UDIMET 901

Special Metals Corp.
Nickel. C 0.05, Cr 12.5, Ni 42.5, Fe 36, Mn 0.1, Si 0.1, Mo 5.7, Ti 2.8, Al 0.2, B 0.015.
At 70°F: 175 ksi TS; 130 ksi YS; 14 El. For turbine discs and parts; for applications in the 1000-1400°F temperature range.

UDIMET A-286

Special Metals Corp.
Nickel. C 0.05, Al 0.2, Ti 2, Mo 1.3, Cr 15, B 0.015, Ni 26, Mn 1.35, Fe 54, V 0.2, Si 0.5.
At 70°F: 146,000 psi TS; 105,000 psi YS. Vacuum melted, precipitation hardened superalloy.

UDIMET C-101

Special Metals Corp.
Nickel. C 0.12, Cr 12.4, Ni 61, Co 9, Mo 1.9, W 3.8, Ti 4.5, Al 3.1, B 0.02, Zr 0.1, Ta 3.9.
At 70°F: 170 ksi TS; 154 ksi YS; 4 El. Cast. Somewhat similar to Inconel 792.

UDIMET D-979

Special Metals Corp.
Nickel. C 0.05, Mn 0.25, Si 0.2, Cr 15, Ni 45, Fe 27, Mo 4, W 4, Ti 3, Al 1, B 0.01.
At 70°F: 204 ksi TS; 146 ksi YS; 15 El. Vacuum melted, precipitation hardened superalloy. Wrought.

UDIMET GMR-235

Special Metals Corp.
Nickel. C 0.15, Cr 15, Ni 69.8, Fe 4, Mo 5, Ti 2.5, Al 3.5, B 0.07.
At 70°F: 112 ksi TS; 103 ksi YS; 3.5 El. Cast.

UDIMET HX

Special Metals Corp.
C 0.05-0.1, Mn 0-1, Si 0-1, Cr 20.5-23, Co 0.5-2.5, W 0.2-1, Mo 8-10, Fe 17-20, bal Ni.
Vacuum melted, high temperature alloy.

UDIMET IN-100

Special Metals Corp.
Nickel. C 0.18, Cr 10, Ni 60, Co 15, Mo 3, Ti 4.7, Al 5.5, B 0.014, Zr 0.06, V 1.
At 70°F: 147 ksi TS; 123 ksi YS; 9 El. Cast.

UDIMET L-605

Special Metals Corp.
C 0.1, Cr 20, Ni 10, W 15, bal Co.
At 1200°F: 103,000 TS; 35,000 YS. Vacuum melted high temperature alloy.

UDIMET M-252

Special Metals Corp.
C 0.15, Al 1, Ti 2.5, Mo 10, Cr 19, Co 10, B 0.007, bal Ni.
At 70°F: 180,000 TS; 120,000 YS; at 1400°F: 130,000 TS; 105,000 YS. Vacuum melted, precipitation hardened superalloy. For turbine buckets, high temperature fasteners.

UDIMET MAR-M-007 (1455)

Special Metals Corp.
Nickel. C 0.08, Cr 8, Ni 65, Co 10, Mo 6, Ti 1, Al 6, B 0.05, Ta 4.
Cast.

UDIMET MAR-M-009 (1422)

Special Metals Corp.
Nickel. C 0.13, Cr 8.5, Ni 59.8, Co 9.5, W 12, Cb 1, Ti 2, Al 5, B 0.015, Hf 2.
Cast.

UDIMET MAR-M-247

Special Metals Corp.
Nickel. C 0.16, Cr 8.2, Ni 60, Co 10, Mo 0.6, W 10, Ti 1, Al 5.5, B 0.02, Zr 0.09, Hf 1.5.
At 70°F: 140 ksi TS; 110 ksi YS; 7 El. Cast.

UDIMET MAR-M-509

Special Metals Corp.
Cobalt. C 0.6, Cr 23.5, Ni 10, Co 55, W 7, Al 0.2, Zr 0.5, Ta 3.5.
At 70°F: 114 ksi TS; 83 ksi YS; 4 El. Cast.

UDIMET MERL 76

Special Metals Corp.
Nickel. C 0.02, Cr 12.5, Ni 54.6, Co 18.5, Mo 3.2, Cb 1.4, Ti 4.3, Al 5, B 0.02, Zr 0.06, Hf 0.4.
At 70°F: 157 ksi YS; 26 El. Powder.

UDIMET N-115

Special Metals Corp.
Nickel. C 0.15, Cr 14.3, Ni 60, Co 13.2, Mo 3.3, Ti 3.7, Al 4.9, B 0.16, Zr 0.04.
At 70°F: 180 ksi TS; 125 ksi YS; 27 El. Wrought.

UDIMET N-155
Special Metals Corp.
C 0.15, Mo 3, Cr 21, Co 20, Ni 10, W 2.5, N 0.15, 1.5 Mn 1.0 Cb, bal Fe.
At 70°F: 118,000 TS; 58,000 YS. At 1200°F: 79,000 TS 43,000 YS. Vacuum melted high temperature alloy.

UDIMET NI-80 A
Special Metals Corp.
C 0.06, Al 1.3, Ti 2.5, Cr 19.5, Co 1.1, bal Ni.
At 70°F: 145,000 psi TS; 90,000 psi YS. At 1200°F: 115,000 psi TS; 80,000 psi YS. Vacuum melted, precipitation hardened superalloy.

UDIMET NICOCRALY
Special Metals Corp.
Nickel. Cr 21, Ni 43.7, Co 22, Al 13, Y 0.3.
Cast.

UDIMET R-41 (CAST)
Special Metals Corp.
Nickel. C 0.09, Cr 19, Ni 55, Co 11, Mo 10, Ti 3.2, Al 1.7, B 0.007.
Cast.

UDIMET R-41 (WROUGHT)
Special Metals Corp.
Nickel. C 0.09, Cr 19, Ni 55, Co 11, Mo 10, Ti 3.1, Al 1.5, B 0.005.
At 70°F: 206 ksi TS; 154 ksi YS; 14 El. Wrought.

UDIMET R-80
Special Metals Corp.
Nickel. C 0.17, Cr 14, Ni 60, Co 9.5, Mo 4, W 4, Ti 5, Al 3, B 0.015, Zr 0.06.
Cast.

UDIMET R-80H
Special Metals Corp.
Nickel. C 0.17, Cr 14, Ni 59.7, Co 9.5, Mo 4, W 4, Ti 4.8, Al 3, B 0.015, Zr 0.01, Hf 0.75.
Cast.

UDIMET S-816
Kulite Tungsten Corp.
Tungsten. C 0.38, Mo 4, Cr 20, Fe 4, Ni 20, W 4, Mn 1.2, Cb 4, bal Co.
At 70°F: 140,000 TS; 67,000 YS. At 1200°F: 112,000 TS; 44,000 YS. Vacuum melted high temperature alloy. *Obsolete*

UDIMET V-57
Special Metals Corp.
Nickel. C 0.08, Mn 0.35, Si 0.75, Al 0.25, Ti 3, Mo 1.25, Cr 14.8, B 0.01, Ni 27, Fe 52.
At 70°F: 170 ksi TS; 120 ksi YS; 26 El. Vacuum melted, precipitation hardened superalloy.

UDIMET WASPALOY
Special Metals Corp.
Nickel. C 0.05, Cr 19.5, Co 13.5, Mo 4.3, Ti 3, Al 1.3, B 0.005, Zr 0.06, Ni 58.
At 70°F: 185 ksi TS; 115 ksi YS; 25 El. For gas turbine blading and discs; high temperature applications up to 1500°F where combined tensile and stress rupture properties are required.

UDIMET X-40
Special Metals Corp.
Cobalt. C 0.5, Mn 0.75, Si 0.75, Cr 25.5, Ni 10.5, Co 54, W 7.5.
At 70°F: 108 ksi TS; 76 ksi YS; 9 El. Cast.

UDIMET X-750
Special Metals Corp.
Nickel. C 0.04, Mn 0.5, Si 0.2, Cr 15.5, Ni 73, Fe 7, Cb 1, Ti 2.5, Al 0.7.
At 70°F: 174 ksi TS; 118 ksi YS; 27 El. Wrought.

UDIMET X-751
Special Metals Corp.
Nickel. C 0.05, Mn 0.5, Si 0.2, Cr 15.5, Ni 72.5, Fe 7, Cb 1, Ti 2.3, Al 1.2.
At 70°F: 200 ksi TS; 150 ksi YS; 50 El. Wrought.

UGICARB
Pechiney/Eurotungstene
WC plus alloy.
For anti-skid tire studs and studded straps; for snow and ice chains.

UGINARC 23 K
Ugine Aciers
C 0-0.02, Ni 9.8, Cr 20, Si 0-0.4, Co 0-0.05, S 0-0.015, P 0-0.02, bal Fe.
Austenitic stainless steel for coated electrodes. AWS E 308 L.

UGINARC 23 S1
Ugine Aciers
C 0-0.025, Ni 10.5, Cr 20, Si 0.9, Co 0-0.05, S 0-0.015, P 0-0.025, bal Fe.
Austenitic stainless steel wire with high silicon for automatic MIG welding. AWS Type ER308LSi.

UGINARC F 36
Ugine Aciers
C 0.12, Ni 9.5, Cr 30, S 0-0.025, P 0-0.03, bal Fe.
Stainless steel for welding. AWS E312 and ER312.

UGINARC MKS
Ugine Aciers
C 0-0.02, Ni 11.3, Cr 18.5, Mo 2.8, bal Fe.
Austenitic stainless steel for coated electrodes. Similar to AISI 316 L.

UGINARC MKS1
Ugine Aciers
C 0-0.025, Ni 12.5, Cr 18.5, Mo 2.7, Si 0.8, bal Fe.
Austenitic stainless steel wire; high silicon grade of AISI 316 L for MIG welding.

UGINARC R 27
Ugine Aciers
C 0.09, Ni 13, Cr 24, S 0-0.015, P 0-0.03, bal Fe.
Austenitic stainless steel for coated electrodes for welding with flux. AWS E 309 and ER 309.

UGINARC R 31
Ugine Aciers
C 0.12, Ni 21, Cr 26, S 0-0.02, P 0-0.03, bal Fe.
Austenitic stainless steel for coated electrodes for welding with flux. AWS E 310 and ER 310.

UGINE 2
Ugine Aciers
C 1.1, bal Fe.
For tools, drills, taps; water hardened. *Obsolete*

UGINE 2P1CS
Ugine Aciers
C 0.85, bal Fe.
For agricultural and edge tools; Type W1, water hardened. *Obsolete*

UGINE 3-E3
Ugine Aciers
C 0.95, bal Fe.
For tools, dies, punches, drills; water hardened. *Obsolete*

UGINE 4-E4
Ugine Aciers
C 0.75, bal Fe.
For tools, springs, punches; water hardening. *Obsolete*

UGINE 5-E5
Ugine Aciers
C 0.75, bal Fe.
For springs, punches; water hardening. *Obsolete*

UGINE AGT
Ugine Aciers
C 0.1, Cr 12.5, Ni 12.5, bal Fe.
Annealed: 68,000-74,000 TS; 28,000-43,000 YS; 45-50 El; 150-160 Brin. Cold drawn: 120,000-128,000 TS; 14 El; 200-210 Brin. For tableware, decorative trim, valve trim; corrosion resistant, good drawability. *Obsolete*

UGINE B 38
Ugine Aciers
C 0.38, Mn 0.7, 40 ppm B, bal Fe.
Boron steel for improved hardenability. Note: B38F grade for cold heading.

UGINE B21
Ugine Aciers
C 0.21, Mn 0.7, 40 ppm B, bal Fe.
For improved hardenability. Boron steel. Note: B21F grade for cold heading.

UGINE BUL
Ugine Aciers
C 1.1, bal Fe.
For chisels, engraving tools, spindles; Type W1, water hardened. *Obsolete*

UGINE C2
Ugine Aciers
C 0.12, bal Fe.
Annealed: 43,000 TS; 30,000 YS; 28 El. Heat treated: 60,000 TS; 35,000 YS; 23 El. For gears, pinions, cams, shafts; SAE 1015; case hardening steel. *Obsolete*

UGINE C4
Ugine Aciers
C 0.4, Mn, bal Fe.
Water hardening steel; for jacks, press columns, shafts. AISI 1040.

UGINE C5
Ugine Aciers
C 0.5, Mn, bal Fe.
Water hardening steel; for heat treated parts. AISI 1050.

UGINE C500
Ugine Aciers
C 0.5, bal Fe.
For gears, shafts, housings; water hardened. *Obsolete*

UGINE C600
Ugine Aciers
C 0.6, bal Fe.
For punches, flat springs, shafts; water hardened. *Obsolete*

UGINE C7
Ugine Aciers
C 0.8, bal Fe.
Annealed: 85,000 TS; 45,000 YS; 13 El. Heat treated: 110,000 TS; 85,000 YS; 10 El. For springs, tools, fixtures, punches; SAE 1080; water hardened. *Obsolete*

UGINE C700
Ugine Aciers
C 0.7, bal Fe.
For punches, flat springs; water hardened. *Obsolete*

UGINE C800
Ugine Aciers
C 0.8, bal Fe.
For tools, dies, punches; Type W1, water hardened. *Obsolete*

UGINE CH110
Ugine Aciers
C 1.1, Cr 1.1, bal Fe.
For blanking and drawing dies, tools; water hardened. *Obsolete*

UGINE CHOC
Ugine Aciers
C 0.9, bal Fe.
For punches, dies, cutters, coining dies; Type W1, water hardened. *Obsolete*

UGINE CU
Ugine Aciers
C 0.1, bal Fe.
Annealed: 35,000-42,000 TS; 20,000-30,000 YS; 26-35 El; 112-140 Brin. For gears, pinions, shafts, cams; SAE 1010; case hardening steel. *Obsolete*

UGINE D100
Ugine Aciers
C, bal Fe.
For machinery parts; extra hard. *Obsolete*

UGINE D36
Ugine Aciers
C 0.12, bal Fe.
Annealed: 40,000 TS; 25,000 YS; 35 El; 110 Brin. For gears, pinions, shafts; SAE 1010; case hardened. *Obsolete*

UGINE DS-TF
Ugine Aciers
C 1.3, W 4.5, bal Fe.
For tools, drills, taps; oil hardened. *Obsolete*

UGINE DY1
Ugine Aciers
C 0.08, Cr 25, N 0.2, bal Fe.
Ferritic type stainless steel; resistant to corrosion at high temperatures. AISI Type 446.

UGINE E1
Ugine Aciers
C 1, bal Fe.
For tools, drills, taps; water hardened. *Obsolete*

UGINE F1
Ugine Aciers
C 0.09, Cr 13, bal Fe.
Martensitic stainless steel. For petroleum industry. Similar to AISI 410.

UGINE F12U
Ugine Aciers
C 0.1, Cr 13, Mo 0.25, S 0.3, bal Fe.
Free machining martensitic stainless steel. Similar to AISI 416.

UGINE F13
Ugine Aciers
C 0.13, Cr 12.8, bal Fe.
Martensitic stainless steel. For cutlery and tablewear; will take high finish. AISI 410.

UGINE F13A
Ugine Aciers
C 0.13, Cr 12, N 0.04, bal Fe.
Martensitic stainless steel. AISI Type 410 for turbines.

UGINE F13B
Ugine Aciers
C 0.15, Cr 12, Cu 0-0.35, bal Fe.
Martensitic stainless steel. For cutlery and tableware; will take high polish.

UGINE F13V
Ugine Aciers
C 0.13, Cr 12.8, bal Fe.
Martensitic stainless steel. For cold heading, bolts and nuts. AISI 410.

UGINE F14PH
Ugine Aciers
C 0-0.06, Ni 4.5, Cr 14.2, Mo 1.4, Cu 3.5, bal Fe.
Precipitation hardening stainless steel; Type 15-5 PH.

UGINE F15
Ugine Aciers
C 0-0.12, Cr 14-18, bal Fe.
Annealed: 70,000 TS; 40,000 YS; 30 El; 55 RA; 150 Brin. Cold drawn: 130,000 TS; 120,000 YS; 2 El; 185 Brin. For oil refinery equipment; bolts, kitchen sinks; Type 430; stainless, ferritic. *Obsolete*

UGINE F17
Ugine Aciers
C 0-0.05, Cr 16.5, bal Fe.
Ferritic type stainless steel; nonhardenable by heat treating. For cold deformation and polishing. AISI 430.

UGINE F17C
Ugine Aciers
C 0.12, Cr 16.8, bal Fe.
Ferritic stainless steel; special grade for cutlery and tableware.

UGINE F17G
Ugine Aciers
C 0.06, Cr 17, S 0-0.015, bal Fe.
Ferritic stainless steel; nonhardenable. For cutlery and tableware. AISI 430.

UGINE F17G1
Ugine Aciers
C 0-0.05, Cr 17, bal Fe.
Ferritic stainless steel; for wire drawing. AISI 430.

UGINE F17M
Ugine Aciers
C 0.05, Cr 16.7, Mo 1, bal Fe.
Ferritic stainless steel; improved corrosion resistance due to molybdenum. AISI Type 434.

UGINE F17U
Ugine Aciers
C 0.09, Cr 16.8, S 0.25, bal Fe.
Free machining, ferritic stainless steel.

UGINE FC
Ugine Aciers
C 0-0.08, Cr 15-18, bal Fe.
Normalized: 78,000-92,000 TS; 15-25 El; 160 Brin. For chemical plant equipment; corrosion resistant. *Obsolete*

UGINE FIA
Ugine Aciers
C 0-0.05, Cr 13, bal Fe.
Normalized: 72,000-85,000 TS; 50,000-65,000 YS; 20-30 El; 150 Brin. For oil refinery equipment; corrosion resistant. *Obsolete*

UGINE FIB
Ugine Aciers
C 0-0.08, Cr 13, Al, bal Fe.
Normalized: 72,000-85,000 TS. For oil refinery equipment; corrosion resistant. *Obsolete*

UGINE FID
Ugine Aciers
C 0.09, Cr 13, bal Fe.
Martensitic stainless steel. For tableware, easily polished to a mirror finish. Similar to AISI 410.

UGINE FJ
Ugine Aciers
C 0.06, Cr 11.5-14.5, Al 0.1-0.3, bal Fe.
Annealed: 70,000 TS; 45,000 YS; 30 El; 160 Brin. For annealing boxes, quenching racks, furnace parts, oil refinery equipment. Type 405 stainless steel, heat and corrosion resistant. *Obsolete*

UGINE FN
Ugine Aciers
C 0.1, Cr 15, Ni 2, bal Fe.
Cast: 85,000 TS; 50,000 YS; 20 El; 40 RA; 180 Brin. For hydraulic turbine wheels; corrosion resistant. *Obsolete*

UGINE FN2
Ugine Aciers
C 0.2, Cr 15-17, Ni 1.25-2.5, bal Fe.
Heat treated: 100,000-135,000 TS; 78,000-112,000 YS; 15-25 El; 45-60 RA; 240-280 Brin. For pump spindles, propeller shafts, aircraft structures; Type 431; stainless, hardenable. *Obsolete*

UGINE G1
Ugine Aciers
C 0.15, Ni 2.8, Cr 0.8, bal Fe.
Annealed: 65,000 TS; 35,000 YS; 179 Brin. Heat treated: 100,000 TS; 90,000 YS; 10 El; 311 Brin. For gears, pinions, shafts, cams, camshafts; SAE 3415; case hardening steel, tough. *Obsolete*

UGINE G11S
Ugine Aciers
C 0.1-0.2, Ni 1.5, Cr 1, Mo 0.2, bal Fe.
Heat treated: 135,000-164,000 TS; 115,000 YS; 10 El. For gears, cams, camshafts, fasteners; case hardened. *Obsolete*

UGINE G12S
Ugine Aciers
C 0.13, Ni 1.4, Cr 1, Mo 0.15, Mn, bal Fe.
Alloy carburizing steel; for case-hardened pinions.

UGINE G14S
Ugine Aciers
C 0.18, Ni 1.3, Cr 1, Mo 0.2, Mn, bal Fe.
Case hardening steel; for large carburized gears, general machinery.

UGINE G1T
Ugine Aciers
C 0.1, Ni 3.5, Cr 0.8, bal Fe.
Annealed: 75,000 TS; 40,000 YS; 187 Brin. Heat treated: 110,000 TS; 90,000 YS; 10 El; 293 Brin. For gears, pinions, shafts, cams, camshafts; SAE 3312; case hardening steel, tough. *Obsolete*

UGINE G2
Ugine Aciers
C 0.14, Ni 2.7, Cr 0.7, Mn, bal Fe.
Carburizing steel; for case hardened gears and pinions, general machinery.

UGINE G4
Ugine Aciers
C 0.17, Ni 3, Cr 0.8, Mn, bal Fe.
Alloy case hardening steel; for gears, pinions, general machinery.

UGINE G5
Ugine Aciers
C 0.15, Ni 3.25, Cr 0.8, bal Fe.
Annealed: 80,000 TS; 42,000 YS; 207 Brin. Heat treated: 145,000 TS; 125,000 YS; 7 El; 370 Brin. For gears, pinions, shafts, cams, camshafts; SAE 3415; case hardening steel, tough. *Obsolete*

UGINE G5T
Ugine Aciers
C 0.1, Ni 4, Cr 0.8, bal Fe.
Annealed: 80,000 TS; 43,000 YS; 207 Brin. Heat treated: 130,000 TS; 100,000 YS; 8 El; 352 Brin. For gears, pinions, shafts, cams, camshafts; SAE 3312; case hardening steel, tough. *Obsolete*

UGINE GB14S
Ugine Aciers
C 0.2, Ni 0.55, Cr 0.5, Mn 0.8, Mo 0.2, 40 ppm B, bal Fe.
Boron steel for improved hardenability. For general machinery and case hardened gears and pinions.

UGINE GB5
Ugine Aciers
C 0.19, Ni 1.3, Cr 1, Mn 0.7, 40 ppm B, bal Fe.
Boron steel for improved hardenability. For large case-hardened parts resistant to overload and fatigue.

UGINE HD20
Ugine Aciers
C 0.21, Ni 0.6, Cr 0.5, Mo 0.2, Mn, bal Fe.
Carburizing steel; pinions for trucks and tractors. AISI 8620.

UGINE IK
Ugine Aciers
C 1, Cr 4, Mo 1.1, V 0.4, bal Fe.
For molds, plugs, gauges; air hardening, non-deforming. *Obsolete*

UGINE K12
Ugine Aciers
C 2.25, Cr 13, bal Fe.
For tools, dies; non-deforming. *Obsolete*

UGINE K2C
Ugine Aciers
C 0.2, Ni 1.3, Cr 1, Mo 0.1, bal Fe.
Annealed: 55,000 TS; 30,000 YS; 179 Brin. Heat treated: 105,000 TS; 88,000 YS; 10 El; 320 Brin. For gears, pinions, shafts, cams, camshafts; case hardening steel, shock resistant. *Obsolete*

UGINE K2D
Ugine Aciers
C 0.15, Ni 1.3, Cr 1, Mo 0.2, bal Fe.
Annealed: 60,000 TS; 30,000 YS; 187 Brin. Heat treated: 125,000 TS; 95,000 YS; 9 El; 340 Brin. For gears, pinions, shafts, cams, camshafts; SAE 4315; case hardening steel, tough. *Obsolete*

UGINE K2F
Ugine Aciers
C 0.2, Ni 1.6, Cr 1, Mo 0.25, bal Fe.
Annealed: 64,000 TS; 35,000 YS; 187 Brin. Heat treated: 140,000 TS; 120,000 YS; 7 El; 401 Brin. For gears, pinions, shafts, cams, camshafts; SAE 4120; case hardening steel, tough. *Obsolete*

UGINE KI17
Ugine Aciers
C 1, Cr 17, Mo 0.5, bal Fe.
Martensitic stainless steel; for stainless ball bearings. AISI 440 C.

UGINE KMD
Ugine Aciers
C 0.2, Cr 1, Mo 0.3, bal Fe.
Normalized: 78,000 TS; 50,000 YS; 12 El; 197 Brin. Heat treated: 100,000 TS; 70,000 YS; 11 El; 277 Brin. For welded structures; requires no heat treatment after welding. *Obsolete*

UGINE KMT
Ugine Aciers
C 0.35, Cr 1, Mo 0.25, Mn, bal Fe.
Oil hardened steel; shafts and gears with high fatigue resistance. AISI 4135.

UGINE KMX
Ugine Aciers
C 0.42, Cr 1, Mo 0.25, Mn, bal Fe.
Oil hardened steel; for petroleum industry. ASTM A193 Gr. B7.

UGINE KMZ
Ugine Aciers
C 0.3, Cr 3, Mo 0.4, Mn, bal Fe.
Oil hardening steel, for large machinery parts. Can be nitrided.

UGINE KN9
Ugine Aciers
C 0.36, Ni 2.8, Cr 2.8, Mo 0.3, Mn, bal Fe.
Air or oil hardened steel; for large shafts and gears, for good fatigue and shock resistance.

UGINE KNA
Ugine Aciers
C 0.35, Ni 3.8, Cr 1.7, Mn, bal Fe.
Oil hardening steel; for large and highly stressed parts.

UGINE KNAMO
Ugine Aciers
Ni 4.25, Cr 1.3, Mo 0.3, C, bal Fe.
Annealed: 100,000 TS; 241 Brin. Heat treated: 190,000 TS; 160,000 YS; 5 El; 477 Brin. For tools, dies, fixtures; air hardened, shock resistant. *Obsolete*

UGINE KNB
Ugine Aciers
C 0.35, Ni 2.8, Cr 0.8, bal Fe.
Annealed: 70,000 TS; 35,000 YS; 201 Brin. Heat treated: 85,000 TS; 70,000 YS; 235 Brin. For gears, shafts, pinions; SAE 3435; shock resistant, oil hardened. *Obsolete*

UGINE KNDMO
Ugine Aciers
C 0.33, Ni 4, Cr 1.8, Mo 0.4, Mn, bal Fe.
Air or oil hardening steel; for large or highly stressed parts requiring high elastic limit and endurance.

UGINE KNE
Ugine Aciers
C 0.35, Ni 3.5, Cr 1, bal Fe.
Annealed: 78,000 TS; 40,000 YS; 217 Brin. Heat treated: 165,000 TS; 142,000 YS; 7 El; 444 Brin. For gears, pinions, crankshafts; SAE 3435; shock resistant, oil hardened. *Obsolete*

UGINE KNH
Ugine Aciers
C 0.35, Ni 2.8, Cr 0.8, bal Fe.
Annealed: 85,000 TS; 229 Brin. Heat treated: 195,000 TS; 165,000 YS; 6 El; 460 Brin. For gears, pinions, shafts; SAE 3435; shock resistant, oil hardened. *Obsolete*

UGINE KNHMO
Ugine Aciers
C 0.3, Ni 4, Cr 1.3, Mo 0.4, Mn, bal Fe.
Oil hardening steel; for highly stressed parts, general machinery and aviation.

UGINE KNMO
Ugine Aciers
C 0.35, Ni 2.8, Cr 0.8, Mo 0.25, bal Fe.
Annealed: 78,000 TS; 40,000 YS; 241 Brin. Heat treated: 165,000 TS; 150,000 YS; 7 El; 444 Brin. For gears, pinions, crankshafts; SAE 4335; shock resistant, oil hardened. *Obsolete*

UGINE KNO
Ugine Aciers
C 0.35, Ni 1.4, Cr 1, Mo 0.23, Mn, bal Fe.
Oil hardened steel; for shafts and gears for good fatigue and shock resistance.

UGINE KOR
Ugine Aciers
C 2.1, Cr 13, V 0.1, Mo, bal Fe.
For blanking and forming dies, shear blades; non-deforming, oil hardened. *Obsolete*

UGINE KORS
Ugine Aciers
C 1.75, Cr 13, V 0.9, Mo, bal Fe.
For shear blades, blanking and forming dies; non-deforming, cold work steel. *Obsolete*

UGINE KS
Ugine Aciers
C 0.95, Cr 1.5, bal Fe.
Annealed: 201 Brin. Heat treated: 600 Brin. For ball and roller bearings; SAE 52100; water or oil hardening. *Obsolete*

UGINE KS2
Ugine Aciers
C 0.9, Cr 2, bal Fe.
For cold rolling rolls; high surface hardness. *Obsolete*

UGINE KT
Ugine Aciers
C 0.4, Cr 0.8, W 1.6, V 0.2, bal Fe.
For punches, headers, crimpers, upsetters; hot work steel, oil hardened. *Obsolete*

UGINE KTD
Ugine Aciers
C 0.5, Cr 0.8, W 1.6, Si 0.7, bal Fe.
For hot punches and shear blades, chisels; high shock resistant, oil hardened. *Obsolete*

UGINE KTH
Ugine Aciers
C 1, Cr 1, W 1, Mn 1, bal Fe.
For milling cutters, taps, drills; water or oil hardened, wear resistant. *Obsolete*

UGINE KVR
Ugine Aciers
C 0.5, Cr 1, V 0.15, bal Fe.
Annealed: 75,000 TS; 38,000 YS; 217 Brin. Heat treated: 200,000 TS; 185,000 YS; 477 Brin. For springs, gears, shafts; SAE 6150; oil hardened. *Obsolete*

UGINE KW
Ugine Aciers
C 1.1, Cr 0.8, W, bal Fe.
For cutting tools, dies, taps, reamers; water hardened, wear resistant. *Obsolete*

UGINE M51S
Ugine Aciers
C 0-0.08, Cr 17-20, Mn 9-12, Ni 1.5-2.5, Ti, bal Fe.
Annealed: 85,000-100,000 TS; 58,000-72,000 YS; 30-40 El; 170-180 Brin. For chemical plant equipment; corrosion resistant. *Obsolete*

UGINE MA
Ugine Aciers
C 0.35, Ni 4.5, Cr 1.5, Mo 0.5, bal Fe.
For hot stamping dies; hot work steel, oil hardened. *Obsolete*

UGINE MARS
Ugine Aciers
C 0.4, Cr 3.5, W 15, V 0.1, bal Fe.
For extrusion dies, rams, punches; hot work steel, oil hardened. *Obsolete*

UGINE MB20
Ugine Aciers
C 0.2, Mn 1.2, 40 ppm B, bal Fe.
Boron steel for improved hardenability. Note: MB20F for cold heading.

UGINE MB38
Ugine Aciers
C 0.38, Mn 1.2, 40 ppm B, bal Fe.
Boron steel for improved hardenability. Note: MB38F for cold heading.

UGINE MBA
Ugine Aciers
C 0.35, Cr 3, W 3, V 0.5, bal Fe.
For extrusion and stamping dies, punches, molds; hot work steel, resists repeated shock. *Obsolete*

UGINE MFC
Ugine Aciers
C 0.9, Cr 0.4, bal Fe.
For punches and dies for bolt work; cold work steel. *Obsolete*

UGINE MN 12
Ugine Aciers
C 1-1.2, Mn 12-13.5, bal Fe.
For wear plates, railroad frogs; wear and abrasion resistant.
Obsolete

UGINE MP14
Ugine Aciers
C 0.3, Cr 13, bal Fe.
Annealed: 95,000 TS; 50,000 YS; 25 El; 55 RA; 200 Brin. For
plastic molding dies; high abrasion resistant, Type 420,
stainless. *Obsolete*

UGINE MS
Ugine Aciers
C 0.8, Si 1.9, Mn 0.7, bal Fe.
Annealed: 85,000 TS; 45,000 YS; 229 Brin. Heat treated:
150,000 TS; 125,000 YS; 7 El; 429 Brin. For springs; water
or oil hardened. *Obsolete*

UGINE MSS
Ugine Aciers
C 0.9, Si 1.3, Mn 0.7, bal Fe.
Annealed: 84,000 TS; 45,000 YS; 229 Brin. Heat treated:
160,000 TS; 140,000 YS; 6 El; 450 Brin. For springs; oil
hardened. *Obsolete*

UGINE MTC
Ugine Aciers
C 0.38, Cr 5, Mo 1.35, W 1.25, V, bal Fe.
For dies, molds, anvils, rams; hot work steel. *Obsolete*

UGINE N2C
Ugine Aciers
C 0.1, Ni 2, bal Fe.
Annealed: 40,000 TS; 25,000 YS; 30 El; 123 Brin. Heat
treated: 75,000 TS; 60,000 YS; 17 El; 155 Brin. For gears,
pinions, cams, camshafts; SAE 2110; case hardening steel,
tough. *Obsolete*

UGINE N3C
Ugine Aciers
C 0.1, Ni 3, bal Fe.
Annealed: 45,000 TS; 35,000 YS; 25 El; 135 Brin. Heat
treated: 110,000 TS; 60,000 YS; 12 El; 187 Brin. For chain
links, gears, pinions, shafts; SAE 2310; case hardening
steel, tough. *Obsolete*

UGINE N5C
Ugine Aciers
C 0.15, Ni 5, bal Fe.
Annealed: 55,000 TS; 38,000 YS; 25 El; 140 Brin. Heat
treated: 140,000 TS; 120,000 YS; 11 El; 375 Brin. For
transmission shafts, gears, pinions, axles; SAE 2515; case
hardening steel, tough. *Obsolete*

UGINE NC16
Ugine Aciers
C 0.15, Ni 1.3, C 1, Mn, bal Fe.
Carburizing steel; automobile pinions.

UGINE NC20
Ugine Aciers
C 0.18, Ni 1.3, Cr 1, Mn, bal Fe.
Carburizing steel; for truck pinions.

UGINE NC32
Ugine Aciers
C 0.3, Ni 1.3, Cr 1, Mo 0.15, bal Fe.
Annealed: 75,000 TS; 40,000 YS; 187 Brin. Heat treated:
185,000 TS; 150,000 YS; 9 El; 350 Brin. For shafts, gears,
crankshafts; SAE 4330; oil hardened. *Obsolete*

UGINE NC35
Ugine Aciers
C 0.35, Cr 1, Ni 1.5, bal Fe.
Heat treated: 228,000-270,000 TS; 200,000 YS; 5 El. For
gears, axles, crankshafts, bolts; oil hardened, shock
resistant. *Obsolete*

UGINE NK8
Ugine Aciers
Ni, Cr.
For electric resistances; heat resistant to 1150 C. *Obsolete*

UGINE NS10
Ugine Aciers
C 0.08-0.16, Cr 12, Ni 12, bal Fe.
For valves, pumps; corrosion resistant. *Obsolete*

UGINE NS190
Ugine Aciers
C 0.1, Cr 17, Ni 13.5, W 3.5, bal Fe.
For high temperature applications; heat and corrosion
resistant. *Obsolete*

UGINE NS190
Ugine Aciers
C, Ni, Cr, bal Fe.
For gas turbine components; heat resistant. *Obsolete*

UGINE NS20
Ugine Aciers
C 0.1, Cr 18, Ni 10, bal Fe.
Annealed: 85,000-99,000 TS; 36,000-43,000 YS; 45-50 El;
60-65 RA; 150-170 Brin. For springs, wire cloth; Type 302,
stainless, austenitic. *Obsolete*

UGINE NS20C
Ugine Aciers
C 0.1, Cr 18, Ni 10.5, bal Fe.
Annealed: 85,000-100,000 TS; 36,000-43,000 YS; 45-50 El;
150-180 Brin. For chemical plant equipment; oxidation
resistant to 800 C, austenitic. *Obsolete*

UGINE NS20CT
Ugine Aciers
C 0-0.08, Cr 17-19, Ni 8-11, Ti = 5 x C, bal Fe.
Normalized: 93,000 TS; 36,000 YS; 45 El; 60 RA; 165 Brin.
Annealed: 87,000 TS; 33,000 YS; 57 El; 73 RA; 155 Brin. For
chemical plant equipment; Type 321; stainless, austenitic.
Obsolete

UGINE NS20E
Ugine Aciers
C 0.1, Ni 7.5, Cr 17.4, bal Fe.
Austenitic stainless steel; work hardens rapidly; for cold-
rolled springs. AISI 301.

UGINE NS20R
Ugine Aciers
C 0.1, Ni 8.5, Cr 18, Mn 0.9, bal Fe.
Austenitic stainless steel; for work hardened springs. AISI
302.

UGINE NS20S
Ugine Aciers
C 0-0.08, Cr 17-19, Ni 8-11, Ti = 5 x C, bal Fe.
Normalized: 93,000 TS; 36,000 YS; 45 El; 60 RA; 165 Brin.
Annealed: 87,000 TS; 33,000 YS; 57 El; 73 RA; 155 Brin. For
chemical plant equipment; Type 321; stainless, austenitic.
Obsolete

UGINE NS21A
Ugine Aciers
C 0-0.05, Ni 9.4, Cr 18.5, bal Fe.
Austenitic stainless steel; wire drawing grade. AISI 304.

UGINE NS21AS
Ugine Aciers
C 0.06, Ni 8.9, Cr 18.5, bal Fe.
Austenitic stainless steel; bar for machining. AISI 304.

UGINE NS21NB
Ugine Aciers
C 0.06, Ni 10.5, Cr 17.8, B 0-1, Nb = 10 x C min, bal Fe.
Stabilized austenitic stainless steel for welding and elevated
temperature. Similar to AISI 347.

UGINE NS21R/NS20P
Ugine Aciers
C 0.07, Ni 8.5, Cr 18.5, bal Fe.
Austenitic stainless steel. For springs, cables, parts in
polished wire. AISI 302/304.

UGINE NS21S
Ugine Aciers
C 0-0.12, Cr 17-19, Ni 9-12, Ti 0.5, bal Fe.
Annealed: 85,000-100,000 TS; 35,000-43,000 YS; 45-50 El.
For chemical plant equipment, welded tanks; austenitic,
stainless. *Obsolete*

UGINE NS21T
Ugine Aciers
C 0.06, Ni 10, Cr 17.8, B 0-0.6, Ti = 5 x C min, bal Fe.
Stabilized austenitic stainless steel for welding and high
temperature. AISI 321.

UGINE NS22
Ugine Aciers
C 0 0.05, Cr 18, Ni 10, bal Fe.
Annealed: 85,000 TS; 36,000 YS; 55 El; 70 RA; 150 Brin.
Cold drawn: 180,000 TS; 125,000 YS; 10 El; 330 Brin. For
hardware, decorative parts, welding rods; Type 304,
stainless, austenitic. *Obsolete*

UGINE NS22S
Ugine Aciers
C 0-0.03, Ni 9.5, Cr 18.5, bal Fe.
Austenitic stainless steel bar; low carbon for resistance to
intergranular corrosion. AISI 304 L.

UGINE NS22SV
Ugine Aciers
C 0-0.03, Ni 11, Cr 18.5, bal Fe.
Austenitic stainless steel wire for cold heading of screws.
AISI 304 L.

UGINE NS24
Ugine Aciers
C 0.16, Ni 13.5, Cr 23, bal Fe.
Austenitic stainless steel, resistant to oxidation to 1050°C.
AISI 309.

UGINE NS30
Ugine Aciers
C 0.1, Ni 20, Cr 25, bal Fe.
Austenitic stainless steel, resistant to oxidation to 1100°C.
AISI 310.

UGINE NS30C
Ugine Aciers
C 0.06, Ni 21.5, Cr 25, Si 2.5, bal Fe.
Austenitic stainless steel for elevated temperature operation.
Low carbon grade of AISI 314.

UGINE NS30Z
Ugine Aciers
C 0.1, Ni 20, Cr 25, Si 2, bal Fe.
Austenitic stainless steel, for high temperature operation
AISI 314.

UGINE NS80C
Ugine Aciers
C 0.1, Cr 18, Ni 10.5, W, bal Fe.
Annealed: 78,000-93,000 TS; 36,000-50,000 YS; 45-50 El;
55-60 RA; 150-180 Brin. For chemical plant equipment;
oxidation resistant to 800 C, austenitic. *Obsolete*

UGINE NS95
Ugine Aciers
C 0.06, Cr 17-19, Ni 9-13, Cb 0.6, bal Fe.
Annealed: 85,000 TS; 35,000 YS; 60 El; B 82 Rock. For
chemical plant equipment, welded structures, evaporators.
Type 347 stainless steel, austenitic, stabilized, welding
grade. *Obsolete*

UGINE NSCD
Ugine Aciers
C 0-0.03, Ni 16.5, Cr 17, Mo 5.25, Cu 2.65, bal Fe.
Austenitic stainless steel; excellent corrosion resistance in strong acid environment.

UGINE NSF
Ugine Aciers
C, Ni, Cr, bal Fe.
Cast. For corrosion resistant castings; stainless. *Obsolete*

UGINE NSM
Ugine Aciers
C 0.1, Cr 17, Ni 11, Mo 2, bal Fe.
Cast: 90,000 TS; 45,000 YS; 40 El; 50 RA; 160 Brin. For castings for welded equipment; stainless, austenitic, Type 316. *Obsolete*

UGINE NSM20
Ugine Aciers
C 0.1, Cr 18, Ni 10, Mo 2, bal Fe.
Annealed: 85,000-99,000 TS; 36,000-43,000 YS; 45-50 El; 60-65 RA; 160-190 Brin. For springs, chemical plant equipment; Type 316, stainless, austenitic. *Obsolete*

UGINE NSM21
Ugine Aciers
C 0.06, Ni 11, Cr 17.2, Mo 2.25, bal Fe.
Austenitic stainless steel. AISI 316.

UGINE NSM21S
Ugine Aciers
C 0-0.03, Ni 11.5, Cr 17.2, Mo 2.25, bal Fe.
Austenitic stainless steel; improved resistance to intergranular corrosion. AISI 316L.

UGINE NSM21SV
Ugine Aciers
C 0-0.03, Ni 12, Cr 17.3, Mo 2.25, bal Fe.
Austenitic stainless steel wire for cold-heading of screws. AISI 316L.

UGINE NSM22
Ugine Aciers
C 0.06, Cr 17, Ni 11, Mo, bal Fe.
Cold drawn: 78,000-93,000 TS; 36,000-43,000 YS; 45-50 El; 60-65 RA; 150-180 Brin. For welded structures, chemical plant equipment; Type 316, austenitic, stainless. *Obsolete*

UGINE NSM22S
Ugine Aciers
C 0.03, Cr 16-18, Ni 10-14, Mo 2-3, bal Fe.
Annealed: 85,000 TS; 35,000 YS; 50 El; 82 Rock B. For chemical plant equipment, welded tanks, digesters, evaporators, valve trim. Type 316L stainless steel, austenitic, acid resistant.

UGINE NSM23S
Ugine Aciers
C 0-0.03, Ni 14.5, Cr 18.5, Mo 3.25, bal Fe.
Austenitic stainless steel. AISI 317L.

UGINE NSMB
Ugine Aciers
C 0-0.12, Cr 19, Ni 10-12, Mo 2.5-3.5, bal Fe.
Annealed: 90,000 TS; 40,000 YS; 45 El; 60 RA; 170 Brin. For acid resistant chemical plant equipment, tanks; Type 316 and 317; stainless, austenitic. *Obsolete*

UGINE NSMC
Ugine Aciers
C 0.06, Ni 11, Cr 17, Mo 2.25, B 0-0.6, Ti = 5 x C min, bal Fe.
Stabilized austenitic stainless steel for chemical and mechanical resistance at high temperature. Modified AISI 316.

UGINE NSMC (CU, TI)
Ugine Aciers
C 0-0.1, Cr 16-18, Ni 10-14, Mo 1.7-2.7, Cu, Ti, bal Fe.
Annealed: 85,000-95,000 TS; 35,000-45,000 YS; 50-60 El; 60-75 RA; 150-190 Brin. For acid resistant, chemical plant equipment; Type 316 Ti; stainless, austenitic. *Obsolete*

UGINE NSV
Ugine Aciers
C 0.05, Ni 12.5, Cr 18.2, bal Fe.
Austenitic stainless steel for mass produced screws and bolts. AISI 305.

UGINE NSV3
Ugine Aciers
C 0-0.02, Ni 9.5, Cr 17.8, Cu 3.5, bal Fe.
Low carbon austenitic stainless steel. UNS S30430.

UGINE NSZ1AV
Ugine Aciers
C 0-0.05, Ni 10, Cr 18.7, bal Fe.
Austenitic stainless steel; wire for cold heading of screws. AISI 304.

UGINE ORH
Ugine Aciers
C 2, Cr 12, Co 3, Mo 1, bal Fe.
For cutters for hard wood at high speeds; nondeforming, wear and abrasion resistant. *Obsolete*

UGINE P
Ugine Aciers
Cr 12-14, 0.15% C min, bal Fe.
Annealed: 88,000 TS; 40,000 YS; 32 El; 68 RA; 170 Brin. Heat treated: 256,000 TS; 190,000 YS; 6 El; 10 RA; 540 Brin. For cutlery, valve trim, turbine blades, springs; Type 420; stainless, hardenable. *Obsolete*

UGINE P12U
Ugine Aciers
C 0.3, Cr 13.2, Mo 0.25, S 0.3, Mn 0-1.25, bal Fe.
Free machining martensitic stainless steel. AISI 420 F.

UGINE PCV
Ugine Aciers
C 0.5, Cr 1, V 0.15, Mn 0.8, bal Fe.
For screw drivers, lathe tongs, chuck jaws; tough and wear resistant. *Obsolete*

UGINE PM
Ugine Aciers
C 0.1, Cr 5, Mo 0.5, bal Fe.
Annealed: 64,000-78,300 TS; 35,000-40,000 YS; 175 Brin. Heat treated: 92,500-107,000 TS; 78,300-92,500 YS; 14-20 El; 225 Brin. For oil refinery equipment; resists sulphurous attack. *Obsolete*

UGINE PM1
Ugine Aciers
C 0-0.15, Cr 4-6, Mo, bal Fe.
Annealed: 65,000 TS; 140 Brin. Heat treated: 108,000 TS; 14 El; 225 Brin. For oil refinery equipment; corrosion resistant. *Obsolete*

UGINE PM2
Ugine Aciers
C 0.1, Cr 5, Mo 0.5, bal Fe.
Annealed: 75,000 TS; 38,000 YS; 175 Brin. Heat treated: 105,000 TS; 92,000 YS; 22 El; 225 Brin. For oil refinery equipment; resists sulphurous attack. *Obsolete*

UGINE PMV
Ugine Aciers
C 0.1, Cr 5, Mo 0.5, V 0.2, bal Fe.
Annealed: 75,000 TS; 38,000 YS; 175 Brin. Heat treated: 110,000 TS; 95,000 YS; 20 El; 230 Brin. For oil refinery equipment; resists sulphurous attack. *Obsolete*

UGINE PY
Ugine Aciers
C, Cr, bal Fe.
For wood working tools; water hardened. *Obsolete*

UGINE QA18
Ugine Aciers
C, Ni, Cr, bal Fe.
For gas turbine components; heat resistant. For furnace parts, heat treat boxes; oxidation resistant to 1150 C. *Obsolete*

UGINE QA2
Ugine Aciers
C 0.2, Cr 18, Ni 38, bal Fe.
Annealed: 71,000-86,000 TS; 36,000-50,000 YS; 25-30 El; 180-200 Brin. For furnace parts, heat treat boxes; heat resistant to 1150 C. *Obsolete*

UGINE QA2D
Ugine Aciers
C 0-0.03, Ni 21.5, Cr 33.5, Ti 0.45, Al 0.3, bal Fe.
Stabilized stainless steel for high temperature operation. UNS N08800.

UGINE QA6
Ugine Aciers
C 0.1, Cr 20, Ni 67, bal Fe.
Annealed: 48,000-65,000 TS; 43,000-57,000 YS. For heat resistant castings; heat resistant to 1200 C. *Obsolete*

UGINE QA8
Ugine Aciers
Cr 20, Ni 80.
Annealed: 95,000 TS; 25-35 El; 55 RA; Rockwell B 90. For heat resistant castings, resistors; high heat resistance. *Obsolete*

UGINE QMS
Ugine Aciers
C 0-0.25, Cr 17, Ni 2, bal Fe.
Annealed: 125,000 TS; 90,000 YS; 20 El; 55 RA; 260 Brin. For pumps, marine hardware, valves; Type 431; corrosion resistant. *Obsolete*

UGINE R
Ugine Aciers
Cr 12-14, 0.15% C min, bal Fe.
Annealed: 88,000 TS; 40,000 YS; 32 El; 68 RA; 170 Brin. Heat treated: 256,000 TS; 190,000 YS; 6 El; 10 RA; 540 Brin. For cutlery, valve trim, springs, turbine blades, Type 420; stainless, hardenable. *Obsolete*

UGINE R12
Ugine Aciers
C 0.35, Cr 14, bal Fe.
Annealed: 95,000 TS; 50,000 YS; 25 El; 55 RA; 195 Brin. Heat treated: 250,000 TS; 215,000 YS; 8 El; 25 RA; 512 Brin. For cutlery, gauges, surgical instruments; Type 420, corrosion resistant. *Obsolete*

UGINE R18
Ugine Aciers
C 0.35, Cr 14, bal Fe.
Annealed: 95,000 TS; 50,000 YS; 25 El; 55 RA; 195 Brin. Heat treated: 250,000 TS; 215,000 YS; 8 El; 25 RA; 512 Brin. For cutlery, gauges, surgical instruments; Type 420, corrosion resistant. *Obsolete*

UGINE RC
Ugine Aciers
C 0.3, W 5, Cr 3, V 0.5, bal Fe.
For extrusion dies and punches, mandrels; hot work steel, oil hardened. *Obsolete*

UGINE RD
Ugine Aciers
C 0.5, W 9.5, Cr 2.5, V 0.1, bal Fe.
For hot work dies and tools; hot work steel. *Obsolete*

UGINE RDS
Ugine Aciers
C 0.25, W 10, Cr 2.5, V 0.5, bal Fe.
For extrusion dies, punches, rams; hot work steel, oil hardened. *Obsolete*

UGINE ROC
Ugine Aciers
C 0.7, W 18, Cr 4.5, Mo 1, V 1.5, Co 10, bal Fe.
For tools, cutters, hobs, reamers, lathe tools; high speed steel, oil hardened, Type T5. *Obsolete*

UGINE ROV
Ugine Aciers
C 1.5, Cr 4.5, W 13, V 5, Co 5, bal Fe.
For permanent magnets; high permeability. *Obsolete*

UGINE RS
Ugine Aciers
C 0.7, W 18-20, Cr 4, V 1, bal Fe.
For drills, reamers, saws, shaving and planer tools; high speed steel, oil hardened, Type T1. *Obsolete*

UGINE RSA
Ugine Aciers
C 0.7, W 6, Cr 4.5, V 1.5, Mo 6, bal Fe.
For turning and slotting tools, milling cutters; Type H42, high speed steel. *Obsolete*

UGINE RSE
Ugine Aciers
C, W, Mo, bal Fe.
For drills, taps, reamers; high speed steel, oil hardened. *Obsolete*

UGINE RSK
Ugine Aciers
C 0.7, Cr 4, W 18, Mo 0.5, V 2, bal Fe.
For drills, taps, reamers, broaches; high speed steel. *Obsolete*

UGINE RSO
Ugine Aciers
C 0.7, W 18, Co 6, Cr 4, V 0.7, Mo 1, bal Fe.
For lathe and planer tools, reamers, taps, hobs, drills; high speed steel, oil hardened. *Obsolete*

UGINE RSO
Ugine Aciers
C 0.7, Cr 4, W 18, Mo 1, V 0.7, Co 6, bal Fe.
For cutting tools, lathe and planer tools; high speed steel. *Obsolete*

UGINE RSS
Ugine Aciers
C 0.7, W 19, Cr 4, V 2.2, bal Fe.
For tools, dies, cutters; high speed steel. *Obsolete*

UGINE S
Ugine Aciers
Cr 12-14, 0.15% C min, bal Fe.
Annealed: 88,000 TS; 40,000 YS; 32 El; 68 RA; 170 Brin.
Heat treated: 256,000 TS; 190,000 YS; 6 El; 10 RA; 540 Brin.
For cutlery, valve trim, turbine blades; Type 420; stainless, hardenable. *Obsolete*

UGINE S12
Ugine Aciers
C 0.35, Cr 14, bal Fe.
Annealed: 95,000 TS; 50,000 YS; 25 El; 55 RA; 195 Brin.
Heat treated: 250,000 TS; 215,000 YS; 8 El; 25 RA; 512 Brin.
For cutlery, gauges, surgical instruments; Type 420, corrosion resistant. *Obsolete*

UGINE SPECIAL 400
Ugine Aciers
C 0.35, Ni 4.5, Cr 1.5, bal Fe.
For cold stamping and deep drawing punches and dies; tough and hard. *Obsolete*

UGINE SPECIAL 460
Ugine Aciers
C 1, Cr 0.5, bal Fe.
For drills, spindles for clock industry; drill rod. *Obsolete*

UGINE TP
Ugine Aciers
C 1.3, bal Fe.
For wood working tools; water hardened. *Obsolete*

UGINE U12
Ugine Aciers
C 0.2, Cr 13, bal Fe.
Martensitic stainless steel; for machinery parts which are in contact with steam, water, wine, beer, etc. AISI 420.

UGINE U12A
Ugine Aciers
C 0.2, Cr 13, bal Fe.
Martensitic stainless steel; very low sulfur grade, for turbines. AISI Type 420.

UGINE U17N
Ugine Aciers
C 0.16, Ni 2.7, Cr 15.6, Mo 0.15, bal Fe.
Martensitic stainless steel; high strength. Similar to AISI 431.

UGINE UGV 182
Ugine Aciers
C 0.06, Cr 18, Mo 1.6, Mn 1.5, S 0.18, bal Fe.
Free machining ferritic stainless steel. Improved corrosion resistance. ASTM A582 XM34; UNS S18200.

UGINE UM3
Ugine Aciers
C 0.18, Cr 1, Mo 0.22, Mn, bal Fe.
Case hardening steel; for bolts and nuts, parts for automobile steering gear. Similar to AISI 4118.

UGINE UM5
Ugine Aciers
C 0.25, Cr 1, Mo 0.25, Mn, bal Fe.
For carbonitrided automobile pinions.

UGINE UM6
Ugine Aciers
C 0.35, Cr 1, Mo 0.25, bal Fe.
Annealed: 64,000 TS; 35,000 YS; 197 Brin. Heat treated: 95,000 TS; 75,000 YS; 12 El; 285 Brin. For transmission shafts, gears, crankshafts; SAE 4135; oil hardened. *Obsolete*

UGINE UM7
Ugine Aciers
C 0.35, Cr 1, Mo 0.25, bal Fe.
For shafts, axles, gears, pinions, crankshafts; SAE 4135; oil hardened.

UGINE VR
Ugine Aciers
C 0.5, Cr 1, V 0.15, bal Fe.
Heat treated: 170,000-200,000 TS; 160,000 YS; 7 El; 375 Brin. For springs; oil hardening. *Obsolete*

UGINE W
Ugine Aciers
C 1.2, W 1, bal Fe.
For milling cutters, drills, reamers, boring blades; water hardened, high surface hardness. *Obsolete*

UGINE WF
Ugine Aciers
C 0.4, Ni 4.3, Cr 0.6, Mo 1.2, bal Fe.
For extrusion tools, wear resistant parts; hot work steel, air hardened. *Obsolete*

UGINE XLX
Ugine Aciers
C, Ni, Cr, Cu, bal Fe.
For heat resistant castings; austenitic, stainless. *Obsolete*

UGINE YB4B
Ugine Aciers
C 0.2, Mn 1.3, Cr 1.2, 40 ppm B, bal Fe.
Boron steel for improved hardenability. Excellent formability; for case hardened gears and pinions for gear boxes.

UGINE YBB
Ugine Aciers
C 0.16, Mn 1.1, Cr 1, 40 ppm B, bal Fe.
Boron steel for improved hardenability. Excellent formability; for case hardened cams, and pinions for gear boxes.

UGINIUM
Ugine Aciers
C 0.1, Cr 12, Ni 12, bal Fe.
For tableware, cutlery, valves, valve trim. Type 12-12 stainless steel. Does not tarnish in air. *Obsolete*

UGINOX F 12 T
Pechiney/Les Toles Inoxydables & Spec.
C 0-0.05, Cr 11, Ti 0.5, bal Fe.
Cold rolled: 380-580 N/mm^2 TS; 195 N/mm^2 YS min; 20 El min. AFNOR Z3CT11; W.Nr. 4512.

UGINOX F 17
Pechiney/Les Toles Inoxydables & Spec.
C 0.07, Cr 17, bal Fe.
Hot or cold rolled: 460-610 N/mm^2 TS; 275 N/mm^2 YS min; 17 El. AFNOR Z8C17; AISI 430; W.Nr. 4016.

UGINOX F 17 MO
Pechiney/Les Toles Inoxydables & Spec.
C 0.06, Cr 17, Mo 1, bal Fe.
Cold rolled: 490-610 N/mm^2 TS; 310 N/mm^2 YS min; 17 El min. AFNOR Z8CD17-01; AISI 434; W.Nr. 4113.

UGINOX FIA
Pechiney/Les Toles Inoxydables & Spec.
C 0.03, Cr 13, bal Fe.
Hot or cold rolled: 430-580 N/mm^2 TS; 250 N/mm^2 YS min; 20 El min. AFNOR Z6C13; AISI 410S; W.Nr. 4000.

UGINOX NG 30
Pechiney/Les Toles Inoxydables & Spec.
C 0.07, Cr 25, Ni 20, Mo 0-1, bal Fe.
Cold rolled: 580-730 N/mm^2 TS; 250 N/mm^2 YS min; 35 El min. For high temperature operation. AFNOR Z12CN 25-20; AISI 310; W.Nr. 4845.

UGINOX NS 20
Pechiney/Les Toles Inoxydables & Spec.
C 0.11, Cr 17, Ni 7.5, bal Fe.
Hot or cold rolled: 650-850 N/mm^2 TS; 265 N/mm^2 YS min; 40 El min. AFNOR Z12CN17-07; AISI 301; W.Nr. 4319.

UGINOX NS 20 E
Pechiney/Les Toles Inoxydables & Spec.
C 0.1, Cr 18, Ni 7.5, bal Fe.
Normally supplied as hard rolled sheet or strip. AFNOR Z12 CN18-07. Similar to AISI 301; Type 4310.

UGINOX NS 20 P
Pechiney/Les Toles Inoxydables & Spec.
C 0-0.09, Cr 18, Ni 8.5, bal Fe.
Cold rolled: 580-730 N/mm^2 TS; 245 N/mm^2 YS min. AFNOR Z10CN18-09; AISI 302.

UGINOX NS 21 A
Pechiney/Les Toles Inoxydables & Spec.
C 0-0.07, Cr 18, Ni 9, bal Fe.
Cold rolled: 560-710 N/mm^2 TS; 235 N/mm^2 YS min; 40 El min. AFNOR Z6CN 8-09. Similar to AISI 304; W.-Nr. 4301.

UGINOX NS 21 AN
Pechiney/Les Toles Inoxydables & Spec.
C 0-0.06, Cr 18, Ni 9.5, bal Fe.
Cold rolled: 540-690 N/mm^2 TS; 210 N/mm^2 YS min; 40 El min. AFNOR Z6CN18-09; BS 304 S15.

UGINOX NS 21 AR
Pechiney/Les Toles Inoxydables & Spec.
C 0-0.07, Cr 18, Ni 9.5, bal Fe.
Cold rolled: 540-690 N/mm^2 TS; 205 N/mm^2 YS min; 45 El min. AFNOR Z6CN18-09. Similar to AISI 304; W.Nr. 4301.

UGINOX NS 21 AS
Pechiney/Les Toles Inoxydables & Spec.
C 0-0.08, Ni 9.5, 18.0 Cr min, bal Fe.
Cold rolled: 560-710 N/mm^2 TS; 215 N/mm^2 min YS; 40 min El. AFNOR Z6CN18-09; AISI 304; W.Nr. 4301.

UGINOX NS 21 C
Pechiney/Les Toles Inoxydables & Spec.
C 0-0.08, Cr 18, Ni 10, Ti = 5 x C min, bal Fe.
Cold rolled: 540-690 N/mm^2 TS; 225 N/mm^2 YS; 35 El min. AFNOR Z6CNT18-10; AISI 321; W.Nr. 4541.

UGINOX NS 22 L
Pechiney/Les Toles Inoxydables & Spec.
C 0-0.03, Ni 9.5, 18.0 Cr min, bal Fe.
Cold rolled: 500-560 N/mm^2 TS; 205 N/mm^2 YS min; 40 El min. AFNOR Z2CN18-10; AISI 304L; W.Nr. 4306.

UGINOX NS 22 S
Pechiney/Les Toles Inoxydables & Spec.
C 0-0.03, Cr 18, Ni 10, bal Fe.
Cold rolled: 500-650 N/mm^2 TS; 205 N/mm^2 YS; 40 El. AFNOR Z2CN18-10. Similar to AISI 304L; W.Nr. 4306.

UGINOX NS 24
Pechiney/Les Toles Inoxydables & Spec.
C 0.15, Cr 23, Ni 13.5, Mo 0-1, bal Fe.
Cold rolled: 580-730 N/mm^2 TS; 270 N/mm^2 YS; 30 El min. For elevated temperature operation. AFNOR Z15 CN24-13; AISI 309.

UGINOX NSCD
Pechiney/Les Toles Inoxydables & Spec.
C 0-0.03, Cr 17.5, Ni 16, Cu 3, 5.0 Mo min, bal Fe.
Cold rolled: 590-740 N/mm^2 TS; 255 N/mm^2 YS min; 35 El min. AFNOR Z2CNDU 17-16.

UGINOX NSM 21
Pechiney/Les Toles Inoxydables & Spec.
C 0-0.07, Cr 17, Ni 11.5, 2.0 Mo min, bal Fe.
Cold rolled: 530-680 N/mm^2 TS; 225 N/mm^2 YS min; 40 El. AFNOR Z6CND17-11; AISI 316; W.Nr. 4401.

UGINOX NSM 21 S
Pechiney/Les Toles Inoxydables & Spec.
C 0-0.03, Cr 17, Ni 12, 2.0 Mo min, bal Fe.
Cold rolled: 500-650 N/mm^2 TS; 215 N/mm^2 YS; 40 El min. AFNOR Type Z2CND17-12; AISI 316L; W.Nr. 4404.

UGINOX NSM 22
Pechiney/Les Toles Inoxydables & Spec.
C 0-0.05, Cr 17, Ni 12.2, Mo 2.6, bal Fe.
Cold rolled: 520-670 N/mm^2 TS; 220 N/mm^2 YS; 40 El min. AFNOR Type Z6CND17-12; AISI 316; W.Nr. 4436.

UGINOX NSM 22 S
Pechiney/Les Toles Inoxydables & Spec.
C 0-0.03, Cr 17.5, Ni 13, 2.5 Mo min, bal Fe.
Cold rolled: 500-650 N/mm^2 TS; 215 N/mm^2 YS; 40 El min. AFNOR Type Z2CND17-13; AISI 316L; W.Nr. 4435.

UGINOX NSM 23 S
Pechiney/Les Toles Inoxydables & Spec.
C 0-0.03, Cr 18.5, Ni 14.5, 3.0 Mo min, bal Fe.
Cold rolled: 520-670 N/mm^2 TS; 225 N/mm^2 YS; 40 El min. AFNOR Z2CND19-15; AISI 317L; W.Nr. 4438.

UGINOX NSMC
Pechiney/Les Toles Inoxydables & Spec.
C 0-0.08, Cr 17, Ni 12, 2.0 Mo min, Ti = 5 x C min, bal Fe.
Cold rolled: 540-690 N/mm^2 TS; 235 N/mm^2 YS min; 35 El min. AFNOR Z6CNDT17-12; W.Nr. 4571.

UGINOX NSZ
Pechiney/Les Toles Inoxydables & Spec.
C 0.15, Cr 20, Ni 11, Mo 2, bal Fe.
Hot or cold rolled: 590-740 N/mm^2 TS; 250 N/mm^2 YS min; 30 El. For elevated temperature operation. AFNOR Z 15CNS20-12; W.Nr. 4828.

UGINOX QA 2
Pechiney/Les Toles Inoxydables & Spec.
C 0-0.1, Cr 21, Ni 32, Mo 0-1, Al 0.3, bal Fe.
Cold rolled: 520-670 N/mm^2 TS; 205 N/mm^2 YS min; 30 El min. For high temperature equipment. AFNOR Z8NC 32-21; W.Nr. 4876.

UGIPLUS GP KMT
Ugine Aciers
C 0.36, Cr 1, Mo 0.25, S 0.07, Mn, bal Fe.
Heat-treated free machining alloy steel.

UGIPLUS GP KN 9
Ugine Aciers
C 0.35, Ni 2.8, Cr 2.8, Mo 0.3, S 0.07, Mn, bal Fe.
Heat-treated free machining alloy steel.

UGIPLUS GP KNO
Ugine Aciers
C 0.35, Ni 1.4, Cr 1, Mo 0.23, S 0.07, Mn, bal Fe.
Heat treated free machining alloy steel.

UGIPLUS GP UM 7
Ugine Aciers
C 0.36, Cr 1, Mo 0.25, S 0.07, Mn, bal Fe.
Free machining alloy steel; oil hardenable.

UHB
Uddeholm Corp.
C 0.8, V 0.2, bal Fe.
For dies for cold heading, embossing, trimming and swaging; cold work steel.

UHB "TRI-Z"
Uddeholm Corp.
C 1.5, Cr 12, Mo 0.8, V 0.9, bal Fe.
For blanking and drawing dies; thread rolling and cold forming dies, broaches; air hardened, low impurity count, non-deforming. *Obsolete*

UHB 14
Uddeholm Corp.
C 0.7, Mn 0.35, Si 0.2, bal Fe.
Heat treated: 122,000-174,000 TS; 82,000-128,000 YS; 12-22 El; 37-52 RA; 240-352 Brin. For die blocks, girders, rails, clutch discs; SAE 1070; water hardened. *Obsolete*

UHB 15
Uddeholm Corp.
C 0.75, bal Fe.
For hollow mine drills; oil hardened. *Obsolete*

UHB 151
Uddeholm Corp.
C 1, Mn 0.6, Si 0.25, Cr 5.25, Mo 1.1, V 0.2, bal Fe.
Annealed: 200-228 Brin. For crimpers, gages, blanking and forming dies. Type A2; air hardened, nondeforming. *Obsolete*

UHB 16
Uddeholm Corp.
C 0.8, bal Fe.
For tools, mine drills; water hardened. *Obsolete*

UHB 16 VA
Uddeholm Corp.
C 0.8, V 0.1, bal Fe.
For chisels. *Obsolete*

UHB 19 VA
Uddeholm Corp.
C 0.92, V 0.1, Mn 0.3, Si 0.25, bal Fe.
For cold heading and coining dies; water hardened, cold work steel. *Obsolete*

UHB 19VA
Uddeholm Corp.
C 1, Si 0.2, Mn 0.25, V 0.1, bal Fe.
Heat treated: 120,000-215,000 TS; 85,000-150,000 YS; 10-20 El; 33-48 RA; 240-500 Brin. For drills, taps, hobs, cutters, reamers. Type W2; water hardened. *Obsolete*

UHB 20
Uddeholm Corp.
C 1.05, bal Fe.
For tools; water hardened. *Obsolete*

UHB 20 VA
Uddeholm Corp.
C 1.05, V 0.15, bal Fe.
For dies; for cold work. *Obsolete*

UHB 20C15
Uddeholm Corp.
C 1, Cr 1.45, bal Fe.
For ball bearings; water hardened. *Obsolete*

UHB 21
Uddeholm Corp.
C 1.05, bal Fe.
For solid mine drills; water hardened. *Obsolete*

UHB 21C10
Uddeholm Corp.
C 1.05, Si 0.25, Mn 0.3, Cr 1, bal Fe.
Heat treated: 225,000 TS; 160,000 YS; 7 El; 28 RA; 575 Brin. For bearings, liners, races, sleeves; water hardened. *Obsolete*

UHB 21C5
Uddeholm Corp.
C 1.1, Si 0.2, Mn 0.3, Cr 0.4, bal Fe.
Heat treated: 125,000-220,000 TS; 90,000-155,000 YS; 8-18 El; 30-45 RA; 245-560 Brin. For bearings, liners, races, sleeves; water hardened. *Obsolete*

UHB 22
Uddeholm Corp.
C 1.1, Mn 0.55, Si 0.2, bal Fe.
Annealed: 100,000 TS; 53,000 YS; 21 El; 42 RA; 200 Brin. For springs, cutters, drills, taps, hobs. Type W1; water hardened. *Obsolete*

UHB 24
Uddeholm Corp.
C 1.2, bal Fe.
For tools; water or oil hardened. *Obsolete*

UHB 24C5
Uddeholm Corp.
C 1.15, Si 0.2, Mn 0.3, Cr 0.65, V 0.1, bal Fe.
Heat treated: 230,000 TS; 165,000 YS; 6 El; 25 RA; 580 Brin. For cold header dies, punches, drills, cutters; water hardened. *Obsolete*

UHB 25C
Uddeholm Corp.
C 1.25, Mn 0.35, Si 0.2, Cr 0.17, bal Fe.
For bearings, pivots, bushings, liners; water hardened. *Obsolete*

UHB 27C5
Uddeholm Corp.
C 1.35, Cr 0.3, bal Fe.
For files; water hardened. *Obsolete*

UHB 3
Uddeholm Corp.
C 0.15, Mn 0.7, Si 0.25, bal Fe.
Annealed: 70,000 TS; 40,000 YS; 25 El; 60 RA; 145 Brin. For screws, bolts, bushing, rivets; SAE 1015; case hardened. *Obsolete*

UHB 3 M 15
Uddeholm Corp.
C 0.12, Mn 1.35, bal Fe.
For gears, shafts; case hardened.

UHB 30C5
Uddeholm Corp.
C 1.4, Si 0.2, Mn 0.3, Cr 0.3, V 0.1, bal Fe.
For engravers' tools, cutters, drills, taps; water hardened.
Obsolete

UHB 34L
Uddeholm Corp.
C 0-0.03, Cr 17, Ni 13.5, Mo 4.3, bal Fe.
Austenitic stainless steel; for marine operations, pulp and paper. *Obsolete*

UHB 3S
Uddeholm Corp.
C 0.15, Si 0.25, Mn 0.7, bal Fe.
Annealed: 70,000 TS; 40,000 YS; 25 El; 60 RA; 145 Brin. For gears, cams, camshafts; case hardened. *Obsolete*

UHB 46
Uddeholm Corp.
C 0.9, Si 0.3, Mn 1.1, Cr 0.5, W 0.5, V 0.1, bal Fe.
For gages, reamers, taps, hobs, knurling tools. Type O1; non-deforming, oil hardened. *Obsolete*

UHB 7
Uddeholm Corp.
C 0.35, Mn 0.35, Si 0.2, bal Fe.
Hot rolled: 85,000 TS; 54,000 YS; 30 El; 53 RA; 185 Brin. For gears, shafts, axles, bolts, screws; SAE 1035; water hardened. *Obsolete*

UHB 711
Uddeholm Corp.
C 0.5, Mn 0.2, Si 0.75, Cr 1.25, W 2.5, V 0.2, bal Fe.
Annealed: 180-220 Brin. For rivet sets, upsetters, punches, dies. Type O1; oil hardened. *Obsolete*

UHB 725LN
Uddeholm Corp.
C 0-0.02, Mn 1.7, Cr 25, Ni 22, Mo 2.1, N 0.12, bal Fe.
Austenitic stainless; for chemical plants. *Obsolete*

UHB 8 M 10
Uddeholm Corp.
C 0.4, Mn 1.25, bal Fe.
For gears, shafts; tough. *Obsolete*

UHB 904L
Uddeholm Corp.
C 0-0.02, Cr 20, Ni 25, Mo 4.5, Cu 1.5, bal Fe.
Austenitic stainless steel; for chemical plant, pulp and paper. *Obsolete*

UHB BORE 2
Uddeholm Corp.
C 1.1, Cr 0.25, W 1, V 0.1, bal Fe.
For twist drills, taps; water hardened. *Obsolete*

UHB CARBON TOOL STEEL
Uddeholm Corp.
C 1.05, Mn 0.3, Si 0.25, bal Fe.
For cold heading dies, punches, gages; water hardened.
Obsolete

UHB COLD HEADER DIE
Uddeholm Corp.
C 0.91-0.98, bal Fe.
For cold heading dies; water hardened. *Obsolete*

UHB CROWN
Uddeholm Corp.
C 0.9, bal Fe.
Annealed: 84,000 TS; 24 El. For cold heading dies, stamping dies, punches; water hardened. *Obsolete*

UHB EXTRA
Uddeholm Corp.
C 1.05, Si 0.25, Mn 0.3, bal Fe.
For blanking dies, drills, hobs, reamers. Type W1; water hardened. *Obsolete*

UHB NO. 46 CROWN
Uddeholm Corp.
C 0.9, Mn 1.2, W 0.5, Cr 0.5, bal Fe.
Annealed: 104,000 TS; 21 El. For tools, dies, gauges, reamers; oil hardened; non-deforming. *Obsolete*

UHB PILGRIM
Uddeholm Corp.
C 0.4, Cr 0.5, Ni 4, Mo 1, W 0.5, bal Fe.
For forging dies, upsetters; hot work steel, oil hardened.
Obsolete

UHB SPECIAL
Uddeholm Corp.
C 0.35, Si 1.05, Cr 5, V 0.4, W 1.5, Mo 1.65, bal Fe.
For blanking and forming dies; non-deforming. AISI 11-12.
Obsolete

UHB STAINLESS 1
Uddeholm Corp.
C 0.1, Cr 13.5, bal Fe.
For corrosion resistant parts; corrosion resistant. *Obsolete*

UHB STAINLESS 15
Uddeholm Corp.
C 0-0.07, Cr 18.5, Ni 20.5, bal Fe.
For chemical plant equipment; austenitic, heat and corrosion resistant. *Obsolete*

UHB STAINLESS 16
Uddeholm Corp.
C 0.27, Cr 13.5, Mo 1.5, bal Fe.
For surgical instruments, cutlery; hardenable, corrosion resistant. *Obsolete*

UHB STAINLESS 17H
Uddeholm Corp.
C 0.25, Cr 17.5, Ni 12.5, W, Cb, bal Fe.
For chemical plant equipment; corrosion and heat resistant.
Obsolete

UHB STAINLESS 18
Uddeholm Corp.
C 0.2, Cr 18, Ni 8, bal Fe.
Annealed: 80,000 TS; 35,000 YS; 55 El; 70 RA; 150 Brin. For chemical plant equipment, tanks, filters, mixers. Type 302; stainless, austenitic. *Obsolete*

UHB STAINLESS 1H
Uddeholm Corp.
C 0.13-0.2, Cr 12-15, bal Fe.
For corrosion resistant parts; corrosion resistant. *Obsolete*

UHB STAINLESS 2
Uddeholm Corp.
C 0.1, Cr 17.5, Si 1, bal Fe.
Annealed: 75,000 TS; 40,000 YS; 30 El; 50 RA; 140 Brin. For chemical plant and oil refinery equipment. Type 430; corrosion resistant. *Obsolete*

UHB STAINLESS 20
Uddeholm Corp.
C 0.2, Cr 20, Ni 10, bal Fe.
Annealed: 90,000 TS; 40,000 YS; 50 El; 60 RA; 170 Brin. For chemical plant equipment; austenitic, stainless. *Obsolete*

UHB STAINLESS 21
Uddeholm Corp.
C 0-0.06, Cr 14, bal Fe.
For stainless parts; nonhardenable, corrosion resistant parts. *Obsolete*

UHB STAINLESS 22
Uddeholm Corp.
C 0.18, Cr 17, Ni 1.5, bal Fe.
Heat treated: 135,000 TS; 112,000 YS; 15 El; 45 RA; 280 Brin. For propeller shafts, pump spindles. Type 431; corrosion resistant. *Obsolete*

UHB STAINLESS 24
Uddeholm Corp.
C 0-0.08, Cr 17, Ni 11.5, Mo 3, bal Fe.
Annealed: 95,000 TS; 45,000 YS; 50 El; 60 RA; 190 Brin. For chemical plant equipment, acid resistant parts. Type 317; stainless, austenitic.

UHB STAINLESS 24L
Uddeholm Corp.
C 0.08, Cr 17.5, Ni 13, Mo 2.8, bal Fe.
Annealed: 85,000 TS; 35,000 YS; 50 El; 65 RA; 160 Brin. For acid resistant chemical plant equipment. Type 316; stainless, austenitic. *Obsolete*

UHB STAINLESS 25
Uddeholm Corp.
C 0-0.07, Cr 23.5, Ni 21.5, bal Fe.
Annealed: 95,000 TS; 45,000 YS; 55 El; 65 RA; 200 Brin. For furnace parts, chemical plant equipment. Type 310; corrosion and heat resistant. *Obsolete*

UHB STAINLESS 26
Uddeholm Corp.
C 0.5, Cr 14, bal Fe.
Annealed: 100,000 TS; 55,000 YS; 200 Brin. For cutlery, valves, surgical instruments. Type 420; stainless, hardenable. *Obsolete*

UHB STAINLESS 27
Uddeholm Corp.
C 0.12, Cr 16, Ni 13.5, W, bal Fe.
For chemical plant equipment; corrosion resistant.
Obsolete

UHB STAINLESS 2M
Uddeholm Corp.
C 0.08, Cr 17, bal Fe.
Annealed: 80,000 TS; 45,000 YS; 30 El; 140 Brin. For chemical plant equipment, kitchen sinks, automotive trim. Type 430; stainless steel, ductile, non-hardenable.
Obsolete

UHB STAINLESS 3
Uddeholm Corp.
C 0.1, Cr 18, Ni 8, bal Fe.
Annealed: 85,000 TS; 35,000 YS; 60 El; 70 RA; 150 Brin. For chemical plant equipment. Type 302; austenitic, stainless.
Obsolete

UHB STAINLESS 3 (P)
Uddeholm Corp.
C 0-0.15, Cr 18, Ni 8.5, bal Fe.
Annealed: 80,000 TS; 35,000 YS; 55 El; 70 RA; 150 Brin. Cold drawn: 180,000 TS; 150,000 YS; 10 El; 300 Brin. For chemical plant equipment, tanks, mixers, agitators. Type 302; stainless, austenitic.

UHB STAINLESS 31
Uddeholm Corp.
C 0.18, Cr 13.5, Ni 0.7, bal Fe.
Annealed: 95,000 TS; 50,000 YS; 25 El; 55 RA; 196 Brin. For cutlery, surgical instruments, valves, turbine blades. Type 420; stainless, hardenable. *Obsolete*

UHB STAINLESS 32
Uddeholm Corp.
C 0.15, Cr 16.5, Ni 1, Mo 0.5, bal Fe.
Heat treated: 130,000 TS; 110,000 YS; 16 El; 48 RA; 250 Brin. For chemical plant equipment, pump parts; corrosion resistant, hardenable. *Obsolete*

UHB STAINLESS 33
Uddeholm Corp.
C 0.1, Cr 14, Ni 13, bal Fe.
Annealed: 75,000 TS; 30,000 YS; 35 El; 200 Brin. For furnace parts, chemical plant equipment; stainless, austenitic. *Obsolete*

UHB STAINLESS 33MM
Uddeholm Corp.
C 0.06, Cr 12.5, Ni 12.5, bal Fe.
For valves, pumps; corrosion and heat resistant. *Obsolete*

UHB STAINLESS 34
Uddeholm Corp.
C 0.06, Cr 17.4, Ni 14, Mo 4.5, Cb, bal Fe.
Annealed: 100,000 TS; 55,000 YS; 40 El; 55 RA; 200 Brin. For welded chemical plant equipment, tanks. Type 317Cb; stainless, austenitic. *Obsolete*

UHB STAINLESS 35
Uddeholm Corp.
C 0-0.2, Cr 24, Ni 12, bal Fe.
Annealed: 90,000 TS; 40,000 YS; 50 El; 65 RA; 170 Brin. For furnace parts and equipment, salt pots. Type 309; heat resistant. *Obsolete*

UHB STAINLESS 3H
Uddeholm Corp.
C 0.15, Cr 18, Ni 8.5, bal Fe.
For heat and corrosion resistant parts; stainless, heat and corrosion resistant. *Obsolete*

UHB STAINLESS 3L
Uddeholm Corp.
C 0.08, Cr 18, Ni 10, bal Fe.
Annealed: 85,000 TS; 35,000 YS; 60 El; 70 RA; 150 Brin. For chemical plant equipment, tanks, mixers. Type 304; stainless, austenitic. *Obsolete*

UHB STAINLESS 3M
Uddeholm Corp.
C 0.06, Cr 19, Ni 10, bal Fe.
Annealed: 85,000 TS; 40,000 YS; 60 El; 150 Brin. For welded structures, chemical plant equipment, evaporators. Type 304 stainless steel, austenitic. *Obsolete*

UHB STAINLESS 3MM
Uddeholm Corp.
C 0-0.07, Ni 9.5, Cr 18, bal Fe.
Annealed: 85,000 TS; 35,000 YS; 60 El; 70 RA; 150 Brin. Cold drawn: 180,000 TS; 125,000 YS; 10 El; 330 Brin. For chemical plant equipment, tanks, mixers, agitators. Type 304; stainless, austenitic. *Obsolete*

UHB STAINLESS 4
Uddeholm Corp.
C 0.1, Cr 18, Ni 8, Mo 1.3, bal Fe.
For stainless parts; stainless, austenitic. *Obsolete*

UHB STAINLESS 41
Uddeholm Corp.
C 0.13, Cr 13.5, bal Fe.
Annealed: 75,000 TS; 40,000 YS; 35 El; 70 RA; 155 Brin. For valves, cutlery, turbine blades. Type 410; corrosion resistant. *Obsolete*

UHB STAINLESS 43
Uddeholm Corp.
C 0.08, Cr 18, Ni 9.5, Mo 0.5, bal Fe.
Annealed: 85,000 TS; 55,000 YS; 60 El; 70 RA; 150 Brin. Cold drawn: 120,000 TS; 95,000 YS; 25 El; 55 RA; 275 Brin. For chemical plant equipment; stainless, austenitic. *Obsolete*

UHB STAINLESS 44
Uddeholm Corp.
C 0.13, Cr 25, Ni 4.5, Mo 1.5, bal Fe.
Normalized: 103,000 TS; 78,000 YS; 18 El; 45 RA; 235 Brin. For valves, pumps, chemical plant equipment. Type 329; corrosion resistant. *Obsolete*

UHB STAINLESS 45
Uddeholm Corp.
C 0.13, Cr 26, Ni 4.5, bal Fe.
For chemical plant equipment. Type 327; corrosion resistant. *Obsolete*

UHB STAINLESS 4H
Uddeholm Corp.
C 0.13-0.2, Cr 16-23, Ni 7-11, Mo, bal Fe.
For heat and corrosion resistant parts; stainless, heat and corrosion resistant. *Obsolete*

UHB STAINLESS 4M
Uddeholm Corp.
C 0-0.07, Cr 16-23, Ni 7-11, Mo, bal Fe.
For heat and corrosion resistant parts; stainless, heat and corrosion resistant. *Obsolete*

UHB STAINLESS 4MM
Uddeholm Corp.
C 0-0.07, Cr 18, Mo 2.1, Ni 11.5, bal Fe.
Annealed: 85,000 TS; 35,000 YS; 50 El; 65 RA; 160 Brin. For acid resistant chemical plant equipment, tanks. Type 316; stainless, austenitic. *Obsolete*

UHB STAINLESS 5
Uddeholm Corp.
C 0.2, Cr 24, Si 1, bal Fe.
For corrosion resistant parts; corrosion resistant. *Obsolete*

UHB STAINLESS 51
Uddeholm Corp.
C 0.1, Cr 14, Mo 1, bal Fe.
For stainless parts; hardenable, corrosion resistant. *Obsolete*

UHB STAINLESS 51 H
Uddeholm Corp.
C 0.15, Cr 14, Mo, bal Fe.
Annealed: 95,000 TS; 46,000 YS; 24 El; 92 Rock B. Heat treated: 240,000 TS; 210,000 YS; 10 El; 50 Rock C. For cutlery, surgical instruments, needle valves, gages. Type 420 stainless steel, hardenable. *Obsolete*

UHB STAINLESS 52
Uddeholm Corp.
C 0.1, Cr 17, Mo 1.5, bal Fe.
Annealed: 80,000 TS; 50,000 YS; 25 El; 50 RA; 150 Brin. For oil refinery equipment, oil heaters. Type 430Mo; corrosion resistant. *Obsolete*

UHB STAINLESS 524
Uddeholm Corp.
C 0.08, Cr 17.5, Ni 13, Mo 2.8, Ti, bal Fe.
Annealed: 90,000 TS; 40,000 YS; 45 El; 60 RA; 180 Brin. For acid resistant chemical plant equipment. Type 316Ti; stainless, austenitic. *Obsolete*

UHB STAINLESS 525
Uddeholm Corp.
C 0.1, Cr 24, Ni 18, Ti, bal Fe.
Annealed: 100,000 TS; 45,000 YS; 50 El; 65 RA; 185 Brin. For furnace parts, valves, pumps. Type 310; stainless. *Obsolete*

UHB STAINLESS 53
Uddeholm Corp.
C 0-0.08, Cr 18, Ni 9.5, Ti, bal Fe.
Annealed: 85,000 TS; 33,000 YS; 58 El; 75 RA; 150 Brin. Cold drawn: 95,000 TS; 60,000 YS; 40 El; 60 RA; 185 Brin. For welded structures, chemical plant equipment. Type 321; stainless, austenitic. *Obsolete*

UHB STAINLESS 54C
Uddeholm Corp.
C 0.06, Cr 17.5, Ni 11.5, Mo 2.2, Ti, bal Fe.
Annealed: 90,000 TS; 40,000 YS; 45 El; 60 RA; 180 Brin. For welded acid resistant equipment. Type 316Ti; stainless, austenitic. *Obsolete*

UHB STAINLESS 55
Uddeholm Corp.
C 0-0.2, Cr 24, Ni 12, bal Fe.
Annealed: 85,000 TS; 35,000 YS; 60 El; 70 RA; 150 Brin. For chemical plant equipment, tanks, filters, mixers. Type 304; stainless, austenitic. *Obsolete*

UHB STAINLESS 6
Uddeholm Corp.
C 0.35, Cr 13.5, bal Fe.
For corrosion resistant parts; corrosion resistant. *Obsolete*

UHB STAINLESS 62
Uddeholm Corp.
C 0-0.1, Cr 17.5, Ti = 7 x C, bal Fe.
Annealed: 80,000 TS; 50,000 YS; 25 El; 50 RA; 150 Brin. For welded oil refinery equipment, sinks. Type 430Ti; austenitic, stainless. *Obsolete*

UHB STAINLESS 624
Uddeholm Corp.
C 0.06, Cr 17.5, Ni 13, Mo 2.8, Cb, bal Fe.
Annealed: 90,000 TS; 40,000 YS; 45 El; 60 RA; 180 Brin. For acid resistant chemical plant equipment. Type 316Cb; stainless, austenitic. *Obsolete*

UHB STAINLESS 63
Uddeholm Corp.
C 0-0.12, Cr 18, Ni 9.5, Cb = 8 x C, bal Fe.
Annealed: 90,000 TS; 45,000 YS; 56 El; 65 RA; 160 Brin. For welded chemical plant equipment, tanks, mixers. Type 347; austenitic, stainless. *Obsolete*

UHB STAINLESS 63 H
Uddeholm Corp.
C 0.06, Cr 16, Ni 13, Cb 0.6, bal Fe.
Annealed: 90,000 TS; 40,000 YS; 50 El; 82 Rock B. For exhaust manifolds, superheaters, low temperature processing equipment. Austenitic, non-hardenable; Type 16-13Cb; stainless steel. *Obsolete*

UHB STAINLESS 64
Uddeholm Corp.
C 0.06, Cr 18, Ni 18, Mo 2, 2 Cb + Cu, bal Fe.
For chemical plant equipment; stainless, austenitic. *Obsolete*

UHB STAINLESS 6H
Uddeholm Corp.
C 0.36-0.5, Cr 12-15, bal Fe.
For corrosion resistant parts; corrosion resistant. *Obsolete*

UHB STAINLESS 7
Uddeholm Corp.
C 0.95, Cr 14, Mn 1, bal Fe.
For corrosion resistant parts; corrosion resistant. *Obsolete*

UHB STAINLESS 703
Uddeholm Corp.
C 0.09, Cr 17.5, Ni 10.5, bal Fe.
Annealed: 85,000 TS; 35,000 YS; 60 El; 70 RA; 150 Brin. For chemical plant equipment, tanks, mixers. Type 304; stainless, austenitic. *Obsolete*

UHB STAINLESS 716
Uddeholm Corp.
C 0.35, Cr 13.5, Mo 1, bal Fe.
Annealed: 100,000 TS; 55,000 YS; 20 El; 50 RA; 210 Brin. For valves, cutlery, surgical and dental instruments. Type 420; corrosion resistant. *Obsolete*

UHB STAINLESS 72
Uddeholm Corp.
C 0.13, Cr 17, bal Fe.
Annealed: 80,000 TS; 50,000 YS; 25 El; 50 RA; 150 Brin. For oil refinery equipment, oil burners and heaters. Type 430; corrosion resistant. *Obsolete*

UHB STAINLESS 731

Uddeholm Corp.
C 0.2, Cr 12-14, bal Fe.
Annealed: 95,000 TS; 45,000 YS; 25 El; 92 Rock B. Heat treated: 250,000 TS; 215,000 YS; 8 El; 52 Rock C. For cutlery, surgical instruments, gauges, needle valves. Type 420; stainless steel, hardenable. *Obsolete*

UHB STAINLESS 734

Uddeholm Corp.
C 0.07, Cr 18-20, Ni 11-15, Mo 3-4, bal Fe.
Annealed: 80,000 TS; 35,000 YS; 70 El; 70 Rock B. Cold rolled: 150,000 TS; 125,000 YS; 15 El; 30 Rock C. For paper and pulp equipment, textile and dye equipment, acid tanks, evaporators, digesters. Type 317; stainless steel, austenitic, acid resistant. *Obsolete*

UHB STAINLESS 75

Uddeholm Corp.
C 0.25, Cr 25, bal Fe.
For corrosion and heat resistant parts; nonhardenable, corrosion resistant. *Obsolete*

UHB STANDARD

Uddeholm Corp.
C 1, bal Fe.
For tools, drills, taps; water hardened. *Obsolete*

UHB STAVAX

Uddeholm Corp.
Now STAVAX ESR.

UHB SUPER

Uddeholm Corp.
C, Mo, W, bal Fe.
For chisels; shock resistant. *Obsolete*

UHB WATER

Uddeholm Corp.
C 1.05, Mn 0.3, Si 0.25, bal Fe.
Annealed: 100,000 TS; 53,000 YS; 21 El; 42 RA; 200 Brin. For blanking and cutting and forming dies; water hardened. *Obsolete*

UHB-AEB

Uddeholm Corp.
C 1, Cr 13.5, bal Fe.
For cutting, forming and blanking dies; air hardened, non-deforming.

UHB-DECOL LEDLOY

Uddeholm Corp.
C 0.97, Si 0.2, Mn 0.45, bal Fe.
Annealed: 100,000 TS; 53,000 YS; 21 El; 42 RA; 200 Brin. For drills, hobs, reamers, taps; Type W1; water hardened. *Obsolete*

UHB-MN-MO

Uddeholm Corp.
C 0.9, Mn 1.4, Si 0.3, Mo 0.25, bal Fe.
For tools, dies, punches; oil hardened, ground flat stock. *Obsolete*

UHB-NICRO 33

Uddeholm Corp.
C 0.13, Cr 0.75, Ni 3, bal Fe.
For gears, cams, camshafts; case hardened, oil hardened. *Obsolete*

UHB-NICRO 43

Uddeholm Corp.
C 0.18, Cr 0.75, Ni 3, bal Fe.
For gears, cams, camshafts; case hardened, oil hardened. *Obsolete*

UHB-VA

Uddeholm Corp.
C 1.05, Mn 0.3, Si 0.25, V 0.2, bal Fe.
For drills, taps, reamers, blanking dies. Type W2; water hardened. *Obsolete*

UHB-X1

Uddeholm Corp.
alloy, bal Fe.
For dies; tools; oil hardened. *Obsolete*

UKI

Pechiney/Eurotungstene
WC plus alloy.
For anti-skid tire studs and studded straps for snow and ice chains.

ULBRASEAL 36

Ulbrich Stainless Steels/Special Metals
C 0-0.05, Cr 0-0.25, Mn 0-0.5, Si 0-0.25, 35.5-36.5 Ni nominal, bal Fe.
Low coefficient of expansion. Used in television picture tubes as mask material, bimetallic thermostats; per MIL I-23011 C.

ULBRASEAL 42

Ulbrich Stainless Steels/Special Metals
Ni 41, C 0-0.05, Mn 0-0.8, Si 0-0.3, Al 0-0.1, Co 0-0.5, bal Fe.
Used for sealing of leads into light bulbs, automotive sealed beam headlights.

ULBRASEAL 42-8

Ulbrich Stainless Steels/Special Metals
Ni 42, Si 0-0.3, Al 0-0.2, 5.6 Cr nominal, bal Fe.
Used in television picture tubes.

ULBRASEAL 46

Ulbrich Stainless Steels/Special Metals
Ni 46, C 0-0.05, Mn 0-0.8, Si 0-0.3, Al 0-0.1, Co 0-0.5, bal Fe.
Used for electrical resistors.

ULBRASEAL 52

Ulbrich Stainless Steels/Special Metals
Ni 50.5, C 0-0.05, Mn 0.6, Si 0-0.3, Al 0-0.1, Co 0-0.5, bal Fe.
Used for cores in mechanical rectifiers and magnetic amplifiers requiring rectangular hysteresis.

ULBRAVAR 29-17

Ulbrich Stainless Steels/Special Metals
Ni 29, C 0-0.06, Mn 0-0.5, Si 0-0.2, Co 17, bal Fe.
Used for hermetic seals in vacuum tubes. AMS 7728.

ULCOMETAL

NL Industries
1-2 Ba+Ca, bal Pb.
14,000 TS; 5 El; 1 RA. For bearings. *Obsolete*

ULCONY

Manufacturer not listed.
Cu 65, Pb 35.
For heavy duty bearings; for poor lubricating conditions.

ULMINIUM S

Wieland-Werke AG Metallwerke
Cu 4.2, Mg 1.6, Mn 0.6, bal Al
Aged: 65,000 TS; 43,000 YS; 14 El; 120 Brin. For transportation and aircraft construction; high strength, age hardened. *Obsolete*

ULTIMIUM N 112

Eutectic Corp.
Tungsten carbide containing electrodes for super hard overlays on steel; AC-DC. For surfacing conveyor screws, crusher liners, muller blades, and wear plates. 68-72 Rock C.

ULTIMIUM N 113

Eutectic Corp.
Tungsten carbide type super abrasion resistant overlays; AC-DC; for all steels and cast irons. For wear plates, mining teeth, crusher liners. Hardness C 60 Rock. *Obsolete*

ULTIMIUM N 114 EXPRESS

Eutectic Corp.
Tungsten carbide containing electrodes for super hard overlays on steel. AC-DC; for conveyor wear surfaces, plows, wear plates. Hardness C 67-69 Rock. *Obsolete*

ULTIMO 200

Rio Algom Corp.
C 0.3, Mn 0.8, Cr 1.65, Mo 0.4, bal Fe.
Annealed: 114,000 TS; 90,000 YS; 25 El; 229 Brin. Hardened: 150,000 TS; 135,000 YS; 22 El; 302 Brin. For heavy duty shafts, gears, crusher rolls, spindles, couplings, clamps. Oil hardening, shock resisting.

ULTIMO 200

Atlas Steels Ltd.
C 0.3, Mn 0.8, Cr 1.65, Mo 0.4, bal Fe.
Annealed: 114,000 TS; 90,000 YS; 25 El; 229 Brin. Hardened: 150,000 TS; 135,000 YS; 22 El; 302 Brin. For heavy duty shafts, gears, crusher rolls, spindles, couplings, clamps. Oil hardening, shock resisting.

ULTIMO 4

Atlas Specialty Steels
C 0.45, Mn 0.75, Cr 1.25, Mo 0.35, bal Fe.
Normalized: 110,000 TS; 85,000 YS; 25 El; 60 RA; 228 Brin. For shafts, gears, arbors; heavy duty.

ULTIMO 6

Atlas Specialty Steels
C 0.55, Mn 55, Si 0.8, Cr 1, Ni 1.6, Mo 0.75, bal Fe.
For shear blades, lead and zinc extrusion tools; oil hardening cold work tool steel.

ULTRA B

Saarstahl AG
C 0.1, Ni 65, Mo 28, bal Fe.
Corrosion resistant alloy. W.Nr. 2.4482. *Obsolete*

ULTRA C

Saarstahl AG
Now SAARSTAHL 2.4537.

ULTRA CAPITAL

Darwins Alloy Castings
C 0.75, Cr 4, W 18, V 1, bal Fe.
For broaches, drills, reamers, hobs, lathe tools; high-speed steel. *Obsolete*

ULTRA CAPITAL + 1

Eagle & Globe Steel Ltd.
Tool material. C 0.76, Cr 4.2, W 18, V 1.15, Co 5, bal Fe. High speed steel. Gives long life in general machine shop work and heavy duty cutting. AS1239 T4A; AISI T4; Werkstoff 1.0255.

ULTRA CAPITAL 22

Darwins Alloy Castings
C 0.75, Cr 4.1, W 22, V 1.25, bal Fe.
For broaches, drills, reamers, milling cutters, hobs; high-speed steel. *Obsolete*

ULTRA CAPITAL HIGH SPEED STEEL

Eagle & Globe Steel Ltd.
Tool material. C 0.75, Cr 4.2, W 18, V 1.1, bal Fe.
Annealed: 255 Brin max. Standard general purpose steel for most cutting tool applications. Low susceptibility to decarburization combined with good red hardness and toughness. AS1239 T1A; AISI T1; BS4659 BT1; Werkstoff 1.3355.

ULTRA CAPITAL PLUS 2

Darwins Alloy Castings
C 0.76, Cr 4.2, W 20, V 1.5, bal Fe.
For broaches, reamers, drills, hobs, cutters, taps; high-speed steel, oil hardened. *Obsolete*

ULTRA CAPITAL PLUS ONE
Darwins Alloy Castings
C 0.85, Cr 4.25, V 1.3, Co 5, W 18, bal Fe.
For lathe and planer tools, hobs, taps, reamers, drills; high-speed steel, oil hardened. *Obsolete*

ULTRA COBALT
Latrobe Steel Co.
C 0.8, V 2, Co 12, W 18, Cr 4, bal Fe.
For tools, high speed cutting tools; high speed steel. *Obsolete*

ULTRA SUPERIOR 25
Stahlwerk Stahlschmidt GmbH & Co.
C 1.3, Cr 4.3, Mo 0.85, V 3.8, W 12, bal Fe.
For forming and blanking dies, cutters; high speed steel.

ULTRA SUPERIOR 25Z
Stahlwerk Stahlschmidt GmbH & Co.
C 0.86, Co 2.8, Cr 4.3, Mo 0.85, V 2.1, W 12, bal Fe.
For lathe and planer tools, reamers, broaches; high speed steel.

ULTRA-CAPITAL STEEL
Adams & Osgood Steel Co.
W 17, Cr 3.4, V 1, C 0.7, bal Fe.
For turning tools and cutters; high speed steel.

ULTRA-CAPITAL STEEL
Balfour Darwins Ltd.
W 17, Cr 3.4, V 1, C 0.7, bal Fe.
For turning tools and cutters; high speed steel.

ULTRA-CUT
Bliss & Laughlin Steel Co.
C 0.08-0.12, S 0.3, bal Fe.
Cold finish: 90,000 TS; 80,000 YS; 15 El; 45 RA; 174-212 Brin. For automatic screw machine products; free machining. *Obsolete*

ULTRA-GOLD
J.F. Jelenko & Co.
Precious metal. Au 87.5, Pt 10, Pd 1.
70,000 psi TS; 60,000 psi YS; 5 El; 150 Brin. Dental alloy for fusing porcelain to metal.

ULTRADIE 1
Cytemp Specialty Steel Div.
C 2.25, Cr 12, Mo 0.8, V 0.2, bal Fe.
For dies; for punching hard, brittle material, drills, reamers, shear blades, lamination dies; non-deforming; cold work die steel; AISI D3. *Obsolete*

ULTRADIE 1M
Cytemp Specialty Steel Div.
C 2.25, Cr 12, V 2, Mo 0.8, bal Fe.
For lamination and drawing dies; punches, gauges; Type D3; oil hardened, non-deforming. *Obsolete*

ULTRADIE 2
Cytemp Specialty Steel Div.
C 1.5, Cr 12, Mo 0.8, bal Fe.
For tools, dies, reamers, taps; non-deforming. *Obsolete*

ULTRADIE 3 (D2)
Cytemp Specialty Steel Div.
Tool material. C 1.5, Cr 12, V 1, Mo 0.8, bal Fe.
For lamination and blanking dies, punches, shear knives; non-deforming; cold work die steel; AISI D2.

ULTRADUR
Plansee, Metallwerk Gesellschaft
Carbide.
For hard facing electrodes; sintered alloy.

ULTRAFORT 6355, ETC.
Thyssen Edelstahlwerke AG
See Werkstoff Nr. 1.6355.
High tensile steels; 5 grades.

ULTRALLOY
Ultraloy Corp.
Sn 48.5, Zn 48.5, Cu 2, Si 0.6, Ag 0.03.
Cast: 12,250 TS; 20 El; 23 Brin. Solder for aluminum; melting point: 700-750°F.

ULTRALOY
Solar Basic Industries
Cr 37.5, Al 7.5, bal Fe.
For furnace parts; see Smith No. 10.

ULTRALOY 10611
Eutectic Corp.
Obsolete

ULTRALUMIN
Ultralumin Leichtmetall AG
Cu 4.7, Mn 0.75, Ni 0.2, Ce, bal Al.
Heat treated: 54,000-60,000 TS; 37,000-43,000 YS; 14-24 El.
For light alloy parts, forgings. *Obsolete*

ULTRAPERM 10
Vacuumschmelze GmbH
Soft magnetic 72 to 83% nickel alloy for circuit breakers, transformers and relays. 300,000 microns permeability max.

ULTRAPERM 200
Vacuumschmelze GmbH
Soft magnetic 72 to 83% nickel alloy for circuit breakers, transformers and relays. 350,000 microns permeability max.

ULTRAPERM F
Vacuumschmelze GmbH
Ni 77, bal Fe.
Soft magnetic alloy with flat hysteresis loop for power electronics; low remanence and low coercivity. Saturation flux density 0.80 T.

ULTRAPERM F 80
Vacuumschmelze GmbH
Ni 77, bal Fe.
Soft magnetic alloy with flat hysteresis loop for power electronics; low remanence and low coercivity. Saturation flux density 0.80 T.

ULTRAPERM Z
Vacuumschmelze GmbH
Ni 77.
Soft magnetic alloy with square hysteresis loop for sensitive magnetic amplifiers.

ULTRATHERM 13
J.C. Soding & Halbach
C 0.45, Si 2.25, Cr 13, bal Fe.
Cast, annealed: 95,000-135,000 TS; 4 El; 200-300 Brin. Ferrite-pearlite; weldable. For parts used at temperatures up to 900°C. DIN G-X45CrSi13.

ULTRATHERM 17
J.C. Soding & Halbach
C 0.45, Si 2.25, Cr 17, bal Fe.
Cast, annealed: 95,000-135,000 TS; 2 El; 200-300 Brin. Ferrite-carbide, weldable. For parts used in oxidizing atmospheres at temperatures up to 950°C. DIN G-X40CrSi 17.

ULTRATHERM 18-9
J.C. Soding & Halbach
C 0.18, Si 1.5, Cr 18, Ni 9, bal Fe.
Cast, annealed: 75,000-110,000 TS; 15 El; 130-200 Brin. Austenitic, weldable. For parts used at elevated temperatures. DIN G-X25CrNiSi189.

ULTRATHERM 19
J.C. Soding & Halbach
C 1.6, Si 1.5, Cr 18, bal Fe.
Cast, annealed: 250-300 Brin. Ferrite-carbide structure, machinability only fair. For parts used at temperatures up to 950°C. DIN G-X160CrSi18.

ULTRATHERM 20-40
J.C. Soding & Halbach
C 0.35, Si 1.35, Cr 17.5, Ni 37.25, bal Fe.
Cast: 70,000 TS; 40,000 YS; 9 El; 170 Brin. Aged: 73,000 TS; 43,000 YS; 5 El; 200 Brin. For retorts, carburizing containers, rails, fixtures, furnace parts. Corrosion and oxidation resistance. Heat resistant, Type HU, austenitic.

ULTRATHERM 23
J.C. Soding & Halbach
C 0.45, Si 1.5, Cr 23, bal Fe.
Cast, annealed: 200-300 Brin. Ferrite-carbide structure, weldable. For parts to be used in oxidizing atmospheres at temperatures up to 1050°C. DIN G-X40CrSi22.

ULTRATHERM 23-11
J.C. Soding & Halbach
C 0.35, Si 1.75, Cr 22, Ni 10, bal Fe.
Cast, annealed: 75,000-110,000 TS; 12 El; 150-220 Brin. Austenitic, weldable. For parts used at elevated temperatures. DIN G-X40CrNiSi229.

ULTRATHERM 23-15
J.C. Soding & Halbach
C 0.25, Si 1.75, Cr 20, Ni 14, bal Fe.
Annealed: 90,000 TS; 40,000 YS; 50 El; 83 Rock B. For heat treating equipment, heat exchangers, furnace parts and fixtures, salt pots, oil burners. Non-hardenable. Heat resistant, austenitic.

ULTRATHERM 25-12
J.C. Soding & Halbach
C 0.35, Si 1.75, Cr 25, Ni 12, bal Fe.
Heat treated: 92,000 TS; 45,000 YS; 8 El; 200 Brin. At 1400°F: 35,000 TS; 18,000 YS; 12 El. For furnace parts, conveyor screws, salt pots, dampers, grate bars. High-corrosion and oxidation resistance. Heat resistant, Type HH, austenitic.

ULTRATHERM 26-21
J.C. Soding & Halbach
C 0.35, Si 1.75, Cr 25.5, Ni 20, bal Fe.
Cast: 75,000 TS; 50,000 YS; 17 El; 170 Brin. Aged: 85,000 TS; 50,000 YS; 10 El; 190 Brin. For salt pots, heat treating boxes, furnace parts, skids, rabble arms. Corrosion and oxidation resistant. Heat resistant, austenitic.

ULTRATHERM 26-21 W
J.C. Soding & Halbach
C 0.15, Si 1.75, Cr 25.5, Ni 20, bal Fe.
Annealed: 95,000 TS; 45,000 YS; 50 El; 89 Rock B. For furnace parts and equipment, heat treating boxes and fixtures, boiler baffles, retorts, kiln linings. High- corrosion and oxidation resistance. Heat resistant, Type 310, austenitic.

ULTRATHERM 28-4H
J.C. Soding & Halbach
C 0.4, Si 1.5, Cr 27, Ni 4, bal Fe.
Cast: 85,000 TS; 48,000 YS; 16 El; 190 Brin. At 1400°F: 36,000 TS; 14 El. For ore roasting furnaces, rabble arms, sintering bars, salt pots, furnace blowers, recuperators. High oxidation resistance in sulfur atmospheres. Heat resistant. Type HD. DIN G-X40CrNi274.

ULTRATHERM 30
J.C. Soding & Halbach
C 0.4, Si 1.5, Cr 28.5, bal Fe.
Cast: 110,000 TS; 75,000 YS; 19 El; 223 Brin. For ore roasting furnaces, grate bars, salt pots, soot blowers, kiln parts. Oxidation resistant to high sulfur atmospheres. Heat resistant. DIN G-X40CrSi29.

ULTRATHERM 30 H
J.C. Soding & Halbach
C 1.3, Si 1.5, Cr 28.5, bal Fe.
Cast, annealed: 250-350 Brin. Ferrite, carbide structure. For parts used in oxidizing atmospheres at temperatures up to 1100°C. DIN G-X130CrSi29.

ULTRATHERM 7
J.C. Soding & Halbach
C 0.25, Si 2.25, Cr 6.5, bal Fe.
Cast, annealed: 95,000-130,000 TS; 4 El; 200-280 Brin.
Ferrite-pearlite, weldable. For parts used at temperatures up
to 850°C. DIN G-X30CrSi6.

ULTRAVAN
Now ELECTRITE ULTRAVAN.

ULTRAVAN
Latrobe Steel Co.
C 1.5, W 6.3, Cr 4.25, V 4.75, Mo 5, Co 5, bal Fe.
For inserted blade cutting tools; high speed steel. *Obsolete*

UMCO-50
SERMAG
Co 48-52, C 0.05-0.12, Cr 27-29, Mn 0.5-1, Si 0.5-1, bal Fe.
As cast, RT: 80,000 TS; 45,600 YS; 8 El. 1830°F: 11,400 TS;
10,000 YS; 18 El. Forged, RT: 134,000 TS; 88,500 YS; 10 El.
1830°F: 11,400 TS; 8,600 YS; 18 El. Excellent strength and
corrosion resistance at elevated temperatures; for furnaces,
heat treat equipment, gas and pulverized coal burners.

UMCO-51
SERMAG
Co 48-52, C 0.25-0.4, Cr 27-29, Nb 2-2.2, Mn 0.5-1, Si 0.5-1,
bal Fe.
As cast, RT: 91,500 TS; 72,000 YS; 2.2 El. 1650°F: 30,400
TS; 26,200 YS; 13.8 El. Properties vary with carbon content.
Excellent strength and corrosion resistance at elevated
temperatures; for furnace parts, heat treat equipment, gas
and pulverized coal burners.

UNA
Una Welding Inc.
C 0.1, bal Fe.
For welding rod.

UNA HIGH SPEED NO. 44
Una Welding Inc.
C 0-0.08, Mn 0.2-0.3, bal Fe.
For welding rod; flux coated for high speed.

UNA HIGH SPEED NO. 45
Una Welding Inc.
C 0.12-0.18, Mn 0.4-0.6, bal Fe.
For welding rod; flux coated for high speed.

UNA HIGH TENSILE NO. 65
Una Welding Inc.
C 0.13-0.18, Mn 0.4-0.6, bal Fe.
For welding rod; flux coated.

UNA NO. 1500
Una Welding Inc.
C 0.08, bal Fe.
For welding rod.

UNA NO. 156
Una Welding Inc.
C 0.13-0.18, bal Fe.
For welding rod; flux coated.

UNA NO. 158
Una Welding Inc.
C 0.08, Mn 0.2-0.3, bal Fe.
For welding rod; flux coated.

UNA NO. 160
Una Welding Inc.
C 0.13-0.18, Mn 0.4-0.6, bal Fe.
For welding rod; flux coated.

UNA NO. 200
Una Welding Inc.
C 0.07-0.12, Mn 0.3-0.5, bal Fe.
For welding rod; flux coated.

UNA NO. 2150
Una Welding Inc.
C 0.03, Mn 0.13, Cu 0.46, Mo 0.07, bal Fe.
For welding rod; for copper bearing steel.

UNA NO. 300
Una Welding Inc.
C 0.13-0.18, Mn 0.4-0.6, bal Fe.
For welding rod; flux coated.

UNA NO. 3175
Una Welding Inc.
C 0.08, Ni 2, Cu 1, bal Fe.
For welding rod; shielded coating.

UNA NO. 3200
Una Welding Inc.
C 0.1-0.14, Mn 0.35-0.55, bal Fe.
For welding rod.

UNA NO. 325
Una Welding Inc.
C 0.85-1.1, Mn 0.3-0.6, bal Fe.
For welding rod; hard and abrasion resisting.

UNA NO. 350
Una Welding Inc.
C 0.17, Ni 4.75-5.25, bal Fe.
For welding rod.

UNA NO. 425
Una Welding Inc.
C 0.08, Mn 0.2-0.3, bal Fe.
Welding rod; protected arc.

UNA NO. 470
Una Welding Inc.
Ni-Cu.
For welding rod; for machinable welds on cast iron.

UNA NO. 560
Una Welding Inc.
Cu.
For welding rod for copper to steel; flux coated.

UNA NO. 601
Una Welding Inc.
Si 0.5, bal Al.
For welding rod; for aluminum.

UNA NO. 712
Una Welding Inc.
C 0.15-0.25, Mn 0.3-0.6, Ni 3.25-3.75, bal Fe.
For welding rod; flux coated.

UNA NO. 76
Una Welding Inc.
C 0.13-0.018, Mn 0.4-0.6, bal Fe.
For welding rod; flux coated.

UNA NO. 911
Una Welding Inc.
C 0.07, Mn 0.5, Cr 18-20, Ni 8-10, bal Fe.
For welding rod; for 18-8 stainless steel.

UNA NO. 912
Una Welding Inc.
C 0.2, Mn 2, Cr 22-26, Ni 12-14, bal Fe.
For welding rod; stainless steel.

UNA NO. 914
Una Welding Inc.
C 0.1, Cr 12-14, Ni 0.25, bal Fe.
For welding rod; for corrosion resistant steel.

UNA NO. 916
Una Welding Inc.
C 0.1, Cr 16-18, Ni 0.25, bal Fe.
For welding rod; for stainless steel.

UNA NO. 917
Una Welding Inc.
C 0.1, Cr 25-30, Ni 0.25, bal Fe.
For welding rod; for stainless steel.

UNA NO. 918
Una Welding Inc.
C 0.15, Cr 4-6, Mo 0.4-0.6, bal Fe.
For welding rod; for 4-6 Cr steel.

UNAMO 1
U.N. Alloy Steel Corp.
C 0.8, Mn 0.3, Si 0.3, Cr 4, V 1, W 1.5, Mo 8, bal Fe.
Hardness: 64-68 Rock C. For cutters, lathe and planer tools,
drills, reamers, broaches, taps. Type M1 high speed steel,
high red-hardness.

UNAMO 10
U.N. Alloy Steel Corp.
C 0.85, Cr 4, V 2, W 1.5, Mo 8, bal Fe.
For twist drills, reamers, broaches, lathe and planer tools.
Type M10 high speed steel, high red-hardness.

UNAMO 2
U.N. Alloy Steel Corp.
C 0.8, Si 0.3, Mn 0.3, Cr 4, V 2, W 6, Mo 5, bal Fe.
Hardened: 64-66 Rock C. For lathe and planer tools,
reamers, broaches, chasers, hobs, taps, form cutters. Type
M2 high-speed steel, high red-hardness.

UNAMO 35
U.N. Alloy Steel Corp.
C 0.8, Cr 4, V 2, W 6, Mo 5, Co 5, bal Fe.
Hardened: 65-67 Rock C. For reamers, broaches, lathe and
planer tools, chasers, drills, hobs, reamers, form cutters.
Type M35 high speed steel, high red-hardness.

UNAMO 7
U.N. Alloy Steel Corp.
C 1, Mn 0.3, Si 0.3, Cr 3.75, V 2, W 1.7, Mo 8.75, bal Fe.
Hardened: 64-67 Rock C. For broaches, chasers, reamers,
hobs. Type M7 high speed steel, high red-hardness.

UNARAPID T-1
U.N. Alloy Steel Corp.
C 0.7, Cr 4, V 1, W 18, bal Fe.
Hardened: 64-66 Rock C. For drills, saws, chasers, reamers,
broaches, lathe and planer tools. Type T1 high speed steel,
high red-hardness.

UNARAPID T-15
U.N. Alloy Steel Corp.
C 1.5, Cr 4.5, V 4.75, W 13.5, Mo 0.5, Co 5, bal Fe.
Hardened: 64-66 Rock C. For textile needles, form cutters,
broaches, reamers, taps. Type T15 high speed steel, high
red-hardness.

UNARAPID T-2
U.N. Alloy Steel Corp.
C 0.8, Cr 4, V 2, W 18, bal Fe.
Hardened: 64-66 Rock C. For lathe and planer tools,
forming tools, drills, broaches, reamers, taps. Type T2 high
speed steel, high red-hardness.

UNARAPID T-4
U.N. Alloy Steel Corp.
C 0.75, Cr 4, V 1, W 18, Co 5, bal Fe.
Hardened: 64-66 Rock C. For lathe and planer tools, drills,
chasers, reamers, broaches, hobs, taps; Type T4 high
speed steel, high red-hardness.

UNARAPID T-5
U.N. Alloy Steel Corp.
C 0.8, Cr 4, V 2, W 18, Co 8, bal Fe.
Hardened: 64-67 Rock C. For lathe and planer tools, form
cutters, drills, reamers, broaches, hobs. Type T5 high speed
steel, high red-hardness.

UNARAPID T-6
U.N. Alloy Steel Corp.
C 0.8, Cr 4.5, V 1.5, W 20, Co 12, bal Fe.
Hardened: 64-68 Rock C. For lathe and planer tools, chasers, form cutters, reamers, broaches, taps, hobs. Type T6 high speed steel, high red-hardness.

UNAVAN 3 TYPE 1
U.N. Alloy Steel Corp.
C 0.95-1.02, Cr 4.5, Ni 0-0.25, V 2.4-2.7, W 5.8-6.3, Mo 5.7-6.2, Co 0-0.25, bal Fe.
Hardened: 64-68 Rock C. For lathe and planer tools, form cutters, reamers, broaches, taps. Type M3 high speed steel, high red-hardness.

UNAVAN 4
U.N. Alloy Steel Corp.
C 1.3, Cr 4, V 4, W 5.5, Mo 4.5, bal Fe.
Hardened: 64-68 Rock C. For chasers, taps, reamers, form cutters, broaches, textile needles. Type M4 high speed steel, high red-hardness.

UNBREAKABLE METAL
American Smelting & Refining Co.
Al, Cu, bal Zn.
For lamp bases, toys, novelties; slush or permanent mold castings.

UNI-DIE
Columbia Tool Steel Co.
Tool material. C 0.71, Mn 2, Mo 1.3, Cr 1, bal Fe.
Air hardened; cold work tool steel. For blanking dies, punches, plug gages. AISI A6.

UNI-DIE SMOOTHCUT
Columbia Tool Steel Co.
C 0.67-0.75, Mn 2-2.25, Mo 1.2-1.4, Cr 0.9-1.1, S, bal Fe.
Heat treated: 292,000 S; 264,000 YS; 1 El; C 55 Rock. For coining and notching dies, shear blades, blanking and forming dies, plastic mold dies. Air hardening, cold work tool steel Type A6. *Obsolete*

UNIBRAZE 1000
J.W. Harris Co., Inc.
Now SAFETY-SILV HARRIS 40 NI 2.

UNIBRAZE 1010, 1010 FC
J.W. Harris Co., Inc.
Low melting point silver brazing rod. For electrical, air conditioning, heating and ventilating work; joins dissimilar metals. Bare and flux coated. *Obsolete*

UNIBRAZE 110 (BARE) 110 FC (FLUX COATED)
J.W. Harris Co., Inc.
Now NICKEL SILVER BARE & FLUX COATED.

UNIBRAZE 1110
J.W. Harris Co., Inc.
Lowest melting point silver brazing rod. General purpose repair and salvage. *Obsolete*

UNIBRAZE 130, 130 FC
J.W. Harris Co., Inc.
Now NICKEL SILVER BARE.

UNIBRAZE 15 PHOSPHOR-COPPER-SILVER
J.W. Harris Co., Inc.
Now STAY-SILV 15.

UNIBRAZE 1550, 1550 FC
J.W. Harris Co., Inc.
Now SAFETY-SILV 56 BARE & FLUX COATED.

UNIBRAZE 160
J.W. Harris Co., Inc.
Low alloy steel electrode, AC-DC, reverse polarity. For welding mild steel and low alloy steel to cast iron and higher carbon steel. *Obsolete*

UNIBRAZE 1640
J.W. Harris Co., Inc.
Silver brazing rod for tool tipping with carbide. *Obsolete*

UNIBRAZE 2 PHOSPHOR-COPPER-SILVER
J.W. Harris Co., Inc.
Now STAY-SILV 2.

UNIBRAZE 210
J.W. Harris Co., Inc.
Phosphor copper brazing rod. Maintenance repair on copper tubing, wire, sheet. *Obsolete*

UNIBRAZE 220
J.W. Harris Co., Inc.
Now SILICON BRONZE ELECTRODE.

UNIBRAZE 230
J.W. Harris Co., Inc.
Phosphor copper silver brazing rod. For brazing copper alloys. *Obsolete*

UNIBRAZE 240
J.W. Harris Co., Inc.
Now PHOS BRONZE C ELECTRODE.

UNIBRAZE 250
J.W. Harris Co., Inc.
Now 310 STAINLESS ELECTRODE.

UNIBRAZE 277
J.W. Harris Co., Inc.
Now 312 SUPER BLUE.

UNIBRAZE 30
J.W. Harris Co., Inc.
Copper cast iron rod.
For torch brazing and build up maintenance on cast iron. *Obsolete*

UNIBRAZE 308 ELC STAINLESS WELD WIRE
J.W. Harris Co., Inc.
Now HARRIS 308 L STAINLESS WELD WIRE.

UNIBRAZE 308 STAINLESS WELD WIRE
J.W. Harris Co., Inc.
Now HARRIS 308 STAINLESS WELD WIRE.

UNIBRAZE 309 STAINLESS WELD WIRE
J.W. Harris Co., Inc.
Now HARRIS 309 STAINLESS WELD WIRE.

UNIBRAZE 31
J.W. Harris Co., Inc.
Ag 31.5, Cu 34, Zn 15.5, Cd 19.
Melting range: 1165-1390°F. Brazing: 1165-1390°F. For torch or induction brazing ferrous and non-ferrous metals. *Obsolete*

UNIBRAZE 310
J.W. Harris Co., Inc.
Now 718 ALUM.

UNIBRAZE 310 STAINLESS WELD WIRE
J.W. Harris Co., Inc.
Now HARRIS 310 STAINLESS WELD WIRE.

UNIBRAZE 312 STAINLESS WELD WIRE
J.W. Harris Co., Inc.
Now HARRIS 312 STAINLESS WELD WIRE.

UNIBRAZE 316 ELC STAINLESS WELD WIRE
J.W. Harris Co., Inc.
Now HARRIS 316 L STAINLESS WELD WIRE.

UNIBRAZE 316 STAINLESS WELD WIRE
J.W. Harris Co., Inc.
Now HARRIS 316 STAINLESS WELD WIRE.

UNIBRAZE 345
J.W. Harris Co., Inc.
Now HARRIS 404 ALUM ELECTRODE.

UNIBRAZE 347 STAINLESS WELD WIRE
J.W. Harris Co., Inc.
Now HARRIS 347 STAINLESS WELD WIRE.

UNIBRAZE 35
J.W. Harris Co., Inc.
Ag 35, Cu 26, Zn 21, Cd 18.
Melting range: 1125-1295°F. Brazing: 1125-1295°F. For torch or induction brazing ferrous and non-ferrous metals where fit-up is poor. AWS A5.8-62T BAg-2; ASTM B260-62T BAg-2; AMS 4768. *Obsolete*

UNIBRAZE 350
J.W. Harris Co., Inc.
Ag 50, Cu 15.5, Zn 15.5, Cd 16, Ni 3.
Melting range: 1195-1270°F. Brazing: 1195-1270°F. For torch or induction brazing ferrous and non-ferrous, including stainless steel; for brazing on carbide tips. AWS A5.8-62T BAg-3; ASTM B260-62T BAg-3; AMS 4771. *Obsolete*

UNIBRAZE 40
J.W. Harris Co., Inc.
Now NI 99 ELECTRODE.

UNIBRAZE 410 STAINLESS WELD WIRE
J.W. Harris Co., Inc.
Now HARRIS 410 STAINLESS WELD WIRE.

UNIBRAZE 410, 410 FC
J.W. Harris Co., Inc.
Now HARRIS AMERICAN LOW FUMING BRONZE. FLUX COATED & BARE.

UNIBRAZE 430 STAINLESS WELD WIRE
J.W. Harris Co., Inc.
Now HARRIS 430 STAINLESS WELD WIRE.

UNIBRAZE 45
J.W. Harris Co., Inc.
Ag 45, Cu 15, Zn 16, Cd 24.
Melting range: 1125-1145°F. Brazing: 1125-1145°F. For torch or induction brazing ferrous and non-ferrous metals where fit-up is good. AWS A5.8-62T BAg-1; ASTM B260-62T BAg-1; AMS 4769. *Obsolete*

UNIBRAZE 5 PHOSPHOR-COPPER-SILVER
J.W. Harris Co., Inc.
Now STAY-SILV 5.

UNIBRAZE 50
J.W. Harris Co., Inc.
Ag 50, Cu 15.5, Zn 16.5, Cd 18.
Melting range: 1160-1175°F. Brazing: 1160-1175°F. For torch or induction brazing ferrous and non-ferrous metals. AWS A5.8-62T BAg-1a; ASTM B260-62T BAg-1a; AMS 4770B. *Obsolete*

UNIBRAZE 533
J.W. Harris Co., Inc.
Die cast rod, Zinc base. For joining zinc to itself or to other alloys. *Obsolete*

UNIBRAZE 555
J.W. Harris Co., Inc.
Now ALSOLDER 720.

UNIBRAZE 6 PHOSPHOR-COPPER-SILVER
J.W. Harris Co., Inc.
Now STAY-SILV 6.

UNIBRAZE 60
J.W. Harris Co., Inc.
Cast iron non machinable electrode; AC-DC, reverse polarity. For repairing gears, motor housings, farm equipment, cams, levers, where machining is not necessary. *Obsolete*

UNIBRAZE 610
J.W. Harris Co., Inc.
Now AZ 92A.

UNIBRAZE 6F PHOSPHOR-COPPER-SILVER
J.W. Harris Co., Inc.
Now STAY-SILV 6HP.

UNIBRAZE ALUMINUM BRONZE
J.W. Harris Co., Inc.
Now ALUMINUM BRONZE AZ.

UNIBRAZE ALUMINUM WELDING ROD
J.W. Harris Co., Inc.
Now HARRIS ALUMINUM WELDING ROD.

UNIBRAZE ALUMINUM WELDING WIRE
J.W. Harris Co., Inc.
Now HARRIS ALUMINUM WELDING WIRE. *Obsolete*

UNIBRAZE CUT ROD
J.W. Harris Co., Inc.
Now HARRIS CUT ROD.

UNIBRAZE DEOXIDIZED COPPER
J.W. Harris Co., Inc.
Now DEOXIDIZED COPPER.

UNIBRAZE FABWELD
J.W. Harris Co., Inc.
All purpose mild steel electrode; AC-DC; reverse polarity. For maintenance and sheet metal welding. *Obsolete*

UNIBRAZE GROOVING ROD
J.W. Harris Co., Inc.
For chamfering, routing, channeling, removing excess metal, gouging out old welds. AC-DC; reverse polarity on DC. *Obsolete*

UNIBRAZE MAGNESIUM WELDING WIRE
J.W. Harris Co., Inc.
Now HARRIS MAGNESIUM WELDING WIRE.

UNIBRAZE NAVAL BRONZE
J.W. Harris Co., Inc.
Now HARRIS NAVAL BRONZE.

UNIBRAZE NICKEL BRONZE
J.W. Harris Co., Inc.
Now HARRIS NICKEL BRONZE.

UNIBRAZE NICKEL SILVER
J.W. Harris Co., Inc.
Now HARRIS NICKEL SILVER.

UNIBRAZE NO. 716 ALUMINUM BRAZING WIRE
J.W. Harris Co., Inc.
Melting range: 970-1085°F. For torch brazing aluminum. AWS and ASTM: BA1Si-3. *Obsolete*

UNIBRAZE NO. 718 ALUMINUM BRAZING WIRE
J.W. Harris Co., Inc.
High silicon, aluminum base wire. Melting range: 1070-1080°F. For torch brazing aluminum. AWS and ASTM: BA1Si-4.

UNIBRAZE O PHOSPHOR-COPPER
J.W. Harris Co., Inc.
Now STAY-SILV O.

UNIBRAZE PHOSPHOR BRONZE A
J.W. Harris Co., Inc.
Now HARRIS PHOSPHOR BRONZE A.

UNIBRAZE PHOSPHOR BRONZE C
J.W. Harris Co., Inc.
Now HARRIS PHOSPHOR BRONZE C.

UNIBRAZE PHOSPHOR BRONZE D
J.W. Harris Co., Inc.
Cu 89.75, Sn 10, P 0.25.
Melting point: 1832°F approx. For inert gas and carbon arc braze welding of ferrous and non-ferrous alloys and for wear resistant surfacing and build-up. AWS A5.13-56T Class RCuSn-D; ASTM A399-56T. *Obsolete*

UNIBRAZE SILICON BRONZE
J.W. Harris Co., Inc.
Now HARRIS SILICON BRONZE.

UNIBROACH
Cytemp Specialty Steel Div.
C 0.8, Cr 4, W 6.25, Mo 6.25, V 2.4, bal Fe.
For finishing tools, cutters, hobs, reamers; high speed steel. *Obsolete*

UNICO
Hidalgo Steel Co. Inc.
C 0.7, Mn 0.4, Si 0.4, Cr 0.7, W 1.4, bal Fe.
For pneumatic and flogging tools, shears; oil hardened, shock resistant.

UNICUT (M3)
Cytemp Specialty Steel Div.
C 1, W 6.25, Cr 4, V 2.4, Mo 6.25, bal Fe.
For lathe and planer tools, drills, taps, reamers, hobs; Type M3; high speed steel.

UNICUT-2
Cytemp Specialty Steel Div.
C 1.2, Mo 6, W 6, Cr 4, V 3, bal Fe.
Hardened: 64-68 Rock C. For cutters, drills, taps, end mills, reamers, hobs, form tools, lathe and planer tools, slitting saws, punches, drawing dies. Abrasion resistant, good red-hardness. AISI M3, Type 2. High speed steel.

UNIDAL
Aluminium Industrie Aktiengesellschaft
Zn 4-6, Mg 0.5-1.5, bal Al.
Extruded: 40,000 TS; 24,000 YS; 17 El; 85 Brin. For architecture; heat treatable.

UNIDUR-100
Aluminium Walzwerke Singen GmbH
Aluminum. Zn 4.5, Mg 1.3, bal Al
Solution heat treated: 350 MPa TS; 10 El. For welded construction of vehicles and equipment.

UNIDUR-100
Swiss Aluminium Ltd.
Aluminum. Si 0.35, Fe 0.4, Cu 0.2, Mn 0.05-0.5, Mg 1-1.4, Cr 0.1-0.35, Zn 4-5, Ti 0.05, Zr 0.08-0.2, bal Al.
Extrusion age hardenable alloy in billet form. 350-420 N/mm^2 TS; 290-370 N/mm^2 YS; 8-10 El; 110 Brin.

UNIFLUX 70 AND V 70
Unicore Inc.
C 0.07, Mn 1.25-1.4, Si 0.45-0.55, bal Fe.
Typical analysis and properties of deposited metal; 88,000 TS, 77,000 YS; 25 El; 65 RA. Flux cored electrodes for shielded arc welding. AWS A5.20 (69); ASME SFA 5.20.

UNIFLUX 75 AND V75
Unicore Inc.
C 0.07, Mn 1.25-1.4, Si 0.45-0.55, bal Fe.
Typical analysis and properties of deposited metal: 82,000-88,000 TS; 72,000 77,000 YS; 25 El; 65 RA. Flux cored electrodes for shielded arc welding. AWS A5.20 (69); ASME SFA 5.20.

UNIFLUX 90 AND V90
Unicore Inc.
C 0.07, Mn 1, Si 0.45, Ni 2.4, bal Fe.
Typical analysis and properties of deposited metal: 85,000-90,000 TS; 78,000-82,000 YS; 26 El; 67 RA. Good low temperature impact values. Flux cored electrodes for shielded arc welding.

UNIFONT
Aluminium Industries, AG
Zn 4-6, Mg 0.5-1.5, bal Al.
Sand cast: 28,000 TS; 20,000 YS; 4.5 El; 65 Brin. Chill cast: 33,000 TS; 22,000 YS; 6 El; 75 Brin. For architecture, ship and auto construction; not heat treatable.

UNIFONT
Aluminium Industrie Aktiengesellschaft
Zn 4-6, Mg 0.5-1.5, bal Al.
Sand cast: 28,000 TS; 20,000 YS; 4.5 El; 65 Brin. Chill cast: 33,000 TS; 22,000 YS; 6 El; 75 Brin. For architecture, ship and auto construction; not heat treatable.

UNIFONT
Swiss Aluminium Ltd.
Zn 4-6, Mg 0.5-1.5, bal Al.
Sand cast: 28,000 TS; 20,000 YS; 4.5 El; 65 Brin. Chill cast: 33,000 TS; 22,000 YS; 6 El; 75 Brin. For architecture, ship and auto construction; not heat treatable.

UNIFONT 5
Swiss Aluminium Ltd.
Aluminum. Mg 0.8-1, Zn 4.8-5.2, bal Al.
Sand cast and aged: 37,000 TS; 24,000 YP; 3-6 El; 70 Brin. Die cast and aged: 40,000 TS; 31,000 YP; 4-8 El; 80 Brin. For architecture, ship and auto components. Self-hardening. Sand and die castings. *Obsolete*

UNIFONT-14
Swiss Aluminium Ltd.
Aluminum. Si 0.4, Fe 0.4, Cu 0.01, Mn 0.1-0.4, Mg 3.5-4.5, Zn 0.9-1.4, Ti 0.05-0.2, bal Al.
Primary foundry alloy in ingot form. 160-230 N/mm^2 TS; 70-100 N/mm^2 YS; 4-9 El; 50-70 Brin.

UNIFONT-50
Swiss Aluminium Ltd.
Aluminum. Si 0.15, Fe 0.15, Cu 0.01, Mn 0.05, Mg 0.6-0.8, Zn 4.3-4.6, Ti 0.15, Cr 0.01-0.2, bal Al.
Primary foundry alloy in ingot form. 200-280 N/mm^2 TS; 130-180 N/mm^2 YS; 5-12 El; 70-80 Brin.

UNIFONT-90
Swiss Aluminium Ltd.
Aluminum. Si 8.5-9, Fe 0.15, Cu 0.01, Mn 0.02, Mg 0.3-0.5, Zn 9.5-10, Ti 0.05, bal Al.
Primary foundry alloy in ingot form. 220-320 N/mm^2 TS; 200-250 N/mm^2 YS; 1-6 El; 90-120 Brin.

UNIFONT-91 DV
Swiss Aluminium Ltd.
Aluminum. Si 8.5-9, Fe 0.15, Cu 0.01, Mn 0.02, Mg 0.3-0.5, Zn 9.5-10, Ti 0.05, bal Al.
Primary foundry alloy in ingot form. 220-250 N/mm^2 TS; 200-230 N/mm^2 YS; 1-2 El; 90-100 Brin.

UNIFONT-94
Swiss Aluminium Ltd.
Aluminum. Si 8.5-9, Fe 0.4, Cu 0.01, Mn 0.4, Mg 0.3-0.5, Zn 9.5-10, Ti 0.05, bal Al.
Primary foundry alloy in ingot form. 300-350 N/mm^2 TS; 230-280 N/mm^2 YS; 2-4 El; 110-120 Brin.

UNIFORM 100
Cold Metal Products Company, Inc.
C 0-0.01, Mn 0-0.3, P 0-0.02, Si 0-0.03, S 0-0.02, Al 0-0.06, N 0-0.005, bal Fe.
Ultra-low carbon strip steel for application in magnetic materials. Soft: 20,000 psi YS; 40,000 psi TS; 40 El.

UNIFORM 200
Cold Metal Products Company, Inc.
C 0-0.02, Mn 0-0.5, P 0-0.025, S 0-0.025, Al 0.02-0.08, bal Fe.
Ultra-low carbon cold rolled strip steel for snap fasteners, eyelets, cylinders, gaskets, and other complex stampings. Low hardness, deep drawability. 25,000 psi YS; 50,000 psi TS; 38 El.

UNIFORM 300
Cold Metal Products Company, Inc.
C 0-0.1, Mn 0-0.5, P 0-0.025, S 0-0.015, Al 0.02-0.08, bal Fe.
Low carbon cold rolled strip steel for eyelets, drawn cylinders and ferrules. Annealed: 45,000 psi YS; 58,000 psi TS; 38 El.

UNIFORM 500
Cold Metal Products Company, Inc.
C 0.47-0.8, Mn 0.5-0.9, P 0-0.025, S 0-0.015, Si 0.15-0.3, Al 0.02-0.08, bal Fe.
As annealed, high carbon cold rolled strip steel for brackets, spring clips, automotive fasteners and roll pins. Low hardness and yield strength. 40,000 psi YS; 65,000 psi TS; 28 El.

UNIFORM 700
Cold Metal Products Company, Inc.

Pre-hardened high carbon cold rolled strip steel for shallow stampings and formed parts. Applications such as springs, putty knives, saw blades, and chain saw guide bars.

UNIFORM 800
Cold Metal Products Company, Inc.
C 0-0.15, Mn 0-1.65, P 0-0.025, S 0-0.015, Si 0-0.6, 0.02 Al min, Cb, V, N, bal Fe.
High strength, low alloy cold rolled strip steel for stamped or formed parts where moderate strength is required. 80,000 psi YS min; 18-24 El.

UNIFORM OIL HARDENING
Jessop Steel Co.
C 0.89, Mn 1.24, Cr 0.55, W 0.47, bal Fe.
For tools, dies; non-deforming. *Obsolete*

UNILOU 24-11
Cytemp Specialty Steel Div.
C 0.15-0.3, Mn 0.5-0.8, Si 0.9-1.2, Cr 22-25, Ni 10-12, bal Fe.
At 70°F: 112,500 TS; 33 El; 236 Brin. At 1200°F: 67,500 TS; 2 El. For heat resisting purposes; cannot be hardened by heat treatment; resists heat up to 2100°F. *Obsolete*

UNILOY 13-8
Cytemp Specialty Steel Div.
Stainless steel. C 0-0.05, Mn 0-0.1, Si 0-0.1, S 0-0.008, P 0-0.01, Cr 12.25-13.25, Ni 7.5-8.5, Mo 2-2.5, Al 0.9-1.35, bal Fe.
For use in aircraft structural components, landing gear parts, shafts, valve parts, fittings, cold headed and machined fasteners. Martensitic, precipitation hardenable. Bars and shapes, grade XM-13: ASTM A564. Plate, sheet, strip, grade XM-13: ASTM A693. Forgings, grade XM-13: ASTM A705. Bars, wire, tubing, rings, forgings and forging stock; vacuum induction, vacuum arc, melted: AMS 5269.

UNILOY 14 CMV
Now UNITEMP 14 CMV.

UNILOY 14 HV
Now UNITEMP 14 HV.

UNILOY 14 HW
Now UNITEMP 14 HW.

UNILOY 14 MV
Now UNITEMP 14 MV.

UNILOY 1409
Now UNILOY 410.

UNILOY 1409 AL
Cytemp Specialty Steel Div.
Al 3.25-4.5, Cr 12-14, bal Fe.
For electrical resistance rod and wire; oxidation resistant to 1600°F. *Obsolete*

UNILOY 1409 M
Now UNILOY 416.

UNILOY 1409 NH
Now UNILOY 405.

UNILOY 1409 NI
Now UNILOY 414.

UNILOY 1409 TB
Now UNILOY 403.

UNILOY 1415 NW
Now UNITEMP 1415 NW.

UNILOY 1420 WM
Now UNITEMP 1420 WM.

UNILOY 1430 MV
Now UNITEMP 1430 MV.

UNILOY 1435
Now UNILOY 420.

UNILOY 1435 M
Now UNILOY 420 F.

UNILOY 15-35
Now UNITEMP 330.

UNILOY 15-5
Cytemp Specialty Steel Div.
Stainless steel. C 0-0.07, Mn 0-1, Si 0-1, S 0-0.03, P 0-0.04, Cr 14-15.5, Ni 3.5-5.5, Cu 2.5-4.5, bal Fe.
For applications requiring high transverse strength and toughness, including valves and fittings, fasteners, gears, cams, shafting, mixing equipment, instrument bourdon tubes, springs, chains, bearings, forgings. Martensitic, precipitation hardenable. Bars and shapes, grade XM-12: ASTM A564. Sheet, plate, strip, grade XM-12: ASTM A693. Forgings, grade XM-12: ASTM A705. Bars, forgings, consumable electrode melted: AMS 5659. Sheet, plate strip, consumable electrode, vacuum arc remelted: AMS 5862.

UNILOY 15100 MO
Cytemp Specialty Steel Div.
C 1.05, Cr 14.5, Mo 4, bal Fe.
Martensitic, corrosion resistant, high strength and hardness to 1000°F. For bearings, gears, shafts, loaded stress parts operating up to 1000°F. AISI 618; modified 440 C + Mo. *Obsolete*

UNILOY 16-18
Cytemp Specialty Steel Div.
C 0-0.08, Cr 15-17, Ni 17-19, Mn 0-2, Si 0-1, bal Fe.
Wire: 75,000 TS; 35,000 YS; 55 El. For cold headed fasteners and parts. Stainless, austenitic; excellent cold headability. *Obsolete*

UNILOY 16-8
Now UNILOY 301.

UNILOY 17-4
Cytemp Specialty Steel Div.
Stainless steel. C 0-0.07, Mn 0-1, Si 0-1, S 0-0.03, P 0-0.04, Cr 15-17.5, Ni 3-5, Cu 3-5, 0.15-0.45 Cb + Ta, bal Fe.
For aerospace, power generation, valve apparatus, fasteners, and miscellaneous applications. Martensitic and magnetic; can be precipitation or age hardened to various levels of hardness and strength.

UNILOY 17-4 MO
Cytemp Specialty Steel Div.
C 0.1, Cr 17, Ni 4, Mo 3, bal Fe.
At RT: 191,900 TS; 146,900 YS; 13.5 El. At 1200°F: 45,240 TS; 29,230 YS; 42.0 El. For knife blades, valves, springs, high strength structural applications up to 1000°F. Precipitation hardening stainless. Name changed to Unitemp 350. *Obsolete*

UNILOY 18-12S
Now UNILOY 305.

UNILOY 18-14S-MO
Now UNILOY 316.

UNILOY 18-8
Now UNILOY 302.

UNILOY 18-8M
Now UNILOY 303.

UNILOY 18-8S
Now UNILOY 304.

UNILOY 18-8TI
Now UNILOY 321.

UNILOY 1809
Now UNILOY 430.

UNILOY 1809 M
Now UNILOY 430 F.

UNILOY 1809 NI
Now UNILOY 431.

UNILOY 18100
Now UNILOY 440 C.

UNILOY 18100 F SE
Now UNILOY 440 F SE.

UNILOY 18110 BM
Now UNILOY 440 C.

UNILOY 1860
Now UNILOY 440 A.

UNILOY 1890
Now UNILOY 440 B.

UNILOY 19-14 SM
Now UNILOY 317.

UNILOY 19-9 DL
Now UNITEMP 19-9 DL.

UNILOY 19-9 W MO
Now UNITEMP 19-9 W MO.

UNILOY 19-9 WX
Now UNITEMP 19-9 WX.

UNILOY 20-10S
Now UNILOY 308.

UNILOY 20-25
Now UNILOY 31.

UNILOY 2009
Now UNILOY 442.

UNILOY 201
Cytemp Specialty Steel Div.
C 0-0.15, Cr 16-18, Ni 3.5-5.5, Mn 5.5-7.5, Si 0-1, N 0-0.25, bal Fe.
Annealed: 115,000 TS; 55,000 min YS; 55 El; 90 Rock B. Good corrosion resistance, nonmagnetic (annealed), good resistance to scaling to 1450°F; good impact strength. For parts requiring corrosion resistance. AISI Type 201 stainless steel; SAE 30201. *Obsolete*

UNILOY 202
Cytemp Specialty Steel Div.
Stainless steel. C 0-0.15, Cr 17-19, Ni 4-6, Mn 7.5-10, Si 0-1, N 0-0.25, bal Fe.
Annealed: 105,000 TS; 55,000 YS; 55 El; 90 Rock B. Good corrosion resistance, non-magnetic (annealed), good resistance to scaling to 1500°F; good impact strength. For parts requiring corrosion resistance. AISI Type 202 stainless steel; SAE 30202.

UNILOY 21-6-9
Cytemp Specialty Steel Div.
Stainless steel.
For billet and plate mill products; 300 series austenitic.

UNILOY 25-12
Now UNILOY 309.

UNILOY 25-12S
Now UNILOY 309 S.

UNILOY 25-20
Now UNILOY 310.

UNILOY 25-20H
Now UNILOY 314.

UNILOY 2525
Now UNILOY 446.

UNILOY 301
Cytemp Specialty Steel Div.
Stainless steel. C 0-0.15, Cr 16-18, Ni 6-8, Mn 0-2, Si 0-1, bal Fe.
Annealed: 110,000 TS; 40,000 YS; 60 El; 85 Rock B. Corrosion resistant; work hardens rapidly to a maximum of about 180,000 psi TS. Cold worked materials for springs and strong, stainless stampings. AISI Type 301 stainless steel; SAE 30301; AMS 5519.

UNILOY 302
Cytemp Specialty Steel Div.
Stainless steel. C 0-0.15, Cr 17-19, Ni 8-10, Mn 0-2, Si 0-1, bal Fe.
Annealed: 85,000 TS; 35,000 YS; 60 El; 150 Brin. Austenitic, non-magnetic (annealed), corrosion resistant, ductile. For stainless steel parts and assemblies, particularly stampings; food industry. AISI Type 302 Stainless; SAE 30302.

UNILOY 303 MA
Cytemp Specialty Steel Div.
C 0-0.15, Mn 0-2, Si 0-1, S 0.11-0.16, Cr 17-19, Ni 8-10, Mo 0.4-0.6, Al 0.6-1, bal Fe.
Cold drawn: 100,000 TS; 60,000 YS; 40 El, 55 RA, 228 Brin. Annealed: 90,000 TS; 40,000 YS; 50 El, 60 RA, 160 Brin. For stainless screw machine products, switch gears, pump and valve parts, fittings. Free machining, stainless, non-magnetic. AMS 5638. *Obsolete*

UNILOY 303, 303 SE
Cytemp Specialty Steel Div.
Stainless steel. C 0-0.15, Cr 17-19, Ni 8-10, Mn 0-2, Si 0-1, Mo 0-0.6, 0.15 min S or Se, bal Fe.
Annealed: 90,000 TS; 35,000 YS; 50 El; 160 Brin. Corrosion resistant, austenitic, free machining. For stainless parts made on automatic screw machines, screws, studs, nuts. AISI Type 303; SAE 30303; AMS 5640.

UNILOY 304
Cytemp Specialty Steel Div.
Stainless steel. C 0-0.08, Cr 18-20, Ni 8-10, Mn 0-2, Si 0-1, bal Fe.
Annealed: 85,000 TS; 35,000 YS; 60 El; 150 Brin. Austenitic, corrosion resistant. For stainless stampings, structural parts, cold formed parts, rivets, beer barrels. AISI Type 304 stainless; SAE 30304.

UNILOY 304 L
Cytemp Specialty Steel Div.
Stainless steel. C 0-0.03, Mn 0-2, Si 0-1, Cr 18-20, Ni 8-10, bal Fe.
Annealed, RT: 85,000 TS; 37,000 YS; 57 El. -320°F: 235,000 TS; 56,000 YS; 40 El. 1400°F: 29,000 TS; 11,000 YS; 36 El. Austenitic, stainless, weldable. For welded assemblies for cryogenic to room temperature to elevated temperature operation, tanks, etc. AISI Type 304l; SAE 30304 L.

UNILOY 305
Cytemp Specialty Steel Div.
C 0-0.12, Cr 17-19, Ni 10-13, Mn 0-2, Si 0-1, bal Fe.
Annealed: 85,000 TS; 38,000 YS; 50 El; 8 Rock B. Austenitic, corrosion resistant; less tendency to work harden. For stainless stampings, formed parts. AISI Type 305 stainless; SAE 30305. *Obsolete*

UNILOY 308
Cytemp Specialty Steel Div.
C 0-0.08, Mn 0-2, Si 0-1, Cr 19-21, Ni 10-12, bal Fe.
Wire, annealed: 115,000 TS; 80,000 YS; 40 El. For weld rod, industrial furnaces, equipment for hot sulfite liquor. AISI Type 308 stainless. *Obsolete*

UNILOY 308L
Cytemp Specialty Steel Div.
C 0-0.03, Mn 0-2, Si 0-1, Cr 19-21, Ni 10-12, bal Fe.
Main use is weld rod for welding austenitic and ferritic stainless steels. *Obsolete*

UNILOY 309
Cytemp Specialty Steel Div.
C 0-0.2, Mn 0-2, Si 0-1, Cr 22-24, Ni 12-15, bal Fe.
Annealed, 70°F: 90,000 TS; 40,000 YS; 45 El; 160 Brin. 1600°F: 21,000 TS; 17,500 YS; 50 El. Austenitic, corrosion resistant, good resistance to scaling to 1800°F, weldable. For furnace parts, aircraft and jet engine parts, heat exchangers, chemical equipment. AISI Type 309 Stainless; SAE 30309. *Obsolete*

UNILOY 309 S
Cytemp Specialty Steel Div.
C 0-0.08, Mn 0-2, Si 0-1, Cr 22-24, Ni 12-15, bal Fe.
Annealed, 70°F: 90,000 TS; 40,000 YS; 45 El; 160 Brin. 1600°F: 21,000 TS; 17,500 YS; 50 El. Austenitic, corrosion resistant, good resistance to scaling to 1800°F, preferred for welding. For furnace parts, welded assemblies for aircraft and jet engine parts, heat exchangers. AISI Type 309S; SAE 30309S; AMS 5650. *Obsolete*

UNILOY 309S-CB
Cytemp Specialty Steel Div.
C 0-0.08, Mn 0-2, Si 0-1, Cr 22-24, Ni 12-15, Cb + Ta = 10 x C, bal Fe.
Low carbon, columbium stabilized, austenitic weld rod for welding Type 30 and similar metals for high temperature service. *Obsolete*

UNILOY 310
Cytemp Specialty Steel Div.
Stainless steel. C 0-0.25, Mn 0-2, Si 0-1.5, Cr 24-26, Ni 19-22, bal Fe.
Annealed, RT: 95,000 TS; 45,000 YS; 50 El; 89 Rock B. 1600°F: 25,000 TS; 17,000 YS; 33 El. For furnace parts, aircraft and jet engine parts, heat exchangers, petroleum refining, weld rods, chemical equipment. AISI Type 310 stainless; SAE 30310; AMS 5651, 5521.

UNILOY 310S
Cytemp Specialty Steel Div.
Stainless steel. C 0-0.08, Mn 0-2, Si 0-1.5, Cr 24-26, Ni 19-22, bal Fe.
Similar to Uniloy 310, but preferred for weld rods and some welded assemblies. AISI Type 310S.

UNILOY 314
Cytemp Specialty Steel Div.
C 0-0.25, Mn 0-2, Si 1.5-3, Cr 23-26, Ni 19-22, bal Fe.
Annealed, RT: 95,000 TS; 45,000 YS; 50 El; 89 Rock B. 1600°F: 25,000 TS; 17,000 YS; 33 El. For furnace parts, heat exchangers, oil refining; particularly for carburizing atmospheres as carburizing boxes and retorts. AISI Type 314; SAE 30314; AMS 5652. *Obsolete*

UNILOY 316
Cytemp Specialty Steel Div.
Stainless steel. C 0-0.08, Mn 0-2, Si 0-1, Cr 16-18, Ni 10-14, Mo 2-3, bal Fe.
Annealed: 80,000 TS; 30,000 YS; 60 El; 149 Brin. Annealed, cold drawn: 90,000 TS; 60,000 YS; 45 El; 190 Brin. Excellent corrosion resistance, weldable. For photographic equipment, pulp and paper, pharmaceutical and textile equipment. AISI Type 316; SAE 30316; AMS 5648.

UNILOY 316F
Cytemp Specialty Steel Div.
Stainless Steel. C 0-0.08, Mn 0-2, Si 0-1, Cr 17-19, Ni 10-14, Mo 1.75-3, 0.10 S min, 0.10 P min, bal Fe.
Free machining grade of Uniloy 316. For stainless parts requiring appreciable machining. AMS 5649. *Obsolete*

UNILOY 316L
Cytemp Specialty Steel Div.
Stainless steel. C 0-0.03, Mn 0-2, Si 0-1, Cr 16-18, Ni 10-14, Mo 2-3, bal Fe.
Preferred for weld wire and welded assemblies for high temperature operation and for equipment for chemical plants. AISI Type 316L; SAE 30316 L; AMS 5653.

UNILOY 317
Cytemp Specialty Steel Div.
Stainless steel. C 0-0.08, Mn 0-2, Si 0-1, Cr 18-20, Ni 11-15, Mo 3-4, bal Fe.
Annealed: 80,000 TS; 30,000 YS; 60 El; 149 Brin. Austenitic, excellent corrosion resistance, maximum resistance to pitting. For boat trim, chemical processing equipment, exhaust manifolds. AISI Type 317; SAE 30317.

UNILOY 317L
Cytemp Specialty Steel Div.
C 0-0.03, Mn 0-2, Si 0-1, Cr 18-20, Ni 11-15, Mo 3-4, bal Fe.
For weld rod and for welded assemblies for corrosive conditions. *Obsolete*

UNILOY 318
Cytemp Specialty Steel Div.
C 0-0.08, Mn 0-2, Si 0-1, Cr 17-19, Ni 13-15, Mo 2-3, Cb + Ta = 10 x C, bal Fe.
Annealed, RT: 80,000 TS; 30,000 YS; 60 El; 149 Brin. 1000°F: 24,000 TS; 16,000 YS; 38 El. For weld wire, welded assemblies and for elevated temperature operation; furnace parts, heat exchangers, exhaust manifolds. *Obsolete*

UNILOY 321
Cytemp Specialty Steel Div.
Stainless steel. C 0-0.08, Mn 0-2, Si 0-1, Cr 17-19, Ni 9-12, Ti = 5 x C min, bal Fe.
Annealed, RT: 85,000 TS; 35,000 YS; 55 El; 150 Brin. 1600°F: 17,000 TS; 13,500 YS; 57 El. For parts and welded assemblies operating intermittently in 900-1600°F range. Aircraft exhaust manifolds, jet engine parts, pressure vessels, heat treat equipment. AISI Type 321; SAE 30321; AMS 5510, 5645.

UNILOY 325

Cytemp Specialty Steel Div.
C 0.4, Si 1.2, Mn 0.6, Cr 7-10, Ni 19-23, bal Fe.
Forged: 141,000-108,000 TS; 27-50 El; 248-167 Brin. For applications where superheated steam, hot oils, sulfides, hot sulfur are encountered; non-corrosive and heat resistant; resists scaling up to 1500°F. *Obsolete*

UNILOY 326

Cytemp Specialty Steel Div.
C 0-0.05, Mn 0-1, Si 0-0.6, Cr 25-27, Ni 6-7, Ti 0-0.25, bal Fe.
About 30% austenitic but ferromagnetic. Annealed: 90,000-100,000 TS; 75,000 YS; 30 El. Hardenable to about 180,000 TS by cold work; resistant to corrosion and stress corrosion. For cold headed bolts and cold headed fasteners for paper pulp, food and chemical industries. *Obsolete*

UNILOY 330

Cytemp Specialty Steel Div.
Stainless steel. C 0-0.2, Mn 0-2, Si 0-1.5, Cr 14.5-16.5, Ni 34-37, bal Fe.
Annealed, RT: 85,000 TS; 42,000 YS; 43 El. 2000°F: 5100 TS; 3900 YS; 72 El. Good strength and scaling resistance at high temperature. For annealing fixtures, carburizing boxes, furnace parts operating up to 2100°F.

UNILOY 347

Cytemp Specialty Steel Div.
Stainless steel. C 0-0.08, Mn 0-2, Si 0-1, Cr 17-19, Ni 9-13, Cb + Ta = 10 x C min, bal Fe.
Stabilized stainless for weld rod, welded assemblies, and for operation 900-1600°F. For aircraft exhaust manifolds, fire walls, pressure vessels, high temperature chemical handling equipment. AISI Type 347; SAE 30347; AMS 5512, 5646.

UNILOY 347F-SE

Cytemp Specialty Steel Div.
C 0-0.08, Mn 0-2, Si 0-1, S 0-0.03, Cr 17-19, Ni 9-12, 0.10 P min, 0.15 Se min,Cb + Ta = 10 x C min, bal Fe.
Annealed, RT: 90,000 TS; 35,000 YS; 50 El, 160 Brin. 1600°F: 20,000 TS; 15,000 YS; 84 El. Free machining, stabilized, austenitic, stainless. For machined parts to operate intermittently to 900-1600°F range. AMS 5642. *Obsolete*

UNILOY 348

Cytemp Specialty Steel Div.
Stainless steel. C 0-0.08, Mn 0-2, Si 0-1, Cr 17-19, Ni 9-13, Ta 0-0.1, Co 0-0.2, Cb + Ta = 10 x C min, bal Fe.
Stabilized, austenitic, weldable stainless, similar to Uniloy 347 except restricted Tantalum and Cobalt content. For nuclear applications. AISI Type 348; SAE 30348; AMS 5646.

UNILOY 36

Cytemp Specialty Steel Div.
Ni 35-36, Mn 0.35, Si 0.12, C 0.04, bal Fe.
Bar: 68,000 TS; 36,000 YS; 44 El, 78 RA. At -100°F: 87,000 TS; 60,000 YS; 42 El, 73 RA. For thermostats, temperature control devices, bimetals, length standards. Low coefficient of expansion. *Obsolete*

UNILOY 403

Cytemp Specialty Steel Div.
Stainless steel. C 0-0.15, Mn 0-1, Si 0-0.5, Cr 11.5-13, bal Fe.
Annealed: 75,000 TS; 40,000 YS; 35 El; 155 Brin. Heat treatable, martensitic, weldable, stainless. Hardenable to 100,000-195,000 psi TS. For turbine wheels, valve parts, cutlery, sporting goods, rifle barrels. AISI Type 403; SAE 51403; AMS 5613.

UNILOY 405

Cytemp Specialty Steel Div.
C 0-0.08, Mn 0-1, Si 0-1, Cr 11.5-14.5, Al 0.1-0.3, bal Fe.
Annealed, cold drawn: 85,000 TS; 70,000 YS; 20 El, 185 Brin. Weldable, not readily hardenable; good corrosion resistance. Designed for welded assemblies that harden only slightly in the welded area. AISI 405; SAE 51405.

UNILOY 410

Cytemp Specialty Steel Div.
Stainless steel. C 0-0.15, Mn 0-1, Si 0-1, Cr 11.5-13.5, bal Fe.
Annealed: 75,000 TS; 40,000 YS; 35 El; 155 Brin. Hardenable to 100,000-190,000 TS; weldable, magnetic, martensitic, corrosion resistant. For turbine blades, cutlery, pump parts, mining machinery, fishing poles. AISI Type 410; SAE 51410; AMS 5504, 5613.

UNILOY 410 CB

Cytemp Specialty Steel Div.
Stainless steel.
For bar, billet, sheet, PRP mill products; 400 series martensitic.

UNILOY 414

Cytemp Specialty Steel Div.
C 0-0.15, Mn 0-1, Si 0-1, Cr 11.5-13.5, Ni 1.25-2.5, bal Fe.
Hardenable, martensitic, stainless. Air or oil hardenable to approximately 200,000 TS max. For beater bars, fasteners, gauge parts, springs, mining equipment cutlery, shafts, valve seats. AISI 414; SAE 51414; AMS 5615. *Obsolete*

UNILOY 416

Cytemp Specialty Steel Div.
Stainless steel. C 0-0.15, Mn 0-1.25, P 0-0.06, Si 0-1, Cr 12-14, Mo 0-0.6, 0.15 S min, bal Fe.
Annealed, cold drawn: 100,000 TS; 85,000 YS; 15 El; 215 Brin. Heat treated: 100,000-190,000 TS; 90,000-145,000 YS; 9-18 El; 26-42 Rock C. Free machining, martensitic, corrosion resistant. For shafts, valves, parts made on automatic screw machines, studs, pump parts. AISI 416; SAE 51416; AMS 5610.

UNILOY 42

Now UNISEAL 42.

UNILOY 420

Cytemp Specialty Steel Div.
Stainless steel. Mn 0-1, Si 0-1, Cr 12-14, 0.15 C min, bal Fe.
Air hardened: 110,000-250,000 TS; 70,000-200,000 YS; 22-7 El; 500-200 Brin. Martensitic, magnetic, corrosion resistant. For cutlery, hand tools, dental and surgical instruments, valve trim and parts, shafts. AISI 420; SAE 51420; AMS 5506, 5621.

UNILOY 420 F

Cytemp Specialty Steel Div.
Mn 0-1, Si 0-1, P 0-0.04, Cr 12-14, Mo 0-0.6, 0.15 C min, 0.15 S min, bal Fe.
Martensitic, free machining, air hardenable, corrosion resistant. Air hardenable to 110,000-250,000 TS. For shafts, hardware, screw machine parts, small turbine wheels, splined shafts. AISI 420 F; SAE 51420 F; AMS 5620.

UNILOY 420 F-SE

Cytemp Specialty Steel Div.
Mn 0-1, Si 0-1, P 0-0.04, S 0-0.03, Cr 12-14, Mo 0-0.6, 0.15 C min, 0.15 Se min, bal Fe.
Same as Uniloy 420 F except that Selenium replaces Sulfur for free machining properties. *Obsolete*

UNILOY 430

Cytemp Specialty Steel Div.
Stainless steel. C 0-0.12, Mn 0-1, Si 0-1, Cr 14-18, bal Fe.
Annealed: 80,000 TS; 55,000 YS; 30 El; 170 Brin. Ferritic, magnetic, non-hardenable by heat treating, corrosion resistant. For architectural and automotive trim and moldings, nitric acid tanks, table ware, plumbing supplies. AISI 430; AMS 51430; AMS 5503, 5627.

UNILOY 430 F

Cytemp Specialty Steel Div.
Stainless steel. C 0-0.12, Mn 0-1.25, Si 0-1, P 0-0.06, Cr 14-18, Mo 0-0.6, 0.15 S min, bal Fe.
Annealed: 80,000 TS; 55,000 YS; 25 El; 170 Brin. Ferritic, magnetic, non-hardenable by heat treating, free machining, corrosion resistant. For corrosion resistant studs, bolts, nuts, etc., being machined on automatic screw machines. AISI 430 F; SAE 51430 F.

UNILOY 430 F-SE

Cytemp Specialty Steel Div.
C 0-0.12, Mn 0-1.25, Si 0-1, P 0-0.06, S 0-0.06, Cr 14-18, 0.15 Se min, bal Fe.
Same as Uniloy 430 F except that Selenium replaces Sulfur for free machining properties. AISI 430 F-Se; SAE 51430 F-Se. *Obsolete*

UNILOY 430 TI

Cytemp Specialty Steel Div.
C 0-0.12, Mn 0-1, Si 0-1, Cr 16-18, Ti = 6 x C min, bal Fe.
Same as Uniloy 430 except that Titanium improves ductility of welds. *Obsolete*

UNILOY 431

Cytemp Specialty Steel Div.
Stainless steel. C 0-0.2, Mn 0-1, Si 0-1, Cr 15-17, Ni 1.25-2.5, bal Fe.
Annealed: 125,000 TS; 95,000 YS; 20 El; 260 Brin. Oil hardened: 150,000-210,000 TS; 110,000-165,000 YS; 20-14 El; 31-45 Rock C. Martensitic, weldable, corrosion resistant; good strength to 900°F. Aircraft fittings, pump shafts, marine hardware, valve parts, conveyor parts. AISI 431; SAE 51431; AMS 5628.

UNILOY 434

Cytemp Specialty Steel Div.
C 0-0.12, Mn 0-1, Si 0-1, Cr 14-18, Mo 0.75-1.25, bal Fe.
Same as Uniloy 430 except that added Molybdenum improves resistance to pitting from de-icing chemicals. AISI 434. *Obsolete*

UNILOY 435

Cytemp Specialty Steel Div.
C 0-0.12, Mn 0-1, Cr 14-18, Si 0-1, 0.40-0.60 Cb + Ta, bal Fe.
Same as Uniloy 430 except that Cb + Ta improves surface smoothness when metal is stretch bent or deep drawn. *Obsolete*

UNILOY 436

Cytemp Specialty Steel Div.
C 0-0.12, Mn 0-1, Si 0-1, Cr 14-18, Mo 0.75-1.25, 0.40-0.60 Cb + Ta, bal Fe.
Similar to Uniloy 430 but produces smoother surfaces on deep drawing and improved resistance to pitting from de-icing chemicals. AISI 436. *Obsolete*

UNILOY 440 A

Cytemp Specialty Steel Div.
C 0.6-0.75, Mn 0-1, Si 0-1, Cr 16-18, Mo 0-0.75, bal Fe.
Annealed: 115,000 TS; 90,000 YS; 12 El; 240 Brin. Heat treated: 260,000 TS; 240,000 YS; 5 El; 510 Brin. Martensitic, magnetic, corrosion resistant. For cutlery, bearings, valves, surgical and dental instruments. AISI 440 A; SAE 51440 A; AMS 5631.

UNILOY 440 B

Cytemp Specialty Steel Div.
C 0.75-0.95, Mn 0-1, Si 0-1, Cr 16-18, Mo 0-0.75, bal Fe.
Air or oil hardenable to 270,000-280,000 TS. Holds high hardness to 900°F. Corrosion resistant, magnetic. For bearings, cutlery, spatula blades, food processing knives. AISI 440 B; SAE 51440 B. *Obsolete*

UNILOY 440 C

Cytemp Specialty Steel Div.
C 0.95-1.2, Mn 0-1, Si 0-1, Cr 16-18, Mo 0-0.75, bal Fe.
Air or oil hardenable to 275,000-285,000 TS and 580 Brin; retains hardness after tempering to 900°F of 55-57 Rock C; corrosion resistant. For cutlery, bearings, nozzles, balls and seats for oil well pumps. AISI 440 C; SAE 51440 C; AMS 5630.

UNILOY 440 F

Cytemp Specialty Steel Div.
C 0.95-1.2, Mn 0-1, Si 0-1, P 0-0.04, Cr 16-18, Mo 0-0.75, 0.05 S min, bal Fe.
Free machining grade of Uniloy 440 C. For parts requiring considerable machining before heat treating. SAE 51440 F; AMS 5632. *Obsolete*

UNILOY 440 F-SE

Cytemp Specialty Steel Div.
C 0.95-1.2, Mn 0-1, Si 0-1, P 0-0.04, S 0-0.03, Cr 16-18, Mo 0-0.75, 0.10 Se min, bal Fe.
Similar to Uniloy 440 F except that Selenium replaces Sulfur for free machining properties. SAE 51440 F-Se. *Obsolete*

UNILOY 442

Cytemp Specialty Steel Div.
C 0-0.2, Mn 0-1, Si 0-1, Cr 18-23, bal Fe.
Annealed, RT: 75,000 TS; 45,000 YS; 30 El; 160 Brin.
1600°F: 4500 TS; 3500 YS; 87 El. Ferritic, non- hardenable stainless, weldable. For annealing boxes, oil burner parts, nozzles, heating elements. AISI 442. *Obsolete*

UNILOY 443

Cytemp Specialty Steel Div.
C 0-0.2, Mn 0-1, Si 0-1, Cr 18-23, Cr 0.9-1.25, bal Fe.
Same properties and characteristics as Uniloy 442, but more resistant to action of dilute sulfuric acid. *Obsolete*

UNILOY 446

Cytemp Specialty Steel Div.
C 0-0.2, Mn 0-1.5, Si 0-1, Cr 23-27, N 0-0.25, bal Fe.
Annealed, RT: 80,000 TS; 50,000 YS; 25 El; 170 Brin.
1800°F: 2500 TS; 2000 YS; 135 El. Ferritic, non-hardenable, stainless, weldable. Oxidation resistant at high temperatures. For annealing boxes, salt bath electrodes, oil burner parts, metal to glass seals. AISI 446; SAE 51446. *Obsolete*

UNILOY 49

Now UNISEAL 52.

UNILOY 50 CR/50 NI

Cytemp Specialty Steel Div.
Cr 50, Ti 1, C 0.06, bal Ni.
Annealed: 125,000 TS; 72,000 YS; 18 El, 20 RA, 28 Rock C. For equipment in the pulp and paper industry, power production, waste incinerator, petroleum refining, fuel-oil fired heaters. Excellent resistance to oil and acid corrosion. Resists sulfidation. *Obsolete*

UNILOY 501

Cytemp Specialty Steel Div.
C 0.1-0, Cr 4-6, bal Fe.
Annealed: 70,000 TS; 30,000 YS; 28 El; 65 RA; 160 Brin.
Heat treated: 175,000 TS; 135,000 YS; 15 El; 50 RA; 370 Brin. For oil refinery equipment, fittings, grate bars; Type 501; corrosion resistant. *Obsolete*

UNILOY 502

Cytemp Specialty Steel Div.
C 0-0.1, Cr 4-6, Mo 0.5, bal Fe.
Annealed: 70,000 TS; 32,000 YS; 28 El; 65 RA; 160 Brin.
Heat treated: 180,000 TS; 140,000 YS; 13 El; 45 RA; 380 Brin. For oil refinery equipment, fittings, grate bars; Type 502; corrosion resistant. *Obsolete*

UNILOY A286

Now UNITEMP A286.

UNILOY D319

Cytemp Specialty Steel Div.
C 0-0.07, Mn 0-2, Si 0-1, Cr 17.5-19.5, Ni 11-15, Mo 2.25-3, bal Fe.
Austenitic, corrosion resistant. For parts, and assemblies for chemical industry; certain improvements over Uniloy 316. *Obsolete*

UNILOY EB 26-1

Cytemp Specialty Steel Div.
Cr 25-27, Mn 0-0.4, Mo 0.75-1.25, Si 0-0.4, bal Fe.
Annealed: 65,000-78,000 TS; 53,000-61,000 YS; 24-30 El; 83-86 Rock B. Hardenable to about 116,000 TS by cold work. Corrosion and stress corrosion resistant, weldable, magnetic. For marine equipment, chemical equipment, food and beverage equipment, pulp and paper. *Obsolete*

UNILOY HV

Now UNITEMP HV.

UNILOY M-252

Now UNITEMP M252.

UNILOY N-155

Now UNITEMP N-155.

UNILOY S 590

Now UNITEMP S 590.

UNILOY S-817

Cytemp Specialty Steel Div.
C 0.32-0.42, Mn 1-2, Cr 19-21, W 3.5-4.5, Ni 19-21, Mo 4, bal Co.
Rolled: 125,000-145,000 TS; 55,000-75,000 YS; 32-22 El; 31-20 RA. For turbine rotors, buckets, shafts, bolts; age-hardenable, heat and corrosion resistant. *Obsolete*

UNILOY SPECIAL NO. 12-12

Universal Cyclops
Cr 11-13, Ni 11-12, C 0.2, Si 0.25-0.5, bal Fe.
For corrosion resistant parts; corrosion resistant. *Obsolete*

UNILOY-HC

Cytemp Specialty Steel Div.
C 0.06, Cr 15, Co 1.5, Mo 16, W 3.75, Fe 5.75, V 0.2, bal Ni.
Heat treated: 116,000 TS; 56,000 YS; 52 El, 95 Rock B. At 1600°F: 57,000 TS; 38,000 YS; 49 El. For chemical processing equipment, process vessels, pump parts, valves, aircraft and aerospace components. Corrosion resistant. Resists repeated thermal shock at 1600 to 1800°F. *Obsolete*

UNIMACH 9-4-20

Cytemp Specialty Steel Div.
For ingot, bar, billet, PRP mill products; high performance alloy.

UNIMACH 9-4-30

Cytemp Specialty Steel Div.
For ingot, bar, billet, PRP mill products; high performance alloy.

UNIMACH AF 1410

Cytemp Specialty Steel Div.
C 0.15, Mn 0.02, Si 0.02, S 0.002, P 0.004, Cr 2, Ni 10, Co 14, Mo 1, Al 0.005, O 0.001, N 0.001, bal Fe.
For highly stressed, fracture critical applications, including airframe structural components, high strength fasteners, landing gear components, solid fuel rocket motor cases, light armor plate, and other applications requiring excellent fracture toughness at high strength. Steel plate: AMS6522. Steel bar, forgings: AMS6527.

UNIMACH NO. 1, H11

Cytemp Specialty Steel Div.
C 0.4, Mn 0.4, Si 1, Cr 5, Mo 1.4, V 1, bal Fe.
Heat treated: 225,000-305,000 TS; 195,000-255,000 YS; 15.0-8.5 El; 50.2-29.5 RA; 445-560 Brin. For die casting dies, hot work punches, hot shears, dummy blocks; hot work steel, air hardened, non deforming.

UNIMAG 50

Cytemp Specialty Steel Div.
C 0.05, Mn 0.05, Si 0.35, Ni 48, bal Fe.
After hydrogen anneal has high saturation flux of 15,000 gausses, and low hysteresis loss in DC, or in AC circuits with frequencies less than 400 cycles per second. For laminated cores, magnetic shielding, relays, solenoid cores and electric motor rotors. *Obsolete*

UNIMAR 250

Cytemp Specialty Steel Div.
Alloy steel. C 0.02, Mn 0.05, Si 0.05, Ni 18, Co 8, Mo 4.8, Ti 0.4, Al 0.1, bal Fe.
Heat treated: 268,000 TS; 258,000 YS; 11 El, 52 Rock C. For jet engine, gas turbine and spacecraft components, pressure vessels, rocket motor cases. Maraging steel, tough, ductile and shock resistant. Precipitation hardened.

UNIMAR 300

Cytemp Specialty Steel Div.
Alloy steel. C 0.02, Mn 0.05, Si 0.05, Ni 18.5, Co 9, Mo 4.8, Ti 0.6, Al 0.1, bal Fe.
Annealed: 50,000 TS; 110,000 YS; 18 El; 32 Rock C. Heat treated: 315,000 TS; 310,000 YS; 10 El, 55 Rock C. For aluminum die casting dies, pressure vessels, rocket motor cases, bolts and fasteners. Maraging steel, vacuum melted. Precipitation hardened.

UNIMAR 350

Cytemp Specialty Steel Div.
C 0.02, Mn 0.05, Si 0.05, Ni 17.5, Co 12, Mo 4.8, Ti 1.5, Al 0.1, bal Fe.
Heat treated 360,000 TS; 350,000 YS; 10 El, 58 Rock C. For jet engine, gas turbine and spacecraft components. Maraging steel, tough, ductile and shock resistant. Precipitation hardened. *Obsolete*

UNIMATIC 6000

Eutectic Corp.
Mild steel electrode for AC DC welding of mild steel;, 68,000 psi TS.
Obsolete

UNIMET

Gilby-Fodor S.A.
Al 1.57, W 1.35, U 0.42, bal Ni.
For electron tube elements. Heat resistant.

UNIMET U-10

TRW Inc./United Greenfield Div.
Sintered, carbide tool material. For roughing cuts on cast iron, non-ferrous and non-metallic materials.

UNIMET U-110

TRW Inc./United Greenfield Div.
Sintered, carbide tool material. For heavy duty and interrupted cutting of cast iron and non-ferrous materials. Also for wear parts

UNIMET U-130

TRW Inc./United Greenfield Div.
Sintered, carbide tool material. For shock resistant tools; withstands medium impact.

UNIMET U-135

TRW Inc./United Greenfield Div.
Sintered, carbide tool material. For fabrication of wear parts; resistant to impact, wear and abrasion.

UNIMET U-140

TRW Inc./United Greenfield Div.
Sintered, carbide tool material. Resists heavy impact.

UNIMET U-20

TRW Inc./United Greenfield Div.
Sintered, carbide tool material. For general purpose roughing and finishing on cast iron, non-ferrous and non-metallic materials.

UNIMET U-20F

TRW Inc./United Greenfield Div.
Sintered, carbide tool material. For use on plastics, glass and highly abrasive materials and for drilling plastic laminates.

UNIMET U-30

TRW Inc./United Greenfield Div.
Sintered, carbide tool material. For light roughing and average finishing of cast iron, non-ferrous and non-metallic materials. Medium high wear resistance and medium low shock resistance.

UNIMET U-40

TRW Inc./United Greenfield Div.
Sintered, carbide tool material. For precision boring of cast iron, non-ferrous and non-metallic materials.

UNIMET U-53
TRW Inc./United Greenfield Div.
Sintered, carbide tool material. For heavy duty roughing cuts on steel and steel alloys at extremely heavy feeds.

UNIMET U-60
TRW Inc./United Greenfield Div.
Sintered, carbide tool material. For general purpose machining and roughing on steel and steel alloys.

UNIMET U-70
TRW Inc./United Greenfield Div.
Sintered, carbide tool material. For average finishing and light roughing cuts on steel and steel alloys.

UNIMET U-73
TRW Inc./United Greenfield Div.
Sintered, carbide tool material. For light roughing and general finishing cuts on steel and steel alloys; fine feeds.

UNIMET U-80
TRW Inc./United Greenfield Div.
Sintered, carbide tool material. For precision boring of steel and steel alloys. High wear resistance coupled with better than average shock resistance. For precision turning, facing and high speed finishing.

UNIMETAL
Unimetal Co.
Sn, Sb, bal Pb.
For welding of white metal. White metal.

UNION
Glacier Metal Co.
Sb, Pb, bal Sn.
For bearings; Babbitt.

UNION 20-CB
Metals & Controls and General Plate
C 0-0.07, Cr 20, Ni 29, 2.0 Mo min, 3.0 Cu min, Cb = 8 x C min, 1.0 Cb + Ta max,, bal Fe.
Annealed: 90,000 TS; 50,000 YS; 45 El; 65 RA; 90 Rock B. For chemical plant equipment to resist sulfuric acid. Stainless, austenitic, heat resistant.

UNION HYMO
Republic Steel Corp.
C 0.3-0.4, Mn 1.35-1.65, bal Fe.
Water quenched: 103,000-195,000 TS; 75,000-178,000 YS; 6-24 El; 28-65 RA; 200-400 Brin. For gears, shafts; machining steel. *Obsolete*

UNION HYMO STEEL
Republic Steel Corp.
C 0.15-0.25, Mn 1-1.3, bal Fe.
Cold drawn: 66,000 TS; 55,000 YS; 15 El; 35 RA. Cold rolled: 75,000-85,000 TS; 55,000-65,000 YS; 15-25 El; 45-55 RA; 143-170 Brin. For case hardened parts, shafts, pinions; free machining case hardening steel. *Obsolete*

UNION MAXCUT
Republic Steel Corp.
C 0.08-0.15, bal Fe.
For machinery and case hardened parts; SAE-X-1112. *Obsolete*

UNION MC QUAID-EHN
Republic Steel Corp.
C 0.15-0.25, Mn 0.3-0.6, P 0.045, S 0.055, bal Fe.
Cold rolled: 70,000-85,000 TS; 60,000-70,000 YS; 15-25 El; 45-55 RA; 149-170 Brin. For carburizing steel for quality carburized parts. *Obsolete*

UNION METAL
Union Metal Mfg. Co.
C 0.3, bal Fe.
Rolled: 60,000-70,000 TS. For bracers, stands, supports; seamless tubing. *Obsolete*

UNION MULTICUT
Republic Steel Corp.
C 0.1-0.2, bal Fe.
For case hardened parts; SAE-1115. *Obsolete*

UNION SPECIAL CARBURIZING STEEL
Republic Steel Corp.
C 0.1-0.2, Mn 0.6-0.9, Si 0.2-0.3, bal Fe.
Turned: 75,000 TS; 55,000 YS; 25 El; 55 RA; 170 Brin. Cold drawn: 90,000 TS; 75,000 YS; 15 El; 50 RA; 187 Brin. For bearings, gears, pinions, shafts, piston pins; case hardening steel. *Obsolete*

UNION-LARSSEN STEEL "RESISTA"
Dortmund-Hoerder Huttenverein A.G.
Cu 0.4, C, bal Fe.
71,300-86,000 TS; 55,000-61,000 YS; 22-26 El. For storage bins, retaining walls for canals; corrosion resistant. *Obsolete*

UNION-LARSSEN STEEL "RESISTA"
Vereinigte Stahlwerke
Cu 0.4, C, bal Fe.
71,300-86,000 TS; 55,000-61,000 YS; 22-26 El. For storage bins, retaining walls for canals; corrosion resistant. *Obsolete*

UNIONALOY
Duraloy Blaw-Knox/Union Steel Casting
C 3.2, Si 2.5, Mn 0.6, bal Fe.
For mill guides, tube mill plugs, hopper liners; resistant to abrasion; cast iron.

UNISEAL 281 ON
Cytemp Specialty Steel Div.
C 0.1, Mn 0.6, Si 0.4, Ni 0.4, Cr 28, bal Fe.
For glass-to-metal seal. Thermal expansion at 77-1472°F: 6.48 in./in.·°F x 10^{-6}. *Obsolete*

UNISEAL 42
Cytemp Specialty Steel Div.
Ni 42, bal Fe.
For glass-to-metal and ceramic-to-metal seals. Controlled thermal expansion matching glasses. *Obsolete*

UNISEAL 52
Cytemp Specialty Steel Div.
Ni 52, bal Fe.
For glass-to-metal and ceramic-to-metal seals. Controlled thermal expansion matching 0120 and 9010 glass. *Obsolete*

UNISIL
British Steel plc
For high efficiency electrical transformers.

UNISIL H (HI-B)
British Steel plc
Similar to Unisil.

UNISON
Erie Steel Co.
C, alloy, bal Fe.
For well bits, arbors, punches, crimpers; nontempering, shock resistant.

UNISPAN 36
Cytemp Specialty Steel Div.
C 0.04, Mn 0.35, Si 0.12, Ni 36, bal Fe.
Annealed: 71,400 TS; 40,000 YS; 41 El; 80 Rock B. Thermal expansion coefficient: -1.07 to +1.8 in./in.·°F x 10^{-6} up to +500°F, depending on previous treatment of metal and temperature range measured. For precision instruments. *Obsolete*

UNISPAN LR 35
Cytemp Specialty Steel Div.
C 0-0.1, Mn 0-0.05, Si 0-0.05, Ni 0-36.5, bal Fe.
Annealed: 64,000-68,000 TS; 37,000-39,000 YS; 33-34 El; 74 Rock B. Thermal expansion coefficient: -8 to 212°F: 0.312 in./in.·°F x 10^{-6}. For precision instruments.

UNISTEEL 410
Reinforcement Steel Services
C 0-0.25, bal Fe.
Weldable, hot rolled reinforcing bar.

UNITED MAGNET
Republic Steel Corp.
C 0.6-0.9, Mn 0.25-0.55, Cr 0.25-2.1, W 0-6, bal Fe.
For magnets, electrical machinery. *Obsolete*

UNITEMP 14 HV
Cytemp Specialty Steel Div.
C 0.45, Cr 1, Mo 0.5, V 0.3, bal Fe.
Heat treated: 110,000-140,000 TS. High temperature, high strength alloy. For truck bodies, transportation equipment. AISI 601. *Obsolete*

UNITEMP 14 HW
Cytemp Specialty Steel Div.
C 0.45, Cr 14, Ni 14, W 2.4, Mo 0.35, bal Fe.
For parts for high temperature operation, heat resistant. *Obsolete*

UNITEMP 14 MHV
Cytemp Specialty Steel Div.
C 0.25-0.3, Mn 0.6-0.9, Si 0.7, Cr 1-1.5, Mo 0.4-0.6, V 0.75-0.95, bal Fe.
At RT: 150,000 TS; 140,000 YS; 16 El; 302 Brin. At 1000°F: 86,000 TS; 74,000 YS; 28 El, 82 RA. For special bolting in steam turbines, hot dies, brake disks. Heat resistant. Useful to 1000°F. *Obsolete*

UNITEMP 14 MV
Cytemp Specialty Steel Div.
C 0.28-0.33, Mn 0.5, Si 0.6, Cr 1-1.5, Mo 0.4-0.6, V 0.2-0.3, bal Fe.
Heat treated: 167,000-298,000 TS; 164,000-272,000 YS; 16.5-8 El; 341-534 Brin. For steam and gas turbine bolting up to 1050°F, brake disks; resists scaling and oxidation to 1200°F. *Obsolete*

UNITEMP 14-14 W
Cytemp Specialty Steel Div.
C 0.45, Cr 13-15, Ni 13-15, W 1.75-3, Mo 0.2-0.5, bal Fe.
Annealed: 110,000 TS; 48,500 YS; 38 El; 51 RA. Aged: 130,000 TS; 68,500 YS; 28 El, 46 RA. For parts requiring corrosion resistance to combustion products at elevated temperatures. Austenitic, stainless. *Obsolete*

UNITEMP 1415 NW
Cytemp Specialty Steel Div.
C 0.16, Mn 0.4, Cr 12.5, Ni 2, Mo 0.2, W 2.75, bal Fe.
Hardenable to 140,000-180,000 TS. Martensitic, corrosion resistant, good high temperature strength to 1200°F. For compressor blades, turbine discs, hardware for jet engines and gas turbine engines. AISI 615 (Greek Ascoloy).

UNITEMP 1420 WM
Cytemp Specialty Steel Div.
C 0.22, Mn 0.75, Cr 11.5, Ni 0.8, Mo 1, W 1, V 0.25, bal Fe.
Hardenable to 190,000-240,000 TS; good strength to 1000°F. For parts for missiles, jet engines, gas turbines, aircraft, steam turbines; corrosion resistant. AISI 616; AISI Type 422.

UNITEMP 1430 MV
Cytemp Specialty Steel Div.
C 0.25-0.35, Cr 11-12, Ni 0-0.5, Mo 2.5-3, V 0.2, bal Fe.
Heat treated: 125,000-210,000 TS; 85,000-120,000 YS; 12-20 El; 32-40 RA; 260-470 Brin. For steam and gas turbine buckets, disks, bolts. Heat resistant to 1200°F. *Obsolete*

UNITEMP 14CMV
Cytemp Specialty Steel Div.
C 0.2, Cr 1, Mo 1, V 0.1, Si 0.75, Mn 0.5, bal Fe.
Heat treated: 138,000 TS; 117,000 YS; 7 El; 45 RA. For high temperature welded structures; good strength and weldability.

UNITEMP 1753

Cytemp Specialty Steel Div.
C 0.24, Cr 16.2, Co 7.2, Mo 1.6, W 8.4, Ti 3.2, Al 1.9, B 0.008, Zr 0.06, Fe 9.5, bal Ni.
Heat treated: 195,000 TS; 130,000 YS; 20 El; 23 RA. For high temperature applications, fasteners, turbine wheels; oxidation resistant to 1675°F. *Obsolete*

UNITEMP 188

Cytemp Specialty Steel Div.
C 0.08, Mn 0.75, Si 0.4, Cr 22, W 14, Ni 23, La 0.05, Fe 1.5, bal Co.
For applications requiring high strength coupled with oxidation and corrosion resistance in high temperature environments up to 2100 °F. Air frame and skin for reentry vehicles, containers for heat transfer media in the nuclear industry, and exhaust fans and reaction vessels for molten salts and corrosive oxidizing gases in the chemical process industry. Austenitic.

UNITEMP 19-9 DL

Cytemp Specialty Steel Div.
C 0.26-0.36, Ni 8-10, Cr 18-22, Mo 1-1.5, W 1-1.5, Cb 0.2-0.8, Ti 0.2-0.6.
For jet and gas engine parts, blades, buckets, wells, rockets; heat resistant.

UNITEMP 19-9 DL

Cytemp Specialty Steel Div.
C 0.3, Cr 19.2, Ni 9, W 1.25, Mo 1.2, Cb 0.4, Ti 0.3, bal Fe.
Hot rolled: 118,000 TS; 69,000 YS; 55 El. For jet engine components, exhaust manifolds; corrosion and heat resistant, austenitic.

UNITEMP 19-9 W-MO

Cytemp Specialty Steel Div.
C 0.00-0.12, Mn 0.4-1, Cr 18-20, W 1-1.75, Ni 8-10, Mo 0.2-0, Cb 0.3-0.6, Ti 0.2-0.5, bal Fe.
For supercharger wheels, blades, casings, turbine blades; heat resistant. *Obsolete*

UNITEMP 19-9 WX

Cytemp Specialty Steel Div.
C 0.07-0.13, Mn 1-2, Cr 19-22, Ni 8-9.5, Mo 0.5, W 1.5, Cb 1.2, Ti 0.2, bal Fe.
For jet engine components, welding rod; heat and corrosion resistant. *Obsolete*

UNITEMP 212

Cannon-Muskegon Corp.
C 0.05-0.15, Cr 15-17, Ni 23-27, Ti 4, Cb 0.6, B 0.1, Zr 0.07, bal Fe.
At RT: 187,000 TS; 135,000 YS; 20 El; 32 RA. At 1200°F: 145,000 TS; 123,000 YS; 15 El; 23 RA. At 1400°F: 108,000 TS; 103,000 YS; 16 El; 25 RA. For aircraft and guilded missile components; age-hardenable, stainless, good to 1400°F service.

UNITEMP 212

Universal Cyclops
C 0.05-0.15, Cr 15-17, Ni 23-27, Ti 4, Cb 0.6, B 0.1, Zr 0.07, bal Fe.
At RT: 187,000 TS; 135,000 YS; 20 El; 32 RA. At 1200°F: 145,000 TS; 120,000 YS; 15 El; 23 RA. At 1400°F: 108,000 TS; 103,000 YS; 16 El; 25 RA. For aircraft and guilded missile components; age-hardenable, stainless, good to 1400°F service.

UNITEMP 350

Cytemp Specialty Steel Div.
C 0-0.1, Cr 16.5-17.5, Ni 4-4.5, Mo 2.5-3, Mn 0.5-0.75, Si 0.2-0.5, bal Fe.
Heat treated: 200,000 TS; 118,000 YS; 10 El; 43 Rock C. Annealed: 164,000 TS; 45,000 YS; 22 El; 43 Rock B. For valves, springs, structures. Heat treatable, stainless. *Obsolete*

UNITEMP 355

Cytemp Specialty Steel Div.
C 0.13, Mn 0.75, Si 0.3, Cr 15.5, Ni 4.25, Mo 2.75, N 0.1, bal Fe.
Solution treated and aged: 180,000-220,000 TS; 170,000-185,000 YS; 12-19 El; 40-48 Rock C. Corrosion resistant, good strength to 1000°F. For jet engine parts, parts for high speed airplanes. AISI 634 (AM-355).

UNITEMP 41

Cytemp Specialty Steel Div.
Nickel. C 0.09, Cr 19, Co 11, Mo 10, Ti 3.1, Al 1.5, Fe 1.8, B 0.005, bal Ni.
For highly stressed parts at high temperature; afterburner parts, turbine casings, nozzle parts, gas turbine parts, high temperature fasteners. AISI 683 (Rene 41).

UNITEMP 450

Cytemp Specialty Steel Div.
For bar mill products; high performance alloy.

UNITEMP 455

Cytemp Specialty Steel Div.
For bar and billet mill products; high performance alloy.

UNITEMP 500

Cytemp Specialty Steel Div.
C 0.12, Cr 17.5, Co 18.5, Mo 4.2, Ti 3.1, Al 3, Zr 0.06, B 0.005, Fe 1.75, Cu 0.03, bal Ni.
Heat treated: 195,000 TS; 125,000 YS; 31 El; 54 RA. For jet engine components, turbine buckets, discs; corrosion and heat resistant. *Obsolete*

UNITEMP 625

Cytemp Specialty Steel Div.
C 0.045, Mn 0.15, Si 0.15, Cr 22, Mo 9, Cb 3.75, Al 0.25, Ti 0.25, Fe 3, bal Ni.
For diverse applications ranging from gas turbine engines to flue gas scrubbers. Austenitic, solution strengthened high performance alloy. Bars, billets, rings, forgings: AMS 5666. Sheet, plate: AMS 5599. Tubing: AMS 5581. Weld wire: AMS 5837.

UNITEMP 718

Cytemp Specialty Steel Div.
C 0.05, Mn 0.1, Si 0.18, Cr 18.25, Ni 53, Mo 3.05, Al 0.5, Ti 1, Co 0.25, B 0.004, Fe 18.6, 5.25 Cb + Ta.
For use in critical jet engine parts such as compressor discs, stators, rotors and shafts, turbine discs; good performance at cryogenic temperatures permits use for rocket engine subassemblies such as bellows, flanges, tubing in compensator lines for liquid propellants. Austenitic, precipitation hardenable high performance alloy.

UNITEMP 750

Cytemp Specialty Steel Div.
Nickel. Cr 15, Ni 73, Cb 0.85, Ti 2.5, Al 0.8, Fe 6.75.
Austenitic, age hardenable, Nickel base alloy. For springs operating up to 1300°F, for aircraft gas turbine parts. AISI 688 (Inconel X).

UNITEMP 901

Cytemp Specialty Steel Div.
Nickel. C 0.05, Cr 12.5, Ni 42.5, Mo 6, Ti 2.5, Al 0.2, B 0.015, bal Fe.
High temperature alloy. For parts requiring high strength and corrosion resistance at 1000-1400°F. AISI 601 or 602 (Incoloy 901).

UNITEMP A286

Cytemp Specialty Steel Div.
C 0-0.08, Cr 13-16, Ni 24-27, Mo 1-2, Ti 1.7-2.2, V 0.4, Al 0-0.35, bal Fe.
Rolled: 135,000-160,000 TS; 85,000-115,000 YS; 30-20 El; 50-30 RA. For jet engine gas turbine buckets, discs and bolts; austenitic, corrosion and heat resistant.

UNITEMP AF2-1DA

Cytemp Specialty Steel Div.
C 0.33-0.37, Zr 0.05-0.15, B 0.015, Al 4.6, Ti 2.8-3.2, Cr 11.5-12.5, W 5.8-6.2, Mo 2.8-3.2, Co 9.5-10.5, Ta 1.3-1.7, bal Ni.
For gas turbine wheels and buckets. High heat and oxidation resistance.

UNITEMP BHT

Cytemp Specialty Steel Div.
C 0.8, Mn 0.3, Cr 4, Mo 4.25, V 1, bal Fe.
For good strength at elevated temperature, bearings, races, gears, cams, and constant speed drive components; dies. AISI 613.

UNITEMP CMV

Cytemp Specialty Steel Div.
C 0.2, Cr 1, V 0.12, bal Fe.
For machine tool parts, gears, bolts; case hardened, tough. *Obsolete*

UNITEMP D6C

Cytemp Specialty Steel Div.
C 0.45, Mn 0.7, Si 0.3, Cr 1.1, Ni 0.6, Mo 1, V 0.1, bal Fe.
Heat treated: 280,000 TS; 235,000 YS; 10 El; 38 RA. For rocket motor cases; high strength and stability. *Obsolete*

UNITEMP EME

Cytemp Specialty Steel Div.
C 0.1-0.2, Cr 18-20, Ni 11-13, W 3-3.5, Cb 0.85-1.2, N 0.1-0.2, bal Fe.
Rolled: 135,000 TS; 100,000 YS; 20 El; 45 RA; 280 Brin. For gas turbine components, high temperature bolts; good strength and ductility to 1200°F, austenitic.

UNITEMP HV

Cytemp Specialty Steel Div.
C 0.45, Cr 1, Mo 0.55, V 0.3, bal Fe.
Heat treated: 167,000-298,000 TS; 164,000-272,000 YS; 8-17 El; 346-534 Brin. Oil hardened; for die casting dies, high temperature bolts, brake discs. *Obsolete*

UNITEMP HX

Cytemp Specialty Steel Div.
Nickel. C 0.1, Mn 0.65, Cr 21.5, Co 1.5, Mo 9, W 0.6, Fe 18.5, bal Ni.
High temperature alloy for furnace parts and fixtures, jet engine tail pipes, afterburner components, shroud rings, turbine blades; forgeable, weldable, machinable. AISI 680 (Hastelloy X).

UNITEMP L 605

Cytemp Specialty Steel Div.
Refractory. C 0.08, Ni 10.5, Fe 2, Cr 20.5, W 15, bal Co.
Rolled: 160,000 TS; 85,000 YS; 55 El. For gas turbine buckets, afterburners; heat resistant.

UNITEMP M-308

Cytemp Specialty Steel Div.
C 0.08, Ni 32.5, Cr 13.75, Mo 4.1, W 6.5, Ti 2.15, Al 0.3, B 0.004, Zr 0.25, bal Fe.
For high temperature applications; heat and corrosion resistant. *Obsolete*

UNITEMP M252

Cytemp Specialty Steel Div.
C 0.15, Cr 18-20, Co 9-11, Mo 9-11, Ti 2.5, Al 0.8, bal Ni.
At 70°F: 173,000 TS. At 1000°F: 155,000 TS; 93,000 YS.
For jet engine and gas turbine buckets; high temperature strength, heat resistant. *Obsolete*

UNITEMP N-155

Cytemp Specialty Steel Div.
Refractory. C 0.1, Mn 1.5, Si 0.5, Cr 21.5, Ni 20.5, Co 20, Mo 3, W 2, Cb 1, N 0.14, bal Fe.
At 70°F: 118,000 TS; 58,000 YS; 49 El. At 1200°F: 73,500 TS; 37,600 YS; 28 El. At 1800°F: 39,000 TS; 30,000 YS; 15 El. For jet engine components, heat and oxidation resistant.

UNITEMP N-155
Cytemp Specialty Steel Div.
C 0.1, Mn 1.5, Cr 20.7, Ni 19.8, Co 19.5, Mo 3, W 2.3, Cb 1.2, N 0.13, Cu 2, bal Fe.
At 70°F: 119,000 TS; 57,000 YS; 43 El; 50 RA. At 1200°F: 80,000 TS; 43,000 YS; 31 El; 42 RA. For afterburners, engine tail pipes, combustion chambers; oxidation resistant to 1800°F.

UNITEMP S-816
Cytemp Specialty Steel Div.
C 0.38, Mn 1.35, Cr 20, Ni 20, Mo 3.75, W 4.2, Cb 4.1, Fe 0-5, bal Co.
Heat treated: 140,000 TS; 55,000 YS; 35 El; 29 RA. For jet engine components, turbine blades; high heat resistance. *Obsolete*

UNITEMP S590
Cytemp Specialty Steel Div.
C 0.38-0.48, Cr 19-22, Ni 18-21, Co 18-21, Mo 4, W 4, Cb 4, bal Fe.
Rolled: 152,00 TS; 80,000 YS; 19 El; 25 RA; 300 Brin. Cast: 84,000 TS; 38,000 YS; 4 El; 10 RA. For jet engine tail pipes, after burners; austenitic, corrosion and heat resistant. *Obsolete*

UNITEMP V-57
Cytemp Specialty Steel Div.
C 0.04, Cr 14.5, Ni 27, Mo 1.25, Ti 3, Al 0.25, B 0.008, V 0.25, bal Fe.
For ingot, billet, PRP mill products; high performance alloy.

UNITEMP VIRGO
Cytemp Specialty Steel Div.
C 0.6, Si 1.2, Mn 0.3, Cr 5, V 0.6, Mo 5.25, bal Fe.
For punches, hot heading dies; hot work steel. *Obsolete*

UNITEMP WASPALOY
Cytemp Specialty Steel Div.
Nickel. C 0.07, Cr 19.75, Co 13.5, Mo 4.45, Ti 3, Al 1.4, Fe 0.75, B 0.005, Zr 0.04, bal Ni.
High strength at high temperature. For turbine buckets for jet engines, discs, engine components. AISI 685 (Waspaloy).

UNITO DRILL ROD
Kidd Drawn Steel
C, bal Fe.
For drills, tools, water hardened.

UNIVAN
U.S. Steel Corp.
C, Mn, V, bal Fe.
For fixtures, housings; tough.

UNIVAN
Duraloy Blaw-Knox/Union Steel Casting
C 0.28-0.32, Ni 1.5, V 0.12, Mn 0.9-1.1, bal Fe.
Cast: 90,000 TS; 60,000 YS; 25 El; 50 RA; 175 Brin. For locomotive frames, wheel centers, crossheads, coupling boxes, spindles, gears, pinions; castings to resist severe shocks and stresses.

UNIVAN C
American Clad Metals Inc.
C 0.3-0.35, Mn 1.3-1.7, Cr 0.4-0.6, Mo 0.27-0.33, bal Fe.
Cast: 85,000 TS; 55,000 YS; 22 El; 45 RA. For locomotive frames, wheel centers, gears, axles, and coupling boxes. Wear resistant.

UNIVERSAL
SWB Stahlformguss Gesellschaft mbH
C 1.25, Si 1.15, Mn 0.7, bal Fe.
For cold work tools, punches; oil hardened, tough. *Obsolete*

UNIVERSAL
Krupp Stahl AG
tool material. C 0.45, Si 0.7, Cr 1.5, Mo 0.5, V 0.8, W 0.5, bal Fe.
For tools, dies, punches; tough, oil hardening. *Obsolete*

UNIVERSAL
J.C. Soding & Halbach
C 0.9, Mn 1.9, V 0.1, bal Fe.
For punches, blanking and forming dies; oil hardened, shock resistant. *Obsolete*

UNIVERSAL ALLOY NO. 10
Universal Alloys Inc.
Cu alloy.
Welded: 65,000 psi TS; 60,000 psi YS; 20 El; 125 Brin. For electrode tips, seam welding rolls; 85% conductivity.

UNIVERSAL NO. 10 BRASS
Universal Castings Corp.
Cu 57-62, Sn 1-2, Pb 0-1.5, Al 0.4, bal Zn.
Cast: 50,000 TS; 25,000 YS; 15 El; 83 Brin. For bearings, gears, retainers, hardware; high strength, free cutting.

UNIVERSAL NO. 20, ALUMINUM BRONZE
Universal Castings Corp.
Fe 0-5, Mn 0-2, 82.0 Cu min, bal Al.
Cast: 83,000 TS; 30,000 YS; 9 El; 150 Brin. For gears, sprockets, bearings, pumps, marine parts; heat treatable, corrosion and wear resistant.

UNIVERSAL NO. XX, MANGANESE BRONZE
Universal Castings Corp.
Cu 60-68, Al 3-7.5, Fe 2-4, Mn 2.5-5, Pb 0-0.2, bal Zn.
Cast: 95,000 TS; 60,000 YS; 5 El; 200 Brin. For valve bodies, gears, boat fittings; resists sea water corrosion.

UNIVERSAL UC101
Universal Castings Corp.
Si 6.5-7.5, Mg 0.3, Fe 0-0.5, Mn 0-0.3, bal Al.
Heat treated: 36,000 TS; 25,000 YS; 3 El; 70 Brin. For fuel pump impellers; heat treatable. *Obsolete*

UNIVERSUM HARTE 1
Hover, Gebruder, Edelstahlwerk
C 0.35, Si 0.9, Cr 1.05, V 0.18, W 1.85, bal Fe.
For shear blades, pneumatic tools, punches; oil hardened, shock resistant.

UNIX
Stahlwerk Stahlschmidt GmbH & Co.
C 1.05, W 1.15, Mn 0.9, bal Fe.
For cutters, blanking and forming dies; water or oil hardened.

UNMAGNETIZABLE WATCH WHEEL ALLOY
English manufacture
Pt 65.75, Ni 18, Cu 18, Cd 1.25.
For watch wheels; corrosion resistant.

UNNAMESSING MWU
Unna Messingwerk, A.G.
Al alloy.
For light alloy parts.

URANIUM (DEPLETED AND NATURAL)
Atomergic Chemetals Corp.
U.
Purities: dendritic 99.96+%, 99.7/8%. Forms: ingot, rod, powder, turnings, plate, sheet, foil.

URANIUM B
Latrobe Steel Co.
C, alloy, bal Fe.
For tools, dies; oil hardened. *Obsolete*

URANIUM R
Wieland-Werke AG Metallwerke
Mg 0.2-2, Cu 3.5-5.5, Si 0.2-1, Mn 0.1-1.2, bal Al.
For aircraft structures; age-hardenable. *Obsolete*

URANIUM STEELS
Manufacturer not listed.
U 3, C 0.2-0.7, Mn, Si, V, bal Fe.
For machinery parts; water hardening.

URANUS 10
Creusot-Loire
C 0.08-0.2, Mn 0-2, Cr 17-19, Ni 8-10, bal Fe.
Annealed: 85,000 TS; 35,000 YS; 60 El; 70 RA; 150 Brin.
Cold drawn: 125,000 TS; 95,000 YS; 25 El; 55 RA; 277 Brin.
For oil refinery and chemical plant equipment; Type 302; stainless, austenitic. *Obsolete*

URANUS 10M
Creusot-Loire
C 0.17-0.25, Cr 18, Ni 8, bal Fe.
Annealed: 90,000 TS; 45,000 YS; 50 El; 70 RA; 180 Brin. For chemical plant equipment, mixers, filters; Type 302; stainless, austenitic. *Obsolete*

URANUS 10SI
Creusot-Loire
C 0-0.08, Cr 17-19, Ni 8-11, Ti = 5 x C, bal Fe.
Normalized: 93,000 TS; 36,000 YS; 45 El; 60 RA; 165 Brin.
Annealed: 87,000 TS; 33,000 YS; 57 El; 73 RA; 155 Brin. For chemical plant equipment; Type 321; stainless, austenitic. *Obsolete*

URANUS 10ST
Creusot-Loire
C 0-0.08, Cr 18, Ni 8, Ti, bal Fe.
Annealed: 85,000 TS; 35,000 YS; 55 El; 65 RA; 150 Brin. For welded chemical plant equipment, tanks, mixers; Type 321; stainless, austenitic. *Obsolete*

URANUS 11W
Dorrenberg Edelstahl GmbH
C 1.3, Si 0-0.25, Mn 0-0.25, bal Fe.
For engravers' tools, reamers, broaches, hobs; Type W1; water hardened.

URANUS 15
Creusot-Loire
C 0-0.08, Cr 18-20, Ni 8-11, Mn 0-2, bal Fe.
Annealed: 90,000 TS; 45,000 YS; 60 El; 135 Brin. Cold drawn: 180,000 TS; 150,000 YS; 10 El; 330 Brin. For chemical plant equipment, welded structures; Type 304; stainless, austenitic. *Obsolete*

URANUS 3
Creusot-Loire
C 0-0.25, Cr 15, Ni 35, bal Fe.
For heat treating boxes, furnace parts; Type 330; heat resistant. *Obsolete*

URANUS 30
Societe des AFY
C 0.2, Ni 7, Cr 4, Cr 20, Si 1.7, bal Fe.
For gas turbine components; heat and creep resistant.

URANUS 4
Compagnie Ateliers et Forges de la Loire
C 0.25, Ni 20, Cr 12, bal Fe.
Annealed: 85,300 TS; 49,800 YS; 30 El. Forged: 107,000 TS; 71,000 YS; 25 El. For pumps, turbines, valves. Heat and corrosion resistant to 1000°C.

URANUS 45 M
Creusot-Loire
C 0-0.07, Mn 0-2, Si 0-1.5, Ni 7-9, Cr 20-22, Mo 2.2-2.8, Cu 0-0.5, bal Fe.
Cast, annealed: 600 N/mm² psi TS min. Austenitic stainless for pumps and piping for nuclear energy equipment. AFNOR Z5 CND 20.8 M.

URANUS 50
Creusot-Loire
C 0-0.06, Cr 21, Ni 7, Mo 2.5, Cu 1.5, bal Fe.
Wrought, annealed: 635 N/mm² psi TS min. Stainless and heat resistant steel; good resistance to stress corrosion. AFNOR Z5 CNDU 21.08.

URANUS 50
Creusot-Loire
C 0-0.1, Cr 16-18, Ni 10-14, Mo 2-3, bal Fe.
Annealed: 85,000-95,000 TS; 35,000-45,000 YS; 50-60 El; 60-75 RA; 150-190 Brin. For acid-resistant chemical plant equipment; Type 316; stainless, austenitic. *Obsolete*

URANUS 50M
Creusot-Loire
C 0-0.07, Mn 0-2, Si 0-1.5, Ni 7-9, Cr 20-22, Mo 2.2-2.8, Cr 1-2, bal Fe.
Cast, annealed: 600 N/mm^2 psi TS min. Austenitic stainless for pump and valve parts in chemical, food, petroleum fertilizer plants. AFNOR Z5 CNDU 20.8 M.

URANUS 50T
Compagnie Ateliers et Forges de la Loire
C 0.03, Cr 17, Ni 13, Cu 1.5, Mo 2.5, bal Fe.
Annealed: 78,000 RS; 32,000 YS; 50 El; 80 Rock B. For chemical and oil refinery plant equipment. Stainless, heat resistant.

URANUS 50T
Societe des AFY
C 0.03, Cr 17, Ni 13, Cu 1.5, Mo 2.5, bal Fe.
Annealed: 78,000 RS; 32,000 YS; 50 El; 80 Rock B. For chemical and oil refinery plant equipment. Stainless, heat resistant.

URANUS 55 M
Creusot-Loire
C 0-0.05, Mn 0-2, Si 0-1.5, Ni 4.5-6, Cr 25-27, Mo 1.5-2.5, Cu 2.5-3.5, bal Fe.
Cast, annealed: 250-290 Brin. Austenitic stainless for pumps, valves that also require abrasion resistance. AFNOR Z3 CNUD 26.5 M; ACI CD 4 M Cu.

URANUS 65
Compagnie Ateliers et Forges de la Loire
C 0-0.03, Cr 25.5, Ni 20, Cb 0.25, bal Fe.
Annealed: 78,200 TS; 31,300 YS; 40 El. For chemical equipment for hot concentrated nitric acid, oil burners, furnace parts, baffle plates. Resists nitric acid, resists intergranular corrosion, austenitic. Oxidation resistant.

URANUS 65
Creusot-Loire
C 0-0.02, Cr 25, Ni 20, Nb, bal Fe.
Wrought, annealed: 490 N/mm^2 psi TS min. Stainless and heat resisting steel; good oxidation resistance at elevated temperature; resists hot nitric acid. AFNOR Z2 CNNb 25.20; similar to AISI 310S.

URANUS 65M
Compagnie Ateliers et Forges de la Loire
C 0.08, Cr 25, Ni 20, Cb 1, bal Fe.
Annealed: 78,000 TS; 31,300 YS; 40 El. For chemical equipment for hot concentrated nitric acid, oil burner parts, baffle plates. Austenitic, corrosion and heat resistant.

URANUS 65M
Creusot-Loire
C 0-0.03, Mn 0-1, Si 0-0.4, Ni 19-22, Cr 23-25, bal Fe.
Cast, annealed: 400 N/mm^2 psi TS min. For nitric acid industries and atomic energy. AFNOR Z2 CN 25.20 M.

URANUS B 6 PM
Creusot-Loire
C 0-0.03, Mn 0-2, Si 0-1, Ni 24-27, Cr 19-22, Mo 4-4.8, Cu 2-3, W 1-5, bal Fe.
Resistant to phosphoric acid. AFNOR Z2 NCDUW 25.20 M.

URANUS B6
Creusot-Loire
C 0-0.02, Cr 20.5, Ni 25.5, Mo 4.5, Cu 1.5, bal Fe.
Wrought, annealed: 550 N/mm^2 psi TS min. Stainless, austenitic grade for equipment for phosphoric and sulfuric acid plants; cellulose, paper and explosives industries. AFNOR Z1 NCDU 25.20.

URANUS B6
Societe des AFY
C 0.03, Cr 20, Ni 25, Cu 1.5, Mo 4.5, bal Fe.
For chemical plant equipment, salt pots; corrosion resistant.

URANUS B6M
Creusot-Loire
C 0-0.04, Mn 0-2, Si 0-1, Ni 24-27, Cr 19-22, Mo 4-4.8, Cu 2-3, bal Fe.
Cast, annealed: 450 N/mm^2 psi TS min. For chemical industries; phosphoric and sulfuric acids, petro-chemicals. AFNOR Z3 NCDU 25.20 M; similar to ACI CN-7M.

URANUS B6SI
Compagnie Ateliers et Forges de la Loire
C 0-0.02, Mn 0.8, Si 2.7, Ni 25, Cr 18, Mo 3.8, Cu 1.5, bal Fe.
Annealed: 136,000 TS; 61,700 YS; 40 El. For chemical plant equipment, furnace parts, conveyor belts, high temperature equipment. Corrosion and heat resistant. Austenitic.

URANUS CH
Compagnie Ateliers et Forges de la Loire
C 0.2, Si 1, Mn 1, Cr 22.5, Ni 3.5, Mo 2.5, Cu 1.5, bal Fe.
Heat treated: 121,000-135,000 TS; 75,000-100,000 YS; 25-35 El. For chemical and petroleum industries, oil refineries. Austenitic, ferritic, corrosion and heat resistant.

URANUS EXTRA
Hidalgo Steel Co. Inc.
C 1, Si 0.2, Mn 0.2, V 0.1, bal Fe.
For cold dies, punches, drills, taps, reamers; water hardened; Type W2.

URANUS F 1
Hidalgo Steel Co. Inc.
C 0.1, Ni 2.5, Cr 20, Cu 1.5, bal Fe.
For furnace parts, chemical plant equipment; heat and corrosion resistant.

URANUS H 3
Now ZCR157.

URANUS R7
Now ATVS MO.

URANUS S
Compagnie Ateliers et Forges de la Loire
C 0-0.02, Mn 0.8, Si 3.7, Ni 14, Cr 17.5, bal Fe.
Annealed: 132,300 TS; 55,200 YS; 32 El; 60 RA. For heat exchangers, chemical and atomic energy plants, nitric acid equipment, cryogenic equipment. Corrosion and heat resistant. Austenitic.

URANUS S1
Creusot-Loire
C 0.015, Cr 17.5, Ni 14.5, Si 3.8, bal Fe.
Wrought, annealed: 540 N/mm^2 psi TS min. Stainless and heat resisting alloy; resistant to oxidizing nitric and sulfuric acids; also used in nuclear operations. AFNOR Z1 CNS 18.15.

URANUS SD
Compagnie Ateliers et Forges de la Loire
C 0-0.02, Mn 0.8, Si 3.7, Ni 15, Cr 17, Mo 2.5, bal Fe.
Annealed: 143,300 TS; 60,000 YS; 35 El. For chemical plant equipment, digesters, acid containers. Heat and corrosion resistant. Austenitic.

URANUS SDM
Creusot-Loire
C 0-0.03, Mn 0-2, Si 3-4, Ni 15-17, Cr 16-18, Mo 2-3, bal Fe.
Cast, annealed: 450 N/mm^2 psi TS min. Austenitic stainless steel. AFNOR Z2 CNSD 17.16 M.

URANUS SM
Creusot-Loire
C 0-0.03, Mn 0-2, Si 3.5-4.5, Ni 13-15, Cr 17-19, bal Fe.
Cast, annealed: 450 N/mm^2 psi TS min. Resistant to various mixed acids, or chlorides. AFNOR Z2 CNS 19.14. M.

URBACH SK10
Carl Urbach & Co., Stahlwerk KG
C 0.76, Co 10, Cr 4.2, Mo 0.8, V 1.8, W 18, bal Fe.
For lathe and planer tools, drills, broaches; high-speed steel.

URBACH SK5
Carl Urbach & Co., Stahlwerk KG
C 0.79, Co 4.75, Cr 4.3, Mo 0.75, V 1.55, W 18, bal Fe.
For lathe and planer tools, drills, reamers, hobs; high-speed steel.

URBACH SS17
Carl Urbach & Co., Stahlwerk KG
C 0.82, Cr 4.1, Mo 0.85, V 1.6, W 8.7, bal Fe.
For lathe and planer tools, form cutters, drills; high-speed steel.

URBACH SS17MO
Carl Urbach & Co., Stahlwerk KG
C 0.95, Cr 4, V, W, Mo, bal Fe.
For lathe and planer tools, form cutters, reamers, taps; high-speed steel.

URBACH SS25
Carl Urbach & Co., Stahlwerk KG
C 0.86, Cr 4.1, Mo 0.85, V 2.5, W 12, bal Fe.
For lathe and planer tools, drills, hobs, reamers; high-speed steel.

URBACH SS35
Carl Urbach & Co., Stahlwerk KG
C 0.85, Cr 4, W, Mo, V, bal Fe.
For lathe and planer tools, drills, reamers; high-speed steel.

URBACH SS45
Carl Urbach & Co., Stahlwerk KG
C 1.3, Cr 4.3, Mo 0.85, V 3.8, W 12, bal Fe.
For blanking and forming dies, cutters; high-speed steel.

URBACH U19
Carl Urbach & Co., Stahlwerk KG
C 0.74, Cr 4.3, V 1.1, W 18.5, bal Fe.
For lathe and planer tools, reamers, broaches, taps; high-speed steel.

URBACH V120
Carl Urbach & Co., Stahlwerk KG
C 0.74, Cr 4.3, V 1.1, W 18.5, bal Fe.
For lathe and planer tools, reamers, taps, drills; high-speed steel.

UROL
AFORA
Alloy steel. C 0.05-0.15, Mn 1-1.5, Si 0-0.06, P 0-0.07, S 0.3-0.4, Pb 0.15-0.3, bal Fe.
Normalized at 900°C (16 mm): 39-52 kgf/mm^2 TS; 22 kgf/mm^2 min YS; 20 min El. Cold drawn (<16 mm): 52-82 kgf/mm^2 TS; 42 kgf/mm^2 min YS; 7 min El. Cold drawn (<40 mm): 47-77 kgf/mm^2 TS; 38 kgf/mm^2 min YS; 8 min El. Cold drawn (<63 mm): 42-72 kgf/mm^2 TS; 31 kgf/mm^2 min YS; 0 min El.

US ULTRA
Raven Steel & Tool Co.
C 0.36, Mn 0.4, Si 1.1, Cr 5, V 0.35, Mo 1.3, bal Fe.
Heat treated: 300,000 TS; 227,000 YS; 9 El; 55 Rock C. For hot dies, fasteners, extrusion and die casting dies, bulkheads. Type H-11 hot-work steel, high toughness and wear resistance.

US ULTRA
U.N. Alloy Steel Corp.
C 0.36, Mn 0.4, Si 1.1, Cr 5, V 0.35, Mo 1.3, bal Fe.
Heat treated: 300,000 TS; 227,000 YS; 9 El; 55 Rock C. For hot dies, fasteners, extrusion and die casting dies, bulkheads. Type H-11 hot-work steel, high toughness and wear resistance.

US ULTRA 2
Now VEW W302.

US ULTRA-4
Now VEW W304.

US ULTRA-4
U.N. Alloy Steel Corp.
C 0.36, Mn 0.4, Si 1, Cr 5, V 0.35, W 1.3, Mo 1.4, bal Fe.
Heat treated: 205,000 TS; 185,000 YS; 12 El; 42 RA; 44 Rock C. For die blocks, extrusion rams, shell piercing tools, upsetters, punches. Type H12 hot work tool steel, air hardening. Resists heat checking.

US ULTRA-4
Raven Steel & Tool Co.
C 0.36, Mn 0.4, Si 1, Cr 5, V 0.35, W 1.3, Mo 1.4, bal Fe.
Heat treated: 205,000 TS; 185,000 YS; 12 El; 42 RA; 44 Rock C. For die blocks, extrusion rams, shell piercing tools, upsetters, punches. Type H12 hot work tool steel, air hardening. Resists heat checking.

USALLOY
United States Pipe & Foundry Co.
C 0-0.2, Mn 0-1.25, S 0-0.05, 0.25 Ni min, 0.20 Cu min (Ni + Cu + Cr = 1.25 min), bal Fe.
65,000 psi min TS; 45,000 psi min YS; 20 min El. For corrosion resistant steel. Tee head bolts and nuts.

USAMET
Elgiloy Limited Partnership
Stainless steel. Cr 17, Ni 7.2, Mn 2, Si 1.4, C 0.1, bal Fe.
Hardened: 280,000 TS min; 265,000 YS. For strip springs.

USCO 13
U.S. Reduction Co.
Now USCO 413.1. A14131

USCO 13A
U.S. Reduction Co.
Cu 0.7, Fe 0.8, Zn 0.1, Si 12, Mg 1, Mn 0.1, Ni 2.5, bal Al.
Permanent mold cast and heat treated: 47,000 TS; 33,000 YS; 0.5 El; 125 Brin. For automotive and diesel pistons; high strength. *Obsolete*

USCO 319.1
U.S. Reduction Co.
Cu 3-4, Fe 0-1, Zn 0-1, Si 5.5-6.5, Mg 0-0.1, Mn 0-0.5, Ni 0-0.5, bal Al.
Sand cast: 28,000 TS; 16,000 YS; 3.5 El; 75 Brin. Heat treated: 36,000 TS; 24,000 YS; 2 El; 80 Brin. For highly stressed castings; heat treatable. A03191

USCO 356.1
U.S. Reduction Co.
Si 6.5-7.5, Mg 0.2-0.4, Fe 0-0.5, bal Al.
A03561

USCO 356.2
U.S. Reduction Co.
Si 7, Mg 0.3, Fe 0-0.12, bal Al.
A13562

USCO 369.1 (SPECIAL K-9)
U.S. Reduction Co.
Si 11-12, Zn 0-1, Cu 0-0.5, Mg 0.25-0.4, Fe 0.4-1, Cr 0.3-0.4, Mn 0-0.35, Ti 0-0.05, Ni 0-0.05, Sn 0-0.1, Pb 0-0.1, bal Al.
Die cast: 23,000-30,000 psi TS; 17,900-23,400 psi YS; 2-5 El. For clutch housings, farm implement parts, electrical equipment.

USCO 413.1
U.S. Reduction Co.
Cu 0-0.6, Fe 0.7, Zn 0-0.5, Si 12, Mg 0.05, Mn 0.3, bal Al.
Sand cast: 26,000 TS; 12,000 YS; 8 El; 60 Brin. Die cast: 35,000 TS; 16,000 YS; 3.5 El. For pressure tight castings; corrosion resistant. A14131

USCO 443.1
U.S. Reduction Co.
Cu 0.6, Fe 0.6, Zn 0.5, Si 4.5-6, bal Al.
Sand cast: 22,000 TS; 10,000 YS; 5 El; 47 Brin. Die cast: 32,000 TS; 15,000 YS; 5 El. For general castings; good corrosion resistance. A04431

USCO 5
U.S. Reduction Co.
Now USCO 443.1. A04431

USCO 5 A
U.S. Reduction Co.
Now USCO A 443.1.

USCO 5-B
U.S. Reduction Co.
Cu 0-0.2, Fe 0-0.4, Zn 0.2, Si 7, Mg 0.3, Mn 0.1, bal Al.
Heat treated: 34,000 TS; 30,000 YS; 2 El; 75 Brin. For highly stressed parts; excellent corrosion resistance. *Obsolete*

USCO 5-W
U.S. Reduction Co.
Cu 1.3, Fe 0-0.6, Zn 0.3, Si 5, Mg 0.5, Mn 0-0.4, bal Al.
Sand cast: 24,000 TS; 21,000 YS; 4.5 El; 55 Brin. Heat treated: 37,000 TS; 29,000 YS; 2.5 El; 85 Brin. For highly stressed parts; good corrosion resistance. *Obsolete*

USCO 535.2 (ALMAG 35)
U.S. Reduction Co.
Mg 6.2-7.5, Mn 0.1-0.2, Ti 0.1-0.2, bal Al.
As cast: 35,000 psi TS min; 18,000 psi YS min; 9.0 El min, 60 Brin min. Weldable casting alloy with good strength and corrosion resistance; dimensionally stable. AA 535.2 sand casting. A05352

USCO 713.1
U.S. Reduction Co.
Cu 0.4-0.8, Fe 0.7-0.8, Zn 7-8, Si 0-0.25, Mg 0.25-0.5, Mn 0-0.6, bal Al.
Sand cast: 29,000 TS; 16,000 YS; 6.5 El; 60 Brin. Aged: 35,000 TS; 25,000 YS; 4.5 El; 74 Brin. For high strength castings; no heat treatment. A07131

USCO 771.2 (PRECEDENT 71 A)
U.S. Reduction Co.
Zn 6.5-7.5, Mg 0.8-1, Cr 0.06-0.2, Ti 0.1-0.2, bal Al.
As cast: 37,000 psi TS; 23,000 psi YS; 10 El; 69 Brin. Aged 3 weeks: 43,000 psi TS; 33,000 psi YS; 4.5 El; 93 Brin. Sand casting alloy with high strength and good stability. AA 771.2. A07712

USCO A
U.S. Reduction Co.
Cu 6.5, Fe 1, Zn 1.4, Si 3.25, Mn 0.4, Mg 0.05, bal Al.
Sand cast: 26,000 TS; 20,000 YS; 2 El; 74 Brin. For general purpose castings; for low stress applications. *Obsolete*

USCO A 242.2
U.S. Reduction Co.
Cu 4, Fe 0.6, Zn 0.1, Si 0.7, Mg 1.5, Mn 0.1, Ni 2, bal Al.
Heat treated: 44,000 TS; 0.5 El; 101 Brin. Permanent mold: 47,000 TS; 42,000 YS; 0.5 El; 110 Brin. For high temperature applications; high properties at elevated temperatures. A02422

USCO A 443.1
U.S. Reduction Co.
Cu 0-0.3, Fe 0-0.6, Zn 0-0.3, Si 4.5-6, Mn 0-0.3, bal Al.
Sand cast: 19,000 TS; 9000 YS; 6 El; 40 Brin. Die cast: 30,000 TS; 14,000 YS; 7 El. For food handling and chemical equipment; sand, permanent mold or die castings.

USCO A-3
U.S. Reduction Co.
Cu 7, Fe 1, Zn 1.5, Si 2, Mg 0.05, Mn 0.4, Ni 0.25, bal Al.
Sand cast: 26,000 TS; 23,000 YS; 1.5 El; 75 Brin. For general purpose castings; for low stress applications. *Obsolete*

USCO B 380.1
U.S. Reduction Co.
Cu 3-4, Fe 0-1, Zn 0-1, Mg 0-0.1, Mn 0.5, Ni 0-0.5, bal Al.
Sand cast: 25,000 TS; 14,000 YS; 2 El; 70 Brin. Die cast: 43,000 TS; 27,000 YS; 2.5 El. For general purpose castings; low shrinkage, very fluid.

USCO B-1
U.S. Reduction Co.
Cu 7, Fe 0.9, Zn 0.5, Si 5.5, Mg 0.3, Mn 0.4, Ni 0.3, bal Al.
Die cast: 28,000 TS; 26,000 YS; 1 El; 95 Brin. Heat treated: 35,000 TS; 26,000 YS; 0.5 El; 100 Brin. For automotive pistons; die and permanent mold castings.

USCO C-1
U.S. Reduction Co.
Cu 4.5, Fe 0.8, Mg 0-0.03, Mn 0.4, Zn 0.3, Si 0-1.2, bal Al.
Sand cast: 25,000 TS; 11,000 YS; 4 El; 65 Brin. Heat treated: 36,000 TS; 26,000 YS; 4.5 El; 74 Brin. For bus wheels, crankcases, transmission housings; for highly stressed parts. *Obsolete*

USCO C-2
U.S. Reduction Co.
Cu 4.5, Fe 0.8, Zn 0-0.3, Si 2.5, Mn 0-0.3, bal Al.
Permanent mold cast: 27,000 TS; 15,000 YS; 2.5 El; 7 Brin. Heat treated: 45,000 TS; 28,000 YS; 5 El; 90 Brin. For automotive and aircraft parts; for highly stressed applications. *Obsolete*

USCO D
U.S. Reduction Co.
Cu 4, Fe 0.7, Zn 0.6, Si 5, Mg 0.5, Ni 0.3, bal Al.
Sand cast: 25,000 TS; 13,000 YS; 2 El; 70 Brin. Die cast: 39,000 TS; 22,000 YS; 3 El. For general purpose castings; heavy secton die castings. *Obsolete*

USCO D-3
U.S. Reduction Co.
Cu 4, Fe 0.7, Zn 0-1, Si 3, Mn 0.4, Ni 0.3, bal Al.
Sand cast: 27,000 TS; 18,000 YS; 2.5 El; 66 Brin. Heat treated: 38,500 TS; 29,000 YS; 3 El; 90 Brin. For cylinder heads, manifold; heat treatable. *Obsolete*

USCO D-4
U.S. Reduction Co.
Cu 2.9, Fe 0-1, Zn 0.7, Si 5.25, Mg 0-0.15, Mn 0.5, Ni 0.3, bal Al.
Sand cast: 28,500 TS; 16,300 YS; 2.7 El; 60 Brin. Heat treated: 41,000 TS; 32,000 YS; 1.5 El; 80 Brin. For general castings; age hardenable. *Obsolete*

USCO D-5
U.S. Reduction Co.
Cu 2.9, Fe 0-1, Zn 0.7, Si 5.25, Mg 0-0.15, Mn 0.5, Ni 0.3, bal Al.
Sand cast: 25,000 TS; 13,000 YS; 2 El; 70 Brin. Aged permanent mold: 33,000 TS: 21,000 YS; 2 El; 79 Brin. For general purpose castings; sand or permanent mold castings. *Obsolete*

USCO K
U.S. Reduction Co.
Now USCO B 380.1.

USCO K-3
U.S. Reduction Co.
Now USCO 319.1. A03191

USCO K-5
U.S. Reduction Co.
Cu 2.5, Fe 0.7, Zn 0.35, Si 6.1, Mg 0-0.13, Mn 0.4, bal Al.
Sand cast: 27,000 TS; 15,500 YS; 3 El; 60 Brin. Permanent mold cast: 32,000 TS; 17,500 YS; 3.5 El; 65 Brin. For pressure tight castings; sand plaster or permanent mold castings. *Obsolete*

USCO K-Z
U.S. Reduction Co.
Zn 0-2, bal Al.
For die castings. *Obsolete*

USCO L
U.S. Reduction Co.
Now USCO A 242.2. A02422

USCO R

U.S. Reduction Co.
Cu 0-0.1, Fe 0.5, Zn 0.3, Si 0-0.3, Mg 3.8, Mn 0-0.6, bal Al.
Sand cast: 25,000 TS; 12,000 YS; 9 El; 50 Brin. For food processing equipment, fittings, hardware; corrosion resistant. *Obsolete*

USCO S-1

U.S. Reduction Co.
Cu 9.5, Fe 0.8, Zn 0-0.5, Si 4, Mg 0.25, Mn 0.3, Ni 0.1, bal Al.
Sand cast: 25,000 TS; 20,000 YS; 0.5 El; 80 Brin. Permanent mold cast: 30,000 TS; 27,000 YS; 1 El; 90 Brin. For high temperature applications; hard wearing surface. *Obsolete*

USCO SPECIAL K-9

U.S. Reduction Co.
Now USCO 369.1 (SPECIAL K-9).

USCO T

U.S. Reduction Co.
Now USCO 713.1. A07131

USCO Z-3

U.S. Reduction Co.
Al, Mg.
For hardener for Zamac 3 alloys; hardener. *Obsolete*

USCO Z-5

U.S. Reduction Co.
Al, Mg.
Hardener. *Obsolete*

USCO-JOBBINS ALMAG 35 (535.2)

U.S. Reduction Co.
Now USCO 535.2 (ALMAG 35). A05352

USCO-JOBBINS PRECEDENT 71A (771.2)

U.S. Reduction Co.
Now USCO 771.2 (PRECEDENT 71 A). A07712

USCO-S-22

U.S. Reduction Co.
Cu 8, Fe 0.8, Zn 0.2, Si 1.5, Mn 4, Ni 0.1, bal Al.
Sand cast: 23,000 TS; 14,000 YS; 2 El; 74 Brin. For general castings; low stress applications. *Obsolete*

USMAC 16

U.S. Magnet & Alloy Corp.
Stainless steel. C 0.1, Mn 0.75, Si 0.65, Cr 12.5, Ni 0.5, S 0.2, bal Fe.
Cast and heat treated: 103,000-203,000 TS; 75,000-173,000 YS; 6-28 El; 9-62 RA; 185-415 Brin. For chemical and food processing equipment, valve bodies, pump castings. Type 416 stainless steel.

USMAC 40C

U.S. Magnet & Alloy Corp.
Stainless steel. C 1.05, Mn 0.5, Si 0.5, Cr 17.5, Ni 0.25, Mo 0.5, bal Fe.
Cast: 110,000 TS; 70,000 YS; 10 El; 210 Brin. For cutlery, valve parts, nozzles, pump parts. Type 440C stainless steel, wear and corrosion resistant.

USMAC 43

U.S. Magnet & Alloy Corp.
Stainless steel. C 0.25, Mn 0.7, Si 0.6, Cr 21, Ni 0.4, Cu 1, bal Fe.
Cast: 85,000 TS; 55,000 YS; 5.0 El; 6.0 RA; 190 Brin. For chemical plant and oil refinery equipment. Type 443 stainless steel.

USMAC 436

U.S. Magnet & Alloy Corp.
C 0.14, Mn 0.5, Si 0.8, Cr 13.25, Ni 1.8, W 2.25, bal Fe.
Cast: 200,000 TS; 140,000 YS; 12.5 El; 40.0 RA; 43-44 Rock C. For jet engine and super charger parts, turbine blades and wheels. AMS 5616. Ultra high strength.

USMAC ML

U.S. Magnet & Alloy Corp.
C 0.18, Mn 1.25, Si 1, Cu 31.75, Fe 2.5, bal Ni.
Cast: 63,000 TS; 31,000 YS; 20 El; 135 Brin. For chemical plant and plastic industry equipment. Similar to Monel Metal. Corrosion resistant.

USO

Hidalgo Steel Co. Inc.
C, alloy, bal Fe.
For striking dies; water hardened.

USO 30CF

Hidalgo Steel Co. Inc.
C 0.3, Cr 3, V 0.25, W 8.8, bal Fe.
For hot dies and tools; hot work steel.

USO 5S

Hidalgo Steel Co. Inc.
C 0-0.1, Cr 4-6, Mo 0.5, bal Fe.
Corrosion resistant.

USO TUNGSTEN

Hidalgo Steel Co. Inc.
Cr 4, V 0.8, W 18, C, bal Fe.
For lathe tools, drills, milling cutters; high speed steel.

USO VANADIUM

Hidalgo Steel Co. Inc.
C 0.9, V 0.2, bal Fe.
For heading and gripping dies, shear blades; water hardened.

USO VANADIUM

Hidalgo Steel Co. Inc.
C 0.8, V 0.2, bal Fe.
For striking dies; water hardened.

USO-AD

Hidalgo Steel Co. Inc.
C 1.55, Cr 11, V 0.2, Mo 0.7, bal Fe.
For blanking and trimming dies; oil hardened.

USO-HE

Hidalgo Steel Co. Inc.
C 1.5, Cr 0.5, V 0.25, W 1.6, bal Fe.
For dies, gauges; oil hardened, non-deforming.

USO-NS

Hidalgo Steel Co. Inc.
C 0.8, Cr 0.2, Mn 1.5, bal Fe.
For broaches, gauges, blanking and forming dies; oil hardened, non-deforming.

USS "MX"

U.S. Steel Corp.
C 0-0.09, Mn 0.7-1, P 0.07-0.12, S 0.24-0.33, bal Fe.
For machinery parts; open hearth or Bessemer free-cutting steel. *Obsolete*

USS 1/2% CR-1/2% MO

U.S. Steel Corp.
C 0.1-0.2, Mn 0.3-0.61, Cr 0.5-0.81, Mo 0.44-0.65, bal Fe.
Annealed: 67,600 TS; 40,700 YS; 30 El; 65 RA. Normalized: 77,000 TS; 45,000 YS; 27 El; 60 RA. At 900 F: 63,000 TS, 31,500 YS; 34 El; 75 RA. For high pressure boilers, superheaters, drying ovens, air preheaters, heat exchangers, cracking still tubes. Reduced graphitization to 950 F. *Obsolete*

USS 1110

United States Steel Corp.
C 0.1, Mn 0.45, S 0.06, bal Fe.
Free machining steel wire and rod.

USS 12 HIGH CARBON

U.S. Steel Corp.
Mn 0-1, Si 0-1, Cr 12-14, 0.15% C min, bal Fe.
Annealed: 95,000 TS; 50,000 YS; 25 El; 55 RA; 195 Brin.
Heat treated: 230,000 TS; 195,000 YS; 8 El; 25 RA; 500 Brin. For cutlery, paper knives, meat blades; heat treatable, corrosion resistant. *Obsolete*

USS 12 MO

United States Steel Corp.
C 0-0.15, Cr 11.5-13.5, Mo 0.4-0.6, bal Fe.
Martensitic stainless steel. Modified AISI 410.

USS 12 NI-5 CR-3 MO

U.S. Steel Corp.
C 0.025, Mn 0.08, P 0.008, S 0.01, Ni 12, Cr 4.75, Mo 3, Ti 0.25, Al 0.3, B 0.002, bal Fe.
Annealed: 135,000 TS; 110,000 YS; 16 El; 70 RA. Aged: 200,000 TS; 190,000 YS; 12 El; 50 RA. For motor cases, solid fuel rockets, hydrospace equipment. Maraging steel. High strength, corrosion resistant. *Obsolete*

USS 12 Z

Manufacturer not listed.
C 0-0.1, Cr 12-14, bal Fe.
For corrosion resistant parts; corrosion resistant. *Obsolete*

USS 12-12

United States Steel Corp.
C 0-0.12, Cr 11.5-13.5, Ni 11.5-13.5, bal Fe.
Austenitic stainless steel.

USS 12-15

United States Steel Corp.
C 0-0.08, Cr 11-13, Ni 14-16, bal Fe.
Austenitic stainless steel. For cold headed parts.

USS 12-2

United States Steel Corp.
C 0-0.15, Cr 11.5-13.5, Ni 1.25-2.5, bal Fe.
Martensitic stainless steel. For valve seats, gage parts, scissors. AISI 414.

USS 12MOV

U.S. Steel Corp.
C 0.2, Cr 12, Mo, V, bal Fe.
Rolled: 100,000 TS; Heat treated: 230,000 TS. For heat and corrosion resistant parts; corrosion resistant. *Obsolete*

USS 15-7 AMV

U.S. Steel Corp.
C 0.24, Mn 0.6, Si 0.8, Ni 7, Cr 15, Mo 2.5, V 0.2, Al 1.4, N, bal Fe.
Long. 140,000 TS; 55,600 YS; 35 El. Trans. 135,400 TS; 56,500 YS; 30.5 El. Heat treated: 264,400 TS; 233,500 YS; 6 El. For aircraft and missile components, valves, fittings. Stainless. Semi-austenitic, precipitation hardening. *Obsolete*

USS 16-2

United States Steel Corp.
C 0-0.2, Cr 15-17, Ni 1.25-2.5, Si 0-1, Mn 0-1, bal Fe.
For springs; corrosion resistant.

USS 17-4 6

United States Steel Corp.
C 0-0.15, Mn 5.5-7.5, Ni 3.5-5.5, Cr 16-18, bal Fe.
Annealed: 115,000 TS; 55,000 YS; 55 El. For stainless parts; stainless, alternate for Type 201.

USS 17-5 MNV

U.S. Steel Corp.
C 0.11, Mn 13.5, Ni 4.6, Cr 16.2, Mo 2, V 0.92, N 0.36, bal Fe.
For corrosion resistant parts; corrosion and heat resistant. *Obsolete*

USS 17-7

United States Steel Corp.
C 0-0.15, Cr 16-18, Ni 6-8, bal Fe.
Annealed: 110,000 TS; 40,000 YS; 60 El; B85 Brin. For aircraft, trailers, car construction. Stainless Type 301.

USS 18-18-2

Molycorp, Inc.
C 0.06, Cr 18, Ni 18, Si 2, bal Fe.
Annealed: 80,000 TS; 36,000 YS; 54 El. At 1000°F: 65,000 TS; 17,400 YS; 47 El. For the chemical, petroleum, electric power and food processing industries, tanks, kettles, pump parts. Resists stress corrosion, austenitic, stainless.

USS 18-5-8

United States Steel Corp.
C 0-0.15, Mn 7.5-10, Ni 4-6, Cr 17-19, bal Fe.
Annealed: 105,000 TS; 55,000 YS; 55 El. For stainless parts; austenitic; alternate for Type 202.

USS 18-8 CB

United States Steel Corp.
C 0.08, Mn 0-2, Si 0-1, Cr 17-19, Ni 9-12, Cb = 10 x C, bal Fe.
88,000-90,000 TS; 30,000 YS; 45-55 El; 70 RA; 165 Brin. For engine manifolds, exhaust stacks, pressure vessels; stainless, austenitic.

USS 18-8 MO CB TA

United States Steel Corp.
C 0-0.1, Cr 16-18, Ni 10-14, Mo 2-3, Cb + Ta = 10 x C min, bal Fe.
Stainless steel. Stabilized type. AISI 316.

USS 18-8 MO-L

United States Steel Corp.
C 0-0.03, Ni 10-14, Cr 16-18, Mo 2-3, bal Fe.
Annealed: 75,000 TS; 32,000 YS; 50 El. For welded stainless parts; resists intergranular corrosion after welding.

USS 18-8-L

United States Steel Corp.
C 0-0.03, Ni 8-12, Cr 18-20, bal Fe.
Annealed: 75,000 TS; 28,000 YS; 50 El; 140 Brin. For welded stainless parts; resists intergranular corrosion after welding.

USS 2090

United States Steel Corp.
C 0.22, Mn 1, bal Fe.
Steel wire and rod for recessed head screws.

USS 25-12 SCBTA

United States Steel Corp.
C 0-0.08, Cr 22-24, Ni 12-15, Cb + Ta = 10 x C min, bal Fe.
Stainless and high temperature alloy; stabilized to improve welding and high temperature properties. Modified AISI 309.

USS 25-12-S

United States Steel Corp.
C 0-0.08, Ni 12-15, Cr 22-24, bal Fe.
Annealed: 90,000 TS; 45,000 YS; 45 El; 217 Brin. For preheaters, furnace parts; corrosion and heat resistant.

USS 25-12SCB

United States Steel Corp.
C 0-0.08, Cr 22-24, Ni 12-15, Cb + Ta = 10 x C min, (0.10 Ta max), bal Fe.
Stainless and high temperature alloy; stabilized to improve welding and high temperature properties. Modified AISI 309.

USS 25-20S

United States Steel Corp.
C 0-0.08, Cr 24-26, Ni 19-22, Si 0-1.5, bal Fe.
Stainless and high temperature alloy. For jet engineering. AISI 310S.

USS 304 LN

U.S. Steel Corp.
C 0-0.03, Ni 8-12, Cr 18-20, N 0.1-0.15, Mn 0-2, Si 0-1, bal Fe.
Rolled: 85,000 TS; 42,000 YS; 59 El; 75 RA. For stainless parts, chemical plant equipment; austenitic, stainless. *Obsolete*

USS 35 N-15 CR

United States Steel Corp.
C 0-0.08, Cr 17-20, Ni 34-37, Si 0.75-1.5, bal Fe.
For high temperature operation; as furnace parts.

USS 41

United States Steel Corp.
C 0.35, Mn 1.5, Si 0.25, bal Fe.
Steel wire and rod for cap screws and automatic wheel bolts. Can be cold headed.

USS 5

U.S. Steel Corp.
Mn 0.3-1, Si 0.1-1, Cr 4-6, Mo 0.4-0.65, 0.1% C min, bal Fe.
Annealed: 65,000-85,000 TS; 25,000-45,000 YS; 25-35 El; 60-75 RA; B 70-80 Rock. For petroleum and chemical plant equipment. Type 501 corrosion resistance. *Obsolete*

USS 5S

U.S. Steel Corp.
C 0-0.1, Mn 0.3-0.6, Si 0-0.5, Cr 4-6, Mo 0.4-0.65, bal Fe.
Annealed: 65,000-85,000 TS; 25,000-45,000 YS; 25-35 El; 60-75 RA; B 70-85 Rock. For petroleum and chemical plant equipment. Heat and corrosion resistant. Good creep and oxidation resistant. *Obsolete*

USS 9% NICKEL STEEL

United States Steel Corp.
C 0-0.13, Mn 0-0.8, Si 0.15-0.3, P 0-0.035, S 0-0.04, bal Fe.
90,000 TS min; 60,000 YS min; 20 El min. For pressure vessels for low temperature operation; cryogenic alloy, good ductility and ease of fabrication. ASTM A353, A553.

USS AIRSTEEL X-200

U.S. Steel Corp.
C 0.43, Mn 0.85, Si 1.5, Cr 2, Mo 0.5, V 0.05, bal Fe.
Heat treated: 280,000 TS; 238,000 YS; 7 El. Annealed: 110,000 TS; 77,400 YS; 19 El; 220 Brin. For missile and rocket components; air hardening, tough. *Obsolete*

USS AMER-LED BESSEMER

U.S. Steel Corp.
C 0-0.09, Mn 0.7-1, P 0.07-0.12, S 0.25-0.35, Pb 0.15-0.35, bal Fe.
For screw machine products; free-cutting. *Obsolete*

USS AMER-LED GR. A

U.S. Steel Corp.
C 0-0.15, Mn 0.8-1.2, P 0.04-0.09, Pb 0.15-0.35, S 0.25-0.35, bal Fe.
For screw machine products; free-cutting. *Obsolete*

USS AMER-LED GR. B

U.S. Steel Corp.
C 0-0.15, Mn 0.85-1.35, P 0.04-0.09, Pb 0.15-0.35, 0.40% min S, bal Fe.
For screw machine products; free-cutting. *Obsolete*

USS AUSTENITIC MANGANESE STEEL

Now MANGANESE GR.

USS B24

United States Steel Corp.
C 0.22, Mn 1, 0.0005 B min, bal Fe.
Steel wire and rod for recessed head screws. Can be cold headed.

USS CHAR-PAC

Now CHAR-PAC.

USS CU-NI-MO

U.S. Steel Corp.
C 0.15, Mn 0.95, Ni 1.35, Mo 0.25, Cu 0.65, bal Fe.
Hot rolled: 65,000 YS. For power shovels, gun carriages, oil rigs; good weldability. *Obsolete*

USS FREE-MACHINING STEEL, LEADED

U.S. Steel Corp.
C 0-0.15, Mn 1, P 0.07, S 0.3, Pb 0.25, bal Fe.
For automatic screw machine products; open hearth, free-cutting. *Obsolete*

USS MARAGING STEEL GR. 200

U.S. Steel Corp.
C 0-0.03, Ni 17-19, Co 7-8.5, Mo 4-4.5, Ti 0.1-0.3, Al 0.05-0.15, B, Zr, Ca, bal Fe.
Maraged: 225,000 TS; 215,000 YS; 12 El; 40 RA; C 48 Rock. For rocket fuel cases, aircraft landing gears, missile and jet engine components, pressure vessels. Maraging, strong and tough. *Obsolete*

USS MARAGING STEEL GR. 250

U.S. Steel Corp.
C 0-0.03, Ni 17-19, Co 7-8.5, Mo 4.6-5.1, Ti 0.3-0.5, Al 0.05-0.15, B, Zr, Ca, bal Fe.
Maraged: 255,000 TS; 245,000 YS; 10 El; 35 RA; C 52 Rock. For rocket fuel cases, aircraft landing gears, missile and jet engine components, pressure vessels. Maraging steel. Strong and tough. *Obsolete*

USS MARAGING STEEL GR. 300

U.S. Steel Corp.
C 0.03, Ni 17-19, Co 8-9.5, Mo 4.6-5.1, Ti 0.6-0.8, Al 0.05-0.15, B, Zr, Ca, bal Fe.
Maraged: 295,000 TS; 290,000 YS; 8 El; 30 RA; C 54 Rock. For rocket fuel cases, aircraft landing gears, missile and jet engine components, pressure vessels. Maraging. Strong and tough. *Obsolete*

USS MX-BESS

U.S. Steel Corp.
C 0-0.09, Mn 0.7-1, P 0.07-0.12, S 0.24-0.33, bal Fe.
For automatic screw machine products; free-cutting. *Obsolete*

USS MX-OH

U.S. Steel Corp.
C 0-0.09, Mn 0.7-1, P 0.07-0.12, S 0.24-0.33, bal Fe.
For automatic screw machine products; free-cutting. *Obsolete*

USS MXI

U.S. Steel Corp.
C 0.09, Mn 0.7-1, P 0.05-0.12, S 0.24-0.33, bal Fe.
Bar: 77,500 TS; 68,000 YS; 17.5 El; 50.2 RA; B 88 Rock; 40,000 fatigue limit. For screw machine products, bolts, fasteners, case-hardened parts, cams, camshafts. Free-cutting, carburizing steel. *Obsolete*

USS PAR-TEN

United States Steel Corp.
C 0-0.12, Mn 0-0.75, V 0-0.07, S 0-0.05, P 0-0.04, Si 0-0.1, Cb 0-0.04, bal Fe.
Hot rolled sheet: 62,000 TS; 45,000 YP; 29 El. For polished and plated components. Good formability and weldability.

USS Q-CORE

United States Steel Corp.
Low core loss motor lamination sheet.

USS Q-TEMP 10B18Q

Now Q-TEMP 10B18Q.

USS Q-TEMP 41BV20Q

United States Steel Corp.
C 0.18-0.23, Mn 0.75-1, Cr 0.25-0.4, Mo 0.15-0.25, V 0.03-0.08, 0.005 B min, bal Fe.
Heat treated: 189,000 TS; 156,000 YS; 15 El; 61 RA. For fasteners, hardware, bolts, extruded parts. Readily cold headed and extruded.

USS RADIO TRANSFORMER 58

United States Steel Corp.
3.8 Si approx, bal Fe.
For audio transformers; high permeability.

USS RADIO TRANSFORMER 65

United States Steel Corp.
3.8 Si approx, bal Fe.
For audio transformers; high permeability.

USS SUPERKORE CC

U.S. Steel Corp.
C 0.2, Mn 0.8, Ni 0.55, Cr 0.5, Mo 0.2, 0.0005% B min, bal Fe.
Heat treated: 170,000 TS; 150,000 YS; 11 El; 45 RA; 340 Brin. For gears, shafts, pinions; case hardened steel.
Obsolete

USS TRANSFORMER 66

United States Steel Corp.
3.25 Si approx, bal Fe.
For high efficiency power and distribution transformers; resistivity 50, watt loss 0.66.

USS TRI-TEN

United States Steel Corp.
C 0-0.22, Mn 0-1.25, Cu 0.2-0.6, 0.02 V min, bal Fe.
Rolled: 63,000-70,000 TS; 43,000-50,000 YS; 25 El. For cranes, shovels, mine cars, truck bodies, derricks. Good resistance to atmospheric corrosion, shock resistant.

USS TYPE 1 MOTOR LAMINATION SHEET

United States Steel Corp.
Low core loss lamination sheet. Similar to AISI Type 1 and ASTM Type 1.

USS TYPE 2-S MOTOR LAMINATION SHEET

United States Steel Corp.
Low core loss sheet. Similar to AISI and ASTM Type 2S.

UT 35

Ugine Aciers
C 0-0.08, N 0-0.05, H 0-0.0125, O 0-0.2, Fe 0-0.2, bal Ti.
Annealed, aged: 290-410 N/mm^2 TS; 195 N/mm^2 YS min; 22 El min. For corrosion resistance and elevated temperature.

UT 35-02

Ugine Aciers
Pd 0.2, C 0-0.08, N 0-0.05, H 0-0.015, O 0-0.2, Fe 0-0.02, bal Ti.
Annealed, aged: 290-410 N/mm^2 TS; 195 N/mm^2 YS min; 22 YS min. Improved corrosion resistance over unalloyed titanium.

UT 40

Ugine Aciers
C 0-0.08, N 0-0.06, H 0-0.0125, O 0-0.25, Fe 0-0.25, bal Ti.
Annealed, aged: 390-540 N/mm^2 TS; 275 N/mm^2 YS min; 20 El min. For corrosion resistance and elevated temperature.

UT 50

Ugine Aciers
C 0-0.08, N 0-0.07, H 0-0.0125, O 0-0.35, Fe 0-0.25, bal Ti.
Annealed, aged: 490-640 N/mm^2 TS; 340 N/mm^2 YS min; 18 El min. For aircraft structural parts.

UT 6242

Ugine Aciers
Al 5.5-6.5, Sn 1.8-2.2, Zr 3.6-4.4, Mo 1.8-2.2, C 0-0.05, H 0-0.0125, O 0.12, Fe 0-0.25, bal Ti.
Annealed, aged: 890 N/mm^2 TS; 820 N/mm^2 YS; 8 El. Weldable; for blades and discs of jet engine components.

UT 651A

Ugine Aciers
Al 6, Zr 5, Sn 2, Mo 1, Si 0.25, bal Ti.
Heat treated: 990 N/mm^2 TS; 850 N/mm^2 YS; 6 El. Heat treatable, high strength alloy.

UT 662

CEZUS
Refractory. Al 6, V 6, Sn 2, bal Ti.
Outstanding corrosion resistance in a wide range of media, including oxidizing acids, chlorine and chlorinated substances, seawater, electrolytes and the intermediates employed in chemical synthesis.

UT 662

Ugine Aciers
Al 5-6, V 5-6, Sn 1.5-2.5, Cu 0.35-1, N 0-0.04, H 0-0.015, O 0.2, Fe 0.35-1, bal Ti.
Annealed: 1000 N/mm^2 TS; 930 N/mm^2 YS; 8 El. High strength, heat treatable alloy.

UT 685

CEZUS
Refractory. Al 6, Zr 5, Mo 0.5, Si 0.3, bal Ti.
Outstanding corrosion resistance in a wide range of media, including oxidizing acids, chlorine and chlorinated substances, seawater, electrolytes and the intermediates employed in chemical synthesis.

UT 685

Ugine Aciers
Al 5.7-6.3, Zr 4-6, Mo 0.25-0.75, Si 0.1-0.4, C 0-0.8, H 0-0.008, O 0-0.2, Fe 0-0.2, bal Ti.
Solution treated, aged: 990 N/mm^2 TS; 850 N/mm^2 YS; 6 El. Aircraft structural parts, blades or disc for jet engine compressors up to 550°C.

UT A 7 D

Ugine Aciers
Al 6.5-7.3, Mo 3.5-4.5, C 0-0.08, N 0-0.5, H 0-0.0125, O 0.2, Fe 0-0.25, bal Ti.
Quenched and aged: 1150 N/mm^2 TS; 1040 N/mm^2 YS; 8 El. High strength alloy, usually as forgings.

UT A6 V

Ugine Aciers
Al 5.5-6.75, V 3.5-4.5, C 0-0.08, H 0-0.015, N 0-0.07, O 0.2, Fe 0-0.3, bal Ti.
Bar, forgings. Annealed: N/mm^2 TS; 830 N/mm^2 YS; 10 El. Most widely used titanium alloy.

UT A8 DV

Ugine Aciers
Al 7.3-8.5, V 0.75-1.25, Mo 0.75-1.25, C 0-0.08, N 0.05, H 0-0.006, O 0-0.12, Fe 0-0.3, bal Ti.
Annealed: 1000 N/mm^2 TS; 950 N/mm^2 YS; 12 El. Forgings for aircraft.

UT40R

Ugine Aciers
Same as UT40 but specially designed as wire for use as rivets for aeronautical industry.

UT60

Ugine Aciers
C 0-0.1, N 0-0.07, H 0-0.125, O 0-0.4, Fe 0-0.35, bal Ti.
Annealed, aged: 540-730 N/mm^2 TS; 440 N/mm^2 YS min; 15 El min. For structural, corrosion resisting parts.

UTA 5E

CEZUS
Refractory. Al 5, Sn 2.5, bal Ti.
Outstanding corrosion resistance in a wide range of media, including oxidizing acids, chlorine and chlorinated substances, seawater, electrolytes and the intermediates employed in chemical synthesis.

UTA 6V

CEZUS
Refractory. Al 6, V 4, bal Ti.
Outstanding corrosion resistance in a wide range of media, including oxidizing acids, chlorine and chlorinated substances, seawater, electrolytes and the intermediates employed in chemical synthesis.

UTA3V

Ugine Aciers
Al 2.5-3.5, V 2-3, C 0-0.05, N 0-0.02, H 0-0.0125, O 0.25, Fe 0-0.25, bal Ti.
Annealed: 640 N/mm^2 TS; 550 N/mm^2 YS; 18 El. Weldable, ductile; for tube, sheet, plate and wire.

UTA5E (L GRADE)

Ugine Aciers
Al 4.5-5.5, Sn 2-3, C 0-0.05, N 0-0.035, H 0-0.0125, O 0-0.12, Fe 0-0.25, 0.32 Fe + O max, bal Ti.
Annealed: 700 N/mm^2 TS; 630 N/mm^2 YS; 10 El. For cryogenic parts and equipment.

UTA5E (NORMAL GRADE)

Ugine Aciers
Al 4.5-5.5, Sn 2-3, C 0-0.15, N 0-0.07, H 0-0.02, O 0-0.2, Fe 0-0.5, bal Ti.
Annealed: 790 N/mm^2 TS; 760 N/mm^2 YS; 10 El. Weldable, not hardenable.

UTALOY

Envirotech Corp.
Ni 35, Cr 12, C 0-0.2, Mo, bal Fe.
Cast: 375 Brin. For heat resisting parts, mining machinery; impact and abrasion resistant. *Obsolete*

UTALOY 41

Envirotech Corp.
C 0.4, Mn 0.95, Si 0.55, Cr 1.1, Mo 0.28, S 0-0.05, P 0-0.05, bal Fe.
Cast, general engineering material.

UTALOY 700

Envirotech Corp.
C 0.66, Mn 1.1, Si 1, Cr 1.5, Mo 0.5, S 0-0.05, P 0-0.05, bal Fe.
Cast, furnished 350-550 Brin. For abrasion resistant applications.

UTALOY 86

Envirotech Corp.
C 0.37, Mn 1.05, Si 0.35, Ni 0.75, Cr 0.65, Mo 0.25, bal Fe.
Cast; general engineering material.

UTC

Ugine Aciers
Cu 2-3, C 0-0.1, N 0-0.05, H 0-0.01, O 0-0.2, Fe 0-0.2, bal Ti.
Sheet, solution treated, aged: 690 N/mm^2 TS; 550 N/mm^2 YS min; 10 El. For aircraft structural parts.

UTEX

Diehl Steel Co.
C 0.95, W 2, Cr 0.6, bal Fe.
For tools; oil hardened.

UTICA

AL Tech Specialty Steel Corp.
C 1.25, W 1.5, Cr 0.4, V 0.2, bal Fe.
Heat treated: 444-740 Brin. For taps, reamers, tools, drills, broaches, punches, dies; non-deforming, non-shrinking steel.

UTILITAS

Hardite Metals Inc.
C 0.6-1.4, bal Fe.
For tools; water hardened. *Obsolete*

UTILITAS

Jessop Steel Co.
C 0.6-1.4, bal Fe.
For tools; water hardened. *Obsolete*

UTILITY

Lehigh Steel Corp.
C 0.6-0.9, Mn 0.25-0.6, bal Fe.
For drills, chisels, blacksmith tools; water hardened.

UTILITY BEARING METAL
Thomas Bolton Ltd.
Cu alloy.
For bearings, bushings; low cost. *Obsolete*

UTILOY 12
Utility Electric Steel Foundry Co.
C 0-0.12, Cr 11-13, Mn 0.4, Si 0.9, bal Fe.
For castings; corrosion resistant.

UTILOY 12N
Utility Electric Steel Foundry Co.
C 0-0.12, Cr 11-13, Ni 1.7-2.5, bal Fe.
For castings; corrosion resistant.

UTILOY 20
Utility Electric Steel Foundry Co.
Ni 29, Cr 20, Mo 2-3, Cu 4, C 0-0.07, Si 1, bal Fe.
Cast: 65,000-75,000 psi TS; 28,000-38,000 psi YS; 50-35 El; 50-40 RA; 120-150 Brin. For chemical plant equipment; resists mixed acids.

UTILOY 3085
Utility Electric Steel Foundry Co.
C 0.28-0.35, Ni 0.8-1.1, bal Fe.
Cast: 82,000 psi TS; 45,000 psi YS; 22 El; 35 RA. For pump parts; water or oil hardened.

UTILOY 46
Utility Electric Steel Foundry Co.
Cr 4-7, C 0-0.25, Mo 0.5, Mn 0.7, Si 0.4, bal Fe.
For castings; corrosion resistant.

UTILOY H
Utility Electric Steel Foundry Co.
C 0.28-0.32, Cr 23-26, Ni 11-13, Si 0.9-1.1, bal Fe.
For castings; corrosion and heat resistant.

UTILOY NH
Utility Electric Steel Foundry Co.
Ni 35, Cr 15, C, bal Fe.
For salt bath pots; heat and corrosion resistant.

UTILOY X
Utility Electric Steel Foundry Co.
C 0-0.15, Cr 18, Ni 8, bal Fe.
For stainless castings; stainless.

UTILOY X-7
Utility Electric Steel Foundry Co.
C 0-0.08, Mn 0-1.5, Si 0-2, Cr 18-21, Ni 8-11, bal Fe.
Cast: 78,000 psi TS; 43,000 psi YS; 45 El; 150 Brin.
Annealed: 78,000 psi TS; 38,000 psi YS; 55 El; 140 Brin. For pumps, valves, autoclaves, mixers, kettles; ACI-CF8; stainless, austenitic.

UTILOY XX
Utility Electric Steel Foundry Co.
C 0-0.07, Cr 18, Ni 8, Mo 3.5-4.5, bal Fe.
For stainless castings; heat and corrosion resistant.

UTP 068 HH
UTP. Welding Materials, Inc.
Nickel. C 0-0.05, Si 0-0.5, Mn 4-6, Cr 18-22, Fe 2-4, Mo 0-2, Ti 0-0.5, Co 0-0.05, Cu 0-0.2, P 0-0.02, S 0-0.01, 67.0 Ni min, 2.0-2.5 Nb/Ta.
Lime coated electrode. Corrosion resistant. Employed in high-grade fabrication for chemical and petrochemical industry, refrigeration and nuclear engineering. 55.0 psi YS min; 97.0 psi TS min; 35.0 El min.

UTP 3127 LC
UTP. Welding Materials, Inc.
Nickel. C 0-0.03, Cr 26-28, Ni 30-32, Mo 3.5-4, Cu 1-1.4, Mn 1-2, S 0-0.01, P 0-0.02, bal Fe.
Low-carbon, fully austenitic electrode, coated as mixed type. High corrosion resistance. 58.0 psi YS min; 87.0 psi TS; 30 El min.

UTP 387
UTP. Welding Materials, Inc.
Copper. C 0-0.03, Si 0-0.5, Mn 1-1.5, Ni 29-32, Fe 0.4-0.75, Ti 0.1-0.5, Cu 67-70, Pb 0-0.02, P 0-0.01, S 0-0.01.
Lime coated cupro-nickel electrode 70/30. 34.8 psi YS min; 56.0 psi TS min; 30.0 El min.

UTP 389
UTP. Welding Materials, Inc.
Copper. C 0-0.03, Si 0-0.4, Mn 1-1.5, Ni 9-12, Fe 1-1.8, Ti 0-0.5, Pb 0-0.02, P 0-0.02, S 0-0.01, bal Cu.
Lime coated cupro-nickel electrode 90/10. 34.8 psi YS min; 56.0 psi TS min; 25.0 El min.

UTP 4225
UTP. Welding Materials, Inc.
Nickel. C 0-0.04, Si 0-0.5, Mn 1.5-3, Cr 24-27, Ni 38-42, Mo 5.5-6.5, Cu 1.5-2.5, Nb 0-0.5, P 0-0.02, S 0-0.01, bal Fe.
Special high-nickel basic coated electrode. Resists reducing media. 50.0 psi YS min; 79.0 psi TS min; 30 El min.

UTP 6123 AL
UTP. Welding Materials, Inc.
Nickel. C 0-0.05, Si 0-1, Mn 0-0.6, Cr 22-24, Fe 10-14, Al 1-1.3, Ti 0-0.5, S 0-0.01, P 0-0.015, bal Ni.
Lime coated electrode.

UTP 6222 MO
UTP. Welding Materials, Inc.
Nickel. C 0-0.05, Si 0-0.6, Mn 0-1, Cr 20-23, Fe 0-4, Mo 8-10, Ti 0-0.4, Co 0-0.05, Cu 0-0.2, P 0-0.02, S 0-0.015, 60.0 Ni min, 3.15-3.70 Nb/Ta.
Lime coated, high nickel electrode. Corrosion resistant. For welding high-strength, highly corrosion-resistant nickel-base alloys of similar nature, for joining ferritic to austenitic steels as well as for surfacing mild and low-alloy steels. 65.0 psi YS min; 110.0 psi TS min; 30.0 El min.

UTP 7013 MO
UTP. Welding Materials, Inc.
Nickel. C 0-0.1, Si 0-0.6, Mn 2.5-4, Cr 12-16, Mo 5-8, Nb 0.5-1.5, Fe 0-10, S 0-0.015, P 0-0.015, W 1-1.5, bal Ni.
High performance lime-type electrode, weldable on AC. For welding cold-tough nickel steels. 60.0 psi YS min; 98.0 psi TS min; 35.0 El min.

UTP 7015
UTP. Welding Materials, Inc.
Nickel. C 0-0.05, Si 0-0.6, Mn 5-7, Cr 15-17, Fe 5-8, Ti 0-0.5, Co 0-0.05, Cu 0-0.2, P 0-0.02, S 0-0.01, 67.0 Ni min, 2.0-2.5 Nb/Ta.
Lime coated electrode for reactor grade materials. 55.0 psi YS min; 90.0 psi TS min; 35.0 El min.

UTP 7015 HL
UTP. Welding Materials, Inc.
Nickel. C 0-0.05, Si 0-0.6, Mn 5-7, Cr 15-17, Fe 5-8, Ti 0-0.5, Co 0-0.05, Cu 0-0.2, P 0-0.02, S 0-0.01, 67.0 Ni min, 2.0-2.5 Nb/Ta.
High performance electrode for reactor engineering.
Recovery 130%. 55.0 psi YS min; 90.0 psi TS min; 35.0 El min.

UTP 7015 MO
UTP. Welding Materials, Inc.
Nickel. C 0-0.05, Si 0-0.6, Mn 2.5-3.5, Cr 15-17, Fe 5-8, Mo 0.5-2, Ti 0-0.5, Co 0-0.05, Cu 0-0.2, P 0-0.02, S 0-0.01, 67.0 Ni min, 1.5-2.5 Nb/Ta.
Lime coated electrode. High-temperature resistant. 55.0 psi YS min; 90.0 psi TS min; 35.0 El min.

UTP 7017 MO
UTP. Welding Materials, Inc.
Nickel. C 0-0.1, Si 0-0.5, Mn 2.5-3.5, Cr 14-17, Mo 2.5-3.5, Nb 2-3, Fe 0-10, S 0-0.015, P 0-0.015, bal Ni.
Lime-type electrode, weldable on AC. For joining cold-tough nickel steels. 56.0 psi YS min; 95.0 psi TS min; 30.0 El min.

UTP 703 KB
UTP. Welding Materials, Inc.
Nickel. C 0-0.02, Si 0-0.2, Mn 0-1.5, Mo 26-30, Fe 0-2, Cr 0-1, P 0-0.02, S 0.015, Cu 0-0.5, Co 0-1, W 0-1, 64.5 Ni min.
Lime coated Ni-Mo electrode. 70.0 psi YS min; 110 psi TS min; 25.0 El min.

UTP 704 KB
UTP. Welding Materials, Inc.
Nickel. C 0-0.02, Mn 0-1, Si 0-0.2, Fe 0-3, Cr 15-17, Mo 15-17, S 0-0.01, Cu 0-0.5, Co 0-2, P 0-0.02, Ti 0-0.4, bal Ni.
Lime coated electrode for high corrosion resistant NiCrMo alloys. 65.0 psi YS min; 100.0 psi TS min; 30.0 El min.

UTP 776 KB
UTP. Welding Materials, Inc.
Nickel. C 0-0.02, Mn 0-1, Si 0-0.2, Fe 4-7, Cr 14.5-16.5, Mo 15-17, W 3-4.5, V 0-0.35, Cu 0-0.5, Co 0-2.5, P 0-0.02, S 0-0.01, bal Ni.
Lime-type coated electrode for high corrosion-resistant NiCrMo alloys. 65.0 psi YS; 100.0 psi TS min; 30.0 El min.

UTP 80 M
UTP. Welding Materials, Inc.
Nickel. C 0-0.05, Si 0-0.8, Mn 2-4, Fe 0.5-2, Ti 0-1, Al 0-0.5, Cu 28-32, P 0-0.02, S 0-0.01, 62.0 Ni min.
Lime coated nickel-copper electrode. 43.0 psi YS; 65.0 psi TS min; 30.0 El min.

UTP 80 NI
UTP. Welding Materials, Inc.
Nickel. C 0-0.03, Si 0-1, Mn 0-0.75, Fe 0-0.6, Ti 1.5-3, Al 0-0.5, Cu 0-0.2, P 0-0.02, S 0-0.01, 93.0 Ni min.
Lime coated pure nickel electrode. Low carbon content. 43.0 psi YS min; 65.0 psi TS min; 30.0 El min.

UTR
Ugine Aciers
Similar to UTC but as wire or bar designed to make rivets for aircraft.

UTRATHERM 27-15
J.C. Soding & Halbach
C 0.35, Si 1.75, Cr 26.5, Ni 14, bal Fe.
Cast: 80,000 TS; 45,000 YS; 12 El; 180 Brin. At 1400°F: 38,000 TS; 6 El. For retorts, skids, brazing fixtures, lead pots, hearth plates, furnace parts. Heat resistant. High-corrosion and oxidation resistance.

UTTER
Uddeholm Corp.
C 0.5, Si 0.9, Cr 0.9, bal Fe.
For springs; oil hardened. *Obsolete*

UVW
W. Ossenberg & Cie Edelstahlwerke
C 1.05, Mn 1, Si 0.2, Cr 1, W 1.2, bal Fe.
Cold work tool steel for knives, shears, punching dies for thin sheets. W.-Nr. 1.2419.

UVW 2
W. Ossenberg & Cie Edelstahlwerke
C 0.9, Mn 2, Si 0.2, V 0.1, bal Fe.
Oil hardening cold work tool steel. For stamping and punching dies. W.-Nr. 1.2842; Similar to AISI O2.

V 25/4 H
Pose-Marre Edelstahlwerk G.m.b.H.
C 1.3, Cr 25, Ni 4, bal Fe.
As cast: 300 HV; 10 El min. For wear resistance against abrasion at elevated temperatures. *Obsolete*

V 28 H
Pose-Marre Edelstahlwerk G.m.b.H.
C 1.3, Cr 28, bal Fe.
As cast: 300 HV; 10 El min. For wear resistance against abrasion.

V 45 STEEL
Bethlehem Steel Corp.
C 0-0.22, Mn 1.25, P 0.04, S 0.05, N 0-0.015, 0.02% V min, bal Fe.
Rolled: 65,000 TS minimum; 45,000 YS minimum; 18 El. For structures, buildings, bridges, booms, derricks, mine cars. Structural plates, tough. *Obsolete*

V 50 STEEL
Bethlehem Steel Corp.
C 0-0.22, Mn 1.25, P 0.04, S 0.05, N 0-0.015, 0.02% V min, bal Fe.
Rolled: 70,000 TS minimum; 50,000 YS minimum; 18 El. For structures, buildings, bridges, booms, derricks, mine cars. Structural plates, tough. *Obsolete*

V 55 STEEL
Bethlehem Steel Corp.
C 0-0.22, Mn 1.25, P 0.04, S 0.05, N 0-0.015, 0.02% V min, bal Fe.
Rolled: 70,000 TS minimum; 55,000 YS minimum; 17 El. For structures, buildings, bridges, booms, derricks, bus bodies. Structural plates, good weldability. *Obsolete*

V 60 STEEL
Bethlehem Steel Corp.
C 0-0.22, Mn 1.25, P 0.04, S 0.05, N 0-0.015, 0.02% V min, bal Fe.
Rolled: 75,000 TS minimum; 60,000 YS minimum; 16 El. For structures, buildings, bridges, derricks, mine cars, bus bodies. Structural plates, good weldability and fabricability. *Obsolete*

V 65 STEEL
Bethlehem Steel Corp.
C 0-0.22, Mn 1.25, P 0.04, S 0.05, N 0-0.015, 0.02% V min, bal Fe.
Rolled: 80,000 TS minimum; 65,000 YS minimum; 15 El. For structures, buildings, bridges, derricks, mine cars, bus bodies. Structural plate, good weldability. *Obsolete*

V 76 (WAS KLOSTER V-76)
Manufacturer not listed.
C 0.6, Si 1.85, Mn 0.7, Mo 0.45, V 0.2, bal Fe.
Oil or water hardening. For tools, dies. AISI S5. Shock resisting tool steel.

V HIGH SPEED STEEL
McInnes Steel Co.
C 0.5-0.9, W 18, Cr 4, V 1.5, bal Fe.
For turning and boring tools, special dies, reamers; tough high speed tool steel. *Obsolete*

V L W 1
Vereinigte Leichtmetallwerke G.m.b.H.
Aluminum. Cu 5, Si 0.4, bal Al.
For light alloy parts; similar to "Lautal."

V L W 14
Vereinigte Leichtmetallwerke G.m.b.H.
Aluminum. Cu 5, Si 0.4, bal Al.
For light alloy parts; same as "Lautal."

V L W 17
Vereinigte Leichtmetallwerke G.m.b.H.
Aluminum. Cu 4-5, bal Al.
For light alloy parts; same as "Duralumin."

V L W 19
Vereinigte Leichtmetallwerke G.m.b.H.
Aluminum. Mg 0.8, Mn 0.8, Si 1.4, bal Al.
For light alloy parts; same as "Pantal."

V L W 2
Vereinigte Leichtmetallwerke G.m.b.H.
Aluminum. Cu 4, Li 0.1, bal Al.
For light alloy parts; similar to "Scleron."

V L W 23
Vereinigte Leichtmetallwerke G.m.b.H.
Aluminum. Li 0.1, Cu 4, bal Al.
For light alloy parts; same as "Scleron."

V L W 3
Vereinigte Leichtmetallwerke G.m.b.H.
Aluminum. Si 11-14, bal Al.
For light alloy parts; similar to "Silumin."

V L W 31
Vereinigte Leichtmetallwerke G.m.b.H.
Aluminum. Si 11-14, bal Al.
For light alloy parts; same as "Silumin."

V L W 41
Vereinigte Leichtmetallwerke G.m.b.H.
Aluminum. Mn 1-2, bal Al.
For light alloy parts; same as "Aluman."

V L W 6
Vereinigte Leichtmetallwerke G.m.b.H.
Aluminum. Mn 1.3, Mg 2.2, Si 0.7, bal Al.
For marine hardware; resists sea water.

V L W 61
Vereinigte Leichtmetallwerke G.m.b.H.
Aluminum. Mn 1.3, Mg 2.2, Sb 0.2, Si 0.7, bal Al.
For marine hardware: same as "K.S. Seewasser."

V L W 63
Vereinigte Leichtmetallwerke G.m.b.H.
Aluminum. Mg 7.5, Mn 0.3, bal Al.
For marine instruments; same as "B.S. Seewasser."

V L W 99
Vereinigte Leichtmetallwerke G.m.b.H.
Aluminum. Al.
For cable; same as pure Al.

V L W LEICHTMETALLE
Vereinigte Leichtmetallwerke G.m.b.H.
Aluminum. Al alloys.
For light alloy parts for airplanes, dirigibles and automobiles; formerly known as "Leichtstahl."

V MOLY HIGH SPEED
McInnes Steel Co.
C 0.8, W 5.5, Cr 4, Mo 4.25, V 1.5, bal Fe.
For tools, dies; high speed steel. *Obsolete*

V PERMANDUR
ITT Components Group Europe
Fe 49.3, Co 48.4, V 2.3.
For telecommunications, diaphragms in telephone receivers; high permeability.

V T-STEEL
Dortmund-Hoerder Huttenverein A.G.
C, Cr, W, Mo, Va, bal Fe.
114,000-135,000 TS; 78,400 YS; 8 El. For steel rails, tires, rims, gears, taps, tough at very low temperature (-20 C); wear resistant. *Obsolete*

V T-STEEL
Vereinigte Stahlwerke
C, Cr, W, Mo, Va, bal Fe.
114,000-135,000 TS; 78,400 YS; 8 El. For steel rails, tires, rims, gears, taps, tough at very low temperature (-20 C); wear resistant. *Obsolete*

V T-STEEL (V T-STAHL)
Dortmund-Hoerder Huttenverein A.G.
C 0.7, bal Fe.
121,000-143,000 TS; 78,400 YS; 8-10 El. For steel rails, tires, rims, gears, taps; wear resistant. *Obsolete*

V T-STEEL (V T-STAHL)
Vereinigte Stahlwerke
C 0.7, bal Fe.
121,000-143,000 TS; 78,400 YS; 8-10 El. For steel rails, tires, rims, gears, taps; wear resistant. *Obsolete*

V-302
VILLARES
Stainless steel. C 0-0.15, Cr 18, Ni 9, bal Fe.
Austenitic. AISI 302; DIN 1.4300.

V-303
VILLARES
Stainless steel. C 0-0.15, Cr 18, Ni 9, S 0.15, bal Fe.
Free cutting austenitic. AISI 303; DIN 1.4305.

V-304
VILLARES
Stainless steel. C 0-0.08, Cr 19, Ni 9.5, bal Fe.
Austenitic. AISI 304; DIN 1.4301.

V-310
VILLARES
Stainless steel. C 0-0.2, Cr 25, Ni 20, bal Fe.
Austenitic. AISI 310; DIN 1.4841.

V-316
VILLARES
Stainless steel. C 0-0.08, Cr 17, Ni 12, Mo 2.5, bal Fe.
Austenitic. AISI 316; DIN 1.4436.

V-347
VILLARES
Stainless steel. C 0-0.08, Cr 18, Ni 11, Nb, bal Fe.
Stabilized; austenitic. AISI 347; DIN 1.4551. *Obsolete*

V-36
Cannon-Muskegon Corp.
C 0.27, Mn 1, Si 0.4, Cr 25, Ni 20, Mo 4, W 2, Cb 2, Fe 3, bal Co.
For high temperature sheet.

V-416
VILLARES
Stainless steel. C 0-0.15, Cr 13, S 0.15, bal Fe.
Free machining; martensitic. AISI 416; DIN 1.4005.

V-5 ALLOY
Foote Mineral Co.
Cr 40, Mn 10, Si 15, bal Fe.
Used in iron manufacturing; for chill reduction. *Obsolete*

V-57
Now CARPENTER, LESCALLOY and UDIMET V-57.

V-7 ALLOY
Foote Mineral Co.
Cr 30, Mn 15, Si 18, bal Fe.
Used in iron manufacturing; for chill reduction. *Obsolete*

V-KUT
Columbia Tool Steel Co.
Tool material. C 0.71, Mn 0.25, Si 0.25, Cr 0.8, V 0.2, bal Fe.
Water or oil hardened tool steel; for rock drills, shear blades, hand stamps.

V-MANG
Abex Corp.
Mn 12-14, C, Mo, bal Fe.
For welding rod; for Mn steel.

V-STAR
Midvale-Heppenstall Co.
Obsolete

V-STAR SPECIAL
Midvale-Heppenstall Co.
C 0.84, Cr 4.15, W 6.4, Mo 5, V 1.9, Co 5.5, bal Fe.
For tools, cutters; high speed steel. *Obsolete*

V-TOOL
Textron Inc.
C 0.9-1, Mn 0.25-0.4, V 0.2-0.3, bal Fe.
For taps, reamers, drills, hobs, broaches; water hardened.

V.A.G. 160
English manufacture
Si 2-5, Mg 0.6, Mn 0.7, bal Al.
For light alloy parts; corrosion resistant.

V.D. TOOL
Joseph T. Ryerson & Son Inc.
C 0.95-1.05, Mn 0.4, V 0.15-0.18, Si 0.25, bal Fe.
For punches, dies, forming tools, shear blades, mandrels;
water hardened; Type W2.

V.H. ATLAS
S.K.F. Industries Inc.
C, alloy, bal Fe.
For tools. *Obsolete*

V.H. NO. 54
S.K.F. Industries Inc.
C, alloy, bal Fe.
For hot punching tools. *Obsolete*

V.M. STEEL
Manufacturer not listed.
Cr 0.7-1, Mo 0.35-0.85, C 0.4, Mn 0.4-0.6, V 0-0.17, bal Fe.
For gears, shafts, blacksmith tools; oil hardening.

V.S.4 STEEL
Edgar Allen Balfour Ltd.
C 0.75-1.05, Mn 0.3, V 0.25, bal Fe.
For cold-heading dies, hammers, pistons; Type W2; water
hardened.

V.S.M.
Now CARPENTER V.S.M.

V/22N
Eagle & Globe Steel Ltd.
Steel. C 0.68-0.75, S 0-0.015, P 0-0.015, Mn 0.3-0.5, Ni 2-2.2,
Mo 0-0.05, bal Fe.
Hardened and tempered strip steel. For wide wood
bandsaws.

V1
Acciaierie Valbruna s.p.a.
C 0.15, Mn 0.3-0.7, bal Fe.
Low carbon steel. Ital. UNI C18; AISI 1015. *Obsolete*

V13
Acciaierie Valbruna s.p.a.
C 0.15, Mn 0.45, Ni 2.75, Cr 0.75, bal Fe.
Ni-Cr carburizing steel. 15 Ni Cr 11. *Obsolete*

V14
Acciaierie Valbruna s.p.a.
C 0.18, Mn 0.7, Cr 0.9, Ni 4.5, bal Fe.
High alloy carburizing steel (18 Ni Cr 18). *Obsolete*

V17
Acciaierie Valbruna s.p.a.
C 0.15, Mn 0.85, Ni 1, Cr 0.8, bal Fe.
Alloy carburizing steel. 15 Cr Ni 4; similar to former SAE 3115.
Obsolete

V174
Acciaierie Valbruna s.p.a.
Stainless Steel. C 0.04, Si 0.5, Mn 0.5, P 0-0.04, S 0-0.01, Cr
15.5, Ni 4.3, Cu 3.3, Nb + Ta = 5 x C, bal Fe.
Precipitation hardening. AISI 630; W. Nr. 1.4542.

V175
Spencer Clark Metal Industries Ltd.
C 0.8, W 18, Cr 4.25, V 1.8, bal Fe.
Vanadium high speed steel.

V18
Acciaierie Valbruna s.p.a.
C 0.18-0.23, Mn 0.8-1.1, Ni 0.9-1.2, Cr 0.9-1.2, bal Fe.
Ni-Cr carburizing steel. 20 Cr Ni 4. *Obsolete*

V19
Keystone Carbon Co.
White bronze. Sintered: 11,000 psi TS; 5 El; 18% porosity;
6.5-6.9 density; 40 Rock H. For bearings; low coefficient of
friction; improved elevated temperature operation.

V1D
Acciaierie Valbruna s.p.a.
C 0.16, Si 0-0.35, Mn 0.65, P 0-0.03, S 0-0.03, Cr 0-0.3, Mo
0.3, bal Fe.
Heat resisting structural steel. W. Nr. 1.5415.

V2
Acciaierie Valbruna s.p.a.
C 0.2, Mn 0.4-0.8, bal Fe.
Low carbon steel. Ital. UNI C20; AISI 1020. *Obsolete*

V2-MD
Acciaierie Valbruna s.p.a.
C 0.17, Si 0-0.6, Mn 1.65, P 0-0.035, S 0-0.035, Mo 0.15, V
0.07, bal Fe.
Special steel.

V20M7K25
Russian manufacture
C 0.8, W 18.9, Co 24.9, Mo 6.6, bal Fe.
For wear resistant cutting tools, wear plates; wear resistant,
high red-hardness.

V22
English manufacture
W alloy.
For gyro and counterweight components; high heat resistant.

V225MN
Acciaierie Valbruna s.p.a.
Stainless Steel. C 0-0.03, Si 0-1, Mn 1.5, P 0-0.03, S 0-0.02,
Cr 22, Mo 3, Ni 5.5, N 0.15, bal Fe.
Heat resisting steel. ASTM F51; W. Nr. 1.4462.

V254
Acciaierie Valbruna s.p.a.
Stainless Steel. C 0.15, Si 1.15, Mn 0-2, P 0-0.045, S 0-0.03,
Cr 25.5, Ni 4.5, bal Fe.
Heat resisting steel. W. Nr. 1.4821.

V274M
Acciaierie Valbruna s.p.a.
Stainless Steel. C 0-0.1, Si 0-1, Mn 0-2, P 0-0.04, S 0-0.03, Cr
27.5, Mo 1.5, Ni 4.5, N 0.07, bal Fe.
Heat resisting steel. AISI 329; W. Nr. 1.4460.

V2AED
German manufacture
C 0.15, Cr 18, Ni 8, W, Ti, Ta, bal Fe.
For turbine blading and nozzles; heat resistant.

V2D
Acciaierie Valbruna s.p.a.
C 0.16, Si 0-0.3, Mn 0.7, P 0-0.03, S 0-0.03, Mo 0.55, bal Fe.
Heat resisting structural steel. ASTM A182-F1; W. Nr. 1.5423.

V2M
Acciaierie Valbruna s.p.a.
C 0.2, Si 0.3, Mn 1.5-2, bal Fe.
Manganese alloy carburizing steel. 20 Mn 8. *Obsolete*

V3
Acciaierie Valbruna s.p.a.
C 0.3, Mn 0.65, bal Fe.
Carbon steel. Ital. UNI C30; AISI 1030. *Obsolete*

V3/1
Acciaierie Valbruna s.p.a.
C 0.35, Mn 0.75, bal Fe.
Medium carbon steel. Ital. UNI C35; AISI 1035. *Obsolete*

V35
Jessop Steel Co.
C 0.32-0.38, Mn 0.5-0.8, Ni 2-2.5, Cr 0.6-0.9, bal Fe.
Ni-Cr structural steel; deep hardening. 35Ni-Cr9.

V4
Acciaierie Valbruna s.p.a.
C 0.4, Mn 0.75, bal Fe.
Medium carbon steel. Ital. UNI C40; AISI 1040. *Obsolete*

V4/1
Acciaierie Valbruna s.p.a.
C 0.43, Mn 0.75, Cr 0-0.25, Ni 0-0.25, bal Fe.
Medium carbon steel. Ital. UNI C43; AISI 1042. *Obsolete*

V41
Acciaierie Valbruna s.p.a.
C 0.27-0.35, Mn 0.5-0.8, Ni 2.6-3.2, Cr 0.6-1, Mo 0.3-0.6, bal
Fe.
Ni-Cr-Mo structural steel; deep hardening. 30 NiCrMo 12.
Obsolete

V44
Acciaierie Valbruna s.p.a.
C 0.3-0.38, Mn 0.3-0.8, Ni 3.5-4, Cr 1.5-1.8, Mo 0.2-0.4, bal
Fe.
Ni-Cr-Mo structural steel, deep hardening. 35 NiCrMo 15.
Obsolete

V444D
Thyssen Edelstahlwerke AG
C 0.4, Si 2.45, Mn 1, Ni 10, Cr 18.5, V 1.5, bal Fe.
For valves; corrosion and heat resistant. *Obsolete*

V5
Acciaierie Valbruna s.p.a.
C 0.5, Si 0-0.4, Mn 0.75, bal Fe.
Carbon structural steel. Ital. UNI C50; AISI 1050. *Obsolete*

V5/1
Acciaierie Valbruna s.p.a.
C 0.48, Mn 0.75, Cr 0-0.25, Ni 0-0.25, Cu 0-0.25, bal Fe.
Carbon steel. Ital. UNI C48; similar to AISI 1050. *Obsolete*

V6
Acciaierie Valbruna s.p.a.
C 0.6, Si 0.15-0.4, Mn 0.7, bal Fe.
Carbon structural steel. Ital. UNI C60; AISI 1060. *Obsolete*

V62
Acciaierie Valbruna s.p.a.
C 0.2, Si 0.5, Mn 1.75, P 0-0.035, S 0-0.035, bal Fe.
Special steel.

V64
Acciaierie Valbruna s.p.a.
C 0.4, Mn 1.35, bal Fe.
Manganese structural steel. 40 Mn 5. *Obsolete*

V68
Acciaierie Valbruna s.p.a.
C 0.95-1.1, Mn 0.3-0.5, Cr 1.4-1.65, bal Fe.
Chromium drill rod. 100 Cr 6; AISI 52100. *Obsolete*

V7
Acciaierie Valbruna s.p.a.
C 0.7, Si 0.35, Mn 0.7, P 0.035, S 0.035, bal Fe.
Carbon steel. Ital. UNI C70; AISI 1074; W. Nr. 1.1231.

V8-C

Acciaierie Valbruna s.p.a.
Tool Material. C 0.8, Si 0-0.4, Mn 0-0.5, Cr 0.55, V 0.2, bal Fe.
Cold work tool steel. W. Nr. 1.2235.

VA-15

VILLARES
C 0.15, Si 0.27, Mn 0.55, Cr 1.2, Ni 3.25, Mo 0.12, bal Fe.
Alloy carburizing steel. AISI 9315.

VA-15

VILLARES
C 0.15, Si 0.27, Mn 0.55, Cr 1.2, Ni 3.25, Mo 0.12, bal Fe.
Alloy carburizing steel. AISI 9315; DIN 14 NiCr 14. *Obsolete*

VA-ALLOY

English manufacture
Al 80.2, Cu 5.01, Zn 13.7, Fe 0.72, V 0.2.
For light alloy parts; non-hardenable.

VA-CRO

Midvale-Heppenstall Co.
C, alloy, bal Fe.
For tools, dies; shock resistant. *Obsolete*

VAC-ARC AGT

Latrobe Steel Co.
C 0.08-0.13, Ni 3-3.5, Cr 1-1.4, Mo 0.08-0.15, bal Fe.
Heat treated: 131,000-179,000 TS; 96,000-143,000 YS; 15-21 El; 59-67 RA; 289-363 Brin. For gears, shafts, pawls, arbors, pneumatic tool parts, ratchets; high core strength, case hardening steel. *Obsolete*

VAC-ARC BG41

Latrobe Steel Co.
C 1.05, Cr 14.5, V 0.12, Mo 4, Mn 0.5, bal Fe.
For bearings for high temperature operation; hardenable, stainless. *Obsolete*

VAC-ARC REGENT

Latrobe Steel Co.
C 1, Mn 0.35, Cr 1.5, bal Fe.
Annealed: 100,000 TS; 81,000 YS; 25 El; 57 RA; 192 Brin.
Heat treated: 237,000 TS; 226,000 YS; 444 Brin. For bearings, liners, bushings; vacuum melted, deep hardening. *Obsolete*

VAC-MELT A

George W. Prentiss & Co.
Ni 74.5, Cr 20, Fe 1.5, Mo 1, Mn 3.
Annealed: 103,000 TS; 27 El. For electrical apparatus; resists heat to 2000 F; resistance alloy. *Obsolete*

VAC-MELT AA

George W. Prentiss & Co.
Ni 77.5, Cr 20, Fe 0.5, Mn 2.
Annealed: 105,000 TS; 30 El. For resistance alloy in electrical equipment; resists heat to 2200 F. *Obsolete*

VAC-MELT B7M

George W. Prentiss & Co.
Ni 60, Cr 15, Mo 7, Mn 2, Fe 16.
Annealed: 114,000 TS; 27 El. For resistance alloy in electrical equipment; resists heat to 1900 F, high resistance to corrosion. *Obsolete*

VAC-MELT C

George W. Prentiss & Co.
Ni 61, Cr 18.5, Fe 16.5, Mn 4.
Annealed: 100,000 TS; 22 El. For resistance alloy in electrical equipment; resists heat to 1900 F. *Obsolete*

VACCUTHERM 5-32

Stahlwerke Sudwestfalen
C 0.15, Cr 11.5, Mo 1, bal Fe.
Annealed: 80,000 TS; 40,000 YS; 40 El; B 92 Rock. Heat treated: 200,000 TS; 160,000 YS; 15 El; C 45 Rock. For chemical plant and oil refinery equipment, springs, table flatware. Corrosion resistant, hardenable. *Obsolete*

VACCUTHERM 5-32 H

Stahlwerke Sudwestfalen
C 0.2, Cr 11.5, Mo 1, bal Fe.
Annealed: 95,000 TS; 40,000 YS; 25 El; 55 RA; B 92 Rock.
Heat treated: 240,000 TS; 210,000 YS; 8 El; 25 RA; C 50 Rock. For gears, shafts, cutlery, knives, surgical instruments, stainless hardware, oil refinery equipment. Corrosion resistant, hardenable. *Obsolete*

VACCUTHERM 5-34

Stahlwerke Sudwestfalen
C 0.2, Cr 11.5, Mo 1, V 0.3, bal Fe.
Annealed: 95,000 TS; 40,000 YS; 25 El; B 92 Rock. Heat treated: 245,000 TS; 210,000 YS; 10 El; C 50 Rock. For gears, shafts, hardware, surgical instruments, oil refinery and equipment. Corrosion resistant, hardenable. *Obsolete*

VACCUTHERM 5-36

Stahlwerke Sudwestfalen
C 0.2, Cr 12, Mo 1, V 0.3, W 0.5, bal Fe.
Annealed: 80,000 TS; 40,000 YS; 24 El; B 94 Rock.
Hardened: 240,000 TS; 205,000 YS; 9 El; C 50 Rock. For cutlery, surgical instruments, valves, bearings, chemical plant and oil refinery equipment. Corrosion resistant, hardenable. *Obsolete*

VACCUTHERM 5-40 H

Stahlwerke Sudwestfalen
C 0.18, Cb 0.2, Cr 11.5, Mo 0.6, Ni 0.6, V 0.25, bal Fe.
Annealed: 85,000 TS; 40,000 YS; 24 El; B 95 Rock.
Hardened: 240,000 TS; 205,000 YS; 10 El; C 50 Rock. For cutlery, surgical instruments, valves, bearings, pivots. Corrosion resistant, hardenable. *Obsolete*

VACODIL 20

Vacuumschmelze GmbH
Ni 20, Mn 6, bal Fe.
Expansion and glass-to-metal sealing alloy for semiconductor techniques.

VACODIL 36

Vacuumschmelze GmbH
Ni 36, bal Fe.
Expansion and glass-to-metal sealing alloy for semiconductor techniques.

VACODIL 42

Vacuumschmelze GmbH
Ni 42, bal Fe.
Expansion and glass-to-metal sealing alloy for semiconductor techniques.

VACODIL 46

Vacuumschmelze GmbH
Ni 46, bal Fe.
Expansion and glass-to-metal sealing alloy for semiconductor techniques.

VACODIL CM 53

Vacuumschmelze GmbH
Mo 53, bal Cu.
Expansion and glass-to-metal sealing alloy for semiconductor techniques.

VACODUR 16

Vacuumschmelze GmbH
Soft magnetic aluminum-iron alloy with high mechanical hardness for wear resistant magnetic heads. *Obsolete*

VACODYM 335

Vacuumschmelze GmbH
Permanent magnet alloy produced from iron and rare earths using powder metallurgy; high field strength with small magnet volume. 280 kJ/m^3 maximum energy product.

VACODYM 370

Vacuumschmelze GmbH
Permanent magnet alloy produced from iron and rare earths using powder metallurgy; high field strength with small magnet volume. 270 kJ/m^3 maximum energy product.

VACOFER S1

Vacuumschmelze GmbH
Pure soft magnetic iron produced by powder metallurgy, for relay components, armatures, pole pieces and sintered shapes; high saturation flux density.

VACOFER S2

Vacuumschmelze GmbH
Pure soft magnetic iron produced by powder metallurgy, for relay components, armatures, pole pieces and sintered shapes; high saturation flux density.

VACOFLEX

Vacuumschmelze GmbH
Thermobimetal expansion alloys; tapes, strips, stamped parts and discs for temperature measurement and control functions, for example, thermostats and appliances. Strip suitable for snap-action discs available as CLICKFLEX. Ten or more standard grades available.

VACOFLUX 48

Vacuumschmelze GmbH
Soft magnetic 50% cobalt-iron alloy for telephone receiver diaphragms, pole pieces, relay components; high saturation flux density.

VACOFLUX 50

Vacuumschmelze GmbH
Soft magnetic 50% cobalt-iron alloy for telephone receiver diaphragms, pole pieces, relay components; high saturation flux density.

VACOFLUX Z

Vacuumschmelze GmbH
Co 50, bal Fe.
Soft magnetic alloy with square hysteresis loop for magnetic amplifiers, converters, chokes and cores. Saturation flux density 2.35 T.

VACOMAX

Vacuumschmelze GmbH
Hard magnetic material, cobalt-iron. Various grades, maximum energy product: VACOMAX 65K: 70 kJ/m^3; VACOMAX 80 T: 80 kJ/m^3; VACOMAX 95 T: 95 kJ/m^3; VACOMAX 145: 160 kJ/m^3; VACOMAX 170: 175 kJ/m^3; VACOMAX 200: 195 kJ/m^3; VACOMAX 165: 180 kJ/m^3; VACOMAX 225: 220 kJ/m^3.

VACOMET CO CR 83/17

Vacuumschmelze GmbH
Cr 17, bal Co.
High purity and homogeneity as target material in production of thin, active layers for magnetic recording and resistors.

VACOMET CO NI 80/20

Vacuumschmelze GmbH
Ni 20, bal Co.
High purity and homogeneity as target material in production of thin, active layers for magnetic recording and resistors; suitable for the deposition of thin films by evaporation and sputtering.

VACOMET NI CR 80/20

Vacuumschmelze GmbH
Cr 20, bal Ni.
High purity and homogeneity as target material in production of thin, active layers for magnetic recording and resistors; anticorrosive protective coating.

VACOMET NI FE 80/20

Vacuumschmelze GmbH
Fe 19, bal Ni.
High purity and homogeneity as target material in production of thin, active layers for magnetic recording and resistors; for the deposition of magnetic films acting as layers with high permeability and low coercivity.

VACOMET S CO
Vacuumschmelze GmbH
Co 100.
High purity and homogeneity as target material in production of coatings to increase the adhesion of functional active layers for magnetic recording and resistors.

VACOMET S CR
Vacuumschmelze GmbH
Cr 100.
High purity and homogeneity as target material in production of coatings to increase the adhesion of functional active layers for magnetic recording and resistors.

VACOMET S FE
Vacuumschmelze GmbH
Fe 100.
High purity and homogeneity as target material in production of coatings to increase the adhesion of functional active layers for magnetic recording and resistors.

VACOMET S NI
Vacuumschmelze GmbH
Ni 100.
High purity and homogeneity as target material in production of coatings to increase the adhesion of functional active layers for magnetic recording and resistors.

VACON 10
Vacuumschmelze GmbH
Ni 29, Co 17, bal Fe.
Expansion and glass-to-metal sealing alloy for electron tubes and semiconductors.

VACON 70
Vacuumschmelze GmbH
Ni 28, Co 23, bal Fe.
Expansion and glass-to-metal sealing alloy for electron tubes and semiconductors.

VACOPERM 100
Vacuumschmelze GmbH
Soft magnetic 72 to 83% nickel alloy for circuit breakers, transformers and relays. 250,000 microns permeability maximum.

VACOPERM 70
Vacuumschmelze GmbH
Soft magnetic 75% nickel-iron alloy for transformer cores, leakage current protective switches, and high grade relays. The same alloy is supplied under the tradename MUMETALL to all countries except the Commonwealth countries, Eire, France and the USA.

VACOPERM BS
Vacuumschmelze GmbH
Soft magnetic 72 to 83% nickel alloy for circuit breakers, transformers and relays. 150,000 microns permeability maximum.

VACOPLUS
Vacuumschmelze GmbH
Cr 10, bal Ni.
For temperature measurement with thermocouples. *Obsolete*

VACOVIT 485
Vacuumschmelze GmbH
Ni 47, Cr 6, bal Fe.
Expansion and glass-to-metal sealing alloy for reed-switches and bushings.

VACOVIT 500
Vacuumschmelze GmbH
Ni 50, bal Fe.
Expansion and glass-to-metal sealing alloy for reed-switches and bushings.

VACOVIT 511
Vacuumschmelze GmbH
Ni 51, Cr 1, bal Fe.
Expansion and glass-to-metal sealing alloy for reed-switches and bushings.

VACOVIT 520
Vacuumschmelze GmbH
Ni 52, bal Fe.
Expansion and glass-to-metal sealing alloy for reed-switches and bushings.

VACOVIT 540
Vacuumschmelze GmbH
Ni 54, bal Fe.
Expansion and glass-to-metal sealing alloy for reed-switches and bushings.

VACOVIT S 505
Vacuumschmelze GmbH
Ni 50, bal Fe.
Sintered expansion and glass-to-metal sealing alloy for reed-switches and bushings.

VACOZET 200
Vacuumschmelze GmbH
Magnetic semi-hard cobalt-iron-nickel alloy for latching reed contacts and remanent magnetism relays.

VACOZET 258
Vacuumschmelze GmbH
Magnetic semi-hard cobalt-iron-nickel alloy for latching reed contacts and remanent magnetism relays.

VACOZET 655
Vacuumschmelze GmbH
Magnetic semi-hard cobalt-iron-nickel alloy for latching reed contacts and remanent magnetism relays.

VACRO 5
Uddeholm Corp.
C 0.27, Mn 1.15, Cr 0.75, V 0.15, bal Fe.
For gears, shafts; shock resistant. *Obsolete*

VACRO 62
Uddeholm Corp.
C 0.3, Cr 2.5, Mo 0.2, V 0.15, bal Fe.
For die casting and plastic mold dies; oil hardened, tough. *Obsolete*

VACROMIUM
Vacuumschmelze GmbH
Nickel-chromium and nickel-chromium-iron alloys for electric heating applications. *Obsolete*

VACRYFLUX
Vacuumschmelze GmbH
Now VACRYFLUX 5001.

VACRYFLUX 5001
Vacuumschmelze GmbH
High-field superconductor made of niobium-titanium in a copper, copper-nickel, or mixed matrix; can be coextruded with aluminum. Provides magnetic fields up to saturation flux density 9 T at 4.2 K. For NMR, MR, fusion research, and high energy physics.

VACRYFLUX NS, HNST
Vacuumschmelze GmbH
High field superconductor made of niobium-tin or niobium-tantalum-tin in a copper-tin matrix. Provides magnetic fields up to saturation flux density 16.5 T at 4.2 K. For NMR, MR, fusion research, and high energy physics.

VACUMET
Vacuumschmelze GmbH
Reference materials for X-ray analysis and spectro-chemical emission analysis. *Obsolete*

VACUMET WASPALLOY
Now CARPENTER WASPALLOY.

VACUMINUS
Vacuumschmelze GmbH
Ni 95.
Nickel alloy for temperature measurement with thermocouples. *Obsolete*

VAL 1
Acciaierie Valbruna s.p.a.
C 0.08-0.12, Si 0-1, Mn 0-1, Cr 12-14, bal Fe.
Chromium stainless steel. W. Nr. 1.4006; similar to AISI 410. *Obsolete*

VAL 106
Acciaierie Valbruna s.p.a.
Tool Material. C 0.5, Si 0.9, Mn 0.5, Cr 8.5, Mo 1.2, W 1.2, bal Fe.
Hot work tool steel. W. Nr. 1.2631.

VAL 41/S
Acciaierie Valbruna s.p.a.
C 0.45-0.6, Si 0-0.4, Mn 0.6-0.9, Cr 0.7-1.2, Mo 0.4-0.6, Ni 1.4-1.8, bal Fe.
Tool steel. W. Nr. 1.2714; U 52 NiCrMo 6 KU. Special purpose type, similar to AISI L6. *Obsolete*

VAL 68
Acciaierie Valbruna s.p.a.
C 1, Si 0-0.3, Mn 0.35, P 0-0.025, S 0-0.025, Cr 1.5, Ni 0-0.3, Cu 0-0.3, bal Fe.
Special steel. AISI 52100; W. Nr. 1.3505.

VAL 68C
Acciaierie Valbruna s.p.a.
C 0.85, Si 0.25, Mn 0.3, P 0-0.035, S 0-0.035, Cr 1.75, bal Fe.
Special steel.

VAL 68M
Acciaierie Valbruna s.p.a.
C 1, Si 0-0.3, Mn 0.7, P 0-0.025, S 0-0.025, Cr 1.85, Mo 0.3, Ni 0-0.3, Cu 0-0.3, bal Fe.
Special steel. W. Nr. 1.3537.

VAL1-S
Acciaierie Valbruna s.p.a.
C 0.16-0.25, Si 0-1, Mn 0-1, Cr 12-14, Ni 0-1, bal Fe.
Martensitic stainless steel. W. Nr. 1.4021; similar to AISI 420. *Obsolete*

VAL1-Z
Acciaierie Valbruna s.p.a.
Stainless Steel. C 0-0.15, Si 0-1, Mn 0-1, Cr 12.5, Mo 0-0.6, Ni 0-0.5, S 0.25, P 0.045, bal Fe.
Free machining martensitic. W. Nr. 1.4005; AISI 416.

VAL102
Acciaierie Valbruna s.p.a.
Tool Material. C 0.4, Si 1, Mn 0-0.4, Cr 5.15, Mo 1.35, V 1, bal Fe.
Hot work tool steel. W. Nr. 1.2344; AISI H 13.

VAL103
Acciaierie Valbruna s.p.a.
Tool Material. C 0.35, Si 1, Mn 0-0.5, Cr 5.25, Mo 1.4, V 0.35, W 1.3, bal Fe.
Hot work tool steel. W. Nr. 1.2608; AISI H12.

VAL104
Acciaierie Valbruna s.p.a.
Tool Material. C 0.36, Si 1, Mn 0-0.5, Cr 5.15, Mo 1.3, V 0.4, bal Fe.
Hot work tool steel. W. Nr. 1.2343; AISI H11.

VAL105
Acciaierie Valbruna s.p.a.
Tool Material. C 0.52, Si 1, Mn 0-0.4, Cr 5.15, Ni 1.4, W 1.2, bal Fe.
Hot work tool steel.

VAL1A

Acciaierie Valbruna s.p.a.
Stainless Steel. C 0-0.11, Si 0-1, Mn 0-1, P 0-0.04, S 0-0.03,
Cr 12.5, Ni 0-0.5, bal Fe.
Martensitic. ASTM A182F6; W. Nr. 4006.

VAL1AL

Acciaierie Valbruna s.p.a.
Stainless Steel. C 0-0.08, Si 0-1, Mn 0-1, P 0-0.04, S 0-0.02,
Cr 12.5, Al 0.2, bal Fe.
Ferritic. AISI 405; W. Nr. 1.4002.

VAL1B

Acciaierie Valbruna s.p.a.
Stainless Steel. C 0-0.15, Si 0-0.5, Mn 0-1, P 0-0.04, S 0-0.03,
Cr 12.5, Ni 0-0.5, bal Fe.
Martensitic. AISI 410; W. Nr. 1.4024.

VAL1MP

Acciaierie Valbruna s.p.a.
Stainless Steel. C 0-0.15, Si 0-0.5, Mn 0.45, Cr 12, Mo 0.5, Ni
0-0.6, P 0.025, S 0.025, bal Fe.
Martensitic.

VAL1NI

Acciaierie Valbruna s.p.a.
Stainless Steel. C 0-0.11, Si 0-1, Mn 0-1, P 0-0.04, S 0-0.03,
Cr 12.5, Ni 1.85, bal Fe.
Martensitic. AISI 414.

VAL1P

Acciaierie Valbruna s.p.a.
Stainless Steel. C 0-0.08, Si 0-0.8, Mn 0-1, Cr 12.5, P 0.04, S
0.03, Ni 0-0.5, bal Fe.
Martensitic. W. Nr. 1.4000; AISI 403.

VAL2

Acciaierie Valbruna s.p.a.
C 0.28-0.35, Si 0-1, Mn 0-1, Cr 12-14, Ni 0-1, bal Fe.
Martensitic stainless steel. W. Nr. 1.4028; X30 Cr 13; similar to
AISI 420. *Obsolete*

VAL2-CS

Acciaierie Valbruna s.p.a.
C 0.36-0.45, Si 0-1, Mn 0-1, Cr 12.5-14.5, Ni 0-1, bal Fe.
Martensitic stainless steel. W. Nr. 1.4034; X40 Cr 13.
Obsolete

VAL2A

Acciaierie Valbruna s.p.a.
Stainless Steel. C 0.19, Si 0-1, Mn 0-1, P 0-0.035, S 0-0.03, Cr
13, Ni 0-1, bal Fe.
Martensitic. AISI 420A; W. Nr. 1.4021.

VAL2AM

Acciaierie Valbruna s.p.a.
Stainless Steel. C 0.19, Si 0-1, Mn 0-1, P 0-0.045, S 0-0.03, Cr
13, Mo 1.1, Ni 0-1, bal Fe.
Martensitic. W. Nr. 1.4120.

VAL2B

Acciaierie Valbruna s.p.a.
Stainless Steel. C 0.3, Si 0-0.8, Mn 0-1, P 0-0.03, S 0-0.03, Cr
13, Ni 0-0.5, bal Fe.
Martensitic. AISI 420B; W. Nr. 1.4028.

VAL2BZ

Acciaierie Valbruna s.p.a.
Stainless Steel. C 0.3, Si 0-1, Mn 0-1.25, P 0-0.04, S 0-0.25,
Cr 13, bal Fe.
Martensitic. AISI 420S.

VAL2C

Acciaierie Valbruna s.p.a.
Stainless Steel. C 0.38, Si 0-1, Mn 0-1, P 0-0.04, S 0-0.03, Cr
13, Ni 0-1, bal Fe.
Martensitic. AISI 420C; W. Nr. 1.4034.

VAL2D

Acciaierie Valbruna s.p.a.
Stainless Steel. C 0.45, Si 0-1, Mn 0-1, P 0-0.045, S 0-0.03, Cr
14.5, Mo 0.3, V 0.1, bal Fe.
Martensitic. W. Nr. 1.4116.

VAL2MV

Acciaierie Valbruna s.p.a.
Stainless Steel. C 0.22, Si 0.35, Mn 0.55, Cr 12, Mo 1, P
0.035, S 0.035, Ni 0.5, W 0.3, bal Fe.
Martensitic. W. Nr. 1.4923.

VAL2W

Acciaierie Valbruna s.p.a.
Stainless Steel. C 0.22, Si 0-0.5, Mn 0.7, P 0-0.025, S 0-0.025,
Cr 12, Mo 1.05, V 0.25, W 1.05, bal Fe.
Martensitic. AISI 616/422; W. Nr. 1.4935.

VAL3

Acciaierie Valbruna s.p.a.
C 0.38, Si 0-1, Mn 0-1, P 0-0.045, S 0-0.03, Cr 16.5, Mo 1.1,
Ni 0-1, bal Fe.
Martensitic. W. Nr. 1.4122.

VAL33

Acciaierie Valbruna s.p.a.
Tool Material. C 0.32, Si 0-0.4, Mn 0-0.4, Cr 2.95, Mo 2.8, V
0.55, bal Fe.
Hot work tool steel. AISI H10; W. Nr. 1.2365.

VAL4

Acciaierie Valbruna s.p.a.
Stainless Steel. C 0.15, Si 0-1, Mn 0-1, Cr 16, Ni 2, P 0.04, S
0.03, bal Fe.
Martensitic. W. Nr. 1.4057; AISI 431.

VAL41D

Acciaierie Valbruna s.p.a.
Tool Material. C 0.42, Si 0-0.4, Mn 0.65, Cr 1.65, Mo 0.5, Ni
3.75, bal Fe.
Hot work tool steel. W. Nr. 1.2766.

VAL41V

Acciaierie Valbruna s.p.a.
Tool Material. C 0.55, Si 0-0.4, Mn 0.8, Cr 1.1, Mo 0.5, Ni
1.65, V 0.1, bal Fe.
Hot work tool steel. AISI L6; W. Nr. 1.2714.

VAL4S

Acciaierie Valbruna s.p.a.
Stainless Steel. C 0-0.16, Si 0-1, Mn 0-1, Cr 19, Ni 2, P 0.04, S
0.03, bal Fe.
Martensitic.

VAL5

Acciaierie Valbruna s.p.a.
Stainless Steel. C 0.45, Si 3.05, Mn 0.6, P 0-0.035, S 0-0.03,
Cr 8.75, Ni 0-0.5, bal Fe.
Valve steel. W. Nr. 1.4718.

VAL5M

Acciaierie Valbruna s.p.a.
Stainless Steel. C 0.4, Si 2.5, Mn 0-0.8, P 0-0.04, S 0-0.03, Cr
10, Mo 1.05, bal Fe.
Valve steel. W. Nr. 1.4731.

VAL6

Acciaierie Valbruna s.p.a.
C 0.6, Si 0.85, Mn 0-0.5, Cr 1, W 2.25, bal Fe.
Shock resisting type tool steel. AISI S1; W. Nr. 1.2550.

VAL66

Acciaierie Valbruna s.p.a.
Tool Material. C 0.3, Si 0-0.4, Mn 0-0.45, Cr 2.5, V 0.55, W
4.8, bal Fe.
Hot work tool steel. W. Nr. 1.2567.

VAL7

Acciaierie Valbruna s.p.a.
C 1.15, Mn 0-0.35, W 1.05, Si 0-0.4, bal Fe.
Special purpose tool steel. AISI F1; W. Nr. 1.2516.

VAL8

Acciaierie Valbruna s.p.a.
Tool Material. C 0.4, Si 0.75, Mn 0-0.5, Cr 1, W 2.05, bal Fe.
Hot work tool steel. W. Nr. 1.2542; AISI S1.

VAL9

Acciaierie Valbruna s.p.a.
Tool Material. C 0.28, Cr 2.6, W 8.75, V 0.3, Si 0-0.4, Mn
0-0.5, bal Fe.
Hot work tool steel. W. Nr. 1.2581; AISI H21.

VALAND 1

Uddeholm Corp.
C 0.3, Ni 1.7, W 9, V 0.3, Cr 3, bal Fe.
For die casting dies, hot working tools; hot work steel, oil
hardened. *Obsolete*

VALAND 2

Uddeholm Corp.
C 0.3, Co 2, Cr 2.4, V 0.25, W 8.5, bal Fe.
For extrusion dies, rams and liners, punches; hot work steel,
oil hardened. *Obsolete*

VALENITE VC-1

Valenite Corp.
Sintered carbide tool material. Primarily for roughing cuts on
cast iron; tough and abrasion resistant; good at withstanding
shock of interrupted cuts.

VALENITE VC-10

Valenite Corp.
Sintered carbide tool material. Excellent general purpose
wear grade; for gage blocks, plug gages.

VALENITE VC-11

Valenite Corp.
Sintered carbide tool material. For wear applications where
some shock is encountered, as planer tools.

VALENITE VC-12

Valenite Corp.
Sintered carbide tool material. High strength and shock
resistant; for punches, blanking dies, lamination dies.

VALENITE VC-125

Valenite Corp.
Sintered carbide tool material. For general purpose and
roughing cuts on steel, including interrupted cuts.

VALENITE VC-13

Valenite Corp.
Sintered carbide tool material. Good shock resistance; for
blanking and punching dies.

VALENITE VC-14

Valenite Corp.
Sintered carbide tool material. For heavy shock resistance as
in blanking dies, swaging, coining dies.

VALENITE VC-2

Valenite Corp.
Sintered carbide tool material. General purpose grade
machining cast iron and nonferrous metals; more wear
resistant than VC-1.

VALENITE VC-28

Valenite Corp.
Sintered carbide tool material. Premium grade for heavy duty
and general purpose machining of cast iron and nonferrous
and nonmetallic materials. Good resistance to edge wear and
chipping.

VALENITE VC-3

Valenite Corp.
Sintered carbide tool material. For finish machining on cast
iron and on nonferrous metals and nonmetallics. Withstands
abrasion but not shock.

VALENITE VC-4
Valenite Corp.
Sintered carbide tool material. Hardest grade, good wear resistance; for finish machining and precision boring of cast iron and nonferrous metals; low shock resistance.

VALENITE VC-55
Valenite Corp.
Sintered carbide tool material. Premium grade for rough and general purpose machining of steel; heavy and interrupted cuts.

VALENITE VC-6
Valenite Corp.
Sintered carbide tool material. For general purpose and light cuts on steel, where cratering is a problem.

VALENITE VC-7
Valenite Corp.
Sintered carbide tool material. High grade for finish cutting steel.

VALENITE VC-8
Valenite Corp.
Sintered carbide tool material. Very hard grade for fine finishing and precision boring steel.

VALENITE VC-83
Valenite Corp.
Sintered carbide tool material. High titanium carbide content; finishing grade for light precision cuts.

VALENITE VC-9
Valenite Corp.
Sintered carbide tool material. Very hard wear resistant grade; for finishing and wear applications where no shock exists.

VALENTINE STEEL
Manufacturer not listed.
C 0.75, Cu 5.25, Mn 1.2, V 1.2, W 12.5, bal Fe.

VALIANT
Osborn Steels Ltd.
C 1.5, W 4, Cr 0.5, bal Fe.
For cutting tools, engraving cutters; water hardened, keen cutting edge. *Obsolete*

VALIMPHY
Creusot-Loire
Cr 8, Mo 1, V 0.5, C, bal Fe.
For petroleum industries, heat transfer tubes; corrosion resistant. *Obsolete*

VALIMPHY 550
Creusot-Loire
C, alloy, bal Fe.
For oil refinery equipment; oxidation resistant. *Obsolete*

VALIMPHY 600
Creusot-Loire
C, alloy, bal Fe.
For oil refinery equipment; oxidation resistant. *Obsolete*

VALIW
Acciaierie Valbruna s.p.a.
C 0.22, Cr 12.5, Mo 1.5, Ni 0.75, V 0.25, W 1, bal Fe.
Annealed: 98,000 TS; 42,000 YS; 22 El; 95 Rock B.
Hardened: 245,000 TS; 210,000 YS; 8 El; 50 Rock C. For surgical instruments, knives, gears, shafts, hardware, pivots. Corrosion resistance, hardenable. *Obsolete*

VALLINOX 301
Vallourec S.A.
C 0.08-0.15, Si 0-1, Mn 0-2, Cr 16-18, Ni 6.5-8.5, bal Fe.
Austenitic stainless steel; work-hardens rapidly. AFNOR Z 12 CN 17-08; AISI 301.

VALLINOX 304
Vallourec S.A.
C 0-0.07, Si 0-1, Mn 0-2, Cr 17-19, Ni 8-10, bal Fe.
Austenitic stainless steel. AFNOR Z 6 CN 18 09; AISI 304.

VALLINOX 316
Vallourec S.A.
C 0-0.07, Si 0-1, Mn 0-2, Cr 16-18, Ni 10-12, Mo 2-2.5, bal Fe.
Austenitic stainless steel; for chemical equipment. AFNOR Z 6 CND 17-11; AISI 316.

VALLINOX 317 L
Vallourec S.A.
C 0-0.03, Si 0-1, Mn 0-2, Cr 18-20, Ni 14-16, Mo 3-4, bal Fe.
Low carbon austenitic stainless steel. For welded stainless chemical equipment. AFNOR Z 2 CND 19-15; similar to AISI 317 L.

VALLINOX 410
Vallourec S.A.
C 0.08-0.15, Si 0-1, Mn 0-1, Cr 11.5-13.5, Mo 0-0.5, bal Fe.
Martensitic stainless steel. AFNOR Z 12 C 13; AISI 410.

VALLINOX 430
Vallourec S.A.
C 0-0.1, Si 0-1, Mn 0-1, Cr 16-18, Mo 0-0.5, bal Fe.
Ferritic type stainless steel. AFNOR Z 8 C 17; AISI 430.

VALLINOX MO
Vallourec S.A.
C 0-0.1, Si 0-1, Mn 0-2, Cr 16-18, Ni 11-13, Mo 2-2.5, Ti = 5 x C (up to 0.60), bal Fe.
Stabilized austenitic stainless steel. AFNOR Z 8 CNDT 17-12. (Stabilized AISI 316).

VALLINOX MO T.B.C.
Vallourec S.A.
C 0-0.03, Si 0-1, Mn 0-2, Cr 16-18, Ni 11-13, Mo 2-2.5, N 0.1-0.2, bal Fe.
Austenitic stainless steel with nitrogen. AFNOR Z 2 CND 17-12.

VALLINOX MONB
Vallourec S.A.
C 0-0.08, Si 0-1, Mn 0-2, Cr 16-18, N 11-13, Mo 2-2.5, Nb + Ta = 10 x C (up to 1.0), bal Fe.
Stabilized austenitic stainless steel. AFNOR Z 8 CNDNb 17-12 (stabilized AISI 316).

VALLINOX NB
Vallourec S.A.
C 0-0.08, Si 0-1, Mn 0-1, Cr 17-19, Ni 10-12, Nb + Ta = 10 x C (up to 1.0), bal Fe.
Stabilized austenitic stainless steel. AFNOR Z 8 CNNb 18-11; AISI 347.

VALLINOX T.B.C.
Vallourec S.A.
C 0-0.03, Si 0-1, Mn 0-2, Cr 17-19, Ni 9-11, N 0.1-0.2, bal Fe.
Austenitic stainless steel with nitrogen. AFNOR Z 2 CN 18-10.

VALLINOX TI
Vallourec S.A.
C 0-0.08, Si 0-1, Mn 0-2, Cr 17-19, Ni 10.12, Ti = 5 x C (up to 0.60), bal Fe.
Stabilized austenitic stainless steel. AFNOR Z 8 CNT 18-11; AISI 321.

VALRAY 1
Henry Wiggin & Co. Ltd.
Ni 80, Cr 20.
For coating automobile exhaust valves; facing alloy, heat resistant. *Obsolete*

VALUTAP DIE STEEL
Teledyne Vasco
C 1.2, W 1.6, Cr 0.7, V 0.2, bal Fe.
For dies, gages, punches. *Obsolete*

VALVE BEARINGS
English manufacture
Sn 71, Sb 24, Cu 5.
For bearings, valve packing; Babbitt.

VALVE BRONZE
American Manganese Bronze Co.
Copper. Cu 85-89, Sn 2-10, Zn 3-7, Pb 3-6.
Cast. For valves; corrosion resistant.

VALVE COPPER
Parker Appliance Co.
Copper. Cu 88, Sn 4, Zn 3, Pb 3, Ni 3.
Cast: 30,000 TS; 15,000 YS; 15 El. For valves and pipe fittings.

VALVE STEEL
American manufacture
Si 5.8, Ti 1.5, V 1.5, bal Fe.
For valves; heat resistant.

VALVE STEEL CHROME I
English manufacture
Cr 11-14, C 0.4-1.2, Si 0.1-0.2, bal Fe.
For valves; heat resistant.

VALVE STEEL CHROME II
Manufacturer not listed.
Cr 6.3, C 0.5-1, Si 0.1-0.3, bal Fe.
For valves; heat resistant.

VALVE STEEL VERY HARD
Manufacturer not listed.
W 60, Fe 26, Ti 5, Cr 4, C 3, Ce 2.
For valves; heat resistant.

VALVE STEEL, TUNGSTEN
Manufacturer not listed.
W 14, Cr 3, C 0.6, bal Fe.
For valves; heat resistant.

VALVE-LOY
Resisto-Loy Company, Inc.
C 0.6, Cr 35, Mo 10, W 8, Co 6, Cu 14, B 0.02, bal Fe.
Welded: 590 Brin. For hard facing electrodes; for valve faces. *Obsolete*

VAMPIRE
Osborn Steels Ltd.
C 0.9, Cr 0.3, Mn 1.5, bal Fe.
For forming and heading dies, punches, crimpers; oil hardened, non-deforming. *Obsolete*

VAN 60
LTV Steel
C 0-0.18, Mn 0-1.4, Si 0-0.4, 0.02 V min or 0.01 Cb, Ce or Zr min, 0.01 Al min, bal Fe.
Good cold forming, weldability, impact toughness. For automobile bumpers and agricultural equipment.

VAN 70
LTV Steel
C 0-0.18, Mn 0-1.5, Si 0-0.5, 0.02 V min or 0.01 Cb, Ce or Zr min, 0.01 Al min, bal Fe.
70,000 psi YS min. Good cold forming, weldability, impact toughness. For automobile bumpers, wheel spiders.

VAN 80
LTV Steel
C 0-0.18, Mn 0-1.6, Si 0-0.6, 0.02 Al min, 0.05 V min, 0.005 N min, bal Fe.
80,000 psi YS min. Good cold forming, weldability, impact toughness. For mobile cranes, transmission towers, and truck frames. ASTM A-656.

VAN ALLEN
English manufacture
Au 64, Ag 18.75, Cu 9, Pd 8, Al 0.25.
For dental alloy; corrosion resistant.

VAN CHIP

Teledyne Firth Sterling
Tool steel. C 1.15, W 6, Cr 4.1, V 3, Mo 5.75, bal Fe.
For plane and lathe tools, reamers, broaches, hobs, taps,
drills; high speed steel, Type M3. *Obsolete*

VAN CUT

Teledyne Vasco
Now VAN CUT TYPE 1.

VAN CUT TYPE 1

Teledyne Vasco
C 1, W 6, Cr 4, V 2.5, Mo 5.5, bal Fe.
For broaches, form tools, chasers; high speed steel. Type
M3-1.

VAN CUT TYPE 2

Teledyne Vasco
C 1.2, W 6.25, Mo 6.25, Cr 4, V 3.1.
65 Rock C. Molybdenum type high speed tool steel. Izod: 31
feet.

VAN DIE CAR

Ziv Steel & Wire Co.
C 1.1, V 0.3, Cr 0.3, bal Fe.
For dies, hobs, knives, punches; wear resistant. *Obsolete*

VAN LOM

Teledyne Vasco
C 0.85, Cr 4, V 2, Mo 8, bal Fe.
For cutting tools, lathe and planer tools; high speed steel.

VAN LOM FM

Teledyne Vasco
C 0.85, Cr 4-4.5, Mo 8-8.5, V 2, bal Fe.
For reamers, drills, punches; high speed steel, free-cutting.
Obsolete

VAN SNAP

Flockton, Tompkin & Co., Ltd.
Low C, Cr, V, M, Mo, bal Fe.
For rivet snaps; oil hardened, nontempering.

VAN-50

LTV Steel
C 0-0.18, Mn 0-1.25, Si 0-0.3, 0.02 V min or 0.01 Cb, Ce, or Zr
min, 0.01 Al min, bal Fe.
50,000 psi YS min. Good cold forming, weldability, impact
toughness. For bumpers, structural tubing, and wheel
spiders. ASTM A-441 A-572.

VAN-CRO 12

St. Lawrence Steel Co.
C 1.5, Cr 12, V 1, Mo 1, bal Fe.
Air or oil hardened cold work tool steel, high carbon/high
chromium type; for shear blades, blanking and trimming
dies, gages. AISI D2.

VAN-LOM

Teledyne Vasco
C 0.85-0.89, Cr 3.8-4.2, Mo 8-8.5, V 2, bal Fe.
For tools, cutters, chasers, reamers, drills, broaches, lathe
and planer tools. AISI M10.

VANADIN 40

SWB Stahlformguss Gesellschaft mbH
C 1.3, Cr 4.3, Mo 0.85, V 3.8, W 12, bal Fe.
For engravers' tools, heading and blanking dies; high speed
steel. *Obsolete*

VANADIN 40 CO

SWB Stahlformguss Gesellschaft mbH
C 1.35, Cr, Co, Mo, V, W, bal Fe.
For engravers' tools, heading and blanking dies; high speed
steel. *Obsolete*

VANADIUM

Teledyne Wah Chang Albany
Refractory. H 0-0.005, C 0-0.02, N 0-0.02, O 0-0.05, Al 0-0.05,
Si 0-0.2, 99.6 V min (by difference), 0.05 Fe + Ni + Cr max,,
0.10 Nb + Mo + Ta max, 0.05 Ti + Zr + Hf max.
Annealed sheet: 29,000-35,000 psi TS; 18,000-35,000 psi YS
(0.2% offset); 35-60 El (in 2 in.). For superconducting wire
and fusion reactor technology.

VANADIUM

Latrobe Steel Co.
C 1.1, W 18, Cr 4, V 3.5, bal Fe.
C 67 Brin. For high speed tools for cutting hard materials;
high speed steel. *Obsolete*

VANADIUM

Atomergic Chemetals Corp.
V.
Purities, zone refined: 99.99+%, dendritic, 99.9%,
99.7-99.8%, 99.3-99.5%, alumothermic 98-99%, 90+%.
Forms: granule, ingot, sheet, rod, foil, wire, powder, single
crystals.

VANADIUM 4

Thyssen Edelstahlwerke AG
C 1.25, Cr 4.15, Mo 0.85, V 3.75, W 12, bal Fe.
For lathe tools, reamers and milling cutters, mainly for finish
cutting; good wear hardness; high speed steel. *Obsolete*

VANADIUM 40

SWB Stahlformguss Gesellschaft mbH
C 1.3, Cr 4.3, Mo 0.85, V 3.8, W 12, bal Fe.
For blanking and forming dies, engravers' tools; high speed
steel. *Obsolete*

VANADIUM 40 CO

SWB Stahlformguss Gesellschaft mbH
C 1.35, Cr, Co, Mo, V, W, bal Fe.
For blanking and forming dies, engravers' tools; high speed
steel. *Obsolete*

VANADIUM 6-6-2

Now VASCO M2.

VANADIUM BRASS

American manufacture
Cu 70, Zn 29.5, V 0.5.
For condenser tubes, sheets, hardware.

VANADIUM BRONZE

American manufacture
Zn 38.5, V 0.5, bal Cu.
For pipes; high strength.

VANADIUM CASTDIE

Columbia Tool Steel Co.
C 0.35, Cr 5, V 1, Mo 1.5, bal Fe.
For hot work tools and dies; hot work steel, oil hardened.
Obsolete

VANADIUM EXTRA

Columbia Tool Steel Co.
Tool material. C 1.06, Mn 0.3, Si 0.25, V 0.2, bal Fe.
Water hardened; tool steel. For trim dies, lathe centers,
mandrels. Type W2-2.

VANADIUM FIREDIE

Columbia Tool Steel Co.
C 0.35, Cr 5, V 1, Mo 1.5, bal Fe.
For Al and Mg die casting dies; Type H13; oil hardened.
Obsolete

VANADIUM GRAINAL NO. 1

Cyprus Foote Mineral Co.
V 25, Ti 15, Al 10, B 0.2, bal Fe.
Used in steel manufacturing; increases hardenability.
Obsolete

VANADIUM GRAINAL NO. 6

Foote Mineral Co.
V 13, Ti 20, Al 12, B 0.2, bal Fe.
Used in steel manufacturing; increases hardenability.
Obsolete

VANADIUM HOT WORK

Manufacturer not listed.
C 0.5, Cr 0.5, V 1, bal Fe.
For hot work tools; hot work steel.

VANADIUM METAL

Foote Mineral Co.
99.5% and 90% V.
For special metallurgical applications; "Vancoram." *Obsolete*

VANADIUM PERMANDUR

Allegheny Ludlum Steel
Co 48, V 2, bal Fe.
For diaphragms, pole pieces; high permeability at high
inductions. *Obsolete*

VANADIUM PERMENDUR

Bell Telephone Laboratories
Fe 49, Co 49, V. 2.
For electrical apparatus working at high flux density; high
permeability at high flux density.

VANADIUM PERMENDUR

Western Electric Co.
Fe 49, Co 49, V. 2.
For electrical apparatus working at high flux density; high
permeability at high flux density.

VANADIUM POTTS BEST

Horace T. Potts Co.
C 0.9-1.2, V 0.2, bal Fe.
For tools, cutters. *Obsolete*

VANADIUM STANDARD

Columbia Tool Steel Co.
Tool material. C 1.06, Mn 0.3, Si 0.25, V 0.2, bal Fe.
For mandrels, cold chisels, blacksmith tools, jigs. Type W2-3;
water hardened; tool steel.

VANADIUM TOOL

English manufacture
C 0.9, Cr 1, V 0.2, bal Fe.
For tools; oil hardening.

VANADIUM TYPE "H"

Teledyne Vasco
C 0.66-0.75, V 0.15-0.25, Cr 0.7-0.9, bal Fe.
Oil treated: 255,000 TS; 222,000 YS; 6 El; 8 RA; 460 Brin. For
punches, dies, tools, axes, hatchets, caulking tools, cold
cutters, swedges; shock resistant. *Obsolete*

VANADIUMSTAHL EXTRA ZH

Stahlwerke Kabel, C.
C 1.15, Cr 0.65, V 0.1, bal Fe.
For blanking and forming dies; oil or water hardened.

VANALIUM

Manufacturer not listed.
Al 80.2, Cu 5.1, Zn 13.7, Fe 0.72, V 0.2.
Cast: 22,000 TS; 16,000 YS; 8 El. For cast parts, for aircraft
and automotive engines; resists corrosion and erosion.

VANASIL

Gillett & Eaton Inc.
Si 21-23, Cu 1-1.5, Ni 2-2.5, Mg 0.75-1.25, V 0.1, Ti 0.15, bal
Al.
Cast: 20,000-36,000 TS; 0.5 El; 90-150 Brin. For pistons,
cylinder sleeves and liners; low coefficient of friction.

VANASIL 77
Gould Inc.
Si 21-23, Ni 2-2.5, Cu 0.9-1.2, Mg 0.75-1.25, bal Al.
T6 at 72°F: 31,000 psi TS. 300°F: 26,000 psi TS. 600°F:
12,000 psi TS. Coefficient of thermal expansion at 72-300°F
9.05 x 10⁻⁶; at 300-600°F 10.45 x 10⁻⁶. For pistons,
compressor blades; good dimensional stability and elevated
temperature strength.

VANCO 5
Darwins Alloy Castings
C 1.5, W 12, Cr 5, V 5, Co 6, bal Fe.
For taps, shear blades, boring and turning tools; high- speed
steel. *Obsolete*

VANCORAM ALUMINUM IRON VANADIUM
Foote Mineral Co.
Al 75, V 2.25, bal Fe.
For use in manufacture of non-ferrous alloys. *Obsolete*

VANCORAM COPPER ALUMINUM VANADIUM
Foote Mineral Co.
25 Cu, 10 Al, 65 V or 35 Cu, 45 Al, 20 V.
For use in manufacture of non-ferrous alloys. *Obsolete*

VANCORAM COPPER MANGANESE VANADIUM
Foote Mineral Co.
Cu 25, Mn 15, V 60.
For use in manufacture of non-ferrous alloys. *Obsolete*

VANCORAM COPPER NICKEL VANADIUM
Foote Mineral Co.
Cu 30, Ni 10, V 60.
For use in manufacture of non-ferrous alloys. *Obsolete*

VANCORAM HIGH CARBON CHROMIUM METAL
Foote Mineral Co.
Cr 86-89, C 10-11.5.
For constituent of alloys; special uses. *Obsolete*

VANCORAM MANGANESE VANADIUM
Foote Mineral Co.
V 40, Mn 60.
For use in manufacture of non-ferrous alloys. *Obsolete*

VANCORAM NICKEL VANADIUM
Foote Mineral Co.
V 40, Ni 60.
For use in manufacture of non-ferrous alloys. *Obsolete*

VANCORAM TITANIUM 70% CR
Foote Mineral Co.
Ti 68-73, Al 10-15, bal Fe.
For alloying element for non-ferrous alloys. *Obsolete*

VANCORUM FERRO VANADIUM
Foote Mineral Co.
V 35-45, C 0-3.5, Si 0-12, bal Fe.
Used in the manufacture of steel; for low V content; open
hearth grade A. *Obsolete*

VANCORUM FERRO VANADIUM CRUCIBLE GRADE B
Foote Mineral Co.
V 50-80, C 0-0.5, Si 0-2.2, bal Fe.
Used in steel manufacturing; for high V content. *Obsolete*

VANCORUM FERRO VANADIUM PRIMOS GRADE C
Foote Mineral Co.
V 50-80, C 0-0.2, Si 1.25, bal Fe.
Used in steel manufacturing; for high V content. *Obsolete*

VANCRO
Hoyland Steel Co.
C 0.45, Mn 0.7, Cr 1, V 0.15, bal Fe.
For dies, cold work tools; water hardening.

VANCRO
Acciaierie Valbruna s.p.a.
Tool Material. C 1, Cr 5.15, Mo 1.15, V 0.25, Si 0.3, Mn 0.55,
bal Fe.
Air hardening cold work tool steel. W. Nr. 1.2363; AISI A2.

VANCRO
Great Western Steel Co.
C 0.45, Cr 1, V 0.15, bal Fe.
For dies, tools; tough, water hardening.

VANCROM
Erie Steel Co.
C, alloy, bal Fe.
For gears, machine parts, pins, spindles; oil hardened,
tough.

VANDERLOY
Van der Horst Corp.
Fe 99.9.
For electroformed molds and dies; electrolytic iron.

VANGUARD
Osborn Steels Ltd.
C 1.3, W 1, Cr 1.5, Mn 0.75, bal Fe.
For heading and forming dies, punches; cold work steel, oil
hardened. *Obsolete*

VANICK
Malleable Iron Fitting Co.
C 2.5, Si 2.5, Ni 0.5, Mn 0.5, trace V, bal Fe.
Cast: 50,000 TS. For general castings; high test gray iron.

VANIDUR
German manufacture
C 0.1, Cr 18, Ni 10, V 1, Ti 0.6, bal Fe.
Annealed: 85,000 TS; 30,000 YS; 55 El; 80 Rock B. For
welded structures, and tanks, chemical and dairy plant
equipment; corrosion and heat resistant, austenitic,
stabilized.

VANIMOLOY
Istituto Sperimentali Metalli Leggeri
C 3, Ni 2, Si, Mn, bal Fe.
For valve lifters; cast iron. *Obsolete*

VANITE
Columbia Tool Steel Co.
Tool material. C 0.82, W 18, Cr 4, V 2, Mo 0.6, bal Fe.
For tools, cutters, reamers, broaches, hobs, milling cutters.
Type T2; high speed steel.

VANQUISH
Osborn Steels Ltd.
C 0.95, V 0.25, bal Fe.
For drills, reamers, punches, taps, crimpers; Type W2; water
hardened. *Obsolete*

VANTRO-S
Teledyne Vasco
C 1.27, Si 2.5, Mn 0.3, W 5.5, Cr 8, V 4, Mo 4.5, Co 5, bal Fe.
For antifriction bearings, high temperature bearings. High
resistance to heat and wear in high temperature service.
Corrosion resistant. *Obsolete*

VAR STEEL GR. 250
Titanium Metals Corp.
C 0-0.03, Ni 18, Mo 4.8, Co 8, Ti 0.4, Al 0.1, B 0.003, Zr, bal
Fe.
Heat treated forging: 254,000 TS; 245,000 YS; 10.5 El; 55 RA.
For missile components, aerospace and jet engine parts,
rocket motor cases. Maraging. High strength and fracture
toughness. *Obsolete*

VAR STEEL GR. 280
Titanium Metals Corp.
C 0-0.03, Ni 18.5, Mo 4.8, Co 9, Ti 0.7, Al 0.1, B 0.003, Zr
0.02, bal Fe.
Annealed: 155,000 TS; 110,000 YS; 15 El; 70 RA; C 30 Rock.
Maraged: 290,000 TS; 280,000 YS; 10 El; 55 RA; C 54 Rock.
For missile components, rocket motor cases, engine parts,
pressure vessels. High strength and fracture toughness.
Maraging. Shock resistant. *Obsolete*

VARIOPERM
English manufacture
Fe 70, Ni 30.
For compensating shunts for electrical equipment;
temperature-sensitive, magnetic.

VAROSS
Associated Spring Co.
Ni 35, bal Fe.
For springs; low expansion, corrosion resistant, non-
magnetic.

VASCO
Teledyne Vasco
C 0.96, Mn 1.3, Cr 0.7, W 0.5, V 0.11, bal Fe.
For tools, dies, cutters. *Obsolete*

VASCO 14-4 CVM
Teledyne Vasco
C 1.05, Si 0.3, Mn 0.5, Cr 14.25, Mo 4, bal Fe.
Heat treated: C 25-61 Rock. For ball and seats for oil well
pumps, cutlery, piston rings and seals, bushings, cryogenic
applications, ball bearings. Hardenable, stainless. Good to
900 F. *Obsolete*

VASCO 155
Teledyne Vasco
C 0-0.07, Si 0-1, Mn 0-1, Ni 4.5, Cr 14.75, Cu 3.5.

VASCO 174
Teledyne Vasco
C 0-0.7, Si 0-1, Mn 0-1, Ni 4, Cr 16.25, Cu 4, Cb + Ta = 5 x C
(0.45 max).
170,000 YS; 190,000 TS; 10 El; 35 RA. Age hardening
stainless steel; for aircraft and missile fittings, fasteners,
gears, jet engine parts, valve parts, chemical process
equipment, pump shafts, paper mill equipment, and aircraft
structural components.

VASCO 300M
Teledyne Vasco
C 0.42, Si 1.65, Mn 0.75, Ni 1.8, Cr 0.8, Mo 0.4, V 0.07.
Heat treated: 230,000 YS; 280,000 TS; 7 El; 25 RA.
Consumable electrode vacuum remelted alloy steels. For
missile and rocket motor cases, landing gear components,
aircraft structural parts, ordnance hardware, high
performance shafting, and fasteners.

VASCO 4330V
Teledyne Vasco
C 0.3, Si 0.27, Mn 0.85, Ni 1.8, Cr 0.85, Mo 0.42, V 0.08.
215,000 YS; 245,000 TS; 18 El; 45 RA. Consumable electrode
vacuum remelted alloy steels. For missile and rocket motor
cases, landing gear components, aircraft structural parts,
ordnance hardware, high performance shafting, and
fasteners.

VASCO 4340
Teledyne Vasco
C 0.4, Si 0.25, Mn 0.75, Ni 1.8, Cr 0.8, Mo 0.25.
Induction vacuum melted followed by consumable electrode
vacuum remelted alloy steels; for helicopter transmission
applications.

VASCO 455
C 0-0.01, Si 0-0.1, Mn 0-0.1, Ni 8.5, Cr 12, Cu 2.
185,000 YS; 200,000 TS; 10 El; 40 RA. Age hardening
stainless steel; for aircraft and missile fittings, fasteners,
gears, jet engine parts, valve parts, chemical process
equipment, pump shafts, paper mill equipment, and aircraft
structural components.

VASCO 6-6-2 DRILL ROD
Teledyne Vasco
C 0.8, Cr 4.25, V 1.7, W 5.7, Mo 5, bal Fe.
For tools; high speed steel. *Obsolete*

VASCO 8N2 FM
Teledyne Vasco
C 0.8, W 1.5, Cr 4, Mo 9, V 1.2, bal Fe.
For cutting tools, reamers, taps, broaches, hobs; high speed steel, free-cutting. *Obsolete*

VASCO 9-4-20
Teledyne Vasco
C 0.2, Mn 0.3, P 0-0.01, S 0-0.01, Ni 9, Co 4.5, Cr 0.75, Mo 1, V 0.09, Si 0-0.1.
Tempered 1025°F: 185,000 YS; 205,000 TS; 17 El; 60 RA; 42 Rock C. Aircraft steel; for applications requiring long time exposure to moderately elevated temperatures.

VASCO 9-4-30
Teledyne Vasco
C 0.3, Mn 0.25, P 0-0.01, S 0-0.01, Si 0-0.2, Ni 7.5, Co 4.5, Cr 1, Mo 1, V 0.09.
Double tempered, 1000°F: 230,000 YS; 195,000 TS; 14 El; 40 RA; 46 Rock C. Aircraft steel; for applications requiring long time exposure to moderately elevated temperatures.

VASCO 9310
Teledyne Vasco
C 0.1, Si 0.25, Mn 0.55, Ni 3.25, Cr 1.2, Mo 0.12.
Induction vacuum melted followed by consumable electrode vacuum remelted alloy steels; for helicopter transmission applications.

VASCO CHROMOLD CVM
Teledyne Vasco
C 0.06-0.12, Si 0.15-0.25, Mn 0.1-0.4, S 0-0.01, P 0-0.01, Cr 2.15-2.45, bal Fe.
Hardened: 120,000 TS; 90,000 YS; C 25 Rock. For plastic molds, glass molds, die cavities. Case hardening steel, easy hubbing for mold cavities. Vacuum melted. *Obsolete*

VASCO CM
Teledyne Vasco
C 1.2-1.3, Si 0.25, Mn 0.6, Cr 0.5, Mo 0.2, bal Fe.
Heat treated: 212,000 TS; 6 El; 7 RA; 480 Brin. Hardened: C 60-65 Rock. For taps, reamers, broaches, punches, arbors, threading dies, drills. Maintains fine cutting edge. *Obsolete*

VASCO D6AC
Teledyne Vasco
C 0.47, Si 0.22, Mn 0.75, Ni 0.55, Cr 1.05, Mo 1, V 0.12.
Heat treated: 205,000 YS; 235,000 TS; 12 El; 35 RA.
Consumable electrode vacuum remelted alloy steels. For missile and rocket motor cases, landing gear components, aircraft structural parts, ordnance hardware, high performance shafting, and fasteners.

VASCO DIE
Teledyne Vasco
C 0.82, Si 1, Mn 0.3, Cr 7.75, V 2.5, Mo 1.55, bal Fe.
Heat treated: 224,000-349,000 TS; 190,000-279,000 YS; 3.8-6.8 El; 11.5-22.7 RA; 47-60 Rock C. For trim dies, thread roll dies, coining dies, plastic molds, slitter knives, shear blades, punches, back up rolls, extrusion dies.

VASCO DIE (6379)
Teledyne Vasco
C 0.82, Si 1, Mn 0.3, Cr 7.75, V 2.5, Mo 1.55, bal Fe.
Heat treated: 200,000-280,000 TS; 170,000-230,000 YS; 6-8 El; 29-31 RA; 47-60 Rock C. For shears, slitters, insert dies, hot and cold punches, forging dies, hot and cold shear blades. Wear resistant, tough with good hot hardness.

VASCO JET X5
Now VASCOJET 1000 VM. 5.

VASCO M-2
Teledyne Vasco
C 0.8-0.85, W 6-6.75, Mo 4.75-5.25, V 1.75-2.05, Cr 3.9-4.4, bal Fe.
High speed tool steel, Mo-W type. For cutting tools, taps, broaches, hobs, milling cutters; AISI M2.

VASCO M-2 DRILL ROD
Teledyne Allvac
C 0.82, Cr 4.25, V 1.9, W 6.4, Mo 5, bal Fe.
For tools, cutters; high speed steel.

VASCO M-33
Teledyne Vasco
C 0.9, W 1.5, Mo 9.5, Cr 3.75, V 1.15, Co 8.
65 Rock C. Molybdenum type high speed tool steel. Izod: 25 feet.

VASCO M10
Teledyne Vasco
Now VAN LOM.

VASCO M50 VM
Teledyne Vasco
C 0.8, Mn 0.3, Si 0.3, Cr 4, V 1, Mo 4, bal Fe.
Annealed: 202 Brin. For gears, valve parts, elevated temperature bearings; cutting tools. AISI M50.

VASCO M7
Teledyne Vasco
C 0.99, W 1.75, Cr 3.75, V 2.05, Mo 8.75, bal Fe.
Heat treated: 22-26 Rock C. For drills, end mills, reamers, hobs, lathe and planer tools, slitting saws, blanking and trimming dies, chasers. Type M-7 high speed steel.

VASCO MARVEL 721
Teledyne Vasco
C 0.33, Cr 3.5, W 9.75, V 0.45, bal Fe.
For extrusion rams and dies, hot punches and shears; hot work steel. *Obsolete*

VASCO MATRIX II-CVM
Teledyne Vasco
C 0.5-0.55, 20% W-Mo-Cr-V-Co total, bal Fe.
Annealed: 95,000 TS; 48,000 YS; 25 El; 55 RA; B 90 Rock.
Heat treated: 406,000 TS; 393,000 YS; 2 El; 2.3 RA; C 63 Rock. For rocket motor cases, gears, pressure vessels, shears, springs, torsion bars. Good high temperature strength. *Obsolete*

VASCO MC
Teledyne Vasco
C 0.32-0.38, Cr 0.7-1, Mo 0.3-0.5, Mn 0.8, bal Fe.
For die casting dies, gears, arbors, spindles; oil hardened, tough, mold and cavity steel. *Obsolete*

VASCO MOMARC CVM
Teledyne Vasco
C 1, Si 0.25, Cr 1.4, Mo 1, bal Fe.
Hardened: C 67-68 Rock. Normalized: 185,000 TS; 139,000 YS; 13 El; 363 Brin. Annealed: 100,000 TS; 81,000 YS; 25 El; 192 Brin. For high temperature bearings, bearing races, special rolls, arbors, dies, gauges. Good wear and heat resistance in high temperature service. *Obsolete*

VASCO NON-SHRINKABLE
Now COLONIAL NO. 6.

VASCO SPECIAL
Teledyne Vasco
C 0.65-1.45, bal Fe.
For chisels, shears, drills. *Obsolete*

VASCO STAINLESS 440C
Teledyne Vasco
C 0.95, Si 0.3, Mn 0.35, Cr 17.5, bal Fe.
For pump parts, wear plates; high corrosion and wear resistance. *Obsolete*

VASCO SUPREME
Teledyne Vasco
C 1.5, W 12.5, V 5, Co 5, bal Fe.
For circular form tools, broaches, cutting tools; high speed steel, free-cutting. Type T-15.

VASCO TUF
Teledyne Vasco
C 0.53, Si 0.9, Mn 0.3, Cr 7.75, V 1.4, Mo 1.35, bal Fe.
Heat treated: 234,000-314,000 TS; 198,000-236,000 YS; 8-9.4 El; 22-34 RA; 48-56 Rock C. For chipper knives, shear blades, heavy duty slitters, punches, gripper dies, plastic molds.

VASCO VANADIUM
Teledyne Vasco
C 0.5-1, Cr 0.8-1.4, V 0.2, bal Fe.
For tools, rock drills, ball-bearing races; shock resistant. *Obsolete*

VASCO VANADIUM TYPE "BB"
Teledyne Vasco
C 0.96-1.1, V 0.15-0.25, Cr 1.3-1.5, bal Fe.
Heat treated: 178,000-213,000 TS; 190,000 YS; 3-8 El; 13-28 RA; 337-400 Brin. For tools, gauges, rolls, ball bearings; oil hardening. *Obsolete*

VASCO VANADIUM TYPE "D"
Teledyne Vasco
C 0.45-0.55, V 0.15-0.25, Cr 0.7-0.9, bal Fe.
Annealed: 60,000-70,000 TS; 34 El; 70 RA. For spindles, gears, arbors, sprockets, wrenches, springs, tools, shear blades, jack screws; tough. *Obsolete*

VASCO VANADIUM TYPE "G"
Teledyne Vasco
C 0.56-0.65, V 0.15-0.25, Cr 0.7-0.9, bal Fe.
Heat treated: 220,000-240,000 TS; 180,000-190,000 YS; 3-5 El; 8-10 RA; 390-412 Brin. For rivet sets, chisels, springs, piston rods, drive shafts, wrist pins, screws; shock resistant. *Obsolete*

VASCO VANADIUM TYPE "K"
Teledyne Vasco
C 0.76-0.85, V 0.15-0.25, Cr 0.7-0.9, bal Fe.
Oil treated: 220,000 TS; 200,000 YS; 5 El; 5 RA; 450 Brin. For cold heading dies, blanking, forming and trimming dies, shear knives, broaches, punches; tough. *Obsolete*

VASCO VANADIUM TYPE "N"
Teledyne Vasco
C 0.86-1, V 0.15-0.25, Cr 0.8, bal Fe.
Oil treated: 195,000 TS; 180,000 YS; 10 El; 15 RA; 395 Brin. For broaches, drills, knives, files, cutlery; oil hardening. *Obsolete*

VASCO WCC 27202
Teledyne Vasco
C 0.4, Mn 0.3, Cr 4.25, Co 4.25, Mo 0.4, W 4.25, V 2.1, bal Fe.
For hot work tools and dies; hot work steel. *Obsolete*

VASCO-VJ AND VASCO VJVM
Now VASCOJET 1000 VM.

VASCODYNE
Teledyne Vasco
C 1, Si 0.85, Mn 0.25, S 0-0.03, P 0-0.03, W 1.6, Cr 3.75, V 1.95, Mo 4.
For the metal cutting industry. High speed steel.

VASCOJET 1000 VM
Teledyne Vasco
C 0.37-0.43, Cr 4.75-5.25, Mo 1.3, V 0.5, Si 0.9, Mn 0.3, bal Fe.
Heat treated: 136,000-311,000 TS; 101,000-240,000 YS; 9-16 El; 29-42 RA; 290-570 Brin. For aircraft landing gear components, engine mounts, fasteners, springs, die blocks; fatigue resistant, high strength.

VASCOJET 2000
Teledyne Vasco
C 0.14, W 1.35, Mo 1.4, Cr 5, V 0.45, Si 0.9, Mn 0.3.
171,000 YS; 226,000 TS; 20 El; 63.1 RA; 46 Rock C.
Tempered: 950°F. Carbon-strengthened ultra-high-strength steel. For aerospace industry; special tooling applications.

VASCOJET 90
Commentyrenne
C 0.12-0.18, Si 0-0.2, Mn 0.8-1.1, Cr 1.25-1.5, Mo 0.8-1, V 0.2-0.3, bal Fe.
Air or oil hardenable. Hardened: 140,000-184,000 TS; 112,000 YS min; 10 El min. High strength, low carbon, low alloy steel. Can be welded; can be case hardened for improved wear resistance. AFNOR 15 CDV 6.

VASCOJET M-A
Teledyne Vasco
C 0.5-0.55, 12.0% W + Mo + Cr + V, bal Fe.
Heat treated: 305,000-362,000 TS; 268,000-293,000 YS; 6-8.5 El; 20-34 RA; C 56-61 Rock. For airframes, fasteners, pressure vessels, engine mounts, gears, axles, rotors. Ultra high-strength to 1000 F. Tough and fatigue resistant. *Obsolete*

VASCOLOY
Now VR/WESSON.

VASCOLOY RAMET
Now VR/WESSON.

VASCOLOY VR-75
Now VR/WESSON VR-75.

VASCOMAX 250
Teledyne Vasco
C 0-0.03, Ni 17-19, Co 7-8.5, Mo 4.6-5.1, Ti 0.3-0.5, Al 0.05-0.15, bal Fe.
Annealed: 140,000 TS; 95,000 YS; 17.0 El; 75.0 RA; 290 Brin. Heat treated: 275,000 TS; 268,000 YS; 10.0 El; 48.0 RA; 520 Brin. For missile and aircraft components, rocket cases. Maraging steel, high strength, tough and ductile.

VASCOMAX 250W, VM
Teledyne Vasco
C 0-0.03, Si 0-0.1, Mn 0-0.1, Ni 17.5-18.5, Co 7.5-8.5, Al 0.05-0.15, Ti 0.4-0.55, bal Fe.
For filler metal in MIG and TIG welding of 18.0 nickel maraging steel. *Obsolete*

VASCOMAX 300
Teledyne Vasco
C 0-0.03, Mo 4.8, Ni 18, Co 9, Al 0.1, Ti 0.6, bal Fe.
Annealed: 150,000 TS; 110,000 YS; 18 El; 72 RA; 30-32 Rock C. Heat treated: 294,000-315,000 TS; 290,000-310,000 YS; 9-12 El; 35-56 RA; 54-55 Rock C. For missiles and aircraft components. High tensile strength and toughness, good corrosion resistance; high notch to smooth tensile ratios.

VASCOMAX 300 W, VM
Teledyne Vasco
C 0-0.03, Si 0-0.1, Mn 0-0.1, Ni 17.5-18.5, Co 9.5-10.5, Mo 4-5, Al 0.05-0.15, Ti 0.55-0.75, bal Fe.
For filler metal in MIG and TIG welding of 18.0 nickel maraging steel. *Obsolete*

VASCOMAX 350
Teledyne Vasco
C 0-0.03, Ni 18, Co 8.5, Mo 5, Ti 0.7, Al 0.1, bal Fe.
Annealed: 165,000 TS; 120,000 YS; 18 El; 35 Rock C. Heat treated: 340,000-360,000 YS; 350,000-370,000 TS; 10 El; 50 RA; 58 Rock C. For shear blades, gripper dies, punches, extrusion tools, torsion bars, collets, missile and aircraft components. Maraging steel, durable and tough. Consumable vacuum melted.

VASCOMAX C250, VM
Teledyne Vasco
Now VASCOMAX 250.

VASCOMAX C300 VM
Teledyne Vasco
Now VASCOMAX 300.

VASCOMAX C350, VM
Teledyne Vasco
Now VASCOMAX 350.

VASCOMAX T-200
Teledyne Vasco
Ni 18.5, Mo 3, Ti 0.7, Al 0.1, Si 0-0.1, Mn 0-0.1, C 0-0.03, S 0-0.01, P 0-0.01.
Annealed: 100,000 YS; 140,000 TS; 21 El; 92 RA; 27-29 Rock C. Referred to as Free-Co alloys. Maraging steels for missile and rocket motor cases, wind tunnel models, recoil springs, flexures, actuators, landing gear components, high performance shafting, gears, and fasteners.

VASCOMAX T-250
Teledyne Vasco
Ni 18.5, Mo 3, Ti 1.4, Al 0.1, Si 0-0.1, Mn 0-0.1, C 0-0.03, S 0-0.01, P 0-0.01.
Annealed: 95,000 YS; 140,000 TS; 16 El; 70 RA; 30-32 Rock C. Referred to as Free-Co alloys. Maraging steels for missile and rocket motor cases, wind tunnel models, recoil springs, flexures, actuators, landing gear components, high performance shafting, gears, and fasteners.

VASCOMAX T300
Teledyne Vasco
Ni 18.5, Mo 4, Ti 1.85, Al 0.1, Si 0-0.1, Mn 0-0.1, C 0-0.03, S 0-0.01, P 0-0.01.
Annealed: 110,000 YS; 150,000 TS; 16 El; 69 RA; 30-32 Rock C. Referred to as Free-Co alloys. Maraging steels for missile and rocket motor cases, wind tunnel models, recoil springs, flexures, actuators, landing gear components, high performance shafting, gears, and fasteners.

VASCOWEAR
Teledyne Vasco
C 1.12, Si 1.2, Mn 0.3, W 1.1, Cr 7.75, Mo 1.6, V 2.4, bal Fe.
Air hardened to 62-64 Rock C. High compressive strength and excellent toughness. For cold finishing rolls, cold forming rolls, cold extrusion dies, trim dies, blanking dies, shear blades.

VATOOL
Disston Inc.
C 0.95, V 0.25, 0.15% Ni and Cr max, bal Fe.
For shear blades, arbors, dies, rivet sets; water hardened. *Obsolete*

VAUCHERS ALLOY
Manufacturer not listed.
Zn 75, Sn 18, Pb 4.5, Sb 2.5.
For bearings.

VAW 41/04
VAW Vereinigte Aluminium-Werke AG
Aluminum. Si 0.5, Fe 0.6, Cu 0.2-0.5, Mn 0.9-1.3, Mg 0.1, Cr 0.03-0.15, Zn 0.1, 0.15 others total, bal Al.
Temper F 17: 150 N/mm^2 YS min; 170-200 N/mm^2 TS; 1 El min. Temper G 15: 125 N/mm^2 YS min; 145-185 N/mm^2 TS; 3 El min. Carbonated screw-caps, pharmaceutica caps.

VAW 41/11
VAW Vereinigte Aluminium-Werke AG
Aluminum. Si 0.5, Fe 0.6, Cu 0.1, Mn 0.5-0.7, Mg 0.15, Cr 0.1, Zn 0.15, 0.15 others total, bal Al.
Temper G 18: 160 N/mm^2 YS min; 180-220 N/mm^2 TS; 1 El min. Wide-neck caps.

VAW 41/20
VAW Vereinigte Aluminium-Werke AG
Aluminum. Si 0.6, Fe 0.7, Cu 0.05-0.2, Mn 1-1.5, Zn 0.1, 0.15 others total, bal Al.
Temper W 10: 100-140 N/mm^2 TS; 10-25 El min (depending on thickness). Temper G 15: 145-185 N/mm^2 TS; 6-13 El min (depending on thickness). Temper G 21: 190 N/mm^2 TS; 1 El min. Semi-rigid food containers.

VAW 61/03
VAW Vereinigte Aluminium-Werke AG
Aluminum. Cu 0.25, Si 0.3, Fe 0.7, Mn 1-1.5, Mg 0.8-1.3, Zn 0.25, 0.1 others total, bal Al.
Temper F 29: 270-275 N/mm^2 YS; 290-330 N/mm^2 TS; 2 El min. Temper G 22: 170 N/mm^2 YS min; 220-260 N/mm^2 TS; 5 El min. Beverage can bodies, deep-drawn can bodies, wide-neck caps.

VAW 61/15
VAW Vereinigte Aluminium-Werke AG
Aluminum. Si 0.6, Fe 0.7, Cu 0.3, Mn 1-1.5, Mg 0.2-0.6, Cr 0.1, Zn 0.25, Ti 0.1, 0.15 others total, bal Al.
Temper G 19: 185-225 N/mm^2 TS; 5 El min. Temper W 13: 125-165 N/mm^2 TS; 10-20 El min (depending on thickness). Semi-rigid food containers.

VAW 63/37
VAW Vereinigte Aluminium-Werke AG
Aluminum. Si 0.2, Fe 0.35, Cu 0.2, Mn 0.2-0.5, Mg 3-4, Cr 0.1, Zn 0.25, Ti 0.05, bal Al.
Temper F 35: 300 N/mm^2 YS; 350-390 N/mm^2 TS; 4 El min. Temper G 32: 290 N/mm^2 YS; 320-360 N/mm^2 TS; 5 El min. Temper G 28: 250 N/mm^2 YS; 280-320 N/mm^2 TS; 6 El min. Beverage can tabs, tear-off can tabs.

VAW 63/45
VAW Vereinigte Aluminium-Werke AG
Aluminum. Si 0.2, Fe 0.35, Cu 0.15, Mn 0.2-0.5, Mg 4-5, Cr 0.1, Zn 0.25, Ti 0.1, 0.1 others total, bal Al.
Temper F 38: 330 N/mm^2 YS; 380-430 N/mm^2 TS; 5 El min. Temper G 37: 320 N/mm^2 YS; 370-410 N/mm^2 TS; 6 El min. Beverage can ends.

VAW 63/52
VAW Vereinigte Aluminium-Werke AG
Aluminum. Si 0.25, Fe 0.4, Cu 0.1, Mn 0.1, Mg 2.2-2.8, Cr 0.15-0.35, Zn 0.1, 0.1 others total, bal Al.
Temper F 32: 280 N/mm^2 YS; 320-360 N/mm^2 YS; 2 El min. Temper G 29: 270 N/mm^2 YS; 290-330 N/mm^2 YS; 7 El min. Beverage can ends, circular tear-off can ends.

VAW 98/50
VAW Vereinigte Aluminium-Werke AG
Aluminum. Si 0.4-0.8, Fe 0.5-1, Cu 0.1, Mn 0.1, Zn 0.1, Ti 0.05, 0.25 others total, bal Al.
Temper W 8: 80-115 N/mm^2 TS; 10-30 El min (depending on thickness). Temper F 13: 110 N/mm^2 YS min; 130-170 N/mm^2 TS; 1.5 El min. Temper G 11: 80 N/mm^2 YS min; 110-150 N/mm^2 TS; 3 El min. Semi-rigid food containers, smoothwall containers, non-carbonated screw caps, pharmaceutical caps.

VAW 99/00
VAW Vereinigte Aluminium-Werke AG
Aluminum. Cu 0.05, Mn 0.05, Mg 0.05, Zn 0.1, Ti 0.05, Si+Fe = 1.0, 0.15 others total, bal Al.
Temper W 6: 60-100 N/mm^2 TS; 8-12 El min (depending on thickness). Closure foil for semi-rigid food containers.

VAW 99/52
VAW Vereinigte Aluminium-Werke AG
Aluminum. Si 0.25, Fe 0.4, Cu 0.05, Mn 0.05, Mg 0.05, Zn 0.07, Ti 0.05, 0.50 others total, bal Al.
Temper F 15: 150-190 N/mm^2 TS; 1 El min. Temper F 17: 150 N/mm^2 YS min; 170-210 N/mm^2 TS; 2 El min. Temper G 15: 140 N/mm^2 YS min; 150-190 N/mm^2 TS; 5 El min. Folded semi-rigid food containers, wide-neck caps.

VB-15
VILLARES
C 0.15, Si 0.27, Mn 0.8, Cr 0.5, Ni 0.55, Mo 0.2, bal Fe.
Alloy carburizing steel. AISI 8615.

VB-17
VILLARES
C 0.17, Si 0.27, Mn 0.8, Cr 0.5, Ni 0.55, Mo 0.2, bal Fe.
Alloy carburizing steel. AISI 8617.

VB-20
VILLARES
C 0.2, Si 0.27, Mn 0.8, Cr 0.5, Ni 0.55, Mo 0.2, bal Fe.
Alloy carburizing steel. AISI 8620.

VB-30
VILLARES
Alloy steel. C 0.3, Si 0.27, Mn 0.8, Cr 0.5, Ni 0.55, Mo 0.2, bal Fe.
Hardenable. AISI 8630.

VB-40
VILLARES
Alloy steel. C 0.4, Si 0.27, Mn 0.87, Cr 0.5, Ni 0.55, Mo 0.2, bal Fe.
Hardenable. AISI 8640.

VB-50
VILLARES
Alloy steel. C 0.5, Si 0.27, Mn 0.87, Cr 0.5, Ni 0.55, Mo 0.2, bal Fe.
Hardenable. AISI 8650.

VB-60
VILLARES
Alloy steel. C 0.6, Si 0.27, Mn 0.87, Cr 0.5, Ni 0.55, Mo 0.2, bal Fe.
Hardenable. AISI 8660.

VC
Now VALENITE VC.

VC 135
Saarstahl AG
Now SAARSTAHL 34 CR 4.

VC-12
Spencer Clark Metal Industries Ltd.
C 0.8, W 20.5, Cr 4.25, V 1.6, Co 12, bal Fe.
Cobalt super high speed steel for heavy duty applications.

VC-13
VILLARES
Tool steel. C 0.85, Cr 1.8, Mo 0.2, V 0.12, bal Fe.
Cold work. AISI L-3; DIN 1.2067. *Obsolete*

VC-130
VILLARES
C 2, Cr 11.5, V 0.2, bal Fe.
Cold work tool steel. Similar to AISI D-3; DIN 1.2080.

VC-131
VILLARES
Tool steel. C 2.1, Cr 11.5, W 0.7, V 0.2, bal Fe.
Cold work tool steel; DIN 1.2436. Similar to AISI D6.

VC-140
VILLARES
Stainless steel. C 0-0.15, Cr 12.5, bal Fe.
Martensitic type stainless steel. AISI 410; similar to DIN 1.4024.

VC-150
VILLARES
Stainless steel. C 0.35, Cr 13, bal Fe.
Martensitic. AISI 420; similar to DIN 1.4034.

VC-52
VILLARES
C 1.05, Si 0.27, Mn 0.35, Cr 1.45, bal Fe.
Steel for bearings. AISI 5210D; W.-Nr. 1.3505.

VC015
Russian manufacture
Co, VC, bal WC.
For hard cutting tools and wear parts; sintered carbides, abrasion resistant.

VC06
Russian manufacture
Co, VC, bal WC.
For hard cutting tools and wear parts; sintered carbides, abrasion resistant.

VC08
Russian manufacture
Co, VC, bal WC.
For hard cutting tools and wear parts; sintered carbides, abrasion resistant.

VC2
Acciaierie Valbruna s.p.a.
C 0.15, Mn 0.85, Cr 0.85, bal Fe.
Chromium carburizing steel (15 Mn Cr 5); similar to former SAE 5115. *Obsolete*

VCA ALLOY
Manufacturer not listed.
V 13, Cr 11, Al 3, bal Ti.
Annealed: 135,000 TS; 131,000 YS; 21 El. Aged: 221,000 TS; 207,500 YS; 4 El. For missile cases, fasteners, welded pressure vessels; recommended for service from -65°F to 600°F.

VCD2
Acciaierie Valbruna s.p.a.
C 0.15, Cr 0.95, Mo 0.55, Si 0-0.035, Mn 0.55, P 0.035, S 0.035, bal Fe.
Low carbon, low alloy steel for carburizing; also for elevated temperature service. W. Nr. 1.7335; ASTM A182-F12.

VCD2S
Acciaierie Valbruna s.p.a.
C 0.15, Mn 0.55, Si 0.75, Cr 1.25, Mo 0.55, P 0.035, S 0.035, bal Fe.
Low alloy steel for valves and parts for high temperature service. ASTM A182-F11.

VCD3
Acciaierie Valbruna s.p.a.
C 0.17, Si 0.35, Mn 0.55, P 0.035, S 0.035, Cr 0.65, Mo 0.55.
Low carbon, low alloy carburizing steel. ASTM A182-F2.

VCD4
Acciaierie Valbruna s.p.a.
C 0.12, Cr 2.25, Mo 1, Si 0.4, Mn 0.5, P 0.035, S 0.035, bal Fe.
Steel for elevated temperature service. W. Nr. 1.7380; ASTM A182-F22.

VCD5
Acciaierie Valbruna s.p.a.
C 0-0.15, Cr 5, Mo 0.55, Si 0.4, Mn 0.4, P 0.03, S 0.03, bal Fe.
Steel for high temperature equipment. ASTM A182-F5; W. Nr. 1.7362.

VCD6
Acciaierie Valbruna s.p.a.
C 0.19, Cr 1.25, Mo 1.05, V 0.25, Si 0.4, Mn 0.65, P 0.03, Ni 0-0.5, bal Fe.
Alloy steel, often used for high pressure hydrogenation vessels.

VCD6-DE
Acciaierie Valbruna s.p.a.
C 0.21, Si 0.45, Mn 0.45, P 0-0.035, S 0-0.035, Cr 1.35, Mo 1.1, Ni 0-0.6, V 0.3, bal Fe.
Heat resisting structural steel. W. Nr. 1.8070.

VCD7
Acciaierie Valbruna s.p.a.
C 0-0.15, Cr 9, Mo 1, Si 0.75, Mn 0.45, P 0.03, S 0.03, bal Fe.
Steel for high temperature equipment. ASTM A182-F9; W. Nr. 1.7386.

VCDE2-DE
Acciaierie Valbruna s.p.a.
C 0.24, Si 0.25, Mn 0.45, P 0-0.035, S 0-0.035, Cr 1.35, Mo 0.55, V 0.2, bal Fe.
Heat resisting structural steel. W. Nr. 1.7733.

VCM
VILLARES
Tool steel. C 0.32, Cr 2.85, Mo 2.9, V 0.5, bal Fe.
Hot work tool steel. DIN 1.2365; similar to AISI H10.

VCMV
Saarstahl AG
Now SAARSTAHL 30 CRMOV 9.

VCO
VILLARES
Tool steel. C 0.5, Cr 1, Ni 3.25, Mo 0.3, bal Fe.
Hot work tool steel. DIN 1.2721.

VD 4 A
Creusot-Loire
C 1.3, Cr 4.5, Mo 4.5, V 4, W 5.5, bal Fe.
High carbon, high speed steel for reamers, broaches, lathe finishing tools, extrusion tooling. AFNOR Type 6-5-4; AISI M4.

VD CHISEL
Joseph T. Ryerson & Son Inc.
C 0.9, Mn 0.4, Si 0.25, bal Fe.
AISI Type W1. Water-hardened tool steel.

VD-2
VILLARES
Tool steel. C 1.5, Cr 12, Mo 0.95, V 0.9, bal Fe.
Cold work tool steel. AISI D2; similar to DIN 1.2379.

VD3
Acciaierie Valbruna s.p.a.
C 0.18-0.23, Mn 0.7-0.9, Ni 0.4-0.7, Cr 0.4-0.6, Mo 0.15-0.25, bal Fe.
Ni-Cr-Mo carburizing steel. 20 NiCrMo 2; AISI 8620. *Obsolete*

VD4
Acciaierie Valbruna s.p.a.
C 0.18, Mn 0.75, Ni 1.3, Cr 0.8, Mo 0.2, bal Fe.
Alloy carburizing steel. 18 NiCrMo 5. *Obsolete*

VD40
Acciaierie Valbruna s.p.a.
C 0.37-0.43, Mn 0.5-0.8, Ni 1.6-1.9, Cr 0.6-0.9, Mo 0.2-0.3, bal Fe.
Ni-Cr-Mo structural steel. 40 NiCrMo 7; AISI 4340. *Obsolete*

VD45
Acciaierie Valbruna s.p.a.
C 0.34-0.42, Mn 0.5-0.8, Ni 0.7-1, Cr 0.7-1, Mo 0.15-0.25, bal Fe.
Ni-Cr-Mo structural steel. 38 NiCrMo 4. *Obsolete*

VD5
Acciaierie Valbruna s.p.a.
C 0.15-0.21, Mn 0.55, Ni 1.5-1.8, Cr 0.4-0.7, Mo 0.2-0.3, bal Fe.
Ni-Cr-Mo carburizing steel. 18 NiCrMo 7; AISI 4320. *Obsolete*

VDC
Now LATROBE VDC.

VDC
Latrobe Steel Co.
C 0.4, Si 1, Mn 0.4, Cr 5.25, Mo 1.35, V 1.
Air hardening. For die casting dies, extrusion dies and backers, shear knives, and forging applications. AISI Type H13.

VDC-RF
Latrobe Steel Co.
C 0.4, Si 1, Mn 0.4, S 0.001, Cr 5.25, Mo 1.35, V 1.
Hot work steel for die casting applications.

VDM LC NICKEL 99.6-ALLOY 205
VDM Nickel-Technologie AG
Nickel. Mn 0-0.3, Fe 0-0.2, Mg 0-0.05, Cu 0-0.1, C 0-0.02, 99.6 Ni min.
Minimum values at 20°C: 340 N/mm² TS; 80 N/mm² YS; 40 El. For chemical and processing plant equipment. Material No. 2.4061. N02205

VDM LC-NICKEL 99.2-ALLOY 201
VDM Nickel-Technologie AG
Nickel. Mn 0-0.35, Fe 0-0.4, Mg 0-0.05, Cu 0-0.25, C 0-0.02, 99.0 Ni min.
Minimum values at 20°C: 340 N/mm² TS; 80 N/mm² YS; 40 El. For chemical and processing plant equipment. Material No. 2.4068. N02201

VDM NICKEL 99.2-ALLOY 200
VDM Nickel-Technologie AG
Nickel. Mn 0-0.3, Fe 0-0.4, Mg 0-0.05, Cu 0-0.25, C 0-0.1, 99.2 Ni min.
Minimum values at 20°C: 370 N/mm² TS; 100 N/mm² YS; 40 El. For chemical and processing plant equipment. Material No. 2.4066. N02200

VDM NICKEL B 9604-FM 61
VDM Nickel-Technologie AG
Fe 0-0.7, Mn 0-0.8, Ti 2-3.5, Si 0-0.7, C 0-0.05, S 0-0.01, Cu 0-0.25, Al 0-1, 93.0 Ni min.
Submerged-arc and electroslag overlay welding on carbon steel or pressure vessel steel for use in the chemical and petrochemical industries. Material No. 2.4155. N02061

VDM NICKEL S 9604-FM 61
VDM Nickel-Technologie AG
Fe 0-0.75, Cu 0-0.25, Mn 0-0.75, Ti 1-4, Si 0-0.75, C 0-0.05, S 0-0.01, Al 0-1, 93.0 Ni min.
Filler metal. Minimum values at RT: 410 N/mm² TS; 200 N/mm² YS; 25 El. Material No. 2.4155. N02061

VDM TITAN 993-TI-GRADE 3
VDM Nickel-Technologie AG
Fe 0-0.3, C 0-0.1, N 0-0.05, H 0-0.013, 0.16 O approx, bal Ti.
Minimum values at 20°C: 460 N/mm² TS; 320 N/mm² YS; 18 El. For heat exchangers and piping systems in seawater desalination equipment, chemical and petrochemical plants and offshore technology, condensers and cooling systems in power plants. Material No. 3.7055. R50550

VDM TITAN 994-TI-GRADE 2
VDM Nickel-Technologie AG
Fe 0-0.25, C 0-0.08, N 0-0.03, H 0-0.013, 0.12 O approx, bal Ti.
Minimum values at 20°C: 390 N/mm² TS; 250 N/mm² YS; 22 El. For heat exchangers and piping systems in seawater desalination equipment, chemical and petrochemical plants and offshore technology, condensers and cooling systems in power plants. Material No. 3.7035. R50400

VDM TITAN 994PD-TI-GRADE 7
VDM Nickel-Technologie AG
Fe 0-0.25, C 0-0.08, N 0-0.03, H 0-0.013, Pd 0.12-0.15, 0.12 O approx, bal Ti.
Minimum values at 20°C: 390 N/mm² TS; 250 N/mm² YS; 22 El. For heat exchangers and piping systems in the chemical industry. Material No. 3.7235. R52400

VDM TITAN 995-TI-GRADE 1
VDM Nickel-Technologie AG
Fe 0-0.2, C 0-0.08, N 0-0.03, H 0-0.013, 0.08 O approx, bal Ti.
Minimum values at 20°C: 290 N/mm² TS; 180 N/mm² YS; 30 El. For heat exchangers and piping systems in seawater desalination equipment, chemical and petrochemical plants and offshore technology, condensers and cooling systems in power plants. Material No. 3.7025. R50250

VDM TITAN 995PD-TI-GRADE 11
VDM Nickel-Technologie AG
Fe 0-0.2, C 0-0.08, N 0-0.03, H 0-0.013, Pd 0.12-0.15, 0.08 O approx, bal Ti.
Minimum values at 20°C: 290 N/mm² TS; 200 N/mm² YS; 30 El. For heat exchangers and piping systems in the chemical industry. Material No. 3.7225. R52250

VDM ZIRKONIUM 702
VDM Nickel-Technologie AG
Hf 0-5, 99.2 Zr + Hf min, 0.16 O min, 0.2 Fe + Cr max.
Minimum values at 20°C: 380 N/mm² TS; 207 N/mm² YS; 16 El. For chemical processes where nickel alloys and stainless steel are inadequate. R60702

VDM ZIRKONIUM 704
VDM Nickel-Technologie AG
Hf 0-5, Sn 1-2, 97.5 Zr + Hf min, 0.16 O min, 0.2-0.4 Fe + Cr.
Minimum values at 20°C: 413 N/mm² TS; 241 N/mm² YS; 14 El. For chemical processes where nickel alloys and stainless steel are inadequate. R60704

VECTOLITE
Crucible Materials Corp.
30 Fe₂O₃, 44 Fe₃O₄, 26 Co₂O₃.

VEDAL
English manufacture
Si, Mg, bal Al.
For light alloy parts; non-hardenable.

VEDAS
Osborn Steels Ltd.
C 0.8-1.2, bal Fe.
For springs, drills, taps, cutters; Type W1; water hardened. *Obsolete*

VEGA
Now CARPENTER VEGA.

VEGA 1
Creusot-Loire
Cr 12-14, 0.15% C min, bal Fe.
Annealed: 88,000 TS; 40,000 YS; 32 El; 68 RA; 170 Brin. Heat treated: 256,000 TS; 190,000 YS; 6 El; 10 RA; 540 Brin. For cutlery, valve trim, turbine blades; Type 420; stainless, hardenable. *Obsolete*

VEGA 10
Creusot-Loire
C 0-0.15, Cr 11.5-13.5, bal Fe.
Heat treated: 120,000-135,000 TS; 110,000-117,000 YS; 15-16 El; 58-63 RA; 220-240 Brin. For cutlery, valves, turbine blades; Type 410; corrosion resistant. *Obsolete*

VEGA 100
Creusot-Loire
C 0.1, Cr 21, Ni 1.5, bal Fe.
For furnace parts, heat treating boxes; heat resistant. *Obsolete*

VEGA 12
Compagnie Ateliers et Forges de la Loire
Stainless steel. C 0.1, Cr 12.5, bal Fe.
Heat treated: 87,000 TS; 64,000 YS; 18 El. Annealed: 64,000 TS; 37,000 YS; 32 El. For turbine parts, cutlery, oil refinery equipment, valves, pivots. Martensitic, stainless steel, hardenable.

VEGA 12VS
Compagnie Ateliers et Forges de la Loire
Stainless steel. C 0.2, Cr 12.5, Mo 1.1, W 1.1, V 0.25, bal Fe.
Heat treated: 114,000-143,000 TS; 85,000-114,000 YS; 12-17 El. For gas and steam turbine blades and rotors, surgical instruments. Martensitic, stainless steel. Useful to 650°C, hardenable.

VEGA 15
Compagnie Ateliers et Forges de la Loire
Stainless steel. C 0.05, Cr 12.5, bal Fe.
Annealed: 78,000 TS; 57,000 YS; 30 El. For oil refinery and food industry equipment, tanks, vessels, trim. Ferritic, stainless steel. Non-hardenable.

VEGA 16
Creusot-Loire
C 0-0.1, Cr 17, Mo 1.5, bal Fe.
Annealed: 80,000 TS; 50,000 YS; 25 El; 50 RA; 150 Brin. For oil refinery equipment, oil burners and heaters, blowers; Type 430 Mo; corrosion resistant. *Obsolete*

VEGA 2
Creusot-Loire
Cr 12-14, 0.15% C min, bal Fe.
Annealed: 88,000 TS; 40,000 YS; 32 El; 68 RA; 170 Brin. Heat treated: 256,000 TS; 190,000 YS; 6 El; 10 RA; 540 Brin. For cutlery, valve trim, turbine blades; Type 420; stainless, hardenable. *Obsolete*

VEGA 20
Creusot-Loire
C 0.2, Cr 15-17, Ni 1.25-2.5, bal Fe.
Heat treated: 100,000-135,000 TS; 78,000-112,000 YS; 15-25 El; 45-60 RA; 240-280 Brin. For pump spindles, propeller shafts, aircraft structures; Type 431; stainless, hardenable. *Obsolete*

VEGA 3
Creusot-Loire
Cr 12-14, 0.15% C min, bal Fe.
Annealed: 88,000 TS; 40,000 YS; 32 El; 68 RA; 170 Brin. Heat treated: 256,000 TS; 190,000 YS; 6 El; 10 RA; 540 Brin. For cutlery, valve trim, turbine blades; Type 420; stainless, hardenable. *Obsolete*

VELINVAR
Japanese manufacture
Co 55-63, V 7-13, bal Fe.
For instruments, chronometers; low coefficient of expansion.

VELOCITAS SPEZIAL HI
Dorrenberg Edelstahl GmbH
C 0.48, Cr, Si, bal Fe.
For gears, pinions, bolts, shafts; oil hardened; tough.

VELOCITAS SPEZIAL HII
Dorrenberg Edelstahl GmbH
C 0.45, Cr, Si, V, bal Fe.
For gears, shafts, crankshafts, fasteners; oil hardened; tough.

VELODAL
Now WIELAND A42.

VELODUR
Wieland-Werke AG Metallwerke
Zn 6.7, Mg 2.3, Cu 2, Mn 0.1, Cr 0.3, bal Al.
Aged: 85,000 TS; 74,000 YS; 7 El; 160 Brin. For transportation and aircraft construction; age-hardened, high strength. *Obsolete*

VELOS
Spencer Clark Metal Industries Ltd.
C 0.85, W 6.4, Mo 5.2, Cr 4.2, V 1.9, bal Fe.
Tungsten-molybdenum high speed steel.

VELOS 42
Spencer Clark Metal Industries Ltd.
C 1.05, W 1.6, Mo 9.5, Cr 4, V 1.2, Co 8.5, bal Fe.
Cobalt-molybdenum high speed steel.

VELOS UR
Spencer Clark Metal Industries Ltd.
C 0.75, W 18, Cr 4.25, V 1.1, bal Fe.
Tungsten high speed steel.

VELVET
English manufacture
C, W, bal Fe.
For tools, drills, taps; water hardened.

VELVET
Union Bronze Co.
Sn 10, Pb 2, bal Cu.
For bushings, Babbitts, bar; hard.

VELVET ANTIFRICTION METAL
McKechnie Metals Ltd.
Pb, Sn, Sb.
For bearings; Babbitt. *Obsolete*

VELVETOUCH
S.K. Wellman Corp.
Cu-Pb-Sn-graphite.
For clutch and brake discs, linings, bearings, facings; sintered.

VENANGO
Now CYCLOPS S2.

VENANGO SPECIAL SHOCK RESISTING STEEL
Cytemp Specialty Steel Div.
C 0.65, Mn 0.5, Si 1.1, V 0.2, Mo 0.5, bal Fe.
Oil hardened: 250,000-340,000 TS; 7-10 El; 10-14 Izod (notched); 50-59 Rock C. Good strength, hardness, shock resistance. For power shock tools, driver bits, knock out pins. *Obsolete*

VENIVICI
Stahlwerke Sudwestfalen
C 0.82, Cr 4, Mo 0.85, V 1.6, W 8.7, bal Fe.
For lathe and planer tools, reamers, broaches, taps; high speed steel. *Obsolete*

VENTOS 4631
Thyssen Edelstahlwerke AG
C 0.05, Cr 20, Ti 2.4, Al 1.4, bal Ni.
High temperature alloy, for valves. DIN NiCr20TiAl; Werkstoff Nr. 2.4631. *Obsolete*

VENTOS 4718
Thyssen Edelstahlwerke AG
C 0.45, Si 3, Cr 9, bal Fe.
For exhaust valves in hot going accelerator, and fuel oil motors. Martensitic, oil or air hardenable. Din X45CrSi93; Werkstoff Nr. 1.4718; formerly Ventos 11C. *Obsolete*

VENTOS 4718, ETC.
Thyssen Edelstahlwerke AG
See Werkstoff Nr. 1.4718.
Valve steels; 9 grades.

VENTOS 4721
Thyssen Edelstahlwerke AG
C 2.1, Cr 11.5, bal Fe.
For valves, valve seat rings in hot running motors. DIN X210Cr12; Werkstoff Nr. 1.4721. *Obsolete*

VENTOS 4732
Thyssen Edelstahlwerke AG
C 0.8, Si 2, Cr 15, Mo 1, W 1, bal Fe.
Valve steel. High temperature strength: 11 kg/mm2 at Martensitic, oil hardenable. DIN X80CrSiMoW152; Werkstoff Nr. 1.4732; formerly Ventos 16 CMW. *Obsolete*

VENTOS 4747
Thyssen Edelstahlwerke AG
C 0.8, Si 2.2, Cr 20, Ni 1.5, bal Fe.
For highly stressed exhaust valves in high speed motors. Martensitic, air hardenable. DIN X80CrNiSi20; Werkstoff Nr. 1.4747; formerly Ventos 23 C. *Obsolete*

VENTOS 4748
Thyssen Edelstahlwerke AG
C 0.85, Mn 17.5, Mo 2.2, V 0.45, bal Fe.
Valve steel. High temperature strength: 11 kg/mm2 at 700 C Martensitic, oil hardenable. DIN X85CrMoV182; Werkstoff Nr. 1.4748; formerly Ventos 20 CMV. *Obsolete*

VENTOS 4780
Thyssen Edelstahlwerke AG
C 0.45, Mn 7.5, Cr 23, Mo 2, Cu 0.7, N 0.3, Co 2, bal Fe.
Austenitic valve steel. DIN X45CrMnN238; Werkstoff Nr. 1.4780; formerly Ventos K. *Obsolete*

VENTOS 4790
Thyssen Edelstahlwerke AG
C 0.7, Mn 6, Cr 21, Ni 1.7, N 0.25, S 0.05, bal Fe.
Austenitic valve steel; formerly Ventos 25 CM. *Obsolete*

VENTOS 4871
Thyssen Edelstahlwerke AG
C 0.5, Mn 9, Cr 21, Ni 4, N 0.45, S 0.05. *Obsolete*

VENTOS 4873
Thyssen Edelstahlwerke AG
C 0.45, Si 2.5, Cr 18, Ni 9, W 1, bal Fe.
For highly stressed valves in flying and racing motors; austenitic. DIN X45CrNiW189; Werkstoff Nr. 1.4873; formerly Ventos CNW. *Obsolete*

VENTOS 4875
Thyssen Edelstahlwerke AG
C 0.55, Mn 8.5, Cr 20.5, Ni 2.2, N 0.3, S 0.05, bal Fe.
Austenitic valve steel. DIN X55CrMnNi208; Werkstoff Nr. 1.4875; formerly Ventos 22 N. *Obsolete*

VENTOS 4971
Thyssen Edelstahlwerke AG
C 0.12, Cr 20, Ni 20, Mo 3, Co 20, W 2.5, Nb 1.3, N 0.15, bal Fe.
High temperature alloy; for valves. DIN X12CrCoNi2120; Werkstoff Nr. 1.4971. *Obsolete*

VENTURELOY II
Venture Corp.
Ag alloy.
For electrical contacts; 75.0% electrical conductivity.

VEP
VILLARES
C 0.04, Cr 3.8, Mo 0.4, bal Fe.
Hot work or mold steel. Similar to AISI P-4; DIN 1.2341.

VERALOY GR. V
Veraloy Products Ltd.
WC.
For cutting tools; sintered carbide.

VERESTA
Thyssen Edelstahlwerke AG
C 1, Cr 1, W 1, bal Fe.
For cutters, form tools, heading dies; oil hardened, wear resistant. *Obsolete*

VERESTA SPECIAL
Thyssen Edelstahlwerke AG
C 1.4, Cr 1.4, bal Fe.
For thread cutters, chasers, taps, engravers tools; oil or water hardened, wear resistant. *Obsolete*

VERESTA-V
Marathon Specialty Steels Inc.
C 0.9, Mn 1.2, V 0.2, W 0.5, Cr 0.5, bal Fe.
Annealed: 84,000 TS; 60,000 YS; 26 El; 185 Brin. Heat treated: 145,000-280,000 TS; 125,000-272,000 YS; 2-8 El; 2-31 RA; 290-535 Brin. For blanking and bending dies, master tools, knurling tools, punches, cutters, master dies, gauges. Type O1, oil hardening, cold work tool steel, shock resistant. *Obsolete*

VERESTA-V
Thyssen Edelstahlwerke AG
C 0.9, Mn 1.2, V 0.2, W 0.5, Cr 0.5, bal Fe.
Annealed: 84,000 TS; 60,000 YS; 26 El; 185 Brin. Heat treated: 145,000-280,000 TS; 125,000-272,000 YS; 2-8 El; 2-31 RA; 290-535 Brin. For blanking and bending dies, master tools, knurling tools, punches, cutters, master dies, gauges. Type O1, oil hardening, cold work tool steel, shock resistant. *Obsolete*

VERIBEST DRILL ROD
Diehl Steel Co.
C 0.9, Mn 1.1, Cr 0.6, W 0.9, V 0.24, bal Fe.
For dies, tools, punches; oil hardening drill rod.

VERILITE-1
Verlite Metals Co.
Aluminum. Ni 0.3, Cu 2.5, Fe 0.7, Si 0.4, bal Al.
Cast: 16,000 TS; 4 El; 4 RA. For aircraft cylinder heads; age-hardenable.

VERILITE-2
Verlite Metals Co.
Aluminum. Ni 1.5, Cu 1, Cr 1.5, Mn 0.5, bal Al.
For aircraft cylinder heads; age-hardenable.

VERILOY
Driver Harris Co.
C 0.35-0.45, Ni 20-22, Cr 10-12, bal Fe.
For chemical engineering equipment, furnace parts; heat and corrosion resistant to 1500 F. *Obsolete*

VERITAS
J.M. Ney Co.
Au 40, Pd 44.7, Ag 5.
White dental casting alloy for metal-ceramic reconstructions. 107,220 psi TS, 60,858 psi YS (0.1% offset); 40 El; 232 HV.

VERNALLOY
Mt. Vernon Furnace & Mfg. Co.
C 0.2, Cr 20, Ni 10, bal Fe.
For heat resistant parts; heat resistant.

VERNIAL
Vereinigte Deutsche Nickel-Werke AG
Mg 0.6-1.4, Si 0.6-1.6, Mn 0.6-1, Cr 0-0.3, bal Al.
Annealed: 21,000 TS; 8,000 YS; 24 El. For window frames, fan blades, gutters, boats; good forming and welding properties. *Obsolete*

VERNIAL
Vereinigte Deutsche Nickel-Werke AG
Aluminum.
Aluminum alloy AlMgSi1.

VERNICON
Vereinigte Deutsche Nickel-Werke AG
Resistance alloy (CuNi 44). *Obsolete*

VERNICORR
Vereinigte Deutsche Nickel-Werke AG
Aluminum.
Aluminum alloy AlMn.

VERNIDUR
Vereinigte Deutsche Nickel-Werke AG
Cu 2.5-5, Mg 0.2-1.8, Mn 0.3-1.5, bal Al.
Annealed: 27,000 TS; 11,000 YS; 22 El; 47 Brin. Heat treated: 72,000 TS; 57,000 YS; 130 Brin. For aircraft structures and fittings, fasteners; age-hardenable. *Obsolete*

VERNIDUR
Vereinigte Deutsche Nickel-Werke AG
Aluminum.
Aluminum alloy AlCuMg.

VERNIKORR
Vereinigte Deutsche Nickel-Werke AG
Mn 0.5-1.5, Cr 0-0.3, bal Al.
Soft: 16,000 TS; 6,000 YS; 40 El. Hard: 29,000 TS; 27,000 YS; 10 El. For cooking utensils, heat exchangers, tanks, furniture; good forming and welding properties. *Obsolete*

VERNIMAG M
Vereinigte Deutsche Nickel-Werke AG
Mg 1.5-3, Mn 0.5-1.5, Cr 0-0.3, bal Al.
Soft: 26,000 TS; 10,000 YS; 20 El; 45 Brin. Hard: 41,000 TS; 36,000 YS; 5 El; 77 Brin. For roofing, hydraulic tubing, architectural trim; good forming and welding properties. *Obsolete*

VERNISIL 12
Vereinigte Deutsche Nickel-Werke AG
Copper. Cu 04, 12 Ni + Co, bal Zn.
Wire and bar for electronics, optical parts, fasteners and household cooking products. C75700

VERNISIL 15
Vereinigte Deutsche Nickel-Werke AG
Ni 14-16, Cu 62-64, bal Zn.
For hardware, cutlery. Nickel silver, corrosion resistant.
Obsolete

VERNISIL 18
Vereinigte Deutsche Nickel-Werke AG
Copper. Cu 62, 18 Ni + Co, bal Zn.
Wire and bar for electronics, optical parts, fasteners and
household cooking products. C75900

VERNISIL 183F
Vereinigte Deutsche Nickel-Werke AG
Ni 17-19, Cu 54-56, bal Zn.
For hardware, cutlery. Nickel silver, corrosion resistant.
Obsolete

VERNISIL 25
Vereinigte Deutsche Nickel-Werke AG
Ni 25, Zn 15, bal Cu.
Annealed: 65,000 TS; 40,000 YS; 42 El; 130 Brin. For
bathroom fixtures, marine hardware. Nickel silver, corrosion
resistant. *Obsolete*

VERONICA B 85
Kind & Co. Edelstahlwerk
C 1, Cr 4, W 1.8, Mo 8.5, V 1.8, Co 5, bal Fe.
High speed steel. W. Nr. 3248.

VERONICA B 88
Kind & Co. Edelstahlwerk
C 1, Cr 4, W 1.8, Mo 8.5, V 1.8, Co 8, bal Fe.
High speed steel. W. Nr. 3249.

VERONICA B 92
Kind & Co. Edelstahlwerk
C 0.8, Cr 4, W 1.7, Mo 8.8, V 1.2, bal Fe.
High speed steel. W. Nr. 3346.

VERONICA B 94
Kind & Co. Edelstahlwerk
C 1, Cr 4, W 1.8, Mo 8.8, V 1.8, bal Fe.
High speed steel. W. Nr. 3348.

VERONICA B 96
Kind & Co. Edelstahlwerk
C 1.15, Cr 4.2, W 2, Mo 9, V 3, bal Fe.
High speed steel.

VERONICA B 98
Kind & Co. Edelstahlwerk
C 1.1, Cr 4, W 1.5, Mo 9.5, V 1.2, Co 8, bal Fe.
High speed steel. W. Nr. 3247.

VERONICA C 33
Kind & Co. Edelstahlwerk
C 1, Cr 4.2, W 3, Mo 2.7, V 2.4, bal Fe.
For lathe and planer tools, hobs, drills, taps. High speed
steel. W. Nr. 3337.

VERONICA C 65
Kind & Co. Edelstahlwerk
C 0.9, Cr 4.2, W 6.5, Mo 5, V 1.9, bal Fe.
For lathe and planer tools, hobs, reamers. High speed steel.
W. Nr. 3343.

VERONICA C 65 H
Kind & Co. Edelstahlwerk
C 0.98, Cr 4.2, W 6.5, Mo 5, V 1.9, bal Fe.
High speed steel. W. Nr. 3342.

VERONICA C 653 VA
Kind & Co. Edelstahlwerk
C 1.25, Cr 4.2, W 6.5, Mo 5, V 3, bal Fe.
High speed steel. W. Nr. 3344.

VERONICA C 655 CO
Kind & Co. Edelstahlwerk
C 0.92, Cr 4.2, W 6.5, Mo 5, V 1.8, Co 5, bal Fe.
High speed steel. W. Nr. 3243.

VERONICA C 655 CO H
Kind & Co. Edelstahlwerk
C 1.1, Cr 4.2, W 6.8, Mo 5, V 1.8, Co 5, bal Fe.
High speed steel.

VERONICA C 658 CO H
Kind & Co. Edelstahlwerk
C 1.45, Cr 4.2, W 6.5, Mo 5, V 2.3, Co 8, bal Fe.
High speed steel. W. Nr. 3222.

VERONICA C 745 CO H
Kind & Co. Edelstahlwerk
C 1.1, Cr 4.2, W 7, Mo 4, V 1.8, Co 5, bal Fe.
High speed steel. W. Nr. 3246.

VERONICA EXTRA
Kind & Co. Edelstahlwerk
C 0.75, Cr 4.2, V 1.2, W 18, bal Fe.
For lathe and planer tools, reamers, broaches, taps. High
speed steel. W. Nr. 3355.

VERONICA EXTRA V 2
Kind & Co. Edelstahlwerk
C 0.85, Cr 4.2, W 18, V 2, bal Fe.
High speed steel. W. Nr. 3357.

VERONICA FSR
Kind & Co. Edelstahlwerk
C 1.2, Cr 11.5, W 2.4, Mo 1.4, V 1.7, bal Fe.
High speed steel.

VERONICA HOCHLEISTUNG 1000
Kind & Co. Edelstahlwerk
C 0.95, Cr 4.2, Mo 1, V 2.5, W 12, bal Fe.
For lathe and planer tools, reamers, hobs, drills. High speed
steel. W. Nr. 3318.

VERONICA HOCHLEISTUNG DIAMANT
Kind & Co. Edelstahlwerk
C 0.65, Cr 4.2, W 18, Mo 0.8, V 1.5, Co 16, bal Fe.
High speed steel. W. Nr. 3257.

VERONICA HOCHLEISTUNG GOLD
Kind & Co. Edelstahlwerk
C 0.75, Cr 4.2, Mo 0.7, V 1.6, W 18, Co 10, bal Fe.
For lathe and planer tools, hobs, broaches, taps. High speed
steel. W. Nr. 3265.

VERONICA HOCHLEISTUNG K 3
Kind & Co. Edelstahlwerk
C 0.85, Co 3, Cr 4.2, Mo 1, V 1.9, W 12, bal Fe.
For lathe and planer tools, reamers, taps, drills. High speed
steel. W. Nr. 3211.

VERONICA HOCHLEISTUNG SILBER
Kind & Co. Edelstahlwerk
C 0.8, Co 5, Cr 4.2, V 1.6, W 18, Mo 0.7, bal Fe.
For lathe and planer tools, reamers, broaches, hobs. High
speed steel. W. Nr. 3255.

VERONICA HOCHLEISTUNG SILBER 4 V
Kind & Co. Edelstahlwerk
C 1.35, Cr 4.2, W 12, Mo 0.9, V 3.8, Co 5, bal Fe.
For engravers' tools, blanking dies. High speed steel. W. Nr.
3202.

VERONICA HOCHLEISTUNG VS 10
Kind & Co. Edelstahlwerk
C 1.25, Cr 4.2, Mo 1, V 4, W 12, bal Fe.
For engravers' tools, blanking and forming dies. High speed
steel. W. Nr. 3302.

VERONICA SPEZIAL
Kind & Co. Edelstahlwerk
C 0.8, Cr 4.2, Mo 0.8, V 1.7, W 9, bal Fe.
For lathe and planer tools, reamers, hobs, drills. High speed
steel. W. Nr. 3316.

VERONICA SPEZIAL RG
Kind & Co. Edelstahlwerk
C 0.85, Cr 4.2, W 12, Mo 1, V 1.7, bal Fe.
High speed steel.

VERONICA SUPER
Kind & Co. Edelstahlwerk
C 1.3, Cr 4.2, W 10, Mo 3.8, V 3.5, Co 10.5, bal Fe.
High speed steel. W. Nr. 3207.

VERONICA SUPER SPEZIAL
Kind & Co. Edelstahlwerk
C 1.45, Cr 4.2, W 10, Mo 3.5, V 4, Co 12, bal Fe.
High speed steel.

VEROTEC 19666
Eutectic Corp.
Alloy powder for metal spraying thick build-ups; will accept
final coat.

VERSAL
Versevorder Metallwerk
Aluminum.
Aluminum alloy AlMg.

VERSALLOY
Atlas Specialty Steels
C 0.11, Cr 17, Ni 8, bal Fe.
Austenitic stainless steel. *Obsolete*

VERSASTEEL
Crucible Materials Corp.
Tool material. C 1, Mn 0.3, Si 2, Cr 4.25, V 1.15, W 0.3, Mo
2.5, bal Fe.
Air hardening tool steel.

VERSATILE
Osborn Steels Ltd.
C 0.4, Cr 1, Ni 1.5, bal Fe.
For gears, bolts, crankshafts, axles; oil hardened, shock
resistant. *Obsolete*

VERTOMAR 1
Arcos Alloys
C 0-0.015, bal Fe (ingot iron).
For electroslag welding wire; extra low carbon. *Obsolete*

VERTOMAR 10
Arcos Alloys
C 0.07-0.13, Mn 0.4-0.7, bal Fe.
For electroslag welding wire. *Obsolete*

VERTOMAR 15
Arcos Alloys
C 0.13-0.19, Mn 0.95-1.3, Si 0.15-0.3, bal Fe.
For electroslag welding wire; 1% manganese. *Obsolete*

VERTOMAR 2M
Arcos Alloys
C 0.11-0.17, Mn 1.75-2.1, bal Fe.
For electroslag welding wire; 2% manganese steel. *Obsolete*

VERTOMAR 2MM
Arcos Alloys
C 0.11-0.17, Mn 1.75-2.1, Mo 0.4-0.6, bal Fe.
For electroslag welding wire; 2% manganese, 0.5%
molybdenum steel. *Obsolete*

VERTOMAR 410 NI
Arcos Alloys
C 0.04, Mn 0.6, Si 0.4, Cr 12, Ni 4, Mo 0.8, bal Fe.
For electroslag welding wire; welding CA6NM castings.
Obsolete

VERTOMAR 6
Arcos Alloys
C 0-0.06, Mn 0-0.15, bal Fe.
For electroslag welding wire. *Obsolete*

VERTOMAR 60
Arcos Alloys
C 0.55-0.66, Mn 0.9-1.25, Si 0.1-0.2, bal Fe.
For electroslag welding wire; medium carbon steel. *Obsolete*

VERTOMAR 8
Arcos Alloys
C 0-0.1, Mn 0.35-0.65, Si 0.1-2, bal Fe.
For electroslag welding wire; silicon killed. *Obsolete*

VERTOMAR CM30
Arcos Alloys
C 0.28-0.33, Mn 0.4-0.6, Si 0.2-0.35, Cr 0.8-1.1, Mo 0.15-0.25, bal Fe.
For electroslag welding wire; AISI type 4130 steel. *Obsolete*

VERTOMAR CNM20
Arcos Alloys
C 0.18-0.23, Mn 0.7-0.9, Si 0.2-0.35, Cr 0.4-0.6, Ni 0.4-0.7, Mo 0.15-0.25, bal Fe.
For electroslag welding wire; AISI type 8620. *Obsolete*

VERTOMAR CV50
Arcos Alloys
C 0.48-0.53, Mn 0.7-0.9, Si 0.2-0.35, Cr 0.8-1.1, 0.15 V min, bal Fe.
For electroslag welding wire; AISI type 6150. *Obsolete*

VERY BEST
Boyd-Wagner Co.
C 1.05, Mn 0.3, Cr 0.5, V 0.1, bal Fe.
Annealed: 185 Brin. For dies and tools, drawing and forming dies, taps, drills, mandrels; tough and hard; water hardened.

VERY BEST
Bissel Steel Co.
C 0.8-1.2, V 0.2, bal Fe.
For cutting tools, drills; water hardened. *Obsolete*

VES
Stahlwerke Sudwestfalen
C 1, V 0.1, Mn 0.25, Si 0.2, bal Fe.
For drills, taps, springs, tools, cutters; Type W2; water hardened. *Obsolete*

VESTALIN
Manufacturer not listed.
Ni 28, C, bal Fe.
For electrical resistances; resistance alloy.

VESTEEL
British Steel plc
Mild steel. For vitreous enameling.

VESUVIUS
Firth-Vickers Stainless Steels Ltd.
C 0.1, Cr 30, Ni 1.7, bal Fe.
Oil treated: 78,000-101,000 TS; 58,000-85,000 YS; 15-25 El; 30-40 RA; 175-225 Brin. For furnace parts, fire bars, stokers, grids; heat and corrosion resistant; Firth Brown J-182.

VET-3
VILLARES
Tool steel. C 0.7, bal Fe.
Water hardening tool steel. AISI W 1-7; similar to DIN 1.1820.

VETD
VILLARES
Tool steel. C 1, V 0.1, bal Fe.
Water hardening tool steel. AISI W 2-9; similar to DIN 1.1640.

VEW 505G
Bohler Gesellschaft M.B.H.
Stainless steel. C 0.1, Cr 18, Ni 8.5, bal Fe.
Austenitic stainless steel casting; for chemical plant equipment. W. Nr. 1.4312. *Obsolete*

VEW A100G
Bohler Gesellschaft M.B.H.
Stainless steel. C 0-0.07, Cr 17.5, Mo 2.75, Ni 12.5, bal Fe.
Austenitic stainless steel casting for corrosion resistant containers up to 300°C. W. Nr. 1.4437. *Obsolete*

VEW A100R
Bohler Gesellschaft M.B.H.
Stainless steel.
Now BOHLER A101.

VEW A102R
Now VEW A102.

VEW A114
Now VEW A120.

VEW A120 G
Bohler Gesellschaft M.B.H.
Stainless steel. C 0-0.07, Si 0-2, Mn 0-1.5, Cr 18.5, Mo 2.25, Ni 11, bal Fe.
Austenitic stainless casting; for pumps, centrifuges. W. Nr. 1.4408. *Obsolete*

VEW A120R
Now VEW A120.

VEW A121
Now VEW A120G.

VEW A350 G
Bohler Gesellschaft M.B.H.
Stainless steel. C 0.08, Si 0-1.5, Mn 0-1.5, Cr 18.5, Mo 2.25, Ni 11.5, Nb = 8 x C, bal Fe.
Austenitic stainless casting; weldable; for armatures, pumps, housings. W. Nr. 1.4581; ACI CF-8M. *Obsolete*

VEW A354G
Bohler Gesellschaft M.B.H.
Stainless steel. C 0-0.1, Si 0-1, Mn 0-2, Cr 17.5, Mo 2.75, Ni 13, Nb = 8 x C, bal Fe.
Austenitic stainless casting; weldable; for containers in chemical industries. W. Nr. 1.4583. *Obsolete*

VEW A500G
Bohler Gesellschaft M.B.H.
Stainless steel. C 0-0.07, Si 0-2, Mn 0-1.5, Cr 18.5, Ni 10, bal Fe.
Austenitic stainless steel casting; for armatures, housings, pumps. W. Nr. 1.4308. *Obsolete*

VEW A750G
Bohler Gesellschaft M.B.H.
Stainless steel. C 0-0.08, Si 0-1.5, N 0-1.5, Cr 18.5, Ni 10, Nb = 8 x C, bal Fe.
Welded austenitic stainless steel casting for equipment for food, paper and textile industries. W. Nr. 1.4552. *Obsolete*

VEW ALLOY
Bohler Gesellschaft M.B.H.
Stainless steel.
VEW alloys are now known as BOHLER alloys.

VEW B110
Bohler Gesellschaft M.B.H.
Tool material. C 0.85, Mn 0.65, Si 0.4, Cr 0.4, bal Fe.
Cold work tool steel. For cutting tools, gages, machine knives. W. Nr. 1.2004. *Obsolete*

VEW B112
Bohler Gesellschaft M.B.H.
Tool material. C 0.75, Mn 0.6, Cr 0.35, bal Fe.
Cold work tool steel for small tools such as punches, mandrels, stamping tools. W. Nr. 1.2003. *Obsolete*

VEW B114
Bohler Gesellschaft M.B.H.
Tool material. C 1.25, Mn 0.35, Cr 0.35, bal Fe.
Water hardening tool steel for punches, cold chisels, mandrels. W. Nr. 1.2002. *Obsolete*

VEW B116
Bohler Gesellschaft M.B.H.
Tool material. C 0.85, Mn 0.4, Cr 0.4, bal Fe.
Cold work tool steel; for gauges, knives. W. Nr. 1.2004. *Obsolete*

VEW B306
Bohler Gesellschaft M.B.H.
Tool material. C 1.15, Cr 0.2, W 2, bal Fe.
Cold work tool steel; for metal cutting saws and hack saw blades. W. Nr. 1.2442. *Obsolete*

VEW B404
Bohler Gesellschaft M.B.H.
Tool material.
Now BOHLER B406.

VEW B535
Bohler Gesellschaft M.B.H.
Tool material. C 0.75, Mn 0.4, Cr 0.25, Ni 0.55, bal Fe.
Water hardening cold work tool steel for hand tools. W. Nr. 1.2703. *Obsolete*

VEW B908
Bohler Gesellschaft M.B.H.
Tool material. C 0.78, Mn 0.7, bal Fe.
Water hardening tool steel for mandrels, collets. W. Nr. 1.1750. *Obsolete*

VEW D330G
Bohler Gesellschaft M.B.H.
C 0.22, Mn 0.7, Cr 1, Mo 0.45, bal Fe.
Oil or water hardening; for elevated temperature operation W. Nr. 1.7354. *Obsolete*

VEW F14
Bohler Gesellschaft M.B.H.
C 0.45, Si 1.65, Mn 0.7, bal Fe.
Silicon-manganese spring steel; oil hardening. Laminated springs, or elliptical springs for vehicles. W. Nr. 1.0902. *Obsolete*

VEW H120G
Bohler Gesellschaft M.B.H.
Stainless steel. C 0.45, Si 2, Mn 0-1, Cr 17, bal Fe.
Stainless cast steel for elevated temperature operations. W. Nr. 1.4740. *Obsolete*

VEW H160G
Bohler Gesellschaft M.B.H.
C 0.3, Si 2, Mn 0-1, Cr 7, bal Fe.
Cast steel for elevated temperature operation as burner parts, tempering furnaces. W. Nr. 1.4710. *Obsolete*

VEW H300G
Bohler Gesellschaft M.B.H.
Stainless steel. C 0.4, Si 1.5, Mn 0-1.5, Cr 27, Ni 4.5, bal Fe.
Stainless casting for elevated temperature operations. W. Nr. 1.4823. *Obsolete*

VEW H520G
Bohler Gesellschaft M.B.H.
Stainless steel. C 0.3, Si 2, Mn 0-1.5, Cr 18, Ni 37, bal Fe.
Stainless casting for high temperature parts such as furnace parts. W. Nr. 1.4865. *Obsolete*

VEW H521
Now VEW H520G.

VEW H527
Bohler Gesellschaft M.B.H.
Stainless steel. C 0.15, Si 1.5, Mn 0-1.5, Cr 25, Ni 20, bal Fe.
Stainless casting for elevated temperature operation. W. Nr. 1.4849. *Obsolete*

VEW H529
Bohler Gesellschaft M.B.H.
Stainless steel. C 0.35, Si 2, Mn 0-1.5, Cr 25, Ni 20, bal Fe.
Stainless casting for elevated temperature operation. W. Nr.
1.4848; ACI HK. *Obsolete*

VEW H537
Bohler Gesellschaft M.B.H.
Stainless steel. C 0.4, Si 2, Cr 25, Ni 13, bal Fe.
Stainless casting for furnace parts, heat treat boxes, pump
bodies. W. Nr. 1.4837; ACI HI. *Obsolete*

VEW H539
Bohler Gesellschaft M.B.H.
Stainless steel. C 0.35, S 2, Mn 0-1.5, Cr 26, Ni 14, bal Fe.
Stainless casting for elevated temperature operation. W. Nr.
1.4846. *Obsolete*

VEW H551
Bohler Gesellschaft M.B.H.
Stainless steel. C 0.4, Si 2, Cr 22, Ni 9.5, bal Fe.
Stainless steel casting for elevated temperature operation. W.
Nr. 1.4826; ACI HF. *Obsolete*

VEW K707
Bohler Gesellschaft M.B.H.
C 1.2, Mn 12.5, bal Fe.
Manganese abrasion resisting steel casting, for rails,
crushers, shovel teeth, drag lines. W. Nr. 1.3401. *Obsolete*

VEW K708
Now K707.

VEW N350G
Bohler Gesellschaft M.B.H.
C 0.23, Cr 17, Ni 1.5, bal Fe.
Corrosion resistant steel castings; for valve guides,
springfaces, spindles. W. Nr. 1.4059. *Obsolete*

VEW N358
Now VEW N359.

VEW N691
Bohler Gesellschaft M.B.H.
Stainless steel. C 1.1, Si 0-2, Mn 0-1, Cr 28, Mo 2.25, bal Fe.
Stainless steel casting; for paper and photo industries. W. Nr.
1.4138. *Obsolete*

VEW N693
Bohler Gesellschaft M.B.H.
C 1.1, Si 0-2, Mn 0-1, Cr 28, bal Fe.
Corrosion resistant casting; good resistance to heat and
abrasion. W. Nr. 1.4086. *Obsolete*

VEW P752
Bohler Gesellschaft M.B.H.
C 0-0.05, Ni 26, Al 14, bal Fe.
Cast permanent magnet. W. Nr. 1.3728. *Obsolete*

VEW P754
Bohler Gesellschaft M.B.H.
C 0.05, Co 15, Ni 20, Al 9.5, Cu 3.5, bal Fe.
Cast permanent magnet. W. Nr. 1.3743; ALNICO 160.
Obsolete

VEW P756
Bohler Gesellschaft M.B.H.
C 0.05, Co 15, Ni 22, Al 10, Cu 3, bal Fe.
Cast permanent magnet. W. Nr. 1.3745; ALNICO 190.
Obsolete

VEW R102
Bohler Gesellschaft M.B.H.
Similar to VEW R100. *Obsolete*

VEW R104
Bohler Gesellschaft M.B.H.
Similar to VEW R100. *Obsolete*

VEW S601, S602
Now VEW S600.

VEW S609
Now VEW S610.

VEW V322
Now VEW 320.

VEW V340G
Bohler Gesellschaft M.B.H.
Alloy steel. C 0.25, Mn 0.65, Cr 1, Mo 0.2, bal Fe.
Alloy steel casting. W. Nr. 1.7218. *Obsolete*

VEW V924
Now VEW V924.

VEW V937
Now VEW 936.

VEW Z904
Bohler Gesellschaft M.B.H.
C 0-0.13, Mn 0.9, P 0.1, S 0.22, bal Fe.
Free machining carbon steel for automatic screw machine
operations. W. Nr. 1.0711; similar to AISI 1212. *Obsolete*

VEW Z906
Bohler Gesellschaft M.B.H.
Carbon steel. C 0-0.14, Mn 1.1, P 0.1, S 0.24-0.32, bal Fe.
Free machining carbon steel for automatic screw machine
operations. W. Nr. 1.0715; similar to AISI 1213. *Obsolete*

VEW Z908
Bohler Gesellschaft M.B.H.
Carbon steel. C 0-0.15, Mn 1.25, P 0.1, S 0.36, bal Fe.
Free machining carbon steel for automatic screw machine
operations. W. Nr. 1.0736; DIN 9 SM 36; similar to old SAE
B1113. *Obsolete*

VEW Z952
Bohler Gesellschaft M.B.H.
C 0-0.15, Mn 1.25, P 0.1, S 0.36, Pb 0.22, bal Fe.
Leaded, free machining steel. W. Nr. 1.0737; similar to AISI
12L14. *Obsolete*

VEW Z980
Bohler Gesellschaft M.B.H.
C 0.1, Mn 0.7, P 0.06, S 0.2, bal Fe.
Free machining steel, for case hardening. W. Nr. 1.0721;
similar to AISI 1212. *Obsolete*

VEW Z982
Bohler Gesellschaft M.B.H.
C 0.36, Mn 0.7, P 0-0.06, S 0.2, bal Fe.
Free machining, heat treatable steel. W. Nr. 1.0726. *Obsolete*

VEW Z984
Bohler Gesellschaft M.B.H.
C 0.36, Mn 0.7, P 0.07, S 0.2, Pb 0.22, bal Fe.
Leaded, free machining steel. W. Nr. 1.0756. *Obsolete*

VEW Z986
Bohler Gesellschaft M.B.H.
C 0.46, Mn 0.7, P 0-0.06, S 0.2, bal Fe.
Free machining, heat treatable steel. W. Nr. 1.0727. *Obsolete*

VEW Z988
Bohler Gesellschaft M.B.H.
C 0.6, Mn 0.7, P 0-0.06, S 0.2, bal Fe.
Free machining, heat treatable steel. W. Nr. 1.0728. *Obsolete*

VFC
VILLARES
C 0.15, Si 0.25, Mn 0.75, bal Fe.
Low carbon steel. AISI 1015; DIN CK 15. *Obsolete*

VH 13
VILLARES
C 0.4, Cr 5, Mo 1.5, V 1, Si 1.
AISI H13; similar to DIN 1.2344.

VH-63
S.K.F. Industries Inc.
C 2.25, Cr 12, V 0.15, Mn 1.2, Si 1.2, bal Fe.
For cold trimming and blanking dies, forming dies, gauges;
oil hardened, non-deforming. *Obsolete*

VI CHROME
Ackerlind Steel Co., Inc.
C 2, Mn 0.3, Cr 12-14, V 0.2, bal Fe.
For dies. Non-deforming; oil hardening; AISI D4.

VI-CHROME "W"
Ackerlind Steel Co., Inc.
C 2, Cr 13, W 1.2, bal Fe.
Air or oil hardening cold work tool and die steel; chromium
type; AISI D6.

VIADUCT 15
Darwins Alloy Castings
C 0.45, Cr 1, W 2, V 0.3, Si 0.55, Mn 0.45, bal Fe.
Annealed: 220 Brin. For pneumatic chisels, shear blades,
cutters; tough, oil hardened. *Obsolete*

VIAG
VAW Vereinigte Aluminium-Werke AG
Aluminum. Cu 4.5-5, bal Al.
For light alloy parts; age-hardenable.

VIBRAC STEEL
British Steel Corp.
C 0.4, Ni 1.5, Cr 0.9, bal Fe.
Heat treated: 135,000 TS; 23 El; 59 RA. For steel cylinders oil
hardened. *Obsolete*

VIBRAC V-30
British Steel Corp.
C 0.27-0.35, Mn 0.6, Ni 2.3-2.8, Cr 0.6, Mo 0.6, bal Fe.
Oil treated: 123,000-230,000 TS; 195,000 YS; 12-22 El; 40-60
RA; 477 Brin. For tool holders, engine parts, die blocks, crank
shafts; shock resisting; free from temper brittleness. *Obsolete*

VIBRAC V-45
British Steel Corp.
C 0.45, Cr 0.6, Ni 2.5, Mo 0.6, bal Fe.
Oil treated: 130,000-270,000 TS; 225,000 YS; 11-22 El; 37-60
RA; 514 Brin. For tool holders, gear wheels, axles, die blocks,
shafts; shock resisting IZ 70. *Obsolete*

VIBRALLOY
Arnold Engineering Co.
Mo 9, Ni 38-42, bal Fe.
For instruments, diaphragms, mechanical filters, vibrating
reeds; constant modulus, high permeability.

VIBRALLOY
Western Electric Co.
Mo 9, Ni 38-42, bal Fe.
For instruments, diaphragms, mechanical filters, vibrating
reeds; constant modulus, high permeability.

VIBRALOY
Audubon Metalwove Belt Corp.
C 0.9, Cr 0.8, bal Fe.
For vibrating screens; abrasion resistant.

VIBRESIST
Atlas Specialty Steels
C 1, V 1.25, Mo 0.3, bal Fe.
For mining drills; oil hardened, hollow drill.

VIBRO
CCS Braeburn Alloy Steel
C 0.5, Cr 1.4, V 0.3, W 1.9, bal Fe.
For tools, dies. Hot work steel. *Obsolete*

VICALLOY

Wilbur B. Driver Co.
React and refract. V 10, Co 52, bal Fe.
For hysteresis motors, magnetic clutches, record tape, magnetic memory, magnets. Saturation induction 10,000 Gausses. coercive force 200 oersteds, magnetically semihard.

VICALLOY

Teledyne Vasco
V 9, Fe 38.1, Mn 0.7, bal Co.
Ductile Co-Fe-V permanent magnet material. 9000 Brin; coercive force 300 oersted, energy product 1.0 x 10^6 max. Designed for magnetic recording tapes.

VICALLOY 1

Telcon Metals Ltd.
Co-Fe-V.
Permanent magnet alloy for rotor assemblies, etc.

VICALLOY 2

Telcon Metals Ltd.
Co-Fe-V
Permanent magnet alloy for rotor assemblies, etc.

VICALLOY I

Bell Telephone Laboratories
V 10, Co 52, Mn 0.3, W 1.9, Si 0.6, bal Fe.
For permanent magnets, cold workable. 900 Brin; 300 H$_c$; 5500 Bo; 60 Rock C.

VICALLOY I

Arnold Engineering Co.
V 10, Co 52, bal Fe.
Permanent magnet material. Residual flux density 7500 gauss; coercive force 250 oersted; peak energy product 0.80 MGO max.

VICALLOY VS30

Krupp Stahl AG
iron, Co 60, Cr 15, bal Fe.
Coercive force-25, remanence 18,000. For electrical and magnetic equipment. Permanent magnet. High magnetic permeability. *Obsolete*

VICKERS B.C.T.

British Steel Corp.
C 0.55-0.65, Si 0.2-0.3, Mn 0.5-0.8, Cr 0.43-0.7, bal Fe.
Annealed: 55,000-100,000 TS; 7-15 El. Hardened: 120,000-140,000 TS; 80,000 YS; 15 El; 40 RA; 255-341 Brin. *Obsolete*

VICKERS B.C.T.

British Steel Corp.
C 0.55-0.65, Si 0.2-0.3, Mn 0.5-0.8, Cr 0.43-0.7, bal Fe.
Annealed: 55,000-100,000 TS; 7-15 El. Hardened: 120,000-140,000 TS; 80,000 YS; 15 El; 40 RA; 255-341 Brin. *Obsolete*

VICTOR BRONZE

American manufacture
Zn 39, Al 1.5, Fe 1, V 0.03, bal Cu.
For pipes; corrosion resistant.

VICTOR DRILL ROD

Crucible Materials Corp.
Tool material. C 1.05, bal Fe.
For shafts, rollers, pins, dowels, push rods; water hardened.

VICTOR HI-CHROME A

Ackerlind Steel Co., Inc.
C 2.25, Mn 0.4, Cr 12-14, V 0.2, bal Fe.
For dies, broaches, lamination and drawing dies, reamers; oil hardening. *Obsolete*

VICTOR HI-CHROME B

Ackerlind Steel Co., Inc.
C 1.5, Cr 13, bal Fe.
For tools, cutters, dies, punches; nondeforming. *Obsolete*

VICTOR METAL

Manufacturer not listed.
Cu 50, Zn 35, Ni 15.
For cast fittings; corrosion resisting.

VICTORIA ALUMINUM

English manufacture
Cu, Zn, Si, Fe, bal Al.
For light alloy parts; v. "Partinium."

VICTORIEUX SAINT JUERY

Societe Nouvelle du Saut-du-Tarn
C 0.7, W 9.5, Cr 2.5, V 0.1, bal Fe.
For hot work dies and tools; hot work steel. *Obsolete*

VICTORY

Lehigh Steel Corp.
C 0.8, Cr 4, V 2, W 6, Mo 5, bal Fe.
For drills, hobs, taps, broaches, reamers; Type M2; high speed steel.

VICTORY COBALT

Teledyne Vasco
C 0.85, W 6, Mo 5, Cr 4, V 2, Co 8.5, bal Fe.
For cutting tools; high speed steel. Type M36.

VICULOY NO. 1

Akron Bronze & Aluminum Inc.
Be 2-2.2, bal Cu.
Cast: 85,000-90,000 TS; 60,000-70,000 YS; 4-12 El; 9 RA; 200-220 Brin. For gears, shafting welding dies and welding electrodes; corrosion resistant, age hardening.

VICULOY NO. 2

Akron Bronze & Aluminum Inc.
Be 1.6-1.8, bal Cu.
Cast: 45,000-50,000 TS; 30,000-35,000 YS; 17-22 El; 40 RA; 120-140 Brin. For circuit breakers, contacts, electrode jaws; corrosion resistant, age hardenable.

VICULOY NO. 3

Akron Bronze & Aluminum Inc.
Be 1.8-2.2, bal Cu.
Heat treated: 160,000-170,000 TS; 120,000 YS; 2-8 El; 4 RA; 370-400 Brin. For gears, die molds, rocker arms, non-sparking tools; corrosion resistant, age hardenable.

VIENNESE ORNAMENTS

Austrian manufacture
Cu 55, Zn 25, Ni 20.
For brass ornaments; corrosion resistant.

VIENNESE SHEET

Austrian manufacture
Cu 60, Zn 20, Ni 20.
For ornaments, hardware; corrosion resistant.

VIENNESE TABLEWARE

Manufacturer not listed.
Cu 50, Zn 25, Ni 25.
For cutlery, ornaments; corrosion resistant.

VIGILANT

Joseph Beardshaw & Son Ltd.
C 0.5-1, bal Fe.
For chisels, hard tools, springs; water hardened.

VIKING

CCS Braeburn Alloy Steel
C 1.05, Cr 1.05, Mo 0.45, bal Fe.
For mandrels, rams. Oil hardened. *Obsolete*

VIKING

Uddeholm Corp.
C 0.5, Si 1, Mn 0.5, Cr 8, Mo 1.5, V 0.5, bal Fe.
300,000 psi TS; 250,000 psi YP; 6 El; 15 RA; 58 Rock C. High alloy tool steel. For blanking and forming, dies, rolls and tools for tube drawing.

VIKING EXTRA

CCS Braeburn Alloy Steel
C 1.05-1.15, Cr 1.25-1.4, Mo 0.3-0.5, bal Fe.
For tools, cutters, rollers for roller bearings, thread rolling dies; oil hardening. *Obsolete*

VIKMANSHYTTAN VH217

Vikmanshyttan AG
C 0.19-0.25, Cr 13, bal Fe.
Annealed: 05,000 TS; 50,000 YS; 25 El; 55 RA; 195 Brin. For valves, cutlery, surgical and dental instruments; Type 420; stainless.

VIKMANSHYTTAN VH273

Vikmanshyttan AG
C 0-0.08, Cr 13, bal Fe.
Annealed: 75,000 TS; 40,000 YS; 35 El; 70 RA; 155 Brin. For valves, cutlery, turbine blades; Type 403; stainless.

VIKMANSHYTTAN VH274

Vikmanshyttan AG
C 0.1, Cr 18, Ni 8, bal Fe.
Annealed: 80,000 TS; 35,000 YS; 55 El; 75 RA; 150 Brin. For chemical plant equipment, tanks, mixers; Type 302; stainless, austenitic.

VIKMANSHYTTAN VH276

Vikmanshyttan AG
C 0.07-0.12, Cr 18, Ni 9, Mo 1.5, bal Fe.
Annealed: 85,000 TS; 35,000 YS; 50 El; 65 RA; 160 Brin. For acid resistant equipment, tanks, mixers; Type 316; stainless, austenitic.

VIKMANSHYTTAN VH280

Vikmanshyttan AG
C 0.14, Cr 13, Mo 1, bal Fe.
For chemical plant and oil refinery equipment; corrosion resistant.

VIKMANSHYTTAN VH295

Vikmanshyttan AG
C 0.75-0.95, Cr 16-18, Mo 0-0.75, bal Fe.
Annealed: 207,000 TS; 62,000 YS; 18 El; 35 RA; 220 Brin. Heat treated: 280,000 TS; 270,000 YS; 3 El; 15 RA; 555 Brin. For cutlery, valves, ball bearings; Type 440 B; corrosion resistant.

VIKMANSHYTTAN VH376

Vikmanshyttan AG
C 0-0.06, Cr 18, Ni 9, Mo 1.5, bal Fe.
Annealed: 85,000 TS; 35,000 YS; 50 El; 65 RA; 160 Brin. For acid resistant equipment, tanks, mixers; Type 316; stainless, austenitic.

VIKMANSHYTTAN VH399

Vikmanshyttan AG
C 0-0.1, Cr 17, Mo 1.5, bal Fe.
Annealed: 80,000 TS; 50,000 YS; 25 El; 50 RA; 150 Brin. For oil refinery equipment, oil burners and heaters; corrosion resistant.

VIKMANSHYTTAN VH406

Vikmanshyttan AG
C 0.13-0.18, Cr 13, bal Fe.
Annealed: 95,000 TS; 50,000 YS; 25 El; 55 RA; 195 Brin. For valves, cutlery, surgical and dental instruments; Type 420; stainless.

VIKMANSHYTTAN VH407

Vikmanshyttan AG
C 0-0.12, Cr 20, Ni 20, bal Fe.
For furnace parts and equipment; corrosion and heat resistant.

VIKMANSHYTTAN VH417

Vikmanshyttan AG
Cr 12-14, 0.15 C min, bal Fe.
Annealed: 95,000 TS; 50,000 YS; 25 El; 55 RA; 195 Brin. For valves, cutlery, surgical and dental instruments; Type 420; stainless.

VIKMANSHYTTAN VH94

Vikmanshyttan AG
C 0-0.15, Cr 11.5-13.5, Ni 1.25-2.5, bal Fe.
For chemical plant and oil refinery equipment; corrosion resistant.

VIKMANSHYTTAN WH605

Vikmanshyttan AG
C 0.2, Cr 12-14, bal Fe.
Annealed: 95,000 TS; 50,000 YS; 25 El; 196 Brin. Hardened: 250,000 TS; 190,000 YS; 10 El; 48 Rock C. For cutlery, surgical instruments, valves, gears, shafts. Type 420 stainless steel, hardenable.

VIKMANSHYTTAN WH620

Vikmanshyttan AG
C 0.1, Cr 14-18, bal Fe.
Annealed: 75,000 TS; 45,000 YS; 28 El; 140 Brin. Cold rolled: 110,000 TS; 90,000 YS; 10 El; 195 Brin. For automotive trim, kitchen sinks, oil burners, fasteners. Type 430 stainless steel, nonhardenable.

VIKMANSHYTTAN WH627

Vikmanshyttan AG
C 0.3, Cr 23-30, bal Fe.
Annealed: 85,0000 TS; 55,000 YS; 25 El; 83 Rock B. For oil burners, heat treating equipment, valves and fittings, furnace parts. Type 446 stainless steel. High heat and corrosion resistance.

VIKMANSHYTTAN WH638

Vikmanshyttan AG
C 0.06, Cr 19, Ni 10, bal Fe.
Annealed: 85,000 TS; 35,000 YS; 60 El; 150 Brin. For chemical and dairy plant equipment, agitators, tanks, digesters. Type 304 stainless steel, austenitic.

VIKMANSHYTTAN WH639

Vikmanshyttan AG
C 0.06, Cr 18-20, Ni 9-11, bal Fe.
Annealed: 85,000 TS; 35,000 YS; 60 El; 150 Brin. For chemical plant equipment, digesters, agitators, vessels. Type 304 stainless steel, austenitic.

VIKMANSHYTTAN WH640

Vikmanshyttan AG
C 0.06, Cr 17-19, Ni 9-12, Cb 0.5, bal Fe.
Annealed: 85,000 TS; 35,000 YS; 60 El; 83 Rock B. For welded chemical plant equipment, tanks, vessels, agitators. Type 347 stainless steel, austenitic.

VIKMANSHYTTAN WH642

Vikmanshyttan AG
C 0.12, Cr 17-19, Ni 8-10, bal Fe.
Annealed: 90,000 TS; 40,000 YS; 50 El; 85 Rock B. For chemical plant equipment, tanks, vessels, digesters, trim, agitators. Type 302 stainless steel, austenitic.

VIKMANSHYTTAN WH643

Vikmanshyttan AG
C 0.06, Cr 18, Ni 11, Ti 0.4, bal Fe.
Annealed: 85,000 TS; 35,000 YS; 55 El; 80 Rock B. For welded chemical plant equipment, tanks, vessels, agitators, mixers. Type 321 stainless steel, austenitic, stabilized.

VIKMANSHYTTAN WH645

Vikmanshyttan AG
C 0.05, Cr 18, Ni 8, Mo 2, bal Fe.
Annealed: 85,000 TS; 35,000 YS; 50 El; 80 Rock B. For acid mixers, chemical plant equipment, agitators, tanks, valve trim. Type 18-8 Mo steel, austenitic.

VIKMANSHYTTAN WH649

Vikmanshyttan AG
C 0.08, Cr 18, Ni 12, Mo 3, bal Fe.
Annealed: 80,000 TS; 40,000 YS; 40 El; 85 Rock B. For dairy and chemical plant equipment, mixers, digesters, tanks. Type 18-12 Mo stainless steel, austenitic, acid resistant.

VIKMANSHYTTAN WH650

Vikmanshyttan AG
C 0.06, Cr 17, Ni 12, Mo 2, bal Fe.
Annealed: 80,000 TS; 30,000 YS; 60 El; 80 Rock B. For textile, pharmaceutical and chemical plant equipment. Type 316-319 stainless steel, austenitic.

VIKMANSHYTTAN WH651

Vikmanshyttan AG
C 0.08, Cr 17, Ni 13, Mo 3, Cb 0.8, bal Fe.
Annealed: 85,000 TS; 40,000 YS; 50 El; 82 Rock B. For chemical plant equipment, mixers, agitators, welded tanks. Type 318 stainless steel, austenitic, welding grade.

VIKMANSHYTTAN WH662

Vikmanshyttan AG
C 0.15, Cr 18, Ni 8, bal Fe.
Annealed: 80,000 TS; 35,000 YS; 60 El; 80 Rock B. For chemical and pharmaceutical plant equipment, tanks, digesters, mixers, agitators. Type 302 stainless steel, austenitic.

VIKMANSHYTTAN WH662D

Vikmanshyttan AG
C 0.06, Cr 19, Ni 11, bal Fe.
Annealed: 70,000 TS; 30,000 YS; 60 El; 80 Rock B. For chemical plant equipment, mixers, tanks, vessels, digesters. Type 304 stainless steel, austenitic.

VINCO

CCS Braeburn Alloy Steel
C 0.7, Mn 0.25, Cr 3.75-4.25, V 1, W 17.5-18.5, bal Fe.
For cutters, hot nut dies, shear knives. Type T1 hot work steel.
Obsolete

VINCO HOT WORK

CCS Braeburn Alloy Steel
C 0.5, W 18, Cr 4, V 1, bal Fe.
Hot work die steel. For hot nut dies, hot extrusion dies, press die insert gripper dies, punches, shear knives. Type H-26.
Obsolete

VINCO HW

Now VINCO HOT WORK.

VIOLA AT13

Ilssa-Viola SpA
C 0.15, Cr 23, Ni 14, S 0.15, bal Fe.
Annealed: 90,000 TS; 40,000 YS; 50 El; 83 Rock B. For furnace parts, skids, incinerators, heat treating boxes, permanent molds. Free-machining. Type 309S stainless steel, corrosion and heat resistant.

VIOLA AT15

Ilssa-Viola SpA
C 0.2, Cr 25, Ni 21, bal Fe.
Annealed: 95,000 TS; 45,000 YS; 50 El; 89 Rock B. For furnace parts, skids, heat treating fixtures, pumps, valves. Free machining. Type 310S stainless steel, corrosion and heat resistant.

VIOLA ICS

Ilssa-Viola SpA
C 0.1, Cr 14-18, bal Fe.
Annealed: 70,000 TS; 40,000 YS; 30 El; 140 Brin. Cold rolled: 120,000 TS; 110,000 YS; 5 El; 195 Brin. For oil burners, oil refinery equipment, heaters, kitchen sinks, furnace parts, septic tanks, fasteners. Type 430 stainless steel, nonhardenable, ferritic.

VIOLA IN

Ilssa-Viola SpA
C 0.12, Cr 18, Ni 9, bal Fe.
Annealed: 90,000 TS; 40,000 YS; 50 El; 85 Rock B. For food processing and chemical plant equipment, tanks, valve trim. Type 302 stainless steel, austenitic, nonhardenable.

VIOLA INC

Ilssa-Viola SpA
C 0.06, Cr 18, Ni 12, Cb 0.6, bal Fe.
Annealed: 85,000 TS; 35,000 YS; 60 El; 80 Rock B. For welded structures, tanks, mixers, agitators, chemical plant equipment. Type 347 stainless steel, stabilized, austenitic, welding grade.

VIOLA IND

Ilssa-Viola SpA
C 0.1, Cr 16-18, Ni 6-8, bal Fe.
Annealed: 110,000 TS; 40,000 YS; 60 El; 85 Rock B. For aircraft and railroad car structures, trailer bodies, springs, automotive and architectural trim. Type 301 stainless steel, austenitic, good ductility.

VIOLA INF25

Ilssa-Viola SpA
C 0.18, Cr 24, bal Fe.
Annealed: 88,000 TS; 56,000 YS; 24 El; 83 Rock B. For heat treating boxes, fixtures and furnace parts, oil burners, conveyor exhaust manifolds. Type 446 stainless steel, corrosion and heat resistant.

VIOLA INF30

Ilssa-Viola SpA
C 0.2, Cr 25, bal Fe.
Annealed: 90,000 TS; 58,000 YS; 23 El; 85 Rock B. For heat treating boxes, fixtures and furnace parts, oil burners, conveyor exhaust manifolds. Type 446 stainless steel, corrosion and heat resistant.

VIOLA INI

Ilssa-Viola SpA
C 0.06, Cr 19, Ni 10, bal Fe.
Annealed: 85,000 TS; 35,000 YS; 60 El; 80 Rock B. For welded structures, chemical and pharmaceutical plant equipment, tanks, mixers. Type 304 stainless steel, austenitic, nonmagnetic.

VIOLA INI/BC

Ilssa-Viola SpA
C 0.05, Cr 20, Ni 12, bal Fe.
Annealed: 88,000 TS; 38,000 YS; 58 El; 82 Rock B. For chemical and pharmaceutical plant equipment, welded structures, tanks, mixers. Type 304 stainless steel, austenitic, nonmagnetic.

VIOLA INMI/-BC

Ilssa-Viola SpA
C 0.8, Cr 18, Mo 2, Ni 8, bal Fe.
Annealed: 80,000 TS; 30,000 YS; 60 El; 80 Rock B. For chemical plant equipment, mixers, digesters, valve trim. Type 18-8 Mo stainless steel. Acid resistant, austenitic.

VIOLA INMS

Ilssa-Viola SpA
C 0.06, Cr 18, Ni 12, Mo 2, bal Fe.
Annealed: 85,000 TS; 35,000 YS; 50 El; 80 Rock B. For chemical plant equipment, agitators, mixers, digesters, valve trim. Type 18-12 Mo stainless steel, acid resistant, austenitic.

VIOLA INS

Ilssa-Viola SpA
C 0.06, Cr 18, Ni 11, Ti 0.3, bal Fe.
Annealed: 85,000 TS; 30,000 YS; 55 El; 80 Rock B. For welded structures, exhaust systems, engine manifolds, refinery equipment, radiant superheaters. Type 321 stainless steel, austenitic, welding grade.

VIOLA MIC

Ilssa-Viola SpA
C 0.12, Cr 12-14, bal Fe.
Annealed: 95,000 TS; 50,000 YS; 25 El; 92 Rock B. For cutlery, surgical instruments, gears, shafts, springs, bearings. Type 420 stainless steel, hardenable.

VIOLET LABEL
Ackerlind Steel Co., Inc.
C 0.9, Cr 0.5, V 0.2, W 1.5, bal Fe.
For cutting dies, tools, gauges, drawing dies; nondeforming.
Obsolete

VIPER
Osborn Steels Ltd.
C 1, Cr 1, Mn 0.45, bal Fe.
For bearings, liners, sleeves; oil or water hardened. *Obsolete*

VIRGINIA SILVER
Manufacturer not listed.
Ni, Zn, bal Cu.
For ornaments; Nickel silver.

VIRGINIA STEEL
Manufacturer not listed.
C, bal Fe.
For construction work.

VIRGO
Now UNITEMP VIRGO.

VIRGO 1
Creusot-Loire
C 0.09-0.12, Cr 13, bal Fe.
For molten salt containers, pots; Type 410; corrosion
resistant. *Obsolete*

VIRGO 1.0
Creusot-Loire
C 0-0.15, Cr 13, Mo 0.5, bal Fe.
Annealed: 75,000 TS; 40,000 YS; 35 El; 70 RA; 155 Brin. For
steam turbines, rotor blades, turbine blades; corrosion
resistant; Type 410. *Obsolete*

VIRGO 104
Creusot-Loire
C 0-0.06, Mn 0-1, Si 0-1, Ni 3.5-5, Cr 12-13.5, Mo 0.4-0.7, bal
Fe.
Stainless casting. Parts for pumps, turbines, nuclear energy.
AFNOR Z4 CND 13.4 M; ACI CA 6NM.

VIRGO 105D
Creusot-Loire
C 0-0.07, Mn 0-1, Si 0-1, Ni 1.5-2.5, Cr 11-13.5, Mo 0-0.75, bal
Fe.
Stainless casting, ferritic type; 200-250 Brin. Parts for steam
turbines. AFNOR Z 4 CN 13.2 M.

VIRGO 106
Creusot-Loire
C 0-0.08, Cr 12-14, bal Fe.
Annealed: 75,000 TS; 40,000 YS; 35 El; 155 Brin. For
equipment in the petroleum industry, flat springs, table
flatware, furnace parts. Corrosion resistant. Non-hardenable.
AFNOR-Z6C13 Spec; AISI 410S. *Obsolete*

VIRGO 112
Creusot-Loire
C 0-0.15, Cr 12-14, bal Fe.
Annealed: 80,000 TS; 43,000 YS; 30 El; 160 Brin. For furnace
parts below 700 C, flat springs, table flatware, oil refinery
equipment. Corrosion resistant, hardenable. AFNOR-Z12C13
Spec; AISI 410. *Obsolete*

VIRGO 11S
Creusot-Loire
C 0-0.08, Cr 17-20, Ni 9-12, bal Fe.
Annealed: 85,000 TS; 30,000 YS; 60 El; B 80 Rock. For
chemical and oil and food industry equipment, welded
structures, tanks, mixers. Stainless, austenitic. France
Z6CN18-10 spec; AISI 304. *Obsolete*

VIRGO 11SA
Creusot-Loire
C 0-0.08, Cr 16-19, Ni 11-14, bal Fe.
Annealed: 85,000 TS; 32,000 YS; 60 El; B 80 Rock. For
chemical and food processing equipment, mixers, agitators,
digesters. Stainless, austenitic, non-magnetic. *Obsolete*

VIRGO 11SB
Creusot-Loire
C 0-0.08, Cr 16-19, Ni 10-13, bal Fe.
Annealed: 83,000 TS; 30,000 YS; B 80 Rock. For chemical
and food processing equipment, digesters, mixers.
Austenitic, stainless. *Obsolete*

VIRGO 11SS
Creusot-Loire
C 0-0.04, Cr 17-19, Ni 9-11, bal Fe.
Annealed: 75,000 TS; 35,000 YS; 50 El; 60 RA; 180 Brin. For
chemical plant equipment, tanks, vessels; stainless,
austenitic; Type 304. *Obsolete*

VIRGO 11SSB
Creusot-Loire
C 0-0.03, Cr 16-19, Ni 10-13, bal Fe.
Annealed: 80,000 TS; 30,000 YS; 60 El; B 78 Rock. For
welded structures, tanks, mixers, chemical and food
processing equipment. Austenitic, stainless. *Obsolete*

VIRGO 12
Creusot-Loire
C 0.17-0.25, Cr 18, Ni 8, bal Fe.
Annealed: 85,000 TS; 40,000 YS; 53 El; 72 RA; 160 Brin. For
chemical plant equipment, tanks, mixers, filters; Type 302;
stainless, austenitic. *Obsolete*

VIRGO 120
Creusot-Loire
C 0-0.25, Cr 12-14, bal Fe.

VIRGO 13
Creusot-Loire
C 0.08-0.2, Cr 18, Ni 9, Si 2.5, bal Fe.
Annealed: 85,000 TS; 40,000 YS; 50 El; 70 RA; 160 Brin. For
furnace parts, heat treating boxes; Type 302B; corrosion and
heat resistant. *Obsolete*

VIRGO 14
Creusot-Loire
C 0-0.12, Cr 18, Ni 10, Mo 2.5, bal Fe.
Annealed: 85,000 TS; 35,000 YS; 50 El; 65 RA; 160 Brin. For
acid resistant chemical plant equipment, mixers; Type 316;
stainless, austenitic. *Obsolete*

VIRGO 14 SB 1
Now ICL 164 FLUAGE.

VIRGO 14S
Creusot-Loire
C 0-0.12, Cr 18, Ni 10, Mo 2.5, Ti, bal Fe.
Annealed: 90,000 TS; 40,000 YS; 45 El; 60 RA; 170 Brin. For
welded acid resistant chemical plant equipment; Type 316 Ti;
stainless, austenitic. *Obsolete*

VIRGO 14SB
Creusot-Loire
C 0-0.08, Cr 16-19, Ni 11-14, Mo 2-2.7, bal Fe.
Annealed: 74,000 TS; 30,000 YS; 40 El. For superheater
tubes, blades, discs, steam pipes. Austenitic, corrosion
resistant. Type 316 stainless. *Obsolete*

VIRGO 14SR
Creusot-Loire
C 0-0.07, Ni 12.5, Cr 18, Mo 2.5, bal Fe.
Annealed: 80,000 TS; 40,000 YS; 40 El; 50 RA; 180 Brin. For
acid resistant equipment in chemical plants; Type 316;
stainless, austenitic. *Obsolete*

VIRGO 14SR1
Creusot-Loire
C 0-0.08, Cr 17-20, Ni 11-14, Mo 3-3.7, bal Fe.
Annealed: 80,000-90,000 TS; 35,000-55,000 YS; 50-70 El;
60-75 RA; B 70-90 Rock. For textile and dye equipment,
paper and pulp plant equipment. Austenitic, stainless. AISI
317, acid resistant. *Obsolete*

VIRGO 14SR2
Creusot-Loire
C 0-0.08, Cr 17-20, Ni 12-15, Mo 3.5-4.2, bal Fe.
Annealed: 80,000 TS; 35,000 YS; 35 El. For chemical and
textile plant equipment, agitators. Type 317 stainless,
austenitic. *Obsolete*

VIRGO 14SS
Creusot-Loire
C 0-0.03, Cr 17-20, Ni 10-13, Mo 2-2.7, bal Fe.
Annealed: 80,000 TS; 30,000 YS; 60 El; B 78 Rock. For paper
mill and die works equipment, acid tanks, mixers. Resists sea
water corrosion. France Z3CND18-12; AISI 316L. *Obsolete*

VIRGO 14SSB
Creusot-Loire
C 0-0.03, Cr 16-19, Ni 11-14, Mo 2-2.7, bal Fe.
Annealed: 85,000 TS; 35,000 YS; 50 El; B 80 Rock. For
chemical plant equipment, mixers, digesters, valve trim,
kettles. Corrosion resistant, austenitic. *Obsolete*

VIRGO 14SSR
Creusot-Loire
C 0-0.03, Cr 17-20, Ni 10-13, Mo 2.5-3.2, bal Fe.
Annealed: 80,000 TS; 30,000 YS; 60 El; B 78 Rock. For
chemical and food processing equipment, mixers, digesters.
Corrosion resistant, austenitic. France spec. Z3CND 18-12;
AISI 316L. *Obsolete*

VIRGO 14SSR1
Creusot-Loire
C 0-0.03, Cr 17-20, Ni 11-14, Mo 3-3.7, bal Fe.
Annealed: 85,000 TS; 45,000 YS; 60 El; B 80 Rock. For acid
tanks, mixers, agitators, chemical plant equipment. Stainless,
austenitic; Type 317L. *Obsolete*

VIRGO 14SSR2
Creusot-Loire
C 0-0.03, Cr 17-20, Ni 12-15, Mo 3.5-4.2, bal Fe.
Annealed: 85,000 TS; 45,000 YS; 60 El; B 80 Rock. For acid
tanks, agitators, mixers, chemical plant equipment. Stainless,
austenitic. AISI 317L, acid resistant. *Obsolete*

VIRGO 15
Creusot-Loire
C 0-0.08, Cr 18, Ni 8, Ti, bal Fe.
Annealed: 85,000 TS; 35,000 YS; 55 El; 65 RA; 150 Brin. For
welded chemical plant equipment, tanks; Type 321; stainless,
austenitic. *Obsolete*

VIRGO 15S
Creusot-Loire
C 0-0.08, Cr 17-19, Ni 8-10, 0.45% min Ti, bal Fe.
Annealed: 75,000 TS; 35,000 YS; 45 El; 55 RA; 180 Brin. For
welded chemical plant equipment, tanks, vessels; Type 321;
stainless, austenitic. *Obsolete*

VIRGO 15SA
Creusot-Loire
C 0-0.07, Ni 13, Cr 18, Ti = 5 x C, bal Fe.
Annealed: 86,000 TS; 31,000 YS; 45 El; 160 Brin. For welded
chemical plant equipment; stainless, austenitic; Type 321.
Obsolete

VIRGO 15SB
Jodots Freres, Ets.
C 0.06, Si 0.4, Mn 1.6, Cr 18, Ni 12, Ti 0.4, bal Fe.
Annealed: 85,000 TS; 35,000 YS; 60 El; 70 RA; 150 Brin. Cold
drawn: 180,000 TS; 125,000 YS; 10 El; 330 Brin. For chemical
plant equipment; Type 304; stainless, austenitic.

VIRGO 17
Creusot-Loire
C 0.08, Ni 13, Cr 17, Mo 2.5, 0.50 Ti or 0.80 Cb, bal Fe.
Annealed: 100,000 TS; 72,000 YS; 18 El. For rotors and rotor
discs; stainless, austenitic. *Obsolete*

VIRGO 17S
Creusot-Loire
C 0-0.08, Cr 17-20, Ni 11-14, Mo 2-2.7, 0.4% Ti min, bal Fe.
Annealed: 85,000 TS; 30,000 YS; B 82 Rock. For welded chemical plant equipment, mixers, digesters. Corrosion resistant, non-magnetic. France spec. Z8CNDT18-12. Austenitic. *Obsolete*

VIRGO 17SA
Creusot-Loire
C 0.08, Ni 13, Cr 18, Mo 2.5, Ti, bal Fe.
Annealed: 85,000 TS; 43,000 YS; 50 El; 190 Brin. For chemical plant equipment, welded structures; Type 316 Ti; stainless, austenitic. *Obsolete*

VIRGO 17SB
Jodots Freres, Ets.
C 0.08, Cr 17, Ni 13, Mo 2.5, 0.5 Ti or Cb, bal Fe.
Annealed: 80,000 TS; 30,000 YS; 60 El; 80 RA; 135 Brin. Cold drawn: 150,000 TS; 135,000 YS; 6 El; 320 Brin. For acid resistant equipment, mixers, evaporators; Type 316 Ti; stainless, austenitic.

VIRGO 17SR
Creusot-Loire
C 0-0.08, Cr 17-20, Ni 11.5-14.5, Mo 2.5-3.2, Ti 0-0.4, bal Fe.
Annealed: 90,000 TS; 45,000 YS; B 90 Rock. For welded structures, tanks, mixers, chemical plant equipment. France spec. Z8CNDT18-12, austenitic, stainless, stabilized. *Obsolete*

VIRGO 18S
Creusot-Loire
C 0-0.08, Cr 17-20, Ni 9-12, 0.9% Cb + Ta max, bal Fe.
Annealed: 90,000 TS; 40,000 YS; 50 El; B 82 Rock. For chemical and textile plant equipment, welded tanks, mixers. Corrosion resistant, austenitic. AISI 347. Welding grade. *Obsolete*

VIRGO 18SB
Creusot-Loire
C 0-0.08, Cr 16-19, Ni 11-14, 0.9% Cb + Ta max, bal Fe.
Annealed: 74,000 TS; 30,000 YS; 40 El. For superheater tubes, blades, discs, steam pipes, welded structures. Austenitic, stainless, nonmagnetic, stabilized. *Obsolete*

VIRGO 19S
Creusot-Loire
C 0-0.08, Cr 17-20, Ni 11-14, Mo 2-2.7, 0.8% Ta + Cb max, bal Fe.
Annealed: 80,000 TS; 30,000 YS; 60 El; B 80 Rock. For chemical and textile plant equipment, mixers, digesters. Stainless, nonmagnetic. Type 316 stabilized, non-hardenable. *Obsolete*

VIRGO 19SB
Creusot-Loire
C 0.06, Ni 12, Cr 18, Ti 0.4, bal Fe.
Annealed: 70,000 TS; 35,000 YS; 25 El. For welded structures, chemical plant equipment; Type 321; stainless, austenitic. *Obsolete*

VIRGO 19SR
Creusot-Loire
C 0-0.08, Cr 17-20, Ni 11.5-14.5, Mo 2.5-3.2, 0.8% Cb + Ta max, bal Fe.
For welded structures, tanks, mixers, chemical plant equipment. Austenitic, stainless, welding grade. *Obsolete*

VIRGO 19SR1
Creusot-Loire
C 0-0.08, Cr 17-20, Ni 12.5-15.5, Mo 3-3.7, 0.8% Cb + Ta max, bal Fe.
Annealed: 85,000 TS; 40,000 YS; 60 El; B 80 Rock. For chemical plant equipment, welded structures, mixers, agitators. Welding grade, austenitic, stainless. *Obsolete*

VIRGO 19SR2
Creusot-Loire
C 0-0.08, Cr 17-20, Ni 13.5-16.5, Mo 3.5-4.2, 0.8% Cb + Ta max, bal Fe.
Annealed: 83,000 TS; 38,000 YS; B 80 Rock. For welded structures, mixers, tanks, kettles, digesters, agitators. Acid resistant, stabilized, austenitic, non-magnetic. *Obsolete*

VIRGO 1B
Creusot-Loire
C 0.07-0.12, Cr 17, bal Fe.
Annealed: 80,000 TS; 50,000 YS; 25 El; 50 RA; 150 Brin. For oil refinery equipment, oil burners and heaters; Type 430; corrosion and heat resistant. *Obsolete*

VIRGO 1S
Creusot-Loire
C 0.1, Cr 11-13, bal Fe.
Annealed: 70,000 TS; 35,000 YS; 30 El; B 80 Rock. For springs, trim, knives, tableware. Type 410 stainless steel, magnetic. *Obsolete*

VIRGO 1SA
Creusot-Loire
C 0-0.08, Cr 12, Al 0.2, bal Fe.
Annealed: 75,000 TS; 40,000 YS; 35 El; 70 RA; 155 Brin. For turbine blades, oil refinery equipment; Type 405; stainless. *Obsolete*

VIRGO 2
Creusot-Loire
C 0.19-0.25, Cr 13, bal Fe.
Heat treated: 115,000 TS; 92,000 YS; 10 El. For turbine blades, cutlery, valves; Type 420; corrosion resistant. *Obsolete*

VIRGO 2.0
Creusot-Loire
C 0-0.25, Cr 13, Mo 0.5, bal Fe.
Annealed: 95,000 TS; 50,000 YS; 25 El; 55 RA; 195 Brin. For cutlery, valves, pumps, turbine blades; Type 420; stainless, hardenable. *Obsolete*

VIRGO 2/3
Creusot-Loire
C 0.26-0.37, Cr 13, bal Fe.
Annealed: 100,000 TS; 55,000 YS; 20 El; 50 RA; 210 Brin. For valves, cutlery, surgical and dental instruments; Type 420; corrosion resistant. *Obsolete*

VIRGO 20
Creusot-Loire
C 0.14, Cr 13, Mo 1, bal Fe.
Annealed: 80,000 TS; 45,000 YS; 30 El; 65 RA; 170 Brin. For oil refinery equipment; heat and creep resistant. *Obsolete*

VIRGO 23
Creusot-Loire
C 0-0.05, Cr 18, Ni 18, Mo 2, Cu 2, Ti = 5 x C, bal Fe.
Annealed: 80,000 TS; 45,000 YS; 40 El; 180 Brin. For autoclaves, chemical plant equipment; stainless, austenitic. *Obsolete*

VIRGO 2S
Creusot-Loire
C 0.08, Cr 11-13, Al 0.2, bal Fe.
Annealed: 60,000-75,000 TS; 32,000-50,000 YS; 20-40 El; 140-180 Brin. For equipment in the petroleum industry, furnace parts in sulfur bearing atmospheres, heat exchangers, tank linings. Corrosion resistant. Nonhardenable. AFNOR-Z6CA13; AISI 405. *Obsolete*

VIRGO 3
Creusot-Loire
C 0-0.25, Cr 12-14, Mo 0-0.5, bal Fe.
Heat treated: 240,000 TS; 200,000 YS; 8 El; C 50 Rock. For valves, cutlery, bearings, surgical instruments, pivots. Hardenable, corrosion resistant. *Obsolete*

VIRGO 31
Creusot-Loire
C 0.28, Cr 24-26, Ni 5, bal Fe.
Annealed: 100,000 TS; 65,000 YS; 15 El; 200 Brin. For furnace parts, grills, heat treating boxes; heat resistant to 1050 C in oxygen and 900 C in sulfur atmosphere. *Obsolete*

VIRGO 31S
Creusot-Loire
C 0-0.08, Cr 15.5-18.5, bal Fe.
Cold rolled: 120,000 TS; 110,000 YS; 5 El; 185 Brin. For nitric acid containers, furnace parts in sulfur-bearing atmospheres. Type 430 stainless steel, good formability and weldability. *Obsolete*

VIRGO 32
Creusot-Loire
C 0.18, Cr 24-26, Ni 5, bal Fe.
Annealed: 97,000 TS; 71,000 YS; 6 El; 200 Brin. For furnace parts, grills, heat treating boxes; heat resistant to 1050 C in O 2 and 900 C in S atmosphere. *Obsolete*

VIRGO 38
Creusot-Loire
C 0-0.06, Mn 0-1, Si 0-1, Ni 4-5.5, Cr 15.5-17.5, bal Fe.
Stainless casting. Parts for water turbines. AFNOR Z 6 CN 16 5M.

VIRGO 39 (CASTING)
Creusot-Loire
C 0-0.06, Mn 0-1, Si 0-1, Ni 4-5.5, Cr 15.5-17.5, Mo 0.5-1.5, bal Fe.
Stainless casting. For water turbines, pumps and other parts used in seawater. AFNOR Z 4 CND 17.4 M.

VIRGO 39 (WROUGHT)
Bohler Gesellschaft M.B.H.
Stainless steel. C 0-0.06, Cr 16, Ni 5, Mo 0.9, bal Fe.
Martensitic stainless: air or oil hardenable up to 880-1080 N/mm^2 TS. Plate: good strength and corrosion resistance to sea water, organic acids. AFNOR Z4 CND 16-05; see also SOLEIL C4. *Obsolete*

VIRGO 3S
Creusot-Loire
C 0-0.15, Cr 12-14, Mo 0-0.5, bal Fe.
Heat treated: 100,000 TS; 71,000 YS; 15 El. Annealed: 80,000 TS; 40,000 YS; 25 El; B 92 Rock. Hardened: 200,000 TS; 160,000 YS; 10 El; C 45 Rock. For steam turbine blades, chemical plant and oil refinery equipment. Corrosion resistant, hardenable. AFNOR-Z12CD13. *Obsolete*

VIRGO 4
Creusot-Loire
C 0-0.15, Cr 11.5-14, Ni 0-1, Mo 0-1, bal Fe.
Annealed: 75,000 TS; 38,000 YS; 28 El; B 82 Rock. For springs, hardware, shafts, tableware. Hardenable, corrosion resistant. *Obsolete*

VIRGO 41
Creusot-Loire
C 0-0.15, Cr 6, Mo 0.5, bal Fe.
Rolled: 110,000 TS; 90,000 YS; 12 El. For oil refinery equipment, pipes, fasteners; heat and creep resistant. *Obsolete*

VIRGO 41
Jodots Freres, Ets.
C 0-0.15, Cr 6, Mo 0.5, bal Fe.
Rolled: 110,000 TS; 90,000 YS; 12 El. For oil refinery equipment, pipes, fasteners; heat and creep resistant. *Obsolete*

VIRGO 41B
Creusot-Loire
C 0.2, Si 0.5, Mn 0.5, Cr 6, Mo 0.5, bal Fe.
Rolled: 114,000 TS; 92,000 YS; 10 El. For oil refinery equipment, pipes, fasteners; heat and creep resistant. *Obsolete*

VIRGO 41B

Jodots Freres, Ets.
C 0.2, Si 0.5, Mn 0.5, Cr 6, Mo 0.5, bal Fe.
Rolled: 114,000 TS; 92,000 YS; 10 El. For oil refinery
equipment, pipes, fasteners; heat and creep resistant.
Obsolete

VIRGO 41S

Creusot-Loire
C 0-0.1, Cr 4-6, Mo 0.4-0.7, bal Fe.
Annealed: 65,000 TS; 25,000 YS; 30 El; B 75 Rock. For oil
refinery and chemical plant equipment, valve trim, oil pump
impellers. AISI 502, corrosion and heat resistant. *Obsolete*

VIRGO 42

Creusot-Loire
C 0-0.17, Cr 6, Mo 0.5, Al 1, bal Fe.
Rolled: 110,000 TS; 90,000 YS; 12 El. For oil refinery
equipment, pipes, fasteners; heat and creep resistant.
Obsolete

VIRGO 42

Jodots Freres, Ets.
C 0-0.17, Cr 6, Mo 0.5, Al 1, bal Fe.
Rolled: 110,000 TS; 90,000 YS; 12 El. For oil refinery
equipment, pipes, fasteners; heat and creep resistant.
Obsolete

VIRGO 43

Creusot-Loire
C 0-0.17, Cr 6, Mo 0.5, V 0.3, bal Fe.
Rolled: 110,000 TS; 90,000 YS; 12 El. For oil refinery
equipment, pipes, fasteners; heat and creep resistant.
Obsolete

VIRGO 43

Jodots Freres, Ets.
C 0-0.17, Cr 6, Mo 0.5, V 0.3, bal Fe.
Rolled: 110,000 TS; 90,000 YS; 12 El. For oil refinery
equipment, pipes, fasteners; heat and creep resistant.
Obsolete

VIRGO 5

Creusot-Loire
C 0-0.15, Cr 12-14, Al 0-1, Si 0-1, bal Fe.
Annealed: 70,000 TS; 40,000 YS; 35 El; 160 Brin. For oil
refinery and chemical plant equipment, furnace parts.
Corrosion and heat resistant. *Obsolete*

VIRGO 51

Creusot-Loire
C 0-0.2, Cr 23-27, 0.2% N 2 max, bal Fe.
Annealed: 88,000 TS; 56,000 YS; 24 El; B 83 Rock. For heat
treating boxes, fixtures and furnace parts, oil burners,
conveyors, exhaust manifolds. AISI 446. Heat and corrosion
resistant. *Obsolete*

VIRGO 52

Creusot-Loire
C 0.5, Cr 24-26, Si 1.5, bal Fe.
For recuperators, furnace parts, grills; heat resistant to 1050
C in oxygen and 950 C in sulfur atmosphere. *Obsolete*

VIRGO 60

Creusot-Loire
C 0-0.25, Cr 25, Ni 20, Si 1.5-3, bal Fe.
For salt pots, heat treating boxes; Type 314; heat resistant
Obsolete

VIRGO 60A

Creusot-Loire
C 0.15, Si 2, Cr 26, Ni 12, bal Fe.
Normalized: 83,000 TS; 43,000 YS; 25 El. For recuperators,
turbines, furnace parts; resists heat to 1100 C in oxygen
atmosphere. *Obsolete*

VIRGO 61

Creusot-Loire
C 0-0.2, Cr 24-26, Ni 21, bal Fe.
Annealed: 96,000 TS; 46,000 YS; 35 El. For heat exchangers,
furnace parts; heat resistant to 1150 C in oxygen
atmosphere. *Obsolete*

VIRGO 61SS

Creusot-Loire
C 0-0.03, Cr 23.5-26.5, Ni 19.5-21.5, bal Fe.
Annealed: 90,000 TS; 40,000 YS; 55 El; B 85 Rock. For
pumps, valves, furnace and heat treating equipment. AISI
310S, corrosion and heat resistant. *Obsolete*

VIRGO 63

Creusot-Loire
C 0.12, Cr 23-27, Ni 20, Si 2, bal Fe.
Annealed: 80,000 TS; 37,000 YS; 32 El. For heat exchangers,
furnace parts; heat resistant to 1200 C in oxygen
atmosphere. *Obsolete*

VIRGO 7

Creusot-Loire
C 0-0.25, Cr 10.5-12.5, Ni 0-0.55, Mo 0-1.2, V 0-0.35, bal Fe.
Heat treated: 130,000 TS; 100,000 YS; 10 El. For valve parts,
pumps, fasteners, gears, shafts, steam turbine blades.
Corrosion and heat resistant. *Obsolete*

VIRGO 70

Creusot-Loire
C 0.08-0.16, Cr 12, Ni 12, bal Fe.
For valves, pumps; corrosion resistant. *Obsolete*

VIRGO 74

Creusot-Loire
C 0-0.1, Cr 16, Mo 17, W 5, Fe 0-5, bal Ni.
Annealed: 75,000 TS; 46,000 YS; 12 El; 14 RA; 200 Brin. For
chemical and oil refinery equipment, valves, pumps,
combustion chambers. Similar to Hastelloy-C. Corrosion and
heat resistant. *Obsolete*

VIRGO 7A

Jodots Freres, Ets.
C 0.22, Cr 11.5, Mo 1.1, V 0.4, Ni 0.4, bal Fe.
Heat treated: 100,000 TS; 72,000 YS; 10 El. For turbine
blades, bolts; corrosion resistant.

VIRGO 84

Creusot-Loire
C 0.2, Cr 22, Ni 12, W 3, bal Fe.
For heat resistant equipment; heat resistant to 1100 C in
oxygen atmosphere. *Obsolete*

VIRGO 86

Creusot-Loire
C 0.45-0.55, Cr 24-28, Ni 8-12, W 6 0, Fe 2, bal Co.
Cast: 113,000 TS; 79,000 YS; 8 El; 9 RA. Aged: 128,000 TS;
113,000 YS; 2 El; 3 RA. For jet engine and gas turbine
blades. Similar to Haynes Stellite No. 31. Heat and corrosion
resistant. High creep strength. *Obsolete*

VIRGO 87B

Creusot-Loire
C 0.14, Cr 12.5, Mo 4.5, Ti 0.6, Cb 2.1, Al 6, Fe 2.5, B, Zr, bal
Ni.
Cast: 117,000 TS; 102,000 YS; 5 El; 400 Brin. For jet engine
and gas turbine blades. High heat and corrosion resistance.
Similar to Inconel 713. *Obsolete*

VIRGO 92

Creusot-Loire
C 0.5, Ni 13-16, Cr 13-16, W 2.2, Mo 0.4, bal Fe.
For furnace parts and equipment, valves. Heat and oxidation
resistant. *Obsolete*

VIRGO 94

Jodots Freres, Ets.
C 0.1, Cr 16-20, Ni 40-48, Ti 1.9-2.6, Co 20-24.
Cold drawn: 127,000 TS; 85,000 YS; 20 El. For turbine
components, valve and pump parts. Austenitic, oxidation and
heat resistant to 1150°C.

VIRGO L 16

Creusot-Loire
C 0.5, Si 14.5-17, Mn 0.5, bal Fe.

VIRGO L 16

Creusot-Loire
C 0.5, Si 14.5-17, Mn 0.5, bal Fe.

VIRGO-11DS

Creusot-Loire
C 0.08, Cr 18, Ni 8, bal Fe.
Annealed: 85,000 TS; 35,000 YS; 60 El; 70 RA; 150 Brin. For
chemical plant equipment, tanks; Type 304; stainless,
austenitic. *Obsolete*

VIRILLIUM

English manufacture
Co 67.9, Cr 24.1, Ni 1.4, Mo 5.3, bal Fe.
Cast: 96,000-98,000 TS; 64,000-69,000 YS; 10-11 El; 295
Brin. For dental alloy; corrosion resistant.

VISCOTHERM 1612

Special Metals Corp.
Cr, Co, Mo, bal Ni.

VISCOTHERM 4

Special Metals Corp.
Cr, Co, Mo, bal Ni.

VISCOTHERM 5

Special Metals Corp.
Cr, Co, Mo, bal Ni.

VISCOTHERM 6

Special Metals Corp.
Cr, Co, Mo, bal Ni.

VISCOTHERM 7

Special Metals Corp.
Cr, Co, Mo, bal Ni.

VISCOUNT 20

Latrobe Steel Co.
C 0.4, Cr 5.25, V 1, Si 1, Mn 0.8, Mo 1.35, S + alloy sulfides,
bal Fe.
For white metal extrusion dies; Type H13; hot work steel.

VISCOUNT 44

Latrobe Steel Co.
C 0.4, Si 1, Mn 0.8, Cr 5.25, Mo 1.35, V 1.
Heat treated AISI Type H13 hot work steel. For dies, backers,
bolsters, dummy blocks, etc.

VISCOUNT 44

Latrobe Steel Co.
C 0.4, Cr 5, V 1, Si 1, Mn 0.3, Mo 1.2, S, bal Fe.
Heat treated: 420-460 Brin. For die casting dies, extrusion
tools, forging die blocks; prehardened, air-hardened, resists
heat checking. AISI H13.

VISTA

Time Steel Service Inc.
C 0.23, Mn 0.6, Si 1.25, Cr 10, Ni 0.75, V 1, W 0.45, Mo 1.2, N
0.1, bal Fe.
Air or oil hardened hot work tool and die steel, chromium
type.

VISTA 11

Time Steel Service Inc.
C 0.4, Mn 0.3, Si 0.9, Cr 5, V 0.5, Mo 1.3, bal Fe.
Chromium type hot work tool and die steel. AISI H11

VISTA 12

Time Steel Service Inc.
C 0.32, Mn 0.3, Si 0.9, Cr 5, V 0.25, W 1.25, Mo 1.5, bal Fe.
Chromium type hot work tool and die steel. AISI H12.

VISTA 13

Time Steel Service Inc.
C 0.4, Mn 0.35, Si 1, C 5, V 1, Mo 1.2, bal Fe.
Chromium type hot work tool and die steel. AISI H13.

VISTA 21

Time Steel Service Inc.
C 0.3, Mn 0.3, Cr 3, V 0.5, W 9, bal Fe.
Hot work tool and die steel, tungsten type. AISI H21.

VISTA 24

Time Steel Service Inc.
C 0.42, Cr 3.5, V 0.3, W 14, bal Fe.
Hot work tool and die steel, tungsten type. AISI H24.

VISTA 26

Time Steel Service Inc.
C 0.52, Cr 3.75, V 0.9, W 17.5, bal Fe.
Hot work tool and die steel, tungsten type. AISI H26.

VISTA 43

Time Steel Service Inc.
C 0.5, Cr 4, V 1.95, Mo 8, bal Fe.
Hot work tool and die steel, molybdenum type. AISI W43.

VISTO

Hidalgo Steel Co. Inc.
C, Cr, V, W, bal Fe.
For tools, cutters; high speed steel.

VISTO

Dorrenberg Edelstahl GmbH
C 0.7, w 18, Cr 4, V 1, bal Fe.
For cutters, tools; high speed steel.

VISTO A50

Dorrenberg Edelstahl GmbH
C 0.56, Ni, Cr, Mo, V, bal Fe.
For gears, shafts, crankshafts, bolts, fasteners; oil hardened; shock resistant.

VISTO ABC III

Dorrenberg Edelstahl GmbH
C 0.95, W, Mo, bal Fe.
For cutters, dies, bearings, liners; water hardened; wear resistant.

VISTO DM1

Dorrenberg Edelstahl GmbH
C 0.45, Cr 1.4, Mo 0.7, V 0.3, bal Fe.
For gears, bolts, crankshafts; oil hardened; tough.

VISTO KD13

Dorrenberg Edelstahl GmbH
C 0.86, Co 2.8, Cr 4.3, Mo 0.85, V 2.1, W 12, bal Fe.
For lathe and planer tools, drills, reamers, hobs; high speed steel.

VISTO KD15

Dorrenberg Edelstahl GmbH
C 0.79, Co 4.7, Cr 4.3, Mo 0.75, V 1.5, W 18, bal Fe.
For lathe and planer tools, broaches; high speed steel.

VISTO KD16

Dorrenberg Edelstahl GmbH
C 0.76, Co 10, Cr 4.2, Mo 0.8, V 1.8, W 18, bal Fe.
For lathe and planer tools, form cutters; high speed steel.

VISTO KD22

Dorrenberg Edelstahl GmbH
C 1.3, Cr 4.3, Mo 0.85, V 3.8, W 12, bal Fe.
For engravers' tools, blanking and forming dies; high speed steel.

VISTO KD25

Dorrenberg Edelstahl GmbH
C 1.35, Cr, W, Co, V, bal Fe.
For engravers' tools, blanking and forming dies; high speed steel.

VISTO KD6

Dorrenberg Edelstahl GmbH
C 0.86, Cr 4.1, Mo 0.85, V 2.5, W 12, bal Fe.
For lathe and planer tools, drills, reamers, taps; high speed steel.

VISTO PL

Dorrenberg Edelstahl GmbH
C 0.45, W, Cr, V, bal Fe.
For punches, crimpers, upsetters; hot work steel; oil hardened.

VISTO REKORD

Dorrenberg Edelstahl GmbH
C 0.74, Cr 4.1, V 1.1, W 18.5, bal Fe.
For lathe and planer tools, reamers, broaches, hobs; high speed steel.

VISTO SUPERIOR

Hidalgo Steel Co. Inc.
C 0.8, Cr 4, Mo 0.8, V 1.6, W 8.7, bal Fe.
For shear blades, dies, drills, reamers, broaches, taps; high speed steel, oil hardened.

VISTO W11C

Dorrenberg Edelstahl GmbH
C 0.3, Co 2, Cr 2.4, V 0.25, W 8.5, bal Fe.
For extrusion tools, punches, shears; hot work steel; oil hardened.

VISTO W5

Dorrenberg Edelstahl GmbH
C 0.3, Cr 2.3, V 0.6, W 4.2, bal Fe.
For punches, riveters, extrusion tools; hot work steel; oil hardened.

VISTO W9

Dorrenberg Edelstahl GmbH
C 0.3, Cr 2.65, V 0.35, W 8.5, bal Fe.
For extrusion rams and liners, punches; hot work steel; oil hardened.

VISTO WC SPEZIAL

Dorrenberg Edelstahl GmbH
C 0.38, Cr, Si, Mo, V, bal Fe.
For upsetters, riveters, punches, shears; hot work steel; oil hardened.

VISTO WC5

Dorrenberg Edelstahl GmbH
C 0.4, Cr, W, V, Si, bal Fe.
For hot work punches and shears; hot work steel; oil hardened.

VISTO Z652

Dorrenberg Edelstahl GmbH
C 0.85, Cr, V, W, Mo, bal Fe.
For lathe and planer tools, drills, taps, hobs; high speed steel.

VITA M-1

LeVita Metal Alloy Co.
C 0.7, Mo 8.5, Cr 4, V 1, W 1.5, bal Fe.
For lathe and planer tools; high speed steel.

VITA M-10

LeVita Metal Alloy Co.
C 0.7, Mo 8, Cr 4, V 2, bal Fe.
For lathe and planer tools; high speed steel.

VITA M-15

LeVita Metal Alloy Co.
C 0.7, Mo 3, W 6.7, V 5, Co 5, bal Fe.
For lathe and planer tools; high speed steel.

VITA M-2

LeVita Metal Alloy Co.
C 0.7, Mo 5, Cr 4, V 2, W 6, bal Fe.
For lathe and planer tools; high speed steel.

VITA M-20

LeVita Metal Alloy Co.
C 0.7, Mo 8, Cr 4, V 1, Co 2.5, bal Fe.
For lathe and planer tools; high speed steel.

VITA M-3

LeVita Metal Alloy Co.
C 0.7, Mo 6, Cr 4, V 2.4, W 6, bal Fe.
For lathe and planer tools; high speed steel.

VITA M-30

LeVita Metal Alloy Co.
C 0.7, Mo 8, Cr 4, V 1, W 2, Co 5, bal Fe.
For lathe and planer tools; high speed steel.

VITA M-34

LeVita Metal Alloy Co.
C 0.7, Mo 8.5, Cr 4, V 2, W 2, Co 8, bal Fe.
For lathe and planer tools; high speed steel.

VITA M-36

LeVita Metal Alloy Co.
C 0.7, Mo 6, Cr 4, V 2, W 6, Co 8, bal Fe.
For lathe and planer tools; high speed steel.

VITA M-4

LeVita Metal Alloy Co.
C 0.7, Mo 4.5, Cr 4.5, V 4, W 5.5, bal Fe.
For lathe and planer tools; high speed steel.

VITA M-40

LeVita Metal Alloy Co.
C 0.7, Mo 8, Cr 4, V 1.5, Co 8, bal Fe.
For lathe and planer tools; high speed steel.

VITA M-50

LeVita Metal Alloy Co.
C 0.7, Mo 4.2, Cr 4, V 1, bal Fe.
For lathe and planer tools; high speed steel.

VITA M-52

LeVita Metal Alloy Co.
C 0.7, Mo 4.2, Cr 4, V 2, bal Fe.
For lathe and planer tools; high speed steel.

VITA M-54

LeVita Metal Alloy Co.
C 0.7, Mo 4.2, Cr 4, V 3, bal Fe.
For lathe and planer tools; high speed steel.

VITA M-56

LeVita Metal Alloy Co.
C 0.7, Mo 4.2, Cr 4, V 4, bal Fe.
For lathe and planer tools; high speed steel.

VITA M-6

LeVita Metal Alloy Co.
C 0.7, Mo 5, W 4, Co 12, Cr 4.5, V 1.5, bal Fe.
For lathe and planer tools; high speed steel.

VITA M-8

LeVita Metal Alloy Co.
C 0.7, Mo 4.5, Cr 4.2, V 1.5, W 5.5, Cb 1.2, bal Fe.
For lathe and planer tools; high speed steel.

VITA T-1

LeVita Metal Alloy Co.
C 0.7, W 18, Cr 4, V 1, bal Fe.
For lathe and planer tools, reamers, taps; high speed steel.

VITA T-12

LeVita Metal Alloy Co.
C 0.7, W 14, Cr 4, V 3, bal Fe.
For lathe and planer tools; high speed steel.

VITA T-15

LeVita Metal Alloy Co.
C 0.7, W 13, Cr 4.2, V 5, Co 5, bal Fe.
For lathe and planer tools; high speed steel.

VITA T-2
LeVita Metal Alloy Co.
C 0.7, W 18, Cr 4, V 2, bal Fe.
For lathe and planer tools; high speed steel.

VITA T-20
LeVita Metal Alloy Co.
C 0.7, W 18.5, Cr 4, V 4, bal Fe.
For lathe and planer tools; high speed steel.

VITA T-3
LeVita Metal Alloy Co.
C 0.7, W 18, Cr 4, V 3.2, bal Fe.
For lathe and planer tools; high speed steel.

VITA T-4
LeVita Metal Alloy Co.
C 0.7, W 18, Cr 4, V 1, Co 5, bal Fe.
For lathe and planer tools; high speed steel.

VITA T-5
LeVita Metal Alloy Co.
C 0.7, W 18, Cr 4, V 2, Co 8, bal Fe.
For lathe and planer tools; high speed steel.

VITA T-6
LeVita Metal Alloy Co.
C 0.7, W 22, Cr 4.5, V 1.5, Co 12, bal Fe.
For lathe and planer tools; high speed steel.

VITA T-7
LeVita Metal Alloy Co.
C 0.7, W 14, Cr 4, V 2, bal Fe.
For lathe and planer tools; high speed steel.

VITA T-8
LeVita Metal Alloy Co.
C 0.7, W 14, Cr 4, V 2, Co 5, bal Fe.
For lathe and planer tools; high speed steel.

VITAL
Osborn Steels Ltd.
C 2, Cr 13, bal Fe.
For blanking and forming dies, punches; oil hardened, non-deforming. *Obsolete*

VITAL
German manufacture
Si 0.6, Cu 0.9, Zn 1, bal Al.
60,000 TS; 5 El. For light alloy parts; non-hardenable.

VITAL-X
Osborn Steels Ltd.
C 1.5, Cr 12, V 0.8, Mo 0.25, bal Fe.
For blanking and forming dies; air hardened, non-deforming. *Obsolete*

VITALLIUM
Howmet Corp.
C 0-0.5, Si 0-0.6, Mn 0-0.75, Mo 5-7, Cr 28-32, bal Co.
Cast: 97,000-130,000 TS; 58,000-80,000 YS; 4-15 El; 2-20 RA; 45-250 Brin. For orthopedic and dental appliances; corrosion resistant.

VITALU
English manufacture
Si 5, bal Al.
For light alloy parts; die castings.

VITINOX 18 CU
Creusot-Loire
C 0-0.06, Cr 18, Ni 10, Cu 3.5, bal Fe.
Wrought, annealed: 490 N/mm^2 psi TS min. Free machining austenitic stainless steel. AFNOR Z4 CNU 18.9.

VITINOX 18X
Creusot-Loire
C 0-0.12, Cr 18, Ni 9, Mo 0-0.6, 0.15 S min, bal Fe.
Wrought, annealed: 510 N/mm^2 psi TS min. Free machining austenitic stainless steel. AFNOR Z10 CNF 18.9.

VITRAL 045/050
Aluminium Laufen AG
Aluminum.
Aluminum alloy AlMgSi0.5.

VITRENAMEL 1
United States Steel Corp.
Low C, Mn, bal Fe.
For porcelain enameled steel sheet for stoves, kitchenware.

VITRENAMEL 2
United States Steel Corp.
Low C, Mn, bal Fe.
For porcelain enameled steel sheet for stoves, kitchenware.

VITRIFORM
Alliance Wall Corp.
Sheet steel with colored porcelain coating. For interior walls.

VITRIX
Westa-Westdeutsche
C 0.35, Cr 1.3, Ni 4.5, Mn 0.6, bal Fe.
For gears, bolts, crankshafts, forging dies; oil hardened, shock resistant.

VITROVAC 6025
Vacuumschmelze GmbH Co.
Amorphous nonmagnetostrictive soft magnetic alloy with high permeability. Saturation flux density 0.55 T.

VITROVAC 6030
Vacuumschmelze GmbH Co.
Amorphous nonmagnetostrictive soft magnetic alloy like VITROVAC 6025 with high saturation flux density and low loss at frequencies 20-100 kHz. Saturation flux density 0.80 T.

VITROVAC 7505
Vacuumschmelze GmbH Fe.
Amorphous nonmagnetostrictive soft magnetic alloy with high saturation flux density and low cores losses. Saturation flux density 1.50 T.

VIVAL
Cie Francais des Metaux
Si 0.5, Mg 1, Mn 0.5, bal Al.
Rolled: 36,000 TS; 34,000 YS; 5 El. For light alloy parts.

VIZOR
Osborn Steels Ltd.
C 0.8-1.2, bal Fe.
For springs, drills, taps, cutters; Type W1; water hardened. *Obsolete*

VK-10E
VILLARES
C 1.3, Cr 4.2, Mo 4.5, W 8, V 2.7, Co 10, bal Fe.
High speed steel. DIN 1.3207.

VK15
Russian manufacture
Co 15, bal WC.
Sintered: 90 Rock A. For hard cutting tools, dies; sintered tungsten carbide.

VK2
Russian manufacture
Co 2, bal WC.
Sintered: 90-93 Rock A. For hard cutting tools, dies; sintered carbide of tungsten.

VK20
Russian manufacture
Co 20, bal WC.
Sintered: 86-88 Rock A. For hard cutting tools, dies; sintered tungsten carbide.

VK25
Russian manufacture
Co 25, bal WC.
Sintered: 86-88 Rock A. For hard cutting tools, dies; sintered tungsten carbide.

VK3
Russian manufacture
Co 3, bal WC.
Sintered: 92 Rock A. For cutting tools, dies; cemented carbides.

VK4
Russian manufacture
Co 4, bal WC.
Sintered: 92 Rock A. For cutting tools, dies; sintered carbide.

VK5E
VILLARES
C 0.92, Cr 4.15, Mo 5, W 6.3, V 1.85, Co 4.8.
Similar to AISI M35; DIN 1.3243.

VK6
Russian manufacture
Co 6, bal WC.
Sintered: 92 Rock A. For hard cutting tools, dies; sintered carbide of tungsten.

VK6M
Russian manufacture
Co 6, bal WC.
Sintered: 92 Rock A. For hard cutting tools; sintered tungsten carbide, low porosity.

VK8
Russian manufacture
Co 8, bal WC.
Sintered: 91.5 Rock A. For hard cutting tools, dies; cemented carbides of tungsten.

VK8V
Russian manufacture
Co 8, bal WC.
Sintered: 91.8 Rock A. For hard cutting tools, dies; sintered carbide.

VKM42
VILLARES
C 1.1, Cr 3.85, Mo 9.5, V 1.15, W 1.5, Co 8.25.
AISI M42; similar to DIN 1.3247.

VL-30
VILLARES
Alloy steel. C 0.3, Si 0.27, Mn 0.5, Cr 0.95, Mo 0.2, bal Fe.
Hardenable. AISI 4130; DIN 25 CrMo 4.

VL-40
VILLARES
Alloy steel. C 0.4, Si 0.27, Mn 0.87, Cr 0.95, Mo 0.2, bal Fe.
Hardenable. AISI 4140; DIN 42 CrMo 4.

VL3
Acciaierie Valbruna s.p.a.
C 0.51, Mn 0.85, Cr 1, V 0.12, Si 0-0.35, P 0.035, S 0.035, bal Fe.
Cr-V structural steel. AISI 6150; W. Nr. 1.8159.

VL4
Acciaierie Valbruna s.p.a.
C 0.45, Si 1.45, Mn 0.6, P 0-0.035, S 0-0.035, Cr 1.45, V 0.1, bal Fe.
Spring steel. W. Nr. 1.2249.

VL5
Acciaierie Valbruna s.p.a.
C 0.61, Si 0.85, Mn 0.75, P 0-0.035, S 0-0.035, Cr 1.15, V 0.1, bal Fe.
Spring steel. W. Nr. 1.2243.

VL7-45U

Russian manufacture
C 0.16, Cr 20, W 8, B 0.06, Fe 25, Ni 46.
Nickel-iron base superalloy. For nozzle guide vanes.

VLX 25-20

Vallourec S.A.
C 0.1, Cr 25, Ni 20, bal Fe.
Oxidation resistant up to 1050°C. Furnace tubes, refinery piping, burners. Similar to AISI 310.

VLX MOBTBC 3

Vallourec S.A.
C 0.03, Cr 17, Ni 13, Mo 2.7, bal Fe.
Good resistance to organic acids and chlorinated media. For petrochemical, textile, paper mill, pharmacy industries. AISI 316L.

VLX SA

Vallourec S.A.
C 0.03, Cr 25, Ni 20, Nb, bal Fe.
Intergranular corrosion resistance to nitric media. For chemical, nuclear industries. Similar to AISI 310.

VLX SL

Vallourec S.A.
C 0.03, Cr 20, Ni 25, 5.5 Mo + Cu, Nb, bal Fe.
Intergranular corrosion resistance to sulfuric, phosphoric and hydrochloric media. For chemical, petrochemical industries; bleaching.

VLX ST

Vallourec S.A.
C 0.04, Cr 12, Ti, bal Fe.
Resistant to stress corrosion by Cl ions and to oxidation up to 700°C. For heat exchangers, exhaust pipes.

VM 118

Saarstahl AG
Now SAARSTAHL ST E 355.

VM DREADNAUGHT

Hawkridge Bros. Co.
C 0.86, Cr 4, V 2, Mo 8, bal Fe.
For tools, cutters; high speed steel. *Obsolete*

VM-20

VILLARES
C 0.2, Si 0.27, Mn 0.55, Cr 0.5, Ni 1.8, Mo 0.25, bal Fe.
Alloy carburizing steel. AISI 4320.

VM-40

VILLARES
Alloy steel. C 0.4, Si 0.27, Mn 0.7, Cr 0.8, Ni 1.8, Mo 0.25, bal Fe.
Hardenable. AISI 4340.

VMARILITE

English manufacture
Al-Mg.
For light alloy parts.

VMC

Acciaierie Valbruna s.p.a.
C 0.37-0.44, Mn 0.5-0.8, Cr 0.9-1.2, bal Fe.
Chromium structural steel. 40 Cr 4; similar to AISI 5140. *Obsolete*

VMC1

Acciaierie Valbruna s.p.a.
C 0.25, Si 0.3, Mn 1.5-2, Cr 0.6-0.9, bal Fe.
Mn-Cr structural steel (25 Mn Cr 8). *Obsolete*

VMC2

Acciaierie Valbruna s.p.a.
C 0.32-0.39, Mn 0.8-1.1, Cr 1-1.3, bal Fe.
Cr-Mn structural steel. 35 CrMn 5; similar to AISI 5135. *Obsolete*

VMK

Pose-Marre Edelstahlwerk G.m.b.H.
C 2.5, Cr 28, Ni 3, W 2, bal Fe.
As cast: 42-50 Rock C. For wear resistance against abrasion.

VML

VILLARES
Tool steel. C 0.55, Cr 1, Mo 0.45, V 0.07, Mn 0.85, bal Fe.
Hot work tool steel.

VMO

VILLARES
Tool steel. C 0.55, Cr 1.1, Ni 1.65, Mo 0.5, V 0.06, bal Fe.
Hot work tool steel. Similar to DIN 1.2713.

VMS 135

Saarstahl AG
Now SAARSTAHL 37 MNSI 5.

VN-32

VILLARES
C 0.32, Si 0.3, Mn 0.4, Cr 0.6, V 0.1, bal Fe.
Wr.-N. 1.2208; DIN 31 CrV 3.

VN-50

VILLARES
C 0.5, Si 0.27, Mn 0.8, Cr 0.95, V 0.15, bal Fe.
Steel for springs. AISI 6150; similar to DIN 50 CrV 4.

VND

VILLARES
Tool steel. C 0.95, Cr 0.5, W 0.5, V 0.12, Mn 1.25, bal Fe.
Oil hardening tool steel. AISI O1; similar to DIN 1.2419.

VNT

Now ARMCO VNT.

VO3, 7.5-7.99 DENSITY

Keystone Carbon Co.
Sintered nickel silver. 7.5-7.99 density; 32,000 psi TS; 80 Rock H. Pleasant appearing structural parts; corrosion resistant.

VO3, 8.0 MIN DENSITY

Keystone Carbon Co.
Sintered nickel silver. 8.0 density min; 36,000 psi TS; 10.0 El; 75 Rock H. Pleasant appearing structural parts; corrosion resistant.

VOELKLINGEN DOCO

Forges et Acieries de Voelkingen
C 1.6, Cr 13, Mo 1, V 1, Co 2, bal Fe.
For blanking and forming dies, punches; air hardened, nondeforming.

VOELKLINGEN NGSA

Forges et Acieries de Voelkingen
C 0.45, Cr 1.5, W 0.4, Mo 0.5, V 0.8, bal Fe.
For punches, shear blades; hot work steel, oil hardened.

VOELKLINGEN RCC

Forges et Acieries de Voelkingen
C 2, Cr 13, W, bal Fe.
For blanking and forming dies; oil or air hardened, nondeforming.

VOELKLINGEN RCW1

Forges et Acieries de Voelkingen
C 0.6, Cr 3.8, W 10, Mo 1, V 0.8, bal Fe.
For shears, punches, upsetters; hot work steel, oil hardened.

VOELKLINGEN RCW2

Forges et Acieries de Voelkingen
C 0.3, Cr 3, W 10, Mo 0.3, V 0.4, Ni 2, bal Fe.
For extrusion dies, liners, punches; hot work steel, oil hardened.

VOELKLINGEN RT10 EXTRA

Forges et Acieries de Voelkingen
C 1, Si 0.15, Mn 0.3, bal Fe.
Heat treated: 185,000 TS; 120,000 YS; 10 El; 30 RA; 390 Brin. For springs, taps, drills, reamers; Type W1; water hardened.

VOELKLINGEN RTWK

Forges et Acieries de Voelkingen
C, alloy, bal Fe.
For machine tool parts; oil hardened.

VOELKLINGEN RUS

Forges et Acieries de Voelkingen
C 0.9, Mn 2, Si 0.3, bal Fe.
For punches, form dies, rolls; oil hardened, nondeforming.

VOELKLINGEN RUS4

Forges et Acieries de Voelkingen
C, alloy, bal Fe.
For machine tool parts; oil hardened.

VOELKLINGEN RWA

Forges et Acieries de Voelkingen
C 0.3, Cr 2.5, W 4.5, Mo 0.3, V 0.5, bal Fe.
For extrusion press liners, mandrels, dies; hot work steel, oil hardened.

VOELKLINGEN RWS

Forges et Acieries de Voelkingen
C 0.4, Cr 3.3, W 3, Mo 0.5, Co 0.3, bal Fe.
For pneumatic shears, punches, chisels; hot work steel, oil hardened.

VOELKLINGEN RWS2

Forges et Acieries de Voelkingen
C 0.25, Cr 1.3, W 4.5, Mo 0.5, V 0.3, bal Fe.
For extrusion press tools, liners; hot work steel, oil hardened.

VOELKLINGEN SGM2

Forges et Acieries de Voelkingen
C 0.3, Cr 2.5, W 8, V 0.5, Co 2.5, bal Fe.
For extrusion press tools; hot work steel, oil hardened.

VOELKLINGEN SGM5

Forges et Acieries de Voelkingen
C 0.35, Cr 5, W 4, Mo 0.4, V 0.4, Co 0.5, bal Fe.
For extrusion press tools; hot work steel, oil hardened.

VOELKLINGEN T7

Forges et Acieries de Voelkingen
C, alloy, bal Fe.
For forging and die casting dies; hot work steel.

VOIZIT

German manufacture
Fe 96.5, 3.50 graphite.
Sintered: 25-50 Brin. For antifriction metal, bearings; pressed powders, 30-40% porosity.

VOKAR

Russian manufacture
C 8-15, W 78-86, Mn, bal Fe.
For cutting tools; sintered hard alloy.

VOLCANO

Lehigh Steel Corp.
C 0.7, W 18, Cr 4, V 1, Co 5, bal Fe.
For high temperature die work; high speed steel.

VOLCO NO. 10

Volco Brass & Copper Co.
Cu 85, Zn 15.
Hard: 69,000 psi TS; 7 El. Soft: 40,000 psi TS; 45 El. For jewelry, hardware; 37% conductivity.

VOLCO NO. 11

Volco Brass & Copper Co.
Cu 65, Zn 25, Ni 10.
Hard: 88,000 psi TS; 7 El. Soft: 55,000 psi TS; 42 El. For hardware; 8.4% conductivity.

VOLCO NO. 12
Volco Brass & Copper Co.
Cu 65, Zn 23, Ni 12.
Hard: 83,000 psi TS; 6 El. Soft: 54,000 psi TS; 41 El. For hardware; 6.3% conductivity.

VOLCO NO. 14
Volco Brass & Copper Co.
Cu 65, Zn 21, Ni 14.
Hard: 84,000 psi TS; 5 El. Soft: 45,000 psi TS; 40 El. For hardware; 6.2% conductivity.

VOLCO NO. 30
Volco Brass & Copper Co.
Cu 90, Zn 10.
Hard: 62,000 psi TS; 6 El. Soft: 57,000 psi TS; 4 El. For screen wire, jewelry, screw stock; 43% conductivity.

VOLCO NO. 5
Volco Brass & Copper Co.
Cu 63, Zn 32, Ni 5.
Hard: 80,000 psi TS; 7 El. Soft: 50,000 psi TS; 50 El. For plumbing fixtures, jewelry; 12% conductivity.

VOLCO NO. 65
Volco Brass & Copper Co.
Cu 65, Zn 35.
Hard: 73,000 psi TS; 10 El. Soft: 45,000 psi TS; 60 El. For rivets, screws, eyelets, reflectors; brass.

VOLCO NO. 68
Volco Brass & Copper Co.
Cu 70, Zn 30.
Hard: 105,000 psi TS; 1 El. Soft: 49,000 psi YS; 48 El. For cartridges, clips; brass.

VOLCO NO. 8
Volco Brass & Copper Co.
Cu 65, Zn 27, Ni 8.
Hard: 82,000 psi TS; 7 El. Soft: 53,000 psi TS; 45 El. For hardware; 9% conductivity.

VOLCO NO. 80
Volco Brass & Copper Co.
Cu 80, Zn 20.
Hard: 73,000 psi TS; 5 El. Soft: 43,000 psi TS; 8 El. For flexible hose, stampings; 32.5% conductivity.

VOLCO NO. 95
Volco Brass & Copper Co.
Cu 95, Zn 5.
Hard: 55,000 psi TS; 5 El. Soft: 35,000 psi TS; 38 El. For jewelry, coins, stampings; 56% conductivity.

VOLKLINGEN NH22
Forges et Acieries de Voelkingen
C 0.2, Cr 19-22, Ni 19-22, bal Fe.
Annealed: 95,000 TS; 45,000 YS; 50 El; 90 Rock B. For furnace parts and equipment, valves, pumps, boiler baffles, turbine and jet engine components. Type 310 stainless steel, heat and corrosion resistant.

VOLKLINGEN NH4
Forges et Acieries de Voelkingen
C 0.16, Cr 23, Ni 14, bal Fe.
Annealed: 75,000 TS; 30,000 YS; 40 El; 220 Brin. For heat exchangers, combustion chambers, salt pots, oil burners. Type 309 stainless steel, austenitic, heat and corrosion resistant.

VOLKLINGEN NH8G
Forges et Acieries de Voelkingen
C 0.15, Cr 24, Ni 13, bal Fe.
Annealed: 75,000 TS; 32,000 YS; 210 Brin. For furnace parts, salt pots, heat exchangers, combustion chambers. Type 309 stainless steel, corrosion and heat resistant.

VOLKLINGEN RNO
Forges et Acieries de Voelkingen
C 0.2, Cr 12-14, bal Fe.
Annealed: 95,000 TS; 50,000 YS; 25 El; 92 Rock B. For pump and valve parts, surgical instruments, cutlery, gauges. Type 420 stainless steel, heat and corrosion resistant, hardenable.

VOLKLINGEN RNOF
Forges et Acieries de Voelkingen
C 0.25, Cr 12-14, bal Fe.
Annealed: 98,000 TS; 52,000 YS; 20 El; 95 Rock B. For cutlery, surgical instruments, pump and valve parts. Type 420 stainless steel, heat and corrosion resistant, hardenable.

VOLKLINGEN RNOW
Forges et Acieries de Voelkingen
C 0.12, Cr 12-14, bal Fe.
Annealed: 80,000 TS; 40,000 YS; 30 El; 85 Rock B. For furnace parts, oil refinery equipment, springs, tableware. Type 410 stainless steel, corrosion and heat resistant.

VOLKLINGEN RNOWW
Forges et Acieries de Voelkingen
C 0.1, Cr 14-18, bal Fe.
Annealed: 75,000 TS; 45,000 YS; 30 El; 80 Rock B. For automotive trim, kitchen sinks, fasteners, heat treating boxes. Type 430 stainless steel, corrosion resistant, nonhardenable.

VOLOMIT
Manufacturer not listed.
WC + MoC.
For cutting tools, dies; similar to "Stellite."

VOLTAL
Busch-Jager Ludenscheider Metallwerke
Cu 4.7, Si 2-4, Zn 0-2.5, Fe 0-1.1, bal Al.
For light alloy parts; age-hardenable.

VOLUMIT
Manufacturer not listed.
Tungsten carbide plus molybdenum. For hard cutting tools and dies; sintered alloy.

VOLVIC
Aubert & Duval
C 0.25, Cr 3.1, W 9, V 0.4, bal Fe.
For rivet sets, extrusion press parts, punches; hot work steel, oil hardened.

VOLVIT
KM-kabelmetal AG
Sn 9, Cu 91.
For bearings. *Obsolete*

VPC
VILLARES
Tool steel. C 0.38, Cr 5, Mo 1.35, V 0.4, Si 1, bal Fe.
Hot work tool steel. AISI H11; similar to DIN 1.2343.

VPCW
VILLARES
Tool steel. C 0.35, Cr 5, Mo 1.5, W 1.35, V 0.25, Si 0.9, bal Fe.
Hot work tool steel. AISI H12; similar to DIN 1.2606.

VPE
VILLARES
Tool steel. C 1, Cr 3.75, bal Fe.
Tool steel. *Obsolete*

VR-15
VILLARES
C 0.15, Si 0.27, Mn 0.8, Cr 0.8, bal Fe.
Alloy carburizing steel. AISI 5115; similar to DIN 15 Cr 3.

VR-30
VILLARES
Alloy steel. C 0.3, Si 0.27, Mn 0.8, Cr 0.95, bal Fe.
Hardenable. AISI 5130.

VR-35
VILLARES
Alloy steel. C 0.35, Si 0.27, Mn 0.7, Cr 0.9, bal Fe.
Hardenable alloy steel. AISI 5135; similar to DIN 34 Cr 4.

VR-40
VILLARES
Alloy steel. C 0.4, Si 0.27, Mn 0.8, Cr 0.8, bal Fe.
Hardenable. AISI 5140; similar to DIN 41 Cr 4.

VR-50
VILLARES
C 0.5, Si 0.27, Mn 0.8, Cr 0.8, bal Fe.
Spring steel. AISI 5150.

VR-60
VILLARES
C 0.6, Si 0.27, Mn 0.87, Cr 0.8, bal Fe.
Spring steel. AISI 5160.

VR/WESSON 26
Fansteel
Co 7, 83 WC, 10 TiC.
Sintered; 225,000 Transverse Strength; 91.8 Rock A; density 12.40. For general purpose and finish machining of steel and non-ferrous metals. Good shock and wear resistance. C-6 Type.

VR/WESSON 2A1
Fansteel
Wc 86, Co 14.
Sintered: 88.5 RA; 375,000 Transverse Strength; 14.1 density For cutting tools requiring medium shock resistance; and for light impact applications. C-11, C-12 type.

VR/WESSON 2A3
Fansteel
Co 11, 89 WC.
Sintered: 325,000 Transverse Strength; 88.3 Rock A; density 14.40. For heavy roughing cuts and interrupted machining of ferrous and non-ferrous metals, and for non-metallics. Excellent shock resistance.

VR/WESSON 2A5
Fansteel
Co 6, 94 WC.
Sintered: 250,000 Transverse Strength; 91.8 Rock A; density 14.90. For general purpose machining, including hardened ferrous and non-ferrous. Very good wear resistance.

VR/WESSON 2A6
Fansteel
Co 8, 92 WC.
Sintered: 300,000 Transverse Strength; 89.5 Rock A; density 14.70. For heavy roughing and interrupted cuts, also for large wire drawing dies. Good shock resistance. C-10 Type.

VR/WESSON 2A68
Fansteel
Co 6, 94 WC.
Sintered: 275,000 Transverse Strength; 90.8 Rock A; density 14.85. For rough machining and interrupted cuts on ferrous and non-ferrous metals. Good shock and wear resistance.

VR/WESSON 2A7
Fansteel
Co 4.5, 95 WC.
Sintered: 225,000 Transverse Strength; 92.2 Rock A; density 15.10. For fine finishing of ferrous and non-ferrous alloys. Excellent wear resistance. C-3, 3-4 Type.

VR/WESSON 630
Fansteel
TiC-base coating on cemented carbide. For cutting tools, excellent abrasion resistance, good toughness and wear resistance. For finishing and general purpose machining. C-2 and C-3 type applications.

VR/WESSON 650
Fansteel

TiC-base coating on alloyed cemented carbide. For cutting tools, excellent shock and wear resistance to cutting temperature, especially for milling and interrupted cuts.

VR/WESSON 660
Fansteel

TiC-base coating on alloyed cemented carbide. For cutting tools, good shock and wear resistance.

VR/WESSON 670
Fansteel

TiC-base coating on alloyed cemented carbide. For cutting tools, mainly semi-finishing and high speed finish machining of steel.

VR/WESSON RAMET I
Fansteel

WC 89.5, Co 10, 0.5 Cr_3C_2.
Sintered: 91.5 RA; 375,000 Transverse Strength; 14.5 density For machining in problem areas, particularly slow speeds; also for dies. C-O, C-1, C-9, C-10, C-11 type.

VR/WESSON VR 14
Fansteel

Co 11.5, 88.5 WC.
Sintered: 350,000 Transverse Strength; 88.2 Rock A; density 14.40. Cutting tools for mining, quarrying and percussion drilling. C-13 Type.

VR/WESSON VR-100
Fansteel

Al_2O_3, TiC ceramic. Hot pressed: 95.0 RA; 115,000 Transverse Strength; 4.3 density. For cutting tools, turning and boring cast irons and steels up to 70 Rock C, non-metallic materials, and milling various cast irons and steels at high speeds.

VR/WESSON VR-13
Fansteel

Co 10, 90 WC.
Sintered: 350,000 Transverse Strength; 88.8 Rock A; density 14.50. Cutting tools for mining, quarrying and percussion drilling. C-13 Type.

VR/WESSON VR-15
Fansteel

Co 13, 87 WC.
Sintered: 375,000 Transverse Strength; 87.5 Rock A; density 14.30. Cuttiing tools for mining, quarrying and percussion drilling. C-14 Type.

VR/WESSON VR-54
Fansteel

Co 7, 93 WC.
Sintered: 275,000 Transverse Strength; 91.8 Rock A; density 14.60. For milling, broaching, hobbing, cast ferrous and non-ferrous alloys. Very good shock resistance.

VR/WESSON VR-65
Fansteel

TiC, Mo_2C, Ni.
Sintered: 92.3 RA; 150,000 Transverse Strength; 5.85 density For high speed fine finishing and boring. C-4, C-8 types.

VR/WESSON VR-71
Fansteel

Co 6, 10 TaC + CbC, 66 WC, 18.0 TiC.
Sintered: 200,000 Transverse Strength; 92,5 Rock A; density 10.90. For high speed finishing and precision boring of steel. Excellent wear resistance. High resistance to cutting temperature. C-7 type.

VR/WESSON VR-73
Fansteel

Co 6.5, 71.5 WC, 12.0 TiC, 10 TaC + CbC.
Sintered: 200,000 Transverse Strength; 92.0 Rock A; density 11.90. For finish machining of steel. Very good wear resistance; resistance to high cutting temperature. C-7 Type.

VR/WESSON VR-75
Fansteel

Co 7.5, 74.5 WC, 8.0 TiC, 10 TaC + CbC.
Sintered: 250,000 Transverse Strength; 91.5 Rock A; density 12.70. For general purpose machining steel. Good shock and wear resistance; resistance to high cutting temperature.

VR/WESSON VR-77
Fansteel

Co 8.5, 73.5 WC, 8.0 TiC, 10 TaC + CbC.
Sintered: 265,000 Transverse Strength; 91.3 Rock A; density 12.60. For heavy roughing and interrupted cuts on steel. Very good shock resistance.

VR/WESSON VR-82
Fansteel

WC 92, Co 5.5, 2.5 TaC/CbC.
Sintered: 92.5 RA; 250,000 Transverse Strength; 14.95 density. For straight machining of high temperature alloys where chips are tough. Fair shock resistance; good wear resistance. C-2, C-3, C-9 types.

VR/WESSON VR-85
Fansteel

Sintered: 91.5 Rock D, 260,000 Transverse Strength; 14.9 density. For straight machining of high temperature alloys where the chips are tough.

VR/WESSON VR-87
Fansteel

Co 17, 55.0 WC, 28.0 TaC.
Sintered: 350,000 Transverse Strength; 85.0 Rock A; density 13.50. Good hot hardness, for machining hot flash from heavy weld in steel pipe. Excellent resistance to shock and to cutting temperatures.

VR/WESSON VR-89
Fansteel

Co 11, 71.0 WC, 18.0 TaC.
Sintered: 275,000 Transverse Strength; 89.9 Rock A; density 14.15. For turning and milling tough alloy steels as high manganese steels; excellent shock resistance.

VR/WESSON VR-97
Fansteel

Al_2O_3, ceramic. Sintered: 93.5 RA; 100,000 Transvers Strength; 3.98 density. For cutting tools, continuous cuts a extremely high speeds, and hard metals, up to 60 Rock C, at lower speeds.

VR/WESSON WH
Fansteel

Co 7, 78.0 WC, 13.0 TiC, 2.0 TaC.
Sintered: 220,000 Transverse Strength; 92.0 Rock A; density 11.75. For general purpose and finish machining of steel. Good wear resistance. C-7 Type.

VR/WESSON WM
Fansteel

Co 10, 75.0 WC, 13.0 TiC, 2 TaC + CbC.
Sintered: 250,000 Transverse Strength; 91.0 Rock A; density 11.50. For roughing and general machining steel. Good shock resistance.

VR/WESSON WS
Fansteel

Co 10.5, 83.5 WC, 4.0 TiC, 2.0 TaC + CbC.
Sintered: 275,000 Transverse Strength; 90.5 Rock A; density 13.20. For rough machining steel; interrupted cuts. Excellent shock resistance.

VS SPEZIAL EXTRA
Thyssen Edelstahlwerke AG
C, W, V, bal Fe.
For steel trimming and weld steel. *Obsolete*

VS-30
Krupp Stahl AG
react and refract. Cr 15, Co 52, bal Fe.
Remanence 18,000; coercive force 25. For hysteresis motors, electro and magnetic devices. Vicalloy type permanent magnet. *Obsolete*

VS-60
VILLARES
C 0.6, Si 2, Mn 0.87, bal Fe.
Spring steel. AISI 9260; similar to DIN 65 Si 7.

VSC
W. Ossenberg & Cie Edelstahlwerke
C 2.25, Mn 0.3, Si 0.3, Cr 14.5, Mo 1, V 0.1, bal Fe.
Oil or air hardenable to 64-66 Rock C. For cold forming and precision rollers.

VT-1-O
Russian manufacture
Unalloyed titanium.

VT-20
VILLARES
C 0.2, Si 0.25, Mn 0.45, bal Fe.
Low carbon steel; AISI 1020; similar to DIN CK 22.

VT-21
VILLARES
C 0.2, Si 0.25, Mn 0.75, bal Fe.
Low carbon steel. AISI 1021. *Obsolete*

VT-30
VILLARES
Carbon steel. C 0.3, Si 0.25, Mn 0.75, bal Fe.
Carbon steel. AISI 1030.

VT-38
VILLARES
Carbon steel. C 0.38, Si 0.25, Mn 0.75, bal Fe.
Carbon steel. AISI 1038; similar to DIN CK 35.

VT-40
VILLARES
Carbon steel. C 0.4, Si 0.25, Mn 0.75, bal Fe.
AISI 1040; similar to DIN CK 40.

VT-45
VILLARES
Carbon steel. C 0.45, Si 0.25, Mn 0.75, bal Fe.
AISI 1045.

VT-5
Russian manufacture
Al 4-5.5, bal Ti.

VT-5-1
Russian manufacture
Al 4-5.5, Sn 2-3, bal Ti.
Annealed: 125,000 TS; 120,000 YS; 15 El. For compressor blades, engine cowlings, support rings; Alpha alloy; corrosion resistant.

VT-50
VILLARES
Carbon steel. C 0.5, Si 0.25, Mn 0.75, bal Fe.
AISI 1050; DIN CK 50.

VT-6
Russian manufacture
Al 5-6.5, V 3.5-4.5, bal Ti.
Annealed: 150,000 TS; 140,000 YS; 15 El; 30 Rock C. For jet engine components, airframe parts, fasteners; Alpha-beta alloy, good hot strength.

VT-60
VILLARES
Carbon steel. C 0.6, Si 0.25, Mn 0.75, bal Fe.
AISI 1060; similar to DIN CK 60.

VT-6S

Russian manufacture
Al 4.5, V 3.5, bal Ti.
Alpha-beta alloy.

VT-70

VILLARES
Carbon steel. C 0.7, Si 0.25, Mn 0.75, bal Fe.
AISI 1070; similar to DIN CK 69.

VT-80

VILLARES
Carbon steel. C 0.8, Si 0.25, Mn 0.75, bal Fe.
AISI 1080.

VT-95

VILLARES
Carbon steel. C 0.95, Si 0.25, Mn 0.4, bal Fe.
AISI 1095.

VTG 129

Engelhard Corp.
Pd 59.01-61, bal Ni.
Uses: Brazing electron tubes. 2260°F MP.

VTG 238

Engelhard Corp.
Au 79.5-80.5, Ag 0-0.05, bal Cu.
For brazing electron tubes. 1630°F MP.

VTG 255

Engelhard Corp.
Au 81.6-82.5, bal Ni.
For brazing electron tubes. 1742°F MP.

VTG 260

Engelhard Corp.
Au 34.5-35.5, Ag 0-0.05, bal Cu.
For brazing electron tubes. 1832-1870°F MP.

VTG 261

Engelhard Corp.
Au 74.5-75.5, Ag 4.5-5.5, bal Cu.
For brazing electron tubes. 1625-1640°F MP.

VTG 301

Engelhard Corp.
Ag 71-73, bal Cu.
For brazing of electron tubes. 1435°F MP.

VTG 428

Engelhard Corp.
Pd 4.5-5.5, bal Ag.
For brazing electron tubes. 1780-1850°F MP.

VTG 431

Engelhard Corp.
Pd 9.5-10.5, bal Ag.
For brazing electron tubes. 1835-1950°F MP.

VTG 447

Engelhard Corp.
Pd 19.5-20.5, bal Ag.
For brazing electron tubes. 1960-2150°F MP.

VTG 478

Engelhard Corp.
Pd 4.5-5.5, Ag 67.69, bal Cu.
For brazing electron tubes. 1480-1490°F MP.

VTG 490

Engelhard Corp.
Pd 14.5-15.5, Ag 64.66, bal Cu.
For brazing electron tubes. 1565-1650°F MP.

VTG 491

Engelhard Corp.
Pd 9.5-10.5, Ag 57.59, bal Cu.
For brazing electron tubes. 1520-1565°F MP.

VTG 492

Engelhard Corp.
Pd 24.5-25.5, Ag 53-56, bal Cu.
For brazing electron tubes. 1650-1740°F MP.

VTG T51

Engelhard Corp.
Ag 74-76, Ni 0.25-0.75, bal Cu.
For brazing electron tubes. 1435-1475°F MP.

VULC-IRON H

Valley-Vulcan Mold Co.
C 3.8-4.2, Si 0.55-0.85, Mn 0.6-0.9, S 0.02-0.05, P 0-0.12, bal Fe.
For ingot molds, stools and related products.

VULC-IRON L

Valley-Vulcan Mold Co.
C 3.4-4, Si 1.4-1.9, Mn 0.6-1.1, S 0.03-0.09, P 0-0.12, bal Fe.
For ingot molds, stools and related products.

VULCAN

Phosphor Bronze Co. Ltd.
87.0 Sn white metal.
High speed. For thin wall linings, locomotive diesel engines applications.

VULCAN "SUPERIOR"

H.K. Porter Co. (Vulcan-Kidd Div.)
C 1, V 0.2, bal Fe.
For chisels, cutters, shop tools, rock drills; for general tool purposes. *Obsolete*

VULCAN CWD

H.K. Porter Co. (Vulcan-Kidd Div.)
C alloy, bal Fe.
For tools. *Obsolete*

VULCAN EXTRA DRILL

H.K. Porter Co. (Vulcan-Kidd Div.)
C 1-1.2, bal Fe.
For miners drills. *Obsolete*

VULCAN HEAVY DUTY

H.K. Porter Co. (Vulcan-Kidd Div.)
C 0.7, Cr 1.5, V 0.2, bal Fe.
For flying shears; tough. *Obsolete*

VULCAN METAL

English manufacture
Cu 80.5, Al 11.7, Fe 4.4, Zn 0.25, Ni 1.5, Sn 0.7, Cr 0.7.
For corrosion resistant structural parts; corrosion resistant.

VULCAN SELF-HARDENING

H.K. Porter Co. (Vulcan-Kidd Div.)
C 0.9, Cr 5, bal Fe.
For tools; self-hardening. *Obsolete*

VULKAN

SWB Stahlformguss Gesellschaft mbH
C 0.56, Ni, Cr, Mo, V, bal Fe.
For gears, crimpers, crankshafts, bolts, studs; oil hardened, shock resistant. *Obsolete*

VULKAN

Hufnagel GmbH
C 0.8, Cr 4.25, Mo 0.6, W 18, V 1.6, Co 5, bal Fe.
High speed steel; for rough machining.

VULKAN 1

Robert-Leyer-Pritzkow & Co.
C 2.1, Cr 11.5, W 0.7, bal Fe.
For blanking, piercing and forming dies, punches; oil or air hardened, non-deforming.

VULKAN 2B

Robert-Leyer-Pritzkow & Co.
C 1.65, Cr 11.5, V 0.1, bal Fe.
For blanking, piercing and forming dies, punches; air hardened, non-deforming.

VULKAN 2MC

Robert-Leyer-Pritzkow & Co.
C 1.5, Co, Mo, Cr, bal Fe.
For blanking and piercing dies, punches; air hardened, non-deforming.

VULKAN 2MW

Robert-Leyer-Pritzkow & Co.
C 1.65, Cr, Mo, W, V, bal Fe.
For blanking and piercing dies, punches; air hardened, non-deforming.

VULKAN 3

Robert-Leyer-Pritzkow & Co.
C 2.1, Cr 11.5, bal Fe.
For blanking, piercing and forming dies, punches; oil or air hardened, non-deforming.

VULKAN 3MW

Robert-Leyer-Pritzkow & Co.
C 1.5, Cr, Mo, V, bal Fe.
For blanking and piercing dies; air hardened, non-deforming.

VULKAN 3V

Robert-Leyer-Pritzkow & Co.
C 2.34, Cr, V, bal Fe.
For blanking, piercing, forming dies, punches; oil hardened, wear resistant.

VULKAN 4

Robert-Leyer-Pritzkow & Co.
C 1.35, Cr, bal Fe.
For bearings, liners, cutters, engravers' tools; water hardened, wear resistant.

VULKAN 55

SWB Stahlformguss Gesellschaft mbH
C 0.55, Si 0.25-0.5, Mn 0.3-0.8, bal Fe.
Annealed: 100,000 TS; 55,000 YS; 14 El; 20 RA; 175 Brin. For axes, axles, shafts, gears, crankshafts; water hardened. *Obsolete*

VULKAN 65

SWB Stahlformguss Gesellschaft mbH
C 0.6, Si 0.25-0.5, Mn 0.3-0.8, bal Fe.
Heat treated: 160,000 TS; 113,000 YS; 12 El; 40 RA; 325 Brin. For rails, hammers, tools, shafts, crankshafts; water hardened. *Obsolete*

VULKAN CM

SWB Stahlformguss Gesellschaft mbH
C 0.4, Cr, Mn, Mo, bal Fe.
For gears, bolts, fasteners, shafts, punches; water hardened. *Obsolete*

VULKAN MS

SWB Stahlformguss Gesellschaft mbH
C 0.53, Si 0.9, Mn 0.9, bal Fe.
Annealed: 98,000 TS; 55,000 YS; 15 El; 22 RA; 180 Brin. For punches, crimpers, gears, shafts; water hardened. *Obsolete*

VULKAN N

SWB Stahlformguss Gesellschaft mbH
C 0.55, Cr 0.7, Mo 0.18, Ni 1.65, V 0.1, bal Fe.
For gears, shafts, crankshafts, bolts, studs; oil hardened, shock resistant. *Obsolete*

VULKAN SPEZIAL

SWB Stahlformguss Gesellschaft mbH
C 0.35, Cr 1.35, Mo 0.25, Ni 3.9, bal Fe.
For gears, bolts, crankshafts; oil hardened, shock resistant. *Obsolete*

VV

Acciaierie Valbruna s.p.a.
C 0.1, Mn 0.3-0.7, bal Fe.
Low carbon steel. Ital. UNI C10; AISI 1010. *Obsolete*

VV-20
VILLARES
C 0.2, Cr 21, Ni 11.5, Si 1, Mn 1.3, N 0.17, bal Fe.
SAE EV4.

VV-35
VILLARES
C 0.35, Cr 19, Ni 8, Si 3, bal Fe.
Valve steel; similar to SAE EV-5.

VV-45
VILLARES
C 0.45, Cr 8.5, Si 3.3, Mn 0.4, bal Fe.
Valve steel. SAE HNV-3; similar to DIN 1.4718.

VV-53
VILLARES
C 0.53, Cr 21, Ni 4, Mn 9, N 0.42, bal Fe.
Valve steel. SAE EV-8.

VV-56
VILLARES
C 0.55, Cr 20, Ni 2, Mn 8, N 0.3, bal Fe.
Similar to SAE EV-12.

VV-73
VILLARES
C 0.73, Cr 21, Ni 1.7, Mn 6.3, Ni 0.2, bal Fe.
Valve steel. SAE EV-11. *Obsolete*

VV-80
VILLARES
C 0.8, Cr 19.5, Ni 1.2, Si 2.3, Mn 0.4, bal Fe.
Valve steel. SAE HNV-6; DIN 1.4747.

VW MAGNESIUM
English manufacture
Al 8, Zn 0.6, Mn 0.3, Be 0.005, bal Mg.
Permanent mold: 20,000-21,500 TS; 14,000-18,500 YS; 3.25
El; 54-56 Brin. Die cast: 21,500-24,000 TS; 17,000-20,000 YS;
2 El; 60-64 Brin. For light alloy parts.

VW-1
VILLARES
Tool steel. C 1.15, Cr 0.2, W 1.1, V 0.12, bal Fe.
Oil hardening tool steel. For cold work. Similar to AISI O7; DIN
1.2516.

VW-3
VILLARES
Tool steel. C 0.45, Cr 1.4, Mo 0.2, W 2, V 0.2, Si 1, bal Fe.
Shock resistant tool steel. Similar to AISI S1; similar to DIN
1.2542.

VW-9
VILLARES
Tool steel. C 0.3, Cr 2.65, W 8.5, V 0.35, bal Fe.
Hot work tool steel. Similar to AISI H20; DIN 1.2581.

VWK-10
VILLARES
C 0.75, Cr 4.25, Mo 0.65, W 18, V 1.6, Co 9.5, bal Fe.
High speed steel. AISI T5; similar to DIN 1.3255.

VWK-5
VILLARES
C 0.8, Cr 4.15, W 14, V 2.1, Co 5, bal Fe.
High speed steel. AISI T8; similar to DIN 1.3251.

VWM-1
VILLARES
C 0.8, Cr 4.25, Mo 9, W 2, V 1.2, bal Fe.
High speed steel. AISI M1; DIN 1.3346. *Obsolete*

VWM-2
VILLARES
C 0.83, Cr 4.1, Mo 5, W 6.1, V 2, bal Fe.
High speed steel. AISI M2; similar to DIN 1.3343.

VWM-7
VILLARES
C 1, Cr 3.75, Mo 8.75, W 1.75, V 2, bal Fe.
High speed steel. AISI M7; similar to DIN 1.3348.

VZH 36-L1
Russian manufacture
Cr 10, Al 5, W 8, Mo 4, B 0.3, bal Ni.
Cast nickel-base superalloy.

VZH 36-L2
Russian manufacture
C 0-0.06, Cr 19-22, Ti 2.3-2.7, Al 3.5-4, B 0-0.03, Fe, bal Ni.
Cast nickel-base superalloy. For automotive turbine blades.

VZH 98
Russian manufacture
C 0-0.1, Cr 25.5, W 15, bal Ni.
High temperature alloy for tailpipes, after burner liners,
combustion cans.

VZHL8
Russian manufacture
C 0.1-0.2, Cr 4-17, Ti 1.8-2.5, Al 2.5-3.5, Mo 4.5-6, B 0.06, Fe
8-12, bal Ni.
Cast nickel-base superalloy. For nozzle guide vanes.

VZKG
Czechoslovakian manufacture
C 0.2, Cr 11, Mo, V, W, bal Fe.
Annealed: 95,000 TS; 40,000 YS; 25 El; 92 Rock B.
Hardened: 240,000 TS; 205,000 YS; 8 El; 50 Rock C. For
valves, cutlery, gears, bearings, surgical instruments.
Corrosion resistant, hardenable.

W 2

English manufacture
C 0.7, Cr 3.3, Cr 0.8, W 16, Co 4, Ni 0.4, Mo 0.6, bal Fe.
For cutters, taps, drills; high speed steel.

W 311

Bergische Stahl Industrie
C 0.35-0.45, Si 0.2-0.4, Mn 1.3-1.6, Cr 1.8-2.1, Mo 0.15-0.25, bal Fe.
Hot work tool steel. W.-Nr. 1.2311.

W 367

Bergische Stahl Industrie
C 0.37-0.43, Si 0.3-0.5, Mn 0.3-0.6, Cr 4.7-5.2, Mo 2.7-3.3, V 0.8-1, bal Fe.
Deep hardening structural steel. W.-Nr. 1.2367.

W 376

Bergische Stahl Industrie
C 0.9-1, Si 0.2-0.4, Mn 0.2-0.4, Cr 11-12, Mo 0.85-0.9, V 0.85-0.95, bal Fe.
Cold work tool steel. W.-Nr. 1.2376.

W 5 ALLOY

Henry Wiggin & Co. Ltd.
Mn 0.5, Si 4, bal Ni.
Annealed: 81,000 TS; 24,000 YS; 56 El; 77 RA; 130 Brin. For spark plug electrodes; resists engine fuels. *Obsolete*

W 565

Bergische Stahl Industrie
C 0.35-0.45, Si 0.8-1.1, Mn 0.3-0.5, Cr 5.1-5.3, W 3.5-4, V 0.15-0.2, bal Fe.
Hot work tool steel. W.-Nr. 1.2565.

W 6 ALLOY

Inco Alloys International Inc.
Mn 0.5, Si 2, bal Ni.
Annealed: 81,000 TS; 24,000 YS; 56 El; 77 RA; 130 Brin. For spark plug electrodes; resists engine fuels. *Obsolete*

W 622

Bergische Stahl Industrie
C 0.55-0.65, Si 0.2-0.4, Mn 0.2-0.4, Cr 3-4.1, W 8.5-9.5, V 0.6-0.8, bal Fe.
Hot work tool steel. W.-Nr. 1.2622.

W ALLOY NO. 6

Wall Colmonoy Corp.
C 0.9-1.4, Cr 25-31, W 3.5-6, Si 0.75-1.7, bal Co.
Welded: 400-440 Brin. For hard facing electrode, valve trim, for oxyacetylene welding; wear and corrosion resistant. *Obsolete*

W BRAND

Now DARWIN W BRAND.

W DIE STEEL

Flockton, Tompkin & Co., Ltd.
C 0.5, W 2.2, Cr 1, V 0.35, Mn 0.5, bal Fe.
For punches, piercing dies, chisels; oil or water hardened, tough.

W HYDROGEN-REDUCED

Kennametal Inc.
W 99.2-99.7, Fe 0.1-0.3, TC 0.05-0.15, 0.01-0.03 free C.
Produced by the conventional hydrogen-reduction process; used in hot-pressed matrices as a filler, wear-rate modifier, and bond-system hardener.

W O O

W. Ossenberg & Cie Edelstahlwerke
C 0.15, Si 0.25, Mn 0.4, bal Fe.
Carbon steel for case-hardened tools. Werkstoff Nr. 1.1705.

W O 10 EXTRA

W. Ossenberg & Cie Edelstahlwerke
C 1, Si 0.25, Mn 0.25, bal Fe.
Carbon tool steel. Werkstoff Nr. 1.1540.

W O 10 PRIMA

W. Ossenberg & Cie Edelstahlwerke
C 1, Si 0.3, Mn 0.35, bal Fe.
Carbon tool steel. Werkstoff Nr. 1.1640.

W O 11 EXTRA

W. Ossenberg & Cie Edelstahlwerke
C 1.1, Si 0.25, Mn 0.25, bal Fe.
Carbon tool steel. Werkstoff Nr. 1.1550.

W O 11 PRIMA

W. Ossenberg & Cie Edelstahlwerke
C 1.15, Si 0.3, Mn 0.35, bal Fe.
Carbon tool steel. Werkstoff Nr. 1.1650.

W O 120

W. Ossenberg & Cie Edelstahlwerke
C 1.4, Si 0.3, Mn 0.3, Cr 0.3, V 0.1, bal Fe.
Cold work tool steel for punches, stamps, centering punch. Werkstoff Nr. 1.2206.

W O 13 PRIMA

W. Ossenberg & Cie Edelstahlwerke
C 1.3, Si 0.3, Mn 0.35, bal Fe.
Carbon tool steel. Werkstoff Nr. 1.1660.

W O 2

W. Ossenberg & Cie Edelstahlwerke
C 0.35, Si 0.4, Mn 0.5, bal Fe.
Carbon tool steel. Werkstoff Nr. 1.1720.

W O 3

W. Ossenberg & Cie Edelstahlwerke
C 0.45, Si 0.35, Mn 0.7, bal Fe.
Carbon tool steel. Werkstoff Nr. 1.1730.

W O 4

W. Ossenberg & Cie Edelstahlwerke
C 0.6, Si 0.4, Mn 0.7, bal Fe.
Carbon tool steel. Werkstoff Nr. 1.1740.

W O 5

W. Ossenberg & Cie Edelstahlwerke
C 0.75, Si 0.4, Mn 0.7, bal Fe.
Carbon tool steel. Werkstoff Nr. 1.1740.

W O 7 EXTRA

W. Ossenberg & Cie Edelstahlwerke
C 0.7, Si 0.25, Mn 0.25, bal Fe.
Carbon tool steel. Werkstoff Nr. 1.1520.

W O 7 PRIMA

W. Ossenberg & Cie Edelstahlwerke
C 0.7, Si 0.3, Mn 0.35, bal Fe.
Carbon tool steel. Werkstoff Nr. 1.1620.

W O 9 EXTRA

W. Ossenberg & Cie Edelstahlwerke
C 0.85, Si 0.25, Mn 0.25, bal Fe.
Carbon tool steel. Werkstoff Nr. 1.1530.

W O 9 PRIMA

W. Ossenberg & Cie Edelstahlwerke
C 0.85, Si 0.3, Mn 0.35, bal Fe.
Carbon tool steel. Werkstoff Nr. 1.1630.

W TAP

Latrobe Steel Co.
C 1.25, W 1.5, Cr 0.4, Mo 0.3, bal Fe.
For taps, drills, reamers; oil hardened. *Obsolete*

W-0.33

Manufacturer not listed.
Al 85, Cu 14, Mn 1.
For pistons, light alloy castings; non-hardenable.

W-25% RE

Now WAH CHANG W-25 RE.

W-5

Jessop Steel Co.
C 1.15, Mn 0.25, Si 0.25, Cr 0.5, bal Fe.
AISI Type W5; water hardening tool steel.

W-545

Now WESTINGHOUSE W-545.

W-545

Cannon Muskegon Corp.
C 0-0.08, Mn 1.5, Si 0.4, Cr 13.5, Ni 26, Mo 1.5, B 0.08, Ti 2.85, Al, bal Fe.
For gas turbine parts, high temperature bolting.

W-722

Now NICKELVAC W-722.

W-AL-CO TYPE 2

Welding Alloy Mfg. Co.
Si 5, bal Al.
For gas welding rod; for aluminum.

W-AL-CO TYPE 1

Welding Alloy Mfg. Co.
Si 5, bal Al.
For arc welding electrodes; for aluminum.

W-AL-CO TYPE E-2S

Welding Alloy Mfg. Co.
Si 5, bal Al.
For welding rod; for aluminum arc welding.

W-AL-CO TYPE G-2S

Welding Alloy Mfg. Co.
Si 5, bal Al.
For welding rod; for aluminum gas welding.

W-ALLOY

Manufacturer not listed.
Al 82, Cu 12, Zn 4.5, W 1.
For light alloy castings; leak proof, hard.

W-DECARB

Kennametal Inc.
W 99.6-99.9, Fe 0.02-0.1.
Used in hot-pressed bond systems as a wear-rate modifier, as a bond hardener, and sometimes as a filler for free-cutting.

W. 3 TUNGSTEN

Hidalgo Steel Co. Inc.
C, W, bal Fe.
For hot dies; heat resistant.

W.9 TUNGSTEN

Hidalgo Steel Co. Inc.
C, W, bal Fe.
For hot dies; heat resistant.

W.K.Z. HOT WORK

Houghton & Richards Inc.
C 0.6, W 8.5, Cr 2.8, V 4, bal Fe.
For hot work tools, dies; hot work steel. AISI H 21.

W.M. 1735

Billiton International Metals B.V.
Sn 10, Sb 15, Cu 1.5, Pb 73.5.
For bearings; white metal.

W.M. 802

Billiton International Metals B.V.
Sn 80, Sb 12, Cu 6, Pb 2.
For bearings; white metal.

W.M. 810

Billiton International Metals B.V.
Sn 80, Sb 10, Cu 10.
For bearings; white metal.

W.M. 855

Billiton International Metals B.V.
Sn 85, Sb 10, Cu 5, Pb 0-0.35.
For bearings; white metal.

W.M. 903

Billiton International Metals B.V.
Sn 90, Sb 7, Cu 3, Pb 0-0.35.
For bearings; white metal.

W.M.S.

British Steel Corp.
C 0.4, Ni 1.5, bal Fe.
Annealed: 74,000 TS; 27 El; 50 RA. Heat treated: 100,000 TS; 27 El; 60 RA. For car axles; shock resisting. *Obsolete*

W100

Thyssen Edelstahlwerke AG
C 0.9, Si 0.25-0.5, Mn 0.3-0.8, bal Fe.
Heat treated: 190,000 TS; 145,000 YS; 10 El; 30 RA; 400 Brin. For springs, drills, taps, punches, reamers; Type W1; water hardened. *Obsolete*

W12MOCR4V4

China Metallurgical Import&Export Corp.
Tool steel. bal Fe.
Quenched and tempered: 63-65 Rock C. Used for simple cutting tools requiring less grinding.

W12MOCR4V4

China Metallurgical Import&Export Corp.
Tool steel. C 1.2-1.4, W 11.5-13, Mo 0.9-1.2, Cr 3.8-4.4, V 3.8-4.4, Si 0.2-0.4, Mn 0.2-0.4, S 0-0.03, P 0-0.03, bal Fe.

W18CR4V

China Metallurgical Import&Export Corp.
Tool steel. bal Fe.
Quenched and tempered: 63-65 Rock C. Used for gears, broaches, twist drills, machine screw taps, machine reamers, lathe tools, saw blades and other cutting tools.

W18CR4V

China Metallurgical Import&Export Corp.
Tool steel. C 0.7-0.8, W 17.5-19, Mo 0-0.3, Cr 3.8-4.4, V 1-1.4, Si 0.2-0.4, Mn 0.2-0.4, S 0-0.03, P 0-0.03, bal Fe.
AISI T1.

W20S

Thyssen Edelstahlwerke AG
C 0.3, Cr 2.5, Mo 0.1, V 0.1, bal Fe.
For extrusion press tools and mandrels, bakelite molds; nitriding steel. *Obsolete*

W22V

Thyssen Edelstahlwerke AG
C 0.45, Ni 1.7, Cr 1.4, Mn, V, bal Fe.
For extrusion press tools, die holders; hot work steel, oil hardened. *Obsolete*

W2MO9CR4V2

China Metallurgical Import&Export Corp.
Tool steel. bal Fe.
Quenched and tempered: 65-67 Rock C. Used for machine screw taps.

W2MO9CR4V2

China Metallurgical Import&Export Corp.
Tool steel. C 0.97-1.05, W 1.4-2.1, Mo 8.2-9.2, Cr 3.5-4, V 1.75-2.25, Si 0.35-0.55, Mn 0.2-0.4, S 0-0.03, P 0-0.03, bal Fe.
AISI M7.

W4 ALLOY

Thyssen Edelstahlwerke AG
C 1, Cr 1.5, Mn 0.3, bal Fe.
For bearings, cutters, cold work tools; oil hardened, abrasion resistant. *Obsolete*

W4 ALLOY

Henry Wiggin & Co. Ltd.
Si, Mn, bal Al.
For spark plug electrodes. *Obsolete*

W4X

Now FINKL W4X.

W6 AND 4

Wardlows Ltd.
W 5.5, Mo 4, Cr 4, V 1.5, C, bal Fe.
For tools, cutters; high speed steel.

W65

Thyssen Edelstahlwerke AG
C 0.45, Si 0.25-0.5, Mn 0.3-0.8, bal Fe.
Hot rolled: 98,000 TS; 59,000 YS; 24 El; 45 RA; 212 Brin. For axles, gears, bolts, bushings, crankshafts; water hardened. *Obsolete*

W6MO5CR4V2

China Metallurgical Import&Export Corp.
Tool steel. C 0.8-0.9, W 5.5-6.75, Mo 4.5-5.5, Cr 3.8-4.4, V 1.6-2.2, Si 0.2-0.4, Mn 0.2-0.4, S 0-0.03, P 0-0.03, bal Fe.
AISI M2.

W6MO5CR4V2

China Metallurgical Import&Export Corp.
Tool steel. bal Fe.
Quenched and tempered: 64-66 Rock C. Suitable for hot forming drills and cutting tools and applications requiring good toughness.

W6MO5CR4V3

China Metallurgical Import&Export Corp.
Tool steel. C 1.15-1.25, W 5-6.75, Mo 4.75-6.5, Cr 3.8-4.4, V 2.75-3.25, Si 0.2-0.4, Mn 0.2-0.4, S 0-0.03, P 0-0.03, bal Fe.
Quenched and tempered: 64-66 Rock C. Used for good wear-resistant cutting tools which need definite toughness for hard-to-cut materials. AISI M3-2.

W6MO5CR4V4

China Metallurgical Import&Export Corp.
Tool steel. C 1.25-1.4, W 5.25-6.5, Mo 4.75-6.5, Cr 3.8-4.4, V 3.75-4.5, Si 0.2-0.45, Mn 0.2-0.4, S 0-0.03, P 0-0.03, bal Fe.
AISI M4.

W7 ALLOY

Henry Wiggin & Co. Ltd.
Mn 2.75, Si 1, bal Ni.
For spark plug electrodes; resists engine fuels. *Obsolete*

W8 ALLOY

Henry Wiggin & Co. Ltd.
Si, Mn, bal Al.
For spark plug electrodes. *Obsolete*

W85

Thyssen Edelstahlwerke AG
C 0.67, Si 0.25-0.5, Mn 0.3-0.8, bal Fe.
Heat treated: 175,000 TS; 130,000 YS; 12 El; 35 RA; 360 Brin. For springs, wheels, die blocks, hammers; water hardened. *Obsolete*

W95-CR

Thyssen Edelstahlwerke AG
C 0.8, Si 0.25-0.5, Mn 0.3-0.8, bal Fe.
Heat treated: 188,000 TS; 143,000 YS; 12 El; 35 RA; 390 Brin. For drills, punches, springs, taps, reamers; Type W1; water hardened. *Obsolete*

WA-1

GTE Valenite Corp.
WC 94, Co 6.
290,000 psi transverse strength; 91.0 Rock A. Density: 14.9. For rough machining of cast iron and non-ferrous materials. Straight WC grade; impact resistant. Code C-1.

WA-10

GTE Valenite Corp.
WC 90, Co 10.
325,000 psi transverse strength; 89.5 Rock A. Density: 14.5 g/cm^3. For punches, and light to medium steel blanking dies. Code C-10.

WA-11

GTE Sylvania
WC 90, Co 10.
Sintered: 360,000 TrS; A 88.9 Rock. Density: 14.55. Abrasion resistance: 36%. Used for coal mining tools, rock bits, blanking dies. Code C-11. *Obsolete*

WA-110

GTE Valenite Corp.
375,000 psi transverse strength; 91.8 Rock A. Density: 14.3 g/cm^3. For rotary and form tools, machining of high-temperature, high-strength nickel-base alloys used in the aircraft industry. Sub-micron grade; non-ferrous machining. Code C-1.

WA-114

GTE Valenite Corp.
400,000 psi transverse strength; 90.0 Rock A. Density: 14.1 g/cm^3. For header dies, punches, dies and mining tools. Sub-micron grade. Code C-10.

WA-119

GTE Valenite Corp.
450,000 psi transverse strength; 88.1 Rock A. Density: 13.5 g/cm^3. For metal forming operations, die punches and bushings. Heavy impact application. Code C-13.

WA-12

GTE Valenite Corp.
WC 87, Co 13.
420,000 psi transverse strength; 88.2 Rock A. Density: 14.2 g/cm^3. For medium blanking dies. Code C-12.

WA-13

GTE Valenite Corp.
WC 80, Co 20.
490,000 psi transverse strength; 86.1 Rock A. Density: 13.5 g/cm^3. For heavy percussive applications. Code C-13.

WA-13

GTE Sylvania
W 84, Co 16.
Sintered: 380,000 TrS; A 86.0 Rock. Density: 13.85. Abrasion resistance: 14%. Blanking die material with good chip and break resistant qualities. Code C-13. *Obsolete*

WA-14

GTE Valenite Corp.
WC 76, Co 24.
480,000 psi transverse strength; 84.8 Rock A. Density: 13.2 g/cm^3. High strength and impact loading. Cold extrusion and cold heading dies. Cold C-14.

WA-2

GTE Valenite Corp.
WC 94, Co 6.
280,000 psi transverse strength; 92.0 Rock A. Density: 14.9 g/cm^3. For machining of cast iron and non-ferrous materials for solid carbide rotary tools. General purpose grade, high hardness and wear resistant. Code C-2.

WA-25

GTE Valenite Corp.
WC 92, Co 6, 2.0 TaC.
250,000 psi transverse strength; 93.1 Rock A. Density: 14.9 g/cm^3. For machining of cast iron and other non-ferrous materials. High temperature alloys and refractory materials. General purpose grade, wear resistant.

WA-3
GTE Valenite Corp.
WC 95.7, Co 4.3.
250,000 psi transverse strength; 92.6 Rock A. Density: 15.1 g/cm^3. For semi-finishing and finishing applications, solid carbide round tools on abrasive materials such as fiberglass, etc. High hardness, wear resistant. Code C-3.

WA-301
Wieland-Werke AG Metallwerke
Si 1, Mg 0.8, Mn 0.8, 2% Pb + Sn + Bi, bal Al.
For screw machine products, hardware, fasteners; free-cutting. *Obsolete*

WA-35
GTE Valenite Corp.
WC 94, Co 6.
Sintered: 260,000 transverse strength; 92.6 Rock A. Density: 14.95. For light finishing cuts on cast iron and non-ferrous materials. Code C-3. *Obsolete*

WA-4
GTE Valenite Corp.
WC 97, Co 3.
260,000 psi transverse strength; 93.0 Rock A. Density: 15.2 g/cm^3. For precision finishing applications on cast iron and non-ferrous materials. Abrasion resistant. Code C-4.

WA-40
GTE Sylvania
WC 80, Co 20.
Sintered: 390,000 TrS; A 85.0 Rock. Density: 13.50. For maximum strength as cold heading or heavy blanking die. Code C-14. *Obsolete*

WA-41
GTE Valenite Corp.
WC 92, Co 8.
Sintered: 310,000 transverse strength; 91.0 Rock A. Density: 14.75. For rough machining on cast iron; also for light percussion tools as vibrating masonry drills, percussion bits, etc. Code C-1. *Obsolete*

WA-47
GTE Valenite Corp.
WC 78.1, Co 8, 8.5 TiC, 5.4 TaC.
250,000 psi transverse strength; 92.0 Rock A. Density: 12.5 g/cm^3. For medium roughing and finishing operations on alloyed steels. High performance semi-finishing grade. Code C-6/C-7.

WA-5
GTE Valenite Corp.
WC 72, Co 8.5, 8.0 TiC, 11.5 TaC.
270,000 psi transverse strength; 91.5 Rock A. Density: 12.5 g/cm^3. For alloy steel forgings and castings. General purpose steel cutting grade. Code C-5/C-6.

WA-50
GTE Valenite Corp.
200,000 psi transverse strength; 91.1 Rock A. Density: 12.9 g/cm^3. For general purpose machining of all types of steel under a wide range of applications. Code C-5/C-6.

WA-510
GTE Valenite Corp.
Sintered: 91.0 Rock A. Non-magnetic. Main use is wear parts for magnetic tape handling equipment. *Obsolete*

WA-54
GTE Valenite Corp.
WC 78.3, Co 11.5, 3.0 TiC, 7.2 TaC.
300,000 psi transverse strength; 90.2 Rock A. Density: 13.5 g/cm^3. For heavy feeds and slow speeds associated with roughing cuts, rough milling. Tough; good edge strength. Code C-5.

WA-57
GTE Valenite Corp.
WC 77, Co 8, 6.0 TiC, 9.0 TaC.
250,000 psi transverse strength; 91.9 Rock A. Density: 13.1 g/cm^3. For large steel mill rolls and alloyed forgings. General purpose grade, good edge strength and crater resistance. Code C-6/C-7.

WA-59
GTE Valenite Corp.
WC 9, Co 8, 1.0 TiC.
Sintered: 300,000 transverse strength; 91.8 Rock A. Density: 14.75. Abrasion resistance: 85%. For roughing cuts on cast iron and non-ferrous materials; for general machining, including stainless. Code C-1. *Obsolete*

WA-6
GTE Sylvania
Co 10, 82% WC, 8% TiC.
Sintered: 250,000 TrS; A 91.2 Rock. Density: 12.5. Abrasion resistance: 39%. For roughing cuts on alloy steels, and general purpose machining. Code C-6. *Obsolete*

WA-623
GTE Valenite Corp.
WC 94, Co 6.
300,000 psi transverse strength; 93.0 Rock A. Density: 14.9 g/cm^3. For circuit board drilling, solid end milling on tough-to-machine metals such as stainless and hardened steels. Micrograin grade, high strength and abrasion resistant. Code C-2.

WA-63
GTE Sylvania
Co 6, 73% WC, 1% TiC, 20% TaC.
Sintered: 265,000 TrS; A 91.4 Rock. Density: 14.6. For general purpose cutting on cast iron, particularly pearlitic malleable brake drums; for lapping and burnishing rolls; resists cratering. Code C-2. *Obsolete*

WA-68
GTE Valenite Corp.
WC 79.5, Co 6, 8.5 TiC, 6.0 TaC.
Sintered: 200,000 transverse strength; 93.0 Rock A. Density: 12.6. High hardness; for finishing cuts on steel and special alloys; precision boring. Code C-7. *Obsolete*

WA-69
GTE Valenite Corp.
WC 91, Co 5, 4.0 TaC.
225,000 psi transverse strength; 91.9 Rock A. Density: 15.0 g/cm^3. For milling applications on cast iron, alloyed cast iron and 300 series stainless steel. Heat resistant. Code C-2.

WA-7
GTE Valenite Corp.
WC 76.5, Co 7.5, 12.0 TiC, 4.0 TaC.
Sintered: 225,000 transverse strength; 92 Rock A. Density: 11.70. Abrasion resistance: 41%. For light roughing cuts or heavy finishing cuts on steel. Code C-7. *Obsolete*

WA-8
GTE Valenite Corp.
WC 75, Co 5, 16.0 TiC, 4.0 TaC.
Sintered: 150,000 transverse strength; 93.0 Rock A. Density: 11.30. Abrasion resistance: 53%. For precision boring or turning of steel. Code C-8. *Obsolete*

WA-800
GTE Valenite Corp.
TiC base.
Sintered: 225,000 transverse strength; 92.7 Rock A. Finish machining pearlitic irons and steels. Code C-7 or C-8. *Obsolete*

WA-870
GTE Sylvania
TiC base.
Sintered: 250,000 TrS; A 92.1 Rock. For medium to light roughing cuts on pearlitic irons and steels. Code C-7 or C-8. *Obsolete*

WA-9
GTE Sylvania
WC 87, Co 13.
Sintered: 300,000 TrS; A 90.0 Rock. Density: 13.0. Abrasion resistance: 70%. Essentially for wear applications where there is only light shock load. Code C-9. *Obsolete*

WABCOLOY
Westinghouse Air Brake Co.
C 2.6, Si 2.5, Mn 0.6, Ni 1.1, Mo 0.8, bal Fe.
Cast: 55,000 TS; 0 El; 255-321 Brin. For cylinders of steam driven compressors, crankshafts. Cast iron, wear resistant.

WABIK METAL
Empire Sheet & Tin Plate Co.
Si 4, bal Fe.
For armatures, motors. *Obsolete*

WAGNER & GUHRS ALUMINUM
Manufacturer not listed.
Sn 80, Zn 20.
For aluminum solder; soft.

WAGNERS FORMULA
Manufacturer not listed.
Cu 50-66, Zn 19-31, Ni 13-18.
For base for plated tableware; corrosion resistant.

WAH CHANG C-120
Teledyne Wah Chang Albany
Mo 5, Zr 1, W 15, bal Cb.
At 70°F: 116,000 TS; 103,000 YS; 4 El. At 2500°F: 27,000 TS; 20,000 YS; 50 El. For space vehicles, nuclear reactors. Good combination of density, strength and oxidation resistant at high temperatures. *Obsolete*

WAH CHANG CB-33ZR
Teledyne Wah Chang Albany
Zr 33, bal Cb.
Uses: superconductors. *Obsolete*

WAH CHANG W-25RE
Teledyne Wah Chang Albany
Re 24-26, bal W.
Sintered: 177,000 TS; 175,000 YS; 33 RA. Recrystallized: 178,000 TS; 164,000 YS; 1-2 El. For electronic tubes, cathodes, heat exchangers, furnace parts. High temperature properties, heat resistant.

WAH CHANG XB-88
Teledyne Wah Chang Albany
W 28, Hf 2, C 0.067, bal Cb.
For gas turbine blades and buckets. Low resistance to shock, good fatigue properties. Heat resistant.

WAH CHANG XB-88
Westinghouse Electric Corp.
W 28, Hf 2, C 0.067, bal Cb.
For gas turbine blades and buckets. Low resistance to shock, good fatigue properties. Heat resistant.

WAKEFIELD 11
Wakefield Corp.
Sn 10, C 4.4, bal Cu.
Metal powder; properties are typical values. 87 Rock H; 9,400 psi TS; 6,500 psi YS; 6 El.

WAKEFIELD 18
Wakefield Corp.
Cu 23.5, C 1.5, bal Fe.
Metal powder; properties are typical values. 65 Rock B; 52,400 psi TS; 48,500 psi YS; 0 El.

WAKEFIELD 19
Wakefield Corp.
Cu 8, C 0.4, bal Fe.
Metal powder; properties are typical values. 39-55 Rock B; 38,100-54,400 psi TS; 32,400-46,900 psi YS; 1 El max.

WAKEFIELD 23
Wakefield Corp.
Sn 10, C 1, bal Cu.
Metal powder; properties are typical values. 45 Rock H; 15,100 psi TS; 9,600 psi YS; 9.7 El.

WAKEFIELD 35
Wakefield Corp.
Cu 20, C 1, bal Fe.
Metal powder; properties are typical values. 93 Rock B; 72,300 psi TS; 57,700 psi YS; 2 El.

WAKEFIELD 39-C
Wakefield Corp.
Ni 4, Cu 1-2, C 0.75, bal Fe.
Metal powder; properties are typical values. 64-84 Rock B; 37,500-63,200 psi TS; 29,600-47,300 psi YS; 1-1.6 El.

WAKEFIELD 41
Wakefield Corp.
Fe 100.
Metal powder; properties are typical values. 26 Rock F; 15,700 psi TS; 11,000 psi YS; 5.7 El.

WAKEFIELD 41-M
Wakefield Corp.
C 0.4, S 0.6, bal Fe.
Metal powder; properties are typical values. 52-74 Rock F; 22,600-34,200 psi TS; 16,500-22,800 psi YS; 2-4.3 El.

WAKEFIELD 76
Wakefield Corp.
Ni 4, C 0.5, bal Fe.
Metal powder; properties are typical values. 73-82 Rock B; 47,900-50,100 psi TS; 38,500-40,400 psi YS; 1.0 El max.

WAKEFIELD 77
Wakefield Corp.
Ni 2, C 0.8, bal Fe.
Metal powder; properties are typical values. 50-70 Rock B; 47,100-57,800 psi TS; 34,700-50,800 psi YS; 1.0 El max.

WAKEFIELD 78
Wakefield Corp.
Ni 2, C 0.5, bal Fe.
Metal powder; properties are typical values. 51-61 Rock B; 35,000-45,200 psi TS; 27,100-31,100 psi YS; 2-2.5 El.

WAKEFIELD 79
Wakefield Corp.
Ni 2, C 0.25, bal Fe.
Metal powder; properties are typical values. 29 Rock B; 25,800 psi TS; 19,000 psi YS; 1.3 El.

WAKEFIELD 80
Wakefield Corp.
Cu 2, C 0.75, bal Fe.
Metal powder; properties are typical values. 68-72 Rock B; 56,300-60,700 psi TS; 51,000-54,200 psi YS; 1 El max.

WAKEFIELD 81
Wakefield Corp.
Cu 20, C 1, bal Fe.
Metal powder; properties are typical values. 45 Rock B; 47,800 psi TS; 40,000 psi YS; 1.3 El.

WAKEFIELD 82
Wakefield Corp.
Ni 1.8, C 0.6, Cu 1.6, Mo 0.5, bal Fe.
Metal powder; properties are typical values. 85 Rock B; 64,400 psi TS; 43,800 psi YS; 1.3 El.

WAKEFIELD 83
Wakefield Corp.
Ni 4, C 0.6, Cu 1.6, Mo 0.5, bal Fe.
Metal powder; properties are typical values. 79 Rock B; 54,600 psi TS; 42,000 psi YS; 2 El.

WAKEFIELD 84
Wakefield Corp.
Ni 1.8, C 0.5, Mo 0.5, bal Fe.
Metal powder; properties are typical values. 52 Rock B; 43,100 psi TS; 36,400 psi YS; 1 El max.

WAKEFIELD 85
Wakefield Corp.
Cr 17, Ni 13, Mo 2.2, Si 0.9, bal Fe.
Metal powder; properties are typical values. 50 Rock B; 43,400 psi TS; 36,000 psi YS; 1.7 El.

WAKEFIELD 85-X
Wakefield Corp.
Cr 17, Ni 13, Mo 2.2, Si 0.9, Cu 15-20, bal Fe.
Metal powder; properties are typical values. 66 Rock B; 59,200 psi TS; 49,700 psi YS; 4.3 El.

WAKEFIELD 86
Wakefield Corp.
Cr 13, Si 0.8, bal Fe.
Metal powder; properties are typical values. 92 Rock B; 66,700 psi TS; 56,900 psi YS; 0 El.

WAKEFIELD 87
Wakefield Corp.
Cu 2, Ni 3.8, C 0.9, Mo 0.75, bal Fe.
Metal powder; properties are typical values. 24 Rock C; 55,800 psi TS; 46,500 psi YS; 1.5 El.

WALCALOY NO. 1
Wall Colmonoy Corp.
General purpose 18-8 stainless wire for use with wirespray gun for build-up on 18-8 stainless and similar steels. *Obsolete*

WALCALOY NO. 2
Wall Colmonoy Corp.
Martensitic stainless wire for use with wirespray gun for build-up on carbon, alloy and 400 series stainless for wear and low shrink requirements.

WALCO METAL
NL Industries
Pb alloy.
For anode for Cr plating, tank lining; resists chromic acid. *Obsolete*

WALCOLOY NO. 1
Wall Colmonoy Corp.
Stainless 18-8 type wire for metallizing gun. For metal spraying austenitic 18-8 stainless steels. *Obsolete*

WALCOLOY NO. 2
Wall Colmonoy Corp.
Martensitic stainless wire for metallizing gun. For metal spraying worn parts of martensitic stainless steel. *Obsolete*

WALCOLOY NO. 4
Wall Colmonoy Corp.
18-10 (316 type) stainless (with Mo) wire for use with wirespray gun for build-up with improved corrosion resistance.

WALCOLOY NO. 4
Wall Colmonoy Corp.
Stainless 18-10 (type 316) wire for metallizing gun. *Obsolete*

WALCOLOY NO. 5
Wall Colmonoy Corp.
Austenitic stainless wire for use with wirespray gun for build-up of good work-hardening and wear surface.

WALDE-LOY 100 METAL
J.M. Pratt & Co.
Pb 87.5, Sn 5.5, Sb 5.3, Cu 1.6, As 0.1.
For storage battery cable terminals and lugs; resists H_2SO_4.

WALLEX NO. 1
Wall Colmonoy Corp.
Cr 30, W 12.5, C 2.25, Si 1.25, 6.0 others max, bal Co.
2355°F MP. Bare rods or castings: 50-55 Rock C; Good corrosion resistance and low coefficient of friction gives good metal-to-metal wear.

WALLEX NO. 1
Wall Colmonoy Corp.
C 2-3, Cr 25-33, W 11-14, Si 0.75-1.75, 43% Co min, bal Fe.
500-550 Brin. For hard facing electrodes, dies, guides; MP 2375 F, abrasion and impact resistant. *Obsolete*

WALLEX NO. 40
Wall Colmonoy Corp.
Ni 24, Cr 16, W 7, Fe 2, Si 1.5, C 0.5, B 2, bal Co.
2080°F MP. Atomized powder for spraywelder, fusewelder. Deposited metal: 41-46 Rock C.

WALLEX NO. 50
Wall Colmonoy Corp.
Ni 18, Cr 19, W 10, Si 2.75, Fe 1, C 0.8, B 3.5, bal Co.
2050°F MP. Bare rods, atomized powder, castings. Deposited metal: 56-61 Rock C.

WALLEX NO. 505
Wall Colmonoy Corp.
Wallex No. 50 with tungsten carbide particles added. Cast: 55-60 Rock C; 2050°F MP (approx). Atomized and crushed powder for hard facing areas requiring extreme abrasion resistance; good resistance to heat, galling and corrosion. Applied by fusewelder torch.

WALLEX NO. 55
Wall Colmonoy Corp.
Wallex No. 50 with tungsten carbide particles added. Cast: 55-60 Rock C; 2050°F MP (approx). Atomized and crushed powder for hard facing areas requiring extreme abrasion resistance; good resistance to heat, galling and corrosion. Applied by spraywelder.

WALLEX NO. 6
Wall Colmonoy Corp.
C 0.9-1.4, Cr 25-31, W 3.5-6, Si 0.75-1.75, 55 Co min, 0.0-6.5 others.
Cast: 39-44 Rock C. For hard facing electrode; corrosion, heat and wear resistant.

WALLRAM H1
Mentah Co.
WC.
For cutting tools, dies; sintered carbides.

WALLRAM H1P
Wallram Hartmetall GmbH
WC 94, Co 6.
For cutters and dies. Wear and abrasion resistant. Sintered carbides.

WALLRAM Z10
Wallram Hartmetall GmbH
MoC, WC, Ni, Co, V.
For extrusion dies and nibs. Sintered carbides. High wear and abrasion resistance.

WALMANG NO. 3
Wall Colmonoy Corp.
C 0.75, Ni 3.5, Si 0.6, Mn 14, bal Fe.
Weld metal: 55 Rock C. Electrode for hard facing; abrasion and impact resisting deposits. *Obsolete*

WALMANG NO. 8
Wall Colmonoy Corp.
C 0.75, Mo 1, Si 0.6, Mn 14, bal Fe.
Weld metal: 50 Rock C. Electrode wire for hard facing; abrasion and impact resisting deposits. *Obsolete*

WALRAMITE
Manufacturer not listed.
Tungsten carbide.
For cutting tools, dies; sintered.

WANDO
Cytemp Specialty Steel Div.
C 0.95, Mn 1.2, Cr 0.5, V 0.2, W 0.5, bal Fe.
For dies, gauges, taps, reamers, punches, slitting saws;
Type O1; non-deforming. *Obsolete*

WAPRESTA
Vereinigte Edelstahlwerke
C 0.3, V 0.18, Cr 1.1, W 3.75, bal Fe.
For pneumatic tools, chisels, punches, dies; oil hardened,
shock resistant. *Obsolete*

WAR BABBITT
Duquesne Smelting Corp.
Sn, Pb, Cu.
Cast: 10,750 TS; 5,630 YS; 2 El; 22 Brin. For bearings;
Babbitt.

WAR BRONZE
German manufacture
Al 2.19, Cu 4.85, Pb 0.92, Sn 0.15, Fe 0.03, bal Zn.
For shell fuses; cast alloy.

WARALOY 20X
Hewitt Metals Corp.
Sn 20, Ag 1.25, bal Pb.
Cast: 7,000 TS; 15 El. For solders: MP 440-490 F. *Obsolete*

WARALOY 30X
Hewitt Metals Corp.
Sn 30, Ag 1.2, bal Pb.
For solder; corrosion resistant. *Obsolete*

WARDLOWS
S. & C. Wardlow
C 0.7-1.4, bal Fe.
For tools. Water hardened.

WARDLOWS TOUGH
S. & C. Wardlow
C 0.9, Mn 1.2, bal Fe.
For tools, dies. Tough, oil hardened.

WARDLOWS UUT
Wardlows Ltd.
C 0.5, Cr 1.5, W 2.25, V 0.25, bal Fe.
For hot work dies; hot work steel.

WARDLOWS WSV
Wardlows Ltd.
C 0.9, V 0.1, bal Fe.
For tools, taps, drills, punches; water hardened.

WARM WORKED 321
British Steel plc
C 0-0.08, Cr 17-19, Ni 9-11, Ti = 5 X C (0.70), bal Fe.
High proof stress version of 321, obtained by controlled low
temperature hot working. *Obsolete*

WARMAN 13
Warman Steel Casting Co.
Cr 11-14, 0.12 C min, bal Fe.
Heat treated: 90,000 TS; 60,000 YS; 24 El; 50 RA. For
corrosion resisting parts; corrosion resistant.

WARMAN 13M
Warman Steel Casting Co.
C 0-0.12, Cr 11.5-14, Mo 0.5-0.7, bal Fe.
Heat treated: 95,000 TS; 65,000 YS; 25 El; 54 RA. For
corrosion resisting castings; corrosion resistant.

WARMAN 5
Warman Steel Casting Co.
C 0.13-0.2, Cr 5, bal Fe.
For corrosion resistant parts; corrosion resistant.

WARMAN 5 M
Warman Steel Casting Co.
C 0.13-0.2, Cr 5, bal Fe.
For corrosion resistant parts; corrosion resistant.

WARMAN 6
Warman Steel Casting Co.
C 0.15, Cr 5-7, bal Fe.
Cast: 90,000 TS; 65,000 YS; 15 El; 35 RA. For corrosion
resisting parts; corrosion resisting.

WARMAN 6 M
Warman Steel Casting Co.
Cr 5-7, Mo 0.5-0.7, 0.18 C min, bal Fe.
Heat treated: 125,000 TS; 100,000 YS; 18 El; 42 RA. For
corrosion resisting parts; corrosion resistant.

WARMAN CALDURO 13-2
Warman Steel Casting Co.
C 0.2, Cr 12.5-15, Ni 1.5-2.5, bal Fe.
Cast: 80,000-110,000 TS; 60,000-90,000 YS; 10-25 El; 30-40
RA. For corrosion resisting parts; corrosion and abrasion
resisting; formerly "Nirosta Calduro KM-1."

WARMAN CALMAR 18-8
Warman Steel Casting Co.
Stainless Steel. C 0.16, Cr 17-20, Ni 7-10, bal Fe.
Cast: 70,000 TS; 30,000 YS; 50 El; 180 Brin. For stainless
parts; corrosion resisting; formerly "Nirosta Calmar KA2."

WARMAN CALMAR 18-8 M
Warman Steel Casting Co.
Stainless Steel. Ni 7-10, Mo 3-4.5, Cr 17-20, 0.16 C min, bal
Fe.
For stainless, heat and corrosion resisting parts; stainless,
corrosion and heat resistant.

WARMAN CALOXO 15-25 M
Warman Steel Casting Co.
Cr 14-17, Ni 23-27, Mo 3-4.5, 0.20 C min, bal Fe.
For heat and corrosion resisting parts; heat and corrosion
resistant.

WARMAN CALOXO 15-35
Warman Steel Casting Co.
Cr 14-17, Ni 33-37, 0.20 C min, bal Fe.
For heat and corrosion resisting parts; heat and corrosion
resistant.

WARMAN CALOXO 18-2
Warman Steel Casting Co.
Cr 16-20, Ni 1.5-2.5, 0.20 C min, bal Fe.
For corrosion resisting parts; corrosion resistant.

WARMAN CALOXO 25-20
Warman Steel Casting Co.
C 0.25, Cr 23-27, Ni 19-21, bal Fe.
Cast: 75,000 TS; 35,000 YS; 25 El; 200 Brin. For furnace
and heat resisting parts; corrosion and abrasion resisting;
formerly "Nirosta Caloxo KNC-3."

WARMAN CALOXO 25-20 M
Warman Steel Casting Co.
Cr 23-27, Ni 17-21, Mo 3-4.5, 0.20 C min, bal Fe.
For heat and corrosion resisting parts; heat and corrosion
resistant.

WARMAN CALOXO 8-18
Warman Steel Casting Co.
Stainless Steel. Cr 7-10, Ni 17-20, 0.16 C min, bal Fe.
For stainless, corrosion resisting parts; corrosion resistant,
stainless.

WARMAN COLOXO 15-25
Warman Steel Casting Co.
Cr 14-17, Ni 23-27, Mo 3-4.5, 0.20 C min, bal Fe.
For heat and corrosion resisting parts; heat and corrosion
resistant.

WARMANS CALOXO 18
Warman Steel Casting Co.
C 0.16, Cr 16-20, bal Fe.
Cast: 90,000 TS; 45,000 YS; 15 El; 30 RA. For corrosion
resisting parts; formerly "Warman
Chrome Stainless."

WARMBRONZE
Waterbury Rolling Mills Inc.
Cu 88, Sn 2, Zn 10.
1/4 H-temper: 53,000 TS; 37 El; 107 Brin. H-temper: 95,000
TS; 3 El; 172 Brin. For hardware, plumbing; corrosion
resistant.

WARMPRESSTAHL M43W
Stahlwerke Kabel, C.
C 0.3, Cr 2.65, V 0.35, W 8.5, bal Fe.
For extrusion dies, rams and liners; oil hardened, hot work
steel.

WARNES METAL
Manufacturer not listed.
Sn 37, Ni 26, Bi 26, Co 11.
Used as substitute for Ag in making ornamental articles;
corrosion resistant.

WARPLIS DRILL ROD
Teledyne Pittsburgh Tool Steel
C 0.9, Cr 0.5, Mn 1.1, W 0.5, V 0.15, bal Fe.
Oil hardenable to 64 Rock C max. For tools, dies, gages,
jigs. Type O1; oil hardening tool steel.

WARRANTED 50-50
Hewitt Metals Corp.
Sn 50, Pb 50.
Block solder: melt range 358-411 F. *Obsolete*

WARRANTED BEST
Jessop-Saville Ltd.
C 0.7-1, bal Fe.
For tools; water hardened.

WARRANTED BEST CAST STEEL
Leadbeater & Scott Ltd.
High C, bal Fe.
For taps, reamers, broaches; water hardened, general
purpose.

**WARRANTED BEST DOUBLE SHEAR
STEEL**
Leadbeater & Scott Ltd.
High C, bal Fe.
For dies, machine knives; water hardened.

WARRANTED C 16.A.
Burys & Co. Ltd.
C 0.5, Si 1, Cr 1.5, bal Fe.
For pneumatic tools, chisels, beading tools; oil hardened,
shock resistant.

WARRANTED CAST STEEL
Darwins Alloy Castings
C 1, bal Fe.
For tools, taps, reamers; water hardened. *Obsolete*

WASHCONITE
Washington Iron Works, Inc.
Si 1.6, Ni 1.5, Cr 0.25, TC 2.0, bal Fe.
Cast: 46,000 TS; 46,000 YS; 0 El; 0 RA; 240 Brin. For gas
and diesel engine cylinders, liners, pistons; alloy cast iron.
Obsolete

WASHCOTE
Harnischfeger Corp.
C 0.2, bal Fe.
Welded: 55,000-60,000 TS; 8-10 El. For welded electrodes;
general purpose. *Obsolete*

WASHER BRASS
American manufacture
Cu 62, Zn 38.
For washers, hardware, yellow brass.

WASHINGTON SPECIAL
Jessop Steel Co.
C 0.6-1.4, V 0.2, Mn 0.35, Si 0.25, bal Fe.
For drills, reamers, springs, punches, taps, cutters; AISI W2; water hardened.

WASHINGTON. (AISI W1 GR. 1)
Hardite Metals Inc.
C 0.6-1.4, bal Fe.
For tools, general tools, broaches, cold heading dies, cutters; water hardened.

WASHINGTON. (AISI W1 GR. 1)
Jessop Steel Co.
C 0.6-1.4, bal Fe.
For tools, general tools, broaches, cold heading dies, cutters; water hardened.

WASPALLOY
Driver Harris Co.
Nickel. Cr 20, Co 13, Mo 4, Ti 3, Al 1, bal Ni.
Welding wire. *Obsolete*

WASPALLOY
Enpar Sonderwerkstoffe GmbH
Nickel.
Alternate manufacturer.

WASPALOY
Now ALLVAC, CARPENTER; CRUCIBLE, UDIMET and UNITEMP WASPALOY.

WASPALOY
Crucible Specialty Steel
Ni 59, Cr 19.5, Co 13.5, Mo 4.2, Ti 3, Mn 0.7, Fe 0-2, Al 1.2, C 0.07.
Heat treated: 188,000 TS; 115,000 YS; 28 El; 25 RA; 375 Brin. At 1400°F; 117,000 TS; 99,000 YS; 28 El; 41 RA. For jet engine turbine buckets and discs, high temperature bolts; heat resistant, high strength, high stress rupture strength up to 1400°F.

WASPALOY
Special Metals Corp.
Ni 59, Cr 19.5, Co 13.5, Mo 4.2, Ti 3, Mn 0.7, Fe 0-2, Al 1.2, C 0.07.
Heat treated: 188,000 TS; 115,000 YS; 28 El; 25 RA; 375 Brin. At 1400°F; 117,000 TS; 99,000 YS; 28 El; 41 RA. For jet engine turbine buckets and discs, high temperature bolts; heat resistant, high strength, high stress rupture strength up to 1400°F.

WASPALOY
Allegheny Ludlum Steel
Ni 59, Cr 19.5, Co 13.5, Mo 4.2, Ti 3, Mn 0.7, Fe 0-2, Al 1.2, C 0.07.
Heat treated: 188,000 TS; 115,000 YS; 28 El; 25 RA; 375 Brin. At 1400°F; 117,000 TS; 99,000 YS; 28 El; 41 RA. For jet engine turbine buckets and discs, high temperature bolts; heat resistant, high strength, high stress rupture strength up to 1400°F.

WATCH ALLOY-1
Manufacturer not listed.
Pd 70, Cu 25, Ag 4, Ni 1.
For watch cases; corrosion resistant.

WATCH ALLOY-2
Manufacturer not listed.
Cu 50, Ni 47.2, Cd 2.8.
For watch cases; corrosion resistant.

WATCH ALLOY-3
Manufacturer not listed.
Au 37.5, Cu 27, Ag 23, Pd 12.5.
For watch cases; corrosion resistant.

WATCH CASE BEZEL
Manufacturer not listed.
Cu 60-63, Zn 21-24, Ni 16.
For watch case bezel; corrosion resistant.

WATCH CASE METAL
Manufacturer not listed.
Cu 55-65, Zn 16-30, Ni 10-28.
For watch cases; corrosion resistant.

WATCH NICKEL
Waltham Precision Instruments
Ni 12, Cu 64, Pb 1, Zn 23.

WATCHMAKERS ALLOY
Manufacturer not listed.
Cu 59, Zn 40, Pb 1.2.
For watch parts; free-cutting.

WATERBURY A10
Waterbury Rolling Mills Inc.
Cu 66, Ni 10, Zn 24.
1/4 H-temper: 65,000 TS; 116 Brin. H-temper: 86,000 TS; 180 Brin. Spring temper: 102,000 TS; 205 Brin. For hardware, cutlery, ornaments; nickel silver, corrosion resistant. *Obsolete*

WATERBURY A15
Waterbury Rolling Mills Inc.
Cu 65, Ni 15, Zn 20.
1/4 H-temper: 65,000 TS; 125 Brin. H-temper: 85,000 TS; 176 Brin. Spring temper: 97,000 TS; 195 Brin. For hardware, cutlery, ornaments; nickel silver, corrosion resistant. *Obsolete*

WATERBURY A18
Waterbury Rolling Mills Inc.
Cu 65, Ni 18, Zn 17.
1/4 H-temper: 60,000 TS; 107 Brin. H-temper: 85,000 TS; 165 Brin. For hardware, ornaments, cutlery, nickel silver, corrosion resistant. *Obsolete*

WATERBURY A5
Waterbury Rolling Mills Inc.
Cu 65, Ni 5, bal Zn.
1/4 H-temper: 72,000 TS; 130 Brin. H-temper: 86,000 TS; 165 Brin. For hardware, cutlery, ornaments; nickel silver, corrosion resistant. *Obsolete*

WATERBURY B-12-S
Waterbury Rolling Mills Inc.
Cu 56.5, Ni 12, Zn 31.5.
1/4 H-temper: 74,000 TS; 130 Brin. H-temper: 96,000 TS; 200 Brin. Spring temper: 116,000 TS; 234 Brin. For hardware, cutlery, ornaments; nickel silver, corrosion resistant. *Obsolete*

WATERBURY B-18-S
Waterbury Rolling Mills Inc.
Cu 55, Ni 18, Zn 27.
1/4 H-temper: 78,000 TS; 147 Brin. H-temper: 100,000 TS; 200 Brin. For hardware, cutlery, ornaments; nickel silver, corrosion resistant. *Obsolete*

WATERBURY K-12
Waterbury Rolling Mills Inc.
Cu 65, Ni 12, Zn 21, Pb 2.
For hardware, screw machine products; nickel silver, free-cutting. *Obsolete*

WATERBURY K-8
Waterbury Rolling Mills Inc.
Cu 65, Ni 8, Zn 25, Pb 2.
For hardware, screw machine products; nickel silver, free-cutting. *Obsolete*

WATERBURY PBA
Waterbury Rolling Mills Inc.
Sn 3.5-5.8, P 0.03-0.35, bal Cu.
Soft: 40,000 TS; 56 Brin. Spring: 105,000 TS; 216 Brin. For springs, clips, diaphragms; P-bronze, wear resistant. *Obsolete*

WATERBURY PBC
Waterbury Rolling Mills Inc.
Sn 7-9, P 0.03-0.35, bal Cu.
Soft: 53,000 TS; 62 Brin. Spring: 119,000 TS; 250 Brin. For springs, clips, diaphragms; P-bronze, wear resistant. *Obsolete*

WATERBURY PBD
Waterbury Rolling Mills Inc.
Sn 9-11, P 0.03-0.35, bal Cu.
Soft: 58,000 TS; 64 Brin. Spring: 129,000 TS; 260 Brin. For springs, clips, diaphragms; P-bronze, wear resistant. *Obsolete*

WATERCRAT
Marshall Steel Co.
C 1.05, Mn 0.35, Si 0.2, Cr 0.5, bal Fe.
For tools, dies, shear blades, punches; precision ground flat stock.

WATERDIE EXTRA
Columbia Tool Steel Co.
Tool material. C 1, Mo 0.35, Si 0.25, Cr 0.5, bal Fe.
Water hardened tool steel. For bushings, forming dies, burnishing rolls. Type W5-2.

WATERDIE STANDARD
Columbia Tool Steel Co.
Tool material. C 1, Mn 0.35, Si 0.25, Cr 0.5, bal Fe.
Water hardened tool steel. For automotive tools, wear plates, mandrels. Type W5-3.

WAUKESHA B
Waukesha Foundry, Inc.
C 0-0.12, Mn 0-1, Si 0-1, Cr 0-1, Mo 26-33, Co 0-2.5, V 0.2-0.6, Fe 4-6, bal Ni.
Cast corrosion resistant alloy. ACI N-12 M.

WAUKESHA C
Waukesha Foundry, Inc.
C 0-0.12, Mn 0-1, Si 0-1, Cr 15-20, Mo 16-20, Co 0-2.5, V 0.2-0.4, Fe 4.5-7.5, W 3.75-5.25, bal Ni.
Cast corrosion resistant alloy. ACI CW-12 M.

WAUKESHA NO. 0
Waukesha Foundry, Inc.
Cu 47-53, Pb 4-6, Sn 10-13, Ni 28-32, Sb 2.5-3.5.
Cast: 60,000 TS; 60,000 YS; 1 El; 230 Brin. For dairy and food processing equipment; corrosion and wear resistant. *Obsolete*

WAUKESHA NO. 1
Waukesha Foundry, Inc.
Cu 48-54, Pb 5-7, Sn 1.5-2.5, Ni 20-23, Zn 16-20.
Cast: 47,000 TS; 25,000 YS; 26-30 El; 105 Brin. For dairy and food processing equipment; corrosion resistant.

WAUKESHA NO. 11
Waukesha Foundry, Inc.
Cu 61-67, Ni 23-27, Fe 6-9, Mn 1-2, Si 0-0.5.
Cast: 62,000-80,000 TS; 50,000 YS; 16-9 El; 158-164 Brin. For dairy and food processing equipment; corrosion resistant.

WAUKESHA NO. 118
Waukesha Foundry, Inc.
Cu 63-69, Pb 3.5-4.5, Sn 2.5-3.5, Ni 18-21, Fe 2.5-3.5, Zn 3-5.
Cast: 52,000 TS; 30,000 YS; 20 El; 123 Brin. For dairy and food processing equipment; corrosion resistant.

WAUKESHA NO. 120
Waukesha Foundry, Inc.
Cu 58-61, Sn 5.5-6.5, Pb 3.5-4.5, Zn 3-5, Ni 23-26, Fe 1.5-2.5.
Cast: 60,000 TS; 45,000 YS; 5 El; 160 Brin. For food and dairy.

WAUKESHA NO. 13
Waukesha Foundry, Inc.
Cu 28-32, Pb 4-6, Ni 55-59, Fe 1, Zn 2.5-4.5, Si 1, Mn 1.
Cast: 72,000-75,000 TS; 30-34 El; 123 Brin. For dairy and food processing equipment; corrosion resistant. *Obsolete*

WAUKESHA NO. 18
Waukesha Foundry, Inc.
Cu 54-60, Pb 5-7, Sn 4-6, Ni 20-24, Zn 7.5-11.
Cast: 68,000-72,000 TS; 2-5 El; 187 Brin. For dairy and food processing equipment; corrosion resistant. *Obsolete*

WAUKESHA NO. 20
Waukesha Foundry, Inc.
C 0-0.07, Ni 29, Cr 20, Mo 3, Cu 4, Si 1, bal Fe.
Cast: 65,000 TS; 30,000 YS; 30-45 El; 40-50 RA; 130-150 Brin. For chemical and pharmaceutical plant equipment; stainless, austenitic. *Obsolete*

WAUKESHA NO. 21
Waukesha Foundry, Inc.
Cu 65-71, Pb 4-6, Sn 9-11, Ni 13-16, Mn 1.5-2.5.
Cast: 61,000-65,000 TS; 2-5 El; 171 Brin. For dairy and food processing equipment; corrosion resistant. *Obsolete*

WAUKESHA NO. 22
Waukesha Foundry, Inc.
Cu 47-53, Sn 13-15, Ni 31-35, Sb 2.5-3.5.
Cast: 50,000-54,000 TS; 1 El; 264 Brin. For dairy and food processing equipment; corrosion resistant. *Obsolete*

WAUKESHA NO. 23
Waukesha Foundry, Inc.
Pb 3.5-4.5, Sn 7-9, Ni 76-80, Zn 6-9, C 0.05-0.15.
Cast: 70,000 TS; 45,000 YS; 10 El; 165-190 Brin. For dairy and food processing equipment, bearings; corrosion and galling resistant. *Obsolete*

WAUKESHA NO. 23C
Waukesha Foundry, Inc.
Pb 3-4.5, Sn 7-9, Zn 6-9, C 0.1-0.25, Mn 1.5-2.5, bal Ni.
Cast: 75,000 TS; 65,000 YS; 10 El; 200-230 Brin. For dairy and food processing equipment, bearings, bushings, pumps. Nongalling, corrosion resistant.

WAUKESHA NO. 3
Waukesha Foundry, Inc.
Cu 61-67, Ni 28-32, Fe 2-3, Si 1-2, Mn 1-2.
Cast: 105,000 TS; 65,000 YS; 2-5 El; 270 Brin. For dairy and food processing equipment; corrosion resistant.

WAUKESHA NO. 35
Waukesha Foundry, Inc.
C 0.02, Cr 27, Ni 5, Mo 2.5, Cu 3, Si 1, Mn 0.75, bal Fe.
Heat treated: 140,000 TS; 90,000 YS; 25 El; 300 Brin. For high strength stainless castings; low distortion in hardening. *Obsolete*

WAUKESHA NO. 4
Waukesha Foundry, Inc.
Cu 54-60, Sn 4-6, Ni 23-27, Fe 2.5-3.5, Zn 4.5-7, Si 0.75-1.25, Mn 2-3.
Cast: 77,000 TS; 67,000 YS; 1-3 El; 240 Brin. For dairy and food processing equipment; corrosion resistant.

WAUKESHA NO. 50
Waukesha Foundry, Inc.
Sn 5-7, Ni 56-60, Bi 2.5, Si 2, Cr 13-17, Mo 3.
Cast: 50,000 TS; 36,000 YS; 7 El; 170 Brin. For bearings, food and chemical plant equipment; non-galling. *Obsolete*

WAUKESHA NO. 54C
Waukesha Foundry, Inc.
Sn 7-9, Zn 7-9, Ag 5-7, C 0.15-0.25, Mn 1.5-2.5, Si 0-0.5, bal Ni.
Cast: 85,000 TS; 60,000 YS; 10 El; 220 Brin. For pumps, valve parts, bushings, bearings, dairy and food processing equipment. High wear and corrosion resistance. Nongalling.

WAUKESHA NO. 7
Waukesha Foundry, Inc.
Cu 54-60, Pb 4-7, Sn 2.5-3.5, Ni 23-27, Fe 2-4, Zn 4.5-7.
Cast: 48,000 TS; 27,000 YS; 18-22 El; 104 Brin.

WAUKESHA NO. 8
Waukesha Foundry, Inc.
Cu 57-63, Pb 4-6, Sn 1, Ni 20-23, Fe 1-2, Zn 8-12.
Cast: 42,000-46,000 TS; 25-29 El; 77 Brin. For dairy and food processing equipment; corrosion resistant. *Obsolete*

WAUKESHA NO. 88
Waukesha Foundry, Inc.
C 0-0.05, Sn 3-5, Bi 3-4.5, Mo 2.5-3.5, Cr 11-14, Fe 0-2, Mn 0-1.5, Si 0-0.5, bal Ni.
Cast: 45,000 TS; 35,000 YS; 6 El; 7 RA; 150 Brin. For food and chemical industry equipment; nongalling against stainless steel.

WAUSAU
Wausau Motor Parts
Mo 1.2, Si 2.2, Mn 0.8, Ni 0.5, Cr 0.15, TC 3.1, CC 0.6, bal Fe.
Cast: 250-270 Brin. For valve seats. Alloy cast iron.

WAXIT
Degussa AG
Wetting agent for the jewelry industry.

WAZ-16
Cannon-Muskegon Corp.
W 16, Al 7, Mo 2, Cb 2, Zr 0.5, C 0.2, bal Ni.
Improved strength at 2200°F (1205°C).

WAZ-20
Cannon-Muskegon Corp.
C 0.15, W 18.5, Al 6.2, Zr 1.5, bal Ni.
Directionally solidified. Cast alloy for jet engine discs and vanes.

WAZ-D
American manufacture
C 0.06, W 16.5, Al 7, Zr 0.8, Fe 4.3, 2.0 Y_2O_3, bal Ni.
Good sress rupture at 2200°F (1200°C).

WBD 200
Wilbur B. Driver Co.
Nickel. 99.5 Ni min.
Commercially pure nickel. ASTM B160.

WBD 205
Wilbur B. Driver Co.
Nickel. 99.5 Ni min.
Commercially pure nickel for electronic applications. ASTM F-9.

WBD 400
Wilbur B. Driver Co.
Nickel. Ni 66, Cu 31, Fe 1.
For corrosion resistance; marine and chemical processing. FED QQN 281 (class A).

WBD 600
Wilbur B. Driver Co.
Nickel. Ni 76, Cr 16, Fe 7.
For high temperature heat resistant applications. ASTM B-166.

WBD WELD 55
Wilbur B. Driver Co.
Nickel. Ni 55, bal Fe.
Core wire for electrodes for welding cast iron. AWS 5.15E NiFe-1.

WBD WELD 60
Wilbur B. Driver Co.
Nickel. Ni 65, Cu 27, Mn 3.5, Ti 2, Si 1.
For welding nickel-copper alloys. AWS 5.14 ERNiCu-7.

WBD WELD 61
Wilbur B. Driver Co.
Nickel. Ni 96, Ti 3.
For arc welding pure nickel. AWS A5.11 ENi-1 and AWS A5.14 ERNi-3.

WBD WELD 62
Wilbur B. Driver Co.
Nickel. Ni 74, Cr 16, Fe 7.5, Cb 2.
For welding nickel base alloys. AWS A5.14 ERNiCrFe-5.

WBD WELD 67
Wilbur B. Driver Co.
Copper. Cu 67, Ni 31, Mn 0.75.
For welding copper-nickel alloys. AWS A5.7 RCuNi, and A5.6 ECuNi.

WBD WELD 82
Wilbur B. Driver Co.
Nickel. Ni 72, Cr 20, Mn 3, Cb 2.5.
For welding nickel base alloys. Corrosion and heat resistant overlays on steel. AWS A5.14 ERNiCr-3,

WBD WELD 92
Wilbur B. Driver Co.
Nickel. Ni 71, Cr 16, Fe 7, Ti 3.
For welding dissimilar alloys. AWS A5.14 ERNiCrFe-6.

WBZ 6 BRONZE
German manufacture
Cu 94, Sn 6.
Cast: 35,600 TS; 15,600 YS; 32 El; 50 Brin. Annealed: 36,900 TS; 15,650 YS; 40 El; 51 Brin.

WC 10
Walsingham Steel Co., Ltd.
C 0.08-0.12, Si 0.3-0.6, Mn 0.3-0.5, Ni 0-0.4, Cr 0-0.25, Mo 0-0.15, Cu 0-0.3, bal Fe.
Meets BS 1617 Gr. A.

WC 18
Walsingham Steel Co., Ltd.
C 0.16-0.2, Si 0.3-0.6, Mn 0.7-0.9, Ni 0-0.4, Cr 0-0.25, Mo 0-0.15, Cu 0-0.3, bal Fe.
Machinable, weldable. Meets BS 1617 Gr. B; BS 592 Gr. A.

WC 20
Walsingham Steel Co., Ltd.
C 0-0.25, Si 0-0.5, Mn 0-0.9, Ni 0-0.4, Cr 0-0.25, Cu 0-0.4, 0.80 Ni-Cr-Mo-Cu max, bal Fe.
Admiralty general purpose castings. Meets D.G.S. 8081 A.

WC 25
Walsingham Steel Co., Ltd.
C 0.22-0.3, Si 0.3-0.6, Mn 0.6-0.8, Ni 0-0.4, Cr 0-0.25, Mo 0-0.15, Cu 0-0.3, bal Fe.
Meets BS 592 Gr. B.

WC 33
Walsingham Steel Co., Ltd.
C 0.31-0.35, Si 0.3-0.6, Mn 0.6-0.9, Ni 0-0.4, Cr 0-0.25, Mo 0-0.15, Cu 0-0.3, bal Fe.
Meets ASTM A358-68 Grade 1.

WC 40
Walsingham Steel Co., Ltd.
C 0.36-0.45, Si 0.3-0.6, Mn 0.7-1, bal Fe.
19 tsi YS min; 35 tsi TS min. Meets BS 592 Gr. C.

WC 48
Walsingham Steel Co., Ltd.
C 0.46-0.5, Si 0.3-0.5, Mn 0.6-1, Ni 0-0.4, Cr 0-0.25, Mo 0-0.15, Cu 0-0.3, bal Fe.
Meets BS 1760 Gr. A.

WC 55
Walsingham Steel Co., Ltd.
C 0.51-0.6, Si 0.3-0.5, Mn 0.6-1, Ni 0-0.4, Cr 0-0.25, Mo 0-0.15, Cu 0-0.3, bal Fe.
Meets BS 1760 Gr. B.

WC-103
Teledyne Wah Chang Albany
Refractory. Ti 0.7-1.3, Hf 9-11, Zr 0.7, Ta 0.5, W 0.5, bal Nb.
54,000 psi TS; 38,000 psi YS; 20 El. For space vehicles, nuclear reactors. Good combination of strength, density, and oxidation resistance at high temperatures.

WC-103
Now COLUMBIUM C-103.

WC-129Y
Teledyne Wah Chang Albany
Refractive. W 13-11, Hf 9-11, Y 0.05-0.3, bal Nb.
80,000 psi TS; 60,000 psi YS; 20 El. For high temperature service, space vehicles, reactors, missiles. Recrystallized at 1900-2200°F. Heat resistant.

WC-1ZR
Teledyne Wah Chang Albany
Refractory. Zr 0.8-1.2, bal Nb.
Recrystallized: 42,000 TS; 23,000 YS; 42 El. For thermal barriers, high temperature parts. High heat and oxidation resistant.

WC-3015
Teledyne Wah Chang Albany
Hf 28-30, W 13-16, Ta 0.4, Zr 1-2, C 0.07-0.3, bal Cb.
Extruded: 140,000-147,000 TS; 135,000-140,000 YS; 5-20 El. For turbines, machine guns, high temperature components. Resists high temperature oxidation. Develops its own surface coating.

WC-3015
Now WAH-CHANG WC-3015.

WC-3015
Teledyne Wah Chang Albany
Hf 28-30, Zr 1-2, W 13-16, Ta 0-4, C 0.07-0.33, bal Cb.
Extruded: 147,000 TS; 140,000 YS; 5-20 El. For turbine components. Good creep and stress-rupture properties. High oxidation and heat resistance. *Obsolete*

WC-752
Teledyne Wah Chang Albany
Refractive. W 9-11, Zr 2-3, bal Nb.
75,000 psi TS; 55,000 psi YS; 20 El. For space vehicles and nuclear reactors. High strength at elevated temperatures. Insensitive to notch stress concentration.

WCC
Teledyne Vasco
C 0.38-0.43, Cr 4-4.5, W 4-4.5, Co 4-4.5, V 2.1, Mo 0.4, bal Fe.
For punches, dies, extrusion dies, permanent molds; hot work steel, wear and thermal fatigue resistant. Type H-10.

WCV
Thyssen Edelstahlwerke AG
C 0.5, Cr 1.05, V 0.1, Mn 0.95, bal Fe.
For gears, bolts, crankshafts, springs; oil hardened, shock resistant. *Obsolete*

WCVH
Thyssen Edelstahlwerke AG
C 0.58, Mn 0.95, Cr 1.05, V 0.1, bal Fe.
For gears, bolts, springs, machine tool parts; oil hardened, shock resistant. *Obsolete*

WE
Thyssen Edelstahlwerke AG
C 0.15, Si 0.25-0.5, Mn 0.3-0.8, bal Fe.
Annealed: 70,000 TS; 40,000 YS; 25 El; 60 RA; 145 Brin. For gears, bolts, nails, rivets, camshafts; case hardened, water hardened. *Obsolete*

WE EXTRA
Thyssen Edelstahlwerke AG
C 0.1, bal Fe.
For bakelite molds; case hardening. *Obsolete*

WE III
Wheeling-Pittsburgh Steel Corp.
Si 1, bal Fe.
Sheet: 60,000 TS; 40,000 YS; 15 El. For electrical equipment, chokes, radio transformers; high permeability. *Obsolete*

WE IV
Wheeling-Pittsburgh Steel Corp.
Si 1-3.5, bal Fe.
Sheet: 50,000-65,000 TS; 32,000-53,000 YS; 12-22 El. For electrical equipment, motors, radio transformers; high permeability. *Obsolete*

WE IX
Wheeling-Pittsburgh Steel Corp.
Si 3-4.2, bal Fe.
Sheet: 67,000-70,000 TS; 55,000-65,000 YS; 4-8 El. For power and distribution transformers; high permeability. *Obsolete*

WE V
Wheeling-Pittsburgh Steel Corp.
Si 1-3.5, bal Fe.
Sheet: 50,000-65,000 TS; 32-000-53,000 YS; 12-22 El. For electrical equipment, motors, radio transformers; high permeability. *Obsolete*

WE VI
Wheeling-Pittsburgh Steel Corp.
Si 1.03-3.5, bal Fe.
Sheet: 50,000-65,000 TS; 32,000-53,000 YS; 12-22 El. For electrical equipment, motors, radio transformers; high permeability. *Obsolete*

WE VII
Wheeling-Pittsburgh Steel Corp.
Si 3-4.2, bal Fe.
Sheet: 67,000-70,000 TS; 55,000-65,000 YS; 4-8 El. For power and distribution transformers; high permeability. *Obsolete*

WE VIII
Wheeling-Pittsburgh Steel Corp.
Si 3-4.2, bal Fe.
Sheet: 67,000-70,000 TS; 55,000-65,000 YS; 4-8 El. For power and distribution transformers; high permeability. *Obsolete*

WE-II
Wheeling-Pittsburgh Steel Corp.
Si 0.5, bal Fe.
Annealed: 45,000 TS; 25,000 YS; 25 El. For armatures, electrical equipment; high permeability. *Obsolete*

WE-X
Wheeling-Pittsburgh Steel Corp.
Si 3-4.2, bal Fe.
Rolled: 70,000 TS; 65,000 YS; 4 El. For motors, electrical equipment; high permeability. *Obsolete*

WE5
Thyssen Edelstahlwerke AG
C 0.06, Cr 5, Mo 0.9, V 0.3, bal Fe.
For bakelite molds; case hardening. *Obsolete*

WEAR DEVIL A
Champion Rivet Co.
C 0.65, Cr 3.7, V 0.2, Mo 0.5, bal Fe.
550 Brin. For hard facing welding electrodes; abrasion resistant.

WEAR DEVIL B
Champion Rivet Co.
C 1, Cr 5, Mo 1.7, bal Fe.
500 Brin. For hard facing welding electrodes; abrasion and erosion resistant.

WEAR DEVIL C-1
Champion Rivet Co.
Co-Cr-Mo-W.
560 Brin. For hard facing welding electrodes. *Obsolete*

WEAR DEVIL C-2
Champion Rivet Co.
Co-Cr-Mo-W.
500 Brin. For hard facing welding electrodes. *Obsolete*

WEAR DEVIL C-3
Champion Rivet Co.
Co-Cr-Mo-W.
For hard facing welding electrodes. *Obsolete*

WEAR GARD
Certified Alloy Products Inc.
C 3.6, Ni 3.2, Cr 1.3, bal Fe.
As cast: 50,000 TS; 600 Brin. For abrasion resisting liners. *Obsolete*

WEAR-ARC
Now WEAR-ARC 40.

WEAR-ARC 12 IP
Chemetron Corp.
Mn 2.7, Cr 13, Si 1.8, Mo 1.1, weld metal: 3.5 C, bal Fe.
AC-DC, straight or reverse polarity, covered electrode for build-up for heavy impact and good abrasion resistance. 54-56 Rc.

WEAR-ARC 3 IP
Chemetron Corp.
Mn 0.9, Cr 2.3, Mo 1.1, Si 0.7, weld metal: 0.20 C, bal Fe.
AC-DC, reverse polarity, covered electrode for build-up to resist wear, impact and compressive loads. 101,000 psi TS; 24 El; 29 Rc.

WEAR-ARC 4 IP
Chemetron Corp.
Mn 0.9, Si 1.3, Cr 2.2, Mo 1, weld metal: 0.45 C, bal Fe.
AC-DC, straight or reverse polarity, covered electrode for hard surfacing. 54-56 Rc.

WEAR-ARC 40
Chemetron Corp.
Mn 0.3, Cr 30, Si 1.8, weld metal: 4.5 C, bal Fe.
AC-DC, reverse polarity, covered electrode for build-up having extreme abrasion resistance with medium impact, even at elevated temperatures. 57 Rc as welded.

WEAR-ARC 5 IP
Chemetron Corp.
Mn 1, Si 0.8, Cr 5.75, Mo 0.65, weld metal: 0.65 C, bal Fe.
AC-DC, reverse polarity, covered electrode for hard surfacing. 58-60 Rc.

WEAR-ARC 6 IP
Chemetron Corp.
Mn 0.8, Cr 6.5, Si 1.8, weld metal: 3.0 C, bal Fe.
AC-DC, straight or reverse polarity, covered electrode for build-up having high abrasion resistance but light impact. 56-59 Rc.

WEAR-ARC CHROME-BORIDE
Chemetron Corp.
C 2, Cr 11, Si 1.3, B 0.9, bal Fe.
Welded: 620 Brin. For hard facing electrodes; wear resistant. *Obsolete*

WEAR-ARC NICKEL MANGANESE
Chemetron Corp.
Mn 14, Si 0.55, Ni 4, weld metal: 0.60 C, bal Fe.
AC-DC, reverse polarity, covered electrode for welding or build-up on high manganese steel equipment, as bucket teeth. Soft, austenitic, as welded; work hardens to about 48 Rc.

WEAR-ARC SUPER WH
Chemetron Corp.
C 0.35, 33% alloying elements, bal Fe.
For hard facing electrode for Mn steel; high strength and wear resistant. *Obsolete*

WEAR-ARC WH
Chemetron Corp.
C 0.5, Mn 4, Cr 19, Ni 9.5, bal Fe.
Welded: 127,000 TS; 19 El; 320 Brin. For hard facing electrode; austenitic, stainless.

WEAR-EX
Pyramid Steel Company
C, alloy, bal Fe.
For hammer mills, mine cars, crushers, pulverizers; tough and abrasion resistant.

WEAR-FLAME 40
Chemetron Corp.
C 4.5, Mn 5, Cr 30, Si 1.8, bal Fe.
Weld: 600 Brin. For hard facing electrodes. Wear and abrasion resistant.

WEAR-O-MATIC 12
Chemetron Corp.
Mn 0.3, Cr 17, Si 1, Mo 0.8, weld metal: 2.3 C, bal Fe.
Semi-automatic, open-arc, DC, reverse polarity electrode for hard surface build-up for heavy impact and severe abrasion resistance. As deposited: 50 Rc.

WEAR-O-MATIC 12A
Chemetron Corp.
C 3, Mn 4, Cr 14.5, Si 1.3, Mo 0.8, bal Fe.
Weld: 48 Rock C. For hard surfacing crushing equipment, muller tires, steel mill pinch roll scraper blades. Submerged arc weld wire. Resists severe abrasion and impact.

WEAR-O-MATIC 12B
Chemetron Corp.
C 2.3, Mn 0.3, Cr 17, Si 1, Mo 0.8, bal Fe
Weld: 50 Rock C. For hard surfacing, heavy rock handling equipment, crushers, impactors, power shovel and dragline bucket parts. Open arc wire. Heavy impact and abrasion resistance.

WEAR-O-MATIC 14
Chemetron Corp.
C 0.25, Mn 1.3, Ni 5, Si 0.3, Mo 4.8, W 0.5, bal Fe.
Weld: 48 Rock C. For hard surfacing steel mill rolls, blast furnace bells and hoppers, roll necks, bearings, journals. Resists heat and abrasion.

WEAR-O-MATIC 15
Chemetron Corp.
C 4, Mn 0.3, Cr 5.5, Si 0.6, Mo 5, bal Fe.
Weld: 60 Rock C. For hard surfacing plug mill knives and augers, cement pump screws, conveyors. Wear and abrasion resistant.

WEAR-O-MATIC 15
Chemetron Corp.
Mn 0.3, Cr 5.5, Si 0.6, Mo 5, weld metal: 4.0 C, bal Fe
Semi-automatic, open-arc, DC, straight or reverse polarity electrode for hard surfacing for extreme abrasion resistance. 60 Rc.

WEAR-O-MATIC 16
Chemetron Corp.
C 0.05, Mn 1.6, Si 0.3, Ni 6.2, Mo 5.3, bal Fe.
Weld: 42-44 Rock C. For hard surfacing of blast furnace bells and hoppers, steel mill twist rolls, pinch rolls. For heat and abrasion resistance without heat checking.

WEAR-O-MATIC 3
Chemetron Corp.
Mn 2, Si 2, Cr 0.5, Mo 0.5, weld metal: 0.07 C, bal Fe.
Semi-automatic, open-arc, DC, straight or reverse polarity electrode for machinable build-up, usually prior to hard surfacing. 30-36 Rc.

WEAR-O-MATIC 40
Chemetron Corp.
C 4, Mn 1.5, Cr 27, Mo 1, Si 1.5, bal Fe.
Welded: 58 Rock C. For hard surfacing compression type crusher parts and hammer mills, mill guides. Wear and abrasion resistant.

WEAR-O-MATIC 40
Chemetron Corp.
Mn 1.5, Cr 27, Mo 1, Si 1.5, weld metal: 4.0 C, bal Fe.
Semi-automatic, open-arc, DC, reverse polarity electrode for build-up for severe abrasion and compression. 58 Rc.

WEAR-O-MATIC 420
Chemetron Corp.
C 0.3, Mn 1.5, Cr 14, Si 1, bal Fe.
Welded: 52 Rock C. For hard surfacing paper mill rolls, pipe forming rolls, hot bar mill guides. Corrosion and abrasion resistance.

WEAR-O-MATIC 6
Chemetron Corp.
C 0.65, Mn 2.6, Cr 3, Si 0.2, Mo 0.5, bal Fe.
Weld: 48 Rock C. For hard surfacing conveyor buckets, dragline and power shovel lips and sides, scraper blades. Severe impact and abrasion resistant. Good compressive strength and high hardness.

WEAR-O-MATIC 6
Chemetron Corp.
Mn 2.6, Cr 3, Si 0.2, Mo 0.5, weld metal: 0.65 C, bal Fe.
Semi-automatic, open-arc, DC, straight or reverse polarity electrode for hard surfacing. High impact resistance; 48 Rc.

WEAR-O-MATIC 7
Chemetron Corp.
C 0.4, Mn 3, Cr 5.3, Si 1.2, Mo 1.5, W 1.4, bal Fe.
Weld: 55 Rock C. For hard surfacing cable sheaves, crane wheels, pinch and pipe forming rolls. Resists severe abrasion and compression.

WEAR-O-MATIC BR
Chemetron Corp.
Si 0.37, Mn 1.6, Cr 2.5, Mo 0.55, weld metal: 0.12 C, bal Fe.
Semi-automatic, open-arc, DC, reverse polarity electrode designed for repair of railroad freight car bolster bowls using 98 Argon - 2 Oxygen. 35-40 Rc.

WEAR-O-MATIC NICKEL MANGANESE
Chemetron Corp.
Mn 13.5, Ni 3.9, Si 0.6, weld metal: 0.30 C, bal Fe.
Semi-automatic, open-arc, DC, reverse polarity electrode for build-up on austenitic manganese steel as in bucket teeth. Soft as welded; work hardens to about 48 Rc.

WEAR-O-MATIC RAIL-ARC
Chemetron Corp.
Mn 3.3, Si 0.5, Cr 17.8, Ni 8.1, weld metal: 0.06 C, bal Fe.
Semi-automatic, open-arc, DC, reverse polarity electrode for multipass build-up on all weldable carbon and austenitic manganese steel rails. Soft as deposited; work hardens to 35 Rc.

WEAR-O-MATIC SUPER WH
Chemetron Corp.
Mn 15, Si 0.65, Cr 17, Ni 1.4, weld metal: 1.1 C, bal Fe.
Semi-automatic, open-arc, DC, reverse polarity electrode for build-up having severe impact resistance with some abrasion. As welded: 30 Rc; work hardens to 47-49 Rc.

WEAR-O-MATIC WH
Chemetron Corp.
C 0.38, Mn 4.23, Si 0.47, Cr 20.2, Ni 9.65, bal Fe.
Weld: 103,000 TS; 70,000 YS; 36 El. For welding manganese steel to carbon steel; 18 Rock C (47 Rock C after work hardening). Tough, resilient, work hardenable.

WEAR-O-MATIC WH
Chemetron Corp.
Mn 4.23, Si 0.47, Cr 20.2, Ni 9.65, weld metal: 0.38 C, bal Fe.
Semi-automatic, open-arc, DC, reverse polarity electrode for weld or build-up for tough surface. As welded: 36 Rc; work hardens to about 47 Rc.

WEAR-PROOFT MAC HEMPITE
Gulf & Western Mfg. Co.
C 0.45, Mn 1.25, Mo 0.35, bal Fe.
Hardened: 700 Brin. For gears, pinions, mill couplings, boxes, cams; wear and shock resistant.

WEARALOY
Adirondack Steel Casting Co.
C 0.5, Ni 1.5, Cr 0.9, Mo 0.2, bal Fe.
For steam shovels, dipper teeth, and sheaves. Wear and abrasion resistant.

WEARCLAD 40, 43, 45
British Steel plc
High C, Cr & Nb.
Abrasion resistant clad plate steel.

WEAREX
Darwins Alloy Castings
C 0.7, W 14.5, Cr 4, V 1, bal Fe.
For lathe and planer tools, woodworking tools, milling cutters; high-speed steel. *Obsolete*

WEARGARD
Centrifugal Products Inc.
C 3.6, Ni 3.2, Cr 1.3, Mo 0.2, bal Fe.
As cast: 555 Brin. Abrasion resistant cast iron. Resin bonded waffle patterns for overlays.

WEARITE 4-11
Mueller Brass Co.
Al 10-11.2, Fe 0-4, 0.5 others, bal Cu.
Plate: 100,000 TS; 50,000 YS; 7 El; 210 Brin. For slides, cams, bushings, liners, bearings, chuck jaws, lathe beds. Extruded aluminum bronze, tough.

WEARITE 4-11
Peninsular Steel Co.
Al 10-11.2, Fe 3-4, 0.5 others, bal Cu.
Plate: 100,000 TS; 50,000 YS; 7 El; 210 Brin. For slides, cams, bushings, liners, bearings, chuck jaws, lathe beds. Extruded aluminum bronze, tough.

WEARITE 4-13
Mueller Brass Co.
Al 12.5-13.6, Fe 3.5-5, bal Cu.
Plate: 105,000 TS; 65,000 YS; 1.5 El; 313 Brin. For slides, cams, bushings, liners, bearings, chuck jaws, lathe beds. Extruded aluminum bronze, tough.

WEARITE 4-13
Peninsular Steel Co.
Al 12.5-13.6, Fe 3.5-5, bal Cu.
Plate: 105,000 TS; 65,000 YS; 1.5 El; 313 Brin. For slides, cams, bushings, liners, bearings, chuck jaws, lathe beds. Extruded aluminum bronze, tough.

WEARLOY
Frank Foundries Corp.
C 0.7, Ni 1.5, Cr 0.6, bal Fe.
For brake drums; for buses and trucks. *Obsolete*

WEARMANG
Atlas Specialty Steels
C 0.3, Mn 1.5, Si 0.25, Mo 0.2, bal Fe.
Rolled: 125,200 TS; 94,000 YS; 20 El; 285 Brin. For wear plates, mining equipment; high wear and abrasion resistance. *Obsolete*

WEARPACT

American Steel Foundries
C 0.23-0.33, Mn 1.3-1.8, Si 0.4, Cr 0.4-1, Mo 0.4-0.6, Ce 0.5, B, bal Fe.
Heat treated: 120,000-260,000 TS; 100,000-195,000 YS; 10-20 El; 23-58 RA; 200-520 Brin. For crushers, dipper teeth; high impact and abrasion resistant. *Obsolete*

WEARTUF STEEL

Horace T. Potts Co.
C 0.9, Mn 1.2, Si 0.6, bal Fe.
Hot rolled: 115,000-145,000 TS; 235-300 Brin. For scraper blades, conveyor buckets, liners, mixers, hoppers, chutes, screens, dipper lips; wear and abrasion resistant. *Obsolete*

WEARWELD

Lincoln Electric Co.
C 0.37, Mn 2.2, Si 0.15, Cr 3.3, bal Fe.
Hard surfacing, arc welding electrodes; resistance to metal to metal wear.

WEARWELL

Stelco Steel
C 0.3-0.35, Mn 1.3-1.65, Si 0.15-0.3, bal Fe.
For riveted and bolted applications requiring abrasion resistance.

WEBB BLUE LABEL

Webb Wire Works
C 0.8-0.9, bal Fe.
For springs; music wire. *Obsolete*

WEBBITE (TAM)

NL Industries
Ti 5-7, bal Al.
To add Ti to Al alloys; for improving aluminum alloys, hardener. *Obsolete*

WEBERT ALLOY

Anaconda Co.
Copper. Zn 14, Si 4, Mn, bal Cu.
For pressure die castings.

WEFAHUTTE HECN

Westfalenhutte Dortmund A.G.
C 0.13, Cr 0.7, Ni 2.5, Mn 0-0.5, bal Fe.
For gears, bolts, plastic mold dies, shafts; case hardened, shock resistant.

WEFAHUTTE HECN15

Westfalenhutte Dortmund A.G.
C 0.15, Ni 1.55, Cr 1.55, bal Fe.
For camshafts, cams, bolts, gears; case hardened, shock resistant.

WEFAHUTTE HECN35

Westfalenhutte Dortmund A.G.
C 0.13, Cr 0.7, Ni 3.5, Mn 0-0.5, bal Fe.
For gears, bolts, plastic mold dies, shafts; case hardened, shock resistant.

WEFAHUTTE HK60

Westfalenhutte Dortmund A.G.
C 0.3, Si 0.25, Mn 1.35, bal Fe.
For punches, gears, crankshafts; water or oil hardened.

WEFAHUTTE HK75

Westfalenhutte Dortmund A.G.
C 0.37, Si 1.2, Mn 1.25, bal Fe.
For punches, upsetters, crimpers; oil hardened, shock resistant.

WEFAHUTTE HKCN15W/H

Westfalenhutte Dortmund A.G.
C 0.28-0.35, Cr 0.5, Ni 1.5, bal Fe.
For gears, bolts, crankshafts, fasteners; oil hardened, shock resistant.

WEFAHUTTE HKCN25W/H

Westfalenhutte Dortmund A.G.
C 0.18-0.35, Cr 0.7, Ni 2.5, bal Fe.
For gears, bolts, machine tool parts; oil hardened, shock resistant.

WEFAHUTTE HKCN35W/H

Westfalenhutte Dortmund A.G.
C 0.22-0.3, Cr 0.7, Ni 3.5, Mn 0.6, bal Fe.
For gears, bolts, crankshafts; oil hardened, shock resistant.

WEFAHUTTE HKCNX25

Westfalenhutte Dortmund A.G.
C 0.36, Ni 1, Cr 1, bal Fe.
For gears, bolts, machine tool parts; oil hardened, tough.

WEFAHUTTE HLMF

Westfalenhutte Dortmund A.G.
C 0.46, Mn, bal Fe.
For machine tool parts, gears, shafts; water hardened.

WEFAHUTTE HLSFH

Westfalenhutte Dortmund A.G.
C 0.65, Si 1.7, Mn 0.7, bal Fe.
For springs, punches, crimpers; oil hardened, shock resistant.

WEFAHUTTE HLSFW

Westfalenhutte Dortmund A.G.
C 0.55, Si 1.7, Mn 0.7, bal Fe.
For springs, punches, upsetters; oil hardened, shock resistant.

WEFAHUTTE HUK10

Westfalenhutte Dortmund A.G.
C 0.34, V 0.1, Mn 0.65, bal Fe.
For gears, bolts, shafts, fasteners; water hardened.

WEFAHUTTE KWF

Westfalenhutte Dortmund A.G.
C 0.46, Si 1.7, Mn 0.65, bal Fe.
For springs, rivet sets, punches; oil hardened, tough.

WEGNER

Manufacturer not listed.
Sn 80, Zn 20.
For bearings; anti-friction.

WEHRALLOY NO. 1

Wehr Steel Corp.
C 0-0.07, Cr 16-23, Ni 7-11, bal Fe.
For heat and corrosion resistant parts. Stainless; heat and corrosion resistant. *Obsolete*

WEHRALLOY NO. 10

Wehr Steel Corp.
C 0.21-0.5, Ni 58-66, Cr 16-23, bal Fe.
For heat and corrosion resistant parts. Heat and corrosion resistant. *Obsolete*

WEHRALLOY NO. 12

Wehr Steel Corp.
C 0-0.12, Cr 12-15, bal Fe.
For corrosion resistant parts. Corrosion resistant. *Obsolete*

WEHRALLOY NO. 13

Wehr Steel Corp.
C 0.13-0.2, Cr 12-15, bal Fe.
For corrosion resistant parts. Corrosion resistant. *Obsolete*

WEHRALLOY NO. 14

Wehr Steel Corp.
C 0.21-0.35, Cr 12-15, bal Fe.
For corrosion resistant parts. Corrosion resistant. *Obsolete*

WEHRALLOY NO. 16

Wehr Steel Corp.
C 0.21-0.35, Cr 16-23, bal Fe.
For heat and corrosion resistant parts. Heat and corrosion resistant. *Obsolete*

WEHRALLOY NO. 17

Wehr Steel Corp.
C 0.81-1.1, Cr 16-23, bal Fe.
For heat and corrosion resistant parts. Heat and corrosion resistant. *Obsolete*

WEHRALLOY NO. 18

Wehr Steel Corp.
C 0.13-0.2, Cr 24-30, bal Fe.
For heat and corrosion resistant parts. Heat and corrosion resistant. *Obsolete*

WEHRALLOY NO. 19

Wehr Steel Corp.
C 0.13-0.2, Cr 24-30, Si, bal Fe.
For heat and corrosion resistant parts. Heat and corrosion resistant. *Obsolete*

WEHRALLOY NO. 2

Wehr Steel Corp.
C 0-0.12, Cr 16-23, Ni 7-11, bal Fe.
For heat and corrosion resistant parts. Stainless; heat and corrosion resistant. *Obsolete*

WEHRALLOY NO. 21

Wehr Steel Corp.
C 0.21-0.35, Cr 24-30, bal Fe.
For heat and corrosion resistant parts. Heat and corrosion resistant. *Obsolete*

WEHRALLOY NO. 22

Wehr Steel Corp.
C 0.13-0.2, Cr 16-23, Ni 7-11, bal Fe.
For heat and corrosion resistant parts. Stainless; heat and corrosion resistant. *Obsolete*

WEHRALLOY NO. 23

Wehr Steel Corp.
C 0.21-0.35, Cr 16-23, Ni 7-11, bal Fe.
For heat and corrosion resistant parts. Stainless; heat and corrosion resistant. *Obsolete*

WEHRALLOY NO. 24

Wehr Steel Corp.
C 0.21-0.35, Cr 16-23, Ni 7-11, bal Fe.
For heat and corrosion resistant parts. Stainless; heat and corrosion resistant. *Obsolete*

WEHRALLOY NO. 26

Wehr Steel Corp.
C 0.36-0.5, Ni 31-39, Cr 16-23, Si, bal Fe.
For heat and corrosion resistant parts. Heat and corrosion resistant. *Obsolete*

WEHRALLOY NO. 27

Wehr Steel Corp.
C 0.36-0.5, Ni 31-39, Cr 16-23, Si, bal Fe.
For heat and corrosion resistant parts. Heat and corrosion resistant. *Obsolete*

WEHRALLOY NO. 28

Wehr Steel Corp.
C 0.51-0.8, Ni 67-75, Cr 16-23, Si, bal Fe.
For heat and corrosion resistant parts. Heat and corrosion resistant. *Obsolete*

WEHRALLOY NO. 29

Wehr Steel Corp.
C 0.21-0.35, Cr 5-7, bal Fe.
For corrosion resistant parts. Corrosion resistant. *Obsolete*

WEHRALLOY NO. 3

Wehr Steel Corp.
C 0.21-0.35, Cr 24-30, Ni 12-15, bal Fe.
For heat and corrosion resistant parts. Stainless; heat and corrosion resistant. *Obsolete*

WEHRALLOY NO. 31

Wehr Steel Corp.
C 0.13-0.2, Cr 24-30, Ni 12-15, bal Fe.
For heat and corrosion resistant parts. Heat and corrosion resistant. *Obsolete*

WEHRALLOY NO. 4

Wehr Steel Corp.
C 0.13-0.35, Cr 24-30, Ni 7-11, bal Fe.
For heat and corrosion resistant parts. Stainless; heat and corrosion resistant. *Obsolete*

WEHRALLOY NO. 43

Wehr Steel Corp.
C 0.13-0.2, Cr 24-30, Ni 12-15, bal Fe.
For heat and corrosion resistant parts. Heat and corrosion resistant. *Obsolete*

WEHRALLOY NO. 5

Wehr Steel Corp.
C, Cr, Ni, bal Fe.
For heat and corrosion resistant parts. Stainless; heat and corrosion resistant. *Obsolete*

WEHRALLOY NO. 6

Wehr Steel Corp.
C 0.13-0.35, Cr 24-30, Si, bal Fe.
For heat and corrosion resistant parts. Heat and corrosion resistant. *Obsolete*

WEHRALLOY NO. 7

Wehr Steel Corp.
C 0.21-0.5, Cr 12-23, Si, bal Fe.
For heat and corrosion resistant parts. Heat and corrosion resistant. *Obsolete*

WEHRALLOY NO. 8

Wehr Steel Corp.
C 0.21-0.5, Ni 24-30, Cr 16-23, Si, bal Fe.
For heat and corrosion resistant parts. Heat and corrosion resistant. *Obsolete*

WEHRALLOY NO. 9

Wehr Steel Corp.
C 0.36-0.5, Ni 31-39, Cr 16-23, Si, bal Fe.
For heat and corrosion resistant parts. Heat and corrosion resistant. *Obsolete*

WEHRALLOY STAINLESS 18-8

Wehr Steel Corp.
C 0.13-0.2, Cr 18, Ni 8, bal Fe.
For stainless parts. Stainless. *Obsolete*

WEHRALLOY STAINLESS 18-8 SPECIAL

Wehr Steel Corp.
C 0-0.07, Cr 18, Ni 8, bal Fe.
For stainless parts. Stainless. *Obsolete*

WEIDRIUM

German manufacture
Ni, Cu, Zn, Sn, Fe, Mg.

WEIGER

Manufacturer not listed.
Ag 77-80, Cu 18-20, Pt 2-5.
For silver solder; corrosion resistant.

WEIGHTS

Manufacturer not listed.
Cu 90, Sn 8, Zn 2.
For weights; corrosion resistant.

WEINGARTNER EW 1540 EXTRA

Emil Weingartner & Co.
C 1, bal Fe.
Water hardening tool steel. Werkstoff Nr. 1.1540; AISI W1 tool steel.

WEINGARTNER EW 2343

Emil Weingartner & Co.
C 0.38, Si 1, Mn 0.4, Cr 5.3, Mo 1.1, V 0.4, bal Fe.
Air or oil hardening hot work tool steel. Werkstoff Nr. 1.2343; similar to AISI H11.

WEINGARTNER EW 2601

Emil Weingartner & Co.
V 1.65, Si 0.3, Mn 0.3, C 12, Mo 0.6, bal Fe.
Air or oil hardening cold work tool steel. Werkstoff Nr. 1.2601; AISI D2.

WEINGARTNER EW 3343

Emil Weingartner & Co.
C 0.78-0.86, Cr 4, Mo 5, V 1.75, W 6, bal Fe.
High speed tool steel. Werkstoff Nr. 1.3343; AISI M2.

WEINGARTNER EW 4024

Emil Weingartner & Co.
C 0.12-0.17, Si 0-1, Mn 0-1, Cr 13, bal Fe.
Air or oil hardenable martensitic stainless steel. Werkstoff Nr. 1.4024; similar to AISI 410 stainless steel.

WEINGARTNER EW 4301

Emil Weingartner & Co.
C 0-0.07, Si 0-1, Mn 0-2, Cr 18, Ni 10, bal Fe.
Austenitic stainless steel. Werkstoff Nr. 1.4301; AISI 304; stainless.

WEINGARTNER EW 4704

Emil Weingartner & Co.
C 0.45, Si 0.4, Mn 0.45, Cr 2.65, bal Fe.
Oil hardening valve steel. Werkstoff Nr. 1.4704; DIN X 45 SiCr4.

WEINGARTNER EW 4841

Emil Weingartner & Co.
C 0-0.2, Si 2.05, Mn 0-2, Cr 25, Ni 20, bal Fe.
Austenitic high temperature alloy. Werkstoff Nr. 1.4841; similar to AISI 310 stainless steel.

WEINGARTNER EW 5919

Emil Weingartner & Co.
C 0.15, Mn 0.4, Cr 1.55, Ni 1.55, bal Fe.
Oil hardenable carburizing steel. Werkstoff Nr. 1.5919; DIN 15 CrNi6.

WEINGARTNER EW 8159

Emil Weingartner & Co.
C 0.5, Si 0.25, Mn 0.95, Cr 1.05, V 0.1, bal Fe.
Oil hardening steel; for springs, shafts. Werkstoff Nr. 1.8159; DIN 50 CrV4; AISI 6150.

WEINGARTNER EW 8507

Emil Weingartner & Co.
C 0.34, Mn 0.6, Cr 1, Mo 0.2, Al 1, bal Fe.
Nitriding steel. Werkstoff Nr. 1.8507.

WEIRALEAD

National Steel Corp.
Pb coated steel.
For structural sheets; hot dip. *Obsolete*

WEIRCOLOY

National Steel Corp.
C 0.2, Cu 0.5, bal Fe.
For galvanized structures; copper bearings; sheet steel. *Obsolete*

WEIRITE

National Intergroup Inc.
Sn coated soft steel.
For cans, roofing; tin plate, electrolytic.

WEIRZIN

National Intergroup Inc.
Zn coated steel.
For structural sheets; electrolytic plate.

WEISSPUNKT

SWB Stahlformguss Gesellschaft mbH
C 0.9, Mo, Cr, V, W, bal Fe.
For lathe and planer tools, reamers, broaches; high speed steel. *Obsolete*

WEISSPUNKT BM09

SWB Stahlformguss Gesellschaft mbH
Cr 4, C, W, Mo, V, bal Fe.
For lathe and planter tools, reamers; high speed steel. *Obsolete*

WEL-MET BRONZE

Wel-Met Co.
Copper. Cu 84-86, Sn 7-8, Pb 5-6, 1.0-2.0 graphite.
Sintered: 12,000 TS; 5 El; 30 Brin. For bearings. Self-lubricating, sintered, and oil impregnated.

WEL-MET STEEL

Wel-Met Co.
Fe 89-92, Cu 7-10, 0.5-2.0 graphite.
Sintered: 27,000 TS; 1-2 El; 00-00 Brin. For bearings. Sintered and self-lubricating.

WEL-TEN 100N

Yawata Iron & Steel Co., Ltd.
C 0-0.18, Ni 0-1.5, Cu 0.15-0.5, Cr 0.4-0.8, Mo 0-0.6, V 0-0.1, Mn 0.6-1.2, bal Fe.
Quenched and tempered plate: 140,000-163,000 TS; 130,000 YS; 15 El. For heavy duty welded structures and equipment, pressure vessels, bridges. Conforms to ASTM-A300 Cl.1. Good weldability, high strength.

WEL-TEN 50

Yawata Iron & Steel Co., Ltd.
C 0-0.18, Mn 0.9-1.5, Si 0.25-0.45, bal Fe.
Rolled: 71,000-82,000 TS; 47,000 YS min; 20 El min. For pressure vessels, bridges, penstocks, construction machinery. Constructional steel. Good weldability.

WEL-TEN 55

Yawata Iron & Steel Co., Ltd.
C 0-0.18, Mn 1.2-1.5, Si 0.35-0.55, bal Fe.
Normalized: 78,000-90,000 TS; 51,000 YP min; 20.0 El min. For pressure vessels, bridges, penstocks, construction machinery, booms, derricks. Readily fabricated, tough, construction steel. Good weldability.

WEL-TEN 60

Yawata Iron & Steel Co., Ltd.
C 0-0.16, Mn 0-1.3, Si 0-0.55, Cr 0-0.4, V 0-0.15, Ni 0-0.6, bal Fe.
Heat treated: 85,000 TS min; 65,000 YP min; 16 El min. For railroad and mine cars, pressure vessels, agricultural equipment, derricks. Good weldability, tough. High-strength low-alloy constructional steel.

WEL-TEN 60

Nippon Steel USA Inc.
C 0.13, Si 0.35, Mn 1.24, P 0.012, S 0.008, V 0.04, bal Fe.
6-50 mm plate, quenched and tempered: 92,900 psi TS; 82,200 psi YS; 30 El. High strength steel for welded structures of oil storage tanks, bridges, earth moving and offshore structures.

WEL-TEN 60 CF

Nippon Steel USA Inc.
C 0.07, Si 0.26, Mn 1.3, P 0.015, S 0.005, Cr 0.21, Mo 0.15, bal Fe.
6-50 mm plate, quenched and tempered: 92,000 psi TS; 81,400 psi YS; 27 El. High strength steel for welded structures as spherical pressure vessels; low susceptibility to weld cracking.

WEL-TEN 60-LT

Yawata Iron & Steel Co., Ltd.
C 0-0.16, Mn 0.9-1.4, Ni 0-0.6, Cr 0-0.3, V 0.12, bal Fe.
Plate: 85,300-102,000 TS; 64,500-71,100 YS. For pressure vessels, cryogenic equipment. High notch toughness. Good weldability.

WEL-TEN 60H
Yawata Iron & Steel Co., Ltd.
C 0-0.18, Si 0.15-0.75, Mn 1-1.5, Ni 0-1, 0.15 Nb + V max, bal Fe.
Normalized: 85,000-102,000 TS; 64,000 YP min; 20 El min.
For pressure vessels, bridges, penstocks, construction machinery. Good weldability and formability.

WEL-TEN 62
Yawata Iron & Steel Co., Ltd.
C 0-0.18, Si 0.15-0.55, Mn 0.9-1.4, Ni 0-0.6, Cr 0-0.3, V 0-0.12, bal Fe.
Quenched and tempered: 88,000-107,000 TS; 71,000 YS min; 19 El min. For pressure vessels, penstocks, bridges, construction machinery. Good formability and weldability.

WEL-TEN 62
Nippon Steel USA Inc.
C 0.14, Si 0.46, Mn 1.18, P 0.017, S 0.008, Cr 0.15, V 0.04, bal Fe.
6-50 mm plate, quenched and tempered: 89,900 psi TS; 78,200 psi YS; 29 El. High strength steel for pressure vessels.

WEL-TEN 62 CF
Nippon Steel USA Inc.
C 0.07, Si 0.28, Mn 1.36, P 0.014, S 0.003, Cr 0.25, Mo 0.22, V 0.03, bal Fe.
6-50 mm plate, quenched and tempered: 93,900 psi TS; 82,500 psi YS; 26 El. High strength steel for welded structures as spherical pressure vessels; low susceptibility to weld cracking.

WEL-TEN 70
Nippon Steel USA Inc.
HSLA steel. C 0.11, Si 0.45, Mn 1, P 0.01, S 0.003, Cu 0.02, Ni 0.9, Cr 0.3, Mo 0.4, V 0.04, bal Fe.
50 mm thick: 112,800 TS; 103,000 YS; 44 El. High strength low alloy steel for structural purposes.

WEL-TEN 80
Yawata Iron & Steel Co., Ltd.
C 0.18, Mn 0.6-1.2, Si 0.15-0.35, B 0.006, Cr 0.4-0.8, Cu 0.15-0.5, Mo 0.6, Ni 1.5, V 0.1, bal Fe.
Plate: 114,000-135,000 TS; 100,000 YS min; 18-20 El (quenched and tempered). For pressure vessels, bridges, mine cars, power shovels, cranes, trucks, trailers. Shock and wear resistant. Good weldability.

WEL-TEN 80
Nippon Steel USA Inc.
C 0.11, Si 0.21, Mn 0.85, P 0.015, S 0.006, Cu 0.22, Ni 0.97, Cr 0.53, Mo 0.43, V 0.05, B 0.0008, bal Fe.
6-100 mm plate, quenched and tempered: 119,000 psi TS; 112,000 psi YS; 23 El. High strength steel for welded structures as long span bridge, penstock, earth moving equipment.

WEL-TEN 80 C
Nippon Steel USA Inc.
C 0.12, Si 0.26, Mn 0.86, P 0.01, S 0.004, Cu 0.29, Cr 0.74, Mo 0.42, V 0.04, B 0.0008, bal Fe.
6-50 mm plate, quenched and tempered: 118,000 psi TS; 110,000 psi YS; 23 El. High strength steel for welded structures as pressure vessels, earth moving equipment.

WEL-TEN 80 C-LT
Yawata Iron & Steel Co., Ltd.
C 0-0.18, Mn 0.6-1.2, Cr 0.7-1.3, Mo 0-0.6, V 0.15-0.5, B 0-0.006, bal Fe.
Plate: 113,800-135,100 YS; 99,600 YS min. For pressure vessels, cryogenic equipment. High notch toughness, good weldability.

WEL-TEN 80 E
Nippon Steel USA Inc.
C 0.19, Si 0.35, Mn 1.43, P 0.016, S 0.01, B 0.0008, bal Fe.
6-20 mm plate, quenched and tempered: 121,000 psi TS; 110,000 psi YS; 36 El. High strength steel for welded structures as earth moving equipment.

WEL-TEN 80C
Yawata Iron & Steel Co., Ltd.
C 0-0.18, Si 0.15-0.35, Mn 0.6-1.2, Cu 0.15-0.5, Cr 0.7-1.3, Mo 0-0.6, B 0-0.0006, bal Fe.
Quenched and tempered: 114,000-135,000 TS; 100,000 YP min. For pressure vessels, bridges, penstocks, construction machinery, bulldozers, power shovels. Good formability and weldability.

WEL-TEN 80P
Nippon Steel USA Inc.
HSLA steel. C 0.1, Si 0.31, Mn 1.46, P 0.01, S 0.01, Cu 0.07, Mo 0.59, B 0.003, Cb 0.049, bal Fe.
25 mm thick plate: 116,000 TS; 111,000 YS; 14.3 El. High strength low alloy steel for structural purposes.

WEL-TEN 80S
Nippon Steel USA Inc.
HSLA steel. C 0.08, Si 0.54, Mn 1.36, P 0.016, S 0.0008, Cu 0.14, Ni 0.97, Cr 1.03, Mo 0.56, V 0.05, Ti 0.015, bal Fe.
25 mm thick plate: 118,000 TS; 108,000 YS; 26 El. High strength low alloy steel for structural purposes.

WELCHS ALLOY
Manufacturer not listed.
Sn 52, Ag 48.
For dental alloy, solder; corrosion resistant.

WELCON 2H
Japan Steel Works Ltd.
C 0.15, Si 0.5, Mn 1.2, bal Fe.
Heat treated: 85,000-100,000 TS; 71,000 YP min; 16 El min. For pressure vessels, bridges, mine cars, cranes, truck bodies. Good fabricability and weldability. Constructional steel plates, tough.

WELCON 2H SUPER
Japan Steel Works Ltd.
C 0.08-0.16, Mn 0.6-1.2, Si 0.55, Cr 0.5, Mo 0.4, Ni 1, bal Fe.
Plate: 100,000-114,000 TS; 90,000 YS min; 18 El. For pressure vessels, bridges, bus and trailer bodies, cranes, mine cars. Wear and shock resistant. Heat treated by mill.

WELCON 2H ULTRA
Japan Steel Works Ltd.
C 0.08-0.16, Mn 0.6-1.2, B 0.006, Cr 0.8, Mo 0.7, Cu 0.15-0.5, Ni 1.5, V 0.1, bal Fe.
Plate: 114,000-135,000 TS; 100,000 YS min; 18-22 El. For pressure vessels, bridges, mine cars, power shovels, cranes, trucks, trailers. Shock and wear resistant. Heat treated by mill.

WELCON 2H-100
Japan Steel Works Ltd.
C 0.18, Si 0.45, Mo 0.6, V 0.1, bal Fe.
Plate: 138,000-164,000 TS; 128,000 YP. For pressure vessels, bridges, bus and trailer bodies, derricks, booms. Wear and shock resistant.

WELCON 50
Japan Steel Works Ltd.
C 0.18, Mn 0-1.35, Si 0-0.55, bal Fe.
Rolled plate: 71,000-82,000 TS; 47,000 YP min; 26 El min. For ship plate, oil tankers, thermal power generating equipment, mine cars, bus bodies. Constructional steel plate, tough.

WELCON-2H CR
Japan Steel Works Ltd.
C 0.11, Si 0.35, Mn 1.01, P 0.014, Si 0.008, Ni 0.14, Cr 0.5, Mo 0.34, bal Fe.
29 mm thick plate: 97,000 TS; 86,100 YS; 45 El. High strength low alloy steel for structural purposes.

WELD-ARC
Champion Rivet Co.
C 0.1, bal Fe.
Welded. For welding rod for mild steel; arc welding. *Obsolete*

WELD-FAST
Revere Copper Products, Inc.
Sn, bal Cu.
For welding rod; for bronze. *Obsolete*

WELDANKA
Dunford Hadfields Ltd.
C 0.2, Cr 18-25, 7-12% Ni + alloys, bal Fe.
78,500 TS. For turbine blading, chemical and aircraft parts, paper and pulp industry; high resistance to corrosion and heat. *Obsolete*

WELDBEST ALBRONZE 100
Weldwire Co., Inc.
Copper. Al, bal Cu.
Welded: 100 Brin. Aluminum bronze for welding electrodes for copper alloys; all purpose.

WELDBEST ALBRONZE 250
Weldwire Co., Inc.
Copper. Al, bal Cu.
Welded: 250 Brin. Aluminum bronze for welding electrodes for copper alloys; all purpose.

WELDEX
Pittsburgh Brass Mfg. Co.
C 2.2, Si 2.4, P 0.03, S 0.01, Mn 0.25, 0.08 combined C, bal Fe.
Malleable iron. 65,000-68,000 TS; 47,000-51,000 YS; 12-16 El.

WELDFIL 525 ATOMIZED STEEL POWDER
Hoeganaes Corp.
Ni 1.8, Mo 0.5, Mn 0.2, C 0.02, 0.5 H_2 (loss), bal Fe.
For one-pass shielded arc welding of plate up to 2 in. thick.

WELDMACO
Welding Material Co.
C 0.2, bal Fe.
For welding rod.

WELDOLOY NO. 685
Park Sales Co.
Cd 95, Ag 5.
For solder on steel, bronze and cast iron; flows at 750°F.

WELDOLOY NO. 690
Park Sales Co.
Si 8, bal Al.
Cast: 35,000 TS. For harder solder for Al; flows at 930°F.

WELDRAWN 18-8
Superior Tube Company
C 0.2, Cr 18, Ni 8, bal Fe.
For tubing; welded, stainless. *Obsolete*

WELL-CAST A
Wellman Dynamics Corp.
Mg 6.2-7.5, Mn 0.1-0.25, bal Al.
F-temper: 40,000 TS; 21,000 YS; 13 El; 70 Brin. For light alloy parts; Almag 35. *Obsolete*

WELL-CAST E
Wellman Dynamics Corp.
Si 4.5-6, Fe 0-0.8, bal Al.
F-temper: 19,000 TS; 8,000 YS; 8 El; 40 Brin. For light alloy parts; Alloy 43, corrosion resistant. *Obsolete*

WELL-CAST J
Wellman Dynamics Corp.
Cu 3.5-4.5, Si 2.5-3.5, Fe 0-1.2, bal Al.
F-temper: 21,000 TS; 14,000 YS; 2.5 El; 55 Brin. For light alloy parts; Alloy 108. *Obsolete*

WELL-CAST K
Wellman Dynamics Corp.
Mg 9.5-10.6, Ti 0.2, bal Al.
T4-temper: 46,000 TS; 25,000 YS; 14 El; 75 Brin. For light alloy parts; Alloy 220, age-hardened. *Obsolete*

WELL-CAST P

Wellman Dynamics Corp.
Cu 4-5, Si 0-1.2, bal Al.
T4-temper: 32,000 TS; 16,000 YS; 8.5 El; 60 Brin. T6-temper: 36,000 TS; 24,000 YS; 5 El; 75 Brin. T62-temper: 40,000 TS; 34,000 YS; 1.5 El; 95 Brin. For light alloy parts; Alloy 195, age-hardened. *Obsolete*

WELL-CAST R

Wellman Dynamics Corp.
Zn 2-4, Zr 0.5-1, 2.5-4% rare earths, bal Mg.
T5-temper: 23,000 TS; 16,000 YS; 50 Brin. For light alloy parts; Alloy EZ33A, age-hardened. *Obsolete*

WELL-CAST T

Wellman Dynamics Corp.
Th 2.5-4, Zr 0.4-1, bal Mg.
T6-temper: 32,000 TS; 18,000 YS; 6 El. For light alloy parts; age-hardened. *Obsolete*

WELL-CAST V

Wellman Dynamics Corp.
Cu 3-4.5, Si 5.5-7, bal Al.
F-temper: 27,000 TS; 18,000 YS; 2 El; 70 Brin. T6-temper: 36,000 TS; 24,000 YS; 2 El; 80 Brin. For light alloy parts; Alloy 319, age-hardened. *Obsolete*

WELL-CAST W

Wellman Dynamics Corp.
Si 6.5-7.5, Mg 0.2-0.4, bal Al.
T4-temper: 28,000 TS; 16,000 YS; 6 El; 55 Brin. T5-temper: 25,000 TS; 20,000 YS; 2 El; 60 Brin. T6-temper: 33,000 TS; 24,000 YS; 3.5 El; 70 Brin. For light alloy parts; Alloy 356, age-hardened. *Obsolete*

WELL-CAST X

Wellman Dynamics Corp.
Al 5.3-6.7, Zn 2.5-3.5, bal Mg.
F-temper: 29,000 TS; 14,000 YS; 6 El; 50 Brin. T4-temper: 40,000 TS; 14,000 YS; 12 El; 55 Brin. T6-temper: 40,000 TS; 18,000 YS; 5 El; 73 Brin. For light alloy parts; Alloy AZ63A, age-hardened. *Obsolete*

WELL-CAST Y

Wellman Dynamics Corp.
Al 8.1-9.3, Zn 0.4-1, bal Mg.
F-temper: 24,000 TS; 14,000 YS; 2 El; 52 Brin. T4-temper: 40,000 TS; 14,000 YS; 11 El; 55 Brin. T6-temper: 40,000 TS; 19,000 YS; 4 El; 73 Brin. For light alloy parts; Alloy AZ91C, age-hardened. *Obsolete*

WELL-CAST Z

Wellman Dynamics Corp.
Al 8.3-9.7, Zn 1.8-2.4, bal Mg.
F-temper: 24,000 TS; 14,000 YS; 2 El; 65 Brin. T4-temper: 40,000 TS; 14,000 YS; 10 El; 63 Brin. T6-temper: 40,000 TS; 23,000 YS; 2 El; 84 Brin. For light alloy parts; Alloy AZ92A, age-hardened. *Obsolete*

WELLCAST NO. 116

Wellman Dynamics Corp.
Cu 64, Ni 16, Zn 10, Pb 6, Sn 4.
Cast: 28,000-30,000 TS; 18-22 El; 70 Brin. For dairy industry castings; acid resisting. *Obsolete*

WELLCAST NO. 13A

Wellman Dynamics Corp.
Cu 2-3, Fe 0.6, Zn 11-13, bal Al.
Cast: 24,000-26,000 TS; 1-3 El; 60-80 Brin. For bearing caps, castings; non-hardenable. *Obsolete*

WELLCAST NO. 16S

Wellman Dynamics Corp.
Si 12-13, bal Al.
Cast: 20,000-24,000 TS; 10-15 El; 40-50 Brin. For light alloy castings; wear resisting. *Obsolete*

WELLCAST NO. 17S

Wellman Dynamics Corp.
Cu 1.4, Zn 2, Mg 0.3, Mn 1.4, Ti 0.15, bal Al.
Cast: 28,000-30,000 TS; 4-6 El; 50-60 Brin. For light alloy parts; non-hardenable. *Obsolete*

WELLCAST NO. 3S

Wellman Dynamics Corp.
Cu 6-8, Si 1.5-2, Zn 1.5, bal Al.
Cast: 18,000-21,000 TS; 1-3 El; 60-80 Brin. For crankcases, oil pans, transmission cases; good machinability. *Obsolete*

WELLCAST NO. 4A

Wellman Dynamics Corp.
Cu 4-5, Si 1, Fe 1, Mg 1, bal Al.
Cast: 20,000-24,000 TS; 1-2 El; 60-70 Brin. For light alloy castings; heat treatable. *Obsolete*

WELLCAST NO. 5S

Wellman Dynamics Corp.
Cu 0.6, Si 4.5-6, bal Al.
Cast: 17,000-20,000 TS; 2-4 El; 40-55 Brin. For marine fittings, aircraft engine parts, pressure tight. *Obsolete*

WELLCAST NO. 65

Wellman Dynamics Corp.
Cu 55-58, Zn 38-40, Mn 0.2-0.5, Fe 1-1.5, Al 1.
Cast: 65,000-75,000 TS; 20 El; 110 Brin. For gears, valves; resists sea water corrosion. *Obsolete*

WELLCAST NO. 73

Wellman Dynamics Corp.
Cu 70, Pb 25, Sn 5.
Cast: 22,000-25,000 TS; 19 El; 40 Brin. For high speed bearings, antifriction castings; heavy duty. *Obsolete*

WELLCAST NO. 78

Wellman Dynamics Corp.
Cu 77-79, Zn 11-12, Pb 6-8, Sn 2-3.
Cast: 25,000-27,000 TS; 8-10 El; 55-65 Brin. For castings; free-cutting. *Obsolete*

WELLCAST NO. 80

Wellman Dynamics Corp.
Cu 80, Sn 10, Pb 10, Trace P.
Cast: 25,000-30,000 TS; 6-8 El; 60-70 Brin. For bearings; heavy loads, severe service. *Obsolete*

WELLCAST NO. 85

Wellman Dynamics Corp.
Cu 85, Sn 5, Zn 5, Pb 5.
Cast: 25,000-27,000 TS; 15 El; 50 Brin. For pressure castings; free-cutting, pressure tight. *Obsolete*

WELLCAST NO. 88

Wellman Dynamics Corp.
Cu 88, Sn 10, Zn 2, Trace P.
Cast: 30,000-35,000 TS; 16-20 El; 70 Brin. For valves, nuts, propellers, gears; Navy "G" Bronze. *Obsolete*

WELLCAST NO. 99C

Wellman Dynamics Corp.
Cu 98-99, Sn 1-2.
Cast: 20,000-22,000 TS; 25 El; 50 Brin. For electrode holders; high electical conductivity *Obsolete*

WELMET 20

Welland Electric Steel Foundry
C 0-0.07, Ni 29, Cr 20, Mo 3-4, Cu 4, bal Fe.
Cast: 65,000-75,000 psi TS; 28,000-38,000 psi YS; 35-50 El; 45-50 RA; 120-160 Brin. For chemical plant equipment, valves, agitators, mixers; resists mixed acids, austenitic.

WELMET NO. 1

Welland Electric Steel Foundry
C 0-0.15, Cr 25, Ni 35, Mo 5, bal Fe.
For pipes, valves, fittings, steam nozzles; heat and corrosion resistant in SO_2 atmosphere.

WENDT-SONIS

Now CQ and CY.

WESGO 35 AU-65 CU

Western Gold & Platinum Co.
Au 35, Cu 65.
For brazing Cu, Kovar and nickel brass. Melting point: 970-1005°C.

WESGO 40 AU-60 CU

Western Gold & Platinum Co.
Au 40, Cu 60.
For brazing Cu, Kovar and nickel brass.

WESGO 50 AU-50 CU

Western Gold & Platinum Co.
Au 50, Cu 50.
For brazing Cu, Kovar, nickel brass. Melting point: 930-950°C.

WESGO AL-300

GTE Products Corp./Wesgo Div.
97.6 Al_2O_3.
Dense alumina for application in corrosive, heat, and wear environments. 75 Rock 45N.

WESGO AL-500

GTE Products Corp./Wesgo Div.
94 Al_2O_3.
Dense alumina for application in corrosive, heat, and wear environments. 78 Rock 45N.

WESGO AL-600

GTE Products Corp./Wesgo Div.
96 Al_2O_3.
Dense alumina for application in corrosive, heat, and wear environments. 79 Rock 45N.

WESGO AL-995

GTE Products Corp./Wesgo Div.
99.5 Al_2O_3.
Dense alumina for application in corrosive, heat, and wear environments. 81 Rock 45N.

WESGO DECARBONIZED

Western Gold & Platinum Co.
Cu 28.1, bal Ag.
For brazing. Melting point: 780°C.

WESGO SNW-1000

GTE Products Corp./Wesgo Div.
Silicon nitride ceramic for laser component and aerospace applications. 91 Rock A.

WESGO SNW-2000

GTE Products Corp./Wesgo Div.
Silicon nitride ceramic for laser component and aerospace applications. 90 Rock A.

WESSEL'S SILVER

Manufacturer not listed.
Cu 51-66, Ni 19-32, Zn 12.5-17, Fe 0-0.5, Ag 0-2.
For ornamental parts, architectural trim; corrosion resistant.

WESSLING

Gusstahl-Handels GmbH
C 0.38, Si 17.6, Mn 0.8, bal Fe.
Cast: 15,000 TS, 50 Rock C. For pumps, valves, drains, fittings, agitators, and castings. Acid resistant. Brittle as cast.

WESSONMETAL 26

Wesson Co.
WC 83, Co 7, 10.0 TiC.
Sintered: 225,000 tranverse strength; 91.7 Rock A. For cutting tools, dies; high hardness and strength; wear and crater resistant.

WESSONMETAL GR. 900

Wesson Co.
TiC, MoC, Ni.
Sintered: 92.2-92.7 Rock A; 200,000 tranverse strength. For cutters in high speed finish machining. Excellent wear resistance with moderate shock resistance.

WESSONMETAL GR. GA
Wesson Co.
WC 95.5, Co 4.5.
Sintered: 92.5 Rock A; 210,000 tranverse strength. For cutting tools to machine cast iron and non-ferrous materials. Extreme abrasion resistance.

WESSONMETAL GR. GF
Wesson Co.
WC 97, Co 3.
Sintered: 92.7 Rock A; 195,000 tranverse strength. For cutting tools to machine cast iron and non-ferrous materials. High wear resistance, hard and strong.

WESSONMETAL GR. GI
Wesson Co.
WC 94, Co 6.
Sintered: 91.5-92.5 Rock A; 250,000 tranverse strength. For cutting tools to machine cast iron, non-ferrous materials. High strength and hardness, abrasion resistant.

WESSONMETAL GR. GS
Wesson Co.
WC 93, Co 7.
Sintered: 91.2 Rock A; 260,000 tranverse strength. For cutting tools to machine cast iron and non-ferrous materials. Shock resistant, high strength.

WESSONMETAL GR. HR
Wesson Co.
W, TiC, TaC, CbC, Co.
Sintered: 91.2 Rock A; 300,000 tranverse strength. For cutting tools for roughing and interrupted cuts. High heat resistance.

WESSONMETAL GR. HV
Wesson Co.
WC 83, Mo 1, Co 3, 13.0 TiC.
Sintered: 94.0 Rock A; 185,000 tranverse strength. For cutting tools for high speed machining. Extreme high hardness.

WESSONMETAL GR. M
Wesson Co.
WC 83.5, Co 13, 3.5 TiC.
Sintered: 90.2 Rock A; 290,000 tranverse strength. For cutting tools in heavy or interrupted cuts on steel or cast iron. High strength and hardness.

WESSONMETAL GR. WH
Wesson Co.
WC 80, Co 7, 13.0 TiC.
Sintered: 92.4 Rock A; 212,000 tranverse strength. For cutting tools in medium and light machining of steel at high speeds. High crater resistance.

WESSONMETAL GR. WM
Wesson Co.
WC 77.25, Co 10, 12.75 TiC.
Sintered: 91.2 Rock A; 225,000 tranverse strength. For general purpose cutters.

WESSONMETAL GR. WP
Wesson Co.
WC 87, Co 13.
Sintered: 88.5-89.5 Rock A; 312,000 tranverse strength. For cutting tools and mining tools. High impact resistance.

WESSONMETAL GR. WS
Wesson Co.
WC 83, Co 13, 4.0 TiC.
Sintered: 90.5 Rock A; 280,000 tranverse strength. For cutting tools in heavy, intermittent machining of tough steel. High shock resistance.

WEST 110 SUPERCAST
West Steel Casting Co.
C, alloy, bal Fe.
Cast: 110,000 TS; 90,000 YS; 15 El; 25 RA. For die blocks, car wheels, skip buckets; oil hardening.

WEST 60 STANCAST
West Steel Casting Co.
C, alloy, bal Fe.
Cast: 60,000 TS; 35,000 YS; 32 El; 54 RA. For annealing boxes, pots, fittings, autoclaves; heat resistant.

WEST 75 HY-CAST
West Steel Casting Co.
C, alloy, bal Fe.
Cast: 75,000 TS; 42,000 YS; 28 El; 45 RA. For anvil blocks, car wheels, pump impellers; oil hardening.

WEST 90 DURACAST
West Steel Casting Co.
C, alloy, bal Fe.
Cast: 90,000 TS; 58,000 YS; 22 El; 40 RA. For gears, pipe fittings, clutch jaws; oil hardening.

WEST NO. 1
West Steel Casting Co.
C 0.2-0.28, Mn 0.6-0.8, Si 0.3-0.4, bal Fe.
Annealed: 38,000-65,000 TS; 112-149 Brin. For machinery parts, shafts. SAE 1225.

WEST NO. 10 (CUMLOY)
West Steel Casting Co.
C 0.4-0.5, Mn 0.6-0.8, Si 0.3-0.4, Cr 0.8-1.1, Mo 0.2-0.3, bal Fe.
Heat treated: 60,000-100,000 TS; 196-514 Brin. For gears, shafts, machinery parts; resists wear, cannot be welded. SAE 4145.

WEST NO. 11 (CUMLOY)
West Steel Casting Co.
C 0.3-0.4, Mn 0.8-1.1, Si 0.3-0.4, Cr 0.8-1.1, Mo 0.2-0.3, bal Fe.
Heat treated: 75,000-110,000 TS; 196-514 Brin. For high strength forgings; wear resistant. SAE 4153.

WEST NO. 12 (CUMLOY)
West Steel Casting Co.
C 0.2-0.3, Mn 0.6-0.8, Si 0.7-0.8, Cr 0.9-1.25, Mo 0.4-0.6, bal Fe.
Heat treated: 70,000-90,000 TS; 170-402 Brin. For machinery parts; tough.

WEST NO. 13 (CUMLOY)
West Steel Casting Co.
C 0.35-0.45, Mn 1-1.25, Si 0.7-0.8, Cr 0.8-1.1, Mo 0.2-0.3, bal Fe.
Heat treated: 80,000-120,000 TS; 217-600 Brin. For dies; cannot be welded.

WEST NO. 14
West Steel Casting Co.
C 0.5-0.6, Mn 0.6-0.8, Si 0.3-0.4, Ni 1-1.25, Mo 0.4-0.5, bal Fe.
Heat treated: 80,000-120,000 TS; 217-512 Brin. For dies, blocks; Cumloy.

WEST NO. 14A
West Steel Casting Co.
C 0.3-0.4, Mn 0.6-0.8, Si 0.3-0.4, Ni 1.25-1.5, Mo 0.3-0.4, bal Fe.
Heat treated: 70,000-100,000 TS; 196-514 Brin. For cams; Cumloy.

WEST NO. 2
West Steel Casting Co.
C 0.4-0.5, Mn 0.6-0.8, Si 0.3-0.4, bal Fe.
Annealed: 45,000-80,000 TS; 156-196 Brin. For machinery parts, gears, shafts, axles; tough. SAE 1245.

WEST NO. 3
West Steel Casting Co.
C 0.15-0.2, Mn 0.3-0.5, Si 2-3, bal Fe.
70,000 TS; 170 Brin. For heat treating boxes; cannot be welded.

WEST NO. 3A
West Steel Casting Co.
C 0.15-0.2, Mn 0.3-0.5, Si 1.5-2, bal Fe.
Special: 70,000 TS; 156 Brin. For heat resistant parts; heat resistant.

WEST NO. 4
West Steel Casting Co.
C 0-0.2, Mn 0.3-0.5, Si 0.3-0.4, bal Fe.
Annealed: 30,000-55,000 TS; 103-131 Brin. For electrical and magnetic parts. SAE 1215.

WEST NO. 4A
West Steel Casting Co.
C 0-0.2, Mn 0.6-0.8, Si 0.3-0.4, bal Fe.
Annealed: 35,000-60,000 TS; 103-137 Brin. For machinery parts, case-hardened parts; carburizing steel. SAE 1215.

WEST NO. 5
West Steel Casting Co.
C 0.28-0.35, Mn 0.6-0.8, Si 0.3-0.4, Ti, bal Fe.
Annealed: 42,000-75,000 TS; 137-156 Brin. For machinery parts. SAE 1230.

WEST NO. 6
West Steel Casting Co.
C 0.3-0.4, Mn 1.25-1.5, Si 0.3-0.4, V 0.12, bal Fe.
Annealed: 60,000-95,000 TS; 126-228 Brin. For cylinders, housings, valves; tough, close grained.

WEST NO. 6A (HY-CAST)
West Steel Casting Co.
C 0.3-0.4, Mn 0.8-1.1, Si 0.3-0.4, bal Fe.
Annealed: 55,000-90,000 TS; 170-208 Brin. For shafts, gears, axles; water hardening.

WEST NO. 7 (DURACAST)
West Steel Casting Co.
C 0.3-0.4, Mn 0.6-0.8, Si 0.3-0.4, Mo 0.15-0.2, bal Fe.
Annealed: 50,000-80,000 TS; 170-277 Brin. For tools, dies; tough.

WEST NO. 7A (DURACAST)
West Steel Casting Co.
C 0.3-0.4, Mn 0.6-0.8, Si 0.3-0.4, Mo 0.3-0.4, bal Fe.
Heat treated: 60,000-90,000 TS; 170-302 Brin. For tools, dies; tough.

WEST NO. 8 (DURACAST)
West Steel Casting Co.
C 0.4-0.5, Mn 0.6-0.8, Si 0.3-0.4, V 0.15-0.18, bal Fe.
Heat treated: 60,000-90,000 TS; 170-402 Brin. For tools, dies, shafts, gears, anvils; tough.

WEST NO. 9 (DURACAST)
West Steel Casting Co.
C 0.6-0.8, Mn 0.6-0.8, Si 0.3-0.4, V 0.15-0.18, bal Fe.
Heat treated: 80,000-110,000 TS; 228-600 Brin. For tools, dies, punches; water hardened.

WEST NO. 9A (CUMLOY)
West Steel Casting Co.
C 0.3-0.4, Mn 1.25-1.5, Si 0.3-0.4, Mo 0.2-0.3, V 0.12, bal Fe.
Heat treated: 70,000-100,000 TS; 196-514 Brin. For shafts, gears, machinery parts; tough.

WESTA 003A
Westa-Westdeutsche
C 1.15, Si 0-0.25, bal Fe.
Annealed: 110,000 TS; 58,000 YS; 18 El; 38 RA; 210 Brin. For drills, taps, reamers, broaches; Type W1; water hardened.

WESTA 2
Westa-Westdeutsche
C 1.3, Si 0-0.25, Mn 0-0.25, bal Fe.
For engravers' tools, drills, cutters, reamers; Type W1; water hardened.

WESTA 2002
Westa-Westdeutsche
C 2.1, Cr 11.5, bal Fe.
For blanking and forming dies, punches; oil hardened, nondeforming.

WESTA 2002 SPEZIAL
Westa-Westdeutsche
C 2.1, Cr 11.5, W 0.7, bal Fe.
For blanking and forming dies, gages, punches; oil hardened, nondeforming.

WESTA 2002W
Westa-Westdeutsche
C 1.65, Cr 11.5, V 0.1, bal Fe.
For blanking and forming dies, gages, punches; air hardened, nondeforming.

WESTA 212
Westa-Westdeutsche
C 0.3, Cr 2.6, V 0.35, W 8.5, bal Fe.
For extrusion rams, liners and dies; hot work steel, oil hardened.

WESTA 212D
Westa-Westdeutsche
C 0.3, Cr 2.3, V 0.6, W 4.25, bal Fe.
For hot and cold work tools and dies; oil hardened, tough.

WESTA 3
Westa-Westdeutsche
C 1.15, Si 0-0.25, Mn 0-0.25, bal Fe.
Annealed: 110,000 TS; 58,000 YS; 18 El; 38 RA; 210 Brin.
For reamers, taps, form and milling cutters, drills; Type W1; water hardened.

WESTA 301
Westa-Westdeutsche
C 0.3, Co 2, Cr 2.4, V 0.25, W 8.5, bal Fe.
For extrusion rams and liners, punches, shears; hot work steel, oil hardened.

WESTA 4
Westa-Westdeutsche
C 1, Si 0-0.25, Mn 0-0.25, bal Fe.
Heat treated: 120,000-215,000 TS; 85,000-150,000 YS; 10-20 El; 48-33 RA; 240-550 Brin. For taps, drills, reamers, hobs, Type W1; water hardened.

WESTA 425
Westa-Westdeutsche
C 0.3, Cr 1.1, V 0.18, W 3.75, Si 1, bal Fe.
For cold work tools and dies, upsetters; oil hardened, tough.

WESTA 465M
Westa-Westdeutsche
C 0.9, Mn 1.9, V 0.1, bal Fe.
For punches, crimpers, blanking dies; oil hardened, nondeforming.

WESTA 5
Westa-Westdeutsche
C 0.85, Si 0-0.25, Mn 0-0.25, bal Fe.
Heat treated: 130,000-190,000 TS; 88,000-145,000 YS; 12-20 El; 50-35 RA; 255-390 Brin. For tools, cutters, springs; Type W1; water hardened.

WESTA 546
Westa-Westdeutsche
C 0.45, Cr 1.3, Mo 0.45, V 0.8, W 0.45, bal Fe.
For cold work tools and dies; oil hardened, tough.

WESTA 6
Westa-Westdeutsche
C 0.7, Si 0-0.25, Mn 0-0.25, bal Fe.
Heat treated: 122,000-175,000 TS; 82-000-128,000 YS; 12-22 El; 52-37 RA; 240-350 Brin. For punches, rolls, crimpers, springs; Type W1; water hardened.

WESTA 725
Westa-Westdeutsche
C 0.65, Cr 3.7, Mo 0.85, V 0.7, W 8.5, bal Fe.
For lathe and planer tools, drills, reamers, hobs; high speed steel, oil hardened.

WESTA AK SPEZIAL
Westa-Westdeutsche
C 0.4, Cr 13, Si 0.4, Mn 0.3, bal Fe.
Annealed: 95,000 TS; 50,000 YS; 20 El; 50 RA; 200 Brin.
Cold drawn: 110,000 TS; 90,000 YS; 15 El; 45 RA; 240 Brin.
For turbine blades, cutlery, gauges; Type 420; corrosion resistant.

WESTA AK2 SPEZIAL
Westa-Westdeutsche
C 0.2, Si 0.4, Mn 0.3, Cr 13, bal Fe.
Annealed: 95,000 TS; 50,000 YS; 26 El; 55 RA; 195 Brin.
Cold drawn: 105,000 TS; 85,000 YS; 17 El; 50 RA; 215 Brin.
For turbine blades, surgical instruments; Type 420; corrosion resistant.

WESTA AL 14
Westa-Westdeutsche
C 0.33, Al 1.1, Cr 1.7, Ni 1, bal Fe.
For oil refinery equipment; creep resistant.

WESTA AL16
Westa-Westdeutsche
C 0.27, Al 1.1, Cr 1.4, bal Fe.
For oil refinery equipment; creep resistant.

WESTA BE
Westa-Westdeutsche
C 0.13, Cr 0.7, Ni 2.5, bal Fe.
For gears, fasteners, camshafts, cams, machine tool parts; case hardened, shock resistant.

WESTA BOW
Westa-Westdeutsche
C 0.28, Cr 0.7, Ni 2.5, bal Fe.
For gears, bolts, fasteners, machine tool parts; oil hardened, shock resistant.

WESTA CA
Westa-Westdeutsche
C 0.13, Cr 0.5, bal Fe.
For gears, bolts, machine tool parts; case hardened.

WESTA CE
Westa-Westdeutsche
C 0.15, Mn 0.5, Cr 0.65, bal Fe.
For gears, bolts, machine tool parts; case hardened.

WESTA CE SPEZIAL
Westa-Westdeutsche
C 0.15, Cr 0.65, Mn 0.5, bal Fe.
For gears, bolts, machine tool parts; case hardened.

WESTA CE2 SPEZIAL
Westa-Westdeutsche
C 0.2, Cr 1.15, Mn 1.25, bal Fe.
For gears, cams, camshafts, fasteners; case hardened, tough.

WESTA CM1
Westa-Westdeutsche
Cr 1.2, Mo 0.2, C 0.15, bal Fe.
For gears, bolts, machine tool parts; case hardened, tough.

WESTA CM2
Westa-Westdeutsche
C 0.2, Cr 1.2, Mo 0.2, bal Fe.
For gears, bolts, camshafts, fasteners; case hardened, tough.

WESTA CM3
Westa-Westdeutsche
C 0.25, Cr 1, Mo 0.2, Mn 0.65, bal Fe.
For gears, bolts, machine tool parts; oil hardened, tough.

WESTA CM4
Westa-Westdeutsche
C 0.33, Mn 0.65, Cr 1, Mo 0.2, bal Fe.
For gears, bolts, machine tool parts; oil hardened, tough.

WESTA CM5
Westa-Westdeutsche
C 0.42, Mn 0.65, Cr 1, Mo 0.2, bal Fe.
For gears, bolts, crankshafts, studs; oil hardened, tough.

WESTA CN SPEZIAL
Westa-Westdeutsche
C 0.28-0.35, Cr 0.7, Ni 2.5, bal Fe.
For gears, bolts, crankshafts, machine tool parts; oil hardened, shock resistant.

WESTA CNBD
Westa-Westdeutsche
C, alloy, bal Fe.
For machine tool parts; oil hardened.

WESTA CNE
Westa-Westdeutsche
C 0.13, Ni 2.5, Cr 0.7, bal Fe.
For gears, bolts, cams, camshafts; case hardened, shock resistant.

WESTA CNH SPEZIAL
Westa-Westdeutsche
C 0.5, Cr 1, Ni 3.5, Mn 0.5, bal Fe.
For gears, bolts, crankshafts, axles, shafts; oil hardened, shock resistant.

WESTA CNL
Westa-Westdeutsche
C 0.35, Mn 0.6, Cr 1.3, Ni 4.5, bal Fe.
For gears, bolts, crankshafts, machine tool parts; oil hardened, shock resistant.

WESTA CNL EXTRA
Westa-Westdeutsche
C 0.5, Cr 1.05, Ni 3.25, bal Fe.
For gears, bolts, crankshafts, axles; oil hardened, shock resistant.

WESTA CNS 95
Westa-Westdeutsche
C 0.22-0.3, Cr 0.7, Ni 3.5, bal Fe.
For gears, bolts, shafts, studs, fasteners; oil hardened, shock resistant.

WESTA CNS SPEZIAL
Westa-Westdeutsche
C 0.22-0.3, Cr 0.7, Ni 3.5, bal Fe.
For gears, bolts, crankshafts, machine tool parts; oil hardened, shock resistant.

WESTA CNSH
Westa-Westdeutsche
C 0.22-0.3, Cr 0.7, Ni 3.5, bal Fe.
For gears, bolts, crankshafts; oil hardened, shock resistant.

WESTA CR
Westa-Westdeutsche
C 0.9, Cr 0.8, Mn 0.3, Si 0.25, bal Fe.
For bearings, cutters, springs, liners, bushings; water hardened.

WESTA CR1
Westa-Westdeutsche
C 1, Cr 1.1, Mn 0.07, Si 0.25, bal Fe.
For bearings, liners, bushings; water hardened, wear resistant.

WESTA CR1/W
Westa-Westdeutsche
C 0.9, Si 0.25, Mn 0.3, Cr 0.8, bal Fe.
For bearings, liners, sleeves, bushings; water hardened, wear resistant.

WESTA CR2
Westa-Westdeutsche
C 0.85, Si 0.25, Mn 0.35, Cr 1.75, bal Fe.
For bearings, bushings, liners; water hardened, wear resistant.

WESTA CRK
Westa-Westdeutsche
C 1.05, Cr, bal Fe.
For bearings, liners, bushings; water hardened, wear resistant.

WESTA CVM
Westa-Westdeutsche
C 0.5, Mn 0.65, Cr 1, Mo 0.2, bal Fe.
For gears, bolts, crankshafts, axles; oil hardened, shock resistant.

WESTA CX(2)
Westa-Westdeutsche
C 0.38, Si, Cr, V, bal Fe.
For gears, bolts, machine tool parts; oil hardened, shock resistant.

WESTA DS SPEZIAL
Westa-Westdeutsche
C 1.15, Cr 0.65, V 1, bal Fe.
For bearings, cutters, liners; oil or water hardened.

WESTA E
Westa-Westdeutsche
C 0.55, Si 0.1-0.4, Mn 0.5-0.7, bal Fe.
For gears, bolts, shafts, fasteners; water hardened.

WESTA ECN
Westa-Westdeutsche
C 0.13, Si 0-0.35, Mn 0.5, Cr 0.7, Ni 2.5, bal Fe.
For gears, bolts, camshafts, cams; case hardened, shock resistant.

WESTA EK
Westa-Westdeutsche
C 1.45, Cr 1.4, Mn 0.6, bal Fe.
For bearings, liners, sleeves, bushings; water hardened, wear resistant.

WESTA EKA
Westa-Westdeutsche
C 1.45, Cr 1.4, Mn 0.6, bal Fe.
For bearings, liners, sleeves, bushings; water hardened, wear resistant.

WESTA ES EXTRA
Westa-Westdeutsche
C 0.53, Si 0.9, Mn 0.9, bal Fe.
For machine tool parts, gears; water hardened.

WESTA EXTRA
Westa-Westdeutsche
C 1.3, Mn 0.3, Cr 0-0.2, W 4.7, bal Fe.
For cutters; water hardened.

WESTA EXTRA ZH
Westa-Westdeutsche
C 1, Si 0-0.25, Mn 0-0.25, bal Fe.
Annealed: 100,000 TS; 53,000 YS; 21 El; 42 RA; 200 Brin.
For springs, drills, taps, reamers; Type W1; water hardened.

WESTA EZH
Westa-Westdeutsche
C 1, Mn 0-0.25, S 0-0.25, bal Fe.
For springs, tools, drills, taps, reamers; Type W1; water hardened.

WESTA EZH SPEZIAL
Westa-Westdeutsche
C 1, V 0.1, Mn 0.5, Si 0.2, bal Fe.
For drills, taps, cutters, springs; Type W2; water hardened.

WESTA F
Westa-Westdeutsche
C 1.1, Si 0-0.25, Mn 0-0.25, bal Fe.
Annealed: 110,000 TS; 56,000 YS; 18 El; 40 RA; 200 Brin.
For drills, taps, cutters, hobs, springs; Type W1; water hardened.

WESTA KZA
Westa-Westdeutsche
C 1.45, Si 0.25, Mn 0.6, Cr 1.4, bal Fe.
For bearings, bushings, cutters; water hardened, wear resistant.

WESTA MK
Westa-Westdeutsche
C 0.76, Co 4.75, Cr 4.3, Mo 0.75, V 1.5, W 18, bal Fe.
For lathe and planer tools, drills; high speed steel.

WESTA ORI
Westa-Westdeutsche
C 1.25, Si 1.15, Mn 0.7, Cr 1.2, bal Fe.
For cutters, forming dies; oil hardened.

WESTA R
Westa-Westdeutsche
C 1.05, Cr 1, W 1.15, Mn 0.9, bal Fe.
For cutters, tools; water or oil hardened.

WESTA R-SPEZIAL
Westa-Westdeutsche
C 1.42, W, V, bal Fe.
For bearings, cutters, forming dies; oil hardened, wear resistant.

WESTA RCR1
Westa-Westdeutsche
C 1.4, Cr 0.3, V 0.1, Mn 0.3, bal Fe.
For forming and blanking dies, cutters; oil or water hardened, wear resistant.

WESTA SC
Westa-Westdeutsche
C 0.67, Si 1.3, Mn 0.5, Cr 0.5, bal Fe.
For springs, chisels, upsetters; oil hardened, shock resistant.

WESTA SP
Westa-Westdeutsche
C 1.2, Cr 0.2, V 0.1, W 1, bal Fe.
For cutters, reamers, forming dies; water or oil hardened.

WESTA T3
Westa-Westdeutsche
C 1.1, Si 0-0.25, Mn 0-0.25, bal Fe.
Heat treated: 200,000 TS; 125,000 YS; 8 El; 27 RA; 400 Brin.
For springs, reamers, drills, taps; water hardened; Type W1.

WESTA T4
Westa-Westdeutsche
C 1, Si 0-0.25, Mn 0-0.25, bal Fe.
Heat treated: 190,000 TS; 120,000 YS; 10 El; 30 RA; 390 Brin. For drills, taps, reamers, broaches; Type W1; water hardened.

WESTA T5 EXTRA
Westa-Westdeutsche
C 0.75, Si 0.25-0.5, Mn 0.3-0.8, bal Fe.
Heat treated: 175,000 TS; 130,000 YS; 12 El; 37 RA; 360 Brin. For springs, clutch discs, girders, rails; Type W1; water hardened.

WESTA T5 HART
Westa-Westdeutsche
C 0.61, Si 0.25-0.65, Mn 0.65, bal Fe.
Heat treated: 160,000 TS; 112,000 YS; 12 El; 40 RA; 330 Brin. For wheels, die blocks, rails, girders; water hardened.

WESTA T5W
Westa-Westdeutsche
C 0.6, Si 0.25-0.5, Mn 0.3-0.8, bal Fe.
Heat treated: 160,000 TS; 112,000 YS; 12 El; 40 RA; 330 Brin. For wheels, die blocks, rails, girders; water hardened.

WESTA T6H
Westa-Westdeutsche
C 0.45, Si 0.25, Mn 0.65, bal Fe.
Hot rolled: 98,000 TS; 58,000 YS; 24 El; 45 RA; 212 Brin. For axles, gears, bolts, crankshafts; water hardened.

WESTA T6H EXTRA
Westa-Westdeutsche
C 0.45, Si 0.25-0.5, Mn 0.3-0.8, bal Fe.
Hot rolled: 98,000 TS; 58,000 YS; 24 El; 45 RA; 212 Brin. For axles, gears, bolts, crankshafts; water hardened.

WESTA T6W
Westa-Westdeutsche
C 0.35, Si 0.25, Mn 0.55, bal Fe.
Hot rolled: 85,000 TS; 54,000 YS; 30 El; 53 RA; 180 Brin. For gears, shafts, axles, bolts; water hardened.

WESTA T6W EXTRA
Westa-Westdeutsche
C 0.35, Si 0.25-0.5, Mn 0.3-0.8, bal Fe.
Hot rolled: 85,000 TS; 54,000 YS; 30 El; 53 RA; 180 Brin. For gears, shafts, axles, bolts; water hardened.

WESTA T7
Westa-Westdeutsche
C 0.22, Si 0.25, Mn 0.45, bal Fe.
Annealed: 73,000 TS; 40,000 YS; 22 El; 58 RA; 140 Brin. For fan blades, bolts, screws, gears; water hardened.

WESTA TBM1
Westa-Westdeutsche
C 0.55, Mn 0.6, Cr 0.7, Mo 0.18, Ni 1.65, V 0.1, bal Fe.
For forging and heading dies, shear blades; oil hardened, tough.

WESTA TBM1 EXTRA
Westa-Westdeutsche
C 0.56, Cr 0.85, Mo 0.2, Ni 2, V 0.1, bal Fe.
For forging and heading dies; oil hardened, tough.

WESTA TE SPEZIAL
Westa-Westdeutsche
C 0.13, Cr 0.7, Ni 3.5, bal Fe.
For gears, cams, camshafts, fasteners; case hardening steel.

WESTA TEJ
Westa-Westdeutsche
C 0.13, Ni 4.5, Cr 1.1, bal Fe.
For gears, bolts, machine tool parts; case hardened, shock resistant.

WESTA TEM
Westa-Westdeutsche
C 0.13, Cr 0.7, Ni 3.5, bal Fe.
For gears, bolts, machine tool parts; case hardened, shock resistant.

WESTA TH
Westa-Westdeutsche
C 1.05, Cr 1, Mn 0.3, Si 0.25, bal Fe.
For bearings, liners, sleeves; water hardened, wear resistant.

WESTA TPA
Westa-Westdeutsche
C 0.35, Cr 1.3, Ni 4.5, bal Fe.
For gears, bolts, crankshafts, axles; oil hardened, shock resistant.

WESTA TY 1 W
Westa-Westdeutsche
C, alloy, bal Fe.
For machine tool parts; oil hardened, tough.

WESTA TY 2 W
Westa-Westdeutsche
C 0.13, Cr 0.2, Ni 1.5, bal Fe.
For gears, bolts, shafts, cams, camshafts; case hardened, shock resistant.

WESTA VCN
Westa-Westdeutsche
C 0.28-0.35, Cr 0.7, Ni 2.5, bal Fe.
For gears, bolts, crankshafts, fasteners; oil hardened, shock resistant.

WESTA W8
Westa-Westdeutsche
C 0.15, Si 0.25, Mn 0.37, bal Fe.
For gears, bolts, machine tool parts; case hardened.

WESTA WM4
Westa-Westdeutsche
C 0.3, Si 1, Cr 1.1, V 0.18, W 3.75, bal Fe.
For pneumatic tools, chisels, upsetters, punches; oil hardened, tough.

WESTA WP
Westa-Westdeutsche
C 0.45, Cr 1.4, Mo 0.7, V 0.3, Mn 0.7, bal Fe.
For forging and die casting dies; oil hardened, tough.

WESTEECO NO. 35-15
Western Crucible Steel Casting Co.
Cr 35, Ni 15, C, bal Fe.
For furnace parts, heat resisting castings. Heat resistant.

WESTEECO NO. 4
Western Crucible Steel Casting Co.
C 0.3, Ni, Cr, Mo, bal Fe.
Heat treated: 110,000 TS; 80,000 YS; 20 El; 38 RA; 230 Brin.
For gears, shafts, machinery parts. Oil hardened, tough.

WESTEECO NO. 5
Western Crucible Steel Casting Co.
C 0.15, Ni 55-60, Cr 15-18, bal Fe.
For heat resisting castings. Heat resistant.

WESTEECO NO. 6
Western Crucible Steel Casting Co.
C 0.25, Ni 1-1.25, Mo 0.25, bal Fe.
Heat treated: 95,000 TS; 70,000 YS; 24 El; 42 RA; 195-215 Brin. For lift arms for dump trucks. Oil hardened, shock resistant.

WESTERN ALLOY NO. 10
Olin Corp.
Zn 10, bal Cu.
Hard: 61,000 TS; 54,000 YS; 5 El. Soft: 37,000 TS; 10,000 YS; 45 El. For jewelry, hardware, bullet jackets, screwshells; good strength and corrosion resistance. *Obsolete*

WESTERN ALLOY NO. 116
Olin Corp.
Cu 89.25, Zn 2.75, Pb 4, Sn 4.
Hard: 75,000 TS; 7 El. Soft: 46,000 TS; 45 El. For thrust washers, clutch plates, wrist pin bearings; excellent bearing material. *Obsolete*

WESTERN ALLOY NO. 13
Olin Corp.
Cu 87, Zn 13.
Hard: 66,000 TS; 56,000 YS; 5 El. Soft: 39,000 TS; 11,000 YS; 46 El. For decorative items, lipstick cases; gold color. *Obsolete*

WESTERN ALLOY NO. 133
Olin Corp.
Cu 61.5, Zn 24, Ni 13, Pb 1.5.
Hard: 91,000 TS; 4 El. For keys, hardware; nickel silver, corrosion resistant. *Obsolete*

WESTERN ALLOY NO. 15
Olin Corp.
Cu 85, Zn 15.
Hard: 70,000 TS; 57,000 YS; 5 El. Soft: 40,000 TS; 12,000 YS; 47 El. For belt buckles, escutcheons, pencil caps; red brass, resists stress cracking. *Obsolete*

WESTERN ALLOY NO. 178
Olin Corp.
Cu 56.5, Ni 12, Zn 31.5.
Hard: 98,000 TS; 2 El. For spring parts in optical goods; corrosion resistant, nickel silver. *Obsolete*

WESTERN ALLOY NO. 180
Olin Corp.
Cu 66.5, Zn 23.5, Ni 10.
Hard: 86,000 TS; 75,000 YS; 4 El. Soft: 51,000 TS; 19,000 YS; 46 El. For builders' hardware; corrosion resistant, nickel silver. *Obsolete*

WESTERN ALLOY NO. 222
Olin Corp.
Cu 93, Zn 4.7, Fe 2.3.
Hard: 67,000 TS; 6 El. Soft: 49,000 TS; 35 El. For furnace brazing of evaporators; good joining properties. *Obsolete*

WESTERN ALLOY NO. 34
Olin Corp.
Zn 34, bal Cu.
Hard: 74,000 TS; 60,000 YS; 8 El. Soft: 46,500 TS; 15,000 YS; 62 El. For contacts, fasteners, ferrules; yellow brass, good deep drawing. *Obsolete*

WESTERN ALLOY NO. 51
Olin Corp.
Cu 99.9, Ag 0.028.
Hard: 51,000 TS; 6 El. Soft: 33,000 TS; 45 El. For automobile radiator fins; retains hardness to 600 F. *Obsolete*

WESTERN ALLOY NO. 52
Olin Corp.
Cu 99.98, P 0.02.
Hard: 50,000 TS; 45,000 YS; 6 El. Soft: 32,000 TS; 10,000 YS; 45 El. For headers for heat exchangers; for brazed and welded parts. *Obsolete*

WESTERN ALLOY NO. 53
Olin Corp.
Cu 99.92, Ag.
Hard: 50,000 TS; 45,000 YS; 6 El. Soft: 32,000 TS; 10,000 YS; 45 El. For metal to glass seals, electronic parts; severe drawability. *Obsolete*

WESTERN ALLOY NO. 55
Olin Corp.
Cu 99.96, P 0.04.
Hard: 50,000 TS; 45,000 YS; 6 El. Soft: 32,000 TS; 10,000 YS; 45 El. For cooking utensils, bus bars, heat exchangers; high heat and electrical conductivity. *Obsolete*

WESTERN ALLOY NO. 62
Olin Corp.
Cu 64, Zn 35, Pb 1.
Hard: 74,000 TS; 60,000 YS; 7 El. Soft: 47,000 TS; 15,000 YS; 60 El. For key stock; good machinability. *Obsolete*

WESTERN ALLOY NO. 63
Olin Corp.
Cu 66.5, Zn 32.5, Pb 1.
Hard: 74,000 TS; 60,000 YS; 7 El. Soft: 47,000 TS; 15,000 YS; 60 El. For gears, plates, washers; good blanking properties. *Obsolete*

WESTERN ALLOY NO. 68
Olin Corp.
Cu 61.5, Zn 36.65, Pb 1.85.
Hard: 74,000 TS; 60,000 YS; 7 El. Soft: 47,000 TS; 15,000 YS; 60 El. For clock and watch parts; free-cutting. *Obsolete*

WESTERN ALLOY NO. 86
Olin Corp.
Cu 95, Zn 4, Sn 1.
Hard: 62,000 TS. Soft: 38,000 TS; 46 El. For electrical spring contacts; good electrical conductivity. *Obsolete*

WESTERN ALLOY NO. 98
Olin Corp.
Sn 0.75, Zn 39.25, bal Cu.
Hard: 88,000 TS; 66,000 YS; 18 El. Soft: 58,000 TS; 26,000 YS; 41 El. For condenser tubes, bolts, marine hardware, spindles; resists sea water corrosion. *Obsolete*

WESTERN ALUMINUM BRONZE
Olin Corp.
Cu 87, Al 4, Fe 1, Zn 2.
For bearings, bushings, clutch plates; wear and corrosion resistant. *Obsolete*

WESTERN BEARING BRONZE
Olin Corp.
Cu 88, Zn 4, Sn 4, Pb 4.
Hard: 78,000 TS; 7 El. Soft: 45,000 TS; 40 El. For bushings, bearings, clutch plates; heavy duty. *Obsolete*

WESTERN BEST QUALITY BRASS
Olin Corp.
Cu 75, Zn 25.
Hard: 73,500 TS; 8 El. Soft: 46,000 TS; 51 El. For springs, radio tube sockets; good for drawing and spinning. *Obsolete*

WESTERN BRAZING BRASS #340
Olin Corp.
Cu 60.5, P 0.17, bal Zn.
For brazing sheets; low melting temperature. *Obsolete*

WESTERN BRONZE NO. 222
Olin Corp.
Cu 92, Fe 2.3, bal Zn.
Hard: 68,000 TS; 5 El. Soft: 48,000 TS; 35 El. For refrigeration evaporators; resists grain growth. *Obsolete*

WESTERN COPPER #55
Olin Corp.
Cu 99.96, 0.04% oxygen.
Hard: 50,000 TS; 45,000 YS; 6 El. Soft: 32,000 TS; 10,000 YS; 45 El. For bus bars, terminals, auto gaskets; high electrical conductivity. *Obsolete*

WESTERN CUPRO NICKEL 7%-#142
Olin Corp.
Cu 93, Ni 7.
Hard: 59,000 TS. Soft: 44,000 TS. For fuse components, resistance strips; high electrical resistivity. *Obsolete*

WESTERN CUPRO NICKEL-7%
Olin Corp.
Cu 93.2, Ni 6-8.
Hard: 58,000 TS; 6 El. Soft: 40,000 TS; 38 El. Corrosion resistant. *Obsolete*

WESTERN CUPRO-NICKEL-9%
Olin Corp.
Cu 91, Ni 9.
Hard: 63,000 TS; 5 El. For corrosion resistant parts; corrosion resistant. *Obsolete*

WESTERN CUPRONICKEL 1290
Olin Corp.
Cu 88, Ni 12.
Hard: 65,000 TS; 4 El. For corrosion resistant parts; corrosion resistant. *Obsolete*

WESTERN DEOXIDIZED COPPER #52
Olin Corp.
Cu 99.98, P 0.02.
Hard: 50,000 TS; 45,000 YS; 6 El. Soft: 32,000 TS; 10,000 YS; 45 El. For headers, deep drawn articles; resists hydrogen embrittlement. *Obsolete*

WESTERN GILDING ALLOY NO. 5
Olin Corp.
Zn 5, bal Cu.
Hard: 56,000 TS; 50,000 YS; 5 El. Soft: 34,000 TS; 10,000 YS; 45 El. For jewelry, coins, spark plug gaskets, name plates; brass, high luster. *Obsolete*

WESTERN HIGH BRASS
Olin Corp.
Cu 67.5, Zn 38.5.
Hard: 74,000 TS; 9 El. Soft: 46,500 TS; 52 El. For hub caps, reflectors, hardware; good cold workability. *Obsolete*

WESTERN HIGH CONDUCTIVITY BRONZE #86
Olin Corp.
Cu 95, Zn 4, Sn 1.
Hard: 60,000 TS; 7 El. Soft: 38,000 TS; 45 El. For electrical spring contacts; good electrical conductivity. *Obsolete*

WESTERN HIGH LEADED BRASS #64
Olin Corp.
Cu 63.5, Zn 34.65, Pb 1.85.
Hard: 75,000 TS; 60,000 YS; 7 El. Soft: 52,000 TS; 20,000 YS; 50 El. For gear wheels, keys, dials, nuts; good machining and blanking properties. *Obsolete*

WESTERN HIGH LEADED BRASS #68
Olin Corp.
Cu 61.5, Zn 36.65, Pb 1.85.
Hard: 74,000 TS; 60,000 YS; 7 El. Soft: 47,000 TS; 15,000 YS; 60 El. For clock and watch parts, gears, levers, cams; best machining and blanking properties. *Obsolete*

WESTERN HIGH LEADED BRASS 64
Olin Corp.
Zn 34.6, Pb 1.8, bal Cu.
Hard: 75,000 TS; 60,000 YS; 7 El. Soft: 52,000 TS; 20,000 YS; 50 El. For keys, hardware, gears, instrument plates; good blanking and machining. *Obsolete*

WESTERN JEWELRY BRONZE #13
Olin Corp.
Cu 87, Zn 13.
Hard: 66,000 TS; 56,000 YS; 5 El. Soft: 39,000 TS; 11,000 YS; 46 El. For jewelry, medals, slide fasteners; good cold workability. *Obsolete*

WESTERN LEADED BEARING BRONZE #116
Olin Corp.
Cu 89.25, Pb 4, Sn 4, Zn 2.75.
Hard: 77,000 TS; 72,000 YS; 7 El. Soft: 46,000 TS; 17,000 YS; 45 El. For wrist pin bearings, thrust washers; best bearing properties. *Obsolete*

WESTERN LEADED NICKEL SILVER 10%. #131
Olin Corp.
Cu 61.5, Zn 27, Ni 10, Pb 1.5.
Hard: 92,000 TS; 83,000 YS; 2 El. 1/4 H-temper: 68,000 TS; 41,000 YS; 16 El. For keys, watch plates, wheels and lock parts; corrosion resistant, free-cutting. *Obsolete*

WESTERN LEADED NICKEL SILVER 12%, #133
Olin Corp.
Cu 61.5, Zn 24, Ni 13, Pb 1.5.
Hard: 92,000 TS; 4 El. Soft: 57,000 TS; 41 El. For keys, watch plates, lock parts; corrosion resistant, free-cutting. *Obsolete*

WESTERN LEADED NICKEL SILVER 8%, #130
Olin Corp.
Cu 63.5, Zn 27.9, Ni 7.9, Pb 1.5.
Hard: 86,000 TS; 2 El. 1/4 H-temper: 64,000 TS; 16 El. For keys, watch wheels; corrosion resistant, free-cutting. *Obsolete*

WESTERN LOW BRASS 20
Olin Corp.
Zn 20, bal Cu.
Hard: 74,000 TS; 59,000 YS; 7 El. Soft: 44,000 TS; 14,000 YS; 50 El. For flexible hose, floats, bellows; good workability. *Obsolete*

WESTERN LUBALOY #84
Olin Corp.
Cu 90, Zn 9.5, Sn 0.05.
Hard: 64,000 TS; 58,000 YS; 5 El. Soft: 40,000 TS; 11,000 YS; 46 El. For bushings, bearings, thrust washers; good wear resistance and weldability. *Obsolete*

WESTERN LUBALOY-X #80
Olin Corp.
Cu 90, Zn 8.25, Sn 1.75.
Hard: 70,000 TS; 59.000 YS; 7 El. Soft: 45,000 TS; 19,000 YS; 50 El. For weatherstrip, jewelry, fuse clips, contact springs; good fatigue strength and resilience. *Obsolete*

WESTERN LUBRONZE #83
Olin Corp.
Cu 87, Zn 12, Sn 1.
Hard: 66,000 TS; 56,000 YS; 5 El. Soft: 42,000 TS; 12,000 YS; 48 El. For flat link chains and springs; good bending properties. *Obsolete*

WESTERN MANGANESE BRASS #277
Olin Corp.
Cu 70, Zn 28.75, Mn 1.25.
Hard: 77,000 TS; 7 El. Soft: 48,000 TS; 59 El. For evaporators, resistance strips; high electrical resistance. *Obsolete*

WESTERN MEDIUM LEADED BRASS #62
Olin Corp.
Cu 64, Zn 35, Pb 1.
Hard: 74,000 TS; 60,000 YS; 7 El. Soft: 47,000 TS; 15,000 YS; 60 El. For keystock, hose couplings; good machinability and formability. *Obsolete*

WESTERN MEDIUM LEADED BRASS #63
Olin Corp.
Cu 66.5, Zn 32.5, Pb 1.
Hard: 74,000 TS; 60,000 YS; 7 El. Soft: 47,000 TS; 15,000 YS; 60 El. For sink strainers, gears, cams, hose couplings; good blanking and deep drawing properties. *Obsolete*

WESTERN MUNTZ METAL
Olin Corp.
Cu 60, Zn 40.
Cast: 48,000 TS; 54 El. Hard: 80,000 TS; 9.5 El. For condenser plates, tanks; high strength. *Obsolete*

WESTERN NICKEL SILVER 55-18, #168
Olin Corp.
Cu 55, Ni 18, Zn 27.
Hard: 100,000 TS; 85,000 YS; 3 El. Soft: 60,000 TS; 27,000 YS; 40 El. For relay springs, contact arms, street car tokens; spring qualities, tarnish resistant. *Obsolete*

WESTERN NICKEL SILVER 56-12, #178
Olin Corp.
Cu 56.5, Zn 31.5, Ni 12.
Hard: 98,000 TS; 91,000 YS; 2 El. 1/4 H-temper: 74,000 TS; 44,000 YS; 14 El. For surgical and dental instruments, spring contacts; high strength and durability. *Obsolete*

WESTERN NICKEL SILVER 65-18, #164
Olin Corp.
Cu 65, Ni 18, Zn 17.
Hard: 86,000 TS; 74,000 YS; 3 El. Soft: 54,000 TS; 25,000 YS; 40 El. For fishing tackle, surgical instruments, terminals; corrosion resistant. *Obsolete*

WESTERN NICKEL SILVER 66-10, #180
Olin Corp.
Cu 66.5, Zn 23.5, Ni 10.
Hard: 86,000 TS; 75,000 YS; 4 El. Soft: 51,000 TS; 19,000 YS; 46 El. For fishing reels, hardware; corrosion resistant. *Obsolete*

WESTERN NICKEL SILVER 70-12, #176
Olin Corp.
Cu 70, Zn 18, Ni 12.
Hard: 79,000 TS; 4 El. Soft: 52,000 TS; 50 El. For arch supports, builders hardware; corrosion resistant. *Obsolete*

WESTERN NICKEL SILVER 72-18, #163
Olin Corp.
Cu 72, Ni 18, Zn 10.
Hard: 78,000 TS; 4 El. Soft: 50,000 TS; 45 El. For musical instruments, trophies, ferrules; deep drawing properties. *Obsolete*

WESTERN NO. 30
Olin Corp.
Zn 30, bal Cu.
Hard: 76,000 TS; 62,000 YS; 8 El. Soft: 47,000 TS; 15,000 YS; 62 El. For cartridge cases, radiator tanks; high ductility, good workability. *Obsolete*

WESTERN OFHC #53
Olin Corp.
99.92% min Cu.
Hard: 50,000 TS; 45,000 YS; 6 El. Soft: 32,000 TS; 10,000 YS; 45 El. For electronic parts; severe drawing properties. *Obsolete*

WESTERN PHOSPHOR BRONZE 1.25%, #101
Olin Corp.
Cu 98.7, Sn 1.3, P 0.12.
Hard: 64,000 TS; 59,000 YS; 6 El. Soft: 41,000 TS; 14,000 YS; 55 El. For welded tubing, metal hose, clutch plates; good weldability and wear resistance. *Obsolete*

WESTERN PHOSPHOR BRONZE 5%, #106
Olin Corp.
Cu 95, Sn 5, P 0-0.35.
Hard: 85,000 TS; 75,000 YS; 10 El. Soft: 47,000 TS; 19,000 YS; 64 El. For diaphragms, contact springs, bellows; excellent spring qualities. *Obsolete*

WESTERN PHOSPHOR BRONZE 8%, #113
Olin Corp.
Cu 92, Sn 8, P 0-0.25.
Hard: 95,000 TS; 91,000 YS; 13 El. Soft: 55,000 TS; 21,000 YS; 68 El. For contact springs, diaphragms; best spring properties, resistant to wear and fatigue. *Obsolete*

WESTERN RED BRASS #15
Olin Corp.
Cu 85, Zn 15.
Hard: 70,000 TS; 57,000 YS; 5 El. Soft: 40,000 TS; 12,000 YS; 47 El. For coffee pots, novelties, gift ware, hardware; resists corrosion and season cracking. *Obsolete*

WESTERN SILVER BEARING COPPER #51
Olin Corp.
Ag 0.028, 99.9% min Cu.
Hard: 51,000 TS; 45,000 YS; 6 El. Soft: 33,000 TS; 10,000 YS; 45 El. For heat exchangers; high softening point *Obsolete*

WESTFALIA RF 1
W. Ossenberg & Cie Edelstahlwerke
C 0.12, Cr 13, bal Fe.
Martensitic, corrosion resistant steel for pump shafts, armature shafts, spindles. Werkstoff Nr. 1.4001; X 7 Cr 14; AISI 410.

WESTFALIA RF 2

W. Ossenberg & Cie Edelstahlwerke
C 0.15, Cr 13, bal Fe.
Martensitic stainless steel; for valves, surgical instruments, armatures. Werkstoff Nr. 1.4024; similar to AISI 410.

WESTFALIA RF 3

W. Ossenberg & Cie Edelstahlwerke
C 0.2, Cr 13, bal Fe.
Martensitic stainless steel; for instrument shafts, valves, surgical instruments, cutlery. Werkstoff Nr. 1.4021; similar to AISI 420.

WESTFALIA RF 4

W. Ossenberg & Cie Edelstahlwerke
C 0.4, Cr 13, bal Fe.
Martensitic stainless steel; for pump parts, instrument shafts and gears, cutlery. Werkstoff Nr. 1.4034; similar to AISI 420.

WESTFALIA RF 5

W. Ossenberg & Cie Edelstahlwerke
C 0.2, Cr 13, Mo 1.2, bal Fe.
Martensitic stainless steel; for blades and parts for steam turbines. Werkstoff Nr. 1.4120; similar to AISI 420.

WESTFALIA RFS 1

W. Ossenberg & Cie Edelstahlwerke
C 0.08, Cr 16.5, bal Fe.
Ferritic type stainless steel; for lightly loaded structural parts in water or steam. Not hardenable by heat treat. Werkstoff Nr. 1.4016; AISI 430.

WESTFALIA RFS 2

W. Ossenberg & Cie Edelstahlwerke
C 0.1, Cr 17, Ti, bal Fe.
Stabilized ferritic type stainless steel; for welded corrosion resistant parts. Werkstoff Nr. 1.4510.

WESTFALIA RFS 3

W. Ossenberg & Cie Edelstahlwerke
C 0.15, Cr 16.5, Mo 0.25, S, bal Fe.
Free-machining martensitic stainless steel for screws, nuts, bolts. Werkstoff Nr. 1.4104.

WESTFALIA RFS 4

W. Ossenberg & Cie Edelstahlwerke
C 0.22, Cr 17, Ni 1.5, bal Fe.
Heat treated: 80-95 kp/mm^2 TS; 60 kp/mm^2 YS. For strong shafts, arbors, pump parts to resist corrosion. Werkstoff Nr. 1.4057.

WESTFALIA RFS H1

W. Ossenberg & Cie Edelstahlwerke
C 1.1, Cr 15, Mo 0.5, V, bal Fe.
Martensitic stainless steel; hardenable to about 59 Rock C. For stainless ball and roller bearings, valves, spray nozzles. Werkstoff Nr. 1.4111.

WESTFALIA RFS H2

W. Ossenberg & Cie Edelstahlwerke
C 0.9, Cr 18, Mo 1, V 0.1, bal Fe.
Martensitic stainless steel; hardenable to about 57 Rock C. For stainless parts subject to high wear as ball and roller bearings. Werkstoff Nr. 1.4112.

WESTFALIA RFS H3

W. Ossenberg & Cie Edelstahlwerke
C 0.9, Cr 16.5, Mo 0.5, Co 1.5, V 0.25, bal Fe.
Martensitic stainless steel; hardenable to about 60 Rock C. For ball and roller bearings, valves, cutlery. Werkstoff Nr. 1.4535.

WESTFALIA RFS H4

W. Ossenberg & Cie Edelstahlwerke
C 0.35, Cr 16.5, Mo 1.2, bal Fe.
Martensitic stainless steel; hardenable to about 450 Brin. For cutlery; for arbors, spindles operating at elevated temperatures. Werkstoff Nr. 1.4122.

WESTFALIA S B 1

W. Ossenberg & Cie Edelstahlwerke
C 0.15, Cr 18, Ni 8, bal Fe.
Austenitic stainless steel. For food and dairy equipment. Werkstoff Nr. 1.4300; AISI 302.

WESTFALIA S B 2

W. Ossenberg & Cie Edelstahlwerke
C 0.07, Cr 18, Ni 10, bal Fe.
Austenitic stainless steel. For food and dairy equipment. Werkstoff Nr. 1.4301; AISI 304.

WESTFALIA S B 3

W. Ossenberg & Cie Edelstahlwerke
C 0.1, Cr 18, Ni 10, Ti, bal Fe.
Stabilized austenitic stainless steel. For apparatus and parts for photo, paper, soap and textile industries. Werkstoff Nr. 1.4541.

WESTFALIA S B 4

W. Ossenberg & Cie Edelstahlwerke
C 0.1, Cr 18, Ni 9, Nb, bal Fe.
Stabilized austenitic stainless steel. For apparatus and parts for photo, paper, soap, and textile industries. Werkstoff Nr. 1.4550.

WESTFALIA S B 5

W. Ossenberg & Cie Edelstahlwerke
C 0.07, Cr 18, Ni 11.5, Mo 2, bal Fe.
Austenitic stainless steel. For equipment for chemical industry, photo, rubber, die and textile industries Werkstoff Nr. 1.4401; AISI 316.

WESTFALIA S B 6

W. Ossenberg & Cie Edelstahlwerke
C 0.1, Cr 18, Ni 10, Mo 2, Ti, bal Fe.
Titanium stabilized austenitic stainless steel for welded assemblies in chemical textile, die and photo industries. Werkstoff Nr. 1.4571.

WESTFALIA S B 7

W. Ossenberg & Cie Edelstahlwerke
C 0.1, Cr 18, Ni 10, Mo 2, Nb, bal Fe.
Niobium stabilized austenitic stainless steel for welded assemblies in photo, chemical and textile industries. Werkstoff Nr. 1.4580.

WESTFALIA S B 9

W. Ossenberg & Cie Edelstahlwerke
C 0.1, Cr 27, Mo 1.5, Ni 5, bal Fe.
Ferrite-austenite stainless. 65-80 kp/mm^2 TS; 50 Kp/mm^2 YS min. Stainless parts for high chemical and mechanical stress. Werkstoff Nr. 1.4460.

WESTFALISCHE CDM

Westfalische Stahlgesellschaft A.G.
C 1.1, Mn 0.3, Cr 0.4, bal Fe.
For cutters, bearings, tools; water hardened, wear resistant.

WESTFALISCHE CEZ

Westfalische Stahlgesellschaft A.G.
C 1, Mn 0.07, Cr 1.1, bal Fe.
For bearings, cutters, bushings; water hardened, wear resistant.

WESTFALISCHE CFSS

Westfalische Stahlgesellschaft A.G.
C 1.05, Cr, bal Fe.
For bearings, bushings, liners; water hardened, wear resistant.

WESTFALISCHE EH10

Westfalische Stahlgesellschaft A.G.
C 1, Si 0-0.25, Mn 0-0.25, bal Fe.
Annealed: 100,000 TS; 53,000 YS; 21 El; 42 RA; 200 Brin. For tools, cutters, drills, taps, reamers; Type W1; water hardened.

WESTFALISCHE EM07

Westfalische Stahlgesellschaft A.G.
C 0.7, Si 0-0.25, Mn 0-0.25, bal Fe.
Heat treated: 175,000 TS; 130,000 YS; 12 El; 36 RA; 355 Brin. For springs, tools, punches, rails, cutters, axes; Type W1; water hardened.

WESTFALISCHE EPS

Westfalische Stahlgesellschaft A.G.
C 0.3, Cr 2.65, V 0.35, W 8.5, bal Fe.
For punches, shears, extrusion rams and liners; hot work steel, oil hardened.

WESTFALISCHE EWP970

Westfalische Stahlgesellschaft A.G.
C 0.3, Co 2, Cr 2.4, V 0.25, W 8.5, bal Fe.
For punches, shears, extrusion rams and liners; hot work steel, oil hardened.

WESTFALISCHE EZ13

Westfalische Stahlgesellschaft A.G.
C 1.0, Si 0-0.25, Mn 0-0.25, bal Fe.
For engravers' tools, forming dies, reamers; Type W1; water hardened.

WESTFALISCHE HSS

Westfalische Stahlgesellschaft A.G.
C 1.65, Cr 11.5, V 0.1, Mn 0.3, bal Fe.
For blanking and forming dies, punches; air hardened, nondeforming.

WESTFALISCHE ISS

Westfalische Stahlgesellschaft A.G.
C 0.9, Mn 1.9, V 0.1, bal Fe.
For punches, dies, shears, cutters; oil hardened, nondeforming.

WESTFALISCHE OPS

Westfalische Stahlgesellschaft A.G.
C 0.45, Si, Cr, V, bal Fe.
For springs, gears, bolts, fasteners; oil hardened, tough.

WESTFALISCHE OWS

Westfalische Stahlgesellschaft A.G.
C 0.38, Si, Cr, V, bal Fe.
For gears, bolts, crankshafts, fasteners; oil hardened, tough.

WESTFALISCHE PM06

Westfalische Stahlgesellschaft A.G.
C 0.6, Si 0.25-0.5, Mn 0.3-0.8, bal Fe.
Heat treated: 115,000-160,000 TS; 77,000-113,000 YS; 12-23 El; 40-54 RA; 230-320 Brin. For wheels, die blocks, rails, girders; water hardened.

WESTFALISCHE PM45

Westfalische Stahlgesellschaft A.G.
C 0.45, Si 0.25-0.5, Mn 0.3-0.8, bal Fe.
Hot rolled: 98,000 TS; 59,000 YS; 24 El; 45 RA; 212 Brin. For axles, gears, bolts, bushings, crankshafts; water hardening.

WESTFALISCHE PW10

Westfalische Stahlgesellschaft A.G.
C 1.1, Si 0-0.25, Mn 0-0.25, bal Fe.
For springs, taps, cutters, hobs, reamers; Type W1; water hardened.

WESTFALISCHE RZ5

Westfalische Stahlgesellschaft A.G.
C 0.5, Mn 0.95, Cr 1.05, V 0.1, bal Fe.
For springs, gears, bolts, crankshafts; oil hardened, shock resistant.

WESTFALISCHE SP07

Westfalische Stahlgesellschaft A.G.
C 0.75, Si 0.25-0.5, Mn 0.3-0.8, bal Fe.
Heat treated: 185,000 TS; 140,000 YS; 14 El; 38 RA; 370 Brin. For springs, rails, clutch discs, girders; Type W1; water hardened.

WESTFALISCHE SS BLAU
Westfalische Stahlgesellschaft A.G.
C 0.86, Cr 4.1, Mo 0.85, V 2.5, W 12, bal Fe.
For drills, taps, reamers, broaches; high speed steel.

WESTFALISCHE SS GELB
Westfalische Stahlgesellschaft A.G.
C 0.86, Co 2.8, Cr 4.3, Mo 0.85, V 2.1, W 12, bal Fe.
For broaches, taps, cutters, drills, reamers; high speed steel.

WESTFALISCHE SS GOLD
Westfalische Stahlgesellschaft A.G.
C 0.76, Co 10, Cr 4.2, Mo 0.85, V 1.8, W 18, bal Fe.
For cutters, taps, drills, hobs, reamers; high speed steel.

WESTFALISCHE SS GRUN
Westfalische Stahlgesellschaft A.G.
C 0.82, Cr 4.1, Mo 0.85, V 1.6, W 8.7, bal Fe.
For reamers, drills, broaches, taps, cutters; high speed steel.

WESTFALISCHE SS KUPFER
Westfalische Stahlgesellschaft A.G.
C 0.74, Cr 4.1, V 1.1, W 18.5, bal Fe.
For reamers, drills, broaches, hobs, taps; high speed steel.

WESTFALISCHE SS ROT
Westfalische Stahlgesellschaft A.G.
C 1.3, Cr 4.3, Mo 0.85, V 3.8, W 12, bal Fe.
For blanking and forming dies, engravers' tools; high speed steel.

WESTFALISCHE SS SCHWARTZ
Westfalische Stahlgesellschaft A.G.
C 1.35, Cr 4, V, W, Mo, bal Fe.
For blanking and forming dies, reamers, taps; high speed steel.

WESTFALISCHE SS SILBER
Westfalische Stahlgesellschaft A.G.
C 0.79, Co 4.75, Cr 4.3, Mo 0.75, V 1.5, W 18, bal Fe.
For taps, chasers, drills, hobs, reamers; high speed steel.

WESTFALISCHE SS WEISS
Westfalische Stahlgesellschaft A.G.
C 0.85, Cr 4, V, W, Mo, bal Fe.
For lathe and planer tools, reamers; high speed steel.

WESTFALISCHE SS09
Westfalische Stahlgesellschaft A.G.
C 0.85, Si 0-0.25, Mn 0-0.25, bal Fe.
Heat treated: 190,000 TS; 145,000 YS; 10 El; 30 RA; 400 Brin. For springs, drills, taps, reamers, hammers; Type W1; water hardened.

WESTFALISCHE SS10
Westfalische Stahlgesellschaft A.G.
C 1, Si 0-0.25, Mn 0-0.25, bal Fe.
Annealed: 100,000 TS; 53,000 YS; 21 El; 42 RA; 200 Brin. For springs, taps, reamers, drills, hobs; Type W1; water hardened.

WESTFALISCHE SZ09
Westfalische Stahlgesellschaft A.G.
C 0.85, Si 0-0.25, Mn 0-0.25, bal Fe.
Heat treated: 190,000 TS; 145,000 YS; 10 El; 30 RA; 400 Brin. For springs, taps, reamers, drills; Type W1; water hardened.

WESTFALISCHE TSS
Westfalische Stahlgesellschaft A.G.
C 1.15, Cr 0.65, V 0.1, bal Fe.
For blanking and forming dies, bearings; water hardened, wear resistant.

WESTFALISCHE USS
Westfalische Stahlgesellschaft A.G.
C 1.25, Si 1.15, Mn 0.7, Cr 1.2, bal Fe.
For blanking and forming dies, bearings, liners; oil hardened.

WESTIG 100 CR 6
Westig (U.K.) Ltd.
C 1, Si 0.25, Mn 0.35, P 0-0.03, S 0-0.025, Cr 1.5, bal Fe.
Ball bearing steel equivalent to BS 534 A 99; JIS SUS 2; AISI 52100.

WESTIG 100 K6
Now WESTIG 100 CR 6; WESTIG SSRM5 now WESTIG S 6-5-2; WESTIG SSR SPEZIAL now WESTIG 18-0-1.

WESTIG 105 CR 4
Semi-Alloys
C 1-1.1, Cr 0.9-1.15, bal Fe.
Ball bearing steel. Werkstoff Nr. 1.3503; similar to AISI 52100.

WESTIG 105WCR6
Westig (U.K.) Ltd.
Alloy steel. C 1.05, Si 0.22, Mn 0.95, P 0-0.03, S 0-0.03, Cr 1, W 1.15, bal Fe.
Alloyed tool steel equivalent to JIS SK S 31; SKS 2; SKS 3.

WESTIG 110 W CR V 5
Semi-Alloys
C 1.1, Si 0.15-0.3, Mn 0.2-0.4, Cr 1.1-1.3, W 1.2-1.4, V 0.15-0.25, bal Fe.
Cold work tool steel; for leather and rubber machine knives, staybolt taps. Werkstoff Nr. 1.2519.

WESTIG 115 CR V 3
Semi-Alloys
C 1.1-1.25, Si 0.15-0.3, Mn 0.2-0.4, Cr 0.5-0.8, V 0.07-0.12, bal Fe.
Cold work tool steel; for drills, taps, reamers, scraping tools. Werkstoff Nr. 1.2210.

WESTIG 115 W 8
Semi-Alloys
C 1.15, Si 0.15-0.3, Mn 0.2-0.4, Cr 0.15-0.25, W 1.8-2.1, bal Fe.
Cold work tool steel; for metal-cutting saws and hack saw blades. Werkstoff Nr. 1.2442.

WESTIG 120 W 4
Semi-Alloys
C 1.2, Si 0.15-0.3, Mn 0.2-0.35, W 0.9-1.1, bal Fe.
Cold work tool steel; for taps, center drills, twist drills. Werkstoff Nr. 1.2414.

WESTIG 120 WV 4
Semi-Alloys
C 1.2, Si 0.15-0.3, Mn 0.2-0.35, Cr 0.15-0.25, W 0.9-1.1, V 0.07-0.12, bal Fe.
Cold work tool steel; for twist drills, rotary files, countersinks. Werkstoff Nr. 1.2516.

WESTIG 125 CR 1
Westig (U.K.) Ltd.
Alloy steel. C 1.25, Si 0.22, Mn 0.32, P 0-0.03, S 0-0.03, Cr 0.35, bal Fe.
Alloyed tool steel.

WESTIG 125 CR 1
Semi-Alloys
C 1.25, Si 0.15-0.3, Mn 0.25-0.4, Cr 0.3-0.4, bal Fe.
Cold work tool steel; for mandrels, punches, taps, drawing dies. Werkstoff Nr. 1.2002.

WESTIG 125 CR 2
Westig (U.K.) Ltd.
Alloy steel. C 1.25, Si 0.22, Mn 0.32, P 0-0.03, S 0-0.03, Cr 0.5, bal Fe.
Alloyed tool steel.

WESTIG 140 CR 3
Semi-Alloys
C 1.35-1.5, Si 0.15-0.3, Mn 0.25-0.4, Cr 0.4-0.7, bal Fe.
Cold work tool steel; for files, burnishing tools, cutters. Werkstoff Nr. 1.2008.

WESTIG 15CR3
Westig (U.K.) Ltd.
C 0.16, Si 0.25, Mn 0.5, P 0-0.035, S 0-0.035, Cr 0.7, bal Fe.
Case hardening structural steel equivalent to BS 523 M 15; JIS SCr415(H); AISI 5015.

WESTIG 16MNCR5
Westig (U.K.) Ltd.
C 0.16, Si 0.3, Mn 1.15, P 0-0.035, S 0-0.035, Cr 0.95, bal Fe.
Case hardening structural steel equivalent to BS 527 M 20; AISI 5115.

WESTIG 20MNCR5
Westig (U.K.) Ltd.
C 0.2, Si 0.3, Mn 1.25, P 0-0.035, S 0-0.035, Cr 1.15, bal Fe.
Case hardening structural steel equivalent to AISI 5120.

WESTIG 2390 SPEZIAL
Westig (U.K.) Ltd.
C 0.31, Si 0.28, Mn 0.95, P 0-0.03, S 0-0.015, Cr 3.85, Mo 1.1, Ni 0.7, V 0.35, bal Fe.
Alloyed high grade steel (backing material).

WESTIG 31 CR V 3
Semi-Alloys
C 0.28-0.35, Si 0.25-0.4, Mn 0.4-0.6, Cr 0.4-0.7, V 0.07-0.12, bal Fe.
Cold work tool steel; for screw drivers, wrenches. Werkstoff Nr. 1.2208.

WESTIG 4005
Westig (U.K.) Ltd.
C 0.1, Si 0-1, Mn 0-1, P 0-0.045, S 0.4, Cr 12.5, Mo 0.6, bal Fe.
Stainless free cutting steel equivalent to BS 416s21; AISI 416.

WESTIG 4005 SE
Westig (U.K.) Ltd.
C 0.1, Si 0-1, Mn 0-1, P 0-0.045, S 0.4, Cr 12.5, Mo 0.6, Se 0.07, bal Fe.
Stainless free cutting steel.

WESTIG 4006
Semi-Alloys
C 0.08-0.12, Cr 12-14, bal Fe.
Chrome stainless steel. Werkstoff Nr. 1.4006. (Was CHRONIFER F 14).

WESTIG 4016
Westig (U.K.) Ltd.
Stainless steel. C 0-0.08, Si 0-1, Mn 0-1, P 0-0.045, S 0-0.03, Cr 16.5, bal Fe.
Ferritic stainless steel equivalent to BS 430 S 15; SUS 430; AISI 430.

WESTIG 4016
Semi-Alloys
C 0-0.1, Cr 15.5-17.5, bal Fe.
Ferritic stainless steel, nonhardenable. Werkstoff Nr. 1.4016; similar to AISI 430. (Was CHRONIFER F-17).

WESTIG 4021
Westig (U.K.) Ltd.
Stainless steel. C 0.2, Si 0-1, Mn 0-1, P 0-0.04, S 0-0.03, Cr 13, bal Fe.
Stainless steel equivalent to BS 420 S 37; JIS SUS 420 J 1; AISI 421.

WESTIG 4021
Semi-Alloys
C 0.17-0.22, Cr 12-14, bal Fe.
Martensitic stainless steel. Werkstoff Nr. 1.4021; similar to AISI 420. (Was CHRONIFER V-13).

WESTIG 4028
Westig (U.K.) Ltd.
Stainless steel. C 0.32, Si 0.17, Mn 0.3, P 0-0.025, S 0-0.01, Cr 13.8, Mo 0-0.5, Ni 0-0.5, bal Fe.
Martensitic stainless steel equivalent to BS 430 S 45; SUS 420 J2.

WESTIG 4028 MO
Westig (U.K.) Ltd.
Stainless steel. C 0.32, Si 0.17, Mn 0.3, P 0-0.025, S 0-0.01, Cr 13.8, Mo 0-0.5, Ni 0-0.5, bal Fe.
Martensitic stainless steel equivalent to BS 430 S 45; SUS 420 J2.

WESTIG 4034
Westig (U.K.) Ltd.
Stainless steel. C 0.45, Si 0-1, Mn 0-1, P 0-0.04, S 0-0.03, Cr 13, Mo 0.6, bal Fe.
Stainless steel equivalent to BS 420 S 45; JIS SUS 420 J 2.

WESTIG 4034 C
Westig (U.K.) Ltd.
Stainless steel, C 0.62, Si 0.35, Mn 0.6, P 0-0.025, S 0-0.02, Cr 12.3, Ni 0-0.5, bal Fe.
Martensitic stainless steel.

WESTIG 4034 R
Westig (U.K.) Ltd.
Stainless steel. C 0.62, Si 0.35, Mn 0.6, P 0-0.025, S 0-0.02, Cr 12.3, Ni 0-0.5, bal Fe.
Martensitic stainless steel.

WESTIG 4034 S
Westig (U.K.) Ltd.
C 0.45, Si 0-1, Mn 0.75, P 0-0.045, S 0.2, Cr 13.5, Mo 0.6, bal Fe.
Stainless free cutting steel equivalent to AISI 420 F.

WESTIG 4057
Westig (U.K.) Ltd.
Stainless steel, C 0.2, Si 0-1, Mn 0-1, P 0-0.04, S 0-0.03, Cr 16.5, Ni 2, bal Fe.
Stainless steel equivalent to BS 431 S 29; JIS SUS 431; AISI 431.

WESTIG 4104
Westig (U.K.) Ltd.
C 0.15, Si 0-1, Mn 1.25, P 0-0.045, S 0.25, Cr 16.5, Mo 0.25, bal Fe.
Stainless free cutting steel equivalent to JIS SUS 430F; AISI 430 F.

WESTIG 4112
Westig (U.K.) Ltd.
Stainless steel. C 0.9, Si 0-1, Mn 0-1, P 0-0.045, S 0-0.03, Cr 18, Mo 1.1, V 0.1, bal Fe.
Stainless steel equivalent to JIS SUS 440 B; AISI 440 B.

WESTIG 4122
Westig (U.K.) Ltd.
Stainless steel. C 0.42, Si 0-1, Mn 0-1, P 0-0.025, S 0-0.025, Cr 16, Mo 1.1, Ni 0-1, bal Fe.
Martensitic stainless steel.

WESTIG 4301
Westig (U.K.) Ltd.
Stainless steel. C 0-0.07, Si 0-1, Mn 0-2, P 0-0.045, S 0-0.03, Cr 18, Ni 9, bal Fe.
Austenitic stainless steel equivalent to BS 304 S 15; JIS SUS 304; AISI 304.

WESTIG 4301
Semi-Alloys
C 0-0.07, Cr 18.5, Ni 9.5, bal Fe.
Austenitic stainless steel. Werkstoff Nr. 1.4301; AISI 304. (Was CHRONIFER SPEZIAL SUPRA).

WESTIG 4301 SG
Westig (U.K.) Ltd.
Stainless steel. C 0-0.07, Si 0-1, Mn 0-2, P 0-0.045, S 0-0.03, Cr 18, Ni 9, bal Fe.
Austenitic stainless steel.

WESTIG 4303 FR
Westig (U.K.) Ltd.
Stainless steel. C 0-0.03, Si 0-1, Mn 0-2, P 0-0.02, S 0-0.01, Cr 16, Ni 14, bal Fe.
Austenitic stainless steel equivalent to BS 305 S 19; JIS SUS 305; AISI 308; 305.

WESTIG 4303 NI
Westig (U.K.) Ltd.
Stainless steel. C 0-0.05, Si 0-0.8, Mn 0-2, P 0-0.02, S 0-0.01, Cr 18, Ni 12, bal Fe.
Austenitic stainless steel equivalent to BS 305 S 19; JIS SUS 305; AISI 308; 305.

WESTIG 4305
Westig (U.K.) Ltd.
C 0.12, Si 0.3, Mn 1.75, P 0-0.045, S 0.3, Cr 17.5, Mo 0.4, Ni 9, bal Fe.
Stainless free cutting steel equivalent to BS 303 S 21; JIS SUS 303; AISI 303.

WESTIG 4305
Semi-Alloys
C 0-0.15, S 0.15-0.35, Cr 18, Ni 9, bal Fe.
Free machining austenitic stainless steel. Werkstoff Nr. 1.4305; AISI 303. (Was CHRONIFER SPEZIAL D).

WESTIG 4305 SE
Westig (U.K.) Ltd.
C 0.12, Si 0.3, Mn 1.75, P 0-0.045, S 0.25, Cr 17.5, Mo 0.4, Ni 9, Se 0.06, bal Fe.
Stainless free cutting steel.

WESTIG 4310
Westig (U.K.) Ltd.
Stainless steel. C 0.13, Si 0.9, Mn 1.3, P 0-0.03, S 0-0.025, Cr 17, Mo 0-0.5, Ni 6.75, bal Fe.
Austenitic stainless steel equivalent to JIS SUS 301; AISI 301.

WESTIG 4401
Semi-Alloys
C 0-0.07, Cr 17.5, Ni 12, Mo 2-2.5, bal Fe.
Austenitic stainless steel. Werkstoff Nr. 1.4401; similar to AISI 316. (Was CHRONIFER SPEZIAL D SUPRA).

WESTIG 4541
Semi-Alloys
C 0-0.1, Cr 18, Ni 10, Ti = 5 x C, bal Fe.
Stabilized austenitic stainless steel. Werkstoff Nr. 1.4541; AISI 321. (Was CHRONIFER SPEZIAL EXTRA).

WESTIG 46 CR MO V 4
Westig (U.K.) Ltd.
C 0.46, Si 0.22, Mn 0.95, P 0-0.03, S 0-0.015, Cr 1.05, Mo 0.95, Ni 0.5, V 0.3, bal Fe.
Alloyed high grade steel (backing material).

WESTIG 50 CR V 4
Westig (U.K.) Ltd.
C 0.5, Si 0.27, Mn 0.9, P 0-0.03, S 0-0.015, Cr 1.05, V 0.15, bal Fe.
Alloyed high grade steel (backing material) equivalent to BS 735 A 50; JIS SUP 10; AISI 6150.

WESTIG 50 CR V 4
Semi-Alloys
C 0.47-0.55, Mn 0.7-1.1, Cr 0.9-1.2, V 0.1-0.2, bal Fe.
Chromium-manganese steel. Werkstoff Nr. 1.8159; similar to A 6150.

WESTIG 55 NI CR MO V 6
Semi-Alloys
C 0.55, Cr 0.6-0.8, Ni 1.5-1.8, Mo 0.25-0.35, V 0.07-0.12, bal Fe.
Cold work tool steel; forging and stamping dies. Werkstoff Nr. 1.2713.

WESTIG 58 CRV4
Westig (U.K.) Ltd.
Alloy steel. C 0.58, Si 0.27, Mn 0.9, P 0-0.03, S 0-0.015, Cr 1.05, V 0.15, bal Fe.
Alloyed tool steel.

WESTIG 70 SICR5
Westig (U.K.) Ltd.
C 0.7, Si 1.3, Mn 0.7, P 0-0.03, S 0-0.03, Cr 0.7, bal Fe.
Alloyed spring steel.

WESTIG 71 SI 7
Westig (U.K.) Ltd.
C 0.71, Si 1.65, Mn 0.7, P 0-0.03, S 0-0.03, bal Fe.
Alloyed spring steel.

WESTIG 8 NI 6
Westig (U.K.) Ltd.
C 0.08, Si 0.27, Mn 0.4, P 0-0.03, S 0-0.015, Ni 1.45, bal Fe.
Alloyed case hardening steel equivalent to ASTM A 350-LF 5.

WESTIG A 60
Semi-Alloys
C 0.6, Mn 0.5-0.7, P 0-0.07, S 0.15-0.25, bal Fe.
Free machining medium high carbon steel. Werkstoff Nr. 1.0729.

WESTIG A 60 PB
Semi-Alloys
C 0.6, Mn 0.5-0.7, P 0-0.07, S 0.15-0.25, Pb, bal Fe.
Free machining steel.

WESTIG C 100 W
Westig (U.K.) Ltd.
Carbon steel. C 1, Si 0.2, Mn 0.42, P 0-0.03, S 0-0.01, bal Fe.
Carbon tool steel.

WESTIG C 105 W 2
Semi-Alloys
C 1.05, Si 0.1-0.3, Mn 0.1-0.35, bal Fe.
Carbon tool steel; for stone tools, embossing tools, scythes. Werkstoff Nr. 1.1645.

WESTIG C 110 W 2
Semi-Alloys
C 1.1, Si 0-0.3, Mn 0-0.35, bal Fe.
Carbon tool steel; for wood and leather working tools. Werkstoff Nr. 1.1654.

WESTIG C 125 R
Westig (U.K.) Ltd.
Carbon steel. C 1.25, Si 0.2, Mn 0.22, P 0-0.015, S 0-0.006, bal Fe.
Carbon tool steel.

WESTIG C 125 W
Westig (U.K.) Ltd.
Carbon steel. C 1.25, Si 0.2, Mn 0.22, P 0-0.03, S 0-0.01, bal Fe.
Carbon tool steel equivalent to JIS SK2; AISI W 112.

WESTIG C 125 W 2
Semi-Alloys
C 1.2-1.35, Si 0.1-0.3, Mn 0.1-0.35, bal Fe.
Carbon tool steel; for files, scrapers, hard knives. Werkstoff Nr. 1.1663.

WESTIG C 60
Westig (U.K.) Ltd.
Carbon steel. C 0.61, Si 0-0.4, Mn 0.75, P 0-0.045, S 0-0.045, bal Fe.
Carbon tool steel equivalent to BS 080 A 62; AISI 1060.

WESTIG C 67 K
Westig (U.K.) Ltd.
Carbon steel. C 0.68, Si 0.22, Mn 0.42, P 0-0.025, S 0-0.015, bal Fe.
Carbon tool steel.

WESTIG C 75
Westig (U.K.) Ltd.
Carbon steel. C 0.75, Si 0.22, Mn 0.5, P 0-0.03, S 0-0.025, bal Fe.
Carbon tool steel equivalent to BS 070 A 72.

WESTIG C 75 BR
Westig (U.K.) Ltd.
Carbon steel. C 0.75, Si 0.22, Mn 0.5, P 0-0.015, S 0-0.006, bal Fe.
Carbon tool steel.

WESTIG C 75 K
Westig (U.K.) Ltd.
Carbon steel. C 0.76, Si 0.22, Mn 0.57, P 0-0.025, S 0-0.01, bal Fe.
Carbon tool steel.

WESTIG C 75 MN
Westig (U.K.) Ltd.
Carbon steel. C 0.75, Si 0.27, Mn 0.7, P 0-0.03, S 0-0.025, bal Fe.
Carbon tool steel.

WESTIG C 75 S
Westig (U.K.) Ltd.
Carbon steel. C 0.72, Si 0.25, Mn 0.8, P 0-0.03, S 0-0.01, bal Fe.
Carbon tool steel.

WESTIG C 75 W 3
Semi-Alloys
C 0.72-0.82, Si 0.15-0.4, Mn 0.6-0.8, bal Fe.
Carbon tool steel; for mandrels, collet chucks. Werkstoff Nr. 1.1750.

WESTIG C 80 W 2
Westig (U.K.) Ltd.
Carbon steel. C 0.8, Si 0.25, Mn 0.5, P 0-0.03, S 0-0.025, bal Fe.
Plain carbon tool steel.

WESTIG C 85
Westig (U.K.) Ltd.
Carbon steel. C 0.82, Si 0.25, Mn 0.42, P 0-0.03, S 0-0.025, bal Fe.
Carbon tool steel.

WESTIG C 85 F
Westig (U.K.) Ltd.
Carbon steel. C 0.82, Si 0.25, Mn 0.42, P 0-0.015, S 0-0.006, bal Fe.
Carbon tool steel.

WESTIG CK 101
Westig (U.K.) Ltd.
Carbon steel. C 1, Si 0.18, Mn 0.4, P 0-0.03, S 0-0.015, bal Fe.
Plain carbon tool steel.

WESTIG CK 60
Westig (U.K.) Ltd.
C 0.6, Si 0-0.35, Mn 0.75, P 0-0.035, S 0-0.035 bal Fe.
Quenched and tempered structural steel equivalent to BS 080 A 62; JIS S 58 C; AISI 1060.

WESTIG CK 75
Westig (U.K.) Ltd.
Carbon steel. C 0.75, Si 0.25, Mn 0.5, P 0-0.03, S 0-0.025, bal Fe.
Plain carbon tool steel.

WESTIG MK 63
Semi-Alloys
C 0.6-0.64, Mn 0.35-0.55, N 0-0.007, bal Fe.
Structural steel. Werkstoff Nr. 1.1222.

WESTIG N 100
Semi-Alloys
C 1, Si 0.15-0.3, Mn 0.15-0.3, S 0-0.025, P 0-0.025, V 0.1-0.15, bal Fe.
Wire, for needles. Werkstoff Nr. 1.2833.

WESTIG S 12-1-2
Semi-Alloys
C 0.95, Cr 4.2, Mo 0.85, V 2.2, W 12, bal Fe.
High speed steel; for lathe tools, milling cutters. Werkstoff Nr. 1.3318.

WESTIG S 18-0-1
Semi-Alloys
C 0.75, Cr 4.2, V 1, W 18, bal Fe.
High speed steel; for lathe tools, milling cutters, drills. Werkstoff Nr. 1.3355; similar to AISI T1.

WESTIG S 2-10-1-8
Semi-Alloys
C 1.08, Cr 4, Mo 9.5, V 1.2, W 1.5, Co 8, bal Fe.
High speed steel; for engraving milling cutters, lathe tools for hard materials. Werkstoff Nr. 1.3247; similar to AISI M42.

WESTIG S 2-9-2
Semi-Alloys
C 1, Cr 3.8, Mo 8.6, V 2, W 1.8, bal Fe.
High speed steel; for milling tools, drills, lathe tools. Werkstoff Nr. 1.3348; similar to AISI M1.

WESTIG S 3-3-2
Semi-Alloys
C 1, Cr 4.2, Mo 2.7, V 2.4, W 2.85, bal Fe.
High speed steel; for milling cutters, twist drills, broaches. Werkstoff Nr. 1.3333.

WESTIG S 6-5-2
Semi-Alloys
C 0.9, Cr 4.2, Mo 5, V 1.85, W 6.4, bal Fe.
High speed steel; for twist drills, lathe tools, milling cutters. Werkstoff Nr. 1.3343; similar to AISI M2.

WESTIG S 6-5-2-5
Semi-Alloys
C 0.92, Cr 4.1, Mo 5, V 1.9, W 8.4, Co 4.8, bal Fe.
High speed steel; for highly stressed twist drills. Werkstoff Nr. 1.3243.

WESTIG S 6-5-3
Semi-Alloys
C 1.2, Cr 4.2, Mo 5, V 3, W 6.4, bal Fe.
High speed steel; for heavy duty milling cutters, broaches, reamers. Werkstoff Nr. 1.3344.

WESTIG S 7-4-2-5
Semi-Alloys
C 1.1, Cr 4.2, Mo 3.8, W 6.8, V 1.8, Co 5, bal Fe.
High speed steel; for twist drills, taps, reamers, milling cutters. Werkstoff Nr. 1.3246.

WESTIG W 95
Westig (U.K.) Ltd.
Carbon steel. C 1, Si 0.17, Mn 0.4, P 0-0.03, S 0-0.015, bal Fe.
Carbon tool steel.

WESTIG W 95 BR
Westig (U.K.) Ltd.
Carbon steel. C 1, Si 0.17, Mn 0.4, P 0-0.015, S 0-0.006, bal Fe.
Carbon tool steel.

WESTIG W 95 MN
Westig (U.K.) Ltd.
Carbon steel. C 1, Si 0.37, Mn 0.7, P 0-0.03, S 0-0.025, bal Fe.
Carbon tool steel.

WESTIG X 165 CR MO V 12
Semi-Alloys
C 1.55-1.75, Cr 11-12, Mo 0.5-0.7, W 0.4-0.6, V 0.1-0.5, bal Fe.
Cold work tool steel; for coining dies, sheet metal punching dies. Werkstoff Nr. 1.2601.

WESTINGHOUSE 52 NI-FE
Westinghouse Electric Corp.
Ni 52, bal Fe.
For semiconductors and hermetically sealed electronic components. Controlled expansion. *Obsolete*

WESTINGHOUSE B-33
Westinghouse Electric Corp.
V 4, bal Cb.
At 73°F: 81,000 psi TS; 595,000 psi YS; 32 El. At 2000°F: 33,000 psi TS; 30,300 psi YS; 34 El. At -148°F: 91,700 psi TS; 74,200 psi YS; 31 El. For space vehicles, nuclear reactors. Combination of density, strength and oxidation resistance at high temperatures.

WESTINGHOUSE B-66
Westinghouse Electric Corp.
V 5, Mo 5, Zr 1, bal Cb.
At -148°F: 128,000 psi TS; 108,000 psi YS; 12 El. At 73°F: 115,000 psi TS; 91,000 psi YS; 14 El. At 2000°F: 65,000 psi TS; 58,000 psi YS; 28 El. For space vehicles, nuclear reactors. Combination of density, strength and oxidation resistance at high temperatures.

WESTINGHOUSE B-77
Westinghouse Electric Corp.
V 5, W 10, Zr 1, bal Cb.
At -148°F: 159,000 psi TS; 137,000 psi YS; 20 El. At 73°F: 132,000 psi TS; 106,500 psi YS; 18 El. At 2400°F: 30,000 psi TS; 27,000 psi YS; 34 El. For space vehicles, nuclear reactors. Combination of density, strength, and oxidation resistance at high temperatures.

WESTINGHOUSE HI-120
Westinghouse Electric Corp.
Ti, bal Cb.
Ductility of alloy windings enable the magnet to cycle from full strength to zero field without damage. Operates in liquid helium at -452°F. Superconducting magnet. *Obsolete*

WESTINGHOUSE NC155
Westinghouse Electric Corp.
Mo 5, V 5, bal Cb.
Heat resistant for high temperature applications, space vehicles. *Obsolete*

WESTINGHOUSE SR26
Westinghouse Electric Corp.
C 0-0.05, Cr 18, Ni 37, Co 20, Mo 3.5, Ti 4.2, Al 0.2, B 0.05, Zr 0.05, bal Fe.
For high temperature applications; heat resistant. *Obsolete*

WESTINGHOUSE T-111
Westinghouse Electric Corp.
W 7-9, Hf 2-2.8, C 0.003, bal Ta.
Stress relieved sheet: 150,000 psi TS; 144,800 psi YS; 9 El. At 2000°F: 92,100 psi TS; 67,500 psi YS; 8 El. At 3000°F: 16,300 psi TS; 14,100 psi YS; 52 El. For aerospace structural components, containers, for high temperature liquid metals in nuclear reactors. Superior strength at 2000-3500°F. Good ductility for forming.

WESTINGHOUSE T-222
Westinghouse Electric Corp.
C 0.01, W 9.64, Hf 2.4, bal Ta.
At -320°F: 184,600 psi TS; 175,000 psi YS; 28 El. At 75°F:
110,600 psi TS; 105,000 psi YS; 30 El. At 3000°F: 24,200
psi TS; 24,000 psi YS; 26 El. For high temperature, re-entry
space vehicles, rocket reaction chambers, nozzle parts,
liquid metal systems. High temperature strength, good
ductility at low temperatures, good welding.

WESTINGHOUSE W545
Westinghouse Electric Corp.
C 0.05, Ni 26, Cr 12.5, Mo 1.5, B 0.08, Ti 2.6, Al 0.2, bal Fe.
Bar: 162,000 psi TS; 115,000 psi YS; 20 El; 25 RA; 350 Brin.
Sheet: 142,000 psi TS; 108,000 psi YS; 8 El; 320 Brin. For
jet engine components, gas turbine discs; age hardenable,
heat resistant, high tensile and creep strength. *Obsolete*

WESTROHR 17L
Mannesmannrohren-Werke A.G.
C 0.13, Cr 0.85, Mo 0.45, bal Fe.
For cams, gears, camshafts, fasteners; case hardened,
tough. *Obsolete*

WESTROHR N10
Mannesmannrohren-Werke A.G.
C 0.21, Cr 3, Mo 0.4, V 0.8, W 0.37, bal Fe.
For die casting dies, ejector pins, liners; oil hardened.
Obsolete

WESTROHR N5B
Mannesmannrohren-Werke A.G.
C 0.1, Cr 2.8, Mn 0.6, bal Fe.
For oil refinery equipment; creep and heat resistant.
Obsolete

WESTROHR N8
Mannesmannrohren-Werke A.G.
C 0.17, Mo 0.25, V 0.15, Mn 0.6, Cr 2.7, bal Fe.
For oil refinery equipment; creep and heat resistant.
Obsolete

WESTROHR N9
Mannesmannrohren-Werke A.G.
C 0.2, Cr 3.25, Mo 0.5, V 0.5, bal Fe.
For oil refinery equipment; creep and heat resistant.
Obsolete

WEXITE GRADE CR-6
Colt Industries
WC + Co.
For cutting tools; sintered, carbide. *Obsolete*

WF
Now FINKL WF.

WF 11
Now CRUCIBLE WF 11.

WF-11
Cannon-Muskegon Corp.
C 0.1, Mn 1.5, Si 0.5, Cr 20, Ni 10, W 15, bal Co.
For jet engines parts, sheet.

WF-31
Cannon-Muskegon Corp.
C 0.15, Mn 1.42, Cr 20, Ni 10, Mo 2.6, W 10.7, Ti 1, bal Co.
For high temperature operation.

WFF
Wardlows Ltd.
C 1.3, W 4.5, bal Fe.
For tools, cutters; fast finishing steel.

WH 05U
Sandvik Hard Materials Ltd.
Co 3, WC 96.9, 0.1 others.
Cemented carbide. 1950 Vickers; 7780 MNm^{-2} compressive
strength

WH 10
Sandvik Hard Materials Ltd.
Co 6, WC 93.7, 0.3 others.
Cemented carbide. 1750 Vickers; 6870 MNm^{-2} compressive
strength.

WH 25
Sandvik Hard Materials Ltd.
Co 10, WC 89.5, 0.5 others.
Cemented carbide. 1600 Vickers; 6470 MNm^{-2} compressive
strength.

WH 35
Sandvik Hard Materials Ltd.
Co 15, WC 84.25, 0.75 others.
Cemented carbide. 1400 Vickers; 5610 MNm^{-2} compressive
strength.

WHEEL BRAND BABBITT
Manufacturer not listed.
Sn 83, Sb 14, Cu 3.
For bearings, bushings; anti-friction

WHEEL BRASS
American manufacture
Cu 68, Zn 30, Pb 2.
For wheels, hardware; free-cutting.

WHEELERITE
Electro Foundry Co.
Si 1.5-2.5, TC 3-3.5, Mg 0.05-0.15, bal Fe.
Cast: 102,000 TS; 80,000 YS; 9 El; 241 Brin. Annealed:
67,000 TS; 46,000 YS; 26 El; 207 Brin. For gears, housings,
and machine tool castings. Ductile cast iron.

WHEELING METAL
Continental Foundry & Machine Co.
C 0.5, bal Fe.
For rolls; water hardened.

WHEELING SUPER STEEL
Continental Foundry & Machine Co.
C 0.7, bal Fe.

WHELCO A
Metalsource Corp.
C 0.1, Cr 13.5, Ni 0.25, bal Fe.
For corrosion resistant parts; corrosion resistant. *Obsolete*

WHELCO ALLOY TAP
Metalsource Corp.
C 1.1, Cr 0.35, V 0.2, W 0.85, bal Fe.
For tools, taps; oil hardened. *Obsolete*

WHELCO D
Metalsource Corp.
C 0.15, Cr 18, Ni 8, bal Fe.
For stainless parts; stainless. *Obsolete*

WHELCO DIE AND TOOL
Metalsource Corp.
C, Cr, bal Fe.
For punches and dies. *Obsolete*

WHELCO FINISHING
Metalsource Corp.
C 1.3, W 4, Mo 0.4, bal Fe.
For tools and dies, reamers. *Obsolete*

WHELCO HIGH SPEED
Metalsource Corp.
C 0.7, W 18, Cr 3.85, V 1, bal Fe.
For cutters, reamers, punches; high speed steel. *Obsolete*

WHELCO HOT DIE
Metalsource Corp.
C 0.8-1, Cr 3-4, bal Fe.
For bolt and rivet dies, hot piercing punches, forging
mandrels; hot work steel. *Obsolete*

WHELCO HOT WORK
Metalsource Corp.
C, W, bal Fe.
For hot work dies, tools; hot work steel. *Obsolete*

WHELCO M
Metalsource Corp.
C 0.7, Cr 0.75, Mo 0.25, Ni 1.5, bal Fe.
For cams, gears, arbors, punches; Type L6. *Obsolete*

WHELCO NO. 1
Metalsource Corp.
C 0-0.12, Cr 13, Ni 0.25, bal Fe.
For corrosion resistant parts; corrosion resistant. *Obsolete*

WHELCO NO. 2
Metalsource Corp.
C 0-0.25, Cr 18, Ni 8, bal Fe.
For stainless parts; stainless. *Obsolete*

WHELCO NO. 3
Metalsource Corp.
C 0.4, Cr 8, Ni 19, bal Fe.
For corrosion resistant parts; corrosion resistant. *Obsolete*

WHELCO NO. 4
Metalsource Corp.
C 0-0.12, Cr 17-21, Ni 0.25, bal Fe.
For heat and corrosion resistant parts; heat and corrosion
resistant. *Obsolete*

WHELCO NO. 5
U.S. Metalsource
Cr 16-23, C 0-0.12, bal Fe.
For corrosion resistant parts. *Obsolete*

WHELCO NO. 6
Metalsource Corp.
C 0.4, Cr 13.5, Ni 0.25, bal Fe.
For corrosion resistant parts; corrosion resistant. *Obsolete*

WHELCO NO. 7
Metalsource Corp.
C 0.65, Cr 17, Ni 0.25, bal Fe.
For corrosion resistant parts; corrosion resistant. *Obsolete*

WHELCO NO. 8
Metalsource Corp.
C 1.15, Cr 17, Ni 0.25, bal Fe.
For corrosion resistant parts; corrosion resistant. *Obsolete*

WHELCO NO. 9
Metalsource Corp.
C 0.3, Cr 26, Ni 0.25, bal Fe.
For corrosion resistant parts; corrosion resistant. *Obsolete*

WHELCO OIL HARDENING
U.S. Metalsource
C 0.9, Mn 1.2, Cr, V, bal Fe.
For tools and dies, punches; oil hardened. *Obsolete*

WHELCO PISTON
Metalsource Corp.
C, Cr, V, bal Fe.
For pistons, tools. *Obsolete*

WHELCO STANDARD
Metalsource Corp.
C 1, V 0.18, bal Fe.
For tools, taps, reamers; water hardened. *Obsolete*

WHELCO STD
Metalsource Corp.
C 0.95, Mn 0.3, V 0.03, bal Fe.
For cold work dies; water hardening. *Obsolete*

WHELCO SUPERIOR
Metalsource Corp.
C 0.7-1.2, bal Fe.
For tools, cutters, drills; water hardened. *Obsolete*

WHISTLE LAFOND'S
Manufacturer not listed.
Cu 80-81, Sn 17-18, Zn 0-2, Sb 0-2.
For steam whistles; bronze.

WHIT-ALLOY
Whiton Machine Co.
Al alloy.
For body chucks; non-hardenable.

WHITE ALLOY-1
Manufacturer not listed.
Cu 5-64, Sn 32, As 3.5.
For metallic mirrors; high polish.

WHITE ALLOY-2
Manufacturer not listed.
Cu 49-53, Zn 23-25, Ni 22-25, Fe 2-2.4.
For corrosion resistant white metal parts; corrosion resistant.

WHITE BENEDICT
Manufacturer not listed.
Cu 60.4, Zn 18.1, Ni 16.4, Sn 0.7, Pb 4.4.
For restaurant white metal ware; corrosion resistant.

WHITE BRASS
American manufacture
Zn 66, Cu 34.
For ornamental parts, castings.

WHITE BRASS
American manufacture
Cu 65, Zn 32-33, Sn 2-3.
For ornamental parts.

WHITE BRONZE
Belmont Metals Inc.
Zn 20, Mn 20, Al, bal Cu.
65,000 psi TS. For jewelry, hardware, decorative castings; melting point 1505-1550°F.

WHITE BUTTON ALLOY
Manufacturer not listed.
Cu 48.8-53, Zn 23-24.4, Ni 22-24.4, Fe 2-2.4.
For white buttons; corrosion resistant.

WHITE CAST IRON NO. 1
American manufacture
Mn 0.2, Si 0.75, CC 2.5, bal Fe.
Cast: 29,000 TS; 225 Brin. For cylinders, housings; very hard and brittle.

WHITE COPPER
Manufacturer not listed.
Cu 70, Zn 18, Ni 12.
For corrosion resistant white metal parts; corrosion resistant.

WHITE END
Evans Steel Co.
C 0.7, Cr 1.5, V 0.2, bal Fe.
For rolls; oil hardening.

WHITE GOLD SOLDER, 10 CARAT
Manufacturer not listed.
Au 30, Ag 55, Cu 1, Zn 12, Cd 2.
For gold solders; corrosion resistant.

WHITE GOLD SOLDER, 14 CARAT
Manufacturer not listed.
Au 50, Ag 36, Cu 1, Zn 11, Cd 2.
For gold solder; corrosion resistant.

WHITE GOLD SOLDER, 18 CARAT "A"
Manufacturer not listed.
Au 60.8, Ag 14, Cu 4, Ni 6, Zn 5.2.
For gold solder; corrosion resistant.

WHITE GOLD SOLDER, 18 CARAT "B"
Manufacturer not listed.
Au 61, Ag 13.5, Cu 1.5, Ni 7, Zn 17.
For gold solder; corrosion resistant.

WHITE GOLD SOLDER, 18 CARAT "C"
Manufacturer not listed.
Au 82, Ni 10, Zn 6, Cd 2.
For gold solder; corrosion resistant.

WHITE GOLD, 10 CARAT
Manufacturer not listed.
Au 41.6, Cu 25, Ni 25, Zn 8.3.
For jewelry; corrosion resistant.

WHITE GOLD, 14 CARAT "A"
Manufacturer not listed.
Au 58.3, Cu 17, Ni 17, Zn 7.6.
For jewelry; corrosion resistant.

WHITE GOLD, 14 CARAT "B"
Manufacturer not listed.
Au 59, Cu 25.5, Ni 12.3, Zn 3.2.
For jewelry; corrosion resistant.

WHITE GOLD, 18 CARAT
Manufacturer not listed.
Au 75, Ag 18.5, Cu 1, Zn 5.5.
For jewelry; corrosion resistant.

WHITE GOLD-1
Manufacturer not listed.
Au 90, Pd 10.
For jewelry; corrosion resistant.

WHITE GOLD-2
Manufacturer not listed.
Au 75-85, Ni 8-10, Zn 2-9.
For jewelry; corrosion resistant.

WHITE LABEL
Peninsular Steel Co.
C 1.55, Cr 11.5, Mo 0.8, V 0.9, bal Fe.
For blanking, punching, forming and shearing dies, wear plates, thread rollers and drawing dies. AISI D2 cold work tool steel.

WHITE LABEL
Ackerlind Steel Co., Inc.
C 1.05, Mn 0.3, Si 0.25, V 0.2, bal Fe.
AISI Type W2 water hardening tool steel.

WHITE LABEL
Peninsular Steel Co.
C 1.6, Cr 13, V 0.3, Mo 0.8, bal Fe.
For tools, dies. *Obsolete*

WHITE LABEL
Wallace Murray Corp.
C 0.75-0.9, bal Fe.
For blacksmith tools depending on temper, chisels, rock drills; "Heller's Electrical Tool."

WHITE LABEL
Houghton & Richards Inc.
C 0.7, W 18, Cr 4, V 1, bal Fe.
For cutting tools, hobs; high speed steel.

WHITE LABEL
A. Milne & Co.
0.95-1.30 C (as desired).
AISI Type W1; water hardening, tool steel.

WHITE LABEL S
Peninsular Steel Co.
C 2.05, Cr 11.75, V 0.6, bal Fe.
For dies, tools; oil hardening, non-deforming.

WHITE LABEL STYRIAN-1
Skoda Works National Corp.
C 0.77, Mn 0.17, Cr 4.34, W 20.7, bal Fe.
For tools, cutters, drills; high speed steel.

WHITE LABEL STYRIAN-2 NEW RAPID
Skoda Works National Corp.
C 0.7, Mn 0.14, Cr 4.63, W 15.42, V 1, bal Fe.
For tools, cutters, reamers; high speed steel.

WHITE METAL, DUTCH
Manufacturer not listed.
Sn 81.5, Sb 8.8, Cu 9.6.
For bearings, domestic ware; Babbitt.

WHITE METAL, HANOVER
Manufacturer not listed.
Sn 86.8, Sb 7.6, Cu 5.6.
For bearings, domestic ware; Babbitt.

WHITE METAL-1
Manufacturer not listed.
Pb 77, Sb 15, Sn 5, Cu 2.3.
For bearings, bushings; antifriction.

WHITE METAL-2
Manufacturer not listed.
Sn 49-53, Pb 33-34, Sb 11-14, Cu 2.4-3.3, Zn 0.1.
For bearings, bushings; antifriction.

WHITE NICKEL
Manufacturer not listed.
Cu 55-64, Ni 18, Fe 0.35, bal Zn.
For trimmings, control brackets, levers, fittings, plumbing; S.A.E. Spec 42.

WHITE NICKEL ALLOY
Manufacturer not listed.
Cu 65, Ni 32.25, Al 2.75.
For strong heat and corrosion resistant parts; corrosion resistant. For bearings, bushings; antifriction.

WHITE TOMBASIL
Illingworth Steel Co.
Zn, Mn, Ni, Pb, bal Cu.
Cast: 46,000-90,000 TS; 18,000-45,000 YS; 7-50 El; 100-180 Brin. Die cast and permanent mold; low melting temperature white alloy; corrosion resistant; for building hardware, marine hardware, ornamental, drain covers, pumps, swimming pool fixtures.

WHITELEY
Manufacturer not listed.
Au 45-55, Pd 30-35, Pt 15-20.
For dental alloy; corrosion resistant.

WHITELIGHT ALUMINUM
White Metal Rolling & Stamping Corp.
There are 13 aluminum alloys extruded;, see AA listing for properties.

WHITELIGHT AZ10
White Metal Rolling & Stamping Corp.
Al 0.1-1.5, Zn 0.2-0.6, 0.20 Mn min, bal Mg.
Extruded: 33,000 TS; 20,000 YS; 5-10 El. Low cost, good machinability, weldability, workability. For light weight machined parts.

WHITELIGHT AZ31B
White Metal Rolling & Stamping Corp.
Al 2.5-3.5, Zn 0.7-1.5, bal Mg.
Extruded: 38,000 TS; 26,000 YS; 15 El; 49 Brin. Rolled: 42,000 TS; 32,000 YS; 15 El; 73 Brin. For light alloy parts; wrought.

WHITELIGHT AZ51
White Metal Rolling & Stamping Corp.
Al 4.1-5.5, Zn 0.4-1.3, 0.15 Mn min, bal Mg.
Extruded: 38,000-41,000 TS; 23,000-28,000 YS; 10-11 El. Good structural properties, high endurance limit, good weldability, machinability, and corrosion resistance.

WHITELIGHT AZ61A

White Metal Rolling & Stamping Corp.
Al 5.8-7.2, Zn 0.4-1.5, 0.15 Mn min, bal Mg.
Extruded: 45,000 TS; 32,000 YS; 16 El; 60 Brin. For light alloy parts; wrought.

WHITELIGHT AZ80A

White Metal Rolling & Stamping Corp.
Al 8.5, Zn 0.5, bal Mg.
Extruded: 49,000 TS; 36,000 YS; 11 El; 60 Brin. For light alloy parts; wrought.

WHITELIGHT M1A

White Metal Rolling & Stamping Corp.
Mn 1.2, bal Mg.
Extruded: 32,000-39,000 TS; 26,000-29,000 YS; 10-11 El; 44 Brin. For light alloy parts; wrought.

WHITELIGHT ZK20A

White Metal Rolling & Stamping Corp.
Zn 2.3, Zr 0.55, bal Mg.
Extruded: 38,000 TS; 28,000 YS; 4 El. For light weight structures.

WHITELIGHT ZK30A

White Metal Rolling & Stamping Corp.
Zn 3, Zr 0.7, bal Mg.
Extruded: 44,000 TS; 34,000 YS. For light weight structures.

WHITELIGHT ZK60A

White Metal Rolling & Stamping Corp.
Zn 5.7, Zr 0.55, bal Mg.
Extruded: 49,000 TS; 38,000 YS; 14 El; 75 Brin. T5-temper: 52,000 TS; 43,000 YS; 11 El; 82 Brin. For light weight structures; age-hardened.

WHITEX ZINC

White Metal Rolling & Stamping Corp.
Zn.
For trim, molding.

WHITWORTH

English manufacture
C, alloy, bal Fe.
For armor plate.

WHIZ

Horace T. Potts Co.
C, alloy, bal Fe.
For tools. *Obsolete*

WHP

Wardlows Ltd.
C 1.5, Cr 11.5, Mo 0.75, bal Fe.
For dies, tools; non-deforming.

WI 765

Bergische Stahl Industrie
C 0.6-0.75, Si 0-1, Mn 0-1, Cr 13-15, bal Fe.
Martensitic stainless steel. W.-Nr. 1.4109.

WI-301

Wieland-Werke AG Metallwerke
Cu 4, Mg 0.8, Mn 0.6, 2% Pb + Sn + Cd + Bi, bal Al.
For screw machine products, fasteners, free cutting. *Obsolete*

WI-52

Waimet Alloys Co.
Cr 20-22, W 10-12, C 0.4-0.5, Fe 1-2.5, Ni 0-1, 1.5-2.5% Cb + Ta, bal Co.
At 70 F: 125,000 TS; 85,000 YS; 5 El; 5 RA; 380 Brin. At 1500 F: 75,000 TS; 52,000 YS; 8 El; 12.5 RA. For turbine nozzle vanes in jet engines; high heat resistance, investment cast. *Obsolete*

WI-52

Waimet Alloys Co.
Cr 20-22, W 10-12, C 0.4-0.5, Fe 1-2.5, Ni 0-1, 1.5-2.5 Cb + Ta, bal Co.
At 70°F: 125,000 TS; 85,000 YS; 5 El; 5 RA; 380 Brin. At 1500°F: 75,000 TS; 52,000 YS; 8 El; 12.5 RA. For turbine nozzle vanes in jet engines; high heat resistance, investment cast.

WI-52

Cannon-Muskegon Corp.
Cr 20-22, W 10-12, C 0.4-0.5, Fe 1-2.5, Ni 0-1, 1.5-2.5 Cb + Ta, bal Co.
At 70°F: 125,000 TS; 85,000 YS; 5 El; 5 RA; 380 Brin. At 1500°F: 75,000 TS; 52,000 YS; 8 El; 12.5 RA. For turbine nozzle vanes in jet engines; high heat resistance, investment cast.

WICROMAL

J. & A. Erbsloh Aluminium
Aluminum. Mn 0.5-1.5, Cr 0-0.3, bal Al.
Soft: 16,000 TS; 6000 YS; 40 El. Hard: 29,000 TS; 27,000 YS; 10 El. For cooking utensils, heat exchangers, tanks, furniture; good forming and welding properties. *Obsolete*

WIDALOX

Krupp Stahl AG
tool material. Al_2O_3 + additives.
For cutting tools. Ceramic, high hardness. *Obsolete*

WIDDER

German manufacture
Sn 10-25, bal Pb.
For soft solder.

WIDIA "M-68"

Fried. Krupp Huttenwerke AG
WC + Co.
For precision boring, used only for light work; hard grade cemented carbide. *Obsolete*

WIDIA "M-68"

Thomas Prosser & Sons
WC + Co.
For precision boring, used only for light work; hard grade cemented carbide. *Obsolete*

WIDIA F 1

Krupp Stahl AG
tool material. WC 69, Co 8, 25.0 TiC.
For cutting tools, dies; sintered. *Obsolete*

WIDIA F 2

Krupp Stahl AG
tool material. Co 5.5, 34.5 WC; 60 TiC.
For cutting tools, dies; sintered. *Obsolete*

WIDIA G1

Krupp Stahl AG
tool material. WC 94, Co 6.
For cutting tools for cast iron; cemented. *Obsolete*

WIDIA G2

Krupp Stahl AG
tool material. WC 80, Co 11.
For cutting tools for cast iron; cemented. *Obsolete*

WIDIA G3

Krupp Stahl AG
tool material. WC 85, Co 15.
For cutting tools for cast iron; cemented. *Obsolete*

WIDIA H

Krupp Huttenwerke
WC + Co.
For machining hard materials, chilled castings; glass; cemented carbide.

WIDIA H

Prosser & Sons, Thos.
WC + Co.
For machining hard materials, chilled castings; glass; cemented carbide.

WIDIA H1

Krupp Stahl AG
tool material. WC 94, Co 6.
For cutting tools for cast iron; cemented. *Obsolete*

WIDIA H2

Krupp Stahl AG
tool material. WC 91.5, Co 7, VC 0.5, 1.0 TaC.
For cutting tools, dies; sintered. *Obsolete*

WIDIA N

Krupp Huttenwerke
W 88, C 5, Co 6.
For machining gray cast iron, brass, non-ferrous; best combination of hardness and toughness.

WIDIA N

Prosser & Sons, Thos.
W 88, C 5, Co 6.
For machining gray cast iron, brass, non-ferrous; best combination of hardness and toughness.

WIDIA S

Krupp Huttenwerke
WC + Co.
For machining light work, used for light tools as broaches, reamers, forming tools; cemented carbide.

WIDIA S

Prosser & Sons, Thos.
WC + Co.
For machining light work, used for light tools as broaches, reamers, forming tools; cemented carbide.

WIDIA S1

Krupp Stahl AG
tool material. WC 78, Co 8, 16.0 TiC.
For cutting tools for steel; cemented. *Obsolete*

WIDIA S2

Krupp Stahl AG
tool material. WC 78, Co 8, 14.0 TiC.
For cutting tools for steel; cemented. *Obsolete*

WIDIA S3

Krupp Stahl AG
tool material. WC 88, Co 7, 5.0 TiC.
For cutting tools for steel; cemented. *Obsolete*

WIDIA TH10

Krupp Stahl AG
tool material. WC 91, Co 6, 3 (TiC + TaC).
For cutting tools, and dies, high temperature components. Cemented carbides. Hard and abrasion resistant. *Obsolete*

WIDIA TT25

Krupp Stahl AG
tool material. Co 9, 20 TiC + TaC, bal WC.
For high temperature applications, cutters, dies. Oxidation and corrosion resistant, wear and abrasion resistant. *Obsolete*

WIDIA TT40

Krupp Stahl AG
tool material. Co 13, 12 TiC + TaC, bal WC.
For cutting tools and drills, dies, wear parts. Cemented carbides, hard and abrasion resistant. *Obsolete*

WIDIA TT50

Krupp Stahl AG
tool material. Co 17, 15 TiC + TaC, bal WC.
For cutting tools and drills, dies, wear parts. Cemented carbides. Hard and abrasion resistant. *Obsolete*

WIDIA X
Krupp Huttenwerke
WC, Ti, C, Co.
For machining alloy steel of over 275 Brinell; cemented carbide.

WIDIA X
Prosser & Sons, Thos.
WC, Ti, C, Co.
For machining alloy steel of over 275 Brinell; cemented carbide.

WIDIA XX
Krupp Huttenwerke
WC, Ti, C, Co.
For machining mild and soft steels up to 275 Brinell; cemented carbide.

WIDIA XX
Prosser & Sons, Thos.
WC, Ti, C, Co.
For machining mild and soft steels up to 275 Brinell; cemented carbide.

WIDIA-1
Thomas Prosser & Sons
W 84, C 3, Co 13.
Sintered: 215,000 TS. For tools, dies, tool bits, lathe and planer tool cutters, shaper tools, reamers, twist drills; sintered alloy.

WIDIA-2
Krupp Stahl AG
tool material. W 87.4, C 5.68, Co 6.1.
Sintered: 250,000 TS. For tools, dies, tool bits, lathe and planer tool cutters, shaper tools, reamers, twist drills; sintered alloy. *Obsolete*

WIDIE
German manufacture
Sn 30, bal Pb.
For soft solder.

WIEGOLD
Now WIELAND-S10.

WIELAND ALBZ4
Wieland-Werke AG Metallwerke
Al 4, Cu 96.
Soft: 50,000 TS; 17,000 YS; 55 El; 90 Brin. Hard: 71,000 TS; 50,000 YS; 15 El; 150 Brin. For paper and chemical industries. Aluminum bronze. Wear and corrosion resistant. *Obsolete*

WIELAND ALBZ9
Wieland-Werke AG Metallwerke
Al 9, Cu 91.
Soft: 54,000 TS; 28,000 YS; 30 El; 100 Brin. Hard: 78,000 TS; 60,000 YS; 8 El; 160 Brin. For paper and chemical industries. Aluminum bronze. Wear and corrosion resistant. *Obsolete*

WIELAND EK2
Wieland-Werke AG Metallwerke
Cu 76, Al 2, P 0.03, bal Zn.
For condenser tubes; resists sea water corrosion. *Obsolete*

WIELAND FW
Wieland-Werke AG Metallwerke
Sn 3-5, Zn 3-5, P 0-0.1, bal Cu.
Soft: 50,000 TS; 50 El; 75 Brin. Hard: 70,000 TS; 11 El; 140 Brin. For paper and chemical industries. Corrosion resistant. *Obsolete*

WIELAND FW444
Wieland-Werke AG Metallwerke
Sn 3-5, Zn 3-5, Pb 3-5, P 0-0.1, bal Cu.
Soft: 47,000 TS; 40 El; 70 Brin. Hard: 65,400 TS; 8 El; 135 Brin. For paper and chemical industries. Corrosion resistant. *Obsolete*

WIELAND FW6
Now WIELAND-B06 and B16.

WIELAND FW8
Wieland-Werke AG Metallwerke
Sn 7.5-9, P 0-0.4, Zn 0-0.3, bal Cu.
Soft: 57,000 TS; 60 El; 85 Brin. Hard: 80,000 TS; 18 El; 155 Brin. For paper and chemical industries; corrosion resistant. *Obsolete*

WIELAND HLOS
Now WIELAND-K10.

WIELAND PK1
Now WIELAND-S28.

WIELAND SOMS68
Now WIELAND-S31.

WIELAND T
Now WIELAND-M20.

WIELAND T15
Now WIELAND-M05.

WIELAND T5
Now WIELAND-M15.

WIELAND TF
Wieland-Werke AG Metallwerke
Sn 5-7, Zn 5-7, P 0-0.1, bal Cu.
Soft: 55,500 TS; 52 El; 85 Brin. Hard: 78,200 TS; 13 El; 160 Brin. For marine hardware, fasteners, clips, terminals. Corrosion resistant. *Obsolete*

WIELAND WB206
Now WIELAND-S37.

WIELAND WB444
Wieland-Werke AG Metallwerke
Cu 88, Zn 4, Sn 4, Pb 4.
For bearings; free-cutting. *Obsolete*

WIELAND WB681
Now WIELAND-S31.

WIELAND WB800
Now WIELAND-B09.

WIELAND WBT II
Now WIELAND-W20.

WIELAND WS 100
Now WIELAND-N30.

WIELAND WS 120
Now WIELAND-N32.

WIELAND WS 125
Now WIELAND-N12.

WIELAND WS 131
Wieland-Werke AG Metallwerke
Cu 45-49, Ni 11-13, bal Zn.
Soft: 64,000 TS; 40 El; 100 Brin. Hard: 92,000 TS; 25 El; 180 Brin. For swivels in tool holders, optical goods, hardware, diaphragms, fasteners. Nickel bronze, corrosion resistant. *Obsolete*

WIELAND WS 175
Now WIELAND-N37.

WIELAND WS 180
Now WIELAND-N18.

WIELAND WS 80
Wieland-Werke AG Metallwerke
Cu 68, Ni 8, bal Zn.
Soft: 50,000 TS; 40 El; 70 Brin. Hard: 71,000 TS; 12 El; 130 Brin. For hardware, optical goods, novelties. Nickel silver. Corrosion resistant. *Obsolete*

WIELAND-A05
Wieland-Werke AG Metallwerke
99.5 Al min.
Extruded only: 75 MPa TS; 30 MPa YS; 25 El; 19 Brin. For construction of containers, vehicles and machinery. *Obsolete*

WIELAND-A08
Wieland-Werke AG Metallwerke
99.8 Al min.
Extruded only: 65 MPa TS; 30 MPa YS; 27 El; 18 Brin. For construction of containers, vehicles and machinery. *Obsolete*

WIELAND-A11
Wieland-Werke AG Metallwerke
Mg 1, bal Al.
Extruded only: 120 MPa TS; 50 MPa YS; 18 El; 40 Brin. For construction of containers, railroad cars, and chemical process equipment. *Obsolete*

WIELAND-A13
Wieland-Werke AG Metallwerke
Mg 3, bal Al.
Extruded only: 170 MPa TS; 90 MPa YS; 17 El; 48 Brin. For shipbuilding, vehicle, construction, chemical and food, optical, precision mechanics; resistant to corrosion. *Obsolete*

WIELAND-A15
Wieland-Werke AG Metallwerke
Mg 5, bal Al.
Extruded only: 265 MPa TS; 135 MPa YS; 15 El; 60 Brin. For chemical industry, optical, water and land craft, building trade; corrosion resistant. *Obsolete*

WIELAND-A22
Wieland-Werke AG Metallwerke
Mg 0.5, Si 0.5, bal Al.
Naturally age-hardened: 140 MPa TS; 70 MPa YS; 23 El; 45 Brin. Artificially age-hardened: 225 MPa TS; 170 MPa YS; 20 El; 75 Brin. For building industry; age-hardenable. *Obsolete*

WIELAND-A25
Wieland-Werke AG Metallwerke
Mg 1, Si 0.5, bal Al.
Naturally age-hardened: 160 MPa TS; 80 MPa YS; 23 El; 47 Brin. Artificially age-hardened: 255 MPa TS; 230 MPa YS; 13 El; 75 Brin. For building industry; age-hardenable. *Obsolete*

WIELAND-A28
Wieland-Werke AG Metallwerke
Mg 1, Si 0.8, bal Al.
Naturally age-hardened: 210 MPa TS; 110 MPa YS; 16 El; 70 Brin. Artificially age-hardened: 285 MPa TS; 250 MPa YS; 12 El; 92 Brin. For building industry, construction of machinery and chemical process equipment. *Obsolete*

WIELAND-A30
Wieland-Werke AG Metallwerke
Mg 0.6, Si 0.9, Mn 0.25, bal Al.
Naturally age-hardened: 210 MPa TS; 110 MPa YS; 16 El; 70 Brin. Artificially age-hardened: 285 MPa TS; 250 MPa YS; 12 El; 92 Brin. For building industry, chemical process equipment; age-hardenable material. *Obsolete*

WIELAND-A32
Wieland-Werke AG Metallwerke
Mg 1, Si 1, Mn 1, bal Al.
Naturally age-hardened: 250 MPa TS; 190 MPa YS; 15 El;
68 Brin. Artificially age-hardened: 340 MPa TS; 310 MPa YS;
12 El; 100 Brin. Age-hardenable material for construction.
Obsolete

WIELAND-A42
Wieland-Werke AG Metallwerke
Zn 4.5, Mg 1, bal Al.
Naturally age-hardened: 340 MPa TS; 230 MPa YS; 17 El;
90 Brin. Artificially age-hardened: 420 MPa TS; 360 MPa YS;
12 El; 110 Brin. For vehicles, machinery and chemical
process equipment; no loss of material strength after
welding. *Obsolete*

WIELAND-A61
Wieland-Werke AG Metallwerke
Mn 1.3, Fe 0.5, bal Al.
Extruded only: 110 MPa TS; 60 MPa YS; 30 El; 29 Brin.
Tubing for heat exchangers and irrigating plants;
containers, chemical process equipment; corrosion
resistant. *Obsolete*

WIELAND-A90
Wieland-Werke AG Metallwerke
Cu 4, Mg 0.5, Mn 0.5, bal Al.
Similar to AA 2014. *Obsolete*

WIELAND-B06
Wieland-Werke AG Metallwerke
Copper. Sn 6, bal Cu.
Soft: 360 MPa TS; 180 MPa YS; 55 El; 90 Brin. Hard: 490
MPa TS; 420 MPa YS; 24 El; 155 Brin. Tin bronze; for paper
and chemical industry; good spring properties, resistant to
corrosion; solderable. Similar to CDA and UNS C90200.
C51900

WIELAND-B08
Wieland-Werke AG Metallwerke
Sn 8, bal Cu.
Soft: 390 MPa TS; 180 MPa YS; 65 El; 90 Brin. Hard: 550
MPa TS; 510 MPa YS; 26 El; 165 Brin. For slide bearings;
resistant to wear. Similar to CDA and UNS C90200.
Obsolete

WIELAND-B09
Wieland-Werke AG Metallwerke
Copper. Sn 8, bal Cu.
Soft: 390 MPa TS; 180 MPa YS; 65 El; 90 Brin. Hard: 550
MPa TS; 510 MPa YS; 26 El; 165 Brin. For slide bearings;
resistant to wear and to corrosion. C52100

WIELAND-B10
Wieland-Werke AG Metallwerke
Copper. Sn 8, bal Cu.
Used only for Bourdon tubes. C52100

WIELAND-B14
Wieland-Werke AG Metallwerke
Copper. Sn 4, bal Cu.
For connectors, switches and similar electrical components;
good cold-forming, soldering properties, spring properties,
resistant to corrosion. C51100

WIELAND-B15
Wieland-Werke AG Metallwerke
Copper. Sn 5, bal Cu.
For connectors, switches and similar electrical components;
good cold-forming, solderable properties, spring properties,
resistant to corrosion. C51000

WIELAND-B16
Wieland-Werke AG Metallwerke
Copper. Sn 6, bal Cu.
Soft: 375 MPa TS; 180 MPa YS; 55 El; 85 Brin. Hard: 530
MPa TS; 490 MPa YS; 25 El; 160 Brin. For electrical
switches, connectors, relays, springs; resistant to stress
corrosion. C51800

WIELAND-B18
Wieland-Werke AG Metallwerke
Copper. Sn 8, bal Cu.
Soft: 390 MPa TS; 180 MPa YS; 65 El; 90 Brin. Hard: 550
MPa TS; 490 MPa YS; 25 El; 180 Brin. Relay springs, parts
for electrical, shipbuilding, paper industry; solderable,
resistant to stress corrosion. C52100

WIELAND-B66
Wieland-Werke AG Metallwerke
Copper. Sn 6, Zn 6, bal Cu.
Soft: 400 MPa TS; 195 MPa YS; 60 El; 100 Brin. Spring
hard: 670 MPa TS; 620 MPa YS; 14 El; 200 Brin. Leaf
springs of switches and electrical equipment, diaphragms.

WIELAND-B92
Wieland-Werke AG Metallwerke
Copper. Sn 5, Ni 2, Zn 10, bal Cu.
Nickel-tin bronze.

WIELAND-B98
Wieland-Werke AG Metallwerke
Sn 4, Zn 4, bal Cu.
Soft: 345 MPa TS; 140 MPa YS; 50 El; 80 Brin. Hard: 490
MPa TS; 430 MPa YS; 12 El; 145 Brin. For diaphragms, leaf
springs of switches and electrical equipment. *Obsolete*

WIELAND-G05
Wieland-Werke AG Metallwerke
Copper. Sn 5, Zn 5, Pb 4.5, bal Cu.
Cast: 250 MPa TS min; 110 MPa YS (approx); 13 El; 65 Brin.
For bearing bushes and bushings. C83600

WIELAND-G07
Wieland-Werke AG Metallwerke
Copper. Sn 6.5, Pb 6, Zn 3.8, bal Cu.
Cast: 320 MPa TS min; 170 MPa YS; 30 El; 85 Brin. For
bearing bushes and bushings. C93200

WIELAND-G10
Wieland-Werke AG Metallwerke
Copper. Sn 9.5, Ni 1, Pb 0.7, bal Cu.
Cast: 350 MPa TS min; 210 MPa YS; 18 El; 100 Brin. For
bearing bushes and bushings. C92700

WIELAND-G12
Wieland-Werke AG Metallwerke
Copper. Sn 11.5, Ni 1, Pb 0.7, bal Cu.
Cast: 350 MPa TS; 230 MPa YS; 15 El; 105 Brin. For bearing
bushes and bushings. C92500

WIELAND-G14
Wieland-Werke AG Metallwerke
Copper. Sn 13.5, Ni 1, Pb 0.7, bal Cu.
Cast: 330 MPa TS; 230 MPa YS; 8 El; 120 Brin. For bearing
bushes and bushings. C90900

WIELAND-G21
Wieland-Werke AG Metallwerke
Copper. Sn 10, Pb 10, bal Cu.
Cast: 290 MPa TS; 180 MPa YS; 16 El; 85 Brin. For bearing
bushes and bushings. C93700

WIELAND-G22
Wieland-Werke AG Metallwerke
Copper. Sn 8, Pb 15, bal Cu.
Cast: 260 MPa TS; 160 MPa YS; 15 El; 75 Brin. For bearing
bushes and bushings. C93800

WIELAND-G66
Wieland-Werke AG Metallwerke
Copper. Ni 6, Sn 6, bal Cu.
Cast. Similar to UNS C94700. For bearing bushes and
bushings; suitable for high duty applications.

WIELAND-G91
Wieland-Werke AG Metallwerke
Copper. Sn 12, Pb 1.5, bal Cu.
Cast. C92500

WIELAND-G92
Wieland-Werke AG Metallwerke
Copper. Pb 7, Sn 7, bal Cu.
Cast. Similar to CDA and UNS C93400. C93200

WIELAND-G93
Wieland-Werke AG Metallwerke
Copper. Sn 7.5, Pb 7.5, Zn 1.2, Ni 1.3, bal Cu.
Cast. Similar to CDA and UNS C93400.

WIELAND-G94
Wieland-Werke AG Metallwerke
Sn 9.5, Pb 5, Ni 4, bal Cu.
Obsolete

WIELAND-G96
Wieland-Werke AG Metallwerke
Copper. Sn 13.5, Ni 3, bal Cu.
Cast.

WIELAND-GB2
Wieland-Werke AG Metallwerke
Copper. Pb 5, Sn 5, Zn 5, bal Cu.
Cast. For bearing bushes and bushings. C83600

WIELAND-GB4
Wieland-Werke AG Metallwerke
Copper. Pb 6.5, Sn 6.5, Zn 4, bal Cu.
Cast. Similar to UNS C93200. For bearing bushes and
bushings.

WIELAND-GB5
Wieland-Werke AG Metallwerke
Copper. Sn 11.5, Pb 0.7, bal Cu.
Cast. Similar to UNS C92500. For bearing bushes and
bushings.

WIELAND-GB6
Wieland-Werke AG Metallwerke
Copper. Pb 22, Sn 2, bal Cu.
Cast. Similar to UNS C98200. For bearing bushes and
bushings.

WIELAND-HGA7
Wieland-Werke AG Metallwerke
Copper. Sn 10, Zn 2, bal Cu.
Cast. C90500

WIELAND-HGA8
Wieland-Werke AG Metallwerke
Copper. Sn 9.5, Pb 5, bal Cu.
Cast: 300 MPa TS; 180 MPa YS; 16 El; 95 Brin. For bearing
bushes and bushings.

WIELAND-HGA9
Wieland-Werke AG Metallwerke
Copper. Sn 4.5, Pb 20.5, bal Cu.
Cast: 200 MPa TS; 120 MPa YS; 14 El; 55 Brin. For bearing
bushes and bushings. C94100

WIELAND-HGB1
Wieland-Werke AG Metallwerke
Copper. Sn 11.5, Ni 2, bal Cu.
Cast. Also CDA and UNS C91700. C91600

WIELAND-HK65
Wieland-Werke AG Metallwerke
Copper. Fe 2.4, P 0.03, Zn 0.12, bal Cu.
Soft: 320-380 MPa TS; 200 MPa YS; 27 El; 85 Brin. Hard:
410-480 MPa TS; 370 MPa YS; 8 El; 130 Brin. Spring hard:
480-530 MPa TS; 430 MPa YS min; 4 El; 140 Vickers. Extra
spring: 530-580 MPa TS; 470 MPa YS min; 5 El; 150 Vickers.
Connector and switch parts, lead frames for integrated
circuits in semi-conductor engineering; high electrical and
thermal conductivity, good temperature resistance, very
good cold-forming and excellent soldering properties.
C19400

WIELAND-K10
Wieland-Werke AG Metallwerke
Copper. 99.99 Cu min.
Oxygen free. Electrical and electronic industry, vacuum technology (especially suitable); very high purity, high electrical conductivity. C10100

WIELAND-K12
Wieland-Werke AG Metallwerke
Copper. 99.95 Cu min.
Oxygen free. Electrical industry; high electrical conductivity. C10300

WIELAND-K20
Wieland-Werke AG Metallwerke
Copper. P 0.015-0.04, 99.9 Cu min.
Oxygen free. For heating and sanitary installations, refrigeration and air-conditioning equipment. C12200

WIELAND-K21
Wieland-Werke AG Metallwerke
Copper. P 0.015-0.04, 99.9 Cu min.
Oxygen free. For heat exchangers in refrigeration and air-conditioning plants; very good cold-forming properties. C12200

WIELAND-K30
Wieland-Werke AG Metallwerke
Copper. 99.9 Cu min.
Oxygen containing. Electrical industry; high electrical conductivity. C11000

WIELAND-K60
Wieland-Werke AG Metallwerke
Copper. Cr 0.3-1.2, Zr 0.04-0.1, bal Cu.
Age-hardened: 360 MPa TS; 290 MPa YS; 18 El; 120 Brin. Cold work-hardened: 520 MPa TS; 470 MPa YS; 8 El; 155 Brin. Electrical engineering, welding electrodes; high electrical conductivity, high strength, good tempering properties. C18200

WIELAND-K62
Wieland-Werke AG Metallwerke
Copper. Cu 98.5, bal Cr, Ni, Sn, Ti.
400-600 MPa TS; 300-550 MPa YS; 5-30 El; 120-220 Vickers. For flat springs and connector parts; high electrical and thermal conductivity, good bending properties.

WIELAND-K72
Wieland-Werke AG Metallwerke
Copper. Cu 98.5, bal Cr, Ni, Sn, Ti.
Temper 1: 450 MPa TS max; 300 MPa YS min; 18 El; 130 Vickers. Temper 2: 450-550 MPa TS; 350 MPa YS min; 12 El; 150 Vickers. Temper 3: 550 MPa TS min; 460 MPa YS min; 9 El; 170 Vickers. For leadframes.

WIELAND-K75
Wieland-Werke AG Metallwerke
Copper. Cu 99, bal Cr, Si, Ti.
330-580 MPa TS; 300-500 MPa YS; 3-25 El; 105-160 Vickers; in tempers soft to hard. Connector and switch parts, leadframes; high electrical and thermal conductivity, good temperature resistance, good stamping properties, little die wear, good bending properties. C18070

WIELAND-L05
Wieland-Werke AG Metallwerke
Ni 5, Fe 1, bal Cu.
Soft: 260 MPa TS; 100 MPa YS; 45 El; 65 Brin. Tubing for heat exchangers, chemical process equipment; resistant to corrosion. *Obsolete*

WIELAND-L10
Wieland-Werke AG Metallwerke
Copper. Ni 10, Fe 1, bal Cu.
Soft: 330 MPa TS; 120 MPa YS; 35 El; 70 Brin. Tubing for heat exchangers, chemical process equipment, shipbuilding industry; resistant to corrosion and salt water. C70600

WIELAND-L30
Wieland-Werke AG Metallwerke
Copper. Ni 30, Fe 0.5, Mn 1, bal Cu.
Soft: 420 MPa TS; 160 MPa YS; 35 El; 95 Brin. Tubing for heat exchangers and conduits in ships, sea water desalination plants; very resistant to corrosion and to salt water. C71500

WIELAND-L49
Wieland-Werke AG Metallwerke
Copper. Cu 89, Ni 9, Sn 2.
Soft: 390 MPa TS; 180 MPa YS; 40 El; 85 Brin. Hard: 530 MPa TS; 520 MPa YS; 5 El; 150 Brin. Contact elements, connectors, relays; good spring qualities, minimal stress relaxation, resistant to tarnishing, corrosion, stress corrosion; solderable. C72500

WIELAND-L50
Wieland-Werke AG Metallwerke
Copper. Cu 89, Ni 9, Sn 2.
Soft: 410 MPa TS max; 250 MPa YS max; 35 El; 75 Vickers. Halfhard: 410-500 MPa TS; 320 MPa YS min; 10 El; 110 Vickers. Hard: 500-570 MPa TS; 450 MPa YS min; 3 El max; 160 Vickers. Spring: 560 MPa TS min; 500 MPa YS min; 3 El max; 180 Vickers. Contact elements, connectors, relays; good spring qualities, minimal stress relaxation. C72500

WIELAND-M05
Wieland-Werke AG Metallwerke
Copper. Cu 95, Zn 5.
Soft: 250 MPa TS; 100 MPa YS; 45 El; 65 Brin. Hard: 330 MPa TS; 300 MPa YS; 10 El; 110 Brin. For jewelry, art objects, watch and clock making industry, electrical industry; resistant to stress corrosion, good for enameling. C21000

WIELAND-M10
Wieland-Werke AG Metallwerke
Copper. Cu 90, Zn 10.
Soft: 270 MPa TS; 100 MPa YS; 45 El; 65 Brin. Hard: 360 MPa TS; 330 MPa YS; 10 El; 115 Brin. For jewelry, art objects, watch and clock making industry, electrical industry; resistant to stress corrosion. C22000

WIELAND-M15
Wieland-Werke AG Metallwerke
Copper. Cu 85, Zn 15.
Soft: 290 MPa TS; 110 MPa YS; 45 El; 65 Brin. Hard: 420 MPa TS; 390 MPa YS; 12 El; 120 Brin. For jewelry, art objects, electrical industry. C23000

WIELAND-M20
Wieland-Werke AG Metallwerke
Copper. Cu 80, Zn 20.
Soft: 310 MPa TS; 120 MPa YS; 50 El; 70 Brin. Hard: 420 MPa TS; 350 MPa YS; 20 El; 125 Brin. For jewelry, art objects, Bourdon tubes, electric equipment for automotive corrugation tubes. C24000

WIELAND-M28
Wieland-Werke AG Metallwerke
Copper. Cu 72, Zn 28.
Soft: 330 MPa TS; 110 MPa YS; 55 El; 70 Brin. Hard: 450 MPa TS; 410 MPa YS; 20 El; 130 Brin. For metal goods and musical instruments; very good cold forming properties.

WIELAND-M30
Wieland-Werke AG Metallwerke
Copper. Cu 70, Zn 30.
Soft: 320 MPa TS; 130 MPa YS; 55 El; 70 Brin. Hard: 450 MPa TS; 410 MPa YS; 20 El; 130 Brin. For jewelry, dials, contact springs; very good cold forming properties, good etching. C26000

WIELAND-M32
Wieland-Werke AG Metallwerke
Copper. Cu 69, Zn 31.
Soft: 320 MPa TS; 130 MPa YS; 55 El; 70 Brin. Hard: 450 MPa TS; 410 MPa YS; 20 El; 130 Brin. For jewelry, dials, contact springs; very good cold-forming properties. C26000

WIELAND-M33
Wieland-Werke AG Metallwerke
Copper. Cu 67, Zn 33.
Soft: 330 MPa TS; 130 MPa YS; 55 El; 70 Brin. Hard: 460 MPa TS; 410 MPa YS; 20 El; 135 Brin. For musical instruments, watch and clock parts, electrical goods; good cold-forming properties and mechanical polishing. C26800

WIELAND-M36
Wieland-Werke AG Metallwerke
Copper. Cu 64, Zn 36.
Soft: 340 MPa TS; 130 MPa YS; 55 El; 70 Brin. Hard: 490 MPa TS; 470 MPa YS; 13 El; 145 Brin. Musical instruments, watch and clock parts; good for deep-drawing, forming and spinning. C27000

WIELAND-M37
Wieland-Werke AG Metallwerke
Copper. Cu 63, Zn 37.
Soft: 340 MPa TS; 130 MPa YS; 55 El; 70 Brin. Hard: 490 MPa TS; 470 MPa YS; 13 El; 145 Brin. Metal goods, electrical parts; good spring properties. C27200

WIELAND-M38
Wieland-Werke AG Metallwerke
Copper. Cu 63, Zn 37.
Soft: 340 MPa TS; 130 MPa YS; 55 El; 70 Brin. Hard: 490 MPa TS; 470 MPa YS; 13 El; 145 Brin. For musical instruments, contact springs; good cold-forming properties, good spring properties. C27400

WIELAND-N12
Wieland-Werke AG Metallwerke
Copper. Cu 64, Ni 12, bal Zn.
Soft: 380 MPa TS; 160 MPa YS; 45 El; 85 Brin. Hard: 510 MPa TS; 470 MPa YS; 13 El; 155 Brin. Strip, sheet. For tableware optical and precision instruments; can be deep-drawn; resistant to tarnishing and corrosion. C75700

WIELAND-N17
Wieland-Werke AG Metallwerke
Copper. Cu 55, Ni 18, bal Zn.
Soft: 440 MPa TS; 200 MPa YS; 45 El; 95 Brin. Hard: 580 MPa TS; 550 MPa YS; 15 El; 170 Brin. Strip. For relay springs; good punching and soldering properties, resistant to tarnishing. C77000

WIELAND-N18
Wieland-Werke AG Metallwerke
Copper. Cu 61, Ni 18, bal Zn.
Soft: 410 MPa TS; 170 MPa YS; 40 El; 90 Brin. Hard: 550 MPa TS; 520 MPa YS; 10 El; 170 Brin. Strip, sheet. For cutlery, relay springs, ornaments; soft temper grade can be deep-drawn; solderable, resistant to tarnishing. C76400

WIELAND-N19
Wieland-Werke AG Metallwerke
Copper. Cu 61, Ni 18, bal Zn.
380-700 MPa TS; 230-650 MPa YS; 40 El max; 85-230 Vickers; in tempers soft to extra spring. Strip, sheet. For cutlery, ornaments, connector parts, relay springs; resistance to tarnishing. C76400

WIELAND-N22
Wieland-Werke AG Metallwerke
Copper. Cu 64, Ni 12, bal Zn.
340-650 MPa TS; 250-600 MPa YS; 45 El max; 90-200 Brin; in tempers soft to spring. Rod, bar, wire, tube; resistant to tarnishing and corrosion. C75700

WIELAND-N28
Wieland-Werke AG Metallwerke
Copper. Cu 61, Ni 18, bal Zn.
430-700 MPa TS; 300-600 MPa YS; 40 El max; 120-220 Brin; in tempers soft to spring. Rod, bar, wire, tube; resistant to tarnishing and corrosion. C76400

WIELAND-N30
Wieland-Werke AG Metallwerke
Cu 46, Ni 10, Pb 1, bal Zn.
Hard: 660 MPa TS; 490 MPa YS; 16 El; 180 Brin. Machined parts for watches, clocks, optical and precision instruments; good machining and hot forming properties. *Obsolete*

WIELAND-N31
Wieland-Werke AG Metallwerke
Copper. Cu 47, Ni 8, Mn 5, Pb 3, bal Zn.
Halfhard: 510-630 MPa TS; 370 MPa YS min; 12 El; 160 Brin. Hard: 630 MPa TS min; 440 MPa YS min; 5 El; 180 Brin. Rod, wire, sections.

WIELAND-N32
Wieland-Werke AG Metallwerke
Copper. Cu 57, Ni 12, Pb 1, bal Zn.
Hard: 530 MPa TS; 480 MPa YS; 20 El; 150 Brin. Machined parts for watches, clocks, optical and precision instruments; good machining and hot forming properties. C79300

WIELAND-N37
Wieland-Werke AG Metallwerke
Copper. Cu 60, Ni 17.5, Pb 1, bal Zn.
Hard: 550 MPa TS; 510 MPa YS; 10 El; 160 Brin. For optical and precision instrument parts, drawing instruments; good machining; resistant to tarnishing. C76300

WIELAND-N90
Wieland-Werke AG Metallwerke
Copper. Cu 62, Ni 23, Sn 2, bal Zn.
Used in optical industry.

WIELAND-S10
Wieland-Werke AG Metallwerke
Copper. Sn 1, Ni 1, Zn 11, bal Cu.
Soft: 290 MPa TS; 120 MPa YS; 40 El; 70 Brin. Hard: 420 MPa TS; 340 MPa YS; 15 El; 125 Brin. For art objects; color similar to gold.

WIELAND-S12
Wieland-Werke AG Metallwerke
Copper. Cu 88, Sn 2, bal Zn.
300-710 MPa TS; 130-690 MPa YS; 50 El max; 70-225 Vickers; in tempers soft to extra spring. Strip for connector parts, switches, fuses. C42500

WIELAND-S23
Wieland-Werke AG Metallwerke
Copper. Cu 73, Al 3.5, Co 0.5, bal Zn.
Soft: 560-600 MPa TS; 430 MPa YS max; 40 El; 125 Brin. Hard: 790-840 MPa TS; 670 MPa YS min; 3 El min; 210 Brin. Cold work-hardened special brass alloy. Spring contacts for electrical engineering. C68800

WIELAND-S28
Wieland-Werke AG Metallwerke
Copper. Cu 71, Sn 1.2, As 0.03, bal Zn.
Soft: 340 MPa TS; 120 MPa YS; 55 El; 70 Brin. 1/2 hard: 390 MPa TS; 190 MPa YS; 45 El; 90 Brin. Tubes for condensers, heat exchangers in land plants (not sea water); corrosion resistant. C44300

WIELAND-S31
Wieland-Werke AG Metallwerke
Copper. Cu 68, Si 1, Pb 0.2, bal Zn.
Drawn: 370-570 MPa TS; 130-470 MPa YS; 15 El; 160 Brin. For slide bearings, valve guides, fasteners, bolts; corrosion resistant. C69800

WIELAND-S35
Wieland-Werke AG Metallwerke
Copper. Cu 59, Ni 2.5, Mn 2, Al 0.7, Pb 0.6, bal Zn.
Half hard: 490-540 MPa TS; 290 MPa YS min; 18 El; 130 Brin. Hard: 540 MPa TS min; 390 MPa YS min; 12 El; 160 Brin. Rod, bar, sections, wire.

WIELAND-S37
Wieland-Werke AG Metallwerke
Copper. Cu 60, Al 1, Mn 1, Pb 0.3, bal Zn.
Soft: 420 MPa TS; 170 MPa YS; 32 El; 110 Brin. Hard: 570 MPa TS; 370 MPa YS; 18 El; 145 Brin. For slide bearings and parts subject to heavy wear.

WIELAND-S39
Wieland-Werke AG Metallwerke
Copper. Cu 58, Mn 1.4, Fe 0.6, Pb 0.4, bal Zn.
Half hard: 490-600 MPa TS; 290 MPa YS min; 12 El; 130 Brin. Rod and sections. For forgings and use in architecture.

WIELAND-S40
Wieland-Werke AG Metallwerke
Copper. Cu 58, Mn 2, Al 1.5, Pb 0.7, bal Zn.
Extruded: 640 MPa TS; 340 MPa YS; 18 El; 155 Brin. Drawn: 690 MPa TS; 390 MPa YS; 15 El; 170 Brin. For slide bearings and parts subject to heavy wear.

WIELAND-S41
Wieland-Werke AG Metallwerke
Copper. Cu 58, Mn 2, Al 1.5, Pb 0.7, Fe 0.3, bal Zn.
540-650 MPa TS; 240-350 MPa YS; 10-20 El; 160 Brin. Rod for gear parts and valve guide bearings.

WIELAND-S91
Molycorp, Inc.
Cu 60, Pb 0.7, Sn 0.7, bal Zn.

WIELAND-SH76
Wieland-Werke AG Metallwerke
Copper. Cu 76, Al 2, As 0.03, bal Zn.
Soft: 370 MPa TS; 130 MPa YS; 55 El; 80 Brin. 1/2 hard: 410 MPa TS; 180 MPa YS; 45 El; 90 Brin. Tubes for evaporators, condensers, seawater desalination plants, ship equipment; salt water resistant. C68700

WIELAND-U20
Wieland-Werke AG Metallwerke
Al 10, Fe 2, Mn 2, bal Cu.
Extruded: 670 MPa TS; 260 MPa YS; 20 El; 150 Brin. Drawn: 730 MPa TS; 460 MPa YS; 16 El; 175 Brin. For bushes, worm wheels, spindle nuts, pump parts in ships and seawater desalination plants; high strength, good temperature stability, resistant to both corrosion and to wear. *Obsolete*

WIELAND-U90
Wieland-Werke AG Metallwerke
Al 10, Fe 3, Ni 2, bal Cu.
Similar to WIELAND-U20. *Obsolete*

WIELAND-Z10
Wieland-Werke AG Metallwerke
Copper. Cu 63, Pb 0.3, bal Zn.
Soft: 340 MPa TS; 130 MPa YS; 55 El; 75 Brin. Hard: 410 MPa TS; 310 MPa YS; 12 El; 140 Brin. For jewelry, connector parts; for spinning and embossing; good brazing and welding. C33500

WIELAND-Z11
Wieland-Werke AG Metallwerke
Copper. Cu 63, Pb 1, bal Zn.
Soft: 330 MPa TS; 120 MPa YS; 55 El; 70 Brin. Hard: 480 MPa TS; 450 MPa YS; 12 El; 140 Brin. Carbon brush holders, electric terminals; good machining and embossing. C34000

WIELAND-Z12
Wieland-Werke AG Metallwerke
Copper. Cu 63, Pb 2, bal Zn.
290-460 MPa TS; 200-360 MPa YS; 14-50 El; 70-140 Brin. Rod, bar, wire, sections in tempers soft to hard. C34200

WIELAND-Z14
Wieland-Werke AG Metallwerke
Copper. Cu 61.5, Pb 2, bal Zn.
290-460 MPa TS; 230-360 MPa YS; 14-50 El; 70-140 Brin. Rod, bar, wire, sections in tempers soft to hard. C35300

WIELAND-Z20
Wieland-Werke AG Metallwerke
Copper. Cu 61, Pb 0.2, bal Zn.
Soft: 360 MPa TS; 140 MPa YS; 45 El; 75 Brin. Hard: 490 MPa TS; 430 MPa YS; 18 El; 140 Brin. For fixtures, hardware and lock parts, watch and clock casings; suitable for embossing and hot stamping. C28000

WIELAND-Z21
Wieland-Werke AG Metallwerke
Copper. Cu 61, Pb 1.7, bal Zn.
Soft: 350 MPa TS; 120 MPa YS; 50 El; 75 Brin. Hard: 480 MPa TS; 450 MPa YS; 10 El; 140 Brin. For watch and clock parts, valve parts, taps and fittings, brass rules; good machining and riveting. C35000

WIELAND-Z22
Wieland-Werke AG Metallwerke
Copper. Cu 61, Pb 0.5, bal Zn.
340-550 MPa TS; 230-500 MPa YS; 45 El max; 75-170 Vickers. Strip in tempers soft to spring. C36500

WIELAND-Z23
Wieland-Werke AG Metallwerke
Copper. Cu 61, Pb 3, bal Zn.
Soft: 360 MPa TS; 200 MPa YS; 37 El; 85 Brin. Hard: 480 MPa TS; 440 MPa YS; 12 El; 140 Brin. Turned parts for watch, clock, electrical work, screws, bolts, nuts; excellent for machining on automatic lathes. C36000

WIELAND-Z30
Wieland-Werke AG Metallwerke
Copper. Cu 59, Pb 1.7, bal Zn.
Soft: 380 MPa TS; 160 MPa YS; 45 El; 84 Brin. Hard: 530 MPa TS; 490 MPa YS; 12 El; 150 Brin. For side plates and movements for watches and clocks, instruments; good hot forming and machining properties. C37700

WIELAND-Z31
Wieland-Werke AG Metallwerke
Copper. Cu 58, Pb 2, bal Zn.
Hard: 530 MPa TS; 460 MPa YS; 13 El; 145 Brin. For lock parts, electric terminals, hardware, movements and side plates for clocks, watches. C38000

WIELAND-Z32
Wieland-Werke AG Metallwerke
Copper. Cu 58.5, Pb 3, bal Zn.
360-520 MPa TS; 230-410 MPa YS; 11-35 El; 90-150 Brin. Rod, bar, wire, and sections in tempers soft to hard. C38500

WIELAND-Z33
Wieland-Werke AG Metallwerke
Copper. Cu 58, Pb 3, bal Zn.
Hard: 540 MPa TS; 480 MPa YS; 12 El; 150 Brin. Machined parts (gears) for clocks, watches and precision instruments; screws, nuts, bolts; excellent for machining on automatic lathes. C38500

WIELAND-Z34
Wieland-Werke AG Metallwerke
Copper. Cu 58, Pb 4, bal Zn.
For machining on automatic lathes. C38510

WIELAND-Z35
Wieland-Werke AG Metallwerke
Copper. Zn 35, Pb 2, As 0.04, bal Cu.
Soft: 340 MPa TS; 230 MPa YS; 45 El; 75 Brin. Half hard: 370 MPa TS; 250 MPa YS; 20 El; 120 Brin. Hard: 440 MPa TS; 340 MPa YS; 10 El; 140 Brin. Rod, bar, tubes, with improved resistance against dezincification. C35330

WIELAND-Z40
Wieland-Werke AG Metallwerke
Copper. Cu 55, Pb 2, bal Zn.
Extruded only: 580 MPa TS; 250 MPa YS; 16 El; 130 Brin.
For extruded sections and shapes; good machining; not
suitable for cold forming.

WIELAND-Z90
Wieland-Werke AG Metallwerke
Copper. Cu 67, Pb 0.6, bal Zn.
290-500 MPa TS; 160-350 MPa YS; 15-55 El; 60-150 Brin.
Tube in tempers soft to hard. C33000

WIELAND-ZB1
Wieland-Werke AG Metallwerke
Copper. Zn 35, Pb 2, As 0.04, bal Cu.
Half hard: 370 MPa TS; 250 MPa YS; 20 El; 120 Brin. Hard:
440 MPa TS; 340 MPa YS; 10 El; 140 Brin. Rod for free
machining; improved resistance against dezincification
certified. C35330

WIESILBER
Wieland-Werke AG Metallwerke
Cu 46-67, Ni 7-19, Pb 0-2, bal Zn.
For marine hardware; nickel silver. *Obsolete*

WIG METAL
Steinvertriebs Aktiengesellschaft
tungsten carbide.
For hard tools, dies, drill bits; high wear resistance.

WIGGIN ALLOY C263
Mond Nickel Co. Ltd.
Cr 20, Co 20, Mo 5.9, Ti 2.2, Al 0.5, bal Ni.
For gas turbine jet pipes, thrust reversers, noise
suppressors, flame tubes. Creep resistant for service to
850°C.

WIKMAN'S CRU-2
Wikmanshytte Bruks A.B.
C 1, Si 0.2, Mn 0.3, bal Fe.
Annealed: 100,000 TS; 53,000 YS; 21 El; 42 RA; 200 Brin.
For springs, cutters, reamers, drills, taps, and hobs. Type
W1; water hardened.

WIKMAN'S CRU-EX49
Wikmanshytte Bruks A.B.
C 0.8, Cr 4, Mo 5, W 6.2, V 2, bal Fe.
For lathe and planer tools, hobs, and reamers. High speed
steel.

WIKMAN'S VH10KA
Wikmanshytte Bruks A.B.
C 1-1.5, Si 0.2, Mn 0.3, Cr 1, bal Fe.
For cutting tools, files, and bearings. Water hardened; wear
resistant.

WIKMANS CRU-02
Wikmanshytte Bruks A.B.
C 1.5, Si 0.3, Mn 0.3, bal Fe.
For engravers' tools and blanking tools. Water hardened.

WIKMANS CRU-03
Wikmanshytte Bruks A.B.
C 1.4, Si 0.3, Mn 0.3, bal Fe.
For engravers' tools and blanking and forming dies. Type
W1; water hardened.

WIKMANS CRU-1
Wikmanshytte Bruks A.B.
C 1.2, Si 0.2, Mn 0.3, bal Fe.
For springs, cutters, drills, reamers, and broaches. Type W1;
water hardened.

WIKMANS CRU-3
Wikmanshytte Bruks A.B.
C 0.8, Si 0.2, Mn 0.3, bal Fe.
Heat treated: 130,000-188,000 TS; 87,000-145,000 YS;
12-21 El; 35-50 RA; 235-390 Brin. For drills, punches,
reamers, and hobs. Type W1; water hardened.

WIKMANS CRU-4
Wikmanshytte Bruks A.B.
C 0.7, Si 0.2, Mn 0.3, bal Fe.
Heat treated: 122,000-174,000 TS; 82,000-128,000 YS;
12-22 El; 37-52 RA; 240-355 Brin. For wheels, die blocks,
girders, rails, and springs. Type W1; water hardened.

WIKMANS CRU-EX12
Wikmanshytte Bruks A.B.
C 0.7, Cr 4.5, W 18.5, Co 0-0.6, bal Fe.
For drills, reamers, broaches, lathe and planer tools; high
speed steel.

WIKMANS CRU-EX20
Wikmanshytte Bruks A.B.
C 0.8, Mo 1.2, Cr 4.5, W 18.5, Co 2.5, V 1.6, bal Fe.
For milling cutters, planer and shaper tools, and broaches.
High speed steel.

WIKMANS CRU-EX22
Wikmanshytte Bruks A.B.
C 0.8, Mo 1.2, Cr 4.5, W 18.5, Co 5.5, V 1.6, bal Fe.
For milling cutters, hobs, and planer and shaper tools. High
speed steel.

WIKMANS CRU-EX26
Wikmanshytte Bruks A.B.
C 0.8, Mo 1, W 18.5, Cr 4.5, Co 10, V 1.6, bal Fe.
For milling cutters, planer, and shaper cutters. High speed
steel.

WIKMANS CRU-X42
Wikmanshytte Bruks A.B.
C 0.8, Cr 4, Mo 3, W 6.2, V 2, bal Fe.
For lathe and planer tools, hobs, drills, and taps. High
speed steel.

WIKMANS VH 341
Wikmanshytte Bruks A.B.
C 0.35, Mn 0.55, Cr 1.15, Ni 2.6, bal Fe.
For gears and shafts. Tough; oil hardened.

WIKMANS VH VULC
Wikmanshytte Bruks A.B.
Cr 12-14, 0.15 C min, bal Fe.
Annealed: 88,000 TS; 40,000 YS; 32 El; 68 RA; 170 Brin. For
cutlery, valve trim, and pump parts. Type 420; stainless;
hardenable.

WIKMANS VH-312
Wikmanshytte Bruks A.B.
C 0.25, Mn 0.65, Cr 1.05, Mo 0.2, bal Fe.
For gears, shafts, and machinery parts. Oil or water
hardened; tough.

WIKMANS VH-313
Wikmanshytte Bruks A.B.
C 0.3, Mn 0.55, Cr 1, Ni 3.2, Mo 0.25, bal Fe.
For gears, shafts, axles, mandrels, and machinery parts. Oil
hardened; shock resistant.

WIKMANS VH-35
Wikmanshytte Bruks A.B.
C 0.4, Mn 1.15, Cr 1.15, bal Fe.
For machinery parts and gears. Oil hardened.

WIKMANS VH-364
Wikmanshytte Bruks A.B.
C 0.12, Mn 0.55, Cr 0.75, Ni 3, bal Fe.
For gears, camshafts, and crankshafts. Case hardened.

WIKMANS VH-392
Wikmanshytte Bruks A.B.
C 0.2, Mn 0.55, Ni 1.8, Mo 0.25, bal Fe.
For gears, pinions, and machinery parts. Case hardened;
shock resistant.

WIKMANS VH-68
Wikmanshytte Bruks A.B.
C 0.3, Mn 0.55, Cr 1.25, Ni 4.25, bal Fe.
For gears and shafts. Tough; oil hardened.

WIKMANS VH-CN2
Wikmanshytte Bruks A.B.
C 0.25, Cr 0.7, Ni 3, bal Fe.
For gears, shafts, and machinery parts. Shock resistant.

WIKMANS VH-CN3
Wikmanshytte Bruks A.B.
C 0.18, Si 0.3, Mn 0.55, Cr 0.75, Ni 3, bal Fe.
For gears, camshafts, axles, and machinery parts. Case
hardened; shock resistant.

WIKMANS VH-EX1939
Wikmanshytte Bruks A.B.
C 0.8, Mo 0.2, Cr 4, W 12.5, Co 0.8, V 2.5, bal Fe.
For drills, planer and shaper tools, hobs, and reamers. High
speed steel.

WIKMANS VH-FS
Wikmanshytte Bruks A.B.
C 0.55, Si 1.75, Mn 0.75, bal Fe.
For springs; oil hardened.

WIKMANS VH10
Wikmanshytte Bruks A.B.
C 1, Si 0.2, Mn 0.3, bal Fe.
Annealed: 100,000 TS; 53,000 YS; 21 El; 42 RA; 200 Brin.
For springs, cutters, reamers, drills, taps, and hobs. Type
W1; water hardened.

WIKMANS VH10CV
Wikmanshytte Bruks A.B.
C 1, Si 0.2, Mn 0.3, Cr 0.5, V 0.1, bal Fe.
For bearings, heading, and blanking dies. Oil or water
hardened.

WIKMANS VH11
Wikmanshytte Bruks A.B.
C 1.1, Si 0.2, Mn 0.3, bal Fe.
Annealed: 100,000 TS; 53,000 YS; 21 El; 42 RA; 200 Brin.
For springs, cutters, drills, reamers, and hobs. Type W1;
water hardened.

WIKMANS VH12
Wikmanshytte Bruks A.B.
C 1.2, Si 0.2, Mn 0.3, bal Fe.
For springs, header dies, punches, cutters, and drills. Type
W1; water hardened.

WIKMANS VH13C
Wikmanshytte Bruks A.B.
C 1.35, Cr 0.3, Mn 0.25, Si 0.25, bal Fe.
For files, razors, and finishing tools. Water hardened.

WIKMANS VH214
Wikmanshytte Bruks A.B.
Cr 12-14, 0.15 C min, bal Fe.
Annealed: 88,000 TS; 40,000 YS; 32 El; 68 RA; 170 Brin. For
cutlery, valve trim, and pump parts. Type 420; stainless;
hardenable.

WIKMANS VH223
Wikmanshytte Bruks A.B.
C 1.8, Cr 12, Mo 0.6, Co 3.25, bal Fe.
For cutting and cold work tools. Air hardened; non-
deforming.

WIKMANS VH225
Wikmanshytte Bruks A.B.
C 0.45, Si 0.9, Mn 0.3, Cr 1.2, Mo 0.25, W 2.2, V 0.15, bal
Fe.
For pneumatic tools, chisels, shears, and punches. Hot
work steel; oil hardened.

WIKMANS VH233
Wikmanshytte Bruks A.B.
C 0.8, Si 0.3, V 0.1, bal Fe.
Heat treated: 190,000 TS; 145,000 YS; 12 El; 32 RA; 400
Brin. For blanking and forming dies, and punches. Water
hardened.

WIKMANS VH236
Wikmanshytte Bruks A.B.
C 0.55, Cr 1.5, Ni 3, Mn 0.4, bal Fe.
For bakelite dies and cold work tools. Oil hardened; shock resistant.

WIKMANS VH238
Wikmanshytte Bruks A.B.
C 0.9, Mn 1.2, Cr 0.5, W 0.5, V 0.1, bal Fe.
For drawing and forming dies and punches. Oil hardened; non-deforming.

WIKMANS VH239
Wikmanshytte Bruks A.B.
Cr 12-14, 0.15 C min, bal Fe.
Annealed: 88,000 TS; 40,000 YS; 32 El; 68 RA; 170 Brin. For cutlery, valve trim, and turbine blades; Type 420; stainless; hardenable.

WIKMANS VH240
Wikmanshytte Bruks A.B.
C 0.35, Cr 23-27, bal Fe.
Annealed: 90,000 TS; 60,000 YS; 20 El; 45 RA; 180 Brin. For furnace parts and heat treating boxes. Type 446; heat resistant.

WIKMANS VH244
Wikmanshytte Bruks A.B.
C 1.35, Cr 0.3, W 6, Mn 0.3, bal Fe.
For drawing and forming dies and finishing tools. Oil hardened; high hardness.

WIKMANS VH266
Wikmanshytte Bruks A.B.
C 0.55, Cr 1, Ni 3, Mo 0.3, Mn 0.4, Si 0.3, bal Fe.
For drop forging dies and cold work tools. Oil or air hardened; shock resistant.

WIKMANS VH267
Wikmanshytte Bruks A.B.
C 0.4, Si 1.3, Cr 5, Mo 0.75, W 5, bal Fe.
For Al die casting dies and punches. Oil or air hardened; hot work steel.

WIKMANS VH273
Wikmanshytte Bruks A.B.
C 0-0.12, Cr 11.5-13, Al 0.1-0.3, bal Fe.
Annealed: 71,000 TS; 42,600 YS; 22 El; 70 RA; 150 Brin. Heat treated: 175,000 TS; 145,000 YS; 21 El; 64 RA; 352 Brin. For oil refinery and chemical plant equipment. Type 405; corrosion resistant.

WIKMANS VH274
Wikmanshytte Bruks A.B.
C 0-0.1, Mn 0-2, Cr 17-19, Ni 8-10, bal Fe.
Annealed: 85,000 TS; 35,000 YS; 60 El; 70 RA; 150 Brin. Cold drawn: 125,000 TS; 95,000 YS; 22 El; 55 RA; 277 Brin. For oil refinery and chemical plant equipment. Type 304; stainless; austenitic.

WIKMANS VH276
Wikmanshytte Bruks A.B.
C 0-0.1, Cr 18, Ni 10, Mo 1.3, bal Fe.
Annealed: 85,000-95,000 TS; 35,000-45,000 YS; 50-60 El; 60-75 RA; 150-190 Brin. For acid resistant chemical plant equipment. Type 316; stainless; austenitic.

WIKMANS VH277
Wikmanshytte Bruks A.B.
C 0.65, Ni 1.4, Mo 0.3, Cr 0.7, V 0.1, bal Fe.
For forging and cold working dies. Oil hardened; shock resistant.

WIKMANS VH282
Wikmanshytte Bruks A.B.
C 0.4, Cr 15, Ni 15, Si 1.5, W 2.1, bal Fe.
For valves for large diesel engines; corrosion and heat resistant.

WIKMANS VH300
Wikmanshytte Bruks A.B.
C 0.3, Cr 3, Ni 1.7, W 9, V 0.3, Mn 0.3, bal Fe.
For hot work tools, Cu die casting dies. Hot work steel; oil hardened.

WIKMANS VH320
Wikmanshytte Bruks A.B.
C 1.5, Cr 12, Mo 0.8, V 0.2, bal Fe.
For drawing and blanking dies and cold work tools. Air hardened; non-deforming.

WIKMANS VH322
Wikmanshytte Bruks A.B.
C 0.4, Cr 12, Mo 1, Si 2.5, bal Fe.
For valves for large diesel engines; corrosion and heat resistant.

WIKMANS VH33
Wikmanshytte Bruks A.B.
C 0.95, Si 1.5, Mn 0.75, Cr 1, bal Fe.
For drawing and forming dies, and punches. Oil hardened; shock resistant.

WIKMANS VH357
Wikmanshytte Bruks A.B.
C 1.2, Si 0.2, Mn 0.3, W 0.55, V 0.1, bal Fe.
For drills, taps, cutters, and heading dies. Oil or water hardened.

WIKMANS VH376
Wikmanshytte Bruks A.B.
C 0-0.08, Cr 18, Ni 9, Mo 1.3, bal Fe.
Annealed: 85,000 TS; 35,000 YS; 50 El; 65 RA; 160 Brin. For chemical plant and oil refinery equipment. Type 316; stainless; austenitic.

WIKMANS VH388
Wikmanshytte Bruks A.B.
C 1, Si 0.2, Mn 0.6, Cr 5.2, Mo 1.1, V 0.2, bal Fe.
For forming and drawing dies, punches, and shears. Oil or air hardened; non-deforming.

WIKMANS VH407
Wikmanshytte Bruks A.B.
C 0-0.12, Cr 20, Ni 19, bal Fe.
For heat treat boxes, furnace parts and equipment. Stainless; austenitic.

WIKMANS VH410
Wikmanshytte Bruks A.B.
C 0.35, Si 1, Cr 5, Mo 1.5, V 0.4, bal Fe.
For extrusion dies for Al. Oil or air hardened; non-deforming.

WIKMANS VH417
Wikmanshytte Bruks A.B.
C 0.3-0.36, Cr 13, bal Fe.
Annealed: 100,000 TS; 55,000 YS; 25 El; 55 RA; 195 Brin. For surgical instruments, valves, and cutlery. Type 420; corrosion resistant.

WIKMANS VH54
Wikmanshytte Bruks A.B.
C 0.3, Cr 1, Ni 3.5, W 5.5, Mn 0.25, Si 0.2, bal Fe.
For mandrels and extrusion press dies and liners. Hot work steel; oil hardened.

WIKMANS VH60
Wikmanshytte Bruks A.B.
C 1.25, Cr 2, V 0.15, Si 0.7, Mn 0.3, bal Fe.
For finishing tools, and cutters. Oil or water hardened.

WIKMANS VH602
Wikmanshytte Bruks A.B.
C 0.09-0.13, Cr 12, Ni 0.6, Mo 0.5, bal Fe.
For turbine blades, and valves. Corrosion resistant.

WIKMANS VH604
Wikmanshytte Bruks A.B.
C 0-0.12, Cr 15.5, bal Fe.
Annealed: 80,000 TS; 50,000 YS; 25 El; 50 RA; 150 Brin. For hardware, kitchen sinks, and fasteners. Type 430; corrosion resistant.

WIKMANS VH611
Wikmanshytte Bruks A.B.
C 0.53-0.58, Cr 14, bal Fe.
For butchers' knives, and surgical instruments. Corrosion resistant; hardenable.

WIKMANS VH620
Wikmanshytte Bruks A.B.
C 0-0.1, Cr 18, bal Fe.
Annealed: 80,000 TS; 50,000 YS; 25 El; 50 RA; 150 Brin. For oil refinery equipment, soot blowers and bolts. Corrosion and heat resistant.

WIKMANS VH63
Wikmanshytte Bruks A.B.
C 2.2, W 1.2, Cr 13, Mn 0.75, Si 0.3, bal Fe.
For forming and drawing dies and cold work tools. Oil or air hardened; non-deforming.

WIKMANS VH650
Wikmanshytte Bruks A.B.
C 0-0.07, Cr 18, Ni 10, Mo 2.6, bal Fe.
Annealed: 85,000 TS; 35,000 YS; 50 El; 65 RA; 160 Brin. For acid resistant chemical plant equipment. Type 316; stainless; austenitic.

WIKMANS VH7
Wikmanshytte Bruks A.B.
C 0.7, Si 0.2, Mn 0.3, bal Fe.
Heat treated: 122,000-174,000 TS; 82,000-128,000 YS; 12-22 El; 37-52 RA; 240-355 Brin. For wheels, die blocks, girders, rails, and springs. Type W1; water hardened.

WIKMANS VH78
Wikmanshytte Bruks A.B.
C 1.1, Si 0.2, Mn 0.3, Cr 0.3, W 1, V 0.1, bal Fe.
For twist drills, punches, reamers, and cutting tools. Oil or water hardened.

WIKMANS VH8
Wikmanshytte Bruks A.B.
C 0.8, Si 0.2, Mn 0.3, bal Fe.
Heat treated: 130,000-188,000 TS; 87,000-145,000 YS; 12-21 El; 35-50 RA; 235-390 Brin. For drills, punches, reamers, and hobs. Type W1; water hardened.

WIKMANS VH94
Wikmanshytte Bruks A.B.
Cr 12-14, 0.15 C min, bal Fe.
Annealed: 88,000 TS; 40,000 YS; 32 El; 68 RA; 170 Brin. For cutlery, valve trim, and turbine blades. Type 420; stainless; hardenable.

WIKMANS WH602
Wikmanshytte Bruks A.B.
C 0.08, Cr 14.5, Mo 1, bal Fe.
Annealed: 80,000 TS; 35,000 YS; 30 El; 92 Rock B. For springs, table flatware, oil refinery and chemical plant equipment. Corrosion resistant.

WILCO AMPLEX
Engelhard Corp.
For thermal cutouts in circuit breakers; bi-metal, temperature range 0-350°F. *Obsolete*

WILCO C-5
Engelhard Corp.
Ag alloy.
For electrical contacts; sintered. *Obsolete*

WILCO DA
Engelhard Corp.
Pd, bal Pt.
For electrical relays, voltage regulators; low contact resistance. *Obsolete*

WILCO H T CONSTANT
Engelhard Corp.
For thermostats, temperature controls, bi-metal, temperature range 250-600°F. *Obsolete*

WILCO H.T. SPECIAL
Engelhard Minerals & Chemicals Corp.
For thermostats, temperature controls. *Obsolete*

WILCO HIGH HEAT
Engelhard Corp.
For thermostats, temperature controls; bi-metal, temperature range 300-650°F. *Obsolete*

WILCO HIGHFLEX
Engelhard Corp.
Ni 18, Cr 11, C 0.25, 36 Ni + Co, bal Fe.
For temperature indicating instruments, contact making devices; bi-metal, temperature range 0-350°F. *Obsolete*

WILCO HIGHFLEX 45
Engelhard Corp.
Ni 19.4, C 0.55, Mn 0.9, Cr 2.2, Co 0.1, C 0.12, 36 Ni + Co, bal Fe.
For temperature indicating instruments, contact making devices; bi-metal, temperature range 0-350°F. *Obsolete*

WILCO HIGHHEAT 47
Engelhard Corp.
For thermostats, temperature controls; bi-metal, temperature range 300-650°F. *Obsolete*

WILCO NO. 10
Engelhard Corp.
Al alloy.
For electric contacts; hard, wear resistant. *Obsolete*

WILCO NO. 14
Engelhard Corp.
Pt alloy.
For electric contacts; arc-resisting. *Obsolete*

WILCO NO. 19
Engelhard Corp.
Pt, bal Ag.
For electric contacts; low contact resistance. *Obsolete*

WILCO NO. 20
Engelhard Corp.
Ag 80, Cd 20.
For relay contacts; high conductivity. *Obsolete*

WILCO NO. 28
Engelhard Corp.
Zn, bal Ag.
For electric contacts; resists film formation. *Obsolete*

WILCO NO. 29
Engelhard Corp.
Pd, bal Ag.
For electric contacts; for relays. *Obsolete*

WILCO NO. 3
Engelhard Corp.
Au, bal Pt.
For electric contacts; low corrosion resistance. *Obsolete*

WILCO NO. 31
Engelhard Corp.
Ag alloy.
For electric contacts; high electrical conductivity. *Obsolete*

WILCO NO. 33
Engelhard Corp.
Au, bal Ag.
For electric contacts; corrosion resistant. *Obsolete*

WILCO NO. 34
Engelhard Corp.
Pd, bal Pt.
For electric contacts; for low current applications. *Obsolete*

WILCO NO. 35
Engelhard Corp.
Pd, bal Ag.
For electric contacts. *Obsolete*

WILCO NO. 38
Engelhard Corp.
Ag, Cu.
For electric contacts; high conductivity. *Obsolete*

WILCO NO. 39
Engelhard Corp.
Ag alloy.
For electric contacts; for voltage regulators. *Obsolete*

WILCO NO. 4
Engelhard Corp.
Au, bal Pt.
For electric contacts; free from atmospheric films. *Obsolete*

WILCO NO. 40
Engelhard Corp.
Ag alloy.
For electric contacts. *Obsolete*

WILCO NO. 43
Engelhard Corp.
Ru, bal Pt.
For electric contacts. *Obsolete*

WILCO NO. 50
Engelhard Corp.
Rh 33, Os 67.
For voltage regulators; corrosion resistant. *Obsolete*

WILCO NO. 6
Engelhard Corp.

For electrical contacts; resists oxidation and corrosion. *Obsolete*

WILCO NO. 7
Engelhard Corp.
Au, bal Pt.
For electric contacts. *Obsolete*

WILCO P-10
Engelhard Corp.
Ir, bal Pt.
For electrical contacts, thermostats, relays; hard, wear resistant. *Obsolete*

WILCO P-15
Engelhard Corp.
Ir, bal Pt.
For electrical contacts, thermostats, relays; hard, wear resistant. *Obsolete*

WILCO P-150
Engelhard Corp.
Ir, Ru, Os, Rh, Pd, bal Pt.
For electrical contacts, voltage regulators, speed controls; oxidation and wear resistant. *Obsolete*

WILCO P-20
Engelhard Corp.
Ir, bal Pt.
For electrical contacts, thermostats, relays; hard, wear resistant. *Obsolete*

WILCO P-21
Engelhard Corp.
Ir, bal Pt.
For electrical contacts, thermostats, relays; hard, wear resistant. *Obsolete*

WILCO P-34
Engelhard Corp.
Ag, bal Pd.
For electrical contacts, thermostats, relays; low surface contact resistance. *Obsolete*

WILCO P-35
Engelhard Corp.
Pd, bal Ag.
For electrical contacts, relays, radio vibrators; resists sulfide formation. *Obsolete*

WILCO P-43
Engelhard Corp.
Ru, bal Pt.
For electrical contacts, thermostats, relays; hard, wear resistant. *Obsolete*

WILCO P-5
Engelhard Corp.
Ir, bal Pt.
For electrical contacts, relays, thermostats; hard, wear resistant. *Obsolete*

WILCO P-50
Engelhard Corp.
Ir, Ru, Os, Rh, Pd, bal Pt.
For voltage regulators, electrical contacts, speed governors; high corrosion and wear resistance. *Obsolete*

WILCO P-53
Engelhard Corp.
Ru, bal Pt.
For electrical contacts, thermostats, relays; hard, wear resistant. *Obsolete*

WILCO P-54
Engelhard Corp.
Ag, bal Pd.
For electrical contacts, thermostats, relays; high hardness and wear resistance. *Obsolete*

WILCO P-58
Engelhard Corp.
Ru, bal Pd.
For electrical contacts, thermostats, relays; hard, and wear resistant. *Obsolete*

WILCO P-6
Engelhard Corp.
Ru, bal Pt.
For electrical contacts, thermostats, relays; hard, wear resistant. *Obsolete*

WILCO P-60
Engelhard Corp.
Pd, bal Pt.
For electrical contacts, relays, magnetos; wear and impact resistant. *Obsolete*

WILCO P-61
Engelhard Corp.
Ag, Ni, bal Pd.
For electrical contacts, thermostats, relays; low surface contact resistance. *Obsolete*

WILCO P-62
Engelhard Corp.
Ag, Ni, bal Pd.
For electrical contacts, voltage regulators; low surface contact resistance. *Obsolete*

WILCO P-63
Engelhard Corp.
Pd, Ni, bal Ag.
For electrical contacts, relays, voltage regulators; low surface contact resistance. *Obsolete*

WILCO P-68
Engelhard Corp.
Ag, Ni, bal Pd.
For electrical contacts, thermostats, relays; low surface contact resistance. *Obsolete*

WILCO SCOFLEX
Engelhard Corp.

For temperature indicating instruments, contact making devices; bi-metal, temperature range 0-350°F. *Obsolete*

WILCO SILVER SOLDER NO. 18
Engelhard Corp.
Ag 49.5, Cu 34.5, Zn 16.
For silver solder; corrosion resistant. *Obsolete*

WILCO SILVER SOLDER NO. 2
Engelhard Corp.
Now ENGELHARD ALLOY NO. 6833.

WILCO SILVER SOLDER NO. 25
Engelhard Corp.
Now ENGELHARD ALLOY NO. 257.

WILCO SILVER SOLDER NO. 27
Engelhard Corp.
Now ENGELHARD ALLOY NO. 273.

WILCO STANDARD
Engelhard Corp.
For temperature indicating instruments, contact making devices; bi-metal; temperature range 0-300°F. *Obsolete*

WILCO THERMO-METAL CIRFLEX
Engelhard Corp.
Thermostatic bimetal; operating temperature 50-300°F. *Obsolete*

WILCO THERMO-METAL MIDFLEX 46
Engelhard Corp.
Thermostatic bimetal; operating temperature 150-450°F. *Obsolete*

WILCO THERMOMETAL RUFLEX
Engelhard Corp.
Thermometal for steam traps, gas pilot controls; bimetal. *Obsolete*

WILCO W METAL
Engelhard Corp.
For thermostatic bimetals. *Obsolete*

WILCOLOY
Engelhard Corp.
WC.
For telegraph relays; tungsten carbide. *Obsolete*

WILCOLOY T-1
Engelhard Corp.
Co, Be, bal Cu.
Heat treated: 120,000 psi TS; 90,000 psi YS; 222 Brin. For contact backing, spring arms, reeds, switches; age-hardenable. *Obsolete*

WILCOLOY-B
Engelhard Corp.
Be, bal Cu.
Heat treated: 190,000 psi TS; 150,000 psi YS; 400 Brin. For contact backing, spring arms, circuit vibrators; age-hardened, maximum service temperature to 700°F. *Obsolete*

WILCOLOY-C
Engelhard Corp.
Cr, Si, bal Cu.
Heat treated: 76,000 psi TS; 61,000 psi YS; 144 Brin. For contact backing, circuit breakers; heat treatable, high electrical conductivity. *Obsolete*

WILCOLOY-E
Engelhard Corp.
Cd, bal Cu.
Cold drawn: 64,000 psi TS; 55,000 psi YS; 137 Brin. For contact backing springs, reeds, relays; high electrical conductivity and strength. *Obsolete*

WILLEYS 10A
Now EXCELL-O 10A.

WILLEYS 509
Now EXCELL-O 509.

WILLEYS 606
Now EXCELL-O 606.

WILLEYS 6A
Now EXCELL-O 6A.

WILLEYS 6AX
Now EXCELL-O 6AX.

WILLEYS 710
Ex-Cell-O
Co 11, WC, 10% TiC.
For cutters for general machining; sintered. *Obsolete*

WILLEYS 8A
Now EXCELL-O 8A.

WILLEYS 945
Ex-Cell-O
Co 10, 4% TiC, bal WC.
For tools, dies, cutters; sintered; Tr.S. 240000. *Obsolete*

WILLEYS E-10
Ex-Cell-O
Co 10, bal WC.
For tools, dies, cutters; sintered; Tr.S. 280,000. *Obsolete*

WILLEYS E-12
Ex-Cell-O
WC.
For cutters, rest blades, centers; sintered, tough. *Obsolete*

WILLEYS E-13
Ex-Cell-O
Co 13, bal WC.
For tools, dies, cutters; sintered; Tr. S. 325000. *Obsolete*

WILLEYS E-18
Ex-Cell-O
Co 18, bal WC.
For tools, dies, cutters; sintered; Tr.S. 400,000. *Obsolete*

WILLEYS E-25
Now EXCELL-O E25.

WILLEYS E-3
Now EXCELL-O E3.

WILLEYS E-4.5
Ex-Cell-O
Co 4.5, bal WC.
For tools, dies, cutters; sintered; Tr.S. 190,000. *Obsolete*

WILLEYS E-5
Now EXCELL-O E5.

WILLEYS E-6
Now EXCELL-O E6.

WILLEYS E-8
Now EXCELL-O E8.

WILLEYS E13
Ex-Cell-O
Co 13, 87 WC.
Sintered: Rockwell A89. For wear surfaces; heavy shock, carbides. *Obsolete*

WILLEYS E16
Now EXCELL-O E16.

WILLEYS E9
Now EXCELL-O E9.

WILLEYS W12C
Ex-Cell-O
Co 12, 88 WC.
Sintered: Rockwell A88. For mining and percussion tools; carbides. *Obsolete*

WILLEYS W9C
Ex-Cell-O
Co 9, 91 WC.
Sintered: Rockwell A89. For mining and percussion tools; carbides. *Obsolete*

WILLEYS X3
Ex-Cell-O
Co 6.5, 3 TaC, 90.5 WC.
Sintered: Rockwell A92. For general purpose cutting tools; carbides. *Obsolete*

WILLEYS X620
Ex-Cell-O
Co 6, 20 TaC, 74 WC.
Sintered: Rockwell A92. For special applications; carbides. *Obsolete*

WILLIAMS S1000
Williams Precious Metals
Low melting. Cu 60, Au 40.
Solder alloy. 1000°C FP; 980°C MP.

WILLIAMS S1005
Williams Precious Metals
Low melting. Cu 62.5, Au 37.5.
Solder alloy. 1005°C FP; 985°C MP.

WILLIAMS S1010
Williams Precious Metals
Low melting. Cu 65, Au 35.
Solder alloy. 1010°C FP; 990°C MP.

WILLIAMS S1020
Williams Precious Metals
Low melting. Au 50, Ag 50.
Solder alloy. 1020°C FP; 1000°C MP.

WILLIAMS S1030
Williams Precious Metals
Low melting. Au 99.4, Sb 0.6.
Solder alloy. 1030°C FP; 360°C MP.

WILLIAMS S1063
Williams Precious Metals
Low melting. Au 100.
Solder alloy.

WILLIAMS S1063A
Williams Precious Metals
Low melting. Au 99.9-99.98, P 0.02-0.1.
Solder alloy.

WILLIAMS S1063B
Williams Precious Metals
Low melting. Au 99.9999, Be 0.0001.
Solder alloy.

WILLIAMS S1063C
Williams Precious Metals
Low melting. Au 99.9, B 0.1.
Solder alloy.

WILLIAMS S118
Williams Precious Metals
Low melting. In 52, Sn 48.
Solder alloy. 118°C FP; 118°C MP.

WILLIAMS S1238
Williams Precious Metals
Low melting. Pd 60, Ni 40.
Solder alloy. 1238°C FP; 1238°C MP.

WILLIAMS S125
Williams Precious Metals
Low melting. In 50, Sn 50.
Solder alloy. 125°C FP; 118°C MP.

WILLIAMS S131
Williams Precious Metals
Low melting. Sn 52, In 48.
Solder alloy. 131°C FP; 118°C MP.

WILLIAMS S138
Williams Precious Metals
Low melting. Bi 58, Sn 42.
Solder alloy. 138°C FP; 138°C MP.

WILLIAMS S143
Williams Precious Metals
Low melting. In 97, Ag 3.
Solder alloy. 143°C FP; 143°C MP.

WILLIAMS S145
Williams Precious Metals
Low melting. Sn 58, In 42.
Solder alloy. 145°C FP; 118°C MP.

WILLIAMS S149
Williams Precious Metals
Low melting. In 80, Pb 15, Ag 5.
Solder alloy.

WILLIAMS S154
Williams Precious Metals
Low melting. In 99.5, Ga 0.5.
Solder alloy.

WILLIAMS S1552
Williams Precious Metals
Low melting. Pd 100.
Solder alloy.

WILLIAMS S157
Williams Precious Metals
Low melting. In 100.
Solder alloy.

WILLIAMS S174
Williams Precious Metals
Low melting. In 70, Pb 30.
Solder alloy. 174°C FP; 160°C MP.

WILLIAMS S1769
Williams Precious Metals
Low melting. Pt 100.
Solder alloy.

WILLIAMS S179
Williams Precious Metals
Low melting. Sn 62.5, Pb 36.1, Ag 1.4.
Solder alloy. 179°C FP; 179°C MP.

WILLIAMS S181
Williams Precious Metals
Low melting. Pb 37.5, Sn 37.5, In 25.
Solder alloy. 181°C FP; 134°C MP.

WILLIAMS S183
Williams Precious Metals
Low melting. Sn 63, Pb 37.
Solder alloy. 183°C FP; 183°C MP.

WILLIAMS S183A
Williams Precious Metals
Low melting. Sn 62, Pb 38.
Solder alloy. 183°C FP; 183°C MP.

WILLIAMS S184
Williams Precious Metals
Low melting. Sn 65, Pb 35.
Solder alloy. 184°C FP; 183°C MP.

WILLIAMS S185
Williams Precious Metals
Low melting. In 60, Pb 40.
Solder alloy. 184°C FP; 183°C MP.

WILLIAMS S186
Williams Precious Metals
Low melting. Sn 70, Pb 30.
Solder alloy. 186°C FP; 183°C MP.

WILLIAMS S188
Williams Precious Metals
Low melting. Sn 60, Pb 40.
Solder alloy. 188°C FP; 183°C MP.

WILLIAMS S192
Williams Precious Metals
Low melting. Sn 75, Pb 27.
Solder alloy. 192°C FP; 183°C MP.

WILLIAMS S199
Williams Precious Metals
Low melting. Sn 80, Pb 20.
Solder alloy. 199°C FP; 183°C MP.

WILLIAMS S199A
Williams Precious Metals
Low melting. Sn 91, Zn 9.
Solder alloy. 199°C FP; 199°C MP.

WILLIAMS S200
Williams Precious Metals
Low melting. Sn 55, Pb 45.
Solder alloy. 200°C FP; 183°C MP.

WILLIAMS S205
Williams Precious Metals
Low melting. Sn 85, Pb 15.
Solder alloy. 205°C FP; 183°C MP.

WILLIAMS S209
Williams Precious Metals
Low melting. In 50, Pb 50.
Solder alloy. 209°C FP; 180°C MP.

WILLIAMS S210
Williams Precious Metals
Low melting. Pb 55, Sn 44, Ag 1.
Solder alloy. 210°C FP; 177°C MP.

WILLIAMS S212
Williams Precious Metals
Low melting. Sn 50, Pb 50.
Solder alloy. 212°C FP; 183°C MP.

WILLIAMS S213
Williams Precious Metals
Low melting. Sn 90, Pb 10.
Solder alloy. 213°C FP; 183°C MP.

WILLIAMS S216
Williams Precious Metals
Low melting. Sn 50, Pb 49.5, Sb 0.5.
Solder alloy. 216°C FP; 183°C MP.

WILLIAMS S217
Williams Precious Metals
Low melting. Sn 90, Au 10.
Solder alloy. 217°C FP; 217°C MP.

WILLIAMS S218
Williams Precious Metals
Low melting. Pb 52, Sn 48.
Solder alloy. 218°C FP; 183°C MP.

WILLIAMS S221
Williams Precious Metals
Low melting. Sn 96.5, Ag 3.5.
Solder alloy. 221°C FP; 221°C MP.

WILLIAMS S222
Williams Precious Metals
Low melting. Sn 95, Pb 5.
Solder alloy. 222°C FP; 183°C MP.

WILLIAMS S225
Williams Precious Metals
Low melting. Pb 60, In 40.
Solder alloy. 225°C FP; 195°C MP.

WILLIAMS S226
Williams Precious Metals
Low melting. Pb 95, Ag 3.5, Sb 1.5.
Solder alloy. 226°C FP; 218°C MP.

WILLIAMS S226A
Williams Precious Metals
Low melting. Sn 97.5, Ag 2.5.
Solder alloy. 226°C FP; 221°C MP.

WILLIAMS S227
Williams Precious Metals
Low melting. Pb 55, Sn 45.
Solder alloy. 227°C FP; 183°C MP.

WILLIAMS S231
Williams Precious Metals
Low melting. Pb 58, Sn 40, Sb 2.
Solder alloy. 231°C FP; 185°C MP.

WILLIAMS S232
Williams Precious Metals
Low melting. Pb 60, Sn 37, Ag 3.
Solder alloy. 232°C FP; 179°C MP.

WILLIAMS S232A
Williams Precious Metals
Low melting. Sn 100.
Solder alloy.

WILLIAMS S235
Williams Precious Metals
Low melting. Sn 99, Sb 1.
Solder alloy.

WILLIAMS S237
Williams Precious Metals
Low melting. In 90, Ag 10.
Solder alloy. 237°C FP; 141°C MP.

WILLIAMS S238
Williams Precious Metals
Low melting. Pb 60, Sn 40.
Solder alloy. 238°C FP; 183°C MP.

WILLIAMS S238A
Williams Precious Metals
Low melting. Sn 97, Sb 3.
Solder alloy. 238°C FP; 232°C MP.

WILLIAMS S240
Williams Precious Metals
Low melting. Sn 95, Ag 5.
Solder alloy. 240°C FP; 221°C MP.

WILLIAMS S240A
Williams Precious Metals
Low melting. Sn 95, Sb 5.
Solder alloy. 240°C FP; 232°C MP.

WILLIAMS S243
Williams Precious Metals
Low melting. Pb 63.2, Sn 35, Sb 1.8.
Solder alloy. 243°C FP; 185°C MP.

WILLIAMS S247
Williams Precious Metals
Low melting. Pb 65, Sn 35.
Solder alloy. 247°C FP; 183°C MP.

WILLIAMS S248
Williams Precious Metals
Low melting. Sn 61.5, Pb 35.5, Ag 3.
Solder alloy. 248°C FP; 179°C MP.

WILLIAMS S250
Williams Precious Metals
Low melting. Pb 68.4, Sn 30, Sb 1.5.
Solder alloy. 250°C FP; 185°C MP.

WILLIAMS S253
Williams Precious Metals
Low melting. Pb 70, Sn 27, Ag 3.
Solder alloy. 253°C FP; 179°C MP.

WILLIAMS S257
Williams Precious Metals
Low melting. Pb 70, Sn 30.
Solder alloy. 257°C FP; 183°C MP.

WILLIAMS S260
Williams Precious Metals
Low melting. Sn 50, Pb 47, Ag 3.
Solder alloy. 260°C FP; 179°C MP.

WILLIAMS S260A
Williams Precious Metals
Low melting. Pb 90, Sb 10.
Solder alloy. 260°C FP; 252°C MP.

WILLIAMS S263
Williams Precious Metals
Low melting. Pb 73.7, Sn 25, Sb 1.3.
Solder alloy. 263°C FP; 184°C MP.

WILLIAMS S264
Williams Precious Metals
Low melting. Pb 75, In 25.
Solder alloy. 264°C FP; 250°C MP.

WILLIAMS S268
Williams Precious Metals
Low melting. Pb 75, Sn 25.
Solder alloy. 268°C FP; 183°C MP.

WILLIAMS S270
Williams Precious Metals
Low melting. Pb 79, In 20, Sb 1.
Solder alloy. 270°C FP; 184°C MP.

WILLIAMS S271
Williams Precious Metals
Low melting. Bi 100.
Solder alloy.

WILLIAMS S280
Williams Precious Metals
Low melting. Pb 80, Sn 20.
Solder alloy. 280°C FP; 100°C MP.

WILLIAMS S280A
Williams Precious Metals
Low melting. Pb 81, In 19.
Solder alloy. 280°C FP; 270°C MP.

WILLIAMS S280B
Williams Precious Metals
Low melting. Au 80, Sn 20.
Solder alloy. 280°C FP; 280°C MP.

WILLIAMS S285
Williams Precious Metals
Low melting. Pb 92, Sn 5, Sb 3.
Solder alloy. 285°C FP; 239°C MP.

WILLIAMS S288
Williams Precious Metals
Low melting. Pb 85, Sn 15.
Solder alloy. 288°C FP; 183°C MP.

WILLIAMS S289
Williams Precious Metals
Low melting. Pb 57, Sn 40, Ag 3.
Solder alloy. 289°C FP; 179°C MP.

WILLIAMS S292
Williams Precious Metals
Low melting. Pb 90, Ag 5, Sn 5.
Solder alloy.

WILLIAMS S295
Williams Precious Metals
Low melting. Sn 90, Ag 10.
Solder alloy. 295°C FP; 221°C MP.

WILLIAMS S295A
Williams Precious Metals
Low melting. Pb 95, Sb 5.
Solder alloy. 295°C FP; 252°C MP.

WILLIAMS S296
Williams Precious Metals
Low melting. Pb 92.5, Sn 5, Ag 2.5.
Solder alloy. 296°C FP; 287°C MP.

WILLIAMS S300
Williams Precious Metals
Low melting. Pb 92.5, In 5, Ag 2.5.
Solder alloy.

WILLIAMS S302
Williams Precious Metals
Low melting. Pb 90, Sn 10.
Solder alloy. 302°C FP; 275°C MP.

WILLIAMS S303
Williams Precious Metals
Low melting. Pb 97.5, Ag 2.5.
Solder alloy. 303°C FP; 303°C MP.

WILLIAMS S304
Williams Precious Metals
Low melting. Pb 95.5, Ag 2.5, Sn 2.
Solder alloy. 304°C FP; 299°C MP.

WILLIAMS S309
Williams Precious Metals
Low melting. Pb 97.5, Ag 1.5, Sn 1.
Solder alloy. 309°C FP; 309°C MP.

WILLIAMS S310
Williams Precious Metals
Low melting. Pb 90, In 5, Ag 5.
Solder alloy. 310°C FP; 290°C MP.

WILLIAMS S314
Williams Precious Metals
Low melting. Pb 95, In 5.
Solder alloy. 314°C FP; 292°C MP.

WILLIAMS S314A
Williams Precious Metals
Low melting. Pb 95, Sn 5.
Solder alloy. 314°C FP; 310°C MP.

WILLIAMS S315
Williams Precious Metals
Low melting. Pb 98, Sb 1.2, Ga 0.8.
Solder alloy.

WILLIAMS S320
Williams Precious Metals
Low melting. Pb 98, Sn 2.
Solder alloy. 320°C FP; 300°C MP.

WILLIAMS S322
Williams Precious Metals
Low melting. Pb 98.5, Sb 1.5.
Solder alloy. 322°C FP; 310°C MP.

WILLIAMS S327
Williams Precious Metals
Low melting. Pb 100.
Solder alloy.

WILLIAMS S356
Williams Precious Metals
Low melting. Au 88, Ge 12.
Solder alloy. 356°C FP; 356°C MP.

WILLIAMS S363
Williams Precious Metals
Low melting. Au 96.85, Si 3.15.
Solder alloy. 363°C FP; 363°C MP.

WILLIAMS S364
Williams Precious Metals
Low melting. Pb 95, Ag 5.
Solder alloy. 364°C FP; 305°C MP.

WILLIAMS S365
Williams Precious Metals
Low melting. Sn 65, Ag 25, Sb 10.
Solder alloy. 365°C FP; 232°C MP.

WILLIAMS S495
Williams Precious Metals
Low melting. Au 81, In 19.
Solder alloy. 495°C FP; 480°C MP.

WILLIAMS S577
Williams Precious Metals
Low melting. Al 88.3, Si 11.7.
Solder alloy. 577°C FP; 577°C MP.

WILLIAMS S650
Williams Precious Metals
Low melting. Ag 56, Cu 22, Zn 17, Sn 5.
Solder alloy. 650°C FP; 620°C MP.

WILLIAMS S660
Williams Precious Metals
Low melting. Al 100.
Solder alloy.

WILLIAMS S705
Williams Precious Metals
Low melting. Cu 80, Ag 15, P 5.
Solder alloy. 705°C FP; 640°C MP.

WILLIAMS S705A
Williams Precious Metals
Low melting. Ag 61.5, Cu 24, In 14.5.
Solder alloy. 705°C FP; 630°C MP.

WILLIAMS S720
Williams Precious Metals
Low melting. Ag 60, Cu 30, Sn 10.
Solder alloy. 720°C FP; 600°C MP.

WILLIAMS S730
Williams Precious Metals
Low melting. Ag 63, Cu 27, In 10.
Solder alloy. 730°C FP; 685°C MP.

WILLIAMS S780
Williams Precious Metals
Low melting. Ag 72, Cu 28.
Solder alloy. 780°C FP; 780°C MP.

WILLIAMS S785
Williams Precious Metals
Low melting. Ag 71.5, Cu 28, Ni 0.5.
Solder alloy. 785°C FP; 775°C MP.

WILLIAMS S790
Williams Precious Metals
Low melting. Ag 90, Ge 10.
Solder alloy. 790°C FP; 651°C MP.

WILLIAMS S795
Williams Precious Metals
Low melting. Ag 71.15, Cu 28.1, Ni 0.75.
Solder alloy. 795°C FP; 780°C MP.

WILLIAMS S800
Williams Precious Metals
Low melting. Au 98, Si 2.
Solder alloy. 800°C FP; 370°C MP.

WILLIAMS S800A
Williams Precious Metals
Low melting. Ag 63, Cu 28.5, Sn 6, Ni 2.5.
Solder alloy. 800°C FP; 690°C MP.

WILLIAMS S810
Williams Precious Metals
Low melting. Ag 68, Cu 27, Pd 5.
Solder alloy. 810°C FP; 807°C MP.

WILLIAMS S830
Williams Precious Metals
Low melting. Ag 68.8, Cu 26.7, Ti 4.5.
Solder alloy. 830°C FP; 850°C MP.

WILLIAMS S840
Williams Precious Metals
Low melting. Ag 97, Si 3.
Solder alloy. 840°C FP; 840°C MP.

WILLIAMS S845
Williams Precious Metals
Low melting. Au 60, Ag 20, Cu 20.
Solder alloy. 845°C FP; 835°C MP.

WILLIAMS S852
Williams Precious Metals
Low melting. Ag 58, Cu 32, Pd 10.
Solder alloy. 852°C FP; 824°C MP.

WILLIAMS S890
Williams Precious Metals
Low melting. Au 80, Cu 20.
Solder alloy. 890°C FP; 890°C MP.

WILLIAMS S895
Williams Precious Metals
Low melting. Au 75, Cu 20, Ag 5.
Solder alloy. 895°C FP; 885°C MP.

WILLIAMS S900
Williams Precious Metals
Low melting. Ag 65, Cu 20, Pd 15.
Solder alloy. 900°C FP; 850°C MP.

WILLIAMS S925
Williams Precious Metals
Low melting. Au 81.5, Cu 16.5, Ni 2.
Solder alloy. 925°C FP; 910°C MP.

WILLIAMS S950
Williams Precious Metals
Low melting. Au 82, Ni 18.
Solder alloy. 950°C FP; 950°C MP.

WILLIAMS S950A
Williams Precious Metals
Low melting. Ag 54, Pd 25, Cu 21.
Solder alloy. 950°C FP; 900°C MP.

WILLIAMS S96
Williams Precious Metals
Low melting. Bi 52, Pb 30, Sn 18.
Solder alloy. 96°C FP; 96°C MP.

WILLIAMS S961
Williams Precious Metals
Low melting. Ag 100.
Solder alloy.

WILLIAMS S965
Williams Precious Metals
Low melting. Cu 87.75, Ge 12, Ni 0.25.
Solder alloy. 965°C FP; 820°C MP.

WILLIAMS S970
Williams Precious Metals
Low melting. Au 50, Cu 50.
Solder alloy. 970°C FP; 955°C MP.

WILMIL
William Mills & Co. Ltd.
Si 10-13, Mn 0.5, Cu 0.1, Mg 0.1, bal Al.
Sand cast: 28,000 TS; 9,000 YS; 5-14 El; 50 Brin. For housings, ship construction; corrosion resistant, good castability. *Obsolete*

WILMIL M
William Mills & Co. Ltd.
Cu 0.1, Mg 0.4, Mn 0.5, Fe 0.6, bal Al.
Cast: 25,000 TS; 16,000 YS; 1.5 El; 65 Brin. Aged: 35,000 TS; 29,000 YS; 95 Brin. For highly stressed castings; age-hardenable. *Obsolete*

WILMOTT'S ALUMINUM
Manufacturer not listed.
Sn 86, Bi 14.
For aluminum solder.

WILRICH 142
Eaton Corp.
C 0.4, Cr 13, Ni 13, W 3, Si 1.5, bal Fe.
For supercharger buckets; high heat resistant.

WILRICH 300
Eaton Corp.
Bimetal.
For hard facing electrodes; corrosion resistant.

WILRICH 301
Eaton Corp.
Ni, Cr.
Cast: 300-400 Brin. For hard facing electrodes; corrosion resistant.

WILRICH 350
Eaton Corp.
Ni, Cr-Cu.
For hard facing electrodes; corrosion resistant.

WILRICH 600
Eaton Corp.
Cr, W.
Cast: 620 Brin. For hard facing electrodes; corrosion resistant.

WILRICH 625
Eaton Corp.
Cr, W, Co.
Cast: 650 Brin. For hard facing electrodes; heat, abrasion and corrosion resisting.

WILSON
Wilson Welder & Metals Co.
C, bal Fe.
For welding rod.

WILSON
Manufacturer not listed.
C, alloy, bal Fe.
For armor plate.

WILSON 109
Airco Welding Products
C 0.07, Mn 0.28, P 0.078, S 0.02, Si 0.15, bal Fe.
Welded: 70,000-80,000 TS; 60,000-68,000 YS; 21-25 El; 36-42 RA. For welding rod; mild steel welding. *Obsolete*

WILSON 512
Airco Welding Products
C 0.1, Mn 0.5, P 0.028, S 0.023, Si 0.49, bal Fe.
Welded: 75,000-80,000 TS; 60,000-65,000 YS; 25-30 El; 52-57 RA. For welding rod for high S steels; low H_2 electrode. *Obsolete*

WILSON 520 AND 520A
Airco Welding Products
C 0.1, Mn 0.4, Si 0.35, S 0.003, bal Fe.
Welded: 70,000-80,000 TS; 58,000-66,000 YS; 18-25 El; 40-60 RA. For welding rod; mild steel fabrication. *Obsolete*

WILSON 575
Airco Welding Products
C 0.83, Mn 0.72, P 0.14, S 0.003, Ni 92.66, Fe 5.7.
Welded: 58,250-59,800 TS; 5.0-6.5 El. For welding rod; for welding cast iron. *Obsolete*

WILSON NO. 10LC
Wilson Welder & Metals Co.
C 0-0.17, Ni 4.5-5.2, bal Fe.
For welding rod; wear resistant.

WILSON NO. 10SA
Wilson Welder & Metals Co.
C 0.15, Ni 5, bal Fe.
For welding rod; shielded.

WILSON NO. 512
Wilson Welder & Metals Co.
C, bal Fe.
For welding electrodes; mild steel.

WILSON NO. 520
Wilson Welder & Metals Co.
C, bal Fe.
For welding rod for mild steel; shielded, arc.

WILSON NO. 520
Engelhard Corp.
C, bal Fe.
For welding electrodes. *Obsolete*

WILSON NO. 520A
Wilson Welder & Metals Co.
C, bal Fe.
For welding rod Cr-Mo steel; shielded, arc.

WILSON NO. 575
Wilson Welder & Metals Co.
High Ni core.
For welding electrodes for cast iron; coated.

WIMET 10F
Sandvik Hard Materials Ltd.
Co 10, bal WC (micrograin).
Sintered: 1650 Vickers. For cutting aerospace materials; special forming and extrusion tools. *Obsolete*

WIMET 110B
Wimet Ltd.
Co 11, bal WC.
Sintered: 1225 VDH. General purpose rock drilling.
Obsolete

WIMET 15F
Sandvik Hard Materials Ltd.
Co 15, bal WC (micrograin).
Sintered: 1400 Vickers. For forming, blanking, cold
extrusion. *Obsolete*

WIMET 3F
Sandvik Hard Materials Ltd.
Co 3, bal WC (micrograin).
Sintered: 1950 Vickers. For special wire dies. *Obsolete*

WIMET 90B
Sandvik Hard Materials Ltd.
Co 9, bal WC.
Sintered: 1260 Vickers. For rock drilling and general mining
applications. *Obsolete*

WIMET BP1
Sandvik Hard Materials Ltd.
Co 16, bal WC.
Sintered: 1150 Vickers. For pressing and blanking tools,
forming tools, heavy duty dies. *Obsolete*

WIMET CM
Sandvik Hard Materials Ltd.
Co 9.5, bal WC.
Sintered: 1400 Vickers. For wood working tools. ISO K30.
Obsolete

WIMET CT
Sandvik Hard Materials Ltd.
Co 9, bal WC.
Sintered: 1300 Vickers. For rock drilling; rotary tools and
percussive drills for abrasive rocks. *Obsolete*

WIMET CW540
Sandvik Hard Materials Ltd.
TiCN coated carbide for steel cutting. *Obsolete*

WIMET CW620
Sandvik Hard Materials Ltd.
TiCN coated carbide for cutting cast iron. *Obsolete*

WIMET CXT
Sandvik Hard Materials Ltd.
Co 9, bal WC.
Sintered: 1225 Vickers. For tools for rock drilling. *Obsolete*

WIMET G
Sandvik Hard Materials Ltd.
Co 11, bal WC.
Sintered: 1325 Vickers. For dies, forming tools, cutting tools
for wood and plastics. ISO K40 (Am. Code C-1). *Obsolete*

WIMET GW520
Sandvik Hard Materials Ltd.
TiCN coated carbide for steel cutting. *Obsolete*

WIMET H
Sandvik Hard Materials Ltd.
Co 6, bal WC.
Sintered: 1750 Vickers. For cutting iron at high speeds;
glass, non-ferrous metals. ISO K10. *Obsolete*

WIMET HH
Wimet Ltd.
Co 7, 1% TaC, 0.5% VC, bal WC.
Sintered: 1775 VDH. For cutting hard cast irons; high speed
precision turning and boring. ISO KO1 (Am. Code C-4).
Obsolete

WIMET N
Sandvik Hard Materials Ltd.
Co 6, bal WC.
Sintered: 1575 Vickers. For turning and milling cast iron and
non-ferrous metals; wire drawing dies, wear parts. ISO K20
(Am. Code C-2). *Obsolete*

WIMET NC
Wimet Ltd.
Co 9.5, bal WC.
Sintered: 1400 VDH. Tough cutting grade for low hardness
cast iron, soft steels, and steel castings, with sand
inclusions; dies. ISO K30 (Am. Code C-1). *Obsolete*

WIMET NH
Wimet Ltd.
Co 6, bal WC.
Sintered: 1650 VDH. For cutting cast iron and non-ferrous
materials. ISO K15 (Am. Code C-3). *Obsolete*

WIMET R11
Sandvik Hard Materials Ltd.
Co 11, bal WC.
Sintered: 1140 Vickers. Heavy duty percussive rock drilling.
Obsolete

WIMET T
Sandvik Hard Materials Ltd.
Co 20, bal WC.
Sintered: 1050 Vickers. For cold heading and extrusion
dies. *Obsolete*

WIMET TT
Sandvik Hard Materials Ltd.
Co 25, bal WC.
Sintered: 950 Vickers. For cold heading and extrusion dies.
Obsolete

WIMET TTX
Wimet Ltd.
Co 30, bal WC.
Sintered: 850 VDH. For high duty cold heading and
extrusion dies. *Obsolete*

WIMET X
Wimet Ltd.
WC + Co.
For tipped cutting tools for aluminum; sintered carbide.
Obsolete

WIMET X L 2
Sandvik Hard Materials Ltd.
Co, TiC, TaC, WC.
Sintered: 1575 Vickers. For steel cutting at high speeds and
light loads. ISO P20. *Obsolete*

WIMET XL1
Wimet Ltd.
Co 5, 16% TiC, bal WC.
Sintered: 1700 VDH. High speed turning on steels; finishing
cuts and precision boring. ISO PO1 (Am. Code C-8).
Obsolete

WIMET XL2B
Sandvik Hard Materials Ltd.
Co 10, 15 (Ta + Nb) C, 19.0 TiC, bal WC.
Sintered: 1525 Vickers. Steel turning at medium to high
speeds, semi-finish and finishing cuts. ISO P10 (Am. Code
C-7). *Obsolete*

WIMET XL3
Sandvik Hard Materials Ltd.
Co 9, 12 (Ta + Nb) C, 9.0 TiC, bal WC.
Sintered: 1450 Vickers. General purpose steel cutting with
good resistance to shock and impact. ISO P30 (Am. Code
C-6). *Obsolete*

WIMET XL35
Sandvik Hard Materials Ltd.
Co 10, 2 (Ta + Nb) C, 10.0 TiC, bal WC.
Sintered: 1550 Vickers. For turning and milling steel having
interrupted cuts. ISO P25 (Am. Code C-6). *Obsolete*

WIMET XL4
Wimet Ltd.
Co 9, 5% TiC, bal WC.
Sintered: 1475 VDH. For heavy duty planing and turning
steel at low speeds; roughing cuts. ISO P30 (Am. Code
C-5). *Obsolete*

WIMET XL45
Sandvik Hard Materials Ltd.
Co 11, 8 (Ta + Nb) C, 4.0 TiC, bal WC.
Sintered: 1500 Vickers. For rough turning and planing
operations on steel. *Obsolete*

WINCHESTER DRAWING DIE
Manufacturer not listed.
C 1.54, Mn 0.25, W 5.11, bal Fe.
For tools, dies; oil hardening.

WINDSOR
Jessop Steel Co.
C 0.98, Cr 5.25, V 0.25, Mo 1, bal Fe.
For dies; air hardened. AISI A2.

WINDSOR (CAST TO SHAPE)
Jessop Steel Co.
C 1, Mn 0.45, Si 0.25, Cr 5.25, V 0.25, Mo 1.1, bal Fe.
For heavy duty production loads; machinery cams, etc. Air
hardening tool steel.

WINNS SUPERHEAT
Manufacturer not listed.
Cu 62, Zn 35, Sn 2, Pb 0.4, As 0.13, Fe 0.2.
For valves, fittings for super heated steam.

WINSTON
Time Steel Service Inc.
C 1, Mn 0.6, Cr 5.25, V 0.25, Mo 1.1, bal Fe.
Medium alloy air hardened tool steel. AISI A2.

WINSTON V
Time Steel Service Inc.
C 2.4, Cr 5.25, V 4.25, Mo 1.1, bal Fe.
Air hardened tool steel. AISI A7.

WIPING SOLDER
American manufacture
Sn 40, Pb 60.
For commercial lead solder; melting point 446°F.

WIPLA METAL
German manufacture
C 0.2, Cr 18, Ni 8, bal Fe.
For machinery parts; non-rusting.

WIPTAM
Krupp Stahl AG
superalloy. Co 45.5, Cr 28.3, Ni 24.4, C 0.1, Si 1.1, Mn 0.7.
Rolled: 90,000 TS; 74,000 YS; 1.2 El; 378 Brin. For dental
alloy, dentures. Corrosion and wear resistant. *Obsolete*

WIRE BRASS
Manufacturer not listed.
Cu 65-72, Zn 27-35, Sn 0.17, Pb 0.28.
For wire applications; high ductility.

WIRE DRAWING ALLOY
Columbia Tool Steel Co.
C 2.35, Cr 1.9, Mn 1.65, W 10.5, Mo 0.55, bal Fe.
For tools, wire drawing dies; abrasion resisting. *Obsolete*

WIRESPRAY ALUMINUM
Wall Colmonoy Corp.
Aluminum wire for wirespray gun build-up, mainly on
aluminum parts.

WIRESPRAY ALUMINUM S
Wall Colmonoy Corp.
Silicon-aluminum wire for wirespray gun coating and build-up. *Obsolete*

WIRESPRAY BABBITT "A"
Wall Colmonoy Corp.
High tin, lead free Babbitt wire for spray build-up of bearings.

WIRESPRAY BRASS "Y"
Wall Colmonoy Corp.
Yellow brass wire for spray build-up and repair of brass parts. *Obsolete*

WIRESPRAY BRONZE "A"
Wall Colmonoy Corp.
Widely used bronze (containing Al) wire for spray build-up and repair. Machinable, produces good finish. *Obsolete*

WIRESPRAY BRONZE "C"
Wall Colmonoy Corp.
Commercial bronze wire for spray build-up and repair for light wear applications. *Obsolete*

WIRESPRAY BRONZE "M"
Wall Colmonoy Corp.
Manganese bronze wire for spray build-up and repair of manganese bronze parts. *Obsolete*

WIRESPRAY BRONZE "P"
Wall Colmonoy Corp.
Phosphor bronze wire for spray build-up and repair of phosphor bronze parts. *Obsolete*

WIRESPRAY BRONZE "T"
Wall Colmonoy Corp.
Tobin bronze wire for spray build-up and repair. *Obsolete*

WIRESPRAY COPPER
Wall Colmonoy Corp.
High purity copper wire for spray build-up. For electrical applications.

WIRESPRAY CUPRO-NICKEL
Wall Colmonoy Corp.
Cu 70, Ni 30.
Wire for spray build-up and repair. *Obsolete*

WIRESPRAY MOLYBDENUM
Wall Colmonoy Corp.
Molybdenum wire for spray coating and build-up. Usually used as thin undercoat between base metal and final coating material.

WIRESPRAY MONEL
Wall Colmonoy Corp.
Monel wire for spray build-up and repair. Good corrosion protection where wear is not severe.

WIRESPRAY NICHROME 62-16
Wall Colmonoy Corp.
Ni-Cr-Fe high temperature wire for spray build-up. Used alone or in combination with and where aluminum is the top coating. *Obsolete*

WIRESPRAY NICHROME 80-20
Wall Colmonoy Corp.
Ni-Cr alloy wire used usually for spray undercoat for ceramics. Heat and oxidation resistant. *Obsolete*

WIRESPRAY NICKEL
Wall Colmonoy Corp.
Nickel wire for spray build-up and repair. *Obsolete*

WIRESPRAY NO. 10
Wall Colmonoy Corp.
General purpose carbon steel wire for metallizing gun. Deposits are machinable.

WIRESPRAY NO. 25
Wall Colmonoy Corp.
Carbon steel wire for metallizing gun. Deposits harder than No. 10, but still machinable.

WIRESPRAY NO. 3 LO-SHRINK
Wall Colmonoy Corp.
High chrome stainless wire for metallizing gun. For producing heavy, machinable stainless overlays.

WIRESPRAY NO. 80
Wall Colmonoy Corp.
Low shrink steel wire for metallizing gun. Too hard to machine except with carbide; good wear resistance.

WIRESPRAY TIN
Wall Colmonoy Corp.
Tin wire for spray build-up or repair. Used for dairy equipment, decorative purposes, and to prepare surfaces for soldered joints. *Obsolete*

WIRESPRAY ZINC
Wall Colmonoy Corp.
Zinc wire for spray build-up for corrosion protection in outdoor atmospheres. *Obsolete*

WIRONET 2012 EN
Thyssen Edelstahlwerke AG
C 0-0.1, Cr 18, Ni 13, Mo 2.8, Nb = 10 x C min, bal Fe.
Annealed (quenched): 71,000-106,000 TS; 33,000 YS minimum; 40 El minimum. Minimum yield strength 20,000 psi at 550 C. Weldable austenitic stainless. For welded assemblies for operation in chemical plants and for up to 550 C. DIN X10CrNiMoNb1812. *Obsolete*

WIRONIT 140
Thyssen Edelstahlwerke AG
C 0.05, Cr 17, Ni 4, Cu 4, Nb 0.4, bal Fe.
Precipitation hardenable stainless steel, good strength below 400 C. DIN X7CrNiCuNb174; Werkstoff Nr. 1.4540. *Obsolete*

WIRONIT 150
Thyssen Edelstahlwerke AG
C 0.05, Cr 17, Ni 7, Al 1, bal Fe.
Precipitation hardenable stainless steel, good strength below 400 F. DIN X7CrNiAl177; Werkstoff Nr. 1.4568. *Obsolete*

WIRONIT 151
Thyssen Edelstahlwerke AG
C 0.05, Cr 15, Ni 7, Mo 3, Al 1.2, bal Fe.
Precipitation hardenable stainless steel, good strength below 800 F. DIN X7CrNiMoAl157; Werkstoff Nr. 1.4532. *Obsolete*

WIRONIT 151 L
Thyssen Edelstahlwerke AG
C 0-0.09, Si 0-1, Mn 0-1, Cr 15, Mo 2.5, Ni 7, Al 1.1, bal Fe.
Solution treat and age type stainless. DIN X7CrNiMoAl157; Werkstoff Nr. 1.4574; AMS 5657; PH 15-7 Mo. *Obsolete*

WIRONIT 2001
Thyssen Edelstahlwerke AG
C 0-0.02, Cr 17, Ni 13, Mo 2.2, bal Fe.
Annealed (quenched): 64,000-100,000 TS; 28,000 YS minimum; 45 El minimum. Austenitic, excellent corrosion resistance, non-hardenable except by cold work. For equipment for food, beverage, dairy, textile, and chemical industries. *Obsolete*

WIRONIT 2003
Thyssen Edelstahlwerke AG
C 0-0.03, Cr 17, Ni 12, Mo 2.2, bal Fe.
Annealed (quenched): 64,000-100,000 TS; 28,000 YS minimum; 45 El minimum. Austenitic, excellent corrosion resistance, non-hardenable except by cold work. For equipment for food, beverage, chemical industries. DIN X2CrNiMo1810. *Obsolete*

WIRONIT 2004
Thyssen Edelstahlwerke AG
C 0-0.03, Cr 17, Ni 13.5, Mo 2.8, bal Fe.
Annealed (quenched): 64,000-100,000 TS; 28,000 YS minimum; 45 El minimum. Austenitic, excellent corrosion resistance, non-hardenable except by cold work. For equipment for paper, food, textile, chemical industries. DIN X2CrNiMo1812. *Obsolete*

WIRONIT 2005
Thyssen Edelstahlwerke AG
C 0-0.03, Cr 18, Ni 14, Mo 2.6, N, bal Fe.
Annealed (quenched): 92,000-113,000 TS; 42,500 YS minimum; 35 El minimum. Austenitic stainless, improved strength. For stainless parts requiring better strength. DIN XCrNiMoN1814. *Obsolete*

WIRONIT 2007
Thyssen Edelstahlwerke AG
C 0-0.07, Cr 18, Ni 10.5, Mo 2, bal Fe.
Annealed: 85,000 TS; 35,000 YS; 50 El; 65 RA; 160 Brin. Cold drawn: 150,000 TS; 135,000 YS; 6 El; 300 Brin. For acid resistant chemical plant equipment, tanks; Type 316; stainless, austenitic. *Obsolete*

WIRONIT 2008
Thyssen Edelstahlwerke AG
C 0-0.07, Cr 18, Ni 13, Mo 2.8, bal Fe.
Annealed (quenched): 71,000-100,000 TS; 30,000 YS minimum; 45 El minimum. Austenitic, excellent corrosion resistance, non-hardenable except by cold work. For equipment for food, paper, textile and chemical industries. DIN X5CrNiMo1812. *Obsolete*

WIRONIT 2010E
Thyssen Edelstahlwerke AG
C 0-0.12, Cr 18, Ni 10.5, Mo 2, Ti = 4 x C, bal Fe.
Annealed: 85,000 TS; 40,000 YS; 45 El; 60 RA; 170 Brin. For welded acid resistant equipment, tanks, mixers; Type 316Ti; stainless, austenitic. *Obsolete*

WIRONIT 2010EN
Thyssen Edelstahlwerke AG
C 0-0.12, Cr 18, Ni 10.5, Mo 2, Cb = 8 x C, bal Fe.
Annealed: 85,000 TS; 40,000 YS; 45 El; 60 RA; 170 Brin. For welded acid resistant equipment, tanks, mixers; Type 316Cb; stainless, austenitic. *Obsolete*

WIRONIT 2012 E
Thyssen Edelstahlwerke AG
C 0-0.1, Cr 18, Ni 13, Mo 2.8, Ti = 6 x C min, bal Fe.
Annealed (quenched): 71,000-106,000 TS; 33,000 YS minimum; 40 El minimum. Austenitic, excellent corrosion resistance, weldable, non-hardenable except by cold work. For welded assemblies in food and chemical industries; for operation up to 400 C. DIN X10CrNiMoTi1812. *Obsolete*

WIRONIT 2020
Thyssen Edelstahlwerke AG
C 0-0.07, Cr 18, Ni 19.5, Mo 2.3, Cu 2, Nb = 8 x C min, bal Fe.
Annealed (quenched): 71,000-106,000 TS; 31,000 YS minimum; 40 El minimum. Austenitic, weldable stainless, non-hardenable except by cold work. For welded assemblies in corrosive environments and up to 350 C. DIN X5CrNiMoCuNb1818. *Obsolete*

WIRONIT 2107
Thyssen Edelstahlwerke AG
C 0-0.07, Cr 17, Mo 4.7, Ni 13, bal Fe.
Annealed: 90,000 TS; 42,000 YS; 42 El; 58 RA; 180 Brin. For acid resistant chemical plant equipment, tanks; Type 317; stainless, austenitic. *Obsolete*

WIRONIT 4113
Thyssen Edelstahlwerke AG
C 0.07, Cr 17, Mo 1.1, bal Fe.
Ferritic type corrosion resistant steel for automobile trim. DIN X6CrMo17; Werkstoff Nr. 1.4113. *Obsolete*

WIRONIT 4460

Thyssen Edelstahlwerke AG
C 0.1, Cr 27, Ni 4.5, Mo 1.5, bal Fe.
Ferritic austenitic type corrosion resistant steel with good elevated temperature strength. DIN X8CrNiMo275; Werkstoff Nr. 1.4460. *Obsolete*

WIRONIT 599

Thyssen Edelstahlwerke AG
C 0-0.08, Cr 13, bal Fe.
Annealed: 64,000-92,000 TS; 35,000 YS minimum; 20 El minimum. 28,000 YS minimum at 400 F. Ferritic, magnetic. For rust and acid-resisting parts. DIN X7Cr13. *Obsolete*

WIRONIT 599 AL

Thyssen Edelstahlwerke AG
C 0-0.08, Cr 13, Al 0.1-0.3, bal Fe.
Annealed: 64,000-92,000 TS; 35,000 YS minimum; 20 El minimum. 28,000 YS minimum at 400 F. Ferritic, magnetic, weldable, non-hardenable. For rust and acid-resisting parts. DIN X7CrAl13; AISI 405. *Obsolete*

WIRONIT 600

Thyssen Edelstahlwerke AG
C 0-0.12, Si 0.4, Cr 13, bal Fe.
Annealed: 75,000 TS; 40,000 YS; 35 El; 70 RA; 155 Brin. For turbine blades, valves, cutlery, surgical instruments; Type 410; corrosion resistant. *Obsolete*

WIRONIT 600 A

Thyssen Edelstahlwerke AG
C 0.1, Cr 13, S 0.2, bal Fe.
Heat treated: 85,000-112,000 TS; 64,000 YS minimum; 12 El minimum. Martensitic, magnetic, free machining, hardenable, corrosion resistant. For stainless parts of intermediate hardness; cutlery, tableware. DIN X12CrS13; AISI 416. *Obsolete*

WIRONIT 601

Thyssen Edelstahlwerke AG
C 0.15, Cr 13, bal Fe.
Heat treated: 92,000-112,000 TS; 71,000 YS minimum; 16 El minimum. 47,000 YS minimum at 400 C. Martensitic, magnetic, weldable. For turbine wheels and blades, cutlery, tableware, surgical and dental instruments. DIN X15Cr13; AISI 403, 410. *Obsolete*

WIRONIT 602

Thyssen Edelstahlwerke AG
C 0-0.08, Cr 14.5, bal Fe.
Annealed: 64,000-92,000 TS; 35,000 YS minimum; 20 El minimum. Ferritic, magnetic, weldable, non-hardenable. For tableware, welded assemblies for rust resisting applications. DIN X7Cr14. *Obsolete*

WIRONIT 610

Thyssen Edelstahlwerke AG
C 0.08, Cr 18, bal Fe.
Annealed: 80,000 TS; 50,000 YS; 25 El; 50 RA; 150 Brin. For chemical plant and oil refinery equipment; corrosion and heat resistant; Type 430. *Obsolete*

WIRONIT 610 EN

Thyssen Edelstahlwerke AG
C 0-0.1, Cr 17, Nb = 12 x C, bal Fe.
Annealed: 64,000-85,000 TS; 35,000 YS minimum; 20 El minimum. Ferritic, magnetic, weldable, non-hardenable. For welded assemblies in corrosive conditions. DIN X8CrNb17. *Obsolete*

WIRONIT 610E

Thyssen Edelstahlwerke AG
C 0-0.1, Cr 17.5, Ti = 7 x C, bal Fe.
Annealed: 85,000 TS, 50,000 YS; 25 El; 50 RA; 160 Brin. For welded chemical plant and oil refinery equipment; corrosion and heat resistant. *Obsolete*

WIRONIT 615

Thyssen Edelstahlwerke AG
C 0.2, Si 0.4, Cr 13, bal Fe.
Annealed: 95,000 TS; 50,000 YS; 25 El; 55 RA; 195 Brin. For turbine blades, cutlery, valves, surgical instruments; Type 420; corrosion resistant. *Obsolete*

WIRONIT 620

Thyssen Edelstahlwerke AG
C 0-0.1, Cr 20, Ni 0.8, bal Fe.
Annealed: 57,000-85,000 TS; 42,000 YS minimum; 20 El minimum. Ferritic, magnetic, weldable, non-hardenable. For parts subject to corrosive or elevated temperature conditions. DIN X10CrNi20. *Obsolete*

WIRONIT 632

Thyssen Edelstahlwerke AG
C 0.22, Cr 17, Ni 1.5, bal Fe.
Annealed: 125,000 TS; 95,000 YS; 20 El; 55 RA; 260 Brin. Cold drawn: 130,000 TS; 110,000 YS; 15 El; 35 RA; 270 Brin. For pumps, valves, marine hardware; Type 431; corrosion resistant. *Obsolete*

WIRONIT 632 L

Thyssen Edelstahlwerke AG
C 0.2, Si 0-1, Mo 0-1, Cr 16.15, Ni 1.7, bal Fe.
Hardenable corrosion resistant steel. For highly stressed bolts, shafts, studs at temperatures to 400 C. DIN X22CrNi17; Werkstoff Nr. 1.4044; AMS 5628; AISI 431. *Obsolete*

WIRONIT 645

Thyssen Edelstahlwerke AG
C 0.4, Si 0.4, Cr 13, bal Fe.
Annealed: 95,000 TS; 50,000 YS; 25 El; 55 RA; 200 Brin. For cutlery, valves, pumps, surgical instruments; Type 420; corrosion resistant. *Obsolete*

WIRONIT 696

Thyssen Edelstahlwerke AG
C 0.85, Cr 18, V 1, bal Fe.
Annealed: 107,000 TS; 62,000 YS; 18 El; 35 RA; 220 Brin. Heat treated: 280,000 TS; 270,000 YS; 3 El; 15 RA; 550 Brin. For bearings, liners, cutlery, valves; Type 440B; corrosion resistant. *Obsolete*

WIRONIT 706E

Thyssen Edelstahlwerke AG
C 0-0.1, Cr 17, Mo 1.8, Ti = 7 x C, bal Fe.
Annealed: 80,000 TS; 50,000 YS; 25 El; 50 RA; 150 Brin. For welded furnace and pump parts; corrosion and heat resistant. *Obsolete*

WIRONIT 710A

Thyssen Edelstahlwerke AG
C 0.12, Cr 16.5, Mo 0.25, S 0.2, bal Fe.
Annealed: 80,000 TS; 50,000 YS; 25 El; 50 RA; 160 Brin. For screw machine products, shafts, fasteners; Type 430f. *Obsolete*

WIRONIT 720

Thyssen Edelstahlwerke AG
C 0.2, Si 0.4, Cr 13, bal Fe.
Annealed: 95,000 TS; 50,000 YS; 25 El; 55 RA; 195 Brin. Cold drawn: 105,000 TS; 85,000 YS; 17 El; 50 RA; 215 Brin. For turbine blades, cutlery, valves, surgical instruments; Type 420; corrosion resistant. *Obsolete*

WIRONIT 735

Thyssen Edelstahlwerke AG
C 0.35, Cr 16.5, Mo 1.15, bal Fe.
For cutlery, valves, pumps, furnace parts; corrosion and heat resistant. *Obsolete*

WIRONIT 745

Thyssen Edelstahlwerke AG
C 0.45, Cr 14.5, Mo 0.55, bal Fe.
Air or oil hardenable to 56-58 Rock C. Martensitic, magnetic, stainless. For cutlery, kitchen knives, dental and surgical instruments, hardened shafts for non rusting conditions. DIN X48CrMoV15. *Obsolete*

WIRONIT 755

Thyssen Edelstahlwerke AG
C 0.55, Cr 14, Mo 0.55, bal Fe.
Air or oil hardenable to 50-57 Rock C. Martensitic, magnetic, corrosion resistant. For ball and roller bearings, shafts, cutlery, surgical instruments. DIN X52CrMo13. *Obsolete*

WIRONIT 765

Thyssen Edelstahlwerke AG
C 0.62, Cr 14, Mo 0.55, bal Fe.
Air or oil hardenable to 56-58 Rock C. Martensitic, magnetic, stainless. For cutlery, kitchen knives, dental and surgical instruments, hardened shafts for non-rusting conditions. DIN X65CrMo14. *Obsolete*

WIRONIT 770

Thyssen Edelstahlwerke AG
C 0.65, Cr 14, Mo 1.1, Co 1.4, V 0.4, bal Fe.
Air or oil hardening to 52-60 Rock C. Martensitic, corrosion resistant. Holds good sharp edge, for cutlery, surgical and dental instruments. *Obsolete*

WIRONIT 785

Thyssen Edelstahlwerke AG
C 0.9, Cr 17, Mo 0.5, V 0.25, Co 1.5, bal Fe.
Air or oil hardenable to 58-60 Rock C. Martensitic, magnetic, stainless. For hardened parts in food, beverage and textile industries. DIN X90CrCoMoV17. *Obsolete*

WIRONIT 795

Thyssen Edelstahlwerke AG
C 0.9, Cr 18, V 1, bal Fe.
Annealed: 107,000 TS; 62,000 YS; 18 El; 35 RA; 220 Brin. For bearings, valves, cutlery, surgical instruments; Type 440B; corrosion resistant. *Obsolete*

WIRONIT 798

Thyssen Edelstahlwerke AG
C 1, Cr 13, Mo 0.5, bal Fe.
Air or oil hardenable to 59-61 Rock C. Martensitic, magnetic, stainless. For cutlery, dental and surgical instruments, stainless bearings and bearing races. DIN X100CrMo13. *Obsolete*

WIRONIT 801

Thyssen Edelstahlwerke AG
C 0-0.02, Cr 18, Ni 12, bal Fe.
Annealed (quenched): 71,000-100,000 TS; 26,000 YS minimum; 50 El minimum. Austenitic, stainless, weldable, non-hardenable except by cold work. For food, dairy, textile, chemical plant equipment. *Obsolete*

WIRONIT 803

Thyssen Edelstahlwerke AG
C 0-0.03, Cr 18, Ni 11, bal Fe.
Annealed (quenched): 71,000-100,000 TS; 26,000 YS minimum; 50 El minimum. Austenitic, stainless, weldable, non-hardenable except by cold work. For food, dairy, textile, chemical plant equipment. DIN X2CrNi189. *Obsolete*

WIRONIT 805

Thyssen Edelstahlwerke AG
C 0.02, Cr 18, Ni 11, N, bal Fe.
Annealed (quenched): 85,000-106,000 TS; 37,000 YS minimum; 40 El minimum. Austenitic, stainless, non-hardenable except by cold work. Stainless parts for special conditions. DIN XCrNiN1812. *Obsolete*

WIRONIT 807

Thyssen Edelstahlwerke AG
C 0-0.07, Cr 18, Ni 9.5, bal Fe.
Annealed: 80,000 TS; 35,000 YS; 55 El; 75 RA; 150 Brin. For chemical plant equipment, tanks, mixers, filters; Type 304; stainless, austenitic. *Obsolete*

WIRONIT 807 L

Thyssen Edelstahlwerke AG
C 0-0.07, Si 0-1, Mn 0-2, Cr 18, Ni 10, bal Fe.
Austenitic stainless steel. DIN X5CrNi189; Werkstoff Nr.
1.4314; AISI 304. *Obsolete*

WIRONIT 808

Thyssen Edelstahlwerke AG
C 0-0.1, Cr 12.5, Ni 12, bal Fe.
For valves, pumps, cutlery; corrosion and heat resistant.
Obsolete

WIRONIT 812

Thyssen Edelstahlwerke AG
C 0-0.15, Cr 18, Ni 8.5, bal Fe.
Annealed: 80,000 TS; 35,000 YS; 55 El; 75 RA; 150 Brin. For
chemical plant equipment, tanks, mixers, filters; Type 302;
stainless, austenitic. *Obsolete*

WIRONIT 812 A

Thyssen Edelstahlwerke AG
C 0-0.15, Cr 18, Ni 9, S 0.15, bal Fe.
Annealed (quenched): 71,000-100,000 TS; 31,000 YS
minimum; 50 El minimum. Austenitic, stainless, free-
machining, non-hardenable except by cold work. For
machined parts for food, dairy, textile, chemical plant
equipment. DIN X12CrNiS188. *Obsolete*

WIRONIT 812 EL

Thyssen Edelstahlwerke AG
C 0-0.08, Si 0-1, Mn 0-2, Cr 18, Ni 10, Ti = 6 x C min, bal
Fe.
Austenitic stainless, weldable; usable for parts at 400-700 C.
DIN X10CrNiTi189; Werkstoff Nr. 1.4544; AISI 321. *Obsolete*

WIRONIT 812 L

Thyssen Edelstahlwerke AG
C 0-0.12, Si 0-1, Mn 0-2, Cr 18, Ni 9, bal Fe.
Austenitic stainless steel. DIN X12CrNi188; Werkstoff Nr.
1.4304; AMS 5637; AISI 302. *Obsolete*

WIRONIT 812E

Thyssen Edelstahlwerke AG
C 0-0.12, Cr 18, Ni 9.5, Ti = 4 x C, bal Fe.
Annealed: 85,000 TS; 35,000 YS; 55 El; 65 RA; 150 Brin. For
welded structures, chemical plant equipment; Type 321;
stainless, austenitic. *Obsolete*

WIRONIT 812EN

Thyssen Edelstahlwerke AG
C 0-0.12, Cr 18, Ni 9.5, Cb = 8 x C, bal Fe.
Annealed: 90,000 TS; 45,000 YS; 56 El; 65 RA; 160 Brin.
Cold drawn: 100,000 TS; 65,000 YS; 45 El; 60 RA; 205 Brin.
For welded chemical plant equipment, tanks, mixers; Type
347; stainless, austenitic. *Obsolete*

WIRONIT 815

Thyssen Edelstahlwerke AG
C 0-0.1, Cr 18, Ni 8.5, bal Fe.
Annealed: 85,000 TS; 35,000 YS; 60 El; 70 RA; 150 Brin.
Cold drawn: 180,000 TS; 125,000 YS; 10 El; 330 Brin. For
chemical plant equipment, tanks, vessels; Type 304;
stainless, austenitic. *Obsolete*

WIRONIT B

Thyssen Edelstahlwerke AG
C 0.03, Fe 5, Mo 28, bal Ni.
Good corrosion resistant alloy with special resistance to
reducing media. DIN NiMo30; Werkstoff Nr. 2.4810.
Obsolete

WIRONIT C

Thyssen Edelstahlwerke AG
C 0-0.03, Cr 20, Si 0-0.1, V 0.3, Mo 15, bal Ni.
Corrosion resistant and high temperature alloy. DIN
NiCr20Mo15; Werkstoff Nr. 2.4811. *Obsolete*

WISIL

Krupp Stahl AG
react and refract. Co 66.2, Cr 27, Mo 4.5, C 0.35, Si 0.4, Mn
1.
Cast: 123,000 TS; 87,000 YS; 10 El; 362 Brin. For dental
alloy; corrosion resistant. *Obsolete*

WISSCO NO. 1

Wickwire Spencer Steel Co.
Stainless Steel. C 0-0.12, Cr 12-14, bal Fe.
130,000-233,000 TS. For stainless wires; stainless, can be
forged and gas welded.

WISSCO NO. 1 M

Wickwire Spencer Steel Co.
Stainless Steel. C 0.12, Cr 12-15, S, Se or Mo, bal Fe.
For stainless wire; free machining.

WISSCO NO. 2

Wickwire Spencer Steel Co.
Stainless Steel. C 0.12-0, Cr 12-15, bal Fe.
For stainless wires; stainless, oil hardened.

WISSCO NO. 3

Wickwire Spencer Steel Co.
Stainless Steel. C 0-0.12, Cr 16-18, bal Fe.
For stainless wires.

WISSCO NO. 4

Wickwire Spencer Steel Co.
Stainless Steel. Cr 18, Ni 8, C 0.2, bal Fe.
170,000-300,000 TS. For stainless wires; stainless, can be
forged and welded.

WISSCO NO. 4 H T

Wickwire Spencer Steel Co.
Cr 18, Ni 8, C 0.07, bal Fe.
For high strength hard drawn wires; stainless. *Obsolete*

WISSCO NO. 4 M

Wickwire Spencer Steel Co.
Stainless Steel. C 0.08-0.2, Cr 17-19, Ni 7-9, P, S or Se, bal
Fe.
For stainless wire; free machining.

WISSCO NO. 4 S

Wickwire Spencer Steel Co.
Stainless Steel. C 0-0.11, Cr 17-19, Ni 7-9, bal Fe.
For welding rods; stainless.

WISSCO NO. 5 A

Wickwire Spencer Steel Co.
C 0-0.2, Cr 22-26, Ni 11-13, bal Fe.
For heat resistant wire; heat resistant.

WISSCO NO. 5 C

Wickwire Spencer Steel Co.
Stainless Steel. Cr 27, C 0.3, bal Fe.
For high heat resistant wires; stainless.

WISSCO NO. 5 C N

Wickwire Spencer Steel Co.
Stainless Steel. C 0-0.25, Cr 19-21, Ni 24-26, Mn, Si, bal Fe.
For high heat resistant wires, springs for high temperature
service; stainless.

WISSLER HIGH SPEED

English manufacture
Cr 15-35, W 15-40, B 0.5-2.5, C 0.75-2.5, 15-50 Ni or Co.
For cutting tools, dies; cast nonferrous.

WITHERM 10

Thyssen Edelstahlwerke AG
C 0-0.12, Cr 6.5, Al 0.75, Si 0.7, bal Fe.
For oil refinery equipment; creep and heat resistant.
Obsolete

WITHERM 15

Thyssen Edelstahlwerke AG
C 0-0.12, Si 1.15, Cr 13, Al 0.95, bal Fe.
For oil refinery equipment; corrosion and heat resistant.
Obsolete

WITHERM 20

Thyssen Edelstahlwerke AG
C 0-0.12, Si 0.9, Cr 18, Al 0.95, bal Fe.
For oil refinery equipment; corrosion and heat resistant.
Obsolete

WITHERM 25

Thyssen Edelstahlwerke AG
C 0-0.15, Si 1.4, Cr 24, Al 1.45, bal Fe.
For oil refinery equipment; creep and heat resistant.
Obsolete

WITHERM 29

Thyssen Edelstahlwerke AG
C 0.2, Si 1.2, Cr 25, Ni 4, bal Fe.
Annealed: 90,000 TS; 65,000 YS; 20 El; 212 Brin. Aged:
97,000 TS; 65,000 YS; 18 El; 210 Brin. For valves, pumps,
retorts, heat treat boxes; corrosion and heat resistant.
Obsolete

WITHERM 32

Thyssen Edelstahlwerke AG
C 0.08, Si 0.5, Cr 18, Ni 10, Ti, bal Fe.
Austenitic Cr-Ni stainless steel for high temperature
operation. Useful to 800 C. DIN 12CrNiTi189; Werkstoff Nr.
4878. *Obsolete*

WITHERM 36

Thyssen Edelstahlwerke AG
C 0.15, Si 2, Cr 19.5, Ni 9.5, bal Fe.
Annealed: 80,000 TS; 35,000 YS; 55 El; 75 RA; 160 Brin. For
chemical plant equipment, tanks, mixers, filters; Type 302;
austenitic, stainless. *Obsolete*

WITHERM 48

Thyssen Edelstahlwerke AG
C 0.15, Si 2, Cr 24, Ni 19, bal Fe.
Annealed: 100,000 TS; 45,000 YS; 50 El; 65 RA; 185 Brin.
For furnace parts, valves, pumps, heat treat boxes; Type
310; austenitic, stainless. *Obsolete*

WITHERM 48 L

Thyssen Edelstahlwerke AG
C 0-0.2, Si 2, Mn 1.3, Cr 25, Ni 20, bal Fe.
Stainless steel for high temperature operation, weldable.
DIN X15CrNiSi2520; Werkstoff Nr. 1.4844; AISI 310.
Obsolete

WITHERM 52

Thyssen Edelstahlwerke AG
C 0.12, Si 1.9, Cr 16, Ni 36, bal Fe.
Austenitic Ni-Cr steel for high temperature operation; resists
scaling and carburizing; useful to 1100 C. DIN
X12NiCrSi3616; Werkstoff Nr. 4864. *Obsolete*

WITHERM 61

Thyssen Edelstahlwerke AG
C 0.1, Si 1, Ni 32, Cr 20, bal Fe.
Austenitic Ni-Cr steel for high temperature operation; resists
scaling and carburization; useful to 1200 C. DIN
X10NiCr3220; Werkstoff Nr. 4861. *Obsolete*

WITHERM 8

Thyssen Edelstahlwerke AG
C 0.1, Si 2.4, Cr 6, bal Fe.
Annealed: 78,000-100,000 TS; 57,000 YS; 18 El. For high
temperature operation, usable in air from 800-900 C. DIN
X10CrSi6; Werkstoff Nr. 4712. *Obsolete*

WITTEN 2260

Thyssen Edelstahlwerke AG
C 0.07, Cr 22, Mo 9, Fe 0-5, Nb 0.4, bal Ni.
Nickel base precipitation hardening alloy, corrosion
resistant. DIN NiCr20Mo9Nb; Werkstoff Nr. 2.4856.
Obsolete

WITTEN 5H

Thyssen Edelstahlwerke AG

C 0.53, Si 0.9, Mn 0.9, bal Fe.

For gears, crimpers, punches, axes, axles; water hardened. *Obsolete*

WITTEN AZ 35 M

Thyssen Edelstahlwerke AG

C 0.35, Mn 0.8, Al 1, Cr 1.1, Mo 0.2, bal Fe.

Nitriding steel for plastic molds, particularly for impressions possessing thin webs. DIN 33AlCrMo4. *Obsolete*

WITTEN AZ 38 M

Thyssen Edelstahlwerke AG

C 0.35, Mn 0.5, Al 1, Cr 1.7, Mo 0.2, bal Fe.

Nitriding steel for plastic molds, particularly for impressions possessing thin webs. *Obsolete*

WITTEN BCNO 18

Thyssen Edelstahlwerke AG

C 0.21, Mn 0.8, Cr 0.5, Mo 0.2, Ni 0.6, B, bal Fe.

For case hardening fairly heavy sections as drive gears, splined couplings; automotive and heavy transport equipment. DIN 21NiCrMoB2. *Obsolete*

WITTEN BCNO 22

Thyssen Edelstahlwerke AG

C 0.2, Mn 0.7, Cr 0.5, Mo 0.45, Ni 1.6, B, bal Fe.

For case hardening heavily loaded driving gears, spline shafts, drive couplings. DIN 20NiCrMoB6. *Obsolete*

WITTEN BCO 23

Thyssen Edelstahlwerke AG

C 0.23, Mn 0.65, Cr 1.05, Mo 0.15, B, bal Fe.

For case hardening intermediate thick sections; high hardness in core area. For drive gears, spline shafts, transmission gears. DIN 23CrMoB4; Werkstoff Nr. 1.7211. *Obsolete*

WITTEN BCO 24

Thyssen Edelstahlwerke AG

C 0.23, Mn 0.8, Cr 0.8, Mo 0.35, B, bal Fe.

For case hardening heavy sections to produce strong core properties. For drive gears, splined couplings, axles, shafts for heavy transport equipment. DIN 23CrMoB33; Werkstoff Nr. 1.7271. *Obsolete*

WITTEN BL EXTRA

Thyssen Edelstahlwerke AG

C 0.45, Cr 1.4, Ni 4, W 0.5, bal Fe.

Oil hardening die steel for hobbing tools, die sinking dies, embossing tools for table knives, spoons. *Obsolete*

WITTEN C 023 L

Thyssen Edelstahlwerke AG

C 0.25, Si 0.25, Mn 0.65, Cr 1, Mo 0.2, bal Fe.

Alloy steel. Werkstoff Nr. 1.7214. *Obsolete*

WITTEN C 33

Thyssen Edelstahlwerke AG

C 0.34, Mn 0.75, Cr 1.05, bal Fe.

Water or oil hardening carbon steel. Annealed, cold finished: 223 Brin. maximum. Hardenable to 115,000-135,000 psi TS in sections up to 40 mm diameter. For shafts, axles, splined couplings, drive pins. DIN 34Cr4; Werkstoff Nr. 1.7033. *Obsolete*

WITTEN C 37

Thyssen Edelstahlwerke AG

C 0.37, Mn 0.75, Cr 1.05, bal Fe.

Water or oil hardening low alloy steel. Annealed, cold finished: 235 maximum Brin. Hardenable to 115,000-142,000 psi TS in sections up to 40 mm diameter. For cog wheels, bolts, piston rods, shafts. DIN 37Cr4; Werkstoff Nr. 1.7034. *Obsolete*

WITTEN C 38

Thyssen Edelstahlwerke AG

C 0.38, Mn 0.65, Cr 0.5, bal Fe.

Water or oil hardening carbon steel. Annealed: 207 Brin. maximum. Hardenable to 115,000-135,000 psi TS in sizes not over 16 mm diameter. For shafts, spline couplings. DIN 38Cr2; Werkstoff Nr. 1.7003. *Obsolete*

WITTEN C 46

Thyssen Edelstahlwerke AG

C 0.46, Mn 0.65, Cr 0.5, bal Fe.

Water or oil hardening carbon steel. Annealed: 207 Brin maximum. Cold finished: 223 Brin maximum. Hardenable to 115,000-135,000 psi TS up to 40 mm diameter sections. For shafts, axles, splined couplings, drive pins, induction hardened parts. DIN 46Cr2; Werkstoff Nr. 1.7006. *Obsolete*

WITTEN C40

Thyssen Edelstahlwerke AG

C 0.41, Cr 1.1, Mn 0.7, bal Fe.

For gears, bolts, machine tool parts; water or oil hardened. *Obsolete*

WITTEN CN 15

Thyssen Edelstahlwerke AG

C 0.15, Mn 0.5, Cr 1.5, Ni 1.5, bal Fe.

For case hardening; intermediate depth of hardening of core area. For drive gears and splines, axles, shafts, gauges, wear parts. DIN 15CrNi6; Werkstoff Nr. 1.5919. *Obsolete*

WITTEN CN 15 L

Thyssen Edelstahlwerke AG

C 0.15, Si 0.25, Mn 0.5, Cr 1.5, Ni 1.5, bal Fe.

Alloy carburizing steel; core strength after hardening: 128,000-170,000 psi TS. For gears, splined shafts. DIN 15CrNi6; Werkstoff Nr. 1.5924. *Obsolete*

WITTEN CN 18 L

Thyssen Edelstahlwerke AG

C 0.18, Si 0.25, Mn 0.5, Cr 2, Ni 2, bal Fe.

Alloy carburizing steel; core strength after heat treating: 140,000-205,000 psi TS. DIN 18CrNi8; Werkstoff Nr. 1.5934. *Obsolete*

WITTEN CN18

Thyssen Edelstahlwerke AG

C 0.18, Cr 2, Ni 2, Mn 0.5, bal Fe.

For gears, shafts, fasteners; case hardening steel. *Obsolete*

WITTEN CNE3

Thyssen Edelstahlwerke AG

C 0.13, Cr 0.7, Ni 3.5, bal Fe.

For gears, bolts, camshafts, cams, fasteners; case hardening steel, shock resistant. *Obsolete*

WITTEN CNE4

Thyssen Edelstahlwerke AG

C 0.13, Cr 1.1, Ni 4.5, bal Fe.

For gears, cams, camshafts, machine tool parts; case hardening steel, shock resistant. *Obsolete*

WITTEN CNL

Thyssen Edelstahlwerke AG

C 0.45, Ni, Cr, W, bal Fe.

For upsetters, extrusion press parts; oil hardened, tough. *Obsolete*

WITTEN CNO 17

Thyssen Edelstahlwerke AG

C 0.17, Mn 0.5, Cr 1.6, Mo 0.35, Ni 1.5, bal Fe.

For case hardening medium thick sections. For drive gears, axles, cams, spline shafts, tractor gears. DIN 17CrNiMo6; Werkstoff Nr. 1.6587. *Obsolete*

WITTEN CNO 18 F

Thyssen Edelstahlwerke AG

C 0.21, Mn 0.8, Cr 0.5, Mo 0.2, Ni 0.6, bal Fe.

For case hardening. For gears, pinions, splined shafts, cams. DIN 21NiCrMo2; AISI 8620; Werkstoff Nr. 1.6523. *Obsolete*

WITTEN CNO 40

Thyssen Edelstahlwerke AG

C 0.34, Mn 0.55, Cr 1.55, Mo 0.2, Ni 1.55, bal Fe.

Oil hardenable to 142,000-170,000 psi TS in sections 40-100 mm diameter. For large and highly stressed shafts, axles, and drive couplings in heavy equipment. DIN 34CrNiMo6; Werkstoff Nr. 1.6582. *Obsolete*

WITTEN CNO36

Thyssen Edelstahlwerke AG

C 0.36, Cr 1, Mo 0.2, Ni 1, bal Fe.

For gears, bolts, machine tool parts; oil hardened, shock resistant. *Obsolete*

WITTEN CNOV 30

Thyssen Edelstahlwerke AG

C 0.3, Mn 0.45, Cr 2, Mo 0.4, Ni 2, bal Fe.

Oil hardenable to 142,000-170,000 psi TS in sections 100-160 mm diameter. For very large and highly stressed axles, shafts and drive couplings in earth moving equipment, cranes, ordnance. DIN 30CrNiMo8; Werkstoff Nr. 1.6580. *Obsolete*

WITTEN CNOV 30 L

Thyssen Edelstahlwerke AG

C 0.31, Si 0.25, Mn 0.45, Cr 2, Mo 0.4, Ni 2, bal Fe.

Deep hardening alloy steel; weldable. Werkstoff Nr. 1.6604. *Obsolete*

WITTEN CNOV 37 L VA

Thyssen Edelstahlwerke AG

C 0.37, Si 0.25, Mn 0.65, Cr 0.8, Mo 0.35, Ni 1.8, V 0.1, bal Fe.

Alloy steel, hardenable to 250,000 psi TS. Werkstoff Nr. 1.6944; similar to AISI 4340. *Obsolete*

WITTEN CNV1

Thyssen Edelstahlwerke AG

C 0.28-0.35, Cr 0.5, Ni 1.5, Mn 0.6, bal Fe.

For gears, pinions, crankshafts, fasteners; oil hardened, shock resistant. *Obsolete*

WITTEN CNV2

Thyssen Edelstahlwerke AG

C 0.28-0.35, Cr 0.7, Ni 2.5, bal Fe.

For gears, pinions, studs, bolts; oil hardened, shock resistant. *Obsolete*

WITTEN CNV3

Thyssen Edelstahlwerke AG

C 0.22-0.3, Cr 0.7, Ni 3.5, bal Fe.

For gears, bolts, crankshafts, fasteners; oil hardened, shock resistant. *Obsolete*

WITTEN CNV4

Thyssen Edelstahlwerke AG

C 0.35, Ni 4.5, Cr 1.3, bal Fe.

For gears, bolts, machine tool parts; oil hardened, shock resistant. *Obsolete*

WITTEN CO 20 AF

Thyssen Edelstahlwerke AG

C 0.2, Mn 0.75, Cr 0.35, Mo 0.5, bal Fe.

For case hardening; good core strength and hardness up to about 30 mm cross section thickness by proper heat treating. For gears, pinions, shafts, axles, splined couplings. DIN 20MoCr4; Werkstoff Nr. 1.7321. *Obsolete*

WITTEN CO 20 AL

Thyssen Edelstahlwerke AG

C 0.2, Si 0.25, Mn 0.7, Cr 0.45, Mo 0.4, bal Fe.

Low alloy carburizing steel. DIN 20CrMo2; Werkstoff Nr. 1.7334. *Obsolete*

WITTEN CO 20 L

Thyssen Edelstahlwerke AG

C 0.2, Si 0.25, Mn 0.7, Cr 0.45, Mo 0.45, bal Fe.

Low alloy carburizing steel. Werkstoff Nr. 1.7324. *Obsolete*

WITTEN CO 23

Thyssen Edelstahlwerke AG
C 0.25, Mn 0.65, Cr 1, Mo 0.2, bal Fe.
For gears, pinions, cams, bolts, camshafts; case hardening steel, tough. *Obsolete*

WITTEN CO 25 F

Thyssen Edelstahlwerke AG
C 0.25, Mn 0.7, Cr 0.45, Mo 0.5, bal Fe.
For case hardening; higher core hardness through heavier sections. For drive gears, heavy pinions, axles, shafts, splined couplings. DIN 25MoCr4; Werkstoff Nr. 1.7325. *Obsolete*

WITTEN CO 32

Thyssen Edelstahlwerke AG
C 0.33, Mn 0.65, Cr 1, Mo 0.2, bal Fe.
For gears, bolts, studs, shafts, fasteners; oil hardened, tough. *Obsolete*

WITTEN CO 38

Thyssen Edelstahlwerke AG
C 0.43, Mn 0.65, Cr 1, Mo 0.2, bal Fe.
For gears, bolts, studs, shafts, fasteners; oil hardened, tough. *Obsolete*

WITTEN CO 46

Thyssen Edelstahlwerke AG
C 0.5, Mn 0.65, Cr 1, Mo 0.2, bal Fe.
For gears, bolts, studs, shafts, fasteners; oil hardened, tough. *Obsolete*

WITTEN COL 30

Thyssen Edelstahlwerke AG
C 0.32, Mn 0.55, Cr 3, Mo 0.4, bal Fe.
Oil hardenable to 142,000-170,000 psi TS in sections 40-100 mm diameter. For large, heavily loaded shafts, drive couplings. DIN 32CrMo12; Werkstoff Nr. 1.7361. *Obsolete*

WITTEN COV 13 L

Thyssen Edelstahlwerke AG
C 0.15, Si 0-0.2, Mn 0.9, Cr 1.4, Mo 0.9, V 0.25, bal Fe.
Alloy carburizing steel. Werkstoff Nr. 1.7734. *Obsolete*

WITTEN COV 30

Thyssen Edelstahlwerke AG
C 0.3, Mn 0.55, Cr 2.5, Mo 0.2, V 0.15, bal Fe.
Oil hardenable to 142,000-170,000 psi TS in sections 100-160 mm diameter. For large heavily loaded shafts for tractor, railroad, ordnance equipment. DIN 30CrMoV9; Werkstoff Nr. 1.7707. *Obsolete*

WITTEN COV 30 L

Thyssen Edelstahlwerke AG
C 0.3, Si 0.25, Mn 0.5, Cr 2.5, Mo 0.2, V 0.15, bal Fe.
Alloy steel, fairly deep hardening. DIN 30CrMoV9; Werkstoff Nr. 1.7704. *Obsolete*

WITTEN CRL

Thyssen Edelstahlwerke AG
C 1.05, Cr 1.2, bal Fe.
For bearings, sleeves, liners, bushings; water hardened, wear resistant. *Obsolete*

WITTEN CRW1

Thyssen Edelstahlwerke AG
C 0.35, Cr 1.05, V 0.18, W 1.85, bal Fe.
For header dies, upsetters, crimpers; oil hardened, tough. *Obsolete*

WITTEN CRW2

Thyssen Edelstahlwerke AG
C 0.45, Cr 1.05, V 0.2, W 1.85, bal Fe.
For header dies, upsetters, punches, shears; oil hardened, tough. *Obsolete*

WITTEN CRW3

Thyssen Edelstahlwerke AG
C 0.55, Si 0.9, Cr 1.05, V 0.18, W 1.85, bal Fe.
For header dies, upsetters; oil hardened, tough. *Obsolete*

WITTEN CV 48

Thyssen Edelstahlwerke AG
C 0.5, Mn 0.9, Cr 1.05, V 0.12, bal Fe.
Oil hardening low alloy steel. Annealed: 235 Brin maximum. Hardenable to 142,000-170,000 psi TS in sections up to 40 mm diameter. For axles, shafts, gears, connecting rods, spline couplings, hatchet heads. DIN 50CrV4; Werkstoff Nr. 1.8159; AISI 6150. *Obsolete*

WITTEN D 1414

Thyssen Edelstahlwerke AG
C 0.32-0.4, Si 0.15-0.35, Mn 0.4-0.7, Cr 0.5, bal Fe.
For forgings, bolts, screws, nuts; for operation up to 500 C. DIN Ck35; Werkstoff Nr. 1.1181. *Obsolete*

WITTEN D 4415

Thyssen Edelstahlwerke AG
C 0.12-0.2, Si 0.15-0.5, Mn 0.5-0.8, Mo 0.45-0.65, bal Fe.
For forgings, stampings, flanges for operation up to 530 C. DIN 16Mo5; Werkstoff Nr. 1.5423. *Obsolete*

WITTEN D 4519 S

Thyssen Edelstahlwerke AG
C 0.1-0.18, Si 0.15-0.35, Mn 0.3-0.6, Cr 0.3-0.6, Mo 0.5-0.65, V 0.25-0.35, bal Fe.
For forgings, stampings, flanges for operation up to 560 C. DIN 14MoV63; Werkstoff Nr. 1.7715. *Obsolete*

WITTEN D 5325

Thyssen Edelstahlwerke AG
C 0.08-0.12, Si 0.15-0.35, Mn 0.3-0.5, Cr 2.7-3, Mo 0.2-0.3, bal Fe.
For parts to be used up to about 520 C, particularly in hydrogenation plants. DIN 10CrMo11; Werkstoff Nr. 1.7276. *Obsolete*

WITTEN D 5509

Thyssen Edelstahlwerke AG
C 0.15-0.2, Si 0.15-0.35, Mn 0.3-0.5, Cr 2.7-3, Mo 0.2-0.3, V 0.1-0.2, bal Fe.
For forgings, stampings, vessels for steam lines and similar operations at temperatures up to about 520 C. DIN 17CrMoV10; Werkstoff Nr. 1.7766. *Obsolete*

WITTEN D 5514 S

Thyssen Edelstahlwerke AG
C 0-0.15, Si 0.3-0.5, Mn 0.3-0.6, Cr 4.5-5.5, Mo 0.45-0.65, bal Fe.
For forgings and parts for high pressure hydrogenation plants for use at temperatures up to about 520 C. DIN 12CrMo195; Werkstoff Nr. 1.7362. *Obsolete*

WITTEN D 5535

Thyssen Edelstahlwerke AG
C 0.17-0.25, Si 0.3-0.6, Mn 0.3-0.6, Cr 1.2-1.5, Mo 1-1.2, V 0.25-0.35, Ni 0-0.6, bal Fe.
For forgings for turbine parts, flanges, nuts, bolts and similar hardware for operation up to 550 C. DIN 21CrMoV511; Werkstoff Nr. 1.8070. *Obsolete*

WITTEN D 5607

Thyssen Edelstahlwerke AG
C 0-0.15, Si 0.15-0.5, Mn 0.4-0.6, Cr 2-2.5, Mo 0.9-1.1, bal Fe.
For forgings for boiler plants, flanges, and similar parts for operation up to 580 C. DIN 10CrMo910; Werkstoff Nr. 1.7380. *Obsolete*

WITTEN D 6 G

Thyssen Edelstahlwerke AG
C 0.65, Si 0.25, Mn 0.4, bal Fe.
Water hardening tool steel for small and medium size forging dies, swaging dies, trimming dies, woodworking tools. *Obsolete*

WITTEN D 7 G

Thyssen Edelstahlwerke AG
C 0.75, Si 0.3, Mn 0.3, bal Fe.
Water hardening steel for small and medium plain carbon forging dies. DIN C75W3. *Obsolete*

WITTEN D 7 S

Thyssen Edelstahlwerke AG
C 0.75, Si 0.25, Mn 0.3, bal Fe.
Water hardening tool steel for hot shears, shear blades, screw drivers. AISI W1. *Obsolete*

WITTEN D 8512

Thyssen Edelstahlwerke AG
C 0.19, Cr 12, Mo 1, bal Fe.
Heat treated: 100,000-120,000 TS; 71,000 YS; 14-16 El. At 500 C: 38,800 YS minimum. For turbine parts, high temperature bolts and nuts. DIN X19CrMo121; Werkstoff Nr. 1.4921. *Obsolete*

WITTEN D 8514

Thyssen Edelstahlwerke AG
C 0.17, Cr 11, Ni 0.6, Mo 0.6, V 0.25, Nb 0.2, bal Fe.
Heat treated: 135,000-162,000 TS; 109,000 YS minimum; 6-8 El. At 600 C: 61,000 YS minimum. For turbine shafts and wheels. DIN X17CrMoVNb121; Werkstoff Nr. 4913. *Obsolete*

WITTEN D 8516

Thyssen Edelstahlwerke AG
C 0.2, Cr 12, Ni 0.55, Mo 1, V 0.3, bal Fe.
Heat treated: 100,000-120,000 TS; 71,000 YS; 14-16 El. At 500 C: 38,000 YS minimum. For shafts and hardware for turbines and boilers operating at temperatures up to 600 C. DIN X20CrMoV121; Werkstoff Nr. 1.4922. *Obsolete*

WITTEN D 8518

Thyssen Edelstahlwerke AG
C 0.22, Cr 12, Ni 0.55, Mo 1, V 0.3, bal Fe.
Heat treated: 112,000-134,000 TS; 85,000 YS minimum; 12-14 El. At 500 C: 49,500 YS minimum. For turbine shafts and other parts operating up to 600 C. DIN X22CrMoV121; Werkstoff Nr. 1.4923. *Obsolete*

WITTEN D 8518 L

Thyssen Edelstahlwerke AG
C 0.22, Si 0.3, Mn 0.5, Cr 11.5, Mo 1, Ni 0.5, V 0.3, bal Fe.
Martensitic stainless, forgeable, heat treatable. DIN X22CrMoV121; Werkstoff Nr. 1.4934. *Obsolete*

WITTEN D 8518 W

Thyssen Edelstahlwerke AG
C 0.2, Cr 12, Ni 0.55, Mo 1, W 0.5, V 0.3, bal Fe.
Properties same as Witten DA 8518. For turbine parts, shafts, blades, hardware, operating up to 600 C. DIN X20CrMoWV121; Werkstoff Nr. 1.4935. *Obsolete*

WITTEN D 9 S

Thyssen Edelstahlwerke AG
C 0.9, Si 0.25, Mn 0.3, bal Fe.
Water hardening tool steel for punches, mandrels, drifts, scissors, woodworking tools. AISI W1. *Obsolete*

WITTEN D2417

Thyssen Edelstahlwerke AG
C 0.19, Si 0.5, Mn 1.15, bal Fe.
For gears, bolts, cams, machine tool parts; case hardened. *Obsolete*

WITTEN D4

Thyssen Edelstahlwerke AG
C 0.45, Si 0.25-0.5, Mn 0.3-0.8, bal Fe.
Hot rolled: 98,000 TS; 60,000 YS; 24 El; 55 RA; 212 Brin. For gears, bolts, fasteners, shafts; water hardened. *Obsolete*

WITTEN D4412S

Thyssen Edelstahlwerke AG
C 0.15, Si 0.25, Mn 0.6, Mo 0.3, bal Fe.
For gears, bolts, machine tool parts; case hardened, tough. *Obsolete*

WITTEN D5512

Thyssen Edelstahlwerke AG
C 0.24, Cr 1.15, Mo 0.25, Mn 0.55, bal Fe.
For gears, bolts, machine tool parts; oil or water hardened, tough. *Obsolete*

WITTEN D5515S

Thyssen Edelstahlwerke AG
C 0.13, Cr 0.85, Mo 0.45, Mn 0.55, bal Fe.
For gears, bolts, machine tool parts; case hardened, tough.
Obsolete

WITTEN D5517

Thyssen Edelstahlwerke AG
C 0.2, Cr, Mo, V, bal Fe.
For gears, bolts, machine tool parts; case hardened, tough.
Obsolete

WITTEN D5520

Thyssen Edelstahlwerke AG
C 0.24, Cr 1.35, Mo 0.55, V 0.2, bal Fe.
For gears, bolts, crankshafts, axles; oil hardened, shock resistant. *Obsolete*

WITTEN DA 1525 LVA

Thyssen Edelstahlwerke AG
C 0-0.08, Si 0-1, Mn 1.5, Cr 15, Mo 1.3, Ni 26, V 0.3, Ti 2.1, B 0.006, bal Fe.
High temperature alloy, forgeable, age hardenable. DIN X5NiCrTi2615; Werkstoff Nr. 1.4944; AMS 5734; A-286.
Obsolete

WITTEN DA 1525 VU

Thyssen Edelstahlwerke AG
C 0.05, Cr 15, Ni 25, Mo 1.3, V 0.3, Ti 2.1, Al 0.3, B, bal Fe.
Hardened: 125,000-170,000 TS; 92,000 YS min; 12 El. At 700 C: 45,000 YS min. For stainless parts requiring good strength. Usable to about 700 C. DIN X5NiCrTi2615; Werkstoff Nr. 1.4980. *Obsolete*

WITTEN DA 1573

Thyssen Edelstahlwerke AG
C 0-0.03, Cr 15, Fe 5, bal Ni.
Ni-Cr corrosion resistant and high temperature alloy. DIN LC-NiCr15Fe; Werkstoff Nr. 2.4814. *Obsolete*

WITTEN DA 1613

Thyssen Edelstahlwerke AG
C 0.06, Cr 16, Ni 13, Nb, bal Fe.
Solution annealed: 74,000-100,000 TS; 30,000 YS min; 22-35 El. Niobium stabilized austenitic stainless steel. For turbine parts and welded assemblies for use to 750 C. DIN X8CrNiNb1613; Werkstoff Nr. 1.4961. *Obsolete*

WITTEN DA 1613 V

Thyssen Edelstahlwerke AG
C 0.06, Cr 16, Ni 13, Mo 1.3, V 0.7, N 0.1, Nb, bal Fe.
Annealed: 78,000-106,000 TS; 37,000 YS min; 20-30 El. Hot-cold work: 92,000-120,000 TS; 71,000 YS min; 12-16 El. At 600 C: 30,000 YS min. Parts in hot-cold worked condition usable in turbines to 650 C. DIN X8CrNiMoVNb1613; Werkstoff Nr. 1.4988. *Obsolete*

WITTEN DA 1616

Thyssen Edelstahlwerke AG
C 0.07, Cr 16, Ni 16, Mo 1.8, Nb, bal Fe.
Solution annealed: 75,000-100,000 TS; 31,000 YS min; 22-35 El. Nobium stabilized austenitic stainless. For parts to be used at temperatures up to 750 C. DIN X8CrNiMoNb1616; Werkstoff Nr. 1.4981. *Obsolete*

WITTEN DA 1616 B

Thyssen Edelstahlwerke AG
C 0.07, Cr 16, Ni 16, Mo 1.8, B 0.06, Nb, bal Fe.
Annealed: 78,000-106,000 TS; 40,000 YS min; 20-30 El. Hot-cold work: at 600 C: 37,000 YS min. Parts in hot-cold work condition have good tensile properties at 650-700 C. DIN X8CrNiMoBNb1616; Werkstoff Nr. 1.4986. *Obsolete*

WITTEN DA 1616 L

Thyssen Edelstahlwerke AG
C 0-0.1, Si 0.4, Mn 1.2, Cr 16.5, Mo 1.8, Ni 16.5, Nb = 10 x C min, bal Fe.
High temperature alloy, weldable. DIN X8CrNiMoNb1616; Werkstoff Nr. 1.4984. *Obsolete*

WITTEN DA 1616 W

Thyssen Edelstahlwerke AG
C 0.07, Cr 16, Ni 16, W 3, N 0.1, Nb, bal Fe.
Solution annealed: 78,000-106,000 TS; 37,000 YS min; 20-30 El. At 600 C: 21,000 YS min. Stabilized austenitic stainless. For parts to be used at temperatures up to 750 C. DIN X8CrNiWNb1616; Werkstoff Nr. 1.49495. *Obsolete*

WITTEN DA 1713

Thyssen Edelstahlwerke AG
C 0.06, Cr 17, Ni 13, Mo 2.2, bal Fe.
Solution annealed: 71,000-100,000 TS; 30,000 YS min; 34-45 El. At 500 C: 15,500 YS min. Austenitic, good corrosion resistance. For parts and assemblies operating at temperatures up to 750 C. DIN X6CrNiMo1713; Werkstoff 1.4919. *Obsolete*

WITTEN DA 1713 W

Thyssen Edelstahlwerke AG
C 0.12, Cr 16, Ni 13.5, W 3, Ti, bal Fe.
Annealed: 71,000-106,000 TS; 34,000 YS min. Hot-cold work: 92,000-112,000 TS; 64,000-78,000 YS; 25 El. At 600 C: 24,000 YS min. In hot-cold worked condition it is usable for turbine parts up to 650 C. DIN X12CrNiWTi1613; Werkstoff Nr. 1.4962. *Obsolete*

WITTEN DA 1811

Thyssen Edelstahlwerke AG
C 0.06, Cr 18, Ni 11, bal Fe.
Solution annealed: 71,000-100,000 TS; 27,000 YS min; 37-50 El. Austenitic stainless. DIN X6CrNi1811; Werkstoff Nr. 1.4948. *Obsolete*

WITTEN DA 2019

Thyssen Edelstahlwerke AG
C 0.12, Cr 21, Co 20, Ni 20, Mo 3.2, W 2.5, Nb 1, N 0.15, bal Fe.
Hardened: 100,000-142,000 TS; 50,000 YS min; 25 El. At 600 C: 34,000 YS min. For gas turbine parts operating up to 730 C. DIN X12CrCoNi2120; Werkstoff Nr. 1.4971. *Obsolete*

WITTEN DA 2019 L

Thyssen Edelstahlwerke AG
C 0.12, Si 0-1, Mn 1.5, Cr 21, Mo 3, Ni 20, Co 20, Nb 1, N 0.15, W 2.5, bal Fe.
High temperature alloy. DIN X12CrCoNi2120; Werkstoff Nr. 1.4974. AMS 5768; Multimet N-155. *Obsolete*

WITTEN DA 2020

Thyssen Edelstahlwerke AG
C 0.4, Cr 20, Co 20, Ni 20, Mo 4, W 4, Nb 4, bal Fe.
Hardened: 112,000-142,000 TS; 57,000 YS min; 20 El. At 700 C: 30,000 YS min. For gas turbine parts operating up to 900 C. DIN X40CoCrNi2020; Werkstoff Nr. 1.4977. *Obsolete*

WITTEN DA 2040

Thyssen Edelstahlwerke AG
C 0.1, Cr 20, Ni 10, Fe 0-3, W 15, bal Co.
Cobalt base high temperature alloy, for high temperature springs. DIN CoCr20W15Ni; Werkstoff Nr. 2.4967. *Obsolete*

WITTEN DA 2040 L

Thyssen Edelstahlwerke AG
C 0.1, Si 0-1, Mn 1.5, Cr 20, Ni 10, W 15, bal Co.
Cobalt base high temperature alloy. DIN CoCr20W15Ni; Werkstoff Nr. 2.4964. *Obsolete*

WITTEN DA 2060 TI

Thyssen Edelstahlwerke AG
C 0.07, Cr 20, Co 10, Fe 0-5, Ti 2.4, Al 1.4, bal Ni.
Nickel base precipitation hardening alloy, corrosion resistant. DIN NiCr20Co18Ti; Werkstoff Nr. 2.4969. *Obsolete*

WITTEN DA 2060 TIL

Thyssen Edelstahlwerke AG
C 0.05, Si 0.35, Mn 0.5, Cr 19.5, Co 17.5, Ti 2.3, Al 1.4, Fe 3, bal Ni.
Nickel base high temperature alloy. DIN NiCr20Co18Ti; Werkstoff Nr. 2.4632. *Obsolete*

WITTEN DA 2080

Thyssen Edelstahlwerke AG
C 0.12, Cr 20, Fe 0-5, Ti 0.4, bal Ni.
Nickel base high temperature alloy. DIN NiCr20Ti; Werkstoff Nr. 2.4951. *Obsolete*

WITTEN DA 2080 L

Thyssen Edelstahlwerke AG
C 0.1, Si 0.35, Mn 0.5, Cr 19.5, Ti 0.5, Fe 3, bal Ni.
Nickel base high temperature alloy. DIN NiCr20Ti; Werkstoff 2.6430. *Obsolete*

WITTEN DA 2080 TI

Thyssen Edelstahlwerke AG
C 0.07, Cr 20, Fe 0-5, Ti 2.4, Al 1.4, bal Ni.
Nickel base precipitation hardening alloy, corrosion resistant. DIN NiCr20TiAl; Werkstoff Nr. 2.4952. *Obsolete*

WITTEN DA 2080 TIL

Thyssen Edelstahlwerke AG
C 0.05, Si 0.35, Mn 0.5, Cr 19.5, Ti 2.2, Al 1.2, Fe 3, bal Ni.
Nickel base high temperature alloy. DIN NiCr20TiAl; Werkstoff Nr. 2.4631. *Obsolete*

WITTEN DMO

Thyssen Edelstahlwerke AG
C 0.45, Cr, V, Si, bal Fe.
For gears, shafts, crankshafts, springs; oil hardened, shock resistant. *Obsolete*

WITTEN DMW

Thyssen Edelstahlwerke AG
C 0.38, Cr, V, Si, bal Fe.
For gears, shafts, bolts, studs; oil hardened, shock resistant. *Obsolete*

WITTEN EM 3

Thyssen Edelstahlwerke AG
C 0.15, Cr 0.7, Ni 3.5, bal Fe.
Case hardening steel, designed largely for high-duty plastic molds. DIN 15NiCr14. *Obsolete*

WITTEN EML

Thyssen Edelstahlwerke AG
C 0.19, Cr 1.25, Mo 0.2, Ni 3.75, bal Fe.
For gears, bolts, camshafts, fasteners; case hardening steel, shock resistant. *Obsolete*

WITTEN EMW

Thyssen Edelstahlwerke AG
C 0.04, Cr 3.8, Mo 0.5, bal Fe.
Cold hobbing steel to be subsequently case hardened for plastic molds. DIN X6CrMo4. *Obsolete*

WITTEN G 110

Thyssen Edelstahlwerke AG
C 1.15, Si 0.25, Mn 0.25, bal Fe.
Water hardening tool steel, for chisels, hammers, sculptor tools. *Obsolete*

WITTEN G 65

Thyssen Edelstahlwerke AG
C 0.7, Si 0.25, Mn 0.25, bal Fe.
Water hardening tool steel, hollow or solid for drills, trepanning and core drilling tools for rock or soil. *Obsolete*

WITTEN G75

Thyssen Edelstahlwerke AG
C 0.85, Si 0-0.25, Mn 0-0.25, bal Fe.
Heat treated: 190,000 TS; 145,000 YS; 10 El; 30 RA; 400 Brin. For drills, taps, reamers, springs, tools; Type W1; water hardened. *Obsolete*

WITTEN GB

Thyssen Edelstahlwerke AG
C 1, Mn 0.9, Cr 1, W 1.15, bal Fe.
For cutters, tools, forming dies; water or oil hardened. *Obsolete*

WITTEN GBA

Thyssen Edelstahlwerke AG
C 0.95, Mn 1.1, Cr 0.6, V 0.1, W 0.6, bal Fe.
Oil hardenable to 60-64 Rock C. Non-deforming, for dies, gauges, milling cutters, reamers, forming tools. *Obsolete*

WITTEN GKM

Thyssen Edelstahlwerke AG
C 1, Cr 1.55, Mn 0.35, bal Fe.
For bearings, liners, bushings, cutters; water hardened, wear resistant. *Obsolete*

WITTEN GLHA

Thyssen Edelstahlwerke AG
C 0.45, Mn 1.4, Cr 1.9, Mo 0.2, bal Fe.
Oil hardening hot work steel for drop forging dies. Available as hardened to 100 Kp/mm 2 TS for plastic molds. DIN 40CrMnMo7. *Obsolete*

WITTEN GSO

Thyssen Edelstahlwerke AG
C 1.45, Cr 1.4, Mn 0.6, bal Fe.
For bearings, bushings; water hardened, wear resistant. *Obsolete*

WITTEN HBL

Thyssen Edelstahlwerke AG
C 1, Cr 1.2, Mo 0.3, bal Fe.
Oil hardening tool steel for drills, counter bores, rock drills. *Obsolete*

WITTEN HCE

Thyssen Edelstahlwerke AG
C 0.15, Mn 0.5, Cr 0.65, bal Fe.
For case hardening; hardness of core after case hardening: 143-187 Brin. For small machine parts as studs, screws, bolts, cap screws, levers, pins. DIN 15Cr3; Werkstoff Nr. 1.1141. *Obsolete*

WITTEN HFC 5 LVA

Thyssen Edelstahlwerke AG
C 0.4, Si 0.9, Mn 0.3, Cr 5, Mo 1.3, V 0.5, bal Fe.
Air hardening hot work tool steel. Werkstoff Nr. 1.7784; AMS 6487; AISI H11. *Obsolete*

WITTEN HFC. 16

Thyssen Edelstahlwerke AG
C 0-0.03, Cr 13.5, Co 8, Ni 4, Mo 4, Ti 0.3, bal Fe.
High temperature corrosion resistant steel. *Obsolete*

WITTEN HFN 18

Thyssen Edelstahlwerke AG
C 0-0.03, Co 7.5, Ni 18, Mo 5, Ti 0.5, bal Fe.
Maraging steel, high strength. DIN X2NiCoMo1885; Werkstoff Nr. 1.6359. *Obsolete*

WITTEN K 1 L

Thyssen Edelstahlwerke AG
C 1.05, Si 0.25, Mn 0.3, Cr 0.5, bal Fe.
Water or oil hardening high carbon steel for ball bearings, balls, races. DIN 105Cr2; Werkstoff Nr. 1.3501; AISI 50100. *Obsolete*

WITTEN K 2 L

Thyssen Edelstahlwerke AG
C 1.05, Si 0.25, Mn 0.3, Cr 1, bal Fe.
Water or oil hardening high carbon steel. For ball and roller bearings and races. DIN 105Cr4; Werkstoff Nr. 1.3503; AISI 51100. *Obsolete*

WITTEN K 3 L

Thyssen Edelstahlwerke AG
C 1, Si 0.25, Mn 0.3, Cr 1.5, bal Fe.
Water or oil hardening high carbon steel. For ball and roller bearings and races. DIN 100Cr6; Werkstoff Nr. 1.3505; AMS 6440; AISI E 52100. *Obsolete*

WITTEN KOBALT 10

Thyssen Edelstahlwerke AG
C 0.76, Cr 4.15, Mo 0.65, Co 9.5, V 1.6, W 18, bal Fe.
For lathe, planer and milling tools for heavy cuts at high speeds; high speed steel with good red hardness. *Obsolete*

WITTEN KOBALT 105 M

Thyssen Edelstahlwerke AG
C 1.23, Cr 4.2, Mo 3.15, V 3.2, W 9.5, Co 10.2, bal Fe.
Air or oil hardenable to 65-68 Rock C. For cutting tools requiring good wearing properties and hot strength; well adopted to free cutting and engraving work. DIN S10-4-3-10. *Obsolete*

WITTEN KOBALT 5

Thyssen Edelstahlwerke AG
C 0.8, Cr 4.1, Mo 0.6, V 1.6, W 18, Co 4.7.
For lathe, planer, milling cutters to operate at higher temperatures; good red hardness. DIN S18-1-2-5; AISI T4. *Obsolete*

WITTEN KOBALT 54 V

Thyssen Edelstahlwerke AG
C 1.35, Cr 4.1, Mo 0.85, V 3.75, W 12, Co 4.65, bal Fe.
Air or oil hardenable to 65-67 Rock C. For lathe tools and other cutting tools requiring good wear resistance and good red hardness. DIN S12-1-4-5; AISI T15 high speed steel. *Obsolete*

WITTEN KOBALT 55 H

Thyssen Edelstahlwerke AG
C 1.12, Cr 4.1, Mo 5, V 1.75, W 6.4, Co 4.75, bal Fe.
Air or oil hardenable to 67-69 Rock C. For twist drills, lathe tools requiring good wear resistance at high speeds. DIN S7-4-2-5; AISI M41 high speed steel. *Obsolete*

WITTEN KOBALT 55M

Thyssen Edelstahlwerke AG
C 0.92, Cr 4.1, Mo 5.1, V 1.9, W 6.4, Co 4.75, bal Fe.
Air or oil hardenable to 64-67 Rock C. For taps, reamers, milling cutters, lathe tools requiring good cutting efficiency, red hardness and toughness. DIN S6-5-2-5; AISI M35 high speed steel. *Obsolete*

WITTEN KOBALT 810 M

Thyssen Edelstahlwerke AG
C 1.08, Cr 4, Mo 9.3, V 1.2, W 1.5, Co 8, bal Fe.
For finish turning, boring tools, requiring sharp edge at high speed. DIN S2-10-1-8; AISI M42. *Obsolete*

WITTEN KV 32 COL

Thyssen Edelstahlwerke AG
C 0.34, Si 0.25, Mn 0.65, Cr 1, Mo 0.2, bal Fe.
Alloy steel. DIN 34CrMo4; Werkstoff Nr. 1.7220; AISI 4135. *Obsolete*

WITTEN LDM

Thyssen Edelstahlwerke AG
C 1.1, Mn 0.35, Cr 1.2, V 0.15, W 1.3, bal Fe.
Oil hardening tool steel; for cutting light metals and non-metallics. *Obsolete*

WITTEN M 28

Thyssen Edelstahlwerke AG
C 0.28, Mn 1.5, bal Fe.
Water or oil hardening carbon tool steel. Hardenable to 115,000-135,000 TS in sizes not over 16 mm diameter. For shafts, connecting rods, spline couplings, lever arms. DIN 28Mn6; Werkstoff Nr. 1.5065. *Obsolete*

WITTEN M41

Thyssen Edelstahlwerke AG
C 0.4, Mn 0.95, Si 0.37, bal Fe.
For gears, bolts, shafts; water hardened. *Obsolete*

WITTEN MC14

Thyssen Edelstahlwerke AG
C 0.16, Cr 0.95, Mn 1.15, Si 0.25, bal Fe.
For gears, cams, machine tool parts; case hardening steel. *Obsolete*

WITTEN MC18

Thyssen Edelstahlwerke AG
C 0.2, Mn 1.25, Cr 1.15, bal Fe.
For gears, cams, machine tool parts; case hardening steel, tough. *Obsolete*

WITTEN MO 5

Thyssen Edelstahlwerke AG
C 0.85, W, Mo, bal Fe.
For tools, dies; oil hardened. *Obsolete*

WITTEN MO 53 V

Thyssen Edelstahlwerke AG
C 1.22, Cr 4.1, Mo 5.1, V 3.2, W 6.4, bal Fe.
Air or oil hardenable to 64-67 Rock C. Lathe tools and other cutting tools requiring extra high wear resistance. DIN S6-5-3; AISI M3 Type 2 HSS. *Obsolete*

WITTEN MO 5H

Thyssen Edelstahlwerke AG
C 0.98, Cr 4.1, Mo 5.1, V 1.9, W 6.4, bal Fe.
Air or oil hardenable to 65-67 Rock C. For drills, reamers, lathe tools; good wear resistance. DIN SC6-5-2; AISI M2 (Mod) high speed steel. *Obsolete*

WITTEN MO 9

Thyssen Edelstahlwerke AG
C 0.88, Cr 3.8, Mo 8.75, V 1.1, W 1.7, bal Fe.
Air or oil hardenable to 63-66 Rock C. For drills, lathe tools for roughing cuts, milling cutters and other tools requiring toughness. DIN S2-9-1; AISI M1 high speed steel. *Obsolete*

WITTEN MO 9V

Thyssen Edelstahlwerke AG
C 1.01, Cr 4, Mo 8.75, V 1.9, W 1.7, bal Fe.
Air or oil hardenable to 64-66 Rock C. Lathe tools, reamers, drills requiring improved wear resistance. DIN S2-9-2; AISI M7 high speed steel. *Obsolete*

WITTEN MS 33

Thyssen Edelstahlwerke AG
C 0.45, Mn 1.1, Cr 2.3, Mo 0.5, V 0.1, bal Fe.
Air hardening tool steel for various types of hand chisels. *Obsolete*

WITTEN MS 5

Thyssen Edelstahlwerke AG
C 0.5, Cr 1, V 0.1, bal Fe.
Oil hardening tool steel for cold chisels, punches, center punches. *Obsolete*

WITTEN MS 5 P (HT)

Thyssen Edelstahlwerke AG
C 0.5, Mn 0.9, Cr 1, V 0.1, bal Fe.
Normally supplied heat treated. Heat treated: 100 kg/mm 2 TS (Rehardening to higher values possible). For molds and frames for processing plastic material. DIN 50CrV4. *Obsolete*

WITTEN MSL

Thyssen Edelstahlwerke AG
C 0.45, Mn 0.9, Cr 1.8, Mo 0.3, bal Fe.
Oil or air hardening tool steel for chisels and upsetting tools. *Obsolete*

WITTEN PCVWL

Thyssen Edelstahlwerke AG
C 0.5, Si 0.25, Mn 0.9, Cr 1, V 0.1, bal Fe.
Alloy steel; hardened and tempered 190,000 psi TS. DIN 50CrV4; Werkstoff Nr. 1.8154. AMS 6448; AISI 6150. *Obsolete*

WITTEN PW SPEZIAL

Thyssen Edelstahlwerke AG
C 0.95, Si 0.3, Mn 0.4, V 0.4, bal Fe.
Oil or water hardening tool steel for embossing tools, dies, grippers and headers for making bolts and nuts. *Obsolete*

WITTEN PWV
Thyssen Edelstahlwerke AG
C 1.45, V 3.3, bal Fe.
Water hardening steel for header dies and striking jaws in
screw and rivet industries, hardenable to above 60 Rock C.
DIN 145V33. *Obsolete*

WITTEN RMH
Thyssen Edelstahlwerke AG
C 0.4, Cr 13, Mn 0.3, bal Fe.
Annealed: 100,000 TS; 55,000 YS; 20 El; 50 RA; 200 Brin.
For valves, cutlery, surgical and dental instruments; Type
420; corrosion resistant. *Obsolete*

WITTEN RMH 735
Thyssen Edelstahlwerke AG
C 0.4, Cr 16, Mo 1.1, bal Fe.
Corrosion resistant steel, hardenable to 54 Rc. For working
corrosive plastic materials. DIN X36CrMo17. *Obsolete*

WITTEN S 1525 LVA
Thyssen Edelstahlwerke AG
C 0-0.08, Si 0-1, Mn 1.5, Cr 15, Mo 1.3, Ni 26, V 0.0, Ti 2.1,
B 0.006, bal Fe.
Welding wire. DIN X5NiCrTi2615; Werkstoff Nr. 1.4954; AMS
5805; A-286. *Obsolete*

WITTEN S 48 L
Thyssen Edelstahlwerke AG
C 0-0.15, Si 0-1.5, Mn 0-2.5, Cr 25, Ni 20, bal Fe.
Stainless steel for deep drawing and for high temperation
operation; weldable; welding rod. DIN X10CrNi2520;
Werkstoff Nr. 1.4854; AISI 310. *Obsolete*

WITTEN S 87 NL
Thyssen Edelstahlwerke AG
C 0-0.08, Si 0-2, Mn 0-1.5, Cr 19, Ni 9, Nb = 12 x C min, bal
Fe.
Austenitic stainless steel, weldable. For welded stainless
assemblies, welding wire. DIN X8CrNiNb199; Werkstoff Nr.
1.4554; AISI 347. *Obsolete*

WITTEN S2
Thyssen Edelstahlwerke AG
C 0.9, Mn 1.9, V 0.1, bal Fe.
For punches, shears, dies, upsetters; oil hardened, non-
deforming. *Obsolete*

WITTEN SCS
Thyssen Edelstahlwerke AG
C 0.67, Si 1.3, Cr 0.5, bal Fe.
For springs, punches; oil hardened, tough. *Obsolete*

WITTEN SOZ
Thyssen Edelstahlwerke AG
C 1.25, Si 1.15, Mn 0.7, Cr 1.2, bal Fe.
For bearings, header dies, punches; oil hardened.
Obsolete

WITTEN SS 105
Thyssen Edelstahlwerke AG
C 1, Cr 5.25, Mo 1.1, V 0.1, bal Fe.
Oil hardening tool steel; for punching, forming and cold
heading dies; used also for hot working dies. *Obsolete*

WITTEN SS 110
Thyssen Edelstahlwerke AG
C 1.05, Cr 13.5, Mo 0.5, bal Fe.
Oil hardening cold work tool steel; for punching and
forming dies. *Obsolete*

WITTEN SS 116
Thyssen Edelstahlwerke AG
C 1.6, Cr 12, Mo 0.65, W 0.5, V 1.2, bal Fe.
Air or oil hardening tool steel; for thread rolling tools, deep
drawing dies with fine impressions. DIN X165CrVMo121;
Werkstoff Nr. 2609. *Obsolete*

WITTEN SS 117 M
Thyssen Edelstahlwerke AG
C 1.6, Cr 12, Mo 0.65, V 0.45, W 0.5, bal Fe.
Air or oil hardenable to 60-64 Rc. Non-deforming; for
cutting and forming dies, nail and tack making dies, hobs,
wood working tools. AISI D2. *Obsolete*

WITTEN SS 125
Thyssen Edelstahlwerke AG
C 2.5, Cr 12, bal Fe.
Oil hardening cold work tool steel; for punching and
forming dies. *Obsolete*

WITTEN SS 213 C
Thyssen Edelstahlwerke AG
C 2.1, Cr 12, Mo 0.5, Co 1, W 0.7, bal Fe.
Air or oil hardening tool steel; cold or hot work. For
punching and forming dies. *Obsolete*

WITTEN SS 65 M
Thyssen Edelstahlwerke AG
C 0.62, Cr 14, Mo 0.55, bal Fe.
Air or oil hardening tool steel; for cutting tools, dies and
molds. *Obsolete*

WITTEN SS115
Thyssen Edelstahlwerke AG
C 1.65, Cr 11.5, V 0.1, Mn 0.3, bal Fe.
For forming and blanking dies, punches; air hardened,
non-deforming. *Obsolete*

WITTEN SS115C
Thyssen Edelstahlwerke AG
C 1.65, Cr 11.5, V 0.1, Co, bal Fe.
For forming and blanking dies; air hardened, non-
deforming. *Obsolete*

WITTEN SS120
Thyssen Edelstahlwerke AG
C 2.1, Cr 11.5, Mn 0.3, bal Fe.
For forming and blanking dies, punches; oil hardened,
non-deforming. *Obsolete*

WITTEN SS131
Thyssen Edelstahlwerke AG
C 2.1, Cr 11.5, W 0.7, bal Fe.
For forming and blanking dies, punches; oil hardened,
non-deforming. *Obsolete*

WITTEN SSO
Thyssen Edelstahlwerke AG
C 1.05, Mn 1.1, Cr 0.9, V 0.1, bal Fe.
Oil hardenable to 60-64 Rock C. Non-deforming; for
complicated cutting punches, blanking dies, blanking
plates; also cutting tools as thread milling cutters. *Obsolete*

WITTEN SW
Thyssen Edelstahlwerke AG
C 1.15, Cr 0.65, V 0.1, Mn 0.3, bal Fe.
For header dies, trimming dies; oil hardened. *Obsolete*

WITTEN UAM
Thyssen Edelstahlwerke AG
C 0.55, Cr 0.7, Ni 1.7, Mo 0.3, V 0.1, bal Fe.
Oil hardening hot work steel for forging dies, hot trimming
dies and hot shear blades. DIN 55NiCrMoV6. *Obsolete*

WITTEN UAM EXTRA
Thyssen Edelstahlwerke AG
C 0.55, Cr 1.1, Ni 1.7, Mo 0.8, V 0.1, bal Fe.
Air or oil hardening hot work steel for forging dies, extrusion
dies for production of light metal shapes. DIN
57NrCrMoV77. *Obsolete*

WITTEN UAM SPEZIAL
Thyssen Edelstahlwerke AG
C 0.56, Cr 0.9, Mo 0.2, Ni 2, bal Fe.
For forging and heading dies; oil hardened, tough.
Obsolete

WITTEN UM 100
Thyssen Edelstahlwerke AG
C 0-0.1, Cr 18, Mo 9, Ni 6, bal Fe.
Annealed: 100,000 TS; 42,500 YS; 15 El. Austenitic, non-
magnetic, corrosion resistant; hardenable by cold work to
about 175,000 TS. For non-magnetic, corrosion resistant
structural parts; weldable. DIN X8CrMnNi189; Werkstoff Nr.
3965. *Obsolete*

WITTEN UM 110
Thyssen Edelstahlwerke AG
C 0.6, Mn 9.5, Cr 22, Ni 4.5, bal Fe.
Annealed: 140,000 TS; 85,000 YS; 30 El. Austenitic, non-
magnetic, corrosion resistant. For highly stressed non-
magnetic, stainless parts. DIN X50CrMnNi229; Werkstoff Nr.
3967. *Obsolete*

WITTEN UM 115
Thyssen Edelstahlwerke AG
C 0.5, Mn 9, Cr 21, Ni 4, V 1, bal Fe.
Solution annealed: 120,000-155,000 TS; 78,000 YS; 30 El.
Aged: 150,000-175,000 TS; 120,000 YS; 12 El. For structural
parts requiring good strength and corrosion resistance. DIN
X50CrMnNiV229; Werkstoff Nr. 3947. *Obsolete*

WITTEN UM 125
Thyssen Edelstahlwerke AG
C 0.3, Mn 3.5, P 0.4, Cr 18, Ni 10, bal Fe.
Solution annealed: 92,000 TS; 42,000 YS; 40 El. Aged:
112,000-127,000 TS; 67,000-78,000 YS; 20 El. DIN
X25CrNiMnP1810; Werkstoff Nr. 3961. *Obsolete*

WITTEN UM 30
Thyssen Edelstahlwerke AG
C 0.45, Mn 19, Cr 4, N 0.08-0.12, bal Fe.
Annealed: 112,000 TS; 54,000 YS; 45 El. Austenitic, non-
magnetic, corrosion resistant; hardenable by cold work to
about 200,000 TS; weldable. For shipbuilding, instruments,
electrical equipment. DIN X40MnCrN19; Werkstoff Nr. 3813.
Obsolete

WITTEN UM 40
Thyssen Edelstahlwerke AG
C 0.1, Mn 18, Cr 12, Ni 2, N 0.12, bal Fe.
Annealed: 100,000 TS; 43,000 YS; 45 El. Austenitic, non-
magnetic, corrosion resistant; hardenable by cold work to
about 190,000 TS; weldable. For shipbuilding, instruments
and electrical work. DIN X12MnCr1812; Werkstoff Nr. 3968.
Obsolete

WITTEN UM 45
Thyssen Edelstahlwerke AG
C 0.5, Mn 20, Cr 14, V 1, bal Fe.
Solution annealed: 127,000-141,000 TS; 71,000 YS; 40 El.
Aged: 141,000-170,000 TS; 112,000 YS; 30 El. For structural
parts requiring good strength and corrosion resistance. DIN
X50MnCrV2014; Werkstoff Nr. 3819. *Obsolete*

WITTEN UM 50
Thyssen Edelstahlwerke AG
C 0-0.05, Mn 18, Cr 13, Ni 2.5, N 0.12, bal Fe.
Annealed: 92,000 TS; 37,000 YS; 45 El. Austenitic, non-
magnetic, corrosion resistant; hardenable only by cold
work, weldable. For structural parts to resist corrosion. DIN
X5MnCr1813; Werkstoff Nr. 3949. *Obsolete*

WITTEN UM 60
Thyssen Edelstahlwerke AG
C 0-0.04, Cr 18, Ni 13, bal Fe.
Annealed: 70,000 TS; 31,000 YS; 45 El. Austenitic, non-
magnetic, corrosion resistant, hardenable only by cold
work, weldable. For parts to resist corrosion and elevated
temperature scaling. DIN X4CrNi1813; Werkstoff Nr. 3941.
Obsolete

WITTEN UM 70

Thyssen Edelstahlwerke AG
C 0-0.1, Cr 18, Ni 12, bal Fe.
Annealed: 78,000 TS; 31,000 YS; 45 El. Austenitic, non-magnetic, corrosion resistant; hardenable only by cold work, weldable. For shipbuilding, non-magnetic electrical equipment, structural parts operating in corrosive media. DIN X10CrNi1812; Werkstoff Nr. 3956; AISI 305. *Obsolete*

WITTEN UM 80

Thyssen Edelstahlwerke AG
C 0-0.03, Cr 17.5, Ni 14, Mo 2.3, bal Fe.
Annealed: 78,000 TS; 31,000 YS; 45 El. Austenitic, non-magnetic, corrosion resistant; hardenable only by cold work, weldable. For corrosion resisting parts in food and chemical plants, electrical equipment. DIN X2CrNiMo1813; Werkstoff Nr. 3953. *Obsolete*

WITTEN UM 85

Thyssen Edelstahlwerke AG
C 0.05, Cr 18, Ni 14, Mo 2.5, N 0.15, bal Fe.
Annealed: 71,000-100,000 TS; 42,500 YS; 40 El. Austenitic, non-magnetic, corrosion resistant; hardenable by cold work to 200,000 TS. For corrosion resistant parts in electric equipment; weldable. DIN X4CrNiMoN1814; Werkstoff Nr. 3952. *Obsolete*

WITTEN UM 90

Thyssen Edelstahlwerke AG
C 0.03, Cr 20, Ni 15, Mo 2.8, N 0.25, bal Fe.
Annealed: 100,000-120,000 TS; 58,000 YS; 40 El. Austenitic, non-magnetic, corrosion resistant, hardenable by cold work to about 190,000 TS. For non-magnetic, corrosion resistant structural parts; weldable. DIN X2CrNiMoN2015. *Obsolete*

WITTEN UMK

Obsolete

WITTEN VM 03

Thyssen Edelstahlwerke AG
C 0.45, Cr 1.6, Mo 0.6, V 0.8, W 0.6, bal Fe.
Oil hardening steel for pressure discs, mandrels, inner liners for extruding heavy metals, and for water cooled hot shear blades. DIN 45CrVMo58. *Obsolete*

WITTEN VM23

Thyssen Edelstahlwerke AG
C 0.45, Cr 1.4, Mo 0.7, V 0.3, bal Fe.
For forging and header dies; oil hardened, tough. *Obsolete*

WITTEN W 01 E

Thyssen Edelstahlwerke AG
C 0.1, Mn 0.4, bal Fe.
For case hardening; core remains soft. For small parts of machines, levers, links, bolts, pins. DIN Ck10; Werkstoff Nr. 1.1121. *Obsolete*

WITTEN W 10 L

Thyssen Edelstahlwerke AG
C 0.1, Si 0.25, Mn 0.4, bal Fe.
For carburizing purposes, soft core. DIN Ck15; Werkstoff Nr. 1.1144; AMS-5060. *Obsolete*

WITTEN W 25 E

Thyssen Edelstahlwerke AG
C 0.25, Mn 0.55, bal Fe.
Annealed: 64,000-78,000 TS; 35,000 minimum YS; 25 minimum El. Hardened: (16 mm maximum cross section): 78,000-100,000 TS; 64,000 minimum YS; 19 minimum El. For structural purposes on automotive, containers, small machines. DIN Ck25; Werkstoff Nr. 1.1158. *Obsolete*

WITTEN W 30 L

Thyssen Edelstahlwerke AG
C 0.35, Si 0.25, Mn 0.6, bal Fe.
Water hardenable carbon steel. DIN Ck35; Werkstoff Nr. 1.1174; AMS 5080. *Obsolete*

WITTEN W 34

Thyssen Edelstahlwerke AG
C 0.3, Cr 3, Mo 2.7, V 0.5, bal Fe.
Air or oil hardening hot work steel for light and heavy metal pressure casting tools, die inserts, brass forging dies and dies and punches for screws and nuts; annealed material designed for cold hubbing. DIN X32CrMoV33. *Obsolete*

WITTEN W 40 L

Thyssen Edelstahlwerke AG
C 0.45, Si 0.25, Mn 0.7, bal Fe.
Water hardening carbon steel. DIN Ck45; Werkstoff Nr. 1.1194; AISI C 1045. *Obsolete*

WITTEN W 50 E

Thyssen Edelstahlwerke AG
C 0.55, Mn 0.75, bal Fe.
Water hardening carbon steel. Normalized Annealed: 95,000-120,000 TS; (16 mm maximum); 52,000 minimum YS; 15 minimum El, 229 Brin. Hardened (16 mm maximum thickness): 114,000-125,000 TS; 78,000 minimum YS; 12 minimum El. For shafts, connecting rods, lever arms, induction hardened parts. DIN Ck55; Werkstoff Nr. 1.1203. *Obsolete*

WITTEN W 53 SPECIAL

Thyssen Edelstahlwerke AG
C 0.4, Cr 5, Mo 3.2, V 0.9, bal Fe.
Air or oil hardening hot work steel for die inserts, dies, mandrels in forging presses, piercing mandrels and tube extrusion. DIN X40CrMoV53. *Obsolete*

WITTEN W 66

Thyssen Edelstahlwerke AG
C 0.4, Si 1, Cr 5.2, Mo 1.3, V 0.4, bal Fe.
Air or oil hardening hot work steel for pressure casting tools for light metals, extrusion rams, mandrels, trimming dies and hot shear blades. DIN X38CrMoV51. *Obsolete*

WITTEN W 66 SPECIAL

Thyssen Edelstahlwerke AG
C 0.4, Si 1, Cr 5.2, Mo 1.3, V 1, bal Fe.
Air or oil hardened to 48-54 Rock C. Q + Temp 600 C: 225,000-240,000 psi TS. Hot work tool steel for ejector pins, die casting dies, pressure casting models. DIN X40CrMoV51; AISI H 13. *Obsolete*

WITTEN W 77

Thyssen Edelstahlwerke AG
C 0.35, Si 1, Cr 5.2, Mo 1.5, V 0.3, W 1.3, bal Fe.
Air or oil hardening hot work steel for light metal extrusion dies, forging press dies, trimming tools and hot shear blades. DIN X37CrMoW51. *Obsolete*

WITTEN W10E

Thyssen Edelstahlwerke AG
C 0.15, Si 0.25, Mn 0.37, bal Fe.
Annealed: 70,000 TS; 40,000 YS; 25 El; 60 RA; 145 Brin. For gears, bolts, machine tool parts; case hardened. *Obsolete*

WITTEN W1W7

Thyssen Edelstahlwerke AG
C 0.7, Si 0-0.25, Mn 0-0.25, bal Fe.
Heat treated: 175,000 TS; 130,000 YS; 12 El; 37 RA; 355 Brin. For springs, tools, rails, hammers, axes, cutters; Type W1; water hardened. *Obsolete*

WITTEN W1W8

Thyssen Edelstahlwerke AG
C 0.85, Si 0-0.25, Mn 0-0.25, bal Fe.
Heat treated: 190,000 TS; 145,000 YS; 10 El; 30 RA; 400 Brin. For springs, tools, drills, taps, hobs, reamers; Type W1; water hardened. *Obsolete*

WITTEN W30E

Thyssen Edelstahlwerke AG
C 0.35, Mn 0.55, Si 0.25, bal Fe.
Hot rolled: 85,000 TS; 54,000 YS; 30 El; 53 RA; 185 Brin. For gears, bolts, machine tool parts; water hardened. *Obsolete*

WITTEN W40E

Thyssen Edelstahlwerke AG
C 0.45, Si 0.25, Mn 0.55, bal Fe.
Hot rolled: 98,000 TS; 59,000 YS; 24 El; 45 RA; 212 Brin. For gears, bolts, machine tool parts; water hardened. *Obsolete*

WITTEN W44

Thyssen Edelstahlwerke AG
C 0.3, Si 1, Cr 1.1, V 0.18, W 3.7, bal Fe.
For header and upsetter dies, rivet sets; oil hardened, tough. *Obsolete*

WITTEN W60E

Thyssen Edelstahlwerke AG
C 0.6, Si 0.25, Mn 0.55, bal Fe.
Heat treated: 160,000 TS; 113,000 YS; 12 El; 40 RA; 325 Brin. For springs, axes, hammers, shafts; water hardened. *Obsolete*

WITTEN W88A

Thyssen Edelstahlwerke AG
C 0.3, Cr 2.35, V 0.6, W 4.25, bal Fe.
For extrusion rams and liners, punches; hot work steel, oil hardened. *Obsolete*

WITTEN W99

Thyssen Edelstahlwerke AG
C 0.3, Cr 2.65, V 0.35, W 8.5, Mn 0.3, bal Fe.
For extrusion rams and liners, punches; hot work steel, oil hardened. *Obsolete*

WITTEN WIB

Thyssen Edelstahlwerke AG
C 1.2, Cr 0.2, W 1, bal Fe.
Water hardening tool steel; for drills, boring tools, and other cutting tools. *Obsolete*

WITTEN WIB100

Thyssen Edelstahlwerke AG
C 1.2, Cr 0.2, V 0.1, W 1, Mn 0.28, bal Fe.
For header and upsetting dies, punches; oil hardened, wear resistant. *Obsolete*

WITTEN WIB100

Thyssen Edelstahlwerke AG
C 1.2, Cr 0.2, V 0.1, W 1, bal Fe.
For header dies, punches; oil hardened. *Obsolete*

WITTEN WIW 10

Thyssen Edelstahlwerke AG
C 1.1, Si 0-0.25, Mn 0-0.25, bal Fe.
Annealed: 105,000 TS; 55,000 YS; 20 El; 40 RA; 210 Brin. For springs, cutters, drills, taps, reamers, hobs; Type W1; water hardened. *Obsolete*

WITTEN WIW 9

Thyssen Edelstahlwerke AG
C 1, Si 0.2, Mn 0.2, bal Fe.
Water hardening tool steel, hardenable to high hardness and strength. For shears, punching and trimming dies, cutters. *Obsolete*

WITTEN WIZ 11

Thyssen Edelstahlwerke AG
C 1.15, Si 0.25, Mn 0.3, bal Fe.
Water hardenable to 60-66 Rock C. For hand tools, woodworking tools. DIN C110W2; AISI W1, tool steel. *Obsolete*

WITTEN WIZ 11 L

Thyssen Edelstahlwerke AG
C 1.1, Si 0.25, Mn 0.25, bal Fe.
Water hardening tool steel; drill rod. DIN C110W2; Werkstoff Nr. 1.1654; AISI 1095. *Obsolete*

WITTEN WIZ 12
Thyssen Edelstahlwerke AG
C 1.15, Si 0-0.25, Mn 0-0.25, bal Fe.
Annealed: 110,000 TS; 56,000 YS; 18 El; 40 RA; 210 Brin.
For springs, tools, reamers, broaches; Type W1; water hardened. *Obsolete*

WITTEN WIZ 7
Thyssen Edelstahlwerke AG
C 0.7, Si 0.25, Mn 0.3, bal Fe.
Water hardening tool steel for drop forging dies, swaging dies, riveting sets, shear blades, knife blades, scissors, wood working tools. AISI W1. *Obsolete*

WITTEN WIZ 8
Thyssen Edelstahlwerke AG
C 0.85, Si 0-0.25, Mn 0-0.25, bal Fe.
Heat treated: 190,000 TS; 145,000 YS; 12 El; 35 RA; 390 Brin. For springs, tools, taps, drills, reamers; Type W1; water hardened. *Obsolete*

WITTEN WIZ 9
Thyssen Edelstahlwerke AG
C 1, Mn 0-0.25, Si 0-0.25, bal Fe.
Annealed: 100,000 TS; 53,000 YS; 20 El; 42 RA; 200 Brin.
For springs, tools, taps, drills; Type W1; water hardened. *Obsolete*

WITTEN WOLFRAM 184
Thyssen Edelstahlwerke AG
C 0.75, Cr 4.1, V 1.1, W 18, bal Fe.
Air or oil hardenable to 63-66 Rock C. Lathe tools, hobs, milling cutters, taps, broaches for cutting soft and medium-hard materials. DIN S18-0-1; AISI T1, high speed steel. *Obsolete*

WITTEN WS 45
Thyssen Edelstahlwerke AG
C 0.45, Cr 4.5, Mo 0.5, V 2, W 4.5, Co 4.5, bal Fe.
Air hardening hot work tool steel for impression and die inserts, mandrels and hot pressure die casting dies. DIN X45CoCrWV555. *Obsolete*

WITTEN ZR3
Thyssen Edelstahlwerke AG
C 1.42, W, V, bal Fe.
For bearings, bushings, header dies; oil hardened, wear resistant. *Obsolete*

WITTEN ZR5
Thyssen Edelstahlwerke AG
C 1.3, W, bal Fe.
For engraver's tools, cutters, dies; water or oil hardened. *Obsolete*

WITTEN ZW
Thyssen Edelstahlwerke AG
C 1.4, Cr 0.3, V 0.1, bal Fe.
For engraving tools, drills, taps; Type W2; water hardened. *Obsolete*

WIZARD
Agawam Tool Co.
C 0.45, Cr 0.8, W 0.85, bal Fe.
For pneumatic chisels, shock tools; tough.

WIZARD
Worthington Steel & Annealing Co.
C, alloy, bal Fe.
For high speed tools. High speed steel.

WIZARD
Ziv Steel & Wire Co.
C 0.45, Mn 0.3, Cr 0.9, W 1, Mo 0.2, bal Fe.
For pneumatic tools, hand chisels, punches, crimpers; shock resistant, oil or water hardened.

WK 249
Bergische Stahl Industrie
C 0.4-0.5, Si 1.3-1.6, Mn 0.5-0.7, Cr 1.3-1.6, V 0.07-0.12, bal Fe.
Cold work tool steel. W.-Nr. 1.2249.

WK 500
Bergische Stahl Industrie
C 0.36-0.42, Si 0.9-1.2, Mn 0.3-0.5, Cr 4.8-5.8, Mo 0.8-1.4, V 0.25-0.5, bal Fe.
Hot work tool steel. W.-Nr. 1.2343.

WK 575
Bergische Stahl Industrie
C 0.28-0.35, Si 0.2-0.4, Mn 0.2-0.4, Cr 2.7-3.2, Mo 2.6-3, V 0.4-0.7, bal Fe.
Hot work tool steel. W.-Nr. 1.2365.

WK 600
Bergische Stahl Industrie
C 0.32-0.4, Si 0.9-1.2, Mn 0.3-0.6, Cr 5-5.6, Mo 1.3-1.6, W 1.2-1.4, V 0.15-0.4, bal Fe.
Hot work tool steel. W.-Nr. 1.2606.

WKE 4
Fagersta Stainless AB
Tool material. C 1.25, Cr 4.1, Mo 3.1, V 3.1, W 9, Co 9, bal Fe.
Cobalt high speed steel for lathe tools, drills, reamers, milling cutters. Good red hardness. *Obsolete*

WKE 45
Fagersta Stainless AB
Tool material. C 1.4, Cr 4.2, Mo 3.5, V 3.5, W 9, Co 11, bal Fe.
Cobalt high speed steel for lathe tools, milling cutters, reamers. Good abrasion resistance and red hardness. *Obsolete*

WKE-4
Now FAGERSTA WKE-4.

WKL
Thyssen Edelstahlwerke AG
C 1, Cr 1.5, bal Fe
For cutting tools, bearings; water hardened. *Obsolete*

WKZV
Bergische Stahl Industrie
C 0.25-0.35, Si 0.15-0.3, Mn 0.2-0.4, Cr 2.5-2.8, W 8-9, V 0.3-0.4, bal Fe.
Hot work tool steel. W.-Nr. 1.2581.

WM 344
Bergische Stahl Industrie
C 0.37-0.42, Si 0.9-1.2, Mn 0.3-0.5, Cr 5-5.5, Mo 1.2-1.5, V 0.9-1.1, bal Fe.
Hot work tool steel. W.-Nr. 1.2344.

WM 362
Bergische Stahl Industrie
C 0.6-0.65, Si 1-1.2, Mn 0.3-0.5, Cr 5-5.5, Mo 1-1.3, V 0.25-0.35, bal Fe.
Cold work tool steel. W.-Nr. 1.2362.

WM 559
Thyssen Edelstahlwerke AG
C 0.45, Cr 1.5, Mo 0.5, V 0.8, W 0.5, bal Fe.
For screw dies, hot work tools; hot work steel, oil hardened. *Obsolete*

WM 631
Bergische Stahl Industrie
C 0.45-0.55, Si 0.8-1, Mn 0.4-0.6, Cr 8-9, Mo 1.1-1.3, W 1.1-1.3, bal Fe.
Cold work tool steel. W.-Nr. 1.2631.

WM 80F
German manufacture
Sn 80, Sb 11, Cu 9.
For bearings; anti-friction.

WM10
Friedr. Lohmann GmbH
C 0.3, Cr 2.65, V 0.35, W 8.5, bal Fe.
For extrusion press rams and liners, punches, shears. Type H21; hot work steel, oil hardened. T20821

WMD
Kasle Steel Co.
C 0.33, Mn 0.3, Si 0.3, Cr 2.9, V 0.5, Mo 2.9, bal Fe.
Oil or air hardening hot work tool steel. *Obsolete*

WMD. EXTRA
Kasle Steel Co.
C 0.33, Mn 0.3, Si 0.3, Cr 2.9, V 0.5, Mo 2.9, Co 3, bal Fe.
Air or oil hardening hot work tool steel. *Obsolete*

WN 20
Sandvik Hard Materials Ltd.
Co 6, WC 94.
Cemented carbide. 1580 Vickers; 5890 MNm^{-2} compressive strength.

WN 30
Sandvik Hard Materials Ltd.
Co 9.5, WC 90.5.
Cemented carbide. 1400 Vickers; 5200 MNm^{-2} compressive strength.

WN 4
Bergische Stahl Industrie
C 0.25-0.35, Si 0.8-1.1, Mn 0.3-0.5, Cr 0.9-1.2, W 3.5-4, V 0.15-0.2, bal Fe.
Hot work tool steel. W.-Nr. 1.2564.

WN 40
Sandvik Hard Materials Ltd.
Co 11, WC 89.
Cemented carbide. 1290 Vickers; 4600 MNm^{-2} compressive strength.

WN 45
Sandvik Hard Materials Ltd.
Co 13, WC 87.
Cemented carbide. 1240 Vickers; 4400 MNm^{-2} compressive strength.

WN 5
Bergische Stahl Industrie
C 0.25-0.35, Si 0.15-0.3, Mn 0.2-0.4, Cr 2.2-2.5, W 4-4.5, V 0.5-0.7, bal Fe.
Hot work tool steel. W.-Nr. 1.2567.

WN 50
Sandvik Hard Materials Ltd.
Co 15, WC 85.
Cemented carbide. 1150 Vickers; 4100 MNm^{-2} compressive strength.

WN-102
Now SYLVANIA WN-102.

WN-103
Now SYLVANIA WN-103.

WNS
Wardlows Ltd.
C 0.95, Mn 1.2, W 0.5, Cr 0.5, bal Fe.
For tools, dies; non-deforming.

WOCO
Bethlehem Steel Corp.
C 0.45, Cr 1.5, W 2.25, V 0.25, bal Fe.
For hot work dies; hot work steel. *Obsolete*

WOCO
Bisset Steel Co.
C 0.45, Cr 1.5, W 2.25, V 0.25, bal Fe.
For pneumatic and rivet busting tools, shear blades; shock tools, tough. *Obsolete*

WODURIT

SWB Stahlformguss Gesellschaft mbH
C 1.42, W, V, bal Fe.
For blanking and forming dies, punches; oil hardened, wear resistant. *Obsolete*

WOLFF CC16

Otto Wolff Handelgesellschaft
C 1, Mn 0.35, Cr 1.55, Si 0.3, bal Fe.
For bearings, liners, bushings; water or oil hardened, wear resistant.

WOLFF CC5

Otto Wolff Handelgesellschaft
C 1.1, Cr 0.4, Mn 0.3, bal Fe.
For bearings, liners, sleeves, cutters; water hardened, wear resistant.

WOLFF CM SPEZIAL

Otto Wolff Handelgesellschaft
C 0.45, Mn 0.7, Cr 1.4, Mo 0.7, V 0.3, bal Fe.
For gears, bolts, crankshafts, fasteners; oil hardened, tough.

WOLFF CMV12

Otto Wolff Handelgesellschaft
C 0.5, Mn 0.95, Cr 1.05, V 0.1, bal Fe.
For gears, springs, bolts, shafts; oil hardened, shock resistant.

WOLFF CR-08

Otto Wolff Handelgesellschaft
C 0.9, Si 0.25, Mn 0.3, Cr 0.8, bal Fe.
For bearings, liners, bushings; water hardened, wear resistant.

WOLFF CR-10

Otto Wolff Handelgesellschaft
C 1.05, Si 0.25, Mn 0.3, Cr 1.2, bal Fe.
For bearings, liners, bushings; water hardened, wear resistant.

WOLFF CR-12

Otto Wolff Handelgesellschaft
C 2.1, Cr 11.5, Si 0.35, Mn 0.3, bal Fe.
For blanking and forming dies, punches; oil hardened, nondeforming.

WOLFF CSC 12

Otto Wolff Handelgesellschaft
C 1.25, Si 1.15, Mn 0.7, Cr 1.2, bal Fe.
For bearings, bushings, liners, sleeves; water hardened.

WOLFF CV 60

Otto Wolff Handelgesellschaft
C 0.55, Si 0.1-0.4, Mn 0.5-0.7, bal Fe.
For gears, pinions, machine tool parts; water hardened.

WOLFF EC 15

Otto Wolff Handelgesellschaft
C 0.15, Si 0.25, Mn 0.25, bal Fe.
Annealed: 70,000 TS; 40,000 YS; 25 El; 60 RA. For gears, bolts, machine tool parts; case hardened.

WOLFF EW

Otto Wolff Handelgesellschaft
C 0.15, Si 0.15-0.35, Mn 0.25-0.5, bal Fe.
For gears, pinions, machine tool parts; case hardened.

WOLFF EXTRA HART

Otto Wolff Handelgesellschaft
C 1, Si 0-0.25, Mn 0-0.25, bal Fe.
For drills, taps, reamers, broaches, springs; Type W1; water hardened.

WOLFF EXTRA ZAH

Otto Wolff Handelgesellschaft
C 0.85, Si 0-0.25, Mn 0-0.25, bal Fe.
Heat treated: 190,000 TS; 145,000 YS; 10 El; 30 RA; 400 Brin. For springs, tools, drills, taps; Type W1; water hardened.

WOLFF MCVII

Otto Wolff Handelgesellschaft
C 0.58, Cr 1, V 0.09, Mn 0.95, bal Fe.
For springs, gears, bolts, shafts; oil hardened, shock resistant.

WOLFF MS 10

Otto Wolff Handelgesellschaft
C 0.53, Si 0.9, Mn 0.9, bal Fe.
For punches, crimpers, bolts, gears; water hardened.

WOLFF MS 15

Otto Wolff Handelgesellschaft
C 0.7, Si 1.7, Mn 0.7, bal Fe.
For springs; oil hardened.

WOLFF SCRV 15

Otto Wolff Handelgesellschaft
C 0.45, Si, Cr, V, bal Fe.
For springs, gears, bolts, shafts; oil hardened, shock resistant.

WOLFF SVC 13H

Otto Wolff Handelgesellschaft
C 0.61, Cr 1.18, V 0.1, Mn 0.75, bal Fe.
For springs, crankshafts, punches; oil hardened, shock resistant.

WOLFRAM

Joseph Beardshaw & Son Ltd.
C 0.55, W 2.75, Cr 1.25, V 0.25, bal Fe.
For hot stamping dies; shock resistant, oil hardened.

WOLFRAM 184

Thyssen Edelstahlwerke AG
C 0.74, Cr 4.1, V 1.1, W 18.5, bal Fe.
For lathe and planer tools; reamers; high speed steel.

WOLFRAMANT

Plansee, Metallwerk Gesellschaft
WC, Co.
For inserts for boring crowns; sintered carbides.

WOLFRAMANT

Otto Wolff Handelgesellschaft
C 1.42, W, V, bal Fe.
For engraving tools, fast finishing cutters; Type W2; water hardened.

WOLFRAMANT

Thyssen Edelstahlwerke AG
C 0.7, W 18, Cr 4, V 1, Co 5, bal Fe.
For cutters, flutters, hobs; high speed steel.

WOLFRAMINIUM

Manufacturer not listed.
Al 98, Sb 1.4, Cu 0.36, Sn 1, W 0.05.
Cast: 20,000 TS; 6 El. Rolled: 48,000 TS; 6 El. Annealed: 24,000 TS; 16 El. For motor car body work; non-tarnishing.

WOLFRAMSTAHL, EXTRA SPEZIAL

Stahlwerke Kabel, C.
C 1.42, W, V, bal Fe.
For engravers' tools, cutters, blanking dies; oil hardened, wear resistant.

WOLVERINE 1060 ALUMINUM

Wolverine Tube, Inc.
Aluminum.
For condenser and heat exchanger tubes. Meets ASTM B-234. H 14 temper: 12 ksi TS min; 10 ksi YS min. A91060

WOLVERINE 3003 ALUMINUM

Wolverine Tube, Inc.
Aluminum.
For condenser and heat exchanger tubes. Meets ASTM B-234. H 14 temper: 20 ksi TS min; 17 ksi YS min; 10 El min. A93003

WOLVERINE 5052 ALUMINUM

Wolverine Tube, Inc.
Aluminum.
For condenser and heat exchanger tubes. Meets ASTM B-234. H 32 temper: 31 ksi TS min; 23 ksi YS min; 10 El min. H 34 temper: 34 ksi TS min; 26 ksi YS min; 10 El min. A95052

WOLVERINE ADMIRALTY, TYPE B

Wolverine Tube, Inc.
Copper.
For condenser and heat exchanger tubes. Meets ASTM B-111. Annealed: 45 ksi TS min; 15 ksi YS min; 35 El min. C44300

WOLVERINE ADMIRALTY, TYPE C

Wolverine Tube, Inc.
Copper.
For condenser and heat exchanger tubes. Meets ASTM B-111. Annealed: 45 ksi TS min; 15 ksi YS min; 35 El min. C44400

WOLVERINE ADMIRALTY, TYPE D

Wolverine Tube, Inc.
Copper.
For condenser and heat exchanger tubes. Meets ASTM B-111. Annealed: 45 ksi TS min; 15 ksi YS min. C44500

WOLVERINE ALCLAD 3003

Wolverine Tube, Inc.
Aluminum.
For condenser and heat exchanger tubes. Meets ASTM B-234. H 14 temper: 19 ksi TS min; 16 ksi YS min; 10 El min. A93003

WOLVERINE ALLOY 01

Wolverine Tube, Inc.
Copper. Cu 99.9, P 0.025.
Annealed: 34,000 TS; 45 El. Ductile copper with good corrosion resistance; used in air conditioners, plumbing tubes, and heat exchanger tubes. *Obsolete*

WOLVERINE ALLOY 02

Wolverine Tube, Inc.
Copper. Cu 99.4, As 0.3, P 0.025.
Annealed: 34,000 TS; 45 El. For heat exchanger applications. *Obsolete*

WOLVERINE ALLOY 07

Wolverine Tube, Inc.
Copper. Cu 99.9, P 0.007.
Annealed: 34,000 TS; 45 El. High conductivity copper for electrical conductivity applications such as waveguide and bus tubes. *Obsolete*

WOLVERINE ALLOY 08

Wolverine Tube, Inc.
Copper. Cu 98.7, P 0.025, Fe 1.
Annealed: 40,000 TS; 40 El. For air conditioners, plumbing tubes and heat exchangers. *Obsolete*

WOLVERINE ALLOY 21

Wolverine Tube, Inc.
Copper. Cu 66, Pb 0.5, bal Zn.
Cold drawn: 68,000 TS; 29 El. Improved machinable brass used in plumbing brass goods, pump cylinders. *Obsolete*

WOLVERINE ALLOY 22

Wolverine Tube, Inc.
Copper. Cu 66.5, Pb 1.5, Zn 32.
Annealed: 52,000 TS; 50 El. Hard drawn: 75,000 TS; 7 El. For applications requiring extensive machining; free cutting material. *Obsolete*

WOLVERINE ALLOY 24

Wolverine Tube, Inc.
Copper. Cu 70, Zn 30.
Annealed: 52,000 TS; 55 El. Hard drawn: 78,000 TS; 8 El. Highly ductile brass for miscellaneous fabrications, plumbing brass goods and pump cylinders. *Obsolete*

WOLVERINE ALLOY 26

Wolverine Tube, Inc.
Copper. Cu 71, Zn 28, Sn 1, As, 0.02 Sb or P.
Annealed: 53,000 TS; 65 El. Used as heat exchanger tubes for steam condensers in circulating seawater or brackish water. *Obsolete*

WOLVERINE ALLOY 28

Wolverine Tube, Inc.
Copper. Cu 85, Zn 15.
Annealed: 44,000 TS; 45 El. Condenser and heat exchanger alloy for use in fresh waters, water pipes. *Obsolete*

WOLVERINE ALLOY 29

Wolverine Tube, Inc.
Copper. Cu 90, Zn 10.
Annealed: 38,000 TS; 50 El. Drawn: 52,000 TS; 20 El. For use in rotating bands, marine hardware. *Obsolete*

WOLVERINE ALLOY 30

Wolverine Tube, Inc.
Copper. Cu 76, Zn 22, Al 2, 0.05 As or P.
Annealed: 60,000 TS; 55 El. For condenser and heat exchanger applications in seawater; resistant to impingement attack. *Obsolete*

WOLVERINE ALLOY 32

Wolverine Tube, Inc.
Copper. Cu 95, Al 5, As 0.02.
Annealed: 52,000 TS; 60 El. For condenser and heat exchanger tube applications, especially in seawater, brackish water, and polluted seawater. *Obsolete*

WOLVERINE ALLOY 40

Wolverine Tube, Inc.
Aluminum. Fe 0.4, Si 0.15, Cu 0.1, 99.0 Al min.
O Temper: 13,000 TS; 45 El. H-18: 24,000 TS; 15 El. For refrigeration tube. *Obsolete*

WOLVERINE ALLOY 41

Wolverine Tube, Inc.
Aluminum. Mn 1.25, Fe 0.5, Si 0.3, 97.0 Al min.
O Temper: 16,000 TS; 40 El. H-18: 29,000 TS; 10 El. Air conditioning and refrigeration tubing, condensers and heat exchangers in sulfur dioxide and hydrogen sulfide environments. *Obsolete*

WOLVERINE ALLOY 42

Wolverine Tube, Inc.
Aluminum. Mg 2.5, Cr 0.25, 0.4 Si + Fe, bal Al.
O Temper: 28,000 TS; 30 El. Good fatigue resistance; for hydraulic lines. *Obsolete*

WOLVERINE ALLOY 43

Wolverine Tube, Inc.
Aluminum. Mg 0.8, Si 0.5, Fe 0.25, bal Al.
O Temper: 13,000 TS. T6 Temper: 35,000 TS; 12 El. Heat treatable high strength alloy used for furniture tubes. *Obsolete*

WOLVERINE ALLOY 44

Wolverine Tube, Inc.
Aluminum. 99.45 Al min.
O Temper: 12,000 TS; 35 El. H-19 Temper: 27,000 TS; 5 El. For electrical connectors and bus tubing. *Obsolete*

WOLVERINE ALLOY 45

Wolverine Tube, Inc.
Aluminum. Si 0.68, Mn 0.11, Mg 1.1, Cu 0.3, Cr 0.2, bal Al.
O Temper: 18,000 TS; 25 El. T6 Temper: 45,000 TS; 12 El. Heat treatable alloy, high strength structural applications. *Obsolete*

WOLVERINE ALLOY 51

Wolverine Tube, Inc.
Copper. Cu 68.7, Ni 30, Mn 0.7, Fe 0.5.
Annealed: 60,000 TS; 45 El. Condenser and heat exchanger tube alloy for use in sea or brackish waters, resistant to impingement attack. *Obsolete*

WOLVERINE ALLOY 52

Wolverine Tube, Inc.
Nickel. Ni 64, Mn 1, Fe 1, Cu 34.
Annealed: 78,000 TS; 43 El. Heat exchanger tube alloy with high strength and corrosion resistance. Also known by tradename "CUNEL". *Obsolete*

WOLVERINE ALLOY 53

Wolverine Tube, Inc.
Copper. Cu 89, Ni 10, Fe 1.
Annealed: 44,000 TS; 42 El. Condenser and heat exchanger tube alloy for use in sea or brackish waters, resistant to impingement attack. *Obsolete*

WOLVERINE ALUMINUM BRASS

Wolverine Tube, Inc.
Copper.
For condenser and heat exchanger tubes. Meets ASTM B-111. Annealed: 50 ksi TS min; 18 ksi YS min; 35 El min. C68700

WOLVERINE ALUMINUM BRONZE

Wolverine Tube, Inc.
Copper.
For condenser and heat exchanger tubes. Meets ASTM B-111. Annealed: 50 ksi TS min; 19 ksi YS min; 35 El min. C60800

WOLVERINE ARSENICAL COPPER

Wolverine Tube, Inc.
Copper.
For condenser and heat exchanger tubes. Meets ASTM B-111, B-75. Light drawn: 36 ksi TS min; 30 ksi YS min; 20 El min. Hard drawn: 45 ksi TS min; 40 ksi YS min; 6 El min. C14200

WOLVERINE COPPER

Wolverine Tube, Inc.
Copper.
For condenser and heat exchanger tubes. Meets ASTM B-111, B-75. Light drawn: 36 ksi TS min; 30 ksi YS min; 20 El min. Hard drawn: 45 ksi TS min; 40 ksi YS min; 6 El min. C12200

WOLVERINE COPPER-NICKEL, 10%

Wolverine Tube, Inc.
Copper.
For condenser and heat exchanger tubes. Meets ASTM B-111. Annealed: 40 ksi TS min; 15 ksi YS min; 35 El min. Light drawn: 45 ksi TS min; 35 ksi YS min; 20 El min. C70600

WOLVERINE COPPER-NICKEL, 20%

Wolverine Tube, Inc.
Copper.
For condenser and heat exchanger tubes. Meets ASTM B-111. Annealed: 45 ksi TS min; 16 ksi YS min; 35 El min. C71000

WOLVERINE COPPER-NICKEL, 30%

Wolverine Tube, Inc.
Copper.
For condenser and heat exchanger tubes. Meets ASTM B-111. Annealed: 52 ksi TS min; 18 ksi YS min; 35 El min. Drawn, stress relieved: 72 ksi TS min; 50 ksi YS min. C71500

WOLVERINE COPPER-NICKEL, 5%

Wolverine Tube, Inc.
Copper.
For condenser and heat exchanger tubes. Meets ASTM B-111. Annealed: 38 ksi TS min; 12 ksi YS min; 35 El min. Light drawn: 40 ksi TS min; 30 ksi YS min; 20 El min. C70400

WOLVERINE RED BRASS 85%

Wolverine Tube, Inc.
Copper.
For condenser and heat exchanger tubes. Meets ASTM B-111. Annealed: 40 ksi TS min; 12 ksi YS min; 35 El min. C23000

WOMPCO

Worthington Pump Inc.
S 0.115, Mn 0.74, Si 2.01, 0.90% combined C, 2.90% total C, bal Fe.
39,600-50,000 TS; 0 El; 0 RA; 213-230 Brin. For cast iron castings, cylinders, gears; cast iron; S.S. 55000. *Obsolete*

WONICO

American manufacture
W 80, Ni 13, Co 5.
For alloy for sealing in glass; coefficient of expansion 55 x 10^{-7}

WOOD BAND SAW

Colt Industries
C 0.7, Cr 2, bal Fe.
For wood band saws; water or oil hardening. *Obsolete*

WOOD FUSIBLE ALLOY-1

American manufacture
Bi 50, Sn 12.5, Pb 25, Cd 12.5.
For safety plugs, fuses; melting point 71°C.

WOOD FUSIBLE ALLOY-2

American manufacture
Bi 52.5, Pb 31.5, Sn 16.
For safety plugs, fuses; melting point 98°C.

WOOD SAW

Disston Inc.
C 0.75, Ni 0.7, Cr 0.2, bal Fe.
For saws for wood. *Obsolete*

WOOD'S METAL

Cerro Metal Products Co.
Bi 40.5, Pb 24.95-34.5, Sn 9.3-22.4, Cd 6.2-12.5.

WORKWEAR 14

Ludlow Steel Corp.
Mn 14, C, bal Fe.
Plate: 145,000-155,000 TS; 73-75 El; 54-56 RA; 200 Brin as rolled; 500+ Brin as work hardened. For chute liners, crushers, shovel and dragline buckets, quarry and mine skips. Austenitic, wear and abrasion resistant, shock and impact resistant.

WORTHITE

Worthington Pump Inc.
Stainless steel. C 0-0.07, Mn 0-1, Si 2.5-3.5, Cr 18-20, Cu 1.5-2, Ni 22-25, Mo 2.5-3, bal Fe.
Commercially available as ASTM A744 Grade CN7MS, but no longer produced by Worthington Pump. Cast: 70,000 psi TS min; 30,000 psi YS min; 35 El; 137-183 Brin. Stainless steel for pumps requiring great resistance to corrosion. *Obsolete*

WORTLE DIE

Colt Industries
C 1.2-1.3, W 2.5-3.5, Mn 0.5-0.75, bal Fe.
For drawing dies; oil hardened. *Obsolete*

WORTLE DIE

John A. Crowley Inc.
C 1.2, W 3, Mn 0.6, bal Fe.
For tools, dies; oil hardened.

WORTLE DIE NO. 1

Colt Industries
C 2.1, Cr 1.9, W 4.3, bal Fe.
For cold drawing dies; oil hardening. *Obsolete*

WORTLE DIE NO. 4

Colt Industries
C 2.1, W 3.2, bal Fe.
For cold drawing dies; for hard abrasive metals. *Obsolete*

WORTLE DIE NO. 5

Colt Industries
C 2, Mn 2, Cr 2, W 12, bal Fe
For cold drawing dies; for abrasive metals. *Obsolete*

WORTLE DIE STEEL
Latrobe Steel Co.
C 1.2, W 2.5-3.5, Mn 0.7, bal Fe.
For dies, draw dies. *Obsolete*

WORTLE NO. 4 DRAWING DIE
Hawkridge Bros. Co.
C 1.95, W 3, bal Fe.
For drawing dies, tools; oil hardening. *Obsolete*

WOTAN AKS-98
Hans Kanz Metallwerke
Al, bal Cu.
At 20°C: 71,100 TS; 34,100 YS; 44 El. At 600°C: 8500 TS; 2850 YS; 2 El. For autoclaves, chemical equipment, pumps, valves; aluminum-bronze, corrosion resistant, tough.

WOTAN COBALT 10
Henckels Zwillingwerke, G.A.
C 0.76, Co 10, Cr 4, Mo 0.8, V 1.8, W 18, bal Fe.
For lathe and planer tools, taps, reamers, hobs; high speed steel; 76WCoV7240.

WOTAN COBALT 3
Henckels Zwillingwerke, G.A.
C 0.86, Co 2.8, Cr 4.3, Mo 0.85, V 2.1, W 12, bal Fe.
For lathe and planer tools, reamers, broaches, drills; high speed steel; 91WCoV3811.

WOTAN COBALT 5
Henckels Zwillingwerke, G.A.
C 0.79, Co 4.75, Cr 4.3, Mo 0.75, V 1.5, W 18, bal Fe.
For reamers, hobs, drills, taps, broaches; high speed steel; 79WCo7419.

WOTAN COBALT 513
Henckels Zwillingwerke, G.A.
C 0.8, Cr, Co, W, V, Mo, bal Fe.
For lathe and planer tools, taps, reamers, hobs; high speed steel; 80WCoCrVMo50.22.

WOTAN DRILLING
Henckels Zwillingwerke, G.A.
C 1.3, Cr 4.3, Mo 0.85, V 3.8, W 12, bal Fe.
For reamers, drills, lathe and planer tools, broaches; high speed steel; 130WV3838.

WOTAN DRILLING CO
Henckels Zwillingwerke, G.A.
C 1.35, Cr, W, V, bal Fe.
For cutters, form tools; high speed steel; 135WCo4619.

WOTAN EXTRA F
Henckels Zwillingwerke, G.A.
C 0.74, Cr 4.1, V 1.1, W 18.5, bal Fe.
For lathe and planer tools, drills, hobs, reamers; high speed steel; 74WV74.

WOTAN III/3
Henckels Zwillingwerke, G.A.
C 0.95, Cr 4, V 1, Mo, W, bal Fe.
For lathe and planer tools, reamers, hobs, taps; high speed steel; 95WMo1126.

WOTAN ILLING
Henckels Zwillingwerke, G.A.
C 0.82, Cr 4.1, Mo 0.85, V 1.6, W 8.7, bal Fe.
For lathe and planer tools, reamers, drills, hobs; high speed steel; 82WV3419.

WOTAN MO 5
Henckels Zwillingwerke, G.A.
C 0.85, Cr, W, Mo, bal Fe.
For lathe and planer tools, hobs, broaches, taps; high speed steel; 82WV3419.

WOTAN MO 5 CO
Henckels Zwillingwerke, G.A.
C, Cr, W, Mo, V, bal Fe.
For cutters, hobs, broaches, drills; high speed steel.

WOTAN MO 5 V 3
Henckels Zwillingwerke, G.A.
C, Cr, W, Mo, V, bal Fe.
For cutters, hobs, broaches, drills; high speed steel.

WOTAN MO 9
Henckels Zwillingwerke, G.A.
C, Cr, W, Mo, V, bal Fe.
For cutters, hobs, broaches, drills; high speed steel.

WOTAN ZWILLING
Henckels Zwillingwerke, G.A.
C 0.86, Cr 4.1, Mo 0.85, V 2.5, W 12, bal Fe.
For lathe and planer tools, broaches, reamers, hobs, taps; high speed steel; 86WV3826.

WP300, WP500
British Steel plc
Medium C, Cr, Mo.
Quenched and tempered. High impact and wear resistance.

WPS
W. Ossenberg & Cie Edelstahlwerke
C 0.55, Si 0.3, Mn 0.6, Cr 0.7, Mo 0.3, Ni 1.7, V 0.1, bal Fe.
Hot work tool steel. For forging and pressing dies. Werkstoff Nr. 1.2713.

WPS 1
W. Ossenberg & Cie Edelstahlwerke
C 0.55, Si 0.3, Mn 0.7, Cr 1, Mo 0.5, Ni 1.7, V 0.1, bal Fe.
Hot work tool steel; forging dies. Werkstoff Nr. 1.2714.

WPS 2
W. Ossenberg & Cie Edelstahlwerke
C 0.45, Si 0.3, Mn 0.7, Cr 1.5, Mo 0.7, V 0.3, bal Fe.
Hot work tool steel. Molds for casting or forming light metals, lead, tin and zinc. Werkstoff Nr. 1.2323.

WRM 100
Western Reserve Manufacturing Co., Inc.
Copper. Sn 9-11, Pb 0-0.3, Zn 1-3, Fe 0-0.15, P 0-0.05, 0.25 others, bal Cu.
As cast: 30,000 TS; 25,000 YS; 18 El. CDA 905.

WRM 100B
Western Reserve Manufacturing Co., Inc.
Copper. Cu 87-90.5, Sn 9-11, Pb 0-0.3, Zn 0-0.5, Fe 0-0.1, Ni 0-0.5, P 0.75-1.5, Sb 0-0.1.
As cast: 50,000 TS; 22,000 YS; 15 El; 110 Brin. CDA 907.

WRM 1010
Western Reserve Manufacturing Co., Inc.
Copper. Sn 9-11, Pb 8-11, Zn 0-0.75, Fe 0-0.15, Ni 0-0.75, P 0-0.25, Sb 0-0.5, bal Cu.
As cast: 40,000 TS; 15 El. CDA 937.

WRM 1010B
Western Reserve Manufacturing Co., Inc.
Copper. Sn 9-11, Pb 8-11, Zn 0-0.75, Fe 0-0.15, Ni 0-0.75, P 0-1.5, Sb 0-0.5, bal Cu.
As cast: 50,000 TS; 27,000 YS; 10 El; 93 Brin min. CDA 937.

WRM 1010N
Western Reserve Manufacturing Co., Inc.
Copper. Sn 9-11, Pb 9-11, Ni 2-3, bal Cu.
As cast: 39,000 TS; 24,000 YS; 6 El. CDA 937.

WRM 102
Western Reserve Manufacturing Co., Inc.
Copper. Sn 9-11, Pb 1-2.5, Zn 0-0.75, Ni 0-1, P 0-0.25, bal Cu.
As cast: 40,000 TS; 10 El. CDA 927.

WRM 102N
Western Reserve Manufacturing Co., Inc.
Copper. Sn 9-11, Pb 2-3.25, Ni 2.75-4, P 0-0.75, bal Cu.
As cast: 53,000 TS; 31,000 YS; 16 El; 105 Brin. CDA 929.

WRM 110
Western Reserve Manufacturing Co., Inc.
Copper. Sn 10-12, Ni 0-0.05, P 0.1-0.3, bal Cu.
As cast: 50,000 TS; 27,000 YS; 15 El. CDA 907.

WRM 110N
Western Reserve Manufacturing Co., Inc.
Copper. Sn 10-12, Pb 0-0.5, Zn 0-0.5, Fe 0-0.2, Ni 1-3, P 0.1-0.3, Sb 0-0.2, bal Cu.
As cast: 40,000 TS; 15 El. CDA 907.

WRM 111N
Western Reserve Manufacturing Co., Inc.
Copper. Sn 10-12, Pb 1-1.5, Fe 0-0.3, Ni 0.75-1.25, P 0.2-0.3, Sb 0-0.2, 0.1 others, bal Cu.
As cast: 40,000 TS; 1 El. CDA 925.

WRM 119N
Western Reserve Manufacturing Co., Inc.
Copper. Cu 75-79, Sn 10-14, Pb 8-10, Zn 0-0.1, Ni 1.5-4.5.
As cast: 32,000 TS; 20,000 YS; 10 El; 86-100 Brin. CDA 937.

WRM 150
Western Reserve Manufacturing Co., Inc.
Copper. Cu 84-86, Sn 13-15, Pb 0-0.2, Zn 0-1.5, Fe 0-0.1, Ni 0-0.75, P 0-0.05.
As cast: 31,000 TS; 1 El. CDA 910.

WRM 160
Western Reserve Manufacturing Co., Inc.
Copper. Cu 82-85, Sn 15-17, Pb 0-0.25, Zn 0-0.25, Fe 0-0.25, P 0-1.
As cast: 31,000 TS; 135 Brin. CDA 911.

WRM 165
Western Reserve Manufacturing Co., Inc.
Copper. Cu 78-82, Sn 15-17, Pb 4-6, 0.50 others.
As cast: 72-82 Brin. CDA 928.

WRM 190
Western Reserve Manufacturing Co., Inc.
Copper. Cu 79-82, Sn 18-20, Pb 0-0.25, Zn 0-0.25, Fe 0-0.25, P 0-1.
As cast: 31,000 TS; 1 El; 160 Brin. CDA 913.

WRM 210
Western Reserve Manufacturing Co., Inc.
Copper. Cu 79-82, Sn 2-3, Pb 9-11, Zn 6.5-8.5, Fe 0-0.4, Ni 0-1, P 0-0.04.
As cast: 28,000 TS; 15 El. CDA 844.

WRM 223
Western Reserve Manufacturing Co., Inc.
Copper. Sn 1-3, Pb 22-25, Zn 0-0.5, Fe 0-0.2, Ni 0-1, P 0-0.05, bal Cu.
As cast. CDA 943.

WRM 230
Western Reserve Manufacturing Co., Inc.
Copper. Sn 1.5-2.5, Pb 27-34, Ni 0.25-0.75, bal Cu.
As cast: 25-50 Brin.

WRM 310N
Western Reserve Manufacturing Co., Inc.
Copper. Sn 1.5-4, Pb 8-12, Zn 6-11, Fe 0-0.3, Ni 1-3, P 0-1.5, Sb 0-0.2, bal Cu.
As cast: 80 Brin. CDA 844.

WRM 37
Western Reserve Manufacturing Co., Inc.
Copper. Sn 2.25-3.5, Pb 6-8, Zn 7-10, Fe 0-0.4, Ni 0-1, P 0-0.05, bal Cu.
As cast: 30,000 TS; 16 El. CDA 844.

WRM 44N
Western Reserve Manufacturing Co., Inc.
Copper. Cu 63-65, Sn 3.5-4.5, Pb 3.5-5, Fe 0-1, Ni 19.5-21, P 0-0.05, Sb 0-0.25, bal Zn.
As cast: 30,000 TS; 8 El. CDA 976.

WRM 46
Western Reserve Manufacturing Co., Inc.
Copper. Cu 82-83.8, Sn 3.3-4.2, Pb 5-7, Zn 5-8, Fe 0-0.3, Ni 0-1, P 0-0.03, Sb 0-0.25.
As cast. CDA 838.

WRM 50N H.T.
Western Reserve Manufacturing Co., Inc.
Copper. Sn 5-6, Pb 0-0.01, Zn 1-3, Fe 0-0.15, Ni 4.5-5.5, P 0-0.03, bal Cu.
As cast: 75,000 TS; 50,000 YS; 12 El; 165 Brin. CDA 947.

WRM 514 N
Western Reserve Manufacturing Co., Inc.
Copper. Sn 4-6, Pb 12-15, Zn 0-1, Fe 0-0.4, Ni 1-2.5, P 0-0.1, Sb 0-0.4, bal Cu.
As cast. CDA 938.

WRM 51N
Western Reserve Manufacturing Co., Inc.
Copper. Sn 4.5-6, Pb 0-1, Zn 0-2.5, Fe 0-0.25, Ni 4.5-6, P 0-0.03, Sb 0-0.15, 0.25 others, bal Cu.
As cast: 35,000 TS; 20 El. CDA 948.

WRM 520
Western Reserve Manufacturing Co., Inc.
Copper. Sn 4.25-5.5, Pb 18-22, Zn 0-0.5, Ni 0-0.5, Fe 0-0.2, P 0-0.05, bal Cu.
As cast: 28,000 TS; 19,000 YS; 15 El; 60 Brin. CDA 941.

WRM 525
Western Reserve Manufacturing Co., Inc.
Copper. Sn 4.5-6, Pb 23-26, Zn 0-0.5, Fe 0-0.15, Ni 0-0.5, P 0-0.05, Sb 0-0.75, bal Cu.
As cast: 21,000 TS; 15 El; 38-41 Brin. CDA 943.

WRM 55
Western Reserve Manufacturing Co., Inc.
Copper. Sn 4-6, Pb 4-6, Zn 4-6, Fe 0-0.3, Ni 0-1, P 0-0.05, bal Cu.
As cast: 35,000 TS; 20 El. CDA 836.

WRM 55B
Western Reserve Manufacturing Co., Inc.
Copper. Cu 82-85, Sn 4-6, Pb 4-6, Zn 4-6, Fe 0-0.3, Ni 0-1, P 0-1.5.
As cast: 50,000 TS; 83 Brin min. CDA 836.

WRM 55N
Western Reserve Manufacturing Co., Inc.
Copper. Sn 4-6, Pb 4-6, Zn 4-6, Fe 0-0.2, Ni 4-6, P 0-0.05, bal Cu.
As cast. CDA 836.

WRM 59
Western Reserve Manufacturing Co., Inc.
Copper. Sn 4.5-6, Pb 8-10, Zn 0-2, Fe 0-0.2, Ni 0-0.5, 0.30 others, bal Cu.
As cast: 35,000 TS; 10,000 YS. CDA 935.

WRM 61
Western Reserve Manufacturing Co., Inc.
Copper. Sn 5.5-6.5, Pb 1-2, Zn 3-5, Fe 0-0.25, Ni 0-1, P 0-0.05, bal Cu.
As cast: 45,000 TS; 20,000 YS; 30 El. CDA 922.

WRM 616
Western Reserve Manufacturing Co., Inc.
Copper. Sn 5-7, Pb 14-18, Zn 0-1.5, Fe 0-0.4, Ni 0-0.75, P 0-0.05, Sb 0-0.5, bal Cu.
As cast: 20,000 TS; 10 El. CDA 939.

WRM 619
Western Reserve Manufacturing Co., Inc.
Copper. Sn 5-6.5, Pb 18-21, Zn 0-0.75, Fe 0-0.15, Ni 0-1, P 0-0.05, Sb 0-0.75, bal Cu.
As cast: 22,000 TS; 10 El. CDA 941.

WRM 773
Western Reserve Manufacturing Co., Inc.
Copper. Sn 6.25-7.5, Pb 6-8, Zn 2-4, Fe 0-0.2, Ni 0-0.5, P 0-0.15, Sb 0-0.35, bal Cu.
As cast: 35,000 TS; 10 El. CDA 932.

WRM 80
Western Reserve Manufacturing Co., Inc.
Copper. Sn 7.5-9, Pb 0-0.3, Zn 3-5, Fe 0-0.15, Ni 0-1, P 0-0.05, bal Cu.
As cast: 44,000 TS; 20,000 YS; 25 El. CDA 903.

WRM 810
Western Reserve Manufacturing Co., Inc.
Copper. Sn 7-9, Pb 9-11, Zn 1.5-3.5, Fe 0-0.2, Ni 0-0.75, P 0-0.1, Sb 0-0.5, bal Cu.
As cast. CDA 937.

WRM 822
Western Reserve Manufacturing Co., Inc.
Copper. Sn 7.5-8.5, Pb 19.5-24, Zn 0-0.75, Fe 0-0.2, Ni 0-1, P 0-1.5, Sb 0-0.8, bal Cu.
As cast: 32,000 TS; 18,000 YS; 20 El. CDA 945.

WRM 88
Western Reserve Manufacturing Co., Inc.
Copper. Sn 7-9, Pb 7-9, Zn 0-0.75, Fe 0-0.15, Ni 0-1, P 0-0.5, bal Cu.
As cast: 30,000 TS; 16,000 YS; 20 El. CDA 934.

WRM ALLOY A-10
Waterbury Rolling Mills Inc.
Cu 66, Ni 10, Zn 24.
Annealed: 49,000 TS; 18,000 YS; 49 El; 63 Brin. Hard: 86,000 TS; 75,000 YS; 4 El; 180 Brin. For rivets, screws, fasteners, optical parts; ASTM-B122 alloy 3; nickel silver. *Obsolete*

WRM ALLOY A-12
Waterbury Rolling Mills Inc.
Cu 65, Ni 12, Zn 23.
Annealed: 48,000 TS; 17,000 YS; 62 El; 114 Brin. Hard: 105,000 TS; 101,000 YS; 2 El; 210 Brin. For holloware, screen cloth, tableware, camera parts; nickel silver; ASTM-B151 alloy D. *Obsolete*

WRM ALLOY A-15
Waterbury Rolling Mills Inc.
Cu 65, Ni 15, Zn 20.
Annealed: 53,000 TS; 18,000 YS; 43 El; 63 Brin. Hard: 85,000 TS; 75,000 YS; 3 El; 172 Brin. For optical goods, camera parts, jewelry; nickel silver, corrosion resistant. *Obsolete*

WRM ALLOY A-18
Waterbury Rolling Mills Inc.
Cu 65, Ni 18, Zn 17.
For hardware, flatware; nickel silver; ASTM-B122 alloy 2. *Obsolete*

WRM ALLOY A-5
Waterbury Rolling Mills Inc.
Cu 65, Ni 5, bal Zn.
For hardware, trim, fasteners; nickel silver. *Obsolete*

WRM ALLOY B-12-S
Waterbury Rolling Mills Inc.
Cu 56.5, Ni 12, Zn 31.5.
For hardware, flatware, fasteners; nickel silver; ASTM -B122 alloy 8. *Obsolete*

WRM ALLOY B-18-S
Waterbury Rolling Mills Inc.
Cu 55, Ni 18, Zn 27.
Annealed: 60,000 TS; 27,000 YS; 40 El; 100 Brin. Hard: 100,000 TS; 85,000 YS; 3 El; 190 Brin. For switches, springs, hardware; nickel silver; ASTM B122 alloy 4. *Obsolete*

WRM ALLOY K-12
Waterbury Rolling Mills Inc.
Cu 65, Ni 12, Zn 21, Pb 2.
For hardware, flatware, fasteners; leaded nickel silver, free-cutting. *Obsolete*

WRM ALLOY K-8
Waterbury Rolling Mills Inc.
Cu 65, Ni 8, Zn 25, Pb 2.
For keys; leaded nickel silver, free-cutting. *Obsolete*

WROUGHT IRON 4D
A.M. Byers Co.
C 0.02, S 0.02, P 0.12, Mn 0.06, Si 0.12, 2.5 silicates, bal Fe.
Rolled: 48,000 TS; 27,000 YS; 14 El; 97-105 Brin. For condensers, fasteners, tanks, heaters; easy to work. *Obsolete*

WS 353
Bergische Stahl Industrie
C 0.24-0.3, Si 0.4-0.6, Mn 0.3-0.7, Cr 1.3-1.5, Mo 1.1-1.4, V 0.35-0.45, bal Fe.
Oil hardening structural steel. W.-Nr. 1.2353.

WS 718
Bergische Stahl Industrie
C 0.5-0.57, Si 0.15-0.3, Mn 0.4-0.5, Cr 0.5-0.7, Ni 2.5-3, bal Fe.
Cold work tool steel. W.-Nr. 1.2718.

WS1 EX
Thyssen Edelstahlwerke AG
C 1.2, V 0.1, W 1, bal Fe.
For cutting tools; water hardened. *Obsolete*

WSCR
Thyssen Edelstahlwerke AG
C 0.67, Si 1.3, Mn 0.5, Cr 0.5, bal Fe.
For springs, chisels, punches; oil hardened, shock resistant. *Obsolete*

WT 70
Sandvik Hard Materials Ltd.
Co 25, WC 75.
Cemented carbide. 950 Vickers; 3400 MNm^{-2} compressive strength.

WT DRILL ROD
Latrobe Steel Co.
C 1.23, W 1.27, Cr 0.4, V 0.3, bal Fe.
For drill rod, tools; oil hardening. *Obsolete*

WUEST NO. 1
Manufacturer not listed.
Zn 50, Al 30, Cu 20.
For aluminum solder.

WUEST NO. 2
Manufacturer not listed.
Zn 65, Al 20, Cu 15.
For aluminum solder.

WUNDUS
Sanderson Kayser Ltd.
C 0.7, Co 6, W 22, bal Fe.
For milling cutters, twist drills, turning, planing, shaping and slotting tools; for cutting hardest metals.

WW-1
W.W. Alloys Inc.
Cu alloy.
65,000 TS; 60,000 YS; 20 El; 60 RA; 125 Brin. For resistance welding electrode; 85% conductivity, for coated metals. *Obsolete*

WW-100W
W.W. Alloys Inc.
Cu-W.
Bars: 200,000 TS; Rockwell A76. For resistance welding electrode; 30% conductivity, for Red Brass and Cu. *Obsolete*

WW-10W3
W.W. Alloys Inc.
Cu 23, W 77.
Bars: 160,000 TS; 205 Brin. For resistance welding electrode; 28% conductivity, for inserts and facings. *Obsolete*

WW-120
W.W. Alloys Inc.
Al 9.5-10.5, Fe 1.5, bal Cu.
Rolled: 70,000 TS; 40,000 YS; 12 El; 14 RA; 120 Brin. For pickling equipment, gears; Al-bronze, corrosion resistant. *Obsolete*

WW-125
W.W. Alloys Inc.
Cu 88, Al 9, Fe 3.
Cast: 80,000 TS; 29,000 YS; 25 El; 25 RA; 125 Brin. For bushings, bearings, sleeves; corrosion and wear resistant. *Obsolete*

WW-150
W.W. Alloys Inc.
Cu 86.2, Al 10.1, Fe 3.4, 0.3% special elements.
Cast: 80,000 TS; 35,000 YS; 22 El; 20 RA; 150 Brin. For bushings, gears, slides, liners; corrosion and shock resistant, Al-bronze. *Obsolete*

WW-175
W.W. Alloys Inc.
Cu 85.2, Al 10.8, Fe 3.7, 0.3% special elements.
Rolled: 90,000 TS; 35,000 YS; 12 El; 8 RA; 175 Brin. Heat treated: 95,000 TS; 43,000 YS; 12 El; 12 RA; 190 Brin. For gears, aircraft parts, bushings, acid equipment; Al bronze, shock and corrosion resistant. *Obsolete*

WW-190
W.W. Alloys Inc.
Al 12, bal Cu.
Cast: 70,000 TS; 45,000 YS; 3 El; 190 Brin. For welding dies and welder jaws; good arc resistance characteristics. *Obsolete*

WW-1W3
W.W. Alloys Inc.
Cu-W.
Bars: 135,000 TS; 130 Brin. For resistance welding electrode; 35% conductivity, for stainless steel and brass. *Obsolete*

WW-2
W.W. Alloys Inc.
Cr, bal Fe.
Cast: 50,000 TS; 35,000 YS; 15 El; 50 RA; 116 Brin. Rolled: 75,000 TS; 65,000 YS; 15 El; 50 RA; 150 Brin. For electrical contacts, resistance welding electrode; 75% conductivity, for carbon steel and brass. *Obsolete*

WW-200
W.W. Alloys Inc.
Al 12, Fe 4.1, 0.4% special elements, bal Cu.
Cast: 82,000 TS; 37,000 YS; 3 El; 200 Brin. Heat treated: 95,000 TS; 40,000 YS; 5 El; 210 Brin. For cams, gibs, slides, forming rolls; Al-bronze, corrosion and wear resistant. *Obsolete*

WW-20W3
W.W. Alloys Inc.
Cu-W.
Bars: 170,000 TS; 228 Brin. For resistance welding electrode; 27% conductivity, for heavy projection welds. *Obsolete*

WW-230
W.W. Alloys Inc.
Al 10.5-11.5, Fe 4-5, Ni 4-5, bal Cu.
Cast: 100,000 TS; 50,000 YS; 8 El; 230 Brin. Heat treated: 105,000 TS; 60,000 YS; 10 El; 240 Brin. For heavy duty wear parts, gears; Al-bronze, corrosion and wear resistant. *Obsolete*

WW-275
W.W. Alloys Inc.
Al 13, Fe 4.5, 0.5% special elements, bal Cu.
Cast: 80,000 TS; 55,000 YS; 1.5 El; 275 Brin. For forming dies, die inserts, wear strips; corrosion and abrasion resistant. *Obsolete*

WW-3
W.W. Alloys Inc.
Co, Be, bal Cu.
Rolled: 110,000 TS; 100,000 YS; 10 El; 210 Brin. Cast: 95,000 TS; 65,000 YS; 6 El; 210 Brin. For resistance welding electrode; 50% conductivity, for stainless steel. *Obsolete*

WW-325
W.W. Alloys Inc.
Cu 81, Al 14, Fe 5.
Cast: 85,000 TS; 65,000 YS; 0.5 El; 325 Brin. For cams, forming and drawing dies; corrosion and abrasion resistant. *Obsolete*

WW-375
W.W. Alloys Inc.
Al 14.5-15.5, 4.5-5.5% special elements, bal Cu.
Cast: 80,000 TS; 60,000 YS; 375 Brin. For forming and drawing dies; C.S. 225,000. *Obsolete*

WW-5
W.W. Alloys Inc.
Cu-alloy.
Cast: 80,000 TS; 10 El; 185 Brin. Rolled: 100,000 TS; 20 El; 210 Brin. For flash welding dies, bushings, structural parts; 45% conductivity; good strength. *Obsolete*

WW-6
W.W. Alloys Inc.
Cu alloy.
Rolled: 115,000 TS; 95,000 YS; 10 El; 10 RA; 210 Brin. For resisting welding wheels, dies, electrodes; for stainless st *Obsolete*

WW-7
W.W. Alloys Inc.
Be, Co, bal Fe.
Cast: 110,000 TS; 60,000 YS; 2 El; 380 Brin. Rolled: 195,000 TS; 170,000 YS; 4 El; 380 Brin. For resistance welding electrode; corrosion and abrasion resistant. *Obsolete*

WW-8
W.W. Alloys Inc.
Cu alloy.
Cast: 65,000-75,000 TS; 12,000-16,000 YS; 2-10 El; 116-165 Brin. For resistance welding electrode. *Obsolete*

WW105F
W.W. Alloys Inc.
Al 5, Fe 1.5, bal Cu.
Forged: 60,000 TS; 25,000 YS; 38 El; 100 Brin. For machine tool slides, gibs, wear strips; corrosion and wear resistant. *Obsolete*

WW110F
W.W. Alloys Inc.
Al 7, Fe 2, bal Cu.
Forged: 70,000 TS; 32,000 YS; 30 El; 110 Brin. For machine tool parts, liners, wear plates; corrosion resistant, Al-bronze. *Obsolete*

WW140F
W.W. Alloys Inc.
Al 9, Fe 3, bal Cu.
Forged: 85,000 TS; 40,000 YS; 20 El; 140 Brin. For gears, worm wheels; good weldability, corrosion resistant. *Obsolete*

WW175A
W.W. Alloys Inc.
Al 10.5, Fe 4, bal Cu.
Cast: 85,000 TS; 32,000 YS; 16 El; 175 Brin. For bearings, gears, worm wheels, shifter forks, Al-bronze, wear and shock resistant. *Obsolete*

WW175B
W.W. Alloys Inc.
Al 10.5, Fe 4, bal Cu.
Cast: 90,000 TS; 50,000 YS; 4 El; 220 Brin. For bushings, bearings; Al-bronze, wear and shock resistant. *Obsolete*

WWS
Saarstahl AG
C 0.6, Si 0.9, Mn 1, bal Fe.
Hot or cold work tool steel; for hammer dies, trimming tools, punches; shock resistant. Werkstoff Nr. 1.2826; AFNOR 60 MS 4. *Obsolete*

WYCKOFF 1215
AMCO
C 0-0.09, Mn 0.75-1.05, P 0.04-0.09, S 0.26-0.35, bal Fe.
Cold drawn: 79,000 TS; 61,000 YS; 10 El; 35 RA; 167 Brin. For gears, shafts, hardware, bolts, fasteners. Free-cutting. Case hardening. *Obsolete*

WYCKOFF 12L14X
AMCO
C 0-0.09, Mn 0.85-1.15, P 0.04-0.09, S 0.26-0.35, Pb 0.15-0.35, 0.035% Te min, bal Fe.
Cold drawn: 75,000 TS; 56,000 YS; 15 El; 45 RA; 156 Brin. For gears, shafts, hardware, bolts, fasteners. Free-cutting, case-hardening steel. *Obsolete*

WYCKOFF WXB-1
AMCO
C 0.08-0.13, Mn 0.8-0.9, S 0.08-0.13, P 0.07-0.12, bal Fe.
For nails, screws, machinery parts, screw machine products; B 1111 Steel; cold drawn, free-cutting. *Obsolete*

WYCKOFF WXB-2
AMCO
C 0-0.13, Mn 0.7-1, P 0.07-0.12, S 0.16-0.23, bal Fe.
For screw machine products; B 1112 Steel; cold drawn, free-cutting. *Obsolete*

WYCKOFF WXB-3
AMCO
C 0-0.13, Mn 0.7-1, P 0.07-0.12, S 0.24-0.33, bal Fe.
For screw machine products; B 1113 Steel; cold drawn, free-cutting. *Obsolete*

WYCLIFFE BLACK HEART MALLEABLE IRON
Follsain-Wycliffe Foundries, Ltd.
Castings for agriculture, automobile, ship building, textile, commercial vehicle. Meets British Standard BS 310/1958, Gr. B 22-14.

WYCO 100-44
AMCO
C 0.4-0.48, Mn 1.35-1.65, P 0-0.04, S 0.24-0.33, bal Fe.
Cold drawn: 125,000 TS; 100,000 YS minimum; 12 El; C 25 Rock. For gears, shafts, hardware, bolts, fasteners. Free-cutting, oil or water hardening. *Obsolete*

WYCO 100-50
AMCO
C 0.48-0.55, Mn 0.6-0.9, P 0-0.04, S 0-0.05, bal Fe.
Cold drawn: 125,000 TS; 100,000 YS minimum; 11 El; C 25 Rock. For gears, crankshafts, machinery parts, bolts, fasteners, hardware. Oil or water hardening. *Obsolete*

WYCO 125-44
AMCO
C 0.4-0.48, Mn 1.35-1.65, S 0.24-0.33, bal Fe.
Cold drawn: 140,000 minimum TS; 125,000 YS minimum; 5 El; 15 RA; C 30 Rock. For gears, crankshafts, bolts, fasteners, hardware. Free machining, water or oil hardening. *Obsolete*

WYCO COLD FINISHED STEEL
AMCO
C 0.3, Ni 0.5, bal Fe.
For cold drawn steel products, machine tool parts; superior surface and machinability. *Obsolete*

WYNDALOY
Wyndale Mfg. Co.
Cu 60, Ni 20, Mn 20.
Hardened: 200,000 psi TS; 170,000 psi YS; 475 Brin.
Annealed: 98,000 psi TS; 80,000 psi YS; 140 Brin. For
valves, pistons, pump rods, screws, bolts, nuts, corrosion
resistant, hardenable, non-magnetic.

WYNITE HIGH DUTY NI-CR CAST IRON
Follsain-Wycliffe Foundries, Ltd.
Cast: 10-18 tsi TS. For pumps, jigs. Meets British Standard
BS 1452, Gr. 14 or Gr. 17.

WZ-1
American Electro Metals Corp.
TiC.
For gas turbine parts, jet engine and nuclear power plants;
sintered, high heat resistant.

WZ-12A
Plansee, Metallwerk Gesellschaft
Titanium. Ti 75, Ni 1.5, Co 5, Cr 5.
For gas turbine blades. Sintered, oxidation resistant.

WZ-12B
Plansee, Metallwerk Gesellschaft
Ni 24, Co 8, Cr 8, 60.0 TiC.
For gas turbines, jet engines, and nuclear power plants.
Sintered; oxidation resistant.

WZ-12B
American Electro Metals Corp.
Ni 24, Co 8, Cr 8, 60.0 TiC.
For gas turbines, jet engines, nuclear power plants;
sintered, oxidation resistant.

WZ-12C
American Electro Metals Corp.
Ni 30, Co 10, Cr 10, 50.0 TiC.
For gas turbines, jet engines, nuclear power plants;
sintered, oxidation resistant.

WZ-12C
Plansee, Metallwerk Gesellschaft
Ni 30, Co 10, Cr 10, 50.0 TiC.
For gas turbines, jet engines, and nuclear power plants.
Sintered; oxidation resistant.

WZ-1B
American Electro Metals Corp.
Ni 32, Cr 8, 60.0 TiC.
For gas turbines, jet engines, nuclear power plants,
sintered, oxidation resistant.

WZ-1B
Plansee, Metallwerk Gesellschaft
Ni 32, Cr 8, 60.0 TiC.
For gas turbines, jet engines, and nuclear power plants.
Sintered; oxidation resistant.

WZ-1C
Plansee, Metallwerk Gesellschaft
Ni 40, Cr 10, 50.0 TiC.
For gas turbine blades. Sintered; oxidation resistant.

WZ-1D
Plansee, Metallwerk Gesellschaft
Nickel. Ni 52, Cr 13, 35.0 TiC.
For high temperature applications; sintered cermet.

WZ-2
Plansee, Metallwerk Gesellschaft
Co 28, Cr 12, 60.0 TiC.
For gas turbine blades. Sintered, oxidation resistant.

WZ3
Austrian manufacture
Ni 32, Cr 8, 50.0 TiC, 10.0 TaC.
Sintered: 1070 Vickers. For jet engine components;
sintered, carbides.

WZM
American manufacture
W 25, Zr 0.1, C 0.03, bal Mo.

X

Now BETHLEHEM X.

X 8 NICOCRTI55-20-20

German manufacture
C 0.1, Si 0-1.5, Mn 0-1, Al 0.8-1.8, Fe 0-5, Ti 1.8-2.7, Cr 18-21, Co 15-21, bal Ni.
At 20°C: 155,000 TS; 90,000 YS; 39 El; 20 RA. At 800°C: 78,000 TS; 57,000 YS; 7 El; 4 RA. For jet engine and gas turbine components; high heat, corrosion and oxidation resistant, Nimonic 90, heat treatable.

X 8 NICRALTI 75-20

German manufacture
C 0.04, Cr 21, Ti 2.5, Al 0.7, Mn 0.6, bal Ni.
At 20°C: 132,000 TS; 80,000 YS; 45 El; 36 RA. At 800°C: 62,000 TS; 53,000 YS; 8 El; 10 RA. For gas turbine blades, jet engine components; high heat, corrosion and oxidation resistant, Nimonic 80A, creep resistant.

X A R 15

National Intergroup Inc.
C 0.1-0.21, Mn 0.6-1.1, Si 0.4 0.8, B 0.0025, Cr 0.4-0.8, Mo 0.18-0.28, Zr 0.05-0.15, bal Fe.
Heat treated plate: 360 Brin min; 190,000 TS; 175,000 YS; 16 El; 55 RA. For bridges, booms, dipper sticks, penstocks, chutes, wear plates, truck bodies, fan blades, floor plates. Quenched and tempered at mill. Tough. Wear and abrasion resistant.

X A R 30

National Intergroup Inc.
C 0-0.3, Mn 0.6-1.1, Si 0.4-0.8, B 0.0025, Cr 0.4-0.8, Mo 0.18-0.28, Zr 0.05-0.15, bal Fe.
Heat treated plate: 360 Brin min; 190,000 TS; 175,000 YS; 16 El; 55 RA. For bridges, booms, dipper sticks, penstocks, chutes, wear plates, truck bodies, fan blades, floor plates. Quenched and tempered at mill. Tough. Wear and abrasion resistant.

X WIPING SOLDER

Hewitt Metals Corp.
Sn 40, bal Pb.
Block solder, for plumbers, wiping solder. *Obsolete*

X-11

Teledyne Firth Sterling
Tungsten. W 97, bal Ni, Fe, Cu.
Used in ordnance products. 880-1160 MPa TS; 1-12 El; 30-43 Rock C; 18.5 g/cm^3 density.

X-110

Now COLUMBIUM D-43.

X-15C

Teledyne Firth Sterling
Tungsten. W 95, Ni 2.5, Fe 1, Co 1.5.
Used in ordnance products. 970-1430 MPa TS; 6-35 El; 31-47 Rock C; 17.3-18.6 g/cm^3 density.

X-15R

Teledyne Firth Sterling
Tungsten. W 95, Ni 3.5, Fe 1.5.
For counterweights, inertial masses, radiation shielding, X-ray collimation, and ordnance products. 850-1400 MPa TS; 2-35 El; 29-44 Rock C; 14.0-18.6 g/cm^3 density

X-1900

Now RENE 85.

X-2

Teledyne Vasco
C 0.24, Si 0.89, Mn 0.32, Cr 4.9, V 0.54, Mo 1.3, bal Fe.
Annealed: 85,000 TS; 35,000 YS; 12 El; 50 RA; 512 Brin. For gears and shafts, strength structures; air or oil hardened, hot work steel. *Obsolete*

X-21C

Teledyne Firth Sterling
Tungsten. W 93, Ni 3.4, Fe 1.5, Co 2.1.
Used in ordnance products. 970-1430 MPa TS; 6-35 El; 31-47 Rock C; 17.3-18.6 g/cm^3 density.

X-21R

Teledyne Firth Sterling
Tungsten. W 93, Ni 4.9, Fe 2.1.
For counterweights, inertial masses, radiation shielding, X-ray collimation, and ordnance products. 850-1400 MPa TS; 2-35 El; 29-44 Rock C; 14.0-18.6 g/cm^3 density.

X-27C

Teledyne Firth Sterling
Tungsten. W 91, Ni 4.4, Fe 1.9, Co 2.7.
Used in ordnance products. 970-1430 MPa TS; 6-35 El; 31-47 Rock C; 17.3-18.6 g/cm^3 density.

X-27R

Teledyne Firth Sterling
Tungsten. W 91, Ni 6.3, Fe 2.7.
For counterweights, inertial masses, radiation shielding, X-ray collimation, and ordnance products. 850-1400 MPa TS; 2-35 El; 29-44 Rock C; 14.0-18.6 g/cm^3 density.

X-27X

Teledyne Firth Sterling
Tungsten. W 91.
Used in high performance ordnance applications. 1030-1730 MPa TS; 8-39 El; 30-50 Rock C; 17.4 g/cm^3 density.

X-30R

Teledyne Firth Sterling
Tungsten. W 90, Ni 7, Fe 3.
For counterweights, inertial masses, radiation shielding, X-ray collimation, and ordnance products. 850-1400 MPa TS; 2-35 El; 29-44 Rock C; 14.0-18.6 g/cm^3 density.

X-40

Special Metals Corp.
C 0.5, Mn 0.5, Si 0.5, Cr 22, Ni 10, W 7.5, Fe 1.5, bal Co.
High temperature alloy. (Haynes Stellite Alloy 31).

X-40

Teledyne Allvac
C 0.5, Mn 0.5, Si 0.5, Cr 22, Ni 10, W 7.5, Fe 1.5, bal Co.
High temperature alloy. (Haynes Stellite Alloy 31).

X-41

Manufacturer not listed.
C 0.5, Mn 0.5, Cr 25, Ni 8, W 7.5, Cr 1.75, B 2, Fe 1, bal Co.
For high temperature applications; heat resistant.

X-42-W

Gulf States Steel, Inc.
HSLA steel. C 0.21, Mn 1.25, P 0.04, S 0.05, Si 0.3, 0.01 V min, bal Fe.
High strength low alloy steel with good formability and weldability. 42,000 YS min; 60,000 TS min; 24 El.

X-42-W, ETC.

Now REPUBLIC X-42-W, ETC.

X-45

General Electric Co.
C 0.25, Cr 25, Ni 10, W 7, Fe 1.5, bal Co.
High temperature alloy.

X-45

Cannon-Muskegon Corp.
C 0.25, Mn 0-1, Cr 25.5, Ni 10.5, W 7.5, B 0.01, Fe 0-2, bal Co.
Cobalt base superalloy; casting. For marine gas turbines, nozzle vanes.

X-45 W, ETC.

Now REPUBLIC X-45 W, ETC.

X-45-W

Gulf States Steel, Inc.
HSLA steel. C 0.22, Mn 1.25, P 0.04, S 0.05, Si 0.3, 0.01 V min, bal Fe.
High strength low alloy steel with good formability and weldability. 45,000 YS min; 60,000 TS min; 22 El.

X-50

General Electric Co.
C 0.76, Mn 0.6, Cr 22, Ni 20, Co 40, W 12, Fe 1.
For high temperature applications; heat resistant.

X-50

General Electric Co.
C 0.75, Cr 22, Ni 20, W 12, Fe 2.5, bal Co.
Cobalt base superalloy.

X-50

Manufacturer not listed
C 0.76, Mn 0.6, Cr 22, Ni 20, Co 40, W 12, Fe 1.
For high temperature applications; heat resistant.

X-50-W

Gulf States Steel, Inc.
HSLA steel. C 0.23, Mn 1.35, P 0.04, S 0.05, Si 0.3, 0.01 V min, bal Fe.
High strength low alloy steel with good weldability and formability. 50,000 YS min; 65,000 TS min; 22 El.

X-55-W

Gulf States Steel, Inc.
HSLA steel. C 0.25, Mn 1.35, P 0.04, S 0.05, Si 0.3, 0.01 V min, bal Fe.
High strength low alloy steel with good formability and weldability. 55,000 YS min; 70,000 TS min; 20 El.

X-60-W

Gulf States Steel, Inc.
HSLA steel. C 0.26, Mn 1.35, P 0.04, S 0.05, Si 0.3, 0.01 V min, bal Fe.
High strength low alloy steel with good formability and weldability. 60,000 YS min; 75,000 TS min; 18 El.

X-63

General Electric Co.
C 0.4, Cr 22, Ni 10, Co 58.5, Mo 6, Al 1.25, Fe 2.
Cobalt base superalloy; casting.

X-65-W

Gulf States Steel, Inc.
HSLA steel. C 0.26, Mn 1.35, P 0.04, S 0.05, Si 0.3, 0.01 V min, bal Fe.
High strength low alloy steel with good formability and weldability. 65,000 YS min; 80,000 TS min.

X-70-W

Gulf States Steel, Inc.
HSLA steel. C 0.26, Mn 1.65, P 0.04, S 0.05, Si 0.3, 0.01 V min, bal Fe.
High strength low alloy steel with good formability and weldability. 70,000 YS min; 85,000 TS min.

X-750

Now INCONEL ALLOY X-750 and NICKELVAC X-750.

X-75R

Teledyne Firth Sterling
Tungsten. W 75, Ni 17.5, Fe 7.5.
For counterweights, inertial masses, radiation shielding, X-ray collimation, and ordnance products. 850-1400 MPa TS; 2-35 El; 29-44 Rock C; 14.0-18.6 g/cm^3 density.

X-76

American manufacture
Cu 0.5-2, Mn 0.4-0.8, Mg 1.4-3, Zn 5-8, bal Al.
For supercharger casings, diffusers, impellers; age hardenable.

X-782
American manufacture
C 2, Mn 0.5, Si 0.5, Cr 26, Fe 4, W 8.7, bal Ni.
For engine valves.

X-80-W
Gulf States Steel, Inc.
HSLA steel. C 0.26, Mn 1.6, Si 0.3, 0.10 V min, 0.01 Cb min, bal Fe.
High strength steel with good formability and weldability. 80,000 YS min; 95,000 TS min.

X-90R
Teledyne Firth Sterling
Tungsten. W 70, Ni 21, Fe 9.
For counterweights, inertial masses, radiation shielding, X-ray collimation, and ordnance products. 850-1400 MPa TS; 2-35 El; 29-44 Rock C; 14.0-18.6 g/cm^3 density.

X-9R
Teledyne Firth Sterling
Tungsten. W 97, Ni 2.1, Fe 0.9.
For counterweights, inertial masses, radiation shielding, X-ray collimation, and ordnance products. 850-1400 MPa TS; 2-35 El; 29-44 Rock C; 14.0-18.6 g/cm^3 density.

X-ALLOY
English manufacture
Cu 3.6, Mg 0.6, Ni 0.7, Fe 1.2, Si 0.7, bal Al.
For pistons; cast.

X-B
Alloy Engineering & Casting Co.
Ni 33-37, Cr 13-17, bal Fe.
Cast: 64,000 TS; 36,000 YS; 5 El; 196 Brin. For castings and furnace parts. Resists heat. *Obsolete*

X-CEL SUPER X
Hewitt Metals Corp.
Block solder: melt range 358-424 F. *Obsolete*

X-ITE
Alloy Engineering & Casting Co.
Mn 1.12, Ni 38, Cr 18, Si 2, C 0.5, bal Fe.
65,000-75,000 TS; 53,000-56,000 YS; 2.5 El; 4.0 RA. For furnace parts, hearth plates, and skid and roller rails. Heat, corrosion, and abrasion resistant. *Obsolete*

X-ITE CB
Alloy Engineering & Casting Co.
C 0.35-0.75, Cr 17-21, Ni 37-41, Cb, bal Fe.
Cast: 70,000 TS; 40,000 YS; 10 El; 180 Brin. For furnace parts, heat treating boxes, salt pots, retorts, and burners. Heat and corrosion resistant. *Obsolete*

X-SUPERMAL
Dresser Industries
TC 1-1.6, Si 1.1-1.3, Mn 0.3, bal Fe.
Heat treated: 60,000-70,000 TS; 50,000-60,000 YS; 6-8 El; 179-201 Brin. For chainlinks, buckets, conveyor equipment; wear resistant.

X134
Acciaierie Valbruna s.p.a.
Stainless Steel. C 0-0.06, Si 0-0.5, Mn 0-1, P 0-0.04, S 0-0.04, Cr 13, Ni 4, bal Fe.
Martensitic. W. Nr. 1.4313.

X13RG
Aubert & Duval
C 0.3, Cr 12, Mo 0.3, Ni 1, bal Fe.
Annealed: 95,000 TS; 40,000 YS; 25 El; 55 RA; B 92 Rock.
Heat treated: 240,000 TS; 215,000 YS; 8 El; 25 RA; C 52 Rock. For cutlery, surgical instruments, gears, knives, hardware, scissors. Corrosion resistant, hardenable. *Obsolete*

X154MU
Acciaierie Valbruna s.p.a.
Stainless Steel. C 0.07, Si 1, Mn 1, P 0-0.04, S 0-0.03, Cr 14.5, Mo 1.6, Ni 4, Cu 4, bal Fe.
Precipitation hardening.

X17
Acciaierie Valbruna s.p.a.
Stainless Steel. C 0-0.1, Si 0-0.8, Mn 0-1, P 0-0.04, S 0-0.03, Cr 16.5, Mo 0-0.6, Ni 0-0.5, bal Fe.
Ferritic. AISI 430; W. Nr. 1.4016.

X17AL
Acciaierie Valbruna s.p.a.
Stainless Steel. C 0-0.12, Si 1, Mn 0-1, P 0-0.04, S 0-0.03, Cr 18, Al 1, bal Fe.
Ferritic. W. Nr. 1.4742.

X17C
Acciaierie Valbruna s.p.a.
C 0-0.1, Si 0-1, Mn 0-1, Cr 16-18, Ni 0-0.5, bal Fe.
Ferritic stainless steel. W. Nr. 1.4016; X 8 Cr 17; AISI 430. *Obsolete*

X17M
Acciaierie Valbruna s.p.a.
Stainless Steel. C 0-0.07, Si 0-1, Mn 0-1, P 0-0.045, S 0-0.03, Cr 17, Mo 1, bal Fe.
Ferritic. AISI 434; W. Nr. 1.4113.

X17MC
Acciaierie Valbruna s.p.a.
Stainless Steel. C 0-0.1, Si 0-1, Mn 0-1, P 0-0.045, S 0-0.03, Cr 18, Mo 2.1, Nb + Ta = 12 x C, bal Fe.
Ferritic. W. Nr. 1.4525.

X17Z
Acciaierie Valbruna s.p.a.
Stainless Steel. C 0-0.12, Si 0-1, Mn 0-1.25, Cr 17, Mo 0.25, Ni 0-0.5, S 0.25, P 0.06, bal Fe.
Free machining ferritic type. AISI 430 F.

X17ZDE
Acciaierie Valbruna s.p.a.
Stainless Steel. C 0.13, Si 0-1, Mn 0-1.5, P 0-0.045, S 0-0.25, Cr 16.5, bal Fe.
Ferritic. W. Nr. 1.4104.

X25
Acciaierie Valbruna s.p.a.
Stainless Steel. C 0-0.2, Si 0-1.5, Mn 0-1, P 0-0.04, S 0-0.03, Cr 25, bal Fe.
Ferritic. AISI 446; W. Nr. 1.3810.

X25AL
Acciaierie Valbruna s.p.a.
Stainless Steel. C 0-0.12, Si 0-1, Mn 0-1, P 0-0.04, S 0-0.03, Cr 25, Al 1.5, bal Fe.
Ferritic. W. Nr. 1.4762.

X4037
Eagle & Globe Steel Ltd.
Alloy steel. C 0.37, Mn 1, Mo 0.15, S 0.1, bal Fe.
600-700 MPa TS; medium tensile strength and good impact toughness.

X40COCRNIW 45-20
German manufacture
C 0.4, Mn 1.2, Cr 20, Ni 20, Co 43, W 4, Fe 3, Cb 4.
Heat treated: 140,000 TS; 67,000 YS; 31 El; 300 Brin. For turbine blades, rotors, hardware, jet engine components; high heat and oxidation resistance to 1500°F.

X4150 HIGH TENSILE STEEL
Eagle & Globe Steel Ltd.
Alloy steel. C 0.5, Si 0.25, Mn 1.3, Cr 0.65, Mo 0.18, S 0.08, bal Fe.
Heat treated. Free machining high tensile steel, welded by standard methods, preheating and postheating required. For machine tools, transportation equipment, excavating and road machinery, draw bench chain, oil drilling, nuts and bolts. AS1444 X4150.

X7AL
Acciaierie Valbruna s.p.a.
Stainless Steel. C 0-0.12, Si 0.75, Mn 0-1, P 0-0.03, S 0-0.03, Cr 7, Al 0.75, bal Fe.
Ferritic. W. Nr. 1.4713.

XALOY 100
Xaloy Inc.
Cast iron. C 3.5, B, Ni, Si, Mn, bal Fe.
For extruder and injection molding barrels, cylinder liners, pump parts; hard, wear resistant white cast iron. Formerly DI-HARD.

XALOY 101
Xaloy Inc.
Cast iron. C 3.5, B, Ni, Cr, Si, Mn, bal Fe.
For extruder and injection molding barrels, cylinder liners, pump parts; hard, wear resistant white cast iron.

XALOY 1525
Xaloy Inc.
Cr 12, B 2.5, Si 3.5, Fe, Mo, Cu, W, bal Ni.
For hard surfacing, plastic processing, chemical plant and other industrial equipment; wear and corrosion resistant. *Obsolete*

XALOY 306
Xaloy Inc.
Cr 7, Ni 40, B, Si, Mn, bal Co.
For extruder and injection molding barrels, cylinder liners, pump parts; wear and corrosion resistant.

XALOY 800
Xaloy Inc.
WC dispersion in Ni-Si-B alloy matrix. Wear and corrosion resistant bore coating for extruder and injection molding barrels, cylinder liners, pump parts.

XANTAL A
Alluminio S.A.
Cu 87.2, Al 9, Fe 3, Mn 0.8.
Sand cast: 71,000-85,000 TS; 15-25 El. For castings, propellers; corrosion resistant.

XANTAL B
Alluminio S.A.
Cu 81, Al 11, Fe 4, Ni 4.
Cast: 101,000 TS; 1-2 El. Heat treated: 114,000 TS; 2-4 El. For castings, marine hardware; corrosion resistant.

XANTAL M
Alluminio S.A.
Cu 89.5, Al 9, Ni 1.5.
Sand cast: 64,000-78,000 TS; 18-28 El. Tough.

XANTAL S
Alluminio S.A.
Cu 90.4, Al 7.8, Ni 0.5, Mn 0.3, Zn 1.
Annealed: 71,000 TS; 50 El. Forged: 65,000 TS; 50 El. For marine hardware castings; corrosion resistant.

XB
Now CARPENTER C-XB VALVE STEEL and FIRTH-BROWN XB.

XB CAST
TRW Inc.
C 1.45, Mn 0.4, Si 2.25, Cr 20, Ni 1.3, bal Fe.
For diesel engine valves.

XB-88

Now WESTINGHOUSE XB-88, WAH CHANG XB-88, and COLUMBIUM XB-88.

XB-STEEL

Italian manufacture
C 0.8, Cr 19.5, Ni 1.45, Si 2, Mn 0.4, bal Fe.
For exhaust valves operating up to 750°C; heat and corrosion resistant.

XB75S

Aluminum Company of America
Zn 5.1-6.1, Mg 2.1-2.9, Cu 1.2-2, Cr 0.15-0.4, bal Al.
O-temper: 33,000 TS; 15,000 YS; 17 El; 60 Brin. T6-temper: 82,000 TS; 72,000 YS; 11 El; 150 Brin. For aircraft parts; extrusions. *Obsolete*

XC-60 STEEL

National Steel Corp.
C 0-0.26, Mn 0-1.3, Cb 0-0.05, bal Fe.
For gas line pipes; combines high strength and weldability. *Obsolete*

XCR

American manufacture
Cr 23.75, Ni 4.75, Si 0-1, Mo 2.75, Mn 1, C 0.45, bal Fe.
For exhaust valves; heat resistant.

XCV GRADE 140

English manufacture
C 1.15, Mn 0.24, Cr 0.3, V 0.15, bal Fe.
For tools, dies, drills, taps; water hardened.

XH-80

Russian manufacture
C 0.1, Cr 20, Ni 80.
Resistance wire for electric toasters, heaters.

XHD N102

Eutectic Corp.
For hard, wear resistant overlays on steels and cast iron; AC-DC. For plow shares, cement grinder rings, mixers, and excavator teeth; 50-55 Rock C.

XK1147

Eagle & Globe Steel Ltd.
Alloy steel. C 0.4-0.47, Mn 1.6-1.9, P 0-0.05, S 0.07-0.12, Si 0.1-0.35, bal Fe.
750 MPa TS; 7 El. Free cutting carbon steel for stub axles, pins and bolts requiring high strength and wear resistance. Resulfurized to improve machinability.

XK1320

Eagle & Globe Steel Ltd.
Alloy steel. C 0.18-0.23, Mn 1.4-1.7, P 0-0.05, S 0-0.05, Si 0.1-0.35, bal Fe.
590 MPa TS; 10 El. Pearlitic manganese steel. High core strength; for automotive parts, heat treated bolts and nuts. Fatigue resistant steel with good mechanical properties. BSS970 EN Series EN14A/1.

XK1320

Eagle & Globe Steel Ltd.
Carbon steel. C 0.18-0.23, Mn 1.4-1.7, Si 0.1-0.35, P 0-0.05, S 0-0.05, bal Fe.
For automotive components, nuts and bolts, gears and spline shafts. AS1442 SK1320; BS970 150M19 (EN14A); SAE 1524.

XK1335

Eagle & Globe Steel Ltd.
Alloy steel. C 0.33-0.38, Mn 1.4-1.7, P 0-0.05, S 0-0.05, Si 0.1-0.35, bal Fe.
680 MPa TS; 9 El. Pearlitic manganese steel. High core strength; for automotive parts, heat treated bolts and nuts. Fatigue resistant steel with good mechanical properties. BSS970 EN Series EN15; BSS970 Part 1 1972 150M36.

XK1340

Eagle & Globe Steel Ltd.
Alloy steel. C 0.38-0.43, Mn 1.4-1.7, P 0-0.05, S 0-0.05, Si 0.1-0.35, bal Fe.
740 MPa TS; 9 El. Pearlitic manganese steel. High core strength; for automotive parts, heat treated bolts and nuts. Fatigue resistant steel with good mechanical properties.

XK1340

Eagle & Globe Steel Ltd.
Carbon steel. C 0.38-0.43, Mn 1.4-1.7, Si 0.1-0.35, P 0-0.05, S 0-0.05, bal Fe.
For bolts, shafts and forgings. AS1442 XK1340; SAE 1541.

XK1345

Eagle & Globe Steel Ltd.
Alloy steel. C 0.43-0.48, Mn 1.4-1.7, P 0-0.05, S 0-0.05, Si 0.1-0.35, bal Fe.
775 MPa TS; 8 El. Pearlitic manganese steel. High core strength; for automotive parts, heat treated bolts and nuts. Fatigue resistant steel with good mechanical properties.

XK5150S

Eagle & Globe Steel Ltd.
Alloy steel. C 0.48-0.55, P 0-0.05, S 0-0.05, Mn 0.7-1, Si 0.1-0.35, Cr 0.7-0.9, bal Fe.
Spring steel bars. 840 MPa TS; 248 Brin.

XK5155S

Eagle & Globe Steel Ltd.
Alloy steel. C 0.5-0.6, P 0-0.05, S 0-0.05, Mn 0.7-1, Si 0.1-0.35, Cr 0.7-0.9, bal Fe.
Spring steel bars. 865 MPa TS; 255 Brin.

XK5160S

Eagle & Globe Steel Ltd.
Alloy steel. C 0.55-0.65, P 0-0.05, S 0-0.05, Mn 0.7-1, Si 0.1-0.35, Cr 0.7-0.9, bal Fe.
Spring steel bars. 880 MPa TS; 260 Brin.

XK9258S

Eagle & Globe Steel Ltd.
Carbon steel. C 0.5-0.65, P 0-0.05, S 0-0.05, Mn 0.7-1.05, Si 1.6-2.2, bal Fe.
Spring steel bars. 865 MPa TS; 255 Brin.

XK9261S

Eagle & Globe Steel Ltd.
Alloy steel. C 0.55-0.65, P 0-0.05, S 0-0.05, Mn 0.7-1, Si 1.8-2.2, Cr 0.1-0.25, bal Fe.
Spring steel bars. 865 MPa TS; 255 Brin.

XL CHISEL

Latrobe Steel Co.
C 0.55, W 2, Cr 1.35, V 0.25, Si 0.25, Mn 0.25, bal Fe.
For chisels, punches; oil hardening.

XL CUT

United Engineering Steels Ltd.
C 0-0.09, Si 0-0.02, Mn 0.97, P 0.05, S 0.3, bal Fe.
Low carbon resulfurized grade. For automatic screw machines.

XL CUT PB

United Engineering Steels Ltd.
C 0-0.09, Si 0-0.02, Mn 0.97, P 0.06, S 0.3, 0.15 Pb min, bal Fe.
Low carbon resulfurized, leaded steel. For automatic screw machines.

XL CUT SPB

United Engineering Steels Ltd.
C 0-0.08, Si 0-0.01, Mn 0.97, P 0.09, S 0.3, N 0.008, 0.2 Pb min, bal Fe.
Low carbon resulfurized, leaded steel. For automatic screw machines.

XL METAL

Koppers Co. Inc.
TC 3-3.5, Si 1.5-2, Mn 0-1, Cr 0-0.5, bal Fe.
150-200 Brin. For piston rings for marine steam engines, steam piston rings; castings. *Obsolete*

XL PLATE

Teledyne Pittsburgh Tool Steel
C 0.2, Mn, bal Fe.
Stress relieved. AISI C1020.

XLO

Atlas Specialty Steels
C 0.6, Cr 0.75, Mo 0.3, Ni 1.75, bal Fe.
For drop forge die blocks; oil hardening.

XS1112

Eagle & Globe Steel Ltd.
Alloy steel. C 0.08-0.15, Mn 1.1-1.4, P 0-0.05, S 0.2-0.3, bal Fe.
450 MPa TS; 9 El. Machinability and free cutting properties. For automatic and semiautomatic machines for repetition parts. Rephosphorized, resulfurized, leaded steel can be case hardened. BSS970 EN Series EN1A; BSS970 Part 1 1972 220M07.

XS11L12

Eagle & Globe Steel Ltd.
Alloy steel. C 0.08-0.15, Mn 1.1-1.4, P 0-0.05, S 0.2-0.3, Pb 0.15-0.35, bal Fe.
450 MPa TS; 9 El. Machinability and free cutting properties. For automatic and semiautomatic machines for repetition parts. Rephosphorized, resulfurized, leaded steel can be case hardened. BSS970 EN Series EN1A.

XTRA TOUGH

Now DARWIN EXTRA TOUGH.

XTROCUT-250

Joseph T. Ryerson & Son Inc.
C 0-0.09, Mn 0.75-1.05, P 0.04-0.09, S 0.26-0.35, bal Fe.
Extruded hot: 60,000 TS; 40,000 YS; 30 El; B 65-75 Rock. Extruded cold drawn: 85,000 TS; 75,000 YS; 15 El; B 90-100 Rock. For rings, couplings, rolls, sleeves. Free-machining tubing. *Obsolete*

XUPER 1020 XFC

Eutectic Corp.
Cadmium free silver solder type for brazing stainless steel, copper alloys. Bonding temperature 1050°F; 85,000 psi TS.

XUPER 146 XFC

Eutectic Corp.
Bronze type alloy for torch brazing ferrous and copper base alloys. Bonding temperature 1400-1600°F; 65,000 psi TS.

XUPER 16 XFC

Eutectic Corp.
Electrode for torch brazing all steels. Bonding temperature 1400-1600°F; 100,000 psi TS.

XUPER 18 XFC

Eutectic Corp.
Electrode for metallic arc joining copper base alloys, steel, and cast iron. Bonding temperature 1400-1600°F; 70,000 psi TS.

XUPER 185 XFC

Eutectic Corp.
Bronze electrode for torch build-up; machinable and wear resistant. Bonding temperature 1400-1600°F; 130 Brin; work hardens to 200 Brin.

XUPER 2100

Eutectic Corp.
For DC metallic arc welding of aluminum alloys containing magnesium; 30,000-35,000 psi TS. *Obsolete*

XUPER 22-33N
Eutectic Corp.
For machinable welds on cast iron; AC-DC reverse; for pump housings and pressure chambers. 55,000-60,000 psi TS.

XUPER 2240
Eutectic Corp.
For machinable welds on cast iron; AC-DC; strength: 50,000 psi TS.

XUPER 2240
Eutectic Corp.
For AC-DC metallic arc to penetrate contaminated cast iron surfaces; nodular deposits; machinable; 55,000-60,000 psi TS.

XUPER 680 CGS
Eutectic Corp.
Electrode for AC-DC metallic arc welding all steels, dies, and tools; 120,000 psi TS.

XUPER 9080
Eutectic Corp.
Now EUTECTRODE N9080. *Obsolete*

XUPER BRONZTEC 19868
Eutectic Corp.
High hardness aluminum bronze powder for metal spraying.

XUPER DIAMAX 10999
Eutectic Corp.
Powder for spray coating that has good abrasion resistance. 80-85 RA.

XUPER DRIL-TEC 8800
Eutectic Corp.
Flux coated electrode containing hard particles for torch build-up on metals; bonding temperature 1400-1600°F. Matrix 200 Brin, particles 89-91 Brin. Good abrasion resistance.

XUPER ELASTODUR R8811
Eutectic Corp.
Electrode for torch build-up; has hard particles in matrix for abrasion resistance.

XUPER EUTECBOR 9000
Eutectic Corp.
Electrode for torch build-up on metals for high hardness and corrosion resistance. Bonding temperature 1800°F; 55-62 Rock C.

XUPER EXOTRODE
Eutectic Corp.
Electrode for AC-DC to chamfer, gouge, cut and pierce all metals.

XUPER MODULTEC 2250
Eutectic Corp.
For AC-DC nodular graphite deposits on nodular cast iron for improved ductility; machinable; 55,000-60,000 psi TS. *Obsolete*

XUPER NUCLEOTEC 2222
Eutectic Corp.
Electrode for AC-DC metallic arc welding all steels and nickel alloys; massive sections; good shock resistance; 100,000 psi TS.

XUPER ULTRABOND 5000
Eutectic Corp.
Alloy powder for metal spraying base coat on all metals except pure copper. Maximum adherence.

XUPERBOND
Eutectic Corp.
Now XUPER ULTRABOND 5000. *Obsolete*

XX
Bethlehem Steel Corp.
C 0.9-1.3, bal Fe.
For gripper and header dies, punches; Type W1; cold heading. *Obsolete*

XX METAL
Kinite Corp.
C, alloy, bal Fe.
For machinery parts; oil hardening.

XX SUPERIOR
English manufacture
C, Cr, bal Fe.
For tools; water hardened.

XYRON 2-23
Eutectic Corp.
Now XUPER 22-33N. *Obsolete*

XYRON 2-24
Eutectic Corp.
Now XUPER 2240. *Obsolete*

XYRON 2-25
Eutectic Corp.
For machinable welds on cast iron; Meehanite and Ni-Resist; AC-DC. 56,000 psi TS. *Obsolete*

XYRON 2-26
Eutectic Corp.
Electrode for repair and salvage of defective cast iron castings; AC-DC; good color match. Strength 62,000 psi. *Obsolete*

XYRON 244
Eutectic Corp.
Electrode for machinable welds and repairs on gray and alloyed castings; can join cast iron to steel; AC-DC. 53,000 psi TS.

Y ALLOY
Sterling International Technology Ltd.
Cu 3.5-4.5, Mg 1.2-1.7, Ni 1.8-2.3, bal Al.
Sand cast: 31,000-36,000 TS; 28,000-30,000 YS; 0-1 El;
95-105 Brin; 0.6 Izod. Sand or chill cast aluminum. For high
temperature parts such as automobile pistons. BS 1490 LM
14-WP, 16-W, 16-WP.

Y W A STEEL
Edgar Allen Balfour Ltd.
C 0.28, Si 0.45, Cr 1.3, Ni 3.5, W 5.8, V 0.25, bal Fe.
For extrusion dies, hot punches. High temperature and
abrasion resistance, tough.

Y-ALLOY
Birmingham Aluminum Casting Co.
Aluminum. Cu 3.5-4.5, Mg 1.2-1.7, Ni 1.8-2.3, Ti 0-0.2, bal
Al.
34,000 TS; 28,000 YS; 100 Brin. For pistons, cylinder heads;
hardenable.

Y-ALLOY
Soc. Alluminio Veneto per Azioni
Aluminum. Cu 0.8-4.2, Mg 1.3-1.7, Ni 1.8-2.3, bal Al.
For pistons, cylinder heads; high temperature properties.

Y-PHOSPHOR BRONZE NO. 207A
Victor Mfg. Co.
Cu 70, Zn 29, P 1.
For wire for weaving purposes. Great ductility.

Y11
Keystone Carbon Co.
Sintered high conductivity copper. 8.3 density; 25,000 psi
TS; 4.0 El; 80 Rock H. For parts requiring high electrical
conductivity.

YALE BRONZE
American manufacture
Zn 7-8, Sn 0.5-1.5, Pb 0.7-1.5, bal Cu.
For screw machine products, bolts, bushings; free-cutting.

YAW-TEN 41
Yawata Iron & Steel Co., Ltd.
C 0-0.18, Si 0-0.35, Mn 0-1.2, Cu 0.25-0.5, Cr 0.4-0.65, bal
Fe.
Normalized: 74,000 TS; 31,000 YP. Rolled: 58,000 TS;
34,000 YP. For buildings, bridges, industrial machinery.
Good fabricability and weldability. Resists atmospheric
corrosion.

YAW-TEN 50
Yawata Iron & Steel Co., Ltd.
C 0-0.12, Si 0.15-0.35, Mn 0-0.9, P 0.06-0.12, Cu 0.25-0.5, Ti
0-0.15, bal Fe.
Annealed: 67,000 TS min; 50,000 YP min. Rolled: 71,000 TS
min; 57,000 YP min. For bridges, structures, mine cars, bus
bodies, railroad cars, ships, farm implements. Good
weldability and impact resistance. Resists atmospheric
corrosion.

YAW-TEN 60
Yawata Iron & Steel Co., Ltd.
C 0-0.18, Si 0.15-0.55, Mn 0.8-1.4, Cu 0.25-0.5, Cr 0.4-0.65,
Ti 0-0.15, bal Fe.
Heat treated: 82,000-100,000 TS; 63,000-67,000 YP. For
buildings, bridges, industrial machinery. Good fabricability
and weldability. Resists atmospheric corrosion.

YAW-TEN M
Yawata Iron & Steel Co., Ltd.
C 0-0.12, Si 0.15-0.35, Mn 0-0.9, S 0.25-0.5, Cr 0.5-1, Ti
0-0.15, bal Fe.
Rolled: 64,000 TS min; 47,000 YP min. For air preheaters,
flues, stacks, gas exhaust systems. For applications
exposed to corrosive attack from sulfuric acid gases and
other corrosive chemicals.

YB-TEN
Youngstown Steel
S 0-0.12, Mn 0-0.75, P 0-0.04, S 0-0.05, Si 0-0.1, 0.01 min
Cb or V, bal Fe.
Sheet: 62,000 TS; 45,000 YS; 28 El. For automobile
bumpers. High strength low alloy steel. Good weldability
and fabricability.

YC135A
American manufacture
Sn 82, Sb 13, Cu 5.
Die casting alloy. As cast: 10,000 psi TS; 11 El.

YCB
American manufacture
Cr 15, Ni 25, Mo 4, Si 1, Cb 2, bal Fe.
For supercharger wheels for jet engine; heat resistant.

YD NICRAL
Special Metals Corp.
Cr 16, Al 4.8, 1.0 Y_2O_3, bal Ni.
Yttrium oxide dispersion nickel-chromium alloy, for high
temperature operation. At 70°F: 170 ksi TS, 12 El. At
2000°F: 11 ksi TS, 10-20 El. Good high temperature stress
rupture strength.

YD-NI
American manufacture
Cr 16, Al 4, 1.5 Y_2O_3, bal Ni.
Cast superalloy.

YELLOW BRASS 274
Anaconda Co.
Copper. Cu 63, Zn 37.
Hard: 65,000 TS; 57,000 YS; 20 El; 75 Rock B. Soft: 46,000
TS; 17,000 YS; 60 El; 20 Rock B. For cold headed screws,
bolts, rivets, studs and accessories, novelties, fixtures,
appliances. Electrical conductivity: 26%. High strength.

YELLOW BRASS 59
Anaconda Co.
Copper. Zn 34, bal Cu.
Soft: 45,000 TS; 60 El; 52 Brin. Hard: 76,000 TS; 4 El; 153
Brin. For bead chains, eyelets, grommets, lamp fixtures.
Yellow brass.

YELLOW BRASS 66%
Chase Brass & Copper Co., Inc.
Copper. Cu 66, Zn 34.
Annealed: 49,000 psi TS; 17,000 psi YS; 57 El. Rolled:
74,000 psi TS; 40,000 psi YS; 8 El. For ornaments, lamp
fixtures, automobile parts; deep drawing.

YELLOW BRASS, 65%
Chase Brass & Copper Co., Inc.
Copper. Cu 65, Zn 35.
Annealed: 50,000 psi TS; 60 El. Drawn (84%): 128,000 psi
TS; 3 El. For springs, pins, rivets, screws. CDA 270.

YELLOW BRASS-61
Anaconda Co.
Copper. Cu 63, Zn 37.
Soft: 46,000 TS; 17,000 YS; 60 El. Hard: 85,000 TS; 50,000
YS; 20 El. For pins, screws, rivets. High strength yellow
brass.

YELLOW END
Sanderson Kayser Ltd.
C 1-1.15, bal Fe.
For tools, taps, drills; water hardened. *Obsolete*

YELLOW GOLD-1
American manufacture
Au 53, Ag 25, Cu 22.
Cast: 205 Brin. For jewelry; corrosion resistant.

YELLOW GOLD-2
American manufacture
Au 50, Ag 25, Cu 25.
Heat treated: 159 Brin. For jewelry; corrosion resistant.

YELLOW LABEL
Peninsular Steel Co.
C 0.9, Mn 1.2, Cr 0.5, W 0.5, V 0.2, bal Fe.
For tools, dies, punches; non-deforming, oil hardening.

YELLOW LABEL
Wallace Murray Corp.
C 0.8-1.1, V 0.15-0.25, bal Fe.
For tools, dies, boring and shaping tools, milling cutters;
"Heller's Alloy Die."

YELLOW LABEL CAST
Jessop-Saville Ltd.
C 0.7-1.2, bal Fe.
For tools and dies, drills, taps, punches; water hardened.

YELLOW TUBE BRASS-218
Anaconda Co.
Copper. Cu 66.5, Zn 33, Pb 0.5.
Soft: 45,000 TS; 17,000 YS; 55 El. Hard: 73,000 TS; 60,000
YS; 10 El. For plumbing goods, flashlight shells; tubes.

YIDOR
Uddeholm Corp.
C 1.45, Cr 1.4, Mn 0.6, bal Fe.
For blanking and forming dies, cutters; water or oil
hardened, abrasion resistant. *Obsolete*

YND-30
Yawata Iron & Steel Co., Ltd.
C 0-0.14, Mn 1-1.5, bal Fe.
Plate: 42,700 YS min; 61,200 TS min. For storage tanks,
transportable containers for liquefied petroleum gas,
pressure vessels. High notch toughness for cryonic
temperatures. Good weldability.

YND-33
Yawata Iron & Steel Co., Ltd.
C 0.14, Mn 1-1.5, 0.035 S & P max, bal Fe.
Plate: 64,000 TS min; 46,900 YS min. For storage tanks and
transportable containers for liquefied petroleum gas,
chemical industry apparatus, refrigerated ships. Low
temperature steel, high notch toughness.

YND-37
Yawata Iron & Steel Co., Ltd.
C 0.14, Mn 1-1.5, Ni 0-0.7, bal Fe.
Plate: 71,000 TS min; 52,600 YS min. For storage tanks and
transportable containers for liquefied petroleum gas
chemical industry apparatus, refrigerated ships. High notch
toughness at low temperatures.

YND-58
Yawata Iron & Steel Co., Ltd.
C 0-0.14, Ni 2-2.7, Cr 0-0.5, Mo 0-0.55, bal Fe.
Plate: 96-700-16,000 TS; 82,500 YS min. For storage tanks
and transportable containers for liquefied petroleum gas
chemical equipment, pressure vessels. High notch
toughness, low temperature steel, good weldability.

YO-FLEX 60
Youngstown Steel
C 0.13, Mn 0.45, P 0.01, S 0.025, bal Fe.
60,000 YS. For high strength hot dipped galvanized sheet
steel applications.

YO-FLEX 70
Youngstown Steel
C 0.13, Mn 0.45, P 0.01, S 0.025, bal Fe.
70,000 YS. For high strength hot dipped galvanized sheet
steel applications.

YO-FLEX 80
Youngstown Steel
C 0.13, Mn 0.45, P 0.01, S 0.025, bal Fe.
80,000 YS. For high strength hot dipped galvanized sheet
steel applications.

YO-FLEX 90
Youngstown Steel
C 0.13, Mn 0.45, P 0.01, S 0.025, bal Fe.
90,000 YS. For high strength hot dipped galvanized sheet steel applications.

YO-LEAD TYPE A
Youngstown Steel
C 0.13, Mn 1, P 0.06, S 0.3, Pb 0.25, bal Fe.
For free machining bar applications.

YO-LEAD TYPE B
Youngstown Steel
C 0.15, Mn 1.2, P 0.6, S 0.4, Pb 0.25, bal Fe.
For free machining bar applications.

YO-MAN
Youngstown Steel
C 0-0.25, Mn 1.1-1.6, Si 0-0.3, Cu 0.2, bal Fe.
Plate: 70,000 TS min; 50,000 YS min; 20 El. For shovels, structural members, shovel blades. High strength low alloy steel.

YO-NAMEL
Youngstown Steel
C 0.005, Mn 0.3, P 0.003, S 0.02, bal Fe.
For porcelain enameling.

YO-OH TYPE 1
Youngstown Steel
C 0.09, Mn 0.85, P 0.1, S 0.3, bal Fe.
For free machining bar applications.

YO-OH TYPE 2
Youngstown Steel
C 0.09, Mn 0.9, P 0.06, S 0.32, bal Fe.
For free machining bar applications.

YOCOMITE
James Yocum & Son Inc.
Cast Iron. Ni 3, Cr 0.75, high C, bal Fe.
50,000 TS; 450 Brin. For dies. Cast iron.

YOLOY "E" ACR
Youngstown Steel
C 0-0.1, Ni 0-0.6, Cr 0-0.35, Cu 0.2-0.5, bal Fe.
Rolled: 60,000 maximum TS; 45,000 maximum YS; 30 El. For coal cars, dump trucks, bodies; good formability and weldability. *Obsolete*

YOLOY "E" HSX
Youngstown Steel
C 0-0.18, Ni 0.4-1, Cr 0.2-0.35, Cu 0.2-0.5, bal Fe.
Rolled: 60,000 TS; 45,000 YS; 25 El. For coal cars, dump trucks, bodies; good formability and weldability. *Obsolete*

YOLOY 45-W
Youngstown Steel
C 0.1, Mn 0.8, Si 0.2, Cb 0.03, bal Fe.
Rolled: 60,000 TS minimum; 45,000 YS minimum; 25 El minimum. For trucks, plow frames, pressure vessels. Good weldability and fabricability. *Obsolete*

YOLOY ACR
Now YOLOY YSW-50 STEEL.

YOLOY ACR
Youngstown Steel
C 0-0.1, Mn 0-0.6, Ni 0.5-1, Cu 0.25-0.5, bal Fe.
Rolled: 60,000 TS; 45,000 YS; 30 El; 60-75 Brin. For caskets, refrigeration cases; good drawability. *Obsolete*

YOLOY C
Youngstown Steel
C 0-0.1, Cr 0-0.3, Mn 0-0.6, Cu 0.25-0.5, bal Fe.
Rolled: 60,000 TS; 45,000 YS; 30 El. For caskets, tanks, truck trailers, trim; good welding and forming.

YOLOY E-HS
Youngstown Steel
C 0-0.18, Ni 0.4-1, Cr 0.2-0.35, Mn 0-0.9, Cu 0.2-0.5, bal Fe.
Rolled: 70,000 TS; 50,000 YS; 22 El. For coal cars, dump truck bodies; good welding and forming. *Obsolete*

YOLOY HS
Youngstown Steel
C 0-0.15, Mn 0-0.75, P 0-0.1, Si 0-0.3, Cu 0.75-1.25, bal Fe.
Rolled: 70,000 TS; 50,000 YS; 22 El; 62 RA; 85 Brin. For trucks, trailers, cars, bodies, bridges; good weldability and formability.

YOLOY HSK
Youngstown Steel
C 0-0.12, Mn 0-0.6, Ni 1.5-2, Cu 0.75-1.25, bal Fe.
Rolled: 75,000 TS; 45,000 YS; 27 El; 75-85 Brin. For railroad cars, trucks, bridges, mine equipment; good weldability and formality. *Obsolete*

YOLOY HSX
Youngstown Steel
C 0-0.15, Mn 0-1, Cu 0.75, Ni 1, bal Fe.
Rolled: 65,000 TS; 45,000 YS; 25 El. For railroad cars, trucks, bridges, mine equipment; good weldability and formability.

YOLOY M GR. B
Youngstown Steel
C 0-0.23, Mn 0-1.4, Si 0.15-0.2, 0.20% min Cu, bal Fe.
Rolled: 70,000 TS; 45,000 YS; 22 El. For road machinery; high strength. *Obsolete*

YOLOY MGRA
Youngstown Steel
C 0-0.25, Cu 0-0.3, Mn 0-1.6, bal Fe.
Rolled: 70,000 TS; 50,000 YS; 20 El. For coal cars, dump trucks, trailer bodies; high strength, low alloy constructio steel. *Obsolete*

YOLOY S
Youngstown Steel
C 0-0.12, Cu 0.75-1.25, Ni 1.65-2, Mn 0-0.6, bal Fe.
Rolled: 60,000 TS; 45,000 YS; 25 El. For transportation equipment, mine cars, truck bodies; good weldability, precipitation hardenable.

YOLOY T-50
Youngstown Steel
C 0.1, Mn 0.45, P 0.01, S 0.025, Si 0.4, Cu 0.25, Cr 0.55, Ti 0.1, Al, bal Fe.
50,000 YS. Hot rolled and cold rolled sheet for high strength weathering applications.

YOLOY T-60
Youngstown Steel
C 0.1, Mn 0.45, P 0.01, S 0.025, Si 0.4, Cu 0.25, Cr 0.55, Ti 0.1, Al, bal Fe.
60,000 YS. Hot rolled and cold rolled sheet for high strength weathering steel applications.

YOLOY T-70
Youngstown Steel
C 0.1, Mn 0.45, P 0.01, S 0.025, Si 0.4, Cu 0.25, Cr 0.55, Ti 0.2, Al, bal Fe.
70,000 YS. Hot rolled and cold rolled sheet for high strength weathering steel applications.

YOLOY T-80
Youngstown Steel
C 0.1, Mn 0.45, P 0.01, S 0.025, Si 0.4, Cu 0.25, Cr 0.55, Ti 0.2, Al, bal Fe.
80,000 TS. Hot rolled and cold rolled sheet for high strength weathering steel applications.

YOLOY YSW 42
Youngstown Steel
C 0.16, Mn 0.75-1, V 0.02, Cb 0.02, bal Fe.
Plate: 62,000 TS min; 42,000 YS min; 19 El min. For trucks, crane booms, mine cars, bridges, pressure vessels, trailers. Low alloy high strength steel. High weldability.

YOLOY YSW 45
Youngstown Steel
C 0.16, Mn 0.75-1, V 0.02, Cb 0.02, bal Fe.
Plate: 60,000 TS min; 45,000 YS min; 19 El min. For trucks, crane booms, bridges, mine cars, trailers, pressure vessels. Low alloy high strength steel. High weldability.

YOLOY YSW 50 STEEL
Youngstown Steel
C 0-0.22, Mn 0-1.25, 0.01 Cb or V min, bal Fe.
Rolled: 65,000 TS; 50,000 YS; 22 El. For structurals, auto and truck frames, railroad construction; high strength low alloy steel.

YOLOY YSW 55
Youngstown Steel
C 0.16, Mn 1, V 0.02, Cb 0.02, bal Fe.
Plate: 70,000 TS; 55,000 YS; 17 El min. For bridges, trucks, crane booms, mine cars, pressure vessels, trailers. Low alloy high strength steel. High weldability.

YOLOY YSW 60
Youngstown Steel
C 0.16, Mn 0.75-1, V 0.02, Cb 0.02, bal Fe.
Plate: 75,000 TS min; 60,000 YS min; 16 El min. For mine cars, trailers, trucks, crane booms, bridges, power shovels. Low alloy high strength steel. Good weldability.

YOLOY YSW 65
Youngstown Steel
C 0.16, Mn 1, V 0.02, Cb 0.02, bal Fe.
Plate: 80,000 TS min; 65,000 YS min; 15 El min. For power shovels, crane booms, railroad cars, pressure vessels, mine cars. Low alloy high strength steel. Good weldability.

YOLOY YSW 70
Youngstown Steel
C 0.16, Mn 1, V 0.02, bal Fe.
Plate: 85,000 TS min; 70,000 YS min; 12 El min. For mine cars, crane booms, buildings, pressure vessels, bridges, power shovels. Low alloy high strength steel. High weldability.

YOLOY YSW-A441
Youngstown Steel
C 0-0.22, Mn 0-1.25, Cu 0.2, V 0.02, bal Fe.
Sheet: 60,000 TS; 45,000 YS; 25 El min. Bar: 67,000 TS; 46,000 YS; 19 El min. For railroad cars, bridges, buildings, crane booms, power shovels, mining machinery. High strength low alloy steel.

YOONSTEEL A1
Yoonsteel (Malaysia) Sdn. Bhd.
Carbon steel. C 0.1-0.25, Si 0.3-0.6, Mn 0.5-0.9, P 0-0.06, S 0-0.06, bal Fe.
Carbon steel that meets BS 3100 A1; ASTM A27 U60-30; JIS SC 42; DIN GS 45. 230 N/mm^2 YS; 430 N/mm^2 TS; 22 El.

YOONSTEEL A2
Yoonsteel (Malaysia) Sdn. Bhd.
Carbon steel. C 0.25-0.35, Si 0.3-0.6, Mn 0.5-1, P 0-0.06, S 0-0.06, bal Fe.
Carbon steel that meets BS 3100 A2; ASTM A27 70-36; JIS SC 49; DIN GS 52. 260 N/mm^2 YS; 490 N/mm^2 TS; 18 El.

YOONSTEEL A3
Yoonsteel (Malaysia) Sdn. Bhd.
Carbon steel. C 0.35-0.45, Si 0.3-0.6, Mn 0.5-1.1, P 0-0.06, S 0-0.06, bal Fe.
Carbon steel that meets BS 3100 A3; ASTM A148 80-40; DIN GS 60. 295 N/mm^2 YS; 540 N/mm^2 TS; 14 El.

YOONSTEEL A4
Yoonsteel (Malaysia) Sdn. Bhd.
Carbon steel. C 0.18-0.25, Si 0.3-0.6, Mn 1.2-1.6, P 0-0.05, S 0-0.05, bal Fe.
Carbon steel that meets BS 3100 A4; ASTM A148 80-54; JIS SCMn 1A. 320 N/mm^2 YS; 540 N/mm^2 TS; 16 El; 152-207 Brin.

YOONSTEEL A5

Yoonsteel (Malaysia) Sdn. Bhd.
Carbon steel. C 0.25-0.33, Si 0.3-0.6, Mn 1.2-1.6, P 0-0.05, S 0-0.05, bal Fe.
Carbon steel that meets BS 3100 A5; ASTM A148 90-60; JIS SCMn 2A; DIN GS 62. 370 N/mm² YS; 620 N/mm² TS; 13 El; 179-229 Brin.

YOONSTEEL B1

Yoonsteel (Malaysia) Sdn. Bhd.
C 0.2-0.45, Si 0.4-0.8, Mn 0.4-0.8, P 0-0.04, S 0-0.04, Ni 0.5-2.5, Cr 1.5-4.5, Mo 0.2-0.5, bal Fe.
Wear resistant steel. 1600 N/mm² TS; 2 El; 45-52 Rock C.

YOONSTEEL B2

Yoonsteel (Malaysia) Sdn. Bhd.
C 0.45-0.55, Si 0.35-0.75, Mn 0.5-1, P 0-0.04, S 0-0.04, Cr 0.8-1.2, bal Fe.
Low alloy steel that meets BS 3100 BW2; DIN 1.7145. 540 N/mm² YS; 850 N/mm² TS; 8 El; 14-27 Rock C.

YOONSTEEL B2

Yoonsteel (Malaysia) Sdn. Bhd.
C 0.45-0.55, Si 0.35-0.75, Mn 0.5-1, P 0-0.04, S 0-0.04, Cr 0.8-1.2, bal Fe.
Low alloy steel that meets BS 3100 BW2; DIN 1.7145. 540 N/mm² YS; 850 N/mm² TS; 8 El; 14-27 Rock C.

YOONSTEEL B3

Yoonsteel (Malaysia) Sdn. Bhd.
P 0-0.04, S 0-0.04, bal Fe.
Low alloy steel that meets BS 3100 BT2; ASTM A148 105-85. 540 N/mm² YS; 850 N/mm² TS; 12 El; 23-28 Rock C.

YOONSTEEL B4

Yoonsteel (Malaysia) Sdn. Bhd.
C 0.55-0.65, Si 0.35-0.75, Mn 0.5-1, P 0-0.04, S 0-0.04, Ni 0.8-1.5, Cr 0.2-0.4, bal Fe.
Low alloy steel that meets BS 3100 BW4; DIN 1.7266. 1050 N/mm² TS; 6 El; 36-45 Rock C.

YOONSTEEL B5

Yoonsteel (Malaysia) Sdn. Bhd.
C 0.5-0.7, Si 0.3-0.6, Mn 0.6-1, P 0-0.04, S 0-0.04, Cr 2-5, Mo 0.2-0.5, bal Fe.
Wear resistant steel. 1100 N/mm² TS; 1.2 El; 50-58 Rock C.

YOONSTEEL C1

Yoonsteel (Malaysia) Sdn. Bhd.
Cast iron. C 3-3.5, Si 2-2.4, Mn 0.4-0.8, P 0-0.3, S 0-0.1, bal Fe.
Gray cast iron that meets BS 1452 180; JIS FC 20; DIN GG 18. 180 N/mm² TS.

YOONSTEEL C2

Yoonsteel (Malaysia) Sdn. Bhd.
Cast iron. C 3-3.3, Si 1.5-2.2, Mn 0.5-1, P 0-0.2, S 0-0.1, bal Fe.
Gray cast iron that meets BS 1452 220; JIS FC 25; DIN GG 22. 220 N/mm² TS.

YOONSTEEL C3

Yoonsteel (Malaysia) Sdn. Bhd.
Cast iron. C 2.4-3.6, Si 0.3-1, Mn 0.5-1.5, P 0-0.1, S 0-0.06, Cr 14-18, Mo 1-3.5, bal Fe.
Alloyed white cast iron that meets ASTM 532 II B, C; DIN GX 315 CrMo 18.2. 54-62 Rock C.

YOONSTEEL C4

Yoonsteel (Malaysia) Sdn. Bhd.
Cast iron. C 2.3-3.2, Si 0.3-1, Mn 0.5-1.5, P 0-0.1, S 0-0.06, Ni 0-1.5, Cr 23-30, Mo 0-1.5, bal Fe.
Alloyed white cast iron that meets ASTM A 532 III. 55-62 Rock C; 620 N/mm² TS.

YOONSTEEL D1

Yoonsteel (Malaysia) Sdn. Bhd.
Cast iron. C 3.4-4, Si 1.75-4, Mn 0.1-0.5, P 0-0.1, S 0-0.01, bal Fe.
Ductile cast iron that meets DIN GGG-42; JIS FCD 40; BS 2789 420/12. 390 N/mm² TS; 255 N/mm² YS; 12 El.

YOONSTEEL D2

Yoonsteel (Malaysia) Sdn. Bhd.
Cast iron. C 3.4-4, Si 1.75-3.75, Mn 0.2-0.5, P 0-0.1, S 0-0.02, bal Fe.
Ductile cast iron that meets DIN GGG-50; JIS FCD 50; BS 2789 500/7. 490 N/mm² TS; 340 N/mm² YS; 7 El.

YOONSTEEL D3

Yoonsteel (Malaysia) Sdn. Bhd.
Cast iron. C 3.4-4.1, Si 1.5-3.5, Mn 0.2-0.5, P 0-0.1, S 0-0.02, Cu 0.5-2, bal Fe.
Ductile cast iron that meets DIN GGG-60; JIS FCD 60; BS 2789 600/3. 580 N/mm² TS; 390 N/mm² YS; 2 El.

YOONSTEEL H1

Yoonsteel (Malaysia) Sdn. Bhd.
Cast iron. C 3-3.5, Si 1.8-2.8, Mn 0.5-0.9, P 0-0.2, S 0-0.1, Cr 0.2-2.2, bal Fe.
Heat and corrosion resistant iron.

YOONSTEEL H2

Yoonsteel (Malaysia) Sdn. Bhd.
Cast iron. C 2.7-3.1, Si 1.5-2.2, Mn 0.8-1.2, P 0-0.2, S 0-0.1, Ni 14-22, Cr 1.5-2.5, Cu 0-6.5, bal Fe.
Heat and corrosion resistant iron. 155 N/mm² TS.

YOONSTEEL M1

Yoonsteel (Malaysia) Sdn. Bhd.
C 1-1.35, Si 0.4-0.8, Mn 11-14, P 0-0.07, S 0-0.05, bal Fe.
Wear resistant steel that meets BS 3100 BW 10; ASTM 128 A; JIS SCMnH 1; DIN GX120Mn13. 620 N/mm² TS; 35 El; 200-220 Brin.

YOONSTEEL M2

Yoonsteel (Malaysia) Sdn. Bhd.
C 1-1.35, Si 0.4-0.8, Mn 11-14, P 0-0.07, S 0-0.05, Cr 1.5-2.5, bal Fe.
Wear resistant steel that meets ASTM 128 C; JIS SCMnH 11; DIN GX120MnCr13.2. 736 N/mm² TS; 20 El; 210-230 Brin.

YOONSTEEL M3

Yoonsteel (Malaysia) Sdn. Bhd.
C 0.9-1.3, Si 0.4-0.8, Mn 11.5-14, P 0-0.07, S 0-0.04, Cr 0-2.5, Mo 0.5-1.2, bal Fe.
Wear resistant steel that meets ASTM 128 E1; DIN GX120MnCrMo13.2. 736 N/mm² TS; 20 El; 210-230 Brin.

YOONSTEEL M4

Yoonsteel (Malaysia) Sdn. Bhd.
C 1-1.35, Si 0.4-0.8, Mn 6-8, P 0-0.07, S 0-0.05, Mo 0.9-1.2, bal Fe.
Wear resistant steel that meets ASTM 128 F. 540 N/mm² TS; 220-240 Brin.

YOONSTEEL M5

Yoonsteel (Malaysia) Sdn. Bhd.
bal Fe.
Wear resistant manganese steel. 210-230 Brin.

YOONSTEEL M6

Yoonsteel (Malaysia) Sdn. Bhd.
bal Fe.
Wear resistant manganese steel. 900 N/mm² TS; 36 El; 230-255 Brin.

YOONSTEEL N1

Yoonsteel (Malaysia) Sdn. Bhd.
Cast iron. C 2.5-3.6, Si 0.4-0.7, Mn 0.4-0.7, P 0-0.3, S 0-0.15, Ni 3.5-5, Cr 1.4-3.5, bal Fe.
Alloyed white cast iron that meets DIN GH 280 NiCr18.8. 54-61 Rock C; 370 N/mm² TS.

YOONSTEEL N2

Yoonsteel (Malaysia) Sdn. Bhd.
Cast iron. C 2.5-3.6, Si 1-2.2, Mn 0.4-1, P 0-0.1, S 0-0.15, Ni 5-7, Cr 7-11, bal Fe.
Alloyed white cast iron that meets DIN GH 290 CrNi34.22; ASTM ID. 54-61 Rock C; 370 N/mm² TS.

YOONSTEEL S1

Yoonsteel (Malaysia) Sdn. Bhd.
Stainless steel. C 0.1-0.2, Si 0.3-1, Mn 0.5-1, P 0-0.04, S 0-0.04, Ni 0-1, Cr 11.5-13.5, bal Fe.
Stainless steel that meets BS 3100 420C29; ASTM A296 CA 15; JIS SCS 2 TYPE 2. 460 N/mm² YS; 690 N/mm² TS; 11 El.

YOONSTEEL S2

Yoonsteel (Malaysia) Sdn. Bhd.
Stainless steel. C 0.05-0.08, Si 0.5-2, Mn 0.5-1.5, P 0-0.04, S 0-0.04, Ni 8-11, Cr 18-21, bal Fe.
Stainless steel that meets BS 3100 304C15; ASTM A743 CF 8; JIS SCS 13 TYPE 13. 200 N/mm² YS; 485 N/mm² TS; 30 El.

YOONSTEEL S3

Yoonsteel (Malaysia) Sdn. Bhd.
C 0.2-0.6, Si 0.5-2, Mn 0.5-2, P 0-0.04, S 0-0.04, Ni 19-23, Cr 23-28, Mo 0-0.5, bal Fe.
Heat resistant steel that meets BS 3100 310C40; ASTM A297 HK; JIS SCH 21 TYPE 21; DIN 1.4848. 240 N/mm² YS; 450 N/mm² TS; 10 El.

YOONSTEEL S4

Yoonsteel (Malaysia) Sdn. Bhd.
C 0.2-0.5, Si 0.5-2, Mn 0.5-2, P 0-0.04, S 0-0.04, Ni 11-14, Cr 24-28, Mo 0-0.5, bal Fe.
Heat resistant steel that meets BS 3100 309C35; ASTM A297 HH; JIS SCH 13 TYPE 13; DIN 1.4837. 240 N/mm² YS; 510 N/mm² TS; 10 El.

YOONSTEEL S5

Yoonsteel (Malaysia) Sdn. Bhd.
C 0.2-0.5, Si 0.5-2, Mn 0.5-2, P 0-0.04, S 0-0.04, Ni 23-27, Cr 19-23, Mo 0-0.5, bal Fe.
High temperature heat resistant steel that meets BS 3100 311C11; ASTM A297 HN; JIS SCH 19 TYPE 19. 430 N/mm² TS; 8 El.

YOR CASTAN

Yorkshire Imperial Metals Ltd.
Cu 88, Sn 12.
As Drawn: 106,000-121,000 TS; 10-15 El; 194-242 Brin. For condenser and heat exchanger tubes. Resists sea water corrosion.

YORCALBRO (ALUMINUM BRASS)

IMI Yorkshire Alloys Ltd.
As 0.02-0.035, Al 1.8-2.3, Cu 76-78, bal Zn.
Annealed: 47,800 psi TS; 23,200 psi YS; 70 El; 70-75 DPN (68-73 Brin). Tube for condensers and seawater pipelines.

YORCALBRO/ALUMBRO

Yorkshire Imperial Metals Ltd.
Cu 76, Zn 22, Al 2, As 0.04.
For tubes in steam condensers and heat exchangers, oil refineries, sea water pipe lines. Resists corrosion and erosion of sea water. Aluminum brass.

YORCORON

Yorkshire Imperial Metals Ltd.
Ni 31, Fe 2, Mn 2, bal Cu.
Plates and tubes. Annealed: 62,700-71,700 TS; 22,400-36,000 YS; 40-50 El; 97-116 Brin. Hard: 98,600-110,000 TS; 90,000-107,000 YS; 4-8 El; 175-213 Brin. Very resistant to abrasion by such as sand.

YORCUNIC/KUNIFER 10

Yorkshire Imperial Metals Ltd.
Ni 10, Fe 2, Mn 1, bal Cu.
For evaporator tubes in sugar industry, feed water heaters. Resists sea water corrosion.

YORCUNIFE/KUNIFER 5
Yorkshire Imperial Metals Ltd.
Ni 5.5, Fe 1.2, Mn 0.5, bal Cu.
For sea water pipelines. Resists sea water corrosion.

YORK
Time Steel Service Inc.
C 0.75, Cr 1, Mo 1.35, Mn 2, Si 0.3, bal Fe.
Air hardened tool steel. AISI A6.

YORKSHIRE 70/30
Yorkshire Imperial Metals Ltd.
Ni 30, Fe 0.7, Mn 0.8, bal Cu.
For condenser tubes, evaporator tubes, feed water heater tubes, heat exchangers. Resists sea water corrosion. Cupro-nickel.

YOUNGSTOWN ARMATURE
Youngstown Steel
Si 0.5, bal Fe.
For armatures, electrical equipment. *Obsolete*

YOUNGSTOWN ELECTRICAL
Youngstown Steel
Si 1, bal Fe.
For electrical equipment. *Obsolete*

YOUNGSTOWN FIELD
Youngstown Steel
Si 0.25, bal Fe.
For electrical apparatus. *Obsolete*

YOUNGSTOWN GALVANNEALED
Youngstown Steel
Zn coated low carbon steel.
For roofing, siding, culverts, fence wire.

YOUNGSTOWN SPECIAL DYNAMO
Youngstown Steel
Si 2.5, bal Fe.
For dynamos. *Obsolete*

YOUNGSTOWN SPECIAL MOTOR
Youngstown Steel
Si 2, bal Fe.
For electrical equipment. *Obsolete*

YOUNGSTOWN TRANSFORMER
Youngstown Steel
Si 3.5, bal Fe.
For transformers. *Obsolete*

YOUNGSTOWN TRANSFORMER EXTRA SPECIAL
Youngstown Steel
Si 3.5, bal Fe.
For transformers. *Obsolete*

YOUNGSTOWN TRANSFORMER SPECIAL
Youngstown Steel
Si 4, bal Fe.
For transformers. *Obsolete*

YS-T100
Youngstown Steel
C 0.09, Mn 0.9, P 0.01, S 0.025, Al 0.04, Ti 0.25, bal Fe.
100,000 YS. For high strength cold rolled steel sheet applications.

YS-T120
Youngstown Steel
C 0.09, Mn 0.9, P 0.01, S 0.025, Al 0.04, Ti 0.25, bal Fe.
120,000 YS. For high strength cold rolled steel sheet applications.

YS-T140
Youngstown Steel
C 0.09, Mn 0.9, P 0.01, S 0.025, Al 0.04, Ti 0.25, bal Fe.
140,000 YS. For high strength cold rolled steel sheet applications.

YS-T50
Youngstown Steel
C 0.09, Mn 0.45, P 0.01, S 0.025, Al 0.04, Ti 0.1, bal Fe.
Hot rolled sheet and plate, cold rolled: 50,000 YS. For high strength steel applications.

YS-T60
Youngstown Steel
C 0.09, Mn 0.45, P 0.01, S 0.025, Al 0.04, Ti 0.15, bal Fe.
Hot rolled sheet and plate, cold rolled sheet: 60,000 YS. For high strength steel applications.

YS-T70
Youngstown Steel
C 0.09, Mn 0.45, P 0.01, S 0.025, Al 0.04, Ti 0.2, bal Fe.
Hot rolled sheet and plate, cold rolled sheet: 70,000 YS. For high strength steel applications.

YS-T80
Youngstown Steel
C 0.09, Mn 0.45, P 0.01, S 0.025, Al 0.04, Ti 0.25, bal Fe.
Hot rolled sheet and plate, cold rolled sheet: 80,000 YS. For high strength steel applications.

YS-T90
Youngstown Steel
C 0.09, Mn 0.9, P 0.01, S 0.025, Al 0.04, Ti 0.25, bal Fe.
Hot rolled sheet: 90,000 YS. For high strength steel applications.

YSW 42, ETC.
Now YOLOY YSW 42, ETC.

YSW-50 F
Youngstown Steel
C 0.09, Mn 0.9, P 0.01, S 0.025, Cb 0.04, bal Fe.
Hot rolled and cold rolled high strength steel sheet: 50,000 YS.

YSW-55 F
Youngstown Steel
C 0.09, Mn 0.9, P 0.01, S 0.025, Cb 0.05, bal Fe.
Hot rolled and cold rolled high strength steel sheet: 55,000 YS.

YSW-60 F
Youngstown Steel
C 0.09, Mn 0.9, P 0.01, S 0.025, Cb 0.06, bal Fe.
Hot rolled and cold rolled high strength steel sheet: 60,000 YS.

YSW-65 F
Youngstown Steel
C 0.09, Mn 1, P 0.01, S 0.025, Cb 0.07, bal Fe.
Hot rolled and cold rolled high strength steel sheet: 65,000 YS.

YTTERBIUM
Atomergic Chemetals Corp.
Yb.
Purities: 99.9% special distilled grade, 99.5+%. Forms: ingot, lump, sheet, foil, wire, turnings, rod.

YTTRIUM
Atomergic Chemetals Corp.
Y.
Purities: 99.9%, 99.5%, 90%, 70%. Forms: ingot, sponge, lump, sheet, foil, rod, wire, powder, turnings.

YUMA CHROME MAGNET STEEL
Allegheny Ludlum Steel
C 1, Cr 4, bal Fe.
For magnets. *Obsolete*

YUNDK 24
Russian manufacture
Ni 14, Co 24, Al 9, Cu 3, bal Fe.
Sintered: 45 Rockwell C. For electrical and magnetic equipment; permanent magnet, sintered alloy, high external energy and permeability.

Z 178

Bergische Stahl Industrie
C 0-0.07, Si 0-1, Mn 0-2, Cr 17-20, Ni 8.5-10, bal Fe.
Austenitic stainless steel. W.-Nr. 1.4301; similar to AISI 304.

Z 178 E

Bergische Stahl Industrie
C 0-0.1, Si 0-1, Mn 0-2, Cr 17-19, Ni 9-11.5, Ti = 5 x C min, bal Fe.
Stabilized austenitic stainless steel. W.-Nr. 1.4541; similar to AISI 321.

Z 178 H

Bergische Stahl Industrie
C 0-0.12, Si 0-1, Mn 0-2, Cr 17-19, Ni 8-10, bal Fe.
Austenitic stainless steel. W.-Nr. 1.4300; AISI 302.

Z 5 Z

English manufacture
Zn 3.5-5.5, Zr 0.4-1, bal Mg.
Sand cast and precipitation treated: 230 MPa TS; 145 MPa YS; 5 El Brin. Higher strength sand or chill casting. BS 2970 MAG 4.

Z A

Hanson-Van Winkle-Munning Co.
Zn, Al.
For electroplating anode.

Z A M

Hanson-Van Winkle-Munning Co.
Al 0.5, Hg 0.25, bal Zn.
For anodes for Zn plating; acid resistant.

Z-1R

Zinkberatungsstelle GmbH
Al 0.8-1, Cu 0.3-0.4, bal Zn.
Rolled: 31,000-42,500 TS; 40-70 Brin. For structures.

Z-2

C.M. Grey Mfg. Co.
Al 4, Cu 3, Mg 0.03, bal Zn.
Die cast: 50,000 TS; 4-8 El; 82-87 Brin. For gears, pumps, motor frames, hardware, die castings.

Z-3

Sterling International Technology Ltd.
Cu 6-8, Si 2-4, Zn 2-4, bal Al.

Z-3

C.M. Grey Mfg. Co.
Al 4, Mg 0.04, bal Zn.
Die cast: 37,000 TS; 4-6 El; 60-65 Brin. For gears, frames, hardware, die castings.

Z-3 ALLOY

Birmingham Aluminum Casting Co.
Aluminum. Cu 6-8, Zn 2-4, Si 2-4, Fe 0.7-1.1, bal Al.
Cast: 23,000-28,000 TS; 1-2 El; 90 Brin. For crankcases, gear boxes, flywheel housings, brake shoes; die and sand cast.

Z-4M, MAGNETIC GRADE

Keystone Carbon Co.
Fe.
Sintered: 6.0-6.5 density; coercive force 2.8-2.9 oersteds. Remanence: 6,500-8,100 Gausses. Soft magnetic material for pole pieces, rotor and stator cores for small motors. *Obsolete*

Z-ALLOY

Engelhard Corp.
Now ENGELHARD ALLOY NO. 24737 (Z-ALLOY).

Z-GUSS

Aluminiumwerke Maulbronn
Al 6, Zn 3, bal Mg.
33,000-40,000 TS; 1-3 El; 100 Brin. For light alloys; age hardenable.

Z-METAL

Alloy Foundries
C 2-2.6, Si 1.1, Mn 0.7-1.2, bal Fe.
Cast: 70,000-90,000 TS; 40,000-60,000 YS; 8-18 El; 155-225 Brin. For castings, camshafts, gears, nozzles; spheroidized cast iron.

Z-METAL

Castings Corp.
C 2-2.6, Si 1.1, Mn 0.7-1.2, bal Fe.
Cast: 70,000-90,000 TS; 40,000-60,000 YS; 8-18 El; 155-225 Brin. For castings, camshafts, gears, nozzles; spheroidized cast iron.

Z-METAL

Industrial Furnace Corp.
TC 2-2.6, CC 0.3-0.8, Si 0.9-1.1, Mn 0.75-1.25, bal Fe.
Cast: 75,000-85,000 TS; 50,000-55,000 YS; 18-12 El; 155-225 Brin. For gears, sprockets, wrenches, bolts, pipe clamps, tool bits, brake drums, air drills; a malleable cast iron; resistance to wear and corrosion; spheroidized.

Z-METAL

Buffalo Brake Beam Co.
TC 2-2.6, CC 0.3-0.8, Si 0.9-1.1, Mn 0.75-1.25, bal Fe.
Cast: 75,000-85,000 TS; 50,000-55,000 YS; 18-12 El; 155-225 Brin. For gears, sprockets, wrenches, bolts, pipe clamps, tool bits, brake drums, air drills; a malleable cast iron; resistance to wear and corrosion; spheroidized.

Z-METAL M-17

Alloy Foundries
TC 2-2.6, C 0.3-0.8, Si 0.9-1.1, Mn 0.75-1.2, bal Fe.
Heat treated: 80,000-100,000 TS; 60,000-75,000 YS; 6-14 El; 197-241 Brin. Pearlitic malleable iron for gears, crankshafts, axle housings, hammer heads.

Z-METAL M-50

Alloy Foundries
TC 2-2.6, CC 0.3-0.8, Si 0.9-1.1, Mn 0.75-1.2, bal Fe.
Heat treated: 70,000-90,000 TS; 45,000-60,000 YS; 3-9 El; 179-223 Brin. Pearlitic malleable iron for gears, crankshafts, axle housings, shears.

Z. N.

CCS Braeburn Alloy Steel
C 0.45-0.5, Cr 0.85, Mn 0.7, bal Fe.
For die casting dies; hot work steel. *Obsolete*

Z.T. DIE STEEL

Latrobe Steel Co.
C 0.4, Cr 0.8, bal Fe.
For dies. *Obsolete*

Z02, 6.8-7.2 DENSITY

Keystone Carbon Co.
Sintered steel manufactured to AISI 4630 composition. As sintered: 65 Rock B. Heat treated: 25 Rock C; 110,000 psi TS. For structural parts.

Z05, 5.6-6.0

Keystone Carbon Co.
5.6-6.0 g/cm^3 density. Graphite bearing material that offers good bearing characteristics coupled with quiet operation. ASTM B782-88, Grade 1.

Z07, 5.6-6.0

Keystone Carbon Co.
5.6-6.0 g/cm^3 density; 25,000 psi K-value min; 75 Rock H. Iron graphite bearing material for applications requiring heavy leads and high speeds. ASTM B782-88, Grade 2.

Z23

Keystone Carbon Co.
Fe-Cu-P alloy. Sintered: 6.1-6.5 density; 60,000 psi TS; 45 Rock B. Standard grade for structural parts.

Z24

Keystone Carbon Co.
Sintered electrolytic iron, high density: 7.3 density min; 40,000 psi TS; 40 Rock B. Can be hardened or case hardened, and plated. For many types of structural parts. ASTM B310-83A, Type V, Class A.

Z28

Keystone Carbon Co.
Diluted bronze 40-60 (40 bronze, 60 Fe). Sintered: 14,000 psi TS; 18% porosity; 5.8-6.2 density; 40 Rock H. Lower priced bearing.

Z29M

Keystone Carbon Co.
P 0.45, bal Fe.
Sintered: 7.0-7.6 density; coercive force 1.0-1.2 oersteds. Remanence: 1100-1300 Gausses. Soft magnetic material for cores for AC relays, rotor and stator cores for small AC motors and generators.

Z33, 5.7-6.1 DENSITY

Keystone Carbon Co.
Sintered carbon steel, Class C: 5.7-6.1 density; 26,000 psi TS; 35 Rock B; hardenable to 40,000 psi TS; 55 RA. For structural parts. ASTM B310-83A, Type I; SAE 852.

Z33, 6.11-6.50 DENSITY

Keystone Carbon Co.
Sintered carbon steel, Class C: 6.11-6.50 density; 34,000 psi TS; 55 Rock B; hardenable to 50,000 psi TS; 60 Rock A. For structural parts. ASTM B310-83A, Type II; SAE 855.

Z33, 6.51-6.90 DENSITY

Keystone Carbon Co.
Sintered carbon steel, Class C: 6.51-6.90 density; 44,000 psi TS; 70 Rock B; hardenable to 64,000 psi TS; 64 Rock A. For structural parts. ASTM B310-83A, Type III.

Z33, 6.91-7.30 DENSITY

Keystone Carbon Co.
Sintered carbon steel, Class C: 6.91-7.30 density; 60,000 psi TS; 85 Rock B; hardenable to 80,000 psi TS; 68 Rock A. For structural parts. ASTM B310-83A, Type IV.

Z34, 5.70-6.10 DENSITY

Keystone Carbon Co.
Sintered carbon copper steel, Class C: 5.7-6.1 density; 30,000 psi TS; 35 Rock B; hardenable to 40,000 psi TS; 55 Rock A. For structural parts; recommended for parts requiring induction hardening. ASTM B426-38A, Type I, Gr. 1; SAE 846A.

Z34, 6.11-6.50 DENSITY

Keystone Carbon Co.
Sintered carbon copper steel, Class C: 6.11-6.50 density; 41,000 psi TS; 50 Rock B; hardenable to 50,000 psi TS; 60 Rock A. For structural parts; recommended for parts requiring induction hardening. ASTM B426-83A, Type II, Gr. 1; SAE 864B.

Z46, 5.7-6.1 DENSITY

Keystone Carbon Co.
Cu 3-6, bal Fe.
Sintered carbon copper steel, Class C: 5.70-6.10 density; 35,000 psi TS; 45 Rock B; hardenable to 50,000 psi TS; 55 Rock A. All purpose grade for structural parts. ASTM B426-83A, Type I, Gr.2; SAE 865A.

Z46, 6.11-6.5 DENSITY

Keystone Carbon Co.
Cu 3-6, bal Fe.
Sintered carbon copper steel, Class C: 6.11-6.5 density; 50,000 psi TS; 60 Rock B; hardenable to 75,000 psi TS; 80 Rock A. All purpose grade for structural parts. ASTM B426-83A, Type II, Gr.2; SAE 865B.

Z46, 6.51-6.90 DENSITY

Keystone Carbon Co.
Sintered carbon copper steel, Class C, 3-6 Cu: 6.51-6.90 density; 60,000 psi TS; Rock B 65; hardenable to 85,000 psi TS; Rock A 64. All purpose grade for structural parts. ASTM B426-65 Gr.2, except density.

Z63, 6.4-6.8 DENSITY

Keystone Carbon Co.
Sintered nickel steel: 6.3-6.7 density; 40,000 psi TS; 45 Rock B. Hardened and tempered: 65,000 psi TS; 60 Rock A. For structural parts; may be oil impregnated. ASTM B484-83A, Grade II, Class C, Type 1.

Z63, 6.81-7.20 DENSITY

Keystone Carbon Co.
Sintered nickel steel: 6.81-7.20 density; 55,000 psi TS; 60 Rock B. Hardened and tempered: 110,000 psi TS; 65 Rock A. For structural parts. ASTM B484-83A, Grade II, Class C, Type 1.

Z63, 7.21-7.60 DENSITY

Keystone Carbon Co.
Sintered nickel steel, special process: 7.21-7.60 density; 100,000 psi TS; 80 Rock B. Hardened and tempered: 180,000 psi TS; 70 Rock A. For structural parts. ASTM B484-83A, Grade II, Class C, Type 2.

Z64, 6.4-6.8

Keystone Carbon Co.
Sintered nickel steel: 6.4-6.8 g/cm^3; 48,000 psi TS; 55 Rock B. Hardened and tempered: 110,000 psi TS; 25 Rock C. For structural parts. ASTM B484-83A, Grade 1, Class C, Type 1.

Z64, 6.81-7.20

Keystone Carbon Co.
Sintered nickel steel: 6.81-7.20 g/cm^3 density; 55,000 psi TS; 75 Rock B. Hardened and tempered: 125,000 psi TS; 30 Rock C. For structural parts. ASTM B484-83A, Grade 1, Class C, Type 2.

Z64, 7.21 MIN

Keystone Carbon Co.
Sintered nickel steel: 7.21 g/cm^3 density min; 80,000 psi TS; 85 Rock B. Hardened and tempered: 160,000 psi TS; 35 Rock C. For structural parts. ASTM B484-83A, Grade 1, Class C, Type 3.

Z72, 6.8-7.2

Keystone Carbon Co.
Sintered 4650 alloy steel: 6.8-7.2 g/cm^3 density; 80,000 psi TS; 80 Rock B. Hardened and tempered: 140,000 psi TS; 30 Rock C. For high strength gear applications.

Z90

Friedr. Lohmann GmbH
C 1, Mn 0.25, V 0.1, bal Fe.
For springs, cutters, drills, reamers, taps. Type W2; water hardened. T72302

Z91, 6.6-7.0

Keystone Carbon Co.
Air hardening: 70,000 psi TS; 95 Rock B. Wear resistant. For room and high temperature applications.

ZA-12

Bunting Bearings Corp.
Zinc/aluminum. Al 10.5-11.5, Cu 0.5-1, Mg 0.01-0.03, bal Zn. Permanent mold process.

ZA-12

Aluminum Smelting & Refining Co., Inc.
Zinc. Al 10.5-11.5, bal Zn.
Cold chamber die cast: 60,000 psi TS max; 48,000 psi YS max; 4-7 El.

ZA-27

Aluminum Smelting & Refining Co., Inc.
Zinc. Al 25-28, Cu 2-2.5, bal Zn.
64,000 psi TS max; 55,000 psi YS max; 116-122 Brin. Die casting alloy.

ZA-27

Bunting Bearings Corp.
Zinc/aluminum. Al 25-28, Cu 2-2.5, Mg 0.01-0.02, bal Zn. Permanent mold process.

ZA-8

Aluminum Smelting & Refining Co., Inc.
Zinc. Al 8-8.8, bal Zn.
54,000 psi TS; 42,000 psi YS; 6.5 El max. Die casting alloy; can be hot chamber die cast.

ZA-8

Bunting Bearings Corp.
Zinc/aluminum. Al 8-8.8, Cu 0.8-1.3, Mg 0.015-0.03, bal Zn. Permanent mold process. Centrifugally cast process.

ZADUR 5 BR

Dusseldorf-Heerdt GmbH & Co., KG
C 1.42, W, V, bal Fe.
For engravers' tools, bearings, cutters; water or oil hardened.

ZADUR B 25

Dusseldorf-Heerdt GmbH & Co., KG
C, Cr, Ni, V, bal Fe.
For molds, dies.

ZADUR B25S

Dusseldorf-Heerdt GmbH & Co., KG
C, Cr, Ni, V, bal Fe.
For headers, upsetters, punches; cold work steel.

ZADUR BC1

Dusseldorf-Heerdt GmbH & Co., KG
C 1.05, Cr 1, Si 0.25, Mn 0.3, bal Fe.
For bearings, liners, sleeves, forming dies; water hardened.

ZADUR BC12

Dusseldorf-Heerdt GmbH & Co., KG
C 2.1, Cr 11.5, bal Fe.
For blanking and forming dies; oil or air hardened, nondeforming.

ZADUR BC12W

Dusseldorf-Heerdt GmbH & Co., KG
C 1.65, Cr 11.5, V 0.1, bal Fe.
For blanking and forming dies, punches; air hardened, nondeforming.

ZADUR BC13

Dusseldorf-Heerdt GmbH & Co., KG
C 2.1, Cr 11.5, W 0.7, bal Fe.
For blanking and forming dies, punches; oil or air hardened, nondeforming.

ZADUR BC13M EXTRA

Dusseldorf-Heerdt GmbH & Co., KG
C, Cr, Mn, bal Fe.
For upsetters, headers, dies, punches; cold work steel.

ZADUR BC13W

Dusseldorf-Heerdt GmbH & Co., KG
C 1.65, Cr 11.5, V 0.1, Mo 0.2, bal Fe.
For blanking and forming dies, punches; air hardened, nondeforming.

ZADUR BCN

Dusseldorf-Heerdt GmbH & Co., KG
C 0.45, Ni, Cr, W, bal Fe.
For upsetters, heading dies, crimpers; oil hardened, tough.

ZADUR BDK

Dusseldorf-Heerdt GmbH & Co., KG
V 0.2, W 1.85, C 0.45, Cr 1.05, bal Fe.
For cold heading dies, headers, punches; oil hardened, tough.

ZADUR BDK EXTRA

Dusseldorf-Heerdt GmbH & Co., KG
C 0.45, Si 0.9, Cr 1.05, V 0.2, W 1.85, bal Fe.
For cold heading dies, headers, punches; oil hardened, tough.

ZADUR BF2

Dusseldorf-Heerdt GmbH & Co., KG
C 1.05, Mn 1.25, Cr 1.15, bal Fe.
For bearings, liners, sleeves; water hardened, water resistant.

ZADUR BF2 EXTRA

Dusseldorf-Heerdt GmbH & Co., KG
C 0.9, Mn 1.9, V 0.1, bal Fe.
For punches, dies, upsetters, shears, tools; oil hardened, nondeforming.

ZADUR BFC

Dusseldorf-Heerdt GmbH & Co., KG
C 1.05, Cr 1, W 1.15, Mn 0.9, bal Fe.
For cutting tools, bearings, liners; oil or water hardened, wear resistant

ZADUR BGM

Dusseldorf-Heerdt GmbH & Co., KG
C 0.34, Cr 1.55, Ni 1.55, Mo 0.2, bal Fe.
For gears, shafts, crankshafts, bolts, fasteners; oil hardened, shock resistant.

ZADUR BGS

Dusseldorf-Heerdt GmbH & Co., KG
C 0.45, Cr, Ni, bal Fe.
For gears, bolts, fasteners; oil hardened, shock resistant.

ZADUR BKM

Dusseldorf-Heerdt GmbH & Co., KG
C 1, V 0.1, Mn 0.25, Si 0.25, bal Fe.
For cutters, hobs, drills, taps, reamers; Type W2; water hardened.

ZADUR BLS

Dusseldorf-Heerdt GmbH & Co., KG
C 0.55, Cr 1.05, V 0.18, W 1.85, bal Fe.
For cold heading dies, punches, headers; oil hardened, tough.

ZADUR BMR

Dusseldorf-Heerdt GmbH & Co., KG
C 0.4, Cr, Mo, bal Fe.
For gears, pinions, shafts, bolts; oil hardened, tough.

ZADUR BP5

Dusseldorf-Heerdt GmbH & Co., KG
C 0.45, Si, Cr, V, bal Fe.
For springs, bolts, gears; oil hardened, tough.

ZADUR BPK

Dusseldorf-Heerdt GmbH & Co., KG
C 0.4, Cr, V, Mn, bal Fe.
For gears, bolts, crankshafts; oil hardened, tough.

ZADUR BRK

Dusseldorf-Heerdt GmbH & Co., KG
C 0.5, Cr 1.05, Ni 3.25, Mn 0.5, bal Fe.
For gears, bolts, crankshafts, fasteners; oil hardened, shock resistant.

ZADUR BST1

Dusseldorf-Heerdt GmbH & Co., KG
C, Cr, W, Ni, V, Co, bal Fe.

ZADUR BWF2

Dusseldorf-Heerdt GmbH & Co., KG
C, Cr, Ni, V, bal Fe.
For tools, dies; oil hardened, tough.

ZADUR BWM
Dusseldorf-Heerdt GmbH & Co., KG
C 0.3, Cr 2.65, V 0.35, W 8.5, bal Fe.
For extrusion press dies and rams, upsetters; hot work steel, oil hardened.

ZADUR BWM EXTRA
Dusseldorf-Heerdt GmbH & Co., KG
C, Cr, W, V, Ni, bal Fe.
For header and upsetter dies, shears, punches; hot work steel, oil hardened.

ZADUR BWM100
Dusseldorf-Heerdt GmbH & Co., KG
C 0.3, Cr 1.1, Si 1, V 0.18, W 75, bal Fe.
For header dies, upsetters, crimpers; oil hardened, tough.

ZADUR BWM5
Dusseldorf-Heerdt GmbH & Co., KG
C 0.3, Cr 2.35, V 0.6, W 4.25, bal Fe.
For extrusion press dies and rams, upsetters; hot work steel, oil hardened.

ZADUR BWMCO
Dusseldorf-Heerdt GmbH & Co., KG
C 0.3, Co 2, Cr 2.4, V 0.25, W 8.5, bal Fe.
For extrusion press dies and rams, upsetters; oil hardened, tough.

ZADUR CCK
Dusseldorf-Heerdt GmbH & Co., KG
C 1.45, Cr 1.4, Mn 0.6, bal Fe.
For bearings, bushings, liners, sleeves; water hardened, wear resistant.

ZADUR CCW
Dusseldorf-Heerdt GmbH & Co., KG
C, Cr, Ni, W, V, bal Fe.
Oil hardened; tough.

ZADUR D0
Dusseldorf-Heerdt GmbH & Co., KG
C 0.28, Ni, Mo, bal Fe.
For gears, bolts, machine tool parts; oil hardened, tough.

ZADUR D1
Dusseldorf-Heerdt GmbH & Co., KG
C 0.28, Ni, Cr, Mo, V, bal Fe.
For header dies, machine tool parts; oil hardened, tough.

ZADUR D3
Dusseldorf-Heerdt GmbH & Co., KG
C 0.26, Ni, Cr, Mo, bal Fe.
For machine tool parts, gears, bolts; oil hardened, tough.

ZADUR DS10
Dusseldorf-Heerdt GmbH & Co., KG
C 0.65, Cr 3.75, Mo 0.85, W 8.5, V 0.7, bal Fe.
For lathe and planer tools, reamers, drills, taps; high speed steel.

ZADUR EDH11
Dusseldorf-Heerdt GmbH & Co., KG
C 0.37, Si 1.25, Mn 1.25, bal Fe.
For gears, shafts, crankshafts; oil hardened, tough.

ZADUR EDH2
Dusseldorf-Heerdt GmbH & Co., KG
C 0.42, Cr 1, Mo 0.2, bal Fe.
For gears, bolts, shafts, fasteners; oil hardened, tough.

ZADUR EDH3
Dusseldorf-Heerdt GmbH & Co., KG
C 0.4, Cr, Mo, bal Fe.
For gears, bolts, shafts, fasteners; oil hardened, tough.

ZADUR EDHV
Dusseldorf-Heerdt GmbH & Co., KG
C 0.5, Mn 0.85, Cr 1, V 0.09, bal Fe.
For gears, springs, bolts, shafts; oil hardened, shock resistant.

ZADUR EH
Dusseldorf-Heerdt GmbH & Co., KG
C 0.15, Cr 0.65, Mn 0.5, bal Fe.
For gears, machine tool parts, shafts; case hardening steel.

ZADUR EHC125
Dusseldorf-Heerdt GmbH & Co., KG
C 0.16, Mn 1.15, Cr 0.95, bal Fe.
For gears, bolts, fasteners, cams; case hardening steel, tough.

ZADUR EHC135
Dusseldorf-Heerdt GmbH & Co., KG
C 0.2, Mn 1.15, Cr 0.95, bal Fe.
For bolts, gears, camshafts, cams, pinions; case hardening steel.

ZADUR FE501
Dusseldorf-Heerdt GmbH & Co., KG
C 0.7, V, W, Cr, bal Fe.
For lathe and planer tools, drills, reamers, hobs; high speed steel.

ZADUR FE501R
Dusseldorf-Heerdt GmbH & Co., KG
C 0.7, Cr 4, V, W, bal Fe.
For lathe and planer tools, drills, reamers, taps; high speed steel.

ZADUR FE601
Dusseldorf-Heerdt GmbH & Co., KG
C 0.86, Cr 4.1, Mo 0.85, V 2.5, W 12, bal Fe.
For lathe and planer tools, reamers, drills, taps; high speed steel.

ZADUR FEX
Dusseldorf-Heerdt GmbH & Co., KG
C 0.74, Cr 4.1, V 1.1, W 18.5, bal Fe.
For lathe and planer tools, reamers, drills, taps; high speed steel.

ZADUR FN
Dusseldorf-Heerdt GmbH & Co., KG
C 0.82, Cr 4.1, Mo 0.85, V 1.6, W 8.7, bal Fe.
For lathe and planer tools, reamers, broaches, taps; high speed steel.

ZADUR GC15
Dusseldorf-Heerdt GmbH & Co., KG
C 0.25, Cr 14.5, Ni 0-1, bal Fe.
For oil refinery equipment; creep and heat resistant.

ZADUR GC17
Dusseldorf-Heerdt GmbH & Co., KG
C 0.25, Cr 17, Ni 1.8, bal Fe.
Annealed: 125,000 TS; 95,000 YS; 20 El; 55 RA; 260 Brin.
Cold drawn: 130,000 TS; 110,000 YS; 15 El; 35 RA; 270 Brin.
For pumps, marine hardware, cutlery; Type 431; stainless and heat resistant.

ZADUR GC30H
Dusseldorf-Heerdt GmbH & Co., KG
C 1.2, Si 1.3, Cr 29, bal Fe.
For wear plates, grates, grids, liners, valves; heat and abrasion resistant.

ZADUR GC30W
Dusseldorf-Heerdt GmbH & Co., KG
C 0.4, Si 1.3, Cr 29, bal Fe.
Cast: 90,000 TS; 65,000 YS; 2 El; 212 Brin. For cylinder liners, valve seats and bodies, bushings; heat and abrasion resistant.

ZADUR GCM30H
Dusseldorf-Heerdt GmbH & Co., KG
C 1.2, Si 1.3, Cr 29, Mo 2, bal Fe.
For cylinder liners, valve seats and bodies, bushings; heat and abrasion resistant.

ZADUR GD15
Dusseldorf-Heerdt GmbH & Co., KG
C 0.38, Si, Cr, V, bal Fe.
For springs, bolts, gears, machine tool parts; oil hardened, shock resistant.

ZADUR GD2
Dusseldorf-Heerdt GmbH & Co., KG
C 0.7, Cr, W, V, bal Fe.
For lathe and planer tools, drills, reamers; high speed steel.

ZADUR GD22
Dusseldorf-Heerdt GmbH & Co., KG
C 0.35, Cr 1.05, V 0.18, W 1.85, bal Fe.
For upsetter dies, header dies, crimpers, punches; oil hardened, tough.

ZADUR GRMN
Dusseldorf-Heerdt GmbH & Co., KG
C 0.15, Cr 18, Ni 8.5, bal Fe.
Annealed: 80,000 TS; 35,000 YS; 55 El; 75 RA; 150 Brin. For chemical plant equipment, tanks; Type 302; stainless, austenitic.

ZADUR GRMP
Dusseldorf-Heerdt GmbH & Co., KG
C 0.15, Cr 18, Ni 9.5, Mo 2, bal Fe.
Annealed: 85,000 TS; 35,000 YS; 50 El; 65 RA; 160 Brin. For acid resistant chemical plant equipment, tanks, mixers, filters; Type 316; stainless, austenitic.

ZADUR HDL
Dusseldorf-Heerdt GmbH & Co., KG
C, W, V, bal Fe.
For machine tool parts; oil hardened.

ZADUR MSA1
Dusseldorf-Heerdt GmbH & Co., KG
C 0.53, Si 0.9, Mn 0.9, bal Fe.
For gears, bolts, shafts, fasteners; water hardened.

ZADUR MSA2
Dusseldorf-Heerdt GmbH & Co., KG
C 0.55, Cr 0.7, Ni 1.65, V 0.1, Mo 0.18, bal Fe.
For gears, bolts, crankshafts, fasteners; oil hardened, shock resistant.

ZADUR MSA3
Dusseldorf-Heerdt GmbH & Co., KG
C 0.56, Cr 0.8, Mo 0.2, Ni 1.75, V 0.1, bal Fe.
For gears, bolts, crankshafts, fasteners; oil hardened, shock resistant.

ZADUR RBM
Dusseldorf-Heerdt GmbH & Co., KG
C 0.45, Cr 1.4, Mo 0.7, V 0.3, bal Fe.
For forging and heading dies, upsetters, die casting dies; oil hardened, tough.

ZADUR RMA
Dusseldorf-Heerdt GmbH & Co., KG
C 0.9, Mo 1.15, V 1, Cr 18, bal Fe.
For cutlery, valves, ball bearings; hardenable, corrosion resistant.

ZADUR RMC
Dusseldorf-Heerdt GmbH & Co., KG
C 0.2, Cr 13, Mo 1.15, bal Fe.
Annealed: 95,000 TS; 50,000 YS; 25 El; 55 RA; 195 Brin. For turbine blades, cutlery, valves; hardenable, corrosion resistant.

ZADUR RME
Dusseldorf-Heerdt GmbH & Co., KG
C 0.35, Cr 16.5, Mo 1.15, bal Fe.
For oil refinery equipment; corrosion and heat resistant.

ZADUR RMH
Dusseldorf-Heerdt GmbH & Co., KG
C 0.4, Cr 13, Mn 0.3, bal Fe.
Annealed: 100,000 TS; 55,000 YS; 20 El; 50 RA; 200 Brin. For turbine blades, valves, cutlery, surgical instruments; Type 420; stainless.

ZADUR RMI
Dusseldorf-Heerdt GmbH & Co., KG
C 0-0.1, Si 2, Cr 18, Ni 8.5, bal Fe.
Annealed: 85,000 TS; 35,000 YS; 60 El; 70 RA; 150 Brin. For chemical plant equipment, tanks, agitators; Type 304; stainless, austenitic.

ZADUR RMK
Dusseldorf-Heerdt GmbH & Co., KG
C 0-0.12, Cr 18, Ni 9.5, Ti = 4 x C, bal Fe.
Annealed: 85,000 TS; 35,000 YS; 55 El; 65 RA; 150 Brin. For welded chemical plant equipment; Type 321; stainless, austenitic.

ZADUR RML
Dusseldorf-Heerdt GmbH & Co., KG
C 0-0.12, Cr 18, Ni 10.5, Mo 2, Ti = 4 x C, bal Fe.
Annealed: 85,000 TS; 35,000 YS; 50 El; 65 RA; 160 Brin. For welded acid resistant equipment; Type 316 Ti; stainless, austenitic.

ZADUR RMM
Dusseldorf-Heerdt GmbH & Co., KG
C 0.2, Cr 13, Si 0.4, bal Fe.
Annealed; 95,000 TS; 50,000 YS; 25 El; 55 RA; 195 Brin. For turbine blades, valves, cutlery, surgical instruments; Type 420 stainless, hardenable.

ZADUR RMN
Dusseldorf-Heerdt GmbH & Co., KG
C 0-0.15, Cr 18, Ni 8.5, bal Fe.
Annealed: 80,000 TS; 35,000 YS; 55 El; 75 RA; 150 Brin. For chemical plant equipment, tanks, mixers, filters; Type 302; stainless austenitic.

ZADUR RMO
Dusseldorf-Heerdt GmbH & Co., KG
C 0.22, Cr 17, Ni 1.5, Si 0.4, bal Fe.
Annealed: 125,000 TS; 90,000 YS; 20 El; 55 RA; 260 Brin. For pumps, marine hardware, valves; Type 431; corrosion resistant.

ZADUR RMP
Dusseldorf-Heerdt GmbH & Co., KG
C 0-0.1, Cr 18, Mo 2, Ni 9.5, bal Fe.
Annealed: 85,000 TS; 35,000 YS; 50 El; 65 RA; 160 Brin. For acid resistant chemical plant equipment; Type 316; stainless, austenitic.

ZADUR RMR
Dusseldorf-Heerdt GmbH & Co., KG
C 0-0.07, Cr 18, Ni 9.5, bal Fe.
Annealed: 85,000 TS; 35,000 YS; 60 El; 70 RA; 150 Brin. For chemical plant equipment, tanks, mixers; Type 304; stainless, austenitic.

ZADUR RMT
Dusseldorf-Heerdt GmbH & Co., KG
C 0-0.07, Cr 18, Ni 10.5, Mo 2, bal Fe.
Annealed: 85,000 TS; 35,000 YS; 50 El; 65 RA; 160 Brin. For acid resistant chemical plant equipment; Type 316; stainless, austenitic.

ZADUR RMW
Dusseldorf-Heerdt GmbH & Co., KG
C 0-0.12, Si 0.4, Cr 13, bal Fe.
Annealed: 75,000 TS; 40,000 YS; 35 El; 70 RA; 155 Brin. For turbine blades, valves, cutlery; Type 410; stainless.

ZADUR SO EXTRA
Dusseldorf-Heerdt GmbH & Co., KG
C 1.3, Cr 4.3, Mo 0.85, V 3.8, W 12, bal Fe.
For blanking and forming dies; high speed steel.

ZADUR SS CO SPEZIAL
Dusseldorf-Heerdt GmbH & Co., KG
C 1.35, Co, W, Mo, Cr, V, bal Fe.
For blanking and forming dies, engravers' tools; high speed steel.

ZADUR SS CO10
Dusseldorf-Heerdt GmbH & Co., KG
C 0.76, Co 10, Cr 4.2, Mo 0.8, V 1.8, W 18, bal Fe.
For lathe and planer tools, form cutters, hobs; high speed steel.

ZADUR SS CO15
Dusseldorf-Heerdt GmbH & Co., KG
C 0.7, Cr 4, W, V, Co, bal Fe.
For lathe and planer tools, drills, taps, hobs; high speed steel.

ZADUR SS CO3
Dusseldorf-Heerdt GmbH & Co., KG
C 0.86, Co 2.8, Cr 4.3, Mo 0.85, V 2.1, W 12, bal Fe.
For lathe and planer tools, reamers, drills; high speed steel.

ZADUR SS CO5
Dusseldorf-Heerdt GmbH & Co., KG
C 0.79, Co 4.75, Cr 4.3, Mo 0.75, V 1.5, W 18, bal Fe.
For lathe and planer tools, cutters, reamers; high speed steel.

ZADUR SSB50N
Dusseldorf-Heerdt GmbH & Co., KG
C 0.8, W, Co, Cr, V, Mo, bal Fe.
For lathe and planer tools, reamers, broaches; high speed steel.

ZADUR ZRS
Dusseldorf-Heerdt GmbH & Co., KG
C 1.4, Cr 0.3, V 0.1, bal Fe.
For blanking and heading dies, engravers' tools; water or oil hardened, wear resistant.

ZAL-DIE
Osborn Steels Ltd.
C 0.35, Mn 0.5, Cr 1, Ni 3, Mo 0.3, bal Fe.
For shafts, dies, gears, axles, crankshafts; oil hardened, shock resistant. *Obsolete*

ZALUTITE
British Steel plc
Hot-dipped zinc-aluminum alloy coated steel. Mild steel.

ZAMA
Italian manufacture
Al, Mg, bal Zn.
For die castings, instrument cases; corrosion resistant.

ZAMAK ALPHA
Zinkberatungsstelle GmbH
Al 4, Cu 0.5-1, Mg 0.3, bal Zn.
Rolled: 44,000-71,000 TS; 75-140 Brin. For die castings. Rolled or cast.

ZAMAK BETA
Zinkberatungsstelle GmbH
Al 10, Cu 0.7, Mg 0.03, bal Zn.
Rolled: 45,000-78,000 TS; 75-160 Brin. For housings, hardware.

ZAMAK ETA
Zinkberatungsstelle GmbH
Al 10, Cu 0.3, bal Zn.
Rolled: 31,000-45,000 TS; 55-65 Brin. For housings and hardware.

ZAMAK ETA H
Zinkberatungsstelle GmbH
Al 10, Cu 0.3, Mg 0.01, bal Zn.
Rolled: 56,700-70,000 TS; 100-130 Brin. For housings and hardware.

ZAMAK GAMMA
German manufacture
Al 0.8, Cu 0.4, Mg 0-0.2, bal Zn.
For castings or drawn wire.

ZAMAK NO. 2
New Jersey Zinc Co., Inc.
Al 4, Cu 3, Mg 0.03, bal Zn.
Die cast: 50,000 TS; 4-8 El; 82-87 Brin. For gears, motor frames, carburetors, hardware, pumps. *Obsolete*

ZAMAK NO. 6
New Jersey Zinc Co., Inc.
Al 4, Cu 1.25, bal Zn.
Die cast: 42,000 TS; 5-7 El; 65-75 Brin. For die castings for maximun fluidity and exceptional toughness. *Obsolete*

ZAMAK Z 400
Metallgesellschaft Reuterweg
Al 3.7-4.3, Cu 0-0.6, Mg 0.02-0.05, bal Zn.
Die cast: 36,000-43,000 TS; 1.5-4 El; 70-80 Brin. For housings, cases, ornaments, general die castings; impact resistant, shrinks on aging.

ZAMAK Z 410
Metallgesellschaft Reuterweg
Al 3.7-4.3, Cu 0.6-1, Mg 0.02-0.05, bal Zn.
Die cast: 39,000-46,000 TS; 2-5 El; 80-90 Brin. For housings, cases, ornaments, general die castings; good corrosion resistance.

ZAMAK Z 430
Metallgesellschaft Reuterweg
Zn alloy.
For mold inserts, tools, dies.

ZAMIUM
Manufacturer not listed.
Ni 60, Cr 40, Mn, W.
For heat resistant parts; corrosion and heat resistant.

ZAPP 120
Robert Zapp Werkstofftechnik GmbH
C 0-0.08, Cr 13, Ni 13, bal Fe.
For chemical plant equipment; corrosion resistant. W. Nr. 4307.

ZAPP 13
Robert Zapp Werkstofftechnik GmbH
C 0-0.08, Cr 13, bal Fe.
Ferritic type stainless steel. Annealed: 130-180 Brin. Werkstoff Nr. 4000.

ZAPP 135 P
Robert Zapp Werkstofftechnik GmbH
C 0-0.07, Cr 17, Mo 4.5, Ni 13.5, bal Fe.
Austenitic stainless steel; for chemical equipment. Werkstoff Nr. 4449; AISI 317.

ZAPP 135T
Bochum Stahlwerke AG
C 0.05, Cr 17, Ni 13, Mo 2, Ta, Cb, bal Fe.
Annealed: 85,000 TS; 35,000 YS; 50 El; 60 RA; 150 Brin. Cold drawn: 150,000 TS; 135,000 YS; 6 El; 300 Brin. For welded acid resistant equipment, tanks, mixers; Type 316Cb; stainless, austenitic. *Obsolete*

ZAPP 13A1
Robert Zapp Werkstofftechnik GmbH
C 0-0.08, Cr 13, Al 0.1-0.3, bal Fe.
Ferritic type stainless steel. Annealed: 130-180 Brin. Werkstoff Nr. 4002; AISI 405.

ZAPP 13B
Robert Zapp Werkstofftechnik GmbH
C 0-0.08, Cr 14, bal Fe.
Ferritic type stainless steel. Annealed: 130-180 Brin. Werkstoff Nr. 4001.

ZAPP 17
Robert Zapp Werkstofftechnik GmbH
C 0-0.1, Cr 17, bal Fe.
Ferritic type stainless steel. Annealed: 130-170 Brin. Werkstoff Nr. 4016; AISI 430.

ZAPP 17 F
Robert Zapp Werkstofftechnik GmbH
C 0-0.07, Cr 17, Mo 1, bal Fe.
Ferritic type stainless steel. Annealed: 130-180 Brin. Werkstoff Nr. 4113; similar to AISI 434.

ZAPP 17 NB
Robert Zapp Werkstofftechnik GmbH
C 0-0.1, Cr 17, Nb = 12 x C min, bal Fe.
Ferritic type stainless steel. Annealed: 130-170 Brin. Werkstoff Nr. 4511.

ZAPP 17 T
Robert Zapp Werkstofftechnik GmbH
C 0-0.1, Cr 17, Ti = 7 x C min, bal Fe.
Ferritic type stainless steel. Annealed: 130-170 Brin. Werkstoff Nr. 4510.

ZAPP 17 U
Robert Zapp Werkstofftechnik GmbH
C 0-0.15, Cr 17, Mo 0.25, S, bal Fe.
Ferritic type stainless steel. Annealed: 190-230 Brin. Werkstoff Nr. 4104; AISI 430 F.

ZAPP 182 RNB
Robert Zapp Werkstofftechnik GmbH
C 0-0.07, Cr 18, Mo 2.3, Ni 20, Cu/Ti/Ta, bal Fe.
Austenitic stainless steel. Werkstoff Nr. 4505.

ZAPP 182RT
Bochum Stahlwerke AG
C 0-0.07, Cr 17.5, Ni 17.5, Mo 2, Cu 2, Ti = 7 x C, bal Fe.
Annealed: 85,000 TS; 35,000 YS; 50 El; 65 RA; 160 Brin. For welded acid resistant equipment, tanks, mixers; stainless, austenitic, stabilized. *Obsolete*

ZAPP 223 RNB
Robert Zapp Werkstofftechnik GmbH
C 0-0.07, Cr 17-18, Mo 3-3.5, Ni 22-23, Nb, bal Fe.
Austenitic stainless steel. Werkstoff Nr. 4586.

ZAPP 252 T
Robert Zapp Werkstofftechnik GmbH
C 0-0.06, Cr 25, Mo 2.2, Ni 25, N, bal Fe.
Austenitic stainless steel. Werkstoff Nr. 4577.

ZAPP 26 NL
Robert Zapp Werkstofftechnik GmbH
C 0-0.1, Cr 26, Mo 1.5, Ni 4.5, bal Fe.
Ferritic type stainless steel. Annealed: 190-230 Brin. Werkstoff Nr. 4460.

ZAPP 28
Bochum Stahlwerke AG
C 1.2, Si 1.3, Cr 29, bal Fe.
For wear plates, crushers, furnace equipment; corrosion and heat resistant. *Obsolete*

ZAPP 28L
Bochum Stahlwerke AG
C 1.2, Si 1.3, Cr 29, Mo 2, bal Fe.
For wear plates, crushers, furnace equipment; corrosion and heat resistant. *Obsolete*

ZAPP 28N
Bochum Stahlwerke AG
C 0.4, Si 1.3, Cr 27, Ni 4, bal Fe.
For furnace parts, heat treat boxes; heat and oxidation resistant. *Obsolete*

ZAPP 28W
Bochum Stahlwerke AG
C 0.4, Cr 27, Si, bal Fe.
For furnace parts, heat treat boxes; heat and oxidation resistant. *Obsolete*

ZAPP 42MNV7
Bochum Stahlwerke AG
C 0.42, Mn 1.75, V 0.1, bal Fe.
For punches, crimpers, dies; oil hardened, shock resistant. *Obsolete*

ZAPP 58 M
Robert Zapp Werkstofftechnik GmbH
C 0-0.1, Cr 18, Ni 5.5, Mn 8.5, N, bal Fe.
Manganese austenitic stainless. Werkstoff Nr. 4371; AISI 202.

ZAPP 80
Bochum Stahlwerke AG
C 0.15, Si 1.5, Cr 18, Ni 8.5, bal Fe.
Annealed: 80,000 TS; 35,000 YS; 55 El; 75 RA; 150 Brin. For chemical plant equipment, tanks, mixers, agitators; Type 302; stainless, austenitic. *Obsolete*

ZAPP 80 ELC
Robert Zapp Werkstofftechnik GmbH
C 0-0.03, Cr 18, Ni 11, bal Fe.
Low carbon austenitic stainless steel. Solution annealed: 130-180 Brin. Werkstoff Nr. 4306; AISI 304 L.

ZAPP 80 FH
Robert Zapp Werkstofftechnik GmbH
C 0-0.15, Cr 17, Ni 7.5, bal Fe.
Austenitic type stainless steel. Solution treated: 170-210 Brin. Work hardens rapidly. Werkstoff Nr. 4310; AISI 301.

ZAPP 80 K
Robert Zapp Werkstofftechnik GmbH
C 0-0.12, Cr 18, Ni 9, bal Fe.
Austenitic type stainless steel. Solution treated: 130-180 Brin. Werkstoff Nr. 4300; AISI 302.

ZAPP 80 NB
Robert Zapp Werkstofftechnik GmbH
C 0-0.1, Cr 18, Ni 10.5, Nb/Ta, bal Fe.
Stabilized austenitic stainless steel. For welded structures and elevated temperatures. Werkstoff Nr. 4550; AISI 347.

ZAPP 80P
Robert Zapp Werkstofftechnik GmbH
C 0-0.07, Cr 18, Ni 9.5, bal Fe.
Annealed: 85,000 TS; 35,000 YS; 60 El; 70 RA; 150 Brin. Cold drawn: 180,000 TS; 125,000 YS; 10 El; 330 Brin. For chemical plant equipment, welded structures, tanks; Type 304; stainless, austenitic. W. Nr. 4301.

ZAPP 80T
Robert Zapp Werkstofftechnik GmbH
C 0-0.12, Cr 18, Ni 9.5, Ti = 4 x C, bal Fe.
Annealed: 85,000 TS; 35,000 YS; 55 El; 65 RA; 150 Brin. Cold drawn: 95,000 TS; 60,000 YS; 40 El; 60 RA; 185 Brin. For welded structures, chemical plant equipment; Type 321; stainless, austenitic. W. Nr. 4541.

ZAPP 80U
Robert Zapp Werkstofftechnik GmbH
C 0.12, Ni 9.5, Cr 18, S 0.2, bal Fe.
Annealed: 80,000 TS; 35,000 YS; 40 El; 55 RA; 160 Brin. For screw machine products, bolts; Type 302; stainless, free-cutting. W. Nr. 4305.

ZAPP 80X
Bochum Stahlwerke AG
C 0-0.12, Cr 18, Ni 9.5, Cb = 8 x C, bal Fe.
Annealed: 90,000 TS; 35,000 YS; 56 El; 65 RA; 160 Brin. Cold drawn: 100,000 TS; 65,000 YS; 40 El; 60 RA; 202 Brin. For welded structures, chemical plant equipment; Type 347; stainless, austenitic. *Obsolete*

ZAPP 82
Bochum Stahlwerke AG
C 0.15, Cr 18, Ni 9.5, Mo 2, bal Fe.
Annealed: 85,000 TS; 35,000 YS; 50 El; 65 RA; 160 Brin. Cold drawn: 150,000 TS; 135,000 YS; 6 El; 300 Brin. For acid resistant equipment, mixers, agitators; Type 316; stainless, austenitic. *Obsolete*

ZAPP 82 NB
Robert Zapp Werkstofftechnik GmbH
C 0-0.1, Cr 17.5, Mo 2.3, Ni 12, Nb/Ta, bal Fe.
Stabilized austenitic stainless steel. For welded chemical equipment. Werkstoff Nr. 4580.

ZAPP 82 T
Robert Zapp Werkstofftechnik GmbH
C 0-0.1, Cr 17.5, Mo 2.3, Ni 12, Ti, bal Fe.
Stabilized austenitic stainless steel. Werkstoff Nr. 4571; similar to AISI 316.

ZAPP 82P
Robert Zapp Werkstofftechnik GmbH
C 0-0.07, Cr 18, Ni 10.5, Mo 2, bal Fe.
Annealed: 80,000 TS; 35,000 YS; 55 El; 70 RA; 150 Brin. Cold drawn: 140,000 TS; 130,000 YS; 7 El; 280 Brin. For acid resistant equipment, mixers, agitators; Type 316L; stainless, austenitic. W. Nr. 4401.

ZAPP 82RT
Bochum Stahlwerke AG
C 0.05, Cr 18, Ni 10, Cu, bal Fe.
Annealed: 80,000 TS; 35,000 YS; 55 El; 70 RA; 150 Brin. For chemical plant equipment; stainless, austenitic. *Obsolete*

ZAPP 82S
Bochum Stahlwerke AG
C 0-0.1, Cr 18, Ni 9.5, Mo 2, bal Fe.
Annealed: 80,000 TS; 35,000 YS; 55 El; 70 RA; 150 Brin. Cold drawn: 140,000 TS; 130,000 YS; 7 El; 280 Brin. For acid resistant equipment, mixers, agitators; Type 316; stainless, austenitic. *Obsolete*

ZAPP 82X
Bochum Stahlwerke AG
C 0-0.12, Cr 18, Ni 10.5, Mo 2, Cb = 8 x C, bal Fe.
Annealed: 85,000 TS; 35,000 YS; 50 El; 65 RA; 160 Brin. Cold drawn: 150,000 TS; 135,000 YS; 6 El; 300 Brin. For welded acid resistant equipment; Type 316Cb; stainless, austenitic. *Obsolete*

ZAPP 83 ELC
Robert Zapp Werkstofftechnik GmbH
C 0-0.03, Cr 18, Mo 2.8, Ni 13.5, bal Fe.
Low carbon austenitic steel; for chemical equipment. Werkstoff Nr. 4435; AISI 316 L.

ZAPP 83 NB
Robert Zapp Werkstofftechnik GmbH
C 0-0.1, Cr 17.5, Mo 2.8, Ni 13, Nb/Ta, bal Fe.
Stabilized austenitic stainless steel. For welded chemical equipment. Werkstoff Nr. 4583.

ZAPP 83 P
Robert Zapp Werkstofftechnik GmbH
C 0-0.07, Cr 18, Mo 2.8, Ni 13, bal Fe.
Austenitic stainless steel. For chemical equipment. Werkstoff Nr. 4436; AISI 316.

ZAPP 83 T
Robert Zapp Werkstofftechnik GmbH
C 0-0.1, Cr 17.5, Mo 2.8, Ni 13, Ti, bal Fe.
Stabilized austenitic stainless steel; for welded chemical equipment. Werkstoff Nr. 4573.

ZAPP AC14
Robert Zapp Werkstofftechnik GmbH
C 0.34, Al 1.1, Cr 1.4, bal Fe.
For oil refinery equipment; heat and creep resistant.

ZAPP BC115

Robert Zapp Werkstofftechnik GmbH
C 1.55-1.75, Si 0.25-0.4, Mn 0.2-0.4, Co 1.2-1.4, Cr 11-12, Mo 0.5-0.6, bal Fe.
Cold work tool steel. Werkstoff Nr. 1.2880.

ZAPP BC115S

Robert Zapp Werkstofftechnik GmbH
C 2-2.25, Si 0.2-0.4, Mn 0.2-0.4, Co 0.8-1.1, Cr 11-12.5, Mo 0.3-0.5, W 0.6-0.8, bal Fe.
Cold work tool steel. Werkstoff Nr. 1.2884. *Obsolete*

ZAPP BC120

Robert Zapp Werkstofftechnik GmbH
C 1.65, Cr 11.5, Co, bal Fe.
For blanking and forming dies, punches; air hardening, nondeforming. *Obsolete*

ZAPP BCVW85

Robert Zapp Werkstofftechnik GmbH
C 0.3, Cr 2.4, V 0.25, W 8.5, Co 2, bal Fe.
For hot work tools, punches, shears, extrusion rams; oil hardened, tough. *Obsolete*

ZAPP C10

Robert Zapp Werkstofftechnik GmbH
C 1.05, Mn 0.3, Cr 1, bal Fe.
For bearings, cutters, punches; water hardened, wear resistant.

ZAPP C120

Robert Zapp Werkstofftechnik GmbH
C 2.1, Cr 11.5, W 0.7, bal Fe.
For blanking and forming dies, gauges, punches; oil or air hardened, non-deforming.

ZAPP C130W

Robert Zapp Werkstofftechnik GmbH
C 0.2, Cr 13, Mn 0.3, Si 0.4, bal Fe.
Annealed: 95,000 TS; 50,000 YS; 25 El; 55 RA; 196 Brin. For valve trim, turbine blades, surgical instruments; Type 420; corrosion resistant.

ZAPP C135M

Robert Zapp Werkstofftechnik GmbH
C 0.4, Si 0.4, Mn 0.3, Cr 13, bal Fe.
Annealed: 100,000 TS; 55,000 YS; 22 El; 50 RA; 200 Brin. For valves, turbine blades, cutlery, knives; Type 420; stainless, hardenable.

ZAPP C14

Robert Zapp Werkstofftechnik GmbH
C 1.05, Cr 1.2, Mn 0.3, bal Fe.
For bearings, cutters, punches; water hardened, wear resistant.

ZAPP C16

Robert Zapp Werkstofftechnik GmbH
C 1, Cr 1.55, Mn 0.35, bal Fe.
For bearings, liners, sleeves; water hardened, wear resistant. W. Nr. 1.2067.

ZAPP C3

Robert Zapp Werkstofftechnik GmbH
C 1.4, Cr 0.3, V 0.1, bal Fe.
For blanking and forming dies, engravers' tools; water hardened, wear resistant. *Obsolete*

ZAPP C5

Bochum Stahlwerke AG
C 1.1, Cr 0.4, Mn 0.3, bal Fe.
For bearings, cutters, forming dies; water hardened, wear resistant. *Obsolete*

ZAPP C8

Robert Zapp Werkstofftechnik GmbH
C 0.9, Cr 0.8, Mn 0.3, bal Fe.
For bearings, cutters, punches; water hardened, wear resistant.

ZAPP CLWV8

Robert Zapp Werkstofftechnik GmbH
C 0.45, Cr 1.35, V 0.8, Mo 0.45, W 0.45, bal Fe.
For forging dies, upsetters, header dies; oil hardened, tough.

ZAPP CM12

Robert Zapp Werkstofftechnik GmbH
C 1-1.1, Si 0.15-0.3, Mn 1-1.2, Cr 0.7-1, bal Fe.
Cold work tool steel. Werkstoff Nr. 1.2127.

ZAPP CN14

Robert Zapp Werkstofftechnik GmbH
C 0.4-0.5, Si 0.15-0.35, Mn 0.5-0.8, Cr 1.2-1.5, Ni 1.5-1.8, bal Fe.
Cold work tool steel. Werkstoff Nr. 1.2710.

ZAPP CN15

Robert Zapp Werkstofftechnik GmbH
C 0.12-0.17, Mn 0.4-0.6, Ni 1.4-1.7, Cr 1.4-1.7, bal Fe.
Alloy carburizing steel. Werkstoff Nr. 1.2712. *Obsolete*

ZAPP CN35

Robert Zapp Werkstofftechnik GmbH
C 0.5, Cr 1.05, Ni 3.25, Mn 0.5, bal Fe.
For gears, bolts, machine tool parts; oil hardened, shock resistant.

ZAPP CN58

Robert Zapp Werkstofftechnik GmbH
C 0.1-0.17, Si 0.2-0.35, Mn 0.3-0.5, Cr 0.65-0.85, Ni 3.3-3.6, bal Fe.
Cold work tool steel. Werkstoff Nr. 1.2735.

ZAPP CN59

Robert Zapp Werkstofftechnik GmbH
C 0.1-0.17, Si 0.2-0.3, Mn 0.3-0.5, Cr 0.9-1.2, Ni 4.2-4.7, bal Fe.
Cold work tool steel. Werkstoff Nr. 1.2745.

ZAPP CNL3

Robert Zapp Werkstofftechnik GmbH
C 0.55, Cr 0.7, Mo 0.18, Ni 1.65, V 0.1, bal Fe.
For gears, bolts, crankshafts; oil hardened, shock resistant.

ZAPP CNL5

Robert Zapp Werkstofftechnik GmbH
C 0.57, Cr 0.8, Mo 0.2, Ni 1.75, V 0.1, bal Fe.
For gears, bolts, crankshafts; oil hardened, shock resistant.

ZAPP CPM 10 V

Robert Zapp Werkstofftechnik GmbH
C 2.45, Cr 5.25, Mo 1.3, V 9.75, S 0.07, bal Fe.
For blanking and forming, powder pressing tools, etc. Cold work steel; PM produced.

ZAPP CPM 9 V

Robert Zapp Werkstofftechnik GmbH
C 1.78, Cr 5.25, Mo 1.3, V 9, bal Fe.
For blanking and forming dies, powder pressing tools, etc. Cold work steel; PM produced.

ZAPP CPM REX 76

Robert Zapp Werkstofftechnik GmbH
C 1.5, Cr 3.75, W 10, Mo 5.25, V 3.1, Co 9, S 0.06, bal Fe.
For cutting tools. High speed steel; PM produced.

ZAPP CPM REX M 4

Robert Zapp Werkstofftechnik GmbH
C 1.35, Cr 4.25, W 5.75, Mo 4.5, V 4, S 0.06, bal Fe.
For cutting tools, cold extrusion tools. Type M4 high speed steel; PM produced.

ZAPP CPM T 440 V

Robert Zapp Werkstofftechnik GmbH
C 2.15, Cr 17.5, Mo 0.5, V 5.75, bal Fe.
For plastic molds, machine components. Type corrosion resistance, martensitic steel; PM produced.

ZAPP CVL 10 EXTRA

Robert Zapp Werkstofftechnik GmbH
C 0.32-0.4, Si 0.9-1.2, Mn 0.3-0.5, Cr 5-5.6, Mo 1.3-1.6, W 1.2-1.4, V 0.15-0.4, bal Fe.
Hot work tool steel. Werkstoff Nr. 1.2606.

ZAPP CVL 30

Robert Zapp Werkstofftechnik GmbH
C 0.28-0.35, Si 0.2-0.4, Mn 0.2-0.4, Cr 2.7-3.2, Mo 2-3, V 0.4-0.7, bal Fe.
Hot work tool steel. Werkstoff Nr. 1.2365.

ZAPP CVL10

Robert Zapp Werkstofftechnik GmbH
C 0.38-0.42, Si 0.9-1.2, Mn 0.3-0.5, Cr 4.8-5.8, Mo 0.8-1.4, V 0.25-0.5, bal Fe.
Hot work tool steel. Werkstoff Nr. 1.2343.

ZAPP CVL10V

Robert Zapp Werkstofftechnik GmbH
C 0.37-0.42, Si 0.9-1.2, Mn 0.3-0.5, Cr 5-5.5, Mo 1.2-1.5, V 0.9-1.1, bal Fe.
Hot work tool steel. Werkstoff Nr. 1.2344.

ZAPP CVL6

Robert Zapp Werkstofftechnik GmbH
C 0.45, Mn 0.7, Cr 1.4, Mo 0.7, V 0.3, bal Fe.
For gears, springs, bolts, shafts; oil hardened, tough.

ZAPP CVW35

Robert Zapp Werkstofftechnik GmbH
C 1.42, W, V, bal Fe.
For cutters, forming dies, tools; water or oil hardened.

ZAPP CVW40

Robert Zapp Werkstofftechnik GmbH
C 0.3, V 0.18, W 3.75, Cr 1.1, bal Fe.
For extrusion dies, liners, rams, upsetters; oil hardened, tough.

ZAPP CVW45

Robert Zapp Werkstofftechnik GmbH
C 0.3, Cr 2.35, V 0.6, W 4.25, bal Fe.
For extrusion dies, liners, rams, upsetters; oil hardened, tough.

ZAPP CVW50

Robert Zapp Werkstofftechnik GmbH
C 1.25-1.35, Si 0.2-0.3, Mn 0.2-0.4, Cr 0-0.2, W 4.7-5.2, bal Fe.
Cold work tool steel. Werkstoff Nr. 1.2453. *Obsolete*

ZAPP CVW85

Robert Zapp Werkstofftechnik GmbH
C 0.3, V 0.35, W 8.5, Cr 2.65, bal Fe.
For extrusion dies, rams, liners; oil hardened, tough.

ZAPP CVW95

Robert Zapp Werkstofftechnik GmbH
C 0.65, Mo 0.85, W 8.5, Cr 3.75, V 0.7, bal Fe.
For extrusion dies, rams, liners, cutters; high speed steel.

ZAPP CWN130

Robert Zapp Werkstofftechnik GmbH
C 0.45, Ni, Cr, W, bal Fe.
For forging dies, upsetters; oil hardened, tough. *Obsolete*

ZAPP GUSS 1.4740

Robert Zapp Werkstofftechnik GmbH
C 0.3-0.6, Si 1-2.5, Mn 0-1, Cr 16-18, bal Fe.
Heatproof steel castings. DIN G-X40 CrSi 17.

ZAPP GUSS 1.4776

Robert Zapp Werkstofftechnik GmbH
C 0.3-0.6, Si 1-2.5, Mn 0-1, Cr 27-30, bal Fe.
Heatproof steel castings. DIN G-X40 CrSi 29.

ZAPP GUSS 1.4777
Robert Zapp Werkstofftechnik GmbH
C 1.2-1.4, Si 1-2.5, Mn 0-1, Cr 27-30, bal Fe.
Heatproof steel castings. DIN G-X130 CrSi 29.

ZAPP GUSS 1.4823
Robert Zapp Werkstofftechnik GmbH
C 0.3-0.5, Si 1-2, Mn 0-1.5, Cr 26-28, Ni 3.5-5.5, bal Fe.
Heatproof steel castings. DIN G-X40 CrNiSi 27 4.

ZAPP GUSS 1.4826
Robert Zapp Werkstofftechnik GmbH
C 0.3-0.5, Si 1-2.5, Mn 0-1.5, Cr 21-23, Ni 9-11, bal Fe.
Heatproof steel castings. DIN G-X40 CrNiSi 22 9.

ZAPP GUSS 1.4837
Robert Zapp Werkstofftechnik GmbH
C 0.2-0.5, Si 1-2.5, Mn 0-1.5, Cr 24-26, Ni 11-14, bal Fe.
Heatproof steel castings. DIN G-X35 CrNiSi 25 12.

ZAPP GUSS 1.4846
Robert Zapp Werkstofftechnik GmbH
C 0.2-0.5, Si 1-2.5, Mn 0-1.5, Cr 25-28, Ni 13-16, bal Fe.
DIN G-X40 CrNiSi 26 14.

ZAPP GUSS 1.4848
Robert Zapp Werkstofftechnik GmbH
C 0.2-0.5, Si 1-2.5, Mn 0-1.5, Cr 24-27, Ni 19-21, bal Fe.
Heatproof steel castings. DIN G-X40 CrNiSi 25 20; AISI 310.

ZAPP GUSS 1.4865
Robert Zapp Werkstofftechnik GmbH
C 0.2-0.5, Si 1-2.5, Mn 0-1.5, Cr 16-19, Ni 36-39, bal Fe.
Heatproof steel castings. DIN G-X40 CrNiSi 36 16.

ZAPP K 10
Robert Zapp Werkstofftechnik GmbH
C 0.1, Cr 13, bal Fe.
Martensitic type stainless steel. Heat treated: 170-210 Brin
(700-750°C temper). Werkstoff Nr. 4006; AISI 403; 410.

ZAPP K 15
Robert Zapp Werkstofftechnik GmbH
C 0.15, Cr 13, bal Fe.
Martensitic type stainless steel. Heat treated: 190-240 Brin
(700-750°C temper). Werkstoff Nr. 4024; similar to AISI 410.

ZAPP K 15 L
Robert Zapp Werkstofftechnik GmbH
C 0.15, Cr 13, Mo 1.2, bal Fe.
Martensitic type stainless steel. Heat treated: 220-260 Brin
(600-700°C temper). Werkstoff Nr. 4119.

ZAPP K 20
Robert Zapp Werkstofftechnik GmbH
C 0.2, Cr 13, bal Fe.
Martensitic type stainless steel. Heat treated: 180-250 Brin
(700-750°C temper). Werkstoff Nr. 4021; AISI 420.

ZAPP K 40
Robert Zapp Werkstofftechnik GmbH
C 0.45, Cr 13, bal Fe.
Martensitic type stainless steel. Heat treatable to 50 Rock C
min. Werkstoff Nr. 4034.

ZAPP K 50
Robert Zapp Werkstofftechnik GmbH
C 0.5, Cr 13.5, Mo 0.5, bal Fe.
Martensitic type stainless steel. Annealed: 180-225 Brin;
hardenable to 50 Rock C. Werkstoff Nr. 4110.

ZAPP K 90 B
Robert Zapp Werkstofftechnik GmbH
C 0.9, Cr 17, Mo 0.5, Co, V, bal Fe.
Martensitic type stainless steel. Werkstoff Nr. 4535.

ZAPP K 90 L
Robert Zapp Werkstofftechnik GmbH
C 0.9, Cr 18, Mo 1.2, V, bal Fe.
Martensitic type stainless steel. Hardenable to 52 Rock C
min. Werkstoff Nr. 4112; AISI 440 B.

ZAPP K15
Robert Zapp Werkstofftechnik GmbH
C 0.15, Si 0.3-0.5, Mn 0.25-0.5, bal Fe.
Cold drawn: 72,000 TS; 60,000 YS; 22 El; 58 RA; 145 Brin.
For gears, shafts, machine tool parts; case hardening steel.

ZAPP K2 PRIMA
Bochum Stahlwerke AG
C 1.3, Si 0-0.25, Mn 0-0.25, bal Fe.
For engravers' tools, form cutters; Type W1; water hardened.
Obsolete

ZAPP K20
Robert Zapp Werkstofftechnik GmbH
C 0.2, Si 0.4, Cr 13, bal Fe.
Annealed: 95,000 TS; 50,000 YS; 25 El; 55 RA; 195 Brin. Cold
drawn: 105,000 TS; 85,000 YS; 17 El; 50 RA; 215 Brin. For
valves, cutlery, turbine blades, surgical instruments; Type
420; stainless.

ZAPP K20L
Robert Zapp Werkstofftechnik GmbH
C 0.2, Mo 1.15, Cr 13, bal Fe.
Annealed: 95,000 TS; 50,000 YS; 25 El; 55 RA; 195 Brin. For
valves, cutlery, pumps, turbine blades; Type 420 Mo;
stainless. W. Nr. 4120.

ZAPP K20N
Robert Zapp Werkstofftechnik GmbH
C 0.22, Cr 17, Ni 1.5, bal Fe.
Annealed: 125,000 TS; 95,000 YS; 20 El; 55 RA; 260 Brin. For
pumps, marine hardware, valves; Type 431; heat and
corrosion resistant. W. Nr. 4057.

ZAPP K3 EXTRA
Bochum Stahlwerke AG
C 1.1, Si 0-0.25, Mn 0-0.25, bal Fe.
Annealed: 110,000 TS; 56,000 YS; 18 El; 40 RA; 210 Brin. For
springs, drills, taps, cutters; Type W1; water hardened.
Obsolete

ZAPP K3 PRIMA
Bochum Stahlwerke AG
C 1.1, Si 0-0.25, Mn 0-0.25, bal Fe.
Annealed: 110,000 TS; 56,000 YS; 18 El; 40 RA; 210 Brin. For
springs, tools, cutters, drills, taps; Type W1; water hardened.
Obsolete

ZAPP K35L
Robert Zapp Werkstofftechnik GmbH
C 0.35, Mo 1.15, Cr 16.5, bal Fe.
Annealed: 130,000 TS; 95,000 YS; 18 El; 50 RA; 270 Brin. For
oil refinery equipment, marine hardware; heat and corrosion
resistant. W. Nr. 4122.

ZAPP K4 EXTRA
Robert Zapp Werkstofftechnik GmbH
C 1, Si 0-0.25, Mn 0-0.25, bal Fe.
Annealed: 110,000 TS; 56,000 YS; 18 El; 40 RA; 210 Brin. For
springs, tools, cutters, taps, reamers; Type W1; water
hardened. W. Nr. 1.1543.

ZAPP K4 PRIMA
Robert Zapp Werkstofftechnik GmbH
C 1, Si 0-0.25, Mn 0-0.25, bal Fe.
Annealed: 110,000 TS; 56,000 YS; 18 El; 40 RA; 210 Brin. For
springs, tools, cutters, taps, reamers; Type W1; water
hardened. W. Nr. 1.1645.

ZAPP K45
Robert Zapp Werkstofftechnik GmbH
C 0.45, Si 0.3-0.5, Mn 0.3-0.8, bal Fe.
Hot rolled: 98,000 TS; 60,000 YS; 24 El; 45 RA; 212 Brin. For
gears, bolts, machine tool parts; water hardened.

ZAPP K5 EXTRA
Robert Zapp Werkstofftechnik GmbH
C 0.85, Si 0-0.25, Mn 0-0.25, bal Fe.
Heat treated: 190,000 TS; 145,000 YS; 10 El; 30 RA; 400 Brin.
For springs, tools, cutters, drills, taps, hobs; Type W1; water
hardened. W. Nr. 1.1525.

ZAPP K5 EXTRA
Bochum Stahlwerke AG
C 0.85, Si 0-0.25, Mn 0-0.25, bal Fe.
Heat treated: 190,000 TS; 145,000 YS; 10 El; 30 RA; 400 Brin.
For springs, tools, cutters, drills, taps, hobs; Type W1; water
hardened. *Obsolete*

ZAPP K5 PRIMA
Robert Zapp Werkstofftechnik GmbH
C 0.85, Si 0-0.25, Mn 0-0.25, bal Fe.
Heat treated: 190,000 TS; 145,000 YS; 10 El; 30 RA; 400 Brin.
For springs, tools, cutters, drills, taps, hobs; Type W1; water
hardened. W. Nr. 1.1625. *Obsolete*

ZAPP K5 PRIMA
Bochum Stahlwerke AG
C 0.85, Si 0-0.25, Mn 0-0.25, bal Fe.
Heat treated: 190,000 TS; 145,000 YS; 10 El; 30 RA; 400 Brin.
For springs, tools, cutters, drills, taps, hobs; Type W1; water
hardened. *Obsolete*

ZAPP K55S
Robert Zapp Werkstofftechnik GmbH
C 0.55, Si 0.1-0.4, Mn 0.5-0.7, bal Fe.
Annealed; 100,000 TS; 55,000 YS; 15 El; 22 RA; 180 Brin. For
gears, bolts, machine tool parts; water hardened. *Obsolete*

ZAPP K6 EXTRA
Robert Zapp Werkstofftechnik GmbH
C 0.7, Si 0-0.25, Mn 0-0.25, bal Fe.
Heat treated: 175,000 TS; 128,000 YS; 12 El; 37 RA; 355 Brin.
For rails, springs, hammers, axes, punches; Type W1; water
hardened. *Obsolete*

ZAPP K6 PRIMA
Robert Zapp Werkstofftechnik GmbH
C 0.7, Si 0-0.25, Mn 0-0.25, bal Fe.
Heat treated: 175,000 TS; 128,000 YS; 12 El; 37 RA; 355 Brin.
For rails, hammers, springs, axes, punches; Type W1; water
hardened. *Obsolete*

ZAPP K60
Robert Zapp Werkstofftechnik GmbH
C 0.6, Si 0.1-0.4, Mn 0.5-0.7, bal Fe.
Heat treated: 160,000 TS; 115,000 YS; 12 El; 40 RA; 325 Brin.
For gears, shafts, fasteners; water hardened.

ZAPP K60G
Bochum Stahlwerke AG
C 0.6, Si 0.1-0.4, Mn 0.5-0.7, bal Fe.
Heat treated: 160,000 TS; 115,000 YS; 12 El; 40 RA; 325 Brin.
For gears, shafts, fasteners, machine tool parts; water
hardened. *Obsolete*

ZAPP K90G
Bochum Stahlwerke AG
C 0.9, Si 0.25-0.5, Mn 0.3-0.8, bal Fe.
Heat treated: 190,000 TS; 145,000 YS; 10 El; 30 RA; 400 Brin.
For drills, taps, punches, cutters, reamers; Type W1; water
hardened. *Obsolete*

ZAPP LC 120
Robert Zapp Werkstofftechnik GmbH
C 1.65, Cr, Mo, V, bal Fe.
For blanking and piercing dies, punches; air hardened.

ZAPP LC115S
Robert Zapp Werkstofftechnik GmbH
C 0.9-1, Si 0.2-0.4, Mn 0.2-0.4, Cr 11-12, Mo 0.85-0.95, V
0.85-0.95, bal Fe.
Cold work tool steel. Werkstoff Nr. 1.2376.

ZAPP LC12

Robert Zapp Werkstofftechnik GmbH
C 0.9-1.1, Si 0.15-0.3, Mn 0.2-0.4, Cr 1.1-1.3, bal Fe.
Cold work tool steel. Werkstoff Nr. 1.2303. *Obsolete*

ZAPP LC120S

Robert Zapp Werkstofftechnik GmbH
C 1.5-1.6, Si 0.3-0.5, Mn 0.3-0.5, Cr 11.5-12.5, Mo 0.6-0.8, V 0.9-1.1, bal Fe.
Cold work tool steel. Werkstoff Nr. 1.2379.

ZAPP LC40

Robert Zapp Werkstofftechnik GmbH
C 0-0.07, Si 0-0.2, Mn 0-0.2, Cr 3.5-4, Mo 0.3-0.6, bal Fe.
Cold work tool steel. Werkstoff Nr. 1.2341. *Obsolete*

ZAPP LCN

Robert Zapp Werkstofftechnik GmbH
C 0.55, Cr 0.7, Mo 0.18, Ni 1.65, V 0.1, bal Fe.
For header and forging dies; oil hardened, tough.

ZAPP LCN EXTRA

Robert Zapp Werkstofftechnik GmbH
C 0.5, Cr 0.8, Mo 0.2, Ni 1.8, V 0.1, bal Fe.
For header and forging dies, punches; oil hardened, tough.

ZAPP LCN SUPRA

Robert Zapp Werkstofftechnik GmbH
C 0.5-0.6, Si 0.15-0.35, Mn 0.6-0.8, Cr 0.9-1.2, Mo 0.7-0.9, Ni 1.5-1.8, V 0.7-0.12, bal Fe.
Hot work tool steel. W.-Nr. 1.2744. *Obsolete*

ZAPP LCN35E

Robert Zapp Werkstofftechnik GmbH
C 0.16-0.22, Si 0.15-0.3, Mn 0.3-0.5, Cr 1.1-1.4, Ni 3.8-4.3, Mo 0.15-0.25, bal Fe.
Cold work tool steel. W.-Nr. 1.2764.

ZAPP LCN40

Robert Zapp Werkstofftechnik GmbH
C 0.32-0.38, Si 0.15-0.3, Mn 0.4-0.6, Cr 1.2-1.5, Mo 0.2-0.4, Ni 3.8-4.3, bal Fe.
Hot work tool steel. W.-Nr. 1.2766.

ZAPP LCN45

Robert Zapp Werkstofftechnik GmbH
C 0.4-0.5, Si 0.15-0.3, Mn 0.3-0.5, Cr 1.2-1.5, Ni 3.8-4.3, Mo 0.15-0.35, bal Fe.
Cold work tool steel. W.-Nr. 1.2767 Mo.

ZAPP LCW 50

Robert Zapp Werkstofftechnik GmbH
C 0.4-0.5, Si 0.3-0.5, Mn 0.3-0.5, Cr 4-5, Co 4-5, Mo 0.4-0.6, W 4-5, V 1.8-2.1, bal Fe.
Hot work tool steel. W.-Nr. 1.2678. *Obsolete*

ZAPP LVC25

Robert Zapp Werkstofftechnik GmbH
C 0.26-0.34, Si 0.15-0.35, Mn 0.4-0.7, Cr 2.3-2.7, Mo 0.15-0.25, V 0.1-0.2, bal Fe.
W.-Nr. 1.2307.

ZAPP LVC50

Robert Zapp Werkstofftechnik GmbH
C 0.9-1.05, Si 0.2-0.4, Mn 0.4-0.7, Cr 4.8-5.5, Mo 0.9-1.25, V 0.1-0.3, bal Fe.
Cold work tool steel. W.-Nr. 1.2363.

ZAPP M 180

Robert Zapp Werkstofftechnik GmbH
C 0-0.15, Cr 12, Mo 0.5, Ni 2, Mn 18, bal Fe.
Manganese austenitic stainless steel. W.-Nr. 4451.

ZAPP MC13E

Robert Zapp Werkstofftechnik GmbH
C 0.2, Mn 1.25, Cr 1.15, bal Fe.
For gears, cams, camshafts; case hardening steel.

ZAPP MC9

Bochum Stahlwerke AG
C 1, Cr 1.1, Mn 0.07, Si 0.25, bal Fe.
For bearings, bushings, liners; water hardened. *Obsolete*

ZAPP MCL3

Robert Zapp Werkstofftechnik GmbH
C 0.4, Cr, Mn, Mo, bal Fe.
For gears, bolts, crankshafts, fasteners; oil hardened, tough.

ZAPP MCW10

Robert Zapp Werkstofftechnik GmbH
C 1.05, Cr 1, W 1.15, Mn 0.9, bal Fe.
For bearings, cutters, form tools; water hardened.

ZAPP MCWV

Robert Zapp Werkstofftechnik GmbH
C 0.9-1.05, Si 0.15-0.35, Mn 1-1.2, Cr 0.5-0.7, W 0.5-0.7, V 0.05-0.15, bal Fe.
Cold work tool steel. Werkstoff Nr. 1.2510.

ZAPP MS10

Robert Zapp Werkstofftechnik GmbH
C 0.53, Si 0.9, Mn 0.9, bal Fe.
For gears, bolts, crankshafts, shears, punches; oil hardened, tough.

ZAPP MVC12

Robert Zapp Werkstofftechnik GmbH
C 0.61, Si 0.85, Mn 0.75, Cr 1.18, V 0.1, bal Fe.
For springs, dies, upsetters, shears; oil hardened, shock resistant.

ZAPP P300

Robert Zapp Werkstofftechnik GmbH
C 0.5-0.57, Si 0.15-0.3, Mn 0.4-0.5, Cr 0.5-0.7, Ni 2.5-3, bal Fe.
Cold work tool steel. Werkstoff Nr. 1.2718.

ZAPP REX T 15

Robert Zapp Werkstofftechnik GmbH
C 1.55, Cr 4, W 12.25, V 5, Co 5, S 0.06, bal Fe.
For cutting tools, cold extrusion tools, Type T15 high speed steel; PM produced.

ZAPP SC10

Robert Zapp Werkstofftechnik GmbH
C 0.9, Cr 1.2, Si 1.15, Mn 0.7, bal Fe.
For blanking and forming dies; oil hardened, wear resistant.

ZAPP SC12

Robert Zapp Werkstofftechnik GmbH
C 1.25, Si 1.15, Mn 0.7, Cr 1.2, bal Fe.
For blanking and forming dies; oil hardened, wear resistant.

ZAPP SC5

Robert Zapp Werkstofftechnik GmbH
C 0.67, Cr 0.5, Mn 0.5, Si 1.3, bal Fe.
For springs, shears, chisels, punches; oil hardened, tough.

ZAPP SCW 20M

Robert Zapp Werkstofftechnik GmbH
C 0.45, W, Cr, V, bal Fe.
For extrusion liners and rams, upsetters; hot work tools, oil hardened.

ZAPP SCW20H

Robert Zapp Werkstofftechnik GmbH
C 0.55, Si 0.9, Cr 1.05, V 0.2, W 1.85, bal Fe.
For forging and die casting dies; oil hardened, tough.

ZAPP SSB 100

Robert Zapp Werkstofftechnik GmbH
C 0.76, Co 10, Cr 4.2, Mo 0.8, V 1.8, W 18, bal Fe.
For lathe and planer tools, form tools; high speed steel.

ZAPP SSB30

Robert Zapp Werkstofftechnik GmbH
C 0.86, Co 2.8, Cr 4.3, Mo 0.85, V 2.1, W 12, bal Fe.
For lathe and planer tools, reamers, drills, taps; high speed steel.

ZAPP SSB50

Robert Zapp Werkstofftechnik GmbH
C 0.79, Co 4.75, Cr 4.3, Mo 0.75, V 1.55, W 18, bal Fe.
For lathe and planer tools, reamers, broaches, taps; high speed steel.

ZAPP SSL25

Robert Zapp Werkstofftechnik GmbH
C 0.95, Cr 4, W, Mo, V, bal Fe.
For lathe and planer tools, reamers, broaches; high speed steel.

ZAPP SSV25

Robert Zapp Werkstofftechnik GmbH
C 0.86, Cr 4.1, Mo 0.85, V 2.5, W 12, bal Fe.
For lathe and planer tools, drills, reamers, taps; high speed steel.

ZAPP SSV50

Robert Zapp Werkstofftechnik GmbH
C 1.3, Cr 4.3, Mo 0.85, V 3.8, W 12, bal Fe.
For blanking and forming dies, cutters, reamers; high speed steel.

ZAPP SSVB50

Robert Zapp Werkstofftechnik GmbH
C 1.35, Cr 4, W, Co, V, Mo, bal Fe.
For blanking and forming dies; high speed steel.

ZAPP SSW180

Robert Zapp Werkstofftechnik GmbH
C 0.74, Cr 4.3, V 1.1, W 18.5, bal Fe.
For lathe and planer tools, drills, reamers, taps; high speed steel.

ZAPP SSW85

Robert Zapp Werkstofftechnik GmbH
C 0.82, Cr 4.1, Mo 0.85, V 1.6, W 8.7, bal Fe.
For lathe and planer tools, reamers, broaches; high speed steel.

ZAPP SSWL50

Robert Zapp Werkstofftechnik GmbH
C 0.85, Cr 4, W, Mo, V, bal Fe.
For lathe and planer tools, reamers, drills, taps; high speed steel.

ZAPP SVC14H

Robert Zapp Werkstofftechnik GmbH
C 0.45, Si, Cr, V, bal Fe.
For springs, bolts, gears, crankshafts; oil hardened, shock resistant.

ZAPP SVC14W

Robert Zapp Werkstofftechnik GmbH
C 0.45, Si, Cr, V, bal Fe.
For springs, bolts, gears, crankshafts; oil hardened, shock resistant.

ZAPP SVC25

Robert Zapp Werkstofftechnik GmbH
C 0.38, Si, Cr, V, bal Fe.
For springs, bolts, gears, crankshafts; oil hardened, shock resistant.

ZAPP VC10

Robert Zapp Werkstofftechnik GmbH
C 0.5, Cr 1.05, V 0.1, Mn 0.95, bal Fe.
For gears, bolts, crankshafts; oil hardened, shock resistant.

ZAPP VC12

Robert Zapp Werkstofftechnik GmbH
C 0.55-0.62, Si 0.15-0.35, Mn 0.8-1.1, Cr 0.9-1.2, bal Fe.
Cold work tool steel. Werkstoff Nr. 1.2242.

ZAPP VC120
Robert Zapp Werkstofftechnik GmbH
C 1.65, Cr 11.5, V 0.1, bal Fe.
For blanking and forming dies, punches; air hardened, nondeforming.

ZAPP VC120S
Robert Zapp Werkstofftechnik GmbH
C 2.1-2.3, Si 0.2-0.4, Mn 0.2-0.4, Cr 11.5-12.5, Mo 0.85-0.95, V 2.1-2.3, bal Fe.
Cold work tool steel. Werkstoff Nr. 1.2378. *Obsolete*

ZAPP VC4
Robert Zapp Werkstofftechnik GmbH
C 0.75-0.85, Si 0.25-0.4, Mn 0.3-0.5, Cr 0.4-0.7, V 0-0.15, bal Fe.
Cold work tool steel. Werkstoff Nr. 1.2235. *Obsolete*

ZAPP VC5
Robert Zapp Werkstofftechnik GmbH
C 1.15, Cr 0.65, V 0.1, Mn 0.3, bal Fe.
For bearings, blanking and forming dies; oil hardened, wear resistant.

ZAPP VK2 SUPRA
Robert Zapp Werkstofftechnik GmbH
C 1.4-1.5, Si 0.2-0.35, Mn 0.3-0.5, V 3-3.5, bal Fe.
Cold work tool steel. Werkstoff Nr. 1.2838. *Obsolete*

ZAPP VK4 EXTRA
Robert Zapp Werkstofftechnik GmbH
C 1, V 0.1, Mn 0.25, Si 0.2, bal Fe.
For heading and blanking dies, bearings, cutters; Type W2; water hardened.

ZAPP VK4 SUPRA
Robert Zapp Werkstofftechnik GmbH
C 0.9-1, Si 0.2-0.35, Mn 0.3-0.5, V 0.35-0.45, bal Fe.
Cold work tool steel. W.-Nr. 1.2835.

ZAPP VM20
Robert Zapp Werkstofftechnik GmbH
C 0.9, Mn 1.9, V 0.1, Si 0.25, bal Fe.
For punches, forming dies, shear blades, upsetters; oil hardened, nondeforming.

ZAPP VW10
Robert Zapp Werkstofftechnik GmbH
C 1.2, Cr 0.2, V 0.1, W 1, bal Fe.
For forming and blanking dies; water hardened, tough.

ZAPP WC120
Robert Zapp Werkstofftechnik GmbH
C 2.1, Cr 11.5, W 0.7, bal Fe.
For blanking and forming dies, punches; oil hardened, nondeforming.

ZAPP WCN45
Robert Zapp Werkstofftechnik GmbH
C 0.4-0.5, Si 0.15-0.3, Mn 0.3-0.5, Cr 1.2-1.5, Ni 3.8-4.3, W 0.5, bal Fe.
Cold work tool steel. Werkstoff Nr. 1.2767W.

ZCR 157 (URANUS R3)
Creusot-Loire
C 0.07, Cr 15, Ni 7, Mo 2.2, Al 1.15, bal Fe.
Precipitation hardenable. Stainless: 1460-1760 N/mm² TS max; 7 El. For mechanical structures in off-shore drilling equipment. AFNOR Z8CNDA.

ZCR 173
Creusot-Loire
C 0-0.06, Mn 0-1, Si 0-1, Ni 3.5-5, Cr 15-17, Cu 2.3-3.3, Nb, bal Fe.
Stainless casting; heat treated: 365-420 Brin. Good corrosion resistance at elevated temperature. AFNOR Z4CNV 16.4.

ZCR 174
Creusot-Loire
C 0-0.06, Mn 0-1, Si 0-1, Ni 3.5-5, Cr 15-17.5, Cu 3-4.5, 0.15 Nb min, bal Fe.
Stainless casting; heat treated: 365-410 Brin. Good corrosion resistance at elevated temperature. AFNOR Z4CNVN6 16.4M.

ZE 10 A
Now MAGNESIUM ZE 10 A.

ZE41A
Fansteel/Wellman Dynamics
Magnesium. Zn 3.5-5, RE 0.75-1.75, Zr 0.4-1, bal Mg.
Magnesium casting alloy. Cast: 29,000 psi TS; 19,500 psi YS; 2.5 El; 62 Brin.

ZE41A
Various foundries
Zn 4.2, Zr 0.7, 1.2 Rare Earths, bal Mg.
T5-temper: 28,000-30,000 TS; 16,000-20,000 YS; 4-5 El. At 600°F: 11,000 TS; 8,000 YS; 43 El. Magnesium sand castings with good pressure tightness and elevated temperature properties. For parts operating at 350-600°F. ASTM B80-57T; British RZ5.

ZE41A (STANDARD ALLOY)
Dow Chemical Co.
Magnesium. Zn 3.5-5, Zr 0.4-1, 0.75-1.75 rare earths, bal Mg.
Heat treated: 30,000-32,500 psi TS; 20,000-22,000 psi YS; 5-3 El; 65-75 Brin. For structures up to 400°F service; age hardenable; pressure tight. *Obsolete*

ZEDABRONZE ZA4
Le Bronze Industriel
Zn 23, Al 5, Ni 1, Fe 2, Mn 2.5, bal Cu.
Zinc-manganese bronze, cast or wrought: 160-210 Brin. AFNOR Z23/A4; SAE 430A.

ZEDABRONZE ZA7
Le Bronze Industriel
Zn 19-25, Al 5-5.5, Ni 3, Mn 4-5, bal Cu.
Zinc-manganese bronze, cast or wrought: 200-240 Brin. SAE 430B; ASTM B138; 670.

ZEDABRONZE ZA9
Le Bronze Industriel
Zn 18-21, Al 6.5, Fe 3.5, Mn 4.5-5.5, bal Cu.
Zinc-manganese bronze, cast or wrought: 225-275 Brin. AFNOR UZ19 A6.

ZELCO
Manufacturer not listed.
Zn 83, Al 15, Cu 2.
For aluminum solder.

ZELNICKER BABBITTS
W.A. Zelnicker
Pb, Sb, bal Sn.
For bearings; Babbitt.

ZENITE
Zenith Foundry Co.
C 2.85-3.15, Si 1.6-1.8, Mn 0.8, Ni 0.65-0.85, Cu 0.15-0.35, Mo 0.2-0.4, Cr 0-0.2, bal Fe.
Cast: 45,000 TS min; 210-260 Brin. For gears, shafts, housings. High strength alloy cast iron.

ZENITH HIGH SPEED
Manufacturer not listed.
C 0.61, Mn 0.12, Cr 3.09, W 19.2, bal Fe.
For tools, high speed cutting tools; high speed steel.

ZENITH SPECIAL
Zenith Foundry Co.
Si 2-2.3, C 3.2-3.5, S 0-0.12, P 0-0.15, Mn 0.5-0.8, Ni 0.2-0.4, Cr 0.15-0.35, Mo 0.5-0.7, bal Fe.
Cast: 40,000 TS min; 212-262 Brin. For gears, shafts, housings. High strength alloy cast iron.

ZENO
Zenith Foundry Co.
C 3.3, Si, Mn, alloy, bal Fe.
For gears, shafts, machinery housings; cast iron.

ZEPHOR BRONZE
Bridgeport Rolling Mills Co.
Cu 95, Sn 4.2-5.8, P 0.03-0.35.
1/2 Hard: 68,000 TS; 55,000 YS; 28 El; Rock B 78. Hard: 81,000 TS; 75,000 YS; 10 El; Rock B 87. Spring: 100,000 TS; 80,000 YS; 4 El; Rock B 95. For diaphragms, electrical switches, washers, springs, fuse clips, fasteners. Phosphor Bronze, corrosion resistant. *Obsolete*

ZEPHYR
J.M. Ney Co.
Au 80, Pd 5.
Alloy for retention loops for acrylic facings; nonoxidizing. Fusing temperature: 2000°F.

ZERANIN
Isabellenhuette
Copper. Cu 88, Mn 6, Ge 6.
Annealed: 360 N/mm² TS. For electrical equipment and instruments. Resistance alloy. Maximum working temperature to 140°C.

ZERGAL-X3
Italian manufacture
Zn 5.1-6.1, Mg 2.1-2.9, Cu 1.2-2.2, Zr 0.18-0.4, Mn 0.1-0.3, Ti 0-0.2, bal Al.
O Temper: 40,000 TS; 24,000 YS; 10 El. T6 Temper: 80,000 TS; 72,000 YS; 7 El. For aircraft structures, mobile equipment, hydraulic systems; age-hardenable, high strength.

ZERON METAL
Manufacturer not listed.
Mn 0.8-1, C, bal Fe.
Heat treated: 70,000-80,000 TS; 55,000-60,000 YS; 16-14 El; 170-180 Brin. For wrenches, hardware, fittings; heat treated malleable cast iron.

ZETONIA
German manufacture
Pb, Sb, bal Sn.
For bearings, bushings; white antifriction.

ZEUS
German manufacture
Cu 80, Ag 20.
For safety fuses.

ZEUS EV37T
Krupp Stahl AG
alloy steel. C 0.12, Mn 0.3-0.5, P 0.04, S 0.035, Ti 0.05, bal Fe.
For electrodes; welding. *Obsolete*

ZEUS GV37 T
Krupp Stahl AG
alloy steel. C 0.12, Si 0.05, Mn 0.5, Ti 0.05, bal Fe.
For electrodes; welding. *Obsolete*

ZEVESCAL
Calumet Steel Castings Corp.
C, Cr, Mo, bal Fe.
Cast: 700 Brin. For gravel mixers and pug mill paddles. Abrasion resistant.

ZEVESCAL W
Calumet Steel Castings Corp.
C 1.8-2.4, Cr 15-17, Mo 1.5-2, bal Fe.
Annealed: 120,000 TS; 95,000 YS; 2 El; 2 RA; 200 Brin. For chemical plant equipment. Wear and abrasion resistant.

ZGS PLATINUM
Johnson Matthey
Pt 99.95, 600 ppm zirconia.
Annealed: 186.2 MPa TS; 40 El; 60 HV. Higher strength and creep resistance of platinum at higher temperatures.

1384 / WOLDMAN'S ENGINEERING ALLOYS

ZGS RHODIUM-PLATINUM
Johnson Matthey
Pt 89.95, Rh 10, 600 ppm zirconia.
Annealed: 344.7 MPa TS; 30 El; 110 HV. Higher strength and creep resistance at elevated temperatures.

ZH62A
Various foundries
Zn 5.7, Zr 0.7, Th 1.8, bal Mg.
T5-temper: 35,000-40,000 TS; 22,000-25,000 YS; 4-6 El. At 500°F: 14,000 TS; 10,000 YS; 30 El. Magnesium sand castings with good pressure tightness and freedom from porosity. For parts operating at 300-500°F. ASTM B80-69; AMS 4438; QQ-M-56; SAE 508.

ZH62A (STANDARD ALLOY)
Dow Chemical Co.
Magnesium. Th 1.4-2.2, Zn 4.8-6.2, 0.5 Zr min, bal Mg.
Aged: 43,500 psi TS; 26,500 psi YS; 4 El; 75 Brin. At 500°F: 14,000 psi TS; 10,000 psi YS; 30 El. For airframe castings, gear boxes, engine components; fatigue resistant to 400°F. *Obsolete*

ZHS-3
Russian manufacture
C 0.16, Cr 15, Mo 4, W 5, Ti 2, Al 2, B 0.02, bal Ni.
For gas turbine engines, buckets, and discs; heat and oxidation resistant.

ZHS-6
Russian manufacture
C 0.15, Cr 12.5, Mo 4.7, W 7, Ti 2.6, Al 5, B 0.01, bal Ni.
For gas turbine engines, buckets and discs; heat and oxidation resistant.

ZHS-6K
Russian manufacture
C 0.17, Cr 11.5, Co 5, Mo 4, W 5, Ti 2.7, Al 5.5, B 0.01, bal Ni.
For gas turbine engines, buckets and discs; heat and oxidation resistant.

ZHS3
Russian manufacture
C 0.14, Cr 15, Ti 2, Al 2, W 5.5, B 0-0.02, bal Ni.
Cast nickel-base superalloy for nozzle guide vanes.

ZI-184
American manufacture
C 0.8-1, Cr 7-9, W 4-5, V 1.1-1.5, bal Fe.
For tools, dies; oil hardening.

ZICRAL
Pechiney/Cegedur
Al alloy.
For light alloy parts; non-hardenable.

ZICRAL A-28GU
French manufacture
Cu, Mg, Zn, bal Al.
For light alloy parts; heat treatable.

ZIERAL
KM-kabelmetal AG
Mg 0.6-1.4, Si 0.6-1.6, Mn 0.6-1, Cr 0-0.3, bal Al.
Annealed: 21,000 TS; 8000 YS; 24 El. For window frames, gutters, fan blades, boats; good welding and forming properties. *Obsolete*

ZILLOY 40
Now ZILLOY 45.

ZIMAL
Birmingham Aluminum Casting Co.
Zinc. Al 4, Cu 3, bal Zn.
Cast die: 45,000 TS; 36,000 YS; 1 El; 100 Brin. For carburetors, gears, corrosion resistant.

ZIMALIUM
German manufacture
Mg 3.7-7.1, Zn 2.8-4.5, bal Al.

ZINC
Atomergic Chemetals Corp.
Zn.
Purities, zone refined: 99.9999%, 99.9995%, 99.999%, 99.99%. Forms: rod, ingot, shot, powder, sheet, foil, wire, splatters, semicircular bar, single crystals.

ZINC BABBITT
English manufacture
Zn 69, Sn 26, Cu 5, Sb 3.
For bearings; anti-friction.

ZINC DURALUMIN ("E" ALLOY)
English manufacture
Zn 20, Cu 2.5, Mg 0.5, Mn 0.5, bal Al.
Heat treated: 91,000 TS. For strong light alloy parts; non-hardenable.

ZINC-RICH PRIMED
United States Steel Corp.
Low C, Mn, bal Fe.
Prepainted steel sheet.

ZINCDIE
Columbia Tool Steel Co.
C 0.3, Mn 0.8, Si 0.5, Mo 0.4, Cr 1.65, bal Fe.
Hardened: C 40-48 Rock. For die casting dies, plastic molds, compression and plunger type molds. Type P20, mold and die steel. *Obsolete*

ZINCGRIP
Armco
See ARMCO GAINEX and FORMABLE series.

ZINELL
Gulf & Western Industries
C 0.4, bal Fe.
For forgings; water hardened.

ZINKAL M
Zinkberatungsstelle GmbH
Mn 0-1, bal Zn.
Rolled: 21,300-35,500 TS; 25-40 Brin. For structures.

ZINKALIUM
English manufacture
Mg 0.8-8.3, Zn 0.8-8.3, bal Al.
For light alloy parts; non-hardenable.

ZINNAL
Vereinigte Silberhammerwerke Hetzel
Al sheet coated on both sides with Sn and rolled so as to form a firmly welded whole. For instrument cases; corrosion resistant.

ZINNBRONZE SNBZ4
VDM Nickel-Technologie AG
Sn 4, P 0.3, bal Cu.
Hard: 71,600-92,000 TS; 20-35 El; 140-170 Brin. Soft: 46,000-60,000 TS; 55-65 El; 70-80 Brin. For paper and chemical plant equipment; corrosion resistant P-bronze. *Obsolete*

ZINNBRONZE SNBZ6
VDM Nickel-Technologie AG
Sn 6, P 0.3, bal Cu.
Hard: 71,600-92,000 TS; 20-35 El; 150-175 Brin. Soft: 50,000-64,000 TS; 50-65 El; 75-85 Brin. For paper and chemical plant equipment; corrosion resistant P-bronze. *Obsolete*

ZINNBRONZE SNBZ8
VDM Nickel-Technologie AG
Sn 8, P 0.2, bal Cu.
Hard: 78,000-100,000 TS; 20-35 El; 160-190 Brin. Soft: 58,000-72,000 TS; 45-60 El; 80-90 Brin. For paper and chemical plant equipment; corrosion resistant P-bronze. *Obsolete*

ZINTEC
British Steel plc
Electro-zinc coated steel; mild steel.

ZIP
Agawam Tool Co.
C 0.65, Cr 4, Co 7.5, V 1.65, W 17.5, Mo 0.85, bal Fe.
For lathe, planer and shaper tools; high speed steel.

ZIP
Ziv Steel & Wire Co.
C 0.75, W 19, Co 7, Cr 4.5, V 2, Mo 1, bal Fe.
For reamers, broaches, lathe and planer cutters, hobs; Type T5; high speed steel.

ZIP ZIP
Hardite Metals Inc.
C 0.6, Cr 0.75, W 1.8, bal Fe.
For tools, chisels, screw drivers, pneumatic tools; oil hardened. *Obsolete*

ZIP ZIP
Jessop Steel Co.
C 0.6, Cr 0.75, W 1.8, bal Fe.
For tools, chisels, screw drivers, pneumatic tools; oil hardened. *Obsolete*

ZIPPALLOY
Seymour Products Co.
Cu 87, bal Zn.
1/2 H-temper: 54,000 TS; 116 Brin. Hard temper: 65,000 TS; 137 Brin. Spring: 75,000 TS; 160 Brin. For electrical components, hardware, jewelry, watches, lamps; high brass, corrosion resistant.

ZIRCADYNE 702
Teledyne Wah Chang Albany
Hf 4.5, H 0.005, N 0.025, C 0.05, O 0.16, 99.2 Zr + Hf min, 0.20 Fe + Cr max.
379 MPa TS min; 207 MPa YS min; 16 El min (0.2% offset). Unalloyed; for chemical processing industry; corrosion resistant. ASTM R60702.

ZIRCADYNE 704
Teledyne Wah Chang Albany
Hf 4.5, Sn 1-2, H 0.005, N 0.025, C 0.05, O 0.18, 97.5 Zr + Hf min, 0.20-0.40 Fe + Cr.
413 MPa TS min; 241 MPa YS min; 14 El min (0.2% offset). Similar to nuclear grade alloys. ASTM R60704.

ZIRCADYNE 705
Teledyne Wah Chang Albany
Hf 4.5, H 0.005, N 0.025, C 0.05, Nb 2-3, O 0.18, 95.5 Zr + Hf min, 0.20 Fe + Cr max.
552 MPa TS min; 379 MPa YS min; 16 El min (0.2% offset). Similar corrosion resistance to ZIRCADYNE 702, but higher strength. ASTM R60705.

ZIRCADYNE 706
Teledyne Wah Chang Albany
Hf 4.5, H 0.005, N 0.025, C 0.05, Nb 2-3, O 0.16, 95.5 Zr + Hf min, 0.20 Fe + Cr max.
510 MPa TS min; 345 MPa YS min; 20 El min (0.2% offset). For severe forming applications such as panel type heat exchangers. ASTM R60706.

ZIRCAL
French manufacture
Zn 7-8.5, Mg 2.5, Cu 1.5, Cr 0.25, bal Al.
Rolled: 78,000-92,000 TS; 64,000-71,000 YS. Wire: 97,000 TS; 92,000 YS; 6 El; 180 Brin. For light alloy parts; age-hardenable.

ZIRCALLOY-2
Chase Brass & Copper Co., Inc.
Sn 1.5, O 0.12, Fe 0.13, Cr 0.1, Ni 0.05, bal Zr.
Extruded and drawn: 65,000 TS; 47,000 YS; 25 El. Low neutron cross section. For pressure tubes in nuclear reactors.

ZIRCALOY 2

CEZUS
Refractory. Sn 1.2-1.7, Fe 0.07-0.2, Cr 0.05-0.15, Ni 0.03-0.08,
bal Zr.

ZIRCALOY 3

Westinghouse Electric Corp.
C 0-0.05, Sn 0.2-0.3, Fe 0.2-0.3, bal Zr.
Room temperature: 61,100 psi TS; 44,200 psi YS; 36.5 RA;
165 Brin. At 500°F: 26,400 psi TS; 16,700 psi YS; 43.0 RA.
For high temperature applications; good strength to 750°F.
Obsolete

ZIRCALOY 4

CEZUS
Refractory. Sn 1.2-1.7, Fe 0.18-0.24, Cr 0.07-0.13, bal Zr.

ZIRCALOY-2

Westinghouse Electric Corp.
Sn 1.2-1.7, Fe 0.07-0.2, Cr 0.05-0.15, Ni 0.03-0.08, Fe + Ni +
Cr = 0.18 to 0.38.

ZIRCALOY-2

Sandvik/Coromant
Sn 1.5, Fe 0.12, Cr 0.1, Ni 0.05, bal Zr.
At 70°F: 76,000 TS; 49,000 YS; 28 RA; 185 Brin. At 500°F:
42,000 TS; 24,000 YS; 42 RA. For nuclear reactors; low
neutron absorption.

ZIRCALOY-2

Westinghouse Electric Corp.
Sn 1.5, Fe 0.12, Cr 0.1, Ni 0.05, bal Zr.
At 70°F: 76,000 TS; 49,000 YS; 28 RA; 185 Brin. At 500°F:
42,000 TS; 24,000 YS; 42 RA. For nuclear reactors; low
neutron absorption.

ZIRCALOY-4

Chase Brass & Copper Co., Inc.
Sn 1.5, O 0.12, Fe 0.21, Cr 0.1, bal Zr.
Extruded and drawn: 65,000 TS; 47,000 YS; 25 El. Low
neutron cross section. For pressure tubes in nuclear reactors.

ZIRCALOY-4

Westinghouse Electric Corp.
Sn 1.5, Fe 0.2, Cr 0.1, bal Zr.
Bar: 80,000-115,000 TS; 40,000-105,000 YS; 3-20 El; 82-100
Rock B. For reactors, fuel sheathing, structural components
in pressurized water reactors. Corrosion resisting, high
strength.

ZIRCALOY-4

Universal Cyclops
Sn 1.5, Fe 0.2, Cr 0.1, bal Zr.
Bar: 80,000-115,000 TS; 40,000-105,000 YS; 3-20 El; 82-100
Rock B. For reactors, fuel sheathing, structural components
in pressurized water reactors. Corrosion resisting, high
strength.

ZIRCALOY-4

Martin Marietta Aluminum Inc.
Sn 1.5, Fe 0.2, Cr 0.1, bal Zr.
Bar: 80,000-115,000 TS; 40,000-105,000 YS; 3-20 El; 82-100
Rock B. For reactors, fuel sheathing, structural components
in pressurized water reactors. Corrosion resisting, high
strength.

ZIRCALOY-4

Westinghouse Electric Corp.
Sn 1.2-1.7, Fe 0.18-0.24, Cr 0.07-0.13, Fe + Cr = 0.28 to
0.37.

ZIRCOGRAF

Pechiney Electrometallurgie
Miscellaneous nonferrous. Al 1.3, Ca 1.5, Mn 6, Si 66, Zr 6,
bal Fe.
Cast iron inoculant.

ZIRCONIUM

Chase Brass & Copper Co., Inc.
Hf 4, bal Zr.
Annealed: 60,000 TS; 35,000 YS; 20 El. Corrosion resistant.
Tube for chemical processing.

ZIRCONIUM

Atomergic Chemetals Corp.
Zr.
Purities, zone refined: 99.995+%, crystal bar 99.95%, reactor
grade 99.6+%, commercial grade. Forms: sponge, rod, bar,
powder, wire, plate, sheet, foil, tubing, single crystals.

ZIRCONIUM

Vallourec S.A.
95.0 Zr min.
Resistant to acids. For chemical industries.

ZIRCONIUM COPPER 150

Anaconda Co.
Copper. Cu 99.83, Zr 0.17.
Heat treated: 60,000 TS; 50,000 YS; 12 El; 75 Rock B. For
commutators, collector copper rings, canned motor windings,
rectifier bases. Heat treatable, corrosion resistant. 93%
electrical conductivity.

ZIRCONIUM DISILICIDE

NL Industries
Zr 52.7, Si 35.5, Al 8, Fe 1.5.
For special alloy applications; cermet. *Obsolete*

ZIRCONIUM GR. 11

Scientific Products Inc.
N 0.01, 0.2 Fe + Cr, 99.5 Zr + Hf min.
165 Brin max. For chemical process applications. Corrosion
resistant.

ZIRCONIUM GR. 12

Scientific Products Inc.
N 0.01, 0.05 Fe + Cr, 99.5 Zr + Hf min
150 Brin. For chemical plant process equipment. High
corrosion resistance.

ZIRCONIUM GR. 32

Scientific Products Inc.
Sn 1.5, Fe 0.14, Cr 0.1, Ni 0.05, bal Zr.
Annealed: 70,000 TS; 43,000 YS; 28 El; 150 Brin. At 1000°F:
18,000 TS; 10,000 YS; 45 El. Cold reduced 60%: 110,000 TS;
102,000 YS; 35 RA. For nuclear reactors, fuel sheathing,
boiling and pressurized water reactors. Resists irradiated
water corrosion.

ZIRCONIUM GR. 34

Scientific Products Inc.
Sn 1.5, Fe 0.21, Cr 0.1, bal Zr.
Annealed: 70,000 TS; 43,000 YS; 28 El. At 500°F: 35,000 TS;
120 DPH. For nuclear reactors, fuel sheathing, structural
components. Resists irradiated water corrosion.

ZIRCONIUM STEEL

English manufacture
Zr 0.1-0.6, C 0.2-0.6, Mn, Si, P, S, bal Fe.
For machinery parts, shafts; water hardened.

ZIRCONIUM-2.5 NIOBIUM

Chase Brass & Copper Co., Inc.
Nb 2.6, O 0.11, bal Zr.
Extruded and drawn: 100,000 TS; 70,000 YS; 12 El. Low
neutron cross section. For pressure tubes in nuclear reactors.

ZIRCONIUM-AJR

Metropolitan-Vickers Electrical Co. Ltd.
Cu 1, Mo 1.5, bal Zr.
For nuclear reactors. Resists CO_2 corrosion at high
temperatures.

ZIRCONIUM-ATR

Metropolitan-Vickers Electrical Co. Ltd.
Cu 0.5, Mo 0.5, bal Zr.
At 600°F: 45,000 TS; 42,000 YS; 24 El; 84 Rock B. For
nuclear reactors, fuel sheathing, reactor structural
components. Resists CO_2 corrosion at high temperatures.

ZIRCONIUM-COPPER 992

Olin Brass, Indianapolis
Zr 0.13, bal Cu.
Annealed: 30,000 psi TS; 10,000 psi YS; 45 El; 40 Brin. Aged:
60,000 psi TS; 55,000 psi YS; 15 El; 116 Brin. For resistance
welding electrodes, grid wires, and contacts; heat treatable;
high conductivity. *Obsolete*

ZIRCONIUM-COPPER N-4 (NIPPERT ALLOY N-4)

Nippert Co.
Zr 0.25, bal Cu.
At 75°F: 53,000 TS; 50,500 YS; 10 El; 54 RA; 115 Brin. At
550°F: 42,000 TS; 40,500 YS; 8 El; 37 RA; 63 Brin. For
commutators; high strength and conductivity at high
temperature. Same as "AMZIRC".

ZIRKONAL

German manufacture
Si 0.5, Cu 15, Mn 8, bal Al.
Rolled: 170-190 Brin. For light alloy parts; not hardenable.

ZIRMET

Foote Mineral Co.
Zr 99.9.
Cold rolled: 111,300 TS; 77,500 YS. Annealed: 73,000 TS;
29,800 YS. For purifying gases; ductile, absorbs gases,
corrosion resistant. *Obsolete*

ZIRTEN

Japan Steel Works Ltd.
C 0-0.16, Si 0.35-0.65, Mn 0.3-1.2, P 0-0.12, Cu 0.25-0.55, Ni
0-0.5, Cr 0.4-0.8, Zr 0-0.15, bal Fe.
Plate: 70,000 TS min; 50,000 YP min; 18 El min. For railroad
and mine cars, bridges, booms, chutes. Corrosion resisting
to atmosphere. High strength low alloy steel plates.

ZIRTUNG

GTE Products Corp.
W, Zr.
For welding rod for inert gas shielded arc welding.

ZISIUM

English manufacture
Al 82-83, Cu 1-3, Zn 15, Sn 0-1.
For strong light alloy parts; non-hardenable.

ZISKON

Manufacturer not listed.
Al 60, Zn 40.
For light alloy parts; non-hardenable.

ZISKON

Manufacturer not listed.
Al 25-33, Zn 67-75.
For castings; non-hardenable.

ZIV EXTRA

Ziv Steel & Wire Co.
C 1.1, Mn 0.3, Si 0.2, V 0.2, bal Fe.
For forming dies, shear blades, reamers, taps; water
hardened; Type W2.

ZIV SPECIAL

Ziv Steel & Wire Co.
C 1.1, Mn 0.3, V 0.2, Cr 0.2, bal Fe.
For tools, taps, reamers, shear blades; water hardened.

ZIV SUPER

Ziv Steel & Wire Co.
C 0.7, W 18, Cr 4, V 1, bal Fe.
Hardened: 64-66 Rock C. For drills, hobs, reamers, broaches,
taps, lathe and planer tools. Type T1 high-speed steel.

ZIV'S COBALT
Ziv Steel & Wire Co.
C 0.7, W 19, Cr 4.5, V 2, Co 5, bal Fe.
For broaches, reamers, gear cutters, hobs, taps; Type T4; high speed steel.

ZIV'S REGULAR
Ziv Steel & Wire Co.
C 0.95, Si 0.2, Mn 0.25, bal Fe.
For cold chisels, shear knives, punches, drills; water hardened; Type W1.

ZIV'S SOLID DRILL
Ziv Steel & Wire Co.
C, bal Fe.
For mining drills; water hardened.

ZIVAN 45
Ziv Steel & Wire Co.
C 0.45, Mn 0.75, Cr 0.95, V 0.16, bal Fe.
For machinery parts; oil hardening.

ZIVCO
Ziv Steel & Wire Co.
C 0.95-1.05, Mn 0.5, Si 0.2, bal Fe.
For sledges, picks, drills, taps; Type W1; water hardened.

ZIVCO VANADIUM
Ziv Steel & Wire Co.
Tool material. C 1-1.1, V 0.15-0.25, bal Fe.
AISI Type W2 water hardening tool steel.

ZIVS BLOCK TESTED HOLLOW DRILL
Ziv Steel & Wire Co.
C 0.7-0.9, bal Fe.
For mining drills; water hardened. *Obsolete*

ZK 51
French manufacture
Zn 4.5, Zr 0.7, bal Mg.
Cast: 34,000-38,000 TS; 23,000-26,000 YS; 6-8 El. AMS 4443; SAE 509.

ZK-NICKEL
Vereinigte Deutsche Nickel-Werke AG
Mn, Cu, bal Ni.
For spark plug electrodes. *Obsolete*

ZK51A
Various foundries
Zn 4.5, Zr 0.7, bal Mg.
T5-Temper: 34,000-40,000 TS; 20,000-24,000 YS; 5-8 El. At 600°F: 8,000 TS; 6,000 YS; 16 El. Magnesium sand castings with good corrosion resistance; recommended for simple, highly stressed parts of uniform cross section. ASTM B80-69; AMS 4443; SAE 509; QQ-M-56.

ZK51A (STANDARD ALLOY)
Dow Chemical Co.
Magnesium. Zn 3.6-5.5, Mn 0-0.15, 0.55 Zr min, bal Mg.
Cast: 33,000 psi TS; 19,000 psi YS; 11 El; 60 Brin. For airframe structures, landing wheels, high yield strength casting alloy. *Obsolete*

ZK60A
Dow Chemical Co.
Magnesium. Zn 4.8-6.2, 0.45 Zr min, bal Mg.
F-temper: 43,000-49,000 psi TS; 31,000-38,000 psi YS; 5-14 El; 75 Brin. T5-temper: 45,000-53,000 psi TS; 36,000-44,000 psi YS; 4-14 El; 82 Brin. Extrusions for highly stressed parts, primarily in aircraft and military; also used as forgings. ASTM B107-69, B91-68; AMS 4352, 4362.

ZK61A
Dow Chemical Co.
Zn 6, Zr 0.8, bal Mg.
Casting alloy.

ZL 80 (1)
Pechiney Electrometallurgie
Miscellaneous nonferrous. Al 1.4, Ca 2.6, Si 66, Zr 1.6, bal Fe.
Cast iron inoculant.

ZL 80 (2)
Pechiney Electrometallurgie
Miscellaneous nonferrous. Al 1.4, Ca 2.6, Si 75, Zr 1.6, bal Fe.
Cast iron inoculant.

ZN AL-1 ALLOY
German manufacture
Al 0.7-0.9, Cu 0.35-0.5, bal Zn.
For conductors for power transmission; high conductivity.

ZN-FE ALLOY
German manufacture
Fe 0.13, bal Zn.
Rolled: 23,000-25,000 TS; 25-40 Brin. For conductors for power transmission.

ZNCUBE
Manufacturer not listed.
Cu, Be, bal Zn.
For hardware, lamp and fuse sockets; high strength. *Obsolete*

ZO2, 7.2 MIN DENSITY
Keystone Carbon Co.
Sintered steel manufactured to AISI 4630 composition. As sintered: 75 Rock B. Heat treated: 35 Rock C; 160,000 psi TS. For highly stressed structural parts; hydraulic pump gears.

ZO4, HIGH DENSITY
Keystone Carbon Co.
Sintered iron. 6.91-7.29 density; 30,000 psi TS; 35 Rock B; hardenable to 51,000 psi TS; 65 RA. For heavier loaded structural parts. ASTM B310-83A, Type IV, Class A.

ZO4, LOW DENSITY
Keystone Carbon Co.
Sintered iron. 5.7-6.1 density; 16,000 psi TS; 50 Rock H. For lightly loaded structural parts. ASTM B310-83A, Type I Class A; SAE 850.

ZO4, MEDIUM DENSITY
Keystone Carbon Co.
Sintered iron. 6.11-6.50 density; 20,000 psi TS; 60 Rock H. For lightly loaded structural parts. ASTM B310-83A, Type II, Class A; SAE 853.

ZO4, MEDIUM HIGH DENSITY
Keystone Carbon Co.
Sintered iron. 6.51-6.90 density; 26,000 psi TS; 30 Rock B; hardenable to 60 RA. For medium loaded structural parts. ASTM B310-83A, Type III, Class A.

ZODIAC
Henry Wiggin & Co. Ltd.
Ni 20, Cu 64, Zn 16.
84,000 TS. For shunts, rheostats, resistance wire; nickel silver. *Obsolete*

ZOLLNER Z132
Manufacturer not listed.
Si 11-13, Cu 0.5-2.75, Mg 0.7-1.3, Ti 0-0.15, Fe 0-1.3, bal Al.
Cast: 31,000 min TS; 90-120 Brin. For pistons; age hardenable.

ZORITE
Michiana Products Corp.
Ni 35, Cr 15-17, Mn 1.75, C 0.5, Si 1, bal Fe.
Cast: 65,000-55,000 TS; 45,000-40,000 YS; 2-5 El; 4-6 RA; 190-180 Brin. For furnaces, heaters, electrical resistances; heat and corrosion resisting; maximum working temperature 1900°F.

ZOVM
Thyssen Edelstahlwerke AG
C, Cr, Mo, V, bal Fe.
12 El. For machinery parts; oil hardened. *Obsolete*

ZR CU
CEZUS
Refractory. Cu 1.6-2, bal Zr.

ZR CU MO (ATR)
CEZUS
Refractory. Cu 0.5, Mo 0.5, bal Zr.

ZR NB
CEZUS
Refractory. Nb 2.4-2.8, bal Zr.

ZRE 1
English manufacture
Zn 0.8-3, Zr 0.4-1, 2.5-4.0 Rare Earths, bal Mg.
Sand cast and precipation treated: 140 MPa TS; 95 MPa YS; 3 El; 50 Brin. BS 2970 MAG 6.

ZT 1
English manufacture
Zn 1.7-2.5, Zr 0.4-1, Th 2.5-4, bal Mg.
Sand cast and precipitation treated: 185 MPa TS; 85 MPa YS; 5 El; 50 Brin. BS 2970 MAG 8.

ZUNIT 13
Bochum Stahlwerke AG
C 0-0.12, Si 2.2, Cr 13, bal Fe.
Annealed: 75,000 TS; 40,000 YS; 35 El; 70 RA; 155 Brin. For valves, cutlery, oil refinery equipment; corrosion resistant. *Obsolete*

ZUNIT 13A
Robert Zapp Werkstofftechnik GmbH
C 0-0.12, Si 1.2, Al 1, Cr 13, bal Fe.
Annealed: 75,000 TS; 40,000 YS; 35 El; 70 RA; 155 Brin. For oil refinery equipment, valves; creep and heat resistant.

ZUNIT 17
Now ZAPP GUSS 1.4740.

ZUNIT 18
Bochum Stahlwerke AG
C 0-0.12, Si 2, Cr 18, bal Fe.
Annealed: 80,000 TS; 50,000 YS; 25 El; 50 RA; 150 Brin. For valves, pumps, marine hardware, furnace parts; corrosion and heat resistant. *Obsolete*

ZUNIT 18A
Robert Zapp Werkstofftechnik GmbH
C 0-0.12, Si 1, Al 1, Cr 18, bal Fe.
Annealed: 80,000 TS; 50,000 YS; 25 El; 50 RA; 150 Brin. For valves, oil refinery equipment; heat and creep resistant.

ZUNIT 24A
Robert Zapp Werkstofftechnik GmbH
C 0-0.12, Al 1.5, Cr 24, bal Fe.
Annealed: 85,000 TS; 50,000 YS; 30 El; 55 RA; 180 Brin. For oil refinery equipment, valves, pumps; creep and heat resistant.

ZUNIT 25N
Robert Zapp Werkstofftechnik GmbH
C 0.2, Si 1.2, Cr 25, Ni 4, bal Fe.
Cast: 90,000 TS; 65,000 YS; 2 El; 212 Brin. Heat treated: 97,000 TS; 65,000 YS; 18 El; 210 Brin. For cylinder liners, bushings, valve seats and bodies; corrosion and heat resistant; Type CC-20.

ZUNIT 28
Now ZAPP GUSS 1.4776.

ZUNIT 28K
Now ZAPP GUSS 1.4777.

ZUNIT 28N
Now ZAPP GUSS 1.4873.

ZUNIT 6
Bochum Stahlwerke AG
C 0-0.12, Si 2.3, Cr 6, bal Fe.
For oil refinery equipment; creep and heat resistant.
Obsolete

ZUNIT N10
Robert Zapp Werkstofftechnik GmbH
C 0.15, Cr 19.5, Ni 9.5, Si 2, bal Fe.
Cast: 85,000 TS; 45,000 YS; 35 El; 165 Brin. For heat treating boxes, baskets, burner tips, conveyors; Type HF; corrosion and heat resistant.

ZUNIT N10
Now ZAPP GUSS 1.4826.

ZUNIT N11
Now ZAPP GUSS 1.4837.

ZUNIT N15
Now ZAPP GUSS 1.4846.

ZUNIT N20
Robert Zapp Werkstofftechnik GmbH
C 0.15, Si 2, Cr 24, Ni 19, bal Fe.
Cast: 75,000 TS; 50,000 YS; 17 El; 170 Brin. Aged: 85,000 TS; 50,000 YS; 10 El; 190 Brin. For furnace parts, retorts, rabble arms, dampers; Type HK; corrosion and heat resistant.

ZUNIT N20
Now ZAPP GUSS 1.4848.

ZUNIT N30
Now ZAPP GUSS 1.4865.

ZUNIT N8
Robert Zapp Werkstofftechnik GmbH
C 0-0.15, Si 0-1, Mn 0-20, Cr 17-19, Ni 9-11, bal Fe.
Austenitic stainless; heat resisting. W.-Nr. 1.4878.

ZUNIT ST
Robert Zapp Werkstofftechnik GmbH
C 0-0.1, Si 1, Mn 0-1, Ti = 5 x C, bal Fe.
Ferritic low carbon rolling and forging steel; heat resisting. W.-Nr. 1.5310.

ZUNIT-GUSS 17
Robert Zapp Werkstofftechnik GmbH
C 0.5, Si 1.5, Cr 17, bal Fe.
For furnace parts, heat treating boxes, retorts; corrosion and heat resistant.

ZUNIT-GUSS 22
Bochum Stahlwerke AG
C 0.6, Si 1.5, Cr 22, bal Fe.
For grates, baffles, support skids, furnace parts; heat resistant. *Obsolete*

ZUNIT-GUSS 28
Robert Zapp Werkstofftechnik GmbH
C 0.6, Si 1.5, Cr 29, bal Fe.
For grates, baffles, support skids, furnace parts; heat resistant.

ZUNIT-GUSS 28K
Robert Zapp Werkstofftechnik GmbH
C 1.3, Si 1.5, Cr 29, bal Fe.
For dies, crusher parts; heat resistant.

ZUNIT-GUSS 28N
Robert Zapp Werkstofftechnik GmbH
C 0.4, Si 1.3, Cr 27, Ni 4, bal Fe.
Cast: 70,000 TS; 65,000 YS; 2 El; 190 Brin. Aged: 115,000 TS; 80,000 YS; 18 El. For furnace parts, salt pots, soot blowers, grate bars; Type HC; heat resistant.

ZUNIT-GUSS N10
Robert Zapp Werkstofftechnik GmbH
C 0.4, Si 1.3, Cr 27, Ni 4, bal Fe.
Cast: 70,000 TS; 65,000 YS; 2 El; 190 Brin. Aged: 115,000 TS; 80,000 YS; 18 El. For furnace parts, salt pots, soot blowers, grate bars; Type HC; heat resistant.

ZUNIT-GUSS N11
Robert Zapp Werkstofftechnik GmbH
C 0.4, Si 2, Cr 28, Ni 11, bal Fe.
Cast: 70,000 TS; 65,000 YS; 2 El; 190 Brin. Aged: 115,000 TS; 80,000 YS; 15 El. For furnace parts, salt pots, baffles, grids; Type HE; corrosion and heat resistant.

ZUNIT-GUSS N15
Robert Zapp Werkstofftechnik GmbH
C 0.4, Si 2, Cr 26, Ni 14, bal Fe.
Cast: 75,000 TS; 47,000 YS; 17 El; 25 RA; 200 Brin. For heat treating boxes, furnace parts, fixtures, grates; Type HH; corrosion and heat resistant.

ZUNIT-GUSS N20
Robert Zapp Werkstofftechnik GmbH
C 0.4, Si 2, Cr 26, Ni 4, bal Fe.
Cast: 70,000 TS; 65,000 YS; 2 El; 190 Brin. Aged: 115,000 TS; 80,000 YS; 15 El. For furnace parts, salt pots, baffles, blowers; Type HC; corrosion and heat resistant.

ZUNIT-GUSS N30
Robert Zapp Werkstofftechnik GmbH
C 0.5, Si 1.8, Cr 25, Ni 30, bal Fe.
For furnace parts, heat treating boxes, retorts; corrosion and heat resistant.

ZURCO GALVACOTE
Zurbach Steel Co.
C 0.2, bal Fe.
For structures; galvanized sheet.

ZURCO NO. 18
Zurbach Steel Co.
Cr 16-18, C 0.1, bal Fe.
Annealed: 80,000 TS; 55,000 YS. For non-corrosive parts; rustless steel; for continuous heat up to 1200°F.

ZURCO NO. 25
Zurbach Steel Co.
Cr 25-30, C 0.1, bal Fe.
For non-corrosive parts; rustless steel; for continuous heat up to 2200°F.

ZURCO RUSTLESS IRON
Zurbach Steel Co.
Cr 16-18, C, bal Fe.
For non-corrosive parts; rustless.

ZURCO RUSTLESS STEEL
Zurbach Steel Co.
C 0.1, Cr 12-14, bal Fe.
Annealed: 75,000 TS; 44,000 YS. For corrosion resisting parts; rustless, maximum operating temperature 1000°F.

ZURCO SILVATEX
Zurbach Steel Co.
C 0.2, bal Fe.
For structures; coated sheet.

ZURCO STEEL "A"
Zurbach Steel Co.
C 0.08-0.12, bal Fe.
For cold rolled or annealed pickled sheets; deep drawing.

ZURCO STEEL "AAA"
Zurbach Steel Co.
C 0.08-0.12, bal Fe.
For annealed pickled sheets for Ni and Cr plating; extra deep drawing.

ZURCO STEEL "B"
Zurbach Steel Co.
C 0.08-0.12, bal Fe.
For furniture sheets, filing cabinets, lockers; deep drawing.

ZURCO STEEL "C"
Zurbach Steel Co.
C 0.08-0.12, bal Fe.
For sheet; deep drawing.

ZW.1
Birmetals Ltd.
Zn 1.2, Zr 0.4-0.8, bal Mg.
Plate, sheet, strip, extruded sections. 28,000-34,000 TS; 14,000-22,000 Proof Stress (0.1%); 8-10 El; 55-75 VDH. Medium strength, good corrosion resistance; for structural components. *Obsolete*

ZW.3
Birmetals Ltd.
Zn 3, Zr 0.4-0.8, bal Mg.
Plate, sheet, extrusions, forgings. 30,000-40,000 TS; 16,000-28,000 Proof Stress (0.1%); 8-10 El; 60-80 VDH. High strength with excellent corrosion resistance for structural components. *Obsolete*

ZW.6
Birmetals Ltd.
Zn 5.5, Zr 0.4-0.8, bal Mg.
Extruded: 38,000-40,000 TS; 26,000-28,000 Proof Stress (0.1%); 8-10 El. High strength with good corrosion resistance, not easily formed. *Obsolete*

ZYKLON 10
Otto Wolff Handelgesellschaft
C 0.3, Cr 2.65, V 0.35, W 8.5, bal Fe.
For extrusion press parts, rams, and liners; hot work steel, oil hardened.

ZYKLON 122
Otto Wolff Handelgesellschaft
C 0.35, Cr 1.05, V 0.18, W 1.85, bal Fe.
For punches, shears, upsetters, cold headers; oil hardened, hot work steel.

ZYKLON 12C
Otto Wolff Handelgesellschaft
C 0.3, Co 2, Cr 2.4, V 0.25, W 8.5, bal Fe.
For extrusion press parts, hot punches and shears; oil hardened, hot work steel.

ZYKLON 5
Otto Wolff Handelgesellschaft
C 0.3, Cr 2.35, V 0.6, W 4.25, bal Fe.
For upsetters, extrusion press parts; oil hardened, hot work steel.

ZYKLON 5SN
Otto Wolff Handelgesellschaft
C 0.3, Cr 1.1, V 0.18, W 3.75, bal Fe.
For upsetters, extrusion press parts, rams; hot work steel, oil hardened.

ZYKLON CMV211
Otto Wolff Handelgesellschaft
C 0.4, Cr, Mn, Mo, bal Fe.
For gears, bolts, crankshafts; oil hardened.

ZYKLON CSM155
Otto Wolff Handelgesellschaft
C 0.67, Si 1.3, Mn 0.5, Cr 0.5, bal Fe.
For springs; oil hardened.

ZYKLON MNU35
Otto Wolff Handelgesellschaft
C 0.56, Cr 0.7, Mo 0.18, Ni 1.65, V 0.1, bal Fe.
For gears, pinions, bolts, studs, crankshafts; oil hardened, shock resistant.

ZYKLON MNV15
Otto Wolff Handelgesellschaft
C 0.55, Cr 0.7, Mo 0.18, Ni 1.65, V 0.1, bal Fe.
For gears, pinions, bolts, studs, crankshafts; oil hardened,
shock resistant.

MANUFACTURER LISTING

Note: Some of the addresses listed below may not be current or complete because of the inclusion of obsolete alloys from previous editions.

Abex Corp.
American Brake Shoe Co.

Abex Corp.
AMSCO Division
485 Frontage Road
Burr Ridge, IL 60521
(312)789-7700

Abex Corp.
Electro Alloys Division
Taylor Street & Abbe Road
P. O. Box 4012
Elyria, OH 44035
(216)323-3202

Acciaierie Valbruna s.p.a.
Viale della Scienza n. 25 (zona ind.)
36100 Vicenza
Italy
19000 Valbruna

Accurate Brass Co.
Bristol, CT 06010

ACENOR, S.A.
Planta de Basauri
Formerly called:
Echevarria S.A. (Aceros HEVA)
Apartado 55
Basauri (Vizcaya) 48970
Spain
416 30 00

Achorn Steel Co.
Cambridge, MA

Acieral Co. of America
New York, NY

Acieries de Champagnole
4 a 18 Rue Jules Ferry
Boite Postale No. 92
93123 Courneuve 93123
France

Acieries de Sambra & Meuse
France

Acieries du Forez
17 Rue Pierre Copel
F-42010 St. Etienne (Loire)
France

Acieries et Forges d'Anor
5 Av. de la Republique
Paris 11e
France

Acieries S. A. Bedel
F-42 Saint Etienne
Paris
France

ACIPCO Steel Products Division
American Cast Iron Pipe Co.
1501 N. 31st Avenue
Birmingham, AL 35207
(205)325-7701

Ackerlind Steel Co., Inc.
45-15 Barnett Ave.
Long Island City, NY 11104
(212)361-1545

Acme Aluminum Alloys, Inc.
Dayton, OH 45401

Acme Electric Welder Co.
Los Angeles, CA 90058

Acme Foundry Co.
Detroit, MI 48233

Acme Foundry & Machine Co.
P.O. Box 563
Coffeyville, KS 67337

Acme Steel Co.
13500 S. Perry Ave.
Riverdale, IL 60627-1182
(708)849-2500

Adamas Carbide Corp.
Market & Passaic Sts.
Kenilworth, NJ 07033
(201)241-1000

J.D. Adams & Co.
Indianapolis, IN

Adams & Osgood Steel Co.
Boston, MA 02109

Adams Hardfacing Co.
Alloy Rod Division
P.O. Box 47
Wakita, OK 73771

Adirondack Steel Casting Co.
(Reported to be out of business)
Watervliet, NY 12189

Advance Aluminum Casting Corp.
Chicago, IL 60607

Advance Foundry Co.
107 Seminary Avenue
Dayton, OH 45401
(513)253-4148

Aetna Standard Engineering Co.
321 First Street
Ellwood City, PA 16117

AFORA
Carretera de Zumarraga
Apartado 2
Azcoitia (Guipuzcoa)
Spain
81 02 00

Agawam Tool Co.
Springfield, MA 01101

Agil Chemie
Berlin
Germany

Aiken Industries Inc./Carbide Div.
1024 E. Smithfield St.
McKeesport, PA 15135

Airco Vacuum Metals
Formerly called:
Airco Welding Products
100 Mountain Ave.
New Providence, NJ 07974
(201)464-2400

Airco Welding Products
See Airco Vacuum Metals

Ajax Metal
See H. Kramer & Co.

Ajax Steel & Forge Co.

Akron Bronze & Aluminum Inc.
579 Wolf Ledges Pkwy.
Akron, OH 44309

Aktiebolaget Svenska Metallverken
Stockholm
Sweden

AL Tech Specialty Steel Corp.
P.O. Box 152
Dunkirk, NY 14048
(716)366-1000

Aladdin Welding Products Inc.
1300 Burton St., S.E.
Grand Rapids, MI 49507
(616)243-2531

Alais Forges et Camargue
Paris
France

Alamo Iron Works
P.O. Box 231
San Antonio, TX 78291
(512)223-6161

Alan Wood Steel Co.

Alar, Ltd.
London
England

Alcaloy Inc.
Cleveland, OH

Alcan Canada Products Ltd.
Aluminium Goods Div.
158 Sterling Road
Toronto 154 M6R 2B8
Canada
(416)531-9911

Alcan Metal Powders Division
Alcan Aluminum Corp.
P.O. Box 290
Elizabeth, NJ 07207
(201)353-4600

Alcan-Booth Industries, Ltd.
1 Mount Street
Berkeley Square
London W1Y 6HP
England

Alco Products
Schenectady, NY

Alcoa of Great Britain
Alcoa House, P.O. Box 15
Droitwich
Worcestershire WR9 7BG
England
09057 3441

Alcoa/Massena Operations
Massena Operations
P.O. Box 150
Massena, NY 13662

Alfa Romeo
Milan
Italy

Algoma Steel Corp. Ltd.
Sault Ste. Marie
Ontario P6A 5P2
Canada
(705)945-2351

A. Allan & Son
325 Washington Ave.
P.O. Box 54
Carlstadt, NJ 07072
(201)438-1317

Allard, Soc. Anon. Usines et Acieries
Mont-sur-Marchienne
Belgium

Allegheny Ludlum Corp./Special Materials
Special Materials Division
695 Ohio St.
Lockport, NY 14094
(716)439-5327

Allegheny Ludlum Steel
Allegheny Ludlum Corp.
1000 Six PPG Place
Pittsburgh, PA 15222-5479
(412)394-2800

L.B. Allen Co.
9329 Bernice Ave.
Shiller Park, IL 60176

Allen Mfg. Co.
Drawer 570
Hartford, CT 06101
(203)242-8511

Allen-Sherman-Hoff Pump Co.
Wynnewood, PA

Allgemeines Deutsches Metallwerk GmbH
Berlin
Germany

Allgemeine Elektrizitats-Gesellschaft
AEG Telefunken AG.
Hohenzollerndamn 150
Berlin 1000
Germany

Alliance Wall Corp.
Box 247
Alliance, OH 44601

Allied Die Casting Corp.
New York, NY

Allied Process Brass Corp.
(Reported to be out of business)
New York, NY

Allied Products Corp.
5700 W. Roosevelt Dr.
Chicago, IL 60650

Allied Steel & Chemical Co.
New York, NY

Allied Steel & Tractor Products Inc.
Amalgamated Steel Division
5800 Harper Rd.
Solon, OH 44139

Allis-Chalmers Mfg. Co.
Box 512
Milwaukee, WI 53201

Allmetal Screw Products Co., Inc.
821 Stewart Ave.
Garden City, NY 11530
(516)741-1200

Alloy Cast Products, Inc.
700 Swenson Dr.
Kenilworth, NJ 07033
(201)245-2255

Alloy Engineering & Casting Co.
1700 West Washington
P.O. Box 3066
Champaign, IL 61821
(217)398-3200

Alloy Foundries
Eastern Co.
Eastern Co.
112 Bridge St.
P.O. Box 460
Naugatuck, CT 06770
(203)729-2255

Alloy Metal Products Inc.
626 Schmidt Rd.
Davenport, IA 52808
(319)324-3511

Alloy Metals Inc.
2807 Elliott
Troy, MI 48084
(313)585-2400

Alloy Technology International, Inc.
Formerly called:
Chromalloy American Corp./Sintercast Div.
169 Western Highway
West Nyack, NY 10994

Alloys & Products, Inc.
(Reported to be out of business)
See Belmont Metals Inc.

Alluminio SA
Milan
Italy

Allyne-Ryan Foundry Co.
Cleveland, OH

Alpha Metals Inc.
600 Route 440
Jersey City, NY 07304

Alten Foundry & Machine Works
2000 W. Wheeling St.
Lancaster, OH 43130

Alumetal
Montecatini Edison
Via Alserio 22
Milano 1-20159
Italy

Aluminium Belge, S.A.
Luttich
Belgium

Aluminium Co. of Canada, Ltd.
1188 Sherbrooke St. West
Montreal, Quebec H3A 362
Canada
(514)848-8000

Aluminium Francais
23 rue Balzac
Paris 8
France

Aluminium GmbH
Rheinfelden/Baden
Germany

Aluminium Industrie Aktiengesellschaft
See Alusuisse, Zurich

Aluminium Industries, AG
Chippis
Switzerland

Aluminium Laufen AG
Switzerland

Aluminium Norf GmbH
Draht und Verseilwerk Lunen
Postfach 100353
Koblenzer Strasse
D-4040 Neuss 1
Germany

Aluminium Press-und Walzwerk
Munchenstein AG
CH-4142 Muchenstein
Switzerland

Aluminium Union Ltd.
London
England

Aluminium Walzwerke Singen GmbH
Alusingen-Platz 1
77 Singen/HTWL
Germany
0 77 31-80 0

Aluminium Wire & Cable Co., Ltd.
See British Alcan Wire Ltd.

Aluminium-Werke Wutoschingen GmbH
7896 Wutoschingen/Baden
Germany
07 746 81 0

Aluminium-Zentrale e.V.
Beratung und Information
Konigsallee 30
Postfach 12 07
D-4000 Dusseldorf 1
Germany
0211 32 08 21

Aluminiumwerke Maulbronn
Maulbronn
Wurtt
Germany

Aluminiumwerke Nurnberg, G.m.b.H.
Nurnberg
Germany

Aluminiumwerke Rorschach AG
Rorschach CH-9400
Switzerland

Aluminum Company of America
1501 Alcoa Building
425 Sixth Avenue
Pittsburgh, PA 15219
(412)553-4545

Aluminum Industries Inc.
Milwaukee Ave.
Chicago, IL

Aluminum Smelting & Refining Co., Inc.
Certified Alloys Co.
5463 Dunham Rd.
Maple Heights, OH 44137
(216)662-3100

Aluminum Solder Corp. (ALSOCO)
New York, NY

Alusuisse Deutschland GmbH
Seestr. 1
Postfach 5040
D-7750 Konstanz
Germany
07531/8 06 0

Alusuisse France SA
Division SFRM
4 Rue du Parc
F-93150 Blanc-Mesnil
France

Alusuisse Metals, Inc.
21-00 Route 208
Fairlawn, NJ 07410
(201)791-1360

Alusuisse, Zurich
Alusuisse Metals, Inc.
Formerly called:
Aluminium Industrie Aktiengesellschaft

AMAX Corp.
Climax Division
See Climax Performance Materials Corp.

AMAX Inc.
AMAX Specialty Metals Corp. Division
See Climax Specialty Metals

Ambo-Stahl-Gesellschaft
Gerhard Sevenich GmbH & Co.
Postfach 300120
D-5000 Koln 30
Germany

Ambolt Machine Tool Co.
New York, NY

AMCO

American Abrasive Metals Co.
460 Coit St.
Irvington, NJ 07111

American Alloys Corp.
3227 Gardner Avenue
Kansas City, MO 64120

American Art Alloys Inc.
Kokomo, IN

American Art Metals Inc.
New York, NY

American Brazing Alloys Co.
Pelham, NY

American Carbide Corp.
Union City, NJ

American Chain & Cable
Page Steel & Wire Div.
See Page Aluminized Steel
100 Monongahela St.
Monessen, PA 15062

American Clad Metals Inc.
Pawtucket, RI 02862

American Crucible Products Co.
1305 Oberlin Ave.
Lorain, OH 44052
(216)245-6826

American Cutting Alloys Inc.
New York, NY

American Electro Metals Corp.
Yonkers, NY

American Gage & Machine Co.
See Elgiloy Limited Partnership

American Hoist & Derrick Co.
St. Paul Foundry
57 South Robert Street
St. Paul, MN 55107
(612)293-4149

American Injector Co.
Detroit, MI

American Light Alloys Co.
1265 McBride Ave.
Little Falls, NJ 07424

American Magnesium Co.
(Reported to be out of business)
Cleveland, OH

American Manganese Bronze Co.
Philadelphia, PA

American Manganese Mfg. Co.
(Reported to be out of business)
Dunbar, PA

American Marsh Pumps Inc.
Novo Pump & Machine Division
Lansing, MI 48905

American Metallurgical Products Co.
9800 McKnight Road
Pittsburgh, PA 15239
(412)931-5040

American Meter Co.
Reliance Foundry Division
Pittsburgh, PA

American Nickeloid Co.
127 Westacre Drive
Peru, IL 61354

American Non-Gran Bronze Co.
Berwyn, PA

American Radiator Co.
New York, NY

American Saw & Mfg. Co.
301 Chestnut Street
East Longmeadow, MA 01028
(413)525-3961

American Silver Co.
Flushing, NY

American Smelting & Refining Co.
Federated Metals Division
P.O. Box 151
Perth Amboy, NJ 08862
(201)826-6200

American Solder Co.
Philadelphia, PA

American Stainless Steel Co.
Pittsburgh, PA

American Steel Co.
600 Spring St.
Ellwood City, PA 16117

American Steel Foundries
1005 Prudential Plaza
Chicago, IL 60601
(312)644-4080

American Tank & Fabricating Co.
12314 Elmwood Ave.
Cleveland, OH 44111
(216)252-1500

Amesbury Brass & Foundry Co.
Amesbury, MA

AMF Inc.
777 Westchester Ave.
White Plains, NY 10604

Ampal, Inc.
P.O. Box 2270
Flemington, NJ 08822
(201)782-5454

Ampco Metal
Ampco Pittsburgh Corp.
P.O. Box 2004
1745 South 38th Street
Milwaukee, WI 53201
(414)645-3750

Ampco Pittsburgh Corp.
Wyckoff Steel Division
700 Porter Building
Pittsburgh, PA 15219
(412)456-4400

Anaconda Co.
Brass Division
See Atlantic Richfield
414 Meadow St.
P.O. Box 830
Waterbury, CT 06720
(203)574-8500

Ancast, Inc.
3194 Townline Road
Sodus, MI 49126
(616)927-1985

Anchor Alloys Inc.
968 Meeker Ave.
Brooklyn, NY 11222

Anchor Drawn Steel Co.
See Teledyne Allvac

Thomas Andrews & Co. Ltd.
Royds Works
Sheffield S4 7WZ
England

Andrews Toledo Ltd.
(Reported to be out of business)
London
England

Antaciron, Inc.
Wellsville, NY

Apex Bronze Foundry Co.
Oakland, CA

Apex International Alloys, Inc.
Subsidiary of Alumax,Inc.
2340 Des Plaines Avenue
Des Plaines, IL 60018

Apex Steel Co.
(Reported to be out of business)
Sheffield
England

Apollo Metals Inc.
6894 S. Oak Park Ave.
Chicago, IL 60638

Apollo Steel Co.
Apollo, PA

Apothecaries Hall Co.
Waterbury, CT

Arcade Malleable Iron Co.
Worcester, MA

Arcos Alloys
Hoskins Mfg.
P.O. Box 308
Mount Carmel, PA 17851
(717)339-5200

Ardal Ltd.
Worle
Weston-Super-Mare
England

Armco
Eastern Steel Division
ARMCO Int'l
P.O. Box 600
Middletown, OH 45043
(513)425-6541

Armco Steel Co.
Sheffield Steel Division
(Reported to be out of business)
Houston, TX

Armco Steel Corp.
Rustless Iron Division
See Armco Steel
Baltimore, MD

Armstrong Bros. Tool Co.
5273 W. Armstrong Ave.
Chicago, IL 60646
(312)763-3333

Arnold Engineering Co.
Subsidiary of Allegheny Ludlum Steel Corp.
P.O. Box G
300 West St.
Marengo, IL 60152
(815)568-2000

Arpocalloy Co.
14012 Bannister Road
Kansas, MO 64139
(816)524-3437

Arwood Corp.
Main Office and Technical Center
Subsidiary of Interlake, Inc.
Rockleigh, NJ 07647
(201)767-0600

N.C. Ashton Ltd.
St. Andrew's House
P.O. Box B.24
Huddersfield HD1 6RE
England
0484-26263

Associated Electrical Industries Ltd.
Manchester
England

Associated Spring Co.
Manross Div.
18 Main St.
Bristol, CT 06010

Associated Steel Corp.
P.O. Box 7045
18200 Miles Ave.
Cleveland, OH 44128
(216)475-8000

Astro
P.O. Drawer 520
Wooster, OH 44691
(216)264-8639

Ateliers de la Gironde
Paris
France

E.C. Atkins & Co.
Indianapolis, IN

Atkinson Co.
Rochester, NY

E. Frank Atkinson & Sons Ltd.
Sheffield S18 6NS
England

Atlantic Casting & Engineering Co.
810 Bloomfield Ave.
Clifton, NJ 07012
(201)779-2450

Atlantic Richfield
Formerly called:
Anaconda Co.
414 Meadow St.
P.O. Box 830
Waterbury, CT

Atlantic Steel Casting Co.
Sixth & Lloyd Sts.
Chester, PA 19016

Atlantic Steel Co.
P.O. Box 1714
Atlanta, GA 30301
(404)897-4505

Atlantic Steel Corp.
P.O. Box 1714
Atlanta, GA 30301
(404)897-4505

Atlantic Zinc Works
New York, NY

Atlas Brass Foundry
Santa Fe at 11th St.
Los Angeles, CA 90021

Atlas Foundry & Machine Co.
3021 South Wilkeson Street
Tacoma, WA 98409-7857
(206)475-4600

Atlas Foundry Co.
(Reported to be out of business)
Irvington, NJ

Atlas Metal & Alloys Co. Ltd.
Chicago, IL

Atlas Pattern & Model Works
179 Morgan Ave.
Brooklyn, NY 11237

Atlas Specialty Steels
Sammi Atlas Inc.
Box 1000
One Centre St.
Welland, Ontario L3B 5R7
Canada
(416)735-5661

Atlas Steels Ltd.

Atomergic Chemetals Corp.
222 Sherwood Ave.
Farmingdale, NY 11735-1527
(516)694-9000

Atrax Cemented Carbide
Wallace Murray Corp.
P.O. Box 486
McKeesport, PA 15134
(412)466-8300

Aubert & Duval
Acierie des Ancizes
41 rue de Villiers
92202 Neuilly-sur-Seine
France
40 88 20 00

Audubon Metalwove Belt Corp.
P.O. Box 14722-T
Philadelphia, PA 19134

August Thyssen Co.
Berlin
Germany

Aurora Industries
Aurora Metals Division
Formerly called:
Aurora Metal Co.
1002 Greenfield Road
Montgomery, IL 60538
(708)844-4900

Auto Specialties Mfg. Co.
601 Graves St.
St. Joseph, MI 49085

Avesta AB
Formerly called:
Avesta Jernverks AB
Avesta S-774 01
Sweden
46 226 810 00

Avesta Jernverks AB
See Avesta AB

G.A. Avril Co.
Brass and Bronze Ingot Division
4445 Kings Run Drive
P.O. Box 32066
Cincinnati, OH 45232-0066

Axelson Mfg. Co.
Los Angeles, CA

Babcock & Wilcox
Power Generation Group
McDermott
20 South Van Buren Ave.
Barberton, OH 44203-0351
(216)753-4511

Babcock & Wilcox Co.
P.O. Box 401
Beaver Falls, PA 15010
(412)846-0100

Bachite Development Corp.

Badell Co. Inc.
4904 Calumet Blvd.
Hammond, IN 46327

Baker, Perkins & Co. Ltd.
Peterborough
England

BALCO
Bahrain Saudi Aluminum Marketing Co.
B.S.C.
P.O. Box 20079
Bahrain

Baldwin Steel Co.
7255 Division St.
Oakwood, OH 44146

Baldwin-Lima-Hamilton Corp.
Eddystone Division
Philadelphia, PA

Arthur Balfour & Co., Ltd.
See Darwins Alloy Castings.

E.A. Balfour Steel
Balfour Darwins Ltd.
Sheffield
England

Edgar Allen Balfour Ltd.
Imperial Steel Works
P.O. Box 10
Manchester M21 2ER
England
061 223 13

Balfour Darwins Ltd.
See Darwins Alloy Castings

F.J. Ballard & Co.
Tividale
Tipton Staff
England

Barber Asphalt Co.
Philadelphia, PA

Barium Stainless Steel Corp.
Canton, OH

Barker & Allen Ltd.
Nickel Silver Works
Heath Street South
Springhill
Birmingham B18 7PZ
England
021 452 1199

Barronia Metals, Ltd.
London
England

Barrow Haematite Steel Co., Ltd.
Barrow-in-Furness
Lancashire
England

Barworth Flockton Ltd.
Ecclesfield
Sheffield S30 3XH
England

Battelle Memorial Institute
Cobalt Information Center
505 King Ave.
Columbus, OH 43201

Batterium Metal & Vislok, Ltd.
Harborough
England

Bausch Machine Tool Co.
Springfield, MA

Joseph Beardshaw & Son Ltd.
(Reported to be out of business)
Acme Steel Works
30-36 Burton Road
Sheffield, Yorkshire S3 8DJ
England

Bearium Metals Corp.
1171 Chili Ave.
Rochester, NY 14624
(716)235-5360

Beaumont Birch Co.
1505 Race St.
Philadelphia, PA 19102

Becker Bros. Carbon Corp.
3450 S. Laramie St.
Cicero, IL 60650

Becker Stahlwerk AG
Krefeld
Rhineland
Germany

Bedford Tool & Forge Co.
P.O. Box 46096-T
Bedford, OH 44146

Bell Telephone Laboratories
600 Mountain Ave.
Murray Hill, NJ 07974
(201)582-3000

Belle City Malleable Co.
(Reported to be out of business)
Racine, WI

Belmont Metals Inc.
Formerly called:
Belmont Smelting & Refining Works, Inc.
200 Expressway South
Brooklyn, NY 11207

Bendix Steel Co.
1612 Enterprise Pkwy.
Twinsburg, OH 44087

Alexander Benecke, Inc.
New York, NY

Benedict-Miller, Inc.
Marin Avenue & Orient Way
P.O. Box 912, Dept. T/R
Lyndhurst, NJ 07071
(201)438-3000

Bergische Stahl Industrie
5630 Remscheid 1
West Germany

Bergmann Elektrizitawerke
Wibmersdorfer Strasse 39
Charlottenburg
Berlin D-1000
Germany

Paul Bergsoe & Son
Glostrup
Denmark

Bergstrom Alloys Corp.
New York, NY

Berry Metal Co.
Harmony, PA 16037

Berted Foundry Co.
Columbiana, OH

Bethlehem Foundry & Machine Co.
Bethlehem, PA

Bethlehem Steel Corp.
Bethlehem, PA 18016
(215)694-2424

Bidault-Elion SA
Paris F-75000
France

Billington & Newton Ltd.
Longport
Staffordshire
England

Billiton International Metals B.V.
PO Box 436
2260 Akleidschendam
Netherlands

Binney Castings Co.
2745 Avondale Ave.
Toledo, OH 43607

Birdsboro Corp.
Pennsylvania Engineering Corp.
Furnace St.
Birdsboro, PA 19508
(215)582-2011

Birkett, Billington & Newton Ltd.
Hanley
England

Birmetals Ltd.
Birmabright Works
Clapgate Lane, Quinton
Birmingham B32 3BX
England
021-422-2253

Birmetals Ltd., Birmabright Ltd.
Woodgate Works
Quinton
Birmingham, 32
England

Birmingham Aluminium Casting Co.
Birmid Works, Dartmouth Road
Smethwick, Warley
West Midlands B66 1BW
England
021-558-1431

Birmingham Battery & Metal Co., Ltd.
P.O. Box 320
Selly Oak
Birmingham, 29
England

Bishop Tube Co.
Route 30 & Main Rd.
Frazer, PA 19355
(215)647-3450

Bissett Steel Co.
945 E. 67th St.
Cleveland, OH 44103
(216)391-2000

Black Drill Co., Inc.
1400 E. 222 St.
Cleveland, OH 44117
(216)531-2202

Black-Clawson Co.
605 Clark St.
Middletown, OH 45042

Blackalloy Company of America, Inc.
10 Vesper St.
P.O. Box 2157
Paterson, NJ 07509
(201)742-1234

Blackor Co.
Los Angeles, CA

Blackstone Corp.
Jamestown Malleable Iron Division
1111 Allen St.
Jamestown, NY 14701
(716)665-2620

Blackwell's Metallurgical Works
26 Park Road
Lytham St. Annes
Lancashire
England

Bleiwerk Goslar GmbH & Co. KG
Besserer & Ernst
Postfach 1220
Im Schleeke 8
D-3380 Goslar
Germany
05321 754-0

Bliss & Laughlin Steel Co.
281 East 155th Street
Harvey, IL 60426
(312)264-1800

BNF Metals Technology Centre
Wantage Business Park
Denchworth Road
Oxfordshire, Wantage OX12 9BJ
England
0235 772992

Bochum Stahlwerk AG
See SWB Stahlformguss Gesellschaft mbH

Bochumer Verein
See Krupp Stahl AG
Bochum
Germany

Bofors AB
Nobel Industries
S-69180 Bofors
Sweden
46 586-810 00

Bohler Gesellschaft m.b.H.
Formerly called:
Vereinigte Edelstahlwerke
Mariazeller Strasse 25
Postfach 96
A-8605 Kapfenberg
Austria
03862 20 0

Bohn Aluminum & Brass Corp.
Gulf & Western Mfg. Co.
23100 Providence Drive
P.O. Box 999
Southfield, MI 48075

Bohn Bearing Div.
BOHN/ADEC Group
Wickes Mfg. Co.
P.O. Box 319
P.O. Box 999
Greensburg, IN 47240
(812)663-3401

Bohn Extruded Products Div.
Gulf and Western Mfg. Corporation
365 W. 24th Street
Holland, MI 49423

H. Boker & Co.
(Reported to be out of business)
New York, NY

Boiler Development Corp.
Boiler Industries Div.
720 East Fourth St.
Marion, IN 46952

Thomas Bolton Ltd.
Formerly called:
Thomas Bolton & Sons Ltd.
P.O. Box 1
Froghall
Stoke-on-Trent ST10 2HF
England
053-84-2241

Bonney-Floyd Co.
See Shenango Co., Columbus Plant
Columbus, OH 43207

Booth Aluminium Ltd., James
See Alcan Booth
Kitts Green
Birmingham 33
England

Borg-Warner Automotive Powdered Metals Corp.
Formerly called:
Michigan Powdered Metal Products Inc.
32059-T Schoolcraft Road
Livonia, MI 48150

Borolite Corp.
Niagara Falls, NY

Robert Bosch Metallwerk GmbH
Postfach 10 60 50
7000 Stuttgart 10, Stuttgart
West Germany
0711 811-0

Bowsteel Distributors Corp.
P.O. Box 164
Linden, NJ 07037

Boyd-Wagner Co.
Chicago, IL

Bradley & Foster Ltd.
Staffordshire
England

Bradley Laboratories
London
England

Braeburn Alloy Steel
See CCS Braeburn Alloy Steel
River Rd.
Lower Burrell, PA 15068
(412)224-6000

Braun-Steeples Co. Ltd.
San Francisco, CA

Breda Co.
Rome
Italy

Bresciana Metallurgica
Rome
Italy

Bridgeport Brass Corp.
See Olin Brass, Indianapolis

Bridgeport Rolling Mills Co.
Subsidiary of Atco Industries, Inc.
1 Bruce Ave.
Stratford, CT 06497
(203)375-4434

Brighton Electric Steel Casting
510 45th Street
P.O. Box 206
Beaver Falls, PA 15010
(412)846-7377

British Alcan Wire Ltd.
British Alcan Aluminium Plc
Formerly called:
Aluminium Wire & Cable Co., Ltd.
Port Tennant
Swansea SA1 8PS
Wales
0792-52251

British Aluminium Co., Ltd.
Chalfont Technological Center
Chalfont Park, Gerralds Cross
Buckinghamshire SL9 0QB
England
87373

British Driver-Harris Co. Ltd.
Manchester
England

British General Electric Co. Ltd.
London
England

British Insulated Callender's Cables Ltd.
(Reported to be out of business)
Prescot
England

British Metal Corp., Ltd.
Princes House
93 Gresham St.
London EC 2
England

British Oxygen Co., Ltd.
London
England

British Piston Ring Co.
Coventry
England

British Rolling Mills Ltd.
Tipton
Staffordshire
England

British Standards Institution
101 Pentonville Road
London N1 9ND
England
01-837-8801

British Steel Corp.
P.O. Box 35
Bridge Street
Sheffield S3 8AZ
England
0742 26401

British Steel plc
9 Albert Enbankment
London SE1 7SN
England
44 01735 7654

British Thomson Houston Co., Ltd.
Rugby
Warwickshire
England

Brockhouse Casting Co.
Wolverhampton
England

Bronze Die Casting Co.
Franklin St. at Ohio River
Pittsburgh, PA 15233

Brooks & Perkins Inc.
Advanced Structures Division
12633 Inkster Road
Livonia, MI 48150
(313)522-2000

Brown & Sharpe Mfg. Co.
140 Waterman Ave.
Precision Park
North Kingstown, RI 02852
(401)884-3000

Brown Alloy Works
Detroit, MI

David Brown Foundries Co.
Penistone
Near Sheffield
England

Brown-Wales Co.
Formerly called:
Horace Potts
165 Rindge Ave.Ext.
Cambridge, MA 02140
(617)864-4300

Brukokoncernen AB
Fagersta
Sweden

Brunswick Technetics, ECS
2000 Brunswick Lane
DeLand, FL 32724-9990
(904)736-1700

Brush Wellman
Engineered Materials
17876 St. Clair Ave.
Cleveland, OH 44110
(216)486-4200

Brush Wellman Corp.
Pennrold Division
P.O. Box 973
Reading, PA 19603

Buckeye Brass & Mfg. Co.
6410 Hawthorne Ave.
Cleveland, OH 44103

Buffalo Brake Beam Co.
Acme Steel & Malleable Iron Works Div.
39 Chandler Street
Buffalo, NY 14207

Buffalo Bronze Die Casting Co.
(Reported to be out of business)
Buffalo, NY

Buffalo Wire Works Co. Inc.
P.O. Box 129
Buffalo, NY 14240
(716)826-4666

Bull's Metal & Marine Ltd.
Glasgow
Scotland

Bundy Corp.
Bundy Tubing Div.
12345 E. 9 Mile Road
Warren, MI 48090
(313)758-4511

Bunting Bearings Corp.
Formerly called:
Markey Bronze Corp.
200 Van Buren St.
Delta, OH 43515
(419)822-3456

Burden Iron Co.
(Reported to be out of business)
Troy, NY

Burdett Oxygen Co.
See Burdox Inc.
3300 Lakeside Ave.
Cleveland, OH 44114

Burdox Inc.
Formerly called:
Burdett Oxygen Co.
3300 Lakeside Ave.
Cleveland, OH 44114

Burgess-Parr Inc.
(Reported to be out of business)
Exchange & Liberty St.
Freeport, IL 61032

Burndy Corp.
5 Richards Ave.
Norwalk, CT 06856
(203)838-4444

Burys & Co. Ltd.
Sheffield
England

Busch-Jager Ludenscheider Metallwerke
Ludenscheid
Germany

A.M. Byers Co.
(Reported to be out of business)
Ambridge, PA

Byrwill Co.
Cincinnati, OH

Cabot Corporation
See Haynes International, Inc.

A.W. Cadman Mfg. Co.
2816 Smallman St.
Pittsburgh, PA 15222
(412)281-6683

Calloy Ltd.
Avonmouth
England

Caloriz Corp. of Great Britain Ltd.
London
England

Calorizing Co.
Pittsburgh, PA

Calumet & Hecla, Universal Oil Products
Wolverine Tube Division
Allen Park, MI 48101

Calumet Steel Castings Corp.
1636 Summer Street
P.O. Box 727
Hammond, IN 46325
(219)932-4406

Cambridge Wire Cloth Co.
P.O. Box 399
Cambridge, MD 21613
(301)228-3000

Cameron & Son Ltd.
(Reported to be out of business)
Sheffield
England

Canada Electric Steel Castings Ltd.
Orillia, Ontario
Canada

Canadian Quebec Metallurgical Corp.
Quebec
Canada

Cannon-Muskegon Corp.
2875 Lincoln St.
Box 506
Muskegon, MI 49443
(616)755-1681

Cannon-Stein Steel Co.
(Reported to be out of business)
Syracuse, NY

Capito & Klein

Capitol Castings Inc.
Formerly called:
Midland Ross Corp.
P.O. Box 27328
Tempe, AZ 85285-7328
(602)839-3000

Capper Pass & Son Ltd.
North Ferriby
North Humberside
England

Carbidie
Formerly called:
Aiken Industries Inc./Carbide Div.
P.O. Box 509
Arona Rd.
Irwin, PA 15642-0509
(412)864-5900

Carbone Loraine Industries Corporation
Pechiney Corp.
P.O. Box 89
Boonton, NJ 07005
(201)334-0700

Carborundum Co.
345 Third St.
P.O. Box 156
Niagara Falls, NY 14303
(716)278-2000

G.O. Carlson Inc.
P.O. Box 526
350 Marshallton Rd.
Thorndale, PA 19372
(215)384-2800

Carmet Materials
Allegheny Ludlum Industries
1100 Mandoline
Madison Heights, MI 48071
(313)566-3190

Carobronze Ltd.
School Road
Belmont Road
London W4
England

Carondelet Foundry Co.
2101 South Kingshighway Blvd
St. Louis, MO 63110
(314)771-0906

Carpenter Technology Corp.
P.O. Box 14662
Reading, PA 19612-4662
(215)371-2000

Richard W. Carr & Co., Ltd.
Pluto Works
Wadsley Bridge
Sheffield S6 1LL
England
349451 STD0742

Charles Carr Ltd.
Birmingham
England

Donald Carroll Metals, Inc.
(Reported to be out of business)
201 N. Division St.
Bensenville, IL 60106

Castalloy Co., Inc.
West Central Street
Natick, MA

Castings Corp.
Brewster, NY 10509

Castings Development Co.
Philadelphia, PA

CCS Braeburn Alloy Steel
River Road
Lower Burrell, PA 15068
(412)224-6900

Cegedur Pechiney
Pechiney
66 Avenue Marceau
Paris Cedex 08 F-75361
France

Central Brass & Aluminum Foundry Co.
1020 Woodward St.
Cincinnati, OH 45203

Central Engineering & Supply Co.
834-T Main Ave.
Passaic, NJ 07055

Central Foundry Co.
Joplin, MO

Central Institute of Metals
Leningrad
USSR

Central Iron & Steel Co.
Harrisburg, PA

Central Pattern & Foundry Co.
Chicago, IL

Centrifugal Products Inc.
3245 Cherry Ave.
Long Beach, CA 90801

Century Brass Products Inc.
59 Mill Street
Waterbury, CT 06720
(203)574-7700

Cercast Group
Pechiney Corporation
3905 Industrial Blvd.
Montreal-North, Quebec H1H 2Z2
Canada
(514)322-2371

Cerro Copper Products
See Cerro Metal Products Co.

Cerro Metal Products Co.
Bellefonte Works
Marmon Group, Inc.
P.O. Box 388
Rt. 144 South
Bellefonte, PA 16823
(814)355-6217

Cerro Wire & Cable Company
See Cerro Metal Products

Certanium Alloys & Research Co.

Certified Alloy Products, Inc.
3245 Cherry Ave.
P.O. Box 90
Long Beach, CA 90801
(213)595-6621

CEZUS
Cedex 21 Tour Manhattan
Paris, La Defense 92087
France
47 62 86 71

W.M. Chace Co.
GTE Sylvania
1600 Beard Ave.
Detroit, MI 48209
(313)842-7400

Chain Belt Co.
Milwaukee, WI

Chambersburg Engineering Co.
150 Derbyshire St.
P.O. Box 359
Chambersburg, PA 17201
(717)264-7151

Champion Rivet Co.
5137 Indianapolis Blvd.
P.O. Drawer 398
East Chicago, IN 46312
(210)397-8750

Champion Steel Co.
P.O. Box 97
8247 Penniman Road
Orwell, OH 44076
(216)437-5161

Chapman Valve Mfg. Co.
Indian Orchard, MA

Chase Brass & Copper Co.
Cleveland Refractory Metals Div.
29855 Aurora Rd.
Solon, OH 44139

Chase Brass & Copper Co., Inc.
Kennecott Copper Corp.
20600 Chagrin Blvd.
Cleveland, OH 44122
(216)283-3900

Chateaugay Ore & Iron Co.
Standish, NY

Chemalloy Co., Inc.
P.O. Box 350
County Line Road
Bryn Mawr, PA 19010-0350
(215)527-3700

Chemalloy Electronics Corp.
Gillespie Airport
Santee, CA 92071

Chemetron Corp.
ARC Products Division
P.O. Box 517
Hanover, PA 17331
(717)637-8911

Chemetron Welding Products
See Chemetron Corp.

Chicago Development Corp.
#1 Hwy. N.
P.O. Box 266
Ashland, VA 23005

Chicago Hardware Foundry Co.
North Chicago, IL

Chicago Malleable Casting Co.
Chicago, IL

Chicago Steel Foundry
Chicago, IL

Chiers-Chatillon
33-35 Rue d'Alsace
Boite Postale 64
92 302 Levallois Cedex T759 7000
France
280 6311

China Metallurgical Import & Export Corp.
46 Dongsixi Dajie
Beijing
China
550197

Chromalloy American Corp./Sintercast Div.
See Alloy Technology International, Inc.

Chrome Alloys Mfg. Co.
Oakland, CA

Chrome Steel Corp.
(Reported to be out of business)
Carteret, NJ

Chromium Corp. of America
8701 Union Ave.
Cleveland, OH 44105

Chromium Mining & Smelting Corp. Ltd.
3720 Pl. Victoria
Montreal
Canada

Chrysler Corp.
P.O. Box 118
CIMS 418-19-30
Detroit, MI 48288
(313)956-3973

Cie Francais de Metaux
Paris
France

Cinaudagraph Corp.
Stamford, CT

Cincinnati Steel Castings Co.
Cincinnati, OH

Cindal Aluminium Ltd.
Birmingham
England

Cleveland Automatic Machinery Co.
Cleveland, OH

Cleveland Brass Corp.
P.O Drawer C
Kinsman, OH 44428

Cleveland Twist Drill Co.
1242 E. 49th St.
P.O. Box 6656
Cleveland, OH 44101

Charles Clifford Industries Ltd.
Dog Pool Mills Stirchley
Birmingham Works
Birmingham, B 30
England

Climax Performance Materials Corp.
7666 W. 63rd. St.
Summit, IL 60501

Climax Specialty Metals
21803 Tungsten Rd.
Cleveland, OH 44117

CMW Inc.
3029 E. Washington St.
P.O. Box 2266
Indianapolis, IN 46206
(317)634-891

R.W. Coan Ltd.
London
England

Coast Metals, Inc.
201 Redneck Avenue
P.O. Box 7
Little Ferry, NJ 07643
(201)641-1200

Cobalt Information Center
Columbus, OH

Cochrane Corp.
Philadelphia, PA

Cogne Co.
Rome
Italy

A. Cohn Ltd.
London
England

Sigmund Cohn Corp.
121 So. Columbus Ave.
Mount Vernon, NY 10553
(914)664-5300

Cold Metals Products Company, Inc.
P.O. Box 6078
45 S. Montgomery Avenue
Youngstown, OH 44501
(216)746-6311

Colonial Alloys Co.
Ridge & West Crawford Sts.
Philadelphia, PA 19129

Colonial Metals Co.
P.O. Box 311
Columbia, PA 17512
(717)684-2311

Colorado Fuel & Iron Co.
Denver, CO

Colt Industries
Stellite Division/Victor Mfg. Co.
London
England

Columbia Bronze Corp.
Freeport, NY

Columbia Tool Steel Co.
500 Lincoln Ave.
Chicago Hts.,, IL 60411
(708)757-5353

Columbus McKinnon Corp.
Audubon and Sylvan Pkwys.
Amherst, NY 14228

Comalco Ltd.
Melbourne
Australia

Cominco Metals
1500 120 Adelaide St. W.
Toronto, Ontario M5H 1T1
Canada
(416)869-1850

Cominco Trail
Box 1000
Trail, British Columbia BIV 1R 428
Canada
(604)364-4222

Commentryenne
35 rue d'Alsace
92531 Levallois-Perret
France
47 59 67 30

Commerce Pattern Co.
New York, NY

Commercial Alloys Co.
San Francisco, CA

Commonwealth Aircraft Corp.
Pty. Ltd., Box 779 H GPO
Melbourne, 3001
Australia

Compagnie Ateliers et Forges de la Loire
12, rue de la Rochefoucauld
F-75, Paris 9
France

Compagnie d'Orleans
Paris
France

Compagnie des Alliages
Speciaux d'Aluminium
Asnieres
France

Compagnie des Mines
Fonderies et Forges d'Alais
Temaris
France

Compagnie Francaise de l'Etain
Paris 8e
France

Compound Electro Metals Ltd.
London
England

Compressed Industrial Gases Co.
Chicago, IL

Connecticut Metals Corp.
Hampden, CT

Consolidated Ashcroft Hancock Co. Inc.
Bridgeport, CT

Consolidated Car Heating Co.
Albany, NY

Constrictor Ltd.
(Reported to be out of business)
London
England

Contact Technologies, Inc.
Formerly called:
Stackpole Carbon Co.
112 Curry Ave.
P.O. Box 149
St. Marys, PA 15857-0149
(814)781-1234

Continental Copper & Steel Industries Inc.
American Boron Products Division
Buffalo, NY

Continental Foundry & Machine Co.
10060 Lamar Ave.
Olive Branch, MS 38654

Continental Industries Corp.
New York, NY

Continental Steel Corp.
Formerly called:
Penn-Dixie Corp.
1111 South Main Street
Kokomo, IN 46902
(317)457-3211

George Cook & Co., Ltd.
Uniformity Steel Works
Corporation St.
Sheffield S3 8RE
England
737603

Cooper Alloy Corp.
Shellcast Corporation Division
100 Bloy Street
Hillside, NJ 07205
(201)688-4120

Cooper Metallurgical Corp.
3560 Ridge Road
Cleveland, OH 44102

Cop-Sil-Loy Inc.
Hollywood, CA

Copper Development Assoc., Inc.
405 Lexington Ave.
New York, NY 10017
England
(212)953-7300

Copper Range Co.
630 Fifth Ave.
New York, NY 10020

Copperweld Steel Co.
4000 Mahoning Ave., N.W.
Warren, OH 44483
(216)841-7100

Corning Glass Co.
Corning, NY 14830

Coshocton Stainless
Cyclops Corporation
P.O. Box 548
Cochocton, OH 43812
(614)829-2341

COSIPA-Companhia Siderurgica Paul
Estrada de Piacaguera, km 6
Cubatao, SP
Brazil

Coulter Steel & Forge Company
1494 67th Street
P.O. Box 8008
Emeryville, CA 94662
(415)653-2512

Crane Co.
Valve Division
104 N. Chicago St.
Joliet, IL 60431
(815)727-2600

Torrey S. Crane Co.
492 Summer St.
Plantsville, CT 06479

Crescent Tool Co.
Jamestown, NY

Creusot-Loire
Des Operations de Marche
B.P. 218-09
75428 Paris Cedex 09
France
1 280-65-77

Criterion Metals Inc.
Attleboro Industrial Park
35 Extension Street
Attleborro, MA 02703
(617)226-0800

Crobalt Inc.
2800 S. State
Ann Arbor, MI 48106

Cronite Foundry Co. Ltd.
Lawrence Road
Tottingham
London N 15
England

John A. Crowley Inc.
See Kloster Steel Corp.
New York, NY

Crucible Electric Steel Co.
Homestead, PA

Crucible Materials Corp.
Formerly called:
Colt Industries/Crucible Specialty Metals
Box 977
Syracuse, NY 13201-0877
(315)487-4111

Crucible Specialty Metals
See Crucible Materials Corp.

Crucible Specialty Steel
Sanderson Steel Division
See Crucible Materials Corp.

Crucible Steel
Halcomb Steel Division
See Crucible Materials Corp.

Crucible Steel Castings Co.
(Reported to be out of business)
W. 84th and Elmira Ave.
Cleveland, OH 44102
(216)631-4255

M.E. Cunningham Co.
Box 9996-C
Pittsburgh, PA 15233

Curtis Bay Copper & Iron Works
(Reported to be out of business)
Baltimore, MD

Curtis-Wright Corp.
Quehanna, PA

Herbert Cutanit Ltd.
(Reported to be out of business)
Cliff Lane, Grappenhall
Warrington
Lancashire WA4 3JX
England

Cutler Hammer Inc.
See Eaton Cutler Hammer
4201 N. 27th St.
Milwaukee, WI 53216
(414)449-6000

Cyclops Industries/Eastern Stainless
Formerly called:
Eastern Stainless Steel Co.
P.O. Box 1975
Baltimore, MD 21203
(301)522-6200

Cyprus Foote Mineral Co.
Cyprus Minerals Co.
Formerly called:
Foote Mineral Co.
Suite 301
301 Lindenwood Drive
Malvern, PA 19355-1740
(215)889-9605

Cytemp Specialty Steel Division
Cyclops Corporation
P.O. Box 606
Titusville, PA 16354
(814)827-3641

D-M-E Company
Fairchild Industries
29111 Stephenson Highway
Madison Heights, MI 48071

D.A.B. Industries Inc.
466 Stevenson Hwy.
Troy, MI 48084
(313)585-2000

Daido Steel Co. Ltd.
200 Park Ave., Suite 342 East
New York, NY 10017

Damascus Steel Casting Co.
Blockhouse Run Road
New Brighton, PA 15066
(412)846-2770

Darbyshire Steel Co. Inc.
P.O. Box 12903
El Paso, TX 79912

Darwin & Milner Inc.
Cleveland, OH

Darwin Tools Ltd.

Darwin's Ltd.
See Balfour-Darwins
Sheffield
England

Darwin-Toledo Ltd.
Toledo Steel Works
Nepsend Lane
Sheffield 3
England

Darwins Alloy Castings
Edgar Allen Balfour Steels Ltd.
Formerly called:
Balfour Darwins Ltd.
Sheffield Road
Tinsley
Sheffield S9 1RL
England
0742 448421

Davis & Geck Inc.
Pearl River, NY 10965

Dayton Malleable Iron Co.
G.H.R. Foundry Division
Dayton, OH

Degefors Iron & Steel Works
Degefors
Sweden

Degussa AG
Geschaftsbereich Dental
Weibfraunstr. 9
Postfach 110533
D-6000 Frankfurt 11
Germany
069/2 18 01

Delloy Metals
Philadelphia, PA

Deloro Stellite, Inc.
Thermadyne Industries Inc.
101 S. Hanley Road
St. Louis, MO 63105
(314)746-2104

Deloro Stellite Ltd.
Thermadyne
Stratton St. Margaret
Swindon
Wiltshire SN3 4QA
England
0793 82 2451

Delsteel Inc.
P.O. Box 1268
Wilmington, DE 19899
(302)764-2200

Delta (Manganese Bronze) Ltd.
Handford Works
Formerly called:
Manganese Bronze & Brass Co., Ltd.
P.O. Box 22
Hadleigh Road
Ipswich, Suffolk IP2 0EG
England
0473 252127

Delta Enfield Metals Ltd.
Formerly called:
Enfield Rolling Mills Ltd.
Millmarsh Lane
Brimsdown
Enfield, Middlesex EN3 7QB
England
081-804-1255

Delta Extruded Metals Company Ltd.
Delta PLC
Greets Green Road
West Bromwich, West Midlands B70 9ER
England
021 500 6188

Delta Metal (BW) Ltd.
See Delta Extruded Metals Company Ltd.

Denman & Davis Co.
1 Broad St.
Clifton, NJ 07011

Detroit Alloy Steel

Detroit Gray Iron Foundry
Detroit Alloy Steel Co.
Detroit, MI 48201

Detroit Mold Engineering Co.
Newark, NJ

Detroit Steel Co.
Craine-Schrage Steel Division

Detroit Tap & Tool Co.
8615 E. Eight Mile Road
Warren, MI 48090

Deutro GmbH
Berlin
Germany

Deutsche Delta Metallgesellschaft
Dusseldorf-Grafenberg
Germany

Deutsche Gold und Silber Scheide
Hanau
Germany

Deutsche Industrie
Nemen
Germany

Deutsche Messingwerke
Berlin
Germany

Dewramet Ltd.
Formerly called:
Dewrance & Co. Ltd.
York house
Mossland Road
Glasgow G52 4UD
Scotland
041 882 9001

Dewrance & Co. Ltd.
See Dewramet Ltd.

Diamond Metal Alloys
4051 Hollister
Houston, TX 77080
(713)460-5332

Die Casting Appliance Corp., Ltd.
London
England

Die Castings Ltd.
Birmingham
England

Karl Diederichs Stahl-, Walz- und Hammerwerk
Postfach 12 01 20
D-5630 Remscheid 11
Germany
02191 593-0

Diehl Steel Co.
Cincinnati, OH

Dirigold Corp.
Kokomo, IN 46901

Dirilyte Co. of America
Kokomo, IN 46901

Dirostahl Karl Diederichs
Postfaach 120120
D-5630 Remscheid-Luttringhausen
Germany
0 21 91 593-0

Disston Inc.
1030 W. Market St.
Greensboro, NC 27401

Division Lead Co.
7767 W. 61st Street
Summit, IL 60501

Dixon Sintaloy Inc.
535 Hope St.
Stamford, CT 06906

DoAll Co.
254 N. Laurel Ave.
Des Plaines, IL 60016
(312)824-1122

Dodge Foundry & Machine Co.
6501 State Road
Philadelphia, PA 19135
(215)335-7800

Doehler-Jarvis
P.O. Box 902
Toledo, OH 43691
(419)470-8020

Doelger & Kirsten Inc.
3015 W. Chambers St.
Milwaukee, WI 53210

Dofasco Inc.
Formerly called:
Dominion Foundries & Steel, Ltd.
1330 Burlington Street
P.O. Box 2460
Hamilton, ONT L8N 3J5
Canada
(416)544-3761

Dohlen-Stahl Gusstahl-Handels GmbH
Wetter/Ruhr
Germany

Dominion Foundries & Steel Ltd.(Canada)
See Dofasco Inc.

Dominion Magnesium Ltd.
Toronto, Ontario
Canada

Dominion Wheel & Foundry Co.
Toronto
Canada

Donegal Steel Foundry Co.
Marietta, OH

Dorman, Long & Co.
See British Steel Corp.

Dorrenberg Edelstahl GmbH
Runderoth Postfach 2164
D-5250 Engelskirchen
Germany

Dortmund-Hoerder Huttenverein A.G.
(Reported to be out of business)
Dortmund
Germany

Dow Chemical Co.
Texas Operations
Dow Chemical Co.
Freeport, TX 77541

Dresser Industries
Marion Power Shovel Division
1207 Cheney Avenue
P.O. Box 505
Marion, OH 43302
(614)383-5211

Dresser Pump Division
See Worthington Pump Inc.

Driver Harris Co.
Driver-Harris Alloys, Inc. Division
See Harrison Alloys Inc.
308 Middlesex Street
Harrison, NJ 07029
(201)483-4800

Wilbur B. Driver Co.
GTE Sylvania
1875 McCarter Highway
Newark, NJ 07104
(201)481-3100

Duke Steel Co. Inc.
New York, NY

Dunford Hadfields Ltd.
East Hecla Works
Vulcan Rd.
Sheffield S9 1TZ
England
0742 40353

DuPont Co.
Metal Cladding Section
17 North Seventh Avenue
Coatesville, PA 19320
(215)384-1764

E.I. DuPont de Nemours & Co.
1007 Market St.
Wilmington, DE 19898

Duquesne Smelting Corp.
Pittsburgh, PA

Dural Aluminum Composites Corp.
10326 Roselle st.
San Diego, CA 02121
(619)587-1415

Duraloy
Blaw Knox Corp.
P.O. Box 81
Scottdale, PA 15683
(412)887-5100

Duraloy Blaw-Knox
Blaw-Knox Food & Chemical Division
Union Steel Casting Division
See Duraloy

Duraweld Metal Products Corp.
Long Island, NY

Durener Metallwerke
Duren
Germany

Duriron Co. Inc.
Foundry Division
425 N. Findlay St.
P.O. Box 1145
Dayton, OH 45401
(513)226-4000

Dursar Corp.
Newark, NJ

Dusseldorf-Heerdt GmbH & Co., KG
4000 Dusseldorf-Heerdt
Germany

Duval et Poulain
Paris
France

Dymonhard Corp. of America
New York, NY

Dyn-Metal Ltd.
25-29 Chase Road
North Acton
London NW10 6TA
England

Dynamet Technology, Inc.
Eight A Street
Burlington, MA 01803
(617)272-5967

E.M.F. Electric Co. Ltd.
Victoria
Australia

Eagle & Globe Steel Ltd.
50 Orchardleigh Street
Guilford, NSW 2161
Australia
02/681 3600

Eastern Stainless Steel Co.
See Cyclops Industries/Eastern Stainless

Eastmet Corp.
P.O. Box 507
Cockeysville, MD 21030
(301)666-1500

Easton Metal Powder Inc.
900 Line St.
Easton, PA 18042

Eaton Corp.
Eaton Center
Cleveland, OH 44114-2584
(216)523-4400

Eaton Cutler Hammer
Formerly called:
Cutler Hammer Inc.
4201 N. 27th St.
Milwaukee, WI 53216

Eccles & Davis Machinery Co.
Los Angeles, CA

Eclipse-Pioneer Foundries
Formerly called:
Bendix Foundries
Teterboro, NJ

Edelstahlwerk Rochling AG
Volklingen
Saar
Germany

Edgcomb Metals Co.
1 Williams Center
P.O. Box 770
Tulsa, OK 74103

Edgcomb Steel Co.
Hillside, NJ

Egal Metal Products Co.
Baltimore, MD

E.N. Egge Co.
Los Angeles, CA

Eisenwerke Neubrandenburg G.m.b.H.
Berlin
Germany

Eisler Electric Co.
Union, NJ

Ekatit Stahl GmbH
Morikestrasse 30-32
Postfach 136
7120 Bietigheim-Bissingen
Germany
07142-5105356

Ekstrand & Tholand Co.
New York, NY

Electric Auto-Lite Co.
Woodstock, IL

Electro Foundry Co.
Troutdale, OR

Electro Metal Corp.
Yonkers, NY

Electro-Steel Co.
Pittsburgh, PA

Electroloy Co. Inc.
Bridgeport, CT 06605

Electrometal SA Metals Especials
Av. Marginal Direita do Tiete, 952
CEP 05118
Sao Paulo
Brazil
011 261 1900

Electronic Memories & Magnetics Corp.
Indiana General Division
405 Elm St.
Valparaiso, IN 46383
(219)462-3131

Electronic Specialties Co.
H. & S. Metals Co.
Pomona, CA

Elgiloy Limited Partnership
Formerly called:
American Gage & Machine Co.
1565 Fleetwood Drive
Elgin, IL 60123
(312)695-1900

Elgin Watch Co.
570 W. Jackson Blvd.
Chicago, IL 80606

Empire Sheet & Tin Plate Co.
Mansfield, OH

Empire Steel Castings Co.
P.O. Box 139
Reading, PA 19603
(215)921-8101

Empire Steel Corp.
Syracuse, NY

Emsco Derrick & Equipment Co.
Los Angeles, CA

Enfield Rolling Mills Ltd.
See Delta Enfield Metals Ltd.

Engelhard Corp.
Formerly called:
Engelhard Minerals & Chemical Corp.
700 Blair Road
Carteret, NJ 07008

Engelhard Industries
Engelhard Plainville Division
Engelhard
Attleboro, MA 19095

Engelhard Minerals & Chemicals Corp.
American Platinum & Silver Division
See Engelhard Corp.
700 Blair Road
Carteret, NJ 07008

Enpar Sonderwerkstoffe GmbH
Postfach 1144
Overather Strasse 37
D-5250 Engelskirchen
Germany
02263/30 71

Envirotech Corp.
Eimco Foundry Division
870 S. Fifth West
P.O. Box 1740
Salt Lake, UT 84110
(801)521-2000

J. & A. Erbsloh Aluminium
Press und Veredlungswerk
Siebneicker Strasse 235
P.O. Box 150160
D-5620 Velbert 15 (Neviges)
Germany
02053 4910

Ergst, Stahlwerke
Kommanditgesellschaft
See Stahlwerk Ergste GmbH & Co. KG
5840 Schwerte 1
West Germany

Erie EMI Company
Formerly called:
Erie Malleable Iron Co.
603 W. 12th St.
Erie, PA 16501
(814)452-6431

Erie Malleable Iron Co.
See Erie EMI Company

Erie Steel Co.
Erie, PA

Erkenzweig & Schwemann
Hagen
Germany

ESB Inc.
Exide Power Systems Div.
Rising Sun & Adams Aves.
Philadelphia, PA 19120

Escaut & Meuse
Paris
France

ESCO Corp.
2141 NW 25th Avenue
Portland, OR 97210
(503)228-2141

Eureka Welding Alloys, Inc.
Formerly called:
Welding Equipment & Supply Co.
5225 East Davison
Detroit, MI 48212
(313)366-5757

Europa Metalli-LMI SpA
Formerly called:
Societa Metallurgica Italia
50121 Firenze
Borgo Pinti 97/99
Italy

Eurotungstene
Pechiney
65X-38041 Grenoble
Cedex
France

Eutectic Corp.
40-40 172 St.
Flushing, NY 11358-9981
(800)221-1433

Evans Steel Co.
Boston, MA

Everard Tap & Die Corp.

Ex-Cell-O
(Reported to be out of business)
40 Westminster St.
Providence, RI 02903
(401)421-2800

Excelite Co.
Woodbridge, NJ

F.A.C.A.
90 rue de Villiers
Levallois-Perret (92)
France

F.M.C. Corp.
Link-Belt Div.
Prudential Plaza
Chicago, IL 60601

Fagersta Bruks Aktiebolag
Fagersta
Sweden

Fagersta Inc.
2 Henderson Drive
West Caldwell, NJ 07006
(201)575-1600

Fagersta Stainless AB
Box 508
S-77301 Fagersta
Sweden
46 223 445 00

Fahralloy Co.
150th & Lexington
Harvey, IL 60426
(312)333-2060

Fairbanks, Morse & Co.
701 Lawton Ave.
Beloit, WI 53511

Jas. Fairley & Sons Ltd.
Birmingham
England

Fairmount Foundry Inc.
Front & Pine Streets
Hamburg, PA 19526

Faitout Iron & Steel Co.
182 Frelinghuysen Ave.
Newark, NJ 07114

Fakirstahl Hoffmanns GmbH & Co. KG
Oberhutzer Str. 33
D-5630 Remscheid 1
Germany
02191 8561

The Falk Corp.
Subsidiary of Sunstrand Corp.
Box 492
3002 Canal St.
Milwaukee, WI 53201
(414)342-3131

Fansteel
Wellman Dynamics
U.S. Rte. 34
P.O. Box 147
Creston, IA 50801
(515)782-8521

Fansteel Metals
1 Tantalum Pl.
North Chicago, IL 60064
(312)689-4900

Farbenindustrie Aktiengesellschaft, I.G.
Abtielung Elektronmetall
Frankfurt a.M.
Germany

Farrell Co.
See USM Corp.

Farrell-Cheek Steel Co.
706 Lane St.
Sandusky, OH 44870
(419)625-2340

Farrelloy Co.
American Solder & Flux Co., Inc.
Industrial Blvd.
Paoli, PA 19301
(215)647-3575

Federal Foundries & Steel Co., Ltd.
London
England

Federal-Mogul Corp.
Engine & Transmission Product Group
P.O. Box 1966
Detroit, MI 48235
(313)354-7700

Felten & Guilleaume Calswerk AG
Schanzen Strasse 24
D-5000 Koln (Mohlheim)
Germany

Feltrina Societa Metallurgica
Milan
Italy

Ferranti Ltd.
Hollinwood
Lancashire 0L9 7JS
England

FERRO Corp.
Electro Refractories & Abrasives Division
666 Willet Road
Buffalo, NY 14218

Ferry-Capitain, Usines De Bussy
Boite Postale 33-52
Joinville
France

B.A. Field Co.
Newark, NJ

Fillmore Foundry Inc.
Buffalo, NY

A. Finkl & Sons Co.
Republic Steel Corp.
2011 Southport Ave.
Chicago, IL 60614
(312)975-2500

John Finn Metal Works
San Francisco, CA

Firth Brown Ltd.
Atlas Works
See Sheffield Forgemasters Ltd.
Savile St.
Sheffield S4 7US
England
0742-20081

Firth-Vickers Stainless Steels Ltd.
Staybrite Works
Weedon St.
P.O. Box 131
Sheffield S9 2FU
England
49955

Georg Fisher AG
Germany

Fisher Scientific Co.
711 Forbes Ave.
Pittsburgh, PA 15219
(412)562-8300

Fitzsimmons Steel Co.
1629 Wilson Ave.
Youngstown, OH 44501

Flockton, Tompkin & Co., Ltd.
Sheffield
England

Follansbee Steel Co.
P.O. Box 610
State St.
Follansbee, WV 26037
(304)527-4500

Follsain-Wycliffe Foundries, Ltd.
Clayton Dewandre Holdings Co.
Lutterworth
Leics LE17 4HA
England
3551

Fonderie de Precision S.A.
15 rue de Port
92003 Nanterre
Paris
France

Foote Bros. Gear & Machine Corp.
Chicago, IL

Foote Mineral Co.
See Cyprus Foote Mineral Co.

Ford Motor Co.
Casting Division
3001 Miller Road
Dearborn, MI 48121
(313)337-9713

Forges et Acieries de Bonpertuis
38140 Rives-sur-Fure
France

Forges et Acieries de Voelklingen
Voelklingen (Sarre)
France

Forjas Alavesas S.A.
Portal de Gamarra
22 Vitoria
Spain

Fort Pitt Steel Castings

Frank Foundries Corp.
2020 3rd Ave.
Moline, IL 61265
(309)757-8800

Peter A. Frasse & Co.
3 Dakota Drive
Lake Success, NY 11040
(516)775-4800

Fromson Co. Inc.
Rockville, CT 06066

Frontier Bronze Corp.
New and Packard Roads
Niagara Falls, NY 14304

Otto Fuchs Metallwerke
Postfach 1261
D-5882 Meinerzhagen
Germany
02354/7 31

Fugi Iron & Steel Co., Ltd.
Formerly called:
Yawata Iron & Steel Co., Ltd.
Chou-Ku
Tokyo
Japan

Fujikoshi Steel Industry Co., Ltd.
Tokyo
Japan

Fuller & Basche Co.
New York, NY

Fulton Gold Refineries Corp.
New York, NY

Fulton Iron Works Co.
Subsidiary of KATY Industries
3844 Walsh St.
St. Louis, MO 63116
(314)752-2400

Fusion Inc.
4658 E. 355th St.
P.O. Box 390
Willoughby, OH 44094
(216)946-3300

A.B. Gabriel & Co., Ltd.
Row
Birmingham
England

Wm. Gallimore & Sons Ltd.
Sheffield
England

Galv-Weld, Inc.
Dayton, OH

Garford Engineering Co.
Garfield, IN

Garrett Brass & Aluminum Foundry Co.
Garrett, IN

D.G. Gautier & Co.
New York, NY

A. Gayer Co.
Paris
France

General Aircraft Equipment Co.
(Reported to be out of business)

General Alloys Co.
See Alloy Engineering & Casting Co.
(Reported to be out of business)
405 W. First St.
Boston, MA 02127

General Cable Corp.
Cornish Wire Products Division
101 Water St.
Williamstown, MA 01267

General Cerium Co.
See Ronson Metals
(Reported to be out of business)

General Communications Co.
Boston, MA

General Electric Co.
Carboloy Systems Dept.
Box 237 G.P.O.
Detroit, MI 48232
(313)536-9100

General Electric Ltd.
England

General Malleable Corp.
Waukesha, WI

General Metals Powder Co.
130 Elinor Ave.
Akron, OH 44305

General Motors Corp./Central Foundry
P.O. Box 1629
1805 Veterans Memorial Pkwy.
Saginaw, MI 48605-5073

General Steel Industries
Ludlow-Saylor Wire Cloth Division
8474 Delport Dr.
St. Louis, MO 63114

General Thermoelectric Corp.
Princeton, NJ

General Tool & Die Co.
East Orange, NJ

Georgia Iron Works Co.
5000 Wrightsboro Road
Grovetown, GA 30813
(404)863-1011

Georgsmarienwerke Selesiastahl GmbH
Georgsmarienhutte
Germany

Gerhardi & Co.
Ludenscheid
Germany

Gibson Electric Co.
Wesgo Division
GTE Products Corp.
(Reported to be out of business)
477 Harbor Blvd.
Belmont, CA 94002
(415)592-0440

Gilby-Fodor S.A.
29, quai du halage
Rueil-Maimaison F-92
(Seine-et-Oise)
France

Gillett & Eaton Inc.
Lake, MN

F.F. Gilmore & Co.
69 Westford
Needham Heights, MA 02194

Gilson Ltd.
Brussels
Belgium

Gimo-Osterby Bruks, AB
Gimo
Sweden

Giulini Werke A.G.
Rohrsbach
Germany

GKN Powder Met Inc.
See Presmet Corp.

Glacier Metal Co.
Richmond, VA

Glacier Metal Co., Ltd.
Alpertone
England

K.C. Glader Co.
Niles, IL 60648

Globe Metallurgical Inc.
P.O. Box 157
Beverly, OH 45715
(614)984-2361

Glyco-Metallwerke Daelen & Hofmann KG
Postfach 13 03 35
Stielstrasse 11
Wiesbaden 13 D-6200
Germany
0 61 21-201-0

Gold und Silber Scheide Anstalt
Hanau
Germany

Goldschmidt AG
Chem Fabrik
Postfach 10 14 61
D-4300 Essen
Germany

Goodlass Wall & Lead Industries Ltd.
London
England

Gorham Tool Industries Inc.
14400 Woodrow Wilson
Detroit, MI 48238

Gottingen Aluminiumwerke GmbH
Gottingen
Germany

Gould Inc.
Engine Parts Division
35129 Curtis Blvd.
Eastlake, OH 44094
(216)953-5000

Graham Chemical Corp.
Minimax Corp.
5905 N. Clark
Chicago, IL 60660

Grammer, Dempsey & Hudson Co.
Newark, NJ

Grand Northern Products Ltd.
Formerly called:
Resisto-Loy Company, Inc.
1330 Phillips, S.W.
Grand Rapids, MI 49507-1589
(616)243-6833

Grant & West Ltd.
London
England

Graphite Metallizing Corp.
1000 Nepperhan Ave.
Yonkers, NY 10702
(914)968-8400

Graphitized Alloy Corp.
New York, NY

Grasselli Chemical Co.
Cleveland, OH

Grayburn Steel Co.
New York, NY

Great Western Steel Co.
Metalsource Corp.
2310 W. 58 St.
Chicago, IL 60636
(312)434-5800

Grede Foundries, Inc.
9898 W. Blue Mound Rd.
P.O. Box 26499
Milwaukee, WI 53226
(414)257-3600

Greenleaf Corp.
1 Greenleaf Dr.
Saegertown, PA 16433
(814)763-2915

Greenleaf Technical Ceramics
25019 Viking St.
Hayward, CA 94545
(415)783-0120

G.K.M. Greuter & Kerscher GmbH & Co. KG
Boxdorfer Str. 2
Postfach 62 62
D-8510 Furth
Germany
0911/30 20 07

C.M. Grey Mfg. Co.
East Orange, NJ

Griesogen Griesheimer Autogen
Verkaufs, G.m.b.H.
Greisheim a.M.
Germany

Gustav Grimm Edelstahl-Werk GmbH
Postfach 15 01 36
5630 Remscheid 1
Germany
02191 435-0

GTE Products Corp.
Chemical & Metallurgical Division
Hawes St.
Towanda, PA 18848
(717)265-2121

GTE Products Corp.
Wesgo Division
477 Harbor Boulevard
Belmont, CA 94002
(415)592-9440

GTE Sylvania
Chemical & Metallurgical Division
See GTE Valenite Corp.
1989 Hawes St.
Towanda, PA 18848

GTE Valenite Corp.
Walmet Division
Formerly called:
GTE Sylvania
404 E. 10 Mile Road
Pleasant Ridge, MI 48069

William Guertler GmbH
Berlin
Germany

Gulf & Western Industries
New Jersey Zinc Co. Division
65 E. Elizabeth Ave.
Bethlehem, PA 18018
(215)691-5000

Gulf & Western Mfg. Co.
Mackintosh-Hemphill Division
901 Bingham St.
Pittsburgh, PA 15203

Gulf States Steel, Inc.
174 S. 26th Street
Gadsden, AL 35904-1935
(205)543-6700

Gulf Steel Corp.
Dallas, TX

Guronitwerke Vervoort GmbH
Dusseldorf
Germany

Gusstahl-Handels GmbH
van Dohlen-Stahl, Wetter
Ruhr
Germany

Hackett Brass Foundry
Lillibridge St. & Edlie
Detroit, MI 48214

Haeckerstahl GmbH
Postfach 10 02 51
D-5609 Huckeswagen
Germany
02192 40 44

Hallamshire Steel Co.
Sheffield, 3
England

Hamilton Allied Corp.
See Hamilton Foundry

Hamilton Die Cast
999 East Ave.
Hamilton, OH 45011
(513)856-9500

Hamilton Foundry
Formerly called:
Hamilton Allied Corp.
200 Industrial Lane
Harrison, OH 45030
(513)367-6900

Hamilton Technology Inc.
1780 Roherstown Rd.
P.O. Box 3014
Lancaster, PA 17604
(717)569-7061

Hammond & Irving Inc.
254 North St.
P.O. Box 656
Auburn, NY 13201
(315) 253-6265

Hammond Brass Works
Hammond, IN

Handler GmbH
Dusseldorf
Germany

Handy & Harmon
Fairfield Plant
1770 Kings Highway
P.O. Box 610
Fairfield, CT 06430
(203)259-8321

Hanford Foundry Co.
P.O. Box 192
119 S. Arrowhead Ave.
San Bernardino, CA 92402
(714)889-9555

Hans-Heinrich Hutte GmbH
Langelsheim a.Harz
Germany

Hansell-Elcock Co.
Chicago, IL

Hanson-Van Winkle-Munning Co.
9000 Precision Drive
Indianapolis, IN 46236

Hardenite Steel Co. Ltd.
Hardenite Steel Works & Royal Works
Attercliffe Rd.
Sheffield S4 7WZ
England
0742-22131

Hardite Metals Inc.
New York, NY

Charles Hardy & Co.
New York, NY

Harnischfeger Corp.
Milwaukee, WI

Arthur Harris & Co.
202 N. Aberdeen St.
Chicago, IL 60607

J.W. Harris Co., Inc.
10930 Deerfield Rd.
Cincinnati, OH 45242
(513)473-2006

Harrison Alloys Inc.
308 Middlesex St.
Harrison, NJ 07029
(201)483-4800

Harrison Steel Castings Co.
900 Mound St.
P.O. Box 60
Attica, IN 47918

Harrison, Fischer & Co., Ltd.
Eye Witness Works
Sheffield S3 7WJ
England

Harsco Corp.
Taylor-Wharton Division
P.O. Box 229
2900 William Penn Highway
Easton, PA 18042

Harville Machine Inc.
24201 Orange Ave.
Perris, CA 92370

Wm. Hassall & Sons
Manchester
England

Hawkridge Bros Co.
One Wesley St.
Malden, MA 02148

Haynes International, Inc.
Formerly called:
Cabot Corp.
1020 West Park Ave.
P.O. Box 9013
Kokomo, IN 46902-9013
(317)456-6049

Haywood Tyler of Canada, Ltd.
Kitchener, Ontario
Canada

Haywoods NCA Metal Ltd.
London
England

Arthur E. Heckford Ltd.
Birmingham Metal Works
Frederick St. & Regent St.
Birmingham B1 3HH
England
021 236 0525

Oscar W. Hedstrom Corp.
Chicago, IL

Heinkel-Herth Co.
Berlin
Germany

Heinrich Lanz A.G.
Mannheim
Germany

Heinrich Reining GmbH
Postfach 6529
4000 Dusseldorf 1
West Germany
0211/44081

Hellefors Bruks Aktiebolag
See Ovako Steel Hellefors AB

Heller Bros. Co.
See Heller Tool Division, Wallace Murray Corp.
Newark, NJ

Heller Tool
See Wallace Murray Corp.

Hemmings & Co.
Rotherham
England

Henckels Zwillingwerke, G.A.
Solingen
Germany

Henning Bros. & Smith Inc.
99 Scott Ave.
Brooklyn, NY 11237

Heppenstall Co.
4620 Hatfield St.
Pittsburgh, PA 15201

Hetzel & Co. Metallhuttenwerk GmbH
Postfach 2242
Rotterdamer Strasse 135
D-8500 Nurnberg 60
Germany
0911 64 19 0

Hewitt Metals Corp.
Stanley Ave. at Twelfth St.
Detroit, MI 48126
(313)895-3846

John Hewson Co.
New York, NY

Hidalgo Steel Co. Inc.
New York, NY

Theodore Hiertz Metal Co.
St. Louis, MO

High Duty Alloys Ltd.
154 Edinburgh Ave.
Slough Berkshire SLI 4PA
England
0753 72778

Hills-McCanna Co.
400 Maple Ave.
Carpentersville, IL 60110
(708)426-4100

Hobart Bros. Co.
See Hobart Welding Products

Hobart Welding Products
Hobart Square
Troy, OH 45373
(513)332-5123

Hobson, Houghton & Co.
New York, NY

Hochfrequenz-Tiegelstahl GmbH
Bochum
Germany

Hoeganaes Corp.
Interlake Inc.
River Rd. & Taylors Ln.
Riverton, NJ 08077
(609)829-2220

Hoesch Hohenlimburg AG
Postfach 53 08
Langenkampstrasse 14
5800 Hagen 5
Germany
02323 88-1

Hoesch Stahl AG
Formerly called:
Hoesch Huttenwerke
Postfach 9 02
4600 Dortmund 1
Germany
0231 844-1

Hoffman Elektrogusstahlwerk, Alb.
Eschweiler
Germany

Hoffmann & Co. KG
Postfach 1125
D-6369 Schoeneck 2
Germany

Hofors Steel Works
Hofors
Sweden

Holcroft Castings & Forgings Ltd.
P.O. Box 24
Whitehall St.
Rochdale OL12 OLL
England
40911-7

Hollup Corp.
Chicago, IL

Holt Equipment Co.
Independence, OR

Homogeneous Metals Inc.
P.O. Box 294
Clayville, NY 13322
(315)839-5421

Honsel-Werke Aktiengesellschaft
Postfach 1369
D-5778 Meschede 1
Germany
0291 6161

Hoover Ball & Bearing Co.

Hopkinsons Ltd.
Brittania Works
P.O. Box B27
Huddersfield HD2 2UR
England
STD 0484 22171

Horbach & Schmitz GmbH
5000 Koln
Hansaring 49-51
West Germany

Hoskins Mfg. Co.
10776 Hall Road
Hamburg, MI 48139
(313)231-1900

Houghton & Richards Inc.
73 A. Bartlett St.
Marlboro, MA 01752
(508)485-6200

Chr. Hover & Sohn, Edelstahlwerk
5251 Berghausen
uber Engelskirchen
West Germany

Hover, Gebruder, Edelstahlwerk
5251 Kaiserau
uber Engelskirchen
West Germany

Howard Foundry Co.
(Reported to be out of business)
1700 N. Klostner Ave.
Chicago, IL 60639

Howmet Corp.
Alloy Operations
Roy Street
Dover, NJ 07801
(201)361-2310

Howmet Turbine Component Corp.
Pechiney Corporation
475 Steamboat Rd.
Greenwich, CT 06836-1960
(203)661-4600

Hoyland Steel Co.
New York, NY

Hoyt Darchem Ltd.
Garth Rd., Lower Morden
Surrey SM4 4LT
England
081 337 7744

Hoyt Metal Co.
New York, NY

Hoyt Metal Co. of London Ltd.
See Hoyt Darchem Ltd.

Hubbard Steel Co.
East Chicago, IN

Eduard Hueck
Metallwalz und Presswerk
Postfach 1868
D-5880 Ludenscheid
West Germany

Hufnagel GmbH
Schweinauer Haupstrasse 1
8500 Nurnberg
West Germany

F.A. Hughes & Co., Ltd.
London
England

Huntington Alloys Inc.
P.O. Box 1958
Huntington, WV 25720
(304)696-2150

Hurbenium Co. of America
Detroit, MI

Husquarna Vapenfabrik Aktiebolag
Stockholm
Sweden

Hybnickel Alloys Co.
Wilmington, DE

Hyde Park Foundry & Machine Co.
Hyde Park, PA 15641
(412)842-1941

Hydril Co.
1060 Vignes St.
Los Angeles, CA 90012
(213)680-1910

Hydrometals Inc.
1400 Expressway Tower
Dallas, TX 75206

Hytensil Aluminum Co.
Chicago, IL

Idealstahl Breidenbach KG
Postfach 31 02 07
5270 Gummersbach 31
Germany

Illingworth Steel Co.
Philadelphia, PA

Illinois Zinc Co.
Chicago, IL

Illium Corp.
(Reported to be out of business)
Freeport, IL

ILSSA-VIOLA SpA
Via C. Farini 47
I-20159 Milano
Italy
0039 28137451

IMI Knoch Ltd.
IMI PLC
P.O. Box 216
Birmingham, Witton B6 7BA
England

IMI Rod & Wire
See Imperial Metal Industries
P.O. Box 216
Birmingham
England

IMI Rolled Metals Ltd.
See Imperial Metal Industries
P.O. Box 703
Witton
Birmingham B6 7UP
England
021 356 33 44

IMI Titanium Ltd.
P.O. Box 704
Witton
Birmingham B6 7UR
England
021-356-1155

IMI Yorkshire Alloys Ltd.
Formerly called:
Yorkshire Imperial Metals Ltd.
P.O. Box 166
Leeds LS1 1RD
England
0532 701 107

Imperial Aluminium Co., Ltd.
Birmingham
England

Imperial Brass Mfg. Co.
Chicago, IL

Imperial Metal Industries
Formerly called:
IMI Rolled Metals Ltd.
P.O. Box 703
Witton
Birmigham B6 7UP
England

Imperial Smelting Co. Ltd.
See Pasminco Europe (Mazak) Ltd.

Inco Alloys International Inc.
P.O. Box 1958
3200 Riverside Dr.
Huntington, WV 25720
(304)526-5100

Inco Alloys International Ltd.
Wiggin Works
Formerly called:
Henry Wiggin & Co., Ltd.
Holmer Road
Hereford HR4 9SL
England
0432 272777

Indiana General
Valparaiso, IN 46383

Indium Corp. of America
1676 Lincoln Ave.
P.O. Box 269
Utica, NY 13503
(315)768-6400

Induction Steel Castings Co. Inc.
18021 E. Nine Mile Road
P.O. Box 151
East Detroit, MI 48021
(313)775-4545

Industria Nazionale Alluminio, Mori
(Trento)
E. Bolzano
Italy

Industrial Furnace Corp.
Buffalo, NY

Industrial Overlay Metals Corp.
New York, NY

Industrial Steels Inc.
3540 Industrial Rd.
P.O. Box 346
Mims, FL 32754
(407)267-2341

Industries Trading Co.
New York, NY

Ingersoll Steel Co.
Avesta Inc. Division
State Road 38 West
P.O. Box 370
New Castle, IN 47362
(317)529-0124

Ingersoll-Rand Co.
Athens Foundry Division
N. Main Street
Athens, PA 18810

Inland Electronics Products Corp.
Pasadena, CA

Inland Steel Co.
30 West Monroe Street
Chicago, IL 60603
(312)346-0300

T. Inman & Co. Ltd.
Brittania Steel Works
Furnival Road
Sheffield S4 7YA
England
21153/4

INSILCO Corp.
1000 Research Pky.
Meriden, CT 06450

Interlake Inc.
Iron and Steel Division
135th St. & Perry Ave.
Chicago, IL 60627
(312)849-2500

International Alloys Ltd.
Bicester Road, Aylesbury
Buckinghamshire
England
4242

International Development Corp.
New York, NY

International Harvester Co.
Wisconsin Steel Division
401 N. Michigan Blvd.
Chicago, IL 60611

International Lead-Zinc Research Assoc.
292 Madison Ave.
New York, NY 10017

International Nickel Inc.
Park 80 West-Plaza Two
Saddle Brook, NY 07662

International Wire Products
Wyckoff, NJ 07481

IPM Corp.
Allegheny Ludlum Industries Company
2100 Advance Ave.
Columbus, OH 43207
(614)443-4833

Isabellenhuette
Heusler GmbH KG
Postfach 1453
D-6340 Dillenburg
Germany
02771 23031

Iscar Ltd.
P.O. Box 34
Nahariya
Israel

Ishikawajima-Harima Heavy Industries Co.
One World Trade Center, Suite 1101
New York, NY 10048

Isteg Steel Products Co.
Westminster
England

Istituto Sperimentali Metalli Leggeri
Milan
Italy

ITT Components Group Europe
Standard Telephones & Cables Ltd.
(Reported to be out of business)
190, Stranday
London WC2
England

ITT Corp.
(Reported to be out of business)
320 Park Ave.
New York, NY 10022

ITT Electronics Components Corp.
World Headquarters
Hans-Bunte-Strasse 19
P.O. Box 840
D-7800 Freiburg
Germany

Joseph Jackman & Co. Ltd.
Sheffield
England

Jackson Iron & Steel Co.
Jackson, OH 45640

Jamison Steel Corp.
San Francisco, CA

Janney Cylinder Co.
7401 State Road
Philadelphia, PA 19136

Japan Metal Industry Co. Ltd.
Kawasaki
Japan

Japan Special Steel Co.
Tokyo
Japan

Japan Steel & Tube Co. Ltd.
Tokyo
Japan

Japan Steel Works Ltd.
200 Park Ave.
New York, NY 10017

J.F. Jelenko & Co.
99 Business Park Dr.
Armonk, NY 10504
(914)273-8600

Jelliff Corp.
P.O. Box 758
354 Pequot Ave.
Southport, CT 06490
(203)259-1615

Jessop Steel Co.
Atholone Industries Inc.
500 Green St.
Washington, PA 15301
(412)222-4000

Jessop-Saville Ltd.
High Speed Tool Division
Staybrite Works
Weedon St.
Sheffield, 9
England

W.F. Jobbins Inc.
(Reported to be out of business)
Aurora, IL 60506

Jodots Freres, Ets.
Boloeil
Belgium

A. Johnson & Co.
New York, NY

Johnson Bronze Co.
500 S. Mill St.
New Castle, PA 16103
(412)654-6631

Johnson Matthey
Materials Technology Division
Johnson Matthey
Formerly called:
Matthey Bishop
1401 King Rd.
West Chester, PA 19380
(800)523-7792

Johnson Matthey plc
Materials Technology Division Europe
New Garden House
78 Hatton Garden
London
England
01 430 0011

Johnson Mfg. Co.
114 Lost Grove Road
Princeton, IA 52768
(319)289-5123

Johnston Steel & Wire Co. Inc.
53 Wiser Avenue
Worcester, MA 01607
(617)756-8301

Johnstown Corp.
545 Central Avenue
Johnstown, PA 15902
(814)535-6741

Jonas & Culver Ltd.
Neepsend Ltd.
Sheffield, 9
England

Jones & Laughlin Steel Co.
See LTV Steel
3-P Gateway Center
Pittsburgh, PA 15263

Jones & Laughlin Steel Corp.
See LTV Steel

Jones & Rooke Ltd.
Lawmet Slidex
86-92 Northwood St.
Birmingham
England

Jones, Rd., Ltd.
Garrison Lane
Birmingham
England

Earle M. Jorgensen
10650 Alameda St.
P.O. Box 54633
Los Angeles, CA 90054
(213)567-1122

C. Joselin & Co.
New York, NY

Joslyn Stainless Steels Co.
See Slater Steels

Otto Junker GmbH
5101 Lammersdorf uber Aachen
West Germany

Kabel-und Metallwerke AG
3000 Hanover
Germany

Kahl-Holt Co.
Baltimore, MD

Kaiser Aluminum & Chemical Corp.
300 Lakeside Drive
Kaiser Center
Oakland, CA 94643
(415)271-2211

Kaiser Steel
Oakland, CA 94643
(415)271-3300

Kalif Corp.
Emeryville, CA

Kanthal A.B.
Hallstammer
Sweden

Kanthal Corp.
Bulten-Kanthal A.B., Hallstammer, Sweden
119 Wooster St.
Bethel, CT 06801
(203)744-1440

Hans Kanz Metallwerke
Zurich-Albisrieden
Switzerland

Kapfenberg
Steiermark
Austria

Kasle Steel Co.
Box 536
Detroit, MI 48232
(313)846-4200

Kawasaki Steel Corp.
280 Park Ave.
New York, NY 10017
(212)661-8430

Kawecki Berylco Industries Inc.
220 E. 42nd St.
New York, NY 10017

Kay-Brunner Steel Products Inc.
999 Meridian Ave.
Alhambra, CA 91803

E. & E. Kaye Ltd.
See Pechiney
Middlesex EN3 4SS
England

Keasby & Matteson Co.
Ambler, PA

M.W. Kellog
Pullman Kellogg Division
Formerly called:
Pullman Inc.
3 Grainway Plaza
E. Houston, TX 77046
(713)960-2000

Kelly Foundry & Machine Co.
P.O. Drawer 1789
Elkins, WV 26241
(304)636-0390

Kelsey Hayes Co.
Gunite Division
302 Peoples Ave.
Rockford, IL

Kencroft Malleable Co.
Buffalo, NY

Kennametal Inc.
P.O. Box 231
Latrobe, PA 15650
(800)222-9327

Keokuk Electro-Metals Co.
Keokuk, IA

Kerr-McGee Chemical Corp.
Kerr-McGee Center
Oklahoma, OK 73125

Keystone Carbon Co.
St. Marys, PA 15857
(814)781-1591

Kidd Drawn Steel Co.
See Vulcan Kidd Division, H.K. Porter Co.
Aliquippa, PA 15857

Kind & Co. Edelstahlwerk
Postfach 21 80
5276 Wiehl-Bielstein
Germany
02262 840

Kinite Corp.
Milwaukee, WI

Kinney Iron Works
Los Angeles, CA

Kling Bros. Engineering Works
Chicago, IL

Klockner-Werke AG
Mulheimer Str. 50
Duisberg NRW 4100
West Germany

KM-kabelmetal AG
Formerly called:
Osnabruck Kupfer und Drahtwerke
Postfach 33 20
Klosterstrasse 29
Osnabruck D-4500
Germany
0541 321 410

Knapp Mills, Inc.
Long Island, NY

Knowsley Cast Metal Co., Ltd.
Manchester
England

Kobe Steel Ltd.
299 Park Ave.
New York, NY 10022

Koch Light Alloys Ltd.
St. Leonards Road
Willesden
London, NW 10
England

Koerver & Nehring GmbH
Postfach 1640
4150 Krefeld
West Germany
021 51 75 1081

Koppers Co. Inc.
Metal Products Division
3700 Koppers Street
P.O. Box 298
Baltimore, MD 21203
(301)547-7715

H. Kramer & Co.
Ajax Metal Division
44 Richmond St.
Philadelphia, PA 19123

Kreidler Werke G.m.b.H.
Postfach 40 04 80
Schwieberdinger Strasse 3-9
7000 Stuttgart 40
West Germany
0711 82821

Henry A. Kries & Sons Co.
Baltimore, MD

Kropp Forge Co.
5400 West Roosevelt Rd.
Chicago, IL 60650

Fried. Krupp Huttenwerke AG
Postfach 10 13 70
4630 Bochum 1
Germany
0234-63-4990

Krupp Huttenwerke
See Krupp Stahl AG

Krupp Stahl AG
Formerly called:
Stahlwerke Sudwestfalen/Bochumer Verein
Postfach 101220
D-5900 Siegen
Germany

C. Kuhbier & Sohn
See Thyssen Edelstahlwerk AG.
Deutschen Edelstahlwerke G.m.b.H.
5885 Schalksmuhle 2
West Germany

Kulite Tungsten Corp.
160 E. Union Ave.
East Rutherford, NJ 07073
(201)438-9000

L & R Mfg. Co.
Arlington, NJ

La Bour Pump Company
Process Metals Co. Div.
1607 Sterling Ave.
P.O. Box 1187
Elkhart, IN 46515
(219)293-6578

LaClede Brass (or Steel) Works
See Laclede Steel Co.
St. Louis, MO 63102

Ladish Co.
See Tri-Clover

Ladish Co., Inc. (developer)
Cudahy, WI 53110
(414)747-2611

Lake & Elliot Ltd.
Braintree
England

Lake Erie Engineering Corp.
Buffalo, NY

Lakeside Malleable Castings Co.
1333 23rd St.
Racine, WI 53404

Laminoire de la Rochette
Chaudfontaine
Belgium

Lancaster Steel Co. Inc.
New York, NY

Langley Alloys Ltd.
Langley, Slough
Buckinghamshire SL3 6EA
England
44131

LaSalle Steel Co.
Quanex Corporation, Bar Group
PO Box 6800-A
Chicago, IL 60680
(312)734-7800

Latrobe Steel Co.
Timken Co.
2626 Ligonier St.
Latrobe, PA 15650
(412)537-7711

R. Lavin & Sons, Inc.
3426 S. Kedzie Ave.
Chicago, IL 60623
(312)847-1800

Lavorazione Leghe Leggere SpA
Via Agnello 6/L
Lombardi
1-20121 Milano
Italy

Le Bronze Industriel
Rene Loiseau & Cie
49, rue de Paris;B.P.9
93001 Bobigny
France
830 11 37

Le Moyne Steel Co.
Pittsburgh, PA

Leadbeater & Scott Ltd.
(Reported to be out of business)
Sheffield
England

Lebanon Foundry and Machine Co.
Formerly called:
Lebanon Steel Foundry
101 E. Lehman St.
P.O. Box 390
Lebanon, PA 17042
(717)273-1611

Lebanon Steel Foundry
See Lebanon Foundry and Machine Co.

Lehigh Babbitt Co.
Allentown, PA

Lehigh Steel Corp.
16-18 Bethune St.
New York, NY

Les Toles Inoxydables et Specials
Pechiney
Ugine-Guegnon
100, rue de Villiers
92307 Levallois-Perret
France
757 01 20

Lesjofors Aktiebolag
Lesjefors
Sweden

Walker M. Levett
New York, NY

LeVita Metal Alloy Co.
Detroit, MI

Lewin Metals Corp.
East St. Louis, IL

Ley's Malleable Castings Co. Ltd.
Derby DE3 8LY
England
0332 45671

LFM Mfg. Co.
Atchison, KA

Light Alloys Products Co. Ltd.
Munworth
England

Light Metal Works
Tokyo
Japan

Light Metals Inc.
1200 E. 24th St.
Indianapolis, IN 46205

Lincoln Electric Co.
22801 St. Clair Ave.
Cleveland, OH 44117
(216)481-8100

Lithaloys Corp.
New York, NY

Little Bros. Foundry Co.
Port Huron, MI

Little Falls Alloy Inc.
189 Caldwell Ave.
Paterson, NJ 07501
(201)525-1014

Lobdell Co.
Wilmington, DE

Logan Iron & Steel Co.
Philadelphia, PA

Friedr. Lohmann GmbH
Werk fur Spezial & Edelstahle
Postfach 3262
Ruhrtal 2
D 5810 Witten-Herbede
Germany

Lowmoor Best Yorkshire Iron Ltd.
Lowmoor
England

LTV Steel
Formerly called:
Jones & Laughlin Steel Corp.
25 W. Prospect Ave.
Room 902-L
Cleveland, OH 44115
(216)622-5000

Ludlow Steel Corp.
5439 Perkins Road
Bedford, OH 44146

Lukens Steel
Arc Bldg.
Madena Rd.
Coatesville, PA 19320
(215)383-2000

Lumen Bearing Co.
Buffalo, NY

Lunkenheimer Co.
Beekman St. at Waverly Ave.
Cincinnati, OH 45214
(513)921-3400

E. Lunn & Co.
Glasgow
Scotland

Luria Steel & Trading Co.
New York, NY

Lynchburg Foundry Co.
Drawer 411
Lynchburg, VA 24505

Lyon, Conklin & Co.
Race & McComas Sts.
Baltimore, MD 21230

Macauley Foundry Co.
811 Carleton St.
Berkeley, CA 94710
(415)845-2911

Machinenbau AG
See VEB Chemieanlagenbau Grimma

Machinery & Machine Supplies Co. Inc.
New York, NY

Mackenzie's Sons Co. Inc.
Trenton, NJ

Madison Foundry Co.
Cleveland, OH

Madison-Kipp Corp.
222 Waubesa St.
Madison, WI 53704

Magnacast Corp.
1117 East Algonquin Road
Arlington Heights, IL 60005
(312)437-6000

Magnequench
Delco Remy of GM
6435 S.Scatterfield Rd.
Anderson, IN 46013
(317)646-5050

Magnesium Castings & Products Ltd.
Slough
England

Magnesium Elektron Ltd.
Norfolk House
St. James Square
London, SW 14 4JS
England
01 839-8888

Magnetic Metals Corp.
21st St. & Hayes Ave.
P.O. Box 351
Camden, NJ 08101
(609)964-7842

Magnetic Shield Corp.
Perfection Mica Co.
740 N. Thomas Dr.
Bensenville, IL 60106
(312)766-7800

Magnolia Anti-Friction Metal Co.
34 Victoria St.
London SW 1
England

Magnolia Metal Corp.
RTE 2 Box 13M
Auburn, NE 68305
(402)274-3152

Magotteaux SA
Societe Anonyme
B-4601 Vaux-sous-Chevremont
Belgium

Mahle GmbH
Postfach 50 07 69, 50 07 80
Pragstasse 26-46
D-7000 Stuttgart 50
West Germany
0711 5011

Main Metal Ltd.
Monmouth St.
London, WC 2
England

Major Engineering Co.
1021 N. Columbia Place
P.O. Box 700
Tulsa, OK 74101

Malleable Iron Fitting Co.
50 Maple St.
Branford, CT 06405

Manco Products, Inc.
2401 Schaefer Road
Melvindale, MI 48122
(313)928-7411

Manganese Bronze & Brass Co. Ltd.
See Delta (Manganese Bronze) Ltd.
Ipswich
England

Manganese Steel Forge Co.
Harsco Corp
Philadelphia, PA 19134

Mannesmann-Huttenwerke A.G.
Duisberg
West Germany

Mannesmannrohren-Werke AG
Rohrpresswerk Reisholz
Postfach 1104
Henkelstrasse 209
4000 Dusseldorf 13
West Germany
0211 8751

Manning, Maxwell & Moore Co.
Handcock Valve Div.
618 West Ave.
Norwalk, CT 06852
(203)366-8764

Manos Ltd.
Zurich
Switzerland

Marathon Specialty Steels Inc.
See Thysson Deutsche Edelstahlwerke A.G.
Empire State Bldg.
350 Fifth Ave.
New York, NY 10001

Markey Bronze Corp.
See Bunting Bearings Corp.

Marquette Corp.
Minneapolis, MN

Marrel Freres, S.A.
Boite Postale 46
F-42800 Rive-de-Gier
France

Marsh Bros. & Co. Ltd.
Ponds Steel Works
Shude Lane
Sheffield
England

Marshall Steel Co.
P.O. Box 340
LaGrange, IL 60525
(312)447-7420

Martin Marietta Aluminum Inc.
P.O. Box 711
The Dalles, OR 97058

Martin-Marietta Corp.
Martin Metals Co.
1450 S. Rolling Road
Baltimore, MD 21227

Martinel Steel Co., Ltd.
Sheffield
England

Massey-Harris Ltd.
Toronto, ONT
Canada

Massillon Steel Casting Co.
Massillon, OH 44646

Materials Development Corp.
81 Hicks Avenue
Medford, MA 02155
(617)391-0400

Materials Research Corporation
Rte. 303 & Glenshaw Street
Orangeburg, NY 10962
(914)359-4200

Mather & Platt Ltd.
Manchester
England

Mathieson-Heglar Co.
La Salle, IL 61301

Matthey Bishop
See Johnson Matthey

Maulbronn Aluminiumwerke
Maulbronn-Wurtt
Germany

Maywood Chemical Works
Maywood, NJ

McCalls Special Products
P.O. Box 71
Hawke Street
Sheffield S9 2LN
England

McCallum-Hatch Bronze Co. Inc.
Buffalo, NY

McCauley Alloy Sales Co.
New York, NY

P.F. McDonald & Co.
Boston, MA

McGean Chemical Corp.
See McGean-Rohco, Inc.

McGean-Rohco, Inc.
McGean Group
Formerly called:
McGean Chemical Corp.
2910 Harvard Ave.
P.O. Box 09087
Cleveland, OH 44109-0087
(216)441-4900

McGill Mfg. Co. Inc.
1100 N. Lafayette St.
Valparaiso, IN 46383

McInnes Steel Co.
441 East Main Street
Corry, PA 16407
(814)664-9664

McKechnie Metals Ltd.
McKechnie Group
Middlemore Lane Aldridge
Walsall WS9 8DN
England
0922 53321

McLouth Steel Corp.
300 S. Livernois Ave.
Detroit, MI 48209
(313)843-3000

Wm. McPhail & Sons
Glasgow
Scotland

McQuay-Norris Mfg. Co.
St. Louis, MO

Medart Engineering & Equipment Co.
1213 Hadley St.
St. Louis, MO 63106

Meech Foundry Inc.
4730 Warner Rd.
Cleveland, OH 44125

Meehanite Metal Corp.
Meehanite Worldwide Division
112 Carolina Cove
Beaufort, SC 29902
(803)525-1700

Meigh Castings Co. Ltd.
Uckington Foundry
Near Swindon
Cheltenham
England

Mentah Co.
Vorgtlinder
West Berlin
Germany

Merco Nordstrom Valve Co.
Pittsburgh, PA

Meridian Steel Co. Inc.
50 Jericho Turnpike
Jericho, L.I., NY 11753
(516)334-8250

Mesta Machine Co.
P.O. Box 1466
Pittsburgh, PA 15230
(412)464-4040

Metal & Thermit Corp.
New York, NY

Metal Carbides Corp.
6001 Southern Blvd.
P.O. Box 3749
Youngstown, OH 44513
(216)788-6541

Metal Castings Ltd.
Worcester
England

Metal Sales Corp.
Jersey, NJ

Metal Specialties Inc.
515 Commerce Dr.
Fairfield, CT 06430
(203)384-0331

Metallbank AG
Heddernheimer Copper Works
Heddernheim
Germany

Metallgesellschaft A.G.
Postfach 10 15 01
D-6000 Frankfort 1
West Germany

Metallgesellschaft Reuterweg
14 Postfach 3724
D-6000 Frankfurt AMI
Germany

Metallhuttenwerke Schaefer und Schael, A.G.
Postfach 56 04
D-4000 Dusseldorf
Germany

Metallo-Chemische Fabrik
Postfach 1527
D-3012 Langenhagen (Hann.)
Germany

Metalloy Products Co.
1341 Sussex Lane
Newport Beach, CA 92660

Metalltechnik Schmidt GmbH & Co.
Schulstrasse 41
D-7024 Filderstadt 4
Germany
0711 77 10 84

Metallwerk Olsberg GmbH
Emscherstr. 63
D-4300 Essen 12
Germany
0201 36 40 50

Metallwerk Plansee Gesellschaft
See Plansee Metallwerk Gesellschaft

Metallwerke, A.G.
Vienna
Austria

Metals & Controls and General Plate

Metals & Controls, Inc.
See Texas Instruments Inc./
Material & Controls Group

Metalsource Corp.
Wheelock, Lovejoy & Co. Division
3 Commerce Park Square
23200 Chagrin Blvd.
Cleveland, OH 44122
(216)464-2835

METCO Inc.
Perkin-Elmer
1101 Prospect Ave.
Westbury, Long Island, NY 11590
(516)ED4-1300

Metro Cutanit Ltd.
(Reported to be out of business)
Grappenhall
Warrington
Lancashire
England

Metropolitan-Vickers Electrical Co. Ltd.
Manchester
England

MG Industries
Welding Products
N94 W14355 Garwin Mace Dr.
Menomonee Falls, WI 53051
(414)255-5520

Michiana Products Corp.
Michigan, IN

Michigan Powdered Metal Products Inc.
See Borg-Warner Automotive Powdered Metals Corp.
32059-T Schoolcraft Road
Livonia, MI 48150
(313)261-5322

Michigan Smelting & Refining Co.
Detroit, MI

Michigan Standard Alloy Inc.
See Standard Alloy, Inc.
P.O. Box 688
Benton Harbor, MI 49022
(616)926-1161

Michigan Steel Casting Co.
Detroit, MI

Michigan Tool Co.
Detroit, MI

Michigan Valve & Foundry Co.
Detroit, MI

Midland Motor Cylinder Co.
Smithwick
Birmingham
England

Midland Ross Corp.
See Capitol Castings Inc.

Midvale-Heppenstall Co.
Nicetown
Philadelphia, PA 19140

Millbury Steel Foundry Co.
Millbury, MA

Miller Company
Mill Division
200 Center Street
Meriden, CT 06450
(203)235-4474

Miller Steel Co.
Newark, NJ

Miller Thermal, Inc.
P.O. Box 1081
Appleton, WI 54912
(414)731-6884

William Mills & Co. Ltd.
Friar Park Foundry
British Aluminum Co.
Friar Park Road
Wednesbury
Staffordshire WS10 0JY
England
021 556 1281

A. Milne & Co.
787 Bedford St., N.W.
Atlanta, GA 30318
(404)876-0391

Milwaukee Steel Foundry Co.
Milwaukee, WI

Mir-o-Col Alloy Co.
Los Angeles, CA

Miralite Ltd.
(Reported to be out of business)
Mortlake
England

Robert Mitchell Co. Inc.
350 Decarie Blvd.
St. Laurent, Quebec H4L 3K5
Canada
(514)747-2471

Mitsubishi Metal America Corp.
520 Madison Avenue, 17th Floor
New York, NY 10022
(212)688-9550

Molecu-Wire Corp.
Superior Tube Co.
P.O. Box 495
Farmingdale, NJ 07727
(201)922-9400

Molybdenum Co., N.O.
Reutte-Tyrol
Austria

Molybdenum Corp. of America
See Molycorp, Inc.
280 Park Ave.
New York 10017

Molycorp, Inc.
6 Corporate Park Dr.
White Plains, NY 10604

Monarch Alloy Co.
Ravenna, OH

Monarch Steel Corp.
Indianapolis, IN

Mond Nickel Co. Ltd.
London
England

Montecatini Settore Alluminio
Milan
Italy

Moraine Mfg., Inc.
1212 W. 2nd St.
P.O. Box 678
Oconomowoc, WI 53066

Morgan Crucible Co. Ltd.
London
England

Morris Ashby Ltd.
London
England

Morris Machine Works
See Morris Pumps, Inc.
Baldwinsville, NY
(315)635-3931

Morse Cutting Tools
Formerly called:
Super Tool Co.
First and Bridge St.
Elk Rapids, MI 49629

Motor Castings Co.
1323 S. 65th St.
Milwaukee, WI 53214
(414)476-1434

Mt. Vernon Furnace & Mfg. Co.
Mt. Vernon, IL

Mueller Brass Co.
1925 LaPeer Avenue
Port Huron, MI 48060
(313)987-4000

Multicore
Cantiague Rock Rd.
Westbury, NY 11590
(516)334-7997

P.H. Muntz & Co., Ltd.
West Bromwich
Birmingham
England

Murex Ltd.
Rainham
Essex RM13 9DP
England

Muskegon Piston Ring Co.
1839 Sixth St.
Muskegon, MI 49443

Myrens Verkstec A/S
Oslo
Norway

Nagle/Sybron Corp.
Agile Division
Rochester, NY

Napraloy Co.
Jersey, NJ

Nassau Smelting & Refining Co. Inc.
Staten Island, NY 10307

National Airoil Burner Co.
1264 E. Sedgley Ave.
Philadelphia, PA 19134

National Alloys Ltd.
London
England

National Broach & Machine
Detroit, MI

National Bronze & Aluminum Foundry
Cleveland, OH

National Cable & Metal Co.
Glendale, CA

National Castings, Inc.
Formerly called:
Surface Combustion
1414 East Broadway St.
Toledo, OH 43605
(419)698-4341

National Cleveland Corp.
National Tool Division
Cleveland, OH

National Cylinder Gas
Hollup Division
Chicago, IL

National Erie Corp.
Erie, PA

National Forge Co.
One Front St.
Irvine, PA 16329
(814)563-7522

National Intergroup Inc.
Formerly called:
National Steel Corp.
20 Stanwix St.
Pittsburgh, PA 15222
(412)394-4100

National Physical Laboratory
Teddington
England

National Radiator Co.
Plastic Metals Division
Johnstown, PA

National Research Corp.
Metals Division
Newton, MA 02158

National Standard Co.
Athenia Steel Division
Clifton, NJ

National Steel Corp.
Granite City Division
See National Intergroup, Inc.
20th & State Sts.
Granite, IL 62040
(618)451-3456

National Supply Co.
Spang Chalfont Division
Pittsburgh, PA

Navan Inc.
P.O. Box 3105
Anaheim, CA 92803

Nesaloy Products Inc.
New York, NY

Neubrandenburg Eisenwerke GmbH
Berlin
Germany

Neumayer Kabel u. Metallwerke A.G.
Nurnberg
Germany

New England Brass Company
18 Park Ave.
Taunton, MA 02780
(617)824-5821

New England Collapsible Tube Co.
New London, CT

New England High Carbon Wire Co.
Millbury, MA

New Jersey Zinc Co., Inc.
Palmerton, PA 18071
(215)826-2111

Newcomer Products Inc.
Box 272
Latrobe, PA 15650

Newport Steel Corp.
Interlake, Inc.
Ninth and Lowell Streets
Newport, KY 41072
(606)292-6820

J.M. Ney Co.
Ney Industrial Park
Bloomfield, CT 06002
(203)242-2281

NGK Metals
P.O. Box 13367
Reading, PA 19612
(215)921-5000

Niborium Industries Inc.
960 Broad St.
Providence, RI 02905
(401)467-2840

Nicralium Co.
Jackson, MI

Nihon Jyokiko Seikosho Goshi
Koisha
Japan

Nippert Co.
801 Pittsburgh Dr.
Delaware, OH 43015
(614)363-1981

Nippon Kokan K.K.
See NKK Corp.

Nippon Stainless Steel Co. Ltd.
c/o Sumitomo Metal Ind. Ltd.
420 Lexington Ave.
New York, NY 10017

Nippon Steel Corp.
6-3 Otemachi 2-chome
Chiyoda-ku
Tokyo 100
Japan
03 242 4111

Nippon Steel U.S.A. Inc.
345 Park Ave.
New York, NY 10022
(212)826-6250

Nitralloy Corp.
New York, NY 10022

NKK Corp.
Formerly called:
Nippon Kokan K.K.
1-2-1 Marunouchi
Chiyoda-ku
Tokyo 100
Japan
03 212-7111

NL Industries
Bearings Division
5461 Southwyck Blvd.
Toledo, OH 43614
(419)385-9911

Non-Corrodal Alloys Ltd.
London
England

Non-Ferrous Castings Co. Ltd.
Smethwick
England

Noranda Metal Industries Ltd.
P.O. Box 1158, Sta. "A"
Montreal, QUE H3C 2Y4
Canada

Nordisk Aluminium Industry, A/S
Holmestrand
Norway

Norsk Aluminium Co. A/S
Hoyanger
Norway

Norsk Hydro
Magnesium Division
P.O. Box 2594, Solli
Bygdoy alle 2
N-Oslo 2
Norway

North American Phillips Co.
Ferroxcube Corp. Division
Mt. Marion Rd.
Saugerties, NY 12477

North American Steel Corp.
18030 Riatto
Melvindale, MI 48122
(313)383-7100

Northeast Metals Co.
Philadelphia, PA

Northfield Iron Co.
Northfield, MN

Northwest Alloys, Inc.
P.O. Box 115
Addy, WA 99101

Norton Co.
1 New Bond St.
Worcester, MA

Nova Petrochemicals, Inc.
Formerly called:
Polymer Corp. Ltd.
Box 3001
Sarnia, ONT N7T 7M2
Canada

NRC, Inc.
45 Industrial Place
Newton, MA 02164
(617)969-7690

Nuclear Metals Inc.
2229 Main St.
Concord, MA 01742
(508)236-3119

Nurnberg Aluminiumwerke GmbH
Nurnberg
Germany

Nusite Steel Process Co.
Warsaw, IN

Nyby, Granges, AB
S-644 80 Torshalla
Sweden

Oakes Bronze & Aluminum Co.
P.O. Box 871
Warren, OH 44482

Oederlin & Co. Ltd.
Baden
Switzerland

Ohio Brass Co.
North Main St.
Mansfield, OH 44902
(419)522-711

Ohio Ferro Alloys Corp.
See SiMETCO

Ohio Stainless & Commercial Steel Co.
Cleveland, OH

Oil Well Supply Co.
Houston, TX

Oklahoma Steel Castings Co.
P.O. Box 2709
Tulsa, OK 74101

F.B. Oldham & Co.
Buffalo, NY

Olds Alloys Co.
South Gate, CA 90280

Olin Brass
Olin Brass Division
427 Shamrock St.
East Alton, IL 62024
(618)258-2000

Olin Brass, Indianapolis
Formerly called:
Bridgeport Brass Co.
P.O. Box 51519-T
South Holt Rd. & Minnesota St.
Indianapolis, IN 46251
(317)244-2461

Olin Corp.
Longridge Rd.
Stamford, CT 06904

Olin Metals Research Labs
91 Shelton Ave.
P.O. Drawer 906
New Haven, CT

Oman Non-Friction Metal Co.
El Paso, TX

Oregon Steel Mills
P.O. Box 2760
Portland, OR 97208
(503)286-9651

Osborn Steels Ltd.
Osborn Group
Nether Lane Ecclesfield
Sheffield S30 3ZU
England
3111 0741 5

Oscap Mfg. Co.
New York, NY

Osnabruck Kupfer und Drahtwerke
See KM-kabelmetal AG

W. Ossenberg & Cie Edelstahlwerk
Postfach 7044
5990 Altena 7
Germany
02352 71033

Ostermann GmbH & Co.
Postfach 301029
D-5000 Koln 30
Germany

Otis Elevator Co.
521 5th Ave.
New York, NY 100175

Outokumpu Metals (USA) Inc.
6 Parklane Blvd., Suite 420
Dearborn, MI 48126
(313)271-3210

Outokumpu Oy
Central Mgmt.
Toolonkatu 4
P.O. Box 280
00100 Helsinki
Finland
90 4031

Outokumpu Oy Terasteollisuus
Tornio Works
SF-95400 Tornio
Finland
00358 6984521

Ovako Steel Hellefors AB
Formerly called:
Hellefors Bruks Aktiebolag
Box 77
Hallefors S-712 80
Sweden
591 60 000

William Oxley & Co.
Rotherham
England

P.L. & M. Co.
Los Angeles, CA

Pacific Foundry Co.
San Francisco, CA

Pacific Metal Co.
Portland, OR

Pacific Welding Alloys Mfg. Co.
Los Angeles, CA

Page Aluminized Steel
Formerly called:
American Chain & Cable
100 Monongahela St.
Monessen, PA 15062

Paragon Steel Co.
339 Bergen Ave.
Kearny, NJ 07032
(201)997-1676

Paraloy Co.
Chicago, IL

Park Sales Co.
New York, NY

Parker Appliance Co.
Cleveland, OH

Parker Pen Co.
Janesville, WI 53545

Parker-Kalon Corp.
1140 Rte. 46
Clifton, NJ 07013
(201)471-3700

Parkersburg Steel Co.
Superior Sheet Steel Division
Box 268
Parkersburg, WV 26101

F.M. Parkin Ltd.
Wardsend Road North
Wadsley Bridge
Sheffield S6 1LL
England

Pasminco Europe (Mazak) Ltd.
Formerly called:
Imperial Smelting Co. Ltd.
P.O. Box 237
1 Redcliff Street
Bristol, BS99 7EA
England
0272 215491

Patriarche & Bell
New York, NY

Thomas Paulson & Sons Inc.
Union & Bond St.
Brooklyn, NY 11231

Pechiney
23, rue Balzac
Paris 75008
France

Pechiney Corporation
475 Steamboat Rd.
Greenwich, CT 06836-1960
(203)661-4600

Pechiney Electrometallurgie
Tour Manhattan-Cedex 21
92087 Paris La Defense
France

Pechiney World Trade (USA) Inc.
Pechiney Corporation
500 Plaza Dr.
Secaucus, NJ 07094
(201)863-2400

Peckovers, Ltd.
2195 Langstaff
Concord, Ontario L4K 2B2
Canada
(416)661-2276

Pelton Casteel, Inc.
148 West Dewey Place
Milwaukee, WI 53207
(414)481-3400

Peninsular Steel Co.
Box 05053, Parkgrove Station
Detroit, MI 48205
(313)371-9400

Penn-Dixie Corp.
See Continental Steel Corp.

Pennsylvania Steel Corp.
Detroit, MI

Pennsylvania Steel Foundry & Machine Co.
Third & Arch Sts.
P.O. Box 128
Hamburg, PA 19526

Permanent Magnet Association
(Reported to be out of business)
London
England

Permanente Magnesium Inc.
(Reported to be out of business)
Oakland, CA

Permo Inc.
Chicago, IL

Permold Inc.
P.O. Box P
Medina, OH 44256
(216)723-3251

Perry Barr Metal Co., Ltd.
Birmingham
England

Perry Equipment Corp.
Mount Laurel Road
Hainesport, NJ 08036
(609)267-1600

Peterson Steels Inc.
P.O. Box 157
Union, NJ 07083

Pettibone Corp.
Beardsley & Piper Division
5503 W. Grand Ave.
Chicago, IL 60639

Pfizer Inc.
New York, NY 10842

Phelps Dodge Industries
Phelps Dodge Copper Products Co.
300 Park Ave.
New York, NY 10022
(212)751-3200

N.V. Philips Co.
Eindhover
Netherlands

Philips Electronic & Associated Industries Ltd.
Philips, Netherlands
11/12 Hanover Square
London, W 1
England

C.E. Phillips & Co.
2748 Poplar at Charles
Detroit, MI 48208

Phoenix Steel Corp.
4001 Philadelphia Pike
Claymont, DE 19703
(302)798-1411

Phosphor Bronze Co. Ltd.
GKN Group
P.O. Box 74
Birchall Street
Birmingham B12 0NA
England
021 622 45252

Pickands Mather & Co.
See Pickands Mather Sales Co., Inc.

Pickands Mather Sales Co., Inc.
Formerly called:
Pickands Mather & Co.
1100 Superior Ave.
Cleveland, OH 44114

Pioneer Alloy Products Co.
Cleveland, OH

Pioneer Metals, Inc.
Pioneer Aluminum Inc. Division
3800 East 26th Street
Los Angeles, CA 90023

Pipe Machinery Corp.
(P.M.C. Industries)
Cleveland, OH

Pittsburgh Brass Mfg. Co.
Sandy Hill Road
RD 6, Box 387A
Irwin, PA 15642

Pittsburgh Metallurgical Co. Inc.
Niagara Falls, NY

Pittsburgh Steel Foundry Corp.
Glassport, PA

Plansee Metallwerk Gesellschaft
Postfach 74
6600 Reutte-Tirol
Austria
05672/70 0

Plattawerke GmbH
West Berlin
Germany

Plessey Inc.
Metals Division
320 Long Island Expressway South
Melville, NY 11746
(516)694-7910

Plettenberger Gusstahlfabrik
Postfach 3340
5970 Plettenberg 2
West Germany

Plew Tool Co.
Columbia, IN 46725

Plouff Metallographic Institute
Boston, MA

Plykrome Corp.
New York, NY

Poldi Steel Works
Prague
Czechoslovakia

Polymer Corp. Ltd.
See Nova Petrochemicals, Inc.
Box 3001
Sarnia, ONT N7T 7M2
Canada
(519)337-8251

Pont-St.-Martin
Officine Metallurgiche d'
France

Poro Metals Ltd.
London
England

H.K. Porter Co.
Alloy Metal Wire Works
Porter Bldg.
Pittsburgh, PA 15219
(412)391-1800

H.K. Porter Co.
Vulcan-Kidd Division
Aliquippa, PA 15001

Pose-Marre Edelstahlwerk G.m.b.H.
Postfach 140
Gerberstrasse 26
4006 Erkrath
Germany
0211 24030

Horace T. Potts Co.
Erie Ave. at D Street
Philadelphia, PA 19134
(215)426-4600

Powder Alloy Corporation
12036 Cooperwood Ln.
Cincinnati, OH 45242
(513)984-4016

Powder Alloys Corp.
Clifton, NJ

Powder Metal Products Co.
Franklin Park, IL

Powder Metals Inc.
Long Island, NY

Power Products Inc.
Hunt-Spiller Division
180 Sylvester Rd.
P.O. Box 2607
South San Francisco, CA 94080

J.M. Pratt & Co.
Cincinnati, OH

Pratt & Whitney Cutting Tool & Gage
Colt Industries; 1 Charter Oak Blvd.
West Hartford, CT 06101

Precious Metals Research Works Inc.
320 Washington St.
Mt. Vernon, NY 10553

Precision Casting Co.
Fayetteville, NY 13066

Precision Castparts Corp.
4600 S.E. Harney Drive
Portland, OR 97206
(503)777-3881

Precision Rolled Products, Inc.
14255 Mt. Bismark St.
P.O. Box 60010
Reno, NV 89506
(702)972-0272

Precision-Kidd Steel Co.
Drawer D
Aliquippa, PA 15001
(412)378-0084

George W. Prentiss & Co.
Holyoke, MA

Prescott Co.
Menominee, MI

Presmet Corp.
Mass. Steel Treating Division
Formerly called:
GKN Powder Met Inc.
112 Harding St.
Worcester, MA 01604
(508)792-6400

Pressco Casting & Mfg. Corp.
Chesterton, IN

Pressed Steel Car Co.
Chicago, IL

Thos. Prosser & Sons
New York, NY

William Prym-Werke GmbH & Co. KG
Postfach 1740
Zweifaller Strasse 130
D-5190 Stolberg
Germany

Puget Sound Metal Works
Tacoma, WA

Pullman Inc.
See M.W. Kellog

A.R. Purdy Co. Inc.
2400 W. 95th St.
Chicago, IL 60642
(312)239-4200

Pyramid Steel Company
18400 Miles
Cleveland, OH 44128

PYRON Corp.
Zemex Corp.
Box E, LaSalle Station
Niagara Falls, NY 14304-0305

Q. & C. Co.
New York, NY

Qit-Fer et Titane
79 West Monroe St.
Ste. 1200
Chicago, IL 60603
(312)236-1761

Quaker Alloy Casting Co.
720 S. Cherry St.
Myerstown, PA 17067
(717)866-6511

R-S Products Corp.
Philadelphia, PA

Rajo Motor Company
Belle City Malleable Cast Iron Co. Division
(Reported to be out of business)
Racine, WI 53403

Randall Graphite Products Co.
Chicago, IL

Rankin Mfg. Co.
Mark K. Industries
9720 Distribution Ave.
San Diego, CA 92121
(800)854-2159

Rasmussen Mfg. Co.
Hollydale, CA

J.F. Ratcliff Metals Ltd.
Birmingham
England

Rautaruukki Oy
Raahe Steel Works
SF-92170 Raahensalo
Finland
358/82 301

Raven Steel & Tool Co.
Valley Stream, NY

**RCA Corporate Standards
Engineering**
David Sarnoff Research Center
Camden, NJ 08101

Reactor Experiments, Inc.
963 Terminal Way
San Carlos, CA 94070-3278
(415)592-3355

Reading Alloys, Inc.
P.O. Box 53
Robesonia, PA 19551

Reading Iron Co.
Reading, PA

Redhard Metals Inc.
Hatboro, PA

Regie Nationale des Usines Renault
St. Michael de Maurienne,
55 bd Charonne
Paris 11e
France

Reid-Avery Co.
Cleveland & Chesapeake Ave.
Baltimore, MD 21222

Reinforcement Steel Services
P.O. Box 14
Meadowhall Road
Sheffield S9 1ED
England

Reliance Steel Casting Co.
Pittsburgh, PA

Rely Metal Works
Cape Town
South Africa

Remanit GmbH
Berlin
West Germany

Remington Arms Co. Inc.
Hi-Dense Parts Division
14 Hoefler Ave.
Ilion, NY 13357
(315)894-9961

Remystahl
P.O. Box 1340
D-58 Hagen 1
West Germany
2331 3870

Renewal Services Inc.
17th St. & Lehigh Ave.
Philadelphia, PA 19132

Renfrew Foundries Ltd.
Hillingdon
Glasgow
Scotland

Rennie Tool Co. Ltd.
Manchester
England

Republic Steel Corp.
See LTV Steel Corp.
P.O. Box 6778
Cleveland, OH 44101

Resisto-Loy Company, Inc.
Stanwood Corporation Division
See Grand Northern Products Ltd.
1251 Phillips Avenue S.W.
Grand Rapids, MI 49507
(616)243-7920

Revere Copper Products, Inc.
P.O. Box 300
Rome, NY 13440
(315)338-2554

Rex Buckeye Co.
7508 Associate Ave.
Cleveland, OH 44144

Reynolds Aluminiumwerke G.m.b.H.
Postfach 9, Hegerer Strasse 147
D-5992 Nachrodt/Westf.
West Germany

Reynolds Aluminum
Reynolds Metals Co.
P.O. Box 27003
6601 West Broad St.
Richmond, VA 23261
(804)281-2000

Reynolds Metals Co.
Reduction Division, Listerhill Reduction Plant
E. 2nd Street
P.O. Box 191
Sheffield, AL 35660

**Rheinische Rohrenwerke
Aktiengesellschaft**
Mulheim
West Germany

R.W. Rhodes Metaline Co.
Long Island, NY

Rieger Gebr.
Aalen
West Germany

J.J. Rieter & Co.
New York, NY

Riken Metal Mfg. Co.
Ube
Japan

Rio Algom Corp.
Atlas Steels Division
Cleveland, OH

Riverside Foundry & Galvanizing Co.
Kalamazoo, MI

Riverside Metals Corp.
Riverside, NJ 08075

RMI Company
1000 Warren Ave.
Niles, OH 44446
(216)544-7633

Robbins & Myers, Inc.
1311 Lagonda Ave.
Springfield, OH 45503
(513)328-5100

Robert-Leyer-Pritzkow & Co.
Solingen-Ohligs
West Germany

**Robins Engineers & Constructors
Inc.**
Formerly called:
Robins Conveyors Inc.
711 Union Blvd.
Totowa, NJ 07511
(201)256-7600

Rochling Burbach GmbH
See Saarstahl AG

Rolle Mfg. Co.
Lansdale, PA

Rolled Alloys
125 W. Sterns Rd.
Box 310
Temperance, MI 48182
(313)847-0561

Rolls-Royce Mfg. Co.
Derby
England

Rollschen Eisenwerke, A.G.
Gesellschaft der Ludv. von
Gerlafingen
West Germany

Ronson Metals Corp.
45-65 Manufacturers Place
Newark, NJ 07105
(201)624-1380

Rorschach Aluminiumwerke
Roschach
Switzerland

Ross-Meehan Foundries
P.O. Box 1258
Chattanooga, TN 37401

Rudolf Schmidt Stahlwerke
Vienna
Austria

Ruhrstahl A.G.
Henrichshutte
Hattingen/Ruhr
West Germany

H. Russell & Co. Ltd.
Sheffield, 4
England

Ryer Inc. Ltd.
Los Angeles, CA

Joseph T. Ryerson & Son Inc.
Inland Steel Co.
P.O. Box 8000-A
Chicago, IL 60680
(312)762-2121

S-M Metal Works
London
England

S.I.P.I. Metals Corp.
1722 N. Elston Ave.
Chicago, IL 60622

S.K.F. Industries Inc.
1100 First St.
King of Prussia, PA 19406

Saarstahl AG
DHS
Formerly called:
Rochling Burbach GmbH
Bismarckstrasse 57-59
Postfach 10 19 80
Volklingen D-6620
West Germany
0 68 98 10-1

**SAFE (Societe des Aciers Fins de
L'est)**
40 rue de Paris
Boite Postale 133
92106 Boulogne-Billancourt
France
1 605 89 10

Saginaw Bearing Co.
1400 Agricola Drive
Saginaw, MI 48605

Sanderson Kayser Ltd.
P.O. Box No. 6
Newhall Road
Sheffield S9 2SD
England
0742 449 994

Sandusky Foundry & Machine Co.
615 West Market St.
P.O. Box 5012
Sandusky, OH 44871-8012
(419)626-5340

Sandvik Hard Materials Ltd.
Formerly called:
Wimet Ltd.
P.O. Box 89
Torrington Ave.
Coventry CV4 9XG
England
0203 465 911

Sandvik Steel Co.
P.O. Box 1220
Scranton, PA 18501
(717)587-5191

Sandvik/Coromant
1702 Nevins Rd.
P.O. Box 428
Fairlawn, NJ 07410-0428
(201)794-5000

Sankey & Sons Ltd.
London
England

Sargent & Co.
100 Sargent Drive
New Haven, CT 06509

J.J. Saville & Co. Ltd.
Sheffield
England

Schaeferun Schael, A.G.

Schenk Leichtgusswerke
Maulbronn
West Germany

Karl Schmidt Co.
Metall Schmelzwerk
Kupferstrasse 4
D-7000 Stuttgart 80, Baden-Wurttemberg
West Germany
0711 78 020 77

Schmidt & Clemens Edelstahlwerke
5251 Kaiserau Bez
Koln uber Engelskirchen
Germany

Schmole GmbH
See KM-kabelmetal AG
Postfach 6 20
5750 Menden 1
West Germany

Schneider & Cie
See Creusot-Loire
42 Rue d'Anjou
Paris 75008
France

Schulte Eisenhandlung GmbH
Postfach 201/202
4600 Dortmund
West Germany

Schwaebische Huttenwerke, GmbH
7080 Aslen-Wasseralfingen
West Germany

Scientific Alloys Inc.
379 Old Hopkinton Rd.
P.O. Box 523
Westerly, RI 02891
(401)596-4947

Scientific Products Inc.
1546 E. Grand Blvd.
Detroit, MI 48211

SCM Metal Products Inc.
1468 W. 9th St.
Cleveland, OH 44113
(216)344-8800

SCM Pigments
Formerly called:
Glidden Metals
3901 Glidden Road
Baltimore, MD 21226
(301)355-3600

A.C. Scott & Co. Ltd.
Manchester
England

Scovill Mfg. Co.
See Century Brass Products, Inc.
Waterbury, CT 06720

Seaboard Steel Co. of America
New York, NY

Secon Metals Corp.
7 Intervale St.
White Plains, NY 10606
(914)949-4757

Seibel Metallhuttenwerke Mettmann
Mettmann D-4020
West Germany

W. Seibel Ag
Mettmann Rhld
Germany

Seitzinger's Inc.
Atlanta, GA 30318

Semi-Alloys
888 S. Columbus Ave.
Mt. Vernon, NY 10550
(914)664-2800

Senaca Wire & Mfg. Co.
335 South Vine Street
P.O. Box B
Fostoria, OH 44830
(419)435-9261

SERMAG
Societe d'Etudes et de Recherches
Magnitiques
38 St.-Martin-d'Heres
France

Sessions Foundry Co.
Bristol, CT

Seymour Brass Turning Co.
Formerly called:
Seymour Products Co.
105 Day St.
P.O. Box 110
Seymour, CT 06483

Seymour Products Co.
See Seymour Brass Turning Co.

Shanks & Co. Ltd.
Barrhead
Scotland

Sharon Steel Corp.
P.O. Box 270
Farrell, PA 16121
(412)983-6000

J. Shaw, Son & Greenhalgh, Ltd.
Albert Works
Huddersfield
England

Shawinigan Chemicals, Ltd.
Montreal
Canada

Sheepbridge Alloy Castings Ltd.
Sheepbridge Engineering Ltd.
Hamilton Road
Sutton-in-Ashfield, NG17 5LL
England
0623 511152

Sheepbridge Engineering Ltd.
Sheepbridge Works
Chesterfield
Derbyshire S4I 9QD
England

Sheepbridge Equipment Ltd.
Chesterfield
England

Sheffield Forgemaster Ltd.
Formerly called:
Firth Brown Ltd.
Don Valley House
Saville St.
Sheffield S4 7US
England
44 0742-20081

Sheffield Smelting Co. Ltd.
P.O. Box 28
Windsor St.
Sheffield S4 7WD
England
0742 20966

Sherman & Co.
New York, NY

Siemans & Halske AG.
Berlin
West Germany

Siemens AG
1000 Berlin-8520 Erlangen
Germany

Siemens-Schuckert AG.
Berlin
West Germany

Sifbronze
Suffolk Works
Birmad Qualcast Ltd.
Formerly called:
Suffolk Lawn Mowers
Stowmarket
Suffolk IP14 1EY
England
0449 612183

Sight Feed Generator Co.
Richmond, IN

Silicum Pistons Ltd.
London
England

Silver Creek Precision Co.
Metal & Alloy Division
See TIFA Ltd.

SiMETCO
Formerly called:
Ohio Ferro Alloys Corp.
P.O. Box 8228
Canton, OH 44711
(216)492-5110

Simonds Worden White Co.
Dayton, OH

Simpson Bros. Machine Works
2204 Gallia St.
Portsmouth, OH 45662

Sintered Products Ltd.
Sheepbridge Engineering Group
Hamilton Road
Sutton-in-Ashfield
Nottinghamshire NG17 5LL
England
0623 511152

Sintox Corp. of America
Allentown, PA

Sioux Tools Inc.
2802 Floyd Blvd.
Sioux, IA 51102

Sitkin Smelting & Refining Co.
P.O. Box 708
Lewistown, PA 17044

Sivyer Steel Corp.
Riverside Division
225 South 33rd Street
Bettendorf, IA 52722
(319)355-1811

Skandinaviska Armaturfabriken AG.
Stockholm
Sweden

Skoda Works National Corp.
316000 Plzen
Czechoslovakia

Slater Steels
Fort Wayne Specialty Alloys Division
Joslyn Mfg. & Supply Co.
Formerly called:
Joslyn Stainless Steels Co.
2400 Taylor Street West
Ft. Wayne, IN 46801
(219)432-2561

SMC (Shieldalloy Metallurgical Corp.)
West Boulevard
Newfield, NJ 08344
(609)692-4200

A.O. Smith Co.
P.O. Box 584
Milwaukee, WI 53201

A.O. Smith-Inland Inc.
(Reported to be out of business)
Milwaukee, WI 53201

SMS Corp.
Detroit, MI

J.L. Snowbar
New York, NY

Soc. Alluminio Veneto per Azioni
Porto Marghera
Milan
Italy

Societa Metallurgica Italia
See Europa Metalli-LMI SpA

Societe Alsia
Paris
France

Societe Anon. de Commentry
See Creusot-Loire
Fourchambault et Decazeville

Societe de Vente de l'Aluminium Pechiney
Pechiney
23 bis, rue Balzac
75008 Paris
France

Societe des Acieries de Longwy
Mont-Saint-Martin
France

Societe des AFC
Paris
France

Societe des AFY
Paris
France

Societe des Brevets Berthelmy
Paris
France

Societe du Duralumin
Paris
France

Societe Electro-Cables
Paris
France

Societe Francais d'Electrometallurgie (SOFREM)
Pechiney
10 rue du General Foy
75361 Paris Cedex 08
France

Societe Metallurgique de Gerzat (SMG)
See Pechiney
63360 Gerzat
France

Societe Nouvelle des Acieres de Pompey
47 rue de Villiers
Boite Postale
92202 Neuilly-sur-Seine
France
758 12 15

Societe Nouvelle du Saut-du-Tarn
b avenue de Messine
Forges et Acieries
75008 Paris
France

J.C. Soding & Halbach
Postfach 4241
58 Hagen/Westfalen
West Germany
02331 3961

Solar Basic Industries
Lindberg Hevi Duty Division
2450 W. Hubbard St.
Chicago, IL 60612

Soldering Specialties Co.
Summit, NJ

Sonken-Galamba Corp.
Kansas, KS

Sorcery Metals
1045 E. Atlantic Ave., Suite 313
Delray Beach, FL 33444
(305)276-3851

Southern Malleable Iron Co.
East St. Louis, IL

Southern Metals Co.
St. Louis, MO

Southwire Co.
P.O. Box 1000
Carrollton, GA 30019
(404)832-4242

Sowers Mfg. Co.
Buffalo, NY

Spang Industries/Magnetics Div.
See Spang Specialty Metals

Spang Specialty Metals
Formerly called:
Spang Industries/Magnetics Div.
P.O. Box 391
Butler, PA 16003-0391

Spartan Redheugh Ltd.
Teams
Gateshead
Tyne & Wear NE8 1RD
England

Spear & Jackson (Industrial) Ltd.
Aetna Works
Saville Street East
Sheffield S4 7UR
England
0742 20202

Special Metals Corp.
Middle Settlement Rd.
New Hartford, NY 13413
(315)797-3575

Specialloy Inc.
4025 S. Keeler Ave.
Chicago, IL 60632
(312)376-2612

Specialty Steel Co. of America
Box 7080
Warrensville Heights, OH 44128

Spencer Clark Metal Industries Ltd.
Warren Street
Sheffield S4 7WR
England

Sperry Gyroscope Co.
Great Neck, NY 11020

SPS Industries Inc.
4485 Rt. 8
Alliston Park, PA 15101

SPS Technologies
Formerly called:
Standard Pressed Steel Co.
Highland Ave.
Jenkintown, PA 19046
(215)884-7300

Spuck Iron & Foundry Co.
St. Louis, MO

St. Jacques, Societe des Usines
Montlucon
France

St. Joseph Lead Co.
Monaca, PA

St. Lawrence Steel Co.
18200 S. Miles Parkway
Cleveland, OH 44128

Stackpole Carbon Co.
See Contact Technologies, Inc.

Stackpole Magnet Division
700 Elk Ave.
Kane, PA 16735
(814)837-7000

Stahlwerk Ergste GmbH & Co.
Letmather Str. 69
Schwerte-Ergste
Germany
0 23 04 7 90

Stahlwerk Stahlschmidt GmbH & Co.
Postfach 19 01 20
4000 Dusseldorf 11
West Germany
0211-5601-0

Stahlwerke Kabel, C.
Pouplier
Hagen-Kabel
West Germany

Stahlwerke R. & H. Plate
5880Ludenscheld-Platehof
West Germany
02351 439-0

Stahlwerke Sudwestfalen
See Krupp Stahl AG

Stainless Foundry & Engineering Inc.
5150 N. 35th Street
Milwaukee, WI 53209
(414)462-7220

Stanadyne
Western Steel Division
P.O. Box 4007
Elyria, OH 44036-2007
(216)323-5471

Standard Alloy, Inc.
Formerly called:
Michigan Standard Alloy Inc.
P.O. Box 688
Benton Harbor, MI 49022

Standard Brake Shoe & Foundry Co.
P.O. Box 7677
Pine Bluff, AR 71602
(501)534-0141

Standard Pressed Steel Co.
See SPS Technologies

Stanley Works
1000 Stanley Drive
New Brittain, CT 06053
(203)225-5111

L.S. Starrett Co.
121 Crescent St.
Athol, MA 01331
(617)249-3551

Stauffer Chemical Corp.
Stauffer Metals Division
Nyala Farm Rd.
Westport, CT 06880

Stavanger Co.
Oslo
Norway

Ste d'Escaut & Meuse
Paris 17e
France

Ste de Produits Metallurgiques
Paris
France

Ste des Acieries de Micheville
Paris
France

Ste Metallurgique de Knutange
Paris
France

Stedman Foundry & Machine Co.
P.O. box 209
Aurora, IN 47001

Steel Co. of Canada Ltd.
Hamilton 23
Montreal
Canada

Steel Co., Ltd.
Motherwell
England

Steinvertriebs Aktiengesellschaft
Berlin
Germany

Steirische Gusstahlwerke, A.G.
See Vereinigte Edelstahlwerke
Vienna
Austria

Stelco Steel
P.O. Box 2030
Hamilton, L8N 3T1 ONT
Canada
(416)528-2511

Stellite Div. Colt Industries
See Crucible Materials Corp.
Formerly called:
Victor Mfg. Co.

Stellite Division
Cabot Corp.
1020 W. Park Ave.
Kokomo, IN 46901
(317)457-8411

Stellram Ltd.
Formerly called:
Tungsten & Molybdenum Ltd.
Case Postale 266
1260 Nyon-Suisse
Switzerland
022 61 31 01

Sterling Alloy Casting Corp.
Formerly called:
Sterling Alloys Inc.
302 Miller St.
Sterling, IL 61081
(815)625-3585

Sterling Alloys Inc.
See Sterling Alloy Casting Corp.

Sterling International Technology Ltd.
Birmid Qualcast Ltd.
Formerly called:
Sterling Metals Ltd.
Gypsy Lane
Nuneaton, Warwickshire CV11 4TX
England
44 203 384221

Sterling Metals Ltd.
See Sterling International Technology Ltd.

Sterlite Foundry & Mfg. Co.
Auburn, IN

Stewart Warner Corp.
Alemite Division
1826 Diversey Pkwy.
Chicago, IL 60614

Stewart-Warner
Die Casting Division
1826 Diversey Pkwy.
Chicago, IL 60614

Stone Manganese - J. Stone & Co. Ltd.
Anchor and Hope Lane
London, SE 17
England

Stoody Company
Wrap Division
16425 Gale Avenue
P.O. Box 1901
City of Industry, CA 91744
(818)968-0717

Strasser Co.
E. Rorschache
Switzerland

Stroh Process Steel Co.
See SPS Industries Inc.

Strong, Carlise & Hammond Co.
11760 Berea
Cleveland, OH

Studebaker Chemical Co.
Elyria, OH 44035

Stulz Sickles Steel Co.
P.O. Box 273
930 Julia St.
Elizabeth, NJ 07207
(800)351-1776

Stupakoff Laboratories Inc.
Pittsburgh, PA

Styria-Stahl Steirische Gusstahlwerke AG.
See Vereingte Edelstahlwerke
Vienna
Austria

Sudwestfalen Stahlwerke AG.
See Krupp Stahl AG

Suffolk Iron Foundry Ltd.
Stowmarket
England

Suffolk Lawn Mowers
See Sifbronze

Sulzer Brothers, Inc.
Zurcherstrasse
9 Winterthur CH-8410
Switzerland
052/8111 22

Sumet Corp.
Buffalo, NY

Sumitomo Metal America Inc.
420 Lexington Ave
New York, NY 10017
(212)949-4760

Sumitoms Trading Co.
Osaka
Japan

Super Tool Co.
See Morse Cutting Tools
First and Bridge St.
Elk Rapids, MI 49629
(616)264-8151

Superform Metals Ltd.
British Alcan Aluminium plc
P.O. Box 150
Worcester
England

Superior Die Casting Corp.
Advanced Technology Die Casting Group
1001 London Road
Cleveland, OH 44110
(216)481-3050

Superior Foundry Co.
Cleveland, OH

Superior Metal Co.
Chicago, IL

Superior Steel Corp.
Pittsburgh, PA

Superior Tube Company
P.O. Box 191
Norristown, PA 19404-0191
(215)489-5200

Supersteels Inc.
Cleveland, OH

Surahammars Bruk
Box 201
S-73500 Surahammar
Sweden
46 220 345 00

Surface Combustion
See National Castings, Inc.
Formerly called:
Midland Ross Corp.

Sutcliff, Speakman & Co.
Leigh
Lancaster
England

Svenska Metallverken A.B.
Granges Aluminium
Vasteras S-72188
Sweden

SWB Stahlformguss Gesellschaft mbH
Formerly called:
Bochum Stahlwerk AG
Postfach 10 24 10
Bochum 1
Germany

Swedish American Steel Corp.
Brooklyn, NY

Swedish Crucible Steel Co.
Detroit, MI

Swedish Iron & Steel Corp.
112 Wood Ave.
Middlesex, NJ 08846

Swedish Steel Mills, A.A.
New York, NY

Swiss Aluminium Ltd.
Feldeggstrasse 4
CH-8034 Zurich
Switzerland

Swiss Laboratory Inc.
P.O. Box 1352
Akron, OH 44309

Sybry, Searls & Co. Ltd.
Cannon Steel Works
Sheffield
England

Symington-Gould Corp.
Depew, NY

T.L.M. Co.
Paris
France

Tacony Steel Co.
(Reported to be out of business)
Philadelphia, PA

TAFA Inc.
146 Pembroke Rd.
P.O. Box 1157
Concord, NH 03301
(603)224-9585

Tata Iron & Steel Co.
43 Chowringhee Road
Calcutta 16
India
44 8301

Robert Taylor & Co.
Labert
Scotland

Samuel Taylor Ltd.
Birmingham
England

S.G. Taylor Chain Co.
3 141st Street
Hammond, IN 46320

Taylor-Wilson Mfg. Co.
950-300 Sixth Ave. Bldg.
Pittsburgh, PA 15222

Techalloy Co. Inc.
04 Business Park Dr.
Armonk, NY 10504
(914)273-4500

Techni-Cast Corporation
11220 South Garfield Avenue
South Gate, CA 90280
(213)923-4585

Telcon Metals Ltd.
P.O. Box 12, Manor Royal
Crawley
Sussex RH10 2QH
England
28800

Teledyne Allvac
Formerly called:
Anchor Drawn Steel Co.
P.O. Box 5030
Monroe, NC 28110
(704)289-4511

Teledyne Firth - Sterling
McKeesport, PA 15134

Teledyne McKay
850 Grantley Rd.
P.O. Box 1509
York, PA 17405-1509
(717)845-7581

Teledyne Ohiocast
1075 James St.
Springfield, OH 45501

Teledyne Pittsburgh Tool Steel
Monaca, PA 15061

Teledyne Rodney Metals
1357 E. Rodney French Blvd.
P.O. Box E-915
New Bedford, MA 02742
(617)996-5691

Teledyne Ryan Aeronautical
2701 Harbor Drive
San Diego, CA 92112

Teledyne Vasco
P.O. Box 151
Latrobe, PA 15650
(412)537-5551

Teledyne Wah Chang Albany
P.O. Box 460
Albany, OR 97321-0136
(503)926-4211

Texas Electric Steel Casting Co.
Houston, TX

Texas Instruments Inc.
Material & Controls Group
Formerly called:
Metals & Controls, Inc.
34 Forest St.
Attleboro, MA 02703
(508)699-1621

Textron Inc.
CWC Castings Division
2672-S Henry St.
Dept. TR
Muskegon, MI 49441
(616)733-1331

Thermadyne Industries Inc.
101 S. Hanley Road
St. Louis, MO 63105

Thomas & Skinner Inc.
P.O. Box 150-B
1120 East 23rd St.
Indianapolis, IN 46206
(317)923-2501

Thomas Foundries, Inc.
1007 N. 37th Place
P.O. Box 96
Birmingham, AL 35201
(205)595-4691

Thomas Steel Strip Co.
Delaware Ave. N.W.
Warren, OH 44485
(216)841-6222

Thomson Industries Inc.
Port Washington, NY 11050

3M Co.
Electrical Products Division
3M Center Bldg.
225-4N-05
St. Paul, MN 55144-1000
(612)733-5966

Thyssen Edelstahlwerke AG
Oberschieslenstrasse 16
Postfach 730
4150 Krefeld 1
West Germany
0215/831

Thyssen Specialty Steels, Inc.
365 Village Dr.
Carol Stream, IL 60188
(312)682-3900

Thysson Deutsche Edelstahlwerke A.G.
Formerly called:
Marathon Specialty Steels Inc.
Empire State Bldg.
350 Fifth Ave.
New York, NY 10001

Time Steel Service Inc.
5200 West 164th St.
Cleveland, OH 44142
(216)267-2500

TIMET
P.O. Box 2824
400 Rouser Rd.
Pittsburgh, PA 15230
(412)262-4200

Timken Co.
1835 Dueber Avenue S.W.
Canton, OH 44706-2798
(216)438-4154

Timken-Detroit Axle Co.
Detroit, MI

Tin Research Institute
Greenford
England

Tipaloy Inc.
1435 E. Milwaukee Ave.
Detroit, MI 48211
(313)875-5145

Titan Corp.
Indiana General
Formerly called:
Electronic Memories & Magnetics Corp.
9191 Towne Ctr. Drive
Suite 600
San Diego, CA 92122
(619)453-9592

Titan-Aluminium-Feinguss GmbH
Postfach 13 63
D-5780 Bestwig
Germany

Titanite Alloys Corp.
Cleveland, OH

Titanium Metals Corp.
Standard Steel Division
500 N. Walnut
Burnham, PA 17009
(717)4911

Titanium Metals Corp. of America
Timet Division
See TIMET
400 Rouser Rd.
P.O. Box 2824
Pittsburgh, PA 15230
(412)263-4200

Titusville Forge Co.
Titusville, PA

Tokoshu Seiko Co. Ltd.
24-1, 3-chome
Shiohama, Kawasaki-ku
Kawasaki
Japan

Toledo Steel Works
(Reported to be out of business)
Sheffield
England

Tonawanda Electric Steel Casting
200 River Road
North Tonawanda, NY 14120
(716)693-3090

Toshiba Tungaloy Co., Ltd.
c/o Mitsui & Co.
200 Park Ave.
New York, NY 10017

Toto Steel Co. Ltd.
Tokyo
Japan

Toughite Process Co.
Chicago, IL

Trade ARBED Deutschland G.m.b.H.
Avenue de la Liberte
B.P. 1802
Luxembourg
0221 57291

TradeARBED Inc.
825 Third Ave.
New York, NY 10022

Transformer Steels Ltd.

Transleteur & Co.
Berlin
Germany

Trefileries & Laminoirs du Havre
Antony (Seine)
France

Trefimetaux
Pechiney
6 Boulevard du General Leclerc
F-92115 Clichy
France

Trenite Foundry Corp.
Trenton, NJ

Trent Tube
Crucible
2188 Church St.
P.O. Box 77
East Troy, WI 53120
(414)642-7321

Trethaway Associates
New York, NY

Tri-Clover
Formerly called:
Ladish Co.
9201 Wilmot Rd.
Kenosha, WI 53141
(414)694-5511

Tri-Lok Co.
Pittsburgh, PA

Triangle Conduit & Cable Co.
Copper Rod & Brass Products Division
6900 Jersey Ave.
New Brunswick, NJ 08903

Trimout Mfg. Co.
Boston, MA

True Alloys Inc.
Detroit, MI

TRW Inc.
Metals Division
Minerva, OH 44657

TRW Wendt-Sonis Division
Formerly called:
TRW Inc./United Greefield Div.
900 Skokie Road
Northbrook, IL 60062

TRW, Inc.
1900 Richmond Rd.
Cleveland, OH 44124

TRW Inc./United Greenfield Div.
See TRW Wendt-Sonis Division

Tube Reducing Corp.
Wallington, NJ

Tuff-Hard Corp.
Detroit, MI

D. & J. Tullis Ltd.
London
England

Tungsit Electro-Metals Works Ltd.
London
England

Tungsten & Molybdenum Ltd.
See Stellram Ltd.
Nyon
Switzerland

Tungsten Alloy Mfg. Co., Inc.
306 Sussex St.
Harrison, NJ 07029
(201)484-1323

Tungsten Electric Corp.
Union STATE
NJ

Tungsten Widia Tool Corp.
New York, NY

Tungum Alloy Co.
See Tungum Hydraulics Ltd.

Tungum Hydraulics Ltd.
Formerly called:
Tungum Alloy Co.
The White House, Arle
Cheltenham, Gloucestershire Gl 51 0AD
England
0242 245521

W.S. Tyler Inc.
Combustion Engineering Inc.
8500 Tyler Blvd.
Mentor, OH 44060
(216)255-9131

U.N. Alloy Steel Corp.
106 Finnell St.
Weymouth, MA 02188
(617)335-1420

U.S. Graphite Inc.
Formerly called:
Wickes Engineered Materials Division
1621 Holland Ave.
Saginaw, MI 48601
(517)755-0441

U.S. Magnet & Alloy Corp.
266-T Glenwood Ave.
Bloomfield, NJ 07003

U.S. Metalsource
Formerly called:
Wheelock, Lovejoy & Co.
200 Corporate Center Dr.
Coraopolis, PA 15108
(412)269-4495

U.S. Reduction Co.
9200 Calumet Ave.
Munster, IN 46321
(219)836-0555

U.S. Spring & Bumper Co.
Los Angeles, CA

U.S. Steel Corp.
American Sheet & Tin Plate Division
600 Grant St.
Pittsburgh, PA 15219
(412)433-1121

Uddeholm Corp.
Formerly called:
Uddeholm Steel Corp.
2 Crossroads of Commerce
Rolling Meadows, IL 60008

Uddeholm Steel Corp.
See Uddeholm Corp.

Ugine Aciers
Steel Division
See Pechiney ugine Kuhlmann Corp.
10, rue de General Foy
75361 Paris Cedex 08
France
292 31 00

Ulbrich Stainless Steels & Special Metals Inc.
57 Dodge Avenue
North Haven, CT 06473
(203)239-4481

A.E. Ullman & Associates
New York, NY

Ultraloy Corp.
Chicago, IL

Ultralumin Leichtmetall AG
(Reported to be out of business)
Germany

Una Welding Inc.
Cleveland, OH

Unexcelled Mfg. Co.
New York, NY

Unibraze Corp.
See J.W. Harris Co., Inc.

Unicore Inc.
United Nuclear
11 Defco Park
North Haven, CT 06473
(203)239-3365

Unimetal Co.
Franklin, PA

Union Bronze Co.
Reading, PA

Union Carbide Corp.
Linde Division
39 Old Ridgebury Rd.
Danbury, CT 06817
(203)794-5300

Union Metal Mfg. Co.
1432 Maple Ave. N.E.
Canton, OH 44705

United American Metals Corp.
2246 W. Hubbard St.
Chicago, IL 60612
(312)733-6700

United Engineering Steels Ltd.
Sheffield Road
Rotherham S60 1DQ
England

United States Bronze Powders Inc.
P.O. Box 31
Flemington, NJ 08822

United States Pipe & Foundry Co.
3300 First Avenue North
Birmingham, AL 35202

United States Steel Corp.
600 Grant St.
Pittsburgh, PA 15230
(412)433-1121

United Wire & Supply Co.
1497 Elmwood Ave.
Providence, RI 02910
(401)781-3000

Universal Alloys Inc.
Newark, NJ

Universal Castings Corp.
5821 West 66th St.
Chicago, IL 60638
(312)767-7120

Universal Cyclops
Specialty Steel Division
850 Washington Road
Pittsburgh, PA 15228

Universal Power Corp.
Cleveland, OH

Uniworld Corp. of America
12000 Shaker Blvd.
Cleveland, OH 44120

Unna Messingwerk, A.G.
Unna, Westf.
Germany

Carl Urbach & Co., Stahlwerk KG
Postfach 88
22-2-1974 Konkurs-Anmeldung
5809 Huckeswagen
West Germany

Usines de A. Manoir, Pitres (Eure)
Paris
France

Usines de Jadot
Brussels
Belgium

Usines Emile Henricot, S.A.
Court-Saint-Etienne
Belgium
010 61 22 05

USM Corp.
Farrell Co. Division
148 Maple St.
Ansonia, CT 06401

Utility Electric Steel Foundry Co.
Los Angeles, CA

UTP Welding Materials, Inc.
10535 South Wilcrest Drive, Suite 130
Houston, TX 77099
(713)530-1555

Vacuum Metals Corp.
Cambridge, MA

Vacuumschmelze
Siemens Components, Inc.
186 Wood Avenue South
Iselin, NJ 08830
(201)494-3530

Valenite Corp.
Venite Metals Division
Madison Heights, MI 48071
See also GTE Valenite Corp.

Valley-Vulcan Mold Co.
Formerly called:
Vulcan Mold & Iron Co.
P.O. Box 70
Latrobe, PA 15650
(412)537-3355

Vallourec S.A.
7 Place du Chancellier Adenauer
B.P. 180
75764 Paris Cedex 16
France
1 502-3000

Van der Horst Corp.
Los Angeles, CA 94103

VAW Vereinigte Aluminium-Werke AG
Georg von Boeselager Strasse 25
D-5300 Bonn 1
Germany
0228/5 52 02

VDM Aluminium GmbH
6000 Frankfort 50
Germany

VDM Nickel-Technologie AG
Plettenberger Str. 2
Postfach 1820
D-5980 Werdohl/Westf.
Germany
02392/55 0

VDM Technologies Corp.
10 Sylvan Way
Parsippany, NJ 07054
(201)267-8545

VEB Chemieanlagenbau Grimma
Formerly called:
Machinenbau AG
Bahnhofstrasse 3-5
7240 Grimma
Germany

Venture Corp.
Newark, NJ

Veraloy Products Ltd.
Birmingham
England

Vereinigte Chemische Fabriken
Leopoldshall
Germany

Vereinigte Deutsch Nickel-Werke AG.
Postfach 1340
Rosenweg 15
5840 Schwerte 1
West Germany
023 04 108-1

Vereinigte Deutsche Metallwerke AG
Halbzeuge und Systemtechnik
Postfach 50 11 78
D-6000 Frankfurt 50
Germany
0203 603-1

Vereinigte Deutsche Nickel-Werke AG
Postfach 1840
D-5840 Schwerte 1
Germany

Vereinigte Edelstahlwerke
See Bohler Gesellschaft M.B.H.

Vereinigte Leichtmetallwerke, G.m.b.H.
Hannover-Linden
Germany

Vereinigte Metallwerke Ranshofen-Berndorf AG
A-5282 Braunau-Ranshofen
Austria

Vereinigte Silberhammerwerke Hetzel
Nurnburg
Germany

Vereinigte Stahlwerke
(Reported to be out of business)
Dusseldorf

Verilite Metals Co.
New York, NY

Versevorder Metallwerk
5980 Werdohl
Germany

Vesevorder Metallwerke GmbH
Werdohl/Westf.
Germany

Vickers-Armstrong Co.
See British Steel Corp.

Victor Mfg. Co.
See Stellite Div. Colt Industries
London
England

Vikmanshyttan AG
Vikmanshyttan
Sweden

VILLARES
Formerly called:
Acos Villares S.A.
Avenida do Estado 6116
01510 Sao Paulo
Brazil
112 5276

Voest International, Inc.
919 Third Ave.
New York, NY 10022

Volco Brass & Copper Co.
801 Boulevard
Kenilworth, NJ 07033

Vulcan Foundry Co.
4401 San Leandro St.
Oakland, CA 94601

Vulcan Iron Works
1050 United Penn Bank Bldg.
Wilkes-Barre, PA 18701
(717)822-2161

Vulcan Mold & Iron Co.
See Valley-Vulcan Mold Co.

W.W. Alloys Inc.
(Reported to be out of business)
Detroit, MI

Wabash Alloys, Inc.
Ogden Corp.
4365 Bradley Rd.
Cleveland, OH 44109
(219)563-7461

Waimet Alloys Co.
Dearborn, MI

Wakefield Corp.
P.O. Box 151
29 Foundry St.
Wakefield, MA 01880
(617)245-1828

Walker Metal Products Ltd.
Walkerville, ONT
Canada

Wall Colmonoy Corp.
30261 Stephenson Highway
Madison Heights, MI 48071-1650
(313)585-6400

Wallace Murray Corp.
Heller Tool Division
Formerly called:
Heller Tool
Box 28
Newcomerstown, OH 43832

Wallram Hartmetall GmbH
Essen
Germany

Walsingham Steel Co., Ltd.
Walsingham
Bishop Auckland
Co. Durham DL13 3HX
England

Waltham Precision Instruments
Waltham, MA 02154

Walworth Co.
Aloyco Plant
1400 W. Elizabeth Ave.
Linden, NJ 07036
(201)862-4600

Edgar T. Ward's Sons Co.
Newark, NJ

S. & C. Wardlow
Nutley, NJ

Wardlows Ltd.
Congress Works
Carlisle St.
Sheffield S4 7LJ
England

Warman Steel Casting Co.
Huntington Park, CA

Washburn Wire Co.
New York, NY

Washington Steel Corp.
Blount Inc.
P.O. Box 494
Washington, PA 15301
(412)222-8000

Washington Iron Works, Inc.
400-T E. Lamar
Sherman, TX 75090

Waterbury Rolling Mills Inc.
P.O. Box 550
Waterbury, CT 06720
(203)754-0151

Watsontown Foundry
P.O. Box 126
Watsontown, PA 17777
(717)538-5546

Waukesha Foundry, Inc.
ABEX
1300 Lincoln Ave.
Waukesha, WI 53186
(412)542-0741

Wausau Motor Parts
Schofield, WI

Weatherly Casting & Machine Co.
Formerly called:
Weatherly Foundry & Mfg.
Commerce & Laurel Sts.
P.O. Box 21
Weatherly, PA 18255
(717)427-8611

Weatherly Foundry & Mfg.
See Weatherly Casting & Machine Co.

Webb Wire Works
New Brunswick, NJ

Wehr Steel Corporation
Wehr Corp.
(Reported to be out of business)
10201 W. Lincoln Ave.
Milwaukee, WI 53227
(414)671-2100

Emil Weingartner & Co.
Edelstahlgrosshandel
Postfach 540269
2000 Hamburg 54
West Germany
0 40 8540-0

Wel-Met Co.
Kent, OH

Welded Carbide Co. Inc.
66 Colfax Ave.
Clifton, NJ 07013
(201)779-1678

Welding Alloy Mfg. Co.
Malden, MA

Welding Equipment & Supply Co.
See Eureka Welding Alloys, Inc.

Welding Material Co.
New Orleans, LA

Welding Sales & Engineering Co.
Detroit, MI

Weldwire Co., Inc.
P.O. Box 340
King of Prussia, PA 19406
(215)242-1100

Welland Electric Steel Foundry
Welland, ONT
Canada

S.K. Wellman Corp.
200 Egbert Rd.
Bedford, OH 44146

Wellman Dynamics Corp.
Formerly called:
Fansteel
U.S. Rte. 34
P.O. Box 147
Creston, IA 50801

Wesson Co.
Ferndale, MI

West Steel Casting Co.
Cleveland, OH

Westa-Westdeutsche
Edelstahlhandelsgesellschaft
Von Poldi-hutte, Dusseldorf
West Germany

Western Crucible Steel Casting Co.
Minneapolis, MN

Western Electric Co.
Chicago, IL

Western Gold & Platinum Co.
GTE Sylvania
477 Harbor Blvd.
Belmont, CA 94002
(415)592-9440

Western Reserve Manufacturing Co., Inc.
5311 West River Rd. N.
Lorain, OH 44055
(216)277-1226

Westfalenhutte Dortmund A.G.
Dortmund
Germany

Westfalische Kupfer und Messingwerke A.G.
Postfach 2960
D-5880 Ludenscheid
Germany

Westfalische Leichtmetallwerke GmbH
Nachrodt
Krs. Altena, Westf.
Germany

Westfalische Stahlgesellschaft mbH
F.W. Postfach 52
5970 Plettenberg 1
West Germany
02391 813-0

Westig (U.K.) Ltd.
Provincial House
Solly Street
Sheffield S1 48A
England

Westig GmbH
Postfach 2040
4750 Unna
Germany
02303 203-0

Westinghouse Air Brake Co.
P.O. Box 1
Station St.
Wilmerding, PA 15148

Westinghouse Electric Corp.
Specialty Metals Division
1310 Beaulah Rd.
Pittsburg, PA 15235
(412)374-2231

Wheeling-Pittsburgh Steel Corp.
1134 Market St.
Wheeling, WV 26003
(304)234-2400

Wheelock, Lovejoy & Co.
See U.S. Metalsource

Whipple & Choate Co.
Bridgeport, CT

White Metal Rolling & Stamping Corp.
80-88 Moultrie St.
Brooklyn, NY 11222

Whiton Machine Co.
New London, CT

Whittaker Corp.
Nuclear Metals Division
2229 Main St.
Concord, MA 01742
(508)369-5410

Wickes Engineered Materials Division
See U.S. Graphite Inc.

Wickwire Spencer Steel Co.
New York, NY

Wieland-Werke AG Metallwerke
Postfach 4240
Graf-Arco Strasse 38
D-7900 Ulm
Germany
0731 4 97 0

Henry Wiggin & Co. Ltd.
See Inco Alloys International Ltd.

Wikmanshytte Bruks AB
Wikmanshyttan
Sweden

Friedrich Wilhelms-Hutte
Mulheim
Ruhr
Germany

Wm. Wilkinsen
Shustoke
England

E.A. Williams & Sons
325 Washington Ave.
P.O. Box 6600
Carlstad, NJ 07072
(201)438-0800

Williams Precious Metals
Brush Wellman Inc.
2978 Main Street
Buffalo, NY 14214
(716)837-1000

Williamson Bros. Inc.
Bridgeport, CT

Willworthy Piston Ring Ltd.
London
England

G. Robert Wilms A.G.
Remscheid-Hasten
Germany

Wilson Welder & Metals Co.
New York, NY

Wimet Ltd.
See Sandvik Hard Materials Ltd.

J.T. Wing & Co.
Detroit, MI

W. Martin Winn Ltd.
S. Staffordshire
England

Wintershall, A.G.
Heringer
Germany

Otto Wolff Handelgesellschaft
Postfach 102010
5000 Koln 1
West Germany

Wolverine Tube, Inc.
P.O. Box 2202
Decatur, AL 35609-2202
(205)353-1310

Woodworkers Tool Works
222-T S. Jefferson
Chicago, IL 60606

Worthington Pump Inc.
Dresser Industries
Formerly called:
Dresser Pump Division
401 Worthington Ave.
Harrision, NJ 07029-2097

Worthington Steel & Annealing Co.
Sheffield
England

Wrought Bearing Metals Inc.
New York, NY

Wyndale Mfg. Co.
Indianapolis, IN

Xaloy Inc.
P.O. Box 1441-A
Pulaski, VA 24301
(703)980-7560

Xeram Systems/Technical Components, Inc.
Pechiney Corporation
80 Commerce Drive
Warwick, RI 02886
(401)737-2410

Yawata Iron & Steel Co., Ltd.
See Fugi Iron & Steel Co., Ltd.
Tokyo
Japan

James Yocum & Son Inc.
Philadelphia, PA

Yoonsteel (Malaysia) Sdn. Bhd
Site 14 Tasek Industrial Estate
Ipoh, Perak 31400
Malaysia

York Machine & Supply Co.
York, Pa

Yorkshire Imperial Metals Ltd.
Leeds LS1 1RD
England

Youngstown Alloy Castings Co.
Youngstown, OH

Youngstown Foundry & Machine Co.
Youngstown, OH

Youngstown Steel
Youngstown Sheet & Tube Co. Div.
P.O. Box 900
Youngstown, OH 44501
(216)758-6411

Robert Zapp Werkstofftechnik GmbH
Zapp-Haus
Postfach 9020
4000 Dusseldorf 1
West Germany
0211 35500

W.A. Zelnicker
St. Louis, MO

Zenith Foundry Co.
West Allis, WI

Zinkberatungsstelle GmbH
Berlin
Germany

Ziv Steel & Wire Co.
Detroit, MI

Zurbach Steel Co.
P.O. 1108
Willow St.
Salem, NH 03079
(603)898-2381

Zweigbetrieb der Carp & Hones
Deutschen Edelstahlwerke G.m.b.H.
Postfach 730
4150 Krefeld 1
Germany

TABLES

Chemical elements and symbols

Name	Symbol	Name	Symbol
Aluminum	Al	Mercury	Hg
Antimony	Sb	Molybdenum	Mo
Argon	A	Neodymium	Nd
Arsenic	As	Neon	Ne
Barium	Ba	Nickel	Ni
Beryllium	Be	Nitrogen	N
Bismuth	Bi	Osmium	Os
Boron	B	Oxygen	O
Bromine	Br	Palladium	Pd
Cadmium	Cd	Phosphorus	P
Calcium	Ca	Platinum	Pt
Cesium	Cs	Potassium	K
Carbon	C	Praseodymium	Pr
Cerium	Ce	Radium	Ra
Chlorine	Cl	Radon	Rn
Chromium	Cr	Rhodium	Rh
Cobalt	Co	Rubidium	Rb
Columbium (a)	Cb	Ruthenium	Ru
Copper	Cu	Samarium	Sa
Dysporsium	Dy	Scandium	Sc
Erbium	Er	Selenium	Se
Europium	Eu	Silicon	Si
Fluorine	F	Silver	Ag
Gadolinium	Gd	Sodium	Na
Gallium	Ga	Strontium	Sr
Germanium	Ge	Sulfur	S
Gold	Au	Tantalum	Ta
Hafnium	Hf	Tellurium	Te
Helium	He	Terbium	Tb
Holmium	Ho	Thalium	Tl
Hydrogen	H	Thorium	Th
Indium	In	Thulium	Tm
Iodine	I	Tin	Sn
Iridium	Ir	Titanium	Ti
Iron	Fe	Tungsten (b)	W
Krypton	Kr	Uranium	U
Lanthanum	La	Vanadium	V
Lead	Pb	Xenon	Xe
Lithium	Li	Ytterbium	Yb
Lutecium	Lu	Yttrium	Y
Magnesium	Mg	Zinc	Zn
Manganese	Mn	Zirconium	Zr

(a) Columbium = Niobium, Nb. (b) Tungsten = Wolfram.

Physical constants of principal alloy-forming elements

Element	Atomic weight	Melting point, °F	Boiling point, °F	Density, Mg/m³	Atomic volume	Linear coefficient of thermal expansion per °C x 10⁻⁶, 0-100 °C	Specific heat, cal/g · °C at RT	Thermal conductivity, cal/cm³ · °C at RT	Electrical resistivity, μΩ · cm	Crystallization shrinkage, %	Young's modulus, psi x 10⁶
Al	26.93	1220	4442	2.70	10.0	23.6 (20-100 °C)	0.21	0.53	2.65 (20 °C)	6.6	9
Sb	121.76	1166	2516	6.65	18.0	10.5	0.05	0.043	38.6 (0 °C)	1.4	11
As	74.96	1503	1135	5.73	13.1	4.7	0.08	...	33.3 (20 °C)	...	...
Be	9.01	2336	5020	1.85	4.9	11.6	0.425	0.35	4.3 (20 °C)	...	42.7
Bi	209.0	519.8	2840	9.75	21.4	13.2	0.03	0.019	106.8 (0 °C)	-3.3	4.6
Cd	112.4	609.6	1409	8.64	13.0	29.8	0.055	0.22	6.83 (18 °C)	4.7	8
Ca	40.07	1540	2625	1.54	26.0	22.3	0.149	0.3	...	...	...
C	12.005	...	6512	...	...	...	...	...	...	...	...
Diamond	...	...	...	3.52	3.42	1.18	0.11	...	...	...	...
Graphite	...	6740	8730	2.25	5.35	2.5	0.16	0.06	1375	...	0.7
Ce	140.25	1479	6280	6.79	20.6	8.0	0.045	0.026	75	...	6
Cr	52.0	3407	4829	7.19	7.5	6.2	0.105	0.16	12.9 (0 °C)	...	36
Co	58.97	2723	5250	8.85	6.8	13.8	0.10	0.165	6.24 (20 °C)	...	30
Cu	63.57	1981	4703	8.93	7.15	16.8 (25-100 °C)	0.091	0.94	1.67 (20 °C)	4.0	16
Ge	72.60	1719	5125	5.32	...	5.75	.073	0.14	46	...	...
Au	197.2	1945	5380	19.32	10.2	14.2 (25-100 °C)	0.032	0.70	2.35 (20 °C)	5.2	11.6
In	114.8	313	3632	7.28	15.8	33	0.057	0.057	8.37 (20 °C)	...	1.6
Ir	192.2	4449	9570	22.65	8.6	6.8	0.030	0.14	5.3 (20 °C)	...	76
Fe	55.84	2797	5430	7.88	7.1	11.7	0.102	0.18	9.71 (25 °C)	...	28.5
Pb	207.20	621	3137	11.34	18.3	29.3	0.030	0.083	20.6 (20 °C)	3.4	2
Li	6.94	357	2426	0.534	13.0	56	0.79	0.017	8.55 (0 °C)	...	...
Mg	24.32	1202	2025	1.74	14.0	27.1	0.24	0.38 (0-100 °C)	4.35 (0 °C)	4.2	6.25
Mn	54.93	2273	3900	7.42	7.4	22	0.11	...	185	...	23
Hg	200.6	-37.9	357	13.6	14.7	...	0.033	0.019 (0 °C)	95.74 (20 °C)	3.75	...
Mo	96.0	4730	10,040	10.3	9.3	4.9 (20 °C)	0.063	0.35	5.5 (27 °C)	...	47
Ni	58.68	2645.6	4950	8.9	6.7	13.2 (25-100 °C)	0.102	0.22	6.93 (20 °C)	...	30
Os	190.9	4892	9950	22.5	8.5	4.6 (50 °C)	0.031	...	9.5 (20 °C)	...	81
Pd	106.7	2822	7200	12.16	8.8	11.76	0.055	0.17	10.2 (0 °C)	...	16.3
P	31.04	111	536	1.83	17.0	125.3 (0-40 °C)	0.18	...	1×10^{17} (11 °C)	...	...
Pt	195.2	3217	8185	21.41	9.1	8.99	0.030	0.17	10.96 (0 °C)	...	23.5
Rh	102.9	3571	8130	12.44	8.3	8.3	0.058	0.21 (17 °C)	4.51 (20 °C)	...	42.5
Ru	101.7	4530	8850	12.45	8.4	9.1	0.057	...	7.6 (0 °C)	...	60
Si	28.1	2570	4860	2.33	11.6	5	0.17	0.20	10 (0 °C)	...	16.3
Ag	107.88	1761	4010	10.53	10.2	19.21	0.054	1.00	1.63 (18 °C)	5.0	11
Ta	180.9	5425	9800	16.6	10.9	6.5	0.035	0.13	12.5 (25 °C)	...	27
Th	232.15	3182	7000	11.5	19.1	12.5	0.034	0.09 (100 °C)	13 (0 °C)	...	...
Sn	118.7	449	4118	7.3	16.3	22.96	0.053	1.5	11.0 (0 °C)	2.8	6.3
Ti	47.9	3035	5900	4.5	10.7	8.41	0.12	6.6 (50 °C)	42 (20 °C)	...	16.8
W	184.0	6170	10,706	19.3	9.5	4.44 (27 °C)	0.034	0.397	5.6 (20 °C)	...	50
U	238.07	2070	6904	19.1	12.7	10	0.029	0.07 (70 °C)	30	...	24
V	51.0	3450	6150	6.11	8.5	8.3	0.11	0.074 (-2 °C)	25 (20 °C)	...	19
Zn	65.37	787	1663	7.14	9.1	39.7	0.091	0.27	5.75 (0 °C)	6.5	...
Zr	91.2	3366	6470	6.4	14.1	5.85	0.066	0.05	40	...	13.7

Densities of alloys and metals

Alloys and metals	g/cm^3	lb/in.^{3}
Aluminum	2.69	0.0975
Aluminum and tin:		
Al 91%, Sn 9%	2.85	0.103
Aluminum, copper, and tin:		
Al 85%, Cu, 7.5%, Sn 7.5%	0.301	0.1087
Al 6.25%, Cu 87.5%, Sn 6.25%	7.35	0.2656
Al 5%, Cu 5%, Sn 90%	6.81	0.2459
Aluminum and magnesium:		
Al 70%, Mg 30%	2.00	0.0723
Aluminum and zinc:		
Al 91%, Zn 9%	2.80	0.1012
Antimony	6.71	0.2424
Babbitt alloy	7.27	0.2627
Bismuth	9.80	0.354
Bismuth, lead, and tin:		
Bi 53%, Pb 40%, Sn 7%	10.5	0.3813
Wood's metal:		
Bi 50%, Pb 25%, Cd 12.5%, Sn 12.5%	9.69	0.3501
Brass:		
Cu 90%, Zn 10%	8.58	0.3101
Cu 70%, Zn 30%	8.53	0.308
Cu 60%, Zn 40%	8.34	0.3015
Cu 50%, Zn 50%	8.19	0.2957
Bronze:		
Cu 90%, Sn 10%	8.78	0.3171
Cu 85%, Sn 15%	8.89	0.3211
Cu 80%, Sn 20%	8.73	0.3153
Cu 75%, Sn 25%	8.82	0.3188
Cu 90%, Al 10%	7.69	0.2777
Cu 95%, Al 5%	8.36	0.3020
Cu 97%, Al 3%	8.68	0.3136
Bronze, phosphorus, average	8.60	0.3107
Bronze, tobin, average	8.06	0.291
Cadmium and tin:		
Cd 32%, Sn 68%	7.69	0.2777
Chromium	6.98	0.2523
Cobalt	8.54	0.3084
Copper	8.96	0.324
Copper and nickel:		
Cu 60%, Ni 40%	8.87	0.3206
German silver:		
Cu 60%, Zn 20%, Ni 20%	9.98	0.3607
Cu 52%, Zn 26%, Ni 22%	8.44	0.3049
Cu 59%, Zn 30%, Ni 11%	8.33	0.3009
Cu 63%, Zn 30%, Ni 7%	8.30	0.2997
Gold	19.35	0.699
Gold and copper:		
Au 98%, Cu 2%	18.83	0.6805
Au 90%, Cu 10%	17.15	0.6197
Au 86%, Cu 14%	16.45	0.5943
Gun metal, average	8.71	0.3148
Iridium	22.36	0.8078
Iron, cast	7.15	0.258
Iron, wrought	7.75	0.28
Lead	11.34	0.4097
Lead and antimony:		
Pb 30%, Sb 70%	7.21	0.2604
Pb 37%, Sb 63%	7.37	0.2662
Pb 44%, Sb 56%	7.61	0.2748
Pb 63%, Sb 37%	8.23	0.2974
Pb 83%, Sb 17%	9.55	0.3449
Pb 90%, Sb 10%	10.54	0.3807
Lead and bismuth:		
Bi 67%, Pb 33%	10.23	0.3697
Bi 50%, Pb 50%	10.51	0.3796
Bi 33%, Pb 67%	10.92	0.3946
Bi 25%, Pb 75%	11.16	0.4033
Bi 17%, Pb 83%	11.24	0.4062
Bi 12%, Pb 88%	11.26	0.4068
Lead and tin:		
Pb 87.5%, Sn 12.5%	10.59	0.3825
Pb 84%, Sn 16%	10.31	0.3726
Pb 63.7%, Sn 36.3%	9.42	0.3402
Pb 46.7%, Sn 53.3%	8.73	0.3153
Pb 30.5%, Sn 69.5%	8.23	0.2974
Magnesium	1.738	0.06279
Manganese	7.47	0.270
Manganese, copper, and nickel:		
Mn 12%, Cu 84%, Ni 4%	8.49	0.3067
Mercury	13.54	0.489
Nickel	8.78	0.317
Osmium	22.58	0.816
Palladium	12.02	0.4343
Platinum	21.45	0.775
Platinum and iridium:		
Pt 90%, Iridium 10%	21.53	0.778
Rhodium	12.37	0.447
Ruthenium	12.25	0.4427

Densities of alloys and metals (continued)

Alloys and metals	g/cm^3	lb/in.^{3}
Silver	10.49	0.379
Steel, cast	7.85	0.2835
Tin	7.31	0.264
Tin and antimony:		
Sn 50%, Sb 50%	6.79	0.2453
Sn 75%, Sb 25%	7.08	0.2557
Tin and bismuth:		
Bi 78%, Sn 22%	9.40	0.3396
Bi 63%, Sn 37%	9.13	0.3298
Bi 50%, Sn 50%	8.74	0.3159
Bi 37%, Sn 63%	8.49	0.3067
Bi 22%, Sn 78%	8.07	0.2916
Tin and lead:		
Sn 97%, Pb 3%	7.30	0.2638
Sn 89%, Pb 11%	7.61	0.2748
Sn 80%, Pb 20%	7.80	0.2818
Sn 67%, Pb 33%	8.20	0.2962
Sn 50%, Pb 50%	8.81	0.3182
Titanium	4.51	0.1628
Tungsten	19.3	0.697
Zinc	7.13	0.2577

Energy conversion factors

Foot pounds to joules (1 ft · lb = 1.355818 joules)										
	0	1	2	3	4	5	6	7	8	9
0	0	1	3	4	5	7	8	9	11	12
10	14	15	16	18	19	20	22	23	24	26
20	27	28	30	31	33	34	35	37	38	39
30	41	42	43	45	46	47	49	50	52	53
40	54	56	57	58	60	61	62	64	65	66
50	68	69	71	72	73	75	76	77	79	80
60	81	83	84	85	87	88	89	91	92	94
70	95	96	98	99	100	102	103	104	106	107
80	108	110	111	113	114	115	117	118	119	121
90	122	123	125	126	127	129	130	132	133	134

Joules to foot pounds (1 joule = 0.737561 ft · lb)										
	0	1	2	3	4	5	6	7	8	9
0	0	1	1	2	3	4	4	5	6	7
10	7	8	9	10	10	11	12	13	13	14
20	15	15	16	17	18	18	19	20	21	21
30	22	23	24	24	25	26	27	27	28	29
40	30	30	31	32	32	33	34	35	35	36
50	37	38	38	39	40	41	41	42	43	44
60	44	45	46	46	47	48	49	49	50	51
70	52	52	53	54	55	55	56	57	58	58
80	59	60	60	61	62	63	63	64	65	66
90	66	67	68	69	69	70	71	72	72	73

Force conversion factors

Kilograms per square millimeter to pounds per square inch (1 kg/mm^2 = 1422.334 psi)

	0	1	2	3	4	5	6	7	8	9
0	0	1422	2845	4267	5689	7112	8534	9956	11379	12801
10	14223	15646	17068	18490	19913	21335	22757	24180	25602	27024
20	28447	29869	31291	32714	34136	35558	36981	38403	39825	41248
30	42670	44092	45515	46937	48359	49782	51204	52626	54049	55471
40	56893	58316	59738	61160	62583	64005	65427	66850	68272	69694
50	71117	72539	73961	75384	76806	78228	79651	81073	82495	83918
60	85340	86762	88185	89607	91029	92452	93874	95296	96719	98141
70	99563	100986	102408	103830	105253	106675	108097	109520	110942	112364
80	113787	115209	116631	118054	119476	120898	122321	123743	125165	126588
90	128010	129432	130855	132277	133699	135122	136544	137966	139389	140811

Thousands of pounds per square inch to kilograms per square millimeter (1000 psi = 0.7030696 kg/mm^2)

	0	1	2	3	4	5	6	7	8	9
0	0.00	0.70	1.41	2.11	2.81	3.52	4.22	4.92	5.62	6.33
10	7.03	7.73	8.44	9.14	9.84	10.55	11.25	11.95	12.66	13.36
20	14.06	14.76	15.47	16.17	16.87	17.58	18.28	18.98	19.69	20.39
30	21.09	21.80	22.50	23.20	23.90	24.61	25.31	26.01	26.72	27.42
40	28.12	28.83	29.53	30.23	30.94	31.64	32.34	33.04	33.75	34.45
50	35.15	35.86	36.56	37.26	37.97	38.67	39.37	40.07	40.78	41.48
60	42.18	42.89	43.59	44.29	45.00	45.70	46.40	47.11	47.81	48.51
70	49.21	49.92	50.62	51.32	52.03	52.73	53.43	54.14	54.84	55.54
80	56.25	56.95	57.65	58.35	59.06	59.76	60.46	61.17	61.87	62.57
90	63.28	63.98	64.68	65.39	66.09	66.79	67.49	68.20	68.90	69.60

Kilograms per square millimeter to tons per square inch (1 kg/mm^2 = 0.7111672 short tsi)

	0	1	2	3	4	5	6	7	8	9
0	0.00	0.71	1.42	2.13	2.84	3.56	4.27	4.98	5.69	6.40
10	7.11	7.82	8.53	9.25	9.96	10.67	11.38	12.09	12.80	13.51
20	14.22	14.93	15.65	16.36	17.07	17.78	18.49	19.20	19.91	20.62
30	21.34	22.05	22.76	23.47	24.18	24.89	25.60	26.31	27.02	27.74
40	28.45	29.16	29.87	30.58	31.29	32.00	32.71	33.42	34.14	34.85
50	35.56	36.27	36.98	37.69	38.40	39.11	39.83	40.54	41.25	41.96
60	42.67	43.38	44.09	44.80	45.51	46.23	46.94	47.65	48.36	49.07
70	49.78	50.49	51.20	51.92	52.63	53.34	54.05	54.76	55.47	56.18
80	56.89	57.60	58.32	59.03	59.74	60.45	61.16	61.87	62.58	63.29
90	64.01	64.72	65.43	66.14	66.85	67.56	68.27	68.98	69.69	70.41

Tons per square inch to kilograms per square millimeter (1 short tsi = 1.406139 kg/mm^2)

	0	1	2	3	4	5	6	7	8	9
0	0.00	1.41	2.81	4.22	5.62	7.03	8.44	9.84	11.25	12.66
10	14.06	15.47	16.87	18.28	19.69	21.09	22.50	23.90	25.31	26.72
20	28.12	29.53	30.94	32.34	33.75	35.15	36.56	37.97	39.37	40.78
30	42.18	43.59	45.00	46.40	47.81	49.21	50.62	52.03	53.43	54.84
40	56.25	57.65	59.06	60.46	61.87	63.28	64.68	66.09	67.49	68.90
50	70.31	71.71	73.12	74.53	75.93	77.34	78.74	80.15	81.56	82.96
60	84.37	85.77	87.18	88.59	89.99	91.40	92.81	94.21	95.62	97.02
70	98.43	99.84	101.24	102.65	104.05	105.46	106.87	108.27	109.68	111.08
80	112.49	113.90	115.30	116.71	118.12	119.52	120.93	122.33	123.74	125.15
90	126.55	127.96	129.36	130.77	132.18	133.58	134.99	136.40	137.80	139.21

Force conversion factors (continued)

Thousands of pounds per square inch to megapascals (1000 psi = 6.894757 MPa)

	0	1	2	3	4	5	6	7	8	9
0	0	7	14	21	28	34	41	48	55	62
10	69	76	83	90	97	103	110	117	124	131
20	138	145	152	159	165	172	179	186	193	200
30	207	214	221	228	234	241	248	255	262	269
40	276	283	290	296	303	310	317	324	331	338
50	345	352	359	365	372	379	386	393	400	407
60	414	421	427	434	441	448	455	462	469	476
70	483	490	496	503	510	517	524	531	538	545
80	552	558	565	572	579	586	593	600	607	614
90	621	627	634	641	648	655	662	669	676	683

Kilograms per square millimeter to megapascals (1 kg/mm^2 = 9.806650 MPa)

	0	1	2	3	4	5	6	7	8	9
0	0	10	20	29	39	49	59	69	78	88
10	98	108	118	127	137	147	157	167	177	186
20	196	206	216	226	235	245	255	265	275	284
30	294	304	314	324	333	343	353	363	373	382
40	392	402	412	422	431	441	451	461	471	481
50	490	500	510	520	530	539	549	559	569	579
60	588	598	608	618	628	637	647	657	667	677
70	686	696	706	716	726	735	745	755	765	775
80	785	794	804	814	824	834	843	853	863	873
90	883	892	902	912	922	932	941	951	961	971

Force unit conversion factors

Multiply \ To Get	MPa	psi	kg-f/mm²	N/mm²	tons/in.²
MPa (megapascals)	1	145	0.102	1	64.7×10^{-3}
psi (pounds/inch²)	6.895×10^{-3}	1	0.703×10^{-3}	6.895×10^{-3}	0.446×10^{-3}
kg-f/mm² (kilograms force/millimeters²)	9.806	1.422×10^3	1	9.806	0.634
N/mm² (Newtons/millimeter²)	1	145	0.102	1	64.7×10^{-3}
tons/in.² (British tons/inch²; 2240 pounds/ton)	15.45	2240	1.576	15.45	1

Other conversion factors

1 joule	0.7376 ft · lb
1 ft · lb	1.355 joule
1 ksi$\sqrt{}$in.	0.9101 MPa$\sqrt{}$m
1 MPa$\sqrt{}$m	1.098 ksi$\sqrt{}$in.

Temperature conversion °C to °F (°F = [°C · 1.8] + 32)

°C	0	10	20	30	40	50	60	70	80	90
−200	−328	−346	−364	−382	−400	−418	−436	−454	−472	−490
−100	−148	−166	−184	−202	−220	−238	−256	−274	−292	−310
−0	32	14	−4	−22	−40	−58	−76	−94	−112	−130
0	32	50	68	86	104	122	140	158	176	194
100	212	230	248	266	284	302	320	338	356	374
200	392	410	428	446	464	482	500	518	536	554
300	572	590	608	626	644	662	680	698	716	734
400	752	770	788	806	824	842	860	878	896	914
500	932	950	968	986	1004	1022	1040	1058	1076	1094
600	1112	1130	1148	1166	1184	1202	1220	1238	1256	1274
700	1292	1310	1328	1346	1364	1382	1400	1418	1436	1454
800	1472	1490	1508	1526	1544	1562	1580	1598	1616	1634
900	1652	1670	1688	1706	1724	1742	1760	1778	1796	1814
1000	1832	1850	1868	1886	1904	1922	1940	1958	1976	1994
1100	2012	2030	2048	2066	2084	2102	2120	2138	2156	2174
1200	2192	2210	2228	2246	2264	2282	2300	2318	2336	2354
1300	2372	2390	2408	2426	2444	2462	2480	2498	2516	2534
1400	2552	2570	2588	2606	2624	2642	2660	2678	2696	2714
1500	2732	2750	2768	2786	2804	2822	2840	2858	2876	2894
1600	2912	2930	2948	2966	2984	3002	3020	3038	3056	3074
1700	3092	3110	3128	3146	3164	3182	3200	3218	3236	3254
1800	3272	3290	3308	3326	3344	3362	3380	3398	3416	3434
1900	3452	3470	3488	3506	3524	3542	3560	3578	3596	3614
2000	3632	3650	3668	3686	3704	3722	3740	3758	3776	3794
2100	3812	3830	3848	3866	3884	3902	3920	3938	3956	3974
2200	3992	4010	4028	4046	4064	4082	4100	4118	4136	4154
2300	4172	4190	4208	4226	4244	4262	4280	4298	4316	4334
2400	4352	4370	4388	4406	4424	4442	4460	4478	4496	4514
2500	4532	4550	4568	4586	4604	4622	4640	4658	4676	4694
2600	4712	4730	4748	4766	4784	4802	4820	4838	4856	4874
2700	4892	4910	4928	4946	4964	4982	5000	5018	5036	5054
2800	5072	5090	5108	5126	5144	5162	5180	5198	5216	5234
2900	5252	5270	5288	5306	5324	5342	5360	5378	5396	5414
3000	5432	5450	5468	5486	5504	5522	5540	5558	5576	5594
3100	5612	5630	5648	5666	5684	5702	5720	5738	5756	5774
3200	5792	5810	5828	5846	5864	5882	5900	5918	5936	5954
3300	5972	5990	6008	6026	6044	6062	6080	6098	6116	6134
3400	6152	6170	6188	6206	6224	6242	6260	6278	6296	6314
3500	6332	6350	6368	6386	6404	6422	6440	6458	6476	6494
3600	6512	6530	6548	6566	6584	6602	6620	6638	6656	6674
3700	6692	6710	6728	6746	6764	6782	6800	6818	6836	6854
3800	6872	6890	6908	6926	6944	6962	6980	6998	7016	7034
3900	7052	7070	7088	7106	7124	7142	7160	7178	7196	7214

Temperature conversion °F to °C ($°C = {}^5/_9 \cdot$ [°F - 32])

°F	0	10	20	30	40	50	60	70	80	90
−400	−240	−246	−251	−257	−262	−268	−273	−279	−284	−290
−300	−184	−190	−196	−201	−207	−212	−218	−223	−229	−234
−200	−129	−134	−140	−146	−151	−157	−162	−168	−173	−179
−100	−73	−79	−84	−90	−96	−101	−107	−112	−118	−123
−0	−18	−23	−29	−34	−40	−46	−51	−57	−62	−68
0	−18	−12	−7	−1	4	10	16	21	27	32
100	38	43	49	54	60	66	71	77	82	88
200	93	99	104	110	116	121	127	132	138	143
300	149	154	160	166	171	177	182	188	193	199
400	204	210	216	221	227	232	238	243	249	254
500	260	266	271	277	282	288	293	299	304	310
600	316	321	327	332	338	343	349	354	360	366
700	371	377	382	388	393	399	404	410	416	421
800	427	432	438	443	449	454	460	466	471	477
900	482	488	493	499	504	510	516	521	527	532
1000	538	543	549	554	560	566	571	577	582	588
1100	593	599	604	610	616	621	627	632	638	643
1200	649	654	660	666	671	677	682	688	693	699
1300	704	710	716	721	727	732	738	743	749	754
1400	760	766	771	777	782	788	793	799	804	810
1500	816	821	827	832	838	843	849	854	860	866
1600	871	877	882	888	893	899	904	910	916	921
1700	927	932	938	943	949	954	960	966	971	977
1800	982	988	993	999	1004	1010	1016	1021	1027	1032
1900	1038	1043	1049	1054	1060	1066	1071	1077	1082	1088
2000	1093	1099	1104	1110	1116	1121	1127	1132	1138	1143
2100	1149	1154	1160	1166	1171	1177	1182	1188	1193	1199
2200	1204	1210	1216	1221	1227	1232	1238	1243	1249	1254
2300	1260	1266	1271	1277	1282	1288	1293	1299	1304	1310
2400	1316	1321	1327	1332	1338	1343	1349	1354	1360	1366
2500	1371	1377	1382	1388	1393	1399	1404	1410	1416	1421
2600	1427	1432	1438	1443	1449	1454	1460	1466	1471	1477
2700	1482	1488	1493	1499	1504	1510	1516	1521	1527	1532
2800	1538	1543	1549	1554	1560	1566	1571	1577	1582	1588
2900	1593	1599	1604	1610	1616	1621	1627	1632	1638	1643
3000	1649	1654	1660	1666	1671	1677	1682	1688	1693	1699
3100	1704	1710	1716	1721	1727	1732	1738	1743	1749	1754
3200	1760	1766	1771	1777	1782	1788	1793	1799	1804	1810
3300	1816	1821	1827	1832	1838	1843	1849	1854	1860	1866
3400	1871	1877	1882	1888	1893	1899	1904	1910	1916	1921
3500	1927	1932	1938	1943	1949	1954	1960	1966	1971	1977
3600	1982	1988	1993	1999	2004	2010	2016	2021	2027	2032
3700	2038	2043	2049	2054	2060	2066	2071	2077	2082	2088
3800	2093	2099	2104	2110	2116	2121	2127	2132	2138	2143
3900	2149	2154	2160	2166	2171	2177	2182	2188	2193	2199

Composition ranges and limits for AISI-SAE standard low-alloy steel plate applicable for structural applications

Boron or lead can be added to these compositions. Small quantities of certain elements not required may be found. These elements are to be considered incidental and are acceptable to the following maximum amounts: copper to 0.35%, nickel to 0.25%, chromium to 0.20%, and molybdenum to 0.06%.

AISI-SAE designation	UNS designation	Heat composition ranges and limits, %(a)					
		C	Mn	Si(b)	Cr	Ni	Mo
1330............	G13300	0.27–0.34	1.50–1.90	0.15–0.30	· · ·	· · ·	· · ·
1335............	G13350	0.32–0.39	1.50–1.90	0.15–0.30	· · ·	· · ·	· · ·
1340............	G13400	0.36–0.44	1.50–1.90	0.15–0.30	· · ·	· · ·	· · ·
1345............	G13450	0.41–0.49	1.50–1.90	0.15–0.30	· · ·	· · ·	· · ·
4118............	G41180	0.17–0.23	0.60–0.90	0.15–0.30	0.40–0.65	· · ·	0.08–0.15
4130............	G41300	0.27–0.34	0.35–0.60	0.15–0.30	0.80–1.15	· · ·	0.15–0.25
4135............	G41350	0.32–0.39	0.65–0.95	0.15–0.30	0.80–1.15	· · ·	0.15–0.25
4137............	G41370	0.33–0.40	0.65–0.95	0.15–0.30	0.80–1.15	· · ·	0.15–0.25
4140............	G41400	0.36–0.44	0.70–1.00	0.15–0.30	0.80–1.15	· · ·	0.15–0.25
4142............	G41420	0.38–0.46	0.70–1.00	0.15–0.30	0.80–1.15	· · ·	0.15–0.25
4145............	G41450	0.41–0.49	0.70–1.00	0.15–0.30	0.80–1.15	· · ·	0.15–0.25
4340............	G43400	0.36–0.44	0.55–0.80	0.15–0.30	0.60–0.90	1.65–2.00	0.20–0.30
E4340(c).......	G43406	0.37–0.44	0.60–0.85	0.15–0.30	0.65–0.90	1.65–2.00	0.20–0.30
4615............	G46150	0.12–0.18	0.40–0.65	0.15–0.30	· · ·	1.65–2.00	0.20–0.30
4617............	G46170	0.15–0.21	0.40–0.65	0.15–0.30	· · ·	1.65–2.00	0.20–0.30
4620............	G46200	0.16–0.22	0.40–0.65	0.15–0.30	· · ·	1.65–2.00	0.20–0.30
5160............	G51600	0.54–0.65	0.70–1.00	0.15–0.30	0.60–0.90	· · ·	· · ·
6150(d)	G61500	0.46–0.54	0.60–0.90	0.15–0.30	0.80–1.15	· · ·	· · ·
8615............	G86150	0.12–0.18	0.60–0.90	0.15–0.30	0.35–0.60	0.40–0.70	0.15–0.25
8617............	G86170	0.15–0.21	0.60–0.90	0.15–0.30	0.35–0.60	0.40–0.70	0.15–0.25
8620............	G86200	0.17–0.23	0.60–0.90	0.15–0.30	0.35–0.60	0.40–0.70	0.15–0.25
8622............	G86220	0.19–0.25	0.60–0.90	0.15–0.30	0.35–0.60	0.40–0.70	0.15–0.25
8625............	G86250	0.22–0.29	0.60–0.90	0.15–0.30	0.35–0.60	0.40–0.70	0.15–0.25
8627............	G86270	0.24–0.31	0.60–0.90	0.15–0.30	0.35–0.60	0.40–0.70	0.15–0.25
8630............	G86300	0.27–0.34	0.60–0.90	0.15–0.30	0.35–0.60	0.40–0.70	0.15–0.25
8637............	G86370	0.33–0.40	0.70–1.00	0.15–0.30	0.35–0.60	0.40–0.70	0.15–0.25
8640............	G86400	0.36–0.44	0.70–1.00	0.15–0.30	0.35–0.60	0.40–0.70	0.15–0.25
8655............	G86550	0.49–0.60	0.70–1.00	0.15–0.30	0.35–0.60	0.40–0.70	0.15–0.25
8742............	G87420	0.38–0.46	0.70–1.00	0.15–0.30	0.35–0.60	0.40–0.70	0.20–0.30

(a) Indicated ranges and limits apply to steels made by the open hearth or basic oxygen processes; maximum content for phosphorus is 0.035% and for sulfur 0.040%. For steels made by the electric furnace process, the ranges and limits are reduced as follows: C—0.01%; Mn—0.05%; Cr—0.05% (<1.25%), 0.10% (>1.25%); maximum content for either phosphorus or sulfur is 0.025%. (b) Other silicon ranges may be negotiated. Silicon is available in ranges of 0.10–0.20%, 0.20–0.30%, and 0.35% maximum (when carbon deoxidized) when so specified by the purchaser. (c) Prefix "E" indicates that the steel is made by the electric furnace process. (d) Contains 0.15% V minimum

ASTM specifications of chemical composition for structural plate made of low-alloy steel or carbon steel

ASTM specification	Material grade or type	Composition, %(a) C	Mn	P	S	Si	Cr	Ni	Mo	Cu	Others
Low-alloy steel											
A 514	A	0.15–0.21	0.80–1.10	0.035	0.04	0.40–0.80	0.50–0.80	· · ·	0.18–0.28	· · ·	Zr, 0.05–0.15; B, 0.0025
	B	0.12–0.21	0.70–1.00	0.035	0.04	0.20–0.35	0.40–0.65	· · ·	0.15–0.25	· · ·	V, 0.03–0.08; Ti, 0.01–0.03; B, 0.0005–0.005
	C	0.10–0.20	1.10–1.50	0.035	0.04	0.15–0.30	· · ·	· · ·	0.15–0.30	· · ·	B, 0.001–0.005
	E	0.12–0.20	0.40–0.70	0.035	0.04	0.20–0.40	1.40–2.00	· · ·	0.40–0.60	· · ·	Ti, 0.01–0.10 (b); B, 0.001–0.005
	F	0.10–0.20	0.60–1.00	0.035	0.04	0.15–0.35	0.40–0.65	0.70–1.00	0.40–0.60	0.15–0.50	V, 0.03–0.08; B, 0.0005–0.006
	H	0.12–0.21	0.95–1.30	0.035	0.04	0.20–0.35	0.40–0.65	0.30–0.70	0.20–0.30	· · ·	V, 0.03–0.08; B, 0.0005–0.005
	J	0.12–0.21	0.45–0.70	0.035	0.04	0.20–0.35	· · ·	· · ·	0.50–0.65	· · ·	B, 0.001–0.005
	M	0.12–0.21	0.45–0.70	0.035	0.04	0.20–0.35	· · ·	1.20–1.50	0.45–0.60	· · ·	B, 0.001–0.005
	P	0.12–0.21	0.45–0.70	0.035	0.04	0.20–0.35	0.85–1.20	1.20–1.50	0.45–0.60	· · ·	B, 0.001–0.005
	Q	0.14–0.21	0.95–1.30	0.035	0.04	0.15–0.35	1.00–1.50	1.20–1.50	0.40–0.6	· · ·	V, 0.03–0.08
	R	0.15–0.80	0.85–1.15	0.035	0.04	0.20–0.35	0.35–0.65	0.90–1.10	0.15–0.25	· · ·	V, 0.03–0.08
	S	0.10–0.20	1.10–1.50	0.035	0.04	0.15–0.35	· · ·	· · ·	0.10–0.35	· · ·	B, 0.001–0.005; Nb, 0.06 max(c)
	T	0.08–0.14	1.20–1.50	0.035	0.010	0.40–0:60	· · ·	· · ·	0.45–0.60	· · ·	V, 0.03–0.08; B, 0.001–0.005

(continued)

Note: See Table 4 for the compositions of structural plate made of HSLA steel. (a) When a single value is shown, it is a maximum limit, except for copper, for which a single value denotes a minimum limit. (b) Vanadium can be substituted for part or all of the titanium on a one-for-one basis. (c) Titanium may be present in levels up to 0.06% to protect the boron additions. (d) Specification covers many AISI/SAE grades and chemistries. (e) Limiting values vary with plate thickness. (f) Minimum value applicable only if copper-bearing steel is specified. (g) Plates over 13 mm (½ in.) in thickness shall have a minimum manganese content not less than 2.5 times carbon content. (h) The upper limit of manganese may be exceeded provided C + 1/6 Mn does not exceed 0.40% based on heat analysis.

Composition of high-strength low-alloy steel plate

ASTM specification	Material grade or type	Composition, %(a) C	Mn	P	S	Si	Cr	Ni	Mo	Cu	V	Nb	Others
Structural quality													
A 131	AH32, DH32, EH32, AH36, DH36, EH36	0.18	0.90–1.60	0.04	0.04	0.10–0.50	0.25	0.40	0.08	0.35	0.10	0.05	· · ·
A 242	1	0.15	1.00	0.15	0.05	· · ·	· · ·	· · ·	· · ·	0.20 min	· · ·	· · ·	(b)(e)
A 572	42	0.21	1.35	0.04	0.05	0.40(d)	· · ·	· · ·	· · ·	· · ·	(e)	(e)	(e)
	45	0.22	1.35	0.04	0.05	0.40(d)	· · ·	· · ·	· · ·	(e)	(e)	(e)	(e)
	50	0.23	1.35	0.04	0.05	0.40(d)	· · ·	· · ·	· · ·	(e)	(e)	(e)	(e)
	60	0.26	1.35	0.04	0.05	0.40(d)	· · ·	· · ·	· · ·	(e)	(e)	(e)	(e)
	65	0.26(d)	1.65(d)	0.04	0.05	0.40	· · ·	· · ·	· · ·	(e)	(e)	(e)	(e)
A 588	A	0.19	0.80–1.25	0.04	0.05	0.30–0.65	0.40–0.65	0.40	· · ·	0.25–0.40	0.02–0.10	· · ·	· · ·
	B	0.20	0.75–1.35	0.04	0.05	0.15–0.50	0.50–0.70	0.50	· · ·	0.20–0.40	0.01–0.10	· · ·	· · ·
	C	0.15	0.80–1.35	0.04	0.05	0.15–0.40	0.30–0.50	0.25–0.50	· · ·	0.20–0.50	0.01–0.10	· · ·	· · ·
	D	0.10–0.20	0.75–1.25	0.04	0.05	0.50–0.90	0.50–0.90	· · ·	· · ·	0.30	· · ·	0.04	Zr, 0.05-0.15
	K	0.17	0.50–1.20	0.04	0.05	0.25–0.50	0.40–0.70	0.40	0.10	0.30–0.50	0.005–0.05(f)	· · ·	· · ·
A 633	A	0.18	1.00–1.35	0.04	0.05	0.15–0.50	· · ·	· · ·	· · ·	· · ·	0.05	· · ·	· · ·
	B	0.18	1.00–1.35	0.04	0.05	0.15–0.50	· · ·	· · ·	· · ·	0.10	· · ·	· · ·	· · ·
	C	0.20	1.15–1.50	0.04	0.05	0.15–0.50	· · ·	· · ·	· · ·	· · ·	· · ·	0.01–0.05	· · ·
	D	0.20	0.70–1.60(d)	0.04	0.05	0.15–0.50	0.25	0.25	0.08	0.35	· · ·	· · ·	· · ·
	E	0.22	1.15–1.50	0.04	0.05	0.15–0.50	· · ·	· · ·	· · ·	· · ·	0.04–0.11	(g)	N, 0.01–0.03(h)
A 656	3	0.18	1.65	0.025	0.035	0.60	· · ·	· · ·	· · ·	0.08	0.005–0.15	· · ·	N, 0.020
	7	0.18	1.65	0.025	0.035	0.60	· · ·	· · ·	· · ·	0.005–0.015(i)	0.005–0.015(i)	· · ·	N, 0.020
A 678	D	0.22	1.15–1.50	0.04	0.05	0.15–0.50	· · ·	· · ·	· · ·	0.2 min(j)	0.04–0.11	(g)	N, 0.001–0.03
A 709	50	0.23	1.35(d)	0.04	0.05	0.15–0.40(d)	· · ·	· · ·	· · ·	· · ·	(e)	(e)	(e)
	50W					Identical to A 588 type A, B, or C (as specified)							
A 808	· · ·	0.12	1.65	0.04	0.05	0.15–0.50	· · ·	· · ·	· · ·	0.10	· · ·	0.02–0.10	(Nb + V), 0.15
A 852	· · ·	0.19	0.80–1.35	0.04	0.05	0.20–0.65	0.40–0.70	0.50	· · ·	0.20–0.40	0.02–0.10	· · ·	· · ·
A 871	· · ·	1.20	1.50	0.04	0.05	0.90	0.90	1.25	0.25	1.00	0.10	0.05	Zr, 0.15; Ti, 0.05
Pressure vessel quality													
A 734	B	0.17	1.60	0.035	0.015	0.40	· · ·	0.35	0.25	0.25(j)	0.11	(k)	Al, 0.06; N, 0.030
A 737	B	0.20	1.15–1.50	0.035	0.030	0.15–0.50	· · ·	· · ·	· · ·	· · ·	0.05	(k)	· · ·
	C	0.22	1.15–1.50	0.035	0.030	0.15–0.50	· · ·	· · ·	· · ·	0.04–0.11	(k)		N, 0.03
A 841	· · ·	0.20	0.70–1.60(d)	0.030	0.030	0.15–0.50	0.25	0.25	0.08	0.35	0.06	0.03	Al, 0.020 min

(a) Except as noted, when a single value is shown, it is a maximum limit. (b) Choice and amount of other alloying elements added to give the required mechanical properties and atmospheric corrosion resistance are made by the producer and reported in the heat analysis. (c) Elements commonly added include silicon, chromium, nickel, vanadium, titanium, and zirconium. (d) Limiting values vary with plate thickness. (e) For type 1, 0.005–0.05% Nb; for type 2, 0.01–0.15% V; for type 3, 0.05% Nb max + V = (0.02–0.15%); for type 4, N (with V) 0.015% max. (f) For plates under 13 mm (½ in.) thickness, the minimum niobium limit is waived. (g) Niobium may be present in the amount of 0.01–0.05%. (h) The minimum total aluminum content shall be 0.018% or the vanadium:nitrogen ratio shall be 4:1 minimum. (i) Niobium, or vanadium, or both, 0.005% min. When both are added, the total shall be 0.20% max. (j) Applicable only when specified. (k) 0.05% max Nb may be present.

Carbon steel compositions

Applicable to semifinished products for forging, hot-rolled and cold-finished bars, wire rods, and seamless tubing

UNS number	SAE-AISI number	Cast or heat chemical ranges and limits, %(a)			
		C	Mn	P max	S max
G10050	1005	0.06 max	0.35 max	0.040	0.050
G10060	1006	0.08 max	0.25–0.40	0.040	0.050
G10080	1008	0.10 max	0.30–0.50	0.040	0.050
G10100	1010	0.08–0.13	0.30–0.60	0.040	0.050
G10120	1012	0.10–0.15	0.30–0.60	0.040	0.050
G10130	1013	0.11–0.16	0.50–0.80	0.040	0.050
G10150	1015	0.13–0.18	0.30–0.60	0.040	0.050
G10160	1016	0.13–0.18	0.60–0.90	0.040	0.050
G10170	1017	0.15–0.20	0.30–0.60	0.040	0.050
G10180	1018	0.15–0.20	0.60–0.90	0.040	0.050
G10190	1019	0.15–0.20	0.70–1.00	0.040	0.050
G10200	1020	0.18–0.23	0.30–0.60	0.040	0.050
G10210	1021	0.18–0.23	0.60–0.90	0.040	0.050
G10220	1022	0.18–0.23	0.70–1.00	0.040	0.050
G10230	1023	0.20–0.25	0.30–0.60	0.040	0.050
G10250	1025	0.22–0.28	0.30–0.60	0.040	0.050
G10260	1026	0.22–0.28	0.60–0.90	0.040	0.050
G10290	1029	0.25–0.31	0.60–0.90	0.040	0.050
G10300	1030	0.28–0.34	0.60–0.90	0.040	0.050
G10350	1035	0.32–0.38	0.60–0.90	0.040	0.050
G10370	1037	0.32–0.38	0.70–1.00	0.040	0.050
G10380	1038	0.35–0.42	0.60–0.90	0.040	0.050
G10390	1039	0.37–0.44	0.70–1.00	0.040	0.050
G10400	1040	0.37–0.44	0.60–0.90	0.040	0.050
G10420	1042	0.40–0.47	0.60–0.90	0.040	0.050
G10430	1043	0.40–0.47	0.70–1.00	0.040	0.050
G10440	1044	0.43–0.50	0.30–0.60	0.040	0.050
G10450	1045	0.43–0.50	0.60–0.90	0.040	0.050
G10460	1046	0.43–0.50	0.70–1.00	0.040	0.050
G10490	1049	0.46–0.53	0.60–0.90	0.040	0.050
G10500	1050	0.48–0.55	0.60–0.90	0.040	0.050
G10530	1053	0.48–0.55	0.70–1.00	0.040	0.050
G10550	1055	0.50–0.60	0.60–0.90	0.040	0.050
G10590	1059	0.55–0.65	0.50–0.80	0.040	0.050
G10600	1060	0.55–0.65	0.60–0.90	0.040	0.050
G10640	1064	0.60–0.70	0.50–0.80	0.040	0.050
G10650	1065	0.60–0.70	0.60–0.90	0.040	0.050
G10690	1069	0.65–0.75	0.40–0.70	0.040	0.050
G10700	1070	0.65–0.75	0.60–0.90	0.040	0.050
G10740	1074	0.70–0.80	0.50–0.80	0.040	0.050
G10750	1075	0.70–0.80	0.40–0.70	0.040	0.050
G10780	1078	0.72–0.85	0.30–0.60	0.040	0.050
G10800	1080	0.75–0.88	0.60–0.90	0.040	0.050
G10840	1084	0.80–0.93	0.60–0.90	0.040	0.050
G10850	1085	0.80–0.93	0.70–1.00	0.040	0.050
G10860	1086	0.80–0.93	0.30–0.50	0.040	0.050
G10900	1090	0.85–0.98	0.60–0.90	0.040	0.050
G10950	1095	0.90–1.03	0.30–0.50	0.040	0.050

(a) When silicon ranges or limits are required for bar and semifinished products, the values in Table 1 apply. For rods, the following ranges are commonly used: 0.10 max; 0.07–0.15%; 0.10–0.20%; 0.15–0.35%; 0.20–0.40%; and 0.30–0.60%. Steels listed in this table can be produced with additions of lead or boron. Leaded steels typically contain 0.15–0.35% Pb and are identified by inserting the letter L in the designation (10L45); boron steels can be expected to contain 0.0005–0.003% B and are identified by inserting the letter B in the designation (10B46). Source: "Chemical Compositions of SAE Carbon Steels," SAE J403, *1989 SAE Handbook*, Vol 1, *Materials*, Society of Automotive Engineers, p 1.08–1.10

High-manganese carbon steel compositions

Applicable only to semifinished products for forging, hot-rolled and cold-finished bars, wire rods, and seamless tubing

UNS number	SAE-AISI number	Cast or heat chemical ranges and limits, %			
		C	Mn	P max	S max
G15130	1513	0.10–0.16	1.10–1.40	0.040	0.050
G15220	1522	0.18–0.24	1.10–1.40	0.040	0.050
G15240	1524	0.19–0.25	1.35–1.65	0.040	0.050
G15260	1526	0.22–0.29	1.10–1.40	0.040	0.050
G15270	1527	0.22–0.29	1.20–1.50	0.040	0.050
G15360	1536	0.30–0.37	1.20–1.50	0.040	0.050
G15410	1541	0.36–0.44	1.35–1.65	0.040	0.050
G15480	1548	0.44–0.52	1.10–1.40	0.040	0.050
G15510	1551	0.45–0.56	0.85–1.15	0.040	0.050
G15520	1552	0.47–0.55	1.20–1.50	0.040	0.050
G15610	1561	0.55–0.65	0.75–1.05	0.040	0.050
G15660	1566	0.60–0.71	0.85–1.15	0.040	0.050

Applicable only to structural shapes, plates, strip, sheets, and welded tubing

UNS number	SAE-AISI number	Cast or heat chemical ranges and limits, %				Former SAE number
		C	Mn	P max	S max	
G15240	1524	0.18–0.25	1.30–1.65	0.040	0.050	1024
G15270	1527	0.22–0.29	1.20–1.55	0.040	0.050	1027
G15360	1536	0.30–0.38	1.20–1.55	0.040	0.050	1036
G15410	1541	0.36–0.45	1.30–1.65	0.040	0.050	1041
G15480	1548	0.43–0.52	1.05–1.40	0.040	0.050	1048
G15520	1552	0.46–0.55	1.20–1.55	0.040	0.050	1052

Source: "Chemical Compositions of SAE Carbon Steels," SAE J403, *1989 SAE Handbook,* Vol 1, *Materials,* Society of Automotive Engineers, p 1.08-1.10

Free-cutting (rephosphorized and resulfurized) carbon steel compositions

Applicable to semifinished products for forging, hot-rolled and cold-finished bars, wire rods, and seamless tubing

UNS number	SAE-AISI number	Cast or heat chemical ranges and limits, %				
		C max	Mn	P	S	Pb
G12110	1211	0.13	0.60–0.90	0.07–0.12	0.10–0.15	· · ·
G12120	1212	0.13	0.70–1.00	0.07–0.12	0.16–0.23	· · ·
G12130	1213	0.13	0.70–1.00	0.07–0.12	0.24–0.33	· · ·
G12150	1215	0.09	0.75–1.05	0.04–0.09	0.26–0.35	· · ·
G12144	12L14	0.15	0.85–1.15	0.04–0.09	0.26–0.35	0.15–0.35

Source: "Chemical Compositions of SAE Carbon Steels," SAE J403, *1989 SAE Handbook,* Vol 1, *Materials,* Society of Automotive Engineers, p 1.08-1.10

Free-cutting (resulfurized) carbon steel compositions

Applicable to semifinished products for forging, hot-rolled and cold-finished bars, wire rods, and seamless tubing

UNS number	SAE-AISI number	Cast or heat chemical ranges and limits, %(a)			
		C	Mn	P max	S
G11080	1108	0.08–0.13	0.50–0.80	0.040	0.08–0.13
G11100	1110	0.08–0.13	0.30–0.60	0.040	0.08–0.13
G11170	1117	0.14–0.20	1.00–1.30	0.040	0.08–0.13
G11180	1118	0.14–0.20	1.30–1.60	0.040	0.08–0.13
G11370	1137	0.32–0.39	1.35–1.65	0.040	0.08–0.13
G11390	1139	0.35–0.43	1.35–1.65	0.040	0.13–0.20
G11400	1140	0.37–0.44	0.70–1.00	0.040	0.08–0.13
G11410	1141	0.37–0.45	1.35–1.65	0.040	0.08–0.13
G11440	1144	0.40–0.48	1.35–1.65	0.040	0.24–0.33
G11460	1146	0.42–0.49	0.70–1.00	0.040	0.08–0.13
G11510	1151	0.48–0.55	0.70–1.00	0.040	0.08–0.13

(a) When lead ranges or limits are required, or when silicon ranges or limits are required for bars or semifinished products, the values in Table 1 apply. For rods, the following ranges and limits for silicon are commonly used: up to SAE 1110 inclusive, 0.10% max; SAE 1117 and over, 0.10% max, 0.10–0.20%, or 0.15–0.35%. Source: "Chemical Compositions of SAE Carbon Steels," SAE J403, *1989 SAE Handbook,* Vol 1, *Materials,* Society of Automotive Engineers, p 1.08-1.10

SAE potential standard steel compositions

SAE PS number(a)	Ladle chemical composition limits, wt%								
	C	Mn	P max	S max	Si	Ni	Cr	Mo	B
PS 10	0.19–0.24	0.95–1.25	0.035	0.040	0.15–0.35	0.20–0.40	0.25–0.40	0.05–0.10	...
PS 15	0.18–0.23	0.90–1.20	0.035	0.040	0.15–0.35	...	0.40–0.60	0.13–0.20	...
PS 16	0.20–0.25	0.90–1.20	0.035	0.040	0.15–0.35	...	0.40–0.60	0.13–0.20	...
PS 17	0.23–0.28	0.90–1.20	0.035	0.040	0.15–0.35	...	0.40–0.60	0.13–0.20	...
PS 18	0.25–0.30	0.90–1.20	0.035	0.040	0.15–0.35	...	0.40–0.60	0.08–0.15	0.0005–0.003
PS 19	0.18–0.23	0.90–1.20	0.035	0.040	0.15–0.35	...	0.40–0.60	0.13–0.20	...
PS 20	0.13–0.18	0.90–1.20	0.035	0.040	0.15–0.35	...	0.40–0.60	0.13–0.20	...
PS 21	0.15–0.20	0.90–1.20	0.035	0.040	0.15–0.35	...	0.45–0.65	0.20–0.30	...
PS 24	0.18–0.23	0.75–1.00	0.035	0.040	0.15–0.35	...	0.45–0.65	0.45–0.60	...
PS 30	0.13–0.18	0.70–0.90	0.035	0.040	0.15–0.35	0.70–1.00	0.45–0.65	0.45–0.60	...
PS 31	0.15–0.20	0.70–0.90	0.035	0.040	0.15–0.35	0.70–1.00	0.45–0.65	0.45–0.60	...
PS 32	0.18–0.23	0.70–0.90	0.035	0.040	0.15–0.35	0.20 min	0.20 min	0.05 min	...
PS 33(b)	0.17–0.24	0.85–1.25	0.035	0.040	0.15–0.35	...	0.40–0.60	0.13–0.20	...
PS 34	0.28–0.33	0.90–1.20	0.035	0.040	0.15–0.35	...	0.45–0.65	0.13–0.20	...
PS 36	0.38–0.43	0.90–1.20	0.035	0.040	0.15–0.35	...	0.45–0.65	0.13–0.20	...
PS 38	0.43–0.48	0.90–1.20	0.035	0.040	0.15–0.35	...	0.45–0.65	0.13–0.20	...
PS 39	0.48–0.53	0.90–1.20	0.035	0.040	0.15–0.35	...	0.45–0.65	0.13–0.20	...
PS 40	0.51–0.59	0.90–1.20	0.035	0.040	0.15–0.35	...	0.40–0.70	0.05 min	...
PS 54	0.19–0.25	0.70–1.05	0.035	0.040	0.15–0.35	...	0.45–0.65	0.65–0.80	...
PS 55	0.15–0.20	0.70–1.00	0.035	0.040	0.15–0.35	1.65–2.00	0.45–0.65	0.65–0.80	...
PS 56	0.080–0.13	0.70–1.00	0.035	0.040	0.15–0.35	1.65–2.00	17.00–19.00	1.75–2.25	...
PS 57	0.08 max	1.25 max	0.040	0.15–0.35	1.00 max	...	0.45–0.65	...	...
PS 58	0.16–0.21	1.00–1.30	0.035	0.040	0.15–0.35	...	0.70–0.90	...	...
PS 59	0.18–0.23	1.00–1.30	0.035	0.040	0.15–0.35	...	0.70–0.90	...	...
PS 61	0.23–0.28	1.00–1.30	0.035	0.040	0.15–0.35	...	0.45–0.65	...	0.0005–0.003
PS 63	0.31–0.38	0.75–1.10	0.035	0.040	0.15–0.35	...	0.70–0.90	...	...
PS 64	0.16–0.21	1.00–1.30	0.035	0.040	0.15–0.35	...	0.70–0.90	...	...
PS 65	0.21–0.26	1.00–1.30	0.035	0.040	0.15–0.35	1.65–2.00	0.45–0.75	0.08–0.15	...
PS 66(c)	0.16–0.21	0.40–0.70	0.035	0.040	0.15–0.35	...	0.85–1.20	0.25–0.35	...
PS 67	0.42–0.49	0.80–1.20	0.035	0.040	0.15–0.35	...	...	...	...

(a) Some PS steels may be supplied to a hardenability requirement. (b) Supplied to a hardenability requirement of 15 HRC points within the range of 23-43 HRC at J4 ($^4/_{16}$ in. distance from quenched end), subject to agreement between producer and user. (c) PS66 has a vanadium content of 0.10—0.15%. Source: "Potential Standard Steels," SAE J1081, *1989 SAE Handbook,* Vol 1, *Materials,* Society of Automotive Engineers, p 1.14 - 1.15

Low-alloy steel compositions applicable to billets, blooms, slabs, and hot-rolled and cold-finished bars

Slightly wider ranges of compositions apply to plates.

UNS number	SAE number	Corresponding AISI number	Ladle chemical composition limits, %(a)								
			C	Mn	P	S	Si	Ni	Cr	Mo	V
G13300	1330	1330	0.28–0.33	1.60–1.90	0.035	0.040	0.15–0.35	...	...	...	...
G13350	1335	1335	0.33–0.38	1.60–1.90	0.035	0.040	0.15–0.35	...	...	...	...
G13400	1340	1340	0.38–0.43	1.60–1.90	0.035	0.040	0.15–0.35	...	...	...	...
G13450	1345	1345	0.43–0.48	1.60–1.90	0.035	0.040	0.15–0.35	...	...	...	...
G40230	4023	4023	0.20–0.25	0.70–0.90	0.035	0.040	0.15–0.35	...	...		...
G40240	4024	4024	0.20–0.25	0.70–0.90	0.035	0.035–0.050	0.15–0.35	...	...		
G40270	4027	4027	0.25–0.30	0.70–0.90	0.035	0.040	0.15–0.35	...	...	0.20–0.30	
G40280	4028	4028	0.25–0.30	0.70–0.90	0.035	0.035–0.050	0.15–0.35	...	...	0.20–0.30	
G40320	4032	...	0.30–0.35	0.70–0.90	0.035	0.040	0.15–0.35	...	...	0.20–0.30	
G40370	4037	4037	0.35–0.40	0.70–0.90	0.035	0.040	0.15–0.35	...	...	0.20–0.30	
G40420	4042	...	0.40–0.45	0.70–0.90	0.035	0.040	0.15–0.35	...	...	0.20–0.30	
G40470	4047	4047	0.45–0.50	0.70–0.90	0.035	0.040	0.15–0.35	...	...	0.20–0.30	
G41180	4118	4118	0.18–0.23	0.70–0.90	0.035	0.040	0.15–0.35	...	0.40–0.60	0.08–0.15	...
G41300	4130	4130	0.28–0.33	0.40–0.60	0.035	0.040	0.15–0.35	...	0.80–1.10	0.15–0.25	
G41350	4135	...	0.33–0.38	0.70–0.90	0.035	0.040	0.15–0.35	...	0.80–1.10	0.15–0.25	
G41370	4137	4137	0.35–0.40	0.70–0.90	0.035	0.040	0.15–0.35	...	0.80–1.10	0.15–0.25	
G41400	4140	4140	0.38–0.43	0.75–1.00	0.035	0.040	0.15–0.35	...	0.80–1.10	0.15–0.25	
G41420	4142	4142	0.40–0.45	0.75–1.00	0.035	0.040	0.15–0.35	...	0.80–1.10	0.15–0.25	
G41450	4145	4145	0.41–0.48	0.75–1.00	0.035	0.040	0.15–0.35	...	0.80–1.10	0.15–0.25	
G41470	4147	4147	0.45–0.50	0.75–1.00	0.035	0.040	0.15–0.35	...	0.80–1.10	0.15–0.25	
G41500	4150	4150	0.48–0.53	0.75–1.00	0.035	0.040	0.15–0.35	...	0.80–1.10	0.15–0.25	
G41610	4161	4161	0.56–0.64	0.75–1.00	0.035	0.040	0.15–0.35	...	0.70–0.90	0.25–0.35	
G43200	4320	4320	0.17–0.22	0.45–0.65	0.035	0.040	0.15–0.35	1.65–2.00	0.40–0.60	0.20–0.30	
G43400	4340	4340	0.38–0.43	0.60–0.80	0.035	0.040	0.15–0.35	1.65–2.00	0.70–0.90	0.20–0.30	
G43406	E4340(b)	E4340	0.38–0.43	0.65–0.85	0.025	0.025	0.15–0.35	1.65–2.00	0.70–0.90	0.20–0.30	
G44220	4422	...	0.20–0.25	0.70–0.90	0.035	0.040	0.15–0.35	...	...	0.35–0.45	
G44270	4427	...	0.24–0.29	0.70–0.90	0.035	0.040	0.15–0.35	...	...	0.35–0.45	
G46150	4615	4615	0.13–0.18	0.45–0.65	0.035	0.040	0.15–0.25	1.65–2.00	...	0.20–0.30	
G46170	4617	...	0.15–0.20	0.45–0.65	0.035	0.040	0.15–0.35	1.65–2.00	...	0.20–0.30	
G46200	4620	4620	0.17–0.22	0.45–0.65	0.035	0.040	0.15–0.35	1.65–2.00	...	0.20–0.30	
G46260	4626	4626	0.24–0.29	0.45–0.65	0.035	0.04 max	0.15–0.35	0.70–1.00	...	0.15–0.25	
G47180	4718	4718	0.16–0.21	0.70–0.90	...	...	...	0.90–1.20	0.35–0.55	0.30–0.40	
G47200	4720	4720	0.17–0.22	0.50–0.70	0.035	0.040	0.15–0.35	0.90–1.20	0.35–0.55	0.15–0.25	
G48150	4815	4815	0.13–0.18	0.40–0.60	0.035	0.040	0.15–0.35	3.25–3.75	...	0.20–0.30	
G48170	4817	4817	0.15–0.20	0.40–0.60	0.035	0.040	0.15–0.35	3.25–3.75	...	0.20–0.30	
G48200	4820	4820	0.18–0.23	0.50–0.70	0.035	0.040	0.15–0.35	3.25–3.75	...	0.20–0.30	
G50401	50B40(c)	...	0.38–0.43	0.75–1.00	0.035	0.040	0.15–0.35	...	0.40–0.60	...	...
G50441	50B44(c)	50B44	0.43–0.48	0.75–1.00	0.035	0.040	0.15–0.35	...	0.40–0.60	...	
G50460	5046	...	0.43–0.48	0.75–1.00	0.035	0.040	0.15–0.35	...	0.20–0.35	...	
G50461	50B46(c)	50B46	0.44–0.49	0.75–1.00	0.035	0.040	0.15–0.35	...	0.20–0.35	...	
G50501	50B50(c)	50B50	0.48–0.53	0.75–1.00	0.035	0.040	0.15–0.35	...	0.40–0.60	...	
G50600	5060	...	0.56–0.64	0.75–1.00	0.035	0.040	0.15–0.35	...	0.40–0.60	...	
G50601	50B60(c)	50B60	0.56–0.64	0.75–1.00	0.035	0.040	0.15–0.35	...	0.40–0.60	...	
G51150	5115	...	0.13–0.18	0.70–0.90	0.035	0.040	0.15–0.35	...	0.70–0.90	...	...
G51170	5117	5117	0.15–0.20	0.70–0.90	0.040	0.040	0.15–0.35	...	0.70–0.90	...	
G51200	5120	5120	0.17–0.22	0.70–0.90	0.035	0.040	0.15–0.35	...	0.70–0.90	...	
G51300	5130	5130	0.28–0.33	0.70–0.90	0.035	0.040	0.15–0.35	...	0.80–1.10	...	
G51320	5132	5132	0.30–0.35	0.60–0.80	0.035	0.040	0.15–0.35	...	0.75–1.00	...	
G51350	5135	5135	0.33–0.38	0.60–0.80	0.035	0.040	0.15–0.35	...	0.80–1.05	...	
G51400	5140	5140	0.38–0.43	0.70–0.90	0.035	0.040	0.15–0.35	...	0.70–0.90	...	
G51470	5147	5147	0.46–0.51	0.70–0.95	0.035	0.040	0.15–0.35	...	0.85–1.15	...	
G51500	5150	5150	0.48–0.53	0.70–0.90	0.035	0.040	0.15–0.35	...	0.70–0.90	...	
G51550	5155	5155	0.51–0.59	0.70–0.90	0.035	0.040	0.15–0.35	...	0.70–0.90	...	
G51600	5160	5160	0.56–0.64	0.75–1.00	0.035	0.040	0.15–0.35	...	0.70–0.90	...	
G51601	51B60(c)	51B60	0.56–0.64	0.75–1.00	0.035	0.040	0.15–0.35	...	0.70–0.90	...	
G50986	50100(b)	...	0.98–1.10	0.25–0.45	0.025	0.025	0.15–0.35	...	0.40–0.60	...	...
G51986	51100(b)	E51100	0.98–1.10	0.25–0.45	0.025	0.025	0.15–0.35	...	0.90–1.15	...	
G52986	52100(b)	E52100	0.98–1.10	0.25–0.45	0.025	0.025	0.15–0.35	...	1.30–1.60	...	
G61180	6118	6118	0.16–0.21	0.50–0.70	0.035	0.040	0.15–0.35	...	0.50–0.70	...	0.10–0.15
G61500	6150	6150	0.48–0.53	0.70–0.90	0.035	0.040	0.15–0.35	...	0.80–1.10	...	0.15 min
G81150	8115	8115	0.13–0.18	0.70–0.90	0.035	0.040	0.15–0.35	0.20–0.40	0.30–0.50	0.08–0.15	...
G81451	81B45(c)	81B45	0.43–0.48	0.75–1.00	0.035	0.040	0.15–0.35	0.20–0.40	0.35–0.55	0.08–0.15	...

(continued)

UNS number	SAE number	Corresponding AISI number	Ladle chemical composition limits, %(a)								
			C	Mn	P	S	Si	Ni	Cr	Mo	V
G86150	8615	8615	0.13–0.18	0.70–0.90	0.035	0.040	0.15–0.35	0.40–0.70	0.40–0.60	0.15–0.25	...
G86170	8617	8617	0.15–0.20	0.70–0.90	0.035	0.040	0.15–0.35	0.40–0.70	0.40–0.60	0.15–0.25	...
G86200	8620	8620	0.18–0.23	0.70–0.90	0.035	0.040	0.15–0.35	0.40–0.70	0.40–0.60	0.15–0.25	...
G86220	8622	8622	0.20–0.25	0.70–0.90	0.035	0.040	0.15–0.35	0.40–0.70	0.40–0.60	0.15–0.25	...
G86250	8625	8625	0.23–0.28	0.70–0.90	0.035	0.040	0.15–0.35	0.40–0.70	0.40–0.60	0.15–0.25	...
G86270	8627	8627	0.25–0.30	0.70–0.90	0.035	0.040	0.15–0.35	0.40–0.70	0.40–0.60	0.15–0.25	...
G86300	8630	8630	0.28–0.33	0.70–0.90	0.035	0.040	0.15–0.35	0.40–0.70	0.40–0.60	0.15–0.25	...
G86370	8637	8637	0.35–0.40	0.75–1.00	0.035	0.040	0.15–0.35	0.40–0.70	0.40–0.60	0.15–0.25	...
G86400	8640	8640	0.38–0.43	0.75–1.00	0.035	0.040	0.15–0.35	0.40–0.70	0.40–0.60	0.15–0.25	...
G86420	8642	8642	0.40–0.45	0.75–1.00	0.035	0.040	0.15–0.35	0.40–0.70	0.40–0.60	0.15–0.25	...
G86450	8645	8645	0.43–0.48	0.75–1.00	0.035	0.040	0.15–0.35	0.40–0.70	0.40–0.60	0.15–0.25	...
G86451	86B45(c)	...	0.43–0.48	0.75–1.00	0.035	0.040	0.15–0.35	0.40–0.70	0.40–0.60	0.15–0.25	...
G86500	8650	...	0.48–0.53	0.75–1.00	0.035	0.040	0.15–0.35	0.40–0.70	0.40–0.60	0.15–0.25	...
G86550	8655	8655	0.51–0.59	0.75–1.00	0.035	0.040	0.15–0.35	0.40–0.70	0.40–0.60	0.15–0.25	...
G86600	8660	...	0.56–0.64	0.75–1.00	0.035	0.040	0.15–0.35	0.40–0.70	0.40–0.60	0.15–0.25	...
G87200	8720	8720	0.18–0.23	0.70–0.90	0.035	0.040	0.15–0.35	0.40–0.70	0.40–0.60	0.20–0.30	...
G87400	8740	8740	0.38–0.43	0.75–1.00	0.035	0.040	0.15–0.35	0.40–0.70	0.40–0.60	0.20–0.30	...
G88220	8822	8822	0.20–0.25	0.75–1.00	0.035	0.040	0.15–0.35	0.40–0.70	0.40–0.60	0.30–0.40	...
G92540	9254	...	0.51–0.59	0.60–0.80	0.035	0.040	1.20–1.60	...	0.60–0.80	...	...
G92600	9260	9260	0.56–0.64	0.75–1.00	0.035	0.040	1.80–2.20	...	...	...	...
G93106	9310(b)	...	0.08–0.13	0.45–0.65	0.025	0.025	0.15–0.35	3.00–3.50	1.00–1.40	0.08–0.15	...
G94151	94B15(c)	...	0.13–0.18	0.75–1.00	0.035	0.040	0.15–0.35	0.30–0.60	0.30–0.50	0.08–0.15	...
G94171	94B17(c)	94B17	0.15–0.20	0.75–1.00	0.035	0.040	0.15–0.35	0.30–0.60	0.30–0.50	0.08–0.15	...
G94301	94B30(c)	94B30	0.28–0.33	0.75–1.00	0.035	0.040	0.15–0.35	0.30–0.60	0.30–0.50	0.08–0.15	...

(a) Small quantities of certain elements that are not specified or required may be found in alloy steels. These elements are to be considered as incidental and are acceptable to the following maximum amount: copper to 0.35%, nickel to 0.25%, chromium to 0.20%, and molybdenum to 0.06%. (b) Electric furnace steel. (c) Boron content is 0.0005—0.003%. Source: "Chemical Compositions of SAE Alloy Steels," SAE J404, *1989 SAE Handbook,* Vol 1, *Materials,* Society of Automotive Engineers, p 1.10 - 1.12

Composition limits of principal types of tool steels

Designation AISI	UNS	C	Mn	Si	Cr	Ni	Mo	W	V	Co
Molybdenum high-speed steels										
M1	T11301	0.78–0.88	0.15–0.40	0.20–0.50	3.50–4.00	0.30 max	8.20–9.20	1.40–2.10	1.00–1.35	...
M2	T11302	0.78–0.88; 0.95–1.05	0.15–0.40	0.20–0.45	3.75–4.50	0.30 max	4.50–5.50	5.50–6.75	1.75–2.20	...
M3, class 1	T11313	1.00–1.10	0.15–0.40	0.20–0.45	3.75–4.50	0.30 max	4.75–6.50	5.00–6.75	2.25–2.75	...
M3, class 2	T11323	1.15–1.25	0.15–0.40	0.20–0.45	3.75–4.50	0.30 max	4.75–6.50	5.00–6.75	2.75–3.75	...
M4	T11304	1.25–1.40	0.15–0.40	0.20–0.45	3.75–4.75	0.30 max	4.25–5.50	5.25–6.50	3.75–4.50	...
M7	T11307	0.97–1.05	0.15–0.40	0.20–0.55	3.50–4.00	0.30 max	8.20–9.20	1.40–2.10	1.75–2.25	...
M10	T11310	0.84–0.94; 0.95–1.05	0.10–0.40	0.20–0.45	3.75–4.50	0.30 max	7.75–8.50	...	1.80–2.20	...
M30	T11330	0.75–0.85	0.15–0.40	0.20–0.45	3.50–4.25	0.30 max	7.75–9.00	1.30–2.30	1.00–1.40	4.50–5.50
M33	T11333	0.85–0.92	0.15–0.40	0.15–0.50	3.50–4.00	0.30 max	9.00–10.00	1.30–2.10	1.00–1.35	7.75–8.75
M34	T11334	0.85–0.92	0.15–0.40	0.20–0.45	3.50–4.00	0.30 max	7.75–9.20	1.40–2.10	1.90–2.30	7.75–8.75
M35	T11335	0.82–0.88	0.15–0.40	0.20–0.45	3.75–4.50	0.30 max	4.50–5.50	5.50–6.75	1.75–2.20	4.50–5.50
M36	T11336	0.80–0.90	0.15–0.40	0.20–0.45	3.75–4.50	0.30 max	4.50–5.50	5.50–6.50	1.75–2.25	7.75–8.75
M41	T11341	1.05–1.15	0.20–0.60	0.15–0.50	3.75–4.50	0.30 max	3.25–4.25	6.25–7.00	1.75–2.25	4.75–5.75
M42	T11342	1.05–1.15	0.15–0.40	0.15–0.65	3.50–4.25	0.30 max	9.00–10.00	1.15–1.85	0.95–1.35	7.75–8.75
M43	T11343	1.15–1.25	0.20–0.40	0.15–0.65	3.50–4.25	0.30 max	7.50–8.50	2.25–3.00	1.50–1.75	7.75–8.75
M44	T11344	1.10–1.20	0.20–0.40	0.30–0.55	4.00–4.75	0.30 max	6.00–7.00	5.00–5.75	1.85–2.20	11.00–12.25
M46	T11346	1.22–1.30	0.20–0.40	0.40–0.65	3.70–4.20	0.30 max	8.00–8.50	1.90–2.20	3.00–3.30	7.80–8.80
M47	T11347	1.05–1.15	0.15–0.40	0.20–0.45	3.50–4.00	0.30 max	9.25–10.00	1.30–1.80	1.15–1.35	4.75–5.25
M48	T11348	1.42–1.52	0.15–0.40	0.15–0.40	3.50–4.00	0.30 max	4.75–5.50	9.50–10.50	2.75–3.25	8.00–10.00
M62	T11362	1.25–1.35	0.15–0.40	0.15–0.40	3.50–4.00	0.30 max	10.00–11.00	5.75–6.50	1.80–2.10	...
Tungsten high-speed steels										
T1	T12001	0.65–0.80	0.10–0.40	0.20–0.40	3.75–4.50	0.30 max	...	17.25–18.75	0.90–1.30	...
T2	T12002	0.80–0.90	0.20–0.40	0.20–0.40	3.75–4.50	0.30 max	1.00 max	17.50–19.00	1.80–2.40	...
T4	T12004	0.70–0.80	0.10–0.40	0.20–0.40	3.75–4.50	0.30 max	0.40–1.00	17.50–19.00	0.80–1.20	4.25–5.75
T5	T12005	0.75–0.85	0.20–0.40	0.20–0.40	3.75–5.00	0.30 max	0.50–1.25	17.50–19.00	1.80–2.40	7.00–9.50
T6	T12006	0.75–0.85	0.20–0.40	0.20–0.40	4.00–4.75	0.30 max	0.40–1.00	18.50–21.00	1.50–2.10	11.00–13.00
T8	T12008	0.75–0.85	0.20–0.40	0.20–0.40	3.75–4.50	0.30 max	0.40–1.00	13.25–14.75	1.80–2.40	4.25–5.75
T15	T12015	1.50–1.60	0.15–0.40	0.15–0.40	3.75–5.00	0.30 max	1.00 max	11.75–13.00	4.50–5.25	4.75–5.25
Intermediate high-speed steels										
M50	T11350	0.78–0.88	0.15–0.45	0.20–0.60	3.75–4.50	0.30 max	3.90–4.75	...	0.80–1.25	...
M52	T11352	0.85–0.95	0.15–0.45	0.20–0.60	3.50–4.30	0.30 max	4.00–4.90	0.75–1.50	1.65–2.25	...
Chromium hot-work steels										
H10	T20810	0.35–0.45	0.25–0.70	0.80–1.20	3.00–3.75	0.30 max	2.00–3.00	...	0.25–0.75	...
H11	T20811	0.33–0.43	0.20–0.50	0.80–1.20	4.75–5.50	0.30 max	1.10–1.60	...	0.30–0.60	...
H12	T20812	0.30–0.40	0.20–0.50	0.80–1.20	4.75–5.50	0.30 max	1.25–1.75	1.00–1.70	0.50 max	...
H13	T20813	0.32–0.45	0.20–0.50	0.80–1.20	4.75–5.50	0.30 max	1.10–1.75	...	0.80–1.20	...
H14	T20814	0.35–0.45	0.20–0.50	0.80–1.20	4.75–5.50	0.30 max	...	4.00–5.25	...	...
H19	T20819	0.32–0.45	0.20–0.50	0.20–0.50	4.00–4.75	0.30 max	0.30–0.55	3.75–4.50	1.75–2.20	4.00–4.50
Tungsten hot-work steels										
H21	T20821	0.26–0.36	0.15–0.40	0.15–0.50	3.00–3.75	0.30 max	...	8.50–10.00	0.30–0.60	...
H22	T20822	0.30–0.40	0.15–0.40	0.15–0.40	1.75–3.75	0.30 max	...	10.00–11.75	0.25–0.50	...
H23	T20823	0.25–0.35	0.15–0.40	0.15–0.60	11.00–12.75	0.30 max	...	11.00–12.75	0.75–1.25	...
H24	T20824	0.42–0.53	0.15–0.40	0.15–0.40	2.50–3.50	0.30 max	...	14.00–16.00	0.40–0.60	...
H25	T20825	0.22–0.32	0.15–0.40	0.15–0.40	3.75–4.50	0.30 max	...	14.00–16.00	0.40–0.60	...
H26	T20826	0.45–0.55(b)	0.15–0.40	0.15–0.40	3.75–4.50	0.30 max	...	17.25–19.00	0.75–1.25	...
Molybdenum hot-work steels										
H42	T20842	0.55–0.70(b)	0.15–0.40	...	3.75–4.50	0.30 max	4.50–5.50	5.50–6.75	1.75–2.20	...
Air-hardening, medium-alloy, cold-work steels										
A2	T30102	0.95–1.05	1.00 max	0.50 max	4.75–5.50	0.30 max	0.90–1.40	...	0.15–0.50	...
A3	T30103	1.20–1.30	0.40–0.60	0.50 max	4.75–5.50	0.30 max	0.90–1.40	...	0.80–1.40	...
A4	T30104	0.95–1.05	1.80–2.20	0.50 max	0.90–2.20	0.30 max	0.90–1.40	...	...	...
A6	T30106	0.65–0.75	1.80–2.50	0.50 max	0.90–1.20	0.30 max	0.90–1.40	...	...	...
A7	T30107	2.00–2.85	0.80 max	0.50 max	5.00–5.75	0.30 max	0.90–1.40	0.50–1.50	3.90–5.15	...
A8	T30108	0.50–0.60	0.50 max	0.75–1.10	4.75–5.50	0.30 max	1.15–1.65	1.00–1.50	...	...
A9	T30109	0.45–0.55	0.50 max	0.95–1.15	4.75–5.50	1.25–1.75	1.30–1.80	...	0.80–1.40	...
A10	T30110	1.25–1.50(c)	1.60–2.10	1.00–1.50	...	1.55–2.05	1.25–1.75	...	...	...
High-carbon, high-chromium, cold-work steels										
D2	T30402	1.40–1.60	0.60 max	0.60 max	11.00–13.00	0.30 max	0.70–1.20	...	1.10 max	...
D3	T30403	2.00–2.35	0.60 max	0.60 max	11.00–13.50	0.30 max	...	1.00 max	1.00 max	...
D4	T30404	2.05–2.40	0.60 max	0.60 max	11.00–13.00	0.30 max	0.70–1.20	...	1.00 max	...
D5	T30405	1.40–1.60	0.60 max	0.60 max	11.00–13.00	0.30 max	0.70–1.20	...	1.00 max	2.50–3.50
D7	T30407	2.15–2.50	0.60 max	0.60 max	11.50–13.50	0.30 max	0.70–1.20	...	3.80–4.40	...
Oil-hardening cold-work steels										
O1	T31501	0.85–1.00	1.00–1.40	0.50 max	0.40–0.60	0.30 max	...	0.40–0.60	0.30 max	...
O2	T31502	0.85–0.95	1.40–1.80	0.50 max	0.50 max	0.30 max	0.30 max	...	0.30 max	...
O6	T31506	1.25–1.55(c)	0.30–1.10	0.55–1.50	0.30 max	0.30 max	0.20–0.30	...	...	...
O7	T31507	1.10–1.30	1.00 max	0.60 max	0.35–0.85	0.30 max	0.30 max	1.00–2.00	0.40 max	...
Shock-resisting steels										
S1	T41901	0.40–0.55	0.10–0.40	0.15–1.20	1.00–1.80	0.30 max	0.50 max	1.50–3.00	0.15–0.30	...
S2	T41902	0.40–0.55	0.30–0.50	0.90–1.20	...	0.30 max	0.30–0.60	...	0.50 max	...
S5	T41905	0.50–0.65	0.60–1.00	1.75–2.25	0.50 max	...	0.20–1.35	...	0.35 max	...
S6	T41906	0.40–0.50	1.20–1.50	2.00–2.50	1.20–1.50	...	0.30–0.50	...	0.20–0.40	...
S7	T41907	0.45–0.55	0.20–0.90	0.20–1.00	3.00–3.50	...	1.30–1.80	...	0.20–0.30(d)	...
Low-alloy special-purpose tool steels										
L2	T61202	0.45–1.00(b)	0.10–0.90	0.50 max	0.70–1.20	...	0.25 max	...	0.10–0.30	...
L6	T61206	0.65–0.75	0.25–0.80	0.50 max	0.60–1.20	1.25–2.00	0.50 max	...	0.20–0.30(d)	...

(continued)

AISI	UNS	C	Mn	Si	Cr	Ni	Mo	W	V	Co
Low-carbon mold steels										
P2	T51602	0.10 max	0.10–0.40	0.10–0.40	0.75–1.25	0.10–0.50	0.15–0.40	...	...	...
P3	T51603	0.10 max	0.20–0.60	0.40 max	0.40–0.75	1.00–1.50	...	...	...	...
P4	T51604	0.12 max	0.20–0.60	0.10–0.40	4.00–5.25	...	0.40–1.00	...	...	...
P5	T51605	0.10 max	0.20–0.60	0.40 max	2.00–2.50	0.35 max	...	...	...	...
P6	T51606	0.05–0.15	0.35–0.70	0.10–0.40	1.25–1.75	3.25–3.75	...	...	...	...
P20	T51620	0.28–0.40	0.60–1.00	0.20–0.80	1.40–2.00	...	0.30–0.55	...	...	...
P21	T51621	0.18–0.22	0.20–0.40	0.20–0.40	0.50 max	3.90–4.25	...	...	0.15–0.25	1.05–1.25Al
Water-hardening tool steels										
W1	T72301	0.70–1.50(e)	0.10–0.40	0.10–0.40	0.15 max	0.20 max	0.10 max	0.15 max	0.10 max	...
W2	T72302	0.85–1.50(e)	0.10–0.40	0.10–0.40	0.15 max	0.20 max	0.10 max	0.15 max	0.15–0.35	...
W5	T72305	1.05–1.15	0.10–0.40	0.10–0.40	0.40–0.60	0.20 max	0.10 max	0.15 max	0.10 max	...

(a) All steels except group W contain 0.25 max Cu, 0.03 max P, and 0.03 max S; group W steels contain 0.20 max Cu, 0.025 max P, and 0.025 max S. Where specified, sulfur may be increased to 0.06 to 0.15% to improve machinability of group A, D, H, M, and T steels. (b) Available in several carbon ranges. (c) Contains free graphite in the microstructure. (d) Optional. (e) Specified carbon ranges are designated by suffix numbers.

Compositions of tool steels no longer in common use

Type	C	W	Mo	Cr	V	Others
High-speed steels						
M6	0.80	4.25	5.00	4.00	1.50	12.00 Co
M8	0.80	5.00	5.00	4.00	1.50	1.25 Nb
M15	1.50	6.50	3.50	4.00	5.00	5.00 Co
M45	1.25	8.00	5.00	4.25	1.60	5.50 Co
T3	1.05	18.00	...	4.00	3.00	...
T7	0.75	14.00	...	4.00	2.00	...
T9	1.20	18.00	...	4.00	4.00	...
Hot-work steels						
H15	0.40	...	5.00	5.00	...	...
H16	0.55	7.00	...	7.00	...	...
H20	0.35	9.00	...	2.00	...	...
H41	0.65	1.50	8.00	4.00	1.00	...
H43	0.55	...	8.00	4.00	2.00	...
Cold-work steels						
D1	1.00	...	1.00	12.00	...	...
A5	1.00	...	1.00	1.00	...	3.00 Mn
Shock-resisting steels						
S3	0.50	1.00	...	0.74	...	...
S4	0.55	...	...	...	...	2.00 Si, 0.80 Mn
Mold steel						
P1	0.10	...	...	...	...	...
Special-purpose steels						
L1	1.00	...	...	1.25	...	...
L3	1.00	...	...	1.50	0.20	...
L4	1.00	...	...	1.50	0.25	0.60 Mn
L5	1.00	...	0.25	1.00	...	1.00 Mn
L7	1.00	...	0.40	1.40	...	0.35 Mn
F1	1.00	1.25	...	...	...	...
F2	1.25	3.50	...	...	...	...
F3	1.25	3.50	...	0.75	...	...
Water-hardening tool steels						
W3	1.00	...	...	...	0.50	...
W4	0.60/1.40(b)	...	...	0.25	...	...
W6	1.00	...	...	0.25	0.25	...
W7	1.00	...	...	0.50	0.20	...

(b) Various carbon contents were available.

Compositions of standard stainless steels

Type	UNS designation	C	Mn	Si	Cr	Ni	P	S	Other
Austenitic types									
201	S20100	0.15	5.5–7.5	1.00	16.0–18.0	3.5–5.5	0.06	0.03	0.25 N
202	S20200	0.15	7.5–10.0	1.00	17.0–19.0	4.0–6.0	0.06	0.03	0.25 N
205	S20500	0.12–0.25	14.0–15.5	1.00	16.5–18.0	1.0–1.75	0.06	0.03	0.32–0.40 N
301	S30100	0.15	2.00	1.00	16.0–18.0	6.0–8.0	0.045	0.03	. . .
302	S30200	0.15	2.00	1.00	17.0–19.0	8.0–10.0	0.045	0.03	. . .
302B	S30215	0.15	2.00	2.0–3.0	17.0–19.0	8.0–10.0	0.045	0.03	. . .
303	S30300	0.15	2.00	1.00	17.0–19.0	8.0–10.0	0.20	0.15 min	0.6 Mo(b)
303Se	S30323	0.15	2.00	1.00	17.0–19.0	8.0–10.0	0.20	0.06	0.15 min Se
304	S30400	0.08	2.00	1.00	18.0–20.0	8.0–10.5	0.045	0.03	. . .
304H	S30409	0.04–0.10	2.00	1.00	18.0–20.0	8.0–10.5	0.045	0.03	. . .
304L	S30403	0.03	2.00	1.00	18.0–20.0	8.0–12.0	0.045	0.03	. . .
304LN	S30453	0.03	2.00	1.00	18.0–20.0	8.0–12.0	0.045	0.03	0.10–0.16 N
302Cu	S30430	0.08	2.00	1.00	17.0–19.0	8.0–10.0	0.045	0.03	3.0–4.0 Cu
304N	S30451	0.08	2.00	1.00	18.0–20.0	8.0–10.5	0.045	0.03	0.10–0.16 N
305	S30500	0.12	2.00	1.00	17.0–19.0	10.5–13.0	0.045	0.03	. . .
308	S30800	0.08	2.00	1.00	19.0–21.0	10.0–12.0	0.045	0.03	. . .
309	S30900	0.20	2.00	1.00	22.0–24.0	12.0–15.0	0.045	0.03	. . .
309S	S30908	0.08	2.00	1.00	22.0–24.0	12.0–15.0	0.045	0.03	. . .
310	S31000	0.25	2.00	1.50	24.0–26.0	19.0–22.0	0.045	0.03	. . .
310S	S31008	0.08	2.00	1.50	24.0–26.0	19.0–22.0	0.045	0.03	. . .
314	S31400	0.25	2.00	1.5–3.0	23.0–26.0	19.0–22.0	0.045	0.03	. . .
316	S31600	0.08	2.00	1.00	16.0–18.0	10.0–14.0	0.045	0.03	2.0–3.0 Mo
316F	S31620	0.08	2.00	1.00	16.0–18.0	10.0–14.0	0.20	0.10 min	1.75–2.5 Mo
316H	S31609	0.04–0.10	2.00	1.00	16.0–18.0	10.0–14.0	0.045	0.03	2.0–3.0 Mo
316L	S31603	0.03	2.00	1.00	16.0–18.0	10.0–14.0	0.045	0.03	2.0–3.0 Mo
316LN	S31653	0.03	2.00	1.00	16.0–18.0	10.0–14.0	0.045	0.03	2.0–3.0 Mo; 0.10–0.16 N
316N	S31651	0.08	2.00	1.00	16.0–18.0	10.0–14.0	0.045	0.03	2.0–3.0 Mo; 0.10–0.16 N
317	S31700	0.08	2.00	1.00	18.0–20.0	11.0–15.0	0.045	0.03	3.0–4.0 Mo
317L	S31703	0.03	2.00	1.00	18.0–20.0	11.0–15.0	0.045	0.03	3.0–4.0 Mo
321	S32100	0.08	2.00	1.00	17.0–19.0	9.0–12.0	0.045	0.03	5 × %C min Ti
321H	S32109	0.04–0.10	2.00	1.00	17.0–19.0	9.0–12.0	0.045	0.03	5 × %C min Ti
330	N08330	0.08	2.00	0.75–1.5	17.0–20.0	34.0–37.0	0.04	0.03	. . .
347	S34700	0.08	2.00	1.00	17.0–19.0	9.0–13.0	0.045	0.03	10 × %C min Nb
347H	S34709	0.04–0.10	2.00	1.00	17.0–19.0	9.0–13.0	0.045	0.03	8 × %C min − 1.0 max Nb
348	S34800	0.08	2.00	1.00	17.0–19.0	9.0–13.0	0.045	0.03	0.2 Co; 10 × %C min Nb; 0.10 Ta
348H	S34809	0.04–0.10	2.00	1.00	17.0–19.0	9.0–13.0	0.045	0.03	0.2 Co; 8 × %C min − 1.0 max Nb; 0.10 Ta
384	S38400	0.08	2.00	1.00	15.0–17.0	17.0–19.0	0.045	0.03	. . .
Ferritic types									
405	S40500	0.08	1.00	1.00	11.5–14.5	. . .	0.04	0.03	0.10–0.30 Al
409	S40900	0.08	1.00	1.00	10.5–11.75	0.50	0.045	0.045	6 × %C min − 0.75 max Ti
429	S42900	0.12	1.00	1.00	14.0–16.0	. . .	0.04	0.03	. . .
430	S43000	0.12	1.00	1.00	16.0–18.0	. . .	0.04	0.03	. . .
430F	S43020	0.12	1.25	1.00	16.0–18.0	. . .	0.06	0.15 min	0.6 Mo(b)
430FSe	S43023	0.12	1.25	1.00	16.0–18.0	. . .	0.06	0.06	0.15 min Se
434	S43400	0.12	1.00	1.00	16.0–18.0	. . .	0.04	0.03	0.75–1.25 Mo
436	S43600	0.12	1.00	1.00	16.0–18.0	. . .	0.04	0.03	0.75–1.25 Mo; 5 × %C min − 0.70 max Nb
439	S43035	0.07	1.00	1.00	17.0–19.0	0.50	0.04	0.03	0.15 Al; 12 × %C min − 1.10 Ti
442	S44200	0.20	1.00	1.00	18.0–23.0	. . .	0.04	0.03	. . .
444	S44400	0.025	1.00	1.00	17.5–19.5	1.00	0.04	0.03	1.75–2.50 Mo; 0.025 N; 0.2 + 4 (%C + %N) min − 0.8 max (Ti + Nb)
446	S44600	0.20	1.50	1.00	23.0–27.0	. . .	0.04	0.03	0.25 N
Duplex (ferritic-austenitic) type									
329	S32900	0.20	1.00	0.75	23.0–28.0	2.50–5.00	0.040	0.030	1.00–2.00 Mo
Martensitic types									
403	S40300	0.15	1.00	0.50	11.5–13.0	. . .	0.04	0.03	. . .
410	S41000	0.15	1.00	1.00	11.5–13.5	. . .	0.04	0.03	. . .
414	S41400	0.15	1.00	1.00	11.5–13.5	1.25–2.50	0.04	0.03	. . .
416	S41600	0.15	1.25	1.00	12.0–14.0	. . .	0.06	0.15 min	0.6 Mo(b)
416Se	S41623	0.15	1.25	1.00	12.0–14.0	. . .	0.06	0.06	0.15 min Se
420	S42000	0.15 min	1.00	1.00	12.0–14.0	. . .	0.04	0.03	. . .
420F	S42020	0.15 min	1.25	1.00	12.0–14.0	. . .	0.06	0.15 min	0.6 Mo(b)
422	S42200	0.20–0.25	1.00	0.75	11.5–13.5	0.5–1.0	0.04	0.03	0.75–1.25 Mo; 0.75–1.25 W; 0.15–0.3 V
431	S43100	0.20	1.00	1.00	15.0–17.0	1.25–2.50	0.04	0.03	. . .
440A	S44002	0.60–0.75	1.00	1.00	16.0–18.0	. . .	0.04	0.03	0.75 Mo
440B	S44003	0.75–0.95	1.00	1.00	16.0–18.0	. . .	0.04	0.03	0.75 Mo
440C	S44004	0.95–1.20	1.00	1.00	16.0–18.0	. . .	0.04	0.03	0.75 Mo
Precipitation-hardening types									
PH 13-8 Mo	S13800	0.05	0.20	0.10	12.25–13.25	7.5–8.5	0.01	0.008	2.0–2.5 Mo; 0.90–1.35 Al; 0.01 N
15-5 PH	S15500	0.07	1.00	1.00	14.0–15.5	3.5–5.5	0.04	0.03	2.5–4.5 Cu; 0.15–0.45 Nb
17-4 PH	S17400	0.07	1.00	1.00	15.5–17.5	3.0–5.0	0.04	0.03	3.0–5.0 Cu; 0.15–0.45 Nb
17-7 PH	S17700	0.09	1.00	1.00	16.0–18.0	6.5–7.75	0.04	0.04	0.75–1.5 Al

(a) Single values are maximum values unless otherwise indicated. (b) Optional

Compositions of ACI heat-resistant casting alloys

ACI designation	UNS number	ASTM specifications(a)	Composition, %(b)			
			C	Cr	Ni	Si (max)
HA	...	A 217	0.20 max	8–10	...	1.00
HC	J92605	A 297, A 608	0.50 max	26–30	4 max	2.00
HD	J93005	A 297, A 608	0.50 max	26–30	4–7	2.00
HE	J93403	A 297, A 608	0.20–0.50	26–30	8–11	2.00
HF	J92603	A 297, A 608	0.20–0.40	19–23	9–12	2.00
HH	J93503	A 297, A 608, A 447	0.20–0.50	24–28	11–14	2.00
HI	J94003	A 297, A 567, A 608	0.20–0.50	26–30	14–18	2.00
HK	J94224	A 297, A 351, A 567, A 608	0.20–0.60	24–28	18–22	2.00
HK30	...	A 351	0.25–0.35	23.0–27.0	19.0–22.0	1.75
HK40	...	A 351	0.35–0.45	23.0–27.0	19.0–22.0	1.75
HL	J94604	A 297, A 608	0.20–0.60	28–32	18–22	2.00
HN	J94213	A 297, A 608	0.20–0.50	19–23	23–27	2.00
HP	...	A 297	0.35–0.75	24–28	33–37	2.00
HP-50WZ(c)	...	...	0.45–0.55	24–28	33–37	2.50
HT	J94605	A 297, A 351, A 567, A 608	0.35–0.75	13–17	33–37	2.50
HT30	...	A 351	0.25–0.35	13.0–17.0	33.0–37.0	2.50
HU	...	A 297, A 608	0.35–0.75	17–21	37–41	2.50
HW	...	A 297, A 608	0.35–0.75	10–14	58–62	2.50
HX	...	A 297, A 608	0.35–0.75	15–19	64–68	2.50

(a) ASTM designations are the same as ACI designations. (b) Rem Fe in all compositions. Manganese content: 0.35 to 0.65% for HA, 1% for HC, 1.5% for HD, and 2% for the other alloys. Phosphorus and sulfur contents: 0.04% (max) for all but HP-50WZ. Molybdenum is intentionally added only to HA, which has 0.90 to 1.20% Mo; maximum for other alloys is set at 0.5% Mo. HH also contains 0.2% N (max). (c) Also contains 4 to 6% W, 0.1 to 1.0% Zr, and 0.035% S (max) and P (max)

Typical room-temperature properties of ACI heat-resistant casting alloys

Alloy	Condition	Tensile strength		Yield strength		Elongation, %	Hardness, HB
		MPa	ksi	MPa	ksi		
HC	As-cast	760	110	515	75	19	223
	Aged(a)	790	115	550	80	18	...
HD	As-cast	585	85	330	48	16	90
HE	As-cast	655	95	310	45	20	200
	Aged(a)	620	90	380	55	10	270
HF	As-cast	635	92	310	45	38	165
	Aged(a)	690	100	345	50	25	190
HH, type 1	As-cast	585	85	345	50	25	185
	Aged(a)	595	86	380	55	11	200
HH, type 2	As-cast	550	80	275	40	15	180
	Aged(a)	635	92	310	45	8	200
HI	As-cast	550	80	310	45	12	180
	Aged(a)	620	90	450	65	6	200
HK	As-cast	515	75	345	50	17	170
	Aged(b)	585	85	345	50	10	190
HL	As-cast	565	82	360	52	19	192
HN	As-cast	470	68	260	38	13	160
HP	As-cast	490	71	275	40	11	170
HT	As-cast	485	70	275	40	10	180
	Aged(b)	515	75	310	45	5	200
HU	As-cast	485	70	275	40	9	170
	Aged(c)	505	73	295	43	5	190
HW	As-cast	470	68	250	36	4	185
	Aged(d)	580	84	360	52	4	205
HX	As-cast	450	65	250	36	9	176
	Aged(c)	505	73	305	44	9	185

(a) Aging treatment: 24 h at 760 °C (1400 °F), furnace cool. (b) Aging treatment: 24 h at 760 °C (1400 °F), air cool. (c) Aging treatment: 48 h at 980 °C (1800 °F), air cool. (d) Aging treatment: 48 h at 980 °C (1800 °F), furnace cool

Compositions and typical microstructures of ACI corrosion-resistant cast steels

ACI type	Wrought alloy type(a)	ASTM specifications	Most common end-use microstructure	C	Mn	Si	Cr	Ni	Others(c)
Chromium steels									
CA-15	410	A 743, A 217, A 487	Martensite	0.15	1.00	1.50	11.5–14.0	1.0	0.50Mo(d)
CA-15M	···	A 743	Martensite	0.15	1.00	0.65	11.5–14.0	1.0	0.15–1.00Mo
CA-40	420	A 743	Martensite	0.40	1.00	1.50	11.5–14.0	1.0	0.5Mo(d)
CA-40F	···	A 743	Martensite	0.2–0.4	1.00	1.50	11.5–14.0	1.0	···
CB-30	431, 442	A 743	Ferrite and carbides	0.30	1.00	1.50	18.0–22.0	2.0	···
CC-50	446	A 743	Ferrite and carbides	0.30	1.00	1.50	26.0–30.0	4.0	···
Chromium-nickel steels									
CA-6N	···	A 743	Martensite	0.06	0.50	1.00	10.5–12.5	6.0–8.0	···
CA-6NM	···	A 743, A 487	Martensite	0.06	1.00	1.00	11.5–14.0	3.5–4.5	0.4–1.0Mo
CA-28MWV	···	A 743	Martensite	0.20–0.28	0.50–1.00	1.00	11.0–12.5	0.50–1.00	0.9–1.25Mo; 0.9–1.25W; 0.2–0.3V
CB-7Cu-1	···	A 747	Martensite, age hardenable	0.07	0.70	1.00	15.5–17.7	3.6–4.6	2.5–3.2Cu; 0.20–0.35Nb; 0.05N max
CB-7Cu-2	···	A 747	Martensite, age hardenable	0.07	0.70	1.00	14.0–15.5	4.5–5.5	2.5–3.2Cu; 0.20–0.35 Nb; 0.05N max
CD-4MCu	···	A 351, A 743, A 744, A 890	Austenite in ferrite, age hardenable	0.04	1.00	1.00	25.0–26.5	4.75–6.0	1.75–2.25Mo; 2.75–3.25Cu
CE-30	312	A 743	Ferrite in austenite	0.30	1.50	2.00	26.0–30.0	8.0–11.0	···
CF-3(e)	304L	A 351, A 743, A 744	Ferrite in austenite	0.03	1.50	2.00	17.0–21.0	8.0–12.0	···
CF-3M(e)	316L	A 351, A 743, A 744	Ferrite in austenite	0.03	1.50	2.00	17.0–21.0	8.0–12.0	2.0–3.0Mo
CF-3MN	···	A 743	Ferrite in austenite	0.03	1.50	1.50	17.0–21.0	9.0–13.0	2.0–3.0Mo; 0.10–0.20N
CF-8(e)	304	A 351, A 743, A 744	Ferrite in austenite	0.08	1.50	2.00	18.0–21.0	8.0–11.0	···
CF-8C	347	A 351, A 743, A 744	Ferrite in austenite	0.08	1.50	2.00	18.0–21.0	9.0–12.0	Nb(f)
CF-8M	316	A 351, A 743, A 744	Ferrite in austenite	0.08	1.50	2.00	18.0–21.0	9.0–12.0	2.0–3.0Mo
CF-10	···	A 351	Ferrite in austenite	0.04–0.10	1.50	2.00	18.0–21.0	8.0–11.0	···
CF-10M	···	A 351	Ferrite in austenite	0.04–0.10	1.50	1.50	18.0–21.0	9.0–12.0	2.0–3.0Mo
CF-10MC	···	A 351	Ferrite in austenite	0.10	1.50	1.50	15.0–18.0	13.0–16.0	1.75–2.25Mo
CF-10SMnN	···	A 351, A 743	Ferrite in austenite	0.10	7.00–9.00	3.50–4.50	16.0–18.0	8.0–9.0	0.08–0.18N
CF-12M	316	···	Ferrite in austenite or austenite	0.12	1.50	2.00	18.0–21.0	9.0–12.0	2.0–3.0Mo
CF-16F	303	A 743	Austenite	0.16	1.50	2.00	18.0–21.0	9.0–12.0	1.50Mo max; 0.20–0.35Se
CF-20	302	A 743	Austenite	0.20	1.50	2.00	18.0–21.0	8.0–11.0	···
CG-6MMN	···	A 351, A 743	Ferrite in austenite	0.06	4.00–6.00	1.00	20.5–23.5	11.5–13.5	1.50–3.00Mo; 0.10–0.30Nb; 0.10–30V; 0.20–40N
CG-8M	317	A 351, A 743, A 744	Ferrite in austenite	0.08	1.50	1.50	18.0–21.0	9.0–13.0	3.0–4.0Mo
CG-12	···	A 743	Ferrite in austenite	0.12	1.50	2.00	20.0–23.0	10.0–13.0	···
CH-8	···	A 351	Ferrite in austenite	0.08	1.50	1.50	22.0–26.0	12.0–15.0	···
CH-10	···	A 351	Ferrite in austenite	0.04–0.10	1.50	2.00	22.0–26.0	12.0–15.0	···
CH-20	309	A 351, A 743	Austenite	0.20	1.50	2.00	22.0–26.0	12.0–15.0	···
CK-3MCuN	···	A 351, A 743, A 744	Ferrite in austenite	0.025	1.20	1.00	19.5–20.5	17.5–19.5	6.0–7.0V; 0.18–0.24N; 0.50–1.00Cu
CK-20	310	A 743	Austenite	0.20	2.00	2.00	23.0–27.0	19.0–22.0	···
Nickel-chromium steel									
CN-3M	···	A 743	Austenite	0.03	2.00	1.00	20.0–22.0	23.0–27.0	4.5–5.5Mo
CN-7M	···	A 351, A 743, A 744	Austenite	0.07	1.50	1.50	19.0–22.0	27.5–30.5	2.0–3.0Mo; 3.0–4.0Cu
CN-7MS	···	A 743, A 744	Austenite	0.07	1.50	3.50(g)	18.0–20.0	22.0–25.0	2.5–3.0Mo; 1.5–2.0Cu
CT-15C	···	A 351	Austenite	0.05–0.15	0.15–1.50	0.50–1.50	19.0–21.0	31.0–34.0	0.5–1.5V

(a) Type numbers of wrought alloys are listed only for nominal identification of corresponding wrought and cast grades. Composition ranges of cast alloys are not the same as for corresponding wrought alloys; cast alloy designations should be used for castings only. (b) Maximum unless a range is given. The balance of all compositions is iron. (c) Sulfur content is 0.04% in all grades except: CG-6MMN, 0.030% S (max); CF-10SMnN, 0.03% S (max); CT-15C, 0.03% S (max); CK-3MCuN, 0.010% S (max); CN-3M, 0.030% S (max), CA-6N, 0.020% S (max); CA-28MWV, 0.030% S (max); CB-7Cu-1 and -2, 0.03% S (max). Phosphorus content is 0.04% (max) in all grades except: CF-16F, 0.17% P (max); CF-10SMnN, 0.060% P (max); CT-15C, 0.030% P (max); CK-3MCuN, 0.045% P (max); CN-3M, 0.030% P (max); CA-6N, 0.020% P (max); CA-28MWV, 0.030% P (max); CB-7Cu-1 and -2, 0.035% P (max). (d) Molybdenum not intentionally added. (e) CF-3A, CF-3MA, and CF-8A have the same composition ranges as CF-3, CF-3M, and CF-8, respectively, but have balanced compositions so that ferrite contents are at levels that permit higher mechanical property specifications than those for related grades. They are covered by ASTM A 351. (f) Nb, 8 × %C min (1.0% max); or Nb + Ta × %C (1.1% max). (g) For CN-7MS, silicon ranges from 2.50 to 3.50%.

Summary of specification requirements for various carbon steel castings

Unless otherwise noted, all the grades listed in this table are restricted to a phosphorus content of 0.040% max and a sulfur content of 0.045% max.

Class or grade	Tensile strength(a) MPa	ksi	Yield strength(a) MPa	ksi	Minimum elongation in 50 mm (2 in.), %	Minimum reduction in area, %	Chemical composition(b), % C	Mn	Si	Other requirements	Condition or specific application
ASTM A 27: carbon steel castings for general applications											
N-1	...	...	...	...	...	...	0.25(c)	0.75(c)	0.80	0.06% S, 0.05% P	Chemical analysis only
N-2	...	...	...	...	...	...	0.35(c)	0.60(c)	0.80	0.06% S, 0.05% P	Heat treated but not mechanically tested
U60-30	415	60	205	30	22	30	0.25(c)	0.75(c)	0.80	0.06% S, 0.05% P	Mechanically tested but not heat treated
60–30	415	60	205	30	24	35	0.30(c)	0.60(c)	0.80	0.06% S, 0.05% P	Heat treated and mechanically tested
65–35	450	65	240	35	24	35	0.30(c)	0.70(c)	0.80	0.06% S, 0.05% P	Heat treated and mechanically tested
70–36	485	70	250	36	22	30	0.35(c)	0.70(c)	0.80	0.06% S, 0.05% P	Heat treated and mechanically tested
70–40	485	70	275	40	22	30	0.25(c)	1.20(c)	0.80	0.06% S, 0.05% P	Heat treated and mechanically tested
ASTM A 148: carbon steel castings for structural applications(d)											
80–40	550	80	275	40	18	30	(e)	(e)	(e)	0.06% S, 0.05% P	Composition and heat treatment necessary to achieve specified mechanical properties
80–50	550	80	345	50	22	35	(e)	(e)	(e)	0.06% S, 0.05% P	Composition and heat treatment necessary to achieve specified mechanical properties
90–60	620	90	415	60	20	40	(e)	(e)	(e)	0.06% S, 0.05% P	Composition and heat treatment necessary to achieve specified mechanical properties
105–85	725	105	585	85	17	35	(e)	(e)	(e)	0.06% S, 0.05% P	Composition and heat treatment necessary to achieve specified mechanical properties
SAE J435c: see Table 2 for alloy steel castings specified in SAE J435c											
0022	...	...	...	...	...	...	0.12–0.22	0.50–0.90	0.60	187 HB max	Low-carbon steel suitable for carburizing
0025	415	60	207	30	22	30	0.25(c)	0.75(c)	0.80	187 HB max	Carbon steel welding grade
0030	450	65	241	35	24	35	0.30(c)	0.70(c)	0.80	131–187 HB	Carbon steel welding grade
0050A	585	85	310	45	16	24	0.40–0.50	0.50–0.90	0.80	170–229 HB	Carbon steel medium-strength grade
0050B	690	100	485	70	10	15	0.40–0.50	0.50–0.90	0.80	207–255 HB	Carbon steel medium-strength grade
080	550	80	345	50	22	35	...	...	...	163–207 HB	Medium-strength low-alloy steel
090	620	90	415	60	20	40	...	...	...	187–241 HB	Medium-strength low-alloy steel
HA, HB, HC(f)	...	...	...	...	...	...	0.25–0.34	(f)	(f)	See Fig. 2.	Hardenability grades (Fig. 2)
ASTM A 216: carbon steel castings suitable for fusion welding and high-temperature service											
WCA	415–585	60–85	205	30	24	35	0.25	0.70(c)	0.60	(g)	Pressure-containing parts
WCB	485–655	70–95	250	36	22	35	0.30	1.00(c)	0.60	(g)	Pressure-containing parts
WCC	485–655	70–95	275	40	22	35	0.25	1.20(c)	0.50	(g)	Pressure-containing parts
Other ASTM cast steel specifications with carbon steel grades(h)											
A 352-LCA	415–585	60–85	205	30	24	35	0.25	0.70(c)	0.60	(g)(i)(j)	Low-temperature applications
A 352-LCB	450–620	65–90	240	35	24	35	0.30	1.00	0.60	(g)(j)(k)	Low-temperature applications
A 356-grade 1	485	70	250	36	20	35	0.35	0.70(c)	0.60	0.035% P max, 0.030% S max	Castings for valve chests, throttle valves, and other heavy-walled components for steam turbines
A 757-A1Q	450	65	240	35	24	35	0.30	1.00	0.60	(j)(k)(l)	Castings for pressure-containing applications at low temperatures

(a) Where a single value is shown, it is a minimum. (b) Where a single value is shown, it is a maximum. (c) For each reduction of 0.01% C below the maximum specified, an increase of 0.04% Mn above the maximum specified is permitted up to the maximums given in the applicable ASTM specifications. (d) Grades may also include low-alloy steels; see Table 2 for the stronger grades of ASTM A 148. (e) Unless specified by purchaser, the compositions of cast steels in ASTM A 148 are selected by the producer in order to achieve the specified mechanical properties. (f) Purchased on the basis of hardenability, with manganese and other elements added as required. (g) Specified residual elements include 0.30% Cu max, 0.50% Ni max, 0.50% Cr max, 0.20% Mo max, and 0.03% V max, with the total residual elements not exceeding 1.00%. (h) These ASTM specifications also include alloy steel castings for the general type of applications listed in the Table. (i) Testing temperature of −32 °C (−25 °F). (j) Charpy V-notch impact testing at the specified test temperature with an energy value of 18 J (13 ft · lbf) min for two specimens and an average of three. (k) Testing temperature of −46 °C (−50 °F). (l) Specified residual elements of 0.03% V, 0.50% Cu, 0.50% Ni, 0.40% Cr, and 0.25% Mo, with total amount not exceeding 1.00%. Sulfur and phosphorus content, each 0.025% max

Summary of specification requirements for various alloy steel castings with chromium contents up to 10%

Material class(a)	Tensile strength(b) MPa	ksi	Yield strength(b) MPa	ksi	Minimum elongation in 50 mm (2 in.), %	Minimum reduction in area, %	C	Mn	Si	Composition(c), % Cr	Ni	Mo	Other
ASTM A 148: steel castings for structural applications(d)													
115–95	795	115	655	95	14	30	. . .	. . .	. . .	. . .	. . .	. . .	(e)
135–125	930	135	860	125	9	22	. . .	. . .	. . .	. . .	. . .	. . .	(e)
150–135	1035	150	930	135	7	18	. . .	. . .	. . .	. . .	. . .	. . .	(e)
160–145	1105	160	1000	145	6	12	. . .	. . .	. . .	. . .	. . .	. . .	(e)
165–150	1140	165	1035	150	5	20	. . .	. . .	. . .	. . .	. . .	. . .	(f)
165–150L	1140	165	1035	150	5	20	. . .	. . .	. . .	. . .	. . .	. . .	(f)
210–180	1450	210	1240	180	4	15	. . .	. . .	. . .	. . .	. . .	. . .	(f)
210–180L	1450	210	1240	180	4	15	. . .	. . .	. . .	. . .	. . .	. . .	(f)
260–210	1795	260	1450	210	3	6	. . .	. . .	. . .	. . .	. . .	. . .	(f)
260–210L	1795	260	1450	210	3	6	. . .	. . .	. . .	. . .	. . .	. . .	(f)
SAE J435c: see Table 1 for the carbon steel castings specified in SAE J435c													
0105	725	105	586	85	17	35	. . .	. . .	. . .	. . .	. . .	. . .	(h)
0120	827	120	655	95	14	30	. . .	. . .	. . .	. . .	. . .	. . .	(h)
0150	1035	150	862	125	9	22	. . .	. . .	. . .	. . .	. . .	. . .	(h)
0175	1207	175	1000	145	6	12	. . .	. . .	. . .	. . .	. . .	. . .	(h)
ASTM A 217: alloy steel castings for pressure-containing parts and high-temperature service													
WC1	450–620	65–90	240	35	24	35	0.25	0.50–0.80	0.60	0.35(i)	0.50(i)	0.45–0.65	(i)(j)
WC4	485–655	70–95	275	40	20	35	0.20	0.50–0.80	0.60	0.50–0.80	0.70–1.10	0.45–0.65	(j)(k)
WC5	485–655	70–95	275	40	20	35	0.20	0.40–0.70	0.60	0.50–0.90	0.60–1.00	0.90–1.20	(j)(k)
WC6	485–655	70–95	275	40	20	35	0.20	0.50–0.80	0.60	1.00–1.50	0.50(i)	0.45–0.65	(i)(j)
WC9	485–655	70–95	275	40	20	35	0.18	0.40–0.70	0.60	2.00–2.75	0.50(i)	0.9–1.20	(i)(j)
WC11	550–725	80–105	345	50	18	45	0.15–0.21	0.50–0.80	0.30–0.60	1.00–1.75	0.50(i)	0.45–0.65	(i)(l)
C5	620–795	90–115	415	60	18	35	0.20	0.40–0.70	0.75	4.00–6.50	0.50(i)	0.45–0.65	(i)(j)
C12	620–795	90–115	415	60	18	35	0.20	0.35–0.65	1.00	8.00–10.00	0.50(i)	0.90–1.20	(i)(j)
ASTM A 389: alloy steel castings (NT) suitable for fusion welding and pressure-containing parts at high temperatures													
C23	485	70	275	40	18	35	0.20	0.30–0.80	0.60	1.00–1.50	. . .	0.45–0.65	(h)(m)
C24	550	80	345	50	15	35	0.20	0.30–0.80	0.60	1.00–1.25	. . .	0.90–1.20	(h)(m)
ASTM A 487: alloy steel castings (NT or QT) for pressure-containing parts at high temperatures													
1A (NT)	585–760	85–110	380	55	22	40	0.30	1.00	0.80	0.35(n)	0.50(n)	0.25(n)(o)	0.5 Cu(h)(n)
2B (QT)	620–795	90–115	450	65	22	45	0.30	1.00	0.80	0.35(n)	0.50(n)	0.25(n)(o)	0.5 Cu(h)(n)
1C (NT or QT)	620	90	450	65	22	45	0.30	1.00	0.80	0.35(n)	0.50(n)	0.25(n)(o)	0.5 Cu(h)(n)
2A (NT)	585–760	85–110	365	53	22	35	0.30	1.10–1.40	0.80	0.35(i)	0.50(i)	0.10–0.30	(i)(p)
2B (QT)	620–795	90–115	450	65	22	40	0.30	1.10–1.40	0.80	0.35(i)	0.50(i)	0.10–0.30	(i)(p)
2C (QT)	620	90	450	65	22	40	0.30	1.10–1.40	0.80	0.35(i)	0.50(i)	0.10–0.30	(i)(p)
4A (NT or QT)	620–795	90–115	415	60	20	40	0.30	1.00	0.80	0.40–0.80	0.40–0.80	0.15–0.30	(k)(p)
4B (QT)	725–895	105–130	585	85	17	35	0.30	1.00	0.80	0.40–0.80	0.40–0.80	0.15–0.30	(k)(p)
4C (NT or QT)	620	90	415	60	20	40	0.30	1.00	0.80	0.40–0.80	0.40–0.80	0.15–0.30	(k)(p)
4D (QT)	690	100	515	75	17	35	0.30	1.00	0.80	0.40–0.80	0.40–0.80	0.15–0.30	(k)(p)
4E (QT)	795	115	655	95	15	35	0.30	1.00	0.80	0.40–0.80	0.40–0.80	0.15–0.30	(k)(p)
6A (NT)	795	115	550	80	18	30	0.38	1.30–1.70	0.80	0.40–0.80	0.40–0.80	0.30–0.40	(k)(p)
6B (QT)	825	120	655	95	15	35	0.38	1.30–1.70	0.80	0.40–0.80	0.40–0.80	0.30–0.40	(k)(p)
7A (QT)(p)	795	115	690	100	15	30	0.20	0.60–1.00	0.80	0.40–0.80	0.70–1.00	0.40–0.60	(k)(p)(r)
8A (NT)	585–760	85–110	380	55	20	35	0.20	0.50–0.90	0.80	2.00–2.75	. . .	0.90–1.10	(k)(p)
8B (QT)	725	105	585	85	17	30	0.20	0.50–0.90	0.80	2.00–2.75	. . .	0.90–1.10	(k)(p)
8C (QT)	690	100	515	75	17	35	0.20	0.50–0.90	0.80	2.00–2.75	. . .	0.90–1.10	(k)(p)
9A (NT or QT)	620	90	415	60	18	35	0.33	0.60–1.00	0.80	0.75–1.10	0.50(i)	0.15–0.30	(i)(p)
9B (QT)	725	105	585	85	16	35	0.33	0.60–1.00	0.80	0.75–1.10	0.50(i)	0.15–0.30	(i)(p)
9C (NT or QT)	620	90	415	60	18	35	Composition same as 9A (NT or QT) but with a slightly higher tempering temperature						
9D (QT)	690	100	515	75	17	35	0.33	0.60–1.00	0.80	0.75–1.10	0.50(i)	0.15–0.30	(i)(p)
10A (NT)	690	100	485	70	18	35	0.30	0.60–1.00	0.80	0.55–0.90	1.40–2.00	0.20–0.40	(k)(p)
10B (QT)	860	125	690	100	15	35	0.30	0.60–1.00	0.80	0.55–0.90	1.40–2.00	0.20–0.40	(k)(p)
11A (NT)	485–655	70–95	275	40	20	35	0.20	0.50–0.80	0.60	0.50–0.80	0.70–1.10	0.45–0.65	(p)(s)
11B (QT)	725–895	105–130	585	85	17	35	0.20	0.50–0.80	0.60	0.50–0.80	0.70–1.10	0.45–0.65	(p)(s)
12A (NT)	485–655	70–95	275	40	20	35	0.20	0.40–0.70	0.60	0.50–0.90	0.60–1.00	0.90–1.20	(p)(s)
12B (QT)	725–895	105–130	585	85	17	35	0.20	0.40–0.70	0.60	0.50–0.90	0.60–1.00	0.90–1.20	(p)(s)
13A (NT)	620–795	90–115	415	60	18	35	0.30	0.80–1.10	0.60	0.40(t)	1.40–1.75	0.20–0.30	(p)(t)
13B (QT)	725–895	105–130	585	85	17	35	0.30	0.80–1.10	0.60	0.40(t)	1.40–1.75	0.20–0.30	(p)(t)
14A (QT)	825–1000	120–145	655	95	14	30	0.55	0.80–1.10	0.60	0.40(t)	1.40–1.75	0.20–0.30	(p)(t)
16A (NT)(u)	485–655	70–95	275	40	22	35	0.12(v)	2.10(v)	0.50	0.20(s)	1.00–1.40	0.10(s)	(s)(w)

(a) NT, normalized and tempered; QT, quenched and tempered. (b) When a single value is shown, it is a minimum. (c) When a single value is shown, it is a maximum. (d) Unless specified by the purchaser, the compositions of cast steels in ASTM A 148 are selected by the producer and therefore may include either carbon or alloy steels; see Table 1 for the lower-grade steels specified in ASTM A 148. (e) 0.06% S (max), 0.05% P (max). (f) 0.020% S (max), 0.020% P (max). (g) Similar to the cast steel in ASTM A 148. (h) 0.045% S (max), 0.040% P (max). (i) When residual maximums are specified for copper, nickel, chromium, tungsten, and vanadium, their total content shall not exceed 1.00%. (j) 0.50% Cu (max), 0.10% W (max), 0.045% S (max), 0.04% P (max). (k) When residual maximums are specified for copper, nickel, chromium, tungsten, and vanadium, their total residual content shall not exceed 0.60%. (l) 0.35% Cu (max), 0.03% V (max), 0.015% S (max), 0.020% P (max). (m) 0.15–0.25% V. (n) The specified residuals of copper, nickel, chromium, and molybdenum (plus tungsten), shall not exceed a total content of 1.00%. (o) Includes the residual content of tungsten. (p) 0.50% Cu (max), 0.10% W (max), 0.03% V (max), 0.045% S (max), 0.04% P (max). (q) Material class 7A is a proprietary steel and has a maximum thickness of 63.5 mm (2½ in.). (r) Specified elements include 0.15–0.50% Cu, 0.03–0.10% V, and 0.002–0.006% B. (s) When residual maximums are specified for copper, nickel, chromium, tungsten, molybdenum, and vanadium, their total content shall not exceed 0.50%. (t) When residual maximums are specified for copper, nickel, chromium, tungsten, and vanadium, their total content shall not exceed 0.75%. (u) Low-carbon grade with double austenitization. (v) For each reduction of 0.01% C below the maximum, an increase of 0.04% Mn is permitted up to a maximum of 2.30%. (w) 0.20% Cu (max), 0.10% W (max), 0.02% V (max), 0.02% S (max), 0.02% P (max)

Wrought copper alloys (Composition, percent maximum unless otherwise indicated)

Copper No.	Designation	Description	Cu (incl Ag), % min	Ag, % min	Troy oz, min	As	Sb	P	Te	Others
C10100 (a)	OFE	Oxygen free electronic	99.99 (b)	...	...	(c)	(c)	0.0003	0.0010	(c)
C10200 (a)	OF	Oxygen free	99.95	...	...	...	...		...	...
C10300	OFXLP		99.95 (d)	...	...	...	...	0.001-0.005	...	...
C10400 (a)	OFS	Oxygen free with Ag	99.95	0.027	8	...	...		...	...
C10500 (a)	OFS	Oxygen free with Ag	99.95	0.034	10	...	...		...	...
C10700	OFS	Oxygen free with Ag	99.95	0.085	25	...	...		...	...
C10800	OFLP		99.95 (d)	...	...	...	...	0.005-0.012	...	...
C10920	...		99.90	...	...	...	...		...	0.02 O$_2$
C10930	...		99.90	0.044	13	...	...		...	0.02 O$_2$
C10940			99.90	0.085	25	...	...		...	0.02 O$_2$
C11000 (a)	ETP	Electrolytic tough pitch	99.90	...	...	...	...		...	(e)
C11010 (a)	"RHC"	Remelted high conductivity	99.90	...	...	...	...		...	(e)
C11020 (a)	FRHC	Fire-refined high conductivity	99.90	...	...	...	...		...	(e)
C11030 (a)	CRTP	Chemically refined tough pitch	99.90	...	...	...	...		...	(e)
C11100 (a)	...	Electrolytic tough pitch, anneal resistant	99.90	...	...	...	...		...	(f)
C11300 (a)	STP	Tough pitch with Ag	99.90	0.027	8	...	...		...	(e)
C11400 (a)	STP	Tough pitch with Ag	99.90	0.034	10	...	...		...	(e)
C11500 (a)	STP	Tough pitch with Ag	99.90	0.054	16	...	...		...	(e)
C11600 (a)	STP	Tough pitch with Ag	99.90	0.085	25	...	...		...	(e)
C11700	...		99.9 (g)	...	...	...	...	0.04	...	0.004-0.02 B
C11904 (a,h)	...	...	99.9	0.027	8	...	...		...	...
C11905 (a,h)	...	...	99.9	0.034	10	...	...		...	...
C11906 (a,h)	...	...	99.9	0.054	16	...	...		...	...
C11907 (a,h)	...	...	99.9	0.085	25	...	...		...	...
C12000	DLP	Phosphorus deoxidized, low residual phosphorus	99.90	...	...	...	...	0.004-0.012	...	...
C12100	...		99.90	0.014	4	...	...	0.005-0.012	...	...
C12200 (j)	DHP	Phosphorus deoxidized, high residual phosphorus	99.9	...	...	...	...	0.015-0.040	...	...
C12210	...	...	99.90	...	...	...	...	0.015-0.025	...	...
C12220	...	...	99.9	...	...	...	...	0.040-0.065	...	...
C12300	...	...	99.90	...	...	...	...	0.015-0.040	...	...
C12500	FRTP	Fire refined, tough pitch	99.88	...	...	0.012	0.003		0.025 (k)	0.050 Ni, 0.003 Bi, 0.004 Pb
C12700	FRSTP	Fire refined, tough pitch with Ag	99.88	0.027	8	0.012	0.003		0.025 (k)	0.050 Ni, 0.003 Bi, 0.004 Pb
C12800	FRSTP	Fire refined, tough pitch with Ag	99.88	0.034	10	0.012	0.003		0.025 (k)	0.050 Ni, 0.003 Bi, 0.004 Pb
C12900	FRSTP	Fire refined, tough pitch with Ag	99.88	0.054	16	0.012	0.003		0.025 (k)	0.050 Ni, 0.003 Bi, 0.004 Pb
C13000	FRSTP	Fire refined, tough pitch with Ag	99.88	0.085	25	0.012	0.003		0.025 (k)	0.050 Ni, 0.003 Bi, 0.004 Pb
C14180	...	...	99.90	...	...	...	...	0.075	...	0.02 Pb, 0.01 Al
C14181	...	...	99.90	...	...	...	...	0.002	...	0.002 Cd, 0.005 C, 0.002 Pb, 0.002 Zn
C14200	DPA	Phoshorus deoxidized, arsenical	99.4	...	...	0.15-0.50	...	0.015-0.040	...	...
C14210	...	...	99.20	...	...	0.30-0.50	...	0.013-0.050	...	...
C14300	...	Cadmium copper, deoxidized	99.90 (m)	...	...	...	...		...	0.05-0.15 Cd
C14310	...	...	99.90 (m)	...	...	...	...		...	0.10-0.30 Cd
C14400	...	...	99.90 (n)	...	...	...	0.003	0.013-0.025	...	0.10-0.20 Sn, 0.02 Se + Te, 0.03 Fe, 0.05 Ni, 0.05 Zn
C14410	...	...	99.90 (p)	...	...	...	...	0.005-0.020	...	0.05 Fe, 0.05 Pb, 0.10-0.20 Sn
C14420	...	...	99.90 (q)	...	...	...	...		0.02-0.05	0.05-0.15 Sn
C14500 (h)	...	Tellurium bearing	99.90 (q)	...	...	...	...	0.004-0.012	0.40-0.7	...
C14510	...	Tellurium bearing	99.90 (q)	...	...	...	...	0.010-0.030	0.30-0.7	0.05 Pb
C14520	DPTE	Phosporus deoxidized, tellurium bearing	99.40 (q)	...	...	...	...	0.004-0.020	0.40-0.7	...
C14700	...	Sulfur bearing	99.90 (r)	...	...	...	...		...	0.20-0.50 S
C14710	...	...	99.90 (h,r)	...	...	...	...	0.010-0.030	...	0.05-0.15 S, 0.05 Pb
C14720	...	...	99.90 (h,r)	...	...	...	...	0.010-0.030	...	0.20-0.50 S, 0.10 Pb
C15000	...	Zirconium copper	99.80	...	...	...	...		...	0.10-0.20 Zr
C15100	...		99.82	...	...	...	...		...	0.005 Al, 0.005 Mn, 0.05-0.15 Zr
C15500	...	...	99.75	0.027-0.10	8-30	...	...	0.040-0.080	...	0.08-0.13 Mg
C15600	...	...	99.6	...	...	...	...	0.06-0.09	...	0.20-0.30 Co, 0.02 Mg

(a) High conductivity coppers that have a minimum conductivity of 100% IACS when annealed. (b) Cu is exclusive of Ag. (c) Total of Se, Te, Bi, As, Sb, Sn, and Mn not to exceed 40 ppm (0.0040%). Hg 1 ppm max (0.0001%); Zn 1 ppm max (0.0001%); Cd max (0.0001%); S 18 ppm max (0.0018%); Pb 10 ppm max (0.0010%); Se 10 ppm max (0.0010%); Bi 10 ppm max (0.0010%); oxygen 10 ppm max (0.0010%). (d) Includes P. (e) Oxygen and trace elements may vary with process. (f) Small amounts of Cd or other elements may be added by agreement to improve resistance to softening at elevated temperatures. (g) Includes B. (h) Includes oxygen-free or deoxidized grades with deoxidizers (such as P, B, Li, or others) in an amount agreed upon. (j) Includes oxygen-free copper which contains P in an amount agreed upon. (k) Includes Te + Se. (m) Includes Cd, deoxidized with Li or other elements as agreed upon. (n) Includes Cu + Sn + P. (p) Includes Cu + Ag + Sn. (q) Includes Te. (r) Includes Ag + S + P + Pb. Source: *1988 Standards Handbook*, Copper Development Association Inc., p 5-19.

Wrought copper alloys (Composition, percent maximum unless otherwise indicated)

Copper No.	Cu (incl Ag), % min	Al (a)	Fe	Pb	Oxygen (a)
C15710	99.71	0.08-0.12	0.01	0.01	0.07-0.15
C15715	99.62	0.13-0.17	0.01	0.01	0.12-0.19
C15720	99.52	0.18-0.22	0.01	0.01	0.16-0.24
C15725	99.43	0.23-0.27	0.01	0.01	0.20-0.28
C15735	99.24	0.33-0.37	0.01	0.01	0.29-0.37
C15760	98.77	0.58-0.62	0.01	0.01	0.52-0.59

(a) All aluminum present as Al_2O_3; 0.04% oxygen present as Cu_2O with a negligible amount in solid solution with copper. Source: *1988 Standards Handbook*, Copper Development Association Inc., p 5-19.

Wrought high-copper alloys (Composition, percent maximum unless shown as a range or minimum)

Copper No.	Previous trade name	Cu (incl Ag)	Fe	Sn	Ni	Co	Cr	Si	Be	Pb	Others
C16200	Cadmium copper	rem (a)	0.02	...	...	...	...	...	...	...	0.7-1.2 Cd
C16210	...	rem (a)	...	...	...	...	...	...	...	...	0.50-1.2 Cd
C16500	...	rem (a)	0.02	0.50-0.7	...	...	...	...	...	...	0.6-1.0 Cd
C17000	Beryllium copper	rem (a)	(b)	...	(b)	(b)	...	0.20	1.60-1.79	...	0.20 Al
C17200	Beryllium copper	rem (a)	(b)	...	(b)	(b)	...	0.20	1.80-2.00	...	0.20 Al
C17300	...	rem (a)	(b)	...	(b)	(b)	...	0.20	1.80-2.00	0.20-0.6	0.20 Al
C17400	...	rem (a)	0.20	...	...	0.15-0.35	...	0.20	0.15-0.50	...	0.20 Al
C17410	...	rem (a)	0.20	...	...	0.35-0.6	...	0.20	0.15-0.50	...	0.20 Al
C17420	...	rem (a)	0.20	...	...	0.05-0.6	...	0.20	0.05-0.15	...	0.20 Al
C17500	Beryllium copper	rem (a)	0.10	...	...	2.4-2.7	...	0.20	0.40-0.7	...	0.20 Al
C17510	...	rem (a)	0.10	...	1.4-2.2	0.30	...	0.20	0.20-0.6	...	0.20 Al
C17520	...	rem (a)	...	...	0.50-1.5	...	...	...	0.10-0.30	...	0.10-0.30 Zr, 0.03-0.06 Mg
C17600	...	rem (a)	0.10	...	...	1.4-1.7	...	0.20	0.25-0.50	...	0.9-1.1 Ag, 0.20 Al
C17700	...	rem (a)	0.10	...	...	2.4-2.7	...	0.20	0.40-0.7	...	0.40-0.6 Te, 0.20 Al
C18000	...	rem (a)	0.15	...	2.0-3.0 (c)	...	0.10-0.6	0.40-0.8	...	...	...
C18030	...	rem (d)	...	0.08-0.12	...	...	0.10-0.20	...	...	...	0.005-0.15 P
C18070	...	99.0 (e)	...	...	...	...	0.15-0.40	0.02-0.07	...	...	0.01-0.40 Ti
C18090	...	96.0 min (f)	...	0.7-1.2	0.7-1.2	...	0.30-1.0	...	...	...	0.40-0.8 Ti
C18100	...	98.7 min (a)	...	...	...	...	0.40-1.0	...	...	...	0.03-0.06 Mg, 0.08-0.20 Zr
C18135	...	rem (a)	...	...	...	...	0.20-0.6	...	...	...	0.20-0.6 Cd
C18150	...	rem (23)	...	...	...	...	0.50-1.5	...	...	...	0.05-0.25 Zr
C18200	Chromium copper	rem (a)	0.10	...	...	...	0.6-1.2	0.10	...	0.05	...
C18400	Chromium copper	rem (a)	0.15	...	...	...	0.40-1.2	0.10	...	...	0.005 As, 0.005 Ca, 0.05 Li, 0.05 P, 0.7 Zn
C18500	Chromium copper	rem (a)	...	...	...	...	0.40-1.0	...	...	0.015	0.04 P, 0.08-0.12 Ag
C18700	...	rem (a)	...	...	...	...	...	...	...	0.8-1.5	...
C18900	...	rem (a)	...	0.6-0.9	...	...	...	0.15-0.40	...	0.02	0.05 P, 0.01 Al, 0.10-0.30 Mn, 0.10 Zn
C18980	...	98.0 (a)	...	1.0	...	...	...	0.50	...	0.02	0.50 Mn, 0.15 P
C18990	...	rem (d)	...	1.8-2.2	...	...	0.10-0.20	...	...	...	0.005-0.015 P
C19000	...	rem (a)	0.10	...	0.9-1.3	...	...	...	...	0.05	0.8 Zn, 0.15-0.35 P
C19010	...	rem (a)	...	...	0.8-1.8	...	...	0.15-0.35	...	...	0.01-0.05 P
C19020	...	rem (e)	...	0.30-0.9	0.50-3.0	...	...	...	...	...	0.01-0.20 P, 0.35 Mn + Si
C19100	...	rem (a)	0.20	...	0.9-1.3	...	...	...	...	0.10	0.50 Zn, 0.35-0.6 Te, 0.15-0.35 P

(a) Cu + sum of named elements = 99.5% min. (b) Ni + Co, 0.20% min; Ni + Fe + Co, 0.6% max. (c) Includes Co. (d) Cu + sum of named elements = 99.9% min. (e) Cu + sum of named elements = 99.8% min. (f) Cu + sum of named elements = 99.85% min. Source: *1988 Standards Handbook*, Copper Development Association Inc., p 5-19.

Wrought high-copper alloys (Composition, percent maximum unless shown as a range or minimum)

Copper No.	Cu	Fe	Sn	Zn	Al	Pb	P	Others
C19200	98.5 min (a)	0.8-1.2	...	0.20	...	...	0.01-0.04	...
C19210	rem (a)	0.05-0.15	...	...	...	...	0.025-0.040	...
C19220	rem (a)	0.10-0.30	0.05-0.10	...	...	...	0.03-0.07	0.005-0.015 B, 0.10-0.25 Ni
C19260	98.5 min (b)	0.40-0.8	...	...	...	...	...	0.20-0.40 Ti, 0.02-0.15 Mg
C19280	rem (a)	0.50-1.5	0.30-0.7	0.30-0.7	...	...	0.005-0.015	...
C19400	97.0 min (a)	2.1-2.6	...	0.05-0.20	...	0.03	0.015-0.15	...
C19500	96.0 min (a)	1.0-2.0	0.10-1.0	0.20	0.02	0.02	0.01-0.35	0.30-1.3 Co
C19520	96.6 min (a)	0.50-1.5	0.50-1.5	...	...	...	0.01-0.35	...
C19600	rem (a)	0.9-1.2	...	0.35	...	...	0.25-0.35	...
C19700	rem (a)	0.30-1.2	0.20	0.20	...	0.05	0.10-0.40	0.01-0.20 Mg, 0.05 Ni, 0.05 Co, 0.05 Mn
C19750	rem (a)	0.35-1.2	0.05-0.40	0.20	...	0.05	0.10-0.40	0.01-0.20 Mg, 0.05 Ni, 0.05 Co, 0.05 Mn

(a) Cu + sum of named elements = 99.8% min. (b) Cu + sum of named elements = 99.9% min. Source: *1988 Standards Handbook*, Copper Development Association Inc., p 5-19.

Wrought copper-zinc alloys (brasses) (Composition, percent maximum unless a range or minimum)

Copper No.	Previous trade name	Cu	Pb	Fe	Zn	Others
C20500	...	97.0-98.0 (a)	0.02	0.05	rem	...
C21000	Gilding, 95%	94.0-96.0 (a)	0.03	0.05	rem	...
C22000	Commercial bronze, 90%	89.0-91.0 (a)	0.05	0.05	rem	...
C22600	Jewelry bronze, 87.5%	86.0-89.0 (a)	0.05	0.05	rem	...
C23000	Red brass, 85%	84.0-86.0 (a)	0.05	0.05	rem	...
C23030	...	83.5-85.5 (a)	0.05	0.05	rem	0.20-0.40 Si
C23400	...	81.0-84.0 (a)	0.05	0.05	rem	...
C24000	Low brass, 80%	78.5-81.5 (a)	0.05	0.05	rem	...
C24080	...	78.0-82.0 (a)	0.20	...	rem	0.10 Al
C25000	...	74.0-76.0 (b)	0.05	0.05	rem	...

(a) Cu + sum of named elements = 99.8% min. (b) Cu + sum of named elements = 99.7% min. Source: *1988 Standards Handbook*, Copper Development Association Inc., p 5-19.

Wrought copper-zinc alloys (brasses) (Composition, percent maximum unless shown as a range or minimum)

Copper No.	Previous trade name	Cu (a)	Pb	Fe	Zn	Others
C26000	Cartridge brass, 70%	68.5-71.5	0.07	0.05	rem	...
C26100	...	68.5-71.5	0.05	0.05	rem	0.02-0.05 P
C26130	...	68.5-71.5	0.05	0.05	rem	0.02-0.08 As
C26200	...	67.0-70.0	0.07	0.05	rem	...
C26380	...	68.0-72.0	0.30	...	rem	0.10 Al
C26800	Yellow brass, 66%	64.0-68.5	0.15	0.05	rem	...
C27000	Yellow brass, 65%	63.0-68.5	0.10	0.07	rem	...
C27200	...	62.0-65.0	0.07	0.07	rem	...
C27400	Yellow brass, 63%	61.0-64.0	0.10	0.05	rem	...
C28000	Muntz metal, 60%	59.0-63.0	0.30	0.07	rem	...
C28580	...	49.0-52.0	0.50	0.10	rem	...

(a) Cu + sum of named elements = 99.7% min. Source: *1988 Standards Handbook*, Copper Development Association Inc., p 5-19.

Wrought copper-zinc-lead alloys (leaded brasses) (Composition, percent maximum unless shown as a range or minimum)

Copper No.	Previous trade name	Cu	Pb	Fe	Sn	Zn	Others
C31200	...	87.5-90.5 (a)	0.7-1.2	0.10	...	rem	0.25 Ni
C31400	Leaded commercial bronze	87.5-90.5 (a)	1.3-2.5	0.10	...	rem	0.7 Ni
C31600	Leaded commercial bronze (nickel bearing)	87.5-90.5 (a)	1.3-2.5	0.10	...	rem	0.7-1.2 Ni, 0.04-0.10 P
C32000	Leaded red brass	83.5-86.5 (a)	1.5-2.2	0.10	...	rem	0.25 Ni
C32510	...	69.0-72.0 (a)	0.30-0.7	...	...	rem	0.02-0.06 As
C33000	Low leaded brass (tube)	65.0-68.0 (a)	0.25-0.7 (b)	0.07	...	rem	
C33100		65.0-68.0 (a)	0.8-1.5	0.06	...	rem	...
C33200	High leaded brass (tube)	65.0-68.0 (a)	1.5-2.5	0.07	...	rem	...
C33500	Low leaded brass	62.0-65.0 (a)	0.25-0.7	0.15 (c)	...	rem	...
C33530	...	62.5-66.5 (a)	0.30-0.8	0.10	...	rem	0.02-0.06 As
C34000	Medium leaded brass, 64½%	62.0-65.0 (a)	0.8-1.5	0.15 (c)	...	rem	...
C34200	High leaded brass, 64½%	62.0-65.0 (a)	1.5-2.5	0.15 (c)	...	rem	...
C34500	...	62.0-65.0 (a)	1.5-2.5	0.15	...	rem	...
C35000	Medium leaded brass, 62%	60.0-63.0 (a,d)	0.8-2.0	0.15 (c)	...	rem	...
C35300	High leaded brass, 62%	60.0-63.0 (e,d)	1.5-2.5	0.15 (c)	...	rem	...
C35330	...	60.5-64.0 (e)	1.5-3.5 (f)	...	...	rem	0.02-0.25 As
C35340	...	60.0-63.0 (e)	1.5-2.5	0.10-0.30	...	rem	...
C35600	Extra high leaded brass	60.0-63.0 (e)	2.0-3.0	0.15 (c)	...	rem	...
C36000	Free cutting brass	60.0-63.0 (e)	2.5-3.7	0.35	...	rem	...
C36200	...	60.0-63.0 (e)	3.5-4.5	0.15	...	rem	...
C36500	Leaded muntz metal, uninhibited	58.0-61.0 (a)	0.25-0.7	0.15	0.25	rem	...
C36600	Leaded muntz metal, arsenical	58.0-61.0 (a)	0.25-0.7	0.15	0.25	rem	0.02-0.06 As
C36700	Leaded muntz metal, antimonial	58.0-61.0 (a)	0.25-0.7	0.15	0.25	rem	0.02-0.10 Sb
C36800	Leaded muntz metal, phosphorized	58.0-61.0 (a)	0.25-0.7	0.15	0.25	rem	0.02-0.10 P
C37000	Free cutting muntz metal	59.0-62.0 (a)	0.8-1.5	0.15	...	rem	...
C37100	...	58.0-62.0 (a)	0.6-1.2	0.15	...	rem	...
C37700	Forging brass	58.0-61.0 (e)	1.5-2.5	0.30	...	rem	...
C37710	...	56.5-60.0 (e)	1.0-3.0	0.30	...	rem	...
C37800	...	56.0-59.0 (e)	1.5-2.5	0.30	...	rem	...
C38000	Architectural bronze, low leaded	55.0-60.0 (e)	1.5-2.5	0.35	0.30	rem	0.50 Al
C38010	...	55.0-60.0 (e)	1.5-3.0	0.30	...	rem	0.10-0.6 Al
C38500	Architectural bronze	55.0-59.0 (e)	2.5-3.5	0.35	...	rem	...
C38510	...	56.0-60.0 (e)	2.5-4.5	...	...	rem	...
C38590	...	56.5-60.0 (e)	2.0-3.5	0.35	...	rem	...

(a) Cu + sum of named elements = 99.6% min. (b) For tube over 5 in. O.D., the Pb may be less than 0.20%. (c) For flat products, the iron shall be 0.10% max. (d) Cu, 61.0% min for rod. (e) Cu + sum of named elements = 99.5% min. (f) Pb may be reduced to 1.0% by agreement.
Source: *1988 Standards Handbook*, Copper Development Association Inc., p 5-19.

Wrought copper-zinc-tin alloys (tin brasses) (Composition, percent maximum unless a range or minimum)

Copper No.	Previous trade name	Cu	Pb	Fe	Sn	Zn	P	Others
C40400	...	rem (a)	...	...	0.35-0.7	2.0-3.0	...	...
C40500	...	94.0-96.0 (a)	0.05	0.05	0.7-1.3	rem	...	...
C40800	...	94.0-96.0 (a)	0.05	0.05	1.8-2.2	rem	...	...
C41000	...	91.0-93.0 (a)	0.05	0.05	2.0-2.8	rem	...	...
C41100	...	89.0-92.0 (a)	0.10	0.05	0.30-0.7	rem	...	...
C41300	...	89.0-93.0 (a)	0.10	0.05	0.7-1.3	rem	...	...
C41500	...	89.0-93.0 (a)	0.10	0.05	1.5-2.2	rem	...	...
C42000	...	88.0-91.0 (a)	...	...	1.5-2.0	rem	0.25	...
C42100	...	87.5-89.0 (a)	0.05	0.05	2.2-3.0	rem	0.35	0.15-0.35 Mn
C42200	...	86.0-89.0 (a)	0.05	0.05	0.8-1.4	rem	0.35	...
C42500	...	87.0-90.0 (a)	0.05	0.05	1.5-3.0	rem	0.35	...
C43000	...	84.0-87.0 (a)	0.10	0.05	1.7-2.7	rem	...	...
C43200	...	85.0-88.0 (a)	0.05	0.05	0.40-0.6	rem	0.35	...
C43400	...	84.0-87.0 (a)	0.05	0.05	0.40-1.0	rem	...	...
C43500	...	79.0-83.0 (a)	0.10	0.05	0.6-1.2	rem	...	...
C43600	...	80.0-83.0 (a)	0.05	0.05	0.20-0.50	rem	...	...
C44300	Admiralty, arsenical	70.0-73.0 (b)	0.07	0.06	0.8-1.2 (c)	rem	...	0.02-0.06 As
C44400	Admiralty, antimonial	70.0-73.0 (b)	0.07	0.06	0.8-1.2 (c)	rem	...	0.02-0.10 Sb
C44500	Admiralty, phosphorized	70.0-73.0 (b)	0.07	0.06	0.8-1.2 (c)	rem	0.02-0.10	...
C45450	...	65.0-66.0 (d)	...	...	0.10-0.30	rem	0.10-0.30	0.20-0.40 Al
C46200	Naval brass, 63½%	62.0-65.0 (b)	0.20	0.10	0.50-1.0	rem	...	...
C46210	...	61.0-64.0 (b)	0.05	...	1.0	rem	...	0.03 Al, 0.50 Si
C46400	Naval brass, uninhibited	59.0-62.0 (b)	0.20	0.10	0.50-1.0	rem	...	...
C46420	...	61.0-63.5 (b)	0.20	0.20	1.0-1.4	rem	...	...
C46500	Naval brass, arsenical	59.0-62.0 (b)	0.20	0.10	0.50-1.0	rem	...	0.02-0.06 As
C46600	Naval brass, antimonial	59.0-62.0 (b)	0.20	0.10	0.50-1.0	rem	...	0.02-0.10 Sb
C46700	Naval brass, phosphorized	59.0-62.0 (b)	0.20	0.10	0.50-1.0	rem	0.02-0.10	...
C47000	Naval brass welding and brazing rod	57.0-61.0 (b)	0.05	...	0.25-1.0	rem	...	0.01 Al
C47600	...	86.0-88.0 (b)	1.8-2.2	0.05	1.8-2.2	rem	0.03-0.07	0.05-0.15 Mn
C47940	...	63.0-66.0 (b)	1.0-2.0	0.10-1.0	1.2-2.0	rem	...	0.10-0.50 Ni (incl. Co)
C48200	Naval brass, medium leaded	59.0-62.0 (b)	0.40-1.0	0.10	0.50-1.0	rem	...	...
C48500	Naval brass, high leaded	59.0-62.0 (b)	1.3-2.2	0.10	0.50-1.0	rem	...	...
C48600	...	59.0-62.0 (b)	1.0-2.5	...	0.8-1.5	rem	...	0.02-0.25 As
C49080	...	49.0-52.0 (b)	0.50	...	3.0-4.0	rem	...	0.10 Al

(a) Cu + sum of named elements = 99.7% min. (b) Cu + sum of named elements = 99.6% min. (c) For tubular products, the minimum Sn content may be 0.9%. (d) Cu + sum of named elements = 99.8% min. Source: *1988 Standards Handbook*, Copper Development Association Inc., p 5-19.

Wrought copper-tin alloys (phosphor bronzes) (Composition, percent maximum unless shown as a range or minimum)

Copper No.	Previous trade name	Cu (a)	Pb	Fe	Sn	Zn	P	Others
C50100	...	rem	0.05	0.05	0.50-0.8	...	0.01-0.05	...
C50200	...	rem	0.05	0.10	1.0-1.5	...	0.04	...
C50500	Phosphor bronze, 1.25% E	rem	0.05	0.10	1.0-1.7	0.30	0.03-0.35	...
C50700	...	rem	0.05	0.10	1.5-2.0	...	0.30	...
C50710	...	rem	...	...	1.7-2.3	...	0.15	0.10-0.40 Ni
C50715	...	rem	0.02	0.05-0.15	1.7-2.3	...	0.025-0.04	(b)
C50800	...	rem	0.05	0.10	2.6-3.4	...	0.01-0.07	...
C50900	...	rem	0.05	0.10	2.5-3.8	0.30	0.03-0.30	...
C51000	Phosphor bronze, 5% A	rem	0.05	0.10	4.2-5.8	0.30	0.03-0.35	...
C51100	...	rem	0.05	0.10	3.5-4.9	0.30	0.03-0.35	...
C51800	Phosphor bronze	rem	0.02	...	4.0-6.0	...	0.10-0.35	0.01 Al
C51900	...	rem	0.05	0.10	5.0-7.0	0.30	0.03-0.35	...
C52100	Phosphor bronze, 8% C	rem	0.05	0.10	7.0-9.0	0.20	0.03-0.35	...
C52400	Phosphor bronze, 10% D	rem	0.05	0.10	9.0-11.0	0.20	0.03-0.35	...

(a) Cu + sum of named elements = 99.5% min. (b) Cu + Sn + Fe + P = 99.5% min. Source: *1988 Standards Handbook*, Copper Development Association Inc., p 5-19.

Wrought copper-tin-lead alloys (leaded phosphor bronzes) (Composition, percent maximum, unless shown as a range or minimum)

Copper No.	Previous trade name	Cu (a)	Pb	Fe	Sn	Zn	P
C53200	Phosphor bronze B	rem	2.5-4.0	0.10	4.0-5.5	0.20	0.03-0.35
C53400	Phosphor bronze B-1	rem	0.8-1.2	0.10	3.5-5.8	0.30	0.03-0.35
C54400	Phosphor bronze B-2	rem	3.5-4.5	0.10	3.5-4.5	1.5-4.5	0.01-0.50
C54800	...	rem	4.0-6.0	0.10	4.0-6.0	0.30	0.03-0.35

(a) Cu + sum of named elements = 99.5% min. Source: *1988 Standards Handbook*, Copper Development Association Inc., p 5-19.

Wrought copper-phosphorus and copper-silver-phosphorus alloys (Composition, percent maximum unless shown as a range or minimum)

Copper No.	Cu (a)	Ag	P
C55180	rem	...	4.8-5.2
C55181	rem	...	7.0-7.5
C55280	rem	1.8-2.2	6.8-7.2
C55281	rem	4.8-5.2	5.8-6.2
C55282	rem	4.8-5.2	6.5-7.0
C55283	rem	5.8-6.2	7.0-7.5
C55284	rem	14.5-15.5	4.8-5.2

(a) Cu + sum of named elements = 99.85% min. Source: *1988 Standards Handbook*, Copper Development Association Inc., p 5-19.

Wrought copper-aluminum alloys (aluminum bronzes) (Composition, percent maximum unless shown as a range or minimum)

Copper No.	Cu (a) (incl Ag)	Pb	Fe	Sn	Zn	Al	Mn	Si	Ni (incl Co)	Others
C60600	rem	...	0.50	...	...	4.0-7.0	...	...	...	...
C60700	rem	0.01	...	1.7-2.0	...	2.3-2.9	...	...	...	...
C60800	rem	0.10	0.10	...	...	5.0-6.5	...	...	...	0.02-0.35 As
C61000	rem	0.02	0.50	...	0.20	6.0-8.5	...	0.10	...	...
C61300	rem (b)	0.01	2.0-3.0	0.20-0.50	0.10 (c)	6.0-7.5	0.20	0.10	0.15	0.015 P(c)
C61400	rem	0.01	1.5-3.5	...	0.20	6.0-8.0	1.0	...	...	0.015 P
C61500	rem	0.015	...	...	...	7.7-8.3	...	...	1.8-2.2	...
C61550	rem (d)	0.05	0.20	0.05	0.8	5.5-6.5	1.0	...	1.5-2.5	0.8 Zn
C61800	rem	0.02	0.50-1.5	...	0.02	8.5-11.0	...	0.10	...	...
C61900	rem	0.02	3.0-4.5	0.6	0.8	8.5-10.0	...	...	...	...
C62200	rem	0.02	3.0-4.2	...	0.02	11.0-12.0	...	0.10	...	...
C62300	rem	...	2.0-4.0	0.6	...	8.5-10.0	0.50	0.25	1.0	...
C62400	rem	...	2.0-4.5	0.20	...	10.0-11.5	0.30	0.25	...	...
C62500	rem	...	3.5-5.5	...	...	12.5-13.5	2.0	...	...	...
C62580	rem	0.02	3.0-5.0	...	0.02	12.0-13.0	...	0.04	...	...
C62581	rem	0.02	3.0-5.0	...	0.02	13.0-14.0	...	0.04	...	...
C62582	rem	0.02	3.0-5.0	...	0.02	14.0-15.0	...	0.04	...	...
C62730	rem	0.05	4.0-6.0	0.10	0.40	8.5-11.0	0.50	0.10	4.0-6.0	0.05 Mg
C63000	rem	...	2.0-4.0	0.20	0.30	9.0-11.0	1.5	0.25	4.0-5.5	...
C63010	78.0 min (b)	...	2.0-3.5	0.20	0.30	9.7-10.9	1.5	...	4.5-5.5	...
C63020	74.5 min	0.03	4.0-5.5	0.25	0.30	10.5-11.5	1.5	...	4.2-6.0	0.20 Co, 0.05 Cr
C63200	rem	0.02	3.5-4.3 (e)	...	...	8.7-9.5	1.2-2.0	0.10	4.0-4.8 (e)	...
C63230	75.9-84.5	0.02	3.0-5.0 (c)	...	...	8.5-9.5	3.5	0.10	4.0-5.5 (e)	...
C63280	rem	0.02	3.0-5.0	...	...	8.5-9.5	0.6-3.5	...	4.0-5.5	...
C63300	rem	0.02	2.0-6.0	...	...	5.0-7.5	11.0-13.0	1.5	1.0-2.5	...
C63380	rem	0.02	2.0-4.0	...	0.15	7.0-8.5	11.0-14.0	0.10	1.5-3.0	...
C63400	rem	0.05	0.15	0.20	0.50	2.6-3.2	...	0.25-0.45	0.15	0.15 As
C63600	rem	0.05	0.15	0.20	0.50	3.0-4.0	...	0.7-1.3	0.15	0.15 As
C63800	rem	0.05	0.20	...	0.8	2.5-3.1	0.10	1.5-2.1	0.20 (f)	0.25-0.55 Co
C64110	rem	1.0-2.0	...	...	...	8.0-11.0	0.50	...	...	...
C64200	rem	0.05	0.30	0.20	0.50	6.3-7.6	0.10	1.5-2.2	0.25	0.15 As
C64210	rem	0.05	0.30	0.20	0.50	6.3-7.0	0.10	1.5-2.0	0.25	0.15As
C64250	rem	...	1.0	...	...	5.5-7.5	0.50	1.5-3.0	...	...
C64400	rem	0.03	0.05	0.10	0.20	3.5-4.5	...	0.8-1.3	4.2-5.0	...

(a) Cu + sum of named elements = 99.5% min. (b) Cu + sum of named elements = 99.8% min. (c) When the product is for subsequent welding applications and is so specified by the purchaser, Cr, Cd, Zr, and Zn shall each be 0.05% max. (d) For tubular products, the minimum Sn content may be 0.9%. (e) Fe content shall not exceed Ni content. (f) Not including Co. Source: *1988 Standards Handbook*, Copper Development Association Inc., p 5-19.

Wrought copper-silicon alloys (silicon bronzes) (Composition, percent maximum unless shown as a range or minimum)

Copper No.	Previous trade name	Cu (a) (incl Ag), % min	Pb	Fe	Sn	Zn	Al	Mn	Si	Ni (incl Co)	Others
C64700	...	rem	0.10	0.10	...	0.50	...	...	0.40-0.8	1.6-2.2	...
C64710	...	95.0	...	...	...	0.20-0.50	...	0.10	0.50-0.9	2.9-3.5	...
C64900	...	rem	0.05	0.10	1.2-1.6	0.20	0.10	...	0.8-1.2	0.10	...
C65100	Low silicon bronze B	rem	0.05	0.8	...	1.5	...	0.7	0.8-2.0	...	...
C65300	...	rem	0.05	0.8	...	...	...	...	2.0-2.6	...	...
C65400	...	rem	0.05	...	1.2-1.9	0.50	...	...	2.7-3.4	...	0.01-0.12 Cr
C65500	High silicon bronze A	rem	0.05	0.8	...	1.5	...	0.50-1.3	2.8-3.8	0.6	...
C65600	...	rem	0.02	0.50	1.5	1.5	0.01	1.5	2.8-4.0	...	0.10 P
C65620	...	90.0	...	1.0-2.0	...	1.5-4.0	...	...	2.4-4.0	...	...
C65800	...	rem	0.05	0.25	...	...	...	0.50-1.3	2.8-3.8	0.6	...
C66100	...	rem	0.20-0.8	0.25	...	1.5	...	1.5	2.8-3.5	...	...

(a) Cu + sum of named elements = 99.5% min. Source: *1988 Standards Handbook*, Copper Development Association Inc., p 5-19.

Wrought miscellaneous copper-zinc alloys (Composition, percent maximum, unless a range or minimum)

Copper No.	Previous trade name	Cu (a) (incl Ag)	Pb	Fe	Sn	Zn	Ni (incl Co)	Al	Mn	Si	Others
C66400	...	rem	0.015	1.3-1.7 (b)	0.05	11.0-12.0	0.05(c)	0.05	0.05	0.05	0.02 P, 0.30-0.7 Co(b), 0.05 Ag
C66410	...	rem	0.015	1.8-2.3	0.05	11.0-12.0	0.05	0.05	0.05	0.05	0.02 P, 0.05 As
C66700	Manganese brass	68.5-71.5	0.07	0.10	...	rem	...	...	0.8-1.5	...	...
C66800	...	60.0-63.0	0.50	0.35	0.30	rem	0.25	0.25	2.0-3.5	0.50-1.5	...
C66900	...	62.5-64.5	0.05	0.25	...	rem	...	...	11.5-12.5	...	...
C67000	Manganese bronze B	63.0-68.0	0.20	2.0-4.0	0.50	rem	...	3.0-6.0	2.5-5.0	...	...
C67130	...	56.0-59.0	0.50-1.5	...	0.50-1.5	rem	0.50-1.5	0.10-1.0	0.50-1.5	...	...
C67300	...	58.0-63.0	0.40-3.0	0.50	0.30	rem	0.25	0.25	2.0-3.5	0.50-1.5	...
C67400	...	57.0-60.0	0.50	0.35	0.30	rem	0.25	0.50-2.0	2.0-3.5	0.50-1.5	...
C67410	...	55.5-59.0	0.8	1.0	0.50	rem	2.0	1.3-2.3	1.0-2.4	0.7-1.3	...
C67500	Manganese bronze A	57.0-60.0	0.20	0.8-2.0	0.50-1.5	rem	...	0.25	0.05-0.50	...	...
C67600	...	57.0-60.0	0.50-1.0	0.40-1.3	0.50-1.5	rem	...	...	0.05-0.50	...	...
C67620	...	55.0-57.0	0.50-1.5	0.50-1.2	...	rem	...	...	1.0-2.0	...	...
C67700	...	55.0-58.0	0.50-1.0	0.7-1.5	...	rem	1.5-2.3	...	0.05-0.30	...	0.40-0.8 As
C67800	...	56.0-59.0	0.30	0.7-1.5	0.20	rem	...	0.50-1.5	0.20-0.6	...	...
C67810	...	56.5-59.5	1.0	1.0	0.50	rem	1.5	0.40-1.6	0.40-1.8	0.6	...
C67820	...	56.0-60.0	0.10	0.50-1.2	0.30-1.0	rem	...	0.30-1.3	0.30-2.0	...	...
C68000	Bronze, low fuming (nickel)	56.0-60.0	0.05	0.25-1.25	0.75-1.10	rem	0.20-0.8	0.01	0.01-0.50	0.04-0.15	...
C68100	Bronze, low fuming	56.0-60.0	0.05	0.25-1.25	0.75-1.10	rem	...	0.01	0.01-0.50	0.04-0.15	...
C68200	...	58.0-60.0	...	0.6-1.0	...	rem	...	...	0.6-1.0	0.07-0.15	...
C68600	...	56.0-60.0	0.50-1.5	0.50-1.2	0.20-1.0	rem	...	0.30-1.5	0.30-2.0	...	...
C68700	Aluminum brass, arsenical	76.0-79.0	0.07	0.06	...	rem	...	1.8-2.5	...	...	0.02-0.06 As
C68800	...	rem	0.05	0.20	...	21.3-24.1 (d)	...	3.0-3.8 (d)	...	...	0.25-0.55 Co
C69000	...	72.0-74.5	0.025	0.05	...	rem	0.50-0.8	3.0-3.8 (d)	...	...	...
C69100	...	81.0-84.0	0.05	0.25	0.10	rem	0.8-1.4	0.7-1.2	0.10 min	0.8-1.3	...
C69400	Silicon red brass	80.0-83.0	0.30	0.20	...	rem	...	...	...	3.5-4.5	...
C69430	...	80.0-83.0	0.30	0.20	...	rem	...	...	...	3.5-4.5	0.03-0.06 As
C69440	...	80.0-83.0	0.30	0.20	...	rem	...	...	...	3.5-4.5	0.03-0.06 Sb
C69450	...	80.0-83.0	0.30	0.20	...	rem	...	...	0.40	3.5-4.5	0.03-0.06 P
C69700	...	75.0-80.0	0.50-1.5	0.20	...	rem	...	...	0.40	2.5-3.5	...
C69710	...	75.0-80.0	0.50-1.5	0.20	...	rem	...	...	0.40	2.5-3.5	0.03-0.06 As
C69720	...	75.0-80.0	0.50-1.5	0.20	...	rem	...	...	0.40	2.5-3.5	0.03-0.06 Sb
C69730	...	75.0-80.0	0.50-1.5	0.20	...	rem	...	...	0.40	2.5-3.5	0.03-0.06 P
C69800	...	66.0-70.0	0.8	0.4	...	rem	0.50	...	...	0.7-1.3	...
C69900	Incramute	rem	0.02	0.10	...	0.14	0.10	1.4-2.3	40.0-48.0	...	0.20 Co, 0.05 C, 0.05 Ag, 0.05 Cd, 0.01 As
C69910	...	rem	0.01	1.0-1.4	...	3.0-5.0	...	0.25-0.8	28.0-32.0	...	...
C69950	...	51.0-54.0	...	...	...	...	8.5-10.5	...	36.0-40.0	...	...

(a) Cu + sum of named elements = 99.5% min. (b) Fe + Co shall be 1.8-2.3%. (c) Not including Co. (d) Al + Zn shall be 25.1-27.1%. Source: *1988 Standards Handbook*, Copper Development Association Inc., p 5-19.

Wrought copper-nickel alloys (Composition, percent maximum unless shown as a range or minimum)

Copper No.	Previous trade name	Cu (a) (incl Ag), % min	Pb	Fe	Zn	Ni (incl Co)	Sn	Mn	Others
C70100	...	rem	...	0.05	0.25	3.0-4.0	...	0.50	...
C70200	...	rem	0.05	0.10	...	2.0-3.0	...	0.40	...
C70250	...	rem	0.05	0.20	1.0	2.2-4.2	...	0.10	0.05-0.30 Mg, 0.25-1.2 Si
C70320	...	rem	...	...	...	2.5-5.0	...	...	0.20-1.2 Al, 0.20-1.2 Si, 0.18-0.50 Cr
C70400	Copper nickel, 5%	rem	0.05	1.3-1.7	1.0	4.8-6.2	...	0.30-0.8	...
C70440	...	rem	0.05	1.0-1.8	1.0	4.5-6.0	0.10	0.50-1.5	0.05 C, 0.05 S, 0.35-0.45 Si
C70500	Copper nickel, 7%	rem	0.05	0.10	0.20	5.8-7.8	...	0.15	...
C70600	Copper nickel, 10%	rem	0.05(b)	1.0-1.8	1.0(b)	9.0-11.0	...	1.0	(b)
C70610	...	rem	0.01	1.0-2.0	...	10.0-11.0	...	0.50-1.0	0.05 S, 0.05 C
C70690	...	rem	0.001	0.005	0.001	9.0-11.0	...	0.001	(c)
C70700	...	rem	...	0.05	...	9.5-10.5	...	0.50	...
C70800	Copper nickel, 11%	rem	0.05	0.10	0.20	10.5-12.5	...	0.15	...
C70900	...	rem	0.05	0.6	1.0	13.5-16.5	...	0.6	...
C71000	Copper nickel, 20%	rem	0.05	1.0	1.0	19.0-23.0	...	1.0	...
C71100	...	rem	0.05	0.10	0.20	22.0-24.0	...	0.15	...
C71110	...	rem	...	...	...	21.5-23.5	...	0.35	(b) 0.05 Ti
C71300	...	rem	0.05	0.20	1.0	23.5-26.5	...	1.0	...
C71500	Copper nickel, 30%	rem	0.05(b)	0.40-1.0	1.0(b)	29.0-33.0	...	1.0	(b)
C71580	...	rem	0.05	0.50	0.05	29.0-33.0	...	0.30	(d)
C71581	...	rem	0.02	0.40-0.7	...	29.0-32.0	...	1.0	(e)
C71590	...	rem	0.001	0.005	0.001	29.0-33.0	...	0.001	(c)
C71630	...	rem	0.01	0.40-1.0	...	30.0-32.0	...	0.50-1.5	0.06 C, 0.08 S
C71640	...	rem	0.01	1.7-2.3	...	29.0-32.0	...	1.5-2.5	0.03 S, 0.06 C
C71700	...	rem	...	0.40-1.0	...	29.0-33.0	...	...	0.30-0.7 Be
C71900	...	rem	0.015	0.50	0.05	28.0-33.0	...	0.20-1.0	2.2-3.0 Cr, 0.02-0.35 Zr, 0.01-0.20 Ti, 0.04 C, 0.25 Si, 0.015 S, 0.02 P
C72150	...	rem(a)	0.05	0.10	0.20	43.0-46.0	...	0.05	0.10 C, 0.50 Si
C72200	...	rem(f)	0.05(b)	0.50-1.0	1.0(b)	15.0-18.0	...	1.0	(b) 0.30-0.7 Cr, 0.03 Si, 0.03 Ti
C72400	...	rem(a)	0.05	0.10	0.50	11.0-15.0	0.05	1.0	1.5-2.5 Al, 0.05-0.40 Mg
C72420	...	rem(g)	0.02	0.7-1.2	0.20	13.5-16.5	0.10	3.5-5.5	1.0-2.0 Al, 0.50 Cr, 0.15 Si, 0.05 Mg, 0.15 S, 0.01 P, 0.05 C
C72500	...	rem(f)	0.05	0.6	0.50	8.5-10.5	1.8-2.8	0.20	...
C72600	...	91.0-93.0 (g)	0.05	0.20	0.50	3.5-4.5	3.5-4.5	0.20	0.05 P
C72700	...	rem(g)	0.02(h)	0.50	0.50	8.5-9.5	5.5-6.5	0.05-0.30	0.10 Nb, 0.15 Mg
C72800	...	rem(g)	0.005	0.50	1.0	9.5-10.5	7.5-8.5	0.05-0.30	0.10 Al, 0.001 B, 0.001 Bi, 0.10-0.30 Nb, 0.005-0.15 Mg, 0.005 P, 0.0025 S, 0.02 Sb, 0.05 Si, 0.01 Ti
C72900	...	rem(g)	0.02(h)	0.50	0.50	14.5-15.5	7.5-8.5	0.30	0.10 Nb, 0.15 Mg
C72950	...	rem(g)	0.05	0.6	...	20.0-22.0	4.5-5.7	0.6	...

(a) Cu + sum of named elements = 99.5% min. (b) When the product is for subsequent welding applications and so specified by the purchaser, Zn shall be 0.50% max, P 0.02% max, Pb 0.02% max, S 0.02% max (0.008% max for Alloy C71110) and C 0.05% max. (c) The following maximum limits shall apply: 0.03 C, 0.02 Si, 0.003 S, 0.002 Al, 0.001 P, 0.0005 Hg, 0.001 Ti, 0.001 Sb, 0.001 As, 0.001 Bi, and 0.001 Sn. For Copper Alloy No. C70690, Co shall be 0.02 max and for Copper Alloy No. C71590, Co shall be 0.05 max. (d) The following maximum limits shall apply: 0.07 C, 0.15 Si, 0.024 S, 0.05 Al, and 0.03 P. (e) 0.02 P max, 0.25 Si max, 0.01 S max, 0.20-0.50 Ti. (f) Cu + sum of named elements = 99.8% min. (g) Cu + sum of named elements = 99.7% min. (h) 0.005% Pb max for hot rolling. Source: *1988 Standards Handbook*, Copper Development Association Inc., p 5-19.

Wrought copper-nickel-zinc alloys (nickel silvers) (Composition, percent maximum unless shown as a range or minimum)

Copper No.	Previous trade name	Cu (a)	Pb	Fe	Zn	Ni (incl Co)	Mn	Others
C73150	...	rem(b)	0.10	0.25	9.0-15.0	4.0-7.0	0.50	...
C73200	...	rem	0.05	0.6	3.0-6.0	19.0-23.0	1.0	...
C73500	...	70.5-73.5	0.10	0.25	rem	16.5-19.5	0.50	...
C73800	Nickel silver, 70-12	68.5-71.5	0.05	0.25	rem	11.0-13.0	0.50	...
C74000	...	69.0-73.5	0.10	0.25	rem	9.0-11.0	0.50	...
C74300	...	63.0-66.0	0.10	0.25	rem	7.0-9.0	0.50	...
C74500	Nickel silver, 65-10	63.5-66.5	0.10(c)	0.25	rem	9.0-11.0	0.50	...
C75200	Nickel silver, 65-18	63.5-66.5	0.05	0.25	rem	16.5-19.5	0.50	...
C75400	Nickel silver, 65-15	63.5-66.5	0.10	0.25	rem	14.0-16.0	0.50	...
C75700	Nickel silver, 65-12	63.5-66.5	0.05	0.25	rem	11.0-13.0	0.50	...
C75720	...	60.0-65.0	0.04	0.25	rem	11.0-13.0	0.05-0.30	...
C75900	...	60.0-63.0	0.10	0.25	rem	17.0-19.0	0.50	...
C76000	...	60.0-63.0	0.10	0.25	rem	7.0-9.0	0.50	...
C76100	...	59.0-63.0	0.10	0.25	rem	9.0-11.0	0.50	...
C76200	...	57.0-61.0	0.10	0.25	rem	11.0-13.5	0.50	...
C76300	...	60.0-64.0	0.50-2.0	0.50	rem	17.0-19.0	0.50	...
C76390	...	59.0-63.0	0.8-1.1	0.25	rem	23.0-26.0	0.50	0.40-0.6 Sn
C76400	...	58.5-61.5	0.05	0.25	rem	16.5-19.5	0.50	...
C76600	...	55.0-58.0	0.10	0.25	rem	11.0-13.5	0.50	...
C76700	Nickel silver, 56.5-15	55.0-58.0	...	...	rem	14.0-16.0	0.50	...
C77000	Nickel silver, 55-18	53.5-56.5	0.05	0.25	rem	16.5-19.5	0.50	...
C77010	...	54.0-56.0	0.03	0.30	rem	17.0-19.0	0.05-0.35	...
C77300	...	46.0-50.0	0.05	...	rem	9.0-11.0	...	0.01 Al, 0.25 P, 0.04-0.25 Si
C77310	...	46.0-56.0	0.05	...	rem	9.0-11.0	0.50	0.01 Al, 0.25 P, 0.04-0.25 Si
C77400	...	43.0-47.0	0.20	...	rem	9.0-11.0	...	...
C77600	Nickel silver, 43.5-13	42.0-45.0	0.25	0.20	rem	12.0-14.0	0.25	0.15 Sn
C78200	...	63.0-67.0	1.5-2.5	0.35	rem	7.0-9.0	0.50	...
C78800	...	63.0-67.0	1.5-2.0	0.25	rem	9.0-11.0	0.50	...
C79000	...	63.0-67.0	1.5-2.2	0.35	rem	11.0-13.0	0.50	...
C79200	...	59.0-66.5	0.8-1.4	0.25	rem	11.0-13.0	0.50	...
C79300	...	55.0-59.0	0.50-2.0	0.50	rem	11.0-13.0	0.50	...
C79600	Leaded nickel silver, 10%	43.5-46.5	0.8-1.2	...	rem	9.0-11.0	1.5-2.5	...
C79620	...	46.0-48.0	0.50-2.0	...	rem	8.0-11.0	0.50	...
C79800	...	45.5-48.5	1.5-2.5	0.25	rem	9.0-11.0	1.5-2.5	...
C79810	...	46.0-48.0	2.0-3.5	...	rem	8.0-11.0	0.50	...
C79900	...	47.5-50.5	1.0-1.5	0.25	rem	6.5-8.5	0.50	...

(a) Cu + sum of named elements = 99.5% min. (b) Cu + sum of named elements = 99.7% min. (c) 0.05% Pb max for rod and wire. Source: *1988 Standards Handbook*, Copper Development Association Inc., p 5-19.

Cast copper (Composition, percent maximum unless a range or minimum)

Copper No.	Cu (incl Ag), % min	% min	Troy oz, min	B	P
C80100	99.95	...	...	...	...
C80300	99.95	0.034	10	...	...
C80500	99.75	0.034	10	...	...
C80700	99.75	...	...	0.02	...
C80900	99.70	0.034	10	0.02	...
C81100	99.70	...	...	...	...
C81200	99.9	...	...	...	0.045-0.065

Source: *1988 Standards Handbook*, Copper Development Association Inc., p 5-19.

Cast high-copper alloys (Composition, percent maximum unless shown as a range or minimum)

Copper No.	Cu (a)	Ag	Be	Co	Si	Ni	Fe	Al	Sn	Pb	Zn	Cr
C81300	98.5 min	...	0.02-0.10	0.6-1.0	...	...	...	...	...	...	...	...
C81400	98.5 min	...	0.02-0.10	...	...	...	...	...	...	...	...	0.6-1.0
C81500	98.0 min	...	...	...	0.15	...	0.10	0.10	0.10	0.02	0.10	0.40-1.5
C81540	95.1 min (b)	...	...	...	0.40-0.8	2.0-3.0 (c)	0.15	0.10	0.10	0.02	0.10	0.10-0.6
C81700	94.2 min	0.8-1.2	0.30-0.55	0.25-1.5	...	0.25-1.5	...	...	...	...	...	...
C81800	95.6 min	0.8-1.2	0.30-0.55	1.4-1.7	...	...	...	...	...	...	...	...
C82000	95.0 min	...	0.45-0.8	2.4-2.7 (c)	0.15	0.20	0.10	0.10	0.10	0.02	0.10	0.10
C82100	95.5 min	...	0.35-0.8	0.25-1.5	...	0.25-1.5	...	...	...	...	...	...
C82200	96.5 min	...	0.35-0.8	...	...	1.0-2.0	...	...	...	...	...	...
C82400	96.4 min	...	1.65-1.75	0.20-0.40	...	0.10	0.20	0.15	0.10	0.02	0.10	0.10
C82500	95.5 min	...	1.90-2.15	0.35-0.7 (c)	0.20-0.35	0.20	0.25	0.15	0.10	0.02	0.10	0.10
C82510	95.5 min	...	1.90-2.15	1.0-1.2	0.20-0.35	0.20	0.25	0.15	0.10	0.02	0.10	0.10
C82600	95.2 min	...	2.25-2.45	0.35-0.7	0.20-0.35	0.20	0.25	0.15	0.10	0.02	0.10	0.10
C82700	94.6 min	...	2.35-2.55	...	0.15	1.0-1.5	0.25	0.15	0.10	0.02	0.10	0.10
C82800	94.8 min	...	2.50-2.75	0.35-0.7 (c)	0.20-0.35	0.20	0.25	0.15	0.10	0.02	0.10	0.10

(a) Cu + sum of named elements = 99.5% min. (b) Includes Ag. (c) Ni + Co. Source: *1988 Standards Handbook*, Copper Development Association Inc., p 5-19.

Cast copper-tin-zinc and copper-tin-zinc-lead alloys (red brasses and leaded red brasses) (Composition, percent maximum unless shown as a range or minimum)

Copper No.	Cu (a,b)	Sn	Pb	Zn	Fe	Sb	As	Ni(a) (incl Co)	S	P (c)	Al	Si	Bi
C83300	92.0-94.0	1.0-2.0	1.0-2.0	2.0-6.0	...	...	...	...	...	...	...	...	...
C83400	88.0-92.0	0.20	0.50	8.0-12.0	...	...	...	...	...	...	...	...	...
C83410	88.0-91.0	1.0-2.0	0.10	rem	0.05	...	...	0.05	...	...	0.005	0.005	...
C83420	88.0-92.0	0.25-0.7	0.50	rem	0.10	...	...	...	...	...	...	...	...
C83450	87.0-89.0	2.0-3.5	1.5-3.0	5.5-7.5	0.30	0.25	...	0.8-2.0	0.08	0.03	0.005	0.005	...
C83500	86.0-88.0	5.5-6.5	3.5-5.5	1.0-2.5	0.25	0.25	...	0.50-1.0	0.08	0.03	0.005	0.005	...
C83520	rem	3.5-4.5	3.5-4.5	1.5-4.0	0.30	0.25	...	1.0	...	...	...	...	...
C83600	84.0-86.0	4.0-6.0	4.0-6.0	4.0-6.0	0.30	0.25	...	1.0	0.08	0.05	0.005	0.005	...
C83700	83.0-88.0	1.0	0.50	rem	0.30	...	0.05-0.20	0.30	...	...	0.005	0.005	...
C83800	82.0-83.8	3.3-4.2	5.0-7.0	5.0-8.0	0.30	0.25	...	1.0	0.08	0.03	0.005	0.005	...
C83810	rem	2.0-3.5	4.0-6.0	7.5-9.5	0.50 (d)	(d)	(d)	2.0	...	...	0.005	0.005	0.10

(a) In determining copper min, copper may be calculated as Cu + Ni. (b) Cu + sum of named elements = 99.3% min. (c) For continuous castings, 1.5% P max. (d) Fe + Sb + As shall be 0.50% max. Source: *1988 Standards Handbook*, Copper Development Association Inc., p 5-19.

Cast semi-red brasses and leaded semi-red brasses (Composition, percent maximum unless shown as a range or minimum)

Copper No.	Cu (a,b)	Sn	Pb	Zn	Fe	Sb	Ni(a) (incl Co)	S	P (c)	Al	Si	Bi
C84200	78.0-82.0	4.0-6.0	2.0-3.0	10.0-16.0	0.40	0.25	0.8	0.08	0.05	0.005	0.005	...
C84400	78.0-82.0	2.3-3.5	6.0-8.0	7.0-10.0	0.40	0.25	1.0	0.08	0.02	0.005	0.005	...
C84410	rem	3.0-4.5	7.0-9.0	7.0-11.0	(d)	(d)	1.0	...	...	0.01	0.20	0.05
C84500	77.0-79.0	2.0-4.0	6.0-7.5	10.0-14.0	0.40	0.25	1.0	0.08	0.02	0.005	0.005	...
C84800	75.0-77.0	2.0-3.0	5.5-7.0	13.0-17.0	0.40	0.25	1.0	0.08	0.02	0.005	0.005	...

(a) In determining copper min, copper may be calculated as Cu + Ni. (b) Cu + sum of named elements = 99.2% min. (c) For continuous castings, 1.5% P max. (d) Fe + Sb + As shall be 0.8% max. Source: *1988 Standards Handbook*, Copper Development Association Inc., p 5-19.

Cast yellow brasses and leaded yellow brasses (Composition, percent maximum unless shown as a range or minimum)

Copper No.	Cu (a)	Sn	Pb	Zn	Fe	Sb	Ni (a) (incl Co)	Mn	As	S	P	Al	Si
C85200	70.0-74.0(b)	0.7-2.0	1.5-3.8	20.0-27.0	0.6	0.20	1.0	...	...	0.05	0.02	0.005	0.05
C85210	70.0-75.0(c)	1.0-3.0	2.0-5.0	rem	0.8	...	1.0	...	0.02-0.06	...	...	0.005	0.005
C85310	68.0-73.0(c)	1.5	2.0-5.0	rem	0.8(d)	(d)	1.0	...	0.02-0.06 (d)	...	...	0.005	...
C85400	65.0-70.0(c)	0.50-1.5	1.5-3.8	24.0-32.0	0.7	...	1.0	...	...	...	...	0.35	0.05
C85500	59.0-63.0(c)	0.20	0.20	rem	0.20	...	0.20	0.20	...	...	...	...	...
C85700	58.0-64.0(e)	0.50-1.5	0.8-1.5	32.0-40.0	0.7	...	1.0	...	...	...	...	0.8	0.05
C85710	58.0-63.0(e)	1.0	1.0-2.5	rem	0.8	...	1.0	0.50	...	...	...	0.20-0.8	0.05
C85800	57.0 min(e)	1.5	1.5	31.0-41.0	0.50	0.05	0.50	0.25	0.05	0.05	0.01	0.55	0.25

(a) In determining copper min, copper may be calculated as Cu + Ni. (b) Cu + sum of named elements = 99.1% min. (c) Cu + sum of named elements = 99.0% min. (d) Fe + Sb + As shall be 0.8% max. (e) Cu + sum of named elements = 98.7% min. Source: *1988 Standards Handbook*, Copper Development Association Inc., p 5-19.

Cast manganese bronze and leaded manganese bronze alloys (Composition, percent maximum unless shown as a range or minimum)

Copper No.	Cu (a,b)	Sn	Pb	Zn	Fe	Ni (a) (incl Co)	Al	Mn
C86100	66.0-68.0	0.20	0.20	rem	2.0-4.0	...	4.5-5.5	2.5-5.0
C86200	60.0-66.0	0.20	0.20	22.0-28.0	2.0-4.0	1.0	3.0-4.9	2.5-5.0
C86300	60.0-66.0	0.20	0.20	22.0-28.0	2.0-4.0	1.0	5.0-7.5	2.5-5.0
C86400	56.0-62.0	0.50-1.5	0.50-1.5	34.0-42.0	0.40-2.0	1.0	0.50-1.5	0.10-1.0
C86500	55.0-60.0	1.0	0.40	36.0-42.0	0.40-2.0	1.0	0.50-1.5	1.0-1.5
C86700	55.0-60.0	1.5	0.50-1.5	30.0-38.0	1.0-3.0	1.0	1.0-3.0	1.0-3.5
C86800	53.5-57.0	1.0	0.20	rem	1.0-2.5	2.5-4.0	2.0	2.5-4.0

(a) In determining copper min, copper may be calculated as Cu + Ni. (b) Cu + sum of named elements = 99.0% min. Source: *1988 Standards Handbook*, Copper Development Association Inc., p 5-19.

Cast copper-silicon alloys (silicon bronzes and silicon brasses) (Composition, percent maximum unless shown as a range or minimum)

Copper No.	Cu	Sn	Pb	Zn	Fe	Al	Si	Mn	Mg	Ni (incl Co)	S	Others
C87300	94.0 min(a)	...	0.20	0.25	0.20	...	3.5-4.5	0.8-1.5	...	...	...	...
C87400	79.0 min(b)	...	1.0	12.0-16.0	...	0.8	2.5-4.0	...	...	...	...	...
C87410	79.0 min(b)	...	1.0	12.0-16.0	...	0.8	2.5-4.0	...	...	...	...	0.03-0.06 As
C87420	79.0 min(b)	...	1.0	12.0-16.0	...	0.8	2.5-4.0	...	...	...	...	0.03-0.06 Sb
C87430	79.0 min(b)	...	1.0	12.0-16.0	...	0.8	2.5-4.0	...	...	...	...	0.03-0.06 P
C87500	79.0 min(a)	...	0.50	12.0-16.0	...	0.50	3.0-5.0	...	...	...	...	...
C87510	79.0 min(a)	...	0.50	12.0-16.0	...	0.50	3.0-5.0	...	...	...	...	0.03-0.06 As
C87520	79.0 min(a)	...	0.50	12.0-16.0	...	0.50	3.0-5.0	...	...	...	...	0.03-0.06 Sb
C87530	79.0 min(a)	...	0.50	12.0-16.0	...	0.50	3.0-5.0	...	...	...	...	0.03-0.06 P
C87600	88.0 min(a)	...	0.50	4.0-7.0	0.20	...	3.5-5.5	0.25	...	...	...	...
C87610	90.0 min(a)	...	0.20	3.0-5.0	0.20	...	3.0-5.0	0.25	...	...	...	...
C87800	80.0 min(a)	0.25	0.15	12.0-16.0	0.15	0.15	3.8-4.2	0.15	0.01	0.20	0.05	0.01 P, 0.05 As, 0.05 Sb
C87900	63.0 min(a)	0.25	0.25	30.0-36.0	0.40	0.15	0.8-1.2	0.15	...	0.50	0.05	0.01 P, 0.05 As, 0.05 Sb

(a) Cu + sum of named elements = 99.5% min. (b) Cu + sum of named elements = 99.2% min. Source: *1988 Standards Handbook*, Copper Development Association Inc., p 5-19.

Cast copper-tin alloys (tin bronzes) (Composition, percent maximum unless shown as a range or minimum)

Copper No.	Cu (a,b)	Sn	Pb	Zn	Fe	Sb	Ni (a) (incl Co)	S	P (c)	Al	Si	Mn
C90200	91.0-94.0(b)	6.0-8.0	0.30	0.50	0.20	0.20	0.50	0.05	0.05	0.005	0.005	...
C90250	89.0-91.0	9.0-11.0	0.30	0.50	0.25	0.20	0.8	0.05	0.05	0.005	0.005	0.10
C90300	86.0-89.0	7.5-9.0	0.30	3.0-5.0	0.20	0.20	1.0	0.05	0.05	0.005	0.005	...
C90500	86.0-89.0(d)	9.0-11.0	0.30	1.0-3.0	0.20	0.20	1.0	0.05	0.05	0.005	0.005	...
C90700	88.0-90.0	10.0-12.0	0.50	0.50	0.15	0.20	0.50	0.05	0.30	0.005	0.005	...
C90710	rem	10.0-12.0	0.25	0.05	0.10	...	0.10	...	0.50-1.2	0.005	0.005	...
C90800	85.0-89.0	11.0-13.0	0.25	0.25	0.15	0.20	0.50	0.05	0.30	0.005	0.005	...
C90810	rem	11.0-13.0	0.25	0.30	0.15	0.20	0.50	0.05	0.15-0.8	0.005	0.005	...
C90900	86.0-89.0	12.0-14.0	0.25	0.25	0.15	0.20	0.50	0.05	0.05	0.005	0.005	...
C91000	84.0-86.0	14.0-16.0	0.20	1.5	0.10	0.20	0.8	0.05	0.05	0.005	0.005	...
C91100	82.0-85.0	15.0-17.0	0.25	0.25	0.25	0.20	0.50	0.05	1.0	0.005	0.005	...
C91300	79.0-82.0	18.0-20.0	0.25	0.25	0.25	0.20	0.50	0.05	1.0	0.005	0.005	...
C91600	86.0-89.0	9.7-10.8	0.25	0.25	0.20	0.20	1.2-2.0	0.05	0.30	0.005	0.005	...
C91700	84.0-87.0	11.3-12.5	0.25	0.25	0.20	0.20	1.2-2.0	0.05	0.30	0.005	0.005	...

(a) In determining copper min, copper may be calculated as Cu + Ni. (b) Cu + sum of named elements = 99.4% min. (c) For continuous castings, 1.5% P max. (d) Cu + sum of named elements = 99.7% min. Source: *1988 Standards Handbook*, Copper Development Association Inc., p 5-19.

Cast copper-tin-lead alloys (leaded tin bronzes) (Composition, percent maximum unless otherwise indicated)

Copper No.	Cu (a,b)	Sn	Pb	Zn	Fe	Sb	Ni (a) (incl Co)	S	P (c)	Al	Si	Bi
C92200	86.0-90.0	5.5-6.5	1.0-2.0	3.0-5.0	0.25	0.25	1.0	0.05	0.05	0.005	0.005	...
C92300	85.0-89.0	7.5-9.0	0.30-1.0	2.5-5.0	0.25	0.25	1.0	0.05	0.05	0.005	0.005	...
C92310	rem	7.5-8.5	0.30-1.5	3.5-4.5	...	...	1.0	...	...	0.005	0.005	0.03
C92400	86.0-89.0	9.0-11.0	1.0-2.5	1.0-3.0	0.25	0.25	1.0	0.05	0.05	0.005	0.005	...
C92410	rem	6.0-8.0	2.5-3.5	1.5-3.0	0.20	0.25	0.20	...	...	0.005	0.005	0.05
C92500	85.0-88.0	10.0-12.0	1.0-1.5	0.50	0.30	0.25	0.8-1.5	0.05	0.30	0.005	0.005	...
C92600	86.0-88.5	9.3-10.5	0.8-1.5	1.3-2.5	0.20	0.25	0.7	0.05	0.03	0.005	0.005	...
C92610	rem	9.5-10.5	0.30-1.5	1.7-2.8	0.15	...	1.0	...	...	0.005	0.005	0.03
C92700	86.0-89.0	9.0-11.0	1.0-2.5	0.7	0.20	0.25	1.0	0.05	0.25	0.005	0.005	...
C92710	rem	9.0-11.0	4.0-6.0	1.0	0.20	0.25	2.0	0.05	0.10	0.005	0.005	...
C92800	78.0-82.0	15.0-17.0	4.0-6.0	0.8	0.20	0.25	0.8	0.05	0.05	0.005	0.005	...
C92900	82.0-86.0	9.0-11.0	2.0-3.2	0.25	0.20	0.25	2.8-4.0	0.05	0.50	0.005	0.005	...

(a) In determining copper min, copper may be calculated as Cu + Ni. (b) Cu + sum of named elements = 99.3% min. (c) For continuous castings, 1.5% P max. Source: *1988 Standards Handbook*, Copper Development Association Inc., p 5-19.

Cast copper-tin-lead alloys (high-leaded tin bronzes) (Composition, percent maximum unless shown as a range or minimum)

Copper No.	Cu (a)	Sn	Pb	Zn	Fe	Sb	Ni (a) (incl Co)	S	P (b)	Al	Si
C93100	rem (c)	6.5-8.5	2.0-5.0	2.0	0.25	0.25	1.0	0.05	0.30	0.005	0.005
C93200	81.0-85.0 (c)	6.3-7.5	6.0-8.0	1.0-4.0	0.20	0.35	1.0	0.08	0.15	0.005	0.005
C93400	82.0-85.0 (c)	7.0-9.0	7.0-9.0	0.8	0.20	0.50	1.0	0.08	0.50	0.005	0.005
C93500	83.0-86.0 (d)	4.3-6.0	8.0-10.0	2.0	0.20	0.30	1.0	0.08	0.05	0.005	0.005
C93600	79.0-83.0 (e)	6.0-8.0	11.0-13.0	1.0	0.20	0.55	1.0	0.08	0.15	0.005	0.005
C93700	78.0-82.0 (f)	9.0-11.0	8.0-11.0	0.8	0.15 (g)	0.55	1.0	0.08	0.15	0.005	0.005
C93720	83.0 min (f)	3.5-4.5	7.0-9.0	4.0	0.35	0.50	0.50	...	...	...	...
C93800	75.0-79.0 (h)	6.3-7.5	13.0-16.0	0.8	0.15	0.8	1.0	0.08	0.05	0.005	0.005
C93900	76.5-79.5 (h)	5.0-7.0	14.0-18.0	1.5	0.40	0.50	0.8	0.08	1.5	0.005	0.005
C94000	69.0-72.0 (j)	12.0-14.0	14.0-16.0	0.50	0.25	0.50	0.50-1.0	0.08 (k)	0.05	0.005	0.005
C94100	72.0-79.0 (j)	4.5-6.5	18.0-22.0	1.0	0.25	0.8	1.0	0.08 (k)	0.05	0.005	0.005
C94300	68.5-73.5 (f)	4.5-6.0	22.0-25.0	0.8	0.15	0.8	1.0	1.0 (k)	0.08	0.005	0.005
C94310	rem (f)	1.5-3.0	27.0-34.0	0.50	0.50	0.50	0.25-1.0	...	0.05	...	...
C94320	rem (f)	4.0-7.0	24.0-32.0	...	0.35	...	...	...	...	...	...
C94330	68.5-75.5 (f)	3.0-4.0	21.0-25.0	3.0	0.35	0.50	0.50	...	...	...	...
C94400	rem (f)	7.0-9.0	9.0-12.0	0.8	0.15	0.8	1.0	0.08	0.50	0.005	0.005
C94500	rem (f)	6.0-8.0	16.0-22.0	1.2	0.15	0.8	1.0	0.08	0.05	0.005	0.005

(a) In determining copper min, copper may be calculated as Cu + Ni. (b) For continuous castings, 1.5% P max. (c) Cu + sum of named elements = 99.2% min. (d) Cu + sum of named elements = 99.4% min. (e) Cu + sum of named elements = 99.3% min. (f) Cu + sum of named elements = 99.0% min. (g) Iron shall be 0.35% max when used for steel backed bearings. (h) Cu + sum of named elements = 98.9% min. (j) Cu + sum of named elements = 98.7% min. (k) For continuous castings, sulfur shall be 0.25% max. Source: *1988 Standards Handbook*, Copper Development Association Inc., p 5-19.

Cast copper-tin-nickel alloys (nickel-tin bronzes) (Composition, percent maximum unless shown as a range or minimum)

Copper No.	Cu	Sn	Pb	Zn	Fe	Sb	Ni (incl Co)	Mn	S	P	Al	Si
C94700	85.0-90.0(a)	4.5-6.0	0.10(b)	1.0-2.5	0.25	0.15	4.5-6.0	0.20	0.05	0.05	0.005	0.005
C94800	84.0-89.0(a)	4.5-6.0	0.30-1.0	1.0-2.5	0.25	0.15	4.5-6.0	0.20	0.05	0.05	0.005	0.005
C94900	79.0-81.0(c)	4.0-6.0	4.0-6.0	4.0-6.0	0.30	0.25	4.0-6.0	0.10	0.08	0.05	0.005	0.005

(a) Cu + sum of named elements = 99.3% min. (b) The mechanical properties of Copper Alloy No. C94700 (heat treated) may not be attained if the lead content exceeds 0.01%. (c) Cu + sum of named elements = 99.2% min. Source: *1988 Standards Handbook*, Copper Development Association Inc., p 5-19.

Cast copper-aluminum-iron and copper-aluminum-iron-nickel alloys (aluminum bronzes) (Composition, percent maximum unless shown as a range or minimum)

Copper No.	Cu	Pb	Fe	Ni (incl Co)	Al	Mn	Mg	Si	Zn	P	Sn
C95200	86.0 min(a)	...	2.5-4.0	...	8.5-9.5	...	...	...	...	...	...
C95210	86.0 min(a)	0.05	2.5-4.0	1.0	8.5-9.5	1.0	0.05	0.25	0.50	...	0.10
C95220	rem(b)	...	2.5-4.0	2.5	9.5-10.5	0.50	...	...	...	...	...
C95300	86.0 min(a)	...	0.8-1.5	...	9.0-11.0	...	...	...	...	...	...
C95400	83.0 min(b)	...	3.0-5.0	1.5	10.0-11.5	0.50	...	...	...	...	...
C95410	83.0 min(b)	...	3.0-5.0	1.5-2.5	10.0-11.5	0.50	...	...	...	...	...
C95420	83.5 min(b)	...	3.0-4.3	0.50	10.5-12.0	0.50	...	...	...	...	...
C95500	78.0 min(b)	...	3.0-5.0	3.0-5.5	10.0-11.5	3.5	...	...	...	...	...
C95510	78.0 min(c)	...	2.0-3.5	4.5-5.5	9.7-10.9	1.5	...	...	0.30	...	0.20
C95600	88.0 min(a)	...	...	0.25	6.0-8.0	...	...	1.8-3.3	...	...	...
C95700	71.0 min(b)	0.03	2.0-4.0	1.5-3.0	7.0-8.5	11.0-14.0	...	0.10	...	...	...
C95710	71.0 min(b)	0.05	2.0-4.0	1.5-3.0	7.0-8.5	11.0-14.0	...	0.15	0.50	0.05	1.0
C95800	79.0 min(b)	0.05	3.5-4.5(d)	4.0-5.0 (d)	8.5-9.5	0.8-1.5	...	0.10	...	...	0.10
C95810	79.0 min(b)	0.10	3.5-4.5(d)	4.0-5.0 (d)	8.5-9.5	0.8-1.5	0.05	0.10	0.50	...	...
C95900	rem(b)	...	3.0-5.0	0.50	12.0-13.5	1.5	...	...	...	...	...

(a) Cu + sum of named elements = 99.0% min. (b) Cu + sum of named elements = 99.5% min. (c) Cu + sum of named elements = 99.8% min. (d) Fe content shall not exceed Ni content. Source: *1988 Standards Handbook*, Copper Development Association Inc., p 5-19.

Cast copper-nickel-iron alloys (copper-nickels) (Composition, percent maximum unless shown as a range or minimum)

Copper No.	Cu (a)	Pb	Fe	Ni (incl Co)	Mn	Si	Nb	C	Be	Others
C96200	rem	0.01	1.0-1.8	9.0-11.0	1.5	0.50	0.50-1.0	0.10	...	0.02 S, 0.02 P
C96300	rem	0.01	0.50-1.5	18.0-22.0	0.25-1.5	0.50	0.50-1.5	0.15	...	0.02 S, 0.02 P
C96400	65.0-69.0	0.03(b)	0.25-1.5	28.0-32.0	1.5	0.50	0.50-1.5	0.15	...	...
C96600	rem	0.01	0.8-1.1	29.0-33.0	1.0	0.15	...	...	0.40-0.7	...
C96700	rem	0.01	0.7-1.0	29.0-33.0	0.7	0.15	...	...	1.1-1.2	0.10-0.20 Zr, 0.10-0.20 Ti
C96800	rem	0.005	0.50	9.5-10.5	0.05-0.30	0.05	0.10-0.30	...	...	(c)

(a) Cu + sum of named elements = 99.5% min. (b) For welding grades, Pb may not exceed 0.01%. (c) The following additional impurity limits shall apply: Al, 0.10 max; B, 0.001 max; Bi, 0.001 max; Mg, 0.005-0.15; P, 0.005 max; S, 0.0025 max; Sb, 0.02 max; Sn, 7.5-8.5; Ti, 0.01 max; Zn, 1.0 max. Source: *1988 Standards Handbook*, Copper Development Association Inc., p 5-19.

Cast copper-nickel-zinc alloys (nickel silvers) (Composition, percent maximum unless a range or minimum)

Copper No.	Cu	Sn	Pb	Zn	Fe	Sb	Ni (incl Co)	S	P	Al	Mn	Si
C97300	53.0-58.0(a)	1.5-3.0	8.0-11.0	17.0-25.0	1.5	0.35	11.0-14.0	0.08	0.05	0.005	0.50	0.15
C97400	58.0-61.0(a)	2.5-3.5	4.5-5.5	rem	1.5	...	15.5-17.0	...	...	...	0.50	...
C97600	63.0-67.0(b)	3.5-4.5	3.0-5.0	3.0-9.0	1.5	0.25	19.0-21.5	0.08	0.05	0.005	1.0	0.15
C97800	64.0-67.0(c)	4.0-5.5	1.0-2.5	1.0-4.0	1.5	0.20	24.0-27.0	0.08	0.05	0.005	1.0	0.15

(a) Cu + sum of named elements = 99.0% min. (b) Cu + sum of named elements = 99.7% min. (c) Cu + sum of named elements = 99.6% min. Source: *1988 Standards Handbook*, Copper Development Association Inc., p 5-19.

Cast copper-lead alloys (leaded coppers) (Composition, percent maximum unless shown as a range or minimum)

Copper No.	Cu	Sn	Pb	Ag	Zn	P	Fe
C98200	73.0-79.0	0.50(a)	21.0-27.0	...	...	...	0.35
C98400	67.0-74.0	0.25	25.0-32.0	1.5	0.10	0.02	0.35
C98600	60.0-70.0	0.50	30.0-40.0	1.5	...	...	0.35
C98800	56.5-62.5(b)	0.25	37.5-42.5(c)	5.5(c)	0.10	0.02	0.35
C98820	rem	1.0-5.0	40.0-44.0	...	...	...	0.35
C98840	rem	1.0-5.0	44.0-58.0	...	...	...	0.35

(a) Sn max may be raised to 1.3% by agreement. (b) Includes Ag. (c) Pb and Ag may be adjusted to modify the alloy hardness. Source: *1988 Standards Handbook*, Copper Development Association Inc., p 5-19.

Cast copper special alloys (Composition, percent maximum unless shown as a range or minimum)

Copper No.	Other designations	Cu (a)	Sn	Pb	Ni	Fe	Al	Co	Si	Mn	Others
C99300	Incramet 800	rem	0.05	0.02	13.5-16.5	0.40-1.0	10.7-11.5	1.0-2.0	0.02	...	...
C99400	...	rem	...	0.25	1.0-3.5	1.0-3.0	0.50-2.0	...	0.50-2.0	0.50	0.50-5.0 Zn
C99500	...	rem	...	0.25	3.5-5.5	3.0-5.0	0.50-2.0	...	0.50-2.0	0.50	0.50-2.0 Zn
C99600	Incramute 1	rem	0.10	0.02	0.20	0.20	1.0-2.8	0.20	0.10	39.0-45.0	0.20 Zn, 0.05 C
C99700	...	54.0 min	1.0	2.0	4.0-6.0	1.0	0.50-3.0	...	...	11.0-15.0	19.0-25.0 Zn
C99750	...	55.0-61.0	0.50-2.5	...	5.0	1.0	0.25-3.0	...	...	17.0-23.0	17.0-23.0 Zn

(a) Cu + sum of named elements = 99.7% min. Source: *1988 Standards Handbook*, Copper Development Association Inc., p 5-19.

Wrought Aluminum
chemical composition limits
(Only composition limits which are identical to those listed herein or are registered with The Aluminum Association should be designated as "AA" alloys.)

AA number	Registered date	Si	Fe	Cu	Mn	Mg	Cr	Ni	Zn	Ga	V		Ti	Others(c) Each	Others(c) Total	Aluminum min(d,e)
1030	...	0.35	0.6	0.10	0.05	0.05	...	...	0.10	...	0.05	...	0.03	0.03	...	99.30(e)
1035	8/22/78	0.35	0.6	0.10	0.05	0.05	...	...	0.10	...	0.05	...	0.03	0.03	...	99.35(e)
1040	8/22/73	0.30	0.50	0.10	0.05	0.05	...	...	0.10	...	0.05	...	0.03	0.03	...	99.40(e)
1045	8/22/73	0.30	0.45	0.10	0.05	0.05	...	...	0.05	...	0.05	...	0.03	0.03	...	99.45(e)
1050	...	0.25	0.40	0.05	0.05	0.05	...	...	0.05	...	0.05	...	0.03	0.03	...	99.50(e)
1055	8/22/73	0.25	0.40	0.05	0.05	0.05	...	...	0.05	...	0.05	...	0.03	0.03	...	99.55(e)
1060	...	0.25	0.35	0.05	0.03	0.03	...	...	0.05	...	0.05	...	0.03	0.03	...	99.60(e)
1065	8/22/73	0.25	0.30	0.05	0.03	0.03	...	...	0.05	...	0.05	...	0.03	0.03	...	99.65(e)
1070	...	0.20	0.25	0.04	0.03	0.03	...	...	0.04	...	0.05	...	0.03	0.03	...	99.70(e)
1075	...	0.20	0.20	0.04	0.03	0.03	...	...	0.04	...	0.05	...	0.03	0.03	...	99.75(e)
1080	...	0.15	0.15	0.03	0.02	0.02	...	...	0.03	0.03	0.05	...	0.03	0.02	...	99.80(e)
1085	...	0.10	0.12	0.03	0.02	0.02	...	...	0.03	0.03	0.05	...	0.02	0.01	...	99.85(e)
1090	...	0.07	0.07	0.02	0.01	0.01	...	...	0.03	0.03	0.05	...	0.01	0.01	...	99.90(e)
1095	...	0.030	0.040	0.010	0.010	0.010	...	...	0.010	...	...	...	0.005	0.005	...	99.95(d)

AA number	By	Date	Si	Fe	Cu	Mn	Mg	Cr	Ni	Zn	Ga	V		Ti	Others(c) Each	Others(c) Total	Aluminum min(d,e)
1100	...	...	1.0 Si + Fe	...	0.05-0.20	0.05	...	...	...	0.10	...	...	(t)	...	0.05	0.15	99.00(e)
1200	AATD	9/23/66	1.0 Si + Fe	...	0.05	0.05	...	...	...	0.10	...	...	...	0.05	0.05	0.15	99.00(e)
1230(j)	...	...	0.7 Si + Fe	...	0.10	0.05	0.05	...	...	0.10	...	0.05	...	0.03	0.03	...	99.30(e)
1135	Alcoa	4/23/57	0.65 Si + Fe	...	0.05-0.20	0.04	0.05	...	...	0.10	...	0.05	...	0.03	0.03	...	99.35(e)
1235(g)	...	...	0.65 Si + Fe	...	0.05	0.05	0.05	...	...	0.10	...	0.05	...	0.03	0.03	...	99.35(e)
1435	Kaiser	3/5/58	0.15	0.30-0.50	0.02	0.05	0.05	...	...	0.10	...	0.05	...	0.03	0.03	...	99.35(e)
1145(g)	...	...	0.55 Si + Fe	...	0.05	0.05	0.05	...	...	0.05	...	0.05	...	0.03	0.03	...	99.45(e)
1345(v)	Alcoa	10/8/56	0.30	0.40	0.10	0.05	0.05	...	...	0.05	...	0.05	...	0.03	0.03	...	99.45(e)
1250		1/3/56	0.20	0.40	0.10	0.01	0.01	0.01	...	0.05	...	...	(ff)	...	0.03	...	99.50(e)
1350(w)	AATD	1/24/75	0.10	0.04	0.05	0.01	...	0.01	...	0.05	0.03	...	0.05 B, (ff)	...	0.03	0.10	99.50(e)
1170	Alcoa	1/16/58	0.30 Si + Fe	...	0.03	0.03	0.02	0.03	...	0.04	...	0.05	...	0.03	0.03	...	99.70(e)
1175(h)	...	...	0.15 Si + Fe	...	0.10	0.02	0.02	...	...	0.04	0.03	0.05	...	0.02	0.02	...	99.75(e)
1180(i)	...	...	0.09	0.09	0.01	0.02	0.02	...	...	0.03	0.03	0.05	...	0.02	0.02	...	99.80(e)
1185	...	...	0.15 Si + Fe	...	0.01	0.02	0.02	...	...	0.03	0.03	0.05	...	0.02	0.01	...	99.85(e)
1285	...	...	0.08(k)	0.08(k)	0.02	0.01	0.01	...	...	0.03	0.03	0.05	...	0.02	0.01	...	99.85(e)
1188(i)	...	...	0.06	0.06	0.005	0.01	0.01	...	...	0.03	0.03	0.05	...	0.01	0.01	...	99.88(e)
1199(i)	Alcoa	3/12/56	0.006	0.006	0.006	0.002	0.006	...	...	0.006	0.005	0.005	...	0.002	0.002	...	99.99(d)
2011	...	...	0.40	0.7	5.0-6.0	...	...	...	...	0.30	...	...	(l)	...	0.05	0.15	Remainder
2014	...	...	0.50-1.2	0.7	3.9-5.0	0.40-1.2	0.20-0.8	0.10	...	0.25	...	...	0.20 Zr + Ti	0.15	0.05	0.15	Remainder
2214	...	...	0.50-1.2	0.30	3.9-5.0	0.40-1.2	0.20-0.8	0.10	...	0.25	...	...	0.20 Zr + Ti	0.15	0.05	0.15	Remainder
2017	...	...	0.20-0.8	0.7	3.5-4.5	0.40-1.0	0.40-0.8	0.10	...	0.25	...	...	0.20 Zr + Ti	0.15	0.05	0.15	Remainder
2117	...	...	0.8	0.7	2.2-3.0	0.20	0.20-0.50	0.10	...	0.25	...	...	...	...	0.05	0.15	Remainder
2018	...	...	0.9	1.0	3.5-4.5	0.20	0.45-0.9	0.10	1.7-2.3	0.25	...	...	...	...	0.05	0.15	Remainder
2218	...	...	0.9	1.0	3.5-4.5	0.20	1.2-1.8	0.10	1.7-2.3	0.25	...	...	...	...	0.05	0.15	Remainder
2618	...	...	0.10-0.25	0.9-1.3	1.9-2.7	...	1.3-1.8	...	0.9-1.2	0.10	...	...	...	0.04-0.10	0.05	0.15	Remainder
2219	Alcoa	8/13/54	0.20	0.30	5.8-6.8	0.20-0.40	0.02	...	...	0.10	...	0.05-0.15	(ll)	0.02-0.10	0.05	0.15	Remainder
2319(u)	Alcoa	6/5/58	0.20	0.30	5.8-6.8	0.20-0.40	0.02	...	...	0.10	...	0.05-0.15	(ll, t)	0.10-0.20	0.05	0.15	Remainder
2419	Kaiser	10/12/72	0.15	0.18	5.8-6.8	0.20-0.40	0.02	...	...	0.10	...	0.05-0.15	(ll)	0.02-0.10	0.05	0.15	Remainder
2024	...	...	0.50	0.50	3.8-4.9	0.30-0.9	1.2-1.8	0.10	...	0.25	...	...	0.20 Zr + Ti	0.15	0.05	0.15	Remainder
2124	AATD	10/2/70	0.20	0.30	3.8-4.9	0.30-0.9	1.2-1.8	0.10	...	0.25	...	...	0.20 Zr + Ti	0.15	0.05	0.15	Remainder
2025	...	...	0.50-1.2	1.0	3.9-5.0	0.40-1.2	0.05	0.10	...	0.25	...	...	...	0.15	0.05	0.15	Remainder
2036	Reynolds	8/13/70	0.50	0.50	2.2-3.0	0.10-0.40	0.30-0.6	0.10	...	0.25	...	...	...	0.15	0.05	0.15	Remainder
2048	Reynolds	8/2/72	0.15	0.20	2.8-3.8	0.20-0.6	1.2-1.8	...	...	0.25	...	...	...	0.10	0.05	0.15	Remainder
3002	Alcoa	7/3/61	0.08	0.10	0.15	0.10-0.25	0.05-0.20	...	...	0.05	...	0.05	...	0.03	0.03	0.10	Remainder
3102	Alcoa	3/1/72	0.40	0.7	0.10	0.05-0.40	...	...	...	0.30	...	...	...	0.10	0.05	0.15	Remainder
3003	...	...	0.6	0.7	0.05-0.20	1.0-1.5	...	...	...	0.10	...	...	...	...	0.05	0.15	Remainder
3303	Alcoa	5/31/74	0.6	0.7	0.05-0.20	1.0-1.5	...	...	...	0.30	...	...	...	...	0.05	0.15	Remainder
3004	...	...	0.30	0.7	0.25	1.0-1.5	0.8-1.3	...	...	0.25	...	...	...	...	0.05	0.15	Remainder
3005	...	...	0.6	0.7	0.30	1.0-1.5	0.20-0.6	0.10	...	0.25	...	...	...	0.10	0.05	0.15	Remainder
3105	Alcoa	5/27/60	0.6	0.7	0.30	0.30-0.8	0.20-0.8	0.20	...	0.40	...	...	...	0.10	0.05	0.15	Remainder
3006	Alumax	8/24/73	0.50	0.7	0.10-0.30	0.50-0.8	0.30-0.6	0.20	...	0.15-0.40	...	...	...	0.10	0.05	0.15	Remainder
3007	Nat. Alum.	3/17/76	0.50	0.7	0.10-0.30	0.30-0.8	0.6	0.20	...	0.40	...	...	...	0.10	0.05	0.15	Remainder
4002	AATD	6/11/68	3.5-4.5	0.35	0.05-0.15	0.03	0.05-0.15	...	...	0.15	...	...	0.8-1.4 Cd	0.02	0.05	0.15	Remainder
4004(hh)	Reynolds	10/5/71	9.0-10.5	0.8	0.25	0.10	1.0-2.0	...	...	0.20	...	...	...	...	0.05	0.15	Remainder
X4104	Reynolds	2/26/74	9.0-10.5	0.8	0.25	0.10	1.0-2.0	...	...	0.20	...	...	0.02-0.20 Bi	...	0.05	0.15	Remainder
X4005(f)	Reynolds	10/5/71	9.5-11.0	0.8	0.25	0.10	0.20-1.0	...	...	0.20	...	...	...	...	0.05	0.15	Remainder
4032	...	...	11.0-13.5	1.0	0.50-1.3	...	0.8-1.3	0.10	0.50-1.3	0.25	...	...	...	...	0.05	0.15	Remainder
4043(u)	...	...	4.5-6.0	0.8	0.30	0.05	0.05	...	...	0.10	...	...	(t)	0.20	0.05	0.15	Remainder
4343(o)	...	...	6.8-8.2	0.8	0.25	0.10	...	...	...	0.20	...	...	...	...	0.05	0.15	Remainder
4543	...	...	5.0-70	0.50	0.10	0.05	0.10-0.40	0.05	...	0.10	...	...	...	0.10	0.05	0.15	Remainder
4643(u)	Alcoa	8/14/63	3.6-4.6	0.8	0.10	0.05	0.10-0.30	...	...	0.10	...	...	(t)	0.15	0.05	0.15	Remainder
4044(bb)	Reynolds	7/15/69	7.8-9.2	0.8	0.25	0.10	...	...	...	0.20	...	...	...	...	0.05	0.15	Remainder
4545(o)	...	...	9.0-11.0	0.8	0.30	0.05	0.05	...	...	0.10	...	...	...	0.20	0.05	0.15	Remainder
4045	...	...	10.0	...	...	...	...	...	...	...	...	...	...	...	...	...	Remainder

Registered Compositions (continued)

AA number	By	Date	Si	Fe	Cu	Mn	Mg	Cr	Ni	Zn	Ga	V		Ti	Each	Total	Aluminum min (d,e)
4145(o,u)	Alcoa	4/30/57	9.3-10.7	0.8	3.3-4.7	0.15	0.15	0.15	...	0.20	...	...	(t)	...	0.05	0.15	Remainder
4047(o,u)	...	...	11.0-13.0	0.8	0.30	0.15	0.10	...	...	0.20	...	...	(t)	...	0.05	0.15	Remainder
5005	...	...	0.30	0.7	0.20	0.20	0.50-1.1	0.10	...	0.25	...	...	...	...	0.05	0.15	Remainder
5205	AATD	5/29/67	0.15	0.7	0.03-0.10	0.10	0.6-1.0	0.10	...	0.05	...	...	...	...	0.05	0.15	Remainder
5010	Alumax	10/3/61	0.40	0.7	0.25	0.10-0.30	0.20-0.6	0.15	...	0.30	...	...	...	0.10	0.05	0.15	Remainder
X5020	Alcoa	7/12/72	0.30	0.7	1.3-1.7	0.10-0.50	2.4-3.2	0.20	...	0.20	...	...	...	0.10	0.05	0.15	Remainder
5040	Alcoa	2/24/61	0.30	0.7	0.25	0.9-1.4	1.0-1.5	0.10-0.30	...	0.25	...	...	...	...	0.05	0.15	Remainder
5050	...	...	0.40	0.7	0.20	0.10	1.1-1.8	0.10	...	0.25	...	...	...	...	0.05	0.15	Remainder
5051	Alumax	3/1/67	0.40	0.7	0.25	0.20	1.7-2.2	0.10	...	0.25	...	...	...	0.10	0.05	0.15	Remainder
5151	Alcoa	9/25/70	0.20	0.35	0.15	0.10	1.5-2.1	0.10	...	0.15	...	...	...	0.10	0.05	0.15	Remainder
5052	...	...	0.25	0.40	0.10	0.10	2.2-2.8	0.15-0.35	...	0.10	...	...	...	...	0.05	0.15	Remainder
5252	Kaiser	2/24/61	0.08	0.10	0.10	0.10	2.2-2.8	...	...	0.05	...	0.05	...	...	0.03	0.10	Remainder
5352	Alcoa	9/23/71	0.45 Si + Fe	...	0.10	0.10	2.2-2.8	0.10	...	0.10	...	...	...	0.10	0.05	0.15	Remainder
5652	...	...	0.40 Si + Fe	...	0.04	0.01	2.2-2.8	0.15-0.35	...	0.10	...	...	...	...	0.05	0.15	Remainder
5154(n)	...	...	0.25	0.40	0.10	0.10	3.1-3.9	0.15-0.35	...	0.20	...	...	...	0.20	0.05	0.15	Remainder
5254	...	...	0.45 Si + Fe	...	0.05	0.01	3.1-3.9	0.15-0.35	...	0.20	...	...	...	0.20	0.05	0.15	Remainder
5454	Alcoa	7/8/57	0.25	0.40	0.10	0.50-1.0	2.4-3.0	0.05-0.20	...	0.25	...	...	...	...	0.05	0.15	Remainder
5554(u)	Alcoa	3/5/58	0.25	0.40	0.10	0.50-1.0	2.4-3.0	0.05-0.20	...	0.25	...	...	(t)	0.05-0.20	0.05	0.15	Remainder
5654(u)	AATD	5/29/68	0.45 Si + Fe	...	0.05	0.01	3.1-3.9	0.15-0.35	...	0.20	...	...	(t)	0.05-0.15	0.05	0.15	Remainder
5056	...	...	0.30	0.40	0.10	0.05-0.20	4.5-5.6	0.05-0.20	...	0.10	...	...	...	...	0.05	0.15	Remainder
5356(u)	...	...	0.25	0.40	0.10	0.05-0.20	4.5-5.5	0.05-0.20	...	0.10	...	...	(t)	0.06-0.20	0.05	0.15	Remainder
5456	Alcoa	10/4/56	0.25	0.40	0.10	0.50-1.0	4.7-5.5	0.05-0.20	...	0.25	...	...	...	0.20	0.05	0.15	Remainder
5556(u)	Alcoa	10/9/56	0.25	0.40	0.10	0.50-1.0	4.7-5.5	0.05-0.20	...	0.25	...	...	(t)	0.05-0.20	0.05	0.15	Remainder
5357	...	...	0.12	0.17	0.20	0.15-0.45	0.8-1.2	...	...	0.05	...	...	...	...	0.05	0.15	Remainder
5457	...	12/24/57	0.08	0.10	0.20	0.15 0.45	0.8-1.2	...	...	0.05	...	0.05	...	...	0.03	0.10	Remainder
5657	Reynolds	2/26/60	0.08	0.10	0.10	0.03	0.6-1.0	...	...	0.05	0.03	0.05	...	...	0.02	0.05	Remainder
5082	Alcoa	8/14/63	0.20	0.35	0.15	0.15	4.0-5.0	0.15	...	0.25	...	...	...	0.10	0.05	0.15	Remainder
5182	Alcoa	11/10/67	0.20	0.35	0.15	0.20-0.50	4.0-5.0	0.10	...	0.25	...	...	...	0.10	0.05	0.15	Remainder
5083	...	...	0.40	0.40	0.10	0.40-1.0	4.0-4.9	0.05-0.25	...	0.25	...	...	...	0.15	0.05	0.15	Remainder
5183(u)	Kaiser	6/7/57	0.40	0.40	0.10	0.50-1.0	4.3-5.2	0.05-0.25	...	0.25	...	...	(t)	0.15	0.05	0.15	Remainder
X5085	Alcoa	7/12/72	0.20	0.40	0.15	0.20	5.8-6.8	0.20	...	0.20	...	...	...	0.10	0.05	0.15	Remainder
5086	...	...	0.40	0.50	0.10	0.20-0.7	3.5-4.5	0.05-0.25	...	0.25	...	...	...	0.15	0.05	0.15	Remainder
X5087(u)	Alcoa	7/10/74	0.20	0.35	0.10	0.50-1.0	4.7-5.5	0.05-0.20	...	0.25	...	...	(t, jj)	0.05-0.20	0.05	0.15	Remainder
X5090	Conalco	3/19/70	0.50	0.50	0.25	0.35	6.0-8.0	0.05-0.30	...	0.20	...	...	(gg)	0.02	0.05	0.15	Remainder
6101(p)	Reynolds, Revere	7/8/55	0.30-0.7	0.50	0.10	0.03	0.35-0.8	0.03	...	0.10	...	...	0.06 B	...	0.03	0.10	Remainder
6201(z)	Kaiser	9/7/60	0.50-0.9	0.50	0.10	0.03	0.6-0.9	0.03	...	0.10	...	...	0.06 B	...	0.03	0.10	Remainder
6301	Kaiser	4/27/70	0.50-0.9	0.7	0.10	0.15	0.6-0.9	0.10	...	0.25	...	...	...	0.15	0.05	0.15	Remainder
6003(q)	...	...	0.35-1.0	0.6	0.10	0.8	0.8-1.5	0.35	...	0.20	...	...	...	0.10	0.05	0.15	Remainder
6004	Kaiser	2/9/73	0.30-0.6	0.10-0.30	0.10	0.20-0.6	0.40-0.7	...	...	0.05	...	...	...	...	0.05	0.15	Remainder
6005	AATD	12/20/62	0.6-0.9	0.35	0.10	0.10	0.40-0.6	0.10	...	0.10	...	...	...	0.10	0.05	0.15	Remainder
6105	Revere	11/23/65	0.6-1.0	0.35	0.10	0.10	0.45-0.8	0.10	...	0.10	...	...	...	0.10	0.05	0.15	Remainder
6205	Conalco	3/10/70	0.6-0.9	0.7	0.20	0.05-0.15	0.40-0.6	0.05-0.15	...	0.25	...	...	0.05-0.15 Zr	0.15	0.05	0.15	Remainder
6006	Kaiser	10/20/71	0.20-0.6	0.35	0.15-0.30	0.05-0.20	0.45-0.9	0.10	...	0.10	...	...	...	0.10	0.05	0.15	Remainder
6007	Conalco	4/4/75	0.9-1.4	0.7	0.20	0.05-0.25	0.6-0.9	0.05-0.25	...	0.25	...	...	0.05-0.20 Zr	0.15	0.05	0.15	Remainder
6011	...	...	0.6-1.2	1.0	0.40-0.9	0.8	0.6-1.2	0.30	0.20	1.5	...	...	...	0.20	0.05	0.15	Remainder
6151	...	...	0.6-1.2	1.0	0.35	0.20	0.45-0.8	0.15-0.35	...	0.25	...	...	...	0.15	0.05	0.15	Remainder
6351	Kaiser	12/16/58	0.7-1.3	0.50	0.10	0.40-0.8	0.40-0.8	...	...	0.20	...	...	...	0.20	0.05	0.15	Remainder
6951	...	...	0.20-0.50	0.8	0.15-0.40	0.10	0.40-0.8	...	...	0.20	...	...	...	...	0.05	0.15	Remainder
6053	...	...	(r)	0.35	0.10	...	1.1-1.4	0.15-0.35	...	0.10	...	...	...	...	0.05	0.15	Remainder
6253(x)	...	...	(r)	0.50	0.10	...	1.0-1.5	0.15-0.35	...	1.6-2.4	...	...	...	...	0.05	0.15	Remainder
6061	...	...	0.40-0.8	0.7	0.15-0.40	0.15	0.8-1.2	0.04-0.35	...	0.25	...	...	...	0.15	0.05	0.15	Remainder
6261	Alcan	4/23/68	0.40-0.7	0.40	0.15-0.40	0.20-0.35	0.7-1.0	0.10	...	0.20	...	...	...	0.10	0.05	0.15	Remainder
6162	Reynolds	3/26/59	0.40-0.8	0.50	0.20	0.10	0.7-1.1	0.10	...	0.25	...	...	...	0.10	0.05	0.15	Remainder
6262	Alcoa	1/14/60	0.40-0.8	0.7	0.15-0.40	0.15	0.8-1.2	0.04-0.14	...	0.25	...	...	(aa)	0.15	0.05	0.15	Remainder
6063	...	...	0.20-0.6	0.35	0.10	0.10	0.45-0.9	0.10	...	0.10	...	...	...	0.10	0.05	0.15	Remainder
6463	Alcoa	4/15/57	0.20-0.6	0.15	0.20	0.05	0.45-0.9	...	...	0.05	...	...	...	...	0.05	0.15	Remainder
6763	National	12/4/72	0.20-0.6	0.08	0.04-0.16	0.03	0.45-0.9	...	...	0.03	...	0.05	...	...	0.03	0.10	Remainder
6066	...	...	0.9-1.8	0.50	0.7-1.2	0.6-1.1	0.8-1.4	0.40	...	0.25	...	...	...	0.20	0.05	0.15	Remainder
6070	Alcoa	1/18/62	1.0-1.7	0.50	0.15-0.40	0.40-1.0	0.50-1.2	0.10	...	0.25	...	...	...	0.15	0.05	0.15	Remainder
7001	...	...	0.35	0.40	1.6-2.6	0.20	2.6-3.4	0.18-0.35	...	6.8-8.0	...	...	...	0.20	0.05	0.15	Remainder
7004	Alcan	3/19/64	0.25	0.35	0.05	0.20-0.7	1.0-2.0	0.05	...	3.8-4.6	...	...	0.10-0.20 Zr	0.05	0.05	0.15	Remainder
7104	Alcan	3/19/64	0.25	0.40	0.03	...	0.50-0.9	...	...	3.6-4.4	...	...	...	0.05	0.05	0.15	Remainder
7005(dd)	Alcoa	8/13/62	0.35	0.40	0.10	0.20-0.7	1.0-1.8	0.06-0.20	...	4.0-5.0	...	...	0.08-0.20 Zr	0.01-0.06	0.05	0.10	Remainder
7008(ee)	Alcoa	11/15/68	0.10	0.10	0.05	0.05	0.7-1.4	0.12-0.25	...	4.5-5.5	...	...	...	0.05	0.05	0.10	Remainder
7011(ee)	Reynolds	12/2/68	0.15	0.20	0.05	0.10-0.30	1.0-1.6	0.05-0.20	...	4.0-5.5	...	...	...	...	0.05	0.15	Remainder
7013(f)	Alcoa	1/29/76	0.6	0.7	0.10	1.0-1.5	...	...	...	1.5-2.0	...	...	...	...	0.05	0.15	Remainder
X7016	Reynolds	6/29/72	0.10	0.12	0.45-1.0	0.03	0.8-1.4	...	...	4.0-5.0	...	0.05	...	0.03	0.03	0.10	Remainder
X7116	Reynolds	6/12/75	0.15	0.30	0.50-1.1	0.05	0.8-1.4	...	...	4.2-5.2	0.03	0.05	...	0.05	0.03	0.10	Remainder
X7029	Reynolds	12/8/75	0.10	0.12	0.50-0.9	0.03	1.3-2.0	...	...	4.2-5.2	...	0.05	...	0.03	0.03	0.10	Remainder
7039	Kaiser	7/16/62	0.30	0.40	0.10	0.10-0.40	2.3-3.3	0.15-0.25	...	3.5-4.5	...	...	...	0.10	0.05	0.15	Remainder
X7046	Alcoa	5/16/73	0.20	0.40	0.25	0.05-0.30	1.0-1.6	0.06-0.20	...	6.6-7.6	...	...	0.06-0.18 Zr	0.06	0.05	0.15	Remainder
7049	Kaiser	5/10/68	0.25	0.35	1.2-1.9	0.20	2.0-2.9	0.10-0.22	...	7.2-8.2	...	...	...	0.10	0.05	0.15	Remainder
7149	Kaiser	10/28/75	0.15	0.20	1.2-1.9	0.20	2.0-2.9	0.10-0.22	...	7.2-8.2	...	...	...	0.10	0.05	0.15	Remainder
7050	Alcoa	2/1/71	0.12	0.15	2.0-2.6	0.10	1.9-2.6	0.04	...	5.7-6.7	...	...	0.08-0.15 Zr	0.06	0.05	0.15	Remainder
7070(f)	Alcan	12/20/72	0.15	0.25	0.05	...	...	...	...	1.3-1.8	...	...	...	...	0.05	0.15	Remainder
7072(y)	...	...	0.7 Si + Fe	...	0.10	0.10	0.10	...	...	0.8-1.3	...	...	...	...	0.05	0.15	Remainder
7472	Alcoa	12/19/60	0.25	0.6	0.05	0.05	0.9-1.5	...	...	1.3-1.9	...	...	...	...	0.05	0.15	Remainder
7075	...	...	0.40	0.50	1.2-2.0	0.30	2.1-2.9	0.18-0.28	...	5.1-6.1	...	...	0.25 Zr + Ti	0.20	0.05	0.15	Remainder
7175	Alcoa	11/8/57	0.15	0.20	1.2-2.0	0.10	2.1-2.9	0.18-0.28	...	5.1-6.1	...	...	...	0.10	0.05	0.15	Remainder
7108...	...	...	...	...	...	...	1.0	...	...	5.0	...	...	0.18Zr	...	...	...	Remainder

Registered Compositions (continued)

AA number	Registered By	Registered Date	Si	Fe	Cu	Mn	Mg	Cr	Ni	Zn	Ga	V		Ti	Others(c) Each	Others(c) Total	Aluminum min (d,e)
7475	Alcoa	9/15/69	0.10	0.12	1.2-1.9	0.06	1.9-2.6	0.18-0.25	...	5.2-6.2	...	...	...	0.06	0.05	0.15	Remainder
7076	...	...	0.40	0.6	0.30-1.0	0.30-0.8	1.2-2.0	...	...	7.0-8.0	...	...	...	0.20	0.05	0.15	Remainder
7277(n)	...	...	0.50	0.7	0.8-1.7	...	1.7-2.3	0.18-0.35	...	3.7-4.3	...	...	...	0.10	0.05	0.15	Remainder
7178(n)	...	...	0.40	0.50	1.6-2.4	0.30	2.4-3.1	0.18-0.35	...	6.3-7.3	...	...	...	0.20	0.05	0.15	Remainder
7079	...	...	0.30	0.40	0.40-0.8	0.10-0.30	2.9-3.7	0.10-0.25	...	3.8-4.8	...	...	...	0.10	0.05	0.15	Remainder
7179	Alcoa	11/8/57	0.15	0.20	0.40-0.8	0.10-0.30	2.9-3.7	0.10-0.25	...	3.8-4.8	...	...	...	0.10	0.05	0.15	Remainder
8001	Alcoa	9/5/57	0.17	0.45-0.7	0.15	...	...	...	0.9-1.3	0.05	...	...	(s)	...	0.05	0.15	Remainder
8112(n)	...	...	1.0	1.0	0.40	0.6	0.7	0.20	...	1.0	...	...	...	0.20	0.05	0.15	Remainder
8020	Conalco	6/13/73	0.10	0.10	0.005	0.005	...	...	...	0.005	...	0.05	(kk)	...	0.03	0.10	Remainder
X8030	A.E.I.	9/29/75	0.10	0.30-0.8	0.15-0.30	...	0.05	...	...	0.05	...	...	0.001-0.04 B	...	0.03	0.10	Remainder
X8130	Reynolds	3/31/76	0.15(mm)	0.40-1.0(mm)	0.05-0.15	...	...	...	...	0.10	...	...	...	...	0.03	0.10	Remainder
8040	Conalco	11/15/62	1.0 Si + Fe	...	0.20	0.05	...	...	...	0.20	...	...	0.10-0.30 Zr	...	0.05	0.15	Remainder
8076	Alcoa	7/24/72	0.10	0.6-0.9	0.04	...	0.08-0.22	...	...	0.05	...	...	0.04 B	...	0.03	0.10	Remainder
X8176	Southwire	1/21/76	0.03-0.15	0.40-1.0	...	...	...	...	...	0.10	0.03	...	...	...	0.05	0.15	Remainder
X8077	Alcan	5/20/75	0.10	0.10-0.40	0.05	...	0.10-0.30	...	...	0.05	...	...	(m)	...	0.03	0.10	Remainder
8079(g)	Reynolds	1/9/69	0.05-0.30	0.7-1.3	0.05	...	...	...	...	0.10	...	...	...	...	0.05	0.15	Remainder
8280(n)	...	...	1.0-2.0	0.7	0.7-1.3	0.10	...	...	0.20-0.7	0.05	...	...	5.5-7.0 Sn	0.10	0.05	0.15	Remainder
8081	Alcoa	2/8/65	0.7	0.7	0.7-1.3	0.10	...	...	...	0.05	...	...	18.0-22.0 Sn	0.10	0.05	0.15	Remainder
8017	...	...	0.7	0.15	...	0.03	...	...	...	...	...		...	...	...	...	Remainder
8177	...	...	0.35	...	...	0.06	...	...	...	...	...		...	...	...	...	Remainder

(a) Composition in percent maximum unless shown as a range or a minimum. Standard limits for alloying elements and impurities are expressed to the following places: less than 1/1000 percent, 0.000X; 1/1000 to 1/100 percent, 0.00X; 1/100 to 1/10 percent unalloyed aluminum made by a refining process, 0.0XX; 1/100 to 1/10 percent alloys and unalloyed aluminum not made by a refining process, 0.0X; 1/10 through 1/2 percent, 0.XX; over 1/2 percent 0.X, X.X, etc. (b) For purposes of determining conformance to these limits, an observed value or a calculated value obtained from analysis is rounded off to the nearest unit in the last right-hand place of figures used in expressing the specified limit, in accordance with the following: American National Standard Rules for Rounding Off Numerical Values (ANSI Z25.1). When the figure next beyond the last figure or place to be retained is less than 5, the figure in the last place retained should be kept unchanged. When the figure next beyond the last figure or place to be retained is greater than 5, the figure in the last place retained should be increased by 1. When the figure next beyond the last figure or place to be retained is 5 and (1) there are no figures, or only zeroes, beyond this 5, if the figure in the last place to be retained is odd, it should be increased by 1; if even, it should be kept unchanged; (2) if the 5 next beyond the figure in the last place to be retained is followed by any figures other than zero, the figure in the last place retained should be increased by 1, whether odd or even. (c) Analysis is regularly made only for the elements for which specific limits are shown, except for unalloyed aluminum. If, however, the presence of other elements is suspected to be, or in the course of routine analysis is indicated to be in excess of the specified limits, further analysis is made to determine that these other elements are not in excess of the amount specified. (d) The aluminum content for unalloyed aluminum made by a refining process is the difference between 100.00 percent and the sum of all other metallic elements present in amounts of 0.0010 percent or more each, expressed to the third decimal before determining the sum, which is rounded to the second decimal before subtracting. (e) The aluminum content for unalloyed aluminum not made by a refining process is the difference between 100.00 percent and the sum of all other metallic elements present in amounts of 0.010 percent or more each, expressed to the second decimal before determining the sum. (f) Cladding alloy. (g) Foil. (h) Cladding on clad 1100 and clad 3003 reflector sheet. (i) Capacitor alloy. (j) Cladding on alclad 2024. (k) Silicon plus iron, 0.14 max. (l) Lead, bismuth 0.20-0.6 each. (m) Boron 0.05 max.; zirconium 0.02-0.08. (n) Consider as original alloy. (o) Brazing alloy. (p) Bus conductor. (q) Cladding on alclad 2014. (r) Silicon 45-65 percent of magnesium, (s) Boron, cobalt 0.001 max. each; cadmium 0.003 max.; lithium 0.008 max. (t) Beryllium 0.0008 max. for welding electrode only. (u) Welding electrode. (v) Mechanical wire. (w) Formerly designated EC. (x) Cladding on alclad 5056. (y) Cladding on alclad 2219, alclad 3003, alclad 3004, alclad 5050, alclad 5155, alclad 6061, alclad 7075, alclad 7475 and alclad 7178. (z) Conductor alloy. (aa) Lead, bismuth 0.40-0.7 each. (bb) Cladding on brazing sheet. (cc) Cadmium 0.05-0.20; tin 0.03-0.08. (dd) Extruded products. (ee) High strength cladding for sheet and plate products. (ff) Vanadium plus titanium, 0.02 max. (gg) Beryllium 0.001-0.02; boron 0.001-0.05. (hh) Cladding on X7, X8, X13 and X14 brazing sheet. (ii) These designations may be used for registration of new compositions after all unregistered numbers in the same series are used. Thereafter, the appropriate number with the oldest cancellation date shall be the next number re-used. (jj) Cobalt 0.30-0.7. (kk) Bismuth 0.10-0.50; tin 0.10-0.25. (ll) Zirconium 0.10-0.25. (mm) Silicon plus iron, 1.0 max.

Chemical Composition Limits (a, b)

(Only composition limits which are identical to those listed herein or are registered with The Aluminum Association should be designated as "AA" alloys.)

Registered alloys in the form of XXX.0 castings, XXX.1 ingot and XXX.2 ingot

AA number	Former designation	Registered By	Registered Date	Product(c)	Si	Fe	Cu	Mn	Mg	Cr	Ni	Zn	Sn	Ti	Others(d) Each	Others(d) Total	Al min (e)
100.1(f)	...	AATL(g)	6/30/70	Ingot	0.15	0.6-0.8	0.10	(h)	...	(h)	...	0.05	...	(h)	0.03(h)	0.10	99.00
130.1(f)	...	AATD	6/30/70	Ingot	(i)	(i)	0.10	(h)	...	(h)	...	0.05	...	(h)	0.03(h)	0.10	99.30
150.1(f)	...	AATD	6/30/70	Ingot	(j)	(j)	0.05	(h)	...	(h)	...	0.05	...	(h)	0.03(h)	0.10	99.50
160.1	...	Alcoa	1/28/76	Ingot	0.10(j)	0.25(j)	...	(h)	...	(h)	...	0.05	...	(h)	0.03(h)	0.10	99.60
170.1(f)	...	AATD	6/30/70	Ingot	(k)	(k)	...	(h)	...	(h)	...	0.05	...	(h)	0.03(h)	0.10	99.70
201.0	...	Conalco	4/17/68	S	0.10	0.15	4.0-5.2	0.20-0.50	0.15-0.55	...	...	...	...	0.15-0.35	0.05(l)	0.10	Remainder
201.2	...	Conalco	4/17/68	Ingot	0.10	0.10	4.0-5.2	0.20-0.50	0.20-0.55	...	...	...	...	0.15-0.35	0.05(l)	0.10	Remainder
A201.0	...	Conalco	10/9/70	S	0.05	0.10	4.0-5.0	0.20-0.40	0.15-0.35	...	...	...	...	0.15-0.35	0.03(l)	0.10	Remainder
A201.1	A201.2	Conalco	10/9/70	Ingot	0.05	0.07	4.0-5.0	0.20-0.40	0.20-0.35	...	...	...	...	0.15-0.35	0.03(l)	0.10	Remainder
B201.0	...	...	9/21/84	S	0.05	...	4.5-5.0	0.20-0.50	...	...	...	...	...	...	0.05(30)	0.15	Remainder
203.0	Hiduminium 350	M&A Co.	12/2/72	S	0.30	0.50	4.5-5.5	0.20-0.30	0.10	...	1.3-1.7	0.10	...	0.15-0.25(m)	0.05(n)	0.20	Remainder
203.2	Hiduminium 350	M&A Co.	12/2/72	Ingot	0.20	0.35	4.8-5.2	0.20-0.30	0.10	...	1.3-1.7	0.10	...	0.15-0.25(m)	0.05(n)	0.20	Remainder
204.0	A-U5GT	Howmet	10/1/74	S&P	0.20	0.35	4.2-5.0	0.10	0.15-0.35	...	0.05	0.10	0.05	0.15-0.30	0.05	0.15	Remainder
204.2	A-U5GT	Howmet	10/1/74	Ingot	0.15	0.10-0.20	4.2-4.9	0.05	0.20-0.35	...	0.03	0.05	0.05	0.15-0.25	0.05	0.15	Remainder
206.0	...	Trialco	4/23/76	S&P	0.10	0.15	4.2-5.0	0.20-0.50	0.15-0.35	...	0.05	0.10	0.05	0.15-0.30	0.05	0.15	Remainder
206.2	...	Trialco	4/23/76	Ingot	0.10	0.10	4.2-5.0	0.20-0.50	0.20-0.35	...	0.03	0.05	0.05	0.15-0.25	0.05	0.15	...
A206.0	...	Trialco	4/23/76	S&P	0.05	0.10	4.2-5.0	0.20-0.50	0.15-0.35	...	0.05	0.10	0.05	0.15-0.30	0.05	0.15	Remainder
A206.2	...	Trialco	4/23/76	Ingot	0.05	0.07	4.2-5.0	0.20-0.50	0.20-0.35	...	0.03	0.05	0.05	0.15-0.25	0.05	0.15	Remainder
208.0	108	AATD	...	S	2.5-3.5	1.2	3.5-4.5	0.50	0.10	...	0.35	1.0	...	0.25	...	0.50	Remainder
208.1	108	AATD	...	Ingot	2.5-3.5	0.9	3.5-4.5	0.50	0.10	...	0.35	1.0	...	0.25	...	0.50	Remainder
208.2	108	AATD	...	Ingot	2.5-3.5	0.8	3.5-4.5	0.30	0.03	...	...	0.20	...	0.20	...	0.30	Remainder
213.0	C113	AATD	...	S&P	1.0-3.0	1.2	6.0-8.0	0.6	0.10	...	0.35	2.5	...	0.25	...	0.50	Remainder
213.1	C113	AATD	...	Ingot	1.0-3.0	0.9	6.0-8.0	0.6	0.10	...	0.35	2.5	...	0.25	...	0.50	Remainder
222.0	122	AATD	...	S&P	2.0	1.5	9.2-10.7	0.50	0.15-0.35	...	0.50	0.8	...	0.25	...	0.35	Remainder
222.1	122	AATD	...	Ingot	2.0	1.2	9.2-10.7	0.50	0.20-0.35	...	0.50	0.8	...	0.25	...	0.35	Remainder
224.0	...	Alcoa	4/2/69	S&P	0.06	0.10	4.5-5.5	0.20-0.50	...	...	...	...	...	0.35	0.03(o)	0.10	Remainder
224.2	...	Alcoa	4/2/69	Ingot	0.02	0.04	4.5-5.5	0.20-0.50	...	...	...	...	...	0.25	0.03(o)	0.10	Remainder
240.0	A240.0 (A140)	Alcoa	...	S	0.50	0.50	7.0-9.0	0.30-0.7	5.5-6.5	...	0.30-0.7	0.10	...	0.20	0.05	0.15	Remainder
240.1	A240.1 (A140)	Alcoa	...	Ingot	0.50	0.40	7.0-9.0	0.30-0.7	5.6-6.5	...	0.30-0.7	0.10	...	0.20	0.05	0.15	Remainder
242.0	142	AATD	...	S&P	0.7	1.0	3.5-4.5	0.35	1.2-1.8	0.25	1.7-2.3	0.35	...	0.25	0.05	0.15	Remainder
242.1	142	AATD	...	Ingot	0.7	0.8	3.5-4.5	0.35	1.3-1.8	0.25	1.7-2.3	0.35	...	0.25	0.05	0.15	Remainder

(continued)

Registered alloys in the form of XXX.0 castings, XXX.1 ingot and XXX.2 ingot (continued)

AA number	Former designation	Registered By	Date	Prod-uct(e)	Si	Fe	Cu	Mn	Mg	Cr	Ni	Zn	Sn	Ti	Others(d) Each	Total	Al min (e)
242.2	142	AATD	...	Ingot	0.6	0.6	3.5-4.5	0.10	1.3-1.8	...	1.7-2.3	0.10	...	0.20	0.05	0.15	Remainder
A242.0	A142	AATD	...	S	0.6	0.8	3.7-4.5	0.10	1.2-1.7	0.15-0.25	1.8-2.3	0.10	...	0.07-0.20	0.05	0.15	Remainder
A242.1	A142	AATD	...	Ingot	0.6	0.6	3.7-4.5	0.10	1.3-1.7	0.15-0.25	1.8-2.3	0.10	...	0.07-0.20	0.05	0.15	Remainder
A242.2	A142	AATD	...	Ingot	0.35	0.6	3.7-4.5	0.10	1.3-1.7	0.15-0.25	1.8-2.3	0.10	...	0.07-0.20	0.05	0.15	Remainder
243.0	ML	...	4/21/78	S	0.35	0.40	3.5-4.5	0.15-0.45	1.8-2.3	0.20-0.40	1.9-2.3	0.05	...	0.06-0.20	0.05(26)	0.15	Remainder
243.1	ML	...	4/21/78	Ingot	0.35	0.30	3.5-4.5	0.15-0.45	1.9-2.3	0.20-0.40	1.9-2.3	0.05	...	0.06-0.20	0.05(26)	0.15	Remainder
295.0	195	AATD	...	S	0.7-1.5	1.0	4.0-5.0	0.35	0.03	...	...	0.35	...	0.25	0.05	0.15	Remainder
295.1	195	AATD	...	Ingot	0.7-1.5	0.8	4.0-5.0	0.35	0.03	...	...	0.35	...	0.25	0.05	0.15	Remainder
295.2	195	AATD	...	Ingot	0.7-1.2	0.8	4.0-5.0	0.30	0.03	...	...	0.30	...	0.20	0.05	0.15	Remainder
296.0	B295.0 (B195)	AATD	...	P	2.0-3.0	1.2	4.0-5.0	0.35	0.05	...	0.35	0.50	...	0.25	0.05	0.35	Remainder
296.1	B295.1 (B195)	AATD	...	Ingot	2.0-3.0	0.9	4.0-5.0	0.35	0.05	...	0.35	0.50	...	0.25	0.05	0.35	Remainder
296.2	B295.2 (B195)	AATD	...	Ingot	2.0-3.0	0.8	4.0-5.0	0.30	0.03	...	...	0.30	...	0.20	0.05	0.15	Remainder
305.0	...	Reynolds	11/5/74	S&P	4.5-5.5	0.6	1.0-1.5	0.50	0.10	0.25	...	0.35	...	0.25	0.05	0.15	Remainder
305.2	...	Reynolds	9/24/73	Ingot	4.5-5.5	0.14-0.25	1.0-1.5	0.05	...	...	...	0.05	...	0.20	0.05	0.15	Remainder
A305.0	...	Reynolds	11/5/74	S&P	4.5-5.5	0.20	1.0-1.5	0.10	0.10	...	...	0.10	...	0.20	0.05	0.15	Remainder
A305.1	...	Kaiser	6/4/74	Ingot	4.5-5.5	0.15	1.0-1.5	0.05	...	...	...	0.05	...	0.20	0.05	0.15	Remainder
A305.2	...	Reynolds	9/24/73	Ingot	4.5-5.5	0.13	1.0-1.5	0.05	...	...	...	0.05	...	0.20	0.05	0.15	Remainder
308.0	A108	AATD	...	S&P	5.0-6.0	1.0	4.0-5.0	0.50	0.10	...	...	1.0	...	0.25	...	0.50	Remainder
308.1	A108	AATD	...	Ingot	5.0-6.0	0.8	4.0-5.0	0.50	0.10	...	...	1.0	...	0.25	...	0.50	Remainder
308.2	A108	AATD	...	Ingot	5.0-6.0	0.8	4.0-5.0	0.30	0.10	...	...	0.50	...	0.20	...	0.50	Remainder
319.0	319, AllCast	AATD	...	S&P	5.5-6.5	1.0	3.0-4.0	0.50	0.10	...	0.35	1.0	...	0.25	...	0.50	Remainder
319.1	319, AllCast	AATD	...	Ingot	5.5-6.5	0.8	3.0-4.0	0.50	0.10	...	0.35	1.0	...	0.25	...	0.50	Remainder
319.2	319, AllCast	AATD	...	Ingot	5.5-6.5	0.6	3.0-4.0	0.10	0.10	...	0.10	0.10	...	0.20	...	0.20	Remainder
A319.0	...	AATD	8/28/70	S&P	5.5-6.5	1.0	3.0-4.0	0.50	0.10	...	0.35	3.0	...	0.25	...	0.50	Remainder
A319.1	...	AATD	8/28/70	Ingot	5.5-6.5	0.8	3.0-4.0	0.50	0.10	...	0.35	3.0	...	0.25	...	0.50	Remainder
B319.0	...	SAE 329	10/30/81	S&P	5.5-6.5	12	3.0-4.0	0.8	0.10-0.50	...	0.50	1.0	...	0.25	...	0.50	Remainder
B319.1	...	...	10/30/81	Ingot	5.5-6.5	0.9	3.0-4.0	0.8	0.15-0.50	...	0.50	1.0	...	0.25	...	0.50	Remainder
B320.0	...	...	4/8/82	S&P	5.0-8.0	12	2.0-4.0	0.8	0.05-0.60	...	0.35	3.0	...	0.25	...	0.50	Remainder
B320.1	...	...	4/8/82	Ingot	5.0-8.0	12	2.0-4.0	0.8	0.10-0.60	...	0.35	3.0	...	0.25	...	0.50	Remainder
324.0	324	Alcoa	...	P	7.0-8.0	1.2	0.40-0.6	0.50	0.40-0.7	...	0.30	1.0	...	0.20	0.15	0.20	Remainder
324.1	324	Alcoa	...	Ingot	7.0-8.0	0.9	0.40-0.6	0.50	0.45-0.7	...	0.30	1.0	...	0.20	0.15	0.20	Remainder
324.2	324	Alcoa	1/26/72	Ingot	7.0-8.0	0.6	0.40-0.6	0.10	0.45-0.7	...	0.10	0.10	...	0.20	0.05	0.15	Remainder
328.0	Red X-8	AATD	...	S	7.5-8.5	1.0	1.0-2.0	0.20-0.6	0.20-0.6	0.35	0.25	1.5	...	0.25	...	0.50	Remainder
328.1	Red X-8	AATD	...	Ingot	7.5-8.5	0.8	1.0-2.0	0.20-0.6	0.25-0.6	0.35	0.25	1.5	...	0.25	...	0.50	Remainder
332.0	F332.0 (F132)	AATD	...	P	11.0-13.0	1.2	0.50-1.5	0.35	0.7-1.3	...	2.0-3.0	0.35	...	0.25	0.05	...	Remainder
332.1	F332.1 (F132)	AATD	...	Ingot	11.0-13.0	0.9	0.50-1.5	0.35	0.8-1.3	...	2.0-3.0	0.35	...	0.25	0.05	0.15	Remainder
332.2	F332.2 (F132)	AATD	...	Ingot	11.0-13.0	0.9	0.50-1.5	0.10	0.9-1.3	...	2.0-3.0	0.10	...	0.20	0.05	0.15	Remainder
333.0	333	AATD	...	P	8.0-10.0	1.0	3.0-4.0	0.50	0.05-0.50	...	0.50	1.0	...	0.25	...	0.50	Remainder
333.1	333	AATD	...	Ingot	8.0-10.0	0.8	3.0-4.0	0.50	0.10-0.50	...	0.50	1.0	...	0.25	...	0.50	Remainder
A333.0	...	AATD	8/28/70	P	8.0-10.0	1.0	3.0-4.0	0.50	0.05-0.50	...	0.50	3.0	...	0.25	...	0.50	Remainder
A333.1	...	AATD	8/28/70	Ingot	8.0-10.0	0.8	3.0-4.0	0.50	0.10-0.50	...	0.50	3.0	...	0.25	...	0.50	Remainder
336.0	A332.0 (A132)	AATD	...	P	8.5-10.5	1.2	2.0-4.0	0.50	0.50-1.5	...	0.50	1.0	...	0.25	...	0.50	Remainder
336.1	A332.1 (A132)	AATD	...	Ingot	8.5-10.5	0.9	2.0-4.0	0.50	0.6-1.5	...	0.50	1.0	...	0.25	...	0.50	Remainder
336.2	A332.2 (A132)	AATD	...	Ingot	8.5-10.0	0.6	2.0-4.0	0.10	0.9-1.3	...	0.10	0.10	...	0.20	...	0.30	Remainder
339.0	Z332.0 (Z132)	AATD	...	P	11.0-13.0	1.2	1.5-3.0	0.50	0.50-1.5	...	0.50-1.5	1.0	...	0.25	...	0.50	Remainder
339.1	Z332.1 (Z132)	AATD	...	Ingot	11.0-13.0	0.9	1.5-3.0	0.50	0.6-1.5	...	0.50-1.5	1.0	...	0.25	...	0.50	Remainder
343.0	X443Z	MRCI	10/27/72	D	6.7-7.7	1.2	0.50-0.9	0.50	0.10	0.10	...	1.2-2.0	0.5	...	0.10	0.35	Remainder
343.1	X443Z	MRCI	10/27/72	Ingot	6.7-7.7	0.50-0.9	0.50-0.9	0.50	0.10	0.10	...	1.2-1.9	0.5	...	0.10	0.35	Remainder
354.0	354	AATD	...	P	8.6-9.4	0.20	1.6-2.0	0.10	0.40-0.6	...	...	0.10	...	0.20	0.05	0.15	Remainder
354.1	354	AATD	...	Ingot	8.6-9.4	0.15	1.6-2.0	0.10	0.45-0.6	...	...	0.10	...	0.20	0.05	0.15	Remainder
355.0	355	AATD	...	S&P	4.5-5.5	0.6(p)	1.0-1.5	0.50(p)	0.40-0.6	0.25	...	0.35	...	0.25	0.05	0.15	Remainder
355.1	355	AATD	...	Ingot	4.5-5.5	0.50(p)	1.0-1.5	0.50(p)	0.45-0.6	0.25	...	0.35	...	0.25	0.05	0.15	Remainder
355.2	355	AATD	...	Ingot	4.5-5.5	0.14-0.25	1.0-1.5	0.05	0.50-0.6	...	...	0.05	...	0.20	0.05	0.15	Remainder
A355.0	...	...	9/17/81	S&P	4.5-5.5	0.09	...	0.05	0.45-0.6	...	...	0.05	...	0.04-0.2	0.05	0.15	Remainder
A355.2	...	...	9/17/81	Ingot	4.5-5.5	0.06	1.0-1.5	0.03	0.50-0.6	...	...	0.03	...	0.04-0.2	0.03	0.10	Remainder
C355.0	C355	AATD	...	S&P	4.5-5.5	0.20	1.0-1.5	0.10	0.40-0.6	...	...	0.10	...	0.20	0.05	0.15	Remainder
C355.1	...	Kaiser	6/4/74	Ingot	4.5-5.5	0.15	1.0-1.5	0.10	0.40-0.6	...	...	0.10	...	0.20	0.05	0.15	Remainder
C355.2	C355	AATD	...	Ingot	4.5-5.5	0.13	1.0-1.5	0.05	0.50-0.6	...	...	0.05	...	0.20	0.05	0.15	Remainder
356.0	356	AATD	...	S&P	6.5-7.5	0.6	0.25	0.35	0.20-0.40	...	...	0.35	...	0.25	0.05	0.15	Remainder
356.1	356	AATD	...	Ingot	6.5-7.5	0.50	0.25	0.35	0.25-0.40	...	...	0.35	...	0.25	0.05	0.15	Remainder
356.2	356	AATD	...	Ingot	6.5-7.5	0.13-0.25	0.10	0.05	0.30-0.40	...	...	0.05	...	0.20	0.05	0.15	Remainder
A356.0	A356	AATD	...	S&P	6.5-7.5	0.20	0.20	0.10	0.20-0.40	...	...	0.10	...	0.20	0.05	0.15	Remainder
A356.1	...	Kaiser	6/4/74	Ingot	6.5-7.5	0.15	0.20	0.10	0.20-0.40	...	...	0.10	...	0.20	0.05	0.15	Remainder
A356.2	A356	AATD	...	Ingot	6.5-7.5	0.12	0.10	0.05	0.30-0.40	...	...	0.05	...	0.20	0.05	0.15	Remainder
B356.0	...	...	9/17/81	S&P	6.5-7.5	0.09	0.05	0.05	0.25-0.45	...	...	0.05	...	0.04-0.20	0.05	0.15	Remainder
B356.2	...	...	9/17/81	Ingot	6.5-7.5	0.06	0.03	0.03	0.30-0.45	...	...	0.03	...	0.04-0.20	0.03	0.10	Remainder
C356.0	...	...	5/30/85	S&P	6.5-7.5	0.07	0.05	0.05	0.25-0.45	...	...	0.05	...	0.04-0.20	0.05	0.15	Remainder
C356.2	...	...	5/30/85	Ingot	6.5-7.5	0.15	0.05	0.03	0.45-0.6	...	...	0.05	...	0.20	0.05	0.15	Remainder
F356.0	...	Reynolds	10/20/71	S&P	6.5-7.5	0.20	0.20	0.10	0.17-0.25	...	...	0.10	...	0.20	0.05	0.15	Remainder
F356.2	...	Reynolds	10/20/71	Ingot	6.5-7.5	0.12	0.10	0.05	0.17-0.25	...	...	0.05	...	0.20	0.05	0.15	Remainder
357.0	357	AATD	...	S&P	6.5-7.5	0.15	0.05	0.03	0.45-0.6	...	...	0.05	...	0.20	0.05	0.15	Remainder
357.1	357	AATD	...	Ingot	6.5-7.5	0.12	0.05	0.03	0.45-0.6	...	...	0.05	...	0.20	0.05	0.15	Remainder
A357.0	A357	AATD	...	S&P	6.5-7.5	0.20	0.20	0.10	0.40-0.7	...	...	0.10	...	0.10-0.20	0.05(q)	0.15	Remainder
A357.2	A357	AATD	...	Ingot	6.5-7.5	0.12	0.10	0.05	0.45-0.7	...	...	0.05	...	0.10-0.20	0.03(q)	0.10	Remainder
B357.0	...	...	9/17/81	S&P	6.5-7.5	0.09	0.05	0.05	0.40-0.6	...	...	—0.05	...	0.04-0.2	0.05	0.15	Remainder
D357.2	...	...	9/17/81	Ingot	6.5-7.5	0.06	0.03	0.03	0.50-0.7	...	...	0.03	...	0.04-0.20	0.03(6)	0.10	Remainder
C357.0	...	...	9/17/81	S&P	6.5-7.5	0.09	0.05	0.05	0.45-0.6	...	...	0.03	...	0.04-0.20	0.03	0.10	Remainder
C357.2	...	...	9/17/81	Ingot	6.5-7.5	0.06	0.03	0.03	0.40-0.7	...	...	—0.05	...	0.04-0.2	0.05(6)	0.15	Remainder
D357.2	...	...	9/21/84	S	6.5-7.5	0.20	...	0.10	0.55-0.6	...	...	...	...	0.10-0.20	0.05(6)	0.15	Remainder
B358.0	Tens-50	AATD	...	S&P	7.6-8.6	0.30	0.20	0.20	0.40-0.6	0.20	...	0.20	...	0.10-0.20	0.05(r)	0.15	Remainder
B358.2	Tens-50	AATD	...	Ingot	7.6-8.6	0.20	0.10	0.10	0.45-0.6	0.05	...	0.10	...	0.12-0.20	0.05(s)	0.15	Remainder

Registered alloys in the form of XXX.0 castings, XXX.1 ingot and XXX.2 ingot (continued)

AA number	Former designation	Registered By	Date	Prod-uct(e)	Si	Fe	Cu	Mn	Mg	Cr	Ni	Zn	Sn	Ti	Others(d) Each	Total	Al min (e)
359.0	359	AATD	...	S&P	8.5-9.5	0.20	0.20	0.10	0.50-0.7	...	...	0.10	...	0.20	0.05	0.15	Remainder
359.2	359	AATD	...	Ingot	8.5-9.5	0.12	0.10	0.10	0.55-0.7	...	...	0.10	...	0.20	0.05	0.15	Remainder
360.0(t)	360	AATD	...	D	9.0-10.0	2.0	0.6	0.35	0.40-0.6	...	0.50	0.50	0.15	...	...	0.25	Remainder
360.2	360	AATD	...	Ingot	9.0-10.0	0.7-1.1	0.10	0.10	0.45-0.6	...	0.10	0.10	0.10	...	...	0.20	Remainder
A360.0(t)	A360	AATD	...	D	9.0-10.0	1.3	0.6	0.35	0.45-0.6	...	0.50	0.50	0.15	...	...	0.25	Remainder
A360.1(t)	A360	AATD	...	Ingot	9.0-10.0	1.0	0.6	0.35	0.45-0.6	...	0.50	0.40	0.15	...	...	0.25	Remainder
A360.2	A360	AATD	...	Ingot	9.0-10.0	0.6	0.10	0.05	0.45-0.6	...	...	0.05	...	...	0.05	0.15	Remainder
361.0	...	...	6/30/78	D	9.5-10.5	1.1	0.50	0.25	0.40-0.6	0.20-0.30	0.20-0.30	0.50	0.10	0.20	0.05	0.15	Remainder
361.1	...	...	6/30/78	Ingot	9.5-10.5	0.8	0.50	0.25	0.45-0.6	0.20-0.30	0.20-0.30	0.40	0.10	0.20	0.05	0.15	Remainder
363.0	363	Kaiser	1/16/70	S&P	4.5-6.0	1.1	2.5-3.5	(u)	0.15-0.40	(u)	0.25	3.0-4.5	0.25	0.20	(v)	0.30	Remainder
363.1	363	Kaiser	1/16/60	Ingot	4.5-6.0	0.8	2.5-3.5	(u)	0.20-0.40	(u)	0.25	3.0-4.5	0.25	0.20	(v)	0.30	Remainder
364.0	364	AATD	...	D	7.5-9.5	1.5	0.20	0.10	0.20-0.40	0.25-0.50	0.15	0.15	0.15	...	0.05(w)	0.15	Remainder
364.2	364	AATD	...	Ingot	7.5-9.5	0.7-1.1	0.20	0.10	0.25-0.40	0.25-0.50	0.15	0.15	0.15	...	0.05(w)	0.15	Remainder
369.0	Special K-9	...	4/4/78	D	11.0-12.0	1.3	0.50	0.35	0.25-0.45	0.30-0.40	0.05	1.0	0.10	...	0.05	0.15	Remainder
369.1	Special K-9	...	4/4/78	Ingot	11.0-12.0	1.0	0.50	0.35	0.30-0.45	0.30-0.40	0.05	0.9	0.10	...	0.05	0.15	Remainder
380.0(t)	380	AATD	...	D	7.5-9.5	2.0	3.0-4.0	0.50	0.10	...	0.50	3.0	0.35	...	...	0.50	Remainder
380.2	380	AATD	...	Ingot	7.5-9.5	0.7-1.1	3.0-4.0	0.10	0.10	...	0.10	0.10	0.10	...	...	0.20	Remainder
A380.0(t)	A380	AATD	...	D	7.5-9.5	1.3	3.0-4.0	0.50	0.10	...	0.50	3.0	0.35	...	...	0.50	Remainder
A380.1(t)	A380	AATD	...	Ingot	7.5-9.5	1.0	3.0-4.0	0.50	0.10	...	0.50	2.9	0.35	...	...	0.50	Remainder
A380.2	A380	AATD	...	Ingot	7.5-9.5	0.6	3.0-4.0	0.10	0.10	...	0.10	0.10	...	...	0.05	0.15	Remainder
B380.0	A380	AATD	...	D	7.5-9.5	1.3	3.0-4.0	0.50	0.10	...	0.50	1.0	0.35	...	...	0.50	Remainder
B380.1	A380	AATD	...	Ingot	7.5-9.5	1.0	3.0-4.0	0.50	0.10	...	0.50	0.9	0.35	...	...	0.50	Remainder
383.0	...	AATD	...	D	9.5-11.5	1.3	2.0-3.0	0.50	0.10	...	0.30	3.0	0.15	...	...	0.50	Remainder
383.1	...	AATD	...	Ingot	9.5-11.5	0.6-1.0	2.0-3.0	0.50	0.10	...	0.30	2.9	0.15	...	...	0.50	Remainder
383.2	...	AATD	...	Ingot	9.5-11.5	0.6-1.0	2.0-3.0	0.10	0.10	...	0.10	0.10	0.10	...	...	0.20	Remainder
384.0	384	AATD	...	D	10.5-12.0	1.3	3.0-4.5	0.50	0.10	...	0.50	3.0	0.35	...	...	0.50	Remainder
384.1	384	AATD	...	Ingot	10.5-12.0	1.0	3.0-4.5	0.50	0.10	...	0.50	2.9	0.35	...	...	0.50	Remainder
384.2	384	AATD	...	Ingot	10.5-12.0	0.6-1.0	3.0-4.5	0.10	0.10	...	0.10	0.10	0.10	...	...	0.20	Remainder
A384.0	384	AATD	...	D	10.5-12.0	1.3	3.0-4.5	0.50	0.10	...	0.50	1.0	0.35	...	...	0.50	Remainder
A384.1	384	AATD	...	Ingot	10.5-12.0	1.0	3.0-4.5	0.50	0.10	...	0.50	0.9	0.35	...	...	0.50	Remainder
385.0	B384.0 (384)	Alcoa	1/21/70	D	11.0-13.0	2.0	2.0-4.0	0.50	0.30	...	0.50	3.0	0.30	...	...	0.50	Remainder
385.1	B384.1 (384)	Alcoa	1/21/70	Ingot	11.0-13.0	0.7-1.1	2.0-4.0	0.50	0.30	...	0.50	2.9	0.30	...	...	0.50	Remainder
390.0	390	AATD	...	D	16.0-18.0	1.3	4.0-5.0	0.10	0.45-0.65	...	...	0.10	...	0.20	0.10	0.20	Remainder
390.2	390	AATD	...	Ingot	16.0-18.0	0.6-1.0	4.0-5.0	0.10	0.50-0.65	...	...	0.10	...	0.20	0.10	0.20	Remainder
A390.0	A390	AATD	...	S&P	16.0-18.0	0.50	4.0-5.0	0.10	0.45-0.65	...	...	0.10	...	0.20	0.10	0.20	Remainder
A390.1	A390	AATD	...	Ingot	16.0-18.0	0.40	4.0-5.0	0.10	0.50-0.65	...	...	0.10	...	0.20	0.10	0.20	Remainder
B390.0	...	...	3/29/79	D	16.0-18.0	1.3	4.0-5.0	0.45-0.65	...	...	0.10	1.5	...	0.20	0.10	0.20	Remainder
B390.1	...	...	3/29/79	Ingot	16.0-18.0	1.0	4.0-5.0	0.50-0.65	...	...	0.10	1.4	...	0.20	0.10	0.20	Remainder
392.0	392	AATD	...	D	18.0-20.0	1.5	0.40-0.8	0.20-0.60	0.8-1.2	...	0.50	0.50	0.30	0.20	0.15	0.50	Remainder
392.1	392	AATD	...	Ingot	18.0-20.0	1.1	0.40-0.8	0.20-0.6	0.9-1.2	...	0.50	0.40	0.30	0.20	0.15	0.50	Remainder
393.0	Vanasil	USCO	...	SP&D	21.0-23.0	1.3	0.7-1.1	0.10	0.7-1.3	...	2.0-2.5	0.10	...	0.10-0.20	0.05(x)	0.15	Remainder
393.1	Vanasil	USCO	...	Ingot	21.0-23.0	1.0	0.7-1.1	0.10	0.8-1.3	...	2.0-2.5	0.10	...	0.10-0.20	0.05(x)	0.15	Remainder
393.2	Vanasil	USCO	...	Ingot	21.0-23.0	0.8	0.7-1.1	0.10	0.8-1.3	...	2.0-2.5	0.10	...	0.10-0.20	0.05(x)	0.15	Remainder
408.2(y)	...	Reynolds	9/24/73	Ingot	8.5-9.5	0.6-1.3	0.10	0.10	...	...	...	0.10	...	...	0.10	0.20	Remainder
409.2(y)	...	Reynolds	9/24/73	Ingot	9.0-10.0	0.6-1.3	0.10	0.10	...	...	...	0.10	...	...	0.10	0.20	Remainder
411.2(y)	...	Reynolds	9/24/73	Ingot	10.0-12.0	0.6-1.3	0.20	0.10	...	...	...	0.10	...	...	0.10	0.20	Remainder
413.0(t)	13	AATD	...	D	11.0-13.0	2.0	1.0	0.35	0.10	...	0.50	0.50	0.15	...	...	0.25	Remainder
413.2	13	AATD	...	Ingot	11.0-13.0	0.7-1.1	0.10	0.10	0.07	...	0.10	0.10	0.10	...	...	0.20	Remainder
A413.0(t)	A13	AATD	...	D	11.0-13.0	1.3	1.0	0.35	0.10	...	0.50	0.50	0.15	...	...	0.25	Remainder
A413.1(t)	A13	AATD	...	Ingot	11.0-13.0	1.0	1.0	0.35	0.10	...	0.50	0.40	0.15	...	...	0.25	Remainder
A413.2	A13	AATD	...	Ingot	11.0-13.0	0.6	0.10	0.05	0.05	...	0.05	0.05	0.05	...	...	0.10	Remainder
B413.0	...	...	11/6/84	Ingot	11.0-13.0	0.40	0.10	0.35	0.05	...	0.05	0.10	...	0.25	0.05	0.20	Remainder
B413.1	...	...	11/6/84	S&P	11.0-13.0	0.50	0.10	0.35	0.05	...	0.05	0.10	...	0.25	0.05	0.20	Remainder
435.2(27)	...	...	12/18/81	Ingot	3.3-3.9	0.40	0.05	0.05	0.05	...	...	0.10	...	...	0.05	0.20	Remainder
443.0	43	AATD	...	S	4.5-6.0	0.8	0.6	0.50	0.05	0.25	...	0.50	...	0.25	...	0.35	Remainder
443.1	43	AATD	...	Ingot	4.5-6.0	0.6	0.6	0.50	0.05	0.25	...	0.50	...	0.25	...	0.35	Remainder
443.2	43	AATD	...	Ingot	4.5-6.0	0.6	0.10	0.10	0.05	...	...	0.10	...	0.20	0.05	0.15	Remainder
A443.0	43(0.30 max Cu)	AATD	...	S	4.5-6.0	0.8	0.30	0.50	0.05	0.25	...	0.50	...	0.25	...	0.35	Remainder
A443.1	43(0.30 max Cu)	AATD	...	Ingot	4.5-6.0	0.6	0.30	0.50	0.05	0.25	...	0.50	...	0.25	...	0.35	Remainder
B443.0	43(0.15 max Cu)	AATD	...	S&P	4.5-6.0	0.8	0.15	0.35	0.05	...	...	0.35	...	0.25	0.05	0.15	Remainder
B443.1	43(0.15 max Cu)	AATD	...	Ingot	4.5-6.0	0.6	0.15	0.35	0.05	...	...	0.35	...	0.25	0.05	0.15	Remainder
C443.0	A43	AATD	...	D	4.5-6.0	2.0	0.6	0.35	0.10	...	0.50	0.50	0.15	...	...	0.25	Remainder
C443.1	A43	AATD	...	Ingot	4.5-6.0	1.0	0.6	0.35	0.10	...	0.50	0.40	0.15	...	...	0.25	Remainder
C443.2	A43	AATD	...	Ingot	4.5-6.0	0.7-1.1	0.10	0.10	0.05	...	...	0.10	...	...	0.05	0.15	Remainder
444.0	...	Reynolds	11/5/74	S&P	6.5-7.5	0.6	0.25	0.35	0.10	...	...	0.35	...	0.25	0.05	0.15	Remainder
444.2	...	Reynolds	9/24/73	Ingot	6.5-7.5	0.13-0.25	0.10	0.05	0.05	...	...	0.05	...	0.20	0.05	0.15	Remainder
A444.0	A344	AATD	...	P	6.5-7.5	0.20	0.10	0.10	0.05	...	...	0.10	...	0.20	0.05	0.15	Remainder
A444.1	...	Kaiser	6/4/74	Ingot	6.5-7.5	0.15	0.10	0.10	0.05	...	...	0.10	...	0.20	0.05	0.15	Remainder
A444.2	A344	AATD	...	Ingot	6.5-7.5	0.12	0.05	0.05	0.05	...	...	0.05	...	0.20	0.05	0.15	Remainder
445.2(y)	B444.2	Reynolds	9/24/73	Ingot	6.5-7.5	0.6-1.3	0.10	0.10	...	...	...	0.10	...	...	0.10	0.20	Remainder
511.0	F514(F214)	AATD	...	S	0.35	0.50	0.15	0.35	3.5-4.5	...	...	0.15	...	0.25	0.05	0.15	Remainder
511.1	F514.1(F214)	AATD	...	Ingot	0.35	0.40	0.15	0.35	3.6-4.5	...	...	0.15	...	0.25	0.05	0.15	Remainder
511.2	F514.2(F214)	AATD	...	Ingot	0.30	0.30	0.10	0.10	3.6-4.5	...	...	0.10	...	0.20	0.05	0.15	Remainder
512.0	B514.0(B214)	AATD	...	P	0.30	0.40	0.10	0.30	3.5-4.5	...	...	1.4-2.2	...	0.20	0.05	0.15	Remainder
512.2	B514.2(B214)	AATD	...	Ingot	0.30	0.30	0.10	0.10	3.6-4.5	...	...	1.4-2.2	...	0.20	0.05	0.15	Remainder
513.0	A514.0(A214)	AATD	...	S	1.4-2.2	0.6	0.35	0.8	3.5-4.5	0.25	...	0.35	...	0.25	0.05	0.15	Remainder
513.2	A514.2(A214)	AATD	...	Ingot	1.4-2.2	0.30	0.10	0.10	3.6-4.5	...	...	0.10	...	0.20	0.05	0.15	Remainder
514.0	214	AATD	...	S	0.30-0.7	0.50	0.15	0.35	3.5-4.5	...	...	0.15	...	0.25	0.05	0.15	Remainder
514.1	214	AATD	...	Ingot	0.30-0.7	0.40	0.15	0.35	3.6-4.5	...	...	0.15	...	0.25	0.05	0.15	Remainder
514.2	214	AATD	...	Ingot	0.30-0.7	0.30	0.10	0.10	3.6-4.5	...	...	0.10	...	0.20	0.05	0.15	Remainder

(continued)

Registered alloys in the form of XXX.0 castings, XXX.1 ingot and XXX.2 ingot (continued)

AA number	Former designation	Registered By	Date	Product(e)	Si	Fe	Cu	Mn	Mg	Cr	Ni	Zn	Sn	Ti	Others(d) Each	Total	Al min (e)
515.0	L514.0(L214)	Reynolds	1/2/70	D	0.50-1.0	1.3	0.20	0.40-0.6	2.5-4.0	...	...	0.10	...	...	0.05	0.15	Remainder
515.2	L514.2(L214)	Reynolds	1/2/70	Ingot	0.50-1.0	0.6-1.0	0.10	0.40-0.6	2.7-4.0	...	...	0.05	...	...	0.05	0.15	Remainder
516.0	...	...	9/30/83	D	0.30-1.5	0.35-1.0	0.30	0.15-0.40	2.5-4.5	...	0.25-0.40	0.20	0.10	0.10-0.20	0.05[28]	...	Remainder
516.1	...	...	9/30/83	Ingot	0.30-1.5	0.35-0.7	0.30	0.15-0.40	2.6-4.5	...	0.25-0.40	0.20	0.10	0.10-0.20	0.05[28]	...	Remainder
518.0	218	AATD	...	D	0.35	1.8	0.25	0.35	7.5-8.5	...	0.15	0.15	0.15	...	...	0.25	Remainder
518.1	218	AATD	...	Ingot	0.35	1.0	0.25	0.35	7.6-8.5	...	0.15	0.15	0.15	...	...	0.25	Remainder
518.2	218	AATD	...	Ingot	0.25	0.7	0.10	0.10	7.6-8.5	...	0.05	...	0.05	...	...	0.10	Remainder
520.0	220	AATD	...	S	0.25	0.30	0.25	0.15	9.5-10.6	...	...	0.15	...	0.25	0.05	0.15	Remainder
520.2	220	AATD	...	Ingot	0.15	0.20	0.20	0.10	9.6-10.6	...	...	0.10	...	0.25	0.05	0.15	Remainder
535.0	Almag 35	AATD	...	S	0.15	0.15	0.05	0.10-0.25	6.2-7.5	...	...	...	...	0.10-0.25	0.05(z)	0.15	Remainder
535.2	Almag 35	AATD	...	Ingot	0.10	0.10	0.05	0.10-0.25	6.6-7.5	...	...	...	...	0.10-0.25	0.05(z)	0.15	Remainder
A535.0	A218	AATD	...	S	0.20	0.20	0.10	0.10-0.25	6.5-7.5	...	...	...	...	0.25	0.05	0.15	Remainder
A535.1	A218	AATD	...	Ingot	0.20	0.15	0.10	0.10-0.25	6.6-7.5	...	...	...	...	0.25	0.05	0.15	Remainder
B535.0	B218	AATD	...	S	0.15	0.15	0.10	0.05	6.5-7.5	...	...	...	...	0.10-0.25	0.05	0.15	Remainder
B535.2	B218	AATD	...	Ingot	0.10	0.12	0.05	0.05	6.6-7.5	...	...	...	...	0.10-0.25	0.05	0.15	Remainder
705.0	603, Ternalloy 5	AATD	...	S&P	0.20	0.8	0.20	0.40-0.6	1.4-1.8	0.20-0.40	...	2.7-3.3	...	0.25	0.05	0.15	Remainder
705.1	603, Ternalloy 5	AATD	...	Ingot	0.20	0.6	0.20	0.40-0.6	1.5-1.8	0.20-0.40	...	2.7-3.3	...	0.25	0.05	0.15	Remainder
707.0	607, Ternalloy 7	AATD	...	S&P	0.20	0.8	0.20	0.40-0.6	1.8-2.4	0.20-0.40	...	4.0-4.5	...	0.25	0.05	0.15	Remainder
707.1	607, Ternalloy 7	AATD	...	Ingot	0.20	0.6	0.20	0.40-0.6	1.9-2.4	0.20-0.40	...	4.0-4.5	...	0.25	0.05	0.15	Remainder
710.0	A712.0 (A612)	AATD	...	S	0.15	0.50	0.35-0.65	0.05	0.6-0.8	...	...	6.0-7.0	...	0.25	0.05	0.15	Remainder
710.1	A712.1 (A612)	AATD	...	Ingot	0.15	0.40	0.35-0.65	0.05	0.65-0.8	...	...	6.0-7.0	...	0.25	0.05	0.15	Remainder
711.0	C712.0 (C612)	Alcoa	...	P	0.30	0.7-1.4	0.35-0.65	0.05	0.25-0.45	...	...	6.0-7.0	...	0.20	0.05	0.15	Remainder
711.1	C712.1 (C612)	Alcoa	...	Ingot	0.30	0.7-1.1	0.35-0.65	0.05	0.30-0.45	...	...	6.0-7.0	...	0.20	0.05	0.15	Remainder
712.0	D712.0 (D612, 40E)	AATD	...	S	0.30	0.50	0.25	0.10	0.50-0.65	0.40-0.6	...	5.0-6.5	...	0.15-0.25	0.05	0.20	Remainder
712.2	D712.2 (D612, 40E)	AATD	...	Ingot	0.15	0.40	0.25	0.10	0.50-0.65	0.40-0.6	...	5.0-6.5	...	0.15-0.25	0.05	0.20	Remainder
713.0	613, Tenzaloy	AATD	...	S&P	0.25	1.1	0.40-1.0	0.6	0.20-0.50	0.35	0.15	7.0-8.0	...	0.25	0.10	0.25	Remainder
713.1	613, Tenzaloy	AATD	...	Ingot	0.25	0.8	0.40-1.0	0.6	0.25-0.50	0.35	0.15	7.0-8.0	...	0.25	0.10	0.25	Remainder
771.0	Precedent 71A	USCO	...	S	0.15	0.15	0.10	0.10	0.8-1.0	0.06-0.20	...	6.5-7.5	...	0.10-0.20	0.05	0.15	Remainder
771.2	Precedent 71A	USCO	...	Ingot	0.10	0.10	0.10	0.10	0.85-1.0	0.06-0.20	...	6.5-7.5	...	0.10-0.20	0.05	0.15	Remainder
772.0	B771.0 (Precedent 71B)	USCO	...	S	0.15	0.15	0.10	0.10	0.6-0.8	0.06-0.20	...	6.0-7.0	...	0.10-0.20	0.05	0.15	Remainder
772.2	B772.2 (Precedent 71B)	USCO	...	Ingot	0.10	0.10	0.10	0.10	0.65-0.8	0.06-0.20	...	6.0-7.0	...	0.10-0.20	0.05	0.15	Remainder
850.0	750	AATD	...	S&P	0.7	0.7	0.7-1.3	0.10	0.10	...	0.7-1.3	...	5.5-7.0	0.20	...	0.30	Remainder
850.1	750	AATD	...	Ingot	0.7	0.50	0.7-1.3	0.10	0.10	...	0.7-1.3	...	5.5-7.0	0.20	...	0.30	Remainder
851.0	A850.0 (A750)	AATD	...	S&P	2.0-3.0	0.7	0.7-1.3	0.10	0.10	...	0.30-0.7	...	5.5-7.0	0.20	...	0.30	Remainder
851.1	A850.1 (A750)	AATD	...	Ingot	2.0-3.0	0.50	0.7-1.3	0.10	0.10	...	0.30-0.7	...	5.5-7.0	0.20	...	0.30	Remainder
852.0	B850.0 (B750)	AATD	...	S&P	0.40	0.7	1.7-2.3	0.10	0.6-0.9	...	0.9-1.5	...	5.5-7.0	0.20	...	0.30	Remainder
852.1	B850.1 (B750)	AATD	...	Ingot	0.40	0.50	1.7-2.3	0.10	0.7-0.9	...	0.9-1.5	...	5.5-7.0	0.20	...	0.30	Remainder
853.0	XC850.0 (XC750)	Alcoa	...	S&P	5.5-6.5	0.7	3.0-4.0	0.50	...	...	...	...	5.5-7.0	0.20	...	0.30	Remainder
853.2	XC850.2 (XC750)	Alcoa	...	Ingot	5.5-6.5	0.50	3.0-4.0	0.10	...	...	...	...	5.5-7.0	0.20	...	0.30	Remainder

(a) Composition in percent maximum unless shown as a range or a minimum. Standard limits for alloying elements and impurities are expressed to the following places: Less than 1/100 percent, 0.000X; 1/1000 to 1/100 percent, 0.00X; 1/100 to 1/10 percent alloys and unalloyed aluminum not made by a refining process, 0.0X; 1/10 through 1/2 percent, 0.XX; over 1/2 percent, 0.X, X.X, etc. (Magnesium percent for some alloys are exceptions to this rule.) (b) For purposes of determining conformance to these limits, an observed value or a calculated value obtained from analysis is rounded off to the nearest unit in the last right-hand place of figures used in expressing the specified limit, in accordance with the following: American National Standard Rules for Rounding Off Numerical Values (ANSI Z25.1). When the figure next beyond the last figure or place to be retained is less than 5, the figure in the last place retained should be kept unchanged. When the figure next beyond the last figure or place to be retained is greater than 5, the figure in the last place retained should be increased by 1. When the figure next beyond the last figure or place to be retained is 5 and (1) there are no figures, or only zeroes, beyond this 5, if the figure in the last place to be retained is odd, it should be increased by 1; if even, it should be kept unchanged; (2) if the 5 next beyond the figure in the last place to be retained is followed by any figures other than zero, the figure in the last place retained should be increased by 1, whether odd or even. (c) D = die casting; P = permanent mold; S = sand. (d) Analysis is regularly made only for the elements for which specific limits are shown, except for minimum purities of 99.00 percent. If, however, the presence of other elements is suspected to be, or in the course of routine analysis is indicated to be in excess of the specified limits, further analysis is made to determine that these other elements are not in excess of the amount specified. (e) The aluminum content for unalloyed aluminum not made by a refining process is the difference between 100.00 percent and the sum of all other metallic elements present in amounts of 0.010 percent or more each, expressed to the second decimal before determining the sum. (f) Rated minimum conductivities: 100.1 ingot, 54 percent IACS; 130.1 ingot, 55 percent IACS; 150.1 ingot, 57 percent IACS; 170.1 ingot, 59 percent IACS. The rating of ingot metal for minimum conductivity characteristic is based on established relations between electrical conductivity and metal composition. (g) Aluminum Association Technical Division. (h) Manganese + chromium + titanium + vanadium 0.025 max. (i) Iron/silicon ratio 2.5 min. (j) Iron/silicon ratio 2.0 min. (k) Iron/silicon ratio 1.5 min. (l) Silver 0.40-1.0. (m) Titanium + zirconium 0.50 max. (n) Antimony 0.20-0.30; cobalt 0.20-0.30; zirconium 0.10-0.30. (o) Vanadium 0.05-0.15; zirconium 0.10-0.25. (p) If iron exceeds 0.45, manganese content shall not be less than one-half iron content. (q) Beryllium 0.04-0.07. (r) Beryllium 0.10-0.30. (s) Beryllium 0.15-0.30. (t) A360.1, A380.1 and A413.1 ingot is used to produce 360.0 and A360.0; 380.0 and A380.0; 413.0 and A413.0 castings, respectively. (u) Manganese and chromium 0.8 max. total. (v) Lead 0.25 max. (w) Beryllium 0.02-0.04. (x) Vanadium 0.08-0.15. (y) B444.2, 408.2, 409.2 and 411.2 are used to coat steel. (z) Beryllium 0.003-0.007, boron 0.002 max. Source: The Aluminum Association Inc.